ENCYCLOPEDIA *of* GEOMAGNETISM AND PALEOMAGNETISM

Encyclopedia of Earth Sciences Series

ENCYCLOPEDIA OF GEOMAGNETISM AND PALEOMAGNETISM

Volume Editors

David Gubbins is Research Professor of Earth Sciences in the School of Earth and Environment, University of Leeds, UK. He did his PhD on geomagnetic dynamos in Cambridge, supervised by Sir Edward Bullard and has worked in the USA, and in Cambridge before moving to Leeds in 1989. His work has included dynamo theory and its connection with the Earth's thermal history, modeling the Earth's magnetic field from historical measurements, and recently the interpretation of paleomagnetic data. He is a fellow of the Royal Society and has been awarded the gold medal of the Royal Astronomical Society and the John Adam Fleming Medal of the American Geophysical Union for original research and leadership in geomagnetism.

Emilio Herrero-Bervera is Research Professor of Geophysics at the School of Ocean and Earth Science and Technology (SOEST) within the Hawaii Institute of Geophysics and Planetology (HIGP) of the University of Hawaii at Manoa, where he is the head of the Paleomagnetics and Petrofabrics Laboratory. During his career he has published over 90 papers in professional journals including Nature, JGR, EPSL, JVGR. He has worked in such diverse fields as volcanology, sedimentology, plate tectonics, and has done field work on five continents.

Aim of the Series

The *Encyclopedia of Earth Sciences Series* provides comprehensive and authoritative coverage of all the main areas in the earth sciences. Each volume comprises a focused and carefully chosen collection of contributions from leading names in the subject, with copious illustrations and reference lists.

These books represent one of the world's leading resources for the Earth sciences community. Previous volumes are being updated and new works published so that the volumes will continue to be essential reading for all professional Earth scientists, geologists, geophysicists, climatologists, and oceanographers as well as for teachers and students.

See the back of this volume for a current list of titles in the *Encyclopedia of Earth Sciences Series*. Go to http://www.springerlink.com/reference-works/ to visit the "Earth Sciences Series" on-line.

About the Editors

Professor Rhodes W. Fairbridge[†] has edited 24 encyclopedias in the Earth Sciences Series. During his career he has worked as a petroleum geologist in the Middle East, been a World War II intelligence officer in the SW Pacific, and led expeditions to the Sahara, Arctic Canada, Arctic Scandinavia, Brazil, and New Guinea. He was Emeritus Professor of Geology at Columbia University and was affiliated with the Goddard Institute for Space Studies.

Professor Michael Rampino has published more than 100 papers in professional journals including *Science*, *Nature*, and *Scientific American*. He has worked in such diverse fields as volcanology, planetary science, sedimentology, and climate studies, and has done field work on six continents. He is currently Associate Professor of Earth and Environmental Sciences at New York University and a consultant at NASA's Goddard Institute for Space Studies.

ENCYCLOPEDIA OF EARTH SCIENCES SERIES

ENCYCLOPEDIA *of* GEOMAGNETISM AND PALEOMAGNETISM

edited by

DAVID GUBBINS
University of Leeds

and

EMILIO HERRERO-BERVERA
University of Hawaii at Manoa

A C.I.P. Catalogue record for this book is available from the Library of Congress.

ISBN-13: 978-1-4020-3992-8

This publication is available also as
Electronic publication under ISBN 978-1-4020-4423-6 and
Print and electronic bundle under ISBN 978-1-4020-4866-1

Published by Springer
PO Box 17, 3300 AA Dordrecht, The Netherlands

Printed on acid-free paper

Cover photo: Part of "A Digital Age Map of the Ocean Floor", by Mueller, R.D., Roest, W.R., Royer, J.-Y., Gahagan, L.M., and Sclater, J.G., SIO Reference Series 93-30, Scripps Institution of Oceanography (map downloaded courtesy of NGDC).

Dedication

Jack A. Jacobs (1916–2003)

This encyclopedia is dedicated to the memory of Jack Jacobs. He made contributions across the whole spectrum of geomagnetism and paleomagnetism throughout a long and productive career. His books *Micropulsations*, *Geonomy, The Earth's Core and Geomagnetism, Reversals of the Earth's Magnetic Field*, and the four volumes of *Geomagnetism* intended to replace *Chapman & Bartels'* work of the same name, cover the field. For this encyclopedia he completed articles for both editors, on *disc dynamo* and *geomagnetic excursions*, and was working on *superchrons and changes in reversal frequency* at the end.

Contents

List of Contributors xv

Preface xxv

Aeromagnetic Surveying 1
Mark Pilkington

Agricola, Georgius (1494–1555) 3
Allan Chapman

Alfvén Waves 3
Christopher Finlay

Alfvén, Hannes Olof Gösta (1908–1995) 6
Carl-Gunne Fälthammar and David Gubbins

Alfvén's Theorem and the Frozen Flux Approximation 7
Paul H. Roberts

Anelastic and Boussinesq Approximations 11
Stanislav I. Braginsky and Paul H. Roberts

Anisotropy, Electrical 20
Karsten Bahr

Antidynamo and Bounding Theorems 21
Friedrich Busse and Michael Proctor

Archeology, Magnetic Methods 23
Armin Schmidt

Archeomagnetism 31
Donald D. Tarling

Auroral Oval 33
Stephen Milan

Baked Contact Test 35
Kenneth L. Buchan

Bangui Anomaly 39
Patrick T. Taylor

Barlow, Peter (1776–1862) 40
Emmanuel Dormy

Bartels, Julius (1899–1964) 42
Karl-Heinz Glaßmeier and Manfred Siebert

Bauer, Louis Agricola (1865–1932) 42
Gregory A. Good

Bemmelen, Willem van (1868–1941) 44
Art R.T. Jonkers

Benton, E. R. 44
David Loper

Bingham Statistics 45
Jeffrey J. Love

Biomagnetism 48
Michael D. Fuller and Jon Dobson

Blackett, Patrick Maynard Stuart, Baron of Chelsea (1887–1974) 53
Michael D. Fuller

Bullard, Edward Crisp (1907–1980) 54
David Gubbins

Carnegie Institution of Washington, Department of Terrestrial Magnetism 56
Gregory A. Good

Carnegie, Research Vessel 58
Gregory A. Good

Champ 59
Stefan Maus

Chapman, Sydney (1888–1970) 61
Stuart R.C. Malin

Coast Effect of Induced Currents 61
Ted Lilley

Compass 66
Art R.T. Jonkers

Conductivity Geothermometer 69
Ted Lilley

Conductivity, Ocean Floor Measurements 71
Steven Constable

Convection, Chemical 73
David Loper

Convection, Nonmagnetic Rotating 74
Andrew Soward

Core Composition 77
William F. McDonough

Core Convection 80
Keke Zhang

Core Density 82
Guy Masters

Core Motions 84
Kathryn A. Whaler

Core Origin 89
Francis Nimmo

Core Properties, Physical 91
Frank D. Stacey

Core Properties, Theoretical Determination 94
David Price

Core Temperature 98
David Price

Core Turbulence 101
Bruce Buffett and Hiroaki Matsui

Core Viscosity 104
Lidunka Vočadlo

Core, Adiabatic Gradient 106
Orson L. Anderson

Core, Boundary Layers 111
Emmanuel Dormy, Paul H. Roberts, and Andrew Soward

Core, Electrical Conductivity 116
Frank D. Stacey

Core, Magnetic Instabilities 117
David R. Fearn

Core, Thermal Conduction 120
Frank D. Stacey

Core-Based Inversions for the Main Geomagnetic Field 122
David Gubbins

Core-Mantle Boundary Topography, Implications for Dynamics 124
Andrew Soward

Core-Mantle Boundary Topography, Seismology 125
Andrea Morelli

Core-Mantle Boundary, Heat Flow Across 127
Stéphane Labrosse

Core-Mantle Coupling, Electromagnetic 130
Richard Holme

Core-Mantle Coupling, Thermal 132
Jeremy Bloxham

Core-Mantle Coupling, Topographic 135
Dominique Jault

Cowling, Thomas George (1906–1990) 137
Leon Mestel

Cowling's Theorem 138
Friedrich Busse

Cox, Allan V. (1926–1987) 139
Kenneth P. Kodama

Crustal Magnetic Field 140
Dhananjay Ravat

D″ and F-Layers 145
David Gubbins

D″ as a Boundary Layer 145
David Loper

D″, Anisotropy 146
Michael Kendall

D″, Composition 149
Quentin Williams

D″, Seismic Properties 151
Thorne Lay

Della Porta, Giambattista (1535–1615) 156
Allan Chapman

Demagnetization 156
Jaime Urrutia-Fucugauchi

Depth to Curie Temperature 157
Mita Rajaram

Dipole Moment Variation 159
Catherine Constable

Dynamo Waves 161
Graeme R. Sarson

Dynamo, Backus 163
Ashley P. Willis

Dynamo, Braginsky 164
Graeme R. Sarson

Dynamo, Bullard-Gellman 166
Graeme R. Sarson

Dynamo, Disk 167
Jack A. Jacobs

Dynamo, Gailitis 169
Agris Gailitis

Dynamo, Herzenberg 170
Paul H. Roberts

Dynamo, Lowes-Wilkinson 173
Frank Lowes

Dynamo, Model-Z 174
Rainer Hollerbach

Dynamo, Ponomarenko 175
Paul H. Roberts

Dynamo, Solar 178
Eugene N. Parker

Dynamos, Experimental 183
Andreas Tilgner

Dynamos, Fast 186
Michael Proctor

Dynamos, Kinematic 188
Philip W. Livermore

Dynamos, Mean-Field 192
Karl-Heinz Raedler

Dynamos, Periodic 200
David Gubbins

Dynamos, Planetary and Satellite 203
David J. Stevenson

Earth Structure, Major Divisions 208
Brian Kennett

Elsasser, Walter M. (1904–1991) 214
Eugene N. Parker

EM Modeling, Forward 215
Dmitry B. Avdeev

EM Modeling, Inverse 219
Gary D. Egbert

EM, Industrial Uses 223
Graham Heinson

EM, Lake-Bottom Measurements 227
Adam Schultz

EM, Land Uses 228
Louise Pellerin

EM, Marine Controlled Source 231
Nigel Edwards

EM, Regional Studies 242
Oliver Ritter

EM, Tectonic Interpretations 245
Malcolm Ingham

Environmental Magnetism 248
Barbara A. Maher

Environmental Magnetism, Paleomagnetic Applications 256
Andrew P. Roberts

Equilibration of Magnetic Field, Weak- and Strong-Field Dynamos 262
Keke Zhang

Euler Deconvolution 263
Alan B. Reid

First-Order Reversal Curve (FORC) Diagrams 266
Adrian R. Muxworthy and Andrew P. Roberts

Fisher Statistics 272
Jeffrey J. Love

Fleming, John Adam (1877–1956) 273
Shaun J. Hardy

Fluid Dynamics Experiments 274
Jonathan M. Aurnou

Galvanic Distortion 277
Karsten Bahr

Gauss' Determination of Absolute Intensity 278
Stuart R.C. Malin

Gauss, Carl Friedrich (1777–1855) 279
Karl-Heinz Glaßmeier

Gellibrand, Henry (1597–1636) 280
Stuart R.C. Malin

Geocentric Axial Dipole Hypothesis 281
Michael W. McElhinny

Geodynamo 287
Chris Jones

Geodynamo, Dimensional Analysis and Timescales 297
David Gubbins

Geodynamo, Energy Sources 300
Stéphane Labrosse

Geodynamo: Numerical Simulations 302
Gary A. Glatzmaier

Geodynamo, Symmetry Properties 306
David Gubbins

Geomagnetic Deep Sounding 307
Roger Banks

Geomagnetic Dipole Field 310
Frank Lowes

Geomagnetic Excursion 311
Jack A. Jacobs

Geomagnetic Field, Asymmetries 313
Phillip L. McFadden and Ronald T. Merrill

Geomagnetic Hazards 316
Alan W.P. Thomson

Geomagnetic Jerks 319
Susan Macmillan

Geomagnetic Polarity Reversals 320
Alain Mazaud

Geomagnetic Polarity Reversals, Observations 324
Bradford M. Clement

Geomagnetic Polarity Timescales 328
William Lowrie

Geomagnetic Pulsations 333
Karl-Heinz Glaßmeier

Geomagnetic Reversal Sequence, Statistical Structure 335
Phillip L. McFadden

Geomagnetic Reversals, Archives 339
Jean-Pierre Valet and Emilio Herrero-Bervera

Geomagnetic Secular Variation 346
Ingo Wardinski

Geomagnetic Spectrum, Spatial 350
Frank Lowes

Geomagnetic Spectrum, Temporal 353
Catherine Constable

Geomagnetism, History of 355
Art R.T. Jonkers

Gilbert, William (1544–1603) 360
Allan Chapman

Gravitational Torque 362
Jean-Louis Le Mouel

Gravity-Inertio Waves and Inertial Oscillations 364
Keith Aldridge

Grüneisen's Parameter for Iron and Earth's Core 366
Orson L. Anderson

Halley, Edmond (1656–1742) 375
Sir Alan Cook

Hansteen, Christopher (1784–1873) 376
Johannes M. Hansteen

Harmonics, Spherical 377
Denis Winch

Harmonics, Spherical Cap 395
G.V. Haines

Hartmann, Georg (1489–1564) 397
Allan Chapman

Helioseismology 398
Michael J. Thompson

Higgins-Kennedy Paradox 401
Friedrich Busse

Humboldt, Alexander Von (1759–1859) 402
Friedrich Busse

Humboldt, Alexander Von and Magnetic Storms 404
G.S. Lakhina, B.T. Tsurutani, W.D. Gonzalez, and S. Alex

IAGA, International Association of Geomagnetism and Aeronomy 407
David Kerridge

Ideal Solution Theory 408
Dario Alfè

IGRF, International Geomagnetic Reference Field 411
Susan Macmillan

Induction Arrows 412
Oliver Ritter

Induction from Satellite Data 413
Steven Constable

Inhomogeneous Boundary Conditions and the Dynamo 416
Keke Zhang

Inner Core Anisotropy 418
Xiaodong Song

Inner Core Composition 420
Lidunka Vočadlo

Inner Core Oscillation 422
Keith Aldridge

Inner Core Rotation 423
Paul G. Richards and Anyi Li

Inner Core Rotational Dynamics 425
Michael G. Rochester

Inner Core Seismic Velocities 427
Annie Souriau

Inner Core Tangent Cylinder 430
Rainer Hollerbach and David Gubbins

Inner Core, PKJKP 433
Hanneke Paulssen

Instrumentation, History of 434
Gregory A. Good

Interiors of Planets and Satellites 439
Gerald Schubert

Internal External Field Separation 448
Denis Winch

Ionosphere 452
Arthur D. Richmond

Iron Sulfides 454
Leonardo Sagnotti

Jesuits, Role in Geomagnetism 460
Agustín Udías

Kircher, Athanasius (1602–1680) 463
Oriol Cardus

Langel, Robert A. (1937–2000) 465
Michael E. Purucker

Laplace's Equation, Uniqueness of Solutions 466
David Gubbins

Larmor, Joseph (1857–1942) 468
David Gubbins

Lehmann, Inge (1888–1993) 468
David Gubbins

Length of Day Variations, Decadal 469
Richard Holme

Length of Day Variations, Long-Term 471
L.V. Morrison and F.R. Stephenson

Lloyd, Humphrey (1808–1881) 472
Deanis Weaire and J.M.D. Coey

Magnetic Anisotropy, Sedimentary Rocks and Strain Alteration 475
Peter D. Weiler

Magnetic Anomalies for Geology and Resources 477
Colin Reeves and Juha V. Korhonen

Magnetic Anomalies, Long Wavelength 481
Michael E. Purucker

Magnetic Anomalies, Marine 483
James R. Heirtzler

Magnetic Anomalies, Modeling 485
Jafar Arkani-Hamed

Magnetic Domains 490
Susan L. Halgedahl

Magnetic Field of Mars 502
Jafar Arkani-Hamed

Magnetic Field of Sun 505
Steven M. Tobias

Magnetic Indices 509
Jeffrey J. Love and K.J. Remick

Magnetic Mineralogy, Changes due to Heating 512
Bernard Henry

Magnetic Properties, Low-Temperature 515
Andrei Kosterov

Magnetic Proxy Parameters 525
Mark J. Dekkers

Magnetic Remanence, Anisotropy 535
Ann M. Hirt

Magnetic Shielding 540
Gary R. Scott

Magnetic Surveys, Marine 542
Maurice A. Tivey

Magnetic Susceptibility, Anisotropy 546
František Hrouda

Magnetic Susceptibility, Anisotropy, Effects of Heating 560
Jaime Urrutia-Fucugauchi

Magnetic Susceptibility, Anisotropy, Rock Fabric 564
Edgardo Cañón-Tapia

Magnetic Susceptibility (MS), Low-Field 566
Brooks B. Ellwood

Magnetization, Anhysteretic Remanent 572
Bruce M. Moskowitz

Magnetization, Chemical Remanent (CRM) 580
Shaul Levi

Magnetization, Depositional Remanent 588
Jaime Urrutia-Fucugauchi

Magnetization, Isothermal Remanent 589
Mike Jackson

Magnetization, Natural Remanent (NRM) 594
Mimi J. Hill

Magnetization, Oceanic Crust 596
Julie Carlut and Hélène Horen

Magnetization, Piezoremanence and Stress Demagnetization 599
Stuart Alan Gilder

Magnetization, Remanent, Ambient Temperature and Burial Depth from Dyke Contact Zones 603
Kenneth L. Buchan

Magnetization, Remanent, Fold Test 607
Jaime Urrutia-Fucugauchi

Magnetization, Thermoremanent 609
Özden Özdemir

Magnetization, Thermoremanent, in Minerals 616
Gunther Kletetschka

Magnetization, Viscous Remanent (VRM) 621
David J. Dunlop

Magnetoconvection 630
Keke Zhang and Xinhao Liao

Magnetohydrodynamic Waves 632
Christopher Finlay

Magnetohydrodynamics 639
Paul H. Roberts

Magnetometers, Laboratory 654
Wyn Williams

Magnetosphere of the Earth 656
Stanley W.H. Cowley

Magnetostratigraphy 664
William Lowrie

Magnetotellurics 670
Martyn Unsworth

Magsat 673
Michael E. Purucker

Main Field Maps 674
Mioara Mandea

Main Field Modeling 679
Mioara Mandea

Main Field, Ellipticity Correction 683
Stuart R.C. Malin

Mantle, Electrical Conductivity, Mineralogy 684
Tomoo Katsura

Mantle, Thermal Conductivity 688
Frank D. Stacey

Matuyama, Motonori (1884–1958) 689
Masaru Kono

Melting Temperature of Iron in the Core, Experimental 689
Guoyin Shen

Melting Temperature of Iron in the Core, Theory 692
David Price

Microwave Paleomagnetic Technique 694
John Shaw

Nagata, Takesi (1913–1991) 696
Masaru Kono

Natural Sources for Electromagnetic Induction Studies 696
Nils Olsen

Néel, Louis (1904–2000) 700
Pierre Rochette

Nondipole Field 701
Catherine Constable

Nondynamo Theories 704
David J. Stevenson

Norman, Robert (flourished 1560–1585) 707
Allan Chapman

Observatories, Overview 708
Susan Macmillan

Observatories, Instrumentation 711
Jean L. Rasson

Observatories, Automation 713
Lawrence R. Newitt

Observatories, Intermagnet 715
Jean L. Rasson

Observatories, Program in Australia 717
Peter A. Hopgood

Observatories, Program in the British Isles 720
David Kerridge

Observatory Program in France 721
Mioara Mandea

Observatories, Program in USA 722
Jeffrey J. Love and J.B. Townshend

Observatories in Antarctica 723
Jean-Jacques Schott and Jean L. Rasson

Observatories in Benelux Countries 725
Jean L. Rasson

Observatories in Canada 726
Lawrence R. Newitt and Richard Coles

Observatories in China 727
Dongmei Yang

Observatories in East and Central Europe 728
Pavel Hejda

Observatories in Germany 729
Hans-Joachim Linthe

Observatories in India 731
Gurbax S. Lakhina and S. Alex

Observatories in Italy 733
Massimo Chiappini

Observatories in Japan and Asia 733
Toshihiko Iyemori and Heather McCreadie

Observatories in Latin America 734
Luiz Muniz Barreto

Observatories in New Zealand and the South Pacific 735
Lester A. Tomlinson

Observatories in Nordic Countries 736
Truls Lynne Hansen

Observatories in Russia 737
Oleg Troshichev

Observatories in Southern Africa 739
Pieter Kotzé

Observatories in Spain 739
Miquel Torta and Josep Batlló

Ocean, Electromagnetic Effects 740
Stefan Maus

Oldham, Richard Dixon (1858–1936) 742
Johannes Schweitzer

Ørsted 743
Nils Olsen

Oscillations, Torsional 746
Mathieu Dumberry

Paleointensity: Absolute Determinations Using Single Plagioclase Crystals 749
John A. Tarduno, Rory D. Cottrell, and Alexei V. Smirnov

Paleointensity, Absolute, Techniques 753
Jean-Pierre Valet

Paleointensity, Relative, in Sediments 758
Stefanie Brachfeld

Paleomagnetic Field Collection Methods 765
Edgardo Cañón-Tapia

Paleomagnetic Secular Variation 766
Steve P. Lund

Palcomagnctism 776
Ronald T. Merrill and Phillip L. McFadden

Paleomagnetism, Deep-Sea Sediments 781
James E.T. Channell

Paleomagnetism, Extraterrestrial 788
Michael D. Fuller

Paleomagnetism, Orogenic Belts 801
John D.A. Piper

Parkinson, Wilfred Dudley 807
Ted Lilley

Peregrinus, Petrus (flourished 1269) 808
Allan Chapman

Periodic External Fields 809
Denis Winch

Plate Tectonics, China 816
Xixi Zhao and Robert S. Coe

Pogo (OGO-2, -4 and -6 Spacecraft) 828
Joseph C. Cain

Polarity Transition, Paleomagnetic Record 829
Kenneth A. Hoffman

Polarity Transitions: Radioisotopic Dating 834
Brad S. Singer

Pole, Key Paleomagnetic 839
Kenneth L. Buchan

Potential Vorticity and Potential Magnetic Field Theorems 840
Raymond Hide

Precession and Core Dynamics 842
Philippe Cardin

Price, Albert Thomas (1903–1978) 844
Bruce A. Hobbs

Principal Component Analysis in Paleomagnetism 845
Jeffrey J. Love

Project Magnet 850
David G. McMillan

Proudman-Taylor Theorem 852
Raymond Hide

Radioactive Isotopes, Their Decay in Mantle and Core 854
V. Rama Murthy

Reduction to Pole 856
Dhananjay Ravat

Repeat Stations 858
Susan Macmillan

Reversals, Theory 859
Graeme R. Sarson

Rikitake, Tsuneji (1921–2004) 862
Y. Honkura

Ring Current 863
Thomas E. Moore

Robust Electromagnetic Transfer Functions Estimates 866
Gary D. Egbert

Rock Magnetism 870
Ronald T. Merrill

Rock Magnetism, Hysteresis Measurements 874
David Krása and Karl Fabian

Rock Magnetometer, Superconducting 883
William S. Goree

Runcorn, S. Keith (1922–1995) 886
Neil Opdyke

Runcorn's Theorem 888
Andrew Jackson

Sabine, Edward (1788–1883) 890
David Gubbins

Seamount Magnetism 891
James R. Heirtzler and K.A. Nazarova

Secular Variation Model 892
Christopher G.A. Harrison

Sedi 902
David Loper

Seismic Phases 903
Brian Kennett

Seismo-Electromagnetic Effects 908
Malcolm J.S. Johnston

Shaw and Microwave Methods, Absolute Paleointensity Determination 910
John Shaw

Shock Wave Experiments 912
Thomas J. Ahrens

Spinner Magnetometer 920
Jiří Pokorný

Statistical Methods for Paleovector Analysis 922
Jeffrey J. Love

Storms and Substorms, Magnetic 926
Mark Lester

Superchrons, Changes in Reversal Frequency 928
Jack A. Jacobs

Susceptibility 931
Eduard Petrovsky

Susceptibility, Measurements of Solids 933
Z.S. Teweldemedhin, R.L. Fuller, and M. Greenblatt

Susceptibility, Parameters, Anisotropy 937
Edgardo Cañón-Tapia

Taylor's Condition 940
Rainer Hollerbach

Thellier, Émile (1904–1987) 942
David J. Dunlop

Thermal Wind 945
Peter Olson

Time-Averaged Paleomagnetic Field 947
David Gubbins

Time-Dependent Models of the Geomagnetic Field 948
Andrew Jackson

Transfer Functions 953
Martyn Unsworth

Transient Em Induction 954
Maxwell A. Meju and Mark E. Everett

True Polar Wander 956
Vincent Courtillot

ULVZ, Ultra-Low Velocity Zone 970
Ed J. Garnero and M. Thorne

Units 973
David Gubbins

Upward and Downward Continuation 974
Dhananjay Ravat

Variable Field Translation Balance 977
David Krása, Klaus Petersen, and Nikolai Petersen

Verhoogen, John (1912–1993) 979
Peter Olson

Vine-Matthews-Morley Hypothesis 980
Maurice A. Tivey

Volcano-Electromagnetic Effects 984
Malcolm J.S. Johnston

Voyages Making Geomagnetic Measurements 987
David R. Barraclough

Watkins, Norman David (1934–1977) 992
Brooks B. Ellwood

Westward Drift 993
Richard Holme

Color Plates 997

Subject Index 1013

Contributors

Thomas J. Ahrens
CALTECH
MS 252-21
Pasadena, CA 91125, USA
email: tja@caltech.edu

Keith Aldridge
Department of Earth & Atmospheric Sciences
York University
4700 Keele Street
Toronto, ON M3J 1P3, Canada
email: keith@yorku.ca

Sobhana Alex
Indian Institute of Geomagnetism
New Panvel (W)
Navi Mumbai-410 218, India
email: salex@iigs.iigm.res.in

Dario Alfè
Department of Earth Sciences
University College London
Gower Street
London, WC1E 6BT, UK
email: d.alfe@ucl.ac.uk

Orson L. Anderson
Institute of Geophysics and Planetary Physics
F83 Department of Earth and Space Sciences
University of California, Los Angeles, CA 90095, USA
email: ola@ess.ucla.edu

Jafar Arkani-Hamed
Earth & Planetary Sciences
McGill University
3450 University St
Montreal, QC H3A 2A7, Canada
email: jafar@eps.mcgill.ca

Jonathan M. Aurnou
Department of Earth and Space Sciences
University of California, Los Angeles
595 Charles Young Drive East
Los Angeles, CA 90095-1567, USA
email: jona@ess.ucla.edu/aurnou@ucla.edu

Dmitry B. Avdeev
School of Cosmic Physics
Dublin Institute for Advanced Studies
5 Merrion Square
Dublin 2, Ireland
email: davdeev@cp.dias.ie

Karsten Bahr
Geophysical Institute
Universität Göttingen
Herzberger Landstr. 180
37075 Göttingen, Germany
email: kbahr@uni-geophys.gwdg.de

Roger Banks
Fernwood, Rogerground
Hawkshead, Ambleside
Cumbria, LA22 0QG, UK
email: rbanks@beeb.net

David Barraclough
49 Liberton Drive
Edinburgh, EH16 6NL, UK
email: drbarraclough@hotmail.com

Luiz Muniz Barreto
Observatorio Nacional
Rua general Jose Cristino, 77
Rio de Janeiro, Brazil
email: barreto@on.br

Josep Batlló
Department Matematica Aplicada 1
Universitat Polytecnica de Catalunya
Spain
email: Josep.batllo@ups.es

Jeremy Bloxham
Department of Earth and Planetary Sciences
Harvard University
20 Oxford Street
Cambridge, MA 02138, USA
email: bloxham@geophysics.harvard.edu/jeremy_bloxham@harvard.edu

Stefanie Brachfeld
Department of Earth and Environmental Studies
Montclair State University
Montclair, NJ 07043, USA
email: brachfelds@mail.montclair.edu

Stanislav I. Braginsky
Institute of Geophysics and Planetary Physics
UCLA
405 Hilgard Avenue
Los Angeles, CA 90024-1567, USA
email: sbragins@ucla.edu

Kenneth L. Buchan
Geological Survey of Canada
Natural Resources Canada
601 Booth Street
Ottawa, Ontario, K1A 0E8, Canada
email: KBuchan@nrcan.gc.ca

Bruce Buffett
Department of Geophysical Sciences
University of Chicago
5734 S. Ellis Avenue
Chicago, IL 60637, USA
email: buffett@geosci.uchicago.edu

Friedrich Busse
Institute of Physics
University of Bayreuth
95440 Bayreuth, Germany
email: busse@uni-bayreuth.de

Joseph C. Cain
Department of Geology
Florida State University
Tallahassee, FL 32306-3026, USA
email: cain@geomag.gfdi.fsu.edu

Edgardo Cañón-Tapia
Department of Geology
CICESE
P.O. Box 434843
San Diego, CA 92143, USA
email: ecanon@cicese.mx

Philippe Cardin
Universite Joseph Fourier de Grenoble
Laboratoire de Geophysique interne et Tectonophysique
1381 Rue de la Piscine, BP 53
Grenoble Cedex 9, 38041, France
email: philippe.cardin@obs.ujf-grenoble.fr

Oriol Cardus
Observatori de l'Ebre
Roquetes
Tarragona, 43520, Spain
email: jocardus@obserbre.es

Julie Carlut
Laboratoire de Geologie
Ecole Normale Superieure
24 rue Lhomond
Paris, 75235, France
email: jcarlut@mailhost.geologie.ens.fr

James E.T. Channell
University of Florida
Department of Geological Sciences
P.O. Box 112120
Gainesville, FL 32611-2120, USA
email: jetc@nersp.nerdc.ufl.edu

Allan Chapman
Modern History Faculty Office
University of Oxford
Broad St.
Oxford, OX1 3BD, UK
email: rachel.chapman@classics.ox.ac.uk

Massimo Chiappini
Instituto Nazionale di Geofisica e Vulcanologia
Vigna Murata 605
Rome, 00143, Italy
email: massimo.chiappini@ingv.it

Bradford M. Clement
Florida International University
Department of Earth Science
SW 8th St & 107th Ave
Miami, FL 33199, USA
email: clementb@fiu.edu

Robert S. Coe
Institute of Geophysics and Planetary Physics
University of California Santa Cruz
1156 High Street
Santa Cruz, CA 95064, USA
email: rcoe@es.ucsc.edu

J. Michael D. Coey
Physics Dept.
Trinity College
College Green
Dublin 2, Ireland
email: jcoey@tcd.ie

Richard Coles
Geomagnetism Laboratory
Natural Resources Canada
7 Observatory Crescent
Ottawa, Ontario K1A0Y3, Canada

Catherine Constable
Institute of Geophysics and Planetary Physics
Scripps Institution of Oceanography
University of California at San Diego
La Jolla, CA 92093 0225, USA
email: cconstable@ucsd.edu

Steven Constable
Scripps Institution of Oceanography
La Jolla, CA 920930225, USA
email: sconstable@ucsd.edu

Sir Alan Cook (deceased)

Rory D. Cottrell
Department of Earth and Environmental Sciences
University of Rochester
Hutchison Hall 227
Rochester, NY 14627, USA
email: rory@earth.rochester.edu

Vincent Courtillot
Institut de Physique du Globe de Paris
4 place Jussieu
Paris Cedex 05, 75252, France
email: courtil@ipgp.jussieu.fr

Stanley W.H. Cowley
Department of Physics & Astronomy
University of Leicester
Leicester, LE1 7RH, UK
email: swhc1@ion.le.ac.uk

Mark J. Dekkers
Department of Earth Sciences
Utrecht University
Budapestlaan 17
Utrecht, 3584 CD, The Netherlands
email: dekkers@geo.uu.nl

Jon Dobson
Centre for Science & Technology in Medicine
Keele University
Thornburrow Drive
Hartshill, Stoke-on-Trent ST4 7QB, UK

Emmanuel Dormy
C.N.R.S./I.P.G.P./E.N.S.
Département de Physique
Ecole Normale Supérieure
24, rue Lhomond
75231 Paris Cedex 05, France
email: dormy@phys.ens.fr

Mathieu Dumberry
School of Earth and Environment
University of Leeds
Leeds, LS2 9JT, UK
email: dumberry@earth.leeds.ac.uk

David J. Dunlop
Department of Physics
University of Toronto
Mississauga, Ontario L5L 1C6, Canada
email: dunlop@physics.utoronto.ca

Nigel Edwards
Department of Physics
University of Toronto
60 St George Street
Toronto, Ontario M5S 1A7, Canada
email: edwards@core.physics.utoronto.ca

Gary D. Egbert
College Oceanography
Oregon State University
Oceanography Admin Bldg 104
Corvallis, OR 97331-5503, USA
email: egbert@coas.oregonstate.edu

Brooks B. Ellwood
Department of Geology and Geophysics
Louisiana State University
Baton Rouge, LA 70803, USA
email: ellwood@geol.lsu.edu

Mark E. Everett
Department of Geology & Geophysics
Texas A & M University
College Station, TX 77843-3114, USA
email: everett@geo.tamu.edu/colt45@beerfrdg.tamu.edu

Karl Fabian
Department of Earth and Environmental Sciences
University of Munich
Theresienstr. 41
80333 München, Germany
email: karl.fabian@geophysik.uni-muenchen.de

Carl-Gunne Fälthammar
Dept. of Plasma Physics
Royal Institute of Technology
Stockholm, SE-10044, Sweden
email: carl-gunne.falthammar@alfvenlab.kth.se

David R. Fearn
Department of Mathematics
University of Glasgow
Glasgow G12 8QW, UK
email: D.Fearn@maths.gla.ac.uk

Christopher Finlay
ETH-Hönggerberg
Institute of Geophysics
Schaftmattstrasse 30
CH-8093 Zürich, Switzerland
email: efinlay@erdw.ethz.ch

Michael D. Fuller
Paleomagnetics and Petrofabrics Laboratory
1680 East West Rd
Honolulu, Hawaii, 96822, USA
email: MFU3961215@aol.com

Robert L. Fuller
909 River Rd.
Colgate Palmolive co.
Piscataway, NJ 08854, USA

Agris Gailitis
Institute of Physics
University of Latvia
Miera iela 32
Salaspils, LV 2169, Latvia
email: gailitis@sal.lv

Edward J. Garnero
Dept Geological Sciences
Arizona State University
Box 871404
Tempe, AZ 85287-1404, USA
email: eddie@seismo.berkeley.edu

Stuart Alan Gilder
Geophysics Section
Ludwig Maximillians University
Theresienstrasse 41
80333 München, Germany
email: gilder@geophysik.uni-muenchen.de

Karl-Heinz Glaßmeier
Institute of Geophysics and Extraterrestrial Physics
Technical University of Braunschweig
Mendelssohnstr. 3
38106 Braunschweig, Germany
email: kh.glassmeier@tu-bs.de

Gary A. Glatzmaier
Department of Earth Sciences
University of California
Santa Cruz, CA 95064, USA
email: glatz@es.ucsc.edu

Walter Demétrio Gonzalez
INPE-Caixa Postal 515
2200 Sao Jose Dos Campos
Sao Paulo, Brazil

Gregory A. Good
History Department
West Virginia University
Morgantown, WV 26506-6303, USA
email: Greg.good@mail.wvu.edu

William S. Goree
Inc. and 2 G Enterprises
2040 Sunset Drive
Pacific Grove, CA 93950, USA
email: billgoree@earthlink.net
wgoree@2gsuper.com

Martha Greenblatt
Department of Chemistry and Chemical Biology
Rutgers University
610 Taylor Road
Piscataway, NJ 08854-8087, USA
email: greenbla@rci.rutgers.edu

David Gubbins
School of Earth and Environment
University of Leeds
Leeds LS2 9JT, UK
email: gubbins@earth.leeds.ac.uk

G.V. Haines
69 Amberwood Cr
Ottawa, ON K2E 7C2, Canada
email: haines@geolab.emr.ca
er309@ncf.ca

Susan L. Halgedahl
Department of Geology and Geophysics
University of Utah
Salt Lake City, Utah 84112, USA
email: shalg@mail.mines.utah.edu

Truls Lynne Hansen
Tromso Geophysical Laboratory
University of Tromso
Tromso, N-9037, Norway
email: truls.hansen@tgo.uit.no

Johannes M. Hansteen (deceased)

Shaun J. Hardy
Carnegie Institution of Washington
5241 Broad Branch Rd., N.W.
Washington, DC 20015, USA
email: hardy@dtm.ciw.edu

Christopher G.A. Harrison
Rosenstiel School of Marine and Atmospheric Science
University of Miami
4600 Rickenbacker Causeway
Miami, FL 33149, USA
email: harrison@mail.rsmas.miami.edu

Graham Heinson
Department of Geology and Geophysics
Adelaide University
Adelaide, SA 5005, Australia
email: gheinson@geology.adelaide.edu.au
Graham.Heinson@adelaide.edu.au

James R. Heirtzler
NASA, Goddard Space Flight Center
MC 920
Greenbelt, MD 20771-0001, USA
email: jamesh@ltpmail.gsfc.nasa.gov

Pavel Hejda
Geophysical Institute
Academy of Sciences of the Czech Republic
Bocni II/1401
Prague 4, 141 31, Czech Republic
email: ph@ig.cas.cz

Bernard Henry
Géomagnétisme et Paléomagnétisme
IPGP and CNRS
4 avenue de Neptune
Saint-Maur Cedex, 94107, France
email: henry@ipgp.jussieu.fr

Emilio Herrero-Bervera
Paleomagnetics and Petrofabrics Laboratory
University of Hawaii at Manoa
Honolulu, Hawaii 96822, USA
email: herrero@soest.hawaii.edu

Raymond Hide
17 Clinton Ave
East Molesey, Surrey, KT8 0HS, UK
email: r.hide@ic.ac.uk

Mimi J. Hill
Department of Earth and Ocean Sciences
University of Liverpool
Oxford Street
Liverpool, L69 7ZE, UK
email: mimi@liv.ac.uk
m.hill@liverpool.ac.uk

Ann M. Hirt
ETH-Hönggerberg
Institute of Geophysics
Zürich, 8093, Switzerland
email: hirt@mag.ig.erdw.ethz.ch

Bruce A. Hobbs
Department of Geology and Geophysics
University of Edinburgh
West Main Road
Edinburgh, EH9 3JW, UK
email: b.a.hobbs@ed.ac.uk

Kenneth A. Hoffman
Physics Department
California Polytechnic State University
San Luis Obispo, CA 93407, USA
email: khoffman@calpoly.edu

Rainer Hollerbach
School of Mathematics
University of Leeds
Leeds, LS2 9JT, UK
email: rh@maths.leeds.ac.uk

Richard Holme
Department of Earth and Ocean Sciences
University of Liverpool
4 Brownlow Street
Liverpool, L69 3GP, UK
email: holme@liv.ac.uk

Yoshimori Honkura
Tokyo Institute of Technology
Department of Earth & Planetary Sciences
2-12-1-I2-6 Ookayama
Meguro-ku, Tokyo, 152-8551, Japan
email: yhonkura@geo.titech.ac.jp

Peter A. Hopgood
Geoscience Australia
GPO Box 378
Canberra, ACT, 2601, Australia
email: peter.hopgood@ga.gov.au

Hélène Horen
Laboratoire de Geologie
Ecole Normale Superieure
24 rue Lhomond, Paris
75235, France
email: horen@geologie.ens.fr

František Hrouda
AGICO, Inc.
Advanced Geoscience Instruments Company
Jecna 29a
Brno, CZ 621 00, Czech Republic
email: fhrouda@agico.cz

Malcolm Ingham
Department of Physics
University of Victoria
PO Box 600
Wellington, New Zealand
email: malcolm.ingham@vuw.ac.nz

Toshihiko Iyemori
Graduate School Science
Kyoto University
Data Analysis Center
Kyoto, 606-8502, Japan
email: iyemori@kugi.kyoto-u.ac.jp

Andrew Jackson
School of Earth and Environment
University of Leeds
Leeds LS2 9JT, UK
email: a.jackson@earth.leeds.ac.uk

Mike Jackson
Department of Geology and Geophysics
Institute for Rock Magnetism
University of Minnesota
100 Union Street SE
Minneapolis, MN 55455, USA
email: irm@umn.edu

Jack A. Jacobs (deceased)

Dominique Jault
LGIT, CNRS and University Joseph-Fourier
BP 53
38041Grenoble Cedex9, France
email: Dominique.Jault@obs.ujf-grenoble.f.

Malcolm J.S. Johnston
USGS
345 Middlefield Rd MS 977
Menlo Park, CA 94025, USA
email: mal@usgs.gov

Chris Jones
Department of Applied Mathematics
University of Leeds
Leeds LS2 9JT, UK
email: c.a.jones@maths.leeds.ac.uk

Art R.T. Jonkers
Department of Earth and Ocean Sciences
The Jane Herdman Laboratories
University of Liverpool
4 Brownlow St
Liverpool, L69 3GP, UK
email: jonkers@liv.ac.uk

Tomoo Katsura
Institute for Study of the Earth's Interior
Okayama University
Misasa, Tottori-ken, 682-0193, Japan
email: tkatsura@misasa.okayama-u.ac.jp

Michael Kendall
Dept of Earth Sciences
University of Bristol
Queen's Rd.
Bristol, BS8 1RJ, UK
email: gljmk@bristol.ac.uk

Brian Kennett
Research School of Earth Sciences
Australian National University
GPO Box 4
Canberra, ACT 0200, Australia
email: brian@rses.anu.edu.au

David Kerridge
Geomagnetism Group
British Geological Survey West Mains Road
Edinburgh, EH9 3LA, UK
email: djk@bgs.ac.uk

Gunther Kletetschka
NASA Goddard Space Flight Center
Greenbelt, Maryland, 20771, USA
email: gkletets@pop600.gsfc.nasa.gov

Kenneth P. Kodama
Department of Earth and Environmental Sciences
Lehigh University
31 Williams Drive, Bethlehem, PA 18015-3188, USA
email: kpk0@lehigh.edu

Masaru Kono
Institute for Study of the Earth's Interior
Okayama, University of Misasa
Yamada 827
Misasa, Tottori Prefecture, 682 0193, Japan
email: mkono@misasa.okayama-u.ac.jp

Juha V. Korhonen
Geological Survey of Finland
POB 96
Espo, 02151, Finland
email: juha.korhonen@gtk.fi

Andrei Kosterov
Nikolaeva 5-56, Kiev, Ukraine
email: andrei_kosterov@mail.ru/irm@umn.edu

Pieter Kotzé
Geomagnetism Group
Hermanus Magnetic Observatory
PO Box 32, Hermanus, 7200, South Africa
email: pkotze@hmo.ac.za

David Krása
School of GeoSciences
University of Edinburgh
King's Buildings
Edinburgh, EH9 3JW, UK
email: david.krasa@ed.ac.uk

Stéphane Labrosse
Departement des Geomateriaux
Institut de Physique du Globe de Paris
4 place Jussieu
Paris Cedex 05, 75252, France
email: labrosse@ipgp.jussieu.fr

Gurbax S. Lakhina
Indian Institute of Geomagnetism
New Panvel, Navi Mumbai, 410218, India
email: lakhina@iigs.iigm.res.in

Thorne Lay
Earth Sciences Dept.
University California Santa Cruz
Santa Cruz, CA 95064-1077, USA
email: tlay@es.ucsc.edu/thorne@pmc.ucsc.edu

Jean-Louis Le Mouel
Institut de Physique du Globe de Paris
4 place Jussieu
Paris Cedex 05, 75252, France
email: lemouel@ipgp.jussieu.fr

Mark Lester
Dept. of Physics and Astronomy
University of Leicester
Leicester, LE1 7RH, UK
email: mle@ion.le.ac.uk

Shaul Levi
College of Oceanic and Atmospheric Sciences
Oregon State University
Corvallis, OR 97331-5503, USA
email: slevi@coas.oregonstate.edu

Anyi Li
Lamont-Doherty Earth Observatory
and Department of Earth and Environmental Sciences
Columbia University
Palisades, NY 10964, USA
email: anyili@ldeo.columbia.edu

Xinhao Liao
Shanghai Astronomical Observatory
80 Nandan Road
Shanghai, 200030, China

Ted Lilley
Research School of Earth Sciences
Australian National University
GPO Box 4
Canberra, ACT 0200, Australia
email: ted.lilley@anu.edu.au

Hans-Joachim Linthe
Geomagnetic Adolf Schmidt Observatory Niemegk
Geoforschungszentrum Potsdam
Lindenstr. 7
14823 Niemegk, Germany
email: linthe@gfz-potsdam.de

Philip W. Livermore
School of Mathematics
University of Leeds
Leeds, LS2 9JT, UK
email: livermor@maths.leeds.ac.uk

David Loper
Florida State University
GFDI-4360
Tallahassee, FL 32306-0000, USA
email: loper@gfdi.fsu.edu

Jeffrey J. Love
USGS Golden
Box 25045 MS966 DFC
Denver, CO 80227, USA
email: jlove@usgs.gov

Frank Lowes
Department of Physics
The University of Newcastle-upon-Tyne
Newcastle upon Tyne, NE1 7RU, UK
email: f.j.lowes@ncl.ac.uk

William Lowrie
Institute of Geophysics
ETH-Hönggerberg
CH-8093 Zürich, Switzerland
email: lowrie@mag.ig.erdw.ethz.ch

Steve P. Lund
Department of Earth Sciences
University of Southern California
Los Angeles, CA 90089-0740, USA
email: slund@usc.edu

Susan Macmillan
British Geological Survey
West Mains Road
Edinburgh, EH9 3LA, UK
email: smac@bgs.ac.uk

Barbara A. Maher
Centre for Environmental Magnetism and Palaeomagnetism
Lancaster University
Lancaster, LA1 4YB, UK
email: b.maher@lancs.ac.uk

Stuart R.C. Malin
30 Wemyss Road
Blackheath
London, SE3 0TG, UK
email: smalin@dialstart.net

Mioara Mandea
GeoForschungZentrum
Telegrafenberg
14473 Potsdam, F269, Germany
email: mioara@gfz-potsdam.de

Guy Masters
IGPP Scripps Institute of Oceanography
University of California, San Diego
9500 Gilman Drive
La Jolla, CA 92093-0225, USA
email: gmasters@ucsd.edu

Hiroaki Matsui
Dept. Geophysical Sciences
University of Chicago
5734 S. Ellis Ave.
Chicago, IL 60637, USA
email: matsui@geosci.uchicago.edu

Stefan Maus
National Geophysical Data Center
NOAA E/GC1
325 Broadway
Boulder, CO 80305-3328, USA
email: Stefan.Maus@noaa.gov

Alain Mazaud
Laboratoire des Sciences du Climat et de l'Environnement (LSCE)
CEA-CNRS
Avenue de la Terrasse
Gif-sur-Yvette Cedex, 91198, France
email: mazaud@lsce.cnrs-gif.fr

Heather McCreadie
World Data Centre for Geomagnetism
Kyoto University
Kyoto, 606-8502, Japan

William F. McDonough
Department of Geology
University of Maryland
College Park, Maryland, 20742, USA
email: mcdonoug@geol.umd.edu

Michael W. McElhinny
Gondwana Consultants
31 Laguna Place
Port Macquarie, NSW 2444, Australia
email: mikemce@midcoast.com.au
mmcelhinny@optusnet.com.au

Phillip L. McFadden
Geoscience Australia
GPO Box 378
Canberra, ACT 2601, Australia
email: pmcfadde@pcug.org.au

David G. McMillan
Department of Earth & Space Science and Engineering
York University
Toronto, Ontario, Canada
email: dgm@yorku.ca

Maxwell A. Meju
Dept Environmental Sci.
Lancaster University
University Rd.
Bailrigg, LA1 4YQ, UK
email: m.meju@lancaster.ac.uk

Ronald T. Merrill
Geophysics Program, AK50
University of Washington
P.O. Box 433934
Seattle, Washington 98195, USA
email: ron@geops.geophys.washington.edu

Leon Mestel
Astronomy Centre
University of Sussex
Brighton BN1 9QH, UK
email: lmestel@sussex.ac.uk

Stephen Milan
Radio and Space Plasma Physics Group
Department of Physics and Astronomy
University of Leicester
Leicester, LE1 7RH, UK
email: steve.milan@ion.le.ac.uk

Thomas E. Moore
Laboratory for Solar and Space Physics
Mail Code 612
Greenbelt, MD 20771, USA
email: thomas.e.moore@nasa.gov

Andrea Morelli
Istituto Nazionale di Geofisica e Vulcanologia
Via Donato Creti 12
40128 Bologna, Italy
email: morelli@ingv.it

Leslie V. Morrison
28 Pevensley Park Road
Westham, Pevensley
East Sussex, BN24 5HW, UK
email: lmorr49062@aol.com

Bruce M. Moskowitz
Department of Geology and Geophysics
University of Minnesota
310 Pillsbury Dr. SE
Minneapolis, MN 55455, USA
email: bmosk@umn.edu
bmosk@bmosk.email.umn.edu

V. Rama Murthy
Department of Geology and Geophysics
University of Minnesota
310 Pillsbury Drive SE
Minneapolis, MN55455, USA
email: vrmurthy@umn.edu

Adrian R. Muxworthy
National Oceanography Centre
School of Ocean and Earth Science
University of Southampton
Southampton, SO14 3ZH, UK
email: adrian.muxworthy@soton.ac.uk
arob@soc.soton.ac.uk

Katherine A. Nazarova
ITSS/NASA Goddard Space Flight Center
Greenbelt, MD 20771, USA
email: katianh@core2.gsfc.nasa.gov

Lawrence R. Newitt
1 Observatory Crescent
Geological Survey Canada
Ottawa, ON K1A 0Y3, Canada
email: newitt@geoLAB.nrcan.gc.ca

Francis Nimmo
Dept. Earth Sciences
University of California
Santa Cruz, CA 95064, USA
email: fnimmo@es.ucsc.edu

Nils Olsen
Danish National Space Center
Juliane Maries Vej
Copenhagen, 2100, Denmark
email: nio@spacecenter.dk

Peter Olson
Earth & Planetary Sciences
The Johns Hopkins University
Baltimore, MD 21218-2681, USA
email: olson@jhu.edu

Neil Opdyke
Department of Geology
University of Florida
1112 Turlington Hall
Gainsville, FL 32611, USA
email: drno@ufl.edu

Özden Özdemir
Department of Physics
University of Toronto
Mississauga, Ontario L5L 1C6, Canada
email: ozdemir@physics.utoronto.ca

Eugene N. Parker
1323 Evergreen Rd
Homewood, IL 60430, USA
email: parker@odysseus.uchicago.edu

Hanneke Paulssen
Universiteit Utrecht
Institute of Earth Sciences
P O Box 80021
Utrecht, 3508 TA, The Netherlands
email: paulssen@geo.uu.nl

Louise Pellerin
Green Engineering, Inc.
6543 Brayton Drive
Anchorage, AK 99507, USA
email: pellerin@ak.net

Klaus Petersen
Petersen Instruments
Torstr. 173
10115 Berlin, Germany
email: petersen@vftb.com

Nikolai Petersen
Department of Earth and Environmental Sciences
University of Munich
Theresienstr. 41
80333 München, Germany
email: petersen@geophysik.uni-muenchen.de

Eduard Petrovsky
Geophysical Institute
Bocni II/1401
Prague 4, 141 31, Czech Republic
email: edp@ig.cas.cz

Mark Pilkington
Geological Survey of Canada
615 Booth Street
Ottawa, ON, Canada, K1A 0E9
email: mpilking@nrcan.gc.ca

John D.A. Piper
Department of Earth Sciences
Geomagnetism Laboratory
University of Liverpool
Liverpool, L69 7ZEm, UK
email: sg04@liverpool.ac.uk

Jiří Pokorný
AGICO, Inc.
Advanced Geoscience Instruments Company
Jecna 29a
Brno, CZ 621 00, Czech Republic
email: agico@agico.cz

David Price
Department of Geological Science
University College London
Gower Street
London, WC1E 6BT, UK
email: d.price@ucl.ac.uk

Michael Proctor
University of Cambridge
D.A.M.T.P, F1.07 CMS
Wilberforce Rd.
Cambridge, CB3 0WA, UK
email: M.R.E.Proctor@damtp.cam.ac.uk
mrep@damtp.cam.ac.uk

Michael E. Purucker
Goddard Space Flight Centre
Geodynamics Branch
Hughes-STX
Greenbelt, MD 20771, USA
email: purucker@geomag.gsfc.nasa.gov

Karl-Heinz Raedler
Astrophysikalisches Institut Potsdam
Andersternwarte 16
14482 Potsdam, Germany
email: khraedler@aip.de
khraedler@arcor.de

Mita Rajaram
Indian Institute of Geomagnetism
New Panvel, Navi Mumbai, 410218, India
email: mita@iigs.iigm.res.in

Jean L. Rasson
Centre de Physique du Globe
Institut Royal Météorologique
Dourbes, 5670, Belgium
email: jr@oma.be

Dhananjay Ravat
Department of Geology 4324
Southern Illinois University Carbondale
Carbondale, IL 62901-4324, USA
email: ravat@geo.siu.edu
ravat@siu.edu

Colin Reeves
Earthworks
Achterom 41a
Delft, 2611PL, The Netherlands
email: reeves.earth@planet.nl

Alan B. Reid
49 Carr Bridge Drive
Leeds, LS16 7LB, UK
email: alan@reid-geophys.co.uk

Karen J. Remick
USGS Golden
Box 25045 MS966 DFC
Denver, CO 80227, USA
email: kremick@usgs.gov

Paul G. Richards
Lamont-Doherty Earth Observatory
Columbia University
61 Route 9W
Palisades, NY 10964-1000, USA
email: richards@ldeo.columbia.edu

Arthur D. Richmond
NCAR
High Altitude Observatory
POB 3000
Boulder, CO 80307-3000, USA
email: richmond@hao.ucar.edu

Oliver Ritter
GeoForschungsZentrum
Telegrafenberg A45
14473 Potsdam, Germany
email: oritter@gfz-potsdam.de

Andrew P. Roberts
National Oceanography Centre
School of Ocean and Earth Science
University of Southampton
Southampton, SO14 3ZH, UK
email: arob@noc.soton.ac.uk
arob@soc.soton.ac.uk

Paul H. Roberts
Institute of Geophysics and Planetary Physics
UCLA
405 Hilgard Avenue
Los Angeles, CA 90095, USA
email: roberts@math.ucla.edu

Michael G. Rochester
Dept of Earth Sciences
Memorial University of Newfoundland
St. John's, N.L., A1B 3X5, Canada
email: mrochest@mun.ca

Pierre Rochette
CNRS-Université d'Aix-Marseille 3
CEREGE BP80 Europole de l'Arbois
Aix en Provence Cedex 4, 13545, France
email: rochette@cerege.fr

Leonardo Sagnotti
Istituto Nazionale di Geofisica e Vulcanologia
Via di Vigna Murata 605
Roma, 00143, Italy
email: sagnotti@ingv.it

Graeme R. Sarson
School of Mathematics and Statistics
University of Newcastle
Newcastle upon Tyne, NE1 7RU, UK
email: g.r.sarson@ncl.ac.uk

Armin Schmidt
Department of Archaeological Sciences
University of Bradford
BD7 1DP, UK
email: A.Schmidt@Bradford.ac.uk

Jean-Jacques Schott
Ecole et Observatoire des Sciences de la Terre
5, rue Descartes
Strasbourg Cedex, 67084, France
email: jeanJacques.Schott@eost.u-strasbg.fr

Gerald Schubert
Department of Earth & Space Sciences
University of California
2707 Geology Building
Los Angeles, CA 90024-1567, USA
email: gschubert@geovax.ess.ucla.edu
schubert@ucla.edu

Adam Schultz
College of Oceanic and Atmospheric Sciences
Oregon State University
Corvallis, OR 97331-5503, USA
email: adam@coas.oregonstate.edu

Johannes Schweitzer
NORSAR
Instituttveien 25
POB 53
Kjeller, 2027, Norway
email: johannes.schweitzer@norsar.no

Gary R. Scott
Berkeley Geochronology Center
2455 Ridge Road
Berkeley, CA 95709, USA
email: gscott@bgc.org

John Shaw
Department of Earth Sciences
University of Liverpool
Oxford Street, P.O. Box 147
Liverpool, L69 3BX, UK
email: shaw@liverpool.ac.uk

Guoyin Shen
Center for Advanced Radiation Sources
University of Chicago
Chicago, IL 60439, USA
email: shen@cars.uchicago.edu

Manfred Siebert
Institute of Geophysics
University of Göttingen
Friedrich-Hund-Platz 1
37077 Göttingen, Germany

Brad S. Singer
Department of Geology and Geophysics
University of Wisconsin-Madison
1215 West Dayton Street
Madison, WI 53706, USA
email: bsinger@geology.wisc.edu

Alexei V. Smirnov
Department of Earth and Environmental Sciences
University of Rochester
Hutchison Hall 227
Rochester, NY 14627, USA
email: alexei@earth.rochester.edu

Xiaodong Song
Dept. of Geology
University of Illinois
1301 W. Green St. 245NHB
Urbana, IL 61801, USA
email: xsong@uiuc.edu

Annie Souriau
CNRS
Observatoire Midi-Pyrenees
14 Ave. Edouard Belin
Toulouse, 31400, France
email: annie.souriau@cnes.fr

Andrew Soward
School of Mathematical Sciences
University of Exeter
Exeter, EX4 4QE, UK
email: A.M.Soward@exeter.ac.uk

Frank D. Stacey
CSIRO Exploration and Mining
PO Box 883
Kenmore, Queensland 4069, Australia
email: Frank.Stacey@csiro.au

F. Richard Stephenson
Dept. of Physics
University of Durham
South Road
Durham, DH1 3LE, UK
email: frstephenson@durham.ac.uk

David J. Stevenson
CALTECH
Div Geology & Planetary Sci, 150-21
Pasadena, CA 91125, USA
email: djs@gps.caltech.edu

John A. Tarduno
Department of Earth and Environmental Sciences
University of Rochester
Hutchison Hall 227
Rochester, NY 14627, USA
email: john@earth.rochester.edu
john@volterra.earth.rochester.edu

Donald D. Tarling
Department of Geological Sciences
Plymouth Polytechnic
Drake Circus
Plymouth, Devon, PL4 8AA, UK
email: d.tarling@plymouth.ac.uk

Patrick T. Taylor
Laboratory for Planetary Geodynamics
NASA/Goddard Space Flight Center
Greenbelt, MD 20771, USA
email: patrick.taylor@nasa.gov

Z.S. Teweldemedhin (no address)

Michael J. Thompson
Dept. of Applied Mathematics
University of Sheffield
Sheffield, S3 7RH, UK
email: michael.thompson@sheffield.ac.uk

Alan W.P. Thomson
British Geological Survey
West Mains Road
Edinburgh, EH9 3LA, UK
email: a.thomson@bgs.ac.uk

Michael Thorne
Department of Geological Sciences
Arizona State University
Tempe, AZ 85287-1404, USA

Andreas Tilgner
Institute of Geophysics
University of Gottingen
Herzberger Landstr. 180
37075 Göttingen, Germany
email: andreas.tilgner@geo.physik.uni-goettingen.de

Maurice A. Tivey
Dept Geology & Geophysics
WHOI
360 Woods Hole Rd
Woods Hole, MA 02543-1542, USA
email: mtivey@whoi.edu

Steven M. Tobias
Department of Applied Mathematics
University of Leeds
Leeds, LS2 9JT, UK
email: s.m.tobias@leeds.ac.uk

Lester A. Tomlinson
Geoscience, Electronics & Data Services
30 Kirner St.
Christchurch, 8009, New Zealand
email: geoserve@xtra.co.nz

Miquel Torta
Observatori de l'Ebre
Roquetes (Tarragona), 43520, Spain
email: ebre.jmtorta@readysoft.es

John B. Townshend
USGS
Box 25046 MS 966
Golden, CO 80225, USA

Oleg Troshichev
Arctic and Antarctic Research Institute
38 Bering St.
St. Petersburg, 199397, RUSSIA
email: olegtro@aari.nw.ru

Bruce T. Tsurutani
Jet Propulsion Laboratory
California Institute of Technology
4800 Oak Grove Drive
Pasadena, CA 91009, USA
email: BTSURUTANI@jplsp.jpl.nasa.gov

Agustín Udías
Facultad de Ciencias Físicas
Departamento de Geofísica
Universidad Complutense
Ciudad Universitaria
Madrid, 28040, Spain
email: audiasva@fis.ucm.es

Martyn Unsworth
University of Alberta
Edmonton, Alberta, T6G 2J1, Canada
email: unsworth@phys.ualberta.ca

Jaime Urrutia-Fucugauchi
Instituto de Geofisica, Laboratorio de Paleomagnetismo y Paleoambiente
Universidad Nacional Autonoma de Mexico
Mexico D.F., 04510, Mexico
email: juf@geofisica.unam.mx

Jean-Pierre Valet
Institut de Physique du Globe de Paris
4 place Jussieu
Paris Cedex 05, 75252, France
email: valet@ipgp.jussieu.fr

Lidunka Vočadlo
Dept. Earth Sciences
University College London
Gower St.
London, WC1E 6BT, UK
email: l.vocadlo@ucl.ac.uk

Ingo Wardinski
GeoForschungsZentrum Potsdam
Sektion 2.3 Geomagnetische Felder
Telegrafenberg
14473 Potsdam, Germany
email: ingo@gfz-potsdam.de

Deanis Weaire
Department of Physics
Trinity College
College Green
Dublin 2, Ireland
email: dweaire@tcd.ie

Peter D. Weiler
Baseline Environmental Consulting
5900 Hollis Street
Emeryville, CA 94608, USA
email: peter.weiler@lfr.com

Kathryn A. Whaler
Grant Institute of Earth Science
The University of Edinburgh
West Mains Road
Edinburgh, EH9 3JW, UK
email: kathy.whaler@ed.ac.uk

Quentin Williams
Earth Sciences
UCSC
Santa Cruz, CA 95064, USA
email: qwilliams@es.ucsc.edu

Wyn Williams
Department of Geology & Geophysics
University of Edinburgh
West Mains Road
Edinburgh, EH9 3JW, UK
email: wyn.williams@ed.ac.uk

Ashley P. Willis
Dept of Mathematics
University of Bristol
Bristol, BS8 1TW, UK
email: a.willis@bristol.ac.uk

Denis Winch
School of Mathematics & Statistics F07
University of Sydney
Sydney, NSW 2006, Australia
email: winch-d@maths.su.oz.au
denisw@maths.usyd.edu.au

Dongmei Yang
Institute of Geophysics
China Earthquake Administration, No. 5
Minzudaxuenanlu
Haidan District, Beijing, 100081, China
email: ydmgeomag@263.net

Keke Zhang
Department of Mathematical Sciences
University of Exeter
Exeter, EX4 4QE, UK
email: kzhang@ex.ac.uk

Xixi Zhao
Institute of Geophysics and Planetary Physics
University of California Santa Cruz
1156 High Street
Santa Cruz, CA 95064, USA
email: xzhao@emerald.ucsc.edu
xzhao@es.ucsc.edu

Preface

Geomagnetism is the study of the earth's magnetic field: its measurement, variation in time and space, origins, and its use in helping us to understand more about our Earth. Paleomagnetism is the study of the record left in the rocks; it has contributed much to our understanding of the geomagnetic field's past behavior and many other aspects of geology and earth history. Both have applications, pure and applied: in navigation, in the search for minerals and hydrocarbons, in dating rock sequences, and in unraveling past geologic movements such as plate motions. The entire subject is a small subdiscipline of earth science, and our goal has been to cover it in fine detail at a level accessible to anyone with a general scientific education. We envisage the encyclopedia to be of greatest use to those starting in the subject and those needing to know something of the field for their own application, but the topic is broad and demanding—as we have become increasingly aware while editing the huge variety of contributions—and we also expect it to be of use to experts in geomagnetism or paleomagnetism who need to stray outside their own area of expertise, for nobody is an expert in the whole field.

The scope of the encyclopedia is defined by the "GP" section of the American Geophysical Union: the magnetic field of internal origin. Over 25% of the membership of GP has contributed to this book. External sources of magnetic field are included insofar as they are used in solid earth geomagnetism—for example periodic external fields, because they induce electric currents in the earth that are useful in mapping deep electrically conducting regions—and articles are included on the ionosphere, magnetosphere, Sun, and planets. External geomagnetism as such is a separate discipline in most research establishments as well as the AGU, and is therefore not treated.

Geomagnetism is the oldest earth science, having its origins in simple human curiosity in the lodestone's ability to point north. It claims what most believe to be the first scientific treatise, *William Gillbert*'s (*q.v.*) De Magnete published in 1601, the claim being founded on its use of deduction from experimental measurement. These innocent beginnings were soon to give way to the intensely practical business of finding one's way at sea, and during the European age of discovery understanding the geomagnetic field and using it for navigation became a burning challenge for early scientists. A century after the publication of De Magnete saw *Edmond Halley* (*q.v.*) in charge of a Royal Navy vessel making measurements throughout the Atlantic Ocean. Halley's plans to fix position more accurately by using the departures of magnetic north from geographic north were dashed by the geomagnetic field's rapid changes in time, and the longitude problem was of course finally solved by Harrison and his accurate clock, but the compass remains an essential aid to navigation to this day.

Almost a century after Halley's voyages James Cook was making even more accurate measurements throughout the oceans, and in the 19th century, *Alexander von Humboldt* (*q.v.*) and *Carl Friedrich Gauss* (*q.v.*) set up a network of magnetic observatories, the first example of international cooperation in a scientific endeavor. The data compilation continued throughout the 19th century, with typical Victorian tenacity, led by *Edward Sabine* (*q.v.*), and detailed magnetic measurements were made by James Clark Ross' expedition to the poles and the voyage of HMS Challenger. Impressive though these data collections were, with hindsight they yielded rather little in the way of pure scientific discovery or useful application. True, Sabine was to identify the source of magnetic storms with activity on the Sun and they left us a wonderful record of the geomagnetic field in the 19th century, laying the foundation for modern surveying, but the real prize of discovering the geomagnetic field's origin eluded them.

Developments in the early 20th century were to catapult geomagnetism into the limelight yet again, this time in the quest for minerals. Metal ores, base and noble, are concentrated in rocks rich in magnetites that are intensely magnetic. Geomagnetism provided a cheap and simple prospecting tool for exploration, and magnetic surveys proliferated on land as never before. Geomagnetism provides the cheapest geophysical exploration tool, and while it may lack the precision of seismic methods it continues to produce economic returns—a year's profit from one of the larger mines would probably pay for all the mapping in the last century. The discovery of electromagnetic induction provided yet another technique for exploring the earth's interior and even more significantly it changed prevailing views on the origin of geomagnetic fluctuations and the earth's main dipole field. The many theories for the origin proposed around the turn of the 20th century are reviewed here by David Stevenson (see *Nondynamo theories*). The only one to survive the test of time was *Joseph Larmor*'s (*q.v.*) self-exciting dynamo theory, but even this was to suffer a half-century setback from *Thomas Cowling*'s (*q.v.*) proof that no dynamo could sustain a magnetic field with symmetry about an axis, which the earth's dipole has to a good approximation.

Spectacular progress was being made at about this time by French physicists such as Bernard Brunhes and *Motonori Matuyama* (*q.v.*) from Japan trying to understand the magnetic properties of rocks. In founding the science of paleomagnetism they discovered they were able to determine the direction of the earth's magnetic field at the time of the rock's formation, and made the astonishing discovery that the magnetic field had reversed direction in the past. This discovery, like the dynamo theory, suffered a setback when, in the late 1950s, Seiya Uyeda and Takesi Nagata from Japan found that some minerals reverse spontaneously: this providing an alternative but rather mundane explanation that appealed to some in a skeptical scientific community. Evidence for polarity reversals mounted, thanks in great part to the efforts of *Keith Runcorn* (*q.v.*) and colleagues in England, but it required precise radiometric dating and access to a suite of rocks younger than 5 million years to establish a complete chronology and put the question beyond any doubt: this was finally achieved by *Allan Cox* (*q.v.*) and colleagues in the USA in 1960, on the eve of the plate tectonic revolution.

It is hard to comprehend the rapidity of scientific developments in earth science in the 1960s and impossible to underestimate the importance of the role played by paleomagnetism and geomagnetism in the development of plate tectonics. The establishment of polarity reversals came together with H. Hess' ideas on seafloor spreading and the discovery of magnetic stripes on the ocean floor to provide a means to map the age of the oceans (see *Vine-Matthews-Morley hypothesis*) and confirm once and for all Wegener's ancient ideas of continental drift. Even today, almost all our quantitative knowledge of plate movements in the geological past comes from paleomagnetism and geomagnetism and the development of a *reversal timescale* (*q.v.* see also Bill Lowrie's article *Geomagnetic polarity timescales*). 1958 saw at last the removal of Cowling's objection to the dynamo theory (see *Dynamo, Backus* and *Dynamo, Herzenberg*) and work done in Eastern Europe and the former Soviet Union provided the mathematical foundation and physical insight needed to understand how the geomagnetic field could be generated in the earth's liquid iron core (see *Dynamos, mean-field* and *Dynamo, Braginsky*). Improvements in instrumentation accelerated magnetic surveys: electromagnetic surveys became routine, proton magnetometers could be towed behind ships, and Sputnik ushered in the satellite era with a magnetometer on board.

So where does geomagnetism and paleomagnetism stand now? The dynamo theory of the origin of the main field still presents one of the most difficult challenges to classical physics, but computers are now fast enough to solve the equations of *magnetohydrodynamics* (*q.v.*, see also *Geodynamo, numerical simulations*) in sufficient complexity to reproduce many of the observed phenomena; we have just entered a decade of magnetic observation with two satellites operational at the time of writing and two more launches planned in the near future; there are more aircraft devoted to industrial magnetic surveying than ever before; electromagnetic methods have found application in the search for hydrocarbon reserves and have moved into the marine environment (see *EM, marine controlled source*). Paleomagnetists have begun to map details of the magnetic field during polarity transition (see *Geomagnetic polarity reversals*), discover many examples of excursions (aborted reversals), are mapping systematic departures from the simple dipole structure, and have automated laboratory techniques to the point where, in a single day, they can make more measurements of the absolute paleointensity of magnetic materials such as ceramics and basaltic rocks than pioneer *Emile Thellier* (*q.v.*) could do in a lifetime (see articles on *Absolute paleointensity*).

The subject divides naturally into the studies of magnetic fields with different origins—indeed these differences often make it difficult even for experts to understand other branches of their own subject! Those studying the permanent magnetization of the earth's crustal rocks deal with magnetic fields that owe their origin to permanent magnetism at the molecular or crystal grain level; those in electromagnetic induction study magnetic fields caused by electric currents induced in solid rocks deep inside the Earth; while dynamo theorists deal with induction by a fluid, and have to deal with the additional complexity of advection by a moving conductor. Paleomagnetism naturally separates into studies of the magnetism of rocks, or *rock magnetism* (*q.v.*), laboratory methods for determining the ancient field, and the history of the ancient field itself. This classification dictated our choice of topics. Special effort has been made to represent the activities of the global network of permanent magnetic observatories. These rarely feature in scientific papers and most practicing scientists are unaware of the meticulous nature of the work and the dedication of those unsung heroes charged with maintaining standards over decades—a persistence rarely experienced in modern science. The observatory section represents, to our knowledge, the first attempt to draw together into one place this rather loosely connected international endeavor.

Our subject relates to many other disciplines, either because the geomagnetic field is a vital part of our environment and provides a surprising range of useful techniques to others, from stratigraphy through navigation to radio communication. Partly because of this, and partly in an effort to provide a self-contained volume, we have strayed outside the strict remit of GP. We have included articles on earth structure, particularly those esoteric regions (see for example articles on *D"*) important for geomagnetism, and have covered the fascinating magnetic fields of other planets and satellites.

Our main thanks must go to our contributors, who have so willingly and energetically contributed to make this a truly community effort: we have received very few refusals to our requests to contribute. Alan Jones and Kathryn Whaler advised on electromagnetic induction, a difficult subject for both editors. Thanks go to Stella Gubbins for her unstinting work in organizing the geomagnetism articles and presenting them to the publishers in good order.

March 2007
David Gubbins and Emilio Herrero-Bervera

A

AEROMAGNETIC SURVEYING

Introduction

Magnetic surveying is one of the earliest geophysical methods ever used, with a magnetic compass survey being used in Sweden in 1640 to detect magnetic iron ores. Once instruments were developed in the 1880s to measure the magnitude of the Earth's magnetic field, magnetic surveys applied to mineral exploration became widespread (Hanna, 1990). These early surveys, comprising magnetometer measurements taken on or close to the Earth's surface, were limited in extent and only small areas could be covered in any great detail. With the advent of an aircraft-mounted magnetometer system (Muffly, 1946), developed primarily for submarine detection during World War II, the number and areal coverage of magnetic surveys expanded rapidly. The first airborne magnetic or aeromagnetic survey for geological purposes was flown in 1945 in Alaska by the U.S. Geological Survey and the U.S. Navy. By the end of the 1940s, aeromagnetic surveys were being flown worldwide.

Survey objectives

Aeromagnetic surveys are flown for a variety of reasons: geologic mapping, mineral and oil exploration, and environmental and groundwater investigations. Since variations in the measured magnetic field reflect the distribution of magnetic minerals (mainly magnetite) in the Earth's crust and human-made objects, surveys can be used to detect, locate, and characterize these magnetic sources. Most surveys are flown to aid in surface geologic mapping, where the magnetic effects of geologic bodies and structures can be detected even in areas where rock outcrop is scarce or absent, and bedrock is covered by glacial overburden, bodies of water, sand, or vegetation. Magnetic surveys can also detect magnetic sources at great depth (tens of kilometers) within the Earth's crust, being limited only by the depth at which magnetic minerals reach their Curie point and cease to be ferrimagnetic. Broad correlations can be made between rock type and magnetic properties, but the relationship is often complicated by temperature, pressure, and chemical changes that rocks are exposed to (Grant, 1985). Nevertheless, determining the location, shape, and attitude of magnetic sources and combining this with available geologic information can result in a meaningful geological interpretation of a given area. Certain kinds of ore bodies may produce magnetic anomalies that are desirable targets for mineral exploration surveys. Although hydrocarbon reservoirs are not directly detectable by aeromagnetic surveys, magnetic data can be used to locate geologic structures that provide favorable conditions for oil/gas production and accumulation (Gibson and Millegan, 1998). Similarly, mapping the magnetic signatures of faults and fractures within water-bearing sedimentary rocks provides valuable constraints on the geometry of aquifers and the framework of groundwater systems.

Survey design

Aeromagnetic surveys are usually flown in a regular pattern of equally spaced parallel lines (flightlines). A series of control or tie-lines is also flown perpendicular to the flightline direction to assist in the processing of the magnetic field data set. The ratio of control to flightline spacing generally ranges from 3 to 10 depending on the desired data quality. Flightline spacing is dependent on the primary aim of the survey and governs the amount of detail in the measured magnetic field. For reconnaissance mapping of large regions (e.g., states, countries), where little or no knowledge of the magnetic field to be mapped is available, typical line spacings are 1.5 km in Australia and 0.8 km in Canada. Reconnaissance survey lines are usually oriented north-south or east-west. For smaller areas with similar financial resources available, these spacings can be reduced (e.g., 0.2 km in Sweden and Finland). For smaller regions, the line spacing is governed by the size of the target (geological structure, oil prospect, ore body) being investigated. For example, mineral exploration surveys generally have spacings in the range 50–200 m and are flown perpendicular to the dominant geologic strike direction, although directions no more than 45° from magnetic north are preferred because N-S wavelengths are shorter than E-W wavelengths at low and intermediate latitudes. For areas with distinct regions of differing geologic strike, costs may permit splitting the survey into several flight directions or a single compromise direction must be assigned.

The survey flying height is intimately related to flightline spacing with lower heights being more appropriate for closer line spacings. This is due to the rapid decrease in magnetic field intensity and increase in wavelength as a function of the distance from a magnetic source. Theoretical arguments suggest a line spacing to height ratio of 2 or less is desirable to accurately sample variations in the magnetic field (Reid, 1980). The majority of surveys, however, have ratios higher than this, i.e., 2.5 to 8. The minimum flight height is limited by government restrictions, safety considerations, and ruggedness of the terrain. Topographic variations also affect the mode of flying used in a survey. For geologic mapping and mineral exploration surveys, measurements are desirable at levels as close to the magnetic sources as possible, hence surveys are flown at a constant height (mean terrain clearance) above the ground surface. Prior to the availability of real-time global positioning system (GPS) navigation, surveys over mountainous regions were flown at a constant altitude. Postsurvey processing of the collected magnetic field data could then be applied to artificially "drape" the measurements onto a surface at some specified mean terrain clearance. This can improve the resolution of magnetic anomalies that is degraded by flying at the higher constant altitude. However, current usage of more accurate navigational systems (GPS) now permits fixed-wing surveys to be flown along a preplanned artificial drape surface in regions of rugged topography. The drape surface is set to a constant height above ground level but is modified to take into consideration the climb and descent rate of the survey aircraft. Helicopters can also offer an alternative (usually costlier) to fixed-wing aircraft in mountainous areas.

Survey data acquisition

Along with measurements of the magnetic field, information relating to the accurate location and time of these recorded values is gathered. A modern data acquisition system (usually computer-based) records and archives magnetic, navigational, temporal, altitude, and perhaps aircraft attitude data. In addition, video recordings of the flight-path of the aircraft are made that may be used later to identify magnetic effects of human-made sources and to provide checks on navigation. Accurate synchronization of all the data streams is crucial for subsequent processing of the raw magnetometer data. Magnetometer measuring rates of 10 samples per second and navigational information at one sample per second are current norms.

Magnetometers

Compared to the first magnetometers used in the 1940s, the resolution and accuracy of magnetic field measurements have increased significantly. The early fluxgate magnetometers had resolutions of about 1 nT and noise envelopes of 2 nT (Horsfall, 1997). These instruments only produced relative measurements of the field and suffered from appreciable drift, possibly 10 nT/hr. Proton precession magnetometers followed with a resolution of 0.1 nT and noise envelope of 1 nT. These instruments measured the magnetic field intensity and had minimal drift. These have been superseded by the optically pumped, generally caesium-vapor magnetometers which have a resolution of 0.001 nT and a noise envelope of 0.005 nT. Such levels of measurement accuracy are not met in practice since errors arise through unremoved aircraft magnetic effects, navigation effects, and postsurvey data processing. Standard sampling rates of 10 samples per second coupled with aircraft speeds of 220–280 km/hr result in measurements at an interval of 6–8 meters.

Single magnetometer aircraft configurations simply produce measurements of the magnetic field intensity in the direction of the Earth's magnetic field. By adding extra sensors, various other quantities can be measured. Two vertically separated sensors can provide vertical magnetic gradient information while two wing-tip sensors allow measurement of the transverse horizontal magnetic gradient. Vertical gradient measurements have the advantage in effectively suppressing longer wavelength magnetic anomalies and providing a higher resolution definition of shorter wavelength features. This is crucial in surveys flown for mapping of near-surface geology where the longer wavelength effects of deep sources are of secondary interest. In theory, having two sensors also implies that a simple subtraction of the two measurements removes any temporal effects in the data. However, in practical applications, temporal effects may still be present in the data. Transverse (perpendicular to the flight path) gradient measurements are not generally used as a final interpretation product but can be exploited when the magnetic data are interpolated between the flight lines to produce a regular grid of magnetic field values (Hogg, 1989).

Aircraft noise

Magnetic noise caused by the survey aircraft arises from permanent and induced magnetization effects and from the flow of electrical currents. The permanent magnetization of the aircraft results in a heading error which is a function of the flightline direction. Induced magnetizations occur due to the motion of the aircraft in the Earth's magnetic field. These effects are partially reduced by mounting the magnetometer as far away as possible from the source of the noise, either on a boom mounted on the aircraft's tail or in a towed "bird" attached by a cable. Although the bird can be located several tens of meters below the aircraft, this may introduce orientation and positional errors if the sensor does not maintain a fixed location relative to the aircraft. For magnetometers mounted on the aircraft, a technique called compensation is needed to reduce remaining magnetic noise effects to an acceptable level. Modern aircraft are compensated by establishing and modeling the magnetic response of the aircraft to changes in its orientation. This is usually done prior to surveying by flying lines in each of the four local survey directions in an area of low magnetic field gradient and at high altitude, while carrying out several roll, pitch, and yaw maneuvers. The compensation may be done digitally in real time through the data acquisition system, or later as part of data processing. Information on the aircraft's orientation and direction is used to predict the resultant magnetic effect, which is then removed as the survey is flown, or later. The figure of merit, consisting of the sum of the absolute values of the magnetic effects of the roll, pitch, and yaw maneuvers for each of the four headings (12 values in all), can be as low as one nanotesla. Once a survey is started, the aircraft's noise level should be assessed repeatedly to ensure that the same level of data quality is maintained.

Positioning

No matter what accuracies are achieved in the magnetic field measurements, the value of the final survey data is dependent on accurate location of the measurement points. Traditional methods of navigation relied heavily on visual tracking using aerial photographs. The actual location of the flight path would be recovered manually by comparing these photographs with images from onboard video cameras. Synchronizing the data recording with the cameras would allow positioning of the magnetic field data. Electronic navigation systems were also used either to interpolate between visually located points or provide positioning information for offshore surveys. Positional errors of the order of hundreds of meters would not be unexpected. These techniques have been superseded by the introduction in the 1990s of GPS. Navigation using GPS relies on the information sent from an array of satellites whose locations are known precisely. Signals from a number of satellites are used to triangulate the position of the receiver in the aircraft so that its position is known for navigational purposes and to locate the magnetic field measurements. GPS systems also provide highly accurate time information that forms the basis for synchronizing each component of the data acquisition system. Positional accuracies of ± 5 m horizontally and ± 10 m vertically can be achieved with a single GPS receiver. These values can be improved upon by using another receiver and refining positions by making differential corrections. The added receiver is set up at the survey base and simultaneously collects data from the satellites as the survey is flown. Postsurvey processing of all the positional data reduces errors in information provided by the satellites and accuracies of ± 2–4 m can be achieved. Current surveys generally use real-time differential GPS navigation where the raw positional information is corrected as the data is being collected. The GPS location and altitude are also used to calculate (and subtract) the expected normal field for the location, as predicted by the International Geomagnetic Reference Field (IGRF). This generally removes an unhelpful regional component arising from the Earth's core field.

Temporal effects

Since the Earth's magnetic field varies temporally as well as spatially, time variations that occur over the period of surveying must be determined and removed from the raw measurements. The dominant effect for airborne surveys is the diurnal variation, which generally has an amplitude of tens of nanotesla. Shorter time-period variations due to magnetic storms can be much larger (hundreds of nanotesla) and severe enough to prevent survey data collection. Data can also be degraded my micropulsations with short periods and amplitudes of several nanotesla. Monitoring of the Earth's field is an essential component of a survey in order to mitigate these time-varying effects. One or more base station magnetometers is used to track changes in the field during survey operations. When variations are unacceptably large, surveying is suspended and any flightlines recording during disturbed periods are reflown. The smoothly changing diurnal variation is removed from the data using tie-line leveling. Simply subtracting this variation from the measured data is not sufficient since diurnal changes may vary significantly over the survey area. Nonetheless, the recorded diurnal can be used as a guide in the leveling process. Tie-line leveling is based on the differences in the measured field at the intersection of flightlines and tie-lines. If the distance and hence the time taken to fly between these intersection points is small enough then it can be assumed that the diurnal varies approximately linearly and can be corrected for (Luyendyk, 1997).

Data display and interpretation

The final product resulting from an aeromagnetic survey is a set of leveled flightline data that are interpolated onto a regular grid of magnetic field intensity values covering the survey region. These values can be displayed in a variety of ways, the most common being a color map or image, where the magnetic field values, based on their magnitude, are assigned a specific color. Similarly, the values can be represented as a simple line contour map. Both kinds of representation can be used in a qualitative fashion to divide the survey area into subregions of high and low magnetizations. Since the data are available digitally, it is straightforward to use computer-based algorithms to modify and enhance the magnetic field image for the specific purpose of the survey. Transformation and filtering allows certain attributes of the data to be enhanced, such as the effects due to magnetic sources at shallow or deep levels, or occurring along a specified strike direction. More sophisticated methods may estimate the depths, locations, attitudes, and the magnetic properties of magnetic sources.

Mark Pilkington

Bibliography

Gibson, R.I., and Millegan, P.S. (eds.), 1998. *Geologic Applications of Gravity and Magnetics: Case Histories.* Tulsa, OK, U.S.A.: Society of Exploration Geophysicists and American Association of Petroleum Geologists.

Grant, F.S., 1985. Aeromagnetics, geology and ore environments, I. Magnetite in igneous, sedimentary and metamorphic rocks: An overview. *Geoexploration*, **23**: 303–333.

Hanna, W.F., 1990. Some historical notes on early magnetic surveying in the U.S. Geological Survey. In Hanna, W.F., (ed.), *Geologic Applications of Modern Aeromagnetic Surveys.* United States Geological Survey Bulletin 1924, pp. 63–73.

Hogg, R.L.S., 1989. Recent advances in high sensitivity and high resolution aeromagnetics. In Garland, G.D. (ed.), *Proceedings of Exploration '87, Third Decennial International Conference on Geophysical and Geochemical Exploration for Minerals and Groundwater.* Ontario Geological Survey, Special Volume 3, pp. 153–169.

Horsfall, K.R., 1997. Airborne magnetic and gamma-ray data acquisition. *Australian Geological Survey Organization Journal of Australian Geology and Geophysics*, **17**: 23–30.

Luyendyk, A.P.J., 1997. Processing of airborne magnetic data. *Australian Geological Survey Organization Journal of Australian Geology and Geophysics*, **17**: 31–38.

Muffly, G., 1946. The airborne magnetometer. *Geophysics*, **11**: 321–334.

Reid, A.B., 1980. Aeromagnetic survey design. *Geophysics*, **45**: 973–976.

Cross-references

Compass
Crustal Magnetic Field
Depth to Curie Temperature
IGRF, International Geomagnetic Reference Field
Magnetic Anomalies for Geology and Resources
Magnetic Anomalies, Modeling

AGRICOLA, GEORGIUS (1494–1555)

Agricola, which seems to have been the classical academic pen-name for Georg Bauer, was born at Glauchau, Germany, on March 24, 1494 as the son of a dyer and draper. He studied at Leipzig University and trained as a doctor, and became Town Physician to the Saxon mining community of Chemnitz. His significance derives from his great treatise on metal mining, *De Re Metallica* (1556), or "On Things of Metal." This sumptuously illustrated work was a masterly study of all aspects of mining, including geology, stratigraphy, pictures of mineshafts, machinery, smelting foundries, and even the diseases suffered by miners.

Book III of *De Re Metallica* contains a description and details of how to use the miner's compass which, by 1556, seems to have been a well-established technical aid to the industry. Agricola drew a parallel between the miner's and the mariner's compass, saying that the direction cards of both were divided into equidistant divisions in accordance with a system of "winds." Agricola's compass is divided into four quadrants, with each quadrant subdivided into six: 24 equidistant divisions around the circle. And just as a mariner named the "quarters" of his compass from the prevailing winds, such as "Septentrio" for north and "Auster" for south, so the miner likewise named his underground metal veins, depending on the directions in which they ran.

Agricola also tells us that when prospecting for ore, a mining engineer would use his compass to detect the veins of metal underground and use it to plot their direction and the likely location of their strata across the landscape.

Agricola, it was said, always enjoyed robust health and vigor. He served as Burgomaster of Chemnitz on several occasions and worked tirelessly for the sick during the Bubonic Plague epidemic that ravaged Saxony during 1552–1553. Then, on November 21, 1555, he suddenly died from a "four days" fever.

Allan Chapman

Bibliography

Agricola, G., *De Re Metallica* (Basilae, 1556), translated by Hoover, H.C., and Hoover, L.H. (1912; Dover Edition, New York, 1950).

ALFVÉN WAVES

Introduction and historical details

Alfvén waves are transverse magnetic tension waves that travel along magnetic field lines and can be excited in any electrically conducting fluid permeated by a magnetic field. *Hannes Alfvén* (*q.v.*) deduced their existence from the equations of electromagnetism and hydrodynamics (Alfvén, 1942). Experimental confirmation of his prediction was found seven years later in studies of waves in liquid mercury (Lundquist, 1949). Alfvén waves are now known to be an important mechanism for transporting energy and momentum in many geophysical and astrophysical hydromagnetic systems. They have been observed in Earth's magnetosphere (Voigt, 2002), in interplanetary plasmas (Tsurutani and Ho, 1999), and in the solar photosphere (Nakariakov *et al.*, 1999). The ubiquitous nature of Alfvén waves and their role in communicating the effects of changes in electric currents and magnetic fields has ensured that they remain the focus of increasingly detailed laboratory investigations (Gekelman, 1999). In the context of geomagnetism, it has been suggested that Alfvén waves could be a crucial aspect of the dynamics of Earth's liquid outer core and they have been proposed as the origin of *geomagnetic jerks* (*q.v.*) (Bloxham *et al.*, 2002). In this article a description is given of the Alfvén wave mechanism, the Alfvén wave equation is derived and the consequences of Alfvén waves for geomagnetic observations are discussed. Alternative introductory perspectives on Alfvén waves can be found in the books by Alfvén and Fälthammar (1963), Moffatt (1978), or Davidson (2001). More technical details concerning Alfvén waves in Earth's core can be found in the review article of Jault (2003).

The Alfvén wave mechanism

The restoring force responsible for Alfvén waves follows from two simple physical principles:

1. Lenz's law applied to conducting fluids: "Electrical currents induced by the motion of a conducting fluid through a magnetic field give rise to electromagnetic forces acting to oppose that fluid motion."
2. Newton's second law for fluids: "A force applied to a fluid will result in a change in the momentum of the fluid proportional to the magnitude of the force and in the same direction."

The oscillation underlying Alfvén waves is best understood via a simple thought experiment (Davidson, 2001). Imagine a uniform magnetic field permeating a perfectly conducting fluid, with a uniform flow initially normal to the magnetic field lines. The fluid flow will distort the magnetic field lines (see *Alfvén's theorem and the frozen flux approximation*) so they become curved as shown in Figure A1 (part (b)). The curvature of magnetic field lines produces a magnetic (Lorentz) force on the fluid, which opposes further curvature as predicted by Lenz's law. By Newton's second law, the Lorentz force changes the momentum of the fluid, pushing it (and consequently the magnetic field lines) in an attempt to minimize field line distortion and restore the system toward its equilibrium state. This restoring force provides the basis for transverse oscillations of magnetic fields in conducting fluids and therefore for Alfvén waves. As the curvature of the magnetic field lines increases, so does the strength of the restoring force. Eventually the Lorentz force becomes strong enough to reverse the direction of the fluid flow. Magnetic field lines are pushed back to their undistorted configuration and the Lorentz force associated with their curvature weakens until the field lines become straight again. The sequence of flow causing field line distortion and field line distortion exerting a force on the fluid now repeats, but with the initial flow (a consequence of fluid inertia) now in the opposite direction. In the absence of dissipation this cycle will continue indefinitely. Figure A1 shows one complete cycle resulting from the push and pull between inertial acceleration and acceleration due to the Lorentz force.

Consideration of typical scales of physical quantities involved in this inertial-magnetic (Alfvén) oscillation shows that the strength of the magnetic field will determine the frequency of Alfvén waves. Balancing inertial accelerations and accelerations caused by magnetic field curvature, we find that $U/T_{\mathrm{A}} = B^2/L_{\mathrm{A}}\rho\mu$ where U is a typical scale of the fluid velocity, T_{A} is the time scale of the inertial-magnetic (Alfvén) oscillation, L_{A} is the length scale associated with the oscillation, B is the scale of the magnetic field strength, ρ is the fluid density, and μ is the magnetic permeability of the medium. For highly electrically conducting fluids, magnetic field changes occur primarily through advection (see *Alfvén's theorem and the frozen flux approximation*) so we have the additional constraint that $B/T_{\mathrm{A}} = UB/L_{\mathrm{A}}$ or $U = L_A/T_A$. Consequently $L_{\mathrm{A}}^2/T_{\mathrm{A}}^2 = B^2/\rho\mu$ or $v_A = B/(\rho\mu)^{1/2}$. This is a characteristic velocity scale associated with Alfvén waves and is referred to as the Alfvén velocity. The Alfvén velocity will be derived in a more rigorous manner and its implications discussed further in the next section.

Physical intuition concerning Alfvén waves can be obtained through an analogy between the response of a magnetic field line distorted by fluid flow across it and the response of an elastic string when plucked. Both rely on tension as a restoring force, elastic tension in the case of the string and magnetic tension in the case of the magnetic field line and both result in transverse waves propagating in directions perpendicular to their displacement. When visualizing Alfvén waves it can be helpful to think of a fluid being endowed with a pseudoelastic nature by the presence of a magnetic field, and consequently supporting transverse waves. Nonuniform magnetic fields have similar consequences for Alfvén waves as nonuniform elasticity of solids has for elastic shear waves.

The Alfvén wave equation

To determine the properties of Alfvén waves in a quantitative manner, we employ the classical technique of deriving a wave equation and then proceed to find the relationship between frequency and wavelength necessary for plane waves to be solutions. Consider a uniform, steady, magnetic field $\boldsymbol{B_0}$ in an infinite, homogeneous, incompressible, electrically conducting fluid of density ρ, kinematic viscosity ν, and magnetic diffusivity $\eta = 1/\sigma\mu$ where σ is the electrical conductivity.

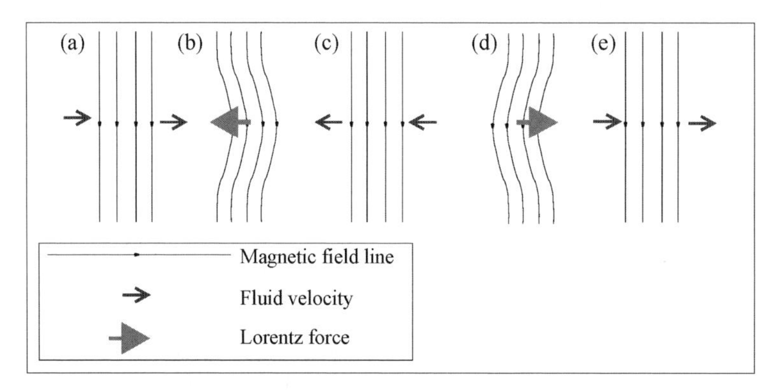

Figure A1 The Alfvén wave mechanism. In (a) an initial fluid velocity normal to the uniform field lines distorts them into the curved lines shown in (b) giving rise to a Lorentz force which retards and eventually reverses the fluid velocity, returning the field lines to their undisturbed position as shown in (c). The process of field line distortion is then reversed in (d) until the cycle is completed with the return to the initial configuration in (e).

We imagine that the fluid is perturbed by an infinitesimally small flow $\boldsymbol{u}$ inducing a perturbation magnetic field $\boldsymbol{b}$. Ignoring terms that are quadratic in small quantities, the equations describing Newton's second law for fluids and the evolution of the magnetic fields encompassing Lenz's law are

$$\underbrace{\frac{\partial \boldsymbol{u}}{\partial t}}_{\substack{\text{Inertial}\\\text{acceleration}}} = \underbrace{-\frac{1}{\rho}\nabla p}_{\substack{\text{Combined mechanical}\\\text{and magnetic}\\\text{pressure gradient}}} + \underbrace{\frac{1}{\rho\mu}(\boldsymbol{B_0}\cdot\nabla)\boldsymbol{b}}_{\substack{\text{Lorentz acceleration}\\\text{due to magnetic tension}}} + \underbrace{\nu\nabla^2\boldsymbol{u}}_{\substack{\text{Viscous}\\\text{diffusion}}}, \quad \text{(Eq. 1)}$$

$$\underbrace{\frac{\partial \boldsymbol{b}}{\partial t}}_{\substack{\text{Change in the}\\\text{magnetic field}}} = \underbrace{(\boldsymbol{B_0}\cdot\nabla)\boldsymbol{u}}_{\substack{\text{Stretching of magnetic}\\\text{field by fluid motion}}} + \underbrace{\eta\nabla^2\boldsymbol{b}}_{\substack{\text{Magnetic}\\\text{diffusion}}} \quad \text{(Eq. 2)}$$

Taking the curl $(\nabla\times)$ of equation 1, we obtain an equation describing how the fluid vorticity $\xi = \nabla\times\boldsymbol{u}$ evolves

$$\frac{\partial\xi}{\partial t} = \frac{1}{\rho\mu}(\boldsymbol{B_0}\cdot\nabla)(\nabla\times\boldsymbol{b}) + \nu\nabla^2\xi. \quad \text{(Eq. 3)}$$

$\nabla\times$ equation 2 gives

$$\nabla\times\frac{\partial\boldsymbol{b}}{\partial t} = (\boldsymbol{B_0}\cdot\nabla)\xi + \eta\nabla^2(\nabla\times\boldsymbol{b}). \quad \text{(Eq. 4)}$$

To find the wave equation, we take a further time derivative of equation 3 so that

$$\frac{\partial^2\xi}{\partial t^2} = \frac{1}{\rho\mu}(\boldsymbol{B_0}\cdot\nabla)\left(\nabla\times\frac{\partial\boldsymbol{b}}{\partial t}\right) + \nu\left(\nabla^2\frac{\partial\xi}{\partial t}\right), \quad \text{(Eq. 5)}$$

and then eliminate $\boldsymbol{b}$ using an expression for $\frac{1}{\rho\mu}(\boldsymbol{B_0}\cdot\nabla)(\nabla\times\partial\boldsymbol{b}/\partial t)$ in terms of ξ obtained by operating with $\frac{1}{\rho\mu}(\boldsymbol{B_0}\cdot\nabla)$ on (4) and substituting for $\frac{1}{\rho\mu}(\boldsymbol{B_0}\cdot\nabla)(\nabla\times\boldsymbol{b})$ from equation 3 which gives

$$\frac{1}{\rho\mu}(\boldsymbol{B_0}\cdot\nabla)\left(\nabla\times\frac{\partial\boldsymbol{b}}{\partial t}\right) = \frac{1}{\rho\mu}(\boldsymbol{B_0}\cdot\nabla)^2\xi - \nu\eta\nabla^4\xi + (\nu+\eta)\nabla^2\left(\frac{\partial\xi}{\partial t}\right), \quad \text{(Eq. 6)}$$

which when substituted into equation 5 leaves the Alfvén wave equation

$$\frac{\partial^2\xi}{\partial t^2} = \frac{1}{\rho\mu}(\boldsymbol{B_0}\cdot\nabla)^2\xi - \nu\eta\nabla^4\xi + (\nu+\eta)\nabla^2\left(\frac{\partial\xi}{\partial t}\right). \quad \text{(Eq. 7)}$$

The first term on the right hand side is the restoring force which arises from the stretching of magnetic field lines. The second term is the correction to the restoring force caused by the presence of viscous and ohmic diffusion, while the final term expresses the dissipation of energy from the system due to these finite diffusivities.

Dispersion relation and properties of Alfvén waves

Substituting a simple plane wave solution of the form $\xi = Re\{\hat{\xi}e^{i(\boldsymbol{k}\cdot\boldsymbol{r}-\omega t)}\}$ where $\boldsymbol{k} = k_x\hat{\boldsymbol{x}} + k_y\hat{\boldsymbol{y}} + k_z\hat{\boldsymbol{z}}$ and $\boldsymbol{r} = x\hat{\boldsymbol{x}} + y\hat{\boldsymbol{y}} + z\hat{\boldsymbol{z}}$ into the Alfvén wave equation 7, we find that valid solutions are possible provided that

$$\omega^2 = \left(\frac{B_0^2(\boldsymbol{k}\cdot\hat{\boldsymbol{B}}_0)^2}{\rho\mu} - \nu\eta k^4\right) - i(\nu+\eta)k^2\omega. \quad \text{(Eq. 8)}$$

where $\hat{\boldsymbol{B}}_0 = \boldsymbol{B_0}/|B_0|$.

Equation 8 is the dispersion relation, which specifies the relationship between the angular frequency ω and the wavenumber k of Alfvén waves. It is a complex quadratic equation in ω, so we can use the well known formula to find explicit solutions for ω, which are

$$\omega = \frac{-i(\nu+\eta)k^2}{2} \pm \sqrt{\frac{B_0^2(\boldsymbol{k}\cdot\hat{\boldsymbol{B}}_0)^2}{\rho\mu} - \frac{(\nu-\eta)^2k^4}{4}}. \quad \text{(Eq. 9)}$$

In an idealized medium with $\nu = \eta = 0$ there is no dissipation and the dispersion relation simplifies to

$$\omega = \pm v_A(\boldsymbol{k}\cdot\hat{\boldsymbol{B}}_0), \quad \text{(Eq. 10)}$$

where v_A is the Alfvén velocity

$$v_A = \frac{B_0}{(\rho\mu)^{1/2}}. \quad \text{(Eq. 11)}$$

This derivation illustrates that the Alfvén velocity is the speed at which an Alfvén wave propagates along magnetic field lines. Alfvén waves are nondispersive because their angular frequency is independent of $|k|$ and the phase velocity and the group velocity (at which energy and information are transported by the wave) are equal. Alfvén waves are however anisotropic, with their properties dependent on the angle between the applied magnetic field and the wave propagation direction.

An idea of Alfvén wave speeds in Earth's core can be obtained by inserting into equation 11 the seismologically determined density of the outer core fluid $\rho = 1\times10^4$ kg, the magnetic permeability for a metal above its Curie temperature $\mu = 4\pi\times10^{-7}$ T^2 $mkg^{-1}s^2$ and a plausible value for the strength of the magnetic field in the core (which we take to be the typical amplitude of the radial field strength observed at the core surface $B_0 = 5\times10^{-4}$ T) giving an Alfvén velocity of $v_A = 0.004$ m s^{-1} or 140 km yr^{-1}. The time taken for such a wave to travel a distance of order of the core radius is around 25 years. It should also be noted that Alfvén waves in the magnetosphere (where they also play an important role in dynamics) travel much faster because the density of the electrically conducting is very much smaller.

Considering Alfvén waves in Earth's core, Ohmic dissipation is expected to dominate viscous dissipation but for large scale waves will still be a small effect so that $\nu k^2 \ll \eta k^2 \ll v_A^2$. Given this assumption the dispersion relation reduces to

$$\omega = \frac{-i\eta k^2}{2} \pm v_A(\boldsymbol{k}\cdot\hat{\boldsymbol{B}}_0) \quad \text{(Eq. 12)}$$

and the wave solutions have the form of simple Alfvén waves, damped on the Ohmic diffusion timescale of $T_{ohm} = 2/\eta k^2$

$$\xi = Re\left\{\hat{\xi}e^{i(\boldsymbol{k}\cdot\boldsymbol{r} \pm v_A(\boldsymbol{k}\cdot\hat{\boldsymbol{B}}_0)t)}e^{-t/T_{ohm}}\right\} \quad \text{(Eq. 13)}$$

Smaller scale waves are rapidly damped out by Ohmic diffusion, while large scale waves will be the longer lived, so we expect these to have the most important impact on both dynamics of the core and the observable magnetic field.

Observations of Alfvén waves and their relevance to geomagnetism

The theory of Alfvén waves outlined above is attractive in its simplicity, but can we really expect such waves to be present in Earth's core? In the outer core, rotation will have a strong influence on the fluid dynamics (see *Proudmann-Taylor theorem*). Alfvén waves cannot exist when the Coriolis force plays an important role in the force balance; in this case, more complex wave motions arise (see *Magnetohydrodynamic waves*). In addition convection is occurring (see *Core convection*) and gives rise to a dynamo generated magnetic field (see *Geodynamo*), which is both time dependent and spatially nonuniform, while the boundary conditions imposed by the mantle are heterogeneous—these factors combine to produce a formidably complex system.

Braginsky (1970) recognized that, despite all these complications, a special class of Alfvén waves is likely to be the mechanism by which angular momentum is redistributed on short (decadal) timescales in Earth's core. He showed that when Coriolis forces are balanced by pressure forces, Alfvén waves involving only the component of the magnetic field normal to the rotation axis can exist. The fluid motions in this case consist of motions of cylindrical surfaces aligned with the rotation axis, with the Alfvén waves propagating along field lines threading these cylinders and being associated with east-west oscillations of the cylinders. Similarities to torsional motions familiar from classical mechanics led Braginsky to christen these geophysically important Alfvén waves torsional oscillations (see *Oscillations, torsional*). Although the simple Alfvén wave model captures the essence of torsional oscillations and leads to a correct order of magnitude estimate of their periods, coupling to the mantle and the nonaxisymmetry of the background magnetic field should be taken into account and lead to modifications of the dispersion relation given in equation 9. A detailed discussion of such refinements can be found in Jault (2003).

The last 15 years have seen a rapid accumulation of evidence suggesting that Alfvén waves in the form of torsional oscillations are indeed present in Earth's core. The transfer of angular momentum between the mantle and torsional oscillations in the outer core is capable of explaining decadal changes in the rotation rate of Earth (see *Length of day variations, decadal*). Furthermore, core flows determined from the inversion of global magnetic and secular variation data show oscillations in time of axisymmetric, equatorially symmetric flows which can be accounted for by a small number of spherical harmonic modes with periodic time dependence (Zatman and Bloxham, 1997). The superposition of such modes can produce abrupt changes in the second time derivative of the magnetic field observed at Earth's surface, similar to geomagnetic jerks (Bloxham *et al.*, 2002). Interpreting axisymmetric, equatorially symmetric core motions with a periodic time dependence as the signature of torsional oscillations leads to the suggestion that geomagnetic jerks are caused by Alfvén waves in Earth's core. Further evidence for the wave-like nature of the redistribution of zonally averaged angular momentum derived from core flow inversions has been found by Hide *et al.* (2000), with disturbances propagating from the equator towards the poles. The mechanism exciting torsional oscillations in Earth's core is presently unknown, though one suggestion is that the time dependent, nonaxisymmetric magnetic field could give rise to a suitable fluctuating Lorentz torque on geostrophic cylinders (Dumberry and Bloxham, 2003).

Future progress in interpreting and understanding Alfvén waves in Earth's core will require the incorporation of more complete dynamical models of torsional oscillations (see, for example, Buffett and Mound, 2005) into the inversion of geomagnetic observations for core motions.

Christopher Finlay

Bibliography

Alfvén, H., 1942. Existence of electromagnetic-hydrodynamic waves. *Nature*, **150**: 405–406.

Alfvén, H., and Fälthammar, C.-G., 1963. *Cosmical Electrodynamics, Fundamental Principles.* Oxford: Oxford University Press.

Bloxham, J., Zatman, S., and Dumberry, M., 2002. The origin of geomagnetic jerks. *Nature*, **420**: 65–68.

Braginsky, S.I., 1970. Torsional magnetohydrodynamic vibrations in the Earth's core and variations in day length. *Geomagnetism and Aeronomy*, **10**: 1–10.

Buffett, B.A., and Mound, J.E., 2005. A Green's function for the excitation of torsional oscillations in Earth's core. *Journal of Geophysical Research,* Vol. 110, B08104, doi: 10.1029/2004JB003495.

Davidson, P.A., 2001. *An introduction to Magnetohydrodynamics.* Cambridge: Cambridge University Press.

Dumberry, M., and Bloxham, J., 2003. Torque balance, Taylor's constraint and torsional oscillations in a numerical model of the geodynamo. *Physics of Earth and Planetary. Interiors*, **140**: 29–51.

Gekelman, W., 1999. Review of laboratory experiments on Alfvén waves and their relationship to space observations. *Journal of Geophysical Research*, **104**: 14417–14435.

Hide, R., Boggs, D.H., and Dickey, J.O., 2000. Angular momentum fluctuations within the Earth's liquid core and solid mantle. *Geophysical Journal International*, **125**: 777–786.

Jault, D., 2003. Electromagnetic and topographic coupling, and LOD variations. In Jones, C.A., Soward, A.M., and Zhang, K., (eds.), Earth's core and lower mantle. *The Fluid Mechanics of Astrophysics and Geophysics,* **11**: 56–76.

Lundquist, S., 1949. Experimental investigations of magnetohydrodynamic waves. *Physical Review*, **107**: 1805–1809.

Moffatt, H.K., 1978. *Magnetic Field Generation in Electrically Conducting Fluids.* Cambridge: Cambridge University Press.

Nakariakov, V.M., Ofman, L., DeLuca, E.E., Roberts, B., and Davila, J.M., 1999. TRACE observation of damped coronal loop oscillations: Implications for coronal heating. *Science*, **285**: 862–864.

Tsurutani, B.T., and Ho, C.M., 1999. A review of discontinuities and Alfvén waves in interplanetary space: Ulysses results. *Reviews of Geophysics*, **37**: 517–541.

Voigt, J., 2002. Alfvén wave coupling in the auroral current circuit. *Surveys in Geophysics*, **23**: 335–377.

Zatman, S., and Bloxham, J., 1997. Torsional oscillations and the magnetic field within the Earth's core. *Nature*, **388**: 760–763.

Cross-references

Alfvén's Theorem and the Frozen Flux Approximation
Alfvén, Hannes Olof Gösta (1908–1995)
Core Convection
Geodynamo
Length of Day Variations, Decadal
Magnetohydrodynamic Waves
Oscillations, Torsional
Proudman-Taylor Theorem

ALFVÉN, HANNES OLOF GÖSTA (1908–1995)

Hannes Alfvén is best known in geomagnetism for the "frozen flux" theorem that bears his name and for the discovery of magnetohydrodynamic waves. He started research in the physics department at the University of Uppsala, where he studied radiation in triodes. His early work on electronics and instrumentation was sound grounding for his later discoveries in cosmic physics. When his book *Cosmical Electrodynamics* (Alfvén, 1950) was published, the author was referred to by one of the reviewers—*T.G. Cowling (q.v.)*—as "an electrical engineer in Stockholm." All of Hannes Alfvén's scientific work reveals a profound physical insight and an intuition that allowed him to derive results of great generality from specific problems.

Hannes Alfvén is most widely known for his discovery (Alfvén, 1942) of a new kind of waves now generally referred to as *Alfvén waves* (*q.v.*). These are a transverse mode of magnetohydrodynamic waves, and propagate with the Alfvén velocity, $B/(\mu_0\rho)^{1/2}$. In the Earth's core they occur as torsional oscillations as well as other Alfvén-type modes that are altered by the Coriolis force and have quite a different character (see *Magnetohydrodynamic waves*). Before Alfvén, electromagnetic theory and hydrodynamics were well developed but as separate scientific disciplines. By combining them, Alfvén founded the new discipline of *magnetohydrodynamics* (*q.v.*). Magnetohydrodynamics is of fundamental importance for the physics of the fluid core of the Earth and other planets but has also much wider significance. It is indispensable in modern plasma physics and its applications, both in the laboratory (e.g., in fusion research) and in space (ionospheres, magnetospheres, the sun, stars, stellar winds and interstellar plasma). A famous theorem in magnetohydrodynamics, sometimes called the *Alfvén theorem* (*q.v.*), is that of frozen flux, which says that if the conductivity of a magnetized fluid is high enough, magnetic field and fluid motion are coupled in such a way that the fluid appears to be frozen to the magnetic field lines. This is a powerful tool in many applications, for example in studying motions in the Earth's core (see *Core motions*). The degree to which the frozen condition holds in a fluid is characterized by a dimensionless parameter called the Lundquist number, $\mu_0\sigma B\lambda/\rho^{1/2}$ (see *Geodynamo, dimensional analysis and timescales*). It was derived by Alfvén's student Stig Lundqvist (1952), who also was the first to prove the existence of magnetohydrodynamic waves experimentally in liquid metal (Lundqvist, 1949). In plasmas, especially space plasmas, there are important exceptions to the validity of the frozen field concept, and Alfvén himself vigorously warned against its uncritical use.

The second fundamental contribution by Alfvén was the guiding center theory, in which the average motion of a charged particle gyrating in a magnetic field is represented by the motion of its center of gyration. This theory dramatically simplified many plasma physics problems and laid the foundation of the adiabatic theory of charged particle motion.

Specific fields where Hannes Alfvén contributed were the theory of aurora and *magnetic storms* (*q.v.*) (Alfvén, 1939), evolution of the solar system, and cosmology (he was a vocal opponent to the Big Bang theory). His auroral theory was disputed, in particular by *S. Chapman* (*q.v.*), and was generally disregarded. But when direct measurements in space became possible, many of Hannes Alfvén's ideas, especially about the auroral acceleration process, were vindicated. A similar fate was shared by many of Hannes Alfvén's ideas. Even his discovery of the magnetohydrodynamic waves was not taken seriously until Enrico Fermi acknowledged the possibility of their existence. His early work in cosmic ray physics led Alfvén to predict the existence of a galactic magnetic field, but his prediction was universally rejected, and by the time the galactic field was discovered, his prediction was long forgotten.

Hannes Alfvén was active in public affairs throughout his life, particularly in opposition to nuclear proliferation. He was president of the Pugwash Conference from 1970 to 1975. He wrote several popular science books, including a work of science fiction under the pseudonym O. Johannesson. Further details of his life and a complete list of publications may be found in Pease and Lindqvist (1998); more details of his life are on http://public.lanl.gov/alp/plasma/people/alfven.html.

Carl-Gunne Fälthammar and David Gubbins

Bibliography

Alfvén, H., 1939. A theory of magnetic storms and of the aurorae (I), *Kungliga Svenska Vetenskapsakademiens Handlingar, Tredje Serien*, **18**: 1–39; Partial reprint EOS Transactions of the American Geophysical Union, 1970, 51: 181–193.

Alfvén, H., 1942. Existence of electromagnetic-hydrodynamic waves. *Nature*, **150**: 405–406.

Alfvén, H., 1950. *Cosmical Electrodynamics*, Oxford: Clarendon Press.

Lundqvist, S., 1949. Experimental demonstrations of magneto-hydrodynamic waves. *Nature*, **164**: 145–146.

Lundqvist, S., 1952. Studies in magnetohydrodynamics. *Arkiv för Fysik*, **5**: 297–347.

Pease, R.S., and Lindqvist, S., 1998. Hannes Olof Gösta Alfvén. *Biographical Memoirs of the Royal Society*, **44**: 3–19.

Cross-references

Alfvén's Theorem and the Frozen Flux Approximation
Alfvén Waves
Chapman, Sydney (1888–1970)
Core Motions
Cowling, Thomas George (1906–1990)
Geodynamo, Dimensional Analysis and Timescales
Magnetohydrodynamics
Magnetohydrodynamic Waves
Oscillations, Torsional
Storms and Substorms, Magnetic

ALFVÉN'S THEOREM AND THE FROZEN FLUX APPROXIMATION

History

In 1942, *Hannes Alfvén* (*q.v.*) published a paper that announced the discovery of the wave that now bears his name (Alfvén, 1942; see *Alfvén waves*), and that is now often regarded as marking the birth of magnetohydrodynamics or "MHD" for short (see *Magnetohydrodynamics*). In interpreting the waves in an associated paper (Alfvén, 1943, §4), he enunciated a result that has become known as "Alfvén's theorem":

> *Suppose that we have a homogeneous magnetic field in a perfectly conducting fluid.... In view of the infinite conductivity, every motion (perpendicular to the field) of the liquid in relation to the lines of force is forbidden because it would give infinite eddy currents. Thus the matter of the liquid is "fastened" to the lines of force...*

Here, one understands "in relation" to mean "relative." Also the term "frozen" instead of "fastened" has been thought more appealing, and the theorem is frequently referred to as "the frozen flux theorem." It is now more often stated in terms such as

> *Magnetic flux tubes move with a perfectly conducting fluid as though frozen to it.*

The reason why it is a little more accurate to refer to "magnetic flux tubes" rather than "magnetic field lines" will be explained in section "Formal demonstrations of the theorem."

According to Alfvén's theorem, a perfect conductor cannot gain or lose magnetic flux. The theorem therefore has no bearing on field generation by dynamo action although, to understand how the conductor initially acquired the magnetic flux threading it, one must recognize the imperfect conductivity of the fluid and reopen the dynamo question. Similarly, the theorem shows that it is impossible to change the topology of the field lines passing through the conductor in any way whatever and, to understand how a real fluid allows the field lines to sever and reconnect, one must recognize that its resistivity is finite.

Underlying physics

Although MHD phenomena usually require the coupled equations of fluid dynamics and electromagnetism (em) to be solved together, all that is needed to establish Alfvén's theorem is the "kinematic" part of the relationship; the fluid velocity **V** is regarded as given and the magnetic field **B** is found by solving

$$\nabla \cdot \mathbf{B} = 0 \qquad \text{(Eq. 1)}$$

and the em induction equation:

$$\partial_t \mathbf{B} = \nabla \times (\mathbf{V} \times \mathbf{B}) + \eta \nabla^2 \mathbf{B}. \qquad \text{(Eq. 2a)}$$

Here

$$\eta = 1/\mu_0\sigma \quad \text{(Eq. 2b)}$$

is the "magnetic diffusivity" (assumed uniform), σ being the electrical conductivity, and μ_0 the permeability of free space: SI units are used here. In kinematic dynamo theory, **V** is sought for which (Eq. 2a) has self-sustaining solutions **B** (satisfying the appropriate boundary conditions). There is also considerable geophysical interest in the inverse problem of inferring **V** from the **B** observed at the Earth's surface; see section "The inverse problem."

Alfvén's theorem follows from (Eq. 2a) in the zero diffusion limit:

$$\partial_t \mathbf{B} = \nabla \times (\mathbf{V} \times \mathbf{B}) \quad \text{for} \quad \sigma = \infty \quad \text{(Eq. 3)}$$

A solution to this equation can be regarded as the first term in an expansion of the field in inverse powers of a magnetic Reynolds number. Such an expansion is, in the parlance of perturbation theory, "singular" since the term $\eta\nabla^2\mathbf{B}$ involving the highest spatial derivatives of **B** has been ejected from (Eq. 2a). The highest derivatives reassert themselves in the structure of boundary layers; see *Core, boundary layers* and the section "The inverse problem".

Equation (3) has the same form as the equation that governs the vorticity $\boldsymbol{\omega} = \nabla \times \mathbf{V}$ in an inviscid fluid of uniform density driven by conservative forces:

$$\partial_t \boldsymbol{\omega} = \nabla \times (\mathbf{V} \times \boldsymbol{\omega}). \quad \text{(Eq. 4)}$$

Kelvin's theorem follows from (Eq. 4):

Vortex tubes move with an inviscid fluid of uniform density as though frozen to it.

Clearly, Alfvén's theorem is the MHD analog of Kelvin's theorem. There is however a significant difference: (Eq. 3) is a linear equation that determines the evolution of **B** from an initial state for given **V**; (Eq. 4) is a nonlinear relationship between $\nabla \times \mathbf{V}$ and **V**.

Formal demonstrations of the theorem

Some readers, with as firm a grasp of em theory as Alfvén, will find his explanation of the theorem, given in the section "History", sufficiently persuasive. Others, who have a grounding in classical fluid mechanics, may be satisfied by the analogy with Kelvin's theorem, although that theorem too requires proof. Yet others, with a more mathematical bent, may prefer a direct demonstration, such as the one given below.

Let $\Gamma(t)$ be an arbitrary curve "frozen" to the fluid as it moves, and let $\mathbf{ds}(t)$ be an infinitesimal element of Γ whose ends are situated at $\mathbf{s}(t)$ and $\mathbf{s}(t) + \mathbf{ds}(t)$. The fluid velocities at these points are $\mathbf{V}(\mathbf{s}(t),t)$ and $\mathbf{V}(\mathbf{s}(t) + \mathbf{ds}(t), t) = \mathbf{V}(\mathbf{s}(t), t) + \mathbf{ds}(t) \cdot \nabla\mathbf{V}(\mathbf{s}(t), t)$. At a time $\mathrm{d}t$ later, the ends are therefore situated at $\mathbf{s} + \mathbf{V}\mathrm{d}t$ and $\mathbf{s} + \mathbf{ds} + [\mathbf{V} + \mathbf{ds} \cdot \nabla\mathbf{V}]\mathrm{d}t$. It follows that

$$d_t(\mathbf{ds}) = \mathbf{ds} \cdot \nabla\mathbf{V}, \quad \text{(Eq. 5a)}$$

where

$$d_t = \partial_t + \mathbf{V} \cdot \nabla \quad \text{(Eq. 5b)}$$

is the "motional" or "Lagrangian" derivative, i.e., the time derivative following the fluid motion.

In the first application of this kinematic result, we combine it with (Eq. 3), written in the form

$$d_t\mathbf{B} = \mathbf{B} \cdot \nabla\mathbf{V} - \mathbf{B}\nabla \cdot \mathbf{V}, \quad \text{(Eq. 6a)}$$

where we have used (Eq. 1) and a vector identity. The result is

$$d_t(\mathbf{B} \times \mathbf{ds}) = -(\mathbf{B} \times \mathbf{ds})_j \nabla V_j, \quad \text{for} \quad \sigma = \infty, \quad \text{(Eq. 6b)}$$

where the summation convention applies to the repeated suffix j. It follows that

$$\mathbf{B} \times \mathbf{ds} = \mathbf{0} \quad \text{at} \quad t = 0 \quad \Rightarrow \quad \mathbf{B} \times \mathbf{ds} = \mathbf{0} \quad \text{for all } t. \quad \text{(Eq. 6c)}$$

A *magnetic field line* is a curve every element **ds** of which is parallel to **B**. According to (Eq. 6c), if Γ is initially a field line, it is always a field line. This establishes that magnetic field lines move with a perfectly conducting fluid, the weak form of Alfvén's theorem. The conclusion would still follow if a source term parallel to **B** were added to the right-hand side of (Eq. 3). The absence of this term leads to the stronger form of the theorem given below.

For the second application of (Eq. 5a), let $\Sigma(t)$ be an arbitrary (open) surface "frozen" to the fluid as it moves and let $\mathbf{dS}(t)$ be an infinitesimal element of Σ having the shape of a parallelogram, two adjacent sides being along curves $\Gamma^{(1)}$ and $\Gamma^{(2)}$, so that

$$\mathbf{dS} = \mathbf{ds}^{(1)} \times \mathbf{ds}^{(2)}, \quad \text{(Eq. 7a)}$$

where $\mathbf{ds}^{(1)}$ and $\mathbf{ds}^{(2)}$ are elements of $\Gamma^{(1)}$ and $\Gamma^{(2)}$ to which (Eq. 5a) applies, leading to the result

$$d_t(\mathbf{dS}) = \mathbf{dS}(\nabla \cdot \mathbf{V}) - \mathrm{d}S_j\nabla V_j. \quad \text{(Eq. 7b)}$$

It follows from (Eq. 6a) and (Eq. 7b) that

$$d_t(\mathbf{B} \cdot \mathbf{dS}) = 0, \quad \text{for} \quad \sigma = \infty, \quad \text{(Eq. 8a)}$$

and therefore

$$\mathbf{B} \cdot \mathbf{dS} = \mathbf{0} \quad \text{at} \quad t = 0 \quad \Rightarrow \quad \mathbf{B} \cdot \mathbf{dS} = \mathbf{0} \quad \text{for all } t. \quad \text{(Eq. 8b)}$$

A *magnetic surface*, $M(t)$, is a surface composed of field lines, so that every surface element, **dS**, is perpendicular to **B**. According to (Eq. 8b), if M is initially a magnetic surface, it is always a magnetic surface. This conclusion also follows from (Eq. 6c).

To obtain the stronger result, let M be a *magnetic flux tube*. This is a bundle of magnetic field lines and is therefore bounded by a magnetic surface for all t. Let $\Sigma(t)$ be a cross-section of the tube, i.e., a surface that cuts across every magnetic field line within the tube. The net magnetic flux through Σ,

$$m = \int_{\Sigma(t)} \mathbf{B} \cdot \mathbf{dS}, \quad \text{(Eq. 9)}$$

is called the "strength" of the tube. It is easily seen from (Eq. 1) that m is the same for every cross-section Σ. Equation (8b) shows further that m is time independent:

$$d_t \int_{\Sigma(t)} \mathbf{B} \cdot \mathbf{dS} = 0, \quad \text{for} \quad \sigma = \infty. \quad \text{(Eq. 10a)}$$

This conservation of flux is the strong form of Alfvén's theorem.

It is worth observing that, according to Eqs. (1), (3), and Stokes' theorem, (Eq. 10a) may also be written in Eulerian terms as

$$\int_{\Sigma(t)} (\partial_t\mathbf{B}) \cdot \mathbf{dS} + \oint_{\Gamma(t)} \mathbf{B} \cdot (\mathbf{V} \times \mathbf{ds}) = 0, \quad \text{for} \quad \sigma = \infty, \quad \text{(Eq. 10b)}$$

where $\Gamma(t)$ is the perimeter of $\Sigma(t)$. If Σ were fixed in space, the first term on the left-hand side of (Eq. 10b) would be the rate of increase of the flux through Σ. The $\mathbf{V} \times \mathbf{ds}$ in the second term is the rate at which the vector area of Σ increases as $\mathbf{ds}$, on the perimeter Γ of Σ, is advected by the fluid motion $\mathbf{V}$. The associated rate at which Σ gains magnetic flux is therefore $\mathbf{B} \cdot (\mathbf{V} \times \mathbf{ds})$.

The inverse problem

Even though no material, apart from superconductors, can transmit electricity without ohmic loss, Alfvén's theorem is often useful in visualizing MHD processes. For example, in rapidly rotating convective systems such as the Earth's core, the Coriolis force deflects the rising and falling buoyant streams partially into the zonal directions, and creates zonal field. This "ω-effect" is readily pictured with the help of Alfvén's theorem: The field lines of the axisymmetric part of $\mathbf{B}$, for example, are dragged along lines of latitude by the zonal shear, i.e., the shear adds a zonal component to $\mathbf{B}$. Alfvén's theorem is also helpful in attacking the problem of inferring unobservable fluid motions from observed magnetic field behavior, and it is this application that will be considered now: $\mathbf{B}$ and $\partial_t\mathbf{B}$ will be assumed known and $\mathbf{V}$ will be sought.

The electrical conductivity of the mantle is so low that the toroidal electric currents induced in the mantle by the changing MHD state of the core are small. If they are neglected, the magnetic field created by the core is, everywhere in the mantle and above, a potential field that can be expressed as a sum of internal spherical harmonics with time-varying Gauss coefficients whose values are obtained by analyzing observatory and satellite data (see *Time-dependent models of the geomagnetic field*). This sum can be used to compute $\mathbf{B}$ at the core-mantle boundary (CMB) but the larger the order n of the spherical harmonic the more rapidly it increases with depth into mantle. The energy spectrum, which is dominated by the dipolar components at the Earth's surface, is almost flat at the CMB (see *Geomagnetic spatial spectrum*). Clearly, small errors that arise in the large n coefficients when the data at the Earth's surface is analyzed can have serious consequences when $\mathbf{B}$ is extrapolated to the CMB.

Although $\mathbf{B}$ is continuous across the CMB, its radial derivative is not. Extrapolation cannot therefore provide information about $\nabla^2\mathbf{B}$ in the core, even at its surface. If the unknown $\eta\nabla^2\mathbf{B}$ in (Eq. 2a) is significant, it is impossible to learn anything about $\mathbf{V}$ from the observed $\mathbf{B}$. If however $\eta\nabla^2\mathbf{B}$ is small enough, (Eq. 3) may be a good first approximation to (Eq. 2a), and the extrapolated $\mathbf{B}$ and $\partial_t\mathbf{B}$ may yield information about $\mathbf{V}$ (Roberts and Scott, 1965).

The first task is to assess the importance of $\eta\nabla^2\mathbf{B}$ in (Eq. 2a), restricting attention to the large scales L of $\mathbf{B}$ corresponding to the Gauss coefficients ($n \leq 12$) that are accessible. The three terms appearing in (Eq. 2a) are respectively of order B/T, V_SB/L, and $\eta B/L^2$, where T is the timescale of the large-scale $\mathbf{B}$. The third term is small compared with one or both of the other terms if

$$V_S > \eta/L \approx 5 \times 10^{-6}\,\mathrm{m\,s^{-1}} \quad \text{and/or} \quad T << L^2/\eta \approx 2 \times 10^3\,\mathrm{yr}, \qquad \text{(Eq. 11a, b)}$$

where we have taken $\eta = 2\,\mathrm{m^2\,s^{-1}}$ and $L = R_1/10$ as representative, where R_1 is the core radius. Many scales of geophysical interest satisfy (Eq. 11), and the adoption of (Eq. 3) in place of (Eq. 2a) seems plausible for these scales. In geomagnetic data analysis this is known as "the frozen flux hypothesis" or "the frozen flux approximation."

The plethora of geodynamo simulations that have been performed over the past decade have afforded many opportunities of comparing different methods of solving the inverse problem, and in particular of testing the frozen flux hypothesis, without having to incur the inaccuracies of downward extrapolation. Roberts and Glatzmaier (2000) used one of their highly numerically resolved simulations to evaluate the unsigned flux,

$$U = \int_{\mathrm{CMB}} |B_r|\mathrm{d}S, \qquad \text{(Eq. 12)}$$

at three epochs separated by about 150 years; see also Glatzmaier and Roberts (1996). It follows from (Eq. 10a) that U should be constant. They found instead that U varied, though by less than 4% over the $\sim$300 years. In deriving this result, they pretended that, as for the real Earth, only the harmonics of degree n less than 13 were available in calculating U but, even when all harmonics for $n \leq 36$ were used, U changed by only about 8% over the 300 year interval. (The slightly increased variability is expected since (Eq. 11) are less well obeyed for smaller L, i.e., larger n.)

It was already noticed earlier that (Eq. 3) governs the leading (i.e., largest) term in a singular expansion of $\mathbf{B}$ in powers of parameters such as η/VL and $\eta T/L^2$, which are small according to (Eq. 11). In fluid mechanics, such expansions are often said to be "mainstream expansions," since they apply in the main body of the fluid, away from thin diffusive "boundary layers." The nature of the boundary layers was discussed by Roberts and Scott (1965) but more realistically by Backus (1968) and by Hide and Stewartson (1972) (see *Core, boundary layers*). See also Braginsky and Le Mouël (1993) and Gubbins (1996). The $\eta\nabla^2\mathbf{B}$ term in (Eq. 2a) is important in the boundary layers since it, and it alone, can adjust the mainstream $\mathbf{B}$ to conditions imposed by the mantle. The nature of the singular expansion of $\mathbf{B}$ becomes important in interpreting what is meant by "the core surface velocity."

In the most convenient frame of reference, the one used here that rotates with the mantle, $\mathbf{V} = \mathbf{0}$ on the CMB (the no-slip condition). But both $\mathbf{V}$ and $\mathbf{B}$ change rapidly across the diffusive boundary layer at the CMB. This layer adjusts the (zero) $\mathbf{V}$ and the extrapolated $\mathbf{B}$ on the CMB to the mainstream $\mathbf{V}$ and $\mathbf{B}$ governed by core MHD. It is the velocity, $\mathbf{V}_S$, on the lower surface S of this boundary layer that is the core surface velocity sought. The thickness, δ, of the boundary layer is small compared with the scales L of $\mathbf{B}$ that can be extracted from the geomagnetic data. The normal components of these large-scale $\mathbf{B}$ and $\mathbf{V}$ cannot change substantially within the distance δ. This has two consequences. First, since $V_r = 0$ on the CMB,

$$V_r = 0, \quad \text{on} \quad S. \qquad \text{(Eq. 13)}$$

Second, by (Eq. 1), the B_r and $\partial_t B_r$ extrapolated to the CMB are also the B_r and $\partial_t B_r$ on S. These are related since, according to (Eq. 3),

$$\partial_t B_r + \nabla_\mathrm{H} \cdot (B_r\mathbf{V}_S) = 0, \quad \text{on } S \text{ for} \quad \sigma = \infty, \qquad \text{(Eq. 14)}$$

where ∇_H is the horizontal gradient operator.

Once (Eq. 14) is accepted, instead of the corresponding equation derived from (Eq. 2a), questions of nonexistence and nonuniqueness arise. By (Eq. 10a)

$$d_t \int_{\Sigma(t)} B_r\, \mathrm{d}S = 0, \qquad \text{(Eq. 15a)}$$

for every co-moving surface area $\Sigma(t)$ on S. Lacking knowledge of $\mathbf{V}_S$, it is impossible to test (Eq. 15a) for every $\Sigma(t)$, but a more limited objective can be reached. One may transform (Eq. 15a), as in (Eq. 10b), into

$$\int_{\Sigma(t)} (\partial_t B_r)\mathrm{d}S + \oint_{\Gamma(t)} B_r(\hat{\mathbf{n}} \cdot \mathbf{V}_S)\mathrm{d}s = 0, \quad \text{for } \sigma = \infty, \qquad \text{(Eq. 15b)}$$

where $\hat{\mathbf{n}}$ is the (horizontal) normal vector to $\Gamma(t)$, directed outwards along S from the interior of $\Sigma(t)$. [Equation (15b) also follows by integrating (Eq. 14) over $\Sigma(t)$ and applying the two-dimensional divergence theorem.] Let $\Gamma(t)$ be a *null flux curve* (NFC), which is a closed curve on S defined by

$$B_r = 0, \quad \text{on an NFC.} \qquad \text{(Eq. 16a)}$$

Equation (15b) gives (Backus, 1968)

$$\int_{\Sigma(t)} (\partial_t B_r)\mathrm{d}S = 0, \quad \text{on an NFC.} \qquad \text{(Eq. 16b)}$$

This, in contrast to (Eq. 15a), can easily be tested for the given $\partial_t B_r$. By (Eq. 1), there is at least one NFC on S, the magnetic equator, which separates the southern magnetic "hemisphere," where most (or conceivable all) of the magnetic flux leaves the core, from the northern magnetic "hemisphere" where it re-enters the core. Other smaller NFC can, and apparently do exist within these hemispheres. Because their scale L is smaller, (Eq. 11a,b) and therefore (Eq. 16b) are less well satisfied.

One must expect that observations will be inconsistent with (Eq. 16b) for three reasons. First, the analysis of geomagnetic data into spherical harmonic components is not completely accurate; second, the downward extrapolation to the CMB introduces errors that increase with the spherical harmonic number n of the Gauss coefficients; third, because σ is not infinite in the core, the magnetic flux through $\Sigma(t)$ is not perfectly conserved anyway! At first sight, the third reason might seem to be the most significant, in view of the current rapid decrease in the axial dipole moment of the Earth (about 8% over the last 150 years). One should not forget however that the frozen flux hypothesis implies Alfvén's theorem, according to which the secular variation is caused by the rearrangement of the existing field lines already entering and leaving the core. Such a rearrangement can easily accommodate such a rapid change in dipole moment without contradicting the hypothesis; see Bondi and Gold (1950). The possibility that the other two reasons are mainly responsible for violations of (Eq. 16b) remains open. This led Roberts and Scott (1965) to suggest that the frozen field constraint should be turned into an advantage. They proposed that future spherical harmonic analyses of the geomagnetic field should impose the constraint both to improve the spherical harmonic representation of **B** and to obtain a finite $\mathbf{V}_S$ as a by-product. For implementations of this idea, see Benton *et al.* (1987), Constable *et al.* (1993) and O'Brien *et al.* (1997); see also *Benton E.R.*

When (Eq. 16b) is satisfied, a finite $\mathbf{V}_S$ exists but it is not unique. This was noted by Roberts and Scott (1968) and is most easily seen from the Backus (1968) representation of $\mathbf{V}_S$:

$$B_r\mathbf{V}_S = \nabla_{\mathrm{H}}\phi - \hat{\mathbf{r}}\times\nabla_S\psi, \qquad \text{(Eq. 17a)}$$

where $\hat{\mathbf{r}}$ is the radial unit vector. It follows from (Eq. 14) that

$$\nabla^2_{\mathrm{H}}\phi = -\partial_t B_r, \qquad \text{(Eq. 17b)}$$

from which ϕ can be determined uniquely from the given $\partial_t B_r$, but ψ remains largely undetermined, though Backus (1968) showed how a little information about ψ might be extracted.

The possibility of using the horizontal components of **B** to obtain information about $\mathbf{V}_S$ was mooted by Roberts and Scott (1965) and was the subject of a penetrating analysis by Backus (1968). There is a fundamental difficulty: when integrated across the boundary layer at the CMB, the electric currents (density **J**), which are horizontal on the CMB, are equivalent to a surface current **I**, and

$$\mathbf{B}_S = \mathbf{B}_{\mathrm{CMB}} + \mu_0\hat{\mathbf{r}} \times \mathbf{I}; \qquad \text{(Eq. 18)}$$

(see *Core, boundary layers*). Evidently, it is impossible to determine $\mathbf{B}_S$ from $\mathbf{B}_{\mathrm{CMB}}$ without knowledge of **I**. Nevertheless Barraclough *et al.* (1989), making use of an interpretation of Backus' analysis due to Moffatt, undertook a study that led them to believe that **I** is small enough to be neglected. The resulting $\mathbf{V}_S$ was, however, unrealistic.

It was mentioned above that the secular variation is, according to the frozen flux hypothesis, created by the rearrangement of preexisting magnetic flux to and from the core. The alternative process of "flux creation" relies on ohmic diffusion. The rapid decrease in the Earth's axial dipole has been attributed by Bloxham and Gubbins (1985) to flux creation, more specifically to growing reversed flux patches within the southern magnetic hemisphere. They proposed to model this process by determining **V** from (Eq. 2a) rather than (Eq. 3).

One mechanism that has attracted particular attention is conversion of the otherwise undetectable toroidal field in the core into poloidal field that escapes into the mantle. The idea is analogous to the ejection of the large subsurface zonal magnetic field in the sun into the solar atmosphere, with the creation of a sunspot pair on the solar surface. This can be modeled directly by Alfvén's theorem since the solar surface is a "free" surface on which V_r may be nonzero; after a flux tube erupts, the fluid within the tube can fall back under gravity to the solar surface, leaving the erupted field in the solar atmosphere. This is not possible for the Earth's core by (Eq. 13). The flux can be ejected only by diffusion, i.e., by violating Alfvén's theorem through finite σ. An upwelling crowds field lines close to the CMB, and the enhanced **J** that results promotes field diffusion between the CMB and mantle. This idea was modeled in several publications, starting with Coulomb (1955), Allen and Bullard (1958, 1966), Nagata and Rikitake (1961), Hide and Roberts (1961), and Rikitake (1967), and was more recently revived by Bloxham (1986) and Drew (1993). It was originally motivated first by the belief that the ω-effect (see above) is so potent that the toroidal field dominates the poloidal field everywhere in the core (except at the CMB), and second by the thought that even the slow diffusion of a sufficiently strong toroidal field could create significant secular variation. Recent geodynamo simulations have suggested however that the Lorentz force so strongly opposes the ω-effect that the toroidal and poloidal fields have similar magnitudes. This makes toroidal flux expulsion models less attractive.

Paul H. Roberts

Bibliography

Alfvén, H., 1942. Existence of electromagnetic-hydrodynamic waves. *Nature*, **150**: 405–406.

Alfvén, H., 1943. On the existence of electromagnetic-hydrodynamic waves. *Arkiv foer Matematik, Astronomi och Fysik*, **39**: 2.

Allen, D.W., Bullard, E.C., 1958. Distortion of a toroidal field by convection. *Reviews of Modern Physics*, **30**: 1087–1088.

Allen, D.W., Bullard, E.C., 1966. The secular variation of the earth's magnetic field. *Proceedings of the Cambrige Philosophical Society*, **62**: 783–809.

Backus, G.E., 1968. Kinematics of geomagnetic secular variation in a perfectly conducting core. *Philosophical Transactions of the Royal Society of London*, **A263**: 239–266.

Barraclough, D., Gubbins, D., Kerridge, D., 1989. On the use of the horizontal components of the magnetic field in determining core motions. *Geophysical Journal International*, **98**: 293–299.

Benton, E.R., Estes, R.H., Langel, R.A., 1987. Geomagnetic field modeling incorporating constraints from frozen-flux electromagnetism. *Physics of Earth and Planetary Interiors*, **48**: 241–264.

Bloxham, J., 1986. The expulsion of magnetic flux from the Earth's core. *Geophysical Journal of the Royal Astronomical Society*, **87**: 669–678.

Bloxham, J., Gubbins, D., 1985. The secular variation of the Earth's magnetic field. *Nature*, **317**: 777–781.

Bondi, H., Gold, T., 1950. On the generation of magnetism by fluid motion. *Monthly Notices of the Royal Astronomical Society*, **110**: 607–611.

Braginsky, S.I., Le Mouël, J.-L., 1993. Two-scale model of a geomagnetic field variation. *Geophysical Journal International*, **112**: 147–158.

Constable, C.G., Parker, R.L., Stark, P.B., 1993. Geomagnetic field models incorporating frozen flux constraints. *Geophysical Journal International*, **113**: 419–433.
Coulomb, J., 1955. Variation seculaire par convergence ou divergence à la surface du noyau. *Annals of Geophysics*, **11**: 80–82.
Drew, S.J., 1993. Magnetic field expulsion into a conducting mantle. *Geophysical Journal International*, **115**: 303–312.
Glatzmaier, G.A., Roberts, P.H., 1996. On the magnetic sounding of planetary interiors. *Physics of Earth and Planetary Interiors*, **98**: 207–220.
Gubbins, D., 1996. A formalism for the inversion of geomagnetic data for core motions with diffusion. *Physics of Earth and Planetary Interiors*, **98**: 193–206.
Hide, R., Roberts, P.H., 1961. The origin of the main geomagnetic field. *Physics and Chemistry of the Earth*, **4**: 27–98.
Hide, R., Stewartson, K., 1972. Hydromagnetic oscillations of the earth's core. *Reviews of Geophysics Space Physics*, **10**: 579–598.
Nagata, T., Rikitake, T., 1961. Geomagnetic secular variation and poloidal magnetic fields produced by convectional motion in the Earth's core. *Journal of Geomagnetism and Geoelectricity*, **13**: 42–53.
O'Brien, M.S., Constable, C.G., Parker, R.L., 1997. Frozen-flux modeling for epochs 1915 and 1980. *Geophysical Journal International*, **128**: 434–450.
Rikitake, T., 1967. Non-dipole field and fluid motion in the Earth's core. *Journal of Geomagnetism and Geoelectricity*, **19**: 129–142.
Roberts, P.H., Glatzmaier, G.A., 2000. A test of the frozen-flux approximation using a new geodynamo model. *Philosophical Transactions of the Royal Society of London*, **A358**: 1109–1121.
Roberts, P.H., Scott, S., 1965. On analysis of the secular variation I. A hydromagnetic constraint: theory. *Journal of Geomagnetism and Geoelectricity*, **17**: 137–151.

Cross-references

Alfvén Waves
Alfvén, Hannes Olof Gösta (1908–1995)
Benton, Edward R. (1934–1992)
Core, Boundary Layers
Geomagnetic Spectrum, Spatial
Magnetohydrodynamics
Time-dependent Models of the Geomagnetic Field

ANELASTIC AND BOUSSINESQ APPROXIMATIONS

Introduction

Background

The anelastic and Boussinesq approximations are simplifications of the system of equations governing buoyantly driven flows. The Boussinesq equations were the first to appear in print. In an influential paper on thermal convection, Rayleigh (1916) attributed them, without reference, to Boussinesq and the name has stuck, even though they had previously been employed by Oberbeck (1888). Although variations in density are the very essence of buoyancy, they are neglected everywhere in the Boussinesq equations "except in so far as they modify the action of gravity" (Rayleigh, 1916), i.e., the density is assumed to be constant except in the buoyancy force. This retains the essential physics with a minimum of complexity. The Boussinesq approximation usually performs well in modeling convection in laboratory conditions in which variations in pressure scarcely affect the density of the fluid. It is, however, unsatisfactory for large systems like the Earth's core, the lower part of which is significantly compressed by the overlying material. The Boussinesq approximation may then introduce unacceptable errors. A pressure dependence of density was first incorporated into the thermal convection equations by Spiegel and Veronis (1960) and, for atmospheric convection, by Ogura and Phillips (1962) (see also Gough, 1969). Braginsky (1964) introduced the anelastic approximation (calling it "the inhomogeneous model") for studying convection in a two-component fluid modeling the Earth's core. In what follows, "AA" is frequently used as an abbreviation for the "anelastic approximation" (see the section "The anelastic approximation (AA)"), "BA" for the "Boussinesq approximation" (see the section "The Boussinesq approximation (BA)"), and "CA" as a general term to describe these "convective approximations" and others to be described below.

Although none of the CAs can be used if the convective velocities are comparable with the speed of sound, one or another is applicable to convection in the laboratory, the oceans, the atmosphere, and the Earth's mantle and fluid core (see *Core convection*).

The following sections aim to describe the essence of the CAs with a minimum of inessentials. There are several simplifications:

a. Only a one-component fluid is considered. In many geophysical systems, this is too restrictive; convection is complicated by the effects of salinity in the ocean, water vapor in the atmosphere, phase transitions in the mantle, and light admixture in the core (see *Convection, chemical*).
b. The system under consideration is supposed to be driven steadily, by constant heat sources within it and/or constant temperature differences between its upper and lower boundaries (both assumed horizontal and stationary); this does not imply that the convection is steady.
c. Variations in the gravitational acceleration, **g**, created by the density differences associated with the convection are ignored; **g** is assumed to be time-independent and specified in terms of a potential U so that $\mathbf{g} = -\nabla U$. The more general self-gravitational case, in which **g** is created by the mass of the fluid itself and varies as the convection redistributes mass, is not treated here.
d. The inertial frame is used; there are no Coriolis or centripetal accelerations. There are no magnetic fields and associated Lorentz forces. For more general cases, see *Convection, nonmagnetic rotating; Magnetoconvection; Magnetohydrodynamics; Geodynamo*.

For the Earth's core, restrictions (a)-(d) are applied in papers by Braginsky (1964) and Braginsky and Roberts (1995, 2003). In particular, the slow evolution of the Earth, which has significant implications for core dynamics, is allowed for by a minor extension of the methods presented here. For mantle convection, see Davis and Richards (1992) and Schubert *et al.* (2001); phase transitions are considered in Tackley (1997). For oceanic and atmospheric convection, see Kamenkovich (1977), Gill (1982), and Monin (1990). For astrophysical convection, see Lantz and Fan (1999).

Basic equations

In their primitive, unapproximated form, the laws governing conservation of mass, momentum, and energy may be written as (e.g., Landau and Lifshitz, 1987)

$$\partial_t \rho = -\nabla \cdot (\rho \mathbf{V}), \qquad \text{(Eq. 1)}$$

$$\rho d_t \mathbf{V} = -\nabla P + \rho \mathbf{g} + \mathbf{F}^{\mathrm{v}}, \qquad \text{(Eq. 2)}$$

$$\rho d_t \mathcal{E}^{\mathrm{I}} + \nabla \cdot \mathbf{I}^q = -P \nabla \cdot \mathbf{V} + Q^{\mathrm{R}} + Q^{\mathrm{v}}, \qquad \text{(Eq. 3)}$$

where $\partial_t = \partial/\partial t$ and $d_t = \partial_t + \mathbf{V} \cdot \nabla$ is the motional derivative; ρ is the density; P is the pressure; $\mathbf{V}$ is the fluid velocity, $\mathbf{F}^{\mathrm{v}}$ is the viscous force per unit volume; $\mathcal{E}^{\mathrm{I}}(\rho, S)$ is the internal energy per unit mass (i.e., the "specific" internal energy); S is the specific entropy; $\mathbf{I}^q$ is the heat flux; Q^{R} is a constant heat production per unit volume from internal heat sources (such as dissolved radioactivity); and Q^{v} is the heating rate per unit volume through viscous friction. The first term on the right-hand side of (Eq. 3) represents reversible $P\,\mathrm{d}\rho^{-1}$ work that, for example, is exchanged between the kinetic and internal energy when fluid motions compress or decompress the fluid; from (Eq. 1), $\nabla \cdot \mathbf{V} = -\rho d_t \rho^{-1}$, where ρ^{-1} is the specific volume.

The first law of thermodynamics for a moving fluid is

$$\mathrm{d}\mathcal{E}^{\mathrm{I}} = -P\mathrm{d}\rho^{-1} + T\mathrm{d}S, \qquad \text{(Eq. 4)}$$

where T is temperature. It therefore follows from Eqs. (1) and (3) that

$$\rho T d_t S + \nabla \cdot \mathbf{I}^q = Q^{\mathrm{R}} + Q^{\mathrm{v}}. \qquad \text{(Eq. 5)}$$

This equation, which may be called the "heat equation," is sometimes more convenient than (Eq. 3). It may also be written as

$$\rho d_t S + \nabla \cdot \mathbf{I}^S = \sigma^S, \qquad \text{(Eq. 6)}$$

and is then often called the "entropy equation" because it shows how entropy is transported and created. It is transported by the entropy flux, $\mathbf{I}^S$, which is related to the heat flux, $\mathbf{I}^q$, by

$$\mathbf{I}^S = \mathbf{I}^q / T. \qquad \text{(Eq. 7)}$$

It is created at a rate σ^S per unit volume which, according to (Eq. 5), is

$$\sigma^S = \sigma^{\kappa} + \sigma^{\mathrm{v}} + \sigma^{\mathrm{R}}, \qquad \text{(Eq. 8)}$$

the three contributions to which arise respectively from thermal conduction, viscosity, and the internal heat sources:

$$\sigma^{\kappa} = -T^{-1}\mathbf{I}^S \cdot \nabla T = \mathbf{I}^q \cdot \nabla T^{-1}, \quad \sigma^{\mathrm{v}} = T^{-1}Q^{\mathrm{v}}, \quad \sigma^{\mathrm{R}} = T^{-1}Q^{\mathrm{R}}. \qquad \text{(Eq. 9)}$$

The second law of thermodynamics states that evolution is irreversible. This requires that

$$\sigma^S \geq 0. \qquad \text{(Eq. 10)}$$

Since this must hold for all possible evolving $\mathbf{V}$ and T, the three contributions (Eq. 9) to σ^S must be individually nonnegative.

It follows from Eqs. (1) and (2) that the evolution of the specific kinetic energy density, $\mathcal{E}^{\mathrm{K}} = \frac{1}{2}V^2$, is governed by

$$\rho d_t \mathcal{E}^{\mathrm{K}} = -\mathbf{V} \cdot \nabla P + \rho \mathbf{V} \cdot \mathbf{g} + \mathbf{V} \cdot \mathbf{F}^{\mathrm{v}}. \qquad \text{(Eq. 11)}$$

Conservation of the total energy density,

$$\partial_t u^{\mathrm{total}} + \nabla \cdot \mathbf{I}^{\mathrm{total}} = Q^{\mathrm{R}}, \qquad \text{(Eq. 12)}$$

follows from (3) and (11). Here

$$u^{\mathrm{total}} = \rho(\mathcal{E}^K + \mathcal{E}^{\mathrm{I}} + U), \qquad \text{(Eq. 13)}$$

$$\mathbf{I}^{\mathrm{total}} = \rho(\mathcal{E}^K + \mathcal{E}^H + U)\mathbf{V} + \mathbf{I}^q + \mathbf{I}^{\mathrm{v}}, \qquad \text{(Eq. 14)}$$

are the total energy density per unit volume and the total energy flux, $\mathcal{E}^H = \mathcal{E}^I + P/\rho$ is the specific enthalpy, and $\mathbf{I}^{\mathrm{v}}$ is the energy flux due to viscosity. The derivation of (Eq. 12) makes use of alternative expressions for the last two terms in (Eq. 11). First, from (Eq. 1) and the time independence of U, the rate of working of the buoyancy force is

$$\rho \mathbf{V} \cdot \mathbf{g} = -[\partial_t(\rho U) + \nabla \cdot (\rho U \mathbf{V})]. \qquad \text{(Eq. 15)}$$

Second, the rate of working of the frictional force is

$$\mathbf{V} \cdot \mathbf{F}^{\mathrm{v}} = -\nabla \cdot \mathbf{I}^{\mathrm{v}} - Q^{\mathrm{v}}. \qquad \text{(Eq. 16)}$$

More explicitly, the viscous force is the divergence of the viscous stress tensor, $\overleftrightarrow{\boldsymbol{\pi}}^{\mathrm{v}}$:

$$F_i^{\mathrm{v}} = \nabla_j \pi_{ji}^{\mathrm{v}}, \quad \text{where} \quad \pi_{ij}^{\mathrm{v}} = \pi_{ji}^{\mathrm{v}}, \qquad \text{(Eq. 17)}$$

so that (Eq. 16) holds with

$$I_i^{\mathrm{v}} = -V_j \pi_{ji}^{\mathrm{v}}, \quad Q^{\mathrm{v}} = \pi_{ij}^{\mathrm{v}} \nabla_j V_i. \qquad \text{(Eq. 18)}$$

Fourier's law will be employed below, i.e.,

$$\mathbf{I}^q = -K^T \nabla T, \qquad \text{(Eq. 19)}$$

$$\sigma^{\kappa} = K^T (\nabla T)^2 / T^2, \qquad \text{(Eq. 20)}$$

where $K^T (\geq 0)$ is the thermal conductivity.

For later convenience, (Eq. 4) is augmented here by further thermodynamic relations (e.g., Landau and Lifshitz, 1980):

$$\rho^{-1}\mathrm{d}\rho = (\rho v_T^2)^{-1}\mathrm{d}P - \alpha \mathrm{d}T, \qquad \text{(Eq. 21)}$$

$$\rho^{-1}\mathrm{d}\rho = (\rho v_S^2)^{-1}\mathrm{d}P - \alpha^S \mathrm{d}S, \qquad \text{(Eq. 22)}$$

$$T\mathrm{d}S = C_p(\mathrm{d}T - \alpha^S \mathrm{d}P/\rho), \qquad \text{(Eq. 23)}$$

$$T\mathrm{d}S = C_v(\mathrm{d}T - \alpha^S v_S^2 \mathrm{d}\rho/\rho), \qquad \text{(Eq. 24)}$$

where v_T and v_S are the isothermal and adiabatic sound speeds, C_p and C_v are the specific heats at constant pressure and volume, and α and α^S are the thermal and entropic expansion coefficients:

$$v_T^2 = \left(\frac{\partial P}{\partial \rho}\right)_T, \quad v_S^2 = \left(\frac{\partial P}{\partial \rho}\right)_S, \quad C_p = T\left(\frac{\partial S}{\partial T}\right)_P, \quad C_v = T\left(\frac{\partial S}{\partial T}\right)_\rho, \qquad \text{(Eq. 25)}$$

$$\alpha = -\frac{1}{\rho}\left(\frac{\partial \rho}{\partial T}\right)_P = -\rho\left(\frac{\partial S}{\partial P}\right)_T, \quad \alpha^S = -\frac{1}{\rho}\left(\frac{\partial \rho}{\partial S}\right)_P = \rho\left(\frac{\partial T}{\partial P}\right)_S = \frac{\alpha T}{C_p}. \qquad \text{(Eq. 26)}$$

The equivalence of the two expressions for α follow from the Gibbs relation $(\partial \rho^{-1}/\partial T)_P = -(\partial S/\partial P)_T$. When written in Jacobian form as $\partial(\rho^{-1}, P)/\partial(T, P) = -\partial(T, S)/\partial(T, P)$, this relation also leads to a result that will be useful in section "A more general CA":

$$\frac{\partial(\rho^{-1}, P)}{\partial(S, T)} = 1. \qquad \text{(Eq. 27)}$$

In the same way, the Gibbs relation $(\partial \rho^{-1}/\partial S)_P = (\partial T/\partial P)_S$ confirms (Eq. 27) and also establishes the equivalence of two expressions for α^S in (Eq. 26). In what follows, the coefficients (Eq. 25) and (Eq. 26) are evaluated in reference states denoted by $_0$, $_a$ and $_b$; to avoid a proliferation of suffixes, these suffixes will be omitted, and also from K^T, the thermal diffusivity $\kappa = K^T/\rho C_p$, and the kinematic viscosity ν.

Other useful thermodynamic relations are

$$C_p/C_v = v_S^2/v_T^2, \quad C_p - C_v = C_v \alpha T \gamma, \quad v_S^2 - v_T^2 = v_T^2 \alpha T \gamma, \qquad \text{(Eq. 28)}$$

where

$$\gamma = \frac{\rho}{T}\left(\frac{\partial T}{\partial \rho}\right)_S = \frac{\alpha v_S^2}{C_p} \qquad \text{(Eq. 29)}$$

is the Grüneissen parameter which, in making numerical estimates below, we shall assume is O(1). We shall frequently need to distinguish between cases in which the working fluid is a liquid and in which it is a gas. For a liquid, it is usually true that $\alpha T \ll 1$; then (Eq. 28) shows that $C_v \approx C_p$ and $v_T \approx v_S$. For a gas, αT is close to unity.

Finally, another consequence of (Eq. 4) will be needed later:

$$d\mathcal{E}^H = \rho^{-1}dP + TdS. \qquad \text{(Eq. 30)}$$

The Boussinesq approximation (BA)

The governing equations Eqs. (1)–(5) describe a multitude of processes having widely different characteristic time scales, τ, e.g., fast processes (acoustic/seismic) and slow processes, usually called "convective." In the case of the Earth, the short timescales range from seconds to hours and the long timescales from years to millions of years. To model convection, it is advantageous to "filter out" the fast processes; this is the goal of the CAs. These approximations regard a convecting fluid as being, from a thermodynamic point of view, a small departure from a suitably defined, time-independent "reference state." This allows thermodynamic relations to be linearized. There is, however, no suggestion that other nonlinearities, such as $\rho\mathbf{V}\cdot\nabla\mathbf{V}$ in the inertial term of (Eq. 2), can be ignored.

The simplest form of CA is the Boussinesq approximation (BA), which was originally devised for liquid systems for which (i) the density ρ and temperature T have nearly constant values ρ_0 and T_0, and (ii) the deviation $\rho_1 = \rho - \rho_0$ is small compared with ρ_0 and depends only on $T_1 = T - T_0$, its dependence on the pressure P being neglected. The BA is widely and unquestioningly used; see, for example, Chandrasekhar (1961). It is employed in most, but not all, simulations of core dynamics as, for instance, in the geodynamo model of Glatzmaier and Roberts (1995). The aim of this section is to provide a simple motivation for the BA; the section "The generalized Boussinesq approximation (GBA)" presents a less direct but more satisfactory alternative.

Define a uniform reference state in which all thermodynamic quantities have constant values, indicated by the suffix $_0$. (We omit $_0$ from $C_p, \alpha, v_S^2, \ldots$ to avoid a proliferation of suffixes). The small deviations from the reference values are denoted by the suffix $_1$:

$$\rho = \rho_0 + \rho_1, \quad T = T_0 + T_1,$$
$$P = P_0 + P_1 = P_0 - \rho_0 U + \rho_0\Pi_1, \quad S = S_0 + S_1, \ldots \qquad \text{(Eq. 31)}$$

The variables carrying the suffix $_1$ have both static and time-varying parts. In particular, P_1 contains the hydrostatic term $-\rho_0 U$, which plays no essential role in the convective process.

The condition for the applicability of the BA may be illustrated through values typical of laboratory experiments in a layer of water of depth $H \sim 10$ cm:

$$\rho_0 \approx 1\text{ cm}^{-3} = 10^3\text{ kg m}^{-3}, \quad T_0 \sim 300\text{ K}, \quad P_0 = 1\text{ atm},$$
$$v_S \approx v_T \approx 1.5\text{ km s}^{-1}, \quad \alpha \sim 10^{-4}\text{K}^{-1},$$
$$\Delta T \sim 10\text{ K}, \quad g \approx 10\text{ m s}^{-2}, \qquad \text{(Eq. 32)}$$

where ΔT is typical of the temperature differences: $T_1 \sim \Delta T$. The derivation rests on "thermodynamic linearization" in which $dT, dP, d\rho, \ldots$ are replaced by $T_1, P_1, \rho_1, \ldots$ For example, (Eq. 21) gives

$$\rho_1/\rho_0 = P_1/\rho_0 v_T^2 - \alpha T_1. \qquad \text{(Eq. 33)}$$

The fractional change in ρ created by the temperature difference $T_1 \sim \Delta T$ is of order

$$\varepsilon_c \equiv (\Delta\rho/\rho)_T = \alpha\Delta T. \qquad \text{(Eq. 34)}$$

The main part of the fractional change in ρ created by the pressure difference P_1 is generally due to the hydrostatic force $-\rho_0 U$; this is usually much greater than the part due to the small correction $\rho_0\Pi_1$ associated with the convection, which is therefore neglected. According to (Eq. 33), it is quantified by $\varepsilon_a \sim P_1/\rho v_S^2$, i.e.,

$$\varepsilon_a \equiv (\Delta\rho/\rho)_P = H/H_\rho, \quad \text{where} \quad H_\rho = v_S^2/g \qquad \text{(Eq. 35)}$$

is the "density scale-height." [Despite (Eq. 33), ε_a is here defined, for later convenience, using v_S rather than v_T; for liquids, the main concern of this section, $v_S \approx v_T$.] From (Eq. 32), we see that $\varepsilon_c \sim 10^{-3} \ll 1$ and that $H_\rho \sim 200$ km is very large, so that $\varepsilon_a \sim 5 \times 10^{-7}$. Therefore both conditions (i) and (ii) are well satisfied: the density change is small and the contribution made by the P variation is much smaller than that due to T:

$$\varepsilon_a \ll \varepsilon_c \ll 1. \qquad \text{(Eq. 36)}$$

In other words, "the BA always applies in a thin enough layer."

By (Eq. 26), the thermodynamically linearized Eqs. (23) and (4) are

$$T_0 S_1 = C_p T_1 - \alpha T_0 P_1/\rho_0, \quad \mathcal{E}_1^{\mathrm{I}} = T_0 S_1 + (P_0/\rho_0^2)\rho_1. \qquad \text{(Eq. 37)}$$

By Eqs. (35) and (29), $P_1/\rho_0 \sim \varepsilon_a v_S^2 \sim \varepsilon_a(\gamma C_p/\alpha)$, so that $T_0 S_1 = C_P T_1[1 + \alpha T_0 \mathrm{O}(\varepsilon_a/\varepsilon_c)]$. i.e., $T_0 S_1$ may be replaced by $C_p T_1$ with an error that is even smaller than $\varepsilon_a/\varepsilon_c$ because αT_0 is small for a liquid. In terms of the "height of an equivalent column," $H_P = P_0/\rho_0 g$, the second of (Eq. 37) gives $\mathcal{E}_1^{\mathrm{I}} = C_p T_1[1 + \gamma H_P/H_\rho]$, where we have substituted αT_1 for ρ_1/ρ_0. According to (Eq. 32), $H_P \sim 10$ m, and hence $H_P/H_\rho \sim 5 \times 10^{-5} \ll 1$. Therefore, when conditions (Eq. 36) are satisfied, the thermodynamics gives, to high accuracy,

$$\rho_1/\rho_0 = \alpha T_1, \qquad \text{(Eq. 38)}$$

$$T_0 S_1 = C_p T_1, \qquad \text{(Eq. 39)}$$

$$\mathcal{E}_1^{\mathrm{I}} = C_p T_1. \qquad \text{(Eq. 40)}$$

With these simplifications, Eqs. (1), (2), and (3) or (5) give the BA:

$$\nabla\cdot\mathbf{V} = 0, \qquad \text{(Eq. 41)}$$

$$d_t\mathbf{V} = -\nabla\Pi_1 - \alpha T_1\mathbf{g} + \mathbf{F}^{\mathrm{v}}/\rho_0, \qquad \text{(Eq. 42)}$$

$$d_t T_1 = \kappa\nabla^2 T_1 + (Q^{\mathrm{R}} + Q^{\mathrm{v}})/\rho_0 C_p. \qquad \text{(Eq. 43)}$$

It may be seen that every term in Eqs. (1)–(5) has been evaluated on the assumption that $\rho = \rho_0$, apart from the buoyancy force, where ρ has been replaced by $\rho_0(1 - \alpha T_1)$.

A steadily convecting system may be thought of as a heat engine that does no useful work (though useful work would be done if a "windmill" were inserted into the fluid, driven by the convection). It does no useful work because, though thermal energy is converted into a higher form (kinetic energy), this returns as thermal energy through the viscous dissipation Q^{v}. The Boussinesq layer is a heat engine of very low efficiency, defined here as the ratio of the rate of working of the buoyancy forces to the rate at which the internal energy changes through convection:

$$\alpha T_1\mathbf{V}\cdot\mathbf{g}/C_p d_t T_1 \sim \alpha g H/C_p = \gamma\varepsilon_a \ll 1. \qquad \text{(Eq. 44)}$$

Here $d_t T_1$ has been estimated as $T_1/\tau_c \sim T_1 V/H$, where V and τ_c are velocity and timescales typical of the convection, τ_c being the timescale of convective overturning.

In the BA, the kinetic energy density, $\mathcal{E}^K = \frac{1}{2}V^2$, is an insignificant part of the total energy density, $\mathcal{E}^{total}$. The heat balance Eq. (43) is in fact a first, and rather good, approximation to the total energy balance Eq. (12). But, because the (small) rate of working of the buoyancy force, $-\rho_0 \alpha T_1 \mathbf{V} \cdot \mathbf{g}$, has no equal but opposite term in Eq. (43) to remove it from the total energy balance, the illusion is created that the gravitational field provides the energy necessary to drive the convection. In reality, the mean gravitational energy of a steadily convecting system does not change; the gravitational field on average does no work.

The dimensionless parameter that appears in virtually all convection studies is the Rayleigh number,

$$Ra = g\alpha\Delta T H^3/\nu\kappa. \qquad \text{(Eq. 45)}$$

For buoyancy to be influential, Ra must typically be greater than O(1). The Rayleigh number is the product of two nondimensional numbers, ε_c, and $gH^3/\nu\kappa$. Unless $\varepsilon_c \ll 1$ and $gH^3/\nu\kappa \gg 1$, it is not obvious that the BA is useful. Stated another way, although Eq. (36) should hold, the product $\varepsilon_a\varepsilon_c$ must be much larger than $\nu\kappa/(Hv_S)^2$, which in the case of (32) is of order 10^{-12}.

Returning to convection in the Earth's fluid core, and taking $\alpha = 10^{-5}\text{K}^{-1}$, $\nu = 10^{-6}\ \text{m}^2\ \text{s}^{-1}$, $\kappa = 3\times10^{-5}\ \text{m}^2\ \text{s}^{-1}$, $H^3 = 10^{19}\text{m}^3$, and $\Delta T = 1500$ K as typical, it is seen that Ra is enormously large ($Ra \sim 10^{29}$). This strongly indicates that a fresh approach is needed for this geophysical application. This will be provided in the next section, where it is shown that, by taking ΔT to be the temperature difference between the inner core boundary (ICB) and the core-mantle boundary (CMB), as was done above, Ra is overestimated by a factor of order 10^8.

The adiabatic (isentropic) reference state

In the section "The Boussinesq approximation (BA)," the basic state was uniform; its density, pressure, and all other thermodynamic variables were constant in space and time. This is appropriate only for a thin layer for which $\varepsilon_a \ll 1$. More generally the thermodynamic variables in the reference state depend on position, and here they will be distinguished by the suffix, $_a$. An expansion like Eq. (31) is envisaged:

$$\rho = \rho_a + \rho_c, \quad P = P_a + P_c, \quad T = T_a + T_c, \quad S = S_a + S_c, \ldots, \qquad \text{(Eq. 46)}$$

in which the suffix "c" distinguishes the convective parts of thermodynamic variables. As in the section "The Boussinesq approximation (BA)," it is unnecessary to add a suffix to $\mathbf{V}$ since $\mathbf{V}_a = \mathbf{0}$. Also, since by assumption the reference state is steady, $\partial_t f_a = 0$ for all adiabatic variables, f_a.

For the convective motions to be slow, it is necessary that the reference state be in "quasiequilibrium," i.e., it must be a steady state in which the dominant terms in the mechanical and thermal equations (2) and (6) balance. The mechanical quasiequilibrium is hydrostatic balance:

$$\nabla P_a = \rho_a \mathbf{g} = -\rho_a \nabla U. \qquad \text{(Eq. 47)}$$

There is more than one way of enforcing thermal "quasiequilibrium"; see also "A more general CA." In this section we focus on one extreme, though extremely useful, choice that is particularly appropriate for a strongly convecting system: the *adiabatic reference state*. (This also motivated the choice $_a$ for reference state variables.)

Within a fluid parcel of dimensions ℓ, entropy S is carried by heat conduction, at a speed of order κ/ℓ. When the velocity V of the parcel is large compared with κ/ℓ, i.e., when the Péclet number $Pe = V\ell/\kappa$ is large, S is transported far more effectively by convection. If, with the Earth's core in mind, we take $\kappa \sim 10^{-5}\ \text{m}^2\ \text{s}^{-1}$, $V \sim 10^{-4}\ \text{m s}^{-1}$ and $\ell \sim 10$ km, it is quickly seen that $\kappa/\ell \sim 10^{-5}V$, i.e., $Pe \sim 10^5$. When $Pe \gg 1$, the entropy in the parcel is carried over a long distance, of order $Pe\ \ell$, virtually unchanged. This is confirmed by (Eq. 6) which for large V is approximately $d_t S = 0$ though, over short distances, thermal diffusion smoothes out S. In this way, S is thoroughly homogenized by convection. The net result is a state that is nearly isentropic (except in boundary layers; see "Boundary conditions for CA"). It is therefore appropriate to complete the specification of the adiabatic reference state by supplementing (Eq. 47) with

$$\nabla S_a = \mathbf{0}. \qquad \text{(Eq. 48)}$$

Trivially, all thermodynamic variables defined by Eqs. (47) and (48) are functions of U in the reference state.

It is well known that (Eq. 48) is the condition for the neutral stability of a compressible fluid; see e.g., Landau and Lifshitz (1987), "The anelastic approximation (AA)." A state in which $\mathbf{g}\cdot\nabla S$ is large and positive would be top-heavy and would set up fast overturning motions; if $\mathbf{g}\cdot\nabla S$ is large and negative, the state would be bottom-heavy and overturning would be inhibited. It is only when $\nabla S \approx \mathbf{0}$ that slow convective motions are possible, and the system is then close to a basic state for which $\nabla S_a = \mathbf{0}$.

As noted by Braginsky and Roberts (2003), Eqs. (30) and (48) imply that the hydrostatic balance (Eq. 47) can be re-expressed as:

$$\nabla(E_a^H + U) = \mathbf{0}. \qquad \text{(Eq. 49)}$$

This parallels a similar statement ($\mathcal{E}^G + U =$ constant, where $\mathcal{E}^G = \mathcal{E}^H - TS$ is the specific Gibbs free energy) for hydrostatic equilibrium in an isothermal fluid; see, for example, Landau and Lifshitz (1987), "The adiabatic (isentropic) reference state."

Similar arguments apply to a chemically inhomogeneous fluid; the mass fractions of the constituents in the moving parcel are virtually unchanged during its displacement, but are smoothed out over short distances. We emphasize that zero gradients arise only for "miscible" quantities that, in the absence of diffusion, are advected without change by the flow. Prominent amongst quantities that do not become constant through mixing, are ρ_a, T_a, and P_a. For example, P_a in the Earth's core varies from about 136 GPa at the CMB to 329 GPa at the ICB. This creates the nonzero "adiabatic gradients" in ρ_a and T_a that, by Eqs. (47), (48), (22) and (23), are

$$\nabla\rho_a = \left(\frac{\partial\rho}{\partial P}\right)_S \nabla P_a = \frac{\rho_a}{v_S^2}\mathbf{g}, \qquad \text{(Eq. 50)}$$

$$\nabla T_a = \left(\frac{\partial T}{\partial P}\right)_S \nabla P_a = \alpha^S \mathbf{g}. \qquad \text{(Eq. 51)}$$

Equation (51) is often written as $T_a^{-1}\nabla T_a = \gamma\mathbf{g}/v_S^2$. As in (Eq. 35), we introduce

$$\varepsilon_a \equiv \Delta\rho_a/\rho_a = H/H_\rho, \quad \text{where} \quad H_\rho = v_S^2/g. \qquad \text{(Eq. 52)}$$

This quantifies the fractional change in ρ_a across the fluid; for the Earth's core, $H_\rho \sim 10^4$ km so that $\varepsilon_a \sim 0.2$. The fractional change, $\Delta T_a/T_a$, in temperature is of the same order, but the fractional change in P_a is much greater.

Often little error is made in assuming that large bodies like the Earth are spherically symmetric, and (Eq. 51) then gives

$$\Gamma \equiv -\frac{dT_a}{dr} = \frac{g\alpha T_a}{C_p} = g\alpha^S, \qquad \text{(Eq. 53)}$$

where r is distance from the center of symmetry. In the Earth's core, $\Gamma \sim 1$ K km^{-1}, so that the *adiabatic heat flux*, $K^T\Gamma$, is about 0.04 W m^{-2} near the CMB, accounting for a heat loss to the mantle of order 6 TW.

The more vigorous the convection, the more appropriate is (Eq. 48) as choice of reference state. This is because the process of smoothing S becomes stronger. The precision of a result such as

$$\rho_c/\rho_a = P_c/\rho_a v_S^2 - \alpha^S S_c, \quad \text{(Eq. 54)}$$

obtained from (Eq. 22), can be assessed by comparing a force such as $\rho_a g$ in the hydrostatic balance (Eq. 47) with a typical force appearing in the equations governing convection such as, in the case of Earth's core, the Coriolis force $2\rho_a \mathbf{\Omega} \times \mathbf{V}$. The ratio of these forces is

$$\varepsilon_c = 2\Omega V/g, \quad \text{(Eq. 55)}$$

which is about 10^{-8}, if we assume, as is suggested by observations, that the typical flow speed, V, is at most 10^{-3} m s^{-1}. Since $\rho_c \sim \varepsilon_c\rho_a$ and $\varepsilon_c \ll 1$, it follows not only that the terms discarded during thermodynamic linearization are negligible, but also that, whenever ρ appears, it can be safely replaced by ρ_a, except in the buoyancy force. (This parallels an analogous step in "The Boussinesq approximation (BA)," where ρ_1 was neglected in comparison with ρ_0 except in the buoyancy force.) The smallness of ε_c underscores the wisdom of using Eqs. (47) and (48) in defining the reference states for vigorously convecting systems; in the case of the Earth's core, the pressure gradient and buoyancy force in (Eq. 2) are greater than the remaining terms by a factor of order 10^8, and by adopting (Eq. 47) they are "removed," so exposing the small terms that control convection.

The anelastic approximation (AA)

As emphasized in "The Boussinesq approximation (BA)," the prime aims of a CA are to "filter out" fast processes and to make use of thermodynamic linearization which, in the "anelastic approximation" or "AA," is linearization about the adiabatic state. The reduced forms of Eqs. (1), (2) and (6) are

$$\nabla \cdot (\rho_a \mathbf{V}) = 0, \quad \text{(Eq. 56)}$$

$$d_t\mathbf{V} = -\nabla\Pi_c - \alpha^S S_c \mathbf{g} + \mathbf{F}^v/\rho_a, \quad \text{(Eq. 57)}$$

$$\rho_a d_t S_c + \nabla \cdot \mathbf{I}^S = \sigma^S, \quad \text{(Eq. 58)}$$

where $\Pi_c = P_c/\rho_a$ and

$$\mathbf{I}^S = T_a^{-1}\mathbf{I}^q, \quad \mathbf{I}^q = \mathbf{I}^q_a + \mathbf{I}^q_c, \quad \sigma^S = T_a^{-1}(Q^R + Q^v) - \mathbf{I}^S \cdot \nabla T_a. \quad \text{(Eq. 59)}$$

It is sometimes useful to re-express $\alpha^S S_c g$ in (Eq. 57) as $S_c \nabla T_a$; see (Eq. 51).

The arguments leading to Eqs. (56)–(59) are now described. To filter out the fast time scales, the term $\partial_t\rho_c$ on the left-hand side of (Eq. 1) must be omitted. When convection occurs, this term is of order $\rho_c/\tau_c \sim \rho_c V/H$. The right-hand side of (Eq. 1) is $-\nabla \cdot (\rho\mathbf{V}) \sim \rho_a V/H$. On the grounds (see above) that terms of order $\rho_c = O(\varepsilon_c\rho_a) \ll \rho_a$ are tiny, we may therefore simplify (Eq. 1) as (Eq. 56), and we have also, for the same reason, taken the opportunity of replacing $\nabla \cdot (\rho\mathbf{V})$ by $\nabla \cdot (\rho_a\mathbf{V})$. Equation (56) states that the curl-free part of the momentum flux, $\rho_a\mathbf{V}$, is negligible compared with its rotational part. Once the term $\partial_t\rho$ has been omitted from (Eq. 1), the system loses all solutions of acoustic/seismic type that depend on the "elasticity" of the medium. This is the essence of the AA and gives it its name. The "filtering out" of the elastic waves is justified because their timescale is short compared with that of convective overturning. The elasticity of the medium is, however, responsible also for the non-uniform density of the basic state created by the pressure gradient, and this is recognized by the AA.

The replacement of ρ by ρ_a on the left-hand side of (Eq. 2) is similarly justified, but the two dominant terms on the right-hand side, $-\nabla P_a$ and $\rho_a\mathbf{g}$, have already canceled out by (Eq. 47), so that (Eq. 2) becomes

$$\rho_a d_t\mathbf{V} = -\nabla P_c + \rho_c\mathbf{g} + \mathbf{F}^v. \quad \text{(Eq. 60)}$$

A revealing simplification of (Eq. 60) was noticed by Braginsky and Roberts (1995, 2003). The density perturbation ρ_c resulting from P_c and S_c create, according to Eqs. (50) and (54), the buoyancy force

$$\rho_c\mathbf{g} = (P_c/\rho_a)\nabla\rho_a - \rho_a\alpha^S S_c\mathbf{g}, \quad \text{(Eq. 61)}$$

so that the two first terms on the right-hand side of (60) are

$$-\nabla P_c + \rho_c\mathbf{g} = -\rho_a\nabla(P_c/\rho_a) - \rho_a\alpha^S S_c\mathbf{g}. \quad \text{(Eq. 62)}$$

Equation (57) follows.

The resemblance of (Eq. 57) to the Boussinesq momentum equation (Eq. 42) is so striking that it is worth stressing that (Eq. 57) is a consequence of the assumption of an adiabatic, well–mixed reference state. It should also be emphasized that the density perturbation due to P_c is fully taken into account in (Eq. 57). The "elastic" part of the density perturbation ρ_c has not been ignored in the AA; it has been absorbed into Π_c and therefore cannot supply the restoring force necessary for acoustic/seismic wave propagation (see above). Equation (57) shows clearly that the buoyancy force associated with deviations from a well-mixed adiabatic state is created *only* by the variations in entropy; the buoyancy force associated with pressure variations, though it may be just as large, does not contribute because, as for the BA, it is conservative and can be absorbed into the effective pressure to create a potential term, here $\nabla\Pi_c$. [Braginsky and Roberts (1995) show that these simplifications apply even when $U(= U_a + U_c)$ is created by the self-gravitation of the fluid mass and satisfies the appropriate Poisson equation.]

Because $\rho_c \sim \varepsilon_c\rho_a \ll \rho_a$, $T \sim \varepsilon_c T_a \ll T_a$ and $d_t S_a = 0$, the first term of the entropy equation (6) may be replaced by $\rho_a d_t S_c$, and (Eq. 58) results. Since $\partial_t T_a = 0$, (Eq. 58) may be re-written, with help from (Eq. 51), as the heat equation

$$\rho_a d_t(T_a S_c) + \nabla \cdot \mathbf{I}^q - \rho_a\alpha^S S_c\mathbf{V} \cdot \mathbf{g} = Q^R + Q^v. \quad \text{(Eq. 63)}$$

On taking the scalar product of (57) with $\rho_a\mathbf{V}$ and using (Eq. 56) and (Eq. 16), it is found that the evolution of kinetic energy is governed by

$$\rho_a d_t E^K + \nabla \cdot (P_c\mathbf{V} + \mathbf{I}^v) + \rho_a\alpha^S S_c\mathbf{V} \cdot \mathbf{g} = -Q^v. \quad \text{(Eq. 64)}$$

By combining (63) and (64), it follows that

$$\rho_a d_t(\mathcal{E}^K + T_a S_c) + \nabla \cdot (\mathbf{I}^q + P_c\mathbf{V} + \mathbf{I}^v) = Q^R. \quad \text{(Eq. 65)}$$

It may be particularly noted that the rate of working of the buoyancy force, $-\rho_a\alpha^S S_c\mathbf{V} \cdot \mathbf{g}$, appearing in (Eq. 64) is canceled out by the same term with an opposite sign in (Eq. 63). To achieve this cancelation, it is necessary to use the fact that $\nabla T_a = \alpha^S\mathbf{g}$, a step that was impossible in "The Boussinesq approximation (BA)" since there $\nabla T_0 = \mathbf{0}$.

The total energy equation that follows from Eqs. (56) and (65) is

$$\partial_t u_c^{\text{total}} + \nabla \cdot \mathbf{I}_c^{\text{total}} = Q^R, \quad \text{(Eq. 66)}$$

where

$$u_c^{\text{total}} = \rho_a(\mathcal{E}^K + T_a S_c), \quad \text{(Eq. 67)}$$

$$\mathbf{I}_c^{\text{total}} = \rho_a(\mathcal{E}^K + \mathcal{E}_c^H)\mathbf{V} + \mathbf{I}_a^q + \mathbf{I}_c^q + \mathbf{I}^v, \quad \mathcal{E}_c^H = \rho_a^{-1}P_c + T_a S_c. \quad \text{(Eq. 68)}$$

Here E_c^H follows from (Eq. 30) by thermodynamic linearization.

The expressions Eqs. (67) and (68) for u_c^{total} and $\mathbf{I}_c^{\text{total}}$ are different from Eqs. (13) and (14). To clarify, we shall temporarily call the latter $\tilde{u}_c^{\text{total}}$ and $\tilde{\mathbf{I}}_c^{\text{total}}$, and introduce the next term in the ε_c–expansion of $\mathbf{V}$, writing $\mathbf{V} = \mathbf{V}_1 + \mathbf{V}_2 + \ldots$, where $\mathbf{V}_1$ has until now been written as $\mathbf{V}$. Equation (1) gives

$$\nabla \cdot (\rho_a \mathbf{V}_1) = 0, \quad \text{(Eq. 69)}$$

$$\partial_t \rho_c + \nabla \cdot (\rho_c \mathbf{V}_1 + \rho_a \mathbf{V}_2) = 0. \quad \text{(Eq. 70)}$$

It can be readily shown from Eqs. (13) and (14) that

$$\tilde{u}_c^{\text{total}} = u_c^{\text{total}} + (E_a^H + U)\rho_c, \quad \text{(Eq. 71)}$$

$$\tilde{\mathbf{I}}_c^{\text{total}} = \mathbf{I}_c^{\text{total}} + (E_a^H + U)[\rho_a \mathbf{V}_1 + (\rho_c \mathbf{V}_1 + \rho_a \mathbf{V}_2)]. \quad \text{(Eq. 72)}$$

According to Eqs. (49), (69), and (70), the additional terms cancel out when (Eq. 71) is substituted into Eqs. (12), and (66)–(68) are recovered.

As was noted in "The adiabatic (isentropic) reference state," convection is driven by the superadiabatic temperature difference across the layer rather than by the full temperature difference (adiabatic plus superadiabatic), as was used in crudely estimating the Rayleigh number, *Ra*, for the Earth's core at the end of "The Boussinesq approximation (BA)." This means that a more realistic *Ra* is smaller by a factor of order ε_c so that $Ra \sim 10^{21}$. This is still so large that the core must be in a well mixed, turbulent state, and this confirms *a posteriori* the suitability of the AA in describing core motions. It might seem that when, thermodynamically, the convective state is a small perturbation away from the adiabatic state, the heat flux, must also be a small perturbation away from the adiabatic heat flux, $\mathbf{I}_a^q$. This however is not necessarily the case. When *Ra* is large, the same motions that mix the fluid and lead to (Eq. 48) also transport a large amount of heat. Even with a small T_c, this may be of the same order as, or even larger than, the heat flux, $K^T\Gamma$, down the adiabat. This question and the role of turbulence is re-opened in the section "The role of turbulence."

The AA generalizes easily to convection in multicomponent fluids, such as the binary alloy often used to describe the dynamics of Earth's core. As for *S*, the mass fraction $\xi(= \xi_a + \xi_c)$ of the light component of the alloy is almost homogenized by convection when *S* is, so that $\nabla\xi_a = \mathbf{0}$ supplements (Eq. 48). Compositional differences create buoyancy forces, so that $-\alpha^S S_c \mathbf{g}$ in (Eq. 57) is replaced by $C\mathbf{g}$, where $C = -\alpha^S S_c - \alpha^\xi \xi_c$ is the *co-density* and $\alpha^\xi = -\rho^{-1}(\partial\rho/\partial\xi)_{P,S}$; cf. (Eq. 26). Also, ξ_c is governed by an equation similar to (Eq. 58), although S_c obeys a different boundary condition. Full details of this generalized AA are given by Braginsky and Roberts (1995, 2003). It was basic in the geodynamo simulation of Glatzmaier and Roberts (1996).

Alternative forms of the AA and BA

The temperature-based anelastic approximation (TAA)

The central thermodynamic variable describing convection in the AA is S_c; see Eqs. (56)–(58). This is convenient when studying core convection, because the nonadiabatic heat flux, $\mathbf{I}_c^q$, is mainly due to turbulence and is therefore best expressed in terms of the inhomogeneities in the entropy (see "The role of turbulence."). For systems such as the Earth's mantle and oceans however, the temperature, T_c, is a more convenient variable to employ. One is then led to a "temperature-based anelastic approximation" or "TAA" in which S_c has been replaced by T_c in Eqs. (56), (57), and (63):

$$\nabla \cdot (\rho_a \mathbf{V}) = 0, \quad \text{(Eq. 73)}$$

$$d_t \mathbf{V} = -\nabla\Pi_c - \alpha T_c \mathbf{g} + \mathbf{F}^v/\rho_a, \quad \text{(Eq. 74)}$$

$$\rho_a d_t(C_p T_c) + \nabla \cdot \mathbf{I}^q - \rho_a \alpha T_c \mathbf{V} \cdot \mathbf{g} = Q^R + Q^v, \quad \text{(Eq. 75)}$$

where

$$\mathbf{I}^q = -K^T\nabla T_a - K^T\nabla T_c, \quad \text{and} \quad \Pi_c = P_c/\rho_0. \quad \text{(Eq. 76)}$$

As before, the presence of $\rho_a \alpha T_c \mathbf{V} \cdot \mathbf{g}$ in (Eq. 75) assures conservation of total energy.

The derivation of Eqs. (73)–(76) requires only

$$T_a S_c = C_p T_c - (\alpha T_a/\rho_a)P_c, \quad \text{(Eq. 77)}$$

which follows from (Eq. 23) and thermodynamic linearization. If we use (Eq. 77) in full to remove reference to S_c in Eqs. (56)–(58), we obtain a theory in every way equivalent to the AA, though complicated by P_c terms from (Eq. 77). In order of magnitude, however, $P_c/\rho_a H \sim \Pi_c/H \sim \alpha T_c g$, so that (Eq. 77) implies

$$\alpha^S S_c = \alpha T_c[1 + \mathrm{O}(\varepsilon_a \alpha T_a \gamma)], \quad \text{(Eq. 78)}$$

$$T_a S_c = C_p T_c[1 + \mathrm{O}(\varepsilon_a \alpha T_a \gamma)]. \quad \text{(Eq. 79)}$$

Whenever $\alpha T_a \ll 1$, as is the case for fluids and solids (but not for gases, where $\alpha T_a \approx 1$), the product $\varepsilon_a \alpha T_a \gamma$ in Eqs. (78) and (79) is negligibly small, even when $\varepsilon_a \sim 1$. We may then replace $\alpha^S S_c \mathbf{g}$ in (Eq. 57) by $\alpha T_c \mathbf{g}$ and $d_t(T_a S_c)$ in (Eq. 63) by $d_t(C_p T_c)$. [Variations in C_p may be small, and $d_t(C_p T_c)$ in (Eq. 75) may then be replaced simply by $C_p d_t T_c$, as is often done. We may also recall (see "Introduction") that, since $\alpha T_a \ll 1$, the differences between v_T and v_S and between C_v and C_p may be ignored.] Since the density change, $\Delta\rho_a$, across the system is not small when $\varepsilon_a \sim 1$, it is necessary to retain ρ_a within the divergence in (Eq. 73).

Slow, creeping motions occur in the Earth's mantle through a process called "subsolidus convection." This is well described by the basic equations of "Introduction," the viscosity defining $\mathbf{F}^v$ being of order 10^{21} Pa s ($\nu \sim 2 \times 10^{17}$ m^2 s^{-1}). The resulting motions are so slow ($V \sim 1$ cm year^{-1}) that the inertial term on the left-hand side of (Eq. 74) is utterly insignificant. Nevertheless, $Ra \sim 10^7 - 10^8$ through much the mantle's vertical extent, in which therefore the temperature is close to an abiabat. The AA is then appropriate but, because $\alpha T_a \sim 3 \times 10^{-2}$, the approximate form Eqs. (73)–(75) of the AA applies almost equally well; see Schubert *et al.* (2001).

The generalized Boussinesq approximation (GBA)

The approximation just described becomes even simpler when $\varepsilon_a \equiv gH/v_S^2 \ll 1$. For example, for the oceans, $H \sim 5$ km and $v_S \sim 1.5$ km s^{-1}, gives $\varepsilon_a \sim 2 \times 10^{-2}$. In such a case, we may substitute

$$\rho_a = \rho_0 + \rho_{a1}, \quad T_a = T_0 + T_{a1}, \quad \ldots\ldots \quad \text{(Eq. 80)}$$

into Eqs. (73) to (76), where $\rho_0, T_0, \ldots$ are constants. Because $\rho_{a1}/\rho_0 \sim T_{a1}/T_0 \sim \varepsilon_a \ll 1$, we may omit ρ_{a1} and T_{a1} everywhere except where $\nabla\rho_a$ and ∇T_a are required, as in (Eq. 76). It follows that

$$\nabla \cdot \mathbf{V} = 0, \quad \text{(Eq. 81)}$$

$$d_t\mathbf{V} = -\nabla\Pi_c - \alpha T_c\mathbf{g} + \mathbf{F}^{v}/\rho_0, \quad \text{(Eq. 82)}$$

$$\rho_0 C_p d_t T_c + \nabla\cdot\mathbf{I}^q - \rho_0\alpha T_c\mathbf{V}\cdot\mathbf{g} = Q^R + Q^v, \quad \text{(Eq. 83)}$$

where C_p and the other coefficients are assumed to be constants and

$$\mathbf{I}^q = -K^T\nabla T_a - K^T\nabla T_c, \quad \text{and} \quad \Pi_c = P_c/\rho_0. \quad \text{(Eq. 84)}$$

Since $\varepsilon_a \ll 1$, the term $-\rho_0\alpha T_c\mathbf{V}\cdot\mathbf{g}$ required to maintain energy conservation is small, but it has been retained in (Eq. 83), as it was for the AA and TAA.

Equations (81)–(84) define what Braginsky (1964) and Braginsky and Roberts (1995) termed the "homogeneous model." The density appearing in Eqs. (82) and (83) is the constant reference density ρ_0 and, because of this similarity with the BA, we call the present theory the "generalized Boussinesq approximation" or "GBA," but often it is simply called the "Boussinesq approximation." As its name suggests, the GBA is more widely applicable than the BA. The BA requires that (Eq. 36) holds, but the GBA may be used when

$$\varepsilon_a \sim \varepsilon_c \ll 1. \quad \text{(Eq. 85)}$$

This extension in the range of validity comes about because, whereas the BA involves the departure T_1 from a constant reference temperature T_0, the GBA employs the deviation T_c from an adiabatic T_a.

The GBA is especially appropriate for the oceans. The fractional changes in density created by the pressure and temperature are of order ε_a and ε_c, respectively and, as in "The Boussinesq approximation (BA)," they are small. The pressure in the ocean changes from 1 atm at its surface to several times 10^5 atm at a depth of several kilometers. Hence P_1 cannot be treated as a small correction to a constant P_0; the changes in pressure and temperature are of the same order so that (Eq. 85) holds and not (Eq. 36). Nevertheless, as in the TAA above, it is only the temperature-created change in density, $\rho_0\alpha T_c\mathbf{g}$, that drives convection. Although the GBA is relevant to oceanic convection, compositional buoyancy, arising from differences in salinity, is also a significant factor that requires Eqs. (81)–(84) to be appropriately generalized; see Kamenkovich (1977).

Despite the fact that $\varepsilon_a \approx 0.2$ in the Earth's core, so that $\rho_a \approx \rho_0$ and $T_a \approx T_0$ are poor approximations, the GBA is often used in studying core convection. Since $\alpha T_a \sim 5\times10^{-2}$ however, the approximations $T_a S_c \approx C_p T_c$ and $\alpha^S S_c \approx \alpha T_c$ are well satisfied, with an error only of order 10^{-3}, so that the TAA is applicable.

The role of turbulence

It was observed in "The anelastic approximation (AA)" that, when Ra is large, there is every likelihood that the fluid motions are turbulent and therefore inaccessible to accurate numerical simulation either now or in the foreseeable future. This formidable difficulty besets the natural sciences. Progress can be made only by abandoning all hopes of describing $\mathbf{V}, S_c, \ldots$ in detail, and by seeking instead the evolution of their averages, $\overline{\mathbf{V}}, \overline{S}_c, \ldots$; these may be smooth enough to be numerically resolvable. The remaining "turbulent" fields, $\mathbf{V}^t, S_c^t, \ldots$, are often termed the "sub-grid scale" fields, and are numerically unobtainable. Since however small-scale motions may diffuse the mean fields far more effectively than the molecular diffusivities, their effect on the mean fields cannot be ignored. It is absolutely necessary to incorporate a description of the small-scale fields into the mean field theory.

The fields are separated into mean and turbulent parts:

$$\mathbf{V} = \overline{\mathbf{V}} + \mathbf{V}^t, \quad S_c = \overline{S}_c + S^t, \quad \ldots\ldots. \quad \text{(Eq. 86)}$$

(We omit the $_c$ from S_c^t because turbulence automatically implies convection.) Consider first the transport of entropy. The ensemble average of the entropy equation, in the form (Eq. 58), is

$$\rho_a \overline{d}_t \overline{S}_c + \nabla\cdot\overline{\mathbf{I}}^S = \overline{\sigma}^S, \quad \text{(Eq. 87)}$$

where $\overline{d}_t = \partial_t + \overline{\mathbf{V}}\cdot\nabla$ and

$$\mathbf{I}^S = \mathbf{I}_a^S + \overline{\mathbf{I}}_c^{Sm} + \overline{\mathbf{I}}^{St}, \quad \overline{\sigma}^S = T_a^{-1}(Q^R + \overline{Q}^{vm} + \overline{Q}^{vt}) - \mathbf{I}^S\cdot\nabla T_a. \quad \text{(Eq. 88)}$$

Here

$$\overline{\mathbf{I}}^{St} = \rho_a\overline{S^t\mathbf{V}^t} \quad \text{(Eq. 89)}$$

is the "turbulent entropy flux," from which the "turbulent heat flux," $\overline{\mathbf{I}}^{qt} = T_a\overline{\mathbf{I}}^{St}$ can be obtained. This may be comparable with the molecular heat flux down the adiabat, $\mathbf{I}_a^q = -K^T\nabla T_a$, and is large compared with $\overline{\mathbf{I}}_c^{qm} = -K^T\nabla\overline{T}_c$, which is of order $\varepsilon_c\mathbf{I}_a^q$ where $\varepsilon_c \ll 1$; the superfix ^{m} stands for "molecular." The viscous dissipation rate has a molecular part, $\overline{Q}^{vm} = \overline{\pi}_{ij}^{vm}\nabla_j\overline{V}_i$, and a turbulent part, $\overline{Q}^{vt} = \overline{\pi_{ij}^{vt}\nabla_j V_i^t}$, that is usually much larger.

Similar arguments apply to (Eq. 57), the average of which may be written as

$$\rho_a\overline{d}_t\overline{\mathbf{V}} - \overline{\mathbf{F}}^{vt} = \rho_a\nabla\overline{\Pi}_c - \rho_a\alpha^S\overline{S}_c\mathbf{g} + \overline{\mathbf{F}}^{vm}, \quad \text{(Eq. 90)}$$

where

$$\overline{\mathbf{F}}^{vt} = \nabla\cdot\overline{\overset{\leftrightarrow}{\boldsymbol{\pi}}}^{vt}, \quad \text{and} \quad \overline{\pi}_{ij}^{vt} = -\rho_a\overline{V_i^t V_j^t} \quad \text{(Eq. 91)}$$

is the "Reynolds stress tensor"; $\overline{\mathbf{F}}^{vt}$ may be combined with the viscous force, $\overline{\mathbf{F}}^{vm}$, of molecular origin created by $\overline{\mathbf{V}}$, previously denoted by $\nabla\cdot\overset{\leftrightarrow}{\boldsymbol{\pi}}^{v}$. It is however generally much greater (except perhaps, in the case of the Earth's core, in Ekman layers). The rate of working of the turbulent buoyancy force is

$$\overline{Q}^t = -\rho_a\alpha^S\overline{S^t\mathbf{V}^t}\cdot\mathbf{g} = -\overline{\mathbf{I}}^{St}\cdot\nabla T_a, \quad \text{(Eq. 92)}$$

and may be comparable with the rate of working, $-\rho_a\alpha^S\overline{S}_c\overline{\mathbf{V}}\cdot\mathbf{g}$, of the mean buoyancy force. By using (Eq. 90) to remove the averaged fields from (Eq. 57), an equation governing $\mathbf{V}^t$ is derived. Taking the scalar product of this with $\rho_a\mathbf{V}^t$ and averaging, it is found that, in statistically steady turbulence ($\mathcal{E}^{Kt} = \frac{1}{2}\overline{(V^t)^2} = \text{constant}$),

$$\overline{Q}^t = \overline{Q}^{vt}, \quad \text{(Eq. 93)}$$

i.e., the buoyant energy released on the turbulent scales supplies on average the kinetic energy necessary to maintain the turbulent motions against viscous decay. It should be emphasized however that, in magnetohydrodynamic systems such as the Earth's core, the right-hand side of (Eq. 93) is replaced by the sum of the viscous and ohmic dissipations, and the ohmic dissipation may dominate; see Braginsky and Roberts (1995, 2003).

In summary, the mean fields are governed by

$$\nabla\cdot(\rho_a\overline{\mathbf{V}}) = 0, \quad \text{(Eq. 94)}$$

$$\overline{d}_t\overline{\mathbf{V}} = -\nabla\overline{\Pi}_c - \alpha^S\overline{S}_c\mathbf{g} + \overline{\mathbf{F}}^{\text{vtotal}}/\rho_a, \quad \text{(Eq. 95)}$$

$$\rho_a T_a\overline{d}_t\overline{S}_c + \nabla\cdot\overline{\mathbf{I}}^{q\text{total}} = Q^R + \overline{Q}^{\text{vtotal}}, \quad \text{(Eq. 96)}$$

where $\overline{\Pi}_c = \overline{P}_c/\rho_a$ and

$$\overline{\mathbf{F}}^{\text{vtotal}} = \overline{\mathbf{F}}^{vm} + \overline{\mathbf{F}}^{vt}, \quad \overline{Q}^{\text{vtotal}} = \overline{Q}^{vm} + \overline{Q}^{vt},$$
$$\overline{\mathbf{I}}^{q\text{total}} = \mathbf{I}_a^q + \overline{\mathbf{I}}^{qm} + \overline{\mathbf{I}}_c^{qt}. \quad \text{(Eq. 97)}$$

Equations (94)–(96) have the same form as Eqs. (56)–(58), but involve mean quantities, $\overline{F}^{vt}, \overline{Q}^{vt}, \overline{\overleftrightarrow{\pi}}^{vt}$ and $\overline{\mathbf{I}}_c^{qt}$, that have to be supplied by experimental measurement or by turbulence theory. Until this is done, the turbulent entropy and momentum fluxes must be parameterized.

"Local turbulence theory" provides one possible parameterization; this rests on the assumption of "scale separation," i.e., that the length and time scales on which the mean fields evolve are so widely separated from turbulent scales that the effects of the latter are, as far as the mean fields are concerned, only "local." The diffusive action of local turbulence on the averaged motion can be parameterized by turbulent diffusivities, K^t and v^t, that are much greater than their molecular counterparts, e.g., $\overline{\mathbf{I}}^{St} = -K^t\nabla\overline{S}_c$. [This parameterization is also commonly used in describing astrophysical convection.] Such forms are appropriate when the turbulence is isotropic but when, as for the Earth's core, this is probably not the case, anisotropic (tensor) diffusivities are more realistic, e.g., $\overline{\mathbf{I}}^{St} = -\overleftrightarrow{\mathbf{K}}^t \cdot \nabla\overline{S}_c$. In either case, it is $\nabla\overline{S}_c$ and not $\nabla\overline{T}_c$, that is significant. For this reason "the anelastic approximation (AA)" is more convenient for core convection than the alternative "temperature-based anelastic approximation (TAA)," even though the conditions required for the validity of the latter are fairly well satisfied. For the compositionally generalized AA of "the anelastic approximation (AA)," the turbulent flux $\overline{\mathbf{I}}^{\xi t} = \rho_a\overline{\xi^t\mathbf{V}^t}$ of ξ analogous to (Eq. 89) may be parameterized by $\overline{\mathbf{I}}^{\xi t} = -\overleftrightarrow{\mathbf{K}}^t \cdot \nabla\overline{\xi}_c$, with the same turbulent diffusivity $\overleftrightarrow{\mathbf{K}}^t$ as in $\overline{\mathbf{I}}^{St} = -\overleftrightarrow{\mathbf{K}}^t \cdot \nabla\overline{S}_c$; see Braginsky and Roberts (1995, 2003).

More detailed discussions about the role of turbulence may be found in Braginsky (1964), Braginsky and Meytlis (1990), and Braginsky and Roberts (1995, 2003). See also *Core turbulence*.

Final remarks

An alternative expression of the BA

Formulations of the BA, different from "the Boussinesq approximation (BA)" and "the generalized Boussinesq approximation (GBA)," are often preferred; see for example Spiegel and Veronis (1960). In the convecting state, (Eq. 31) is replaced by

$$\rho = \rho_b + \rho', \quad T = T_b + T', \quad P = P_b + P', \quad \ldots, \qquad \text{(Eq. 98)}$$

where the primed variables arise from convection and variables carrying the suffix "b" refer to a "basic conductive state" that is in mechanical and thermal equilibrium: $\mathbf{V}_b = \mathbf{0}$. Evidently, unlike "The Boussinesq approximation (BA)," the basic state is nonuniform. Mechanical equilibrium requires

$$\nabla P_b = \rho_b\mathbf{g} = -\rho_b\nabla U. \qquad \text{(Eq. 99)}$$

This implies that ρ_b, P_b, and therefore T_b are constant on equipotential surfaces, i.e., surfaces on which U is constant. Thermal equilibrium requires that T_b satisfies the imposed thermal boundary conditions together with

$$\nabla \cdot (K^T\nabla T_b) + Q^R = 0, \qquad \text{(Eq. 100)}$$

obtained from Eqs. (3) and (19).

According to Eqs. (46) and (98), $T_c = T' + T_b - T_a$. On substituting this into Eqs. (81)–(83) and assuming that $T_c \sim T' \sim T_b - T_a \ll T_b$, the equations governing the convective motions become

$$\nabla \cdot \mathbf{V} = 0, \qquad \text{(Eq. 101)}$$

$$d_t\mathbf{V} = -\nabla\Pi' - \alpha T'\mathbf{g} + \mathbf{F}^v/\rho_0, \qquad \text{(Eq. 102)}$$

$$d_tT' + \mathbf{V}\cdot\nabla(T_b - T_a) = [\nabla\cdot(K^T\nabla T') + Q^v]/\rho_0C_p. \qquad \text{(Eq. 103)}$$

The term in (Eq. 103) that is proportional to the superadiabatic gradient is included implicitly in the GBA and is significant when $\varepsilon_a \sim \varepsilon_c (\ll 1)$.

One of the disadvantages of the present formulation is apparent in the case of "free convection," in which boundary conditions prevent P_b and T_b from being constant on equipotential surfaces. Simultaneous solution of Eqs. (99) and (100) is then impossible, and the assumption $\mathbf{V}_b = \mathbf{0}$ is untenable; convection occurs for all nonzero Ra. By adopting Eqs. (31) or (46) in preference to (Eq. 98), free convection motions do not have to be treated separately; they are on the same footing as convective instabilities and are included with them.

A more general CA

In some contexts, convection is weak or nonexistent in some regions of the system but is strong in other regions. For example, consider the thermal boundary layer at a no-slip wall confining a strongly convecting system, or consider a transition layer via which convection in a superadiabatic region penetrates into an adjacent subadiabatic region. In such situations it is useful to select the basic state in a more general way than has so far been contemplated. The basic state (denoted by subscript "b") is again hydrostatic, but satisfies the heat equation (5) in the form

$$\rho_bT_bd_tS_b + \nabla\cdot\mathbf{I}_b^q = Q^R, \qquad \text{(Eq. 104)}$$

which reduces to (Eq. 48) in a strongly convecting region but to (Eq. 100) in an adjacent weakly or nonconvecting region. Since $d_tS_b (= \mathbf{V}\cdot\nabla S_b)$ involves $\mathbf{V}$, (Eq. 104) imposes a consistency requirement on the solution of the convection equations. Despite this complication, the theory uses thermodynamic linearization and is therefore easier to apply than the full panoply of "Basic equations."

Denoting departures from the basic state by $'$, convection is governed by

$$\nabla\cdot(\rho_b\mathbf{V}) = 0, \qquad \text{(Eq. 105)}$$

$$d_t\mathbf{V} = -\nabla\Pi' + (T'\nabla S_b - S'\nabla T_b) + \mathbf{F}^v/\rho_b, \qquad \text{(Eq. 106)}$$

$$\rho_bd_t(T_bS') - \rho_b\mathbf{V}\cdot(S_b\nabla T' - T_b\nabla S') + \nabla\cdot\mathbf{I}_b^q = Q^v, \qquad \text{(Eq. 107)}$$

where $\Pi' = P'/\rho_b$. In deriving (Eq. 106), the relations $\nabla\rho_b = v_S^{-2}\nabla P_b - \rho_b\alpha^S\nabla S_b$ and $\rho' = v_S^{-2}P' - \rho_b\alpha^S S'$ obtained from (Eq. 22) have been used. It follows from these that $\rho'\nabla P_b - P'\nabla\rho_b = \rho_b^2(T'\nabla S_b - S'\nabla T_b)$, by (Eq. 27). The buoyancy term $T'\nabla S_b - S'\nabla T_b$ in (Eq. 106) can also be written as $-\alpha^S S'\mathbf{g} + \Pi'\alpha^S\nabla S_b$ or as $-\alpha T'\mathbf{g} + \Pi'\alpha\nabla T_b$; cf. Eqs. (62) and (74).

A generalized CA of the type considered here has been developed by Rogers and Glatzmaier (2005).

Boundary conditions for CA

Boundary conditions are specific to the system under consideration and are imposed on the solutions of the basic equations of "Basic equations." They translate directly into corresponding conditions on solutions of the CA equations. Their application to the AA, however, raises questions that are briefly addressed here, using the Earth's core as an illustration.

The core and mantle form one, coupled system that must obey the correct thermal conditions at the CMB. The first of these is

$$T_{\text{Core}} = T_{\text{Mantle}}, \quad \text{on the CMB}, \qquad \text{(Eq. 108)}$$

where each side is the sum of adiabatic and convective parts, e.g., $T_{\text{Core}} = T_{\text{Core,a}} + T_{\text{Core,c}}$. The second expresses continuity of the normal component of the heat flux

$$I^q_{\text{Core},n} = I^q_{\text{Mantle},n}, \quad \text{on the CMB}, \qquad \text{(Eq. 109)}$$

where each side is the sum of adiabatic and convective parts, e.g., $I^q_{\text{Core},n} = I^q_{\text{Core,a}} + I^q_{\text{Core},cn}$. In the case of the core, these were denoted earlier simply by $I^q_{\text{a}} = -K^T \partial_n T_{\text{a}} = K^T \Gamma$ and the normal component of $\mathbf{I}^q_{\text{c}}$.

It is clearly inconvenient to have to solve for core and mantle convection simultaneously but, because $V_{\text{mantle}} \ll V_{\text{core}}$, the two systems can be considered separately. Each of the four fluxes comprising (Eq. 109) are a few TW and, since $I^q_{\text{Core},cn} \sim (\rho C_p V_n T_{\text{c}})_{\text{Core}}$ and $I^q_{\text{Mantle},cn} \sim (\rho C_p V_n T_{\text{c}})_{\text{Mantle}}$, it follows that $T_{\text{Core,c}} \ll T_{\text{Mantle,c}}$. To a very good approximation therefore, (Eq. 108) reduces to

$$T_{\text{Mantle}} = T_{\text{Core,a}}, \quad \text{on the CMB}. \qquad \text{(Eq. 110)}$$

This is the thermal condition to apply to mantle convection. It would be wrong from a physical point of view to assign T_{CMB} when solving core convection. Having solved the mantle convection equations with the T_{CMB} prescribed, $I^q_{\text{Mantle},n}$ on the CMB becomes known (in principle), and this determines the right-hand side of (Eq. 109), which becomes the condition to be applied to core convection:

$$I^q_{\text{Core},cn} = I^q_{\text{CMB}}, \text{ where } I^q_{\text{CMB}} = (Nu - 1) I^q_{\text{Core},a}, \text{ on the CMB}. \qquad \text{(Eq. 111)}$$

Here $Nu = I^q_{\text{Mantle},n} / I^q_{\text{Core,a}}$ is the Nusselt number, a nondimensional measure of the importance of convective heat transport in the core relative to the conductive; in general it is a function of horizontal position on the CMB (see *Core-mantle coupling, thermal*).

In the bulk of the core, turbulent motions are responsible for the flux I^q_{cn} but V_n, falls to zero as the CMB is approached, so that thermal conduction must increasingly carry the necessary heat flux until, on the CMB itself, $I^q_{cn} = -K^T \nabla_n T_{\text{c}}$. Since I^q_{cn} is typically large, $\nabla_n T$ must be large too, both at the CMB and in a thin, nonadiabatic, boundary layer abutting the CMB. We shall not discuss the structure of the boundary layer here (see *Core, boundary layers*).

A note on terminology

All CAs filter out elastic waves (seismic/acoustic) and they could therefore all be described as "anelastic," i.e., not elastic. Paradoxically, the AA and TAA are the only CAs that recognize elasticity at all, in the sense that they include the slow elastic response of the density to changes in pressure. If these are termed "anelastic," the BA is *totally* anelastic! A CA is commonly called "Boussinesq" if it uses $\nabla \cdot \mathbf{V} = 0$ but "anelastic" if it uses $\nabla \cdot (\rho_{\text{a}} \mathbf{V}) = 0$ instead.

Stanislav I. Braginsky and Paul H. Roberts

Bibliography

Braginsky, S.I., 1964. Magnetohydrodynamics of the Earth's core. *Geomagnetism and Aeronomy*, **4**: 698–712.

Braginsky, S.I., and Meytlis, V.M., 1990. Local turbulence in the Earth's core. *Geophysical and Astrophysical Fluid Dynamics*, **55**: 71–87.

Braginsky, S.I., and Roberts, P.H., 1995. Equations governing convection in the Earth's core and the geodynamo. *Geophysical and Astrophysical Fluid Dynamics*, **79**: 1–97.

Braginsky, S.I., and Roberts, P.H., 2003. On the theory of convection in the Earth's core. In Ferriz-Mas, A., and Núñez, M., (eds.), *Advances in Nonlinear Dynamos*. London: Taylor and Francis, pp. 60–82.

Chandrasekhar, S., 1961. *Hydrodynamic and hydromagnetic stability*, Oxford UK: Clarendon Press; reissued in 1981 by Dover Publications Inc., New York.

Davis, G.F., and Richards, M.A., 1992. Mantle convection. *Journal of Geology*, **100**: 151–206.

Gill, A.E., 1982. *Atmosphere-Ocean Dynamics*, London: Academic Press.

Glatzmaier, G.A., and Roberts, P.H., 1995. A three-dimensional convective dynamo solution with rotating and finitely conducting inner core and mantle. *Physics of the Earth and Planetary Interiors*, **91**: 63–75.

Glatzmaier, G.A., and Roberts, P.H., 1996. An anelastic evolutionary geodynamo simulation driven by compositional and thermal convection. *Physica D*, **97**: 81–94.

Gough, D.O., 1969. The anelastic approximation for thermal convection. *Journal of the Atmospheric Sciences*, **26**: 448–456.

Kamenkovich, V.M., 1977. *Fundamentals of Ocean Dynamics*. Amsterdam: Elsevier.

Landau, L.D., and Lifshitz, E.M., 1980. *Statistical Physics, Part 1*. 3rd edn. Oxford: Pergamon.

Landau, L.D., and Lifshitz, E.M., 1987. *Fluid Mechanics*, 2nd edn. Oxford: Pergamon.

Lantz, S.R., and Fan, Y., 1999. Anelastic magnetohydrodynamic equations for modeling solar and stellar convection zones. *Astrophysical Journal Supplement*, **121**: 247–264.

Monin, A.S., 1990. *Theoretical Geophysical Fluid Dynamics*, Dordrecht: Kluwer.

Oberbeck, A., 1888. On the phenomena of motion in the atmosphere. *Sitz. König. Preuss. Akad. Wiss.*, 261–275. English translation in Saltzman (1962).

Ogura, Y., and Phillips, N.A., 1962. Scale analysis of deep and shallow convection in the atmosphere. *Journal of the Atmospheric Sciences*, **19**: 173–179.

Rayleigh, Lord. 1916. On convection currents in a horizontal layer of fluid, when the higher temperature is on the under side. *Philosophical Magazine* (6), **32**: 529–546. Reprinted in Saltzman (1962).

Rogers, T.M., and Glatzmaier, G.A., 2005. Penetrative convection within the anelastic approximation. *Astrophysics Journal*, **620**: 432–441.

Saltzman, B., (ed.) 1962. *Selected Papers on the Theory of Thermal Convection with Special Application to the Earth's Planetary Atmosphere*, New York: Dover.

Schubert, G., Turcotte, D.L., and Olson, P., 2001. *Mantle Convection in the Earth and Planets*, Cambridge: University Press.

Spiegel, E.A., and Veronis, G., 1960. On the Boussinesq approximation for a compressible fluid. *Astrophysics Journal*, **131**: 442–497.

Tackley, P.J., 1997. Effects of phase transition on three-dimensional mantle convection. In Crossley, D.J., (ed.), *Earth's Deep Interior*. London: Gordon and Breach.

Cross-references

Convection, Chemical
Convection, Nonmagnetic Rotating
Core Convection
Core-mantle Coupling, Thermal
Core Turbulence
Core, Boundary Layers
Geodynamo, Energy Sources
Magnetoconvection
Magnetohydrodynamics

ANISOTROPY, ELECTRICAL

In the geosciences, the term "electrical anisotropy" describes the directional dependence of the electrical conductivity of Earth materials. In the case of intrinsic anisotropy, this directional dependence occurs on all scales, while in the case of macroscopic anisotropy composite media are formed from isotropic components in such a way that the bulk conductivity is anisotropic. Intrinsic anisotropy can be described with numerical modeling of the electromagnetic (see *EM*) induction process (e.g., Pek and Verner, 1997). An example of macroscopic anisotropy is presented in Figure A2.

Field observations: magnetotellurics

Fournier (1963) first realized that differences between amplitudes of orthogonal electromagnetic fields, measured with *magnetotellurics* (MT) (*q.v.*), can be associated with electrical anisotropy. In modern MT, anisotropic Earth models are derived from differences between the phases (phase lags between electric and magnetic fields) of two polarizations of the electromagnetic field (Kellett *et al.*, 1992)—see Figure A2. In contrast, amplitude differences are rarely evaluated now due to *galvanic distortion* (*q.v.*) (Simpson and Bahr, 2005). Tests are available which allow to distinguish between anisotropic Earth models and models of conductivity anomalies that only generate anisotropic data (Bahr *et al.*, 2000). In these tests the phase differences obtained from many MT sites in a target area are compared, and anisotropic Earth models are considered if all sites have similar phase differences.

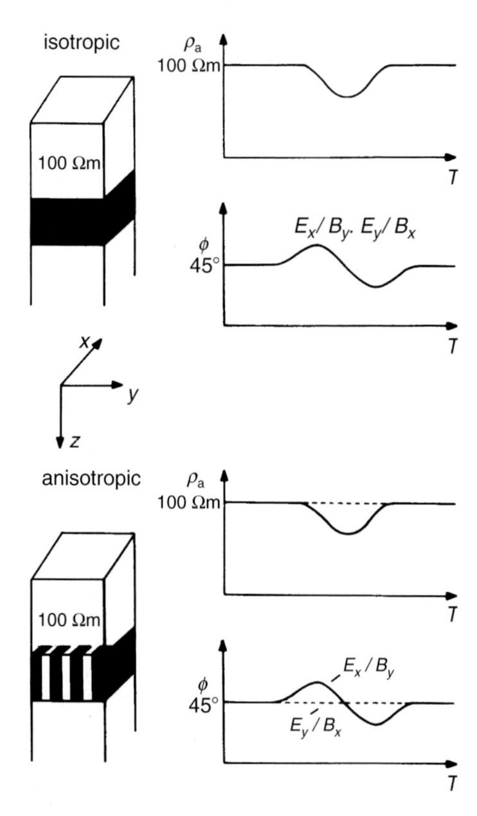

Figure A2 Isotropic and anisotropic MT Earth models and the corresponding transfer functions, displayed as apparent resistivity ρ_a (amplitude of the E/B transfer function) and phase φ (phase lag between the electric field E and the magnetic field B), as a function of period T. Due to the skin effect, electromagnetic fields with longer periods penetrate deeper into the Earth than shorter periods. The conductive (black) intermediate layer of the isotropic model generates a minimum in the apparent resistivity curve at intermediate periods as well as a variation in the phase curve, regardless of the polarization of the electromagnetic field, E_x/B_y or E_y/B_x. In the anisotropic model, this intermediate layer is conductive only in the x direction and therefore affects only the transfer function for the E_x/B_y polarization, resulting in a phase difference between the E_x/B_y and the E_y/B_x polarization in the intermediate period band.

Laboratory measurements and scaling

The ratio between the highest and the lowest conductivity is referred to as anisotropy factor. The electrical anisotropy factors found in laboratory studies of crustal rocks (e.g., Rauen and Lastovickova, 1995) are smaller than those found in MT earth models, and this discrepancy has been explained with scaling techniques in composite media (Bahr, 1997) (Figure A3). In contrast, the anisotropy factor of hydrogen diffusity in olivine can be as large as 40 (Kohlstedt and Mackwell, 1998) and this can result in a large electrical anisotropy of "wet" olivine (Lizzaralde *et al.*, 1995).

Geodynamic implications

MT data from Northern Bavaria have been interpreted with electrical anisotropy in the middle crust with the direction of highest conductance (*electromagnetic strike*) aligned parallel to paleoterrane boundaries (Bahr *et al.*, 2000). Anisotropic conductive structures in the continental crust can therefore be fossil remainings of terrane accretion. Marine-controlled electromagnetic data from the Pacific show that the oceanic crust can be electrically anisotropic, with the direction of highest conductance aligned parallel to the fossil spreading direction (Everett and Constable, 1999). Strain-aligned mineralogical fabric with connected accumulations of trace conductors such as graphite or sulfides can account for this anisotropy. Graphite is also the most likely cause of electrical anisotropy in the North American lithosphere (Mareschal *et al.*, 1995).

With long period MT, electrical anisotropy has been found at the base of the Australian lithosphere with the direction of highest conductance rotated 30° relative to the direction of the present day absolute

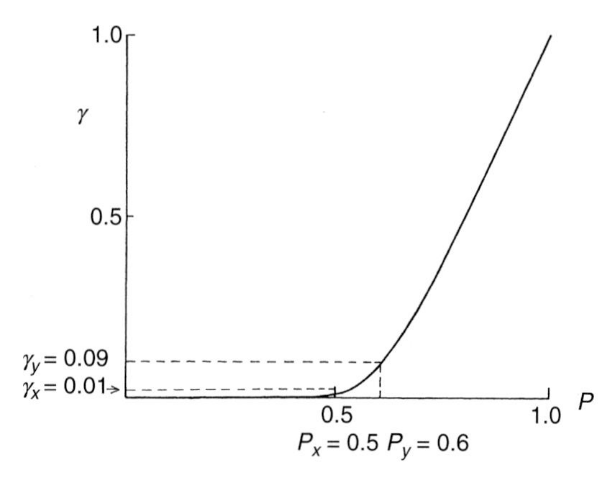

Figure A3 Electrical connectivity γ of a random resistor network representing a two-component medium as function of the fraction p of the high conductive component. At the percolation threshold, a small increase of this fraction results in a relatively large increase of the connectivity. Suppose the high conductive component is distributed in microcracks: If these cracks occur slightly more often in the y direction than in the x direction, then a large difference between the connectivities γ_y and γ_x in the y and x direction is created. This difference results in anisotropic bulk conductivity.

plate motion (APM) that is determined relative to the hotspot reference frame (Simpson, 2001). But the APM direction that is determined relative to a reference frame defined by requiring no net rotation of the lithosphere matches the electrical anisotropy direction better (Simpson, 2002). Mantle flow models suggest that shear of the mantle imparted by the drag of an overriding plate favors alignment of olivine [100] axes parallel to the direction of plate motion (McKenzie, 1979), and this has been confirmed by seismological studies with surface waves (e.g., Debayle and Kennett, 2000). Hydrogen diffusivities enhance the conductivity of olivine crystals anisotropically (e.g., Lizzaralde *et al.*, 1995), providing a possible explanation for the MT anisotropy if hotspots are not stationary relative to the deep mantle (Simpson, 2002). However, electrical anisotropy has also been found below the European lithosphere where current plate motion is slow, and therefore paleoflow can also explain the electrical anisotropy of the sublithospheric mantle (Leibecker *et al.*, 2002; Gatzemeier and Moorkamp, 2005).

Karsten Bahr

Bibliography

Bahr, K., 1997. Electrical anisotropy and conductivity distribution functions of fractal random networks and of the crust: the scale effect of connectivity. *Geophysical Journal International*, **30**: 649–660.

Bahr, K., Bantin, M., Jantos, Chr., Schneider, E., and Storz, W., 2000. Electrical anisotropy from electromagnetic array data: implications for the conduction mechanism and for distortion at long periods. *Physics of the Earth and Planetary Interiors*, **119**: 237–257.

Debayle, E., and Kennett, B.L.N., 2000. The Australian continental upper mantle: structure and deformation inferred from surface waves, *Journal of Geophysical Research*, **105**: 25423–25450.

Fournier, H., 1963. La spectrographic directionelle magnéto-tellurique a Garchy (Niève). *Annales Geophysicae*, **19**: 138–148.

Gatzemeier, A., and Moorkamp, M., 2005. 3D modelling of electrical anisotropy from electromagnetic array data: hypothesis testing for different upper mantle conduction mechanisms. *Physics of the Earth and Planetary Interiors*, **149**: 225–242.

Kellett, R.L., Mareschal, M., and Kurtz, R.D., 1992. A model of lower crustal electrical anisotropy for the Pontiac Subprovince of the Canadian Shield. *Geophysical Journal International*, **111**: 141–150.

Kohlstedt, D.L., and Mackwell, S., 1998. Diffusion of hydrogen and intrinsic point defects in Olivine. Z. Phys. Chemie, **207**: 147–162.

Leibecker, J., Gatzemeier, A., Hönig, M., Kuras, O., and Soyer, W., 2002. Evidence of electrical anisotropic structures in the lower crust and the upper mantle beneath the Rhenish Shield. *Earth Planet Science Letter*, **202**: 289–302.

Lizzaralde, D., Chave, A., Hirth, G., and Schultz, A., 1995. Northeastern Pacific mantle conductivity profile from long-period magnetotelluric sounding using Hawaii-to-California submarine cable data. *Journal of Geophysical Research*, **100**: 17837–17854.

Mareschal, M., Kellett, R.L., Kurtz, R.D., Ludden, J.N., and Bailey, R.C., 1995. Archean cratonic roots, mantle shear zones and deep electrical anisotropy. *Nature*, **375**: 134–137.

McKenzie, D., 1979. Finite deformation during fluid flow. *Geophysical Journal of the Royal Astronomical Society*, **58**: 689–715.

Pek, J., and Verner, T., 1997. Finite-difference modelling of magnetotelluric fields in two-dimensional anisotropic media. *Geophysical Journal International*, **128**: 505–521.

Rauen, A., and Lastovickova, M., 1995. Investigation of electrical anisotropy in the deep borehole KTB. *Surveys in Geophysics*, **16**: 37–46.

Simpson, F., 2001. Resistance to mantle flow inferred from the electromagnetic strike of the Australian upper mantle. *Nature*, **412**: 632–635.

Simpson, F., 2002. Intensity and direction of lattice-preferred orientation of olivine: are electrical and seismic anisotropies of the Australian mantle reconcilable? *Earth Planet Science Letter*, **203**: 535–547.

Simpson, F., and Bahr, K., 2005. *Practical Magnetotellurics.* Cambridge: Cambridge University Press.

Cross-references

Galvanic Distortion
Induction
Magnetotellurics
Mantle, Electrical Conductivity, Mineralogy

ANTIDYNAMO AND BOUNDING THEOREMS

Dynamos are devices that convert mechanical energy into electromagnetic energy. While technical dynamos depend on a suitable arrangement of multiply connected regions (usually wires) of high electrical conductivity within an insulating space, the generation of magnetic fields in planets and stars must occur in simply connected domains of essentially uniform finite conductivity. For a long time after Larmor (1919) (see *Larmor, J.*) first proposed this homogeneous dynamo process as the origin of magnetic fields in sunspots it has been doubtful whether it is possible. Mathematicians and geophysicists have proved antidynamo theorems in order to determine the conditions under which homogeneous dynamos are possible. The first and most famous theorem has been formulated and proved by Cowling (1934) (see *Cowling, T.G.*). He demonstrated that axisymmetric or two-dimensional magnetic fields cannot be generated by the homogeneous dynamo process (see *Cowling's Theorem*). Since the Earth's magnetic field is nearly axisymmetric the dynamo hypothesis of the origin of geomagnetism remained in doubt for a quarter of a century until Backus (1958) and Herzenberg (1958) could show in a mathematically convincing way that nonaxisymmetric fields could indeed be generated (see *Dynamo, Herzenberg* and *Dynamo, Backus*).

Antidynamo and bounding theorems can be divided into the following categories:

1. Lower bounds on magnetic Reynolds numbers or related quantities that are necessary for amplifying magnetic fields
2. Structures and symmetries of velocity fields that cannot be dynamos (e.g., Toroidal theorem)
3. Structures and symmetries of magnetic fields that can not be generated by the dynamo process (e.g., Cowling's theorem)

Lower bounds for magnetic Reynolds numbers

Nearly all theorems that have appeared in the literature are based on the equation of magnetic induction for the magnetic flux density $\boldsymbol{B}$,

$$\left(\frac{\partial}{\partial t} + \boldsymbol{v} \cdot \nabla\right)\boldsymbol{B} + \nabla \times (\lambda \nabla \times \boldsymbol{B}) = \boldsymbol{B} \cdot \nabla \boldsymbol{v}, \qquad \text{(Eq. 1)}$$

which can be derived from Maxwell's equations in the magnetohydrodynamic approximation in which the displacement current is neglected and from Ohm's law for a moving conductor,

$$\lambda \nabla \times \boldsymbol{B} = \boldsymbol{v} \times \boldsymbol{B} + \boldsymbol{E} \qquad \text{(Eq. 2)}$$

where $\boldsymbol{E}$ is the electric field, $\boldsymbol{v}$ is the velocity vector field, and λ is the magnetic diffusivity. The latter is defined as the inverse of the product of magnetic permeability and electrical conductivity, $\lambda = (\sigma \cdot \mu)^{-1}$. In writing equation 1, a solenoidal velocity field $\boldsymbol{v}$ has also been assumed.

Antidynamo theorems of type (i) can be obtained by multiplying equation 1 by $\boldsymbol{B}$ and integrating it over a finite domain V of finite conductivity

$$\frac{1}{2}\frac{\mathrm{d}}{\mathrm{d}t}\int_{V+V'}|\boldsymbol{B}|^2\mathrm{d}^3V = \int_V \boldsymbol{B}\cdot(\boldsymbol{B}\cdot\nabla)\boldsymbol{v}\mathrm{d}^3V - \int_V \lambda|\nabla\times\boldsymbol{B}|^2\mathrm{d}^3V \qquad \text{(Eq. 3)}$$

Here the integral on the left hand side is extended over the entire space $V + V'$ where the integral over the electrically insulating complement V' of V arises from the partial integration of the term $\nabla\times(\lambda\nabla\times\boldsymbol{B})$ (for details see Backus, 1958). Using the maximum $m(t)$ of the rate of strain tensor and the minimum decay rate s_d, which are defined by

$$m(t) = \max\left[\frac{1}{2}(\partial_i v_j + \partial_j v_i)\right], \quad s_\mathrm{d} \equiv \min\left(\frac{\int_V \lambda|\nabla\times\boldsymbol{B}|^2\mathrm{d}^3V}{\int_{V+V'}|\boldsymbol{B}|^2\mathrm{d}^3V}\right) \qquad \text{(Eq. 4)}$$

the inequality

$$\frac{1}{2}\frac{\mathrm{d}}{\mathrm{d}t}\int_{V+V'}|\boldsymbol{B}|^2\mathrm{d}^3V \le (m(t) - s_\mathrm{d})\int_{V+V'}|\boldsymbol{B}|^2\mathrm{d}^3V \qquad \text{(Eq. 5)}$$

can be obtained (Backus, 1958). This relationship indicates the rate of stretching of the magnetic field by the velocity field must be sufficiently strong in order to overcome the effect of Ohmic dissipation. In the special case when V is a sphere of constant λ with radius r_0 and when the velocity field is steady such that $m(t) = v_0/r_0$ can be written with a constant v_0 of the dimension of a velocity, inequality (Eq. 5) can be obtained in the form

$$\frac{1}{2}\frac{\mathrm{d}}{\mathrm{d}t}\int_{V+V'}|\boldsymbol{B}|^2\mathrm{d}^3V \le (Rm - \pi^2)\frac{\lambda}{r_0^2}\int_{V+V'}|\boldsymbol{B}|^2\mathrm{d}^3V \qquad \text{(Eq. 6)}$$

where the magnetic Reynolds number $Rm \equiv v_0 r_0/\lambda$ has been introduced and the expression $s_\mathrm{d} = \pi^2\lambda/r_0^2$ for the smallest rate of decay of a magnetic field in a conducting motionless sphere has been used. This inequality states that unless the magnetic Reynolds number Rm exceeds the value π^2 a dynamo cannot exist in a conducting sphere. Another version of this theorem was obtained by Childress (1969), which assumes the same form (Eq. 6) except that v_0 in the definition of Rm denotes the maximum of $|\boldsymbol{v}|$, which is the more common definition of the magnetic Reynolds number. For an improvement of Backus' bound, see Proctor (1977). A related criterion has been derived by Proctor (1979). It states for

$$\int_V |\nabla\boldsymbol{v}|^2\mathrm{d}^3V \le \frac{\lambda^2\pi}{16 r_0} \qquad \text{(Eq. 7)}$$

no growing magnetic field can be obtained. The expression on the left hand side denotes, of course, the viscous dissipation of the velocity field $\boldsymbol{v}$ divided by the viscosity. It should be noted that all equations and derivations used in this article remain valid in a rotating system, which makes the results especially useful for geophysical and astrophysical applications.

Velocity fields that cannot act as dynamos

For the discussion of other dynamo theorems it is convenient to introduce the general representation for solenoidal vector fields such as $\boldsymbol{B}$ and $\boldsymbol{v}$ (in the case when $\nabla\cdot\boldsymbol{v} = 0$ can be assumed),

$$\boldsymbol{B} = \nabla\times(\nabla h\times\boldsymbol{r}) + \nabla g\times\boldsymbol{r},$$
$$\boldsymbol{v} = \nabla\times(\nabla\Phi\times\boldsymbol{r}) + \nabla\Psi\times\boldsymbol{r} \equiv \boldsymbol{v}_\mathrm{pol} + \boldsymbol{v}_\mathrm{tor} \qquad \text{(Eq. 8)}$$

where $\boldsymbol{r}$ is the position vector. The poloidal scalar functions h and Φ and the toroidal scalar functions g and Ψ are uniquely defined when the condition is imposed that their averages over surfaces $|\boldsymbol{r}| = const.$ vanish (Backus, 1958). A famous theorem of type (ii) that was already known to Elsasser (1946) (see *Elsasser W.*) and been discussed in detail by Bullard and Gellman (1956) is the Toroidal theorem, which states that a purely toroidal velocity field for which $\Phi \equiv 0$ holds cannot generate a magnetic field when $\lambda = \lambda(r)$ is a function of the radius $r = |\boldsymbol{r}|$ only. The validity of this theorem can easily be seen when equation 1, multiplied by $\boldsymbol{r}$,

$$\left(\frac{\partial}{\partial t} + \boldsymbol{v}\cdot\nabla\right)\boldsymbol{B}\cdot\boldsymbol{r} - \lambda(r)\nabla^2\boldsymbol{B}\cdot\boldsymbol{r} = \boldsymbol{B}\cdot\nabla\boldsymbol{v}\cdot\boldsymbol{r} \qquad \text{(Eq. 9)}$$

is considered. Since the right hand side vanishes for a purely toroidal velocity field, equation 9 becomes the heat equation for the "temperature" $\boldsymbol{B}\cdot\boldsymbol{r}$ without heat sources for which only decaying solutions exist. After the r-component of $\boldsymbol{B}$ has decayed it is easy to see that the other components must decay as well. The question could be raised whether a purely poloidal velocity field can generate a magnetic field. The answer is "yes." Gailitis (1970) provides an example in that he used an axisymmetric purely poloidal velocity field for his ring dynamo (see *Dynamo, Gailitis*).

Since a finite radial component of the velocity field is required for dynamo action according to the Toroidal theorem, it is of interest to see whether bounds on its magnitude can be obtained. The condition

$$\max|\boldsymbol{v}\cdot\boldsymbol{r}| \ge \lambda M_\mathrm{pol}/M \qquad \text{(Eq. 10)}$$

derived on the basis of equation 9 for a sphere of constant diffusivity λ provides a necessary condition for a dynamo (Busse, 1975). Here M denotes the total energy of the magnetic field, while M_pol refers to the energy of its poloidal component.

A related criterion that is independent of magnetic energies has recently been derived by Proctor (2004). It states that the energies of poloidal and toroidal components of the magnetic field decay exponentially if

$$\frac{1}{2}|\boldsymbol{v}_\mathrm{pol}|_\mathrm{m}|\boldsymbol{v}_\mathrm{tor}|_\mathrm{m} + \sqrt{2}|\boldsymbol{v}_\mathrm{pol}|_\mathrm{m}\frac{\lambda}{r_0} < \frac{\lambda^2}{r_0^2} \qquad \text{(Eq. 11)}$$

is satisfied where $|\boldsymbol{v}_\mathrm{pol}|_\mathrm{m}$ and $|\boldsymbol{v}_\mathrm{tor}|_\mathrm{m}$ indicate the maximum values of $|\boldsymbol{v}_\mathrm{pol}|$ and $|\boldsymbol{v}_\mathrm{tor}|$ in the conducting sphere of radius r_0. For $|\boldsymbol{v}_\mathrm{tor}|_\mathrm{m} > \pi\lambda/r_0$ this criterion is more severe than Childress' criterion mentioned above in connection with relationship (Eq. 6).

Antidynamo theorems have also been derived for velocity fields with only a radial component (Namikawa and Matsushita, 1970) in which case the property $\nabla\cdot\boldsymbol{v} = 0$ must be dropped, of course. For a discussion of antidynamo theorems for a combination of a toroidal velocity field and a purely radial velocity we refer to Ivers (1995) and earlier papers mentioned therein.

Magnetic fields that cannot be generated by the dynamo process

We have already mentioned the fact that equation 1 does not permit growing axisymmetric magnetic fields as solutions as stated by Cowling's theorem. But it is also generally believed that neither a purely toroidal nor a purely poloidal magnetic field can be generated by the dynamo process in a sphere. The assumption of spherical symmetry is essential since the boundary conditions for poloidal and toroidal components separate on a spherical surface. The only other configuration with this property appears to be the planar layer, which may be regarded as the limit of a thin spherical shell. Proofs in the spherical and planar cases are thus essentially identical. A proof that any physically reasonable purely toroidal field must decay has been obtained by

Table A1 Existence of homogeneous dynamos

Properties	Of magnetic field	Of velocity field
Axisymmetry	No	Yes
Purely toroidal	No	No
Purely poloidal	No (?)	Yes
Helical symmetry	Yes	Yes

Kaiser *et al.* (1994). In the poloidal case a proof depending on a physical plausibility argument has been given by Kaiser (1995).

Concluding remarks

The search for antidynamo theorems has led to some interesting mathematical theorems in the past decades. Just as important are the physical insights into the dynamo process that have been gained. The interactions of poloidal and toroidal components of the magnetic field are evidently essential for the operation of a dynamo. To summarize the main results of antidynamo theorems the Table A1 has been composed. In it we have included the case of helical symmetry, which is known to be dynamo friendly. Indeed Lortz (1968) was able to derive a dynamo with helical symmetry.

Friedrich Busse and Michael Proctor

Bibliography

Backus, G., 1958. A class of self-sustaining dissipative spherical dynamos, *Annals of Physics*, **4**: 372–447.

Bullard, E., and Gellman, H., 1954. Homogeneous dynamos and terrestrial magnetism. *Philosophical Transactions of the Royal Society of London*, **A247**: 213–278.

Busse, F.H., 1975. A necessary condition for the geodynamo. *Journal of Geophysical Research*, **80**: 278–280.

Childress, S., 1969. Théorie magnétohydrodynamique de l'effet dynamo, Lecture Notes. *Département Méchanique de la Faculté des Sciences, Paris.*

Cowling, T.G., 1934. The magnetic field of sunspots. *Monthly Notices of the Royal Astronomical Society*, **34**: 39–48.

Elsasser, W.M., 1946. Induction effects in terrestrial magnetism. *Physical Review*, **69**: 106–116.

Gailitis, A., 1970. Magnetic field excitation by a pair of ring vortices. *Magnetohydrodynamics (N.Y.)*, **6**: 14–17.

Herzenberg, A., 1958. Geomagnetic Dynamos. *Philosophical Transactions of the Royal Society of London*, **A250**: 543–585.

Ivers, D.J., 1995. On the antidynamo theorem for partly symmetric flows. *Geophysical and Astrophysical Fluid Dynamics*, **80**: 121–128.

Kaiser, R., 1995. Towards a poloidal magnetic field theorem. *Geophysical and Astrophysical Fluid Dynamics*, **80**: 129–144.

Kaiser, R., Schmitt, B.J., and Busse, F.H., 1994. On the invisible dynamo. *Geophysical and Astrophysical Fluid Dynamics*, **77**: 91–109.

Larmor, J., 1919. How could a rotating body such as the sun become a magnet? *Reports of the British Association for the Advancement of Science*, 159–160.

Lortz, D., 1968. Exact solutions of the hydromagnetic dynamo problem. *Plasma Physics*, **10**: 967–972.

Namikawa, T., and Matsushita, S., 1970. Kinematic dynamo problem. *Geophysical Journal of the Royal Astronomical Society*, **19**: 319–415.

Proctor, M.R.E., 1977. On Backus' necessary condition for dynamo action in a conducting sphere. *Geophysical and Astrophysical Fluid Dynamics*, **9**: 89–93.

Proctor, M.R.E., 1979. Necessary conditions for the magnetohydrodynamic dynamo. *Geophysical and Astrophysical Fluid Dynamics*, **14**: 127–145.

Proctor, M.R.E., 2004. An extension of the toroidal theorem. *Geophysical and Astrophysical Fluid Dynamics*, **98**: 235–240.

Cross-references

Cowling's Theorem
Cowling, Thomas George (1906–1990)
Dynamo, Backus
Dynamo, Gailitis
Dynamo, Herzenberg
Elsasser, Walter M. (1904–1991)
Larmor, Joseph (1857–1942)

ARCHEOLOGY, MAGNETIC METHODS

Magnetism and archeology

Introduction

Magnetic methods have become important tools for the scientific investigation of archeological sites, with magnetic prospection surveys and archeomagnetic dating being the most prominent ones. The principles behind these techniques were initially applied to larger and older features, for example prospecting for ore deposits (see *Magnetic anomalies for geology and resources*) or paleomagnetic dating (see *Paleomagnetism*). When these techniques were adapted for archeological targets it was soon established that very different methodologies were required. Archeological features are relatively small and buried at shallow depth and the required dating accuracy is in the order of tens of years. More importantly, the relationship between archeological features and magnetism is often difficult to predict and the planning of investigations can hence be complicated. Related is the problem of interpretation. Geophysical results on their own are only of limited use to resolve an archeological problem. It is the archeological interpretation of the results using all possible background information (site conditions, archeological background knowledge, results from other investigations, etc.), which provides useful new insights. If the relationship between magnetic properties and their archeological formation is unknown, such interpretation may become speculative.

All magnetic investigations depend on the contrast in a magnetic property between the feature of interest and its surrounding environment, for example the enclosing soil matrix. The most important magnetic properties for archeological studies are magnetization and magnetic susceptibility.

Remanent magnetism

Thermoremanent magnetization is probably the best understood magnetic effect caused by past human habitation. If materials that are rich in iron oxides are heated above their Curie temperature and then allowed to cool in the ambient Earth's magnetic field they have the potential to acquire a considerable thermoremanence that is fixed in the material until further heating. Typical archeological examples are kilns and furnaces, often built of clay, which during their heating cycles often exceed the Curie temperatures of magnetite (Fe_3O_4) and maghaemite (γ-Fe_2O_3) (578°C and 578–675°C, respectively). Such iron oxides are commonly found in the clay deposits that were used for the construction of these features. Even if the clays only contained weakly magnetic haematite or goethite, the heating and cooling cycles may have converted these into ferrimagnetic iron oxides (see *Magnetic mineralogy, changes due to heating*). Similarly, fired bricks and pottery can exhibit thermoremanence but when the finished bricks are used as building material their individual vectors of magnetization will point into many different directions producing an overall weakened magnetic signature (Bevan, 1994). The same applies to heaps of pottery shards. The strong magnetic remanence of kilns led to their discovery with magnetometers in 1958 (Clark, 1990), which triggered the widespread use of archeological magnetometer surveys today.

Kilns also helped to establish the archeomagnetic dating technique as a new tool for chronological studies in archeology (Clark *et al.*, 1988).

Detrital remanence is caused when the Earth's magnetic field aligns magnetized particles that are suspended in solution and gradually settle (see *Magnetization, depositional remanent*). An example is stagnant water loaded with "magnetic sediments". The resulting deposits can exhibit a weak but noticeable remanence, which could be used for dating and prospection. It is suspected that similar effects have produced a small remanence and hence a noticeable positive signal in Egyptian mud-bricks, which were made from wet clay pushed into moulds and dried in the sun (Herbich, 2003). However, other magnetometer surveys over mud-brick structures (Becker and Fassbinder, 1999) have shown a negative magnetic contrast of these features against the surrounding soil. Although it is possible to consider burning events that could lead to such results during the demolition of buildings, it is more likely that these bricks were made from clay that has lower magnetic susceptibility than the soil on the building site.

Induced magnetism

Any material with a magnetic susceptibility will acquire an induced magnetization in the Earth's magnetic field (see *Magnetic susceptibility*). Hence, if past human habitation has led to enhanced levels of magnetic susceptibility in the soils, magnetic measurements can be used for their detection. The relationship between human activities and the enhancement of magnetic susceptibility was investigated by Le Borgne (1955, 1960), distinguishing thermal and bacterial enhancement. When soil is heated in the presence of organic material (for example during bush- or camp-fires), oxygen is excluded and the resulting reducing conditions lead to a conversion of the soil's hematite (α-Fe_2O_3, antiferromagnetic) to magnetite (Fe_3O_4, ferrimagnetic) with a strong increase of magnetic susceptibility. On cooling in air, some of the magnetite may be reoxidized to maghemite (γ-Fe_2O_3, ferrimagnetic), thereby preserving the elevated magnetic susceptibility. In contrast, the "fermentation effect" refers to the reduction of hematite to magnetite in the presence of anerobic bacteria that grew in decomposing organic material left by human habitation, either in the form of rubbish pits ("middens") or wooden building material. This latter effect requires further research but it is reported that changes in pH/Eh conditions as well as the bacteria's use of iron as electron source are responsible for the increase of magnetic susceptibility (Linford, 2004). The level of magnetic susceptibility that can be reached through anthropogenic enhancement also depends on the amount of iron oxides initially available in the soil for conversion. The level of enhancement can hence be quantified by relating a soil's current magnetic susceptibility to the maximum achievable value. This ratio is referred to as "fractional conversion" and is determined by heating a sample to about 700°C to enhance its magnetic susceptibility as far as possible (Crowther and Barker, 1995; Graham and Scollar, 1976). Whether the initial magnetic susceptibility was enhanced by pedogenic or anthropogenic effects can however not be distinguished with this method. It is also worth remembering that magnetite and maghemite have the highest magnetic susceptibility of the iron oxides commonly found in soils, and in the absence of elemental iron, a sample's magnetic susceptibility is hence a measure for the concentration of these two minerals.

More recent investigations have indicated additional avenues for the enhancement of magnetic susceptibility. One of the most interesting is a magnetotactic bacterium that thrives in organic material and grows magnetite crystals within its bodies (Fassbinder *et al.*, 1990) (see also *Biomagnetism*). Their accumulation in the decayed remains of wooden postholes led to measurable magnetic anomalies and is probably responsible for the detection of palisade walls in magnetometer surveys (Fassbinder and Irlinger, 1994). Other causes include the low-temperature thermal dehydration of lepidocrocite to maghemite (e.g., Özdemir and Banerjee, 1984) and the physical alteration of the constituent magnetic minerals, especially their grain size (Linford and Canti, 2001; Weston, 2004). Magnetic susceptibility can also be enhanced by the creation of iron sulfides in perimarine environments with stagnant waters (Kattenberg and Aalbersberg, 2004). These may fill geomorphological features, like creeks, that were used for settlements and can therefore be indirect evidence for potential human activity.

Archeological prospection

Archeological prospection refers to the noninvasive investigation of archeological sites and landscapes for the discovery of buried archeological features. To understand past societies it is of great importance to analyze the way people lived and interacted and the layout of archeological sites gives vital clues; for example the structure of a Roman villa's foundations or the location of an Iron-aged ditched enclosure within the wider landscape. Such information can often be revealed without excavation by magnetic surveys. These techniques have therefore become a vital part of site investigation strategies. Buried archeological features with magnetic contrast will produce small anomalies in measurements on the surface and detailed interpretation of recorded data can often lead to meaningful archeological interpretations. The techniques are not normally used to "treasure-hunt" for individual ferrous objects but rather for features like foundations, ditches, pits, or kilns (Sutherland and Schmidt, 2003).

Magnetic susceptibility surveys

Since human habitation can lead to increased magnetic susceptibility (see above), measurements of this soil property are used for the identification of areas of activity. Such surveys can either be carried out *in situ* (i.e., with nonintrusive field measurements) or by collecting soil samples for measurements in a laboratory. These two methods are often distinguished as being volume- and mass-specific, respectively, although such labeling only vaguely reflects the measured properties.

Most instruments available for the measurement of low field magnetic susceptibility internally measure the "total magnetic susceptibility" (k_t with units of m^3), which is proportional to the amount of magnetic material within the sensitive volume of the detector: the more material there is, the higher will be the reading. For field measurements, the amount of investigated material is usually estimated by identifying a "volume of sensitivity" (V) for the employed sensor (e.g., a hemisphere with the sensor's diameter for the Bartington MS2D field coil). The "volume specific magnetic susceptibility" is then defined as $\kappa = k_t/V$ (dimensionless). In contrast, laboratory measurements normally relate instrument readings to the weight of a sample (m), which can be determined more accurately. The "mass specific magnetic susceptibility" is then $\chi = k_t/m$ (with units of $m^3\ kg^{-1}$). Accordingly, it is possible to calculate one of these quantities from the other using the material's bulk density (ρ): $\chi = \kappa/\rho$.

The main difference between field and laboratory measurements, however, is the treatment of samples. It is common practice (Linford, 1994) to dry and sieve soil samples prior to measuring their magnetic susceptibility in the laboratory. Drying eliminates the dependency of mass specific magnetic susceptibility on moisture content, which affects the bulk density. Sieving removes coarse inclusions (e.g., pebbles) that are magnetically insignificant. In this way, laboratory measurements represent the magnetic susceptibility of a sample's soil component and can therefore be compared to standard tables. For field measurements, however, results can be influenced by nonsoil inclusions and the conversion of volume-specific measurements to mass-specific values is affected by changes in environmental factors (e.g., soil moisture content).

The measured magnetic susceptibility depends on the amount of iron oxides available prior to its alteration by humans (mostly related to a soil's parent geology), and also on the extent of conversion due to the anthropogenic influences. As a consequence, the absolute value of magnetic susceptibility can vary widely between different sites and "enhancement" can only be identified as a contrast between areas of higher susceptibility compared to background measurements. There is no predetermined threshold for this contrast and it is therefore justifiable to use the more qualitative measurements of field-based magnetic volume susceptibility (see above) for archeological prospection.

The instrument most commonly used for field measurements of volume specific magnetic susceptibility is the Bartington MS2D field coil consisting of a 0.2 m diameter "loop," which derives 95% of its signal from the top 0.1 m of soil (Schmidt *et al.*, 2005). Since most archeological features are buried deeper than this sensitivity range, the method relies on the mixing of soil throughout the profile, mostly by ploughing.

Magnetic susceptibility surveys, either performed as *in situ* field measurements or as laboratory measurements of soil samples, can be used in three different ways: (i) as primary prospection method to obtain information about individual buried features, (ii) to complement magnetometer surveys and help with their interpretation by providing data on underlying magnetic susceptibility variations, or (iii) for a quick and coarse "reconnaissance" survey using large sampling intervals to indicate areas of enhancement instead of outlining individual features. Figure A4a shows the magnetic susceptibility survey (MS2D) of a medieval charcoal burning area. Since a detailed study of this feature was required, the data were recorded with a spatial resolution of 1 m in both x- and y-directions. Corresponding magnetometer data (FM36) are displayed in Figure A4b. The magnetic susceptibility results outline the burnt area more clearly and are on this site well suited for the delineation of features. On former settlement sites where ploughing has brought magnetically enhanced material to the surface and spread it across the area, magnetic susceptibility measurements at coarse intervals of 5, 10, or 20 m can be used to identify areas of enhancement that can later be investigated with more detailed sampling, for example with a magnetometer (see below). Even where ploughing has mixed the soil, magnetic susceptibility measurements can vary considerably over short distances of about 2 m. It is hence not normally possible to interpolate coarsely sampled data and settlement sites can only be identified if several adjacent measurements have consistently high levels compared to the surrounding area. Figure A5 shows data from a survey at Kirkby Overblow, North Yorkshire, in search for a lost medieval village. The dots mark the individual measurements and their size represents the strength of the magnetic susceptibility, highlighting the singularity of each measurement. Nevertheless, areas of overall enhancement can be identified and these were later investigated with magnetometer surveys. The gray shading in this diagram visualizes the data in a contiguous way, coloring the Voronoi cell around each individual measurement according to its magnetic susceptibility. The limitations of such diagrams for widely separated measurements should be considered carefully.

Magnetometer surveys

In contrast to surveys that measure magnetic susceptibility directly, magnetometer surveys record the magnetic fields produced by a contrast in magnetization, whether it is induced as a result of a magnetic susceptibility contrast, or remanent, for example from thermoremanent magnetization. If the shape and magnetic properties of a buried archeological feature were known, the resulting magnetic anomaly could be calculated (Schmidt, 2001) (see also *Magnetic anomalies, modeling*). The inverse process, however, of reconstructing the archeological feature from its measured anomaly, is usually not possible due to the nonuniqueness of the magnetic problem and the complex shape and heterogeneous composition of such features. Some successful inversions were achieved when archeologically informed assumptions were made about the expected feature shapes (e.g., the steepness of ditches) and magnetic soil properties (for example from similar sites) (Neubauer and Eder-Hinterleitner, 1997; Herwanger *et al.*, 2000). If surveys are conducted with sufficiently high spatial resolution, the mapped data often already provide very clear outlines of the buried features (Figure A6), allowing their archeological interpretation even without data inversion. However, when interpreting measured data directly, the typical characteristics of magnetic anomalies have to be taken into consideration. For example, in the northern hemisphere

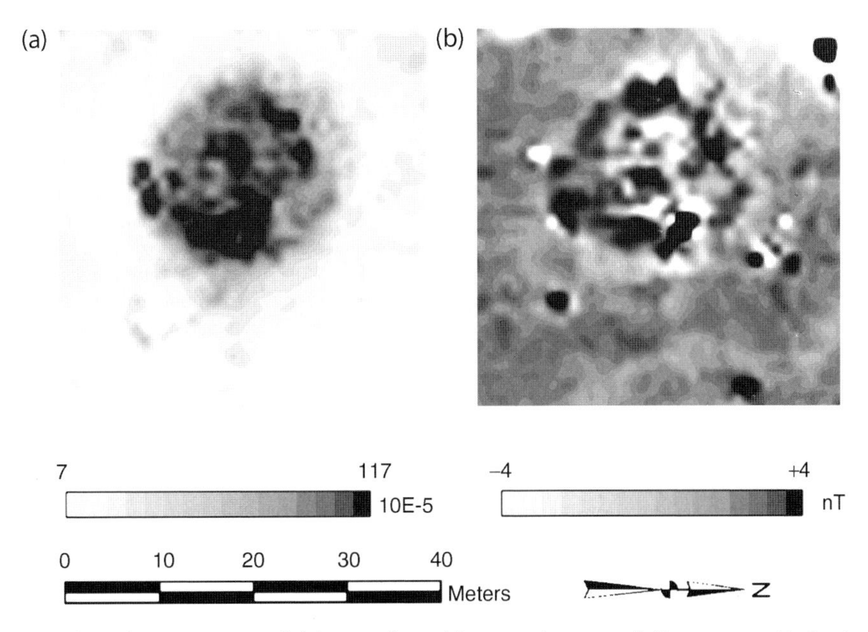

Figure A4 Medieval charcoal production site in Eskdale, Cumbria. (a) Magnetic susceptibility survey with Bartington MS2D field coil. (b) Fluxgate gradiometer survey with Geoscan FM36. Both surveys were conducted over an area of 40 m × 40 m with a spatial resolution of 1 m × 1 m.

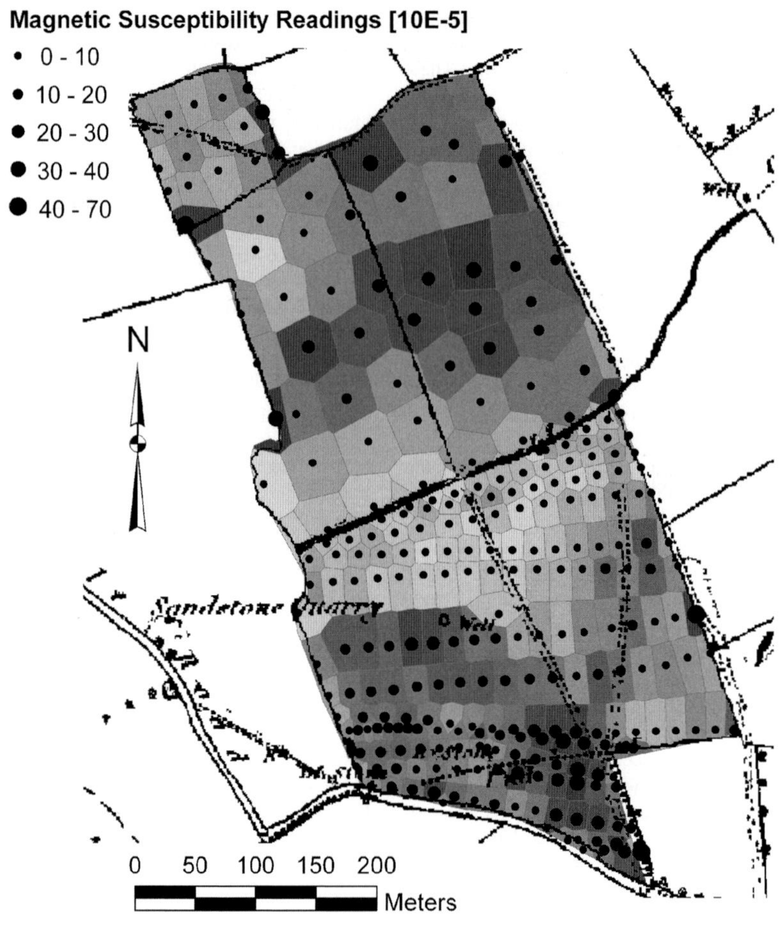

Figure A5 Kirkby Overblow, North Yorkshire. To search for a deserted medieval village, a magnetic susceptibility survey was undertaken with the Bartington MS2D field coil. The measurements are represented as scaled dots and as shaded Voronoi cells. Basemap from 1st edition Ordnance Survey data.

anomalies created by soil features with induced magnetization have a negative trough to the north of the archeological structure and a slight shift of the positive magnetic peak to the south of its center (Figure A7).

To estimate the strength of a typical archeological response it is possible to approximate the anomaly with a simplified dipole field: $B = \mu_0 m/r^3 = \Delta\kappa V B_{\mathrm{Earth}}/r^3$, where μ_0 is the magnetic permeability of free space ($4\pi \times 10^{-7}$ Tm A^{-1}), m is the total magnetic moment of the feature, r the distance between measurement position and sample, $\Delta\kappa$ its volume-specific magnetic susceptibility contrast, V its volume, and B_{Earth} the Earth's magnetic flux density. For a buried pit, the following values can be used: $\Delta\kappa = 10 \times 10^{-5}$, $V = 1$ m^3, $B_{\mathrm{Earth}} = 48{,}000$ nT, and $r = 1$ m. This yields an anomaly strength of only 4.8 nT, which is typical for archeological soil features (e.g., pits or ditches). Anomalies created by the magnetic enrichment of soils through magnetotactic bacteria, for example in palisade ditches, can be as low as 0.3 nT (Fassbinder and Irlinger, 1994) and are therefore only detectable with very sensitive instruments and on sites where the signals caused by small variations in the undisturbed soil's magnetic properties are very low (low "soil noise"). Peak values higher than approximately 50 nT are normally only measured over ferrous features with very high magnetic susceptibility, or over features with thermoremanent magnetization, like furnaces or kilns. As shown by the numerical approximation above, the anomaly strength depends both on the magnetization and the depth of a buried feature and can therefore not be used for the unambiguous characterization of that feature. More indicative is the spatial variation of an anomaly (see Figure A7) since deeper features tend to create broader anomalies.

Surveys are normally undertaken with magnetometers on a regular grid. The required spatial resolution obviously depends on the size of the investigated features, but since these are often unknown at the outset, a high resolution is advisable. Recommendations by English Heritage have suggested a minimum resolution of 0.25 m along lines with a traverse spacing of 1 m or less (0.25 m × 1 m) (David, 1995). However, to improve the definition of small magnetic anomalies even denser sampling is required. Such high-definition data can even show peaks and troughs of bipolar anomalies from small, shallowly buried iron debris (e.g., farm implements) and therefore help to distinguish these from archeological features with wider anomaly footprints. More recently, it has become possible to collect randomly sampled data with high accuracy (Schmidt, 2003) that can either be gridded to a predefined resolution or visualized directly with Delaunay triangulation (Sauerländer *et al.*, 1999).

Due to the often small anomaly strength caused by archeological features, very sensitive magnetometers are required for the surveys. The first investigations (Aitken *et al.*, 1958) were made with proton-free precession magnetometers (see also *Observatories, instrumentation*) but due to the slow speed of operation and their relative insensitivity, these are rarely used for modern surveys. In Britain, the most commonly used magnetometers use fluxgate sensors (see also *Observatories, instrumentation*), which can achieve noise levels as low as 0.3 nT if sensors and electronics

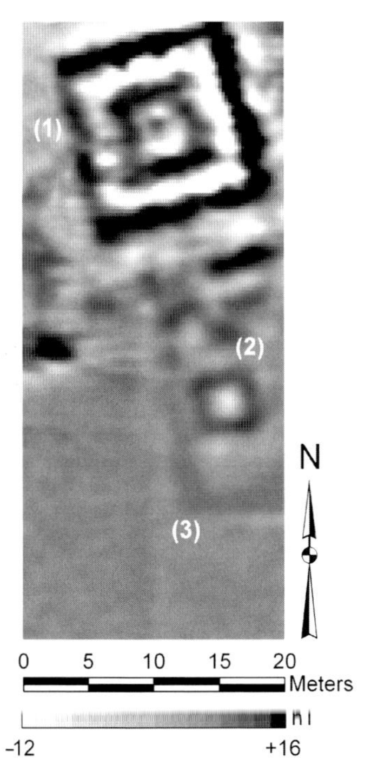

Figure A6 Fluxgate gradiometer data from Ramagrama, Nepal. The outlines of a small Buddhist temple complex are clearly visible, including the outer and inner walls of the courtyard building (1). In addition, the foundations of a small shrine (2) within an enclosure wall (3) can be discerned.

are carefully adjusted. In Austria and Germany, some groups use cesium vapor sensors that are built to the highest possible specifications and have reported sensitivities of 0.001 nT (Becker, 1995). On loess soils, which produce very low magnetic background variations, the weak anomalies caused by magnetotactic bacteria in wood (see above) were detected with cesium magnetometers, revealing palisade trenches of Neolithic enclosures (Neubauer and Eder-Hinterleitner, 1997).

Magnetometer sensors measure the combination of the archeological anomaly and the Earth's magnetic field. Hence, to reveal the archeological anomalies, readings have to be corrected carefully for changes in the ambient field, caused by diurnal variations (see *Geomagnetic secular variation*) or magnetic storms (see *Storms and substorms, magnetic*). Proton free-precession instruments are often used in a differential arrangement ("variometer") by placing a reference sensor in a fixed location to monitor the Earth's magnetic field. The survey is then carried out with an additional sensor, which is affected by the same ambient field. Subtracting the data from both sensors cancels out the Earth's field. The same can be achieved with a gradiometer arrangement where the second sensor is usually rigidly mounted 0.5–1 m above the first so that both are carried together. Their measurements can be subtracted instantly to form the gradiometer reading. Despite the heavier weight of such an arrangement, it allows for improved instrument design. Especially for fluxgate sensors a signal feedback system, linked to the upper sensor, can be used to enhance the sensitivity of the instrument. Archeological features that can be detected with magnetometers are often buried at shallow depth and a gradiometer's sensor separation is then similar to the distance between the feature and the instrument. Therefore, the gradiometer reading is not an approximation for the field gradient and is better recorded as the difference in magnetic flux density (nT) between the sensors and not as a gradient (nT m^{-1}). The magnetic field created by the feature of interest and the ambient magnetic field combine at the sensor and the recorded signal depends on the vector component measured with the particular instrument. For example, fluxgate sensors measure a single component of the magnetic field (usually the vertical component) and a gradiometer therefore records exactly this component of the archeological anomaly. In contrast, gradiometers built from field intensity sensors, like proton free-precession and cesium vapor sensors, measure the component of the anomaly in the direction of the ambient Earth's magnetic field, since this is much larger than the anomaly. Its direction governs the vector addition of the two contributions and the gradiometer reading can be approximated as $\left|\mathbf{B}_{\mathrm{Earth}} + \mathbf{B}_{\mathrm{anomaly}}\right| - \left|\mathbf{B}_{\mathrm{Earth}}\right| \approx \mathbf{B}_{\mathrm{anomaly}} \cdot \mathbf{e}_{\mathrm{Earth}}$ where $\mathbf{e}_{\mathrm{Earth}}$ is the unit-vector in the direction of the Earth's magnetic field (Blakely, 1996). As a consequence, vertical fluxgate sensors record weaker signals in areas of low magnetic latitude (i.e., near the equator) and data from the two sensor types cannot be compared directly.

The "detection range" of archeological geophysical surveys depends on the magnetization and depth of the features (see above) as well as the sensitivity of the used magnetometers and can therefore not easily be specified. Based on practical experience with common instruments, it is estimated that typical soil features, like pits or ditches, can be detected at depths of up to 1–2 m, while ferrous and thermoremanent features can be identified even deeper. Weak responses were recorded from paleochannels that are buried by more than 3 m of alluvium (Kattenberg and Aalbersberg, 2004).

Prior to their final interpretation, magnetometer data often have to be treated with computer software for the improvement of survey deficiencies and the processing of resulting data maps (Schmidt, 2002). Data improvement can reduce some of the errors introduced during the course of a survey, such as stripes and shearing between adjacent survey lines. To reduce the time of an investigation, adjacent lines of a magnetometer survey are often recorded walking up and down a field ("zigzag" recording). However, since most magnetometers have at least a small heading error, a change in sensor alignment resulting from this data acquisition method can lead to slightly different offsets for adjacent lines, which are then visible as stripes in the resulting data (Figure A8a, middle). A common remedy for this effect is the subtraction of the individual mean or median value from each survey line. This helps to balance the overall appearance but also removes anomalies running parallel to the survey lines. If the sensor positions for the forward and backward survey direction are systematically offset from the desired recording position (e.g., always 0.1 m "ahead") anomalies will be sheared and data will appear "staggered" (Figure A8a, left). If this defect is sufficiently consistent, it can be removed (Figure A8b) by fixed or adaptive shifting of every other data line (Ciminale and Loddo, 2001). Another common problem found with some instruments is the "drift" of their offset value, mostly due to temperature effects. If regular measurements are made over dedicated reference points the effects of drift can be reduced numerically. All these methods of data improvement require detailed survey information, for example about the length of each survey line, the direction of the lines, the size of data blocks between reference measurements, etc. It is hence essential that metadata are comprehensively recorded (see below). Once all data have been corrected for common problems and all survey blocks balanced against each other, they can be assembled into a larger unit (often referred to as "composite") and processed further. Typical processing steps may include low- and high-pass filtering and reduction-to-the-pole. However, most processing can introduce new artifacts into the data (Schmidt, 2003) and should therefore only be used if the results help with the archeological interpretation.

Many archeological magnetometer surveys are commissioned to resolve a clearly defined question, for example to find archeological remains in a field prior to its development for housing. However, most data are also of potential benefit beyond their initial intended use and should therefore be archived. For example, the removal of

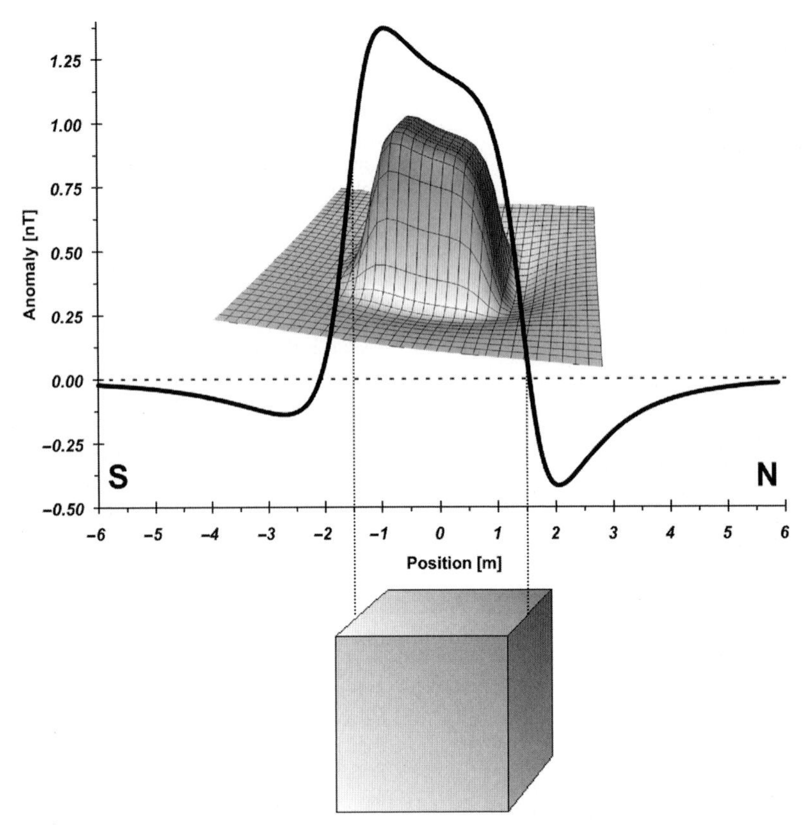

Figure A7 Calculated shape of the magnetic anomaly caused by the induced magnetization of a buried cubic feature with 3 m side length. The anomaly is measured with a vertical fluxgate gradiometer (0.5 m sensor separation), 0.5 m above the cube's top surface. The strength of the anomaly was calculated for an inclination of 70° and a magnetic susceptibility contrast of $\kappa = 1 \times 10^{-8}$.

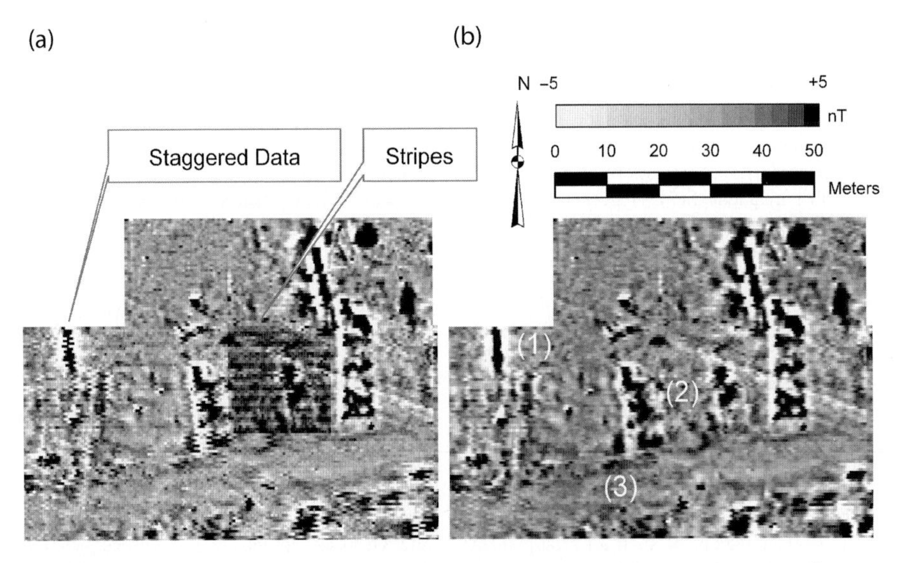

Figure A8 Fluxgate gradiometer data from Adel Roman Fort, West Yorkshire. (a) Field measurements with staggered data (*left*) and stripes (*center*). (b) After their improvement the data clearly show a ditch (1) and the soldiers' barracks (2) to the north of the road through the Fort (3).

all archeological remains during the development of a building site may mean that the collected geophysical data are the most important record of an ancient settlement. It is therefore essential that data archiving is undertaken according to recognized standards. In particular, detailed information describing data collection procedures and the layout of site and survey is important. Such information is usually referred to as "metadata" (Schmidt, 2002) and complements the numerical instrument readings as well as processed results. Related to the archiving of data from geophysical surveys is the recommendation that at least a brief report should be provided and archived, whenever possible.

Archeomagnetic dating

As with paleomagnetic dating (see *Paleomagnetism*), the magnetic remanence preserved in archeological structures can be used for their dating. Although some research has been undertaken on depositional ("detrital") remanence in sediments (Batt and Noel, 1991), archeomagnetic dating is mainly applied to thermoremanent magnetization. By firing archeological structures that are rich in iron oxides above their Curie temperature (ca 650–700°C) they become easily magnetized in the direction of the ambient field, which is usually the Earth's magnetic field. On subsequent cooling below the blocking temperature, this acquired magnetization will form a magnetic remanence. When archeological features that were exposed to such heating and cooling are excavated, oriented samples can be recovered and their thermoremanent magnetization measured. By comparing these data with an archeomagnetic calibration curve that charts the variation of magnetic parameters with time, a date can be determined for the last firing of the archeological feature.

Most archeomagnetic dating methods use two or three components of the remanent magnetization vector (inclination, declination, and sometimes intensity). It is therefore a pre-requisite that samples are collected from structures that have not changed their orientation since the last firing. Typical features include kilns, hearths, baked floors, and furnaces. Unfortunately, it is not always possible to assess whether an archeological feature is found undisturbed and *in situ*. For example, due to instabilities following the abandonment of a kiln, the walls may have moved slightly or the area around a fire place may have been disturbed by modern agricultural activities. Only the final statistical analysis (see below) can ascertain the validity of results. In recent years, advances in archeointensity dating have been made using only the magnitude of the magnetization for age determinations (see *Shaw and microwave methods, absolute paleointensity determination*). It is therefore possible to magnetically date materials that are no longer in their original position, like fired bricks that were used in buildings or even nonoriented pottery fragments (Shaw *et al.*, 1999; Sternberg, 2001).

A variety of different sampling methods exist, which all have their respective benefits. Some groups extract samples with corers, others encase the selected samples in Plaster of Paris before lifting them together with the plaster block, and in Britain plastic disks are commonly glued to the samples before extraction. As it is important to accurately record the orientation of the sample while still *in situ*, plaster and disks are usually leveled horizontally and the north direction is marked with a compass (either conventional, digital, or sun-based). Determining the right sampling locations within a feature is important as the effect of magnetic refraction often causes magnetic field lines to follow the shape of heated features similar to the demagnetization effects observed in grains and elongated objects. Soffel (1991) reports an approximately sinusoidal dependence of both inclination and declination from the azimuthal angle in a hollow cylindrical feature while Abrahamsen *et al.* (2003) have found that sample declinations from a hemisphere of solid iron slag with 0.5 m diameter vary throughout 360°. It is therefore essential that several samples from different parts of a feature are compared to statistically assess whether a consistent magnetization vector can be determined.

After the last firing of an archeological structure, magnetically soft materials may have acquired a viscous remanent magnetization that gradually followed the changing direction of the Earth's magnetic field (see *Geomagnetic secular variation*). Its contribution to a sample's overall magnetization can lead to wrong estimates for the age, and it therefore has to be assessed with a Thellier experiment and then removed. This removal can either be accomplished through stepwise thermal demagnetization (see *Demagnetization*) or with stepwise alternating field (AF) demagnetization (Hus *et al.*, 2003). The subsequent measurement of a sample's magnetization vector, for example in a spinner magnetometer, is then a good approximation of its thermoremanence. Fisher statistics (see *Fisher statistics*) is used to assess the distribution of all the measured magnetization vectors of an archeological feature by calculating a mean value for the direction together with its angular spread α_{95}. The latter describes the distribution of all measured directions around the mean and is half of the opening angle of a cone (hence appearing as a "radius" on a stereographic plot) that contains the mean vector direction with a probability of 95%. The spread of the individual vector directions can be due to errors in sample marking, measurement errors, and the distribution of different directions within a single feature (see above). It has therefore become common practice to expect α_{95} to be less than 5° for a reliable investigation (Batt, 1998).

Once the magnetization of an archeological feature has been established it can be compared to a calibration curve to derive the archeological age. The construction and use of such calibration data has been a matter of recent research. In Britain, a calibration curve was compiled by Clark *et al.* (1988) using 200 direct observations (since 1576 AD) and over 100 archeomagnetic measurements from features that were dated by other means, as far back as 1000 BC. All data were corrected for regional variations of the Earth's magnetic field and converted to apparent values for Meriden (52.43°N, 1.62°W). After plotting results on a stereographic projection, the authors manually drew a connection line that was annotated with the respective dates. Measurements from any new feature could be drawn on the same diagram and the archeological date was determined by visual comparison. This approach is compatible with the accuracy of the initial calibration curve but has clear limitations (Batt, 1997). Similar calibration curves exist for other countries and due to short-scale variations of the Earth's magnetic field's nondipole component (see *Nondipole field*), they are all slightly different and have to be constructed from individually dated archeological materials. The reference curve for Bulgaria, for example, now extends back to nearly 6000 BC (Kovacheva *et al.*, 2004). Although the calibration curve by Clark *et al.* (1988) has been a useful tool for archeomagnetic dating, improvements are now being made. Batt (1997) used a running average to derive the calibration curve more consistently from the existing British data. Kovacheva *et al.* (2004) calculated confidence limits for archeological dates using Bayesian statistics (Lanos, 2004) for the combination of inclination, declination, and paleointensity of measured samples. To improve the accuracy and reliability of the archeomagnetic method, more dated archeological samples are required and comprehensive international databases are currently being compiled.

Some researchers have attempted to use magnetometer surveys over furnaces to derive archeomagnetic dates for their last firing. For this, the magnetization causing the recorded magnetic anomaly has to be estimated and can then be used with a calibration curve for the dating of the buried archeological feature. To accommodate the complex shape and inhomogeneous fill of partly demolished iron furnaces in Wales, Crew (2002) had to build models with up to five dipole sources to approximate the measured magnetic anomaly maps. The dipole parameters were chosen to achieve the best possible fit between measured and modeled data and their relationship with the magnetization of the furnaces' individual components is not entirely clear. In addition to the sought after thermoremanence, this magnetization also has contributions from acquired viscous remanence and from the induced magnetization in the current Earth's magnetic field. Even after estimates for these two sources have been taken into consideration, the results were in poor agreement with the archeomagnetic calibration curve. The magnetic anomalies from slag-pit furnaces in Denmark were approximated with individual single dipole sources by Abrahamsen *et al.* (2003). They found that the distribution of 32 adjacent furnaces produced an unacceptably high α_{95} value of 18° and concluded that the method is therefore unsuitable for the dating of these features.

Conclusion

There are many ways in which magnetic methods can be used in archeological research and this application has made them popular with the public. Magnetometer surveys have become a tool for archeologists, nearly as important as a trowel. In an excavation, many archeological remains are only revealed by their contrast in color or texture compared with the surrounding soil. Searching for a contrast in magnetic properties is therefore only an extension of a familiar archeological concept. The science explaining the magnetic properties of buried features may be complex but the application of the techniques has become user-friendly. Similarly, archeomagnetic dating is an important part of an integrated archeological dating strategy. For archeological sites, dates are often derived with many different methods simultaneously, ranging from conventional archeological typological determinations over radiocarbon dating to luminescence methods. Combining these different data, for example with Bayesian statistics, allows a significant reduction of each method's errors and leads to improved results. The wealth of information stored in the magnetic record has certainly made an important contribution to modern archeology.

Armin Schmidt

Bibliography

Abrahamsen, N., Jacobsen, B.H., Koppelt, U., De Lasson, P., Smekalova, T., and Voss, O., 2003. Archaeomagnetic investigations of iron age slags in Denmark. *Archaeological Prospection*, **10**: 91–100.

Aitken, M.J., Webster, G., and Rees, A., 1958. Magnetic prospecting. *Antiquity*, **32**: 270–271.

Batt, C.M., and Noel, M., 1991. Magnetic studies of archaeological sediments. In Budd *et al.* (eds.), *Archaeological Science 1989*, pp. 234–241.

Batt, C.M., 1997. The British archaeomagnetic calibration curve: an objective treatment. *Archaeometry*, **39**: 153–168.

Batt, C.M., 1998. Where to draw the line? The calibration of archaeomagnetic dates. *Physics and Chemistry of the Earth*, **23**: 991–995.

Becker, H., 1995. From Nanotessla to Picotessla—a new window for magnetic prospecting in archaeology. *Archaeological Prospection*, **2**: 217–228.

Becker, H., and Fassbinder, J.W.E., 1999. In search for piramesses — the lost capital of Ramses II. the Nile delta (Egypt) by caesium magnetometry. In Fassbinder, J.W.E., and Irlinger, W.E. (eds.), *Archaeological Prospection*. München: Bayerisches Landesamt für Denkmalpflege, 146–150.

Bevan, B.W., 1994. The magnetic anomaly of a brick foundation. *Archaeological Prospection*, **1**: 93–104.

Blakely, R.J., 1996. *Potential Theory in Gravity and Magnetic Applications*. Cambridge: Cambridge University Press.

Ciminale, M., and Loddo, M., 2001. Aspects of magnetic data processing. *Archaeological Prospection*, **8**: 239–246.

Clark, A., 1990. *Seeing Beneath the Soil*. London: Batsford.

Clark, A.J., Tarling, D.H., and Noel, M., 1988. Developments in archaeomagnetic dating in Britain. *Journal of Archaeological Science*, **15**: 645–667.

Crew, P., 2002. Magnetic mapping and dating of prehistoric and medieval iron-working sites in Northwest Wales. *Archaeological Prospection*, **9**: 163–182.

Crowther, J., and Barker, P., 1995. Magnetic susceptibility: distinguishing anthropogenic effects from the natural. *Archaeological Prospection*, **2**: 207–216.

David, A., 1995. Geophysical survey in archaeological field evaluation, *English Heritage Research and Professional Services Guideline*, Vol. 1.

Fassbinder, J.W.E., and Irlinger, W.E., 1994. Aerial and magnetic prospection of an eleventh to thirteenth century Motte in Bavaria. *Archaeological Prospection*, **1**: 65–70.

Fassbinder, J.W.E., Stanjek, H., and Vali, H., 1990. Occurrence of magnetic bacteria in soil. *Nature*, **343**: 161–163.

Graham, I., and Scollar, I., 1976. Limitations on magnetic prospection in archaeology imposed by soil properties. *Archaeo-Pysika*, **6**: 1–125.

Herbich, T., 2003. Archaeological geophysics in Egypt: the Polish contribution. *Archaeologia Polona*, **41**: 13–56.

Herwanger, J., Maurer, H., Green, A.G., and Leckebusch, J., 2000. 3-D inversions of magnetic gradiometer data in archeological prospecting: Possibilities and limitations. *Geophysics*, **65**: 849–860.

Hus, J., Ech-Chakrouni, S., Jordanova, D., and Geeraerts, R., 2003. Archaeomagnetic investigation of two mediaeval brick constructions in north Belgium and the magnetic anisotropy of bricks. *Geoarchaeology*, **18**: 225–253.

Kattenberg, A.E., and Aalbersberg, G., 2004. Archaeological prospection of the Dutch perimarine landscape by means of magnetic methods. *Archaeological Prospection*, **11**: 227–235.

Kovacheva, M., Hedley, I., Jordanova, N., Kostadinova, M., and Gigov, V., 2004. Archaeomagnetic dating of archaeological sites from Switzerland and Bulgaria. *Journal of Archaeological Science*, **31**: 1463–1479.

Lanos, P., 2004. Bayesian inference of calibration curves: application to archaeomagnetism. In Buck, C.E., and Millard, A.R. (eds.), *Tools for Constructing Chronologies: Crossing Discipline Boundaries*. Lecture Notes in Statistics. Berlin: Springer-Verlag, pp. 43–82.

Le Borgne, E., 1955. Susceptibilité magnétique anormale du sol superficiel, *Annales de Géophysique*, **11**: 399–419.

Le Borgne, E., 1960. Influence du feu sur les propriétés magnétiques du sol et sur celles du schiste et du granite. *Annales de Géophysique*, **16**: 159–195.

Linford, N., 1994. Mineral magnetic profiling of archaeological sediments. *Archaeological Prospection*, **1**: 37–52.

Linford, N.T., 2004. Magnetic ghosts: mineral magnetic measurements on Roman and Anglo-Saxon graves. *Archaeological Prospection*, **11**: 167–180.

Linford, N.T., and Canti, M.G., 2001. Geophysical evidence for fires in antiquity: preliminary results from an experimental study. *Archaeological Prospection*, **8**: 211–225.

Neubauer, W., and Eder-Hinterleitner, A., 1997. 3D-Interpretation of postprocessed archaeological magnetic prospection data. *Archaeological Prospection*, **4**: 191–205.

Özdemir, O., and Banerjee, S.K., 1984. High temperature stability of maghemite. *Geophysical Research Letters*, **11**: 161–164.

Sauerländer, S., Kätker, J., Räkers, E., Rüter, H., and Dresen, L., 1999. Using random walk for on-line magnetic surveys, *European Journal of Environmental and Engineering Geophysics*, **3**: 91–102.

Schmidt, A., 2001. Visualisation of multi-source archaeological geophysics data. In Cucarzi, M., and Conti, P. (eds.), *Filtering, Optimisation and Modelling of Geophysical Data in Archaeological Prospecting*. Rome: Fondazione Ing. Carlo M. Lerici, pp. 149–160.

Schmidt, A., 2002. *Geophysical Data in Archaeology: A Guide to Good Practice,* ADS series of Guides to Good Practice. Oxford: Oxbow Books.

Schmidt, A., 2003. Remote Sensing and Geophysical Prospection. *Internet Archaeology*, 15 (http://intarch.ac.uk/journal/issue15/schmidt_index.html).

Schmidt, A., Yarnold, R., Hill, M., and Ashmore, M., 2005. Magnetic Susceptibility as Proxy for Heavy Metal Pollution: A Site Study. *Geochemical Exploration*, **85**: 109–117.

Shaw, J., Yang, S., Rolph, T.C., and Sun, F.Y., 1999. A comparison of archaeointensity results from Chinese ceramics using microwave and conventional Thellier's and Shaw's methods. *Geophysical Journal International*, **136**: 714–718.

Soffel, H. Chr., 1991. *Paläomagnetismus und Achäomagnetismus*. Berlin, Heidelberg, New York: Springer-Verlag.

Sternberg, R.S., 2001. Magnetic properties and archaeomagnetism. In Brothwell, D.R., and Pollard, A.M. (eds.), *Handbook of Archaeological Sciences*, Chichester: John Wiley & Sons, Ltd.

Sutherland, T., and Schmidt, A. 2003. Towton, 1461: An Integrated Approach to Battlefield Archaeology. *Landscapes*, **4**: 15–25.

Weston, D.G., 2004. The influence of waterlogging and variations in pedology and ignition upon resultant susceptibilities: A series of laboratory reconstructions. *Archaeological Prospection*, **11**: 107–120.

Cross-references

Biomagnetism
Demagnetization
Fisher Statistics
Geomagnetic Secular Variation
Magnetic Anomalies for Geology and Resources
Magnetic Anomalies, Modeling
Magnetic Mineralogy, Changes Due to Heating
Magnetic Susceptibility
Magnetization, Depositional Remanent (DRM)
Magnetization, Thermoremanent (TRM)
Nondipole Field
Observatories, Instrumentation
Paleomagnetism
Shaw and Microwave Methods, Absolute Paleointensity Determination
Storms and Substorms, Magnetic

ARCHEOMAGNETISM

The science and utilization of the magnetization of objects associated with archeological sites and ages. In practice, it is mostly concerned with the magnetic dating of archeological materials, but can also be used for reconstruction, magnetic sourcing, environmental analysis, and exploration. It depends on (1) the property of particular archeological materials containing magnetic impurities (particularly, the iron oxides, magnetite, and hematite) that can retain a record of the past direction and strength of the Earth's magnetic field, usually at the time that they cooled after being heated, but also when deposited, fluidized, or chemically altered; (2) the Earth's magnetic field gradually changes in both direction and intensity (*secular variation, q.v.*). As archeological materials normally record events, such as firing, at a specific point and time, most archeomagnetic records are intermittent in both space and time. The establishment of regional records of directions and intensities therefore require spatial corrections to be applied to individual determinations. Such "Master Curves" for directional studies are now mostly based on the assumption that the geomagnetic field, over an area of some 10^6 km^2, can be represented by an inclined geocentric dipole model of the field and have been constructed for several areas, particularly in Europe, the Middle East, Japan, and Central and southwestern North America. Such corrections appear justified by studies of the regional variation of the present geomagnetic field but may not be valid for all areas and times. Spatial variations in intensity are usually modeled using an axial geocentric dipole model of the geomagnetic field.

History

Bricks were shown to be magnetic by Boyle, in 1691, and that they lose their magnetization when heated and acquire it while cooling. Volcanic rocks used in the amphitheatre of Pompeii were demonstrated to have retained their original magnetization for at least 2000 years by Melloni in 1853. Folgerhaiter examined the magnetization of pottery in 1894. Chevallier, in 1925, constructed a record of historical changes in the direction of the geomagnetic field using lavas from Mt Etna. Actual dating of archeological fired clays, involving incremental thermal demagnetization of their direction and intensity of magnetization, were initiated by Thellier in 1936.

Procedures

(a) Directional Studies. Oriented samples are selected from part of the structure or object considered to have been magnetized at the same time. Their initial directions of remanence are measured and the samples subjected to demagnetization in incremental steps. This demagnetization can by heating, alternating magnetic fields or tuned microwaves. The vectors at successive steps are then analyzed to determine the characteristic vector associated with the time when the sample had originally cooled. These vectors are then combined, giving each determination equal weight, to determine the mean direction for the site. When the age of the firing is already known, then this direction is incorporated into a "Master Curve" for that region. When the age is unknown, the direction is compared with the appropriate "Master Curve" to assess the likely age of firing.

(b) Paleointensity Studies. As the intensity of magnetization is a scalar quantity, it is unnecessary that the sample should be unoriented. This also enables the technique to be used for isolated objects, such as pottery shards, that are no longer in their original firing position. It is therefore a far more widely applicable method. Until 2000, most paleointensity determinations were based on comparing the observed intensity of magnetization of a sample with the intensity it acquired in a known external magnetic field strength. Assuming no chemical changes occurred, i.e., the bulk magnetic susceptibility (K) remains constant, there is a simple relationship in which the ancient geomagnetic field strength (A) equals the intensity of natural remanence (NRM) acquired in that field multiplied by the strength of laboratory field (F) divided by the intensity of thermal remanence (TRM) acquired in that field (F), i.e., $K = \mathrm{NRM}/A = \mathrm{TRM}/F$. The problems are to ensure there is no change in susceptibility and that the NRM thermal component that was actually acquired during the original cooling has been isolated from all other NRM components. Conventionally, various tests, such as monitoring the susceptibility, repeat readings of the NRM component at successive temperatures and testing for the linear NRM/TRM relationship during incremental heating are all used. However, even when these are all satisfied, there can still be discrepancies between paleointensity determinations on specimens taken from the same sample. While such discrepancies are few, they suggest that thermally induced changes in the magnetic properties are occurring during laboratory heating even when all such tests are satisfied. The origin of these effects remains unclear and possibly relates to thermally induced physical changes as well as chemical changes in the magnetic properties. While other tests are still being developed, the more recent microwave method offers a radically different technique of demagnetizing both the natural and laboratory remanences. The microwaves are tuned to preferentially affect the quanta of spin wave energy in magnetic materials, magnons. The microwave energy is imposed briefly at different power levels, resulting in incremental demagnetization at successive power increments. In the current systems, little of the microwave energy leaks into the thermal band and the consequent rise in temperature is less than 120°C at the current maximum power of 120 W at 14 GHz. Such a temperature rise could still affect some minerals, but the hydroxide minerals that are most likely to be affected will only be present in weathered samples that are, in any case, not suitable for paleointensity determinations. This form of energy could also affect the magnetic domain structures but, as for thermally induced chemical changes, the brevity of the energy application appears to inhibit all such changes. A major advantage is that it can be applied rapidly, currently enabling paleointensity estimates to be made in a fraction of the

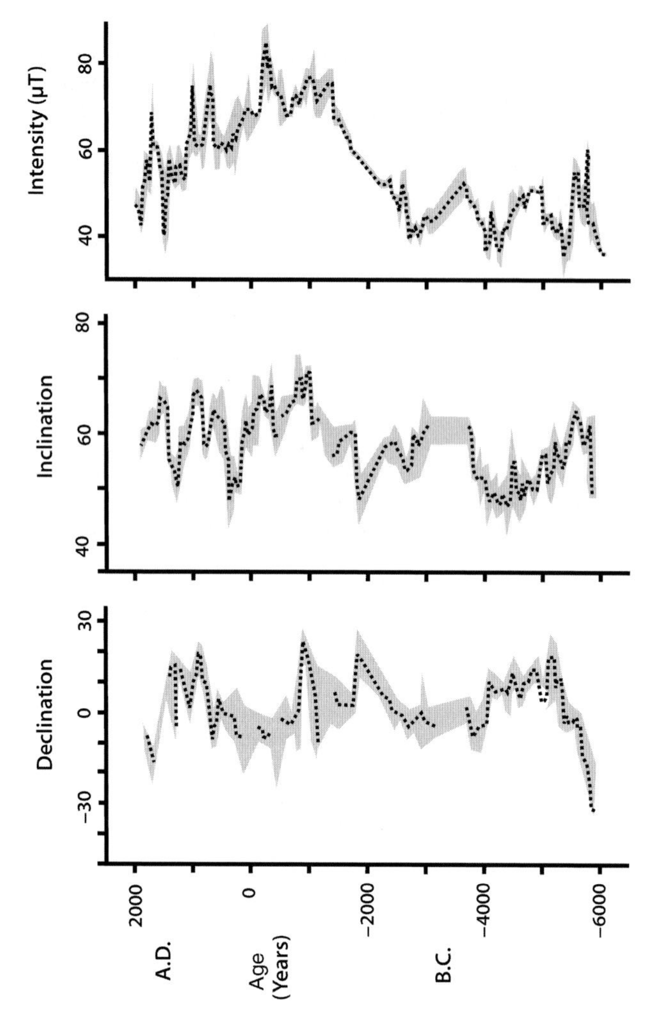

Figure A9 The "Master Curve" of geomagnetic secular variation in Bulgaria. The curves are discontinuous where there is little or no data. The 95% probability error is shaded. Modified from M.Kovacheva, Bulgarian Academy of Sciences (personal communication).

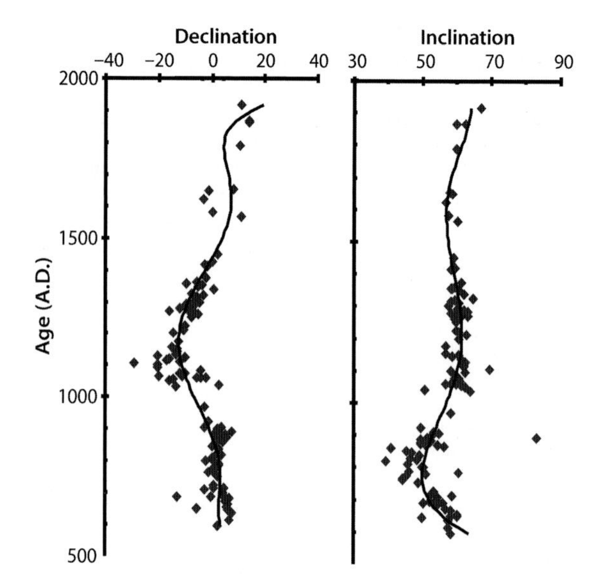

Figure A10 The "Master Curve" of geomagnetic secular variation in SW USA. This shows only the best-dated, well defined declinations and inclinations, after spatial correction. Modified from J.Eighmy, University of Colorado State University (personal communication).

time needed for standard thermal paleointensity methods. Nonetheless, several aspects of this method must still be considered experimental and it is mostly designed for materials containing magnetite rather than hematite.

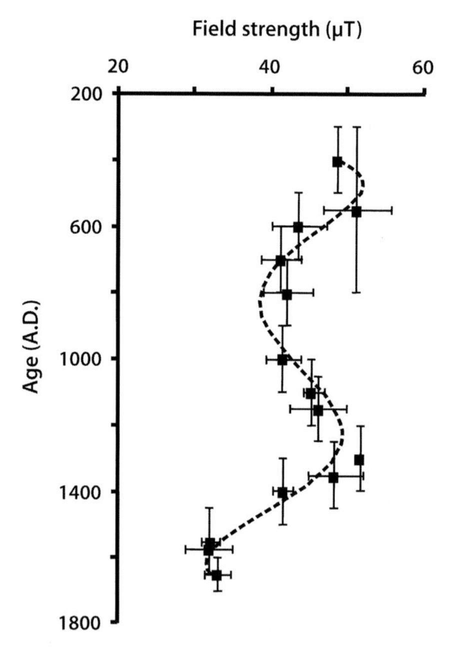

Figure A11 The paleointensity secular variation in Peru. These data are determined using microwave demagnetization techniques. Modified from J. Shaw, University of Liverpool (personal communication).

Errors

In directional studies, the largest, non-Gaussian errors in regional "Master Curves" arise from original age inaccuracies. Individual determinations can be affected by magnetic anisotropy, magnetic refraction, but mainly from physical disturbances of the site since firing. The precision of measurement and inhomogeneity remain problems in paleointensity studies, although the former is reducing.

Present Situation: The longest, continuous, and most systematic record that is available is for Bulgaria (FigureA9) and includes both directional and intensity data. The most readily accessible data are for England and Wales, France, southwestern USA (FigureA10), and potentially for Japan. Standard methods of curve fitting are being applied to local sequences of directional and intensity determinations, mostly assuming the errors to be Gaussian. This assumption may not always be valid and can result in smoothing fluctuations in the readings that may be of geomagnetic origin. Paleointensities using microwave techniques are now available for certain periods in Egypt and Peru (FigureA11).

Donald D. Tarling

Cross-references

Geomagnetic Secular Variation
Paleomagnetism
Secular Variation Model
Shaw and Microwave Methods, Absolute Paleointensity Determination

AURORAL OVAL

The auroras have fascinated humans since prehistory, but it was not until 1770 that Captain James Cook reported that the *aurora borealis* in the Northern Hemisphere had a counterpart in the Southern Hemisphere, the *aurora australis*. It was later noted that the frequency of observation of the aurora did not increase all the way to the poles, but maximized near latitudes of $\sim 70°$ and decreased again as the poles were approached. This allowed Elias Loomis in 1860 to draw a map of greatest auroral frequency as an irregular oval encircling the northern pole, though with a centroid somewhat displaced toward the northern coast of Greenland, such that it ran through the northern reaches of Scandinavia, Canada, and Siberia. Subsequently, photographic surveys have shown that the auroral observations are organized by the geomagnetic field, the ovals being roughly centered on the geomagnetic poles, though displaced away from the Sun, such that a magnetic latitude and magnetic local time coordinate system is the most appropriate for describing auroral morphology. The auroras are located at higher magnetic latitudes ($\sim 75°$) on the dayside than on the nightside ($\sim 70°$); the oval has a typical radius of 1500–2000 km and a latitudinal width of some 200–1000 km (Feldstein and Starkov, 1967). The auroras tend to be most luminous and of greatest latitudinal extent near local midnight. Before the advent of the Space Age, the ovals could only be considered as a statistical phenomenon. However, with the launch of auroral imagers onboard spacecraft such as Dynamics Explorer 1, it was found that the ovals were visible as continuous rings surrounding the pole (Frank and Craven, 1988). In recent years, near-continuous, high-temporal resolution auroral imagery from the ground and space has provided a wealth of information regarding the nature of the auroral oval and its relationship to the dynamic *magnetosphere* (*q.v.*) (e.g., Milan *et al.*, 2006).

Auroral processes

The auroras are formed by precipitation of charged particles, electrons, and protons, following magnetic field lines from the magnetosphere into the atmosphere. Here, collisions with neutral atmospheric atoms or molecules can lead to the promotion of electrons from their ground state or, in cases where the collisional energy exceeds the ionization potential, produce ion-electron pairs to supplement the pre-existing *ionosphere* (*q.v.*) produced mainly by solar photoionization (e.g., Vallance Jones, 1974). Excited atoms and molecules emit photons as they relax back to their ground states, to form the luminous aurora; approximately 1% of the incoming kinetic energy is converted to visible light, the power output of a typical 10×1000 km auroral arc being of the order of 1 GW. The initial kinetic energy of the precipitating particles determines the altitude at which most collisions occur, higher energy particles penetrating deeper into the atmosphere before being slowed significantly (Rees, 1989). Early triangulation of auroral altitudes (Størmer, 1955) showed that peak auroral emission occurred at 100–150 km altitude (the ionospheric E region), though significant luminosity could be observed higher, even above 250 km (F region). The auroral wavelengths produced depend on the atoms or molecules involved in the collisions, the atmospheric concentrations of which themselves depend on altitude. Although many spectral lines can be identified in the aurora, a few dominate. In the lower altitude regime, the main spectral line observed is the forbidden green oxygen line at 557.7 nm (^1S to ^1D), produced by incoming electrons with energies of ~ 10 keV. At higher altitudes, the red doublet at 630 and 636.4 nm (^1D to ground state) dominates, generated by ~ 1 keV electrons. Weak hydrogen lines are also observed, as incoming protons are excited by their collisions with the atmosphere. In addition, emission outside of the visible spectrum, the N_2 Lyman-Birge-Hopfield lines in the ultraviolet for instance, are important in auroral imaging from space. Other emission mechanisms include bremsstrahlung X-ray emission from highly energetic electrons penetrating to altitudes below 90 km.

The forms of aurora vary greatly, from auroral curtains or arcs, to curls, corona, and rayed aurora. Auroras can also be separated into *diffuse* and *discrete* forms, the latter often appearing embedded within the former. Discrete auroras are thought to arise due to acceleration processes taking place at relatively low altitudes (below 1 Earth radius). Low-Earth orbit observations of the characteristics of precipitating particles are shedding new light on these accelerating mechanisms (e.g., Carlson *et al.*, 1998). Spatial scales also vary from the global scale of the auroral ovals themselves, to meso- and micro-scale features within individual arcs; indeed, the closer the auroras are examined, the smaller the spatial scales that become apparent. One of the many open questions within auroral physics is the mechanism by which such diverse features, some as small as a few meters across, arise within the oval (Borovsky, 1993).

Auroral characteristics and the magnetosphere

The ultimate source of the precipitating electrons and protons is the solar wind plasma constantly streaming away from the Sun. Observing the morphology and dynamics of the auroral oval provides a great deal of information about the mechanisms by which solar wind particles gain entry to the magnetosphere, are distributed within it, and are subsequently accelerated to impact the atmosphere. The auroral ovals are not static, uniform features but display local time asymmetries and temporal variations on timescales ranging from subsecond, through minutes, hours, the seasons, and up to the 11-year solar cycle. In general, shorter timescale variations reflect internal magnetospheric dynamics such as magnetohydrodynamic wave activity, whereas longer changes are caused by magnetospheric responses to external stimuli, such as interactions between the terrestrial field with the *Sun's magnetic field* (*q.v.*) carried frozen-in (*Alfvén's Theorem and the frozen flux approximation, q.v.*) to the solar wind, where it is known as the interplanetary or heliospheric magnetic field. Solar cycle variations arise due to long-term changes in the characteristics of the solar wind and its magnetic field.

Magnetic reconnection between the interplanetary field and the magnetospheric field at the dayside magnetopause causes the two to become interconnected, such that the magnetosphere is no longer *closed*. The *open* field lines are eventually disconnected from the solar wind again by reconnection occurring in the magnetotail. It is this constant topological reconfiguration of the magnetospheric field that results in a general circulation and structuring of the magnetospheric plasma, as first proposed by James Dungey (Dungey, 1961), now known as the Dungey cycle. The open field lines allow solar wind plasma to enter the magnetosphere, in general streaming towards the Earth, funneled towards auroral latitudes in the noon sector by the dipolar nature of the *geomagnetic dipole field* (*q.v.*). Particles, which penetrate to sufficiently low altitudes that they can collide with atmospheric neutrals, give rise to *cusp* aurora in the dayside auroral oval, so-called after the magnetospheric cusp topology through which particles enter the magnetosphere. These particles tend to be relatively unaccelerated and so excite red line aurora. Those which do not collide are said to *mirror* on entering the higher field strength at low altitudes, return upwards along the magnetic field and are carried antisunward to populate the magnetotail. Such particles come eventually to reside near the equatorial plane where they are heated to form a dense hot *plasma sheet*. Confined in general to move along the near-dipolar magnetic field lines, these *trapped* particles approach the Earth most closely at auroral latitudes as they mirror backwards and forwards between conjugate hemispheres. At each bounce there is a possibility of an atmospheric collision and the generation of aurora; plasma sheet particles, having been accelerated in the magnetotail, give rise to green line aurora. The plasma sheet extends from the magnetotail around the Earth to the dayside, forming the familiar rings of aurora around the poles. The lower latitude boundary of the auroral oval is governed by the distance of the inner edge of the plasma sheet from the Earth in the equatorial plane, usually of order 7 Earth radii. The higher latitude boundary of the oval is closely related to the transition from closed dipolar field lines to open field lines which map deep into the magnetotail and out into the solar wind. Such nondipolar field lines do not provide the trapping geometry that requires particles to constantly return to the atmosphere, and so auroral luminosities tend to be negligible poleward of 80° latitude. On occasions, Sun-aligned arcs can be observed at high latitudes inside the oval, also known as *transpolar arcs* or *theta aurora*, in the latter case due to the characteristic shape they form together with the oval. There is still much debate regarding the source of these auroras and the magnetospheric configuration that gives rise to them.

Seasonal variations have been found in the distribution of auroral luminosity within the ovals (Newell *et al.*, 1998) and the overall level of auroral activity (Russell and McPherron, 1973). The latter is in part a consequence of the differing reconnection geometries available at the magnetopause as the tilt of the rotational and magnetic axes of the Earth change the orientation of the magnetosphere with respect to the Sun-Earth line. However, variations in auroral distribution controlled by the extent of solar illumination of the polar regions point toward complex feedback mechanisms between ionosphere and magnetosphere.

The substorm cycle

The auroral oval is not static with time, but undergoes a 2–4 hr cycle of growth and decay known as the *substorm cycle* (intimately related to the Dungey cycle), first adequately described by Akasofu (1968). During periods of strong reconnection at the dayside magnetopause, generally when the interplanetary magnetic field points southwards, an increasing proportion of the terrestrial magnetic flux becomes opened. As a consequence, the auroral oval moves to lower latitudes as the open field line region expands, during what is known as the *growth phase*, which can last several tens of minutes. At some point the magnetosphere can no longer support continued opening of flux and reconnection is initiated in the magnetotail to close flux once again. This flux closure is associated with vivid nightside auroral displays, termed *substorm break-up aurora*, as plasma is accelerated earthwards into the midnight sector atmosphere. The nightside aurora expands rapidly polewards to form a *substorm auroral bulge* as the dim open flux region contracts, leading to this phase of the substorm cycle being known as *expansion phase*, which can last from 30 min to a few hours. The auroral bulge expands not only polewards but also rapidly westwards, spear-headed by the most luminous feature of the substorm, the *westwards-travelling surge*. Eventually, the substorm enters the *recovery phase* as the magnetosphere relaxes back to a more quiescent state. This phase is accompanied by large-scale wave-like perturbations of the poleward border of the postmidnight oval, termed *omega bands* after their characteristic appearance.

Associated with substorms are elevated levels of geomagnetic activity caused by currents flowing horizontally in the E region ionosphere, facilitated by enhanced ionospheric conductivity associated with the auroral ionization. These currents flow to close pairs of upward and downward field-aligned currents (Iijima and Potemra, 1978), also known as Birkeland currents, which transmit stress from the outer magnetosphere into the ionosphere and drive ionospheric convection, the counterpart of the magnetospheric circulation described by the Dungey/substorm cycle. The main components of the ionospheric current systems are the *westward* and *eastward electrojets* flowing along the auroral ovals in the dawn and dusk sectors, respectively, and the development of the *westward substorm electrojet* in the midnight sector during expansion phase. These currents are largely responsible for the magnetic deflections detected at the ground and compiled to form various geomagnetic indices that measure the general level of geomagnetic disturbance, such as the planetary K_P index devised by *Julius Bartels* (*q.v.*) (Bartels *et al.*, 1939) and auroral electrojet index *AE* (Davis and Sugiura, 1966).

Stephen Milan

Bibliography

Akasou, S.-I., 1968. *Polar and magnetospheric substorms.* Dordrecht: Reidel.
Bartels, J., Heck, N.H., and Johnston, H.F., 1939. The three-hour-range index measuring geomagnetic activity. *Journal of Geophysical Research*, **44**: 411.
Borovsky, J.E., 1993. Auroral arc thickness as predicted by various theories. *Journal of Geophysical Research*, **9**: 6101.
Carlson, C.W., Pfaff, R.F., and Watzin, J.G., 1998. The Fast Auroral SnapshoT (FAST) mission. *Geophysical Research Letters*, **25**: 2013.
Davis, T.N., and Sugiura, M., 1966. Auroral electrojet activity index AE and its universal time variations. *Journal of Geophysical Research*, **71**: 785.
Dungey, J.W., 1961. Interplanetary magnetic field and the auroral zones. *Physics Review Letters*, **6**: 47.
Feldstein, Y.I., and Starkov, G.V., 1967. Dynamics of the auroral belt and polar geomagnetic disturbances. *Planetary and Space Science*, **15**: 209.
Frank, L.A., and Craven, J.D., 1988. Imaging results from Dynamics Explorer 1. *Reviews of Geophysics and Space Physics*, **26**: 249.
Iijima, T., and Potemra, T.A., 1978. Large-scale characteristics of field-aligned currents associated with substorms. *Journal of Geophysical Research*, **83**: 599.
Milan, S.E., Wild, J.A., Grocott, A., and Draper, N.C., 2006. Space- and ground-based investigations of solar wind-magnetosphere-ionosphere coupling. *Advances in Space Research*, **38**: 1671–1677.
Newell, P.T., Meng, C.-I., and Wing, S., 1998. Relation to solar activity of intense aurorae in sunlight and darkness. *Nature*, **393**: 342.
Rees, M.H., 1989. *Physics and Chemistry of the Upper Atmosphere.* Cambridge: Cambridge University Press.
Russell, C.T., and McPherron, R.L., 1973. Semi-annual variation of geomagnetic activity. *Journal of Geophysical Research*, **78**: 92.
Størmer, C., 1955. *The Polar Aurora.* Oxford: Clarendon Press.
Vallance Jones, A., 1974. *Aurora.* Dordrecht: Reidel.

Cross-references

Alfvén's Theorem and the Frozen Flux Approximation
Bartels, Julius (1899–1964)
Geomagnetic Dipole Field
Ionosphere
Magnetic Field of Sun
Magnetosphere of the Earth
Storms and Substorms, Magnetic

B

BAKED CONTACT TEST

Introduction

At the time of emplacement, an igneous unit (intrusion or volcanic flow) heats adjacent host rocks. Upon cooling in the Earth's magnetic field the igneous unit and adjacent host rocks acquire similar directions of remanent magnetization. The simple case of thermal overprinting of host rocks that have a consistent composition and magnetic mineralogy is illustrated in Figure B1. The host rock magnetization is completely reset in a "baked" zone adjacent to the contact, partially reset in a zone of "hybrid" remanence farther away, and unaffected in the "unbaked" zone at even greater distances.[1] On the other hand, a later regional metamorphic event would, if sufficiently intense, produce a consistent remanence direction throughout the profile.

Two types of baked contact tests utilize the paleomagnetic remanence across the contact of igneous units in order to establish whether the remanence in the igneous unit is primary. The standard *baked contact test* (or *contact test*) compares remanence directions in the igneous unit, the baked zone, and the unbaked zone (Figure B1). Although relatively robust, difficulties can arise with the interpretation of this test if chemical changes have occurred in the host rocks. A more rigorous test, herein called the *baked contact profile test*, includes a detailed study of overprinting in the hybrid zone (Figure B1). It can demonstrate conclusively that the overprint remanence was acquired as a thermoremanent magnetization (TRM) at the time of emplacement of the igneous unit. Both the standard baked contact and the more detailed baked contact profile tests are described and discussed below.

Importance of baked contact tests

Well-dated paleomagnetic poles (see *Pole, key paleomagnetic*) are a prerequisite to defining reliable apparent polar wander paths, tracking the drift of continents, and establishing continental reconstructions. In general, key paleopoles should be demonstrated primary and the rock unit from which they are derived should be precisely and accurately dated (Buchan *et al.*, 2001). In the Precambrian most key paleopoles are derived from igneous rocks because fossil evidence for the age of sedimentary rocks is usually lacking. Although there are several field tests of the primary nature of the magnetic remanence of igneous rocks (e.g., Buchan and Halls, 1990), the baked contact test is the most widely used.

[1] In some publications the "baked", "hybrid" and "unbaked" zones are referred to as the "contact", "hybrid" and "host (magnetization)" zones, respectively.

In a recent review of the worldwide database of paleomagnetic poles for the 1700–500 Ma period, Buchan *et al.* (2001) concluded that only 18 paleopoles, of many hundreds that are available, are sufficiently well-dated to be used for robust continental reconstructions. All 18 are from well-dated igneous rocks. Ten of the 18 have a baked contact test or a baked contact profile test demonstrating that the remanence is primary, whereas eight are shown primary using other types of tests. This emphasizes the importance of igneous rocks and baked contact tests in paleomagnetic studies, especially in the Precambrian.

Baked contact test

Brunhes (1906) first proposed a comparison of the remanence of an igneous unit with that of adjacent baked host rocks, concluding that if the remanence direction is similar in the two units it is stable and primary. Everitt and Clegg (1962) pointed out that the Brunhes test was incomplete because regional metamorphism will result in simultaneous remagnetization of both igneous and baked rocks. Therefore, they proposed an extension of the Brunhes test in which unbaked host rocks farther from the intrusion are also sampled. Everitt and Clegg used their test to look for a difference in remanence characteristics between baked and unbaked sedimentary host rocks. Today, the Everitt and Clegg (1962) test is commonly referred to as the baked contact test (or contact test) and applied mainly to a comparison of the direction of remanent magnetization in an igneous unit and its igneous or sedimentary host rocks.

Method

The standard baked contact test requires collection of oriented paleomagnetic samples from the igneous unit, from baked host rocks and from unbaked host rocks in relatively close proximity to the igneous unit. In addition, the baked and unbaked host rocks should be of similar composition so that a direct comparison can be made of their rock properties.

The baked contact test is "positive" (Figure B2a) and the remanence considered primary if the igneous unit and baked host rocks carry a consistent, stable direction of magnetization, whereas the unbaked host rocks have a stable but different direction. It is important to reiterate that the unbaked host rock must be stably magnetized. An example of a positive baked contact test is shown in Figure B3. Here a Marathon dyke crosscuts an older Matachewan dyke in the Superior Province of the Canadian Shield. The Matachewan dyke far from the Marathon contact carries a stable SSW remanence direction, whereas the Marathon dyke and adjacent baked Matachewan dyke have similar SE remanence

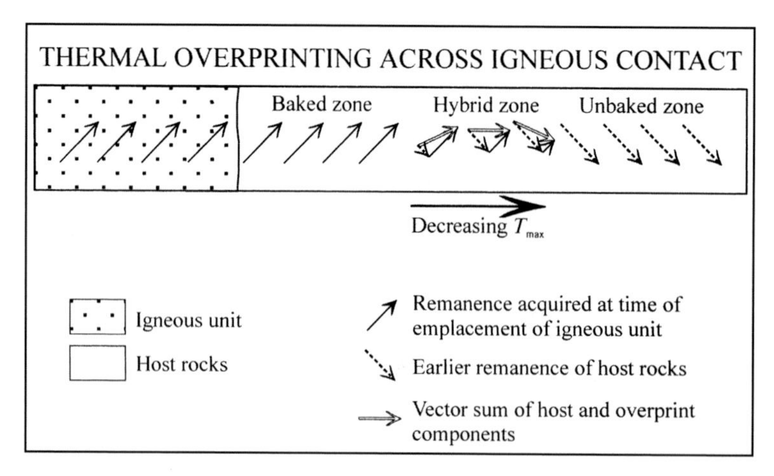

Figure B1 Direction of remanent magnetization with distance from the contact of an igneous intrusion. It is assumed that the magnetic mineralogy of the host rock is similar throughout the profile and that overprinting in the baked and hybrid zones is a thermoremanent magnetization.

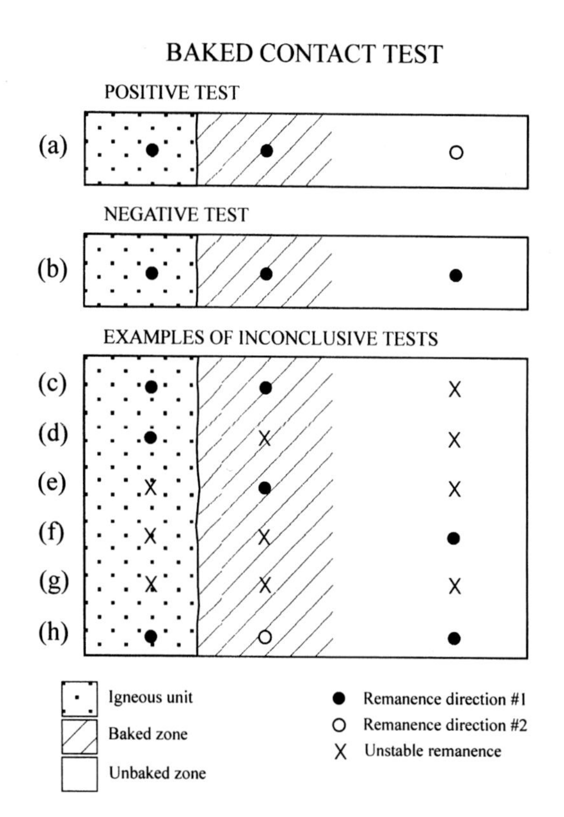

Figure B2 Baked contact tests. (a) Positive test. (b) Negative test. (c-h) Examples of inconclusive tests.

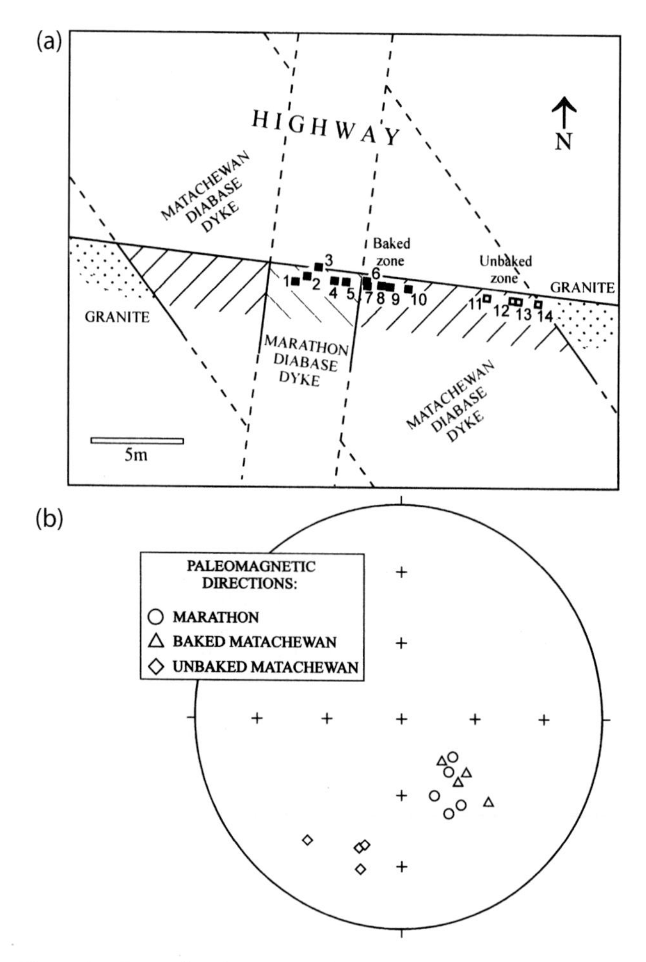

Figure B3 Example of a positive baked contact test for a ca. 2.10 Ga Marathon dyke which crosscuts a ca. 2.45 Ga Matachewan dyke in a roadcut at a locality in the Superior Province of the Canadian Shield. (a) Patterns indicate area of outcrop. Open and closed squares refer to samples for which typical Marathon and Matachewan remanence directions were obtained, respectively. (b) Paleomagnetic directions are plotted on an equal-area net. All magnetizations are directed up. Data are discussed in the text and in Buchan *et al.* (1996).

directions. This positive test indicates that the Marathon dyke carries a primary remanence from the time of its intrusion (Buchan *et al.*, 1996).

Comparing the magnetic properties of the unbaked and baked host rocks may provide information on whether chemical alteration has occurred in the baked rocks at the time of emplacement of the igneous unit. The presence of chemical changes can make the test less reliable. For example, consider the case of an igneous unit with unblocking temperatures that are lower than those of host rocks into which it is intruded. Suppose that the unblocking temperatures of host rocks adjacent to the igneous unit are lowered as a result of chemical changes that occur due to heating by the igneous unit. Then a later mild regional reheating event can reset the remanence of the igneous unit and adjacent host rocks without resetting the remanence of distant host rocks that have retained their higher unblocking temperatures. This situation would yield a positive baked contact test even though the remanence of the igneous unit is secondary. (Note that the more detailed baked contact profile test described below can eliminate any uncertainty about chemical changes by clearly establishing that the remanence in the baked zone is a TRM and that it was acquired *at the time when the igneous unit was emplaced*).

A positive test also demonstrates that the unbaked host rocks carry a remanence that significantly predates the emplacement of the igneous unit.

The baked contact test is "negative" (Figure B2b) if the igneous unit and the baked and unbaked host all carry a similar stable direction of magnetization. A negative test usually indicates that the remanence of the igneous unit is secondary, having been acquired significantly after emplacement. However, there are specific situations in which the igneous unit carries a primary remanence, but the baked contact test is negative. These include cases where (a) the igneous unit and its host rocks are of similar age, (b) the host rocks acquired a regional overprint shortly before emplacement of the igneous unit, and (c) the direction of the Earth's magnetic field was similar at the time of emplacement of the igneous unit and at some earlier time when the host rocks were magnetized. Therefore, although a remanence can usually be established as primary with a positive baked contact test, it cannot be demonstrated as secondary with the same degree of certainty by a negative baked contact test.

Many baked contact tests do not meet the criteria described above for a positive or a negative test. In a few specific situations, some limited information concerning the age of the remanence in the baked zone can still be obtained. For example, if the igneous unit is unstably magnetized, but the baked and unbaked host rocks are stably magnetized and can be shown to have similar magnetic properties, similar remanence directions in the baked and unbaked host rocks likely indicate a secondary overprint, whereas dissimilar directions likely indicate that the remanence of the baked zone was acquired at the time of emplacement of the igneous unit. However, in most cases in which the criteria for a positive or negative test are not achieved, the test is "inconclusive" (e.g., Figure B2c-h) and does not provide information concerning the primary or secondary nature of the remanence.

Incorrect application of the baked contact test

The baked contact test is often applied incorrectly or misinterpreted in the literature. In addition, terminology applied to the test is often confusing or misleading. Examples of these problems are discussed below.

In some studies only the igneous unit and its baked zone are sampled and the remanence is interpreted to be primary based on similar remanence directions in these two units. However, as noted above, Everitt and Clegg (1962) described how this result would also be obtained if both units were overprinted at some later date. Such tests are incomplete and should not be referred to as baked contact tests because all three elements (igneous, baked host, and unbaked host rocks) that are necessary for a baked contact test have not been sampled.

There are also many examples in the literature where an igneous unit and baked host rocks carry a consistent direction of magnetization but unbaked host is unstably magnetized (Figure B2c). Terms such as "semipositive test," "not fully positive test," or "partial test" are sometime used in such cases. However, these terms are misleading and should not be used, because they imply incorrectly that the test gives some information on whether the remanence of the igneous unit is primary. For example, host rocks that are unstably magnetized before emplacement of the igneous unit can acquire a stable remanence in the baked zone at the time of emplacement as a result of chemical alterations that occur only in the baked zone. During a subsequent regional metamorphism the igneous unit and baked host will be remagnetized (with similar remanence directions), while the unbaked host remains unstably magnetized. This is especially likely if temperatures during the metamorphic event remain below that necessary to induce chemical changes in the unbaked host rocks. Thus, when unbaked host rocks carry an unstable remanence, the baked contact test is inconclusive and the age of the remanence in the igneous unit is uncertain.

In other studies, the unbaked host rocks are collected at great distance from the igneous unit and its baked host rocks, often as a result of a lack of suitable outcrop in the intervening area. This can cause serious problems in the interpretation of the results. For example, overprinting of the local area near the igneous unit may have occurred after its emplacement, resulting in remagnetization of the igneous unit and its baked zone, whereas the distant unbaked site may have escaped overprinting at that time. In addition, it is more likely that magnetic mineralogy will differ between such widely separated baked and unbaked host rocks, making the comparison of the magnetic directions between the two locations more problematic.

Baked contact profile test

Although the baked contact test described above is generally considered to be a robust test of primary remanence, there are some circumstances involving chemical alterations in the host rocks when it can be misleading. Buchan *et al.* (1993) and Hyodo and Dunlop (1993) pointed out that analyzing magnetic overprinting along a profile that includes the hybrid zone (Figure B1), following the procedure of Schwarz (1977), yields a more rigorous test of primary remanence than the standard baked contact test. In particular, such a baked contact profile test can demonstrate that the overprint component is a TRM and that it was acquired at the time of emplacement of the igneous unit (Buchan and Schwarz, 1987; Schwarz and Buchan, 1989; Hyodo and Dunlop, 1993).

The baked contact profile test is the most powerful method for establishing that remanence in igneous rocks is primary. However, it is not widely used because of the necessity of sampling a continuous profile, the difficulty in locating the often-narrow hybrid zone, and the difficulty in analyzing the hybrid zone magnetizations. To date it has only been applied in a few instances to the study of dyke contacts (e.g., Schwarz, 1977; Buchan *et al.*, 1980; Symons *et al.*, 1980; Buchan and Schwarz, 1981; McClelland Brown, 1981; Schwarz and Buchan, 1982; Schwarz *et al.*, 1985; Hyodo and Dunlop, 1993; Oveisy, 1998; Wingate and Giddings, 2000), usually as part of studies to determine ambient host rock temperatures and depth of burial (see *Magnetization, Remanent, Ambient Temperature and Burial Depth from Dyke Contact Zones*).

Method

A continuous profile is sampled from the igneous unit, through the baked and hybrid zones into the unbaked host zone (see Figure B1). The distance of each sample from the contact is recorded, as are the dimensions of the igneous unit.

Samples are thermally demagnetized in stepwise fashion in order to determine the maximum temperature (T_{max}) to which each hybrid sample was reheated at the time of emplacement of the igneous unit. If the overprint component is a partial TRM (pTRM) resulting from heating by the igneous unit, the unblocking temperature spectra of the overprint component will occupy a range immediately below that of the earlier host component (Dunlop, 1979; McClelland Brown, 1982; Schwarz and Buchan, 1989; Dunlop and Özdemir, 1997). Any overlap of the two unblocking spectra should be explainable in terms of viscous pTRM. T_{max} is given by the maximum unblocking temperature of the overprint component, most easily determined using an orthogonal component (or Zijderveld) plot of the horizontal and vertical component of the thermal demagnetization data. T_{max} must be corrected for the effects of magnetic viscosity using published curves

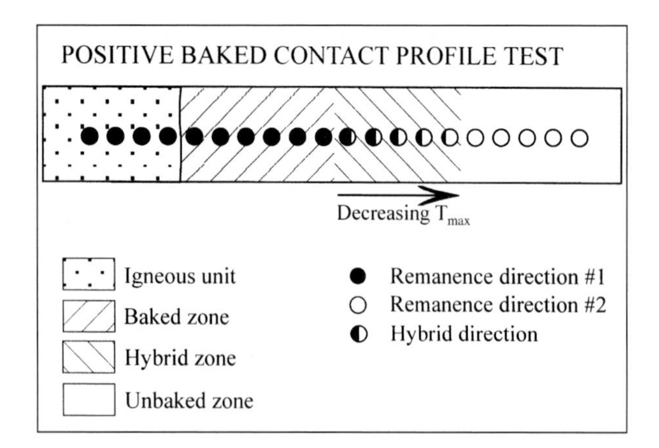

Figure B4 Positive baked contact profile test. T_{max}, the maximum temperature that is obtained at a given location in the host rocks as the result of emplacement of the igneous unit, is determined from orthogonal component projections in the hybrid zone. A systematic decrease in the T_{max} through the hybrid zone with increasing distance from the contact demonstrates that the overprint is a TRM.

for the appropriate magnetic mineral (e.g., Pullaiah *et al.*, 1975). The procedure for analyzing magnetic components in the hybrid zone and determining T_{max} is described in more detail in the article on *Magnetization, remanent, application (q.v.)*.

If the overprint component is a chemical remanent magnetization (CRM), T_{max} cannot be determined. Schwarz and Buchan (1989) discuss how CRM components can be distinguished in hybrid zones.

The baked contact profile test is "positive" (Figure B4) if the criteria for a positive baked contact test are satisfied (i.e., the igneous unit and baked host carry a consistent, stable direction of magnetization, whereas the unbaked host has a stable but different direction) and if T_{max} values in the hybrid zone decrease systematically away from the contact, as predicted by heat conduction theory (Jaeger, 1964). The critical aspect of the test is that T_{max} values decrease systematically across the hybrid zone. The most detailed comparisons of experimental and theoretical T_{max} values across hybrid zones have been described by McClelland Brown (1981).

A positive baked contact profile test demonstrates that the overprint remanence of the hybrid zone is a pTRM and that it was acquired as the igneous unit cooled following emplacement. The similarity between the direction of the overprint in the hybrid zone and that of the remanence in the igneous unit itself indicates that the latter is primary.

The baked contact profile test is "negative" if the igneous unit and all elements of the host profile yield consistent remanence directions. Interpretation of the negative baked contact profile test is similar to that of the negative baked contact test described above.

The test is "inconclusive" if the criteria for "positive" or "negative" test are not met. For example, if T_{max} values cannot be obtained from the hybrid zone or if they do not decrease systematically with distance from the contact, then it cannot be concluded that the overprint is a TRM from the time of emplacement of the igneous unit.

Kenneth L. Buchan

Bibliography

Brunhes, B., 1906. Recherches sur la direction d'aimantation des roches volcaniques. *Journal de Physique*, **5**: 705–724.

Buchan, K.L., and Halls, H.C., 1990. Paleomagnetism of Proterozoic mafic dyke swarms of the Canadian Shield. In Parker, A.J., Rickwood, P.C., and Tucker, D.H., (eds.), *Mafic Dykes and Emplacement Mechanisms*. Balkema: Rotterdam, 209–230.

Buchan, K.L., Mortensen, J.K., and Card, K.D., 1993. Northeast-trending Early Proterozoic dykes of southern Superior Province: multiple episodes of emplacement recognized from integrated paleomagnetism and U-Pb geochronology. *Canadian Journal of Earth Sciences*, **30**: 1286–1296.

Buchan, K.L., Halls, H.C., and Mortensen, J.K., 1996. Paleomagnetism, U-Pb geochronology, and geochemistry of Marathon dykes, Superior province, and comparison with the Fort Frances swarm. *Canadian Journal of Earth Sciences*, **30**: 1286–1296.

Buchan, K.L., and Schwarz, E.J., 1981. Uplift estimated from remanent magnetization: Munro area of Superior Province since 2150 Ma. *Canadian Journal of Earth Sciences*, **18**: 1164–1173.

Buchan, K.L., and Schwarz, E.J., 1987. Determination of the maximum temperature profile across dyke contacts using remanent magnetization and its application. In Halls, H.C., and Fahrig, W.F. (eds.), *Mafic Dyke Swarms*. Geological Association of Canada, Special Paper **34**, pp. 221–227.

Buchan, K.L., Ernst, R.E., Hamilton, M.A., Mertanen, S., Pesonen, L.J., and Elming, S.-Å., 2001. Rodinia: the evidence from integrated palaeomagnetism and U-Pb geochronology. *Precambrian Research*, **110**: 9–32.

Buchan, K.L., Schwarz, E.J., Symons, D.T.A., and Stupavsky, M., 1980. Remanent magnetization in the contact zone between Columbia Plateau flows and feeder dykes: evidence for groundwater layer at time of intrusion. *Journal of Geophysical Research*, **85**: 1888–1898.

Dunlop, D.J., 1979. On the use of Zijderveld vector diagrams in multicomponent paleomagnetic studies. *Physics of the Earth and Planetary Interiors*, **20**: 12–24.

Dunlop, D.J., and Özdemir, Ö., 1997. Rock magnetism: fundamentals and frontiers. Cambridge: Cambridge University Press, 573 pp.

Everitt, C.W.F., and Clegg, J.A., 1962. A field test of paleomagnetic stability. *Journal of the Royal Astronomical Society*, **6**: 312–319.

Hyodo, H., and Dunlop, D.J., 1993. Effect of anisotropy on the paleomagnetic contact test for a Grenville dike. *Journal of Geophysical Research*, **98**: 7997–8017.

Jaeger, J.C., 1964. Thermal effects of intrusions. *Reviews of Geophysics*, **2**(3): 711–716.

McClelland Brown, E., 1981. Paleomagnetic estimates of temperatures reached in contact metamorphism. *Geology*, **9**: 112–116.

McClelland Brown, E., 1982. Discrimination of TRM and CRM by blocking-temperature spectrum analysis. *Physics of the Earth and Planetary Interiors*, **30**: 405–411.

Oveisy, M.M., 1998. Rapakivi granite and basic dykes in the Fennoscandian Shield; a palaeomagnetic analysis. Ph.D. thesis, Luleå University of Technology, Luleå, Sweden.

Pullaiah, G., Irving, E., Buchan, K.L., and Dunlop, D.J., 1975. Magnetization changes caused by burial and uplift. *Earth and Planetary Science Letters*, **28**: 133–143.

Schwarz, E.J., 1977. Depth of burial from remanent magnetization: the Sudbury Irruptive at the time of diabase intrusion (1250 Ma). *Canadian Journal of Earth Sciences*, **14**: 82–88.

Schwarz, E.J. and Buchan, K.L., 1982. Uplift deduced from remanent magnetization: Sudbury area since 1250 Ma ago. *Earth and Planetary Science Letters*, **58**: 65–74.

Schwarz, E.J. and Buchan, K.L., 1989. Identifying types of remanent magnetization in igneous contact zones. *Physics of the Earth and Planetary Interiors*, **68**: 155–162.

Schwarz, E.J., Buchan, K.L., and Cazavant, A., 1985. Post-Aphebian uplift deduced from remanent magnetization, Yellowknife area of Slave Province. *Canadian Journal of Earth Sciences*, **22**: 1793–1802.

Symons, D.T.A., Hutcheson, H.I., and Stupavsky, M., 1980. Positive test of the paleomagnetic method for estimating burial depth using a dike contact. *Canadian Journal of Earth Sciences*, **17**: 690–697.

Wingate, M.T.D., and Giddings, J.W., 2000. Age and paleomagnetism of the Mundine Well dyke swarm, Western Australia: implications for an Australia-Laurentia connection at 755 Ma. *Precambrian Research*, **100**: 335–357.

Cross-references

Magnetization, Remanent, Ambient Temperature and Burial Depth from Dyke Contact Zones
Magnetization, Thermoremanent (TRM)
Paleomagnetism
Pole, Key Paleomagnetic

BANGUI ANOMALY

"Bangui anomaly" is the name given to one of the Earth's largest crustal magnetic anomalies and the largest over the African continent. It covers two-thirds of the Central African Republic and the name derives from the capital city Bangui that is near the center of this feature. From surface magnetic survey data, Godivier and Le Donche (1962) were the first to describe this anomaly. Subsequently high-altitude aerial magnetic surveying (see *Aeromagnetic surveying*) by the US Naval Oceanographic Office (Project Magnet) recorded a greater than 1000 nT dipolar, peak-to-trough anomaly with the major portion being negative (Figure B5). Satellite observations (Cosmos 49) were first reported in 1964 (Benkova *et al.*, 1973); these revealed a −40nT anomaly at 350 km altitude. Subsequently the higher altitude (417–499 km) Polar Orbiting Geomagnetic Observatory (POGO) satellites data recorded peak-to-trough anomalies of 20 nT. These data were added to Cosmos 49 measurements by Regan *et al.* (1973) for a regional satellite altitude map. In October 1979, with the launch of Magsat (see *Magsat*), a satellite designed to measure crustal magnetic anomalies, a more uniform satellite altitude magnetic map was obtained (Girdler *et al.*, 1992). From the more recent CHAMP (see *CHAMP*) satellite mission a map was computed at 400 km altitude, a greater than 16 nT anomaly was recorded (Figure B6/Plate 6c). The Bangui anomaly is elliptically shaped and is approximately 760 by 1000 km centered at 6° N, 18° E (see *Magnetic anomalies, long wavelength*). It is composed of three segments, with positive anomalies north and south of a large central negative anomaly. This displays the classic pattern of a magnetic anomalous body being magnetized by induction in a zero inclination field. This is not surprising since the magnetic equator passes near the center of the body.

While the existence and description of the Bangui anomaly is well-known, what is less established and controversial is the origin or cause that produced this large magnetic feature. It is not possible to discuss its origin without mentioning the other associated geophysical and geologic information. There is a −120 mGal Bouguer gravity anomaly coincident with the magnetic anomaly (Boukeke, 1994) and a putative

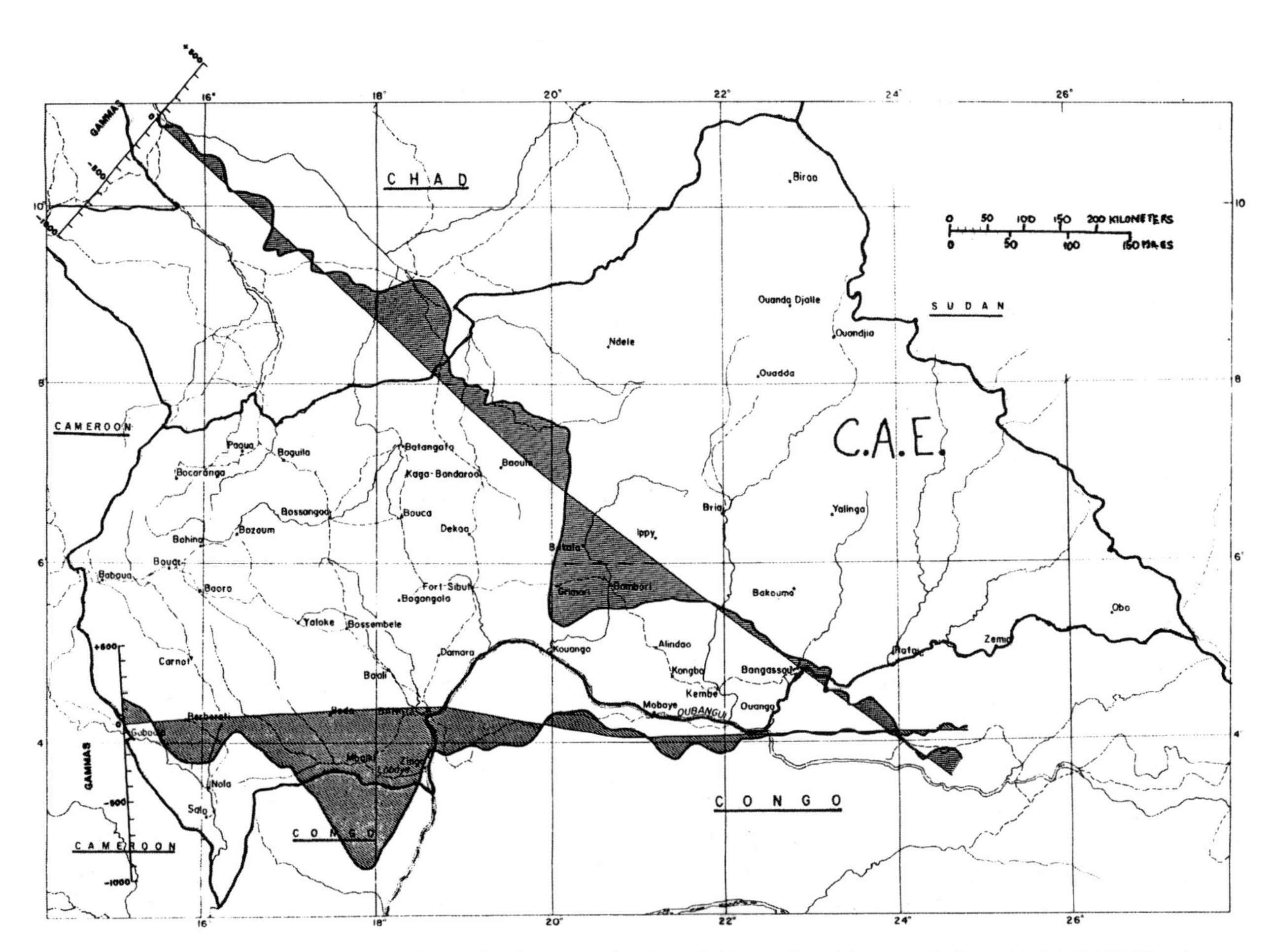

Figure B5 Total field aeromagnetic anomaly profile data over the Central African Republic from Project MAGNET, US Naval Oceanographic Office (from Regan and Marsh, 1982, their figure 3, Reproduced by permission of American Geophysical Union).

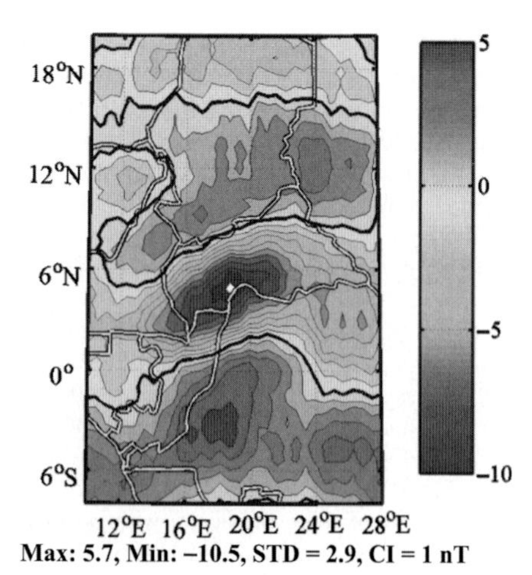

Figure B6/Plate 6c Satellite altitude (400 km) scalar magnetic anomaly map of the Central African Republic region from CHAMP mission data. Anomaly maximum, minimum, standard deviation and contour interval are given on the figure (Hyung Rae Kim, UMBC and NASA/GSFC).

topographic ring some 810 km diameter associated with this feature (Girdler *et al.*, 1992). In Rollin's (1995) recently compiled tectonic/geologic map of the Central African Republic, Late Archean and Early Proterozoic rocks are exposed beneath the central part of the anomaly. Lithologically the area is dominated by granulites and charnockites (a high temperature/pressure granite believed to be part of the lower crust). There are, in addition, significant exposures of greenstone belts and metamorphosed basalts with itabrite (a metamorphosed iron formation). There are several theories for the origin of the anomaly. Regan and Marsh (1982) proposed that a large igneous intrusion into the upper crust became denser on cooling and sank into the lower crust with the resulting flexure producing the overlying large basins of this region (see *Magnetic anomalies for geology and resources*). The intrusion the source of the magnetic anomaly; the sedimentary basin fill the source of the gravity anomaly. Another hypothesis is that it is the result of a large extraterrestrial impact (Green, 1976; Girdler *et al.*, 1992). Ravat *et al.* (2002) applied modified Euler deconvolution techniques to the Magsat data and their analysis supports the impact model of Girdler *et al.* (1992). Unfortunately, it is not possible to discriminate between these theories based solely on geophysical data. However, the key to the solution may lie in the origin of carbonados (microcrystalline diamond aggregates). Carbonados are restricted to the Bahia Province, Brazil and the Central African Republic, with the latter having a greater number. Smith and Dawson (1985) proposed that a meteor impacting into carbon-rich sediment produced these microdiamonds. More recently De *et al.* (1998) and Magee (2001) have failed to confirm this hypothesis. The origin of this large crustal anomaly remains uncertain.

Patrick T. Taylor

Bibliography

Benkova, N.P., Dolginow, S.S., and Simonenko, T.N., 1973. Residual magnetic field from the satellite Cosmos 49. *Journal of Geophysical Research*, **78**: 798–803.

Boukeke, D.B., 1994. Structures crustales D'Afrique Centrale Déduites des Anomalies Gravimétriques et magnétiques: Le domaine précambrien de la République Centrafricaine et du Sud-Cameroun. ORSTOM TDM **129**.

De, S., Heaney, P.J., Hargraves, R.B., Vicenzi, E.P., and Taylor, P.T., 1998. Microstructural observations of polycrystalline diamond: a contribution to the carbonado conundrum. *Earth and Planetary Science Letters*, **164**: 421–433.

Godivier, R., and Le Donche, L., 1962. Réseau magnétique ramené au 1er Janvier 1956: République Centrafricaine, Tchad Méridonial. 19 pages: 6 maps 1:2,500,000, Cahiers ORSTOM/Geophysique, No. **1**.

Girdler, R.W., Taylor, P.T., and Frawley, J.J., 1992. A possible impact origin for the Bangui magnetic anomaly (Central Africa). *Tectonophysics*, **212**: 45–58.

Green, A.G., 1976. Interpretation of Project MAGNET aeromagnetic profiles across Africa. *Journal of the Royal Astronomical Society*, **44**: 203–208.

Magee, C.W. Jr., 2001. Constraints on the origin and history of carbonado diamond. PhD thesis, The Australian National University, Canberra.

Ravat, D., Wang, B., Widermuth, E., and Taylor, P.T., 2002. Gradients in the interpretation of satellite-altitude magnetic data: and example from central Africa. *Journal of Geodynamics*, **33**: 131–142.

Regan, R.D., Davis, W.M., and Cain, J.C., 1973. The detection of "intermediate" size magnetic anomalies in Cosmos49 and 0602.4.6 data. *Space Research*, **13**: 619–623.

Regan, R.D., and Marsh, B.D., 1982. The Bangui magnetic anomaly: Its geological origin. *Journal of Geophysical Research*, **87**: 1107–1120.

Rollin, P., 1995. Carte Tectonique de République Centrafricaine. Université de Besançon.

Smith, J.V., and Dawson, J.B., 1985. Carbonado: diamond aggregates from early impacts of crustal rocks? *Geology*, **13**: 342–343.

Cross-references

Aeromagnetic Surveying
CHAMP
Crustal Magnetic Field
Magnetic Anomalies for Geology and Resources
Magnetic Anomalies, Long Wavelength
Magsat

BARLOW, PETER (1776–1862)

A British mathematician and physicist, born at Norwich, England, Peter Barlow is now remembered for his mathematical tables, the Barlow wheel and Barlow lens. His contributions to science in general and magnetism in particular are most impressive. We will concentrate here chiefly on his contributions in direct relation with geomagnetism, which are too often not given the attention they deserve.

Despite lacking formal education, Peter Barlow became assistant mathematical master at the Royal Military Academy in Woolwich in 1801. He was promoted to a professorship in 1806 and worked in Woolwich until retiring in 1847. His first researches were mainly focused on pure mathematics (his "Theory of Numbers" appeared in 1811), but in 1819 he began to work on magnetism. In May 1823, Peter Barlow was elected fellow of the Royal Society. He later also became a member of several of the leading overseas societies (including correspondant of the French Académie des Sciences in 1828). He worked on problems associated with magnetic mesurements and the issue of deviation in ship compasses caused by iron pieces in the hull. In 1825, he was awarded the Royal Society Copley Medal for his method of correcting the deviation by juxtaposing the compass with a suitably shaped piece of iron used as neutralizing plate.

Guided by a suggestion from John Herschel, Peter Barlow conducted experiments on the influence of rotation upon magnetic and non-magnetic bodies. In a letter to Major Colby dated December 20, 1824, he relates:

> *"Having been lately speculating on the probable causes of the earths magnetic polarity. It occured to me that it might possibly be due to the rotation, and if so the same ought to be the case*

with any revolving mas of iron. I therefore fixed one of our 13 inch shells upon one of the turning lathes in the arsenal driven by the steam engine, and the very few trials were most conclusive and satisfactory."

The next year, in the *Philosophical Transactions*, Peter Barlow describes how the experimental mesurements were made extremely difficult because of the disturbing influence of the lathe and other machinery on the needle. After careful investigations, he reports negative conclusions:

"I have certainly found a stronger effect produced by rotation than I anticipated, yet it does not appear to be of a kind to throw any new light upon the difficult subject of terrestrial magnetism. I think there are strong reasons for assuming, that the magnetism of the earth is of that kind which we call induced magnetism; but at present we have no knowledge of the inductive principle, (...)"

Years later, *Lord Blackett* (*q.v.*) revisited this possibility with similar conclusions (Blackett, 1952).

Following on Öersted's discovery of the magnetism associated with electrical current (Experimenta circa effectum Conflictus Electrici in Acum Magneticam, 1820), the French physicist André-Marie Ampère proposed (Annales de chimie et de physique, 1820) that electrical currents within the Earth could account for the geomagnetic field (these currents were then assumed to be of galvanic origin). Barlow was the first to test the practicability of Ampere's proposal and designed a remarkable experiment to that end. This experiment is presented in the *Philoposphical Transactions* for 1831. Barlow built a wooden hollow globe 16 in. in diameter and cut grooves in it. A copper wire was placed around the sphere along the grooves in the manner of a solenoid. When this globe is connected to a powerful galvanic battery, current passing through the coils sets up a dipolar magnetic field. Barlow describes how, if one turns

"(...) the globe so as to make the pole approach the zenith, the dip will increase, till at the pole itself the needle will become perfectly vertical. Making now this pole recede, the dip will decrease, till at the equator it vanishes, the needle becoming horizontal. (...) Nothing can be expected nor desired to represent more exactly on so small a scale all the phenomena of terrestrial magnetism, than does this artificial globe (...) I may therefore, I trust, be allowed to say, that I have proved the existence of a force competent to produce all the phenomena of terrestrial magnetism, without the aid of any body usually called magnetic."

Figure B7 Teaching instrument, based on Barlow's sphere, used to demonstrate how a current passing through a coil produces a dipolar field similar to that of the Earth. [The Physics Museum of the University of Coimbra, CAT. 1851: 25.O.III, 39 × 25.8 × 41, wood, brass, and copper. Photography: João Pessoa-Divisão de Documentação Fotográfica do Instituto Português de Museus].

This interpretation of the principal geomagnetic field clearly represents the premise of present dynamo theory. Barlow's globe was originally constructed in 1824; this experiment yielded a teaching apparatus still preserved in some universities around the world (see Figure B7).

Peter Barlow also did work with geomagnetic observations, in 1833 he constructed a new declination chart (then called "variation" chart) in which he embraced earlier magnetic observations. This chart is illustrated and described in the *Philoposphical Transactions* in 1833. Barlow notes that the lines of equal variation (following the terminology introduced by E. Halley in his original 1701 chart) are very regular, denoting the deep origin of these structures. Barlow also discusses the evolution of these lines in time by comparison to previous charts (i.e., the secular variation).

Barlow concludes his opus by noting that he shall be most happy if this

"labour should furnish the requisite data for either a present or future development of those mysterious laws which govern the magnetism of the terrestrial globe, an object as interesting in philosophy as it is important in navigation."

Peter Barlow died in March 1862 in Kent, England.

Acknowledgments

Figure reproduced by permission of João Pessoa (Divisão de Documentação Fotográfica do Instituto Português de Museus) and the Physics Museum of the University of Coimbra.

Emmanuel Dormy

Bibliography

Barlow, P., 1824. Letter to Major Colby at the Royal Military Academy dated dec. 20th 1824. *Archives of the Royal Society, HS.*3. 287.

Barlow, P., 1825. On the temporary magnetic effect induced in iron bodies by rotation, In a Letter to J.F.W. Herschel. *Philosophical Transactions*, **115**: 317–327.

Barlow, P., 1831. On the probable electric origin of all the phenomena of terrestrial magnetism; with an illustrative experiment. *Philosophical Transactions*, **121**: 99–108.

Barlow, P., 1833. On the present situation of the magnetic lines of equal variation, and their changes on the terrestrial surface. *Philosophical Transactions*, **123**: 667–673.

Blackett, P.M.S., 1952. A Negative Experiment Relating to Magnetism and the Earth's Rotation. *Philosophical Transactions*, **A245**: 309–370.

Ingenuity and Art, 1997. A collection of Instruments of the Real "Gabinete de Física". *Catalogue of the Physics Museum of the University of Coimbra (Portugal).*

Mottelay, P.F., 1922. *Biographical History of Electricity and Magnetism.* London: Charles Griffin & Company Limited.

Obituary notices of fellows deceased, *Proceedings of the Royal Society*, 1862–1863, **12**: xxxiii–xxxiv.

Cross-references

Blackett, Patrick Maynard Stuart, Baron of Chelsea (1897–1974)
Geodynamo
Geomagnetic Secular Variation
Halley, Edmond (1656–1742)

BARTELS, JULIUS (1899–1964)

Amongst the 20th century scientists working in the field of geomagnetism, Julius Bartels was certainly one of the outstanding contributors. He is known to most geophysicists as the author, along with *Sydney Chapman* (*q.v.*), of the two-volume monograph *Geomagnetism*, which used to be the bible of geomagnetism for several decades. Born on August 17, 1899 in Magdeburg (Germany), his life and scientific career was intimately related with the development of the field of geophysics to become a scientific discipline of its own. Julius Bartels studied mathematics and physics at the University of Göttingen as a student of Emil Wiechert as well as David Hilbert, Max Born, and James Franck. In 1923 he received his Ph.D. with a dissertation on "New methods of the calculation and display of daily pressure variations, especially during strong nonperiodic oscillations," which was supervised by Wilhelm Meinardus. Also his habilitation thesis "On atmospheric tides," which he submitted to the University of Berlin, dealt with the determination of tidal-like oscillations of the atmosphere using long-term pressure observations.

Trained as a mathematician Julius Bartels always had a keen interest in developing and applying statistical methods to geophysical problems. He became interested in geomagnetism when studying daily variations of the geomagnetic field caused by tidal oscillations of the ionosphere. His teacher in this new field was Adolf Schmidt whose assistant he became in 1923 after he graduated from Göttingen.

In 1929 he was offered professorship in meteorology and geophysics at the Forestry College in Eberswalde, close to Berlin. After several years of teaching there he took over the chair of geophysics at the University of Berlin and became the director of the Geomagnetic Observatory in Potsdam, which later became the Observatory of Niemegk. During his years in Potsdam, Julius Bartels laid the foundation for our current quantitative knowledge about geomagnetic variations and their relation to solar activity. With these statistical methods, he was able to disprove many widely accepted geophysical models on the one hand and develop some very useful and easy to apply geophysical indices on the other. In particular, the K-index and its planetary counterpart, K_P, have been developed in these years. The K_P index was, and partially still is, the key index to describe geomagnetic activity.

Eliminating regular variations of the Earth's magnetic field, Bartels demonstrated the impact of solar wave and particle radiation on the geomagnetic field. He recognized that seizing this concept in numbers would form a powerful tool for scientists in the research of solar-terrestrial relations. Bartels realized that his K_P tables would only be accepted by the scientific community if they were presented in a clearly arranged manner. The striking success of the K_P index is also due to its appealing representation as a kind of musical diagram. Using only geomagnetic data and statistical methods, Bartels claimed the existence of M-regions on the Sun that are responsible for geomagnetic activity. The claimed M-regions were later identified as coronal holes, which are the source of the fast solar wind that intensifies geomagnetic activity.

Between 1931 and 1940 Bartels also worked as a research associate at the Department of Terrestrial Magnetism at the Carnegie Institution (Washington, D.C.) for longer periods of time, where he closely cooperated with his friend Sydney Chapman. In 1940, the first edition of their book *Geomagnetism* was printed. After the Second World War, Bartels accepted an offer from the University of Göttingen to take over the chair of geophysics there, succeeding Gustav Angenheister. In 1956 Julius Bartels also became the director of the Institute for Stratospheric Research of the Max-Planck-Institut für Aeronomie, now the Max-Planck-Institute for Solar System Research.

Following the tradition of *Carl-Friedrich Gauss* (*q.v.*), Wilhelm Weber, and *Alexander von Humboldt* (*q.v.*), who initiated the "Göttinger Magnetischer Verein," Julius Bartels strongly favored international cooperation. He was one of the initiators of the International Geophysical Year 1957/1958, served as the president of the *International Association of Geomagnetism and Aeronomy* (*q.v.*) from 1954 to 1957, and became the vice-president of the International Union of Geodesy and Geophysics in 1961. Julius Bartels was an elected member of the Academies in Berlin and Göttingen, the Deutsche Akademie der Naturforscher und Ärzte Leopoldina in Halle, the Royal Astronomical Society in London, and the International Academy of Astronautics in Paris. His outstanding scientific contribution to the field of geomagnetism was honored by the Emil Wiechert Medal of the Deutsche Geophysikalische Gesellschaft, the Charles Chree Medal of the British Physical Society, and the William Bowie Medal of the American Geophysical Union. His published work spans 23 monographs and 143 scientific papers. Julius Bartels died on March 6, 1964 in Göttingen.

Karl-Heinz Glaßmeier and Manfred Siebert

Bibliography

Kertz, W., 1971. Einführung in die Geophysik II, Mannheim: Bibliographisches Institut.
Siebert, M., 1964. Julius Bartels, Gauss-Gesellschaft e. V., Göttingen.

Cross-references

Chapman, Sydney (1888–1970)
Gauss, CarlFriedrich (1777–1855)
Humboldt, Alexander Von (1759–1859)
IAGA, International Association of Geomagnetism and Aeronomy
Magnetosphere of the Earth
Periodic External Fields

BAUER, LOUIS AGRICOLA (1865–1932)

Louis Agricola Bauer (January 26, 1865 to April 12, 1932) founded and directed the Department of Terrestrial Magnetism (DTM) of the Carnegie Institution of Washington from 1904 until the 1920s. (See *Carnegie Institution of Washington, Department of Terrestrial Magnetism*). Seizing the opportunity provided by a large bequest by Andrew Carnegie, Bauer conceived of a tightly structured, worldwide magnetic survey that would be completed in a single generation. This goal was largely achieved by the time he was replaced as director of the DTM by *John Adam Fleming* (*q.v.*).

Bauer was born in Cincinnati, Ohio, as the son of German immigrants Ludwig Bauer and Wilhelmina Buehler. Bauer was the sixth of nine children. He studied civil engineering at the University of Cincinnati, graduating in 1888 and moving directly into a position as an aide in the US Coast and Geodetic Survey's computing division. Here he worked under Charles Anthony Schott, a German immigrant of 1848 who had dedicated his career at the Survey to geomagnetism. Under Schott, Bauer learned how to use magnetic instruments and how to reduce magnetic data. After four years at the Survey, Bauer enrolled in 1892 at the University of Berlin, Germany, to earn a Ph.D. in physics and learn the theoretical background to geomagnetic research. There he studied under Max Planck, Wilhelm von Bezold, and Wilhelm Foerster, and in 1895 successfully defended a dissertation on geomagnetic secular variation. Many geomagnetic researchers are still familiar with "Bauer plots" of declination versus inclination, which he introduced in 1895. These appear to indicate a cyclical behavior of the magnetic field (Bauer, 1895; Good, 1999).

While at the University of Berlin, Bauer worked as an observer at the magnetic observatory at Telegraphenberg, Potsdam. While a student, he met Adolf Schmidt, one of the most important geomagnetic researchers of the time, with whom he collaborated throughout his career.

Bauer returned to the United States as one of the country's first self-proclaimed geophysicists, with a prestigious Ph.D. from Germany. In 1896 he founded the international journal *Terrestrial Magnetism*, which became the *Journal of Geophysical Research* in 1948.

He taught physics and mathematics at the University of Chicago and the University of Cincinnati in the late 1890s and then returned to the US Coast and Geodetic Survey in 1899 to direct its new division of terrestrial magnetism. Although he periodically taught a course on geomagnetism at the Johns Hopkins University in Baltimore, his move to the Survey marked a permanent transition away from academia and toward a life dedicated to researching geomagnetism full time. He was following, as he saw it, the tradition of *Edmond Halley* (*q.v.*), *Alexander von Humboldt* (*q.v.*), and *Carl Friedrich Gauss* (*q.v.*). His goal was to understand Earth's magnetism in all its facets, including the origin of the main field, secular variation, and shorter term variations due to "cosmic" influences. The Coast and Geodetic Survey gave him the opportunity to develop an ambitious magnetic survey across the broad expanse of the United States and also the opportunity to elaborate a chain of magnetic observatories, which would allow him to investigate the variation of the field on different timescales. Here he also learned to manage an instrument shop, a computational division, field operations, cartographers, and publications. He soon deployed these skills on a global scale.

In 1902, Andrew Carnegie endowed the Carnegie Institution of Washington to encourage scientific research beyond the established disciplines. Bauer saw an opportunity in this to move beyond the borders of academic and government science and to address for the first time the global nature of geomagnetism. He established the *Department of Terrestrial Magnetism* (*q.v.*) at the Carnegie Institution in 1904. Bauer intended the DTM to observe Earth's magnetism everywhere that others were not. This meant Asia, Africa, South America, and the polar and ocean regions. The DTM was also to act as a coordinating and standardizing bureau to guarantee that data obtained by all observers would be intercomparable. The DTM under Bauer completed the first "World Magnetic Survey," sending out 200 land-based survey teams and circling the globe numerous times with the ocean-going magnetic observatories, the ships *Galilee* and *Carnegie* (*q.v.*) (see Figures B8 and B9). He also established magnetic observatories

Figure B8 Louis Agricola Bauer, observing the horizontal intensity of Earth's magnetic field onboard the DTM vessel *Galilee* in 1906. Courtesy of Carnegie Institution of Washington, Department of Terrestrial Magnetism Archives. Photo 0.4.

Figure B9 Louis Agricola Bauer, portrait presented to the crew of the *Carnegie* on its maiden voyage in 1909. Bauer was noted for his sense of duty. Photo courtesy of Carnegie Institution of Washington.

at Huancayo, Peru, and Watheroo, Western Australia. The survey was nearly complete by 1929, when the *Carnegie* burned at Apia, Samoa (Good, 1994).

Bauer saw the need to professionalize and coordinate geomagnetic research and wanted to see progress made both in observational programs and in theory. Toward the end of his career, he participated extensively in the formation of the International Union of Geodesy and Geophysics, and served as general secretary (1919–1927) and as president (1927–1930) of the Section of Terrestrial Magnetism and Electricity (now the *International Association of Geomagnetism and Aeronomy or IAGA*, *q.v.*). His retirement from the DTM was gradual, as he experienced increasingly intense periods of depression from 1924 on. Fleming was "acting director" from 1927 until 1932. In 1932, the *Washington Post* reported Bauer's death (by suicide) on the front page. Bauer's influence on geomagnetic research in the early 20th century was immeasurable.

Gregory A. Good

Bibliography

Bauer, L.A., 1895. On the secular motion of a free magnetic needle. *Physical Review*, **2**: 456–465; **3**: 34–48.

Good, Gregory A., 1994. Vision of a global physics: the Carnegie Institution and the first world magnetic survey. *History of Geophysics*, **5**: 29–36.

Good, Gregory A., 1999. Louis Agricola Bauer. In John A. Garraty and Mark C. Carnes (eds.), *American National Biography.* 24 vols. New York: Oxford University Press, Vol. 2, pp. 349–351.

Cross-references

Carnegie Institution of Washington, Department of Terrestrial Magnetism
Carnegie, Research Vessel
Fleming, John Adam (1877–1956)
Gauss, Carl Friedrich (1777–1855)
Halley, Edmond (1656–1742)
Humboldt, Alexander von (1759–1859)
IAGA, International Association of Geomagnetism and Aeronomy

BEMMELEN, WILLEM VAN (1868–1941)

Dutch geophysicist and meteorologist Willem van Bemmelen was born on August 26, 1868 in Groningen (the Netherlands) as the son of Prof. Dr. Jacob Maarten van Bemmelen and Maria Boeke. As a student of physics and mathematics in his early twenties, he became interested in the Earth's magnetic field through the work of *Christopher Hansteen* (*q.v.*) (Hansteen, 1819), who had derived isogonic charts and a geomagnetic hypothesis from historical magnetic measurements gathered from ship's logbooks. Aware of Hansteen's limited coverage of the seventeenth century and the period before 1600 and having encountered an abundance of original manuscript sources in various Dutch archives during an attempt to construct a secular variation curve for the city of Utrecht, Van Bemmelen decided to fill the gap with all useful data he could find from the period 1540–1690. The results he initially laid down in his Ph.D. thesis (Bemmelen, 1893), which tabulated the original declination observations gleaned from 38 ships' logbooks (169 on land, about 1000 at sea). The thesis moreover included secular variation curves for 19 cities and six isogonic reconstructions, at 1540, 1580, 1610, 1640, 1665, and 1680.

After receiving his doctorate, he extended this research during the next three years, covering additional archives in the Netherlands (1893–1896), London (early 1896), and Paris (summer 1897). Later that year he sailed to the East Indies, where he was appointed director of the Dutch Royal Magnetical and Meteorological Observatory at Batavia (Djakarta). There he met his future wife, Soetje Hermanna de Iongh (November 6, 1876, Pankal Pinang to July 5, 1969, Driebergen), whom he married on January 10, 1899. That same year appeared Van Bemmelen's "Die Abweichung der Magnetnadel" (Bemmelen, 1899), a substantial extension of the earlier compilation, which contained 388 observations on land, 5276 at sea (about 60% from original sources), 20 small isogonic charts for the period 1492–1740, and six larger ones on Mercator projection for 1500, 1550, 1600, 1650, and 1700. A decade later, he published a geomagnetic survey of the Dutch East Indies (1903–1907), which included the readings obtained at the Batavia observatory (Bemmelen, 1909). In the next decade, he directed his attention increasingly to meteorological investigations in the Dutch colony (notably rain and wind). He eventually returned to his native country, where he died at The Hague on January 28, 1841. His collected declinations have since become incorporated in the world's largest compilation of historical geomagnetic data, held at the World Data Center C1 at Edinburgh, United Kingdom.

Art R.T. Jonkers

Bibliography

Bemmelen, W. van, 1893. De Isogonen in de XVIde en XVIIe Eeuw. Ph.D. thesis, Leyden University. Utrecht: De Industrie.

Bemmelen, W. van, 1893. Über Ältere Erdmagnetische Beobachtungen in der Niederlanden, *Meteorologische Zeitschrift*, 10 (February 1893), 49–53.

Bemmelen, W. van, 1899. Die Abweichung der Magnetnadel: Beobachtungen, Säcular-Variation, Wert- und Isogonensysteme bis zur Mitte des XVIIIten Jahrhunderts. Supplement to *Observations of the Royal Magnetical and Meteorological Observatory, Batavia*, **21**: 1–109. Batavia: Landsdrukkerij (Government Printing Office).

Bemmelen, W. van, 1909. *Magnetic Survey of the Dutch East Indies, 1903–1907.* Observations made at the Royal Magnetical and Meteorological Observatory at Batavia, 30, appendix 1. Batavia: Landsdrukkerij.

Hansteen, C., 1819. *Untersuchungen über den Magnetimus der Erde.* Christiania: Lehmann & Gröndahl.

Cross-references

Geomagnetism, History of
Hansteen, Christopher (1784–1873)
Voyages Making Geomagnetic Measurements

BENTON, E. R.

Edward R. Benton, known to all as "Ned", is remembered for both his geophysical research and administration. He began his career as an applied mathematician, receiving the PhD degree in that discipline from Harvard University in 1961 under the supervision of Professor Arthur E. Bryson, Jr. Ned's early research focused on aerodynamics, beginning with a dissertation on "Aerodynamic origin of the magnus effect on a finned missile".

Following a year as lecturer in mathematics at the University of Manchester, Ned moved to Boulder, Colorado, taking a position as a staff scientist at the National Center for Atmospheric Research (NCAR) in the fall of 1963. In 1964 he became a lecturer in the Department of Astro-Geophysics (now the Department of Astrophysical, Planetary and Atmospheric Sciences) at the University of Colorado (UC), thus beginning an employment oscillation between NCAR and UC that lasted through 1977. Milestones along the way included academic appointments at UC to assistant professor in 1965, to associate professor in 1967, and Full Professor in 1971, and serving as assistant director of the NCAR Advanced-Study Program from 1967 to 1969, departmental chairman from 1969 to 1974, special assistant to the president for university relations of the University Corporation for Atmospheric Research (the parent organization for NCAR) from 1975 to 1977 and associate director of the Office of Space Science and Technology of UC from 1986 to 1987.

Ned's transition from engineer to geophysicist began with the publication in 1964 of a study of zonal flow inside an impulsively started rotating sphere. This work blossomed into a series of fundamental studies of the combined effects of rotation and magnetic fields on confined fluids, culminating in 1974 with a review of spin up (Benton, 1974). During this period, Ned also had a secondary interest in solutions of Burghers' equation. The next geophysically related area in which Ned took an interest was "kinematic dynamo theory," with a series of three papers published in 1975–1979 on Lortz-type dynamos. These papers developed a systematic categorization of these simple kinematic dynamos and provided some useful insights into their possible structures. This work was not continued as Ned's interest was soon drawn in other directions.

The work up until 1979 was prelude to his major scientific contribution in the area of inversion of the geomagnetic field. In that year he published a burst of papers on this topic, highlighted by papers on "Magnetic probing of planetary interiors" and "Magnetic contour maps at the core-mantle boundary" (Benton 1979a, b). In the mid-60s, workers had peered through this geomagnetic window on the Earth's core, but when it was shown that the view was both incomplete and flawed, interest in this subject had languished. Following Ned's revitalization of the subject, it has flourished, with contour maps of magnetic field and velocity at the top of the core being published by a number of groups. Now, the accurate determination of these features is a vital part of solving the dynamo problem.

Many know Ned best for his service as chairman of Study of the Earth's Deep Interior (**SEDI**) from 1987 to 1991. SEDI began as an idea at the IAGA Scientific Assembly in Prague, Czechoslovakia, in August, 1985, when IAGA Working Group I-2 (on theory of planetary magnetic fields and geomagnetic secular variation) called for the creation of a project to study the Earth's core and lower mantle. Ned cochaired the approach to IAGA and IUGG which led to SEDI's adoption, and served as its first chairman from 1987 to 1991. During his term as chairman, SEDI held symposia at Blanes, Spain, in June 1988, and Santa Fe, New Mexico in August 1990, as well as a large number of special sessions at various meetings of AGU, EGS, IAGA and IASPEI. Also, a number of national SEDI groups were formed, including those in Britain, Canada,

France, Japan and the United States. Several of these groups were successful in instituting national scientific projects with the help of SEDI. Also, SEDI endorsed several new scientific projects, including ISOP, INTERMAGNET, and the Canadian GGP. All these activities were a great boon for the deep-earth geoscientific community and helped these studies maintain their visibility and viability. This success was due in large part to Ned's steady hand at the helm during SEDI's crucial formative years.

Ned received a number of honors for his scientific research and service, including NASA's Group Achievement Award in 1983, a scholarship from the Cecil H. and Ida M. Green Foundation for Earth Sciences, and election as Fellow of the American Geophysical Union shortly before his untimely death. The loss of his scientific talents, leadership ability, and companionship is still felt strongly by his many colleagues.

David Loper

Bibliography

Benton E.R. 1974. Spin-up. *Annual Review on Fluid Mechanics*, **6**: 257–280.

Benton E.R. 1979a. Magnetic probing of planetary interiors. *Physics of the Earth and Planetary Interiors*, **20**: 111–118.

Benton E.R. 1979b. Magnetic contour maps at the core-mantle boundary. *Journal of Geomagnetism and Geoelectricity*, **31**: 615–626.

Cross-references

IAGA, International Association of Geomagnetism and Aeronomy
SEDI, Study of the Earth's Deep Interior

BINGHAM STATISTICS

When describing the dispersion of paleomagnetic directions expected to have antipodal symmetry, it is a standard practice within paleomagnetism to employ Bingham (1964, 1974) statistics. The Bingham distribution that forms the basis of the theory is derived from the intersection of a zero-mean, trivariate Gaussian distribution with the unit sphere. For full-vector Cartesian data $\mathbf{x} = (x_1, x_2, x_3)$ the Gaussian density function is

$$p_g(\mathbf{x}|\mathbf{C}) = \frac{1}{(2\pi)^{3/2}|\mathbf{C}|^{1/2}} \exp\left[-\frac{1}{2}\mathbf{x}^{\mathrm{T}}\mathbf{C}^{-1}\mathbf{x}\right], \quad \text{(Eq. 1)}$$

where $\mathbf{C}$ is a covariance matrix. But for directional data the Cartesian vectors $\hat{\mathbf{x}} = (\hat{x}_1, \hat{x}_2, \hat{x}_3)$ are of unit length, and the corresponding density function is Bingham's distribution

$$p_b(\hat{\mathbf{x}}|\mathbf{EKE}^{\mathrm{T}}) = \frac{1}{F(\kappa_1, \kappa_2, \kappa_3)} \exp\left[\hat{\mathbf{x}}^{\mathrm{T}}\mathbf{EKE}^{\mathrm{T}}\hat{\mathbf{x}}\right]. \quad \text{(Eq. 2)}$$

The matrix $\mathbf{E}$ is defined by the eigenvectors $\mathbf{e}^m$ of the covariance matrix $\mathbf{C}$ (see *Principal component analysis for paleomagnetism*). The diagonal concentration matrix

$$\mathbf{K} = \begin{pmatrix} \kappa_1 & 0 & 0 \\ 0 & \kappa_2 & 0 \\ 0 & 0 & \kappa_3 \end{pmatrix}, \quad \text{(Eq. 3)}$$

is formed from the Bingham concentration parameters κ_m. Normalization of (Eq. 2) is obtained through the confluent hypergeometric function, represented here as a triple sum (see Abramowitz and Stegan, 1965):

$$F(\kappa_1, \kappa_2, \kappa_3) = \sum_{i,j,k=0}^{\infty} \frac{\Gamma(i+\frac{1}{2})\Gamma(j+\frac{1}{2})\Gamma(k+\frac{1}{2})}{\Gamma(i+j+k+\frac{3}{2})} \frac{\kappa_1^i \kappa_2^j \kappa_3^k}{i!j!k!}, \quad \text{(Eq. 4)}$$

Let us now examine some of the properties of the Bingham distribution. The density function (Eq. 2) is clearly antipodally symmetric:

$$p_b(\hat{\mathbf{x}}) = p_b(-\hat{\mathbf{x}}). \quad \text{(Eq. 5)}$$

The distribution is also invariant for any change in the sum of the concentration parameters,

$$\sum_m \kappa_m \rightarrow \sum_m \kappa_m + \kappa', \quad \text{(Eq. 6)}$$

where κ' is an arbitrary real number. Therefore, for specificity, we set the largest parameter equal to zero, and we choose an order for the other two, so that

$$\kappa_1 \leq \kappa_2 \leq \kappa_3 = 0. \quad \text{(Eq. 7)}$$

Thus, the Bingham distribution is a two-parameter distribution. For all possible values of κ_1 and κ_2, the density (Eq. 2) describes a wide range of distributions on the sphere (see Figure B10). In spherical coordinates of inclination I and declination D the Bingham density function reduces to

$$p_b(\hat{\mathbf{x}}|\kappa_1, \kappa_2, \mathbf{e}^1, \mathbf{e}^2) = \frac{1}{F(\kappa_1, \kappa_2)} \exp\left[\kappa_1(\hat{\mathbf{x}} \cdot \mathbf{e}^1)^2 + \kappa_2(\hat{\mathbf{x}} \cdot \mathbf{e}^2)^2\right] \cos I, \quad \text{(Eq. 8)}$$

where the unit data vectors are given by

$$\hat{x}_1 = \cos I \cos D, \quad \hat{x}_2 = \cos I \sin D, \quad \hat{x}_3 = \sin I, \quad \text{(Eq. 9)}$$

where the unit eigenvectors are given by

$$\hat{e}_1^m = \cos I^m \cos D^m, \quad \hat{e}_2^m = \cos I^m \sin D^m, \quad \hat{e}_3^m = \sin I^m, \quad \text{(Eq. 10)}$$

and where the normalization function is

$$F(\kappa_1, \kappa_2) = \sqrt{\pi} \sum_{i,j=0}^{\infty} \frac{\Gamma(i+\frac{1}{2})\Gamma(j+\frac{1}{2})}{\Gamma(i+j+\frac{3}{2})} \frac{\kappa_1^i \kappa_2^j}{i!j!}. \quad \text{(Eq. 11)}$$

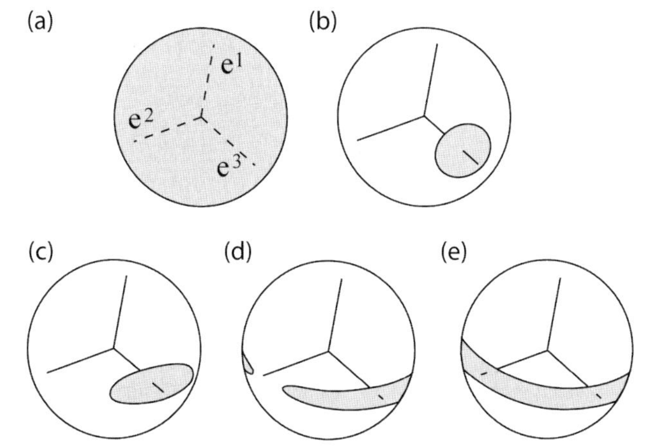

Figure B10 Bingham density function, with representative contours for a (a) uniform density $\kappa_1 = \kappa_2 = 0$, (b) symmetric bipolar density $\kappa_1 < \kappa_2 \ll 0$, (c) asymmetric bipolar density $\kappa_1 < \kappa_2 \ll 0$, (d) asymmetric girdle $\kappa_1 \ll \kappa_2 < 0$, and (e) symmetric girdle density $\kappa_1 \ll \kappa_2 = 0$. (After Collins and Weiss, 1990).

Three important special cases are worthy of attention, which are most clearly illustrated in the coordinate system determined by the eigenvectors $\mathbf{e}^m$. First, for $\kappa_1 = \kappa_2$ we have a axially symmetric bipolar distribution, with density function

$$p_3(\theta|\kappa) = \frac{1}{F(-\kappa)}\exp\left[-\kappa\cos^2\theta\right]\sin\theta, \quad \text{for} \quad \kappa \leq 0, \qquad \text{(Eq. 12)}$$

(see (Eq. 19) of *Statistical methods for paleomagnetic vector analysis*.) The angle θ is defined by the directional datum $\hat{x}$ and the principal axis determined by the eigenvector with the largest eigenvalue $\mathbf{e}^3$

$$\cos\theta = \hat{\mathbf{x}} \cdot \mathbf{e}^3, \qquad \text{(Eq. 13)}$$

normalization is given by

$$F(\kappa) = \sum_{i=0}^{\infty} \frac{\Gamma(i+\frac{1}{2})}{\Gamma(i+\frac{3}{2})}\frac{\kappa^i}{i!}. \qquad \text{(Eq. 14)}$$

Second, for $\kappa_1 < \kappa_2 = 0$ the distribution is a axially symmetric girdle, with density function (Dimroth 1962; Watson 1965)

$$p_1(\theta|\kappa) = \frac{1}{F(\kappa)}\exp\left[\kappa\cos^2\theta\right]\sin\theta, \quad \text{for} \quad \kappa \leq 0 \qquad \text{(Eq. 15)}$$

The angle θ is defined by the directional datum $\hat{x}$ and the principal axis determined by the eigenvector with the smallest eigenvalue $\mathbf{e}^1$

$$\cos\theta = \hat{\mathbf{x}} \cdot \mathbf{e}^1. \qquad \text{(Eq. 16)}$$

These two special cases of the more general Bingham distribution are useful for describing the dispersion (Eq. 12) of bipolar data about some mean pole, and the dispersion (Eq. 15) of bipolar data about some mean plane. And, finally, the third symmetric distribution is uniform, obtained for $\kappa_1 = \kappa_2 = 0$. In spherical coordinates this is just

$$p_0(\theta|\kappa) = \sin\theta, \quad \text{for} \quad \kappa = 0. \qquad \text{(Eq. 17)}$$

The uniform distribution has no preferred direction and so θ can be measured from an arbitrary axis. Of course, if we now allow κ to be positive or negative (or zero), then the full range of axially symmetric density functions is available (Mardia, 1972, p. 234):

$$p_{1,3}(\theta|\kappa) = \frac{1}{F(\kappa)}\exp\left[\kappa\cos^2\theta\right]\sin\theta, \quad \text{for} \quad -\infty \leq \kappa \leq \infty. \qquad \text{(Eq. 18)}$$

The off-axis angle θ is then defined by the axis of symmetry of the distribution. Symmetric density functions, ranging from bipolar to girdle, obtained for a variety of values of κ, are illustrated in Figure B11.

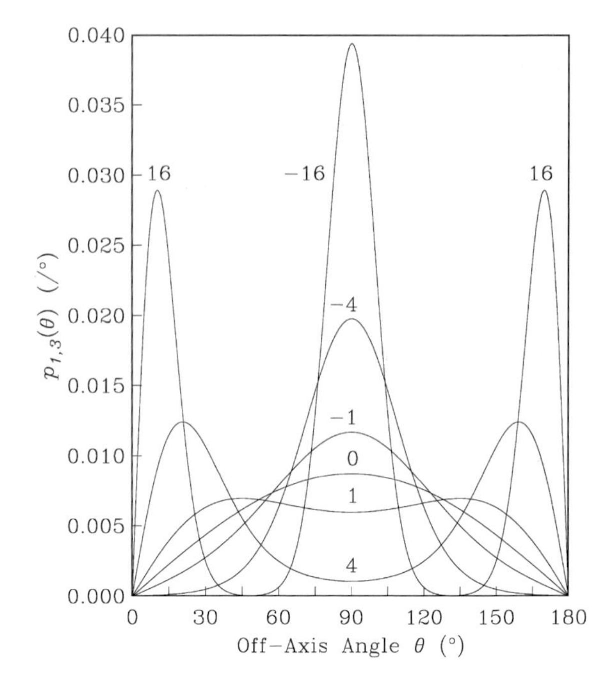

Figure B11 Examples of the axially symmetric Bingham probability density function $p_{1,3}(\theta)$, (Eq. 18) for a variety of κ concentration parameters: $0, \pm1, \pm4, \pm16$. Note that as $|\kappa|$ is increased the dispersion decreases.

Maximum-likelihood estimation

In fitting a Bingham distribution to paleomagnetic directional data a convenient method is that of maximum-likelihood; for a general review see Stuart *et al.* (1999). With this formalism, the likelihood function is constructed from the joint probability-density function for the existing data set. Using the general form of the Bingham density function (Eq. 8) the likelihood for N data is just

$$L(\kappa_1, \kappa_2, \mathbf{e}^1, \mathbf{e}^2) = \prod_{j=1}^{N} pb\left(I_j, D_j|\kappa_1, \kappa_2, \mathbf{e}^1, \mathbf{e}^2\right). \qquad \text{(Eq. 19)}$$

Maximizing L is accomplished numerically (Press *et al.*, 1992), an exercise yielding a pair of eigenvectors $\mathbf{e}^1$ and $\mathbf{e}^2$ and their corresponding concentration parameters κ_1 and κ_2. The third eigenvector $\mathbf{e}^3$ is determined by orthogonality and its concentration parameter by convention (Eq. 7) is zero. Some investigators (e.g., Onstott, 1980; Tanaka, 1999) prefer a two-step estimation method, where the eigenvectors are determined by principal component analysis, but these vectors are identical to those found by maximizing (Eq. 19). In any case, obtaining the eigenvalues of the data set through principal component analysis is still required for establishing confidence limits.

Confidence limits

It is unfortunate that the relationship between the concentration parameters κ_m, determined (usually) by maximum likelihood, and the eigenvalues of the covariance matrix λ_m, determined (usually) by principal component analysis, is very complicated. This fact makes the establishment of confidence limits on the eigenvectors difficult. However, Bingham (1974, p. 1220) has discovered an approximate formula for the confidence limit valid under certain circumstances. The confidence ellipse, within which a specified percentage (%) of estimated eigenvectors $\mathbf{e}^m$ can be expected to be realized from a statistically identical data set, is given by the elliptical axes

$$\alpha_{\%}^{mn} = \left[\frac{\chi_{\%}^2}{2N\mathbf{\Delta}_{mn}}\right]^{1/2}, \qquad \text{(Eq. 20)}$$

for

$$\Delta_{mn} = (\kappa_m - \kappa_n)(\lambda_m - \lambda_n) \ll 1, \quad \text{and} \quad N \to \infty, \qquad \text{(Eq. 21)}$$

Table B1 Principal component and maximum likelihood analysis of the Hawaiian paleomagnetic directional data recording a mixture of normal and reverse polarities over the past 5 Ma

Eigenvalue	Eigen direction		Bingham κ_m	Confidence axes		
	I(o)	D(o)		α_{95}^{m1}	α_{95}^{m2}	α_{95}^{m3}
0.0304	−14.6	87.6	−9.8749		1	0.0765
0.0601	−58.9	−48.4	−9.3490	1		0.0801
0.9088	31.1	−1.5	0.0000	0.0765	0.0801	

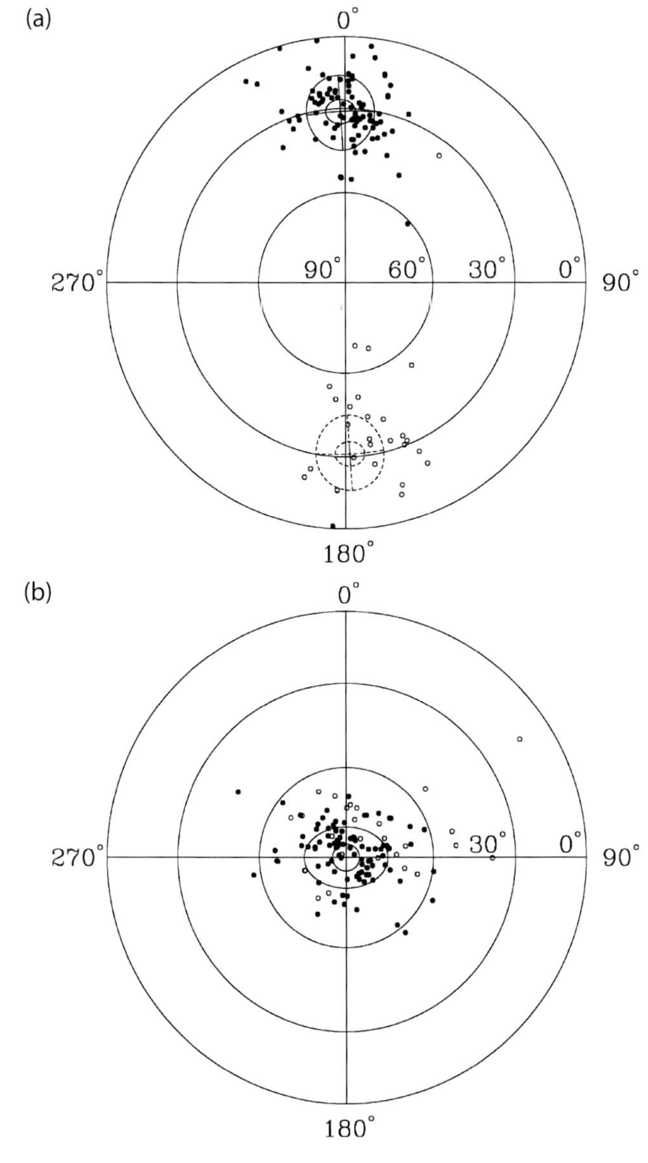

Figure B12 Equal-area projection of Hawaiian directional data, defined in (a) geographic coordinates and (b) eigen coordinates. Also shown are the projections of the variance minor ellipse, defined by λ_1 and λ_2, and, inside of that, the α_{95}^{3n} confidence ellipse. As is conventional, the azimuthal coordinate is declination (clockwise positive, 0° to 360°), and the radial coordinate is inclination (from 90° in the center to 0° on the circular edge).

and where $\chi^2_{\%}$ is the usual chi-squared value for two degrees of freedom. Alternative methods have been proposed for establishing confidence limits, most notably the bootstrap method popularized within paleomagnetism by Tauxe (1998).

Using the mixed-polarities directional from Hawaii covering the past 5 Ma (see *Principal component analysis*), in Table B1 we summarize the statistical parameters, and in Figure B12 we show the 95% confidence limit α_{95}^{3n} about the eigenvector defining the mean bipolar direction ($\mathbf{e}^3$). Note that the confidence limit is much smaller than the variance of the data.

Jeffrey J. Love

Bibliography

Abramowitz, M., and Stegun, I.A., 1965. *Handbook of Mathematical Functions.* New York: Dover.

Bingham, C., 1964. Distributions on the sphere and on the projective plane, PhD Dissertation, Yale University, New Haven, CT.

Bingham, C., 1974. An antipodally symmetric distribution on the sphere. *Annals of Statistics*, **2**: 1201–1225.

Collins, R., and Weiss, R., 1990. Vanishing point calculation as a statistical inference on the unit sphere. International Conference on Computer Vision, December, pp. 400–403.

Dimroth, E., 1962. Untersuchungen zum Mechanismus von Blastesis und syntexis in Phylliten und Hornfelsen des südwestlichen Fichtelgebirges I. Die statistische Auswertung einfacher Gürteldiagramme. *Tscherm. Min. Petr. Mitt.*, **8**: 248–274.

Mardia, K.V., 1972. *Statistics of Directional Data.* New York: Academic Press.

Onstott, T.C., 1980. Application of the Bingham distribution function in paleomagnetic studies. *Journal of Geophysical Research*, **85**: 1500–1510.

Press, W.H., Teukolsky, S.A., Vetterling, W.T., and Flannery, B.P., 1992. *Numerical Recipes.* Cambridge: Cambridge University Press.

Stuart, A., Ord, K., and Arnold, S., 1999. Kendall's advanced theory of statistics, *Classical Inference and the Linear Model.* Volume 2A, London: Arnold.

Tanaka, H., 1999. Circular asymmetry of the paleomagnetic directions observed at low latitude volcanic sites. *Earth, Planets, and Space*, **51**: 1279–1286.

Tauxe, L., 1998. *Paleomagnetic Principles and Practice.* Dordrecht: Kluwer Academic.

Watson, G.S., 1965. Equatorial distributions on a sphere. *Biometrika*, **52**: 193–201.

Cross-references

Fisher Statistics
Principal Component Analysis in Paleomagnetism
Statistical Methods for Paleo Vector Analysis

BIOMAGNETISM

Biomagnetism has emerged from a somewhat checkered earlier history. Indeed it has the dubious honor through Dr. Mesmer of contributing the verb mesmerize to the language and to have been the butt of Da Ponte and Mozart's humor in *Così fan Tutte*. However, it has survived to become a mature science covering a wide range of areas including animal navigation and orientation, possible harmful effects of magnetic fields, and searches for mechanisms explaining these phenomena are underway. In addition, electromagnetic instrumentation plays a major role in medicine and synthetic magnetic particles have been used to measure viscosity in cells and to transport chemicals within the human body, Here we discuss biological sensitivity to magnetic fields, the biogenic synthesis of magnetite, and its possible role in neurodegenerative diseases in humans. We leave the major areas of electromagnetically based medical instrumentation techniques, such as magnetocardiagrams, magnetoencephalograms, and the uses of nanoparticles of magnetite for other reviews.

Sensitivity to magnetic fields

The most readily demonstrated sensitivity to magnetic fields is that of magnetotactic bacteria. Several discussions of the phenomena are given in "Magnetite Biomineralization and Magnetoreception in Organisms" by Kirschvink *et al.* (1985), which provides an excellent source for work in this research area up to that time. As is well known, the bacteria are flagellum driven and the presence of a linear chain of magnetite particles results in their motion being constrained to move along, but not across magnetic lines of force. However, the sense of motion with respect to the field is more complicated than originally recognized and can in some cases change during the course of the day. The simple chain enables the bacterium to navigate between oxygenated and anoxic conditions at different depths in the water column as it swims.

Use of the geomagnetic field as a navigational cue

A wide variety of animals have been shown to sense the geomagnetic field and to use it as a navigational cue, e.g., salmon (Quin, 1980) honeybees (Gould *et al.* 1978), robins (Wiltschkow and Wiltschkow, 1972), homing pigeons (Walcott, 1977, 1978, 1996), field mice (Mather and Baker, 1981) and it has been suggested that human beings can also sense the geomagnetic field (Baker, 1980). There is no longer any doubt that certain animals can sense the magnetic field and indeed detect small changes in that field (e.g., Walker and Bitterman, 1989). The homing pigeons appear to use the gradient of the field (Walker, 1998, 1999), although this work remains controversial. The geomagnetic field intensity is of the order of tens of microteslas and typical gradients are a few nanoteslas per kilometer. To detect the necessary changes in the field to establish the gradients over distances of kilometers would therefore require sensitivity to field differences of the order of nanoteslas. Yet, pigeons appear to be able to do this (Windsor, 1975; Walker, 1998). It has long been thought that a magnetite-based sensory receptor makes this possible.

The nature of the magnetic sense organ has proved elusive, but an example of a magnetite based sensor has been exquisitely demonstrated in the rainbow trout by Walker *et al.* (1997). Behavioral studies established that the animal could discriminate between the presence and the absence of a field comparable to the geomagnetic field. Intracellular magnetite was discovered in its nose. Magnetic force microscopy observations of the magnetite were consistent with the magnetic behavior of single domain magnetite particles (Diebel *et al.*, 2000). Finally, neural signal responses to field changes were recorded in nerves from this region. Other detailed demonstrations of magneto-reception are likely to emerge as various animals, known to be capable of sensing the field, are examined by similar techniques. Indeed demonstrations of changes in firing rates in response to small magnetic field variations have been reported for some time (Figure B13) (for example, see Semm and Beason, 1990; Schiff, 1991). Magnetite in the bee abdomen has been implicated in the orientation of honeybees during their "dance" with a sensitivity to local magnetic field variations many times smaller than the geomagnetic field (Kirschvink *et al.*, 1997).

Having established that certain animals can sense the geomagnetic field, the next question is "how is the sensing used in navigation?" One of the most extensively investigated topics is avian navigation, which has recently been reviewed (Wiltschkow and Wiltschkow, 2003). Here we restrict the discussion to the pigeon (for which some of the most convincing evidence of a magnetic compass is available), although as noted above, the topic remains controversial (e.g., Walcott, 1977, 1978, 1996; Walker, 1998, 1999; Walraff, 1999; Reilly, 2002). Kramer's map and compass model (e.g., Kramer, 1961) is the starting point of most interpretations. In such magnetic navigational systems the two necessary requirements are: (1) some means whereby the animal can locate itself in relation to its loft and (2) some form of compass whereby it can set and maintain a course for its target. These require sensing of the magnetic field and of relatively small changes in the field, as we noted above.

How exactly the ability to sense the field is used to provide a map or compass, remains obscure and may well differ from species to species. A particularly important clue has emerged in the case of the homing pigeon from studies of the direction flown immediately after release (Windsor, 1975; Walker, 1998). This is illustrated in Figure B14, in which we see that initial departure directions are symmetrical about the local field gradient, which is indicated by the heavy solid NW/SE line. If birds are released close to this line, the errors are small, but they increase as release sites depart from it. Moreover, if the birds are released to the SW of the gradient line, the starting direction errs to the SW, and conversely if they are released from the site to the NE they err to the NE. Walker (1998) proposed a vector summation model to account for the directions and final return to the loft. It is clear that the chosen directions for the most part take the bird up or down the local gradient in the appropriate sense to move toward the loft. When the release site is close to the local gradient leading to the loft, errors are small, but when the site is away from this line flying down the gradient gives rise to large errors.

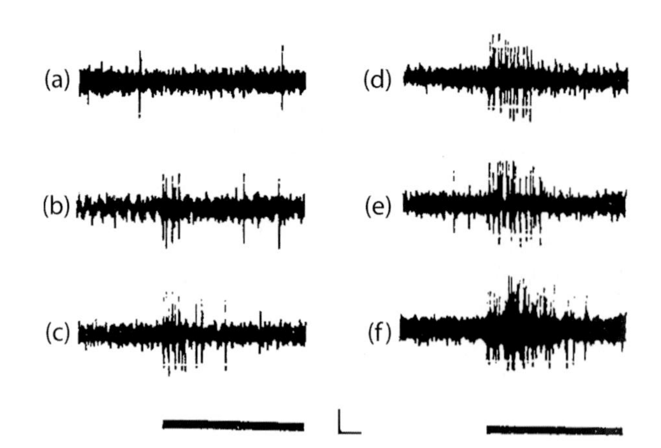

Figure B13 Observation of enhanced firing rates in response to magnetic field changes in the Bobolink. The response of a single ganglion cell to field intensity changes—(a) spontaneous activity, (b) response to 200 nT change, (c) response to 5000 nT change, (d) response to 15000 nT change, (e) response to 25000 nT change, and (f) response to 100000 nT change. Note the geomagnetic field is of the order of 10000 nT, or gammas, or tens of microteslas. The stimulus is indicated by the horizontal heavy line. The vertical bar indicates 2 mV and the horizontal bar 50 ms (From Semm and Beason. 1990).

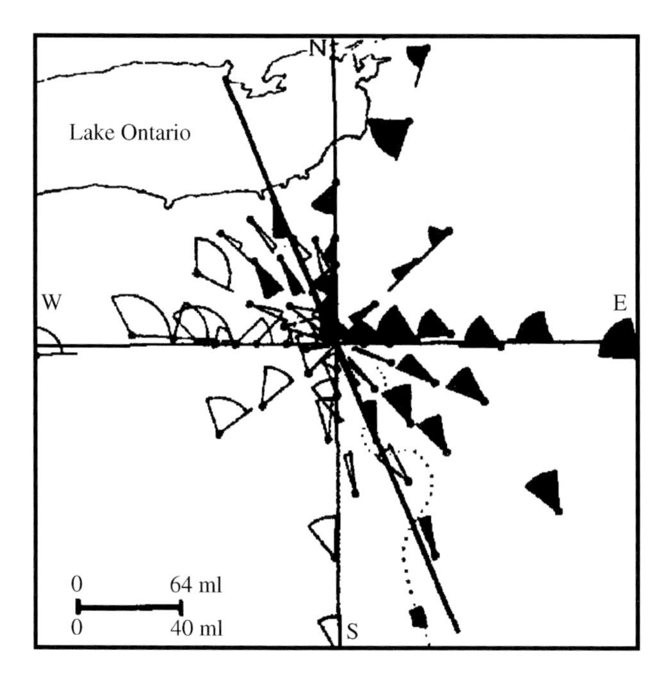

Figure B14 Initial departure directions of pigeons released at various distance and azimuthal direction from their loft. Note that the departure directions are symmetrical about the local field gradient indicated by the heavy NW/SE line (From Windsor, 1975 and Walker, 1998).

Indeed, some of the birds released due west of the loft appear to set off in completely the wrong direction. The overall pattern of directions demonstrates that the birds are able to determine the local field gradient and have a memory of whether the field is higher or lower at the release site than at the loft. The birds appear to use the gradient of the field in their navigation, which is a long standing idea in homing noted by Wiltschkow and Wiltschkow (2003), but it is not clear whether they use nonmagnetic clues or a more sophisticated vector summation magnetic model, such as that suggested by Walker (1998) to complete their return to the loft. These ideas have been strongly criticized by Wallraff (1999) and Reilly (2002), but given the observation of the relation of initial flight directions to the field gradient (Windsor, 1975; Walker, 1998), it is hard to avoid the interpretation that the pigeon is able to sense and follow the field gradient.

Many other animals are able to sense the geomagnetic field. There are examples demonstrating determination of the direction of the horizontal component and of the inclination of the total field vector field. Beason (1989) demonstrated sensitivity to the sign of the vertical component of a magnetic field in the Bobolink (*Dolichonyx orzivorous*), which is a nocturnal migratory species. The experiments were carried out in a planetarium and the initial departure directions were found to reverse with the reversal of the vertical component of the field. The experiment provided convincing evidence of an inclination based compass because the direction of horizontal component of the field was maintained when the vertical component was reversed and yet the initial departure directions reversed. Particularly convincing evidence of the ability to sense the direction of the horizontal component of the field comes from species which orient themselves, e.g., resting termites (Roonwal, 1958) or structures they build, e.g., termite galleries (Becker, 1975), and honey bee hives (Lindauer and Martin, 1972). These and other examples are presented in an excellent review of orientation in Arthropods (Lehrer, 1997). With all the many examples of sensitivity to the geomagnetic field in animals, it is natural to ask whether human beings are sensitive to the Earth's field and claims that we are indeed able to sense the field have been made (Baker, 1980), but the results have not been replicated. A review of human magnetoreception is given in Kirschvink *et al.* (1985).

In addition to useful magnetic field sensing in relation to navigation, there is a broader issue of field sensitivity including possible harmful effects of magnetic fields. Such effects are not likely to be related to the normal field strength of the geomagnetic field because animals have evolved in the presence of this field. Rather, it is the absence of the geomagnetic field, or the presence of much stronger fields that may prove harmful. During the period of low field intensity associated with field reversals, when the simple geometry of the field is also lost, animals relying on the field for navigation will presumably experience difficulty. In a more direct test of the importance of the geomagnetic field, Shibab *et al.* (1987) reported impairment of axonal ensheathment and myelination of peripheral nerves of newborn rats kept in 10 nT and 0.5 mT environments compared with controls.

Sensitivity to ac fields and dc fields larger than the geomagnetic field

Studies of effects of alternating magnetic fields and static fields that are stronger than the geomagnetic field have also tested sensitivity of the nervous system to these fields, but the main emphasis of the studies is on possible harmful effects, such as carcinogenesis, mutagenesis, and developmental effects. Much of this work has been stimulated by epidemiological studies such as that of Feychting and Ahlbom (1995), which suggested links between power line exposure and cancer in children. This has generated an enormous literature, but no conclusive evidence of cause and effect has emerged.

As an example of an effect of a static field only marginally larger than the geomagnetic field, Bell *et al.* (1992) reported changes in electrical activity in the human brain on exposure to fields of 0.78 G, or 78 μT. Power spectra of electroencephalograms (EEG) were found to be modified in all but one case, and in most cases an increase was observed. Interictal (between seizure) firing rates in the hippocampus of epileptic patients are increased by fields of the order of milliteslas. Such fields are some 20 times larger than the geomagnetic field (Dobson *et al.*, 1995; Fuller *et al.*, 1995, 2003). The procedure had some clinical value in the preoperative evaluation of patients with drug resistant epilepsy. In contrast to the generally observed increases in activity, activity was inhibited when the field was applied during periods of strong activity (Dobson *et al.*, 1998). Work on hippocampal slices by Wieraszko (2000) showed a complicated response pattern, with initial inhibition of firing followed by an amplification phase after the field was removed. The amplification was modulated by dantrolene, which is an inhibitor of intracellular calcium channels. In contrast, to the observations of neural responses to field changes observed in animals that use the field sensitivity for navigation, these responses were on a relatively long timescale of seconds and increased firing rates persisted in some case for tens of minutes. This suggests that these phenomena are very different from the magnetite sensor-mediated responses discussed above in the trout and the Bobolink. In those cases, the response took place within milliseconds of the field change and was completed within seconds.

Wikswo and Barach (1980) showed that very large fields of 24 T would be needed to bring about a 10% difference in conduction, in the discussion of the direct effect of magnetic fields on nerve transmission. In this, they noted that they had specifically ignored "microscopic chemical effects as well as those due to induced fields." It is evidently through these more subtle effects that weak magnetic fields affect biological systems.

The direct effect of magnetic fields on the rate of recombination of free radicals is an example of a mechanism that could play a role in field sensitivity. Moreover, given the importance of free radicals in biological systems this could be a significant effect. It arises because the probability of recombination of free radicals depends upon their spin state, which in turn can be affected by a dc field (McLauchlan, 1989) and possibly by ac fields (Hamilton *et al.*, 1988).

Whereas the sensitivity of the trout to the geomagnetic field is clearly through the intermediary of a magnetite based sensory receptor, it may well be that there are also direct effects of magnetic fields on the nervous system. Indeed, it seems that the effects on the epileptic patients are very different from the process in the trout and may result from direct effects on cell processes.

Biogenic magnetite

Biogenic magnetite clearly plays a central role in the sensitivity of the trout to magnetic fields and probably in many other animals. Biogenic magnetite was first discovered in organisms as a capping on the radula (teeth) of the chiton—a marine mollusk (Lowenstam, 1962). Since that time it has been found in many organisms ranging from bacteria to humans. In some species it has been demonstrated to form a part of a geomagnetic field receptor system (e.g., Frankel, 1984; Walker *et al.*, 1997; Wiltschko and Wiltschko, 2002; Kirschvink and Walker, 1995; Lohmann and Lohmann, 1996), but in most it is not obviously linked to magnetic orientation/navigation and its role is unclear.

Magnetite is ferrimagnetic because of the imbalance of Fe^{2+}, which is confined to one of the two opposed sub-lattices, i.e., on the octahedral sites. It has a saturation magnetization of 4.71×10^5 A m^{-1} at room temperature, a Curie (Néel) point of 578°C and a high density of 5.2 g cm^{-3}. It is presumably the high density and hardness that account for its presence in the radula of Chiton. In magnetotactic bacteria and in species that sense the magnetic field, it plays a role in detecting the field. However, magnetite is also a source of ferrous iron, which can generate potentially harmful free radicals. Here we will discuss the synthesis of magnetite, its occurrence in organisms including human beings and its possible role in neurodegenerative diseases.

Synthesis and occurrence of biogenic magnetite

Perhaps the best understood example of biogenic magnetite occurs in the various species of magnetotactic bacteria. These motile, gram-negative bacteria synthesize chains of highly pure and crystallographically perfect magnetite nanoparticles for use as a navigational aid—a phenomenon known as "magnetotaxis" (Blakemore, 1975; Frankel, 1984). The morphology of these particles is characteristic of the various species, however, in all cases, the size of the particles is generally just above the superparamagnetic limit such that they are magnetically blocked. This enables the chain to experience a torque when the geomagnetic field is at an angle to the long axis of the chain.

Recent work from Bertani and others (2001) has led to the identification of genes, which are involved in the synthesis of magnetite in magnetotactic bacteria. Though this gives an indication of the processes involved in magnetite synthesis, the mechanism employed by the bacteria to form perfect crystals of very specific size has remained difficult to identify. Work published recently, however, characterized proteins associated with biogenic magnetite crystals from the bacterium *M. magnetotacticum*, which are believed to regulate crystal growth (Arakakai *et al.*, 2003). The authors were able to use these proteins to direct the synthesis of magnetite nanoparticles in the laboratory with sizes and shapes very similar to those found in *M. magnetotacticum*.

Biogenic magnetite in organisms

Biogenic magnetite is found in many organisms; however, its physiological role in most of these is not well understood (Webb *et al.*, 1990). The mineral is also found in human tissue. Biogenic magnetite, along with maghemite (γFe_2O_3—an oxidation product of magnetite), was first reported in humans in 1992 by a group at the California Institute of Technology led by Joseph Kirschvink (Kirschvink *et al.*, 1992). In that study, the group examined human brain tissue samples taken from cadavers and this work proved somewhat controversial.

In order to allay this controversy and to examine the possibility of contamination and *postmortem* changes in brain iron chemistry, a series of studies on tissue removed from the human hippocampus was undertaken. The analyses were performed on tissue resected during amygdalohippocampectomies (a surgical procedure in which the damaged hippocampus of focal epilepsy patients is removed) as well as from cadaver tissue. Thus it was possible to control *post mortem* artifacts, which may have complicated the interpretation of Kirschvink's earlier results. These studies demonstrated clearly that biogenic magnetite is present in human brain tissue and confirm the results of the Cal Tech group (Dunn *et al.*, 1995; Dobson and Grassi, 1996; Schultheiss-Grassi and Dobson, 1999). In addition, recent results appear to demonstrate that biogenic magnetite concentration increases with age in males; however, this relationship is not seen in female subjects (Dobson, 2002).

While initial analysis of human brain tissue focused on magnetometry studies and the identification of biogenic magnetite by proxy, particles of this material also have been extracted, imaged, and characterized using transmission electron microscopy (TEM) (Kirschvink *et al.*, 1992; Schultheiss-Grassi *et al.*, 1999) (Figure B15). The particles are generally smaller than 200 nm and, in most cases, are on the order of a few tens of nanometers. While some particles exhibit dissolution edges, others preserve pristine crystal faces, and all particles examined thus far are chemically pure (this is common in biogenic magnetite). Morphologically, many of the particles are similar to those observed in magnetotactic bacteria (Schultheiss-Grassi *et al.*, 1999) though other particles are more irregularly shaped. A particularly intriguing transmission electron image has suggested that the iron oxide present in some hippocampal tissue may be concentrated in or near the cell membrane (Dobson, *et al.*, 2001) (Figure B16). Whether this location is related to cell membrane transport, or simply iron storage, is not known. However, the high concentration of iron oxide in this cell suggests a degree of specialization because such a concentration in all hippocampal tissue would give an unrealistically high concentration in this tissue. Unfortunately, magnetic particles have, for the most part, only been observed in tissue extracts, and work is currently underway to develop new techniques for imaging the particles in tissue slices and mapping their distribution to tissue structures.

Biogenic magnetite also has been found in the human heart, spleen, and liver. Again, this was determined by magnetometry studies, and calculation of magnetite concentrations showed that the heart has the highest levels of all organs examined thus far (Schultheiss-Grassi *et al.*, 1997). These investigations are continuing with the aim of determining the origin and role of biogenic magnetite in the human body.

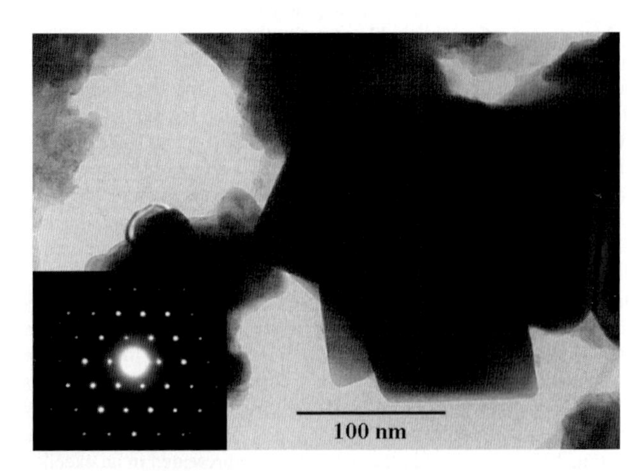

Figure B15 Transmission electron micrograph of biogenic magnetite extracted from the human hippocampus. Inset shows an electron diffraction pattern confirming the presence of magnetite. (From Schultheiss-Grassi et *al.*, 1999).

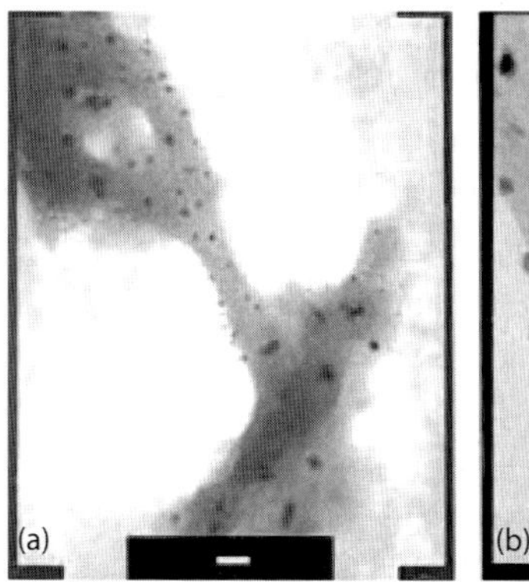

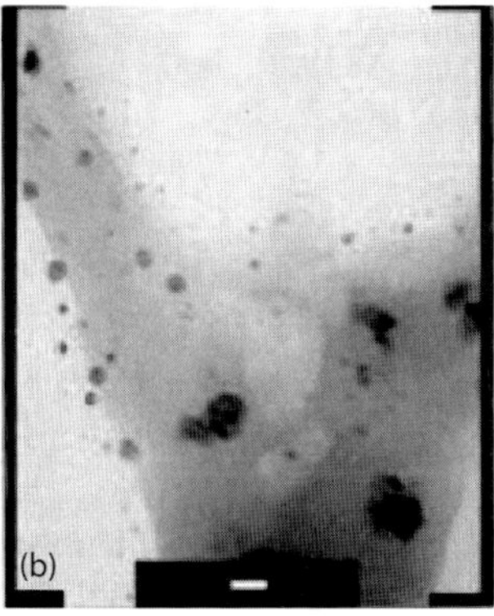

Figure B16 Electron microscope image of cell with iron oxide particles. (a) scale bar 200 nm and (b) scale bar 100 nm (From Dobson *et al.*, 2001).

Biogenic magnetite and neurological disorders

Though the physiological role of magnetite in humans is not yet known, recent studies have demonstrated a tentative link between magnetite biogenesis and neurodegenerative disorders such as Alzheimer's disease (AD) (Dobson, 2001, 2004; Hautot *et al.*, 2003). It has been known for 50 years that neurodegenerative diseases are associated with elevated levels of iron (Goodman, 1953). However, since that discovery, very little progress has been made in identifying the iron compounds present, their role, and their origin until this recent work.

Magnetite was first reported in AD tissue in the original study of human brain tissue by Kirschvink *et al.* (1992). These results did not demonstrate any correlation between elevated levels of biogenic magnetite and AD. Although the results were negative, it appears that this may have been due to the particular measurement technique rather than the absence of a correlation altogether. The isothermal remanent magnetization (IRM) of the tissue was measured using the superconducting quantum interference device (SQUID) magnetometry. One of the drawbacks of this method is that the contribution of very fine (superparamagnetic) particles (less than ~50 nm) is lost from the signal, particularly when measured at 0°C as was the case in this first study. In this case, any concentration which is calculated from the data depends on making some assumptions—one assumption being that the signal is due to particles which are large enough to contribute to the remanent magnetization (i.e., blocked). In this study, that may not have been a true picture of all the biogenic magnetite in the tissue samples.

Recently, Hautot *et al.* (2003) have developed methods for measuring the total biogenic magnetite signal. This is done by measuring the magnetization vs. applied field and then modeling the contributions from material such as the diamagnetic tissue, ferritin, and heme iron. In this way, the total concentration of magnetite—regardless of grain size—can be determined. Using this method, preliminary evidence of a correlation between biogenic magnetite concentration has been demonstrated in female AD patients.

By measuring other parameters, such as magnetic interactions between groups of magnetite particles, it has also been shown that there is preliminary evidence of a correlation between particle packing geometry and focal epilepsy. There was no demonstrable correlation between biogenic magnetite concentration and epileptic tissue, but again, these results are from measurements of remanent magnetization.

New studies using microfocused synchroton x-ray sources have now enabled iron compounds in brain tissue to be located and characterized in situ with subcellular resolution (Mikhaylova *et al.*, 2005; Collingwood *et al.*, 2005). This breakthrough allows, for the first time, the mapping of specific iron compounds including magnetite, to structures and cells within neurodegenerative tissue. Magnetite-rich iron anomalies have been unequivocally identified in AD tissue using this method, and it shows great promise for furthering our understanding of the role of magnetite and other iron compounds in neurological disorders.

Conclusion

Biomagnetism is now a major research area with a healthy mix of biologists, physicists, and chemists involved. We can safely conclude that many organisms sense the geomagnetic field and use it as a navigational cue, or for other purposes, such as nest alignment. Again it is clear that strong magnetic fields can produce mutagenic effects and produce harmful developmental effects in the several animals investigated. It is much less clear how much damage the relatively weak fields of power lines, mobile phones, and other electrical devices of modern society actually harm human beings. Magnetite is synthesized by many organisms and is involved in field detection in some. It is however present in many organisms in which its role is unclear. Indeed, as we noted, magnetite may be harmful as a source of free radicals. Presently, the possibility that magnetite may play a role in neurodegenerative diseases, such as Alzheimer's, Parkinson's, or Huntington's, is being investigated and may explain the association of iron enrichment and these diseases.

Mike Fuller and Jon Dobson

Bibliography

Arakaki, A., Webb, J., and Matsunaga, T., 2003. A novel protein tightly bound to bacterial magnetic particles in *Magnetospirillum magneticum* strain AMB-1. *Journal of Biological Chemistry*, **278**: 8745–8750.

Baker, R.R., 1980. Goal orientation by blindfolded humans after long distance displacement: possible involvement of a magnetic sense. *Science*, **210**: 555–557.

Beason, R.C., 1989. Use of an inclination compass during migratory orientation by the Bobolink (Dolichonyx orzivorous). *Journal of Experimental Biology*, **125**: 49–56.

Becker, G., 1975. Einfluss von mangetischen, elektrischen und Schwerefeldern auf den Galeriebau von Termiten, *Umschau in Wissenschaft und Technik*, **75**: 183–185.

Bell, G.B., Marno, A.A., and Chesson, A.L., 1992. Alterations in electrical activity caused by magnetic fields, detecting the detection process. *Electroencephalography and clinical Neurophysiology*, **83**: 389–392.

Bertani, L.E., Weko, J., Phillips, K.V., Gray, R.F., and Kirschvink, J.L., 2001. Physical and genetic characterization of the genome of Magnetospirillum magnetotacticum, strain MS-1. *Gene*, **264**: 257–263.

Blakemore, R.P., 1975. Magnetotactic bacteria. *Science*, **190**: 377–379.

Collingwood, J.F., Mikhailova, A., Davidson, M., Batich, C., Streit, W.J., Terry, J., and Dobson, J., 2005. *In-situ* characterization and mapping of iron compounds in Alzheimer's tissue. *Journal of Alzheimer's Disease*, **7**: 267–272.

Diebel, C.E., Proksch, R., Green, C.R., Neilson, P., and Walker, M.M., 2000. Magnetite defines a magnetoreceptor. *Nature*, **406**: 299–302.

Dobson, J., 2001. Nannoscale biogenic iron oxides and neurodegenerative disease. *FEBS Letters*, **496**: 1–5.

Dobson, J., 2002. Investigation of age-related variations in biogenic magnetite levels in the human hippocampus. *Experimental Brain Research*, **144**: 122–126.

Dobson, J., 2004. Magnetic iron compounds in neurological disorders. *Annals of the New York Academy of Sciences*, **1012**: 183–192.

Dobson, J., Fuller, M., Moser, S., Dunn, J.R., and Zoeger, J., (1995) Evocation of epileptiform activity by weak DC magnetic fields and iron biomineralization in the human brain. In Decke, L., Baumgartner, C., Stroink, G., Williamson, S.J. Dobson, J., Weiser, H.G., and Fuller, M., (eds.), *Advances in Biomagnetism*, 16–19, 1995.

Dobson, J., St. Pierre, T., Gregor Wieser, H., and Fuller, M., 2000. Changes in paroxysmal brainwave patters of epileptics by weak-field magnetic stimulation, *Bioelectromagnetics*, **21**: 94–96.

Dobson, J., and Grassi, P., 1996. Magnetic properties of human hippocampal tissue: Evaluation of artefact and contamination sources. *Brain Research Bulletin*, **39**: 255–259.

Dobson, J., Gross, B., Exley, C., Mickhailova, A., Batich, C., and Pardoe, H., 2001. Biomedical applications of biogenic and biocompatible magnetic nanoparticles. In Ayrapetyan, S., and North, A. (eds.) *Modern Problems of Cellular and Molecular Biophysics*, Yerevan: Noyan Tapan/UNESCO, 121–130.
Dunn, J.R., Fuller, M., Zoeger, J., Dobson, J., Heller, F., Caine, E., and Moskowitz, B.M., 1995. Magnetic material in the human hippocampus. *Brain Research Bulletin*, **36**: 149–153.
Feychting, M., and Ahlbom, A., 1995. Magnetic fields and cancer in children residing near Swedish high-voltage power lines. *American Journal of Epidemiology*, **138**: 7, 467–481.
Frankel, R.B., 1984. Magnetic guidance of organisms. *Annual Review of Biophysics and Bioengineering*, **13**: 85–103.
Fuller, M., Dobson, J., Wieser, H.G., and Moser, S., 1995. On the sensitivity of the human brain to magnetic fields: Evocation of epileptiform activity. *Brain Research Bulletin*, **36**(2): 155–159.
Fuller, M., Wilson, C.L., Velasco, A.L., Dunn, J.R., and Zoeger, J., 2003. On the confirmation of an effect of magnetic fields on the interictal firing rate of epileptic patients. *Brain Research Bulletin*, **611**: 1–10.
Goodman, L., 1953. Alzheimer's disease—a clinicopathologic analysis of 23 cases with a theory on pathogenesis. *Journal of Nervous and Mental Diseases*, **118**: 97–130.
Gould, J.L., Kirschvink, J.L., and Deffeyes, K.S., 1978. Bees have magnetic remanence. *Science*, **201**: 1026–1028.
Hautot, D., Pankhurst, Q.A., Kahn, N., and Dobson, J., 2003. Preliminary evaluation of nanoscale biogenic magnetite and Alzheimer's disease. *Proceedings of the Royal Society B: Biology Letters*, **270**: S62–S64.
Kirschvink, J.L., Jones, D.S., and MacFadden, B., 1985. *Magnetite Biomineralization and Magnetoreception in Organisms: A New Biomagnetism*. New York: Plenum Press.
Kirschvink, J.L., Kobayashi-Kirschvink, A., and Woodford, B.J., 1992. Magnetite biomineralization in the human brain. *Proceedings of the National Academy of Science USA*, **89**: 7683–7687.
Kirschvink, J.L., and Walker, M.M., 1995. Magnetoreception in honeybees. *Science*, **269**: 1889–1889.
Kirschvink, J.L., Padmanabha, S., Boyce, C.K., and Oglesby, J., 1997. Measurement of the threshold sensitivity of honeybees to weak, extremely low frequency magnetic fields. *Journal of Experimental Biology*, **200**: 1363–1368.
Kramer, G., 1961. Long distance orientation, In Marshall, E.G. (ed.), *Biology and Comparative Physiology of Birds*, London: Academic Press, pp. 341–371.
Lehrer, M. (ed.), 1997. *Orientation and Communication in Arthropods*. Berlin: Birkhäuser.
Lindauer, M., and Martin, M., 1972. Magnetic effects on dancing bees, In Galler, S.R., Schmidt-Koenig, K., and Belleville, R.E. (eds.), *Animal Orientatioin and Navigation*. Washington, D.C.: US Government Printing Office, pp. 559–567.
Lohmann, K.J, and Lohmann, C.M.F., 1996. Detection of magnetic field intensity by sea turtles. *Nature*, **380**: 59–61.
Lowenstam, H.A., 1962. Magnetite in denticle capping in recent chitons (*Polyplacophora*). *Geological Society of America Bulletin*, **73**: 435–438.
Mather, J.G., and Baker, R.R., 1981. A magnetic sense of direction in woodmice for route based navigation, *Nature*, **291**: 152–155.
McLauchlan, K.A., 1992. Are environmental magnetic fields dangerous? *Physics World*, **January**: 41–45.
Mikhailova, A., Davidson, M., Channel, J.E.T., Guyodo, Y., Batich, C., and Dobson, J., 2005. Detection, identification and mapping of iron anomalies in brain tissue using x-ray absorption spectroscopy. *Journal of the Royal Society Interface*, **2**: 33–37.
Quin, T.P., 1980. Evidence for celestial and magnetic compass orientation in lake migrating sockeye salmon fry, *Journal of Comparative Physiology*, **137**: 243–248.
Reilly, WI., 2002. Magnetic position determination by homing pigeons, *Journal of Theoretical Biology*, **218**: 47–54.
Roonwal, M.L., 1958. Recent work on termite research in India (1947–1957) *Transactions of the Bose Research Institute*, **22**: 77–100.
Schiff, H., 1991. Modulation of spike frequencies by varying the ambient magnetic field and magnetite candidates in bees (*Apis mellifera*). *Comparative Biochemistry and Physiology*, **100A**: 975–985.
Schultheiss-Grassi, P.P., Heller, F., and Dobson, J., 1997. Analysis of magnetic material in the human heart, spleen and liver. *Biometals*, **10**: 351–355.
Schultheiss-Grassi, P.P. and Dobson, J., 1999. Magnetic analysis of human brain tissue. *Biometals*, **12**: 67–72.
Schultheiss-Grassi, P.P., Wessiken, R., and Dobson, J., 1999. TEM observation of biogenic magnetite extracted from the human hippocampus. *Biochimica et Biophysica Acta*, **1426/1**: 212–216.
Semm, P., and Beason, R.C., 1990. Responses to small Magnetic Variations by the trigeminal system of the Bobolink, *Brain Research Bulletin*, **25**: 735–740.
Shibab, K., Brock, M., Gosztonyi, G., Erne, S.E., Hahlbohm, H.D., and Schoknecht, G., 1987. The geomagnetic field: a factor in cellular interactions. *Neurological Research*, **9**: 225–235.
Walcott, C., 1977. Magnetic fields and the orientation of homing pigeons under sun, *Journal of Experimental Biology*, **70**: 105–123.
Walcott, C., 1978. Anomalies in the earth's magnetic field increase the scatter of pigeon's vanishing bearings, In Schmidt-Koenig, K., and Keeton, W.T., (eds.), *Animal Migration, Navigation and Homing*, Berlin: Springer, pp. 143–151.
Walcott, C., 1996. Pigeon homing observations, experiments and confusions, *Journal of Experimental Biology*, **199**: 21–27.
Walker, M.M., 1998. On a wing and a vector: a model for magnetic navigation by homing pigeons. *Journal of Theoretical Biology*, **192**: 341–349.
Walker, M.M., 1999. Magnetic position determining by homing positions, *Journal of Theoretical Biology*, **197**: 271–276.
Walker, M.M., and Bitterman, M.E., 1989., Bees can be trained to respond to very small changes in geomagnetic field intensity, *Journal of Experimental Biology*, **141**: 447–451.
Walker, M.M., Diebel, C.E., Haugh, C.V., Pankhurst, P.M., Montgomery, J.C., and Green, C.R., 1997. Structure and function of the vertebrate magnetic sense, *Nature*, **390**: 371–376.
Wallraff, H.G., 1999. The magnetic map of the homing pigeon, an evergreen phantom, *Journal of Theoretical Biology*, **197**: 265–269.
Webb, J., St. Pierre, T.G., Macey, D.J., 1990. Iron biomineralization in invertebrates. In Frankel, R.B., and Blakemore, R., (eds.), *Iron Biominerals*, New York: R.P. Plenum Publishing Corp., 193–220.
Wieraszko, A., 2000. Dantrolene modulates the influence of steady magnetic fields on hippocampal evoked potentials *in vitro, Bioelectromagnetics*, **21**: 175–182.
Wiltschkow, W., and Wiltschkow, R., 1972. Magnetic compass of European Robins, *Science*, **176**: 62–64.
Wiltschko, W., and Wiltschko, R., 2003. Magnetic compass orientation in birds and its physiological basis. *Naturwissenschaften*, **89**: 445–452.
Windsor, D.M., 1975. Regional expression of directional preferences by experienced homing pigeons, *Animal Behaviour*, **23**: 335–343.
Winklhofer, M., Holtkamp-Rotzler, E., Hanzlik, M., Fleissner, G., and Petersen, N., 2001. Clusters of superparamagnetic magnetite particles in the upper-beak skin of homing pigeons: evidence of a magnetoreceptor? *European Journal of Mineralogy*, **13**: 659–669.

Cross-references

Magnetization, Isothermal Remanent
Rock Magnetism

BLACKETT, PATRICK MAYNARD STUART, BARON OF CHELSEA (1887–1974)

Blackett (Figure B17) came to geophysics late in a remarkable career. He had been educated at Osborne Naval College and Dartmouth and joined the Royal Navy, seeing action at the Battles of Jutland and of the Falklands in the Great War, as it was then known. After the war, he resigned from the navy, went to Cambridge to Magdalene College, and read for the Natural Science Tripos. He joined Lord Rutherford's group at the Cavendish as a graduate student in 1921, where he worked on transmutation of atoms using cloud chambers, carrying out in the succeeding two decades much of the work for which he was eventually awarded a Nobel Prize in Physics in 1948.

In 1933 he became professor of physics at Birkbeck College, London and in 1937 he moved to Manchester University as the Langworthy Professor of Physics. Prior to the Second World War, he had been brought into the Air Defense Committee and with the advent of the war was involved in many technical projects and made notable contributions in operations research. After the Second World War, he returned to Manchester and to research and was instrumental in establishing the first chair in radioastronomy and in the development of the Jodrell Bank Research Station that would lead to the famous steerable radio telescope. Meanwhile he became interested in magnetic fields in stars and planets, which led to his research in geophysics. In 1953, he was appointed professor and head of the physics department at Imperial College and built up an influential group in rock magnetism and paleomagnetism. Professor Blackett was elected president of the Royal Society in 1965 and made a life peer in 1969. He died in 1974.

Figure B18 Cloud chamber photograph showing alpha particles traversing the chamber and an example on the left of a collision between an alpha particle and a nitrogen atom, yielding a heavy oxygen isotope and a proton. The proton leaves to the right of the initial track (after Close *et al.*, 1987).

Blackett was a member of the heroic generation of physicists, most of whom were European, who revolutionized physics between the two world wars. As one of Lord Rutherford's students, he was assigned to study the disintegration of elements using the cloud chamber, which had been invented by C.T.R. Wilson at Cambridge nearly a quarter of a century earlier. With it, Wilson had demonstrated the tracks of ionizing alpha particles emitted from a small radioactive source. In his initial studies of thousands of photographs using such a source and with the cloud chamber containing nitrogen, Blackett found just eight, in which something remarkable was recorded. As Figure B18 shows, most of the alpha particles from the radioactive source below the chamber pass through it, not striking the atoms of nitrogen gas in the chamber. However, on the left a track shows that an alpha particle has hit a nitrogen atom. This has yielded a heavy oxygen isotope and a proton that leaves to the right of the original track. The heavy oxygen leaves to the left with a stronger track and is seen to collide again. Blackett had trapped a nuclear transmutation on film! He went on to establish the nature of the scattering of alpha particles from a number of nuclei of varying mass.

Figure B17 P.M.S. Blackett.

A major problem with the cloud chamber used in Blackett's early experiments was that thousands of photographs were needed to have a good chance of seeing an interesting transmutation. This problem became all the more daunting when the cloud chambers were turned to the analysis of cosmic rays. It was then a matter of chance, whether a cosmic ray crossed the chamber close to the time of expansion. The problem was solved by Blackett and a young Italian physicist Giuseppe Occhialini. Occhialini had studied with Rossi at Florence and brought with him knowledge of the work on coincident signals from Geiger counters. Blackett and Occhialini's solution was to place a Geiger counter above and below the cloud chamber so that a simultaneous signal from the two could be used to detect a cosmic ray particle and fire the cloud chamber. Of course by that time, the cosmic ray particle was long gone, but the ionization trail was still there and produced the necessary condensation for a track to be formed when the chamber was expanded. Although Anderson was the first to detect Dirac's antielectron, or positron as it has become known, Blackett and Occhialini confirmed its existence with their technique and demonstrated the production of positrons. To do this they had placed a copper plate above the cloud chamber, which produced gamma rays from the interaction of energetic cosmic ray electrons with the copper. Within the chamber numerous symmetrically divergent tracks were formed, as the particles with opposite electric charge formed and responded to the strong magnetic field imposed. These divergent tracks recorded the generation of matter, in the form of electrons and positrons, from energy, in the form of gamma rays, as predicted by Einstein. The citation for Blackett's 1948 Noble Prize was "for his development of the Wilson cloud chamber method and his discoveries therewith in the fields of nuclear physics and cosmic radiation."

Blackett's Nobel acceptance speech in 1948 reveals his social sensitivity. After expressing his personal feelings of satisfaction in the award, he remarked that he saw the Nobel Prize "as a tribute to the vital school of European Experimental Physics." He then went on to address the problem of avoiding the catastrophe of nuclear war, noting that "technological progress and pure science are but different facets of the same growing mastery by man over the force of nature. It is our task as scientists and citizens to ensure these forces are used for the good of men and not their destruction." In 1948 he published more on these ideas in the book *Military and Political Consequences of Atomic Energy*. His socialist political thought caused difficulty with the America of that time, but his recognition of the dangers of a world divided into very poor and very rich countries remains all too relevant to this day.

Blackett was clearly an excellent experimentalist and it was this ability that led him into the development of a sensitive magnetometer that opened up the field of paleomagnetism of the weakly magnetized sediments. Blackett had very skillfully constructed a sensitive magnetometer to test a theory of his about magnetic fields and the rotation of massive bodies. The idea was in relation to the origin of the magnetic fields of stars and of the Earth as a fundamental property of their rotation. The work developing this system and the negative result were described in a famous paper by Blackett entitled "A Negative Experiment Relating to Magnetism and the Earth's Rotation" and published in the *Transactions of the Royal Society*. There used to be a reprint of this paper in the geophysics library at Madingly Rise, Cambridge, with a note from Blackett, apologizing for the mistaken theoretical basis for the experiment. However, his instrument provided the means of measuring the weak magnetization of sediments, which was essential to test the idea of continental drift. This was one mistake that had a substantial silver lining!

Blackett's rock magnetism group worked initially at Manchester and then moved with him to Imperial College. His laboratory at Imperial College, with its orderly arrangement of instruments and polished wooden floors, was reminiscent of the quarterdeck of a Royal Navy battleship, presumably a legacy of his naval days. Important work was done both in fundamental rock magnetism and in paleomagnetism. Leng and Wilson demonstrated the reality of geomagnetic field reversals. Haigh made critical contributions to the understanding of magnetization acquired by magnetic materials as they grow in a magnetic field. Everitt studied the acquisition of magnetization as magnetic particles cooled in a magnetic field and developed the baked contact test—one of the standard paleomagnetic field tests. The paper by Blackett, Clegg, and Stubbs (1961) was one of the most convincing early demonstrations by paleomagnetism of the veracity of the idea of continental drift. Blackett's lectures on rock magnetism taken from his Second Weizman Memorial Lectures of 1954 is a little gem, combining insight on the grand scale, but also reflecting the pleasure of experimental work (Blackett 1956).

Blackett's main contributions were outside of geophysics. One immediately thinks of his experimental skills and the insights that came from the cloud chamber work; his leadership during World War II; his contributions to the development of radioastronomy. In addition to all this, his experimental skills played an important role in the development of rock magnetism and paleomagnetism. The way in which his group operated was in a direct lineage from Rutherford's Cavendish, which was brought to our field. We were lucky that our subject was touched by Blackett.

Michael D. Fuller

Bibliography

Blackett, P.M.S., 1952. A negative experiment relating to magnetism and the Earth's rotation, *Philosophical Transactions of the Royal Society of London, Series A*, **A245**: 309–370.

Blackett, P.M.S., 1954. *Lectures on Rock Magnetism*. Jerusalem: The Weizmann Science Press of Israel.

Blackett, P.M.S., Clegg, J.A., and Stubbs, P.H.S., 1960. An analysis of rock magnetic data, *Proceedings of the Royal Society of London*, **256**: 291–322.

Close, F., Marten, M., and Sutton, C., 1987. *The Particle Explosion*. Cambridge: Oxford University Press.

BULLARD, EDWARD CRISP (1907–1980)

Edward Crisp Bullard, known to everyone as Teddy, was the greatest geophysicist working in the United Kingdom in the second half of the 20th century (Figure B19). He made seminal contributions to the theory of the origin of the magnetic field, to the analysis of the westward drift and secular variation, and to studies of electromagnetic induction. He championed a number of developments in the theory of plate tectonics, and placed his department at the centre of the plate tectonic revolution of the 1960s. His full biography is in McKenzie (1987).

Teddy was born in Norwich into a wealthy family of brewers. The building that housed the brewery still stands in an attractive part of Norwich, but the business was taken over in the 1960s and Bullard's beer can no longer be found; the name enjoyed a revival in the early days of the campaign for Real Ale and won a prize at the 1974 Cambridge beer festival. The young Bullard was educated at Repton School where his physics teacher, A.W. Barton, initiated his interest in physics. He studied physics in Cambridge, where he found the lectures dull with the exception of *J. Larmor* (*q.v.*), Rutherford, and Pars, and proceeded to research in Rutherford's laboratory. He was in fact supervised by *P.M.S. Blackett* (*q.v.*), although Rutherford was probably the greater influence. His Ph.D. project was on electron scattering, but on taking up a position in the Department of Geodesy and Geophysics he began work on the measurement of gravity with pendulums and made observations with them in East Africa. His Ph.D. thesis, only 26 pages long, includes both the electron scattering and gravity work. He was elected a fellow of the Royal Society in 1941; the citation includes work on seismics, heat flow, and gravity: everything, in fact, except geomagnetism.

On the outbreak of the Second World War he joined the Admiralty Research Laboratory and set to work on reducing the magnetic fields around ships. The Germans had laid magnetic mines in shallow waters around Britain, which sank 60 ships in the "phoney war" in December 1939. The "degaussing" was successful (Bullard, 1946). He subsequently joined Blackett to work on what later became known as operational research. The degaussing of ships was his first work on magnetism, the only previous reference being to the effect of the Earth's magnetic field on the period of invar pendulums (Bullard, 1933), but during his war work he read Chapman and Bartels (1940), and this probably stimulated him to work on geomagnetism after the war was over.

Postwar Cambridge offered little for an ambitious researcher, particularly in geophysics, and Teddy moved to Toronto. There followed the most productive period of his career. He used his knowledge of

Figure B19 "Teddy" as the author knew him in about 1970.

heat flow to develop the convection theory for the dynamo (Bullard, 1950) and analyzed global observations for secular variation and the westward drift (Bullard *et al.*, 1950). Both these studies remain substantially correct today, although his theory for westward drift, that the magnetic field acted on the lower mantle to drive it eastwards by an electric motor effect, is no longer believed because the electrical conductivity of the mantle is thought to be too small.

W.M. Elsasser (*q.v.*) also began working on the origin of the Earth's magnetic field at this time, but he concentrated on the mechanism by which a magnetic field could be generated by flow in the core, whereas Bullard had worked with the observations and on the heat sources driving the convection. Elsasser derived some fundamental results in dynamo theory and constructed a formalism for solving the equations, but did not attempt to solve them. This was a very formidable problem for that time—indeed, it is still a very difficult problem with today's computers. Bullard and a research student at Toronto, H. Gellman, attempted to find a solution with the tools at hand (see *Bullard-Gellman dynamo*). In 1950 Bullard moved back to the United Kingdom as director of the National Physical Laboratory, where he had access to one of the most powerful computers of the day.

Although Bullard and Gellman (1954) is one of the most influential papers ever published in dynamo theory, their solutions had not converged and their claim to have found a working dynamo was subsequently proved incorrect (Gibson and Roberts, 1969). Nonetheless, it was a brave and very early effort to carry out a numerical solution of a partial differential equation governing a problem in fluid mechanics. The dynamo failed because they chose a fluid flow that lacked helicity. It is clear that they intended to design a flow with helicity because their paper contains a discussion of magnetic field lines being pulled and twisted to reinforce the dipole field, but they were restricted by numerical considerations to a very simple flow. Bullard was undoubtedly influenced by Elsasser's ideas, which were later developed by E.N. Parker into what is now known as an α-effect dynamo (Parker, 1955). Kumar and Roberts (1975) eventually found a working dynamo using the Bullard-Gellman formalism and a more complicated fluid flow.

Bullard desperately wanted to return to Cambridge. Eventually a post became available when *S.K. Runcorn* (*q.v.*) moved to Newcastle, and Bullard accepted a very junior position, lower than the one he occupied before the war. He was eventually promoted as reader then professor in 1964, over 20 years since his election to fellowship of the Royal Society and 10 years after his knighthood.

On the question of polarity reversals he sat on the fence, much to Runcorn's annoyance, remaining sceptical because of the discovery of self-reversal in certain minerals. He was finally won over by the classic work of Cox *et al.* (1963), and when Harry Hess visited Cambridge in 1962 and talked about his ideas on seafloor spreading he encouraged Fred Vine, then a research student, to pursue his interpretation of reversely magnetized seafloor in terms of Hess' model. The result is now known as the *Vine-Matthews-Morley hypothesis* (*q.v.*).

In later years he had a succession of successful research students working on electromagnetic induction, among them Bob Parker, Nigel Edwards, and Dick Bailey. He put his name to very few scientific papers written during this period, but his influence was everywhere, both on the group as a whole and on individual students. Bullard worked to the end, finishing his last paper on historical measurements at London with Stuart Malin (Malin and Bullard, 1981) shortly before he died.

David Gubbins

Bibliography

Bullard, E.C., 1933. The effect of a magnetic field on relative gravity determinations by means of an invar pendulum. *Proceedings of the Cambridge Philosopjical Society*, **29**: 288–296.

Bullard, E.C., 1946. The protection of ships from magnetic mines. *Proceedings of the Royal Institution of Great Britain*, **33**: 554–566.

Bullard, E.C., 1950. The transfer of heat from the core of the Earth. *Monthly Notices of the Royal Astronomical Society*, **6**: 36–41.

Bullard, E.C., Freedman, C., Gellman, H., and Nixon, J., 1950. The westward drift of the Earth's magnetic field. *Philosophical Transactions of the Royal Society of London, Series A*, **A243**: 67–92.

Bullard, E.C., and Gellman, H., 1954. Homogeneous dynamos and terrestrial magnetism. *Philosophical Transactions of the Royal Society of London, Series A*, **247**: 213–278.

Chapman, S., and Bartels, J., 1940. *Geomagnetism.* London: Oxford University Press.

Cox, A., Doell, R.R., and Dalrymple, G.B., 1963. Geomagnetic polarity epochs and Pleistocene geochronometry. *Nature*, **198**: 1049.

Gibson, R.D., and Roberts, P.H., 1969. The Bullard and Gellman dynamo. In Runcorn, S.K. (ed.), *The Application of Modern Physics to the Earth and Planetary Interiors.* New York: Wiley Interscience, pp. 577–602.

Kumar, S., and Roberts, P.H., 1975. A three dimensional kinematic dynamo. *Proceedings of the Royal Society*, **344**: 235–258.

Malin, S.R.C., and Bullard, E.C., 1981. The direction of the Earth's magnetic field at London, 1570–1975. *Philophical Transactions of the Royal Society of London, Series A*, **A299**: 357–423.

McKenzie, D.P., 1987. Edward Crisp Bullard. *Biographical Memoirs of Fellows of the Royal Society*, **33**: 67–98.

Parker, E.N., 1955. Hydromagnetic dynamo models. *Astrophysical Journal*, **122**: 293–314.

Cross-references

Blackett, Patrick Maynard Stuart, Baron of Chelsea (1897–1974)
Bullard-Gellman dynamo
Dynamo, Bullard-Gellman
Dynamos, Kinematic
Elsasser, Walter M. (?–1991)
Geomagnetic Polarity Reversals, Observations
Geomagnetic Secular Variation
Larmor, Joseph (1857–1942)
Mantle, Electrical Conductivity, Mineralogy
Runcorn, S. Keith (1922–1995)
Transient EM Induction
Vine-Matthews-Morley Hypothesis
Westward Drift

C

CARNEGIE INSTITUTION OF WASHINGTON, DEPARTMENT OF TERRESTRIAL MAGNETISM

When the Carnegie Institution of Washington (CIW) was established in 1902 with a large bequest from Andrew Carnegie, one of its goals was to encourage research in the "borderlands" between the scientific disciplines. In its first years, the CIW established two departments that became stages for research in the geosciences: the Geophysical Laboratory and the Department of Terrestrial Magnetism (DTM). While the Geophysical Laboratory moved quickly in the direction of geochemical investigation and toward developing laboratory techniques to study how rocks behave at great depth within the Earth, the DTM focused on a series of problems related to geomagnetism and geo-electricity during its first several decades. The DTM branched out starting in the 1920s to investigate the atomic nucleus, cosmic rays, and even cosmology. These two departments have now existed for over a century and today are located on a single campus where their researchers can more easily interact (Trefil and Hazen, 2002).

The geophysicist *Louis Agricola Bauer* (*q.v.*), who had been engaged in geomagnetic research for over a decade, established the DTM in 1904 with the general goal of studying Earth's magnetic and electric phenomena, but with a definite project to prosecute "quickly" a magnetic survey of the planet. This project was to be completed within a generation in order to provide data which could be more easily reduced to a common epoch than the data then available. Magnetic data had only been gathered since the mid-19th century with any degree of consistency, and even then, it was concentrated in just a few regions and was spread over decades. Entire continents were uninvestigated and charts available for magnetic declination in the world's oceans were frequently in error by 1° and 2°, while inclination measurements were 1°–3° off. Moreover, outside the main shipping lanes very few readings had been made at all (Good, 1994b). Bauer proposed a World Magnetic Survey to the Carnegie Institution of Washington because, as a nongovernmental agency, it would enjoy greater credibility for mounting expeditions in Africa, Asia, South America, and other places where the institutions of major governments might be looked upon with suspicion. Moreover, free of any government connection, the DTM could directly approach scientific officials in other countries without going through their supervising bureaucracies. As he put it in his mission statement for the new department, the DTM would survey the magnetism of areas not being studied by others. This was most of the globe.

The DTM was to have a second purpose too: to coordinate and standardize magnetic measurements then being undertaken by dozens of governments and agencies around the world. That is, it would be the international standardizing bureau. It would carefully compare the great variety of geomagnetic instruments (see *History of Instrumentation*) and promote the best methods of investigation and measurement. To this end, for decades the DTM had its traveling "magneticians" visit major magnetic observatories around the world to standardize instruments and to learn and teach different working methods. DTM promoted data exchange and uniform methods of reduction. In these regards the DTM acted in the way that various commissions of the *International Association for Geomagnetism and Aeronomy* or *IAGA* (*q.v.*) and *Intermagnet* do today.

Bauer set the DTM's first task as evaluation of existing data for secular variation studies and he contracted Adolph Schmidt to determine the status of data on magnetic storms. The second over-arching goal was to "establish the facts" and to convince theoreticians that the data and mathematical tools could become available for decisive testing among different theories.

Bauer also brought his journal *Terrestrial Magnetism and Atmospheric Electricity* (since 1948 the *Journal of Geophysical Research*) to the DTM, where he and his associate director *John A. Fleming* (*q.v.*) used it to provide a disciplinary and international identity for geomagnetic researchers. In its pages, Bauer published critical studies of various instruments, theoretical discussions, and "news of the profession," including capsule biographies of geomagnetic role models like *Edmond Halley* (*q.v.*), *Alexander von Humboldt* (*q.v.*), and *Carl Friedrich Gauss* (*q.v.*).

Bauer's most dramatic project, however, was the global magnetic survey. Between 1905 and World War II, the DTM employed two hundred magnetic observers on land expeditions. Counting the number of expeditions is a little difficult, since many of them split and rejoined and they varied from a few weeks to many months in duration. The DTM reported in 1928 that 178 land expeditions had thus far occupied 5685 magnetic stations. This averaged 271 stations per year. By 1928, when Fleming was taking over supervisory duties for the DTM, the main land survey was completed. Dwindling numbers of expeditions from then until the mid-1940s concentrated on secular variation measurements at a smaller number of *repeat stations* (*q.v.*).

These troops of magneticians were a tightly organized, well-trained army of observers. They learned the physics of electricity and magnetism, the use of magnetometers, induction coils, and sine galvanometers (see *History of Instrumentation*) and the techniques of data reduction. They also were resourceful, independent travelers. Unlike other expeditionaries, these scientists often traveled alone and rarely in groups of more than three. They transported hundreds of kilograms of instrumentation and other gear and frequently traveled by camel or canoe, as well as by steamer or locomotive. The magnetic data was hard won and surprisingly, not one observer died of disease or accident in all of those travels into the Polar Regions, Amazonia, and bandit-ridden China (Figure C1).

Figure C1 Magnetic observer Frederick Brown traveled around Asia by camel and sedan chair. This photo is from April 1915. Photo courtesy of Carnegie Institution of Washington. Photo DTM 5769.

Meanwhile, Bauer carefully executed a parallel set of oceanic expeditions. He consulted in 1904 and 1905 with experts in geomagnetism (e.g., Arthur Schuster) and hydrographers (e.g., George W. Littlehales and Ettrick W. Creak) (Good, 1994b). In 1906, 1907, and 1908 the brig *Galilee* followed three, spiraling loops in the northern Pacific Ocean, providing magnetic measurements within each 5° quadrangle in the region, or one station every 200 miles. The DTM constantly improved sea-going magnetic instruments, so that by 1908, magnetic observers read declination to within 0.5°, compared with 0.1° for land observers. By this last cruise, errors due to iron on board ship was a larger source of error than the instruments.

Hence, Bauer obtained funding from the CIW for construction of a nonmagnetic yacht, the *Carnegie* (*q.v.*), which was launched in 1909. This yacht was magnificent, if peculiar. Its dominant peculiarity was the deck house, a glass enclosed galley with a glass observing dome at each end. Protected within, magnetic observers could make measurements even in quite rough seas. From 1909 until its demise in 1929, the *Carnegie* conducted seven cruises, including a celebrated fastest-ever circumnavigation of Antarctica for a sailing vessel in 118 days; all the while the observers continued their scientific work.

A map published by the DTM in 1929 shows the globe enmeshed in the observations of Bauer's remarkable World Magnetic Survey (Figure C2). Bauer had imagined an ambitious plan to make thousands of measurements around the globe in a single generation and he found the funding, personnel, and instruments to do it. The effort required to produce this first geomagnetic "snap-shot" was nothing short of remarkable.

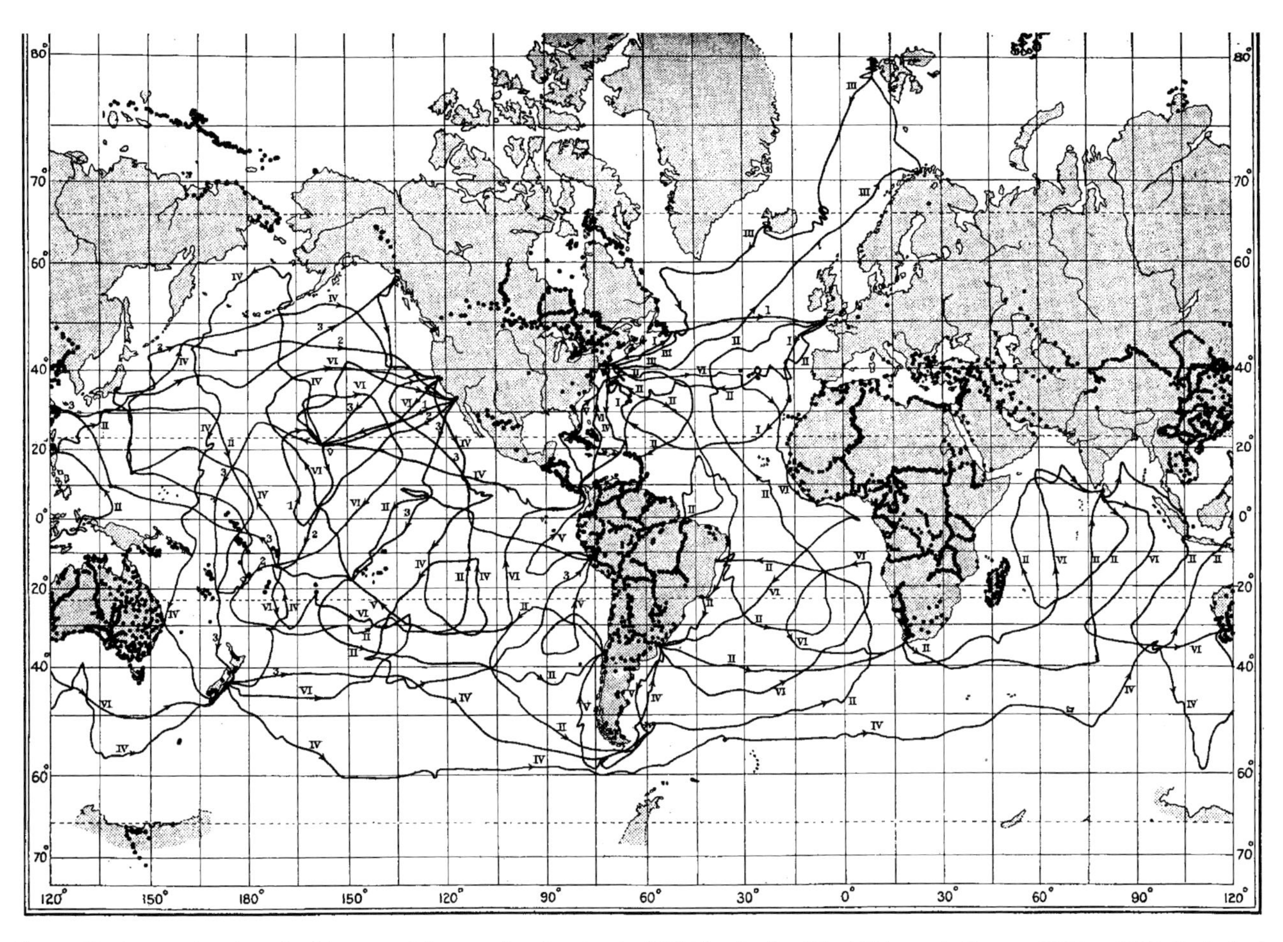

Figure C2 A map showing all of the expeditions undertaken by the DTM for the first World Magnetic Survey, 1904 to 1929. Each dot is a discreet station. The lines on the oceans indicate the tracks of the *Galilee* and the *Carnegie* between 1905 and 1929. Map courtesy of Carnegie Institution of Washington.

Although the DTM still exists, its focus on expeditionary, geomagnetic work ended with World War II. The department published two volumes in 1947, coordinated by E.H. Vestine, in which the data of the enterprise were summarized and analyzed (Vestine *et al.*, 1947a,b). Although the DTM has continued to conduct geophysical research since then, its character has been much more varied, including research in seismology and nuclear physics.

Gregory A. Good

Bibliography

Good, Gregory A. (ed.), 1994a. *The Earth, the Heavens and the Carnegie Institution of Washington.* History of Geophysics, volume 5. Washington, D.C.: American Geophysical Union, 252 pp.

Good, Gregory A., 1994b. Vision of a global physics: The Carnegie Institution and the first world magnetic survey. *History of Geophysics*, **5**: 29–36.

Trefil, J., and Hazen, M.H., 2002. *Good Seeing: a Century of Science at the Carnegie Institution of Washington.* Washington, D.C.: Joseph Henry Press.

Vestine, E.H., Laporte, L., Cooper, C., *et al.*,1947a. Description of the earth's main magnetic field and its secular change, 1905–1945. Carnegie Institution of Washington, Washington, D.C., Publication No. 578. 532 p.

Vestine, E.H., Laporte, L., Lange, I., and Scott, W.E., 1947b. *The Geomagnetic Field, its Description and Analysis.* Carnegie Institution of Washington, Washington, D.C., Publication No. 580.

Cross-references

Bauer, Louis Agricola (1865–1932)
Carnegie, Research Vessel
Fleming, John Adam (1877–1956)
Gauss, Carl Friedrich (1777–1855)
Halley, Edmond (1656–1742)
Humboldt, Alexander von (1759–1859)
IAGA, International Association of Geomagnetism and Aeronomy
Repeat Stations

CARNEGIE, RESEARCH VESSEL

When *Louis Agricola Bauer* (*q.v.*) established the *Department of Terrestrial Magnetism* (DTM) at the Carnegie Institution of Washington (cv) in 1904, he intended to focus the DTM's efforts strongly on an oceanic magnetic survey. The oceans represented both the largest and the most accessible gap in geomagnetic data. Bauer felt that most of the oceanic data available was not of high enough quality and that it was mainly restricted to coastal waters and trade routes. From 1905 to 1908, the DTM leased a wooden sailing vessel, the brig *Galilee.* Bauer had the steel rigging replaced with hemp and stripped away as much other iron as possible, but bolts and nails had to stay. The *Galilee* sailed over 70000 miles and established 442 magnetic stations, but its most important result was this: The *Galilee* demonstrated that a completely nonmagnetic ship was necessary to carry on this research.

Bauer convinced the Carnegie Institution's Board (and Andrew Carnegie) that the DTM needed its own yacht, specially designed for magnetic research. The shipyard in Brooklyn, New York that had built Kaiser Wilhelm's *Meteor* (and the winner of the 1906 Atlantic cup race) designed and built a most peculiar ship: the *Carnegie* (Figure C3). It was 155 feet long and carried 13000 square feet of sails. Bauer stationed a magnetic investigator at the shipyard for 7 months to guarantee that no magnetic materials were used in its construction. The planking was all wood, the nails were of locust tree, and the bolts were copper or bronze. Its bronze anchors each weighed 5500 lb and instead of anchor chains, the ship used 11-in. hemp cable. At $115000 (perhaps ten million dollars today), the *Carnegie* cost more than any other scientific instrument in history.

Figure C3 The *Carnegie* under full sail. Photo courtesy of Carnegie Institution of Washington. Photo H-112r.

Figure C4 A full complement of *Carnegie* observers in the two observing domes. Captain W.J. Peters is on the deck above. The photo was taken in the Indian Ocean in 1911. Courtesy of Carnegie Institution of Washington. Photo H-118.

The most important part of the ship for geomagnetic research was the observation deck, a special glass-enclosed room with a circular observation dome at each end (Figure C4). The domes were of glass panels with a brass framework. Magnetic observations were made in foul weather and fair, in order to obtain the density of measurements that was required.

A crowd of 3500 people turned out to launch the ship in 1909. It sailed the world's oceans for 20 years, collecting many thousands of magnetic measurements, and sailed on seven cruises, the equivalent distance of circumnavigating the globe twelve times. It met its demise in 1929 in the harbor at Apia, Samoa, when a gasoline engine exploded during refueling. The Carnegie Institution never again operated its own research vessels. Although the British government proposed a new "non-magnetic" research vessel in the 1930s, the *Research*, World War Two prevented it being placed in service. The Soviet government did launch the nonmagnetic schooner *Zarya* in the 1950s and continued to use it for several decades. (The name *Zarya* is now attached to the Russian component of the International Space Station.)

Simultaneously with the oceanic surveys of the *Galilee* and the *Carnegie*, the DTM (see Carnegie Institution of Washington, Department of Terrestrial Magnetism) conducted an ambitious magnetic land campaign. The goal was to gather enough magnetic data in a generation to provide a much better "snapshot" of Earth's magnetism than was then available. While this survey was completed by about 1930, the DTM continued to send observers to selected *"repeat stations"* (*q.v.*) to obtain data useful for studying secular variation.

Gregory A. Good

Bibliography

Good, Gregory A., 1994. Vision of a Global Physics: The Carnegie Institution and the First World Magnetic Survey. *History of Geophysics*, **5**: 29–36.

Cross-references

Bauer, Louis Agricola (1865–1932)
Carnegie Institution of Washington, Department of Terrestrial Magnetism
Repeat Stations

CHAMP

Challenging Minisatellite Payload (Figure C5) is a satellite mission dedicated to improving gravity and magnetic field models of the Earth. CHAMP was proposed in 1994 by Christoph Reigber of GeoForschungsZentrum Potsdam in response to an initiative of the German Space Agency (DLR) to support the space industry in the "New States" of the united Germany by financing a small satellite mission. The magnetic part of the mission is lead by Hermann Lühr. CHAMP was launched with a Russian COSMOS vehicle on July 15, 2000 onto a low Earth orbit. Initially planned to last 5 years, the mission is now projected to extend to 2009 (Figure C6). The official CHAMP website is at http://op.gfz-potsdam.de/champ/.

Satellite and orbit

A limiting factor for low Earth satellite missions is the considerable drag of the atmospheric neutral gas below 600 km altitude. This brought down *Magsat* (*q.v.*) within 7 months, despite of its elliptical orbit, and necessitated the choice of a higher altitude orbit for *Ørsted* (*q.v.*). To achieve long mission duration on a low orbit, CHAMP was given high weight (522 kg), a small cross section (Figure C5), and a stable attitude. It was launched onto an almost circular, near polar ($i = 87.3°$) orbit with an initial altitude of 454 km. While Magsat was on a strictly sun synchronous dawn/dusk orbit, CHAMP advances one hour in local time within eleven days. It takes approximately 90 min to complete one revolution at a speed of about 8 km s^{-1}.

Magnetic mission instrumentation

Magnetometers

At the tip of the 4 m-long boom, a proton precession Overhauser *magnetometer* (*q.v.*), measures the total intensity of the *magnetic field* (*q.v.*) once per second. This instrument, which was developed by LETI, Grenoble, has an absolute accuracy of <0.5 nT. Its measurements are used in the absolute calibration of two redundant vector magnetometers, located mid-boom on an optical bench. These fluxgate magnetometers were developed and supplied by DTU Lyngby, Denmark. They sample the field at 50 Hz with a resolution <0.1 nT.

Star cameras

The orientation of the optical bench in space is given by a star camera, which was developed and supplied by DTU Lyngby, Denmark. Attitude uncertainty is the largest source of error in satellite vector magnetic data. Star cameras are often blinded by the sun or moon, and provide unreliable attitude with regard to rotations about their direction of vision (boresight). For this reason, CHAMP was equipped with a dual head star camera, improving relative attitude by an order of magnitude to 3 arc seconds, corresponding to around 0.5 nT accuracy for the vector components. Since this high accuracy is only achieved in dual head mode (62% of CHAMP data), future magnetic field missions (e.g., the European Space Agency mission *swarm,* scheduled for launch in 2010) will have triple head star cameras. A further, redundant dual head star camera on the body of CHAMP is of limited utility for the magnetic field measurements, due to the flexibility of the boom.

Electric field, electron density, and temperature

A digital ion drift meter (DIDM) and a planar Langmuir probe (PLP) were provided by the Air Force Research Laboratory, Wright-Patterson Air Force Base, Ohio. The DIDM, designed to measure the electric field from ion velocities, partly failed due to frictional overheating during the launch phase of CHAMP. The PLP, which was not damaged, provides

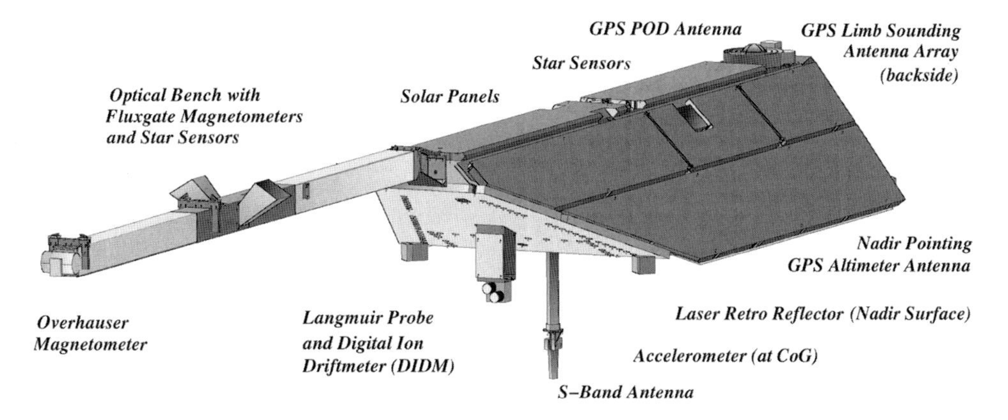

Figure C5 Front view of the CHAMP satellite (courtesy GFZ Potsdam). The acronyms POD and CoG stand for Precise Orbit Determination and Center of Gravity, respectively.

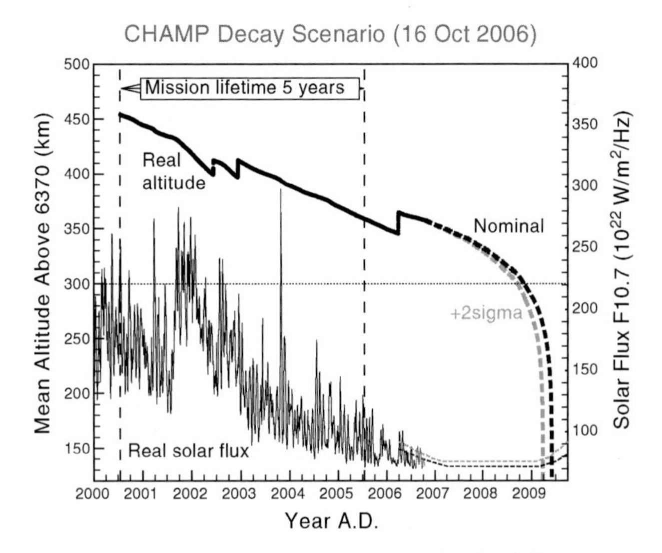

Figure C6 The decay of CHAMP's orbital altitude (*left scale*) depends on the neutral gas density, which is enhanced by solar activity (*right scale*). CHAMP's orbit has been raised thrice to prolong the mission, which is predicted to end after the solar activity minimum in 2009 (courtesy of F.-H. Massmann, GFZ Potsdam).

the spacecraft potential, electron temperature, and electron density, once in 15 s. As discussed below, these quantities can be used to correct magnetic field measurements for the diamagnetic effect of the plasma surrounding the satellite. The PLP measurements have thus, turned out to be essential for accurate geomagnetic field modelling.

GPS receiver

Apart from providing the accurate position of CHAMP, the Black Jack GPS receiver (supplied by NASA) has the important task of providing an absolute time frame. A pulse delivered every second is used to synchronize all of the instruments on board. Furthermore, it provides a stable reference frequency for the proton precession magnetometer readings, giving them absolute accuracy.

Data products

CHAMP's standard science products are labelled from level-0 to level-4, according to the amount of preprocessing applied to the original data. Scientific utility starts with level-2 products, which are calibrated, flagged, and merged with accurate orbits and are supplied as daily files in common data format (CDF). Level-3 products comprise the final processed, edited, and calibrated data, as well as the rapid delivery products. Derived products such as the initial CHAMP main field model CO2 (Holme *et al.*, 2003) are classified as level-4. The level-2 to level-4 products are archived and distributed by the Information System and Data Centre (ISDC) at GFZ Potsdam (http://isdc.gfz-potsdam.de/champ/).

Early science results

Main field models

Combined analysis of measurements from the three magnetic satellite missions (CHAMP, *Ørsted* (*q.v.*) and SAC-C) has to a breakthrough in main field model accuracy (Holme *et al.*, 2003). While models from ground-based observations are limited in resolution to about 5000 km wavelength, the new combined satellite models resolve the *main field* (*q.v.*) and its *secular variation* (*q.v.*) to 3000 km. Shorter wavelengths of the main field are masked by the *crustal magnetic field* (*q.v.*).

Crustal field models

With its low altitude, circular, polar orbit, CHAMP is particularly suited for mapping the field caused by the magnetic minerals in the Earth's crust. The first CHAMP crustal field model (Maus *et al.*, 2002)—extending to spherical harmonic degree 80, corresponding to a wavelength of 500 km—shows that the Earth's crustal magnetic field is weaker than indicated by earlier observations from Magsat.

Magnetic signal of ocean flow

The flow of electrically conducting seawater through the Earth's magnetic field acts as a global dynamo, inducing electric fields which drive electric currents and thus, give rise to secondary magnetic fields. Focussing on the dominant semidiurnal lunar tide, periodic ocean flow magnetic signals were clearly identified in CHAMP satellite data (Tyler *et al.*, 2003).

Diamagnetic effect of ionospheric plasma

A diamagnetic substance responds to an applied magnetic field by circular currents that produce a magnetisation opposing the applied field. Such a diamagnetic response, which is a well-known property of plasma in the outer magnetosphere and the solar wind, was until recently, thought to be negligible in the ionosphere. However, bands of enhanced plasma density on both sides of the magnetic equator have been found to cause depressions of the order of 5 nT in the geomagnetic field intensity (Lühr *et al.*, 2003). With the electron density and temperature acquired by the PLP instrument, CHAMP's magnetic field readings can be corrected for this effect. A related phenomenon is the formation of surface currents on plasma cavities. The CHAMP satellite has made it possible for the first time to directly observe these currents in their magnetic field signature (Lühr *et al.*, 2002).

Stefan Maus

Bibliography

Holme, R., Olsen, N., Rother, M., and Lühr, H., 2003. CO_2—A CHAMP magnetic field model. In Reigber, C., Lühr, H., and Schwintzer, P. (eds.), *CHAMP Mission Results I.* Berlin: Springer, pp. 220–225.

Lühr, H., Maus, S., Rother, M., and Cooke, D., 2002. First *in-situ* observation of night-time F region currents with the CHAMP satellite. *Geophysical Research Letters*, **29**(10):10.1029/2001GL013845.

Lühr, H., Maus, S., Rother, M., Mai, W., and Cooke, D., 2003. The diamagnetic effect of the equatorial Appleton anomaly: its characteristics and impact on geomagnetic field modelling. *Geophysical Research Letters*, **30**(17): 1906, doi:10.1029/2003GL017407.

Maus, S., Rother, M., Holme, R., Lühr, H., Olsen, N., and Haak, V., 2002. First scalar magnetic anomaly map from CHAMP satellite data indicates weak lithospheric field. *Geophysical Research Letters*, **29**(14): 10.1029/2001GL013685.

Tyler, R., Maus, S., and Lühr, H., 2003. Satellite observations of magnetic fields due to ocean flow. *Science*, **299**: 239–241.

Cross-references

Crustal Field
Magnetometer
Magsat
Main Field
Ørsted
Secular Variation

CHAPMAN, SYDNEY (1888–1970)

Sydney Chapman received 17 degrees (mostly honorary doctorates), published seven books and over 400 scientific papers, received nine major medals (including the Royal and Copley medals of the Royal Society), held ten visiting chairs, was president of eight professional societies, and fellow of many more. Rather than try to encompass a career of such breadth and depth, it is mainly his contribution to geomagnetism that is considered here, though he also contributed massively to, *inter alia*, solar-terrestrial physics, ionospheric physics, meteorology, aurorae, and the kinetic theory of gasses.

Chapman was born in Eccles, Lancashire, on January 29, 1888. He was set to enter an engineering firm, but, before doing so, he won the last of 15 university scholarships offered by Lancashire and went on to take an engineering degree at Manchester University. He stayed on for a further year and added a mathematics B.Sc. On the advice of Horace Lamb, he then tried for and obtained a Cambridge mathematics scholarship. He completed the examinations after two years at Trinity College, but was required to stay on for a further year before graduating. During this time he received a visit from Frank Dyson, the Astronomer Royal, offering him the prestigious and well-paid post of Chief Assistant at the Royal Observatory, Greenwich. (The other Chief Assistant at the time was the cosmologist Arthur Eddington, with whom Chapman devised the cycling performance measure n, which is the number of occasions on which n or more miles have been cycled in a day. Eddington achieved $n = 77$, and Chapman rather more.)

Chapman was at Greenwich from 1911 to 1914, and again from 1916 to 1918 to help out when many of the staff were at the war. He was a conscientious objector during the World War I—a far from easy option—but had revised his views by the time of World War II. While at Greenwich, Chapman was introduced to geomagnetism. He organized the re-instrumentation of the magnetic observatory, took part in the routine magnetic observations (on his own admission he was not a good observer), and set about the interpretation of geomagnetic variations.

In 1919 he moved from Trinity College, Cambridge, to succeed Lamb in the chair of mathematics and natural philosophy at Manchester. In the same year he was elected to Fellowship of the Royal Society at the early age of 31. This was largely for his work on the theory of solar and lunar daily geomagnetic variations.

In 1924, Chapman became chief professor of mathematics at Imperial College, London, where he remained until 1946. While there, he won the John Couch Adams essay prize, the subject of which (chosen at the suggestion of Dyson, with Chapman in mind) was the theoretical interpretation of geomagnetic phenomena. One of the conditions of the award was that the work should be published, which Chapman satisfied with a number of papers. But he also felt obliged to turn the essay into a book. In this he collaborated with the young German scientist *Julius Bartels* (*q.v.*) to eventually produce the two-volume *Geomagnetism* in 1940. This became the "bible" for geomagneticians for half a century. Though the instrumental part and the bibliography are now outdated, the mathematical sections are still of sufficient value to make this book an important part of any geomagnetic library.

It was also during his time in London that Chapman started his collaboration with V.C.A. Ferraro on the theory of magnetic storms, which was to be superseded only when satellite observations became available. With another of his students, *A.T. Price* (*q.v.*), he examined the currents induced in the Earth by Sq and their implication for mantle conductivity.

In 1946, Chapman moved to Oxford University to become Sedleian Professor of Natural Philosophy, a post he held until approaching compulsory retirement at 65. Rather than retire, he moved to the United States, where he already held visiting appointments. This was during the lead-up to the 1957/8 International Geophysical Year (IGY) and Chapman, as president of the IGY Special Committee, took the opportunity to travel widely, soliciting support for the project. As well as his theoretical work, he had undertaken many analyses of vast quantities of observational data and was well aware of the need for such data on a world-wide basis.

In the final, though far from inactive, part of his career (more than 150 papers were published after his "retirement"), Chapman divided his time between the Universities of Colorado and Alaska. In Alaska he worked with S.-I. Akasofu on geomagnetic disturbance phenomena and in Colorado with several colleagues, largely updating and extending earlier work in the light of recently acquired data, including IGY data. He was still actively working until a few days before his death on June 16, 1970.

Chapman's personal qualities were as remarkable as his scientific ability. He was slightly built, but very athletic, though his feats were of endurance rather than speed. Some mention has already been made of his cycling prowess. He gave up cycling at about 70, mainly because American highways were not cyclist-friendly, as he and his wife discovered experimentally. He swam half a mile a day throughout his life, going to considerable lengths to avoid missing a day. The nearest he came to owning a car was a half-share in one with Bartels, but only briefly. He much preferred to walk. He was a gentleman with strong moral principles and was always greatly considerate of the feelings of others. His many colleagues and students valued him as much as a friend as for his teaching.

Stuart R.C. Malin

Bibliography

Akasofu, S.-I., Fogle, B., and Haurwitz, B. (eds.), 1968. *Sydney Chapman, Eighty, From his Friends.* Privately published by University of Alaska, University of Colorado and University Corporation for Atmospheric Research.

Chapman, S., and Bartels, J., 1940. *Geomagnetism.* Oxford: Clarendon Press.

Cowling, T.G., 1971. Sydney Chapman 1888–1970. *Biographical Memoirs of Fellows of The Royal Society*, **17**: 53–89.

Cowling, T.G., and Ferraro, V.C.A., 1972. Obituary Sydney Chapman. *Quarterly Journal of the Royal Astronomical Society*, **13**: 464–467.

Malin, S.R.C., 1988. Sydney Chapman 1888–1970: a centennial celebration. *Eos, Transactions, American Geophysical Union*, **69**: 57–67.

Cross-references

Bartels, Julius (1899–1964)
Geomagnetism, History of
Price, Albert Thomas (1903–1978)

COAST EFFECT OF INDUCED CURRENTS

Introduction

When changes in Earth's magnetic field are measured on timescales of about 1 hour, the vertical component near most coastlines is abnormally large and correlates with the onshore horizontal component. This phenomenon is known as the geomagnetic coast effect. It is caused by electromagnetic induction occurring naturally in the Earth's surface layer.

The "primary" currents in the induction process are electric currents, which flow external to the Earth and change with time. As the solid Earth and its oceans conduct electricity, "secondary" currents are induced to flow in them. At Earth's surface, observations of magnetic and electric variations record the effects of the primary and secondary fields combined. Because in many instances the primary fields are uniform and spatially uncorrelated from one induction event to the next, spatial patterns occurring systematically reflect spatial patterns in the secondary fields due to structure in the electrical conductivity of the Earth.

The main electrical conductivity contrast at Earth's surface is between land and seawater. Seawater is a relatively homogeneous material, with its electrical conductivity, σ, depending mainly on salt content and temperature. A typical value for σ throughout the world's oceans is 3.3 S/m, in a range of 3 S/m for cold, less saline water to 6 S/m for warm, salty water (Bullard and Parker, 1970).

Land generally consists of inhomogeneous rock material and its conductivity, in contrast to that of seawater, varies by orders of magnitude (Sheriff, 1999). However the electrical conductivity of rock material generally is less than that of seawater. Further, the electrical conductivity of rock material is generally less than 1 S/m, leading to the common use of electrical resistivity ρ (the reciprocal of conductivity σ) when describing rock material. Thus a resistivity of 1 Ω m is equivalent to a conductivity of 1 S/m and a resistivity of 10 Ω m is equivalent to a conductivity of 0.1 S/m.

Then rock material typically has a resistivity of some hundreds of Ω m in the crust and possibly thousands of Ω m in ancient cratonic rocks. In sedimentary basins the resistivity may be less, say tens of Ω m. In ore bodies, and in crustal concentrations of the good conductor graphite, the resistivity may be only small fractions of an Ω m; however such good conductors form only a minor part of the total volume of crustal rocks.

The land-ocean conductivity contrast exerts a strong effect on electromagnetic induction at Earth's surface. Induction at places of such conductivity contrast is characterized by the generation of a vertical magnetic field component. Near a land-ocean contrast, this vertical component is seen as the geomagnetic coast effect.

Discovery of the coast effect

The coast effect was first observed by Parkinson (1959, 1962) and the discovery demonstrated that conductivity contrasts at and near Earth's surface could be detected by using natural electromagnetic induction at periods of about 1 hour. Parkinson's early results for Australia are shown in Figure C7 (see *Parkinson, Wilfred Dudley*).

At about the same time detailed observations of magnetic fluctuation patterns were also made in other continents, for example by Schmucker (1963); Rikitake (1964); Price (1967). The advent of digital recording and time-series analysis using electronic computers led to numerical analysis methods replacing Parkinson's graphical methods and the phenomenon was investigated as a function of frequency (Everett and Hyndman, 1967; Schmucker, 1970). One consequence of this development was then to demonstrate in-phase and out-of-phase components of the coast effect. Generally the in-phase component is stronger than the out-of-phase component (a result of the high conductivity of seawater) and often the effect is described and interpreted simply in terms of the in-phase component. If the out-of-phase component is being studied, care must be taken with the definition of phase (see Lilley and Arora, 1982). See also *induction arrows* (*q.v.*).

Observation of the coast effect

The review by Parkinson and Jones (1979) gives a global survey of the observation of the coast effect, which is still geographically representative. The authors note that the few coastal locations where the coast effect is absent are tectonically anomalous. In a regular coast effect, the strength of the phenomenon, as reckoned by the response in the vertical component of the fluctuating field, reduces to about half its coastal strength at a distance of some hundreds of kilometers inland.

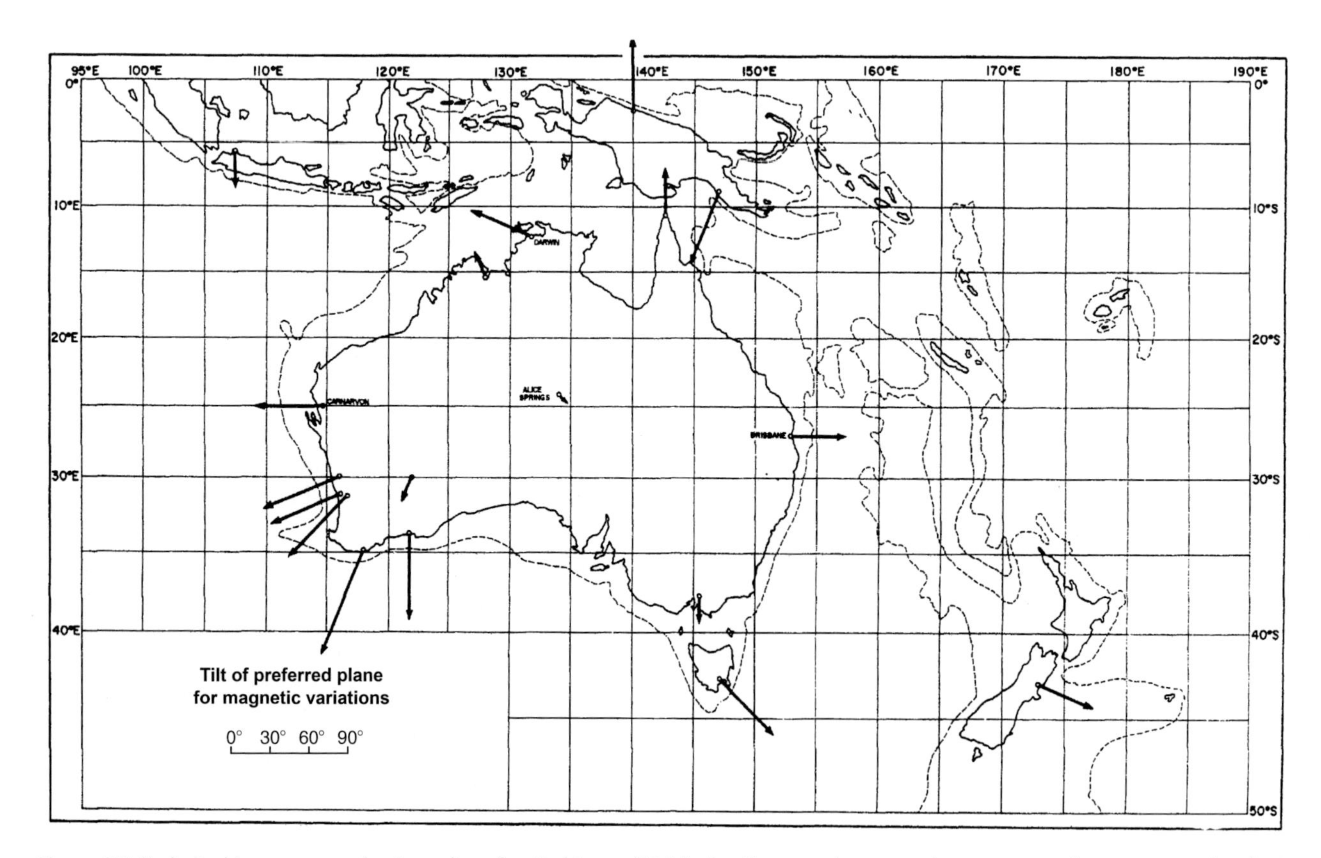

Figure C7 Early Parkinson arrows for Australia, after Parkinson (1964). Small magnetic vector changes at an observatory tend to lie in a "preferred plane." The horizontal projection of the downwards normal to this plane gives arrows as shown which, generally pointing to the high conductivity side of a nearby conductivity contrast, point to the deep ocean.

The authors also note that the coast effect varies only slightly with period, reaching a broad maximum in the period range 30–90 min. An important quantity is the electromagnetic "skin depth," δ,

$$\delta = \sqrt{\frac{2}{\mu\sigma\omega}} \qquad \text{(Eq. 1)}$$

where ω denotes angular frequency and μ denotes the permeability of the material (usually taken as that of free space). For the period range 30–90 min, the skin depth of seawater is some 20 km, which is greater than the ocean depth. Thus an external magnetic fluctuation signal reaches the ocean floor and penetrates the seafloor rock material and the ocean may be regarded as an electromagnetic "thin sheet."

An example of induction arrows for the coast effect of southeast Australia, observed on both land and seafloor, is shown in Figure C8 (White *et al.*, 1990). The site is favorable for a clear demonstration of the effect, as the coastline runs relatively straight for hundreds of kilometers, the continental shelf is narrow, and at the foot of the continental slope the seafloor is deep (almost 5000 m) and flat, comprising the Tasman Abyssal Plain. Included in the compilation are sites spaced down the continental slope. As can be seen from the induction arrows, the coast effect is at a maximum about halfway down the slope into the deep ocean.

An example of the data recorded in this particular case is shown in Figure C9. Displayed are the horizontal and vertical components of magnetic variation at five stations. It is in the vertical components that the coast effect is evident and the signal can be seen to be a maximum at station 4 (the trace marked CS4Z).

Fluctuations in the geomagnetic field occur over a wide frequency band, and the coast effect has also been investigated at shorter and longer periods. At shorter periods, the effect has been demonstrated to occur in "pulsations" of period order 1 min and less. At these short periods the skin depth of seawater no longer is greater than the depth of the deep ocean, and the effect takes place in the upper layer of the ocean. The relatively shallow seawater of the continental shelf becomes important.

At longer periods, a natural frequency band at which to examine the coast effect is that of the magnetic daily variations (termed "Solar quiet," and denoted Sq). However, for these signals different observation and analysis techniques are required due to the presence of a significant vertical component in the Sq source field.

Generally there is agreement that major "electric current gyres" are induced to flow in the oceans by the Sq source fields (Beamish *et al.*, 1983) and these also have a measurable effect at observatories near coast lines. Bennett and Lilley (1973) showed how a coast-effect vertical component could be separated from an observed record by subtracting an inland record observed simultaneously, which was taken to be a "normal" (i.e., background) component.

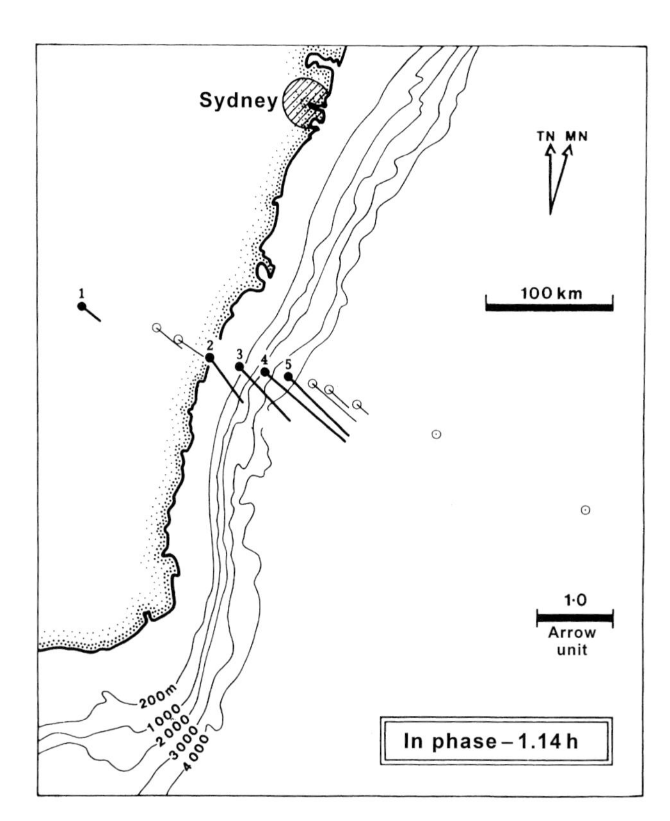

Figure C8 Induction arrows (in-phase, for period 1.14 h) for the coast of southeast Australia, for a line of stations from inland to deep seafloor (Ferguson, 1988; Lilley *et al.*, 1989; White *et al.*, 1990).

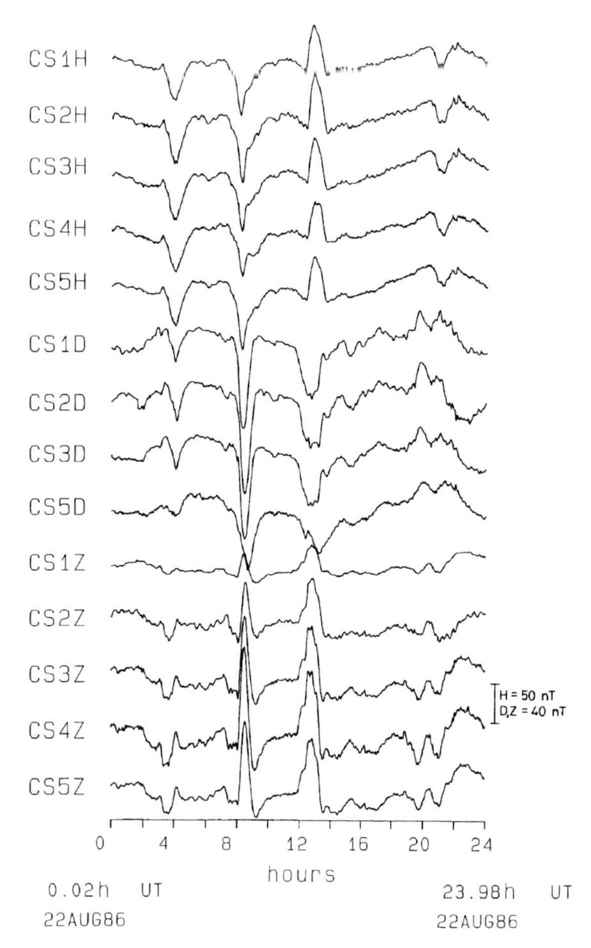

Figure C9 Examples of magnetic fluctuation records for one day, showing the coast effect (White *et al.*, 1990). The top five traces, with suffix H, are the horizontal north components for stations 1 to 5, as marked on Figure C8. The middle set of traces, with suffix D, are the horizontal east components. The lower five traces, with suffix Z, are the vertical components. The strength of the signal in the vertical component characterizes the coast effect, as it is the result of the electrical conductivity structure at the land-sea interface. Note this signal is strongest at station 4, half way down the continental slope.

Analogue modeling of the coast effect

The relative electrical conductivities of rock material and seawater lend themselves to analogue model studies. Following the initial global "terrella" model of Parkinson (1964), model studies have been carried out on geometries seen in coastlines worldwide (Dosso, 1973). These analogue models have the ability to model 3D situations, such as the islands of Tasmania (Dosso *et al.*, 1985) and New Zealand (Dosso *et al.*, 1996).

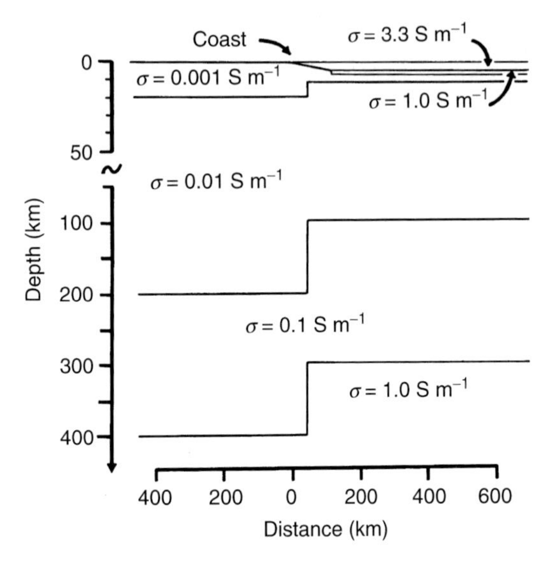

Figure C10 A two-dimensional model for the coast effect shown in Figures C8 and C9, with seawater layer and contrasting electrical conductivity structure beneath continent and ocean (Kellett *et al.*, 1991).

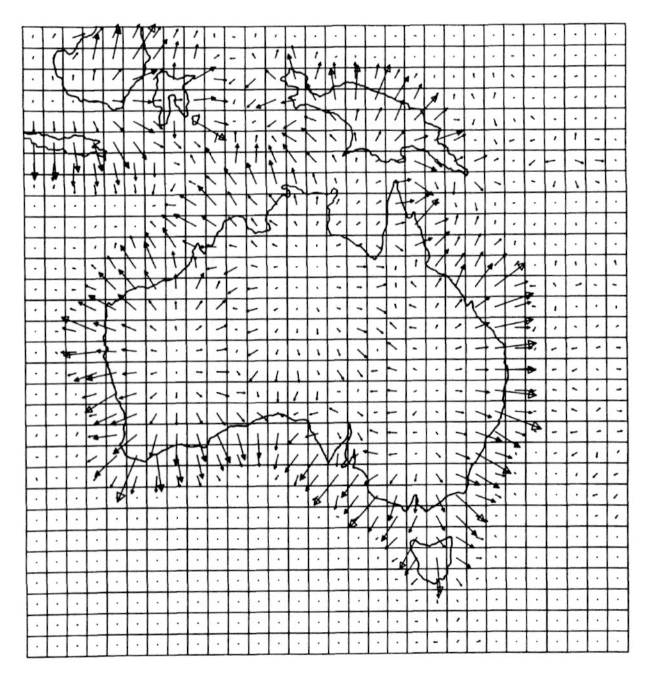

Figure C11 The electromagnetic response of a thin-sheet model of Australia presented as induction arrows, for period 1 h (in-phase) from Corkery and Lilley (1994). (Scale: an arrow which exactly spans a grid unit has a magnitude of 0.3).

Calculation of the coast effect

The basic calculation of the coast effect is a two-dimensional problem and the forward problem was solved in one of the early achievements of numerical modeling of electromagnetic induction in the Earth (Jones and Price, 1970; Weaver, 1994). Given that the ocean bathymetry is generally known well, the calculation becomes an exercise in modeling the rock material beneath the seawater. Figure C10 shows such a 2D model for the observed data shown in Figures C8 and C9.

As abilities in 2D and 3D numerical modelling and inversion have progressed, the coast effect has been a basic case for testing new methods and techniques. As an example, which may be compared with Figures C7 and C8, the induction arrow response for a numerical model of the Australian continent set in its surrounding seas (Corkery and Lilley, 1994) is shown in Figure C11.

Geophysical ocean-coast transects

Geophysical studies now frequently address the tectonic boundaries, which may be present at coastlines. By making observations along a transect, which crosses the coast line from deep ocean to inland, information may be obtained, which characterizes the structure and history of the region. Electromagnetic studies to obtain electrical conductivity information have been foremost amongst such transects, which have included subduction zones (Yukutake *et al.*, 1983; EMSLAB-Group, 1988) and passive margins (Wannamaker *et al.*, 1996). These transects

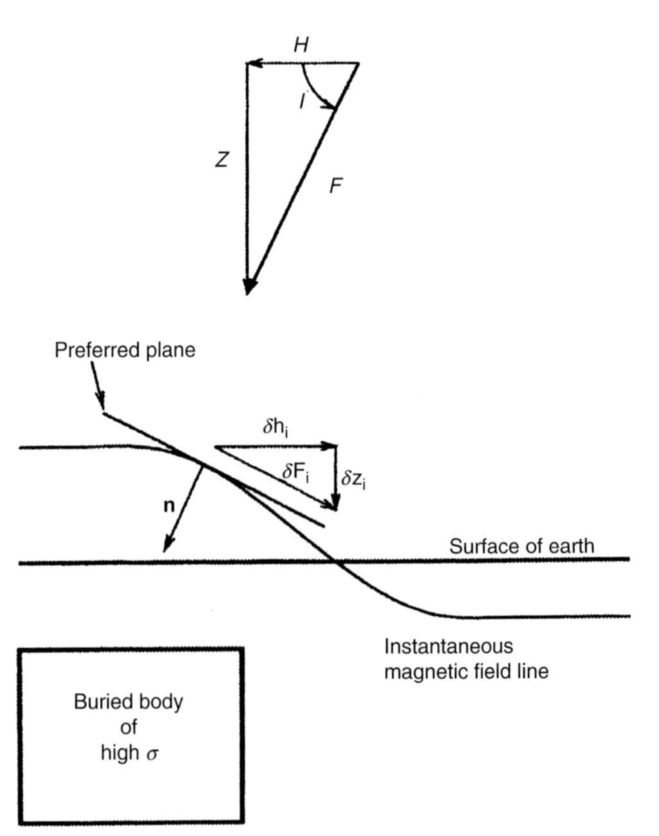

Figure C12 Diagram showing, near a buried body of high electrical conductivity, the "preferred plane" in which small vector magnetic changes tend to lie, and the normal **n** to the preferred plane. When the normal **n** is parallel or antiparallel to the main field **F**, a total-field magnetometer such as a proton-precession instrument will be insensitive to the small vector changes (Lilley *et al.*, 1999).

have given some of the most comprehensive observations of the coast effect, with electric and magnetic field measurements on both land and seafloor. Note also that Hitchman *et al.* (2000) made observations of the coast effect offshore at the sea surface, using a floating total-field magnetometer.

An application of the coast effect

Coast effects cover large regions and their influence may be widespread. An example is the "amphidrome" effect, which, as a result of the coast effect, exists across a wide swath of southern Australia (Lilley *et al.*, 1999). Here the coast effect causes the Parkinson preferred plane to be approximately orthogonal to the main magnetic field vector. Thus fluctuation events, which generally lie in the preferred plane, tend to be orthogonal to the main field and the component of them measured by a "total field" magnetometer (as in an aeromagnetic survey) is much reduced.

Figure C12 illustrates the "amphidrome effect" in which magnetic fluctuation signals, measured by a "total field" magnetometer, are suppressed when the normal to the Parkinson preferred plane is parallel to Earth's main magnetic field vector.

Conclusion

The geomagnetic coast effect is regarded as being well understood, and is primarily due to the electrical conductivity contrast between land and seawater, controlling natural electromagnetic induction in the environs of a coastline.

Ted Lilley

Bibliography

Beamish, D., Hewson-Browne, R.C., Kendall, P.C., Malin, S.R.C., and Quinney, D.A., 1983. Induction in arbitrarily shaped oceans—VI. Oceans of variable depth. *Geophysical Journal of the Royal Astronomical Society*, **75**: 387–396.

Bennett, D.J., and Lilley, F.E.M., 1973. An array study of daily magnetic variations in southeast Australia. *Journal of Geomagnetism and Geoelectricity*, **25**: 39–62.

Bullard, E.C., and Parker, R.L., 1970. Electromagnetic induction in the oceans. In Maxwell, A.E., (ed.), *The Sea.* Wiley-Interscience, Vol. 4, pp. 695–730.

Corkery, R.W., and Lilley, F.E.M., 1994. Towards an electrical conductivity model for Australia. *Australian Journal of Earth Sciences*, **41**: 475–482.

Dosso, H.W., 1973. A review of analogue model studies of the coast effect. *Physics of the Earth and Planetary Interiors*, **7**: 294–302.

Dosso, H.W., Nienaber, W., and Parkinson, W.D., 1985. An analogue model study of electromagnetic induction in the Tasmania region. *Physics of the Earth and Planetary Interiors*, **39**: 118–133.

Dosso, H.W., Chen, J., Chamalaun, F.H., and McKnight, J.D., 1996. Difference electromagnetic induction arrow responses in New Zealand. *Physics of the Earth and Planetary Interiors*, **97**: 219–229.

EMSLAB-Group, 1988. The EMSLAB electromagnetic sounding experiment. *Eos, Transactions of the American Geophysical Union*, **69**: 89–99.

Everett, J.E., and Hyndman, R.D., 1967. Geomagnetic variations and electrical conductivity in south-western Australia. *Physics of the Earth and Planetary Interiors*, **1**: 24–34.

Ferguson, I.J.,1988. *The Tasman Project of Seafloor Magnetotelluric Exploration*, Ph.D. thesis, Australian National University.

Hitchman, A.P., Lilley, F.E.M., and Milligan, P.R., 2000. Induction arrows from offshore floating magnetometers using land reference data. *Geophysical Journal International*, **140**: 442–452.

Jones, F.W., and Price, A.T., 1970. The perturbations of alternating geomagnetic fields by conductivity anomalies. *Geophysical Journal of the Royal Astronomical Society*, **20**: 317–334.

Kellett, R.L., Lilley, F.E.M., and White, A., 1991. A two-dimensional interpretation of the geomagnetic coast effect of southeast Australia, observed on land and seafloor. *Tectonophysics*, **192**: 367–382.

Lilley, F.E.M., and Arora, B.R., 1982. The sign convention for quadrature Parkinson arrows in geomagnetic induction studies. *Reviews of Geophysics and Space Physics*, **20**: 513–518.

Lilley, F.E.M., Filloux, J.H., Ferguson, I.J., Bindoff, N.L., and Mulhearn, P.J., 1989. The Tasman Project of Seafloor Magnetotelluric Exploration: Experiment and observations. *Physics of the Earth and Planetary Interiors*, **53**: 405–421.

Lilley, F.E.M., Hitchman, A.P., and Wang, L.J., 1999. Time-varying effects in magnetic mapping: amphidromes, doldrums and induction hazard. *Geophysics*, **64**: 1720–1729.

Parkinson, W.D., 1959. Directions of rapid geomagnetic fluctuations. *Geophysical Journal of the Royal Astronomical Society*, **2**: 1–14.

Parkinson, W.D., 1962. The influence of continents and oceans on geomagnetic variations. *Geophysical Journal of the Royal Astronomical Society*, **6**: 441–449.

Parkinson, W.D., 1964. Conductivity anomalies in Australia and the ocean effect. *Journal of Geomagnetism and Geoelectricity*, **15**: 222–226.

Parkinson, W.D., and Jones, F.W., 1979. The geomagnetic coast effect. *Reviews of Geophysics and Space Physics*, **17**: 1999–2015.

Price, A.T., 1967. Electromagnetic induction within the Earth. In *Physics of Geomagnetic Phenomena,* Matsushita, S., and Campbell, W.H. (eds.), Vol. 11-I, of International Geophysics Series, London: Academic Press, pp. 235–298.

Rikitake, T., 1964. Outline of the anomaly of geomagnetic variations in Japan. *Journal of Geomagnetism and Geoelectricity*, **15**: 181–184.

Schmucker, U., 1963. Anomalies of geomagnetic variations in the southwestern United States. *Journal of Geomagnetism and Geoelectricity*, **15**: 193–221.

Schmucker, U., 1970. *Anomalies in Geomagnetic Variations in the South-Western United States*. Bulletin of the Scripps Institution of Oceanography, Vol. 13.

Sheriff, R.E.,1999. *Encyclopedic Dictionary of Exploration Geophysics*. Society of Exploration Geophysicists.

Wannamaker, P.E., Chave, A.D., Booker, J.R., Jones, A.G., Filloux, J.H., Ogawa, Y., Unsworth, M., Tarits, P., and Evans, R., 1996. Magnetotelluric experiment probes deep physical state of southeastern United States. *Eos (Transactions of American Geophysical Union)*, **77**: 329–336.

Weaver, J.T., 1994. *Mathematical Methods for Geo-electromagnetic Induction.* Baldock, UK: Research Studies Press.

White, A., Kellett, R.L., and Lilley, F.E.M., 1990. The continental slope experiment along the Tasman Project profile, southeast Australia. *Physics of the Earth and Planetary Interiors*, **60**: 147–154.

Yukutake, T., Filloux, J.H., Segawa, J., Hamano, Y., and Utada, H., 1983. Preliminary report on a magnetotelluric array study in the Northwest Pacific. *Journal of Geomagnetism and Geoelectricity*, **35**: 575–587.

Cross-references

Electromagnetic Induction (EM)
Geomagnetic Deep Sounding
Induction Arrows
Magnetotellurics
Mantle, Electrical Conductivity, Mineralogy
Parkinson, Wilfred Dudley

COMPASS

A freely suspended, magnetized iron needle will attempt to align itself with the ambient magnetic field vector (**F**); when made to float in a bowl of liquid, or balanced horizontally on a pivot, it will orientate itself along the local magnetic meridian (**H**, the horizontal part of the vector). At what specific instant during the Middle Ages this attribute was first discovered remains shrouded in the mists of time; China, the Arab world, Greece, Italy, France, and Scandinavia have all claimed primacy in this respect. The ubiquity of *lodestone* (or *loadstone*, various iron-rich, permanently magnetized minerals) at and near the Earth's surface must have enabled many cultures to become aware of this mysterious, but highly useful, property.

Technically speaking, any contraption incorporating magnetic alignment to indicate direction can be considered a compass. The word itself can be traced back to Mediterranean countries around the mid-13th century, initially referring to sailing directions (both written instructions and nautical charts). Two centuries later, its meaning had expanded to include dividers applied in course plotting on a chart, whereas in German-speaking regions the word additionally became associated with pocket sundials, which contained a small dry-pivoted magnetic needle, inset at the base, for meridional orientation. It would take another century before the term would be predominantly associated with the currently typical magnetized needles bearing a graduated card, often encased and in cardanic suspension. The following discussion will outline some of the most notable aspects of this evolution, focussing in turn on its basic design, the observation compass, technical improvements, and the problem of deviation.

Basic compass design

The earliest unequivocal evidence of a magnetized piece of iron used for orientation stems from China around the turn of the 12th century A.D., in sources describing the floating magnetic "compass fish," and south-pointing needles magnetized by lodestone touch suspended with a silk wire or balanced on a fingernail. The floating needle (in a little water-filled saucer, marked with 12 or 24 wind directions) was probably incidentally used at sea during overcast weather by both Chinese, Korean, and Japanese navigators, until "dry" compasses were introduced by European ships from the 16th century. However, Southeast Asian shipping was primarily coastal and monsoon-driven, limiting reliance on the magnetic pathfinder. By contrast, contemporary Arabian navigators traversing the Indian Ocean set their courses using a 32-point sidereal card without a needle, deriving directions instead from the near-perpendicular rise and set of specific stars, such as Altair (East) and Antares (Southeast). Their 13th-century adoption of the compass on Mediterranean voyages was established through European contacts.

In western Europe, references to floating needles start to appear from the 12th century, in accounts by A. Neckam (1187), P. de Maricourt (Petrus Peregrinus, 1269), and in various ship's inventories. Over the next century, notions of magnetization and compass needles became so commonplace that they were increasingly used in a metaphorical sense in literature. During the 14th century, dry-pivoted instruments came to replace their floating precursors and the needle was attached to a starch-stiffened paper disk depicting the rose of the winds (6–8 in. in diameter); its cardinal points were painted blue, the half cardinals red. The direction of Jerusalem (east) was initially marked with a cross, the others with the first letter of their associated Mediterranean wind. The north-indicating letter "T" (for *Tramontana*) may have mutated into the ornamental *fleur-de-lys* around 1500. Points division grew from 8 through 16–32 in the 16th century. From about 1650, standardized designs engraved in copper plate replaced hand-drawn cards, while during the 18th century, a more austere appearance with rim graduation became the norm. At the time, maximum steering accuracy of square-rigged vessels generally did not exceed one quarter point (2°49′).

With the advent of commercial cartography and triangulation, the surveying compass rose to prominence on land. This instrument was usually tripod-mounted and equipped with a plumb line to be aligned with the sighted object. Its pivoted, straight needle rotated over a card attached to the inside bottom of the case. In the 17th century, the mounting of more elaborate sighting devices (double visors, cross-hairs, small telescopes) allowed more precise readings to be taken. Another compass-like instrument, used on land from the late 16th century, was the *versorium*, an arrow-shaped iron needle, up to a foot long, applied by natural philosophers such as *W. Gilbert* (*q.v.*) and N. Cabeo to investigate electrical and magnetic phenomena, for example, to detect the electrostatic attraction produced by rubbed amber.

At sea, the mariner's needle became attached to the underside of the card, providing instantaneous identification of all directions and a larger moment of inertia. By the late 16th century, the whole became encased in a round or square bowl or box made of wood, copper, or brass, suspended in brass or tin gimbals, inside a larger square box. The box was covered with a glass or crystal lid, sealed with putty or wax to keep out wind, dirt, and moisture, while its bottom was detachable for maintenance purposes, so the watertight top seal did not have to be broken each time the needle required retouching.

The steering compass was housed on deck inside the binnacle, which was horizontally divided into three compartments; the two outer ones each held a compass, the middle one a lantern. When sailing craft grew larger, the fore-and-aft line that indicated the ship's heading became less clearly distinguishable to the helmsman, so a corresponding mark inside the compass bowl, the *lubber's line*, was introduced as a steering aid. Various production centers mushroomed, such as Sicily, Genoa, and Venice (Mediterranean), Danzig (Baltic), and Seville, Lisbon, La Rochelle, Bordeaux, Rouen, and Flanders (west European coasts). From the 17th century, the capitals of the seagoing nations likewise became the seat of compass making firms and individual craftsmen, which soon came to dominate the manufacture of these instruments.

To compensate for magnetic declination in steering, the needle was sometimes affixed to the card at an angle to true north; another physical solution was the *rectifier*, a second, rotatable card on top of the first, with which changing declination values along a voyage could be compensated by applying a variable (usually fixed-step) shift. To compensate for magnetic inclination, which changes with latitude along the voyage and dips the needle, thus increasing friction, a lump of wax was attached to the card; from the 18th century, increasingly sophisticated, adjustable counterweights were used.

Regarding needle shape, at its crudest stage, a single piece of iron wire was bent into rhombic form, the two ends forming an acute angle at one apex, and the middle of the wire forming another at the opposite end, allowing space in the middle for a cap to balance the needle on the pivot. Alternatively, two separate curved pieces of iron could be joined at the ends to form a lozenge or two empty parentheses. Single, straight needles of well-tempered steel (enabling more intense, longer-lasting magnetization), rose to prominence in the first half of the 18th century. Metal caps then became replaced by glass and agate ones, which suffered less wear from pivotal friction. Admiralties and commercial shipping companies adopted standardized, often patented designs (from 1766), for example, by G. Knight and R. Walker (18th century), and by the firms of Stebbing, Weilbach, and Kleman (19th century).

The observation compass

The main difference between a steering compass and an observation compass is a sighting device mounted on top of the box (see Figure C13), to be aligned with a landmark (surveying and charting), the ship's wake (estimating leeway), or a celestial body (determining magnetic declination). Such compasses could be set up on land or anywhere on deck using a wooden stool or special tripod. Despite great

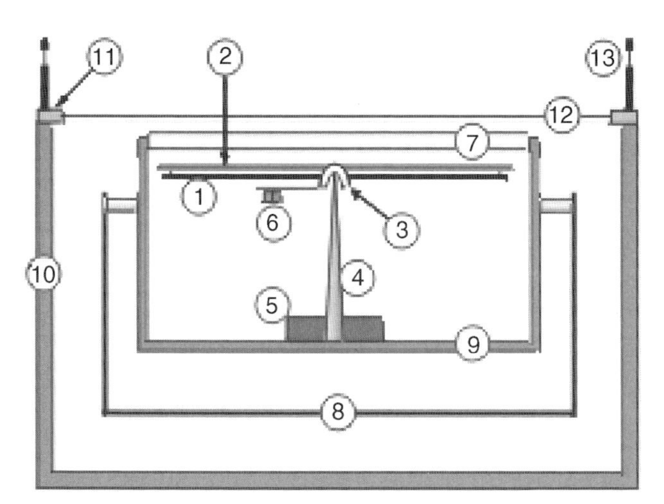

Figure C13 Cross section of an early-modern amplitude compass, as used in oceanic navigation. 1, needle; 2, card; 3, cap; 4, pin; 5, balancing weight; 6, inclination counterweight; 7, glass lid; 8, cardanic suspension; 9, compass bowl (inner box); 10, outer box; 11, rotatable verge ring; 12, fiducial line; 13, crosshair visor.

variability of design, all types had cards graduated in degrees (or less) to measure the horizontal arc between two imaginary great circles perpendicular to the horizon and meeting in the zenith. The arc distance from true east was called *amplitude*; that from true north was called *azimuth*. The most significant practical distinction between *amplitude compass* and *azimuth compass* is that the former, older device only allowed near-horizontal sightings directly over the card, whereas the latter's much taller visors, introduced in many countries in the second half of the 18th century, permitted multiple observations to be taken in short succession, and at a much greater altitude above the horizon. The acceptance of azimuth as a superior measurement was, however, reliant on adequate technology, the implementation of the new, longer calculations, and awareness of the problems associated with amplitude readings (limited opportunity, poor sighting conditions at the horizon due to clouds and haze and atmospheric refraction).

The first description of an observation compass can be found in De Maricourt's "Epistle on the Magnet" (1269). Apart from a silver or brass transverse index, it carried a sighting rule with upright styles at both ends to align with heavenly bodies. Similar 16th-century descriptions are known from the Portuguese J. de Lisboa (1508), Italian L. Sanuto (1588), Dutchman R. Pietersz van Twisk (1595), and Englishman W. Barlowe (1597, with a rotatable verge ring carrying a brass ruler with two upright, triangular sights). In France, two types of amplitude compass existed concurrently, one with small pins fixed to the rim of the box, the other with two small glass windows in opposite sides, furnished with crosshairs. French navigators adopted azimuth visors only from the 19th century onward. Some English 17th-century amplitude compasses had a large, broad semicircle made of brass, which extended beyond the compass rim and could be rotated around it. Opposite this dial was a large foldable style, carrying a sight and a hypotenusal thread from the top of the index down to the plate that could cast a noonday shadow onto the card. This method was rarely used at sea.

In terms of measurement accuracy, by taking the average or (more often) the median of a series of magnetic declination readings obtained in the course of a nautical day, a standard deviation of little under half a degree was maintained by the majority of early-modern navigational practitioners engaged in oceanic shipping. On land, with the aid of more accurate instruments and fixed meridians engraved in stone, this margin could sometimes be reduced to a few arc minutes. Around 1820, a more modern version of the azimuth compass was introduced at sea, allowing celestial measurements near the zenith with a card divided in half degrees.

Compass design improvements

Besides complaints about careless construction, low-quality materials, skewed or warped cards, and the aforementioned wear on pin and cap, the most serious defect of the instrument, mentioned time and again in early-modern documents, was weak magnetization. Needles made of iron wire (in use up to the early 18th century) required several retouchings at sea on long voyages, which is why some vessels still carried lodestones on board as well. Magnetization itself was deemed highly dependent on the proper method; suggestions range from repeatedly striking the needle with a hammer while placed on the lodestone, through rotating its tips inside small holes made at the lodestone's poles, to (1–12) stroking motions of the lodestone along (half or all of) the needle's length. A separate debate focussed on the various types and qualities of lodestone, using color, density, and origin as decisive characteristics. With the 18th-century advent of steel needles and artificial magnets (fully magnetized steel rods, much stronger than natural magnets), the earlier dependence on lodestones, both by compass makers and sailors, was greatly reduced.

A different way to improve directivity was to increase the number of needles inside the instrument; by placing two, four, or even eight straight needles in parallel under the card; their combined directive property would greatly exceed that of any single one. The first controlled experiment that compared the performance of single- and parallel needle compasses dates from 1649, by C.J. Lastman and I. Blaeu, two navigation officials employed by the Dutch East India Company. The superior results of parallel needles were disseminated in leaflets and navigation textbooks, and from 1655 until about 1710, Dutch Eastindiamen routinely carried two or more such instruments.

After incidental descriptions in French and Spanish textbooks without practical implementation, Danish naval officer and hydrographer Lous rekindled interest from 1767, advocating a compass with four flat, steel needles of 150 mm in length, held in a wooden or brass frame. Similar designs were proposed by French mathematician C. de Borda (1772) and astronomer P.-C. Le Monnier (1776), and by English captain C.H. Lane (1788). However, it would take until 1840 before the new British Admiralty's standard compass would carry four straight steel needles attached to a ring beneath the card, itself to be supplanted by the improved card invented by W. Thomson (1878, adopted 1889), which carried eight parallel needles suspended by wire. In subsequent decades, this dry compass fought a losing battle against another design improvement which would be universally adopted in the 20th century: the liquid compass.

The liquid compass has the advantages of damping needle movements and exterior disturbances and supporting the card to reduce friction, but at the cost of introducing such new complications as freezing, leakage, and temperature-induced changes of volume or pressure. After Royal Society experiments with lodestones acting under water in the 1660s, *E. Halley* (*q.v.*) seems to have demonstrated a rudimentary liquid compass there in 1690. The first comprehensive description appeared in 1779, written by Dutch physician J. Ingen-Housz: a large, sensitive needle inside a steel tube with a float, housed in a waterproof casing filled with linseed oil. Several other designs were propounded by instrument makers G. Wright (1781), Barton (1792), and C.C. Lous (1795); liquid compass patents were, for example, issued to inventor Francis Crow (1813), and the Danish compass production firm of Weilbach (1830). The problem of variable fluid volume was finally solved by W.R. Hammersley's system of expansion chambers (1856). A late-19th century liquid compass with two parallel artificial magnets is depicted in Figure C14.

Compass deviation

The deflection of compass needles away from the geomagnetic vector due to the attraction of magnetized iron nearby is known as compass

Figure C14 An American 19th-century liquid steering compass with two parallel needles (on *left*), and a Swedish late 20th-century handheld, dry-pivoted scouting compass (author's collection).

deviation (the term was coined by Arctic explorer J. Ross, ca. 1820). The effect was first recorded by Portuguese navigator J. de Castro in 1538, who ascribed inconsistent readings of magnetic declination to a cast iron gun on deck. From the 17th century onward, regulations regarding compass use warned both mariners and surveyors against wearing clothing attributes made of iron and prompted them to take note of metal structures that might affect the readings. On the largely wooden ships, the disturbance remained limited to incidental aberrations of a few degrees at most, the cause of which was usually identified swiftly, and either removed or avoided thereafter.

The introduction of iron and steel in ship construction, however, radically exacerbated the problem. Despite the advantages of stronger, longer, cheaper, more spacious hulls, the increasing reliance on structural iron parts (late 18th and first half of the 19th century) and extensive armor plating (Crimea War, 1854–56), all introduced substantial sources of magnetic deviation, up to tens of degrees. The period from the first iron-hulled clippers (ca. 1800) up to all-steel ship design (1877) was hallmarked by the scientific investigation of deviational compass error, and the invention of various methods of compensation.

Possibly the oldest method to investigate compass deviation was "swinging the ship," which involved turning the ship about a vertical axis, and recording the incurred deviation on all headings. The procedure was first suggested by American surveyor J. Churchman in 1790. British naval surveyor M. Flinders, while traversing Tasmanian waters in 1798, likewise noticed that the amount of compass error varied as a function of course. On his next assignment in the region, he identified (latitude-related) geomagnetic inclination as another factor to be considered. Subsequent captivity on the French isle of Mauritius (1803–1810) allowed him to develop his theory of compass error further, using his findings to correct and redraw much of his survey. Upon his return to England, he suggested a counter attractor, the "Flinders bar" (an iron rod of specific length placed inside the binnacle, opposite the side of greatest deviation), which was eventually introduced on board ship some 50 years later.

Research intensified from the 1810s, amongst others by explorer W. Scoresby, astronomer F. Arago, Board of Longitude secretary Th. Young, and *Professor P. Barlow* (*q.v.*), who developed a disk ("Barlow's plate," 1819) to quantify deviation aboard ship. Theoretical aspects received attention in a series of treatises (1821–1838) by French mathematician S.D. Poisson, relating geomagnetism and ship magnetism using C.A. de Coulomb's inverse-square law. Soon thereafter, Astronomer Royal G.B. Airy started a series of experiments on iron ships, using combinations of bar magnets and iron chains as counteracting physical correctors. In the second half of the century, hydrographer Frederick Evans worked with mathematician A. Smith on the problem, based on the Hydrographic Office's first isogonic chart (1858), resulting in the "Admiralty Manual for Deviations of the Compass" in 1862.

Notwithstanding these efforts, deviation proved a highly complex phenomenon that was never fully defeated. A different solution ensuring accurate and safe navigation free from interference was eventually found in the gyrocompass, based on the gyroscopic principle, which forces the axis of any sufficiently fast spinning top to keep a fixed orientation, for example, relative to the Earth's axis (described in 1836 by E. Sang in Britain, and in 1852 by J.B.L. Foucault in France). The first fully operational models were patented by H. Anschütz-Kaempfe and E. A. Sperry in the early 20th century, and underwent sea trials from 1907. Over the next few decades, the gyrocompass proved a worthy successor to the magnetic compass, which became relegated to the more modest role of auxiliary safety backup on large ships. Nevertheless, the magnetic compass continues to serve the majority of smaller, private vessels, as well as scouts, hikers, and other travelers on land (see Figure C14). In addition, the historical accumulation of magnetic compass readings recorded in ship's logbooks presently provides an invaluable source of data on geomagnetic secular variation over the last four hundred years.

Art R.T. Jonkers

Bibliography

Bennett, J.A., 1987. The Divided Circle: A History of Instruments for Astronomy, Navigation and Surveying. Oxford: Phaidon.

Blackman, M., 1983. The Lodestone: A Survey of the History and the Physics, *Contemporary Physics*, **24** no. 4: 319–331.

Cotter, C.H., 1997. The Early History of Ship Magnetism: the Airy-Scoresby Controversy, *Annals of Science*, **34**: 589–599.

Day, A., 1967. The Admiralty Hydrographic Service 1795–1919. London: Ministry of Defence.

Fanning, A.E., 1986. Steady as She Goes: A History of the Compass Department of the Admiralty. London: Her Majesty's Stationery Office.

Freiesleben, H.C., 1978. Geschichte der Navigation. Wiesbaden: Steiner.

Hackmann, W.D., 1994. Jan van der Straet (Stradanus) and the Origins of the Mariner's Compass, In Hackmann, W.D., and Turner, A.J. (eds.), Learning, Language and Invention: Essays Presented to Francis Maddison. Aldershot: Variorum, 148–179.

Hewson, J.B., 1951. A History of the Practice of Navigation. Glasgow: Brown, Son and Ferguson.

Hitchins, H.L., and May, W.E., 1955. *From Lodestone to Gyro-Compass.* London: Hutchinson.

Körber, H.-G., 1965. Zur Geschichte der Konstruktion von Sonnenuhren und Kompassen des 16. bis 18. Jahrhunderts, Veröffentlichungen des Staatlichen Mathematisch-Physikalischen Salons 3. Berlin.

Maddison, F.R., 1969. Medieval Scientific Instruments and the Development of Navigational Instruments in the Fifteenth and Sixteenth Centuries. Agrupamento de Estudos de Cartografia Antiga 30. Coimbra: Junta de Investigações do Ultramar.

McConnell, A., 1980. *Geomagnetic Instruments before 1900: An Illustrated Account of their Construction and Use.* London: Harriet Wynter.

Merrifield, F.G., 1963. *Ship Magnetism and the Magnetic Compass, Navigation and Nautical Courses*, Vol. 1, Oxford: Pergamon.

Schück, A., 1911–1915. *Der Kompass.* Hamburg: Schück.

Smith, J.A., 1992. Precursors to Peregrinus: The Early History of Magnetism and the Mariner's Compass in Europe, *Journal of Medieval History.* **18**: 21–74.

Waters, D.W., 1958. *The Art of Navigation in England in Elizabethan and Early Stuart Times.* London: Hollis and Carter.

Wolkenhauer, A., 1982. Der Schiffskompas in 16. Jahrhundert und die Ausgleichung der Magnetischen Deklination, In Köberer, W. (ed), *Das Rechte Fundament der Seefahrt: Deutsche Beitrage zur Geschichte der Navigation.* Hamburg: Hoffmann and Campe.

Cross-references

Barlow, Peter (1776–1862)
Geomagnetism, History of
Gilbert, William (1544–1603)
Halley, Edmond (1656–1742)
Voyages Making Geomagnetic Measurements

CONDUCTIVITY GEOTHERMOMETER

Introduction

The concept of the electrical conductivity geothermometer comes from the laboratory-demonstrated dependence of the electrical conductivity of rock material on temperature (Hinze, 1982). These laboratory results are consistent with theoretical considerations.

Beneath various complications that arise in the upper crust of Earth, rock material is generally considered to conduct electricity according to the physical mechanism of electronic semiconduction. The partial melting of rock material may also increase rock conductivity once connectivity is established for the melt fraction (Shankland *et al.*, 1981).

For electronic semiconduction, the relationship is the Arrhenius one between electrical conductivity σ and absolute temperature T, given by (Parkinson and Hutton, 1989)

$$\sigma = \sigma_e \exp(-E_e/kT) + \sigma_i \exp(-E_i/kT) \qquad \text{(Eq. 1)}$$

Here the first term in Eq. (1) describes extrinsic conduction, operating through thermally activated crystal defects and impurities, and the second term describes intrinsic conduction at higher temperatures. The terms σ_e and σ_i are the high-temperature asymptotes for each type of conduction, E_e and E_i are activation energies, and k is the Boltzmann's constant.

The strategy is to exploit the phenomenon of electromagnetic induction at Earth's surface by natural source fields (Campbell, 1997) to obtain a σ profile. Then, in terms of parameters σ_e, σ_i, E_e and E_i determined experimentally from laboratory studies or on the basis of other laboratory information, to obtain a T profile from the σ profile. The procedure is thus in two parts, which will now be discussed in turn.

Determining electrical conductivity as a function of depth

The observed data are time-series of magnetic and electric fields measured at various places over the surface of Earth. Generally these places are well-established magnetic observatories, sometimes supplemented by more temporary magnetotelluric stations. The primary signals have generally come to the solid Earth from outside it, and because they change with time they induce secondary fields within Earth, as the planet is an electrical conductor. At the surface, observations record both primary and secondary fields combined. A process of data inversion is needed, based on the known physics of induction in the Earth (Weaver, 1994), to give from these observations a profile of electrical conductivity with depth.

Activity in this topic is now such that specialist workshops on Electromagnetic Induction in the Earth have been held in different parts of the world every two years since the first one in Edinburgh, Scotland in 1972. Review and contributed papers are published, for example see Weaver (1999). Data on electrical conductivity structure from electromagnetic observations are used to obtain information on geological structure, composition, and history.

An important point is that the process of electromagnetic induction in the Earth is frequency dependent. An indication of the depth of penetration into a material by an induction process is given by the electromagnetic "skin depth," δ,

$$\delta = \sqrt{\frac{2}{\mu\sigma\omega}} \qquad \text{(Eq. 2)}$$

where ω denotes angular frequency, σ denotes the electrical conductivity of the material, and μ denotes the permeability of the material (usually taken as that of free space).

Thus lower frequency signals penetrate deeper into the Earth and information on deep Earth conductivity depends on them. A natural limit to the lowest frequencies which can be observed is imposed by the spectrum of the source fields. Commonly the longest period studied is 1 year, exploiting a periodicity which has its origin in the seasonal cycle of Earth. A longer period of 11 years, which has its origin in a periodicity seen in solar activity, is also sometimes used. A limit may also be imposed by the length of observatory record needed to obtain data for these lowest frequencies.

If the Earth had ideal spherical symmetry, and its properties (including electrical conductivity) varied with depth only, the analysis of observations to give structure would be more straightforward. This situation, termed one-dimensional (1D), has mathematics which, while far from simple, are not as complicated as the next stages of complexity, which are termed two-dimensional (2D) and three-dimensional (3D) respectively (Weaver, 1994). For a 2D situation, the conductivity varies with two spatial parameters such as depth and one horizontal direction; while in a 3D situation the conductivity may vary in all directions.

Many case histories now demonstrate that departures from 1D are common in the electrical conductivity structure of the crust and upper mantle of Earth (Booker and Chave, 1989; Gough, 1989; Mareschal *et al.*, 1995). In addition to the possibility of "conductivity anomalies" caused by partial melt, other notable causes of increased conductivity are graphite and water (ELEKTB-Group, 1997).

However it is common to model the deeper regions of the Earth's mantle as 1D. This action is taken on the grounds that the material is expected to be more homogeneous than the more easily inspected crust.

An example of global conductivity profiles thus obtained by inverting surface observations are those of Olsen (1999) shown in Figure C15.

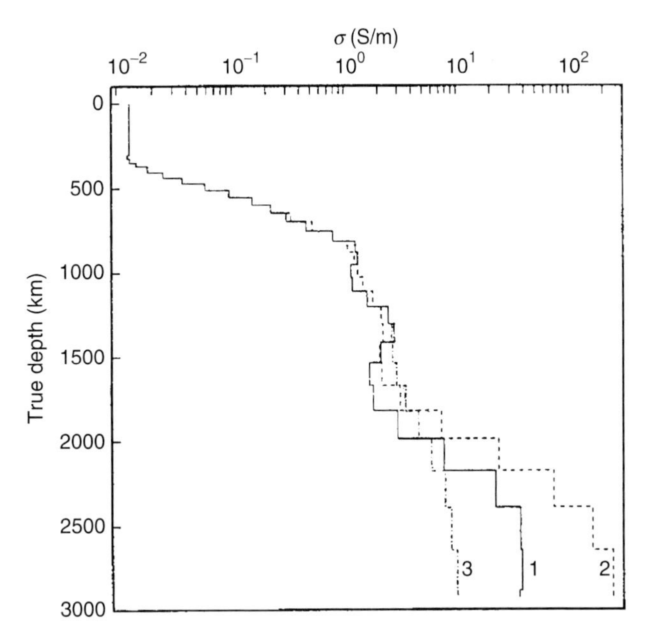

Figure C15 Profiles of electrical conductivity of the Earth as a function of depth, after Olsen (1999). The different profiles result from different data inversion procedures.

In this case, as in others, the mantle-core boundary is taken to be a step to effectively infinite conductivity, due to the molten metallic material of the Earth's core.

Determining temperature from electrical conductivity

When a conductivity profile with depth has been obtained, there is then the question regarding what mineral data should be used to convert the conductivity data to temperature. While there is general agreement about the main mineralogical composition of Earth, some fine points, upon which conductivity is highly dependent, remain uncertain. Prime amongst these is the oxygen state, in Earth's mantle, of the mineral olivine (Shankland and Duba, 1990; Constable and Roberts, 1997).

Figure C16 from Shankland (1981) combines data on rock and mineral conductivities. It shows firstly that electrical conductivity is an Earth quantity which varies by orders of magnitude in common rock materials. Then, within the range of conductivity values shown, the two lines (on the log-log graph) for synthetic "hot-pressed olivine" give an indication of the range of values in Eq. (1) possible when inverting conductivity data to give an "electrogeotherm" (and see Xu and Shankland (1999) for more recent experimental results). Such inversions are made on the basis (generally accepted by mineralogists) that the mantle of Earth is comprised mainly of olivine.

A number of research workers have thus published temperature profiles for the Earth based on analyses of electromagnetic induction at Earth's surface. The early result of Duba (1976), shown in Figure C17, is instructive regarding the general magnitudes of the quantities involved. Note that Constable (1993) explores the possibility that the temperature at 410 km is less than as shown in Figure C17, for consistency with experimental evidence that a phase transition at that depth should occur at 1400°C. The electrogeotherm models of Constable (1993), incorporating results from Constable *et al.* (1992), are shown in Figure C18.

It is important to note also that at the high temperatures near the base of the mantle, Eq. (1) becomes less temperature dependent, as its asymptotic form for high temperatures is approached. Thus electrical conductivity, even if known, becomes less useful in predicting temperature at such high temperatures.

Conclusion

Increasing research into the various elements needed to construct an electrical conductivity geothermometer has produced the result (common in research) that the matter is more complicated than might have at first been expected. While increased computer power enables continued progress in the numerical modeling of complicated conductivity structures and of the inversion of observed data in terms of them

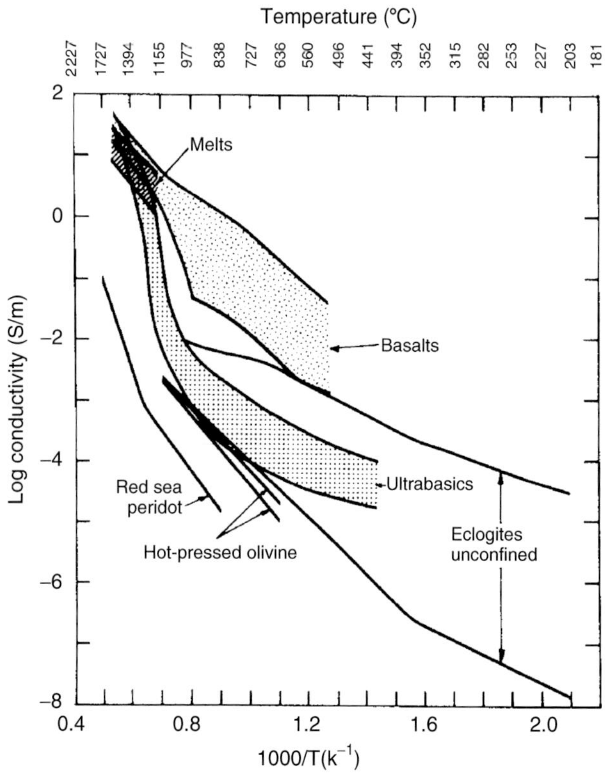

Figure C16 Conductivities of possible upper mantle constituents (after Shankland (1981) and Rai and Manghani (1978)).

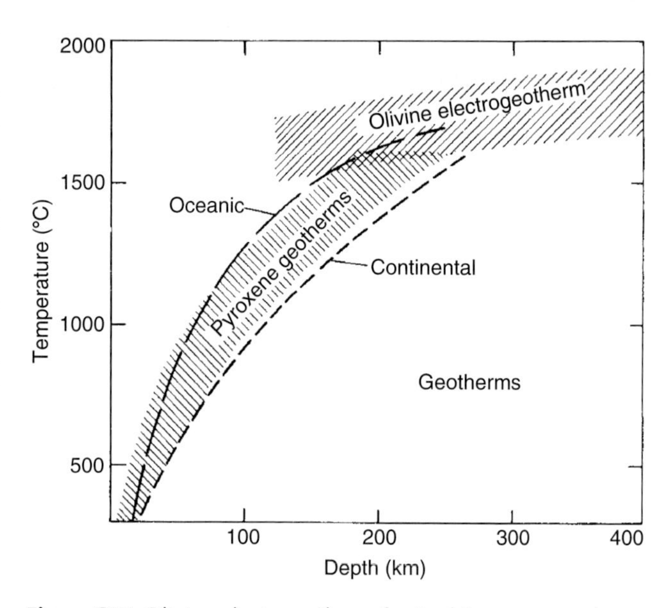

Figure C17 Olivine electrogeotherm for Earth's upper mantle compared with geotherms determined petrologically (after Duba, 1976).

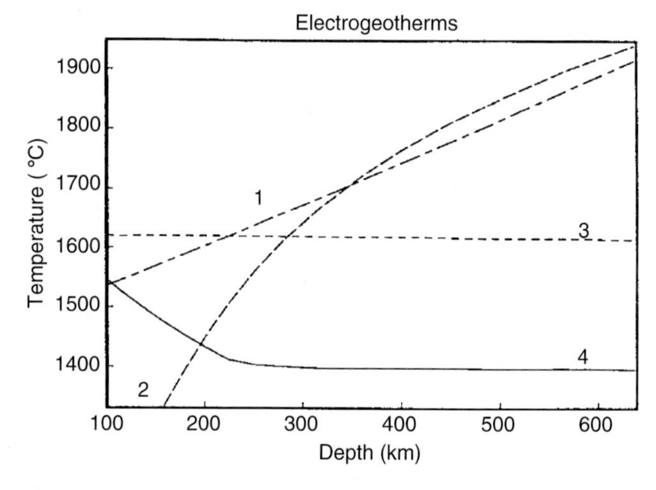

Figure C18 Upper mantle electrogeotherms, after Constable (1993). The different curves result from different data inversion techniques, especially regarding a possible step increase in electrical conductivity at depth 600 km.

(Fujii and Schultz, 2002), uncertainty about the details of the main constituent of the mantle (olivine, and in particular its oxygen fugacity) remain and limit the application of the strategy.

In fact at mantle depths the relationship between electrical conductivity and temperature is increasingly being examined from another view point. Rather than assume that the material in the mantle is known, it is increasingly the case that the temperature is estimated from other considerations. Then, such temperature estimates are used with laboratory measurements of different minerals to determine, from electrical conductivity profiles, the mineralogy of the mantle (Shankland *et al.*, 1993; Constable, 1993).

Ted Lilley

Bibliography

Booker, J.R., and Chave, A.D., 1989. Introduction to special section on the EMSLAB—Juan de Fuca experiment. *Journal of Geophysical Research*, **94**: 14093–14098.

Campbell, W.H., 1997. *Introduction to Geomagnetic Fields.* Cambridge: Cambridge University Press.

Constable, S., Shankland, T.J., and Duba, A., 1992. The electrical conductivity of an isotropic olivine mantle. *Journal of Geophysical Research*, **97**: 3397–3404.

Constable, S.C., 1993. Constraints on mantle electrical conductivity from field and laboratory measurements. *Journal of Geomagnetism and Geoelectricity*, **45**: 707–728.

Constable, S.C., and Roberts, J.J., 1997. Simultaneous modeling of thermopower and electrical conduction in olivine. *Physics and Chemistry of Minerals*, **24**: 319–325.

Duba, A., 1976. Are laboratory electrical conductivity data relevant to the Earth? *Acta Geodaetica, Geophysica et Montanistica of the Academy of Sciences of Hungary*, **11**: 485–495.

ELEKTB-Group, 1997. KTB and the electrical conductivity of the crust. *Journal of Geophysical Research*, **102**: 18289–18305.

Fujii, I., and Schultz, A., 2002. The 3D electromagnetic response of the Earth to ring current and auroral oval excitation. *Geophysical Journal International*, **151**: 689–709.

Gough, D.I., 1989. Magnetometer array studies, earth structure, and tectonic processes. *Reviews of Geophysics*, **27**: 141–157.

Hinze, E., 1982. Laboratory electrical measurements on mantle relevant minerals. *Geophysical Surveys*, **4**: 337–352.

Mareschal, M., Kellett, R.L., Kurtz, R.D., Ludden, J.N., Ji, S., and Bailey, R.C., 1995. Archaen cratonic roots, mantle shear zones and deep electrical anisotropy. *Nature*, **375**: 134–137.

Olsen, N., 1999. Long-period (30 days – 1 year) electromagnetic sounding and the electrical conductivity of the lower mantle beneath Europe. *Geophysical Journal International*, **138**: 179–187.

Parkinson, W.D., and Hutton, V.R.S., 1989. The electrical conductivity of the Earth. In Jacobs, J.A., (ed.), *Geomagnetism.* Vol. 3, Academic Press Inc., pp. 261–321.

Rai, C.S., and Manghani, M.H., 1978. Electrical conductivity of ultramafic rocks to 1820 kelvin. *Physics of the Earth and Planetary Interiors*, **17**: 6–13.

Shankland, T.J., 1981. Electrical conduction in mantle materials. In O'Connell, R.J., and Fyfe, W.S., (eds.), *Evolution of the Earth.* Vol. 5 of Geodynamics Series. Washington, D.C.: AGU, pp. 256–263.

Shankland, T.J., and Duba, A.G., 1990. Standard electrical conductivity of isotropic, homogeneous olivine in the temperature range 1200–1500°C. *Geophysical Journal International*, **103**: 25–31.

Shankland, T.J., O'Connell, R.J., and Waff, H.S., 1981. Geophysical constraints on partial melt in the Upper Mantle. *Reviews of Geophysics and Space Physics*, **19**: 394–406.

Shankland, T.J., Peyronneau, J., and Poirer, J.P., 1993. Electrical conductivity of the Earth's lower mantle. *Nature*, **366**: 453–455.

Weaver, J.T., 1994. *Mathematical Methods for Geo-electromagnetic Induction.* Baldock, UK: Research Studies Press.

Weaver, J.T., 1999. Collective review papers presented at the 14th Workshop on Electromagnetic Induction in the Earth. *Surveys in Geophysics*, **20**: 197–200.

Xu, Y., and Shankland, T.J., 1999. Electrical conductivity of orthopyroxene and its high pressure phases. *Geophysical Research Letters*, **26**: 2645–2648.

Cross-references

Conductivity, Ocean Floor Measurements
Geomagnetic Deep Sounding
Magnetotellurics
Mantle, Electrical Conductivity, Mineralogy

CONDUCTIVITY, OCEAN FLOOR MEASUREMENTS

The ocean floor presents a particularly harsh environment in which to carry out electrical measurements, with pressures of up to 600 atm (60 MPa), temperatures of around 3°C, and no possibility of radio contact with instrumentation. Furthermore, seawater is a corrosive, conductive fluid. Thus, progress in the field of electrical conductivity studies has largely followed the availability of reliable underwater technology and has not become truly routine until recently.

Most electromagnetic methods can be adapted for seafloor use, but the high conductivity of seawater dominates both how data are collected and how they are interpreted. Seawater conductivity depends on salinity and temperature; in practice salinity variations are too small to be significant and so to a good approximation seawater conductivity is given by $(3 + T/10)$ S/m where T is temperature in degree celsius. The bulk of the ocean thus has a resistivity of about 0.3 Ωm, with warmer surface waters 0.2 Ωm.

DC resistivity is only practical for shallow investigations where the seafloor conductivity is similar to or greater than the seawater. The most common EM method applied to seafloor studies is the *magnetotelluric* (MT) method (*q.v.*). First experiments date from the 1960s, and Charles Cox, Jean Filloux, and Jimmy Larson (1971) report an MT response from measurements made in the Pacific in 1965, using 1 km long cables on the seafloor and a new seafloor magnetometer developed by Filloux. Filloux later also developed a system for making electric field measurements using short (about 3 m) pipes acting as salt bridges and a device to reverse the connection of electrodes to the pipes, allowing any electrode self potential to be removed from long period electric field variations (Filloux, 1987). Although this technique is still important for studies of ocean currents using seafloor *E*-field recorders, it has since proved unnecessary for MT studies, and it is now common practice to simply mount electrodes at the ends of four approximately 5 m plastic arms to give 10 m *E*-field dipoles. The torsion fiber magnetometers originally used by Filloux to keep power consumption low have been replaced in more recent instrumentation by fluxgate sensors and induction coils.

There are advantages and disadvantages to the seafloor MT method. On the advantage side, it is easy to make a low impedance, low noise electrical contact with the environment. The sensor of choice for this is silver-silver chloride nonpolarizable electrodes although a new carbon fiber electrode has been developed for short period studies. The seafloor is also free of the cultural noise that can plague land MT surveys. Access is good and permitting, if required, is usually valid for the entire survey area. Arrays of seafloor MT recorders lend themselves well to the new array processing techniques available.

On the other hand, the skin depth (exponential decay length for EM fields) at 1 Hz in seawater is only 270 m and so the overlying ocean removes the short period source fields, a problem exacerbated by the red nature of the geomagnetic spectrum (see *Geomagnetic temporal spectrum*). Depending on water depth and the type of instrumentation, data are limited to periods longer than 10 to 1000 seconds. Water motion produces a Lorenz ($\mathbf{E} = \mathbf{V} \times \mathbf{B}$) field that is prominent at tidal periods and eventually dominates the induced fields (see *Electromagnetic ocean effects*). Thus, seafloor MT is ultimately band limited. The recent use of induction coil *magnetometers* (*q.v.*) on marine MT instruments (Constable *et al.*, 1998), coupled with high gain, AC-coupled electric field amplifiers, optimizes the collection of higher frequency data, but even with this system one is limited to about 10 s period data on the deep seafloor (in 4–6 km water).

The geomagnetic *coast effect* (*q.v.*) is very much more severe on the ocean side. Electric charge accumulating on coastlines from the MT fields leak into the conductive part of the mantle over a boundary layer distance L given approximately by $\sqrt{ST}$ (Cox, 1980), where S is the conductance of the oceans and T is the resistivity-thickness product of the uppermost seafloor rocks. We know from CSEM studies that T is about 10^9 Ωm^2, so long period MT anomalies from the coast effect are attenuated horizontally over an L of 3000 km. Undoubtedly, there is electrical connection between the ocean and conductive mantle at volcanic regions and possibly at subduction zones, but at passive margins the coast effect is so strong that T can be estimated from land MT data and global induction studies. For these reasons, it is important in marine MT interpretations include coastlines and bathymetry during modeling and interpretation.

In the late 1970s, Charles Cox (1981) proposed replacing the high-frequency energy lost to the magnetotelluric signal with a deep towed human-made transmitter, and carried out the first experiments in 1979 (Spiess *et al.*, 1980). This controlled source electromagnetic (CSEM) sounding method complements the MT method, because not only is it sensitive to shallower structure, but also it is preferentially sensitive to resistive seafloor, whereas the MT signal is induced in conductive rocks. In the most common application of this technique, a horizontal electric transmitting dipole is towed close to the seafloor, and the resulting electric fields are measured as a function of range and frequency by a number of seafloor electric field recorders. At a typical transmission frequency of 1 Hz, the skin depth in seawater is so small that receivers placed more than a few kilometers from the transmitter record only signals that have propagated through the more resistive seafloor. For two decades the marine CSEM method remained an exotic technique used to study deep seafloor and mid-ocean ridges, but during the first years of the 21st century it was enthusiastically embraced by the petroleum industry. Hydrocarbon reservoirs are relatively resistive (10–100 Ωm) compared to the marine sediments in which they are often found (around 1 Ωm), and this resistivity contrast combined with CSEM sounding's ability to detect thin resistive layers makes the method useful for mapping oil and gas reservoirs (Constable and Weiss, 2005).

Nigel Edwards (1985) and colleagues developed a marine application of a land technique called magnetometric resistivity (MMR) sounding, in which a bipolar vertical electric field transmitter extends from the ship to near the seafloor. A low-frequency current is transmitted and the resulting magnetic fields are measured by seafloor recorders some distance away. Since the essentially DC current flow from the transmitter is distorted by the resistive seafloor rocks, the horizontal magnetic fields become a measure of seafloor conductivity.

Figure C19 illustrates the three types of measurements discussed here. The recorders are essentially the same in every case, except that for long-period MT measurements a fluxgate magnetometer would supplement or replace the induction coil sensors. Figure C20 illustrates the current understanding of oceanic conductivity structure. Seawater, as noted, is around 0.3 Ωm except above the thermocline near the surface. The upper 400–600 m of oceanic crust is composed of extrusive volcanics, lava flows, and pillow lavas. Corresponding to seismic layer 2, these rocks are highly porous and have resistivities between 1 and 10 Ωm. They are more conductive at mid-ocean ridges (MORs) because pore-water is heated by volcanic action. Any sediments overlaying the basalts will have a resistivity between that of seawater and about 1 Ωm.

At a depth of about 500 m, porosity, permeability, and electrical conductivity all drop as the lithology changes to intrusive basaltic volcanics (dikes and sills). Away from ridges, resistivity will be a few tens of Ωm in this layer. At around 2 km below the seafloor, we find the gabbros that make up most of the crust. Away from the MOR axes, resistivity increases to several tens of thousands of Ωm by the time mantle depths are reached; whether this increase occurs at the top of the gabbroic layer, or deeper in the crust, is hard to determine because resistivity is monotonically increasing below the seafloor and the resolving kernels of EM methods are intrinsically smooth. At fast MOR spreading centers, and at currently active places on slowly spreading ridges, this layer contains the magma chamber that fuels crustal accretion. Resistivity within the magma chamber will be between that of basaltic magma (0.3 Ωm) and about 30 Ωm, depending on melt content.

Off-axis, crustal conductivity is simply a reflection of porosity, and any conductivity contrast at the moho, which would typically be below 600°C, will be small, and the lithospheric mantle, like the lower crust, is many thousands of Ωm. It is possible that the oceanic upper mantle, depleted in volatiles during crustal formation at mid-ocean ridges, is about an order of magnitude more resistive than continental mantle. At the base of the thermal lithosphere (from 6 to 100 km deep, depending on plate age), conductivity increases to between 100 and 500 Ωm consistent with the electrical conductivity of dry subsolidus

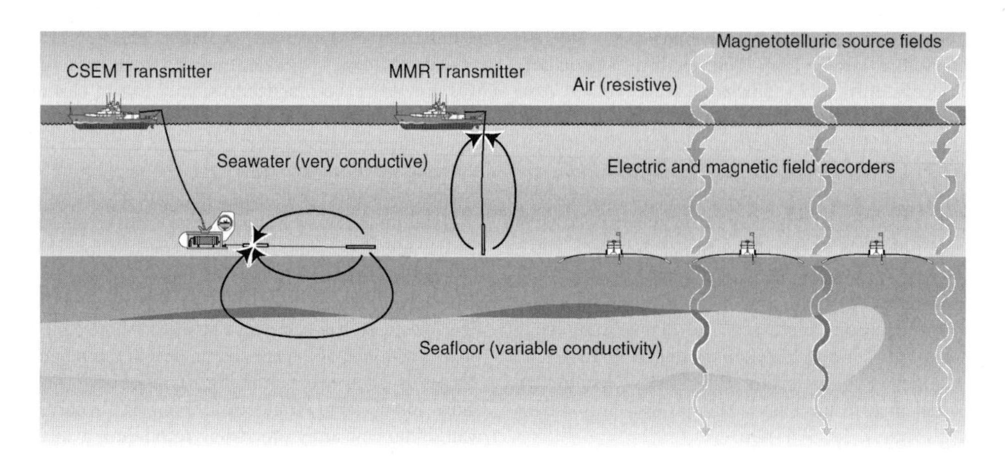

Figure C19 Experimental layout for seafloor-controlled source EM (CSEM) sounding (*left*), magnetometric resistivity (MMR) sounding (*center*), and magnetotelluric (MT) sounding (*right*). In all cases the receiver instruments are very much the same; for MT sounding a CSEM or MMR transmitter is not required.

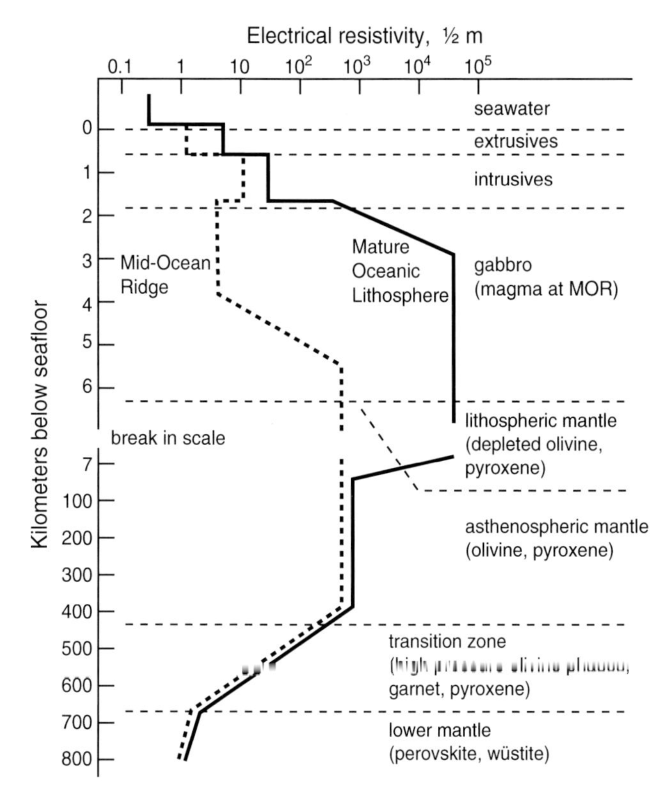

Figure C20 Approximate seafloor conductivity profiles below a mid-ocean ridge (*broken line*) and older, normal oceanic lithosphere (*solid line*).

olivine near its melting point, or perhaps a little more conductive (see *Conductivity geothermometer*). Many early marine MT studies reported a highly conductive layer thought to correspond to the asthenosphere between about 100 and 200 km deep. Although there may well be a conductivity anomaly associated with the base of the lithosphere, the very high conductivities originally modeled are probably an artifact of using one-dimensional interpretation on data that are affected by the *coast effect* (*q.v.*). The controversy that once surrounded this issue seems to be abating, however, and more recent multisite, multidimensional studies often include the coastlines explicitly and tend not to require an asthenospheric conducting layer.

Both laboratory studies of high pressure mantle minerals (see *Mantle, Electrical conductivity, Mineralogy*) and induction studies using long-period magnetic observatory records (see *Geomagnetic deep sounding*) show that the lower mantle (believed to be composed of silicate perovskite and mangnesio-wüstite) is very much more conductive than the upper mantle and is about 0.2–0.5 Ωm. Whether this conductivity increase occurs entirely at 670 km or starts shallower in the transition zone is unclear. Recent laboratory measurements on the high-pressure phases of olivine suggest that conductivity should increase at 440 km and some induction studies support this. However, the jump in conductivity at 670 km is so large that smooth inversions for mantle conductivity inevitably exhibit an increase starting at shallower depths.

Steven Constable

Bibliography

Constable, S., Orange, A., Hoversten, G.M., and Morrison, H.F., 1998. Marine magnetotellurics for petroleum exploration 1. A seafloor instrument system. *Geophysics*, **63**: 816–825.

Constable, S., and Weiss, C.J., 2005. Mapping thin resistors (and hydrocarbons) with marine EM methods: Insights from 1D modeling. *Geophysics*, **71**: 643–651.

Cox, C.S., 1980. Electromagnetic induction in the oceans and inferences on the constitution of the Earth. *Geophysical Survey*, **4**: 137–156.

Cox, C.S., 1981. On the electrical conductivity of the oceanic lithosphere. *Physics of the Earth and Planetary Interiors*, **25**: 196–201.

Cox, C.S., Filloux, J.H., and Larsen, J.C., 1971. Electromagnetic studies of ocean currents and electrical conductivity below the ocean-floor. In A.E. Maxwell (ed.), *The Sea*. New York: Wiley, pp. 637–693.

Edwards, R.N., Law, L.K., Wolfgram, P.A., Nobes, D.C., Bone, M.N., Trigg, D.F., and DeLaurier, J.M., 1985. First results of the MOSES experiment: Sea sediment conductivity and thickness determination, Bute Inlet, British Columbia, by magnetometric off-shore electrical sounding. *Geophysics*, **50**: 153–160.

Filloux, J.H., 1987. Instrumentation and experimental methods for oceanic studies. In J.A. Jacobs (ed.), *Geomagnetism*, London: Academic Press, pp. 143–248.

Spiess, F.N., Macdonald, K.C., Atwater, T., Ballard, R., Carranza, A., Cordoba, D., Cox, C., Diaz Garcia, V.M., Francheteau, J., Guerrero, J., Hawkings, J., Haymon, R., Hessler, R., Juteau, T., Kastner, M., Larson, R., Luyendyk, B., Macdougall, J.D., Miller, S., Normark, W. Orcutt, J., and Rangin, C., 1980. East Pacific Rise: hot springs and geophysical experiments. *Science*, **207**: 1421–1433.

Cross-references

Coast Effect of Induced Currents
Conductivity Geothermometer
EM, Industrial Uses
EM, Marine Controlled Source
Geomagnetic Deep Sounding
Geomagnetic Spectrum, Temporal
Magnetotellurics
Mantle, Electrical Conductivity, Mineralogy
Ocean, Electromagnetic Effects

CONVECTION, CHEMICAL

Convection is the differential motion which occurs due to the action of the gravitational body force on density inhomogeneities within a fluid body. The density of a fluid body is a function of its pressure, temperature, and chemical composition. Pressure variations cannot cause convection; they play a secondary role, at most. "Chemical convection" refers to motions which arise due to spatial variations in density arising from variations in the chemical content of the fluid; an alternate and equivalent name is "compositional convection." Similarly, motions which arise due to spatial variations in temperature are called "thermal convection."

The study of convection is facilitated by assuming that the density, ρ, is independent of pressure and is a linear function of both temperature, T, and chemical content (i.e., mass or volume fraction of one of two chemical constituents), C; for example,

$$\rho = \rho_0 - \rho_0\alpha[T - T_0] - \rho_0\beta[C - C_0],$$

where α is the coefficient of thermal expansion, β is the compositional expansion coefficient, and a subscript 0 denotes a constant (reference) value. For normal fluids $\alpha > 0$ (hot fluid is lighter than cold), while the sign of β depends on whether C measures the fraction of denser or less dense constituent. Commonly, C is taken to quantify the less dense constituent and then $\beta > 0$. In the core, C quantifies the nonmetallic constituent and β is of unit order.

If the surfaces of constant density are not perpendicular to the local acceleration of gravity, convection invariably occurs. On the other hand, if these surfaces are everywhere perpendicular to gravity, then a quiescent state is always possible. An important question is whether convection can and will occur in such a fluid body. The fluid is stable if all perturbations of arbitrary form and amplitude decay with time. On the other hand, if any small perturbation of the quiescent state leads to motions which increase in amplitude with time, the fluid is said to be (convectively) unstable.

The tendency for a fluid body to convect is measured by the Rayleigh number:

$$Ra = \frac{(\Delta\rho)gh^3}{\rho_0 \nu\kappa}$$

where g is the acceleration of gravity, h is the vertical extent of the body, $\Delta\rho$ is a measure of the density perturbations (excluding those due to pressure), ν is the kinematic viscosity of the fluid, and κ is the diffusivity of the factor causing density variations (either temperature or composition). Often the symbol D is used in place of κ when variations are due to composition. Commonly $\Delta\rho$ is defined such that a positive value denotes lighter fluid beneath denser and a negative value denotes denser fluid beneath lighter.

There is a dynamic similarity between the situation in which the density of the fluid depends only on the relative proportions of two chemical constituents of the fluid (with $\Delta\rho/\rho_0 = \beta\Delta C$) and that in which density depends only on the temperature (with $\Delta\rho/\rho_0 = \alpha\Delta T$). In either case the fluid is convectively unstable when the Rayleigh number exceeds a critical value, the magnitude of which depends on the shape and nature of the boundaries of the fluid body (see Chandrasekhar, 1961). If the density is a function of both temperature and composition, then the stability of the fluid body depends on two Rayleigh numbers, measuring the density changes due to these two variables, and is significantly more complicated (see Turner, 1974).

The existence of Earth's magnetic field is very strong evidence that convection occurs in the outer core, as there is no plausible explanation for the origin of this field other than dynamo action driven by convective motions (see *Geodynamo*). The outer core convects because it is cooled and the principal sources of buoyancy are the latent heat and light material released as the inner core grows by solidification of the denser component of the liquid outer core (see *Geodynamo, energy sources* and *Core convection*). That is, convection is likely due to both thermal and chemical (or compositional) differences; both the thermal and compositional Rayleigh numbers are likely to be positive, particularly near the inner-core boundary.

Chemical convection of another sort can occur within the uppermost portion of the inner core. It is likely that the inner core is not completely solid, but rather a mixture of solid and liquid, sometimes called a *mush*. The solid and liquid phases are in phase equilibrium, placing a constraint on the temperature and composition of the liquid phase. Ignoring pressure effects, this may be characterized by a simple linear liquidus relation:

$$T = T_0 - \Gamma[C - C_0],$$

where Γ is a constant quantifying the rate at which the temperature of pure iron is decreased by addition of (nonmetallic) impurities in the core. Using this constraint to eliminate composition from the equation for density,

$$\rho = \rho_0 - \rho_0\left[\alpha - \frac{\beta}{\Gamma}\right][T - T_0].$$

For almost all materials, the change of density with composition is sufficiently large that $\beta > \alpha\Gamma$. In this case, the normal density-temperature relation is reversed; in a mush, cold fluid is less dense than warm.

In the core, pressure effects are important; and in the above formulas, one must replace the temperature by the potential temperature, or equivalently, normalize the temperature with the adiabatic temperature (see *Core, adiabatic gradient*). While the actual temperature in the core increases with depth, the normalized temperature decreases with depth. (This is why the outer core solidifies at the bottom, rather than the top.) The point to be emphasized is that this situation prevails in the mush at the top of the inner core and the liquid phase is prone to convective instability. Within the mush, thermal and chemical effects are no longer independent and it is not proper to consider thermal and chemical convections separately. However, given the dominance of density changes due to chemical differences, convection is properly characterized as chemical.

Chemical convection within mushy zones is commonly observed in the casting of metallic alloys (e.g., see Worster, 1997). The situation in Earth's core is simulated when an alloy is cooled from below and the denser phase freezes first. In this situation, convection within the mush has a curious structure. Upward motion induces melting of the crystals of the solid phase, thereby reducing the resistance to flow. This provides a positive feedback leading to the formation of discrete chimneys in the mush. The cold, chemically buoyant fluid in the mush convects by moving laterally to and rising up these chimneys; mass is conserved by a downward return flow of warm, chemically denser fluid into the mush away from the chimneys. It is uncertain whether chimney convection occurs in the core, where both Coriolis and Lorentz forces act on the moving fluid.

David Loper

Bibliography

Chandrasekhar, S., 1961. *Hydrodynamic and Hydromagnetic Stability*, London: Oxford University Press.

Turner, J.S., 1974. Double-diffusive phenomena, *Annual Review of Fluid Mechanics*, **6**: 37–56.

Worster, M.G., 1997. Convection in mushy layers, *Annual Review of Fluid Mechanics*, **29**: 91–122.

Cross-references

Core Convection
Core, Adiabatic Gradient
D″ and F layers of the Earth
Geodynamo
Geodynamo, Energy Sources

CONVECTION, NONMAGNETIC ROTATING

Various physical processes are responsible for maintaining motion, possibly turbulent, in the Earth's fluid core (see *Core motions*), which is ultimately responsible for the *geodynamo* (*q.v.*). Here, however, we focus attention on thermal convection (see *Core convection*) driven by an unstable temperature gradient. Compressibility will be ignored except where density variations lead to buoyancy forces, the so-called *Boussinesq approximation* (*q.v.*). The basic model (Chandrasekhar, 1961) generally adopted in theoretical studies concerns a self-gravitating sphere, gravitational acceleration $-g\mathbf{r}$, in which fluid is confined to a spherical shell, inner radius r_i, and outer radius r_o. Relative to cylindrical polar coordinates (s, ϕ, z), the system rotates rapidly about the z-axis with constant angular $\boldsymbol{\Omega} = \Omega\hat{\mathbf{z}} \equiv (0, 0, \Omega)$. The Boussinesq fluid, density ρ, kinematic viscosity ν, has coefficient of thermal expansion α and thermal diffusivity κ. The fluid is assumed to contain a uniform distribution of heat sources with thermal boundary conditions, which leads to a spherically symmetric temperature exhibiting an adverse radial temperature gradient $-\beta\mathbf{r}$.

The basic static equilibrium state just described may be buoyantly unstable. The ensuing fluid motion, velocity $\mathbf{u}$, causes temperature perturbations T of the basic state (see *Core temperature*), which satisfy the heat conduction equation

$$D_t T = \beta \mathbf{r} \cdot \mathbf{u} + \kappa \nabla^2 T, \qquad \text{(Eq. 1a)}$$

where $D_t \equiv \partial/\partial t + \mathbf{u} \cdot \nabla$ is the material derivative: here β is a constant for the case of heating by a uniform distribution of heat sources in a full sphere but may be a function of the radius r for more general situations. In turn, the ensuing density variations $-\alpha\rho T$ drive the motion itself, which is governed by

$$D_t \mathbf{u} + 2\boldsymbol{\Omega} \times \mathbf{u} = -\nabla(p/\rho) + g\alpha T \mathbf{r} + \nu \nabla^2 \mathbf{u}$$

$$\text{with} \quad \nabla \cdot \mathbf{u} = 0, \qquad \text{(Eq. 1b)}$$

where p is the fluid pressure, and both ρ and g are constants. The vorticity equation obtained by taking its curl is

$$D_t \boldsymbol{\zeta} - (2\boldsymbol{\Omega} + \boldsymbol{\zeta}) \cdot \nabla \mathbf{u} = -g\alpha \mathbf{r} \times \nabla T + \nu \nabla^2 \boldsymbol{\zeta}, \qquad \text{(Eq. 1c)}$$

where $\boldsymbol{\zeta} \equiv \nabla \times \mathbf{u}$ is the relative vorticity ($2\boldsymbol{\Omega} + \boldsymbol{\zeta}$ is the absolute vorticity). Generally the inner and outer spherical boundaries are regarded as isothermal ($T = 0$) and impermeable ($\mathbf{r} \cdot \mathbf{u} = 0$), whereas both the cases of stress-free and no-slip boundaries are often considered separately.

The physical system is characterized by four dimensionless numbers, namely the aspect ratio η, the Ekman number, the Prandtl number, and the Rayleigh number:

$$\eta = r_\mathrm{i}/r_\mathrm{o}, \quad E = \nu/2\Omega r_\mathrm{o}^2, \quad P = \nu/\kappa, \quad R = g\alpha\beta r_\mathrm{o}^6/\kappa\nu \qquad \text{(Eq. 2)}$$

These parameters arise naturally for the linear convection that occurs at the onset of instability. For a given system such as the Earth's core, we must regard η, E, and P as given. Interest then focuses on what happens with increasing R. The initial bifurcation occurs at the so-called critical Rayleigh number R_c. The strength of the finite amplitude that ensues is measured by some convenient parameter such as the Rossby number $U_0/r_\mathrm{o}\Omega$ based on a typical velocity U_0.

The onset of instability

The linear problem for the onset of instability has a long history. The early studies (Chandrasekhar, 1961) focused on axisymmetric convection. Later it was realized (Roberts, 1968) that, in the geophysically interesting limit of small Ekman number ($E \ll 1$), the onset is characterized by nonaxisymmetric wave-like modes proportional to $\exp \mathrm{i}(m\phi - \omega t)$ with $m = \mathcal{O}(E^{-1/3})$ and $(r_\mathrm{o}^2/\nu)\omega = \mathcal{O}(E^{-2/3})$; they occur when $R = \mathcal{O}(E^{-4/3})$.

Since the rotation (as measured by the smallness of E) is so rapid, the flow is almost geostrophic; $2\boldsymbol{\Omega} \times \mathbf{u} \approx \nabla(p/\rho)$ (see Eq. (1b)). True geostrophic flow satisfies the *Proudman-Taylor theorem* (*q.v.*) and is independent of z, i.e., $2\boldsymbol{\Omega} \cdot \nabla \mathbf{u} = 0$ (see Eq. (1c)). In a spherical shell, geostrophic flow is purely azimuthal and not convective. The system overcomes this difficulty by occurring on the small $\mathcal{O}(E^{1/3} r_\mathrm{o})$ azimuthal length scale upon which it is quasigeostrophic (i.e., independent of z on the $\mathcal{O}(E^{1/3} r_\mathrm{o})$ length scale of the convection but dependent on z on the $\mathcal{O}(r_\mathrm{o})$ length scale of the core radius):

$$p \approx -2\rho\Omega\Psi, \quad \mathbf{u} \approx \nabla \times \Psi\hat{\mathbf{z}} + W\hat{\mathbf{z}}, \quad \boldsymbol{\zeta} \approx -(\nabla_\perp^2 \Psi)\hat{\mathbf{z}} + \nabla \times W\hat{\mathbf{z}}, \qquad \text{(Eq. 3)}$$

where $\nabla_\perp^2 \equiv \nabla^2 - \partial^2/\partial z^2$. Note that $\nabla \cdot \mathbf{u} = 0$ does not imply that $\partial W/\partial z = 0$. Instead it simply means that there are small extra components of $\mathbf{u}$, orthogonal to $\hat{\mathbf{z}}$, of order $E^{1/3} W$ ignored in Eq. (3). Similar remarks apply to the representation of $\boldsymbol{\zeta}$. Within the framework of these approximations, the linearized form of the z-components of the vorticity Eq. (1c) and the equation of motion (1b) are

$$(\partial_t - \nu\nabla_\perp^2)\nabla_\perp^2 \Psi + 2\Omega\partial_z W = g\alpha\partial_\phi T \qquad \text{(Eq. 4a)}$$

$$(\partial_t - \nu\nabla_\perp^2) W - 2\Omega\partial_z \Psi = g\alpha z T, \qquad \text{(Eq. 4b)}$$

respectively, where $\partial_t \equiv \partial/\partial t$ etc. The corresponding form of the heat conduction Eq. (1a) is

$$(\partial_t - \kappa\nabla_\perp^2) T = \beta(\partial_\phi \Psi + z W). \qquad \text{(Eq. 4c)}$$

Correct to leading order it is sufficient to solve the system (4) subject to the impermeable boundary condition

$$\mathbf{r} \cdot \mathbf{u} \approx \partial_\phi \Psi + z W = 0 \quad \text{on} \quad r = r_\mathrm{i} \text{ and } r_\mathrm{o}, \qquad \text{(Eq. 5)}$$

namely the inner and outer boundaries.

In the early solutions of Eq. (4) for a full sphere, it was assumed that motion is localized in the vicinity of some cylinder $s = s_\mathrm{c}$ on a short radial scale, yet long compared to the azimuthal length scale $\mathcal{O}(s_c E^{1/3})$. Accordingly, the approximation $\nabla_\perp^2 = s^{-2}\partial_\phi^2$ was also made (Roberts, 1968; Busse, 1970). The localized convection has a columnar structure, for which the evolution of its axial vorticity $-\nabla_\perp^2 \Psi$ is governed by Eq. (4a). The buoyancy torques $-g\alpha\partial_\phi T$ that drive motion are most effective near the equator, where the radial component of gravity is strongest. Nevertheless, the large tilt of the boundaries at the ends of the column necessarily break the constraints of the Proudman-Taylor theorem and inhibit motion. This effect is quantified by the term $2\Omega\partial_z W$, which describes the stretching of absolute vorticity and is minimized at the poles where the boundary tilt vanishes. In consequence, the preferred location of the convection is at mid latitudes, where $s_\mathrm{c}/r_\mathrm{o}$ is roughly a half.

A heuristic appreciation of the solution is obtained by making the geostrophic ansatz that Ψ and T are independent of z (i.e., assume Eq. (4b) is simply $2\Omega\partial_z \Psi = 0$) and that W is linear in z, which according to the boundary condition (5) implies $W = -(z/h^2)\partial_\phi \Psi$, where $h = (r_\mathrm{o}^2 - s^2)^{1/2}$ is the column half height. Linked to these heuristic assumptions is the neglect of W in Eq. (4c), which then leaves the model equations (Busse, 1970)

$$(\partial_t - \nu\nabla_\perp^2)\nabla_\perp^2 \Psi - 2(\Omega/h^2)\partial_\phi \Psi = g\alpha\partial_\phi T, \qquad \text{(Eq. 6a)}$$

$$(\partial_t - \kappa\nabla_\perp^2) T = \beta\partial_\phi \Psi \qquad \text{(Eq. 6b)}$$

In the absence of viscosity $\nu = 0$ and buoyancy forces $g\alpha\beta = 0$, Eq. (6a) possesses eastward propagating Rossby wave solutions, which under the long radial length assumption $\nabla_\perp^2 = s^{-2}\partial_\phi^2$ have frequency $\omega = 2\Omega s^2/h^2 m$. When dissipation and buoyancy are reinstated, the dispersion relation becomes

$$(P\tilde{\omega} + \mathrm{i}\kappa^2)[(\tilde{\omega} + \mathrm{i}\kappa^2) - E^{-1}\tilde{s}(1 - \tilde{s}^2)^{-1}/k + \tilde{s}^2 R = 0, \qquad \text{(Eq. 7)}$$

where we have introduced the dimensionless parameters $\tilde{\omega} = r_\mathrm{o}^2\omega/\nu$, $k = r_\mathrm{o} m/s$ and $\tilde{s} = s/r_\mathrm{o}$. The onset of convection occurs at the minimum value of R over $\tilde{s}$ and k subject to the constraint that $\tilde{\omega}$ is real. This recovers our earlier critical value estimates $\tilde{s}_\mathrm{c} \approx 0.5$, $\tilde{k}_\mathrm{c} = \mathcal{O}(E^{-1/3})$, $\tilde{\omega}_\mathrm{c} = \mathcal{O}(E^{-2/3})$, and $R_\mathrm{c} = \mathcal{O}(E^{-4/3})$ (Busse, 1970). The marginal waves, like the Rossby waves travel eastward, and indeed have the Rossby wave character (not to be confused with an alternative class of solutions, namely the equatorially trapped inertial waves discussed later) in the small Prandtl number limit. For that reason, the waves that occur at finite P are referred to as thermal Rossby waves.

Though the system (6) only constitutes model equations in relation to the rotating self-gravitating sphere, they are correct for an ingenious rapidly rotating annulus model (Busse, 1970). The idea is that the annulus rotates so fast that the centrifugal acceleration produces an effective gravity in the radial s-direction. The top and bottom boundaries are slightly tilted to produce small axial velocities of exactly the form proposed in the heuristic model. This configuration provides the basis of many laboratory investigations (e.g., Busse and Carrigan, 1976; Carrigan and Busse, 1983; but see also *Fluid dynamics experiments*). The local solutions of both the full-sphere system (4) and annulus system (5) certainly illuminate the physical processes involved but they fail to capture the radial structure, whose resolution is essential for the proper solution of the eigenvalue problem. The correct asymptotic theory was first developed for a model system of annulus type (Yano, 1992) and later for the full-sphere Eqs. (4) and (5) (Jones *et al.*, 2000). The essential difficulty lies in the character of the local dispersion relation $\omega = \omega(s)$ at given m with all other physical parameters fixed. Since ω, now complex in general, varies with position, the angular phase velocity ω/m varies with radius s. As a result the convection rolls, which the theory assumes are elongated in the radial s-direction (to obtain the relatively long radial length scale), are in fact twisted in a prograde sense because of their tendency to propagate eastward faster with increasing s (Busse and Hood, 1982; Carrigan and Busse, 1983). A balance is in fact achieved at a finite angle of twist which asymptotic theory can predict (Jones *et al.*, 2000) together with the true critical Rayleigh number, which is larger than that obtained by simply minimizing over local values as in the Roberts-Busse theory. In the case of a spherical shell, the Jones *et al.* asymptotic theory continues to apply provided the convection is localized outside the tangent cylinder $s = r_i$ of the inner sphere, which it will be for parameter values of the Earth's core. For planets with large inner cores, however, that will no longer be true. Then convection generally occurs outside but adjacent to the inner sphere tangent cylinder at critical values predicted by local theory (Busse and Cuong, 1977). A systematic theory of this *inner core tangent cylinder* (*q.v.*) convection (Dormy *et al.*, 2004) also shows that, with alternative heating profiles exhibiting r-dependent β, such convection may occur irrespective of the inner core size.

The asymptotic theories described effectively apply to stress-free boundaries for which there are no strong additional boundary layers (see *Core, boundary layers*). When the boundaries are rigid, an Ekman layer forms on the boundary and the analysis of the system (4) applies outside the Ekman layer. The Ekman pumping condition on the outer sphere that replaces the impermeable condition is $\mathbf{r}\cdot\mathbf{u} \approx -(Er_0/2h)^{1/2}\mathbf{r}\cdot\boldsymbol{\zeta}$ on $r = r_o$ (Greenspan, 1968) and this leads to a correction to the Rayleigh number smaller by a factor $\mathcal{O}(E^{1/6})$. Numerical results (Zhang and Jones, 1993) suggest that this correction is positive (negative) for small (large) Prandtl number, with a switch over when P is roughly unity. There are other even smaller corrections arising from weak thermal boundary layers.

The validity of the asymptotic theories presented is not so clear in the limit of small Prandtl number and the double limit $E \to 0$, $P \to 0$ is not yet properly resolved. Nevertheless, numerical studies suggest that at fixed small E the onset of instability in the limit $P \to 0$ occurs in the form of equatorially trapped inertial waves with $m = \mathcal{O}(1)$ rather than as finite latitude Rossby waves with $m \gg 1$. Asymptotic theory indicates that, for $\kappa/2\Omega r_o = P^{-1}E \geq \mathcal{O}(1)$, the critical Rayleigh number R_c for this inertial wave convection is sensitive to the kinematic boundary conditions: for stress-free conditions $R_c = \mathcal{O}(1)$ (Zhang, 1994); for no-slip conditions $R_c = \mathcal{O}(E^{-1/2})$ (Zhang, 1995).

The nonlinear development

An early attempt at a nonlinear asymptotic theory (Soward, 1977) suggested that finite amplitude solutions existed close to the local critical Rayleigh number, which is somewhat smaller than the true (or global) critical value. There is recent evidence that such solutions exist but are unstable. The asymptotic theory depends strongly on the large azimuthal wave number m of the convective mode. As a consequence, the nonlinearity only generates large axisymmetric perturbations of T and $\mathbf{u}$, manifest dynamically in large azimuthal geostrophic flows and *thermal winds* (*q.v.*). These interact with the assumed convective mode and the generation of higher harmonics is ignored. Recent investigations, similar in spirit, taken well into the nonlinear regime have adopted Busse's annulus model with constant tilt boundaries (Abdulrahman *et al.*, 2000), as well as nonlinear versions of Eq. (6) adopting spherical boundaries (Morin and Dormy, 2004).

Most of the earlier numerical work on nonlinear convection at small but finite E (as opposed to the perceived asymptotic limit $E \to 0$ of the previous paragraph) was based on Busse's annulus model. The bifurcation sequence has been traced in a series of papers (see Schnaubelt and Busse, 1992, which should be contrasted with Abdulrahman *et al.*, 2000). An important primary instability is the so-called mean flow instability discussed recently in the review (Busse, 2002). The corresponding studies in spherical shells following the pioneering linear investigations (see, for example, Zhang and Busse, 1987; Zhang, 1992) are now extensive and described in review articles (Busse, 1994, 2002). As the Rayleigh number R is increased, a typical scenario is that vacillating convection develops, followed by spatial modulation and a breakdown towards chaos. Surprisingly, with further increase of R more orderly relaxation oscillations are identified (Grote and Busse, 2001). It would seem that the strong shear produced by the convection actually suppresses the convection. There is then a long period over which the shear decays slowly due to viscous damping. After this relaxation, convection begins again becoming vigorous in a relatively short time and the cycle is repeated. There is recent evidence that, for annulus type model systems (Morin and Dormy, 2003), this relaxation oscillation can occur at Rayleigh numbers close to critical in the asymptotic limit $E \to 0$.

Andrew Soward

Bibliography

Abdulrahman, A., Jones, C.A., Proctor, M.R.E., and Julien, K., 2000. Large wavenumber convection in the rotating annulus. *Geophysical and Astrophysical Fluid Dynamics*, **93**: 227–252.

Busse, F.H., 1970. Thermal instabilities in rapidly rotating systems. *Journal of Fluid Mechanics*, **44**: 441–460.

Busse, F.H., 1994. Convection driven zonal flows and vortices in the major planets. *Chaos*, **4**(2): 123–134.

Busse, F.H., 2002. Convective flows in rapidly rotating spheres and their dynamo action. *Physics of Fluids*, **14**(4): 1301–1314.

Busse, F.H., and Carrigan, C.R., 1976. Laboratory simulation of thermal convection in rotating planets and stars. *Science*, **191**: 81–83.

Busse, F.H., and Cuong, P.G., 1977. Convection in rapidly rotating spherical fluid shells. *Geophysical and Astrophysical Fluid Dynamics*, **8**: 17–44.

Busse, F.H., and Hood, L.L., 1982. Differential rotation driven by convection in a rotating annulus. *Geophysical and Astrophysical Fluid Dynamics*, **21**: 59–74.

Carrigan, C.R., and Busse, F.H., 1983. An experimental and theoretical investigation of the onset of convection in rotating spherical shells. *Journal of Fluid Mechanics*, **126**: 287–305.

Chandrasekhar, S., 1961. *Hydrodynamic and Hydromagnetic Stability.* Oxford: Clarendon Press.

Dormy, E., Soward, A.M., Jones, C.A., Jault, D., and Cardin, P., 2004. The onset of thermal convection in rotating spherical shells. *Journal of Fluid Mechanics*, **501**: 43–70.

Greenspan, H.P., 1968. *The Theory of Rotating Fluids.* Cambridge: Cambridge University Press.

Grote, E., and Busse, F.H., 2001. Dynamics of convection and dynamos in rotating spherical fluid shells. *Fluid Dynamics Research*, **28**: 349–368.

Jones, C.A., Soward, A.M., and Mussa, A.I., 2000. The onset of thermal convection in a rapidly rotating sphere. *Journal of Fluid Mechanics*, **405**: 157–179.
Morin, V., and Dormy, E., 2004. Time dependent β-convection in rapidly rotating spherical shells. *Physics of Fluids*, **16**: 1603–1609.
Roberts, P.H., 1968. On the thermal instability of a rotating-fluid sphere containing heat sources. *Philosophical Transactions of Royal Society of London*, **A263**: 93–117.
Schnaubelt, M., and Busse, F.H., 1992. Convection in a rotating cylindrical annulus. Part 3 Vacillating spatially modulated flow. *Journal of Fluid Mechanics*, **245**: 155–173.
Soward, A.M., 1977. On the finite amplitude thermal instability of a rapidly rotating fluid sphere. *Geophysical and Astrophysical Fluid Dynamics*, **9**: 19–74.
Yano, J.-I., 1992. Asymptotic theory of thermal convection in rapidly rotating systems. *Journal of Fluid Mechanics*, **243**: 103–131.
Zhang, K., 1992. Spiralling columnar convection in rapidly rotating spherical fluid shells. *Journal of Fluid Mechanics*, **236**: 535–556.
Zhang, K., 1994. On coupling between the Poincaré equation and the heat equation. *Journal of Fluid Mechanics*, **268**: 211–229.
Zhang, K., 1995. On coupling between the Poincaré equation and the heat equation: Non-slip boundary conditions. *Journal of Fluid Mechanics*, **284**: 239–256.
Zhang, K.-K., and Busse, F.H., 1987. On the onset of convection in rotating spherical shells. *Geophysical and Astrophysical Fluid Dynamics*, **39**: 119–147.
Zhang, K., and Jones, C.A., 1993. The influence of Ekman boundary layers on rotating convection in spherical fluid shells. *Geophysical and Astrophysical Fluid Dynamics*, **71**: 145–162.

Cross-references

Anelastic and Boussinesq Approximations
Core Convection
Core Motions
Core Temperature
Core, Boundary Layers
Core, Magnetic Instabilities
Fluid Dynamics Experiments
Geodynamo
Inner Core Tangent Cylinder
Proudman–Taylor Theorem
Thermal Wind

CORE COMPOSITION

Primary planetary differentiation produced a metallic core and silicate shell surrounded by a thin hydrous and gaseous envelope. Emil Wiechert proposed this simple first order picture of the Earth at the end of the 19th century, while in 1914 Beno Gutenberg, Wiechert's former PhD student, determined that the depth to the core-mantle boundary at 2900 km (c.f., the present day value is 2895 $\pm$ 5 km depth, Masters and Shearer, 1995). Establishing a more detailed picture of the Earth's core has been a considerable intellectual and technological challenge, given the core's remote setting. The composition of the Earth's core is determined by integrating observations and constraints from geophysics, cosmochemistry, and mantle geochemistry; a unilateral approach from any of these perspectives cannot produce a significant compositional model.

Geophysical methods provide the only direct measurements of the properties of the Earth's core. The presence and size of the core and its material properties are revealed by such studies. Foremost among these observations include (1) its seismic wave velocity and the free oscillation frequencies, (2) the moment of inertia (both of these observations plus the Earth's mass collectively define a density profile for the core and mantle that is mutually and internally consistent (Dziewonski and Anderson, 1981), (3) the distribution and secular variation in magnetic field, and (4) laboratory data on mineral physics (e.g., equation of state (EOS) for materials at core appropriate conditions). When combined with the element abundance curve for the solar system and compositional models for the Earth's mantle (McDonough and Sun, 1995) these observations give us constraints on the mineralogical and chemical constituents of the core and mantle. Washington (1925), Birch (1952), and more recent studies (see McDonough (2004) for a recent update and literature review) have used these constraints to develop compositional models for the Earth and the core. These models consistently converge on the result that the core contains approximately 85% Fe, 5% Ni, and $\leq$10% of minor lighter components (in weight %, or about 77% Fe, 4% Ni, and 19% other in atomic proportions).

The minor component in the core is an alloy of lower atomic mass that accounts for the core's lower density when compared with that of liquid Fe at core conditions. Washington (1925), drawing upon analogies with phases in iron meteorites, recognized that the core contained a minor amount of an atomically light component (e.g., sulfide, carbide, phosphide). Birch (1964) suggested that this light component represented some 10% of the core's mass and offered a suite of candidate elements (e.g., H, C, O, Si, or S). Anderson and Isaak (2002) more recently reviewed the relevant literature and concluded that only ~5% of this light component is needed. More recently, however, Masters and Gubbins (2003) show that the density increase for the liquid outer core to the solid inner core is much greater than previously considered and Lin *et al.* (2005) found that Birch's law (a linear relationship between sound velocity and density) does not hold at core pressures. Both findings have implications for the bulk core composition and imply potentially greater amounts of a light component in the outer core. The nature and proportion of the elements that make up this alloy are controlled by three main factors: (1) the behavior of elements during metal-silicate segregation, (2) the integrated pressures and temperatures experienced during core formation, and (3) whether or not there is (or has been) mass transfer across the core-mantle boundary since core formation. Studies of meteorites identify the behavior of the elements in the early solar nebula and during planetismal formation, thus identifying elements that are likely concentrated in the core. Analyses of mantle samples constrain the composition of the Earth's primitive mantle (the combined crust plus mantle) and from this one ascertains the volatile element inventory for the planet. Studies of the secular variation of the mantle composition define the extent of core–mantle mass exchange. However, there is the proviso that we must sample this change; chemical changes occurring at the core-mantle boundary that remain isolated at the base of the mantle can only be speculated upon, but not demonstrated. Collectively, these data establish a bulk planetary composition; subtracting the primitive silicate mantle composition from this reveals the core composition.

The compositional diversity of the planets in the solar system and that of chondritic meteorites (primitive, undifferentiated meteorites) provide a guide to the bulk Earth composition. However, this diversity of samples presents a problem in that there is no unique meteorite composition that characterizes the Earth. The solar system is compositionally zoned from volatile-poor planets closer to the sun to volatile-rich gas giants further out. Relative to the other planets, the Earth has a relatively intermediate size-density relationship and volatile element inventory and is more depleted in volatile components than CI chondrites, the most primitive of all of the meteorites. The bulk Earth's composition is more similar to that of some carbonaceous chondrites and less so the ordinary or enstatite chondrites (Figure C21), especially in regard to the four most abundant elements (Fe, O, Si, and Mg; in terms of atomic proportions these elements represent 95% of the inventory of elements in chondrites and the Earth) and their ratios. Thus, we need to establish the absolute abundances of the refractory elements in the Earth and the signature of the volatile element depletion pattern.

The silicate Earth, or primitive mantle, encompasses the solid Earth minus the core. There is considerable agreement at the major and minor element level for the composition of the primitive mantle. The relative abundances of the lithophile elements (e.g., Ca, Al, Ti, REE (rare earth elements), Li, Na, Rb, B, F, Zn, etc) in the primitive mantle establish both the absolute abundances of the refractory elements in the Earth and the planetary signature of the volatile element depletion pattern. The volatile lithophile elements, those with half-mass condensation temperatures <1200 K (i.e., a reference temperature at which half the mass of a specific element condenses into a mineral from a cooling solar nebular under a given oxygen partial pressure) and excluded from the core, have CI chondrite relative abundances that systematically decrease with condensation temperature.

A compositional model for the Earth, the core, and mantle is presented in Table C1. The model core composition is derived from the above constraints and has a light element component that is consistent with existing geophysical requirements. Significantly, along with Fe and Ni the core contains most of the planet's sulfur, phosphorous, and carbon. It contains little or no gallium, an element widely found in iron meteorites, and is devoid of radioactive elements.

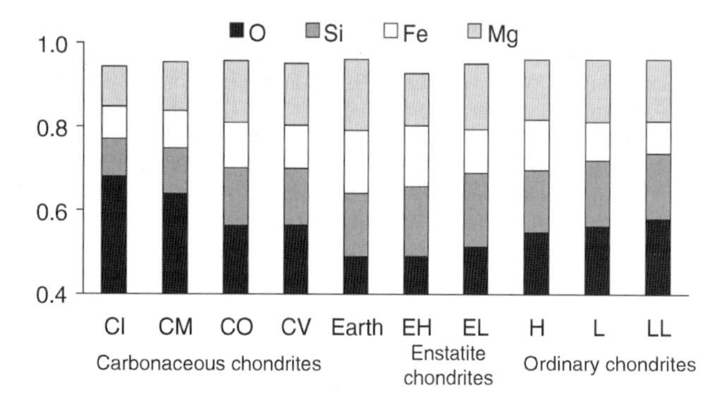

Figure C21 Atomic proportions of oxygen, silicon, iron, and magnesium in carbonaceous chondrites, the Earth, enstatite chondrites, and ordinary chondrites.

Light elements in the core

Given an outer core density deficit of 5%–10%, a host of elements (e.g., H, C, O, Si, P, S) can potentially be present. Commonly, metallurgical or cosmochemical arguments, coupled with meteoritical analogies (e.g., reduced metals, such as silicides in enstatite chondrites) are used to argue for one's preferred candidate element. There is sufficient evidence to suggest that some, but not all, of this density deficiency is due to the presence of sulfur (McDonough, 2004). However, beyond this, we are still working to place further constraints on this issue and a number of elements in the mix (particularly H, C, O, Si) are considered viable for explaining the core's density deficit.

The Earth's volatility curve can be established from the relative abundances of the lithophile elements, which are elements that (1) are excluded from the core and concentrated in the mantle, (2) have different condensation temperatures, and (3) widely vary in their chemical characteristics (McDonough, 2004). A comparison of the Earth's volatility curve with that of chondritic meteorites shows that the sulfur content of the core is restricted to 1.5–2 wt% due to its volatility. In addition, the planetary volatility curve reveals that the proportions of Si and O in the Earth, both of which have half-mass condensation temperatures <1200 K, cannot be readily established from chondritic observations.

There are a number of issues that need to be evaluated when considering mixed alloying elements to account for the core's density deficit. In calculating the density deficit one needs to know the thermal gradient in the deep Earth and the EOS for appropriate alloys at core conditions. Such data are still lacking for many of these alloys at core

Table C1 The composition of the Earth, core, and mantle (silicate Earth)

	Earth	Core	Mantle		Earth	Core	Mantle		Earth	Core	Mantle
H	260	600	100	Zn	40	0	55	Pr	0.17	0	0.25
Li	1.1	0	1.6	Ga	3	0	4	Nd	0.84	0	1.25
Be	0.05	0	0.07	Ge	7	20	1.1	Sm	0.27	0	0.41
B	0.2	0	0.3	As	1.7	5	0.05	Eu	0.10	0	0.15
C	730	2000	120	Se	2.7	8	0.075	Gd	0.37	0	0.54
N	25	75	2	Br	0.3	0.7	0.05	Tb	0.067	0	0.10
O %	29.7	0	44	Rb	0.4	0	0.6	Dy	0.46	0	0.67
F	10	0	15	Sr	13	0	20	Ho	0.10	0	0.15
Na %	0.18	0	0.27	Y	2.9	0	4.3	Er	0.30	0	0.44
Mg %	15.4	0	22.8	Zr	7.1	0	10.5	Tm	0.046	0	0.068
Al %	1.59	0	2.35	Nb	0.44	0	0.66	Yb	0.30	0	0.44
Si %	16.1	6.0	21	Mo	1.7	5	0.05	Lu	0.046	0	0.068
P	715	2000	90	Ru	1.3	4	0.005	Hf	0.19	0	0.28
S	6350	19000	250	Rh	0.24	0.74	0.001	Ta	0.025	0	0.037
Cl	76	200	17	Pd	1	3.1	0.004	W	0.17	0.47	0.029
K	160	0	240	Ag	0.05	0.15	0.008	Re	0.075	0.23	0.0003
Ca %	1.71	0	2.53	Cd	0.08	0.15	0.04	Os	0.9	2.8	0.003
Sc	10.9	0	16	In	0.007	0	0.01	Ir	0.9	2.6	0.003
Ti	810	0	1200	Sn	0.25	0.5	0.13	Pt	1.9	5.7	0.007
V	105	150	82	Sb	0.05	0.13	0.006	Au	0.16	0.5	0.001
Cr	4700	9000	2625	Te	0.3	0.85	0.012	Hg	0.02	0.05	0.01
Mn	800	300	1045	I	0.05	0.13	0.01	Tl	0.012	0.03	0.004
Fe %	32.0	85.5	6.26	Cs	0.035	0.065	0.021	Pb	0.23	0.4	0.15
Co	880	2500	105	Ba	4.5	0	6.6	Bi	0.01	0.03	0.003
Ni	18200	52000	1960	La	0.44	0	0.65	Th	0.055	0	0.08
Cu	60	125	30	Ce	1.13	0	1.68	U	0.015	0	0.02

Concentrations are given in μg/g (p.p.m.), unless stated as "%", which are given in weight %.

Table C2 Competing models for the composition of the Earth and core

Wt%	Si-bearing model		O-bearing model	
	Earth	Core	Earth	Core
Fe %	32.0	85.5	32.9	88.3
O %	29.7	0	30.7	3
Si %	16.1	6	14.2	0
Ni %	1.82	5.2	1.87	5.4
S %	0.64	1.9	0.64	1.9
Cr %	0.47	0.9	0.47	0.9
P %	0.07	0.20	0.07	0.20
C %	0.07	0.20	0.07	0.20
H %	0.03	0.06	0.03	0.06
Mean atomic #		23.5		23.2
		Atomic proportions		
Fe		0.768		0.783
O		0.000		0.093
Si		0.107		0.000
Ni		0.044		0.045
S		0.030		0.029
Cr		0.003		0.003
P		0.009		0.009
C		0.008		0.008
H		0.030		0.029
Total		1.000		1.000

conditions. Likewise, the mutual compatibility of different alloying element pairs has yet to be established for many of these elements. There appears to be good evidence showing that alloys of Si-O and Si-S are not compatible because of contrasting solubility (see recent review of Li and Fei, 2004). Thus, based on a model of heterogeneous accretion for the Earth (i.e., a secular variation in the composition of material that contributed to the growing Earth), the ~2% (by mass) sulfur in the core is mixed with either silicon or oxygen as the other dominant alloying element in the outer core. From this perspective two competing models for the composition of the core are offered in Table C2, which contain either Si or O, but not both. Other light elements contribute to the density deficit in the core, including H, C, and P (see Table C1), but these elements are in significantly lower proportions due to their partitioning into other parts of the Earth and the markedly lower condensation temperatures for H and C. Such a diverse array of light elements may well be mutually compatible in iron liquids, particularly if liquid segregation occurs at high pressure (ca. 10–25 GPa).

Radioactive elements in the core

The existence of radioactive elements in the Earth's core is usually invoked to offer an explanation for the power needed to run the geodynamo (Labrosse *et al.*, 2001) and/or as a way to increase the volatile element content of the Earth. An alternative model for the core's power budget was offered by Christensen and Tilgner (2004) who showed that ohmic losses by the core's geodynamo are sufficiently large enough to remove the need for radioactive heating in the core. Arguments for potassium in the core are based on three lines of evidence: (1) cosmochemical—potassium sulfide is found in some meteorites, (2) high pressure s-d electronic transition, and (3) solubility of K in Fe-S (and Fe-S-O) liquids at high pressure.

The cosmochemical argument for K in the core is based on the observation that a potassium-bearing sulfide (djerfisherite) is found in trace quantities in enstatite chondrites (Lodders, 1995). However, this *ad hoc* hypothesis overlooks the fact that enstatite chondrites also contain a myriad of other sulfides that are more abundant than djerfisherite, including oldhamite (CaS), ferroan alabandite ((Mn,Fe)S), and niningerite ((Mg,Fe)S), and that alkali sulfides like djerfisherite are typically found in trace quantities as replacement or alteration phases (Brearley and Jones, 1998). Incorporating the more common, higher temperature sulfide phases, which contain REE and other refractory lithophile elements in nonchondritic proportions, would grossly and adversely change both the elemental and isotopic abundances of the Sm/Nd, Lu/Hf, and Th/U isotopic systems. Given an absence of fractionation of these isotopic and elemental systems, one can also conclude that there are negligible quantities of K sulfide in the core. At high pressure s-orbital electrons in alkali metal can be compressed into d-orbital electronic states. For K this electronic transition is theoretically possible, but this effect is likely to be even greater for Rb and Cs. Whereas an argument can be made for Cs in the core, it cannot be so for Rb (McDonough, 2004). Moreover, the coherent depletion pattern of the volatile lithophile elements is not restricted to just the alkali metals and thus the pattern cannot be dismissed as "being stored in the core" based on pressure induced s-d electronic transitions. Finally, Gessmann and Wood (EPSL, 2002) demonstrate that K is soluble in Fe-S and Fe-S-O liquids at high pressure and suggested that K can therefore be sequestered into the core. However, Ca is also incorporated into the metallic liquid and having Ca in the core creates problems given that the mantle's Ca content is in chondritic proportions. Other recent models (Rama Murthy *et al.*, 2003, Lee and Jeanloz, 2003) proposing K in the Earth's core, however, also are accompanied by geochemical consequences that limit their acceptability (see McDonough, 2003 for a further review). Overall, it is unlikely that there is any K in the core, although much debate continues.

Timing of core formation

Defining the age and duration of core formation depends on having an isotope system in which the parent-daughter isotope pair are fractionated by core sequestration over a time interval within the functional period of the system's half-life. Fortunately, recent analytical advances in the W-Hf isotope system provide us with a tool to gauge the timing of core formation. Both Hf and W are refractory elements (lithophile and siderophile, respectively) having chondritic relative concentrations in the Earth, with ~90% of the Earth's budget of W being hosted in the core and the entire planet's Hf budget in the silicate Earth. Recent studies (Yin *et al.*, 2002; Kleine *et al.*, 2002; Schoenberg *et al.*, 2002) found that the Earth's W-isotopic composition is higher than that of chondrites, indicating that core separation had to have been completed by ~30 Ma after t_0 (4.56 Ga ago).

The inner core and core-mantle exchange

Based on geophysical data the inner core comprises ~5% of the core's mass. Its existence is not disputed, although there is much speculation about when the inner core began to crystallize, the extent of element fractionation between the inner and outer core and the role it plays in models of core–mantle exchange. Recently, Labrosse *et al.* (2001), examining the power budget for the core, suggested that inner core crystallization began in the latter half of Earth's history (ca. 1–2 Ga) and that some amount of radioactive heating is necessary. In contrast, however, support for early inner core crystallization comes from the Os isotopic compositions of some Hawaiian basalts, which has been explained by parent-daughter isotope fractionation due to core differentiation followed by core contributions to the Hawaiian magma source region. The origin of this Os isotopic signature in these basalts is interpreted to be due to exchange across the core-mantle boundary (with the base of the mantle being the putative source of these basalts) of a radiogenic Os isotopic component that was generated by inner core crystallization, which incorporated Os and to a lesser extent, Re and Pt (Brandon *et al.*, 2002). The extent of core-mantle exchange can be monitored, albeit on a less sensitive scale, from studies of peridotites and basalts. Ratios of Mg/Ni and Fe/Mn in the mantle have been fixed (i.e., $\leq\pm15\%$) for the last 3.8 Ga (McDonough and Sun 1995), which is inconsistent with significant core-mantle exchange, given the proportions of these elements in the core and mantle. Likewise, other sensitive element ratios (e.g., Re/Yb, P/Nd, Mo/Ce, W/Ba) involving siderophile element (e.g., Re, P, Mo, W) and similarly incompatible (i.e., elements readily entering a melt relative to the solid) lithophile elements (e.g., Yb, Ce, Nd, Ba) show that <0.5% of a core contribution can be incorporated into the source regions of mid-ocean ridge basalts or intraplate (plume-derived) basalts. In general, however, models invoking early growth of the inner core are at odds with findings from power budget calculations.

William F. McDonough

Bibliography

Anderson, O.L., and Isaak, D.G., 2002. Another look at the core density deficit of Earth's outer core. *Physics of the Earth and Planetary Interiors*, **131**(1): 19–27.

Birch, F., 1952. Elasticity and Constitution of the Earth's Interior. *Journal of Geophysical Research*, **57**(2): 227–286.

Birch, F., 1964. Density and composition of mantle and core. *Journal of Geophysical Research*, **69**(20): 4377–4388.

Brandon, A.D., Walker, R.J., Puchtel, I.S., Becker, H., Humayun, M., and Revillon, S., 2003. $^{186}Os/^{187}Os$ systematics of Gorgona Island komatiites: implications for early growth of the inner core. *Earth Planetary Science Letters*, **206**: 411–426.

Brearley, A.J., and Jones, R.H., 1998. Chondritic meteorites. In Papike, J.J. (ed.), *Planetary Materials,* Vol. 36, Washington, D.C.: Mineralogical Society of America, pp. 3-01–3-398.

Christensen, U.R., and Tilgner, A., 2004. Power requirement of the geodynamo from ohmic losses in numerical and laboratory dynamos. *Nature*, **429**: 169–171.

Dziewonski, A., and Anderson, D.L., 1981. Preliminary reference Earth model. *Physics of Earth and Planetary Interiors*, **25**: 297–356.

Gessmann, C.K., and Wood, B.J., 2002. Potassium in the Earth's core? *Earth and Planetary Science Letters*, **200**(1–2): 63–78.

Kleine, T., Münker, C., Mezger, K., and Palme, H., 2002. Rapid accretion and early core formation on asteroids and the terrestrial planets from Hf-W chronometry. *Nature*, **418**: 952–955.

Labrosse, S., Poirier, J.P., and LeMouel, J.L., 2001. The age of the inner core. *Earth and Planetary Science Letters*, **190**: 111–123.

Lee, K.K.M., and Jeanloz, R., 2003. High-pressure alloying of potassium and iron: Radioactivity in the Earth's core? *Geophysical Research Letters*, **30**(23): 2212, doi:10.1029/2003GL018515.

Li, J., and Fei, Y., 2003. Experimental constraints on core composition, In Carlson, R.W. (ed.), *The Mantle and Core,* In Holland, H.D., and Turekian, K.K. (eds.), *Treatise on Geochemistry*, Vol. 2, Oxford: Elsevier-Pergamon, pp. 521–546.

Lin, J.-F., Strarhahn, W., Zhao, J., Shen, G., Mao, H.-k., and Hemley, R.J., 2005. Sound velocities of hot dense iron: Birch's Law revisited. *Science*, **308**: 1892–1895.

Lodders, K., 1995. Alkali elements in the Earth's core—evidence from enstatite meteorites. *Meteoritics*, **30**(1): 93–101.

Masters, G., and Gubbins, D., 2003. On the resolution of density within the Earth. *Physics of Earth and Planetary Interiors*, **140**: 159–167.

Masters, T.G., and Shearer, P.M., 1995. Seismic models of the Earth: Elastic and Anelastic. In Ahrens, T.J. (eds.), *Global Earth Physics: a Handbook of Physical Constants, Vol. AGU Reference Shelf.* Washington, D.C.: American Geophysical Union, pp. 88–103.

McDonough, W.F., and Sun, S.-S. 1995. The composition of the Earth. *Chemical Geology*, **120**: 223–253.

McDonough, W.F., 2003. Compositional model for the Earth's core. In Carlson, R.W. (ed.), *The Mantle and Core,* In Holland, H.D., and Turekian, K.K. (eds.), *Treatise on Geochemistry*, Vol. 2, Oxford: Elsevier-Pergamon, pp. 547–568.

Rama Murthy, V., van Westrenen, W., and Fei, Y., 2003. Experimental evidence that potassium is a substantial radioactive heat source in planetary cores. *Nature*, **423**: 163–165.

Schoenberg, R., Kamber, B.S., Collerson, K.D., and Eugster, O., 2002. New W-isotope evidence for rapid terrestrial accretion and very early core formation. *Geochimica et Cosmochimica Acta*, **66**: 3151–3160.

Washington, H.S., 1925. The chemical composition of the Earth. *American Journal of Science*, **9**: 351–378.

Yin, Q., Jacobsen, S.B., Yamashita, K., Blichert-Toft, J., Télouk, P., and Albarede, F., 2002. A short timescale for terrestrial planet formation from Hf-W chronometry of meteorites. *Nature*, **418**: 949–952.

CORE CONVECTION

Types of core convection

There exist primarily two different types of convection taking place in the Earth's fluid core. The first type is driven by thermal instabilities. A distribution of heat sources from radioactive elements in the core can produce a radial temperature gradient $\mathrm{d}T(r)/\mathrm{d}r$, where T is the temperature and r is the distance from the core centre. Thermal convection may occur when the temperature in the core decreases more rapidly than that of the adiabatic gradient,

$$\frac{\mathrm{d}T}{\mathrm{d}r} < \left(\frac{\mathrm{d}T}{\mathrm{d}r}\right)_{\text{adiabatic}} = \frac{\alpha g T}{\rho C_P}, \qquad \text{(Eq. 1)}$$

where ρ is the density of the core, g is the acceleration of gravity, α is the thermal expansion coefficient, and C_P is the specific heat at constant pressure. The adiabatic gradient in the core is about -0.1 K/km. The unstable stratification described by Eq. (1) offers buoyancy forces that may drive thermal convection in the fluid core. This is a necessary condition for thermal instability, which is usually referred to as Schwarzschild's criterion (or the Adams-Williamson condition) (for example, Gubbins and Roberts, 1987). The actual temperature gradient in the core decreases on approaching the core-mantle boundary because the gradient required to conduct a given amount of heat reduces with radius. Except possibly at the top of the core where the adiabatic gradient is steep as a result of higher g and lower ρ, Schwarzschild's instability condition is likely to be satisfied in the whole fluid core. In general, the condition for core convection is that its Rayleigh number R defined as

$$R = \frac{\alpha\beta g r_0^4}{\nu\kappa}, \qquad \text{(Eq. 2)}$$

where ν is the kinematic viscosity of the fluid, r_0 is the radius of the fluid core, κ is the thermal diffusivity, and β is the superadiabatic temperature gradient $(\mathrm{d}T/\mathrm{d}r - (\mathrm{d}T/\mathrm{d}r)_{\text{adiabatic}})$ in the core, must be sufficiently large. In other words, Schwarzschild's criterion does not represent a sufficient condition for thermal convection because of viscous and magnetic damping in the fluid core. Gubbins (2001) estimated that the Rayleigh number for thermal convection in the Earth's core is enormously large, $R = 10^{29}$, when molecular diffusivities are used, while turbulent diffusivities give a much lower value with $R = 10^{12}$. These values of the Rayleigh number exceed the critical value for nonmagnetic thermal convection.

The second type of convection is nonthermal, in connection with slow growth of the solid inner core of the Earth through freezing from the outer core (Gubbins *et al.*, 1979; Labrosse *et al.*, 1997). The density of the outer liquid core is smaller than that of pure iron, being composed of a mixture of iron with lighter elements. During the general cooling of the Earth, the inner core grows and the light constituents excluded from the inner core rise upward buoyantly through the outer core to drive compositional convection (Gubbins *et al.*, 1979; Moffatt and Loper, 1994). Some fraction of the light constituents might survive remixing to create and sustain a stably stratified layer at the top of the core, "the hidden ocean" (Braginsky, 1999). This type of the fluid motion in the core is usually referred to as compositional convection. Gubbins (2001) estimated that the Rayleigh number for compositional convection in the Earth's core is also enormously large, $R = 10^{38}$, when molecular diffusivities are used, while turbulent diffusivities give a much lower value with $R = 10^{15}$. These values of the Rayleigh number exceed the critical value for nonmagnetic compositional convection.

Dynamics of core convection

Core convection, driven by either thermal instability or by the solidification of the inner core or by both, sustains geomagnetic fields through magnetohydrodynamic dynamo processes, which convert the kinetic energy of convection into magnetic energy. The primary dynamics of core convection is controlled by (i) rapid rotation of the Earth, (ii) viscous effects, (iii) thermal or compositional buoyancy, and (iv) convection-generated magnetic fields (Fearn, 1998; Zhang and Gubbins, 2000). Other details such as compressibility, variable rotation, boundary conditions, and the detailed driving mechanism of core convection are of secondary importance. In order that core convection can occur, the dominant Coriolis force must be balanced by five forces at any instant of time:

$$2\boldsymbol{\Omega} \times \mathbf{u} = -\frac{1}{\rho}\nabla p - \alpha\Theta g_0 \mathbf{r} - \frac{\mathrm{D}\mathbf{u}}{\mathrm{D}t} + \frac{1}{\rho\mu}(\nabla \times \mathbf{B}) \times \mathbf{B} + \nu\nabla^2\mathbf{u}, \qquad \text{(Eq. 3)}$$

where $\mathbf{r}$ is the position vector, $\boldsymbol{\Omega}$ is the angular velocity of the Earth, Θ is the deviation from the adiabatic or compositional stratification, μ is the magnetic permeability, $\mathbf{u}$ is the velocity field, and $\mathbf{B}$ is the generated magnetic field. It is important to note that the Coriolis force $2\boldsymbol{\Omega} \times \mathbf{u}$ can only be balanced in major part by the pressure force $-\nabla p/\rho$. All other forces, the buoyancy (thermal or compositional) force $-\alpha\Theta g_0\mathbf{r}$, the magnetic force $(\nabla \times \mathbf{B}) \times \mathbf{B}/\rho\mu$, the inertial force $\mathrm{D}\mathbf{u}/\mathrm{D}t$, and the viscous force $\nu\nabla^2\mathbf{u}$, play significant roles in core convection. As a consequence of the delicate dynamic balance described by Eq. (3), there exist large disparities both in spatial and temporal scales of core convection, dependent largely upon how the balance between the viscous, Coriolis, and magnetic forces is achieved (Zhang and Gubbins, 2000).

Structure of core convection

The spatial structure of core convection is likely to vary at different times in the Earth's history. Suppose at one instant that the magnetic field is weak and plays an insignificant role in dynamics. In this case, while the buoyancy force supplies energy to drive convection, the role of viscosity must be inverted from the usual one of inhibiting convection: it provides the necessary frictional forces to offset the part of the Coriolis force $2\boldsymbol{\Omega} \times \mathbf{u}$ that cannot be balanced by the pressure gradient $-\nabla p/\rho$. Moreover, the corresponding convection must be in the form of nearly two-dimensional columnar rolls selecting the symmetry (Roberts, 1968; Busse, 1970; Jones *et al.*, 2000; Zhang and Liao, 2004)

$$(u_r, u_\theta, u_\phi)(r, \theta, \phi) = (u_r, -u_\theta, u_\phi)(r, \pi - \theta, \phi),$$
$$\Theta(r, \theta, \phi) = \Theta(r, \pi - \theta, \phi), \qquad \text{(Eq. 4)}$$

where (r, θ, ϕ) are spherical polar coordinates with $\theta = 0$ at the axis of rotation. With large viscous forces the Rayleigh number R required to initiate convection has to be extremely large

$$R = O(E^{-4/3}), \qquad \text{(Eq. 5)}$$

and the horizontal scale L of the rolls must be extremely small

$$L = O(E^{1/3} r_0), \qquad \text{(Eq. 6)}$$

where $E = \nu/2\boldsymbol{\Omega} r_0^2$ is the Ekman number and $E \leq O(10^{-9})$ in the Earth's core. An example of such convection, which shows the convective flow in two different sections, is displayed in Figure C22. The convective motions are the columnar rolls aligned with the axis of rotation and located and localized at higher latitudes with radial phase shifts (Busse, 1970; Zhang, 1992; Jones *et al.*, 2000; Sumita and Olson, 2000).

Suppose at another time that the fluid in the core is permeated by a strong dynamo-generated large-scale magnetic field. With the presence of the strong magnetic field, the dynamical role of viscosity is taken up by the magnetic force $(1/\rho\mu)(\nabla \times \mathbf{B}) \times \mathbf{B}$. As a result, core convection becomes more efficient and larger scale, in sharp contrast to

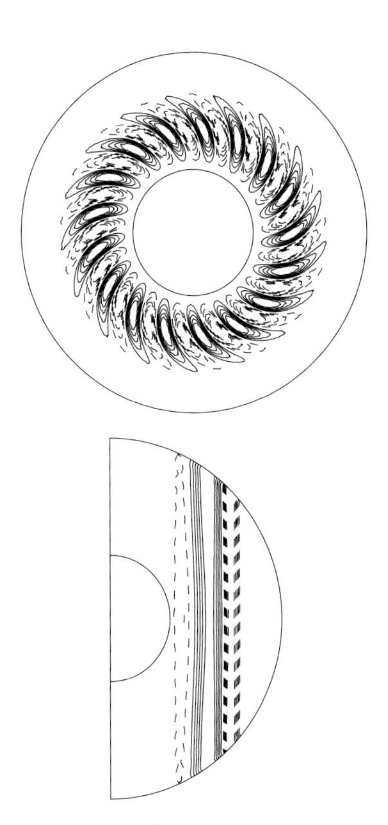

Figure C22 Structure of weakly nonlinear convection in a rapidly rotating spherical shell at $E = /2r_0^2 = 2 \times 10^{-6}$ when the effect of magnetic field is weak. Displayed are contours of the radial velocity u_s in the equatorial plane (*upper panel*) and in a meridional plane (*lower panel*) for the Prandtl number Pr = 1.0. *Solid contours* indicate $u_s > 0$ and *dashed contours* correspond to $u_s < 0$.

nonmagnetic convection shown in Figure C22. By implication, the Rayleigh number R required to initiate convection becomes much smaller:

$$R = O(E^{-1}), \qquad \text{(Eq. 7)}$$

together with the larger scale of convective motions

$$L = O(r_0). \qquad \text{(Eq. 8)}$$

In summary, because core convection is critically affected by both the Coriolis and magnetic forces, the huge contrast between the spatial and temporal scales is an important characteristic of core convection.

Keke Zhang

Bibliography

Braginsky, S.I., 1999. Dynamics of the stably stratified ocean at the top of the core. *Physics of the Earth and Planetary Interiors*, **111**: 21–34.

Busse, F.H., 1970. Thermal instabilities in rapidly rotating systems. *Journal of Fluid Mechanics*, **44**: 441–460.

Fearn, D.R., 1998. Hydromagnetic flow in planetary cores. *Reports on Progress in Physics*, **61**: 175–235.

Gubbins, D., 2001. The Raleigh number for convection in the Earth's core. *Physics of the Earth and Planetary Interiors*, **128**: 3–12.

Gubbins, D., Masters, T.G., and Jacobs, J.A., 1979. Thermal evolution of the Earth's core. *Geophysical Journal of the Royal Astronomical Society*, **59**: 57–100.

Gubbins, D., and Roberts, P.H., 1987. Magnetohydrodynamics of the Earth's core. In Jacobs J.A. (ed.), *Geomagnetism*. Vol. 2, London: Academic Press, pp. 1–183.

Labrosse, S., Poirier, J.-P., LeMouël, J.-L., 1997. On cooling of the Earth's core. *Physics of the Earth and Planetary Interiors*, **99**: 1–17.

Moffatt, H.K., and Loper, D.E., 1994. Hydromagnetics of the Earth's core, I. The rise of a buoyant blob. *Geophysical Journal International*, **117**: 394–402.

Roberts, P.H., 1968. On the thermal instability of a self-gravitating fluid sphere containing heat sources. *Philosophical Transactions of the Royal Society of London: Series A*, **263**: 93–117.

Sumita, I., and Olson, P., 2000. Laboratory experiments on high Rayleigh number thermal convection in a rapidly rotating hemispherical shell. *Physics of the Earth and Planetary Interiors*, **117**: 153–170.

Jones, C.A., Soward, A.M., and Mussa, A.I., 2000. The onset of thermal convection in a rapidly rotating sphere. *Journal of Fluid Mechanics*, **405**: 157–179.

Zhang, K., 1992. Spiralling columnar convection in rapidly rotating spherical fluid shells. *Journal of Fluid Mechanics*, **236**: 535–556.

Zhang, K., and Gubbins, D., 2000. Scale disparities and magnetohydrodynamics in the Earth's core. *Philosophical Transactions of the Royal Society of London A*, **358**: 899–920.

Zhang, K., and Liao, X., 2004. A new asymptotic method for the analysis of convection in a rapidly rotating sphere. *Journal of Fluid Mechanics*, **518**: 319–346.

Cross-references

Core Temperature
Core, Magnetic Instabilities
Geodynamo
Magnetoconvection
Magnetohydrodynamic Waves
Proudman-Taylor Theorem

CORE DENSITY

Introduction

Knowledge of the density distribution inside the core is of interest for several reasons. Perhaps most importantly, the size of the density jump at the inner core boundary controls how important inner core growth is for powering the dynamo. Also of interest is the presence or otherwise of anomalous (e.g., convectively stable) regions in the outer core. This latter problem is more difficult as it requires an evaluation of the density *gradient* and how much this deviates from the Adams-Williamson equation (see below), which is expected to control the density gradient in a vigorously convecting outer core.

The main constraints on density in the core come from measurements of the frequencies of free oscillations of the Earth (Masters and Gubbins, 2003). However, it is also possible to use the sensitivity of reflected body waves to impedance contrasts to infer the density jump at the ICB from the amplitudes of the seismic phase *PKiKP* (Koper and Pyle, 2004; Cao and Romanowicz, 2004). We compare the results of these studies in the following.

Background

For a variety of reasons, we believe that most or all of the outer core is convecting vigorously and so will be adiabatic and homogeneous. In such a region, the density distribution will follow the Adams-Williamson equation closely. The Adams-Williamson equation is

$$\left(\frac{\mathrm{d}\rho}{\mathrm{d}r}\right)_{\mathrm{ad}} = \left(\frac{\partial\rho}{\partial p}\right)_{\mathrm{s}}\frac{\mathrm{d}p}{\mathrm{d}r} = -\frac{\rho g}{\left(\frac{\partial p}{\partial \rho}\right)_{\mathrm{s}}} = -\frac{\rho g}{\phi} \qquad \text{(Eq. 1)}$$

where ρ is density, g is the acceleration due to gravity, s is entropy, and the pressure, p is taken to be hydrostatic. $(\partial p/\partial \rho)_{\mathrm{s}}$ is often given the symbol ϕ and is called the "seismic parameter" because it can be computed from the velocities of propagation of seismic waves in the Earth. In an isotropic solid, we have

$$\phi = \frac{K_{\mathrm{s}}}{\rho} = V_{\mathrm{p}}^2 - \frac{4}{3}V_{\mathrm{s}}^2 \qquad \text{(Eq. 2)}$$

where V_{p} and V_{s} are the compressional and shear velocity respectively and K_{s} is the adiabatic bulk modulus.

The "Bullen parameter," η, is a measure of departure of the density gradient from the Adams-Williamson condition and is a measure of stability of a region (see Masters, 1979 for a thorough discussion). η is simply given by

$$\eta = -\frac{\phi}{\rho g}\frac{\mathrm{d}\rho}{\mathrm{d}r} \qquad \text{(Eq. 3)}$$

and so is 1 in a vigorously convecting region. If $\eta > 1$, the region is stable—i.e., a radially displaced parcel of material will be negatively buoyant and will oscillate around its initial radius. $\eta < 1$ implies that a region is convectively unstable though the effects of viscosity and conduction may inhibit convection from occurring. An order of magnitude calculation shows that η is insignificantly less than one in the outer core before convection will start. Note that η will appear to be less than one in the boundary layers of a convecting region where the temperature gradients are strongly superadiabatic. Such boundary layers would be visible in the mantle, but boundary layers in the outer core are expected to be extremely thin because of the low viscosity of this region.

Bullen (1975) made extensive use of an alternative form for η:

$$\eta = \frac{\mathrm{d}K_{\mathrm{s}}}{\mathrm{d}p} + \frac{1}{g}\frac{\mathrm{d}\phi}{\mathrm{d}r} \qquad \text{(Eq. 4)}$$

by arguing that $dK_s/dp \equiv (dK_s/dr)/(dp/dr)$ is slowly varying in the Earth (and has a value of about 4) so that departures from the Adams-Williamson state can be determined by looking at the radial gradient of the seismic parameter. In fact, there is every reason to believe that dK_s/dp will be strongly affected in regions of strong temperature gradient or rapid variation in chemistry so Bullen's approach is somewhat suspect.

Density models using The Adams-Williamson Equation

Early models of the density within the deep Earth assumed that the Adams-Williamson equation held in the lower mantle and outer core but the development of an extensive data set of degenerate free oscillation frequencies over the last 30 years has given us independent constraints on density, which, in principle, allow us to evaluate departures from the Adams-Williamson state. The resolution of density by this data set has been discussed in detail by Masters and Gubbins (2003). They used a variant of Backus-Gilbert resolving power theory in which linear combinations of the data are taken to localize information about some model property (e.g., density) about some target radius inside the Earth. The resulting "local averages" are relatively precise if averages are sought over large depth intervals but decrease in precision for averages over smaller depth intervals. Some results of this analysis are presented below but first we consider a different approach to constructing density profiles in the core. This approach takes advantage of the fact that the mode data can give precise estimates of the mean density of the whole core and of the outer core alone. Given ϕ, it is then possible to integrate the Adams-Williamson equation in the outer core and inner core and determine a density distribution for the whole core. The result of doing this for five different models of ϕ is shown in Figure C23. Clearly, the resulting density profile is very insensitive to the choice of seismic velocity model and all models have a density jump of about 0.8 Mg/m^3 at the inner core boundary. We also indicate the uncertainty of density in the core which comes primarily from the uncertainty of the estimates of mean density of the outer core and the whole core. The value obtained for the density jump at the inner core boundary agrees with the results of Masters and Gubbins (2003) and Cao and Romanowicz (2004) but is larger than the upper bound of 0.45 Mg/m^3 obtained by Koper and Pyle (2004). The reason for this discrepancy is not yet clear.

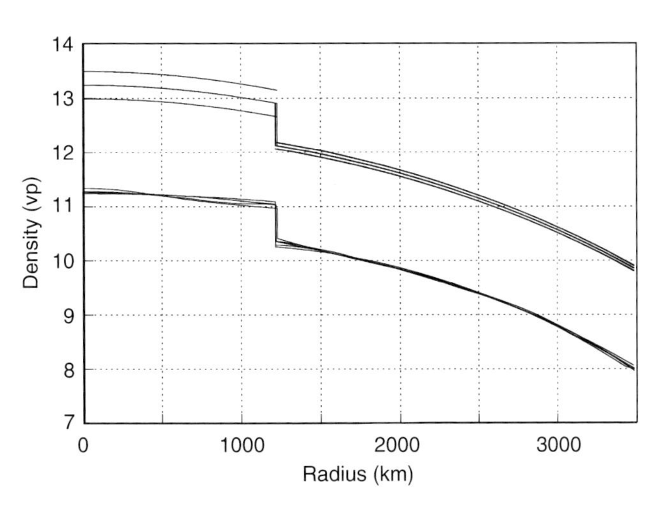

Figure C23 Five models of density in the core (*top line with error corridor*) made by integrating the Adams-Williamson equation using five different velocity models (*plotted beneath the density models*). The density distribution is insensitive to the particular velocity model chosen. The five velocity models are 1066A and 1066B of Gilbert and Dziewonski (1975), PEMA of Dziewonski *et al.* (1975), isotropic PREM of Dziewonski and Anderson (1981), and AK135 of Montagner and Kennett (1996). The error corridor is determined by how well we can resolve the mean density of the whole core and the outer core.

It is now of interest to ask if there are significant departures from the Adams-Williamson condition. Suppose we relax the constraint of adiabaticity (but keep the assumption of homogeneity—see Masters, 1979, for the general case). Density is a function of temperature and pressure:

$$\frac{d\rho}{dr} = \left(\frac{\partial\rho}{\partial T}\right)_p \frac{dT}{dr} + \left(\frac{\partial\rho}{\partial p}\right)_T \frac{dp}{dr} \qquad \text{(Eq. 5)}$$

The temperature gradient in an adiabatic and homogeneous region is given by

$$\left(\frac{dT}{dr}\right)_{ad} = \left(\frac{\partial T}{\partial p}\right)_s \frac{dp}{dr} = -\frac{\alpha g T}{C_p} = -\frac{gT\gamma}{\phi} \qquad \text{(Eq. 6)}$$

where α is the coefficient of volume expansion, C_p is the specific heat, and γ is Grüneisen's ratio, which is a dimensionless number with a value close to 1 for nearly all materials. Let the total temperature gradient be

$$\frac{dT}{dr} = \left(\frac{dT}{dr}\right)_{ad} - \tau \qquad \text{(Eq. 7)}$$

This equation defines the superadiabatic temperature gradient τ. Substituting in the above equation and using some thermodynamic identities gives

$$\frac{d\rho}{dr} = -\frac{\rho g}{\phi} + \alpha\rho\tau \qquad \text{(Eq. 8)}$$

and the Bullen parameter becomes

$$\eta = 1 - \frac{\alpha\phi}{g}\tau \qquad \text{(Eq. 9)}$$

It is extremely unlikely that the bulk of the outer core is significantly superadiabatic but it is possible that a stable (subadiabatic) region exists. The most stable the core could reasonably become in a thermal sense is isothermal, i.e., $dT/dr = 0$. This leads to a value of η:

$$\eta = 1 + \alpha T\gamma \qquad \text{(Eq. 10)}$$

The dimensionless combination $\alpha T\gamma$ is a small quantity inside the Earth (because α rapidly decreases as pressure increases) and has a value on the order of 0.08 in the core. The consequence of this is that a density model for a core which has a thermally stable region is going to be almost identical to those shown in Figure C23. In fact, even if the whole core is isothermal, the maximum deviation in density from the models shown in Figure C23 is only 1%.

The case of a compositionally stable region in the outer core is discussed by Masters (1979) and it is possible that strong compositional gradients would be observable using the free-oscillation data set. To check this possibility, we performed resolution experiments of the type discussed in Masters and Gubbins (2003). Here, we confine attention to the outer core since resolution of the mode dataset in the inner core is not good enough to detect reasonable departures of the density from the Adams-Williamson state. Figure C24 shows "local averages" of density with errors of 0.5%–1%. These correspond to averages over a radial distance of about 400 km. Though there is apparently some structure in the local averages at a radius of about 2000 km, the local averages follow the Adams-Williamson models quite well and are certainly consistent

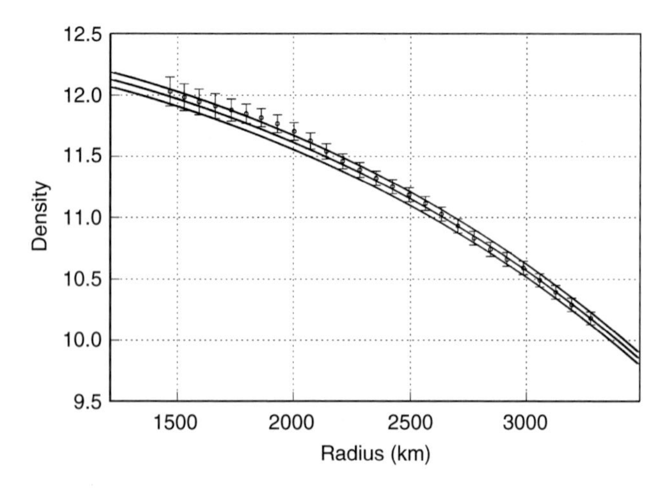

Figure C24 Local averages of density (with 1 sigma error bars) in the outer core computed using the method described by Masters and Gubbins (2003) compared with the models computed using the Adams-Williamson equation. The local averages are for a target error of 0.5% and are averages over about 400 km in radius.

within one standard deviation. Note that we cannot make local averages for target depths closer to the inner core boundary or the outer core boundary since the resulting resolution kernels have significant energy in the inner core or mantle and the local averages are biased.

Summary

Given the current observational constraints on the density in the core, we conclude that use of the Adams-Williamson equation is justified and results in a tightly constrained density profile with uncertainties of about 0.5% in the outer core and about 2% in the inner core. The models have a density jump of about 0.8 Mg/m^3 at the inner core boundary. A resolution analysis of the free-oscillation dataset gives local averages of density, which are consistent with the models derived from the Adams-Williamson equation.

Guy Masters

Bibliography

Bullen, K.E., 1975. *The Earth's Density.* London: Chapman and Hall.

Cao, A., and Romanowicz, B., 2004. Constraints on density and shear velocity contrasts at the inner core boundary. *Geophysics Journal International*, **157**: 1146–1151.

Dziewonski, A.M., and Anderson, D.L., 1981. Preliminary reference Earth model. *Physics of Earth and Planetary Interiors*, **25**: 297–356.

Dziewonski, A.M., Hales, A.L., and Lapwood, E.R., 1975. Parametrically simple earth models consistent with geophysical data. *Physics of Earth and Planetary Interiors*, **10**: 12–48.

Gilbert, F., and Dziewonski, A.M., 1975. An application of normal mode theory to the retrieval of structural parameters and source mechanisms from seismic spectra. *Philosophical Transactions Of the Royal Society of London*, **A278**: 187–269.

Koper, K., and Pyle, M., 2004. Observations of PKiKP/PCP amplitude ratios and implications for Earth structure at the boundaries of the liquid core. *Journal of Geophysical Research*, **109**: B03301, doi:10.1029/2003 JB002750.

Masters, G., 1979. Observational constraints on the chemical and thermal structure of the earth's deep interior. *Geophysical Journal of the Royal Astronomical Society*, **57**: 507–534.

Masters, G., and Gubbins, D., 2003. On the resolution of density within the Earth. *Physics of Earth and Planetary Interiors*, **140**: 159–167.

Montagner, J-P., and Kennett, B.L.N., 1996. How to reconcile body-wave and normal-mode reference Earth models. *Geophysics Journal International*, **125**: 229–248.

Cross-references

Adiabatic Gradient in the Core
Boundary Layers in the Core
Core Temperature
Core, Adiabatic Gradient
Core, Boundary Layers
Grüneisen's Parameter for Iron and Earth' Core
Higgins–Kennedy Paradox

CORE MOTIONS

Introduction

The geomagnetic field is widely agreed to evolve as a result of self-sustaining geodynamo action, consisting of convective movement of the electrically conducting, liquid iron mixture forming the outer core. Thus we can use magnetic field changes to investigate the flow responsible for the dynamo although, for reasons outlined below (see also *Alfvén's Theorem and the frozen flux approximation*), such studies provide only a limited amount of information. More details of all but the most recent work are given in review articles by Bloxham and Jackson (1991), and Whaler and Davis (1997).

The starting point is a combination of Maxwell's equations and Ohm's Law, in the nonrelativistic limit appropriate for the core:

$$\frac{\partial \mathbf{B}}{\partial t} = \nabla \times (\mathbf{v} \times \mathbf{B}) + \lambda \nabla^2 \mathbf{B} \qquad \text{(Eq. 1)}$$

where $\mathbf{B}$ is the magnetic field, $\mathbf{v}$ the velocity, and $\lambda = 1/\mu_0\sigma$ is magnetic diffusivity (see *Magnetohydrodynamics* and *Alfvén's Theorem and the frozen flux approximation*). The first term on the right hand side of this equation represents advection, where magnetic field lines are moved around and re-arranged by the flow, and the second represents their creation and destruction, i.e., diffusion; both contribute to temporal changes of the magnetic field, on the left hand side of the equation.

To use Eq. (1) to determine the flow, we need to know the magnetic field and its spatial and temporal derivatives in the core. What we can actually measure are these quantities at the Earth's surface. Extrapolation to the base of the mantle is straightforward in principle if we assume that the mantle is electrically insulating, since then the magnetic field can be described as the gradient of a scalar potential which satisfies Laplace's equation. This potential represents the poloidal part of the field; the toroidal part is confined to the core and cannot be observed directly at the Earth's surface (see *Magnetohydrodynamics*). The coefficients of the scalar potential, known as spherical harmonic or Gauss coefficients, can be deduced from surface and satellite measurements of the field; techniques for doing so are described in *Main Field Modeling*. In practice, field extrapolation towards the sources is not a stable procedure, since field coefficients representing short wavelengths, and their uncertainties, are preferentially amplified. Fortunately, regularization techniques can be used to produce magnetic field models at the core–mantle boundary (CMB) in which the uncertain short wavelength features are suppressed (*Core-based inversions*). The weakly conducting mantle ($\lesssim$10 Sm^{-1}; see *Mantle, electrical conductivity, mineralogy*) has only a small effect on the downward continued field, so that values extrapolated through an insulator should not be seriously in error. This process determines the field at the base of the mantle, but only its radial component is guaranteed continuous across the electrical conductivity jump at the CMB (see *Alfvén's Theorem and the frozen flux approximation*). By a boundary layer analysis, Hide and Stewartson (1972) argued that the jump in horizontal components across the CMB should be small (see *Alfvén's Theorem and the frozen flux approximation* and *Core, boundary layers*), but there are additional problems in using the horizontal components of Eq. (1). The presence of a toroidal field component in the core means that, although we know the radial (poloidal) field at the CMB, we cannot estimate its spatial derivatives (required to calculate the diffusion term in Eq. (1)) there, and we have no knowledge of **B** in the bulk of the core.

The frozen-flux assumption

Fortunately, the high core conductivity means that its magnetic diffusivity is small. If diffusion is neglected, the radial component of (1) becomes

$$\frac{\partial \mathbf{B}_r}{\partial t} = \mathbf{B}_r \nabla_{\mathrm{H}} \cdot \mathbf{v} + \mathbf{v} \cdot \nabla_{\mathrm{H}} \mathbf{B}_r \qquad \text{(Eq. 2)}$$

where the subscript H indicates tangential derivatives, i.e.,

$$\nabla_{\mathrm{H}} = \nabla - \hat{\mathbf{r}}(\hat{\mathbf{r}} \cdot \nabla).$$

In fact, the situation is slightly more complicated in that there is a boundary layer adjacent to the CMB across which quantities adjust to match the boundary conditions (see *Alfvén's Theorem and the frozen flux approximation* and *Core, boundary layers*); where we refer to CMB flow and magnetic field in what follows, we actually mean those quantities at the top of the free stream immediately below the boundary layer.

Equation (2) provides one constraint on two unknowns, the tangential components of **v**—the CMB is a material boundary and thus the radial flow vanishes there. Hence we are unable to determine the flow uniquely, even at the CMB, without making additional assumptions. This nonuniqueness was first pointed out by Roberts and Scott (1965); Backus (1968) gives a clear exposition of exactly what can be determined from Eq. (2). He noted that, just as the CMB flow can be expanded into its toroidal and poloidal components, represented by scalars that integrate to zero over the CMB, so can the product of the flow and the radial field:

$$\mathbf{v}\mathbf{B}_r = \nabla_{\mathrm{H}}\chi - \hat{\mathbf{r}} \times \nabla_{\mathrm{H}}\psi. \qquad \text{(Eq. 3)}$$

Substituting (3) into (2) gives

$$\frac{\partial \mathbf{B}_r}{\partial t} + \nabla^2_{\mathrm{H}}\chi = 0.$$

This determines χ to within a function whose Laplacian vanishes, but does not constrain ψ at all. Some information on ψ comes from (3) evaluated on null-flux curves (NFCs). NFCs are contours on which B_r vanishes, i.e., they separate patches of the core where flux enters from those where it leaves. Setting $\mathbf{B}_r = 0$ in (3) gives

$$\nabla_{\mathrm{H}}\psi = -\hat{\mathbf{r}} \times \nabla_{\mathrm{H}}\chi,$$

determining ψ to within an arbitrary constant along each NFC. On NFCs, (2) becomes

$$\frac{\partial \mathbf{B}_r}{\partial t} = \mathbf{v}.\nabla_{\mathrm{H}}\mathbf{B}_r$$

from which the component of flow parallel to $\nabla_{\mathrm{H}}\mathbf{B}_r$, i.e., perpendicular to NFCs, v_n, can be deduced:

$$v_n = -\frac{\partial \mathbf{B}_r/\partial t}{|\nabla_{\mathrm{H}}\mathbf{B}_r|} \qquad \text{(Eq. 4)}$$

v_n is the only component of the flow that can be determined without further assumptions, and even this requires the frozen-flux hypothesis. Estimates of the magnetic Reynolds number for the core, and the diffusion rate of the dipole field, both suggest that the hypothesis is reasonable on the decade timescale. The assumption can be tested against the data: integrating (2) over a patch of the core bounded by a NFC, and noting that NFCs are material curves, gives

$$\frac{\partial}{\partial t}\iint_{S_i} \mathbf{B}_r dS = 0, \text{ or } \iint_{S_i} \mathbf{B}_r dS = \text{constant}, \qquad \text{(Eq. 5)}$$

where S_i is a patch of the core bounded by a NFC (see *Alfvén's Theorem and the frozen flux approximation*). Booker (1969) found that nonregularized CMB field models available at that time were inconsistent with these patch integral constraints, but nevertheless produced maps of the component of flow perpendicular to NFCs given by (4), on the basis that failure to satisfy the constraints may be due to temporal changes in unresolved shorter wavelength components of the magnetic field. More recent tests using regularized CMB field models are equivocal, since uncertainty bounds on the integrals are difficult to determine, and their interpretation depends on the philosophical viewpoint of the inversion practitioner (e.g., Backus, 1988).

In the diffusionless limit in which (2) applies, magnetic field lines are tied to fluid parcels (see *Alfvén's Theorem and the frozen flux approximation*). This means they can be used as tracers of the flow, but individual fluid parcels on a contour of $\mathbf{B}_r$ cannot be tracked, i.e., there is no information on circulation around a contour. However, the Navier-Stokes equation provides a constraint on the average circulation around NFCs. For an insulating mantle, the radial component of current density vanishes at the CMB, so on NFCs the Lorentz force is purely radial and the horizontal momentum balance thus geostrophic. Then the average circulation speed, $\bar{v}_c$, around a NFC satisfies

$$\bar{v}_c L + 2\Omega A_P = \text{constant},$$

where L is the length of the NFC and A_p its area projected onto the equatorial plane. A second piece of flow information that can be deduced without making further assumptions is the amount of horizontal convergence and divergence at points on the CMB where $\nabla_{\mathrm{H}}\mathbf{B}_r$ vanishes, i.e., at extrema (maxima, minima, and saddle points) of $\mathbf{B}_r$: from (2)

$$\nabla_H \cdot \mathbf{v} = -\frac{\dot{\mathbf{B}}_r}{\mathbf{B}_r}.$$

Toroidal-poloidal decomposition

Backus' (1968) analysis makes clear the severe nature of the CMB flow nonuniqueness. Several assumptions which reduce it and, like the frozen-flux hypothesis, are testable against the data, have been proposed. Before introducing them, and detailing the extent to which they reduce the ambiguity, it is useful to introduce the standard method for CMB flow modeling. Express the CMB flow as

$$\mathbf{v} = \mathbf{v}_{\mathrm{T}} + \mathbf{v}_{\mathrm{P}} = \nabla \times (T\hat{\mathbf{r}}) + \nabla_{\mathrm{H}}S. \qquad \text{(Eq. 6)}$$

The toroidal part, $\mathbf{v}_{\mathrm{T}}$, represented by the streamfunction, T, is divergence-free and represents motion such as solid body rotation and gyres, whereas the poloidal part, $\mathbf{v}_{\mathrm{P}}$, represented by the velocity potential, S, is curl-free and represents tangentially converging and diverging fluid. Expanding T and S as sums of spherical harmonics (qv),

$$\begin{aligned} T(\theta,\varphi) &= c\sum_{l,m}\left(t_l^{mc}\cos m\varphi + t_l^{ms}\sin m\varphi\right)P_l^m(\cos\theta) \\ &\equiv c\sum_{l,m} t_l^m Y_l^m(\theta,\varphi) \\ S(\theta,\varphi) &= c\sum_{l,m}\left(s_l^{mc}\cos m\varphi + s_l^{ms}\sin m\varphi\right)P_l^m(\cos\theta) \\ &\equiv c\sum_{l,m} s_l^m Y_l^m(\theta,\varphi) \end{aligned} \qquad \text{(Eq. 7)}$$

where θ and φ are colatitude and longitude, respectively, and c is the CMB radius. The flow at any point can be estimated if the coefficients $\{t_l^m\}$ and $\{s_l^m\}$ are known. The usual method for determining the coefficients is to substitute for $\mathbf{B}_r$ and $\partial\mathbf{B}_r/\partial t$ also expressed as spherical harmonic expansions, together with the spherical harmonic expressions (7) for T and S from (6), into (2), multiply by the complex conjugate of an arbitrary spherical harmonic, integrate over the core surface, and interchange integration and summation. The orthogonality of the spherical harmonics (qv) then isolates a secular variation (SV) coefficient on the left-hand side of (2), whilst the right-hand side can be manipulated into sums of Elsasser and Gaunt (sometimes referred to as Adams-Gaunt) integrals involving the coefficients of the main field, multiplied by the coefficients of T and S respectively. Various selection rules govern which Elsasser and Gaunt integrals are nonzero, and (2) reduces to a finite set of linear equations for the flow coefficients $\{t_l^m\}$ and $\{s_l^m\}$ if the sums (7) are terminated at a given spherical harmonic degree and order (see *Bullard-Gellman dynamo; Spherical harmonics*). In matrix form, we therefore have

$$\dot{\mathbf{g}} = (\mathbf{E}(\mathbf{g}) : \mathbf{G}(\mathbf{g}))\begin{pmatrix}\mathbf{t}\\ ..\\ \mathbf{s}\end{pmatrix} \qquad \text{(Eq. 8)}$$

where $\dot{\mathbf{g}}$ is a vector of SV spherical harmonic coefficients, $\mathbf{E}$ and $\mathbf{G}$ are matrices whose elements are linear combinations of main field spherical harmonic coefficients (represented by the vector $\mathbf{g}$) multiplying Elsasser and Gaunt integrals respectively, and $\mathbf{t}$ and $\mathbf{s}$ are vectors of spherical harmonic coefficients of the streamfunction and velocity potential respectively. In early studies such as that of Kahle *et al.* (1967), the flow coefficients were estimated from (8) using standard least squares analysis, without appreciating the ambiguity in the flow. In fact, most of the ambiguity-reducing assumptions discussed below can be incorporated into an estimation of the flow coefficients based on Eq. (8).

Ambiguity-reducing assumptions

The commonly adopted ambiguity-reducing assumptions are that the flow is purely toroidal (Whaler, 1980), tangentially geostrophic (Hills, 1979; Le Mouël, 1984), steady (Voorhies, 1986), or helical (Amit and Olsen, 2004); in several cases, the assumptions have been combined. Perhaps the easiest assumption conceptually is that the CMB flow is purely toroidal. Then the velocity potential is identically zero and (8) reduces to

$$\dot{\mathbf{g}} = \mathbf{E}(\mathbf{g})\mathbf{t}.$$

Under this assumption, the first term on the right-hand side of (2) vanishes, and thus the component of the flow perpendicular to contours of $\mathbf{B}_r$ can be estimated almost everywhere. Equation (4) now applies for all contours of $\mathbf{B}_r$, not just the NFCs. Flow around nonzero contours of $\mathbf{B}_r$ is unconstrained. At extrema of $\mathbf{B}_r$, (2) shows that $\dot{\mathbf{B}}_r$ must vanish, i.e., contours of zero radial SV must pass through them.

A more general test, analogous to that for frozen-flux, is provided by integrating (2) over patches of the CMB, but now those bounded by any contour of $\mathbf{B}_r$. Using the divergence theorem in a plane, with the radial flow at the CMB vanishing, we can show that (5) holds for patches S_i bounded by any nonzero contour of $\mathbf{B}_r$. Tests indicate that this condition is as well satisfied as that for patches bounded by NFCs, so that if we are willing to accept the frozen-flux hypothesis, we should also be willing to accept the purely toroidal flow hypothesis. If the flow is incompressible, horizontal flow convergence and divergence at the CMB, $\nabla_H\cdot\mathbf{v}\equiv\nabla_H\cdot\mathbf{v}_T$, equates to upwelling and downwelling flow which would be absent if the core was stably stratified adjacent to the CMB (see *Higgins–Kennedy paradox*).

Assuming that a toroidal flow is also steady significantly reduces the ambiguity. At two different epochs, the normal to a contour of $\mathbf{B}_r$ will in general point in a different direction, allowing two different components of the steady flow to be determined from (4). By combining them vectorially, using the fact that the flow is identical at the two epochs, the full flow is obtained everywhere except at those locations where the normal to the contour does not change direction. This concept can be extended to a steady general flow; magnetic field information is required at three epochs rather than two, and the flow is unique except where

$$\Delta = \hat{\mathbf{r}}.[\mathbf{B}_0(\mathbf{b}_1\times\mathbf{b}_2)+\mathbf{B}_1(\mathbf{b}_2\times\mathbf{b}_0)+\mathbf{B}_2(\mathbf{b}_0\times\mathbf{b}_1)] = 0 \qquad \text{(Eq. 9)}$$

where $\mathbf{b}_i = \nabla_H\mathbf{B}_r(\theta,\phi,t_i)$, $\mathbf{B}_i$ is the radial field component at t_i, and 0, 1, and 2 are the three epochs between which the flow is assumed steady. However, uncertainties on magnetic field components at the CMB, which are amplified when its spatial derivatives are taken, mean that Δ is not significantly larger than its uncertainty for time spans short enough for the steady flow assumption to be reasonable, i.e., Δ is difficult to distinguish from zero. In most practical determinations of the flow, steadiness is implicit over the period used to derive the SV. Steady flows are normally calculated by solving simultaneously sets of equations of condition (8) at different epochs (at least three) for a single set of flow coefficients, which defines a linear system. An alternative method is to embody the steady flow assumption in a nonlinear system, which can be solved iteratively.

The tangentially geostrophic (TG) flow assumption is a dynamical constraint making use of the Navier–Stokes equation with the approximation that the essential force balance at the CMB is between Coriolis and buoyancy forces, neglecting Lorentz forces (Bloxham and Jackson, 1991, discuss its validity). Curling the Navier–Stokes equation gives the thermal wind equation, whose radial component is

$$\nabla_H(\mathbf{v}\cos\theta) = 0. \qquad \text{(Eq. 10)}$$

This equation provides a second constraint on the flow, which then becomes uniquely determined except in so-called ambiguous patches, which are areas of the CMB bounded by contours of $\mathbf{B}_r/\cos\theta$ that do not cross the equator. Within the ambiguous patches, the component of flow around contours of $\mathbf{B}_r/\cos\theta$ cannot be determined, but the flow is fully resolved over the rest of the CMB. The test of consistency with the TG assumption is that (5) must hold over patches bounded by contours of $\mathbf{B}_r/\cos\theta$. Again, tests indicate that this condition is as well satisfied as that for NFC patch integrals. TG flows can be estimated either by imposing (10) as a side constraint during inversion of (8), or by setting up a geostrophic basis for the flow, and solving for its coefficients. In terms of the geostrophic basis, the flow is

$$\mathbf{v} = \sum_{l,m} z_l^m \boldsymbol{\omega}_l^m \qquad \text{(Eq. 11a)}$$

where the $\{z_l^m\}$ are the unknown coefficients and

$$\boldsymbol{\omega}_l^m = \begin{cases} \mathbf{T}_l^0 & m = 0 \\ \mathbf{S}_l^m + a_{l-1}^m \mathbf{T}_{l-1}^m + b_{l+1}^m \mathbf{T}_{l+1}^m & m \neq 0 \end{cases}. \qquad \text{(Eq. 11b)}$$

In (11b), $\{a_l^m, b_l^m\}$ are known functions, and

$$\mathbf{T}_l^m = \mathbf{r} \times \nabla Y_l^m$$

$$\mathbf{S}_l^m = r\nabla_H Y_l^m.$$

The geostrophic basis is complete, but it requires careful implementation in any numerical scheme where the expansion (11a) must be terminated, due to the coupling between successive spherical harmonics embodied in (11b).

Amit and Olson (2004) assume that CMB flow convergence and divergence, $\nabla_H \cdot \mathbf{v} = \nabla_H^2 S$, and its radial vorticity, $\varsigma = \hat{\mathbf{r}} \cdot (\nabla \times \mathbf{v}) = \nabla_H^2 T$, are proportional, i.e.,

$$\nabla_H^2 S = \mp k_0 \nabla_H^2 T, \qquad \text{(Eq. 12)}$$

where k_0 is a positive constant. The minus sign is associated with k_0 in the northern hemisphere, and the plus sign in the southern hemisphere. Helicity, $H = \varsigma \cdot \mathbf{v}$, vanishes at the CMB, but a nonzero value in the bulk of the core indicates a correlation between vorticity and velocity, and hence (12) is referred to as the helical flow assumption. The size of k_0 determines whether helicity is strong or weak. Amit and Olson (2004) also introduce what they refer to as columnar flow, where the toroidal and poloidal flow components are coupled through the adjustment of "Busse roll" flow to the core's spherical boundaries (see *Convection, nonmagnetic rotating*). The constraint is very similar to that for TG flows: (10) can be rewritten

$$\nabla_H \cdot \mathbf{v} = \frac{\tan\theta}{c} \mathrm{v}_\theta$$

whereas the columnar flow constraint is

$$\nabla_H \cdot \mathbf{v} = \frac{2\tan\theta}{c} \mathrm{v}_\theta. \qquad \text{(Eq. 13)}$$

A general constraint on upwelling is obtained by taking a linear combination of (12) and (13), with specific values of the combination constants giving helical, toroidal, TG, or columnar flows. Helical, TG flows are unique, and a necessary condition for the flow to be of this form is that (5) must be satisfied, where now the patches are bound by contours of Γ which satisfy the elliptic equation

$$\left(\frac{\partial \mathbf{B}_r}{\partial \phi} \pm k_0 \sin\theta \frac{\partial \mathbf{B}_r}{\partial \theta}\right) \frac{\partial \Gamma}{\partial \theta} - \left(\frac{\partial \mathbf{B}_r}{\partial \theta} + \mathbf{B}_r \tan\theta \mp k_0 \frac{1}{\sin\theta} \frac{\partial \mathbf{B}_r}{\partial \phi}\right) \frac{\partial \Gamma}{\partial \phi} = \pm k_0 \mathbf{B}_r c^2 \sin\theta \nabla_H^2 \Gamma.$$

A comparison of flows produced by the steady, toroidal, and TG assumptions calculated from the same data set shows that steady flows are *predominantly* toroidal and TG, although Bloxham and Jackson (1991) argue that the flow cannot be *purely* so. Modeling SV data obtained from observatory records (rather than spherical harmonic SV models) by a purely toroidal flow gives a good, but statistically inadequate, fit. Steady flows also fit the gross features of the SV record, and thus can model it over long periods with a very small number of parameters. Holme and Whaler (2001) extended the concept to calculate steady flows in a reference frame drifting with respect to the mantle. This maintains the numerical advantage of needing to estimate only a small number of parameters, but reproduces more of the detail of the SV. TG flows are attractive because they allow the flow to be deduced in the interior of the core (see below).

Flow construction

Most flows are calculated using a spectral approach, i.e., calculating coefficients of a global expression for **v**. Most commonly, the coefficients of the velocity potential, S, and streamfunction, T, are solved for, but TG flows have been constructed by estimating the geostrophic basis coefficients in (11a). Typically, inversions for the coefficients are regularized to ensure flows are not dominated by small-scale, poorly constrained motion. A number of regularization constraints have been proposed, the most commonly employed being minimizing the root-mean-square (rms) energy in the flow, or minimizing the rms of its second spatial derivatives. Alternatively, the flow minimizing the rms radial SV generated by advection of the main field can be sought; like the constraints on the flow itself, this is a quadratic norm of **v**. Time dependence can be introduced by parameterizing the coefficients of the streamfunction and velocity potential, e.g., through cubic B-splines, in much the same way as time-dependent models of the magnetic field itself have been constructed. During regularized inversion, the quantity minimized to find the flow coefficients is a linear combination of the sum of squares of residuals between the SV spherical harmonic coefficients (or occasionally, the SV values at points on the Earth's surface) and their predictions by the flow, and a quadratic norm of the flow. This imposes a practical uniqueness on the flow, regardless of the inherent ambiguity. In this linear inverse problem, the numerical value of the Lagrange multiplier is chosen to achieve the required balance between fitting the SV coefficients and the roughness of the flow. If required, the TG constraint (10) can be included with a second Lagrange multiplier.

Chulliat and Hulot (2000) and Amit and Olson (2004) chose instead to calculate local solutions. Chulliat and Hulot (2000) calculated what they called the "geostrophic pressure"—the CMB overpressure associated with a TG flow—outside the ambiguous patches. Amit and Olson (2004) used a finite difference method on a regular grid covering the CMB to calculate flows including helicity and columnar constraints. Near the equator, helical flows are dominated by the TG assumption, and poloidal flow sources are located in regions of meridional flow; away from the equator, flows are dominated by helicity, and poloidal flow sources coincide with vortex centers.

The most common features of CMB velocity maps, regardless of the ambiguity-reducing assumptions imposed, are a band of westward flow centered on the equator extending from about 90° E to 90° W, an anticlockwise gyre beneath the south-western Indian Ocean, a jet from the North (and to a lesser extent, South) pole approximately along the 90° E meridian, and slower flow beneath the hemisphere centered on the 180° E meridian (see Figure C25). The band of westward flow produces the westward drift of the geomagnetic field in the hemisphere centered on the Greenwich meridian, known since the time of Halley to be a first-order feature of the SV. Likewise, we would expect slower flow beneath the Pacific because the magnetic field there changes much more slowly, a feature of the palaeomagnetic as well as geomagnetic observational record.

SV can be interpreted in terms of wave motion (see *Magnetohydrodynamic waves*). There is evidence from time dependent CMB field models of westward drifting waves, both at mid-latitudes, and centered on the equator (with wavenumber 5, period 270 years, drifting westward at 17 km/year; Finlay and Jackson (2003)). If a stably stratified layer exists adjacent to the CMB, short-term SV variations could propagate through it by horizontally-polarized MAC waves (rather than be screened by the layer's skin depth).

Torsional oscillations and extrapolation into the core

The response of the core to a torque applied from the mantle is to set up torsional oscillations. These are oscillations of cylinders coaxial with the rotation axis, and thus the flow is constant on their surfaces. Torsional oscillations are responsible for changes in angular momentum of the core thought to maintain the angular momentum budget of the Earth system on the decade timescale (see *Length of day variations, decadal*).

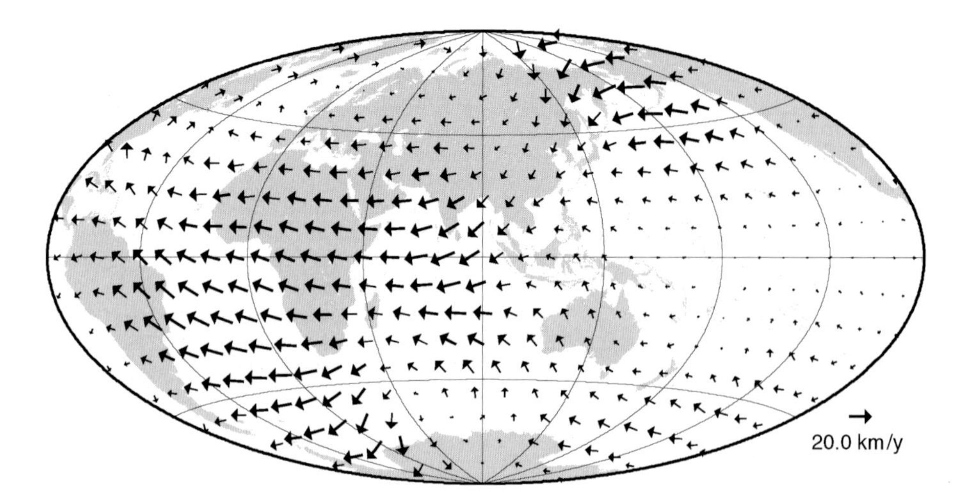

Figure C25 Typical steady flow for the period 1960–1980 inferred from the secular variation at the core–mantle boundary, illustrating features mentioned in the text. Continents are shown only for reference.

For such motions, the CMB flow is symmetric about the equator, and about the center of the Earth. About 80% of the CMB flow energy in unconstrained flows is symmetric about the equator. Assuming the symmetries required for TG flow on these coaxial cylinders reduces the number of independent velocity parameters by a factor of eight, but does not degrade the misfit to the SV significantly. The axial component of the core's angular momentum depends on just two coefficients of (7), t_1^0 and t_3^0, of a TG flow (neglecting the effect of the inner core). If the flows are realistic, its time changes should match the axial component of angular momentum changes in the mantle, which can be deduced from changes in the length-of-day. There is good agreement between the length-of-day signal and that predicted by time-varying TG CMB flow over the 20th century. Holme and Whaler (2001) found that, if they interpreted the drift of the core reference frame with respect to the mantle as representing solid body rotation, its time dependence also matched the observed length-of-day changes. Decadal changes in length-of-day indicate that the core flow is not steady, but the steady flow assumption remains a useful approximation in many circumstances.

Zatman and Bloxham (1997, 1999) used torsional waves as a means of extrapolating into the interior of the core. Modeling the SV by a steady CMB flow plus two torsional oscillations reproduces many features of observed geomagnetic impulses or "geomagnetic jerks" signal (see *Geomagnetic jerks*) well (Bloxham *et al.*, 2002). If s is distance from the rotation axis, the excitation and damping of these oscillations depends on B_s averaged over a cylinder centered on the rotation axis, $\bar{B}_s$. This relationship can be inverted for $\bar{B}_s$ as a function of s.

Summary

CMB flows can be deduced from geomagnetic observations and the frozen-flux induction equation using regularized inversion combined with additional assumptions on the nature of the flow, including that it is steady, toroidal, tangentially geostrophic, helical, and/or columnar. The extent to which the additional assumptions reduce the inherent ambiguity can be characterized, and the validity of the assumptions tested against the data, commonly through examining the time stationarity of surface integrals of B_r over patches of the CMB bounded by contours of specific functions of B_r. The main features of the flows are robust, regardless of the assumptions made concerning the nature of the flow. They have westward drift in an equatorial band in the western hemisphere, an anticlockwise gyre beneath the south-western Indian Ocean, and many have slower flow beneath the Pacific Ocean. Interpreting the flow in terms of torsional oscillations allows some extrapolation into the core. Time-varying flows predict angular momentum changes of the core which compensate for those of the mantle deduced from length-of-day variations.

Kathryn A. Whaler

Bibliography

Amit, H., and Olson, P., 2004. Helical core flow from geomagnetic secular variation. *Physics of Earth and Planetary Interiors*, **147**: 1–25.

Backus, G.E., 1968. Kinematics of geomagnetic secular variation in a perfectly conducting core. *Philosophical Transactions of the Royal Society of London*, **A263**: 239–266.

Backus, G.E., 1988. Comparing hard and soft bounds in geophysical inverse problems. *Geophysical Journal*, **94**: 249–261.

Bloxham, J., and Jackson, A., 1991. Fluid flow near the surface of the Earth's outer core. *Reviews of Geophysics*, **29**: 97–120.

Bloxham, J., Zatman, S., and Dumberry, M., 2002. The origin of geomagnetic jerks. *Nature*, **420**: 65–68.

Booker, J.R., 1969. Geomagnetic data and core motions. *Proceedings of the Royal Society of London*, **A309**: 27–40.

Chulliat, A., and Hulot, G., 2000. Local computation of the geostrophic pressure at the top of the core. *Physics of Earth and Planetary Interiors*, **117**: 309–328.

Finlay, C.C., and Jackson, A., 2003. Equatorially dominated magnetic field change at the surface of Earth's core. *Science*, **300**: 2084–2086.

Hide, R., and Stewartson, K., 1972. Hydromagnetic oscillations of the Earth's core. *Reviews of Geophysics and Space Physics*, **10**: 579–598.

Hills, R.G., 1979. Convection in the Earth's mantle due to viscous shear at the core-mantle interface and due to large-scale buoyancy, PhD thesis, New Mexico State University, Las Cruces.

Holme, R., and Whaler, K.A., 2001. Steady core flow in an azimuthally drifting reference frame. *Geophysical Journal International*, **145**: 560–569.

Kahle, A.B., Vestine, E.H., and Ball, R.H., 1967. Estimated surface motions of the Earth's core. *Journal of Geophysical Research*, **72**: 1095–1108.

Le Mouël, J.-L., 1984. Outer-core geostrophic flow and secular variation of Earth's geomagnetic field. *Nature*, **311**: 734–735.

Roberts, P.H., and Scott, S., 1965. On the analysis of secular variation, 1, A hydromagnetic constraint: Theory. *Journal of Geomagnetism and Geoelectricity*, **17**: 137–151.

Voorhies, C.V., 1986. Steady flows at the top of Earth's core derived from geomagnetic field models. *Journal of Geophysical Research*, **91**: 12444–12466.

Whaler, K.A., 1980. Does the whole of the Earth's core convect? *Nature*, **287**: 528–530.

Whaler, K.A., and Davis, R.G., 1997. Probing the Earth's core with geomagnetism. In Crossley, D.J. (ed.), *Earth's Deep Interior.*, Amsterdam: Gordon and Breach, pp. 114–166.

Zatman, S., and Bloxham, J., 1997. Torsional oscillations and the magnetic field within the Earth's core. *Nature*, **388**: 760–763.

Zatman, S., and Bloxham, J., 1999. On the dynamical implications of models of B_s in the Earth's core. *Geophysical Journal International*, **138**: 679–686.

Cross-references

Alfvén's Theorem and the Frozen Flux Approximation
Convection, Nonmagnetic Rotating
Core, Boundary Layers
Core-based Inversions for the Main Magnetic Field
Core–Mantle Coupling, Electromagnetic
Core–Mantle Coupling, Thermal
Dynamo, Bullard-Gellman
Geodynamo, Symmetry Properties
Geomagnetic Jerks
Geomagnetic Secular Variation
Harmonics, Spherical
Higgins–Kennedy Paradox
Length of Day Variations, Decadal
Magnetohydrodynamic Waves
Magnetohydrodynamics
Main Field Modeling
Mantle, Electrical Conductivity, Mineralogy
Oscillations, Torsional
Time-dependent Models of the Geomagnetic Field
Westward Drift

CORE ORIGIN

All major bodies of the inner solar system, including the Earth, are composed primarily of iron and silicates. Due to its high density, the iron component—the core—is invariably found at the center of the body. In addressing the origin of the Earth's core, four main questions arise. Firstly, composition: of what does the core consist? Secondly, accretion: how and when was this material assembled? Thirdly, differentiation: how and when did the Earth develop a recognizable core at its center? And finally, evolution: how did the core change subsequent to its formation? The first of these questions is dealt with elsewhere (see *Core composition; Inner core composition*), and will be only briefly summarized below. The other questions form the bulk of this article. Although the focus of the article will be the Earth's core, where possible these questions will also be discussed for other terrestrial planets. The answers to these questions are in many cases poorly understood, especially for planets for which few or no samples are currently available.

Composition

Because the solar photosphere and primitive (chondritic) meteorites have similar ratios of most elements, it is a reasonable supposition that these elements are present in the same ratios in the bulk Earth (and other terrestrial planets) (Taylor, 1992). The chondritic Si:Fe atomic ratio of 1.2:1 suggests that terrestrial planets should have roughly comparable core and mantle masses. The actual ratios of core to mantle mass are 85:15, 33:67, 15:85, and 2:98 for Mercury, Earth/Venus, Mars, and the Moon (Lodders and Fegley, 1998). This rather wide range of values is probably a result of late-stage impacts, as discussed below.

The composition of planetary cores depends partly on what elements were present at the time of core formation. Simple models of solar system formation suggest that elements with relatively low condensation temperatures (i.e., volatiles) are likely to have been more abundant at greater distances from the Sun. The terrestrial planets and meteorites appear to be depleted in elements with condensation temperatures $< \sim 1000$ K, compared to the initial solar nebula (Taylor, 1992). Whether there are gradations in volatile content between the terrestrial planets is currently unclear. Potassium isotope data (Humayun and Clayton, 1995) are a strong argument against such gradients occurring as a result of fractionating processes. Furthermore, any initial gradation will have been reduced by mixing of bodies from different solar distances during the accretion process.

For the Earth, more information is available because of seismological constraints on its interior structure, and constraints from samples on its bulk composition. As discussed elsewhere (*Core composition; Inner core composition*), the core consists of Fe plus about 10 wt% of one or more light elements, probably either S, Si, or O. These light elements are more concentrated in the liquid outer core and can significantly reduce the melting temperature of the liquid (see *Melting temperature of iron in the core, theory; Ideal Solution theory*). If core solidification occurs at the center of the planet, expulsion of the light elements during freezing can drive compositional convection (see *Convection, chemical; Geodynamo, energy sources*).

Because alkali element abundances are depleted with respect to the solar nebula, it is likely that potassium was mostly lost as a volatile during Earth's formation. Nonetheless, the remaining potassium is a potentially significant constituent of the core. Recent experimental evidence and theoretical arguments both suggest that potassium may partition into the core, especially if sulfur is a major constituent (see *Radioactive isotopes, their decay in mantle and core*). Radioactive decay of potassium-40 may help to power the terrestrial geodynamo (see *Geodynamo, energy sources*; Nimmo *et al.*, 2004) and could also be important on other planets. The presence of these secondary elements is presumably determined by their overall availability and the conditions that applied the last time the iron was in equilibrium with its surroundings.

There are few constraints on the detailed compositions or states of other planetary cores (see Taylor, 1992 and Nimmo and Alfe, 2006 for summaries). The Moon is very depleted in volatile elements and has a small core; these observations can be explained if it accreted at high temperatures from debris left over after a Mars-sized object collided with the Earth (e.g., Canup and Asphaug, 2001). Analyses of meteorites thought to have come from Mars have been used to deduce a planet that is richer in S and Fe than the Earth. The response of Mars to solar tides has been used to argue that the core of Mars is at least partially liquid. Similar observations have also been made for the Moon, Mercury, and Venus. Since most of these bodies are small and cool rapidly, these observations strongly suggest the presence of substantial amounts ($\sim$10 wt%) of S in the outer cores of these bodies. Because S is not incorporated into the solid core at low pressures, it becomes progressively harder to freeze the remaining liquid as solidification progresses. Whether the presence of large quantities of S is likely at Mercury's orbit, or following a high energy impact, remains an unanswered question.

Solar system accretion

The process of accretion is now reasonably well understood, at least in outline (see Weidenschilling (2000) and Chambers (2004) for summaries of the likely timescales and processes). The solar system began when a nebular cloud of dust and gas became gravitationally unstable, and collapsed in on itself, forming a star surrounded by the orbiting remnants of the original cloud. Models suggest that the accumulation of these cloud remnants into Mars-sized planetary embryos was a rapid process, taking O(10^5 years) in the inner solar system and somewhat longer at greater distances from the Sun. The final stage of accretion, in which these planetary embryos collided in high-energy impacts, was much slower, taking O(10^8 years). The early formation of Jupiter is presumed to have frustrated planet formation within the asteroid belt (see Chambers, 2004).

Presumably, each planetary embryo began as a relatively homogeneous object, similar to the undifferentiated asteroids observed today. However, collisions between different embryos caused these bodies to increase in temperature as they grew (e.g., Stevenson, 1989). Furthermore, bodies, which formed early, may also have been heated by the decay of now-extinct radioactive elements, especially ^{26}Al (half-life 0.73 Myr). Thus, at some critical size, the planetary embryo will have begun to melt and (as discussed below) core formation is likely to have proceeded rapidly thereafter. This critical size probably varied in both time and space; for instance, Ceres, the largest asteroid (present-day diameter $D = 913$ km) appears not to have differentiated at all, while Vesta ($D = 520$ km) underwent widespread melting, which certainly involved core formation.

The process of differentiation is discussed in more detail below. Of more interest, from the point of view of accretion, is the timescale over which differentiation occurs. This timescale may be obtained from hafnium-tungsten (Hf/W) isotopic observations, as follows (see Harper and Jacobsen, 1996 for a summary).

Core differentiation leads to fractionation between Hf and W, with Hf going preferentially into the silicate phase and about 80% of the W into the core. The unstable isotope of Hf, ^{182}Hf, decays to ^{182}W with a half-life of 9 Ma. If differentiation happens before all the ^{182}Hf is exhausted, then the mantle will show an excess of ^{182}W over stable ^{180}W relative to an undifferentiated (chondritic) sample. The amount of the ^{182}W excess depends on the initial ^{182}Hf/^{180}Hf ratio and the timing of differentiation. Since the former is known, observations of the ^{182}W/^{180}W ratio have been used to constrain core formation timescales.

If core formation is assumed to occur instantaneously, then the times after solar system formation at which core formation took place are (Kleine *et al.*, 2002): 4 Ma, 13 Ma, 33 Ma, and 26–33 Ma for Vesta, Mars, the Earth, and the Moon, respectively. These timescales indicate that planetary accretion and differentiation is a relatively rapid process. Core formation, as discussed below, occurs most rapidly if the iron is liquid. Vesta's early core formation age and small size suggest that heating either by ^{26}Al or large impacts was probably important. For larger bodies like the Earth, core melting would have occurred simply due to the gravitational energy of accretion (Stevenson, 1989; see below).

In practice, accretion is likely to occur over a long timescale (tens of million years for an Earth-sized body; Chambers, 2004). Furthermore, for bodies the size of the Earth, it is likely that many of the accreting planetary embryos had themselves already undergone differentiation. Thus, the interpretation of the apparent core age is not straightforward. For instance, the increase in apparent core age with object size (Figure C26) is probably a reflection of a more prolonged accretion process for the larger bodies. An additional problem in the interpretation of Hf/W data is that the degree to which incoming bodies re-equilibrate with the planet's mantle is uncertain, and probably depends on the size of the impactor and the state of the mantle (magma ocean or solid) (Harper and Jacobsen 1996). The Moon-forming impact almost certainly produced a magma ocean on Earth and may also have partially or completely re-set the isotopic Hf/W clock.

Similar issues arise when considering the concentrations of siderophile (iron-loving) elements in the mantle (see Righter (2003) for a review). On Earth, these concentrations are significantly higher than those expected for equilibration at low pressures and temperatures. A possible solution is that equilibration took place at the bottom of a deep ($\sim$1000 km) magma ocean. Interestingly, siderophile abundances estimated for the Martian mantle, based on Martian meteorite compositions, are more consistent with equilibration at lower temperatures and pressures (Righter, 2003). Although highly model-dependent, these observations suggest that Mars lacked a deep magma ocean. These conclusions are consistent with the expectation that larger bodies have higher internal temperatures owing to their greater gravitational potential energy.

Differentiation

As described above, small planetary embryos (or planetesimals) probably consisted of a homogeneous mixture of metallic, silicate, and volatile components. However, as the planetesimals grew, collision velocities and thus the heat energy of each impact increased. Simultaneously, radioactive decay within the planetesimals further increased their internal temperature, especially if short-lived species such as ^{26}Al were still active at this point. At some particular size, probably depending on the rate of growth, the interior of the planetesimal started to melt. As described below, melting permits the denser components of the planetesimal to migrate towards the center of the body, thus achieving the process of differentiation. The following description is based largely on Stevenson (1990) and Solomon (1979) (see also Figure C27).

In general, metallic iron melts before end-member mantle silicates. Thus, differentiation can potentially occur either by the core liquid migrating through a solid silicate matrix or by liquid core droplets settling in a silicate liquid. The latter process is appropriate, for example, to the kind of magma ocean that is formed after giant impacts and results in rapid core formation. Based on the balance between viscous and surface tension forces in a convecting magma ocean, the likely

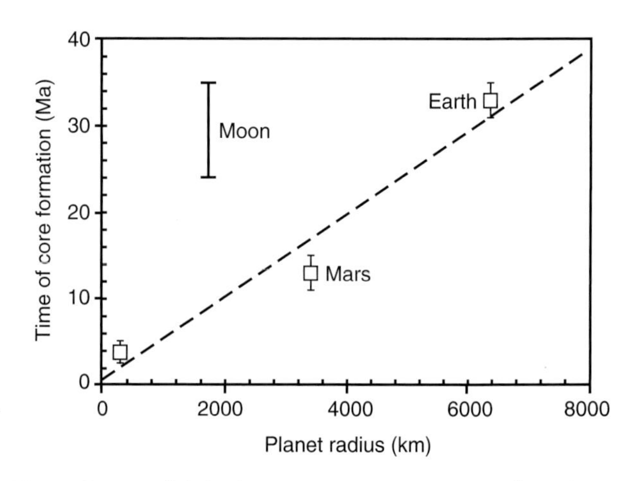

Figure C26 Model single-stage core formation age after solar system formation compared with present-day radius of body. From Kleine *et al.* (2002).

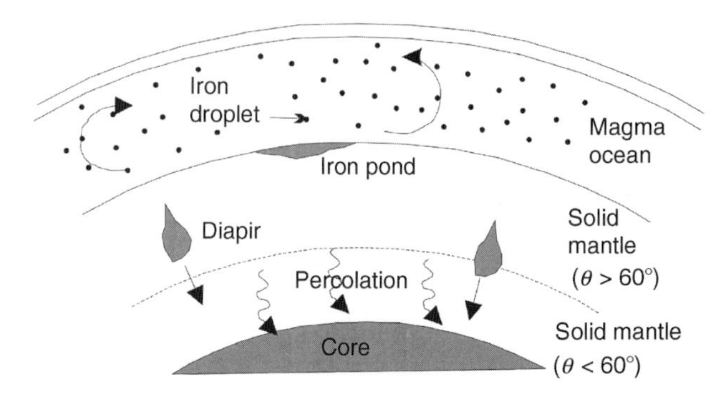

Figure C27 Mechanisms of core differentiation, after Stevenson (1990). is the dihedral angle.

iron droplet size is $\sim$1 cm. At these length scales, chemical equilibration with the surrounding mantle material is likely to be complete.

The former process, core liquid migrating through a solid silicate matrix, is controlled largely by the dihedral angle, the characteristic angle formed at an interface between solid grains and an interstitial melt. For large dihedral angles, typical of iron-silicate interfaces at low pressures, melt percolation is inefficient because the iron droplets accumulate in disconnected pockets. Percolation is much more rapid if the dihedral angle is smaller ($<60°$), the melt fraction is large ($>\sim10\%$), or large shear stresses occur. The dihedral angle typically decreases with pressure, so that large variations in permeability with depth may have occurred within accreting planets. For dihedral angles $<60°$, transit timescales through the mantle for individual iron particles are typically 10^4–10^5 y. For dihedral angles $>60°$, an alternative mechanism for differentiation is the accumulation of large bodies of iron which then migrate downwards as diapirs.

Once differentiation begins, it is likely to go rapidly to completion: the downwards motion of the dense iron results in a release of gravitational potential energy, leading to local heating, viscosity reduction, and an increase in the rate at which differentiation proceeds. This runaway effect is larger for larger bodies, but even for Mars-sized objects the energy of core differentiation is sufficient to raise the temperature of the entire planet by 300 K (Solomon, 1979).

Evolution and inner core growth

Once the bulk of differentiation has ended, subsequent evolution of planetary cores is more leisurely. Core cooling is controlled by the rate at which the overlying mantle can remove heat (see *Core-mantle boundary, heat flow across; Interiors of planets and satellites*). Reactions between the core and mantle may take place, but are likely to be relatively slow (see *D″, Composition*). Partly because of the energy released during differentiation, the cores are initially likely to be both hot and convecting, and in many cases (e.g., Mars, Earth, the Moon?) develop early dynamos (see *Dynamos, planetary and satellite; Magnetic field of Mars;* Nimmo and Alfe, 2006).

Over the longer term, the core will cool and core solidification will set in, acting as another energy source for the dynamo (see *Geodynamo, energy sources*). Only for the Earth is the existence of a solid inner core certain (see *Inner core composition*). Similar inner cores may exist for the other terrestrial planets, depending on the slopes of the core adiabat (see *Core, adiabatic gradient*) and melting curve, and the core sulfur content (Nimmo and Alfe, 2006). The age of the inner core depends on the rate at which the mantle is extracting heat from the core. Current estimates (see *Core-mantle boundary, heat flow across*) suggest an inner core age of only $\sim$1 Ga. Although the uncertainties in this value are still quite large, a cooling rate sufficient to maintain the terrestrial dynamo implies an inner core significantly younger than the age of the Earth (Nimmo *et al.*, 2004). Prior to the formation of the inner core, the terrestrial dynamo was driven mainly by cooling of the liquid core, with possible assistance from radioactive decay (see *Radioactive isotopes, their decay in mantle and core*).

Francis Nimmo

Bibliography

Canup, R.M., and Asphaug, E., 2001. Origin of the Moon in a giant impact near the end of the Earth's formation. *Nature*, **412**: 708–712.

Chambers, J.E., 2004. Planetary accretion in the inner Solar System. *Earth and Planetary Science Letters*, **223**: 241–252.

Harper, C.L., and Jacobsen, S.B., 1996. Evidence for 182Hf in the early solar system and constraints on the timescale of terrestrial accretion and core formation. *Geochimica et Cosmochimica Acta*, **60**: 1131–1153.

Humayun, M., and Clayton, R.N., 1995. Potassium isotope cosmochemistry—genetic implications of volatile element depletion. *Geochimica et Cosmochimica Acta*, **59**: 2131–2148.

Kleine, T., Munker, C., Mezger, K., and Palme, H., 2002. Rapid accretion and early core formation on asteroids and the terrestrial planets from Hf-W chronometry. *Nature*, **418**: 952–955.

Lodders, K., and Fegley, B., 1998. The planetary scientist's companion, Oxford: Oxford University Press.

Nimmo, F., and Alfe, D., 2006. Properties and evolution of the Earth's core and geodynamo. In Sammonds, P.R., and Thompson, J.M.T., (eds.), *Advances in Science: Earth Science.* London: Imperial College Press.

Nimmo, F., Price, G.D., Brodholt, J., Gubbins, D., 2004. The influence of potassium on core and geodynamo evolution. *Geophysical Journal International*, **156**: 363–376.

Righter, K., 2003. Metal-silicate partitioning of siderophile elements and core formation in the early Earth. *Annual Reviews of Earth and Planetary Sciences*, **31**: 135–174.

Solomon, S.C., 1979. Formation, history and energetics of cores in the terrestrial planets. *Physics of Earth and Planetary Interiors*, **19**: 168–182.

Stevenson, D.J., 1989. Formation and early evolution of the Earth. In Peltier, W.R., (ed.), *Mantle Convection and Plate Tectonics.* London: Gordon and Breach, pp. 818–868.

Stevenson, D.J., 1990. Fluid dynamics of core formation. In Newson, H.E., and Jones, J.E., (eds.), *Origin of the Earth.* New York: Oxford University Press, pp. 231–249.

Taylor, S.R., 1992. Solar system evolution, Cambridge: Cambridge University Press.

Weidenschilling, S.J., 2000. Formation of planetesimals and accretion of the terrestrial planets. *Space Science Review*, **92**: 295–310.

Cross-references

Convection, Chemical
Core Composition
Core, Adiabatic Gradient
Core–Mantle Boundary, Heat Flow Across
D″, Composition
Dynamos, Planetary and Satellite
Geodynamo, Energy Sources
Ideal Solution Theory
Inner Core Composition
Interiors of Planets and Satellites
Magnetic Field of Mars
Melting Temperature of Iron in the Core, Theory
Radioactive Isotopes, Their Decay in Mantle and Core

CORE PROPERTIES, PHYSICAL

This is an overview of properties, many of which have individual entries in this encyclopedia [see *Core composition; Core properties, theoretical determination;* D″ *as a boundary layer; Grüneisen's parameter for iron and Earth's core;* and *Core viscosity*]. Some properties are very well determined, notably density and elasticity, but others are hardly more than guesses and for some there is significant disagreement, as in the case of the Grüneisen parameter. Values given here are selected, as far as possible to make a self-consistent set, constrained by relevant thermodynamic relationships. Table C3 summarizes them for the top and bottom of the outer core and the top of the inner core.

Elasticity and the equation of state

The most secure data on the core that we have are obtained from seismology and are presented in the digested form of Earth models. The most widely used of these is PREM (Preliminary Reference Earth Model—Dziewonski and Anderson, 1981). PREM is a parameterized model in the sense that density and the speeds of P and S waves are

Table C3 Physical properties of the core: a summary. Numbers are constrained to give a self-consistent set, retaining digits beyond the level of absolute accuracy

Property	Outer core at CMB	Outer core at ICB	Inner core at ICB
Density, ρ(kgm^{-3})	9902	12163	12983
Bulk modulus, K_S (GPa)	645.9	1301.3	1303.7
K_T (GPa)	588.7	1214.4	1221.4
$K'_S \equiv (\partial K_S/\partial P)_S$	3.513	3.317	3.338
$K_S K''_S$	−1.276	−0.703	−0.759
Rigidity modulus, μ(GPa)	–	–	169.4
$\mu' \equiv (\partial \mu/\partial P)_S$	–	–	0.207
Mean atomic weight $\bar{m}$	48.1	48.1	52.5
Specific heat, C_P (JK^{-1}kg^{-1})	815	794	728
Temperature, T (K)	3739	5000	5000
$\mathrm{d}T/\mathrm{d}z$ (K/km)	0.884	0.286	0.049
Grüneisen parameter, γ	1.443	1.390	1.391
$q = (\partial \ell n\gamma/\partial \ell n V)_T$	0.254	0.129	0.111
Volume expansion coefficient α (10^{-6} K^{-1})	18.0	10.3	9.7
Electrical resistivity, ρ_e (μΩ m)	2.12	2.02	1.6
Electrical conductivity, σ_e (10^5 S m^{-1})	4.7	5.0	6.3
Magnetic diffusivity, η_m (m^2 s^{-1})	1.69	1.61	1.27
Thermal conductivity, κ (W m^{-1} K^{-1})	46	63	79
Thermal diffusivity, η (10^{-6} m^2 s^{-1})	5.7	6.5	8.5
Viscosity (Pa s)	$\sim 10^{-2}$	$\sim 10^{-2}$	–
Latent heat, melting (10^5 J kg^{-1})		9.6	
Density change on melting (kg m^{-3})		−200	

presented as polynomials in radius, over different ranges. In the case of the outer core they are third order polynomials. This makes the model convenient to use for many purposes and it represents well the broad scale of properties, density (and therefore gravity and pressure), as well as elasticity. However, there is a limitation. When we consider derivative properties, especially the pressure derivatives of bulk modulus, K, that is $K' = \mathrm{d}K/\mathrm{d}P$ and $K'' = \mathrm{d}^2K/\mathrm{d}P^2$, the model parameterization gives unrealistic variations. The dimensionless derivatives K' and KK'' are essential parameters of an equation-of-state (EoS) so it is necessary to constrain them by imposing an EoS and using PREM to fit EoS parameters. This means that we decide on the form of the EoS before fitting model data and the bias of this decision influences the conclusions. The result is a divergence of inferences about properties according to the favored equations.

It must be noted that the bulk modulus observed seismologically is the adiabatic value, K_S, which applies to the compression of thermally isolated material. It is related to the isothermal modulus, K_T, describing the compression of material held at constant temperature, by the identity

$$K_S = K_T(1 + \gamma\alpha T)$$

γ being the Grüneisen parameter and α the volume expansion coefficient, which are discussed below and listed in Table C3. In common parlance, "bulk modulus" means K_T unless otherwise specified. In the core K_S and K_T differ by 7%–10%. We must also be careful to specify that in the outer core the depth dependence of K_S is given by its adiabatic derivative, $K'_S \equiv (\partial K_S/\partial P)_S$, because the temperature gradient is adiabatic. Similarly, $K''_S \equiv (\partial K'_S/\partial P)_S$.

The status and acceptability of current rival equations are reviewed by Stacey and Davis (2004). The situation has improved noticeably since about 2000 with new thermodynamic constraints that restrict the acceptable equations to those that are specifically written as equations for K'. One of these, which we refer to as the "reciprocal K-primed equation," was used to obtain the EoS parameters in Table C3:

$$\frac{1}{K'} = \frac{1}{K'_0} + \left(1 - \frac{K'_\infty}{K'_0}\right)\frac{P}{K}$$

where subscripts zero and infinity indicate the zero and infinite pressure extrapolations of K'. A particular advantage of an equation of this form is that it makes use of an identity (common to all of the rival equations, but not recognized)

$$K'_\infty = 1/(P/K)_\infty$$

In the case of the core, K'_∞ is quite tightly constrained by the observation of inner core rigidity, so this identity provides a fixed infinite pressure end point to the EoS.

The need for a theoretical constraint on the EoS is obvious also in consideration of the rigidity modulus, μ, of the inner core, which is not well observed seismologically. PREM gives a satisfactory average value but a gradient such that the ratio μ/K increases with depth, which is incompatible with homogeneous material. We have a relationship first developed for application to the inner core but shown to fit lower mantle data precisely

$$\frac{\mu}{K} = \left(\frac{\mu}{K}\right)_0 - \left[\left(\frac{\mu}{K}\right)_0 - \left(\frac{\mu}{K}\right)_\infty\right] K'_\infty \frac{P}{K}$$

In fact it is not only μ' that is anomalous in the inner core by PREM, but also K'. Both can be adjusted to agree with the above equations without changing the profile of $(K + 4/3\mu)$ which is reliably observed from the P wave speed. Now, μ/K in the inner core is quite small, so that there is little scope for extrapolation of the μ/K equation above. μ/K must fall below the inner core value, but cannot vanish, that is $(\mu/K)_\infty = 0$ gives an extreme upper bound to the P/K scale and therefore a lower bound on K'_∞. The result is a very restricted range of possible values: $K'_\infty = 3.0 \pm 0.1$. When the reciprocal K' equation is applied to the core even this range is seen to be too wide; the value 2.9 is implausible. So, we now have a firm basis for adjusting the PREM tabulation for the core and the values in Table C3 were obtained in this way. Details of fitting and a complete tabulation are given by Stacey and Davis (2004).

Specific heat

At the high temperatures of the core the lattice component of specific heat, C_ℓ, can differ little from the classical (Dulong-Petit) value, $3R$ (per mole). With the estimated mean atomic weights in Table C3, this means 519 $JK^{-1}kg^{-1}$ for the outer core and 475 JK^{-1} kg^{-1} for the inner core. Added to this is an electron component, C_e, arising from the thermal excitation of conduction electrons. At specified volume, C_e is proportional to temperature, T, because both the number of electrons that are thermally excited and the level to which they are excited are proportional to T, making the electron thermal energy proportional to T^2. The proportionality constant depends on the density of electron states from which electrons may be thermally excited and this decreases with compression by an increasing energy spread of the bands of energy levels. As a convenient analytical representation of the results of a numerical band structure calculation, Boness *et al.* (1986) wrote the electron specific heat of iron at density ρ as

$$C_e = (A\rho_0/\rho - B)T$$

ρ_0 being the zero pressure density (at 290 K). Allowing for the dilution by lighter elements, we can use this expression for outer core alloy with $A = 0.12709$ $JK^{-2}kg^{-1}$ and $B = 0.02378$ $JK^{-2}kg^{-1}$, making the total specific heat at constant volume

$$C_V = C_\ell + C_e$$

This is converted to C_P, specific heat at constant pressure, as listed in Table C3, by the identity

$$C_P = C_V(1 + \gamma\alpha T)$$

which is similar to the relationship (above) between K_S and K_T.

Temperature and temperature gradient

Core temperatures in Table C3 are anchored to 5000 K at the inner core boundary. This estimate could be in error by 500 K. It is based on the coexistence at that level of solid and liquid, with the assumption that there is some melting point depression by the solutes, relative to the melting point of pure iron, which is usually estimated to be nearer to 6000 K at ICB pressure. The temperature gradient is calculated from the identity

$$(\partial \ln T/\partial \ln \rho)_s = \gamma$$

and the temperature at the core mantle boundary is obtained by integrating this gradient with a density dependence of γ, as considered below.

The Grüneisen parameter and thermal expansion coefficient

The thermodynamic definition of the Grüneisen parameter is

$$\gamma = \frac{\alpha K_T}{\rho C_V} = \frac{\alpha K_S}{\rho C_P}$$

It is a dimensionless number, of order unity, that is a simple combination of four familiar properties. Three of them are subject to independent estimates. K_S/ρ is obtained from the seismic wave speeds and C_P, calculated as described above, cannot be far from 800 $JK^{-1}kg^{-1}$ in the core (Table C3). The thermal expansion coefficient, α, is the odd one out. Our only effective way of estimating α is by means of γ. Knowledge of one is equivalent to knowledge of the other and equations for thermodynamic properties of the Earth's interior can be written in terms of either, but since α is obtained from γ it is usually more convenient to use γ directly. This is essential to calculation of the adiabatic temperature gradient, as above, and hence the thermodynamic efficiency of convection and much other thermal geophysics.

As has been recognized since the early work of E. Grüneisen and J.C. Slater, in principle γ can be represented in terms of elastic moduli and their pressure derivatives, but the theories that do this are not free of doubts and difficulties. Grüneisen defined the parameter named after him in terms of volume dependence of crystal vibration frequencies, making the connection to elasticities obvious. As Stacey and Davis (2004) showed by an application of Einstein's theory of specific heat, the Grüneisen and thermodynamic definitions are equivalent if the mode frequencies are independent of temperature at constant volume. Then, by assuming that the crystal mode frequencies are adequately represented by the seismologically observed moduli, K and μ, we obtain a formula known as the acoustic γ. This works well for the mantle, but is not applicable to the core. Apart from the problem of applying to the liquid outer core a formula using μ and μ', electrons have an important role in metals that is not properly accounted for in the Grüneisen approach.

Another way of looking at γ is to see it as the ratio of thermal pressure to thermal energy. If we raise the temperature of a unit volume of material by ΔT, by applying heat $\rho C_P \Delta T$ at constant P, causing thermal expansion $\alpha\Delta T$, and then recompress it adiabatically to the original volume, the pressure is increased by $\alpha K_S \Delta T$. This is thermal pressure and we see that the ratio of this to the applied heat coincides with the thermodynamic definition of γ. Still considering only atomic vibrations, we can apply this concept to calculate the mean forces between vibrating atoms (and hence pressure) relative to their thermal energy. This leads to a formula of the form

$$\gamma = \frac{(1/2)(dK/dP) - 1/6 - (f/3)(1 - P/3K)}{1 - (2f/3)P/K}$$

where f is a factor that depends on how the atoms vibrate (correlations, mutual interactions between bonds, and so on). Rival theories give a wide range of alternatives ($f = 0, 1, 2, 2.35$) but none is convincing. However, the general form of the equation appears satisfactory and it can be used with a value of f selected empirically to match a relevant observation, such as a measured zero pressure value of γ. On the basis of a comparison with the acoustic γ for the lower mantle, Stacey and Davis (2004) suggested $f = 1.44$, but hesitated to assume that any constant value was acceptable. There is one situation in which we can be confident of this formula. In the limit $P \to \infty$, so that $K'_\infty \to (K/P)_\infty$, we have $\gamma_\infty = (1/2)K'_\infty - 1/6$, independently of f. We can note that there is no difficulty in applying this equation to a liquid, such as the outer core, and there is no appeal to mode frequencies so that electron pressure is properly included, but that doubt remains about the role of shear modes.

A third approach to the calculation of γ relies on the thermodynamic requirement that its volume derivative, $q = (\partial \ell n\gamma/\partial \ell nV)_T$, must decrease strongly with pressure, vanishing in the infinite pressure limit, so that the next derivative, $\lambda = (\partial \ell nq/\partial \ell nV)_T$, is at least nearly constant. This method also requires calibration by a known value of γ_0, but uses also the fix on γ_∞ (on which all theories agree). Interestingly, Stacey and Davis (2004) found that when applied to the core the second and third of these approaches gave almost identical values. Thus, although the theory is still incomplete and numerical calculations give higher values [see *Core properties, theoretical determination*] thermodynamic arguments lead to the values in Table C3, which reports values of q as well as γ and also the values of α that follow from it.

Electrical and thermal conductivities of the core

As pointed out in the separate entries for these properties, they are closely related. Electrons dominate thermal as well as electrical conduction. Calculation of electrical conductivity, σ_e, is usually a calculation of its inverse, resistivity, $\rho_e = 1/\sigma_e$. This is a measure of the impediment to the free motion of electrons by thermal vibration, crystal irregularities, and impurities. It is the impurity resistivity that gives the greatest difficulty, but the overall conductivity is now almost certainly estimated to $\pm 25\%$. The precise value does not appear crucial to the dynamo, but the consequent thermal conductivity is more critical because of the drain on core heat by thermal conduction. The conductivities of core alloy appear nicely balanced, being high enough for dynamo action but not so high as to exhaust the energy source by conductive heat loss.

Magnetic susceptibility and diffusivity

Iron in any of its forms is strongly paramagnetic, that is it responds positively to a magnetic field, but under core conditions the magnetic susceptibility is only about 5×10^{-6}, that is the permeability, μ, differs from the vacuum value, $\mu_0 = 4\pi \times 10^{-7}$ Hm^{-1}, by about 5 parts in 10^6. It can have no influence on the dynamo or other properties, including the magnetic diffusivity, $\eta_m = 1/(\sigma_e \mu)$.

Viscosity

The molecular viscosity of the outer core, $\sim 10^{-2}$ Pa s, is believed to be little different from that of liquid iron at ordinary pressures. It is low enough to cause no geophysical effect. This is even true of the eddy viscosity, arising from turbulence in the magnetically controlled flow, being 1 to 2×10^4 Pa s, but anisotropic according to the local orientation of the magnetic field.

Frank D. Stacey

Bibliography

Boness, D.A., Brown, J.M., and McMahan, A.K., 1986. The electronic thermodynamics of iron under Earth core conditions. *Physics of Earth and Planetary Interiors*, **42**: 227–240.

Dziewonski, A.M., and Anderson, D.L., 1981. Preliminary reference Earth model. *Physics of Earth and Planetary Interiors*, **25**: 297–356.

Stacey, F.D., and Davis, P.M., 2004. High pressure equations of state and applications to the lower mantle and core. *Physics of Earth and Planetary Interiors*, **142**: 137–184.

Cross-references

Core Composition
Core Density
Core Properties, Theoretical Determination
Core Viscosity
Core, Adiabatic Gradient
Core, Electrical Conductivity
Core, Thermal Conduction
D″ as a Boundary Layer
Grüneisen's Parameter for Iron and Earth's Core
Shock Wave Experiments

CORE PROPERTIES, THEORETICAL DETERMINATION

Introduction

The fact that the core is largely composed of Fe was firmly established as a result of Birch's analysis of mass-density/sound-wave velocity systematics. Today we believe that the outer core is about 6%–10% less dense than pure liquid Fe, while the solid inner core is a few percent less dense than crystalline Fe. Before a full understanding of the chemically complex core is reached, however, it is necessary to understand the properties and behavior at high pressure (P) and temperatures (T) of its primary constituent, namely metallic Fe. Experimental techniques have evolved rapidly in the past years, and today using diamond anvil cells or shock experiments, the study of minerals at pressures up to $\sim$200 GPa and temperatures of a few thousand Kelvin is possible. These studies, however, are still far from routine and results from different groups are often in conflict. As a result, therefore, in order to complement these existing experimental studies and to extend the range of pressure and temperature over which we can model the Earth, computational mineral physics has, in the past decade, become an established and growing discipline.

Within computational mineral physics a variety of atomistic simulation methods (developed originally in the fields of solid state physics and theoretical chemistry) are used. These techniques can be divided approximately into those that use some form of interatomic potential model to describe the energy of the interaction of atoms in a mineral as a function of atomic separation and geometry, and those that involve the approximate solution of Schrödinger's equation to calculate the energy of the mineral species by quantum mechanical techniques. For the Earth sciences, the accurate description of the behavior of minerals as a function of temperature is particularly important and computational mineral physics usually uses either lattice dynamics or molecular dynamics methods to achieve this important step. The relatively recent application of all of these advanced condensed matter physics methods to geophysics has only been made possible by the very rapid advances in the power and speed of computer processing. Techniques, which in the past were limited to the study of structurally simple compounds, with small unit cells, can today be applied to describe the behavior of complex, low symmetry structures (which epitomize most minerals) and liquids. In this entry, we will focus on recent computational studies of Fe, which have been aimed at predicting its geophysical properties and behavior under core conditions. These recent quantum mechanical calculations of the free energy of the solid and liquid phases of Fe have enabled the determination of a number of geophysically significant thermodynamic properties of Fe (Alfè *et al.*, 2001, 2002).

The structure of Fe under core conditions

Under ambient conditions, Fe adopts a body-centered cubic (bcc) structure that transforms with temperature to a face-centered cubic (fcc) form, and with pressure transforms to a hexagonal close packed (hcp) phase, ε-Fe. The high P/T phase diagram of pure iron itself however is still controversial. Various diamond anvil cell (DAC) based studies have been interpreted as showing that hcp Fe transforms at high temperatures to a phase that has variously been described as having a double hexagonal close-packed structure (dhcp) or an orthorhombicly distorted hcp structure. Furthermore, high-pressure shock experiments have also been interpreted as showing a high pressure solid-solid phase transformation, which has been suggested could be due to the development of a bcc phase. Other experimentalists, however, have failed to detect such a post-hcp phase and have suggested that the previous observations were due either to minor impurities or to metastable strain-induced behavior.

Further progress in interpreting the nature and evolution of the core would be severely hindered if the uncertainty concerning the crystal structure of the core's major chemical component remained unresolved. Such uncertainties can be resolved, however, using *ab initio* calculations (e.g., Vočadlo *et al.*, 1999). Spin-polarized simulations have been performed on candidate phases (including a variety of distorted bcc and hcp structures and the dhcp phase) at pressures ranging from 325 to 360 GPa. These reveal that under these conditions only bcc Fe has a residual magnetic moment and all other phases have zero magnetic moments. It should be noted however, that the magnetic moment of bcc Fe disappears at temperatures >1000 K, indicating that even bcc Fe will have no magnetic stabilization energy under core conditions. Static simulations find that at core pressures, the bcc polymorph of iron is mechanically unstable and continuously transforms to the fcc when allowed to relax to a state of isotropic stress. In contrast, hcp, dhcp, and fcc Fe remain mechanically stable at core pressures, and so it is possible to calculate their phonon frequencies and hence free energies. From such an analysis the hcp phase of Fe is found to have the lowest Gibbs free energy over the whole *P-T* space thought to represent inner core conditions.

Despite the fact that bcc Fe is mechanically unstable at high pressure and low temperatures (and so not amenable to lattice dynamical modeling), it has been suggested that it may become entropically stabilized at high temperatures (e.g., Matsui and Anderson, 1997). This behavior is found in some other transition metal phases, such as Ti and Zr. Vočadlo *et al.* (2003a) have performed molecular dynamic simulations of bcc Fe and have found that this structure is indeed mechanically stable above ~1000 K at core densities, but in the absence of S or Si, still remains metastable with respect to hcp-Fe.

Calculated thermodynamic properties of hcp Fe

Alfè *et al.* (2001) have presented a comprehensive set of quantum mechanically derived thermodynamic data for hcp-Fe. Their calculated total constant-volume specific heat per atom C_v (Figure C28) emphasizes the importance of electronic excitations. In a purely harmonic system, C_v would be equal to $3k_B$, and it is striking that C_v is considerably greater than that even at the modest temperature of 2000 K, while at 6000 K it is nearly doubled. The decrease of C_v with increasing pressure evident in Figure C28 comes from the suppression of electronic excitations by high compression and to a smaller extent from the suppression of anharmonicity.

The thermal expansivity α (Figure C29) is one of the few cases where it is possible to compare calculated properties with experimental measurements and both approaches show that α decreases strongly with increasing pressure. Calculation also shows that α increases significantly with temperature. The product αK_T of expansivity and isothermal bulk modulus, which is equal to $(dP/dT)_v$, is important because it is sometimes assumed to be independent of pressure and temperature over a wide range of conditions, and this constancy is used to extrapolate experimental data. Calculated isotherms for αK_T (Figure C30) indicate that its dependence on P is indeed weak, especially at low temperatures, but that its dependence on T certainly cannot be ignored, since it increases by at least 30% as T goes from 2000 to 6000 K at high pressures.

The thermodynamic Grüneisen parameter $\gamma = V(dP/dE)_v = \alpha K_T V/C_v$ plays an important role in high-pressure physics because it relates the thermal pressure and the thermal energy. Assumptions about the value of γ are frequently used in reducing shock data from the Hugoniot to an isothermal state. If one assumes that γ depends only on V, then the thermal pressure and energy are related by

$$P_{th}V = \gamma E_{th}$$

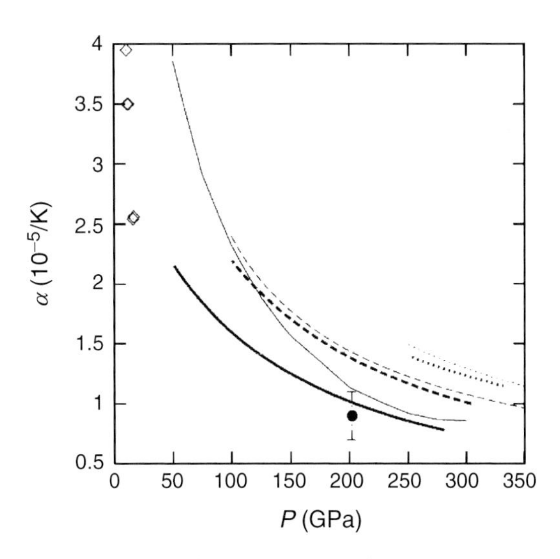

Figure C29 The thermal expansivity (α) of hcp-Fe as a function of pressure on isotherms $T = 2000$ K (*continuous curves*), 4000 K (*dashed curves*), and 6000 K (*dotted curves*). Heavy and light curves show results of Alfè *et al.* (2000) and of Stixrude *et al.* (1997), respectively. The black circle is the experimental value of Duffy and Ahrens (1993) at $T = 5200 \pm 500$ K. Diamonds are data from Boehler (1990) for temperatures between 1500 and 2000 K.

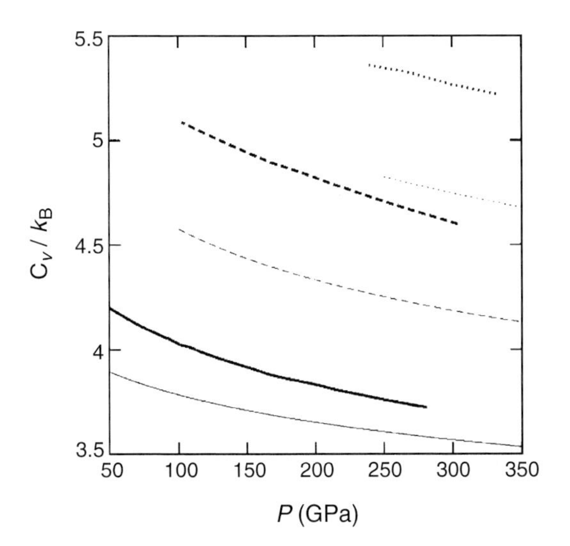

Figure C28 Total constant-volume specific heat per atom (C_V, in units of k_B) of hcp Fe as a function of pressure on isotherms $T = 2000$ K (*continuous curves*), 4000 K (*dashed curves*), and 6000 K (*dotted curves*). Heavy and light curves show results of Alfè *et al.* (2000) and of Stixrude *et al.* (1997), respectively.

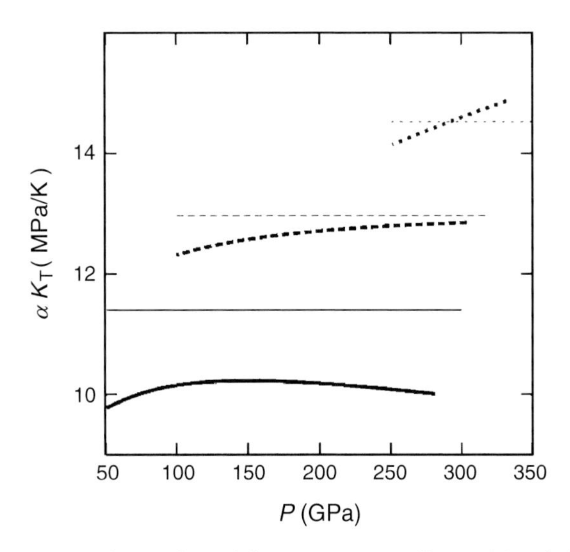

Figure C30 The product of the expansion coefficient (α) and the isothermal bulk modulus (K_T) of hcp-Fe as a function of pressure on isotherms $T = 2000$ K (*continuous curves*), 4000 K (*dashed curves*), and 6000 K (*dotted curves*). Heavy and light curves show results of Alfè *et al.* (2000) and of Stixrude *et al.* (1997), respectively.

the well known Mie-Grüneisen equation of state. At low temperatures, where only harmonic phonons contribute to E_{th} and P_{th}, γ should indeed be temperature independent above the Debye temperature, because $E_{th} = 3k_BT$ per atom, and $P_{th}V = -3k_BT(\mathrm{d}\ln\omega/\mathrm{d}\ln V) = 3k_BT\gamma_{ph}$, so that $\gamma = \gamma_{ph}$ (the phonon Grüneisen parameter), which depends only on V. But at high temperatures, the temperature independence of γ will clearly fail because of electronic excitations and anharmonicity. Calculated results for γ (Figure C31) indicate that it varies rather little with either pressure or temperature in the region of interest. At temperatures below ~4000 K, it decreases with increasing pressure, as expected from the behavior of γ_{ph}. This is also expected from the often-used empirical rule of thumb $\gamma = (V/V_0)^q$, where V_0 is a reference volume and q is a constant exponent usually taken to be roughly unity. Since V decreases by a factor of about 0.82 as P goes from 100 to 300 GPa, this empirical relation would make γ decrease by the same factor over this range, which is roughly what we see. However, the pressure dependence of γ is very much weakened as T increases until, at 6000 K, γ is almost constant.

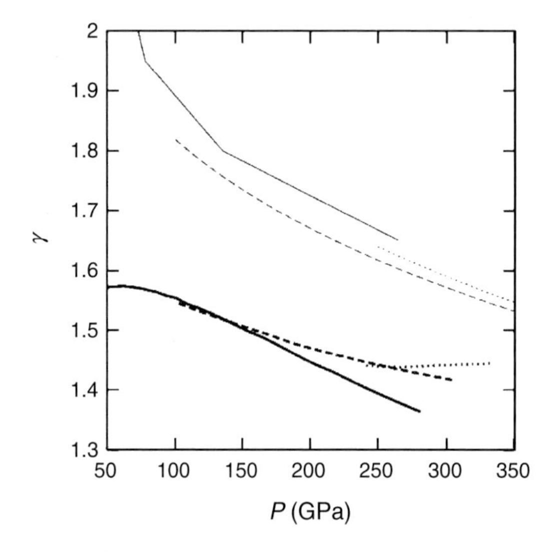

Figure C31 The Grüneisen parameter (γ) of hcp-Fe as a function of pressure on isotherms $T = 2000$ K (*continuous curves*), 4000 K (*dashed curves*), and 6000 K (*dotted curves*). Heavy and light curves results of Alfè *et al.* (2000) and of Stixrude *et al.* (1997), respectively.

The measured and calculated elastic constants of hcp-Fe are presented in Table C4. Although there is some scatter on the reported values of c_{12}, overall, the agreement between the experimental and various *ab initio* studies are excellent.

Rheological and thermodynamical properties of liquid Fe

De Wijs *et al.* (1998) in Alfè *et al.* (2000) have reported the quantum mechanically determined viscosity, diffusion, and thermodynamic properties of Fe in the liquid state. Their simulations cover the temperature range 3000–8000 K and the pressure range 60–390 GPa. Table C5 shows a comparison of the pressures calculated in the simulations with the pressures deduced by Anderson and Ahrens (1994) from a conflation of experimental data, and Figure C32a-e summarizes for liquid Fe the calculated values of density, thermal expansion coefficient, adiabatic and isothermal bulk moduli, heat capacity (C_v), and Grüneisen parameter. These values are in close accord with the estimates available in Anderson and Ahrens (1994). It is worth noting that the systematics of the behavior of heat capacity (C_v), Grüneisen parameter, etc., for liquids is quite different for those of solids.

Table C6 reports the calculated values for the self diffusion of Fe in liquid Fe, and the viscosity of liquid Fe for the same range of P and T conditions reported above. For the entire pressure-temperature domain of interest for the Earth's outer core, the diffusion coefficient, D, and viscosity, η, are comparable with those of typical simple liquids, D being $\sim 5 \times 10^{-9}$ m^{2}s^{-1} and η being in the range 8–15 mPa s, depending on the detailed thermodynamic state.

Conclusion

The past decade has seen a major advance in the application of *ab initio* methods in the solution of high pressure and temperature geophysical problems, thanks to the rapid developments in high-performance computing. The progress made means that it is now

Table C4 A compilation of elastic constants (c_{ij}, in GPa), bulk (K) and shear (G) moduli (in GPa), and longitudinal (v_P) and transverse (v_S) sound velocity (in km s^{-1}) as a function of density (ρ, in g cm^{-3}) and atomic volume (in Å^3 per atom). In this table $K = (\langle c_{11}\rangle + 2\langle c_{12}\rangle)/3$ and $G = (\langle c_{11}\rangle - \langle c_{12}\rangle + 3\langle c_{44}\rangle)/5$, where $\langle c_{11}\rangle = (c_{11} + c_{22} + c_{33})/3$, etc. Previous calculated values are from Stixrude and Cohen (1995), Steinle-Neumann *et al.* (1999), Söderlind *et al.* (1996), and Vočadlo *et al.* (2003b). The experimental data of Mao *et al.* (1999) is also presented

	V	ρ	c_{11}	c_{12}	c_{13}	c_{33}	c_{44}	c_{66}	K	G	v_P	v_S
Stixrude and Cohen	9.19	10.09	747	301	297	802	215	223	454	224	8.64	4.71
	7.25	12.79	1697	809	757	1799	421	444	1093	449	11.50	5.92
Steinle-Neumann *et al.*	8.88	10.45	930	320	295	1010	260	305	521	296	9.36	5.32
	7.40	12.54	1675	735	645	1835	415	470	1026	471	11.49	6.13
	6.66	13.93	2320	1140	975	2545	500	590	1485	591	12.77	6.51
Mao *et al.*	9.59	9.67	500	275	284	491	235	113	353	160	7.65	4.06
	7.36	12.60	1533	846	835	1544	583	344	1071	442	11.48	5.92
Söderlind *et al.*	9.70	9.56	638	190	218	606	178	224	348	200	8.02	4.57
	7.55	12.29	1510	460	673	1450	414	525	898	448	11.03	6.04
	6.17	15.03	2750	893	1470	2780	767	929	1772	789	13.70	7.24
Vocadlo *et al.*	9.17	10.12	672	189	264	796	210	242	397	227	8.32	4.74
	8.67	10.70	815	252	341	926	247	282	492	263	8.87	4.96
	8.07	11.49	1082	382	473	1253	309	350	675	333	9.86	5.38
	7.50	12.37	1406	558	647	1588	381	424	900	407	10.80	5.74
	6.97	13.31	1810	767	857	2007	466	522	1177	500	11.77	6.13
	6.40	14.49	2402	1078	1185	2628	580	662	1592	630	12.95	6.59

Table C5 Pressure (in GPa) calculated as a function of temperature (T) and density for liquid Fe. Experimental estimates are in parentheses, from Anderson and Ahrens (1994)

ρ (kg m^{-3})					
T (K)	9540	10700	11010	12130	13300
3000	60				
4300		132 (135)			
5000		140 (145)			
6000	90	151 (155)	170 (170)	251 (240)	360 (335)
7000		161	181	264 (250)	375 (350)
8000		172	191	275	390 (360)

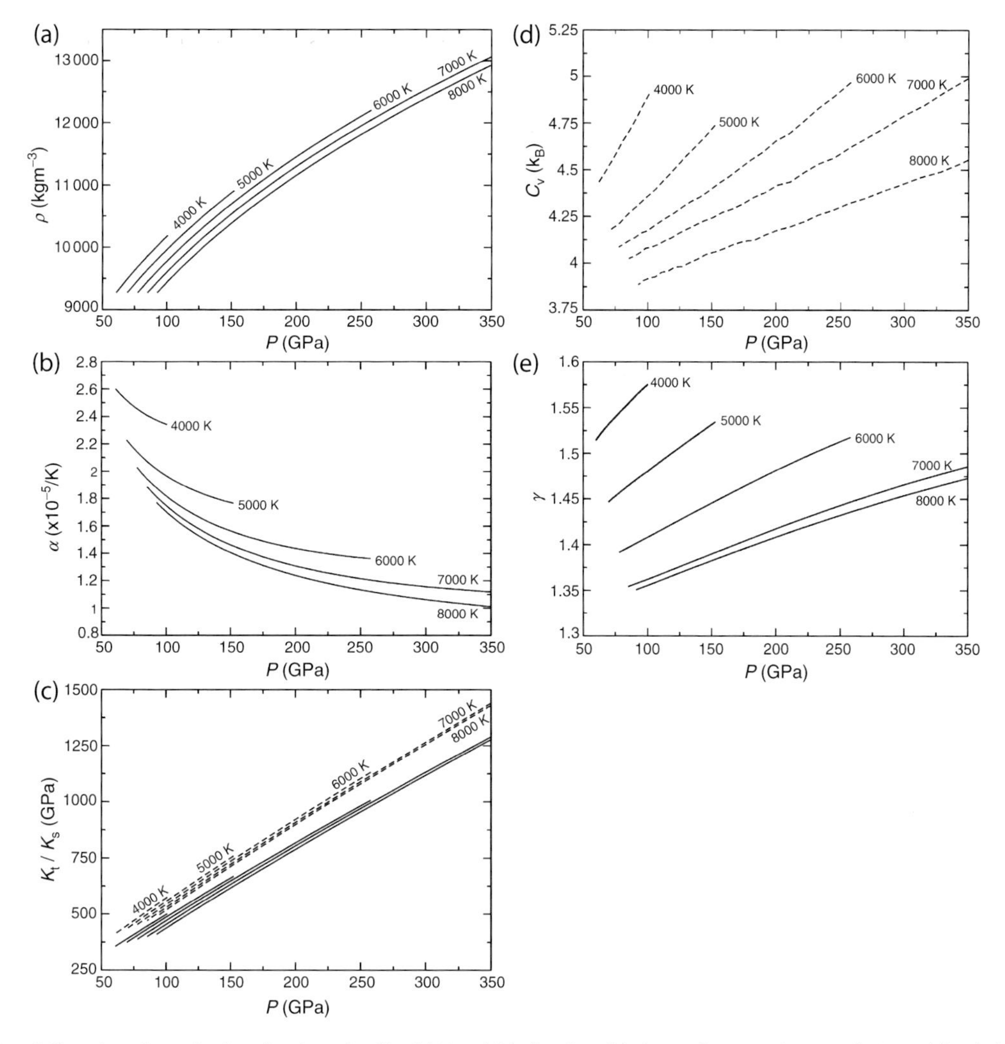

Figure C32 Panel Showing the calculated values for liquid Fe of (a) density, (b) thermal expansion coefficient, (c) adiabatic (*dashed lines*) and isothermal bulk (*soild lines*) moduli, (d) heat capacity (C_v), and (e) Grüneisen parameter, all as a function of P at temperatures between 4000 and 8000 K.

Table C6 The diffusion coefficient (D) and the viscosity (η) from our *ab initio* simulations of liquid Fe at a range of temperatures and densities. The error estimates come from statistical uncertainty due to the short duration of the simulations

	ρ (kg m^{-3})					
	T (K)	9540	10700	11010	12130	13300
D (10^{-9} m^2 s^{-1})	3000	4.0 ± 0.4				
	4300		5.2 ± 0.2			
	5000		7.0 ± 0.7			
	6000	14 ± 1.4	10 ± 1	9 ± 0.9	6 ± 0.6	5 ± 0.5
	7000		13 ± 1.3	11 ± 1.1	9 ± 0.9	6 ± 0.6
η (mPa s)	3000	6 ± 3				
	4300		8.5 ± 1			
	5000		6 ± 3			
	6000	2.5 ± 2	5 ± 2	7 ± 3	8 ± 3	15 ± 5
	7000		4.5 ± 2	4 ± 2	8 ± 3	10 ± 3

possible to calculate from first principles the free energies of solid and liquid phases, and hence to determine both the phase relations and the physical properties of planetary-forming minerals. In the future, with "tera-scale computing," it will be possible to model more complex and larger systems to investigate for example solid-state rheological problems or physical properties such as thermal and electrical conductivity, which are currently beyond current quantum mechanical simulations.

David Price

Bibliography

Alfè, D., Kresse, G., and Gillan, M.J., 2000. Structure and dynamics of liquid Iron under Earth's core conditions. *Physical Review*, **61**: 132–142.

Alfè, D., Price, G.D., and Gillan, M.J., 2001. Thermodynamics of hexagonal-close-packed iron under Earth's core conditions. *Physical Review*, **B64**: 045123.

Alfè, D., Price, G.D., and Gillan, M.J., 2002. Iron under Earth's core conditions: Liquid-state thermodynamics and high-pressure melting curve from *ab initio* calculations. *Physical Review*, **B65**: 165118, 1–11.

Anderson, W.W., and Ahrens, T.J., 1994. An equation of state for liquid iron and implication for the Earth's core. *Journal of Geophysical Research*, **99**: 4273–4284.

Brown, J.M., and McQueen, R.G., 1986. Phase transitions, Grüneisen parameter and elasticity for shocked iron between 77 GPa and 400 GPa. *Journal of Geophysical Research*, **91**: 7485–7494.

Duffy, T.S., and Ahrens, T.J., 1993. Thermal-expansion of mantle and core materials at very high-pressures. *Geophysical Research Letters*, **20**: 1103–1106.

Mao, H.K., Shu, J., Shen, G., Hemley, R.J., Li, B., and Sing, A.K., 1999. Elasticity and rheology of iron above 220GPa and the nature of the Earth's inner core. *Nature*, **399**: 280.

Matsui, M., and Anderson, O.L., 1997. The case for a body-centred cubic phase for iron at inner core conditions. *Physics of Earth and Planetary Interiors*, **103**: 55–62.

Söderlind, P., Moriarty, J.A., and Wills, J.M., 1996. First-principles theory of iron up to earth-core pressures: Structural, vibrational and elastic properties. *Physical Review*, **B53**: 14063–14072.

Steinle-Neumann, G., Stixrude, L., and Cohen, R.E., 1999. First-principles elastic constants for the hcp transition metals Fe, Co, and Re at high pressure. *Physical Review*, **B60**: 791–799.

Stixrude, L., and Cohen, R.E., 1995. Constraints on the crystalline structure of the inner core—mechanical instability of bcc iron at high-pressure. *Geophysical Research Letters*, **22**: 125–128.

Stixrude, L., Wasserman, E., and Cohen, R.E., 1997. Composition and temperature of the Earth's inner core. *Journal of Geophysical Research*, **102**: 24729–24739.

Vočadlo, L., deWijs, G.A., Kresse, G., Gillan, M.J., and Price, G.D., 1997. First principles calculations on crystalline and liquid iron at Earth's core conditions. *Faraday Discussions*, **106**: 205–217.

Vočadlo, L., Brodholt, J., Alfè, D., Price, G.D., and Gillan, M.J., 1999. The structure of iron under the conditions of the Earth's inner core. *Geophysical Research Letters*, **26**: 1231–1234.

Vočadlo, L., Alfè, D., Gillan, M.J., Wood, I.G., Brodholt, J.P., and Price, G.D., 2003a. Possible thermal and chemical stabilisation of body-centred-cubic iron in the Earth's core? *Nature*, **424**: 536–553.

Vočadlo, L., Alfè, D., Gillan, M.J., and Price, G.D., 2003b. The Properties of iron under core conditions from first principles calculations. *Physics of Earth and Planetary Interiors*, **140**: 101–125.

Cross-references

Core Density
Core Properties, Physical
Core Temperature
Core Viscosity
Core, Adiabatic Gradient
Grüneisen's Parameter for Iron and Earth's Core
Inner Core Anisotropy
Inner Core Seismic Velocities
Melting Temperature of Iron in the Core, Experimental
Melting Temperature of Iron in the Core, Theoretical
Shock Wave Experiments

CORE TEMPERATURE

High-pressure experiments combined with geophysical and geodynamic modeling suggest that the temperature near the bottom of the mantle probably lies between 2500 and 3000 K. There is still uncertainty over the heat flow out of the core and the thermal conductivity of the materials in D″ zone, so it is not possible to give a precise estimate of the temperature increase across the thermal boundary layer at the base of the mantle, although it is generally accepted that this may be of the order of 1000 K. Another approach to constraining the temperature of the core is based upon the assumption that the solid inner core is crystallizing from the liquid outer core, and the ICB marks a crystallization interface. Given that the core is iron (alloyed with about 10% of lighter elements), and given that impurities lower the melting point, then the melting temperature (T_m) of iron at the pressure of the ICB can be used to place an upper bound on the core's temperature at this depth. As reviewed elsewhere in this volume (see *Melting temperature of iron in the core, experimental* and *Melting temperature of iron in the core, theory*) estimates for T_m can be obtained from shock melting experiments, by laser-heated diamond-anvil cells studies, and from theory. The experimental measurements are difficult and their interpretation is controversial, with ICB melting temperatures inferred from experiments ranging from ~4850 K (Boehler, 1996) to ~6850 K (Yoo *et al.*, 1993). Values from theory, however, now seem to be more closely clustered, with values of T_m of Fe at 360 GPa of between 6000 and 6400 K (see *Melting temperature of iron in the core, theory*). However, to constrain the temperature at the ICB more tightly, it is essential to infer an estimate of the true composition of the core and to determine the resulting suppression of the melting temperature caused by the light alloying elements.

On the basis of seismology and data for pure Fe, it is considered that the outer and inner core contain some light element impurities (see *Core composition*). Cosmochemical abundances of the elements, combined with models of the Earth's history, limit the possible impurities to a few candidates. Those most often discussed are S, O, and Si (e.g., Poirier, 1994; Allègre *et al.*, 1995), and most studies are, to date, confined to these three alloying elements. Boehler (2000) reviews the high *P* melting studies done on Fe-O and Fe-S systems, and concludes that the *in situ* detection of the solidus in such experiments is difficult because of the small melt fraction produced at the eutectic temperature in a multicomponent system. Nevertheless, he claims that experiments suggest that alloying elements in Fe produce only a small depression in T_m. This contrasts with historical estimates (Stevenson, 1981) of the effect of, for example, S on the T_m of Fe at core pressures, which indicated that a suppression of ~1100 K should be expected.

To address this problem, Alfè *et al.* (2002 a,b) have recently developed a strategy that uses quantum mechanical molecular dynamics to both constrain seismologically and thermodynamically allowed compositions for the outer and inner core, and to determine the magnitude of alloy-induced suppression of the melting temperature of Fe at the ICB. Their approach is based on the supposition that the solid inner core is slowly crystallizing from the liquid outer core and that therefore the inner and outer cores are in thermodynamic equilibrium at the ICB. This implies that the chemical potentials of Fe and of each impurity must be equal on the two sides of the ICB. If the core consisted of pure Fe, equality of the chemical potential (Gibbs free energy in this case) would tell us only that the temperature at the ICB is equal to the melting temperature of Fe at the ICB pressure of 330 GPa. With impurities present, equality of the chemical potentials for each impurity element imposes a relation between the mole fractions in the liquid and the solid, so that with S, O, and Si there are three such relations. But these three relations must be consistent with the inferred values of the mass densities in the inner and outer core deduced from seismic and free-oscillation data.

The chemical potential, μ_x, of a solute *X* in a solid or liquid solution is conventionally expressed as $\mu_x = \mu_x^0 + k_B T \ln a_x$, where μ_x^0 is a constant and a_x is the activity. It is common practice to write $a_x = \gamma_x c_x$, where γ_x is the activity coefficient and c_x the concentration of *X*. The chemical potential can therefore be expressed as

$$\mu_x = \mu_x^0 + k_B T \ln \gamma_x c_x \qquad \text{(Eq. 1)}$$

which can be rewritten as

$$\mu_x = \mu_x^* + k_B T \ln c_x \qquad \text{(Eq. 2)}$$

It is helpful to focus on the quantity μ_x^* for two reasons: first, because it is a convenient quantity to obtain by *ab initio* calculations (Alfè *et al.*, 2002b); second, because at low concentrations the activity coefficient, γ_x, will deviate only weakly from unity by an amount proportional to c_x and by the properties of the logarithm the same will be true of μ_x^*.

The equality of the chemical potentials μ_x^l and μ_x^s in coexisting liquid and solid (superscripts l and s, respectively) then requires that

$$\mu_x^{*l} + k_B T \ln c_x^l = \mu_x^{*s} + k_B T \ln c_x^s \qquad \text{(Eq. 3)}$$

or equivalently

$$c_x^s / c_x^l = \exp\left[(\mu_x^{*l} - \mu_x^{*s})/k_B T\right] \qquad \text{(Eq. 4)}$$

This means that the ratio of the mole fractions c_x^s and c_x^l in the solid and liquid solution is determined by the liquid and solid thermodynamic quantities μ_x^{*l} and μ_x^{*s} . Although liquid-solid equilibrium in the Fe/S and Fe/O systems has been experimentally studied up to pressures of around 60 GPa, there seems little prospect of obtaining experimental data for $\mu_x^{*l} - \mu_x^{*s}$ for Fe alloys at the much higher ICB pressure. However, Alfè *et al.* (2002 b) have recently shown that the fully *ab initio* calculation of μ_x^{*l} and μ_x^{*s} is technically feasible. Thus, the chemical potential, μ_x, of chemical component *X* can be defined as the change of Helmholtz free energy when one atom of *X* is introduced into the system at constant temperature *T* and volume *V*. In *ab initio* simulations, it is awkward to introduce a new atom, but the awkwardness can be avoided by calculating $\mu_X^* - \mu_{Fe}^*$, which is the free energy change, ΔF, when an Fe atom is replaced by an X atom. For the liquid, this ΔF is computed by applying the technique of "thermodynamic integration" to the (hypothetical) process in which an Fe atom is continuously transmuted into an X atom.

Alfè *et al.* find a major qualitative difference between O and the other two impurities. For S and Si, μ_x^* is almost the same in the solid and the liquid, the differences being at most 0.3 eV, i.e., markedly smaller than $k_B T$ ~0.5 eV; but for O the difference of μ_x^* between solid and liquid is ~2.6 eV, which is much bigger than $k_B T$. This means that added O will partition strongly into the liquid, but added S or Si will have similar concentrations in both the solid and the coexisting liquid phase.

Their simulations of the chemical potentials of the alloys can be combined with simulations of their densities to investigate whether the known densities of the liquid and solid core can be matched by any binary Fe/X system, with X = S, O, or Si. Using their calculated partial volumes of S, Si, and O in the binary liquid alloys, one finds that the mole fractions required to reproduce the liquid core density are 16%, 14%, and 18% respectively (Figure C33, panel (a) displays the predicted liquid density as a function of c_x compared with the seismic density). The calculated chemical potentials in the binary liquid and solid alloys then give the mole fractions in the solid of 14%, 14%, and 0.2%, respectively, that would be in equilibrium with these liquids (see Figure C33, panel (b)). Finally, the partial volumes in the binary solids give ICB density discontinuities of 2.7 ± −0.5, 1.8 ± −0.5 and 7.8 ± −0.2%, respectively (Figure C33, panel (c)). As expected, for S and Si, the discontinuities are considerably smaller than the known value of 4.5 ± −0.5%; for O, the discontinuity is markedly greater than the known value. Alfè *et al.* conclude that none of the binary systems can account for the discontinuity quantitatively. The density and seismic character of the core can, however, be accounted for by O together with either or both of S and Si. *Ab initio* calculations on general quaternary alloys containing Fe, S, O, and Si will clearly be feasible in the future, but currently they are computationally too demanding, so for the moment it is assumed that the chemical potential of each impurity species is unaffected by the presence of the others. This led Alfè *et al.* to estimate that the mole fractions needed to account for the ICB density discontinuity were 8.5 ± 2.5 mole % S (and/or Si) and 0.2 ± 0.1% O in the inner core and 10 ± 2.5% S/Si and 8 ± 2.5% O in the liquid outer core. This compositional estimate is based on the value of the density discontinuity at the ICB determined by Shearer and Masters (1990). Since then, Masters and Gubbins (2003) have reassessed the free oscillation data set and have determined the density jump at the ICB to be

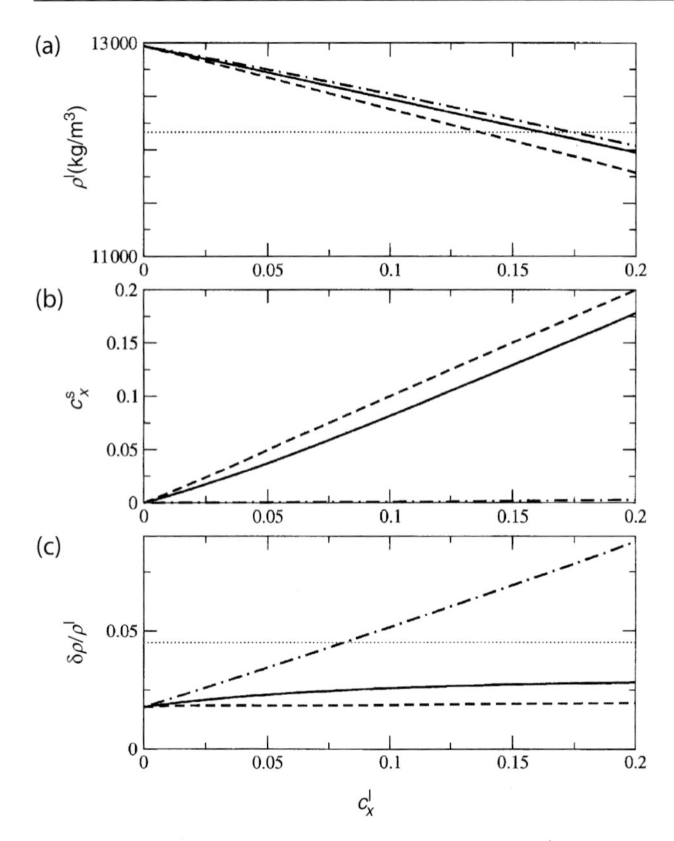

Figure C33 Liquid and solid impurity mole fractions c_x^l and c_x^s of impurities X = S, Si, and O and resulting densities of the inner and outer core predicted by *ab initio* simulations. *Solid, dashed* and *chain curves* represent S, Si, and O, respectively. (a) Liquid density ρ^l (kg m^{-3}); horizontal *dotted line* shows density from seismic data. (b) Mole fractions in solid resulting from equality of chemical potentials in solid and liquid. (c) Relative density discontinuity ($\delta\rho/\rho^l$) at the ICB; horizontal *dotted line* is the value from free oscillation data.

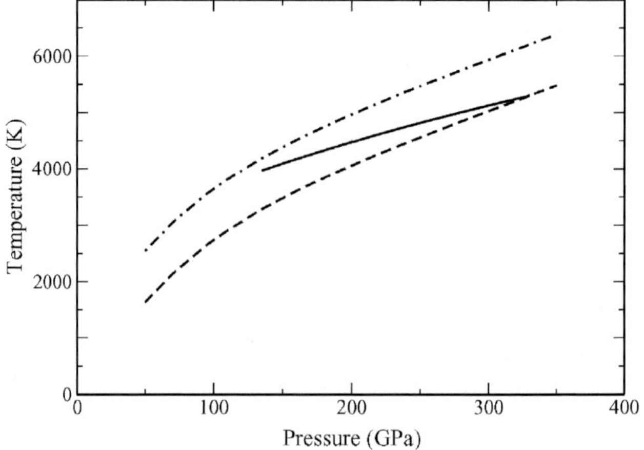

Figure C34 The calculated melting curve of Fe (*chained line*), the suppressed melting curve of Fe-alloy in the core assuming that ΔT_m is independent of depth (*dash line*), and the resulting adiabatic temperature profile (*solid line*) for the outer core obtained using Eq. (6).

0.82 ± 0.18 Mg m^{-3}, which is larger than the previous estimates. Using the new density data of Masters and Gubbins leads to a revised core composition of 7 ± 2.5 mole % S/Si and 0.2 ± 0.1% O for the inner core and 8 ± 2.5% S/Si and 13 ± 2.5% O for the outer core. This sensitivity emphasizes the need for very accurate seismic data if the composition of the core is to be constrained more precisely (see also recent studies by Koper and Pyle (2004) and Cao and Romanowicz (2004)). Finally, we should note that Braginsky and Roberts (1995) have considered the effect of S, Si, and O on core convection and the geodynamo. They concluded on the basis of their dynamical studies, however, that silicon or sulfur would be favored over oxygen as the principal light alloying constituent. The conflict between their conclusions and the atomistic-based deductions of Alfè *et al.* is still to be resolved.

Using however the calculated impurity chemical potentials of Alfè *et al.*, it is possible to use the Gibbs-Duhem relation to compute the change in the Fe chemical potential caused by the impurities in the solid and liquid phases. Alfè *et al.* (2002b) show that the shift of melting temperature (ΔT_m) due to the presence of the impurities is

$$\Delta T_m = k_B T_m (c_x^s - c_x^l)/\Delta S_m \qquad \text{(Eq. 5)}$$

where k_B is Boltzmann's constant, ΔS_m is the entropy of melting of the pure Fe, and c_x^s and c_x^l are the concentration of solute atoms in the solid inner core and liquid outer core, respectively. From their calculated values it is clear that the suppression of the core melting temperature (ΔT_m) is dominated by the effect of O, and is approximately –900 K depending upon the exact value of the density contrast at the ICB. Using their *ab initio* estimate of the melting temperature of pure Fe (see *Melting temperature of iron in the core, theory*) at core pressure, Alfè *et al.* predict that the Earth's temperature at the ICB to be $\sim 5300 \pm 300$ K.

There now appears to be an emerging consensus on the temperature of the ICB, as the value inferred from the work of Alfè *et al.* ($\sim 5300 \pm 300$ K) is close to the value of 5700 K inferred by Steinle-Neumann *et al.* (2001) from independently determined *ab initio* calculations on the elastic properties of the inner core and to the values favored by Poirier and Shankland (1993) and that of 5050 ± 300 K suggested by Anderson (2003). It appears now that only the results inferred from diamond anvil cell (DAC) studies (Boehler, 1996, 2000) lie outside this region of agreement, with these experiments suggesting that the temperature at the ICB is less than 4800 K. The origin for the disagreement between melting temperatures inferred from DAC studies and those obtained from shock data and theory is still to be satisfactorily explained.

This notwithstanding, given the temperature of the ICB, it is possible to calculate the temperature of the core at the CMB, assuming that the outer core temperature profile is adiabatic since

$$\partial T/\partial r = \gamma g T/\Phi \qquad \text{(Eq. 6)}$$

where Φ is the seismic parameter, γ is the Grüneisen parameter in the liquid outer core, and g is the acceleration due to gravity (see also Figure C34). As with T_{ICB}, there is an emerging consensus about the value of γ for the liquid core, with *ab initio* values (e.g., Vočadlo *et al.*, 2003) and those inferred from thermodynamic relations (e.g., Anderson, 2003) both finding γ to be virtually constant along the adiabat, with a value of ~ 1.5 (however, see Stacey and Davis (2004), who favor a lower value of γ). Using the *ab initio* value and the melting behavior of Fe alloy determined by Alfè *et al.* leads to an entirely *ab initio* estimate of the core temperature at the CMB of between 4000 ± 300 K. Again, this is in excellent agreement with that inferred from thermodynamic arguments (Anderson, 2003), and coincidentally identical to the value inferred by Boehler (1996), although in this case the agreement is fortuitous, as Boehler's assumed outer core adiabat is much shallower than that used in other studies.

In conclusion, there seems to be growing agreement over the temperature of the core at the ICB and at the CMB. However, more work is still needed to define the density contrast at the ICB, as it critically effects models for the composition of the core (which in turn determine the value of ΔT_m). Also it is essential that the experimentally and theoretically derived values of the T_m of Fe at ICB conditions be reconciled. Until these issues are resolved, the exact temperature profile in the core will remain contentious.

David Price

Bibliography

Alfè, D., Gillan, M.J., and Price, G.D., 2002a. Composition and temperature of the Earth's core constrained by combining *ab initio* calculations and seismic data. *Earth and Planetary Science Letters*, **195**: 91–98.

Alfè, D., Gillan, M.J., and Price, G.D., 2002b. *Ab initio* chemical potential of solid and liquid solutions and chemistry of the Earth's core. *Journal of Chemical Physics*, **116**: 7127–7136.

Allègre, C.J., Poirier, J.P., Humler, E., and Hofmann, A.W., 1995. The chemical composition of the Earth. *Physics of Earth and Planetary Interiors*, **134**: 515–526.

Anderson, O.L., 2003. The three-dimensional phase diagram of iron. In Dehant, V., Creager, K.C., Karato, S.I., and Zatman, S. (eds.), *Earth Core.* Geodynamics Series, **31**. Washington, D.C.: American Geophysical Union, pp. 83–104.

Boehler, R., 1996. Melting temperature of the Earth's Mantle and Core. *Annual Review of Earth and Planetary Sciences*, **24**: 15–40.

Boehler, R., 2000. High-pressure experiments and the phase diagram of lower mantle and core materials. *Reviews of Geophysics*, **38**: 221–245.

Braginsky, S.I., and Roberts, P.H., 1995. Equations governing convection in earths core and the geodynamo. *Geophysical and Astrophysical Fluid Dynamics*, **79**: 1–97.

Cao, A.M., and Romanowicz, B., 2004. Constraints on density and shear velocity contrasts at the inner core boundary. *Geophysical Journal International.*, **157**(3): 1146–1151.

Koper, K.D., and Pyle, M.L., 2004. Observations of PkiKP/PcP amplitude ratios and implications for Earth structure at the boundaries of the liquid core. *Journal of Geophysical Research-Solid Earth*, **109**(B3): art. no. B03301.

Masters, G., and Gubbins, D., 2003. On the resolution of density within the Earth. *Physics of Earth and Planetary Interiors*, **140**: 159–167.

Poirier, J.P., 1994. Light elements in the Earth's outer core: A critical review. *Phys. Earth Planet. Inter.*, **85**: 319–337.

Poirier, J.P., and Shankland, T.J., 1993. Dislocation melting of iron and the temperature of the inner-core boundary, revisited. *Geophysical Journal International*, **115**: 147–151.

Shearer, P.M., and Masters, G., 1990. The density and shear velocity contrast at the inner core boundary. *Geophysical Journal International*, **102**: 491–498.

Stacey, F.D., and Davis, P.M., 2004. High pressure equations of state with applications to the lower mantle and core. *Physics of the Earth and Planetary Interiors*, **142**: 137–184.

Steinle-Neumann, G., Stixrude, L., Cohen, R.E., and Gülseren, O., 2001. Elasticity of iron at the temperature of the Earth's inner core. *Nature*, **413**: 57–60.

Stevenson, D.J., 1981. Models of the Earth's core. *Science*, **214**: 611–619.

Vočadlo, L., Alfè, D., Gillan, M.J., and Price, G.D., 2003. The Properties of Iron under Core Conditions from First Principles Calculations. *Physics of Earth and Planetary Interiors*, **140**: 101–125.

Yoo, C.S., Holmes, N.C., Ross, M., Webb, D.J., and Pike, C., 1993. Shock temperatures and melting of iron at Earth core conditions. *Physical Review Letters*, **70**: 3931–3934.

Cross-references

Core Density
Core Properties, Physical
Core Properties, Theoretical Determination
Core, Adiabatic Gradient
Grüneisen's Parameter for Iron and Earth's Core
Inner Core Seismic Velocities
Melting Temperature of Iron in the Core, Experimental
Melting Temperature of Iron in the Core, Theory
Shock Wave Experiments

CORE TURBULENCE

Introduction

Turbulence is a ubiquitous feature of geophysical flows. The challenge in dealing with turbulence is the broad range of spatial and temporal scales. For the Earth's core, the largest scales of motion ($L \approx 10^6$ m) are set by the geometry of the core, whereas the smallest dissipative scales ($l < 0.1$ m) are determined by the low viscosity and chemical diffusivity of liquid iron alloys (Dobson, 2000; Vocadlo *et al.*, 2000). The range of time scales is even greater (Hollerbach, 2003). Buoyancy-driven inertial oscillations have diurnal periods (Zhang, 1994), whereas changes in the frequency of magnetic reversals occur on time scales of 10^8 years (McFadden and Merrill, 2000). Such a vast range of spatial and temporal scales prohibits direct simulations of the Earth's *geodynamo* (*q.v.*). Instead, modelers must confine their attention to the largest scales and deal with the problem of limited spatial resolution by parameterizing processes that operate at subgrid scales. Spatially constant eddy diffusivities or hyperdiffusivities are routinely used to represent the effects of small-scale flow in geodynamo models (Glatzmaier and Roberts, 1995, 1996; Kageyama and Sato, 1997; Kuang and Bloxham, 1997; Kida *et al.*, 1997; Christensen *et al.*, 1998, 1999; Grote *et al.*, 1999; Kono and Sakuraba, 2000). Even though many of these models are capable of reproducing features of the Earth's field (dipole dominance, episodic reversals, etc.), there is little doubt that simple models based on scalar diffusivities are inadequate representations of the small-scale processes (Buffett, 2003). Theoretical studies (Braginsky and Meytlis, 1990; Shimizu and Loper, 1997) and numerical simulations (St. Pierre, 1996; Matsushima *et al.*, 1999) have shown that small-scale motion is likely to be highly anisotropic due to the influence of rotation and a strong, large-scale magnetic field. Turbulence in the core is expected to have a plate-like structure, which affects the transport of heat and momentum at small scales. The influence of this anisotropy on the large-scale flow is presently unknown.

The nature of turbulence in the core differs in several important respects from the types of turbulence that are encountered in other areas of fluid mechanics (e.g., Tennekes and Lumley, 1972). The first distinction involves the role of the Reynolds number (Re) in characterizing the flow. The Reynolds number describes the relative importance of inertia and viscous forces in the momentum equation, so high Re usually means that inertia has a leading-order role in the dynamics. This expectation may not be valid when other forces are present. The value of Re in the Earth's core is thought to be quite large (perhaps as large as 10^8) by virtue of the low viscosity of liquid iron and the large size of the core. However, the inertial forces are not expected to have a leading-order role in the momentum equation, where the primary force balance includes the Coriolis, Lorentz, and pressure forces. Large Re in the core means that the effect of inertia in the core is large compared with the effects of viscosity, but inertia is still small compared with other forces. This distinction makes the dynamics of turbulence in the core fundamentally different from that in other high-Re flows.

One consequence of this difference is manifest in the spatial distribution of dissipation in the core. Flow in the core is driven by thermal and compositional anomalies. These density anomalies are probably distributed over a very broad range of spatial scales. We expect the work done by buoyancy forces (e.g., the input of energy to the flow) to be distributed over an equally broad range of scales. Another expectation, based on current geodynamo models, is that much of the input energy is dissipated at the scale where it is supplied (Olson and Christensen, 1999; Buffett and Bloxham, 2001). This view of convection in the core is strikingly different from high-Re turbulence, where energy is input at large scales and cascades through a range of progressively smaller scales until viscous dissipation becomes appreciable. Transfers of energy between scales still occur in the geodynamo problem. In fact, these interactions are essential for the operation of the geodynamo (Buffett and Bloxham, 2001). The subtlety of energy transfers to and from the small scales make core turbulence an important part of the geodynamo problem.

Eddy diffusion

Eddy diffusion is based on an analogy with molecular transport of heat and momentum in an ideal gas. The continuum approximation for molecular transport works well when the average distance between collisions of molecules, λ, is small compared with the scale of the flow. With the usual assumption that the velocity scale of molecular motion is proportional to the speed of sound a, the kinetic theory of gases predicts that the kinematic viscosity of the gas is (Landau and Lifshitz, 1981)

$$\nu \approx a\lambda. \qquad \text{(Eq. 1)}$$

The analogy between molecular and turbulent transport implies that the eddy viscosity (or diffusivity) is described by $\nu_T \approx vl$, where v and l are the velocity and size of the unresolved eddies. Although this analogy is conceptually appealing, it has several serious shortcomings. One of these is a lack of separation between the scale of the turbulent transport and the scale of the resolved motions. Normally, the distinction between the turbulent and resolved scales is imposed by the choice of grid spacing in the numerical calculations. Near the scale of truncation the resolved and unresolved scales are comparable, so the analogy breaks down. Interactions between the resolved and unresolved scales are liable to depend on their relative size, so the effective viscosity would need to vary with the scale of the resolved flow in order to reproduce the effect of interactions between the turbulent and resolved flows.

Another challenge for eddy diffusion models in the geodynamo problem arises from the anisotropic structure of small-scale convection (Braginsky and Metlyis, 1990). Eddies are expected to have longer dimensions in the directions of the rotation axis $\mathbf{\Omega}$ and the local magnetic field $\mathbf{B}$, and a much shorter dimension in the perpendicular direction (e.g., $\mathbf{\Omega} \times \mathbf{B}$). Expectations about the magnitude of the eddy diffusivity (e.g., $\nu_T \approx vl$) could be extended from a scalar quantity to a tensor using information about the dimensions of the eddies in the three defining directions. However, such an approach would require substantial effort because the size and orientation of eddies may change as $\mathbf{B}$ evolves. Moreover, the eddies may not be characterized by a single size or shape. It is more likely that eddies exist over a broad range of scales. The absence of any clear separation between the scale of the eddies and the smallest resolved flows returns us to the problem of requiring the eddy viscosity to vary with the scale of the resolved flow. The most comprehensive attempt to include the effects of anisotropic diffusion in geodynamo models is described in a series of papers by Phillips and Ivers (2000, 2001, 2003).

A variant of the eddy diffusion model replaces the derivatives in the dissipative terms with higher-order operators. This approach is called hyperdiffusion because it enhances the damping of motion or the diffusion of heat as the length scale decreases. Hyperdiffusion addresses the need to make the eddy diffusivity dependent on the scale of the flow. However, the imposed dependence is usually *ad hoc*. One approach to hyperdiffusion replaces the Laplacian operator ∇^2 with $-\nabla^4$ or ∇^6, which is equivalent to making the eddy diffusion depend on the wavenumber of the motion in the spectral domain. On the other hand, adding higher order derivatives introduces the problem of requiring additional boundary conditions to solve the problem. Most implementations of hyperdiffusion in geodynamo models alter only the horizontal part of ∇^2 in spherical coordinates, so that the order of the equations for the radial part of the solution is unchanged (and the number of boundary conditions is unchanged). This means that turbulent diffusion is handled differently in the horizontal and radial directions (Grote *et al.*, 2000). An unintended anisotropy may arise from this implementation of hyperdiffusivity, which is unrelated to the effects of rotation and large-scale magnetic fields.

Eddy diffusion and hyperdiffusion models are computationally simple, but they do a poor job of representing the effects of small-scale turbulence in magnetoconvection calculations (Buffett, 2003). Several alternative strategies have developed in recent years for other applications (see Lesieur and Metais, 1996; Meneveau and Katz, 2000 for recent reviews). Of particular interest is the scale-similarity model, and its variants, because these models have the potential to reproduce the anisotropy of core turbulence. These approaches have grown out of a conceptual framework known as large-eddy simulations. This framework is useful because it defines the modifications to the governing equations when the fields are not resolved. Unfortunately, it does not specify how these additional terms should be evaluated. Another promising approach, based on Lagrangian averaging, is briefly described in the concluding remarks.

Large-Eddy simulations

Large-eddy simulations use spatial filtering to eliminate scales that are smaller than the grid spacing Δ. Each field in the calculation (velocity $\mathbf{V}$, for example) is convolved with a filter function $G(\mathbf{x})$ to define the large-scale fields, e.g.,

$$\tilde{\mathbf{V}}(\mathbf{x}, t) = \int G_\Delta(\mathbf{x} - \mathbf{x}')\mathbf{V}(\mathbf{x}')d\mathbf{x}' \qquad \text{(Eq. 2)}$$

Analogous expressions define the large-scale temperature $\tilde{T}$ and magnetic field $\tilde{\mathbf{B}}$. Applying the filter to the governing equations yields a set of equations for the large-scale fields. To illustrate, we filter the energy equation for an incompressible fluid

$$\frac{\partial T}{\partial T} + \nabla \cdot (\mathbf{V}T) = \kappa\nabla^2 T \qquad \text{(Eq. 3)}$$

to obtain

$$\frac{\partial \tilde{T}}{\partial t} + \nabla \cdot (\tilde{\mathbf{V}}\tilde{T}) = \kappa\nabla^2 \tilde{T} - \nabla \cdot \mathbf{I}^\Delta, \qquad \text{(Eq. 4)}$$

where

$$I_i^\Delta = \widetilde{V_i T} - \tilde{V}_i\tilde{T} \qquad \text{(Eq. 5)}$$

is the subgrid-scale (SGS) heat flux; $\mathbf{I}^\Delta$ represents a correction for the omitted correlations between $\mathbf{V}$ and T when the large-scale heat flux is calculated from $\tilde{V}_i\tilde{T}$ instead of $\widetilde{V_i T}$. Other SGS terms arise from the other nonlinear terms in the relevant governing equations. For the problem of convection in the Earth's core, these additional terms include the SGS momentum flux (or Reynolds stress), the SGS Maxwell stress and the SGS induction term (see Buffett, 2003 for details).

The scale-similarity model makes a direct attempt to estimate (5) using only the resolved part of the fields (Bardina *et al.*, 1980; also see Leonard, 1974). Suppose that we filter the complete (and unknown) fields **T** and **V** using a wider filter, say $G_{2\Delta}(\mathbf{x})$, to define the fields $\overline{\mathbf{T}}$ and $\overline{\mathbf{V}}$. It follows that the SGS heat flux on the coarser grid (with resolution 2Δ) is defined by

$$I_i^{2\Delta} = \overline{V_i T} - \overline{V}_i \overline{T} \qquad \text{(Eq. 6)}$$

When the amplitude of the fields with scales between 2Δ and Δ are large compared with the amplitudes at scales below Δ, we can approximate V_i and T in Eq. (6) using $\tilde{V}_i$ and $\tilde{T}$. This substitution yields an approximation for $I_i^{2\Delta}$. The scale-similarity method supposes that $I_i^{2\Delta}$ is similar to I_i^{Δ}. Specifically, the spatial pattern is the same, but the amplitude may differ. The modeled SGS heat flux becomes [Germano, 1986]

$$I_i^{\mathrm{m}} = C(\overline{\tilde{V}_i \tilde{T}} - \overline{\tilde{V}}_i \, \overline{\tilde{T}}) \qquad \text{(Eq. 7)}$$

where the constant C is used to adjust the amplitude of $\boldsymbol{I}^{\mathrm{m}}$. Automated procedures for evaluating the model constants have been devised by Germano *et al.* (1991). Implementation of this general scheme for a variant of the scale-similarity method (known as the nonlinear gradient method) yields model constants C that are typically closed to unity in both magnetoconvection and plane-layer dynamo calculations (Matsui and Buffett, 2005). These results are consistent with conclusions drawn previously by comparing I_i^{Δ} and I_i^{m} using fully resolved calculations (Buffett, 2003; Matsushima, 2004). A number of technical challenges remain to be solved to deal with the effects of boundaries and nonuniform grid spacing. However, the scale-similarity method provides a framework for making progress in modeling the effects of core turbulence.

Other approaches and considerations

Another method for dealing with resolved motions in numerical simulations is based on Lagrangian averages of fluid parcel trajectories (Holm, 2002). The derivation of the governing equations (known as the alpha model) is based on a variational description of ideal fluid motion (Chen *et al.*, 1999). The current version of the alpha models uses a closure based on Taylor's hypothesis for frozen-in turbulence, although more elaborate descriptions of the small-scale fluctuations could be used, in principle. Extensions of the model to include dissipative effects have been developed using a filtered form of Kelvin's circulation theorem (Foias *et al.*, 2001) and the results have been cast in the form of an LES model (Montgomery and Pouquet, 2002), even though the underlying derivation is fundamentally different. These recent developments offer promising new lines of investigation for the study of core turbulence. It is too early to know whether any of these methods will find widespread use in geodynamo models.

Computational considerations will also govern the choice of suitable turbulence models. The most viable models must be adaptable to computational schemes that can exploit massively parallel computers. Almost all of the existing geodynamo models rely on a spectral expansion of the fields in spherical harmonics. This approach is ideally suited for determining the potential field outside of the core. However, it is poorly suited for parallel computing. The use of spherical harmonics requires excessive communication between processors, and this problem becomes more acute as the number of processors increases. Increasing availability of massively parallel computers has motivated a shift in attention to other methods of solution that are more efficient for parallel computations. New dynamo models based on finite elements (Chan *et al.*, 2001; Matsui and Okuda, 2004), finite differences (Kageyama and Sato, 2004), and spectral elements (Fournier *et al.*, 2004) are now in various stages of development. One or more of these methods may become the future standard. The choice of turbulence models should be guided by these recent trends because the success of turbulence models is often dependent on the method of solution (Meneveau and Katz, 2000).

Bruce Buffett and Hiroaki Matsui

Bibliography

Bardina, J., Ferziger, J.H., and Reynolds, W.C., 1980. Improved subgrid scale models for large-eddy simulations. *American Institute of Aeuronautics and Acoustics Journal*, **34**: 1111–1119.

Braginsky, S.I., and Meytlis, V.P., 1990. Local turbulence in the Earth's core. *Geophysical and Astrophysical Fluid Dynamics*, **55**: 71–87.

Buffett, B.A., 2003. A comparison of subgrid-scale models for large-eddy simulations of convection in the Earth's core. *Geophysics Journal International*, **153**: 753–765.

Buffett, B.A., and Bloxham, J., 2002. Energetics of numerical geodynamo models. *Geophysical Journal International*, **149**: 211–221.

Chan, K.H., Zhang, K., Zou, J., and Schubert, G., 2001. A nonlinear 3-D spherical α^2 dynamo using a finite element method. *Physical Earth Planet Interiors*, **128**: 35–50.

Chen, S., Foias, C., Holm, D.D., Olson, E., Titi, E.S., and Wynne, S., 1999. A connection between the Camassa-Holm equations and turbulent flow in pipes and channels. *Physics of Fluids*, **11**: 2343–2353.

Christensen, U., Olson, P., and Glatzmaier, G.A., 1998. A dynamo interpretation of geomagnetic field structures. *Geophysical Research Letters*, **25**: 1565–1568.

Christensen, U., Olson, P., and Glatzmaier, G.A., 1999. Numerical modeling of the geodynamo: A systematic parameter study. *Geophysics Journal International*, **138**: 393–409.

Clune, T.C., Elliott, J.R., Miesh, M.S., Toomre, J., and Glatzmaier, G.A., 1999. Computational aspects of a code to study rotating turbulent convection in spherical shells. *Parallel Computing*, **25**: 361–380.

Dobson, D.P., 2000. Fe-57 and Co tracer diffusion in liquid Fe-FeS at 2 and 5 GPa. *Physics of the Earth and Planetary Interiors*, **120**: 137–144.

Foias, C., Holm, D.D., and Titi, E.S., 2001. The Navier-Stokes-alpha model of fluid turbulence. *Physica D*, **152**: 505–519.

Germano, M., 1986. A proposal for a redefinition of the turbulent stresses in the filtered Navier-Stokes equations. *Physics of Fluids*, **29**: 2323–2324.

Germano, M., Piomelli, U., Moin, P., and Cabot, W.H., 1991. A dynamic subgrid-scale eddy viscosity model. *Physics of Fluids A*, **3**: 1760–1765.

Glatzmaier, G.A., and Roberts, P.H., 1995. A three-dimensional convective dynamo solution with rotating and finitely conducting inner core and mantle. *Physics of Earth and Planetary Interiors*, **91**: 63–75.

Glatzmaier, G.A., and Roberts, P.H., 1996. An anelastic evolutionary geodynamo simulation driven by compositional and thermal buoyancy. *Physica D*, **97**: 81–94.

Grote, E., Busse, F.H., and Tilgner, A., 1999. Convection-driven quadrupolar dynamos in rotating spherical shells. *Physical Review E*, **60**: R5025–R5028.

Grote, E., Busse, F.H., and Tilgner, A., 2000. Effects of hyperdiffusivities on dynamo simulations. *Geophysical Research Letters*, **27**: 2001–2004.

Hollerbach, R., 2003. The range of timescales on which the geodynamo operates. In Dehant, V., *et al.*, (ed.), *Earth's Core: Dynamics, Structure, Rotation*. Geodynamics Monograph, American Geophysical Union, Vol. 31, 181–192.

Holm, D.D., 2002. Lagrangian averages, averaged Lagrangians and the mean effect of fluctuations in fluid dynamics. *Chaos*, **12**: 518–530.

Kageyama, A., and Sato, T., 1995. The Complexity Simulation Group, Computer simulation of a magnetohydrodynamic dynamo. *Physics of Plasmas*, **2**: 1421–1431.
Kageyama, A., and Sato, T., 1997. Generation mechanism of a dipole field by a magnetohydrodynamic dynamo. *Physical Review Letters E*, **55**: 4617–4626.
Kageyama, A., and Sato, T., 2004. Yin-Yang grid: An overset grid in spherical geometry. *Geochemistry, Geophysics, and Geosystems*, **5**, Q09005.
Katayama, J.S., Matsushima, M., and Honkura, Y., 1999. Some characteristics of magnetic field behavior in a model of MHD dynamo thermally driven in a rotating spherical shell. *Physics of Earth and Planetary Interiors*, **111**: 141–159.
Kono, M., Sakuraba, A., and Ishida, M., 2000. Dynamo simulation and paleosecular variation models. *Philosophical Transactions of the Royal Society of London*, **A358**: 1123–1139.
Kuang, W., and Bloxham, J., 1997. An Earth-like numerical dynamo model. *Nature*, **389**: 371–374.
Landau, L.D., and Lifshitz, E.M., 1981. *Physical Kinetics, Course of Theoretical Physics*, Vol. 10, Oxford: Pergamon Press.
Leonard, A., 1974. Energy cascade in large-eddy simulations of turbulent flow. *Advances In Geophysics*, **18**: 237–248.
Lesieur, M., and Metais, O., 1996. New trends in large-eddy simulations of turbulence. *Annual Review of Fluid Mechanics*, **28**: 45–82.
Matsui, H., and Okuda, H., 2004. Development of a simulation code for MHD dynamo processes using the GEOFEM platform. *International Journal of Computational Fluid Dynamics*, **18**: 323–332.
Matsui, H., and Buffett, B.A., 2005. Subgrid-scale modeling of convection-driven dynamos in a rotating plane layer. *Physics of Earth and Planetary Interiors*, **153**: 108–123.
Matsushima, M., Nakajima, T., and Roberts, P.H., 1999. The anisotropy of local turbulence in the Earth's core. *Earth, Planets and Space*, **51**: 277–286.
Matsushima, M., 2004. Scale similarity of MHD turbulence in the Earth's core. *Earth, Planets, Space*, **56**: 599–605.
McFadden, P.L., and Merrill, R.T., 2000. Evolution of the geomagnetic reversal rate since 160 Ma: Is the process continuous? *Journal of Geophysical Research*, **105**: 28455–28460.
Meneveau, C., and Katz, J., 2000. Scale invariance and turbulence models for large-eddy simulation. *Annual Review of Fluid Mechanics*, **32**: 1–32.
Moffatt, H.K., 1978. *Magnetic Field Generation in Electrically Conducting Fluids*, Cambridge, UK: Cambridge University Press, 343 pp.
Montgomery, D.C., and Pouquet, A., 2002. An alternative interpretation for the Holm alpha model. *Physics of Fluids*, **14**: 3365–3566.
Olson, P., Christensen, U., and Glatzmaier, G.A., 1999. Numerical modeling of the geodynamo: Mechanisms of field generation and equilibration. *Journal of Geophysical Research*, **104**: 10383–10404.
Phillips, C.G., and Ivers, D.J., 2000. Spherical anisotropic diffusion models for the Earth's core. *Physics of Earth and Planetary Interiors*, **117**: 209–223.
Phillips, C.G., and Ivers, D.J., 2001. Spectral interactions of rapidly rotating anisotropic turbulent viscous and thermal diffusion in the Earth's core. *Physics of Earth and Planetary Interiors*, **128**: 93–107.
Phillips, C.G., and Ivers, D.J., 2003. Strong-field anisotropic diffusion models for the Earth's core. *Physics of Earth and Planetary Interiors*, **140**: 13–28.
Piomelli, U., and Zang, T.A., 1991. Large-eddy simulation of transitional channel flow. *Computer Physics Communications*, **65**: 224–230.
Roberts, P.H., and Glatzmaier, G.A., 2000a. Geodynamo theory and simulations. *Reviews of Modern Physics*, **72**: 1081–1123.
Shimizu, H., and Loper, D.E., 1997. Time and length scales of buoyancy-driven flow structures in a rotating hydromagnetic fluid. *Physics of Earth and Planetary Interiors*, **104**: 307–329.
St. Pierre, M.G., 1996. On the local nature of turbulence in Earth's outer core. *Geophysical and Astrophysical Fluid Dynamics*, **83**: 293–306.
Tennekes H., and Lumley, J.L., 1972. *A First Course in Turbulence*, Cambridge, MA: The MIT Press.
Vocadlo, L., Alfe, D., Price, G.D., and Gillan, M.J., 2000. First-principle calculations of the diffusivity and viscosity of liquid Fe-FeS at experimentally accessible conditions. *Physics of Earth and Planet. Interiors*, **120**: 145–152.
Zhang, K.K., 1994. On the coupling between Poincare equation and the heat equation. *Journal of Fluid Mechanics*, **268**: 211–229.

CORE VISCOSITY

Definition of viscosity

The transport properties of fluid materials (either solid or liquid) are determined by their viscosity, which defines the resistance of the material to fluid flow (i.e., how "runny" it is). Viscosity, η, describes the time dependence of material motion through the ratio of applied shear stress, σ, to strain rate, $\dot{\varepsilon}$ via

$$\eta = \frac{\sigma}{\dot{\varepsilon}} \qquad \text{(Eq. 1)}$$

The Earth's liquid outer core will have a relatively low viscosity compared to that of the solid inner core, which deforms on a much longer timescale. In geophysics, the quantity η is referred to as the dynamic viscosity, which, when normalized by the density, ρ, is termed the kinematic viscosity, v:

$$v = \frac{\eta}{\rho} \qquad \text{(Eq. 2)}$$

Viscosity is a quantity dependent on the properties of the fluid at a molecular level. Therefore, both the viscosities defined above may also be referred to as "molecular viscosity." In the outer core, fluid motion occurs over a range of length scales from that at a molecular level to large-scale motion with characteristic distances comparable with the outer core radius. Such large-scale motion does not exist in isolation, but has embedded within it turbulence over a range of length scales, which serve to drain the largest scale flow of its energy. Consequently, this energy dissipation causes an increase in the viscosity of the large-scale fluid, resulting in an "effective viscosity" or "turbulent viscosity" of a much larger magnitude than the molecular viscosity defined earlier. This effective viscosity has been observed experimentally in turbulent fluids yet it is very difficult to quantify and impossible to derive rigorously; however estimates for effective viscosities are often used when modeling the outer core. Viscosity is a very important parameter in geophysics since the viscosity of materials in the Earth's core are a contributory factor in determining overall properties of the core itself, such as core convection (see *Core properties* and *Core convection*); indeed, the fundamental equations governing the dynamics of the outer core and the generation and sustention of the magnetic field are dependent, in part, on the viscosity of the outer core fluid.

Quantifying outer core viscosity

There have been many estimates made for outer core viscosity derived from geodetic, seismological, geomagnetic, experimental, and theoretical studies. However, the values so obtained span 14 orders of magnitude (see Secco, 1995).

Geodetic observations (e.g., free oscillations, the Chandler wobble, length of day variations, nutation of the Earth, tidal measurements, gravimetry) lead to viscosity estimates ranging from 10 mPa s (observations of the Chandler wobble, Verhoogen, 1974) to 10^{13} mPa s (analysis of free oscillation data, Sato and Espinosa, 1967). Theoretical geodetic studies (e.g., viscous coupling of the core and mantle, theory of rotating fluids, *inner core oscillation* (*q.v.*), core nutation) lead to viscosity estimates ranging from 10 mPa s (evaluation of decay time of *inner core oscillation* (*q.v.*), Won and Kuo, 1973) to 10^{14} mPa s (secular deceleration of the core by viscous coupling, Bondi and Lyttleton, 1948). Generally much higher values for viscosity (10^{10}–10^{14} mPa s) are obtained from seismological observations of the attenuation of P- and S-waves through the core (e.g., Sato and Espinosa, 1967; Jeffreys, 1959), and from geomagnetic data (e.g., 10^{10} mPa s; Officer, 1986).

The viscosities of core-forming materials may also be determined experimentally in the laboratory and theoretically through computer simulation. Empirically, viscosity follows an Ahrrenius relation of the form (see Poirier, 2002):

$$\eta \propto \exp\left(\frac{Q_v}{k_B T}\right) \qquad \text{(Eq. 3)}$$

where Q_v is the activation energy. Poirier (1988) analyzed data for a number of liquid metals and found that there is also an empirical relation between Q_v and the melting temperature:

$$Q_v \cong 2.6RT_m \qquad \text{(Eq. 4)}$$

This very important result implies that the viscosity of liquid metals remains constant (i.e., independent of pressure) along the melting curve and therefore equal to that at the melting point at ambient pressure, which is generally of the order of a few mPa s. Furthermore, Poirier went on to state that the viscosity of liquid iron in the outer core would, therefore, be equal to that at ambient pressure (~6 mPa s).

On a microscopic level, an approximation for the viscosity of liquid metals is given by the Stokes-Einstein equation, which provides a relationship between diffusion and viscosity of the form:

$$D\eta = \frac{k_B T}{2\pi a} \qquad \text{(Eq. 5)}$$

where a is an atomic diameter, T is the temperature, k_B is the Boltzmann constant, and D is the diffusion coefficient.

Theoretical values for diffusion coefficients have been obtained from *ab initio* molecular dynamics simulations on liquid iron at core conditions (de Wijs *et al.*, 1998), leading to a predicted viscosity of ~12–15 mPa s using the Stokes-Einstein relation above. However, although the Stokes-Einstein equation has proved successful in establishing a link between viscosity and diffusion for a number of monatomic liquids, it is not necessarily the case that it should be effective for alloys or at high pressures and temperatures. To address the validity of the Stokes-Einstein relation and to assess the effect of impurities on viscosity coefficients, a number of experimental and theoretical studies have been performed on the Fe-FeS system. High-pressure tracer diffusion experiments (Dobson *et al.*, 2001) have been carried out on liquid Fe-FeS alloys at 5 GPa, resulting in high diffusivities (10^{-5} cm^2 s^{-1}) in excellent agreement with *ab initio* molecular dynamics calculations performed at the same conditions (Vočadlo *et al.*, 2000). When incorporated into Eq. 5, these diffusivities lead to values for viscosity of a few mPa s.

Direct viscosity measurements (Dobson *et al.*, 2000) of Fe-FeS alloys by means of the falling-sphere technique have been made at similar pressures and temperatures to those used in the diffusion experiments above; these resulted in values for viscosities in excellent agreement with those derived experimentally using the Stokes-Einstein relation. Furthermore these results are in excellent agreement with *ab initio* molecular dynamics calculations of viscosity based on rigorous Green-Kubo functions of the stresses obtained directly from the simulations (Vočadlo *et al.*, 2000). All of these results thus provide both experimental and theoretical verification of the Stokes-Einstein relation (Eq. 5).

The results from these studies further show not only that the viscosity of liquid iron at core pressures is approximately equal to that at low pressures, verifying Poirier's groundbreaking and insightful result (Poirier, 1988), but also that the introduction of light elements into Fe liquid does not appear significantly to effect the values. Both *ab initio* calculations and experiments consistently give viscosities of the order of a few mPa s. This suggests that viscosity changes little with homologous temperature, and it is now generally accepted that the viscosity of the outer core is likely to be a few mPa s (comparable to that of water on the Earth's surface).

In general, viscosities derived from Earth observations are high (10^7–10^{14} mPa s) and those based on laboratory experiments and theoretical considerations are much lower (~1–10 mPa s). The disparate values for viscosity arise for two main reasons: the type of viscosity being measured (molecular or effective) and the large uncertainties associated with the interpretation of Earth observation data. It is also possible that viscosity measurements made from the direct observations of the Earth are not actually measuring viscosity at all, but some other effects that are being attributed to viscosity; this would certainly be the case if the "true" viscosity were of the order of mPa s, since such a small viscosity is unlikely to be detectable.

A low viscosity leads to viscous forces that are essentially negligible when compared to the Coriolis force, supporting the view of an outer core in a state of small-circulation turbulent convection rather than a more coherent pattern of convection on a much larger scale comparable with the core radius. If this is the case, it is necessary to obtain accurate values for viscosity in order to quantify this turbulence in the outer core.

Viscosity and the inner core

The inner core is not perfectly elastic and has a finite viscosity with deformation occurring over long timescales. Placing numerical constraints on the viscosity of the inner core is fundamental to understanding important core processes such as differential *inner core rotation* (*q.v.*), *inner core oscillation* (*q.v.*), and *inner core anisotropy* (*q.v.*) (see Bloxham, 1988).

High-temperature experiments on solid iron at ambient pressure lead to estimates for viscosities of ~10^{13} Pa s (Frost and Ashby, 1982); however, this is likely to be a lower limit as the value may increase at higher pressures. Seismological and geodetic observations have led to a number of estimates for inner core viscosity ranging from 10^{11} to 10^{20} Pa s (see Dumberry and Bloxham, 2002). In particular, Buffett (1997) modeled the viscous relaxation of the inner core by calculating the relaxation time for the inner core to adjust, as it rotates, back to its equilibrium shape after small distortions due to perturbations in gravitational potential imposed by the overlying mantle. He suggested that the viscosity has to be constrained to be either less than 10^{16} Pa s (if the whole inner core is involved in the relaxation) or greater than 10^{20} Pa s (if there is no relaxation of the inner core), although this latter case may lead to gravitational locking and hence no differential rotation.

Quantifying the viscosity of the phases present in the inner core at a microscopic level is a very difficult problem. At temperatures close to the melting point (as expected in the inner core) viscous flow is likely to be determined either by dislocation creep (Harper-Dorn creep) or diffusion creep (Nabarro-Herring creep).

The overall viscosity of inner core material has diffusion-driven and dislocation-driven contributions:

$$\eta = \left(\left(\frac{1}{\eta_{\text{diff}}}\right) + \left(\frac{1}{\eta_{\text{disl}}}\right)\right)^{-1} \qquad \text{(Eq. 6)}$$

Diffusion-controlled viscosity, whereby the material strain is caused by the motion of lattice defects (e.g., vacancies) under applied stress, is given by

$$\eta_{\text{diff}} = \frac{d^2RT}{\alpha D_{\text{sd}} V} \quad \text{(Eq. 7)}$$

where d is the grain size, R is the gas constant, T is the temperature, α is a geometric constant, and V is the volume. The self-diffusion coefficient, D_{sd}, is given by

$$D_{\text{sd}} = D_0 \exp\left(-\frac{\Delta H}{RT}\right) \quad \text{(Eq. 8)}$$

where D_0 is a preexponential factor and ΔH is the activation enthalpy for self-diffusion.

For simple materials, dislocation-controlled viscosity, whereby material strain is caused by the movement of linear defects along crystallographic planes, is given by

$$\eta_{\text{disl}} = \frac{RT}{\rho D_{\text{sd}} V} \quad \text{(Eq. 9)}$$

where ρ is the dislocation density.

Both dislocation- and diffusion-controlled creep mechanisms are thermally activated and the thermally controlled parameter in both cases is the self-diffusion coefficient, D_{sd}. A commonly used empirical relation for metals assumes that ΔH is linearly proportional to the melting temperature, T_{m}, and hence that

$$D_{\text{sd}} = D_0 \exp\left(-\frac{gT_{\text{m}}}{T}\right) \quad \text{(Eq. 10)}$$

where g is a constant taking a value of $\sim$18 for metals (Poirier, 2002).

Considering iron close to its melting point at core pressures ($\sim$5500 K), and using reasonable estimates for other quantities in Eqs. 7, 9, and 10 ($\alpha \sim 42$, $D_0 \sim 10^{-5}\ \text{m}^2\ \text{s}^{-1}$, $V \sim 5 \times 10^{-6}\ \text{m}^3\ \text{mol}^{-1}$), we obtain values for η_{diff} and η_{disl} of $\sim 10^{21} d^2$ Pa s and $\sim 6 \times 10^{22}/\rho$ Pa s, respectively. Unfortunately, the strong dependence of the viscosity expressions on the completely unknown quantities of grain size and dislocation density means that it is extremely difficult to produce reliable final numerical values. Grain sizes in the inner core could be anything from 10^{-3}–10^{3} m, resulting in diffusion viscosities in the range 10^{15}–10^{27} Pa s; dislocation densities could be as low as $10^{6}\ \text{m}^{-2}$ or nearer to the dislocation melting limit of $10^{13}\ \text{m}^{-2}$, resulting in dislocation driven viscosities of 10^{9}–10^{16} Pa s. Thus, even the relative contributions from dislocation controlled and diffusion controlled viscosity are as yet unknown.

Clearly, inner core viscosity is not a well-constrained property, with estimates varying over many orders of magnitude. Future microscopic simulations, combined with high-resolution seismic and geodetic data, should constrain this quantity further and thereby improve our understanding of inner core dynamics.

Lidunka Vočadlo

Bibliography

Bloxham, J., 1988. Dynamics of angular momentum in the Earth's core. *Annual Reviews in Earth and Planetary Science*, **26**: 501–517.

Bondi, H., and Lyttleton, R.A., 1948. On the dynamical theory of the rotation of the Earth. *Proceedings of the Cambridge Philosophical Society*, **44**: 345–359.

Buffet, B.A., 1997. Geodynamic estimates of the viscosity of the Earth's inner core. *Nature*, **388**: 571–573.

Dobson, D.P., Crichton, W.A., Vočadlo, L., Jones, A.P., Wang, Y., Uchida, T., Rivers, M., Sutton, S., and Brodholt, J., 2000. In situ measurement of viscosity of liquids in the Fe-FeS system at high pressures and temperatures. *American Mineralogist*, **85**: 1838–1842.

Dobson, D.P., Brodholt, J.P., Vočadlo, L., and Chrichton, W., 2001. Experimental verification of the Stokes-Einstein relation in liquid Fe-FeS at 5 GPa. *Molecular Physics*, **99**: 773–777.

Dumberry, M., and Bloxham, J., 2002. Inner core tilt and polar motion. *Geophysical Journal International*, **151**: 377–392.

Frost, H.J., and Ashby, M.F., 1982. *Deformation Mechanism Maps.* Oxford: Pergamon Press.

Jeffreys, H., 1959. *The Earth, its Origin, History and Physical Constitution.* Cambridge: Cambridge University Press.

Officer, C.B., 1986. A conceptual model of core dynamics and the Earth's magnetic field. *Journal of Geophysics*, **59**: 89–97.

Poirier, J.P., 1988. Transport properties of liquid metals and viscosity of the Earth's core, *Geophysical Journal International*, **92**: 99–105.

Poirier, J.P., 2002. Rheology: elasticity and viscosity at high pressure. In Hemley, R.J., Chiarotti, G.L., Bernasconi, M., and Ulivi, L. (eds.), *Proceedings of the International School of Physics "Enrico Fermi".* Amsterdam: IOS Press. Course CXLVII.

Sato, R., and Espinosa, A.F., 1967. Dissipation factor of the torsional mode T_2 for a homogeneous mantle Earth with a soft-solid or viscous-liquid core. *Journal of Geophysical Research*, **72**: 1761–1767.

Secco, R.A., 1995. Viscosity of the outer core. In Ahrens, T.J. (ed.), *Mineral Physics and Crystallography: A Handbook of Physical Constants,* AGU Reference Shelf 2. Washington, D.C.: American Geophysical Union.

Verhoogen, J., 1974. Chandler wobble and viscosity in the Earth's core. *Nature*, **249**: 334–335.

Vočadlo, L., Alfè, D., Price, G.D., and Gillan, M.J., 2000. First principles calculations on the diffusivity and viscosity of liquid Fe-S at experimentally accessible conditions. *Physics of the Earth and Planetary Interiors*, **120**: 145–152.

de Wijs, G.A., Kresse, G., Vočadlo, L., Dobson, D.P., Alfè, D., Gillan, M.J., and Price, G.D., 1998. The viscosity of liquid iron at the physical conditions of the Earth's core. *Nature*, **392**: 805–807.

Won, I.J., and Kuo, J.T., 1973. Oscillation of the Earth's inner core and its relation to the generation of geomagnetic field. *Journal of Geophysical Research*, **78**: 905–911.

Cross-references

Core Convection
Core Properties, Physical
Core Properties, Theoretical Determination
Inner Core Anisotropy
Inner Core Oscillation
Inner Core Rotation

CORE, ADIABATIC GRADIENT

Introduction

The importance of temperature T to physical problems of Earth's core is well understood. The adiabatic gradient dT/dz is also important because it induces the transfer of conductive power (the larger the adiabatic gradient, the more conductive power is transferred). When conduction is the dominant form of heat transfer, as it is in the transfer of heat from the core to the mantle, the relation between power Q (Watts) transferred and the adiabatic gradient is

$$Q = Ak\left(\frac{dT}{dz}\right) \quad \text{(Eq. 1)}$$

where A (m^2) is the area; k ($W\ m^{-1}K^{-1}$) is the thermal conductivity, a physical property of core material; and z (m) is depth. The ratio of Q to A is called energy flux, Q ($W\ m^{-2}$ or $J\ t^{-1}\ m^{-2}$).

In an adiabatic state, such as that found in Earth's outer core, the adiabatic gradient of the temperature profile is at constant entropy S, and dT/dz in Eq. 1 is replaced by $(\partial T/\partial z)_S$. Equation 1 becomes

$$Q = Ak\left(\frac{\partial T}{\partial z}\right)_S \quad \text{(Eq. 2)}$$

To evaluate Eq. 2 for the power transferred from the core to the mantle (at radius 1221 km, the core side of the core-mantle boundary (CMB), or at pressure $P = 135$ GPa), the core's surface area on the core side of the CMB ($15.26 \times 10^{13}\ m^2$) is needed.

Adiabaticity is defined as a condition of no heat input or heat output in a region where heat flow occurs. The core has energy input from various sources, yet remains in an adiabatic state. Because the outer core is in a low-viscosity liquid state and well-mixed and because the flows are fast, any departure from the equilibrium of adiabaticity is immediately compensated for. Masters (see *Density distribution in the core*) found that a density gradient could be estimated from normal mode eigenfrequencies (an advanced seismic technique). His density gradient showed adiabaticity.

Ab-initio calculations by de Wijs *et al.* (1998) and Vočadlo *et al.* (2003) verify that the value of the core's viscosity is close to that at ambient conditions. The low viscosity of the core confirms the applicability of Eq. 2 to the outer core.

In the following section, we derive an equation for the adiabatic gradient of the core in terms of two parameters of Earth from seismology and two parameters (T and γ_ℓ, the liquid Grüneisen parameter) from mineral physics. In the section entitled, "Data for the outer core's adiabatic gradient," we make the calculation for the adiabatic gradient versus pressure for the outer core using two sets of mineral physics data. In the section entitled "Data for the inner core's adiabatic gradient," we make the calculation for the adiabatic gradient versus radius of Earth for the inner core using two sets of mineral physics data. In the last section, we discuss the implications arising from information listed in the previous sections, especially the maximum power transferred from Earth's outer core to its mantle by conduction.

The adiabatic gradient's dependence on the Grüneisen parameter

The adiabatic gradient of Earth's interior depends on the value of the Grüneisen parameter γ. Although the formula for this parameter has many manifestations (see *Grüneisen's parameter for iron and Earth's core*), its general definition is that it measures the change in pressure resulting from a change in energy density. The formula for γ fundamental to derivations in this article is

$$\gamma = -\left(\frac{\partial \ln T}{\partial \ln V}\right)_S = \left(\frac{\partial \ln T}{\partial \ln \rho}\right)_S \quad \text{(Eq. 3)}$$

where T is temperature, V is volume, ρ is density, and S indicates constant entropy or a path along an adiabat. This equation is applicable to the outer core because we have shown that adiabaticity exists there. We rewrite Eq. 3 to show how temperature varies with density on an adiabat

$$\left(\frac{\partial T}{\partial \rho}\right)_S = \frac{\gamma T}{\rho} \quad \text{(Eq. 4)}$$

Using the definition of the adiabatic bulk modulus given by $K_S = \rho(dP/d\rho)$, we have

$$\left(\frac{\partial T}{\partial P}\right)_S = \frac{\gamma T}{K_S} \quad \text{(Eq. 5)}$$

which shows how temperature varies with pressure along an adiabat. Using the equation for hydrostatic equilibrium in a fluid, $(\partial P/\partial z)_S = \rho g$, where g is gravity and z is depth, Eq. 5 is transformed to show how temperature varies with depth along an adiabat

$$\left(\frac{\partial T}{\partial z}\right)_S = \frac{\gamma g T}{\phi} \quad \text{(Eq. 6)}$$

where ϕ, the seismic parameter, $K_S/\rho = v_p^2 - 4/3v_s^2$, is determinable from the seismic velocities. To write Eq. 6 where the depth z is replaced by the radial distance of Earth r, take $z = a - r$ (where a is Earth's radius), so that

$$\left(\frac{\partial T}{\partial r}\right)_S = -\frac{\gamma g T}{\phi} \quad \text{(Eq. 7)}$$

To evaluate Eq. 6 or 7, take data for ϕ and g from seismic tables, in particular, PREM (Dziewonski and Anderson, 1981), and data on γ and T from mineral physics sources such as enumerated below. When considering Earth's interior, it is important to distinguish gamma associated with a liquid phase from gamma associated with a solid phase (γ_l, liquid; γ_s, solid). Although the formula for gamma is often the same for both phases, the values are usually different because the physical properties of the two phases are different. For the outer core, γ_l is the proper choice and γ_s is to be used for the inner core. For low values of pressure, $\gamma_l \gg \gamma_s$, but $\gamma_l \rightarrow \gamma_s$ as the pressure increases (see Figure 4, *Grüneisen's parameter for iron and Earth's core*). At the ICB (inner core boundary) pressure, $\gamma_l \cong \gamma_s$ (see discussion in last section).

Data for the outer core's adiabatic gradient

A major goal in mineral physics for the last two decades has been to find accurate values of the temperature of the outer core along its adiabat and the associated gamma, γ_l. Since the year 2000, advances in theory and experiment have made possible establishment of good values for these two important parameters at core conditions. Three important advances are listed below:

a. Values for physical properties of solids at extreme conditions (calculated using *ab-initio* techniques) have been published for the last two decades, but only in the last six years have good *ab-initio* results on liquid iron at core conditions become available. Calculations on properties of liquid hcp iron and of simulated core material at high pressure and temperature have been made by Dario Alfè and his colleagues at University College, London (Alfè *et al.*, 2001, 2002a,b,c); resulting data, especially on γ_l and T_m are found in Table C7.
b. Stacey and Davis (2004) presented calculations on the physical properties of the core, including γ_l and T, based only on Stacey's K-primed equation of state (Stacey, 2000) and seismic data. Their unique approach does not require the intermediate step of using properties of pure iron followed by adding impurities. These new data are used in this article and compared with those arising from the *ab-initio* method. Data from Stacey and Davis are found in Tables C7 and C8.
c. Properties of solid iron at extreme conditions are needed at the solid-liquid boundary of the core (330 GPa) and throughout the inner core. Experiments on the variation of the Debye constant Θ_D with volume for hcp iron through the pressure range of the core

Table C7 Data for calculation of adiabatic gradient versus pressure of the liquid outer core (Alfè *et al.*, 2001, 2002a,b,c; Stacey and Davis, 2004; see *Thermal conduction in the core*)

P (GPa)	Alfè *et al.*		Stacey and Davis		PREM, seismic table		Adiabatic gradient (CMB)	
	T_m (K)	γ_l	T_m (K)	γ_l	g (m s^{-2})	$\phi = K_s/\rho$ (m^2 s^{-2})	$(\partial T/\partial z)_S$ (K km^{-1}) Alfè *et al.*	$(\partial T/\partial z)_S$ (K km^{-1}) Stacey and Davis
135	4112	1.527	3739.0[a]	1.443[a]	10.680	67.330	0.996	0.886[a]
160	4335	1.526	3932.2	1.4307	10.537	71.191	0.979	0.833
180	4487	1.525	4081.9	1.4232	9.9110	76.016	0.892	0.757
200	4640	1.522	4223.5	1.4163	9.4216	81.98	0.812	0.688
220	4787	1.519	4358.1	1.4105	9.0404	85.04	0.773	0.654
240	4927	1.516	4486.7	1.4055	8.5023	89.28	0.711	0.601
260	5062	1.512	4610.0	1.4011	7.9197	93.34	0.649	0.548
280	5193	1.508	4728.1	1.3973	6.2862	97.33	0.506	0.427
300	5319	1.504	4842.1	1.3939	5.8966	99.98	0.472	0.398
320	5440	1.501	4952.4	1.3908	4.9413	105.49	0.383	0.323
330	5500	1.500	5001.0	1.3908	4.4002	107.24	0.339	0.285

[a]Data from Stacey, see *Thermal conduction in the core.*

Table C8 Data for calculation of adiabatic gradient versus Earth radius of the solid inner core (Stacey and Davis, 2004 [SD], Anderson, 2002a [A])

r (km)	Stacey and Davis		Ander. *et al.*		PREM, seismic data		Adiabatic gradient	
	T (K)	γ_s	T (K)	γ_s	g (m s^{-2})	$\phi = K_s/\rho$ (m^2 s^{-2})	$(\partial T/\partial r)_S$ (K km^{-1}) SD	$(\partial T/\partial r)_S$ (K km^{-1}) A
0	5030	1.3872	5456	1.53	0	108.9	0	0
200	5029.2	1.3874	5450	1.53	0.7311	108.8	− 0.047	− 0.056
400	5026.8	1.3877	5434	1.53	1.4804	108.51	− 0.094	− 0.112
600	5022.8	1.3881	5406	1.53	2.1862	108.02	− 0.141	− 0.167
800	5017.1	1.3889	5367	1.53	2.9068	107.33	− 0.189	− 0.222
1000	5009.9	1.3897	5259	1.53	3.6203	106.45	− 0.237	− 0.274
1200	5001.9	1.3908	5258	1.53	4.3251	105.38	− 0.285	− 0.330
1222	5000.0	1.3908	5250	1.53	4.4002	107.24	− 0.285	− 0.330

(Anderson *et al.*, 2001) provided the foundation for determination of γ_s versus V, the melting temperature (solidus, T_m) and the adiabatic temperature (Anderson, 2002a,b; Anderson *et al.*, 2003; Isaak and Anderson, 2003). This method, called the thermal physics approach, relies heavily on thermodynamics and experimental data. Data from this method are found in Table C8.

The variation of $(\partial T/\partial z)_S$ with respect to pressure and temperature using Eq. 6 is shown in Table C7. It is noted that the adiabatic gradient at the CMB (135 GPa) is high (0.996 K km^{-1} – 0.886 K km^{-1}), comparable to the value used by Nimmo *et al.* (2004), 0.80 K km^{-1}. The Stacey and Davis adiabatic gradient values are somewhat (~15%) lower than those of Alfè *et al.*

Data for the inner core's adiabatic gradient

Table C8 displays the calculated adiabatic gradient values for the solid inner core obtained using adiabatic temperature data and γ_s values taken from thermal physics (Anderson, 2002a, Figure 9) and from the K-primed EoS method (Stacey and Davis, 2004). The thermal physics method adopted 1.53 as the value of γ_s throughout the inner core, since γ_s is virtually independent of volume near the ICB (Anderson, 2002a). When γ_s is independent of volume or density, Eq. 4 can be rewritten as $(\mathrm{d}T/T) = \gamma(\mathrm{d}\rho/\rho)$, yielding

$$\left(\frac{T_1}{T_0}\right) = \left(\frac{\rho_1}{\rho_0}\right)^{\gamma} \qquad \text{(Eq. 8)}$$

which holds for any region in which γ is independent of depth. In the thermal physics approach, the density ratio $(\rho(r)/12\,763.60)$ is used to find the temperature ratio of $T(r)/5250$ for $\gamma = \gamma_s = 1.53$.

Stacey and Davis retain five significant figures in their data, probably in part because PREM tabular data have five significant figures. If their values of γ_s in Table C8 were rounded to three significant figures, a constant value of $\gamma_s = 1.39$ would hold throughout the entire inner core. Therefore, the calculated value of $(\partial T/\partial r)_S$ is only listed to three significant figures. The calculated adiabatic gradient from the K-primed EoS approach is about 15% lower than that arising from the thermal physics approach at the ICB pressure.

Implications of a high adiabatic gradient at the CMB

Equation 2 shows that a high value of $(\partial T/\partial z)_S$ at the CMB, coupled with a high value of k results in a high value of Q, which is the conductive power transferred from the core to the D″ region. The work here implies a large value for the power Q conducted from the core to the mantle. The value of thermal conductivity k to be used in Eq. 1 is needed.

There are three recent estimates of the value of k at the core side of the CMB where $P = 135$ GPa. Labrosse *et al.* (1997) proposed 60 W m^{-1} K^{-1}. Anderson (1998) proposed 44 W m^{-1} K^{-1}. Stacey and Anderson (2001) proposed 46 W m^{-1} K^{-1}. (Older estimates of k are: 40 W m^{-1} K^{-1} (Stevenson, 1981) and 28.6 W m^{-1} K^{-1} (Stacey, 1992).) Many of the estimates referenced above are based on shock-wave work on electrical conductivity measurements of iron-silica alloy done at the Lawrence Livermore National Laboratory (Matassov, 1977). (Matassov's work is recorded in his Ph.D. dissertation filed in the Library of the University of California at Davis, California, and also as an Lawrence Livermore National Laboratory Technical Report (UCRL–52322)).

To find k for a solid, it is necessary to add the lattice conductivity k_l to the electronic conductivity k_e. But k_l is smaller, 3.1 W m^{-1} K^{-1} (Stacey and Anderson, 2001), and less sensitive to P and T than k_e. We take

$$k = (3.1 + k_e) \quad \mathrm{W\,m^{-1}\,K^{-1}} \qquad \text{(Eq. 9)}$$

and obtain k_e from electrical conductivity measurements, σ. We assume that the solid-state conductivity at core pressure can be used for a liquid at core pressure, even though the liquid does not have a periodic lattice. There is a correction to be made in using k with a solid rather than a liquid. Braginsky and Roberts (1995) invoked a 10% reduction in thermal conductivity at the transition.

Electrical conductivity σ can be measured more easily at high pressure than thermal conductivity because the latter is a transport property requiring accurate measurements of gradients, whereas the former requires the simpler measurements of current and field. The Wiedemann-Franz ratio, by which the conduction electrons' contribution k_e to the total thermal conductivity k is calculated from σ, was derived using the classical free electron model of atoms (Joos, 1958)

$$\frac{k_e}{\sigma} = \frac{1}{3}\left(\frac{\pi \boldsymbol{k}}{e}\right)^2 T \qquad \text{(Eq. 10)}$$

It is seen that k_e is proportional to both T and σ, where $\boldsymbol{k}$ is the Boltzmann constant and e is the electron's charge. On an isotherm, k_e is proportional to σ.

The value of k reported by Bi *et al.* (2002) for pure iron at core conditions is 30 W m^{-1} K^{-1}, and a similar value, also for pure iron at core conditions, was found in the work of Matassov (1977); both values were obtained from shock-wave experiments. Our interest here, however, is ultimately in core materials, not pure iron.

Bridgman (1957) found that electrical resistivity is increased as the silicon content is increased in an Fe-Si alloy. Electrical conductivity, being the inverse of electrical resistivity, must therefore decrease as silicon content increases, and from Eq. 10, thermal conductivity must also decrease as silicon content increases. However, pressure increases k.

We follow Stacey's (see *Core, thermal conduction*) derivation of k_e from Matassov's 1977 measured electrical conductivity and use Stacey's derived parameters for CMB conditions, $k_\ell = 3.1$ W m^{-1} K^{-1}, $k_e = 43.21$ W m^{-1} K^{-1}, and $k = 46.31$ W m^{-1} K^{-1}. These values are for a liquid core, of an Fe-Ni-Si composition, in which $T = 3739$ K and $\gamma_\ell = 1.433$ (Stacey and Davis, 2004), from which we find a value of Q(CMB) $= 6.23$ TW (see Table C9).

The value of k_e scales as T at constant P (see equation 10) and is therefore larger at the CMB if the T_m is larger than Stacey's 3739 K. Alfè *et al.* (2002a,b) found T_m to be 4112 K, so in this case the appropriate value of k_e is 47.52 W m^{-1} K^{-1}, larger than Stacey's value (see Table C9). But the value for γ_l (CMB) found by Alfè *et al.* (2002a,b), 1.527, is larger than that found by Stacey, and this induces a larger value of $(\partial T/\partial z)_S$ (0.996). The net result is that for Alfè *et al.*'s data, Q(CMB) $= 7.69$ TW, larger than Stacey's value, 6.23 TW (see Table C9). The bottom row of Table C9 contains experimental (thermal physics) data. In the thermal physics approach, the measured γ (not shown) is for a solid, not a liquid. Since we are interested in a liquid core, this value of γ_s is generally not appropriate for a calculation of Q. However, γ_l, while much larger than γ_s at low pressure, decreases rapidly with pressure. (A plot of γ_l and γ_s versus pressure is shown in Figure 1 of Anderson, 1998). Convergence of γ_s and γ_l appears to occur above the CMB pressure, 135 GPa. At the ICB pressure (330 GPa), the difference between γ_l (from theory) and γ_s (from experiment) is vanishingly small. Thus, we take $\gamma_s = \gamma_l$ at the ICB and use the data on γ_l versus pressure to find T_m (CMB) and Q(CMB), as shown below.

We now give details on the determination of T_m and k_e from the *ab initio* and thermal physics approaches. The value of T_m (CMB) in Table C9 for the K' equation of state is for the core composition itself, since the K' equation of state deals with the real Earth. For the *ab initio* and thermal physics approaches, values of T_m are initially determined at the liquid-solid boundary for pure iron and then changed as corrections for light impurities are invoked. In these two approaches, two corrections must be made. First, there is a correction, resulting from the light impurities, called the temperature depression.

The temperature depression is calculated by assuming chemical equilibrium between the solid and liquid phases. The value $\Delta T_m = T_m - T_m^*$ represents the temperature depression from T_m of pure iron, where the superscript * represents the melting temperature of the solute. The symbols c_x^l and c_x^s represent the mole fraction of the solute in the liquid and solid phases, respectively. The formula for ΔT_m is (Landau and Lifshitz, 1958)

$$\Delta T_m = \frac{\boldsymbol{k} T_m (c_x^s - c_x^l)}{\Delta S} \qquad \text{(Eq. 11)}$$

where ΔS is the entropy of melting. By assuming no impurities in the solid core and ideal solutions, Stevenson (1981) simplified equation 10 to an approximation holding only for the liquid outer core, $\Delta T_m / T_m = -\Sigma \ln(1 - \chi_i)$, where χ_i is the mole fraction of the ith impurity. The approximation is now known to overestimate the value of ΔT_m. Anderson (1998) used this approximation to find ΔT_m for various combinations of light impurities in iron and obtained values near −1000 K.

Alfè *et al.* (2002b) solved the general problem given by equation 11, taking into account that there are light impurities in the inner core as well as the outer core and that partitioning of impurities can occur at

Table C9 Values used in finding Q(CMB), conductive power entering D″

Methods	γ_l(CMB)	T_m(CMB) K	k_e W m^{-1} K^{-1}	K W m^{-1} K^{-1}	$(\partial T/\partial z)_S$ K km^{-1}	Q(CMB) TW
K' equation of state[a]	1.435	3739	43.21	46.31	0.8856	6.23
Theory (*ab-initio*)[b] (Table 1)	1.527	4112	47.52	50.62	0.996	7.69
Experiment (thermal physics)[c]	1.51	3940 ± 100	45.53	48.63	0.944	7.00 ± 0.2

[a]See *Thermal conduction in the core*.
[b]Alfè *et al.* (2002a,b).
[c]Anderson (1998); Anderson *et al.* (2003).

the boundary of the liquid and the solid. Alfè *et al.* (2002b) used the impurities O, S, and Si and assumed concentrations for the core. They found:

a. A large concentration of oxygen impurities ($\sim$8 mole percent) is in the outer liquid core and a much smaller concentration ($\sim$0.3 mole percent) is in the inner solid core (Alfè *et al.*, 2002b). These oxygen concentrations in the inner and outer cores explain the density jump at the ICB (a solid-liquid core interface) found from seismology. Silicon and sulfur impurity concentrations, which are almost evenly distributed between the inner and outer cores, cannot help explain the density jump.
b. ΔT_m due to impurities $= -700$ K (Alfè *et al.*, 2002a), which satisfies the seismically determined density jump at the ICB. Since T_m of iron at the ICB pressure is 6200 K, the value of $T_m(\text{ICB}) = 5500$ K in the *ab initio* approach. Thus, $\Delta T_m / T_m = -0.1129$.

To find the temperature depression due to impurities from the thermal physics approach, we adopt the results of the calculation for $\Delta T_m / T_m$ from the *ab initio* approach, e.g., $\Delta T_m / T_m = -0.1129$. Since T_m for iron at the ICB pressure is 6050 K (Anderson *et al.*, 2003), we have $\Delta T_m = -683$ K and $T_m(\text{ICB}) = 5367$ K.

We now make the calculation for the value of T_m of the core at the CMB pressure (135 GPa). The value of γ_l along the liquidus found using *ab initio* methods goes from 1.488 at the CMB to 1.527 at the ICB, a change of only 3% (Alfè, pers. comm. 2003). These values are sufficiently close such that the value of T_m (CMB) may be calculated using $\gamma = 1.51$ in equation 8, resulting in

$$T_m(\text{CMB}) = T_m(\text{ICB})(9903/12\,166)^{\gamma} \quad \text{(Eq. 12)}$$

For $\gamma = 1.51$ and $T_m(\text{ICB}) = 5500$ K, we find $T_m(\text{CMB}) = 4112$ K from the *ab initio* method. For the thermal physics method, we use equation 12 with $\gamma = 1.51$, obtaining $T_m(\text{CMB}) = 3940 \pm 100$ K, using $\Delta T_m / T_m = -0.1129$ and $T_m(\text{ICB}) = 5367$ K. In the calculations, we have implicitly assumed that the values are for the liquidus, even though the measured properties are for the solidus in the thermal physics approach. We justify this because at the ICB the measured $\gamma_s = 1.50$, whereas from the *ab initio* approach the value is $\gamma_l = 1.53$. Thus, γ_l and γ_s have nearly converged. We assume convergence and therefore can use the liquidus properties for the adiabatic decompression calculation. We assign an error of ± 100 K to T_m(CMB) because of the lack of exact convergence.

The value of thermal conductivity must be found. Since T_m(CMB) is higher for the *ab initio* method and the thermal physics method than for the K' EoS method, the value of k_e is correspondingly higher. Equation 10 shows that k_e is proportional to T, and appropriate values of k_e and k are entered in Table C9. The resulting values of Q(CMB) in Table C9 are found from equation 1. It is seen that the highest value of Q(CMB), 7.7 TW, is from the *ab initio* method and the lowest value, 6.2 TW, is from the K' EoS method. Midway between the extremes is the value of Q(CMB) from the thermal physics method, 7.0 ± 0.2 TW.

How is one to choose between the three values of Q(CMB) shown in Table C9? First, we note that the three values are fairly close, especially when one considers the differences in the three methodologies. Second, all three methods have a history of success, so preferring a value from one method is somewhat arbitrary. Third, perhaps the best way to regard these results is that Q(CMB) should lie between a minimum value, 6.2 TW, and a maximum value, 7.7 TW. It is to be noted that the maximum value, 7.7 TW, is the closest to three recent results arising from theories of core thermal history. The total power passing from the core to D″ was reported as 8.3 TW by Gubbins *et al.* (2003), 9 TW by Gubbins *et al.* (2004), and 7.31 TW by Nimmo *et al.* (2004).

Orson L. Anderson

Bibliography

Alfè, D., Price, G.D., and Gillan, M.J., 2001. Thermodynamics of hexagonal close-packed iron under Earth's core conditions. *Physical Review B*, **64**: 045,123 1–16, doi:10.1103/PhysRevB.64.045,123.

Alfè, D., Gillan, M.J., and Price, G.D., 2002a. Composition and temperature of the Earth's core constrained by combining *ab initio* calculations and seismic data. *Earth and Planetary Science Letters*, **195**: 91–98.

Alfè, D., Gillan, M.J., and Price, G.D., 2002b. *Ab initio* chemical potentials of solid and liquid solutions and the chemistry of the Earth's core. *Journal of Chemical Physics*, **116**: 7127–7136.

Alfè, D., Price, G.D., and Gillan, M.J., 2002c. Iron under Earth's core conditions: Liquid-state thermodynamics and high-pressure melting curve from *ab initio* calculations. *Physical Review B*, **65,** 165,118 1–11, doi:10.1103/PhysRevB.65.165,118.

Anderson, O.L., 1998. The Grüneisen parameter for iron at outer core conditions and the resulting conductive heat and power in the core. *Physics of the Earth and Planetary Interiors*, **109**: 179–197.

Anderson, O.L., 2002a. The power balance at the core-mantle boundary. *Physics of the Earth and Planetary Interiors*, **131**: 1–17.

Anderson, O.L., 2002b. The three-dimensional phase diagram of iron. In Dehant, V., Creager, K.C., Karato, S.I., and Zatman, S., (eds.), *Earth's Core: Dynamics, Structure, Rotation.* vol. 31 of *Geodynamics Series*, Washington, DC: American Geophysical Union, doi: 10.1029/031GD07.

Anderson, O.L., Dubrovinsky, L., Saxena, S.K., and Le Bihan, T., 2001. Experimental vibrational Grüneisen ratio values for ε-iron up to 330 GPa at 300 K. *Geophysical Research Letters*, **28**: 399–402.

Anderson, O.L., Isaak, D.G., and Nelson, V.E., 2003. The high-pressure melting temperature of hexagonal close-packed iron determined from thermal physics. *Journal of Physics and Chemistry of Solids*, **64**: 2125–2131.

Bi, Y., Tan, H., and Jing, F., 2002. Electrical conductivity of iron under shock compression up to 200 GPa. *Journal of Physics:Condensed Matter*, **14**: 10849–10854.

Braginsky, S.I., and Roberts, P.H., 1995. Equations governing convection in Earth's core and the geodynamo. *Geophysical and Astrophysical Fluid Dynamics*, **79**: 1–97.

Bridgman, P.W., 1957. Effects of pressure on binary alloys, V and VI. *Proceedings of the American Academy of Arts and Sciences*, **84**: 131–216.

Dziewonski, A.M., and Anderson, D.L., 1981. Preliminary reference Earth model. *Physics of the Earth and Planetary Interiors*, **25**: 297–356.

Gubbins, D., Alfè, D., Masters, G., Price, G.D., and Gillan, M.J., 2003. Can the Earth's dynamo run on heat alone? *Geophysical Journal International*, **155**: 609–622.

Gubbins, D., Alfè, D., Masters, G., Price, G.D., and Gillan, M., 2004. Gross thermodynamics of two-component core convection. *Geophysical Journal International*, **157**: 1407–1414.

Isaak, D.G., and Anderson, O.L., 2003. Thermal expansivity of hcp iron at very high pressure and temperature. *Physica B*, **328**: 345–354.

Joos, G., 1958. *Theoretical Physics*, 3rd ed. London: Blackie. With the collaboration of I.M. Freeman, 885 pp.

Labrosse, S., Poirier, J., and Mouël, J.L., 1997. On cooling of the Earth's core. *Physics of the Earth and Planetary Interiors*, **99**: 1–17.

Landau, L.D., and Lifshitz, E.M., 1958. *Statistical Physics.* London, UK: Pergamon Press Ltd. Translated from the Russian by E. Peierls and R. F. Peierls.

Matassov, G., 1977. Electrical conductivity of iron-silicon alloys at high pressures and the Earth's core. Technical Report UCRL–52322, Lawrence Livermore National Laboratory. http://www.llnl.gov/tid/lof/documents/pdf/176480.pdf.

Nimmo, F., Price, G.D., Brodholt, J., and Gubbins, D., 2004. The influence of potassium on core and geodynamo evolution. *Geophysical Journal International*, **156**: 363–376.
Stacey, F.D., 1992. *Physics of the Earth,* 3rd ed. Brisbane: Brookfield Press, 513 pp.
Stacey, F.D., 2000. The K-primed approach to high-pressure equations of state. *Geophysical Journal International*, **143**: 621–628.
Stacey, F.D., and Anderson, O.L., 2001. Electrical and thermal conductivities of Fe-Ni-Si alloy under core conditions. *Physics of the Earth and Planetary Interiors*, **124**: 153–162.
Stacey, F.D., and Davis, P.M., 2004. High pressure equations of state with applications to the lower mantle and core. *Physics of the Earth and Planetary Interiors*, **142**: 137–184.
Stevenson, D.J., 1981. Models of the Earth's core. *Science*, **214**: 611–619.
Vočadlo, L., Alfè, D., Gillan, M.J., and Price, G.D., 2003. The properties of iron under core conditions from first principles calculations. *Physics of the Earth and Planetary Interiors*, **140**: 101–125.
de Wijis, G.A, Kresse, G., Vočadlo, L., Dobson, D.P., Alfè, D., Gilan, M.J., and Price, G.D., 1998. The viscosity of liquid iron at the physical conditions of the Earth's core. *Nature*, **392**: 805–807.

Cross-references

Alfvén's Theorem and the Frozen Flux Approximation
Anelastic and Boussinesq Approximations
Core, Adiabatic Gradient
Core Convection
Core, Electrical Conductivity
Core, Magnetic Instabilities
Core-Mantle Boundary Topography, Implications for Dynamics
Core-Mantle Boundary Topography, Seismology
Core-Mantle Boundary, Heat Flow Across
Core-Mantle Coupling, Electromagnetic
Core-Mantle Coupling, Thermal
Core-Mantle Coupling, Topographic
Core Properties, Physical
Core, Thermal Conduction
Core Viscosity
Inner Core Tangent Cylinder

CORE, BOUNDARY LAYERS

Basic ideas

It might be said that the term *boundary layer* means different things to different people. To someone observing fluid flowing past a flat plate, it might seem that there are two distinct regimes of motion. Far from the plate, the flow might seem too fast for the eye to follow, markers carried by the fluid appearing blurred as they speed past the plate; near the plate however the flow is so sluggish that it is easily followed by eye. The observer may call this the boundary layer, and the region beyond the free-stream or the mainstream, and he may feel that the interface between the two is reasonably sharp, so that he can call it the edge of the boundary layer. The theoretician will see no such sharp interface but will employ a mathematical technique, sometimes called matched asymptotics that similarly distinguishes an inner region near the plate from the outer region beyond. For him, the edge of the boundary layer is a region where the two solutions are required to agree with one another, i.e., to match.

To be successful, the asymptotic approach requires that the kinematic viscosity, ν, is small, as measured by an appropriate nondimensional parameter, such as the inverse Reynolds number in the case of the flow past the plate, or the Ekman number, E, for the situations we encounter below. The thickness, δ, of the boundary layer is then small compared with the characteristic scale, L, of the system, so that $\varepsilon = \delta/L$ is small and vanishes with ν; this does not imply that it is proportional to ν.

The inner and outer solutions are developed as expansions in powers of ε. The relative size of successive terms in the expansions is determined by the matching process. Provided ε is sufficiently small, only a few terms in each expansion are needed to obtain useful solutions of acceptable accuracy. In what follows, we shall retain only the first, or "leading" term in the expansion of the inner solution, $\mathbf{u}$, but will require the first and second terms in the expansion of the outer solution, $\mathbf{U}$. The primary (leading order) part, $\mathbf{U}_0$, of the outer solution is independent of ν and can therefore only satisfy one condition at a stationary impermeable boundary Γ, namely $U_{0\perp} \equiv \mathbf{n} \cdot \mathbf{U} = 0$, where $\mathbf{n}$ is the unit normal to Γ, directed into the fluid. The components $\mathbf{U}_{0\|}$ of $\mathbf{U}_0$ that are tangential to Γ will then in general be nonzero on $z = 0$, where z measures distance from Γ in the direction of $\mathbf{n}$. The task of the boundary layer is to reconcile $\mathbf{U}_{0\|}$ at $z = 0$ with the no-slip condition: $\mathbf{u} = \mathbf{0}$ on Γ. This means that $\mathbf{u}_\|$ must, through the action of viscosity, be reduced from $\mathbf{U}_{0\|}$ to zero in the distance δ. To achieve this deceleration, the viscous term in the momentum equation, $-\nu\nabla^2\mathbf{u}_\|$, must be finite and nonzero. Since this force is of order $\nu\mathbf{u}_\|/\delta^2$, it follows that δ is proportional to $\nu^{1/2}$.

To derive the inner expansion, the stretched coordinate, $\zeta = z/\delta$, is introduced to replace the distance z from Γ. The boundary layer is then characterized by $\zeta = \mathcal{O}(1)$ and $\partial_\zeta \equiv \partial/\partial\zeta = \mathcal{O}(1)$, so that $\partial_z \equiv \partial/\partial z = \mathcal{O}(1/\delta) \gg 1/L$. In contrast, $\nabla_\|$ is much smaller; as in the mainstream, it is $\mathcal{O}(1/L)$. The matching principle asserts that the inner and outer expansions should agree with each other at the edge of the boundary layer, which is defined as a region where $z \ll L$ but $\zeta \gg 1$. [For example, if $z = \mathcal{O}(L\delta)^{1/2}$, then $z/L = \mathcal{O}((\delta/L)^{1/2}) \downarrow 0$ and $\zeta = \mathcal{O}((L/\delta)^{1/2}) \uparrow \infty$ as $\varepsilon \downarrow 0$.]

A significant consequence follows from mass conservation and the matching principle. Since variations in density across the thin boundary layer can be ignored at leading order, mass conservation requires that $\nabla \cdot \mathbf{u} = 0$. At leading order, this gives

$$\partial_\zeta u_\perp + \nabla_S \cdot \mathbf{u}_\| = 0. \qquad \text{(Eq. 1a)}$$

Here ∇_S is the surface divergence, which may be defined, in analogy with the better known definition of the three-dimensional divergence, as a limit. For any vector $\mathbf{Q}$ depending on position $\mathbf{x}_\|$ on Γ and directed tangentially to Γ,

$$\nabla_S \cdot \mathbf{Q} = \lim_{A\to 0}\frac{1}{A}\oint_\gamma \mathbf{Q}\cdot\mathbf{N}\mathrm{d}s. \qquad \text{(Eq. 1b)}$$

Here γ is the perimeter of a small "penny-shaped disk" on Γ of area A; ds is arc length and $\mathbf{N}$ is the outward normal to γ lying in Γ. This is shown in Figure C35, which also gives the disk a small thickness

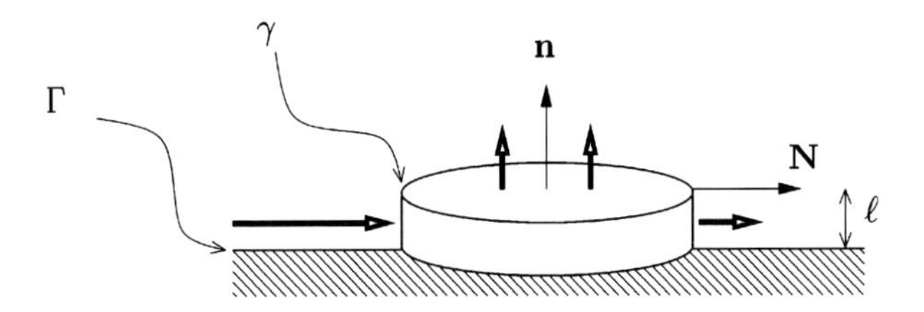

Figure C35 Efflux from the top of a "penny shape" volume encompassing the boundary layer. This compensates the volume flux deficit in the boundary layer through the sides of the volume.

$z = \ell$. Since $\mathbf{N}ds = \mathbf{ds} \times \mathbf{n}$, where the vector element ds of arc length is in the right-handed sense with respect to $\mathbf{n}$, an application of Stokes's theorem to Eq. (1b) gives

$$\nabla_S \cdot \mathbf{Q} = \lim_{A \to 0} \frac{1}{A} \oint_\gamma \mathbf{ds} \cdot (\mathbf{n} \times \mathbf{Q}) = \lim_{A \to 0} \frac{1}{A} \int_A \mathbf{dS} \cdot (\nabla \times (\mathbf{n} \times \mathbf{Q})),$$

where $\mathbf{dS} = \mathbf{n}dS$ is the vector element of surface area on A. From which it follows that

$$\nabla_S \cdot \mathbf{Q} = \mathbf{n} \cdot \nabla \times (\mathbf{n} \times \mathbf{Q}). \qquad \text{(Eq. 1c)}$$

This is generally a more convenient form of the surface divergence than (1b).

Since Γ is impermeable, $u_\perp = 0$ on $\zeta = 0$ and Eq. (1a) gives (on taking $\mathbf{Q} = \int_0^\zeta \mathbf{u}_\parallel d\zeta$)

$$u_\perp(\zeta) = \mathbf{n} \cdot \nabla \times \left(\int_0^\zeta \mathbf{u}_\parallel d\zeta \times \mathbf{n} \right). \qquad \text{(Eq. 1d)}$$

We now choose the "top" face of the penny in Figure C35 to be at the edge of the boundary layer, i.e., $\delta \ll \ell \ll L$. Then, by the asymptotic matching principle, the left-hand side of Eq. (1d) is both $u_\perp(\zeta \uparrow \infty)$ and $U_\perp(z \downarrow 0)$. Since the right-hand side of Eq. (1d) is $\mathcal{O}(\delta)$, this contribution to $\mathbf{U}$ is small compared with $\mathbf{U}_0$, i.e., it refers to the second, or "secondary," term $\mathbf{U}_1$ in the expansion of $\mathbf{U}$. This also establishes that the ratio of the secondary and primary terms in the expansion of $\mathbf{U}$ is $\mathcal{O}(\delta)$ and not, as might have been supposed a *priori*, $\mathcal{O}(\nu) = \mathcal{O}(\delta^2)$.

It follows from (1d) that, for $z = 0$,

$$U_{1\perp} = \mathbf{n} \cdot \nabla \times (\mathbf{Q}_\parallel \times \mathbf{n}), \qquad \text{(Eq. 2a)}$$

where $\mathbf{Q}_\parallel$ is now the volume flux deficit in the boundary layer:

$$\mathbf{Q}_\parallel = \delta \int_0^\infty (\mathbf{u}_\parallel - \mathbf{U}_\parallel) d\zeta, \quad \text{or} \quad \mathbf{Q}_\parallel = \int_0^{z/\delta \uparrow \infty} (\mathbf{u}_\parallel - \mathbf{U}_\parallel) dz. \qquad \text{(Eq. 2b, c)}$$

The second of these equivalent forms, which will be used later, provides a convenient way of reminding us that the z-integration is only across the boundary layer and not across the entire fluid. In each expression, $\mathbf{U}_\parallel$ is evaluated at $z = 0$ and therefore depends on $\mathbf{x}_\parallel$ only.

Equation 2a provides a crucial boundary condition on $\mathbf{U}_1$. Further boundary conditions are not needed and in fact could not be imposed without overdetermining the solution. Not even the term $\nu \nabla^2 \mathbf{U}_0$ appears in the equation governing $\mathbf{U}_1$, since this term is $\mathcal{O}(\nu) = \mathcal{O}(\delta^2)$, i.e., is asymptotically small compared with $\mathbf{U}_1$. For flow past a flat plate, the source (2a) at the boundary has the effect of displacing the effective boundary by an amount of order δ, which is then often called "the displacement thickness"; e.g., see Rosenhead (1963, Chapter V.5). For the Ekman and Ekman-Hartmann layers considered in "The Ekman layer" and "The Ekman-Hartmann layer" below, the effect of the secondary flow is more dramatic.

Ekman layers arise at the boundaries of highly rotating fluids; Ekman–Hartmann layers are their generalization in magnetohydrodynamics (MHD). The Ekman-Hartmann layer must reconcile both the mainstream flow $\mathbf{U}$ and the magnetic field $\mathbf{B}$ to conditions at a boundary Γ, where we assume that $\mathbf{u} = \mathbf{0}$ and $\mathbf{b} = \mathbf{B}_\Gamma$. It follows from $\nabla \cdot \mathbf{b} = 0$ that $b_\perp$ is the same everywhere in the boundary layer. It therefore coincides with $B_{\Gamma\perp}$ at the wall and with $B_{0\perp}$ at the edge of the boundary layer. The remaining components of $\mathbf{b}$ change rapidly through the boundary layer: $\mathbf{b} = \mathbf{B}_0$ at its "upper" edge and $\mathbf{b} = \mathbf{B}_\Gamma$, at its "lower" edge. This occurs because the current density, $\mathbf{j}_\parallel$, in the boundary layer is large. In analogy with Eq. (2a), there is a secondary current flow $J_{1\perp}$ between the boundary layer and the mainstream given by

$$J_{1\perp} - J_{\Gamma\perp} = \mathbf{n} \cdot \nabla \times (\mathbf{I}_\parallel \times \mathbf{n}), \qquad \text{(Eq. 3a)}$$

where $\mathbf{I}_\parallel$ is the electric current deficit:

$$\mathbf{I}_\parallel = \delta \int_0^\infty (\mathbf{j}_\parallel - \mathbf{J}_\parallel) d\zeta, \quad \text{or} \quad \mathbf{I}_\parallel = \int_0^{z/\delta \to \infty} (\mathbf{j}_\parallel - \mathbf{J}_\parallel) dz. \qquad \text{(Eq. 3a, b)}$$

If the exterior region, $z < 0$, is electrically insulating, $\mathbf{J}_\Gamma = \mathbf{0}$, so bringing Eq. (3a) even closer to Eq. (2a). When attention is focused on the mainstream magnetic field alone, $\mathbf{I}$ is the symptom of the boundary layer that has most significance, and it is thought of as a "current sheet." In a similar way, the volumetric deficit $\mathbf{Q}$ may be regarded as a "vortex sheet."

The theory necessary to evaluate $\mathbf{Q}$ and $\mathbf{I}$ will be described in "The Ekman layer." It will be found that, as far as the effects of viscosity are concerned, the relevant boundary layer scale is the Ekman layer thickness, $\delta_E = (\nu/\Omega)^{1/2} = E^{1/2}L$, where $E = \nu/\Omega L^2$ is the Ekman number and Ω is the angular speed of the system.

Geophysical overtones

The mathematical concepts introduced above need to be handled with care when applied to a complex object like the Earth's core. Various intricate physical effects can occur in the core and a mathematical model by essence requires some simplifying assumptions. Let us first note that the dynamics of the Earth's core is often modeled by a set of coupled physical quantities: the magnetic field, the fluid velocity, and a driving mechanism (usually thermal and/or chemical). This introduces (see *Anelastic and Boussinesq approximations*) at least three diffusivities. The magnetic diffusivity largely dominates the two others. It follows that boundary layers (associated with low diffusivities) can develop both based on the smallness of viscosity (as discussed above) and on the smallness of say the thermal diffusivity (compositional diffusivity being even smaller). The thickness of such thermal (compositional) boundary layers, δ_κ, depends on the thermal (compositional) diffusivity (or the relevant turbulent diffusivity), κ, of core material. These layers are conceptually significant in understanding how heat enters and leaves the FOC. We shall not consider them in this article and refer the reader to *Anelastic and Boussinesq approximations*.

Assume that $L = 2 \times 10^6$ m is a typical large length scale of core motions and magnetic fields (see *Core motions*). If the molecular viscosity ν is 10^{-6} m^2/s, then E is about 10^{-15}. This suggests that an asymptotic solution to core MHD is fully justified as far as the effects of viscosity are concerned. A mainstream solution would be expected away from the boundaries and a boundary layer would develop near the mantle and near the solid inner core.

This line of thought is also the basis for the determination of core surface motions from observations of the main geomagnetic field and its secular variation. On the assumption that electric currents generated in the mantle by the geodynamo are negligible, the observed fields can be extrapolated downwards to the CMB and used to provide information about the fluid motions "at the top of the core," using Alfvén's theorem. A difficulty however remains: the fluid in contact with the mantle co-rotates with it by the no-slip condition. There *is* no relative motion! Realistically, one can hope only to determine the mainstream flow at the edge of the boundary layer. The connection between the fields at the CMB to the fields at the edge of the boundary layer was considered by Backus (1968), who presented an analysis of Ekman-Hartmann type. Hide and Stewartson (1972) made further developments of the theory.

The boundary layer thickness, δ_E, obtained from the above estimate of the Ekman number is however only about 10 cm and it is natural to wonder whether such a thin layer can have any effect whatever on the dynamics of a fluid body that is about ten million times thicker!

Turbulence provides amelioration. It is widely accepted that the molecular viscosity is inadequate to transport large-scale momentum in the core, its role being subsumed by small-scale turbulent eddies. These, and the associated momentum flux, are highly anisotropic through the action of Coriolis and Lorentz forces (Braginsky and Meytlis, 1990). Nevertheless, a crude ansatz is commonly employed: the molecular ν is replaced by an isotropic turbulent viscosity having the same order of magnitude as the molecular magnetic diffusivity, that is to say about 1 m^2/s. Then $E \approx 10^{-9}$ and the asymptotic approach still seems secure. Even ignoring the possible intrinsic instability of the boundary layer (see "Stability of the Ekman-Hartmann layer"), it is reasonable to suppose that the small-scale eddies would penetrate the boundary layer so that a better estimate of δ_E would be 100 m. That such a layer would have a dynamical effect on the dynamics of the core then seems less implausible.

There are, however, skeptics who believe that, even if the boundary layers are 100 m thick, they are still too thin to be of geophysical interest. Kuang and Bloxham (1997) removed them by replacing the no-slip conditions on the core-mantle boundary (CMB) and inner core boundary (ICB) by the conditions of zero tangential viscous stress. This step obviously eliminates viscous coupling between the fluid outer core (FOC) and mantle and between the FOC and the solid inner core (SIC). Then, in the absence of other coupling mechanisms, the rotation of the mantle and the SIC does not affect the dynamics of the FOC and spin-up does not occur (see "The Ekman layer"). Geophysical justification for the step rests on estimates indicating that, owing to the large electrical conductivity of the SIC and despite the relatively small electrical conductivity in the mantle, the magnetic couples between the FOC and both the mantle and the SIC still greatly exceed the viscous couples, which can therefore be disregarded by adopting the zero viscous stress conditions. In this article we shall ignore such complications and adopt the "traditional approach" where no-slip conditions apply at both boundaries. See also *Core–mantle coupling*.

Although the smallness of E makes the asymptotic approach to MHD very attractive, purely analytic methods are not powerful enough to provide the solutions needed; numerical methods must be employed. One might nevertheless visualize a semianalytic, seminumerical approach in which the computer finds an inviscid mainstream flow $\mathbf{U}$ and magnetic field $\mathbf{B}$ that satisfy the boundary conditions (2a) and (3a). This too is not straightforward, since solutions for $\nu = 0$ raise important numerical difficulties and to restore ν to the mainstream would be tantamount to repudiating the boundary layer concept.

At this stage it is natural to wonder what use asymptotic methods have in core MHD. In the present state of algorithmic development and computer capability, the answer may seem to be, "Not much!" The boundary layer concept is, however, valuable in locating regions requiring special attention in numerical work, such as free shear layers and boundary layer singularities ("Free shear layers"). Moreover, the only way of verifying that the numerical simulations have reduced the effect of viscosity on core dynamics to a realistic level is by confronting them with the expectations of asymptotic theory.

The Ekman layer

Consider an incompressible fluid of uniform density, ρ, moving steadily with velocity $\mathbf{u}$, relative to a reference frame rotating with constant angular velocity $\boldsymbol{\Omega}$. In this reference frame, the Coriolis force per unit volume, $2\rho\boldsymbol{\Omega} \times \mathbf{u}$, is balanced by the pressure gradient $-\nabla p$ and the viscous stresses $\rho\nu\nabla^2\mathbf{u}$, where ν is the kinematic viscosity. The normal component of this balance gives, to leading order in the boundary layer expansion, $\nabla_\perp p = \mathbf{0}$, which implies that the pressure, p, throughout the boundary layer coincides with the pressure, P, in the mainstream at the edge of the boundary layer. The tangential components give dominantly

$$2\Omega_\perp \mathbf{n} \times \mathbf{u}_\parallel = -\nabla_\parallel(P/\rho) + \nu\partial_{zz}\mathbf{u}_\parallel, \qquad \text{(Eq. 4)}$$

which shows that only the component, $\Omega_\perp \equiv \boldsymbol{\Omega}\cdot\mathbf{n}$, of angular velocity normal to the boundary is significant at leading order. We assume here that $\Omega_\perp \neq 0$.

Inside the *Ekman layer* governed by Eq. (4), the interplay between the additional viscous forces needed to meet the no-slip boundary condition causes the flow to be deflected from the direction of $\mathbf{U}$, leading to the well known Ekman spiral of the velocity $\mathbf{u}(z)$ as z/δ increases from zero to infinity; see Figure C36.

As a consequence of the Ekman spiral there is a transverse mass transport, which is quantified by the volume flux deficit

$$\mathbf{Q}_\parallel = \tfrac{1}{2}\delta_E[-\mathbf{U}_0 + (\mathrm{sgn}\Omega_\perp)\mathbf{n} \times \mathbf{U}_0], \quad \text{where} \quad \delta_E = \sqrt{(\nu/|\Omega_\perp|)} \qquad \text{(Eq. 5a, b)}$$

is the appropriately redefined Ekman layer thickness which, over curved boundaries such as the CMB and ICB, depends on $\mathbf{x}_\parallel$ through $\Omega_\perp$. The velocity $U_{1\perp}$ follows directly from Eq. (2a); see also Greenspan (1968, p. 46):

$$U_{1\perp} = \mathbf{n}\cdot\mathbf{u} = \tfrac{1}{2}\mathbf{n}\cdot\nabla \times \{\delta_E[\mathbf{n} \times \mathbf{U}_0 + (\mathrm{sgn}\Omega_\perp)\mathbf{U}_0]\}, \qquad \text{(Eq. 6a)}$$

which, when the mainstream velocity $\mathbf{U}_0$ is geostrophic and boundary is planar normal to the rotation vector ($\mathbf{n} \times \boldsymbol{\Omega} = \mathbf{0}$), takes the simpler form

$$U_{1\perp} = \mathbf{n}\cdot\mathbf{u} = \tfrac{1}{2}\delta_E(\mathrm{sgn}\Omega_\perp)\mathbf{n}\cdot\nabla \times \mathbf{U}_0. \qquad \text{(Eq. 6b)}$$

This relates the normal flow to the vorticity $\nabla \times \mathbf{U}_0$ of the primary mainstream flow (see also Pedlosky, 1979).

The phenomenon described by Eq. (6) is often called *Ekman pumping* or *Ekman suction* depending on whether $u_\perp >$ or < 0. It should again be stressed that the secondary flow induced by the mainstream boundary condition (6) is scaled by the boundary layer thickness δ_E and is therefore small. It is worth emphasizing however that this modification of the effective boundary conditions introduces dissipation into the mainstream force balance, and this generally provides the dominant dissipation mechanism for a sufficiently large-scale flow. More precisely, for all flows in the mainstream characterized by a length scale larger than $LE^{1/4}$, dissipation within the boundary layers will dominate over the bulk effects of viscosity. Indeed, Ekman pumping/suction is a particularly significant process in determining the evolution of the angular momentum of a rotating fluid, which occurs on the spin-up timescale, $\tau_{s\text{-}u} = E^{-1/2}/\Omega = E^{1/2}L^2/\nu$, i.e., on a timescale intermediate between the rotation period $2\pi/\Omega$ (the day) and the viscous diffusion time L^2/ν (which, for the molecular ν, would exceed the age of the Earth).

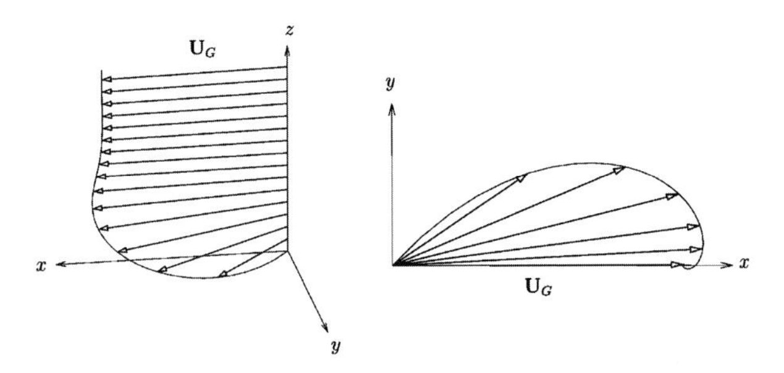

Figure C36 Side (*left*) and top (*right*) views of the Ekman layer profile. The velocity executes a spiral from zero velocity on the boundary to the mainstream velocity **U** at the edge of the boundary layer. Copyright 2007 from 'Mathematical Aspects of Natural Dynamos' by E. Dormy and A.M. Soward (eds.). Reproduced by permission of Routledge/Taylor & Francis Group, LLC.

The theory of the Ekman layer adumbrated above assumes that conditions are steady. It can be adapted to time-dependent situations provided they do not change too rapidly, i.e., provided that they occur on timescales longer than the rotation period $2\pi/\Omega$. When considering the effects on core flow of the lunisolar precession, in which $\mathbf{\Omega}$ changes on the diurnal timescale, the present theory is inapplicable.

The Ekman-Hartmann layer

The situation described in "The Ekman layer" is now generalized to MHD; the fluid is electrically conducting and a magnetic field is present. In the mainstream, the electric current density, **J**, is given by Ohm's law:

$$\mathbf{J} = \sigma(\mathbf{E} + \mathbf{U} \times \mathbf{B}), \qquad \text{(Eq. 7a)}$$

where **E** is the electric field and σ is the electrical conductivity. The Lorentz force per unit volume, $\mathbf{J} \times \mathbf{B}$, is therefore of order $\sigma B^2 U$. In the Earth's core, this is comparable with the Coriolis force $2\rho\mathbf{\Omega} \times \mathbf{U}$. The Elsasser number, $\Lambda = \sigma B^2/\rho\Omega$, is therefore of order unity. The asymptotic limit of interest is therefore $E \downarrow 0$ with $\Lambda = \mathcal{O}(1)$. Since $\Lambda = M^2 E$, where $M = BL(\sigma/\rho\nu)^{1/2}$ is the Hartmann number, it follows that $M \sim E^{-1/2} \uparrow \infty$ as $E \downarrow 0$.

Both **B** and the magnetic field, **b**, in the boundary layer obey Gauss's law, and to dominant order $\nabla \cdot \mathbf{b} = 0$ gives, as before ("Basic ideas"), $\partial_\zeta \mathrm{b}_\perp = 0$, from which $\mathrm{b}_\perp = B_{\Gamma\perp} = \mathrm{B}_{0\perp}$, at the edge of the mainstream. We shall suppose that $\mathrm{B}_{0\perp} \neq 0$. Ohm's law in the boundary layer is essentially the same as (7a) but, in our notation, it is written as

$$\mathbf{j} = \sigma(\mathbf{e} + \mathbf{u} \times \mathbf{b}). \qquad \text{(Eq. 7b)}$$

We again suppose steady conditions so that $\mathbf{E} = -\nabla\Phi$ and $\mathbf{e} = -\nabla\phi$, where Φ and ϕ are the electric potentials in the mainstream and boundary layer. The above expressions in terms of electric potentials correspond to a low magnetic Reynolds number description. It is important to remember we are here concerned with the magnetic Reynolds based on the boundary layer scale. This is a very small quantity in the case of the Earth's core. To leading order, the normal component of Eq. (7b) is $e_\perp = -\partial_\zeta\phi = 0$ and this shows that, throughout the boundary layer, ϕ coincides with Φ both at the edge of the mainstream and on the boundary itself. This generally depends on $\mathbf{x}_\parallel$, so that a current $\mathbf{J}_\Gamma = \sigma_\Gamma \mathbf{E}_\Gamma$ flows in the stationary wall, if its conductivity σ_Γ is nonzero. To leading order, the components of Eq. (7b) parallel to Γ together with Ampère's law, $\mu\mathbf{j} = \nabla \times \mathbf{b}$, where μ is the magnetic permeability, imply

$$\mu^{-1}\mathbf{n} \times \partial_z \mathbf{b}_\parallel = \sigma(-\nabla_\parallel \phi + \mathbf{u}_{0\parallel} \times \mathbf{B}_{0\perp}). \qquad \text{(Eq. 8)}$$

[The last term in Eq. (7b) also contributes $\mathbf{U}_{1\perp} \times \mathbf{b}_{0\parallel}$ but, as this is asymptotically smaller than the term $\mathbf{u}_{0\parallel} \times \mathbf{B}_{0\perp}$ retained, it has therefore been discarded.]

The determination of the boundary layer structure requires the equation of motion to be satisfied too. This differs from Eq. (4) by the addition of the Lorentz force. This may be decomposed into a magnetic pressure gradient and the divergence of the Maxwell stress $\mathbf{bb}/\mu$. The magnetic pressure can be absorbed into p to form a total pressure which, by the same argument as before, is constant across the boundary layer to leading order. The components of the equation of motion parallel to Γ are governed by

$$2\Omega_\perp \mathbf{n} \times (\mathbf{u}_\parallel - \mathbf{U}_{0\parallel}) = -(\mathrm{B}_\perp^2/\rho\mu\eta)(\mathbf{u}_\parallel - \mathbf{U}_{0\parallel}) + \nu\partial_{zz}\mathbf{u}_\parallel \qquad \text{(Eq. 9)}$$

see Gilman and Benton, (1968) and Loper, (1970).

Equations 8 and 9 determine the structure of the Ekman-Hartmann layer which, in the limit $\Lambda \downarrow 0$, becomes the Ekman layer considered in "The Ekman layer" (see Figure C37). For $\Lambda \uparrow \infty$, the Coriolis force is unimportant and the theory reduces to that governing the Hartmann layer, which is a well known boundary layer that arises in the study of MHD duct flow at large Hartmann number; see, e.g., Roberts (1967), Müller and Bühler (2001). It is then found that

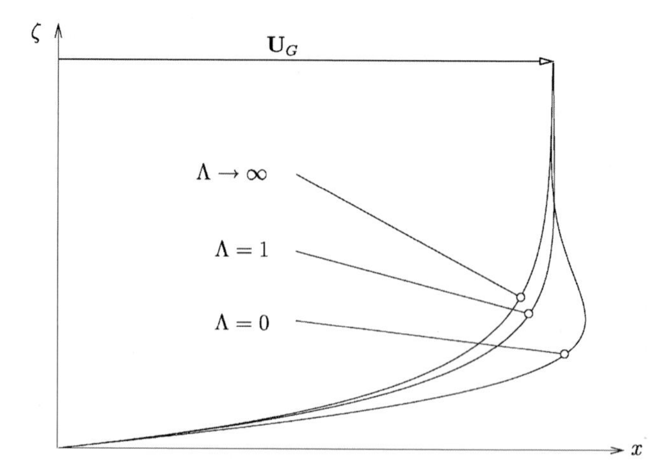

Figure C37 Streamwise component of the velocity in the boundary layer for different values of the Elsasser number, illustrating the smooth transition from an Ekman-type boundary layer to the Hartmann-type boundary layer.

$$\mathbf{u}_\parallel = \mathbf{U}_{0\parallel}[1 - \exp(z/\delta_\mathrm{H})] \qquad \text{(Eq. 10a, b)}$$

where $\delta_\mathrm{H} = \sqrt{(\rho\nu/\sigma)}/|B_\perp|$ is the Hartmann layer thickness. For a flat boundary, there is no flux deficit and $U_{1\perp}$ therefore vanishes.

Equations (8) and (9) also show that

$$\Lambda = B_\perp^2/\rho\mu\eta\Omega_\perp = (\delta_\mathrm{E}/\delta_\mathrm{H})^2 \qquad \text{(Eq. 11)}$$

is a convenient redefinition of the Elsasser number measuring the relative importance of Lorentz and Coriolis forces in determining the boundary layer structure. For small Λ, the Ekman spiral persists together with the associated boundary layer pumping. As E increases through $\mathcal{O}(1)$ values, these effects decrease until at large Λ the flow becomes unidirectional without any associated boundary layer pumping.

Free shear layers

Supposing that the Earth's fluid core occupies the spherical $r_i \leq r \leq r_0$, we redefine the Ekman and Hartmann numbers by

$$E = \nu/r_0^2\Omega = \delta_\mathrm{E}^2/r_0^2 \quad \text{and} \quad M^2 = r_0^2|\mathbf{B}|^2/\rho\nu\mu\eta = r_0^2/\delta_\mathrm{H}^2 \qquad \text{(Eq. 12a, b)}$$

respectively, and redefine the Elsasser number (11) as $\Lambda = M^2 E$. In what follows, (r, θ, φ) will be spherical coordinates in which the colatitude θ is zero at the north pole. We shall also employ cylindrical polar coordinates (s, φ, z).

As explained earlier, Ekman-Hartmann layers are generally present on both the CMB, $r = r_0$ and the ICB $(r = r_i)$. Since horizontal variations are negligible in comparison with the very rapid variation across the layers, the curvature of the spherical boundaries does not affect our earlier results as these depend only on the components of $\mathbf{\Omega}$ and **B** normal to the boundaries. Where these vanish, the theory adumbrated in "The Ekman layer" and "The Ekman-Hartmann layer" breaks down, and boundary layer singularities arise. To focus on these, we at first consider only the nonmagnetic case, $M = 0$. At the equator of the inner sphere, where the Ekman layer is singular, Eq. (2a) ceases to apply and a new force balance is struck, and a new type boundary layer arises called the equatorial Ekman layer. This is intimately linked to a free shear layer, which surrounds the tangent cylinder, i.e., the imaginary cylinder touching the inner sphere at the equator and having generators parallel to the rotation axis (see *Inner core: tangent cylinder*). This shear layer exhibits a complicated asymptotic structure that is best illustrated by the Proudman-Stewartson problem (Proudman, 1956; Stewartson, 1957, 1966), which concerns the slow steady axisymmetric flow induced by rotating the solid inner core at a slightly faster rate than the outer solid mantle. The flow pattern that occurs in this fundamental configuration illustrates the interplay between the various boundary and shear layers. It is important to stress that these layers do not take into account magnetic effects. Such effects are important in the Earth core and lead to a variety of shear layers whose details are not discussed here (see Kleeorin *et al.* 1997).

The mainstream flow is dominantly geostrophic and azimuthal: $\mathbf{U}_G = U_G(s)\widehat{\boldsymbol{\varphi}}$, where $\widehat{\boldsymbol{\varphi}}$ denotes the unit vector in the φ-direction. Outside the tangent cylinder, $s = r_i$, the fluid co-rotates with the outer sphere, and there is no Ekman layer on the CMB. Within the tangent cylinder, $\mathbf{U}_G(s)$ adjusts its value so that the Ekman suction, $-U_r(r_i) = \mathcal{O}(E^{1/2}U_G)$, into the Ekman layer on the ICB equals the Ekman pumping, $U_r(r_o)$, out of the Ekman layer on the CMB. This generates a secondary flow $U_z(s)\widehat{\mathbf{z}}$ in the mainstream that has only a z-component. It depends on s alone; such flows are termed *geostrophic*. In the northern (southern) hemisphere it is negative (positive).

As the tangent cylinder is approached from within, $\mathbf{U}_G$ tends to the inner sphere velocity. The jump in the geostrophic velocity across the tangent cylinder is smoothed out in an exterior quasigeostrophic layer $(s > r_i)$ in which the effects of Ekman pumping on the outer core boundary $\sim \Omega E^{1/2}\partial_s U_G$ is now balanced in the axial vorticity equation by lateral friction $\nu\partial_{sss}U_G$ in a layer of width $\mathcal{O}(r_0 E^{1/4})$, the Stewartson $E^{1/4}$-layer. There is a comparable but thinner $E^{2/7}$–layer inside $(s < r_i)$, whose main function is to smooth out $\partial_s U_G$ and so achieve continuity of stress.

These quasigeostrophic layers do not resolve all the flow discontinuities. The secondary mainstream flow, $U_z(s)\widehat{\mathbf{z}}$, feeds the Ekman layers on the ICB with fluid. In each hemisphere, the associated mass flux deficit, $\mathbf{Q}_\parallel$, is directed towards the equator and builds up as the equator is approached. Mass conservation demands that this fluid be accounted for, and this is one of the main functions of the shear layer at the tangent cylinder. The fluid is ejected towards the CMB as a jet in an inner ageostrophic layer of width $\mathcal{O}(r_0 E^{1/3})$, the Stewartson $E^{1/3}$-layer.

As the equator is approached, the Ekman layer becomes singular: from (5b), $\delta_E \sim \mathcal{O}(|\theta - \pi/2|^{-1/2} E_i^{1/2} r_i) \to \infty$, as $\theta \to \pi/2$, where $E_i = \nu/r_i^2\Omega = E(r_0/r_i)^2$ is the Ekman number based on the inner core radius. When $|\theta - \pi/2| = \mathcal{O}(E_i^{1/5})$ however, this expression for δ_E becomes of the same order as the distance, $\mathcal{O}((\theta - \pi/2)^2 r_i)$, of the point concerned in the Ekman layer from the tangent cylinder. This defines the lateral extent, $\mathcal{O}(E_i^{1/5} r_i)$, of the equatorial Ekman layer, i.e., the distance over which the solution (5a) fails. The radial extent of the equatorial Ekman layer is $\mathcal{O}(E_i^{2/5} r_i)$.

The MHD variant of Proudman-Stewartson problem reveals various other shear layers (Hollerbach, 1994a; Kleeorin *et al.*, 1997; Dormy *et al.*, 2002), while a nonaxisymmetric version exhibits even more structure (Hollerbach, 1994b; Soward and Hollerbach, 2000). Other investigations have been undertaken in plane layer (Hollerbach, 1996) and cylindrical (Vempaty and Loper, 1975, 1978) geometries.

Stability of the Ekman-Hartmann layer

The laminar Ekman-Hartmann layer profiles are determined by the linearized equations 8 and 9. This corresponds to a small Re approximation, where $\mathrm{Re} = UL/\nu$ is the Reynolds number and L is some characteristic length, possibly the core radius r_0, but perhaps smaller. In view of the low viscosity in the Earth's liquid core, and of its large size, Re is naturally huge. In the boundary layers, however, the relevant length scale is the boundary layer width δ. Thus the corresponding Reynolds number $\mathrm{Re}_{BL} = U\delta/\nu$ (often referred to as the "*boundary layer Reynolds number*") is much less than $\mathrm{Re}_0 = Ur_0/\nu$ for the full core. Since the width of the Ekman-Hartmann layer is based on the normal components $\Omega_\perp$ and $B_\perp$ of both the rotation and the magnetic field, Re_{BL} depends on position, $\mathbf{x}_\parallel$. Both these components decrease with latitude (at any rate for a magnetic field having dipole symmetry) and so the boundary layer width δ increases as the equator is approached, just as we explained in connection with the equatorial Ekman layer. Consequently the boundary layer Reynolds number Re_{BL} increases in concert.

Evidently the boundary layer Reynolds number Re_{BL} is much smaller than Re_0 and for geophysical parameter values it may be sufficiently small to justify the linear approach described in the previous sections. In that circumstance a linear stability analysis may be undertaken to determine a critical value of Re_{BL} (say Re_c, which the boundary layer becomes unstable (usually as a traveling wave). It is well known that boundary layer profiles with inflection points are generally prone to instability (e.g., Schlichting and Gersten, 2000). So, on the one hand, the Hartmann layer (10), which lacks an inflection point, is extremely stable to disturbances up to high values of Re_{BL} while, on the other hand, the Ekman and Ekman Hartmann profiles, which spiral, can develop instabilities at moderate Re_c.

Much effort has been devoted identifying the critical Reynolds number Re_c and the associated traveling wave mode of instability, whose orientation is determined by its horizontal wave vector $\mathbf{k}_\parallel$. Comprehensive Ekman layer stability studies have been undertaken in both the case of vertical rotation $\boldsymbol{\Omega}_\parallel = \mathbf{0}$ (Lilly, 1966) and oblique rotation $\boldsymbol{\Omega}_\parallel \neq \mathbf{0}$ (Leibovich and Lele, 1985).

The Ekman-Hartmann layer stability characteristics have been investigated in the context of the Earth's liquid core. For a model with normally directed magnetic field and rotation $(\boldsymbol{\Omega}_\parallel = \mathbf{B}_\parallel = 0)$, it has been shown for Earth core values of Ω and B (Gilman, 1971) that Re_{BL} is less than the critical value Re_0 necessary for an instability to grow. Since Re_{BL} is so small, the linearization leading to Eqs. (8) and (9) is justified. The more general orientation with $\boldsymbol{\Omega}_\parallel \neq \mathbf{0}$ and $\mathbf{B}_\parallel \neq \mathbf{0}$ appropriate to the local analysis of a shell with an axisymmetric dipole magnetic field has also been studied. Desjardins *et al.* (2001) show that, while the Ekman-Hartmann layer is stable in the polar caps, an equatorial band extending some 45° both north and south of the equator could develop instabilities.

Emmanuel Dormy, Paul H. Roberts, and Andrew M. Soward

Bibliography

Backus, G.E., 1968. Kinematics of geomagnetic secular variation in a perfectly conducting core. *Philosophical Transactions of the Royal Society of London*, **A263**: 239–266.

Braginsky, S.I., and Meytlis, V.P., 1992. Local turbulence in the Earth's core. *Geophysical and Astrophysical Fluid Dynamics*, **55**: 71–87.

Desjardins, B., Dormy, E., and Grenier, E., 2001. Instability of Ekman-Hartmann boundary layers, with application to the fluid flow near the core-mantle boundary. *Physics of the Earth and Planetary Interiors*, **124**: 283–294.

Dormy, E., Jault, D., and Soward, A.M., 2002. A super-rotating shear layer in magnetohydrodynamic spherical Couette flow. *Journal of Fluid Mechanics*, **452**: 263–291.

Dormy, E., and Soward, A.M. (eds), 2007. *Mathematical Aspects of Natural Dynamos*. The Fluid Mechanics of Astrophysics and Geophysics, CRC/Taylor & Francis.

Gilman, P.A., 1971. Instabilities of the Ekman-Hartmann boundary layer. *Physics of Fluids*, **14**: 7–12.

Gilman, P.A., and Benton, E.R., 1968. Influence of an axial magnetic field on the steady linear Ekman boundary layer. *Physics of Fluids*, **11**: 2397–2401.
Greenspan, H.P., 1968. *The Theory of Rotating Fluids.*, Cambridge: Cambridge University Press.
Hide, R., and Stewartson, K., 1972. Hydromagnetic oscillations of the earth's core. *Reviews of Geophysics and Space Physics*, **10**: 579–598.
Hollerbach, R., 1994a. Magnetohydrodynamic Ekman and Stewartson layers in a rotating spherical shell. *Proceedings of the Royal Society of London A*, **444**: 333–346.
Hollerbach, R., 1994b. Imposing a magnetic field across a nonaxisymmetric shear layer in a rotating spherical shell. *Physics of Fluids*, **6**(7): 2540–2544.
Hollerbach, R., 1996. Magnetohydrodynamic shear layers in a rapidly rotating plane layer. *Geophysical and Astrophysical Fluid Dynamics*, **82**: 281–280.
Kleeorin, N., Rogachevskii, A., Ruzmaikin, A., Soward, A.M., and Starchenko, S., 1997. Axisymmetric flow between differentially rotating spheres in a magnetic field with dipole symmetry. *Journal of Fluid Mechanics*, **344**: 213–244.
Kuang, W., and Bloxham, J., 1997. An Earth-like numerical dynamo model. *Nature*, **389**: 371–374.
Leibovich, S., and Lele, S.K., 1985. The influence of the horizontal component of the Earth's angular velocity on the instability of the Ekman layer. *Journal of Fluid Mechanics*, **150**: 41–87.
Lilly, D.K., 1966. On the instability of the Ekman boundary layer. *Journal of the Atmospheric Sciences*, **23**: 481–494.
Loper, D.E., 1970. General solution for the linearised Ekman-Hartmann layer on a spherical boundary. *Physics of Fluids*, **13**: 2995–2998.
Müller, U., and Bühler, L., 2001. *Magnetofluiddynamics in Channels and Containers.* Berlin: Springer.
Pedlosky, J., 1979. *Geophysical Fluid Dynamics.* Berlin: Springer.
Proudman, I., 1956. The almost rigid rotation of a viscous fluid between concentric spheres. *Journal of Fluid Mechanics*, **1**: 505–516.
Roberts, P.H., 1967. *An Introduction to Magnetohydrodynamics.* London: Longmans.
Rosenhead, L. (ed.), 1963. *Laminar Boundary Layers.* Oxford: Clarendon Press.
Schlichting, H., and Gersten, K., 2000. *Boundary Layer Theory.* Berlin: Springer.
Soward, A.M., and Hollerbach, R., 2000. Non-axisymmetric magnetohydrodynamic shear layers in a rotating spherical shell. *Journal of Fluid Mechanics*, **408**: 239–274.
Stewartson, K., 1957. On almost rigid rotations. *Journal of Fluid Mechanics*, **3**: 299–303.
Stewartson, K., 1966. On almost rigid rotations. Part 2. *Journal of Fluid Mechanics*, **26**: 131–144.
Vempaty, S., and Loper, D., 1975. Hydromagnetic boundary layers in a rotating cylindrical container. *Physics of Fluids*, **18**: 1678–1686.
Vempaty, S., and Loper, D., 1978. Hydrodynamic free shear layers in rotating flows. *ZAMP*, **29**: 450–461.

Cross-references

Alfvén's Theorem and the Frozen Flux Approximations
Anelastic and Boussinesq Approximations
Core Motions
Core-Mantle Boundary Topography, Implications for Dynamics
Core-Mantle Boundary Topography, Seismology
Core-Mantle Boundary, Heat Flow Across
Core-Mantle Coupling, Electromagnetic
Core-Mantle Coupling, Thermal
Core-Mantle Coupling, Topographic
Geodynamo
Inner Core Tangent Cylinder
Magnetohydrodynamics

CORE, ELECTRICAL CONDUCTIVITY

Core processes responsible for the geomagnetic field dissipate energy by two competing mechanisms that both depend on the electrical conductivity, σ_e, or equivalently the reciprocal quantity, resistivity, $\rho_e = 1/\sigma_e$. The obvious dissipation is ohmic heating. A current of density i (amperes/m^2) flowing in a medium of resistivity ρ_e (ohm m) converts electrical energy to heat at a rate $i^2\rho_e$ (watts/m^3). Thus one requirement for a planetary dynamo is a sufficiently low value of ρ_e (high σ_e) to allow currents to flow freely enough for this dissipation to be maintained. In a body the size of the Earth, this means that the core must be a metallic conductor. However, a metal also has a high thermal conductivity, introducing a competing dissipative process (see *Core, thermal conduction*). The stirring of the core that is essential to dynamo action maintains a temperature gradient that is at or very close to the adiabatic value (see also *Core, adiabatic gradient*) and conduction of heat down this gradient is a drain on the energy that would otherwise be available for dynamo action.

Heat transport by electrons dominates thermal conduction in a metal and the thermal and electrical conductivities are related by a simple expression (the Wiedemann-Franz law; see *Core, thermal conduction*). Thus, while the viability of a dynamo depends on a conductivity that is high enough for dynamo action, it must not be too high. In reviewing planetary dynamos, Stevenson (2003) concluded that high conductivity is a more serious limitation. It is evident that if the Earth's core were copper, instead of iron alloy, there would be no geomagnetic field.

The conductivity of iron

By the standards of metals, iron is a rather poor electrical conductor. At ordinary temperatures and pressures its behavior is complicated by the magnetic properties, but these have no relevance to conduction under conditions in the Earth's deep interior. We are interested in the properties of nonmagnetic iron, which means iron above its Curie point, the temperature of transition from a ferromagnetic to a paramagnetic (very weakly magnetic) state (1043 K), or in one of its nonmagnetic crystalline forms, especially the high pressure form, epsilon-iron (ϵ-Fe). Extrapolations from high temperatures and high pressures both indicate that the room temperature, zero pressure resistivity of nonmagnetic iron would be about 0.21 μΩ m. This is slightly more than twice the value for the familiar, magnetic form of iron and more than ten times that of a good conductor, such as copper. This is one starting point for a calculation of the conductivity of the core. A more secure starting point is the resistivity of liquid iron, just above its zero pressure melting point (1805 K), 1.35 μΩ m, although this is only marginally different from a linear extrapolation from 0.21 μΩ m at 290 K, because melting does not have a major effect on the resistivity of iron.

Effects of temperature and pressure

For a pure metal, resistivity increases almost in proportion to absolute temperature, but increasing pressure has an opposite effect. Phonons, quantized thermal vibrations of a crystal structure, scatter electrons, randomizing the drift velocities that they acquire in an electric field. The number of phonons increases with temperature, shortening the average interval between scattering events and increasing resistivity. Pressure stiffens a crystal lattice, restricting the amplitude of thermal vibration, or, in quantum terms, reducing the number of phonons at any particular temperature. It is convenient to think of the vibrations as transient departures from a regular crystal structure and that electrons are scattered by the irregularities. Although this is a highly simplified view it conveys the sense of what happens. Temperature increases crystal irregularity and pressure decreases it.

The temperature and pressure effects are given a quantitative basis by referring to a theory of melting, due in its original form to F.A. Lindemann. As modified by later discussions, Lindemann's idea was that melting occurs when the amplitude of atomic vibrations reaches a critical fraction of the atomic spacing. Stacey (1992) suggested that this was equivalent to saying that melting occurs at a particular level of the same crystal irregularity as is responsible for electrical resistivity and therefore that the resistivity (of a pure metal) is constant on the melting curve. The suggestion was given a mathematical basis by Stacey and Anderson (2001), simplifying the problem of extrapolating the resistivity of pure iron to core conditions. We need to know only how far the core is from the melting curve of iron and to make a relatively minor adjustment from the resistivity of liquid iron at zero pressure.

To verify the validity of this approach we can examine the available measurements on resistivity of iron and its alloys at relevant pressures and temperatures. The only experiments used the shock-compression method to measure resistivities at pressures up to 140 GPa, just inside the core range. The primary data source is an unpublished thesis by G. Matassov, who made a series of measurements on Fe-Si alloys and assembled earlier measurements on Fe and Fe-Ni. Some care is required in the interpretation because temperature is uncontrolled and must be estimated from insecure thermodynamic data on the Hugoniot or shock-compression curve, but Stacey and Anderson (2001) found that within the 10% uncertainty of the calculation the resistivity of pure solid iron at the melting point and a pressure of 140 GPa agreed with the zero pressure melting point value.

Effect of impurities

Theory is weaker when we consider the effect of alloying ingredients, that is impurity resistivity (see *Core composition*). Measurements on dilute alloys of iron with several other elements at nominal pressures up to 10 GPa (7% of the pressure at the top of the core) were reported by P.W. Bridgman. In this range the impurity effect is reasonably simple. It is more or less independent of both temperature and pressure and the added resistivity increases in a regular way with the total impurity content but is similar for different elements. These observations are helpful to our understanding of impurity resistivity in alloys with higher impurity contents, for which the simple picture breaks down, and to the behavior at much higher pressures, but for direct observations we rely almost entirely on data in the Matassov thesis.

Assuming that the conclusion from measurements on dilute alloys, that the total impurity content is more important than the identities of the impurity atoms, can be extended to higher concentrations, we can use the Fe-Si and Fe-Ni data reported by Matassov as indicative of core resistivity, without knowing what the core impurities are. We simply require sufficient of these core impurities to lower the outer core density by 10% relative to pure iron. On this basis Stacey and Anderson (2001) estimated the core impurity resistivity to be 0.90 μΩ m.

Conductivity variation within the core

The density contrast between the inner, solid and outer, liquid cores substantially exceeds the difference between densities of solid and liquid of the same composition. The inferred partial rejection of light solutes by the solid as the inner core grows (see *Convection, chemical*) means that there is a melting point depression and that the inner core boundary temperature is below the melting point of pure iron (see *Melting temperature of iron in the core*). The difference is less higher up in the core because the melting point gradient is steeper than the adiabat, so that there is a slight increase in resistivity upwards in the core. The temperature corrections are applied to the pure liquid iron melting point resistivity, assumed to coincide with the zero pressure value, 1.35 μΩ m, by taking resistivity to be proportional to absolute temperature and then the constant impurity resistivity, 0.90 μΩ m, is added. In this way Stacey and Anderson (2001) obtained outer core resistivities of 2.12 μΩ m at the core-mantle boundary and 2.02 μΩ m at the inner core boundary. The difference is less than the uncertainties in these values, but the variation is real. With a lower impurity content, plus solidification, the inner core resistivity is about 1.6 μΩ m.

Converting these values to conductivities, we have $\sigma_e = 4.7 \times 10^5$ S m^{-1} at the core-mantle boundary, 5.0×10^5 S m^{-1} at the inner core boundary, and 6.3×10^5 S m^{-1} in the inner core.

Magnetic diffusivity

It is sometimes convenient to refer not to conductivity or resistivity but to magnetic diffusivity. For nonferromagnetic materials, this is

$$\eta_m = \rho_e/\mu_0$$

where $\mu_0 = 4\pi \times 10^{-7}$ H m^{-1} is the permeability of free space. As for any other diffusivity the SI unit is m^{2}s^{-1} and with the resistivity estimates above the outer core value is 1.6 to 1.7 m^{2}s^{-1}.

Frank D. Stacey

Bibliography

Stacey, F.D., 1992. *Physics of the Earth*, 3rd edn. Brisbane: Brookfield Press.

Stacey, F.D., and Anderson, O.L., 2001. Electrical and thermal conductivities of Fe-Ni-Si alloy under core conditions. *Physics of Earth and Planet Interiors*, **124**: 153–162.

Stevenson, D.J., 2003. Planetary magnetic fields. *Earth Planetary Science Letters*, **208**: 1–11.

Cross-references

Convection, Chemical
Core Composition
Core, Adiabatic Gradient
Core, Thermal Conduction
Melting Temperature of Iron in the Core, Experimental
Melting Temperature of Iron in the Core, Theory

CORE, MAGNETIC INSTABILITIES

Introduction

The technique of considering the stability of some basic state to a perturbation is an essential one in fluid dynamics since it can give significant insight into fundamental mechanisms (see for example Chandrasekhar, 1961; Drazin and Reid, 1981). The classic example is that of Rayleigh-Bénard convection, in which fluid is contained between two horizontal flat plates. The lower plate is heated and the upper cooled. Since colder fluid is heavier than warmer fluid, the basic state consists of fluid at rest, with density ρ increasing with height z. This configuration is potentially unstable (heavy fluid lying over light fluid). However, convection (overturning of the fluid) will only take place if the density gradient $d\rho/dz$ is sufficiently large; the buoyancy force arising from the variation in density must be large enough to overcome the viscous drag of the fluid. An alternative way of thinking about this is to consider a small parcel of fluid. Consider a perturbation of the basic state that causes the parcel to move upwards. This results in the displacement of heavier fluid downward, the whole process releasing gravitational potential energy stored in the basic state. If this energy is greater than the work done against viscous drag then the perturbation will grow. Otherwise it will decay. The density gradient that marks the borderline between these two situations is known as the critical density gradient. The nondimensional number commonly used to measure the density gradient is known as the Rayleigh number

$$Ra \equiv g\alpha\beta\ell^4/\nu\kappa. \qquad \text{(Eq. 1)}$$

where g is the gravitational acceleration, α the coefficient of volume expansion $(-\rho^{-1}\partial\rho/\partial T)$, β the temperature gradient $(-\mathrm{d}T/\mathrm{d}z)$, ℓ the distance between the plates, ν the kinematic viscosity, and κ the thermal diffusivity of the fluid.

Much can be learned from linear theory which determines the critical value Ra_c of the Rayleigh number. Of fundamental interest to the geodynamo problem is the dependence of Ra_c on the rotation rate Ω and the magnetic field $\mathbf{B}$ (see *Core convection*).

Here, since the topic of thermal instability (or convection) is dealt with elsewhere, we shall focus on other sources of instability, in particular magnetic instability. The principle is the same; the basic state stores energy and a perturbation may extract that energy. The perturbation grows (i.e., the basic state is unstable) if energy is extracted and is more than enough to overcome any diffusive losses.

Magnetic instability

Associated with any magnetic field $\mathbf{B}$ is its magnetic energy $\int_V (B^2/2\mu)\mathrm{d}V$ where μ is the magnetic permeability and V is the region containing the field. If a rearrangement of field lines would result in a lower magnetic energy (just as the interchange of heavy and light fluid, as discussed above, results in a lower gravitational potential energy), then there is the possibility of instability driven by the magnetic field. Of course, the field cannot be considered in isolation. In the Earth's core, the field permeates a conducting fluid and the evolution of the field and the motion of the core fluid are strongly coupled. This inevitably constrains what rearrangements of field lines are possible. If a permitted rearrangement results in a lower total energy, then instability will result if the energy released exceeds the diffusive losses resulting from the rearrangement.

Mean-field dynamo theory (see *Dynamos, mean field*) focuses on the generation of an axisymmetric (or mean) magnetic field by the action of a mean electromotive force (e.m.f.) and differential rotation. A topic that has received somewhat less attention is that of the stability of the field to nonaxisymmetric perturbations. In mean-field dynamo theory, the field is maintained when the generation effect of the mean e.m.f. and differential rotation balance the decay due to ohmic diffusion. However, if the field is sufficiently strong and it satisfies certain other conditions then the field may be in an unstable configuration. Instability can extract energy from the mean field, so the generation mechanism may have a second sink of energy to counteract. Magnetic instabilities may therefore play an important role in determining what fields are observed and how strong they are. Linear theory has established that the minimum field strength required for instability (though depending on many factors) is comparable with estimates of the Earth's toroidal field strength. Also, a careful analysis (McFadden and Merrill, 1993) of the reversal data has concluded that "reversals are triggered by internal instabilities of the fluid motion of the core." Here, we review the various classes of magnetic instability and the conditions required for instability.

Classes of instability

Energy can be extracted from a basic magnetic field by a rearrangement of field lines in one of two ways; with or without reconnecting field lines. In a perfectly conducting fluid, field lines are frozen into the fluid (see *Alfvén's theorem*) and field lines can neither be broken nor reconnected. This constrains what perturbations are possible. An instability is known as ideal if it can extract energy without reconnecting field lines and can therefore exist in a perfectly conducting fluid. Alternatively, if the existence of an instability depends on reconnecting field lines and is therefore absent in a perfectly conducting fluid, the instability is described as resistive.

A given magnetic field configuration $\mathbf{B}$ in an ideal fluid may be stable. Adding resistive effects can destabilize it. While initially somewhat counter-intuitive, the reason is quite clear; adding the effect of resistivity increases the number of degrees of freedom of the system by allowing field-line reconnection. A very similar situation is familiar in parallel shear flows where an inviscid flow may be stable, but can be destabilized by adding viscosity (see for example Drazin and Reid, 1981).

The key parameter; the Elsasser number

For both ideal and resistive modes of instability, resistive effects (otherwise known as ohmic diffusion) also play a more traditional role; diffusion acts to damp out instability if diffusion is sufficiently strong, just as is the case for viscosity in thermal convection, see Eq. (1). The key parameter in the case of magnetic instability in a rapidly rotating system (such as the Earth's core) is the Elsasser number

$$\Lambda \equiv \frac{B^2}{2\Omega\mu_0\rho_0\eta}\left(=\frac{\sigma B^2}{2\Omega\rho_0}=\frac{\tau_\eta}{\tau_s}\right), \qquad \text{(Eq. 2)}$$

where the ohmic diffusion time

$$\tau_\eta = \frac{\mathcal{L}^2}{\eta} \qquad \text{(Eq. 3)}$$

and the slow MHD time scale

$$\tau_s = \frac{2\Omega}{\Omega_A^2}, \quad \text{where} \quad \Omega_A^2 = \frac{\mathcal{B}^2}{\mu_0\rho_0\mathcal{L}^2}. \qquad \text{(Eq. 4)}$$

In the above, $\mathcal{B}$ is the magnetic field strength, Ω is the rotation frequency of the Earth, μ_0 is the magnetic permeability of free space, ρ_0 is the core density, $\mathcal{L}$ is a characteristic length scale, for example the radius of the core, σ is the electrical conductivity, and $\eta = 1/(\mu_0\sigma)$ is the magnetic diffusivity.

The Elsasser number is a nondimensional measure of the field strength. It can also be thought of as an inverse measure of the strength of magnetic diffusion; $\Lambda \to \infty$ is the perfectly conducting limit. The expression of Λ in terms of the ratio of τ_η and τ_s is instructive; τ_s is the timescale on which diffusionless magnetic waves evolve in a rapidly rotating system for which $\Omega \gg \Omega_A$ (Ω_A is the Alfvén frequency, see *Alfvén waves*). When Λ is large $(\tau_\eta \gg \tau_s)$ the timescale on which magnetic waves evolve is short compared with the timescale on which magnetic diffusion acts and we therefore expect diffusive damping to be negligible. By contrast, when Λ is of order unity we can expect diffusive damping to be important. When Λ is small, the magnetic field is weak; it will have insufficient energy to drive an instability and will not play a dominant role in the dynamics of the fluid.

Conditions for instability

As expected from the above argument, detailed model calculations (based on the simultaneous solution of the Navier-Stokes equations and the magnetic induction equation for a prescribed axisymmetric field in a given geometry, see for example Fearn, 1994) indeed show that a necessary condition for magnetic instability is

$$\Lambda > \Lambda_c \qquad \text{(Eq. 5)}$$

where the exact value of Λ_c of course depends on the choice of field used in the model. Values of order 10 are typical. This is significant, because, for the Earth's core, $\Lambda = 10$ corresponds to a field strength of some 5 mT. Field strengths in the core are believed to be of this order so it is probable that magnetic instabilities are relevant to the dynamics of the Earth's core, and indeed may play an important role in constraining the field strength.

In addition to the condition (5) on the field strength, instability is also dependent on the field geometry (or shape). Condition (5) is about there being sufficient energy stored in the field. The geometric conditions described below are about whether that energy can be extracted.

The specific discussion here is tailored to the case of a rapidly rotating fluid, as is appropriate for application to the Earth's core. Most of the qualitative ideas, though, apply equally well to systems that are not rotating or where rotational effects are much less important, for example in laboratory plasmas (e.g., see Davidson, 2001) and the solar atmosphere (e.g., see Priest, 1982).

Ideal instability

Motivated by the belief that the toroidal part of the core field dominates the poloidal part, early work focussed on purely azimuthal fields of the form

$$\mathbf{B} = B\mathbf{e}_\phi \quad \text{(Eq. 6)}$$

where (s, ϕ, z) are cylindrical polar coordinates and $\mathbf{e}_\phi$ is the unit vector in the ϕ-direction and z is the axial direction. A local stability analysis for $B = B(s)$ (Acheson, 1983) has shown this to be unstable if B increases sufficiently rapidly with s somewhere in the core. The nature of this condition has led to the alternative name field gradient instability. If $B \propto s^\alpha$ then the condition for the field-gradient instability is $\alpha > 3/2$. More generally Acheson (1983) found instability if

$$\Delta \equiv \frac{2s^2}{B}\frac{d}{ds}\left(\frac{B}{s}\right) > m^2, \quad \text{(Eq. 7)}$$

where m is the azimuthal wavenumber of the instability.

Numerical studies of field (6) have confirmed Acheson's prediction and then gone on to consider more complex fields (see Fearn, 1994 for a review). Where Eq. (7) is not satisfied everywhere, there is a tendency for the instability to be concentrated in the region where Eq. (7) is satisfied.

It is worth commenting at this stage on the field $B \propto s(\alpha = 1)$ that results from a uniform current in the z-direction. This choice leads to a particularly simple form of the Lorentz force. For this reason, it has often been used in studies of the effect of a magnetic field on thermally driven convection. It is not ideally unstable in a rapidly rotating system, but *is* ideally unstable in a nonrotating system. Rotation is therefore seen to have an inhibiting effect. There are several studies that identify rather esoteric instabilities of magnetic origin for $B \propto s$. Fearn (1988) was able to link these to the presence of some additional effect, for example stable density stratification, counteracting the inhibiting effect of rotation. The message from this is that while $B \propto s$ is a perfectly adequate choice for the purpose of studying the effect of a magnetic field on convection, it is not a typical field for the study of magnetic instability.

Resistive instability

Resistive instability is usually associated with so-called critical levels $\mathbf{k} \cdot \mathbf{B} = 0$ where $\mathbf{k}$ is the wave vector of the instability. For fields of the form (6), this condition reduces to $B = 0$, i.e., resistive instability is associated with there being a zero of the azimuthal field somewhere in the core. The condition $\mathbf{k} \cdot \mathbf{B} = 0$ is well known in the nonrotating plasma physics literature and the main effect of rotation is to modify the timescale on which the instability operates. Field-line curvature is unimportant; the instability has been found both for curved fields of the form (6) and for straight field lines.

Discussion

Further studies have looked at adding a z-dependence to B and also investigated poloidal fields. These have shown that the basic qualitative understanding derived from studying the field (6) is robust; see discussion and references in Fearn (1998). In applying these ideas to the core, we know that the mean toroidal field vanishes on the axis, so must increase with s somewhere, before decreasing again to zero at the core-mantle boundary. It is also likely that the resistive instability condition $\mathbf{k} \cdot \mathbf{B} = 0$ will be satisfied somewhere. To further investigate fields relevant to the core, Zhang and Fearn (1994, 1995) have investigated the stability of the toroidal and poloidal decay modes of the core (see Figures C38 and C39) and found them to be unstable, with Λ_c typically in the range 10–20. Given all this, it seems highly likely that any dynamo generated field will have a configuration that is unstable somewhere in the core. Whether or not it is actually unstable will then depend on the field strength.

Several studies have investigated the combined effects of thermal convection and magnetic instability. A recent study (Zhang and Gubbins, 2000) builds on the Zhang and Fearn work referred to above, incorporating a basic magnetic field that is a combination of toroidal and poloidal decay modes and also includes density stratification. In general, instability is a result of both thermal and magnetic forcing.

Figure C38 Contours of $\mathbf{B}_\phi$ for two toroidal decay modes (*one on the left and one on the right*) studied by Zhang and Fearn (1994). Reproduced with the permission of Taylor & Francis.

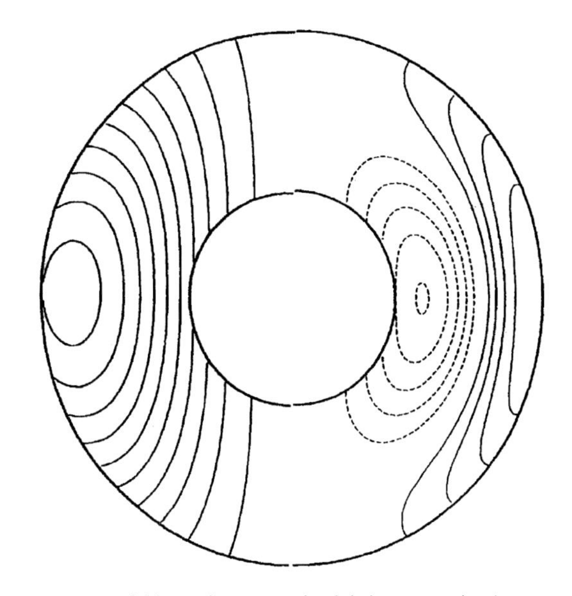

Figure C39 Field lines for two poloidal decay modes (*one on the left and one on the right*) studied by Zhang and Fearn (1995). Reproduced with the permission of Taylor & Francis.

One way of visualizing this is through a graph of Ra_c versus Λ. This has a negative gradient (Ra_c decreases as Λ increases) with the graph cutting the horizontal axis ($Ra = 0$) at $\Lambda = \Lambda_c$. As Λ is decreased from this point the contribution from magnetic energy to the instability decreases and the thermal contribution must increase (Ra_c increases) to compensate. The slope of the graph depends on the Roberts number $q = \kappa/\eta$, approaching the vertical for small q.

Recent developments

The stability analyses leading to an understanding of the necessary conditions for instability (see above) have all been linear. Such analyses can say nothing about how the growth of the instability feeds back on the dynamo process generating **B**.

The effect of the instability can be thought of as twofold. Firstly, and most simply, the instability extracts energy from the field so represents a drain on the field's energy in addition to ohmic diffusion. We therefore expect that for a given energy source driving the geodynamo, the field generated would be weakened by the presence of an instability. The big question is "by how much?" Secondly, there will be a mean e.m.f. associated with the instability. This will feed back on the mechanism generating the field. Recent work (Fearn and Rahman, 2004) has begun to investigate this and found that this feedback effect can be important and that magnetic instability can significantly constrain the strength of the mean field.

Acknowledgments

Figures C38 and C39 are reproduced from Figure 1 of Zhang and Fearn (1994) and Figure 1 of Zhang and Fearn (1995) respectively with permission of the publishers Taylor and Francis (http://www.tandf.co.uk).

David R. Fearn

Bibliography

Acheson, D.J., 1983. Local analysis of thermal and magnetic instabilities in a rapidly rotating fluid. *Geophysical and Astrophysical Fluid Dynamics*, **27**: 123–136.

Chandrasekhar, S., 1961. *Hydrodynamic and Hydromagnetic Stability.* Oxford: Clarendon Press.

Davidson, P.A., 2001. *An Introduction to Magnetohydrodynamics.* Cambridge: Cambridge University Press.

Drazin, P.G., Reid, W.H., 1981. *Hydrodynamic stability.* Cambridge: Cambridge University Press.

Fearn, D.R., 1988. Hydromagnetic waves in a differentially rotating annulus IV. Insulating boundaries. *Geophysical and Astrophysical Fluid Dynamics*, **44**: 55–75.

Fearn, D.R., 1994. Magnetic instabilities in rapidly rotating systems. In Proctor, M.R.E., Matthews, P.C., and Rucklidge, A.M., (eds) *Solar and Planetary Dynamos* Cambridge: Cambridge University Press, pp. 59–68.

Fearn, D.R., 1998. Hydromagnetic flows in planetary cores. *Reports on Progress in Physics*, **61**: 175–235.

Fearn, D.R., Rahman, M.M., 2004. Instability of non-linear α^2-dynamos. *Physics of the Earth and Planetary Interiors*, **142**: 101–112.

McFadden, P.L., Merrill, R.T., 1993. Inhibition and geomagnetic reversals. *Journal of Geophysical Research*, **98**: 6189–6199.

Priest, E.R., 1982. *Solar Magnetohydrodynamics.* Dordrecht: Reidel.

Zhang, K., Fearn, D.R., 1994. Hydromagnetic waves in rapidly rotating spherical shells generated by magnetic toroidal decay modes. *Geophysical and Astrophysical Fluid Dynamics*, **77**: 133–157.

Zhang, K., Fearn, D.R., 1995. Hydromagnetic waves in rapidly rotating spherical shells generated by poloidal decay modes. *Geophysical and Astrophysical Fluid Dynamics*, **81**: 193–209.

Zhang, K., Gubbins, D., 2000. Is the geodynamo process intrinsically unstable? *Geophysical Journal International*, **140**: F1–F4.

Cross-references

Alfvén Waves
Alfvén's Theorem
Core Convection
Core Motions
Dynamos, Mean Field
Geodynamo, Dimensional Analysis and Timescales
Magnetoconvection
Magnetohydrodynamic Waves
Reversals, Theory

CORE, THERMAL CONDUCTION

The three-dimensional stirring of the outer core that is required for dynamo action is very rapid compared with thermal diffusion, ensuring that the temperature gradient is maintained very close to the adiabatic value (see *Core, adiabatic gradient*). Thus there is a steady flux of conducted heat at all levels, regardless of the convected heat transport. Thermal convection requires additional heat, but there is also a possibility of refrigerator action by compositional convection, carrying some of the conducted heat back down (see *Core convection* and *Core composition*). Either way the conducted heat cannot contribute to dynamo action, but is a "base load" on core energy sources that must be provided before anything else can happen. The magnitude of this "base load" depends on the thermal conductivity.

Relationship between thermal and electrical conductivities

Heat transport in the core alloy is dominated by the conduction electrons. We can write the total conductivity, κ, as a sum of two terms, κ_e representing the electron contribution and κ_ℓ the lattice contribution, with $\kappa_e \gg \kappa_\ell$

$$\kappa = \kappa_e + \kappa_\ell$$

An originally empirical but now well documented and theoretically understood relationship connects κ_e to the electrical conductivity, σ_e (see *Core, electrical conductivity*). This is the Wiedemann-Franz law:

$$\kappa_e = LT\sigma_e$$

where T is absolute temperature and $L = 2.45 \times 10^{-8} \mathrm{W}\Omega\,\mathrm{K}^{-2}$, known as the Lorentz number, is the same for all metals. Derivations of the Wiedemann-Franz law (see, for example, Kittel, 1971) assume that the temperature is not too low and that the mechanisms of electron scattering are the same for electrically and thermally energized electrons. These conditions are satisfied in the core. Thus, we estimate thermal conductivity from electrical conductivity, for which there are satisfactory independent estimates.

Lattice conductivity, κ_ℓ, occurs in all materials and is discussed under *Mantle, thermal conductivity* (*q.v.*). It is not well constrained theoretically, but is so much smaller than κ_e that the uncertainty has little influence on the estimated total κ for the core. Following Stacey and Anderson (2001), we assume a uniform value, $\kappa_\ell = 3.1\ \mathrm{Wm^{-1}K^{-1}}$ throughout the core and add this to κ_e, as obtained from the Wiedemann-Franz law.

The variation in $\rho_e(=1/\sigma_e)$ over the depth range of the outer core is slight. Values given under *Core, electrical conductivity* (*q.v.*), 2.12 $\mu\Omega$ m at the core-mantle boundary decreasing to 2.02 $\mu\Omega$ m at the inner core boundary, are used here to represent the radial variation, retaining digits beyond the level of absolute significance. These give the values of κ listed in Table C10. Most of the radial variation is

Table C10 Thermal conduction in the core

Radius (km)	T (K)	dT/dz (K/km)	κ (Wm^{-1}K^{-1})	η (10^{-6}m^{2}s^{-1})	$\dot{Q}/A$ (mWm^{-2})	$\dot{Q}$ (10^{12} W)
0	5030.0	0	80.38	8.57	0	0
200	5029.2	0.0080	80.32	8.56	0.65	0.0003
400	5026.8	0.0161	80.17	8.55	1.29	0.003
600	5022.8	0.0241	79.72	8.54	1.93	0.007
800	5017.1	0.0322	79.57	8.52	2.56	0.021
1000	5009.9	0.0402	79.13	8.49	3.18	0.040
1200	5001.0	0.0483	78.59	8.46	3.79	0.069
1221.5	5000.0	0.0491	78.53	8.46	3.86	0.072
1221.5	5000.0	0.2857	63.74	6.60	18.21	0.342
1400	4945.7	0.3236	62.95	6.56	20.37	0.502
1600	4876.6	0.3677	61.95	6.51	22.78	0.733
1800	4798.5	0.4133	60.83	6.46	25.14	1.024
2000	4711.1	0.4604	59.58	6.40	27.43	1.379
2200	4614.2	0.5091	58.20	6.34	29.63	1.802
2400	4507.4	0.5594	56.70	6.26	31.72	2.296
2600	4390.3	0.6119	55.07	6.18	33.70	2.862
2800	4262.5	0.6668	53.31	6.10	35.55	3.502
3000	4123.4	0.7258	51.42	6.00	37.32	4.221
3200	3972.3	0.7871	49.39	5.90	38.87	5.002
3400	3808.2	0.8548	47.22	5.78	40.36	5.864
3480	3738.7	0.8836	46.31	5.73	40.92	6.227

due to the temperature factor in the Wiedemann-Franz law, with the adiabatic temperature profile also given in the table. The inner core is more nearly isothermal, but both electrical and thermal conductivities are higher on account of the lower content of "impurities." The impurity resistivity is taken to be half of the outer core value of 0.90 μΩ m.

Conducted heat in the outer core

The flux of conducted heat through any radius is

$$\dot{Q} = \kappa A \mathrm{d}T/\mathrm{d}z$$

where $A = 4\pi r^2$ is the area of the surface at radius r and $\mathrm{d}T/\mathrm{d}z = -\mathrm{d}T/\mathrm{d}r$ is taken as the adiabatic gradient and also listed in Table C10. The decrease in $\dot{Q}/A$ with depth is caused by the decreasing temperature gradient, which more than compensates for the increase in κ.

Diffusivity and thermal relaxation of the inner core

To discuss progressive cooling we need to know the thermal diffusivity, defined by

$$\eta = \kappa/\rho C_p$$

with ρ being density and C_P the specific heat (see *Core properties, physical*). This is the ratio of conductivity to heat capacity per unit volume and is a measure of the speed with which the internal temperature of a body may change by conduction of heat into or out of it. The unit is m^{2}s^{-1}, reflecting the fact that the time required for any change depends on the square of the linear dimension of a body. In the case of a sphere there is a simple expression for thermal relaxation in terms of η. Consider a uniform sphere (with no internal heat sources) of total radius R and a surface temperature $T(R)$ that is held constant as the interior loses heat by conduction to the surface. The internal temperature profile, $T(r)$, relaxes towards a radial dependence given by

$$T(r) - T(R) = [T(r=0) - T(R)]\sin(\pi r/R)/(\pi r/R)$$

and decays exponentially with time, t

$$T(r) - T(R) = [T(r) - T(R)]_{t=0}\exp(-t/\tau)$$

where

$$\tau = (1/\eta)(R/\pi)^2$$

is the thermal relaxation time constant.

In applying this to the inner core, we see from Table C10 that it suffices to assume a constant thermal diffusivity, $\eta = 8.5 \times 10^{-6}$ m^{2}s^{-1}. Taking $R = 1.22 \times 10^6$ m, we have $\tau = 1.77 \times 10^{16}$s $= 0.56 \times 10^9$ years. The significance of this number is that it is small enough to ensure that, unless the inner core has more than 5×10^{11}W of radiogenic heat, its temperature gradient is less than adiabatic, disallowing inner core convection and a convective explanation of its anisotropy (see *Inner core anisotropy*). This concentration of radiogenic heat in the core as a whole would give about 10^{13}W, preventing core cooling and causing difficulty with thermal history. Radiogenic heat in the inner core is not allowed for in Table C10.

Competition between heat sources and conduction

Consider the heat loss by the core due to bodily cooling only, with neglect of the gravitational energy of compositional convection and latent heat of inner core formation. The temperature profile remains adiabatic, so that the rate of temperature fall at any depth is proportional to temperature and increases inwards. There is also an inward increase in the heat capacity per unit volume, due to the increasing density. In this situation the total heat flux through any radius would be almost exactly proportional to the conducted heat flux, $\dot{Q}$, as listed in Table C10. Convective instability would arise throughout the outer core at the same value of the total heat flux from the core. Still without considering the gravitational energy release, when we introduce the latent heat release at the inner core boundary, the added flux is the same at all radii, r, so that the additional heat flux per unit area varies as $1/r^2$. Then the total heat flux exceeds the conducted heat by a margin that increases inwards. It follows that the driving force for convection and dynamo action is strongest deep in the outer core and it is possible that the outer part would not convect thermally, but is driven by the compositional convection. But, whatever the process, the stirring action maintains an adiabatic gradient in the outer core, with the consequent conducted heat, $\dot{Q}$, listed in Table C10.

Frank D. Stacey

Bibliography

Kittel, C., 1971. *Introduction to Solid State Physics*, 4th edn. New York: Wiley.

Stacey, F.D., and Anderson, O.L., 2001. Electrical and thermal conductivities of Fe-Ni-Si alloy under core conditions. *Physics of Earth and Planetary Interiors*, **124**: 153–162.

Cross-references

Core Composition
Core Convection
Core Properties, Physical
Core, Adiabatic Gradient
Core, Electrical Conductivity
Inner Core Anisotropy
Mantle, Thermal Conductivity

CORE-BASED INVERSIONS FOR THE MAIN GEOMAGNETIC FIELD

Maps of the Earth's magnetic field have been made for navigational purposes since at least the time of *Edmond Halley* (*q.v.*); for details see *Main field modeling*. This activity has culminated in production of the *International Geomagnetic Reference Field* (*q.v.*) using a representation of the main field in spherical harmonic (or geomagnetic or Gauss) coefficients by least-squares fitting to global data set. The basic mathematical representation and statistical procedure were developed by *C.F. Gauss* (*q.v.*); the current approach using computers is described in Barraclough and Malin (1968).

The geomagnetic field $\boldsymbol{B}$ is represented as the gradient of a potential V satisfying Laplace's equation:

$$\boldsymbol{B} = -\nabla V$$

$$\nabla^2 V = 0,$$

the solution of which is written in terms of spherical harmonics as the sum

$$V(r,\theta,\phi) = a\sum_{l=1}^{\infty}\sum_{m=0}^{l}\left(\frac{a}{r}\right)^{l+1}\left[g_l^m \cos m\phi + h_l^m \sin m\phi\right]P_l^m(\cos\theta) \qquad \text{(Eq. 1)}$$

where r, θ, ϕ are the usual spherical coordinates, a is Earth's radius, and P_l^m is an associated Legendre function (see *Harmonics, spherical*). The measured component of the magnetic field is determined by differentiating V and setting r, θ, ϕ to the coordinates of the measurement point. The unknown geomagnetic coefficients $\{g_l^m, h_l^m\}$ are then determined by truncating the series at some suitable degree L and fitting to the observations by minimizing the sum of squares of the residuals. The truncation point depends not only on the quality and coverage of the measurements but also on the level of magnetization of crustal rocks, which is regarded as noise because the main field is defined as the part of the geomagnetic field that originates in the core. The truncation point has risen from $L = 8$ to $L = 10$ as data quality has improved, and may increase further in the near future. The core field is now thought to dominate below $L = 12$ while the crustal field is thought to dominate above that level, although the cross-over point is difficult to establish.

There are problems with abruptly truncating a spherical harmonic series. These are closely related to the more familiar problem of truncating a Fourier series abruptly—indeed, the expansion (1) is exactly a Fourier series in ϕ. Truncating an infinite series can be thought of as multiplying the coefficients by a "truncation sequence" that is equal to 1 if $l \leq L$ and to 0 if $l > L$. In the space domain this is equivalent to convolution with the spatial function corresponding to the truncation sequence. In ordinary Fourier spectral analysis the truncation function is called a boxcar and its spatial function is the familiar sinc function, or $\sin N\phi/\sin\phi$, where N depends on the width of the boxcar or truncation level. The spherical harmonic problem is two-dimensional but the principle is the same: the spatial representation of the truncated series is the convolution of the spatial representation of the truncation function with the correct function, the one corresponding to the infinite series:

$$V_L(r,\theta,\phi) = \int V(r,\theta',\phi')A(\theta,\phi;\theta',\phi')\mathrm{d}\Omega' \qquad \text{(Eq. 2)}$$

where A is the spatial representation of the trunction function, or averaging function, and the integral is taken over all solid angles Ω'. Symmetry arguments show that A must be only a function of the angular distance Δ between (θ,ϕ) and (θ',ϕ') (Whaler and Gubbins, 1981):

$$A(\theta,\phi;\theta',\phi') = \frac{1}{4\pi}\sum_{l=2}^{L}(2l+1)P_l(\cos\Delta). \qquad \text{(Eq. 3)}$$

A is also plotted in Figure C40. Note the similarities between A and the sinc function: both have peaks at the estimation point (θ,ϕ) and both have zeroes and oscillations or side lobes well away from the estimation point. These are obviously undesirable because the estimate is influenced by the magnetic field over a very wide area. Side lobes in the spherical case are particularly insidious because of the very large lobe at 180°—the estimate is strongly influenced by behavior at the antipodal point! In Fourier spectral analysis this problem is treated by tapering the Fourier series rather than truncating it abruptly and a whole range of tapers and their properties have been studied. The same can be applied to the spherical problem, and in our case we can design the taper by assuming the field originates in the Earth's core.

The part of the geomagnetic field at the Earth's surface that originates in the core is large scale because short wavelengths are attenuated with distance upward to the Earth's surface (see *Upward and downward continuation*). The magnetic potential at the core surface is obtained from equation 1 by setting $r = c$, the core radius. This increases the degree l spherical harmonic coefficients by a factor $(a/c)^{l+1}$, where $a/c \approx 1.8$. The shorter wavelengths are severely attenuated by upward continuation from the core and are increased by downward continuation to the core. Any field model we produce by fitting data at the Earth's surface may therefore produce a very rough field at the core surface, with a large amount of energy in shorter wavelengths. Indeed, the Earth's surface has no physical significance in this problem at all, it is simply the place where measurements are made. Shure *et al.* (1982) therefore proposed finding the model that was in some sense optimally smooth at the core surface. This needs a mathematical definition of "smooth," the simplest being to minimize the "surface energy," $\int \boldsymbol{B}^2\mathrm{d}\Omega$, where $\boldsymbol{B}$ is evaluated at $r = c$, the surface of the core. Shure *et al.* (1982) prove a number of useful properties of this solution; the most important is uniqueness, which means the core–surface constraint can replace truncation of the spherical harmonic series.

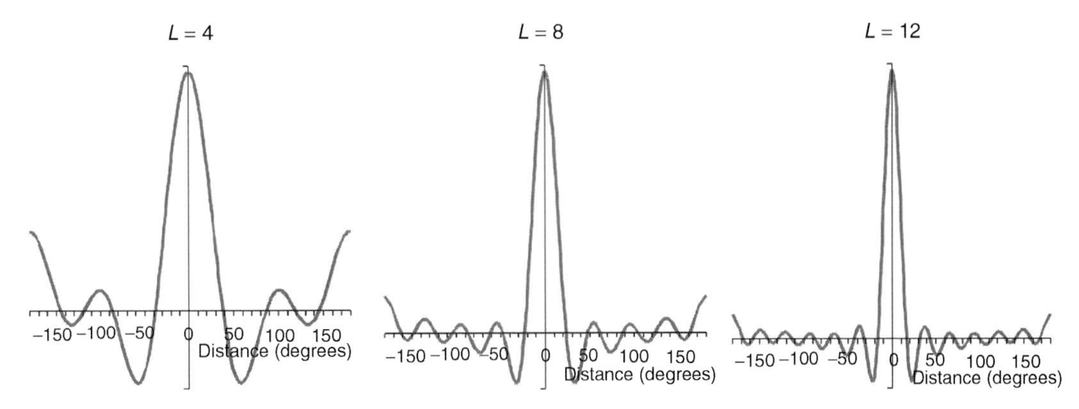

Figure C40 Truncating a spherical harmonic series at degree L is equivalent to convolving the original function of (,) with an averaging function $A_L()$, where is the angular distance from the point (,). Averaging functions are shown for 3 levels of truncation, $L = 4, 8, 12$. Note the side lobes, which have the undesirable effect of allowing the field at considerable distance from the point (,) to influence the result, and the very large side lobe at 180°. Tapering instead of abrupt truncation can reduce or remove the side lobes, improving resolution at the expense of increasing the error.

These core-based inversions have been implemented in two distinct ways, one by developing functions called harmonic splines (Shure *et al.*, 1982) and one by slight modification of the existing IGRF procedure (Gubbins, 1983); uniqueness guarantees the solutions are the same, and here I focus on the latter method. The constraint

$$\int \boldsymbol{B}^2 \mathrm{d}\Omega = 4\pi \sum_{l=1}^{\infty} \sum_{m=0}^{l} (l+1) \left(\frac{a}{c}\right)^{2l+4} (g_l^{m2} + h_l^{m2}) = M \qquad \text{(Eq. 4)}$$

may be incorporated in the least squares fit to the data by the method of Lagrange multipliers. This modifies the normal equations of the least squares method by adding a factor proportional to $(l+1)^2(a/c)^{2l}$ to each diagonal element. This procedure is called "damping," but very specific damping as dictated by the physics of the problem, the requirement that the main field originate in the core. The damping is very powerful because the diagonal elements are increased by an additional factor of $[(l+2)a/(l+1)c]^2$ for each rise in degree and the result is a very smooth magnetic field at the Earth's surface. The factor $[(l+2)/(l+1)]^2$ is relatively unimportant because damping only has a significant effect on the solution at high l when there is little control from the data, and at high l this factor approximates to 1. The downward continuation factor, however, remains at $(a/c)^2 \approx 3.2$. The degree of smoothness is determined by the choice of M, which also determines the fit to the data, larger M allowing a rougher field that will fit the data better. In practice M is adjusted until both the fit and the roughness of the field are satisfactory, usually when the fit to the data matches the error estimates and/or the norm of the model matches an *a priori* estimate. Solution is accomplished numerically again by truncating the spherical harmonic series, but at a much higher level than is needed in the absence of damping. A truncation point of 20 is usually enough to ensure numerical convergence to 5 significant figures. Harmonic splines give the exact solution in principle, but they require numerical evaluation of integrals and are therefore not analytically exact.

A Bayesian formulation allows one to place formal errors on the geomagnetic coefficients obtained from core-based inversions. Replacing the constraint (4) with $\int \boldsymbol{B}^2 \mathrm{d}\Omega < M_1$ defines a volume in parameter space that can be used to derive formal bounds on the geomagnetic coefficients. Little use has been made of these formal errors at the Earth's surface, but some attempts have been made to use them to estimate the accuracy of our estimates of the core field. This requires a stronger constraint than Eq. (4), which imposes no limits on the core field. Constraints based on bounding the Ohmic heating using estimates of the heat flow across the *core-mantle boundary* (*q.v.*) do produce finite limits, but they are too large to be of any interest. Enhancing the constraint has led to some debate; the interested reader is referred to the original articles (Bloxham and Gubbins, 1986; Backus, 1988; Bloxham *et al.*, 1989).

Core-based inversions arose because theoreticians needed self-consistent models of the Earth's main magnetic field for comparing with output from dynamo models, determining *core motions* (*q.v.*) and other studies such as decadal variations in the *length of the day* (*q.v.*), which the IGRF failed to provide. The work on core-based models has culminated in time-dependent models and movies of the core field (see *Time-dependent models of the geomagnetic field*) that have stimulated many new ideas in geomagnetism. The IGRF continues to be derived using traditional methods, as is perhaps appropriate given its use in mapping and the need for continuity of standards, but in the near future we may see some move towards using core-based regularization in the IGRF also.

David Gubbins

Bibliography

Backus, G.E., 1988. Bayesian inference in geomagnetism. *Geophysical Journal*, **92**: 125–142.

Barraclough, D.R., and Malin, S.R.C., 1968. 71/1 Inst. Geol. Sci. HMSO, London.

Bloxham, J., and Gubbins, D., 1986. Geomagnetic field analysis—IV. testing the frozen-flux hypothesis. *Geophysical Journal of the Royal Astronomical Society*, **84**: 139–152.

Bloxham, J., Gubbins, D., and Jackson, A., 1989. Geomagnetic secular variation. *Philosophical Transactions of the Royal Society of London*, **329**: 415–502.

Gubbins, D., 1983. Geomagnetic field analysis I: Stochastic inversion. *Geophysical Journal of the Royal Astronomical Society*, **73**: 641–652.

Shure, L., Parker, R.L., and Backus, G.E., 1982. Harmonic splines for geomagnetic modelling. *Physics of the Earth and Planetary Interiors*, **28**: 215–229.

Whaler, K.A., and Gubbins, D., 1981. Spherical harmonic analysis of the geomagnetic field: An example of a linear inverse problem. *Geophysical Journal of the Royal Astronomical Society*, **65**: 645–693.

Cross-references

Core Motions
Core-Mantle Boundary, Heat Flow Across
Gauss, Carl Fiedrich (1777–1855)
Halley, Edmond (1656–1742)
Harmonics, Spherical
IGRF, International Geomagnetic Reference Field
Laplace's Equation, Uniqueness of Solutions
Length of Day Variations, Decadal
Main Field Maps
Main Field Modeling
Upward and Downward Continuation

CORE-MANTLE BOUNDARY TOPOGRAPHY, IMPLICATIONS FOR DYNAMICS

Ever since it was suggested (Hide and Malin, 1970) that the observed correlation between the variable parts of the Earth's surface gravitational and geomagnetic fields may be explained by the influence of topography (or surface bumps) on the core-mantle interface, there has been considerable interest in the resulting electromagnetic induction and core magnetohydrodynamics (see *Core motions*; *Core-mantle coupling, topographic*). Though the size of the bumps is uncertain, seismic data suggests that the biggest bumps, on the longest length scales, have heights up to the order 10 km (see *Core-mantle boundary, seismology*) Though this is small compared to the core radius, they may have a considerable influence on core motion provided that the horizontal scale is large because of the large Coriolis force and the resulting influences of the *Proudman-Taylor theorem* (*q.v.*). Other Core-mantle couplings, including electromagnetic and thermal, have also been investigated.

Wave motions

The excitation of *magnetohydrodynamic waves* (*q.v.*) at the core-mantle boundary has been investigated within the framework of a 2-D model. Relative to rectangular Cartesian coordinates (x,y,z), the boundary is close to the plane $z = 0$ with small undulations described by $z = h(x)$. The system rotates rapidly about the z-axis, $\mathbf{\Omega} = (0, 0, \Omega)$, and the electrically conducting fluid, constant density ρ, permeability μ, conductivity σ, occupying the region $z > h$ is permeated by the unidirectional magnetic field $\mathbf{B}_0 = (B_0(z), 0, 0)$. Both the linear sheared field $B_0 = B_0' z$ (Anufriev and Braginsky, 1975) and the uniform field $B_0 =$ constant (Moffatt and Dillon, 1976) have been investigated. The aligned uniform flow $\mathbf{U}_0 = (U_0, 0, 0)$ is maintained and the steady linear waves excited due to its motion over the sinusoidal boundary $h = h_0 \sin(x/L)$ are considered.

In both these problems, inertia and viscosity are neglected while ohmic dissipation is retained. The key dimensionless parameters are the Elsasser number $\Lambda = \sigma B_0^2/\rho\Omega$ and the magnetic Reynolds number $R_{\mathrm{m}} = \sigma\mu L U_0$. For the simpler uniform magnetic field problem (Moffatt and Dillon, 1976) solutions are sought proportional to $\exp(-\lambda z/L)$, where λ is a complex constant, which satisfies the dispersion relation

$$4\lambda^2[R_{\mathrm{m}} - \mathrm{i}(1-\lambda^2)]^2 + \Lambda^2(1-\lambda^2) = 0.$$

Three of the roots $\lambda = \lambda_n (n = 1, 2, 3)$ have positive real part. These distinct solutions must be combined to meet the boundary conditions at the CMB. Generally the motion has a boundary layer character. In the limit $R_{\mathrm{m}} \to \infty$, however, wave-like solutions (Re $\lambda = 0$; MC-waves) are possible, when $2L\Omega U_0 > B_0^2/\mu\rho$; otherwise they are evanescent.

Convection

Sphere

The role of core-mantle boundary-bumps on Boussinesq convection (see *Anelastic and Boussinesq approximations*; *Core convection*) in a rapidly rotating self-gravitating sphere, radius r_0, containing a uniform distribution of heat sources (see *Convection, nonmagnetic rotating*) has been investigated (Bassom and Soward, 1996). Relative to cylindrical polar coordinates (s, ϕ, z) (the z-axis and rotation axis are aligned), we denote the core boundary location by $z = z_{\pm}(s, \phi)(z_+$ and z_- in the northern and southern hemispheres respectively) small bumps are considered of z-height $\sqrt{r_0^2 - s^2} \mp z_{\pm}$ proportional to $\sin m\phi$. The main new idea linked to the topography is that the geostrophic contours $s = s_{\mathrm{G}}$, on which the separation $\mathcal{H}(s_{\mathrm{G}}, \phi) = z_+ - z_-$ is constant, are no longer circular $s_{\mathrm{G}} = \bar{s}_{\mathrm{G}} =$ constant but distorted so that they become $s_{\mathrm{G}} - \bar{s}_{\mathrm{G}} \propto \varepsilon(r_0/m)\sin m\phi$ with $\varepsilon \ll 1$.

In rapidly rotating systems fluid motion is constrained by the *Proudman-Taylor theorem* (*q.v.*), which demands that geostrophic flow $\mathbf{U}_{\mathrm{G}}$ follows the geostrophic contours. Since gravity is radial, so is the buoyancy force $\mathbf{F}_{\mathrm{B}} = f_{\mathrm{B}}\mathbf{r}$. In the case of a sphere without bumps $\mathbf{F}_{\mathrm{B}}$ has no component parallel to the geostrophic contour. The buoyancy force can do no work and so no such convective motion is possible. Instead the onset of convection is quasigeostrophic and has the character of thermal Rossby waves (see *Rotating convection, nonmagnetic rotating*; *Core, magnetic instabilities*) with short azimuthal length scale $\mathcal{O}(E^{1/3}r_0)$, where $E = \nu/r_0^2\Omega$ is the Ekman number. On the other hand, with bumps there is a small component $\mathcal{O}(\varepsilon)$ of $\mathbf{F}_{\mathrm{B}}$ aligned to the geostrophic contour, which allows for the possibility that the buoyancy force can do work that drives $\mathbf{U}_{\mathrm{G}}$.

The results for the onset of the geostrophic mode of convection identify the importance of bump height as measured by the parameter

$$\Theta = \varepsilon/m^{1/2}E^{1/4}.$$

On increasing the bump height from zero, the thermal Rossby wave is preferred until $\Theta = \mathcal{O}((mE^{1/3})^{1/4})$. Thereafter a steady geostrophic mode is preferred, for which the adverse density gradient necessary to drive it decreases with increasing bump height (i.e., Θ). The main interest is for large scale bumps with $mE^{1/3} \ll 1$ for which the transition occurs at small Θ, while even order one values of Θ are readily obtained for geophysically realistic parameter values. For an Ekman number $E = \mathcal{O}(10^{-8})$, based on a turbulent viscosity, the estimate of $\Theta = \mathcal{O}(1)$ is achieved by undulations of about 10 km in depth on a horizontal length scale comparable to the core radius length scale, while both the depth and lateral extent length scale estimates are reduced considerably when based on $E = \mathcal{O}(10^{-15})$ appropriate to the kinematic rather than the turbulent viscosity.

To drive the geostrophic flow, diffusive processes must occur which lead to density variations along the geostrophic contours. This leads to an $\mathcal{O}(\varepsilon)$ secondary "convective" motion vital to drive the geostrophic flow determined by a variant of *Taylor's condition* (*q.v.*), which takes account of Ekman boundary layer suction and pumping (see *Core, boundary layers*), first obtained in the dynamo context. When $\Theta \ll 1$ this "convective" motion is confined close to the outer boundary (shallow convection), but when $\Theta \gg 1$ penetrates the full height H of the geostrophic columns (deep convection).

The nature of the geostrophic mode is sensitive to the boundary conditions. In the case of rigid boundaries, motion is localized close to some critical radius $s = s_{\mathrm{c}}$ on a short radial length $\mathcal{O}(E^{1/8}r_0)$ as effected by Ekman pumping at the boundary. In contrast for stress-free boundaries, there is negligible Ekman pumping and geostrophic motion driven by the buoyancy force is diffused by lateral friction (internal viscous stresses) so as to fill the sphere.

Annulus

Since many of the important physical features of rotating spherical convection are reproduced by Busse's annulus model, which provides a model for convection localized in radial extent in the vicinity of $s = s_c$ (see *Convection, nonmagnetic rotating*), it has also been employed for the case of a bumpy boundary. There are some essential differences even in the linear theory between the spherical and annular cases. Since the z-dependence is suppressed in the annulus model, it has no equivalent of the shallow convection mode. Instead, there is a minimum $\mathcal{O}(1)$ value of Θ at which the steady geostrophic mode can occur. Below that value the geostrophic mode is oscillatory.

The theory has been developed and extended into the nonlinear regime from an asymptotic point of view with $E \ll 1$ (Bell and Soward, 1996). That theory focuses exclusively on the geostrophic mode of convection. In contrast, results have been obtained numerically at large but finite E (Hermann and Busse, 1998), which illustrate both the thermal Rossby and geostrophic modes of convection. Both developments address possible long wavelength modulation. Experimental results have been reported (Westerburg and Busse, 2003) and compared with the theoretical predictions.

Andrew Soward

Bibliography

Anufriev, A.P., and Braginsky, S.I., 1975. Influence of irregularities of the boundary of the Earth's core on the velocity of the liquid and on the magnetic field. *Geomagnetism and Aeronomy*, **15**: 754–757.

Bassom, A.P., and Soward, A.M., 1996. Localised rotating convection induced by topography, *Physica D*, **97**: 29–44.

Bell, P.I., and Soward, A.M., 1996. The influence of suerace topography on rotating convection. *Journal of Fluid Mechanics*, **313**: 147–180.

Hermann, J., and Busse, F.H., 1998. Stationary and time dependent convection in the rotating cylindrical annulus with modulated height. *Physics of Fluids*, **10**(7): 1611–1620.

Hide, R., and Malin, S.R.C., 1970. Novel correlations between global features of the Earth's gravitational and magnetic fields. *Nature*, **225**: 605–609.

Moffatt, H.K., and Dillon, R.F., 1976. The correlation between gravitational and geomagnetic fields caused by the interaction of the core fluid motion with a bumpy core-mantle boundary. *Physics of the Earth and Planetary Interiors*, **13**: 67–78.

Westerburg, M., and Busse, F.H., 2003. Centrifugally driven convection in the rotating cylindrical annulus with modulated boundaries. *Nonlinear Processes in Geophysics*, **10**: 275–280.

Cross-references

Anelastic and Boussinesq Approximations
Convection, Nonmagnetic Rotating
Core Convection
Core Motions
Core, Boundary Layers
Core, Magnetic Instabilities
Core-Mantle Boundary Topography, Seismology
Core-Mantle Coupling, Topographic
Magnetohydrodynamic Waves
Proudman-Taylor Theorem
Taylor's Condition

CORE-MANTLE BOUNDARY TOPOGRAPHY, SEISMOLOGY

The main compositional change in the Earth, the core-mantle boundary (CMB), translates into a strong discontinuity in seismic properties—density, bulk modulus, rigidity—that affects the propagation of seismic waves very significantly. If chemical and dynamic processes induce topographic variations of the CMB, as it appears plausible, then seismology should in principle be able to detect and map them. In fact, systematic geographical variations of seismic body wave travel times pointed to the existence of large-scale undulations of the CMB. Also, stochastic analyses of travel times, and detection of scattered seismic energy, have been used to characterize the statistical properties of boundary topography down to spatial scales of a few kilometers. Modeling the relevant seismic waves remains however a challenging task, because the rays also cross the heterogeneous mantle, including the very complicated D″ layer—both must be accounted for—and because the signal to noise ratio in available data is very small. As a result, in spite of a general recognition of the existence of large and small scale "seismic" topography of the CMB, the debate on the actual shape and amplitude range—of the order or hundreds of meters to kilometers—is still under way, as agreement among different models is often unsatisfactory.

Reflection and refraction at the CMB generate a variety of seismic phases, which are coded by seismologists following a simple convention by which each leg of the ray and each topside reflection is represented by a letter (see *Seismic phases*). PcP designates seismic energy that traveled as a P wave from the hypocenter, was reflected at the core and went back up as P again (Figure C41). PKKP is a wave that traveled through mantle, outer core, was reflected at the bottom of the CMB, and went back up to the surface. PKIKP are rays traveled almost straight across the whole Earth. The seismic phase most often used is PcP, because it has good sensitivity to CMB radius, it is easily identified on seismograms (Figure C42), and has traveled through the heterogeneous part of the Earth for the shortest path possible. Additionally, CMB-refracted phases (PKP and PKIKP, Figure C42) and

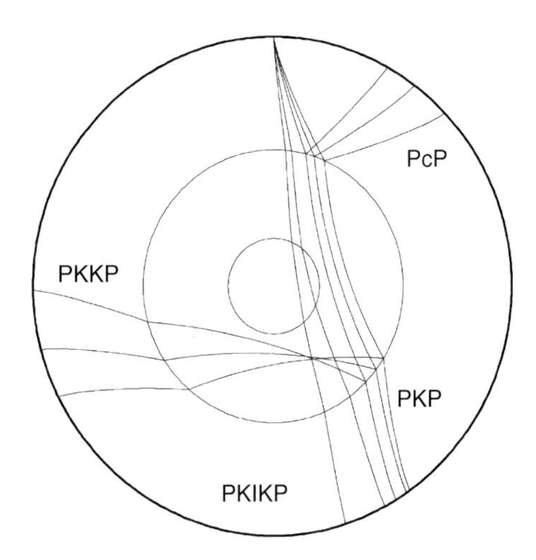

Figure C41 Some seismic ray paths interacting with the core–mantle boundary. Body waves reflected at the mantle (PcP) and the core-side (PKKP) of the CMB, as well as refracted waves—PKP, crossing the outer core, and PKIKP, penetrating to the inner core—have been used to study the shape of the CMB. PcP rays are affected the least by deep earth structure. These ray paths are traced in a spherically symmetric reference earth model.

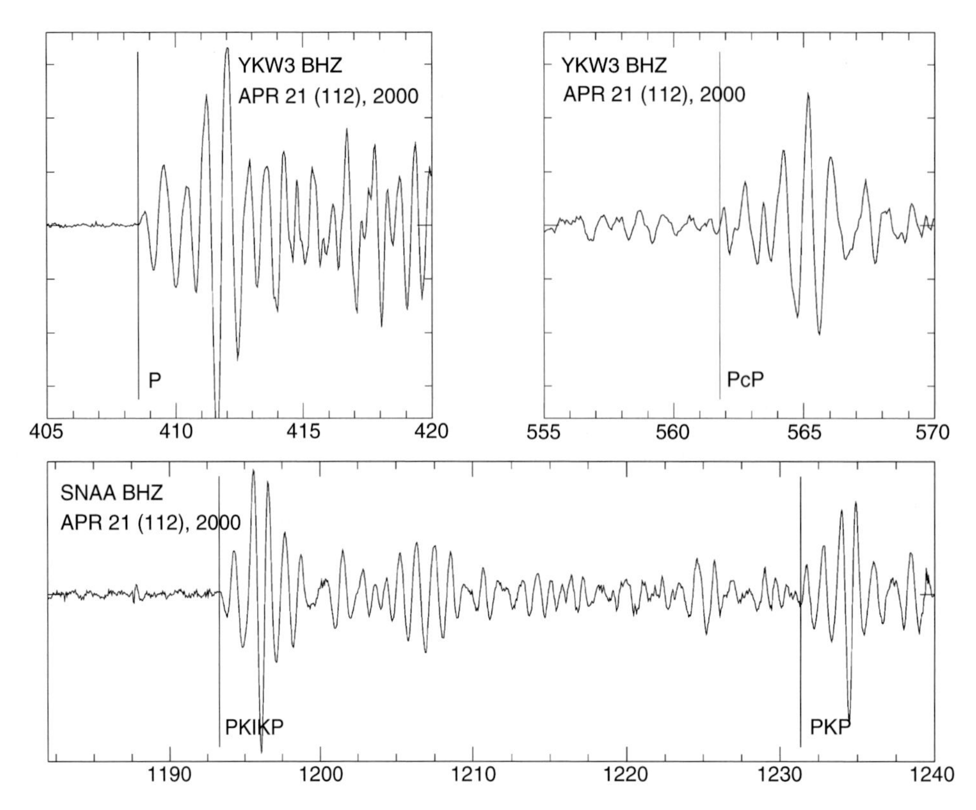

Figure C42 High-pass (1 Hz) filtered seismograms for a magnitude 6.0 earthquake located beneath Aleutian Islands. Top panels show two time windows—with the same, arbitrary, vertical amplification—of the record from seismographic station YKW3 at Yellowknife (N.W.T., Canada) with epicentral distance of 3873 km, or 35°. Horizontal scale is time, in seconds, since event origin. The bottom panel shows the seismogram from station SNAA (Sanae, Antarctica) with epicentral distance of 17773 km, about 160°, for the same earthquake. Both PKIKP and PKP arrivals are visible.

underside reflections (PKKP) have also been used in large-scale tomographic studies. A discrepancy of the travel time of a core phase with respect to the reference value can be interpreted as due to a different radius of the reflector/refractor.

Of course, the theoretical travel times of core phases must necessarily account for the three-dimensional variations of wave speed in the whole mantle. Core phases do not have good sensitivity to the vertical location of heterogeneities in the Earth structure, and are therefore not capable, alone, of discriminating well between small variations of wave speed or boundary radius. Our analysis must then rely on a mantle model previously determined by means of direct P waves. After these critical corrections are made, statistical analyses of travel time "residuals" (Morelli and Dziewonski, 1987; Rodgers and Wahr, 1993; Garcia and Souriau, 2000) have shown the existence of small but geographically coherent signal, that would correspond to boundary undulations with an amplitude of the order of a few kilometers on length scales of several hundred to a few thousand kilometers. For each single seismic phase, boundary undulation would in fact trade-off rather well with velocity perturbations in a thin, highly heterogeneous layer. The presence of this signal on top-side reflections (PcP) would require this layer to be located on the mantle side of the CMB. The original observation in favor of the topographic origin of the travel time anomalies (Morelli and Dziewonski, 1987) consisted of the agreement of the maps derived from separate PcP and PKP analyses, in spite of the opposite signs of their sensitivity to CMB radius perturbation—reflected and refracted waves could be consistently fit by allowing CMB radius to vary, but would otherwise require anticorrelated boundary layer wave speed heterogeneities. Later studies, using different data selection rules and mantle models, were unable to find such a good correlation, but maps derived exclusively from PcP data shared some common traits (Figure C43). Likeness among models derived from the other seismic phases is worse or nonexistent. The unsatisfactory correspondence raises some skepticism on the current ability of seismology to accurately map the shape of the CMB, discriminating it from other structural complexities. Some intriguing correspondence among maps derived from PcP indeed exist (Figure C43): elevated regions appear under the north Pacific, the Indian, and the north Atlantic oceans; depressed regions figure beneath central Africa, southeast Asia, and the Pacific coast of South America. The presence of at least some of these anomalies is also supported by mantle dynamic models (see *Core-mantle boundary topography, implications for dynamics*). The worse performance of PKP and PKKP data may derive from additional complexities in their modeling. Although no compelling evidence of lateral variations of seismic speed in the outer core has ever been found, waves transmitted through the boundary could possibly also be affected by structure of the core-mantle transition zone located beneath the seismic reflector—in the case of PKIKP, also by the seismically heterogeneous and anisotropic inner core. The added complication of each ray sampling the CMB in two (PKP) or three (PKKP) spots, combined with a rarefied sampling, may constitute yet another source of error. A plausible scenario then involves a core-mantle transition zone with a complicated structure, not only involving undulations of a sharp boundary, but also seismic velocity gradients—which we are still unable to model globally, and which can consequently contaminate in various ways the retrieved topography. PcP data may be the least affected, yielding better agreement among different analyses.

Spatial resolution of travel times is limited by the wavelength and is often represented by the Fresnel zone. For P body waves used in travel time studies, the Fresnel zone is about 300 km wide at the CMB. Information on smaller length scales can be obtained by studying high

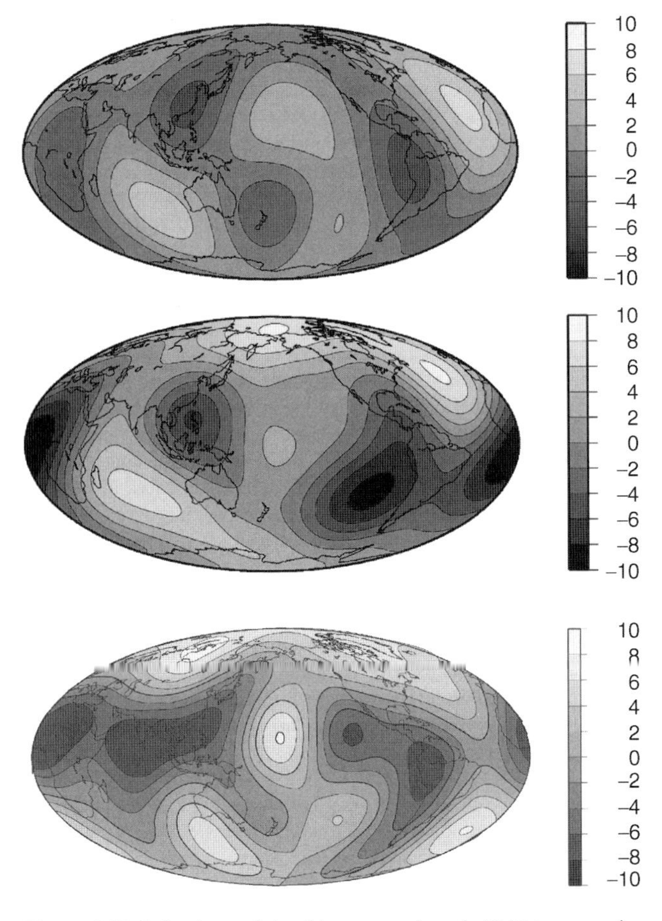

Figure C43 Seismic models of long-wavelength CMB topography (from top to bottom: Morelli and Dziewonski, 1987; Obayashi and Fukao, 1997; Rodgers and Wahr, 1993). All models are derived from analysis of PcP travel times, and are represented by low-degree spherical harmonics (qv) expansions (up to degree 5 for the last model, 4 for the others). The grayscale represents changes of core radius with respect to the reference figure of hydrostatic equilibrium, up to +10 km (white) and −10 km (black).

frequency scattered energy or the pulse distortion of short-period body waves, preferably in situations where the scattered energy precedes, rather than follows, the main seismic arrival, so that it is not obliterated by the coda. (This is possible for seismic phases, such as PKKP, which follow a maximum-time path.) By studying precursors to PKKP it has been possible to infer the existence of CMB topography with root mean square amplitude a few hundred meters with a correlation length of 7–10 km (Earle and Shearer, 1997).

Seismic imaging of the core-mantle boundary topography is in principle quite straightforward, but it is in reality complicated by a number of practical issues. The scanty and irregular global sampling of the CMB, the imperfect account of mantle and core structure, the simplified theoretical modelling of wave propagation, all concur with the small signal to noise ratio to limit the imaging potential of seismic core data. As a result, the different studies on the topic have not quite gone beyond a qualitative agreement, and often disagree. However, the qualitative correspondence found among different models, the stochastic analyses of travel times, and the accordance with inferences from mantle dynamic modeling, point towards the existence of kilometric topography at the CMB. Better knowledge of the seismic structure of the Earth's mantle, including the still elusive D″ layer, and perhaps a better modeling of the full effects of such three-dimensional structures on seismic waves, are still needed to wipe our glasses and clarify our vision.

Andrea Morelli

Bibliography

Earle, P.S., and Shearer, P.M., 1997. Observations of PKKP precursors used to estimate small-scale topography on the core-mantle boundary. *Science*, **277**: 667–670.

Garcia, R., and Souriau, A., 2000. Amplitude of the core-mantle boundary topography estimated by stochastic analysis of core phases. *Physics of the Earth and Planetary Interiors*, **117**: 345–359.

Morelli, A., and Dziewonski, A.M., 1987. Topography of the core-mantle boundary and lateral homogeneity of the outer core. *Nature*, **325**: 678–683.

Obayashi, M., and Fukao, Y., 1997. P and PcP travel time tomography for the core-mantle boundary. *Journal of Geophysical Research*, **102**: 17825–17841.

Rodgers, A., and Wahr, J., 1993. Inference of core-mantle boundary topography from ISC PcP and PKP traveltimes. *Geophysical Journal International*, **115**: 911–1011.

Cross-references

Core-Mantle Boundary Topography, Implications for Dynamics
Core-Mantle Boundary Topography, Seismology
Core-Mantle Boundary, Heat Flow Across
Core-Mantle Coupling, Topographic
D″, Seismic Properties
Inner Core Anisotropy
Inner Core Seismic Velocities
ULVZ, Ultra-low Velocity Zone
Seismic Phases

CORE-MANTLE BOUNDARY, HEAT FLOW ACROSS

The boundary between the core and the mantle is of huge importance for the dynamics of the Earth's interior and the seat of many interesting phenomena: (see *Core-mantle boundary and Core-mantle coupling, thermal*). From the core point of view, this is the surface across which all heat produced must escape. The value of the integrated heat flow across the core-mantle boundary (CMB) is then of primary importance for the thermal evolution of the core and it controls the amplitude of the energy sources available to drive the geodynamo (see *Geodynamo, energy sources*).

All the energy that flows out of the core must be transported upward across the mantle. The mantle is primarily solid (except in the very localized regions of partial melting) but it can creep and is subject to convective movements on time scales of some tens of millions of years (Myr), the surface expression of these motions being plate tectonics. On the other hand, the outer core is liquid, with a viscosity close to that of water (see *Core viscosity*) and its dynamics is then much faster. The secular variation of the magnetic field suggests that the velocity in the fluid core is on the order of 10^{-4} m s^{-1} (e.g., Hulot *et al.*, 2002) which, together with the radius of the core 3.48×10^6 m, gives a typical time scale of about a thousand years.

The mantle, being much more sluggish than the core, constitutes the limiting factor for the thermal evolution of the Earth, and the heat flow across the CMB is the heat flow that mantle convection can accept. Of course, the state of the core affects this value by imposing a temperature difference between the bottom and the top surfaces of the mantle, that is, in nondimensional units, a Rayleigh number for the mantle. Both systems are then coupled in their evolution.

Total heat loss of the core

Energy balance of the mantle

The heat flow across the CMB is controlled by mantle convection and gross estimates of its integrated value can be obtained by use of the global energy balance of the mantle. This balance states very simply that the heat flow at the surface of the Earth is, to first approximation, equal to the sum of three terms: The heat production by decay of radioactive elements in the mantle, the cooling heat associated with the heat capacity of the mantle (this term is called secular cooling of the mantle), and the heat flow across the CMB. The latter of these terms should then be obtained by simple difference, providing all others can be determined accurately. We will see here that this approach can give reasonable estimates but hardly any precise value.

Other energies are involved in the thermodynamics of mantle convection but they are either negligible or internally balanced. Among the first group, one can cite tidal friction (Verhoogen, 1980), latent heat due to the secular movement of phase boundaries upon mantle cooling and adiabatic heating from slow contraction. Another energy term, which is often discussed, viscous dissipation, can be locally important but does not enter the global energy balance because it is internally balanced by the work of buoyancy forces (as is the case for the core: see *Geodynamo, energy sources* and Hewitt *et al.* (1975)). Gravitational energy due to chemical stratification was a very important source of energy during the formation of the core (Stevenson, 1981), early in Earth history, but is likely gone by now. Similar processes like the formation of the continental crust or opposite ones like the slow mixing of a primordial chemical stratification (Davaille, 1999) could still be at work but are likely negligible due to the small masses involved.

The total heat loss of the Earth is reasonably well known to be about 42 TW from heat flux measurements made both on oceans and continents, as well as interpolation based on models for the cooling of oceanic plates (Sclater *et al.*, 1980). The value of total heat production by radioactive decay relies on cosmochemical models for the formation of the Earth, or more precisely on the choice of meteorites that is supposed to have formed the Earth. Most models assume a CI carbonaceous chondrite origin for the Earth, leading to a total heat production in the silicate Earth (mantle plus crust) of about 20 TW (see Javoy, 1999, for a critical discussion). The distribution of the heat-producing isotopes (^{40}K, ^{235}U, ^{238}U, ^{232}Th) in the Earth is clearly nonuniform, the continental crust being largely enriched compared to the mantle, but this matter does not affect the estimates presented here as long as the heat flow at the surface is considered. The secular cooling term is the least well constrained of all terms but we can already see that, unless the mantle is actually heating up, the heat flow across the CMB cannot be greater than 22 TW, a value obtained in the case of a steady mantle temperature.

To get a more useful upper bound one needs a model of thermal evolution of the mantle. Unfortunately, our understanding of mantle convection is still far from complete and no entirely satisfying model of mantle cooling is available at present. In particular, these models usually predict a very fast cooling of the Earth in the early stages of the evolution so that they cannot maintain the present surface heat flow with the concentration in radioactive elements estimated from geochemistry (e.g., Grigné and Labrosse, 2001). Nevertheless, these models all predict present day rate of mantle cooling between 50 and 100 K Ga^{-1} (e.g., Schubert *et al.*, 2001), which is similar to estimates based on the chemistry of basalts (e.g., Abbott *et al.*, 1994). Using a heat capacity of about 10^3 J kg^{-1} K^{-1} and a mass of the mantle of $4\ 10^{24}$ kg, this gives a contribution of secular cooling of about 6–12 TW to the heat flow at the surface. Subtracting this value to the 22 TW estimated above, a heat flow across the CMB of 10–16 TW is obtained. Considering the very approximate nature of this estimate, it should only be taken as a crude but plausible range.

A mechanical point of view

Alternatively to looking at the global balance of the mantle and modeling its temporal evolution, the heat flow at the CMB has been estimated by computing the buoyancy flux of the different mantle hot plumes that are supposed to be responsible for the formation of hotspot tracks, such as Hawaii. Using different hotspot catalogs, values around 3 TW for the total heat coming out of hotspots have been typically obtained (Davies, 1988) and this value has been often quoted as the total heat loss of the core. The idea behind that estimate is that all the heat that comes out of the core is collected in the boundary layer at the bottom of the mantle and transported upward to the surface by mantle plumes, the surface expressions being hotspots.

The value obtained is low compared to the range given previously but could be acceptable (considering the approximate nature of this range), although the reasoning used to get it is questionable. Without even entering the long standing debate regarding the origin of hotspots, two issues regarding the link between the heat transported to the surface by hot mantle plumes and the total heat loss of the core can be raised, both coming from the difficulty of considering any particular dynamical object as a separate entity in a convective flow. First, it is difficult to imagine that hot plumes move straight up from the boundary layer where they originate to the surface without any effect of the large-scale mantle circulation associated with plates. Several models actually show that many (if not most of) hot plumes that form at the bottom of a convective system containing some important ingredients of mantle convection such as temperature-dependent viscosity, volumetric (radiogenic) heating, and large (plate size) scale flow do not rise to the surface (Labrosse, 2002).

Second, there are two modes of cooling of the core, both operating simultaneously. You can take some hot matter and move it upward in the form of plumes or take some cold matter from the surface and spread it on top of the core. This separation is a little arbitrary since in both cases you replace some hot matter by some cold matter, but, in the former, this matter comes from the ambient mantle whereas in the latter it comes from the surface, in the form of plates. The relative importance of these two modes depends on both the mass flux of cold matter brought to the boundary and the temperature difference involved. Both these parameters are more important for the cold down welling currents than for the hot up welling plumes in convection models that include volumetric heating (Labrosse, 2002) and it is easy to think that this difference will be even larger in the case of the mantle, where cold plates represent the dominant feature of lateral heterogeneities. This means that a better estimate of the heat flow across the CMB could be obtained if the mass flux associated with plates were known in the lower mantle. This is unfortunately not the case and no estimate better than the one proposed using the gross energy balance above seems possible for the time being.

Estimates from the core side

Even though mantle convection controls the heat flow across the core mantle boundary, the fact that a dynamo has been operating in the core for most of its history places constraints on this value: It must be large enough to maintain the dynamo action. The thermodynamics of a convective geodynamo has been studied by several authors (see *Geodynamo, energy sources*) and the results will be shortly recalled here.

The decay time of the dipole of the Earth is of the order of 15 ka (see *Geodynamo, dimensional analysis and timescales*), that is much smaller than the age of the Earth magnetic field, which means that a dynamo action must be regenerating it against ohmic dissipation. A particularity of the thermodynamics of convective flows, compared to the classical Carnot engine, is that the work is performed inside the system so that the energy balance only states the equilibrium between the sum of heat sources and the total heat loss, the heat flow across the CMB in the case of interest here. On the other hand, an entropy balance can be used to relate the total ohmic dissipation in the core to the energy sources that drive the dynamo, with the introduction of some efficiency factors. Several combinations of different energy sources are possible, particularly depending on whether one thinks that radioactive elements are present in the core or not, and added to the uncertainties attached to most physical parameters, a total heat flow across the CMB in the range of 5–15 TW can be predicted for a total ohmic dissipation in the core of about 1–2 TW. It must be noted that the uncertainty in this total ohmic dissipation is probably the largest in the whole set of parameters of this type of derivation, owing to our current lack of knowledge concerning an important part of the magnetic field in the core, the toroidal, and the small scale poloidal part.

The estimates discussed above fall in about the same range as those obtained previously from the mantle side, which can be taken as a good sign, but it means that it does not really help in narrowing down this range. Another point of comparison is provided by the average adiabatic temperature gradient in the core, along which about 7 TW is conducted. If the actual heat flow is lower than this value, a stable thermal stratification would tend to be established at the top of the core.

Lateral variations

In addition to the value of the total heat flow across the CMB, the amount and scale of lateral variations of the heat flux may be of equal importance for the dynamics of the core. These lateral variations are imposed by mantle convection and evolve on timescales much longer than those associated with the dynamo. These variations can then be seen as a permanent inhomogeneous boundary condition imposed on the core by the mantle, the effects of which being still not completely understood (see *Inhomogeneous boundary conditions and the dynamo*).

Convection being responsible for the heterogeneities in seismic velocities at the bottom of the mantle, it is tempting to use the images of these heterogeneities as produced by seismic tomography to convert them to maps of lateral variations of the heat flux at the CMB. This assumes that all the lateral variations in seismic velocities are of thermal origin and that a single, so-called, conversion factor can be used for the operation. On the other hand, there is some evidence for a non-purely thermal origin of these heterogeneities, in which case the chemical part of the signal must first be separated, a still formidable task. Several studies about the effect of heterogeneous boundary conditions on core dynamics (e.g., Glatzmaier *et al.*, 1999; Olson and Christensen, 2002) have nevertheless assumed a purely thermal origin of the lateral variation of the seismic velocities and the heat flux pattern obtained is often called the "tomographic heat flux," the amplitude of the lateral variations remaining a largely unknown adjustable parameter. This pattern is essentially controlled by the large-scale structure of tomographic images and, in particular, features a large heat flux ring falling bellow the subduction zones around the Pacific Ocean (Figure C44). This can be optimistically taken as a sign of consistency since the subducting plates, if they reach the bottom of the mantle, will produce a high heat flux.

There is no direct measurement of the amplitude of the lateral variations of heat flux at the CMB, but some argument can be made about their being quite large. First, the contrast between the core and the mantle is as important as the one between the mantle and the ocean, so that variations of the same relative importance can be expected at the CMB than at the surface. The existence of crypto-continents at the CMB has even been proposed (e.g., Stacey, 1991), and such could be the name given to the piles of chemically dense material that are envisioned in some models of mantle convection with chemical heterogeneities (e.g., Tackley, 1998; Davaille, 1999). At the Earth surface, the mean heat flux through oceanic plates is about 100 mW m^{-2}, with a pic-to-pic variation equal to this value (Sclater *et al.*, 1980), and it is reasonable to assume that variations of the same relative amplitude exist at the CMB. One can even defend the case of larger variations, owing to the fact that these variations are controlled by the effect of down welling plates, which are associated with the lateral variations of the heat flux at the surface. One can for example construct a mantle convection model driven by internal heating and a top surface maintained at a low temperature with a zero net heat flow at the bottom surface, although displaying lateral variations of a quite large amplitude. Pic-to-pic variations of the heat flux at the core mantle boundary of the same order as the mean value present then a conceivable picture.

The possibility of thermal stratification that was mentioned above has to be discussed again in the context of the lateral variations of the heat flux at the CMB. Considering the likely importance of these variations and the closeness of the estimates of the actual heat flux at the CMB and of that conducted down the adiabatic temperature gradient, it is likely that some regions of the CMB present a stable

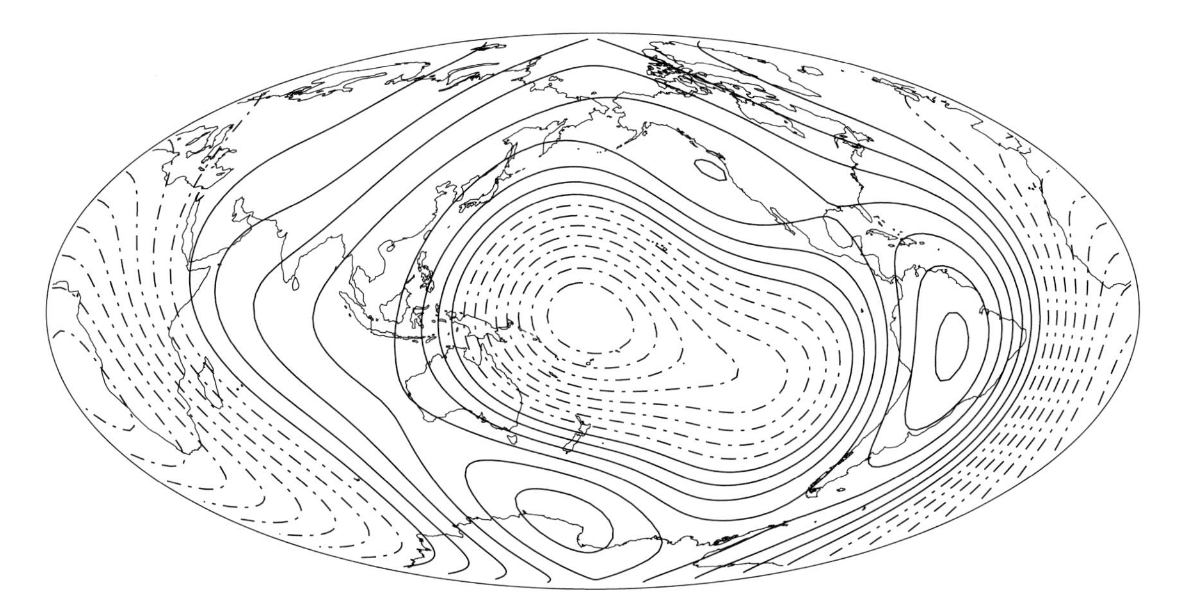

Figure C44 Pattern of heat flux lateral variations at the core mantle boundary in the so-called "tomographic" model of Christensen and Olson (2003). Heat flux higher than averaged is in plain contours and correlate well with the position of subduction zones at the surface. Reproduced by permission of U. Christensen and P. Olson and Elsevier.

thermal stratification. The effects of such patches on the dynamics of the core and the generation of the magnetic field are still largely unknown but have to be investigated.

Variations with time

The heat flux across the CMB is not only varying laterally but also with time, on the timescale of mantle convection. Moreover, the pattern of lateral variations is also evolving on the same timescales and both temporal evolutions are likely to be of great importance to understand the long-term variations of the magnetic field. Unfortunately, the knowledge that one can get for the time evolution of the heat flow across the CMB is even scarcer than about the present day value. In particular, most of what we know about the present heat loss of the Earth comes from the study of the seafloor (Sclater *et al.*, 1980), which is not older than about 200 Ma. However, one can put forward the argument that some ergodicity holds so that the variations of this heat loss with time can be as large as the lateral variations observed at the present day. Most of the heat loss of the Earth comes from the formation of mid-oceanic ridges and the heat budget of the Earth can be considerably changed by modifying the number of active ridges, or equivalently the average size of mantle convection cells. For example, the Atlantic ocean is responsible for about 6 TW of the 30 TW of total oceanic heat flow and it opened between 180 and 120 Ma ago. Before that time, if no other ridge of equivalent importance was active, the heat loss of the Earth must have been 6 TW lower than at present.

From the previous discussion, one can easily imagine (although not prove) variations of the total heat loss of the Earth of the order of 10 TW on timescales of about 200 Ma (Grigné *et al.*, 2005). As for the heat flow across the CMB, it is likely to have undergone variations on a similar timescale, with amplitudes that could be as large as the amplitude of the present day lateral variation, which, unfortunately, is poorly known as discussed above. This means that important variations of the ohmic dissipation in the core must have occurred (see *Geodynamo, energy sources*) and understanding the link between these variations and the variations of the magnetic field characteristics would provide a means of investigating this evolution.

Stéphane Labrosse

Bibliography

Abbott, D., Burgess, L., and Longhi, J., 1994. An empirical thermal history of the Earth's upper mantle. *Journal of Geophysical Research*, **99**: 13835–13850.

Christensen, U.R., and Olson, P., 2003. Secular variation in numerical geodynamo models with lateral variations of boundary heat flow. *Physics of the Earth and Planetary Interiors*, **138**: 39–54.

Davaille, A., 1999. Simultaneous generation of hotspots and superswells by convection in a heterogeneous planetary mantle. *Nature*, **402**: 756–760.

Davies, G.F., 1988. Ocean bathymetry and mantle convection, 1. large-scale flow and hotspots. *Journal of Geophysical Research*, **93**: 10467–10480.

Glatzmaier, G.A., Coe, R.S., Hongre, L., and Roberts, P.H., 1999. The role of the Earth's mantle in controlling the frequency of geomagnetic reversals. *Nature*, **401**: 885–890.

Grigné, C., and Labrosse, S., 2001. Effects of continents on Earth cooling: Thermal blanketing and depletion in radioactive elements. *Geophysical Research Letters*, **28**: 2707–2710.

Grigné, C., Labrosse, S., and Tackley, P.J., 2005. Convective heat transfer as a function of wavelength: Implications for the cooling of the Earth. *Journal of Geophysical Research*, **110**: B03409. Doi:10.1029/2004JB003376.

Hewitt, J.M., McKenzie, D.P., and Weiss, N.O., 1975. Dissipative heating in convective flows. *Journal of Fluid Mechanics*, **68**: 721–738.

Hulot, G., Eymin, C., Langlais, B., Mandea, M., and Olsen, N., 2002. Small-scale structure of the geodynamo inferred from Oersted and Magsat satellite data. *Nature*, **416**: 620–623.

Javoy, M., 1999. Chemical Earth models. *Comptes Rendus de l'Académie des Sciences Series II A Earth and Planetary Science*, **329**: 537–555.

Labrosse, S., 2002. Hotspots, mantle plumes and core heat loss. *Earth and Planetary Science Letters*, **199**: 147–156.

Olson, P., and Christensen, U.R., 2002. The time-average magnetic field in numerical dynamos with nonuniform boundary heat flow. *Geophysical Journal International*, **151**: 809–823.

Schubert, G., Turcotte, D.L., and Olson, P., 2001. *Mantle Convection in the Earth and Planets*. Cambridge: Cambridge University Press.

Sclater, J.G., Jaupart, C., and Galson, D., 1980. The heat flow through oceanic and continental crust and the heat loss of the Earth. *Reviews of Geophysics and Space Physics*, **18**: 269–312.

Stacey, F.D., 1991. Effects on the core of structure within D″. *Geophysical and Astrophysical Fluid Dynamics*, **60**: 157–163.

Stevenson, D., 1981. Models of the Earth core. *Science*, **214**: 611–619.

Tackley, P.J., 1998. Three-dimensional simulations of mantle convection with a thermochemical CMB boundary layer: D″? In Gurnis, M., Wyession, M.E., Knittle, E., and Buffett, B.A., (eds.), *The Core-Mantle Boundary Region*, Washington, DC: American Geophysical Union, pp. 231–253.

Verhoogen, J., 1980. *Energetics of the Earth*. Washington, DC: National Academy Press. 139 pp.

Cross-references

Core Viscosity
Core-Mantle Boundary
Core-Mantle Coupling, Thermal
Geodynamo, Dimensional Analysis and Timescales
Geodynamo, Energy Sources
Inhomogeneous Boundary Conditions and the Dynamo

CORE-MANTLE COUPLING, ELECTROMAGNETIC

Variations in length of day (equivalently, the rate of rotation of the Earth's surface) are observed on timescales from subdaily to millions of years. They result either from external torques (for example, the tidal torques exerted on the Earth by the moon and sun), or from the exchange of angular momentum between different reservoirs within the Earth system. Of these, the fluid core is by far the most massive, and there is convincing evidence that at least *decadal variations in length of day* (ΔLOD) (*q.v.*) result from the exchange of angular momentum between the core and the mantle. Such exchange requires a coupling mechanism between the core and mantle. The mechanisms that have been suggested fall into four classes: viscous coupling, *topographic coupling* (*q.v.*), gravitational coupling, and electromagnetic coupling.

Although the mantle is a poor electrical conductor in comparison with the molten iron core, it is nonetheless not a perfect insulator. Currents flowing in the core can "leak" into the mantle and changing magnetic fields from the core, which penetrate the mantle, can induce electric currents. These currents incur resistive losses, which, as a result of Lenz' law, do work against the processes generating them. Globally, a torque between the core and mantle is generated.

Many attempts have been made to quantify the effect of electromagnetic coupling between the core and the mantle. The underlying theory is presented by Gubbins and Roberts (1987) amongst others. The torque Γ is given by the moment of the Lorentz force integrated over the mantle

$$\Gamma = \int_M \mathbf{r} \times (\mathbf{J} \times \mathbf{B})\, dV$$

where $\mathbf{r}$ is the position vector, $\mathbf{J}$ the current density, and $\mathbf{B}$ the magnetic field, integrated over the volume V of the conducting mantle M. This integral can be reformulated as a surface integral calculated over the core mantle boundary (CMB)

$$\Gamma = -\frac{1}{\mu_0} \int_{\mathrm{CMB}} (\mathbf{r} \times \mathbf{B})\, B_r\, dS$$

where μ_0 is the permeability of free space, and B_r is the radial magnetic field at the CMB. The axial component Γ_z of this torque (which affects the length of day) can be written

$$\Gamma_z = -\frac{c^3}{\mu_0} \int_{\mathrm{CMB}} B_\varphi B_r \sin^2\theta\, d\theta\, d\varphi \qquad \text{(Eq. 1)}$$

where c is the radius of the core, θ is colatitude, φ is longitude, and B_φ is the field in the longitudinal direction.

Because the radial field B_r must be continuous over the core-mantle boundary, it can in principle be inferred from *downward continuation* (*q.v.*) of the field observed at the Earth's surface, but B_φ is less straightforward. A magnetic field can be divided into two parts, written

$$\mathbf{B} = \nabla \times T\mathbf{r} + \nabla \times \nabla \times S\mathbf{r}$$

where T and S are the toroidal and poloidal scalars respectively (see, for example, Gubbins and Roberts, 1987). B_φ has a component from both the toroidal and poloidal fields.

It is convenient to divide the EM torque into three parts. The first, the poloidal torque, depends on the poloidal component of B_φ, which can be inferred by using *time-dependent models of the geomagnetic field* (*q.v.*) for a known mantle conductivity structure, and solving the diffusion equation in the mantle by an iterative, multiple timescale scheme. The toroidal component of B_φ cannot be inferred directly from observations at the Earth's surface. (Attempts have been made to infer the toroidal field from measurements of electric field in submarine telephone cables, but without conclusive results.) The torque which it generates is commonly subdivided into two parts, the advective torque (associated with magnetic induction in the conducting mantle due to changing field at the CMB) and the leakage torque (arising from direct flow of current ("leakage") from the core into the mantle) (e.g., Holme, 1997a and references therein). The advective torque can in principle be inferred from a model of the flow at the top of the core. The radial field at the CMB is advected by the flow and its passage through the conducting mantle leads to magnetic induction. The leakage torque cannot be constrained at all from surface observations, but it has been assumed that the characteristic time scale for changes in diffusive processes at the CMB is likely to be too long for them to be responsible for decadal variations in LOD. (However, results from numerical simulations of the dynamo process suggest that this argument may be weak.)

All components of the torque depend crucially on the structure of the *electrical conductivity of the mantle* (*q.v.*). Downward continuation of the surface field to the CMB assumes an insulating mantle, while EM coupling requires a conducting mantle. However, EM coupling is most strongly enhanced by high conductivity near the base of the mantle, particularly in D″, while secular variation is less sensitive to deep mantle conductivity and more to upper mantle conductivity. Therefore, conductivity located deep in the mantle can give strong EM coupling without greatly affecting downward continuation. A variety of processes have been suggested which might lead to a high conductivity layer at the base of the mantle, including but not limited to core-mantle reaction, partial melt in the lowermost mantle (perhaps associated with *ULVZs* (*q.v.*)), sedimentation of a light component of core fluid at the top of the core, phase transformation of lower mantle silicates, and chemical heterogeneity from a slab graveyard. Any or all of these mechanisms could contribute; for the advective torque, to first order only the conductance (integrated conductivity) of such a layer is important. For EM coupling to be a significant contributor to decadal variations in LOD, a value of this conductance of 10^8 S seems to be necessary. This could arise from, for example, a 200 m thick layer of material of core conductivity, or a conductivity of 500 S m^{-1} over a D″ thickness of 200 km.

There have been many attempts to calculate or estimate the three components of the torque, particularly focusing on the last 150 years or so, to determine whether EM coupling can account for *decadal variations in length of day* (*q.v.*). All studies obtain similar behavior for the poloidal torque: it is found to be westward on the mantle, with only small variations. For deep-mantle conductance of 10^8 S, its value is typically an order of magnitude too small, and unable to explain the oscillatory decadal signal in LOD. There has been much less agreement as to the advective torque, arising because of nonuniqueness in modeling of *core motions* (*q.v.*). The time variation of the radial field is used as a tracer of fluid motion (assuming that diffusion is negligible on decadal time scales—the *frozen-flux approximation* (*q.v.*)), but this provides only one equation to constrain two components of the CMB flow (for example, northwards and westwards). To obtain the other component, other physical assumptions (for example, steadiness, or tangential geostrophy) are adopted, but nevertheless, there are still an infinite number of flows that can explain the secular variation to within a given tolerance. Unfortunately, the advective torque depends on precisely the part of the flow that is not constrained by the observations, but instead by the additional physical constraints. (Magnetic induction from the constrained part of the flow contributes to the poloidal torque.) As a result, it is perhaps not surprising that the torque calculated varies significantly depending on the assumptions adopted. In particular, assuming a steady flow gives an EM torque that does not vary sufficiently to explain the observed ΔLOD. Torques calculated from tangentially geostrophic flows have shown correlation with ΔLOD, particularly if a phase shift between the two signals (due to a delay in transmission of electromagnetic signals through the mantle) is allowed. The confusion is reduced somewhat by considering the inverse problem: is it possible to find a core flow which both explains the observed secular variation and the changes in LOD through EM coupling? The answer is yes, provided a deep mantle conductance of at least 10^8 S is present (Holme, 1997b). Equally, however, flows exist which produce no net torque, or the exact opposite of the observed torque, and still fit the secular variation. While observations do not allow EM coupling to be ruled out as the primary mechanism of core-mantle angular momentum exchange, neither is it possible to confirm its importance (Holme, 1997a).

Most calculations of the torque have assumed a spherically symmetric distribution of mantle conductivity. However, the *core-mantle boundary* (*q.v.*) (CMB) region is known to be very heterogeneous in other physical properties, so it might also be expected to show strong variations in electrical properties. Work is limited on what effect such variations might have on the EM torque, but it seems unlikely that isolated conductivity anomalies could be large enough to either influence secular variation or contribute greatly to coupling. It has been argued that small-scale coupling can enhance magnetic friction, increasing the effectiveness of EM coupling (Buffett, 1996), but on the large scale, only conductance structure of *spherical harmonic* (*q.v.*) degree 1 causes significant changes from the calculations assuming laterally uniform conductance (Holme, 1999), and all evidence is that the deep mantle is dominated by structure that is instead of spherical harmonic degree 2.

This review has concentrated on the effects of electromagnetic coupling on decadal variations in LOD. However, the EM torque also has components perpendicular to the Earth's rotation axis, which contribute to precessional and nutational processes. EM core-mantle coupling plays a particularly important role in the damping of free-core nutations (Buffett, 2002). The limiting value of mantle conductance for which this process is important is again 10^8 S, in agreement with the value at which coupling becomes significant for variations in LOD.

It has been shown that EM coupling can explain variations in LOD, but this is a purely kinematic calculation. The dynamics of the process are less promising. As an idealization of *torsional oscillations* in the core, consider the equation of motion for a simple damped harmonic oscillator

$$\ddot{\theta} + \alpha\dot{\theta} + \omega^2\theta = 0$$

where θ is a measure of angular displacement, and the dots indicate time derivatives. Conservative coupling mechanisms (see *Core-mantle coupling, topographic*) and gravitational contribute primarily to the ω^2 term, allowing the system to support oscillations, while dissipative coupling mechanisms (viscous and electromagnetic) contribute primarily to the α term, leading to damping of these oscillations. The dominant *spherical harmonic* (*q.v.*) component of B_r at the CMB, as at the Earth's surface, is the axial dipole, despite the effects of *downward continuation* (*q.v.*). If changes in core angular momentum are carried by torsional oscillations, then the EM torque for an axial dipole field is directly proportional to the angular momentum relative to the mantle (giving α as a positive constant in the equation above). With only EM coupling, the angular momentum then decays exponentially: a conductance of 10^8 S leads to a damping time of order 90 years, with higher conductance giving a correspondingly shorter timescale. That decadal variations in LOD are observed therefore provides an approximate upper limit on the mean deep mantle conductance: a value of 10^9 S would give a damping time of only 9 years, which is probably too short for free torsional oscillations, which we believe we see. With this insight, we can rule out EM coupling as a primary mechanism to influence paleomagnetic reversal preferred paths, or leading to the observed low level of secular variation in the Pacific over timescales of centuries.

Thus, dynamically, it seems unlikely that EM coupling should be the sole mechanism for explaining decadal variations in length of day—topographic or gravitational coupling are dynamically more favorable to support torsional oscillations. However, a dissipative mechanism is also required for angular momentum transfer, and because the *viscosity of the core* (*q.v.*) is small, EM coupling is the most likely such mechanism. Probably the most promising combination of mechanisms comes from the theory of inner core gravitational coupling to the mantle, in which motions of the inner core excite torsional oscillations in the fluid core, which are then damped by electromagnetic coupling with the mantle. Therefore, although EM coupling is unlikely to be the sole mechanism for explaining *decadal variations in LOD* (*q.v.*), it is nonetheless likely to play an important role.

Richard Holme

Bibliography

Buffett, B.A., 1996. Effects of a heterogeneous mantle on the velocity and magnetic fields at the top of the core. *Geophysical Journal International*, **123**: 303–317.

Buffett, B.A., Mathews, P.M., and Herring, T.A., 2002. Modeling of nutation and precession: Effects of electromagnetic coupling. *Journal of Geophysical Research,* **107**: 10.10292000JB000056.

Gubbins, D., and Roberts, P.H., 1987. Magnetohydrodynamics of the Earth's core. In Jacobs J.A. (eds.), *Geomagnetism,* Vol. 2, London: Academic Press, pp. 1–183.

Holme, R., 1997a. Electromagnetic core-mantle coupling I: Explaining decadal variations in the Earth's length of day. *Geophysical Journal International*, **132**: 167–180.

Holme, R., 1997b. Electromagnetic core-mantle coupling II: probing deep mantle conductance. In Gurnis M., Wysession M.E., Knittle E., and Buffett B.A. (eds.), *The Core-Mantle Boundary Region.*, Washington, D.C.: American Geophysical Union, pp. 139–151.

Holme, R., 1999. Electromagnetic core-mantle coupling III: laterally varying mantle conductance. *Physics of Earth and Planetary Interiors*, **117**: 329–344.

Cross-references

Alfvén's Theorem and the Frozen Flux Approximation
Core Viscosity
Core-Mantle Boundary Topography, Implications for Dynamics
Core-Mantle Boundary Topography, Seismology
Core-Mantle Coupling, Topographic
Harmonics, Spherical
Length of Day Variations, Decadal
Mantle, Electrical Conductivity, Mineralogy
Oscillations, Torsional
Time-dependent Models of the Geomagnetic Field
ULVZ, Ultra-low Velocity Zone
Upward and Downward Continuation

CORE-MANTLE COUPLING, THERMAL

Convection in the Earth's core is driven by the removal of heat from the core by the mantle. In order to remove that heat along with heat from internal heat sources in the mantle, the mantle convects. Thus, at the most fundamental level, convection in the core and in the mantle are coupled. Without mantle convection, the rate of core cooling would be insufficient to maintain core convection. However, the coupling between mantle convection and core convection is not limited to the gross thermodynamics of the core-mantle system: mantle convection results in lateral variations in temperature and composition in the mantle, which in turn affect the pattern of heat flux from the core in to the mantle.

The lateral variations in heat flux from the core that are imposed by the mantle mean that the mantle may exert a long-term influence on the pattern of convection in the core, since the pattern of mantle convection changes on a timescale (say 10^8 years) that is much longer than the timescale of core convection (say 10^3 years).

To understand this further, we begin with a simple physical picture of thermal core-mantle coupling; we then consider the observational evidence and finally we consider results from numerical modeling.

Simple physical model

The heat flux across the core-mantle boundary depends upon the thermal conductivity of and the temperature gradient in the immediately overlying mantle. In this context, by immediately overlying mantle we mean the part of the mantle within the thermal boundary layer at the core-mantle boundary in which the dominant mechanism of heat transport is conduction.

Beneath this thermal boundary layer lies the core-mantle boundary: lateral temperature variations on the core-mantle boundary (assuming that it is an equipotential surface) are of order 10^{-3} K. Above the thermal boundary, the lateral temperature variations in the mantle are of order 10^2 or 10^3 K. As a result, lateral variations in the temperature gradient at the core-mantle boundary are due to the lateral variations in temperature in the overlying mantle, rather than being due to lateral variations in temperature along the core-mantle boundary itself. Thus, where the lowermost mantle is hotter than average, for example in a hot upwelling region, the magnitude of the temperature gradient at the core-mantle boundary will be reduced, resulting in a reduced heat flux from the core to the mantle and, conversely, where the lowermost mantle is colder, for example in a cold downwelling region, the heat flux will be increased.

An alternative way of viewing this is from the perspective of the core. The core responds to lateral temperature variations in the mantle by re-distributing the heat from the core so as to tend to reduce the lateral temperature variations in the mantle. Where the mantle is colder, more heat flows from the core into the mantle.

Mantle convection may also result in lateral variations in composition, which in turn may result in lateral variations in thermal conductivity. These too will result in lateral variations in heat flux: where thermal conductivity in the lowermost mantle is lower than average the heat flux will also be lower than average and vice versa.

A crucial difference between convection in the core and in the mantle, as mentioned above, concerns timescales. In the core, the convective turnover timescale is about 1000 years, while in the mantle it is perhaps as large as 100 Ma. This difference is important because it means that any lateral variations in heat flux imposed on the core by the mantle are stationary on a timescale much greater than the timescale of core convection; in other words, convection in the core may be expected to carry the imprint of lateral variations in heat flux at the core-mantle boundary over a timescale much greater than that on which one might otherwise expect on the basis of considerations of core dynamics in isolation from the mantle.

This provides the key to looking for evidence of core-mantle thermal coupling. Features in the magnetic field that are persistent over timescales comparable or longer than the timescale of core convection may provide evidence of control of the pattern of core convection and hence of the magnetic field by the mantle.

Observational evidence

There are four chief lines of observational evidence for thermal core-mantle interactions: the Pacific dipole window (Fisk, 1931; Cox, 1962), static flux bundles at the core-mantle boundary (Bloxham and Gubbins, 1987), variations in the frequency of geomagnetic reversals (Cox, 1975; Jones, 1977), and the concentration of geomagnetic reversal paths in certain longitude bands (Laj *et al.*, 1991, 1992).

The Pacific dipole window

The central paradigm of geomagnetism is the geocentric axial dipole hypothesis, which maintains that over sufficiently long time intervals the magnetic field averages to that of a geocentric dipole aligned with the rotation axis. Departures from this configuration are referred to as nondipole fields though these must, if the geocentric axial dipole hypothesis holds, be transient. Thus nondipolar fields and secular variation are, on an appropriate timescale, in large part synonymous.

The Pacific dipole window is a region in which the magnetic field around the Pacific region is more nearly dipolar than elsewhere, or equivalently in which the paleomagnetic secular variation is of lower than typical amplitude.

Fisk (1931), from an examination of magnetic observatory data, discovered that secular variation in the Pacific region is anomalously low. Interest in this observation increased following the suggestions of Allan Cox and Richard Doell, based on paleomagnetic studies of Hawaiian lava flows, that nondipole field has been small over the last million years (Doell and Cox, 1961, 1965; Cox and Doell, 1964) and that the low secular variation seen by Fisk dates back at least as far as 1750 (Doell and Cox, 1963). This anomalous behavior in the magnetic field became known as the Pacific dipole window. For the reasons described above, we may consider the Pacific dipole region as one in which the field is more nearly dipolar than elsewhere, or equivalently one in which, on appropriate timescales, the paleomagnetic secular variation is of lower than typical amplitude.

Although some doubt has been since cast on these assertions on paleomagnetic grounds (McElhinny and Merrill, 1975, for example), there is little doubt that the secular variation is different in vigor in the Pacific form elsewhere at least on timescales of several hundred (Bloxham *et al.*, 1989; Jackson *et al.*, 2000) to several thousand years (Johnson and Constable, 1998).

Cox and Doell (1964) suggested two possible causes of the Pacific dipole window: first, that a region of high electrical conductivity in the mantle screens the secular variation; and second that thermal interactions between the mantle and core affected the pattern of convection in the core. Their work was the first to propose that the mantle many influence the morphology of the geomagnetic field.

Static flux bundles at the core-mantle boundary

Using a sequence of maps of the magnetic field at the core-mantle boundary between 1715 and 1980, Gubbins and Bloxham (1987) identified certain features in the field that had drifted little during more than 250 years. Prominent among these static features are a pair of concentrations of magnetic flux at high latitude in each hemisphere, arranged nearly symmetrically about the equator. If westward drift were uniform at the core-mantle boundary, then these patches should have drifted through about 50° during the interval of their study. Even though the magnetic field at the core-mantle boundary is relatively poorly determined in the earliest part of their study, a drift of 50° would have been readily apparent. More recently, using a vastly improved data set, Jackson *et al.* (2000) found substantially the same field morphology over this interval. Furthermore, earlier work, limited to studying the magnetic field at the Earth's surface rather than at the core-mantle boundary, had suggested that while some parts of the magnetic field drift, others are stationary (Yukutake and Tachinaka, 1969; James, 1970).

Three aspects of the high-latitude flux bundles identified by Gubbins and Bloxham (1987) need to be explained: first, their latitude, second, their symmetry about the equator; and, third, their longitude and the fact that they do not drift. The explanation of the first two aspects does not involve core-mantle interactions *per se* and is instead a result of the tangent cylinder in the core (Gubbins and Bloxham, 1987). More pertinently to this discussion, Bloxham and Gubbins (1987) proposed that the longitude and stationarity of the high-latitude flux bundles results from thermal core-mantle interactions on account of the fact that the flux bundles reside beneath regions where the mantle is seismically faster than average. If we assume that seismic heterogeneity in the lowermost mantle is primarily thermal in origin then seismically fast regions correspond to colder regions, in other words to regions of higher heat flux across the core-mantle boundary. Thus, the evidence from the last 300 years is that flux concentrations occur beneath regions of enhanced heat flux from the core.

Of course, 300 years represents less than one convective turnover time for the core, so additional, longer term evidence of the stationarity of high latitude flux bundles is required. A number of studies have examined whether there is evidence for high latitude flux bundles in paleomagnetic measurements of the field. The evidence is mixed: of recent studies of the time-averaged field over the last 5 Ma, Kelly and Gubbins (1997) find flux bundles at approximately the same locations (at least in the northern hemisphere where the paleomagnetic data coverage is reasonable as opposed to the southern hemisphere where it is extremely sparse) as those seen for the last 300 years; Johnson and Constable (1997) find evidence of nonzonal structure, but not unambiguous evidence of flux bundles, while Carlut and Courtillot (1998) fail to find evidence of any nonzonal structure. Given the poor distribution of measurements, high-latitude flux bundles cannot be ruled out based on these studies (Constable *et al.*, 2000). On intermediate timescales, namely from a 3000-year time average of the field, high-latitude flux bundles are observed again in approximately the same locations as recent observations (Johnson and Constable, 1998), though the more recent model (Constable *et al.*, 2000) indicates that while they may exist in the time-average they are not necessarily present at all times.

Despite the somewhat equivocal evidence from studies of longer time span data, any persistent nonzonal structure that is observed is consistent with that seen in the historical record.

Variations in reversal frequency

The chronology of geomagnetic reversals is well-established for roughly the last 150 Ma (see *Geomagnetic polarity reversals*). During that period, the reversal frequency has varied greatly, from as many as eight reversals in a 1-Ma interval around 4 Ma before present to the other extreme of almost 40 Ma between successive reversals during the Cretaceous centered 100 Ma before present. It should be noted, however, that long intervals without reversal are more robustly interpreted as long intervals of predominantly one polarity, since short subintervals of the opposite polarity might be missed. Nonetheless, the record is clear: the frequency of geomagnetic reversal varies greatly, on a timescale, as first pointed out by Cox (1975), of around 50 Ma. As we have seen, this is much longer than the longest timescales commonly associated with magnetic field generation in the core. Cox suggested that these changes in reversal frequency may be due to changes at the core-mantle boundary.

Geomagnetic reversal paths

A fourth line of evidence, first reported by Laj *et al.* (1991, 1992) for thermal core-mantle interactions comes from the tendency of geomagnetic reversal paths, or more precisely of virtual geomagnetic pole paths, to cluster along one of two separate paths, one corresponding in longitude to the Americas and the other to the eastern part of Asia. Such paths correspond roughly to the position of the static flux bundles (discussed above). This work generated considerable controversy; Love (2000), however, in a very careful analysis, found that the clustering of reversal pole paths is statistically significant at about the 95% confidence level. The fact that reversal pole paths are clustered in a way that corresponds geographically to other evidence of core-mantle interaction lends additional credibility.

Numerical models

Following the pioneering work of Glatzmaier and Roberts in the 1990s in producing numerical models of the geodynamo (see *Geodynamo, numerical simulations*), the effect of thermal core-mantle interactions can be simulated numerically. Each of the three sources of observational evidence for thermal core-mantle interactions can be investigated numerically.

The Pacific dipole window

The effect of lateral variations in heat flux at the core-mantle boundary on the long-term secular variation was first studied numerically by Bloxham (2000). By imposing lateral variations in heat flux derived from seismic tomography on a numerical dynamo model he showed that the latitudinal distribution of secular variation was closer to that inferred from observations (than was the case using a homogeneous boundary condition), and that the pattern of secular variation included a Pacific dipole window. In a more recent study, Christensen and Olson (2003) found similar results over a different, but broader, range of parameters.

Static flux bundles at the core-mantle boundary

Sarson *et al.* (1997) were the first to study thermal core-mantle interactions and static flux bundles using a numerical dynamo. They showed that the stationarity of flux bundles may be intermittent, a finding that is consistent with earlier studies of the coupling of purely thermal convection to lateral variations in heat flux (Zhang and Gubbins, 1993, 1997). A later study by Bloxham (2002) found behavior consistent with that seen observationally, namely flux bundles at fixed longitudes in the time-average, but with intervals of intermittency in which the bundles are either moved or absent.

Variations in reversal frequency

Glatzmaier *et al.* (1999) imposed a variety of patterns of heat flux variation on a numerical dynamo model and found differences in the propensity for magnetic reversals, providing compelling evidence that the frequency of magnetic reversal, at least in their numerical dynamo model, is sensitive to the pattern of heat flux variation. One cautionary note should perhaps be mentioned: the changes in pattern of heat flux variation in their study exceeded those that seem reasonable on the timescale on which the reversal frequency changes. Thus, more investigation is needed to investigate the sensitivity of reversal to smaller, more realistic changes in heat flux pattern.

Geomagnetic reversal paths

Two studies are of particular note. First, Coe *et al.* (2000) examined the pole path pattern for four of the geomagnetic reversals reported from the numerical modeling of Glatzmaier *et al.* (1999), finding a correlation between the pattern of heat flux variation and reversal pole paths. In a more extensive study, Kutzner and Christensen (2004) found that pole paths tend to cluster in regions of high heat flux and inferred that regions of high heat flux result in intense magnetic flux bundles, which in turn result in an equatorial dipole component biased in that direction.

Summary

The evidence for thermal core-mantle interactions is compelling: simple physical considerations argue that it will be an important effect; observations provide evidence of thermal core-mantle interactions on a large range of timescales, from change in reversal frequency to the recent secular variation; and numerical dynamo models provide further support. Perhaps the most important finding to emerge from the study of thermal core-mantle interactions is the simple realization that the dynamics of the core cannot be considered in isolation from the mantle: the two convecting systems are fundamentally coupled.

Jeremy Bloxham

Bibliography

Bloxham, J., 2000. The effect of thermal core-mantle interactions on the palcomagnetic secular variation. *Philosophical Transactions of the Royal Society of London A*, **358**: 1171–1179.

Bloxham, J., 2002. Time-independent and time-dependent behaviour of high-latitude flux bundles at the core-mantle boundary. *Geophysical Research Letters,* **29**: 10.1029/2001GL014 543.

Bloxham, J., and Gubbins, D., 1987. Thermal core-mantle interactions. *Nature*, **325**: 511–513.

Bloxham, J., Gubbins, D., and Jackson, A., 1989. Geomagnetic secular variation. *Philosophical Transactions of the Royal Society of London A.*, **329**: 415–502.

Carlut, J., and Courtillot, V., 1998. How complex is the time-averaged geomagnetic field over the last 5 million years? *Geophysical Journal International*, **134**: 527–544.

Christensen, U., and Olson, P., 2003. Secular variation in numerical dynamo models with lateral variations of boundary heat flow. *Physics of the Earth and Planetary Interiors*, **138**: 39–54.

Coe, R., Hongre, L., and Glatzmaier, G., 2000. An examination of simulated geomagnetic reversals from a paleomagnetic perspective. *Philosophical Transactions of the Royal Society of London A.,* **358**: 1141–1170.

Constable, C.G., Johnson, C.L., and Lund, S.P., 2000. Global geomagnetic field models for the past 3000 years: transient or permanent flux lobes? *Philosophical Transactions of the Royal Society of London A.*, **358**: 991–1008.

Cox, A., 1962. Analysis of the present geomagnetic field for comparison with paleomagnetic results. *Journal of Geomagnetism and Geoelectricity*, **13**: 101–112.

Cox, A., 1975. The frequency of geomagnetic reversals and the symmetry of the nondipole field. *Reviews of Geophysics*, **13**: 35–51.

Cox, A., and Doell, R., 1964. Long period variations of the geomagnetic field. *Bulletin of the Seismological Society of America*, **54**: 2243–2270.

Doell, R., and Cox, A., 1961. Palaeomagnetism of Hawaiian lava flows. *Nature*, **192**: 645–646.

Doell, R.R., and Cox, A., 1963. The accuracy of the paleomagnetic method as evaluated from historic Hawaiian lava flows. *Journal of Geophysical Research*, **68**: 1997–2009.

Doell, R.R., and Cox, A., 1965. Paleomagnetism of Hawaiian lava flows. *Journal of Geophysical Research*, **70**: 3377–3405.

Fisk, H., 1931. Isopors and isoporic motion. *Inter. Geodet. Geophys. Un., Terr. Mag. Electr. Sec. Bull.*, 280–292.

Glatzmaier, G., Coe, R., Hongre, L., and Roberts, P., 1999. The role of the Earth's mantle in controlling the frequency of geomagnetic reversals. *Nature*, **401**: 885–890.

Gubbins, D., and Bloxham, J., 1987. Morphology of the geomagnetic field and implications for the geodynamo. *Nature*, **325**: 509–511.

Jackson, A., Jonkers, A.R.T., and Walker, M.R., 2000. Four centuries of geomagnetic secular variation from historical records. *Philosophical Transactions of the Royal Society of London A*, **358**: 957–990.

James, R.W., 1970. Decomposition of geomagnetic secular variation into drifting and non-drifting parts. *Journal of Geomagnetism and Geoelectricity*, **22**: 241–252.

Johnson, C., and Constable, C., 1997. The time-averaged geomagnetic field: Global and regional biases for 0–5 Ma. *Geophysical Journal International*, **131**: 643–666.

Johnson, C., and Constable, C., 1998. Persistently anomalous Pacific geomagnetic fields. *Geophysical Research Letters*, **25**: 1011–1014.

Jones, G., 1977. Thermal interaction of the core and the mantle and long-term behavior of the geomagnetic field. *Journal of Geophysical Research*, **82**: 1703–1709.

Kelly, P., and Gubbins, D., 1997. The geomagnetic field over the past 5 Myr. *Geophysical Journal International*, **128**: 315–330.

Kutzner, C., and Christensen, U., 2004. Simulated geomagnetic reversals and preferred virtual geomagnetic pole paths. *Geophysical Research Letters*, **157**: 1105–1118.

Laj, C., Mazaud, A., Weeks, R., Fuller, M., and Herrero-Brevera, E., 1991. Geomagnetic reversal paths. *Nature*, **351**: 447.

Laj, C., Mazaud, A., Weeks, R., Fuller, M., and Herrero-Brevera, E., 1992. Statistical assessment of the preferred longitudinal bands for recent geomagnetic reversal paths. *Geophysical Research Letters*, **19**: 2003–2006.

Love, J., 2000. Statistical assessment of preferred transitional VGP longitudes based on palaeomagnetic data. *Geophysical Journal International*, **140**: 211–221.

McElhinny, M.W., and Merrill, R.T., 1975. Geomagnetic secular variation over the past 5 my. *Reviews of Geophysics*, **13**: 687–708.

Sarson, G.R., Jones, C.A., and Longbottom, A.W., 1997. The influence of boundary region heterogeneities on the geodynamo. *Physics of the Earth and Planetary Interiors*, **101**: 13–32.

Yukutake, T., and Tachinaka, H., 1969. Separation of the Earth's magnetic field into drifting and standing parts. *Bulletin of the Earthquake Research Institute, Tokyo*, **47**: 65–97.

Zhang, K., and Gubbins, D., 1993. Convection in a rotating spherical fluid shell with an inhomogeneous temperature boundary condition at infinite prandtl number. *Journal of Fluid Mechanics*, **250**: 209–232.

Zhang, K., and Gubbins, D., 1997. Nonlinear aspects of core-mantle interaction. *Geophysical Research Letters*, **20**: 2969–2972.

Cross-references

Geocentric Axial Dipole Hypothesis
Geodynamo, Numerical Simulations
Geomagnetic Polarity Reversals

CORE-MANTLE COUPLING, TOPOGRAPHIC

Topographic core-mantle coupling results from pressure gradients across bumps at the CMB. Hide (1969, 1977) initiated the study of this mechanism. He discussed how pressure lows and highs result from fluid motions below the core surface. Later, dynamic pressure at the core surface ($\simeq 10^3$ Pa from peak to trough) has been estimated from geomagnetic secular variation models through core flow modeling. Subsequent research was focused on the strength of the pressure torque by comparison with electromagnetic, viscous, and gravitational torques acting either at the CMB or at the ICB. These research works have often been controversial because of our poor knowledge of the relief at the core-mantle interface (see *Core-mantle coupling topography, seismology*) and of the motions at the core surface. The axial pressure torque acting on the mantle is:

$$\Gamma_p = \mathbf{k}.\int_{\mathrm{CMB}} p(\mathbf{r}\times\mathbf{n})\,\mathrm{d}S \qquad \text{(Eq. 1)}$$

where $\mathbf{k}$ is the unit vector along the rotation axis, $\mathbf{n}$ is the normal outward to the fluid volume, $\mathbf{r}$ is the position vector, and p is the pressure. It would vanish if the CMB were spherical ($\mathbf{n}$ collinear with $\mathbf{r}$). There is an expression analogous to Eq. (1) for the torque acting on the solid inner core.

There are also equatorial components of the pressure torque. Their calculation is straightforward as the acting pressure is due to the centrifugal acceleration associated with a quasirigid equatorial rotation of the outer core. This is the free core nutation (FCN) problem. The FCN results from a tiny difference between the rotation vectors of, respectively, the mantle and the fluid outer core. Because of this misalignment, centrifugal pressure acts on the CMB, of which the meridional sections are ellipsoidal. As a result, the period of the FCN, which is accurately measured because of resonances with lunar forcing, determines the oblateness of the CMB. The remaining part of this entry is devoted to the axial problem.

In reaction to the pressure torque Γ_p acting from the core on the mantle, there is an exactly opposite torque acting on the core from the mantle. As a result, core motions are influenced by the CMB topography. Less obviously, it can be argued that the core motions are arranged in such a way that the strength of the pressure torque is limited. Thus, estimation of the pressure torque and study of the influence of the CMB topography on core dynamics are two intertwined questions (see *Core-mantle coupling topography, implications for dynamics*).

Geostrophic motions and topographic coupling

Core angular momentum changes in reaction to the torques that act on it. In rapidly rotating and enclosed fluids, angular momentum is carried by motions that are either exactly geostrophic, obeying the leading order force balance between Coriolis and pressure forces, or closely related. Throughout geostrophic circulation, columns of fluid, parallel to the rotation axis, move following geostrophic contours C, which are closed paths on the upper and lower boundaries separated by constant height. Inside the Earth's outer core, the contours C slightly deviate from circles centered on the rotation axis because of topography at the boundaries and generate nearly circular cylinders. For each cylinder, the top and bottom contours are situated either both on the CMB (outside the cylinder tangent to the inner core) or one on the ICB and the other on the CMB (inside this cylinder). Taking the integral over geostrophic cylinders of the momentum equation valid in the frame rotating with the mantle cancels out the leading order terms (Bell and Soward, 1996). Finally, an equation for the geostrophic velocity $\mathbf{u}_g$ is derived:

$$\rho(z_T - z_B)\oint\left(\frac{\partial\mathbf{u}_g}{\partial t}+\rho\frac{\mathrm{d}\Omega}{\mathrm{d}t}(\mathbf{k}\times\mathbf{r})\right)\mathbf{dl} = \int_{z_B}^{z_T}\left(\oint(\mathbf{j}\times\mathbf{B}+\rho\mathbf{g}).\mathbf{dl}\right)\mathrm{d}z, \qquad \text{(Eq. 2)}$$

where **dl** is an element of arc length along C and z_T and z_B denote the z-coordinates (parallel to the rotation axis) along, respectively, the top and bottom geostrophic contours. Neglecting the inertial term on the LHS, the RHS gives a generalized version of *Taylor's condition* (*q.v.*). The magnetic force $\mathbf{j} \times \mathbf{B}$ gives the term that generates *torsional oscillations* (*q.v.*) in reaction to the shearing of the magnetic field lines by the geostrophic motions. It can be shown that Eq. (2) governs the time changes of the kinetic energy of the geostrophic motions, in a frame rotating with the mantle. With spherical boundaries, Eq. (2) can be interpreted also as the equation governing the transport of angular momentum within the fluid outer core and the exchanges with the inner core and the mantle. In presence of topography on the boundaries, Eq. (2) does not, however, give directly the transport of angular momentum (Fearn and Proctor, 1992). This constitutes, arguably, the main difficulty of the studies of topographic core-mantle coupling.

If the solid boundaries were spherical and electrically insulating, core angular momentum would be conserved and the net torque on the core would be zero. This is obviously true but this may also be directly inferred from Eq. (2). Taking into account topography and using a perturbation approach, only the s and z components of the magnetic and buoyancy forces are involved in the terms that are dependent (linearly) on the topography on the RHS of Eq. (2), while a spherical approximation can be used on the LHS. The constraint of conservation of core angular momentum carried by geostrophic motions does not apply any more.

Taking for the sake of the demonstration insulating solid boundaries in a core model, the only torque left to modify the total core angular momentum is the pressure torque Γ_p. Indeed, the buoyancy and magnetic forces that accelerate the geostrophic motions within the core are unable to account for the exchanges of angular momentum with the mantle. Conversely, when there is a nonzero pressure torque, there are time changes of $\mathbf{u}_g$ and Taylor's condition is not exactly obeyed. Thus, in an average state, fulfilling Taylor's condition, the pressure torque would vanish. In the actual Earth, the different contributions to Eq. (1) probably tend to cancel out also except if the electromagnetic torque acting between core and mantle is strong enough to oppose Γ_p. This explains why it will be difficult to estimate Γ_p from geophysical data. If such a cancelation does not take place and if the height of the topography is of the order of 1 km, the pressure torque is much larger than the torque responsible for the observed changes in the Earth's rotation (Jault and Le Mouël, 1990).

From a theoretical viewpoint, it may not be possible to study separately the different torques acting between core and mantle. As an example, addition of a conducting layer at the bottom of the mantle to the model and inclusion of CMB topography cannot be considered in isolation. Both modify the RHS of Eq. (2). Kuang and Chao (2001, 2003) have adapted a numerical geodynamo model to study topographic coupling. Their solutions are calculated in a spherical shell with boundary conditions modified to mimic the role of the CMB topography. With this approach, the main ingredient of topographic coupling, distortion of the geostrophic contours, is included. Kuang and Chao have verified that, in their model, the different contributions to the pressure torque tend to cancel out.

Role of the inner core

The main pressure torque acting on the fluid outer core may result from the action of the hydrostatic pressure on the ICB topography if the ICB does not coincide with an equipotential surface of the gravity field. It is unimportant for the outer core since it cancels exactly a gravity torque acting from the mantle on the outer core. On the other hand, pressure and gravity torques acting on the inner core do not cancel out (see *Inner core oscillation*; *Inner core, rotational dynamics*). Differential rotation between the inner core and the mantle may have some influence on topographic core-mantle coupling because it entails time changes of the geometry of geostrophic contours.

If the height of the topography either on the ICB or on the CMB is not negligible, the cylinder tangent to the inner core equator is not of constant height. Then, there are regions void of closed contours separated by constant height on both sides of the tangent cylinder. There, the role of the geostrophic motions is taken over by low-frequency z-independent inertial waves, known as Rossby waves (Greenspan, 1968, pp. 85–90).

Perspectives

Hopefully, understanding of core-mantle coupling will result from ongoing studies of the propagation of torsional waves within the fluid outer core. Considering the pressure acting on the mantle suffices to calculate the net torque on the mantle but not to estimate how CMB and ICB topographies influence the distribution of geostrophic velocity within the core. Knowledge of the volume forces is required, except in the special case of distortion of the geostrophic contours in the z-direction only, which happens when topography is antisymmetrical with respect to the equatorial plane (Jault *et al.*, 1996). Topographical effects have not yet been adequately taken into account in models of torsional waves.

Perhaps, too much weight has been given in this entry on Taylor's condition. Rapid changes in the differential rotation between geostrophic cylinders are definitely opposed by magnetic forces within the core. On the other hand, only viscous coupling and magnetic coupling with the weakly conducting mantle control global oscillations of the fluid outer core with respect to the mantle.

Dominique Jault

Bibliography

Bell, P.I., and Soward, A.M., 1996. The influence of surface topography on rotating convection. *Journal of Fluid Mechanics*, **313**: 147–180.

Fearn, D.R., and Proctor, M.R.E., 1992. Magnetostrophic balance in nonaxisymmetric, nonstandard dynamo models. *Geophysical and Astrophysical Fluid Dynamics*, **67**: 117–128.

Greenspan, H.P., 1968. *The Theory of Rotating Fluids*. Cambridge: Cambridge University Press.

Hide, R., 1969. Interaction between the Earth's liquid core and solid mantle. *Nature*, **222**: 1055–1056.

Hide, R., 1977. Towards a theory of irregular variations in the length of the day and core-mantle coupling. *Philosophical Transactions of the Royal Society*, **284**: 547–554.

Jault, D., and Le Mouël, J.-L., 1990. Core-mantle boundary shape: constraints inferred from the pressure torque acting between the core and mantle. *Geophysical Journal International*, **101**: 233–241.

Jault, D., Hulot, G., and Le Mouël, J.-L., 1996. Mechanical core-mantle coupling and dynamo modelling. *Physics of the Earth and Planetary Interiors*, **98**: 187–191.

Kuang, W., and Chao, B.F., 2001. Topographic core-mantle coupling in geodynamo modeling. *Geophysical Research Letters*, **28**: 1871–1874.

Kuang, W., and Chao, B.F., 2003: Geodynamo modeling and core-mantle interactions. In Dehant, V., Creager, K., Karato, S.-I., and Zatman, S. (eds.), *Earth's Core Dynamics, Structure, Rotation*. Washington, DC: American Geophysical Union, pp. 193–212.

Cross-references

Core-Mantle Boundary Topography, Implications for Dynamics
Core-Mantle Boundary Topography, Seismology
Core-Mantle Coupling, Electromagnetic
Inner Core Oscillation
Inner Core Tangent Cylinder
Inner Core, Rotational Dynamics
Length of Day Variations, Decadal
Oscillations, Torsional
Taylor's Condition

COWLING, THOMAS GEORGE (1906–1990)

Thomas George Cowling (Figure C45), applied mathematician and astrophysicist, was born on June 17, 1906 at Hackney, London, the second of the four sons of George Cowling, post office engineer, and Edith Eliza Cowling (née Nicholls). His grandparents and their antecedents were nearly all artisans, blue-collar workers, servants, or sailors. Tom was educated at Sir George Monoux Grammar School, from where in 1924 he won an open scholarship in mathematics to Brasenose College, Oxford. He won the University Junior Mathematical Exhibition in 1925 and the Scholarship in 1926. A first class degree was rewarded by a three-year post-graduate scholarship at Brasenose, his research supervisor being the cosmologist Edward A. Milne.

Cowling's acumen was quickly recognized. After gaining his D.Phil., he went by invitation to join *Sydney Chapman* (*q.v.*) at Imperial College as demonstrator. Their collaboration converted Chapman's "skeleton" of a book into a standard text, *The Mathematical Theory of Non-Uniform Gases* (1939), which treats kinetic theory by a systematic attack on the classical Boltzmann equation. Cowling was subsequently assistant lecturer in mathematics at Swansea (1933–1937), and lecturer at Dundee (1937–1938) and at Manchester (1938–1945). From 1945–1948 he was professor of mathematics at Bangor, moving then to his last post as professor of applied mathematics at Leeds, becoming emeritus professor in 1970.

Over the three decades 1928–1958, Cowling made outstanding contributions to stellar structure, cosmical electrodynamics, kinetic theory and plasma physics. Recognition came with the Johnson Memorial Prize at Oxford (1935), election to the Royal Society (1947), the Gold Medal of the Royal Astronomical Society (1956) and its presidency (1965–1967), Honorary Fellowship of Brasenose (1966), the Bruce Medal of the Astronomical Society of the Pacific (1985), and the Hughes Medal of the Royal Society (1990).

Tom Cowling married Doris Marjorie Moffatt on August 24, 1935. There were three children: Margaret Ann Morrison, Elizabeth Mary Offord (deceased 1994), and Michael John Cowling, and six grandchildren.

A few years before Cowling began working on stellar structure, Sir Arthur Eddington had published his celebrated *Internal Constitution of the Stars*, in which he pictured a homogeneous star as a self-gravitating gaseous sphere in radiative equilibrium, with its luminosity fixed by its mass and chemical composition. From comparison of his "standard model" with the masses and radii of observed "main-sequence" stars, Eddington inferred that the as yet unknown stellar energy sources must be highly temperature-sensitive. Cowling's careful analysis strengthened Eddington's case—against the criticisms of Sir James Jeans—by showing that the models would nevertheless almost certainly be vibrationally stable. In parallel with Ludwig Biermann, he also showed that a stellar domain that satisfies the Schwarzschild criterion for convective instability and becomes turbulent needs only a very slightly superadiabatic temperature gradient in order to convect energy of the order of a stellar luminosity.

Figure C45 Thomas George Cowling.

The "Cowling model," with its nearly adiabatic convective core, surrounded by a radiative envelope, is a paradigm for stars somewhat more massive than the Sun, for which energy release by fusion of hydrogen into helium takes place through the highly temperature-sensitive Bethe-Weizsäcker carbon-nitrogen cycle. In low-mass stars like the Sun, hydrogen fusion occurs via the much less temperature-sensitive proton-proton chain, so that the energy-generating core is now radiative; but because of the low surface temperature, there is a deep convective envelope, extending from just beneath the radiating surface. Biermann's calculated depth for the solar convection zone has been confirmed by helioseismology. Biermann also showed that a star of $\sim M_\odot/4$ would be fully convective. Cowling pointed out that for these stars, the luminosity must still satisfy the condition of radiative transport through the nonconvecting surface layers. Earlier, Milne had argued that the luminosity of all stars should be sensitive to conditions in the surface, rather than depending essentially on radiative transfer through the bulk of the star, as in Eddington's theory. Cowling's work vindicated Eddington's approach for main-sequence stars, while implicitly allowing for such sensitivity in giant stars and pre-main sequence stars, with extensive sub-surface convective zones.

Cowling's other contributions to stellar structure include an important paper on binary stars, and two pioneering papers on nonradial stellar oscillations, foreshadowing today's work in helioseismology.

Simultaneous with his work on stellar structure, Cowling published a series of studies on cosmical magnetism and on plasma physics, culminating in his superbly succinct monograph *Magnetohydrodynamics* (1959, 1976). Together with Biermann, Arnulf Schlüter, and Lyman Spitzer Jr., he did much to clarify the concept of "conductivity" for both fully and partially ionized gases. A lively correspondence with Biermann in the 1930s brought out the requirements of a theory of both the structure and the origin of sunspots, and indeed may be thought of as the genesis of the study of magnetoconvection. While recognizing the seminal contributions of *Hannes Alfvén* (*q.v.*) to cosmical magnetism ("Alfvén's best ideas are very good indeed"), he had earlier applied his renowned critical faculties to a demolition of Alfvén's theory of the sunspot cycle and defended the pioneering Chapman-Ferraro, quasi-MHD theory of geomagnetic storms against what he regarded as unjust attacks.

In the geomagnetic and in fact in the general scientific world, Cowling is probably best known for his "antidynamo theorem" (see *Cowling's Theorem*). Sir Joseph Larmor (*q.v.*) had suggested that motion of the conducting gas across a sunspot magnetic field yields an emf, which drives currents against the Ohmic resistance, and these currents could be just those required by Ampère's law to maintain the assumed field. Cowling began what he expected to be a "relatively routine piece of difficult computation," to see what sort of field could be maintained by the Larmor process. He found instead that for an axisymmetric field, the required condition breaks down at the ring of neutral points. In the absence of an energy supply, the Ohmic decay of the axisymmetric field can be pictured as the steady diffusion of closed field lines into the neutral point. The moving gas drags field lines around meridian planes, but does not generate new ones to replace the disappearing loops—that would require gas to emerge from the neutral point with infinite speed. This qualitative argument can be generalized to apply to fields that are topologically similar to the axisymmetric fields.

For a while, Cowling does seem to have surmised that there was waiting to be proved a much more general antidynamo theorem. Later he felt that too much emphasis was laid on his and other antidynamo theorems: their significance is that those fields that can be dynamo-maintained cannot be simple in structure or in mathematical formulation. He illustrated this by showing how one *prima facie* convincing general algorithm, designed for the construction of the velocity field that will maintain a prescribed magnetic field, when applied to a field which is known to be ruled out by an anti-dynamo theorem yields—unsurprisingly—"velocities" that are pure imaginary. He appears to have provisionally accepted the now familiar picture by which nonuniform rotation acting on a given poloidal field generates a toroidal field, which in turn yields the required new poloidal flux when acted upon by nonaxisymmetric convective motions—the "α-effect" in current dynamo terminology. While remaining well aware of the lacunae in our present theoretical knowledge, he concluded (1981): "The dynamo theory of the Sun's magnetic field is subject to a number of unresolved objections, but alternative theories advanced so far are open to much greater objections."

In the absence of dynamo action, the high conductivity of stellar plasma yields an Ohmic decay time for a large-scale primeval stellar field that can exceed the lifetime of the star. This "fossil" field picture is inapplicable to the Sun, in which the poloidal field is now known to reverse as part of the solar cycle, but it remains relevant to the observed strongly magnetic stars, convincingly modelled as oblique magnetic rotators. However, for the Earth, the much smaller length scales yield a decay time that is far too short, so that for both the Sun and the Earth, dynamo maintenance appears to be mandatory.

In his autobiographical essay "Astronomer by accident" (1985), Cowling remarks that he and his brothers inherited the "Puritan work ethic" from their nonconformist parents. He remained an active, non-fundamentalist Baptist, though in his later years he confessed to doubts, speaking of "getting closer to mysticism rather than to a religion of set creeds." His puritanism perhaps showed itself in his very high academic standards, which he applied as much to his own work as to that of others, and earned him the nickname "Doubting Thomas." Everyone knew him by sight: his red hair and his great height made him conspicuous. They also looked up to him metaphorically, because of his combination of kindliness and intellectual power.

Cowling's research activity in his later decades was severely hampered by recurring bouts of ill-health, beginning with a duodenal ulcer in 1954. He died in Newton Green Hospital, Leeds on June 15, 1990.

Leon Mestel

Bibliography

Chapman, S., Cowling, T.G., 1939. *The Mathematical Theory of Non-Uniform Gases.* 1991 Paperback edition. Cambridge: Cambridge University Press.

Cowling, T.G., 1957. *Magnetohydrodynamics.* Interscience (2nd ed., 1976, Adam Hilger).

Cowling, T.G., 1981. The present status of dynamo theory. *Annual Review of Astronomy and Astrophysics*, **19**: 115–135.

Cowling, T.G., 1985. Astronomer by accident. *Annual Review of Astronomy and Astrophysics*, **23**: 1–18.

Mestel, L., 1991. Thomas George Cowling. *Biographical Memoirs of Fellows of the Royal Society*, **37**: 103–125.

Cross-references

Alfvén, Hannes Olof Gösta (1908–1995)
Chapman, Sydney (1888–1970)
Cowling's Theorem
Lamor, Joseph (1857–1942)

COWLING'S THEOREM

Among the theorems of applied mathematics Cowling's theorem on the impossibility of axisymmetric homogeneous dynamos has played a special role because it has cast in doubt for 40 years the dynamo hypothesis of the origin of geomagnetism. This hypothesis goes back to Larmor (1919; see *Larmor, J.*) who argued that the magnetic field of sunspots could be created by motions in an electrically conducting fluid. He envisioned an axisymmetric model where radial motion into a cylindrical tube of magnetic flux would cause an electromotive force $\boldsymbol{v} \times \mathbf{B}$ driving an electric current in the azimuthal direction, which in turn would enhance the magnetic flux in the cylinder. While this model appears to be superficially convincing, Cowling was able to prove that such a dynamo process is not possible.

The proof originally provided by Cowling (1934; see *Cowling, T.G.*) is quite simple since he was only concerned with the possibility of the steady generation of a field confined to meridional planes as envisioned by Larmor. A general axisymmetric magnetic field $\mathbf{B}$ satisfying the condition $\nabla \cdot \mathbf{B} = 0$ can be described by

$$\mathbf{B} = \nabla A_\phi \times \boldsymbol{e}_\phi + B_\phi \boldsymbol{e}_\phi \qquad \text{(Eq. 1)}$$

where e_ϕ is the unit vector in the azimuthal direction and A_ϕ is the ϕ-component of the magnetic vector potential $\mathbf{A}$. Ohm's law for an electrically conducting fluid moving with the velocity field $\boldsymbol{v}$ can be written in the form

$$\lambda \nabla \times \mathbf{B} = \boldsymbol{v} \times \mathbf{B} - \nabla U - \frac{\partial}{\partial t}\mathbf{A} \qquad \text{(Eq. 2)}$$

where the last two terms represent the electric field and where the magnetic diffusivity λ is the inverse of the product of magnetic permeability and electrical conductivity of the fluid, $\lambda = (\sigma \cdot \mu)^{-1}$. Since we are considering an axisymmetric steady process $(\partial/\partial t)\mathbf{A}$ vanishes and ∇U does not contribute to the ϕ-component of Eq. (2). The latter assumes the form

$$v_r \frac{\partial}{\partial r} C + v_z \frac{\partial}{\partial z} C = \lambda \left(\frac{\partial^2}{\partial r^2} C + \frac{\partial^2}{\partial z^2} C - \frac{1}{r}\frac{\partial}{\partial r} C \right) \qquad \text{(Eq. 3)}$$

where a cylindrical coordinate system (r, ϕ, z) has been used and $C(r, z) \equiv rA_\phi$ has been introduced after multiplication of the equation by r. The condition that the magnetic field is generated locally requires that it decays towards infinity at least like a dipolar field, i.e., in proportion to $(r^2 + z^2)^{-3/2}$. The function $C(r, z)$ thus must decay at least like $(r^2 + z^2)^{-1/2}$. Since, on the other hand, C vanishes on the axis, $r = 0$, it must assume at least one maximum or minimum at a point (r_0, z_0) at which $(\partial/\partial r)C = (\partial/\partial z)C = 0$ can be assumed. Such a point in the meridional plane is sometimes called "neutral line." Because at a maximum or minimum usually $(\partial^2/\partial r^2)C + (\partial^2/\partial z^2)C \neq 0$ holds, a contradiction to Eq. (3) is obtained. In the case of "flat" maxima or minima where $(\partial^2/\partial r^2)C + (\partial^2/\partial z^2)C$ may vanish a similar contradiction can be constructed, but this requires a more technical approach. It should be noted that the above argument does not require a solenoidal velocity field, $\nabla \cdot \mathbf{v} = 0$, as Cowling had originally assumed. For a generalization of Cowling's original theorem, which emphasizes its topological nature, see Bullard (1955; see *Bullard, E. C.*). Backus and Chandrasekhar (1956) later refined Cowling's proof by eliminating the possibility of an axisymmetric dynamo with a finite component B_ϕ.

For geophysical and astrophysical applications it is important to consider the possibility of axisymmetric dynamos caused by varying magnetic diffusivity or by time-dependent processes. The former possibility was eliminated by Lortz (1968) while the latter problem motivated Braginsky (1964) to seek a new mathematical approach based on energy arguments. Braginsky also introduced the spatial configuration of a finite axisymmetric domain V of constant electrical conductivity surrounded by an insulating space outside. For his proof of the impossibility of time-dependent axisymmetric dynamos he had however to assume a solenoidal velocity field. For the domain V, Hide (1979) introduced as a new definition for a dynamo that the magnetic flux intersecting the surface ∂V of V remains finite as the time t tends to infinity,

$$\oint_{\partial V} |\boldsymbol{B} \cdot \boldsymbol{d}^2 S| \not\to 0 \quad \text{for} \quad t \to \infty. \qquad \text{(Eq. 4)}$$

Hide and Palmer (1982) tried to prove on this basis Cowling's theorem under rather general conditions. In view of their novel approach it is unfortunate that their proof turned out to be incomplete (Ivers and James, 1984; Núñez, 1996). The efforts of Lortz and coworkers were thus required to arrive at mathematically rigorous conclusions about the impossibility of the generation of axisymmetric magnetic field by time dependent and not necessarily solenoidal velocity fields. For references and details of the mathematical issues we refer to the reviews of Ivers and James (1984) and Núñez (1996).

Cowling (1955) had already realized that the ideas of the proof for the impossibility of an axisymmetric dynamo could easily be applied to prove the impossibility of two-dimensional dynamos for which the magnetic field depends only on two of the three coordinates of a cartesian system. Since the cartesian case is somewhat simpler to deal with many of the mathematical problems have first been explored in this geometry. The 1955 Cowling paper is particularly important in its showing that—because of the presence of non-Hermitian operators—the apparently general kinematic dynamo formalism developed by Elsasser (1947 and earlier papers referred therein; see *Elsasser, W.*) does not guarantee real valued velocities. Cowling illustrates this by showing that the two-dimensional model yields purely imaginary velocities.

It is of interest to note that dynamos with small deviations from axisymmetry can readily be realized. The theory of spherical dynamos developed by Braginsky (1976; see *Dynamo, Braginsky*) is based on a small perturbation of an axisymmetric configuration. A purely axisymmetric dynamo has been obtained by Lortz (1989) by allowing for a small anisotropy of the magnetic diffusivity tensor. It thus appears that in spite of its early negative influence Cowling's theorem has stimulated the understanding of dynamo action and the creation of ingenious solutions for the problem of magnetic field generation.

Friedrich Busse

Bibliography

Backus, G., Chandrasekhar, S., 1956. On Cowling's theorem on the impossibility of self-maintained axisymmetric homogeneous dynamos. *Proceedings of the National Academy of Sciences*, **42**: 105–109.

Braginsky, S.I., 1964. Self-excitation of a magnetic field during the motion of a highly conducting fluid. *Soviet Physics JETP.*, **20**: 726–735.

Braginsky, S.I., 1976. On the nearly axially-symmetrical model of the hydromagnetic dynamo of the Earth. *Physics of the Earth and Planetary Interiors*, **11**: 191–199.

Bullard, E., 1955. A discussion on magneto-hydrodynamics: Introduction. *Proceedings of the Royal Society of London*, **A233**: 289–296.

Cowling, T.G., 1934. The magnetic field of sunspots. *Monthly Notices of the Royal Astronomical Society*, **34**: 39–48.

Cowling, T.G., 1955. Dynamo theories of cosmic magnetic fields. *Vistas in Astronomy*, **1**: 313–323.

Elsasser, W.M., 1947. Induction effects in terrestrial magnetism, Part III Electric modes. *Physical Review*, **72**: 821–833.

Hide, R., 1979. On the magnetic flux linkage of an electrically-conducting fluid. *Geophysical and Astrophysical Fluid Dynamics*, **12**: 171–176.

Hide, R., Palmer, T.N., 1982. Generalization of Cowling's theorem. *Geophysical and Astrophysical Fluid Dynamics*, **19**: 301–309.

Ivers, D.J., James, R.W., 1984. Axisymmetric antidynamo theorems in compressible non-uniform conducting fluids. *Philosophical Transactions of the Royal Society of London*, **A312**: 179–218.

Larmor, J., 1919. How could a rotating body such as the sun become a magnet? *The British Association for the Advancement of Science Report,* 159–160.

Lortz, D., 1968. Impossibility of steady dynamos with certain symmetries. *Physics of Fluids*, **10**: 913–915.

Lortz, D., 1989. Axisymmetric dynamo solutions. *Zeitschrift für Naturforschung*, **44a**: 1041–1045.

Núñez, M., 1996. The decay of axisymmetric magnetic fields: A review of Cowling's theorem. *SIAM Review*, **38**: 553–564.

Cross-references

Bullard, Edward Crisp (1907–1980)
Cowling, Thomas George (1906–1990)
Dynamo, Braginsky
Elsasser, Walter M. (?–1991)
Larmor, Joseph (1857–1942)

COX, ALLAN V. (1926–1987)

Allan Cox was one of the preeminent geophysicists of his generation. He made many important scientific contributions to the field of paleomagnetism and to the study of plate tectonics. His untimely death at the age of 61 was a loss to Stanford University, his home institution, and to the geophysics community.

Cox was born in Santa Ana, California, in 1926 as the son of a house painter. He attended the University of California at Berkeley where he started his education as a chemistry major. He only went to Berkeley for one quarter before he dropped out and joined the merchant marine. His work in the merchant marine allowed him time for one of his great loves, reading over a wide range of topics. After three years in the merchant marine, he returned to Berkeley, again as a chemistry major. One of his life defining experiences at that time was spending summers in Alaska with geologist Clyde Wahrhaftig of the U.S. Geological Survey. These summer jobs gave Cox a strong love of geology. When he returned to Berkeley to continue his education at the end of the summer, he found chemistry dull by comparison and his grades suffered. This led to the loss of his student deferment from the military draft and he ended up in the army for two years. When he came back to Berkeley after the army, strongly motivated to succeed at his studies, he changed his major to geology. In 1955, he earned his bachelor of arts from Berkeley in geology. He continued his education with graduate work in geology at Berkeley. His initial intent was to study ice and rock glaciers, the objects of his summer field work with Wahrhaftig, but instead he decided to work with John Verhoogen on rock magnetism. At that time Verhoogen was one of the few Berkeley faculty members who was receptive to the idea of continental drift. This gave Cox an early exposure and interest in the field that would become plate tectonics.

Cox received a master of arts in 1957 and his PhD from Berkeley in 1959 and went to work at the U.S. Geological Survey in Menlo Park, California. At the Survey he worked with another rock magnetist, Dick Doell, on the paleomagnetic record of geomagnetic field reversals. For this work, Cox and Doell needed some way to accurately date the rocks carrying paleomagnetic records of reversed and normal geomagnetic field polarity, so they convinced the U.S. Geological Survey to hire Brent Dalrymple who could use the radiogenic decay of K to Ar to determine the age of igneous rocks being studied paleomagnetically. The research team of Cox, Doell, and Dalrymple made important contributions to Earth science by unraveling the record of polarity intervals of the geomagnetic field over the past 5 Ma. They traveled to far corners of the globe to collect igneous rocks of various ages, measured their paleomagnetism at their laboratory in Menlo Park, conducted demagnetization experiments to isolate the primary magnetization of the rocks, and dated the rocks radiogenically. There was spirited competition with other researchers conducting similar research at this time, particularly an Australian team consisting of Tarling, Chamalaun, and McDougall, based at the Australian National University. The two teams would make friendly wagers about who would discover the next polarity interval in the quest for the full revelation of the geomagnetic polarity time scale. The bets would be paid off, usually in cocktails, at the next professional meeting. The geomagnetic timescale for the past 5 Ma was completed by 1969 and was crucial to the seafloor spreading interpretation of the seafloor magnetic anomalies that were discovered to be parallel to the mid-ocean ridges at about this time.

In 1967, Allan Cox moved from the U.S. Geological Survey to nearby Stanford University in Palo Alto, California, and became the Cecil and Ida Green Professor of Geophysics. At Stanford he studied various aspects of paleomagnetism with his graduate students, contributing to the understanding of many topics, including the transition of the geomagnetic field between polarity states, the characteristics of geomagnetic field intensity, the statistics of polarity interval lengths, the rock magnetics of seafloor basalts and sedimentary rocks, vertical axis rotations of tectonic blocks, and the structure of a plate's apparent polar wander path. He developed a strong interest in the paleolatitudinal motion of the tectonostratigraphic terranes that were just being recognized by geologists at that time.

In 1979, he became dean of Stanford's School of Earth Sciences and showed a gift for being an administrator. During his stint as dean he kept up his teaching and his research and stayed involved with undergraduates, through undergraduate research projects and helping design the lighting for the student theatre. He was viewed as the paramount example of a true teacher-scholar, excelling equally at teaching, research, and service to Stanford University and his professional societies. He also served his community in the hills to the west of Palo Alto, studying with fellow faculty members and students the effects of logging on soil erosion and watersheds and publishing a small book on the subject, *Logging in Urban Counties*.

Allan Cox had a productive scientific career, publishing over 100 scientific journal articles and two textbooks on plate tectonics, *Plate Tectonics and Geomagnetic Reversals* and *Plate Tectonics, How it Works* with Robert B. Hart. He also received many professional honors. He was elected to the National Academy of Sciences, the American Philosophical Society, and the American Academy of the Arts and Sciences. He was president of the American Geophysical Union from 1978–1980. He received the Day Medal of the Geological Society of America, the Fleming Medal of the American Geophysical Union, and the Vetlesen Medal of Columbia University.

He died in a bicycle accident on January 27, 1987, running off the road in the hills west of Stanford and into a redwood tree. Tragically, his death was ruled a suicide, probably brought on by failings and crises in his personal life.

Kenneth P. Kodama

Bibliography

Cox, A.V., 1973. *Plate Tectonics and Geomagnetic Reversals.* San Francisco, CA: W.H. Freeman and Company.

Cox, A.V., and Hart, R.B., 1986. *Plate Tectonics, How it Works.* Palo Alto, CA: Blackwell Scientific.

Cross-references

Paleomagnetism
Shaw and Microwave Methods, Absolute Paleointensity Determination

CRUSTAL MAGNETIC FIELD

Introduction

Magnetic minerals in the rocks in the upper lithosphere of the Earth and other differentiated planetary bodies having dynamos generate magnetic fields when the minerals are in a cooler environment than their individual Curie temperatures. These magnetic fields are useful for studying crustal materials for a variety of applications. The most widely known example of the utility of the field is its role in deciphering the process of seafloor spreading through the recognition of alternately normally and reversely magnetized rock stripes on the ocean floor (see *Magnetic anomalies, marine*; *Magnetic anomalies for geology and resources*). In continental regions, the use of crustal magnetic field is well-known to Earth scientists because the short-wavelength part of the field closely correlates with the near-surface geologic variations. Crustal magnetic field is thus a proxy for crustal geology and is useful in understanding the structure and composition of the crust as well as its evolution especially when surficial sediments and other geologic formations obscure subsurface geology. Readers interested in the details of the methods on processing, analysis, and interpretation of the entire spectrum of crustal magnetic fields should consult the books by Blakely (1995) and Langel and Hinze (1998).

Magnetization is a complex phenomenon and for rocks it is dependent on a variety of conditions the rocks are exposed to—beginning with their formation, through their tectonic evolution, and exposure to their surroundings and the ambient conditions. A part of the magnetization of the rocks has the memory of these past conditions (based on the strength and direction of the past magnetic field at the time when the rocks are formed or chemically altered, its environmental history/metamorphism, etc.): this memory is retained as components known as "remanent magnetization" (see *Magnetization, remanent* and *Magnetization, natural remanent (NRM)*). The present conditions (magnitude and direction of the surrounding ambient magnetic field, temperature, and mineral chemistry) to which magnetizable minerals in the rocks are exposed are reflected in their "induced magnetization." A vector sum of the induced magnetization and all retained remanent components leads to "total magnetization" of rocks, which causes the magnetic field in the neighborhood of the rocks. This field attenuates with increasing distance from the rocks, adds to the fields originating from the surrounding rocks, and "perturbs" the fields at the observation location caused by the planetary dynamo, electrical currents in the ionosphere and the magnetosphere (the external fields), and induction due to conductivity structure of the planetary interior (see *Core, electrical conductivity* and *Mantle, electrical conductivity, mineralogy*). The perturbing field is traditionally called the "anomalous" field and this is the crustal magnetic field.

The field of the planetary dynamo and the external fields vary spatially as well as in time (see *Geomagnetic secular variation* and *Geomagnetic spectrum, spatial*), and models estimating them at the location and time of observations are required to extract the crustal magnetic field from the raw magnetic field observations. Because the conductivity structure of the Earth is not well-known, its contribution is customarily not explicitly removed when deriving the crustal magnetic field, but efforts are underway to estimate and remove the non-magnetization-related effects from the observed magnetic field in areas where the conductivity variation is large and the geometry is known (e.g., ocean-continent conductivity contrast; see *Coast effect of induced currents*). The typical model describing the core-field dynamo is called the *International Geomagnetic Reference Field (IGRF)* (*q.v.*). However, this does not include the external fields and, therefore, stationary magnetometer basestations tracking the time variations of the external field are employed in a local region of surveys (see *Aeromagnetic surveying*); these time-varying external fields are then removed from the survey observations to construct the magnetic anomaly field. To reduce the effects of imperfect and incomplete reference models and any remaining inconsistencies in the derived anomaly field, it is typical to perform surveys with tracks crossing each other (also called tie-lines) so that the anomaly mismatches at the cross-over locations could be minimized subject to assumptions regarding the nature of the spatiotemporal variations of the external field and the induction effects in the region. Deployment of base stations is not always feasible where surveys are made along tracks longer than a few hundred kilometers (typically over the oceans, but also for continental surveys designed for the determination of the long-wavelength anomaly field). A recently developed specialized model, known as the comprehensive model (CM) (Sabaka *et al.*, 2002), is useful to estimate and remove these noncrustal magnetic field components from such regional survey data. The CM describes the core field and its variation, the long-wavelength variation of the external fields during the magnetically quiet periods, and the first-order induced fields in the Earth using the permanent observatory records beginning 1960s, augmented by fields measured by satellites to fill the large spatial gaps in the magnetic observatory coverage in many parts of the world. Since the CM is a temporally continuous model in comparison to the piece-wise continuous IGRF, it is also effective in minimizing survey base-level mismatches noticeable when neighboring surveys are combined (see examples in Ravat *et al.*, 2003).

On Earth, in the near-surface environment, crustal field generally has magnitude range of a few hundred nanoteslas (nT, an SI unit of magnetic field strength) — much smaller compared to the core-generated magnetic field of 24000–65000 nT (see *Main field maps*). However, over magnetically rich iron formations the magnitude of the crustal field above the Earth's surface can exceed 10000 nT regionally, for wavelengths of a few hundreds of kilometers (e.g., Kursk iron formations in Ukraine). Locally, the magnitude of the anomalies can be stupendously large: for example, ground magnetic surveys over parts of Bjerkreim-Sokndal layered intrusion in Rogaland, Norway, where hemo-ilmenitic rocks having remanent magnetization up to 74 A/m have been measured, a nearly 30000 nT anomaly trough spans the distance of only 500 m (McEnroe *et al.*, 2004). In such places, even compass needles are rendered useless in their usual magnetic north-pointing purpose because they are attracted by the magnetism of the neighboring rocks rather than the Earth's core-generated magnetic field. The sensitivity of commonly available field magnetometers is better than 0.1 nT (see *Aeromagnetic surveying*) and the relative precision of the measurement (which depends partly also on the relative precision of spatial coordinates and the time of measurements) nearly approaches this number in carefully executed surveys. Thus, it is possible to resolve reliably very small spatial changes in magnetic field especially over elongated geologic sources (e.g., dikes or faults). Thus, a magnetic map consists of patterns associated with geology and can be interpreted with fractal magnetization and pattern recognition/characterization concepts (e.g., Pilkington and Todoeschuck, 1993, in Blakely, 1995; Maus and Dimri, 1994; Keating, 1995); this is discussed further in the interpretation methods section.

Examples of the utility of the field

How the crustal magnetic field is used by Earth scientists and the superb subsurface detail reflected in it are best conveyed using simple visual correlations of high-resolution magnetic and geologic maps. Figure C46/Plate 7c shows such a map from the poorly exposed Archean age province of western Australian cratons under the Precambrian and recent cover rocks. In the figure, the magnetic map is superimposed partly by the geological map prepared by Geological Survey of Western Australia and Geoscience Australia from surface outcrops and mining information. Considerable gold workings have been developed in the greenstone formations that form the linear magnetic highs in the shape of an ovoid in the center of the map (this structure is common when domal uplifts have been partly eroded). Aeromagnetics in this case shows clearly the lateral extent of these greenstone formations under the sedimentary cover. In addition, numerous fractures, faults, dykes, as well as textures and patterns that reveal the clues to the origin and evolutionary history of the Archean craton are quite evident in this image. Generally, the correlation of magnetic anomalies with geology requires an intervening step called the *reduction to pole* (*q.v.*) because the anomalies are skewed with respect to their sources and have differently displaced positive and negative lobes depending on the magnetic latitude and the direction of remanent magnetization. Another way of avoiding the skewness issue when comparing with geology is to map the field into magnetic property variation (susceptibility and/or magnetization) as done in the next example.

The second example illustrates the utility of the other end of the crustal magnetic anomaly wavelength spectrum. The greatest advantage of looking at the long-wavelength portion of the crustal magnetic field is that it is uniformly collected by satellites and to a large degree processed similarly over the entire world. One of the most important aspects of the interpretation of the satellite-altitude magnetic field, overlooked by most geologists and geophysicists, leading to its misinterpretation, is that all wavelengths shorter than 500 km are reduced below 1 nT level at the altitude of measurement (about 400 km) and so this field contains virtually no near-surface geologic information unless the near-surface geology itself happens to be a reflection of the structure and composition of the deep crust; therefore, the correlations of these long-wavelength anomalies with geology should not be normally sought. On the other hand, this long-wavelength magnetic field contains an integrated effect of the entire magnetic portion of the lithosphere whereas most geologic/geophysical observations (including near-surface magnetic surveys) are collected over limited areas, processed differently with different methods and then compiled into a regional database—a process that inherently introduces a long-wavelength corruption of the database. Moreover, near-surface geologic samples usually only have limited amount of information about the deep crust. In addition, the present amount and distribution of many data sets that can probe the deep crust (e.g., seismic, electromagnetic, etc.) is not adequate in their spatial resolution to differentiate evolutionary domains of the continents. And finally, all physical properties do not change similarly from one region to the next in the bulk sense and therefore each property has something unique to contribute to our knowledge of the Earth's crust. The main point of this second example is that the bulk magnetization variation is sensitive to a number of changes that occur from one geologic domain to the next and as a result, it is able to delineate some major crustal formation provinces in the United States. This type of knowledge is only possible through a compilation of extensive and time-consuming geochemical analyses of samples throughout the continents. Figure C47a (Plate 6d) shows geologic provinces based on the latest igneous activity in the United States (Van Schmus *et al.*, 1996). In the western portion (for example, the western margin of the Inner Accretionary Belt in Figure C47a and Plate 6d), the geochemical age boundaries well-document the nature in which the continent accreted during the Middle Proterozoic times (1.9–1.6 Ga); however, in the eastern part, later igneous activity of the eastern and southern Granite-Rhyolite provinces (EGR (1470 ± 30 Ma) and SGR (1370 ± 30 Ma)) obscure the fundamental accretionary boundaries. It is only through the analyses of Nd isotope data, the crustal formation age boundary could be defined (shown by a dashed line Nd in Figure C47a and Plate 6d; Van Schmus *et al.*, 1996) and shown by a white dashed line in Figure C47b and Plate 6d). The magnetization variation plotted in Figure C47b (Plate 6d) is the integrated effect of magnetic

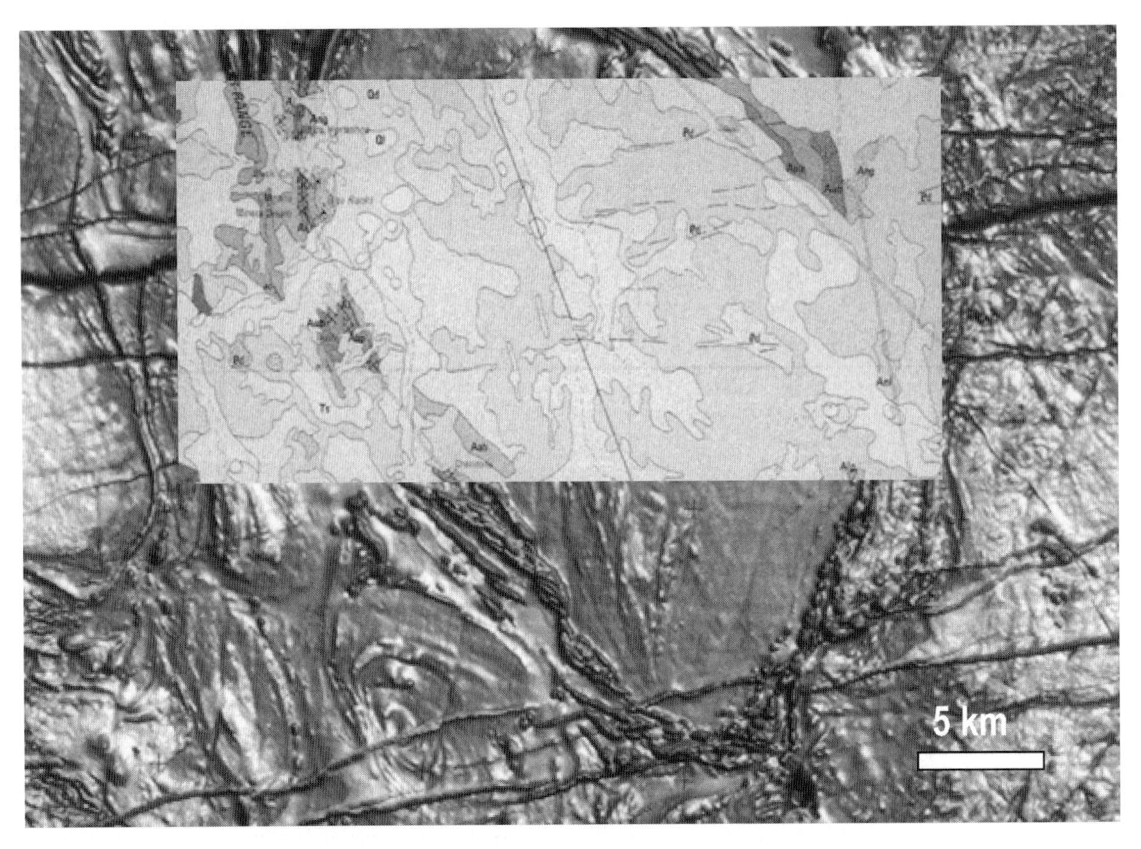

Figure C46/Plate 7c A low-altitude, high-resolution aeromagnetic map of a part of the Archean craton in Western Australia depicting the power of remotely sensed magnetic anomalies in peeling off the sedimentary cover. Pixel size is 25 m which makes it possible to observe subtle variations in the textures of the buried craton. The aeromagnetic image is courtesy of Fugro Airborne Surveys Limited. The example is courtesy of Colin Reeves.

property over the thickness of the magnetic layer; it is based on CHAMP satellite's MF3 crustal magnetic field model (Maus *et al.*, 2004), which is much higher resolution data than its predecessors. The depth integrated magnetic property variation is clearly able to delineate the Nd boundary and reasonably well enclose the Middle Proterozoic province to the west where it is not affected by later tectonic activity reducing the thickness of the magnetic layer (e.g., Rio Grande Rift (R) and Basin and Range Province (BR)). Since the depth integrated magnetic property map is available for the whole world, it could be used to understand better the evolution of the less understood parts of Africa, Antarctica, Asia, Australia, and South America.

In general, the depth integrated magnetic property variation is a good reflection of large changes in magnetic properties of the bulk crust in the continents in locations where the crustal thickness does not vary significantly and also of places where the magnetic crust is thinner compared to the surrounding regions either because the heat flow is high (as in the Rio Grande Rift) or because much of the crust involves nonmagnetic rocks (south of Oklahoma-Alabama transform (T in Figure C47b/Plate 6d) which was the Precambrian continental boundary) and sometimes because of the effect of regional remanent magnetization.

The interpretation methods

The main purpose of the analysis of the crustal magnetic field is to gain knowledge of the crustal geology and its geometry, tectonics, and the evolution of the crust through time. A number of methods have been developed over the past two centuries that allow geophysicists to work toward this goal. The methods fall into two broad categories: qualitative and quantitative methods. Qualitative methods use patterns, magnitude, and wavenumber characteristics of the variation of the magnetic field to identify, group, and contrast adjacent regions into geologic domains, subdomains, and cross-cutting features. For example, referring to Figure C46/Plate 7c, the central region has a significantly different magnetic character than the territories on the west and the east and they can be identified as separate geologic domains on the basis of magnetic pattern. Magnetic highs associated with greenstone belts have significantly different signature than the surrounding magnetic highs, reflecting the structural and compositional variations of the craton. One can also identify, due to their linearity (and high wavenumber nature), the dikes and faults that cut across all the geologic domains and the ones that terminate at the edge of the domains, indicating relative timing of the geologic events. Differences in wavenumber content usually also imply differences in depths to magnetic sources—higher wave numbers nearly always arising from narrow sources reaching closer to the surface. Qualitative correlations among magnetic and other geologic/geophysical information is also immensely useful in the process of interpretation. Many qualitative interpretation tools have their foundation in potential field theory and mathematical computations of various quantities (e.g., regional/residual separation, upward/downward continuation and other filtering techniques), but they are analyzed and interpreted subjectively.

Quantitative methods of analyses can also be subdivided into various categories, but generally the purpose is to generate a physical model resembling geology either through iterative matching of magnetic anomalies from hypothesized magnetic blocks (called "forward modeling" when interpreters judge the changes to be made and "inverse modeling" when matrix inversions and numerical schemes are formulated to judge the adequacy and nature of the changes) (Blakely, 1995) or by coming up with particular characteristics of the

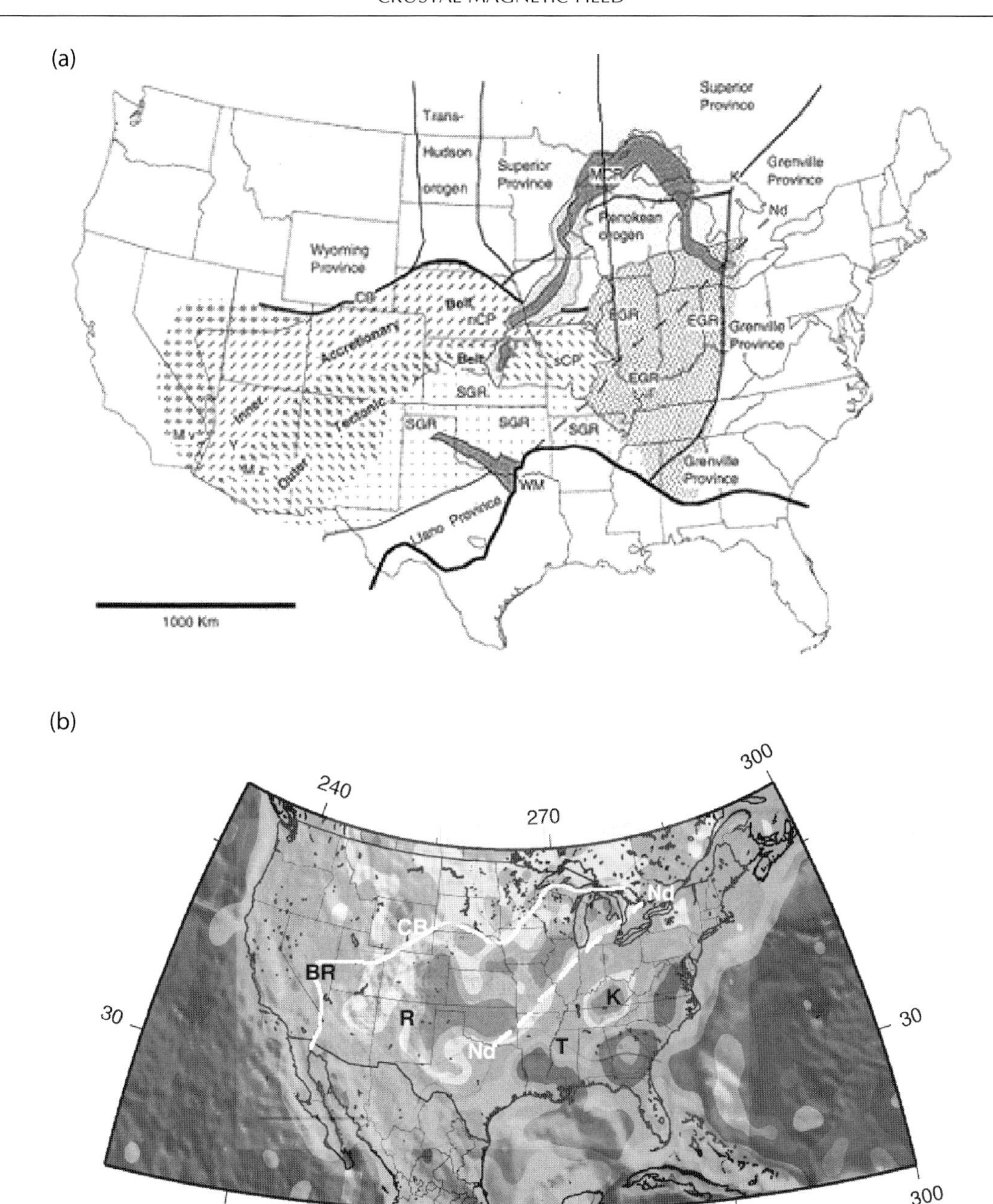

Figure C47/Plate 6d a, Middle Proterozoic (1.9–1.6 Ga) crustal provinces in the United States: bounded on the north and west by Mojave Province (Mv), Cheyenne Belt (CB), and continuing up to the Killarney region (k in the northeast) of Ontario. SGR and EGR are southern Granite-Rhyolite and eastern Granite-Rhyolite provinces. *Dashed line* (Nd) represents inferred southeastern limit as defined by Nd isotopic data which bounds the Middle Proterozoic crustal provinces on its southeastern margin and divides SGR and EGR provinces (from Van Schmus *et al.*, 1996). b, Depth integrated magnetic contrasts over the United States. *Orange/reds* are the relative highs in the susceptibility variation and *blue/light greens* depict relative susceptibility lows. *White continuous* and *dashed lines* (Nd) show the Middle Proterozoic provinces in the United States. Mv from part "a" is excluded from the Middle Proterozoic region because presently this region is governed by the high heat flow related to the Basin and Range province (BR), which has decreased the magnetic thickness of the crust and hence it does not appear as a magnetic high. R, Rio Grande Rift; CB, Cheyenne Belt; K, Kentucky magnetic anomaly; T, Oklahoma-Alabama transform.

nature of the sources (e.g., depths to the top, center or bottom (see *Euler deconvolution*)) through analysis of widths, slopes, and amplitudes of isolated anomalies and their derivative fields from different concentrated geometries (e.g., Analytic Signal method, *Euler deconvolution* (*q.v.*) method in Blakely, 1995; Salem and Ravat, 2004), through the analysis of spectral properties of the field (the Spector and Grant method, the wavenumber domain centroid method of Bhattacharyya and Leu in Blakely, 1995; Fedi *et al.*, 1997) through anomaly attenuation rates and shape factors (see references in Salem *et al.*, 2005), bounds on physical property contrasts (ideal body theory of Parker, 1974, 1975 in Blakely, 1995). Unconstrained modeling is usually not useful in the interpretation of magnetic anomalies because of nonuniqueness of potential fields (which theoretically allows infinite number of different configurations to reproduce the same anomaly) and the only recourse is to employ constraints derived from corollary information (e.g., geologic mapping, borehole information, depths from seismic surveys, etc.) that reduce the ambiguity and lead to meaningful solutions for the particular geologic situation. Nonetheless, when all these aspects of interpretation are carefully applied and limitations assessed, magnetic field provides one of the most valuable tools for insights into the geology and geophysics of the crust.

Dhananjay Ravat

Bibliography

Blakely, R.J., 1995. *Potential Theory in Gravity and Magnetic Applications.* Cambridge: Cambrige University Press.

Keating, P., 1995. A simple technique to identify magnetic anomalies due to kimberlite pipes. *Exploration and Mining Geology*, **4**: 121–125.

Langel, R.A., Hinze, W.J., 1998. *The Magnetic Field of the Earth's Lithosphere: The Satellite Perspective.* Cambridge: Cambridge University Press.

Maus, S., Dimri, V.P., 1994. Scaling properties of potential field due to scaling sources. *Geophysical Research Letters*, **21**: 891–894.

Maus, S., Rother, M., Hemant, K., Lühr, H., Kuvshinov, A., Olsen, N., 2004. Earth's crustal magnetic field determined to spherical harmonic degree 90 from CHAMP satellite measurements. Unpublished manuscript.

McEnroe, S.A., Brown, L.L., Robinson, P., 2004. Earth analog for Martian magnetic anomalies: remanence properties of hemo-ilmenite norites in the Bjerkreim-Sokndal intrusion, Rogaland, Norway. *Journal of Applied Geophysics*, **56**: 195–212.

Ravat, D., Hildenbrand, T.G., Roest, W., 2003. New way of processing near-surface magnetic data: The utility of the comprehensive model of the magnetic field. *The Leading Edge*, **22**: 784–785.

Sabaka, T.J., Olsen, N., Langel, R.A., 2002. A comprehensive model of the quiet-time, near-Earth magnetic field: phase 3. *Geophysical Journal International*, **151**: 32–68.

Salem, A., Ravat, D., 2004. A combined analytic signal and Euler method (AN-EUL) for automatic interpretation of magnetic data. *Geophysics*, **68**: 1952–1961.

Salem, A., Ravat, D., Smith, R., Ushijima, K., 2005. Interpretation of magnetic data using an Enhanced Local Wavenumber (ELW) method. *Geophysics*, **70**: L7–L12.

Van Schmus, W.R, Bickford, M.E., Turek A., 1996. Proterozoic geology of the east-central Midcontinent basement. *Geological Society of America Special Paper*, **308**: 7–32.

Cross-references

Aeromagnetic Surveying
Coast Effect of Induced Currents
Core, Electrical Conductivity
Euler Deconvolution
Geomagnetic Secular Variation
Geomagnetic Spectrum, Spatial
IGRF, International Geomagnetic Reference Field
Magnetic Anomalies for Geology and Resources
Magnetic Anomalies, Marine
Magnetization, Natural Remanent (NRM)
Main Field Maps
Mantle, Electrical Conductivity, Mineralogy
Reduction to Pole
Upward and Downward Continuation

D

D″ AND F-LAYERS

D″ is the rather obscure name for the lowermost mantle adjacent to the liquid core. The encyclopedia contains several articles on the properties of D″ because of its potential influence on the geodynamo. F is the name for the boundary layer at the bottom of the outer core, where liquid core material freezes and differentiates. K.E. Bullen, in 1940–1942, defined the layers of the Earth on the basis of extensive seismological studies linked to probable compositional and mineralogical models. He labelled the layers A to G from the top down, with A the crust, B the upper mantle, C the transition zone, D the lower mantle, E the liquid core, F the layer around the inner core, and G the solid inner core itself (see *Earth structure, major divisions*).

The F-layer was originally proposed to explain a transition region at the bottom of the outer core, introduced by H. Jeffreys in 1939, in which the P-wave velocity decreased with depth. A similar structure was proposed by Gutenberg (1959). Bullen's F-layer was 150 km thick with P-velocity decreasing by about 10% (Bullen and Bolt (1985), p. 317; Lay and Wallace (1995), p. 305). The evidence for Jeffrey's transition zone and the dramatic change in seismic velocity were removed when seismic arrays made possible the measurement of slowness (arrival direction): energy thought to be refracted in the F-layer around the inner core was in fact being scattered from anomalous structure in the lowermost mantle (Doornbos and Husebye, 1972; Haddon and Cleary, 1974).

Region D was therefore separated into D″, the lowest few hundred kilometres of the lower mantle, and D′, the rest of the lower mantle. Region F was abandoned for lack of seismological evidence, but the name remained in use in geomagnetism to denote the boundary layer around the inner core, Braginsky (1963) having already proposed a theory of *energy sources for the geodynamo* (*q.v.*) based on freezing and differentiating of the liquid close to the inner core. Ironically, the only parts of Bullen's original nomenclature in use today is D″, not part of the original list, and F, which no longer exists.

The properties of D″ are described in the next four articles. It is thought to be a thermo-chemical boundary layer with large variations in both temperature and composition that may influence the geodynamo (see *Core-mantle coupling; Electromagnetic; Thermal; and Topographic*). Freezing at the inner core boundary is now thought to result in a thin mushy zone at the top of the inner core (see *Chemical convection*).

David Gubbins

Bibliography

Braginsky, S.I., 1963. Structure of the F layer and reasons for convection in the Earth's core. *Dokl. Akad. Nauk. SSSR English Translation*, **149**: 1311–1314.

Bullen, K.E., and Bolt, B.A., 1985. *An Introduction to the Theory of Seismology.* Cambridge: Cambridge University Press.

Doornbos, D.J., and Husebye, E.S., 1972. Array analysis of *PKP* phases and their precursors. *Physics of the Earth and Planetary Interiors*, **5**: 387–399.

Gutenberg, B., 1959. *Physics of the Earth's Interior.* New York: Academic Press.

Haddon, R.A.W., and Cleary, J.R., 1974. Evidence for scattering of seismic *PKP* waves near the mantle-core boundary. *Physics of the Earth and Planetary Interiors*, **8**: 211–234.

Lay, T., and Wallace, T.C., 1995. *Modern Global Seismology.* New York: Academic Press.

Cross-references

Chemical Convection
Core-Mantle Coupling, Electromagnetic
Core-Mantle Coupling, Thermal
Core-Mantle Coupling, Topographic
Earth Structure, Major Divisions
Energy Sources for the Geodynamo

D″ AS A BOUNDARY LAYER

A *boundary layer* is a thin region (usually situated near a boundary) in a fluid or deformable body that has unusual structural properties. In self-gravitating bodies such as Earth, these regions are characterized by large gradients of temperature or composition or both. The lithosphere is a well-known example of a boundary layer; it is characterized by both a large gradient of temperature and by long-lived chemical heterogeneities (including the continents). The D″ layer near the base of Earth's mantle is believed to have properties complementary to those of the lithosphere, including a large temperature gradient and chemical variations. The focus of this article is on the causes and consequences of its thermal structure; its compositional structure is the subject of a separate entry (see *D″, composition*).

D″ as a thermal boundary layer

Originally the D″ layer was identified as a distinct region immediately above the *Core-mantle boundary* (CMB) by its seismic properties (see *D″, seismic properties*), and these remain our principal source of information on the structure of this region of Earth's interior (Loper and Lay, 1995; see *Earth structure, major divisions*). Its character as a thermal boundary layer follows from the requirement that the core be cooling at a sufficient rate (see *Core-mantle boundary, heat flow across*) so that the outer core is in a state of convective motions sufficiently vigorous to drive the *geodynamo*. In the bulk of the mantle, the upward transfer of heat associated with core cooling is accomplished by convection: upward movement of hot material and downward movement of cold. Since vertical motions are inhibited near the base of the mantle by the strong density contrast between the mantle and the core, vertical transport of heat from the core must be accomplished primarily by thermal conduction in the lowermost part of the mantle. Large temperature gradients are required to conduct the requisite heat (see *Mantle, thermal conductivity*); this is the essence of a thermal boundary layer.

The average temperature contrast across the D″ layer can be estimated by two complementary methods. The first involves extrapolation of the temperatures across the mantle and the core. These extrapolations start from known fixed points and assume that adiabatic temperature gradients exist in the bulk of the mantle and the core. The fixed point in the mantle is the phase transition at 410 km depth, while that in the core is the inner core boundary where the outer core material is at its freezing point. The greatest uncertainty in this calculation is the amount that the freezing temperature is depressed due to impurities in the outer core. Estimates of the temperature contrast obtained from extrapolation range from 200 to 1600°C.

The second method for estimating the temperature contrast across the D″ layer is to use a crude approximation of the equation of heat conduction:

$$q = kA(\Delta T)/(\Delta z)$$

where q is the total heat transferred across the CMB, k is the thermal conductivity, A is the area of the CMB, ΔT is the temperature contrast across the D″ layer, and Δz is its thickness. Knowing the other factors, this may be solved for ΔT. The most uncertain factor in this formula is Δz.

The heat that is transferred from the core has two effects on the material comprising the D″ layer: it makes the material less dense, leading to convective motions which transfer the heat upward through the mantle, and it reduces the viscosity. The thermal and dynamical structures of the D″ layer are governed by the equations of conservation of mass, momentum, and energy, with the latter two being strongly coupled through the dependence of viscosity on temperature. A simple and plausible pattern of convective motion consists of a very slow descent of cold mantle material toward the CMB nearly everywhere and horizontal motion very close to the CMB toward a small number of discrete sites, from which warm mobile material rises into the mantle, forming thermal plumes (Stacey and Loper, 1983). Thermal diffusive thickening of the D″ layer is balanced by the downward motion, leading to a layer of (nearly) steady thickness. Due to the strong variation of viscosity with temperature, the D″ layer has a double structure. Most of the temperature contrast occurs across the thicker portion, in which motions are predominantly downward. Lateral motion is confined to a thinner layer close to the CMB, which is hottest and hence has lowest viscosity.

Lateral (geographic) variations in the temperature of the descending material causes lateral variations in the thermal structure of the D″ layer and consequently lateral variations in the rate of heat transfer from core to mantle (see *Inhomogeneous boundary conditions and the dynamo*). By means of these variations the mantle provides some control on the structure of convection in the core, and of the Earth's magnetic field. In particular, variations in the thermal structure of the D″ layer on the timescale typical of mantle convection (10^8 years) are the most plausible cause of the observed changes in the frequency of reversals on that timescale.

David Loper

Bibliography

Loper, D.E., and Lay, T., 1995. The core-mantle boundary region. *Journal of Geophysical Research*, **100**: 6397–6420.

Stacey, F.D., and Loper, D.E., 1983. The thermal-boundary-layer interpretation of D″ and its role as a plume source. *Physics of the Earth and Planetary Interiors*, **33**: 45–55.

Cross-references

Core-Mantle Boundary, Heat Flow Across
D″, Composition
D″, Seismic Properties
Earth Structure, Major Divisions
Inhomogeneous Boundary Conditions and the Dynamo
Mantle, Thermal Conductivity

D″, ANISOTROPY

Dynamic processes within the Earth can align minerals and inclusions. The resulting rock fabric leads to seismic anisotropy, which means that seismic velocity at a position varies as a function of the direction of wave propagation. In contrast, isotropy refers to the case where velocity is not directionally dependent. For reference, seismic heterogeneity (or inhomogeneity) refers to variation in velocity with position. Conventional seismic imaging techniques (e.g., travel-time tomography) invert data for the isotropic velocity heterogeneity of the mantle, and as such offer insights into thermal and chemical structures. There are though a growing number of methods for estimating seismic anisotropy. As anisotropy results from deformation processes, such analyses offer added insight into the dynamical nature of the Earth. Although we are still in the early stages of mapping anisotropy, it appears that it is most pronounced in the boundary layers of the Earth, regions where deformation is greatest. One such boundary layer is the D″ region (see *D″ as a boundary layer*; *Core-mantle boundary*).

The lowermost mantle boundary layer, referred to as the D″ region or layer, marks the boundary region between the solid silicate (rocky) mantle and the liquid iron core. This region, which extends to a few 100 km above the Core-mantle boundary (CMB), is the site of the most dramatic variations in velocity structure in the lower mantle (see *D″, seismic properties*). Proposed explanations for the anomalous properties of D″ include the effects of a phase change in the region, iron infiltration from the core, primordial material from early mantle differentiation, and subducted material pooling at the CMB. Numerous arguments suggest that mantle plumes may originate in D″. The thermochemical nature of this region will influence thermal, gravitational, topographic, and electromagnetic core-mantle coupling, which in turn influence the dynamo, changes in the length of the Earth's day, and inner core development (see *Core-mantle coupling*; *Core-mantle boundary, heat flow across*). Hence, studies of D″ anisotropy offer insights into the region's influence on large-scale Earth processes.

There are three primary difficulties in studying D″ anisotropy. The first concerns the complexity of seismic wave propagation in anisotropic media. The second involves removing the competing effects of anisotropy from other regions of the mantle. The third lies in the interpretation of anisotropy mechanisms.

Wave propagation in anisotropic media

P- and S-wavefronts emanating from a point source in an isotropic homogeneous medium are spherical. The ray direction (direction of energy transport) is normal to the wavefront, the particle motion of the P-wave is also normal to the wavefront, and the particle motion of the S-wave is tangential to the wavefront and normal to the ray (transverse). The elasticity of an isotropic medium is described by two independent parameters (e.g., the Lamé parameters).

In contrast the elasticity of an anisotropic medium is parameterized by 21 independent elastic parameters. Three independent waves propagate and in a homogeneous anisotropic medium; their wavefronts are not spherical. The ray direction is in general not perpendicular to the wavefront and the particle motion of the P-wave and two S-waves are no longer normal and tangential to the wavefront, respectively. The most unambiguous indicator of anisotropy is the observation of two independent and orthogonally polarized shear waves. As a shear wave in an isotropic medium impinges on an anisotropic region it will split into two shear waves. The two shear waves will continue to propagate separately upon leaving the anisotropic region, retaining a "memory" of the anisotropy. Such splitting is described by the time separation between the two shear waves, δt, and the polarization of the fast shear wave, ϕ (Silver, 1996; Savage, 1999). Such analysis has been used extensively to map anisotropy in the uppermost mantle. Should these two waves encounter another anisotropic medium, they will again split, spawning four independent shear waves. Consequently wave propagation in heterogeneous anisotropic media becomes complicated and it can be difficult to isolate regions of anisotropy.

The number of independent elastic parameters required to describe wave propagation in anisotropic media is further reduced by the symmetry of the medium. For example, a single crystal of olivine is orthorhombic in symmetry and its elasticity is described by nine elastic parameters. A commonly assumed style of anisotropy is transverse isotropy, which is a form of hexagonal symmetry and is described by five elastic parameters. Such a medium with a vertical axis of symmetry is commonly referred to as having vertical transverse isotropy (VTI) symmetry or radial anisotropy. A characteristic of wave propagation in such media is that the fast and slow shear waves are polarized in either the horizontal direction (SH) or the vertical plane (SV). Mechanisms for such anisotropy include the periodic horizontal layering of contrasting materials or the alignment of crystals in a horizontal glide plane with no fixed slip direction.

Evidence for anisotropy in D″

A range of indicators suggest that the lowermost mantle is anisotropic. The decay of phases diffracted along the core-mantle boundary show dramatic regional variations and Doornbos *et al.* (1986) first sought to explain this in terms of VTI anisotropy in the D″ region (see reviews in Lay *et al.*, 1998 and Kendall and Silver, 1998).

D″ anisotropy has been more clearly documented through observations of splitting in shear wave phases that travel hundreds of kilometers through this region (S, ScS and Sdiff; Figure D1) (see *Seismic phases*). The magnitude of the splitting, δt, varies considerably (0 to >7 s) with both region and path length through D″ (Kendall and Silver, 1998; Lay *et al.* 1998). Although this is the site of the largest shear wave splitting magnitudes in the Earth, the splitting is accrued over long path lengths and estimates of anisotropy in the region are 0–3% ($(v_{\text{fast}}-v_{\text{slow}})/v_{\text{ave}}$, where v refers to shear wave velocity). To date, analysis has primarily concentrated on measuring the separation between shear wave arrivals on the transverse and radial components of a three-component seismometer. In other words, the polarization of the fast shear wave, ϕ, is either in the vertical plane or the horizontal direction (SV vs. SH). Such analysis implicitly assumes VTI anisotropy as more general forms of anisotropy would yield fast and slow shear wave energy on both the radial and transverse components. Recently, Panning and Romanowicz (2004) have inverted shear wave data to produce a global map of D″ VTI anisotropy (Figure D2/Plate 11c). Future directions involve developing techniques to invert more general forms of anisotropy (i.e., nonVTI).

A difficultly in measuring D″ anisotropy is ruling out the effects of anisotropy in shallower parts of the mantle. For example, estimating splitting in SKS phases is a now standard technique for studying upper mantle anisotropy (see Silver, 1996; Savage, 1999). Any interpretation of D″ anisotropy must be correct for near-source and near-receiver anisotropy (see Kendall and Silver, 1998). Analysis of SKS splitting at permanent stations provides a correction for near-receiver anisotropy, although care must be taken as this may be directionally dependent. Using very deep earthquakes can mitigate near-source anisotropy, as the transition zone and middle mantle are thought to be largely isotropic (Mainprice *et al.*, 2000). Alternatively, relative changes in shear wave splitting in phases that turn above and below D″ can be used to isolate the anisotropy to the lowermost mantle.

The style of anisotropy in high velocity and hence probably colder regions of D″ is quite consistent. Here shear wave energy on the transverse component leads that on the radial component. Regional studies of D″ beneath the Caribbean, Siberia, Alaska, Australia and the Indian Ocean suggest a roughly VTI anisotropy. Recently, detailed analysis has shown that the anisotropy beneath the Caribbean is not purely VTI: the orientation of the symmetry axis varies up to 20° from the vertical (Garnero *et al.*, 2004). In warmer and more ambient regions the style of D″ anisotropy appears to be more complex and the magnitude more variable. Specifically, the form of anisotropy

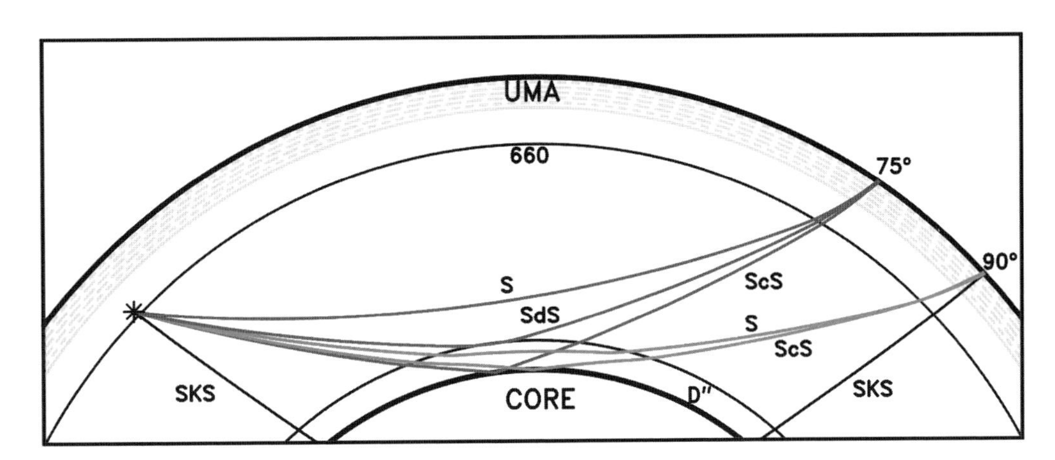

Figure D1 Seismic phases used to analyze D″ anisotropy. UMA refers to upper mantle anisotropy. The SdS phase reflects off a discontinuity that lies at the top of the D″ region. At distances beyond 100°, the phase S and ScS merge and diffract along the CMB (S_{diff}).

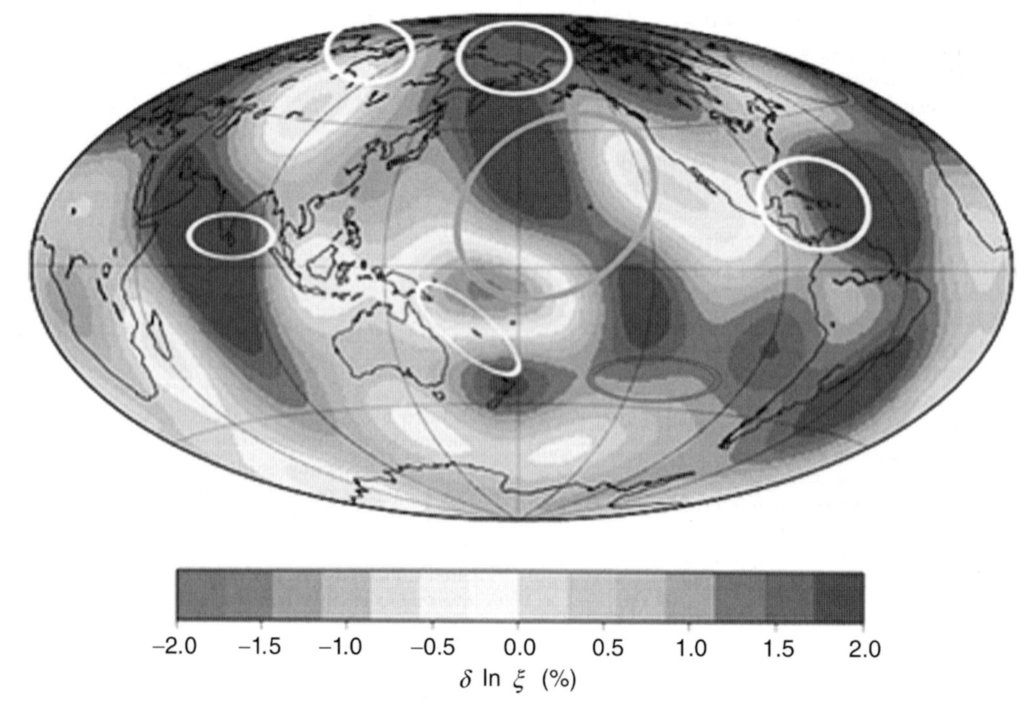

Figure D2/Plate 11c Map of seismic S-wave anisotropy ($\xi = V^2_{SH}/V^2_{SV}$) in D″ based on a global waveform tomography inversion for a 3D model of radial anisotropy (VTI) (from Panning and Romanowicz, 2004). Also shown are areas of detailed regional analysis of D″ anisotropy: white indicates regions wherein D″ horizontally polarized S-waves (SH) are faster than a vertically polarized S-waves (SV); red shows regions where D″ appears to be isotropic; green shows a broad region of D″ beneath the Pacific where the style of anisotropy appears to be complex and quite variable.

beneath the Pacific appears to vary in style and magnitude over relatively short distances. Furthermore, D″ anisotropy is not ubiquitous; regions beneath parts of the Pacific and Atlantic Ocean appear to be isotropic.

Cumulatively, these results suggest that anisotropy correlates with regional variations in other properties of D″ (see *D″, seismic properties*). Although we are still in the early stages of mapping out the magnitude and orientation of such anisotropy, we can loosely group the observations into categories based on regional characteristics. One is associated with regions where slabs are predicted to subduct into the lower mantle. Such regions are characterised by high D″ shear velocities and a more VTI style of anisotropy. Another category involves sites of inferred mantle upwelling, for example, beneath Hawaii. Such regions are characterised by lower than average shear velocities and a more complex style of anisotropy. Finally, initial results suggest that more ambient regions of D″ appear to be isotropic.

Causes of D″ anisotropy

A range of mechanisms can cause anisotropy, including the preferred alignment of crystals, grains, or inclusions, and the fine-scale layering of contrasting materials. As the mantle convects, rock deformation will be accommodated by mechanisms such as dislocation, diffusion, grain boundary migration, and dynamic recrystallization (Karato and Wu, 1993). Dislocation glide and creep mechanisms are effective in aligning crystals and producing a lattice preferred orientation (LPO). In order to evaluate such effects in the lower mantle we need to know the minerals present, their single-crystal elastic properties, and how they deform. We are still a long way from knowing all of this information, but recent advancements in laboratory experiments and theoretical calculations are offering new insights. Candidate minerals for LPO anisotropy include perovskite, magnesiowustite, columbite (Stixrude, 1998), and the recently discovered post-perovskite phase (Murakami *et al.*, 2004) (see *D″, composition*). In high-velocity regions, post-perovskite and magnesiowustite are likely candidates for D″ anisotropy and perovskite is a likely candidate for D″ anisotropy in low-velocity regions.

Alternatively, the anisotropy may be associated with well-ordered subseismic scale heterogeneity. Oriented inclusions or layering can effectively generate anisotropy if there is sufficient contrast in material properties. For example, aligned melt-filled tabular inclusions generate anisotropy very effectively (Kendall and Silver, 1998).

It seems that the physical process responsible for the anisotropy beneath paleoslab regions is different from that beneath regions of mantle upwelling. The infiltration of core material into the D″ region could lead to inclusions of low-velocity material. Should these inclusions be aligned, they may provide an explanation for the anisotropy beneath the central Pacific. This is a region for which there is evidence for an ultralow velocity zone at the base of the mantle and it has been argued that these may be due to the presence of partial melt (see *ULVZ, ultralow velocity zone*). Alternatively, the preferred orientation of perovskite could explain the anisotropy. With either mechanism the magnitude and orientation of the anisotropy must vary over short length scales.

There are a number of arguments that the anisotropy in high-velocity regions is associated with the accumulation of paleoslab material in D″ (Kendall and Silver, 1998). Regions of anisotropy beneath the Caribbean, Alaska, northern Asia, Australia, and the Indian Ocean correlate with predicted CMB locations of the slabs. The high isotropic shear velocities of this region can be explained by the retained slab thermal anomaly and the high strains associated with the slab material colliding with the rigid CMB can align constituent crystals or inclusions.

In summary, there are clear regional variations in the style of D″ anisotropy that are very likely associated with different physical processes. In a sense, these variations are analogous to those observed between continental and oceanic regions in the upper mantle boundary layer. As our knowledge of D″ mineral physics grows and new seismic arrays provide improved coverage, studies of D″ anisotropy will offer new insights into this important but somewhat enigmatic region.

Michael Kendall

Bibliography

Doornbos, D.J., Spiliopoulos, S., and Stacey, F.D., 1986. Seismological properties of D″ and the structure of a thermal boundary layer. *Physics of the Earth and Planetary Interiors*, **57**: 225–239.

Garnero, E.J., Maupin, V., and Fouch, M.J., 2004. Variable azimuthal anisotropy in the Earth's lowermost mantle. *Science*, **306**(5694): 259–261.

Karato, S. and Wu. P., 1993. Rheology of the upper mantle: A synthesis. *Science*, **260**: 771–778.

Kendall J-M. and Silver, P.G., 1998. Investigating causes of D″ anisotropy. In Gurnis, M., Wysession, M.E., Knittle, E., and Buffett, B. A. (eds.), *The Core-Mantle Boundary Region.* Geodynamic Series, **28**, Washington, D.C.: American Geophysical Union, pp. 97–118.

Lay, T., Williams, Q., Garnero, E.J., Kellogg, L., and Wysession, M.E., 1998. Seismic wave anisotropy in the D″ region and its implications. In Gurnis, M., Wysession, M.E., Knittle, E., and Buffett, B.A. (eds.), *The Core-Mantle Boundary Region.* Geodynamic Series, **28**, Washington, D.C.: American Geophysical Union, pp. 299–318.

Mainprice, D., Barruol, G., and Ben Ismail, W., 2000. The seismic anisotropy of the Earth's mantle: From single crystal to polycrystal. In Karato, S., Forte, A.M., Liebermann, R.C., Masters, T.G., and Stixrude, L. (eds.), *Earth's Deep Interior: Mineral Physics and Tomography from the Atomic to the Global Scale.* Geophysical Monograph, 117, Washington, D.C.: American Geophysical Union, pp. 237–264.

Murakami, M., Hirose, K., Kawamura, K., Sato, N., and Ohishi, Y., 2004. Post-perovskite phase transition in $MgSiO_3$. *Science*, **304**: 855–858.

Panning, M., and Romanowicz, B., 2004. Inferences of flow at the base of the Earth's mantle based on seismic anisotropy. *Science*, **303**: 351–353.

Savage, M., 1999. Seismic anisotropy and mantle deformation: What have we learned from shear-wave splitting? *Reviews in Geophysics*, **37**: 65–106.

Silver, P.G., 1996. Seismic anisotropy beneath continents: Probing the depths of geology. *Annual Review Earth Space Sciences*, **24**: 385–432.

Sitxrude, L., 1998. Elastic constants and anisotropy of $MgSiO_3$ perovskite, periclase, and SiO_2 at high pressures. In Gurnis, M., Wysession, M.E., Knittle, E., and Buffett, B.A., (eds.), *The Core-Mantle Boundary Region.* Geodynamic Series, 28, Washington D.C.: American Geophysical Union, pp. 83–96.

Cross-references

Core-Mantle Boundary
Core-Mantle Boundary, Heat Flow Across
Core-Mantle Coupling, Electromagnetic
Core-Mantle Coupling, Thermal
Core-Mantle Coupling, Topographic
D″ as a Boundary Layer
D″, Composition
D″, Seismic Properties
Seismic Phases
ULVZ, Ultralow Velocity Zone

D″, COMPOSITION

Both the mantle and the core of the Earth are fundamentally affected by the properties of the material at the base of the mantle. This region controls the heat flow between the core and the mantle; it may be the source region for mantle plumes; and it is the lowermost window through which the magnetic field of the Earth must propagate to reach the surface. That the base of the mantle differs seismically from the overlying mantle material was recognized nearly a century ago by Gutenberg (1913). The D″ zone was originally defined as a distinct zone by Bullen (1949) as the layer with anomalously low increases in seismic velocity with depth in the lowermost several hundred of kilometers of the mantle. Since the importance of mantle convection was recognized, the chemical composition of this lowermost ~300 km of Earth's 2900 km-thick mantle has widely been viewed as approximately equivalent to that of the overlying lower mantle, with its seismic properties perturbed by a large increase in temperature distributed over the lowermost few hundred kilometers of the mantle. The superadiabatic gradient associated with the basal boundary layer of the Earth's convecting mantle likely involves a gradual temperature increase of ~1000 K to perhaps as much as 2000 K (if the layer is internally stratified) across the D″ zone. Such shifts in temperature can produce decreased gradients of seismic velocities across this zone, a result that generally agrees with laterally averaged seismic models of Earth's mantle, such as the Preliminary Reference Earth Model (PREM: Dziewonski and Anderson, 1981).

However, improvements in seismic observations have driven a reassessment of the possible roles of chemical differentiation and chemical heterogeneity within D″. The specific seismic observations include: a sporadic seismic discontinuity involving an increase of ~1–3% in shear and perhaps in compressional velocity at heights usually 200–300 km above the core-mantle boundary (CMB); a laterally heterogeneous zone 5–40 km above the CMB with decreases in compressional and shear velocity of the order of 10% and 30%, respectively (the ultralow velocity zone, or ULVZ); two large-scale slow structures beneath Africa and the western Pacific spanning lateral dimensions of ~2000 km and extending to heights of ~1000 km above the CMB, with shear decrements of ~3% or larger; and pervasive and laterally varying anisotropy within the D″ layer (see Gurnis *et al.*, 1998; Lay *et al.*, 1998; and *D″, seismic properties*, and *D″, anisotropy*). The chemical characteristics of D″ have thus emerged as a multidisciplinary problem, incorporating the anomalous seismic characteristics of this layer, the somewhat ill-constrained properties of Earth materials at the extreme conditions of the CMB (pressure of 135 GPa and temperature of 3500–5000 K), and the effects of changes in chemistry on the electromagnetic properties of the deep mantle, and thus on the geomagnetic field.

The seismic observations of structural complexity within D″ have driven a broad range of proposals of possible chemical changes that might be present within this region. These possibilities include: descent of melts from the overlying mantle, enriching this zone in incompatible and volatile elements; interaction of material with the underlying core, producing iron enrichment, and contributions from subducted slab material. Additionally, a different chemistry of this region might have been created early in the history of the planet, through early differentiation processes or perhaps even during planetary accretion. All such proposed changes in chemistry must not dramatically alter the seismic velocity or density in this region from that within the lowermost mantle: seismic travel-time and normal mode constraints show that the bulk of D″ can only differ in velocity and density by a few percent from that within the deep mantle (see *Earth structure, major divisions*). This constraint is, however, relaxed in the lowermost 5–40 km of the mantle, where dramatic changes of seismic wave velocities have been observed.

From a geomagnetic perspective, the importance of the chemistry of the lowermost mantle lies primarily in its effect on the propagation of Earth's magnetic field through the basal layer of the mantle. If the electrical conductivity of the lowermost mantle is markedly elevated from that of the overlying material, through enrichment in iron, the presence of partial melt, or both, then temporal variations of the magnetic field could be attenuated within such a layer (see *Mantle, electrical conductivity, mineralogy*). Moreover, any electromagnetic coupling between the core and mantle would generate a torque on the overlying mantle: such a torque has been associated with shifts in the nutation of the Earth and changes in the length of day (e.g., Buffett, 1992). If high conductivity zones were heterogeneously distributed near the base of

Earth's mantle, then the mantle could exercise a structural control on changes in virtual geomagnetic pole positions during reversals of the polarity of Earth's magnetic field, potentially forcing the reversing pole to follow geographically preferred paths (Aurnou *et al.*, 1996).

Core-mantle chemical interactions

The most straightforward way to determine the composition of D″ would be to analyze samples derived from this region. However, the ability to geochemically sample D″ directly via surface volcanism is controversial. For such sampling to occur, an upwelling would have to transit nearly 3000 km of Earth's mantle and retain a chemical signature that could be unambiguously associated with its basal origin. It is possible that a portion of Earth's hot spots may originate from D″, and as such might be expected to carry a geochemical signature of this region or perhaps even of the underlying core. Some evidence exists that some hot spots and flood basalts carry anomalously high $^{187}Os/^{183}Os$ ratios (e.g., Walker *et al.*, 1995). The radiogenic parent of ^{187}Os is ^{187}Re: Re is a strongly siderophilic (iron-loving) element and is thus anticipated to be strongly enriched in Earth's core. Re enrichment is similarly observed in iron meteorites. In contrast, Os is generally lithophilic, partitioning into silicates. Thus, elevated $^{187}Os/^{183}Os$ ratios are associated with the sampling of Re-rich regions. The most parsimonious explanation for such ratios is that some hot spots and flood basalts may record a geochemical signature of interaction with Earth's core. The amount of chemical interaction necessary to generate the observed ratios is not large, but it nevertheless may provide *prima facie* evidence for chemical interaction between Earth's core and mantle.

How might such core-mantle chemical interactions occur? A direct means for producing such interactions is through chemical reactions of the silicate of the lowermost mantle with the liquid iron-rich alloy of the outer core. Such reactions, which involve oxidation of core-derived iron and reduction of mantle silicon, can be described using chemical formulae such as

$$(Mg_xFe_{1-x})\,SiO_3\,(\text{mantle material}) + 3[(1-x)-s]Fe$$
$$= xMgSiO_3 + sSiO_2 + [(1-x)-s]FeSi$$
$$+ [3(1-x)-2s]FeO.$$

Here, s is the amount of silica produced in the reaction and $(1-x)$ is the amount of iron in the reacting mantle material. Such reactions have been both experimentally verified to occur and are thermodynamically driven to the right by a negative volume change of the reaction (e.g., Knittle and Jeanloz, 1991). The intriguing aspect of these reactions is that the products involve both iron-free silicates, which are both likely to be of low density and electrical conductivity, and relatively high density, high electrical conductivity iron alloys. Accordingly, core-mantle reactions can lead to a suite of reaction products with highly heterogeneous physical properties. The key questions associated with such core-mantle reactions are mechanical in nature: over what length-scale such reactions occur at the CMB, how efficiently reaction products are entrained within flow of the overlying mantle or underlying core (exposing fresh mantle material to reaction), and whether the low-density products are able to separate from the high-density products. In short, whether the mantle develops a thin reacted rind at its base, or whether reacted zones can be dynamically entrained into the overlying mantle, producing an enrichment of iron alloys and replenishment of the reacting zone in the lowermost mantle is ill-constrained. The possibility that topographic effects might be important in enhancing such chemical reactions has been suggested (e.g., Jeanloz and Williams, 1998): any topography on the CMB involves either silicate material intruding into iron or *vice versa* (see *Core-mantle boundary topography*). In such cases, the gravitational effects that drive the separation of the core and the mantle material are locally relaxed, with core fluid prospectively being able to infiltrate the mantle material without traveling against gravity.

Subduction and D″

The role of slab (and particularly oceanic crust) enrichment in D″ has frequently been invoked as a possible cause of geochemical heterogeneity in this region. If ancient oceanic crust is sequestered in D″, then this zone would be enriched in Ca, Al, and Si relative to the overlying mantle. Moreover, volatile elements such as hydrogen and carbon that were retained within the slab during its transit through the mantle might also be more abundant in this zone. A compelling case for the existence of such slab-related enrichment would require several separate lines of evidence: first, seismic evidence that slabs penetrate to CMB depths; second, a sufficiently elevated density of subducted material that it would stably reside for long times within D″; and third, a possible mechanism by which oceanic crust can efficiently separate from its underlying depleted mantle. Each of these lines of evidence is not fully developed: seismic tomography shows that slab-related seismic anomalies thicken substantially as they enter into the lower mantle, but these anomalies do not appear to be easily tracked into the lowermost 500–1000 km of the mantle. As the full slab is comprised of oceanic crust overlying depleted mantle material, the chemistry of the slab as a whole lies close to that of the bulk mantle. Therefore, a means by which oceanic crust can delaminate from its depleted complement and accumulate within D″ is required if the dregs of ancient crustal subduction are a major chemical component of the lowermost mantle. Accordingly, while slabs may generate geochemically distinct regions within D″, the available evidence neither requires nor precludes the presence of ancient slabs within D″.

Partial melting

Partial melting and associated melt segregation are well known means of geochemical differentiation in the shallow regions of the planet. Thus, the likely existence of partial melt in the lowermost regions of Earth's mantle may imply that enrichment in incompatible elements—those elements that partition strongly into silicate liquids relative to solids—occurs at these depths. The idea that partial melt exists in the lowermost mantle is driven by the observation that decrements of ~10% in compressional wave velocity and ~30% in shear velocity exist in heterogeneously distributed zones in the lowermost 5–40 km of the Earth's mantle—the ultralow velocity zones (*ULVZ*, *q.v.*). Such decrements likely require 6–30% partial melt (with the amount depending on how the melt is distributed) entrained in parts of the lowermost mantle (Williams and Garnero, 1996). The stable retention of melt at the base of Earth's mantle (rather than it buoyantly rising to the top of Earth's mantle) is probably produced by two tandem, complementary effects. First, the dramatic increase in temperature within D″ relative to the overlying mantle (see *D″ as a boundary layer*) may produce sufficiently high temperatures that the base of the mantle lies above its solidus—the temperature at which melting initiates in multicomponent systems. Second, melts at pressures corresponding to those at the base of the mantle may be denser than their coexisting solids, due to both structural effects associated with highly densified amorphous structures and the preferential partitioning of iron into silicate liquids. Indeed, the available constraints on the ULVZ indicate that this region has an increase in density relative to the overlying mantle material of about 10% (Rost *et al.*, 2005): a result fully consistent with a marked enrichment in iron, with the additional iron most likely being oxidized. Thus, melts likely stably stratify near the base of the mantle, and any partial melt that occurs in the deep lower mantle may descend into D″, enriching this zone in both volatile elements (H and C), and likely incompatible major elements such as Fe, Ca, and Al. Additionally, radiogenic heat-producing elements such as K and U might be enriched within the CMB region.

However, the melting behavior of the mantle material at the extreme pressures and temperatures of the CMB is not yet well-constrained, so the geochemical consequences of partial melting and melt segregation are largely inferred based on experiments at substantially less extreme conditions. The key point about the geochemical effects of partial melting in D″ is that they are not exclusive of either core-mantle reactions or the presence of slab-derived components within D″. Core-mantle reactions may lower the melting point of mantle material and partial melt could facilitate the transfer of reacted material away from the CMB. Similarly, if the melting temperature of ancient oceanic crust (or, indeed, any portion of the slab) is lower than that of normal mantle material, then slab-derived melts could descend into D″. As such, enrichment of D″ in slab-related geochemical components could be driven by partial melting processes. Lastly, the importance of partial melting in D″ is likely to have been far greater earlier in Earth history. If secular cooling of Earth's mantle has occurred over the age of the Earth, then partial melt is anticipated to have been more abundant in the deep mantle early in Earth history, and today's D″ could (at least partially) represent the solidified residue of partial melt.

Primordial chemical layering

A rather different genesis for a chemically distinct D″ was proposed by Ruff and Anderson (1980). They proposed that the composition of the lowermost mantle might differ markedly from that of the overlying material due to effects associated with Earth's accretion. Such primordial layering might have arisen through early accretion of material enriched in Ca, Al, and Ti. These elements are abundant in the initial condensates of the protoplanetary nebula, and the idea was simply that these elements might have heterogeneously accreted to the center of the protoEarth during the earliest stages of planetary accretion. Subsequently, they might have been displaced to the region surrounding the core due to the gravitational instability of subsequently accreted dense, iron-rich material. The dilemma is that this layer must not only persist through impacts of planetary-sized objects on the early Earth, but also endure without mixing throughout Earth history: both the lack of a clearly global discontinuity at the top of D″ and the generally similar seismic properties of the bulk of D″ to the overlying mantle make such broadscale, profoundly chemically distinct, and primordial layering unlikely. However, if primordial layering does persist within D″, it may be less likely to be associated with accretionary effects, and more likely to be produced by differentiation processes within a primordial magma ocean. Such a magma ocean is widely postulated to have existed early in Earth history due to accretional heating associated with giant impacts on the early Earth. In this sense, "primordial" chemical differentiation of D″ could operationally differ little from the effects of partial melting of Earth's deep mantle over the course of Earth history.

However, such early chemical differentiation may also be affected by the presence within the deep mantle of a recently discovered post-perovskite phase [Murakami *et al.*, 2004]. This ultrahigh pressure phase of $(Mg,Fe)(Si,Al)O_3$ may be the dominant mineral within D″, and its pressure of onset appears to correspond with the depth of the seismic discontinuity often viewed as the top of D″ (see *D″, seismic properties*). Little is known about the chemical behavior or affinities of this $CaIrO_3$-structured phase, beyond the likelihood that the transition to this phase has a positive Clapeyron slope. However, depending on the degree to which D″ has equilibrated with the overlying mantle material, D″ could be enriched in elements that are more compatible within the post-perovskite structure than within the silicate perovskites of Earth's lower mantle. Speculatively, if iron preferentially partitions into this phase, then a net iron enrichment of D″ could arise through progressive equilibration between D″ and the overlying mantle. Such equilibration could be generated through selective transport of material across the phase boundary, with progressive fractionation of compatible (and perhaps heavy) elements into the lowermost mantle.

Quentin Williams

Bibliography

Aurnou, J.M., Buttles, J.L., Neumann, G.A., and Olson, P.L., 1996. Electromagnetic core-mantle coupling and paleomagnetic reversal paths. *Geophysical Research Letters*, **23**: 2705–2708.

Buffett, B.A., 1992. Constraints on magnetic energy and mantle conductivity from the forced nutations of the Earth. *Journal of Geophysical Research*, **97**: 19581–19597.

Bullen, K.E., 1949. Compressibility-pressure hypothesis and the Earth's interior. *Monthly Notices of the Royal Astronomical Society, Advances in Geophysics - Supplement*, **5**: 355–368.

Dziewonski, A.M., and Anderson, D.L., 1981. Preliminary reference Earth model. *Physics of the Earth and Planetary Interiors*, **25**: 297–356.

Gurnis, M., Wysession, M.E., Knittle, E., and Buffett, B.A., (eds.) 1998. The core-mantle boundary region, Washington, D.C.: American Geophysical Union.

Gutenberg, B., 1913. Uber die Konstitution des Erdinnern, erschlossen aus Erdbebenbeobachtungen. *Physikalische Zeitschrift*, **14**: 1217–1218.

Jeanloz, R., and Williams, Q., 1998. The core-mantle boundary region. *Reviews in Mineralogy*, **37**: 241–259.

Knittle, E., and Jeanloz, R., 1991. Earth's core-mantle boundary: Results of experiments at high pressures and temperatures. *Science*, **251**: 1438–1443.

Lay, T., Williams, Q., and Garnero, E.J., 1998. The core-mantle boundary layer and deep Earth dynamics. *Nature*, **392**: 461–468.

Murakami, M., Hirose, K., Kawamura, K., Sato, N., and Ohishi, Y., 2004. Post-perovskite phase transition in $MgSiO_3$. *Science*, **304**: 855–858.

Rost, S., Garnero, E.J., Williams, Q., and Manga, M., 2005. Seismic constraints on a possible plume root at the core-mantle boundary. *Nature*, **435**: 666–669.

Ruff, L., and Anderson, D.L., 1980. Core formation, evolution and convection: A geophysical model. *Physics of the Earth and Planetary Interiors*, **21**: 181–201.

Walker, R.J., Morgan, J.W., and Horan, M.F., 1995. Osmium-187 enrichment in some plumes—Evidence for core-mantle interaction. *Science*, **269**: 819–822.

Williams, Q., and Garnero, E.J., 1996. Seismic evidence for partial melt at the base Earth's mantle. *Science*, **273**: 1528–1530.

Cross-references

Core-Mantle Boundary Topography
D″ as a Boundary Layer
D″, Anisotropy
D″, Seismic Properties
Earth Structure, Major Divisions
Mantle, Electrical Conductivity, Mineralogy

D″, SEISMIC PROPERTIES

Across most of Earth's lower mantle, P- and S-wave seismic velocities increase with depth in a fashion consistent with adiabatic self-compression of uniform composition material, and there is relatively little lateral variation in seismological structure (see *Earth structure, major divisions*). However, the lowermost few hundred kilometers of the mantle, called the D″ region, has been recognized since the 1940s as having anomalous seismic properties, with reduced seismic velocity gradients with depth and strong lateral variations in structure. This prompted designation of D″ as a distinct layer in the early labeling of Earth's shells. Seismological characterization of inhomogeneity of the D″ region has advanced dramatically in recent decades, and the unusual seismic properties of D″ are now commonly interpreted as manifestations of complex thermal and chemical boundary layers at the base of the mantle (see *D″ as a boundary layer*). This region is of importance to the Mantle and the core dynamical systems.

At a depth of 2891 km below the surface, the core-mantle boundary (CMB) separates the two principal chemical layers of the Earth, the molten metallic-alloy outer core and the highly viscous silicate- and oxide-mineral mantle (see *Core-mantle boundary*). The contrast in density between the mantle rock and the core alloy exceeds the density change from air to rock at Earth's surface, and the contrast in viscosity is comparable to that between rock and ocean water. The superadiabatic temperature contrast from the core into the mantle is estimated to be as high as 2000°C, possibly exceeding that across the lithospheric thermal boundary layer, and there is heat flow of 2–10 TW across the boundary (44 TW flows through the Earth's surface) (see *Core-mantle boundary, heat flow across*). Over time, dense materials in the mantle are likely to have concentrated in D″, while light dross, expelled from the core may have accumulated at the CMB (see *D″, composition*). The D″ boundary layer is relatively accessible to seismic imaging, and its seismic properties are discussed here (also see *D″, anisotropy*; *D″ as a boundary layer*; *D″, composition*; *ULVZ, ultralow velocity zone*). Garnero (2000), Lay *et al.* (1998a), and many papers in Gurnis *et al.* (1998) provide reviews of ongoing research topics and conceptual models for processes occurring in D″.

One-dimensional seismic velocity models for D″

One-dimensional global seismic velocity models, such as the Preliminary Reference Earth Model (PREM) of Dziewonski and Anderson (1981), usually have reduced velocity gradients in the deepest 150 km of the mantle, reflecting the global departure of D″ velocity structure from that of the overlying lower mantle. For several decades, efforts have been made to interpret the low (or even negative) velocity gradients in such average models of D″ structure as the result of a superadiabatic temperature increase across a thermal boundary layer; large temperature increases above the CMB can be reconciled with the one-dimensional models (see *D″ as a boundary layer*; *Core-mantle boundary, heat flow across*). However, studies over the past half century have established that there is a very substantial heterogeneity in D″ on a wide variety of scale lengths; thus, there is no meaningful "average" structure for the region useful for a robust interpretation of the boundary layer. It appears more useful to discuss the seismic properties of D″ by emphasizing the lateral variations in structure, as is the case for the surface boundary layer.

Large-scale seismic velocity patterns

Seismic tomography has been applied to develop three-dimensional seismic velocity models for the entire mantle for over 15 years. This method extracts three-dimensional P-wave and S-wave velocity fluctuations relative to one-dimensional reference models by inversion of massive data sets with crossing raypaths for which seismic wave arrival times are measured, exploiting the localized sensitivity of seismic waves to velocity structure around the path on which they propagate through the Earth. These three-dimensional velocity inversions indicate that heterogeneity in the mid-lower mantle is relatively weak, but that in the lowermost 300–500 km of the mantle heterogeneity increases with depth, becoming stronger in D″ than anywhere else in the Earth except for the uppermost mantle where major thermal, chemical, and partially molten heterogeneities are known to exist. The most surprising aspect of the D″ seismic velocity heterogeneity is that it is dominated by large-scale patterns, with large continuous volumes of high- or low-velocity material having scale lengths of hundreds to thousands of kilometers. This predominance of large-scale variations is observed in both S-wave and P-wave velocities, as indicated by the representative models for lateral variations in D″ structure shown in Figure D3a/Plate 11a.

S-wave velocity variations (dVs) relative to the global average velocity in D″ (Figure D3a/Plate 11a.) are positive (relatively high

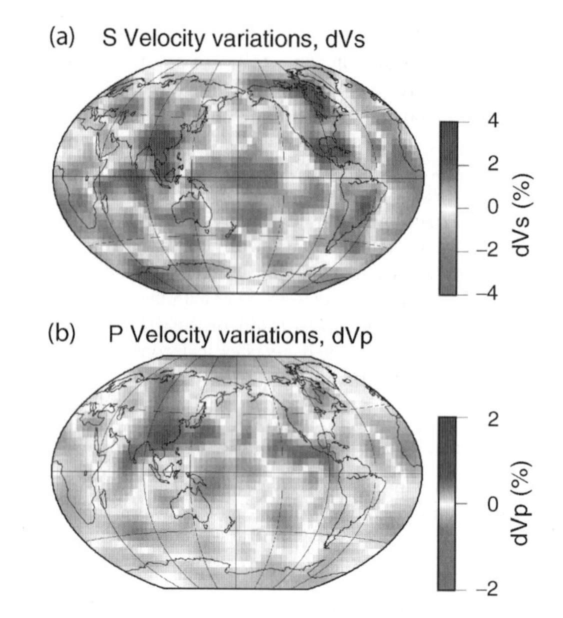

Figure D3/Plate 11a,b Maps of S velocity (a) and P velocity (b) variations relative to the global average values in D″ at a depth near 200 km above the CMB in representative global mantle seismic tomography models. S velocity variations are from Mégnin and Romanowicz (2000) and P velocity variations are from Boschi and Dziewonski (2000). Positive values (blue) indicate higher than average velocities and negative values (red) indicate lower than average velocities. The scale bars differ: S velocity variations are 2–3 times stronger than P velocity variations. Large-scale patterns dominate, suggesting the presence of coherent thermal and chemical structures.

velocity) beneath the circum-Pacific, with relatively low-velocity regions found under the central Pacific and south Atlantic/Africa. High-velocity regions are likely to be at lower temperature than low-velocity regions, although chemical differences may contribute to the variations. Recent tomographic models have ±4% S velocity variations in D″, two or three times stronger than those found in the mid-lower mantle, and comparable to the ±5–8% variations at depths near 150 km in the upper mantle. While mid-mantle variations tend to have more spatially concentrated heterogeneities, there is a significant degree of radial continuity of velocity anomaly patterns throughout the lower mantle, possibly linking subduction zones at the surface to high velocity regions in D″ and hot spot distributions at the surface to low velocity regions in D″.

P-wave velocity variations, dVp, in current tomographic models (Figure D3b/Plate 11b) involve ±1.0–1.5% fluctuations, with low-velocity areas beneath the southern Pacific and south Atlantic/Africa and high velocities under eastern Eurasia. There is not a strong increase in P velocity heterogeneity in D″ relative to that in the overlying mantle as there is for S-wave velocity, which can be attributed to the expected greater sensitivity of S velocity to strong thermal variations in a hot thermal boundary layer.

As apparent in Figure D3/Plate 11a,b, there is a general spatial correlation between S- and P-wave velocity patterns, as expected for temperature-induced variations, but in some regions, such as beneath the northern Pacific, this correlation breaks down. In other regions, such as beneath the central Pacific, the S-wave variations are much stronger than the P-wave variations even though both have the same sign. The decorrelation of dVp and dVs and the variation in their relative strengths provide strong evidence that thermal variations alone cannot explain the large-scale patterns of seismic velocity heterogeneity, so there is likely to be chemical heterogeneity present in D″

(Masters *et al.*, 2000). In fact, if bulk sound velocity variations, dVb, are estimated from P and S arrival times, they are found to have anticorrelation with dVs on very large scale, which requires bulk modulus and shear modulus to have opposite sign perturbations. It is important to recognize that the patterns in Figure D3/Plate 11a,b are relative to the global average velocities at D″ depths, and that these averages do not necessarily define "normal" mantle structure. This results in uncertainty in interpretations of the velocity fluctuations (for example, high-velocity regions may be anomalously low temperature, perhaps due to cool down-wellings that have disrupted an overall hot boundary layer, or low-velocity regions may be anomalously hot regions, perhaps partially melted, within a chemically distinct, high-velocity layer). Resolution of finer scale seismic properties in D″ is pursued in order to overcome this baseline uncertainty of seismic tomography imaging.

Seismic velocity discontinuities in D″

The presence of several percent lateral variations in S and P velocities concentrated within the lowermost mantle suggests the possibility of rapid velocity increases and/or decreases at the top of the D″ region. Various seismic observations suggest that rapid increases in velocity with depth are indeed present at depths from 130–300+ km above the CMB in many regions. Rapid shear velocity increases of 1.5–3% are found over intermediate-scale (500–1000 km) regions, with velocity structures and spatial sampling as indicated in Figure D4. While the PREM model shows a smooth velocity increase with depth in the lower mantle, with D″ being suggested only by a drop in rate of velocity increase with depth 150 km above the CMB, other shear velocity models obtained for localized regions have an abrupt increase in velocity near the top of the D″ region. The velocity discontinuities are required to account for observed S-wave reflections that precede the reflections from the CMB. Similar models are obtained for localized regions for P-waves, usually with a smaller velocity increase of 0.5–1%. The increase in velocity may be distributed over a few tens of kilometers or it may be very sharp, and in some regions it can vary in depth by as much as 100 km over lateral scales of just 200 km. Wysesssion *et al.* (1998) review many observations and models for this D″ seismic velocity discontinuity. While localized one-dimensional models are undoubtedly oversimplified, they do strongly suggest that some regions of D″ appear to effectively be layered, with thermally and compositionally distinct material.

Comparison of Figure D3a/Plate 11a and Figure D4 suggests that the S velocity discontinuity is commonly observed in areas of high S-wave velocity in D″. An abrupt increase in velocity with depth is not expected for a thermal boundary layer structure, so most interpretations of this correlation invoke the notion of localized ponding of cool subducting lithospheric slabs that have sunk to the lowermost mantle, retaining enough thermal and chemical anomalies to account for the high seismic velocity and strong radial gradients. Alternatively, chemically distinct high-velocity material may be present in these regions, perhaps involving ancient accumulated material concentrated during core formation or segregated oceanic crustal materials that have accumulated over Earth history. The observation of a weak and variable S velocity increase beneath the central Pacific, a relatively low-velocity region far from any historical subduction zone, complicates any attempt to interpret the velocity discontinuity as the result of recent slab thermal anomalies.

A possible explanation for the D″ seismic discontinuities has been provided by the discovery in 2004 of a phase change that occurs in magnesium silicate perovskite, $MgSiO_3$, the dominant mineral structure in the lower mantle (Murakami *et al.*, 2004). At pressure and temperature conditions close to those at the top of D″, a slightly denser post-perovskite mineral is expected to form, with a relatively strong increase in shear modulus and a small or even negative change in bulk modulus. This is expected to occur preferentially in lower temperature regions of D″, which may account for the intermittence of the D″ reflections. The effects of chemistry on this phase transition and its implications for boundary layer dynamics are active areas of research.

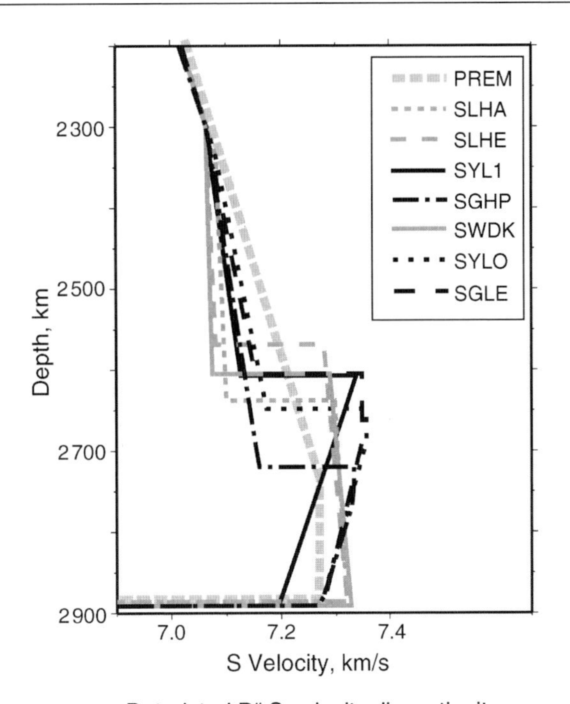

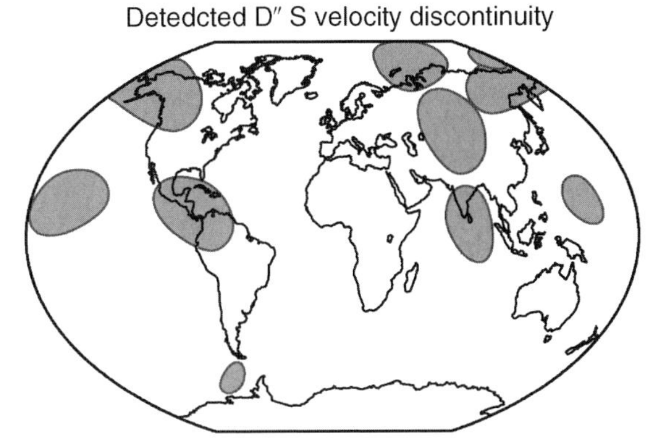

Figure D4 Representative S velocity models obtained from waveform modeling of data from localized regions of the lowermost mantle (top), and a map of regions with large-scale (>500 km) coherent structures having S velocity increases of 2.5–3%, 130–300 km above the CMB (bottom). The S velocity models are one-dimensional approximations of the local structure, and in most regions it appears that there is substantial variability in the depth and strength of the velocity increase on 100–300 km lateral scale-lengths. Most regions with clear S-wave velocity discontinuities have high S-wave velocities in D″ (Figure D3/Plate 11a,b) and underlie regions of extensive subduction of oceanic lithosphere in the past 200 Ma, but the central Pacific is an important exception.

Low-velocity provinces in D″

The central portions of the largest regions with very low shear velocity seen in Figure D3a/Plate 11a appear to have very strong decreases in velocity from above and laterally. For example, the low-velocity region beneath the southern Atlantic and Africa has been modeled as having an abrupt 1 to 3% decrease 250–300 km above the CMB, with average velocities in D″ that are 3–5% lower than for PREM (Figure D5). It appears that these are distinctive low-velocity provinces, comparable in thickness and lateral extent to the high-velocity regions, and having abrupt lateral boundaries (e.g., Wen *et al.*, 2001). The sub-African province appears to thicken to 800 km or so beneath the continent, with a thin

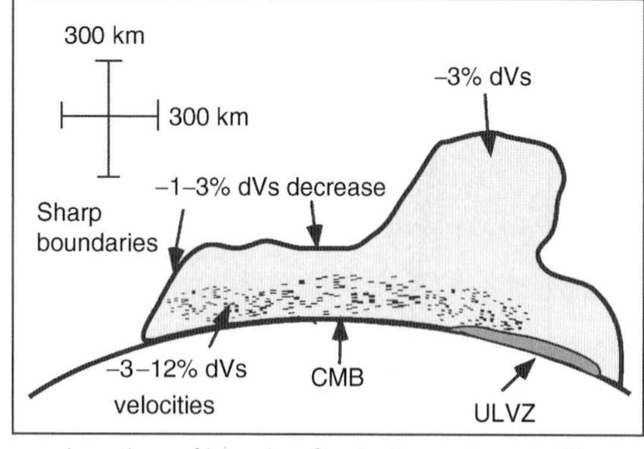

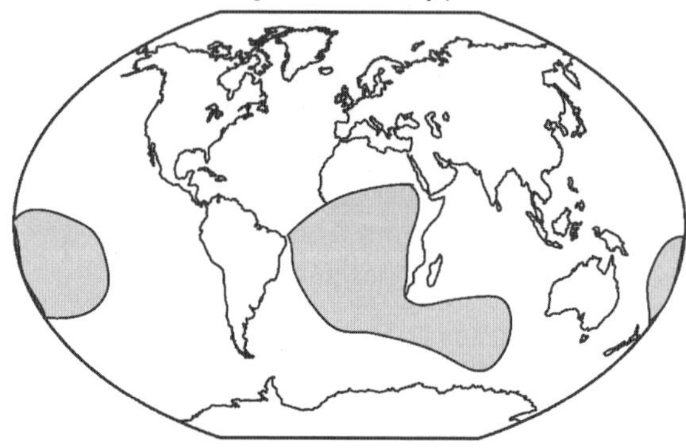

Figure D5 Two massive provinces with thick regions of low S velocity exist in the lower mantle: under the southern Pacific and under the southern Atlantic/Africa/southern Indian Ocean (bottom). A cartoon sketch of the latter feature is shown (top), with a 200–300 km thick layer in D″ having an abrupt decrease in velocity of 1 to 3% overlying a layer with average velocity decreases of 3 to 5%. Under Africa this low velocity body appears to extend upward to as much as 800 km above the CMB, and it may have an ultralow velocity zone (ULVZ) just above the CMB. It has sharp lateral boundaries in D″ and in the mid-mantle. The low S velocity province under the central Pacific is not yet well imaged, but appears comparable in size and strength of velocity reductions.

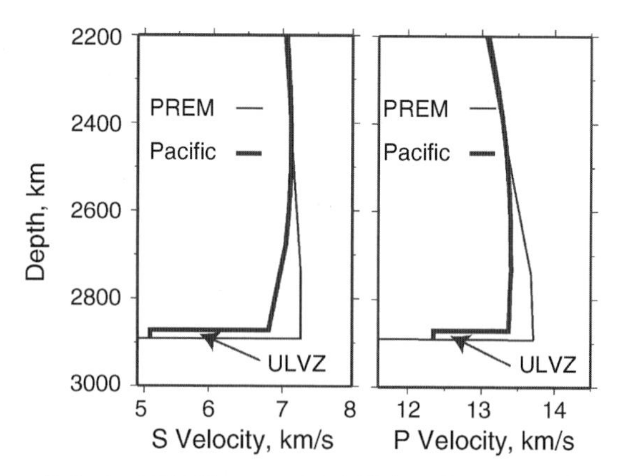

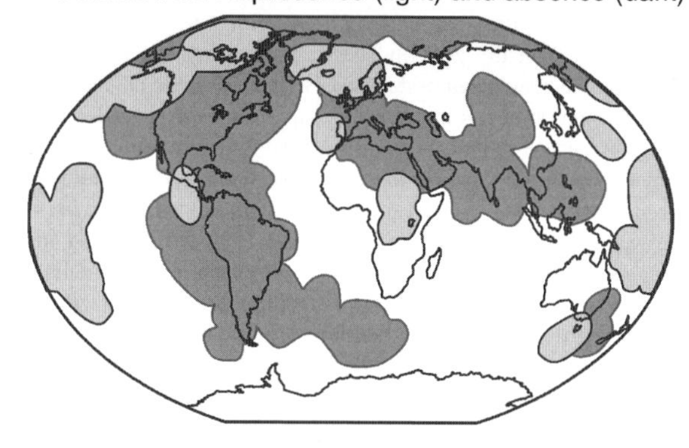

Figure D6 Seismic velocity models for PREM and the central Pacific region, where an ULVZ is detected right above the CMB (top). The ULVZ structures in various regions have P velocity drops of 4 to 10% and S velocity drops of 12 to 30%. The map (bottom) shows regions where ULVZ features have either been observed (light gray) or are not observed (dark gray). In general, regions with ULVZ in D″ tend to be found in regions of low S velocities (Figure D3a/Plate 11a).

region of ultralow velocity right above the CMB (see below). Lateral gradients on the margins of the low-velocity province are very abrupt, on the scale of tens of kilometers, both in D″ and in the upward extension into the lower mantle.

The large low-velocity provinces are commonly assumed to be very hot regions, and have been called Superplume provinces, under the assumption that they are sources of upwelling flow rising from a hot thermal boundary layer at the CMB. Indeed, they do underlie large regions of low velocity in the lower mantle as well as concentrations of hot spot volcanoes at the surface, some having distinctive chemical anomalies. However, the strong lateral gradients and apparent endurance of these structures suggest that there is distinctive chemical composition associated with the low-velocity provinces, and the extent to which thermal and chemical buoyancy trade-off is not yet established. P velocity anomalies of the low S velocity provinces are much smaller, suggesting the possibility of partial melting of these regions, which raises the possibility that these are simply the hottest (and partially molten) regions within a global chemically distinct layer. The strong temperature increase expected across the D″ region may approach or exceed the *eutectic solidus* of lower mantle mineral assemblages, allowing partial melting to take place in the boundary layer (Lay *et al.*, 1998a).

Ultralow velocity zones in D″

The most dramatic seismic velocity reductions that have been detected in D″ are associated with very thin layers or lumps of material just above the CMB that are called ultralow velocity zones (ULVZs) (see *ULVZ, ultralow velocity zone*). Figure D6 shows P and S velocity models characteristic of a region with a ULVZ (the central Pacific). This region, which is lower velocity than PREM over a 200–300 km thick D″ region, has a thin layer 10 to 40 km thick and laterally quite extensive (500–1000 km across), with P velocity reductions of 4 to 10% and S velocity reductions of 12 to 30%. In some regions, such as under Iceland, the ULVZ appears to be in a localized mound a few hundred kilometers across, but with comparably strong velocity reductions to the more extensive areas of ULVZ under the Pacific (Garnero *et al.*, 1998). Comparing Figure D6 and Figure D3a/Plate 11a indicates that ULVZs are commonly found within the margins of large-scale low-velocity regions, including at least portions of the major low S velocity provinces. Existence of a melt component in the ULVZs is the likely explanation for their strong velocity effects. The melt may involve either core-material that has infiltrated into the mantle, or partial melting of mantle material. In either case ULVZs are likely to involve thermochemical anomalies, probably with iron enrichment that causes the melt component to be dense, concentrated near the CMB (see: *D″, composition*; *Core-Mantle boundary*; *ULVZ, ultralow velocity zone*).

Seismic velocity anisotropy in D″

As is the case for upper mantle structure, the velocity structure in D″ is anisotropic, with seismic velocities being dependent on the direction of propagation and the polarization of ground shaking (see *D″, anisotropy*). This results in shear wave splitting, where an S-wave will separate into two components with orthogonal polarization, one traveling slightly faster than the other. By measuring the polarizations and travel-time difference between the fast and slow S-waves, it is possible to determine the anisotropic characteristics of the medium. The results of such analyses of S-wave splitting for phases traversing D″ are summarized in Figure D7. In most cases, the data indicate S velocity models in which horizontally polarized (SH) vibrations travel with 1 to 3% higher velocities than vertically polarized (SV) vibrations for raypaths grazing horizontally through the D″ region. This behavior is consistent with vertical transverse isotropy (VTI), which can result from hexagonally symmetric minerals with vertically oriented symmetry axes or from stacks of thin horizontal layers with periodic velocity fluctuations (Lay *et al.*, 1998b). The regions with the best documented cases for strong VTI in D″ tend to have higher than average S velocities (Figure D7 and Figure D3a/Plate 11a). There are observations favoring slightly tilted (nonvertical) transverse isotropy, which results in weak coupling of the SH and SV signals (the fast wave is close to the SH polarization), as well as limited regions where SV signals are found to propagate faster than SH signals (such as in the low S velocity region of the central Pacific). Obtaining mineralogical and dynamical explanations for anisotropy in the lowermost mantle are areas of active research (see: *D″, anisotropy*). There is great interest in improving the characterization of D″ anisotropy because it has the potential to reveal actual deformation processes occurring in the boundary layer.

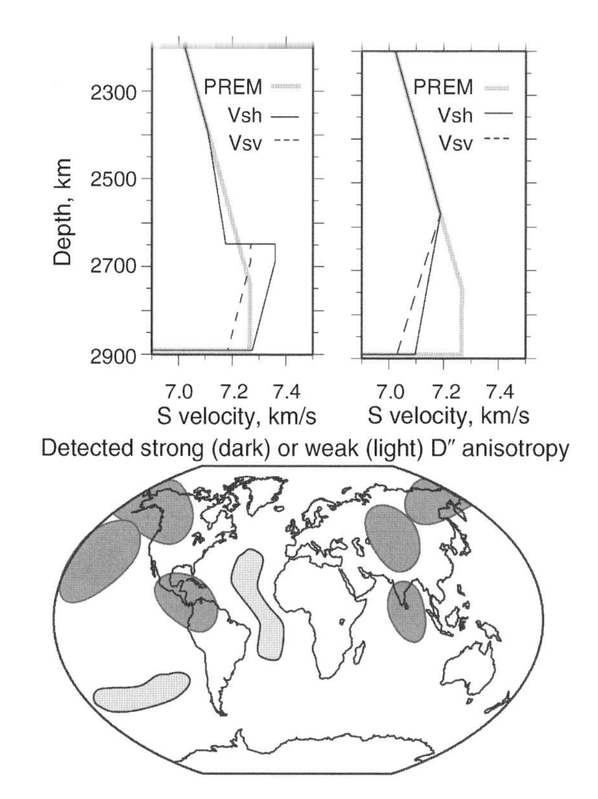

Figure D7 Examples of shear velocity models (top) proposed for various regions of D″ where the velocity structure has been found to be anisotropic, and a map (bottom) showing locations where relatively strong (>1%) vertical transverse isotropy has been detected in D″ (dark shading) or where weak or nonexistent anisotropy has been observed (light shading). The velocity models, which are compared to the PREM reference structure, have a few percent higher velocities for horizontally polarized S-waves (SH) than for vertically polarized S-waves (SV), which is consistent with vertical transverse isotropy. S-wave anisotropy in D″ has been detected in regions with high S velocities (Figure D3a/Plate 11a) and strong D″ S velocity discontinuities (Figure D4), and in regions with low S velocities and weak, or nonexistent S velocity discontinuities.

Discussion and conclusions

The diverse seismic properties of D″ appear to reflect the presence of thermal and chemical boundary layers at the base of the mantle. There is a predominance of large-scale structures and the strength of heterogeneities at intermediate and large-scales appears to be significantly greater than in the overlying lower mantle. Small-scale (<10 km) heterogeneities also appear to be present, but are not clearly more pronounced than in rest of the lower mantle. Interpretation of the seismic properties of D″ involves many uncertainties, and a wide spectrum of scenarios and processes have been advanced to account for the observations. Thermal variations in D″ play an important role in determining the heat flux boundary condition on the core dynamic regime, along with the possible role of D″ as a source of upwelling thermal plumes, so perhaps the greatest current challenge is to identify the relative contributions to the seismic properties of thermal versus chemical heterogeneity.

Thorne Lay

Bibliography

Boschi, L., and Dziewonski, A.M., 2000. Whole Earth tomography from delay times of P, PcP, and PKP phases: Lateral heterogeneities in the outer core or radial anisotropy in the mantle? *Journal of Geophysical Research [solid Earth]*, **105**: 13675–13696.

Dziewonski, A.M., and Anderson, D.L., 1981. Preliminary reference earth model. *Physics of the Earth and Planetary Interiors*, **25**: 297–356.

Garnero, E.J., 2000. Lower mantle heterogeneity. *Annual Reviews of the Earth and Planetary Sciences*, **28**: 509–537.

Garnero, E.J., Revenaugh, J.S., Williams, Q., Lay, T., and Kellogg, L.H., 1998. Ultralow velocity zone at the core-mantle boundary. In Gurnis, M., Wysession, M.E., Knittle, E., and Buffett, B.A. (eds.), *The Core-Mantle Boundary Region*. Washington, D.C.: American Geophysical Union, pp. 319–334.

Gurnis, M., Wysession, M.E., Knittle, E., and Buffett, B.A. (eds.), 1998. *The Core-Mantle Boundary Region*. Washington, D.C.: American Geophysical Union. 334 pp.

Lay, T., Williams, Q., and Garnero, E.J., 1998a. The core-mantle boundary layer and deep Earth dynamics. *Nature*, **392**: 461–468.

Lay, T., Williams, Q., Garnero, E.J., Kellogg, L., and Wysession, M.E., 1998b. Seismic wave anisotropy in the D″ region and its implications. In Gurnis, M., Wysession, M.E., Knittle, E., and Buffett, B.A. (eds.), *The Core-Mantle Boundary Region*. Washington, D.C.: American Geophysical Union, pp. 299–318.

Masters, G., Laske, G., Bolton, H., and Dziewonski, A.M., 2000. The relative behavior of shear velocity, bulk sound speed, and compressional velocity in the mantle: implications for chemical and thermal structure. In Karato, S., Forte, A.M., Liebermann, R.C., Masters, G., and Stixrude, L., (eds.), *Earth's Deep Interior: Mineral Physics and Tomography from the Atomic to the Global Scale*, Washington, D.C.: American Geophysical Union, pp. 63–87.

Mégnin, C., and Romanowicz, B., 2000. The three-dimensional shear velocity structure of the mantle from the inversion of body, surface, and higher-mode waveforms, *Geophysical Journal International*, **143**: 709–728.

Murakami, M., Hirose, K., Kawamura, K., Sata, N., and Ohishi, Y., 2004. Post-perovskite phase transition in $MgSiO_3$, *Science*, **304**: 855–858.

Wen, L., Silver, P., James, D., and Kuehnel, R., 2001. Seismic evidence for a thermo-chemical boundary at the base of the Earth's mantle. *Earth and Planetary Science Letters*, **189**: 141–153.
Wysession, M., Lay, T., Revenaugh, J., Williams, Q., Garnero, E.J., Jeanloz, R., and Kellogg, L., 1998. The D″ discontinuity and its implications. In Gurnis, M., Wysession, M. E., Knittle, E., and Buffett, B.A. (eds.), *The Core-Mantle Boundary Region.* Washington, D.C.: American Geophysical Union, pp. 273–298.

Cross-references

Core-Mantle Boundary
Core-Mantle Boundary Topography: Implications for Dynamics
Core-Mantle Boundary Topography: Seismology
Core-Mantle Boundary, Heat Flow Across
D″ as a Boundary Layer
D″, Composition
Earth Structure, Major Divisions
ULVZ, Ultralow Velocity Zone

DELLA PORTA, GIAMBATTISTA (1535–1615)

Born at Vico, Italy, sometime in October or November 1535, Giambattista Della Porta may well have been self-educated. He was fascinated by the esoteric philosophies of the Renaissance, and his *Magia Naturalis* (1558) became famous for the bewildering mixture of scientific, occult, and classical material that it contained, and was translated into English in 1658.

Magia Naturalis, Book VII, discusses "The Wonders of the Lodestone" in 56 short chapters. Its particular value comes from the rich and abundant stock of classical and medieval references to magnetism contained within it, which in many ways provides us with the first history of this branch of science. Della Porta seems to have taken the innately occult "*vertue*" of the lodestone pretty well as read, though he was always keen to narrate his own experiments, which were usually performed with the intention of testing one legend or another.

Della Porta inquires on a number of occasions in his narrative into why lodestones attract iron. Most remarkably, he mentions that Epicurus in classical Greece ascribed this attractive force to *atoms* coming from the stone (Chapter 2), though most of his reported explanations are mystical and analogous to the power of Orpheus's music, which could allegedly attract and transfix humans.

Of course, *Magia Naturalis* recounts all kinds of tricks that could be performed with magnets, such as "How to make an army of sand fight before you" (Chapter 17), whereby magnetized sand, arranged into ranks and battle array on a table top, could be made to move and collide by the skilful manipulation of lodestones under the table. He also relates stories from classical literature of ancient temples having roofs sheathed in lodestone, so as to make the statue of the goddess within seemingly hang in the air (Chapter 27). On the other hand, Della Porta's experiments exploded the classical myth that onions and garlic destroyed the power of the lodestone, finding that "breathing and belching upon the Loadstone after eating of Garlick, did not stop its vertues" (Chapter 48). He tried similar experiments to test the power of goat's blood and diamonds upon the "*vertue*" of stony and metallic magnets.

However, the lodestone's power was found to be totally destroyed by placing it in a powerful fire, during which the heated lodestone gave off dark blue flames and a stink of sulfur (Chapter 51).

But Della Porta gave lots of advice on practical subjects, such as how to best magnetize the needles of mariners' compasses (Chapter 36), and even suggested how the compass might be employed to find the longitude at sea (Chapter 38) by utilizing the needle's prior recorded deviation from true north on different parts of the Earth's surface. For instance, in the sixteenth century the needle pointed to true north in the Azores, to west of the meridian in the Americas, but to the east of it in Europe.

Della Porta's writings on magnetism must be seen as part of his wider visionary and occult view of nature. For *Magia Naturalis* was immensely influential, and in addition to the tests and experiments which he reported, Book VII is especially important as a history of magnetism from antiquity to the sixteenth century.

Giambattista Della Porta's wider magical and occult beliefs led to an encounter with the Inquisition some time before 1580, but he seems to have otherwise lived as a devout Catholic. He became, being married, a lay brother with the Jesuit Order in 1585 and devoted himself to the movement for the reformation of the Roman Catholic Church. Over the last 30 years of his life, in addition to his scholarly activities he worked in a variety of charitable capacities amongst the poor of Naples. He died in Naples on 4 February 1615.

Allan Chapman

Bibliography

John Baptista Porta (Giambattista Della Porta), *Natural Magick, John Baptista Porta, a Neopolitane: in Twenty Books* (London, 1658). Translated from *Magia Naturalis* (1558) by an unspecified translator.

DEMAGNETIZATION

The multicomponent nature of the remanent magnetization of rocks was recognized early in paleomagnetic research. The natural remanent magnetization (NRM) is the vector sum of several magnetizations generated over the geological history of the rocks. Recognition of the multivectorial nature of NRM, understanding of the magnetization acquisition mechanisms and magnetization types, and development of techniques for separation of magnetization components became part of the fundamental principles for paleomagnetic research and for the success of the multiple applications in geology, geophysics, and related fields (e.g., Irving, 1964; McElhinny, 1973; Tarling, 1983). The various magnetization components are acquired over the rock history since formation by processes related to burial, tectonic deformation, heating, metamorphism, metasomatism, fluid circulation, and weathering, which result in formation of new magnetic minerals, chemical reactions (oxidation, reduction), grain size changes, etc. The magnetization acquired at the time of formation of the rock (i.e., cooling below the blocking temperatures of a volcanic or intrusive rock, deposition of sediments, or a metamorphic event) is referred as primary magnetization, with all other components acquired at later times representing secondary components. Secondary components may be added through formation of new magnetic minerals or through partial or total overprints on existing mineral phases. The acquisition modes and magnetic carriers for the various NRM components, which may be thermal, chemical, or viscous magnetizations residing on iron-titanium oxides, iron sulfides or iron oxyhydroxides, result in characteristic spectra of coercivities, blocking temperatures and solubility. Demagnetization techniques (or cleaning techniques) that use application of alternating magnetic fields, temperature or chemical leaching exploit differences on those characteristics, expecting that each magnetic component is characterized by a particular (and discrete) coercivity, temperature, or solubility spectrum. The temporal changes of direction and intensity at any given location on the Earth s surface resulting form secular variation and polarity reversals of the Earth magnetic field and/or changes of the orientation of a rock unit relative to the ambient field direction cause that in general each generation of NRM components present a distinctive magnetization direction. The separation and interpretation of NRM components make use of laboratory demagnetization techniques, methods for determination of vector components and field and consistency tests.

Demagnetization techniques test the stability of the magnetic minerals that carry the remanent magnetization and provide information on their properties. The standard demagnetization techniques are the thermal and alternating field (AF) demagnetization, which make use of the application or progressively higher temperatures or alternating fields. Chemical demagnetization is a technique used for sedimentary rocks, which exploits the solubility of magnetic minerals. Application of low temperatures has also been employed as a demagnetization technique; it involves successive cooling of samples from room temperature down to liquid nitrogen temperature.

Alternating field demagnetization has been used since early in the development of the paleomagnetism (As and Zijderveld, 1958; Creer, 1959; Kobayashi, 1959). Alternating field demagnetization investigates the coercivity spectrum by application of successively stronger alternating fields to the domain grains of a sample, which is placed in a zero direct field setting. The technique takes the domains around successively larger hysteresis loops at each peak demagnetizing field, which are then slowly reduced as the alternating field strength is decreased to zero (magnetic domains of lower coercivity are locked to different orientations). For single domain grains the relaxation time τ is directly related to the coercive force H_c.

Multidomain grains present low coercivities, and the technique can provide information on the presence of a given assemblage of grains with different domain states (Evans and McElhinny, 1969; Dunlop and West, 1969; Stacey, 1961).

Jaime Urrutia-Fucugauchi

Bibliography

As, J.A., and Zijderveld, J.D.A., 1958. Magnetic cleaning of rocks in palaeomagnetic research. *Geophysical Journal of Royal Astronomical Society*, **1**: 308–319.

Burek, P.J., 1969. Device for chemical demagnetization of red beds. *Journal of Geophysical Research*, **74**: 6710–6712.

Collinson, D.W., 1983. *Methods in Rock Magnetism and Palaeomagnetism*. London: Chapman and Hall, 503 pp.

Collinson, D.W., Creer, K.M. and Runcorn, S.K., 1967. *Methods in Palaeomagnetism*. Amsterdam: Elsevier, 609 pp.

Creer, K.M., 1959. A.C. demagnetization of unstable Triassic Keuper marls from S.W. England. *Geophysical Journal of the Royal Astronomical Society*, **2**: 261–275.

Dunlop, D.J., 2003. Stepwise and continuous low-temperature demagnetization. *Geophysical Research Letters*, **30**(11): Art.No. 1582.

Dunlop, D.J., and West, G.F., 1969. An experimental evaluation of single domaintheories. *Reviews of Geophysics*, **7**: 709–757.

Evans, M.E., and McElhinny, M.W., 1969. An investigation of the origin of stable remanence in magnetite bearing igneous rocks. *Journal of Geomagnetism and Geoelectricity*, **21**: 757–773.

Henry, S.G., 1979. Chemical demagnetization: methods, procedures, and applications through vector analysis. *Canadian Journal of Earth Sciences*, **16**: 1832–1841.

Irving, E., 1964. *Palaeomagnetism and its Application to Geological and Geophysical Problems*. New York: John Wiley, 399 pp.

Kobayashi, K., 1959. Chemical remanent magnetization of ferromagnetic minerals and its application to rock magnetism. *Journal of Geomagnetism and Geoelectricity*, **10**: 99–117.

McElhinny, M.W., 1973. *Palaeomagnetism and Plate Tectonics*. Cambridge: Cambridge University Press.

Park, J.K., 1970. Acid leaching of red beds, and its application to relative stability of the red and black magnetic components. *Canadian Journal of Earth Sciences*, **7**: 1086–1092.

Stacey, F.D., 1961. Theory of the magnetic properties of igneous rocks in alternating magnetic fields. *Phil. Mag.*, **6**: 1241–1260.

Stephenson, A., 1980. Gyroremanent magnetization in anisotropic material. *Nature*, **284**: 49.

Stephenson, A., 1983. Changes in direction of the remanence of rocks produced by stationary alternating-field demagnetization. *Geophysical Journal of the Royal Astronomical Society*, **73**: 213.

Tarling, D.H., 1983. *Palaeomagnetism*. Principles and Applications in Geology, Geophysics and Archaeology. London: Chapman and Hall, 379 pp.

Warnock, A.C., Kodama, K.P., and Zeitler, P.K., 2000. Using thermochronometry and low-temperature demagnetization to accurately date Precambrian paleomagnetic poles. *Journal of Geophysical Research*, **105**(B6): 19435–19453.

Zijderveld, J.D.A., 1967. A.C. demagnetization of rocks: analysis of results. In Collinson, D.W., Creer, K.M., and Runcorn, S.K., (eds.), *Methods in Palaeomagnetism,* Amsterdam: Elsevier, pp. 254–286.

Cross-references

Paleomagnetism
Rock Magnetism

DEPTH TO CURIE TEMPERATURE

Introduction

The Curie temperature or Curie point is the temperature at which a ferromagnetic mineral loses its ferromagnetic properties. At temperatures above the Curie point the thermal agitation causes the spontaneous alignment of the various domains to be destroyed/randomized so that the ferromagnetic mineral becomes paramagnetic. The term Curie point is named after the French scientist Pierre Curie who discovered that ferromagnetic substances lose their ferromagnetic properties above this critical temperature. This discovery has had great utility in providing information about the deep crust of the Earth that lies below direct human access.

The interior of the Earth is considerably hotter than the surface. Depth to the Curie temperature is the depth at which crustal rocks reach their Curie temperature. Although *paramagnetism* and *diamagnetism* contribute to the magnetism of rocks, it is the ferromagnetic minerals that are the dominant carriers of magnetism in rocks (Langel and Hinze, 1998). The dominant ferromagnetic minerals are the *iron-titanium oxide* and *iron sulfides* (*q.v.*). Generally the depth to Curie isotherm is calculated indirectly from analysis of magnetic anomalies; however in a rare case this depth was physically reached in a drilled core. The German Continental Deep Drilling Program drilled a 9.1-km deep core and its petrophysical properties showed that the most abundant ferrimagnetic mineral at this site is monoclinic pyrrhotite (see *iron sulfides*), which disappeared below about 8.6 km, and below this depth, hexagonal pyrrhotite with a Curie temperature of 260°C was the stable phase (Berckhemer *et al.*, 1997). Thus the depth to the Curie temperature of the predominant pyrrhotite was physically reached (as the drilled core reached *in situ* bottom hole temperature of around 265°C).

Utility

Magnetite with a Curie temperature of 580°C is the dominant magnetic mineral in the deep crust within the continental region (Langel and Hinze, 1998). Below the Curie isotherm depth the lithosphere is virtually nonmagnetic. From the analysis of the *crustal magnetic field* (*q.v.*), the depth below which no magnetic sources exist can be estimated. The remotely sensed magnetic (surface, airborne, satellite-borne) measurements thus indirectly provide an isothermal surface within the lower crust that in turn can be translated into geothermal gradients of the region. The nationwide geothermal resources project of Japan, initiated in the early 1980s, was perhaps the first systematic attempt to use Curie point depth estimates from the magnetic data as an integral part of a nationwide geothermal exploration program for such a large area (Okubo *et al.*, 1985). Estimates were made for temperature gradients and utilizing available heat flow data very meaningful average thermal conductivities were determined. The depth to Curie temperature can also provide an understanding of the thermal structure; e.g., in volcanic areas convective heat transfer complicates the determination of the thermal structure from heat flow measurements alone and determination of Curie point depths from magnetic data can prove to be helpful for understanding the thermal structure. From a magnetic analysis of the Tohoku arc, Japan, Okubo, and Matsunaga (1997) find that the Curie isotherm varies from 10 km in the volcanic province of the back arc to 20 km or deeper at the eastern limit of Tohoku. They find that the boundary between the seismic and aseismic zones in the overriding plate correlates with the inferred Curie isotherm, indicating that the seismicity in the overriding plate is related to temperature.

The magnetic method estimates the depth to the bottom of the magnetized layer. Due to the inherent uncertainties of these Curie temperature depths, supporting independent evidence is desirable. Possibilities include the comparison with deep seismic soundings, gravity surveys, and petrological studies. In case the estimated Curie temperature depth correlates with a velocity or density boundary, it is likely to reflect a vertical change in composition. Rajaram *et al.* (2003) applied a series of high pass filters to the aeromagnetic data of Southern India and inferred that below the exhumed crust of the Southern Granulite terrain the magnetic crust is thin (compared to the Dharwar craton) being confined to 22 km; at this depth there is a change in the seismic velocity in the seismic reflection/refraction profiles implying a compositional change being responsible for the lack of magnetic sources below 22 km. The depth extent of magnetic sources has become synonymous with the depth to the Curie temperature though in reality they may represent either a petrological or temperature boundary. Magnetization measurements and petrological studies indicate that ferromagnetic minerals are generally not present in the mantle, at least in the continental region (Langel and Hinze, 1998) implying that the Moho is a magnetic boundary.

Method

Magnetic anomalies contain contributions from the ensemble of sources lying above the Curie isotherm. The depth extent of magnetic sources is often determined by spectral analysis of magnetic data. This is done by either examining the shape of isolated magnetic anomalies or by examining statistical properties of patterns of magnetic anomalies. Spector and Grant (1970) pointed out that the anomaly due to an ensemble of sources, represented by a large number of independent rectangular parallelepipeds, has a power spectrum equivalent to that of a single average source of the ensemble and the depth to the top of the body, z_t, can be estimated from the slope of the log power spectrum. A limited depth extent of the body is predicted to lead to a maximum in the power spectrum, and the wave number of this maximum $k_{\max}$ is related to the bottom of the magnetic source, z_b (depth to the Curie isotherm) thus:

$$k_{\max} = \frac{\log z_b - \log z_t}{z_b - z_t}$$

For a magnetic survey of dimension L the smallest wavenumber is the fundamental wavenumber $k = 2\pi/L$ (Blakely, 1995). Thus, $k_{\max}$ should be at least twice this to be able to resolve a peak in the spectrum and this suggests that the survey dimension must be at least

$$L \geq \frac{4\pi(z_b - z_t)}{\log z_b - \log z_t}$$

For a survey conducted 1 km above the top of magnetic sources, the survey dimension must be at least 160 km for sources extending to 50 km depth. It often turns out, however, that continental scale magnetic compilations do not have a maximum in the power spectrum indicating a flaw in the underlying assumptions. Indeed, it is very difficult to estimate the depth to the bottom of the magnetic sources, as the spectrum is dominated at all wavelengths by the contribution from the shallower parts. To avoid this complication, one calculates the centroid depth, z_0, either from isolated anomalies or by using methods based on Spector and Grant's statistical model and the inferred depth to the Curie temperature is then obtained as $z_b = 2z_0 - z_t$ (Okubo *et al.*, 1985).

The method adopted by Spector and Grant (1970) assumes that the magnetization has no spatial correlation; however there are evidences to the contrary. Pilkington and Todoeschuck (1993) from a statistical analysis of susceptibility and aeromagnetic fields indicate that the magnetization and resulting fields are correlated over large distances and can be described as a self-similar random process. Such processes have a power spectrum proportional to some power of the wavenumber, called the scaling exponent (fractal dimension), and the degree of correlation is indicated by the magnitude of this exponent. Maus *et al.* (1997) find that the limited depth extent of the crustal magnetization is discernible in the power spectra of magnetic anomaly maps of South Africa and Central Asia. From the theoretical power spectrum of a slab using fractal methods, they estimate the Curie temperature depths to lie between 15 and 20 km, with large uncertainties due to the inaccurately known scaling exponent of the crustal magnetization. They conclude that for reliable estimates of the depth to the Curie temperature, magnetic anomalies over an area of at least 1000 km × 1000 km are required.

Magnetic anomalies, long wavelength (*q.v.*), measured by satellites are generally inverted using an equivalent source model (see Langel and Hinze, 1998). The observed magnetic anomaly is inverted by a least squares method for an array of equal area dipole sources at the surface of the Earth. A common assumption is that the magnetization arises from induced magnetization only. The magnetization is proportional to the effective susceptibility and the thickness of the magnetized crust, which in turn is a measure of the depth to the Curie temperature. Using inverse models incorporating *a priori* ocean-continent magnetization contrasts, Purucker *et al.* (1998) generated global integrated susceptibility maps from *Magsat* (*q.v.*), which can be interpreted in terms of thermal characteristics of the crust. Purucker and Ishiara (2005) have used a further refined model using *CHAMP* (*q.v.*) data and find for the subduction region of the 2004 great Sumatran earthquake that the subducting plate is descending so quickly that its temperature does not reach the Curie point, resulting in thickened magnetic crust due to the subducting slab.

Conclusions

Magnetic anomalies derived from *aeromagnetic surveying* (*q.v.*) have been used to estimate the depth and configuration of the Curie isotherm for over half a century. Estimates of depth to the Curie temperature can provide valuable insights in the assessment of geothermal energy, calculation of thermal conductivity, and geodynamic evolution. However, unknown mineralogical and statistical properties of the source magnetization have a strong effect on the accuracy of Curie isotherm depth estimates by magnetic methods. These estimates should therefore be interpreted in the context of independent geological, geophysical, and geothermal information. Estimating depth to Curie temperature on a regional scale from long wavelength anomalies requires that large areas of survey data be used for the calculations. There is still no consensus on the minimum survey area required to arrive at a reliable estimate of the Curie isotherm depth. Also, in the Fourier domain, one works with square grids of data, which can pose a problem due to the shape of the continents e.g., Indian peninsula. However, in the absence of any direct method to observe Curie depth, estimates from magnetic anomalies will continue to provide valuable information. New global magnetic anomaly maps, compiled from all available marine and aeromagnetic data and combined with *CHAMP* (*q.v.*) and upcoming *Swarm* satellite magnetic maps, using *spherical cap harmonics* (*q.v.*), will provide a valuable basis for future global estimates of the Curie isotherm.

Mita Rajaram

Bibliography

Berckhemer, H., Rauen, A., Winter, H., Kern, H., Kontny, A., Lienert, M., Nover, G., Pohl, J., Popp, T., Schult, A., Zinke, J., and Soffel, H.C., 1997. Petrophysical properties of the 9-km deep crustal section at KTB. *Journal of Geophysical Research*, **102**: 18337–18362.

Blakely, R.J., 1995. *Potential Theory in Gravity and Magnetic Applications.* Cambridge University Press, Australia. 441pp.

Langel, R.A., and Hinze, W.J., 1998. *The Magnetic Field of the Lithosphere: The Satellite Perspective.* Cambridge: Cambridge University Press, U.K, 429pp.

Maus, S., Gordon, D., and Fairhead, D., 1997. Curie temperature depth estimation using a self similar magnetization model. *Geophysical Journal International*, **129**: 163–168.

Okubo, Y., Graf, R.J., Hansen, R.O, Ogawa, K., and Tsu, R., 1985. Curie point depths of the island of Kyushu and surrounding areas, Japan. *Geophysics*, **53**: 481–494.

Okubo, Y., and Matsunaga, T., 1997. Curie point depth in northeast Japan and its correlation with regional thermal structure and seismicity. *Journal of Geophysical Research*, **99**: 22363–22371.

Pilkington, M., and Todoeschuck, J.P., 1993. Fractal magnetization of continental crust. *Geophysical Research Letters*, **20**: 627–630.

Purucker, M.E., Langel, R.A., Rajaram, Mita and Raymond, C., 1998. Global magnetization models with apriori information. *Journal of Geophysical Research*, **103**: 2563–2584.

Purucker, M.E., and Ishiara, T., 2005. Magnetic Images of the Sumatran region crust. *EOS*, **86**(10): 101–102.

Rajaram, Mita, Harikumar, P., and Balakrishnan, T.S., 2003. Thin Magnetic Crust in the Southern Granulite Terrain. In Ramakrishnan, M. (ed.), *Tectonics of Southern Granulite Terrain, Kuppam-Palani Geotransect,* Memoirs of the Geological Society of India, Vol. 50, pp. 163–175, Bangalore, India.

Spector, A., and Grant, F.S., 1970. Statistical models for interpreting aeromagnetic data. *Geophysics*, **35**: 293–302.

Cross-references

Aeromagnetic Surveying
CHAMP
Harmonics, Spherical Cap
Iron Sulfides
Magnetic Anomalies for Geology and Resources
Magnetic Anomalies, Long Wavelength
Magnetic Anomalies, Modeling
Magnetic Susceptibility
Magsat

DIPOLE MOMENT VARIATION

The present geomagnetic field is approximately that expected from a dipole located at the center of the Earth and tilted by about 11° relative to the rotation axis. The magnetic moment associated with such a dipole can be inferred from measurements of magnetic field strength. Systematic measurements of relative field strength revealing latitudinal variations were made by De Rossel on the D'Entrecasteaux expedition (1791–1794), but the moment was not evaluated directly until the 1830s when *Gauss* (*q.v.*) carried out the first spherical harmonic analysis using absolute measurements of field strength made at geomagnetic observatories. Although the dipole moment has decreased by about 10% since then, the current value of 7.78×10^{22} A m^2 is close to the average for the past 7 ka (7.4×10^{22} A m^2). A broad range of geomagnetic and paleomagnetic observations indicates that both these values are higher than the longer-term average, but probably not anomalously high, and that the current rate of change seems not atypical for Earth's history. Changes in the dipole moment occur on a wide spectrum of timescales: in general, changes at short periods are small, and the greatest variations are those associated with excursions and full reversals of the geomagnetic field.

Dipole moments and their proxies

How the dipole moment is estimated depends on the information available. The geomagnetic dipole moment m can be computed directly from the degree 1 Gauss coefficients, g_1^0, g_1^1, and h_1^1, of a *spherical harmonic* (*q.v.*) field model via

$$m = \frac{4\pi a^3}{\mu_0}\sqrt{(g_1^0)^2 + (g_1^1)^2 + (h_1^1)^2},$$

with a the average radius of the Earth, and μ_0 the permeability of free space. This clearly distinguishes the dipole from *nondipole field* (*q.v.*) contributions. When there are insufficient data to construct a field model, p can be calculated as a proxy for m using a single measurement of magnetic field strength $\boldsymbol{B}$,

$$p = \frac{4\pi a^3}{\mu_0}\frac{\boldsymbol{B}}{\sqrt{1 + 3\cos^2\theta}}.$$

When θ is the geographic colatitude at which $\boldsymbol{B}$ is measured, p is known as a virtual axial dipole moment (VADM). This is the equivalent moment of a magnetic dipole aligned with the rotation axis that would generate the observed value of $\boldsymbol{B}$. Often the magnetic inclination I is used to derive the magnetic colatitude θ_m via the relationship

$$\tan I = 2\cot\theta_m$$

When θ_m is used in place of θ, the result is known as a virtual dipole moment (VDM). In principle, the VDM might take account of the tilt of the dipole axis, but the effect is complicated by the contribution of the nondipole field present in any instantaneous measurement. VADM and VDM are most often used in comparing paleomagnetic data to account for gross geographic variations in field strength. Temporal and spatial averages of p are sometimes referred to as paleomagnetic dipole moments (PDMs); the hope is that such averages will remove the influence of nondipole field contributions, but recent work indicates the possibility of substantial bias. Paleomagnetic measurements of field strength can be absolute or relative variations with time. Absolute measurements require a thermal origin for the magnetic remanence, while the relative variations commonly acquired from sediments must be tied to an absolute scale. A review of current paleointensity techniques can be found in Valet (2003).

Cosmogenic radioisotopes such as ^{14}C, ^{10}Be, and ^{36}Cl are produced in the stratosphere at a rate that is expected to follow a relationship approximately inversely proportional to the square root of Earth's dipole moment (Elsasser *et al.*, 1956). Thus, cosmogenic isotope records can serve as a proxy for relative dipole moment changes. Production rates are also affected by solar activity, but such changes seem largest on timescales short compared with the dipole moment variations. For ^{14}C, the effects of exchange of CO_2 between atmosphere and ocean must be considered, resulting in substantial smoothing and delay in the recorded signal.

Direct estimates for the period 0–7 ka

Today, the most complete mapping of the magnetic field is achieved using magnetometers on board satellites (e.g., *Ørsted* (*q.v.*) and *CHAMP* (*q.v.*)) that orbit Earth at an altitude of several hundred kilometers. These data are sufficiently dense and accurate that, when combined with observatory measurements, it is possible to obtain spherical harmonic models and detect changes in Earth's dipole moment of the order of 10^{18} A m^2 on timescales as short as a day. Such short-term variations (and at least some of the changes up to periods as long as or longer than the solar cycle) do not reflect changes in the internal part of the magnetic field, but stem from fluctuations in strength of the solar wind. The solar wind modulates the strength of the external ring current that exists between about 3 and 6 Earth radii and is the source of induced magnetic field variations in Earth's moderately electrically conducting mantle. Short-term variations in dipole moment presumably also arise in Earth's core, but will be attenuated by passage through the electrically conducting mantle. On timescales of months to several tens of years, it remains a challenge to separate variations of internal and external origin.

For longer term changes in dipole moment, we turn to models constructed from the historical and paleomagnetic record. An excellent *time-dependent model of the main magnetic field* (*q.v.*) is available for A.D. 1590–1990 (Jackson *et al.*, 2000) but is artificially scaled prior to 1832. Early attempts to extend knowledge of the dipole moment back in time used proxy VDMs and VADMs from archeomagnetic artifacts and lava flows for the past 50 ka (McElhinny and Senanayake, 1982) and indicated that 2000 years ago, the dipole moment was almost 50% higher than today. Time-varying spherical harmonic models now exist for the past 7 ka, and these allow separation of dipole and nondipole variations. Directional data used in these models come from archeomagnetic artifacts, lava flows, and high deposition rate sediments. For intensity, only absolute measurements from archeomagnetic samples and lava flows were included. The resolution depends on the accuracy of the dating and the quality of the observations, which is lower than for direct measurements, and also more heterogeneous, but the dipole moment agrees well with that inferred from historical data in the overlapping time range. Figure D8 shows the dipole moment for the period 0–7 ka, along with VADM calculated for the same data. The estimated dipole moment m is systematically lower than the VADM, but still shows almost a factor of two in variability. The differences are consistent with geographical and temporal bias in sampling of the nondipole field by the available VADMs (Korte and Constable, 2005).

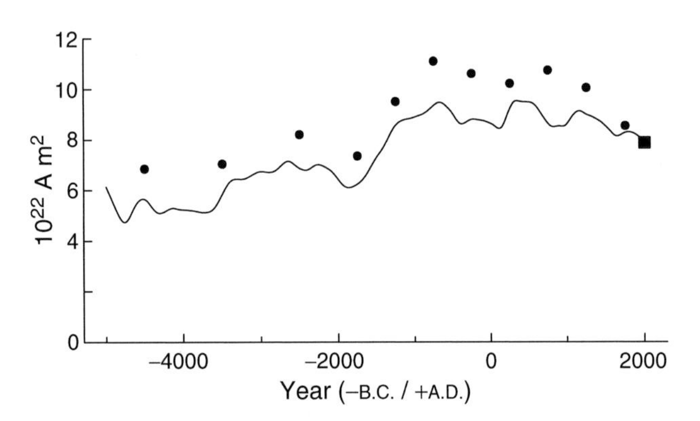

Figure D8 Comparison of estimated dipole moment m (continuous curve) with VADM proxies, p, (circles) for the past 7 ka, each based on the same archeomagnetic intensity data. Square gives the dipole moment from spherical harmonic analysis in 2002.

Proxy records from sediments for 10 ka–1 Ma timescales

On longer timescales, the view of dipole moment variation is far less complete, although the situation is steadily improving. There are no direct estimates of m, so we must rely on VADMs or some other proxy. Relative paleointensity variations in sediments are available from a large number of marine sediment cores (again, see Valet, 2003) at a variety of locations and a range of time intervals. Global stacks and averaging of these records have been used as a proxy for dipole moment variation, initially for the time interval 0–200 ka and more recently for 0–800 ka (the SINT800 record). Longer stacked records will undoubtedly be forthcoming. Detailed analyses of the stacking process have been conducted with simulations of probable noise processes applied to output from numerical dynamo simulations. These show that although individual sedimentary records may be of low quality and inconsistent with others from nearby sites, such techniques can recover long-term variations in dipole moment. The resolution of changes in the record is primarily limited by the quality of the age control and correlations among cores, which depends on sedimentation rate and other environmental factors, but for SINT800 is probably around 20 ka. Regional stacks for the North and South Atlantic regions have better temporal resolution because of higher average sedimentation rates. Some researchers hope to use dipole moment or regional paleointensity variations as a stratigraphic correlation tool, but large local variations in the nondipole field may prevent this from being a useful approach.

The absolute calibration of relative variations in global dipole moment remains difficult, but the general pattern is well understood for the past few million years at a resolution of a few tens of thousand years. For the past 800 ka, a mean value of 5.9×10^{22} A m^2 is estimated for the global dipole moment, about 25% lower than its current value. SINT800 shows unequivocally (Figure D9) that there are repeated large but irregular changes in VADM over the past 800 ka: the standard deviation is a little more than 25% of the mean for the entire record (almost certainly an underestimate of the true field variability because of heavy averaging). Excursional geomagnetic field

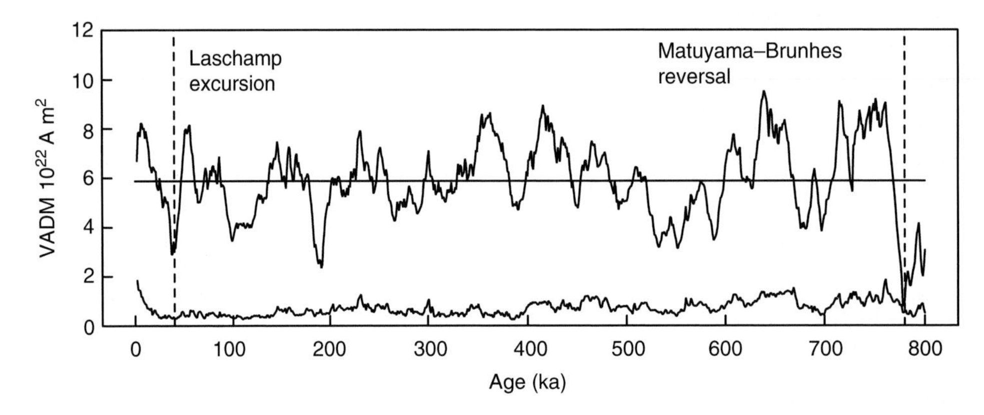

Figure D9 Variations in virtual axial dipole moment inferred from globally distributed marine sediments. Upper curve gives average VADM, lower curve is one standard error in the mean. More details are given in Valet (2003).

behavior in field direction (most recently the Laschamp excursion at about 40 ka), are concurrent with low values of VADM, although the converse is not necessarily true, and the lowest value corresponds to the Matuyama-Brunhes geomagnetic reversal. The exact number of excursions cataloged during the Brunhes normal chron ranges from about 6 to 12 depending on the extent to which one requires global correlations among records. Paleointensity records from volcanics almost exclusively support the notion that excursions are associated with large decreases in geomagnetic intensity, and the idea that geomagnetic reversals are accompanied by a decrease of 80% or 90% in dipole moment is essentially undisputed. Independent support for sedimentary relative paleointensity variations is provided by the general agreement with proxy dipole moment variations derived from the cosmogenic isotopes ^{10}Be and ^{36}Cl.

Proxy records from TRMs and longer timescale variations

In principle, absolute paleointensity measurements on bulk samples and single crystals derived from lava flows and submarine basaltic glass provide the highest quality measurements of the geomagnetic field. The picture that emerges for VADM variations is slowly clarifying as the number of results steadily increases, and it becomes possible to evaluate the quality of the available data through the implementation of consistency checks on laboratory work, and replicate data on specimens from the same sample. Over the time interval 0–160 Ma, the average dipole moment is estimated as 4.5×10^{22} A m^2 with a standard deviation of 1.8×10^{22} A m^2. The data are reasonably approximated by a lognormal distribution (Tauxe, 2006).

It remains hard to make definitive statements about very long-term changes in dipole moment, because of the large geographic (1 standard deviation is about 18% for the present field) and temporal variability in *p*. The relationship between average intensity and reversal rate is still unclear, although at 8.1×10^{22} A m^2 the average VADM for the Cretaceous Normal Superchron is high and also highly variable (1 standard deviation $= 4.3 \times 10^{22}$ A m^2, Tauxe and Staudigel, 2004)). Other times during the Cenozoic show a weak correlation between average field strength and polarity interval length, although for short intervals, this may reflect a significant contribution to the average from the low dipole strength during reversals. Higher average intensities are also correlated with greater variability in VADM, indicating that dipole and nondipole variations may be covariant. It is clear that the average dipole moment depends on the time interval used for the calculation.

Some researchers have inferred correlations between changes in Earth's orbital parameters and geomagnetic field variations. There is to date no undisputed demonstration of such a relationship. Another outstanding question concerns the very ancient geomagnetic field. The field is known to have existed at 3.5 Ga and reversals have been documented from as early as 1.5 Ga (Dunlop and Yu, 2004), but there are so few data that it is not known whether the early field was really dipolar. Many more data are needed to test exciting hypotheses such as whether the nature of geomagnetic field variations changed with the formation and growth of the inner core.

Catherine Constable

Bibliography

Dunlop, D.J., and Yu, Y., 2004. Intensity and polarity of the geomagnetic field during Precambrian time. In Channell, J.E.T.C., Kent, D.V., Lowrie, W., and Meert, J.G. (eds.), *Timescales of the Internal Geomagnetic Field,* Geophysical Monograph Series **145**, Washington, D.C.: American Geophysical Union, pp. 85–100, 10.1029/145GM07.

Elsasser, W.E., Ney, E.P., and Winckler, J.R., 1956. Cosmic ray intensity and geomagnetism. *Nature*, **178**: 1226–1227.

Jackson, A., Jonkers, A.R.T., and Walker, M.R., 2000. Four centuries of geomagnetic secular variation from historical records. *Philosophical Transactions of the Royal Society of London, Series A*, **358**: 957–990.

Korte, M., and Constable, C., 2005. The geomagnetic dipole moment over the last 7000 years—new results from a global model. *Earth and Planetary Science Letters*, **236**: 328–358.

McElhinny, M.W., and Senanayake, W.E., 1982. Variation in the geomagnetic dipole 1: the past 50 000 years. *Journal of Geomagnetism and Geoelectricity*, **34**: 39–51.

Tauxe, L., 2006. Long-term trends in paleointensity. *Physics of the Earth and Planetary Interiors*, **156**: 223–241, doi:10.1016/j.pepi.2005.03.022.

Tauxe, L., and Staudigel, H., 2004. Strength of the geomagnetic field in the Cretaceous Normal Superchron: new data from submarine basaltic glass of the Troodos Ophiolite. *Geochemistry, Geophysics, Geosystems*, **5**: Q02H06, doi: 10.1029/2003GC000635.

Valet, J-P., 2003. Time variations in geomagnetic intensity. *Reviews in Geophysics*, **41**: doi:10.1029/2001/RG000104.

Cross-references

CHAMP
Gauss, Carl Friedrich (1777–1855)
Geomagnetic Spectrum, Temporal
Harmonics, Spherical
Nondipole Field
Ørsted
Reversals, Theory
Time-dependent Models of the Geomagnetic Field

DYNAMO WAVES

Dynamo waves are oscillating solutions of the induction equation of *magnetohydrodynamics* (*q.v.*), which typically involve magnetic field fluctuations on global scales (possibly including global reversals), and fluctuate with periods related to the timescale of magnetic diffusion (i.e., ca. 10^3–10^4 years, for the Earth). They arise as oscillatory solutions of the kinematic *dynamo problem* (*q.v.*): the linear problem for the generation of magnetic field subject to the inductive action of a specified flow. The concept of dynamo waves can be extended beyond the linear regime, however, as the oscillatory behavior is often retained in nonlinear solutions; and in scenarios involving fluctuating velocities, such waves are invoked by several proposed mechanisms for geomagnetic reversals (see *Reversals, theory*).

In this usage, dynamo waves should be distinguished from other forms of *magnetohydrodynamic waves* (*q.v.*), including *Alfvén waves* (*q.v.*), and magnetic *torsional oscillations* (*q.v.*). The behavior of the magnetohydrodynamic waves is determined jointly by the induction equation and the hydrodynamic equation of motion (including the magnetic Lorentz force). As such, the magnetohydrodynamic waves involve various additional timescales (e.g., the inertial and rotational timescales). Whereas magnetohydrodynamic waves are typically considered as perturbations to the basic state of the geodynamo (although they may retain an important role in the basic generation mechanism, as in the *Braginsky dynamo* (*q.v.*)), dynamo waves are part of the fundamental dynamo mechanism.

Axial waves

The idea of dynamo waves goes back to the earliest work on homogeneous dynamos by Parker (1955), where the generative term in the induction equation, later called the "alpha"-effect, was first formulated (see *Dynamos, mean field; Dynamo, Braginsky*). Subsequent early work elucidated the behavior of the resulting equation in axisymmetric systems, and it is in such axial geometries that dynamo waves remain most studied. Given the predominantly axisymmetric character of the geomagnetic field, the importance of the axial modes is clear; and much recent work, despite being fully three-dimensional, continues to focus on these aspects of the solutions. Solving the alpha-effect induction equation in a simplified Cartesian geometry, Parker (1955) found that the most easily excited solutions were normally oscillatory, reversing sign via recurring waves of magnetic activity. This observation was soon corroborated by other authors, and extended to axisymmetric spherical geometries (e.g., Roberts, 1972). The generic behavior of such axisymmetric, mean field dynamos is now well established, and is summarized by Moffatt (1978) and Parker (1979). Different characteristic behavior is obtained if the inductive action of differential fluid rotation—the so-called "omega"-effect—is responsible for part of the field generation (in "alpha-omega" dynamos), or if the alpha effect is the source of dynamo action alone (in "alpha^2" dynamos). The simplest dipole solutions of alpha^2 dynamos tend to be stationary, whereas the corresponding solutions of alpha-omega dynamos tend to be oscillatory (i.e., involving dynamo waves).

It is worth noting, however, that this generic behavior tends to rely upon the spatial distributions of the alpha- and omega-effects being smooth; different behavior is possible with strongly localized forms. In some cases, alpha^2-dynamos can also produce dynamo waves (e.g., Giesecke *et al.*, 2005). There are also various ways in which the oscillations of alpha-omega dynamos can be stabilized; "meridional circulation"—i.e., axisymmetric flow on planes of constant meridian, or longitude—being the most celebrated example (see Braginsky, 1964; Roberts, 1972). The situation is further complicated by the possible competition between distinct symmetries of solutions: dipole and quadrupole (see *Symmetry and the geodynamo*). Oscillatory dipole solutions frequently coexist with stationary quadrupole solutions, and *vice versa*. The nonlinear interaction of the various modes may therefore be rather complex.

The smaller number of three-dimensional kinematic studies which have described dynamo waves are broadly consistent with the characteristic behavior outlined above (e.g., Gubbins and Sarson, 1994; Gubbins and Gibbons, 2002); although no longer purely axisymmetric, these solutions typically retain strong axial symmetry and reverse by a similar mechanism. (Gubbins and Gibbons note, however, that the additional freedom available in three dimensions strongly increases the preference for stationary solutions, rather than dynamo wave solutions.) Willis and Gubbins (2004) also considered the kinematic growth associated with time-periodic velocities in three dimensions; some of their solutions exhibited more complex dynamo wave solutions (see *Dynamos, periodic*).

Much of the early work cited above was more motivated by the *solar dynamo* (*q.v.*) than by the geodynamo, and the concept of dynamo waves remains more developed in the solar context. (Wave solutions are more clearly relevant to the Sun, given the oscillatory nature of the *Sun's magnetic field* (*q.v.*). It was already noted by Braginsky (1964), however, that an isolated dynamo wave might model a geomagnetic reversal; and this insight has been the basis of many kinematic reversal mechanisms (see *Reversals, theory*). The interaction of dynamo waves with more steady states—typically studied using mean field dynamo models—have also been used to model other aspects of the geodynamo; e.g., Hagee and Olson (1991) used nonlinear solutions, combining modes of both steady and oscillatory character, to model aspects of *geomagnetic secular variation* (*q.v.*).

In axisymmetric geometries, the dynamo waves progress via the movement of regions of alternately signed magnetic flux, appearing near the equator and migrating towards the pole (or *vice versa*) to replace the magnetic flux originally there; one such sequence constitutes a field reversal (i.e., half an oscillation). The sense of migration—poleward or equatorward—depends upon the relative signs of the alpha- and omega-effects. At the surface of the dynamo region, this migration is exhibited by bands of oppositely signed radial field; in terms of external field models, the process involves the coupled oscillations of the various zonal multipole terms. In three-dimensional calculations, the migration can also be seen in the motion of local, nonaxisymmetric, surface flux patches. In the solar case, this migration ties in well with the observed sunspot cycle (sunspots being associated with intense magnetic flux). In the case of the Earth, the motion of flux patches of reversed polarity has been proposed as a possible mechanism for a global reversal, on the basis of both observations and theory (e.g., Gubbins, 1987; Gubbins and Sarson, 1994).

Oscillatory features in the output of numerical dynamo models (see *Geodynamo, numerical simulations*) have occasionally been considered with such dynamo wave interpretations in mind (e.g., Glatzmaier and Roberts, 1995); but the three-dimensional simulations are extremely complex, and such identifications remain tentative. The numerical solution analyzed in detail by Wicht and Olson (2004) reversed polarity via a form of dynamo wave; but this solution, which reversed polarity periodically, is not particularly Earth-like. Takahashi *et al.* (2005) analyzed the surface morphology of their numerical reversal in terms of migrating flux patches, but the wave this represents appears to be rather complex.

Nonaxial waves

Following on from the earliest axisymmetric work, most studies of dynamo waves have concentrated on the axial waves discussed above; but other forms of waves are quite generally possible. The most frequently discussed alternatives oscillate via the simple rotation of a nonaxisymmetric field pattern, typically dominated by equatorial field patches (including the equatorial dipole). Such solutions have been studied in nonaxisymmetric solutions of mean field models (e.g., Rädler *et al.*, 1990), and in direct three-dimensional calculations: both kinematic (e.g., Holme, 1997) and dynamic (e.g., Aubert and Wicht, 2004). Such waves may be more relevant to other planets than to the Earth, but they may also have some relevance to the geomagnetic secular variation, where they might explain such features as the *westward drift* (*q.v.*) of field. (Such secular variation may alternatively, however, be explained by other forms of *magnetohydrodynamic waves* (*q.v.*); or else may arise from corotation with the drifting velocity associated with *rotating convection* (*q.v.*).)

Graeme R. Sarson

Bibliography

Aubert, J., and Wicht, J., 2004. Axial vs. equatorial dipolar dynamo models with implications for planetary magnetic fields. *Earth and Planetary Science Letters*, **221**: 409–419.

Braginsky, S.I., 1964. Kinematic models of the Earth's hydrodynamic dynamo. *Geomagnetism and Aeronomy*, **4**: 572–583 (English translation).

Giesecke, A., Rüdiger, G., and Elstner, D., 2005. Oscillating α^2-dynamos and the reversal phenomenon of the global geodynamo. *Astronomische Nachrichten*, **326**: 693–700.

Glatzmaier, G.A., and Roberts, P.H., 1995. A three-dimensional self-consistent computer simulation of a geomagnetic field reversal. *Nature*, **377**: 203–209.

Gubbins, D., 1987. Mechanisms for geomagnetic polarity reversals. *Nature*, **326**: 167–169.

Gubbins, D., and Gibbons, S., 2002. Three-dimensional dynamo waves in a sphere *Geophysical and Astrophysical Fluid Dynamics*, **96**: 481–498.

Gubbins, D., and Sarson, G., 1994. Geomagnetic field morphologies from a kinematic dynamo model. *Nature*, **368**: 51–55.

Hagee, V.L., and Olson, P., 1991. Dynamo models with permanent dipole fields and secular variation. *Journal of Geophysical Research*, **96**: 11673–11687.

Holme, R., 1997. Three-dimensional kinematic dynamos with equatorial symmetry: Application to the magnetic fields of Uranus and Neptune. *Physics of the Earth and Planetary Interiors*, **102**: 105–122.

Moffatt, H.K., 1978. *Magnetic Field Generation in Electrically Conducting Fluids*. Cambridge: Cambridge University Press.
Parker, E.N., 1955. Hydromagnetic dynamo models. *Astrophysical Journal*, **121**: 293–314.
Parker, E.N., 1979. *Cosmical Magnetic Fields*. Oxford: Clarendon Press.
Rädler, K.-H., Wiedemann, E., Brandenburg, A., Meinel, R., and Tuominen, I., 1990. Nonlinear mean-field dynamo models: Stability and evolution of three-dimensional magnetic field configurations. *Astronomy and Astrophysics*, **239**: 413–423.
Roberts, P.H., 1972. Kinematic dynamo models. *Philosophical Transactions of the Royal Society of London, Series A*, **272**: 663–698.
Takahashi, F., Matsushima, M., and Honkura, Y., 2005. Simulations of a quasi-Taylor state geomagnetic field including polarity reversals on the Earth simulator. *Science*, **309**: 459–461.
Wicht, J., and Olson, P., 2004. A detailed study of the polarity reversal mechanism in a numerical dynamo model. *Geochemistry Geophysics Geosystems*, **5**: Q03H10.
Willis, A.P., and Gubbins, D., 2004. Kinematic dynamo action in a sphere: effects of periodic time-dependent flows on solutions with axial dipole symmetry. *Geophysical and Astrophysical Fluid Dynamics*, **98**: 537–554.

Cross-references

Alfvén Waves
Convection, Nonmagnetic Rotating
Dynamo, Braginsky
Dynamo, Solar
Dynamos, Kinematic
Dynamos, Mean Field
Dynamos, Periodic
Geodynamo, Numerical Simulations
Geodynamo, Symmetry Properties
Geomagnetic Secular Variation
Magnetic Field of Sun
Magnetohydrodynamic Waves
Magnetohydrodynamics
Oscillations, Torsional
Reversals, Theory
Secular Variation Model
Westward Drift

DYNAMO, BACKUS

In 1919, *Larmor* (*q.v.*) speculated that the flow of a conducting fluid could generate a magnetic field. At the Earth's surface, the observed magnetic field is largely axisymmetric, but in 1934, however, *Cowling* (*q.v.*) published a proof stating that a two-dimensional magnetic field could not be generated by a two-dimensional flow. For a while it seemed plausible that a generalized antidynamo theorem might exist and another mechanism would have to be sought. Worries were eventually allayed in 1958 with the appearance of two positive examples—the rotor dynamo of *Herzenberg* (*q.v.*), and the stasis dynamo of Backus. Although Bullard and Gellman (1954) had already appeared to have working dynamos, the generated magnetic fields were later shown to be artifacts of insufficient numerical resolution, Gubbins (1973). The existence proof of Backus (1958) bypasses this difficulty by considering a time-dependent flow. Allowing sufficient periods of stasis, the truncation of higher order modes can be rigorously justified.

A magnetic field can be expressed in terms of toroidal and poloidal components $\boldsymbol{B} = \boldsymbol{T} + \boldsymbol{P} = \nabla \wedge (T\hat{\boldsymbol{r}}) + \nabla \wedge \nabla \wedge (P\hat{\boldsymbol{r}})$. Upon substitution into the induction equation

$$\frac{\partial \boldsymbol{B}}{\partial t} = \nabla \wedge (\boldsymbol{u} \wedge \boldsymbol{B}) + \lambda \nabla^2 \boldsymbol{B}, \qquad \text{(Eq. 1)}$$

where λ is the magnetic diffusivity, it becomes apparent that poloidal field must be created by stretching of a toroidal field and *vice versa*. Although originally formulated in terms of topological arguments, Cowling's negative result rested on the impossibility of creating an axisymmetric poloidal field with an axisymmetric flow. The mechanism of toroidal and poloidal exchange, however, formed the basis for the first dynamos shown to exist by analytical methods. Backus (1958) considered the generation of a nonaxisymmetric poloidal field in a sphere of radius R filled with an incompressible fluid. An axisymmetric flow within the sphere is punctuated by periods of stasis; exterior to the sphere the medium is assumed to be electrically insulating.

The proof begins by considering the first few natural decay modes of a stationary conducting sphere. The leading toroidal mode $\boldsymbol{T}_1$ decays more quickly than the leading poloidal mode $\boldsymbol{P}_1$, because boundary conditions confine toroidal fields to the sphere. Therefore, the cycle starts in the situation where only the poloidal mode is present and is normalized by the energy norm, $\| \boldsymbol{P}_1 \| = 1$. A small residual field $\boldsymbol{R}$ may also be present. A large toroidal field is then drawn out from $\boldsymbol{P}_1$ by stretching it with a burst of toroidal flow. Similarly, energy is transfered back from the toroidal field into poloidal field by a poloidal flow. A period of stasis is then required to remove almost all but the leading poloidal mode $\boldsymbol{P}$. By applying a very rapid solid body rotation, it is always possible to align $\boldsymbol{P}$ with the the original, resulting in the field $\gamma(\boldsymbol{P}_1 + \boldsymbol{R}')$. The cycle is sketched in Figure D10.

Backus proceeds to derive a number of criteria that must be satisfied to ensure that the multiplier γ is greater than unity, and such that if the starting residual is no greater than some small value, $\| \boldsymbol{R} \| \leq \varepsilon$, then the proportion of resulting residual field remains small, $\| \boldsymbol{R}' \| \leq \varepsilon$. One of the inequalities that emerges early in the proof, a *bounding theorem* (*q.v.*) for dynamo action is frequently expressed as

$$R'_m \geq \pi^2, \qquad \text{(Eq. 2)}$$

where the magnetic Reynolds number $R'_m = RU'/\lambda$ is defined in terms of the maximum difference in flow velocities over the sphere. To satisfy the remaining criteria, however, the bound (Eq. (2)) is greatly exceeded by Backus' chosen flow.

The flow specified by Backus (the Backus dynamo) is axisymmetric with toroidal and poloidal components

$$\boldsymbol{u}_T = \hat{\boldsymbol{\phi}} \lambda^{-1} r^2 (1 - r^2) \sin\theta \cos\theta, \qquad \text{(Eq. 3)}$$

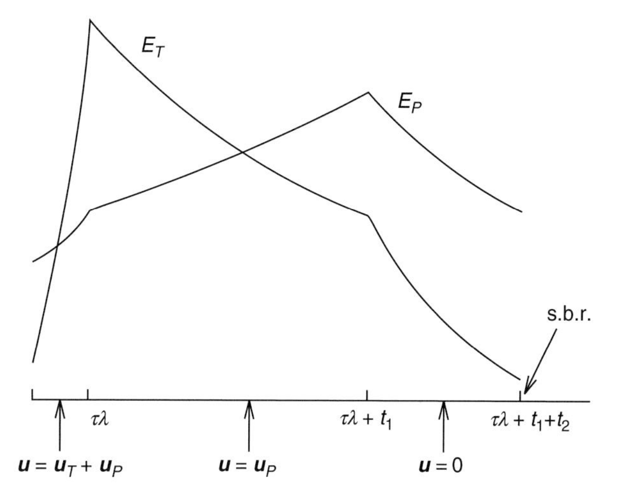

Figure D10 Generation and decay of toroidal and poloidal energies, E_T and E_P, over one cycle of the Backus dynamo.

$$u_P = \hat{r}r^2(1-r^2)^2\sin^2\theta\cos\theta - \hat{\theta}r^2(1-r^2)(1-2r^2)\sin^2\theta. \quad \text{(Eq. 4)}$$

The initial toroidal jerk lasts a short time $\tau\lambda = 0.015$ with $\lambda^{-1} = 8 \times 10^9$. The poloidal component of the flow persists for unit time, so $t_1 = 0.985$, and the stasis period is $t_2 = 0.2105$. After applying a solid body rotation, the flow amplifies the leading poloidal mode initially aligned with the axis. The proportion of residual field at the beginning and end of the cycle is less than $\epsilon = 1.047 \times 10^{-3}$.

Although the flow is axisymmetric, *Cowling's theorem* (*q.v.*) is not violated as the generated field is nonaxisymmetric. The dimensional time- and space-averaged velocity corresponds to approximately 4×10^2 cm s^{-1} for the Earth. The historical record of secular variation suggests that typical velocities near the core-mantle boundary are closer to 4×10^{-2} cm s^{-1}. The constraints of physical plausibility were relaxed in order to establish a successful dynamo, but the rigorous analysis of Backus has stood the test of time. Together with the Herzenberg dynamo, the Backus dynamo demonstrated conclusively that velocity fields are capable of generating a magnetic field.

Ashley P. Willis

Bibliography

Backus, G.E., 1958. A class of self-sustaining dissipative spherical dynamos. *Annals of Physics*, **4**: 372–447.

Bullard, E.C., and Gellman, H., 1954. Homogeneous dynamos and terrestrial magnetism. *Philosophical Transactions of the Royal Society of London, Series A*, **247**: 213–278.

Cowling, T.G., 1934. The magnetic field of sunspots. *Monthly Notices of the Royal Astronomical Society*, **94**: 39–48.

Gubbins, D., 1973. Numerical solutions of the kinematic dynamo problem. *Philosophical Transactions of the Royal Society of London, Series, A*, **274**: 493–521.

Herzenberg, A., 1958. Geomagnetic dynamos. *Philosophical Transactions of the Royal Society of London, Series A*, **250**: 543–583.

Larmor, J., 1919. How could a rotating body such as the Sun become a magnet? *Reports of the—British Association for the Advancement of Science*, 159–160.

Cross-references

Antidynamo and Bounding Theorems
Cowling, Thomas George (1906–1990)
Cowling's Theorem
Dynamo, Herzenburg
Dynamos, Kinematic
Larmor Joseph (1857–1942)

DYNAMO, BRAGINSKY

The Braginsky dynamo—due to Stanislav Iosifovich Braginsky (sometimes spelled Braginskiĭ, or Braginskiy), and first presented in Braginsky (1964a)—was the first model rigorously to show how a dominantly axisymmetric magnetic field could be generated via the homogeneous dynamo mechanism. The key part of Braginsky's analysis elucidates a mechanism for generating dipolar field from azimuthal field; in this respect, the mechanism follows the more heuristic treatment of Parker (1955) and predates the "two-scale" (or turbulent) mechanism developed by Steenbeck *et al.* (1966) (see *Dynamos, mean field*). Unlike these other approaches, Braginsky's analysis was formulated in terms of the magnetic inductive action of large-scale nonaxisymmetric "waves" of a form excited in the Earth's core. The analysis behind this theory produced many additional insights into the *magnetohydrodynamics* (*q.v.*) of the Earth's core.

Nearly axisymmetric magnetic field generation

Although the geomagnetic field is observed to be dominantly axisymmetric (i.e., symmetric about the rotation axis, as exemplified by the dominantly axial dipole field), *Cowling's theorem* (*q.v.*) shows that a purely axisymmetric system cannot maintain a dynamo. While the generation of azimuthal field from dipolar field is relatively straightforward, via the shearing action of nonuniform azimuthal flow (or "differential rotation"), the converse generation of dipolar field from azimuthal field cannot be sustained within such a system. This seemingly rules out a large class of appealingly simple models, and instead requires the consideration of fully three-dimensional systems. Braginsky's analysis circumvented this problem by considering (in a mathematically rigorous way) a "nearly axisymmetric" system, allowing for small deviations in axisymmetry in both flow and magnetic field. Because of the high electrical conductivity of the Earth's core—which is built into the analysis—these asymmetries, although weak, can combine to influence the axisymmetric field significantly. The net effect is to augment the azimuthally averaged equations for the magnetic field with an additional term (arising from the nonaxisymmetric fields); it is this term that allows the sustenance of the axisymmetric field (thus evading Cowling's theorem, as it would apply to the purely axisymmetric system). This term is in the form of what was later to be called an "alpha effect", similar to that obtained from the analysis of *Mean field dynamos* (*q.v.*); it allows the generation of dipolar fields from azimuthal fields. (Braginsky originally denoted this generation term as "Gamma"—as did the earlier, influential, but more schematic analysis of Parker (1955)—but the "alpha" nomenclature has now become the norm.) Together with the so-called "omega" effect, describing the generation of azimuthal field from dipolar field by differential rotation (and thus "completing the cycle"), this allows for a working dynamo.

This analysis required a remarkable appreciation of the magnitudes of several components of the coupled field and flow within the core (with the same scalings applying to both magnetic field and flow field). The azimuthal parts of the axisymmetric fields dominate; the nonaxisymmetric fields are somewhat weaker, and the nonazimuthal parts of the axisymmetric fields (including the dipolar part of the magnetic field) are weaker still. (Technically, these scalings are formulated in powers of the inverse square root of the magnetic Reynolds number; see *Dynamos, kinematic*, for a definition of this nondimensional number.) Although seemingly intricate, this scaling is consistent with the highly conducting nature of the core fluid, and is supported by many later calculations and observations (although the relative dominance of the various components of fields remains somewhat controversial). Notably, this scaling implies that the axial dipole part of the field, which dominates at the Earth's surface, is actually the weakest part of the field within the Earth's core. (This causes no difficulties for the theory, however, as Braginsky's analysis requires the stronger components of field to remain confined within the core. Indeed, one gratifying consequence is that it naturally produces a weak nonaxisymmetric field in the exterior; i.e., it naturally explains the slight tilt observed in the Earth's dipole field.) It is notable that this analysis makes the Braginsky dynamo inescapably of "alpha omega" character, with the "omega" effect mechanism (and the azimuthal fields) inevitably stronger than the "alpha" effect (and the dipolar, or meridional, fields). This need not be a problem regarding applicability to the Earth, as azimuthal differential rotation is relatively easily excited in rotating convective systems (see *Thermal wind*); but it has interesting consequences for the resulting behavior of the dynamo.

Braginsky's original analysis relied on the identification of a number of "effective fields", to simplify the appearance of the azimuthally-averaged equations. These essentially amount to "changes of variables" in the representations of the dipolar magnetic field and of the meridional component of flow (which has a similar geometry to the dipole field, but is confined within the core), to isolate the parts actively involved in the generation mechanism. Such effective variables were found to remain useful when the mathematical analysis of field generation was continued to the next order (Tough, 1967). As Braginsky noted in 1964b, the surprising elegance of this simplification suggests an underlying physical mechanism, and this was later identified by Soward (1972), working in a "pseudo-Lagrangian" framework: the terms needed to give the effective variables can be related to the transformations needed to map streamlines of nearly axisymmetric flow onto axisymmetric, circular, paths about the rotation axis. The development of Soward (1972)—reviewed in Moffatt (1978)—also makes the relation between the Braginsky dynamo mechanism and the mean field dynamo mechanism clearer.

While Braginsky (1964a) initially considered the generation associated with a single wave of nonaxisymmetric flow and field—which might contain more than a single azimuthal mode, but which must rotate at a common angular velocity—the analysis was soon extended to allow for arbitrary combinations of waves, and also to allow for oscillations of the axisymmetric meridional flow and dipolar field on timescales similar to that of the waves (Braginsky, 1964b); the generative effect of the waves effectively sums, once the appropriate new effective quantities are allowed. It is worth noting that to obtain a non-zero generation term, the nonaxisymmetric flow of any of these waves must have some preferred sense of direction; it cannot be mirror symmetric about any plane of constant azimuth. The convective rolls which make up this flow must therefore have a spiral planform. Fortunately such spiraling flows can be excited in convective systems like the Earth's (see *Rotating convection, non-magnetic*). As Braginsky noted, this requirement explains why the *Bullard-Gellman dynamo* (*q.v.*) cannot truly sustain dynamo action, at least in the nearly axisymmetric limit. Sarson and Busse (1998) more explicitly address the importance of spiraling in this limit.

Braginsky (1964a,b) also considered the effect of surfaces where the tangential velocity is discontinuous, or where the axisymmetric azimuthal flow vanishes; he showed that concentrated, or "resonant", generation of field can occur on such surfaces, and analyzed the form of such generation.

Kinematic dynamo action and nonlinear developments

If a particular velocity field is assumed, as in a *kinematic dynamo* (*q.v.*), the detailed dynamo action associated with that flow can be investigated. The mechanics of such kinematic dynamo action were immediately investigated in the axisymmetric system obtained by Braginsky (1964a,b), assuming specific simple forms of the differential rotation, meridional circulation, and generation coefficient "alpha" relevant to that system; Braginsky considered such models both analytically in a plane (Cartesian) geometry approximation (1964b), and numerically in spherical geometry (1964c). Both sets of calculations recover the result, shown by Parker (1955), that "alpha omega" dynamos tend to produce oscillatory magnetic fields; a good model for the solar dynamo, but not for the mainly steady field of the geodynamo. Braginsky found, however, that the presence of meridional circulation could alter the behavior so as to favor stationary solutions. Furthermore, his dynamo equations naturally contain a source of meridional circulation, in terms of the effective meridional circulation described above. Even in the absence of significant "true" meridional circulation—and the meridional circulation in the Earth is thought to be weak—the effective meridional circulation associated with non-axisymmetric waves, or with axisymmetric oscillations, remains present to stabilize the dynamo oscillations. Braginsky (1964b,c) also identified in this effect a possible mechanism for reversals (see *Reversals, theory*): fluctuations in the convective state of the dynamo will naturally lead to fluctuations in the (effective) meridional circulation; and if the latter becomes too weak, the dynamo will become oscillatory, reversing polarity via *dynamo waves* (*q.v.*) until a more normal convective state is resumed. In a notable subsequent corroboration, the relation of the axisymmetric system to the full three-dimensional kinematic dynamo problem was explored by Kumar and Roberts (1975). They prescribed a three-dimensional velocity field (including a nonaxisymmetric wave component), and investigated its kinematic dynamo action, both via direct three-dimensional calculations and via the analogous axisymmetric system obtained from the prescriptions of Braginsky (1964a) (with the generation term and effective meridional circulation being explicitly calculated). In a suite of calculations near to this highly conducting, nearly axisymmetric limit, they found very satisfactory agreement.

The calculations of Braginsky (1964a,b,c) were all concerned with kinematic dynamo action, although Braginsky (1964d) contemporaneously considered the full nonlinear problem of the *magnetohydrodynamics* (*q.v.*) of the Earth's core (requiring the form of the flow to be self-consistently derived). Given the remarkable success of the kinematic development, considerable work was expended in trying to extend the nearly symmetric analysis into the nonlinear regime. Despite some initially promising results—Tough and Roberts (1968) showed that the effective variables described above retain some relevance in the nonlinear problem—such developments proved very difficult. At the level of expansion of the foregoing analysis, the nonlinear modifications did not prove to support dynamos; and attempts (Braginsky *et al.*,) to make progress with modified scalings, theoretically capable of nonlinear dynamo action, also encountered difficulties. Roberts (1987) discusses these efforts.

In the absence of a full nonlinear theory, the Braginsky dynamo model was nevertheless extended into the nonlinear regime in a series of "intermediate" models; these assumed a given form of the generation term "alpha", as motivated by the above, and thus allowed the nonlinear dynamics of the system to be investigated within a numerically tractable axisymmetric system. Notable amongst such models is the "model Z" *dynamo* (*q.v.*), originally presented in Braginsky (1975).

Historical context

The Braginsky dynamo mechanism was of great historical importance. It was the first rigorous elucidation of a mechanism that could sustain a nearly axisymmetric field such as the Earth's, and arguably remains a mechanism more appropriate to the Earth than the two-scale, or turbulent, dynamos frequently studied following the work of Steenbeck *et al.* (1966) While the difficulty of analytic work on the nearly axisymmetric system has limited much further development, the qualitative understanding provided by Braginsky's analysis remains hugely important. Although the current widespread availability of computers capable of simulating three-dimensional dynamo action (see *Geodynamo: numerical simulations*) may naively appear to have marginalized such theories, the results of such large-scale calculations cannot easily be appreciated without the appropriate theoretical guidance.

The importance of Braginsky's contributions to our understanding of the magnetohydrodynamics of the Earth's core is difficult to overstate. The four papers published in 1964 (all were submitted between April and June of that year) constituted a huge advance in the theory of the geodynamo. In addition to the work described here on magnetic field generation, they further advocate the importance of *chemical convection* (*q.v.*), and address the effects of the core conditions on turbulence (see *Core Turbulence*). Later work by Braginsky investigated these varied topics in more detail, and his continued elucidations of the magnetohydrodynamics of geodynamo—most recently in Braginsky and Roberts (1995)—has remained enormously influential.

Graeme R. Sarson

Bibliography

Braginsky, S.I., 1964a. Self-excitation of a Magnetic Field During the Motion of a Highly Conducting Fluid. *Soviet Physics JETP*, **20**: 726–735 (English translation, 1965).

Braginsky, S.I., 1964b. Theory of the Hydromagnetic Dynamo. *Soviet Physics JETP*, **20**: 1462–1471 (English translation, 1965).

Braginsky, S.I., 1964c. Kinematic models of the Earth's hydrodynamic dynamo. *Geomagnetism and Aeronomy*, **4**: 572–583 (English translation).

Braginsky, S.I., 1964d. Magnetohydrodynamics of the Earth's core. *Geomagnetism and Aeronomy*, **4**: 698–712. (English translation).

Braginsky, S.I., 1975. Nearly axially symmetric model of the hydromagnetic dynamo of the Earth, I. *Geomagnetism and Aeronomy*, **15**: 122–128. (English translation)

Braginsky, S.I. and Roberts, P.H., 1995. Equations concerning convection in Earth's core and the geodynamo. *Geophysical and Astrophysical Fluid Dynamics*, **79**: 1–97.

Kumar, S. and Roberts, P.H., 1975. A three-dimensional kinematic dynamo. *Proceedings of the Royal Society of London A*, **344**: 235–258.

Moffatt, H.K. 1978. *Magnetic Field Generation in Electrically Conducting Fluids*. Cambridge: Cambridge University Press.

Parker, E.N., 1955. Hydromagnetic dynamo models. *Astrophysical Journal*, **121**: 293–314.

Roberts, P.H., 1987. Origin of the main field: dynamics. In Jacobs, J.A. (ed.) *Geomagnetism, Volume 2*. London: Academic Press, pp. 251–306.

Sarson, G.R. and Busse, F.H., 1998. The kinematic dynamo action of spiralling convective flows. *Geophysical Journal International*, **133**: 140–158.

Soward, A.M., 1972. A kinematic theory of large magnetic Reynolds number dynamos. *Philosophical Transactions of the Royal Society of London A*, **272**: 431–462.

Steenbeck, M., Krause F., and Rädler, K.-H., 1966. A calculation of the mean electromotive force in an electrically conducting fluid in turbulent motion, under the influence of the Coriolis forces. *Zeitschrift für Naturforschung* **21**a: 369–376 (English translation: Roberts, P.H. and Stix, M. The turbulent dynamo: a translation of a series of papers by Krause, F., Rädler, K.-H., and Steenbeck, M. *Technical Note* 60, NCAR, Boulder CO, 1971).

Tough, J.G., 1967. Nearly symmetric dynamos. *Geophysical Journal of the Royal Astronomical Society*, **13**: 393–396 (Corrigendum: *ibid.*, 1969, 15, 343).

Tough, J.G. and Roberts, P.H., 1968. *Physics of the Earth and Planetary Interiors*, **1**: 288–296.

Cross-references

Convection, Chemical
Convection, Nonmagnetic Rotating
Core Turbulence
Cowling's Theorem
Dynamo Waves
Dynamo, Bullard-Gellman
Dynamo, Model-Z
Dynamos, Kinematic
Dynamos, Mean Field
Geodynamo, Numerical Simulations
Magnetohydrodynamics
Reversals, Theory
Thermal Wind

DYNAMO, BULLARD-GELLMAN

The Bullard-Gellman dynamo, presented by *Edward Crisp Bullard* (*q.v.*) and Harvey Gellman in a pioneering paper of 1954, was the first convincing quantitative model for dynamo action in a fluid sphere. It prescribed a specific velocity, intended as a highly idealized model of the motion in the Earth's core, and considered the kinematic growth of magnetic field produced by this flow (see *Dynamos, kinematic*). Bullard and Gellman considered this problem both theoretically and computationally. In terms of the dynamo explanation for the Earth's magnetic field, this work was ground-breaking in several respects.

The study was formulated using the mathematical framework of vector *spherical harmonics* (*q.v.*). In representing the magnetic field and the velocity in this form—and in analyzing the induction equation determining the time-evolution of magnetic field in the same terms—Bullard and Gellman introduced the formalism most commonly used for dynamo theory today. (A similar treatment had earlier been proposed by Elsasser (1946), but Elsasser had used vector spherical harmonics to represent the magnetic vector potential, rather than the magnetic field itself; and the former option constitutes an unsuitable choice of gauge. Takeuchi and Shimazu (1953) were also engaged in a parallel development.)

At the time, the plausibility of homogeneous fluid dynamos was far from clear; no examples were known, and *Cowling's theorem* (*q.v.*) seemed to shed doubt on the whole concept. The formalism developed by Bullard and Gellman elucidated the practical possibilities of such dynamos. Regardless of their conclusions for the specific velocity model they proposed, their analysis made the mechanisms of kinematic dynamo action the possible subject of detailed study. Indeed, one of the first rigorously proven examples of dynamo action in a sphere—the "stasis" dynamo of Backus (1958)—relies explicitly on the spherical harmonic decomposition developed by Bullard and Gellman.

In choosing to work with a highly idealized flow—which could be represented by a simple, concrete set of vector spherical harmonics—Bullard and Gellman inspired a long series of subsequent investigations into the abstract possibilities of kinematic dynamo action. Many later kinematic dynamos were direct developments of the original Bullard-Gellman flow (e.g., Lilley, 1970; Kumar and Roberts, 1975; Sarson 2003), and many insights into dynamo action were obtained from this "family" of models.

The form of flow they chose—consisting of two distinct parts—was motivated by considerations both of kinematic dynamo action and of the fluid dynamics of the Earth's core. One constituent was an axisymmetric azimuthal flow, of nonuniform angular velocity ("differential rotation"); such a flow, of a form easily excited in a rotating fluid, can shear dipolar magnetic field into azimuthal field. Since Cowling's theorem had shown a purely axisymmetric system to be incapable of sustained dynamo action, a nonaxisymmetric constituent was also included; this involved radial motions, schematically emulating buoyancy-driven convection in the core. Although this component was chosen in a particularly simple form (for reasons of computational convenience), Bullard and Gellman clearly anticipated more "columnar" structures, in line with the *Proudman-Taylor theorem* (*q.v.*); they correctly noted that the dynamo action of such structures would be similar to that of the flow they adopted. Their paper explicitly discusses how this constituent of flow might act to create dipolar field from azimuthal field, along the lines elucidated by Parker (1955) and later formalized in the theory of *mean field dynamos* (*q.v.*).

Even with this relatively simple velocity, the kinematic dynamo problem remains technically daunting (involving the solution of an infinite set of coupled differential equations), and Bullard and Gellman ultimately addressed its solution computationally. In doing so, their work was one of the very first attempts to solve a physical problem governed by partial differential equations numerically. Their calculations were made using the automatic computing engine (ACE) computer of the UK National Physics Laboratory, one of the world's first programmable computers. In this respect, their work is a clear precursor of the large number of numerical simulations now routinely carried out in geodynamo theory (see *Geodynamo: numerical simulations*). Bullard and Gellman's numerical scheme uses a purely spectral treatment of the angular variations (i.e., it considers each term in the vector spherical harmonic decomposition separately, requiring the explicit consideration of all possible mutual interactions) and a finite difference discretization in radius. In this it does not differ greatly from many modern treatments (although the use of purely spectral expansions for nonlinear interactions is now considered less efficient than alternative techniques).

The 1954 paper also contains a number of other significant contributions. Perhaps most famously, it contains an "*antidynamo theorem*" (*q.v.*) stating the impossibility of dynamo action by flows with no radial component (i.e., confined to radial surfaces). The paper also contains significant discussion of the essential magnetohydrodynamics of the problem, and some detailed analysis of the applicability of the work to the Earth.

Despite their many insights and advances, the positive dynamo action Bullard and Gellman tentatively deduced from their calculations was later shown to be an artifact of the low numerical resolution possible at the time (a possibility which the authors themselves had been careful to acknowledge). Nevertheless, the Bullard-Gellman dynamo introduced many methods for solving the induction equation in spherical geometry, took several major steps towards validating the dynamo explanation of the geomagnetic field, and can justly be considered the forerunner of many later computational models. For all these reasons, it stands as an important milestone in the development of dynamo theory.

Graeme R. Sarson

Bibliography

Backus, G.E., 1958. A class of self-sustaining dissipative spherical dynamos. *Annals of Physics*, **4**: 372–447.

Bullard, E.C., and Gellman, H., 1954. Homogeneous dynamos and terrestrial magnetism. *Proceedings of the Royal Society of London, Series A*, **247**: 213–278.

Elsasser, W.M., 1946. Induction effects in terrestrial magnetism: Part I. Theory. *Physical Review*, **69**: 106–116.

Kumar, S., and Roberts, P.H., 1975. A three-dimensional kinematic dynamo. *Proceedings of the Royal Society of London, Series A*, **344**: 235–258.

Lilley, F.E.M., 1970. On kinematic dynamos. *Proceedings of the Royal Society of London, Series A*, **316**: 153–167.

Parker, E.N., 1955. Hydromagnetic dynamo models. *Astrophysical Journal*, **121**: 293–314.

Sarson, G.R., 2003. Kinematic dynamos driven by thermal wind flows. *Proceedings of the Royal Society of London, Series A*, **459**: 1241–1259.

Takeuchi, H., and Shimazu, Y., 1953. On a self-exciting process in magneto-hydrodynamics. *Journal of Geophysical Research*, **58**: 497–518.

Cross-references

Antidynamo and Bounding Theorems
Bullard, Edward Crisp (1907–1980)
Cowling's Theorem
Dynamos, Kinematic
Dynamos, Mean Field
Geodynamo, Numerical Simulations
Harmonics, Spherical
Proudman-Taylor Theorem

DYNAMO, DISK

It is generally believed that the Earth's magnetic field results from some form of electromagnetic induction, producing electric currents flowing in the Earth's fluid, electrically conducting outer core (see *Geodynamo*). The details of the process are not known and several questions have not been answered, such as how the field reverses its polarity. The partial differential equations that govern the generation of the field are complex, nonlinear, and contain physical quantities whose values in the Earth's core are not known with any certainty. Various mathematical models have therefore been proposed, which might simulate reversals of the field.

One of the simplest is the homopolar disk dynamo suggested by Bullard (1955). It is based on the Faraday disk generator and consists of an electrically conducting disk that can be made to rotate on an axle by an applied torque. If the disk rotates in an axial magnetic field, a radial EMF will be produced between the axle and the edge of the disk. If this were all, the EMF would be balanced by an electric charge distribution on the edge and surfaces of the disk and no current would flow. However, if one end of a stationary coil coaxial with the disk and axle is joined to the edge of the disk by a sliding contact (a brush) and the other end is joined to another sliding contact on the axle, a current will flow through the coil and an axial magnetic field will be produced. The system becomes a dynamo when the rotation is fast enough that the induced field becomes equal to the field required to produce it. No external source of field or current is then required and no part of the system needs to be ferromagnetic.

The system can be represented by two nonlinear ordinary differential equations. If the disk is driven by a constant torque G, the equation of motion is

$$C\dot{\omega} - G = MI^2 \qquad \text{(Eq. 1)}$$

where C is the moment of inertia of the disk, ω its angular velocity, I the current and $2\pi M$ the mutual induction of the coil and disk. The equation governing the current is

$$L\dot{I} + RI = M\dot{\omega}\dot{I} \qquad \text{(Eq. 2)}$$

where L and R are the inductance and resistance of the coil, respectively.

Since Eq. (1) is quadratic in I and Eq. (2) linear and homogeneous in I, the dynamo can produce a current in either direction. It cannot, however, switch from one direction to the other, since, if I is zero, so is $\dot{I}$. In order to produce reversals, another term must be introduced into Eq. (2) so that $\dot{I}$ is not proportional to I. This has been achieved in two ways. *Rikitake* (*q.v.*, 1958) suggested using two coupled disk dynamos in series (see Figure D11) and Malkus (1972) suggested an impedance between the brush and the coil and a shunt connected across the coil of a single dynamo to introduce a phase difference. Malkus' suggestion was investigated by Robbins (1977). Bullard and Gubbins (1973) studied the case when the disks oscillated periodically as a model of dynamo action by gravity waves in a stable core and

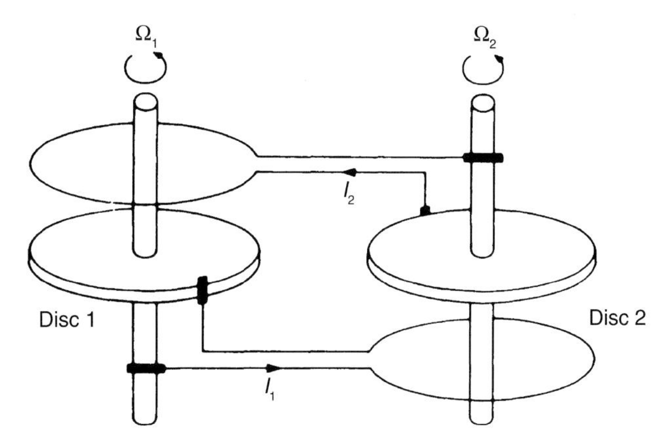

Figure D11 Coupled system of two disk dynamos (after Rikitake, 1958).

found that a single disk dynamo would not generate a magnetic field but that two coupled disk dynamos would, provided the oscillation frequency were not too large.

A number of people have investigated the properties of disk dynamos for values of various parameters and have shown that the system may oscillate randomly between two states, which would correspond to reversals of the geomagnetic field (Figure D12). Such behavior has been found in many other fields and is called chaos. As with other nonlinear systems displaying chaotic motion, the behavior of disk dynamos depends critically on initial conditions. In some models, the amplitude of the oscillations grows monotonically with time—when the amplitude exceeds some particular value, the system flips to the opposite polarity (Figure D12). Hide (1995) questioned the assumption that dissipation due to mechanical friction can be neglected compared with Ohmic effects and suggested that friction may even prevent dynamo action by destroying any initial transients in the system. He demonstrated this for the Rikitake double disk dynamo by adding an extra term $-k\omega_1$, to the right hand side of Eq. (1) for one disk dynamo, and $-k\omega_2$ to the similar equation for the second disk dynamo. The development of the system depends on the values of the parameters and, for some values, chaotic oscillations are inhibited and a dynamo cannot function.

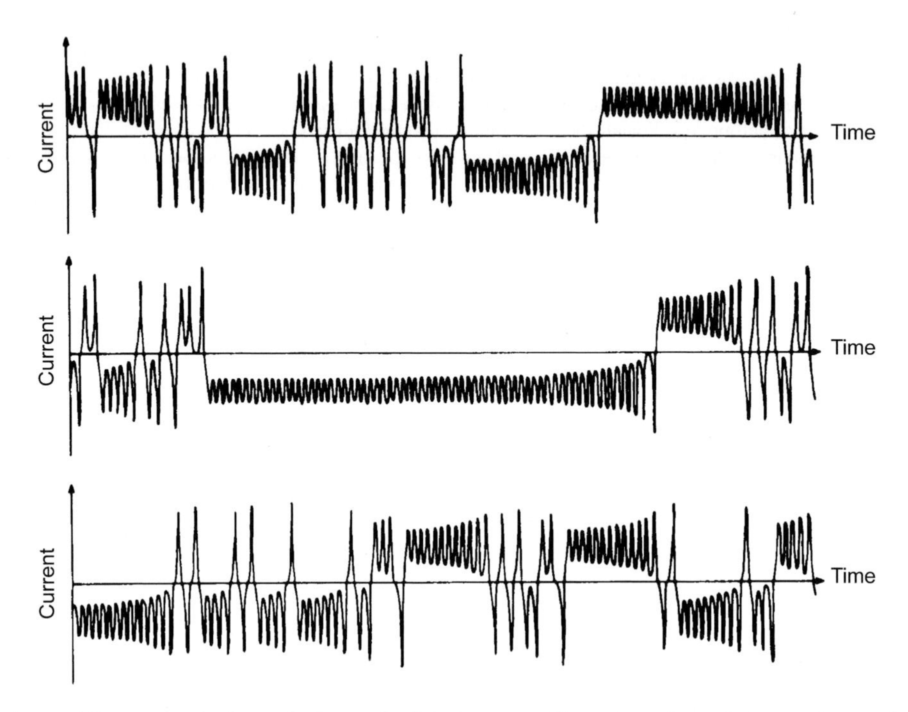

Figure D12 Oscillations of the current in the coil of a single disk dynamo with an impedance between the brush and the coil and a shunt connected across the coil (after Robbins, 1977). The three plots form one continuous time sequence.

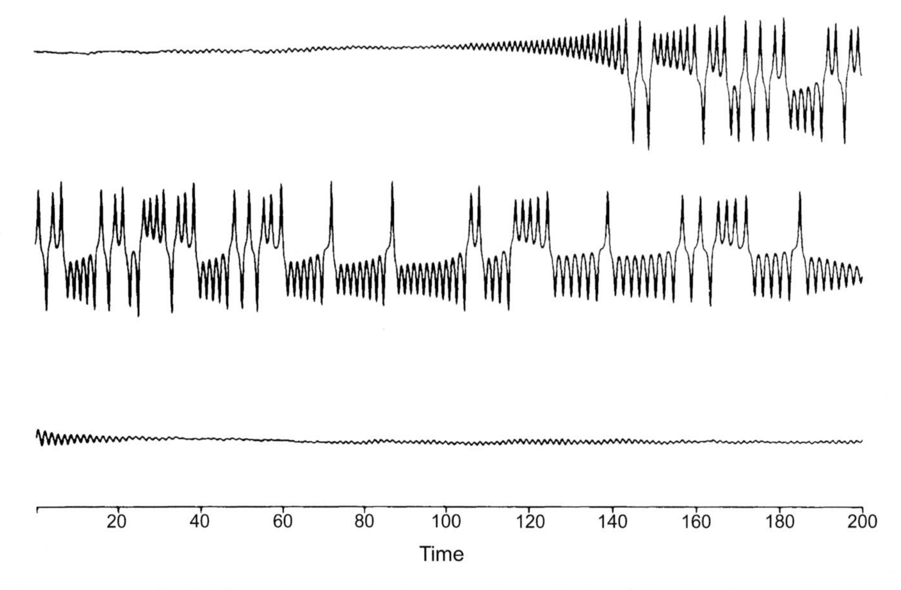

Figure D13 Plot of the current on the Robbins dynamo with brown noise excitation (after Crossley *et al.*, 1986). The three plots form one continuous time sequence.

Crossley *et al.* (1986) investigated the Robbins (1977) dynamo when another component was added into the equations of motion to simulate irregularities in the hydrodynamics of the Earth's core. These irregularities could be due to fluid dynamical turbulence, variations in the material properties, or physical and chemical processes. They represented this "noise" by adding a small time-dependent term to the constant driving torque. Three different types of noise were investigated. Figure D13 shows variations in the current for the Robbins dynamo with the addition of "brown" noise (random walk). An interesting result of their models is that the addition of some noise into the system can either add to or reduce the excursions of the magnetic field attributed to the nonlinear behavior of the equations of motion.

Hide and his coworkers have investigated more sophisticated self-exciting disk dynamo systems. In one study (Hide *et al.*, 1996), they considered a nonlinear electric motor connected in series with the coil of a single disk dynamo. They found that if the torque T on the armature of the motor is proportional to the current I, multiple periodic as well as chaotic persistent fluctuations are possible, but if T is proportional to I^2, persistent fluctuations are completely quenched.

The study of disk dynamos, which began as a search for models of the Earth's magnetic field, has led to the study of sets of nonlinear ordinary differential equations with important applications to other areas of geophysical fluid dynamics.

Jack A. Jacobs

Bibliography

Bullard, E.C., 1955. The stability of a homopolar dynamo. *Proceedings of the Cambridge Philosophical Society*, **51**: 744–760.

Bullard, E.C., and Gubbins, D., 1973. Oscillating disk dynamos and geomagnetism. In *Flow and fracture of rocks*, Monograph 16, American Geophysical Union, p 325–328.

Crossley, D., Jensen, O., and Jacobs, J., 1986. The stochastic excitation of reversals in simple dynamos. *Physics of the Earth and Planetary Interiors*, **42**: 143–153.

Hide, R., 1995. Structural instability of the Rikitake dynamo. *Geophysical Research Letters*, **22**: 1057–1059.

Hide, R., Skeldon, A.C., and Acheson, D.J., 1996. A study of two novel self-exciting single-disk homopolar dynamos: theory. *Proceedings of the Royal Society of London A*, **452**: 1369–1395.

Malkas, W.V.R., 1972. Reversing Bullard's dynamo. *EOS, Transactions of the American Geophysical Union*, **53**: 617.

Rikitake, T., 1958. Oscillations of a system of disk dynamos. *Proceedings of the Cambridge Philosophical Society*, **54**: 89–105.

Robbins, K.A., 1977. A new approach to sub-critical instability and turbulent transitions in a simple dynamo. *Mathematical Proceedings of the Cambridge Philosophical Society*, **82**: 309–325.

Cross-references

Geodynamo
Rikitake, Tsuneji (1921–2004)

DYNAMO, GAILITIS

The Gailitis dynamo is a simple flow pattern in an electrically uniformly conducting (σ) incompressible fluid for which there is a simple mathematical proof that it can generate magnetic field. In that sense, it resembles the *Backus* (*q.v.*), *Herzenberg* (*q.v.*), and *Ponomarenko* (*q.v.*) *dynamos*. Together they supported the belief in dynamo theory at a time when neither experiment nor computation was able to demonstrate dynamo action in uniform fluid conductor.

Geometry

The Gailitis flow is stationary, with axial symmetry, and with velocity **v** in meridian planes only. In the Gailitis (1970) version the whole flow is concentrated in two toroidal eddies. In each of them the fluid rotates round the bent central line like a ring vortex as shown on Figure D14. The exterior is a stationary conductor to infinity.

Self-excitation

As in Herzenberg's dynamo, each eddy alone cannot excite magnetic field but both together can. The lower vortex excites the upper one and vice versa. At an appropriate flow speed, the mutual excitation supports a steady magnetic field. As the flow is axisymmetric, the field has to have angular dependence $B_\varphi \sim \exp(im\varphi)$. The integer m cannot be zero because of *Cowling's anti-dynamo* (*q.v.*) theorem.

Mathematics

For mathematical simplicity, both vortices are assumed to be thin and the motion is assumed to be being fast:

$$R \ll \max(a, z_0) \text{ and } Rm = \mu_0\sigma \int_0^R (r/R)^2 v\,\mathrm{d}r \gg 1.$$

At high Rm, field-lines are semifrozen to the moving fluid, hence the field pattern inside any vortex has the asymptotic form $B_\varphi = \rho b \exp(im\varphi)/a$ (field proportional to the length of field-line) and $B_r = -irmb\exp(im\varphi)/(2a)$ (from div B = 0). As the upper vortex is a mirror reflection of the lower one, at self-excitation the normalization factors b in each vortex have the same magnitude but may differ in sign. Inserting these expressions on the right side of the induction equation written in the integral form gives

$$\mathbf{B}(\mathbf{r}) = \frac{\mu_0\sigma}{4\pi}\int \frac{\mathrm{d}^3r'}{|\mathbf{r}-\mathbf{r}'|}\mathrm{curl}[\mathbf{v}(\mathbf{r}') \times \mathbf{B}(\mathbf{r}')],$$

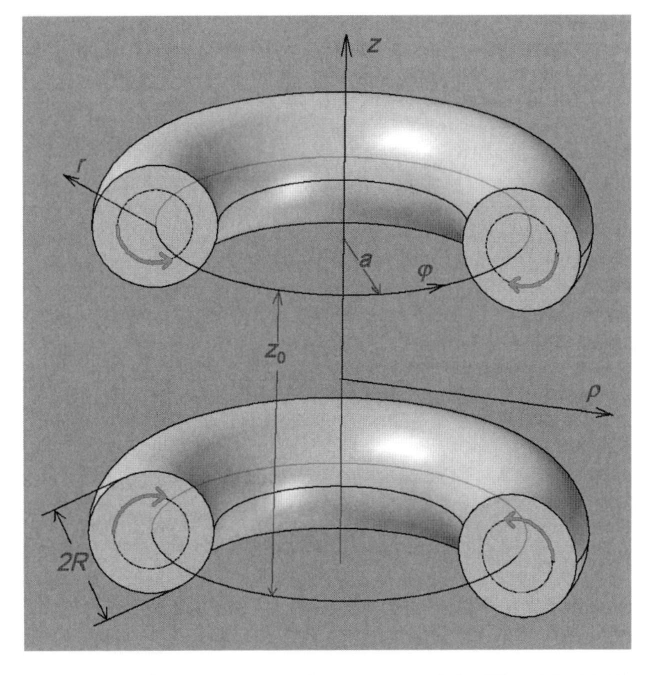

Figure D14 Flow concentrated in two toroidal eddies. The fluid rotates round the bent central line like a ring vortex.

and integrating over the first vortex volume (outside $\mathbf{v} = 0$) with $\mathbf{r}$ fixed at the second vortex, the left side leads to the excitation condition $Rm = \pm(a/R)^2 T_m(z_0/a)$. $T_m(z_0/a)$ is a standard integral. Computed values are given in Figure D15. The dynamo generates a field with different values of m depending on z_0/a. The dominant value of m is the one with lowest $T_m(z_0/a)$. The sign of Rm determines the flow direction. If the flow is directed as on Figure D14, the generated B_φ in both vortices is in the same direction. If the flow direction is opposite to that in Figure D14 the B_φ in the upper vortex is opposite to that in the lower vortex. The absolute value of Rm remains the same.

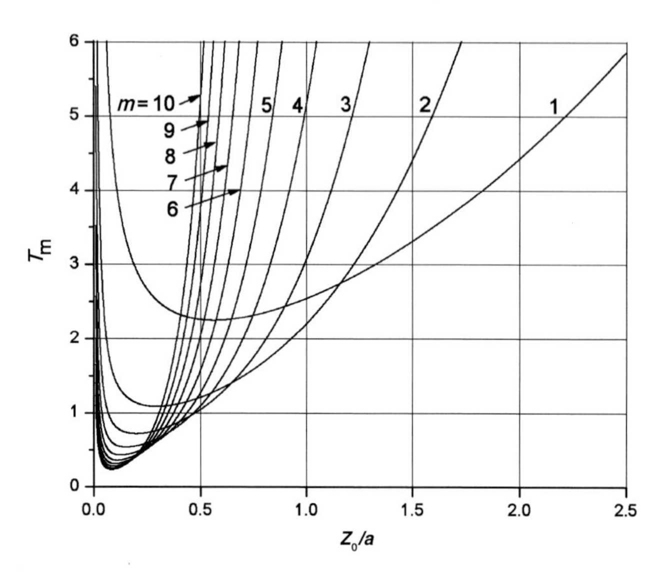

Figure D15 Computed values of the induction equation. The dynamo generates a field with different values of m depending on z_0/a.

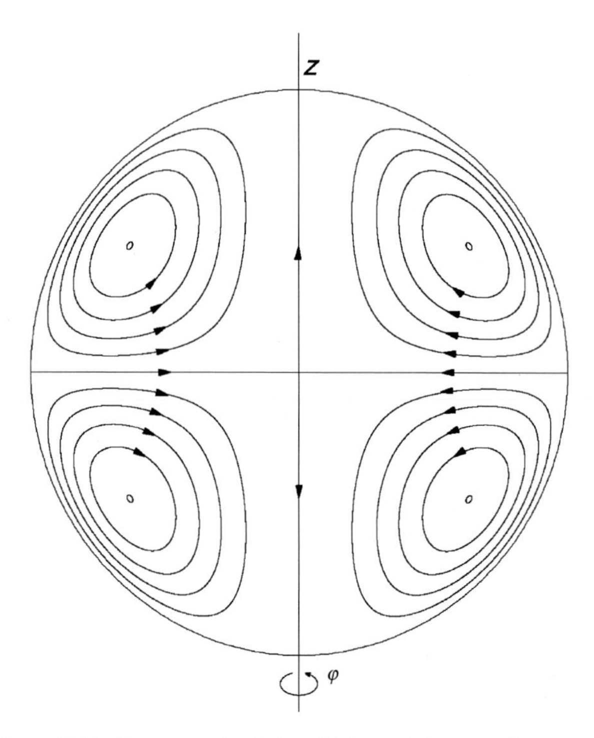

Figure D16 Computer simulation thick vortices in a sphere. Shown is the simplest possible polynomial velocity distribution.

Thick vortices

The thinness of the vortices is not essential for dynamo action. Computer simulation of a pair of thick vortices in a sphere works equally well. For the simplest possible polynomial velocity distribution (Figure D16) the critical $Rm = \mu_0 \sigma v_{max} R$ is 1735 (Gailitis, 1995). As for thin vortices, the absolute magnitude of the critical Rm remains unchanged when the flow direction is altered.

Helicity

The Ponomarenko relies on helicity (**v**.curl **v**). The Gailitis dynamo has zero helicity, like the "second order" spatially periodic dynamos because the flow lies in meridian planes: there is no azimuthal motion. On the other hand, the critical $Rm = 1735$ for Gailitis is two orders of magnitude higher than for Ponomarenko. Adding some azimuthal flow to the thick vortices flow also lowers the critical Rm more than by an order of magnitude. The Gailitis dynamo therefore demonstrates that helicity is not strictly necessary for dynamo action but is helpful.

Agris Gailitis

Bibliography

Gailitis, A., 1970. Self-excitation of a magnetic field by a pair of annular vortices. *Magnetohydrodynamics*, **6**: 14–17.

Gailitis, A., 1995. Magnetic field generation by the axisymmetric conducting fluid flow in a spherical cavity of a stationary conductor. 2. *Magnetohydrodynamics*, **31**: 38–42.

Cross-references

Alfvén's Theorem and the Frozen Flux Approximation
Antidynamo and Bounding Theorems
Cowling's Theorem
Dynamo, Backus
Dynamo, Herzenberg
Dynamo, Ponomarenko
Dynamos, Periodic
Magnetohydrodynamics

DYNAMO, HERZENBERG

Definition of the model

Human-made dynamos are complex multiconnected devices, but those operating in Nature have to function in geometrically simple (usually almost spherical) bodies of electrically conducting fluid, such as Earth's core. During the early development of dynamo theory, it seemed possible that no motion within such a simple "homogeneous" system could maintain a magnetic field, a feeling reinforced by some antidynamo theorems; see *Antidynamo and bounding theorems* and *Cowling's theorem.* Theoreticians sought either to strengthen antidynamo theorems until it could be proved that *homogeneous dynamos* could not exist, or to devise specific examples that worked. In these endeavors, it was simplest to focus on the electrodynamics, assuming that the velocity **V** of the conductor is given, and to ignore the dynamical questions of how **V** is maintained and how a magnetic field **B** would affect it; see *Dynamos, kinematic*. It is then irrelevant whether

V is dynamically possible or not; it is necessary only to find a **V** that can sustain a nonzero **B**.

The first two kinematic dynamos were devised in 1958, one of which is described here; see also *Dynamo, Backus.* These models proved that the search for a general antidynamo theorem was in vain. In 1958, electronic computers were not sufficiently powerful to find homogeneous dynamos. Moreover, numerical integrations could not provide the mathematical "rigor" that the subject demanded in the face of the anti-dynamo theorems. Backus and Herzenberg had to provide demonstrations by rigorous mathematical analysis. In each case, this depended on an ingenious asymptotic argument based on the smallness of a well-chosen parameter.

In essence, the Herzenberg dynamo consists of two spherical rotors, $\mathcal{R}_1$ and $\mathcal{R}_2$, each of radius a, embedded in an infinite conductor of the same magnetic diffusivity, η, with which they are in perfect electrical contact; see Figure D17. The rotors turned steadily with angular velocities, $\boldsymbol{\omega}_1$ and $\boldsymbol{\omega}_2$, different in direction, but having the same magnitude: $\omega = |\boldsymbol{\omega}_1| = |\boldsymbol{\omega}_2|$. Their centers are denoted by O_1 and O_2 where $\overrightarrow{O_1O_2} = \mathbf{R}$. Provided a suitably defined *magnetic Reynolds number*, Rm, is large enough, the rotors can re-enforce one another magnetically in the sense that the magnetic field, $\mathbf{B}_1$, created by $\mathcal{R}_1$ (as it rotates in the field, $\mathbf{B}_2$, produced by $\mathcal{R}_2$) will be large enough for $\mathcal{R}_2$ (rotating in the field $\mathbf{B}_1$) to be able to generate the required $\mathbf{B}_2$. A suitable magnetic Reynolds number is

$$\mathrm{Rm} = \omega a^5/\eta R^3. \qquad \text{(Eq. 1)}$$

The model can sustain a magnetic field if Rm attains a critical value, $\mathrm{Rm_c}$, of order unity. Since **V** is time-independent, **B** is then steady (the DC dynamo) or oscillatory (the AC dynamo). Herzenberg, following the general trend of the times, sought DC dynamos only. The small parameter of Herzenberg's analysis is a/R, or equivalently $\eta/\omega a^2$, since $\eta/\omega a^2 = \mathrm{Rm}^{-1}(a/R)^3$ and $\mathrm{Rm} = \mathrm{O}(1)$.

Dynamos conceptually similar to Herzenberg's have been proposed for binary stellar systems, the magnetic field created by one star providing the field needed by the other star to produce the field required by its partner (Dolginov and Urpin, 1979; Brandenburg *et al.*, 1998).

Ideas behind Herzenberg's evaluation of $\mathrm{Rm_c}$

In 1949, Bullard wrote a seminal paper on induction by spherical rotors, and Herzenberg (1958) made ingenious use of his results; see also Herzenberg and Lowes (1957). Herzenberg's analysis succeeds because of two basic simplifications. First, the rotors are well separated; $R \gg a$.

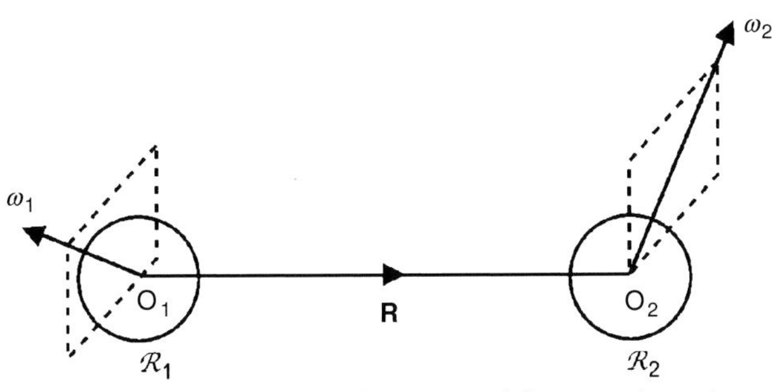

Figure D17 The Herzenberg dynamo model. Two spherical rotors, $\mathcal{R}_1$ and $\mathcal{R}_2$, are embedded in an infinite medium of the same electrical conductivity with which they are in perfect electrical contact. Their centers O_1 and O_2 are separated by **R**, a distance large compared with their radii a. They rotate with angular velocities $\boldsymbol{\omega}_1$ and $\boldsymbol{\omega}_2$ about directions lying in the planes indicated by the *dashed rectangles.*

Second, they are rapidly rotating; $\mathrm{R}_\omega \equiv \omega a^2/\eta \gg 1$. Order one values of Rm arise as the product $\mathrm{R}_\omega (a/R)^3$ of one very large number and one very small one.

To evaluate the field $\mathbf{B}_1$ produced by the rotation of $\mathcal{R}_1$, it is necessary to know, throughout $\mathcal{R}_1$, the field $\mathbf{B}_2$ created by $\mathcal{R}_2$. This field is almost uniform, because of the remoteness of $\mathcal{R}_2$ implied by the first simplification ($R \gg a$). More precisely, $\mathbf{B}_2$ can be expressed in a rapidly convergent Taylor expansion as

$$\mathbf{B}_2(\mathbf{x}) = \mathbf{B}_2(O_1) + \mathbf{x}\cdot(\nabla\mathbf{B}_2)O_1 + \cdots, \qquad \text{(Eq. 2)}$$

where $\mathbf{x}$ is the position vector from O_1. For $r = |\mathbf{x}|$ of order a, the ratio of successive terms in Eq. (2) is of order $a/R \ll 1$, and it might seem at first sight that only the first term in the expansion Eq. (2) is needed, but this is not the case as is shown below.

The second simplification reduces the complications of evaluating $\mathbf{B}_1$. The field $\mathbf{B}_2$ is split into two parts,

$$\mathbf{B}_2 = \overline{\mathbf{B}}_2 + \mathbf{B}_2', \qquad \text{(Eq. 3)}$$

where $\overline{\mathbf{B}}_2$ is symmetric with respect to the direction of $\boldsymbol{\omega}_1$ and $\mathbf{B}_2'$ is the remaining asymmetric part. This separation of $\mathbf{B}_2$ can be achieved by defining cylindrical coordinates (s, ϕ, z), with z-axis parallel to $\boldsymbol{\omega}_1$. The ϕ-independent components of Eq. (2) then define $\overline{\mathbf{B}}_2$:

$$\overline{\mathbf{B}}_2 = A_1^{(1)}\hat{\boldsymbol{\omega}}_1 + a^{-1}A_2^{(1)}[3(\hat{\boldsymbol{\omega}}_1\cdot\mathbf{x})\hat{\boldsymbol{\omega}}_1 - \mathbf{x}] + \cdots, \qquad \text{(Eq. 4)}$$

where $\hat{\boldsymbol{\omega}}_1 = \boldsymbol{\omega}_1/\omega$ and

$$A_1^{(1)} = \hat{\boldsymbol{\omega}}_1\cdot\mathbf{B}_2(O_1), \quad A_2^{(1)} = \frac{1}{2}a\,\hat{\boldsymbol{\omega}}_1\cdot[\hat{\boldsymbol{\omega}}_1\cdot(\nabla\mathbf{B}_2)O_1]. \qquad \text{(Eq. 5)}$$

The advantage of the division in Eq. (3) becomes apparent when the fields B_1' and $\overline{\mathrm{B}}_1$ induced by $\mathbf{B}_2'$ and $\overline{\mathbf{B}}_2$ are compared. In the frame of reference moving with a point on the surface $\mathcal{S}_1$ of $\mathcal{R}_1$, the inducing field is highly oscillatory, and the induced currents are therefore confined to a thin "skin" at $\mathcal{S}_1$. These currents shield the interior of $\mathcal{R}_1$ almost completely from the inducing field $\mathbf{B}_2'$. Since they cancel each other out within $\mathcal{R}_1$, the fields $\mathbf{B}_1'$ and $\mathbf{B}_2'$ must have similar magnitudes everywhere near $\mathcal{R}_1$. This is in sharp contrast to $\overline{\mathbf{B}}_2$ and $\overline{\mathbf{B}}_1$. Since $\overline{\mathbf{B}}_2$ is steady in the frame rotating with $\mathcal{R}_1$, there is no skin of eddy currents at $\mathcal{S}_1$, and the induced field $\overline{\mathbf{B}}_1$ is proportional to ω; see Eq. (6) below. (This is sometimes called "the omega effect"; see *Magnetohydrodynamics.*)

In short, even though $\overline{\mathbf{B}}_2$ and $\mathbf{B}_2'$ are comparable in magnitude, $\overline{\mathbf{B}}_1 = \mathrm{O}(\mathrm{R}_\omega\overline{\mathbf{B}}_2) = \mathrm{O}(\mathrm{R}_\omega\mathbf{B}_2') = \mathrm{O}(\mathrm{R}_\omega\mathbf{B}_1') \gg \mathbf{B}_1'$. The $\mathbf{B}_1'$ part of $\mathbf{B}_1$ can therefore be ignored.

The Bullard (1949) theory shows that the first term in Eq. (4) creates a quadrupolar field; the second produces a combination of dipolar and octupolar fields, of which the latter can be discarded since, for $r \gg a$ (where $\mathbf{B}_1$ is required in order to evaluate the field $\mathbf{B}_2$ generated by the rotation of $\mathcal{R}_2$), the octupolar field is smaller than the dipolar field by a factor of order $(a/R)^2$. For $r \gg a$,

$$\mathbf{B}_1 = \mathrm{R}_\omega\left\{\frac{2}{15}A_2^{(1)}\frac{a^2}{r^3}(\hat{\boldsymbol{\omega}}_1\times\mathbf{x}) - \frac{1}{5}A_1^{(1)}\frac{a^3}{r^5}(\hat{\boldsymbol{\omega}}_1\cdot\mathbf{x})(\hat{\boldsymbol{\omega}}_1\times\mathbf{x}) + \mathrm{O}\left(A_3^{(1)}\frac{a^4}{r^4}\right)\right\}, \qquad \text{(Eq. 6)}$$

$\mathbf{B}_1'$ having been ignored (see above). The necessity of retaining the second term in Eqs. (2) and (4) is now apparent. For $n > 1$ the part of $\mathbf{B}_1$ created by the $A_n^{(1)}$ term is of order $\mathrm{R}_\omega A_n^{(1)}(a/R)^n$ at the rotor $\mathcal{R}_2$, and in particular it is of order $\mathrm{R}_\omega A_2^{(1)}(a/R)^2$ for $n = 2$. The uniform field $A_1^{(1)}$ however makes a contribution to $\mathbf{B}_1$ only of order $\mathrm{R}_\omega A_1^{(1)}(a/R)^3$. It follows that the $A_1^{(1)}$ and $A_2^{(1)}$ are equally effective in inducing field but that, to leading order, all terms not shown in the expansions of Eqs. (2) and (4) can be ignored.

The coordinate-independent way (6) of writing $\mathbf{B}_1$ is due to Gibson and Roberts (1967); see also Moffatt (1978). It eases the task of computing $A_1^{(2)}$ and $A_2^{(2)}$, defined at O_2 in analogy with Eq. (5). For example, on substituting $\mathbf{x} = \mathrm{R}$ in Eq. (6), it is seen that

$$A_1^{(2)} = \Lambda\left[A_1^{(1)}\cos\Theta_1 - \frac{2R}{3a}A_2^{(1)}\right], \qquad \text{(Eq. 7)}$$

where

$$\Lambda = \frac{1}{5}\mathrm{R}_\omega\left(\frac{a}{R}\right)^3\widehat{\mathbf{R}}\cdot(\widehat{\boldsymbol{\omega}}_1\times\widehat{\boldsymbol{\omega}}_2) = \frac{1}{5}\mathrm{Rm_c}\sin\Theta_1\sin\Theta_2\sin\Phi, \qquad \text{(Eq. 8)}$$

and $\widehat{\mathbf{R}} = \mathbf{R}/R$. Here Θ_i is the angle between $\boldsymbol{\omega}_i$ and $\mathbf{R}$, and Φ is the angle between the plane defined by $\boldsymbol{\omega}_1$ and $\mathbf{R}$ and the plane defined by $\boldsymbol{\omega}_2$ and $\mathbf{R}$. Similarly $A_2^{(2)}$ is obtained by setting $\mathbf{x} = \mathbf{R}$ after operating on Eq. (6) with ∇.

Analogous expressions for $A_1^{(1)}$ and $A_2^{(1)}$ are obtained by reversing $\mathbf{R}$ and the roles of $\mathcal{R}_1$ and $\mathcal{R}_2$. In this way four simultaneous equations are obtained for the four constants $A_i^{(j)}(i,j = 1,2)$. In the marginal state, these equations must have a nontrivial solution. This defines the critical values of Λ:

$$[\Lambda^2(\cos\Theta_1\cos\Theta_2 - \sin\Theta_1\sin\Theta_2\cos\Phi) - 3]^2 = 0. \qquad \text{(Eq. 9)}$$

By Eq. (8), this condition clearly cannot be satisfied if Θ_1, Θ_2 or Φ is zero. The configuration shown in Figure 1, in which $\Theta_1 = \Theta_2 = \frac{1}{2}\pi$ and $\Phi = \frac{3}{4}\pi$ satisfies Eq. (9) for $\mathrm{Rm_c} = 5 \times 3^{1/2} \times 2^{3/4} \approx 14.56$. If Eq. (9) is satisfied, it is also satisfied if the sense of rotation of both rotors is reversed but that, if only one of $\boldsymbol{\omega}_1$ and $\boldsymbol{\omega}_2$ is reversed, Eq. (9) is no longer obeyed.

Later developments

The shape of the rotors in Herzenberg's model is not significant, so long as they are axisymmetric (Gailitis, 1973). Also, their solid-body motion can be replaced by a continuously varying internal angular velocity (Gibson, 1969; Moffatt, 1978), so avoiding the unphysical "tearing" of the "fluid" at the surfaces of the rotors. What is significant for such "point vortices" is their angular momenta (Roberts, 1971; Gailitis, 1973), and multirotor models provide a crude representation of field creation by helical turbulence (Kropachev, 1966; Roberts, 1971).

There is a subtlety. In deriving Eq. (9), terms of order $(a/R)^4$ have been discarded. If they were restored, the right-hand side of Eq. (9), though small, might be negative. In that case Eq. (9) would prove nothing, since a complex $\mathrm{Rm_c}$ for a DC dynamo is meaningless. Herzenberg (1958) recognized this difficulty and evaded it in a way relevant to experimental simulations of his dynamo. Instead of assuming that the conductor fills all space, he supposed that the rotors are embedded in a large conducting sphere, surface $\mathcal{S}$, surrounded by an insulator. The electric currents produced by the rotors are "reflected" back by $\mathcal{S}$, so creating new terms on the right-hand side of Eq. (9). He established situations in which these terms are positive. His idea also raises an interesting possibility: perhaps a single rotor can act as a dynamo by re-enforcing the fields reflected by $\mathcal{S}$? This was first suggested by Herzenberg and Lowes (1957); see their case C on p. 568. Kropachev (1966) proposed a specific model of this type. See also Gibson (1968b).

Gibson (1968a) attempted to evade the difficulty raised by the sign of the missing right-hand side of Eq. (9) by taking Herzenberg's analysis to higher order in a/R. Unfortunately the modified Eq. (9) that resulted was again a perfect square, and the difficulty remained. This invited the speculation that, if the analysis could be carried out to *all* orders in a/R, a perfect square would result. Gibson (1968b) did, however, resolve the difficulty in a totally different way. He pointed out that it did not arise for a three-rotor system. He obtained $\mathrm{Rm_c}$ for such a dynamo, again in the asymptotic case of large R_ω and small a/R. See also Kropachev *et al.* (1977a, b); Moffatt (1978).

Gailitis (1973) argued that, if the missing right-hand side of Eq. (9) is positive, there should be two distinct marginal solutions with different $\mathrm{Rm_c}$ but that, if it is negative, an AC dynamo would exist. This conclusion was re-enforced in the analytical work of Brandenburg *et al.* (1998).

Advances in computer capability eventually made it practical to calculate numerically the three-dimensional fields produced by two rotors in the general (nonasymptotic) cases in which a/R is O(1). The integrations of Brandenburg *et al.* (1998) were carried out for a cubical box, with periodic boundaries in x and y, and bounded in z by perfect electrical conductors. As the dimension L of the box was fairly large compared with a and R, the fields reflected from the z-boundaries, and those transmitted through the xy-walls of the box into the adjacent cells of the infinite two-dimensional array of Herzenberg dynamos, were not particularly significant. Brandenburg *et al.* (1998) found that Herzenberg's asymptotic theory gave remarkably accurate values of $\mathrm{Rm_c}$ far outside its apparent range of validity, when the rotors nearly touch one another. One of their main achievements was the discovery of a new family of AC solutions. To be more specific, they considered the case $\Theta_1 = \Theta_2 = \frac{1}{2}\pi$ in which Eq. (9) requires that $\frac{1}{2}\pi < \Phi < \pi$ (or $-\pi < \Phi < -\frac{1}{2}\pi$). But they obtained AC solutions when ;$\Phi < \frac{1}{2}\pi$. It now appears that the Herzenberg dynamo functions, either as a DC dynamo or as an AC dynamo, for nearly all orientations of $\boldsymbol{\omega}_1$ and $\boldsymbol{\omega}_2$ except $\Theta_1 = \Theta_2 = 0$ and $|\Phi| = 0, \frac{1}{2}\pi, \pi$, provided Rm is large enough.

The Herzenberg dynamo provided the inspiration for the first demonstration of homogeneous dynamo action in the laboratory; see *Dynamo, Lowes-Wilkinson.*

Paul H. Roberts

Bibliography

Brandenburg, A., Moss, D., and Soward, A.M., 1998. New results for the Herzenberg dynamo: Steady and oscillatory solutions. *Proceedings of the Royal Society of London*, **A454**: 1283–1300; Erratum p. 3275.

Bullard, E.C., 1949. Electromagnetic induction in a rotating sphere. *Proceedings of the Royal Society of London*, **A199**: 413–443.

Dolginov, A.Z., and Urpin, V.A., 1979. The inductive generation of the magnetic field in binary systems. *Astronomy and Astrophysics*, **79**: 60–69.

Gailitis, A.K., 1973. Theory of the Herzenberg's dynamo. *Magnetohydrodynamics*, **9**: 445–449.

Gibson, R.D., 1968a. The Herzenberg dynamo. I, *The Quarterly Journal of Mechanics and Applied Mathematics*, **21**: 243–255.

Gibson, R.D., 1968b. The Herzenberg dynamo. II, *The Quarterly Journal of Mechanics and Applied Mathematics*, **21**: 257–267.

Gibson, R.D., 1969. The Herzenberg dynamo. In Runcorn, S.K. (ed), *The Application of Modern Physics to the Earth and Planetary Interiors.* London: Wiley, pp. 571–576.

Gibson, R.D., and Roberts, P.H., 1967. Some comments on the theory of homogeneous dynamos. In Hindmarsh, W.R., Lowes, F.J., Roberts, P.H., and Runcorn, S.K. (eds.), *Magnetism and the Cosmos.* Edinburgh: Oliver and Boyd, pp. 108–120.

Herzenberg, A., 1958. Geomagnetic dynamos. *Philosophical Transactions of the Royal Society of London*, **A250**: 543–585.

Herzenberg, A., and Lowes, F.J., 1957. Electromagnetic induction in rotating conductors. *Philosophical Transactions of the Royal Society of London*, **A249**: 507–584.

Kropachev, E.P., 1964. One mechanism of excitation of a stationary magnetic field in a spherical conductor. *Geomagnetism and Aeronomy*, **4**: 281–288.

Kropachev, E.P., 1965. Excitation of magnetic field in a spherical conductor. *Geomagnetism and Aeronomy*, **5**: 744–746.
Kropachev, E.P., 1966. Generation of a magnetic field near the boundary of a conductor, *Geomagnetism and Aeronomy*, **6**: 406–412.
Kropachev, E.P., Gorshkov, S.N., and Serebryanaya, 1977a. Model of the kinematic dynamo of three spherical eddies, I, *Geomagnetism and Aeronomy*, **17**: 343–345.
Kropachev, E.P., Gorshkov, S.N., and Serebryanaya, 1977b. Model of the kinematic dynamo of three spherical eddies, II, *Geomagnetism and Aeronomy*, **17**: 614–616.
Moffatt, H.K., 1978. *Magnetic Field Generation in Electrically Conducting Fluids*. Cambridge UK: Cambridge University Press.
Roberts, P.H., 1971. Dynamo theory. In Reid, W.H. (ed), *Mathematical Problems in the Geophysical Sciences. 2. Inverse Problems, Dynamo Theory and Tides*, Providence RI: American Mathematical Society, pp. 129–206.

Cross-references

Antidynamo and Bounding Theorems
Cowling's Theorem
Dynamo, Backus
Dynamo, Lowes-Wilkinson
Dynamos, Kinematic
Magnetohydrodynamics

DYNAMO, LOWES-WILKINSON

Bullard and Gellman (1954) made the first attempt at numerical solution of a (very crude) mathematical kinematic model of the geomagnetic self-exciting dynamo (see *Dynamo, Bullard-Gellman*). But because of computing limitation, they had to resort to hand-waving arguments as to convergence of the solution. Also, because of misinterpretation of a paper by Cowling (1934), many people believed that such self-excitation was impossible in what was essentially a homogeneous conductor; in effect such people thought that there was an "antidynamo" theorem (see *Cowling's theorem*). Then in 1958 two papers, by Backus and by Herzenberg, each showed algebraically, for quite different systems of motion, that self-excitation was in fact possible. The motions were not at all Earth-like, but these "existence proofs" showed that there could be no such antidynamo theorem.

Herzenberg's approach (see *Dynamo, Herzenberg*) involved the interaction between two rotating solid spherical eddies, embedded in a larger solid sphere, the whole system being of uniform electrical conductivity. To first order the basic interaction was straightforward, being analogous to a two-stage amplifier with positive feedback. However, to prove that successive secondary perturbations converged to a finite result involved a lot of detailed algebra. Also, the paper considered only a steady-state solution, while we knew that the Earth's field reversed.

Because of previous cooperation on a laboratory model of induction in a single "eddy" (Herzenberg and Lowes, 1957), at Newcastle we decided to make a physical laboratory model of the Herzenberg mathematical model, both to confirm the existence of self-excitation and also to investigate its time behavior; Rikitake's (1958) coupled-dynamo extension of Bullard's (1955) mathematical modeling of a lumped-constant self-exciting dynamo (see *Dynamo, disk*) had suggested that a Herzenberg dynamo could spontaneously reverse.

So in 1959 we started to design and build a model. For mechanical convenience we used rotating cylinders rather than spheres, and to improve the efficiency we cut Herzenberg's model in half; mercury was used in the narrow gap between the (half) cylinders and the stationary block to give electrical contact. To give self-excitation, a nondimensional parameter called the magnetic Reynolds number

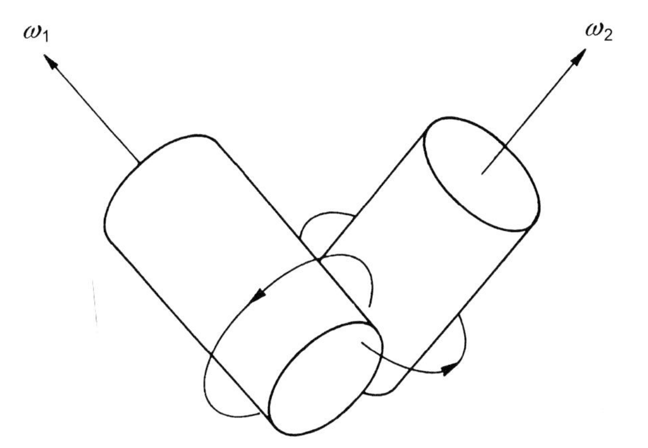

Figure D18 Schematic diagram of the feedback loop in the laboratory dynamo. The arrows ω_1 and ω_2 show the angular velocities, and the circles are the resultant induced magnetic fields.

(see *Geodynamo*) had to be large enough. For our model this number was $\sigma\mu\omega a^2$, where σ and μ were the electrical conductivity and magnetic permeability of the material, ω the angular velocity of the cylinders, and a their radius. The size was limited by material availability and the velocity by turbulent power dissipation in the mercury, so we had to use a material with a large value of $\mu\sigma$. We chose an alloy called Perminvar, which had a high, but fairly constant, initial permeability, and made the radius a 3.5 cm. The cylinders were rotated by electric motors.

Figure D18 shows schematically the intended situation. Cylinder 1 is rotating in the presence of an axial magnetic field (initially applied from outside); this gives radial induced voltage inside the cylinder, leading to current flowing out from the bottom of the cylinder and returning through the side, and hence a "toroidal" magnetic field around the axis. The geometry of the cylinders is chosen so that the induced field of cylinder 1 is along the axis of cylinder 2. Then the same process in cylinder 2 leads to an induced magnetic field which at cylinder 1 is parallel to the original applied field. So there is positive feedback, and if the speeds are large enough, the fed-back field is as large as the original applied field, which is then no longer needed; the system would then be self-exciting.

It turned out that with the original model, we could not go fast enough to reach self-excitation. However we were able to deduce what was going wrong—the spacing between the two cylinders was too small for our approximate calculation to be valid, and the induced fields were more complicated than we expected, giving less feedback. By (effectively) changing the shape of the boundary of the stationary block, we were able to produce the intended simpler induced fields and so achieve self-excitation (Lowes and Wilkinson, 1963). There was now a working self-exciting dynamo on the surface of the Earth, as well as one in its core!

But while the model was self-exciting, it was stable, and did not reverse; this was probably mainly due to the damping coming from the power dissipation in the mercury, which was much larger than the electromagnetic power dissipation. However, with a rather larger steel model, in which the relative geometry of the two cylinders could be varied, we did produce spontaneous reversals (Lowes and Wilkinson, 1967). The reversal waveforms depended on the magnetic properties of the steel and on how the model was powered, so were not directly applicable to the Earth, but the instability which led to the reversals was what would be expected in the Earth. Later investigations by Wilkinson and by D.J. Kerridge are reported by Wilkinson (1984).

Frank Lowes

Bibliography

Backus, G.E., 1958. A class of self-sustaining dissipative spherical dynamos. *Annals of Physics (NY)*, **4**: 372–447.

Bullard, E.C., 1955. The stability of a homopolar dynamo. *Proceedings of the Cambridge Philosophical Society*, **51**: 744–760.

Bullard, E.C., and Gellman, H., 1954. Homogeneous dynamos and terrestrial magnetism. *Philosophical Transactions of the Royal Society of London*, **A 247**: 213–278.

Cowling, T.G., 1934. The magnetic field of sunspots. *Monthly Notes of the Royal Astronomical Society*, **94**: 39–48.

Lowes, F.J., and Wilkinson, I., 1963. Geomagnetic dynamo: a laboratory model. *Nature*, **196**: 1158–1160.

Lowes, F.J., and Wilkinson, I., 1967. Geomagnetic dynamo: an improved laboratory model. *Nature*, **219**: 717–718.

Herzenberg, A., 1958. Geomagnetic dynamos. *Philosophical Transactions of the Royal Society of London*, **A 250**: 543–583.

Herzenberg, A., and Lowes, F.J., 1957. Electromagnetic induction in rotating conductors. *Philosophical Transactions of the Royal Society of London*, **A 249**: 507–584.

Rikitake, T., 1958. Oscillations of a system of disk dynamos. *Proceedings of the Cambrige Philosophical Society*, **54**: 89–105.

Wilkinson, I., 1984. The contribution of laboratory dynamo experiments to our understanding of the mechanism of generation of planetary magnetic fields. *Geophysical Surveys*, **7**: 107–122.

Cross-references

Cowling's Theorem
Dynamo, Bullard-Gellman
Dynamo, Disk
Dynamo, Herzenberg
Geodynamo

DYNAMO, MODEL-Z

Model-Z, first proposed by S.I. Braginsky in 1975, is a dynamical balance in which the poloidal magnetic field in the Earth's core is aligned predominantly along the z-axis (hence the name). Braginsky conjectured that it is through this alignment that the adjustment to Taylor's condition comes about. This article should therefore be read in conjunction with the one on *Taylor's condition* (*q.v.*)

Basic equations

We start with the induction equation

$$\frac{\partial \mathbf{B}}{\partial t} = \nabla^2 \mathbf{B} + \nabla \times (\alpha \mathbf{B}) + \nabla \times (\mathbf{U} \times \mathbf{B}), \qquad \text{(Eq. 1)}$$

with the flow $\mathbf{U}$ given by

$$\mathbf{U} = \mathbf{U}_\mathrm{T} + \mathbf{U}_\mathrm{M} + U_\mathrm{g}(s)\hat{\mathbf{e}}_\phi. \qquad \text{(Eq. 2)}$$

The precise definitions of the thermal wind $\mathbf{U}_\mathrm{T}$, the magnetic wind $\mathbf{U}_\mathrm{M}$, and the geostrophic flow U_g are given in the article on Taylor's condition. See also *Dynamos, mean-field* for a discussion of the α-effect in Eq. (1).

We next apply the toroidal-poloidal decomposition

$$\mathbf{B} = \mathbf{B}_\mathrm{t} + \mathbf{B}_\mathrm{p} = B\hat{\mathbf{e}}_\phi + \nabla \times (A\hat{\mathbf{e}}_\phi) \qquad \text{(Eq. 3)}$$

appropriate for axisymmetric models. The flow may similarly be expressed as $\mathbf{U} = \mathbf{U}_\mathrm{t} + \mathbf{U}_\mathrm{p}$, but the only aspects of this decomposition that we will actually need are that $\mathbf{U}_\mathrm{T}$ and U_g both contribute to $\mathbf{U}_\mathrm{t}$ only. The induction equation then becomes

$$\frac{\partial A}{\partial t} = D^2 A + \alpha B + \hat{\mathbf{e}}_\phi \cdot (\mathbf{U}_\mathrm{p} \times \mathbf{B}_\mathrm{p}), \qquad \text{(Eq. 4)}$$

$$\frac{\partial B}{\partial t} = D^2 B + \hat{\mathbf{e}}_\phi \cdot \nabla \times (\mathbf{U}_\mathrm{t} \times \mathbf{B}_\mathrm{p} + \mathbf{U}_\mathrm{p} \times \mathbf{B}_\mathrm{t}), \qquad \text{(Eq. 5)}$$

where $D^2 = \nabla^2 - 1/s^2$, and (z, s, ϕ) are standard cylindrical coordinates. Note also that α is included only in Eq. (4), but not in Eq. (5). Model-Z is thus an $\alpha\omega$-dynamo, with the poloidal field $\nabla \times (A\hat{\mathbf{e}}_\phi)$ regenerated from B by the α-effect, but the toroidal field $B\hat{\mathbf{e}}_\phi$ regenerated from A only by the ω-effect, which enters Eq. (5) in the form of some prescribed thermal wind $\mathbf{U}_\mathrm{T}$.

From Eq. (11) in *Taylor's condition*, the geostrophic flow becomes

$$U_\mathrm{g}(s) = -E^{-1/2}\frac{(1-s^2)^{1/4}}{4\pi s}\frac{\mathrm{d}}{\mathrm{d}s}\left[s^2 T\right], \qquad \text{(Eq. 6)}$$

where the Taylor integral

$$T = \int_{-z_\mathrm{T}}^{+z_\mathrm{T}} B\frac{\partial A}{\partial z}\mathrm{d}z, \qquad \text{(Eq. 7)}$$

and $z_\mathrm{T} = (1-s^2)^{1/2}$. Note also that by integrating over this particular range $\pm z_\mathrm{T}$ we have implicitly neglected the inner core. See *inner core tangent cylinder* for a discussion of how the Taylor integral becomes considerably more complicated if one includes an inner core.

Braginsky's analysis

We see then that unless $T \le O(E^{1/2})$, the geostrophic flow will increase without bound as $E \to 0$. As discussed in *Taylor's condition*, there are a number of ways of achieving such a scaling. First, we could simply have $\mathbf{B} \le O(E^{1/4})$, the so-called Ekman states. Next, we could have $\mathbf{B} = O(1)$, but with sufficient internal cancellation in the integral. This is the so-called Taylor state. Finally, we could have $\mathbf{B} = O(1)$, but with $\partial A/\partial z \to 0$. This is precisely Braginsky's model-Z, as $\partial A/\partial z \to 0$ implies that the poloidal field lines become aligned with the z-axis. (We note though that $\partial A/\partial z = 0$ does not satisfy the appropriate boundary conditions. There must therefore be a boundary layer in which this alignment breaks down.)

In fact, the difference between the Taylor state and model-Z is more profound than whether $T \to 0$ comes about by having regions of positive and negative $B\partial A/\partial z$ cancel in the integral, or by having $\partial A/\partial z$ essentially zero everywhere. Braginsky argues that T tends to zero *more slowly* than $O(E^{1/2})$, so that $U_g \to \infty$. The fundamental distinction between a Taylor state and model-Z is therefore whether E enters in an essential way (model-Z, where $\mathbf{B} = O(1)$ but $U_g \to \infty$), or only as a perturbation (Taylor state, where $\mathbf{B}$ and U_g are both independent of E at leading order). In other words, are viscous effects important or not?

Now, why might T tend to zero more slowly in model-Z than in a Taylor state? In a Taylor state, any T in excess of $O(E^{1/2})$ immediately induces a series of *torsional oscillations* (*q.v.*), which eventually decay away again, taking the excess with them. However, as Braginsky points out, the efficiency with which torsional oscillations drive T toward zero depends on the extent to which the different cylindrical shells that constitute a torsional oscillation are coupled to one another. In his model, the z-aligned field lines do not thread adjacent cylindrical shells, so the coupling is very weak, so weak that T tends toward zero, but never quite reaches the $O(E^{1/2})$ that it does if the coupling is strong. That is, having $\partial A/\partial z \approx 0$ is the mechanism whereby T tends to zero, but is also the mechanism that prevents it from tending to zero as rapidly as $O(E^{1/2})$.

We can gain further useful insight into the rate at which T tends to zero by considering the energetics associated with the geostrophic flow. We begin by noting that, as part of $\mathbf{U}_t$, U_g affects only B, not A. Specifically, its contribution to Eq. (5) ends up as

$$\frac{\partial B}{\partial t} = -s\frac{\mathrm{d}}{\mathrm{d}s}\left(\frac{U_g}{s}\right)\frac{\partial A}{\partial z}. \qquad \text{(Eq. 8)}$$

Now insert Eq. (6) for U_g, multiply by B, and integrate over the entire volume. The final energy balance one eventually obtains is

$$\frac{1}{2}\frac{\partial}{\partial t}\int B^2 \mathrm{d}V = -\pi E^{-1/2}\int_0^1 \frac{(1-s^2)^{1/4}}{s^3}\left[\frac{\mathrm{d}}{\mathrm{d}s}(s^2T)\right]^2 \mathrm{d}s, \qquad \text{(Eq. 9)}$$

indicating that the geostrophic flow will necessarily act to reduce the energy, unless T is identically equal to zero. Beyond that, we can easily see from Eq. (9) that as long as T tends to zero at least as rapidly as $O(E^{1/4})$, the energy drain associated with the geostrophic flow will still be bounded—as it must be if the energy input via α and $\mathbf{U}_T$ is independent of E—even though U_g itself increases without bound as $E \to 0$.

In fact, in Braginsky's original analysis he has T tending to zero as $E^{1/6}$, even more slowly than $E^{1/4}$. This is energetically sustainable, as he also takes $\mathbf{U}_T$ to scale as $E^{-1/6}$, so an $E^{-1/6}$ energy input via $\mathbf{U}_T$ then balances an $E^{-1/6}$ energy drain via U_g. Taking one of the kinematically prescribed forcing terms to depend on E, and at leading order, is not really desirable though, as it eliminates any possibility of obtaining a solution that does not also depend on E at leading order. That is, by doing so one has eliminated *a priori* the possibility of obtaining a Taylor state. However, as the above analysis demonstrates, this is not an essential feature of model-Z; one can obtain solutions in which U_g scales as some negative power of E even if $\mathbf{U}_T$ (and α) are taken to be independent of E, so that the Taylor state is also accessible.

Finally then, which solution corresponds to the real Earth, Taylor state or model-Z? One weakness of model-Z is that it is a rather special case. That is, *given* that $\partial A/\partial z \approx 0$, one can understand how the other features follow, but why should the field adopt such a special configuration in the first place? And perhaps not surprisingly then, model-Z only arises for certain special choices of α. Most fully three-dimensional numerical models, in which α is not prescribed at all, also show no particular tendency for the poloidal field lines to align with the z-axis (although the Ekman numbers in these models are so large that one is not unambiguously tending toward a Taylor state either). Another drawback of model-Z is that there is no observational evidence (see *Core motions*) for the very large geostrophic flows that it predicts. Walker and Hollerbach (1999) suggest though that Io may have a model-Z dynamo, since there the background Jovian flux is aligned in z, which may prompt the internal field to adopt the same configuration.

Rainer Hollerbach

Bibliography

Braginsky, S.I., 1975. Nearly axially symmetric model of the hydromagnetic dynamo of the Earth. *Geomagnetism and Aeronomy*, **15**: 122–128.

Braginsky, S.I., Roberts, P.H., 1987. A Model-Z geodynamo. *Geophysical and Astrophysical Fluid Dynamics*, **38**: 327–349.

Braginsky, S.I., Roberts, P.H., 1994. From Taylor state to Model-Z. *Geophysical and Astrophysical Fluid Dynamics*, **77**: 3–13.

Walker, M.R., Hollerbach, R., 1999. The adjustment to Taylor's constraint in the presence of an ambient field. *Physics of the Earth and Planetary Interiors*, **114**: 181–196.

Cross-references

Core Motions
Core, Boundary Layers
Dynamos, Mean Field
Dynamos, Planetary and Satellite
Geodynamo, Numerical Simulations
Inner Core Tangent Cylinder
Oscillations, Torsional
Taylor's Condition

DYNAMO, PONOMARENKO

Definition of the model

Human-made dynamos are complex, multiconnected, asymmetric devices, but those operating in Nature have to function in geometrically-simple (usually almost spherical) bodies of electrically conducting fluid, such as Earth's core. During the early development of dynamo theory, it seemed possible that no motion within such a simple "homogeneous" system could maintain a magnetic field, a feeling reenforced by some antidynamo theorems (see *Antidynamo and bounding theorems* and *Cowling's theorem*). Whether *homogeneous dynamos* exist or not is an electrodynamic question, i.e., one that can be addressed without asking how the velocity $\boldsymbol{V}$ of the conductor is maintained and how a magnetic field $\mathbf{B}$ would affect it. It is necessary only to find a $\boldsymbol{V}$ that can sustain a nonzero $\mathbf{B}$, and these are then called "kinematic dynamos", to distinguish them from "MHD dynamos" which satisfy the full equations of magnetohydrodynamics. (See *Dynamos, kinematic* and *Magnetohydrodynamics*.) The first kinematic dynamo models were created in 1958 (see *Dynamo, Backus* and *Dynamo, Herzenberg*). It was clear, especially from the Herzenberg model, that nonmirror-symmetric motions are potentially efficient generators of magnetic field; roughly speaking, the lack of mirror-symmetry provides a homogeneous system with a means of mimicking the asymmetry of human-made dynamos. This added credence to the earlier speculations of Parker (1955) that helical motions could regenerate magnetic field. Ponomarenko (1973) provided the simplest realization of this idea. See also *Geodynamo, symmetry properties*.

As dynamo theory evolved, it became increasingly focused on the MHD dynamo in which $\boldsymbol{V}$ and (nonzero) $\mathbf{B}$ are solutions of the magnetohydrodynamic equations, maintained by appropriate forces such as buoyancy (in the case of convective dynamos). Nevertheless, interest in the Ponomarenko model has been sustained. There are two reasons for this. First, the model provides the simplest illustration of how motions possessing nonzero helicity, $\boldsymbol{V}\cdot\nabla\times\boldsymbol{V}$, can sustain a magnetic field. Second, the model has been the key ingredient in laboratory dynamo experiments (see *Dynamos, experimental*).

Although the Ponomarenko model is one of the simplest homogeneous dynamos, it is not the simplest; a model due to Lortz (1968) can, without recourse to numerical or asymptotic methods, be shown to regenerate magnetic field. Although this model has been developed (see Gilbert, 2003, for references), it has not yet been as influential as the Ponomarenko model, possibly because it does not expose the role of helicity as clearly.

In Ponomarenko's model, the electrical conductor fills all space but motion is confined to the interior of a solid cylinder C, the exterior $\widehat{\mathrm{C}}$ of which is a stationary conductor of the same magnetic diffusivity, η, the electrical contact across the interface S being perfect; see Figure D19. The velocity $\boldsymbol{V}$ of C is a "solid-body" motion, i.e., a motion that the cylinder can execute even if solid; in this respect the model resembles the Herzenberg dynamo in which the moving parts are spherical rotors. The Ponomarenko motion is a combination of a uniform velocity U along the axis Oz of the cylinder and a rotation

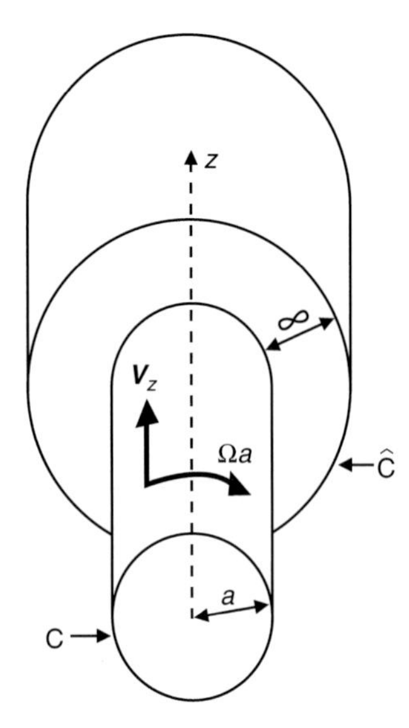

Figure D19 The Ponomarenko dynamo. The cylinder C, radius a, moves as a solid body in a helical motion, a combination of a uniform velocity $U = \mathbf{V}_z$ along the axis Oz of C and a rotation about that axis with angular speed Ω. The ∞ sign indicates that the conductor with which C is in electrical contact fills the rest of space.

about that axis with uniform angular velocity Ω so that, in cylindrical coordinates (s, ϕ, z), its components are

$$\mathbf{V} = \begin{cases} (0, \Omega s, U), & \text{if} \quad s < a \\ 0, & \text{if} \quad s > a \end{cases} \qquad \text{(Eq. 1)}$$

where a is the radius of the cylinder. The motion in C is helical, the conductor at distance s from the axis describing a helix of pitch $\Omega s/U$. The pitch $p = \Omega a/U$ on S serves to single out one model from the complete one-parameter family, $-\infty < p < \infty$, of Ponomarenko dynamos. For $p \neq 0$, the motion Eq. (1) is nonmirror-symmetric, the right-handed screw motion of a model with $p > 0$ being mirrored by the left-handed model of the opposite p.

The theory of Ponomarenko's model; generalizations

The kinematic dynamo problem consists, for a specified motion such as Eq. (1), in solving the magnetic induction equation (see *Magnetohydrodynamics*),

$$\frac{\partial \mathbf{B}}{\partial t} = \nabla \times (\boldsymbol{V} \times \mathbf{B}) + \eta \nabla^2 \mathbf{B}, \qquad \text{(Eq. 2)}$$

in the conductor, which in the Ponomarenko model also includes $\widehat{C}$ (with $\widehat{\mathbf{V}} \equiv \mathbf{0}$). Also, $\nabla \cdot \mathbf{B} = 0$ everywhere. Across the discontinuity, S, the three components of the magnetic field and the two tangential components of the electric field, $\mathbf{E} = -\boldsymbol{V} \times \boldsymbol{B} + \eta \nabla \times \mathbf{B}$, must be continuous, but these conditions are not independent; if four of the five conditions are satisfied, so is the fifth.

A dynamo can legitimately be called a dynamo if two demands are met. The first of these is permanence: if a "seed field" is present initially, then $\mathbf{B}$ must persist for all time. The second is the "dynamo condition": there can be no external sources of magnetic field; the only electric currents flowing are those that are created by the motion itself. For the present model, this requires that $\widehat{\mathbf{B}} \to 0$ as $s \to \infty$.

The kinematic problem just defined governs $\mathbf{B}$ through linear, homogeneous equations. Therefore, $\mathbf{B}$ will either grow without limit (the dynamo succeeds!), or tend to zero as $t \to \infty$ (the dynamo fails), or become constant in amplitude (dynamo action is marginal). Which of these happens depends on whether the creation of field by the $\nabla \times (\boldsymbol{V} \times \mathbf{B})$ term dominates the right-hand side of Eq. (2), or whether it is insufficient to compensate for the ohmic loss due to the $\eta \nabla^2 \mathbf{B}$ term, or whether one term exactly offsets the other. The relative importance of the terms is quantified by a magnetic Reynolds number, such as

$$R_m = aU_{\max}/\eta, \quad \text{where} \quad U_{\max} = \sqrt{\left[U^2 + (\Omega a)^2\right]}, \qquad \text{(Eq. 3)}$$

which is based on the maximum velocity, $U_{\max}$, which occurs on the inner side of S. It is intuitively clear that the motion Eq. (1) cannot function as a dynamo unless R_m *is sufficiently large, but (as the antidynamo theorems show) a large* R_m does not guarantee dynamo action.

The kinematic dynamo defines a *linear stability problem*, of a type familiar in fluid mechanics. The "control parameter" is R_m, and the key questions are

Question 1: What is the critical value, R_{mc}, of R_m, that separates "instability" (i.e., dynamo action) from "linear stability" (no dynamo action)?

Question 2: What is the maximum growth rate of the magnetic field when $R_m > R_{mc}$

A third question becomes pertinent when designing an experiment:

Question 3: Which model of the Ponomarenko family most easily generates magnetic field, i.e., for which value of $|p|$ is $R_{mc}(p)$ smallest, and what is the corresponding value of R_{mc}?

As in linear stability theory, these questions should be answered by solving the general initial value problem, i.e., by determining (perhaps by Laplace transform methods) how a general seed field, specified at time $t = 0$, evolves in time. Because the conductor fills all space, this is a complicated undertaking, but Ponomarenko (1973) concluded that the answers to the three questions can be found by analyzing "normal modes" of the form

$$\mathbf{B} = \mathbf{B}_0(s) \exp[\mathrm{i}(m\phi + kz) + \lambda t]. \qquad \text{(Eq. 4)}$$

Possible values of the complex growth rate, $\lambda = \sigma + \mathrm{i}\omega$, are obtained, for given constant values of the wavenumbers k and m (m being an integer), by solving numerically a certain transcendental "dispersion relationship", obtained from Eq. (2) by Ponomarenko (1973). The mode Eq. (4) travels in the direction of $\nabla(m\phi + kz)$ along the constant s—surfaces as a "wave" that grows, decays, or maintains a constant amplitude, depending on whether σ is positive, negative, or zero. In other words, the phase of the wave travels along helices of pitch ks/m.

For given m, k, and p, Ponomarenko's dispersion relation gives infinitely many R_m for which $\sigma = 0$. The smallest of these is shown in Figure D20 as a function of k for several values of $|m|$ in the case $p = 1.2$. Each curve possesses a single minimum, $R_m = R_{m\,\min}(m, p)$, at one particular value, $k_{\min}(m, p)$, of k. These give (for the m and p concerned) the wavenumber k and the corresponding R_m for which the dynamo first functions when R_m is gradually increased from zero. Figure D21 shows $R_{m\,\min}(m, p)$ as a function of p for $m = 1 - 9$. When $R_{m,\min}$ is minimized over m, the *critical state*, $R_m = R_{mc}(p)$, and the corresponding wavenumbers $m = m_c$ and $k = k_c$ are obtained. Figure D21 shows that $m_c = 1$ when $|p| < 4$, and probably $m_c = 1$ for all p.

The general seed field contains a linear combination of every normal mode Eq. (4) and, if $R_m < R_{mc}(p)$, then $\sigma < 0$ for all of them, so that they all disappear as $t \to \infty$. In the critical state $R_m = R_{mc}(p)$ however, every σ is negative except for a single mode, $(m, k) = (m_c, k_c)$, for which $\sigma = 0$. This mode survives for all time at constant amplitude, and is oscillatory (since $\omega_c \neq 0$). Finding this mode answers Question 1. If $R_m > R_{mc}$, then $\sigma > 0$ in a range of k lying within the $m = 1$ "lobe" in Figure D20 and, as R_m is increased, modes of larger m become successively unstable too.

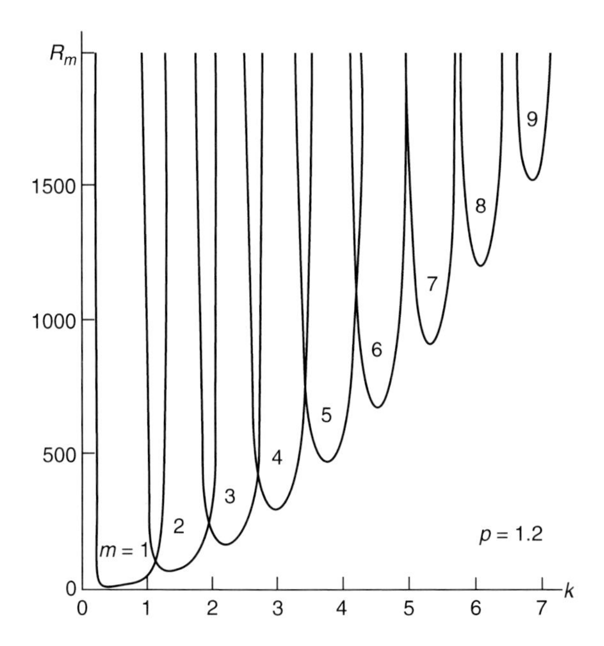

Figure D20 The minimum magnetic Reynolds number R_m as a function of the axial wavenumber k for different values of the angular wavenumber m. For the case shown, $p = 1.2$. The minimum R_m on each constant-m curve is denoted by $R_{m\,\min}(m,p)$. (After Gailitis and Freiberg, 1976.)

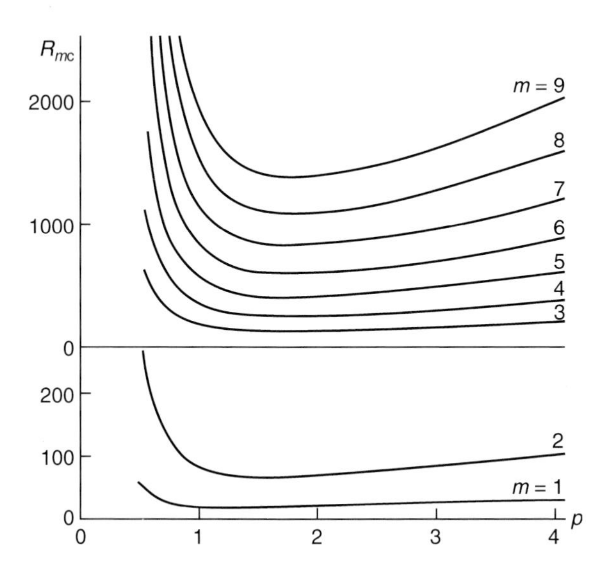

Figure D21 The minimum, $R_{m\,\min}(m,p)$, of the constant-m curves shown in Figure D20 as a function of p for m = 1 – 9. The minimum of $R_{m\,\min}(m,p)$ over m gives the critical magnetic Reynolds number, $R_{mc}(p)$, for the model of the Ponomarenko family having that value of p. The minimum of R_{mc} over p defines the most easily excited member of the family (see Eqs. (5) and (6)). (After Gailitis and Freiberg, 1976.)

Having obtained $R_{mc}(p)$, Question 3 can be answered. As Figure D21 indicates, $R_{mc}(p)$ has a single minimum in p. Numerical work shows that it is

$$R_{m\,c} = 17.7221176\ldots, \quad \text{for} \quad p = \pm 1.31406737\ldots \qquad \text{(Eq. 5)}$$

The corresponding wavenumbers and frequency are

$$m_c = 1, \quad k_c a = \mp 0.387532\ldots, \quad a^2\omega_c/\eta = \mp 0.410300\ldots. \qquad \text{(Eq. 6)}$$

Question 2 has been thoroughly answered but only asymptotically, in the limit of $R_m \to \infty$ (Ponomarenko, 1973; Roberts, 1987; Gilbert, 1988). Nevertheless, the results give insight into how the dynamo functions. The most unstable wave (corresponding to the largest σ for given p) is of small wavelength: $m \gg 1$, $ka \gg 1$. The velocity and field helices are almost orthogonal so that $m\omega + kU \approx 0$. This means that the phase, $m\phi + kz$, of the wave is approximately constant on the helical "streamlines", $\phi = \Omega sz/aU$, of V. Compared with its value on S, the magnetic field becomes exponentially small within a distance of order $1/k$ from S. This highlights the fact that **B** is created by electric currents generated by the potential differences produced by the discontinuity in V at S.

The generation of **B** can be thought of as a two-step process: (i) the creation of the tangential components, B_ϕ and B_z, from the radial component B_s, and (ii) the production of B_s from $\mathbf{B}_{\text{tang}} = (0, B_\phi, B_z)$. Because $R_m \gg 1$, Alfvén's theorem holds approximately (see *Alfvén's theorem and the frozen flux approximation*). The theorem helps in picturing step (i), the action of the differential shear across S. The shear stretches out the radial field B_s crossing S, so creating on and near S a strong field ($\mathbf{B}_{\text{tang}}$) approximately in the direction of the helical motion of C. Since R_m is not infinite, Alfvén's theorem is not strictly enforced and this is significant for step (ii). The field lines of $\mathbf{B}_{\text{tang}}$ are curved and tend to shorten as they diffuse relative to the conductor. As they shorten, they produce B_s so completing the regenerative process.

As $R_m \to \infty$, the maximum of σ tends to a positive value proportional to Ω and, because it is independent of R_m, it is what is known as a *fast dynamo* (see Gilbert, 1988; Childress and Gilbert, 1995 and *Dynamos, fast*). Although a fast dynamo functions only because η is nonzero, the growth rate of magnetic field is, paradoxically, independent of η. The fact that the Ponomarenko dynamo is fast can be traced to the discontinuity in V at S. A real fluid is a continuum in which such "tearing motions" are impossible. When the discontinuity at S is replaced by a thin layer in which V varies rapidly but continuously with s, the dynamo becomes "slow"; σ is proportional to $R_m^{-1/3}\Omega$ and therefore tends to zero as $R_m \to \infty$ (see Gilbert, 1988).

Generalizations

The Ponomarenko model has been generalized in several ways. Gailitis *et al.* (1989) studied several slow dynamos in which V in C is a continuous function of s that vanishes on S. The special case when V is spiral Couette flow was analyzed by Solovyev (1985a,b; 1987). Asymptotic properties of such "screw dynamos" were investigated by Ruzmaikin *et al.* (1988).

Gailitis and Freiberg (1980) generalized the Ponomarenko model by considering models in which C is surrounded by one or more cylindrical shells in which V is nonzero, but different from V in C. One of these provides a theoretical idealization of a successful laboratory experiment based on Ponomarenko's concept. The spatial extent of the electrical conductor in an experiment is necessarily limited in both s- and z-directions. Moreover, the flow along C must return to pumps before being reinjected into C. The efficiency of the dynamo can be improved (i.e., $R_{m\,c}$ can be reduced) by making the conducting fluid return in a sheath surrounding C. In modeling this, Eq. (1) is replaced by

$$\mathbf{V} = \begin{cases} (0, \Omega s, z), & \text{for} \quad s < a; \\ (0, 0, -W), & \text{for} \quad s > a, \end{cases} \qquad \text{(Eq. 7)}$$

the region $s > A$ being an electrical insulator. By taking $W = a^2U/(A^2 - a^2)$, the return flow is equal and opposite to the flow along C. Model Eq. (7) should also be limited in z, but mathematical difficulties arise if the conductor is confined to (say) $|z| < L$, with electrical insulators filling not only $s > A$ but also $|z| > L$. It is hard to solve for the potential field $\widehat{\mathbf{B}}$ outside such a finite cylinder, but Gailitis and Freiberg devised an ingenious way of estimating R_m; see also Gailitis (1990). Another way of limiting the spatial extent of the dynamo led to the models of Dudley and James (1989) which confine the helical motion to the interior of a spherical conductor.

Starting in the mid 1970s, the MHD liquid metals group at the Institute of Physics in Riga set themselves the task of building a laboratory dynamo based on Ponomarenko's model and idealized by Eq. (7) (see Gailitis *et al.*, 1987, 1989 and Gailitis 1996). The difficulty of reaching a magnetic Reynolds number of order 20 (see Eq. (5)) was severe and there were setbacks. Their goal was finally achieved by the turn of the century (see Gailitis *et al.*, 2000, 2001 and *Dynamos, experimental*).

Paul H. Roberts

Bibliography

Childress, S., and Gilbert, A.D., 1995. *Stretch, Twist, Fold: The Fast Dynamo.* Berlin: Springer.

Dudley, M.L., and James, R.W., 1989. Time-dependent kinematic dynamos with stationary flows. *Proceedings of the Royal Sociery of London, Series A*, **425**: 407–429.

Gailitis, A., 1990. In Moffatt, H.K., and Tsinober, A. (eds.), Topological Fluid Mechanics. Cambridge: University Press, pp. 147–156.

Gailitis, A., 1996, Design of a liquid sodium MHD dynamo experiment. *Magnit. Gidro.*, (1), 68–74 (Eng. Trans. *Magnetohydrodynamics*, **32**: 158–162.)

Gailitis, A.K., Freiberg, Ya. G., 1976. Theory of a helical MHD dynamo. *Magnit. Gidro.*, (2), 3–6 (Eng. Trans. *Magnetohydrodynamics*, **12**: 127–130.)

Gailitis, A.K., Freiberg, Ya. G., 1980. Nonuniform model of a helical dynamo. *Magnit. Gidro.*, 15–19 (Eng. Trans. *Magnetohydrodynamics*, (1), **16**: 11–15.)

Gailitis, A.K., Karasev, B.G., Kirillov, I.R., Lielausis, O.A., Luzhanskii, S.M., Ogorodnikov, A.P., Preslitskii, G.V., 1987. Experiment with a liquid-metal model of an MHD dynamo, *Magnit. Gidro.*, (4), 3–8 (Eng. Trans. *Magnetohydrodynamics*, **23**: 349–352.)

Gailitis, A., Lielausis, O., Karasev, B.G., Kirillov, I.R., Ogorodnikov, A.P., 1989. In Lielpeteris, J., and Moreau, R. (eds.), Liquid Metal Magnetohydrodynamics. Dordrecht: Kluwer, pp. 413–419.

Gailitis, A.K., Lielausis, O.A., Dementev, S., Platacis, E., Cifersons, A., Gerbeth, G., Gundrum, T., Stefani, F., Christen, M., Hänel, H., Will, G., 2000. Detection of a flow induced magnetic field eigenmode in the Riga dynamo facility. *Physics Review Letters*, **84**: 4365–4368.

Gailitis, A.K., Lielausis, O.A., Platacis, E., Dementev, S., Cifersons, A., Gerbeth, G., Gundrum, T., Stefani, F., Christen, M., Will, G., 2001. Magnetic field saturation in the Riga dynamo experiment, *Physics Review Letters*, **86**: 3024–3027.

Gailitis, A.K., Lielausis, O.A., and Platacis, E., 2002. Colloquium: Laboratory experiments on hydromagnetic dynamos, *Reviews of Modern Physics*, **74**: 973–990.

Gilbert, A.D., 1988. Fast dynamo action in the Ponomarenko dynamo. *Geophysical and Astrophysical Fluid Dynamics*, **44**: 214–258.

Gilbert, A.D., 2003. In Friedlander, S., and Sevre, D. (eds.), *Handbook of Mathematical Fluid Dynamics*, **2**: 355–441.

Lortz, D., 1968. Exact solutions of the hydromagnetic dynamo problem. *Plasma Physics*, **10**: 967–972.

Parker, E.N., 1955. Hydromagnetic dynamo models. *Astrophysical Journal*, **122**: 293–314.

Ponomarenko, Yu. B., 1973. On the theory of hydromagnetic dynamo. *Zh. Prik. Mech. Tech. Fiz. SSSR* (6): 47–51. (Eng. Trans. *Journal of Applied Mechanics and Technical Physics*, **14**: 775–778.)

Roberts, P.H., 1987. In Nicolis, C., and Nicolis, G. (eds.), Irreversible Phenomena and Dynamical System Analysis in Geosciences. Dordrecht: Reidel, pp. 73–133.

Ruzmaikin, A.A., Sokoloff, D.D., Shukurov, A.M., 1988. A hydromagnetic screw dynamo. *Journal of Fluid Mechanics*, **197**: 39–56.

Solovyev, A.A., 1985a. Magnetic field generation by the axially-symmetric flow of conducting fluid. *Physics of the Solid Earth*, **4**: 101–103.

Solovyev, A.A., 1985b. Couette spiral flows of conducting fluids consistent with magnetic field generation. *Physics of the Solid Earth*, **12**: 40–47.

Solovyev, A.A., 1987. Excitation of a magnetic field by the movement of a conductive liquid in the presence of large values of the magnetic Reynolds number. *Physics of the Solid Earth*, **23**: 420–423.

Cross-references

Alfvén's Theorem and the Frozen Flux
Antidynamo and Bounding Theorems
Cowling's Theorem
Dynamo, Backus
Dynamo, Herzenberg
Dynamos, Experimental
Dynamos, Fast
Dynamos, Kinematic
Geodynamo, Symmetry Properties
Magnetohydrodynamics

DYNAMO, SOLAR

Solar magnetic fields

The strong 1–3 kG magnetic fields of sunspots were discovered by Hale (1908) nearly a hundred years ago. The weaker magnetic fields, appearing all over the surface of the Sun, were discovered, and subsequently mapped and monitored, by Babcock and Babcock (1955) about 50 years ago. The cyclic behavior of the magnetic fields of the Sun, to which the 11-year sunspot cycle is closely tied, indicates a creation and annihilation of magnetic field on an 11-year basis, presumably by the inductive effects of the convective motions within the Sun. For the basic fact is that the Sun is entirely gaseous, and the gas is highly ionized and an excellent conductor of electricity. Even the thin partially ionized layer at the photosphere makes a fairly good conductor, given the 100-km thickness of the layer. Hence there can be no significant electric fields in the frame of reference of the moving gas, from which it follows that the magnetic fields are caught up in the swirling gases and carried bodily with them. The result is, somehow, the complicated cyclic magnetic behavior that we see (see *Magnetic field of Sun*).

The bodily transport of magnetic field **B** is described by the magnetohydrodynamic (MHD) induction equation

$$\frac{\partial \mathbf{B}}{\partial t} = \nabla \times (v \times \mathbf{B}) \qquad \text{(Eq. 1)}$$

in terms of the fluid velocity v (see *Alfvéns theorem and the frozen flux approximation*). In fact, the gas is not a perfect conductor but has some small resistivity. Hence, there is diffusion of the magnetic field relative to the fluid, which can be represented in simplest form by the addition of the term $\eta\nabla^2\mathbf{B}$ on the right-hand side of Eq. (1) (assuming a uniform resistive diffusion coefficient (η). As we shall see, the diffusion is as essential for field production as the bodily transport of magnetic field.

The ongoing deformation of the solar magnetic field takes place largely out of sight, beneath the visible surface, so the first task is to look closely at the fields emerging through the surface that we may infer what is happening below. The outstanding magnetic features on the visible surface of the Sun are the large (2×10^4–2×10^5 km) bipolar magnetic regions, with approximate east-west orientation and field strengths of the general order of 10^2 Gauss in both the positive and negative ends of the bipole (Babcock and Babcock, 1955, see *Magnetic field of Sun*). It is inferred that each bipolar region arises from the upward bulge (a so-called Ω-loop) in a large bundle of east-west (azimuthal) magnetic field lying somewhere below. A bipolar region may endure for months by continuing emergence of fresh Ω-loops at the site (Gaizauskas, *et al.*, 1983), driven to the surface by the intrinsic buoyancy of the magnetic field. A magnetic field **B** provides a pressure $\mathbf{B}^2/8\pi$ without weight, causing buoyancy and a general instability of horizontal fields to form Ω-loops (Parker, 1955a).

Now, the first appearance of the bipolar regions is in the vicinity of 35–40° latitude. Following first appearance, there are soon many bipolar regions distributed around the Sun, forming a broadband of emerging Ω-loops that migrates (1.5 $\mathrm{m\,s^{-1}}$) toward the equator over the next 10 years or so. Note then that the bipolar regions have opposite orientation in the northern and southern hemispheres, and their numbers decline as the two bands of opposite orientation approach each other at the equator. The next wave of bipolar regions has opposite orientation to its predecessor and appears at latitudes of 35–40° at about the time the old bands are fading out at the equator. Thus, the complete magnetic cycle occupies 22 years.

An observational inventory of the magnetic flux emerging over a six month period in a particularly long lived bipolar magnetic region suggests that the subsurface azimuthal field contains a total magnetic flux of at least 2×10^{23} Maxwells in each hemisphere (Gaizauskas, *et al.*, 1983), assuming that the emerging magnetic fields were not recycled in some way and appeared more than once during that time. The fact is that 2×10^{23} Maxwells is a large quantity of magnetic flux, implying a mean azimuthal magnetic field of 2×10^3 Gauss even if spread over 20° in latitude (2×10^{10} cm) and 0.5×10^{10} cm in depth.

There are smaller bipolar regions called *ephemeral regions* ($<2 \times 10^4$ km), appearing at all latitudes and varying but little with the 11-year magnetic cycle. The smaller among them show no tendency to align east and west. The place of the ephemeral regions in the overall magnetic picture is not clear.

In addition to the bipolar regions, there are weaker small-scale fields (with mean intensity of the order of 10 G) in the broad spaces between the bipolar regions. The polar regions, above latitudes of about 55°, exhibit large-scale weak (10 G) fields oriented inward at one pole and outward at the other so that the Sun has a net magnetic dipole moment. This is the principal manifestation of the poloidal field of the Sun. The polar fields reverse at about the time the azimuthal fields reach their peak strength (at latitudes of 15–20°). Some observational analyses of the field reversal show the reversal to be a rotation of the dipole of one half-cycle around across the equator to the opposite orientation of the next half cycle.

Improved telescopic resolution in recent decades shows that all these fields, as well as the stronger fields of the large bipolar regions, are made up of many intense (10^3 Gauss) magnetic fibrils, i.e., isolated bundles of magnetic field, with diameters of the order of 10^2 km. The magnetic field in the wide space between the fibrils is too weak to be detected with existing magnetographs. The mean field in the region is determined by the spacing of the fibrils. In the weak field regions the individual fibrils emerge through the solar surface in the central updrafts of the large (10^4 km) convective cells—the supergranules—subsequently migrating into the weak downdrafts at the supergranule boundaries and the strong downdrafts at the junctions of boundaries. The fibrils interact with each other, producing tiny flares. In the large bipolar regions the emerging Ω-loops (that maintain the regions) are composed of fibrils when they first appear at the surface (Zwaan, 1985).

The observed coordinated cycles of the poloidal and azimuthal (toroidal) magnetic fields suggest that they are both the product of the global internal hydrodynamics of the Sun.

Theory

There have been attempts to explain how the azimuthal magnetic field might come about through torsional oscillations of the Sun interacting with a primordial dipole magnetic field rooted in the central core. Such a dipole would decay with a characteristic time of the order of 10^{10} years, or by a factor of only about $e^{0.5} \approx 1.6$ since the Sun was formed. There are two objections to the concept: The necessary torsional oscillations of the Sun would be readily observed, but have not been seen, and the dipole field observed at the visible surface is not rooted in the central core, because it reverses every 11 years, along with the azimuthal field.

This leaves only the concept of direct MHD dynamo creation and annihilation of the magnetic fields by the internal convection and nonuniform rotation of the Sun (see *Elsasser, W.; Dynamos, kinematic; Dynamos, mean field*). That is to say, the dynamo action evidently takes place in and around the convective zone of the Sun, comprising the outer 2/7 (2×10^5 km) of the solar radius ($R = 7 \times 10^5$ km). The convective velocities increase upward through the convective zone, from essentially zero (1 $\mathrm{m\,s^{-1}}$) at the base to 1 $\mathrm{km\,s^{-1}}$ at the visible surface (see Spruit, 1974). The characteristic scale, or mixing length, of this turbulent convection is believed to be comparable to the local pressure scale height of 5×10^4 km at the bottom (where $T \cong 2 \times 10^6$ K) and 200 km at the top (where $T = 6 \times 10^3$ K).

Now recall that the magnetic field of Earth appears to be generated by the combined nonuniform rotation and the cyclonic convection in the spinning liquid metal core, referred to as an $\alpha\omega$-dynamo and representing the most efficient theoretical convective dynamo mode (see *Elsasser, W.; Geodynamo*). More complicated fluid motions, or, for instance, cyclonic convection alone, can be shown to provide a self-sustaining magnetic field, but they require higher speeds and greater energy input for the same field production (see Steenbeck and Krause, 1966). So consider the basic $\alpha\omega$-dynamo process for the origin of the magnetic fields of the Sun.

The surface of the Sun is observed to rotate nonuniformly, with a 25-day period at the equator and something in excess of 30 days at the poles. Years ago one was inclined, by theoretical considerations on conservation of angular momentum in rising and falling convective columns, to suppose that the rate of rotation depends primarily on distance from the spin axis of the Sun, presumably in the form that is observed at the surface. Fortunately helioseismology (see *Helioseismology*) has stepped in to provide a detailed map of the internal rotation, which turns out to be quite unlike the conjectured form. To put it simply, the angular velocity in the convective zone varies but little with depth below the surface, so at any given depth the angular velocity has much the same latitudinal profile as the visible surface. The radiative zone ($0 < r < 5 \times 10^5$ km) enclosed by the convective zone rotates approximately rigidly with a period of about 28 days. It is evident then that there is a strong shear between the base of the nonuniformly rotating convective zone and the contiguous outer surface of the radiative zone, with the convective zone rotating faster than the radiative zone at low latitudes and slower at high latitudes. The thin shear layer between is called the *tachocline*. It should be appreciated that this remarkable internal rotation has not yet been explained by the theoretical hydrodynamics of the convection, which presumably drives the nonuniform rotation.

The dipole magnetic field of the Sun is caught up in the shearing, producing a strong azimuthal magnetic field at the tachocline—presumably the azimuthal magnetic field that spawns the bipolar magnetic regions observed at the surface. It remains then to understand how the dipole (poloidal) field is generated from the azimuthal (toroidal) field. There is no way to accomplish this within the confines of rotational symmetry about the spin axis of the Sun (see *Elsasser, W.*). Cowling's theorem (see *Cowling's theorem*) points out that no magnetic field with rotational symmetry can be sustained by the inductive effects of fluid motions. The same situation arises in the geodynamo, where the poloidal magnetic field must be generated somehow from the azimuthal field. Note then that the cyclonic convection, arising from the Coriolis force on the many individual convective cells, breaks rotational symmetry around the Sun, and, as it turns out, is the most effective mechanism for producing poloidal field.

The essential point is that cyclonic convection pokes up Ω-shaped loops in the deep azimuthal magnetic field and rotates those Ω-loops into meridional planes. Thus each rotated Ω-loop represents a local magnetic circulation in the meridional plane. Merging many loops through diffusion obliterates the individual loops and leaves the net overall magnetic circulation intact. It is that net circulation of magnetic field in meridional planes in the Sun that is the observed poloidal field. Combining the cyclonic convection with the nonuniform rotation provides the $\alpha\omega$-dynamo process (Parker, 1955b, 1957, 1979).

The azimuthal vector potential **A** provides a convenient description of the poloidal magnetic field, with each rotated Ω-loop contributing a local clump of azimuthal **A**. On average, then,

$$\frac{\partial \mathbf{A}}{\partial t} = \alpha \mathbf{B} \qquad \text{(Eq. 2)}$$

where **B** is the azimuthal magnetic field and α represents the mean effective rotational velocity of the Ω-loops multiplied by their filling factor. Then, allowing for diffusion, the dynamo equation for the generation of the mean poloidal field becomes

$$\frac{\partial \mathbf{A}}{\partial t} - \eta\left(\nabla^2 - \frac{1}{\varpi^2}\right)\mathbf{A} = \alpha \mathbf{B} \qquad \text{(Eq. 3)}$$

where η is again the uniform resistive diffusion coefficient and ϖ represents distance from the spin axis of the Sun (Parker, 1955b, 1970; see *Dynamo, Backus; Dynamos, mean field*). Note that this simple formulation neglects the interaction of the cyclonic convection with the relatively weak poloidal field, which is easily included if desired (see Parker, 1979; Krause and Rädler, 1980; Choudhuri, 1990).

The production of azimuthal field **B** by shearing the poloidal field $\nabla \times A$ follows from the induction equation

$$\frac{\partial \mathbf{B}}{\partial t} - \eta\left(\nabla^2 - \frac{1}{\varpi^2}\right)\mathbf{B} = \{\nabla \times [v \times (\nabla \times \mathbf{A})]\}_\phi \qquad \text{(Eq. 4)}$$

These two equations constitute the dynamo equations to lowest order. They have been derived from first principles in the idealization of intermittent cyclonic motions (Parker, 1955b, 1970, 1979; Backus, 1958) and in the quasilinear approximation (Steenbeck, Krause, and Rädler, 1966). The nature of the dynamo equations and their solutions has been investigated by many authors over the last 50 years (see Moffatt, 1978; Parker, 1979; Krause and Rädler, 1980, and references therein). As already noted, it has not yet been possible to reproduce the fluid motions in the Sun from the hydrodynamic equations, so the dynamo equations are explored by specifying the fluid velocity v—the kinematic dynamo problem (see *Dynamos, kinematic*). Perhaps one day the fluid dynamics can be brought into line and the $\alpha\omega$-dynamo equations solved simultaneously with the fluid equations, as has been accomplished for Earth (Glatzmaier and Roberts, 1996). Note, then, that the quasistationary magnetic field of Earth and the oscillatory field of the Sun arise from the same basic dynamo process, the difference in the fields caused by the different shapes of the dynamo regions and the different distribution of the nonuniform rotation and cyclonic convection within.

For the present brief article, the goal is to display the basic wave properties of the dynamo equations, for which it is sufficient to consider their solution in rectangular coordinates. Consider the simple case that the large-scale fluid motion $v_y(z)$ is in the y-direction and a function only of z, in the form of the uniform shear $G = \mathrm{d}v_y/\mathrm{d}z$. The y-direction corresponds to the azimuthal direction, and $\partial/\partial y = 0$, representing rotational symmetry of the mean fields. It follows that the magnetic field **B** in that direction is given by

$$\frac{\partial \mathbf{B}}{\partial t} - \eta\left(\frac{\partial^2}{\partial x^2} + \frac{\partial^2}{\partial z^2}\right)\mathbf{B} = G\frac{\partial \mathbf{A}}{\partial x}, \qquad \text{(Eq. 5)}$$

where the poloidal field components are

$$\mathbf{B}_x = -\frac{\partial \mathbf{A}}{\partial z}, \mathbf{B}_z = +\frac{\partial \mathbf{A}}{\partial x} \qquad \text{(Eq. 6)}$$

in terms of the vector potential A in the y-direction. The equation for A is

$$\frac{\partial \mathbf{A}}{\partial t} - \eta\left(\frac{\partial^2}{\partial x^2} + \frac{\partial^2}{\partial z^2}\right)\mathbf{A} = \alpha \mathbf{B}. \qquad \text{(Eq. 7)}$$

With uniform α, η, and G, the plane wave solution

$$\mathbf{A} = C \exp(t/\tau + ik_x x + ik_z z), \qquad \text{(Eq. 8)}$$

$$\mathbf{B} = D \exp(t/\tau + ik_x x + ik_z z), \qquad \text{(Eq. 9)}$$

is appropriate, where C and D are constants (see Gubbins and Gibbons, 2002, for 3-D waves in a sphere). Then substituting into the dynamo Eqs. (5) and (7), the result is

$$\frac{D}{C} = \frac{1/\tau + \eta(k_x^2 + k_z^2)}{\alpha} = \frac{ik_x G}{1/\tau + \eta(k_x^2 + k_z^2)}, \qquad \text{(Eq. 10)}$$

so that

$$\frac{1}{\tau} = -\eta(k_x^2 + k_z^2) \pm (\alpha G k_x/2)^{1/2}(1 + i). \qquad \text{(Eq. 11)}$$

It is sufficient for present purposes of illustration to consider the idealized example of an infinitely long train of dynamo waves (real k_x and k_z) with constant amplitude (Re $1/\tau = 0$, Im $1/\tau = \omega$). Supposing αG to be nonnegative, use the upper sign in Eq. (11). The amplitude remains constant, then, when the rate of generation $(\alpha G k_x/2)^{1/2}$ is equal to the rate of dissipation $\eta(k_x^2 + k_z^2)$. The angular frequency of the wave is related to α and G by

$$\omega = (\alpha G k_x/2)^{1/2} = \eta(k_x^2 + k_z^2), \qquad \text{(Eq. 12)}$$

and

$$\frac{1}{\tau} + \eta(k_x^2 + k_z^2) = 2^{1/2}\omega \exp(i\pi/4). \qquad \text{(Eq. 13)}$$

It follows that

$$\alpha = 2\omega^2/Gk_x, \qquad \text{(Eq. 14)}$$

and

$$\frac{D}{C} = \frac{k_x G}{2^{1/2}\omega}\exp(i\pi/4). \qquad \text{(Eq. 15)}$$

The ratio Q of the amplitude of the "azimuthal" field **B** to the vertical "poloidal" z-component $B_z = ik_x\mathbf{A}$ is

$$Q = \frac{D}{ik_x C} = \frac{G}{2^{1/2}\omega}\exp(-i\pi/4). \qquad \text{(Eq. 16)}$$

Application to the Sun

Fitting this simple dynamo wave into the Sun, to represent the migratory bands of azimuthal magnetic field, requires that the vertical half wavelength π/k_z be set equal to something of the order of the characteristic thickness 10^{10} cm of the convective zone, so that $k_z \approx 3 \times 10^{-10}$ cm, with $k_x \approx 10^{-10}$ cm to represent the 3×10^{10} cm horizontal half wavelength from the equator up to latitude 35–40°. The 22-year period of the solar magnetic cycle implies that $\omega \approx 1 \times 10^{-8}$ s^{-1}. It follows from Eq. (12) that the effective diffusion coefficient η must be approximately equal to 1×10^{11} cm^{2}s^{-1}. Then if the shear G arises from a horizontal velocity difference u over a vertical height π/k_z, it follows that $G = k_z u/\pi$. As a rough estimate, put u equal to a twentieth of the 2.0 kms^{-1} equatorial rotation velocity of the Sun, or $u = 1 \times 10^4$ cms^{-1}, with the result that $G \approx 1 \times 10^6$ s^{-1}. It follows from Eq. (14) that $\alpha \approx 2$ cm s^{-1} (more precise models give $\alpha \approx 4$–10 cm s^{-1}).

Consider then the effective values of η and α in the Sun. The diffusion coefficient η is conventionally attributed to turbulent diffusion in the vigorous convection, given by something of the order of 0.1 wh, where w is the characteristic convective velocity and h the associated mixing length, assumed to be comparable to the local pressure scale height. Thus a turbulent convective velocity of 1 kms^{-1} and a scale height of 10^2 km at the visible surface yields η equal to the required 10^{11} cm^{2}s^{-1}. Deeper in the convective zone the convective velocities are smaller (see Spruit, 1974), while the larger-scale height provides a comparable value for η. Turning to the convective rotational velocity α, note that in some ideal way the Coriolis force might produce a rotational velocity as large as $h\Omega\sin\lambda$, where Ω is the angular velocity of the Sun, equal to 3×10^{-6} s^{-1}, and λ is the latitude. With $h = 3 \times 10^9$ cm (comparable to the pressure scale height in the lower convective zone) the result is the ideal rotational velocity $\alpha = 10^4 \sin\lambda$ cms^{-1}, far in excess of the required 2 cms^{-1} estimate. In fact, with the positive and negative contributions to α from the converging and diverging portions of a convective column, the net effect is certainly much smaller than the simple ideal $h\Omega\sin\lambda$.

Now one of the several remarkable features of the magnetic fields in the Sun is the large strength of the azimuthal magnetic field relative to the poloidal field—evidently more than a hundred times stronger. The magnitude of the ratio Q of the amplitudes follows from Eq. (16) as 70, which is on the small side. Several times larger would better fit the present interpretation of the observations of emerging magnetic flux. The total azimuthal flux Φ is given by the integral of $\mathbf{B}$ over the rectangle with sides π/k_x and π/k_z, yielding

$$\Phi = \frac{\pi^2}{k_x k_z}\left(\frac{2}{\pi}\right)^2 Q k_x C \qquad \text{(Eq. 17)}$$

Putting the poloidal field amplitude $k_x C$ equal to 10 G yields the result $\Phi \approx 0.8 \times 10^{23}$ Maxwells. This is rather less than the 2×10^{23} Maxwells inferred from observations of magnetic flux emerging in a large magnetic bipolar region. Nonetheless, it is in the right range of magnitude, and we cannot expect better from the crude plane wave dynamo model.

Conditions in the Sun

A proper theoretical treatment of the solar $\alpha\omega$-dynamo requires solving the dynamo equations in spherical geometry with appropriate distributions of G, α, and η over the region, and then applying the relevant boundary conditions (see Gubbins and Gibbons, 2002). The wave numbers are complex in the general case, with the dynamo wave amplitude growing as the waves progress toward lower latitudes, while Re $1/\tau = 0$. For instance, in the special case of the geodynamo, with more or less static field strength, $1/\tau = 0$, the result is simply

$$k_x = \left(\frac{\alpha G}{\eta^2}\right)^{1/3} \exp(i\pi/6)$$

in the simple case that $k_z = 0$.

Consider then how G and α are distributed in the Sun. The situation is relatively complicated, with the strongest shear G confined mostly to a thin tachocline at the bottom of the convective zone and reversing sign across middle latitudes. The estimation of the magnitude and sign of α as a function of depth and latitude is an obscure process. As already noted, one would expect to find the Coriolis force enhancing the rate of rotation in the region of converging flow at the base of a rising convective column or cell, and retarding the rotation in the expanding rising column above. The effective mean α is the algebraic sum of these two opposing effects. The value of α is certainly a function of height, and only a small net residual α is expected from the large magnitude already estimated from dimensional considerations (see general review by Ossendrijver, 2003 and references therein).

Perhaps the most puzzling parameter is the turbulent diffusion coefficient η, for which mixing length theory provides approximately the theoretically desired value, of the order of 10^{11} cm^{2}s^{-1}. The essential point is that the magnetic field must diffuse over distances comparable to the dimensions of the convective zone in the characteristic five or ten years of the magnetic cycle. Now, the mixing length representation, $\eta = 0.1\ wh$, is based on analogy with the turbulent diffusion of a scalar field, e.g., smoke, where it works very well. One might expect that the mixing length representation would apply to the turbulent diffusion of a vector field in the limit of weak field, but there is no proper theory for this. The big problem is that the azimuthal magnetic field in the Sun is estimated to be 2×10^3 Gauss or more, and that is, in fact, a rather strong field. It is comparable to the equipartition field, for which the tension $\mathbf{B}^2/4\pi$ in the magnetic field is equal to the Reynolds stress ρv^2 in the convection, and it must be appreciated that the magnetic stresses rise rapidly as the convection deforms the magnetic field. So it is not clear how the convection can deform and mix the magnetic field to provide the essential turbulent diffusion. It can only be said that the observed behavior of the magnetic fields in the Sun suggests that there is a way, and that dimensional analysis—which is the essential feature of the application of mixing length theory—somehow leads us to the truth, however much physics we do not understand at the present time.

The puzzles do not stop there, however. For instance, the meridional circulation in the convective zone, poleward and of the order of 20 ms^{-1} at the visible surface, must play a role in the migration of the bands of magnetic field. It is not known how fast or where the equatorward return flow might be deep in the convective zone. The higher gas density suggests that the speed is substantially less than the speed of the poleward flow at the surface. As already noted, the bands of azimuthal magnetic field migrate toward the equator at about 1.5 ms^{-1} deep in the convective zone. The mystery goes back to the fact that the overall hydrodynamics of the convective zone is not yet in hand. Nor is it obvious what improvements in the hydrodynamic theory and modeling are needed. So the investigations have turned to exploring the various plausible theoretical possibilities. Numerical models of the solar dynamo with meridional circulation and overshoot of the convection into the radiative zone below have been constructed to investigate the possible role of meridional circulation above and below the tachocline in the dynamo process (see Gubbins, *et al.*, 2000). Nandy and Choudhuri (2002) explore the consequences of substantial latitudinal transport of azimuthal magnetic field in an equatorward flow below the tachocline.

A few words should be said about boundary conditions for the solar dynamo, essential for constructing quantitative theoretical models. The crucial questions concern the boundary conditions at the under side of the tachocline and at the visible surface. In the simplest picture the bottom surface of the dynamo is rigid and impenetrable. In fact there is overshoot of the convection into the radiative zone, together with the meridional circulation just mentioned. Then it has been suggested that the regions of G and α may not over lap, and communicate only through diffusion η (Parker, 1993; Schmitt, 1993; Tobias, 1996). The idea would be that the cyclonic convection α is important only above the base of the convective zone, while the principal shear G in the tachocline may lie entirely below that level. There is also the possibility that strong azimuthal magnetic field is stored in the convectively stable region beneath the convective zone (Moreno-Insertis, Schüssler, and Ferriz-Mas, 1992).

Boundary conditions at the upper surface of the dynamo, usually identified with the visible surface of the Sun, are even more complicated. It is sometimes assumed in theoretical models that the poloidal field of the dynamo fits smoothly to a vacuum potential field above the surface, with no restriction on the outward diffusion of azimuthal flux at the free surface. However, azimuthal magnetic flux can be lost through the visible surface only if there is some means for unloading the plasma from the magnetic field. Upward bulging Ω-loops do not represent a loss of azimuthal flux by themselves. Only if there is magnetic reconnection between adjacent vertical legs of neighboring Ω-loops around the Sun is there a net loss of flux. In that case the upper portions of the extended Ω-loops are first unloaded when the plasma slides down the vertical legs of the Ω-loops and then cut free of the Sun by the reconnection between neighboring Ω-loops (Parker, 1984). It is not evident from observations that there is much reconnection occurring. In the absence of such universal reconnection the upper boundary is closed and the magnetic loss through the boundary is zero, enhancing the efficiency of the dynamo (Choudhuri, 1984). Finally, it should be noted that the escape through reconnection is not a linear process, so the poloidal field cannot be treated independently of the azimuthal field.

The effects and the causes of the fibril structure of the fields have been speculated on at length (see Schüssler, 1993; Choudhuri, 2003 and references therein). The intense fibril state of the magnetic field at the surface of the Sun, consisting of 100 km diameter magnetic filaments of $1-2 \times 10^3$ G, suggests that the solar dynamo may actually be working with magnetic fibrils rather than the mean field continua for which the dynamo equations are presently formulated. In any case, the mean field treatment represented by the dynamo equations (see *Dynamos, mean field*) is presumed to provide only the large-scale features of the dynamo, but when we inquire into the nature of the diffusion η and the effective cyclonic velocity α, we come face to face with the fibril microstructure, perhaps with rapid reconnection between nonparallel fibrils. Indeed, one wonders if the ephemeral active regions and the individual magnetic fibrils, found all over the surface of the Sun at all times, are related in some way to the subsurface process of turbulent diffusion, magnetic reconnection, and the fibril structure of the magnetic field.

The effects of turbulence may be presumed important, although little is known from formal theory about the detailed interactions of the turbulent convection and the magnetic fibrils. However, one believes that somewhere in hydrodynamics and MHDs, perhaps including effects of radiative transfer in the lower convective zone, and perhaps even extending down into the radiative zone, the generation of the active magnetic fields of the Sun have their rigorous explanation. Theory needs all the guidance from helioseismology and tomography that it can get, because there are too many semiplausible possibilities within the present inadequate understanding of so many complicated processes.

Long-term variation

It must be appreciated, in the development of the theory of the solar dynamo, that direct observations of solar magnetic activity over the last four centuries, combined with radiogenic studies of such nuclei as ^{10}Be and ^{14}C (produced in the terrestrial atmosphere by the cosmic rays that are modulated by the solar wind and the magnetic activity of the Sun) show that the Sun slips off into periods of 50–100 years of greatly reduced activity, e.g., the well-known Maunder Minimum (A.D. 1645–1715), while at other times it exhibits 50–100 year periods of hyperactivity, e.g., the Medieval Maximum (12th to 13th centuries) (Eddy, 1976; Beer, *et al.*, 1990). Sometimes these wide swings can be abrupt, making the transition in a decade or less. One can only conjecture (see Mininni, Gòmez, and Mindlin, 2001; Mininni and Gòmez, 2005 and references therein) on how the hydrodynamics within the Sun, and the consequent MHD dynamo effects, cause this to happen. Observations of other stars show similar variations in their magnetic state (Baliunas, *et al.*, 1995), so there is nothing unique about the Sun, except that we are close enough to it to recognize the general form of the problem, even if not able to see below the surface to resolve the mystery. We must not overlook the important point, however, that the Sun shows activity variations over centuries and perhaps millennia, well beyond the range of contemporary research. Monitoring a large number of other stars—particularly solar-type stars—provides an effective means for viewing in a human lifetime the long-term possibilities for the Sun.

Eugene N. Parker

Bibliography

Babcock, H.W., and Babcock, H.D., 1955. The Sun's magnetic field, 1952–1954. *The Astrophysical Journal*, **121**: 349–366.

Backus, G.E., 1958. The axisymmetric self-excited fluid dynamo. *Annals of Physics*, **4**: 372–447.

Baliunas, S.L., *et al.*, 1995. Chromospheric variations on main-sequence stars II. *The Astrophysical Journal*, **438**: 269–287.

Beer, J., Blinov, J.A., Bonani, G., Finkel, R.C., Hofman, H.J., Lehman, B., Oeschger, H., Sigg, A., Schwander, J., Staffelbach, T., Suter, M., and Wolf, W., 1990. Use of ^{10} Be in polar ice to trace the 11-year cycle of solar activity. *Nature*, **347**: 165–166.

Choudhuri, A.R., 1984. The effect of closed boundary conditions on a stationary dynamo. *The Astrophysical Journal*, **281**: 846–853.

Choudhuri, A.R., 1990. On the possibility of an $\alpha^2\omega$-dynamo in a thin layer in the Sun. *Astrophysical Journal*, **355**: 733–739.

Choudhuri, A.R., 2003. On the connection between mean field dynamo theory and flux tubes. *Solar Physics*, **215**: 31–55.

Eddy, J.A., 1976. The Maunder minimum. *Science*, **192**: 1189–1202.

Gaizauskas, V., Harvey, K.L., Harvey, J.W., and Zwaan, C., 1983. Large-scale patterns formed by solar active regions during the ascending phase of cycle 21. *The Astrophysical Journal*, **265**: 1056–1065.

Glatzmaier, G.A., and Roberts, P.H., 1996. Rotation and magnetism of Earth's inner core. *Science*, **274**: 1887–1892.

Gubbins, D., and Gibbons, S., 2002. Three-dimensional dynamo waves in a sphere. *Geophysical and Astrophysical Fluid Dynamics*, **96**: 481–498.

Gubbins, D., Barber, C.N., Gibbons, S., and Love, J.J., 2000. Kinematic dynamo action in a sphere. I Effects of differential rotation and meridional circulation on solutions with dipole symmetry. *Proceedings of the Royal Society of London, Series A*, **456**: 1333–1353.

Hale, G.E., 1908. On the probable existence of magnetic fields in sunspots. *The Astrophysical Journal*, **28**: 315–343.

Krause, F., and Rädler, K.H., 1980. *Mean-Field Magnetohydrodynamics and Dynamo Theory*. Oxford: Pergamon Press.

Mininni, P.D., Gomez, D.O., and Mindlin, G.B., 2001. Simple model of a stochastically excited solar dynamo, *Solar Physics*, **201**: 203–223.

Moffatt, H.K., 1978. *Magnetic Field Generation in Electrically Conducting Fluids*. Cambridge: Cambridge University Press.

Moreno-Insertis, F., Schüssler, M., and Ferriz-Mas, A., 1992. Storage of magnetic flux tubes in a convective overshoot region. *Astronomy and Astrophysics*, **264**: 686–700.

Nandy, D., and Choudhuri, A.R., 2002. Explaining the latitudinal distribution of sunspots with deep meridional flow. *Science*, **296**: 1671–1673.

Ossendrijver, M., 2003. The solar dynamo. *The Astronomy and Astrophysics Reviews*, **11**: 287–367.

Parker, E.N., 1955a. The formation of sunspots from the solar toroidal field. *The Astrophysical Journal*, **121**: 491–507.
Parker, E.N., 1955b. Hydromagnetic dynamo models. *The Astrophysical Journal*, **122**: 293–314.
Parker, E.N., 1957. The solar hydromagnetic dynamo. *Proceedings of the National Academy of Sciences*, **43**: 8–14.
Parker, E.N., 1970. The generation of magnetic fields in astrophysical bodies. I The dynamo equations. *The Astrophysical Journal*, **162**: 665–673.
Parker, E.N., 1979. *Cosmical magnetic fields*. Chapters 18, 19, and 21. Oxford, Clarendon Press.
Parker, E.N., 1984. Magnetic buoyancy and the excape of magnetic fields from stars. *The Astrophysical Journal*, **281**: 839–845.
Parker. E.N., 1993. A solar dynamo wave at the interface between convection and nonuniform rotation. *The Astrophysical Journal*, **408**: 707–719.
Schmitt, D., 1993. The solar dynamo. *The Cosmic Dynamo,* IAU Symposium No. 157, Dordrecht: Kluwer Academic Publishers, pp. 1–12.
Schüssler, M., 1993. Flux tubes and dynamos. *The Cosmic Dynamo,* IAU Symposium No. 157, Dordrecht: Kluwer Academic Publishers, pp. 27–39.
Spruit, H.C., 1974. A model of the solar convective zone. *Solar Phys.*, **34,** 277–290.
Tobias, S.M., 1996. Diffusivity quenching as a mechanism for Parker's surface dynamo. *The Astrophysical Journal*, **467**: 870–880.
Steenbeck, M., and Krause, F. 1966. Erklärung stellarer und planetarer Magnetfelder durch einen turbulenzbedingten Dynamomechanismus. *Zeit. Naturforsch.*, **21**a: 1285–1296.
Steenbeck, M., Krause, F., and Rädler, K.H., 1966. Berechnung der mittleren Lorentz-Feldstärke **v** × **B** für ein elektrisch leitendes Medium in turbulenter, durch Coriolis-Kräfte beeinflusster Bewegung. *Zeit. Naturforsch.*, **21**a: 369–376.
Zwaan, C., 1985. The emergence of magnetic flux. *Solar Physics*, **100**: 397–414.

Cross-references

Alfvén's Theorem and the Frozen Flux Approximation
Cowling's Theorem
Dynamo, Backus
Dynamos, Kinematic
Dynamos, Mean Field
Elsasser, Walter M. (1904–1991)
Geodynamo
Helioseismology
Magnetic Field of Sun

DYNAMOS, EXPERIMENTAL

Dynamo experiments attempt to reproduce the mechanisms active in planetary dynamos on a laboratory scale. Most researchers would accept convecting liquid metal in a rotating spherical shell as a sensible model of the geodynamo. A straightforward experimental investigation of such a system is technically too demanding, so that driving mechanisms other than convection need to be used. Every experiment focuses on particular features of the dynamo effect, which makes it necessary to build many different experiments in order to obtain a complete picture of the dynamo process. A series of reviews (Busse, 2000; Tilgner, 2000; Gailitis *et al.*, 2002; Nataf, 2003) documents the progress that has been made in the field over the past couple of years.

Fluid flow can only generate a magnetic field if the generating mechanism overcomes ohmic dissipation. This condition translates into the requirement that the magnetic Reynolds number R_m must be large enough. This number is given by $R_m = VL/\lambda$ for a liquid conductor with magnetic diffusivity λ, which fills a volume of extent L and flows with the typical velocity V. If R_m exceeds a critical value, arbitrarily small magnetic fluctuations start to grow. The growth of the magnetic field is eventually stopped because magnetic fields and electric currents in the fluid combine to create the Lorentz force, which modifies the flow so as to saturate field growth. The value of the critical magnetic Reynolds number depends on the flow chosen, but is typically on the order of 10–100 in the experiments.

Lowes and Wilkinson in 1963, built an experiment with solid, electrically conducting, ferromagnetic material in order to keep the magnetic diffusivity small. The motion of the conductor consisted in two cylinders rotating inside a solid block. This experiment was the first laboratory demonstration of a dynamo working in an essentially electrically homogeneous medium (see *Dynamo, Lowes-Wilkinson*).

All other experiments have opted for liquid sodium as working fluid. Among the liquid metals, which are liquid below 500°C, sodium has the lowest magnetic diffusivity. Nonetheless, if an R_m in excess of 10–100 is necessary for dynamo action and if one is willing to build an experiment of size $L = 1$ m, one still needs sodium flowing at velocities 1–10 $\mathrm{m\,s^{-1}}$. Under these conditions, the flow of sodium is fully turbulent. These velocities present a technical challenge even when the sodium is pumped and cannot reasonably be reached in convection-driven flows. For instance, extrapolation of other laboratory experiments predicts flow velocities around 0.1 $\mathrm{m\,s^{-1}}$ if a thermal driving power of 100 kW is applied to a sphere of 1 m radius filled with sodium and spinning at 200 rpm (Nataf, 2003).

Sodium has the pleasant property that its kinematic viscosity at 120°C (the typical working temperature of the experiments) is close to the kinematic viscosity of water at room temperature. Every experiment can thus go through the following design procedures: (1) From analytical or numerical simulations of the induction equation, select a flow and a driving mechanism which are promising (i.e., values of R_m exceeding the critical R_m by a comfortable margin can be achieved with a reasonable driving power). (2) Build a model with water and check whether the selected driving mechanism (usually propellers or pumps) produces the intended flow. Because of turbulence, numerical computation can never tell for sure. (3) Build a sodium experiment. If it is not able to generate a self sustained magnetic field, investigate induction effects with externally applied magnetic fields. For instance, the decay time of the field after the external field has been switched off must go to infinity as R_m approaches the critical value. One can therefore obtain estimates of the critical R_m for the flow under investigation. (4) If necessary, and if the critical R_m estimated in step 3 seems accessible, build a bigger experiment and investigate the properties of the running dynamo.

There are at least two major questions we want these experiments to answer. First, at what magnetic Reynolds numbers does dynamo action set in? In particular, how, if at all, do small turbulent eddies affect the field generation, which mostly occurs at large scales? Second, what exactly determines the strength of the magnetic field at which it stops growing? The role of turbulent eddies is unknown in this context, too. In order to provide answers, experiments monitor the magnetic field as well as the power input into the flow. We have so far only very limited information about the velocity field in dynamo experiments, which reflects the difficulties of anemometry in liquid metals (Brito *et al.*, 2001). The problems of field saturation and of onset of dynamo action are peculiar to the dynamo effect and cannot be answered by MHD experiments with turbulent or rotating liquid metal flows in an externally applied magnetic field. For a recent review of this latter type of experiments relevant to geophysics, see Nataf (2003).

In the following, experiments are classified according to the stage of the design process they are in and are named after their geographic location.

Two different experiments have generated self-sustained magnetic fields. One is located in Riga, Latvia. It is inspired by the arguably simplest known kinematic dynamo (see *Dynamo, Ponomarenko*). The experiment consists of three coaxial pipes. A propeller forces a helical flow in the innermost pipe, the return flow occurs in the second pipe, and the third pipe contains sodium at rest (see Figure D22a). The inner pipe has a length of 3 m and a diameter of 0.25 m. For this extreme aspect ratio, edge effects at the ends of the cylinder are of minor importance, and reasonable design calculations were possible with the computational capabilities of the time at which the design started (Gailitis and Freiberg, 1980). A first experiment in 1987 was unsuccessful (Gailitis, 1987). An improved configuration did produce a self-excited magnetic field in 1999 (Gailitis *et al.*, 2000; Gailitis *et al.*, 2001).

Another experiment in Karlsruhe, Germany, is based on theoretical models dating back to Roberts (1972) (see *Dynamos, periodic*) and Busse (1992). The flow in this experiment is divided into 52 cells arranged in an array (Figure D22b). Each cell contains a right-handed helical flow with alternating directions in adjacent cells. The sodium is guided by pipes and blades made out of stainless steel so that the mechanical structures do not electrically separate the cells. Three independent pumps are necessary to achieve the flow velocities of about 4 $\mathrm{m\,s^{-1}}$ required for dynamo action. As opposed to the Riga dynamo, the helical flow in a single cell of the Karlsruhe dynamo cannot sustain a magnetic field on its own. The Karlsruhe dynamo relies on a cooperative effect: The many small eddies generate a global (large-scale) magnetic field. In the terminology of mean field MHD, the Karlsruhe dynamo is an α^2-dynamo (see *Dynamos, mean field*). Calculations of the onset of dynamo action are based either on a mean field approach (Rädler *et al.*, 2002, 2003) or on direct numerical simulation (Tilgner, 1997, 2002). The flow qualitatively mimics convection in one hemisphere of a rotating planet with the rotation axis aligned with the axis of the eddies. In this context, it is a paradox that the Karlsruhe dynamo excites a field with a dipole moment perpendicular, rather than parallel, to the axis of the eddies. This paradox is dealt with in Tilgner (2004). The Karlsruhe experiment has been running since 1999 and details of the experimental observations are in Stieglitz and Müller (2001), Müller, Stieglitz and Busse (2004), and Stieglitz and Horanyi (2004).

The most important merits of the Riga and Karlsruhe experiments have been to demonstrate the feasibility of fluid dynamo experiments and to reveal a convincing agreement between experimental data and numerical simulations. This last point is not trivial because numerical simulations at present are unable to take into account all geometrical details of an experiment. They also have to ignore turbulence and numerous other real world complications. The dynamo effect appears to be robust enough so that all these features can be safely neglected.

The velocity field cannot change dramatically in response to the magnetic field in both the Riga and Karlsruhe dynamos because of mechanical structures guiding the flow. A series of experiments located in Cadarache (France), Madison (Wisconsin), and College Park (Maryland) are currently in stage 3 of the design process described above and will give more freedom to the fluid. They are based on theoretical work by Dudley and James (1989) and use cylindrical or spherical vessels stirred by one or two propellers (Figure D22c–e). Reports on induction effects in these flows have appeared (Peffley, Cawthorne and Lathrop, 2000; Bourgouin *et al.*, 2002; Pétrélis *et al.*, 2003). Turbulence is a more important ingredient in these experiments than in Riga or Karlsruhe because turbulent eddies are not restricted in size by walls. The importance of turbulence makes predictions of critical magnetic Reynolds numbers more difficult.

Should the experiments just described work as dynamos, there will still be room for exploration. Most importantly, one would like to

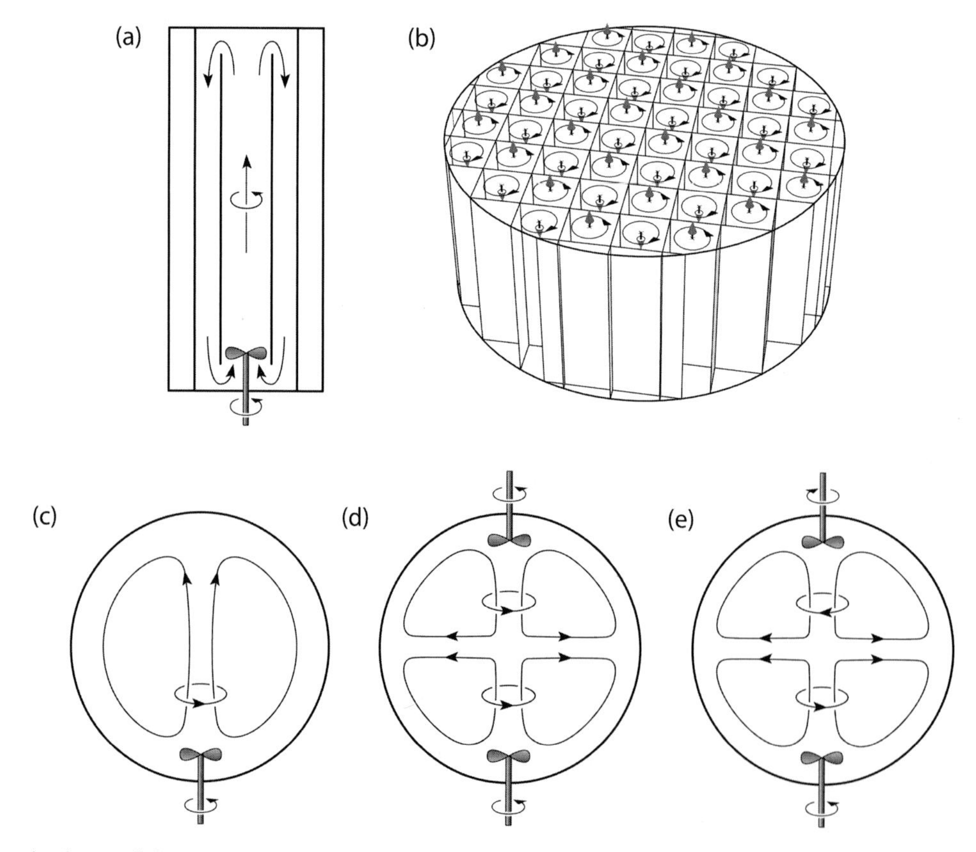

Figure D22 Sketch of several dynamo experiments. (a) the Riga dynamo. The drawing is not to scale. At equal diameter, the pipes should be shown three times as long. (b) The Karlsruhe experiment. The arrows indicate flow directions within each cell. The entire vessel has a diameter of about 2 m and a height of approximately 1 m. (c) to (e) Different flows which can be realized in the experiments in Cadarache, Madison, and Maryland. (e) is the favored flow because it has the smallest critical magnetic Reynolds number according to numerical simulations. The experiment in Cadarache uses a cylindrical instead of a spherical container.

include the effects of rotation since the Coriolis force must play an important role in the equilibrium of forces in real planets. The group in Maryland is building an experiment with two concentric spheres rotating at different angular velocities in order to drive a nontrivial flow in the gap between the spheres. The outer sphere has a diameter of 3 m and magnetic Reynolds numbers of nearly 700 will be accessible. There are several other experiments in stage 2 coming from a variety of backgrounds, but they are all rotating. A configuration studied in New Mexico is motivated by astrophysics and intends to produce an $\alpha\omega$-dynamo (Colgate *et al.*, 2001) (see *Dynamos, mean field*). A group in Perm (Russia) wants to use a torus, rotate it about its axis so that the sodium inside the torus is in solid body rotation, and suddenly stop the torus from rotating. Because of inertia the sodium in the torus continues to rotate and has a nonzero velocity relative to the walls of the torus, so that obstacles placed on the wall can induce in the sodium a flow suitable for a dynamo effect (Dobler *et al.*, 2003). The concept is appealing because of its simplicity but has the disadvantage to generate only a transient flow and not a stationary state.

The possibility exists that the geodynamo is driven by the precession of the rotation axis of the Earth rather than by convection. Following up on earlier experiments by Gans (1970) a group in Meudon (France) explores the feasibility of driving a laboratory dynamo by precession. The apparatus presently in use is a cylinder filled with water rotating about its axis and precessing about an axis perpendicular to the axis of the cylinder (Léorat *et al.*, 2003).

Existing dynamo experiments are not small-scale replica of the geodynamo but can nonetheless provide information directly useful for the analysis of geomagnetic data. For instance, Christensen and Tilgner (2004) extract a scaling from numerical simulations and dynamo experiments which constrains the value of the ohmic dissipation in the Earth's core. The experimental data proves crucial because it is for a magnetic Prandtl number (the ratio of viscosity and magnetic diffusivity) characteristic of liquid metals and presently inaccessible to numerical simulations. Experimental investigation of the dynamo effect is likely to become an increasingly important tool in the study of the geodynamo as more features such as rotation are added to the experiments.

Andreas Tilgner

Bibliography

Bourgouin, M., Marié, L., Pétrélis, F., Gasquet, C., Guigon, A., Luciani, J.-B., Moulin, M., Namer, F., Burguete, J., Chiffaudel, A., Daviaud, F., Fauve, S., Odier, P., and Pinton, J.-F., 2002. Magnetohydrodynamics measurements in the von Karman sodium experiment. *Physics of Fluids*, **14**: 3046–3058.

Brito, D., Nataf, H.-C., Cardin, P., and Masson, J.-P., 2001. Ultrasonic Doppler velocimetry in liquid gallium. *Experiments in Fluids*, **32**: 653–663.

Busse, F.H., 1992. Dynamo theory of planetary magnetism and laboratory experiments. In Friedrich, R., and Wunderlin, A. (eds.), Vol. 69. *Evolution of Dynamical Structures in Complex Systems*. Springer Proceedings in Physics.

Busse, F.H., 2000. Homogeneous dynamos in planetary cores and in the laboratory. *Annual Review of Fluid Mechanics*, **32**: 383–408.

Christensen, U.R., and Tilgner, A., 2004. Power requirement of the geodynamo from ohmic losses in numerical and laboratory dynamos. *Nature*, **429**: 169–171.

Colgate, S.A., Li, H., and Pariev, V., 2001. The origin of the magnetic fields of the universe: the plasma astrophysics of the free energy of the universe. *Physics of Plasmas*, **8**: 2425–2431.

Dobler, W., Frick, P., and Stepanov, R., 2003. Screw dynamo in a time-dependent pipe flow. *Physics Reviews*, **67**: 056309.

Dudley, M.L., and James, R.W., 1989. Time-dependent kinematic dynamos with stationary flows. *Proceedings of the Royal Society of London, Series A*, **425**: 407–429.

Gailitis, A., and Freiberg, Ya., 1980. Nonuniform model of a helical dynamo. *Magnetohydrodynamics*, **16**: 11–15.

Gailitis, A.K., Karasev, B.G., Kirillov, I.R., Lielausis, O.A., Luzhanskii, S.M., Ogorodnikov, A.P., and Presltskii, G.V., 1987. Liquid metal MHD dynamo model experiment. *Magnitnaya Gidrodinamika*, **4**: 3–7.

Gailitis, A., Lielausis, O., Dement'ev, S., Platacis, E., Cifersons, A., Gerbeth, G., Gundrum, Th., Stefani, F., Christen, M., Hänel, H., and Will, G., 2000. Detection of a flow induced magnetic field eigenmode in the Riga dynamo facility. *Physics Review Letters*, **84**: 4365–4368.

Gailitis, A., Lielausis, O., Platacis, E., Dement'ev, S., Cifersons, A., Gerbeth, G., Gundrum, Th., Stefani, F., Christen, M., and Will, G., 2001. Magnetic field saturation in the Riga dynamo experiment. *Physics Review Letters*, **86**: 3024–3027.

Gailitis, A., Lielausis, O., Platacis, E., Stefani, F., and Gerbeth, G., 2002. Laboratory experiments on hydromagnetic dynamos. *Reviews of Modern Physics*, **74**: 973–990.

Gans, R.F., 1970. On hydromagnetic precession in a cylinder. *Journal of Fluid Mechanics*, **45**: 111–130.

Léorat, J., Rigaud, F., Vitry, R., and Herpe, G. 2003. Dissipation in a flow driven by precession and application to the design of a MHD wind tunnel. *Magnetohydrodynamics*, **39**: 321–326.

Müller, U., Stieglitz, R., and Busse, F.H. 2004. On the sensitivity of dynamo action to the system's magnetic diffusivity. *Physics of Fluids*, **16**: L87–L90.

Müller, U., Stieglitz, R., and Horanyi, S. 2004. A two-scale hydromagnetic dynamo experiment. *Journal of Fluid Mechanics*, **498**: 31–71.

Nataf, H.-C., 2003. Dynamo and convection experiments. In Jones, C.A., Soward, A.M., and Zhang, K. (eds.), *Earth's Core and Lower mantle*. Abingdon, UK, Taylor and Francis, pp. 153–179.

Peffley, N.L., Cawthorne, A.B., and Lathrop, D.P., 2000. Toward a self-generating magnetic dynamo: The role of turbulence. *Physics Review*, **61**: 5287–5294.

Pétrélis, F., Bourgouin, M., Marié, L., Burguete, J., Chiffaudel, A., Daviaud, F., Fauve, S., Odier, P., and Pinton, J.-F. 2003. Nonlinear magnetic induction by helical motion in a liquid sodium turbulent flow. *Physics Review Letters*, **90**: 174501.

Rädler, K.-H., and Brandenburg, A., 2003. Contributions to the theory of a two-scale homogeneous dynamo. *Physics Review*, **67**: 026401.

Rädler, K.-H., Rheinhardt, M., Apstein, E., and Fuchs, H., 2002. On the mean-field theory of the Karlsruhe dynamo experiment. I. Kinematic theory. *Magnetohydrodynamics*, **38**: 41–71.

Roberts, G.O., 1972. Dynamo action of fluid motions with two-dimensional periodicity. *Philosophical Transaction of the Royal Society of London, Series A*, **271**: 411–454.

Stieglitz, R., and Müller, U., 2001. Experimental demonstration of a homogeneous two-scale dynamo. *Physics of Fluids*, **13**: 561–564.

Tilgner, A., 1997. A kinematic dynamo with a small scale velocity field. *Physics Letters A*, **226**: 75–79.

Tilgner, A., 2000. Towards experimental fluid dynamos. *Physics of the Earth and Planetary Interiors*, **117**: 171–177.

Tilgner, A., 2002. Numerical simulation of the onset of dynamo action in an experimental two-scale dynamo. *Physics of Fluids*, **14**: 4092–4094.

Tilgner, A., 2004. Small-scale kinematic dynamos: Beyond the α-effect. *Geophysical and Astrophysical Fluid Dynamics*, **98**: 225–234.

Cross-references

Dynamo, Gailitis
Dynamo, Lowes-Wilkinson
Dynamo, Ponomarenko
Dynamos, Kinematic
Dynamos, Mean Field
Dynamos, Periodic
Fluid Dynamics Experiments
Magnetohydrodynamics

DYNAMOS, FAST

As explained in other chapters dynamo action occurs when the effects of advection (Faraday's law) are vigorous enough to overcome the effects of Ohmic diffusion. The ratio between these two effects is expressed by the *magnetic Reynolds number*

$$R_m \equiv \mathcal{U}\mathcal{L}/\eta$$

where $\mathcal{U}$, $\mathcal{L}$ are typical length and timescales, and η is the magnetic diffusivity. If we nondimensionalize the induction equation by scaling lengths with L and time with the turnover time $\mathcal{L}/\mathcal{U}$ we get the following form (assuming η uniform):

$$\frac{\partial \mathbf{B}}{\partial t} = \nabla \times (\mathbf{u} \times \mathbf{B}) - R_m^{-1} \nabla \times \nabla \times \mathbf{B}$$

At the onset of dynamo action, when induction just balances diffusion, R_m is of order unity by definition. However in some astrophysical bodies (though probably not the Earth) R_m is very large, and the question then arises whether magnetic flux can grow at a rate $\mathcal{O}(1)$ (that is on the turnover time) or whether in fact the growth rate is much smaller, depending on some negative power of R_m (the rate R_m^{-1} would imply growth on the Ohmic time $\mathcal{L}^2/\eta$). If the first alternative holds, we say that the dynamo is *fast*, otherwise it is *slow*.

It might be expected to be obvious from Faraday's law (prohibiting any change in the flux linked with a perfect conductor) that all dynamos would be slow, since the growth rate at $R_m = \infty$ would be zero. However if as R_m becomes very large the length scales of the magnetic field become small, then diffusion can never be neglected, even in the limit. Thus the question of the type of dynamo action is intimately bound up with the complexity of the field structure.

It can be important to know whether the dynamo is fast or not; for example the Ohmic time for the Sun is comparable with the age of the universe, whereas the large-scale solar dynamo evolves on scales of a few years. If the solar dynamo were not fast, then it would not be able to generate flux in the way that it does.

While it is probably not the case that the Earth is in a regime appropriate for fast dynamo action, there is no doubt that our understanding of dynamo action in general has been enhanced by the study of fast dynamos, which have proved a valuable interface between MHD and dynamical systems thory; see for example the monograph of Childress and Gilbert (1995).

Examples of slow dynamos

Here we give two rather different examples of slow dynamos. The first is an adaptation of the Faraday disk dynamo introduced by Moffatt (see Figure D23 and *Dynamo, disk*). Then there are simple equations that relate current in the wire I, the current round the disk J, the angular velocity Ω and the fluxes of magnetic field through the wire and the disk Φ_I, Φ_J. These are

$$\Phi_I = LI + MJ, \quad \Phi_J = MI + L'J, \quad RI = \Omega\Phi_J - \frac{d\Phi_I}{dt};$$
$$R'J = -\frac{d\Phi_J}{dt}$$

where M is the mutual inductance, and L, L' the self-inductances of the circuit, and $LL' > M^2$. We can seek solutions $\propto e^{pt}$, and find growth (dynamo action) if $\omega M > R$. The growth rate is

$$p_+ = \left(\sqrt{(RL' + R'L)^2 + 4R'(\Omega M - R)(LL' - M^2)} - (RL' + R'L)\right)/2(LL' - M^2).$$

$p_+ > 0$ for all $\Omega > R/M$ but $p \sim \sqrt{\Omega R'}$ as $\Omega \to \infty$. Identifying R' with η we can see that the growth rate scales as $R_m^{-1/2}$. This is not really surprising since by Faraday's law zero growth rate of the flux through the disk should be expected as $\eta \to \infty$.

A second slow dynamo is the so-called Roberts flow (Robert, 1970) (see also *Dynamos, periodic*). This is a simple steady cellular flow depending on only two space dimensions. The velocity takes the form

$$\mathbf{u}(x,y) = \nabla \times (\psi(x,y)\mathbf{e}_z) + \gamma\psi(x,y)\mathbf{e}_z; \quad \psi = \sin y + \cos x.$$

Thus we can see that the helicity $\langle \nabla \times \mathbf{u} \cdot \mathbf{u} \rangle = \gamma\langle \psi^2 \rangle$. Solutions exist in the form $\mathbf{B} = Re(\tilde{\mathbf{B}}(x,y,t)e^{ikz})$. The optimum growth rate occurs at large k when $R_m \gg 1$, in fact $(k \sim (R_m^{1/2}/\ln R_m)$. These scales, though small, are long compared to the thin boundary layer scale $R_m^{-1/2}$ for field near stagnation points. As $R_m \to \infty$ the optimum growth rate $\sim \mathcal{O}(\ln(\ln R_m)/\ln R_m)$. This is *just* a slow dynamo! What these two rather different examples have in common is that the flows to not act to separate nearby points exponentially in time. That is, they are not stretching flows (in fact the Roberts flow does have exponential stretching at the stagnation points of ϕ, but nowhere else).

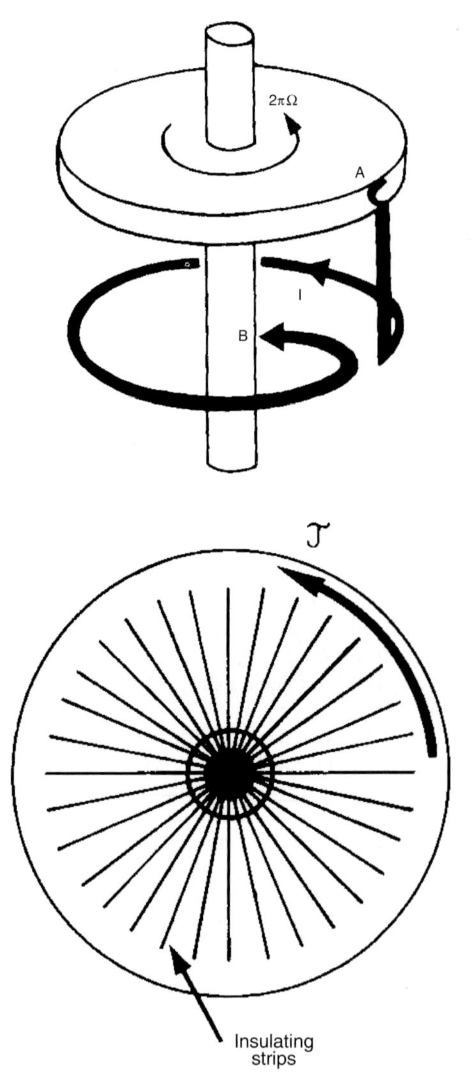

Figure D23 The segmented disk dynamo (Moffatt 1979). Radial strips prevent circumferential current except in the outer part of the disk.

Two examples of fast dynamos

We now turn to two examples of fast dynamos. The first is the stretch-twist-fold (STF) flow of Zel'dovich, which in the absence of diffusion acts to double the magnetic energy by stretching. The diagram in Figure D24 shows how the field has to be moved at large R_m, when the field lines are almost frozen in. Two issues immediately arise. Firstly, the role of diffusion, however small, is not clear near the cross-over points of the field, where gradients may be high. Secondly, the flow field required is certainly not steady, or easily realizable in a finite domain. However, it is possible to manufacture realistic flows that have the required folding properties.

The second example is a development of the Roberts flow introduced by Galloway and Proctor (1992). The velocity field (the "GP-flow") now takes the form: $\mathbf{u}(x,y,t) \propto \nabla \times (\psi(x,y,t)\mathbf{e}_z) + \gamma\psi(x,y,t)\mathbf{e}_z$; $\psi = \sin(y + \epsilon \sin \omega t) + \cos(x + \epsilon \cos \omega t)$ and it can be seen that the Eulerian pattern of the Roberts flow is advected in a circle. However the Lagrangian pattern is completely different; while the Roberts flow has no exponential separation of nearby points (except at corners) the GP-flow has large stretching in large regions of the domain. This can be seen by looking at the Liapunov exponents for the flow with $\epsilon = \omega = 1$: The dynamo behavior of this flow is quite different from that of the Roberts flow. For large R_m, the growth rate levels off at about 0.3, while the critical wavenumber converges to about 0.57. The magnetic field structure becomes very complicated in the limit of large R_m, as can be seen in the right hand panel of Figure D25. In fact diffusion remains important even in the limit, and while the smallest scales $\sim R_m^{-1/2}$, the overall width of the flux structures reaches a finite limit. The contrast between this behavior and that of the Roberts flow can be seen in Figure D26. The key to this complex behavior lies in a delicate balance between the stretching properties of the flow, which tend to increase the energy at a rate independent of diffusion, and the folding of flux elements back on themselves which is inevitable given that the velocity does not lead to unbounded excursions of fluid particles in general. If folding takes place so as to bring oppositely directed field lines close together, the resulting cancellation due to diffusion can result in a net *decay* of the field. This always happens when the field is two-dimensional. In three dimensions, however, with the right flow and the right scales across the folding direction the folding can be *constructive*, leading to net growth of the field even when diffusion is very small.

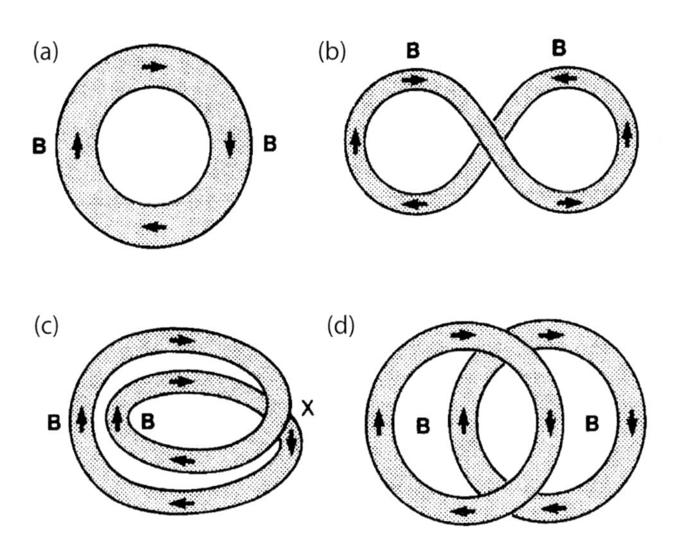

Figure D24 The STF dynamo (after Fearn *et al.*, 1988). The initial loop of flux is made to lie on top of itself but with twice the original energy.

Nonlinear development

The idea of a fast dynamo is essentially a linear one, related to the rate of growth of magnetic fields for a given velocity field **u**. If instead we regard the velocity as being produced by some given force field, subject to the action of Lorentz forces, then as the field grows to finite amplitude the flow field will be modified in such a way that the growth of the field ceases. How is this achieved? One possibility is that the vigor of the flow is decreased while its transport properties remain constant; that is, equilibration is achieved by a reduction of the effective R_m. However, a number of numerical simulations of nonlinear dynamos suggest that instead reduction in dynamo action is achieved by a change in the stretching properties of the flow, with little change in the kinetic energy or other Eulerian properties. When there is no significant field, the rate separation of fluid particles is unconstrained by their initial position. When there is a magnetic field present, and R_m is large on the dominant scales of the flow, then the fluid particles are closely linked to the field lines and so acquire a

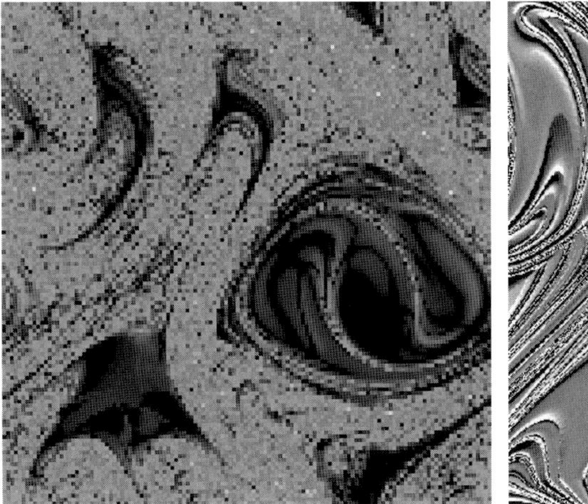
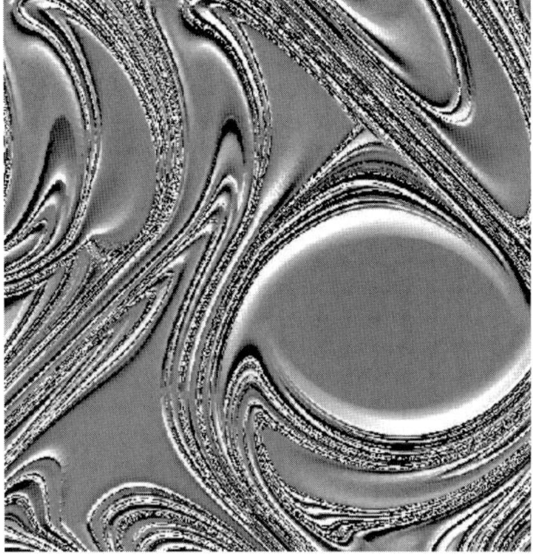

Figure D25 *Left panel*: Liapunov exponents for the GP-flow (courtesy of D.W. Hughes), found by integrating two nearby particle paths for a finite time and measuring the total stretching; dark shades indicate small stretching, lighter shades higher stretching. *Right panel*: intensity graph of the normal component of the dynamo field for this flow (courtesy of F. Cattaneo). Medium gray (red in online encyclopedy) denotes weak, and white strong, field regions. Both plots show a periodic domain in the (x,y) plane. Note the close correlation.

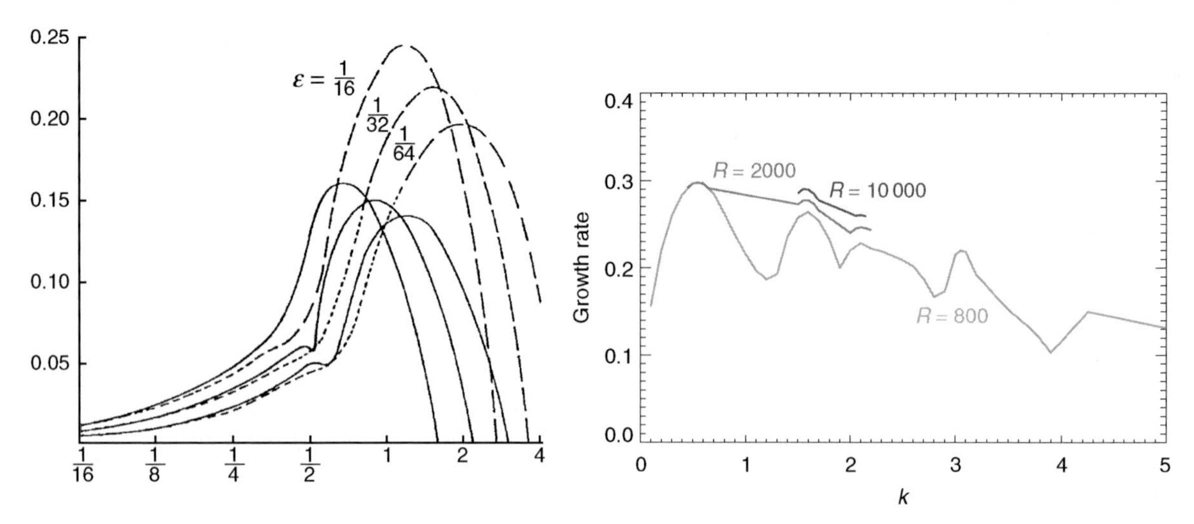

Figure D26 *Left panel*: Growth rates for the Roberts flow as a function of k for various $R_m = 1/\varepsilon$ (solid lines, numerical results, dotted lines, asymptotic theory); *right panel*: Growth rates as a function of k for the GP-flow.

history; unlimited stretching is impossible without inducing large Lorentz forces due to field line folding. When a stationary state is reached the stretching is considerably reduced. The final velocity field has less stretching than the linear flow, and is just such as to achieve a zero growth rate. The resulting dynamo is thus neither slow nor fast, just marginal!

Summary

By means of simple examples I have tried to demonstrate the distinction between fast and slow dynamos. From these it is clear that the limit of infinite R_m is not a simple one. Fast dynamos have become the object of intense study by mathematicians: the outstanding monograph describing this work, and much else concerning fast dynamos, is that of Childress and Gilbert (1995). In the context of the geodynamo, the limit of large R_m is not approached, and the advection (turnover) time and the diffusion time can be considered comparable, so that the limiting behavior discussed here is unlikely to be relevant. Nonetheless, an understanding of dynamos at large R_m has, as explained above, proved very fruitful in advancing our understanding of dynamos in general, including the geodynamo, and so an appreciation of the issues involved is very important.

Michael Proctor

Bibliography

Childress, S., and Gilbert, A.D., 1995. *Stretch, Twist, Fold: The Fast Dynamo*. Berlin: Springer.

Fearn, D.R., Roberts, P.H., and Soward, A.M., 1988. Convection, stability and the dynamo. In Galdi, G.P., and Straughan, B. (ed), *Energy Stability and Convection*. Pitman Research Notes in Mathematics 168, pp. 60–324.

Galloway, D.J., and Proctor, M.R.E., 1992. Numerical calculations of fast dynamos in smooth velocity fields. *Nature*, **356**: 691–693.

Moffatt, H.K., 1979. A self-consistent treatment of simple dynamo systems. *Geophys. Astrophysical Fluid Dynamics*, **14**: 147–166.

Roberts, G.O., 1970. Spatially periodic dynamos. *Philosophical Transaction of the Royal Society of London*, **A266**: 535–558.

Cross-references

Dynamo, Disk
Dynamos, Periodic

DYNAMOS, KINEMATIC

Introduction

In the Earth's fluid outer core, the motion of electrically conducting molten iron is responsible for the generation of the magnetic field. In general, this field reacts back onto the flow by the Lorentz force in a nonlinear fashion, introducing complex behavior into the system which is not well understood. Under the kinematic assumption, we study the simpler problem in which the fluid flow is prescribed (and typically time-independent), ignoring this back reaction. In this case, we are at liberty to analyse the time behavior of the magnetic field in isolation and test whether or not a given flow can act as a dynamo.

Such an analysis is a gross simplification, since the geodynamo is thought to be in the so-called strong-field regime where the flow is significantly affected by the magnetic field and will therefore differ from any fixed kinematic flow structure. Such effects are taken into account in fully dynamical models, however, these are so complex that the behavior of the system often defies simplistic interpretation (see *Geodynamo, numerical simulations*). Additionally, the parameter values at which these models operate are so far from geophysical estimates that it is by no means certain that the results can be extrapolated to the Earth. Bearing this in mind, the analyses of simple kinematic Earth-like models are important in the understanding of the dynamo process.

The induction equation

The magnetic induction equation governs the time evolution of a magnetic field $\mathbf{B}$ under the influence of a moving conductor $\mathbf{u}$ (the outer core) in the reference frame of the mantle and magnetic diffusion (see *Magnetohydrodynamics*). It is written in nondimensional form below, the sole parameter being the magnetic Reynolds number $R_m = \mathcal{U}\mathcal{L}/n$ where $\mathcal{U}$ is a typical value of the flow velocity, $\mathcal{L}$ is the typical length scale taken to be the radius of the outer core and $\eta = (\mu_0 \sigma)^{-1}$ is the (constant) magnetic diffusivity, μ_0 being the permeability of free space, and σ the electrical conductivity;

$$\frac{\partial \mathbf{B}}{\partial t} = R_m \nabla \times (\mathbf{u} \times \mathbf{B}) + \nabla^2 \mathbf{B}. \quad \text{(Eq. 1)}$$

The value of R_m is the ratio of the effects of the flow to magnetic diffusion (associated with Ohmic dissipation of energy), and lies in the geophysical range of 500–1000. Time has been scaled relative to the *a priori* magnetic diffusion timescale $\mathcal{L}^2/n$, which is typically 200000 years for the Earth. A straightforward analysis however shows that the true diffusive decay time for the Earth is more like 20000 years (see Moffatt, 1978). The exterior of the outer core is the mantle, well approximated by an electrical insulator (see *Mantle, electrical conductivity, mineralogy*). The inner core is often ignored for simplicity, but if included, is given the same conductivity as the outer core.

Kinematic dynamo theory is concerned with magnetic field solutions of Eq. (1) that grow in time given a flow structure **u**. It is by no means obvious that these exist, for in a sphere the dynamo mechanism, associated with electric currents, could easily short-circuit prohibiting any magnetic field to grow. Nevertheless, growing solutions have been found, as is outlined in the historical overview below. In general, the kinematic assumption is valid assuming firstly we choose a geophysically motivated flow, plausibly generated by rotating convection (see *Core convection*) and secondly that the Lorentz force is small, usually taken to mean that the initial and subsequent magnetic fields are also small.

The induction equation is typically solved in a spherical geometry using vector spherical harmonics (see *Harmonics, spherical*), in which the divergence-free magnetic field is expanded in toroidal-poloidal form (see Moffatt, 1978).

The outer core is commonly taken to be incompressible and so **u** may also be written in this way:

$$\mathbf{u} = \sum_{l,m} t_l^m + s_l^m = \sum_{l,m} \nabla \times [Y_l^m(\theta,\phi) t_l^m(r)\hat{\mathbf{r}}] + \nabla \times \nabla \times [Y_l^m(\theta,\phi) s_l^m(r)\hat{\mathbf{r}}] \quad \text{(Eq. 2)}$$

where l and m are the degree and order of the spherical harmonic $Y_l^m(\theta,\phi)$ (in spherical polar coordinates), and $\hat{\mathbf{r}}$ is the unit position vector. The toroidal and poloidal scalar functions t_l^m and s_l^m are given. The magnetic field is written in a similar fashion as $\mathbf{B} = \sum_{l,m} \mathbf{T}_l^m + \mathbf{S}_l^m$ (always using upper case). The value $m = 0$ represents a vector independent of ϕ, axisymmetric about the z-axis.

A typical flow comprises several harmonic components, for example, t_1^0 representing differential rotation and s_2^0 representing convection as shown in Figure D27, being nonslip at the boundary $r = 1$. Both of these components are important aspects of core convection.

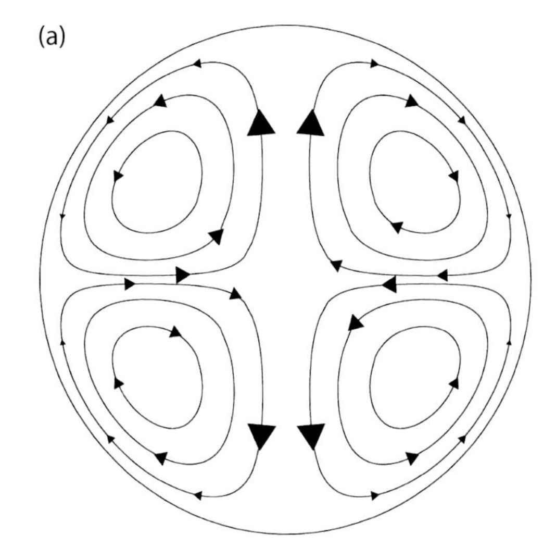

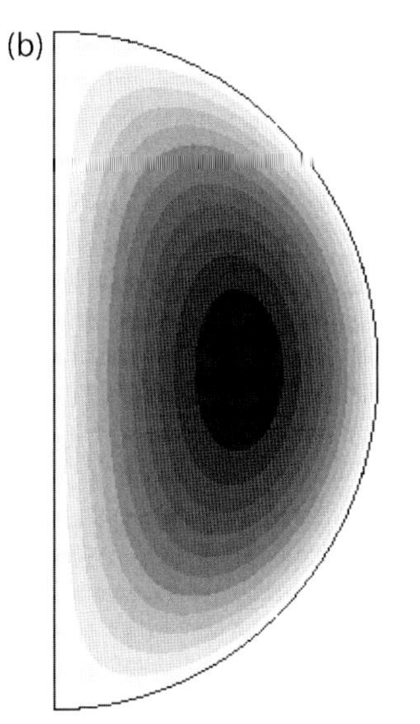

Figure D27 (a) Streamlines of an s_2^0 flow in a meridian plane; (b) contours of u_ϕ in a meridian plane of a t_1^0 flow, giving rise to differential rotation, the flow differing from solid body rotation and is faster midstream than at the boundaries.

An historical overview: after Cowling's theorem

Cowling's theorem (*q.v.*), published in 1933, caused a major setback in the early days of geodynamo theory. It proved that no axisymmetric field could be sustained by dynamo action and since the Earth's field is principally of this type, it was thought at the time that the geodynamo hypothesis was untenable. Indeed, it was not known whether a general antidynamo theorem (see *Antidynamo and bounding theorems*) might exist thus ruling out any kind of generative mechanism altogether. In 1947 Elsasser (see *Elsasser, Walter M.*) was the first to suggest that a complex mechanism might be taking place within the outer core, converting magnetic field between symmetries thus breaking the constraints of Cowling's theorem. He suggested that differential rotation could convert an axisymmetric dipole field, represented by the $\mathbf{S}_1^0$ vector harmonic, into $\mathbf{T}_2^0$ by what is now called the ω-effect (see *Dynamos, mean field*).

The problem remained of how to close the loop, that is, how to change the newly created $\mathbf{T}_2^0$ field back into $\mathbf{S}_1^0$ so that the process might continue indefinitely. Of the two mechanisms he suggested, namely turbulence and tilted flows, the latter was pursued by Bullard (see *Bullard, Edward Crisp*) who argued qualitatively that a flow component s_2^{2c} (describing a tilted convection roll) might be sufficient to regenerate the axial dipole field. Figure D28 shows a schematic picture of this process. Solid arrows show the effect on the field of the t_1^0 flow; dashed

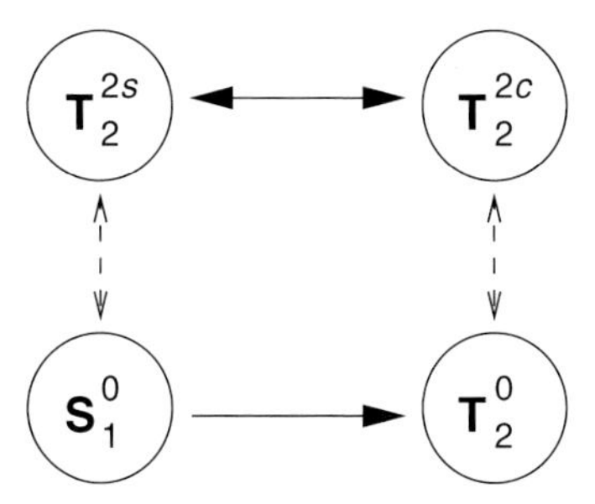

Figure D28 The interactions between the lowest degree harmonics involving the axial dipole $\mathbf{S}_1^0$ that forms a closed loop. Solid arrows denote coupling by t_1^0 motion; dashed by s_2^{2c}. If all links of this chain work sufficiently quickly then the axial dipole harmonic could be infinitely sustained by this process.

arrows show the effect of the flow component s_2^{2c}. Thus, an $\mathbf{S}_1^0$ field, if acted upon in this manner, might be sustained creating byproducts of $\mathbf{T}_2^0$, $\mathbf{T}_2^{2s}$, and $\mathbf{T}_2^{2c}$ fields *en route*. The emphasis at this time was on finding a mechanism whereby the principally observed geophysical field component ($\mathbf{S}_1^0$) could be sustained; no attention was given to other field symmetries. Since this four component field is not axisymmetric, Cowling's theorem does not apply and the idea is plausible.

The Bullard and Gellman model

The first quantitative model of a self-sustaining process taking place inside the Earth's core was proposed by Takeuchi and Shimazu (1953). This work was closely followed by Bullard and Gellman (1954) (see *Dynamo, Bullard-Gellman*); both papers extended the suggestions of Elsasser and Bullard that if a flow was chosen carefully, through a sequence of interactions, the original field might be amplified and lead to a self-sustaining process.

The original procedure used to solve the induction equation was to seek steady fields, which after numerical discretisation, converts Eq. (1) into a generalized eigenvalue problem for R_m; a steady solution was manifested by the existence of a real eigenvalue. The magnetic field was expanded in vector spherical harmonics and each unknown radial scalar function using a finite difference scheme. The velocity chosen was defined by

$$t_1^0(r) = \varepsilon r^2(1-r) \qquad s_2^{2c} = r^3(1-r)^2 \qquad \text{(Eq. 3)}$$

where ε is an adjustable parameter. No inner core was included in the flow pattern to make it as simple as possible. Real eigenvalues were found for R_m, which did not appear to be affected greatly by changing the numerical resolution, at least within the limited computer resources available at that time. Thus it had appeared that a working dynamo had been discovered: the combinations of the differential rotation t_1^0 and the tilted convection roll s_2^{2c} was such that a (nonaxisymmetric) steady field could be maintained with an Earth-like $\mathbf{S}_1^0$ component. However, 13 years later, a repeat of the calculations (Gibson and Roberts, 1967) at higher resolution using vastly more powerful computers showed that the positive dynamo action found was spurious: the results were merely an artifact of the inadequate numerical method used and not of the underlying physics. The question of the possibility of dynamo action in a homogeneous conducting sphere was again reopened.

Other advances

In 1958 both Backus (see *Dynamo, Backus*) and Herzenberg (see *Dynamo, Herzenberg*) published analytic studies that strengthened the belief that dynamo action was possible. The problem with the induction effect is that for any length scale, successively smaller scales will be created by the fluid-field interactions. Numerically it is hoped that on increasing the resolution, the solution converges sufficiently quickly to be captured with the available computing resources. Both of these dynamos use diffusive effects to kill off the small-scale fields: the Backus dynamo by periods of stasis and the Herzenberg dynamo by spatial attenuation, both amenable in analytic treatment. Neither of these solutions would be a possible candidate for the Earth; however, this was evidence that no generalized antidynamo theorem existed.

The study of the dynamo problem in a sphere was also complimented by calculations in other geometries. An analytical solution of a working dynamo in a cylindrical helical flow was published in 1973 by Ponomarenko (see *Dynamo, Ponomarenko*). Roberts (1970) studied infinite spatially periodic dynamos, and found that almost all motions performed as working dynamos. This was real evidence that not only was it possible to find flows that worked as dynamos but that it was relatively easy to do so, finally laying to rest the fear of a generalised antidynamo theorem and gave dynamo theory centre stage for the origin of Earth's magnetic field.

Further work on spherical dynamos

In the 1970s and onward, much progress was made with spherical kinematic dynamos. Lilley (1970) studied the flow $\mathbf{u} = t_1^0 + s_2^{2c} + s_2^{2s}$ and found ostensibly converged growing magnetic field solutions. Unfortunately, yet again, later workers discovered his results to be unconverged; this was to be the last of the spurious discoveries. The first converged solutions were presented by Gubbins (1972), Roberts (1972), and Pekeris *et al.* (1973). The method of solution had switched from seeking steady solutions (leading to a generalised eigenvalue problem for R_m) to time-dependent eigenmode solutions. In this paradigm, the assumption that $\mathbf{B} = \hat{\mathbf{B}}e^{\lambda t}$ is made thereby reducing the induction equation to an eigenvalue problem for the complex growth rate λ. The adjustable parameter R_m was varied and the lowest value giving $\Re(\lambda) = 0$ sought. If such a value could be found (which was by no means certain) this was termed the critical magnetic Reynolds number (R_m^c). It is entirely possible that $\Im(\lambda) \neq 0$ in this critical state so that the dynamo would be oscillatory.

In the various flows studied, these workers found growing magnetic fields although not always of the correct symmetry for the Earth. Nonetheless, this was the first concrete evidence that the geodynamo could work, at least in a simplified case. It was also found that a greater flow complexity introduced by the choice of radial scalar functions favoured dynamo action, i.e., the critical value of R_m was smaller. This characteristic was also found in the study of the nonaxisymmetric flow of Kumar and Roberts (1975) with four components:

$$\mathbf{u} = t_1 + \varepsilon_1 s_2 + \varepsilon_2 s_2^{2c} + \varepsilon_3 s_2^{2s} \qquad \text{(Eq. 4)}$$

where the values of $\varepsilon_1, \varepsilon_2$, and ε_3 are adjustable. The t_1^0 component again represented an azimuthal differential rotation, the s_2^0 a meridional circulation, and the s_2^{2c} and s_2^{2s} terms tilted convection rolls. They found converged field solutions containing the axial dipole, and that the direction of the flow components (i.e., the sign of the values of ε_i) was crucial in determining the symmetry of the growing magnetic field.

In later studies of the Kumar and Roberts flow (e.g., Sarson and Gubbins, 1996; Gubbins *et al.*, 2000) it was found that steady magnetic field solutions were favored with strong meridional circulation. The inclusion of a solid inner core introduced significant changes in the field solutions; the fact that different symmetries adapted in different ways led to the possibility that the inner core might play a role in field symmetry selection. Additionally, it was concluded that whether or not kinematic dynamo action exists depends critically on the exact choice of flow, an issue that makes it difficult to attribute a robust physical interpretation to such a dynamo mechanism. The influence of helicity (the spiralling nature of the flow) was found to be favourable to dynamo action, although was not essential.

Despite the correlation between increasing flow spatial complexity and dynamo efficiency, Dudley and James (1989) found growing magnetic field solutions in very simple axisymmetric convection roll flows, of $t_1^0 s_2^0$, $t_2^0 s_2^0$, and $t_1^0 s_1^0$ type. The fact that a one cell convective flow $t_1^0 s_1^0$ can support dynamo action dispelled beliefs that the flow had to be complex in nature, as previous results suggested.

An alternative methodology to solving the kinematic dynamo problem, instead of seeking eigenvalue solutions as has been popular in most of the previous literature, is to apply an energetic analysis (Livermore and Jackson, 2004). Under this paradigm, instead of only seeking exponentially indefinitely growing solutions, any magnetic field that grows is sought, including those that eventually decay. Crucially, they found that the dynamo action, although not necessarily indefinite, was robust under small changes in the flow, something that previous studies did not find. It is therefore much more straightforward to assign a physical mechanism to such a dynamo process, in this case attributable to magnetic field line stretching by upwelling convective flows, generating Earth-like $\mathbf{S}_1^0$ fields.

The influence of an insulating boundary

Despite many authors finding spherical flows that operated as kinematic dynamos, it was well known that many other choices failed. This was in stark constrast to Roberts (1970) who had showed that almost all infinite spatially periodic flows show dynamo action. One cause has been attributed to the influence of the mantle, introducing electrically insulating boundary conditions on the magnetic field, these making no appearance in a periodic domain. The study of Bullard and Gubbins (1977) showed that the insulating boundary trapped currents near the edge of the flow region and catastrophically increased Ohmic dissipation, inhibiting dynamo action. They suggested that this might not be a problem if the flow near the CMB was quiescent, a topic later taken up by Hutchenson and Gubbins (1994). They found that the addition of a static outer conducting region on top of the fluid core promoted dynamo action: either the critical magnetic Reynolds number was lowered or growing magnetic fields were found when all had previously decayed. The trapped currents could now flow outside the core region facilitating dynamo action, geophysically motivated by the possibility of a stably stratified layer at the top of the outer core.

Motivating the choice of flow

Although one of the great strengths of kinematic theory is the ability to prescribe the flow, the lack of physical self-consistency is one of its great drawbacks. To partially mitigate this issue, Love and Gubbins (1996) studied the problem whereby the dynamo efficiency, defined in terms of the critical magnetic Reynolds number and the Ohmic dissipation produced by the growing field, was minimized over a choice of flow (in this case the the values of ε in the Kumar and Roberts flow). Such a method of choosing a flow has a strong basis on physical grounds because it might be energetically favored. They found dynamo action with much reduced Ohmic dissipation.

A partial solution to the full Navier-Stokes equations (see *Geodynamo*) is another way of choosing the flow. Sarson (2003) considered the geophysically relevant geostrophic approximation, balancing the Coriolis force, buoyancy, and pressure. He derived a fully consistent flow although the driving temperature profile was not geophysically plausible, and found growing field solutions of dipole symmetry in line with the studies of the Kumar and Roberts flow.

Summary

Kinematic dynamo theory has proven very successful in helping scientists understand the complex process of magnetic field generation inside the Earth. Conclusive evidence in the 1970s that the geodynamo could function is of particular historical importance and paved the way for many of the subsequent advances in geodynamo theory. In addition, some links have been made between the generated fields and various flow properties and the presence of a thin stably stratified conducting layer on top of the outer core. Kinematic reversal mechanisms have also been proposed (see *Reversals, theory*).

The main outstanding problem of kinematic dynamos is their sensitivity to the exact choice of flow chosen, making it difficult to attribute physical processes in such cases. Nevertheless, complex dynamical models are still often understood by simple kinematic arguments and this will undoubtedly remain the case for the foreseeable future.

Philip W. Livermore

Bibliography

Bullard, E., and Gellman, H., 1954. Homogeneous dynamos and terrestrial magnetism, *Philosophical Transactions of the Royal Society of London A*, **247**: 213–278.

Bullard, E., and Gubbins D., 1977. Generation of magnetic fields by fluid motions of global scale. *Geophysical and Astrophysical Fluid Dynamics*, **8**: 43–56.

Dudley, M., and James R., 1989. Time-dependent kinematic dynamos with stationary flows. *Proceedings of the Royal Society of London A*, **425**: 407–429.

Gibson, R., and Roberts, P., 1967. Some comments on the theory of homogeneous dynamos. In Hindmarsh *et al.* (ed.), *Magnetism and the Cosmos*. Oliver and Boyd, pp. 108–120.

Gubbins, D., 1973. Numerical solutions of the kinematic dynamo problem. *Philosophical Transactions of the Royal Society of London A*, **274**: 493–521.

Gubbins, D. *et al.*, 2000. Kinematic dynamo action in a sphere: I. Effects of differential rotation and meridional circulation on solutions with axial dipole symmetry. *Proceedings of the Royal Society of London A*, **456**: 1333–1353.

Hutchenson, K., and Gubbins, D., 1994. Kinematic magnetic-field morphology at the core-mantle boundary. *Geophysical Journal International*, **116**: 304–320.

Kumar, S., and Roberts, P., 1975. A three-dimensional dynamo problem. *Proceedings of the Royal Society of London A*, **344**: 235–258.

Lilley, F., 1970. On kinematic dynamos. *Proceedings of the Royal Society of London A*, **316**: 153–167.

Livermore, P., and Jackson, A., 2004. On magnetic energy instability in spherical stationary flows. *Proceedings of the Royal Society of London A*. In press.

Love, J., and Gubbins, D., 1996. Optimized Kinematic Dynamos. *Geophysical Journal International*, **124**: 787–800.

Moffatt, H., 1978. *Magnetic field generation in electrically conducting fluids*. Cambridge University Press.

Pekeris, C. *et al.*, 1973. Kinematic dynamos and the Earth's magnetic field. *Philosophical Transactions of the Royal Society of London A*, **275**: 425–461.

Roberts, G., 1970. Spatially periodic dynamos. *Philosophical Transactions of the Royal Society of London A*, **266**: 535–558.

Roberts, P., 1972. Kinematic dynamo models. *Philosophical Transactions of the Royal Society of London A*, **272**: 663–698.

Sarson, G., 2003. Kinematic dynamos driven by thermal wind flows. *Proceedings of the Royal Society of London A*, **459**: 1241–1259.

Sarson, G., and Gubbins, D., 1996. Three-dimensional kinematic dynamos dominated by strong differential rotation. *Journal of Fluid Mechanics*, **306**: 223–265.

Takeuchi, H., and Shimazu, Y., 1953. On a self-exciting process in magneto-hydrodynamics. *Journal of Geophysical Research*, **58**: 497–518.

Cross-references

Antidynamo and Bounding Theorems
Bullard, Edward Crisp (1907–1980)
Core Convection
Cowling's Theorem
Dynamo, Backus
Dynamo, Bullard-Gellman
Dynamo, Herzenberg
Dynamo, Ponomarenko
Dynamos, Mean Field
Elsasser, Walter M. (1904–1991)
Geodynamo
Geodynamo, Numerical Simulations
Harmonics, Spherical
Magnetohydrodynamics
Mantle, Electrical Conductivity, Mineralogy
Reversals, Theory

DYNAMOS, MEAN-FIELD

Introduction

The mean-field concept was introduced into the dynamo theory of the magnetic fields of the Earth, the Sun, and other cosmic objects in the sixties of the last century. As was known or at least considered as very probable at that time, magnetic fields as well as the motions inside the electrically conducting interiors of these objects show rather complex geometrical structures and time behaviors. In addition antidynamo theorems suggested that dynamo action requires a certain complexity of magnetic field and motion (see *Antidynamo theorems* or *Cowling's theorem*). No solution of the dynamo equations has been found until this time, which could be interpreted as an approximate picture of the situation in the Earth or any cosmic body.

The central idea of the mean-field concept is to define mean magnetic fields, mean velocity fields etc., which reflect essential features of the original fields but show simpler, that is more smooth, geometrical structures and time behaviors, and to derive equations for them, which are easily treatable, in particular numerically solvable with the available tools. Of course, these equations must contain terms accounting for the deviations of the original fields from the mean fields, in which these deviations, however, enter only in the form of averaged quantities.

On the basis of this concept indeed dynamo models for the Earth, the Sun, and many other cosmic objects including galaxies have been developed. A remarkable step in this direction was done by Braginsky in 1964, who considered the Earth and proposed the theory of the "nearly symmetric dynamo" (Braginsky, 1964a,b) (see *Dynamo, Braginsky* and *Dynamo, model Z*). Independent from this, a general mean-field electrodynamics was established by Steenbeck *et al.* in 1966, and it was used to develop a general mean-field dynamo theory, applicable to all objects mentioned (Steenbeck *et al.*, 1966; Krause *et al.*, 1980). The rigorous mathematical formulation of the mean-field theory also covers Parker's ideas on dynamo action described already in 1955 (Parker, 1955, 1957).

A crucial point in mean-field electrodynamics and mean-field dynamo theory is a mean electromotive force resulting from the deviations of motion and magnetic field from mean motion and mean magnetic field. The occurrence of a specific type of this electromotive force, with a component in the direction of the mean magnetic field, is called "α-effect". The mean-field dynamo theory revealed basic dynamo mechanisms working with the α-effect or with related effects. In this way it created a system of ideas which is now of importance in all dynamo theory, also beyond the mean-field theory.

As far as the geodynamo is concerned, a series of direct numerical simulations on very powerful computers has been carried out since the end of the last century (see *Geodynamo, numerical simulations*). The mean-field dynamo models, which, although in some respect problematic, were important for designing the more advanced models investigated in this way, are now no longer of primary interest. Nevertheless, the mean-field concept is still a useful tool for the interpretation of the numerical results and for understanding the basic dynamo mechanisms. In view of the solar and other cosmical dynamos, however, the mean-field approach is till now the only adequate way of describing and investigating them.

In this article a brief outline is given of the main ideas of mean-field electrodynamics and the dynamo theory based on it as well as a few specific applications to the geodynamo. See also more detailed presentations (e.g., Moffatt, 1978; Krause *et al.*, 1980; Zeldovich *et al.*, 1983; Rädler, 1995, 2000).

Mean-field electrodynamics

Basic equations

The standard dynamo theory assumes that the electromagnetic field in an electrically conducting moving fluid is governed by the Maxwell equations and Ohm's law in the form

$$\nabla \times \boldsymbol{E} = -\partial_t \boldsymbol{B}, \quad \nabla \times \boldsymbol{B} = \mu \boldsymbol{j}, \quad \nabla \cdot \boldsymbol{B} = 0 \qquad \text{(Eq. 1)}$$

$$\boldsymbol{j} = \sigma(\boldsymbol{E} + \boldsymbol{u} \times \boldsymbol{B}). \qquad \text{(Eq. 2)}$$

As usual, $\boldsymbol{E}$ is the electric field and $\boldsymbol{B}$ the magnetic flux density, simply called magnetic field in the following, $\boldsymbol{j}$ is the electric current density, $\boldsymbol{u}$ the velocity of the fluid motion, μ the magnetic permeability of the fluid, assumed to be equal to that of free space, and σ the electric conductivity of the fluid. These equations together with proper initial and boundary conditions determine $\boldsymbol{B}$, $\boldsymbol{E}$, and $\boldsymbol{j}$ if $\boldsymbol{u}$ is given. They can be reduced to the induction equation for $\boldsymbol{B}$ alone. For the sake of simplicity it is assumed here that σ does not vary in space. Then it follows that

$$\eta \nabla^2 \boldsymbol{B} + \nabla \times (\boldsymbol{u} \times \boldsymbol{B}) - \partial_t \boldsymbol{B} = 0, \quad \nabla \cdot \boldsymbol{B} = 0, \qquad \text{(Eq. 3)}$$

where η is the magnetic diffusivity defined by $\eta = 1/\mu\sigma$. If $\boldsymbol{B}$ is known from these equations for a given $\boldsymbol{u}$, then both $\boldsymbol{E}$ and $\boldsymbol{j}$ can be determined without further integrations.

Definition of mean fields

Let us focus our attention on situations in which the fluid motion possesses components showing spatial scales that are small compared to the scales of fluid body considered. Typical examples of that are turbulent or convective motions. Then, of course, the electromagnetic field has to show small scales in that sense, too.

In view of the definition of mean fields we consider first a scalar field F. We define the corresponding mean field $\overline{F}$ as an average of F obtained with an averaging procedure which, as a rule, smoothes its variations in space and time. Adopting the terminology of turbulence theory the difference $F' = F - \overline{F}$ is called "fluctuation." To extend the definition to a vector field $\boldsymbol{F}$ we refer to a coordinate system with unit vectors $\mathrm{e}^{(i)}$ and write, using the summation convention, $\boldsymbol{F} = \mathrm{e}^{(i)} F_i$. Then we define the mean-field $\overline{\boldsymbol{F}}$ by $\overline{\boldsymbol{F}} = \mathrm{e}^{(i)} \overline{F}_i$.

Various choices of the averaging procedure may be admitted. The only requirement is that Reynolds' averaging rules apply exactly or at least as an approximation. They read

$$\overline{F + G} = \overline{F} + \overline{G} \qquad \text{(Eq. 4)}$$

$$\overline{\partial F/\partial x} = \partial \overline{F}/\partial x, \quad \overline{\partial F/\partial t} = \partial \overline{F}/\partial t \qquad \text{(Eq. 5)}$$

$$\overline{\overline{F}} = \overline{F} \quad \text{or, what is in this context the same,} \quad \overline{F'} = 0 \qquad \text{(Eq. 6)}$$

$$\overline{\overline{F}G} = \overline{F}\,\overline{G}, \qquad \text{(Eq. 7)}$$

where F as well as G are arbitrary fields, and x stands for any space coordinate. Clearly Eq. (6) is a special case of Eq. (7). An important consequence of Eqs. (4) and (7) is $\overline{FG} = \overline{F}\,\overline{G} + \overline{F'G'}$.

As can easily be seen below, mean vector fields defined on the basis of a Cartesian coordinate system are well different from those defined, e.g., with respect to a cylindrical or a spherical coordinate system.

Giving now a few examples of averages we distinguish between "local" averages, for which $\overline{F}$ in a given point depends only on the values of F in this point or a small neighborhood of it, and "nonlocal" averages.

(i) *Local averages*

(ia) *Statistical or ensemble averages*. In this case we suppose that there is an infinity of copies of the object considered. The individual copies are labeled by a parameter, say p. In that sense any quantity F

to be averaged depends, in addition to the space and time variables, $\boldsymbol{x}$ and t, on this parameter p. The average $\overline{F}(\boldsymbol{x}, t)$ is then defined by averaging F over all p. Averages of this kind clearly ensure the validity of all four rules Eqs. (4)–(7). There is, however, a serious difficulty to relate these averages to observable quantities.

(ib) *Space averages.* A general form of a space average is given by

$$\overline{F}(\boldsymbol{x}, t) = \int_{\infty} F(\boldsymbol{x} + \boldsymbol{\xi}, t) g(\boldsymbol{\xi})\, \mathrm{d}^3\xi, \qquad \int_{\infty} g(\boldsymbol{\xi})\, \mathrm{d}^3\xi = 1. \qquad \text{(Eq. 8)}$$

Here $g(\boldsymbol{\xi})$ is a normalized weight function which is different from zero only in some region around $\boldsymbol{\xi} = \boldsymbol{0}$. The integrations, formally over all $\boldsymbol{\xi}$-space, are in fact over this region only. With such averages the two rules Eqs. (4) and (5) apply exactly but in general Eqs. (6) and (7) are violated. The latter two can be justified as an approximation if there is a gap in the spectrum of the length scales of F, and all large scales are much larger, and all small ones much smaller than the characteristic length of the averaging region. A situation of that kind is sometimes named a "two-scale situation".

(ic) *Time averages.* Similar to space averages, we may define time averages by

$$\overline{F}(\boldsymbol{x}, t) = \int_{\infty} F(\boldsymbol{x}, t + \tau)\, g(\tau)\, \mathrm{d}\tau, \qquad \int_{\infty} g(\tau)\, \mathrm{d}\tau = 1 \qquad \text{(Eq. 9)}$$

with some normalized weight function $g(\tau)$ different from zero in some neighborhood of $\tau = 0$. The comments made under (ib) apply analogously.

(ii) *Nonlocal averages*

(iia) *Azimuthal average.* There are, however, particular space averages to which all averaging rules apply. Consider, for example, a case in which the variation of F in space is properly described by spherical coordinates r, θ, φ, and put

$$\overline{F}(r, \theta, t) = \frac{1}{2\pi} \int_0^{2\pi} F(r, \theta, \varphi, t)\, \mathrm{d}\varphi. \qquad \text{(Eq. 10)}$$

This kind of average is used in particular in Braginsky's theory of the nearly symmetric dynamo. Of course, all mean-fields are by definition axisymmetric. All four rules Eqs. (4)–(7) apply exactly.

(iib) *Averages based on filtering of spectra.* We may, for example, represent F in its dependence on space coordinates by a Fourier integral and define $\overline{F}$ by another integral of this type which covers only the large-scale part of the Fourier spectrum beyond some averaging scale. For averages defined in this way the three rules Eqs. (4)–(6) apply exactly, and with a sufficiently large gap in the spectrum and a proper choice of the averaging scale the remaining rule Eq. (7) can again be justified as an approximation. The azimuthal average defined by Eq. (10) can also be interpreted as one based on filtering a Fourier spectrum with respect to φ. Another interesting possibility consists, e.g., in filtering the multipole spectrum of vector fields so that the mean fields are just dipole fields, or dipole and quadrupole fields, etc.

Basic mean-field equations

Returning to the Eqs. (1), (2), and (3), we understand now $\boldsymbol{B}$, $\boldsymbol{E}$, $\boldsymbol{j}$, and $\boldsymbol{u}$ as superpositions of mean and fluctuating parts. Applying the averaging procedure to Eqs. (1) and (2) we obtain

$$\nabla \times \overline{\boldsymbol{E}} = -\partial_t \overline{\boldsymbol{B}}, \quad \nabla \times \overline{\boldsymbol{B}} = \mu \overline{\boldsymbol{j}}, \quad \nabla \cdot \overline{\boldsymbol{B}} = 0 \qquad \text{(Eq. 11)}$$

$$\overline{\boldsymbol{j}} = \sigma(\overline{\boldsymbol{E}} + \overline{\boldsymbol{u}} \times \overline{\boldsymbol{B}} + \boldsymbol{\mathcal{E}}). \qquad \text{(Eq. 12)}$$

From these equations, or taking the average of Eq. (3), we further obtain

$$\eta \nabla^2 \overline{\boldsymbol{B}} + \nabla \times (\overline{\boldsymbol{u}} \times \overline{\boldsymbol{B}} + \boldsymbol{\mathcal{E}}) - \partial_t \overline{\boldsymbol{B}} = 0, \quad \nabla \cdot \overline{\boldsymbol{B}} = 0. \qquad \text{(Eq. 13)}$$

$\boldsymbol{\mathcal{E}}$ is the mean electromotive force due to the fluctuations of motion and magnetic field,

$$\boldsymbol{\mathcal{E}} = \overline{\boldsymbol{u}' \times \boldsymbol{B}'}. \qquad \text{(Eq. 14)}$$

These equations together with proper initial and boundary conditions determine $\overline{\boldsymbol{B}}$ if $\overline{\boldsymbol{u}}$ and $\boldsymbol{\mathcal{E}}$ are given, and so also $\overline{\boldsymbol{E}}$ and $\overline{\boldsymbol{j}}$.

The crucial point in the elaboration of mean-field electrodynamics is the determination of the mean electromotive force $\boldsymbol{\mathcal{E}}$ for given $\overline{\boldsymbol{u}}$, $\boldsymbol{u}'$, and $\overline{\boldsymbol{B}}$. Using Eqs. (3) and (13), an equation for the magnetic fluctuations $\boldsymbol{B}'$ can be derived. It allows us to conclude that $\boldsymbol{B}'$ can be considered as a functional of $\overline{\boldsymbol{u}}$, $\boldsymbol{u}'$, and $\overline{\boldsymbol{B}}$, which is linear in $\overline{\boldsymbol{B}}$. Thus, $\boldsymbol{\mathcal{E}}$ must show the same dependence on these fields. For the sake of simplicity we restrict our attention to the case in which the magnetic fluctuations $\boldsymbol{B}'$ are due to the interaction of the velocity fluctuations $\boldsymbol{u}'$ with the mean magnetic field $\overline{\boldsymbol{B}}$ only, that is, decay to zero if $\overline{\boldsymbol{B}}$ is zero. In other words, the possibility of a magnetohydrodynamic turbulence with zero $\overline{\boldsymbol{B}}$ is ignored. Apart from some initial time, which is not considered here, $\boldsymbol{\mathcal{E}}$ is then not only linear but also homogeneous in $\overline{\boldsymbol{B}}$, that is, has to vanish if $\overline{\boldsymbol{B}}$ does so. For the sake of simplicity we restrict our attention first to mean fields defined by local averages. Then it can be easily concluded that $\boldsymbol{\mathcal{E}}$ allows the representation

$$\mathcal{E}_i(\boldsymbol{x}, t) = \int_0^{\infty} \int_{\infty} K_{ij}(\boldsymbol{x}, t; \boldsymbol{\xi}, \tau) \overline{B}_j(\boldsymbol{x} + \boldsymbol{\xi}, t - \tau)\, \mathrm{d}^3\xi\, \mathrm{d}\tau, \qquad \text{(Eq. 15)}$$

with some kernel K_{ij} determined by $\overline{\boldsymbol{u}}$ and $\boldsymbol{u}'$. Here, and in what follows, indices like i and j refer to a Cartesian coordinate system and the summation convention is adopted.

Let us accept the further assumption that $\boldsymbol{\mathcal{E}}$ in a given point in space and time depends only on the values of $\boldsymbol{u}$, $\boldsymbol{u}'$, and $\overline{\boldsymbol{B}}$ in some neighborhood of this point. This can easily be justified in the case of turbulent fluid motions but applies for most of the other situations of interest, too. Let us assume further that the variations of $\overline{\boldsymbol{B}}$ in space and time are weak enough so that its behavior in the relevant neighborhood of the considered point is to a good approximation determined by $\overline{\boldsymbol{B}}$ and its first spatial derivatives at this point. This implies that $\boldsymbol{\mathcal{E}}$ can be represented in the form

$$\mathcal{E}_i = a_{ij} \overline{B}_j + b_{ijk} \partial \overline{B}_j / \partial x_k, \qquad \text{(Eq. 16)}$$

where the tensors a_{ij} and b_{ijk} are mean quantities which are determined by $\overline{\boldsymbol{u}}$ and $\boldsymbol{u}'$ but do not depend on $\overline{\boldsymbol{B}}$. It remains of course to be checked in all applications that higher spatial derivatives or the time derivatives of $\overline{\boldsymbol{B}}$ are indeed negligible. Incidentally, as a consequence of $\nabla \cdot \overline{\boldsymbol{B}} = 0$, three elements of b_{ijk} can be arbitrarily fixed.

In the case of a nonlocal, e.g., the azimuthal average in Eqs. (15) and (16) occur primarily in slightly different forms. We may, however, justify Eq. (16) as well.

Let us consider the simple (somewhat academic) example in which the mean motion is zero, $\overline{\boldsymbol{u}} = 0$, and the fluctuations of the velocity field, $\boldsymbol{u}'$, correspond to a homogeneous isotropic turbulence. In this case, no preferred points in space and no preferred directions can be found in the $\boldsymbol{u}'$ field. In other words, all mean quantities depending on $\boldsymbol{u}'$ are invariant under arbitrary translations of the $\boldsymbol{u}'$ field and under arbitrary rotations of this field about arbitrary axes. Simple symmetry considerations allow us then to conclude that a_{ij} and b_{ijk} are isotropic tensors and are independent on position. That is, $a_{ij} = \alpha \delta_{ij}$ and $b_{ijk} = \beta \boldsymbol{\epsilon}_{ijk}$, where the coefficients α and β are independent of position and determined by u' only, and δ_{ij} and ϵ_{ijk} are the Kronecker and the Levi-Civita tensors. This result allows us to write Eq. (16) in the form

$$\mathcal{E} = \alpha\overline{\boldsymbol{B}} - \beta\nabla\times\overline{\boldsymbol{B}}. \quad \text{(Eq. 17)}$$

Together with Eqs. (11) and (12) this yields

$$\overline{\boldsymbol{j}} = \sigma_{\mathrm{m}}(\overline{\boldsymbol{E}} + \alpha\overline{\boldsymbol{B}}), \quad \sigma_{\mathrm{m}} = \frac{\sigma}{1+\beta/\eta}. \quad \text{(Eq. 18)}$$

That is, in Ohm's law for the mean fields occurs a mean-field conductivity σ_{m}, which (if $\beta \neq 0$) differs from the molecular conductivity σ. In addition, there is (if $\alpha \neq 0$) an electromotive force parallel (or antiparallel) to $\overline{\boldsymbol{B}}$. This is remarkable as in the original form (Eq. (2)) of Ohm's law the magnetic field enters only via the term $\boldsymbol{u}\times\boldsymbol{B}$, which has no component in the direction of $\boldsymbol{B}$. The occurrence of a mean electromotive force with a component in the direction of the mean magnetic field $\overline{\boldsymbol{B}}$ is called "α-effect."

The α-effect makes a dynamo possible. As can be easily followed up, Eq. (13) for $\overline{\boldsymbol{B}}$ with $\overline{\boldsymbol{u}} = 0$ and $\mathcal{E}$ specified according to Eq. (17) possesses growing solutions for proper choices of η, α, and β.

Under realistic circumstances an isotropic turbulence is also reflectionally symmetric ("mirror-symmetric") in the sense that there is no preference of left-handed over right-handed helical motions or *vice versa*. More precisely, all mean quantities are invariant under reflections of the $\boldsymbol{u}'$ field, e.g., at the origin of the coordinate system, that is, under exchanging $\boldsymbol{u}'(\boldsymbol{x},t)$ with $-\boldsymbol{u}'(-\boldsymbol{x},t)$. Symmetry considerations show then $\alpha = 0$. Although in a sense unrealistic, the simple example under discussion reveals the fundamental connection between the violation of reflectional symmetry of the turbulence and the α-effect. The turbulence on a rotating body is neither homogeneous nor isotropic. Apart from the fact that reflectional symmetry in the above sense is anyway not compatible with inhomogeneity or anisotropy, it is in particular disturbed by the influence of the Coriolis force. Inhomogeneity and the violation of reflectional symmetry due to the Coriolis force lead again to an α-effect similar to that discussed here and, as a consequence, to dynamo action.

Leaving this simple example and returning to arbitrary $\overline{\boldsymbol{u}}$ and the representation of $\mathcal{E}$ in the form (Eq. (16)) we give an alternative representation of $\mathcal{E}$. The tensor a_{ij} may be split into a symmetric and an antisymmetric part, and the latter can be expressed by a vector. Likewise the gradient tensor $\partial\overline{B}_j/\partial x_k$ can be represented by its symmetric part and a vector, which proves to be proportional to $\nabla\times\overline{\boldsymbol{B}}$. Considering these possibilities we may write

$$\mathcal{E} = -\boldsymbol{\alpha}\circ\overline{\boldsymbol{B}} - \boldsymbol{\gamma}\times\overline{\boldsymbol{B}} - \boldsymbol{\beta}\circ(\nabla\times\overline{\boldsymbol{B}}) - \boldsymbol{\delta}\times(\nabla\times\overline{\boldsymbol{B}}) - \boldsymbol{\kappa}\circ(\nabla\overline{\boldsymbol{B}})^{(s)}. \quad \text{(Eq. 19)}$$

Here $\boldsymbol{\alpha}$ and $\boldsymbol{\beta}$ are symmetric second-rank tensors, $\boldsymbol{\gamma}$ and $\boldsymbol{\delta}$ vectors, and $\boldsymbol{\kappa}$ is a third-rank tensor, all being determined by $\overline{\boldsymbol{u}}$ and $\boldsymbol{u}'$. Further $(\nabla\overline{\boldsymbol{B}})^{(s)}$ is the symmetric part of the gradient tensor of $\overline{\boldsymbol{B}}$, that is, $(\nabla\overline{\boldsymbol{B}})^{(s)}_{ij} = (1/2)(\partial\overline{B}_i/\partial x_j + \partial\overline{B}_j/\partial x_i)$. Of course, $\boldsymbol{\kappa}$ may be assumed to be symmetric in the indices connecting it with $(\nabla\overline{\boldsymbol{B}})^{(s)}$, and because $\nabla\cdot\overline{\boldsymbol{B}} = 0$ three of its elements can be fixed arbitrarily.

Inserting $\mathcal{E}$ according to Eq. (19) into Ohm's law Eq. (12), we may write the latter in the form

$$\overline{\boldsymbol{j}} = \boldsymbol{\sigma}_{\mathrm{m}}\circ(\overline{\boldsymbol{E}} - \boldsymbol{\alpha}\circ\overline{\boldsymbol{B}} + (\overline{\boldsymbol{u}} - \boldsymbol{\gamma})\times\overline{\boldsymbol{B}} - \boldsymbol{\delta}\times(\nabla\times\overline{\boldsymbol{B}}) - \boldsymbol{\kappa}\circ(\nabla\overline{\boldsymbol{B}})^{(s)}). \quad \text{(Eq. 20)}$$

Here, $\boldsymbol{\sigma}_{\mathrm{m}}$ is a conductivity tensor incorporating the $\boldsymbol{\beta}$ term of Eq. (19) and being symmetric. Again, the mean electric current is no longer determined by the molecular conductivity, and the relation between mean current and mean electric field plus mean electromotive force is in general anisotropic. The $\boldsymbol{\alpha}$ term defines some generalization of the α-effect discussed above, that is, an anisotropic α-effect. The $\boldsymbol{\gamma}$ term corresponds to a transport of mean magnetic flux like that by a mean motion, which occurs however even in the absence of any mean motion. Clearly $\overline{\boldsymbol{u}} - \boldsymbol{\gamma}$ is the effective velocity for the transport of mean flux. The $\boldsymbol{\delta}$ term describes an induction effect, which was first found in the special case in which $\boldsymbol{\delta}$ was proportional to an angular velocity $\boldsymbol{\Omega}$ and has been called "$\boldsymbol{\Omega}\times\boldsymbol{j}$-effect". The $\boldsymbol{\kappa}$ term is less easy to interpret. Analogous to the notation α-effect, we speak of "β-effect", "δ-effect" etc. when referring to the induction effects described by the terms with $\boldsymbol{\beta}$, $\boldsymbol{\delta}$ etc. in Eqs. (19) or (20). By the way, we arrive at an alternative form of Ohm's law if we interpret $\boldsymbol{\sigma}_{\mathrm{m}}$ as tensor incorporating both the $\boldsymbol{\beta}$ and $\boldsymbol{\delta}$ terms and cancel the last ones otherwise. Then, however, $\boldsymbol{\sigma}_{\mathrm{m}}$ is no longer symmetric.

It is important to know the dependence of the quantities $\boldsymbol{\alpha}, \boldsymbol{\beta}, \boldsymbol{\gamma}, \boldsymbol{\delta}$, and $\boldsymbol{\kappa}$ on the fluid motion, that is, on $\overline{\boldsymbol{u}}$ and $\boldsymbol{u}'$. Unfortunately there is no simple way to derive general results of that kind. A series of extended calculations have been carried out using specific approximations. Often the "second-order correlation approximation", also called the "first-order smoothing approximation", has been adopted, which, roughly speaking, can be justified only for $\boldsymbol{u}'$ not too large.

Assume for the sake of simplicity that $\overline{\boldsymbol{u}} = 0$ and $\boldsymbol{u}'$ represents a turbulence with a characteristic velocity u'_{c}, a correlation length λ_{c}, and a correlation time τ_{c}. Define then the magnetic Reynolds number $R_m = u'_{\mathrm{c}}\lambda_{\mathrm{c}}/\eta$, the Strouhal number $S = u'_{\mathrm{c}}\tau_{\mathrm{c}}/\lambda_{\mathrm{c}}$, and the quantity $q = \lambda_{\mathrm{c}}^2/\eta\tau_{\mathrm{c}}$, Clearly q is the ratio of the characteristic time $\lambda_{\mathrm{c}}^2/\eta$ for electromagnetic processes in a region with the length scale λ_{c} to the time τ_{c}. In the high-conductivity limit, defined by $q \gg 1$, a sufficient condition for the applicability of the second-order correlation approximation reads $S \ll 1$. In the low-conductivity limit, $q \gg 1$, the corresponding sufficient condition is $R_m \ll 1$.

As an example we give here results for α and β for the simple case in which $\boldsymbol{u}'$ corresponds to a homogeneous isotropic turbulence. In the high-conductivity limit, $q \gg 1$, this approximation yields

$$\alpha = -\frac{1}{3}\int_0^\infty \overline{\boldsymbol{u}'(\boldsymbol{x},t)\cdot(\nabla\times\boldsymbol{u}'(\boldsymbol{x},t-\tau))}\,\mathrm{d}\tau,$$

$$\beta = \frac{1}{3}\int_0^\infty \overline{\boldsymbol{u}'(\boldsymbol{x},t)\cdot\boldsymbol{u}'(\boldsymbol{x},t-\tau)}\,\mathrm{d}\tau, \quad \text{(Eq. 21)}$$

or

$$\alpha = -\frac{1}{3}\overline{\boldsymbol{u}'\cdot(\nabla\times\boldsymbol{u}')}\tau_{\mathrm{c}}^{(\alpha)}, \quad \beta = \frac{1}{3}\overline{\boldsymbol{u}'^2}\tau_{\mathrm{c}}^{(\beta)}. \quad \text{(Eq. 22)}$$

Here $\tau_{\mathrm{c}}^{(\alpha)}$ and $\tau_{\mathrm{c}}^{(\beta)}$ are primarily defined by equating the corresponding right-hand sides of Eqs. (21) and (22). It seems reasonable to assume that they do not differ markedly from τ_{c}. For the low-conductivity limit, $q \ll 1$, it follows that

$$\alpha = -\frac{1}{12\pi\eta}\int_\infty \overline{\boldsymbol{u}'(x,t)\cdot(\nabla\times\boldsymbol{u}'(x+\boldsymbol{\xi},t))}\,\frac{\mathrm{d}^3\xi}{\xi},$$

$$\beta = \frac{1}{12\pi\eta}\int_\infty \overline{u'_\xi(\boldsymbol{x},t)u'_\xi(\boldsymbol{x}+\boldsymbol{\xi},t))}\,\frac{\mathrm{d}^3\xi}{\xi}, \quad \text{(Eq. 23)}$$

where $u'_\xi = (\boldsymbol{u}'\cdot\boldsymbol{\xi})/\xi$. Interestingly enough, if $\boldsymbol{u}'$ is represented in the form $\boldsymbol{u}' = \nabla\times\boldsymbol{a}' + \nabla\phi'$ by a vector potential $\boldsymbol{a}'$ and a scalar potential ϕ' this can be rewritten into

$$\alpha = -\frac{1}{3\eta}\overline{\boldsymbol{a}'\cdot(\nabla\times\boldsymbol{a}')}, \quad \beta = \frac{1}{3\eta}\overline{(\boldsymbol{a}'^2 - \phi'^2)}. \quad \text{(Eq. 24)}$$

With a reasonable assumption on $\overline{u'_\xi(\boldsymbol{x},t)u'_\xi(\boldsymbol{x}+\boldsymbol{\xi},t)}$ it follows that $\beta/\eta = (1/9)R_m^2$. In the high-conductivity limit it is the mean helicity $\overline{\boldsymbol{u}'\cdot(\nabla\times\boldsymbol{u}')}$, in the low-conductivity limit the related quantity $\overline{\boldsymbol{a}'\cdot(\nabla\times\boldsymbol{a}')}$, which are crucial for the α-effect. Both indicate the existence of helical features in the flow pattern and vanish for mirror-symmetric turbulence.

Kinematic mean-field dynamo theory

The kinematic dynamo problem

Let us consider the dynamo problem for a finite simply connected fluid body surrounded by electrically isolating matter (see also *Dynamos, kinematic*). Assume that the electromagnetic fields $\boldsymbol{B}$ and $\boldsymbol{E}$ satisfy the Eqs. (1) and (2) inside this body and the same equations with $\sigma = 0$ and therefore $\boldsymbol{j} = 0$ in all outer space, further that $\boldsymbol{B}$ and the tangential components of $\boldsymbol{E}$ are continuous across the boundary, and finally that $\boldsymbol{B}$ and $\boldsymbol{E}$ vanish at infinity. All this can be reduced to the statement that the magnetic field $\boldsymbol{B}$ satisfies the Eq. (3) inside the body, continues as a potential field in outer space, and vanishes at infinity. The last-mentioned equations and requirements define an initial value problem for $\boldsymbol{B}$. We speak of a dynamo if this problem for $\boldsymbol{B}$ with a given $\boldsymbol{u}$ possesses, for proper initial conditions, solutions which do not decay in the course of time, that is, $\boldsymbol{B} \not\to 0$ as $t \to \infty$. Sometimes the notation "homogeneous dynamo" is used for dynamos as envisaged here in order to stress that they work, in contrast to technical dynamos, in bodies consisting throughout of electrically conducting matter, that is, not containing any electrically insulating parts.

It is well-known that dynamos cannot work with specific geometries of magnetic field or motion. In particular, Cowling's theorem excludes dynamos with magnetic fields $\boldsymbol{B}$ that are symmetric about an axis (see *Cowling's theorem*).

Simple examples of dynamos are those with spatially periodic flows of an infinitely extended fluid as proposed by Roberts already in 1970 (Roberts, 1970, 1972) (see also *Dynamos, periodic*). Assume, for example, that the fluid velocity $\boldsymbol{u}$ is in a Cartesian coordinate system (x, y, z) given by $\boldsymbol{u} = u_\perp a\, \boldsymbol{e} \times \nabla\chi(x,y) + u_\| \boldsymbol{e}\, \chi(x,y)$ with $\chi = \sin(\pi x/a)\sin(\pi y/a)$, where $\boldsymbol{e}$ is the unit vector in z direction and a is some length. This flow possesses helical features. It allows under some condition nondecaying magnetic fields $\boldsymbol{B}$ varying like $\boldsymbol{u}$ periodically in x and y and in addition with a period length, say l, in z. With magnetic Reynolds numbers defined by $R_{m\perp} = u_\perp a/\eta$ and $R_{m\|} = u_\| a/\eta$, this condition reads $R_{m\perp} R_{m\|}\, \phi(R_{m\perp}) \geq 8\pi a/l$, where ϕ is equal to unity in the limit $R_{m\perp} \to 0$ and decays monotonically to zero with growing $R_{m\perp}$.

A useful tool in the investigation of dynamo models is the representation of vector fields like $\boldsymbol{B}$ or $\boldsymbol{u}$ as sums of poloidal and toroidal parts. If a field, say $\boldsymbol{F}$, is symmetric about a given axis the poloidal part $\boldsymbol{F}^{\mathrm{P}}$ and the toroidal one $\boldsymbol{F}^{\mathrm{T}}$ are defined such that $\boldsymbol{F}^{\mathrm{P}}$ lies completely in the meridional planes containing this axis and $\boldsymbol{F}^{\mathrm{T}}$ is everywhere perpendicular to them. This definition can be extended in various ways to the general case, in which $\boldsymbol{F}$ is no longer necessarily axisymmetric. One possibility, which fits best to the situation with spherical objects, is to require that $\boldsymbol{F}^{\mathrm{P}}$ and $\boldsymbol{F}^{\mathrm{T}}$ allow the representations $\boldsymbol{F}^{\mathrm{P}} = \boldsymbol{r}U + \nabla V$ and $\boldsymbol{F}^{\mathrm{T}} = \boldsymbol{r} \times \nabla W$ with $\boldsymbol{r}$ being the radius vector and U, V, and W scalar functions of position. This is indeed a unique definition and generalizes the specific one given for the axisymmetric case (see, e.g., Krause *et al.*, 1980; Rädler, 2000). Then $\boldsymbol{F}^{\mathrm{P}}$ is a specific three-dimensional field but $\boldsymbol{F}^{\mathrm{T}}$ lies completely in spherical surfaces $r = \text{const}$.

The kinematic dynamo problem at the mean-field level

Let us again assume that the fluid motion and so the electromagnetic fields, too, show small-scale parts in the sense explained above. Then it seems reasonable to take the average of all equations applying to fluid body and outer space mentioned in the above formulation of the dynamo problem. This means in particular that the mean magnetic field $\overline{\boldsymbol{B}}$ has to satisfy the Eq. (13) inside the fluid body, to continue as a potential field in outer space and to vanish at infinity. We speak of a "mean-field dynamo" if the problem for $\overline{\boldsymbol{B}}$ posed in this way has nondecaying solutions, $\overline{\boldsymbol{B}} \not\to 0$ as $t \to \infty$. However, the notion "mean-field dynamo" has to be used with care. It does not refer to a real physical object but to a particular model of such an object only, which delivers a simplified picture of the real object. The existence of a mean-field dynamo in the sense of the above definition always implies the existence of a dynamo in the original sense.

It is important to note that mean magnetic fields $\overline{\boldsymbol{B}}$ are not subject to Cowling's theorem. The proofs of this theorem cannot be repeated if Ohm's law Eq. (2) is replaced with its mean-field version Eq. (12). A possible exception is cases with $\boldsymbol{\mathcal{E}} \cdot \overline{\boldsymbol{B}} = 0$. Mean-field dynamos may thus well be axisymmetric. The deviation of $\boldsymbol{B}$ from axisymmetry, which is necessary for a dynamo, need not occur in $\overline{\boldsymbol{B}}$. It is sufficient to have it in $\boldsymbol{B}'$.

A simple illustration of a mean-field dynamo can be given on the basis of the spatially periodic dynamo mentioned above. When defining mean-fields by averaging over all values of x and y, we may derive an equation for $\overline{\boldsymbol{B}}$, which implies an anisotropic α-effect and allows growing solutions. This has been widely discussed in the context of the Karlsruhe dynamo experiment (see below).

Traditional mean-field dynamo models

Many mean-field dynamo models have been developed for various objects like the Earth and the planets, the Sun and several types of stars, or for galaxies. In almost all cases simple symmetries were assumed with respect to the shape of the conducting bodies, to the distributions of the electric conductivity and to the fluid motions.

Let us first formulate general assumptions of that kind. It is always supposed that a rotation axis and an equatorial plane perpendicular to it are defined. We assume that the shape of the fluid body and the distribution of the electric conductivity, or of the magnetic diffusivity, are

- symmetric about the rotation axis,
- symmetric about the equatorial plane,
- steady.

In addition we assume that all averaged quantities depending on the velocity field $\boldsymbol{u}$, that is $\overline{\boldsymbol{u}} + \boldsymbol{u}'$, are invariant under

- rotations of $\boldsymbol{u}$ about the rotation axis,
- reflections of $\boldsymbol{u}$ about the equatorial plane,
- time shifts in $\boldsymbol{u}$.

As the simplest consequence of these last assumptions, we note that the mean velocity $\overline{\boldsymbol{u}}$ is symmetric about both rotation axis and equatorial plane and steady. Another simple consequence is, e.g., that the mean helicity $\overline{\boldsymbol{u}' \cdot (\nabla \times \boldsymbol{u}')}$ of the fluctuating motions as well as the related quantity $\overline{\boldsymbol{a}' \cdot (\nabla \times \boldsymbol{a}')}$ mentioned above, which are of interest for the α-effect, are symmetric about the rotation axis and steady but antisymmetric about the equatorial plane.

The assumptions introduced together with the Eq. (3) governing the magnetic field $\boldsymbol{B}$ also allow us far-reaching conclusions concerning the mean magnetic field $\overline{\boldsymbol{B}}$.

- Firstly, if a field $\overline{\boldsymbol{B}}$ satisfies the relevant equations and conditions formulated above, the field $\overline{\boldsymbol{B}}^{\mathrm{refl}}$, which is generated by reflecting $\overline{\boldsymbol{B}}$ at the equatorial plane, satisfies them, too. The same applies to their sum or their difference, which are symmetric or antisymmetric, respectively, about the equatorial plane.
- Secondly, any field $\overline{\boldsymbol{B}}$ can be decomposed into its Fourier modes $\Re(\hat{\boldsymbol{B}}^{(m)} \exp(im\varphi))$ with respect to the azimuthal coordinate φ, where the $\hat{\boldsymbol{B}}^{(m)}$, with nonnegative integer m, are complex vector fields symmetric about the rotation axis. Each individual Fourier mode of that kind again satisfies the relevant equations and conditions.
- Thirdly, the fields $\overline{\boldsymbol{B}}$ vary with time like $\Re(\hat{\boldsymbol{B}} \exp(pt))$ where $\hat{\boldsymbol{B}}$ is some complex vector field and p a complex constant, or are superpositions of such fields.

Taking these three findings together, we see that it is sufficient to look for solutions of the relevant equations and conditions having the form

$$\overline{\boldsymbol{B}} = \Re\Big(\hat{\boldsymbol{B}}^{(m)} \exp(im\varphi + (\lambda + i\omega)t)\Big). \qquad \text{(Eq. 25)}$$

All further solutions can be gained by superposition of them. Now $\hat{\boldsymbol{B}}^{(m)}$ means a complex vector field being either antisymmetric or symmetric about the equatorial plane, symmetric about the rotation axis and steady; m is a nonnegative integer, and λ and ω are real constants. We denote the solutions of the form Eq. (25) by A or S according to their antisymmetry or symmetry about the equatorial plane, and add the parameter m to characterize the symmetry with respect to the rotation axis. Examples of field patterns of Am and Sm modes are given in Figure D29. The field of a dipole aligned with the rotation axis is of A0 type, that of an quadrupole symmetric about this axis is of S0 type. The field of a dipole perpendicular to this axis is of S1 type. Of course, dynamo-generated magnetic fields of the form Eq. (25) in general do not correspond to single multipoles but to superpositions of several multipoles with the same symmetry properties.

Clearly λ is the growth rate of the solution considered. A mean-field dynamo requires $\lambda \geq 0$. If $\omega = 0$, the solution varies monotonously with time; if $\omega \neq 0$, oscillatory. Axisymmetric modes, $m = 0$, with $\omega \neq 0$ are intrinsically oscillatory. A nonaxisymmetric mode, $m \neq 0$, with $\omega \neq 0$ has the form of a wave traveling in azimuthal direction. Its field configuration rotates rigidly with the angular velocity $-\omega/m$ and is, of course, steady in a corotating frame of reference.

The axisymmetry of the mean velocity $\overline{\boldsymbol{u}}$ allows us to split it into an axisymmetric circulation in meridional planes, $\overline{\boldsymbol{u}}^{\text{circ}}$, and an axisymmetric rotation, $\overline{\boldsymbol{u}}^{\text{rot}}$. The latter has in general the form of a differential rotation, that is $\overline{\boldsymbol{u}}^{\text{rot}} = \omega\hat{\boldsymbol{\omega}} \times \boldsymbol{r}$, where ω is the angular velocity, which is axisymmetric and symmetric about the equatorial plane, $\hat{\boldsymbol{\omega}}$ is the unit vector in axial direction and $\boldsymbol{r}$ again the radius vector.

So far, no assumptions about the structure of $\boldsymbol{\mathcal{E}}$ have been used explicitly. Now we rely again on the assumption that $\boldsymbol{\mathcal{E}}$ in a given point depends on $\overline{\boldsymbol{B}}$ and its first spatial derivatives in this point only, and so on relation Eq. (19). Of course the assumptions made above about the symmetries of the distribution of the magnetic diffusivity and of the motion have consequences for the quantities $\boldsymbol{\alpha}, \boldsymbol{\beta}, \boldsymbol{\gamma}, \boldsymbol{\delta}$, and $\boldsymbol{\kappa}$.

In order to formulate these consequences properly we introduce in addition to the unit vector $\hat{\boldsymbol{\omega}}$ in axial direction another vector, $\hat{\boldsymbol{g}}$, describing another preferred direction in the fluctuating velocity field. We may identify $\hat{\boldsymbol{g}}$, e.g., with the unit vector in the direction opposite to the gravitational force but put it equal to zero where no such a direction can be defined. Whereas $\hat{\boldsymbol{\omega}}$ is independent of position, $\hat{\boldsymbol{g}}$ may vary in space but is symmetric about the rotation axis and the equatorial plane. We write

$$\alpha_{ij} = \alpha_1(\hat{\boldsymbol{\omega}}\cdot\hat{\boldsymbol{g}})\delta_{ij} + \alpha_2(\hat{\boldsymbol{\omega}}\cdot\hat{\boldsymbol{g}})\hat{g}_i\hat{g}_j + \alpha_3(\hat{\boldsymbol{\omega}}\cdot\hat{\boldsymbol{g}})\hat{\omega}_i\hat{\omega}_j + \alpha_4(\hat{\omega}_i\hat{g}_j + \hat{\omega}_j\hat{g}_i) + \alpha_5(\hat{\boldsymbol{\omega}}\cdot\hat{\boldsymbol{g}})(\hat{g}_i\hat{\lambda}_j + \hat{g}_j\hat{\lambda}_i) + \alpha_6(\hat{\omega}_i\hat{\lambda}_j + \hat{\omega}_j\hat{\lambda}_i) \qquad \text{(Eq. 26)}$$

$$\beta_{ij} = \beta_1\delta_{ij} + \beta_2\hat{g}_i\hat{g}_j + \beta_3\hat{\omega}_i\hat{\omega}_j + \beta_4(\hat{\boldsymbol{\omega}}\cdot\hat{\boldsymbol{g}})(\hat{\omega}_i\hat{g}_j + \hat{\omega}_j\hat{g}_i) + \beta_5(\hat{g}_i\hat{\lambda}_j + \hat{g}_j\hat{\lambda}_i) + \beta_6(\hat{\boldsymbol{\omega}}\cdot\hat{\boldsymbol{g}})(\hat{\omega}_i\hat{\lambda}_j + \hat{\omega}_j\hat{\lambda}_i) \qquad \text{(Eq. 27)}$$

$$\gamma_i = \gamma_1\hat{g}_i + \gamma_2(\hat{\boldsymbol{\omega}}\cdot\hat{\boldsymbol{g}})\hat{\omega}_i + \gamma_3\hat{\lambda}_i \qquad \text{(Eq. 28)}$$

$$\delta_i = \delta_1(\hat{\boldsymbol{\omega}}\cdot\hat{\boldsymbol{g}})\hat{g}_i + \delta_2\hat{\omega}_i + \delta_3(\hat{\boldsymbol{\omega}}\cdot\hat{\boldsymbol{g}})\hat{\lambda}_i \qquad \text{(Eq. 29)}$$

where $\hat{\boldsymbol{\lambda}} = \hat{\boldsymbol{\omega}} \times \hat{\boldsymbol{g}}$. As for $\boldsymbol{\kappa}$ we note that $\boldsymbol{\kappa} \circ (\nabla\overline{\boldsymbol{B}})^{(s)}$ can be represented as a sum of the four contributions $\boldsymbol{\beta}^g \circ \boldsymbol{V}^g, \boldsymbol{\beta}^\omega \circ \boldsymbol{V}^\omega, \boldsymbol{\delta}^g \times \boldsymbol{V}^g$, and $\boldsymbol{\delta}^\omega \times \boldsymbol{V}^\omega$ with $\boldsymbol{\beta}^g, \boldsymbol{\beta}^\omega, \boldsymbol{\delta}^g$, and $\boldsymbol{\delta}^\omega$ analogous to $\boldsymbol{\alpha}, \boldsymbol{\beta}, \boldsymbol{\gamma}$, and $\boldsymbol{\delta}$, respectively, where $\boldsymbol{V}^g = (\nabla\overline{\boldsymbol{B}})^{(s)} \circ \hat{\boldsymbol{g}}$ and $\boldsymbol{V}^\omega = (\nabla\overline{\boldsymbol{B}})^{(s)} \circ \hat{\boldsymbol{\omega}}$. Again, three elements of the tensors resulting from $\boldsymbol{\kappa}$ may be fixed arbitrarily. The assumptions introduced above just require that the coefficients $\alpha_1, \alpha_2, \ldots \delta_3$ as well as $\beta_1^g, \beta_2^g, \ldots \delta_3^\omega$ are symmetric about the rotation axis and the equatorial plane and are steady.

Comparing $\boldsymbol{\mathcal{E}}$ as obtained for homogeneous isotropic turbulence and given in Eq. (17) with our results Eq. (19) and Eqs. (26) to (29) we see that the contribution $\alpha\overline{\boldsymbol{B}}$ there, describing the isotropic α-effect, corresponds to $-\alpha_1(\hat{\boldsymbol{\omega}}\cdot\hat{\boldsymbol{g}})\overline{\boldsymbol{B}}$ here, which is however accompanied by other contributions causing an anisotropy of the α-effect. We will use the notation α in the following also in the sense of $\alpha = -\alpha_1(\hat{\boldsymbol{\omega}}\cdot\hat{\boldsymbol{g}})$. Clearly α is then, in contrast to α_1, antisymmetric about the equatorial plane.

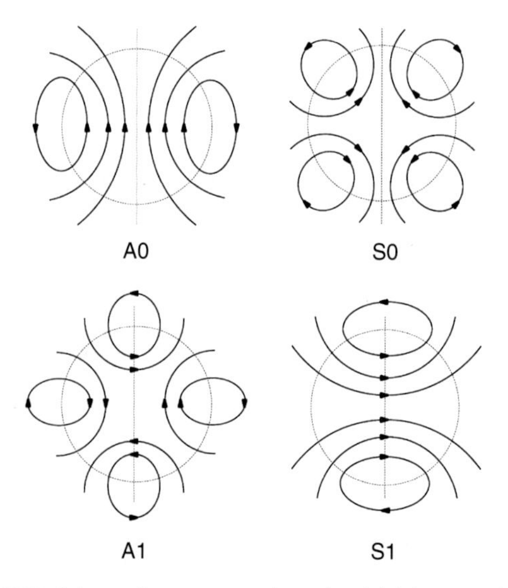

Figure D29 Schematic representation of poloidal magnetic field lines of A0, S0, A1 and S1 modes in meridional planes of spherical bodies. In the case of A0 and S0 modes the patterns agree for all such planes. In the case of the A1 and S1 modes those planes have been chosen which are not crossed by field lines.

Basic dynamo mechanisms

In all dynamo models investigated so far, in which poloidal and toroidal parts of the magnetic field can be defined, dynamo action occurs due to an interplay between these parts. This applies to dynamos in the original sense as well as to mean-field dynamos. So the various mean-field dynamo mechanisms can be characterized by the induction processes, which are dominant in the generation of the poloidal field from the toroidal one and vice versa.

In the following discussion always rotating bodies are considered. In the case of a rigid-body rotation we use a corotating frame of reference in which $\overline{\boldsymbol{u}}^{\text{rot}} = 0$. Nevertheless the rotation occurs in $\boldsymbol{\mathcal{E}}$ via the Coriolis force. It is in particular important for the α-effect. In general we also refer to a somehow fixed rotating frame of reference, in which $\overline{\boldsymbol{u}}^{\text{rot}} = \omega\hat{\boldsymbol{\omega}} \times \boldsymbol{r}$ with an angular velocity ω depending on position. Clearly ω depends on the choice of the frame whereas $\nabla\omega$ is independent of it.

(i) The α^2 and $\alpha\omega$ mechanisms

The α-effect is capable of generating both a poloidal field from a toroidal one and also a toroidal field from a poloidal one. This is the basis of the "α^2 mechanism". Figure D30 demonstrates it for a spherical body and axisymmetric magnetic fields of dipole and quadrupole type, that is, A0 and S0 modes. For the sake of simplicity no other contribution to the electromotive force $\boldsymbol{\mathcal{E}}$ is considered than $\alpha\overline{\boldsymbol{B}}$ with $\alpha > 0$ in the northern and $\alpha < 0$ in the southern hemisphere. As can be readily followed up in the figure, the α-effect with the toroidal field leads to toroidal currents which just support the poloidal field. Likewise the α-effect with the poloidal field results in poloidal currents, which in turn support the toroidal field. In this way, a sufficiently strong α-effect is able to maintain magnetic fields with the configurations envisaged,

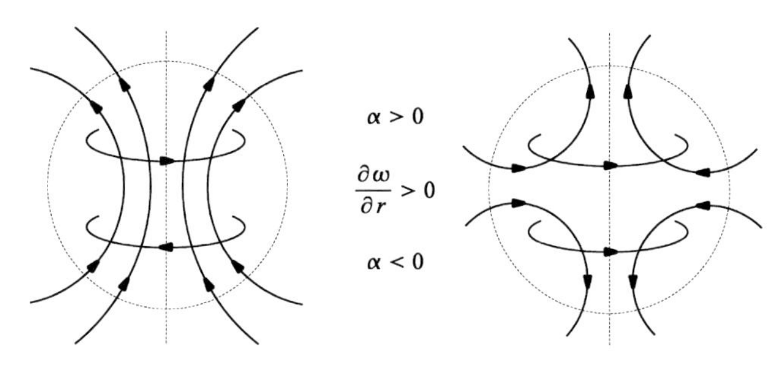

Figure D30 Axisymmetric poloidal and toroidal magnetic field configurations of dipole and quadrupole type as can be maintained by an α^2 mechanism ($\partial\omega/\partial r = 0$) or an $\alpha\omega$ mechanism.

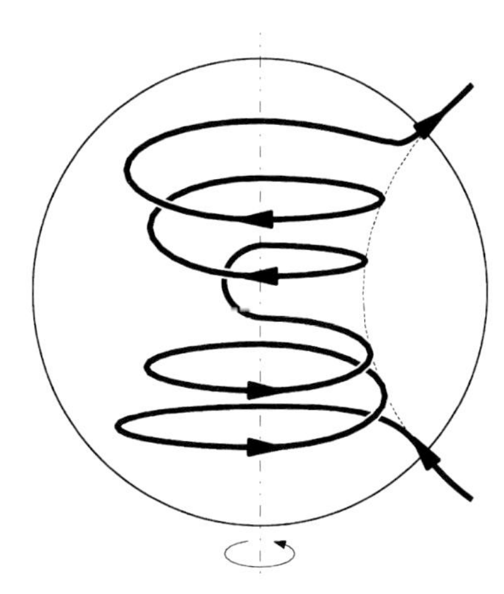

Figure D31 The effect of differential rotation on an axisymmetric poloidal magnetic field. It is assumed that the surface of the fluid body is at rest and the inner parts rotate in the indicated way. A field line given initially by the dotted line in a meridional plane occurs later in the form of the solid line. The magnetic field that results from the action of differential rotation on an axisymmetric poloidal field can be understood as an superposition of the original poloidal field and an additional axisymmetric toroidal field.

or make them grow. If the signs of α are inverted the orientation of either the poloidal or the toroidal fields have to be inverted, too. As a rule, the axisymmetric magnetic fields generated by the α^2 mechanism are nonoscillatory. The α^2 mechanism may work with nonaxisymmetric magnetic fields as well, that is, may support also A1, S1, A2, S2, ... modes. In all cases the poloidal and the toroidal parts of the fields are of the same order of magnitude.

The α^2 mechanism can be modified by differential rotation, that is, by a gradient of ω. As illustrated by Figure D31 a differential rotation generates an axisymmetric toroidal magnetic field from a given axisymmetric poloidal one. The ratio of the magnitude of the toroidal to that of the poloidal field can be arbitrarily high if only the rotational shear is sufficiently strong. So the differential rotation modifies the generation of the toroidal field by the α-effect. A very strong differential rotation can even dominate the generation of this field. In this case we speak of an "$\alpha\omega$ mechanism". It depends on details of the α-effect and the rotational shear whether this mechanism supports preferably an A0 or an S0 mode, and whether this mode is oscillatory or non-oscillatory. The toroidal field is always much stronger than the poloidal one. With nonaxisymmetric fields the effect of differential rotation is more complex. Their structure is changed in such a way that they are subject to dissipation more heavily, and the ratio of the magnitudes of toroidal and poloidal field is therefore bounded. That is why the $\alpha\omega$ mechanism is not effective with non-axisymmetric magnetic fields, that is, with Am or Sm modes with $m \geq 1$.

In general, of course, both α-effect and differential rotation take part in the generation of the toroidal field. This case is sometimes labelled as "$\alpha^2\omega$ mechanism", the case of a negligible influence of the α-effect on the generation of the toroidal field as "pure $\alpha\omega$ mechanism."

Since the end of the 1960s a large number of spherical and other dynamo models working with these mechanisms have been studied, taking into account various contributions to the mean electromotive force $\mathcal{E}$ and various forms of the mean velocity $\bar{\boldsymbol{u}}$, and considering both axisymmetric and nonaxisymmetric magnetic fields. The results have been summarized at several places (e.g., Krause *et al.* 1980; Rädler 1980, 1986, 1995, 2000). We note here only a few facts which are of particular interest for the geodynamo.

In simple models with α^2 mechanism the excitation conditions for A0, S0, A1, and S1 modes are in general close together, whereas the A2, S2, A3, S3, ... modes are less easily excitable. As already mentioned, the axisymmetric modes, A0 and S0, are in almost all cases non-oscillatory. The nonaxisymmetric ones show, depending on the specific form of the α-effect, either eastward or westward migrations. With a fairly isotropic α-effect in general one of the axisymmetric modes is slightly preferred over all others. Anisotropies of the α-effect, also the presence of the γ-effect, lead to preferences of the A1 or S1 mode over all others. In particular the anisotropy of the α-effect due to rapid rotation of the fluid body acts in this sense. That is, under realistic conditions the α^2 mechanism may well favor non-axisymmetric magnetic field structures. Incidentally, the isotropic α-effect together with a weak differential rotation may also lead to a preference of A1 or S1 modes.

As explained above, a pure $\alpha\omega$ mechanism supports only A0 or S0 modes. Which of them is preferably excited, and whether or not it is oscillatory, proved to depend indeed on the distribution of α-effect and rotational shear. For the pure $\alpha\omega$ mechanism, anisotropies of the α-effect play a minor part. In the transition region between the α^2 mechanism and the pure $\alpha\omega$ mechanism, that is, with the $\alpha^2\omega$ mechanism, the situation is even more complex.

(ii) Mechanisms without α-effect

In addition to the α-effect dynamo mechanisms explained so far, other mechanisms due to induction effects covered by the electromotive force $\mathcal{E}$ in combination with differential rotation proved to be possible.

The contribution $-\boldsymbol{\beta} \circ (\nabla \times \overline{\boldsymbol{B}})$ with an anisotropic tensor $\boldsymbol{\beta}$ produces in general also poloidal magnetic fields from toroidal ones and vice versa. Under the reasonable assumption that the mean-field conductivity tensor $\boldsymbol{\sigma}_{\mathrm{m}}$, which is determined by $\boldsymbol{\beta}$, is positive definite, however, a dynamo due to this contribution alone can be excluded. Remarkably enough, a particular contribution to $\boldsymbol{\beta}$, that is, the one with β_5 in Eq. (27), together with differential rotation allows the generation of axisymmetric magnetic fields of both A0 and S0 types. That contribution to $\boldsymbol{\beta}$ occurs due to the Coriolis force.

Likewise the contribution $-\boldsymbol{\delta} \times (\nabla \times \overline{\boldsymbol{B}})$ implies couplings between poloidal and toroidal magnetic fields. Simple energy arguments show that this induction effect alone is not capable of dynamo action. However, the combination of that contribution with a differential rotation may again work as a dynamo for axisymmetric magnetic fields, that is, for such of A0 and S0 types. In the first investigations of dynamos of that kind a specific $\boldsymbol{\delta}$ was considered, given simply by the δ_2 term in Eq. (29). As explained above the induction effect defined by this specific $\boldsymbol{\delta}$ is called "$\boldsymbol{\Omega} \times \boldsymbol{j}$-effect". Therefore the corresponding dynamo is sometimes labeled as "$\boldsymbol{\Omega} \times \boldsymbol{j}$ dynamo."

Incidentally, also the contribution $-\boldsymbol{\kappa} \circ (\nabla\overline{\boldsymbol{B}})^{(s)}$ describes induction effects, which together with differential rotation might lead to dynamo action, or modify the dynamo mechanisms discussed before.

A few studies of these dynamo mechanisms without α-effect have been carried out in spherical models (see, e.g., Krause *et al.*, 1980; Rädler, 1986, 1995, 2000).

Mean-field magnetohydrodynamics and dynamically consistent dynamo models

Mean-field magnetohydrodynamics

The mean-field concept, so far applied to the basic electrodynamic equations, can be extended to the equations of fluid dynamics, too. In the case of incompressible fluids these are the momentum balance, that is the Navier-Stokes equation, with the Lorentz force involved, and the mass balance, that is the continuity equation. In a rotating frame of reference they read

$$\varrho(\partial_t \boldsymbol{u} + (\boldsymbol{u}\cdot\nabla)\boldsymbol{u}) = -\nabla p + \varrho\nu\nabla^2\boldsymbol{u} - 2\varrho\boldsymbol{\Omega}\times\boldsymbol{u} + (1/\mu)(\nabla\times\boldsymbol{B})\times\boldsymbol{B} + \boldsymbol{F}, \quad \nabla\cdot\boldsymbol{u} = 0, \qquad \text{(Eq. 30)}$$

where ϱ is the mass density of the fluid, p the hydrodynamic pressure, ν the kinematic viscosity, $\boldsymbol{\Omega}$ the angular velocity describing the Coriolis force, and $\boldsymbol{F}$ some external force. For compressible fluids, apart from slight changes in these equations, the equation of state and in general also a thermodynamic equation, e.g. the heat conduction equation, have to be added.

Subjecting also the equations of fluid dynamics to averaging we arrive at mean-field magnetohydrodynamics. As we have seen above, in mean-field electrodynamics the basic equations for the mean-fields agree formally with those for the original fields with the exception that an additional mean electromotive force $\boldsymbol{\mathcal{E}}$ occurs, $\boldsymbol{\mathcal{E}} = \overline{\boldsymbol{u}'\times\boldsymbol{B}'}$. In the case of an imcompressible homogeneous fluid, to which attention is restricted in the following, this applies analogously to all mean-field magnetohydrodynamics. Starting from Eq. (30) we find

$$\varrho(\partial_t \overline{\boldsymbol{u}} + (\overline{\boldsymbol{u}}\cdot\nabla)\overline{\boldsymbol{u}}) = -\nabla\overline{p} + \varrho\nu\nabla^2\overline{\boldsymbol{u}} - 2\varrho\boldsymbol{\Omega}\times\overline{\boldsymbol{u}} + (1/\mu)(\nabla\times\overline{\boldsymbol{B}})\times\overline{\boldsymbol{B}} + \overline{\boldsymbol{F}} + \boldsymbol{\mathcal{F}}, \quad \nabla\cdot\overline{\boldsymbol{u}} = 0, \qquad \text{(Eq. 31)}$$

with a mean ponderomotoric force $\boldsymbol{\mathcal{F}}$ given by

$$\boldsymbol{\mathcal{F}} = -\varrho\overline{(\boldsymbol{u}'\cdot\nabla)\boldsymbol{u}'} + (1/\mu)\Big(\overline{(\boldsymbol{B}'\cdot\nabla)\boldsymbol{B}'} - (1/2)\nabla\overline{\boldsymbol{B}'^2}\Big). \qquad \text{(Eq. 32)}$$

An alternative representation of $\boldsymbol{\mathcal{F}}$ is $\mathcal{F}_i = \partial(V_{ij} + M_{ij})/\partial x_j$, where $V_{ij} = -\varrho\overline{u'_i u'_j}$ is the Reynolds stress tensor and $M_{ij} = (1/\mu)(\overline{B'_i B'_j} - (1/2)\overline{B'^2}\delta_{ij})$ the average of the Maxwell stress tensor formed with the magnetic fluctuations.

In mean-field magnetohydrodynamics the two quantities $\boldsymbol{\mathcal{E}}$ and $\boldsymbol{\mathcal{F}}$ play a central role. The fluctuating motion, $\boldsymbol{u}'$, is no longer considered as given but assumptions about its causes, e.g. instabilities, are made. It seems reasonable to evade detailed investigations on these causes by assuming a fluctuating force, say $\boldsymbol{F}'$, that drives these motions. Then $\boldsymbol{u}'$ and also $\boldsymbol{B}'$ are determined by this force and by $\overline{\boldsymbol{u}}$ and $\overline{\boldsymbol{B}}$. So, as a matter of principle, $\boldsymbol{\mathcal{E}}$ and $\boldsymbol{\mathcal{F}}$ can be calculated for a given force $\boldsymbol{F}'$ as functionals of $\overline{\boldsymbol{u}}$ and $\overline{\boldsymbol{B}}$. Eqs. (13) and (31), completed by relations connecting $\boldsymbol{\mathcal{E}}$ and $\boldsymbol{\mathcal{F}}$ with $\overline{\boldsymbol{u}}$ and $\overline{\boldsymbol{B}}$, govern the behavior of $\overline{\boldsymbol{u}}$ and $\overline{\boldsymbol{B}}$.

For sufficiently weak variations of $\overline{\boldsymbol{B}}$ in space and time $\boldsymbol{\mathcal{E}}$ can again be represented in the form of Eq. (16) or of Eq. (19). But the quantities a_{ij}, b_{ijk} or $\boldsymbol{\alpha}, \boldsymbol{\beta}, \ldots$ are then no longer independent of $\overline{\boldsymbol{B}}$. As a consequence of the action of the Lorentz force on the fluctuating motion, apart from an indirect influence via the mean motion, their tensorial structures and the magnitudes of the tensor elements depend on $\overline{\boldsymbol{B}}$ and its derivatives. For example, the α-effect is in general reduced. This fact is called "α-quenching". Likewise corresponding influences on $\boldsymbol{\beta}$, or $\boldsymbol{\sigma}$, are sometimes labeled as "β-quenching", etc.

We refer also to more comprehensive representations of mean-field magnetohydrodynamics (e.g., Rädler, 2000) and more specific results concerning $\boldsymbol{\mathcal{E}}$ or $\boldsymbol{\mathcal{F}}$ (e.g., Rüdiger *et al.*, 1993; Blackman, 2002; Blackman *et al.*, 2002).

Dynamically consistent mean-field dynamo models

When proceeding from a kinematic dynamo model to a dynamically consistent one the electrodynamic Eqs. (1) and (2), or (3), applying inside the fluid body, have to be completed by the momentum balance and the mass balance as given by Eq. (30). These equations, together with proper conditions concerning the continuation of the electromagnetic field in outer space and with boundary conditions for the hydrodynamic quantities, pose a new, more complex initial value problem, which defines in particular $\boldsymbol{B}$ and $\boldsymbol{u}$ if $\boldsymbol{F}$ is given. The problem is nonlinear in both $\boldsymbol{B}$ and $\boldsymbol{u}$. Whereas in the corresponding kinematic dynamo problem, which is linear in $\boldsymbol{B}$, the magnitude of $\boldsymbol{B}$ remains undetermined, now the magnitudes of both $\boldsymbol{B}$ and $\boldsymbol{u}$ are fixed.

At the mean-field level the full dynamo problem has to be formulated on the basis of the mean-field Eqs. (13) and (31) and relations connecting $\boldsymbol{\mathcal{E}}$ and $\boldsymbol{\mathcal{F}}$ with $\overline{\boldsymbol{B}}$ and $\overline{\boldsymbol{u}}$. A first step toward dynamically consistent mean-field dynamo models in that sense are models which consider as in the kinematic case the electrodynamic mean-field equations only but introduce there a dependence of quantities like $\boldsymbol{\alpha}$ or $\boldsymbol{\beta}$ on $\overline{\boldsymbol{B}}$, that is, α or β quenching. As a rule, α-quenching limits the growth of the magnetic field. On this level several studies on the stability of dynamo-generated magnetic field configurations have been carried out (e.g., Rädler *et al.*, 1990). Using indeed the full set of electrodynamic and fluiddynamic equations, in several examples the coupled evolution of the mean magnetic field and the mean motions have been studied (e.g., Hollerbach, 1991).

Mean-field models of the geodynamo

Simple mean-field models

As mentioned above, the great breakthrough in our understanding of the geodynamo came with Braginsky's theory of the nearly symmetric dynamo and the findings of mean-field electrodynamics. The first mean-field dynamo model which was discussed in view of the Earth, an α^2 model, has been proposed by Steenbeck and Krause in 1969 (Steenbeck *et al.*, 1969). A series of similar models were investigated and discussed later on. Whereas the mentioned first mean-field model and some of the following ones considered only axisymmetric magnetic fields, also nonaxisymmetric ones were included in later models (Rädler, 1975; Rüdiger, 1980; Rüdiger *et al.*, 1994). In this way some understanding could be developed not only for the small deviations of the Earth's magnetic field from axisymmetry and for their drifts, but also for the much larger deviations of the magnetic fields of some planets from axisymmetry, in particular of Uranus and Neptune (Rüdiger *et al.*, 1994). The tendency of differential rotation to reduce nonaxisymmetric parts of magnetic fields led to the suggestion that it is not too strong in the interiors of the objects mentioned. A strong differential rotation, however, could be an explanation for the high degree of axisymmetry of the Saturnian magnetic field.

On the applicability of the mean-field concept

For the crude models of the geodynamo and of planetary dynamos addressed so far more or less plausible assumptions were made on the validity of the results of mean-field electrodynamics to the Earth's and planetary interiors. For a more detailed elaboration of such models the applicability of the mean-field concept to these dynamo problems has to be checked carefully.

First of all, a proper averaging procedure has to be adopted which ensures at least the approximate validity of Reynolds' rules. As mentioned above, statistical averages satisfy Reynolds' rules exactly but their relation to measurable quantities is unclear. A spatial average in the sense of Eq. (8) is very problematic. There is no indication of a clear gap in the spectrum of length scales of the motions in the outer core of the Earth, which are relevant for the geodynamo process. That is, there is hardly a possibility to ensure the validity of Reynolds' rules. With the time average in the sense of Eq. (9) the situation is similar.

It allows us only to study the long-term behavior of the magnetic field, that is, the behavior on time scales, which are very long compared with the characteristic time scales of motions in the liquid core. Otherwise the validity of Reynolds' rules is unsure. For several purposes the azimuthal average defined by Eq. (10) can be used, which satisfies these rules exactly. In this case, however, the mean fields are by definition axisymmetric. The investigation of any nonaxisymmetric structures in the geomagnetic field is excluded from the very beginning. In addition the averages are in general not really smooth with respect to the remaining space coordinates and to time. If the fluid motion is of a stochastic nature, the mean electromotive force $\mathcal{E}$ and the mean magnetic field $\overline{\boldsymbol{B}}$ show certain stochastic features too. This applies the more the smaller the number of elements of motion, that is, of eddies or cells, along an averaging circle is. (This aspect is more important for the Earth rather than, e.g., the Sun.) Fairly smooth mean fields would occur only after an additional averaging with respect to the remaining space coordinates or time. The combination of azimuthal averaging with this additional averaging would define a new average, which, of course, can satisfy Reynolds' rules again only approximately.

Already these considerations show that the mean-field concept, although very useful in several respects, is far from being an ideal tool for studying the geodynamo or planetary dynamos. In addition, it remains to be checked whether the standard assumptions of mean-field electrodynamics, in particular the dependence of $\mathcal{E}$ on $\overline{\boldsymbol{B}}$ and its first derivatives only, indeed apply in a given model, or have to be replaced by more general assumptions

The mean-field concept and direct numerical simulations

As already mentioned the mean-field concept was helpful in designing models for direct numerical simulations of the geodynamo process (see *Geodynamo, numerical simulations*). In such models parameters like the Ekman number $E = \nu/\Omega R^2$ or the magnetic Prandtl number $P_{\mathrm{m}} = \nu/\eta$ play an important role; ν and η are again kinematic viscosity and magnetic diffusivity, Ω the angular velocity of rotation and R the radius of the core. With realistic molecular values of ν and η the parameters E and P_{m} are extremely small, typically $E = 10^{-16}$ and $P_{\mathrm{m}} = 10^{-6}$. Such values do not allow to solve the numerical problem with the available computing power. One way out is to understand the underlying equations as mean-field equations based on a "low-level averaging," that is, on averaging over small lengths or short times, and to replace ν and η by the corresponding mean-field quantities. In this way the requirements concerning the computing power are reduced. The direct numerical simulations done so far rest on a specific mean-field concept with "low-level averaging" in the above sense (Roberts *et al.*, 2000; Kono *et al.*, 2002).

At the same time the mean-field concept, now again understood in the usual sense, is a valuable tool for the interpretation of the results of direct numerical simulations. Adopt, e.g., azimuthal averaging. The coefficients which determine the mean electromotive force $\mathcal{E}$, that is $\boldsymbol{\alpha}, \boldsymbol{\beta}, \boldsymbol{\gamma} \ldots$, and the mean velocity $\overline{\boldsymbol{u}}$ can be extracted from the numerical results (Schrinner *et al.*, 2005, 2006). They depend, of course, on the remaining space coordinates and on time (and should perhaps be smoothed in space or time). In this way a mean-field model corresponding to that used for the direct numerical simulation can be constructed. It can tell us, which processes are dominant in the dynamo, whether the dynamo is of α^2 or of $\alpha\omega$ type, to what extent other dynamo mechanisms are important, etc. (Possibly future investigations of this kind will also give some insight in the processes relevant for reversals of the magnetic field.)

Secular variation and reversals

As explained above, when using the azimuthal average and assuming fluid motions of stochastical nature, the electromotive force $\mathcal{E}$ and so the $\boldsymbol{\alpha}, \boldsymbol{\beta}, \boldsymbol{\gamma}, \ldots$, the mean velocity $\overline{\boldsymbol{u}}$ as well as the mean magnetic field $\overline{\boldsymbol{B}}$ show stochastical features, too. On this basis a simple model of the geodynamo has been constructed by Hoyng, Ossendrijver and Schmitt, which is of interest in view of its time behavior (Hoyng *et al.*, 2001; Schmitt *et al.*, 2001; Hoyng *et al.*, 2002). In this model no other induction effect than the α-effect, with a stochastically varying α, is taken into account. The geodynamo occurs then as a bistable oscillator, in which the amplitude of the fundamental nonoscillatory dipolar dynamo mode performs a random walk in a bistable potential. The potential wells represent the normal and reversed polarity states, and the potential hill between the states is due to supercritical excitation. A random transition across the central potential hill corresponds to a reversal. Many features of the secular variation and reversal statistics can be modeled in this way.

Laboratory experiments on dynamos

In 1999 the first two experimental devices aimed at realizing homogeneous dynamos have run successfully, one in Riga, Latvia, and the other in Karlsruhe, Germany (see *Dynamos, experimental*). In the Riga device a dynamo of Ponomarenko type has been realized (Gailitis *et al.*, 2000, 2001). The Karlsruhe device was designed to simulate in a rough way the dynamo process in the Earth's core (Busse, 1975, 1992; Müller *et al.*, 2000; Stieglitz *et al.*, 2001, 2002). The flow pattern was chosen with a view to the convection rolls assumed in the outer core of the Earth (Busse, 1970). It is in fact some modification of a pattern periodic with respect to two Cartesian coordinates, whose capability of dynamo action has been demonstrated by Roberts already in 1970 (Roberts, 1970, 1972). This flow pattern suggests a mean-field formulation of the corresponding dynamo problem. Indeed, a mean-field theory of the Karlsruhe experiment has been developed, the central element of which is an anisotropic α-effect. Its predictions concerning the excitation condition of the dynamo and the geometrical structure of the generated magnetic fields as well as the behavior of the dynamo in the nonlinear regime have been well confirmed by the measured data (Rädler *et al.*, 1998, 2002a,b,c).

Karl-Heinz Raedler

Bibliography

Blackman, E.G., 2002. Recent developments in magnetic dynamo theory. In Falgarone, E., and Passot, T. (ed), *Turbulence and Magnetic Fields in Astrophysics*. Springer Lecture Notes in Physics, pp. 432–463.

Blackman, E.G., and Field, G.B., 2002. New dynamical mean-field dynamo theory and closure approach. *Physical Review Letters*, **89**: 265007/1–4.

Braginsky, S.I., 1964a. Kinematic models of the Earth's hydromagnetic dynamo. *Geomagnetism and Aeron.*, **4**: 732–737.

Braginsky, S.I., 1964b. Theory of the hydromagnetic dynamo. *Sov. Phys. JETP*, **20**: 1462–1471.

Busse, F.H., 1970. Thermal instabilities in rapidly rotating systems. *Journal of Fluid Mechanics*, **44**: 441–460.

Busse, F.H., 1975. A model of the geodynamo. *Geophysical Journal of the Royal Astronomical Society*, **42**: 437–459.

Busse, F.H., 1992. Dynamo theory of planetary magnetism and laboratory experiments. In Friedrich, R., and Wunderlin, A. (ed), *Evolution of Dynamical Structures in Complex Systems*. Berlin: Springer, pp. 197–207.

Gailitis, A., Lielausis, O., Dement'ev, S., Platacis, E., Cifersons, A., Gerbeth, G., Gundrum, T., Stefani, F., Christen, M., Hänel, H., and Will, G., 2000. Detection of a flow induced magnetic field eigenmode in the Riga dynamo facility. *Physical Review Letters*, **84**: 4365–4368.

Gailitis, A., Lielausis, O., Platacis, E., Gerbeth, G., and Stefani, F., 2001. On the results of the Riga dynamo experiments. *Magnetohydrodynamics*, **37**: 71–79.

Hollerbach, R., 1991. Parity coupling in α^2-dynamos. *Geophys Astrophys. Fluid Dyn.*, **60**: 245–260.

Hoyng, P., Ossendrijver, M.A.J.A., and Schmitt, D., 2001. The geodynamo as a bistable oscillator. *Geophys. Astrophys. Fluid Dyn.*, **94**: 263–314.
Hoyng, P., Schmitt, D., and Ossendrijver, M.A.J.H., 2002. A theoretical analysis of the observed variability of the geomagnetic dipole field. *Physics of the Earth and Planetary Interiors*, **130**: 143–157.
Kono, M., and Roberts, P. H., 2002. Recent geodynamo simulations and observations of the geomagnetic field. *Reviews of Geophysics*, **40**(4): 1–53.
Krause, F., and Rädler, K.-H., 1980. *Mean-Field Magnetohydrodynamics and Dynamo Theory.* Berlin: Akademie-Verlag; Oxford: Pergamon Press.
Moffatt, H.K., 1978. *Magnetic Field Generation in Electrically Conducting Fluids.* Cambridge: Cambridge University Press.
Müller, U., and Stieglitz, R., 2000. Can the Earth's magnetic field be simulated in the laboratory? *Naturwissenschaften*, **87**: 381–390.
Parker, E.N., 1955. Hydromagnetic dynamo models. *Astrophysical Journal*, **122**: 293–314.
Parker, E.N., 1957. The solar hydromagnetic dynamo. *Proceedings of the National Academy of Sciences*, **43**: 8–14.
Rädler, K.-H., 1975. Some new results on the generation of magnetic fields by dynamo action. *Memoirs of the Society Royal Society Liege*, **VIII**: 109–116.
Rädler, K.-H., 1980. Mean-field approach to spherical dynamo models. *Astronomische Nachrichten*, **301**: 101–129.
Rädler, K.-H., 1986. Investigations of spherical kinematic mean-field dynamo models. *Astronomische Nachrichten*, **307**: 89–113.
Rädler, K.-H., 1995. Cosmic dynamos. *Reviews of Modern Astronomy*, **8**: 295–321.
Rädler, K.-H., 2000. The generation of cosmic magnetic fields. In Page, D., and Hirsch, J. G. (ed), *From the Sun to the Great Attractor (1999 Guanajuato Lectures in Astrophysics.).* Springer Lecture Notes in Physics, pp. 101–172.
Rädler, K.-H., Apstein, E., Rheinhardt, M., and Schüler, M., 1998. The Karlsruhe dynamo experiment—a mean-field approach. *Studia geophysica et geodaetica*, **42**: 224–231.
Rädler, K.-H., Rheinhardt, M., Apstein, E., and Fuchs, H., 2002a. On the mean-field theory of the Karlsruhe dynamo experiment. *Nonlinear Processes in Geophysics*, **9**: 171–187.
Rädler, K.-H., Rheinhardt, M., Apstein, E., and Fuchs, H., 2002b. On the mean-field theory of the Karlsruhe dynamo experiment. I. Kinematic theory. *Magnetohydrodynamics*, **38**: 41–71.
Rädler, K.-H., Rheinhardt, M., Apstein, E., and Fuchs, H., 2002c. On the mean-field theory of the Karlsruhe dynamo experiment. II. Back-reaction of the magnetic field on the fluid flow. *Magnetohydrodynamics*, **38**: 73–94.
Rädler, K.-H., Wiedemann, E., Brandenburg, A., Meinel, R., and Tuominen, I., 1990. Nonlinear mean-field dynamo models: Stability and evolution of three-dimensional magnetic field configurations. *Astronomy and Astrophysics*, **239**: 413–423.
Roberts, G.O., 1970. Spatially periodic dynamos. *Philosophical Transactions of the Royal Society of London A*, **271**: 411–454.
Roberts, G.O., 1972. Dynamo action of fluid motions with two-dimensional periodicity. *Philosophical Transactions of the Royal Society of London A*, **271**: 411–454.
Roberts, P.H., and Glatzmaier, G.A., 2000. Geodynamo theory and simulations. *Reviews of Modern Physics*, **72**: 1081–1123.
Rüdiger, G., 1980. Rapidly rotating α^2-dynamo models. *Astronomische Nachrichten*, **301**: 181–187.
Rüdiger, G., and Kichatinov, L.L., 1993. Alpha-effect and alpha-quenching. *Astronomics and Astrophysics*, **269**: 581–588.
Rüdiger, G., and Elstner, D., 1994. Non-axisymmetry vs. axisymmetry in dynamo-excited stellar magnetic fields. *Astronomics and Astrophysics*, **281**: 46–50.
Schmitt, D., Ossendrijver, M.A.J.H., and Hoyng, P., 2001. Magnetic field reversals and secular variation in a bistable geodynamo model. *Physics of Earth and Planetary Interiors*, **125**: 119–124.
Schrinner, M., Rädler, K.-H., Schmitt, D., Rheinhardt, M., and Christensen, U., 2005. Mean-field view on rotating magnetoconvection and a geodynamo model. *Astronomische Nachrichten*, **326**: 245–249.
Schrinner, M., Rädler, K.-H., Schmitt, D., Rheinhardt, M., and Christensen, U., 2006. Mean-field view on geodynamo models. *Magnetohydrodynamics*, **42**: 111–122.
Steenbeck, M., and Krause, F., 1969. Zur Dynamotheorie stellarer und planetarer Magnetfelder. II. Berechnung planetenähnlicher Gleichfeldgeneratoren. *Astronomische Nachrichten*, **291**: 271–286.
Steenbeck, M., Krause, F., and Rädler, K.-H., 1966. Berechnung der mittleren Lorentz-Feldstärke $\overline{v \times B}$ für ein elektrisch leitendes Medium in turbulenter, durch Coriolis-Kräfte beeinflußter Bewegung. *Zeitschrift für Naturforschung*, **21a**: 369–376.
Stieglitz, R., and Müller, U., 2001. Experimental demonstration of a homogeneous two-scale dynamo. *Phys. Fluids*, **13**: 561–564.
Stieglitz, R., and Müller, U., 2002. Experimental demonstration of a homogeneous two–scale dynamo. *Magnetohydrodynamics*, **38**: 27–33.
Zeldovich, Ya. B., Ruzmaikin, A.A., and Sokoloff, D.D., 1983. *Magnetic Fields in Astrophysics. The Fluid Mechanics of Astrophysics and Geophysics*, Vol. 3, New York, London, Paris, Montreux, Tokyo: Gordon and Breach Science Publishers.

Cross-references

Antidynamo and Bounding Theorems
Cowling's Theorem
Dynamo, Braginsky
Dynamo, Model-Z
Dynamo, Solar
Dynamos, Experimental
Dynamos, Kinematic
Dynamos, Periodic
Geodynamo
Geodynamo, Numerical Simulations
Magnetohydrodynamics
Westward Drift

DYNAMOS, PERIODIC

A periodic dynamo sustains a magnetic field by fluid flow that repeats periodically either in space or time. Consider *spatially periodic dynamos* first. Their flows are best described in terms of a lattice. A simple, two-dimensional flow that repeats in both y and z coordinates in shown on the left in Figure D32. The flow consists of rolls confined to rectangular cells. The roll structure by itself does not generate a magnetic field (Busse, 1973); dynamo action does result, however, with the addition of a shear flow in the x-direction. This flow was investigated by G.O. Roberts in his PhD thesis of 1969, with supervisor H.K. Moffatt. He showed that most periodic flows are capable of generating magnetic fields. This came at a time when very few examples of homogeneous dynamo action was known, and was an important step forward toward our present view that almost any sufficiently complicated and vigorous flow will generate magnetic fields. Of more lasting consequence, probably, was the use of the mathematical techniques that allowed G.O. Roberts to prove dynamo action.

Dynamo action only occurs for sufficiently large magnetic Reynolds number, $R_m = \mu_0 lV\sigma$ (see *Geodynamo, dimensional analysis and timescales*), where l is the length scale, V a measure of the velocity magnitude, σ the electrical conductivity, and μ_0 the permeability. By their very definition, periodic flows are infinite in extent. The only length scale is therefore the repeat wavelength of the lattice, l. Since dynamo action depends only on R_m, only the product $l\sigma$ is relevant. This allowed Roberts (1972a) to claim dynamo action at almost all values of the conductivity: reduced conductivity can be compensated for by simply increasing the length scale of the flow.

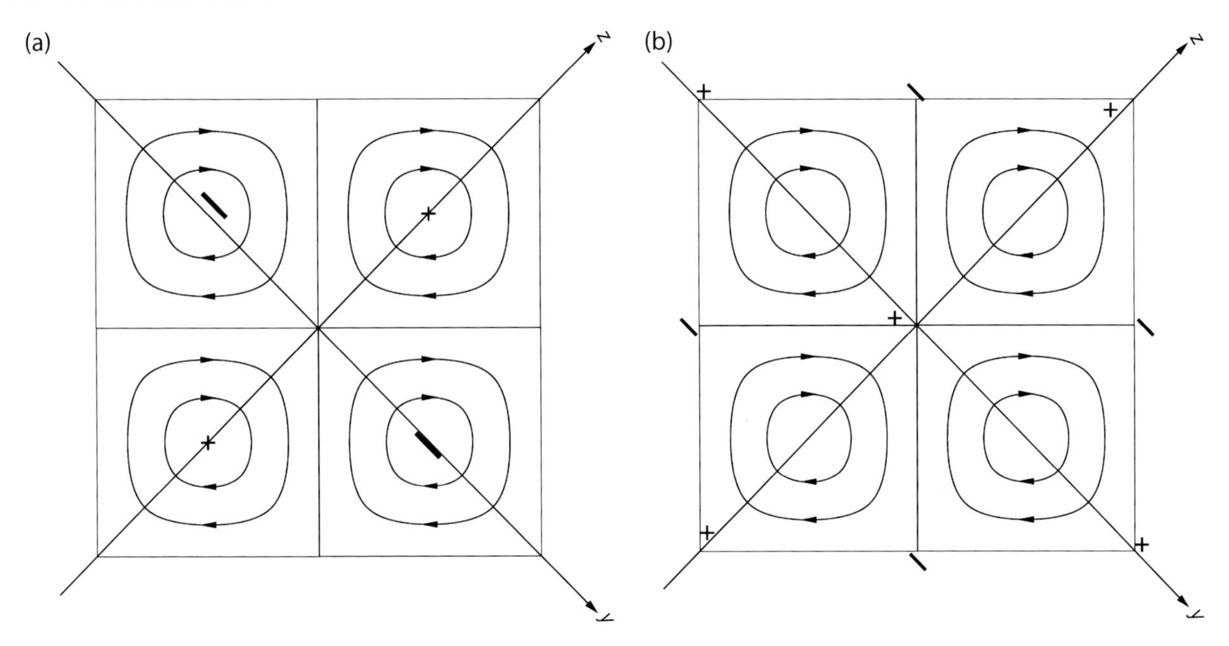

Figure D32 Two simple periodic motions that generate magnetic fields. *Left:* v = (cos*y*−cos*z*, sin*z*, sin*y*); signs +/− indicate sign of *x*-component. The streamlines are all right-handed helices, the flow is maximally helical. *Right:* v = (2 cos*y* cos*z*, sin*z*, sin*y*). The shear flow in the *x*-direction is now maximal at the corners of the cells. This flow has zero net helicity; the streamlines are closed.

Spatially periodic dynamos operate in two steps. Starting from a magnetic field whose length scale is much greater than the period of the flow's lattice, $\boldsymbol{B}_0$, the flow induces a magnetic field $\boldsymbol{B}'$, with the same length scale as the lattice, l. Action of the same flow on $\boldsymbol{B}'$ reinforces the original field by induction. At first it seems surprising that small-scale flow can produce a large-scale field from $\boldsymbol{B}'$; it arises from each roll twisting the field into the same direction in each lattice cell. Mathematically, the induction arises from $\boldsymbol{v} \times \boldsymbol{B}$, and the average $\boldsymbol{v}' \times \boldsymbol{B}'$ contributes to a large-scale field. The simple trigonometric formula $\cos^2 kx = 0.5(1 + \cos 2kx)$ shows that products of small-scale quantities produce large-scale quantities (the 0.5) as well as small-scale ones (the $0.5 \cos 2kx$). Here $k = 2\pi/l$ may be thought of as the wavenumber for the flow.

The mathematical theory for periodic dynamos was initiated by Childress (1969) and developed further by Roberts (1972a). It involves the solution of differential equations with periodic coefficients. Floquet's theorem is central to the treatment of these equations; it states that the solution is a periodic function, with the same period as the coefficients, in this case the period of the flow, multiplied by an exponential factor $\exp iqx$. For a 3-dimensional solution the factor is $\exp i\boldsymbol{j} \cdot x$. The Floquet parameter, q, is a wavenumber ($\boldsymbol{j}$ is a wavevector) that must be found by solving the equations, just as an eigenvalue must be found by solving the characteristic equation of a matrix. It defines a second length scale, $L = 2\pi/q$. In three dimensions the theory is analogous to that of Bloch waves for electrons in solids (Gubbins, 1974). The periodic part of the solution may be expanded in an appropriate Fourier series, which converts the differential equation to a set of algebraic equations. These algebraic equations are then solved for the Fourier coefficients, which provides the complete solution.

The induction term $\boldsymbol{v} \times \boldsymbol{B}$ involves the product of two periodic functions, which transforms into a sum involving all the Fourier coefficients for $\boldsymbol{B}$. This sum is greatly simplified when there is a large disparity of length scales L and l. The general solution is developed as an expansion in the small parameter $\varepsilon = l/L$ (Roberts, 1972a). Further simplification is obtained when the magnetic Reynolds number is small, when each Fourier coefficient is related only to the one corresponding Fourier coefficient in the expansion for the velocity. This is equivalent to the "first order smoothing" or FOST approximation in turbulent dynamo theory (Krause and Rädler, 1980). R_m is based on the small length scale l and is called the *microscale* magnetic Reynolds number. For dynamo action the magnetic Reynolds number must be sufficiently large (see *Antidynamo and bounding "theorems"*). In this case the relevant magnetic Reynolds number is not the microscale R_m but one based on the long length scale, $R_m^L = \mu_0 \sigma V L$. This condition is satisfied provided l/L is sufficiently small.

In his first paper, Roberts (1972a) presents the general theory in three dimensions, including periodicity in time as well as space, and concludes that *in a precisely defined sense, almost all steady spatially periodic motions of a homogeneous conducting fluid will give dynamo action. . . .* The "almost all" here is used in the mathematical sense, that the set of motions that fail to give dynamo action is not empty but has zero measure. Unfortunately, as often seems to be the case in dynamo theory, the set contains many of the simple flows that one thinks of first, such as flow on the right in Figure D32 (see *Antidynamo and bounding theorems*).

In his second paper, Roberts (1972b) concentrates on flows with two-dimensional periodicity (not to be confused with 2D flows) and gives four important illustrative motions. The first is given on the left of Figure D32. The equation governing the mean field contains instead of the velocity the mean helicity, or first Fourier coefficient of $\alpha = \boldsymbol{v} \cdot \nabla \times \boldsymbol{v}$. The mean field equation is analogous with that derived in *mean field dynamo theory* (*q.v.*). The mean field itself is a spiral $\boldsymbol{B} = B_0(0, \cos jx, \sin jx)$. The helical streamlines of the flow twist and push or pull the field lines in adjacent cells in such a way as to reinforce the mean fields at different heights x. Helicity is essential to the process, but what sustains the mean field is the mean helicity averaged over a lattice cell, not the point values of the helicity itself.

The remaining three illustrative flows chosen by Roberts (1972b) all fail as dynamos to first order in the expansion in l/L. These flows have zero mean helicity. One of these is sketched on the right in Figure D32. The shear flow along x is now at a maximum at the corners of the lattice cells rather than in the center. Roberts (1972b) was able to demonstrate numerically that all three flows do give dynamo action. Helicity is sometimes thought to be essential for dynamo action, but Roberts' example shows that this is not the case.

The theory of periodic dynamos is quite similar to that of turbulent dynamo action: it involves small- and large-scale fields, uses approximations that take advantage of the disparity of length scales and weak small-scale fields (FOST), and involves an average over the small-scales. It was developed independently from, but somewhat later than, the turbulence theory of Steenbeck *et al.* (1966). It has the advantage of mathematical rigor because the averaging process is well defined. In periodic dynamo theory the average is taken over the specified lattice cell of the flow, whereas turbulence theory requires an average over an ill-defined volume or measure of an ensemble of realizations of the flow.

Apart from providing an early demonstration of the near-universality of dynamo action, the theory of periodic dynamos provided the basis for two important extensions: application to motions in a sphere, and dynamo action by motions driven by thermal convection.

Childress (1970) considered a flow in a sphere by wrapping many cells of the flow in Eq. (1) around the ϕ-axis. The streamlines are helices wrapped around the axis of the coordinate system. He showed that dynamo action occurred with the same mechanism as for the cartesian flow provided the sphere contained a sufficiently large number of cells, when the boundary conditions at the sphere's surface had little effect. Numerical examples of dynamo action by similar flows were demonstrated by Roberts (1969), Gubbins (1972, 1973), and Dudley and James (1989). These solutions are particularly easy to compute because the flow is axisymmetric and the magnetic field has modes proportional to $\exp \mathrm{i}m\phi$. Cowling's theorem rules out dynamos with $m = 0$ but solutions are possible with $m = 1$ and higher. The numerical problem becomes essentially two-dimensional. Gubbins (1972) was able to investigate the effect of a small number of cells within the sphere, and finally obtained a solution with just one cell in each hemisphere (Bullard and Gubbins, 1972).

Convection in a plane layer, under certain conditions, takes the form of periodically repeating rolls. The theory of periodic dynamos can therefore be used to explore generation of magnetic field by convection. Busse (1973) analyzed dynamo action of Bénard convection between two perfectly conducting parallel plates at infinite Prandtl number. Dynamo action is not possible for simple rolls, and Busse had to introduce a shear flow to mimic G.O. Roberts' periodic flow. He obtained dynamo action and was able to determine the final equilibrium value of the generated field. Meanwhile Childress and Soward (1972) examined dynamo action by Bénard convection with rotation.

Busse (1975) finally developed a model of the geodynamo by considering a set of rolls aligned with the rotation axis in a spherical annulus. The flow relies on the same dynamo properties of the G.O. Roberts example in Figure D32. The flow along the rolls, required to provide helicity for efficient dynamo action, was effected by the sloping spherical boundary at the ends of the rolls. This dynamo model was the earliest attempt to represent the Earth's dynamo with full dynamics, and it succeeded in generating a predominantly-axial dipole field. It is a "weak field dynamo" (see *Equilibration of magnetic field*) in the sense that the generated magnetic field has little effect on the basic form of the flow. This, and the small-scales implied by the low viscosity in the core, make it difficult to reconcile with the Earth's heat flux and other observables. Numerical models of the geodynamo have developed from Busse's original ideas.

The same mathematical approach can be applied to fluid motions that are periodic in time. Roberts (1972a) considered motions periodic in time as well as space. In some circumstances, oscillatory flows can be more effective in generating magnetic field than similar stationary flows (Willis and Gubbins, 2004). In 1971 Higgins and Kennedy (see *Higgins-Kennedy paradox*) raised the possibility that the core might be subadiabatic, in which case convection could not occur. The only allowed radial motions, essential for dynamo action, would be in the form of waves with frequency dominated by the degree of temperature stratification. Just as the theory of spatially periodic dynamos utilized expansions in the ratio of length scales, so the theory of temporally periodic dynamos employed a ratio of timescales. Bullard and Gubbins (1971) showed that dynamo action is less efficient when the flow oscillates with periods short compared with the magnetic decay time. They used a disk dynamo model in which the disks oscillated rather than rotated, but the same general result applies to fluid dynamos. Higgins and Kennedy's stratification required motions with very short periods indeed, a few hours to days and would therefore be very inefficient in generating magnetic field: the core must therefore convect somehow. Current views include the scenario in which part of the core is stratified, with the deeper part convecting (see *Geodynamo, energy sources*).

In summary, the early papers on periodic dynamos provided essential confirmation that the dynamo theory was viable and, because dynamo action seemed almost universal for this particular class of flows, providing confidence that realistic geodynamo models could be found. Perhaps the most lasting legacy is the mathematical technique used for the kinematic studies in an infinite medium, which could be extended to apply to dynamical models, spherical geometry, and periodicity in time.

David Gubbins

Bibliography

Bullard, E.C., and Gubbins, D., 1971. Geomagnetic dynamos in a stable core. *Nature*, **232**: 548–549.

Bullard, E.C., and Gubbins, D., 1972. Oscillating disc dynamos and geomagnetism. In Heard, H.C., Borg, I.Y., Carter, N.L., and Raleigh, C.B. (eds.), *Flow and Fracture of Rocks* American Geophysical Union Geophysical Monograph 16.

Busse, F.H., 1973. Generation of magnetic fields by convection. *Journal of Fluid Mechanics*, **57**: 529–544.

Busse, F.H., 1975. A model of the geodynamo. *Geophysical Journal of the Royal Astronomical Society*, **42**: 437–459.

Childress, S., 1969. *Théorie magnétohydrodynamique de l'effet dynamo*, Dep. Mech. Fac. Sci., Paris, reported in *Roberts and Gubbins* [1987].

Childress, S., 1970. New solutions of the kinematic dynamo problem. *Journal of Mathematical Physics*, **11**: 3063–3076.

Childress, S., and Soward, A.M., 1972. Convection-driven hydromagnetic dynamo. *Physics Review Letters*, **29**: 837–839.

Dudley, M.L., and James, R.W., 1989. Time-dependent dynamos with stationary flows, *Proceedings of the Royal Society of London, Series A*, **425**: 407–429.

Gubbins, D., 1972. Kinematic dynamos and geomagnetism. *Nature*, **238**: 119–121.

Gubbins, D., 1973. Numerical solutions of the kinematic dynamo problem. *Philosophical Transaction of the Royal Society of London, Series A*, **274**: 493–521.

Gubbins, D., 1974. Dynamo action of isotropically driven motions of a rotating fluid. *Studies in Applied Mathematics*, **53**: 157–164.

Krause, F., and Rädler, K.-H., 1980. *Mean-field Magnetohydrodynamics and Dynamo Theory*, Pergammon Press.

Roberts, G.O., 1969. Dynamo waves. In Runcorn, S.K. (ed), *The Application of Modern Physics to the Earth and Planetary Interiors*. Wiley Interscience, pp. 603–628.

Roberts, G.O., 1972. Spatially periodic dynamos. *Philosophical Transaction of the Royal Society of London, Series A*, **266**: 535–558.

Roberts, G.O., 1972. Dynamo action of fluid motions with two-dimensional periodicity. *Philosophical Transaction of the Royal Society of London, Series A*, **271**: 411–454.

Steenbeck, M., Krause, F., and Rädler, K.-H., 1966. A calculation of the mean emf in an electrically conducting fluid in turbulent motion under the influence of coriolis forces, *Z. Naturforsch.*, **21**: 369–376.

Willis, A.P., and Gubbins, D., 2004. Kinematic dynamo action in a sphere: effects of periodic time-dependent flows on solutions with axial dipole symmetry. *Geophysical and Astrophysical Fluid Dynamics*, **98**: 537–554.

Cross-references

Antidynamo and Bounding Theorems
Cowling's Theorem
Dynamos, Kinematic
Dynamos, Mean Field
Equilibration of Magnetic Field, Weak and Strong
Geodynamo
Geodynamo, Dimensional Analysis and Timescales
Geodynamo, Energy Sources
Geodynamo, Numerical Simulations
Higgins-Kennedy Paradox
Magnetohydrodynamics

DYNAMOS, PLANETARY AND SATELLITE

Large magnetic fields in planets and their satellites are thought to arise from dynamo generation, the same process responsible for Earth's magnetic field. A wealth of recent data, mainly from the Galilean satellites of Jupiter and the planet Mars, together with major improvements in our theoretical modeling effort of the dynamo process, have allowed a significant increase in our understanding. However, it is still not possible to state with confidence why only some planets and large satellites have dynamos. A major issue is whether convection can be sustained in a region of adequate electrical conductivity. This depends on the composition and evolution of planets. Even if convection is present, there are criteria that must be satisfied for a dynamo. These have both purely dynamical and energetic aspects. This article concerns all the observed fields, but with emphasis on those that appear to require a dynamo process.

These dynamos arise from thermal or compositional convection in fluid regions of large radial extent in which the electrical conductivity exceeds some value that may be only a few percent that of a typical metal. A significant Coriolis effect on the fluid motions is needed, but all planets rotate sufficiently fast. Although other sources of fluid motion can in principle be relevant, it is thought likely that convection is the dominant source. The maintenance and persistence of convection appears to be easy in gas giants and ice-rich giants, but is not assured in terrestrial planets because of the quite high electrical (and hence thermal) conductivity of iron-rich cores, which allows for a large core heat flow by conduction alone. High electrical conductivity may be unfavorable for a dynamo because it implies high thermal conductivity. If a dynamo operates, the expected field amplitude is plausibly $\sim(2\rho\Omega/\sigma)^{1/2}$ tesla where ρ is the fluid density, Ω is the planetary rotation rate, and σ is the conductivity (SI units). However, dynamo theory may admit solutions with smaller fields. Earth, Ganymede, Jupiter, Saturn, Uranus, and Neptune appear to have fields that are at least roughly consistent with the expectations of dynamo theory. Mercury may also have a dynamo although its field seems anomalously small. Mars has large remanent magnetism from an ancient dynamo, and the Moon might also require an ancient dynamo. Venus is devoid of a detectable global field but may have had a dynamo in the past. The presence or absence of a dynamo in a terrestrial body (including Ganymede) appears to depend mainly on the thermal histories and energy sources of these bodies, especially the convective state of the silicate mantle and the existence and history of a growing inner solid core. Induced fields observed in Europa and Callisto indicate the strong likelihood of water oceans in these bodies.

The significance of these magnetic fields

There are four reasons to be interested in these fields:

1. When a planet or satellite has a large global field, it provides us with insight into the state of matter and the dynamics deep down. There is currently no other way to do this.
2. When a planet has remanent magnetism of near surface rocks, it may tell us about the past behavior of the global field (see *Geomagnetic reversals*) and geological activity. Plate tectonics was deduced primarily from paleomagnetism. The history of the field may also affect climate (by modulating atmospheric escape) and the evolution of life.
3. When a body has an induced field, its magnitude and phase tells us about the body's conductivity structure.
4. Dynamo field generation is a nonlinear chaotic process whose dynamics are of interest in their own right (as a fundamental and very difficult problem in complex systems).

Observations

Magnetic fields, unlike electric fields, do not come from monopoles but from an electrical current, or from the fundamental magnetic moments of elementary particles. In everyday experience, substantial fields arise either from *permanent magnets* where the magnetism arises at the microscopic level and is a thermodynamic property of the material, or through *macroscopic currents* in electrical conductors (e.g., as in a Helmholtz coil). Permanent magnetism is a satisfactory explanation for modest amounts of observed magnetism in solid bodies (e.g., Moon, Mars, and maybe even Mercury), but it requires low temperature (outer regions only of a planet) and it requires an adequate abundance of the minerals that exhibit permanent magnetization (e.g., magnetite, metallic iron). Typically, crustal magnetism has a coherence length small compared to the planet radius, so no large global field arises. On Earth, permanent magnetization accounts for typically 0.1% or less of the observed field. *Localized* fields of up to Earth's global field ($\sim10^{-4}$ T) are possible from permanently magnetized materials; this happens rarely on Earth but may be common in the southern hemisphere of Mars. On Earth, we have a much stronger argument for something else: The field is dynamic (time varying on all timescales from years to billions of years.) This field is generated in Earth's conducting core by a process known as a dynamo and involves very large-scale electrical currents. A similar process operates in many large cosmic bodies including the Sun.

By Faraday's law, a planetary body can also have an "internal" field that is induced by a time-variable external field. These eddy currents and associated fields can be identified by their distinctive time variability, phase, and amplitude. On Earth, these are called *magnetotelluric* currents and fields. They are also observed for the Moon, Europa, and Callisto.

Except for the special case of Jupiter, which is a synchrotron source of radio waves, we learn about planetary magnetic fields by the direct detection of the field (the magnetosphere) from a flyby or orbiter spacecraft. Orbital data are preferred (even for Earth), provided you can measure or get below the effects of an ionosphere. The observations, with likely interpretations, are given below in Table D1. For the large satellites embedded in giant planet magnetospheres, the quoted values have the external field subtracted.

The nature of planets and satellites

Planets and satellites are conveniently categorized according to their primary constituents. Distinguishing between planets and their satellites is artificial, at least for questions pertaining to their evolution and magnetic fields, since satellites are subject to the same planetary processes if they are sufficiently large (>1000 km radius, roughly). *Terrestrial planets* (Mercury, Venus, Earth, Moon, Mars, and Io) consist primarily of materials that condense at high temperatures: oxides and silicates of iron and magnesium, together with metallic iron. The high density and lower melting point of iron alloys relative to silicates generally lead us to expect that these bodies form metallic iron-rich cores. These cores are generally at least partially liquid, even after 4.5 billion years of cooling, because at least one of the core-forming constituents (sulfur) lowers the

Table D1 Observed magnetic fields (see also Russell, 1993 and Connerney, 1993)

Planet or Satellite	Observed Surface Field (in tesla, approximate)	Comments and Interpretation
Mercury	2×10^{-7}	Not well characterized or understood
Venus	$<10^{-8}$ (global); no useful constraint on local fields	No dynamo. Small remanence might exist (but not yet detected).
Earth	5×10^{-5}	Core dynamo
Moon	Patchy; no global fields	Impact generated? Ancient dynamo?
Mars	Patchy but locally strong; no global field	Ancient dynamo, Remanent magnetic lineations and patches.
Jupiter	4.2×10^{-4}	Dynamo (extends to near-surface)
Io[a]	$<10^{-6}$?	Complex (deeply embedded in Jovian field.)
Europa[a]	10^{-7}	Induction response (Salty water ocean)
Ganymede[a]	2×10^{-6}	Dynamo likely
Callisto[a]	4×10^{-9}	Induction response (Salty water ocean)
Saturn	2×10^{-5}	Dynamo
Titan[b]	$<10^{-7}$	No evidence for Ganymede-like dynamo
Uranus	2×10^{-5}	Dynamo
Neptune	2×10^{-5}	Dynamo

[a]All the Galilean satellites are embedded in Jupiter's field and this poses difficulties for determining the fields associated with the satellites, except in the case of Ganymede which has its own magnetosphere. Listed fields for the Galilean satellites are based on downward continuations of the field measured at the spacecraft altitude. In the case of Io there is a major difficulty in this process because the external field from Jupiter is so large. As a consequence, Io's intrinsic field is highly uncertain and even the upper bound is somewhat uncertain.

[b]Titan spends much of its time in the magnetosphere of Saturn. The nearly spin-axisymmetric character of Saturn's magnetic field makes the detection of an induction response more difficult. As of mid-2006, there is no evidence of an induction response in the several flybys of Titan by the Cassini spacecraft.

freezing point of the iron alloy below the operating (convecting) temperature of the overlying mantle. If the sulfur content is small then the fluid region of a core may be thin. *Gas giants* (Jupiter and Saturn) have hydrogen as their major constituent. They may possess "Earthlike" central cores but this may have little bearing on understanding their magnetic fields. *Ice giants* (Uranus and Neptune) contain a hydrogen-rich envelope but their composition is rich in H_2O, CH_4, and NH_3 throughout much of the volume, extending out to perhaps ~80% of their radii. *Large icy satellites and solid icy planets* (Ganymede, Callisto, Titan, Triton, Pluto; also Europa as a special case) contain both ice (predominantly H_2O) and rock. They may be differentiated into an Earthlike structure (silicate rock and possibly an iron-rich core), overlain with varying amounts of primarily water ice, or (as in the case of Callisto) the ice and rock may be partly mixed. Europa is a special case because the water-rich layer is relatively small and may be mostly liquid.

Planets differ from small masses of the same material because of the action of gravity and the difficulty of eliminating heat on billion year timescales. Gravity causes pressure, which can modify the thermodynamic and phase equilibrium behavior of the constituents. This is why bodies rich in materials that are poor conductors at low pressures (e.g., hydrogen, water) may nonetheless have high conductivity at depth. The difficulty of eliminating the heat of formation and subsequent radioactive heat generation leads to unavoidably large internal temperatures, frequently sufficient to guarantee fluidity of a deep conducting region, and often sufficient to guarantee sustained convection.

The geometry of large fields

External to the planet and the large currents responsible for most of the field, the magnetic field **B** can be written as the gradient of a scalar potential that satisfies Laplace's equation. We can identify general solutions to Laplace's equation in terms $\propto Y_{lm}r^{-(l+1)}$ where Y_{lm} is a spherical harmonic, r is the distance from the center of the planet $l = 1$ is the dipole, $l = 2$ is the quadrupole and so on. Terms with $m = 0$ represent spin-axisymmetric components (if we choose the pole of coordinates to be the geographically defined pole of planet rotation), so (for example) $l = 1$ and $m = \pm 1$ represents the tilt of the dipole and the longitude of that tilted dipole. Planetary fields are sometimes described as "tilted, offset dipoles" but this is misleading at best. There is no fundamental significance to a dipole: A current distribution of finite extent will typically produce many additional harmonics. It is nonetheless true that many bodies have fields that are predominantly dipolar, in the sense that the quadrupolar component is significantly smaller than the dipole component, even when evaluated at the core radius. For *Earth, Jupiter, and Saturn* (and probably *Ganymede*, maybe also *Mercury*), the field is predominantly dipolar. The tilt of the dipole relative to the rotation axis is of order 10 degrees for Jupiter and Earth and near-zero for Saturn. For *Uranus and Neptune*, the field is about equally dipole and quadrupole and the tilt of the dipole is 40–60 degrees. Evidently, Uranus and Neptune represent a different class of dynamos.

Large magnetic fields require energy sources

Ohm's law, Ampere's law, and Faraday's law of induction lead to what is often called the *dynamo equation*:

$$\partial \mathbf{B}/\partial t = \lambda \nabla^2 \mathbf{B} + \nabla \mathbf{x}(\mathbf{v} \times \mathbf{B}) \qquad \text{(Eq. 1)}$$

where **B** is the magnetic field, **v** is the fluid motion (relative to a rotating frame of reference) and $\lambda \equiv 1/\mu_0\sigma$ is known as the *magnetic diffusivity* (μ_0 is the permeability of free space, $4\pi \times 10^{-7}$ NA^{-2} (newtons/(Ampere)^2) and σ is the electrical conductivity in S m^{-1} (siemens/meter), and assumed constant). If there is no fluid motion then the field will undergo free ("diffusive") decay on a timescale $\tau \sim L^2/\pi^2\lambda \sim$ (3000 year). $(L/1000\ \text{km})^2 \cdot (1\ \text{m}^2 \cdot \text{sec}^{-1}\ \lambda^{-1})$ where L is some characteristic length scale of the field, no more than the radius of the electrically conducting region (the core). In terrestrial planets, the electrical conductivity corresponds to liquid metallic iron, modified by alloying with other elements (e.g., sulfur). This corresponds to $\sigma \sim 5 \times 10^6$ S m^{-1} and $\lambda \sim 2$ m^2 s^{-1}. In gas giants, shock wave experiments suggest that hydrogen attains the lowest conductivities appropriate to metals ($\sigma \sim 2 \times 10^5$ Sm^{-1}, $\lambda \sim 20$ to 50 m^2s^{-1}) at pressure $P \sim 1.5$ Mbar and $T \sim$ a few thousand degrees. This corresponds to the conditions at 0.8 of Jupiter's radius or 0.5 of Saturn's radius. Shock wave experiments suggest that an "ice" mixture (dominated by water) will reach conductivities of $\sigma \sim 1 \times 10^5$ S m^{-1} ($\lambda \sim 100$ m^2 s^{-1}), conditions met in Uranus and Neptune at around 0.7 of their radii.

In all cases, the free decay time is much less than the age of the solar system. For example, in Earth's core, this timescale is 10,000 years or so. The fact that free decay times are geologically short means that if a planet has a large field *now* then it must have a means of generating the field *now*; it cannot rely on some primordial field or preexisting field.

The dynamo mechanism

The essence of a dynamo lies in electromagnetic induction: The creation of emf and associated currents and field through the motion of conducting fluid across magnetic field lines (Moffatt, 1978; Parker, 1979). Dimensional analysis of the induction equation immediately suggests that the importance of this is characterized by the *magnetic Reynolds number* $R_m \equiv vL/\lambda$ where v is a characteristic fluid velocity and L is a characteristic length scale of the motions or field (e.g., the core radius). Numerical and analytical work suggest that a dynamo will exist if the fluid motions have certain desired features and the magnetic Reynolds number R_m exceeds about 10 or 100. It seems likely that fluid motions of the desired character arise naturally in a convecting fluid (irrespective of the source of fluid buoyancy), provided the Coriolis force has a large effect on the flow, i.e., $v/\Omega L < 1$ where Ω is the planetary rotation rate. Taking into account both criteria, a necessary but not sufficient criterion for a dynamo is that the free decay time of the field be very much larger than the rotation period. This is easily satisfied for any plausible fluid motion of interest, even for slowly rotating planets such as Venus, where the rotation period is at least four orders of magnitude less than the field decay time.

We do not have quantitatively precise sufficient conditions for the existence of a planetary dynamo. Some of the issues can be appreciated by considering the simple case of a generic planet in which the heat flow in the proposed dynamo region arises primarily from cooling, and no phase changes (e.g., freezing or gravitational differentiation) take place. (In terrestrial planets, the dominant source of *surface* heat flow is radioactive decay, but the radioactive elements are not predominantly in the core. In giant planets, cooling from a primordial hot state probably dominates at all levels, though gravitational differentiation may also contribute significantly.) In this approximation, and assuming that the core cools everywhere at the about the same rate, we have

$$F_{\text{total}}(r) = -\rho_c C_p r (\mathrm{d}T_c/\mathrm{d}t)/3 \qquad \text{(Eq. 2)}$$

where $F_{\text{total}}(r)$ is the total heat flow at radius r, ρ_c is the mean core density, C_p is the specific heat, T_c is the mean core temperature, and t is time. In fluid cores, the viscosity is so small that it plays a negligible role in the criterion for convection (totally unlike the case for convection in solid silicate mantles). To an excellent approximation, the condition for convection is that the heat flow must exceed that which can be carried by conduction along an adiabat:

$$F_{\text{total}} > F_{\text{cond, ad}} \equiv k\alpha T g(r)/C_p \Leftrightarrow \text{thermal convection} \qquad \text{(Eq. 3)}$$

where k is the thermal conductivity, α is the coefficient of thermal expansion, and $g(r)$ is the gravitational acceleration at radius r. If the heat flow were less than this value then the core would be stably stratified (vertically displaced fluid elements would tend to oscillate). We can approximate $g(r)$ by $4\pi G\rho_c r/3$ where G is the gravitational constant. Notice that both F_{total} and $F_{\text{cond,ad}}$ are linear in r in this approximation, so the comparison of their magnitudes will be the same independent of planet size and location in the core. From this, we obtain a critical cooling rate that must be exceeded for convection. It is typically about 100 $\mathrm{K\,Ga^{-1}}$ for parameters appropriate to Earth's core and may be as large as 300 or 400 $\mathrm{K\,Ga^{-1}}$ for smaller (but Earth-like) cores. e.g., Ganymede, because α is larger at low pressures. It is substantially lower for giant gas or ice planets, where the conductivity is lower. For Earth's core, a cooling rate like 100 $\mathrm{K\,Ga^{-1}}$ corresponds to a heat flow at the top of the core of around 20 $\mathrm{mW\,m^{-2}}$.

From condensed matter physics, we also have the Wiedemann-Franz "law" that

$$k/\sigma T \equiv L \approx 2 \times 10^{-8}\ \mathrm{W \cdot Ohm\,K^{-2}} \qquad \text{(Eq. 4)}$$

where L is called the Lorenz number. This applies to a metal in which the electrons dominate both the heat and charge transport. Combined with Eq. (3) this implies an *upper* bound to the electrical conductivity in order that thermal convection takes place. For nominal parameter choices, this upper bound is roughly the actual value of the electrical conductivity in Earth's core. This makes the important point that high electrical conductivity may prevent a dynamo! See Figure D33.

Even if convection is possible, it must be sufficiently vigorous. Two possible estimates for convective velocity might be considered. One comes from mixing length theory:

$$V_{ml} \sim 0.3(lF_{\text{conv}}/\rho H_T)^{1/3} \qquad \text{(Eq. 5)}$$

where V_{ml} is the predicted velocity, l is the "mixing length" (plausibly the size of the core), $F_{\text{conv}} = F_{\text{total}} - F_{\text{cond.ad}}$, and $H_T \equiv C_p/\alpha g$ is the temperature scale height, not enormously larger than the core radius except in the limit of small bodies. An alternative estimate, plausibly more relevant if a dynamo is operating, assumes that buoyancy, Coriolis and Lorentz forces are comparable in the fluid flow. In this regime, sometimes referred to as the MAC regime,

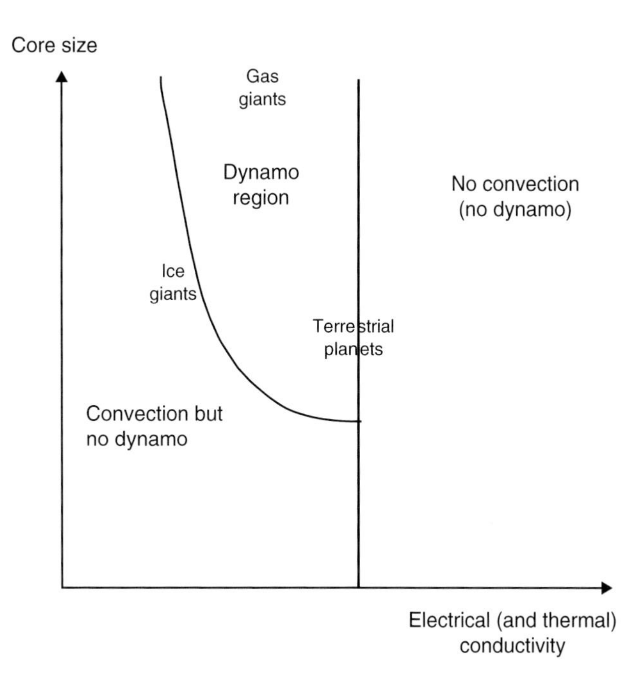

Figure D33 Dependence of dynamo operating region on planet size and electrical conductivity. This schematic diagram focuses on two of the many parameters that determine whether a thermal convection dynamo exists in a planet. At sufficiently low conductivity, there is a size dependent cut-off based on the need to exceed a critical magnetic Reynolds number. As explained in the text, there is also an approximately size-independent criterion for the existence of convection, since if the conductivity is too high then the heat can be carried by conduction. Specific values are omitted from the plot because they depend on many (often poorly known) parameters. However, it is likely that terrestrial planets lie near the right hand dynamo boundary, ice giants lie near the left hand boundary and gas giants are comfortably removed from either boundary. A more complicated but conceptually similar diagram will apply if there is compositional convection.

$$V_{MAC} \sim (F_{conv}/\rho\Omega H_T)^{1/2} \quad \text{(Eq. 6)}$$

and this is typically an order of magnitude or so smaller than V_{ml}. Note that slow rotation is favorable (i.e., increases convective velocity). The acronym MAC stands for magnetic Archimedean Coriolis and refers to a system where the dynamics of the magnetic field (the Lorentz force), the Archimedean force due to buoyancy, and the dynamic effect of rotation (the Coriolis force) are all important for the flow. It should be stressed that the particular form adopted in Eq. (6) is not the only possible way of introducing the effect of rotation, and that there are alternatives in which the dependence on rotation is weaker (though generally the power law is still a negative one, implying less vigorous convection for faster rotation, provided the heat flow is kept constant).

In the dynamo regime, the expected field magnitude *inside the region of field generation* is plausibly given by Elsasser number $\Lambda \equiv \sigma B^2/2\rho\Omega$ of order unity, which implies $B \sim (2\rho\Omega/\sigma)^{1/2}$ T where ρ is the fluid density. As Table D2 shows, this may be roughly satisfied, especially if one allows that the field inside the dynamo region may be larger than at the top of the dynamo region by a factor of a few. There are enough uncertainties that one cannot state with confidence that dynamos operate near an Elsasser number of unity. There appears to be a clear exception for Uranus and Neptune, although downward extrapolation of their fields is difficult because they are not predominantly dipolar.

The cooling rates of giant gas and ice planets can be estimated and predict heat flows that are large compared to conductive transport, and close to the "nominal" values tabulated. However, the situation is far from clear in terrestrial planets, where our understanding of cooling rates is dictated by our (imperfect) understanding of mantle rheologies and heat production. It is possible that the actual heat flow does not exceed the conductive heat flow. This can happen in a convecting fluid provided that there is also compositional buoyancy.

If the core is cooling and the central temperature drops below the liquidus for the core alloy, then an inner core will nucleate. In Earth, we know from seismic evidence that the core is ~10% less dense than pure iron and many suggestions have been offered for the identity of the light elements that are mixed with the iron. As the inner core freezes, it is likely that some or all of these light elements are excluded from the crystal structure. The introduction of light elements into the lowermost core fluid will tend to promote convection and cause mixing throughout all or most of the outer core, provided the cooling is sufficiently fast. Latent heat release at the inner core-outer core boundary will also contribute to the likelihood of convection. However, inner core growth permits outer core convection even when the heat flow through the core-mantle boundary is less (perhaps much less) than the heat carried by conduction along an adiabat. In this regime, the temperature gradient is very slightly less steep than adiabatic and the compositional convection carries heat *downwards*. The total heat flux is still outwards, of course, since the heat carried by conduction is large. This state is possible because the buoyancy release associated with the compositional change exceeds the work done against the unfavorable thermal stratification.

It is possible but not certain that terrestrial planets require inner core growth in order to sustain a dynamo at the present epoch. It does *not* follow that there is a one-to-one correspondence between presence of an inner core and presence of a dynamo. One can have an inner core without a dynamo (conceivably present Mars if the cooling of the core is insufficiently rapid). One can also imagine a dynamo without a growing inner core (conceivably early Earth or other bodies early in their history) if the core were then cooling much more rapidly than now. Partial freeze-out of light material from the core is also a possible dynamo driving mechanism.

Induction fields

The requirement for a significant induction field is much less restrictive than for a dynamo. The conductivity can be much smaller and it does not have to be in differential motion (e.g., it can be a solid). For an external field that varies as exp[$i\omega t$], and a thin, conducting shell of thickness d and radius R, there will be a large induction response if the electromagnetic skin depth $(\lambda/\omega)^{1/2} < (Rd)^{1/2}$. For example, a layer of low-pressure salty water (such as Earth's oceans, with $\lambda \sim 10^6$ m^2 s^{-1}) will satisfy this for a thickness of order 10 km and $\omega \sim 2 \times 10^{-4}$ (corresponding to the frequency of Jupiter's tilted dipole field as it sweeps by Europa). A plausible estimate for R_m in such an ocean is 10^{-3} so there is no significant *internal* induction effect. The observed fields of Europa, Callisto are consistent with an externally induced induction field, and the most likely conductor is salty water.

Summary for each planet

Mercury is likely to have a liquid outer core and some models predict that this core could continue to convect and perhaps sustain a dynamo. Mercury is nonetheless an enigma because the observed field is over an order of magnitude smaller than the field strength predicted for $\Lambda \sim 1$. There are at least four possibilities: permanent magnetism, an exotic nondynamo explanation such as thermoelectric currents, a dynamo that produces much larger internal (e.g., toroidal) fields than the observed external fields (as recent models suggest), or a dynamo that for some reason fails to reach the expected field amplitude Stanley *et al.* (2005). Future missions may test the alternatives through assessment of the harmonic structure of the field. The MESSENGER mission to Mercury was launched in 2004 and should collect data at the end of the decade.

Venus is likely to have a liquid outer core (with or without an inner core) but has no dynamo at present. The predicted dynamo field is over two orders of magnitude larger than the observational upper bound. Slow rotation may be good for dynamos, so if Venus were like Earth in all respects except for its rotation then it would have no difficulty exceeding this upper bound. The most probable interpretation is that the liquid core of Venus does not convect. This could arise because there is no inner core or because the core is currently not cooling. The absence of an inner core is plausible if the inside of Venus

Table D2 Operating conditions for representative dynamos

	Earth	Ganymede	Jupiter	Uranus
Ω, rotation rate (s^{-1})	7×10^{-5}	1×10^{-5}	2×10^{-4}	1.4×10^{-5}
Density (kg m^{-3})	1.1×10^4	6×10^3	1×10^3	1×10^3
Size of dynamo region (m)	3×10^6	7×10^5	3×10^7	$\sim 1 \times 10^7$
H_T, temperature scale height (m)	1×10^7	4×10^7	1×10^8	1×10^7
Conductive heat flow along adiabat (Wm^{-2})	1.5×10^{-2}	1×10^{-3}	$<10^{-1}$	$<10^{-2}$
Nominal convective Heat flow (Wm^{-2})	1×10^{-2}	1×10^{-3}	3	$\sim 10^{-1}$
Magnetic diffusivity (m^{2}s^{-1})	2	4	30	~100
R_m based on V_{ml}	3×10^3	70	3×10^4	700
R_m based on V_{mac}	50	5	400	25
Λ, Elsasser number at top of dynamo	0.3	0.3	0.3	~0.01?

is hotter than the corresponding pressure level of Earth. This can arise because Earth has plate tectonics, which eliminates heat more efficiently than a stagnant lid form of mantle convection. Alternatively (or in addition), Venus' core may not be cooling at present because it is undergoing a transition in convective style following a resurfacing event ~700 Ma ago.

Earth remains imperfectly understood, a humbling reminder of the dangers of claiming an understanding of other planets. Growth of the inner core is thought essential for sustaining convection and sufficient energy to run the dynamo field. Doubts have been expressed about whether Earth's field can be sustained for its known history (at least 3.5 Ga) if the inner core has existed for only of order a couple of billion years or less. An additional energy source may be needed. One possibility is a modest amount of potassium in the core (since potassium has a radioactive isotope, ^{40}K Nimmo *et al.* (2004). Another possibility is that cooling rates of the lower mantle have been underestimated for earlier epochs.

Moon probably has a core that is at least partially liquid. It has patches of strong crustal magnetization that may have been acquired following impacts and compression of conducting plasma at the antipode. It is not known whether the preimpact field was necessarily a global field of the kind that only a dynamo produces. Even if it is a dynamo, it may (uniquely among planets in our solar system) have arisen through mechanical stirring of the inner core. This can arise because of the failure of the core to follow the nutation of the mantle (which currently has a period of 18.6 years). Rapid cooling of a boundary layer immediately above the core-mantle boundary might also conceivably maintain a dynamo for some time.

Mars had an ancient dynamo, probably in the period prior to 4.0 Ga. Connerney *et al.* (2004). There are three possibilities for why this dynamo existed and then died Stevenson (2001): (i) Core cooling decreased to the point where conductive heat loss dominated (but no inner core formed). This is the most well developed hypothesis. (ii) Mars underwent a change in convective style, from an efficient mode (e.g., plate tectonics) to the currently observed stagnant lid mode. This would cause the mantle and core to stop cooling and turn off core convection and the dynamo. This model would work irrespective of whether Mars has an inner core. (iii) The core of Mars froze sufficiently so that the remaining fluid region was too thin to sustain a dynamo.

Jupiter may have dynamo generation out to levels where hydrogen is only a semiconductor, perhaps 80% to 85% of the planet radius. This is compatible with the magnitude and harmonic structure of the field Bagenal *et al.* (2004).

Io exhibits no convincing evidence of a dynamo and no simple inductive response. Although Io has a metallic core, it might not be undergoing much long-term cooling if the mantle is heated steadily by tides.

Europa has a clear signature of an induction field and no evidence of a permanent dipole. The induction field can be explained by a water ocean of similar conductivity to Earth's oceans, provided this ocean has a thickness exceeding ~10 km. No other plausible source of the required conductivity has been suggested.

Ganymede has a clear signature of a permanent dipole. A permanent magnetism explanation is conceivable but unlikely, and the most reasonable interpretation is a dynamo in the metallic core. A liquid Fe-S core is expected in Ganymede. Nonetheless, this dynamo is surprising, partly because of Ganymede's size but mainly because of the difficulty in sustaining convection in such a small body. The presence of large amounts of sulfur and large ^{40}K mantle heating may help. There may also be a much smaller induction signal from a water ocean.

Callisto has a clear induction signal; explained by a salty water ocean that underlies the ice I layer, around 150 to 200 km in depth. This ocean is expected because of radioactive heating alone.

Saturn may have a dynamo very similar to that of Jupiter, but more deep-seated (the reason for the smaller surface field). It may be overlain by a region that greatly reduces the nonspin axisymmetric components, perhaps explaining the small observed dipole tilt.

Titan has an observed upper bound to the field that is less than the field expected for a Ganymede-like dynamo. It may have a water ocean and thus produce an induction signal, potentially detectable by Cassini. However, this will more difficult to detect because Saturn lacks a significant dipole tilt, so the time variable part of Saturn's field is much smaller than that for Jupiter.

Uranus and *Neptune* are very similar in structure and in field strength and geometries. Their very different obliquities are evidently irrelevant to understanding their fields. Although it seems likely that high-pressure water can provide the desired conductivity for a dynamo, it is marginal and the observed field strength seems smaller than expected (see Table D2). This raises the question of whether these planets are actually generating their fields deeper down. Quadrupolar dynamos are permitted by dynamo theory, and the dynamo activity might be limited to a thin shell Stanley and Bloxham (2004).

Triton and Pluto might possibly have water-ammonia oceans and might therefore be capable of induction signals, to the extent that they are subjected to small, time varying external magnetic fields.

Extrasolar giant planets can be expected to be convective at depth, and to have the conductivities sufficient for dynamo action.

David J. Stevenson

Bibliography

Bagenal, F., Dowling, T., and McKinnon, W., 2004. *Jupiter: The Planet, Satellites and Magnetosphere*, Cambridge Planetary Science: Cambridge, UK, Cambridge University Press, 719pp.

Connerney, J.E.P., 1993. Magnetic fields of the outer planets. *Journal of Geophysical Research*, **98**: 18659–18679.

Connerney, J.E.P., Acuna, M.H., Ness, N.F., *et al.*, 2004. Mars crustal magnetism. *Space Science Reviews*, **111**: 1–32.

De Pater, and Lissauer, J.J., 2001. *Planetary Sciences.* New York: Cambridge University Press, 528pp.

Guillot, T., 1999. Interiors of giant planets inside and outside the solar system, *Science*, **286**: 72–77.

Merrill, R.T., McElhinney, M.W., and McFadden, P.L. 1996. *The Magnetic Field of the Earth*, New York: Academic Press, 531pp.

Moffatt, H.K., 1978. *Magnetic Field Generation in Electrically Conducting Fluids.* New York: Cambridge University Press, 336 pp.

Nimmo, F., Price, G.D., Brodholt, J., *et al.*, 2004. The influence of potassium on core and geodynamo evolution. *Geophysical Journal International*, **156**: 363–376.

Parker, E.N., 1979. *Cosmical Magnetic Fields: their Origin and their Activity.* New York: Clarendon Press, Oxford University Press, 841pp.

Poirier, J.-P., 1991. *Introduction to the Physics of the Earth's Interior.* New York: Cambridge University Press, p. 191.

Roberts, P.H., and Glatzmaier, G.A., 2000. Geodynamo theory and simulations, *Reviews of Modern Physics*, **72**: 1081–1123.

Russell, C.T., 1993. Magnetic fields of the terrestrial planets. *Journal of Geophysical Research-Planet*, **98**: 18681–18695.

Showman, A.P., and Malhotra, R., 1999. The Galilean satellites. *Science*, **286**: 77–84.

Stanley, S., Bloxham, J., 2004. Convective-region geometry as the cause of Uranus' and Neptune's unusual magnetic fields. *Nature*, **428**: 151–153.

Stanley, S., and Bloxham, J., Hutchison, W.E., *et al.*, 2005. Thin shell dynamo models consistent with Mercury's weak observed magnetic field. *Earth and Planetary Science Letters*, **234**: 27–38.

Stevenson, D.J., 1983. Planetary magnetic fields. *Reports on Progress in Physics*, **46**: 555–620.

Stevenson, D.J., 2001. Mars' core and magnetism. *Nature*, **412**: 214–219.

Stevenson, D.J., 2003. Planetary Magnetic Fields. *Earth and Planetary Science Letters*, **208**: 1–11.

E

EARTH STRUCTURE, MAJOR DIVISIONS

Although the Earth is a complex body with pervasive three-dimensional structure in its solid portions, the dominant variation in properties is with depth. The figure of the Earth is close to an oblate spheroid with a flattening of 0.003356. The radius to the pole is 6357 km and the equatorial radius is 6378 km; for most purposes, a spherical model of the Earth with a mean radius of 6371 km is adequate. Thus, reference models for internal structure can be used in which the physical properties depend on radius. Three-dimensional variations can then be described by deviations from a suitable reference model.

The main divisions of Earth structure are illustrated in Figure E1. Beneath the thin crustal shell lies the silicate mantle which extends to a depth of 2890 km. The mantle is separated from the metallic core by a major change of seismic properties that has a profound effect on global seismic wave propagation (see *Seismic phases*). The outer core behaves as a fluid at seismic frequencies and does not allow the passage of shear waves, while the inner core appears to be solid. Thus only seismic compressional waves (P) can traverse the outer core.

Much of the evidence for the nature of internal structure comes from the analysis of seismograms and the patterns of propagation of seismic waves.

The existence of a discontinuity at the base of the crust was found by Mohorovičić in the analysis of the Kupatal earthquake of 1909, based on only a limited number of records from permanent seismic stations. Knowledge of crustal structure from seismic methods has developed substantially through the use of controlled sources, e.g., explosions. Indeed, most of the information on the oceanic crust comes from such work. The continental crust varies in thickness from around 20 km in rift zones to 70 km under the Tibetan Plateau. Typical values are close to 35 km. The oceanic crust is thinner with basalt pile about 7 km thick whose structure changes somewhat with the age of the oceanic crust.

The lithosphere continues from the crust into the mantle and also shows significant differences between the oceanic and continental regimes. The entire oceanic lithosphere is generated by the spreading process at mid-ocean ridges and the increase in thickness with age to at least 85 Ma is consistent with thermal cooling. Precambrian shield components of continents have a very thick but lower density lithosphere extending to 200 km (or possibly more in some places). The lithosphere beneath Phanerozoic regions tends to be thinner, about 120 km, with considerable complications in the neighborhood of active tectonic belts.

Beneath the lithosphere lies the asthenosphere that is typically characterized by a decrease in shear wavespeed, increased attenuation and electrical conductivity, and inferred lower viscosity. The transition from lithosphere to asthenosphere appears to be generally sharp in oceanic regions, but probably is more gradational under the continents.

The mantle shows considerable variation in seismic properties with depth, with strong gradients in seismic wavespeed in the top 800 km. The presence of distinct structure in the upper mantle was recognized by Jeffreys in the 1930s who noted a distinct shift in the rate of change of the travel-times of P-waves with distance near 20° from the source. He ascribed this 20° discontinuity to a rapid change in seismic wavespeed with depth. In the 1960s–1970s detailed analysis of the seismic wave field in the distance range from 15° to 25° from earthquake and nuclear explosion sources, using seismic arrays, revealed two clear discontinuities in seismic wavespeed at depths near 410 and 660 km. From mineral physics results, these discontinuities can be associated with major phase changes in the silicate minerals of the mantle. A number of other discontinuities have been proposed. Only a feature near 520 km depth has a widespread occurrence in long-period observations, and appears to represent a small change over a 30–50 km interval in depth (Shearer, 1993). Many different styles of study have shown both the global presence of discontinuities near 410 and 660 km depth, and also significant variations in seismic structure within the upper mantle (for a review see Nolet *et al.*, 1994).

The reduction of seismic wavespeeds in the lowermost mantle led Bullen to identify a subregion D″ of the lower mantle (region D). This zone situated just above the core-mantle boundary has proved to have heterogeneity comparable to that in the uppermost mantle. There are localized discontinuities in shear wavespeed, as first noted by Lay and Helmberger (1983), as well as narrow zones of "ultralow velocity" (Garnero *et al.*, 1998).

The need for a core at depth with greatly reduced seismic wavespeeds was recognized at the end of the last century by Oldham in his analysis of the great Assam earthquake of 1890, because of a zone without distinct P arrivals (a "shadow zone" in PKP, see *Seismic phases*). By 1914, Gutenburg had obtained an estimate for the radius of the core which is quite close to the current value. The presence of the inner core was inferred by Inge Lehmann in 1932 from careful analysis of arrivals within the shadow zone (*PKiKP*), which had to be reflected from some substructure within the core.

The nature of the transition form the outer core to the inner core has been a matter of some debate. Early analysis by Jeffreys, based on a limited number of reports of the travel-times of PKP phases, produced a model with a decrease in P-wavespeed just above the inner core (the

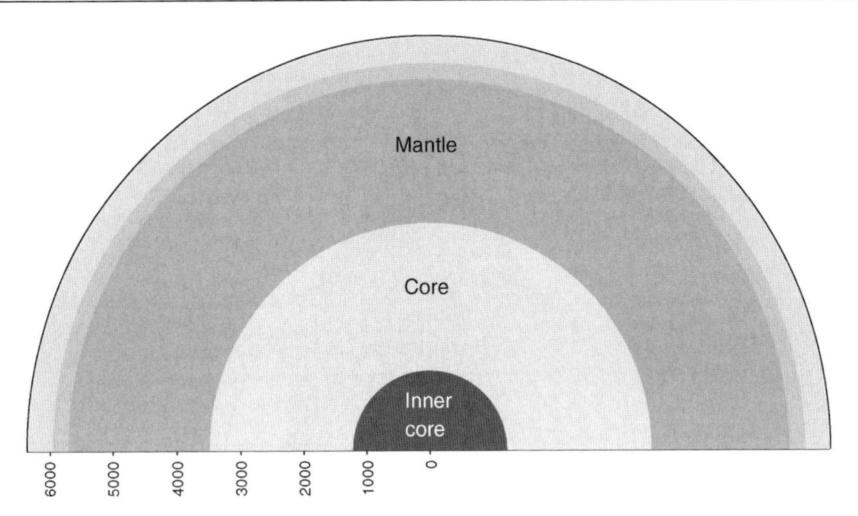

Figure E1 The major divisions of the radial structure of the Earth: the gradations in tone in the upper part of the mantle indicate the presence of discontinuities at 410 and 660 km depth. The diagram is drawn to true scale, so that the Earth's crust appears only as the thin bounding surface.

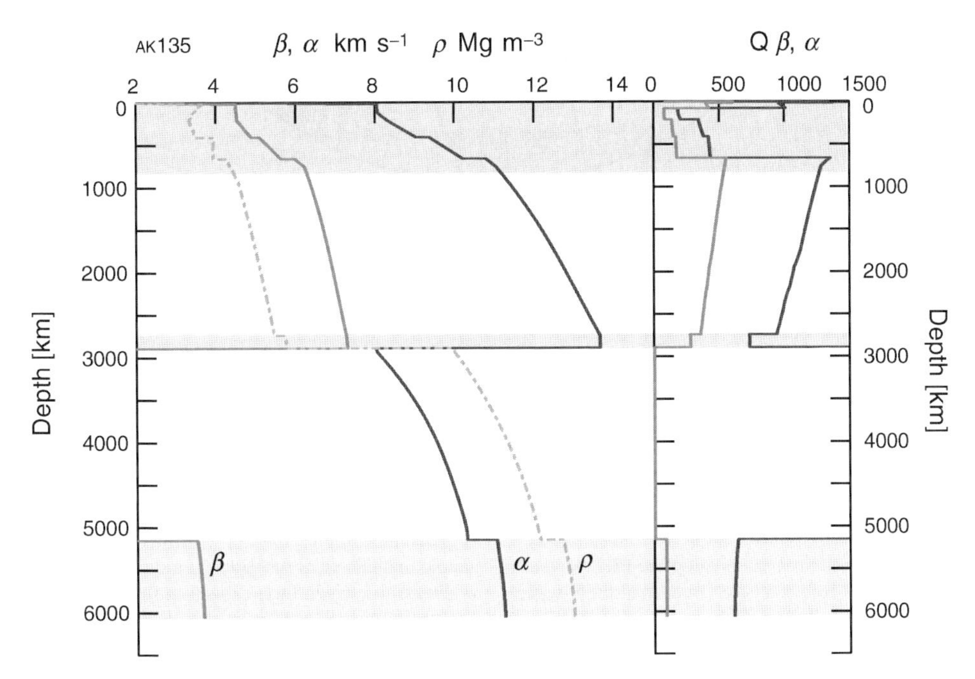

Figure E2 Radial reference earth model AK135, seismic wavespeeds α (P), β (S): Kennett *et al.* (1995), attenuation parameters Q, density (ρ): Montagner and Kennett (1996). Zones marked with gray tones show significant lateral heterogeneity and some level of anisotropy.

F-region in Bullen's notation). In contrast, Gutenburg working with seismograms recorded at Pasadena, California, favored a simple discontinuity with an increase in wavespeed in the inner core. The difficulties arise because of the very complex patterns of arrival expected for the PKP phase (see Kennett, 2002; Chapter 26). Modern analysis using the differential times between arrivals following different paths through the core, or matching observed and computed seismograms, favor a decrease in wavespeed gradient at the base of the outer core, but not a low velocity zone.

The times of arrival of seismic phases on their different paths through the globe constrain the variations in P- and S-wavespeed, and can be used to produce models of the variation with radius. A very large volume of arrival time data from stations around the world have been accumulated by the International Seismological Centre and are available in digital form. This data set has been used to develop high-quality travel-time tables that can in turn be used to improve the locations of events. With reprocessing of the arrival times to improve locations and the identification of the picks for later seismic phases, a set of observations of the relation between travel-time and epicentral distance have been produced for a wide range of phases. The reference model AK135 of Kennett *et al.* (1995) for both P- and S-wavespeeds, illustrated in Figure E2, gives a good fit to the travel-times of mantle and core phases. The reprocessed data set and the AK135 reference model have formed the basis of much recent work

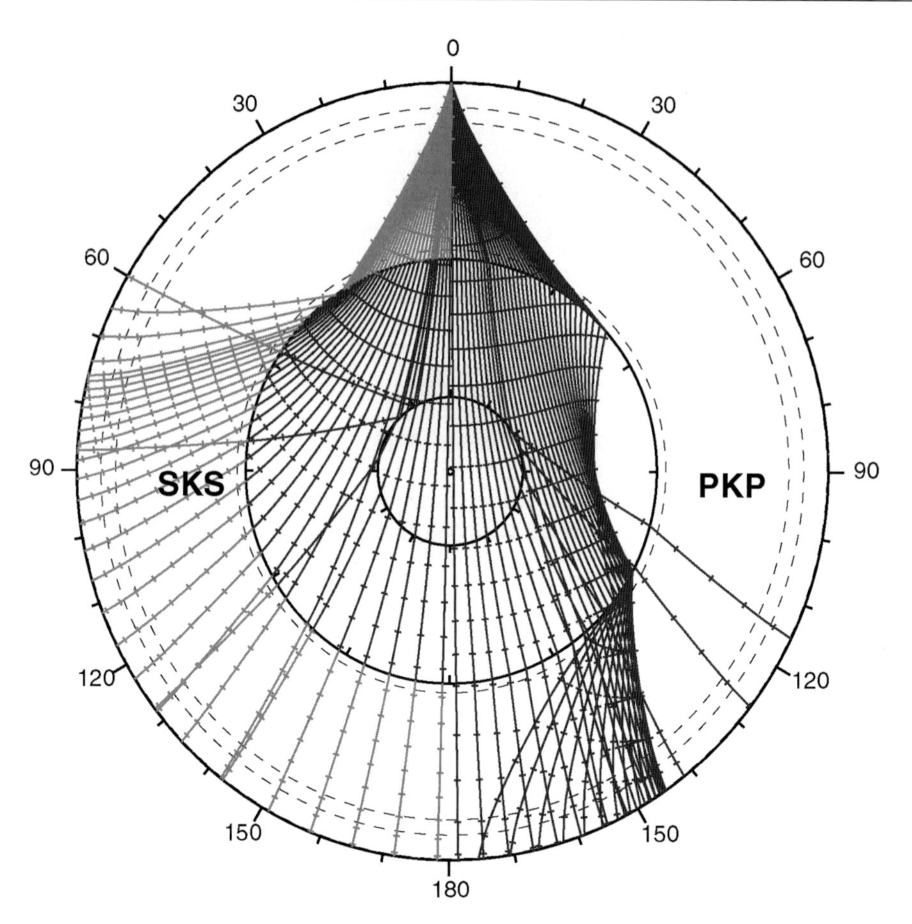

Figure E3 Contrast between the patterns of seismic wave propagation in the core for SKS to the left and PKP to the right. S wavelegs are shown in gray tone. Wavefronts are indicated by ticks on the ray paths at 1 min intervals.

on high-resolution travel-time tomography to determine three-dimensional variations in seismic wavespeed.

One of the difficulties in just using the travel-time relations comes from the sampling of the core by seismic phases. For arrivals traveling solely as P-waves the change in material properties at the core-mantle boundary means that the paths become steeper on entry into the core from Snell's law. Eventually, the effects of sphericity and the gradients in seismic wavespeed are sufficient to turn energy back to the surface but the outermost part of the core is not sampled by the PKP wave. In contrast, the P-wavespeed in the core is a little larger than the S-wavespeed at the base of the mantle and so the converted arrival SKS can sample the whole core. The differing patterns for PKP and SKS are shown in Figure E3 (see also *Seismic phases*). From travel-times alone, the P-wavespeeds at the top of the outer core depend on knowledge of the S-wavespeed throughout the mantle.

The use of the times of arrival of seismic phases enables the construction of models for P- and S-wavespeed, but more information is needed to provide a full model for Earth structure. The density distribution in the Earth has to be inferred from indirect observations and the main constraints come from the mass and moment of inertia. The mean density of the Earth can be reconciled with the moment of inertia if there is a concentration of mass toward the center of the Earth; which can be associated with a major density jump going from the mantle into the outer core and a smaller density contrast at the boundary between the inner and outer cores (Bullen, 1975). With successful observations of the free-oscillations of the Earth following the great Chilean earthquake of 1960, additional information could be extracted from the frequencies of oscillation on both the seismic wavespeeds and the density. Fortunately the inversion of the frequencies of the free-oscillations for a spherically symmetric reference model provides independent constraints on the P-wavespeed structure in the outer core. Even with the additional information from the normal modes the controls on the density distribution are not strong (Kennett, 1998) and additional assumptions such as adiabatic state in the core and lower mantle have often been employed to produce a full model. The reference model PREM of Dziewonski and Anderson (1981) combined the free-oscillation and travel-time information available at the time. A parametric representation of structure was employed in terms of simple mathematical functions to aid the inversion; thus a single cubic was used for seismic wavespeed in the outer core and again for most of the lower mantle. The PREM forms the basis of much current global seismology using quantitative exploitation of seismic waveforms at longer periods (e.g., Dahlen and Tromp, 1998).

In order to reconcile the information derived from the free-oscillations of the Earth and the travel-time of seismic phases, it is necessary to take account of the influence of anelastic attenuation within the Earth. A consequence of attenuation is a small variation in the seismic wavespeeds with frequency, so that waves with frequencies of 0.01 Hz (at the upper limit of free-oscillation observations) travel slightly slower than the 1-Hz waves typical of the short period observations used in travel-time studies. The density and attenuation models shown in Figure E2 were derived by Montagner and Kennett (1996) to link the wavespeed distributions from travel-time analysis to the free-oscillation results.

Figure E4 displays the travel-times for the AK135 model superimposed on the reported phases for a set of 104 test events (83 earthquakes, 21 explosions) assembled by Kennett and Engdahl (1991). These events have well-controlled locations and origin times with a rich set of later phase readings (57655 phases in all). As we can see the reference model provides a good representation of the features

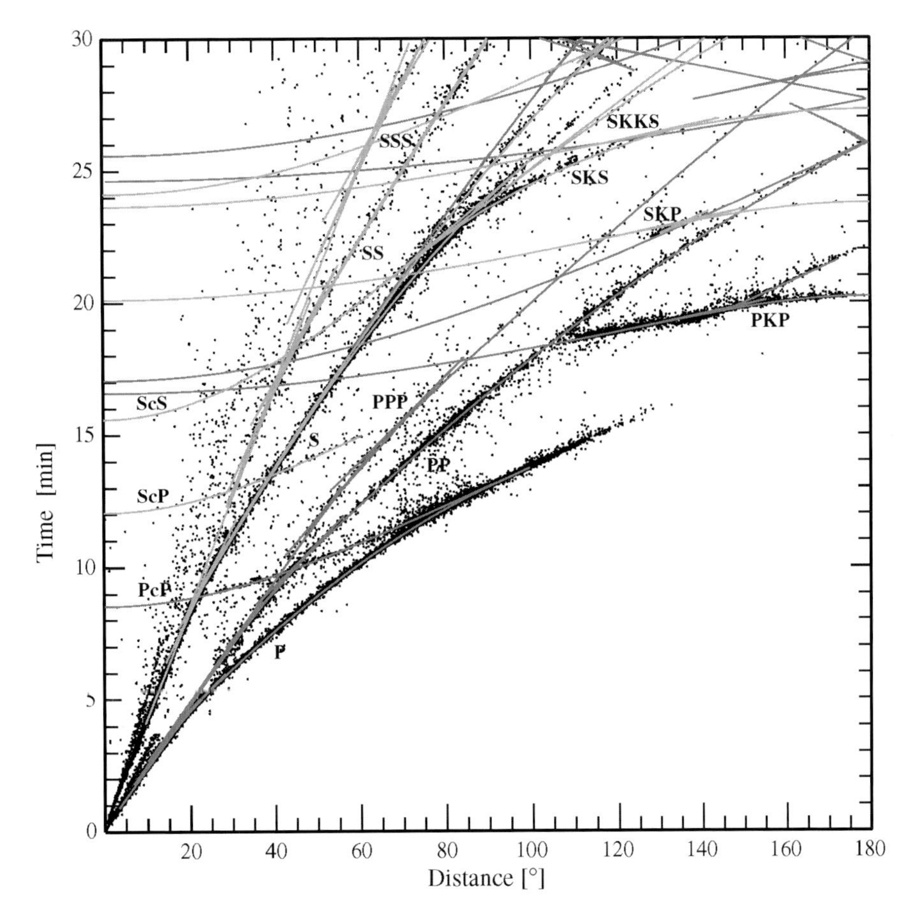

Figure E4 Display of AK135 travel-times superimposed on the times of phases for the test events corrected to surface focus.

of the full set of phase picks (for explanation of the phase codes see *Seismic phases*).

The reference model AK135 (Figure E2) is isotropic and depends solely on radius. However, the real Earth both varies in three-dimensional and displays anisotropy in seismic parameters. The zones of largest variability are indicated in Figure E2 by bands of gray tone. These are also regions in which there is the strongest evidence for anisotropy in seismic properties, from differences in seismic wavespeeds depending on the direction of propagation or the polarization of the seismic waves.

The strongest levels of heterogeneity are found in the outermost 300 km with a strong contrast between the high wavespeeds beneath the ancient cores of continents extending to at least 250 km and the oceanic regime where the fast lithosphere is quite thin. The continental lithosphere also appears to be anisotropic, with evidence from the analysis of seismic surface waves and the shear wave splitting for SKS waves.

The process of subduction brings the cold oceanic lithosphere into the upper mantle and locally there are large contrasts in seismic wavespeeds, well imaged by detailed tomography, that extend down to at least 660 km and in some zones even deeper. Remnant subducted material can have a significant presence in some regions, e.g., above the 660 km discontinuity in the north-west Pacific and in the zone from 660 down to 1000 km beneath Indonesia.

Mantle

The nature of mantle structure varies with depth and it is convenient to divide the mantle up into four major zones (e.g., Jackson and Ridgen, 1998).

Upper mantle (depth $z < 350$ km), with a high degree of variability in seismic wavespeed (exceeding $\pm 4\%$) and relatively strong attenuation in many locations.

Transition zone ($350 < z < 800$ km), including significant discontinuities in P- and S-wavespeeds and generally high velocity gradients with depth.

Lower mantle ($800 < z < 2600$ km) with a smooth variation of seismic wavespeeds with depth that is consistent with adiabatic compression of a chemically homogeneous material.

D″ layer ($2600 < z < 2900$ km) with a significant change in velocity gradient and evidence for strong lateral variability and attenuation.

The two major discontinuities in seismic wavespeeds near depths of 410 and 660 km are associated with phase transformations in silicate minerals induced by the effects of increasing pressure, and represent changes in the organization of the oxygen coordination with the silicon atoms. The change in seismic wavespeed across these discontinuities occurs quite rapidly, and they are seen in both short- and long-period observations. Other minor discontinuities have been proposed, but only one near 520 km appears to have some global presence in long-period stacks but not short-period data. This 520-km transition may occur over an extended zone, e.g., 30–50 km, so that it still appears sharp for long-period waves with wavelengths of 100 km or more. Jackson and Ridgen (1998) provide a broad ranging review of the interpretation of seismological models for the transition zone and their reconciliation with information from mineral physics.

Frequently, the lower mantle is taken to begin below the 660 km discontinuity, but strong gradients in seismic wavespeeds persist to depths of the order of 800 km and it seems appropriate to retain this region in the transition zone. There is increasing evidence for localized sharp transitions in seismic properties at depth around 900 km that appear to be related to the penetration of subducted material into the lower mantle.

Between 800 and 2600 km, the lower mantle has, on average, relatively simple properties which would be consistent with the adiabatic compression of a mineral assemblage of constant chemical composition and phase. Although tomographic studies image some level of three-dimensional structure in this region the variability is much less than in the upper part of the mantle or near the base of the mantle.

The D″ layer from 2600 km to the core-mantle boundary has a distinctive character. The nature of seismic wavespeed distribution changes significantly with a sharp drop in the average velocity gradient. There is a sharp increase in the level of wavespeed heterogeneity near the core-mantle boundary compared with the rest of the lower mantle. The base of the Earth's mantle is a complex zone with widespread indications of heterogeneity on many scales, discontinuities of variable character, and shear wave anisotropy (e.g., Gurnis *et al.*, 1998; Kennett, 2002). The results of seismic tomography give a consistent picture of the long wavelength structure of the D″ region: there are zones of markedly lowered S-wavespeed in the central Pacific and southern Africa, whereas the Pacific is ringed by relatively fast wavespeeds that may represent a "slab graveyard" arising from past subduction. Discordance between P- and S-wave results suggests the presence of chemical heterogeneity rather than just the effect of temperature (e.g., Masters *et al.*, 2000).

There are a number of lines of evidence which suggest the presence of topography on the core-mantle boundary, but the situation is complicated by the influence of the strong heterogeneity in D″. The information from travel-times of seismic phases such as (PcP and PKP, PKKP) is compatible with long wavelength topography on the order of ± 3 km, but such variations are not required by this data. As noted by Buffet (1998), topography of several kilometers on wavelengths of several thousand kilometers is not sufficient to produce the torques between mantle and core inferred from variations in the length of the day. However, such topographic coupling could become important if the core-mantle boundary is rough at smaller wavelengths as suggested from some studies of seismic scattering.

Evidence for a discontinuity at the top of the D″ region comes from the presence of an additional arrival between the direct mantle phase and the core reflection in the distance range 65°–90°. This is not a universal phenomenon, but a D discontinuity is found in many parts of the world. The typical contrast in S-wavespeed is in the range 1.5%–3% at 250 ± 100 km above the core-mantle boundary, with greater variability in P-wavespeed. The nature of the transition varies between different regions and in places may represent a transition zone up to 50 km thick. In some areas the fluctuations in travel-times and amplitude of the reflected phases suggest the presence of strong lateral gradients in discontinuity structure (e.g., Lay *et al.*, 1998).

Beneath the central Pacific Ocean and in a number of other locations, there is evidence for thin zones, around 20 km thick, at the core-mantle boundary with a marked reduction in P-wavespeed of at least 10% (see Garnero *et al.*, 1998). The inferred reduction in S-wavespeed in these ultralow velocity zones somewhat larger.

Since all waves sampling the core have to pass through the D″ region, the complexity of propagation in this region influences inferences made about deeper structure.

Core

The core-mantle boundary at about 2890 km depth marks a substantial change in physical properties associated with a transition from the silicate mantle to the metallic core (see Figure E3). There is a significant jump in density, and a dramatic drop in P-wavespeed from 13.7 to 8.0 km s^{-1}. The major change in wavespeed arises from the absence of

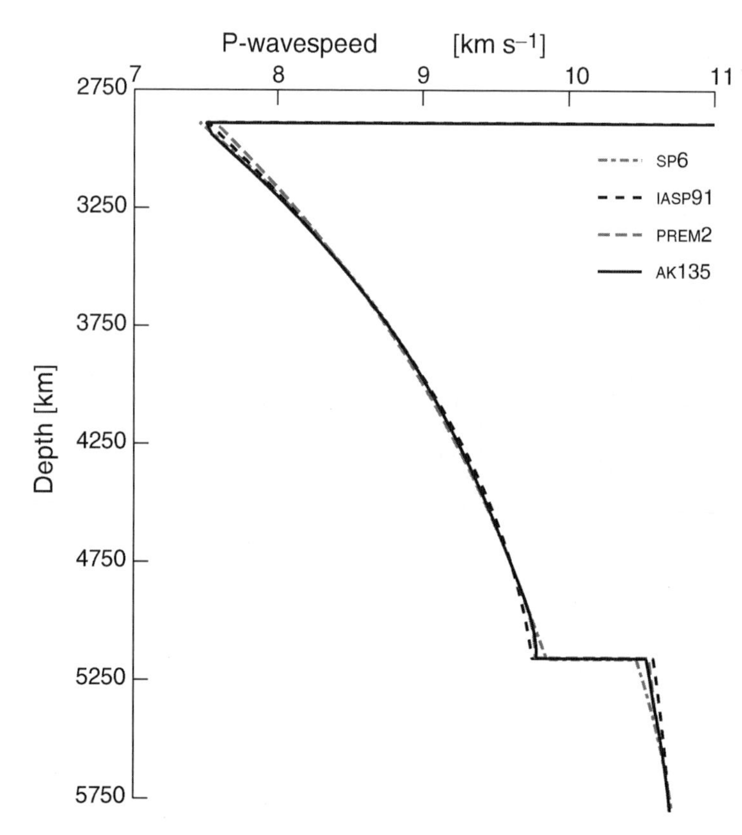

Figure E5 Comparison of P-wavespeed models for the core: IASP91, Kennett and Engdahl, 1991; PREM2, Song and Helmberger, 1992; SP6, Morelli and Dziewonski, 1993; AK135, Kennett *et al.* 1995.

shear strength in the fluid outer core, so that the P-wavespeed depends just on the bulk modulus and density. No shear waves can be transmitted through the outer core so that only the K propagation legs can be present (see *Seismic phases*).

The process of core formation requires the segregation of heavy iron-rich components in the early stages of the accretion of the earth (e.g., O'Neill and Palme, 1998). The core is believed to be largely composed of an iron-nickel alloy, but its density requires the presence of some lighter elemental components. A wide variety of candidates have been proposed for the light components, but it is difficult to satisfy the geochemical constraints on the nature of the bulk composition of the Earth.

The inner core appears to be solid and formed by crystallization of material from the outer core, but it is possible that it could include some entrained fluid in the top 100 km or so. The shear wavespeed for the inner core inferred from free-oscillation studies is very low and the ratio of P- to S-wavespeeds is comparable to a slurry-like material. The structure of the inner core is both anisotropic and shows three-dimensional variation (e.g., Creager, 1999). The variations are complex with some apparent variation between hemispheres, but may be influenced by the passage of phases such as PKIKP through the strong heterogeneity at the base of the mantle in D″.

The fluid outer core is conducting and motions within the core create a self-sustaining dynamo which generates the main component of the magnetic field at the surface of the Earth. The dominant component of the geomagnetic field is dipolar but with significant secondary components. Careful analysis of the historic record of the variation of the magnetic field has lead to a picture of the evolution of the flow in the outer part of the core (e.g., Bloxham and Gubbins, 1989). The presence of the inner core may well be important for the action of the dynamo, and electromagnetic coupling between the inner and outer cores could give rise to differential rotation between the two parts of the core (Glatzmaier and Roberts, 1996). Efforts have been made to detect this differential rotation using the time history of different classes of seismic observations but the results are currently inconclusive.

Even though the main features of the variation in seismic wavespeeds through the core are well established, there are noticeable differences in the details of models proposed by different authors for the regions near the core-mantle boundary and the boundary between the inner core and outer core (Figure E5). The detailed structure just below the core-mantle boundary is primarily controlled by the properties of the multiple reflections SKKS, SKKKS,... from the underside of the core-mantle boundary. The most detailed models of the structure of the boundary region between the inner and outer cores come from matching of observed and calculated waveforms as in the PREM2 model of Song and Helmberger (1992). This work indicates the need for a reduction in seismic wavespeeds just above the inner core boundary, and is supported by studies of the differential times between the branches of the PKP phase that take different paths through the core. However, such studies are susceptible to the influence of heterogeneity and anisotropy in the inner core.

Differential time information was used in the construction of the AK135 model, which is close to PREM2, and differs from SP6 (Morelli and Dziewonski, 1993) that uses a single cubic representation for P-wavespeed throughout the whole core, in a similar way to PREM (Dziewonski and Anderson, 1981).

Brian Kennett

Bibliography

Bloxham, J., and Gubbins, G., 1989. Geomagnetic secular variation. *Philosophical Transactions of the Royal Society of London*, **329A**: 415–502.

Buffet, B.A., 1998. Free oscillations in the length of the day: inferences on physical processes near the core-mantle boundary. In Gurnis, M., Wysession, M.E., Knittle, E., and Buffet, B.A. (eds.), *The Core-Mantle Boundary Region*. Geodynamics Monograph, Vol. 28. Washington, DC: American Geophysical Union.

Bullen, K.E., 1975. *The Earth's Density*. London: Chapman & Hall.

Creager, K.C., 1999. Large-scale variations in inner core anisotropy. *Journal of Geophysical Research*, **104**(23):127–139.

Dahlen, F.A., and Tromp, J., 1998. *Theoretical Global Seismology*. Princeton: Princeton University Press.

Dziewonski, A.M., and Anderson D.L., 1981. Preliminary reference Earth model. *Physics of the Earth and Planetary Interiors*, **25**: 297–356.

Garnero, E.J., Revenaugh, J., Williams, Q., Lay, T., and Kellogg, L.H., 1998. Ultralow velocity zone at the core-mantle boundary. In Gurnis, M., Wysession, M.E., Knittle, E., and Buffet, B.A. (eds.), *The Core-Mantle Boundary Region*. Geodynamics Monograph, Vol. 28. Washington, DC: American Geophysical Union.

Glatzmaier, G.A., and Roberts, P.H., 1996. Rotation and magnetism of Earth's inner core. *Science*, **274**: 1887–1891.

Gurnis, M., Wysession, M.E., Knittle, E., and Buffet, B.A. (eds.), 1998. *The Core-Mantle Boundary Region*, Geodynamics Monograph, Vol. 28. Washington, DC: American Geophysical Union.

Jackson, I., and Rigden, S.M., 1998. Composition and temperature of the Earth's mantle: seismological models interpreted through experimental studies of earth materials. In Jackson, I. (ed.), *The Earth's Mantle: Structure, Composition, and Evolution*. Cambridge: Cambridge University Press, pp. 405–460.

Kennett, B.L.N., 1998. On the density distribution within the Earth. *Geophysical Journal International*, **132**: 374–382.

Kennett, B.L.N., 2002. *The Seismic Wavefield II: Interpretation of Seismograms on Regional and Global Scales*. Cambridge: Cambridge University Press.

Kennett, B.L.N., and Engdahl, E.R., 1991. Travel times for global earthquake location and phase identification. *Geophysical Journal International*, **105**: 429–465.

Kennett, B.L.N., Engdahl, E.R., and Buland, R., 1995. Constraints on seismic velocities in the Earth from travel times. *Geophysical Journal International*, **122**: 108–124.

Lay, T., and Helmberger, D.V., 1983. A lower mantle S-wave triplication and the shear velocity structure of D″. *Geophysical Journal of the Royal Astronomical Society*, **75**: 799–837.

Lay, T., Garnero, E.J., Young, C.J., and Gaherty, J.B., 1997. Scale lengths of shear velocity heterogeneity at the base of the mantle from S-wave differential times. *Journal of Geophysical Research*, **102**: 9887–9910.

Lay, T., Williams, Q., Garnero, E.J., Kellogg, L.H., and Wysession, M.E., 1998. Seismic wave anisotropy in the D″ region and its implications. In Gurnis, M., Wysession, M.E., Knittle, E., and Buffet, B.A. (eds.), *The Core-Mantle Boundary Region*. Geodynamics Monograph, Vol. 28. Washington, DC: American Geophysical Union.

Masters, G., Laske, G., Bolton, H., and Dziewonski, A., 2000. The relative behavior of shear velocity, bulk sound speed, and compressional velocity in the mantle: implications for chemical and thermal structure. In Karato, S.I., Forte, A.M., Liebermann, R.C., Masters, G., and Stixrude, L. (eds.), *Earth's Deep Interior: Mineral Physics and Tomography from the Atomic to the Global Scale,* AGU Geophysical Monograph 117. Washington, DC: American Geophysical Union, pp. 63–87.

Montagner, J-P., and Kennett, B.L.N., 1996. How to reconcile body-wave and normal-mode reference Earth models? *Geophysical Journal International*, **125**: 229–248.

Morelli, A., and Dziewonski, A.M., 1993. Body wave traveltimes and a spherically symmetric P- and S-wave velocity model. *Geophysical Journal International*, **112**: 178–194.

Nolet, G., Grand, S., and Kennett, B.L.N., 1994. Seismic heterogeneity in the upper mantle. *Journal of Geophysical Research*, **99**(23): 753–766.

O'Neill, H.St.C., and Palme, H., 1998. Composition of the silicate Earth: implications for accretion and core formation. In Jackson, I. (ed.), *The Earth's Mantle: Structure, Composition and Evolution*. Cambridge: Cambridge University Press, pp. 3–126.
Shearer, P.M., 1993. Global mapping of upper mantle reflectors from long-period reflectors from long-period SS precursors. *Geophysical Journal International*, **115**: 878–904.
Song, X., and Helmberger, D.V., 1992. Velocity structure near the inner core boundary from waveform modelling. *Journal of Geophysical Research*, **97**: 6573–6586.

Cross-references

Core, Boundary Layers
Core Properties, Physical
Core-Mantle Boundary Topography, Seismology
D″, Anisotropy
D″, as a Boundary Layer
D″, Composition
D″, Seismic Properties
Inner Core Anisotropy
Inner Core Seismic Velocities
Interiors of Planets and Satellites
Lehmann, Inge (1888–1993)
Oldham, Richard Dixon (1858–1936)
Seismic Phases

ELSASSER, WALTER M. (1904–1991)

About 2000 years ago in China it was discovered that a lump of magnetic iron oxide (magnetite), or lodestone, takes up a preferred orientation when freely suspended. Ultimately this discovery led to the magnetic compass for navigation and surveying. Four centuries ago *Gilbert* (1600) (*q.v.*), in England, carried out precise laboratory measurements of the magnetic field around a lodestone. The magnetic records accumulated by the seafaring captains of the time made it clear to Gilbert that the magnetic field of Earth has the same form as that of a lodestone, and he concluded that Earth was itself a huge lodestone.

In 1835, Gauss demonstrated mathematically that the east-west, north-south, and vertical components of the field are observed to vary relative to each other in such a way as to place the origin of the field inside, rather than outside, the body of Earth.

Gilbert's plausible interpretation of the origin of the geomagnetic field stood for about three centuries, until Kelvin in 1862 established the incandescence of the interior of Earth, and Pierre Curie in 1895 studied the abrupt disappearance of magnetic effects in metals and minerals upon heating to hundreds of degrees. The origin of the geomagnetic field became a mystery once again.

Thermoelectric effects in the hot inhomogeneous interior of Earth were a favorite notion, creating electric currents and the associated magnetic fields. However, it was not obvious how to account for a magnetic field as strong as the observed polar 0.6 G, or for the observed secular variations of the field, with characteristic times as short as a few centuries.

Walter M. Elsasser became interested in the problem in the late 30s and investigated the possibility that the thermoelectric effects in the temperature variations in the convective liquid iron and nickel core might provide the observed secular fluctuations of the field observed at the surface of Earth (Elsasser, 1939). Two years later Elsasser (1941) studied the secular variations in some detail and showed that they could be modeled by several randomly oriented dipoles if those dipoles were confined to the liquid core ($r < 0.5R_E$).

The essential role of convection in providing the secular variation turned Elsasser's attention to the possibility that the magnetic field is induced directly by the convective motions in the electrically conducting core. Convinced by then that the thermoelectric effect was too weak to provide the observed magnetic field, Elsasser set about developing the formal mathematical theory of magnetic induction by fluid motions in a sphere (Elsasser, 1946a,b, 1947). Elsasser recognized the fact that the large scale (10^3 km) of the principal convective motions in the liquid iron core, combined with the substantial electrical conductivity $\sigma = 10^{16}\ \mathrm{s}^{-1}$ (resistive diffusion coefficient $\eta = c^2/4\pi\sigma = 10^4 \mathrm{cm\,s}^{-1}$), coupled the magnetic field and the liquid iron over periods of 10^3–10^4 years. This matched the comparable characteristic turnover and nonuniform rotation times of the liquid iron core. So he decomposed the geomagnetic field **B** into what he called the toroidal (azimuthal) and poloidal (meridional) components and expressed the poloidal field in terms of a toroidal vector potential, expanding the whole in spherical harmonics. The fluid velocity **v** in the core was expanded in a similar form so that the magnetohydrodynamic induction equation

$$\frac{\partial \mathbf{B}}{\partial t} - \eta \nabla^2 \mathbf{B} = \nabla \times (\mathrm{v} \times \mathbf{B})$$

providing the growth and decay of each magnetic mode in terms of the field and fluid interaction matrix on the right-hand side.

Elsasser showed that the principal magnetic field in the core is the toroidal magnetic field, created by the interaction of the nonuniform rotation of the convecting core with the main poloidal (dipole) component of the field. Up to this point the poloidal and toroidal fields had rotational symmetry, and Elsasser was aware of Cowling's (1933) theorem (see *Cowling's theorem*) that a magnetic field with rotational symmetry cannot be sustained by fluid motions. Elsasser added quadrupole terms to the fluid motions, thereby breaking the rotational symmetry.

Unfortunately the system of equations for the mode amplitudes showed no signs of converging. The mathematical effort was carried forward, using computers to implement the mathematical analysis, by Bullard (see *Bullard, Edward Crisp* and *Dynamo, Bullard-Gellman*) and others (see *Rikitake, Tsuneji*) but without success. Pekeris, Accad, and Shkoller (1973) got around the difficulties by using a fluid velocity field with strong helicity, so that it required only a small velocity to regenerate the field, and their formal result evidently converges satisfactorily. Gubbins (1973) was able to apply computers to the formal analysis to show that the general method could be made to converge, providing a reliable physical result.

Parker, working as a research associate with Elsasser, noted that the net magnetic circulation in meridional planes (representing the dipole field) is created directly by cyclonic convective cells that push up local Ω-shaped loops in the azimuthal field. The cyclonic cells rotate the many Ω-loops into the meridional planes, where they spread out and merge through resistive diffusion, producing large-scale magnetic circulation, i.e., the dipole component of the poloidal field. The effect is described by the sum of the azimuthal vector potentials $\mathbf{A}_\varphi$ contributed by each small rotated -loop. The net effect is described by the *dynamo equation*

$$\left(\frac{\partial}{\partial t} - \eta \nabla^2\right) \mathbf{A}_\phi = \Gamma(r) \mathbf{B}_\phi$$

where the scalar function $\Gamma(r)$ (nowadays written $\alpha(r)$) is essentially the rotational velocity of the convective cells multiplied by the fraction of space occupied by the cells. Combining this with the Elssaser' equation for the azimuthal field $\mathbf{B}_\phi$ yields a complete set of *dynamo equations* (see *Dynamos, mean field*) easily solved in rectangular geometry in terms of *dynamo waves* (see *Dynamos, kinematic*; *Dynamo waves*), and showing in simple terms how the physics of Elsasser's induction works out in Earth and in the Sun (Parker, 1955).

Elsasser's fundamental dynamo concept, that the magnetic field of Earth is generated by the convection in the liquid iron core, was greeted with outspoken doubt by many eminent physicists. This negative attitude began to weaken only with the publication of his review paper some years later (Elsasser, 1950). The gradual acceptance of Elsasser's conjecture has led to a vast literature over the years (Elsasser, 1956; Moffatt, 1978; Parker, 1979; Krause and Rädler, 1980), exploring a variety of solutions of the dynamo equations in diverse settings (see *Dynamo, Backus*; *Dynamos, planetary and satellite*; *Geodynamo*). It is now believed that the Elsasser's magnetohydrodynamic dynamo concept is the basis for most, if not all, the magnetic fields observed throughout the cosmos. The geomagnetic dynamo has been captured recently in a numerical dynamical model of the convecting core (Glatzmaier and Roberts, 1996), illustrating the progress since Elsasser pointed scientific thinking in the right direction.

In closing, it is interesting to reflect on Elsasser's fundamental contributions to other fields of science. He was the first to point out that electron diffraction is a direct and unavoidable consequence of the de Broglie wavelength of the particle. He was the first to recognize nuclear shell structure. Following his work on geomagnetism, Elsasser turned his attention to the statistical implications of the extreme complexity of biological cells and their reproduction process, publishing his profound conclusions in a book *Reflections on a Theory of Organisms* (Elsasser, 1998). This work was so far ahead of the field at that time that it was ignored until relatively recently, when others began to catch up (see introduction by Harry Rubin to the recent 1998 edition of Elsasser's book). It is perhaps no coincidence that it required a scientist of Elsasser's stature to crack the problem of geomagnetism. The reader is referred to Elsasser (1978) for an autobiographical account of his diverse experiences throughout his remarkable career.

E.N. Parker

Bibliography

Cowling, T.G., 1933. The magnetic field of sunspots. *Monthly Notices of the Royal Astronomical Society*, **94**: 39–48.

Elsasser, W.M., 1939. On the origin of the Earth's magnetic field. *Physical Review*, **55**: 489–498.

Elsasser, W.M., 1941. A statistical analysis of the Earth's internal magnetic field. *Physical Review*, **60**: 876–883.

Elsasser, W.M., 1946a. Induction effects in terrestrial magnetism. *Physical Review*, **69**: 106–116.

Elsasser, W.M., 1946b. Induction effects in terrestrial magnetism. Part II. The secular variation. *Physical Review*, **70**: 202–212.

Elsasser, W.M., 1947. Induction effects in terrestrial magnetism. Part III. Electric modes. *Physical Review*, **72**: 821–833.

Elsasser, W.M., 1950. The Earth's interior and geomagnetism. *Review of Modern Physics*, **22**: 1–35.

Elsasser, W.M., 1956. Hydromagnetic dynamo theory. *Review of Modern Physics*, **28**: 135–163.

Elsasser, W.M., 1978. *Memoirs of a Physicist in the Atomic Age.* New York: Science History Publications.

Elsasser, W.M., 1998. *Reflections on a Theory of Organisms.* Baltimore: Johns Hopkins University Press.

Gilbert, W., 1600. *De Magnete.* London: Chiswick Press.

Glatzmaier, G.A., and Roberts, P.H., 1996. Rotation and magnetism of Earth's inner core. *Science*, **274**: 1887–1892.

Gubbins, D., 1973. Numerical solution of the kinematic dynamo problem. *Philosophical Transactions of the Royal Society London*, Series A, **274**: 493–521.

Krause, F., and Rädler, K.H., 1980. *Mean-field Magnetohydrodynamics and Dynamo Theory.* Oxford: Pergamon Press.

Moffatt, H.K., 1978. *Magnetic Field Generation in Electrically Conducting Fluids.* Cambridge: Cambridge University Press.

Parker, E.N., 1955. Hydromagnetic dynamo models. *Astrophysics Journal*, **122**: 293–314.

Parker, E.N., 1979. *Cosmical Magnetic Fields.* Oxford: Clarendon Press.

Pekeris, C.L., Accad, Y., and Shkoller, B., 1973. Kinetic dynamos and the Earth's magnetic field. *Philosophical Transactions of the Royal Society London Series A*, **275**: 425–461.

Cross-references

Bullard, Edward Crisp (1907–1980)
Cowling's Theorem
Dynamo Waves
Dynamo, Backus
Dynamo, Bullard-Gellman
Dynamos, Kinematic
Dynamos, Mean Field
Dynamos, Planetary and Satellite
Geodynamo
Gilbert William (1544–1603)
Rikitake, Tsuneji (1921–2004)

EM MODELING, FORWARD

Over the last decade, the electromagnetic (EM) induction community had three large international meetings entirely devoted to the theory and practice of three-dimensional (3-D) EM modeling and inversion (Oristaglio and Spies, 1999; Zhdanov and Wannamaker, 2002; Macnae and Liu, 2003; and references therein). As a result, a multitude of various kinds of forward modeling solutions have been revealed to the community. An advanced reader may find it interesting to browse through the above references on his own.

Three-dimensional EM numerical modeling is used today, (i) as an engine for 3-D EM inversion (see *EM modeling, inverse*); (ii) for verification of hypothetical 3-D conductivity models constructed using various approaches; and (iii) as a tool for various feasibility studies and studies to investigate various 3-D effects. During 3-D modeling we solve numerically, Maxwell's equations (here presented in the frequency-domain, with time dependence $\exp(-i\omega t)$ assumed)

$$\nabla \times \mathbf{H} = \tilde{\sigma} \cdot \mathbf{E} + \mathbf{j}^{\text{ext}}, \qquad \text{(Eq. 1a)}$$

$$\nabla \times \mathbf{E} = i\omega\mu \cdot \mathbf{H}, \qquad \text{(Eq. 1b)}$$

where the generalized conductivity $\tilde{\sigma} = \sigma - i\omega\varepsilon$ is a scalar or 3×3 tensor function of x, y, z; σ, ε, and μ are the Earth's electric conductivity, dielectric permittivity, and magnetic permeability, respectively; $\mathbf{j}^{\text{ext}}$ is the impressed current source, $\nabla = (\partial_x, \partial_y, \partial_z)$ stands for the spatial gradient, $\times$ denotes the vector product, $i = \sqrt{-1}$. This allows us to calculate the electric $\mathbf{E}$ and magnetic $\mathbf{H}$ fields that propagate in a 3-D conductive earth within the volume of interest.

Besides *rigorous* 3-D forward problem solutions that solve the problem in three spatial coordinates x, y, z and time t (or frequency ω), there is a variety of so-called *approximate* solutions.

Approximate solutions

Approximate solutions impose constraints on the conductivity models, frequency range, and/or EM field behavior. The reasoning behind the development of such approximate solutions is that until very recently, adequate methods for accurate numerical solutions of 3-D forward problems were not properly developed and computing resources were relatively limited. Even today, a proper solution of some forward problems may take hours or days of computer time. Additionally, approximate solutions may be useful tools at some stages of inversion. Generally, 1-D and 2-D forward problem solutions may be viewed as approximate since the conductivity σ is restricted to be a function of only one or two spatial variables. Both cases are a particular case of the 3-D problem, considered below in detail. To date, numerical solution of 1-D and 2-D forward problems have traditionally been used.

Other solutions imposing additional constraints on the conductivity models are so-called thin sheet solutions, where the 3-D inhomogeneities are collapsed in one or more conductive, or resistive thin sheets.

The calculation within the inhomogeneities is entirely neglected and substituted by known conditions that match the EM field at both sides of the sheets.

Ideally, the rigorous solution for geoelectromagnetic problems has to be able to simulate the EM field in the earth for frequencies ranging from the direct current ($\omega = 0$) up to frequencies of $\omega \approx \sigma\varepsilon^{-1}$, where ε is the Earth's dielectric permittivity. However, accurate simulation for so many decades of frequency is a complex numerical problem. This is why some numerical solutions constrain the frequency range. In geoelectromagnetics the most popular of these are the static limit, or direct current solutions, where only low frequencies are considered.

Lastly, there is also a variety of solutions, based on approximations of the Born-Rytov type, requiring a low contrast assumption. All of the above solutions are particular cases of strict rigorous solutions and require the same approach for their solution as more rigorous ones.

Rigorous solutions

There are three commonly used methods to obtain the numerical solution: finite-difference (FD), finite-element (FE), and volume integral equation (IE). These are now discussed in turn.

Finite-difference method

The most commonly employed is the FD method. The foundation of this method, applied to geoelectromagnetics, is grounded in the work of Yee (1966). Maxwell's equations are reduced to a second-order partial differential equation (PDE) with respect to the total electric, or magnetic field, as

$$\nabla \times \nabla \times \mathbf{E} - i\omega\mu\tilde{\sigma} \cdot \mathbf{E} = i\omega\mu\mathbf{j}^{\text{ext}} \quad \text{(Eq. 2)}$$

An alternative way to write Eq. (1) using the scattered field formulation is

$$\nabla \times \nabla \times \mathbf{E}^{\text{s}} - i\omega\mu\tilde{\sigma} \cdot \mathbf{E}^{\text{s}} = i\omega\mu\mathbf{j}^{\text{s}}, \quad \text{(Eq. 3)}$$

where $\mathbf{E}^{\text{s}} = \mathbf{E} - \mathbf{E}^0$ is the scattered electric field and

$$\mathbf{j}^{\text{s}} = (\tilde{\sigma} - \tilde{\sigma}^0) \cdot \mathbf{E}^0 \quad \text{(Eq. 4)}$$

Here, $\tilde{\sigma}^0$ is the conductivity of a reference model, which is chosen simpler than $\tilde{\sigma}$, $\mathbf{E}^0$ is the electric field excited by $\mathbf{j}^{\text{ext}}$ in the reference model $\tilde{\sigma}^0$. $\mathbf{E}^{\text{s}}$ of Eq. (3) is subject to the Dirichlet boundary condition $\mathbf{E}^{\text{s}} \times \mathbf{n}|_{\partial V} = 0$, where V is the computational volume, and $\mathbf{n}$ is a normal to its boundary ∂V, although other boundary conditions are possible. The preferred forms Eq. (2) or Eq. (3) depend on the specific problem. For numerical modeling one needs to reformulate the problem on a numerical grid. Details of such reformulations are necessary to understand the essence of current FD (and other) numerical methods.

The approximation of Eq. (3) on the FD-staggered grid (SG; Yee, 1966) of $m = n_x \times n_y \times n_z$ rectangular cells yields a linear system of $3m$ equations

$$\nabla'_{\text{SG}} \times \nabla_{\text{SG}} \times \mathcal{E} - i\omega\mu\Sigma \cdot \mathcal{E} = i\omega\mu \cdot \mathcal{J}, \quad \text{(Eq. 5)}$$

where $\mathbf{E}^{\text{s}}$ is sampled at the edges of the cells, while $\mathbf{H}^{\text{s}} = 1/i\omega\mu\nabla \times \mathbf{E}^{\text{s}}$ is sampled at the centers of the cell faces. In Eq. (5), the SG curl-curl operator $\nabla'_{\text{SG}} \times \nabla_{\text{SG}} \times$, Σ, $\mathcal{E}$, and $\mathcal{J}$ are the grid approximations of the continuous curl-curl operator $\nabla \times \nabla \times$, conductivity $\tilde{\sigma}$, electric field $\mathbf{E}^{\text{s}}$, and current $\mathbf{j}^{\text{s}}$, respectively. Conductivity $\tilde{\sigma}$ is uniform within each cell. A matrix form of Eq. (5) is

$$A_{\text{FD}} \cdot E = S, \quad \text{(Eq. 6)}$$

where $A_{\text{FD}} = B - i\omega\mu\Lambda$, E represents $\mathcal{E}$ at the SG nodes, S represents the source $\mathcal{J}$ and the boundary conditions (see Weiss and Newman, 2003; and references therein). The $3m \times 3m$ matrix A_{FD} is large, sparse, with no more than 13 nonzero entries in each row. It has real-valued entries everywhere except those placed at the main diagonal, which are complex-valued. This means that although the matrix A_{FD} is symmetric, it is not Hermitian. A matrix-free variant of the FD method allows for dynamical calculation of the matrix A_{FD} and avoids its storage.

System (6) may be solved by many methods. For 3-D problems the direct methods of matrix inversion become prohibitive in terms of both time and storage. Most of the methods commonly used for the solution of Eq. (6) are variants of Krylov iteration (Greenbaum, 1997). For instance, the quasi-minimum residual (QMR) and biconjugate gradient (BiCG) iterations are applied to solve Eq. (6). However, both methods show similarly slow convergence rates. It is known that the convergence rate is governed by the condition number

$$\kappa(A_{\text{FD}}) = \|A_{\text{FD}}\|\|A_{\text{FD}}^{-1}\|$$

where A_{FD}^{-1} is the inverse matrix, and $\|A_{\text{FD}}\| = \max_u \|A_{\text{FD}}u\|/\|u\|$, where $\|u\| = \sqrt{(u,u)}$. Larger condition numbers produce a slower convergence rate. As a rule, system (6) is ill conditioned with $\kappa(A_{\text{FD}})$ as large as 10^9–10^{12}. These poorly preconditioned systems converge slowly, if at all. In order to obtain a faster solution, Eq. (6) must be transformed to a preconditioned form.

Techniques currently in use to achieve better solution

Today the techniques include: preconditioning, decomposition by potentials, and static divergence correction. In what follows these important issues are briefly addressed.

Preconditioning. A preconditioned form of Eq. (6) is (Greenbaum, 1997)

$$(A_{\text{FD}}M^{-1}) \cdot X = S, \quad \text{(Eq. 7)}$$

where $X = M \cdot E$ is the $3m$-vector of modified unknowns, the $3m \times 3m$ matrix M is called the (right) preconditioner, and M^{-1} is the inverse of M. When Eq. (7) is solved to give an approximate solution X, the solution E of Eq. (6) is resolved from $M \cdot E = X$. The preconditioner M is sought so that $A_{\text{FD}}M^{-1}$ turns out close to the identity matrix. It is desirable to choose M so that Eq. (7) is better preconditioned than Eq. (6), i.e. $1 \approx \kappa(A_{\text{FD}}M^{-1}) \ll \kappa(A_{\text{FD}})$. For moderate to high frequencies system (6) may be effectively solved with the Jacobi preconditioner, $M = diag(A_{\text{FD}})$. An alternative way to precondition Eq. (6) is to use an incomplete Cholesky decomposition, $A_{\text{FD}} = M^tM$, where the preconditioner M is an upper triangular matrix, and the superscript t means its transpose. If $A_{\text{FD}} = A_1 + A_2$, where A_1 dominates over A_2 then $M = A_1$, can be used as a preconditioner. In this case the preconditioned system is

$$(1 + A_2A_1^{-1}) \cdot X = S, \quad \text{(Eq. 8)}$$

where $A_1E = X$. This preconditioning is typically applied to precondition the problem in the static limit.

Static limit. At low frequencies, or at low induction numbers (LIN) λ, i.e., when

$$\lambda = \sqrt{\omega\mu\sigma} \cdot \Delta \ll 1, \quad \text{(Eq. 9)}$$

where Δ stands for the characteristic grid spacing, an iterative solution of Eq. (7) may encounter certain difficulties. A stagnation, or even divergence of the QMR (with Jacobi preconditioner) and spectral Lancsoz decomposition method (SLDM) (see Section SLDM) iterations has been reported. The main reason for the poor convergence is the following. Only one part of Eq. (3) contains the conductivity $\tilde{\sigma}$

and becomes negligibly small as frequencies approach the static limit of Eq. (9). From Eq. (5) and Eq. (9), it follows that $\|i\omega\mu\Sigma\cdot\mathcal{E}\|/\|\nabla'_{SG}\times\nabla_{SG}\times\mathcal{E}\|\approx\lambda^2\ll 1$, since the operator $\nabla'_{SG}\times\nabla_{SG}\times$ is roughly estimated as Δ^{-2}, and Eq. (5) degenerates to

$$\nabla'_{SG}\times\nabla_{SG}\times\mathcal{E}\approx \mathrm{i}\omega\mu\cdot\mathcal{J} \quad \text{(Eq. 10)}$$

The operator of Eq. (10) has a nontrivial null space, since $\mathcal{E}$ can be augmented by the gradient $\nabla'_{SG}\varphi$ of a scalar function φ and still satisfies Eq. (10). To solve Eq. (10), we notice that, when frequencies approach the static limit, the equation

$$\nabla_{SG}\cdot(\Sigma\mathcal{E}+\mathcal{J})=0, \quad \text{(Eq. 11)}$$

which follows from Ampere's law, must be considered as a supplement to Eq. (10) for the Earth's volume and equation $\nabla_{SG}\cdot\mathcal{E}=0$ for the air volume. When these supplements are ignored and a Krylov iteration is applied directly to solve Eq. (10), the null space of Eq. (10) is responsible for poor convergence. To overcome this, the static divergence correction outlined further has been proposed. It dramatically improves convergence and exploits the Helmholtz decomposition.

Helmholtz decomposition. The Helmholtz potentials $\boldsymbol{\Psi}$, φ allow decomposition of the electric field $\mathcal{E}$ into curl-free $\nabla'_{SG}\varphi$ and divergence-free $\boldsymbol{\Psi}$ parts, so that

$$\mathcal{E}=\boldsymbol{\Psi}+\nabla'_{SG}\varphi \quad \text{(Eq. 12)}$$

Imposing the Coloumb gauge condition

$$\nabla_{SG}\cdot\boldsymbol{\Psi}=0 \quad \text{(Eq. 13)}$$

renders the Helmholz's decomposition of Eq. (12) unique. Substituting Eq. (12) in Eq. (5) and Eq. (11) yields

$$\nabla'_{SG}\times\nabla_{SG}\times\boldsymbol{\Psi}-\mathrm{i}\omega\mu\Sigma\cdot(\boldsymbol{\Psi}+\nabla'_{SG}\varphi)=\mathrm{i}\omega\mu\cdot\mathcal{J}, \quad \text{(Eq. 14)}$$

$$\nabla_{SG}\cdot(\Sigma\nabla'_{SG}\varphi)=-\nabla_{SG}\cdot(\Sigma\boldsymbol{\Psi}+J), \quad \text{(Eq. 15)}$$

subject to a mixture of Dirichlet and Neumann conditions: $\mathbf{n}\times\boldsymbol{\Psi}=0$, $(\mathbf{n}\cdot\nabla_{SG})\boldsymbol{\Psi}=0$, and $\varphi=0$ on the computational volume boundary, ∂V. Systems (13)–(15) are strongly elliptic and weakly coupled. It is an alternative and a better form of the traditional system (5) or (6). Again, a Krylov iteration is usually applied to solve this system.

Static divergence correction. When $\lambda\ll 1$, Eq. (14) degenerates to

$$\nabla'_{SG}\times\nabla_{SG}\times\boldsymbol{\Psi}\approx\mathrm{i}\omega\mu\cdot J, \quad \text{(Eq. 16)}$$

or in a matrix form

$$B\Psi=S, \quad \text{(Eq. 17)}$$

where matrix B is a grid representation of the curl-curl operator. Equations (13), (15), and (16) form a complete system of equations in the static limit. Equations (13) and (16) are solved jointly to compute $\boldsymbol{\Psi}$ which is substituted into Eq. (15) to find φ. Given potentials $\boldsymbol{\Psi}$, and φ, the field E is found from Eq. (12). This solution is an effective preconditioner of a type given in Eq. (8), the so-called LIN preconditioner with $M=B$, if the system (13)–(15) is solved in the static limit (Weiss and Newman, 2003). This preconditioner is very effective at low to moderate frequencies and accelerates convergence. At high frequencies, however, the LIN preconditioned Krylov iteration can fail to converge or give much slower solution times than Jacobi preconditioning.

On the basis of these ideas, a multigrid preconditioner for systems (13), (14), and (15) have also been developed (Ariliah and Ascher, 2003). This preconditioner uses single multigrid W-cycles to approximate inversion of matrices of the systems (15) and (16). It gives a preconditioned system with a grid-independent bound on its condition number. This means that for low frequencies the convergence rate does not depend on frequency and conductivity. Once frequency gets larger, the condition number increases quadratically and the convergence rate becomes slower.

The techniques presented in this section are more fundamental than mere mathematical tricks for accelerating the solution convergence. They are deeply rooted in the physics of the EM induction problem and so they allow us to more precisely describe it.

Resume

The main attraction of the FD approach for EM software developers is its relatively simple numerical implementation, especially when compared to other approaches, and as well as its effectiveness for many of EM problems. The drawback is that convergence of an iterative solution of Eq. (6) is not guaranteed even after a preconditioning. The large computational volume needed for stabilization of the EM field at boundaries is another limitation.

The FD method has been extended to fully anisotropic 3-D media (see *Anisotropy, electrical*). Time domain FD solutions have also been developed and implemented (Commer and Newman, 2004; and references therein).

Finite-element method

In the FE method the whole modeling volume V is decomposed into elementary volumes (such as prisms, tetrahedra, or more complex shapes) that specify the geometry of the conductivity model. Accordingly, the electric field of Eq. (3) (or its potentials) is decomposed as

$$\mathbf{E}^s=\sum_{j=1}^{m}\alpha_j(x,y,z)\mathbf{E}_j(\omega), \quad \text{(Eq. 18)}$$

where m is the number of nodes (as a rule, they coincide with the vertices of elementary volumes) and $\mathbf{E}_j$ is the value of electric field at jth node. Functions α_j of Eq. (18) are specified. The coefficients $\mathbf{E}_j$ of decomposition (18) are sought using the Galerkin method: Eq. (3) is multiplied by α_i and integrated over V, Eq. (18) is then substituted into the result to give

$$\sum_{j=1}^{m}\int_{V^s}\alpha_i(\nabla^2-\mathrm{i}\omega\mu\tilde{\sigma})\alpha_j\mathrm{d}r'\mathbf{E}_j=\mathrm{i}\omega\mu\int_{V^s}\alpha_i\mathbf{j}^s\mathrm{d}r', \quad \text{(Eq. 19)}$$

or in matrix form

$$A_{\mathrm{FE}}\cdot E=S, \quad \text{(Eq. 20)}$$

where the $3m\times 3m$ system matrix is $A_{FE}=B-i\omega\mu C$. Here E represents the $3m$-vector of the components of electric field at nodes, $B_{ij}=\int_{V^s}\alpha_i\nabla^2\alpha_j\mathrm{d}r'$, $C_{ij}=\int_{V^s}\alpha_i\tilde{\sigma}\alpha_j\mathrm{d}r'$, and $S_i=i\omega\mu\int_{V^s}\alpha_i j^s\mathrm{d}r'$. To derive Eq. (19) we used the theorem $\nabla\times\nabla\times\mathbf{E}=\nabla\nabla\mathbf{E}-\nabla^2\mathbf{E}$. As in the case of the FD approach, we arrive at a nonsymmetric sparse system of linear equations (20). All numerical techniques to improve the solution accuracy and performance that were outlined above hold for system (20) and generally for the FE approach. The main attraction of the FE approach is that it is believed to be better able to account for geometry (shapes of ore-bodies, topography, cylindrical wells, etc.). This attraction is counterbalanced by a nontrivial and usually time-consuming construction of the finite elements themselves. The FE approach has been implemented by many developers (see Mitsuhata and Uchida, 2004; and references therein).

Integral equation method

With the IE method (Weidelt, 1975; among others), Maxwell's equations (2) are first reduced to a second-kind Fredholm's IE

$$\mathbf{E}(\mathbf{r}) = \mathbf{E}_0(\mathbf{r}) + \int_{V^s} \underline{\underline{G}}_0(\mathbf{r};\mathbf{r}')(\tilde{\sigma} - \tilde{\sigma}_0)\mathbf{E}(\mathbf{r}')\mathrm{d}r' \qquad \text{(Eq. 21)}$$

with respect to the electric field. This is known as the scattering equation (SE). To derive the SE, the Green's function technique is usually applied. In Eq. (21), the free term $\mathbf{E}_0$ is known, $\underline{\underline{G}}_0$ is the 3×3 dyadic for the Green's function of the 1-D reference medium, and V^s is the volume where $(\tilde{\sigma} - \tilde{\sigma}_0)$ differs from zero. A discretization of the SE yields the linear system $A_{\mathrm{IE}} \cdot E = S$, provided that both conductivity $\tilde{\sigma}$ and the unknown electric field $\mathbf{E}$ are constant within each cell. The system matrix A_{IE} is complex and dense, with all entries filled, but more compact than the A_{FD}, or A_{FE} matrices. Again, to get a well preconditioned system matrix A_{IE}, the modified iterative-dissipative method (MIDM) has been successfully developed (Singer, 1995) and implemented (Avdeev *et al.*, 2002; among others). It is surprising that the MIDM-preconditioned system matrix A_{IE} has such a small condition number, $\kappa(A_{\mathrm{IE}}) \leq \sqrt{C_l}$, where C_l is the lateral contrast of conductivity. The main merit of the IE approach is that only the scattering volume V^s is subject to discretization. This reduces the size of the matrix A_{IE} dramatically, as all other methods require a larger volume to be discretized. However, most EM software developers avoid the IE method because accurate computation of the matrix A_{IE} is a tedious and nontrivial problem.

For completeness, I mention the existence of surface IE solutions that assume a constant value of conductivity within the inhomogeneities (Chew, 1999).

Spectral Lancsoz decomposition method

Another efficient FD approach is the SLDM (Davydycheva *et al.*, 2003; see references therein). In order to solve Eq. (6), the Lanczos spectral decomposition, $B = QTQ^{-1}$, is applied. Here Q is the orthogonal matrix, so that $Q^t Q = 1$, and T is a real-valued symmetric three-diagonal matrix. Entries of both matrices Q and T are sought by the Lanczos iterative process (Greenbaum, 1997). Techniques presented in Section *Technique currently in use* ... also hold for the SLDM. SLDM is considered as the method of choice when multifrequency modeling is required because it solves Maxwell's equations at many frequencies for a cost only slightly greater than that for a single frequency. For such numerical effectiveness the SLDM slightly sacrifices its versatility. It assumes that conductivity $\tilde{\sigma}$ and current $\mathbf{j}^{\mathrm{ext}}$ of Eq. (1) are frequency independent, which means that the induced polarization effects cannot be taken easily into account. The SLDM has been extended to anisotropic media. Davydycheva *et al.* (2003) proposed a special conductivity averaging and optimal grid refinement that reduce grid size and accelerate computation. They claim that their new scheme outperforms other FD schemes by an order of magnitude.

Conclusion

Regardless of what method is employed, the initial forward problem (1) is always reduced to a system of linear equations

$$A \cdot X = S \qquad \text{(Eq. 22)}$$

Nowadays, the system is commonly solved iteratively by a preconditioned Krylov iteration. The properties of the matrix A are determined by which method (FD, FE, or IE) is applied to solve the forward problem. In this respect, only two aspects are important, (i) how accurate the system (22) represents Maxwell's equations, and (ii) how well preconditioned the system matrix A is. Competition between various modeling approaches (FD, FE, and IE) is today focused entirely on these two issues. The ultimate goal of 3-D modelers is first, to design a more accurate approximation to Maxwell's equations within a coarser grid discretization. The second important challenge is to find a faster preconditioned linear solver. As a result of this competition between methods, we now have several very effective codes for numerical modeling of 3-D EM fields at our disposal.

The methods described in this article have many important geophysical applications (see *EM, land uses*, *EM, industrial uses*, *Transient EM induction*, and *EM, regional studies*), such as induction logging, airborne EM, magnetotellurics (see *Magnetotellurics*), geomagnetic deep sounding (see *Geomagnetic deep sounding*) and controlled source EM (see *EM, marine controlled source*). While they use Cartesian geometry, some implementations are also available to simulate 3-D spherical earth conductivity models (see *Induction from satellite data*).

Dmitry B. Avdeev

Bibliography

Aruliah, D.A., and Ascher, U.M., 2003. Multigrid preconditioning for Krylov methods for time-harmonic Maxwell's equations in 3D. *SIAM Journal of Scientific Computing*, **24**: 702–718.

Avdeev, D.B., Kuvshinov, A.V., Pankratov, O.V., and Newman, G.A., 2002. Three-dimensional induction logging problems. Part I. An integral equation solution and model comparisons. *Geophysics*, **67**: 413–426.

Chew, W.C., 1999. *Waves and Fields in Inhomogeneous Media.* Piscataway: Wiley-IEEE Press.

Commer, M., and Newman, G., 2004. A parallel finite-difference approach for 3-D transient electromagnetic modeling with galvanic sources. *Geophysics*, **69**: 1192–1202.

Davydycheva, S., Druskin, V., and Habashy, T., 2003. An efficient finite difference scheme for electromagnetic logging in 3-D anisotropic inhomogeneous media. *Geophysics*, **68**: 1525–1536.

Greenbaum, A., 1997. *Iterative Methods for Solving Linear Systems.* Philadelphia, SIAM.

Macnae, J., and Liu, G. (eds.), 2003. *Three-Dimensional Electromagnetics III.* Australia: Australian Society of Exploration Geophysicists.

Mitsuhata, Y., and Uchida, T., 2004. 3-D magnetotelluric modeling using the T-Ω finite-element method. *Geophysics*, **69**: 108–119.

Oristaglio, M., and Spies, B. (eds.), 1999. *Three-Dimensional Electromagnetics.* Tulsa, OK: Society of Exploration Geophysicists.

Singer, B.Sh., 1995. Method for solution of Maxwell's equations in nonuniform media. *Geophysical Journal International*, **120**: 590–598.

Weidelt, P., 1975. Electromagnetic induction in 3-D structures. *Journal of Geophysics*, **41**: 85–109.

Weiss, Ch.J., and Newman,G.A., 2003. Electromagnetic induction in a fully 3-D anisotropic earth. Part 2. The LIN preconditioner. *Geophysics*, **68**: 922–930.

Yee, K.S., 1966. Numerical solution of initial boundary value problems involving Maxwell's equations in isotropic media. *IEEE Transactions of Antennas and Propagations*, **AP-14**: 302–307.

Zhdanov, M.S., and Wannamaker, P.E. (eds.), 2002. *Three-Dimensional Electromagnetics. Methods in Geochemistry and Geophysics*, Vol. 35. Amsterdam: Elsevier.

Cross-references

Anisotropy, Electrical
EM Modeling, Inverse
EM, Industrial Uses
EM, Land Uses
EM, Marine Controlled Source
EM, Regional Studies
Geomagnetic Deep Sounding
Induction from Satellite Data
Magnetotellurics
Transient EM Induction

EM MODELING, INVERSE

The forward problem of electromagnetic (EM) induction involves solution of Maxwell's equations in the electrically conducting Earth, excited by appropriate external sources. Inverse modeling of EM data reverses this process, using data, generally observed on the surface, to image conductivity variations within the Earth. For natural source methods such as magnetotellurics (MT) the frequency-dependent response of the Earth to large-scale sources is determined by statistical estimation of transfer functions (essentially local ratios of field components) from EM time series data (see *Robust EM transfer function estimates*). For controlled source methods source amplitudes and phases are known accurately, and a more direct computation of the Earth response is possible. In either case, at the inversion stage the external sources are taken as fixed and known, and Earth conductivity is adjusted to match the estimated response. In general, lower frequency EM variations penetrate to greater depths, so response functions estimated over a range of frequencies allow depth variations of electrical conductivity in the Earth to be resolved. With sufficient data coverage on the surface, lateral variations of conductivity can also be mapped. In this article we focus on the MT inverse problem. Although active source EM techniques often require more complex modeling codes to account for the greater spatial complexity of sources, theory, and methodology for active source inversion is otherwise quite similar.

Developments in inverse modeling of EM induction data closely parallel those in other subsurface geophysical imaging techniques, such as seismology. Initial efforts focused on one-dimensional (1D) interpretations, with data from one site inverted to obtain information about the local conductivity-depth profile. An extensive and essentially complete theory was developed, including exact analytical solutions to idealized 1D inverse problems. However, it soon became apparent that lateral variations of conductivity must be taken into account for a proper interpretation, and two-dimensional (2D) inversion methods were developed for profiles of data across a dominant geoelectric strike. Finally, as more powerful computational resources have become available, methods for full three-dimensional (3D) inversion have been developed. Although in principle a straightforward generalization of the 2D case, 3D inverse modeling remains a challenging computational problem. Development of new algorithms for 3D inverse modeling is thus still very much an area of active research. Other areas of active research include treatment of conductive anisotropy, and appraisal of nonuniqueness in multidimensional EM inverse problems.

One-dimensional inversion

For uniform external sources with time dependence $e^{i\omega t}$ impinging on a 1D Earth of conductivity $\sigma(z)$, the quasistatic approximation to Maxwell's equations (omitting displacement currents) reduces to the second order linear homogeneous partial differential equation (PDE)

$$E''_x(z,\omega) - i\omega\mu_0\sigma(z)E_x(z,\omega) = 0, \qquad \text{(Eq. 1)}$$

with orthogonal electric and magnetic fields related by

$$E'_x(z,\omega) = i\omega\mu_0 H_y(z,\omega). \qquad \text{(Eq. 2)}$$

In Eqs. (1) and (2), primes denote the derivative with respect to z. The 1D MT inverse problem is to find the conductivity profile $\sigma(z)$ (or equivalently, resistivity $\rho(z) = \sigma(z)^{-1}$), given measurements of the impedance $Z(\omega) = E_x(0,\omega)/H_y(0,\omega)$, or the equivalent inductive length scale $c(\omega) = E_x(0,\omega)/E'_x(0,\omega)$, at the surface $z = 0$. Although the relationship between the data ($Z(\omega)$) and model parameters ($\sigma(z)$) is nonlinear, this 1D inverse problem has an extensive literature and is now well understood. Whittall and Oldenburg (1992) provide a comprehensive review of both the mathematical theory and of practical methods for 1D inversion of MT data. Note that other sorts of EM induction data (e.g., the ratios of internal to external fields for sources of a fixed spherical harmonic degree encountered in global induction studies) can be transformed to equivalent MT responses, so the same theory and methods can be applied more generally.

An apparent resistivity and phase can be defined in terms of the impedance:

$$\rho_a(\omega) = (\omega\mu_0)^{-1}|Z(\omega)|^2 \quad \phi(\omega) = a\tan(Z(\omega)). \qquad \text{(Eq. 3)}$$

If conductivity is independent of depth ($\sigma(z) \equiv \sigma$) then $\rho_a(\omega) \equiv \sigma^{-1}$ and $\phi(\omega) \equiv \pi/2$. In this case computation of the conductivity from the measured impedance is trivial. For the realistic case where conductivity varies with depth, the apparent conductivity ($\sigma_a = \rho_a^{-1}$) is a function of frequency (Figure E6). Conversion from apparent conductivity as a function of frequency ($\sigma_a(\omega)$) to actual conductivity as a function of depth ($\sigma(z)$) is the essence of the 1D EM inverse problem (Figures E6 and E7).

Exact measurements of $Z(\omega)$ at an infinite number of frequencies determines $\sigma(z)$ uniquely. However, as with all practical inverse problems, real data are finite in number and imprecise, so solutions to the MT inverse problem are highly nonunique. In general, MT data constrain only the product of conductivity and layer thickness (the conductance), not the actual conductivity or layer thickness. Furthermore, the resolution of MT data tends to decrease with increasing depth, and the actual resistivity of a section beneath a more conducting layer is typically poorly determined.

A number of approaches for constructing specific solutions to the inverse problem have been proposed. The simplest involve approximate functional transformations of $c(\omega)$ or $Z(\omega)$ to $\sigma(z)$. Although

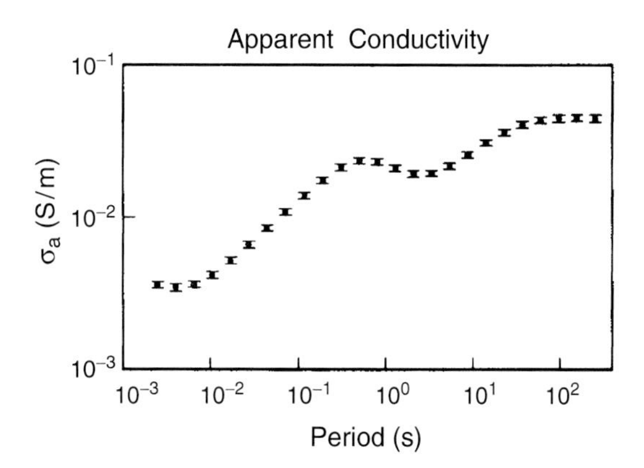

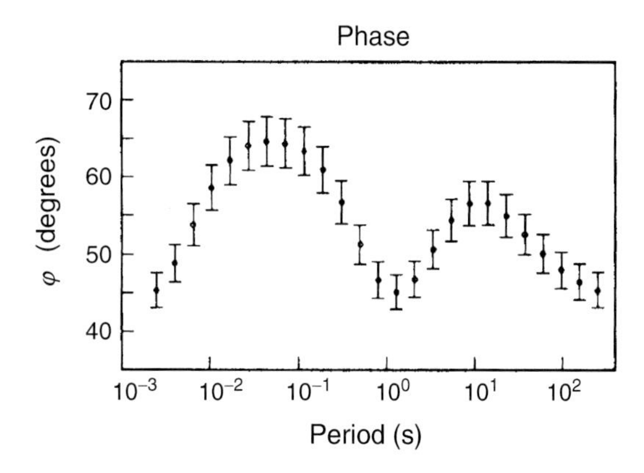

Figure E6 Apparent conductivity (inverse of apparent resistivity), and phase from a five layer 1D model (Figure E7). (After Whittall and Oldenburg, 1992).

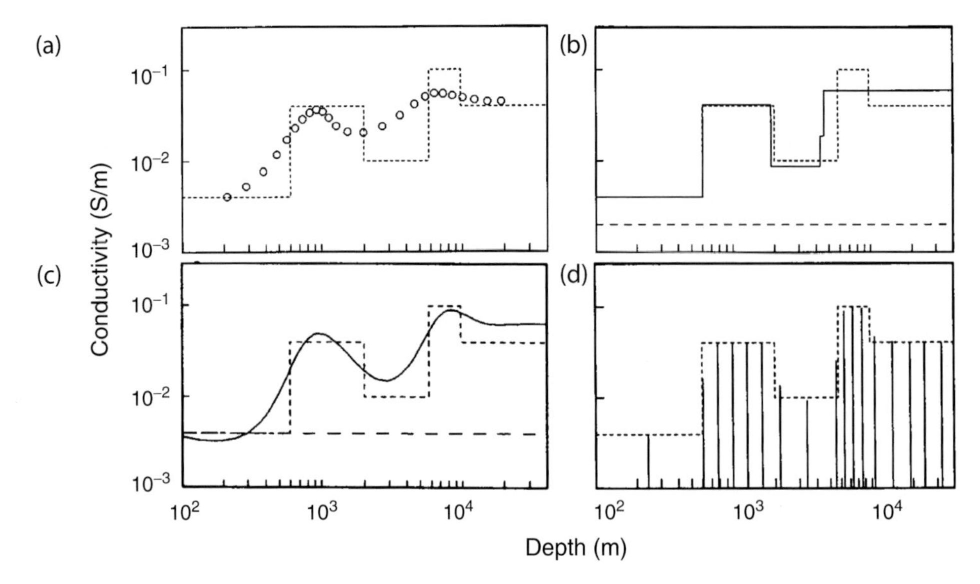

Figure E7 Examples of 1D inversions of the synthetic data of Figure E1. Dashed line gives actual layered model used to generate the data. (a) Heuristic Schmucker inversion of Eq. (4). (b) Layered inversion, with the unknown structure defined by five layers of unknown thickness and conductivity. (c) Minimum structure inversion, obtained by minimizing the norm of $\sigma'(z)$, subject to fitting the data within error bars. (d) D^+ inversion, the model which best fits the synthetic data. Each line corresponds to a thin layer of finite conductance, appropriately scaled to conductivity. (After Whittall and Oldenburg, 1992).

not true inversions (the conductivity profiles generated do not generally reproduce the observed response functions), these transformations provide reasonable rough estimates of conductivity profiles in many situations, and their derivations provide physical insight into the inverse problem. Perhaps the simplest example is the so-called Schmucker inversion. Given $c(\omega_j)$ at a finite set of frequencies, define

$$z_j^* = \Re[c(\omega_j)] \quad \sigma_j^* = (2\omega_j\mu_0)^{-1}\Im[c(\omega_j)]^2. \qquad \text{(Eq. 4)}$$

A plot of σ_j^* vs the discrete set of depths z_j^*, can be shown to provide a reasonable zero order approximation to the conductivity profile. An example for the synthetic data of Figure E6 is given in Figure E7a. Whittall and Oldenburg (1992) provide heuristic justification for this simple approach, and describe a number of similar approximate inversions.

More rigorous inversion of MT data can be accomplished with standard methods of nonlinear geophysical inversion (e.g., Parker, 1994). These methods can generally be formally reduced to minimization of a penalty functional of the form

$$\mathcal{J}(\mathbf{m},\mathbf{d}) = (\mathbf{d}-\mathbf{f}(\mathbf{m}))^{\mathrm{T}}\mathbf{C}_{\mathbf{d}}^{-1}(\mathbf{d}-\mathbf{f}(\mathbf{m})) + (\mathbf{m}-\mathbf{m}_0)^{\mathrm{T}}\mathbf{C}_{\mathbf{m}}^{-1}(\mathbf{m}-\mathbf{m}_0). \qquad \text{(Eq. 5)}$$

Here $\mathbf{d}$ is the data vector of dimension N_{d} (e.g., real and imaginary parts of the complex impedance $Z(\omega_j)$ estimated at a discrete set of frequencies), $\mathbf{C}_{\mathbf{d}}$ is the covariance matrix of data errors (typically diagonal), $\mathbf{m}$ is the N_{m}-dimensional conductivity model parameter vector (e.g., σ_j for a finite number of layers), $\mathbf{f}(\mathbf{m})$ defines the forward mapping, obtained by solving Eq. (1) at the appropriate frequency, $\mathbf{m}_0$ is a prior (or preferred) value of the conductivity parameter $\mathbf{m}$, and $\mathbf{C}_{\mathbf{m}}$ is the covariance of the unknown model parameters used to regularize the inversion (often given as a "roughening operator" $\mathbf{C}_{\mathbf{m}}^{-1}$). The two terms on the right-hand side of Eq. (5) penalize data misfit, and the norm or size (in some general sense) of the model parameter $\mathbf{m}$, respectively.

If the conductivity profile is described in terms of a small number of parameters (e.g., 3–5 layers of unknown thickness and conductivity, for a total of 6–10 unknown parameters; see Figure E7b), the second term explicitly penalizing $\mathbf{m}$ is typically omitted. However, frequently the conductivity model is overparameterized, with a large number of thin layers of unknown conductivity. In this case there are essentially an infinite number of parameter vectors that fit the data equally well, so some sort of penalty on the model size (i.e., $(\mathbf{m}-\mathbf{m}_0)^{\mathrm{T}}\mathbf{C}_{\mathbf{m}}^{-1}(\mathbf{m}-\mathbf{m}_0)$) is required to make the problem well-posed. Once the model parameterization and penalty functional are defined, any of a number of standard search algorithms can be used to seek the minimum of $\mathcal{J}$. Some specific approaches are described in Parker (1994). In general, these involve linearizing the forward mapping $\mathbf{f}(\mathbf{m})$, though for the 1D inverse problem global Monte Carlo search strategies which do not require explicit calculation of derivatives, such as simulated annealing or the genetic algorithm, have been successfully applied.

One very popular linearized approach is the "minimum structure" inversion. In this approach, the model penalty in Eq. (5) is the norm of the first or second derivative of $\sigma(z)$, and the minimizer of $\mathcal{J}$ is the simplest (in the sense of flatness or smoothness of the conductivity profile) consistent with the achieved data misfit (Figure E7c). These minimum structure models are less likely to contain features that are not required by the data. Several different gradient-based optimization algorithms have been used for minimum structure inversions; further details can be found in Whittall and Oldenburg (1992) and Parker (1994).

The 1D MT inverse problem is somewhat unusual in that a number of exact nonlinear inversion algorithms are available. These are important from a theoretical perspective since they provide a constructive proof of existence and uniqueness of inverse solutions. However, these exact nonlinear algorithms all require perfect and complete data, so some adaptation is necessary to apply these to real finite noisy data sets. Generally a two-step procedure is used (Whittall and Oldenburg, 1992): first the incomplete noisy data must be fit to a physically realizable response at all frequencies; then the exact inversion is applied to convert the completed response into the unique corresponding conductivity profile. The most important example of this approach is the so-called D^+ inversion of Parker (1980). Using the representation of $c(\omega)$ in terms of the Sturm-Liouville spectrum of the differential operator in Eq. (1) to complete the estimated response, Parker showed that the best fitting conductivity model for real data sets would generally consist of a finite number of infinitely thin layers of finite conductance, separated by perfectly insulating layers of finite thickness (Figure E7d). Parker and Whaler (1981) developed a stable numerical implementation of this inversion, which has since been widely used in numerous practical applications. Although the conductivity profiles that result from this inversion are obviously not physically realistic, the D^+ inversion is very useful because it provides a rigorous assessment of the degree to which a real data set is consistent with the assumption of a 1D conductivity profile. The nature of the best fitting profile is also instructive with regard to the nonuniqueness of MT data: only conductance, and the rough distribution of conductance with depth, can be resolved by MT data, not the actual value of conductivity at any depth (Figure E7).

Model appraisal

Many particular conductivity profiles will fit any data set equally well, so assessment of which features are common to all acceptable solutions is an important part of the inverse problem. The classical approach to model appraisal for linear problems is Backus-Gilbert theory, based on construction of spatial averages of model parameters that are constrained by the data. For the nonlinear EM inverse problem this resolution analysis requires linearization, and thus only provides a rough characterization of nonuniqueness. Constrained inversion provides an alternative approach to inverse solution appraisal. For example, to test if a particular conductive layer is required, the inversion can be run with conductivity in the appropriate depth range constrained to a low value. If the data cannot be fit adequately subject to this constraint, but can be when the constraint is relaxed, higher conductivity is required on average over this depth range. This general idea can be formalized to construct upper and lower bounds on the total conductance in a given depth range. Details of these, and related appraisal methods for the 1D EM inverse problem are given in Whittall and Oldenburg (1992).

Two-dimensional inversion

For 2D MT modeling a fixed geoelectric strike is assumed, with conductivity $\sigma(y,z)$ varying across the strike (y), and with depth (z), but not along the strike (x). In this case the MT forward problem decouples into two modes: the transverse electric (TE) or E-polarization mode, with electric currents flowing only in the x-direction parallel to the geologic structure, and the transverse magnetic (TM) or B-polarization mode, with electric currents flowing only in the y-direction perpendicular to the strike. In both modes the horizontal magnetic fields are still perpendicular to the electric fields, oriented across strike for TE, and along strike for TM. Now there are distinct impedances for TE and TM modes

$$Z_{xy} = E_x/H_y, \quad Z_{yx} = E_y/H_x, \qquad \text{(Eq. 6)}$$

as well as vertical field transfer functions (or "tippers") in the TE mode $T = H_z/H_y$. As for the 1D case the impedances are often transformed to apparent resistivities and phases, using Eq. (3). Data for 2D interpretation are generally collected along a profile crossing the presumed geoelectric strike. These data are commonly displayed as cross-strike pseudosections, with frequency on the vertical axis (Figure E8). Conversion of these images to conductivity sections, merging information from both modes, and replacing frequency by depth, is the goal of 2D inversion (Figure E8).

For the TE mode, Maxwell's equations can be reduced to a 2D scalar diffusion equation for the along strike electric field

$$\nabla^2 E_x - i\omega\mu_0\sigma E_x = 0. \qquad \text{(Eq. 7)}$$

This is analogous to the 1D equation of Eq. (1). For the TM mode, Maxwell's equations can be reduced to a scalar equation in the along strike magnetic field

$$\nabla^2 B_x + \sigma^{-1}\nabla(\sigma)\cdot\nabla(B_x) - i\omega\mu_0\sigma B_x = 0. \qquad \text{(Eq. 8)}$$

Discretizing Eqs. (7) and (8) with a finite-difference or finite-element approximation results in systems of linear equations that can be solved by a number of approaches, including forming and factoring the banded coefficient matrix, and iterative Krylov space solvers such as biconjugate gradients (e.g., Press *et al.*, 1986).

Although existence and uniqueness of solutions to idealized 2D inverse problems has been proven, theory is far less complete than for the 1D case. In particular, there are no exact analytical procedures for constructing inverse solutions to 2D problems, and there is no general theory to establish consistency with a 2D interpretation for practical datasets.

Some approximate rapid schemes for converting 2D pseudosections to conductivity images have been developed and used. These are mostly based on rewriting the induction equations in integral form, and then using a Born scattering approximation to reduce to an approximate linear inverse problem. Although these methods can be very fast, images constructed are not guaranteed to actually fit the data. With increases in computing power, approaches based on minimizing a penalty functional of the form (5) have become most popular. In principle this can be accomplished with a straightforward extension of the 1D methods, but with the 2D forward, Eqs. (7) and (8) replacing Eq. (1) for computation of the forward mapping. However, for 2D problems there are many more unknown model parameters and data sets are much larger, so issues of computational efficiency become much more central.

Minimization of a penalty functional such as Eq. (5) is most directly accomplished through some sort of gradient based search. The most straightforward scheme is based on computing the Jacobian, i.e., the $N_\mathrm{d} \times N_\mathrm{m}$ matrix of first derivatives of the forward mapping

$$\mathbf{G} = \frac{\partial \mathbf{f}}{\partial \mathbf{m}}, \qquad \text{(Eq. 9)}$$

which can be accomplished most simply by solving nonhomogeneous versions of Eqs. (7) and (8) for every frequency, for each of the N_m columns of $\mathbf{G}$. Using the fact that sources and receivers are symmetric in the governing equations ("reciprocity") computation of $\mathbf{G}$ can generally be accomplished more efficiently by solving one forward problem for each data point, with a source located at the data site. With $\mathbf{G}$ available, the inverse problem can be linearized, and standard methods of linear inverse theory can be applied to find model parameters which more closely fit the data. Using the updated model parameters, $\mathbf{G}$ (which depends on the model conductivity) is recomputed and the inverse model further refined. These steps are iterated until no further reduction in $\mathcal{J}$ is possible. Details on several general schemes for implementing such a linearized iterative search, including the most commonly used for EM inverse problems, are given in Parker (1994), Siripunvaraporn and Egbert (2000), and Rodi and Mackie (2001).

The reciprocity approach improves efficiency of the Jacobian computations considerably, but construction and storage of $\mathbf{G}$ still requires significant computer time and memory. A variety of computational approaches have thus been proposed to make 2D inversion more practical. These include using approximations to the data sensitivities based on simpler 1D models (Smith and Booker, 1991), and constructing only a part of $\mathbf{G}$ to guide the linearized search (Siripunvaraporn and Egbert, 2000). A very common approach used now is to directly minimize J using a nonlinear conjugate gradient (NLCG) approach (Press *et al.*, 1986; Rodi and Mackie, 2001), completely avoiding construction and storage of $\mathbf{G}$. Each step in the NLCG search for the (local) minimum requires computation of the gradient of $\mathcal{J}$; this can be shown to require two TE and TM forward solutions for each frequency. NLCG is attractive because it completely avoids forming or inverting large matrices. However, because many more iterations can be required compared to schemes which make use of the Jacobian, the number of forward solutions may not be significantly reduced,

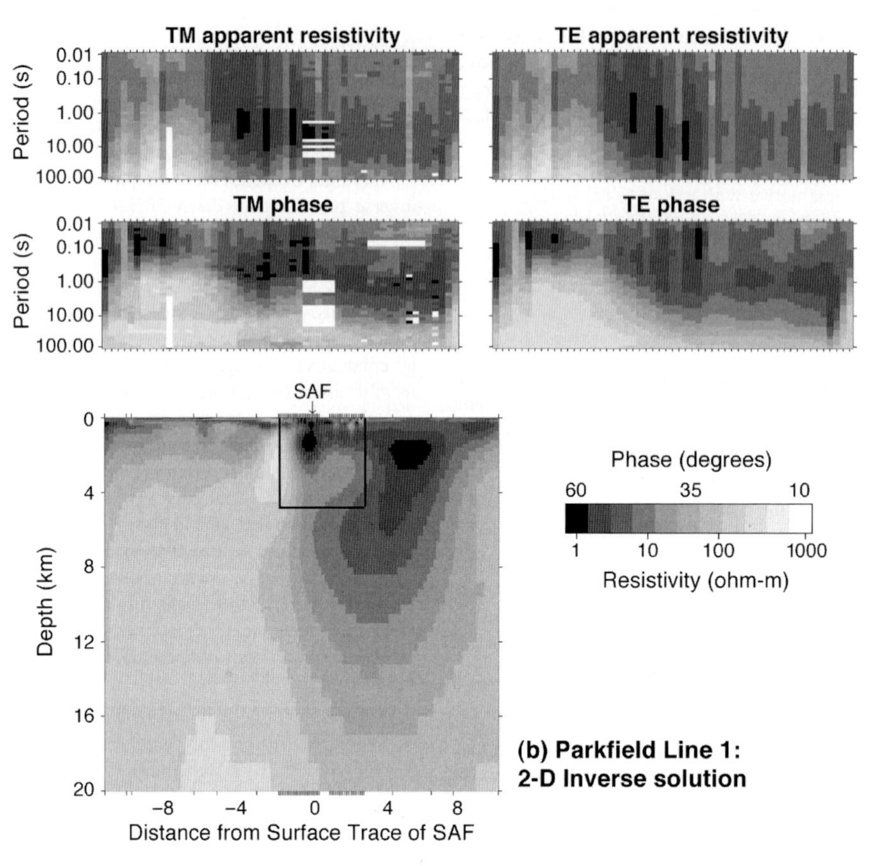

Figure E8 Example of a 2D MT data set and inversion. Pseudosections of (a) TE apparent resistivity; (b) TE phase; (c) TM apparent resistivity; TM phase, from an MT survey near Parkfield California. (d) Resistivity section obtained by inversion of TE and TM data, using a 2D minimum structure inversion. (After Siripunvaraporn and Egbert, 2000).

particularly in comparison with data space search schemes based on a partial sensitivity computation (Siripunvaraporn and Egbert, 2000). Nongradient based search schemes such as the genetic algorithm have been tried for 2D EM inverse problems, but searching the comparatively large parameter space required to represent 2D conductivity variations has not proven practical as yet.

Although a number of computationally feasible 2D MT inversion schemes have been shown to perform quite well on synthetic data, there are a number of complications with real data. Significant expert user intervention is thus often still required for successful 2D inversion of MT data. One challenging problem is that near surface small-scale geology is often very complicated, and can almost never be represented by a single 2D model that matches the regional geoelectric strike. Even if only superficial, these structures can galvanically distort electric fields, resulting in frequency-independent "static shifts" of apparent resistivity curves. Correction for these static shifts has been found to be essential for correct interpretation of deeper structure. In the context of inversion, additional model parameters are required to account for these distortions. Estimation of these surface distortion parameters has been integrated into some 2D inversion codes, but separate distortion analysis is also generally now seen as a critical step in 2D MT interpretation. Because phases are less effected by distortion initial inversions of field data often emphasize fitting of phase data, with apparent resistivities given less weight, and introduced later in a multistage inversion sequence.

Deeper structure can also result in 3D complications. For example, the EM response of a structure elongated along strike, but of finite length, will depend on how close one is to the ends of the structure. In general the TE mode, for which electric currents must flow through the ends of the structure, is most severely effected. As such, TM mode data are also often given higher weight, particularly for initial stages of inversion. Regional structure may also pose difficulties for 2D inversion. For example, a nearby coastline may not be parallel to the local geology. Since regional structure generally has a greater influence at longer periods, this complication may be ameliorated by emphasizing fit to shorter period data, but this of course limits the depth of investigation.

A final, and very important potential complication in 2D interpretation is conductive anisotropy, which may well be ubiquitous in the crust, and perhaps also the upper mantle. Allowing for conductivities that depend on the direction of current flow significantly increases the size of the model space, and as a result ambiguity and nonuniqueness in EM inversions. Some 2D inversion codes now allow for anisotropy, but there is so far little real understanding of resolution and trade-offs when this richer model space is allowed for.

More generally, methods for model appraisal and characterization of nonuniqueness in 2D EM inversion is an area of active research. Linearized resolution analysis can be used if sensitivities are calculated, but the requirement of linearization is perhaps even more of a limitation than for the 1D case. Testing for required features through constrained inversion is also an option for 2D inversion, but methods for constructing rigorous bounds on conductance that have been developed for the 1D EM inverse problem do not generalize in any obvious way to higher dimension.

Three-dimensional inversion

Three-dimensional inversion is a natural extension of 2D inversion, allowing for arbitrary variations of conductivity within the Earth. For the 3D MT problem two possible source polarizations must be considered as in the 2D case, but now there is no preferred coordinate system where the induction equations simplify or decouple. In the general 3D case the MT impedance tensor (i.e., 2×2 frequency domain transfer function matrix relating electric to magnetic fields at a site) has four complex elements: $Z_{xx}, Z_{xy}, Z_{yx}, Z_{yy}$, along with a more general vertical field transfer function giving local ratios of vertical to horizontal magnetic fields (T_x, T_y). Model results for both polarizations are required to form the impedance elements and transfer functions which can be fit in the inversion. The forward problem generally involves solving a vector diffusion equation, formulated variously in terms of electric or magnetic fields, or some sort of potentials. For example, in terms of the 3D vector electric fields **E** the forward problem can be expressed as a PDE

$$\nabla \times \nabla \times \mathbf{E} - i\omega\mu\sigma\mathbf{E} = 0, \quad \text{(Eq. 10)}$$

which is to be solved subject to boundary conditions appropriate to the source assumptions. Integral equation formulations are also used for 3D modeling and inversion. All formulations are readily derived from the usual quasistatic approximation to Maxwell's equations. Factoring the banded coefficient matrix obtained from a finite-difference or finite-element discretization of Eq. (10) is practical for the analogous 2D problem, but prohibitively expensive in 3D. As a result the forward 3D problem is almost always solved with iterative techniques.

Few analytical results are available for the 3D EM inverse problem. As for the 2D problem, efforts have focused on regularized inversion through numerical minimization of a penalty functional such as Eq. (5). Most of the discussion of 2D inversion methods is applicable to the 3D case, but computational challenges are even greater. In particular, the number of model parameters is typically so large that storing the full Jacobian **G** is all but impossible. NLCG, which avoids computing **G** altogether, is thus the most common approach to 3D inversion. Other approaches which have had some success include using very coarse parameterizations, and data space schemes which avoid storing all of **G**.

As of this writing there is relatively little practical experience with 3D inversion. In addition to the severe computational challenges, there are few data sets of sufficient density to truly resolve 3D structures. Data sets collected in industry applications are the most extensive, and much of the relevant experience with 3D inversion is in the private sector, and with active source EM, often for near-surface applications. Even with data sets that are not sufficiently extensive to truly image 3D structure, 3D inversion may be useful for obtaining more reliable interpretations of profile data across approximately 2D structures. Development of more efficient 3D inversion schemes, and learning how to use these for interpretation of diverse EM data sets is an area of active research.

Gary D. Egbert

Bibliography

Parker, R.L., 1980. The inverse problem of electromagnetic induction: existence and construction of solutions based on incomplete data. *Journal of Geophysical Research*, **85**: 4421–4428.

Parker, R.L., and Whaler, K.A., 1981. Numerical methods for establishing solutions to the inverse problem of electromagnetic induction. *Journal of Geophysical Research*, **86**: 9574–9584.

Parker, R.L., 1994. *Geophysical Inverse Theory*. Princeton, NJ: Princeton University Press.

Press, W.H., Flanerry, B.P., Teukolsky, S.A., and Vetterling, W.T., 1986. *Numerical Recipes: The Art of Scientific Computing*. Cambridge: Cambridge University Press.

Rodi, W.L., and Mackie, R.L., 2001. Nonlinear conjugate gradients algorithm for 2-D magnetotelluric inversion. *Geophysics*, **66**: 174–187.

Siripunvaraporn, W., and Egbert, G., 2000. An efficient data-subspace inversion method for two-dimensional magnetotelluric data. *Geophysics*, **65**: 791–803.

Smith, J.T., and Booker, J.R., 1991. Rapid inversion of two- and three-dimensional magnetotelluric data. *Journal of Geophysical Research*, **96**: 3905–3922.

Whittall, K.P., and Oldenburg, D.W., 1992. *Inversion of Magnetotelluric Data for a One-Dimensional Conductivity*. Geophysical Monograph Series No. 5. Tulsa, OK: Society of Exploration Geophysicists.

Cross-references

Anisotropy, Electrical
Conductivity, Ocean Floor Measurements
EM Modeling, Forward
Galvanic Distortion
Geomagnetic Deep Sounding
Induction Arrows
Magnetotellurics
Robust Electromagnetic Transfer Functions Estimates
Transfer Functions

EM, INDUSTRIAL USES

Introduction

Mineral, petroleum, environmental, and geotechnical companies routinely measure electrical conductivity properties of Earth materials to map near-surface geological structures, locate economic resources, and define environmental and engineering problems. Targets may be more electrically conductive than the host environment, for example in the case of many mineral deposits (Ohleoft, 1985) and for the location of groundwater (McNeill, 1990). Alternatively, in the petroleum industry the focus is often on regions that are less conductive than the host, either in the case of a hydrocarbon-filled layer, or for an impermeable formation such as salt domes that trap hydrocarbons (Hoverston *et al.*, 1998). For environmental and geotechnical studies, high-frequency (>1 MHz) EM methods can also be used in a technique called ground-penetrating radar (Daniels *et al.*, 1988). At such high frequencies, electrical conductivity is less important than the dielectric properties of the materials.

To measure the electrical conductivity, the primary methods are either galvanic electrical techniques, in which a constant electrical current is applied to the Earth, or EM induction techniques, in which electrical currents are induced to flow in the Earth by a time-varying external field. The external field can be either magnetic or electrical, and the Earth's response is measured with a *magnetometer* (*q.v.*), an electric dipole coupled to the ground, or a combination of both. The time-varying signals originate either from natural sources, or applied, controlled sources.

EM methods in common use by industry can be broadly categorized in terms of the type of source fields. If the source of the externally varying field is sufficiently distant from the location of measurement, the source is approximately plane-wave, and is said to be one-dimensional (varying in only one direction). On the other hand, if the source and the measurement are close, then the source is said to be three-dimensional (varying in all spatial dimensions). This latter category can be further subdivided into techniques that use a continuously varying controlled source field, usually in the form of a sine wave, known as frequency-domain electromagnetics (FEM), and those that require a source that abruptly switches on and off as a square wave, known as time-domain electromagnetics (TEM) or *transient electromagnetic induction* (*q.v.*).

EM methods have been adapted by industries to a wide range of environments. There is a rapidly growing *marine EM (q.v.)* industrial sector (Chave *et al.*, 1991; Constable *et al.*, 1998) in which the source field is either due to natural changes in Earth's magnetic field or a transmitter is towed behind a ship, and seafloor electrical and magnetic receivers measure the induced signals. Marine EM measurements of the subseabed electrical conductivity are complementary to seismic methods, and help reduce the risk by identifying more suitable drilling targets. Aircraft and helicopters have been modified to carry a wide range of EM sensors, often with a transmitter loop between nose, wingtips, and tail, and with a secondary magnetic receiver towed behind the aircraft (Palacky and West, 1991). Airborne measurements have long been used in mineral exploration by rapid imaging of electrically conductive basement structures, often through many meters of overburden (weathered materials). More recently, these techniques have become important in natural resource management for mapping surface salinity and deeper saline aquifers. Drill-hole EM techniques are widely used in the petroleum industry to determine the electrical conductivity properties of the sequences surrounding the borehole, and in mineral and environmental industries to image the extent of structures away from the borehole (Dyck, 1991).

Plane wave source methods

The most common types of plane wave methods are known as VLF (very low frequency), AMT (audio-*magnetotellurics (q.v.)*), and CSAMT (controlled-source *magnetotellurics (q.v.)*). The VLF method uses a distant vertical electric dipole (denoted E_z) that results from a time-varying current in a long wire orientated vertically above the ground (McNeill and Labson, 1991). Such electric field dipoles are developed as military radio transmitters and operate in the 15–24 kHz range, and are used for global-scale communication with submarines. There are many such transmitters around the world, thus providing a VLF signal in most parts of the globe.

The vertical flow of current generates horizontal magnetic field (denoted B_x and B_y) components, each perpendicular to the direction of propagation, and a small electric field (E_x) in the direction of propagation. When the electric field E_x is incident on a boundary between two materials of different conductivity, electric charges are induced at the interface. This modifies E_x to produce a "current channeling" effect, which produces a secondary magnetic field that has an anomalous vertical component B_z. It is this component that is measured in a VLF survey. It is also possible to measure the ratio of the amplitudes of the horizontal electric and magnetic field components at the Earth's surface to determine the apparent resistivity and phase in the same manner as the magnetotelluric method (Becken and Pedersen, 2003). The primary advantage of the VLF method is that as the source-field is always present and predictable, the anomalous B_z field can be measured quickly with a roving magnetometer to provide a surface map of the electrical conductivity structures.

Mineral and petroleum exploration can be carried out using the *magnetotelluric (q.v.)* method (Vozoff, 1991; Zonge and Hughes, 1991). To provide the highest resolution in the top few kilometers of Earth, frequencies in the bandwidth 1 Hz to 20 kHz are used. At the high-end of the bandwidth, these frequencies are in the audio range, hence the name AMT. Source-fields are either naturally occurring lightning strikes at equatorial latitudes (essentially a large vertical electric field E_z as for the VLF method), or provided by a controlled-source that consists of a distant applied electric or magnetic dipole field. In the AMT or CSAMT method, horizontal and orthogonal components of the electric (E_x and E_y) and magnetic (B_x and B_y) fields are measured to yield an apparent resistivity and phase as a function of frequency. *Forward modeling (q.v.)* and *inverse modeling (q.v.)* provide a depth-dependent image of electrical conductivity. Lateral coverage is slower than the VLF method due to the requirement of measuring the electric fields with grounded electrodes.

Frequency-domain electromagnetic (FEM) methods

In FEM methods, a time-varying current of a known frequency is applied in a transmitter loop of wire at some location, as shown in Figure E9 (Frischknecht *et al.*, 1991). The loop itself may consist of many turns of wire. Electrical currents flowing in a loop generate a primary (p), time-varying magnetic dipole field that will be horizontal (B_x^p) if the loop is orientated vertically, or vertical (B_z^p) is the loop is flat on the ground. The time-varying magnetic dipole fields induce eddy currents to flow in Earth, which in turn generate secondary (s) magnetic fields B^s, as shown in Figure E10. A receiver at a nearby location, which can be a second loop of wire, or a *magnetometer (q.v.)* sensor, measures a combination of the primary fields, B^p, which is known, and the induced secondary fields B^s. The relative contribution of these fields depends on the proximity and orientation of the receiver to the transmitter, and the electrical conductivity of the Earth between the two. The difference in fields is usually measured as a combination of relative amplitude and a phase difference, or as in-phase and out-of-phase (quadrature) components.

In most commercial systems, measurements are made at one or more frequencies, typically in the range 1–100 kHz. Variations in electrical conductivity with depth can be determined by measurements made at different frequencies, and by changing the orientation of the coils. Lateral mapping can be achieved by moving either one or both coils. Depth of penetration is relatively small, limited by the power than can be produced by the transmitter.

FEM methods are widely used in environmental and geotechnical studies, but are also becoming more widespread in *marine EM (q.v.)* exploration for petroleum targets. A particularly common configuration is known as a ground conductivity meter (McNeill, 1980), for which the skin-depth of signal penetration is greater than the loop separation. In this case, the quadrature (or out-of-phase) component

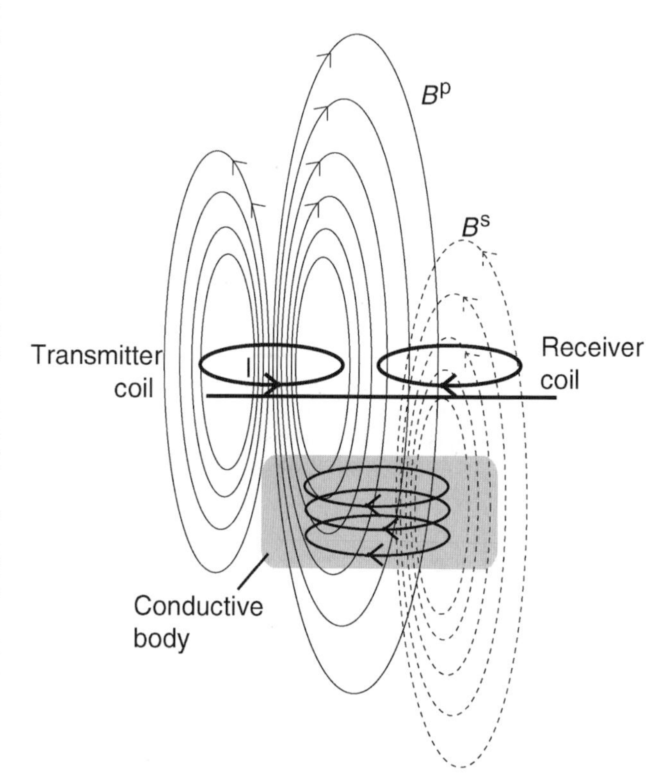

Figure E9 The arrangement of transmitter and receiver loops in frequency-domain electromagnetic (FEM) systems. Current flowing in the transmitter coil generates a primary magnetic (B^p) field that induces eddy currents in a conductor. The resulting secondary magnetic fields (B^s) are measured with a receiver loop.

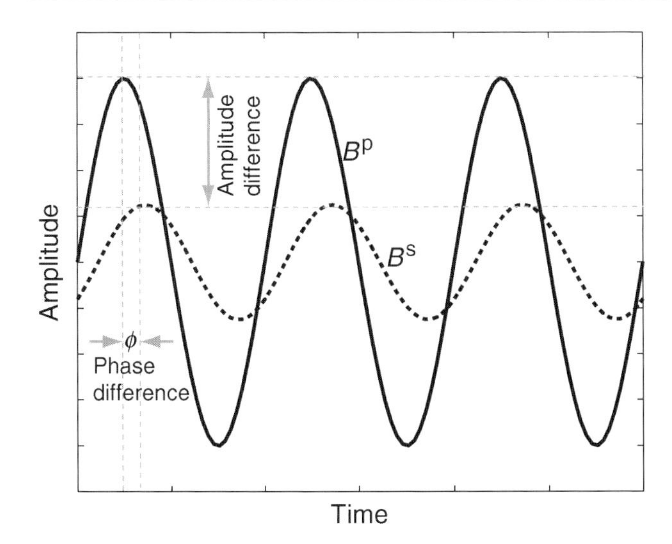

Figure E10 The combination of primary (B^p) and secondary (B^s) fields recorded with a FEM system. The measured response can be determined in terms of an amplitude difference and/or a phase difference ϕ.

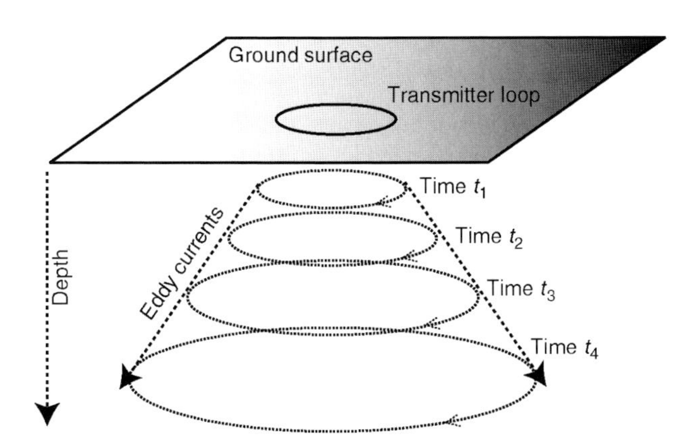

Figure E11 Distribution of eddy currents beneath a time-domain transmitter. At time t_1, immediately after switch-off, eddy currents flow close to the surface; at progressively later times (t_2 to t_4) eddy currents spread out and down.

of the secondary field is linearly proportional to the ground apparent resistivity, which allows rapid spatial mapping of electrical conductivity to be made by foot, vehicle, or helicopter.

Time-domain electromagnetic (TEM) methods

In TEM systems in Figures E11 and E12, the applied electric current in a transmitter loop is switched on and off, in the form of a square wave (Nabighian and Macnae, 1991). In the time intervals when the current is on, a static primary magnetic field B^p is established. When the transmitter current in the loop is extinguished, the primary field collapses almost instantly, and this rapid change in primary field generates large eddy currents in the Earth, close to the surface. Such eddy currents in turn yield a secondary magnetic field B^s that can be measured with a receiver loop, or a magnetometer sensor. Eddy currents continue to flow after switch-off (typically for a few milliseconds) and become broader and deeper. The resulting secondary magnetic field also reduces in amplitude, in Figure E12. Energy is lost as heat, and the rate of decay depends on the electrical conductivity of Earth.

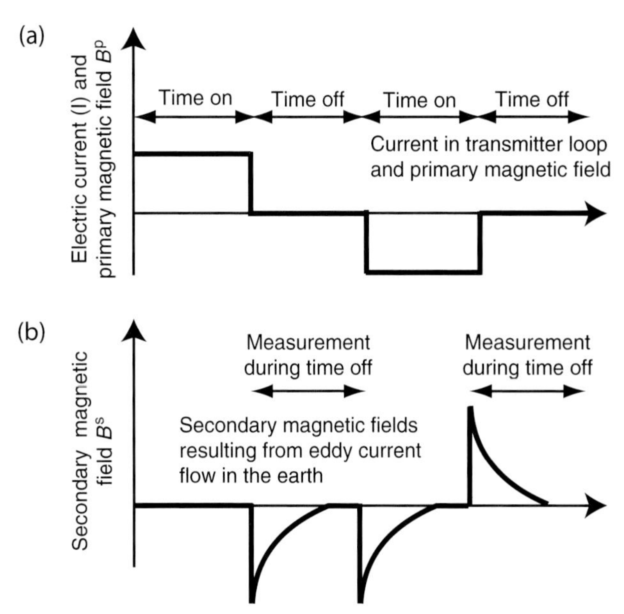

Figure E12 Waveforms encountered in time domain EM systems. (a) The top figure shows the applied current in the transmitter loop and the resulting primary magnetic field. (b) Secondary magnetic fields are measured in the times when the current is switched off, and which are observed to decay with time as eddy currents dissipate.

TEM methods are widely used in the mineral exploration industry, and are increasingly being used for environmental studies. There are two significant advantages of TEM over FEM methods. Firstly, as measurements of signals in the receiver loop or magnetometer sensor need only be made in the intervals when the primary field is switched off (i.e., $B^p = 0$), it is easier to detect the small secondary field B^s decay. Secondly, larger eddy currents can be induced as the primary field changes much more rapidly than in FEM systems. The measured decay in the secondary field can be converted to an apparent resistivity as a function of time. Modeling and inversion techniques can then be used to determine the electrical conductivity of Earth, typically as a function of depth. TEM loops used in the minerals industry are often of dimensions of hundreds of meters, with applied currents of tens of amperes, to provide conductivity models to depth of several hundred meters. For environmental studies the loops are typically 20 m, with targets in the top 50 m.

Grounded wire methods

Methods that involve an injection of time-invariant electrical current into the ground are generally denoted galvanic. Pathways, and resulting current densities, are determined by Earth's electrical conductivity, and are typically measured as a horizontal electric field (E_x and E_y) by a grounded dipole.

A variation on this approach is to measure the resulting secondary magnetic field B^s due to the subsurface current distribution (Edwards *et al.*, 1978; Edwards and Nabighian, 1991). The magnetometric resistivity (MMR) technique is used in mineral exploration and has proved to be useful in imaging basement structure beneath overburden. Two advantages over traditional ground dipole measurements are that data can be acquired rapidly as there is no requirement to have grounded electrodes, and that the secondary magnetic field is the integral sum of the contributions from all the current elements around the measurement's location. The MMR method is therefore less sensitive to near-surface inhomogeneities (Edwards, 1988).

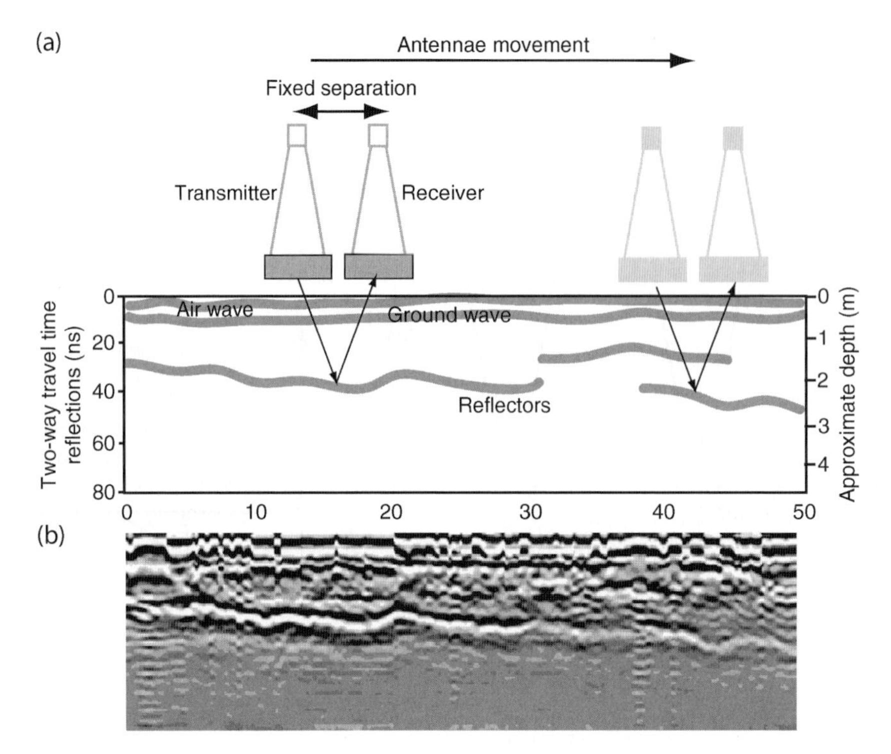

Figure E13 (a) Schematic of a ground-penetrating radar system to detect subsurface structure. The cross section shows interpreted reflections from the radargram in the lower figure. The airwave and groundwave are signals that propagate between the transmitter and receiver through the air and the top few centimeters of the ground, and are therefore almost constant across the profile. (b) Reflections of high frequency (in this case 200 MHz) EM waves shown as a typical grayscale radargram in the lower figure.

Ground-penetrating radar (GPR)

All methods described above are for variations in EM fields that occur at frequencies of 100 kHz, or lower. At these frequencies, EM fields diffuse into the Earth, and are primarily sensitive to electrical conductivity. On the other hand, at sufficiently high frequencies of 1 MHz or more, applied EM fields behave as waves and the main physical parameter of importance is the dielectric permittivity, usually given the symbol ε. The dielectric permittivity determines the reflection and transmission characteristics of interfaces between layers.

In GPR surveys, pairs of interchangeable radar antennas are used as a transmitter and receiver, as shown in Figure E13 (Daniels *et al.*, 1988). A pulse of high-frequency signal is directed into the ground, and the EM echoes are recorded. Depths of penetration are typically about 50 m for 10 MHz, 5 m for 100 MHz, and 0.5 m for 1000 MHz.

In environmental studies, the water content dominates the dielectric properties, as the dielectric permittivity of water is typically ten times greater than that of solid grains. Thus, GPR is used to map depths to water tables and soil moisture content. For geotechnical studies, GPR studies are effective at mapping near-surface engineering structures, including pipes, cavities, and metal objects.

Summary

There continues to be growth in the use of EM methods for resource exploration, and for environmental and engineering applications. These methods are generally rapid and noninvasive, and provide both a lateral and vertical image of electrical conductivity and dielectric permittivity properties of Earth. The industry has adapted these methods to operate in a wide range of environments, particularly in the fields of land, marine, airborne, and downhole studies.

Graham Heinson

Bibliography

Becken, M., and Pedersen, L.B., 2003. Transformation of VLF anomaly maps in apparent resistivity and phase. *Geophysics*, **68**: 497–505.

Chave, A.D., Constable, S.C., and Edwards, R.N., 1991. Electrical exploration methods for the seafloor. In Nabighian, M.N. (ed.), *Electromagnetic Methods in Applied Geophysics.* Tulsa, OK: Society of Exploration Geophysicists, pp. 931–966.

Constable, S., Orange, A., Hoversten, G.M., and Morrison, H.F., 1998. Marine magnetellurics for petroleum exploration. Part 1. A seafloor instrument system. *Geophysics*, **63**: 816–825.

Daniels, D.J., Gunton, D.J., and Scott, H.F., 1988. Introduction to subsurface radar. *IEEE Proceedings*, **135**: 278–320.

Dyck, A.V., 1991. Drill-hole electromagnetic methods. In Nabighian, M.N. (ed.), *Electromagnetic Methods in Applied Geophysics.* Tulsa, OK: Society of Exploration Geophysicists, pp. 881–930.

Edwards, R.N., and Nabighian, M.N., 1991. The magnetometric resistivity method. In Nabighian, M.N. (ed.), *Electromagnetic Methods in Applied Geophysics.* Tulsa, OK: Society of Exploration Geophysicists, pp. 47–104.

Edwards, R.N., 1988. A downhole magnetometric resistivity technique for electrical sounding beneath a conductive surface layer. *Geophysics*, **53**: 528–536.

Edwards, R.N., Lee, H., and Nabighian, M.N., 1978. On the theory of magnetometric resistivity (MMR) method. *Geophysics*, **43**: 1176–1203.

Frischknecht, F.C., Labson, V.F., Spies, B.R., and Anderson, W.L., 1991. Profiling methods using small sources. In Nabighian, M.N. (ed.), *Electromagnetic Methods in Applied Geophysics.* Tulsa, OK: Society of Exploration Geophysicists, pp. 105–270.

Hoversten, G.M., Morrison, H.J., and Constable, S., 1998. Marine magnetellurics for petroleum exploration Part 2. Numerical analysis of subsalt resolution. *Geophysics*, **63**: 826–840.

McNeill, J.D., 1980. Electromagnetic terrain conductivity measurement at low induction numbers. Technical Note TN-6, Geonics Limited, Mississauga, Ontario.

McNeill, J.D., and Labson, V., 1991. Geological mapping using VLF radio fields. In Nabighian, M.N. (ed.), *Electromagnetic Methods in Applied Geophysics*. Tulsa, OK: Society of Exploration Geophysicists, pp. 521–640.

McNeill, J.D., 1990. Use of electromagnetic methods for groundwater studies. In Ward, S.H. (ed.), *Geotechnical and Environmental Geophysics*. Tulsa, OK: Society of Exploration Geophysicists, pp. 191–218.

Nabighian, M.N., and Macnae, J.C., 1991. Time domain electromagnetic prospecting methods. In Nabighian, M.N. (ed.), *Electromagnetic Methods in Applied Geophysics*. Tulsa, OK: Society of Exploration Geophysicists, pp. 427–520.

Olhoeft, G.R., 1985. Low-frequency electrical properties. *Geophysics*, **50**: 2492–2503.

Palacky, G.J., and West, G.F., 1991. Airborne electromagnetic methods. In Nabighian, M.N. (ed.), *Electromagnetic Methods in Applied Geophysics*. Tulsa, OK: Society of Exploration Geophysicists, pp. 811–880.

Vozoff, K., 1991. The magnetotelluric method. In Nabighian, M.N. (ed.), *Electromagnetic Methods in Applied Geophysics*. Tulsa, OK: Society of Exploration Geophysicists, pp. 641–712.

Zonge, K.L., and Hughes L.J., 1991. Controlled source audio-frequency magnetotellurics. In Nabighian, M.N. (ed.), *Electromagnetic Methods in Applied Geophysics*. Tulsa, OK: Society of Exploration Geophysicists, pp. 713–810.

Cross-references

EM Modeling, Forward
EM Modeling, Inverse
EM, Industrial Uses
EM, Land Uses
EM, Marine Controlled Source
Magnetometers, Laboratory
Magnetotellurics
Transient EM Induction

EM, LAKE-BOTTOM MEASUREMENTS

Magnetic fields set up by electrical currents flowing in the Earth's ionosphere and magnetosphere diffuse into the Earth's interior, and induce electrical currents to flow in the oceans and in conductive regions in the Earth's interior. Longer period magnetic field variations have a greater depth of penetration than shorter period variations. By relating the variations in the electric field to those in the magnetic field at different locations and over a range of different periods, it is possible to reconstruct the distribution of electrical conductivity within the Earth, and from that information, to constrain the temperature, composition, fluid and volatile content, and state of Earth materials. This is the magnetotelluric (MT) method (see *Magnetotellurics*; *Galvanic distortion*; *EM, land uses*).

In order to penetrate deep into the lithosphere and below, long-period MT measurements are required. Other methods involving only magnetic field measurements (see *Geomagnetic deep sounding*; *EM, regional studies*) are valid for field variations with periods longer than 2–4 days, and provide information on electrical conductivity of the mantle at depths below 400 km (Schultz *et al.*, 1987) (see *Mantle, electrical conductivity, mineralogy*). Conversely, the MT method is usually employed to investigate periods many decades shorter than this, generally less than 0.2 days. Consequently most MT investigations are restricted to the crust and uppermost mantle. By extending the range of MT investigations to longer periods, finer structural details may be resolved at depth, and information on anisotropy within the crust and mantle may be returned.

The electric field is measured by monitoring the electrical potential across (typically) two orthogonal pairs of metal salt electrodes grounded to Earth and separated by a fixed distance. Electrode noise is a consideration for MT. There is long-period electric field measurement drift due to variations in electrochemical conditions in the soil, moisture content, and thermal effects (Perrier and Morat, 2000), and to electrokinetic effects (Trique *et al.*, 2002). To improve the signal-to-noise ratio, a number of deeply penetrating MT investigations have taken place making use of extremely long electrode separations, by using leased telephone cables on land (Egbert *et al.*, 1992), or abandoned submarine telecommunications cables on the seafloor (see *Conductivity, ocean floor measurements*).

Schultz and Larsen (1987) proposed that the thermally and electrochemically stable environment of lake bottoms might provide an ideal environment to extend MT into periods of 0.2–4 days and below, thus spanning the gap in coverage left by conventional MT and magnetic field methods. While lake-bottom MT shares common features with ocean bottom MT, the high conductance of marine sediments and the conductive seawater overburden at most marine study sites shields the seafloor from the high frequency signals that penetrate to the bottom of lakes, and that can provide higher resolution information on shallow crustal and upper mantle conductivity structure. While contamination of the EM fields by water motions (such as mesoscale eddies) impose a low frequency limit to MT in the oceans, this is generally not a significant consideration in lakes.

A prototype lake-bottom MT station was installed in Lake Washington, in Seattle, Washington, to form an array of Ag-AgCl electrodes, with electrode separations of 800–1300 m. This configuration provided three nonorthogonal components of the electric field, any two of which could be rotated into orthogonal components. A fluxgate magnetometer was buried on the adjacent shore, and high-quality MT data resulted. The MT regional strike direction was seen to be coincident with the strike delineated by regional seismicity and tectonics. The effects of a channeled current were seen, attributed to an Eocene suture zone to the south of the site (Schultz *et al.*, 1987).

A multiyear lake-bottom MT deployment followed in the Archean Superior craton of northern Ontario, Canada (Schultz *et al.*, 1987). More than 2 years of MT data were collected from Carty Lake, following the configuration at Lake Washington. Three zones in the mantle were identified, centered at depths of 280, 456, and 825 km, and roughly coincident with major seismic discontinuities, where the conductivity increased abruptly (Figure E14). The ability to discriminate the jump centered at around 456 km was attributed to MT data with periods of 1 h to 1 week, which were made possible by the stable lake-bottom installation. No other MT experiment to date has produced resolving power sufficient to discriminate the 456 km conductivity increase. The conductance of the upper mantle here was roughly an order of magnitude higher than that predicted by a standard olivine model using a standard geotherm. It was determined that either the upper mantle geotherm was erroneously biased downward, or a dry olivine upper mantle model was inappropriate.

Jones *et al.* (2001a,b, 2003) report on the lake-bottom deployment of compact MT instruments designed for marine use, and of electric dipole receivers inserted through the ice to the bottom of frozen lakes, with magnetometers nearby on shore. These installations in roughly 20 lakes were made to probe to the depth of the top of the asthenosphere beneath the Slave craton in northwestern Canada, in a region of diamondiferous kimberlite pipes. Their investigations serendipitously revealed a highly conductive zone at depths of 80 to below 100 km in the mantle that is interpreted to represent interconnected carbon or graphite along mineral grain faces. Jones *et al.* (2003, 2004) determined that the Superior craton MT data of Schultz *et al.* (1993) and those of the Slave craton indicate that while both provinces differ in shallow structure, on average the deeper mantle sections are nearly identical. Jones *et al.* (2003) compared the lake-bottom MT data against earlier data from the region, and found that these depths had previously been beyond the

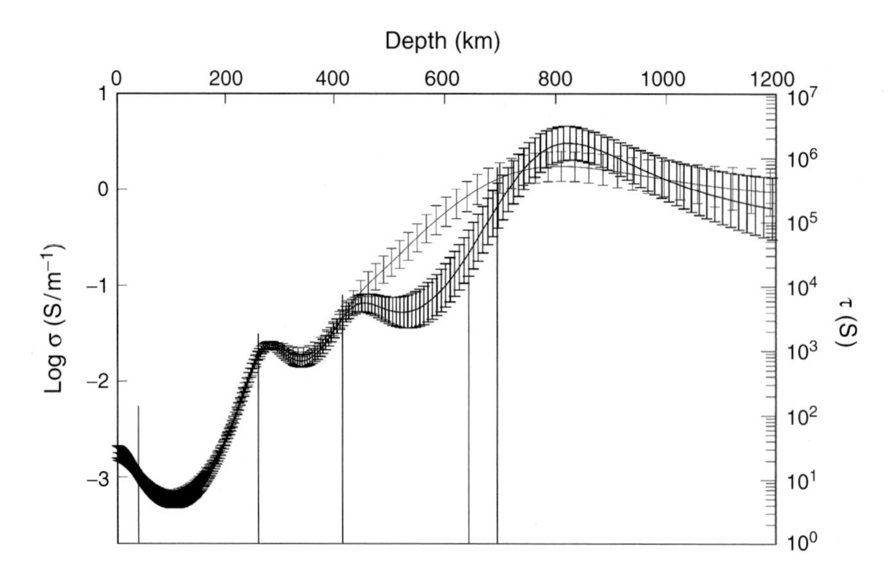

Figure E14 Electrical conductivity model for lake-bottom MT site, Carty Lake, Superior craton, Canada. The thick curve is the model of $\log_{10}$ conductivity vs depth that has the minimum size of jumps in conductivity ($\min_d \log \sigma/dz$) of all models that fit the data. Note the jumps in conductivity centered at 280, 456, and 825 km. The thinner curve is a model with the same properties, but with the lake-bottom MT data with periods of 2 h to 6 days deleted, showing that the peak at 456 km can only be resolved if the lake-bottom data are present. The thin vertical bars represent the best-possible fitting model comprised of infinitely thin zones of finite conductance, τ.

limits of penetration of conventional MT, but were now resolvable because of the extended bandwidth available to lake-bottom MT.

Golden *et al.* (2004a,b) carried out lake-bottom MT measurements in Woelfersheimer lake near Frankfurt am Main, in 2003. These prototype experiments lead to a successful lake-bottom deployment in Iceland. While the analysis of these data is ongoing with the aim of better imaging the putative mantle plume beneath Iceland, a lower noise level is seen for the Icelandic lake-bottom MT data than for nearby conventional MT deployments.

Lake-bottom electromagnetic deployments have now been used at numerous sites in North America, Europe, and Iceland. The experience indicates that this environment provides a means of extending MT observations into the mid-mantle, and for improving the ability to resolve finer structural details than shorter duration, more noise-prone conventional MT installations.

Adam Schultz

Bibliography

Egbert, G.D., Booker, J.R., and Schultz, A., 1992. Very long period magnetotellurics at Tucson observatory—estimation of impedances. *Journal of Geophysical Research*, **97**(B11): 15113–15128.

Golden, S., Björnsson, A., Beblo, M., and Junge, A., 2004a. Project CMICMR: Long-period magnetotellurics on Iceland. American Geophysical Union, Fall Meeting 2004, abstract #GP11A-0819.

Golden, S., Roßberg, R., and Junge, A., 2004b (im Druck)b. Langperiodische MT-Messungen in einem See mit dem Langzeit-Datenlogger Geolore ("Long-periodic MT measurements in a lake with the long-term datalogger Geolore"), *Protocol über das Kolloquium Elektromagnetische Tiefenforschung*, **20**.

Jones, A.G., Ferguson, I.J., Chave, A.D., Evans, R.L., and McNeice, G.W., 2001a. Electric lithosphere of the slave craton. *Geology*, **29**(5): 423–426.

Jones, A.G., Snyder, D., and Spratt, J., 2001b. Magnetotelluric and teleseismic experiments as part of the Walmsley Lake project, northwest territories: experimental designs and preliminary results, *Current Research 2001-C6*, Geological Survey of Canada, 7 pp. Available at: http://gsc.nrcan.gc.ca/bookstore/

Jones, A.G., Lezaeta, P., Ferguson, I.J., Chave, A.D., Evans, R.L., Garcia, X., and Spratt, J., 2003. The electrical structure of the Slave craton, *Lithos*, **71**: 505–527.

Jones, A.G., and Craven, J.A., 2004. Area selection for diamond exploration using deep-probing electromagnetic surveying. *Lithos*, **77**: 765–782.

Perrier, F., and Morat, P., 2000. Characterization of electrical daily variations induced by capillary flow in nonsaturated zone. *Pure and Applied Geophysics*, **157**(5): 785–810.

Schultz, A., Booker, J., and Larsen, J., 1987. Lake bottom magnetotellurics. *Journal of Geophysical Research*, **92**(B10): 10639–10649.

Schultz, A., Kurtz, R.D., Chave, A.D., and Jones, A.G., 1993. Conductivity discontinuities in the upper mantle beneath a stable craton. *Geophysical Research Letters*, **20**(24): 2941–2944.

Trique, M., Perrier, F., Froidefond, T., Avouac, J.-P., and Hautot, S., 2002. Fluid flow near reservoir lakes inferred from the spatial and temporal analysis of the electric potential. *Journal of Geophysical Research*, **107**(B10): 2239, doi:10.1029/2001JB000482.

Cross-references

Conductivity, Ocean Floor Measurements
EM, Land Uses
EM, Regional Studies
Galvanic Distortion
Geomagnetic Deep Sounding
Magnetotellurics
Mantle, Electrical Conductivity, Mineralogy

EM, LAND USES

Land-based electrical and electromagnetic (E&EM) geophysical methods have been applied in mapping the electrical conductivity, or the reciprocal resistivity, structure of the Earth for applications as varied as general geological mapping; waste site characterization; contamination delineation; hydrogeological investigations; exploration for oil and gas, mineral, geothermal, sand, gravel, limestone, and clay;

geotechnical investigations of building and road construction sites; location and identification of subsurface utilities; unexploded ordnance detection; and precision agriculture. Geophysical methodologies are almost as varied as the applications. The electric or magnetic fields or a combination of both are measured for plane wave or controlled sources in either the time or the frequency domain. Imaging of subsurface conductivity structure can extend from simple contouring of the data, to rigorous and computer-intensive, multidimensional inverse modeling. The depths of investigation can be a function of operating frequency, decay time, or the geometric array, ranging from meters to tens of kilometers. Figure E15 shows a schematic of the frequency range for commonly used E&EM methods and approximate depths of investigation for an Earth of 100 Ωm.

The behavior of E&EM fields is governed by Maxwell's equations (Ward and Hohmann, 1988). The physical earth parameters determining the response are the electrical conductivity (σ), or its reciprocal resistivity (ρ), the magnetic permeability (μ), and the dielectric permittivity (ε). The geophysical EM spectrum ranges from ground-penetrating radar (GPR) at hundreds of MHz to low frequencies approximating a direct current (*dc*) as shown in Figure E15.

Land-based EM systems employ frequencies within the quasistatic approximation in which displacement currents are ignored, such that inductive currents predominate and field propagation are diffusive.

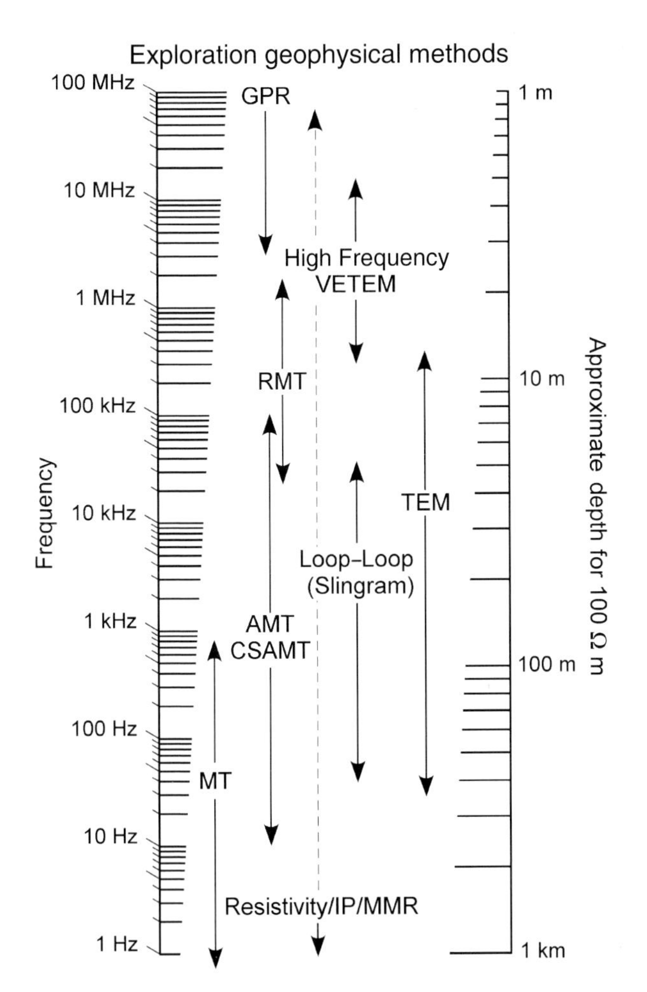

Figure E15 The geophysical EM spectrum showing common exploration methods and the corresponding frequency range, or equivalent time, and approximate depth of investigation for a 100 Ωm half-space. The dashed line related to the resistivity/IP/MMR methods indicate the depth of investigation is related to the array geometry and not the operating frequency.

GPR systems operate in the high-frequency range where displacement currents dominate and field propagation is a wave phenomenon. In the *dc* limit, the electric field is a potential governed by the Poisson equation. In most applications it is assumed $\mu = \mu_0$ (free space), because magnetic materials are often anthropomorphic artifacts, such as drums and pipelines. E&EM techniques can directly map the ionic content of the pore water, an indicator of the subsurface water chemistry and clay content, which is more difficult to achieve with wavefield methods such as GPR.

Electrical resistivity

Electrical current at frequencies near the *dc* limit is injected into the ground over an array of current electrodes using a transmitter, and the resulting voltage is measured over a similar array of voltage electrodes. Current can be injected either galvanically or capacitively. The depth of investigation and resolution are determined by the electrode array and spacing. A description of the wide variety of electrode arrays can be found in Telford *et al.* (1990). Inhomogeneities in the Earth alter the current flow from what would be present with a uniform half-space, and are considered as secondary current distributions. Using modern multielectrode equipment and continuous systems, data are measured in profile or array configurations so that lateral and vertical resistivity variation can be determined through data inversion.

With a multielectrode galvanic system, 20–100 electrodes are usually placed equidistantly in an array, which can be both on and below the surface. A computer-controlled switch box connects four electrodes to input current and measure the resulting voltage. A series of configurations are measured and stored one after the other, and data are stacked until a certain noise level is reached or an upper limit on the number of measurements is attained. Several systems are commercially available and the use of multielectrode geoelectric measurements has increased dramatically over the past 5–10 years. Electrical resistance tomography (ERT) is a term frequently used to describe such systems. Applications are as diverse as precision agriculture, hydrogeological investigations, waste site characterization, road and building foundation investigations, and archaeological studies. Permanent or semipermanent arrays are often used to monitor infiltration and containment migration problems.

Continuous systems make measurements continuously as the instrument is towed over the ground. The Danish pulled array continuous electrical sounding (PACES) system (Sørensen, 1996) uses galvanically coupled steel cylinder electrodes mounted on a tail. The French use spiked wheels as electrodes or capacitively coupled electrodes mounted inside plastic wheels (Panissod *et al.*, 1998). In the United States the OhmMapper system uses a capacitively coupled dipole-dipole array (Pellerin *et al.*, 2004).

The depth of investigation for electrical resistivity systems is limited by the power needed to deliver the required current and the wires needed to support the array. A general estimate for depth of investigation is 40%–60% of the array geometry. Hence for a transmitter-receiver separation of 10 m, the expected depth of investigation would be roughly 5 m. For large depths this entails the use of large amounts of wire, a logistical difficulty. Interpretational complexity increases as different geological structures are crossed. The electrical resistivity method is most practical in the 1–50 m depth range.

A question for production surveys of multielectrode systems is which electrode configurations to use. The choice depends upon optimizing resolution capabilities and the signal-to-noise ratio. Standard electrode configurations include the Wenner and Schlumberger arrays where the potential electrodes are inside the current electrodes along a profile; the dipole-dipole where the current electrodes are offset from the potential electrodes along a profile; and the pole-dipole and pole-pole configurations that utilize a distant electrode away from an electrode array. Practically, the Wenner and Schlumberger arrays are most robust in the presence of cultural noise with a high signal-to-noise ratio, the dipole-dipole array most rapid for deployment over large areas, and the pole-pole array fully general in that all arrays are present.

The continuous systems, such as the PACES and OhmMapper, employ array configurations based on the design of the instrument.

Two-dimensional inversion is the state of the practice for profile-oriented, electrical resistivity data (Loke and Barker, 1996). Fast approximate inversion procedures applicable to large data sets are becoming common along with three-dimensional inversion applicable to large pole-pole arrays and monitoring systems.

Induced polarization

When electrically polarizable minerals are present in the subsurface, induced polarization (IP) at the particle interfaces causes the secondary currents in the vicinity to be out of phase with the applied current (Fink *et al.*, 1990). Although some IP effects are associated with ionic double layers on any silicate surfaces, it is clays and sulfides that tend to exhibit the largest response, thus providing more direct information on lithology than electrical conductivity alone. Historically the IP method has been used successfully in delineating porphyry-copper deposits. More recently IP is used to discriminate between clay-bearing and clean sand units in groundwater prospecting. As shown in many case histories and methodology studies, IP is one of the most powerful techniques for environmental applications. In the 1960s the use of IP was proposed for landfill characterization and after many years of GPR, conductivity meters and electrical resistivity surveys, the use of IP is being shown to be the most accurate tool of the trade.

For accurate IP measurements nonpolarizable electrodes, most often lead/lead chloride or copper/copper sulfate, are needed resulting in high survey costs. However, alternative solutions are emerging as smart electrodes, correction schemes for use of ordinary polarizable electrodes, and measuring strategies allowing for efficient data collection over large areas are being used.

Magnetometric resistivity

Analogous to the electrical resistivity method, the magnetometric resistivity (MMR) method (Edwards and Nabighian, 1991) is based on the measurement of low-frequency magnetic fields associated with noninductive current flow in the ground. Instead of potential electrodes the magnetic field is measured with coils or magnetometers. Though not widely utilized, the technique has been successfully used in mineral exploration, geothermal investigations, and geological mapping related to environmental applications. Strengths of the method include it being relatively insensitive to small conductive bodies, and when employing a vertical electric source, layered structures are not excited and the MMR response is due solely to three-dimensional targets.

Controlled-source frequency-domain electromagnetics

A number of configurations can be used with controlled-source frequency-domain EM systems (Spies and Frischknecht, 1991). The most common is the magnetic dipole-dipole or "Slingram" method that employs a small loop, dipole transmitter and small loop, dipole receiver at multiple-frequencies (Frischknecht *et al.*, 1991). Measurements can be made of the real component, which is in-phase with the transmitted signal, and the out-of-phase, or quadrature, component. The ground conductivity meter (GCM) is a subset of the Slingram method in that it operates where the low frequency inductive approximation is valid and the quadrature component is linearly proportional to the apparent ground conductivity. These methods offer the advantage that ground contact is not necessary, meaning that operation is fast, minimal personnel are required, and continuous data acquisition can be easily implemented. They have been widely used as profiling instruments with the subsequent interpretation based on the simple apparent conductivity representation.

Slingram/GCM data have a number of limitations with regard to quantitative inversion. The secondary field that carries information about the subsurface conductivity is measured in the presence of the primary fields, which can be orders of magnitude larger. This necessitates a compensation of the primary field so that the measured in-phase component only comes from the secondary field. Accuracy of the compensation depends heavily on the transmitter-receiver distance; hence for every transmitter-receiver separation and for every frequency, the instrument must have a compensation circuit. Instruments with coil separations of less than 60 m and a connecting cable between the transmitter and receiver are difficult to calibrate and coil separation errors are detrimental to the accuracy of the in-phase component, unless the coil separation is so large and field conditions favorable that a good relative accuracy is obtainable. The quadrature component is relatively insensitive to close separation geometry.

The in-phase component of magnetic dipole-dipole systems is reliably measured with fixed boom systems where the transmitter and receiver coils are in a rigid frame. For practical reasons these systems are often small. Small coil separations a few meters above a resistive ground effectively constitute a nonconductive environment enabling a primary field zeroing and gain calibration. For coil separations exceeding a few meters it is awkward to get far enough away from the ground. Consequently, calibration is difficult, impeding a quantitative assessment of the reliability of the data. Calibration can be performed with calibration coils, but it is still a time-consuming and cumbersome process.

There is an ongoing debate about the value of having more than one frequency for geological mapping with small fixed-boom systems. One side claims that differences in investigation depths are pronounced enough for multifrequency measurements to be useful. The other side maintains that for conductivity mapping purposes one frequency is sufficient because the skin depth is large compared with the coil separation for all appropriate frequencies, meaning that the sensitivity is controlled by the coil separation. Sounding data can be obtained by using two dipole orientations (horizontal coplanar and vertical coplanar) and measuring at different heights, however using multiple coil separations rather than data at different elevations is preferable. The resolution kernels for soundings at different elevations become almost parallel at relatively shallow depth thus reducing the independence of the data.

For quadrature data the equivalent half-space is linear in conductivity and inversion reduces to a simple transformation. However, the inverse problem is only linear with respect to conductivity at low-induction number, an assumption that can be poor in many near-surface environments. An accurate inversion should take into account the system response: phasing of the instrument, correction for measurement elevation, and calibration factors, as well as the departures from the low-induction number approximation.

Plane wave electromagnetics

Magnetotellurics is arguably the most developed of all EM methods because of the elegance of the plane wave source assumption. Used for deep sounding of the Earth's crust the method has also been used extensively for oil and gas and geothermal exploration. The complex impedance of the Earth is inferred from measurements of the natural electric and magnetic field—or more accurately the voltage measured across orthogonal electric dipoles and induction coils. The MT source below 1 Hz is due to currents in the magnetosphere and above that from worldwide lightning. At about 1 Hz the natural signal strength is low, but schemes have been developed to compensate. The remote reference technique makes use of the simultaneous measurements of the field at multiple sites to removes bias and cultural noise.

The audio magnetotelluric (AMT) and the controlled-source AMT (CSAMT) methods are applied to hydrological and resource exploration targets (Zonge and Hughes, 1991) with depths of investigation from tens to hundreds of meters. The AMT method employs natural signal in the 100 kHz to 10 Hz range; the CSAMT methods uses an artificial source, traditionally a grounded electrical dipole that dominates the natural field although other systems use a magnetic dipole source that augments the low-energy signal at frequencies above 1 kHz. With both approaches the source must be far enough away from the receiver array so that the source has lost its dipolar geometry and is plane wave, but close enough to have an appreciable signal-to-noise ratio. For a grounded electrical dipole source this distance, usually measured in skin depth, can be several kilometers and for a magnetic dipole hundreds of meters.

The RMT method, making use of EM fields from commercial and military radio transmitters and ambient signals, operates in the frequency range between 10 kHz (VLF transmitters) and 1 MHz (AM transmitters) (Tezkan, 1999). The VLF transmitting network will be discontinued in the near future because submarines now are using satellites for communication, hence there will be a lack of signal in the lower frequency range severely limiting the use of the method. It has been used for mapping waste sites and other contaminated areas. The CSAMT and RMT methods have the significant advantage of being able to exploit the multidimensional inversion software that has been well developed in the MT community.

Time-domain electromagnetics

With the time-domain, also known as the transient electromagnetic method (TEM) the Earth is excited by a loop carrying current on the surface of the Earth. The current is abruptly turned off and currents within the Earth are induced to maintain continuity of the magnetic field. The magnetic field decays as the electric currents diffuse through the ground. Magnetic field measurements can be made in, out or coincident with the transmitter loop. Traditionally, only the vertical component is measured, but much information is available in the horizontal components.

The TEM has gained popularity over the past decade. Being an inductive method, it is particularly good at mapping the depth and extent of good conductors, but is relatively poor at distinguishing conductivity contrasts in the low conductivity range. Clay and saltwater intrusion constitute conductive features of special interest in aquifer delineation. Hence, the method has been used extensively in hydrogeophysical investigations.

The very early time EM (VETEM), and equivalent high-frequency domain systems have been constructed in the nanoseconds range (Wright *et al.*, 2000) for depths of investigation of less than a few meters. The response in this range spans the region between diffusion and wave propagation so the data contain information on the conductivity and permittivity in areas of high conductivity where the GPR signal is greatly attenuated.

Louise Pellerin

Bibliography

Edwards, L.E., and Nabighian, M.N., 1991. In Nabighian, M.N. (ed.), *Electromagnetic Methods in Applied Geophysics*, Vol. 2, Part A. Tulsa, OK: Society of Exploration Geophysicists, pp. 47–104.

Fink, J.B., McAlister, E.O., Sternberg, B.B., Wiederwilt, W.G., and Ward, S.H., 1990. *Induced polarization*. Investigations in Geophysics, Vol 4. Tulsa, OK: Society of Exploration Geophysicists.

Frischknecht, F.C., Labson, V.F., Spies, B.R., and Anderson, W.L., 1991. Profiling methods using small sources. In Nabighian, M.N. (ed.), *Electromagnetic Methods in Applied Geophysics,* Vol. 2, Part A. Tulsa, OK: Society of Exploration Geophysicists, pp. 105–270.

Loke, M.H., and Barker, R.D., 1996. Rapid least-squares inversion of apparent resistivity pseudosections by a quasi-Newton method. *Geophysical Prospecting*, **44**: 131–152.

Panissod, C., Michel, D., Hesse, A., Joivet, A., Tabbagh, J., and Tabbagh, A., 1998. Recent developments in shallow-depth electrical and electrostatic prospecting using mobile arrays. *Geophysics*, **63**: 1542–1550.

Pellerin, L., Groom, D., and Johnston, J.M., 2003. Characterization of an old diesel fuel spill? Results of a multireceiver OhmMapper survey. In *Proceedings 73rd Annual International Meeting*. Tulsa, OK: Society of Exploration Geophysicists, pp. 5008–5011.

Sørensen, K.I., 1996. Pulled array continuous electrical profiling. *First Break*, **14**: 85–90.

Spies, B.R., and Frischknecht, F.C., 1991. Electromagnetic sounding. In Nabighian, M.N. (ed.), *Electromagnetic Methods in Applied Geophysics,* Vol. 2, Part A. Tulsa, OK: Society of Exploration Geophysicists, pp. 285–426.

Telford, W.M., Geldart, L.P., and Sheriff, R.E., 1990. *Applied Geophysics*. Cambridge: Cambridge University Press.

Tezkan, B., 1999. A review of environmental applications of quasi-stationary electromagnetic techniques. *Surveys in Geophysics*, **20**: 279–308.

Ward, S.H., and Hohmann, G.W., 1988. Electromagnetic theory for geophysical applications. In Nabighian, M.N. (ed.), *Electromagnetic Methods in Applied Geophysics,* Vol. 1, Tulsa, OK: Society of Exploration Geophysicists, pp. 313–364.

Wright, D.L., Smith, D.V., and Abraham, J.D., 2000. A VETEM survey of a former munitions foundry site at the Denver Federal Center. In *Proceedings Symposium for the Application of Geophysics to Environmental and Engineering Problems (SAGEEP)*. Denver, CO. Society of Environmental and Engineering Geophysics, pp. 459–468.

Zonge, K.L., and Hughes, L.J., 1991. Controlled source audio-frequency magnetotellurics. In Nabighian, M.N. (ed.), *Electromagnetic Methods in Applied Geophysics,* Vol. 2, Part B. Tulsa, OK: Society of Exploration Geophysicists, pp. 713–810.

Cross-references

EM Modeling, Inverse
EM, Industrial Uses
EM, Marine Controlled Source
Magnetotellurics
Robust Electromagnetic Transfer Functions Estimates
Transient EM Induction

EM, MARINE CONTROLLED SOURCE

Background

Controlled source electromagnetic (CSEM) mapping of the electrical conductivity of the seafloor depends on a simple concept of physics. If a time-varying EM field is generated near the seafloor, then eddy currents are induced in the seawater and subjacent crust in accordance with Faraday's law. The outward progress of the currents with time depends on range and electrical conductivity. In particular, the apparent speed in the seawater will be slower than that in the less-conductive crustal zones. Measurements at a remote location of the electric and magnetic fields associated with the eddy currents may be inverted for the crustal resistivity structure, if the resistivity and thickness of the seawater are known.

The concept is easily verified theoretically through an examination of the governing differential equations. The Maxwell interrelationships between the electric field **E** and the magnetic field **B** in an isotropic, homogeneous material may be combined as the damped wave equation

$$-\nabla \times \nabla \times \mathbf{B} = \mu\sigma \frac{\partial \mathbf{B}}{\partial t} + \mu\varepsilon \frac{\partial^2 \mathbf{B}}{\partial t^2}, \qquad \text{(Eq. 1)}$$

where σ, μ, and ε are the conductivity, permeability, and permittivity of the material, respectively. A similar equation may be written for the magnetic field vector **B**. Equation (1) may be rationalized by measuring length and time in units of characteristic length L and characteristic time $\tau = \mu\sigma L^2$. There results

$$-\nabla \times \nabla \times \mathbf{B} = \frac{\partial \mathbf{B}}{\partial t} + \frac{\varepsilon}{\mu\sigma^2 L^2}\frac{\partial^2 \mathbf{B}}{\partial t^2}. \quad \text{(Eq. 2)}$$

The second term on the left-hand side of Eq. (2) usually is neglected in comparison with the first term and the physics simplified to a diffusion process because the scale L of the experiment is large compared with $(\varepsilon/\mu\sigma^2)^{1/2}$ or $1/377\sigma$. The omission is equivalent to the neglect of the magnetic effects of displacement current in comparison with conduction current. The critical scale is largest for very resistive crystalline rock, having a value of the order of few tens of meters. A feel for the time taken for an EM disturbance to diffuse through a uniform medium may be gained by evaluating the characteristic time τ for a few typical cases. If the scale L is set to 1 km and the parameter μ takes its free space value, then τ has value 3.8 s for seawater, typical conductivity 3 S m^{-1}, and 1.2, 0.42, 0.12, 0.042 s for crustal resistivities of 1, 3, 10, and 30 Ωm, respectively. (The characteristic times are approximate and should be treated as upper limits by as much as a factor of 10 for practical systems.)

CSEM is usually used to explore crustal regions. The rocks that are often a simple two-phase system consisting of the resistive grain matrix and conductive pore fluid. Archie's law (Archie, 1942) relates measured bulk resistivities to porosity estimates. In a general form, it is

$$\rho_f = a\rho_w \phi^{-m}, \quad \text{(Eq. 3)}$$

where ρ_f is the measured formation resistivity, ρ_w is the resistivity of seawater, ϕ is the sediment porosity, a is a constant, and m the cementation factor. The latter two parameters can be derived from laboratory measurements and vary between $0.5 < a < 2.5$ and $1.5 < m < 3$. Resistivity values for marine sediments near the seafloor where the porosity is often in excess of 0.5 may be as low as 1 Ωm.

Practical methods: response of a layered earth

A practical CSEM system consists of a transmitter capable of generating an EM disturbance and one or more receivers which detect the disturbance at some later time as it passes nearby. In common with land-based systems used for mineral prospection, the transmitter and the receiver may be electric and/or magnetic dipoles. Usually, for marine exploration, both the transmitter and the receivers are near or on the seafloor, as shown in Figure E16. Further, the system response can be described in frequency domain or time domain. Both use a continuous waveform and are essentially equivalent. In particular, delays in time domain are related to phase changes in frequency domain.

A study of the nature of the response of a layered earth to a time domain controlled source system is a useful learning exercise, as some of the physics is counterintuitive. Disturbances from the transmitter diffuse through the seawater and the seafloor and are seen at the receiver as at least two distinct arrivals separated in time depending on the conductivity contrast between seawater and subjacent crust. If the layered seafloor increases in resistivity with depth, then disturbances propagate laterally more rapidly at depth than immediately beneath the seafloor. The received signal viewed at different ranges in logarithmic time has many of the characteristics of refraction seismic, even though the process is diffusive, and the signal can be processed using seismic methods. We use two basic configurations for marine exploration: the horizontal magnetic dipole-dipole and the horizontal electric dipole-dipole. System geometries can be broadside or in-line. The two geometries yield different information (Yu and Edwards, 1991; Yu *et al.*, 1997). For example, the in-line electric dipole-dipole is sensitive to the vertical resistivity whereas its broadside cousin is sensitive to the horizontal resistivity perpendicular to the line joining the dipoles. The EM fields impressed in the crust by both the horizontal magnetic dipole and the horizontal electric dipole have two polarizations. The polarizations are characterized by the absence of a vertical magnetic field and a vertical electric field, respectively. They average the resistivity in two different ways. If these averages are different, then the medium appears to be only laterally isotropic. Data collected over layered Earth structures must be interpreted using software including anisotropy either as multiple fine isotropic layers or, better, as a few anisotropic layers.

By way of example, a summary of the layered earth responses for the electric dipole-dipole system for the in-line configuration, following Chave and Cox (1982), Edwards and Chave (1986), and Cheesman *et al.* (1987) is presented here. The seawater has a finite thickness d_0 and an electrical conductivity σ_0. The permittivity of the air is ε. The subjacent crust has N layers with thicknesses $d_1, d_2, \ldots, d_{N-1}$ and conductivities $\sigma_1, \sigma_2, \ldots, \sigma_N$, respectively. A current I is switched on at time $t = 0$ and held constant in a transmitting electric dipole of

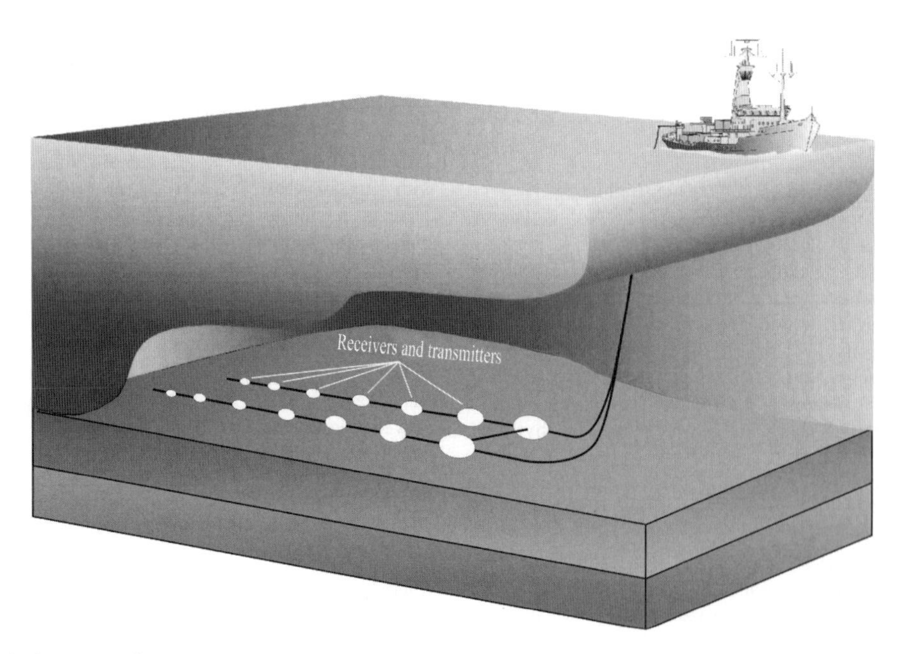

Figure E16 A towed electric seafloor array. Transmitter and receiver electrodes are interchangeable and pairs can be connected to generate many dipole-dipole configurations.

length Δl. The expression for the Laplace transform of the step-on transient electric field at the seafloor for the in-line electric dipole-dipole geometry, dipole separation L, is given in terms of s, the Laplace variable. It is

$$\frac{I\Delta l}{2\pi s}[F(s)+G(s)], \quad \text{(Eq. 4)}$$

where F and G are the Hankel transforms

$$F(s)=-\int_0^\infty \frac{Y_0Y_1}{Y_0+Y_1}\lambda J_1'(\lambda L)\mathrm{d}\lambda, \quad \text{(Eq. 5)}$$

$$G(s)=-(s/L)\int_0^\infty \frac{Q_0Q_1}{Q_0+Q_1}J_1(\lambda L)\mathrm{d}\lambda, \quad \text{(Eq. 6)}$$

and the included Laplace transform of the source dipole moment is $I\Delta l/s$.

If the magnetic effects of displacement currents in the earth are neglected, then the parameters Y_0 and Q_0 are given for the sea layer as

$$Y_0=\frac{\theta_0}{\sigma_0}\left[\frac{\sigma_0 u_a+s\varepsilon\theta_0\tan h(\theta_0 d_0)}{s\varepsilon\theta_0+\sigma_0 u_a\tan h(\theta_0 d_0)}\right] \quad \text{(Eq. 7)}$$

and

$$Q_0=\frac{\mu_0}{\theta_0}\left[\frac{\theta_0+u_a\tan h(\theta_0 d_0)}{u_a+\theta_0\tan h(\theta_0 d_0)}\right], \quad \text{(Eq. 8)}$$

where the EM wavenumbers in the ground and in the air are given by $\theta_0^2=\lambda^2+s\mu\sigma_0$ and $u_a^2=\lambda^2+s^2\mu\varepsilon$.

The parameters Q_1 and Y_1 are evaluated through the downcounting recursion relationships

$$Y_i=\frac{\theta_i}{\sigma_i}\left[\frac{\sigma_i Y_{i+1}+\theta_i\tan h(\theta_i d_i)}{\theta_i+\sigma_i Y_{i+1}\tan h(\theta_i d_i)}\right], \quad \text{(Eq. 9)}$$

and

$$Q_i=\frac{\mu_0}{\theta_i}\left[\frac{\theta_i Q_{i+1}+\mu_0\tan h(\theta_i d_i)}{\mu_0+\theta_i Q_{i+1}\tan h(\theta_i d_i)}\right], \quad \text{(Eq. 10)}$$

where the starting values are $Y_N=\theta_N/\sigma_N$ and $Q_N=\mu_0/\theta_N$, respectively. The wavenumbers θ_i are defined by $\theta_i^2=\lambda^2+s\mu\sigma_i$.

The Y and Q functions denote the polarizations characterized by the absence of a vertical magnetic field and a vertical electric field, respectively. The Y function yields the DC "resistivity" response at late time (Edwards, 1997).

The corresponding form for the broadside electric dipole-dipole geometry is as given in expression (4) but with alternative definitions for F and G.

The methods for inverting the Hankel and Laplace transforms are fully described in the theoretical papers listed above. All other geometries are a linear combination of the in-line and broadside responses.

The frequency domain response of the dipole may be obtained by multiplying the step-on response by s to get the impulse response and then replacing s by $\mathrm{i}\omega$ where ω is the angular frequency of the harmonic current. The inverse Hankel transform becomes a complex function having a phase and an amplitude.

The theory may be extended to include the effects of anisotropy caused by interbedding. The coefficient of anisotropy k is defined by $k=\sqrt{\sigma_h/\sigma_v}$, where σ_h and σ_v are the tangential and normal conductivities in any given layer. The coefficient cannot be smaller than unity. We modify the form of θ only in expression (9) to $\theta^2=k^2\lambda^2+s\mu\sigma$. In both expressions (9) and (10), σ is replaced by σ_h whereever it occurs.

Response curves

The step-on response of a double half-space to the in-line electric dipole-dipole system is shown in Figure E17. The seawater is assigned a conductivity of 3 Sm^{-1} while the subjacent crust has conductivities of 3, 1, 0.1, and 0.3 Sm^{-1}, respectively. The vertical axis has been scaled to yield the DC apparent resistivity at late time. The horizontal axis is in logarithmic time for a transmitter-receiver separation L of 260 m. (Absolute time may be estimated for any system, given a conductivity σ and a value for L, as $\mu_0\sigma L^2/c$, where c is a constant with a value of about 5. Progress of disturbances through several zones of differing conductivity may be obtained by summing such estimators.)

The curves have three characteristics from which the conductivity of the crust may be obtained. There is late-time variation in amplitude (Label 1), which is less sensitive with increasing conductivity contrast. There is an early time change in amplitude which depends on the seafloor conductivity (Label 2). The location in time of this initial change is a strong function of seafloor conductivity (Label 3). A robust estimate of determining apparent resistivity from time-domain curves may be obtained from the maximum in the early time gradient of the response curve. The same information may be gleaned from amplitude and phase curves in the frequency domain.

Electromagnetic refraction

Cross sections of the diffusion of current from a 2D electric dipole at the junction of two half-spaces representing seawater and the subjacent crust may be used to illustrate the origin of the two arrivals. The diffusive process is plotted as a series of contour maps for increasing time steps, as shown in Figure E18a through E18f. The magnetic field is out of the slide, the electric field is in the plane. The model is 500 m^2 and the seawater and crust have conductivities of 3 and 0.5 Sm^{-1}. The contours, originally shown by Edwards (1988), are current streamlines and the shading is proportional to the size of the electric field.

The process can be considered as a form of EM refraction, even though it is diffusive, because of the similarity of the lateral movement of the current flow pattern and the seismic head wave. Consider the direction of the Poynting vector $\vec{P}\sim\vec{E}\times\vec{B}$. Clearly, the first arrival at a receiver dipole is a disturbance that has propagated through the more resistive zone, the crust. In the progress of the diffusion, energy migrates upward from the lower medium through the seafloor. At later times the signal that propagates through the seawater arrives and at the static limit (DC) symmetry is reached about the surface. The maps are consistent with the type curves presented earlier.

Seismic style displays

An impulse response may be obtained as the derivative with respect to logarithmic time of the step-on response. In the displays that follow, the normalized impulse response is plotted as a function of logarithmic time and logarithmic transmitter-receiver separation. The amplitude scaling, which varies from figure to figure, yields the same curve at every offset for a simple double half-space. The curve just moves to later time as the separation increases. The in-line electric dipole-dipole response has two peaks in the double half-space impulse response corresponding to energy that has propagated through the sea and the crust, respectively. The crustal peak is always at earlier time. Its broadside cousin also has two features but of different sign. The crustal peak is positive whereas the seawater trough at later time is negative.

The air wave

Of particular concern to companies that explore in shallow waters such as the shelf seas is the so-called air wave. Some portion of the EM energy travels upward to the sea surface, through the air to the vicinity of the receiver and then downward to the receiver on the seafloor. The up-over-down path can in some instances be faster than any direct path through the seawater or the

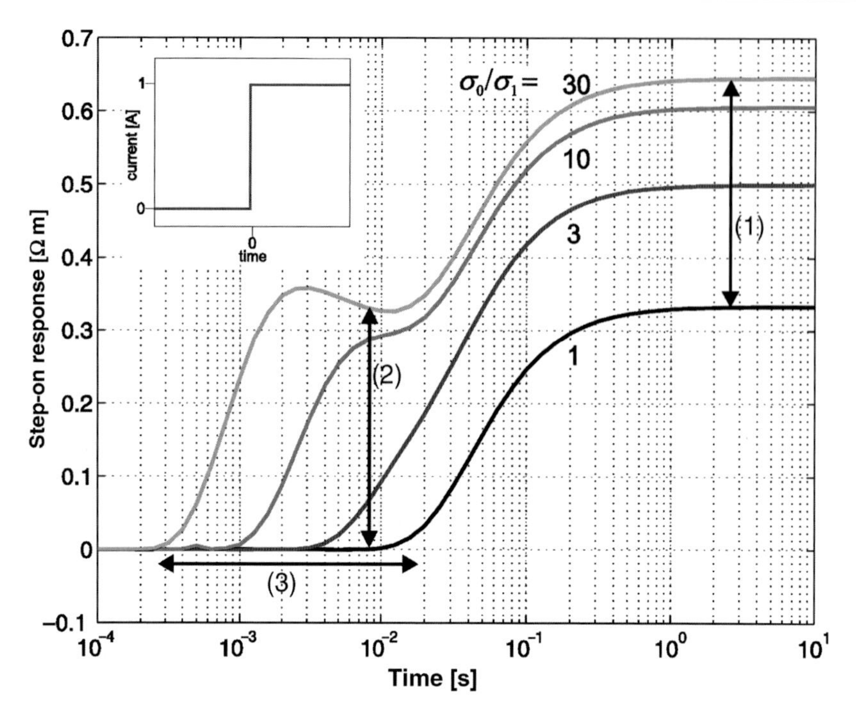

Figure E17 Normalized step-on responses calculated on the interface between two half-spaces to the in-line electric dipole-dipole system. The conductivity of the upper half-space is 3 S m^{-1}, for the lower half-space conductivities of 0.3, 0.1, 1, and 3 S m^{-1} have been assigned. The separation between transmitter and receiver is 260 m. For a conductivity contrast between seawater and subjacent crust larger than 10 the arrival through the signal through the seawater at later times can be clearly separated from the earlier arrival through the crust. In addition, three different effects are noticeable: (1) Amplitude variations at late times depend on the conductivity contrast, but are mainly due to current flow through the seawater; (2) amplitude variations at earlier times depend on the seafloor conductivity; (3) the location in time to the initial change is a function of the seafloor conductivity (after Edwards and Chave, 1986).

subjacent crust. Compare the two models shown in Figure E19a and E19b and the stacked impulse response of these models shown in Figure E19c through E19f, for the in-line and broadside geometries. The sea layer in the first model is infinitely thick while that in the second has a finite thickness of 200 m. As the transmitter-receiver separation increases, the air wave which initially appears at later time appears to move to relatively earlier times and at large separations contaminates the disturbance traveling through the crust. From a practical point of view, the air wave signature is easily removed in the inversion of data provided sufficient dynamic range in the receiver electronics is available to record it properly.

The resistive zone at depth

In the second example, the responses of a double half-space model modified by the inclusion of a rapid increase in resistivity at a depth of 200 m have been computed. They are shown in Figure E20b and compared with the base response of the half-space model, Figure E20a. The stacked impulse responses for the in-line and broadside geometries are shown in Figure E20c through E20f. Notice the distinct refraction visible in the early time crustal response when the EM disturbance sees the resistive buried zone. The location of the refraction in space and its slope on the log time vs log separation diagram may be used to infer the depth to and conductivity of the resistive zone.

Case histories

Gas hydrates in Cascadia

A report of an EM survey for gas hydrate off the west coast of Vancouver Island, Canada, is presented by Schwalenberg *et al.* (2005) using apparatus originally developed by Yuan and Edwards (2000). Natural gas hydrates are ice-like solids found in seafloor sediments. They consist of gas molecules, mainly methane, contained in a cage-like clathrate structure of water molecules. They form under low-temperature and high-pressure conditions, typically in the uppermost few hundreds of meters of sediments in water depth exceeding about 500 m. The global abundance of methane frozen in hydrate exceeds the amount of all other known fossil hydrocarbon resources. Hydrate clearly has a huge potential as a future energy resource. A gas hydrate deposit can be generally identified in a seismic section by the occurrence of a bottom-simulating reflector (BSR) which is associated with the base of the hydrate stability field. The base is a transition zone between hydrate-bearing sediments above it and free gas and water below it. The location of the zone is temperature controlled and depends on the ambient geothermal gradient. The target area is located on the accretionary prism of the Cascadia margin in close vicinity of ODP Site 889, as shown in Figure E21. Here, several seismic blank zones were observed over a vent field which covers an area of roughly 1 km × 3 km. The largest, blank zone 1 is called the Bullseye vent and has a seafloor diameter of about 400 m.

The CSEM experiment was conducted in SW-NE direction, into the prevailing wind and current, along a profile intersecting the Bullseye and approaching the other vent sites, making measurements at 28 sites with a spacing between sites of 250 m. The seafloor array is towed in direct contact with the soft marine sediments, as sketched in Figure E22. At the forward end, a heavy weight (pig) is attached to keep the system in contact with the seafloor. It is followed by a transmitter dipole (TX) 124 m long and, in this experiment, just two receiver dipoles (RX1) and (RX2), each 15 m long located at distances r_1 and r_2 of 174 and 292 m from the transmitter cable, respectively. While the system is operated in the time domain, using a square current waveform with a peak-to-peak amplitude of 20 A and a period

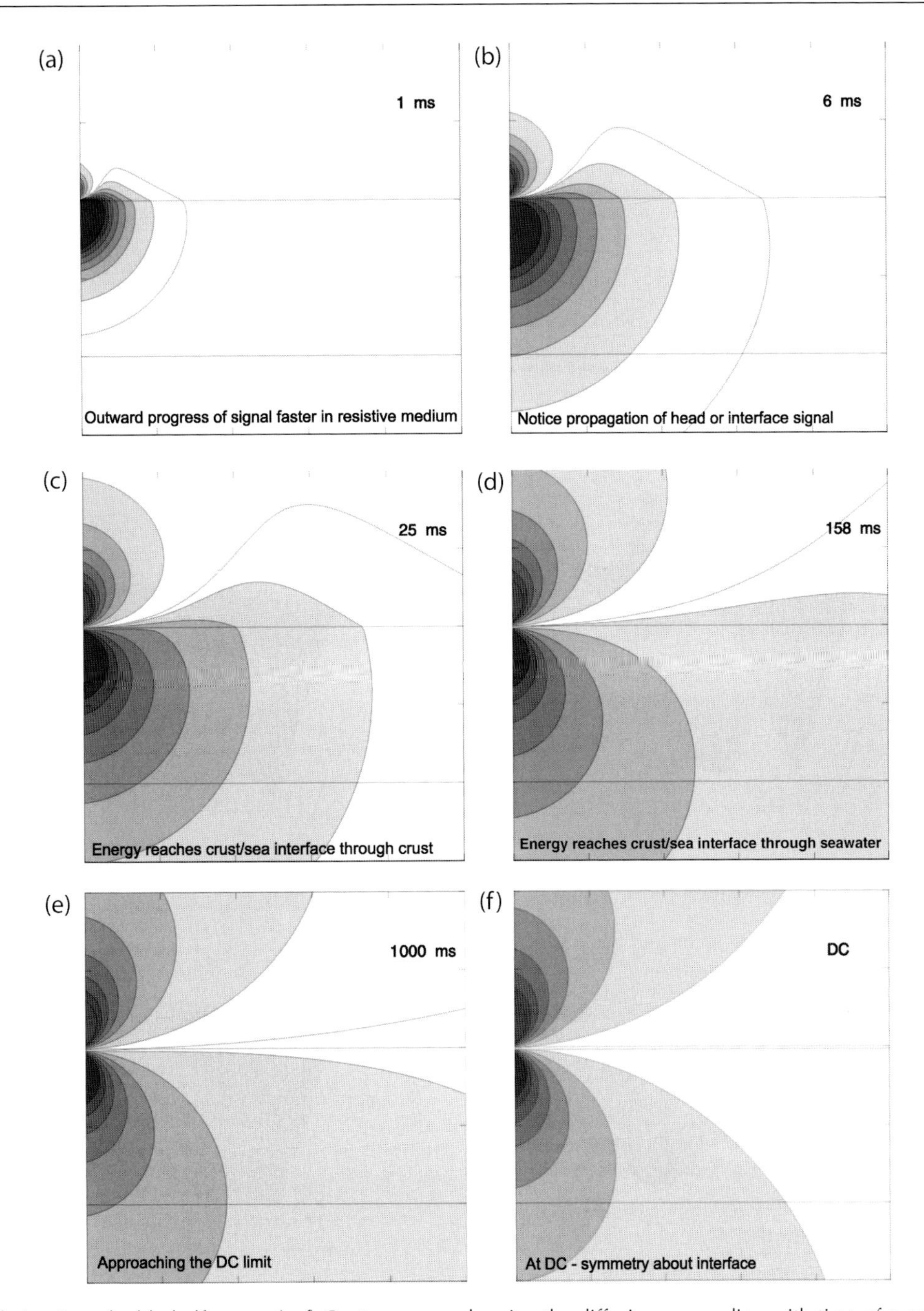

Figure E18 Diffusion in a double half-space. (a–f) Contour maps showing the diffusion proceeding with time of a current from a 2D electric dipole at the interface of two half-spaces representing seawater (3 $\mathrm{S m^{-1}}$) and subjacent crust (0.5 $\mathrm{S m^{-1}}$). The model is 500 $\mathrm{m^2}$. Contours are current streamlines of the electric field and the shading is proportional to the size of the field (after Edwards, 2006).

of 6.6 s, the data analysis and inversion to a multilayered model is completed in frequency domain using phase data only in the band from 0.5 to 100 Hz.

Simplified results for a half-space model are plotted in Figure E23a for both receiver separations over a compatible seismic section, Figure E23b. Two pronounced resistivity anomalies are visible along the profile which are in striking agreement with the seafloor projections of the vent sites from the seismic section. The resistivity values within the anomalous zones are higher for the larger separation RX2 than those for RX1 and rise up locally to more than 5 m over the regional background, which lies between 1.1 and 1.5 Ωm. The gas hydrate concentrations derived from the resistivity profile, shown in Figure E24, over the vent sites exceeds 50% at maximum and about 25% on average of the available pore space.

The resolution of the CSEM method does not permit a detailed analysis of the distribution of the resistive elements within the blank zones but it does provide an integrated value. Assuming that the increase in resistivity is due to a higher hydrate concentration, the latter may be converted to total mass of hydrate and then to total available methane. A rough estimate can be made for the Bullseye vent assuming a cylindrical volume, with diameter and depth of 400 and 200 m, respectively. Twenty five percent of the available pore space corresponds to 3.8 million $\mathrm{m^3}$. With a solid to gas ratio of hydrate of 1:164, the related methane gas volume at STP is 0.62 billion (US) cubic meters.

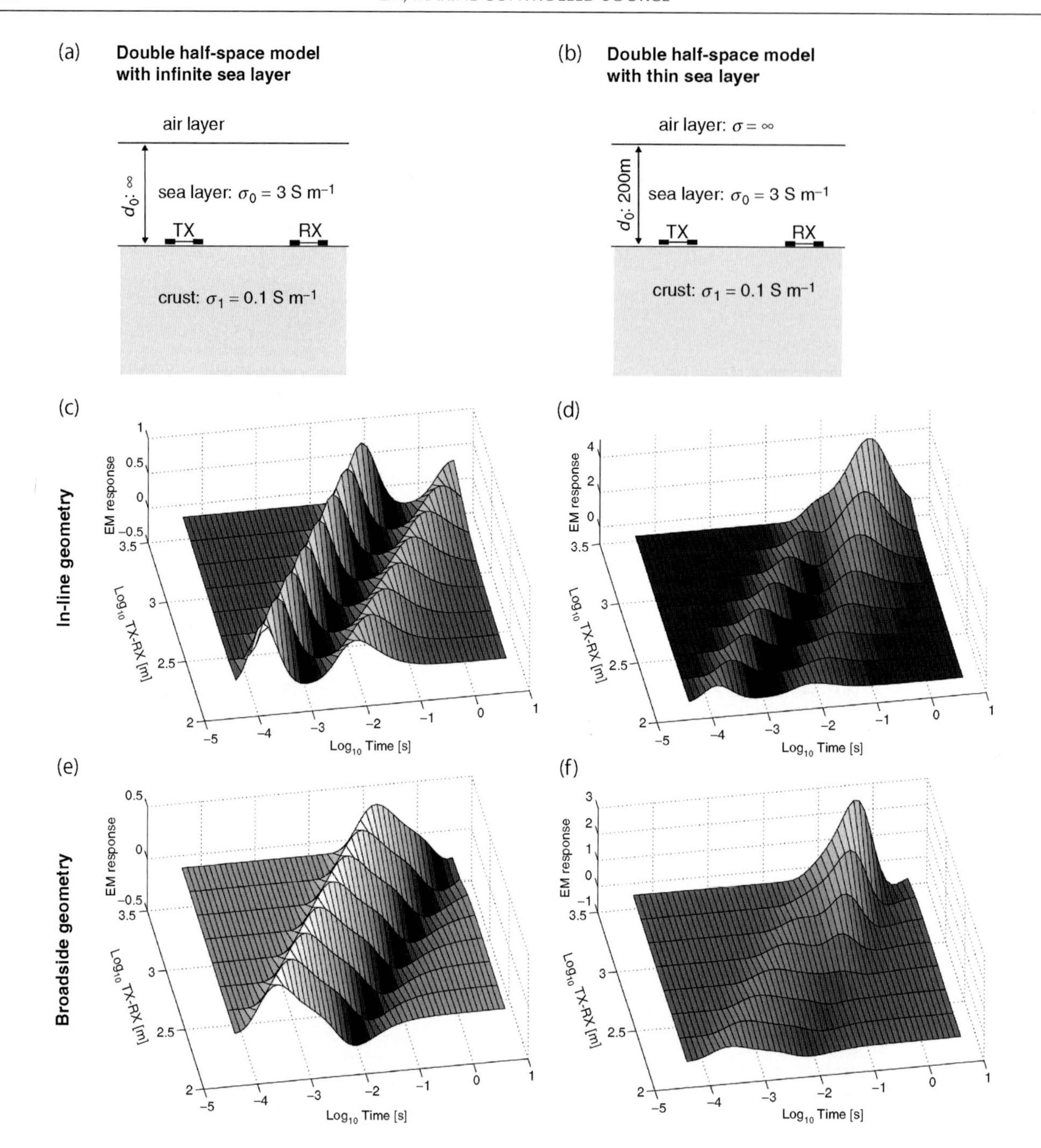

Figure E19 The airwave. The normalized impulse responses of the models in (a) and (b) to an electric dipole-dipole system on the seafloor are shown as functions of logarithmic time and transmitter-receiver separation. Panels (c) and (d) refer to in-line and panels (e) and (f) to broadside geometries (after Edwards, 2006).

The development of software for the interpretation of data like these in 3D is progressing and is clearly needed. Everett and Edwards (1993) and Unsworth *et al.* (1993) have 2.5*D* programs available in both frequency and time domains.

Buried channels, New Jersey

The marine EM group at the Woods Hole Institution of Oceanography and colleagues at the Geological Survey of Canada have built a small-scale coaxial magnetic dipole-dipole system that is contracted for mainly shallow geotechnical surveys. The system is a major improvement of systems described by Cheesman *et al.* (1990, 1991) and Webb and Edwards (1995). Among many other geological problems, they investigated the nature of the infill in a buried channel offshore New Jersey (Evans *et al.*, 2000). The buried channel represents one example of a feature in a shallow section that is analogous to a feature seen in deeper oil-bearing strata. The magnetic dipole system is dragged in contact with the seafloor. The three transmitter-receiver spacings are 4, 13, and 40 m.

Data collected in the frequency domain were processed to give an apparent porosity for each spacing. The maximum depth of investigation was about 20 m. Bounds on physical properties are greatly aided

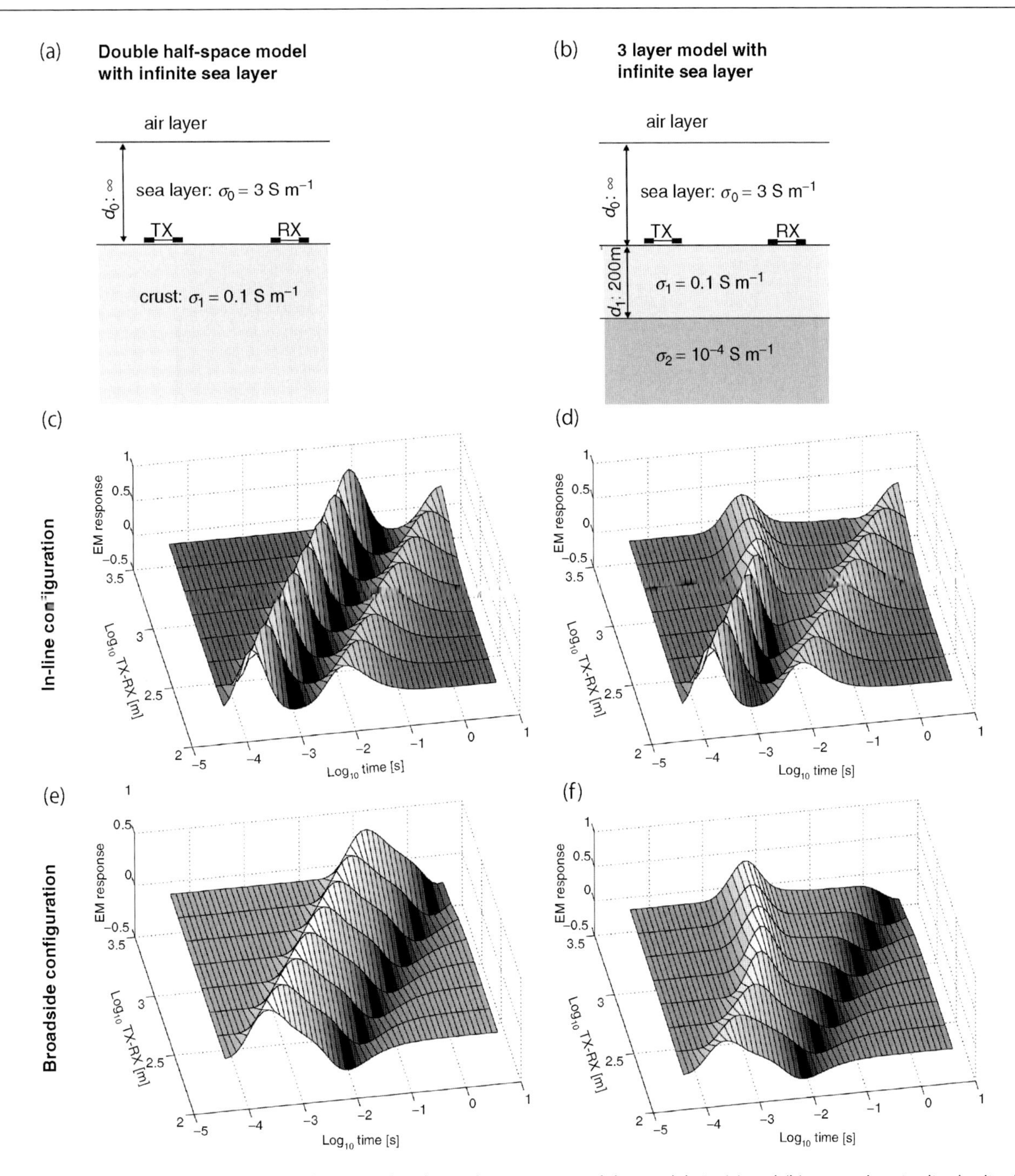

Figure E20 The resistive zone at depth. The normalized impulse responses of the models in (a) and (b) to an electric dipole-dipole system on the seafloor are shown as functions of logarithmic time and transmitter receiver separation. Panels (c) and (d) refer to in-line and panels (e) and (f) to broadside geometries (after Edwards, 2006).

by complementary seismic survey. The latter identified the structure but alone offered no information on nature of the infill. The porosity traces and seismic section are shown in Figure E25a,b. Clearly, there is an excellent correlation between the buried channels visible on the seismic section and an increase of porosity. The channels seem to incise the regional seismic reflector.

Conductivity variations at the mid-ocean ridge

Electrical conductivity varies with the porosity, temperature, degree of partial melt, and the composition of Earth materials. Nowhere are these parameters more variable than at the mid-ocean ridge. There are numerous published case histories of marine CSEM surveys in this environment. They are often combined with a complementary magnetotelluric measurements. The reader is referred to such publications as Sinha *et al.* (1997) or Evans *et al.* (1994). The methodology originated at Scripps Institution of Oceanography, pioneered by C.S. Cox and S.C. Constable.

MacGregor *et al.* (2001) completed a CSEM survey in the vicinity of the Valu Fa Ridge in the Lau Basin. They used fixed receivers and towed a horizontal long wire receiver just above the seafloor. Their survey lines and receiver locations are shown in Figure E26/Plate 15a.

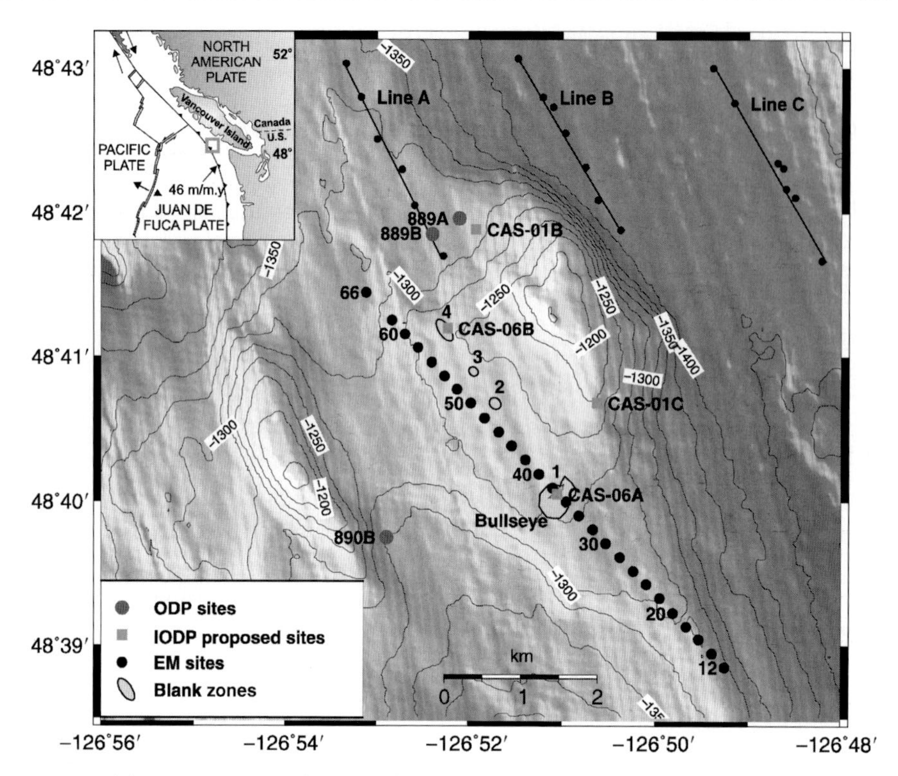

Figure E21 Bathymetry map of the target area on the Cascadia margin. The vent field is located on a bench between two topographic highs in vicinity of ODP sites 889/890. CSEM measurements were conducted along the profile crossing the Bullseye, the largest of 4 vent sites. Lines A, B, and C are EM profiles from a previous survey (after Schwalenberg *et al.*, 2005).

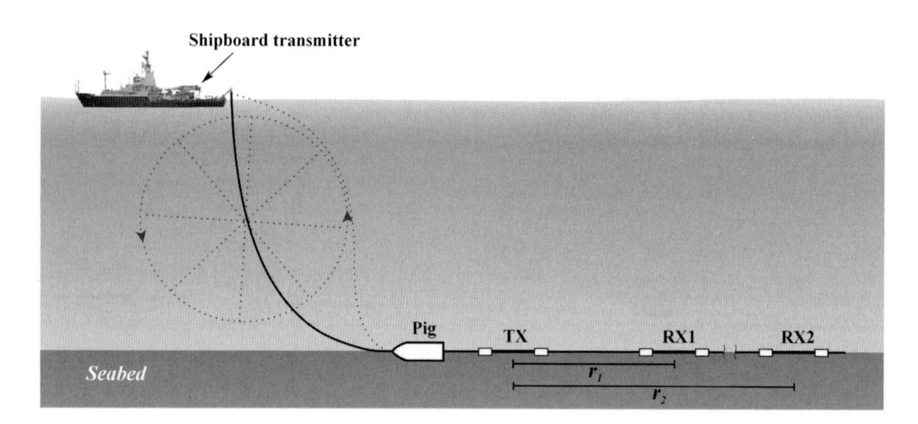

Figure E22 Geometry of the in-line dipole-dipole configuration. A current signal is produced by an onboard transmitter and sent through the coaxial winch cable to the transmitter bipole on the seafloor. Two receiver dipoles at distances r_1 and r_2 record the signal after it passes through the seawater and the sediments. A heavy weight (pig) attached to the front of the system keeps the array on the seafloor while moving along the profile. Moving the ship and taking in the winch cable pulls the array forward and causes a vertical movement of the pig. Solid and dotted line present the winch cable in idle and moving state, respectively. The wheel represents the curve over which the marine cable appears to move while in motion (after Schwalenberg *et al.*, 2005).

The corresponding analyis of the data is displayed in Figure E27/Plate 15b as a crustal cross section of the electrical resistivity anomaly across the axis of the ridge. The colors indicate the deviations (as a logarithmic multiplication factor) from a background model, in which resistivity varies only with depth beneath the seabed. The dark blue layer at the top represents the water, with the top of the crust showing as pale green. A large conductive anomaly can be seen at depths of 1 to 6 km beneath the seafloor, beneath the axis of the ridge, and extending off the axis to the west for nearly 10 km. The anomaly is interpreted as being due to the presence of an axial magma chamber, and an overlying region in which the porosity is occupied by highly conductive hydrothermal fluids. Also shown is a plot of seismic P-wave velocity anomaly (gray contours), calculated with reference to an equivalent background seismic model, and superimposed on the resistivity anomaly. The section was obtained by 2.5*D* regularized inversion of frequency domain CSEM data. The uppermost panel shows the variation in seafloor depth along the profile.

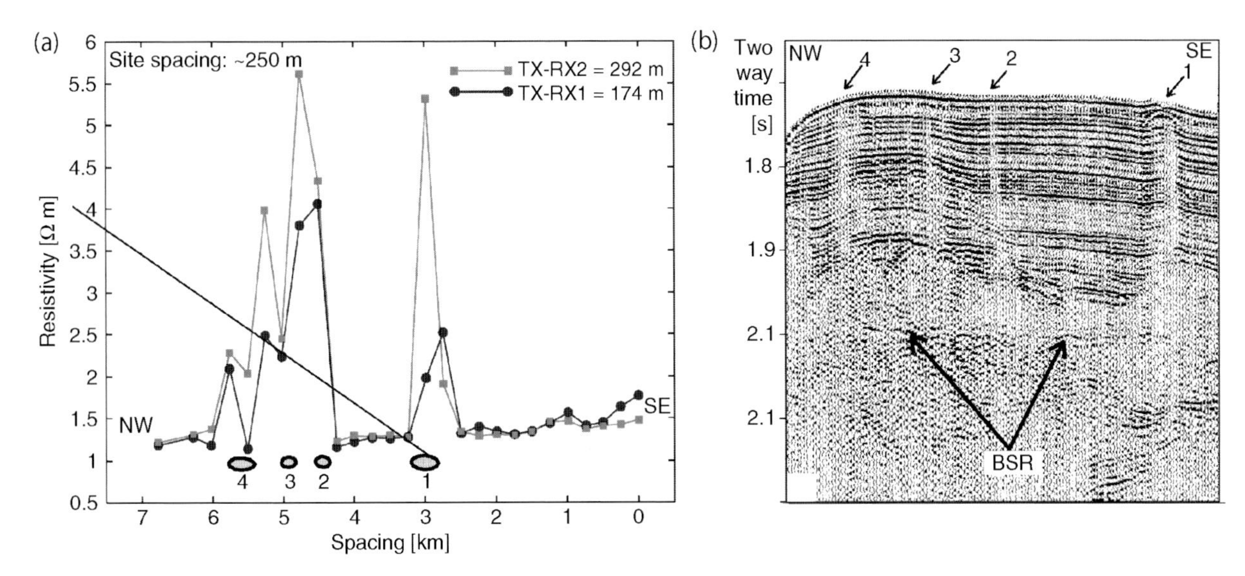

Figure E23 Panel (a) shows the bulk resistivities derived from CSEM data show anomalous resistivities exceeding 5 Ωm over background resistivities between 1.1 and of 1.5 Ωm. The anomalous areas coincide spatially with the surface expression of a series of seismic blank zones displayed in Panel (b) (after Schwalenberg *et al.*, 2005).

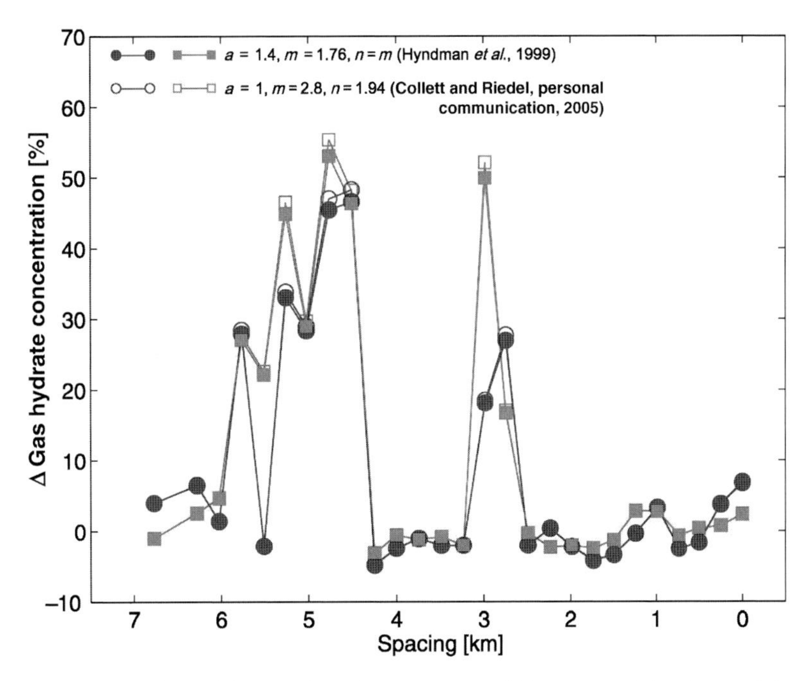

Figure E24 Gas hydrate concentrations derived from Archie's law using two different sets of Archie coefficients. The first set ($a = 1.4$, $m = 1.76$, $n = m$) is based on core data from ODP Leg 146 (Hyndman *et al.*, 2001). A recent reevaluation yielded a second set based on log data ($a = 1$, $m = 2.8$, $n = 1.94$) (Collett and Riedel, personal communication 2005). In this figure a regional gas hydrate concentration profile derived from the baseline resistivities in Figure E20b has been subtracted from the "total" hydrate concentrations. Thus, the profiles represent the additional amount of hydrate and are coincident for both sets of Archie coefficients (after Schwalenberg *et al.*, 2005).

Commercial outlook

In the last 5 years, there been a surge of interest from the exploration community in the use of marine CSEM methods for hydrocarbon exploration (Edwards, 2006). There are marine geological terranes in which the interpretation of seismic data alone is difficult. There are regions dominated by scattering or high reflectivity, such as is found over carbonate reefs, areas of volcanics, and submarine permafrost. Complementary geophysical techniques are required to study these regions. EM was at first not high on the list of alternatives. There was a pervasive belief that the high electrical conductivity of seawater

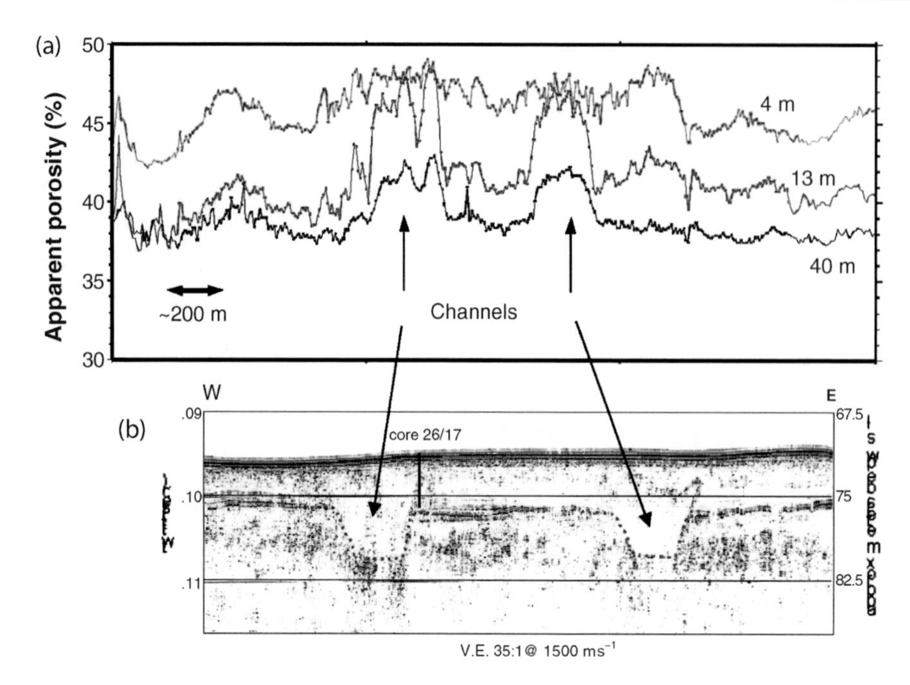

Figure E25 The magnetic dipole-dipole system has been used to find out the nature of the infill of buried paleochannels on the New Jersey continental margin. The apparent resistivities recorded at the three receivers shown in Panel (a) have been converted to apparent porosities using Archie's law. A clear correlation between locally higher porosities and the seismic image of the paleo-channels shown in Panel (b) is evident (after Evans *et al.*, 2000).

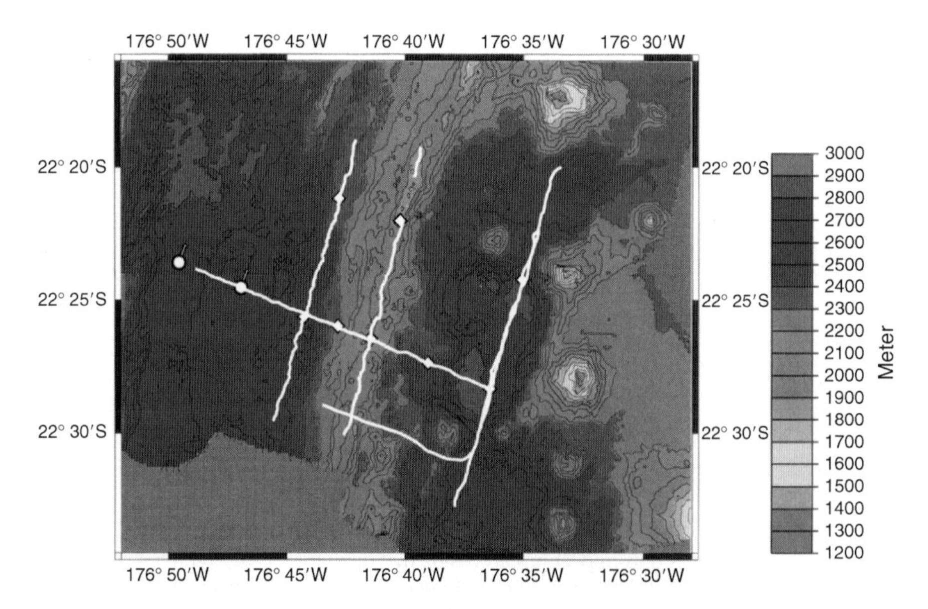

Figure E26/Plate 15a The CSEM transmitter tracks superimposed on a bathymetry map of the Valu Fa Ridge. The receiver locations are shown as white dots (courtesy M.C. Sinha).

precluded the application of EM systems for exploration even though academics had put forward methods specifically designed for the marine environment. The tide turned when a few surveys commissioned from universities proved very successful and over the last 5 years, exploration managers and investors have become aware of the importance of CSEM. Morgan Stanley, a well-known member of the New York Stock Exchange and Investment Manager, reported in August 2004 that in their view the implications of CSEM imaging on offshore drilling, service, and field development activity will be one of the most frequently discussed topics in the oil service industry over the next 12 months. They see a growth in annual revenues from a mere $30 million to $600–900 million in less than 5 years—one quarter of

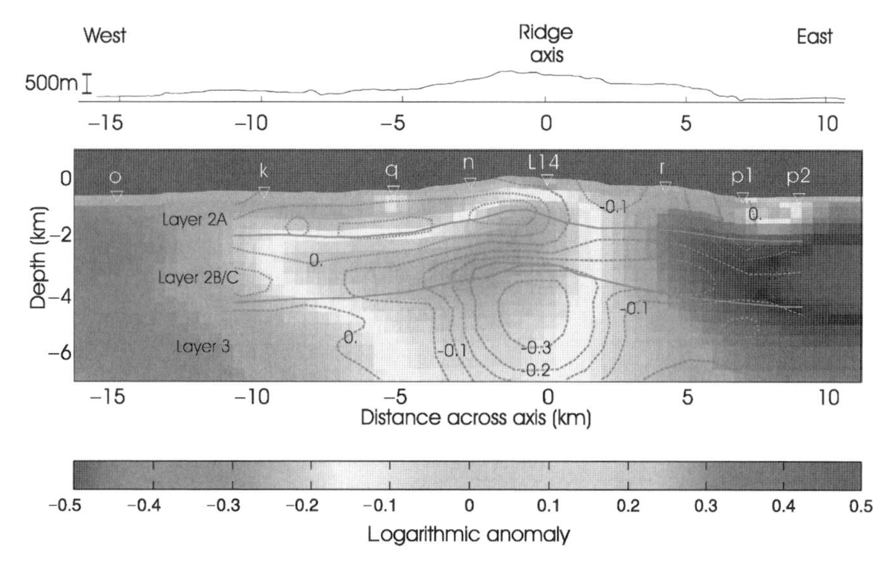

Figure E27/Plate 15b A section through the Valu Fa Ridge showing the logarithmic variation in electrical conductivity and the anomaly in seismic P-wave velocity as color shading and black contours respectively (courtesy M.C. Sinha).

the current spending-on offshore seismic and compare the technological revolution with the growth of 3D seismic in the early 1990s. McBarnet (2004) summarizes recent commercial activity and identifies the players involved.

Nigel Edwards

Bibliography

Archie, G.E., 1942. The electrical resistivity log as an aid in determining some reservoir characteristics. *Journal of Petroleum Technology*, **5**: 1–8.

Chave, A.D., and Cox, C.S., 1982. Controlled electromagnetic sources for measuring the electrical conductivity beneath the oceans. *Journal of Geophysical Research*, **87**: 5327–5338.

Chave, A.D., Constable, S.C., and Edwards, R.N., 1986. Electrical exploration methods for the seafloor. In Nabighian, M.N. (ed.), *Electromagnetic Methods*, Vol. 2: *Applications*. Tulsa, OK: Society of Exploration Geophysicists.

Cheesman, S.J., Edwards, R.N., and Chave, A.D., 1987. On the theory of seafloor conductivity mapping using transient electromagnetic systems. *Geophysics*, **52**: 204–217.

Cheesman, S.J., Edwards, R.N., and Law, L.K., 1990. A short baseline transient electromagnetic method for use on the seafloor. *Geophysical Journal International*, **103**: 431–437.

Cheesman, S.J., Law, L.K., and Edwards, R.N., 1991. Porosity determinations of sediments in Knight Inlet using a transient electromagnetic system. *Geomarine Letters*, **11**: 84–89.

Edwards, R.N., 1988. Two-dimensional modelling of a towed electric dipole-dipole EM system: the optimum time delay for target resolution. *Geophysics*, **53**: 846–853.

Edwards, R.N., 1997. On the resource evaluation of marine gas hydrate deposits using a seafloor transient electric dipole-dipole method. *Geophysics*, **62**: 63–74.

Edwards, R.N., 2006. Marine controlled source electromagnetics: principles, methodologies, future commercial applications. *Surveys in Geophysics*, **26**: 675–700.

Edwards, R.N., and Chave, A.D., 1986. A transient electric dipole-dipole method for mapping the conductivity of the seafloor. *Geophysics*, **51**: 984–987.

Evans, R.L., Sinha, M.C., Constable, S.C., and Unsworth, M.J., 1994. On the electrical nature of the axial melt zone at 13 N on the East Pacific Rise. *Journal of Geophysical Research*, **99**: 77–88.

Evans, R.L., Law, L.K., St. Louis, B., and Cheesman, S.J., 2000. Buried paleochannels on the New Jersey continental margin: channel porosity structures from electromagnetic surveying. *Marine Geology*, **170**: 381–394.

Everett, M.E., and Edwards, R.N., 1993. Transient marine electromagnetics, The 2–5*D* forward problem. *Geophysical Journal International*, **113**: 545–561.

Hyndman, R.D., Spence, G.D., Chapman, R., Reidel, M., and Edwards, R.N., 2001. Geophysical studies of marine gas hydrates in Northern Cascadia. In Paull, C., and Dillon, W.P. (eds.), *Natural Gas Hydrates: Occurrence, Distribution, and Detection*. Geophysical Monograph Series. Washington, DC: American Geophysical Union.

Mac Gregor, L., Sinha, M., and Constable, S., 2001. Electrical resistivity structure of the Valu Fa Ridge, Lau Basin, from marine controlled source electromagnetic sounding. *Geophysical Journal International*, **146**: 217–236.

McBarnet, A., 2004. All at sea with EM. *Offshore Engineer*, **29**: 20–22.

Schwalenberg, K., Willoughby, E., Mir, R., and Edwards, R.N., 2005. Marine gas hydrate electromagnetic signatures in Cascadia and their correlation with seismic blank zones. *First Break*, **23**: 57–63.

Sinha, M.C., Navin, D.A., Mac Gregor, L.M., Constable, S.C., Peirce, C., White, A., Heinson, G., and Inglis, M.A., 1997. Evidence for accumulated melt beneath the slow-spreading Mid-Atlantic Ridge. *Philosophical Transaction of the Royal Society of London*, **A355**: 233–253.

Unsworth, M.J., Travis, B.J., and Chave, A.D., 1993. Electromagnetic induction by a finite electric dipole source over a 2-D earth. *Geophysics*, **58**: 198–214.

Webb, S.C., and Edwards, R.N., 1995. On the correlation of electrical conductivity and heat flow in Middle Valley, Juan de Fuca Ridge. *Journal of Geophysical Research*, **100**: 22,523–22,532.

Yu, L., and Edwards, R.N., 1991. The detection of lateral anisotropy of the ocean floor by electromagnetic methods. *Geophysical Journal International*, **108**: 433–441.

Yu, L., Evans, R.L., and Edwards, R.N., 1997. Transient electromagnetic responses in seafloor with tri-axial anisotropy. *Geophysical Journal International*, **129**: 300–306.

Yuan, J., and Edwards, R.N., 2000. The assessment of marine gas hydrate through electrical remote sounding, hydrate without a BSR? *Geophysical Research Letters*, **27**: 2397–2400.

EM, REGIONAL STUDIES

Introduction

Images of the electrical conductivity can help decipher the architecture of the Earth's interior. The *magnetotelluric* (*q.v.*) method (MT) is one of the few tools capable of imaging from the Earth's surface through to the mantle. Regional studies of EM are conducted to image the conductivity distribution of the subsurface on the scale of a few kilometers to hundreds of kilometers, both in lateral and depth extensions.

The electrical resistivity (ρ) and its inverse the electrical conductivity (σ) characterize charge transport within materials. They are intrinsic material properties, independent of sample size. Rocks and rock-forming minerals vary in their electrical properties, with conductivities ranging from 10^6 to 10^{-14} S m^{-1}.

The MT method is based on the induction of electromagnetic fields in the Earth (see *Natural sources for EM induction studies*). The MT impedance tensor **Z** and the geomagnetic response functions (see *Induction arrows and Geomagnetic Deep sounding*) are the Earth's response to electromagnetic induction (see *Transfer functions; Robust electromagnetic transfer functions estimates*) and thus carry the information about the conductivity distribution of the subsurface.

The word "regional" is not clearly defined. It merely means that the size of the area under investigation is somewhere between "local" (see *EM, industrial uses; EM, land uses*) and "global" (see *Induction from satellite data*). However, most regional studies of EM are initiated to add to our understanding of processes that drive major tectonic and geological events (see also *EM, tectonic interpretations*). Typical examples may be the investigation of large mountain chains, which are formed in subduction or collision zones. Transform faults are expressions of dynamic processes in the Earth's lithosphere. Such large-scale faults can be traced for hundreds to thousands of kilometers on the Earth's surface and there is growing evidence that some of them penetrate the mantle lithosphere. A tectonic regime can be presently active or a fossil. Suture zones, for example, give evidence for past collisions of continents and the closure of former oceans. Most continents consist of Archean nuclei (cratons), which are enclosed by Proterozoic and Phanerozoic tectonic belts and suture zones. Many of the old structures are covered today by huge sedimentary basins. Geophysical deep sounding methods, like seismics or MTs, are arguably the only means to unravel information on the position and structure of such features deep in the Earth's interior.

Tectonic activity generally also involves processes such as formation of new structural fabrics as well as the generation and emplacement of magmas. Chemical reactions due to metamorphism can release vast amounts of fluids or can cause the precipitation of minerals such as ore and graphite. Melt, fluids, and ore deposits are electrically conductive materials. Inclusions of small fractions of these conducting phases can make an entire rock conductive, and if widely enough interconnected, even an entire region. Hence, zones of high electrical conductivity in the Earths crust and mantle hint at present or past traces of a dynamic Earth.

Experiment design

The design of a regional EM study will depend strongly on the target. The geometry of a fault zone, for example, is totally different from that of a subducting slab in a subduction zone. A fault is a narrow, subvertical structure that begins at the surface of the Earth and ends somewhere deep in its interior. Hence, the study of a large fault could have "local" and "regional" components. Locally, a fault may be investigated by one or several short profiles across its surface trace and with a very dense spacing of sites. It is one of the strongholds of the MT method that it is capable of imaging vertical structures (which is difficult with seismics). On the regional scale, a fault zone may widen (distributed shear) or bend to one direction (listric fault) and hence, a much wider area in the vicinity of the fault must be sampled at the surface.

The sounding depth of any electromagnetic method depends on the frequency contents of the induced fields and on the subsurface conductivity (skin effect). High-frequency signals will probe the shallow subsurface while low-frequency fields penetrate a much wider and deeper induction space. Particularly in noisy environments, active EM methods like *transient EM* (*q.v.*) and controlled source EM (*q.v.*) can be a good choice but the sounding depth of these methods is restricted to the first few hundred meters, in favorable conditions perhaps a few kilometers. For the really deep targets we can only rely on natural source MT.

For the more shallow investigation, MT instruments based on induction coil magnetometers will be preferred over fluxgate magnetometers which are more useful for low-frequency work. Electric field sensors for MT are nonpolarizing electrodes (see *Magnetotellurics*). A complete recording system contains some kind of analog signal interface to the sensors, multichannel analog to digital converters, typically with 24-bit accuracy, and digital data storage capacity. An integrated GPS provides accurate timing and position. Modern induction coil magnetometers can be used over a wide frequency range from approximately 10 kHz to 1 mHz and equipment based on these wideband induction coil magnetometers is often called broadband (BB). Fluxgate magnetometers, on the other hand, typically operate in the frequency range from 0.1 Hz to dc and such long-period MT instruments are called LMT. For regional EM studies BB, LMT, or a combination of both types of instruments can be deployed.

Today, there is a clear tendency for MT experiments in which a large number of recording instruments operate simultaneously and with a much denser site spacing than perhaps a few years ago. Another development is toward 3D MT for which instruments must be distributed over an area or aligned in a grid instead of simply following profiles. But a combination of a profile and some wider distributed sites can also be a very useful setup. A dense site spacing, preferably in combination with good areal coverage, enhances model resolution, avoids spatial aliasing, reduces the number of equivalent models, and generally stabilizes inversion results. Any interpretation of anomalies from the deep crust or mantle will be more sound if crustal-scale anomalies are properly resolved instead of having to rely on assumptions.

For practical reasons, it is often necessary however, to make simplifying assumptions when interpreting MT data collected for regional EM studies. A full 3D inversion, is in most cases, not feasible due both to computational requirements and the lack of areal data coverage required to constrain such an inversion (see *EM modeling, inverse*). The complexity reduces considerably if the subsurface can be approximated by a 2D conductivity distribution. In order to determine the 2D conductivity distribution of the subsurface either forward modeling (see *EM modeling, forward*) or inversion is applied to the data. It is at this point that an accurate assessment of the dimensionality of MT data is important. In the case of 2D isotropic Earth structure, the MT equations reduce to fitting two elements of the impedance tensor and one element of the geomagnetic response functions.

Another very important aspect of modern regional EM studies is interdisciplinary work. Electrical conductivity models can provide valuable information but can only add to our knowledge of a more complicated nature. Hence, an interpretation of conductivity models should always consider information from other geoscience disciplines. It is equally important to ensure that colleagues from other fields fully understand the outcome and possible consequences of regional EM studies for their own work.

The following sections give three examples of regional EM work. The case studies are from different tectonic regimes; the observed conductivity anomalies have different causes and the regional extents of the studies (the size of the models) are very dissimilar.

The electrical image of the active Dead Sea transform fault

Figure E28a/Plate 15c may serve as an example for the outcome of a 2D inversion of MT data. The MT data shown were recorded along the innermost 10 km of a much longer profile, centered on the Dead Sea transform in the Arava Valley in Jordan. A dense site spacing of 100 m in the center of the MT profile was supplemented by more widely spaced sites near the profile ends. The derived image is a so-called minimum structure model in a sense that the inversion algorithm attempts to find a trade-off between data misfit and model smoothness. The resistivity values are color-coded. Zones of high conductivity, which are generally the better resolved parameters in any MT model, are shown in red and yellow colors.

For the shallow crust, the inversion model reveals a highly conductive layer from the surface to a depth of ~100 m on the eastern side of the profile. However, the most prominent feature on the MT image is a conductive half-layer confined to west of the fault and beginning at a depth of approximately 1.5 km. The surface trace of the Arava Fault

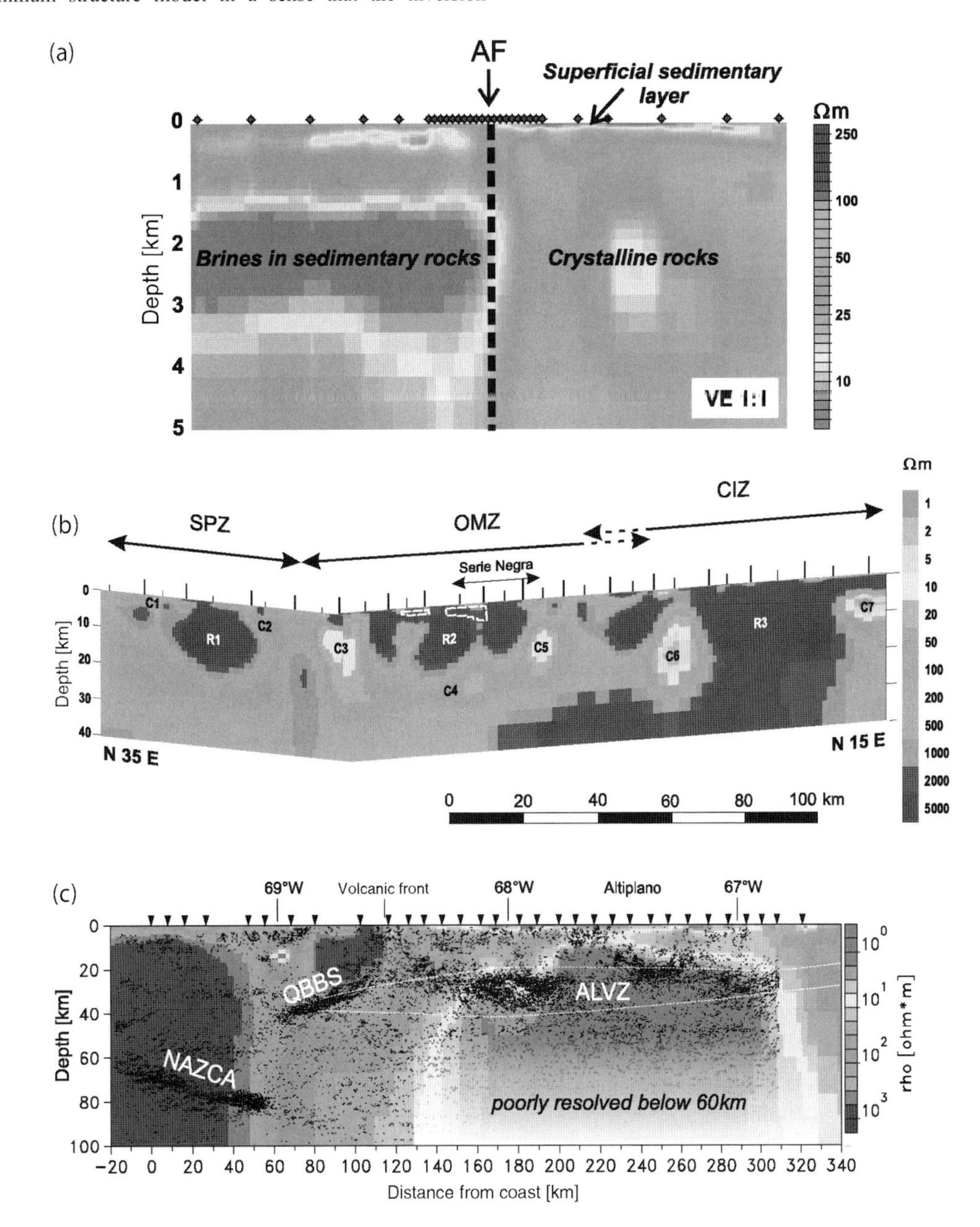

Figure E28/Plate 15c Three examples of regional EM studies. The resistivity values are color coded; red and yellow colors indicate zones of high conductivity: (a) Magnetotelluric profile crossing the Dead Sea Transform Fault, locally known as the Arava Fault (AF), modified after Ritter *et al.* (2003). The most prominent feature in the resistivity model are the sharp lateral contrasts under the surface trace of the AF. (b) Two-dimensional electrical resistivity model of a profile across the western part of the Iberian Peninsula (Iberian Massif), modified after Pous *et al.* (2004). The high conductivity zones coincide with the transitions of suture zones of the Variscan orogen. Labels are R, high-resistivity zones; C, high-conductivity zones; SPZ, South Portuguese Zone; OMZ, Ossa Morena Zone; and CIZ, the Central Iberian Zone. (c) MT model across the Andes, modified after Brasse *et al.* (2002). The most consistent explanation for the broad and deep reaching highly conductive zone is granitic partial melt. Reflection seismic data and the location of the Andean Low-Velocity Zone (ALVZ) are superimposed on the MT model. The Quebrada Blanca Bright Spot (QBBS) marks a highly reflective zone in the middle crust of the forearc.

(AF) correlates with a sharp vertical conductivity boundary at the eastern edge of this feature. The high conductivity may be due to brines in porous sedimentary rocks.

The interpretation that crystalline rocks are situated east of the fault cannot be derived from the MT data. Coincident seismic data, however, reveal a strong increase in P-wave velocities (to values exceeding 5 $\mathrm{km\,s^{-1}}$) east of the AF, where the MT model indicates higher resistivities (Ritter *et al.*, 2003). The seismic velocities are consistent with crystalline basement rocks; however the observed resistivities (50–250 Ωm) are unusually low for unfractured crystalline rocks. Both the seismic and MT observations may be explained by fractured crystalline rocks with interconnected fluid-bearing veins.

A lithological change across the fault may be the cause of the deeper conductivity contrast, however, the near-surface conductors, on opposite sides of the fault, are in similar lithology (alluvial fan deposits). This suggests that an impermeable fault-seal may be arresting cross-fault fluid flow transport at shallow depths. Additionally, the interconnected fluid-bearing veins, posited to exist within the Precambrian basement, do not appear linked to the deep conductor west of the fault. Thus, a fault-seal may be restricting fluid transport at greater depths as well.

The image of the AF in Jordan appears to be an exception, as it shows a distinct lack of an electrically conducting deformation zone at its center. Typical for many of the active faults are subvertical regions of high conductivity. Fluids distributed within the fracture network of the damage zone generally explain the observed high conductivity. Fluid transport within the rock opens pores and cracks, increasing the mobility of solutes such as salts, calcite, or quartz, thereby increasing the bulk conductivity.

The most intensively studied example of any fault is the San Andreas Fault (SAF) in California. Unsworth *et al.* (1997) demonstrated that its internal structure can be imaged with MT measurements. Several short profiles across the SAF image a highly conductive structure down to several kilometers depth. At Parkfield, the anomalous conductivity is confined to a zone centered on the SAF and extending from the surface to 2–5 km depth. At Hollister, the prominent zone of high conductivity is loosely bound between the San Andreas and Calaveras faults, extending to mid-crustal depths beneath the SAF (Bedrosian *et al.*, 2004). Ritter *et al.* (2005) examine how electrical images of different fault zones are linked to specific architectural units, the hydrogeology, and seismicity within the fault.

Fossil faults in suture zones

Exhumed fossil shear zones, in contrast to upper crustal faults, often expose structures which originated below the depth of predominantly brittle deformation (though they may have experienced brittle deformation during reactivation). These shear zones can be similarly conductive, but in these cases, bulk conductivity may be dominated by electron transport, for example in an interconnected graphite network. Graphite coatings lower shear friction and hence add to the mobility of faults. Once created, graphite is stable over very long time spans allowing shear zones to remain conductive long after activity ceases. The critical observation is that the shearing process itself can lead to the interconnection of conductive material (Jödicke *et al.*, 2004; Nover *et al.*, 2005).

The SW Iberian Peninsula constitutes the southern branch of the Iberian Massif, the best exposed fragment of the Variscan Fold Belt in Europe. It was built up by an oblique collision between three continental blocks: the South Portuguese Zone (SPZ), the Ossa Morena Zone (OMZ), and the Central Iberian Zone (CIZ). These blocks are separated by suture zones. In its evolution through time, this area experienced sequences of volcanic rifting, continental collision, and the closure (subduction) of oceans. To unravel traces of these past geodynamic processes by imaging the present Earth's crust is a key question but also a major challenge for geophysical deep sounding methods.

Pous *et al.* (2004) interpret a 200 km long MT profile across this part of the Iberian Massif. The model in Figure E28b (see also Plate 15c) reveals several high conductivity zones which coincide with the transitions of three Variscan terranes (SPZ, OMZ, and CIZ). A general north-dipping trend of the structures appears in the upper crust of the SPZ and OMZ. Fluids in shallow crustal faults could be the cause for the high conductivities observed in C1 and C2. Palaeozoic cover sequences and plutonic intrusions are depicted in the upper to middle crust as high resistivity zones (R1–R3).

The most striking features in the model are the zones of high conductivity which cut across the entire crust (C3 and C6 in Figure E28b and Plate 15c). Most interestingly, these deep reaching zones of enhanced conductivity correlate with the transitions SPZ/OMZ and OMZ/CIZ and it is plausible to conclude that these zones are the (electrical) expressions of suture zones. Another interesting feature of the model in Figure E28b (see also Plate 15c) is the presence of a high conductivity layer (C4) extending over the entire OMZ at middle to lower crustal depth 15–25 km. The top part of this conductive layer is spatially correlated with a broad reflector (derived from reflection seismic data). The preferred explanation of Pous *et al.* (2004) for the high conductivity at greater depth is the presence of interconnected graphite. Mylonitization and shearing along a lower crustal detachment zone could explain both the seismic and the MT observations.

Pous *et al.* (2004) concluded that the enhanced conductivity in the suture zones of the Iberian Massif is generally caused by interconnected graphite enrichment along shear planes. Laboratory measurements on rock samples (Serie Negra)—rocks which were exhumed during Variscian transpression—confirmed the presence of interconnected graphite. Graphite can either accumulate in the schistosity surfaces produced by folding and metamorphism or form metallic films developed in mature faults.

Electrical anisotropy (*q.v.*) is sometimes found in fossil regimes. Typical observations for electrical anisotropy are phase values exceeding 90° (Weckmann *et al.*, 2003) and a stripy appearance (alternating between zones of higher and lower resistivity) of conductive zones. Some parts of the model in Figure E28b (see also Plate 15c) were also found to be electrically anisotropic. These zones are indicated by dashed white lines.

Electrical conductivity image from an active subduction zone

Brasse *et al.* (2002) report on long-period MT studies from the central Andes between latitudes 19.5° and 21° S along two almost parallel profiles of 220 and 380 km lengths. The investigation area extends from the Pacific coast to the southern Altiplano Plateau in the back arc of the South American subduction zone. The main geoelectrical structure resolved is a broad and probably deep reaching highly conductive zone in the middle and deeper crust beneath the high plateau (see the large red block in Figure E28c and Plate 15c). The Andean Continental Research Program (ANCORP) seismic reflection profile revealed highly reflective zones below the Altiplano, in good correlation with the upper boundary of the Altiplano conductor (see Figure E28c and Plate 15c). This highly conductive domain furthermore coincides with low seismic velocities (ALVZ), a zone of elevated V_p/V_s ratios and high heat flow values. Considering all observations, granitic partial melt is the most consistent explanation for this zone of high conductivity.

Schilling and Partzsch (2001) demonstrated that the observed conductivities can be explained with partial melt rates exceeding 14% by carrying out laboratory measurements and theoretical calculations. Assuming a solid rock conductivity of $\sigma_{\mathrm{rock}} = 0.001\ \mathrm{S\,m^{-1}}$ and a melt conductivity of $\sigma_{\mathrm{melt}} = 10\ \mathrm{S\,m^{-1}}$ then the resulting bulk conductivity is $1\ \mathrm{S\,m^{-1}}$. This would correspond to a melt rate of 10% and $10\ \mathrm{S\,m^{-1}}$ represents an upper limit for the conductivity of granitic melt.

The presence of saline fluids, however, cannot the excluded. Particularly for the upper parts of the Altiplano conductor this would also explain the strong seismic reflectivity in the middle crust. An unexpected result of the MT image is the lack of correlation with the down-going slab (NAZCA reflector) or the Quebrada Blanca Bright Spot (QBBS). If the high reflectivity of the QBBS would be caused by large volumes of entrapped fluids ascending through the crust, however, then this should lead to a significant increase in the electrical conductivity. Obviously, the MT data contradict this interpretation, and it is also possible that the reflector is due to some strong structural contrast. The down-going slab and the domains immediately above it could become electrically conductive (i) because of water-rich sediments being transported in the subduction system or (ii) by release of fluids in the blueschist-eclogite transition zone in 80–100 km depth. Modeling studies show that such a conductor is compatible with the data (see Schwalenberg *et al.*, 2002; and H. Brasse, personal communication) but it is not required to fit the data. Hence, if an electrically conductive slab exists, it is obscured by the overwhelming lateral influence of the high conductivity zones associated with the Pacific Ocean in the west and the Altiplano in the east.

Oliver Ritter

Bibliography

Bedrosian, P., Unsworth, M., Egbert, G., and Thurber, C., 2004. Geophysical images of the creeping segment of the San Andreas Fault: implications for the role of crustal fluids in the earthquake process. *Tectonophysics*, **385**: 137–158.

Brasse, H., Lezaeta, P., Schwalenberg, K., Soyer, W., and Haak, V., 2002. The Bolivian Altiplano conductivity anomaly. *Journal of Geophysical Research*, **107**(5): 10.1029/2001JB000391.

Jödicke, H., Kruhl, J.H., Ballhaus, C., Giese, P., and Untiedt, J., 2004. Syngenetic, thin graphite-rich horizons in lower crustal rocks from the Serre San Bruno, Calabria (Italy), and implications for the nature of high-conducting deep crustal layers. *Physics of the Earth and Planetary Interiors*, **141**: 37–58.

Nover, G., Stoll, J.B., and Gönna, J., 2005. Promotion of graphite formation by tectonic stress—a laboratory experiment. *Geophysical Journal International*, **160**: 1059–1067.

Pous, J., Munoz, G., Heise, W., Melgarejo, J.C., and Quesada, C., 2004. Electromagnetic imaging of Variscan crustal structures in SW Iberia: the role of interconnected graphite. *Earth and Planetary Science Letters*, **217**: 435–450.

Ritter, O., Ryberg, T., Weckmann, U., Hoffmann-Rothe, A., Abueladas, A., Garfunkel, Z., and DESERT Research group, 2003. Geophysical images of the Dead Sea transform in Jordan reveal an impermeable barrier for fluid flow. *Geophysical Research Letters*, **30**(14): 1741, doi:10.1029/2003GL017541.

Ritter, O., Hoffmann-Rothe, A., Bedrosian, P.A., Weckmann, U., and Haak, V., 2005. Electrical conductivity images of active and fossil fault zones. In Bruhn, D., and Burlini, L. (eds.), *High-Strain Zones: Structure and Physical Properties,* Vol. 245. Geological Society of London Special Publications, pp. 165–186.

Schilling, F.R., and Partzsch, G., 2001. Quantifying partial melt fraction in the crust beneath the central Andes and the Tibetan Plateau. *Physics and Chemistry of the Earth*, **26**: 239–246.

Schwalenberg, K., Rath, V., and Haak, V., 2002. Sensitivity studies applied to a two-dimensional resistivity model from the Central Andes. *Geophysical Journal International*, **150**(3): 673–686.

Unsworth, M.J., Malin, P.E., Egbert, G.D., and Booker, J.R., 1997. Internal structure of the San Andreas fault at Parkfield, California. *Geology*, **25**(4): 359–362.

Weckmann, U., Ritter, O., and Haak, V., 2003. A magnetotelluric study of the Damara Belt in Namibia 2. MT phases over 90° reveal the internal structure of the Waterberg Fault/Omaruru Lineament. *Physics of the Earth and Planetary Interiors*, **138**: 91–112, doi: 10.1016/S0031-9201(03)00079-1.

Cross-references

Anisotropy, Electrical
Electromagnetic Induction (EM)
EM Modeling, Forward
EM Modeling, Inverse
EM, Industrial Uses
EM, Land Uses
EM, Marine Controlled Source
EM, Tectonic Interpretations
Geomagnetic Deep Sounding
Induction Arrows
Induction from Satellite Data
Magnetotellurics
Natural Sources for EM Induction Studies
Robust Electromagnetic Transfer Functions Estimates
Transfer Functions
Transient EM Induction

EM, TECTONIC INTERPRETATIONS

Studies of electromagnetic induction in the Earth are used to construct models of the variation of electrical conductivity in 1, 2, or 3 dimensions. Although the goal of some studies is the delineation of mineral or geothermal resources, the aim of many such investigations of conductivity structure is an improved understanding of tectonic processes. On a local scale these may include studies of active fault zones or volcanoes, while regional-scale studies include investigations of subduction zones, orogenic belts (regions of mountain building), and even, using seafloor measurements, spreading ocean ridges and mantle hot spots.

Electromagnetic (EM) methods can be used to study active tectonic processes primarily because fluids (molten rock or aqueous solutions) are typically either formed or released during these processes. In comparison to background values of the electrical resistivity of crustal rocks ($\sim$100s–1000s Ωm), both partial melts and aqueous fluids are highly conductive and can have the effect of lowering the bulk resistivity of host rocks, to the extent that they form ideal targets for detection using EM techniques such as magnetotelluric (MT) sounding. The degree to which the bulk resistivity of a rock is reduced by the presence of fluids depends primarily on two factors. The first of these is the resistivity of the fluid itself, but more important is the degree to which the fluid forms an interconnected network and, thus, a continuous electrically conductive path through the host rock. For example, simple numerical calculations of the bulk resistivity of a two-phase system (Figure E29) shows that isolated pockets of fluid (e.g., molten rock) with a resistivity of 0.1–0.25 Ωm have little effect on the bulk resistivity of a solid host rock with a resistivity of 1000 Ωm even if the melt forms 10% by volume. In contrast, a connected melt of as a little as 1% by volume, distributed through the host rock, reduces the bulk resistivity by a factor of nearly 100. The minimum amount of melt or aqueous fluid that is necessary for the fluid to be connected is surprisingly small and possibly less than 1% (Minarik and Watson, 1995).

An excellent example of the manner in which the presence of both partial melt and aqueous fluids facilitates the use of EM methods to investigate tectonic processes is presented by the results of the electromagnetic sounding of the lithosphere and asthenosphere beneath (EMSLAB) study of the Juan de Fuca subduction zone off the Pacific Coast of North America (Wannamaker *et al.*, 1989). The two-dimensional model of electrical resistivity (Figure E30), derived from MT data along a transect of over 300 km in length, shows several regions of anomalously low resistivity. In particular, a thin region of low resistivity dipping to the east below the North American continent is interpreted as being due to the presence of residual oceanic sediments on

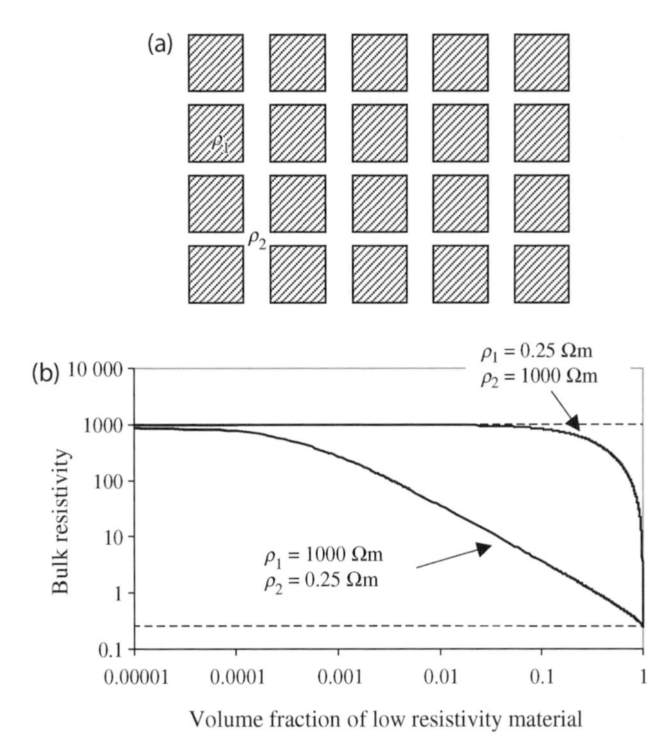

Figure E29 (a) A simple model of a two-phase material consisting of a regular array of cubic cells of resistivity ρ_1 surrounded by a matrix of resistivity ρ_2. (b) The manner in which the bulk resistivity of the material varies with the volume fraction of the low resistivity material.

the surface of the subducted plate. Given the resistivity of seawater ($\approx$0.3 Ωm), a porosity of around 1%–2% is sufficient to account for the observed bulk resistivity of this feature over most of its length. It is also likely that as the plate subducts the increase in temperature with depth leads to the release of additional fluid from hydrated minerals. Low resistivity closer to the surface just to the east of the subduction trench results from the presence of fluids in the accretionary prism formed as the bulk of the oceanic sediments are scraped from the surface of the subducted plate. In the eastern part of the model the large low resistivity region at depths of 20–50 km is interpreted to be due to water rising from the top of the subducted plate as it is released by further dehydration reactions.

The low resistivity region in the western part of the EMSLAB model illustrates the influence of partial melting on resistivity structure. This feature is interpreted to be due to several percent of partial melt associated with regional upwelling connected to the Juan de Fuca Ridge which lies only about 100 km further to the west. An MT study of the Mid-Atlantic Ridge in Iceland (Oskooi *et al.*, 2005) has likewise detected low resistivity inferred to be associated with partial melting at a spreading ridge, and 7% of partial melt has also been calculated as being necessary to explain the low resistivity of the convection plume associated with the Hawaiian islands hot spot (Constable and Heinson, 2004).

Whereas, subduction zones mark the boundaries between oceanic and continental plates, continent-continent collision leads to the rapid uplift of mountain belts. For example, the Southern Alps of New Zealand, which result from the collision of the Pacific Plate with the Australian Plate along the Alpine Fault, are estimated to have undergone 20 km of uplift in the last 26 million years. Similarly, the ongoing collision between the Indian subcontinent and the Asian continent has led to the uplift of the Himalaya and the Tibetan Plateau. Major orogenic events such as these appear to be typified by the presence of crustal low resistivity regions resulting from the presence of

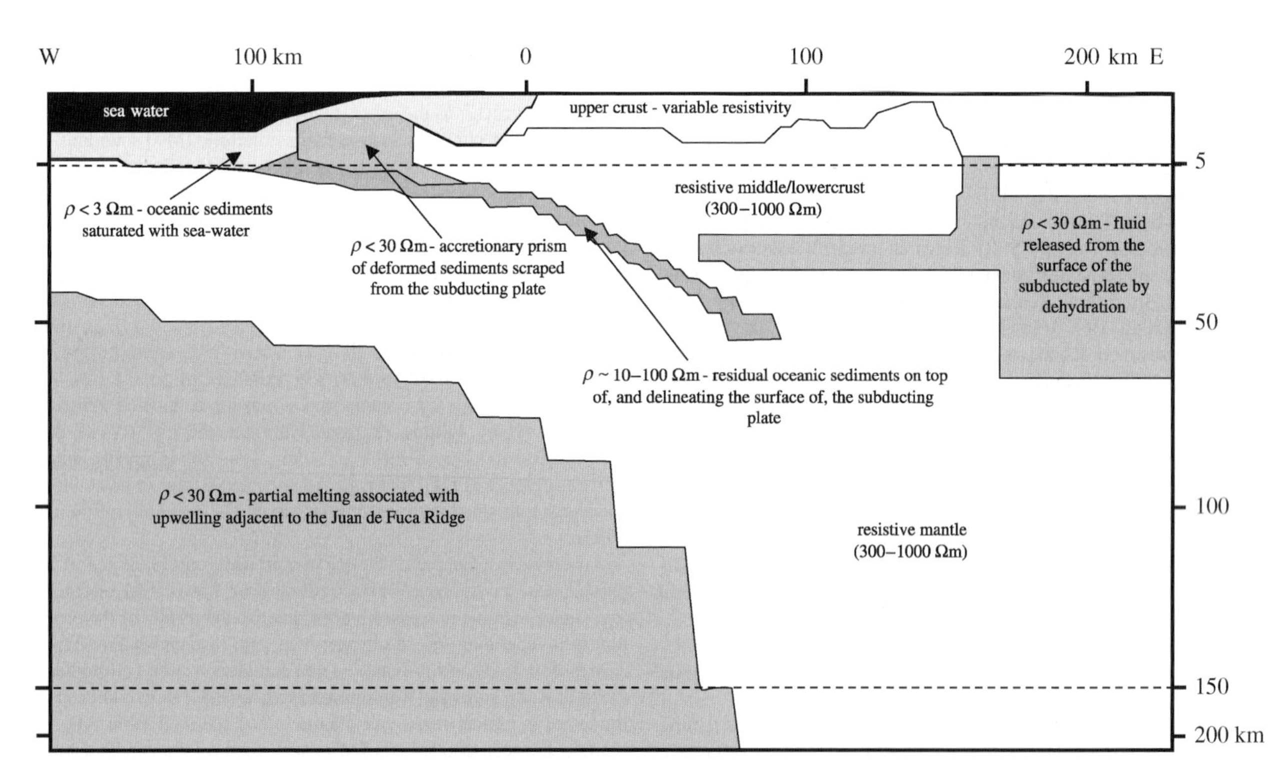

Figure E30 The role played by fluids in influencing the electrical structure associated with tectonic processes illustrated by the electrical resistivity model of the Juan de Fuca subduction zone at latitude 44°–46° N. Note the vertical scale changes at 5 and 150 km. (After Wannamaker *et al.*, 1989).

interconnected fluids. Beneath the Southern Alps low resistivity has been interpreted as being due to the release and upward migration of fluids as the development of a crustal root (i.e., a thickening of the crust beneath the region of uplift) leads to dehydration reactions as cooler material is pushed to greater depth (Ingham, 1996; Wannamaker *et al.*, 2002). A somewhat more complex situation is interpreted to exist beneath the Himalaya and the Tibetan Plateau. Here, significantly enhanced crustal electrical conductivity, detected over a distance of nearly 600 km beyond the original collision zone (Wei *et al.*, 2001; Li *et al.*, 2003), illustrates the difficulty that can occur in differentiating between reduced resistivity due to partial melting and that due to the presence of aqueous fluids. Numerical modeling has been used to place limits on the conductance (the product of the conductivity and thickness) of the region of anomalously low resistivity beneath the Tibetan Plateau. The calculated minimum conductance of this area (6000 S) has then be used, with assumptions concerning the likely resistivity values for partial melt and saline fluids, to estimate the volume fractions of these, and the required thickness of the low resistive layer, which would be necessary to explain both the MT and seismic reflection data. The preferred model of a thin layer containing saline fluids lying above a much thicker layer of partial melt, shows the importance in such situations of synthesis of EM data with data from other geophysical techniques.

Many models that have been proposed to explain the physics of the generation of earthquakes have concentrated on the role played by fluids within fault zones. The formation of a fractured, cataclastic zone along an active fault allows percolation of fluids into the damaged zone, and there has been considerable debate over the possibility that the weakness of many faults may therefore result from the presence of high pressure fluids in the fault zone. As a consequence there have been a number of EM studies aimed at detecting and delineating conductive zones associated with faults, and which have interpreted these in terms of the presence of fluids. For example, a series of MT transects of the San Andreas Fault (Unsworth *et al.*, 1999, 2000; Bedrosian *et al.*, 2002) have suggested a correlation between the presence of an electrically conductive zone along the fault trace and whether the fault zone is locked or not. The existence of a significant fault zone conductor (FZC) on a creeping segment of the San Andreas at Hollister has been associated with a higher fluid content within the fault zone. Whether the presence of fluid in the fault zone facilitates creep of the fault, or is itself a result of creep leading to the opening of fractures, remains somewhat uncertain. However, in contrast, locked segments of the fault zone exhibit only minor FZCs. At Parkfield, the most intensively studied segment of the San Andreas Fault because of its regular recurrence interval for earthquakes, the base of a moderate FZC coincides with the depth beneath which microearthquake activity is observed. Estimates of the fluid resistivity suggest that to achieve the measured bulk resistivity within the FZC at Hollister and Parkfield the porosity must be between 10% and 30%.

Similar conductive zones have been noted to be associated with other major transform faults, although some prominent faults do not exhibit FZCs (e.g., Ritter *et al.*, 2003). Nevertheless, the potential importance of understanding the relationship between fluids, FZCs, and the genesis of earthquakes is illustrated by MT measurements (Tank *et al.*, 2005) on the North Anatolian Fault, which marks the boundary between the Eurasian and Anatolian Plates in northern Turkey. These show that the hypocenters of both the mainshock and aftershocks of the $M_W = 7.4$, 1999 Izmit earthquake were located close to a boundary between resistive and conductive zones.

One other obvious manifestation of tectonism is represented by volcanic activity and there have been numerous EM studies of active volcanoes (e.g., Ogawa *et al.*, 1998; Muller and Haak, 2004). Although, the low resistivity of molten rock ($\sim$0.1 Ωm) suggests that the magma chamber of a volcano ought to present a highly significant electrical target for EM measurements, there are in reality a number of complicating factors which make detection of magma chambers difficult. These include the existence of zones of low resistivity which are the result of either thermal alteration of rocks or the presence of hydrothermal systems associated with volcanic vents. These low resistivity regions normally occur close to the surface and therefore, because EM fields attenuate with depth more rapidly in a low resistivity material (the skin-depth effect), have the effect of decreasing the ability to resolve deeper structural features. Other complicating factors arise from the three-dimensional nature of the electrical structure associated with volcanoes, and the fact that measurements may also be significantly affected by topography.

Malcolm Ingham

Bibliography

Bedrosian, P.A., Unsworth, M.J., and Egbert, G., 2002. Magnetotelluric imaging of the creeping segment of the San Andreas Fault near Hollister. *Geophysical Research Letters*, **29**: doi:10.1029/2001GL014119.

Constable, S., and Heinson, G., 2004. Hawaiian hot spot swell structure from seafloor MT sounding. *Tectonophysics*, **389**: 111–124.

Ingham, M., 1996. Magnetotelluric soundings across the South Island of New Zealand: electrical structure associated with the orogen of the Southern Alps. *Geophysical Journal International*, **124**: 134–148.

Li, S., Unsworth, M.J., Booker, J.R., Wei, W., Tan, H., and Jones, A.G., 2003. Partial melt or aqueous fluid in the mid-crust of southern Tibet? Constraints from INDEPTH magnetotelluric data. *Geophysical Journal International*, **153**: 289–304.

Minarik, W.G., and Watson, E.B., 1995. Interconnectivity of carbonate melt at low melt fraction. *Earth and Planetary Science Letters*, **133**: 423–437.

Muller, A., and Haak, V., 2004. 3-D modelling of the deep electrical conductivity of Merapi volcano (Central Java): integrating magnetotellurics, induction vectors, and the effects of steep topography. *Journal of Volcanology and Geothermal Research*, **138**: 205–222.

Ogawa, Y.N., Matsushima, H., Oshima, H., Takakura, S., Utsugi, M., Hirano, K., Igarashi, M., and Doi, T., 1998. A resistivity cross-section of Usu Volcano, Hokkaido, Japan. *Earth Planets and Space*, **50**: 339–346.

Oskooi, B., Pedersen, L.B., Smirnov, M., Arnason, K., Eysteinsson, H., and Manzella, A., 2005. The deep geothermal structure of the Mid-Atlantic Ridge deduced from MT data in SW Iceland. *Physics of the Earth and Planetary Interiors*, **150**: 183–195.

Ritter, O., Ryberg, T., Weckmann, U., Hoffmann-Rothe, A., Abueladas, A., Garfunkel, Z., and DESERT Research Group, 2003. Geophysical images of the Dead Sea Transform in Jordan reveal an impermeable barrier for fluid flow. *Geophysical Research Letters*, **30**: doi:10.1029/2003GL017541.

Tank, S.B., Honkura, Y., Ogawa, Y., Matsushima, M., Oshiman, N., Tuncer, M.K., Celik, C., Tolak, E., and Isikara, A.M., 2005. Magnetotelluric imaging of the fault rupture area of the 1999 Izmit (Turkey) earthquake. *Physics of the Earth and Planetary Interiors*, **150**: 213–225.

Unsworth, M., Egbert, G., and Booker, J., 1999. High-resolution electromagnetic imaging of the San Andreas fault in Central California. *Journal of Geophysical Research*, **104**: 1131–1150.

Unsworth, M., Bedrosian, P., Eisel, M., Egbert, G., and Siripunvaraporn, W., 2000. Along strike variations in the electrical structure of the San Andreas Fault at Parkfield, California. *Geophysical Research Letters*, **27**: 3021–3024.

Wannamaker, P.E., Booker, J.R., Jones, A.G., Chave, A.D., Filloux, J.H., Waff, H.S., and Law, L.K., 1989. Resistivity cross section through the Juan de Fuca subduction system and its tectonic implications. *Journal of Geophysical Research*, **94**: 14127–14144.

Wannamaker, P.E., Jiracek, G.R., Stodt, J.A., Caldwell, T.G., Gonzalez, V.M., McKnight, J.D., and Porter, A.D., 2002. Fluid generation and pathways beneath an active compressional orogen, the

New Zealand Southern Alps, inferred from magnetotelluric data. *Journal of Geophysical Research*, **107**(6): doi:10.1029/2001JB000186.
Wei, W., Unsworth, M., Jones, A.G., Booker, J., Tan, H., Nelson, D., Chen, L., Li, S., Solon, K., Bedrosian, P., Jin, S., Deng, M., Ledo, J., Kay, D., and Roberts, B., 2001. Detection of widespread fluids in the Tibetan crust by magnetotelluric studies. *Science*, **292**: 716–718.

Cross-references

EM, Regional Studies
Magnetotellurics

ENVIRONMENTAL MAGNETISM

Introduction

Environmental magnetism, which has rapidly developed into an established science from its beginnings some 25 years ago, involves the application of magnetic methods to environmental materials, in order to measure their magnetic properties. The environmental materials in question encompass soils, sediments, dusts, rocks, organic tissues, and man-made materials. Their magnetic properties vary according to the mineralogy, concentration, grain size, morphology, and composition of the magnetic minerals present. These (nondirectional) magnetic properties can act as sensitive recorders of environmental and climatic information, both for the present day and through geological time. Paleo- and rock magnetists use magnetic measurements to identify which magnetic minerals were responsible for acquiring a record of the Earth's magnetic polarity at their time of formation. Environmental magnetists seek to identify the causal links between magnetic properties and ambient climatic, environmental and postdepositional processes. The magnetic properties can then be used, inversely, to identify natural and human-induced changes in climate and environment on both site-specific and regional scales and over current and geological timescales.

Environmental magnetism has become a valued and widely applied methodology because: magnetically distinctive assemblages of magnetic grains—dominantly iron oxides, oxyhydroxides, and sulfides—occur virtually ubiquitously throughout the natural and built environment; these assemblages vary according to their source and depositional history; and sensitive magnetic measurements (i.e., to concentrations of less than 1 part per million for weakly magnetic minerals, like hematite, and less than 1 part per billion for strongly magnetic minerals, like magnetite) can be made relatively cheaply and rapidly compared with other types of mineralogical analysis.

Historical material

As early as 1926, magnetism was used to examine environmental processes, in a glacial setting. Gustav Ising measured the magnetic susceptibility and natural magnetic remanence of annually laminated lake sediments from Sweden. He found that the magnetic properties of the sediments varied seasonally and with distance from their ice margin source. The lake sediment layers deposited in the springtime were much more magnetic than those deposited in winter. Ising interpreted these magnetic changes as reflecting the high specific weight of magnetite grains and thus their response to different rates of flow of glacial meltwater. More magnetite grains were carried at higher rates of (springtime) flow. More than 40 years later, John Mackereth measured magnetic properties of sediments from Windermere, in the English Lake District. His work sparked new interest in the sources of magnetic minerals found in lake sediments and their surrounding catchment soils.

Since the mid-1970s onward, magnetic analyses have been applied to the sediment records of over 100 lakes worldwide. The last 20 years have seen the application of environmental magnetism to solve diverse and global-scale research questions, including: high-resolution correlation of sedimentary units (in marine, lacustrine, and marine-to-terrestrial settings); identification of the flux and timing of detrital inputs (ice-rafted debris, aeolian dust) in the deep-sea sedimentary record; unraveling of postdepositional diagenetic processes in sediments; identification of ferrimagnet formation, and destruction in different types of soils; discovery of sedimentary magnetic mineral contributions by magnetotactic bacteria; identification of natural magnets in a range of different organisms (insects, fish, humans); use of magnetism as a quantitative proxy for paleorainfall in the famous loess/soil sequences of East Europe and East Asia; archaeological prospection; sourcing of fluvial sediments; and mapping of particulate pollution sources and loadings, especially from vehicles (Figure E31). More than 130 magnetic laboratories worldwide are presently engaged in environmental magnetism research.

Essential concepts and applications

The environmental magnetic approach can briefly be summarized as the investigation of magnetic mineralogies in natural and anthropogenic samples by measurement of both their induced and remanent responses to a series of artificially induced magnetic fields. The speed of magnetic measurements, compared with other types of mineralogical analysis, and their nondestructive nature, enables measurement of large numbers of samples, a key advantage in obtaining high-resolution (spatial and/or temporal) environmental data. For any environmental sample, a large number of different magnetic parameters could be measured in order to characterize its magnetic properties and investigate its magnetic mineralogy. In practice, a reasonably small number of magnetic measurements can often provide sufficient information to identify its major magnetic constituents (Figure E32). Table E1 and Figure E33 (Maher *et al.*, 1999) summarise some of the magnetic parameters used routinely in environmental magnetism and typical values for a range of magnetic minerals. Figure E34 shows how magnetic ratio parameters can be used to discriminate between different grain sizes in ferrimagnets. Environmental materials normally contain mixtures of magnetic minerals and magnetic grain sizes. Work is in progress to extend the capability of magnetic methods in order to identify, quantify, and ascribe specific environmental histories/origins to the individual magnetic components within such mixed samples. In the meantime, independent and complementary analyses are also often used, in order to more fully characterize sample mineral properties and compare the magnetic data with other environmental proxies (such as oxygen isotope ratios or sample geochemistry).

The carriers of the magnetic properties of rocks are normally the iron oxides, magnetite, titanomagnetite, maghemite, and hematite. Until recently, because of their significance as paleomagnetic recorders, the titanium-substituted magnetites and maghemites, formed at high temperatures in igneous rocks, were a particular focus of intensive rock magnetic research. Additional rock magnetic attention was also given to hematite, formed at lower temperatures in terrestrial settings by oxidation of iron minerals, during weathering and transport, or postdepositionally by oxidizing pore fluids. Weathering, erosion, and transport of these lithogenic minerals can provide one pathway for the supply of magnetic minerals to a range of environmental materials (soils, sediments, dusts). However, a key advance brought about in environmental magnetism has been the identification of a diverse range of processes and pathways of ferrimagnet formation at or near the Earth's surface, as described below.

Direct bacterial formation of magnetite and greigite, of submicrometer grain size, has been discovered (e.g., Blakemore, 1975; Bazylinski, 1991). A range of microaerobic and anaerobic magnetogenic species actively assimilate iron from the extracellular environment, to form

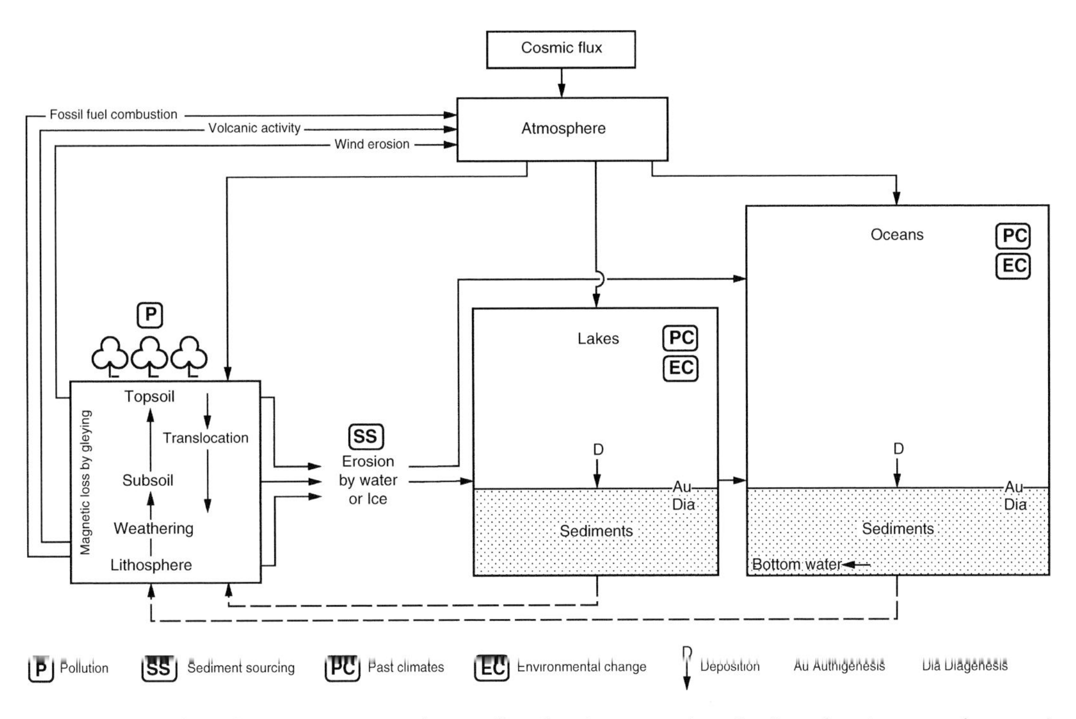

Figure E31 Magnetic minerals in the environment and examples of major areas of application of environmental magnetism (adapted from Thompson and Oldfield, 1986).

distinctive, intracellular chains of ferrimagnetic crystals, some exhibiting unique crystal morphologies (e.g., bullet, boot, and blade shapes). These biogenic crystals normally fall within a narrow grain size range, from ~20 to ~50 nm, and behave as interacting, stable single-domain grains, resulting in strong and stable magnetic behavior. As a result, the torque of the Earth's magnetic field is transferred effectively to the bacterial cell, so that when the bacteria swim, they move along the Earth's field lines. Magnetogenic bacteria, and/or their bacterial magnetofossils (Figure E35), have been found in diverse sedimentary environments, including marine (Petersen *et al.*, 1986), lake (Spring *et al.*, 1993), brackish (Stolz, 1992), and saltmarsh (Sparks *et al.*, 1989) sediments, ocean waters (Bazylinski *et al.*, 1988), soils (Fassbinder *et al.*, 1990), rivers (Bazylinski, 1991), and sewage ponds (Moench and Konetzka, 1978). In terms of population numbers, Petersen *et al.* (1986) estimated for some lake sediments in South Germany a population density of 10^7 magnetogenic bacteria ml^{-1}. In pure culture, the amount of magnetite produced by magnetogenic bacteria has been estimated at about 0.2 g per 10 g (wet weight) of bacteria (equivalent to $\sim10^{11}$ bacterial cells; Frankel, 1987). Other organisms have also been found to synthesize ferrimagnets, including insects (bees, termites, butterflies), fish, birds, and humans.

Nanocrystalline magnetite/maghemite has also been found to form inorganically, *in situ,* and at ambient temperatures (<50°C) within well-drained (generally oxidizing), buffered soils (pH range ~5–8), as a result of release of iron by weathering and subsequent repeated redox cycles (Mullins, 1977; Maher and Taylor, 1988). Magnetite is a mixed Fe^{2+}/Fe^{3+} iron oxide; its *in situ* formation (in the trace amounts responsible for the observed magnetic enhancement in soils) requires the initial presence in the soil matrix of some Fe^{2+}. This pathway of ferrimagnet formation may be mediated by the action of iron-reducing bacteria, which use iron—in the absence of oxygen—as the terminal electron acceptor in their digestion of organic matter. They release the resultant Fe^{2+} into the extracellular environment, where it may partially oxidize to form magnetite (Lovley *et al.*, 1987) and/or by surface adsorption cause other soil iron oxides and oxyhydroxides to dissolve, with possible magnetite formation via a green rust phase (Tamaura *et al.*, 1983). (Action of such Fe-reducing bacteria may also contribute to authigenic formation of magnetite in suboxic sediments of the deep-sea floor, see below.) Unlike bacterial magnetite, which is biologically constrained in crystal size and shape by the intracellular membrane structures into which it is precipitated, extracellular, pedogenically formed magnetite is variable in grain size (1–500 nm) and does not display the unique crystal morphologies observed in bacterial magnetite chains. DNA screening of magnetically enriched topsoils has shown that populations of magnetotactic bacteria are too low (<10^2 bacteria g^{-1}) to account for observed soil ferrimagnetic concentrations (Dearing *et al.*, 2001), whereas the action of one iron-reducing bacterium can mediate the extracellular formation of hundreds of ferrimagnetic grains.

So-called "botanical," nanocrystalline magnetite (1–50 nm) has recently been identified (by transmission electron microscopy and electron diffraction analyses) in grass plant cells. The magnetite is located within the inorganic cores of the plant iron storage compound, phytoferritin (Gajdardziska-Josifovska *et al.*, 2001). Upon breakdown of the organic material, this botanical pathway may constitute another (as yet unquantified) contributor to soil ferrimagnets.

Depending on sedimentation rate and organic matter content, deep-sea sediments can contain authigenic and diagenetic magnetic oxides and/or sulfides, in addition to any detrital magnetic minerals. Generally, as the amount of sedimentary organic matter increases, microorganisms use available oxidants (electron acceptors) in order to metabolize the organic material, with iron oxides being reduced once available nitrate and manganese oxides have been used up. Finally, when the iron oxides too have been reduced, sulfate is used as the oxidant, with the potential to form some magnetic iron sulfides (such as greigite), if this process does not proceed through to formation of pyrite (e.g., Snowball and Thompson, 1988; Tric *et al.*, 1991). Thus, magnetic iron oxides and sulfides can form and dissolve in deep-sea sediments, depending on sedimentary geochemical conditions (Karlin *et al.*, 1987; Langereis and Dekkers, 1999). In turn, the latter are often influenced by changes in climate, especially in generating variations in amounts of organic matter and rate and type of sediment delivery.

In addition to these low-temperature sources of ferrimagnets, two high-temperature pathways have also been identified. Combustion

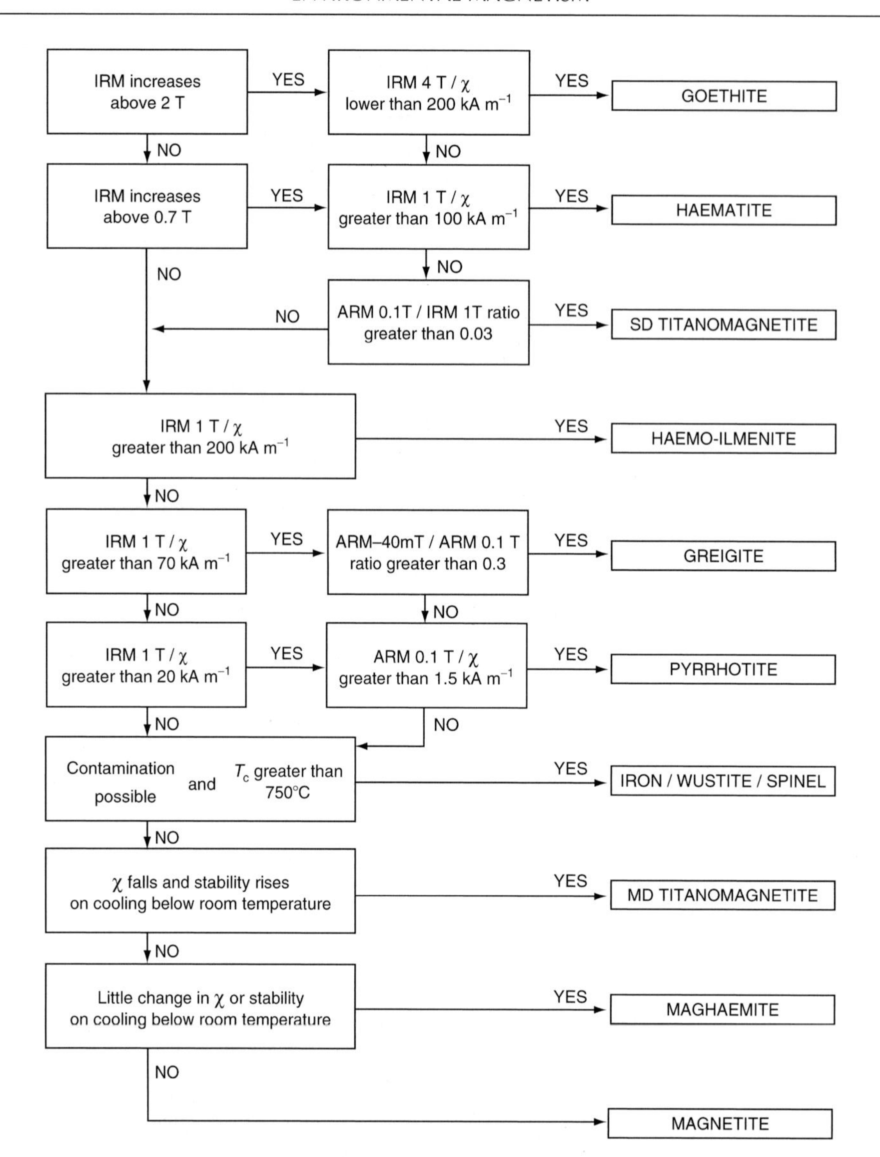

Figure E32 A flowchart indicating how a relatively small number of magnetic measurements can be used to discriminate between the major magnetic minerals found in environmental samples (reproduced with permission from Maher *et al.*, 1999).

Table E1 Typical magnetic properties of natural minerals (reproduced with permission from Maher *et al.*, 1999)

	T_c	M_s	χ	ARM	SIRM
Mineral	(°C)	(Am^2kg^{-1})	(μm^3kg^{-1})	(mAm^2kg^{-1})	(Am^2kg^{-1})
Magnetite (soft)	575	92	560	18	9
Magnetite (hard)	575	92	400	110	22
Titanomagnetite (soft)	200	24	170	80	7
Titanomagnetite (hard)	200	24	200	480	12
Hematite	675	0.5	0.6	0.002	0.24
Ilmenohematite	100	30	25	480	8
Greigite	300	20	120	110	11
Pyrrhotite	300	17	50	80	4.5
Goethite	150	0.5	0.7	0.005	0.05
Iron	770	220	2000	800	80
Paramagnet	1/T	–	~1	0	0
Diamagnet	const	–	−0.006	0	0

Note: T_c, Curie temperature or thermomagnetic behaviour; M_s, saturation magnetization; χ, initial AC susceptibility; ARM, anhysteretic remanence grown in a 100 mT peak alternating field biased with a 0.1 mT DC field; SIRM, isothermal remanence grown in a 1 T DC field.

of organic matter in soils can form ferrimagnets from other soil iron oxides and oxyhydroxides (Longworth *et al.*, 1979), while fossil fuel combustion also releases magnetite-bearing, pollutant particles into the atmosphere (Puffer *et al.*, 1980; Hunt *et al.*, 1984, and see below).

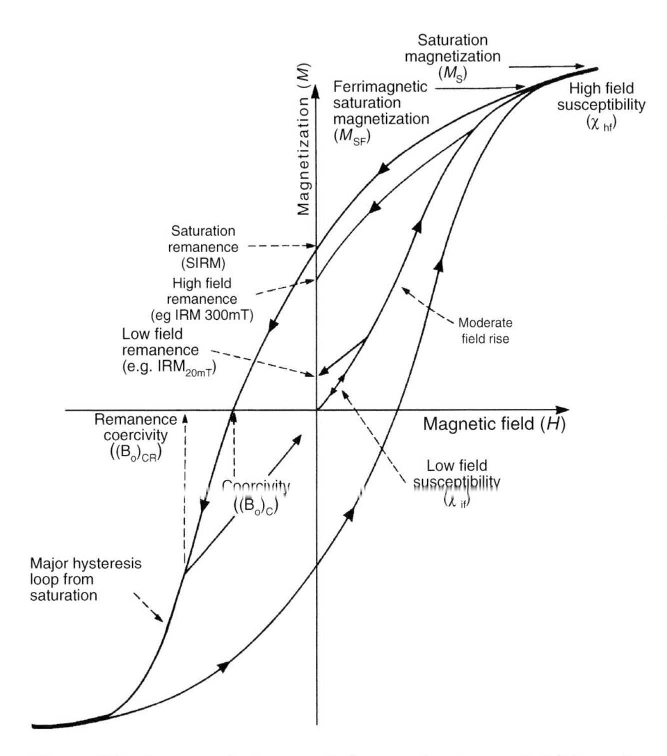

Figure E33 A magnetic hysteresis (magnetization vs field) loop for a typical natural sample containing a mixture of ferrimagnetic and paramagnetic minerals and some of the magnetic parameters routinely used in environmental magnetism (reproduced with permission from Maher *et al.*, 1999).

The magnetic properties of environmental samples are largely dominated by the strongly magnetic ferrimagnets. However, high-field magnetic experiments (i.e., using DC fields of >2 T) can be used to selectively investigate the weakly magnetic minerals goethite and hematite, even when ferrimagnet concentrations are high. In samples deficient both in ferrimagnets and imperfect antiferromagnets, the weak paramagnetism of iron-bearing silicates and the negative, diamagnetic susceptibility of carbonate can be significant (Hounslow and Maher, 1999).

Applications

In modern process studies, the magnetic properties of contemporary soils have been used to identify soil-forming processes (Maher, 1986, 1998), to relatively date weathering surfaces (Pope, 2000), to "fingerprint" different soil types and soil horizons in order to monitor soil erosion and identify sediment sources (e.g., Walling *et al.*, 1979; Dearing *et al.*, 1985; Oldfield *et al.*, 1985), and to identify relationships between modern climate and modern soil magnetic properties (Maher and Thompson, 1995; Han *et al.*, 1996), in order to make paleoclimatic estimates from paleosols ("fossil" or buried soils). This latter approach has been applied to the famous loess/paleosol sequences of the Chinese Loess Plateau, which span quasicontinuously the entire Quaternary period. These alternating windblown loess and soil layers represent colder (and drier) climate stages, when loess deposition proceeded with little soil formation, and warmer, wetter stages, when loess was still deposited but weathering and soil formation also occurred to varying degrees. The magnetic susceptibility values of the soil layers are ~3–5 times higher than the intervening loess units, reflecting *in situ* pedogenic formation of ferrimagnets (Figure E36). To isolate the *in situ*, soil-formed magnetic signal, the pedogenic susceptibility (χ_{ped}) was defined for each soil as the maximum susceptibility value (χ) of the *B* horizon minus the χ of the parent loess. Modern soils, across the Chinese Loess Plateau and the loessic Russian steppe display strong positive correlation between pedogenic χ (and other magnetic parameters, such as frequency-dependent susceptibility, anhysteretic remanence), and annual rainfall (Heller *et al.*, 1993; Maher *et al.*, 1994, 2002). Applying this modern rainfall/magnetism couple to the paleosusceptibility values of the buried soils

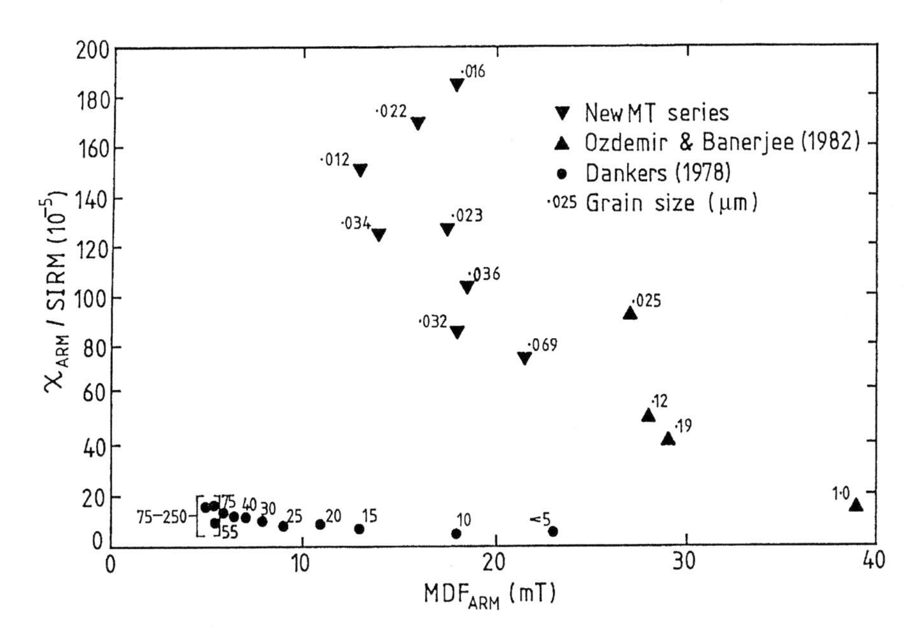

Figure E34 The grain size dependence of selected magnetic properties for magnetite: magnetic remanence ratio (the susceptibility of the anhysteretic remanence/the saturation remanence) and demagnetization (the median destructive field of the anhysteretic remanence) data for discrimination of magnetic grain size in ferrimagnets (from Maher, 1988).

enables calculation of paleorainfall values through the Quaternary. Higher average rainfall is identified for the early to mid-Holocene than at the present day, indicating stronger northward and westward incursion of the rain-bearing summer monsoon winds across the SE Asian continent. Conversely, during glacial stages, rainfall decreased by up to 50%. Kukla *et al.* (1988) contest this magnetism/rainfall relationship, and suggest that the loess/paleosol magnetic variations are controlled by variations in the flux of the weakly magnetic loess—loess flux is higher, and soil magnetic properties weaker, in the northwest of the Loess Plateau. But the loessic Russian steppe, where there is no loess deposition at the present day, also demonstrates the rainfall/magnetism correlation observed in China, indicating the key influence of climate, rather than dust flux, in controlling soil magnetic properties in these areas. Finally, the loess/paleosol magnetic variations correlate with the deep-sea oxygen isotope record, showing that the Asian monsoon systems are interlinked with the processes of midlatitude ice sheet growth and decay.

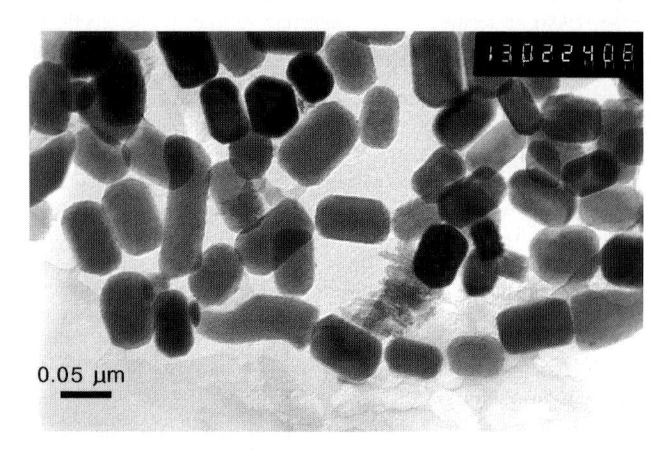

Figure E35 Transmission electron micrograph of bacterial magnetofossils, some with the unique "bullet" and "boot" morphologies, extracted from Cretaceous chalk, Culver Cliff, United Kingdom (photo: M. Hounslow, reproduced with permission from Maher *et al.*, 1999).

Increasingly, magnetic measurements are also being applied to pre-Quaternary paleosols, in order to identify their soil-forming processes and environments. For example, the magnetic properties of paleosols of Permian age appear to record glacial-interglacial fluctuations in low-latitude Pangea (e.g., Tramp *et al.*, 2004).

Soils and weathering bedrock can also be considered as sediment precursors. If they display distinctive magnetic properties, preserved through both erosion and subsequent deposition as sediment, then it may be possible to construct soil-sediment magnetic linkages and quantitatively identify changes in rates of flux of those sources through time. For instance, Caitcheon (1993) demonstrated that magnetism can be used to identify sources of suspended and bedload sediments at stream tributary junctions. The tributary contributions to the resultant downstream mix were also quantifiable. Further developments in quantitative sediment source ascription encompass multivariate statistical unmixing of sediment magnetic properties against potential source magnetic properties, using various types of algorithms, including multiple regression, maximum likelihood unmixing (Shankar *et al.*, 1994) and linear programming (Walden *et al.*, 1997). For example, Shankar *et al.* (1994) estimated that natural catchment sources contributed >97% of stream bedload upstream of an iron ore mine, but that mine waste contributed ~50% of the downstream bedload. An UK-based study of suspended sediment sources (Gruszowski *et al.*, 2003), based on magnetic, geochemical, and radionuclide signatures (in the catchment of the River Leadon), was able to estimate relative contributions of ~43% from subsurface (subsoil and channel bank) sources, and ~27% from topsoil (arable and grassland) sources. The most recent development in tracing of fluvial sediment sources is based on the magnetic signatures of magnetic minerals carried as inclusions within host silicate grains (Hounslow and Morton, 2004). By removing any discrete magnetic grains (through premeasurement acid treatment),

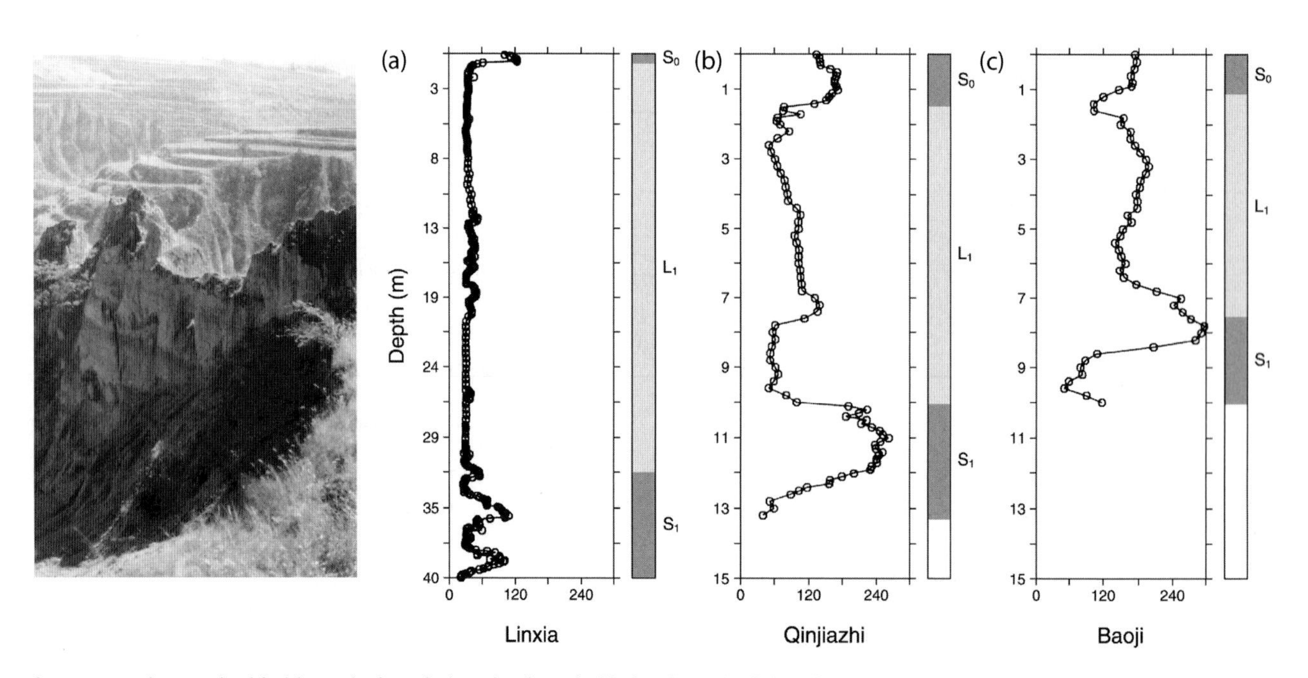

Figure E36 The interbedded loess (pale color) and paleosols (darker layers) of the Chinese Loess Plateau, at Luochuan (central Loess Plateau) together with magnetic susceptibility (in units of 10^{-8} m^{3}kg^{-1}) and sedimentary logs (S_0, modern soil; S_1, Paleosol 1, L_1, Loess 1) for the arid western plateau, and the increasingly humid central and southeastern plateau. Note that the loess/soil sequences are thicker in the west and have increasingly high magnetic susceptibility values toward the southeast.

the sample magnetic properties reflect only magnetic inclusion signatures. As the magnetic inclusions occur within host particles, they are not subject to any hydraulic sorting effects and are protected from any postdepositional, diagenetic alteration/erasure.

Providing a temporal dimension on historical and/or geological timescales, lake, estuarine, and deep-sea sediments can form natural records both of detrital fluxes and productivity in their aquatic communities. The measured magnetic properties of such sediments can thus reflect changes in allogenic inputs through time and/or authigenesis and/or postdepositional diagenesis. Magnetic methods are thus often embedded within other, multiproxy analyses in order to discern the relative significance of these possible inputs. Quantitative "unmixing" of sediment magnetic properties in terms of possible sediment sources, as described above, is increasingly applied in lake sediment studies (e.g., Thompson, 1986; Yu and Oldfield, 1993; Egli, 2004). In the context of allogenic influx, and at the land catchment-scale, much environmental magnetic work has focused on using the magnetic properties of lake sediments to construct source-sediment linkages (as above) but also to identify sedimentation rates through time (e.g., Bloemendal *et al.*, 1979; Dearing and Foster, 1986; Loizeau *et al.*, 1997), often related causally to catchment disturbance, whether natural or anthropogenic. Lake sediments can receive detrital inputs from fluvial supply and from aeolian flux, of dust, volcanic ash, and pollutants. Magnetic oxides can form *in situ*, through the direct bacterial actions of magnetogenic bacteria and/or the extracellular mediation of the iron-reducers (as above). In organic-rich conditions, dissolution of magnetic minerals may occur (e.g., Hilton and Lishman, 1985) while, with a source of sulfate, magnetic iron sulfides may be precipitated (e.g., Snowball and Thompson, 1988; Roberts *et al.*, 1996). Hence, the magnetic properties of lake sediments need to be evaluated on a site-specific basis, and with appropriate age control, in order to identify the key magnetic sources through time, and, inversely, to thus reconstruct past environmental and/or climatic changes. Dearing (1999) provides a review of lake sediment magnetic records for the Holocene period. Where fluvial detrital flux of magnetic minerals can be shown to be of primary importance, changes in hydrological regime, including rainfall (Foster *et al.*, 2003), stream power (e.g., Dearing *et al.*, 1990), snow melt (Snowball *et al.*, 2002), flood frequency and magnitude (Thorndycraft *et al.*, 1998) may be recorded in lake sediment sequences. Conversely, changes in sediment source (e.g., Reynolds *et al.*, 2004) or changes in source magnetic properties, such as results from fire-induced magnetic mineral formation (e.g., Rummery, 1983), may be identifiable. In historical times, anthropogenic pollution particles have been a dominant contributor to the magnetic properties of some lake and estuarine sediments (e.g., Locke and Bertine, 1986; Battarbee *et al.*, 1988; Kodama *et al.*, 1997).

As for lake sediments, the magnetic properties of deep-sea sediments can vary as a result of changes in detrital flux (e.g., aeolian dust, iceberg-rafted debris, and bottom-water transport), bacterially produced or mediated mineral authigenesis of magnetic oxides or sulfides, and/or postdepositional diagenesis. Hence, environmental magnetic properties of marine sediments can also act as proxy indicators of environmental and climatic change, over geological timescales. Sediment redox cycles, influenced by hydrological runoff, sedimentation rates, and organic matter content, can be identified from sediment magnetic properties, as magnetic minerals form, dissolve, and re-form through the diagenetic zone. Sediment magnetic properties in restricted basins like the Mediterranean are very sensitive to even small changes in climate, effectively amplifying the variations in such regions (e.g., Langereis and Dekkers, 1999). Conversely, in the oxic/suboxic sediments of the Atlantic, sediment magnetic properties dominantly reflect differences in detrital sediment sources (although where detrital flux is low, bacterial magnetofossils may constitute a significant contributor to sediment magnetism). Schmidt *et al.* (1999), for instance, identified major sediment sources at the present-day for the South Atlantic, including the riverine runoff from the Amazon Basin and dust flux from the African continent. For the North Atlantic, statistical analysis of a multivariate magnetic data set representing >300 surface sediments across the ocean identified the present spatial distribution of iceberg-rafted debris, dominantly from Greenland and the Labrador Sea/Baffin Bay regions, and two distinctive North African dust plumes (Watkins and Maher, 2003). Mapping and sourcing of iceberg-rafted debris for past glacial stages enables reconstruction of ocean surface currents and testing of ocean circulation models. With regard to aeolian fluxes, robust identification of dust flux to the deep-ocean sedimentary records during past glacial and interglacial stages is important, as dust may play an active role in climate change. Dust can change the radiative properties of the Earth's atmosphere but can also "fertilize" iron-deficient areas of the world's oceans, such as the Southern Ocean, stimulating marine productivity and resulting in drawdown of carbon, with resultant reduction in atmospheric CO_2 levels (Martin, 1990). Dust from the arid/semiarid zone of Africa is characterized by the presence of hematite and/or goethite, weakly magnetic iron oxides with distinctive high-field magnetic properties (Robinson, 1986; Watkins and Maher, 2003). It has thus been possible to identify the flux and provenance of dust by magnetic measurements of suitable deep-sea and lake sediments (e.g., Maher and Dennis, 2001; Kissel *et al.*, 2003; Larrasoaña *et al.*, 2004). Other applications of magnetic measurements to deep-sea sediments include: core correlation; tracing of turbidite flow paths (Abdeldayem *et al.*, 2004), and identification of hydrothermal alteration zones (Urbat and Brandau, 2003).

Current investigations, controversies, and gaps in current knowledge

Recent innovations in environmental magnetism include the use of magnetic measurements of roadside tree leaves and soils as natural pollution monitors (e.g., Matzka and Maher, 1999; Knab *et al.*, 2000; Hanesch *et al.*, 2003); magnetic mapping and *in situ* measurement of magnetic susceptibility in order to identify levels of contaminants in lake sediments (Boyce *et al.*, 2004); magnetic identification of dust flux to ice cores (Lanci *et al.*, 2001); and identification and correlation of volcanic tephras (e.g., Andrews *et al.*, 2002). Development of new measurement techniques continues, such as the use of first-order reversal curves (FORC) diagrams to identify magnetic mineral compositions and grain sizes (e.g., Pike *et al.*, 2001; Carvallo *et al.*, 2004). FORC may be suitable for addressing the problem of investigating the weakly magnetic minerals, hematite and goethite even when magnetite is the dominant mineral (Muxworthy *et al.*, 2004). Remanence measurements at high fields (>2 T) and low temperature (77 K) may also provide a new means of defining the concentration and grain size of these minerals in this context (Maher *et al.*, 2004). Finally, to achieve standardized accuracy as well as precision, interlaboratory calibration is an ongoing research activity internationally (e.g., Sagnotti *et al.*, 2003).

Summary/conclusions

The ubiquity and environmental sensitivity of the magnetic mineralogies present in soils, sediments, and dusts can provide a record of past and present environmental and climatic change and process. The speed and sensitivity of magnetic measurements, and their demonstrated range of environmental applications, has led to their application to a range of key environmental and climatic questions and problems, ranging from site-specific to global in scale. Environmental magnetism has thus become an established discipline, which continues to gain in strength from improved measurement capabilities, quantification, and interaction with other, complementary environmental analyses and proxies.

Barbara A. Maher

Bibliography

Abdeldayem, A.L., Ikehara, K., and Yamazaki, I., 2004. Flow path of the 1993 Hokkaido-Nansei-oki earthquake seismoturbidite, southern margin of the Japan Sea north basin, inferred from anisotropy of magnetic susceptibility. *Geophysics Journal International*, **157**: 15–24.

Andrews, J.T., Geirsdottir, A., Hardardottir J. *et al.*, 2002. Distribution, sediment magnetism, and geochemistry of the Saksunarvatn (10,180 +/– 60 cal. yr BP) tephra in marine, lake, and terrestrial sediments, northwest Iceland. *Journal of Quaternary Science*, **17**: 731–745.

Battarbee, R.W. *et al.*,1988. *Lake Acidification in the United Kingdom 1800–1986.* London: Ensis Publishing.

Bazylinski, D.A., 1991. Bacterial production of iron sulphides. *Material Research Society Symposium Proceedings*, **218**: 81–91.

Bazylinski, D.A., Frankel, R.B., and Jannasch, H.W., 1988. Anaerobic production of magnetite by a marine magnetotactic bacterium. *Nature,* **334**: 518–519.

Blakemore, R.P., 1975. Magnetotactic bacteria. *Science*, **190**: 377–379.

Bloemendal, J., Oldfield, F., and Thompson, R., 1979. Magnetic measurements used to assess sediment influx at Llyn Goddionduon. *Nature*, **280**: 50–53.

Boyce, J.I., Morris, W.A., and Pozza, M.R., 2004. Magnetic mapping and classification of contaminant impact levels in lake sediments. *EOS Transactions American Geophysical Union*, **85**, Joint Assembly Supplement, Abstract NS13A-06.

Caitcheon, G.G., 1993. Sediment source tracing using environmental magnetism—a new approach with examples from Australia. *Hydrological Processes*, **7**: 349–358.

Carvallo, C., Özdemir, O., and Dunlop, D.J., 2004. First-order reversal curve (FORC) diagrams of elongated single-domain grains at high and low temperatures. *Journal of Geophysical Research,* **109**, art no. B04105.

Dearing, J.A., 1999. Holocene environmental change from magnetic proxies in lake sediments. In Maher, B.A., and Thompson, R. (eds.), *Quaternary Climates, Environments and Magnetism.* Cambridge: Cambridge University Press, 231–278.

Dearing, J.A., and Foster, I.D.L., 1986. Limnic sediments used to reconstruct sediment yields and sources in the English Midlands since 1765. In Gardiner, V. (ed.), *International Geomorphology I.* Chichester: John Wiley & Sons.

Dearing, J.A., Maher, B.A., and Oldfield, F., 1985. Geomorphological linkages between soils and sediments: the role of magnetic measurements. In Richards, K.S., Ellis, S., and Arnett, R.R. (eds.), *Soils and Geomorphology.* London: George Allen & Unwin, 245–266.

Dearing, J.A., Alstrom, K., Bergman, A., Regnell, J., and Sandgren, P., 1990. Past and present erosion in southern Sweden. In Boardman, J., Foster, I., and Dearing, J. (eds.), *Soil Erosion on Agricultural Land.* Chichester: John Wiley & Sons, 687 pp.

Dearing, J.A., Hannam, J.A., Anderson, A.S., and Wellington, E.M.H., 2001. Magnetic, geochemical and DNA properties of highly magnetic soils in England. *Geophysical Journal International*, **144**: 183–196.

Egli, R., 2004. Characterization of individual rock magnetic components by analysis of remanence curves. 1. Unmixing natural sediments. *Studia Geophysica et Geodaetica*, **48**: 391–446.

Fassbinder, J.W.E., Stanjek, J., and Vali, H., 1990. Occurrence of magnetotactic bacteria in soil. *Nature*, **343**: 161–163.

Foster, G.C., Dearing, J.A., Jones, R.T. *et al.*,2003. Meteorological and land use controls on past and present hydro-geomorphic processes in the pre-alpine environment: an integrated lake-catchment study at the Petit Lac d'Annecy, France. *Hydrological Processes*, **17**: 3287–3305.

Frankel, R.B., 1987. Anaerobes pumping iron. *Nature*, **330**: 208.

Gajdardziska-Josifovska, M., McClean, R.G., Schofield, M.A. *et al.*, 2001. Discovery of nanocrystalline botanical magnetite. *European Journal of Mineralogy*, **13**: 863–870.

Gruszowski, K.E., Foster, I.D.L., Lees, J.A., and Charlesworth, S.M., 2003. Sediment sources and transport pathways in a rural catchment, Herefordshire, UK. *Hydrological Processes*, **17**: 2665–2681.

Han, J., Lu, H., Wu. N., and Guo, Z., 1996. Magnetic susceptibility of modern soils in China and climate conditions. *Studia Geophysica et Geodaetica*, **40**: 262–275.

Hanesch, M., Maier, G., and Scholger, R., 2003. Mapping heavy metal distribution by measuring the magnetic susceptibility of soils. *Journal of Physics*, **107**: 605–608.

Heller, F., Shen, C.D., Beer, J. *et al.*,1993. Quantitative estimates and palaeoclimatic implications of pedogenic ferromagnetic mineral formation in Chinese loess. *Earth and Planetery Science Letters*, **114**: 385–390.

Hilton, J., and Lishman, J.P., 1985. The effect of redox changes on the magnetic susceptibility of sediments from a seasonally anoxic lake. *Limnology and Oceanography*, **30**: 907–909.

Hounslow, M.W., and Maher, B.A., 1999. Source of the climate signal recorded by magnetic susceptibility variations in Indian Ocean sediments. *Journal of Geophysical Research (Solid Earth)*, **104**: 5047–5061.

Hounslow, M.W., and Morton, A.C., 2004. Evaluation of sediment provenance using magnetic mineral inclusions in clastic silicates: comparison with heavy mineral analysis. *Sedimentary Geology*, **171**: 13–36.

Hunt, A., Jones, J., and Oldfield, F., 1984. Magnetic measurements and heavy metals in atmospheric particulates of anthropogenic origin. *Science of the Total Environment*, **33**: 129–139.

Karlin, R., Lyle, M., and Heath, G.R., 1987. Authigenic magnetite formation in suboxic marine sediments. *Nature*, **326**: 490–493.

Kissel, C., Laj, C., Clemens, S. *et al.*,2003. Magnetic signature of environmental changes in the last 1.2 Myr at ODP site 1146, South China Sea. *Marine Geology*, **201**: 119–132.

Knab, M., Hoffmann, V., and Appel, E., 2000. Magnetic susceptibility as a proxy for heavy metal contamination in roadside soils. *Geologica Carpathica*, **51**: 199.

Kodama, K.P., Lyons, J.C., Siver, P.A. *et al.*,1997. A mineral magnetic and scaled-chrysophyte paleolimnological study of two northeastern Pennsylvanian lakes: records of fly ash deposition, land-use change and paleorainfall variation. *Journal of Paleolimnology*, **17**: 173–189.

Kukla, G., Heller, F., Liu, X. *et al.*,1988. Pleistocene climates in China dated by magnetic susceptibility. *Geology*, **16**: 811–814.

Lanci, L., Kent, D.V., and Biscaye, P.E., 2001. Isothermal remanent magnetization of Greenland ice: preliminary results. *Geophysical Research Letters*, **28**: 1639–1642.

Langereis, C.G., and Dekkers, M.J., 1999. Magnetic cyclostratigraphy: high-resolution dating in and beyond the Quaternary and analysis of periodic changes in diagenesis and sedimentary magnetism. In Maher, B.A., and Thompson, R. (eds.), *Quaternary Climates, Environments and Magnetism.* Cambridge: Cambridge University Press, 352–382.

Larrasoaña, J.C., Roberts, A.P., Rohling, E.J., Winklhofer, M., and Wehausen, R., 2004. Three million years of North Saharan dust supply into the eastern Mediterranean Sea. *Climate Dynamics*, **21**: 689–698.

Locke, G., and Bertine, K.K., 1986. Magnetite in sediments as an indicator of coal combustion. *Applied Geochemistry*, **1**: 345–356.

Longworth, G., Becker, L.W., Thompson, R. *et al.*,1979. Mossbauer effect and magnetic studies of secondary iron oxides in soils. *Journal of Soil Science*, **30**: 93–110.

Lovley, D.R., Stolz, J.F., Nord, G.L., and Phillips, E.J.P., 1987. Anaerobic production of magnetite by a dissimilatory iron-reducing microorganism. *Nature*, **330**: 252–254.

Loizeau, J.L., Dominik, J., Luzzi, T., and Vernet, J.P., 1997. Sediment core correlation and mapping of sediment accumulation rates in Lake Geneva (Switzerland, France) using volume magnetic susceptibility. *Journal of Great Lakes Research*, **23**: 391–402.

Maher, B.A., 1986. Characterisation of soils by mineral magnetic measurements. *Physics of the Earth and Planetary Interiors*, **42**(986): 76–92.

Maher, B.A., 1998. Magnetic properties of modern soils and loessic paleosols: implications for paleoclimate. *Palaeogeography, Palaeoclimatology, Palaeoecology*, **137**: 25–54.

Maher, B.A., and Dennis, P.F., 2001. Evidence against dust-mediated control of glacial-interglacial changes in atmospheric CO_2. *Nature*, **411**: 176–180.

Maher, B.A., and Taylor, R.M., 1988. Formation of ultrafine-grained magnetite in soils. *Nature*, **336**: 368–370.

Maher, B.A., and Thompson, R., 1995. Palaeorainfall reconstructions from pedogenic magnetic susceptibility variations in the Chinese loess and paleosols. *Quaternary Research*, **44**: 383–391.

Maher, B.A., Thompson, R., and Zhou, L.P., 1994. Spatial and temporal reconstructions of changes in the Asian palaeomonsoon: a new mineral magnetic approach. *Earth and Planetary Science Letters*, **125**: 461–471.

Maher, B.A., Thompson, R., and Hounslow, M.W., 1999. Introduction to Quaternary climates, environments and magnetism. In Maher, B.A., and Thompson, R. (eds.), *Quaternary Climates, Environments and Magnetism*. Cambridge: Cambridge University Press, 1–48.

Maher, B.A., Alekseev, A., and Alekseeva, T., 2002. Variation of soil magnetism across the Russian steppe: its significance for use of soil magnetism as a palaeorainfall proxy. *Quaternary Science Reviews*, **21**: 1571–1576.

Maher, B.A., Karloukovski, V.V., and Mutch, T.J., 2004. High-field remanence properties of synthetic and natural submicrometre hematites and goethites: significance for environmental contexts. *Earth and Planetary Science Letters*, **226**: 491–505.

Martin, J.H., 1990. Glacial-interglacial CO_2 change: the iron hypothesis. *Paleoceanography*, **5**: 1–13.

Matzka, J., and Maher, B.A., 1999. Magnetic biomonitoring of roadside tree leaves: identification of spatial and temporal variations in vehicle-derived particulates. *Atmospheric Environment*, **33**: 4565–4569.

Moench, T.T., and Konetzka, W.A., 1978. A novel method for the isolation and study of a magnetotactic bacterium. *Archives of Microbiology*, **119**: 203–212.

Mullins, C.E., 1977. Magnetic susceptibility of the soil and its significance in soil science: a review. *Journal of Soil Science*, **28**: 223–246.

Muxworthy, A.R., King, J.G., Heslop, D., and Williams, W., 2003. Unravelling magnetic mixtures using first-order reveral curve (FORC) diagrams: linear additivity and interaction effects. *EOS Transactions American Geophysical Union*, **84**, Joint Assembly Supplement, Abstract GP31B-0749.

Oldfield, F., Maher, B.A., Donaghue, J., and Pierce, J., 1985. Particle-size related, magnetic source-sediment linkages in the Rhode River catchment. *Journal of the Geological Society (London)*, **142**: 1035–1046.

Petersen, N., von Dobeneck, T., and Vali, H., 1986. Fossil bacterial magnetite in deep-sea sediments from the South Atlantic Ocean. *Nature*, **320**: 611–615.

Pike, C.R., Roberts, A.P., and Verosub, K.L., 2001. First-order reversal curve diagrams and thermal relaxation effects in magnetic particles. *Geophysics Journal International*, **145**: 721–730.

Pope, R.J.J., 2000. The application of mineral magnetic and extractable iron (Fe-d) analysis for differentiating and relatively dating fan surfaces in central Greece. *Geomorphology*, **32**: 57–67.

Puffer, J.H., Russell, E.W., and Rampino, M.R., 1980. Distribution and origin of magnetite spherules in air, waters and sediments of the greater New York city area and the North Atlantic Ocean. *Journal of Sedimentary Petrology*, **50**: 247–256.

Reynolds, R.L., Rosenbaum, J.G., Rapp, J., Kerwin, M.W., Bradbury, J.P., Colman, S., and Adam, D., 2004. Record of Late Pleistocene glaciation and deglaciation in the southern Cascade Range. I. Petrological evidence from lacustrine sediment in Upper Klamath Lake, southern Oregon. *Journal of Paleolimnology*, **31**: 217–233.

Roberts, A.P., Reynolds, R.L., Verosub, K.L., and Adam, D.P., 1996. Environmental magnetic implications of greigite (Fe_3S_4) formation in a 3 myr lake sediment record from Butte Valley, northern California. *Geophysical Research Letters*, **23**: 2859–2862.

Robinson, S.G., 1986. The Late Pleistocene palaeoclimatic record of North Atlantic deep-sea sediments revealed by mineral magnetic measurements. *Physics of the Earth and Planetary Interiors*, **42**: 22–47.

Rummery, T.A., 1983. The use of magnetic measurements in interpreting the fire histories of of lake drainage basins. *Hydrobiologia*, **103**: 53–58.

Sagnotti, L. *et al.*, 2003. Inter-laboratory calibration of low-field magnetic and anhysteretic susceptibility measurements. *Physics of the Earth and Planetary Interiors*, **138**: 25–38.

Schmidt, A.M., von Dobeneck, T., and Bleil, U., 1999. Magnetic characterization of Holocene sedimentation in the South Atlantic. *Paleoceanography*, **14**: 465–481.

Shankar, R., Thompson, R., and Galloway, R.B., 1994. Sediment source modeling—unmixing of artificial magnetization and natural radioactivity measurements. *Earth and Planetary Science Letters*, **126**: 411–420.

Snowball, I.F., and Thompson, R., 1988. The occurrence of greigite in sediments from Loch Lomond. *Journal of Quaternary Science*, **3**: 121–125.

Snowball, I., Zillen, L., and Gaillard, M.J., 2002. Rapid early-Holocene environmental changes in northern Sweden based on studies of two varved lake-sediment sequences. *Holocene*, **12**: 7–16.

Sparks, N.H., Lloyd, J., and Board, R.G., 1989. Saltmarsh ponds—a preferred habitat for magnetotactic bacteria. *Letters in Applied Microbiology*, **8**: 109–111.

Spring, S. *et al.*,1993. Dominating role of an unusual magnetotactic bacterium in the microaerobic zone of a freshwater sediment. *Applied and Environmental Microbiology*, **59**: 2397–2403.

Stolz, J.F., 1992. Magnetotactic bacteria: biomineralization, ecology, sediment magnetism, environmental indicator. In Skinner, H.C.W., and Fitzpatrick, R.W. (eds.), *Biomineralization Processes of Iron and Manganese—Modern and Ancient Environments Catena Supplement No.* 21, pp. 133–145.

Tamaura, Y., Ito, K., and Katsura, T., 1983. Transformation of αFeO (OH) to Fe_3O_4 by adsorption of iron (III) ion on αFeO(OH). *Journal of the Chemical Society, Dalton Transactions*, **1**: 189–194.

Thompson, R., 1986. Modelling magnetization data using SIMPLEX. *Physics of the Earth and Planetary Interiors*, **42**: 113–127.

Thorndycraft, V., Hu, Y., Oldfield, F., Crooks, P.R.J., and Appleby, P.G., 1998. A high-resolution flood event stratigraphy in the sediments of the Petit Lac d'Annecy. *Holocene*, **8**: 741–746.

Tramp, K.L., Soreghan, G.S.L., and Elmore, R.D., 2004. Paleoclimatic inferences from paleopedology and magnetism of the Permian Maroon Formation loessite, Colorado, USA. *Geological Society of America Bulletin*, **116**: 671–686.

Tric, E. *et al.*,1991. High-resolution record of the Upper Olduvai transition from Po Valley (Italy) sediments: support for dipolar transition geometry? *Physics of the Earth and Planetary Interiors*, **65**: 319–336.

Urbat, M., and Brandau, A., 2003. Magnetic properties of marine sediment near the active Dead Dog Mound, Juan de Fuca Ridge, allow to refine hydrothermal alteration zones. *Physics and Chemistry of the Earth*, **28**: 701–709.

Walden, J., Slattery, M.C., and Burt, T.P., 1997. Use of mineral magnetic measurements to fingerprint suspended sediment sources:

approaches and techniques for data analysis. *Journal of Hydrology*, **202**: 353–372.

Walling, D.E., Peart, M.R., Oldfield, F., and Thompson, R., 1979. Suspended sediment sources identified by magnetic measurements. *Nature*, **281**: 110–113.

Watkins, S.J., and Maher, B.A., 2003. Magnetic characterisation of present day deep-sea sediments and sources in the North Atlantic. *Earth and Planetary Science Letters*, **214**: 379–394.

Yu, L., and Oldfield, F., 1993. Quantitative sediment source ascription using magnetic measurements in a reservoir-catchment system near Nijar, S.E. Spain. *Earth Surface Processes and Landforms*, **18**: 441–454.

Cross-references

Archeomagnetism
Biomagnetism
Paleomagnetism
Rock Magnetism

ENVIRONMENTAL MAGNETISM, PALEOMAGNETIC APPLICATIONS

Introduction

Environmental magnetism involves magnetic measurements of environmental materials, including sediments, soils, rocks, mineral dust, anthropogenic pollutants, and biological materials. The magnetic properties of these materials can be highly sensitive to a broad range of environmental processes, which makes mineral magnetic studies widely useful in the environmental sciences. More detailed treatments of the breadth of environmental magnetic research applications, and definitions of magnetic parameters and their interpretation, can be found in recent review articles and books (e.g., Verosub and Roberts, 1995; Maher and Thompson 1999; Evans and Heller 2003; see also *Environmental magnetism*). Most environmental magnetic studies aptly focus on environmental interpretations. Nevertheless, environmental magnetism, whether explicitly stated as such or not, routinely plays an important role in paleomagnetic studies, including tectonic, geochronological, and geomagnetic applications. It is the paleomagnetic applications of environmental magnetism that are in focus here, especially in relation to studies of sediments and sedimentary rocks. These paleomagnetic applications are discussed below following four broad themes:

1. Inter-core correlation and development of relative age models;
2. Development of astronomically calibrated age models;
3. Testing the hypothesis of orbital forcing of the geomagnetic field;
4. Determining the origin of the paleomagnetic signal.

Inter-core correlation and development of relative age models

Sedimentary environments receive inputs of detrital magnetic minerals that vary with time and that are diluted by varying supply of lithogenic mineral components and production of biogenic mineral components, respectively. Stratigraphic variations in magnetic properties within sedimentary basins are usually spatially coherent on scales of tens of meters; in some cases, magnetic property variations can be traced over large portions of an ocean basin. Variations in magnetic properties, such as the low-field magnetic susceptibility, therefore, provide a fundamentally important physical parameter for stratigraphic correlation among cores from a coherent depositional environment. Variations in magnetic susceptibility are routinely used for inter-core correlation and provide the basis for development of relative age models in paleomagnetic studies of geomagnetic field behavior. For example, classic studies of paleosecular variation (see *Paleomagnetic secular variation*) recorded by postglacial lake sediments depended on inter-core correlation based on magnetic susceptibility variations (Turner and Thompson, 1979). The resultant correlation provided independent confirmation that the same paleomagnetic variations were observed at the same time in the respective cores. This procedure yielded a coherent relative age model that was then transformed into a numerical age model by combined use of radiocarbon dating, correlation to historical secular variation measurements from the Greenwich observatory and correlation to archaeomagnetic secular variation data (Turner and Thompson, 1979).

Inter-core correlation, based on magnetic susceptibility variations, is widely performed in paleomagnetic studies of lacustrine and marine sediments. For example, thick marine sedimentary sequences are routinely cored by the Ocean Drilling Program (ODP), and its successor the integrated ODP (IODP), by taking successive ~10-m cores from a single hole. Modern high-resolution studies require acquisition of continuous records, which necessitates coring of multiple holes at the same site with slight depth offsets among cores from adjacent holes to ensure complete stratigraphic recovery. A continuous stratigraphic record is then "spliced" together by correlating high-resolution measurements of physical properties such as magnetic susceptibility from multiple cores (Hagelberg *et al.*, 1992). This routine procedure of using magnetic susceptibility variations for stratigraphic correlation provides the basis for high-resolution paleomagnetic studies of sediment cores as well as for a wide range of other studies. An example of the type of high-resolution inter-core correlation that is possible using magnetic susceptibility variations is shown for two ODP holes from the North Pacific Ocean in Figure E37. The two holes are separated by 00°26′ and were cored at water depths of 2385 m (Hole 883D) and 3826 m (Hole 884D), respectively, on the flanks of Detroit Seamount. The sediments from the two depositional environments have different carbonate and volcanic ash concentrations. The magnetic susceptibility variations for the two holes, measured at 1-cm intervals on u-channel samples, therefore have significant differences (Figure E37a,b). Regardless, the records can still be readily correlated and the depths of correlative horizons from one hole can be transferred into equivalent depths in the other hole (Figure E37c). The physical correlation between ODP holes 883D and 884D shown in Figure E37c enabled Roberts *et al.* (1997) to independently correlate relative geomagnetic paleointensity records for the two holes and to produce a stacked paleointensity record for the North Pacific Ocean for the last 200 ka. This approach to inter-core correlation can also help with analysis of finer details of geomagnetic field behavior. For example, the interval between 8.5 and 9.5 m below seafloor (mbsf) at ODP Site 884 contains a detailed record of a geomagnetic excursion (Roberts *et al.*, 1997). By correlating whole-core magnetic susceptibility records from three holes cored at Site 884 (Figure E38a), in the same way as shown in Figure E37, it is possible to place detailed records of the excursion on the same relative depth scale (Figure E38b).

Inter-core correlation using environmental magnetic parameters that are independent of paleomagnetic variations (Figures E37 and E38) enable powerful direct comparison of details of geomagnetic field behavior in paleomagnetic analyses of the spectrum of geomagnetic field behavior, including studies of paleosecular variation, relative paleointensity of the geomagnetic field, excursions, and polarity transitions. It should be stressed that this technique of inter-core correlation simply provides a relative stratigraphic age model; development of numerical age models depends either on direct dating or on correlation of magnetic properties with an appropriate astronomical target curve, as described below.

Development of astronomically calibrated age models

Stratigraphic variations in magnetic properties are climatically controlled in many depositional environments. This can occur, for

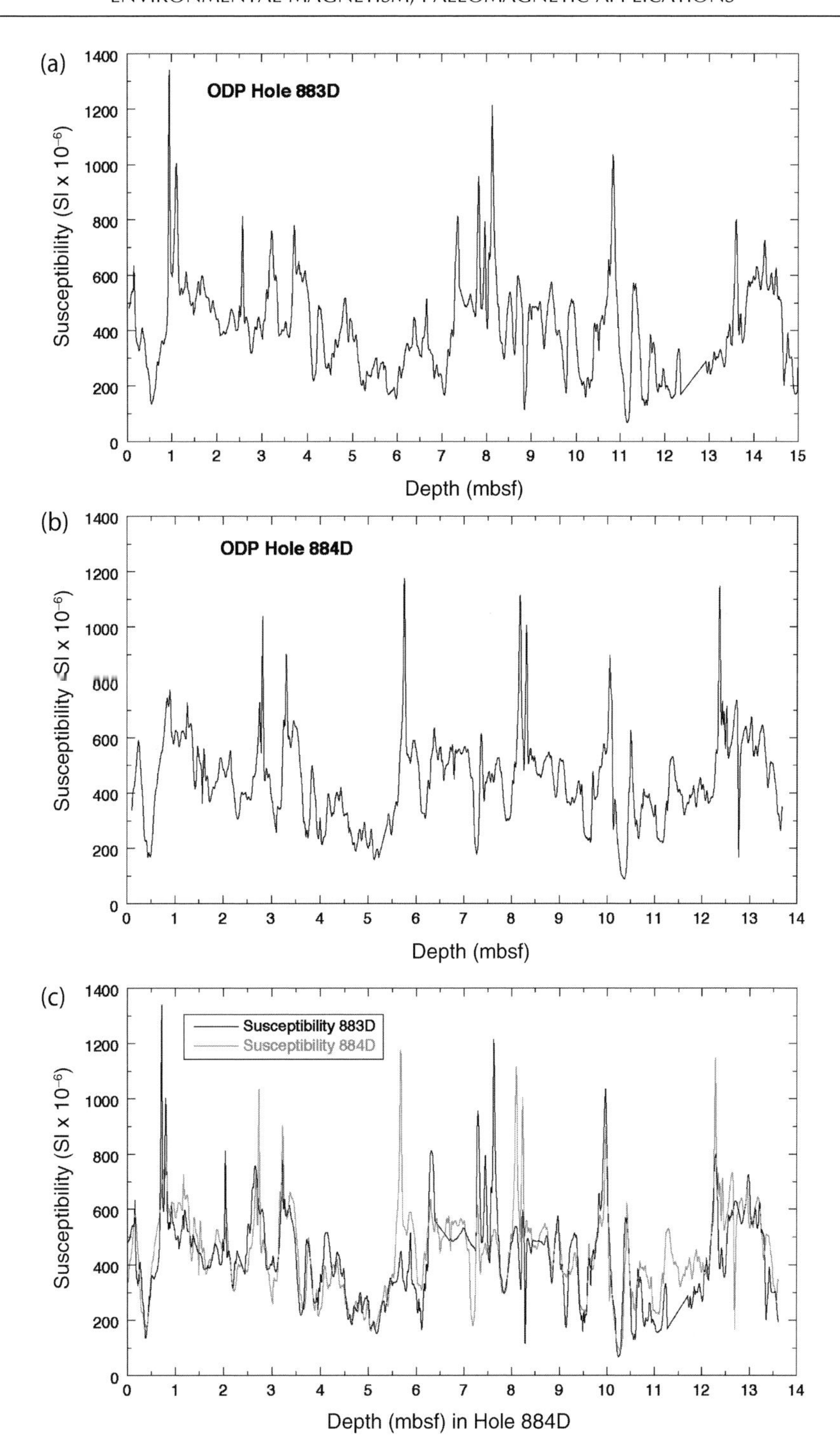

Figure E37 Example of inter-core correlation using low-field magnetic susceptibility. (a) Magnetic susceptibility profile for ODP Hole 883D. (b) Magnetic susceptibility profile for ODP Hole 884D. (c) Correlation of magnetic susceptibility profiles (both for u-channel samples) for ODP holes 883D and 884D, with data from Hole 883D transformed onto the depth scale for Hole 884D (see Roberts *et al.*, 1997 for details). Such correlations based on environmental magnetic parameters provide an independent basis for detailed correlation of paleomagnetic features in cores.

example, as a result of climatic modulation of the supply of lithogenic mineral components eroded from a source rock or through dilution of a roughly constant lithogenic component by a climatically controlled biogenic mineral component. The result is an environmental magnetic record that responds coherently to orbitally forced variations in climate (e.g., Kent, 1982). In some cases, environmental magnetic parameters provide such a good representation of an orbitally controlled forcing parameter (e.g., summer insolation) that the magnetic parameter can be used to directly date the sediment core (e.g., Larrasoaña *et al.*, 2003a). Paleomagnetic studies of relative geomagnetic paleointensity are often conducted on cores where the magnetic properties are climatically controlled. In such cases, age models are routinely developed

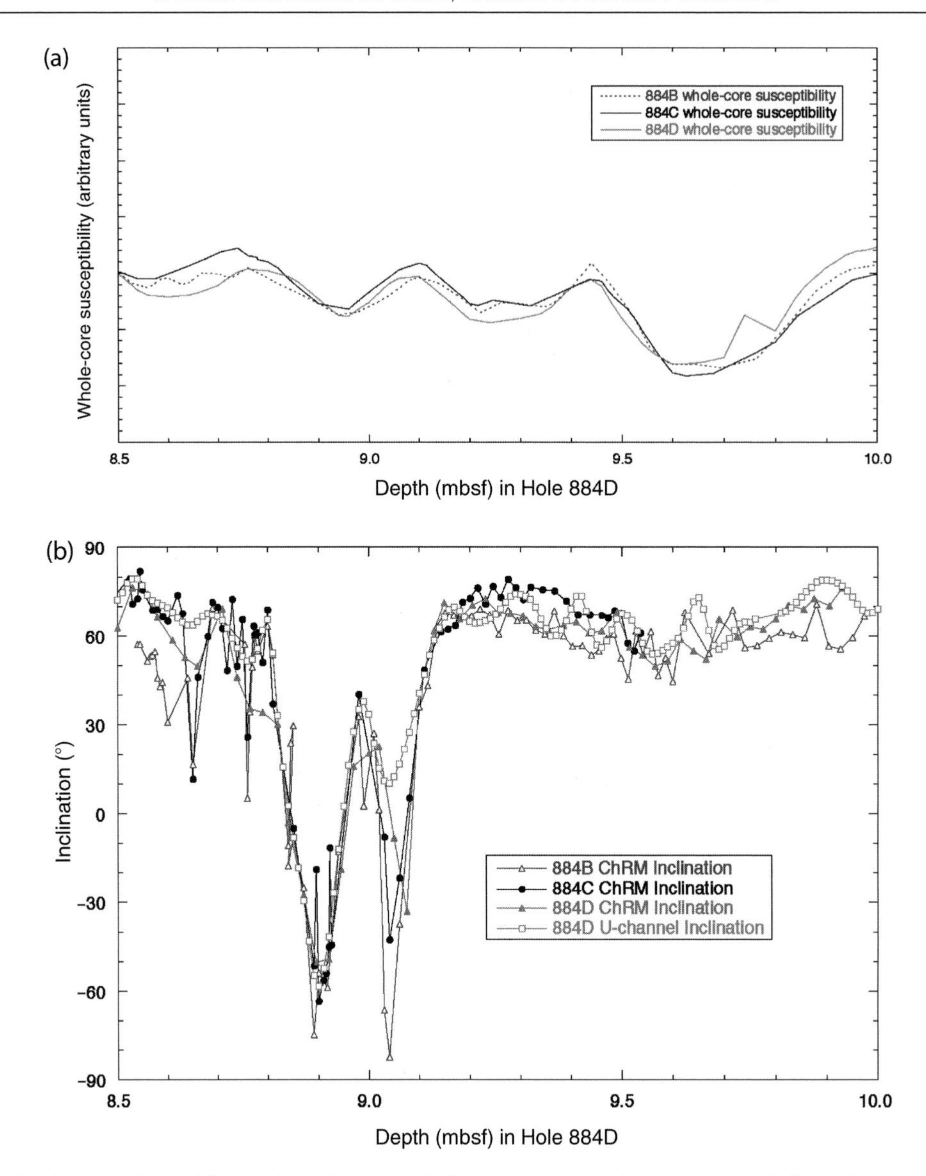

Figure E38 Example of comparison of fine-scale paleomagnetic features based on inter-core correlation using low-field magnetic susceptibility. (a) Correlation of whole-core magnetic susceptibility profiles for ODP holes 884B, 884C, and 884D in the vicinity of a geomagnetic excursion (see Roberts *et al.*, 1997). (b) Paleomagnetic inclination records for the geomagnetic excursion for the characteristic remanent magnetization (ChRM) for discrete samples (holes 884B, C, and D) and for u-channel samples after alternating field demagnetization at 25 mT (Hole 884D). Susceptibility correlation provides a good independent first-order constraint for tying together different high-resolution paleomagnetic records of geomagnetic features.

by correlation of environmental magnetic parameters with astronomical target curves (e.g., Guyodo *et al.*, 1999; Dinarès-Turrell *et al.*, 2002). It is preferable to test such age models against, for example, oxygen isotope records for the same sediment core (e.g., Larrasoaña *et al.*, 2003a); however, age models are often developed solely by correlating environmental magnetic parameters with astronomical target curves.

Environmental magnetic parameters can be routinely used not only for constructing chronologies for individual cores but also for calibrating geological timescales. While development of the astronomical polarity timescale (APTS) (Hilgen, 1991; Langereis and Hilgen, 1991) has been independent of environmental magnetism, extension of the APTS beyond the Miocene-Pliocene boundary has in some cases relied on magnetic properties of sediments. For example, magnetic susceptibility variations have been used to astronomically calibrate the Oligocene and Miocene (Shackleton *et al.*, 1999) via correlation with calculated records of variations in Earth's orbital geometry (Laskar *et al.*, 1993) (see Figure E39).

The APTS was developed by identifying the positions of magnetic reversals in cyclically deposited marine sediments from the Mediterranean Sea, in which organic-rich sapropel deposits provide key tie points to astronomical target curves (Hilgen, 1991; Langereis and Hilgen, 1991). Magnetic minerals are highly sensitive to diagenetic alteration associated with degradation of sedimentary organic matter during early burial, and variations in the burial of organic carbon has produced cycles of nonsteady state diagenesis that has left a characteristic signature in the magnetic properties of these sediments (e.g., Larrasoaña *et al.*, 2003b). In many cases, sapropels that were originally present have been completely removed from the record by postdepositional oxidation, yet magnetic properties can preserve the

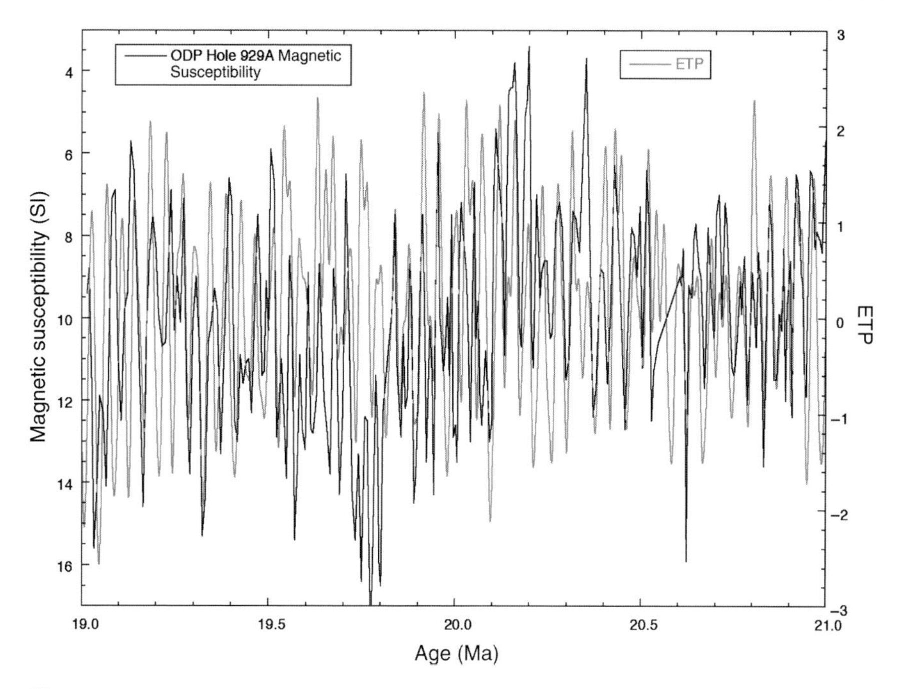

Figure E39 Example of how an environmental magnetic parameter (low-field magnetic susceptibility) can be used to calibrate geological time. The example is from the Early Miocene in ODP Hole 929A from Ceara Rise (Shackleton *et al.*, 1999), with the astronomical target curve being the eccentricity-tilt-precession (ETP) parameter that was extended from the data of Laskar *et al.* (1993). Differences between the curves result from numerous factors (see Shackleton *et al.*, 1999).

characteristic signal associated with the former presence of sapropels (Larrasoaña *et al.*, 2003b). Environmental magnetic analysis of sediments that have undergone orbitally controlled variations of nonsteady state diagenesis can therefore enable development of astronomically calibrated timescales with age tie points in addition to those that are identifiable using standard lithological observations.

Testing the hypothesis of orbital forcing of the geomagnetic field

Long, continuous records of vector variability of the geomagnetic field are needed to understand long-term field evolution and to constrain models of field generation. While it is straightforward to determine the horizontal and vertical components of the paleomagnetic vector in sediments, estimating the intensity of the ancient geomagnetic field has often proved problematical (see *Paleointensity, relative, in sediments*). The problem can be overcome in ideal sediments by normalizing the natural remanent magnetization (NRM) by an appropriate parameter that can remove the effects of variation in magnetic grain size and magnetic mineral concentration. Environmental magnetic parameters such as the low-field magnetic susceptibility, the anhysteretic remanent magnetization, and the isothermal remanent magnetization are routinely used to normalize the NRM to determine the relative geomagnetic paleointensity.

It is accepted that the geomagnetic field is generated by dynamo action within the Earth's electrically conducting fluid outer core (where buoyancy-driven convection generates a self-sustaining dynamo). Calculations indicate that any energization of the dynamo by external orbital forcing is insufficient (by at least an order of magnitude) to control the geomagnetic field (e.g., Rochester *et al.*, 1975). Nevertheless, recent recovery of long, continuous high-resolution vector paleomagnetic records has provided the type of geomagnetic time series needed for testing whether there is any relationship between the field and orbital forcing parameters. Surprisingly, several paleomagnetic records of both the intensity (e.g., Tauxe and Shackleton, 1994; Channell *et al.*, 1998) and direction (e.g., Yamazaki and Oda, 2002) of the field have resurrected the suggestion that orbital energy could drive the geodynamo.

In the case of orbital periodicities embedded within relative paleointensity records, periodograms of these records suggested that orbital periodicities were not present in the environmental magnetic parameters used for relative paleointensity normalization. This was taken to suggest that statistically significant signals at Earth orbital periods in the relative paleointensity records indicate orbital energization of the geomagnetic field (Channell *et al.*, 1998). However, when Guyodo *et al.* (2000) performed wavelet analysis on both the relative paleointensity signals and on the magnetic parameters used for normalization, they found statistically significant periodicities in both the paleointensity estimate and in the normalization parameter during the same time intervals. Guyodo *et al.* (2000) concluded that, overall, there is a subtle climatic contamination of the magnetic parameters that was not removed in the normalization process. The claim of orbital modulation of the geomagnetic field by Channell *et al.* (1998) was therefore retracted.

In another intriguing case, Yamazaki and Oda (2002) reported a link between the paleomagnetic inclination and orbital parameters and explained it by using a model based on long-term regions of nondipole field behavior at the Earth's surface. Horng *et al.* (2003) studied a core that spanned the same age interval from the same region of nondipole field behavior and found no such relationship. Roberts *et al.* (2003) used the data of Yamazaki and Oda (2002), and, while they found the same spectral peaks, they demonstrated that these peaks were not statistically significant. Even when it is considered that sediments might record a forcing signal in a nonlinear manner, comparison of the spectral signal of Yamazaki and Oda (2002) with that of the orbital solution of Laskar *et al.* (1993) demonstrates a lack of coherence between these signals (Figure E40a). In contrast, a demonstrably orbitally controlled environmental magnetic signal (Larrasoaña *et al.*, 2003a) is convincingly coherent with orbital parameters (Figure E40b). Roberts *et al.* (2003) therefore concluded that the hypothesis of (partial) orbital energization of the geomagnetic field remains undemonstrated. Environmental magnetic signals have played a fundamental role in the debate concerning this interesting paleomagnetic application.

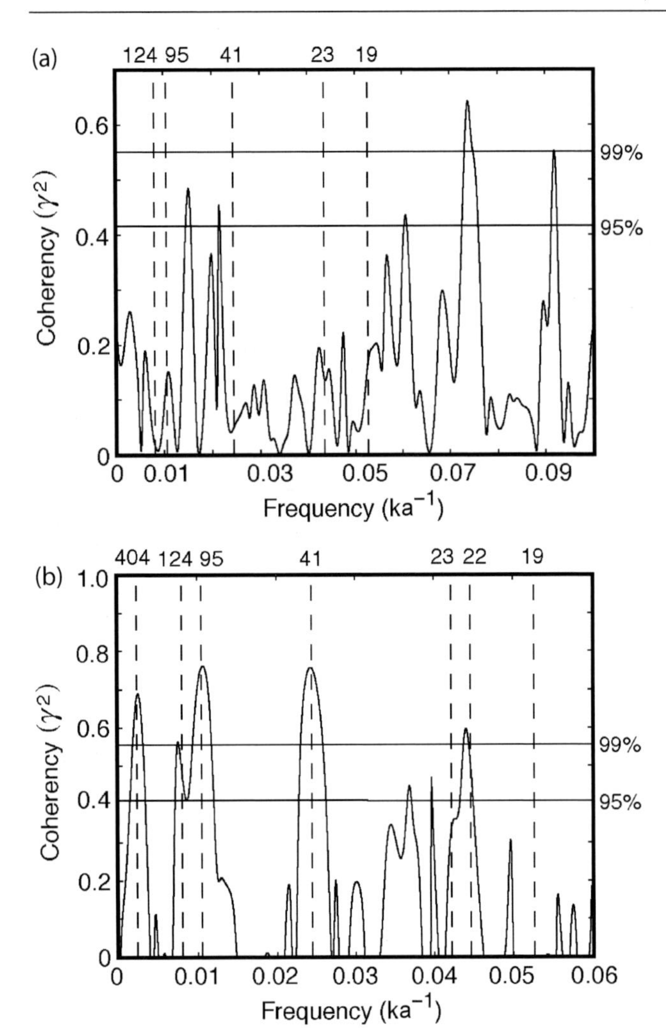

Figure E40 Example of a test of the hypothesis of orbital forcing of the geomagnetic field. (a) Coherency for the ETP parameter from the data of Laskar *et al.* (1993) and the inclination record of Yamazaki and Oda (2002). There is no statistically significant coherency at the 100-ka period (or at other Milankovitch periodicities). (b) Coherency for the ETP parameter from the data of Laskar *et al.* (1993) and the Saharan dust record of Larrasoaña *et al.* (2003a) for the last 3 million years. Orbital forcing of the dust record is clearly demonstrated, with statistically significant coherency with the orbital solutions at the 99% confidence level (see Roberts *et al.*, 2003 for details). This comparison suggests that the hypothesis of orbital forcing of the geomagnetic field is undemonstrated.

Determining the origin of the paleomagnetic signal

Rock magnetic and environmental magnetic investigations constitute a fundamentally important component of most modern paleomagnetic studies. This is because the fidelity of a measured paleomagnetic signal is directly related to the magnetic mineral(s) and the magnetic grain size distribution of the magnetic mineral(s) within a rock. For example, it is generally expected that an assemblage of detrital single-domain or pseudosingle-domain magnetic grains in an undisturbed sediment will faithfully record the geomagnetic field at or near the time of deposition. On the other hand, similar sediments that are magnetically dominated by multidomain grains would not be expected to faithfully record field information dating from the time of deposition. To complicate matters, paleomagnetically important minerals can authigenically grow within some sediments. In such cases, constraining the time of magnetic mineral growth is crucially important for interpretation of paleomagnetic data. Two examples of authigenic iron sulfide minerals, greigite (Fe_3S_4) and monoclinic pyrrhotite (Fe_7S_8), are given here to illustrate the importance of rock magnetic and environmental magnetic studies in support of paleomagnetic studies (see *Iron sulfides*).

Greigite grows within anoxic sedimentary environments that support active bacterial sulfate reduction. Dissolved sulfide forms within pore waters (or within an anoxic water column), which then reacts with detrital iron-bearing minerals, causing their progressive dissolution, resulting in the authigenic formation of pyrite (FeS_2). Pyrite is the end product of sulfidization reactions, with greigite forming as a precursor to pyrite. Greigite can survive in sediments for long periods of geological time and can retain a stable paleomagnetic signal. In early studies involving greigite, it was interpreted to have formed during early diagenesis within tens of centimeters of the sediment-water interface (e.g., Tric *et al.*, 1991; Roberts and Turner, 1993). Subsequently, however, sedimentary greigite has been found to frequently carry a late diagenetic paleomagnetic signal that makes it a highly problematical mineral in many paleomagnetic applications, including magnetostratigraphy (e.g., Florindo and Sagnotti, 1995; Horng *et al.*, 1998; Sagnotti *et al.*, 2005) and tectonics (e.g., Rowan and Roberts, 2005). Paleomagnetic field tests, in particular the fold test, become extremely important for constraining the timing of magnetizations carried by greigite (e.g., Rowan and Roberts, 2005). The reversals test can be misleading and false polarity stratigraphies, resulting from patchy or variably timed remagnetizations (e.g., Florindo and Sagnotti, 1995; Horng *et al.*, 1998; Rowan and Roberts, 2005; Sagnotti *et al.*, 2005), can easily go unidentified if careful rock magnetic and environmental magnetic investigations are not conducted in support of paleomagnetic studies.

Similar to the case with greigite, it is often assumed in paleomagnetic studies that pyrrhotite forms during early diagenesis and that it can carry a stable syndepositional paleomagnetic signal. Geochemical literature, however, indicates that the formation of pyrrhotite is kinetically inhibited at ambient temperatures typical of early burial and that pyrrhotite cannot form during early diagenesis (e.g., Schoonen and Barnes, 1991; Lennie *et al.*, 1995). This presents an apparent conundrum because Horng *et al.* (1998) reported that pyrrhotite carried an identical paleomagnetic signature to detrital magnetite in sedimentary rocks from Taiwan. Horng and Roberts (2006) carried out a source to sink study in southwestern Taiwan and demonstrated that the pyrrhotite in the studied sediments is detrital in origin and that it is sourced from metamorphic rocks in the Central Range of Taiwan. While it is now recognized that pyrrhotite can carry a detrital magnetization (Horng and Roberts, 2006), and while it can also carry late diagenetic remagnetizations (e.g., Jackson *et al.*, 1993; Xu *et al.*, 1998; Weaver *et al.*, 2002), pyrrhotite clearly cannot carry an early diagenetic magnetization. Knowing which type of magnetization is being carried by monoclinic pyrrhotite grains within a sediment is of crucial importance in arriving at a correct paleomagnetic interpretation. These two examples involving different magnetic iron sulfide minerals demonstrate that the assumptions made in many paleomagnetic studies are incorrect. They also make clear the need for careful rock magnetic and environmental magnetic investigations in support of determining the origin, nature, and age of the magnetization in paleomagnetic studies.

Summary

While most environmental magnetic studies appropriately focus on environmental interpretations, environmental magnetism routinely plays an important role in a wide range of paleomagnetic applications. These applications include correlation of paleomagnetic features in suites of sediment cores, development of astronomically calibrated age models, testing the hypothesis of orbital forcing of the geomagnetic field, and determining the origin of the paleomagnetic signal.

The importance of environmental magnetism to this range of paleomagnetic applications therefore makes it fundamentally valuable to paleomagnetism.

Andrew P. Roberts

Bibliography

Channell, J.E.T., Hodell, D.A., McManus, J., and Lehman, B., 1998. Orbital modulation of the Earth's magnetic field intensity. *Nature*, **394**: 464–468.

Dinarès-Turell, J., Sagnotti, L., and Roberts, A.P., 2002. Relative geomagnetic paleointensity from the Jaramillo Subchron to the Matuyama/Brunhes boundary as recorded in a Mediterranean piston core. *Earth and Planetary Science Letters*, **194**: 327–341.

Evans, M.E., and Heller, F., 2003. *Environmental Magnetism: Principles and Applications of Enviromagnetics*. New York: Academic Press, 299 pp.

Florindo, F., and Sagnotti, L., 1995. Palaeomagnetism and rock magnetism in the upper Pliocene Valle Ricca (Rome, Italy) section. *Geophysical Journal International*, **123**: 340–354.

Guyodo, Y., Richter, C., and Valet, J.-P., 1999. Paleointensity record from Pleistocene sediments (1.4–0 Ma) off the California Margin. *Journal of Geophysical Research*, **104**: 22,953–22,964.

Guyodo, Y., Gaillot, P., and Channell, J.E.T., 2000. Wavelet analysis of relative geomagnetic paleointensity at ODP Site 983. *Earth and Planetary Science Letters*, **184**: 109–123.

Hagelberg, T.K., Pisias, N.G., Shackleton, N.J., Mix, A.C., and Harris, S., 1992. Refinement of a high-resolution, continuous sedimentary section for studying equatorial Pacific Ocean paleoceanography, Leg 138. *Proceedings of the Ocean Drilling Program—Scientific Results*, **138**: 31–46.

Hilgen, F.J., 1991. Extension of the astronomically calibrated (polarity) timescale to the Miocene/Pliocene boundary. *Earth and Planetary Science Letters*, **107**: 349–368.

Horng, C.-S., and Roberts, A.P., 2006. Authigenic or detrital origin of pyrrhotite in sediments?: Reso paleomagnetic conundrum, *Earth and Planetary Science Letters*, **241**: 750–762.

Horng, C.S., Torii, M., Shea, K.-S., and Kao, S.-J., 1998. Inconsistent magnetic polarities between greigite- and pyrrhotite/magnetite-bearing marine sediments from the Tsailiao-chi section, southwestern Taiwan. *Earth and Planetary Science Letters*, **164**: 467–481.

Horng, C.-S., Roberts, A.P., and Liang, W.T., 2003. A 2.14-million-year astronomically-tuned record of relative geomagnetic paleointensity from the western Philippine Sea. *Journal of Geophysical Research*, **108**: 2059, doi:10.1029/2001JB001698.

Jackson, M., Rochette, P., Fillion, G., Banerjee, S., and Marvin, J., 1993. Rock magnetism of remagnetized Paleozoic carbonates: low-temperature behavior and susceptibility characteristics. *Journal of Geophysical Research*, **98**: 6217–6225.

Kent, D.V., 1982. Apparent correlation of palaeomagnetic intensity and climate records in deep-sea sediments. *Nature*, **299**: 538–539.

Langereis, C.G., and Hilgen, F.J., 1991. The Rosello composite—a Mediterranean and global reference section for the early to early Late Pliocene. *Earth and Planetary Science Letters*, **104**: 211–225.

Larrasoaña, J.C., Roberts, A.P., Rohling, E.J., Winklhofer, M., and Wehausen, R., 2003a. Three million years of monsoon variability over the northern Sahara. *Climate Dynamics*, **21**: 689–698.

Larrasoaña, J.C., Roberts, A.P., Stoner, J.S., Richter, C., and Wehausen, R., 2003b. A new proxy for bottom-water ventilation based on diagenetically controlled magnetic properties of eastern Mediterranean sapropel-bearing sediments. *Palaeogeography, Palaeoclimatology, Palaeoecology*, **190**: 221–242.

Laskar, J., Joutel, F., and Boudin, F., 1993. Orbital, precessional, and insolation quantities for the Earth from −20 Myr to +10 Myr. *Astronomy and Astrophysics*, **270**: 522–533.

Lennie, A.R., England, K.E.R., and Vaughan, D.J., 1995. Transformation of synthetic mackinawite to hexagonal pyrrhotite: a kinetic study. *American Mineralogist*, **80**: 960–967.

Maher, B.A., and Thompson, R. (eds.), 1999. *Quaternary Climates, Environments and Magnetism*. Cambridge: Cambridge University Press, 390 pp.

Roberts, A.P., and Turner, G.M., 1993. Diagenetic formation of ferrimagnetic iron sulphide minerals in rapidly deposited marine sediments, South Island, New Zealand. *Earth and Planetary Science Letters*, **115**: 257–273.

Roberts, A.P., Lehman, B., Weeks, R.J., Verosub, K.L., and Laj, C., 1997. Relative paleointensity of the geomagnetic field from 0–200 kyr, ODP Sites 883 and 884, North Pacific Ocean. *Earth and Planetary Science Letters*, **152**: 11–23.

Roberts, A.P., Winklhofer, M., Liang, W.-T., and Horng, C.-S., 2003. Testing the hypothesis of orbital (eccentricity) influence on Earth's magnetic field. *Earth and Planetary Science Letters*, **216**: 187–192.

Rochester, M.G., Jacobs, J.A., Smylie, D.E., and Chong, K.F., 1975. Can precession power the geomagnetic dynamo? *Geophysical Journal of the Royal Astronomical Society*, **43**: 661–678.

Rowan, C.J., and Roberts, A.P., 2005. Tectonic and geochronological implications of variably timed remagnetizations carried by authigenic greigite in fine-grained sediments from New Zealand. *Geology*, **33**: 553–556.

Sagnotti, L., Roberts, A.P., Weaver, R., Verosub, K.L., Florindo, F., Pike, C.R., Clayton, T., and Wilson, G.S., 2005. Apparent magnetic polarity reversals due to remagnetization resulting from late diagenetic growth of greigite from siderite. *Geophysical Journal International*, **160**: 89–100.

Schoonen, M.A.A., and Barnes, H.L., 1991. Mechanisms of pyrite and marcasite formation from solution: III. Hydrothermal processes. *Geochimica Cosmochimica Acta*, **55**: 3491–3504.

Shackleton, N.J., Crowhurst, S.J., Weedon, G.P., and Laskar, J., 1999. Astronomical calibration of Oligocene-Miocene time. *Philosophical Transactions of the Royal Society, London, Series A*, **357**: 1907–1929.

Tauxe, L., and Shackleton, N.J., 1994. Relative palaeointensity records from the Ontong-Java Plateau. *Geophysical Journal International*, **117**: 769–782.

Tric, E., Laj, C., Jéhanno, C., Valet, J.-P., Kissel, C., Mazaud, A., and Iaccarino, S., 1991. High-resolution record of the upper Olduvai transition from Po Valley (Italy) sediments: support for dipolar transition geometry? *Physics of the Earth and Planetary Interiors*, **65**: 319–336.

Turner, G.M., and Thompson, R., 1979. Behaviour of the Earth's magnetic field as recorded in sediments of Loch Lomond. *Earth and Planetary Science Letters*, **42**: 412–426.

Verosub, K.L., and Roberts, A.P., 1995. Environmental magnetism: past, present, and future. *Journal of Geophysical Research*, **100**: 2175–2192.

Weaver, R., Roberts, A.P., and Barker, A.J., 2002. A late diagenetic (synfolding) magnetization carried by pyrrhotite: implications for paleomagnetic studies from magnetic iron sulphide-bearing sediments. *Earth and Planetary Science Letters*, **200**: 371–386.

Yamazaki, T., and Oda, H., 2002. Orbital influence on Earth's magnetic field: 100,000-year periodicity in inclination. *Science*, **295**: 2435–2438.

Xu, W., Van der Voo, R., and Peacor, D., 1998. Electron microscopic and rock magnetic study of remagnetized Leadville carbonates, central Colorado. *Tectonophysics*, **296**: 333–362.

Cross-references

Environmental Magnetism
Iron Sulfides
Paleointensity Relative in Sediments
Paleomagnetic Secular Variation

EQUILIBRATION OF MAGNETIC FIELD, WEAK- AND STRONG-FIELD DYNAMOS

The necessity of equilibration of a growing magnetic field

It is enlightening to consider the theory of the Earth's dynamo as two separate constituents: kinematic and dynamic problems. The kinematic problem concerns conditions under which an electrically conducting flow can produce a growing magnetic field. Alternatively, the kinematic dynamo may be regarded as an instability problem where a purely hydrodynamic flow becomes unstable to infinitesimal disturbances of magnetic field. In contrast, the dynamic dynamo problem studies the equilibration of a kinematically growing magnetic field by considering the magnetic feedback on the flow (for example, Bassom and Gilbert, 1997).

In the kinematic dynamo problem, let **u** denote the velocity of nonmagnetic convection in the Earth's fluid core at a given Rayleigh number R, while the magnetic field **B** is described by the dynamo equation

$$\frac{\partial}{\partial t}\mathbf{B} = \nabla \times (\mathbf{u} \times \mathbf{B}) + \lambda \nabla^2 \mathbf{B}, \qquad \text{(Eq. 1)}$$

together with the condition

$$\nabla \cdot \mathbf{B} = 0. \qquad \text{(Eq. 2)}$$

In Eq. (1), λ represents the magnetic diffusivity of the fluid. Equations (1) and (2) admit a solution with the generated magnetic field of the form

$$\mathbf{B}(\mathbf{r}, t) = \mathbf{B}(\mathbf{r}) e^{\sigma t} \qquad \text{(Eq. 3)}$$

together with an appropriate set of boundary conditions for **B** (Moffatt, 1978). When U, the typical speed of the convective flow **u**, is small, any initial magnetic field would decay to zero as $t \to \infty (\mathrm{Real}[\sigma] < 0)$; when U is sufficiently large, $\mathrm{Real}[\sigma] > 0$ and an initial magnetic field would grow without bound, i.e.,

$$|\mathbf{B}| \to \infty, \quad \text{as } t \to \infty. \qquad \text{(Eq. 4)}$$

The critical threshold at the onset of dynamo action takes place at some particular Rayleigh number, R_m, when $\mathrm{Real}[\sigma] = 0$ (for example, Zhang and Busse, 1989). Physically speaking, however, the kinematically growing magnetic field at $R > R_m$ must be brought to an equilibrium, presumably by nonlinear effects.

Equilibration: subcritical and supercritical bifurcation

Mathematically speaking, the equilibration of a kinematically growing magnetic field is described by the momentum equation in a rotating fluid which may be written as

$$2\Omega \times \mathbf{u} = -\frac{1}{\rho}\nabla p + \frac{1}{\rho\mu}(\nabla \times \mathbf{B}) \times \mathbf{B} + \mathbf{f}, \qquad \text{(Eq. 5)}$$

where p is the pressure, μ is the magnetic permeability, ρ is the fluid density, **f** represents forces such as buoyancy, and Ω is the angular velocity of the Earth. For an infinitesimal magnetic field, the magnetic (Lorentz) force $(1/\rho\mu)(\nabla \times \mathbf{B}) \times \mathbf{B}$ is too weak to alter the convective flow **u** and two problems, one pertaining to nonmagnetic convection and the other to kinematic dynamo, are nearly decoupled. A growing magnetic field ultimately leads to a sufficiently large magnetic force which suppresses the convective flow and brings the generated magnetic field to an equilibration.

The equilibration of a kinematically growing magnetic field is associated with a bifurcation from the corresponding nonmagnetic convective state. There exist at least two different sequences of the bifurcation, which are sketched in Figure E41. In the first possible sequence, the linear convective instability occurs at $R = R_c$. When the system is driven harder at larger R, the typical amplitude U of nonmagnetic convection increases. Dynamo action takes place at $R = R_m$ when convection is vigorous enough. The equilibration of the generated magnetic field is achieved when $R > R_m$ as a supercritical bifurcation. In this case, larger values of the Rayleigh number R are required to power the convection-driven dynamo with extra magnetic dissipation.

The second possible sequence of bifurcation is subcritical (or backward), which is also shown in Figure E41. In comparison to a supercritical bifurcation, the nature of equilibration is quite different, which is perhaps more relevant to a rapidly rotating system like the Earth's core dynamo. When R is slightly smaller than R_m, the nonmagnetic convective system is linearly stable but is unstable to finite-amplitude perturbations of a magnetic field. In consequence, the backward branch of the bifurcation (the dashed curve for $R < R_m$ in Figure E41) is unstable and cannot be physically realized. By implication, the Lorentz force resulting from dynamo action has a destabilizing effect which relaxes the inhibiting effect of rapid rotation on nonmagnetic convection. When R is close to R_m, the amplitude of the generated magnetic field suddenly jumps and the equilibrated magnetic field is characterized by a large amplitude. In this case, the usefulness of the kinematic study determining the onset of dynamo action is highly limited.

Equilibration: α-quenching in turbulent dynamos

The equilibration of magnetic field in turbulent dynamos can be modeled through a mechanism called α-quenching. In this process we separate the generated magnetic field **B** into large-scale axisymmetric ($\mathbf{B}_0$) and small-scale nonaxisymmetric ($\hat{\mathbf{B}}$) parts

$$\mathbf{B} = \mathbf{B}_0 + \hat{\mathbf{B}}, \qquad \text{(Eq. 6)}$$

where $\mathbf{B}_0$ is generated by an α-effect in connection with interactions between small-scale turbulent flows and magnetic fields (Moffatt, 1978). The strength of the α-effect would be suppressed when the back-reaction of the generated magnetic field on the flow is significantly strong. This reduced α would prevent the unlimited growth of the mean magnetic field $\mathbf{B}_0$ and bring it to an equilibration. This process is usually described by

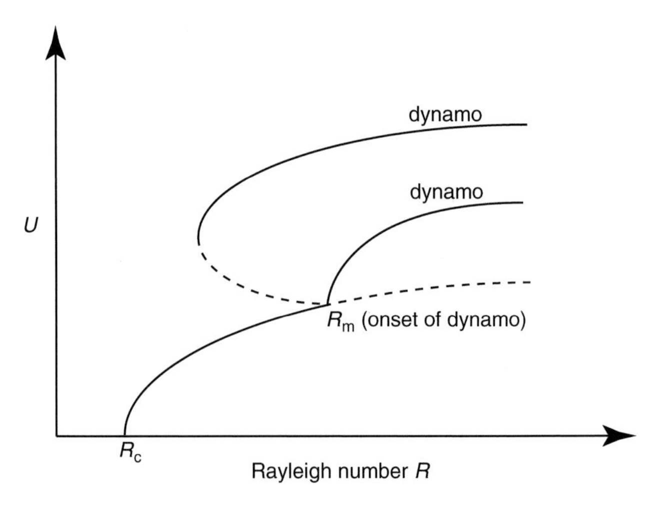

Figure E41 Conjectured structure of bifurcation diagram for the equilibration of convection-driven dynamos. Here R_c is the critical Rayleigh number for the onset of convection and R_m is the another critical Rayleigh number at which dynamo action takes place according to kinematic dynamo theory. Both supercritical (the solid line) and subcritical dynamo (the dashed line) bifurcations are shown.

$$\frac{\partial}{\partial t}\mathbf{B}_0 = \nabla \times \left[\frac{\alpha \mathbf{B}_0}{(1+q|\mathbf{B}_0|^n)}\right] - \lambda_0 \nabla \times \nabla \times \mathbf{B}_0, \qquad \text{(Eq. 7)}$$

where λ_0 is a turbulent magnetic diffusivity, $q > 0$ and n often takes the value 2 (Jepps, 1975; Roberts and Soward, 1992). The α-quenching formulation has been widely used in dynamo simulations and one of its major advantages is that it can simulate the equilibration of the generated magnetic field and the essential dynamo processes without reference to the difficult dynamics of strong nonlinear interaction between the flow and the Lorentz force described by the momentum equation (5). The corresponding bifurcation in the α-quenching dynamo is typically supercritical because the dynamic effect of rotation is not taken into account.

Weak- and strong-field dynamos

In rapidly rotating systems like the Earth's fluid core, magnetohydrodynamic dynamos may be divided into two different categories: weak-field dynamos and strong-field dynamos (for example, Gubbins and Roberts, 1987; Zhang and Schubert, 2000). They are usually defined on the basis of the relative importance of the Lorentz and Coriolis forces in a convection-driven dynamo. In the case of weak-field dynamos, the dynamic role of the Lorentz force plays an insignificant role and the viscous force must be sufficiently strong to offset the part of the Coriolis force that cannot be balanced by the pressure gradient. In the case of strong-field dynamos, the dynamic role of viscosity is taken over by the Lorentz force.

A widely used criterion for the distinction between weak-field and strong-field dynamos is associated with the Elsasser number Λ defined as

$$\Lambda \equiv \frac{B_0^2}{2|\Omega|\mu\rho\lambda},$$

where B_0 is an average or a representative value of the amplitude of the magnetic field generated by a dynamo. Equations (1) and (5) yield an estimate for the relative importance of the Lorentz and Coriolis forces

$$\frac{|(\mu\rho)^{-1}(\nabla \times \mathbf{B}) \times \mathbf{B}|}{|2\Omega \times \mathbf{u}|} = O(\Lambda), \qquad \text{(Eq. 8)}$$

where inertial and viscous terms are neglected. In estimating the value of the Elsasser number Λ, we often take B_0 as a spatially or temporally averaged magnetic field. A strong-field dynamo is one in which $\Lambda \geq O(1)$ while a weak-field dynamo is one in which $\Lambda < O(1)$. However, it should be pointed out that this definition is not entirely satisfactory because a large Elsasser number does not necessarily imply a strong interaction between the convective flow and the generated magnetic field.

An alternative criterion for the distinction between strong- and weak-field dynamos is based on the dynamic effect of the generated magnetic field. A strong-field dynamo is defined as one in which viscosity is dynamically unimportant so that the scale of convection is large as a direct consequence of the dynamic effect of the generated magnetic field. In other words, in a strong-field dynamo the structure and scale of the convective flow are highly sensitive to variations in the generated magnetic field. Similarly, we may define a weak-field dynamo as one in which the dynamic effect of the generated magnetic field can be treated as a perturbation. In other words, in a weak-field dynamo the structure and scale of the convective flow are controlled by the effects of viscosity and rotation and are insensitive to variations in the generated magnetic field. When one turns off dynamo action, convective motions remain largely unchanged for a weak-field dynamo but change almost completely for a strong-field dynamo.

Keke Zhang

Bibliography

Bassom, A.P., and Gilbert, A.D., 1997. Nonlinear equilibration of a dynamo in a smooth helical flow. *Journal of Fluid Mechanics*, **343**: 375–406.

Gubbins, D., and Roberts, P.H. 1987. Magnetohydrodynamics of the Earth's core. In Jacobs, J.A. (ed.), *Geomagnetism,* Vol. 2. London: Academic Press, pp. 1–183.

Jepps, S.A., 1975. Numerical models of hydrodynamic dynamos. *Journal of Fluid Mechanics*, **67**: 625–646.

Moffatt, H.K., 1978. *Magnetic Field Generation in Electrically Conducting Fluids.* Cambridge: Cambridge University Press.

Roberts, P.H., and Soward, A.M., 1992. Dynamo theory. *Annual Review of Fluid Mechanics*, **24**: 459–512.

Zhang, K., and Busse, F., 1989. Convection driven magnetohydrodynamic dynamos in rotating spherical shells. *Geophysical and Astrophysical Fluid Dynamics*, **49**: 97–116.

Zhang, K., and Schubert, G., 2000. Magnetohydrodynamics in rapidly rotating spherical systems. *Ann. Rev. Fluid Mech*, **32**: 409–443.

Cross-references

Core Motions
Core, Magnetic Instabilities
Geodynamo
Magnetoconvection

EULER DECONVOLUTION

Introduction

Magnetic surveys are widely used in mineral and oil exploration to delineate geological structure and locate targets of interest (see *Magnetic anomalies for geology and resources*). They are also used in environmental work to locate buried metal objects such as unexploded ordnance, drums, pipework, and the like. In all of these applications, we need a quick means of turning magnetic field measurements into estimates of magnetic source body location and depth. Euler deconvolution is one such method. Others include Werner deconvolution (Werner, 1955; Hartman *et al.*, 1971) and the Naudy (1971) method.

The method may be applied to survey profile data or gridded data. It operates on a data subset extracted using a moving window. In each window, Euler's equation (see below) is solved for source body coordinates. The source body geometry is specified by a structural index (SI) ranging between 0 (contact of great depth extent) and 3 (dipole or sphere). Possible source bodies and their structural indices are tabulated below. Intermediate bodies have noninteger SIs and the Euler formulation is only approximate. The method does not require any assumptions about direction of exciting field, direction of magnetization, or the relative magnitudes of induced and remanent magnetization of the sources. But it appears to work best on data after *reduction to pole* (*q.v.*).

History

Hood (1963) proposed that Euler's equation might be used with vertical magnetic gradient survey measurements to determine source depth. Ruddock *et al.* (1966) were awarded a US patent describing a graphical method that exploited Euler's equation to determine source depth from magnetic gradient measurements. Thompson (1982) described a fully developed profile method (called EULDPH). It was widely used within the Gulf and Chevron Oil Companies and also by Durrheim (1983) and Corner and Wilsher (1989) to delineate the magnetic markers in the Witwatersrand Basin in the search for gold. Thompson also developed the theory for the case of gridded data. Reid *et al.* (1990)

implemented Thompson's method for gridded data and extended the theory to the case of zero structural index. They also coined the term "Euler deconvolution" by analogy with the established Werner deconvolution technique.

Theory

Strictly speaking, Euler deconvolution is only valid for homogeneous functions. A function $f(\mathbf{v})$ of a set of variables $\mathbf{v} = (v_1, v_2, \ldots)$ is homogeneous of degree n, if

$$f(t\mathbf{v}) = t^n f(\mathbf{v}) \qquad \text{(Eq. 1)}$$

where t is a real number, and n is an integer. This is really a statement of the scaling properties of $f(\mathbf{v})$. If f has a differential at $\mathbf{v}$, then

$$\mathbf{v}\nabla_{\mathbf{v}} f(\mathbf{v}) = nf(\mathbf{v}). \qquad \text{(Eq. 2)}$$

This is Euler's equation, and Euler deconvolution solves it in appropriate cases.

If a field F can be expressed in the form

$$F = A/r^n \qquad \text{(Eq. 3)}$$

F will be homogeneous of degree $-n$. For convenience we define the SI as N ($= -n$, i.e., the negative degree of homogeneity).

Thompson (1982) showed that Euler's equation could usefully be written in the form

$$(x - x_0)\partial T/\partial x + (y - y_0)\partial T/\partial y + (z - z_0)\partial T/\partial z = N(B - T), \qquad \text{(Eq. 4)}$$

where (x_0, y_0, z_0) is the position of a magnetic source whose total field T is detected at (x, y, z). The total field has a regional value of B. N is the structural index.

Strictly speaking, Euler deconvolution is only valid for homogeneous fields. These are only obtained from particular idealized bodies (see Table E2) with integer structural index. The fields of most real bodies do not meet this strict criterion and the method is therefore an approximation in practice. Nevertheless, it has been found useful enough to be applied in a wide variety of cases.

Implementation

The process is implemented as follows.

1. Interpolate line data to a constant spacing, or multiline or point data to a uniform grid. Use an interpolation interval which represents the data well but does not overinterpolate. The interval should be no more than the depth of the shallowest magnetic sources thought to exist in the area.
2. Calculate along line (x) and vertical (z) gradients for profile data and x, y, and z gradients for gridded data, using finite difference or Fourier methods. If measured gradient are available, interpolate them to the same pattern as the field data.
3. Choose a suitable subset size or "window." Grid windows are normally square. The window length should be greater than half the depth to the deepest magnetic sources sought. Since the method assumes that any window contains effects from only one source body, good resolution requires windows to be as small as possible. Typical grid windows might be 7×7 to 10×10 grid points. If necessary, for deep sources, it may be necessary to revisit 1 above, apply an anti-alias low-pass filter and desample the profile or grid.
4. Move the sample window through the profile or grid, using small overlapping steps. At each step, use all the data points in the current window to solve Eq. (4) above for x_0, y_0, z_0, and T for whatever values of structural index N seem relevant to the situation.

The method generates a solution for each window position, regardless of the presence (or lack) of any nearby magnetic source. In addition, the method effectively assumes that the effects of only one source are present in any window. Interference effects give rise to festoons of misleading solutions trailing away from valid solutions that cluster along source body edges. Various strategies have been devised to eliminate unhelpful solutions. They include:

1. Eliminate solutions with high estimation uncertainties (often normalized by depth).
2. Eliminate solutions from poorly conditioned matrices.
3. Identify and select for clusters of solutions.
4. Only solve in windows showing positive field curvature (Fairhead *et al.*, 1994).
5. Eliminate solutions that fall beyond their data window.

Table E2 Possible source bodies and their structural indices

Source body	Magnetic structural index
Dipole (sphere, small, compact body)	3
Thin pipe (vertical—kimberlite; horizontal—pipeline); thin bed fault	2
Thin sheet edge (dyke, sill)	1
Contact with depth extent $\gg$ depth	0 (requires modified equation)

Recent developments and ongoing research

Stavrev (1997) has applied differential similarity transforms to Euler deconvolution. This unconventional approach shows promise of progress in the area of multiple interfering sources. Barbosa *et al.* (1999) developed a means of determining the structural index for an isolated source body by correlating the observed field with the derived "background" field. Mushayandebvu *et al.* (2001, 2003) added rotational invariance to the scaling rule and found it possible to solve for strike, dip, and magnetization contrast or the magnetization-thickness product. Hansen and Suciu (2002) have developed a formulation for multiple sources. But all sources in one solution run must have the same structural index Nabighian and Hansen (2001) applied a Hilbert transform as well as rotational invariance and have found it possible to solve for structural index and dip. They also showed that this form of Euler deconvolution is equivalent to Werner deconvolution (see above).

The following problems are unsolved (this list is almost certainly incomplete):

1. Develop a reliable and robust means of solution selection.
2. Pursue the application of seismic migration approaches to collapsing the festoon (spray) solutions onto their primary points.
3. Establish the bounds of applicability of the method to real data by working on realistic models based on the real Earth.
4. Further explore the potential of differential similarity transforms as applied to Euler deconvolution. They appear to offer scope for advance in multiple-source analysis.

Alan B. Reid

Bibliography

Barbosa, V.C.F., Silva, J.B.C., and Medeiros, W.E., 1999. Stability analysis and improvement of structural index estimation in Euler deconvolution. *Geophysics*, **64**: 48–60.

Corner, B., and Wilsher, W.A., 1989. Structure of the Witwatersrand Basin derived from interpretation of the aeromagnetic and gravity data. In Garland, G.D. (ed.), *Proceedings of Exploration '87: Third Decennial International Conference on Geophysical and Geochemical Exploration for Minerals and Groundwater.* Ontario Geological Survey, Special Volume 3, pp. 532–546.

Durrheim, R.J.,1983. Regional-residual separation and automatic interpretation of aeromagnetic data. Unpublished MSc thesis, University of Pretoria.

Fairhead, J.D., Bennett, K.J., Gordon, D.R.H., and Huang, D., 1994. Euler: beyond the "black box". In *64th Annual International Meeting.* Los Angeles, CA: Society of Exploration Geophysicists, Expanded Abstracts, pp. 422–424.

Hansen, R.O., and Suciu, L., 2002. Multiple-source Euler deconvolution. *Geophysics*, **67**: 525–535.

Hartman, R.R., Teskey, D.J., and Friedberg, J.J., 1971. A system for rapid digital aeromagnetic interpretation. *Geophysics*, **36**: 891–918.

Hood, P.J., 1963. Gradient measurements in aeromagnetic surveying. *Geophysics*, **30**: 891–902.

Mushayandebvu, M.F., van Driel, P., Reid, A.B., and Fairhead, J.D., 2001. Magnetic source parameters of two-dimensional structures using extended Euler deconvolution. *Geophysics*, **66**: 814–823.

Mushayandebvu, M.F., Lesur, V., Reid, A.B., and Fairhead, J.D., 2003. Grid Euler deconvolution with constraints for two-dimensional structures. *Geophysics*, **69**: 489–496.

Nabighian, M.N., and Hansen, R.O., 2001. Unification of Euler and Werner deconvolution in three-dimensions via the generalized Hilbert transform. *Geophysics*, **66**: 1805–1810.

Naudy, H., 1971. Automatic determination of depth on aeromagnetic profiles. *Geophysics*, **36**: 717–722.

Reid, A.B., Allsop, J.M., Granser, H., Millett, A.J., and Somerton, I.W., 1990. Magnetic interpretation in three dimensions using Euler deconvolution. *Geophysics*, **55**: 80–91.

Ruddock, K.A., Slack, H.A., and Breiner, S., 1966. Method for determining depth and falloff rate of subterranean magnetic disturbances utilising a plurality of magnetometers. US Patent 3,263,161, filed March 26, 1963, awarded July 26, 1966, assigned to Varian Associates and Pure Oil Company.

Stavrev, P.Y., 1997. Euler deconvolution using differential similarity transformations of gravity or magnetic anomalies. *Geophysical Prospecting*, **45**: 207–246.

Thompson, D.T., 1982. EULDPH—a new technique for making computer-assisted depth estimates from magnetic data. *Geophysics*, **47**: 31–37.

Werner, S., 1955. Interpretation of magnetic anomalies at sheet-like bodies: Sveriges Geologiska Undersökning. *Årsbok* **43**(1949), No. 6, Series C, No. 508.

Cross-references

Magnetic Anomalies for Geology and Resources
Reduction to Pole

F

FIRST-ORDER REVERSAL CURVE (FORC) DIAGRAMS

Introduction

In many geomagnetic, geological, and environmental magnetic studies, it is important to have reliable methods of characterizing the composition and grain size distribution of magnetic minerals within samples. For example, identification of single domain (SD) magnetic grains is important in absolute paleointensity studies because SD grains produce the most reliable results, while larger multidomain (MD) grains yield the least meaningful results. In paleoclimatic studies, useful environmental information is often revealed by subtle changes in grain size distribution, as revealed by domain state, while the same grain size variations will complicate determination of relative paleointensity from the same sediment. It is therefore crucially important to have reliable methods for determining the magnetic grain size distribution in a wide range of geomagnetic, paleomagnetic, and environmental magnetic applications.

Determining the composition of magnetic minerals in a rock is relatively straightforward, however, identification of the domain state is more difficult. Conventional methods, such as measurement of magnetic hysteresis, can be powerful, but they can also yield ambiguous results because various combinations of mineral composition, grain size, internal stress, and magnetic interactions among the grains can produce the same magnetic behavior. This is particularly true for the commonly used Day plot (Day *et al.*, 1977), which summarizes the bulk magnetic hysteresis properties by plotting ratios of M_{rs}/M_s versus H_{cr}/H_c, where M_{rs} is the saturation remanent magnetization, M_s is the saturation magnetization, H_{cr} is the coercivity of remanence, and H_c is the coercive force (see *Rock magnetism, hysteresis measurements* for a description of how these parameters are determined). For example, numerical studies have shown that magnetostatic interactions among ideal SD grains can cause the bulk hysteresis parameters of samples to plot within the MD region of the Day plot (Muxworthy *et al.*, 2003).

In an attempt to remove some of the ambiguity inherent to conventional hysteresis measurements, Pike *et al.* (1999) and Roberts *et al.* (2000) developed a method of mineral and domain state discrimination using a type of hysteresis curve called a first-order reversal curve (FORC). Measurement of a suite of FORCs provides detailed information from within the major hysteresis loop, which enables determination of the distribution of switching fields and interaction fields for all of the particles that contribute to the hysteresis loop. The ability to measure sufficient FORCs to construct a FORC diagram has only recently become possible with rapid and sensitive vibrating sample magnetometers and alternating gradient magnetometers.

Measuring and constructing FORC diagrams

A FORC is measured by progressively saturating a sample in a field H_{SAT}, decreasing the field to a value H_A, reversing the field and sweeping it back to H_{SAT} in a series of regular field steps (H_B) (Figure F1a). This process is repeated for many values of H_A, which yields a series of FORCs, and the measured magnetization M at each step as a function of H_A and H_B gives $M(H_A, H_B)$ (Figure F1b). $M(H_A, H_B)$ can then be plotted as a function of H_A and H_B in field space (Figure F1c). The field steps are chosen such that H_A and H_B are regularly spaced, which means that $M(H_A, H_B)$ can be plotted on a regular grid. The FORC distribution $\rho(H_A, H_B)$ is defined as the mixed second derivative of the surface shown in Figure F1c:

$$\rho(H_A, H_B) \equiv -\partial^2 M(H_A/H_B)\partial H_A/\partial H_B \quad \text{(Eq. 1)}$$

When $\rho(H_A, H_B)$ is plotted as a contour plot (i.e., as a FORC diagram; Figure F1d), it is convenient to rotate the axes by changing coordinates from (H_A, H_B) to $H_C = (H_B - H_A)/2$ and $H_U = (H_B+H_A)/2$.

As is standard when fitting functions to experimental data, reduction of the effect of noise on FORC distributions is achieved in a piecewise manner. That is, rather than fitting a function to the entire $M(H_A, H_B)$ surface and then directly differentiating this surface to determine $\rho(H_A, H_B)$, the FORC distribution is determined at each point by fitting a mixed second-order polynomial of the form $a_1 + a_2H_A + a_3H_A^2 + a_4H_B + a_5H_B^2 + a_6H_AH_B$ to a local, moving grid. $\rho(H_A, H_B)$ is simply equal to the fitted parameter $-a_6$. As the polynomial is only of second order, it cannot accurately accommodate complex surfaces, however, this will not commonly be a problem for geological or environmental samples, which usually have relatively smooth FORC distributions. The size of the local area is determined by a user-defined smoothing factor (SF), where the size of the grid is simply $(2SF+1)^2$. SF normally takes values between 2 and 5, although ideally it should be 2. Taking the second derivative in Eq. (1) magnifies the noise that is inevitably present in the magnetization data. Increasing SF reduces the contribution of noise to the resulting FORC diagram, but it also removes measured data. Simple tests can be conducted to determine suitable smoothing factors for each sample. These tests determine when all of the noise that has been removed by smoothing; if SF is increased above this value, the effect will be only to remove data (Heslop and Muxworthy, 2005).

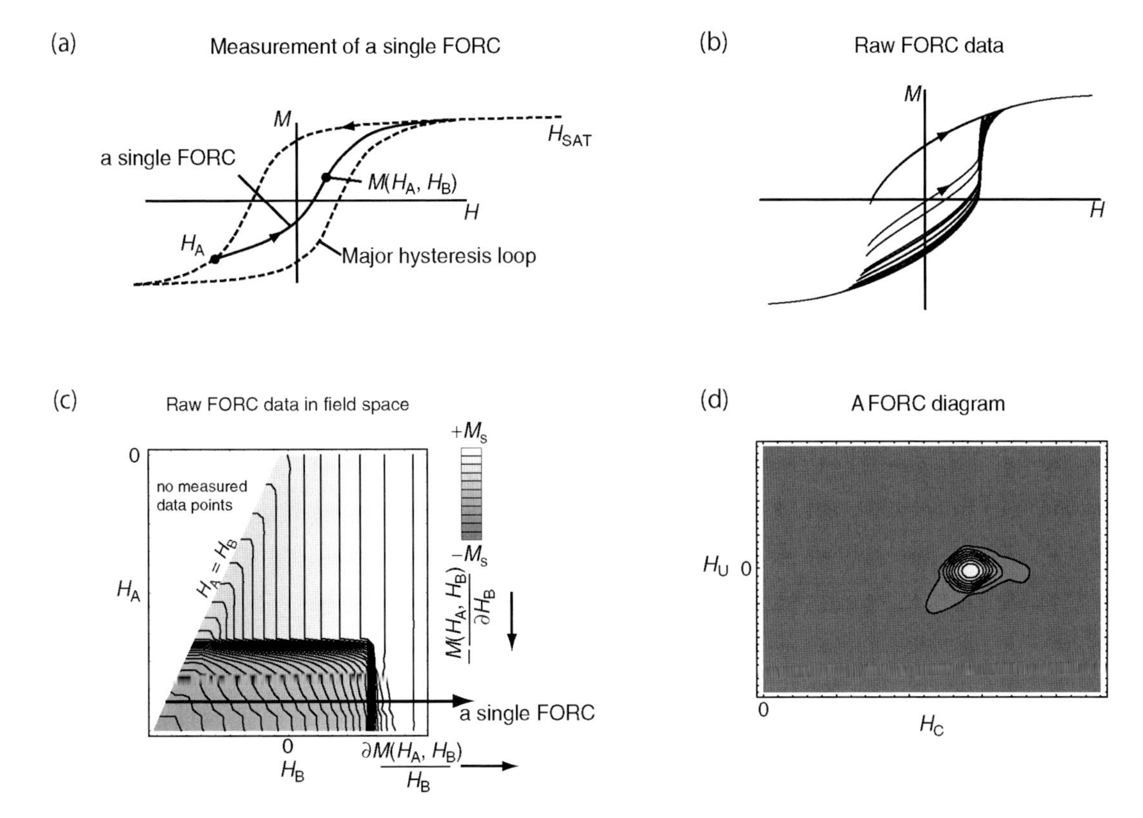

Figure F1 Illustration of how FORC diagrams are constructed. (a) After applying a field at positive saturation (H_{SAT}), the field is reversed to H_A and is then progressively increased at a range of H_B values in the direction of H_{SAT}. The magnetization is denoted by $M(H_A, H_B)$. The *dashed line* represents the major hysteresis loop and the *solid line* represents a single FORC. (b) A set of consecutive FORCs. (c) The $M(H_A, H_B)$ surface plotted in nonrotated field space (H_A, H_B). (d) The resulting FORC diagram for the data shown in (b) and (c) for SF = 4. The FORC data shown in (b) to (d) are for a numerical model of randomly orientated uniaxial SD grains.

A FORC diagram is well described by 100–140 raw FORCs. Recent improvements in system hardware have made it possible to measure 100 FORCs in approximately 1 h. The exact time depends on the field-step size, the aspect ratio of the FORC diagram, H_{SAT}, and the averaging time. The effects of these measurement variables depend as follows:

1. The chosen field-step size depends ultimately on the mineralogy of the sample under investigation. For a simple SD-dominated sample as in Figure F1d, the main peak of the FORC diagram is directly related to the coercive force H_c. It is necessary to choose suitable boundary values for H_U and H_C to fully depict the FORC distribution. The field-step size will be larger for FORC diagrams with larger boundary values, which then increases the measurement time. That is, it will take longer to measure a FORC diagram dominated by high coercivity minerals such as hematite compared to low coercivity minerals such as magnetite.
2. The aspect ratio of the FORC diagram also affects the measurement time. The overall measurement time will decrease as the ratio of the length of the axes, i.e., (H_U axis)/(H_C axis), increases. A square FORC diagram therefore is optimal in terms of measurement time (but the characteristics of the measured sample should dictate the selected axis lengths). Also, in order to rigorously calculate a FORC distribution, it is necessary to measure points outside the limits of the FORC diagram (depending on the value of SF). This will also (marginally) increase the measurement time.
3. H_{SAT} should be sufficient to magnetically saturate the sample. However, making H_{SAT} unnecessarily large will increase measurement time because of the finite time taken to sweep the magnet down from high values of H_{SAT}.
4. The averaging time is the amount of time taken to measure each data point, and is usually 0.1–0.25 s. Typically, the averaging time is set to 0.15 s. Increasing the averaging time increases the total measurement time, but improves signal to noise ratios. In systems that magnetically relax on the same time scale as the averaging time, the averaging time can be critical to the resulting FORC diagram. The averaging time should therefore always be stated in the accompanying text or figure caption.

FORC and Preisach diagrams

The FORC method originates in the phenomenological Preisach-Néel theory of hysteresis (Preisach, 1935; Néel, 1954). There has been much unnecessary confusion in the paleomagnetic community over the relationship between Preisach diagrams and FORC diagrams. Much of this confusion has arisen due to the lack of understanding as to what a Preisach diagram is. Put simply, there are many ways of measuring a Preisach diagram, of which the FORC method is one. FORC diagrams are essentially a new well-defined algorithm or method for rapidly generating a particular class of Preisach diagram.

Characteristic FORC diagrams

Noninteracting SD behavior

To help in the understanding of FORC diagrams, consider a FORC diagram for a SD grain, with uniaxial anisotropy, that is aligned in the direction of the applied field. Such an ideal SD grain will have a perfect square hysteresis loop (Figure F2a). When plotting raw FORC

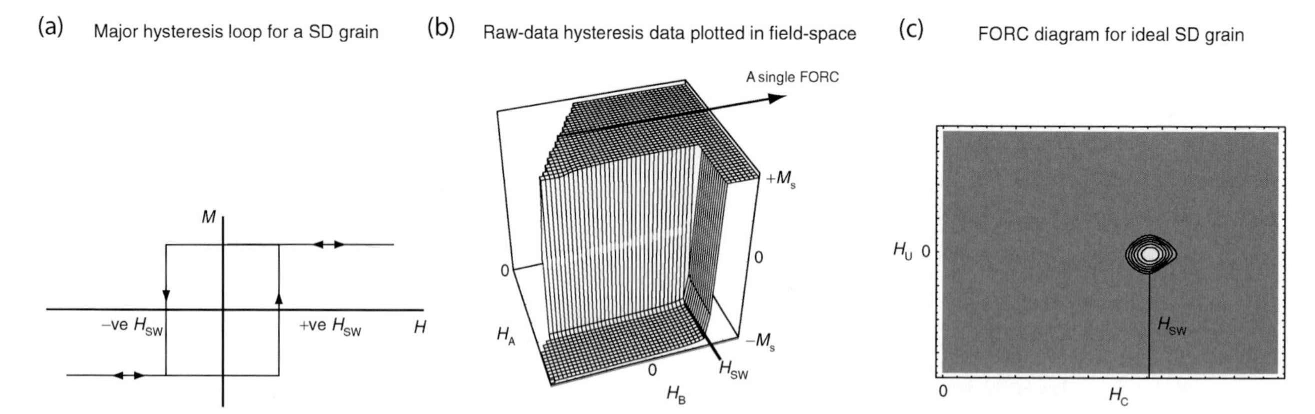

Figure F2 (a) Schematic square hysteresis loop for an ideal noninteracting SD particle, with uniaxial anisotropy, aligned along the direction of the applied field (H_{SW} is the switching field). (b) The raw hysteresis data for a series of FORCs for the grain shown in (a) plotted in nonrotated field space (H_A, H_B). (c) FORC diagram for data shown in (a) and (b). The FORC distribution for this grain lies at $H_c = H_{SW}$ and $H_u = 0$.

data for such a particle in nonrotated field space, $M(H_A, H_B)$ can take one of two values, i.e., $+M_s$ or $-M_s$ (Figure F2b). For $H_A > -H_{SW}$ (where H_{SW} is the switching field), $M(H_A, H_B)$ is $+M_s$, for $H_A < -H_{SW}$ and $H_B < +H_{SW}$, $M(H_A, H_B)$ is $-M_s$, and for $H_A < -H_{SW}$ and $H_B > +H_{SW}$, $M(H_A, H_B)$ is $+M_s$. On differentiating the surface with respect to H_A and H_B, only at $H_A = H_B = H_{SW}$ is $M(H_A,H_B)$ nonzero. That is, the FORC distribution should be a perfect delta function, normal to the FORC plane; however, due to the smoothing factor, $\rho(H_A, H_B)$ will have finite width (Figure F2c). An ideal noninteracting uniaxial SD particle with field applied along the easy axis of magnetization will therefore have a FORC distribution that lies at $H_c = H_{SW}$ and $H_U = 0$.

For assemblages of randomly orientated, noninteracting, identical, uniaxial SD grains, there are three main features in the FORC diagram (Figure F3): first, there is a central peak; second, the main peak displays an asymmetric "boomerang" shape; and third, there is a negative region near the bottom left-hand corner of the FORC diagram. The central peak is due to the switching of the magnetization at H_{SW}. From a more mathematical point of view, the positive peak is associated with the increase in $\partial M/\partial H_B$ with decreasing H_A as highlighted in Figure F3. The lower left-hand arm of the boomerang is related to FORCs near the relatively abrupt positive switching field. The right-hand arm of the boomerang is related to more subtle contours, which are due to the FORCs having different return paths, as highlighted in Figure F3. The shape of the return paths is controlled by the orientation of the grains with respect to the applied field, which results from the fact that the orientation controls the coercivity. Initially, return curve behavior is dominated by grains oriented ~45° to the field. As H_A decreases, grains with orientations closer to 90° and 0° will start contributing to the hysteresis curve, so each time H_A is decreased the return path includes grains with slightly differently shaped hysteresis loops. The right-hand arm is therefore a result of moving from the return path for a 45° assemblage in the first instance, into the return path for a randomly orientated assemblage. This effect is particularly enhanced for an assemblage of identical grains. The origin of the negative region is related to sections of the FORCs where $H_B < 0$ (Newell, 2005). As illustrated in Figure F3, $\partial M/\partial H_B$ decreases with decreasing H_A at H_{B1}, which gives rise to negative values for $\rho(H_A, H_B)$. The decrease in $\partial M/\partial H_B$ with H_A is not as pronounced for $H_B < 0$; consequently the negative region is significantly smaller than the large central peak near H_{B2}. If the SD grains have a distribution of switching fields, this causes the FORC diagram to stretch out in the H_C direction (Figure F4). A distribution of coercivities (e.g., Figure F4) is much more typical of natural samples than an assemblage of identical grains as depicted in Figure F3.

Interacting SD behavior

FORC distributions are highly sensitive to magnetostatic interactions. As a starting point to understanding the contribution of interactions to the FORC diagram, it is simplest to consider Néel's (1954) interpretation of the Preisach (1935) diagram. Néel (1954) showed that for interacting SD grains, H_C corresponds to the coercive force of eachSD loop in the absence of interactions and that H_U is the local interaction field H_I. It follows in the Preisach-Néel interpretation that $\rho(H_a, H_b)$ is the product of two distributions: the coercivity distribution $g(H_C)$ and the interaction field distribution $f(H_U)$. In a simple visualization, when the ideal SD grain in Figure F2a is affected by magnetic interactions, H_I will shift the switching field (Figure F5a). This asymmetry between the forward and backward switching fields gives rise to spreading in the H_U direction on a FORC diagram (Figure F5b). Micromagnetic models for FORC diagrams have shown that Néel's (1954) approximation is correct for moderately interacting systems, but that it breaks down when the grains are separated by less than a grain width (Muxworthy *et al.*, 2004). Highly interacting systems produce FORC diagrams that appear to be essentially the same as those for MD grains (see below and compare Figure F5 with Pike *et al.*, 2001a).

MD and pseudosingle domain grains

FORC diagrams for large MD grains (Pike *et al.*, 2001a) produce a series of contours that run parallel or nearly parallel to the H_U axis (Figure F6a). This spreading of the FORC distribution is essentially the same as for interacting SD grains; instead of the spreading being due to inter-grain magnetostatic interaction fields, it is due to internal demagnetizing fields. Pseudo-single domain (PSD) grains display behavior intermediate between true MD and true SD behavior (Figure F6b). That is, they display both the closed peak structures observed for SD grains and the more open contours that become increasingly parallel to the H_U axis with coarser grain sizes (Roberts *et al.*, 2000).

Superparamagnetic behavior

Superparamagnetic (SP) grains will only manifest themselves on the FORC diagram if their relaxation time is of the same order as the averaging time (Roberts *et al.*, 2000; Pike *et al.*, 2001b). If the grains have shorter relaxation times, i.e., $H_{SW} \rightarrow 0$, then $\rho(H_A, H_B) = 0$ at all values. SP grains with relaxation times of ~0.1–0.25 s have FORC diagrams similar to those for MD grains, i.e., $\rho(H_A,H_B)$ plots as a

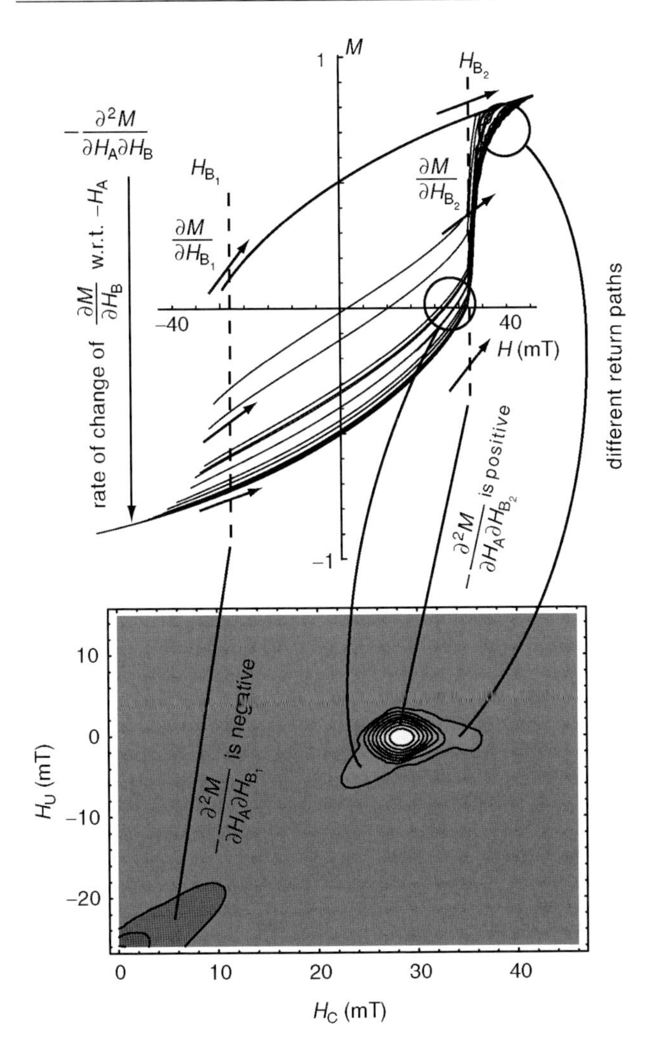

Figure F3 FORCs and FORC diagram for a numerical model of 1000 identical noninteracting SD grains with randomly distributed uniaxial anisotropy (modified after Muxworthy *et al.* (2004)). The origins of the negative and positive regions in the FORC diagram are highlighted (SF = 4). The negative region is due to a decrease in $\partial M/\partial H_B$ with decreasing H_A for negative values of H_B (H_{B_1}). The large positive peak is associated with the increase in $\partial M/\partial H_B$ with decreasing H_A for positive values of H_B, near the switching field (H_{B_2}). The different return paths give rise to the positive region of the FORC distribution to the right of the main peak as illustrated.

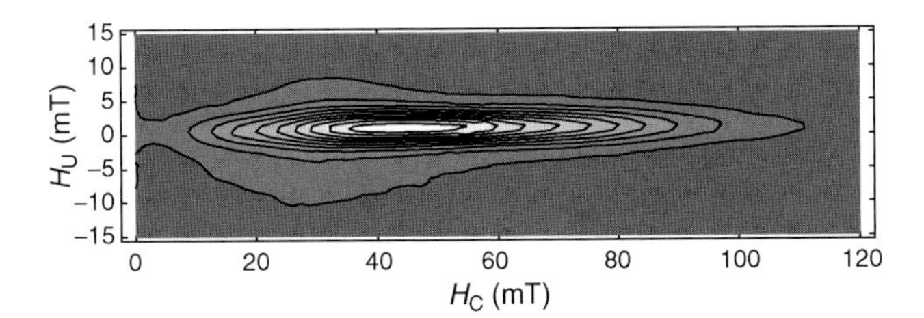

Figure F4 FORC diagram (SF = 3; averaging time = 0.15 s) for an assemblage of noninteracting ideal SD grains for a tuff sample from Yucca Mountain, Nevada (from Roberts *et al.* (2000)).

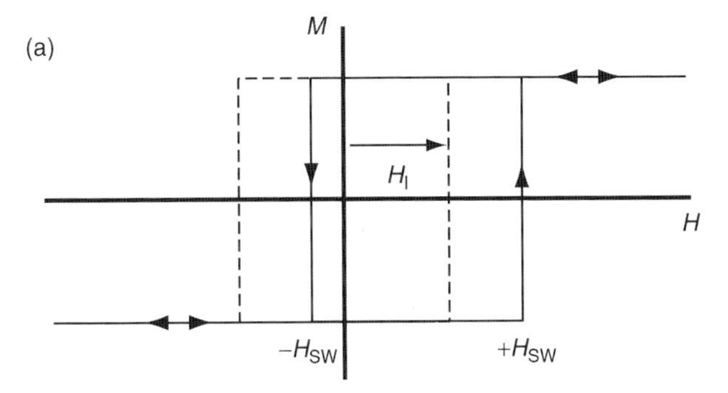

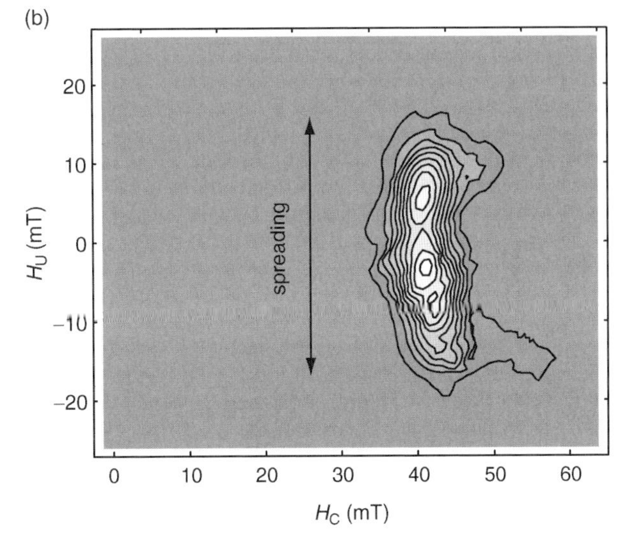

Figure F5 (a) Schematic diagram depicting the effect of a local interaction field H_I on a square hysteresis loop for an ideal SD particle, with uniaxial anisotropy, aligned along the direction of the applied field, as shown in Figure F2a. (b) Numerical simulation for an assemblage of 1000 evenly spaced, magnetite-like ideal SD grains with oriented uniaxial anisotropy. The distance between grain is 1.5 times the grains size.

series of contours running almost parallel to the H_U axis (Figure F6c). Although SP behavior may initially seem similar to MD behavior, the coercivity spectra of MD and SP samples and the shapes of the FORC distributions can be visually distinguished (Figure F6a,c). Cooling a sample by a few degrees can also easily enable identification of thermal relaxation. Due to the exponential nature of SD relaxation times, small temperature variations are sufficient to increase relaxation times, which effectively make the grains thermally stable on the time scale of a measurement. Upon cooling, the FORC diagram for a SP sample would then resemble that of a stable SD sample (cf. Figure F4). In contrast, FORC diagrams for MD samples would not display such variations with temperature.

Further features of FORC diagrams

The H_U axis

Due to the use of the local grid when calculating a FORC diagram for any value of SF, it is necessary to fit the polynomial surface using points outside the plotted FORC diagram. For the upper, lower, and right-hand bounds of the FORC diagram, this is readily achievable by simply measuring extra data points. For the H_U axis, however, this is not possible. This means that points on the H_U axis are not directly

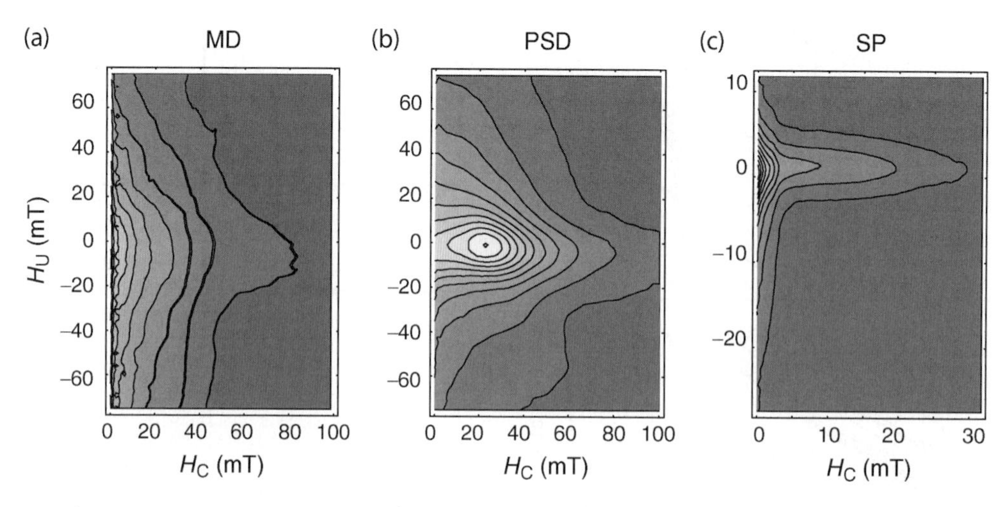

Figure F6 (a) FORC diagram (SF = 4; averaging time = 0.2 s) for a MD magnetite sample (mean grain size of 76 μm). (b) FORC diagram (SF = 3; averaging time = 0.15 s) for a PSD magnetite sample (mean grain size of 1.7 μm). (c) FORC diagram (SF = 3; averaging time = 0.2 s) for tuff sample CS014 from Yucca Mountain, Nevada, containing SP magnetite (from Roberts *et al.* (2000)).

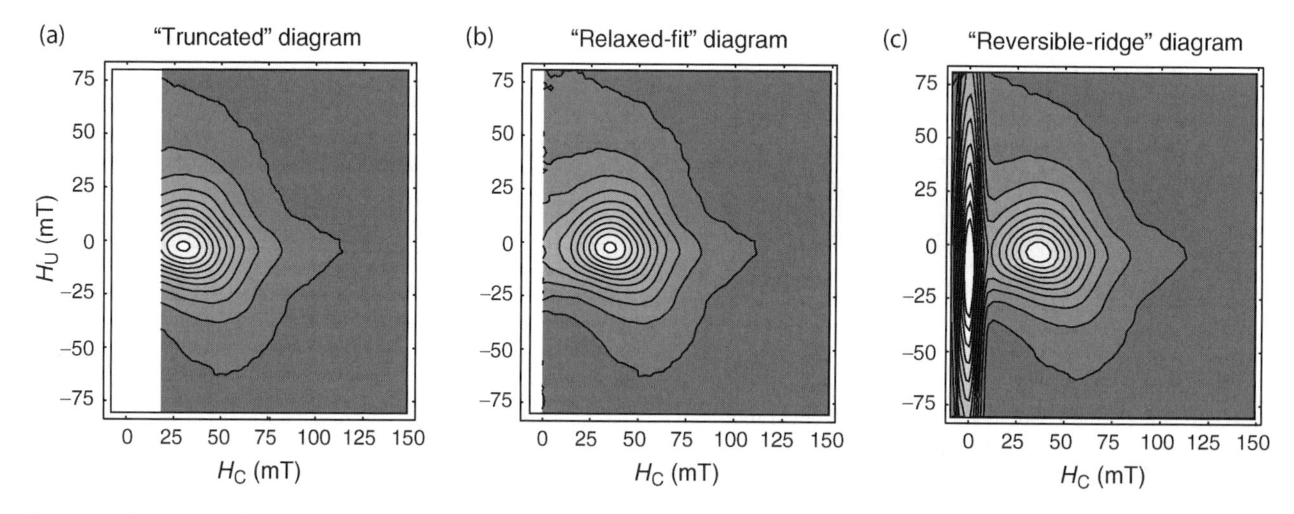

Figure F7 Illustration of three methods for fitting a FORC diagram. (a) A "truncated" FORC diagram, with no extrapolation of data onto the H_U axis; (b) a "relaxed-fit" FORC diagram with extrapolation onto the H_U axis; and (c) "reversible-ridge" fitting (following Pike, 2003). The sample is a PSD magnetite (mean grain size of 0.3 μm; SF = 4; averaging time = 0.1s).

measured, and there will be a gap of size 2SF+1 multiplied by the field spacing between $\rho(H_A, H_B)$ and the H_U axis on the FORC diagram. Three approaches have been made to deal with this problem.

1. Do not plot the FORC distribution for the region where the polynomial cannot be rigorously calculated. This results in a "truncated" FORC diagram (Figure F7a).
2. Relax the calculation of $\rho(H_A, H_B)$ back on to the H_U axis by reducing the smoothing near the H_U axis. Such "relaxed fit" diagrams (Figure F7b) will produce distortions, which can give misleading results if the field step size is too large, but they have the advantage that it is at least possible to observe low coercivity magnetization components. Such components are usually important in natural samples, so relaxing the fit is considered by some workers to be preferable to truncation.
3. Reversible-ridge method (Figure F7c) (Pike, 2003). This approach, which has been developed from Preisach theory, accommodates the reversible component of the magnetization data in addition to the irreversible part. With this approach, FORCs are extrapolated beyond $H_A < 0$, which enables rigorous calculation of $\rho(H_A, H_B)$ back on to the H_U axis. For SD assemblages this approach can be successful, however, for natural samples containing PSD or MD material the reversible ridge can often swamp the signal from the irreversible component of magnetization.

Asymmetry of FORC diagrams

The FORC method is a highly asymmetric method of measuring a Preisach diagram. The asymmetry originates from the measurement method in which each FORC starts from a saturated state. In interacting SD and MD systems, this means that there is bias in the magnetostatic interaction and internal demagnetizing fields in the direction of H_{SAT}. The only FORCs that do not contain a history of the initial H_{SAT} state are those that start from a sufficiently large negative H_A value that saturates the sample in the opposite direction. Asymmetry is therefore inherent to FORC diagrams.

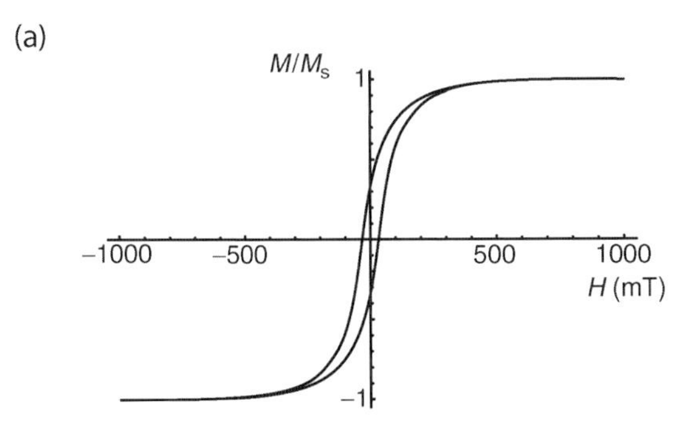

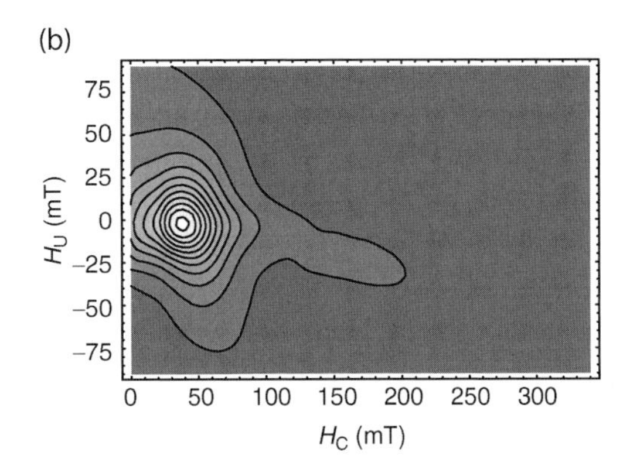

Figure F8 (a) Major hysteresis loop and (b) FORC diagram for a sample containing PSD magnetite and PSD hematite in a ratio of 81:19. On the FORC diagram, the hematite is readily identified, but it is not evident in the major hysteresis loop (SF = 3; averaging time = 0.15 s).

Applications of the FORC diagram

FORC diagrams provide much more detailed information about magnetic assemblages than standard hysteresis measurements. Thermal relaxation, variations in domain state, and magnetostatic interactions all produce characteristic and distinct manifestations on a FORC diagram. Here we illustrate four applications of FORC diagrams in paleomagnetic and environmental magnetic studies.

1. *Unraveling mixed magnetic assemblages*. A key problem in many paleomagnetic and environmental studies is identifying and isolating specific minerals and/or grain size distributions within a sample. This can be difficult in mixed samples if the signal from one mineral is magnetically swamped by that of another mineral. FORC diagrams have been shown to be successful at identifying magnetic signals due to minerals with low intrinsic magnetizations when they are present along with more strongly magnetic minerals (Roberts *et al*., 2000; Muxworthy *et al*., 2005). For example, consider a mixture of PSD magnetite and PSD hematite. The hysteresis loop in Figure F8a provides no evidence for the presence of hematite, whereas in the FORC diagram the presence of hematite is clearly visible by the high-coercivity component that extends to high values of H_C (Figure F8b). Although the FORC diagram is more successful than standard magnetic hysteresis measurements at identifying weaker magnetic phases in mixtures with more strongly magnetic phases, there are still concentration thresholds beyond which the magnetic signal of the strongly magnetic phase dominates and the weakly magnetic phase is no longer detectable in FORC diagrams (Muxworthy *et al*., 2005). In such cases, other magnetic methods may be more suitable than FORC analysis. For example, in mixtures of MD magnetite and hematite, low-temperature techniques can be more sensitive than FORC diagrams at identifying hematite in the presence of magnetite.
2. *Source discrimination*. Due to the detailed nature of FORC diagrams, they can be used to identify subtle differences between source materials that are not so readily observed using magnetic hysteresis data. For example, in a study of a loess/paleosol sequence from Moravia, two different source materials could be clearly distinguished and identified using FORC diagrams (van Oorschot *et al*., 2002). There are many other apparent uses of FORC diagrams in environmental magnetic applications.
3. *Assessing leaching*. Sequential leaching techniques are commonly used by soil chemists to remove iron oxides from samples to facilitate clay mineral analysis. FORC analysis has been shown to help in elucidation of the leaching process where magnetic hysteresis alone was less diagnostic (Roberts *et al*., 2000).
4. *Pre-selection for paleointensity studies*. In absolute paleointensity studies, SD assemblages should provide reliable results, while samples with strong magnetostatic interactions and/or MD grains should yield unreliable results. Domain state and magnetostatic interactions are both readily identifiable on a FORC diagram, which should make them useful for sample screening and selection for absolute paleointensity studies. Likewise, strict criteria apply to the mineralogy, domain state, and concentration of magnetic particles in sedimentary sequences in terms of their suitability for relative paleointensity studies. FORC diagrams therefore also have potential for screening sediments for relative paleointensity studies.

Adrian R. Muxworthy and Andrew P. Roberts

Bibliography

Day, R., Fuller, M., and Schmidt, V.A., 1977. Hysteresis properties of titanomagnetites: grain-size and compositional dependence. *Physics of the Earth and Planetary Interiors*, **13**: 260–267.

Heslop, D., and Muxworthy, A.R., 2005. Aspects of calculating first-order-reversal-curve distributions. *Journal of Magnetism and Magnetic Materials*, **288**: 155–167.

Muxworthy, A.R., Heslop, D., and Williams, W., 2004. Influence of magnetostatic interactions on first-order-reversal-curve (FORC) diagrams: a micromagnetic approach. *Geophysical Journal International*, **158**: 888–897.

Muxworthy, A.R., King, J.G., and Heslop, D., 2005. Assessing the ability of first-order reversal curve (FORC) diagrams to unravel complex magnetic signals. *Journal of Geophysical Research*, **110**: B01105, doi:10.1029/2004JB003195.

Muxworthy, A.R., Williams, W., and Virdee, D., 2003. Effect of magnetostatic interactions on the hysteresis parameters of single-domain and pseudo-single domain grains. *Journal of Geophysical Research*, **108**(B11): 2517, doi:10.1029/2003JB002588.

Néel, L., 1954. Remarques sur la théorie des propriétés magnétiques des substances dures. *Applied Scientific Research B*, **4**: 13–24.

Newell, A.J., 2005. A high-precision model of first-order reversal curve (FORC) functions for single-domain ferromagnets with uniaxial anisotropy. *Geochemistry, Geophysics, Geosystems*, **6**: Q05010, doi:10.1029/2004GC000877.

Pike, C.R., 2003. First-order reversal-curve diagrams and reversible magnetization. *Physics Review B*, 104424.

Pike, C.R., Roberts, A.P., and Verosub, K.L., 1999. Characterizing interactions in fine magnetic particle systems using first order reversal curves. *Journal of Applied Physics*, **85**: 6660–6667.

Pike, C.R., Roberts, A.P., Dekkers, M.J., and Verosub, K.L., 2001a. An investigation of multi-domain hysteresis mechanisms using FORC diagrams, *Physics of the Earth and Planetary Interiors*, **126**: 13–28.

Pike, C.R., Roberts, A.P., and Verosub, K.L., 2001b. First-order reversal curve diagrams and thermal relaxation effects in magnetic particles, *Geophysical Journal International*, **145**: 721–730.

Preisach, F., 1935. Über die magnetische Nachwirkung. *Zeitschrift für Physik*, **94**: 277–302.

Roberts, A.P., Pike, C.R., and Verosub, K.L., 2000. First-order reversal curve diagrams: a new tool for characterizing the magnetic properties of natural samples. *Journal of Geophysical Research*, **105**: 28461–28475.

van Oorschot, I.H.M., Dekkers, M.J., and Havlicek, P., 2002. Selective dissolution of magnetic iron oxides with the acid-ammonium-oxalate/ferrous-iron extraction technique—II. Natural loess and palaeosol samples. *Geophysical Journal International*, **149**: 106–117.

Cross-references

Rock Magnetism, Hysteresis Measurements

FISHER STATISTICS

When describing the dispersion of paleomagnetic directions about some mean direction it is a standard practice within paleomagnetism to employ Fisher (1953) statistics. The theory of Fisher statistics assumes a mean direction, defined by a Cartesian unit vector $\hat{\mathbf{x}}_\mu$, and data, each defined by Cartesian unit vectors $\hat{\mathbf{x}}$. The off-axis angle θ between the mean direction and a datum is defined by

$$\cos\theta = \hat{\mathbf{x}} \cdot \hat{\mathbf{x}}_\mu. \qquad \text{(Eq. 1)}$$

With this, then, the Fisher distribution gives the probability that a particular directional datum $\hat{\mathbf{x}}$ falls between θ and $\theta + \mathrm{d}\theta$ as

$$P_f(\theta|\kappa) = \int_\theta^{\theta+\mathrm{d}\theta} p_f(\theta'|\kappa)\mathrm{d}\theta', \qquad \text{(Eq. 2)}$$

where the probability-density function is

$$p_f(\theta|\kappa) = \frac{\kappa}{2\sinh\kappa}\sin\theta\,\exp(\kappa\cos\theta), \qquad \text{(Eq. 3)}$$

and where κ is a parameter that measures the directional dispersion of the data about the mean direction. If one considers the dispersion of directions over the unit sphere, then the probability of particular a datum falling onto a unit differential area

$$\mathrm{d}A = \sin\theta\mathrm{d}\theta\mathrm{d}\phi, \qquad \text{(Eq. 4)}$$

where ϕ is the azimuthal angle symmetrically-distributed about $\hat{x}_\mu$, is just

$$P_f(A|\kappa) = \int_\phi^{\phi+\mathrm{d}\phi}\int_\theta^{\theta+\mathrm{d}\theta} p_f(A'|\kappa)\sin\theta'\mathrm{d}\theta'\mathrm{d}\phi', \qquad \text{(Eq. 5)}$$

where the corresponding density function is

$$p_f(A|\kappa) = \frac{\kappa}{4\pi\sinh\kappa}\exp(\kappa\cos\theta). \qquad \text{(Eq. 6)}$$

As with all probability-density functions, that a particular datum is realized is certain, we are, after all, describing data that exist, and therefore

$$\int_0^\pi p_f(\theta'|\kappa)\mathrm{d}\theta' = \int_0^{2\pi}\int_0^\pi p_f(A'|\kappa)\sin\theta'\mathrm{d}\theta'\mathrm{d}\phi' = 1. \qquad \text{(Eq. 7)}$$

In Figure F9 we show examples of the Fisher density function.

Application

Let us consider now a set of paleomagnetic data, with the ith paleomagnetic direction $\hat{\mathbf{x}}_i$ defined by an inclination-declination pair (I_i, D_i), such that the Cartesian components are

$$x_i = \cos I_i \cos D_i, \quad y_i = \cos I_i \sin D_i, \quad z_i = \sin I_i. \qquad \text{(Eq. 8)}$$

The mean unit direction $\bar{\mathbf{x}}$ given by N data is

$$\bar{x} = \frac{1}{R}\sum_{i=1}^N x_i, \quad \bar{y} = \frac{1}{R}\sum_{i=1}^N y_i, \quad \bar{z} = \frac{1}{R}\sum_{i=1}^N z_i, \qquad \text{(Eq. 9)}$$

where

$$R^2 = \left(\sum_{i=1}^N x_i\right)^2 + \left(\sum_{i=1}^N y_i\right)^2 + \left(\sum_{i=1}^N z_i\right)^2. \qquad \text{(Eq. 10)}$$

The mean vector $\bar{\mathbf{x}}$ is an estimate of the true mean $\hat{\mathbf{x}}_\mu$ corresponding to the underlying distribution p_f, and with this estimated mean direction we can measure the off-axis angle θ_i of each datum, where

$$\cos\theta_i = \bar{\mathbf{x}} \cdot \hat{\mathbf{x}}_i. \qquad \text{(Eq. 11)}$$

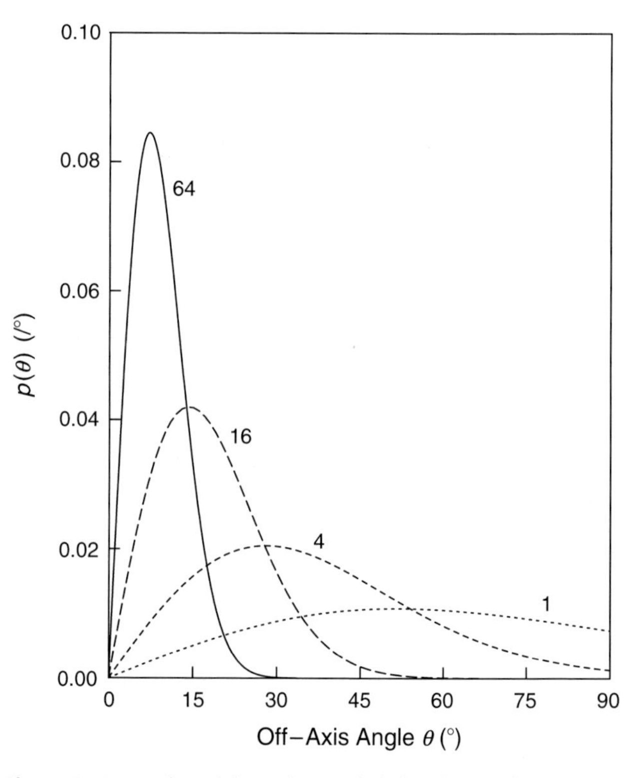

Figure F9 Examples of the Fisher probability-density function $p_f(\theta)$ for a variety of κ dispersion parameters: 1, 4, 16, 64. Note that as κ is increased the dispersion decreases.

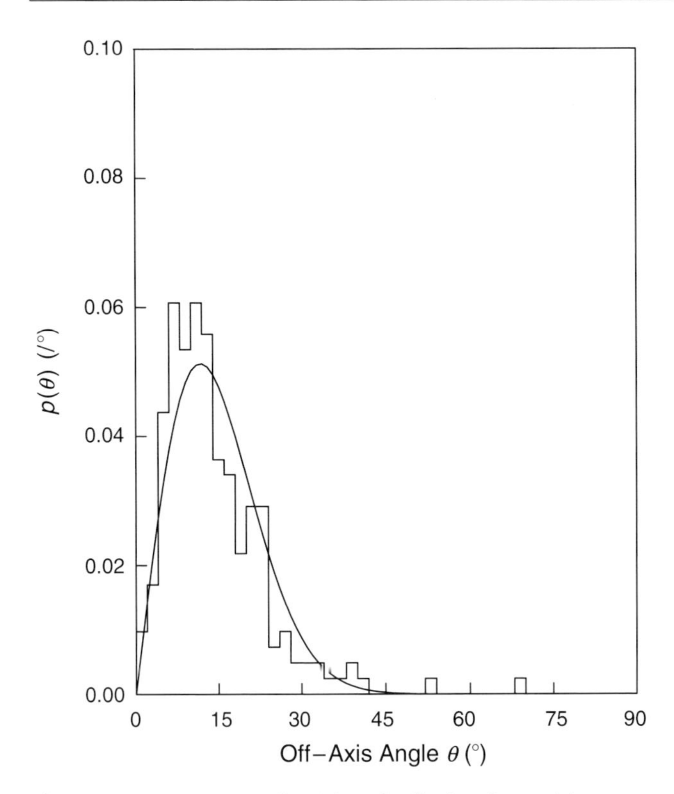

Figure F10 Comparison of a Fisher distribution fit to a histogram of Réunion data recording secular variation during the Brunhes.

The precision parameter κ is given approximately by (McFadden, 1980):

$$\kappa = \frac{N-1}{N-R}, \qquad \text{(Eq. 12)}$$

As $R \to N$ the precision parameter κ increases, and the distribution of directions becomes more tightly clustered about the mean direction.

As an example of a fit of the Fisher distribution to real data, in Figure F10 we show a histogram of off-axis angles corresponding to a compilation of Brunhes-age paleomagnetic data collected at or near the island of Réunion (Love and Constable, 2003). Note that the Fisher distribution fitted using the procedure outlined here captures most of the actual distribution of the data, but that the fit is also not perfect. Indeed, although the Fisher distribution is often used in paleomagnetism, only rarely does paleaomagnetic data actually show a strict Fisher distribution. Both the secular variation of the geomagnetic field and the process by which rocks obtain their paleomagnetic signatures are extremely complicated, and it is, therefore, not too surprising that there is some misfit to a Fisher distribution. Although the Fisher distribution does arise from first principles in the context of the Langevin theory of paramagnetism, more generally, there often is very little reason to expect a set of paleomagnetic data to exhibit perfect Fisher statistics. The real utility of the Fisher distribution comes as a benchmark for comparison, the deviation from its relatively simple mathematical form that is of interest. Further review material on Fisher statistics can be found in the books by Butler (1992) and Tauxe (1998).

Jeffrey J. Love

Bibliography

Butler, R.F., 1992. *Paleomagnetism*, Cambridge, MA: Blackwell Scientific Publications.

Fisher, R.A., 1953. Dispersion on a sphere. *Proceedings of the Royal Society of London, Series A*, **217**: 295–305.

Love, J.J., and Constable, C.G., 2003. Gaussian statistics for palaeomagnetic vectors. *Geophysical Journal International*, **152**: 515–565.

McFadden, P.L., 1980. The best estimate of Fisher's precision parameter. *Geophysical Journal of the Royal Astronomical Society*, **60**: 397–407.

Tauxe, L., 1998. *Paleomagnetic Principles and Practice.* Dordrecht, The Netherlands: Kluwer Academic Publishers.

Cross-references

Magnetization, Natural Remanent (NRM)
Paleomagnetic Secular Variation
Statistical Methods for Paleovector Analysis

FLEMING, JOHN ADAM (1877–1956)

Few individuals influenced geophysics in the 20th century more profoundly than John A. Fleming (Figure F11): geophysicist, engineer, scientific organizer, and administrator. He devoted his life to promoting the study of geomagnetism and building its professional organizations, and played a leading role in organizing magnetic and electric surveys of the Earth during the first half of the 20th century. Yet, as Sydney *Chapman* (*q.v.*) wrote of him, "He was so self-effacing that only those who knew and worked with him can properly assess... what he did for geophysics in USA and in the world at large" (Chapman, 1957).

Fleming was born in Cincinnati, Ohio on January 28, 1877. Educated as a civil engineer at the University of Cincinnati (1895–1899), he worked briefly in construction after receiving his B.S. degree, then joined the U.S. Coast and Geodetic Survey's Division of Terrestrial Magnetism. He advanced steadily at the Survey from 1899 to 1903 and was involved with the planning and construction of magnetic observatories in Alaska, Hawaii, and Maryland. In 1904, he was appointed "Chief Magnetician" at the newly established *Department of Terrestrial Magnetism* (DTM, *q.v.*) of the Carnegie Institution of Washington. He would be associated with the Institution for the next 50 years.

Figure F11 John Adam Fleming (*photograph:* Carnegie Institution, Department of Terrestrial Magnetism).

Working under the direction of DTM's founder, *Louis A. Bauer* (*q.v.*) (with whom he had previously worked at the Coast and Geodetic Survey), Fleming proved himself an able manager of DTM's world magnetic survey—a far-flung enterprise of overland magnetic expeditions and ocean research vessels. He developed new and improved magnetic instruments for use on land and at sea, and made important contributions to establishing and maintaining the world's magnetic standards. He designed the department's geophysical observatories at Huancayo, Peru and Watheroo, Australia. Appointed acting director in 1929, he ran the department for several years during Bauer's incapacitation and after his death. In 1935, bolstered by honorary degrees from the University of Cincinnati (D.Sc., 1933) and Dartmouth College (D.Sc., 1934), Fleming was finally named director of DTM.

Fleming substantially broadened DTM's focus to include studies of the ionosphere, cosmic rays, Earth currents, physical oceanography, marine biology, and atomic physics. He was instrumental in initiating in 1935 the highly productive Washington Conferences on Theoretical Physics. In 1941, in conjunction with the University of Alaska, he established a cooperative research facility for arctic geophysics (College Observatory) in Fairbanks—then later helped lobby successfully for creation of the Geophysical Institute there. During World War II, Fleming oversaw all of DTM's defense contracts, ranging from radio propagation predictions for the armed forces to the development of magnetic compasses and odographs.

E.H. Vestine observed that Fleming possessed "a most unusual gift in judging the future possibilities and capabilities of the young men he brought to DTM" (Vestine, 1956, p. 532). Among these were staff members Merle Tuve, Lloyd Berkner, and Scott Forbush; research associates Julius Bartels and Sydney Chapman; and the department's first postdoctoral fellows, among them James Van Allen.

In addition to authoring more than 130 papers, primarily on magnetic surveys and secular variation, Fleming was a dedicated editor. (A bibliography of Fleming's work was published by Tuve in 1967.) He edited and published the journal *Terrestrial Magnetism and Atmospheric Electricity*—the forerunner of the *Journal of Geophysical Research*—for two decades (1928–1948). He contributed to, and edited, an encyclopedic volume on *Terrestrial Magnetism and Electricity* (1939) in the National Research Council's landmark "Physics of the Earth" series. But Fleming's most enduring legacy was probably as an organizer and administrator of geophysics. In the 1890s Bauer had begun a crusade to professionalize the study of geomagnetism and to set it on an equal footing with established disciplines such as astronomy and meteorology (Good, 1994). Fleming shared Bauer's zeal, and as General Secretary of the American Geophysical Union (AGU) from 1925 to 1947 he devoted himself tirelessly to fostering the community of geophysicists. In the days before AGU had paid administrative staff, Fleming contributed long hours in correspondence, organizing meetings, and editing thousands of pages of the *AGU Transactions*, with only a clerical assistant or two.

In 1941 he was awarded the William Bowie Medal, AGU's highest honor, and when he stepped down as general secretary he was elected "honorary president" (a title created for him) for life. Other noteworthy leadership positions of Fleming's included the presidency of the Association of Terrestrial Magnetism and Electricity (now the *International Association of Geomagnetism and Aeronomy*, *q.v.*) of the International Union of Geodesy and Geophysics (1930–1948), and of the International Council of Scientific Unions (1946–1949). He was a member of the National Academy of Sciences as well as numerous foreign academies and learned societies. In 1945 the Physical Society (London) honored him with its Charles Chree Medal and Prize.

In 1946 Fleming retired from DTM, but continued to serve the Carnegie Institution as Adviser in International Scientific Relations until 1954. His last major project consisted of chronicling the origin and development of the AGU in characteristically detailed and quantitative fashion.

Fleming died in San Mateo, California on July 29, 1956. The AGU posthumously established the John Adam Fleming Medal in his honor. The medal recognizes original research and technical leadership in geomagnetism, atmospheric electricity, aeronomy, space physics, and related sciences. Fleming's colleagues considered him modest, supportive, and scientifically broad-minded, but also demanding, strict, and strong-willed —"a man of enormous energy with a most notable capacity for getting things done. His work was his life and his hobby, but above all he enjoyed companionship and discussions with his many friends and colleagues" (Vestine, 1956, p. 532).

Shaun J. Hardy

Bibliography

Carnegie Institution of Washington, Department of Terrestrial Magnetism. Archives Files, "John A. Fleming" folder. Personal communication, Sydney Chapman to Merle A. Tuve, March 9, 1957.

Good, G.A., 1994. Vision of a Global Physics. In Good, G.A. (ed.) *The Earth, the Heavens, and the Carnegie Institution of Washington.* Washington, DC: American Geophysical Union, pp. 29–36.

Tuve, M.A., 1967. John Adam Fleming, 1877–1956. *Biographical Memoirs, National Academy of Sciences*, **39**: 103–140.

Vestine, E.H., 1956. John Adam Fleming. *Transactions, American Geophysical Union*, **37**: 531–533.

Cross-references

Bartels, Julius
Bauer, Louis Agricola (1865–1932)
Carnegie Institution of Washington, Department of Terrestrial Magnetism
Carnegie, Research Vessel
Chapman, Sydney (1888–1970)
Geomagnetism, History of
IAGA, International Association of Geomagnetism and Aeronomy
Instrumentation, History of

FLUID DYNAMICS EXPERIMENTS

Over the last half century, fluid dynamics experiments have provided researchers with a means of simulating processes relevant to planetary cores. These experiments utilize actual fluids, such as water and liquid metals, to replicate various aspects of core flows, including rotating convection, rotating magnetoconvection, precession, and dynamo generation (Nataf, 2003). In comparison to numerical simulations, laboratory experiments can presently simulate conditions that are closer to those in planetary cores. In addition, experiments can contain small-scale flow structures that are not resolved in numerical models. Thus, experimental studies supply results that are essential for improving our understanding of planetary core dynamics and dynamo action.

In a planetary dynamo, kinetic energy of iron-rich core fluid motions is converted into large-scale magnetic field energy. On the basis of the assumption that buoyancy driven motions dominate precessionally driven motions in Earth's core (see *Precession and core dynamics*), the majority of investigations have focused on understanding how core convection can generate dynamo action. Thus, most laboratory research has recently followed two main tacks: buoyancy-driven convection experiments and mechanically driven dynamo experiments. These approaches provide distinct but complementary ways of studying core processes. The mechanically driven dynamo experiments impose a strong velocity field in order to produce dynamo-generated magnetic fields (see *Dynamos, experimental*). These mechanically forced flows, which are typically driven by pumps or rapidly rotating impellers, may not be realistic analogs to flows in planetary cores. Yet in successful experiments where dynamo generation occurs, the results provide an excellent opportunity to study flows that generate dynamo action and the process by which dynamo-generated magnetic fields equilibrate (see *Dynamo, Gailitis*).

In contrast, buoyancy-driven experiments study the fundamental dynamics and characteristics of core-type flows. Because the laboratory convective flows are far less energetic than flows in the mechanically-driven experiments, the magnetic Reynolds number, which is the ratio of magnetic induction to diffusion, is small. Thus, convection-driven induction processes are weak, and dynamo action cannot be achieved in such experiments. However, these experiments are critical for simulating the details of planetary *core convection* (*q.v.*).

In addition to buoyancy forces, core flows are also strongly affected by Coriolis forces, due to planetary rotation, and Lorentz forces, due to planetary magnetic fields. The Elsasser number, Λ, is the ratio of Lorentz and Coriolis forces. In cases where $\Lambda \ll 1$, Coriolis forces dominate the Lorentz forces and the rotationally constrained fluid motions tend to become two-dimensional along the direction of the rotation axis (see *Proudman-Taylor theorem*). From planetary magnetic field observations, including that of the Earth, Λ is estimated to be $O(1)$ (see *Dynamos, planetary and satellite*). However, numerical simulations of magnetoconvection and dynamo action have found that fluid motions still tend to remain nearly two-dimensional and aligned along the rotation axis for even relatively large Elsasser numbers, i.e., $\Lambda \sim 10$ (see *Magnetoconvection*). Thus, both rotating convection experiments in which Lorentz forces are not present, and rotating magnetoconvection experiments in the $\Lambda \sim 1$ regime can yield critical insights into the dynamics of planetary cores.

Rotating convection experiments

Rotating convection experiments study fluid motions within a spherical shell that mimics the geometry of the Earth's outer core. In the majority of these devices the outer spherical shell, which simulates the Earth's core-mantle boundary, is made of a transparent plastic such as polycarbonate and has a typical radius measuring between 5 and 15 cm. This allows for visualization studies of the convection pattern in transparent fluids like water. By rapidly rotating the shell, a centrifugal acceleration is produced that acts as an effective gravity within the fluid. This centrifugal gravity varies linearly with cylindrical radius, similarly to gravity within a planetary core. However, it points outward, opposite to the direction of a planet's gravity. In order to account for this reversal in gravity's direction, the inner spherical shell, which represents the Earth's inner core boundary, must be held at a lower temperature than the outer shell for thermal convection to occur. Because the direction of gravity and the sign of the temperature difference across the shell are both reversed, the buoyancy forces in the experiment act in the same way as in a planetary core.

The first such experiments studied flows near the onset of *nonmagnetic rotating convection* (*q.v.*) in spherical shells (Busse and Carrigan, 1976). These studies showed that convective motions first manifest as thin cylindrical columns aligned parallel to the axis of rotation (Figure F12a). The two-dimensionality of the flow demonstrates the applicability of the Proudman-Taylor theorem. Further experiments studied stronger convection occurring at higher rotation rates (Cardin and Olson, 1994). In such studies, it was found that strong thermal or compositional buoyancy forces drive fluid motions on columnar structures that fill the entire volume of the spherical shell (Figure F12b). It was also determined that the typical width of the convection columns decreases with the shell's rate of rotation. Sumita and Olson (2000) carried out rotating convection experiments using even stronger thermal buoyancy forces, and found that the convective motions occurred on thin two-dimensional sheets that stretch between the inner and outer spherical boundaries (Figure F12c). Retrograde azimuthal zonal flows were also found to develop in the vast majority of these studies (Aubert *et al.*, 2001; Aurnou *et al.*, 2003; Shew and Lathrop, 2005). *Numerical simulations* (*q.v.*) verify that

(a)

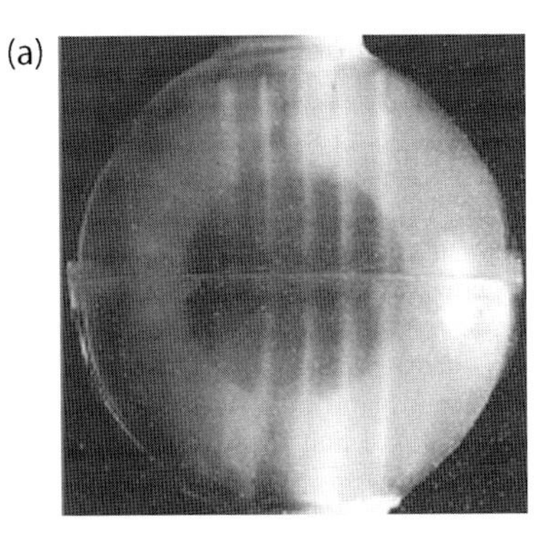

(b)

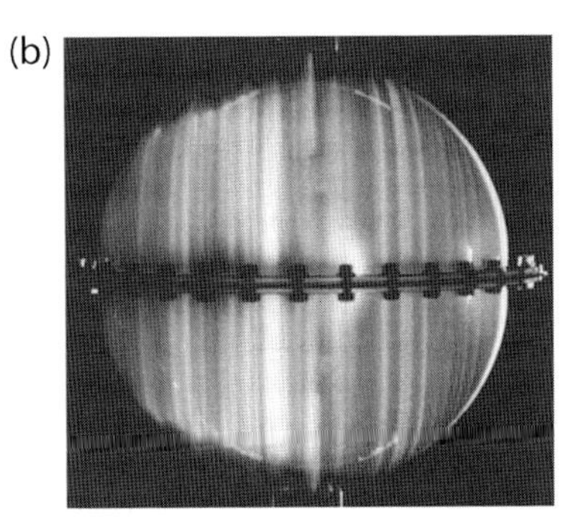

(c)

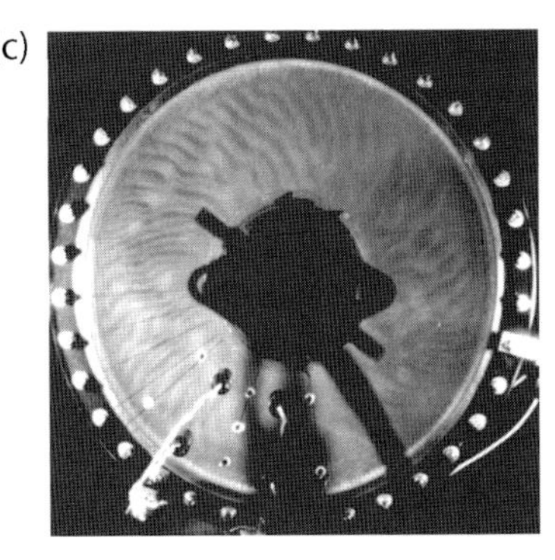

Figure F12 (a) Sideview of the onset of rotating spherical shell convection. Adapted from Busse and Carrigan (1976). (b) Sideview of moderately supercritical rotating spherical shell convection showing convection columns filling the entire shell volume. Adapted from Cardin and Olson (1992). (c) Planview of fully-developed rotating convection in a hemispherical shell showing spiralling sheets that remain aligned with the rotation axis. Adapted from Sumita and Olson (2000).

the well-organized columnar motions and the large-scale zonal flows that are inherent in rotating convection are important components in generating dynamo action.

Recent rotating convection experiments have employed ultrasonic Doppler velocimetry techniques to make precise, quantitative flow velocity measurements (Aubert *et al.*, 2001). They have produced velocity scaling laws as a function of thermal forcing and shell rotation rate, using water and low viscosity liquid gallium as working fluids. Using the same experimental device, Brito *et al.* (2004) have been able to measure changes in water's effective viscosity in a set of turbulent rotating convection experiments. These studies demonstrate how improved experimental diagnostics allow characterizations of the detailed flow properties that are crucial for parametrization and better understanding of core processes.

Rotating magnetohydrodynamics experiments

Rotating magnetoconvection experiments more accurately simulate planetary core physics than rotating convection experiments because they include Lorentz forces, in addition to buoyancy and Coriolis forces. These experiments use liquid metals, such as liquid sodium, liquid gallium, or mercury, as their working fluids in order to simulate the physical properties of the metallic alloys in planetary cores. Until recently, diagnostic techniques were unable to determine flow fields in the opaque liquid metals. Indeed, the paucity of meaningful diagnostics has led to few experimental studies of flows in liquid metals.

Rotating magnetoconvection experiments were first made by Y. Nakagawa (1957, 1958). His experiments studied the onset of thermal convection in a cylindrical, plane layer of liquid mercury subject to a vertical magnetic field and rotation around a vertical rotation axis. These experiments showed that for large rotation rates and magnetic field strengths, thermal convection is facilitated in the $\Lambda \sim 1$ regime, in agreement with the linear stability analysis of S. Chandrasekhar (1961). In experiments with the top lid removed from the convection tank, Nakagawa (1957; 1958) determined the characteristic length scale of the convective motions by observing the motions of sand particles on the surface of the mercury. These observations showed an increase in convective length scale in the $\Lambda \sim 1$ regime, which roughly agree with the predictions of linear stability analysis.

More recently, Aurnou and Olson (2001) studied supercritical heat transfer in rotating magnetoconvection experiments in the $\Lambda \sim 1$ regime but with weaker magnetic fields and rotation rates than in Nakagawa's experiments. For the less extreme parameter values that they investigated, thermal turbulence marked the onset of convection in all their experiments, while no increase in convective heat transfer was measured in the $\Lambda \sim 1$ regime. The difference between their results and Nakagawa's can be explained by the fact that convection is only facilitated in the $\Lambda \sim 1$ regime for sufficiently large magnetic field strengths and rotation rates.

The next generation of rotating *magnetoconvection* and *magnetohydrodynamics* (*q.v.*) experiments will employ the acoustic Doppler techniques that have proven so successful in the studies of Aubert *et al.* (2001) and Brito *et al.* (2004). Such experiments will allow investigators to determine flow patterns, scaling laws and small-scale flow characteristics for convection in liquid metals subject to strong rotational and magnetic forces, as occurs within planetary cores. With experimental information regarding the behavior of small-scale flows (see *Core turbulence*), it may soon be possible for laboratory studies to provide detailed parametrizations of subgrid-scale motions for future numerical models of planetary dynamos.

Jonathan M. Aurnou

Bibliography

Aubert, J., Brito, D., Nataf, H.-C., Cardin, P., and Masson, J.-P., 2001. Scaling relationships for finite amplitude convection in a rapidly rotating sphere, from experiments with water and gallium. *Physics of the Earth and Planetary Interiors*, **128**: 51–74.

Aurnou, J.M., and Olson, P.L., 2001. Experiments on Rayleigh-Benard convection, magnetoconvection, and rotating magnetoconvection in liquid gallium. *Journal of Fluid Mechanics*, **430**: 283–307.

Aurnou, J.M., Andreadis, S., Zhu, L., and Olson, P.L., 2003. Experiments on convection in Earth's core tangent cylinder. *Earth and Planetary Science Letters*, **212**: 119–134.

Brito, D., Aurnou, J., and Cardin, P., 2004. Turbulent viscosity measurements relevant to planetary core-mantle dynamics. *Physics of the Earth and Planetary Interiors*, **141**: 3–8.

Busse, F.H., and Carrigan, C.R., 1976. Laboratory simulation of thermal convection in rotating planets and stars. *Science*, **191**: 81–83.

Cardin, P., and Olson, P., 1994. Chaotic thermal convection in a rapidly rotationg spherical shell: Consequences for flow in the outer core. *Physics of the Earth and Planetary Interiors*, **82**: 235–259.

Chandrasekhar, S., 1961. *Hydrodynamic and Hydromagnetic Stability.* Oxford: Oxford University Press.

Nakagawa, Y., 1957. Experiments on the instability of a layer of mercury heated from below and subject to the simultaneous action of a magnetic field and rotation. *Proceedings of the Royal Society of London, Series A*, **242**: 81–88.

Nakagawa, Y., 1958. Experiments on the instability of a layer of mercury heated from below and subject to the simultaneous action of a magnetic field and rotation. II. *Proceedings of the Royal Society of London, Series A*, **249**: 138–145.

Nataf, H.-C., 2003. Dynamo and convection experiments. In Jones, C.A., Soward, A.M., and Zhang, K. (eds.), *Earth's Core and Lower Mantle.* London: Taylor and Francis, pp. 153–179.

Shew and Lathrop, 2005. A liquid sodium model of convection in Earth's outer core. *Physics of the Earth and Planetary Interiors*, **153**: 136–149.

Sumita, I., and Olson, P., 2000. Laboratory experiments on high Rayleigh number thermal convection in a rapidly rotating hemispherical shell. *Physics of the Earth and Planetary Interiors*, **117**: 153–170.

Cross-references

Convection, Nonmagnetic Rotating
Core Convection
Core Turbulence
Dynamo, Gailitis
Dynamos, Experimental
Dynamos, Planetary and Satellite
Geodynamo, Numerical Simulations
Magnetoconvection
Magnetohydrodynamics
Precession and Core Dynamics
Proudman-Taylor Theorem

G

GALVANIC DISTORTION

The electrical conductivity of Earth materials affects two physical processes: electromagnetic induction which is utilized with *magnetotellurics* (MT) (*q.v.*), and electrical conduction. If electromagnetic induction in media which are heterogeneous with respect to their electrical conductivity is considered, then both processes take place simultaneously: Due to Faraday's law, a variational electric field is induced in the Earth, and due to the conductivity of the subsoil an electric current flows as a consequence of the electric field. The current component normal to boundaries within the heterogeneous structure passes these boundaries continously according to

$$\sigma_1 E_1 = \sigma_2 E_2$$

where the subscripts 1 and 2 indicate the boundary values of conductivity and electric field in regions 1 and 2, respectively. Therefore the amplitude and the direction of the electric field are changed in the vicinity of the boundaries (Figure G1). In electromagnetic induction studies, the totality of these changes in comparison with the electric field distribution in homogeneous media is referred to as galvanic distortion.

The electrical conductivity of Earth materials spans 13 orders of magnitude (e.g., dry crystalline rocks can have conductivities of less than 10^{-6} S m^{-1}, while ores can have conductivities exceeding 10^{6} S m^{-1}). Therefore, MT has a potential for producing well constrained models of the Earth's electrical conductivity structure, but almost all field studies are affected by the phenomenon of galvanic distortion, and sophisticated techniques have been developed for dealing with it (Simpson and Bahr, 2005).

Electric field amplitude changes and static shift

A change in an electric field amplitude causes a frequency-independent offset in *apparent resistivity* curves so that they plot parallel to their true level, but are scaled by a real factor. Because this shift can be regarded as spatial undersampling or "aliasing," the scaling factor or *static shift factor* cannot be determined directly from MT data recorded at a single site. If MT data are interpreted via one-dimensional modeling without correcting for static shift, the depth to a conductive body will be shifted by the square root of the factor by which the apparent resistivities are shifted.

Static shift corrections may be classified into three broad groups:

1. Short period corrections relying on active near-surface measurements such as transient electromagnetic sounding (TEM) (e.g., Meju, 1996).
2. Averaging (statistical) techniques. As an example, electromagnetic array profiling is an adaptation of the magnetotelluric technique that involves sampling lateral variations in the electric field continuously, and spatial low pass filtering can be used to suppress static shift effects (Torres-Verdin and Bostick, 1992).
3. Long period corrections relying on assumed deep structure (e.g., a resistivity drop at the mid-mantle transition zones) or long-period magnetic *transfer functions* (Schmucker, 1973). An equivalence relationship exists between the magnetotelluric impedance Z and Schmucker's C-response:

$$C = \frac{Z}{i\omega\mu_0},$$

which can be determined from the magnetic fields alone, thereby providing an *inductive scale length* that is independent of the distorted electric field. Magnetic *transfer functions* can, for example, be derived from the magnetic daily variation.

The appropriate method for correcting static shift often depends on the target depth, because there can be a continuum of distortion at all scales. As an example, in complex three-dimensional environments near-surface correction techniques may be inadequate if the conductivity of the mantle is considered, because electrical heterogeneity in the deep crust creates additional galvanic distortion at a larger-scale, which is not resolved with near-surface measurements (e.g., Simpson and Bahr, 2005).

Changes in the direction of electric fields and mixing of polarizations

In some target areas of the MT method the conductivity distribution is two-dimensional (e.g., in the case of *electrical anisotropy* (*q.v.*)) and the induction process can be described by two decoupled polarizations of the electromagnetic field (e.g., Simpson and Bahr, 2005). Then, the changes in the direction of electric fields that are associated with galvanic distortion can result in mixing of these two polarizations. The recovery of the undistorted electromagnetic field is referred to as magnetotelluric tensor decomposition (e.g., Bahr, 1988, Groom and Bailey, 1989).

Current channeling and the "magnetic" distortion

In the case of extreme conductivity contrasts the electrical current can be channeled in such way that it is surrounded by a magnetic

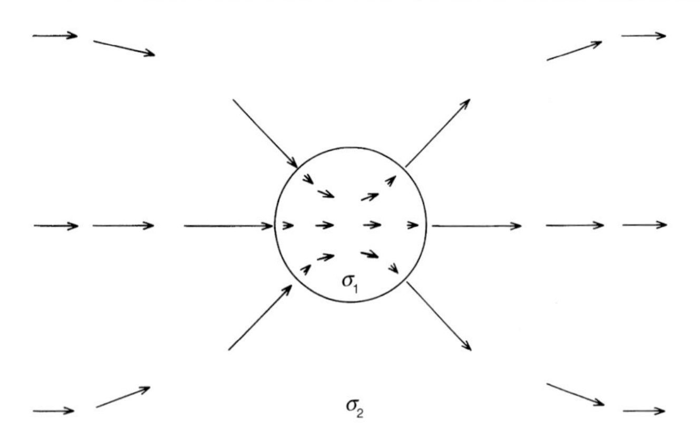

Figure G1 Amplitude and direction of the electric field outside and inside a conductivity anomaly with a conductivity contrast $\sigma_1 : \sigma_2 = 10$ between the anomaly and the host medium.

variational field that has, opposite to the assumptions made in the *geomagnetic deep sounding* (*q.v.*) method, no phase lag with respect to the electric field. The occurrence of such magnetic fields in field data has been shown by Zhang *et al.* (1993) and Ritter and Banks (1998). An example of a magnetotelluric tensor decomposition that includes magnetic distortion has been presented by Chave and Smith (1994).

Karsten Bahr

Bibliography

Bahr, K., 1988. Interpretation of the magnetotelluric impedance tensor: regional induction and local telluric distortion. *Journal of Geophysics*, **62**: 119–127.

Chave, A.D., and Smith, J.T., 1994. On electric and magnetic galvanic distortion tensor decompositions. *Journal of Geophysical Research*, **99**: 4669–4682.

Groom, R.W., and Bailey, R.C., 1989. Decomposition of the magnetotelluric impedance tensor in the presence of local three-dimensional galvanic distortion. *Journal of Geophysical Research*, **94**: 1913–1925.

Meju, M.A., 1996. Joint inversion of TEM and distorted MT soundings: some effective practical considerations. *Geophysics*, **61**: 56–65.

Ritter, P., and Banks, R.J., 1998. Separation of local and regional information in distorted GDS response functions by hypothetical event analysis. *Geophysical Journal International*, **135**: 923–942.

Schmucker, U., 1973. Regional induction studies: a review of methods and results. *Physics of the Earth and Planetary Interiors*, **7**: 365–378.

Simpson, F., and Bahr, K., 2005. *Practical Magnetotellurics.* Cambridge: Cambridge University Press.

Torres-Verdin, C., and Bostick, F.X., 1992. Principles of special surface electric field filtering in magnetotellurics: electromagnetic array profiling (EMAP). *Geophysics*, **57**: 603–622.

Zhang, P., Pedersen, L.B., Mareschal, M., and Chouteau, M., 1993. Channelling contribution to tipper vectors: a magnetic equivalent to electrical distortion. *Geophysical Journal International*, **113**: 693–700.

Cross-references

Anisotropy, Electrical
Geomagnetic Deep Sounding
Magnetotellurics
Mantle, Electrical Conductivity, Mineralogy

GAUSS' DETERMINATION OF ABSOLUTE INTENSITY

The concept of magnetic intensity was known as early as 1600 in *De Magnete* (see *Gilbert, William*). The *relative* intensity of the geomagnetic field in different locations could be measured with some precision from the rate of oscillation of a dip needle—a method used by *Humboldt, Alexander von* (*q.v.*) in South America in 1798. But it was not until Gauss became interested in a universal system of units that the idea of measuring *absolute* intensity, in terms of units of mass, length, and time, was considered. It is now difficult to imagine how revolutionary was the idea that something as subtle as magnetism could be measured in such mundane units.

On 18 February 1832, *Gauss, Carl Friedrich* (*q.v.*) wrote to the German astronomer Olbers:

> "I occupy myself now with the Earth's magnetism, particularly with an *absolute* determination of its intensity. Friend Weber" (Wilhelm Weber, Professor of Physics at the University of Göttingen) "conducts the experiments on my instructions. As, for example, a clear concept of velocity can be given only through statements on time *and* space, so in my opinion, the complete determination of the intensity of the Earth's magnetism requires to specify (1) a weight $= p$, (2) a length $= r$, and then the Earth's magnetism can be expressed by $\sqrt{p/r}$."

After minor adjustment to the units, the experiment was completed in May 1832, when the horizontal intensity (H) at Göttingen was found to be 1.7820 $\mathrm{mg}^{1/2}\,\mathrm{mm}^{-1/2}\,\mathrm{s}^{-1}$ (17820 nT).

The experiment

The experiment was in two parts. In the vibration experiment (Figure G2) magnet A was set oscillating in a horizontal plane by deflecting it from magnetic north. The period of oscillations was determined at different small amplitudes, and from these the period t_0 of infinitesimal oscillations was deduced. This gave a measure of MH, where M denotes the magnetic moment of magnet A:

$$MH = 4\pi^2 I / t_0^2$$

The moment of inertia, I, of the oscillating part is difficult to determine directly, so Gauss used the ingenious idea of conducting the

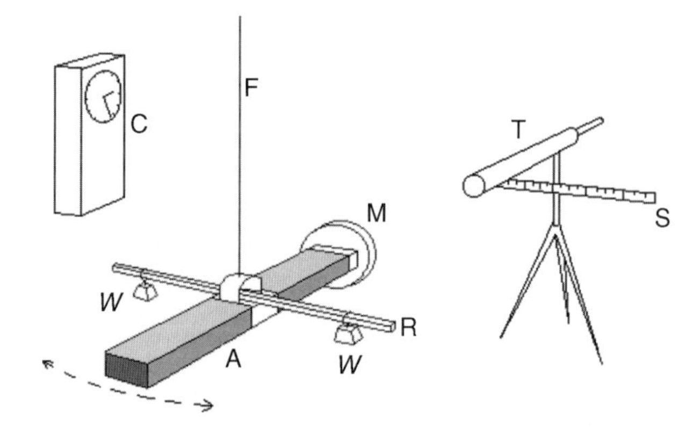

Figure G2 The vibration experiment. Magnet A is suspended from a silk fiber F It is set swinging horizontally and the period of an oscillation is obtained by timing an integral number of swings with clock C, using telescope T to observe the scale S reflected in mirror M. The moment of inertia of the oscillating part can be changed by a known amount by hanging weights W from the rod R.

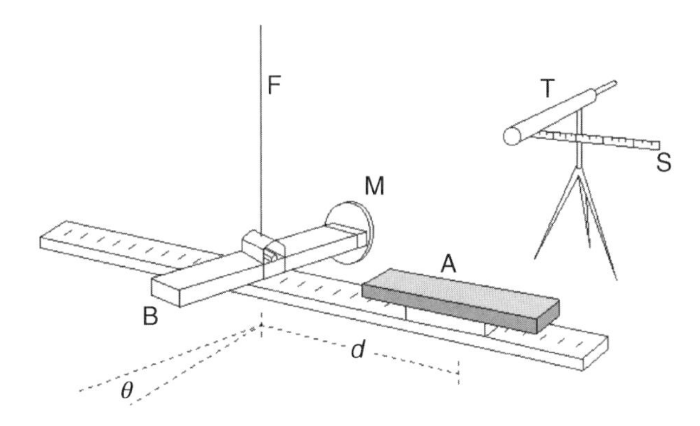

Figure G3 The deflection experiment. Suspended magnet B is deflected from magnetic north by placing magnet A east or west (magnetic) of it at a known distance d. The angle of deflection θ is measured by using telescope T to observe the scale S reflected in mirror M.

experiment for I and then $I + \Delta I$, where ΔI is a known increment obtained by hanging weights at a known distance from the suspension. From several measures of t_0 with different values of ΔI, I was determined by the method of least squares (another of Gauss's original methods).

In the deflection experiment, magnet A was removed from the suspension and replaced with magnet B. The ratio M/H was measured by the deflection of magnet B from magnetic north, θ, produced by magnet A when placed in the same horizontal plane as B at distance d magnetic east (or west) of the suspension (Figure G3). This required knowledge of the magnetic intensity due to a bar magnet. Gauss deduced that the intensity at distance d on the axis of a dipole is inversely proportional to d^3, but that just one additional term is required to allow for the finite length of the magnet, giving $2M\,(1 + k/d^2)/d^3$, where k denotes a small constant. Then

$$M/H = 1/2\,d^3(1 - k/d^2)\tan\theta.$$

The value of k was determined, again by the method of least squares, from the results of a number of measures of θ at different d.

From MH and M/H both M and, as required by Gauss, H could readily be deduced.

Present methods

With remarkably little modification, Gauss's experiment was developed into the Kew magnetometer, which remained the standard means of determining absolute H until electrical methods were introduced in the 1920s. At some observatories, Kew magnetometers were still in use in the 1980s. Nowadays absolute intensity can be measured in seconds with a proton magnetometer and without the considerable time and experimental skill required by Gauss's method.

Stuart R.C. Malin

Bibliography

Gauss, C.F., 1833. *Intensitas vis magneticae terrestris ad mensuram absolutam revocata.* Göttingen, Germany.

Malin, S.R.C., 1982. Sesquicentenary of Gauss's first measurement of the absolute value of magnetic intensity. *Philosophical Transactions of the Royal Society of London*, A**306**: 5–8.

Malin, S.R.C., and Barraclough, D.R., 1982. 150th anniversary of Gauss's first absolute magnetic measurement. *Nature*, **297**: 285.

Cross-references

Gauss, Carl Friedrich (1777–1855)
Geomagnetism, History of
Gilbert, William (1544–1603)
Humboldt, Alexander von (1759–1859)
Instrumentation, History of

GAUSS, CARL FRIEDRICH (1777–1855)

Amongst the 19th century scientists working in the field of geomagnetism, Carl Friedrich Gauss was certainly one of the most outstanding contributors, who also made very fundamental contributions to the fields of mathematics, astronomy, and geodetics. Born in April 30, 1777 in Braunschweig (Germany) as the son of a gardener, street butcher, and mason Johann Friderich Carl, as he was named in the certificate of baptism, already in primary school at the age of nine perplexed his teacher J.G. Büttner by his innovative way to sum up the numbers from 1 to 100. Later Gauss used to claim that he learned manipulating numbers earlier than being able to speak. In 1788, Gauss became a pupil at the Catharineum in Braunschweig, where M.C. Bartels (1769–1836) recognized his outstanding mathematical abilities and introduced Gauss to more advanced problems of mathematics. Gauss proved to be an exceptional pupil catching the attention of Duke Carl Wilhelm Ferdinand of Braunschweig who provided Gauss with the necessary financial support to attend the Collegium Carolinum (now the Technical University of Braunschweig) from 1792 to 1795.

From 1795 to 1798 Gauss studied at the University of Göttingen, where his number theoretical studies allowed him to prove in 1796, that the regular 17-gon can be constructed using a pair of compasses and a ruler only. In 1799, he received his doctors degree from the University of Helmstedt (close to Braunschweig; closed 1809 by Napoleon) without any oral examination and *in absentia*. His mentor in Helmstedt was J.F. Pfaff (1765–1825). The thesis submitted was a complete proof of the fundamental theorem of algebra. His studies on number theory published in Latin language as *Disquitiones arithimeticae* in 1801 made Carl Friedrich Gauss immediately one of the leading mathematicians in Europe. Gauss also made further pioneering contributions to complex number theory, elliptical functions, function theory, and noneuclidian geometry. Many of his thoughts have not been published in regular books but can be read in his more than 7000 letters to friends and colleagues.

But Gauss was not only interested in mathematics. On January 1, 1801 the Italian astronomer G. Piazzi (1746–1820) for the first time detected the asteroid Ceres, but lost him again a couple of weeks later. Based on completely new numerical methods, Gauss determined the orbit of Ceres in November 1801, which allowed F.X. von Zach (1754–1832) to redetect Ceres on December 7, 1801. This prediction made Gauss famous next to his mathematical findings.

In 1805, Gauss got married to Johanna Osthoff (1780–1809), who gave birth to two sons, Joseph and Louis, and a daughter, Wilhelmina. In 1810, Gauss married his second wife, Minna Waldeck (1788–1815). They had three more children together, Eugen, Wilhelm, and Therese. Eugen Gauss later became the founder and first president of the First National Bank of St. Charles, Missouri.

Carl Friedrich Gauss' interest in the Earth magnetic field is evident in a letter to his friend Wilhelm Olbers (1781–1862) as early as 1803, when he told Olbers that geomagnetism is a field where still many mathematical studies can be done. He became more engaged in geomagnetism after a meeting with A. von Humboldt (1769–1859) and W.E. Weber (1804–1891) in Berlin in 1828 where von Humboldt pointed out to Gauss the large number of unsolved problems in geomagnetism. When Weber became a professor of physics at the University of Göttingen in 1831, one of the most productive periods in the field of geomagnetism started. In 1832, Gauss and Weber introduced the well-known Gauss system according to which the magnetic field unit was based on the centimeter, the gram, and the second. The Magnetic Observatory of Göttingen was finished in 1833 and its construction became the prototype for many other observatories all over Europe. Gauss and Weber furthermore developed and improved instruments to measure the magnetic field, such as the unifilar and bifilar magnetometer.

Inspired by A. von Humboldt, Gauss and Weber realized that magnetic field measurements need to be done globally with standardized instruments and at agreed times. This led to the foundation of the *Göttinger Magnetische Verein* in 1836, an organization without any formal structure, only devoted to organize magnetic field measurements all over the world. The results of this organization have been published in six volumes as the *Resultate aus den Beobachtungen des Magnetischen Vereins*. The issue of 1838 contains the pioneering work *Allgemeine Theorie des Erdmagnetismus* where Gauss introduced the concept of the spherical harmonic analysis and applied this new tool to magnetic field measurements. His general theory of geomagnetism also allowed to separate the magnetic field into its externally and its internally caused parts. As the external contributions are nowadays interpreted as current systems in the ionosphere and magnetosphere Gauss can also be named the founder of magnetospheric research.

Publication of the *Resultate* ceased in 1843. W.E. Weber together with such eminent professors of the University of Göttingen as Jacob Grimm (1785–1863) and Wilhelm Grimm (1786–1859) had formed the political group *Göttingen Seven* protesting against constitutional violations of King Ernst August of Hannover. As a consequence of these political activities, Weber and his colleagues were dismissed. Though Gauss tried everything to bring back Weber in his position he did not succeed and Weber finally decided to accept a chair at the University of Leipzig in 1843. This finished a most fruitful and remarkable cooperation between two of the most outstanding contributors to geomagnetism in the 19th century. Their heritage was not only the invention of the first telegraph station in 1833, but especially the network of 36 globally operating magnetic observatories.

In his later years Gauss considered to either enter the field of botanics or to learn another language. He decided for the language and started to study Russian, already being in his seventies. At that time he was the only person in Göttingen speaking that language fluently. Furthermore, he was asked by the Senate of the University of Göttingen to reorganize their widow's pension system. This work made him one of the founders of insurance mathematics. In his final years Gauss became fascinated by the newly built railway lines and supported their development using the telegraph idea invented by Weber and himself.

Carl Friedrich Gauss died on February 23, 1855 as a most respected citizen of his town Göttingen. He was a real genius who was named *Princeps mathematicorum* already during his life time, but was also praised for his practical abilities.

Karl-Heinz Glaßmeier

Bibliography

Biegel, G., and K. Reich, Carl Friedrich Gauss, Braunschweig, 2005.
Bühler, W., Gauss: A Biographical study, Berlin, 1981.
Hall, T., Carl Friedrich Gauss: A Biography, Cambridge, MA, 1970.
Lamont, J., Astronomie und Erdmagnetismus, Stuttgart, 1851.

Cross-references

Humboldt, Alexander von (1759–1859)
Magnetosphere of the Earth

GELLIBRAND, HENRY (1597–1636)

Henry Gellibrand was the eldest son of a physician, also Henry, and was born on 17 November 1597 in the parish of St. Botolph, Aldersgate, London. In 1615, he became a commoner at Trinity College, Oxford, and obtained a BA in 1619 and an MA in 1621. After taking Holy Orders he became curate at Chiddingstone, Kent, but the lectures of Sir Henry Savile inspired him to become a full-time mathematician. He settled in Oxford, where he became friends with Henry Briggs, famed for introducing logarithms to the base 10. It was on Briggs' recommendation that, on the death of Edmund Gunter, Gellibrand succeeded him as Gresham Professor of Astronomy in 1627—a post he held until his death from a fever on 16 February 1636. He was buried at St. Peter the Poor, Broad Street, London (now demolished).

Gellibrand's principal publications were concerned with mathematics (notably the completion of Briggs' *Trigonometrica Britannica* after Briggs died in 1630) and navigation. But he is included here because he is credited with the discovery of geomagnetic secular variation. The events leading to this discovery are as follows (for further details see Malin and Bullard, 1981).

The sequence starts with an observation of magnetic declination made by William Borough, a merchant seaman who rose to "captain general" on the Russian trade route before becoming comptroller of the Queen's Navy. The magnetic observation (Borough, 1581, 1596) was made on 16 October 1580 at Limehouse, London, where he observed the magnetic azimuth of the sun as it rose through seven fixed altitudes in the morning and as it descended through the same altitudes in the afternoon. The mean of the two azimuths for each altitude gives a measure of magnetic declination, D, the mean of which is $11°19'$ E $\pm$ $5'$ rms. Despite the small scatter, the value could have been biased by site or compass errors.

Some 40 years later, Edmund Gunter, distinguished mathematician, Gresham Professor of Astronomy and inventor of the slide rule, found D to be "only 6 gr 15 m" ($6°15'$ E) "as I have sometimes found it of late" (Gunter, 1624, 66). The exact date (ca. 1622) and location (probably Deptford) of the observation are not stated, but it alerted Gunter to the discrepancy with Borough's measurement. To investigate further, Gunter "enquired after the place where Mr. Borough observed, and went to Limehouse with . . . a quadrant of three foot Semidiameter, and two Needles, the one above 6 inches, and the other 10 inches long . . . towards the night the 13 of June 1622, I made observation in several parts of the ground" (Gunter, 1624, 66). These observations, with a mean of $5°56'$ E $\pm$ $12'$ rms, confirmed that D in 1622 was significantly less than had been measured by Borough in 1580. But was this an error in the earlier measure, or, unlikely as it then seemed, was D changing? Unfortunately Gunter died in 1626, before making any further measurements.

When Gellibrand succeeded Gunter as Gresham Professor, all he required to do to confirm a major scientific discovery was to wait a few years and then repeat the Limehouse observation. But he chose instead to go to the site of Gunter's earlier observation in Deptford, where, in June 1633, Gellibrand found D to be "much less than 5°" (Gellibrand, 1635, 16). He made a further measurement of D on the same site on June 12, 1634 and "found it not much to exceed 4°" (Gellibrand, 1635, 7), the published data giving $4°\ 5'$ E $\pm$ $4'$ rms. His observation of D at Paul's Cray on July 4, 1634 adds little, because it is a new site. On the strength of these observations, he announced his discovery of secular variation (Gellibrand, 1635, 7 and 19), but the reader may decide how much of the credit should go to Gunter.

Stuart R.C. Malin

Bibliography

Borough, W., 1581. *A Discourse of the Variation of the Compass, or Magnetical Needle.* (Appendix to R. Norman *The newe Attractive*). London: Jhon Kyngston for Richard Ballard.
Borough, W., 1596. *A Discourse of the Variation of the Compass, or Magnetical Needle.* (Appendix to R. Norman *The newe Attractive*). London: E Allde for Hugh Astley.
Gellibrand, H., 1635. *A Discourse Mathematical on the Variation of the Magneticall Needle. Together with its admirable Diminution lately discovered.* London: William Jones.
Gunter, E., 1624. *The description and use of the sector, the crosse-staffe and other Instruments. First booke of the crosse-staffe.* London: William Jones.
Malin, S.R.C., and Bullard, Sir Edward, 1981. The direction of the Earth's magnetic field at London, 1570–1975. *Philosophical Transactions of the Royal Society of London*, **A299**: 357–423.
Smith, G., Stephen, L., and Lee, S., 1967. *The Dictionary of National Biography.* Oxford: University Press.

Cross-references

Compass
Geomagnetic Secular Variation
Geomagnetism, History of

GEOCENTRIC AXIAL DIPOLE HYPOTHESIS

The time-averaged paleomagnetic field

Paleomagnetic studies provide measurements of the direction of the ancient geomagnetic field on the geological timescale. Samples are generally collected at a number of sites, where each site is defined as a single point in time. In most cases the time relationship between the sites is not known, moreover when samples are collected from a stratigraphic sequence the time interval between the levels is also not known. In order to deal with such data, the concept of the *time-averaged paleomagnetic field* is used. Hospers (1954) first introduced the *geocentric axial dipole hypothesis* (GAD) as a means of defining this time-averaged field and as a method for the analysis of paleomagnetic results. The hypothesis states that the paleomagnetic field, when averaged over a sufficient time interval, will conform with the field expected from a geocentric axial dipole. Hospers presumed that a time interval of several thousand years would be sufficient for the purpose of averaging, but many studies now suggest that tens or hundreds of thousand years are generally required to produce a good time-average.

The GAD model is a simple one (Figure G4) in which the geomagnetic and geographic axes and equators coincide. Thus at any point on the surface of the Earth, the time-averaged paleomagnetic latitude λ is equal to the geographic latitude. If m is the magnetic moment of this time-averaged geocentric axial dipole and a is the radius of the Earth, the horizontal (H) and vertical (Z) components of the magnetic field at latitude λ are given by

$$H = \frac{\mu_0 m \cos\lambda}{4\pi a^2}, \quad Z = \frac{2\mu_0 m \sin\lambda}{4\pi a^2}, \qquad \text{(Eq. 1)}$$

and the total field F is given by

$$F = (H^2 + Z^2)^{1/2} = \frac{\mu_0 m}{4\pi a^2}(1 + 3\sin^2\lambda)^{1/2}. \qquad \text{(Eq. 2)}$$

Since the tangent of the magnetic inclination I is Z/H, then

$$\tan I = 2\tan\lambda, \qquad \text{(Eq. 3)}$$

and by definition, the declination D is given by

$$D = 0^\circ. \qquad \text{(Eq. 4)}$$

The colatitude p (90° minus the latitude) can be obtained from

$$\tan I = 2\cot p (0 \leq p \leq 180^\circ). \qquad \text{(Eq. 5)}$$

The relationship given in Eq. (3) is fundamental to paleomagnetism and is a direct consequence of the GAD hypothesis. When applied to results from different geologic periods, it enables the paleomagnetic latitude to be derived from the mean inclination. This relationship between latitude and inclination is shown in Figure G5.

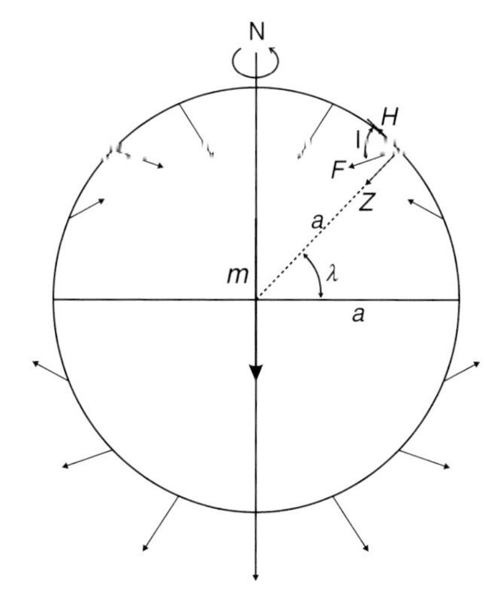

Figure G4 The field of a geocentric axial dipole.

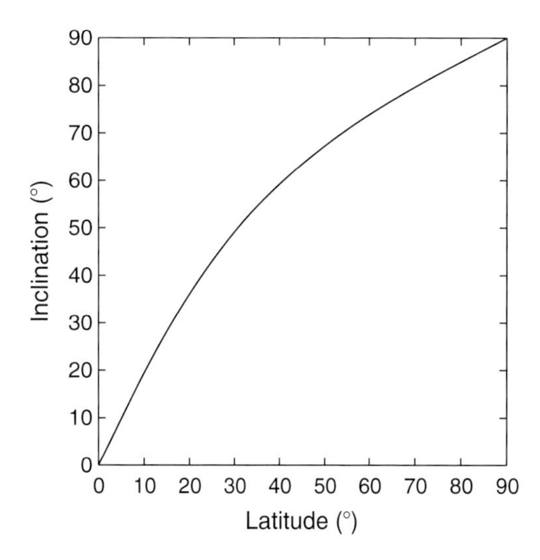

Figure G5 Variation of inclination with latitude for a geocentric dipole.

Paleomagnetic poles

The position where the time-averaged dipole axis cuts the surface of the Earth is called the *paleomagnetic pole* and is defined on the present latitude-longitude grid. Paleomagnetic poles make it possible to compare results from different observing localities, since such poles should represent the best estimate of the position of the geographic pole. These poles are the most useful parameter derived from the GAD hypothesis. If the paleomagnetic mean direction (D_m, I_m) is known at some sampling locality S, with latitude and longitude (λ_s, ϕ_s), the coordinates of the paleomagnetic pole P (λ_p, ϕ_p) can be calculated from the following equations by reference to Figure G6.

$$\sin\lambda_p = \sin\lambda_s \cos p + \cos\lambda_s \sin p \cos D_m \ (-90^\circ \leq \lambda_p \leq +90^\circ) \quad \text{(Eq. 6)}$$

$$\phi_p = \phi_s + \beta, \quad \text{when} \quad \cos p \leq \sin\lambda_s \sin\lambda_p$$

or

$$\phi_p = \phi_s + 180 - \beta, \quad \text{when} \quad \cos p \leq \sin\lambda_s \sin\lambda_p \quad \text{(Eq. 7)}$$

where

$$\sin\beta = \sin p \sin D_m / \cos\lambda_p. \quad \text{(Eq. 8)}$$

The paleocolatitude p is determined from Eq. (5). The paleomagnetic pole (λ_p, ϕ_p) calculated in this way implies that "sufficient" time averaging has been carried out. What "sufficient" time is defined as is a subject of much debate and it is always difficult to estimate the time covered by the rocks being sampled. Any instantaneous paleofield direction (representing only a single point in time) may also be converted to a pole position using Eqs. (7) and (8). In this case the pole is termed a *virtual geomagnetic pole* (VGP). A VGP can be regarded as the paleomagnetic analog of the geomagnetic poles of the present field. The paleomagnetic pole may then also be calculated by finding the average of many VGPs, corresponding to many paleodirections.

Of course, given a paleomagnetic pole position with coordinates (λ_p, ϕ_p), the expected mean direction of magnetization (D_m, I_m) at any site location (λ_s, ϕ_s) may be also calculated (Figure G6). The paleocolatitude p is given by

$$\cos p = \sin\lambda_s \sin\lambda_p + \cos\lambda_s \cos\lambda_p \cos(\phi_p - \phi_s), \quad \text{(Eq. 9)}$$

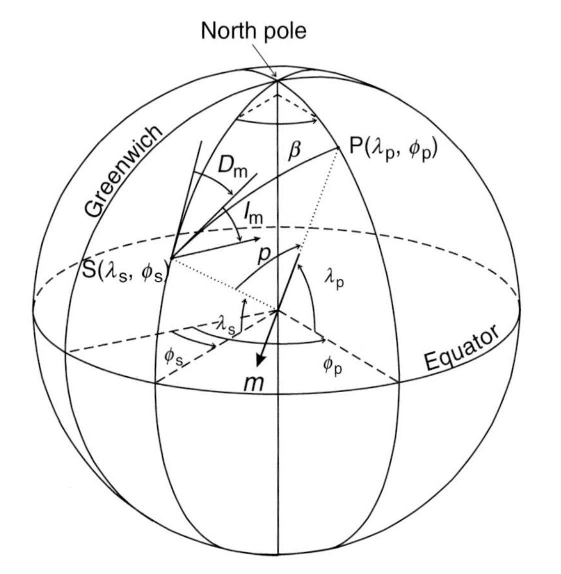

Figure G6 Calculation of the position P (λ_p, ϕ_p) of the paleomagnetic pole relative to the sampling site S (λ_s, ϕ_s) with mean magnetic direction (D_m, I_m).

and the inclination I_m may then be calculated from Eq. (5). The corresponding declination D_m is given by

$$\cos D_m = \frac{\sin\lambda_p - \sin\lambda_s \cos p}{\cos\lambda_s \sin p}, \quad \text{(Eq. 10)}$$

where

$$0^\circ \leq D_m \leq 180^\circ \quad \text{for} \quad 0^\circ \leq (\phi_p - \phi_s) \leq 180^\circ$$

and

$$180^\circ < D_m < 360^\circ \quad \text{for} \quad 180^\circ < (\phi_p - \phi_s) < 360^\circ.$$

The declination is indeterminate (that is any value may be chosen) if the site and the pole position coincide. If $\lambda_s = \pm 90^\circ$ then D_m is defined as being equal to ϕ_p, the longitude of the paleomagnetic pole.

Testing the GAD hypothesis

Timescale 0–5 Ma

On the timescale 0–5 Ma, little or no continental drift will have occurred, so it was originally thought that the observation that worldwide paleomagnetic poles for this time span plotted around the present geographic indicated support for the GAD hypothesis (Cox and Doell, 1960; Irving, 1964; McElhinny, 1973). However, any set of axial multipoles (g_1^0, g_2^0, g_3^0, etc.) will also produce paleomagnetic poles that center around the geographic pole. Indeed, careful analysis of the paleomagnetic data in this time interval has enabled the determination of any second-order multipole terms in the time-averaged field (see below for more detailed discussion of these departures from the GAD hypothesis).

The first important test of the GAD hypothesis for the interval 0–5 Ma was carried out by Opdyke and Henry (1969), who plotted the mean inclinations observed in deep-sea sediment cores as a function of latitude, showing that these observations conformed with the GAD hypothesis as predicted by Eq. (3) and plotted in Figure G5.

Testing the axial nature of the time-averaged field

On the geological timescale it is observed that paleomagnetic poles for any geological period from a single continent or block are closely grouped indicating the dipole hypothesis is true at least to first-order. However, this observation by itself does not prove the *axial* nature of the dipole field. This can be tested through the use of paleoclimatic indicators (see McElhinny and McFadden, 2000 for a general discussion). Paleoclimatologists use a simple model based on the fact that the net solar flux reaching the surface of the Earth has a maximum at the equator and a minimum at the poles. The global temperature may thus be expected to have the same variation. The density distribution of many climatic indicators (climatically sensitive sediments) at the present time shows a maximum at the equator and either a minimum at the poles or a high-latitude zone from which the indicator is absent (e.g., coral reefs, evaporates, and carbonates). A less common distribution is that of glacial deposits and some deciduous trees, which have a maximum in polar and intermediate latitudes. It has been shown that the distributions of paleoclimatic indicators can be related to the present-day climatic zones that are roughly parallel with latitude.

Irving (1956) first suggested that comparisons between paleomagnetic results and geological evidence of past climates could provide a test for the GAD hypothesis over geological time. The essential point regarding such a test is that both paleomagnetic and paleoclimatic data provide independent evidence of past latitudes, since the factors controlling climate are quite independent of the Earth's magnetic field. The most useful approach is to compile the paleolatitude values for a particular occurrence in the form of equal angle or equal area

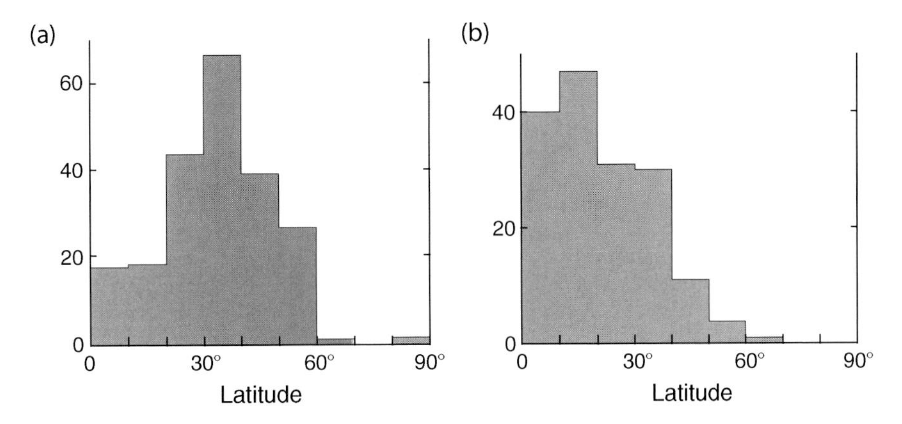

Figure G7 Equal angle latitude histogram for Phanerozoic phosphate deposits. (a) Present latitudes of deposits. (b) Paleolatitudes of those deposits where paleomagnetic data are available. Updated from Cook and McElhinny (1979).

histograms and compare these with the present-day distribution (see Briden and Irving, 1964 for several examples). A more recent example involves the occurrence of phosphate deposits. In this case the distribution of land and sea plays a critical part in their formation. For a phosphate deposit to form, the coastal part of a continent must drift into a low-latitude zone. Thus, the formation of phosphorites is not uniform as a function of time. Cook and McElhinny (1979) have examined the paleolatitudes of Phanerozoic phosphate deposits and have shown that they occur mainly within 40° of the equator as is observed for young phosphorites (Figure G7).

A direct test of the GAD field

If a large continent or block covered a wide latitude range in the past, then the GAD hypothesis can be tested directly by noting the difference in paleolatitude between two sites lying on the same paleomeridian. During the Mesozoic the paleolatitude of Africa covered a range of nearly 60°. Following McElhinny and Brock (1978), McElhinny and McFadden (2000) calculated mean poles for northern Africa and southern Africa. The mean locations for these subsets lie close to a paleomeridian with mean declinations of 340.8° and 338.7° (Figure G8). The present angular difference between these two locations is 50.3° and their paleolatitude difference is 47.4° ± 4.3°. Therefore these data are consistent with the GAD hypothesis during part of the Mesozoic.

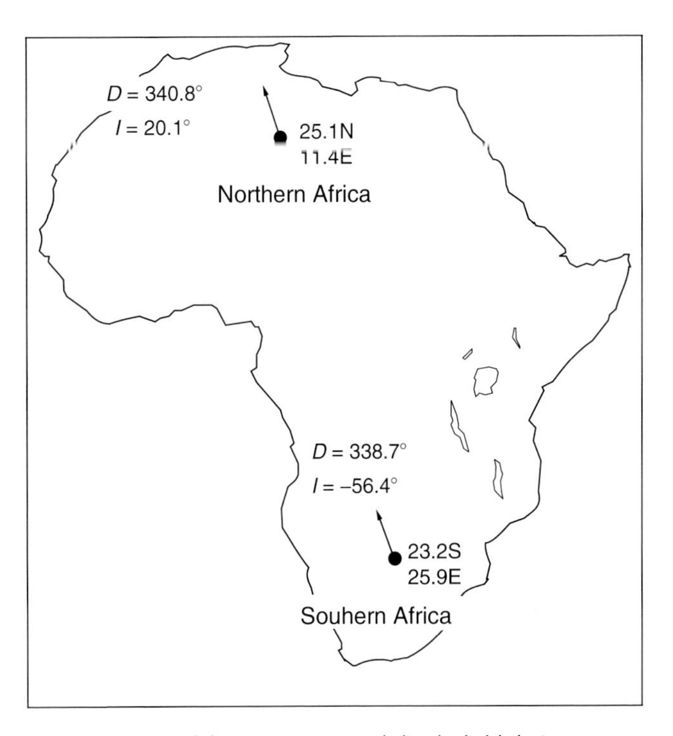

Figure G8 Test of the geocentric axial dipole field during the Mesozoic. Mean values of the paleomagnetic field for northern and southern Africa are shown as calculated by McElhinny and McFadden (2000). The mean sites lie on the same paleomeridian as indicated by the closeness of the mean declinations observed.

Departures from the GAD hypothesis

The far-sided effect

Wilson (1970, 1971) noted that the paleomagnetic poles from Europe and Asia tended to plot too far away from the observation site along the great circle joining the site to the geographic pole (Figure G9). He referred to this as the *far-sided effect* and modeled the effect as originating from a dipole source that is displaced northward along the axis of rotation. Such modeling of the sources of the geomagnetic field is nonunique as there are many possible sources that could equally satisfy the data. It is thus always more useful to use spherical harmonics.

Determining spherical harmonic terms

The potential V (in units of amperes) of the geomagnetic field of internal origin at the surface of the Earth at colatitude θ and longitude ϕ can be represented as a series of spherical harmonics in the form

$$V = \frac{a}{\mu_0}\sum_{l=1}^{\infty}\sum_{m=0}^{l} P_l^m(\cos\theta)(g_l^m \cos m\phi + h_l^m \sin m\phi), \qquad \text{(Eq. 11)}$$

where a is the radius of the Earth and P_l^m is the Schmidt quasinormalized form of the associated Legendre function $P_{l,m}$ of degree l and order m. The coefficients g_l^m and h_l^m are the Gauss coefficients. When averaged over long periods of time such as is required to obtain the time-averaged field when calculating paleomagnetic poles, it might be supposed that the nonzonal components of the magnetic potential will be eliminated leaving only the zonal terms ($m = 0$). Equation (11) then reduces to

$$V = \frac{a}{\mu_0}\sum_{l=1}^{\infty} g_l^0 P_l^0(\cos\theta). \qquad \text{(Eq. 12)}$$

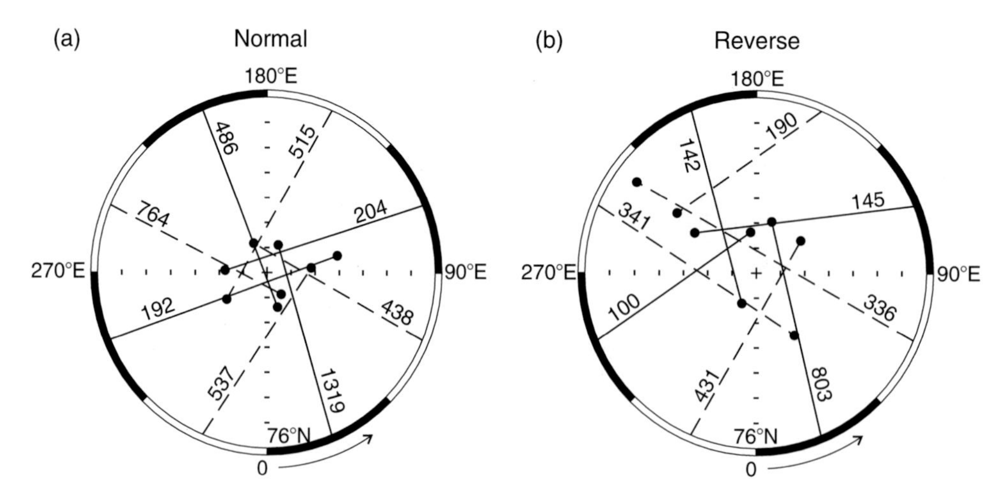

Figure G9 Analysis of global paleomagnetic results for 0–5 Ma from McElhinny *et al.* (1996). Pole positions are averaged over 45° longitude sectors for (a) normal and (b) reverse magnetized data. The paleomagnetic poles tend to plot on the farside of the geographic pole when viewed from each sector. The number of spot readings of the field is indicated for each average. Polar stereographic projection at 76°N.

On the basis of Eq. (12), Merrill and McElhinny (1977) analyzed data for the past 5 Ma by looking at the departure ΔI of the observed inclination I from that expected from a GAD field as a function of latitude. Using a least-squares analysis, they were able to find the best fitting g_2^0 and g_3^0 terms. The most recent least-squares analysis is that of McElhinny *et al.* (1996) in which it was shown that, with present data, only the g_2^0 term is significant and that there are no differences between the normal and reverse field. When the data are restricted to lava flows for the past 5 Ma, several authors have attempted full spherical harmonic analyses. Unlike the least-squares approach, which assumes that nonzonal harmonics are zero, these spherical harmonic analyses are carried out so as to minimize the non-GAD terms (e.g., Johnson and Constable, 1997; Kelly and Gubbins, 1997; Carlut and Courtillot, 1998). It is often claimed in these analyses that significant nonzonal terms are present, but careful analysis of the errors shows that any terms greater than or equal to degree three are not significant (Carlut and Courtillot, 1998). Furthermore, McElhinny and McFadden (1997) and McElhinny (2004) have shown that the data quality is presently not robust enough to determine terms beyond g_2^0, because most of the data were derived during the 1970s when demagnetization techniques were not as robust as those developed during the 1980s. At the present time, the only significant term beyond the GAD field for the interval 0–5 Ma is a persistent geocentric axial quadrupole (g_2^0) of about 4% of the GAD field. Table G1 gives an historical view of the values determined for g_2^0 by various methods over the past 25 years.

Table G1 Values of $G_2 = g_2^0/g_1^0$ for the time-averaged field 0–5 Ma as determined by successive authors based on directional data alone. 95% confidence limits are shown where available. The method used is indicated as least squares (LS) or spherical harmonics (SH)

Authors	Method	Polarity	G_2
Merrill and McElhinny (1977)	LS	N	0.050
	LS	R	0.083
Merrill and McElhinny (1983)	LS	N	0.034 ± 0.006
	LS	R	0.069 ± 0.016
Merrill *et al.* (1990)	LS	N	0.041 ± 0.006
	LS	R	0.080 ± 0.025
	LS	N + R	0.053
Schneider and Kent (1990)	LS	N	0.026 ± 0.010
	LS	R	0.046 ± 0.014
Johnson and Constable (1995)	SH	N	0.044
	SH	R	0.066
McElhinny *et al.* (1996)	LS	N	0.033 ± 0.019
	LS	R	0.042 ± 0.022
	LS	N + R	0.038 ± 0.012
Johnson and Constable (1997)	SH	N	0.036
	SH	R	0.038
Kelly and Gubbins (1997)	SH	N + R	0.042
Carlut and Courtillot (1998)	SH	N	0.036 ± 0.006
	SH	R	0.056 ± 0.018
Hatakeyama and Kono (2002)	SH	N	0.043 ± 0.010
	SH	R	0.080 ± 0.025

Time-averaged field initiative

A program to obtain high-quality data from lava flows in the time interval 0–5 Ma is currently under way and is referred to as the *time-averaged field initiative*. It arose directly as a result of the criticisms put forward by McElhinny and McFadden (1997). New data have been obtained from the Azores (Johnson *et al.*, 1998), Sicily (Zanella, 1998), Hawaii (Laj *et al.*, 1999; Herrero-Bervera and Valet, 2002), the Carribean (Carlut *et al.*, 2000), the Canary Islands (Tauxe *et al.*, 2000), Japan (Morinaga *et al.*, 2000), Possession Island in the southern Indian Ocean (Camps *et al.*, 2001), Easter Island (Miki *et al.*, 1998; Brown, 2002), western Canada (Mejia *et al.*, 2002), southwestern USA (Tauxe *et al.*, 2003), southern Australia (Opdyke and Musgrave, 2004), and Patagonia (Mejia *et al.*, 2004). A significant observation from all of these new data is that the time-averaged field appears to be much simpler than was previously thought. The results appear to indicate that the time-averaged field for 0–5 Ma is very close to that predicted by the GAD hypothesis.

Times > 5 Ma

For times older than 5 Ma, it has been proposed that significant octupole (g_3^0 terms) are present in the time-averaged paleomagnetic field. Axial octupole terms can arise in a variety of ways as a result of poor data quality or data artifacts. A full discussion of these effects is given by McElhinny *et al.* (1996) and McElhinny (2004). When the observed paleomagnetic inclinations appear to be consistently shallower worldwide than those predicted by the GAD field, spherical harmonic analyses interpret this as an axial octupole term. Such inclination shallowing can arise in the following ways:

1. Inclination errors in sediments arising from detrital remanent magnetization (DRM) processes (King and Rees, 1966), or compaction effects (Blow and Hamilton, 1978).
2. Inclination errors in lavas arising from the bulk demagnetizing effect (shape anisotropy) (Coe, 1979). This is unlikely to be a significant effect except in very thin lavas.
3. Creer (1983) has shown that the use of unit vectors in paleomagnetism can cause an artificial shallowing of the observed inclinations.
4. McElhinny *et al.* (1996) have shown that the incomplete magnetic cleaning (demagnetization) of Brunhes-age overprints in reversely magnetized rocks can also create an artificial axial octupole term. Viscous components that would average to the axial dipole field during the Brunhes epoch would have longer relaxation times and would require more careful cleaning procedures.
5. In the least-squares method, the mapping of the inclination anomaly curve ΔI leads to nonorthogonality of the spherical harmonic terms such that the estimates of the coefficients are not entirely independent. However, for small values of the quadrupole and octupole terms, the effect is small.
6. Poor data distribution can cause aliazing between the various axial multipole terms. This problem is exacerbated in the least-squares method where an infinite spherical harmonic representation is truncated to a finite series with only 1, 2, or 3 zonal harmonics. Full spherical harmonic analyses helps overcome this problem, but in this case the individual coefficient estimates should not be overinterpreted as they could arise purely from the poor quality of the data.

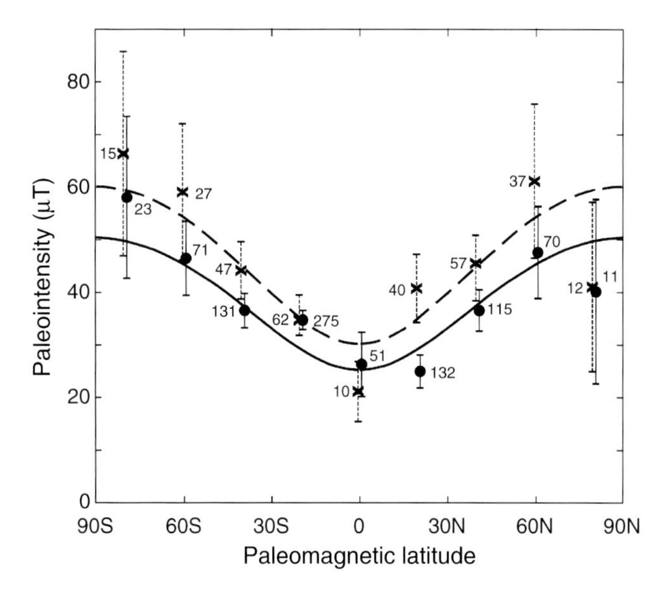

Figure G10 Global paleointensities plotted as a function of paleomagnetic latitude using the mean values averaged over 20° latitude bands calculated by Tanaka *et al.* (1995) for 0–10 Ma (dashed curve) and Perrin and Shcherbakov (1997) for 0–400 Ma (solid curve). The number of units used in each average is indicated with 95% confidence limits for each mean (crosses with dashed error bars for 0–10 Ma, solid circles with solid error bars for 0–400 Ma). The curves represent the best fits for a geocentric axial dipole field.

Random paleogeography test

The random paleogeography test proposed by Evans (1976) has been much used in recent times (e.g., Kent and Smethurst, 1998), purporting to show that significant axial octupole terms were present in the time-averaged field in pre-Cenozoic times. McElhinny and McFadden (2000) surmised that the basic assumption of random paleogeographic sampling required by the Evans (1976) test has not been fulfilled. Meert *et al.* (2003) and McFadden (2004) have both demonstrated that this is indeed the case. On a GAD Earth, sampling over 600 Ma will produce a GAD-like distribution of inclinations as required by the Evans (1976) test only 30% of the time. Inadequate sampling can produce false quadrupole and octupole effects. With the present global paleomagnetic data set, it now appears unlikely that the GAD hypothesis can be tested in this way even using data covering the age of the Earth (McFadden, 2004).

Paleointensities and the GAD hypothesis

The intensity (F) of the GAD field has twice the value at the poles as it does at the equator and varies with latitude (λ) according to the relation

$$F = F_0(1 + 3\sin^2\lambda)^{1/2}, \qquad \text{(Eq. 13)}$$

where F_0 is the intensity of the GAD field at the equator. Tanaka *et al.* (1995) and Perrin and Shcherbakov (1997) have summarized global paleointensity values for the time intervals 0–10 Ma and 0–400 Ma, respectively and calculated mean values over 20° latitude bands. These values should conform with the expected variation in Eq. (13) from the GAD hypothesis. Figure G10 shows these mean values plotted as a function of paleomagnetic latitude and the best-fit curves for a GAD field are drawn through each data set. A chi-square test indicates that the data are consistent with the latitude variation expected for a GAD field. This provides further confirmation that the GAD model is valid, at least to first-order, for the past 400 Ma.

Conclusions

At the present time, the GAD model is a reasonable first-order approximation for the time-averaged field at least for the past 400 Ma and probably for the whole of geological time. A persistent geocentric axial quadrupole term of about 4% of the axial dipole term is present for the interval 0–5 Ma and, with possible variations, can be expected to be a permanent feature of the time-averaged field through time. More precise evaluation of the time-averaged field for 0–5 Ma will become possible when results from the time-averaged field initiative become available and are fully analyzed. Previous analyses were based on poor quality data and new results from the acquisition of new high-quality data worldwide appear to indicate that the time-averaged field is much simpler than was previously thought.

Michael W. McElhinny

Bibliography

Blow, R.A., and Hamilton, N., 1978. Effect of compaction on the acquisition of a detrital remanent magnetization in fine-grained sediments. *Geophysical Journal of the Royal Astronomical Society*, **52**: 13–23.

Briden, J.C., and Irving, E., 1964. Paleolatitude spectra of sedimentary paleoclimatic indicators. In Nairn, A.E.M. (ed.) *Problems in Paleoclimatology*. New York: Wiley-Interscience, pp. 199–250.

Brown, L., 2002. Paleosecular variation from Easter Island revisited: modern demagnetization of a 1970s data set. *Physics of the Earth and Planetary Interiors*, **133**: 73–81.

Camps, P., Henry, B., Prevot, M., and Faynot, L., 2001. Geomagnetic paleosecular variation recorded in plio-Pleistocene volcanic rocks from Possession Island (Crozet Archipelago), southern Indian Ocean. *Journal of Geophysical Research*, **106**: 1961–1971.

Carlut, J., and Courtillot, V., 1998. How complex is the time-averaged geomagnetic field over the past 5 million years? *Geophysical Journal International*, **134**: 527–544.

Carlut, J., Quidelleur, X., Courtillot, V., and Boudon, G., 2000. Paleomagnetic directions and K-Ar dating of 0 to 1 Ma lava flows from La Guadeloupe Island (French West Indies): implications for time-averaged field models. *Journal of Geophysical Research*, **105**: 835–849.

Coe, R.S., 1979. The effect of shape anisotropy on TRM direction. *Geophysical Journal of the Royal Astronomical Society*, **56**: 369–383.

Cook, P.J., and McElhinny, M.W., 1979. A reevaluation of the spatial and temporal distribution of sedimentary phosphate deposits in the light of plate tectonics. *Economic Geology*, **74**: 315–330.

Cox, A., and Doell, R.R., 1960. Review of paleomagnetism. *Geological Society of America Bulletin*, **71**: 645–768.

Creer, K.M., 1983. Computer synthesis of geomagnetic palaeosecular variation. *Nature*, **304**: 695–699.

Evans, M.E., 1976. Test of the nondipolar nature of the geomagnetic field throughout Phanerozoic time. *Nature*, **262**: 676–677.

Hatakeyama, T., and Kono, M., 2002. Geomagnetic field models for the last 5 Myr time-averaged field and secular variation. *Physics of the Earth and Planetary Interiors*, **133**: 181–201.

Herrero-Bervera, E., and Valet, J-P., 2002. Paleomagnetic secular variation of the Honolulu Volcanic Series, (33–700 ka), Oahu (Hawaii). *Physics of the Earth and Planetary Interiors*, **133**: 83–97.

Hospers, J., 1954. Rock magnetism and polar wandering. *Nature*, **173**: 1183.

Irving, E., 1956. Palaeomagnetic and palaeoclimatological aspects of polar wandering. *Geofisica Pura et Applicata*, **33**: 23–41.

Irving, E., 1964. *Paleomagnetism and Its Application to Geological and Geophysical Problems.* New York: John Wiley.

Johnson, C.L., and Constable, C.G., 1995. The time-averaged geomagnetic field as recorded in lava flows over the past 5 Myr. *Geophysical Journal International*, **122**: 489–519.

Johnson, C.L., and Constable, C.G., 1997. The time-averaged geomagnetic field: global and regional biases for 0–5 Ma. *Geophysical Journal International*, **131**: 643–666.

Johnson, C.L., Wijbrans, J.R., Constable, C.G., Gee, J., Staudigal, H., Tauxe, L., Forjaz, V-H., and Salguiero, M., 1998. ^{40}Ar-^{39}Ar ages and paleomagnetism of San Miguel lavas, Azores. *Earth and Planetary Science Letters*, **160**: 637–649.

Kelly, P., and Gubbins, D., 1997. The geomagnetic field over the past 5 million years. *Geophysical Journal International*, **128**: 315–330.

Kent, D.V., and Smethurst, M.A., 1998. Shallow bias of paleomagnetic inclinations in the Paleozoic and Precambrian. *Earth and Planetary Science Letters*, **160**: 391–402.

King, R.F., and Rees, A.I., 1966. Detrital magnetism in sediments: an examination of some theoretical models. *Journal of Geophysical Research*, **71**: 561–571.

Laj, C., Guillou, H., Szeremeta, N., and Coe, R.S., 1999. Geomagnetic secular variation at Hawaii around 3 Ma from a sequence of 107 lavas at Kaena Point (Oahu). *Earth and Planetary Science Letters*, **170**: 365–376.

McElhinny, M.W., 1973. *Palaeomagnetism and Plate Tectonics.* Cambridge: Cambridge University Press.

McElhinny, M.W., 2004. The geocentric axial dipole hypothesis: a least squares perspective. In Channell, J.E.T., Kent, D.V., and Lowrie, W. (eds.), *Timescales of the Internal Geomagnetic Field.* American Geophysical Union Monograph, **145**: 1–12.

McElhinny, M.W., and Brock, A., 1978. A new palaeomagnetic result from East Africa and estimates of the Mesozoic palaeoradius. *Earth and Planetary Science Letters*, **27**: 321–328.

McElhinny, M.W., and McFadden, P.L., 1997. Palaeosecular variation over the past 5 Myr based on a new generalized database. *Geophysical Journal International*, **131**: 240–252.

McElhinny, M.W., and McFadden, P.L., 2000. *Paleomagnetism: Continents and Oceans.* San Diego, CA: Academic Press.

McElhinny, M.W., McFadden, P.L., and Merrill, R.T., 1996. The time-averaged geomagnetic field 0–5 Ma. *Journal of Geophysical Research*, **101**: 25007–25027.

McFadden, P.L., 2004. Is 600 Myr long enough for the random palaeogeographic test of the geomagnetic axial dipole assumption? *Geophysical Journal International*, **158**: 443–445.

Meert, J.G., Tamrat, E., and Spearman, J., 2003. Nondipole fields and inclination bias: insights from a random walk analysis. *Earth and Planetary Science Letters*, **214**: 395–408.

Mejia, V., Barendregt, R.W., and Opdyke, N., 2002. Paleosecular variation of Brunhes age lava flows from British Columbia. *Geochemistry, Geophysics, Geosystems*, **3**(12): 8801, doi:10.1029/2002GC000353.

Mejia, V., Opdyke, N.D., Vilas, J.F., Singer, B.S., and Stoner, J.S., 2004. Plio-Pleistocene time-averaged field in southern Patagonia recorded in lava flows. *Geochemistry, Geophysics, Geosystems*, **5**(3): Q03H08, doi:10.1029/2003GC000633.

Merrill, R.T., and McElhinny, M.W., 1977. Anomalies in the time-averaged paleomagnetic field and their implications for the lower mantle. *Reviews of Geophysics and Space Physics*, **15**: 309–323.

Merrill, R.T., and McElhinny, M.W., 1983. *The Earth's Magnetic Field: Its History, Origin and Planetary Perspective.* London: Academic Press.

Merrill, R.T., McFadden, P.L., and McElhinny, M.W., 1990. Paleomagnetic tomography of the core-mantle boundary. *Physics of the Earth and Planetary Interiors*, **64**: 87–101.

Miki, M., Inokuchi, H., Yamaguchi, S., Matsuda, J., Nagao, K., Isazaki, N., and Yaskawa, K., 1998. Geomagnetic secular variation in Easter Island, southeast Pacific. *Physics of the Earth and Planetary Interiors*, **106**: 93–101.

Morinaga, H., Matsumoto, T., Okimura, Y., and Matsuda, T., 2000. Paleomagnetism of Pliocene to Pleistocene lava flows in the northern part of Hyogo prefecture, northwest Japan and Brunhes chron paleosecular variation in Japan. *Earth, Planets and Space*, **52**: 437–443.

Opdyke, N.D., and Henry, K.W., 1969. A test of the dipole hypothesis. *Earth and Planetary Science Letters*, **6**: 139–151.

Opdyke, N.D., and Musgrave, R., 2004. Paleomagnetic results from the Newer Volcanics of Victoria: contribution to the time averaged field initiative. *Geochemistry, Geophysics, Geosystems*, **5**(3): Q03H09, doi:10.1029/2003GC000632.

Perrin, M., and Shcherbakov, V., 1997. Paleointensity of the Earth's magnetic field for the past 400 Ma: evidence for a dipole structure during the Mesozoic low. *Journal of Geomagnetism and Geoelectricity*, **49**: 601–614.

Schneider, D.A., and Kent, D.V., 1990. The time-averaged paleomagnetic field. *Reviews of Geophysics*, **28**: 71–96.

Tanaka, H., Kono, M., and Uchimura, H., 1995. Some global features of paleointensity in geological time. *Geophysical Journal International*, **120**: 97–102.

Tauxe, L., Staudigal, H., and Wijbrans, J.R., 2000. Paleomagnetism and ^{40}Ar-^{39}Ar ages from La Palma in the Canary Islands. Geochemistry, Geophysics, Geosystems, (9):doi:10.1029/2000GC000063.

Tauxe, L., Constable, C., Johnson, C.L., Koppers, A.A.P., Miller, W.R., and Staudigal, H., 2003. Paleomagnetism of the southwestern U.S.A. recorded by 0–5 Ma igneous rocks. *Geochemistry, Geophysics, Geosystems*, **4**(4): 8802, doi:10.1029/ 2002GC000343.

Wilson, R.L., 1970. Permanent aspects of the Earth's non-dipole magnetic field over Upper Tertiary times. *Geophysical Journal of the Royal Astronomical Society*, **19**: 417–437.

Wilson, R.L., 1971. Dipole offset: the time-averaged palaeomagnetic field over the past 25 million years. *Geophysical Journal of the Royal Astronomical Society*, **22**: 491–504.
Zanella, E., 1998. Paleomagnetism of Pleistocene rocks from Pantelleria Island, (Sicily Channel), Italy. *Physics of the Earth and Planetary Interiors*, **108**: 291–303.

Cross-references

Geocentric Axial Dipole Hypothesis
Pole, Paleomagnetic
Time-averaged Paleomagnetic Field

GEODYNAMO

Introduction

The discovery of the magnetic *compass* (*q.v.*) by the Chinese dates back at least to the A.D. 1st century A.D., and it was known in Europe during the A.D. 12th century. An early idea was that the magnetic needle was attracted to the pole star, but by the year 1600, *William Gilbert* (*q.v.*) had realized that the source of the Earth's magnetic field came from within the Earth rather than outside it. Final confirmation of this was not made until 1838, when *Carl Friedrich Gauss* (*q.v.*) used a *spherical harmonic* (*q.v.*) decomposition of the geomagnetic field to establish that the main field is internal. The important discovery that the field changes with time was made in 1634 by *Henry Gellibrand* (*q.v.*). This showed that the Earth's field cannot be a permanent magnet, as William Gilbert had envisaged. *Edmond Halley* (*q.v.*) investigated the changes in the Earth's field and showed that some magnetic features were drifting westward, and on this basis suggested that the interior of the Earth might be liquid.

Evidence from the study of rock magnetism shows that the Earth's magnetic field is certainly not static, but varies dramatically over long periods of time. Indeed, the Earth has undergone many complete magnetic reversals throughout its history. The most widely accepted theory is that the magnetic field is continually being created and destroyed by fluid motions in the interior of the Earth, as suggested by *Joseph Larmor* (*q.v.*) (1919). Since electricity and magnetism are commonly generated by means of dynamos, the mechanism by which the Earth's magnetic field is created is known as the geodynamo. Permanent magnetism does occur in the *crustal magnetic field* (*q.v.*) of the Earth, and contributes a small and relatively static contribution to the main internally generated magnetic field, the core field.

There are also external components of the magnetic field measured at the Earth's surface. They can be distinguished from the internal core field partly because they increase upward rather than decrease upward, but also because they vary on a much shorter timescale. The origin of these external fields is in the Earth's *ionosphere* (*q.v.*), where charged particles in the solar wind interact with the upper atmosphere. Since solar magnetic activity changes on a timescale of a few days, short bursts of activity known as *magnetic storms and substorms* (*q.v.*) can be detected in magnetic observatories. The external components of the geomagnetic field will not be discussed further here as they are not part of the geodynamo.

Magnetic fields on other planets and stars

The Earth is by no means the only planet to have a strong internal magnetic field. Jupiter has a surface field 10 times larger than the geomagnetic field, and Saturn, Uranus, and Neptune have surface fields at least as strong. Internal fields exist on Mercury (though this field is quite weak) and on Ganymede, one of Jupiter's moons. There are *planetary and satellite dynamos* (*q.v.*) as well as a dynamo in the Earth. Planetary magnetic fields have been reviewed by Stevenson (1982, 2003) and Jones (2003).

Not all the planetary magnetic fields are dominated by an axial dipole, as is the Earth. Uranus and Neptune have more complex fields. Our moon, and the planet Mars, have *crustal magnetic fields* (*q.v.*) which are very likely to have been created by internal dynamos, which operated early in their history, but have now ceased to function. The issue of whether a planet or satellite has an internally generated magnetic field depends on the physical conditions in the *interiors of the planets and satellites* (*q.v.*). The *magnetic field of Sun* (*q.v.*) is much stronger than the geomagnetic field, and some stars have fields that are stronger still. The solar magnetic field is believed to be generated by dynamo processes occurring in the Sun's deep interior, (see the *Solar dynamo*).

The dynamo process

According to dynamo theory, the magnetic field in the Earth is maintained by a system of electrical currents flowing through its liquid metal core. There are four key laws from electromagnetic theory that describe how the magnetic field and the currents behave in terms of concepts from elementary vector calculus. These are discussed in detail in books on *magnetohydrodynamics* (*q.v.*), or MHD for short, such as Roberts (1967), Moffatt (1978), and Davidson (2001). The first is Ampere's law, which can be written

$$\nabla \times \mathbf{B} = \mu \mathbf{j}, \qquad \text{(Eq. 1)}$$

where **B** is the magnetic field, **j** is the current density (the current flowing through a wire is **j** times its cross-sectional area), and μ is the permeability of free space (the high core temperature makes the free space value $4\pi \times 10^{-7}$ Tm A^{-1} appropriate). Maxwell showed that a displacement current term should be added to this equation, but for the Earth's core this extra term can be ignored, so Eq. (1) is sometimes called the pre-Maxwell equation. The second equation is Faraday's law of electromagnetic induction, which is

$$\frac{\partial \mathbf{B}}{\partial t} = -\nabla \times \mathbf{E}, \qquad \text{(Eq. 2)}$$

and says that if a magnetic field is varied in time, an electric field **E** is created. The third result from electromagnetic theory is Ohm's law in a moving conductor, which can be written

$$\mathbf{j} = \sigma(\mathbf{E} + \mathbf{u} \times \mathbf{B}). \qquad \text{(Eq. 3)}$$

Here the conductor (which may be fluid or solid) is moving with velocity **u**. σ is the *electrical conductivity of the core* (*q.v.*), a physical quantity of fundamental importance for the geodynamo. It is estimated to be around 5×10^5 S m^{-1}. Dividing by σ and taking the curl of Eq. (3) allows us to eliminate the electric field **E**, and dividing Eq. (1) by μ and taking the curl allows us to eliminate **j**. The fourth law is simply

$$\nabla \cdot \mathbf{B} = 0, \qquad \text{(Eq. 4)}$$

which is equivalent to saying there are no magnetic monopoles. If μ and σ are constants, we obtain (using the vector identity curl curl = grad div $- \nabla^2$) the induction equation

$$\frac{\partial \mathbf{B}}{\partial t} = \nabla \times (\mathbf{u} \times \mathbf{B}) + \eta \nabla^2 \mathbf{B}, \qquad \text{(Eq. 5)}$$

where $\eta = 1/\mu\sigma$ is called the magnetic diffusivity, and its typical value in the core is $\eta \sim 2$ m^2 s^{-1}. The last diffusive term in Eq. (5) arises because all metals have some electrical resistance.

Seismology has shown that the Earth consists of a solid inner core, the inner core radius being approximately 1100 km (see *Core properties, physical*), surrounded by a fluid outer core of radius approximately 3500 km, surrounded in turn by the mantle. The electrical conductivity of the mantle is very low compared with that of the outer core, so the liquid metal outer core is the most natural seat of the geodynamo.

We first consider what happens if there is no motion, $\mathbf{u} = 0$, so that Eq. (5) becomes the diffusion equation. When the power flowing into a laboratory electromagnet is switched off, the magnetic field usually collapses quite quickly, but the Earth's core is very large and this slows down the decay. Equation (5) with $\mathbf{u} = 0$ can be solved for a spherical conducting core surrounded by an insulating mantle (see e.g., Moffatt, 1978), and the field is found to decay in time as $\exp(-\pi^2 \eta t / a^2)$, where a is the core radius. Putting in the estimates given above, the field reduces by a factor e (the e-folding time) in 20 ka. The core of the Earth is so large that even if the source of the driving were suddenly removed, the field would decay only on this very long timescale. However, the age of the Earth is around 4.5×10^9 y, and paleomagnetic studies show that the Earth's magnetic field has existed for at least 3×10^9 y. This is much longer than the 20 ka magnetic decay time, so an *energy source for the geodynamo* (*q.v.*) is required to maintain the field.

The generation mechanism for the geodynamo is that the field is maintained by *core motions* (*q.v.*) inside the liquid metal outer core of the Earth, so $\mathbf{u}$ is nonzero and the electromagnetic induction term in Eq. (5), $\nabla \times (\mathbf{u} \times \mathbf{B})$, is important. This mechanism has now been demonstrated to work in laboratory *experimental dynamos* (*q.v.*). The same process is used in power stations to generate virtually all our electricity, though there the conducting material is an array of copper wires rather than a core of liquid metal.

The velocity required to generate the magnetic field can be estimated from the size of the induction term relative to the diffusion term. This ratio is approximately $U_* a / \eta = R_m$, the magnetic Reynolds number, where U_* is a typical value of the velocity of the fluid inside the core relative to the frame rotating with the mantle. This can be estimated by observing the *westward drift* (*q.v.*) velocity of the magnetic field itself. Since the inhomogeneities in the field are to a large extent carried along by the fluid motion in the core, this provides a rough estimate of approximately $U_* \sim 2 \times 10^{-4}$ m s^{-1} for the core flow, giving $R_m \sim 300$ (note though that in some places in the core the velocity can be as much as 8×10^{-4} m s^{-1}, Bloxham and Jackson, 1991). Unfortunately, it is not possible to get a complete picture of the flow below the core-mantle boundary (CMB), but *core-based inversions* (*q.v.*) can extract much useful information about core velocities. Since R_m is significantly larger than unity, this ensures that the induction term, which generates the magnetic field, makes good the losses arising from magnetic diffusion. This large value of R_m arising from these natural estimates provides strong support for the dynamo hypothesis in the Earth's core. It is of interest to seek the minimum value of R_m required to overcome diffusion. A number of bounding theorems have been proved to answer this question.

The actual flow $\mathbf{u}$ inside the core that is driving the dynamo will be time-dependent and have a complex structure. However, some idea of the type of flow inside the core suggested by the theory of the dynamics of the core (see section below on "Core dynamics") is shown in Figure G11.

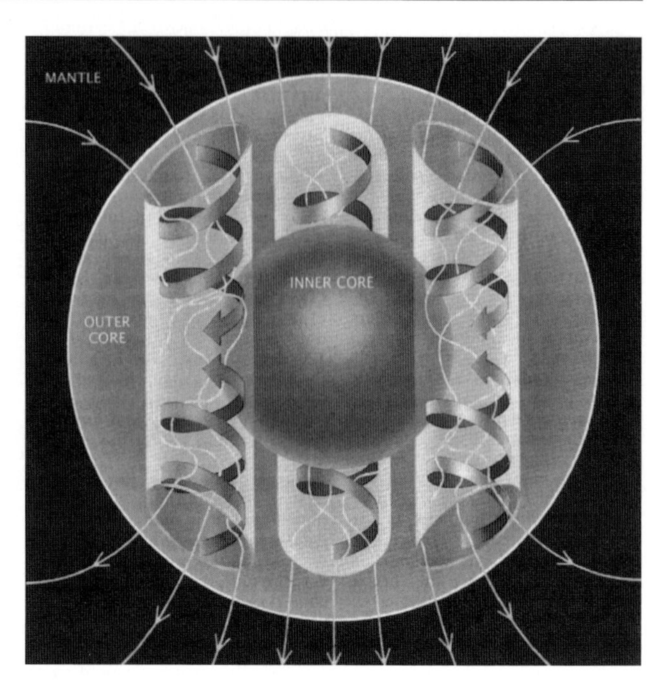

Figure G11 A possible configuration for the fluid flow and the field inside the core (after Bloxham and Gubbins, 1989).

Nondynamo generation mechanisms

Although the dynamo theory is a very reasonable hypothesis, it is necessary to consider the other *nondynamo theories* (*q.v.*) which have been suggested for driving the Earth's magnetic field. Most of these are capable of generating some magnetic field, but are not up to the task of generating the rather strong magnetic field observed. Permanent magnets are demagnetized when raised to temperatures above the Curie point temperature, which for the materials which constitute the core is well below any reasonable estimate of core temperature. Admittedly, the Curie point might be affected by the high pressure in the Earth's core (see *Depth to the Curie temperature*). The strong variability of the geomagnetic field is the best argument against a permanent magnetism explanation. Thermoelectric effects can create currents and hence magnetic fields; indeed Stevenson (1987) has proposed this mechanism for Mercury's field, but it cannot produce enough field to explain geomagnetism. Other possible mechanisms, with references, are listed by Merrill *et al.* (1996), but none has gained any widespread acceptance.

How is the geodynamo driven?

A more controversial issue is the energy source of the fluid flow inside the core. The diffusion of the magnetic field is accompanied by ohmic heating, sometimes called Joule dissipation. This loss of energy requires a source to replace it. *Core convection* (*q.v.*) is the mostly widely accepted source, and the two most developed driving mechanisms are thermal convection and compositional convection. Other possibilities are precession and tidal interaction. For a more detailed discussion on these mechanisms (see *Energy source for the geodynamo*).

The fundamental source of energy for precession and tidal interactions is the rotational energy of the Earth. If these contribute to driving the geodynamo, they would lead to a slowing down of the Earth's rotation, i.e., to a lengthening of the day (see *Decadal length of day variations*). Earth tides and oceanic tides are also slowing down the rotation rate of the Earth, so a precession-driven dynamo would give additional slowing.

Thermal convection derives its energy from the cooling down of the Earth, although if there is radioactive heating in the core, this could contribute too. The issue of whether there is radioactivity in the core has a long history and remains highly controversial (see *Radioactive isotopes and their decay*). There is no doubt that radioactivity is important in the mantle, but whether radioactive elements were carried into the core during formation is a very challenging problem for geochemists. The primordial heating of the Earth occurred as part of the core formation process through gravitational collapse, so the thermal energy of the core may be thought of as originating from gravitational energy.

Another source of energy for driving the dynamo is compositional or *chemical convection* (*q.v.*) (Braginsky, 1963; *J. Verhoogen* (*q.v.*), 1961). The solid inner core of the Earth contains a higher fraction of iron than the fluid outer core (see *Inner core composition*), and so is more dense than the fluid outer core. As the inner core grows, due to core cooling, iron is deposited on the inner core, and light material is released at the moment of freezing. This light material rises due to buoyancy, and stirs up the outer core. Since the net effect of this process is to move heavier material toward the center, gravitational energy is liberated. It is slightly surprising that the core first freezes near the center of the Earth, where the temperature is highest. This happens because the melting temperature of iron in the core increases strongly with pressure.

The question of whether tidal forcing or precessional forcing (see *Precession and core dynamics*) is large enough to drive the geodynamo has been discussed by Malkus (1994). The typical height of an Earth tide inside the core is expected to be about $h \sim 0.15$ m, and precession gives a similar displacement. The typical radial velocity is then $\Omega h \sim 10^{-5}$ m s^{-1}. The azimuthal velocity might be a little larger, so a value not that far off the convective velocity U_* might be achieved. There is a difficulty because this velocity is fluctuating on a timescale of a day, whereas a flow varying only on a much longer timescale (a 1 ka) is needed to sustain a dynamo. Some mechanism for converting oscillatory flows into steady flows is required, and some suggestions for overcoming this difficulty have been proposed (Kerswell, 2002; Aldridge, 2003). A recent example of a precession-driven dynamo has been given by Tilgner (2005), though the conditions under which these dynamos operate are still very far from Earth-like.

Energy balance for a convective dynamo

This naturally leads on to the question of how efficient the geodynamo actually is. In the *solar dynamo* (*q.v.*), only a tiny fraction of the heat energy pouring out of the Sun is converted into the solar magnetic energy: is the Earth more efficient in this respect? To investigate this we need to examine the overall energy and entropy budget in the Earth's core, and to do this we need a model for the *Core temperature of the Earth* (*q.v.*).

Since we believe the fluid outer core is stirred it is reasonable to assume it is approximately adiabatically stratified, so the temperature gradient is given by the *adiabatic gradient in the core* (*q.v.*). This is dependent on a quantity called the *Grüneisen parameter* (*q.v.*) which itself varies somewhat between the inner core boundary (ICB) and the Core-Mantle boundary but if we adopt the currently accepted estimates, we find that the temperature drop from the ICB to the CMB is about 1100 K (Braginsky and Roberts, 1995; Roberts *et al.*, 2003). In principle, the melting temperature of iron at high pressure (see *melting temperature of iron in the core*), should determine the ICB temperature, but unfortunately, the impurities expected in the core significantly depress the melting temperature, so this is somewhat uncertain. A reasonable guess at the present time (Roberts *et al.*, 2003) is $T_{\mathrm{ICB}} = 5100$ K, giving $T_{\mathrm{CMB}} = 4000$ K. An important issue is the amount of *thermal conduction in the core* (*q.v.*). Again there is uncertainty due to the difficulty of the very high core pressures but if we take $\sim$40 W m^{-1} K^{-1} as our estimate, the conducted flux down the adiabat near the CMB is $Q_{\mathrm{cond}} \sim 6$ TW, and only $\sim$0.25 TW near the ICB. There are two reasons for the big difference; first the adiabatic temperature gradient decreases somewhat with depth, so the conducted heat flux per square meter decreases with depth, but secondly the geometry of the much larger surface area near the CMB means that a much larger number of terawatts is conducted there. The critical question is whether the actual heat flux out of the core is greater or less than this conducted flux. If the actual flux exceeds the conducted flux Q_{cond} at any point in the fluid outer core, the excess flux is carried by convection. If however the actual heat flux is less than Q_{cond} then there will be a region where there is a subadiabatic temperature gradient. This is most likely to occur just below the CMB. Here there is either a nonadiabatic stably stratified region where the temperature gradient is significantly subadiabatic, thus reducing the conducted flux, or if compositional convection stirs this subadiabatic region the heat flux produced by compositional stirring is negative (back into the interior). A further possibility is that light material released at the ICB as the inner core forms accumulates just below the CMB. This possibility has been called by Braginsky (1993) the "inverted ocean," since it would consist of a sea of relatively light material floating up against the CMB.

Unfortunately, our knowledge of the *heat flow across the* Core-Mantle Boundary is imprecise, and we cannot yet be certain whether an inverted ocean exists or not. We know that at the present time there are 44 TW of heat coming through the Earth's surface, and that most of that originates in the mantle. Estimates of the CMB heat flux range from 3 TW (Sleep, 1990) to 15 TW (Roberts *et al.*, 2003), and are highly dependent on whether there is core radioactivity or not. In principle, mantle convection simulations could tell us the heat flux across the CMB, since this is one of the bottom boundary conditions to such simulations, and so will affect the structure of mantle convection. However, to date the uncertainties surrounding mantle convection modeling (see e.g., Schubert *et al.*, 2001) preclude this possibility. *Numerical simulations of the geodynamo* (*q.v.*), such as those of Glatzmaier and Roberts (1995, 1997), have typically adopted a compromise value of around 7 TW for the CMB heat flux, which just makes the core fully convective.

The overall energy balance in the core, ignoring any contributions from precession or tides, can be written

$$\mathcal{Q}^{\mathrm{S}} = \mathcal{Q}_{\mathrm{CMB}} - \mathcal{Q}_{\mathrm{ICB}} - \mathcal{Q}^{\mathrm{L}} - \mathcal{Q}^{\mathrm{G}}(-\mathcal{Q}^{\mathrm{R}}). \qquad \text{(Eq. 6)}$$

where $\mathcal{Q}^{\mathrm{S}}$ is the rate of cooling, $\mathcal{Q}_{\mathrm{CMB}}$ is the heat flux passing through the CMB, $\mathcal{Q}_{\mathrm{ICB}}$ is the heat flux passing through the ICB, $\mathcal{Q}^{\mathrm{L}}$ is the latent heat released by the freezing process on the inner core, *Verhoogen* (*q.v.*), 1961 $\mathcal{Q}^{\mathrm{G}}$ is the gravitational energy released by the same process, and $\mathcal{Q}^{\mathrm{R}}$ is the heat produced by radioactivity (if any). For the sake of definiteness, we give reasonable estimates for each of these quantities (see e.g., Braginsky and Roberts, 1995; Roberts *et al.*, 2003), but it cannot be emphasized too strongly that all of these estimates are uncertain, and it would be very surprising if improvements in our understanding of high pressure physics did not lead to radical revision of these estimates over the next decades. Discussions of some of the physics that goes into these estimates can be found in the articles on *core composition* (*q.v.*), *core density* (*q.v.*), and *core properties, physical* (*q.v.*). With this proviso, we take the cooling rate of the core to be $\mathcal{Q}^{\mathrm{S}} \sim 2.3$ TW, the latent heat released at the ICB to be $\mathcal{Q}^{\mathrm{L}} \sim 4.0$ TW, the gravitational energy released at the ICB $\mathcal{Q}^{\mathrm{G}} \sim 0.5$ TW, and $\mathcal{Q}_{\mathrm{ICB}} \sim 0.25$ TW. This gives $\mathcal{Q}_{\mathrm{CMB}} \sim 7$ TW, the value used by Glatzmaier and Roberts (1997). $\mathcal{Q}^{\mathrm{R}}$, the heat released by core radioactivity is extremely uncertain, as mentioned above. Perhaps the best hope is that an improved understanding of the geodynamo might enable us to constrain $\mathcal{Q}^{\mathrm{R}}$, as might a better understanding of the thermal history of planetary interiors.

It might seem surprising that the ohmic dissipation generated by the magnetic field does not appear in the energy balance. This is because it is balanced by the work done by buoyant convection. Some of the heat energy flowing through the core is extracted to drive the fluid motions, only to be returned in full by the ohmic and viscous dissipation.

Entropy balance in the core

To get an estimate which involves the magnetic field we need to consider the rate of entropy production (Hewitt *et al.*, 1975; Gubbins, 1977; Gubbins *et al.*, 1979; Braginsky and Roberts, 1995; Roberts *et al.*, 2003). As magnetic field diffuses, it generates heat through ohmic dissipation,

$$\mathcal{Q}^{\mathrm{D}} = \mu\eta \int \mathbf{j}^2 \mathrm{d}v \qquad \text{(Eq. 7)}$$

Viscous dissipation is believed to be negligible in comparison with ohmic dissipation, because the magnetic energy is much larger than the kinetic energy in the core, and we obtain (Roberts *et al.*, 2003)

$$\mathcal{Q}^{\mathrm{D}} = \frac{T_{\mathrm{D}}}{T_{\mathrm{CMB}}}\left[\left(\mathcal{Q}_{\mathrm{ICB}} + \mathcal{Q}^{\mathrm{L}}\right)\left(1 - \frac{T_{\mathrm{CMB}}}{T_{\mathrm{ICB}}}\right) + \left(\mathcal{Q}^{\mathrm{S}} + \mathcal{Q}^{\mathrm{R}}\right)\right.$$
$$\left.\left(1 - \frac{T_{\mathrm{CMB}}}{T_{\mathrm{M}}}\right)\mathcal{Q}^{\mathrm{G}} - \sum T_{\mathrm{CMB}}\right] \qquad \text{(Eq. 8)}$$

where T_{D} is the average temperature where dissipation occurs, and T_{M} is the average temperature of the fluid outer core. As in a Carnot heat engine, there are "efficiency factors" which mean that thermal convection cannot extract more than $(1 - T_{\mathrm{CMB}}/T_{\mathrm{ICB}}) \sim 20\%$ of the heat flux passing through the core, and for sources which are distributed throughout the core, such as cooling and radioactivity, the efficiency factor may be as low as 10%. Using the estimates above, we find that in the absence of radioactivity the dissipation is around 1 TW, with about 0.5 TW being from compositional convection $\mathcal{Q}^{G}$, and 0.5 TW being from thermal convection (see also Christensen and Tilgner, 2004).

Most current *numerical simulations of the geodynamo* (*q.v.*) have an ohmic dissipation that is less than 1 TW, but this is most probably because they are not yet in the right parameter regime for the geodynamo. Roberts and Glatzmaier (2000) have shown that as the resolution is increased the dissipation rises (see also Kono and Roberts, 2002). Again some geodynamo simulations show a viscous dissipation rate as large as the ohmic dissipation rate, but this again is likely to be a consequence of the numerical difficulties in achieving the right parameter regime.

Kinematic dynamo models

A major difficulty in constructing geodynamo models was noted by Cowling (1934). He showed that it is impossible to maintain an axisymmetric magnetic field by means of dynamo action, a result now known as *Cowling's theorem* (*q.v.*). This was the first of a class of *antidynamo theorems* (*q.v.*) showing that fields with too much symmetry cannot be created by dynamo action. The earliest geodynamo models based on electromagnetic induction appeared in the 1950s (*Bullard* (*q.v.*) and Gellman, 1954), thus showing that although *Cowling's theorem* (*q.v.*) is a mathematical inconvenience (because symmetric solutions of partial differential equations are much easier to find) it is not a fatal objection to the dynamo idea. These models only considered the induction equation (5), the velocity field of the fluid flow being imposed. This is known as the kinematic dynamo problem, the significance of the word kinematic being that the velocity field is simply prescribed, rather than taking the flow to be a solution of an equation of motion. In practice fairly simple flows are chosen, but despite the induction equation (5) being linear in **B**, the nonaxisymmetric three-dimensional nature of the solutions makes it hard to solve, even with today's fast computers. Because of the linearity, the induction equation (5) has either exponentially growing or decaying solutions. We say that a flow is a kinematic dynamo if the solutions grow rather than decay. Fast dynamos, that is dynamos whose growth rate remains finite in the limit of small magnetic diffusion ($R_{\mathrm{m}} \to \infty$) are usually studied in the kinematic dynamo approximation.

It was noticed early on in the study of kinematic dynamos that quadrupolar dynamos, that is dynamos with a dominant quadrupole field (see symmetry and the geodynamo) rather than a dipole field can also be obtained, depending on the chosen flow. The issue of what makes some flows give dipolar dynamos and others quadrupolar dynamos is not yet fully resolved, so we cannot yet give a definite answer to the question why is the Earth dipole dominated rather than quadrupole dominated. *Numerical simulations of the geodynamo* (*q.v.*) usually, but not always, give dipolar fields.

Another issue to arise out of kinematic dynamo studies was the role of meridional circulation, which is the axisymmetric component of flow in the radial and latitudinal direction. (The axisymmetric component of flow in the longitudinal direction is called the azimuthal flow.) The time dependence of a kinematic dynamo is proportional to $\exp(\sigma t)$, and σ is in general complex. If σ happens to real, we say that the dynamo is steady, but if σ has an imaginary part we say it is oscillatory. The *solar dynamo* (*q.v.*) reverses continually in an approximately periodic 22-year cycle, and is therefore naturally modeled as an oscillatory dynamo, but the geodynamo is comparatively steady, although it does reverse (irregularly) occasionally (see *Reversals, theory*). Kinematic dynamo calculations (Roberts, 1972) showed that meridional circulation helps to make a dynamo steady, indicating that meridional circulation could play an important role in the geodynamo. Recent work on kinematic dynamos relevant to the Earth can be found in e.g., Gubbins *et al.* (2000a,b) and Sarson (2003).

Mean field dynamo models

In the 1960s, *mean field dynamos* (*q.v.*) were developed (Steenbeck *et al.*, 1966) see also Moffatt (1978). The basic idea is that small-scale *core turbulence* (*q.v.*) could help to generate the large-scale field. If the magnetic field is $\mathbf{B} + \mathbf{b}$, and the velocity is $\mathbf{U} + \mathbf{u}$ where $\mathbf{b}$ and $\mathbf{u}$ are the small-scale turbulent parts, and $\mathbf{B}$ and $\mathbf{U}$ are the mean parts averaged over the small scales, then the induction equation averaged over the small scales becomes

$$\frac{\partial \mathbf{B}}{\partial t} = \nabla \times (\mathbf{U} \times \mathbf{B}) + \nabla \times (\overline{\mathbf{u} \times \mathbf{b}}) + \nabla^2 \mathbf{B}. \qquad \text{(Eq. 9)}$$

Provided certain criteria are satisfied the new term can be written

$$\nabla \times (\overline{\mathbf{u} \times \mathbf{b}}) = \nabla \times \alpha \mathbf{B} \qquad \text{(Eq. 10)}$$

where α is in general a function of position in the core. More generally, α is a tensor, but it is usually taken to be isotropic so that the comparatively simple form of Eq. (10) can be used. This term in the induction equation is known as the α-effect. The types of turbulent flow which give a nonzero α are those with nonzero mean helicity. The helicity of a flow is $\mathbf{u} \cdot \boldsymbol{\omega}$ where $\boldsymbol{\omega} = \nabla \times \mathbf{u}$ is the vorticity. Flows with a "screw-type" motion have helicity. Parker (1955) pointed out that a blob of hot fluid, rising because of its buoyancy, would twist as it rises due to the action of Coriolis force in a rotating body like the Earth or the Sun. This gives a natural source of helicity. The great advantage of mean field models is that axisymmetric mean fields are now possible solutions of Eqs. (9) and (10). The nonaxisymmetric components of the field which must be there because of *Cowling's theorem* (*q.v.*), do not need to be calculated explicitly. This simplification made it feasible to include dynamics into dynamo models, to obtain what are now known as intermediate dynamo models (see section below on "Intermediate dynamo models").

Mean field models have been extensively used in the *solar dynamo* (*q.v.*) problem, because they give rise to *dynamo waves* (*q.v.*), which can explain many features of the *solar dynamo* (*q.v.*). The main disadvantage of mean field models is that the distribution of α over the core depends on the form of the turbulence in the Earth's core. This cannot be observed, and indeed it is not even known whether such turbulence satisfies the criteria required for a mean field dynamo to be valid. If it were the case that the spatial form of α was not critical, this would be less important, but unfortunately numerical calculations indicated that the α-distribution can make a very big difference to the types of dynamo generated (Hollerbach *et al.*, 1992).

Core dynamics

Some of the most interesting properties of the geodynamo are concerned with the dynamics that is the force balance in the core. This is described by the Navier-Stokes equation with additional forces which are important in the geodynamo. A simplification that is often used is the Boussinesq approximation (see the article on *Boussinesq and anelastic approximations*), in which density variations are ignored except where they are multiplied by g, so that buoyancy forces can be included. This approximation is not strictly valid in the core, as density variations of order 20% can occur, but it is nevertheless a useful simplification of equations which are difficult to solve, and we adopt it here.

The equation of motion for a rotating fluid in which Lorentz (magnetic) forces, buoyancy, and viscosity act can be written

$$\rho\left[\frac{D\mathbf{u}}{Dt} + 2\boldsymbol{\Omega}\times\mathbf{u}\right] = -\nabla p + \mathbf{j}\times\mathbf{B} + \rho\nu\nabla^2\mathbf{u} + \rho g\alpha T\hat{\mathbf{r}}. \quad \text{(Eq. 11)}$$

Here ρ is the density (assumed constant), $\mathbf{u}$ is the velocity, p is the pressure, $\mathbf{j}$ is the current density, $\mathbf{B}$ is the magnetic field, ν is the kinematic viscosity, g is gravity, α is the coefficient of expansion, T is the temperature, and $\hat{\mathbf{r}}$ is the unit vector in the direction of gravity, the radial direction. Not surprisingly, such a complex equation with so many different forces leads to many different physical effects. Note that the Coriolis acceleration, $2\boldsymbol{\Omega}\times\mathbf{u}$ is often thought of as a force (when multiplied by ρ) in a rotating frame. D/Dt is the convective derivative, $\partial/\partial t + \mathbf{u}\cdot\nabla$.

Since the temperature occurs in this equation, another equation determining T is needed,

$$\frac{\partial T}{\partial t} = \kappa\nabla^2 T - \mathbf{u}\cdot\nabla T + Q. \quad \text{(Eq. 12)}$$

Here Q is the heat source and κ is the thermal diffusivity.

A number of dimensionless parameters can be formed from these equations. The Ekman number, $E = \nu/2\Omega d^2$, measures the relative importance of the viscous force to the Coriolis force. To estimate this we need to know the *core viscosity* (*q.v.*), but E is certainly very small, probably of order 10^{-15} in the core. There are two diffusion coefficients, ν and κ, and a third, η, occurs in the induction equation (5), so we have two dimensionless ratios, $\mathrm{Pr} = \nu/\kappa$, known as the Prandtl number, and $\mathrm{P}m = \nu/\eta$ known as the magnetic Prandtl number. Another useful combination involving the magnetic field is the *Elsasser* (*q.v.*) number $\Lambda = B_0^2/2\Omega\mu\rho\eta$, where B_0 is a typical value of the magnetic field. The Elsasser number at the CMB, where the field is typically 0.5 mT, is about 0.25, but inside the core the field is likely to be stronger, and the Elsasser number is believed to be of order unity or a little more.

Rapidly rotating convection

Much of our understanding of convection in rapidly rotating systems comes from the problem of thermally driven *nonmagnetic rotating convection* (*q.v.*), and the problem of convection in an imposed magnetic field (usually either a uniform magnetic field or one with a simple geometry) which is called *magnetoconvection* (*q.v.*). The articles on these topics give a more detailed discussion, but there is a very large literature on both these topics, and some understanding is essential for an appreciation of the geodynamo. Our knowledge has been enhanced by *fluid dynamic experiments* (*q.v.*) and *experimental dynamos* (*q.v.*). It turns out that magnetic field affects the convection considerably, but at least the nonmagnetic case gives us a framework on which to build understanding of the magnetic case.

The onset of instability in a rapidly rotating sphere is now fairly well understood (Roberts, 1968; Busse, 1970; Jones *et al.*, 2000). Rapid rotation means the Ekman number, $E = \nu/2\Omega d^2$ is small, and in this case convection occurs in a columnar form, somewhat as illustrated in Figure G11. However, in nonmagnetic convection the columns are tall but very thin, so that instead of the three columns shown in Figure G11, a much larger number of order $E^{-1/3}$ fit into the sphere. These columns are known as Busse columns (Busse, 1970; Busse and Carrigan, 1976), and their basic structure can be understood in terms of the *Proudman-Taylor theorem* (*q.v.*). The onset of convection in a uniformly heated sphere with stress-free boundaries in the absence of rotation occurs when the Rayleigh number $\mathrm{Ra} = g\alpha\beta r_a^6/\kappa\nu$ exceeds $\mathrm{Ra}_{\mathrm{crit}} = 1\ 546$ (Chandrasekhar, 1961), where $\beta = Q/3\kappa$, so the static gradient is βr. Gravity is taken to be $-gr$, as in a uniform sphere gravity increases linearly from the center. In the small Ekman number limit, i.e., in rapid rotation, the critical Rayleigh number for the onset of convection approaches $4.117E^{-4/3}$ at Prandtl number $\mathrm{Pr} = 1$ (Jones *et al.*, 2000), which gets very large at small E. Although the onset of instability is strongly delayed by the rotation as the Rayleigh number is increased above critical, once the critical value for convection is achieved the flow becomes time-dependent, and then aperiodic, rather quickly. Abdulrahman *et al.* (2000) showed that the bifurcations to chaotic convection occur in a relatively small window in Rayleigh number space $\mathrm{Ra}_{\mathrm{crit}} < \mathrm{Ra} < R_{\mathrm{crit}} + O(E^{-2/3})$.

Convection continues to occur in tall thin columns in the fully nonlinear regime provided the Rossby number $\mathrm{Ro} = U_*/a\Omega$ remains small, although the thickness of the columns does increase somewhat over its very thin linear value, and the columns tend to become rather transient, and occur fairly randomly throughout the sphere. Experiments suggest that in the nonlinear regime, inertial effects take over from viscous effects, and the leading order balance is between inertia, Coriolis force, and buoyancy (for details see Aubert *et al.*, 2001).

When a magnetic field is applied, the *Proudman-Taylor theorem* (*q.v.*) no longer controls the pattern of convection, as the Lorentz force becomes important. An important effect of magnetic field is to increase the thickness of the Busse columns (see e.g., Jones *et al.*, 2003), which occurs even at small Elsasser number $\Lambda \sim O(E^{1/3})$ field strengths. If the Elsasser number is further increased to $O(1)$ values, the columns are still visible, but the number fitting into the sphere is much reduced. Because magnetic field can break the Proudman-Taylor constraint, the critical Rayleigh number is reduced in the presence of a $\Lambda \sim O(1)$ field, so magnetic field can help the system convect, see e.g., Fearn *et al.* (1988).

Dynamical regime in the core

In the magnetic case it is not unreasonable as a first approximation to assume that the large-scale flows and fields have dominant length scales of the size of the core radius. It is then possible to do a useful scale analysis of the dynamics of the Earth's core using Eq. (11) (see e.g., Braginsky and Roberts, 1995; Starchenko and Jones, 2002), although we must keep in mind this can only give order of magnitude estimates, and more exact results require that the geometry of the convection and the spherical domain be taken properly into account.

Some of the physical quantities in the core are quite accurately known, others somewhat less so, but here we only need order of magnitude estimates, to see which terms are large and which small in Eq. (11). We take as a typical value for the velocity in the core $U_* = 2\times10^{-4}$ m s^{-1}, see section "The dynamo process." For D/Dt we take 1/300 years, or 10^{-10} s^{-1}. Significant changes to the field occur on this timescale, though a full reversal of the field might take ten times as long. The density $\rho \sim 10^4$ kg m^{-3} in the core. The Earth's angular velocity $\Omega = 7\times10^{-5}$, and already we can see that the magnitude of the inertial term $|D\mathbf{u}/Dt|$ is very small compared to the Coriolis term $|2\boldsymbol{\Omega}\times\mathbf{u}|$.

We turn next to the Lorentz force term, $\mathbf{j}\times\mathbf{B}$. The field at the CMB is about 5×10^{-4} T. The current is $\nabla\times\mathbf{B}/\mu$, and since the curl is a space derivative, we crudely estimate it at $|\mathbf{B}\pi/a\mu|$ where a is the core radius, 3.5×10^6 m. The Coriolis term $|2\rho\boldsymbol{\Omega}\times\mathbf{u}| \sim 3\times10^{-4}$ N m^{-3} while the Lorentz term $|\mathbf{j}\times\mathbf{B}|$ has magnitude 2.5×10^{-7} N m^{-3}. This is smaller than the Coriolis term, but it is generally believed that the invisible field inside the core is at least 10 times larger than the visible field, which leaves the core, increasing the magnitude of the Lorentz term. Furthermore, the magnetic field pattern at the CMB suggests that it varies on a shorter length scale than the core radius, which will enhance **j**. It is therefore likely that the Lorentz force will be comparable to the other forces at least in some places. The main argument for a significant Lorentz force is that the induction equation (5) is linear in **B** and therefore cannot determine the field strength. This must be limited by the Lorentz force term, so $\mathbf{j} \times \mathbf{B}$ must be significant at least somewhere in the core, and most dynamo models suggest that equilibration of the magnetic field takes place when the Lorentz and Coriolis forces are comparable.

One of the difficulties of numerical simulations is that they require a substantial amount of dissipation, in particular viscous dissipation, for numerical stability. Thus the balance in the models is often between Coriolis force and viscosity, with the Lorentz force playing only a small role in limiting the field strength without greatly affecting the flow. This is known as the weak field regime, in contrast to the strong field regime in which the Lorentz force is comparable to the other large forces in the equation of motion. This is the essential difference between the strong and weak field dynamo regimes. We address this issue further in the section "Taylor's constraint."

The buoyancy force drives the flow, and so we would expect this force to be comparable to the Coriolis and Lorentz force terms. If we estimate g at $10\ \mathrm{ms}^{-2}$, and the coefficient of expansion α at $10^{-5}\ \mathrm{K}^{-1}$, we find that the typical temperature fluctuation is in the range 10^{-3}–10^{-4} K. It is rather remarkable that such tiny temperature fluctuations in the core are driving the whole dynamo process, but we should remember that core flows are very slow; even a lethargic snail could go faster! We also have a consistency check here, because the convective heat flux

$$Q_{\mathrm{conv}} = \rho c_{\mathrm{p}} \int_S u_{\mathrm{r}} T \mathrm{d}S, \quad \text{(Eq. 13)}$$

can be estimated, and when we remember that T here is the very small superadiabatic temperature, which is 10^{-3}–10^{-4} K only, we reassuringly discover that this flux is comparable to that suggested by the energy arguments of the section above "Energy balance for a convective dynamo."

The viscous term in the equation of motion turns out to be very small, and so is not in the primary force balance, except in thin *boundary layers in the core* (*q.v.*). It is however necessary to include the viscous term in the force balance when doing numerical simulations, to ensure numerical stability. Pressure forces are always important in convecting fluids; when buoyant fluid rises, the pressure forces push the ambient fluid aside to make room for the hot rising fluid.

The main balance of forces in the core is therefore between Lorentz force, Coriolis force, buoyancy force, and pressure force. This is known as MAC balance, M standing for magnetic, A for Archimedean force (Buoyancy), and C for Coriolis force (Braginsky, 1967). The word "magnetostrophic" is sometimes used as a synonym for MAC balance, though more commonly magnetostrophic balance refers to situations where buoyancy is absent. Pressure can usually be eliminated by taking the curl of the equation of motion, which converts it into a vorticity equation.

If compositional convection is present, an additional equation of similar form to the temperature equation (12) is required for ξ, the fraction of light material, and the resulting buoyancy must be included in the equation of motion (see e.g., Braginsky and Roberts, 1995). Equations (5), (11), and (12) can be solved numerically in dynamo simulations, but an important issue is whether laminar or turbulent values should be used for the diffusivities. Laminar values of κ and ν are so small it is hard to believe they can be relevant to the core. Turbulent values are therefore generally used, and this raises the interesting question of the nature of turbulence in the Earth's core. The inertial terms, which are responsible for cascading energy down to small scales in "normal" fluids are only effective on tiny length scales in the core, so core turbulence may be quite different from normal turbulence. In particular, the diffusion processes may be quite anisotropic, as suggested by Braginsky and Meytlis (1990).

Waves and instabilities

Calculating how the field and the flow evolves in the core can only be done using sophisticated numerical dynamo models. However, insight can be obtained into the rather complicated nature of the dynamics of the core by considering simple equilibrium flows and fields, and analyzing the *magnetohydrodynamic waves* (*q.v.*) and instabilities that these simple equilibria undergo when they are slightly perturbed. This leads to linear problems which can be investigated mathematically. Perhaps the most fundamental problem is the case of a uniform magnetic field with fluid at rest (or in uniform motion) with a uniform temperature gradient imposed. Linearizing Eqs. (5), (11), and (12) about a uniform magnetic field $\mathbf{B}_0$ and constant temperature gradient β, so that $\mathbf{B} = \mathbf{B}_0 + \mathbf{b}'$, the velocity is $\mathbf{u}'$ and the temperature is $T = \beta z + T'$. The primed quantities are assumed small so that squares and products of primed quantities are neglected, and then wave solutions with $\mathbf{u}', \mathbf{b}'$, and T' proportional to $\exp i(\mathbf{k} \cdot \mathbf{x} - \omega t)$ can be found. Equations (5), (11), and (12) then give the dispersion relation between $\mathbf{k}$ and ω. If the temperature gradient is stabilizing (subadiabatic) then real values of ω are found if diffusion is neglected. If diffusion is retained, ω is complex with a negative imaginary part corresponding to damped oscillations. If the temperature gradient is superadiabatic, growing waves (instabilities) are found provided the diffusion is not too large.

The dispersion relation is actually very complicated, but it can be simplified when as in the Earth's core, different types of waves have very different frequencies (Fearn *et al.*, 1988). The fastest waves are inertial waves (see the article on *gravity-inertio waves and inertial oscillations*), which balance the Coriolis and inertial accelerations, and the inertial wave frequency is given by $\omega_C = 2(\mathbf{\Omega} \cdot \mathbf{k})/|\mathbf{k}|$. The typical period is therefore of the order of a day. These waves have not yet been observed in the core, but they are expected to be driven by tidal forcing. *Alfvén waves* (*q.v.*) result from a balance of inertia and Lorentz force in the equation of motion, when combined with the induction equation. These waves have frequency $\omega_M = (\mathbf{B}_0 \cdot \mathbf{k})/(\mu\rho)^{1/2}$, and travel at the Alfvén speed, $\mathbf{B}_0/(\mu\rho)^{1/2}$, which for a moderate 1 mT core field is around $10^{-2}\ \mathrm{m\ s}^{-1}$, giving around 60 years for the wave to travel round the core. An important class of Alfvén waves are those corresponding to azimuthal motion constant on cylinders, which are called *torsional oscillations* (*q.v.*). These are believed to be important in the core; see "Taylor's constraint" below.

Another timescale comes from the temperature gradient, from the balance of buoyancy and inertia in the equation of motion, combined with the temperature equation (12). The frequency of internal gravity waves is $\omega_A = (g\alpha\beta)^{1/2} k_H/|\mathbf{k}|$, where β is the subadiabatic temperature gradient, and k_H is the component of $\mathbf{k}$ perpendicular to gravity. In a convectively unstable region, the temperature gradient is superadiabatic and β is negative. Then ω_A is imaginary, which corresponds to an exponentially growing unstable mode, with $|\omega_A|$ being the growth rate. With the estimate of 10^{-4} K for a typical superadiabatic temperature fluctuation, $\beta \sim 10^{-4}/a$, giving a typical growth rate of about 1 year.

When all the terms in the equation of motion are present, the dispersion relation gives a fast inertial wave and a slow wave in which only the time-derivative terms in the induction and temperature equations are important, inertia being negligible. These slow waves are known as MAC waves and have frequency

$$\omega_{\mathrm{MAC}} = \omega_{\mathrm{MC}}(1 + \omega_A^2/\omega_M^2)^{1/2}, \quad \text{where } \omega_{\mathrm{MC}} = \omega_M^2/\omega_C. \quad \text{(Eq. 14)}$$

When $|\omega_A| > |\omega_M|$, ω_{MAC} is imaginary, corresponding to a growing convective mode. The rate of growth is slow, since $\tau_{\mathrm{MC}} = 2\pi/\omega_{\mathrm{MC}}$ is of the order of some thousands of years. It is comparable to the magnetic diffusion time, since the ratio $\tau_{\mathrm{diff}}/\tau_{\mathrm{MC}} = B_0^2/2\mu\rho\Omega\eta = \Lambda$ and the Elsasser number Λ has a value of $O(1)$ in the core. The MAC growth rate will be a little larger than $1/\tau_{MC}$ because $|\omega_A| > \omega_M$, but these slow growth times are consistent with the time taken for the typical convective velocity to take fluid across the core, $\tau_{conv} \sim 10^{10}$ s, so the dynamical picture does seem to be self-consistent. The main role of the Coriolis acceleration and the Lorentz force is to constrain the convection and slow it down from an unimpeded growth rate of about a year down to τ_{conv}.

There are, however, some difficulties. Although for general wave vectors **k** the MAC wave timescale is slow, it is much faster if **k** and $\boldsymbol{\Omega}$ are perpendicular. This is the case for motions which are independent of z, the coordinate parallel to the rotation axis. Then ω_{MC} is infinite, which means we have to restore inertia, and we then get the much faster torsional wave frequency. It is also possible for **k** to be perpendicular to both $\boldsymbol{\Omega}$ and $\mathbf{B}_0$. Such waves have motion only along "plates," planes containing the rotation vector and the magnetic field vector. For these waves Eq. (14) is degenerate, with both ω_M and ω_C zero. Such motions are unaffected by the field and the rotation and so will grow on the much shorter ω_A timescale. We therefore expect these modes to dominate small scale convection in the core (Braginsky and Meytlis, 1990).

The above analysis assumes uniform magnetic fields. Actually the field inside the core is likely to have a complicated structure. Such fields often become unstable, and allow magnetic energy to be converted into kinetic energy. In the Sun, this can happen on a very short timescale, and solar flares are the result. Nothing quite as dramatic is expected in the Earth's core, because magnetic diffusion is much more important there. Nevertheless magnetic instabilities could be important in driving shorter timescale motions and in controlling the strength of the magnetic field. Magnetic instabilities generally become important when the Elsasser number is $O(1)$ (Fearn, 1998), (see also *Core magnetic instabilities*) so this may well be an important mechanism in the core.

Taylor's constraint

One interesting consequence of the MAC balance in the core pointed out by J.B. Taylor (1963) was that the geomagnetic field must satisfy a special condition, known as *Taylor's condition* (*q.v.*). Azimuthal flows that are constant on cylinders coaxial with the rotation axis play a rather special role in rotating flows, because the Coriolis acceleration can be balanced entirely by the pressure force. Such flows are called geostrophic flows. The special feature of the cylinders is that a column of fluid moving along such a cylinder has constant height, so no vortex stretching occurs. J.B. Taylor noted that if we integrate the net azimuthal force over a cylinder, the contribution from the Coriolis term is zero, as is the pressure term, and since buoyancy acts radially, not azimuthally, the only force left is Lorentz force. It follows that

$$\int_S (\mathbf{j} \times \mathbf{B})_\phi \mathrm{d}s = 0, \qquad \text{(Eq. 15)}$$

where S is any cylindrical surface not intersecting the inner core. This is known as Taylor's constraint, and provided viscosity is negligible it must be satisfied in the core. Since the current is related to the field by Ampere's law, (1), Taylor's condition is a constraint on the magnetic field. An interesting question is what happens if we solve an initial value problem in which Taylor's constraint is not initially satisfied? We must then restore part of the inertial term, $\rho\partial\mathbf{u}/\partial t$, and the system responds with a fast *torsional oscillation* (*q.v.*), in which Lorentz forces directly accelerate the fluid. These oscillations have a period of some tens of years, and there is evidence that they occur in the core, as they may explain the *decadal variations of the length of day* (*q.v.*) (Jault *et al.*, 1988; Jackson, 1997; Jault, 2003), and possibly shorter period elements of the *geomagnetic secular variation* (*q.v.*), such as the so-called *geomagnetic jerks* (*q.v.*). On the much longer dynamo timescale, these oscillations would then decay away, restoring a field satisfying Taylor's condition. We can view the MAC balance state satisfying Taylor's constraint as a slowly evolving equilibrium state, with the magnetic field slowly changing as the convection moves the core fluid, on a timescale of hundreds of years. Short-term variations of the field on small length scales can be explained by convection, as variations on a short length-scale ℓ would occur on short timescale ℓ/U_*, but the short-term variations on long length scales, such as give rise to the decadal variations of the length of day, could be explained if torsional oscillations were continually being excited, so that the core is always vibrating around its equilibrium MAC state. How these torsional oscillations are excited is not yet understood.

Intermediate dynamo models

The full time-dependent three-dimensional equations (5), (11), and (12) are difficult to solve numerically, so it is only recently that computers have become fast enough to make it feasible to study them. Numerical simulations can also be hard to interpret. A compromise position has been to still retain the α-effect, thus allowing axisymmetric solutions to be found, but to include some dynamics, normally the meridional circulation and the azimuthal flow driven by the magnetic field.

Intermediate dynamo models have often been used to investigate how Taylor's constraint could be met. It was suggested by Malkus and Proctor (1974) that the magnetic field would drive flows of exactly the right type to generate fields satisfying Taylor's constraint. Soward and Jones (1983) looked at an intermediate plane layer model which did indeed lead to solutions satisfying Taylor's constraint, in accord with the Malkus-Proctor scenario. With values of α close to critical, the magnetic field did not satisfy Taylor's constraint. The azimuthal force balance was therefore between the Lorentz force and the Ekman friction in the boundary layer which is $O(E^{1/2})$, so that the magnetic field was limited to a value with Elsasser number $O(E^{1/2})$. This balance is known as an Ekman state. However, as α was increased, a value was reached at which the field did satisfy Eq. (15), and at that point the field dramatically increased, leaving the Ekman state and having a strength with Elsasser number $O(1)$. However, Braginsky (1976, 1994) proposed an almost axisymmetric *model-Z dynamo* (*q.v.*) in which the Taylor constraint was met in an unusual way. In an axisymmetric configuration, Eq. (15) can be written

$$\frac{\mathrm{d}}{\mathrm{d}s}\left(s^2 \int BB_s \mathrm{d}z\right) = 0, \qquad \text{(Eq. 16)}$$

where B_s is the component of magnetic field pointing away from the rotation axis. Taylor's constraint could be satisfied if the meridional part of **B** was purely in the z-direction, so $B_s = 0$, hence the name model Z dynamo. Braginsky proposed a specific distribution of α concentrated near the equatorial region and the CMB, which led to a steady rather than an oscillatory dynamo, as part of the model Z scenario. A more detailed analysis by Jault (1995) suggested however that if the viscosity is reduced to very low values, "Taylorization" (that is the approach to a state in which Eq. (15) holds) occurs in the conventional Malkus-Proctor way, with B_s being small but finite.

Fully three-dimensional simulations usually have to be run with a viscosity too large to achieve "Taylorization" in order to maintain numerical stability. However, Rotvig and Jones (2002) have run fully three-dimensional simulations in plane geometry, which is less demanding computationally, and allows smaller E to be reached. They found that "Taylorization" did occur at the lowest values of E they could reach.

Another way of simplifying the full three-dimensional problem, is the so-called "$2\frac{1}{2}$-dimensional" approach. Here full resolution is used in the r and θ directions, but in the azimuthal ϕdirection a severe truncation is imposed; in the most draconian approximation only the axisymmetric and one nonaxisymmetric $\exp(im\phi)$ modes are retained (e.g., Jones *et al.*, 1995). This still allows nonaxisymmetric convection to occur, so no α-mechanism is needed, and the results are generally qualitatively similar to those obtained from fully three-dimensional simulations, but at much reduced cost. The disadvantage is that unless results are checked by fully three-dimensional simulations, one cannot be certain that the truncation is not affecting the behavior; nevertheless "$2\frac{1}{2}$-dimensional" models are very useful for testing out new ideas in dynamo theory.

Role of the inner core

As mentioned in the energy balance section, the inner core makes an important contribution to the driving of convection in the core. It also significantly affects the dynamics of core convection. The importance of geostrophic motions in rapidly rotating fluids suggests that the solid inner core could play an important role in the dynamics of the Earth's interior. The cylinder, which just touches the inner core, is called the *inner core tangent cylinder* (*q.v.*). Whereas the volume of the inner core itself is only about 4% of the volume of the outer core, the fraction inside the tangent cylinder is much larger.

The flow outside the tangent cylinder can convect heat out in tall thin columns, which although not exactly geostrophic, because some vortex stretching occurs as the columns rotate due to the sloping boundaries, nevertheless minimizes the effect of rotation. Inside the tangent cylinder the convection needs to get the heat out in a direction almost parallel to the rotation axis, and tall thin columns run into the inner core, causing additional friction. In consequence, most models show that convection is more efficient outside the tangent cylinder than inside it. This difference produces a latitudinal temperature gradient, with the poles being slightly hotter than the equator on a sphere of constant radius. Neglecting magnetic forces, the azimuthal component of the curl of Eq. (11), then gives the *thermal wind* (*q.v.*) equation,

$$2\Omega \frac{\partial u_\phi}{\partial z} = \frac{g\alpha}{r} \frac{\partial T}{\partial \theta} \qquad \text{(Eq. 17)}$$

where z is the coordinate parallel to the rotation axis and θ is the meridional direction in spherical polars (r, θ, ϕ). This shows that u_ϕ decreases with z in the northern hemisphere. A similar effect drives the jet stream in the Earth's atmosphere, though the sign is different (the jet stream velocities increase with height, because the pole is colder than the equator, and so the jet stream goes from west to east relative to the surface) and of course temperature gradients and velocities are much smaller in the core. Nevertheless, the thermal wind is sufficient to give a *rotation of the inner core* (*q.v.*) relative to the mantle. The strength of the thermal wind can also be affected by magnetic torques. It has been suggested (Buffett and Glatzmaier, 2000) that the acceleration produced by the thermal wind may be counteracted by a gravitational coupling of the inner core to the mantle, which may be expected if there are slight departures from axisymmetry in the figure of the mantle and the inner core. The rotation of the inner core relative to the mantle can in principle be measured by seismologists, but the actual value of the inner core rotation rate is still controversial (e.g., Collier and Helffrich, 2001).

Reversals and field morphology

Two obvious questions about the geodynamo are (i) why is the field dominated by an axial dipole component? (ii) why does the field periodically reverse its polarity? Our lack of a full understanding of the geodynamo is perhaps highlighted by the fact that neither of these questions can be answered unambiguously. The majority of dynamo simulations do show a dipole dominated field, but by no means all do (e.g., Christensen *et al.*, 1999; Busse, 2002), and the Karlsruhe experimental dynamo was dominated by an equatorial rather than an axial dipole. Quadrupolar dynamos, equatorial dipoles, and axial dipoles can all be produced from flows, which are apparently not that dissimilar. A number of trends are apparent from the simulations; for example, a strong azimuthal flow makes an equatorial dipole less likely, as the shearing motion associated with differential rotation tends to disrupt an equatorial dipole field. Dipolar fields tend to be more common when the convection is driven by a flux of heat from the inner core, whereas quadrupolar dynamos are more common in uniformly heated models, and there is a tendency (Busse, 2002) for dipoles to be preferred over quadrupoles at higher Rayleigh number. However, all these observations are model dependent, and it should not be forgotten that it is not currently possible to run geodynamo models in the correct parameter regime.

A large number of papers have addressed the issue of the *theory of reversals* (*q.v.*) (see e.g., Sarson, 2000). One simple point is that the dynamo equations are invariant under a reversal of the field direction, so if a solution with "normal" polarity exists, an exactly similar one with reversed polarity also exists. To see this note that the transformation $\mathbf{B} \rightarrow -\mathbf{B}$ changes the sign of both $\mathbf{B}$ and $\mathbf{j}$, so the Lorentz force $\mathbf{j} \times \mathbf{B}$ is unchanged, and the induction equation is linear in $\mathbf{B}$ so reversing the sign of $\mathbf{B}$ leaves this equation unchanged. This observation shows that it is reasonable to expect the geodynamo to reverse occasionally, but it doesn't explain why reversals are comparatively infrequent (there are only a few reversals per million years on average) and why they happen relatively quickly (around 5 ka or less) when they occur.

The magnetic field is continually fluctuating, and reversals appear to be part of this process. Magnetic excursions, events where the axis of the dipolar component moves rapidly away from the geographic poles occur much more frequently than full reversals, so it is possible that excursions are in some sense failed reversals, where the fluctuation is large but not large enough to cross a threshold leading to full reversal. Hollerbach and Jones (1993, 1995) suggested that the threshold necessary was that the field in the solid inner core had to be reversed. Since this can only occur by magnetic diffusion, rather than on the somewhat more rapid convective turnover time, only the larger, longer, excursions would allow the inner core field to reverse, and hence establish a full reversal. Numerical simulations of the geodynamo also suggest that the core is in a continually fluctuating state, and that some of these fluctuations occasionally lead to reversals. Glatzmaier *et al.* (1999) found that a spatially inhomogeneous heat flux at the core-mantle boundary could affect reversal frequency (see also *Inhomogeneous boundary conditions and the dynamo*). This is an interesting possibility, as mantle convection models suggest that such inhomogeneities are likely, and indeed seismic measurements of the CMB region also suggest there is such inhomogeneity. This would provide a link between mantle convection and the dynamo, and since mantle convection evolves on a very long million year timescale, this could explain why reversals are very infrequent compared to excursions. For example, on this picture during the 60-million-year Cretaceous superchron, during which there were no significant reversals, mantle convection was in a pattern, which affected the core flow in such a way as to make reversals unlikely. Sarson and Jones (1999) suggested that this might be connected with the meridional circulation, which would be associated with an inhomogeneous CMB heat flux, as in their model the strength of the meridional circulation played an important role in the reversal process.

There is much that we do not yet understand about the dynamics of the Earth's core and the reversal process, but progress is being made on a number of different fronts: better numerical simulations, and deeper understanding of the processes involved; more detailed and more accurate paleomagnetic studies giving information about the past behavior of the geomagnetic field; new results from seismology, probing the structure of the Earth's deep interior ever more thoroughly; a new and better understanding of the physical properties of matter at very high pressure. With all these stimuli, our understanding of the geodynamo will surely radically improve over the coming decades.

Chris Jones

Bibliography

Abdulrahman, A., Jones, C.A., Proctor, M.R.E., and Julien, K., 2000. Large wavenumber convection in the rotating annulus. *Geophysical and Astrophysical Fluid Dynamics*, **93**: 227–252.

Aldridge, K.D., 2003. Dynamics of the core at short periods. In Jones, C.A., Soward, A.M., and Zhang, K. (eds.) *Earth's Core and Lower Mantle*. London: Taylor & Francis, pp. 180–210.

Aubert, J., Brito, D., Nataf, H.C. *et al.*, 2001. A systematic experimental study of rapidly rotating spherical convection in water and liquid gallium. *Physics of the Earth and Planetary Interiors*, **128**: 51–74.

Bloxham, J., and Gubbins, D., 1989. The evolution of the Earth's magnetic field. *Scientific American*, **261**: 30–37.

Bloxham, J., and Jackson, A., 1991. Fluid flow near the surface of the Earth's outer core. *Reviews of Geophysics*, **29**: 97–120.

Braginsky, S.I., 1963. Structure of the F layer and reasons for convection in the Earth's core. *Soviet Physics Doklady*, **149**: 8–10.

Braginsky, S.I., 1967. Magnetic waves in the Earth's core. *Geomagnetism and Aeronomy*, **7**: 851–859.

Braginsky, S.I., 1976. On the nearly axially-symmetrical model of the hydro-magnetic dynamo of the Earth. *Physics of the Earth and Planetary Interiors*, **11**: 191–199.

Braginsky, S.I., 1993. MAC-oscillations of the hidden ocean of the core. *Journal of Geomagnetism and Geoelectricity*, **45**: 1517–1538.

Braginsky, S.I., 1994. The nonlinear dynamo and model-Z. In Proctor, M.R.E., and Gilbert, A.D. (eds.) *Lectures on Solar and Planetary Dynamos*. Cambridge: Cambridge University Press, pp. 267–304.

Braginsky, S.I., and Meytlis, V.P., 1990. Local turbulence in the Earth's core. *Geophysical and Astrophysical Fluid Dynamics*, **55**: 71–87.

Braginsky, S.I., and Roberts, P.H., 1995. Equations governing convection in the Earth's core and the geodynamo. *Geophysical and Astrophysical Fluid Dynamics*, **79**: 1–97.

Buffett, B.A., and Glatzmaier, G.A., 2000. Gravitational braking of inner-core rotation in geodynamo simulations. *Geophysical Research Letters*, **27**: 3125–3128.

Bullard, E.C., and Gellman, H., 1954. Homogeneous dynamos and terrestrial magnetism. *Philosophical Transactions of the Royal Society of London A*, **247**: 213–255.

Busse, F.H., 1970. Thermal instabilities in rapidly rotating systems. *Journal of Fluid Mechanics*, **44**: 441–460.

Busse, F.H., 2002. Convective flows in rapidly rotating spheres and their dynamo action. *Physics of Fluids*, **14**: 1301–1314.

Busse, F.H., and Carrigan, C.R., 1976. Laboratory simulation of thermal convection in rotating planets and stars. *Science*, **191**: 81–83.

Chandrasekhar, S., 1961. *Hydrodynamic and Hydromagnetic Stability*. Oxford: Clarendon Press.

Christensen, U., and Tilgner, A., 2004. Power requirement of the geodynamo from ohmic losses in numerical and laboratory dynamos. *Nature*, **429**: 169–171.

Christensen, U., Olson, P., and Glatzmaier, G.A., 1999. Numerical modelling of the geodynamo: a systematic parameter study. *Geophysical Journal International*, **138**: 393–409.

Collier, J.D., and Helffrich, G., 2001. Estimate of inner core rotation rate from United Kingdom regional seismic network data and consequences for inner core dynamical behaviour. *Earth and Planetary Science Letters*, **193**: 523–537.

Cowling, T.G., 1934. The magnetic field of sunspots. *Monthly Notices of the Royal Astronomical Society*, **94**: 39–48.

Davidson, P.A., 2001. *An Introduction to Magnetohydrodynamics*. Cambridge: Cambridge University Press.

Fearn, D.R., 1998. Hydromagnetic flow in planetary cores. *Reports on Progress in Physics*, **61**: 175–235.

Fearn, D.R., Roberts, P.H., and Soward, A.M., 1988. Convection, stability and the dynamo. In Galdi, G.P., and Straughan, B. (eds.) *Energy, Stability and Convection*. New York: Longmans, pp. 60–324.

Glatzmaier, G.A., and Roberts, P.H., 1995. A three-dimensional self-consistent computer simulation of a geomagnetic field reversal. *Nature*, **377**: 203–209.

Glatzmaier, G.A., and Roberts, P.H., 1997. Simulating the geodynamo. *Contemporary Physics*, **38**: 269–288.

Glatzmaier, G.A., Coe, R.S., Hongre, L., and Roberts, P.H., 1999. The role of the Earth's mantle in controlling the frequency of geomagnetic reversals. *Nature*, **401**: 885–890.

Gubbins, D., 1977. Energetics of the Earth's core. *Journal of Geophysics*, **43**: 453–464.

Gubbins, D., Masters, T.G., and Jacobs, J.A., 1979. Thermal evolution of the Earth's core. *Geophysical Journal of the Royal Astronomical Society*, **59**: 57–99.

Gubbins, D., Barber, C.N., Gibbons, S. *et al.*, 2000a. Kinematic dynamo action in a sphere. I. Effects of differential rotation and meridional circulation on solutions with axial dipole symmetry. *Proceedings of the Royal Society of London A*, **456**: 1333–1353.

Gubbins, D., Barber, C.N., Gibbons, S. *et al.*, 2000b. Kinematic dynamo action in a sphere. II. Symmetry selection. *Proceedings of the Royal Society of London A*, **456**: 1669–1683.

Hewitt, J.M., McKenzie, D.P., and Weiss, N.O., 1975. Dissipative heating in convective flows. *Journal of Fluid Mechanics*, **68**: 721–738.

Hollerbach, R., and Jones, C.A., 1993. Influence of the Earth's inner core on geomagnetic fluctuations and reversals. *Nature*, **365**: 541–543.

Hollerbach, R., and Jones, C.A., 1995. On the magnetically stabilizing role of the Earth's inner core. *Physics of the Earth and Planetary Interiors*, **87**: 171–181.

Hollerbach, R., Barenghi, C.F., and Jones, C.A., 1992. Taylor's constraint in a spherical $\alpha\omega$-dynamo. *Geophysical and Astrophysical Fluid Dynamics*, **67**: 3–25.

Jackson, A., 1997. Time-dependency of tangentially geostrophic core surface motions. *Physics of the Earth and Planetary Interiors*, **103**: 293–311.

Jault, D., 1995. Model Z by computation and Taylor's condition. *Geophysical and Astrophysical Fluid Dynamics*, **79**: 99–124.

Jault, D., 2003. Electromagnetic and topographic coupling, and LOD variations. In Jones, C.A., Soward, A.M., and Zhang, K. (eds.) *Earth's Core and Lower Mantle*. London: Taylor & Francis, pp. 56–76.

Jault, D., Gire, C., and Lemouel, J.L., 1988. Westward drift, core motions and exchanges of angular-momentum between core and mantle. *Nature*, **333**: 353–356.

Jones, C.A., 2003. Dynamos in planets. In Thompson, M.J., and Christensen-Dalsgaard, J.C. (eds.) *Stellar Astrophysical Fluid Dynamics*. Cambridge: Cambridge University Press, pp. 159–176.

Jones, C.A., Longbottom, A.W., and Hollerbach, R., 1995. A self-consistent convection driven geodynamo model, using a mean field approximation. *Physics of the Earth and Planetary Interiors*, **92**: 119–141.

Jones, C.A., Soward, A.M., and Mussa, A.I., 2000. The onset of convection in a rapidly rotating sphere. *Journal of Fluid Mechanics*, **405**: 157–179.

Jones, C.A., Mussa, A.I., and Worland, S.J., 2003. Magnetoconvection in a rapidly rotating sphere: the weak-field case. *Proceedings of the Royal Society of London A*, **459**: 773–797.

Kerswell, R.R., 2002. Elliptical instability. *Annual Review of Fluid Mechanics*, **34**: 83–113.

Kono, M., and Roberts, P.H., 2002. Recent geodynamo simulations and observations of the geomagnetic field. *Reviews of Geophysics*, **40**(4): 1013.

Malkus, W.V.R., 1994. Energy sources for planetary dynamos. In Proctor, M.R.E., and Gilbert, A.D. (eds.) *Lectures on Solar and Planetary Dynamos*. Cambridge: Cambridge University Press, pp. 161–179.

Malkus, W.V.R., and Proctor, M.R.E., 1974. The macrodynamics of α-effect dynamos in rotating fluids. *Journal of Fluid Mechanics*, **67**: 417–443.
Merrill, R.T., McElhinny, M.W., and McFadden, P.L., 1996. *The Magnetic Field of the Earth.* San Diego, CA: Academic Press.
Moffatt, H.K., 1978. *Magnetic Field Generation in Electrically Conducting Fluids.* Cambridge: Cambridge University Press.
Parker, E.N., 1955. The formation of sunspots from the solar toroidal field. *Astrophysics Journal*, **121**: 491–507.
Roberts, P.H., 1967. *An Introduction to Magnetohydrodynamics.* London: Longmans.
Roberts, P.H., 1968. On the thermal instability of a rotating fluid sphere containing heat sources. *Philosophical Transactions of the Royal Society of London A*, **263**: 93–117.
Roberts, P.H., 1972. Kinematic dynamo models. *Philosophical Transactions of the Royal Society of London A*, **272**: 663–703.
Roberts, P.H., and Glatzmaier, G.A., 2000. A test of the frozen flux approximation using geodynamo simulations. *Philosophical Transactions of the Royal Society of London A*, **358**: 1109–1121.
Roberts, P.H, Jones, C.A., and Calderwood, A., 2003. Energy fluxes and ohmic dissipation. In Jones, C.A., Soward, A.M., and Zhang, K. (eds.) *Earth's Core and Lower Mantle.* London: Taylor & Francis, pp. 100–129.
Rotvig, J., and Jones, C.A., 2002. Rotating convection-driven dynamos at low Ekman number. Physical Review E, **66**: 056308-1-15.
Sarson, G.R., 2000. Reversal models from dynamo calculations. *Philosophical Transactions of the Royal Society of London A*, **358**: 921–942.
Sarson, G.R., 2003. Kinematic dynamos driven by thermal-wind flows. *Proceedings of the Royal Society of London A*, **459**: 1241–1259.
Sarson, G.R., and Jones, C.A., 1999. A convection driven geodynamo reversal model. *Physics of the Earth and Planetary Interiors*, **111**: 3–20.
Schubert, G., Turcotte, D.L., and Olson, P., 2001. *Mantle Convection in the Earth and Planets.* Cambridge: Cambridge University Press.
Sleep, N.H., 1990. Hot spots and mantle plumes: some phenomenology. *Journal of Geophysical Research*, **95**: 6715–6736.
Soward, A.M., and Jones, C.A., 1983. α^2-dynamos and Taylor's constraint. *Geophysical and Astrophysical Fluid Dynamics*, **27**: 87–122.
Starchenko, S., and Jones, C.A., 2002. Typical velocities and magnetic field strengths in planetary interiors. *Icarus*, **157**: 426–435.
Steenbeck, M., Krause, F., and Rädler, K-H., 1966. A calculation of the mean electromotive force in an electrically conducting fluid in turbulent motion, under the influence of Coriolis forces. *Zeitschrift für Naturforschung*, **21a**: 369–376.
Stevenson, D.J., 1982. Interiors of the giant planets. *Annual Review of Earth and Planetary Sciences*, **10**: 257–295.
Stevenson, D.J., 1987. Mercury magnetic field—a thermoelectric dynamo. *Earth and Planetary Science Letters*, **82**: 114–120.
Stevenson, D.J., 2003. Planetary magnetic fields. *Earth and Planetary Science Letters*, **208**: 1–11.
Taylor, J.B., 1963. The magneto-hydrodynamics of a rotating fluid and the Earth's dynamo problem. *Proceedings of the Royal Society of London A*, **274**: 274–283.
Tilgner, A., 2005. Precession driven dynamos. Physics of Fluids, 17: 034104-1.
Verhoogen, J., 1961. Heat balance in the Earth's core. *Geophysical Journal*, **4**: 276–281.

Cross-references

Alfvén Waves
Antidynamo and Bounding Theorems
Boussinesq and Anelastic Approximations
Bullard, Edward Crisp (1907–1980)
Compass
Convection, Chemical
Convection, Nonmagnetic Rotating
Core Composition
Core Convection
Core Density
Core Magnetic Instabilities
Core Motions
Core Properties, Physical
Core Temperature
Core Turbulence
Core Viscosity
Core, Adiabatic Gradient
Core, Boundary Layers
Core, Electrical Conductivity
Core, Thermal Conduction
Core-based Inversions for the Main Geomagnetic Field
Core-Mantle Boundary, Heat Flow Across
Cowling's Theorem
Crustal Magnetic Field
Depth to the Curie Temperature
Dynamo Waves
Dynamo, Model-Z
Dynamo, Solar
Dynamos, Experimental
Dynamos, Mean Field
Dynamos, Planetary and Satellite
Elsasser, Walter M. (1904–1991)
Fluid Dynamics Experiments
Gauss, Carl Friedrich (1777–1855)
Gellibrand, Henry (1597–1636)
Geodynamo, Energy Sources
Geodynamo, Numerical Simulations
Geomagnetic Jerks
Geomagnetic Secular Variation
Gilbert William (1544–1603)
Gravito-inertio Waves and Inertial Oscillations
Grüneisen's Parameter for Iron and Earth's Core
Halley, Edmond (1656–1742)
Harmonics, Spherical
Inhomogeneous Boundary Conditions and the Dynamo
Inner Core Composition
Inner Core Rotation
Inner Core Tangent Cylinder
Interiors of Planets and Satellites
Ionosphere
Larmor, Joseph (1857–1942)
Length of Day Variations, Decadal
Magnetic Field of Sun
Magnetoconvection
Magnetohydrodynamic Waves
Magnetohydrodynamics
Melting Temperature of Iron in the Core
Nondynamo Theories
Oscillations, Torsional
Precession and Core Dynamics
Proudman-Taylor Theorem
Radioactive Isotopes and their Decay in Core and Mantle
Reversals, Theory
Storms and Substorms, Magnetic
Taylor's Condition
Thermal Wind
Verhoogen, John (1912–1993)
Westward Drift

GEODYNAMO, DIMENSIONAL ANALYSIS AND TIMESCALES

It is common to put the equations governing a complex physical system like the geodynamo into dimensionless form. This clarifies the importance of each term in the equations and can reduce the number of input parameters to a minimum. Consider the very simple example of the *kinematic dynamo* (*q.v.*), which is governed by a single partial differential equation, the induction equation:

$$\frac{\partial \boldsymbol{B}}{\partial t} = \nabla \times (\boldsymbol{v} \times \boldsymbol{B}) + \frac{1}{\mu_0 \sigma} \nabla^2 \boldsymbol{B}. \qquad \text{(Eq. 1)}$$

The fluid velocity is specified within a sphere and the induction equation solved for a growing magnetic field. It depends on just three input parameters: the amplitude of the velocity V, the radius of the sphere c, and the electrical conductivity σ. In dimensionless form this reduces to just one input parameter, the magnetic Reynolds number $R_m = \mu_0 \sigma V c$, where μ_0 is the permeability of free space (see Tables G2 and G3). Kinematic dynamo behavior therefore depends on the product of the three dimensional input parameter but not on all three independently. This is a great help in exploring the behavior of the system as a function of the input parameters. Incidentally, it means that scaling the Earth down to a laboratory-sized experiment would require a great increase in fluid velocity to compensate for the reduction in length scale (increasing the conductivity is impossible since there are no materials at room temperature with electrical conductivity significantly higher than that of iron in the core).

Nondimensionalization is by no means a unique process: there are many ways in which to scale the dimensional variables, each giving different versions of the same equations. Consider once again the kinematic dynamo problem. The usual approach is to scale length with the radius c and time with the magnetic diffusion time, $\mu_0 \sigma c^2$. This leads to the form

$$\frac{\partial \boldsymbol{B}}{\partial t} = R_m \nabla \times (\boldsymbol{v} \times \boldsymbol{B}) + \nabla^2 \boldsymbol{B}. \qquad \text{(Eq. 2)}$$

so that R_m can be regarded as a dimensionless measure of the fluid velocity. A second way to nondimensionalize the induction equation is to scale time with the overturn, or advection, time c/V, the time it takes for fluid to cross the sphere. This leads to the form

$$\frac{\partial \boldsymbol{B}}{\partial t} = \nabla \times (\boldsymbol{v} \times \boldsymbol{B}) + R_m^{-1} \nabla^2 \boldsymbol{B}. \qquad \text{(Eq. 3)}$$

Now the inverse magnetic Reynolds number is a dimensionless measure of magnetic diffusion. At first sight, Eqs. (2) and (3) appear to contradict each other; the difference arises from the definition of time, t.

In this article, I shall consider the fairly general case of a geodynamo driven by a combination of heat and compositional buoyancy, governed by the equations of momentum, induction, heat, and mass transfer. The full geodynamo problem is usually put into a nondimensional form that leaves just four independent input parameters, the Rayleigh number R_a, Ekman number E, the Prandtl number P_r, and the magnetic Prandtl number P_m. These are defined and estimated in Table G3. Other nondimensional numbers are calculated from the solution of the equation, although they can also be independent input parameters for subproblems. Two examples are the magnetic Reynolds number, which depends on the fluid velocity and is therefore calculated from the convective flow from the full dynamo problem but is

Table G2 Illustrative numerical values of core parameters used

Property	Symbol	Molecular	Turbulent
Density	ρ	10^4 kg m^{-3}	
Gravity	g	0–10 m s^{-2}	
Core radius	c	3484 km	
Inner core radius	r_i	1215 km	
Outer core depth	$d = c - r_i$	2269 km	
Angular velocity	Ω	7.272×10^{-5}	
Kinematic viscosity	ν	10^{-6} m^2 s^{-1}	1 at room temperature
Electrical conductivity	σ	5×10^5 S m^{-1}	
Thermal conductivity	k	50 W m^{-1} K^{-1}	
Specific heat	C_p	700 J kg^{-1} K^{-1}	
Magnetic diffusivity	$\eta = (\mu_0 \sigma)^{-1}$	1.6 m^2 s^{-1}	1.6 m^2 s^{-1}
Thermal diffusivity	$\kappa = k(\rho C_p)^{-1}$	7×10^{-7} m^2 s^{-1}	1.6 m^2 s^{-1}
Molecular diffusion constant	D	10^{-6} m^2 s^{-1}	1.6 m^2 s^{-1}
Thermal expansion	α	5×10^{-6} K^{-1}	
Typical core velocity	V	10^{-4} m^{-1}	
Typical magnetic field	B	1 mT	
Adiabatic gradient	T'_a	0.1 K km^{-1}	
Core heat flux	Q	5 T W	
Temperature gradient	T'	0.5 K km^{-1}	

Those above the line are measured directly; those below the line are inferred from putative core composition, temperature, and pressure T' is the excess temperature gradient over the adiabat T'_a, as defined in Eq. (4). Current dynamo models usually assume some form of turbulence that brings the small diffusivities of heat, momentum (viscosity), and mass up to the larger value of the electrical diffusivity. For illustration I have taken these values equal to 1.6 when calculating turbulent values in Tables G3 and G4.

Table G3 Dimensionless numbers

Name	Definition	Molecular	Turbulent
Rayleigh	$R_a = g\Delta\rho' d^3/\kappa\nu$	10^{30}	7×10^{17}
Modified Rayleigh	$R_a^m = R_a E = g\Delta\rho' d/\kappa\Omega$	10^{15}	8×10^8
Buoyancy	$R_a^B = g\Delta\rho'/\Omega^2 d$	3	3
Ekman	$E = \nu/\Omega c^2$	10^{-15}	10^{-9}
Prandtl	$P_r = \nu/\kappa$	1.4	1
Magnetic Prandtl	$P_m = \mu_0 \sigma \nu = \nu/\eta$	6×10^{-7}	1
Roberts	$P_q = \kappa/\eta$	4×10^{-7}	1
Schmidt	$P_s = \nu/D$	1	1
Magnetic Schmidt	$P_D = \eta/D$	1.6×10^6	1
Lewis	$P_L = D/\kappa$	1.4	1
Magnetic Reynolds	$R_m = \mu_0 \sigma V c$	200	200
Elsasser	$\Lambda = B^2\sigma/\rho\Omega$	1	1
Rossby	$R_o = V/2\Omega c$	2×10^{-7}	2×10^{-7}
Nusselt	$Q/(4\pi c^2 T'_a)$	7	7
Thermal Peclet	$P_e = Vc/\kappa$	5×10^8	200
Mass Peclet	$P_D = Vc/D$	3×10^8	200

Those above the line appear explicitly in formulations of the geodynamo equations; those below are derived from solutions to the geodynamo problem; those in column 4 are for nominal turbulent values of the diffusivities, $\kappa = \nu = D = \eta = 1.6$ m^2 s^{-1}. Rayleigh and buoyancy numbers are calculated for thermal convection by replacing $\Delta\rho'$ with $\alpha\Delta T$, where ρ' is the departure from the adiabatic density gradient, and ΔT is the temperature difference from the adiabat. Values for compositional convection are likely to be higher. Q_a is the heat conducted down the adiabat.

the sole input parameter for the kinematic dynamo as already mentioned; and the Elsasser number (see below), which depends on the magnetic field strength but is the main input parameter to the magnetoconvection problem, in which a magnetic field is imposed rather than being self-generated.

The nondimensional numbers represent ratios of terms appearing in the governing equations: they are ratios of forces in the momentum equation, of magnetic induction in the induction equation, of heat transfer in the heat equation, and of mass transfer in the material diffusion equation. Further physical insight may be obtained by expressing them as ratios of timescales. The Earth's magnetic field varies on an enormous range of timescales, from less than a year for *geomagnetic jerks* (*q.v.*) to a million years between reversals, and its age is comparable with that of the Earth. Timescales contained within the geodynamo equations span an even wider spectrum, from a single day, the timescale of the Coriolis force, to the molecular viscous and heat diffusion times, which exceed the age of the Earth itself (Table G4). This enormous disparity of timescales makes numerical simulation of the geodynamo very difficult, more so than the range of length scales involved, because the computer must resolve details on the shortest timescale, then integrate the equations for sufficient time to demonstrate dynamo generation in order to follow the evolution. The very long viscous and thermal diffusion times are impossible to simulate and are usually shortened by assumed turbulent enhancement of the diffusivities (see *Core turbulence*), often by simply equalizing the diffusivities to the largest, electrical, diffusivity.

A summary of relevant timescales is given in Table G4. Diffusion times are quoted by the formula shown; true diffusion times in a sphere for magnetic, viscous, thermal, and compositional diffusion contain the factor π^2 to give the time taken for a dipole field to fall by a factor of e. Geometrical factors and wavenumbers have been omitted from the formulae for dimensionless numbers in Table G5. The periods of MAC waves depend on the wavenumber and are considerably shorter than the time quoted in the table. Each group of nondimensional parameters is now discussed in turn.

Table G4 Timescales for the Earth's core, definition, and numerical estimates in years

Time	Definition	Molecular	Turbulent
Magnetic diffusion (core)	$\tau_\eta = c^2/\eta\pi^2$	25000	25000
Magnetic diffusion (inner core)	$\tau_i = r_i^2/\eta\pi^2$	3000	
Thermal diffusion	$\tau_\kappa = c^2/\kappa\pi^2$	6×10^{10}	25000
Viscous diffusion	$\tau_\nu = c^2/\nu\pi^2$	4×10^{10}	25000
Mass diffusion	$\tau_D = c^2/D\pi^2$	4×10^{10}	25000
Overturn	$\tau_V = d/V$	700	
Buoyant rise	$\tau_B = \nu/\alpha\Delta Tgd$	2×10^{-19}	3×10^{-13}
Coriolis rise	$\tau_\Omega = c\Omega/g\alpha\Delta T$	10^{-4}	
Day	τ	3×10^{-3}	
MAC wave	$\tau_{MAC} = \Omega\rho\mu_0 c^2/B^2$	3×10^5	

Table G5 Dimensionless numbers and their relationship with timescales

Rayleigh	τ_κ/τ_B
Modified Rayleigh	τ_κ/τ_Ω
Buoyancy	τ/τ_Ω
Ekman	τ/τ_ν
Prandtl	τ_κ/τ_ν
Magnetic Prandtl	τ_η/τ_ν
Roberts	τ_η/τ_κ
Schmidt	τ_D/τ_ν
Magnetic Schmidt	τ_D/τ_η
Lewis	τ_κ/τ_D
Magnetic Reynolds	τ_V/τ_η
Elsasser	τ_η/τ_{MAC}
Rossby	τ/τ_ν
Thermal Péclet	τ_V/τ_κ
Mass Péclet	τ_V/τ_D

Rayleigh, modified Rayleigh, buoyancy numbers—buoyancy: dissipation

The Rayleigh number multiplies the buoyancy force in the dimensionless equation of motion and as such measures the driving force of the convective dynamo. This discussion is restricted to thermal convection but the same remarks apply to compositional convection with concentration replacing temperature, D replacing κ, and a compositional expansion coefficient replacing α. I do not discuss complications involving both thermal and compositional buoyancy since there has been very little discussion of it in the literature to date, and no direct observations.

R_a is a complicated combination of dimensional parameters and is very difficult to estimate numerically. For free convection heated from below, a critical value of the Rayleigh number must be exceeded before convection starts. This is because motion is opposed by viscous forces and heat, the source of the buoyancy, can dissipate away by conduction. Consider the balance of buoyancy and viscous forces for a rising, hot, light blob: $\rho\alpha\Delta Tg = \rho\nu V_S/d^2$. The blob falls with terminal (Stokes) velocity when buoyancy equals viscous drag, of order $V_S = \alpha g\Delta Td^2/\nu$. The time to rise through the core is $\tau_B = d/V_S = \nu/\alpha\Delta Tgd$. Buoyancy is lost by diffusion of heat into the surrounding fluid, restricting the power of the convection. The Rayleigh number is the ratio of the thermal diffusion time $\tau_\kappa = d^2/\kappa$ to the viscous buoyant rise time: $R_a = g\alpha\Delta Td^3/\kappa\nu$. Vigor of convection is measured by the speed with which fluid rises compared to the speed with which heat is lost by conduction.

There are two major difficulties with estimating R_a in the core. First, it is virtually impossible to estimate the temperature difference ΔT driving the convection. The correct boundary condition for core convection is constant heat flux imposed by the convecting and cooling mantle. This gives the temperature gradient at the core surface, from which we must subtract the adiabatic gradient, which would hold in the absence of any convection:

$$T' = \frac{Q}{4\pi c^2 k} - T'_a. \qquad \text{(Eq. 4)}$$

We could then take ΔT to be the product of the temperature gradient T' and the outer core depth d. The problem is that Q is very uncertain and T'_a poorly known. We do know, however, that the core must convect in order to produce a dynamo, and that T' must exceed the adiabat by enough to power the dynamo. Gubbins (2001) estimates R_a in this way to be about 1000 times the critical value for nonmagnetic convection. Jones (2000) gives an independent argument based on the speed of the flow in the core and arrives at a similar value. The critical Rayleigh number depends strongly on both rotation, which increases it, and magnetic field, which in combination with rotation decreases it; R_a in the core seems to be close to the critical value in the absence of any magnetic field (Gubbins, 2001). The Rayleigh number for compositional convection could be much higher.

Studies of magnetoconvection have shown that, at high rotation rates, the critical Rayleigh number varies as E^{-1}, prompting some authors to use the modified Rayleigh number R_aE. This takes account, to some extent, the effect of rotation on the stability of convection. R_aE

contains only the thermal diffusivity and not the fluid viscosity, which plays a negligible role in resisting the flow compared to the magnetic force. The presumed force balance is between Coriolis and buoyancy forces, leading to the scaling $g\alpha\Delta T \sim \Omega V$; the relevant timescale is then

$$\tau_\Omega = c/V = \frac{c\Omega}{g\alpha\Delta T}.$$

The modified Rayleigh number is then seen as the ratio of thermal diffusion time to Coriolis rise time τ_κ/τ_Ω. Others have used a buoyancy parameter that involves no diffusivity. This is the ratio of the day to the Coriolis rise time. Further details may be found in, for example, Gubbins and Roberts (1987).

Nusselt number—convected heat: conducted heat

The Nusselt number is a different dimensionless measure of the heat driving the convection. It is usually quoted as the ratio of convected heat flux to conducted heat flux for convection between boundaries with fixed temperatures. The situation in the Earth's outer core is somewhat different. First, the boundary conditions are fixed temperature at the bottom (melting temperature) and fixed heat flux at the top (as dictated by mantle convection). The Nusselt number is therefore, in some sense, an input parameter for the geodynamo problem. Second, the conduction profile includes the adiabatic gradient, which is steep and responsible for conduction of a large amount of heat in the context of the Earth's thermal history. It is not really an estimate of the vigor of the convection, which depends on the superadiabatic temperature gradient rather than the absolute value. Estimates of N_u in the core are inevitably restricted to low values, from 1 to 10, because of limits on the heat coming from the core.

Ekman number—viscous force: Coriolis force

E is tiny, whether we use molecular or turbulent values of the viscosity, reflecting the enormous disparity between the viscous and diurnal timescales. Smallness of E causes the greatest difficulty in numerical simulations of the geodynamo, even when turbulent diffusivities are used. The smallest values of E achieved in numerical simulations so far are 10^{-5}–10^{-7}, compared with $E = 10^{-9}$ for the Earth's core assuming a turbulent viscosity.

Rossby number—inertial: Coriolis force

R_o is a measure of the importance of inertial forces Dv/Dt in the equation of motion. It is small in the core and many studies have dropped the inertial terms altogether. This may not be appropriate, since inertial terms may play a role in restoring the balance between magnetic, buoyancy, and rotational forces. Inertia also plays an essential role in *torsional oscillations* (*q.v.*).

Prandtl numbers

The Prandtl number measures the ratio of viscous to thermal diffusion. Liquid metals generally have small Prandtl numbers of order 0.1, and many studies have focused on this. Others have taken $P_r = 1$, consistent with the assumption of turbulence equalizing the diffusivities. P_m and P_q also have extremely low values that cause problems. Similar remarks apply to compositional diffusion.

Magnetic Reynolds number—induction: magnetic diffusion

R_m is a dimensionless measure of the fluid velocity, the input parameter for kinematic dynamos, and is the ratio of the overturn and magnetic diffusion times. It must exceed a critical value for dynamo action, for which there are lower bounds (see *Antidynamo and bounding theorems*). Estimates of fluid flow in the core, based on inferences from secular variation, yields $R_m \approx 200$, low but well above the critical values required for dynamo action.

Péclet numbers—advection: diffusion

Like the magnetic Reynolds numbers, P_e and P_D measure the importance of advection in the heat and mass transfer equations, respectively. They are the ratios of the day to viscous diffusion time, and day to compositional diffusion times, respectively.

Elsasser number—magnetic: Coriolis force

Λ is a dimensionless measure of the Lorentz force and an essential input parameter for magnetoconvection. The balance between magnetic, buoyancy, and Coriolis forces (MAC) leads to magnetohydrodynamic waves with periods of centuries to millennia (see *Magnetohydrodynamic waves*). They are highly dispersive and under certain simple conditions have wavespeeds given by $V_{MAC} = kB^2/\Omega\rho\mu_0$, where k is the wavenumber. A "MAC" timescale can then be defined as the time taken for a MAC wave with wavelength c to cross the core,

$$\tau_{MAC} = \frac{c}{V_{MAC}} = \frac{\Omega\rho\mu_0 c^2}{B^2}.$$

The Elsasser number is then $\Lambda = \tau_\eta/\tau_{MAC}$. Small Λ is the condition for large-scale MAC waves to pass through the core with little dissipation. The numerical value for τ_{MAC}in Table G4 is rather long: a typical MAC wave with wavelength 1000 km would have timescale closer to 1 ka; a stronger core field of 10 mT, the value usually taken in MAC wave studies, would reduce it by a factor of 100.

Less common dimensionless numbers

The above list includes those in common usage in geodynamo theory. Other numbers are used occasionally, and some have different names. The Taylor number is the inverse square of the Ekman number, the ratio of centrifugal to viscous force (and for the core is one of the biggest numbers one is ever likely to come across!). The Chandrasekhar number $B^2\sigma d^2$ is a useful measure of the magnetic force in place of the Elsasser number when rotation is unimportant; it measures the relative strengths of magnetic and diffusive forces. The Hartmann number, $B\sigma^{1/2}d/\mu_0^{1/2}$, is important in boundary layer theory; it measures the ratio of magnetic to viscous forces. The *Alfvén* (*q.v.*) number, $V(\rho\mu_0)^{1/2}/B$, is a magnetic Mach number, the ratio of flow speed to the speed of Alfvén waves. Its inverse is the Cowling number. Neither are in common use in geomagnetism, although Merrill and McElhinny (1996) define the Alfvén number as the Cowling number. The degree to which magnetic fields are frozen-in is usefully measured by the Lundqvist number, $\mu_0\sigma B\eta/\rho^{1/2}$, the time for Alfvén waves to cross the core divided by the magnetic diffusion time.

David Gubbins

Bibliography

Gubbins, D., 2001. The Rayleigh number for convection in the Earth's core. *Physics of the Earth and Planetary Interiors*, **128**: 3–12.

Gubbins, D., and Roberts, P.H., 1987. Magnetohydrodynamics of the Earth's core. In Jacobs, J.A. (ed.), *Geomagnetism*, Vol. II, Chapter 1. London: Academic Press, pp. 1–183.

Jones, C.A., 2000. Convection-driven geodynamo models. *Philosophical Transactions of the Royal Society of London, Series A*, **873**: 873–897.

Merrill, R.T., and McElhinny, M.W., 1996. *The Magnetic Field of the Earth*. San Diego, CA: Academic Press.

Cross-references

Alfvén, Hannes Olof Gösta (1908–1995)
Antidynamo and Bounding Theorems
Convection, Chemical
Convection, Nonmagnetic Rotating
Core, Adiabatic Gradient
Core Convection
Core Motions
Core Turbulence
Dynamos, Kinematic
Geodynamo
Geodynamo, Numerical Simulations
Geomagnetic Jerks
Geomagnetic Spectrum, Temporal
Magnetoconvection
Magnetohydrodynamic Waves
Magnetohydrodynamics
Oscillations, Torsional

GEODYNAMO, ENERGY SOURCES

The magnetic field of the Earth is maintained against ohmic losses by dynamo action in the fluid core which takes its energy from several sources of different natures and amplitudes, all being consequences of the thermal evolution of the core. Moreover, each source has a different efficiency in maintaining ohmic dissipation compared to the others. The study of these problems is then of major importance for understanding dynamo theory as well as the thermal evolution of the Earth.

Evolving reference state

The core of the Earth is composed of iron, nickel, and some lighter elements, with concentrations that are still matter of lively debates (see *Core composition*, and Poirier, 1994). The light elements play a very important dynamical role in maintaining the geodynamo since their rejection upon inner core crystallization drives compositional convection in the core (see *Convection, chemical*). The core is therefore usually modeled as a binary alloy of Fe and some light element X. Including a more realistic chemistry in thermal evolution models on the core would add complexity (although no real difficulty) in the derivations without really improving the understanding, unless not yet documented coupled effects occur.

The Earth's core is different from a well-controlled convection experiment in a laboratory, in the fact that it is continually evolving on geological timescales and that the energy for the motion comes from that evolution. Fortunately, the timescales relevant for this evolution and that for the dynamics are very different and can be separated. To get expressions for the different energy sources in the balance, we need to know in what state is the core when averaged on a timescale long compared to the one relevant to the dynamics but short compared to the one associated with the thermal evolution. The short-time dynamics is responsible for maintaining the core close to that reference state whereas the evolution of the reference state provides the energy sources needed by the dynamics.

Convection in the liquid core is assumed to be very efficient so that, outside of tiny boundary layers (see *Core, boundary layers*), all extensive quantities responsible for the movement (entropy s and mass fraction of light elements ξ) are uniform: $\nabla\xi = 0$ and $\nabla s = 0$. In addition, the momentum equation is assumed to average to the hydrostatic balance: $\nabla p = \rho \boldsymbol{g}$, with ρ the density and $\boldsymbol{g}$ the acceleration due to gravity. It is often useful to know what the temperature is in the reference state (see *Core temperature*) and isentropy and isochemistry implies it to be adiabatic (see *Core, adiabatic gradient*) $\nabla T = \alpha \boldsymbol{g} T / C_p$, with α the coefficient of thermal expansion and C_p the heat capacity at constant pressure.

The set of partial differential equations defining the reference state must be supplemented by an equation of state relating the density to the pressure, for example at constant entropy, and by boundary conditions. In terms of temperature, the equilibrium between solid and liquid at the inner core boundary (ICB) provides the required condition, the liquidus of the core alloy being given (see *Melting temperature of iron in the core, Theory and melting temperature of iron in the core, experimental*) as a function of radius in the core. Depending on the choice of equation of state, one can get different expressions for the different profiles in the reference state, usually in the form of some series expansion of powers of the radius.

It should be noted that the actual averaged state cannot be exactly hydrostatic, owing to its being compressible and compressing with time, but the corrections to this balance are negligible for the energetics of the core (Braginsky and Roberts, 1995). The deviations from the isentropic, well mixed state can also be estimated from the amplitude of fluid velocity at the top of the core which also allows an estimate of the typical scale for the size of boundary layers. These are found to be a negligible contribution to averaged quantities (e.g., Braginsky and Roberts, 1995), although they are very important for the dynamics of the core.

Energy equation

The equation for total (internal, kinetic, magnetic, and gravitational) energy conservation can be integrated over the volume of the core and it states that the total heat loss of the core, the heat flow across the core-mantle boundary (CMB), is balanced by the sum of four terms, the secular cooling Q_C associated with the heat capacity of the core, the latent heat Q_L associated with the gradual freezing of the inner core, the gravitational energy E_G associated with the rearrangement in the outer core of the light elements that are released at the ICB by the freezing of the inner core, and possibly the radioactive heating Q_R,

$$Q_{\rm CMB} = Q_{\rm C} + Q_{\rm L} + E_{\rm G} + Q_{\rm R}. \qquad \text{(Eq. 1)}$$

The first three terms are related to the growth of the inner core and can be written as a function of the radius of the inner core c times its time derivative. The exact expression of these three functions depends on the choice of equation of state and parameterization and the expressions of Labrosse (2003) can be written to first order as:

$$Q_{\rm C} = \frac{8\pi}{3} b^3 \rho_0 C_P T_{\rm L}(c)\left(1 - \frac{2}{3\gamma}\right)\frac{c}{L_T^2}\frac{{\rm d}c}{{\rm d}t}, \qquad \text{(Eq. 2)}$$

$$Q_{\rm L} = 4\pi c^2 \rho(c) T_{\rm L}(c) \Delta S \frac{{\rm d}c}{{\rm d}t}, \qquad \text{(Eq. 3)}$$

$$E_{\rm G} = \frac{8\pi^2}{3} G \Delta\rho \rho_0 c^2 b^2 \left(\frac{3}{5} - \frac{c^2}{b^2}\right)\frac{{\rm d}c}{{\rm d}t}, \qquad \text{(Eq. 4)}$$

with b and c the radii of the outer and inner core respectively, ρ_0 the central density, $C_{\rm P}$ the heat capacity, $T_{\rm L}$ the liquidus temperature, γ the Grüneisen parameter (see *Grüneisen's parameter for iron and Earth's core*), L_T the adiabatic length scale, ΔS the entropy of crystallization, $\Delta\rho$ the chemical density jump across the ICB (see *Core density*), and G the gravitational constant. All these physical parameters can be estimated with more or less accuracy using combinations of geophysical data (mostly seismology) and mineral physics (see *Core properties, physical; Core properties, theoretical determination*).

Convective mixing can be assumed to be sufficient to ensure a uniform mass rate $h(t)$ of radioactive heating in the core. Therefore, radioactive heating is not related to inner core growth and is simply equal to $M_N h(t)$, with M_N the mass of the core. This energy source can then be easily computed, provided one knows the concentration in radioactive elements in the core. Among all possible heat producing isotopes, ^{40}K has always been the most popular candidate, owing to its apparent depletion in the mantle compared to Earth forming meteorites and its predicted metalization at high pressure that would allow it to enter the core. However, potassium is also somewhat volatile and its budget in the Earth is influenced by accretion processes. In addition, the concentration of potassium in the core depends strongly on the scenario of core formation, a process still far from perfectly understood (e.g., Stevenson, 1990). The most recent experiments devoted to the partitioning of potassium between iron and silicates (e.g., Gessmann and Wood, 2002; Lee and Jeanloz, 2003; Rama Murthy *et al.*, 2003) favor a concentration of potassium in the core of O(100) ppm, producing less than 1 TW at present but exponentially more in the past. Such a value is too small to affect importantly the thermal evolution of the core (Labrosse, 2003) and radioactivity will not be considered further.

The gravitational energy actually comes in the equations as a compositional energy, due to a change of composition in a gradient of chemical potential (Braginsky and Roberts, 1995; Lister and Buffett, 1995). It is equal to the change of gravitational energy due only to chemical stratification of the core. Other sources of change of gravitational energy do not contribute significantly to this balance and are mostly stored as strain energy.

As can be seen on Eqs. (2–4), for the different energy sources to be estimated, one needs to know the growth rate of the inner core. Since there is no direct way of measuring this number, one usually uses an estimate of the heat flow across the CMB to get this number from the energy balance (1). The heat flow across the CMB is not very well-known (see *Core-mantle boundary, heat flow across*) but for a value of 10 TW (say) and no radioactive elements, one gets approximately (Labrosse, 2003) $Q_C = 5.5$ TW, $Q_L = 2.8$ TW, $E_G = 1.7$ TW.

The energy balance can be used for any time in the history of the Earth to give the growth history of the inner core and the evolution of all energy sources in the balance, provided the heat flow across the CMB is known as a function of time. Moreover, this equation can be integrated between the onset time of inner core crystallization and the present to compute the age of the inner core (Labrosse *et al.*, 2001). A typical example of time evolution of the energy balance is shown on Figure G12, where the onset of inner core crystallization, at an age of 1 Ga, is marked by a qualitative change in the balance: before, only secular cooling is balancing Q_{CMB} (in absence of radioactivity) but this term is greatly decreased when the inner core starts to crystallize and both latent heat and compositional energy come in the balance.

Entropy equation

The energy equation does not involve the magnetic field directly and contains no contribution from the dissipative heating. This is a well-known characteristic of convective engines: in contrary to Carnot engines, dissipation occurs inside the system and is then not lost. In order to relate the energy sources to the magnetic field generation, an entropy balance equation must be written. This equation comes from a combination of the momentum balance equation and the energy balance equation written above (e.g., Braginsky and Roberts, 1995) and states that the entropy that flows out through the CMB is balanced by the sum of the entropy that flows in due to the different heat sources and the entropy that is produced by nonreversible processes, mostly ohmic dissipation and conduction along the adiabatic temperature profile. This equation does not directly involve the gravitational energy, since it is not a heat source. It can however, be brought back into the equation by use of the energy balance to suppress the heat flow across the CMB, giving an efficiency equation,

$$\Phi + T_D \int_{OC} k \left(\frac{\nabla T}{T}\right)^2 dV = \frac{T_D}{T_{CMB}} \left[E_G + \frac{T_{ICB} - T_{CMB}}{T_{ICB}} (Q_{ICB} + Q_L) + \frac{T_R - T_{CMB}}{T_R} Q_R + \frac{T_C - T_{CMB}}{T_C} Q_C \right], \quad \text{(Eq. 5)}$$

where it can be seen that each energy source contributes in maintaining both the total ohmic dissipation Φ and the conduction along the adiabatic gradient, but with different efficiency factors. In particular, this equation shows that the gravitational energy has an efficiency factor that is the ratio of the effective temperature T_D at which ohmic dissipation occurs to the temperature at the CMB, T_{CMB}, whereas all heat sources efficiency factors have a contribution from the classical Carnot engine efficiency factor, $(T_X - T_{CMB})/T_X$, with T_X the temperature at which the heat source X is provided. This shows that compositional energy is more efficiently transformed in dissipation than all heat sources and that the efficiency of each heat source depends on the temperature at which it is provided.

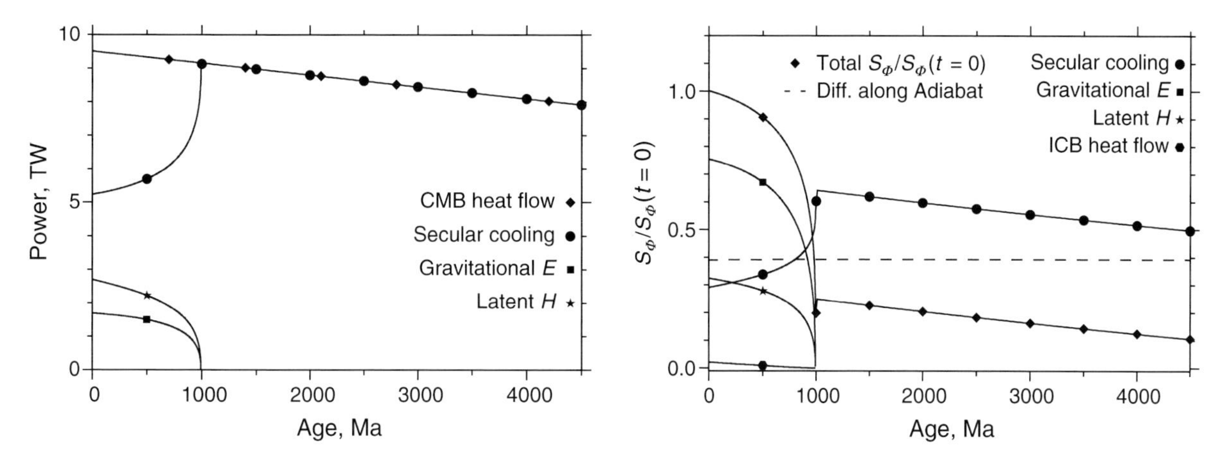

Figure G12 Energy (*left*) and entropy (*right*) balances of the core as a function of time for a typical evolution model without radioactivity. (After Labrosse, 2003).

The energy sources on the right-hand side of Eq. (5) are the same as that appearing in the energy balance Eq. (1) and, except for T_D, their efficiency factors can be expressed using the parameters characterizing the reference state of the core. It can then be proved that all terms, except the radioactive heating one, is a function of the radius of the inner core and is proportional to its growth rate. This means that, if the heat flow across the CMB is known, this growth rate can be computed from the energy equation (1), and the entropy equation (5) then gives the ohmic dissipation that is maintained. Alternatively, one can take the opposite view and compute the growth rate of the inner core that is required to maintain a given ohmic dissipation, the energy balance being then used to get the heat flow across the CMB that makes this growth rate happen. Unfortunately, the ohmic dissipation in the core is no better known than the heat flow across the CMB, since it is dominated by small-scales of the magnetic field and possibly by the invisible toroidal part of it (Gubbins and Roberts, 1987; Roberts *et al.*, 2003). However, the value of $Q_{\mathrm{CMB}} = 10$ TW used above gives a contribution of ohmic dissipation to the entropy balance $\mathbf{\Phi}/T_{\mathrm{D}} = 500\ \mathrm{MWK}^{-1}$. The temperature T_{D} is not well-known, but is bounded by the temperature at the inner core boundary and the CMB and this gives $\mathbf{\Phi} \simeq 2$ TW.

The evolution with time of the entropy balance associated with a given heat flow evolution can be computed and the example shown above gives the result of Figure G12. An interesting feature is the sharp increase of the ohmic dissipation in the core when the inner core starts crystallizing, latent heat and, even more so, gravitational energy being more efficient than secular cooling. Unfortunately, the link between this ohmic dissipation and the magnetic field observed at the surface of the Earth is far from obvious and the detection of such an increase in the paleomagnetic record is unlikely (Labrosse and Macouin, 2003).

Some alternative models for the average structure of the core involving some stratification have been proposed. In particular, the heat conducted along the adiabatic temperature gradient can be rather large (about 7 TW) and might be larger than the heat flow across the CMB (see *Core-mantle boundary, heat flow across*). In this case, two different models have been proposed. In the first one, the adiabatic temperature profile is maintained by compositional convection against thermal stratification, except in a still very thin boundary layer, and this means that the entropy flow across the CMB is less than that required to maintain the conduction along the average temperature profile. In other words, the compositional convection has to fight against thermal stratification to maintain the adiabatic temperature profile in addition to maintaining the dynamo. In the second model (see Labrosse *et al.*, (1997); Lister and Buffett (1998)), a subadiabatic layer of several hundreds of kilometers is allowed to develop at the top of the core and the entropy flow out of the core balances the conduction along the average temperature gradient. In this case, the compositional energy is entirely used for the dynamo. Which of these two options would be chosen by the core is a dynamical question that cannot be addressed by simple thermodynamic arguments as used here.

Stéphane Labrosse

Bibliography

Braginsky, S.I., and Roberts, P.H., 1995. Equations governing convection in Earth's core and the geodynamo. *Geophysical Astrophysical Fluid Dynamics*, **79**: 1–97.

Gessmann, C.K., and Wood, B.J., 2002. Potassium in the Earth's core? *Earth and Planetary Science Letters*, **200**: 63–78.

Gubbins, D., and Roberts, P.H., 1987. Magnetohydrodynamics of the Earth's core. In Jacobs, J.A., (ed.), *Geomagnetism*, Vol. 2. London: Academic Press, pp. 1–183.

Labrosse, S., 2003. Thermal and magnetic evolution of the Earth's core. *Physics of the Earth and Planetary Interiors*, **140**: 127–143.

Labrosse, S., and Macouin, M., 2003. The inner core and the geodynamo. *Comptes Rendus Geosciences*, **335**: 37–50.

Labrosse, S., Poirier, J.-P., and Le Mouël, J.-L., 1997. On cooling of the Earth's core. *Physics of the Earth and Planetary Interiors*, **99**: 1–17.

Labrosse, S., Poirier, J.-P., and Le Mouël, J.-L., 2001. The age of the inner core. *Earth and Planetary Science Letters*, **190**: 111–123.

Lee, K.K.M., and Jeanloz, R., 2003. High-pressure alloying of potassium and iron: radioactivity in the Earth's core? *Geophysical Research Letters*, **30**: 2212, doi:10.1029/2003GL018515.

Lister, J.R., and Buffett, B.A., 1995. The strength and efficiency of the thermal and compositional convection in the geodynamo. *Physics of the Earth and Planetary Interiors*, **91**: 17–30.

Lister, J.R., and Buffett, B.A., 1998. Stratification of the outer core at the core-mantle boundary. *Physics of the Earth and Planetary Interiors*, **105**: 5–19.

Poirier, J.-P., 1994. Light elements in the Earth's core: a critical review. *Physics of the Earth and Planetary Interiors*, **85**: 319–337.

Rama Murthy, V., van Westrenen, W., and Fei, Y., 2003. Radioactive heat sources in planetary cores: experimental evidence for potassium. *Nature*, **423**: 163–165.

Roberts, P.H., Jones, C.A., and Calderwood, A.R., 2003. Energy fluxes and ohmic dissipation in the Earth's core. In Jones, C A., Soward, A.M., and Zhang, K. (eds.) *Earth's Core and Lower Mantle.* London: Taylor & Francis, pp. 100–129.

Stevenson, D.J., 1990. Fluid dynamics of core formation. In Newsom, H.E., and Jones, J.H. (eds.) *Origin of the Earth.* New York: Oxford University Press, pp. 231–249.

Cross-references

Convection, Chemical
Core Composition
Core Density
Core Properties, Physical
Core Properties, Theoretical Determination
Core Temperature
Core, Adiabatic Gradient
Core, Boundary Layers
Core-Mantle Boundary, Heat Flow Across
Grüneisen's Parameter for Iron and Earth's Core
Melting Temperature of Iron in the Core, Experimental
Melting Temperature of Iron in the Core, Theory

GEODYNAMO: NUMERICAL SIMULATIONS

Introduction

The *geodynamo* is the name given to the mechanism in the Earth's core that maintains the Earth's magnetic field (see *Geodynamo*). The current consensus is that flow of the liquid iron alloy within the outer core, driven by buoyancy forces and influenced by the Earth's rotation, generates large electric currents that induce magnetic field, compensating for the natural decay of the field. The details of how this produces a slowly changing magnetic field that is mainly dipolar in structure at the Earth's surface, with occasional dipole reversals, has been the subject of considerable research by many people for many years.

The fundamental theory, put forward in the 1950s, is that differential rotation within the fluid core shears poloidal (north-south and radial) magnetic field lines into toroidal (east-west) magnetic field; and three-dimensional (3D) helical fluid flow twists toroidal field lines into poloidal field. The more sheared and twisted the field structure the faster it decays away; that is, magnetic diffusion (reconnection) continually smooths out the field. The field is self-sustaining if, on average, the generation of field is balanced by its decay. Discovering and understanding the details of how rotating convection in Earth's fluid outer core maintains the observed intensity, structure, and time dependencies requires 3D computer models of the geodynamo.

Magnetohydrodynamic (MHD) dynamo simulations are numerical solutions of a coupled set of nonlinear differential equations that describe the 3D evolution of the thermodynamic variables, the fluid velocity, and the magnetic field. Because so little can be detected about the geodynamo, other than the poloidal magnetic field at the surface (today's field in detail and the paleomagnetic field in much less detail) and what can be inferred from seismic measurements and variations in the length of the day and possibly in the gravitational field, models of the geodynamo are used as much to predict what has not been observed as they are used to explain what has. When such a model generates a magnetic field that, at the model's surface, looks qualitatively similar to the Earth's surface field in terms of structure, intensity, and time-dependence, then it is plausible that the 3D flows and fields inside the model core are qualitatively similar to those in the Earth's core. Analyzing this detailed simulated data provides a physical description and explanation of the model's dynamo mechanism and, by assumption, of the geodynamo.

The first 3D global convective dynamo simulations were developed in the 1980s to study the *solar* dynamo. Gilman and Miller (1981) pioneered this style of research by constructing the first 3D MHD dynamo model. However, they simplified the problem by specifying a constant background density, i.e., they used the Boussinesq approximation of the equations of motion. Glatzmaier (1984) developed a 3D MHD dynamo model using the anelastic approximation, which accounts for the stratification of density within the sun. Zhang and Busse (1988) used a 3D model to study the onset of dynamo action within the Boussinesq approximation. However, the first MHD models of the *Earth's* dynamo that successfully produced a time-dependent and dominantly dipolar field at the model's surface were not published until 1995 (Glatzmaier and Roberts, 1995; Jones *et al.*, 1995; Kageyama *et al.*, 1995). Since then, several groups around the world have developed dynamo models and several others are currently being designed. Some features of the various simulated fields are robust, like the dominance of the dipolar part of the field outside the core. Other features, like the 3D structure and time-dependence of the temperature, flow, and field inside the core, depend on the chosen boundary conditions, parameter space, and numerical resolution. Many review articles have been written that describe and compare these models (e.g., Hollerbach, 1996; Glatzmaier and Roberts, 1997; Fearn, 1998; Busse, 2000; Dormy *et al.*, 2000; Roberts and Glatzmaier, 2000; Christensen *et al.*, 2001; Busse, 2002; Glatzmaier, 2002; Kono and Roberts, 2002).

Model description

Models are based on equations that describe fluid dynamics and magnetic field generation. The equation of mass conservation is used with the very good assumption that the fluid flow velocity in the Earth's outer core is small relative to the local sound speed. The anelastic version of mass conservation accounts for a depth-dependent background density; the density at the bottom of the Earth's fluid core is about 20% greater than that at the top. The Boussinesq approximation simplifies the equations further by neglecting this density stratification, i.e., by assuming a constant background density. An equation of state relates perturbations in temperature and pressure to density perturbations, which are used to compute the buoyancy forces, which drive convection. Newton's second law of motion (conservation of momentum) determines how the local fluid velocity changes with time due to buoyancy, pressure gradient, viscous, rotational (Coriolis), and magnetic (Lorentz) forces. The MHD equations (i.e., Maxwell's equations and Ohm's law with the extremely good assumption that the fluid velocity is small relative to the speed of light) describe how the local magnetic field changes with time due to induction by the flow and diffusion due to finite conductivity. The second law of thermodynamics dictates how thermal diffusion and Joule and viscous heating determine the local time rate of change of entropy (or temperature). Additional equations are sometimes included that account for perturbations in composition and gravitational potential and their effects on buoyancy.

This set of coupled nonlinear differential equations, with a set of prescribed boundary conditions, is solved each numerical time step to obtain the evolution in 3D of the fluid flow, magnetic field, and thermodynamic perturbations. Most geodynamo models have employed spherical harmonic expansions in the horizontal directions and either Chebyshev polynomial expansions or finite differences in radius. The equations are integrated in time typically by treating the linear terms implicitly and the nonlinear terms explicitly. The review articles mentioned above describe the variations on the equations, boundary conditions, and numerical methods employed in the various models of the geodynamo.

Current results

Since the mid-1990s, 3D computer simulations have advanced our understanding of the geodynamo. The simulations show that a dominantly dipolar magnetic field, not unlike the Earth's, can be maintained by convection driven by an Earth-like heat flux. A typical snapshot of the simulated magnetic field from a geodynamo model is illustrated in Figure G13 with a set of field lines. In the fluid outer core, where the field is generated, field lines are twisted and sheared by the flow. The field that extends beyond the core is significantly weaker and dominantly dipolar at the model's surface, not unlike the geomagnetic field. For most geodynamo simulations, the nondipolar part of the surface field, at certain locations and times, propagates westward at about $0.2^\circ\ y^{-1}$ as has been observed in the geomagnetic field over the past couple hundred years.

Several dynamo models have electrically conducting inner cores that on average drift eastward relative to the mantle (e.g., Glatzmaier and Roberts, 1995; Sakuraba and Kono, 1999; Christensen *et al.*, 2001), opposite to the propagation direction of the surface magnetic

Figure G13 A snapshot of the 3D magnetic field simulated with the Glatzmaier-Roberts geodynamo model and illustrated with a set of magnetic field lines. The axis of rotation is vertical and centered in the image. The field is complicated and intense inside the fluid core where it is generated by the flow; outside the core it is a smooth, dipole-dominated, potential field. (From Glatzmaier, 2002.)

field. Inside the fluid core the simulated flow has a "thermal wind" component that, near the inner core, is predominantly eastward relative to the mantle. Magnetic field in these models that permeates both this flow and the inner core tries to drag the inner core in the direction of the flow. This magnetic torque is resisted by a gravitational torque between the mantle and the topography on the inner core surface. The amplitude of the superrotation rate predicted by geodynamo models depends on the model's prescribed parameters and assumptions and on the very poorly constrained viscosity assumed for the inner core's deformable surface layer, which by definition, is near the melting temperature. The original prediction was an average of about 2° longitude per year faster than the surface. Since then, the superrotation rate of the Earth's inner core (today) has been inferred from several seismic analyses, but is still controversial. There is a spread in the inferred values, from the initial estimates of 1° to 3° eastward per year (relative to the Earth's surface) to some that are zero to within an uncertainty of 0.2° per year. More recent geodynamo models that include an inhibiting gravitational torque also predict smaller superrotation rates.

On a much longer timescale, the dipolar part of the Earth's field occasionally reverses (see *Reversals, theory*). The reversals seen in the paleomagnetic record are nonperiodic. The times between reversals are measured in hundreds of thousands of years; whereas the time to complete a reversal is typically a few thousand years, less than a magnetic dipole decay time. Several dynamo simulations have produced spontaneous nonperiodic magnetic dipole reversals (Glatzmaier and Roberts, 1995; Glatzmaier *et al.*, 1999; Kageyama *et al.*, 1999; Sarson and Jones, 1999; Kutzner and Christensen, 2002). Regular (periodic) reversals, like the *dynamo-wave* (*q.v.*) reversals seen in early solar dynamo simulations, have also occurred in recent dynamo simulations.

One of the simulated reversals is portrayed in Figure G14/Plate 16 with four snapshots spanning about 9 ka. The radial component of the field is shown at both the core-mantle boundary (CMB) and the surface of the model Earth. The reversal, as viewed in these surfaces, begins with reversed magnetic flux patches in both the northern and southern hemispheres. The longitudinally averaged poloidal and toroidal parts of the field inside the core are also illustrated at these times. Although when viewed at the model's surface, the reversal appears complete by the third snapshot, another 3 ka is required for the original field polarity to decay out of the inner core and the new polarity to diffuse in.

Small changes in the local flow structure continually occur in this highly nonlinear chaotic system. These can generate local magnetic anomalies that are reversed relative to the direction of the global dipolar field structure. If the thermal and compositional perturbations continue to drive the fluid flow in a way that amplifies this reversed field polarity while destroying the original polarity, the entire global field structure would eventually reverse. However, more often, the local reversed polarity is not able to survive and the original polarity fully recovers because it takes a couple of thousand years for the original polarity to decay out of the solid inner core. This is a plausible explanation for "events," which occur when the paleomagnetic field (as measured at the Earth's surface) reverses and then reverses back, all within about 10 ka.

On an even longer timescale, the frequency of reversals seen in the paleomagnetic record varies. The frequency of nonperiodic reversals in geodynamo simulations has been found to depend on the pattern of outward heat flux imposed over the CMB (presumably controlled in the Earth by mantle convection) and on the magnitude of the convective driving relative to the effect of rotation.

Many studies have been conducted via dynamo simulations to, for example, assess the effects of the size and conductivity of the solid inner core, of a stably stratified layer at the top of the core, of

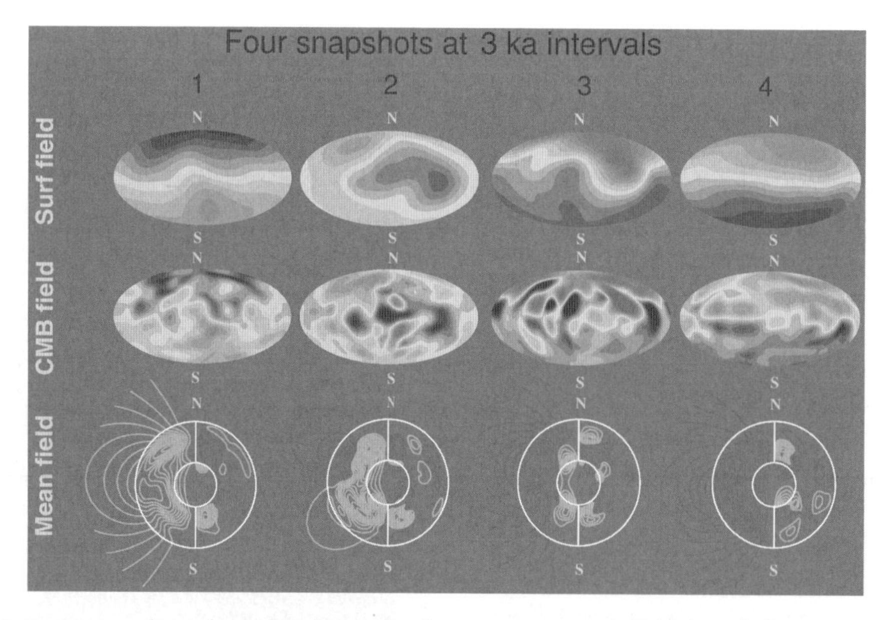

Figure G14/Plate 16 A sequence of snapshots of the longitudinally averaged magnetic field through the interior of the core and of the radial component of the field at the core-mantle boundary and at what would be the surface of the Earth, displayed at roughly 3 ka intervals spanning a dipole reversal from a geodynamo simulation. In the plots of the average field, the small circle represents the inner core boundary and the large circle is the core-mantle boundary. The poloidal field is shown as magnetic field lines on the left-hand sides of these plots (blue is clockwise and red is counterclockwise). The toroidal field direction and intensity are represented as contours (not magnetic field lines) on the right-hand sides (red is eastward and blue is westward). Aitoff-Hammer projections of the entire core-mantle boundary and surface are used to display the radial component of the field (with the two different surfaces displayed as the same size). Reds represent outward directed field and blues represent inward field; the surface field, which is typically an order of magnitude weaker, was multiplied by 10 to enhance the color contrast. (From Glatzmaier *et al.*, 1999.)

heterogeneous thermal boundary conditions, of different velocity boundary conditions, and of computing with different parameters. These models differ in several respects. For example, the Boussinesq instead of the anelastic approximation may be used, compositional buoyancy and perturbations in the gravitational field may be neglected, different boundary conditions, and spatial resolutions may be chosen, the inner core may be treated as an insulator instead of a conductor or may not be free to rotate. As a result, the simulated flow and field structures inside the core differ among the various simulations. For example, the strength of the shear flow on the "tangent cylinder" (the imaginary cylinder tangent to the inner core equator; Figure G13), which depends on the relative dominance of the Coriolis forces, is not the same for all simulations. Likewise, the vigor of the convection and the resulting magnetic field generation tends to be greater outside this tangent cylinder for some models and inside for others. But all the solutions have a westward zonal flow in the upper part of the fluid core and a dominantly dipolar magnetic field outside the core.

When assuming Earth values for the radius and rotation rate of the core, all models of the geodynamo have been forced (due to computational limitations) to use a viscous diffusivity that is at least three to four orders of magnitude larger than estimates of what a turbulent (or eddy) viscosity should be (about 2 m^2s^{-1}) for the spatial resolutions that have been employed. In addition to this enhanced viscosity, one must decide how to prescribe the thermal, compositional, and magnetic diffusivities. One of two extremes has typically been chosen. These diffusivities could be set equal to the Earth's actual magnetic diffusivity (2 m^2s^{-1}), making these much smaller than the specified viscous diffusivity; this was the choice for most of the Glatzmaier-Roberts simulations. Alternatively, they could be set equal to the enhanced viscous diffusivity, making all (turbulent) diffusivities too large, but at least equal; this was the choice of most of the other models. Neither choice is satisfactory.

Future challenges

Because of the large turbulent diffusion coefficients, all geodynamo simulations have produced large-scale *laminar* convection. That is, convective cells and plumes of the simulated flow typically span the entire depth of the fluid outer core, unlike the small-scale turbulence that likely exists in the Earth's core.

The fundamental question about geodynamo models is how well do they simulate the actual dynamo mechanism of the Earth's core? Some geodynamo modelers have argued, or at least suggested, that the large (global) scales of the temperature, flow, and field seen in these simulations should be fairly realistic because the prescribed viscous and thermal diffusivities may be asymptotically small enough. For example, in most simulations, viscous forces (away from the boundaries) tend to be 10^4 times smaller than Coriolis and Lorentz forces. Other modelers are less confident that current simulations are realistic even at the large-scales because the model diffusivities are so large. Only when computing resources improve to the point where we can further reduce the turbulent diffusivities by several orders of magnitude and produce strongly turbulent simulations will we be able to answer this fundamental question.

In the mean time, we may be able to get some insight from very highly resolved 2D simulations of magnetoconvection. These simulations can use diffusivities a thousand times smaller than those of the current 3D simulations. They demonstrate that strongly turbulent 2D rotating magnetoconvection has significantly different spatial structure and time-dependence than the corresponding 2D laminar simulations obtained with much larger diffusivities.

These findings suggest that current 3D laminar dynamo simulations may be missing critical dynamical phenomena. Therefore, it is important to strive for much greater spatial resolution in 3D models in order to significantly reduce the enhanced diffusion coefficients and actually simulate turbulence. This will require faster parallel computers and improved numerical methods and hopefully will happen within the next decade or two. In addition, subgrid scale models need to be added to geodynamo models to better represent the heterogeneous anisotropic transport of heat, composition, momentum, and possibly also magnetic field by the part of the turbulence spectrum that remains unresolved.

Gary A. Glatzmaier

Bibliography

Busse, F.H., 2000. Homogeneous dynamos in planetary cores and in the laboratory. *Annual Review of Fluid Mechanics*, **32**: 383–408.

Busse, F.H., 2002. Convective flows in rapidly rotating spheres and their dynamo action. *Physics of Fluids*, **14**: 1301–1314.

Christensen, U.R., Aubert, J., Cardin, P., Dormy, E., Gibbons, S. *et al.*, 2001. A numerical dynamo benchmark. *Physics of the Earth and Planetary Interiors*, **128**: 5–34.

Dormy, E., Valet, J.-P., and Courtillot, V., 2000. Numerical models of the geodynamo and observational constraints. *Geochemistry, Geophysics, Geosystems*, **1**: 1–42, paper 2000GC000062.

Fearn, D.R., 1998. Hydromagnetic flow in planetary cores. *Reports on Progress in Physics*, **61**: 175–235.

Gilman, P.A., and Miller, J., 1981. Dynamically consistent nonlinear dynamos driven by convection in a rotating spherical shell. *Astrophysical Journal, Supplement Series*, **46**: 211–238.

Glatzmaier, G.A., 1984. Numerical simulations of stellar convective dynamos. I. The model and the method. *Journal of Computational Physics*, **55**: 461–484.

Glatzmaier, G.A., 2002. Geodynamo simulations—how realistic are they? *Annual Review of Earth and Planetary Sciences*, **30**: 237–257.

Glatzmaier, G.A., and Roberts, P.H., 1995. A three-dimensional self-consistent computer simulation of a geomagnetic field reversal. *Nature*, **377**: 203–209.

Glatzmaier, G.A., and Roberts, P.H., 1997. Simulating the geodynamo. *Contemporary Physics*, **38**: 269–288.

Glatzmaier, G.A., Coe, R.S., Hongre, L., and Roberts, P.H., 1999. The role of the Earth's mantle in controlling the frequency of geomagnetic reversals. *Nature*, **401**: 885–890.

Hollerbach, R., 1996. On the theory of the geodynamo. *Physics of the Earth and Planetary Interiors*, **98**: 163–185.

Jones, C.A., Longbottom, A., and Hollerbach, R., 1995. A self-consistent convection driven geodynamo model, using a mean field approximation. *Physics of the Earth and Planetary Interiors*, **92**: 119–141.

Kageyama, A., Ochi, M., and Sato, T., 1999. Flip-flop transitions of the magnetic intensity and polarity reversals in the magnetohydrodynamic dynamo. *Physics Review Letters*, **82**: 5409–5412.

Kageyama, A., Sato, T., Watanabe, K., Horiuchi, R., Hayashi, T. *et al.*, 1995. Computer simulation of a magnetohydrodynamic dynamo. II. *Physics of Plasmas*, **2**: 1421–1431.

Kono, M., and Roberts, P.H., 2002. Recent geodynamo simulations and observations of the geomagnetic field. *Review of Geophysics*, **40**: 41–53.

Kutzner, C., and Christensen, U.R., 2002. From stable dipolar towards reversing numerical dynamos. *Physics of the Earth and Planetary Interiors*, **131**: 29–45.

Roberts, P.H., and Glatzmaier, G.A., 2000. Geodynamo theory and simulations. *Reviews of Modern Physics*, **72**: 1081–1123.

Sakuraba, A., and Kono, M., 1999. Effect of the inner core on the numerical solution of the magnetohydrodynamic dynamo. *Physics of the Earth and Planetary Interiors*, **111**: 105–121.

Sarson, G.R., and Jones, C.A., 1999. A convection driven dynamo reversal model. *Physics of the Earth and Planetary Interiors*, **111**: 3–20.

Zhang, K., and Busse, F.H., 1988. Finite amplitude convection and magnetic field generation in a rotating spherical shell. *Geophysical and Astrophysical Fluid Dynamics*, **41**: 33–53.

Cross-references

Core Convection
Core Turbulence
Core Viscosity
Core-Mantle Boundary Topography, Implications for Dynamics
Core-Mantle Coupling, Electromagnetic
Core-Mantle Coupling, Thermal
Core-Mantle Coupling, Topographic
Dynamo, Solar
Geodynamo
Geomagnetic Dipole Field
Harmonics, Spherical
Inner Core Rotation
Inner Core Seismic Velocities
Inner Core Tangent Cylinder
Magnetohydrodynamics
Reversals, Theory
Thermal Wind
Westward Drift

GEODYNAMO, SYMMETRY PROPERTIES

The behavior of any physical system is determined in part by its symmetry properties. For the geodynamo this means the geometry of a spinning sphere and the symmetry properties of the equations of magnetohydrodynamics. Solutions have symmetry that is the same as, or lower than, the symmetry of the governing equations and boundary conditions. By "symmetry" here we mean a transformation T that takes the system into itself. Given one solution with lower symmetry, we can construct a second solution by applying the transformation T to it. Solutions with different symmetry can evolve independently and are said to be *separable*. If the governing equations are linear separable solutions are also linearly independent: they may be combined to form a more general solution. If the governing equations are nonlinear they may not be combined or coexist but they remain separable. The full *geodynamo* (*q.v.*) problem is nonlinear and separable solutions exist; fluid velocities and magnetic fields with the same symmetry are linearly independent solutions of the linear *kinematic dynamo* (*q.v.*) problem.

Symmetry considerations are important for both theory and observation. For example, solutions with high symmetry are easier to compute than those with lower symmetry and are often chosen for that reason. Time-dependent behavior of nonlinear systems (geomagnetic reversals for example) may be analyzed in terms of one separable solution becoming unstable to one with different symmetry ("symmetry breaking"). Observational applications include detection of symmetries in the geomagnetic field. The axial dipole field has very high symmetry but is not a separable solution of the geodynamo; it does, however, belong to a separable solution with a particular symmetry about the equator. Paleomagnetic data rarely have sufficient global coverage to allow a proper assessment of the spatial pattern of the geomagnetic field, but they can sometimes be used to discriminate between separable solutions with different symmetries.

The sphere is symmetric under any rotation about its centre while rotation is symmetric under any rotation about the spin axis. The symmetries of the spinning sphere are therefore any rotation about the spin axis and reflection in the equatorial plane (Figure G15). This conflict of spherical and cylindrical geometry lies at the heart of many of the properties of rotating convection and the geodynamo. The equations of *magnetohydrodynamics* (*q.v.*) are also invariant under rotation. They are invariant under change of sign of magnetic field B (but not other dependent variables) because the induction equation is linear in B and the magnetic force and ohmic heating are quadratic in B. They are also invariant under time translation.

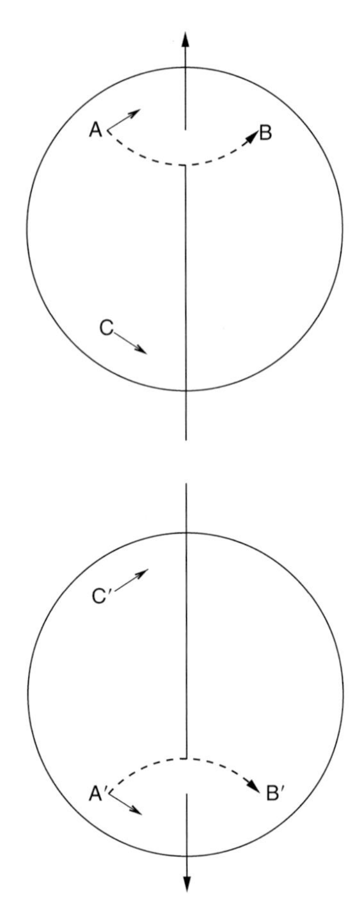

Figure G15 Reflection of a rotating sphere in a plane parallel to the equator. A′ , B′ are the reflections of the points A, B. The reflected sphere turns in the same direction as the original sphere. A vector is equatorial-symmetric (E^S) if its value at C′ appears as a reflection as shown: it is E^A if it appears with a change of sign.

The group of symmetry operations is Abelian because of the infinite number of allowed rotations about the spin axis and time translations. The full set of symmetry operations is found by constructing the group table and using the closure property. The group table, including rotation of π about the spin axis but no higher rotations, is shown in Table G6. Note the additional symmetry operations O; these are combinations of reflection in the equatorial plane and rotation about the spin axis; they amount to reflection through the origin. Note also the subgroups formed by (I,i) and (I,i,E^S,E^A). These are fundamental to some analyses of paleomagnetic data. Arbitrary time translation can be applied to any symmetry to produce steady solutions that are invariant under translation, drifting solutions that are steady in a corotating frame, more complicated time-periodic solutions that may vascillate or have reversing magnetic fields, and solutions that change continually and are sometimes loosely called "chaotic."

A word is needed about the behavior of vectors under reflection. A vector is usually defined by its transformation law under rotation. An axial or pseudovector (or tensor) changes sign on reflection whereas a polar or true vector (or tensor) does not. A true scalar is invariant under reflection, a pseudoscalar changes sign. Examples of true vectors are fluid velocity and electric current. Examples of axial vectors are angular velocity and magnetic field. The cross product changes sign under reflection (to see this consider the simple case of the cross product of two polar vectors); the curl also changes sign under reflection. Vectors v satisfying $\nabla \cdot v = 0$ are often represented in terms of their toroidal and poloidal parts:

Table G6 Multiplication table for a finite subgroup of spatial symmetry operations for a buoyancy-driven dynamo in a rotating sphere

I	i	E^S	E^A	P_2^S	P_2^A	O^S	O^A
i	I	E^A	E^S	P_2^A	P_2^S	O^A	O^S
E^S	E^A	I	i	O^S	O^A	P_2^S	P_2^A
E^A	E^S	i	I	O^A	O^S	P_2^A	P_2^S
P_2^S	P_2^A	O^S	O^A	I	i	E^S	E^A
P_2^A	P_2^S	O^A	O^S	i	I	E^A	E^S
O^S	O^A	P_2^S	P_2^A	E^S	E^A	I	i
O^A	O^S	P_2^A	P_2^S	E^A	E^S	i	I

Note: I, identity; i, field reversal; E^S, reflection in equatorial plane; E^A, reflection in equatorial plane with change of sign of magnetic field; P_2, rotation by π about spin axis; and O, reflection in origin.

$$v = \nabla \times Tr + \nabla \times \nabla \times Pr;$$

if the toroidal part is a true vector the poloidal part will be a pseudovector and vice versa because of the extra curl involved. Helicity, $v \cdot \nabla \times v$, is a pseudoscalar because vorticity changes sign under reflection but velocity does not. Properties of true and pseudovectors are used to determine the symmetry of individual terms in the governing equations and to find separable solutions.

The symmetry of solutions is reflected in their spherical harmonic expansions. Potential fields with E^A symmetry involve only harmonics Y_l^m with $l - m$ odd; E^S fields have $l - m$ even. The symmetries are usually referred to as "dipole" and "quadrupole" families because of their leading terms Y_1^0 (dipole) and Y_2^0 (quadrupole). The terminology is somewhat unsatisfactory for two reasons: first, the equatorial dipole Y_1^1 is a member of the quadrupole family, and second the internal, toroidal, field of each family has the opposite series to that of the external, poloidal, field. Fields with P_2^S symmetry have spherical harmonic series with m even, P_2^A have m odd. Solutions with higher symmetry have series containing m differing by larger integers, for example P_4^S has m multiples of 4.

The main use of symmetry properties in theoretical studies has so far been restricted to reducing the complexity of the solution in order to effect numerical solutions. For example, allowing only a "dipole family," or E^S, solution halves the number of spherical harmonics required to represent the solution (or, equivalently, the solution only need be found in one hemisphere); P_2^S halves it again. In practice solutions with high symmetry may be poor dynamos and be more difficult to compute, despite their apparent lack of spatial complexity, because they require strong driving and involve small-scale magnetic fields. The existence of a separable solution is no guarantee of maintaining a magnetic field: B may still decay to zero. Thus axial symmetry is an allowed symmetry but *Cowling's theorem* (*q.v.*) shows that no axisymmetric magnetic field can be sustained by dynamo action. *Kinematic dynamos* (*q.v.*) with axisymmetric fluid velocities have axisymmetric solutions that decay by Cowling's theorem, but solutions with lower symmetry, each proportional to $expim\phi$, can grow with time. They do not form separable solutions of the full, dynamical, dynamo problem because nonlinear terms in the equations couple the modes to include many values of m. Bullard and Gellman (1954) investigated an $E^S P_2^S$ flow that also possessed a meridional plane of symmetry, again to reduce the computational effort required for the very small, early computer at their disposal. This last symmetry is not an allowed separable solution of the full dynamo problem. The lack of helicity imposed by this symmetry was later found to be the reason for the failure of the *Bullard-Gellman dynamo* (*q.v.*).

The future may see further studies of dynamo behavior making more use of symmetries. For example, reversals may be understood in terms of the "dipole" E^A solution becoming unstable to a "quadrupole" E^S or an oscillatory solution. It was once suggested that reversals may involve only the observed poloidal field, the larger internal toroidal field retaining the same polarity, but this is unlikely because it violates the symmetry properties of the solution.

Symmetry properties have received more attention in observational studies, particularly with paleomagnetic data (see *Geomagnetic field, asymmetries*). The basic observation is that of an E^A field, the dipole, but equatorial symmetry appears to go beyond that of the dipole: the main concentrations of magnetic field on the core-mantle boundary form four lobes, two in the northern hemisphere and two in the southern hemisphere, on closely similar longitudes (see Plate 10c). This basic pattern is close to P_2^S symmetry, but a nonaxisymmetric pattern would only be of interest if it were long term. With homogeneous boundary conditions we would expect the solution to drift without reference to any longitude, averaging to axial symmetry. This has been assumed in many studies in the past; departures from axial symmetry would imply an effect of inhomogeneity on the boundary, such as variable heat flux from the core (see *Core-mantle coupling, thermal*). The lower mantle seismic velocities suggest a P_2^S pattern associated with subduction around the Pacific rim, which could favor P_2^S magnetic fields.

Further discussion of the theory is in Gubbins and Zhang (1993) and of the data analysis in Merrill *et al.* (1996).

David Gubbins

Bibliography

Bullard, E.C., and Gellman, H., 1954. Homogeneous dynamos and terrestrial magnetism. *Philosophical Transactions of the Royal Society of London, Series A*, **247**: 213–278.

Gubbins, D., and Zhang, K., 1993. Symmetry properties of the dynamo equations for paleomagnetism and geomagnetism. *Physics of the Earth and Planetary Interiors*, **75**: 225–241.

Merrill, R.T., McElhinny, M.W., and McFadden, P.L., 1996. *The Magnetic Field of the Earth*. San Diego, CA: Academic Press.

Cross-references

Cowling's Theorem
Core-Mantle Boundary Topography, Implications for Dynamics
Core-Mantle Coupling, Thermal
Dynamo, Bullard-Gellman
Dynamos, Kinematic
Geomagnetic Field, Asymmetries
Magnetohydrodynamics
Paleomagnetic Secular Variation

GEOMAGNETIC DEEP SOUNDING

Geomagnetic deep sounding (GDS) is the use of electromagnetic induction methods to determine the electrical conductivity within the Earth, working from observations of natural geomagnetic variations. It is differentiated from the *magnetotelluric method* (*q.v.*) in employing only the magnetic, and not the electric field. The term GDS is applied both to global and to regional studies. The aim of global investigations is to determine the variation of electrical conductivity with depth. That of regional studies is to map lateral differences in the conductivity of the crust and upper mantle. The book by Rokityansky (1982) covers all aspects of the subject. Weaver (1994) gives a detailed account of the theory.

Sounding the earth using natural geomagnetic variations

A slowly varying magnetic field inside a uniform conductor (conductivity σ and relative magnetic permeability μ) satisfies the induction equation:

$$\nabla^2 \mathrm{B} = \mu\mu_0\sigma\frac{\partial \mathrm{B}}{\partial t}$$

The time-varying field induces eddy currents in the conductor which flow so as to exclude the field from the deeper parts. The amplitude of a spatially uniform field of frequency ω falls to $1/e$ of its surface value at the "skin-depth":

$$z_0 = \sqrt{2/\omega\mu\mu_0\sigma}.$$

This expression provides a rough guide to the "sounding depth" which might be expected of a particular frequency. However, the geometry of the external source also restricts the depth or volume sampled by the field. When induction effects are negligible, a field with spatial wavelength λ falls to $1/e$ of its surface value at depth $\lambda/2\pi$. The basis of the sounding method is to measure the Earth response at a range of frequencies and/or source wavelengths. If the response of a one-dimensional earth is known precisely either at all frequencies or all spatial scales, the radial variation of conductivity is uniquely defined.

The geomagnetic variation spectrum

The frequency range of the externally generated electromagnetic spectrum (Figure G16) is extremely wide. The longest periods available are associated with the solar cycle (11 or 22 years), and penetrate into the lower mantle. Unfortunately, they are difficult to separate from the purely internal variations with periods longer than 3 years generated by the secular behavior of the dynamo. Much of the spectrum between 2 years and 2 days period comes from fluctuations in the total energy of particles in the radiation belts, which drift in the geomagnetic field, creating the *ring current* (*q.v.*). Because the current is located between 3 and 5 Earth radii, the field it creates at the Earth is relatively uniform, and its spatial structure can be represented by a small number of zonal *spherical harmonics* (*q.v.*), of which the first ($n = 1$) is much the most important. Variations with this structure include the semiannual line, the quasiperiodic harmonics at 27, 13.5, 9 days, etc., driven by the Sun's rotation and the persistence both of solar sources and sector structure in the solar wind, and the continuum. The annual variation, however, has a distinct spatial structure that is antisymmetric about the equator, suggesting a seasonal driving force.

The daily variation and its harmonics are created by dynamo action in the ionosphere, where thermal and gravitational tides move plasma through the magnetic field. Their spatial structure can be adequately represented by spherical harmonics up to degree $n = 4$. The continuum from a few days to a few minutes period originates in the current systems produced by geomagnetic storms. Their spatial structure is complex, and in the auroral regions is localized as electrojets in the ionosphere, and field-aligned currents connecting with the outer magnetosphere. Micropulsations are vibrations of geomagnetic field lines, while Schumann resonances are oscillations of the Earth-ionosphere waveguide excited by large-scale thunderstorm activity.

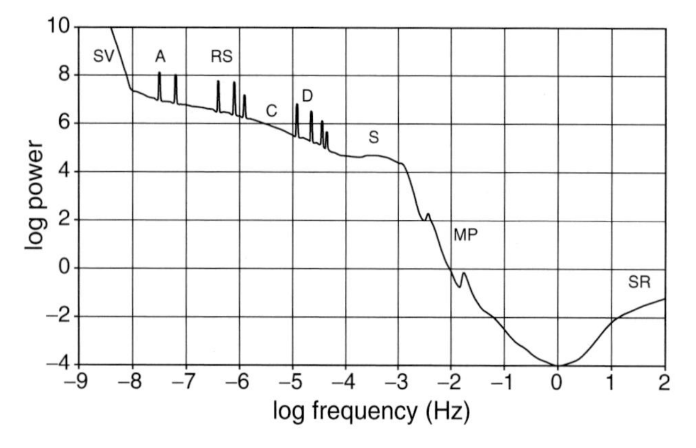

Figure G16 Schematic representation of the natural geomagnetic spectrum. SV, secular variation of the geodynamo; AV, annual variation; RS, recurrent storms; C, continuum; D, daily variation; S, storms and substorms; MP, micropulsations; SR, Schumann resonance.

GDS—the global problem

Only magnetic observatories provide the record lengths required for global sounding to depths of hundreds of kilometers. Because of their poor distribution and insufficient numbers, only the smoother fields are defined adequately. Temporary arrays of magnetometers are deployed to map the more complex fields, and, in the absence of conductivity anomalies, the field gradients can be used for local soundings.

The determination of the vertical variation of conductivity is conveniently divided into two steps. The first is the measurement of the response or *transfer function* (*q.v.*) which links the input—the external part of the magnetic field—to the output—the internal part created by the induced currents. The second is the inversion of the response for the conductivity—discovering what can be inferred about $\sigma(r)$ from the response and its associated errors.

Definition and determination of the response function

In global GDS, *spherical harmonic functions* (*q.v.*) are commonly used to define the spatial structure of the field. The response $Q_n^m(\omega)$ is the ratio of the internal and external parts of the field at frequency ω for a spherical harmonic component of degree n and order m. An alternative is the C response:

$$C = \frac{B_r}{\partial \mathrm{B}_r/\partial r}$$

The radial gradient of the vertical field is replaced by the horizontal gradients of the horizontal components using the condition div $B = 0$. For plane earth geometry, and a smooth external field,

$$C = \frac{B_z}{\partial \mathrm{B}_x/\partial x + \partial B_y/\partial y}$$

The C response has been favored in recent investigations because of its intercomparability between global and local studies, and physical significance as a penetration scale (its dimension is length).

The magnetic observatory network is inadequate for determining all but the largest scale spatial structures. Instead of a full spherical harmonic analysis, a simple spherical harmonic model is usually adopted, based on what is known of the source. For periods between 2 years and 2 days, a single zonal harmonic has been used, with the advantage that the response can be computed from vertical (Z) and horizontal (H) component records at a single observatory. The total potential of internal (g_i) and external (g_e) sources for a spherical harmonic degree $n = 1$ and order $m = 0$, is

$$\Omega = \frac{a}{\mu_0}\left\{g_i\frac{a^2}{r^2} + g_e\frac{r}{a}\right\}\cos\theta$$

The corresponding components of the magnetic induction are

$$Z = -B_r(r = a) = \{-2g_i + g_e\}\cos\theta$$

$$H = -B_\theta(r = a) = -\{g_i + g_e\}\sin\theta$$

The ratio of the vertical and horizontal components of the field, multiplied by $\tan\theta$, is itself a response (W_1^0), which can be computed from the component records by standard response estimation techniques:

$$W_1^0 = \frac{Z}{H}\tan\theta = \frac{\{1-2Q_1^0\}}{-\{1+Q_1^0\}}$$

The major factors limiting the precision of response estimates are the inadequacy of the source model and the influence of lateral variations in conductivity. Attempts to incorporate these effects into response determination and modeling are hampered by the limitations of the observatory network.

Determination of the conductivity

The second step is the transformation of the electromagnetic response data into a conductivity model. In the *forward modeling approach* (*q.v.*), a conductivity distribution is selected either on the basis of independent geophysical constraints, or for mathematical convenience. Its response is computed and compared with the observations, and the model parameters adjusted until a satisfactory fit is achieved. In the *inverse modeling method* (*q.v.*), parameters which define the structure are determined directly from the data. In practice, the two approaches are more alike than at first appears. The class of structure, which is to be the target of an inversion must be selected, and preferences about the smoothness of models incorporated. A preliminary exploration of simple models, for which the relationship between data and structure is well understood, is always worthwhile.

Consider a model in which a perfectly conducting "core," radius $r = Ra$, is surrounded by an insulating shell. The Q_1^0 response (to excitation by ring current-generated fields) can be interpreted using the spherical harmonic model to downward continue the field. The radial component of the field in *any* source-free region is:

$$B_r = -\left\{g_i\frac{-2a^3}{r^3} + g_e\right\}\cos\theta,$$

and it must be zero at the surface of the perfect conductor. If $B_r = 0$ at $r = Ra$,

$$Q_1^0 = g_i/g_e = R^3/2$$

At 27 days period, the value of Q_1^0 is 0.3, which implies $R = 0.84$, corresponding to a depth of 1000 km. This is a strong indication that the conductivity rises steeply in the upper mantle.

The global electromagnetic response and global conductivity distribution

The earliest determinations of the global response are summarized by Chapman and Bartels (1940). Schuster, Chapman, Price, and Lahiri used the daily variation and time-domain analyses of magnetic storms. Conductivity models were restricted to those with analytical solutions—a uniform core surrounded by a uniform insulator, and a power law increase. With the arrival of the digital computer, Fourier transform-based spectral analysis methods were introduced (Currie, 1966). The response could be determined at a continuous range of frequencies, and it was possible to calculate the theoretical response of arbitrary models (Banks, 1969). Recognition of the problems posed by source complexity and lateral variations in conductivity led to more robust response estimation and regionalization of the models (see Constable, 1993). Weidelt and Parker (see Parker, 1983) clarified the inverse problem and demonstrated that the best-fitting model for any set of data was a set of thin conducting sheets. They also showed how to construct more realistic models, which would, however, fit the data less well. Constable applied their techniques to collated response data. Olsen (1998, 1999) further refined both the response and models for the European area.

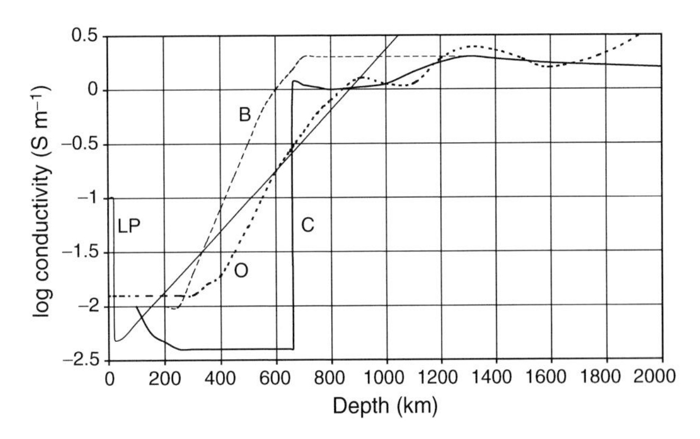

Figure G17 Representative electrical conductivity models. LP, Lahiri and Price model d; B, Banks; C, Constable; O, Olsen.

There was early recognition of a steep rise in conductivity in the 400–800 km depth range to a value of 2 $\mathrm{Sm^{-1}}$ (Figure G17). Later work has made only minor differences. The major factors inhibiting improvement are the inadequacy of the source model and the influence of lateral variations in conductivity. Attempts to incorporate these effects into response determination and modeling (Schultz and Zhang, 1994) are hampered by the nature of the observatory network. *Satellite observations* may be one route to future progress.

GDS—mapping lateral variations in conductivity

Conductivity anomalies and their response

Outside the auroral zones, the externally generated part of the time-varying magnetic field is uniform over hundreds of kilometers. If the electrical conductivity were similarly uniform, the induced currents would double the horizontal component of the external magnetic field but cancel the vertical component. Such a conductivity structure, and the fields associated with it, are referred to as "normal." What additional "anomalous" fields are created when a region of different conductivity—a conductivity "anomaly"—is embedded within the normal structure?

The anomaly's response depends on how its characteristic size (L) relates to the length scale of the normal field, and with a uniform field what matters is its skin-depth z_0, which increases with period. When $z_0 \ll L$ (at high frequencies), the characteristics of the induced currents are controlled by the local structure. When $z_0 \gg L$ (at low frequencies), they are determined by the host body. The pattern of the "normal" induced currents is modified by electric charges set up on the boundaries of the anomaly. A dipolar current system is created which enhances (a conductive anomaly) or opposes (a resistive anomaly) the normal current flow, but which is in phase with it.

Considered as an input/output problem, the input is the normal, spatially uniform horizontal field; the output is the anomalous magnetic field. This is either the entire measured vertical component (since the normal vertical field is zero), or the difference between the local and normal horizontal fields. The output is the sum of the response to two independent inputs—orthogonal directions of the normal horizontal field:

$$B_z(\omega) = Z_{zx}(\omega)B_x^n(\omega) + Z_{zy}(\omega)B_y^n(\omega),$$

where B_z is the vertical field at frequency ω at a site influenced by the anomaly, and B_x^n, B_y^n the north and east components at a normal site.

In practice, recordings at a single site are often used, and the vertical field related to the horizontal field (B_x, B_y) at the same site:

$$B_z(\omega) = T_{zx}(\omega)B_x(\omega) + T_{zy}(\omega)B_y(\omega).$$

Z and T are complex quantities, reflecting the phase shifts between the anomalous currents and the normal fields.

Magnetic variation mapping experiments

An ideal magnetic variation (MV) mapping experiment requires a large array of simultaneously recording magnetometers, spaced sufficiently closely as to avoid aliasing the structure, with at least one instrument at a normal site. Between 1965 and 1985, arrays of up to 50 magnetometers were constructed and deployed (Gough, 1989). They provided valuable initial information on the conductivity structure of the crust and upper mantle in North America, East and Southern Africa, Australia, India, and Europe. However, their limitations were soon clear. Even relatively low frequencies were affected by structures in the upper crust, so spacings less than 10 km were required, limiting the coverage. The analog recording method limited the frequency response and volume of data which could be interpreted. In particular, sampling at 0.1 to 1 Hz for a few weeks restricted the response to the range for which scattering by crustal anomalies was the dominant process. Later experiments with broader-band digital recording were restricted to a small number of instruments, and were forced to revert to the transfer function techniques described above.

The first task in interpretation is to determine the spatial pattern of conductivity without actually modeling the response. With small numbers of irregularly distributed magnetometers, a useful technique is to plot *induction arrows* (*q.v.*). For the T response, the vector is plotted with lengths in the north and east directions proportional to T_{zx} and T_{zy}, respectively. Near two-dimensional bodies, the arrows point toward or away from the structure. Arrows are harder to interpret when the structure is three-dimensional. Once the responses at a network of sites have been determined, the defining equations can be used to predict the spatial pattern of the vertical fields for a selected horizontal field—the "hypothetical event." However, the T response includes the effect of local anomalous horizontal fields, so the predicted vertical field anomalies do not correspond to induction by a uniform horizontal field. Banks (1986) devised a method of determining the Z transfer functions from the spatial variation of T, together with the constraint that the field derives from a potential that satisfies the Laplace equation. Egbert (2002) reviews methods of organizing and displaying MV array data. These include the further step of inverting the MV data for a map of the conductance in a thin sheet. The final step is to apply forward and *inverse modeling techniques* (*q.v.*) to combined MV and MT responses. The latter provide the vital constraints on absolute conductivity values.

General remarks

GDS was the most popular natural source electromagnetic method between 1950 and 1980. The magnetotelluric method was viewed with some suspicion because of the limitations of technology (difficulties in measuring electric fields, the need for high capacity data storage facilities, infield processing to evaluate data quality, etc.—all this came along after 1980 with microprocessor technology), lack of understanding of the effects of very local distortion on the electric field, and inability to compute the electromagnetic fields associated with two- and three-dimensional structures. But it had the huge advantage that measurements at a single site had the potential to define both the vertical and horizontal structure. With the advent of more powerful computing facilities, MT took over from GDS, which was then somewhat ignored. Now there is a realization that measuring and analyzing both electric and magnetic fields adds enormously to the information that can be derived.

Roger Banks

Bibliography

Banks, R.J., 1969. Geomagnetic variations and the electrical conductivity of the upper mantle. *Geophysical Journal of the Royal Astronomical Society*, **17**: 457–487.

Banks, R.J., 1986. The interpretation of the Northumberland trough geomagnetic variation anomaly using two-dimensional current models. *Geophysical Journal of the Royal Astronomical Society*, **87**: 595–616.

Chapman, S., and Bartels, J., 1940. *Geomagnetism.* London: Oxford University Press.

Constable, S., 1993. Constraints on mantle electrical conductivity from field and laboratory measurements. *Journal of Geomagnetism and Geoelectricity*, **45**: 1–22.

Currie, R.G., 1966. The geomagnetic spectrum—40 days to 5.5 years. *Journal of Geophysical Research*, **71**: 4579–4598.

Egbert, G.D., 2002. Processing and interpretation of electromagnetic induction array data. *Surveys in Geophysics*, **23**: 207–249.

Gough, D.I., 1989. Magnetometer array studies, Earth Structure and tectonic processes. *Review of Geophysics*, **27**: 141–157.

Olsen, N., 1998. The electrical conductivity of the mantle beneath Europe derived from C-responses from 3 to 720 h. *Geophysical Journal International*, **133**: 298–308.

Olsen, N., 1999. Long-period (30 days–1 year) electromagnetic sounding and the electrical conductivity of the lower mantle beneath Europe. *Geophysical Journal International*, **138**: 179–187.

Parker, R.L., 1983. The magnetotelluric inverse problem. *Geophysical Surveys*, **6**: 5–25.

Rokityansky, I.I., 1982. *Geoelectromagnetic Investigation of the Earth's Crust and Mantle.* Berlin: Springer-Verlag.

Schultz, A., and Zhang, T.S., 1994. Regularized spherical harmonic analysis and the 3-D electromagnetic response of the earth. *Geophysical Journal International*, **116**: 141–156.

Weaver, J.T., 1994. *Mathematical Methods for Geoelectromagnetic Induction.* Taunton: Research Studies Press.

Cross-references

EM Modeling, Forward
EM Modeling, Inverse
EM, Regional Studies
Harmonics, Spherical
Induction Arrows
Induction from Satellite Data
Internal External Field Separation
Magnetotellurics
Mantle, Electrical Conductivity, Mineralogy
Ring Current
Robust Electromagnetic Transfer Functions Estimates
Transfer Functions

GEOMAGNETIC DIPOLE FIELD

A long thin bar magnet gives a magnetic field, the lines of force of which (in the usual sign convention) leave the magnet near its north magnetic pole, and reenter near its south magnetic pole. If we think of the magnet being physically reduced in size, but keeping the same magnetic moment (see below), then in the limit of infinitesimal size we have, what we call, a dipole field.

Near the Earth, its magnetic field resembles that of a magnetic dipole situated at the geocenter; formally, if we represent the field as a series of spherical harmonics (see *Harmonics, spherical*), then the field given by the $n = 1$, dipole, terms dominates. (Note that these are the fields of *fictitious* dipoles; the real source is electric currents distributed throughout the core.) For many purposes it is adequate to approximate the geomagnetic field as that of a dipole; however, there are several possible definitions of such an approximating dipole.

When averaged over thousands of years, the field is very nearly that of a central axial dipole; i.e., the dipole is at the center of the Earth, and directed along the geographical axis, the spin axis. (This alignment is almost certainly due to the very strong influence the Earth's rotation has on the motions in the liquid core which produce the electric currents—see *Geodynamo*.) At present, the dipole points from north to south (in the sense that the north pole of the fictitious magnet is nearer the south geographic pole), but the direction has reversed many times during geological time (see *Geomagnetic polarity reversals, observations*). This axial dipole corresponds to the coefficient g_1^0 in a spherical harmonic analysis.

Physically, an axial dipole field is produced by a suitable axially symmetric distribution of electric currents, and its magnitude (or strength), called the dipole moment, is given in units of Am^2 (current multiplied by "area turns"); at present its value is about 8×10^{22} A m^2, a value probably rather larger than its average over the last 10^9 y (see *Dipole moment variations*). For the geomagnetic axial dipole, the north-to-south horizontal magnetic field on the equator at radius r is given by

$$B = (\mu_0/4\pi r^3) \quad (\text{Dipole moment}).$$

The spherical harmonic Gauss coefficient g_1^0 gives this value for the Earth's surface, $r = a$, and at present it is about –30000 nT; the negative sign is there because the field is actually directed south-to-north.

But while on average the dipole is axial, at any one time the best-fitting dipole is usually inclined to the spin axis, by an angle of about 10°. A general central dipole can be resolved into three orthogonal components: one along the spin axis (corresponding to the Gauss coefficient g_1^0), plus two in the equatorial plane—one (g_1^1) in the direction of zero longitude (the Greenwich meridian), and the other (h_1^1) in the direction of 90° longitude. The total dipole is called the inclined dipole, and its axis is called the geomagnetic axis; for phenomena (such as the ionosphere) which are controlled by the geometry of the geomagnetic field, it is often convenient to work in a coordinate system which is based on this geomagnetic axis, rather than on the geographic axis.

While this dipole field dominates, the remaining *nondipole field* (*q.v.*) is still significant. This nondipole field corresponds to all the $n > 1$ terms in a spherical harmonic analysis, and at the Earth's surface is typically about 25% that of the dipole field. (At the core-mantle boundary the nondipole field is larger than the dipole field, as its smaller scale fields increase downward more rapidly than the large-scale dipole field.) (See *Harmonics, spherical* and *Geomagnetic spectrum, spatial*.)

Several other planets of the solar system, and many other astronomical bodies, have dipole-like magnetic fields. For a given body, and at a given time, the magnitude and orientation of the dipole are unique. In fact the magnitude and direction of the vector dipole moment are invariant; whatever coordinate system we choose to measure from, we will always get the same dipole moment, i.e., magnitude and direction in space.

The (central) inclined dipole is a reasonable approximation to the observed field. However, if we move the position of this inclined dipole away from the center this introduces three more parameters, so it is possible to get a slightly better fit to the observed field; the displacement reduces the magnitude of what we have called the nondipole field. Conventionally, such a fit is not made to the whole of the nondipole field, but only to the $n = 2$, quadrupole, part of it; while keeping the moment and direction of the inclined dipole constant, the dipole is moved away from the geocenter in such a direction, and by such an amount, as to minimize (in a least-squares sense) the quadrupole field as seen from the displaced origin. Such a displaced dipole is called the eccentric dipole. It should be noted that (like the other dipoles) it is a simply a convenient mathematical fiction, but using it can be a useful arithmetic simplification in, for example, the study of the deflection of cosmic rays. At present the displacement is about 550 km, toward Japan; the displacement is increasing with time, because the dipole field is reducing in magnitude compared with the quadrupole field.

Various authors have suggested other definitions for the "best-fit" dipole (see Lowes, 1994), but those discussed above are the ones currently used.

Frank Lowes

Bibliography

Lowes, F.J., 1994. The geomagnetic eccentric dipole; facts and fallacies. *Geophysical Journal International*, **118**: 671–679.

Cross-references

Dipole Moment Variations
Geodynamo
Geomagnetic Polarity Reversals, Observations
Geomagnetic Spectrum, Spatial
Harmonics, Spherical
Nondipole Field

GEOMAGNETIC EXCURSION

Records of the Earth's magnetic field have shown that on occasions it has reversed its (see *geomagnetic polarity reversals*). Intervals during which the field is predominantly of the same polarity (>1 Ma) have been called chrons. Occasionally within a chron, the magnetic field reverses its polarity for a short time (<0.1 Ma)—these have been called subchrons. The last major change in the Earth's magnetic field occurred ~0.78 Ma ago, when it changed from a predominantly reversed regime (the Matuyama Chron) to its present predominantly normal regime (the Brunhes Chron). A number of cases have also been found when the magnetic field has departed for an even shorter time from its usual near-axial configuration, without establishing, and perhaps not even instantaneously approaching, a reversed direction. Such events have been called excursions (see Figure G18). They have been arbitrarily defined as cases where the colatitude, θ, of the virtual geomagnetic pole is greater than 45°. This definition distinguishes excursions from the *geomagnetic secular variation* (*q.v.*) when θ is less than 45°.

Excursions have been observed in lava flows of various ages in different parts of the world and from deep-sea and lake sediments. A controversial question is whether excursions are worldwide events. This is difficult to decide since the duration of an excursion is short and so accurate dating is essential. It has been achieved for some excursions (e.g., the Laschamp ~45 ka), but in many cases it is still unresolved. The difficulty is further compounded by another controversial issue—does the secular variation, which varies across the world, also reverse its polarity during a reversal.

Measurement of the intensity of the magnetic field at the Earth's surface gives the total field which includes the *nondipole field, ND* (*q.v.*). Unfortunately it is not possible to estimate the contribution of the ND field to the observed value. Excursions are accompanied by a reduction in the strength of the field, perhaps to as much as a tenth of its value. It has been suggested that in such cases, the ND field could dominate over a large portion of the globe, so that local reversals of the field could occur.

In some cases it has been observed that one or more excursions have occurred prior to a major polarity change. Quidelleur *et al.* (2002) found an excursion in lava flows about 40 ka before the Mayuyama/Brunhes transition that correlated with minima observed

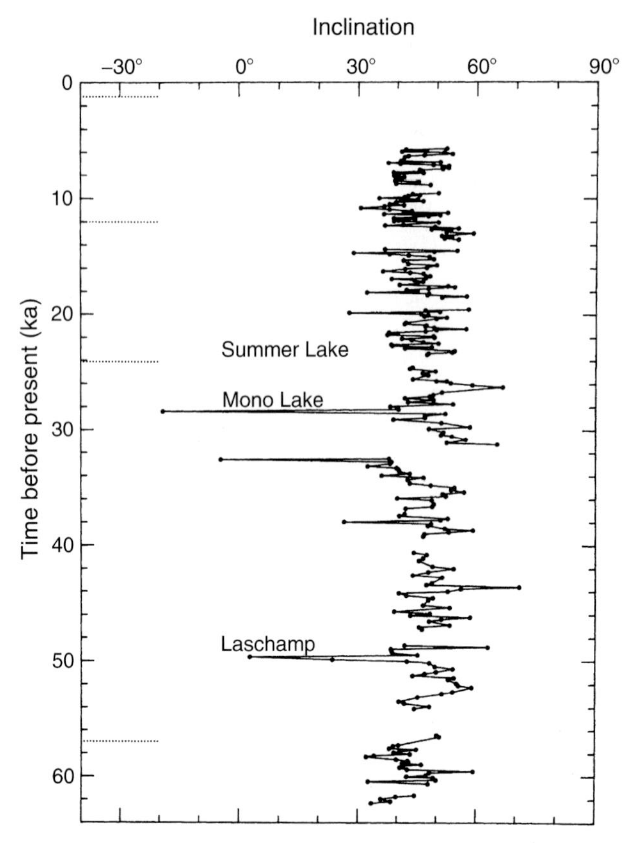

Figure G18 Inclination versus age profile for the paleomagnetically valid section at Hole 480 from 5 to 62 ka (5–49 m subbottom). The horizontal dotted segments at the age ordinate indicate the calibration points for the age estimates: the top of the section and the oxygen isotope stage boundaries. (After Levi and Karlin, 1989.)

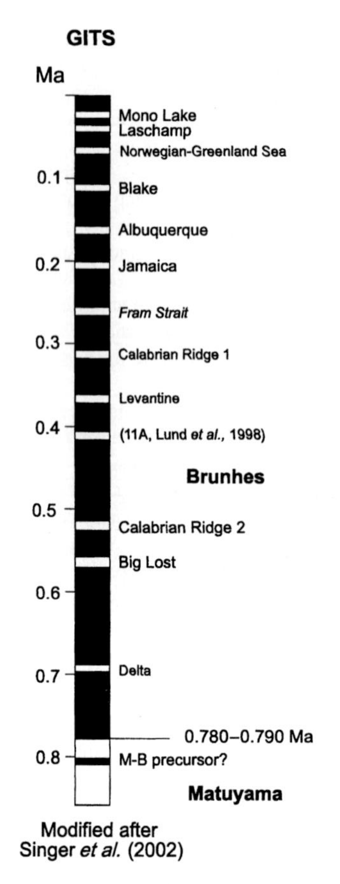

Figure G19 Geomagnetic instability timescale (GITS) for the Brunhes Chron (proposed events which require further verification are labeled with italic letters). Black corresponds to normal polarity and white to reverse polarity. (After Knudsen *et al.*, 2003.)

in the intensity of the field in high-resolution deep-sea records before this transition. The number of subchrons and excursions found in the Brunhes and Matuyama Chrons has increased in the last 4 years (Figure G19). Lund *et al.* (1998) reported 14 such events in the Brunhes Chron. Merrill and McFadden (1994) questioned such a large number of polarity changes, since it would imply that the reversed polarity state was substantially less stable than the normal state during the Brunhes Chron, whereas the dynamo equations show that the two polarities must be statistically the same.

Singer *et al.* (1999) later identified new short-lived polarity events between 1.18 and 0.78 Ma and suggested that during this period of the Matuyama reversed chron there were at least 7, and perhaps more than 11, attempts by the geodynamo to reverse, implying that the geodynamo was equally unstable in this part of the Matuyama Chron as it was in the Brunhes Chron. This was confirmed by Channell *et al.* (2002) who reported four subchrons and seven excursions in the Matuyama Chron.

The large number of subchrons and excursions in the Brunhes and Matuyama Chrons raises a number of unanswered questions on the stability of the Earth's magnetic field. One problem is the lack of reversals observed over geologic time—the oldest known reversal occurred ~3500 Ma ago. There is fair coverage for the last 1 Ma, but comparatively few records older than ~2 Ma. Thus, any conclusions reached about the behavior of the field during the last 2 Ma may not be true for earlier times. Another problem is dating—we cannot give accurate ages for events that last only a few thousand years or less. Possible explanations for excursions are that they are simply the expression of large amplitude changes in the secular variation, that they are aborted attempts to reverse, or that they are the result of chaotic behavior in the nonlinear system of equations that govern the magnetic field.

An interesting further question is what is the distinction, if any, between excursions and reversals. A possible explanation for this has been given by Gubbins (1999) who suggested that the Earth's inner core (IC) could stabilize the magnetic field. The IC is solid and changes in the field can only take place by electrical diffusion. On the other hand, changes in the field in the outer core (OC) can take place much faster since they result from convection in the OC which is fluid. Gubbins proposed that excursions arise when a field reversal takes place in the OC, but not in the IC. A full reversal only occurs when the field reversal in the OC persists long enough for the field in the IC to change polarity as well. If Gubbins' explanation is correct, then excursions would be global. It is also consistent with the approximate ratio of excursions to reversals observed in the Brunhes and Matuyama Chrons.

Jack A. Jacobs

Bibliography

Channell, J.E.T., Mazaud, A., Sullivan, P., Turner, S., and Raymo, M.E., 2002. Geomagnetic excursions and paleointensities in the Matuyama Chron at Ocean Drilling Program sites 983 and 984 (Iceland Basin). *Journal of Geophysical Research*, **107**(B6): doi:10.1029/2001JB000491.

Gubbins, D., 1999. The distinction between geomagnetic excursions and reversals. *Geophysical Journal International*, **137**: F1–F3.

Knudsen, M.F., Abrahamsen, N., and Riisager, P., 2003. Paleomagnetic evidence from Cape Verde Islands basalts for fully reversed excursions in the Brunhes Chron. *Earth and Planetary Science Letters*, **206**: 199–214.

Levi, S., and Karlin, R., 1989. A sixty thousand year paleomagnetic record from Gulf of California sediments: secular variation, late Quaternary excursions and geomagnetic implications. *Earth and Planetary Science Letters*, **92**: 219–233.

Lund, S.P., Acton, G., Clement, B., Hastedt, M., Okada, M., and Williams, T., 1998. Geomagnetic field excursions occurred often during the last million years. *EOS Transactions of the American Geophysical Union*, **79**: 178–179.

Merrill, R.T., and McFadden, P.L., 1994. Geomagnetic field stability: reversals and excursions. *Earth and Planetary Science Letters*, **121**: 57–69.

Quidelleur, X., Carlut, J., Gillot, P-Y., and Soler, V., 2002. Evolution of the geomagnetic field prior to the Matuyama-Brunhes transition: radiometric dating of an 820 ka excursion at La Palma. *Geophysical Journal International*, **151**: F6–F10.

Singer, B.S., Hoffman, K.A., Chauvin, A., Coe, R.S., and Pringle, M.S., 1999. Dating transitionally magnetized lavas of the late Matuyama Chron: toward a new ^{40}Ar/^{39}Ar timescale of reversals and events. *Journal of Geophysical Research*, **104**: 679–693.

Cross-references

Geomagnetic Polarity Reversals
Geomagnetic Secular Variation
Nondipole Field

GEOMAGNETIC FIELD, ASYMMETRIES

The geomagnetic field is produced by dynamo action within the molten iron in the outer core. This is a dynamic process intimately linked to cooling of the core and the rapid spin of the Earth. The very nature of the generative process means that the field is never static but constantly changing, a change referred to as secular variation. Hence, at any given time, there will be asymmetries in the detail of the field. Furthermore, in 1934, Cowling showed that a steady poloidal magnetic field (the part of the Earth's field that we see is a poloidal field) with an axis of symmetry cannot be maintained by motion symmetrical about that axis. This means that at any given time, the magnetic field cannot be symmetric about the rotation axis unless the field is decaying.

The situation is quite different when the field at each point is averaged over a sufficiently long period of time that the statistical variations of the process are averaged out. The resulting field is known as the time-average field and, given a symmetric Earth with symmetric properties, we would expect this time-average field also to be symmetric. Indeed, a fundamental assumption used in paleomagnetic studies is that the time-average field is that of a geocentric axial dipole. Although an asymmetry in the Earth would not necessarily lead to an asymmetry in the magnetic field, we expect that any asymmetry in the time-average geomagnetic field would be a consequence of an asymmetry in the Earth. Hence, asymmetries in the time-average geomagnetic field or in the time-average secular variation are of interest because of the potential to provide information about asymmetries deep within the Earth. Some aspects of the dynamo process are sensitive to boundary conditions and so any asymmetry is likely to be a consequence of asymmetry in the boundary conditions on the core. Hence, the lowermost 150–200 km of the lower mantle, commonly referred to as the D″ layer, probably has the greatest potential to influence dynamo behavior and produce asymmetries. This is a thermal chemical boundary layer that exhibits lateral variation in its structure and composition; analyses of the time-average magnetic field, of the polarity chronology, of magnetic field reversal paths, and of secular variation have all been used to examine the impact of these lateral variations on the magnetic field.

How much time is required to obtain a time-average field?

The magnetic field of internal origin varies on timescales from less than a day to more than 10^8 years. Variations with characteristic times less than a year are screened out by the semiconducting mantle; the very long characteristic times are associated with changes in the rate of magnetic reversals. Most dynamo theorists are of the view that the characteristic times of dynamo processes are less than 10^6 years, and so the changes in the rate of magnetic reversals are usually attributed to changes in boundary conditions of the outer core.

It is well recognized that under ideal conditions an interval of at least 10^4 years is needed to obtain a reasonable estimate of the time-average field. This reflects the characteristic times of geodynamo processes and shows that our record of direct observations is woefully short to provide a reasonable estimate of the time-average field. Furthermore, because the magnetic field occasionally experiences excursions with duration around 10^1 years, much longer times (more like $1 \times 10^6 - 5 \times 10^6$ y) are typically required to obtain a proper time-average field. Consequently it is necessary to appeal to paleomagnetic observations in order to estimate the time-average field. This then brings with it the problems of inadequate and inaccurate recording by the rocks and nonuniform distributions in space and time of the observations. Errors in plate (continent) reconstruction associated with global tectonics can interfere with magnetic field reconstruction when time intervals of order 10^7 y or longer are used and there is potential for the boundary conditions to have changed on these longer time frames. Hence, it is difficult to determine the best time interval to use to estimate the time-average field. Although paleomagnetists involved in tectonic studies often use shorter time spans, most investigators examining the time-average magnetic field properties use a time span around 5×10^6 y.

The issue is then further complicated by the fact that, on occasion, the geomagnetic field reverses its polarity, the two states being referred to as *normal* (the present state) and *reverse* polarity. In the past few million years the average reversal rate has been around 4 to 5 reversals per million years and so the field may well have changed polarity several times during an appropriate averaging interval. If fields of opposite polarity are simply averaged then the result is likely to be close to zero and quite unrepresentative of the actual field. In order to avoid this, the normal and reverse polarity results are either treated separately or combined by reversing the sign of the reverse polarity data. We do not know enough about reverse polarity fields to know if it is actually correct to combine them in this way with normal polarity fields. For example, if only the dipole component changes sign during a reversal then simply reversing the sign of the reverse polarity data will invert all of the nondipole field when in fact that should not be done. This is unlikely to produce a false asymmetry in the analysis but there is a minor possibility that it could incorrectly average out a real asymmetry.

Secular variation asymmetry

In the early 1930s, it was noted that the historical rates of secular variation in the Pacific region were anomalously low. In the 1970s, extensive paleomagnetic investigations of Brunhes-age lava flows from around the world including, especially, data from the Hawaiian Islands, showed that the local dispersion of paleodirections of the magnetic field were anomalously low for Hawaii. This was interpreted as being strong evidence for subdued secular variation in the central Pacific for the past 0.7 Ma, indicating that this was a feature of the time-average field and not just the historical field. The evidence seemed reasonably robust because Hawaii was the most extensively sampled region at that time. Several mechanisms have been suggested, generally invoking lateral or radial inhomogeneity in D″ that suppresses the generation, beneath the Pacific, of nondipole fluctuations. This was interpreted as meaning that the observed variations came from the dipole field, leading to this region being referred to as the *Pacific dipole window* or the Pacific nondipole low.

The claimed low from paleomagnetic evidence has not been sustained by subsequent analyses using much larger data sets. Its origin seems to reflect inadequate time-averaging that occurred because basalt flows in Hawaii typically erupt in clusters. Examination in the 1990s of historical data found an intense focus of the nondipole field in the north Pacific in the 17th century, suggesting that the mere absence of nondipole field and its secular variation in the Pacific is only a recent and therefore transient phenomenon. Nevertheless, Constable and Johnson (1999) proposed there still is evidence for some longitudinal asymmetry in the paleomagnetically estimated secular variation. If these results are confirmed by more data and subsequent analyses, it would indicate an asymmetry that is evidenced in the secular variation data but not in the mean field data.

Polarity asymmetries

Over the years, there have been studies that have concluded that the normal and reverse polarity fields differ in respects other than a simple change of sign. The questions to be addressed are: is there a bias toward one or other of the polarities; do the two polarity states have the same stability; and is the intensity of the field the same for the two polarity states?

The equations governing the geodynamo are complex and nonlinear, making it extraordinarily difficult to obtain a solution. However, the equations are even in **H**, the magnetic field. That is, the equations are insensitive to the sign of **H**, and so if **H** is a solution then so also is –**H**. Hence it is to be expected that the two polarities would have the same statistical properties and any asymmetry would have to be a consequence of some boundary condition.

Polarity bias

Over the years an enormous amount of effort has been put in to determine a reliable chronology for reversals as far back as possible. Such a chronology is now quite well determined for the past 170 Ma from analyses of marine magnetic anomalies. As the reversal timescale developed it became clear that the character of the reversal pattern has changed markedly with time. The most obvious change occurred about 118 Ma ago, when the field apparently locked in to a normal polarity state and remained that way for the next 35 Ma, after which reversals again occurred. Similarly, during the late Paleozoic the field was locked in to a single polarity. This phenomenon of long intervals of time with a single polarity (known as superchrons) has been interpreted as the field exhibiting a clear preference for one polarity state over the other and has been referred to as *polarity bias*. It is interesting to note that during the late Paleozoic this apparent bias was for reverse polarity but during the Cretaceous it was for normal polarity, suggesting that the (interpreted) bias has been a function of time. This perceived polarity bias was attributed to boundary conditions imposed by the D″ layer. Improved data and a better understanding of the reversal process suggest a more simple interpretation. The process producing reversals was evolving in such a way that the reversal rate gradually decreased from about 165 Ma until at about 118 Ma the rate became zero. At that stage, the field was locked in to whatever polarity it had at the time. Some time later the evolution of the process changed and at about 83 Ma reversals again started occurring, the rate gradually increasing from then until the present. Hence, superchrons represent a time when the reversal process ceased rather than representing a time of preference for one polarity state over the other. As such, they do not represent an asymmetry.

Relative stabilities of the two polarity states

Excluding superchrons, when the reversal process was in abeyance, the intervals of time between reversals may be modeled as a gamma distribution. This distribution depends on two parameters: the rate at which the reversals occur; and a parameter k that describes the shape of the distribution. If the gamma parameter, k, equals 1 then the process is the simple Poisson process and the occurrence of an event (reversal) has no impact on the probability of a future event; the process has no memory. For $k > 1$ the probability of a future reversal drops to zero as soon as a reversal occurs and then gradually recovers to its undisturbed value. Thus, the process has a memory of the previous event and this memory temporarily depresses the probability of occurrence of future events. Hence, the parameter k can be interpreted in terms of the stability of the field (against a further reversal) immediately after a reversal. Early studies suggested that k is different for the two polarities, indicating that the relative stabilities of the polarity states are different. A concern with these early studies was that the sense of this asymmetry was sensitive to minor details in the polarity chronology. A major problem is that it is easy to miss a short interval in the polarity record, and this has major consequences for interpretation. Consider the situation if a short reverse interval is missed. The three-interval polarity segment normal-reverse-normal appears as a single-interval-normal polarity segment: the three original intervals have been incorrectly concatenated into a single interval. It is simple to show that if three intervals are drawn at random from a Poisson distribution ($k = 1$) and concatenated into a single interval, then the result is the same as drawing a random interval from a gamma distribution with $k = 3$. Hence, when short intervals are missed it erroneously increases the estimated value of k, and this was not appreciated in the early studies. With this in mind, and with improved data, it is now apparent that the data are consistent with the two polarities having the same value of k. Indeed, for the reversal process, k is close to 1 and so the process is nearly Poisson, if not actually Poisson.

Intensity of the field

Reliable determination of the paleointensity of the geomagnetic field is notoriously difficult. Consequently, conclusions based on paleointensities are generally less robust than those based on paleodirections or on the reversal chronology. Nevertheless, paleointensities are an important source of information. During the 1980s several studies indicated that there were minor but statistically significant differences in the distributions of paleointensity for the normal and reverse polarity fields. However, none of these observations was robust and each suffered from structural problems such as dependence on one or two extreme values or poor spatial distribution of observations. Further work has not supported these apparent differences and there now seems no reason to reject the simple view that the two polarity states have a common time-average paleointensity.

Structure of the normal and reverse polarity fields

By far the most common way to analyze the geomagnetic field is through the use of spherical harmonics. The lower-degree harmonics are probably best known by the terms dipole (degree 2), quadrupole (degree 3), and octupole (degree 4). Zonal harmonics are harmonics that are symmetrical about the spin axis and dominate the structure of the time-average field. In the late 1970s and into the 1980s, analyses of data from the past 5 Ma indicated that the structure of the time-average reverse polarity field has been discernibly different from that of the time-average normal polarity field. Specifically, it appeared that the time-average reverse polarity field had a proportionately larger quadrupole and octupole content than the time-average normal polarity field. However, there were (and remain) significant problems with the data. The data came primarily from lava flows on continents and islands and from deep-sea sedimentary cores, which generally provided only values of the paleomagnetic inclination. The spatial distribution of the data is poor, particularly in the southern hemisphere, and the data are poorly distributed in time. Also, rock magnetic and other effects can produce spurious estimates of odd-degree harmonics, particularly the octupole term. Later work has shown that these early conclusions were not robust and were mainly due to data artifacts.

Polarity transition asymmetry

Our understanding of the structure of the field during transition from one polarity to the other is, at best, rudimentary. Despite the fact that the field at the Earth's surface is unlikely to be dominantly dipolar during a transition, it is standard practice to use the field direction to calculate the position on the Earth's surface where the pole would be if the field were dipolar. This is referred to as a virtual geomagnetic pole, or VGP. In the absence of a dominant dipolar field structure it would be expected that, for a single reversal transition, the VGP transition path would be quite different for observations taken from different locations. Of particular interest here is the fact that in the absence of a persistent asymmetry in the Earth, there should be no consistency in the VGP transition paths from one reversal transition to another.

Several investigators have noted the existence of apparently preferred VGP polarity transition paths, both within individual transitions and persisting across several different transitions. Indeed, there is also some suggestion of similar preferential paths in VGP movement as a consequence of normal secular variation. There are also proponents for periods of VGP stasis during transitions with preferred positions for the clustering of VGPs. Appeals have been made to the existence of persistent regions of relatively high electrical conductivity in D″, anomalous regions of heat flux through D″, or topography at the core-mantle boundary to influence the dynamo behavior and produce these preferred paths. Conversely, there are investigators who feel the interpretation of preferred paths is a consequence of grossly inadequate data. This is currently one of the more controversial topics in paleomagnetism.

Structure of the time-average field

Over the years, several attempts have been made to determine the structure of the time-average field by undertaking spherical harmonic analyses of paleomagnetic data for the past 5 Ma. Of particular interest here is the question of whether any nonzonal harmonic terms are genuinely present in the time-average field. Most of the modelers who have undertaken these investigations do claim the presence of nonzonal terms, but the agreement between different models is poor and this suggests that the conclusions are as yet not robust. As already discussed, there is no reliable evidence for polarity asymmetry and so a comparison of normal and reverse polarity results can be used to assess errors in the estimation of individual harmonic terms. Such a comparison suggests that the actual errors exceed the formal errors assigned by modelers.

Once again there are questions regarding reliability of the data. The data are not well distributed either in space or in time, and it is difficult to detect small rotations in the rocks providing the data. Consequently, our knowledge about the structure of the time-average field remains inconclusive and there is inadequate robust evidence to identify any specific asymmetries.

Field structure from direct observations

Spherical harmonic models have been created using data from magnetic observatories, satellites, and (corrected) ancient mariner logs. There are now models based on these data that extend from approximately 400 years ago to the present. While this time span is well short of that required to obtain a valid time-average magnetic field, some of the results combined with theory have stimulated research on possible long-term magnetic field asymmetry.

In particular, Jeremy Bloxham and David Gubbins found four lobes (in two pairs) in the structure of the field that are placed approximately symmetrically on either side of the equator at 60° latitude and at 120°W and 120°E longitude. Most of the nondipole field drifts westward, but these flux lobes appear to have been fixed for the time interval investigated. A third north-south pair of lobes is required to produce symmetry, and the speculation is that such a third pair has been disrupted by fluid flow near the surface of the outer core. Bloxham and Gubbins hypothesized that the stationary lobes reflect convection rolls found much earlier in weak-field dynamo models by Fritz Busse. ("Weak-field" means that the magnetic field has a negligible effect on fluid motions in the core. Most theory today involves strong-field models.) However, dynamo models indicate that the convection rolls should drift westward or eastward depending on the details of the model. Thus, it was posited that there are thermal anomalies that pin the flux lobes to certain locations in the lowermost mantle. If true, this would lead to long-term asymmetry in the time-average magnetic field. Gubbins and Bloxham also emphasize that the data show a persistent low secular variation in the Pacific hemisphere and, as already discussed, this would also lead to a manifestation of asymmetry in the secular variation data.

Conclusion

Dynamo theory is notoriously complex and even with the computing power available today a working model, based on parameters that actually match those in the Earth, remains elusive. Hence, it is not currently possible to predict that the particular asymmetry in the Earth will manifest as a particular asymmetry in the geomagnetic field. Thus, although there are several mechanisms that it is speculated might lead to geomagnetic asymmetry, such as anisotropy in the inner core, asymmetry in any initial field, thermal-electric and battery effects originating in the mantle, or thermal and chemical heterogeneities in D″, there is no robust evidence that this would actually occur. Certainly, it is not currently possible to invert from an observed asymmetry in the geomagnetic field to the underlying cause of that asymmetry. However, dynamo models do indicate that thermal structure at the base of a mantle does influence the structure of the field generated by the geodynamo.

The case for or against time-average asymmetry in the magnetic field and/or its secular variation remains inconclusive. The case against asymmetry is simple to state: all evidence put forth to advocate asymmetry contains errors, inadequacies, or debatable assumptions. The case for asymmetry is more subtle. Proponents recognize the problems, but point to a variety of data types that each weakly suggest a common location for an asymmetry of some form in the lowermost mantle. For example, the stationary flux lobes evidenced in the direct observations, claimed biases of VGP polarity transition data, and some secular variation data, have all been used to argue for anomalous mantle near the eastern margin of the Pacific hemisphere. This is also a region that appears to exhibit different seismic properties. Although a scientific cliché, we conclude that more data and analyses are required.

Acknowledgments

This article is published with the permission of the Chief Executive Officer, Geoscience Australia.

Phillip L. McFadden and Ronald T. Merrill

Bibliography

Bloxham, J., and Gubbins, D., 1985. The secular variation of the Earth's magnetic field. *Nature*, **317**: 777–781.

Bloxham, J., and Gubbins, D., 1986. Geomagnetic field analysis. IV. Testing the frozen-flux hypothesis. *Geophysical Journal of the Royal Astronomical Society*, **84**: 139–152.

Constable, C., and Johnson, C., 1999. Anisotropic paleosecular variation models: implications for geomagnetic field observables. *Physics of the Earth and Planetary Interiors*, **115**: 35–51.

Cowling, T.G., 1934. The magnetic field of sunspots. *Monthly Notices of the Royal Astronomical Society*, **94**: 39–48.
Merrill, R.T., and McFadden, P.L., 1999. Geomagnetic polarity transitions. *Reviews of Geophysics*, **37**: 201–226.
Merrill, R.T., McElhinny, M.W., and McFadden, P.L., 1996. *The Magnetic Field of the Earth: Paleomagnetism, the Core, and the Deep Mantle*. San Diego, CA: Academic Press.

Cross-references

Cowling's Theorem
D″
Geocentric Axial Dipole Hypothesis
Geomagnetic Polarity Timescales
Geomagnetic Secular Variation
Harmonics, Spherical
Inner Core Anisotropy
Nondipole Field
Reversals, Theory
Westward Drift

GEOMAGNETIC HAZARDS

Introduction

Geomagnetic variations take place over a wide range of timescales. The longer-term variations, typically those occurring over decades to millennia, are predominantly the result of dynamo action in the Earth's core (see *Geodynamo*). However, geomagnetic variations on timescales of seconds to many years also occur due to dynamic processes in the ionosphere (see *Ionosphere*), magnetosphere (see *Magnetosphere of the Earth*), and heliosphere (see *Magnetic field of Sun*). These changes are ultimately tied to variations associated with the solar activity cycle. To a lesser extent geomagnetic variations are also caused by changes in the solar ultraviolet emission that controls the ionization of the Earth's ionosphere (see *Periodic external fields*). However, the largest geomagnetic variations are due to sporadic solar activity in the form of solar coronal mass ejections (CMEs) and the number of CMEs varies with the phase of the solar cycle.

In a CME, the rapid reconfiguration, through magnetic reconnection processes, of the fields in and above solar sunspot regions results in the release of magnetic flux and plasma into the solar wind. Traveling toward Earth, these plasma "bubbles" can interact with the Earth's magnetic field, depositing particles and energy into the magnetosphere, and driving geomagnetic storms (see *Storms and substorms, magnetic*).

Solar activity, and hence geomagnetic storm frequency, varies on a cycle of between about 9 and 14 years (see *Dynamo, solar*). A large-scale dipolar solar magnetic field exists during times of sunspot minimum. This dipolar field progressively diminishes through solar (sunspot) maximum, becoming reestablished with the opposite polarity around the time of the next cycle minimum. Sunspots, flares, and associated CMEs are the means by which the large-scale field is changed and magnetic flux is expelled into the solar wind (see *Magnetic field of Sun*).

The fact that the geomagnetic field does respond to solar conditions can be useful, for example, in investigating Earth structure using magnetotellurics (see *Magnetotellurics*), but it also creates a hazard. This geomagnetic hazard is a risk to technology, rather than to health (for recent reviews see, e.g., Lanzerotti, 2001; Pirjola *et al.*, 2005). Astronauts are certainly at risk from bursts of ionizing solar energetic particle radiation that follow solar flares (Turner, 2001). Astronaut protection involves appropriate spacecraft shielding and positioning relative to the Sun. However, at the Earth's surface, the atmosphere acts as a protective layer equivalent to several meters of concrete. Only in the polar regions and at airline altitudes anywhere may solar radiation storms be a health issue, and major airline operators and aviation authorities are considering the risks (Getley, 2004). In terms of hazard to health, modeling and predicting the changing morphology of the geomagnetic field is important in determining where charged particles may enter into the lower atmosphere (see *Main field modeling*; *Geomagnetic secular variation*).

Conducting networks and geomagnetically induced currents

A time-varying magnetic field external to the Earth induces a secondary magnetic field, internal to the Earth, as a consequence of Faraday's law. Associated with time variations in the induced field is an electric field, which is measurable at the surface of the Earth. The surface electric field causes electrical currents, known as geomagnetically induced currents (GIC), to flow in any conducting structure, for example, a power grid grounded in the Earth. This electric field, measured in $\mathrm{V\,m^{-1}}$, acts as a voltage source across networks, with the different grounding points at different electrical potentials.

Examples of conducting networks are electrical power transmission grids, oil and gas pipelines, undersea communication cables, telephone and telegraph networks, and railways. GIC are often described as being quasi-direct current (DC), although the variation frequency of GIC is governed by the time variation of the electric field. For GIC to be a hazard to technology, the current has to be of a magnitude and frequency that makes the equipment susceptible to either immediate or cumulative damage. The size of the GIC in any network is partly governed by the electrical resistance of the network, relative to the resistance of the underlying Earth. The largest external magnetic field and magnetospheric current variations occur during geomagnetic storms and it is then that the largest GIC occur. Significant variation periods are typically from seconds to about an hour, so the induction process involves the upper mantle and lithosphere (see *Magnetotellurics*; *Geomagnetic deep sounding*). Since the largest magnetic field variations are observed at higher magnetic latitudes, GIC have been regularly measured in Canadian, Finnish, and Scandinavian power grids and pipelines since the 1970s. GIC of tens to hundreds of Amps have been recorded. However, GIC and effects have also been recorded in countries at midlatitudes during major storms. There may even be a risk to low latitude nations during a storm sudden commencement (see *Storms and substorms, magnetic*) because of the high, short-period, rate of change of the field that occurs everywhere on the dayside of the Earth.

GIC have been known since the mid-1800s when it was noted that telegraph systems could run without power during geomagnetic storms, described at the time as operating by means of the "celestial battery" (Boteler *et al.*, 1998). However, technological change and the growth of conducting networks have made the significance of GIC greater and more pervasive in modern society. We therefore describe the GIC hazard in detail in two of the best-studied networks, power grids and pipelines. The technical considerations for undersea cables, telephone and telegraph networks, and railways are similar. However, fewer problems are known, or have been reported in the open literature, about these systems. This suggests that the hazard is less, or that there are reliable methods for equipment protection.

Modern electrical transmission systems consist of generating plant interconnected by electrical circuits that operate at fixed transmission voltages controlled at transformer substations. The grid voltages employed are largely dependent on the path length between these substations and 200–700 kV system voltages are common. There is a trend toward higher voltages and lower line resistances to reduce transmission losses over longer and longer path lengths. However, low line resistances produce a situation favorable to the flow of GIC. Power transformers have a magnetic circuit that is disrupted by the quasi-DC GIC: the field produced by the GIC offsets the operating point of the magnetic circuit and the transformer may go into half-cycle saturation. This produces a harmonic-rich AC waveform, localized heating, and leads to high reactive power demands, inefficient power transmission and possible misoperation of protective measures. Balancing the network in such situations requires significant additional reactive power capacity (Erinmez *et al.*, 2002). The magnitude of GIC that will cause significant problems to transformers varies with transformer type. Modern industry practice is to specify GIC tolerance levels on new transformer purchases.

On 13 March 1989 a severe geomagnetic storm caused the collapse of the Hydro-Quebec power grid in a matter of seconds as equipment protection relays tripped in a cascading sequence of events (Bolduc, 2002). Six million people were left without power for 9 hours, with significant economic loss. Since 1989 power companies in North America, the UK, Northern Europe, and elsewhere have invested time and effort in evaluating the GIC risk and in developing mitigation strategies. GIC risk can, to some extent, be reduced by capacitor blocking systems, maintenance schedule changes, additional on-demand generating capacity, and, ultimately, shedding of load. However, these options are expensive and sometimes impractical. The continued growth of high voltage power networks, for example, in North America and in mainland Europe, is leading to a higher risk. This is partly due to the increase in the interconnectedness at higher voltages; connections to grids in the auroral zone, and commercial considerations that see grids run closer to capacity than was the case historically.

To understand the flow of GIC in power grids and therefore to advise on GIC risk, analysis of the quasi-DC properties of the grid is necessary. This must be coupled with a geophysical model of the Earth that provides the driving surface electric field, determined by combining time-varying ionospheric source fields and a conductivity model of the Earth. Such analyses have been performed for North America, the UK, and in Northern Europe. However, the complexity of power grids, the source ionospheric current systems, and the 3D ground conductivity makes an accurate analysis difficult. By being able to analyze, postevent, major storms, and their consequences we can build a picture of the weak spots in a given transmission system and even run hypothetical event scenarios. An example of a postevent grid analysis for Central Scotland and Northern England during the peak of the 30 October 2003 magnetic storm is shown in Figure G20 (Thomson *et al.*, 2005). At this time the measured GIC was 42 A, a significant, though in this case, manageable level and comparable with the model results.

Grid management is also aided by space weather forecasts of major geomagnetic storms. This allows for mitigation strategies to be implemented. Solar observations provide a 1–3 day warning of an Earth-bound CME, depending on CME speed. Following this, detection of the solar wind shock that precedes the CME in the solar wind, by spacecraft at the Lagrangian L1 point, gives a definite 20–60 min warning of a geomagnetic storm (again depending on local solar wind speed). However, the magnitude and accurate time of arrival of a CME prior to shock detection is unknown, although there is much research and model development within the space weather community. Given the demands on managing power grids more accurate information would have high value.

Major pipeline networks exist at all latitudes and many systems are on a continental scale. Pipeline networks are constructed from steel to contain high-pressure liquid or gas and are covered with special coatings to resist corrosion. Weathering and other damage to the pipeline coating can result in the steel being exposed to moist air or to the ground, causing localized corrosion problems. Cathodic protection rectifiers are used to maintain pipelines at a negative potential with respect to the ground. This minimizes corrosion without allowing any chemical decomposition of the pipe coating and the operating potential is determined from the electrochemical properties of the soil and Earth in the vicinity of the pipeline. The GIC hazard to pipelines is that GIC cause swings in the pipe-to-soil potential, increasing the rate of corrosion during major geomagnetic storms (Gummow, 2002).

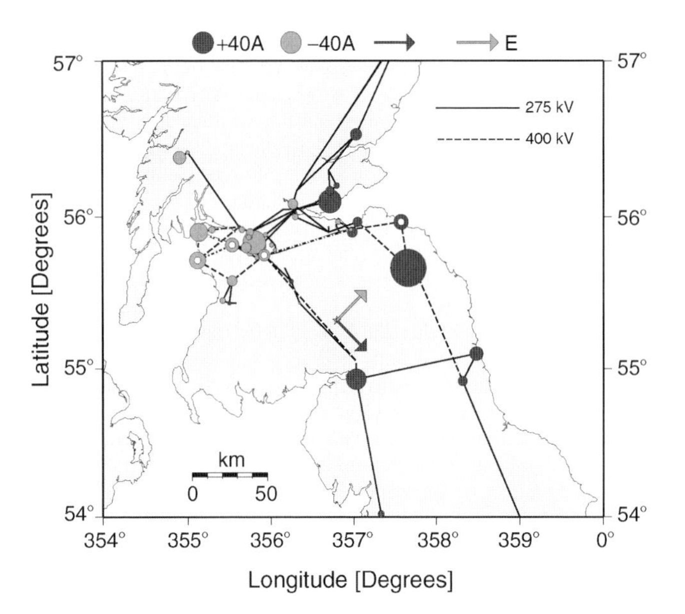

Figure G20 Estimated GIC in a network model of the Central Scotland and Northern England high voltage power grid, at the peak of the 30 October 2003, storm at 21:20 UT. Circle shading denotes GIC flowing to/from earth (dark/light) at transformers in the grid and the arrows denote the instantaneous *E* and *B* fields at Eskdalemuir magnetic observatory. The *E*-field is calculated from a radially varying Earth conductivity model using measured Eskdalemuir geomagnetic variations at this time. GIC amplitudes are proportional to spot size, with 40 A shown for scale. White spots show the locations of permanent GIC measurement sites in the grid.

GIC risk is not, therefore, a risk of catastrophic failure, rather the reduced service lifetime of the pipeline, or parts of it.

Pipeline networks are modeled in a similar manner to power grids, for example, through distributed source transmission line models that provide the pipe-to-soil potential at any point along the pipe (Boteler, 1997). These models need to take into account complicated pipeline topologies that include bends as well as electrical insulators, or flanges, that electrically isolate different sections of the network. From a detailed knowledge of the pipeline response to GIC, pipeline engineers can understand the behavior of the cathodic protection system even during a geomagnetic storm, when pipeline surveying and maintenance may often be suspended.

Satellite operation, navigation, and radio communication

The ionosphere plays a significant role in very low frequency (VLF) through to high-frequency (HF) radio communication, and in navigation systems such as Loran-C and Omega (Lanzerotti, 2001; Schunk, 2001). Ionospheric conductivity is partly affected by geomagnetic storms but more significantly by solar ultraviolet and x-ray control of the ionospheric D, E, and F layers (see *Ionosphere*). Solar flares cause signal-phase anomalies and amplitude variations to occur (fades and enhancements) and conditions can persist for minutes to hours. Solar flares and CMEs are also important as sources of solar energetic particles, affecting radio communications at high latitudes. Disturbed conditions can persist for days to weeks in the Polar Regions during periods of high particle flux into the polar caps. Ultrahigh-frequency (UHF) radio signals are central to the Global Positioning System (GPS) that utilizes satellites in Earth orbit for precise ground position determination. UHF waves pass largely unattenuated through the ionosphere but the system accuracy is sensitive to variations in the total electron content (TEC) in the path between ground and satellite. The TEC determines the signal propagation delay. TEC variations occur during geomagnetic storms and these particularly degrade the accuracy of single-frequency GPS equipment. Geomagnetic storms also produce ionospheric irregularities and scintillations that occur both on the dayside and nightside of the Earth.

In practical terms, radio navigational and communication systems and operators adjust to the prevailing conditions, although accurate forewarning of solar and geomagnetic activity may be useful for planning purposes.

Low Earth orbit satellites and space stations (up to around 1000 km altitude) experience increased air drag during geomagnetic storms. Enhanced ionospheric currents deposit heat in the atmosphere. This causes the atmosphere to expand outward and, at a given altitude, atmospheric density increases. This leads to heightened drag forces, slowing of satellite velocities, and lowering of orbit altitudes. The cumulative effect of geomagnetic storms is therefore to reduce the operational lifetime of satellites, particularly those in low initial orbits, where air density is higher. High-altitude atmospheric density models often parameterize geomagnetic heating effects by geomagnetic activity indices (e.g., Roble, 2001). By predicting geomagnetic indices, on day-to-day and solar cycle timescales, more efficient use can be made of satellite fuel supplies, with judicious orbit reboosts used to extend mission lifetimes. More sophisticated models of the upper atmosphere are being developed, involving near real-time calibration data from many orbiting satellites, as well as a better understanding and modeling of the physics of the atmosphere and its response to solar and geomagnetic forcing.

Satellites also suffer increased surface and internal electrical charging from ionized particles during geomagnetic storms. This can result in system malfunctions as electronic components experience physical damage and logic errors. Satellite manufactures take the solar cycle varying radiation environment into account when designing components, using statistical models of radiation dosages over component lifetimes. Modeling of the geomagnetic field morphology and predicting changes in the field (see *Main field modeling*; *Geomagnetic secular variation*) help to map the radiation environment of the Earth and the response of satellites to that environment. In operation, satellites can be temporarily powered down or placed in an appropriate "safe mode" following warnings of solar and geomagnetic activity.

Geophysical exploration and geomagnetic variations

Aeromagnetic surveys are affected by geomagnetic storm variations (see *Crustal magnetic field*; *Magnetic anomalies for geology and resources*; *Aeromagnetic surveying*). These cause data interpretation problems where external field amplitudes are similar to those of the crustal field in the survey area. Accurate geomagnetic storm warnings, including an assessment of the magnitude and duration of the storm, would allow for an economic use of survey equipment.

For economic and other reasons, oil and gas exploration often involves the directional drilling of well paths many kilometers from a single wellhead in both the horizontal and vertical directions. Target reservoirs may only be a few tens to hundreds of meters across and accurate surveying by gyroscopic methods is expensive since it can involve the cessation of drilling for a number of hours. An alternative is to use magnetic referencing while drilling (Clark and Clarke, 2001). Near real-time magnetic data are used to correct the drilling direction and nearby magnetic observatories prove vital. There is no drilling "down-time" during a magnetic storm and storm forecasts are not normally seen as being important.

Summary and outlook

The geomagnetic hazard to technology results from the strengthening of magnetospheric and ionospheric current systems by the solar wind and by CMEs. These electrical current enhancements cause rapid and high amplitude magnetic variations during geomagnetic storms. Processes internal to the magnetosphere also drive variations, particularly after CME-driven activity. A common theme that emerges from a study of geomagnetic hazards is a need for accurate geomagnetic storm forecasting, in terms of onset time and duration, maximum amplitude, and variation periods. The close connection of geomagnetic hazard with solar activity and space weather is also clear. Some practical applications of geomagnetic variations require only monitoring data, but other applications (will) clearly benefit from a thorough physical understanding of the Sun-Earth magnetic interaction and, in particular, accurate prediction of geomagnetic variations.

Alan W.P. Thomson

Bibliography

Bolduc, L., 2002. GIC observations and studies in the Hydro-Quebec power system. *Journal of Atmospheric and Solar-Terrestrial Physics*, **64**(16): 1793–1802.

Boteler, D.H., 1997. Distributed source transmission line theory for electromagnetic induction studies. In *Supplement of the Proceedings of the 12th International Zurich Symposium and Technical Exhibition on Electromagnetic Compatibility*, February 18 to 20, 1997 at the Swiss Federal Institute of Technology in Zurich, Switzerland, pp. 401–408.

Boteler, D.H., Pirjola, R.J., and Nevanlinna, H., 1998. The effects of geomagnetic disturbances on electrical systems at the Earth's surface. *Advances in Space Research*, **22**(1): 17–27.

Clark, T.D.G., and Clarke, E., 2001. In *Space Weather Workshop: Space weather services for the offshore drilling industry. Looking Towards a Future European Space Weather Programme.* ESTEC, ESA WPP-194.

Erinmez, I.A., Kappenman, J.G., and Radasky, W.A., 2002. Management of the geomagnetically induced current risks on the national grid company's electric power transmission system. *Journal of Atmospheric and Solar-Terrestrial Physics*, **64**(5–6): 743–756.

Getley, I.L., 2004. Observation of solar particle event on board a commercial flight from Los Angeles to New York on October 29, 2003. *AGU Space Weather*, **2**: S05002, doi:10.1029/2003SW000058.

Gummow, R.A., 2002. GIC effects on pipeline corrosion and corrosion-control systems. *Journal of Atmospheric and Solar-Terrestrial Physics*, **64**(16): 1755–1764.

Lanzerotti, L.J., 2001. Space weather effects on technologies. In Song, P., Singer, H.J., and Siscoe, G.L. (eds.) *Space Weather.* Geophysical Monograph 125. Washington, DC: American Geophysical Union, pp. 11–22.

Pirjola, R., Kauristie, K., Lappalainen, H., and Viljanen, A., 2005. Space weather risk. *AGU Space Weather*, **3**: S02A02, doi:10.1029/2004SW000112.

Roble, R.G., 2001. On forecasting thermospheric and ionospheric disturbances in space weather events. In Song, P., Singer, H.J., and Siscoe, G.L. (eds.) *Space Weather.* Geophysical Monograph 125. Washington, DC: American Geophysical Union, pp. 369–376.

Schunk, R.W., 2001. Ionospheric climatology and weather disturbances: a tutorial. In Song, P., Singer, H.J., and Siscoe, G.L. (eds.) *Space Weather.* Geophysical Monograph 125. Washington, DC: American Geophysical Union, pp. 359–368.

Thomson, A.W.P., McKay, A.J., Clarke, E., and Reay, S.J., 2005. Surface electric fields and geomagnetically induced currents in the Scottish power grid during the October 30, 2003, geomagnetic storm. *AGU Space Weather*, **3**: S11002, doi:10.1029/2005SW000156.

Turner, R., 2001. What we must know about solar particle events to reduce the risk to astronauts. In Song, P., Singer, H.J., and Siscoe, G.L. (eds.) *Space Weather.* Geophysical Monograph 125. Washington, DC: American Geophysical Union, pp. 39–44.

Cross-references

Aeromagnetic Surveying
Crustal Magnetic Field
Dynamo, Solar
Geodynamo
Geomagnetic Deep Sounding
Geomagnetic Secular Variation
Ionosphere
Magnetic Anomalies for Geology and Resources
Magnetic Field of Sun
Magnetosphere of the Earth
Magnetotellurics
Main Field Modeling
Periodic External Fields
Storms and Substorms, Magnetic

GEOMAGNETIC JERKS

Observations

Geomagnetic jerks are abrupt changes in the second-time derivative, or secular acceleration, of the magnetic field that arises from sources inside the Earth. They delineate intervals of oppositely signed, near-constant secular acceleration. The first observed geomagnetic jerk was that around 1969 (Courtillot *et al.*, 1978), and since the late 19th century when direct and continuous measurements of the Earth's magnetic field have been made at a number of observatories around the world, geomagnetic jerks have also been observed to occur around 1925, 1978, 1991, and 1999 and possibly also around 1901 and 1913 (e.g., Malin and Hodder, 1982; Courtillot and Le Mouël, 1984; Macmillan, 1996; Alexandrescu *et al.*, 1996; Mandea *et al.*, 2000). These jerks are most readily observed in the first-time derivative of the east component at European observatories (Figure G21). Understanding their origin is important, not only because they result from interesting dynamical processes in the core and may help determine the conductivity of the mantle, but also for improving *time-dependent models of the geomagnetic field* (*q.v.*) and for the strictly practical purpose of forecasting its future behavior, for example, as used in navigation.

Analysis of jerks

Many studies have been undertaken to assess jerk characteristics, such as origin, times of occurrence, and spatial patterns. At observatories the measured magnetic field is the vector sum of fields arising from three primary sources, namely the main field generated in the Earth's core, the crustal field from local rocks, and a combined disturbance field from electrical currents flowing in the upper atmosphere and magnetosphere, which also induce electrical currents in the sea and the ground. As the long-period variations of the disturbance field associated with the 11-year solar cycle are in the same frequency band as variations arising from sources inside the Earth, the separation of the sources is an important part of any analysis of geomagnetic jerks. Spherical harmonic analysis applied to monthly or yearly mean geomagnetic observatory data has shown that jerks are internal in origin and, where sufficient data exist, are global phenomena (e.g., Malin and Hodder, 1982, McLeod, 1985).

Various analysis techniques have been applied to jerks to investigate specific aspects of their temporal and spatial characteristics. Methods such as optimal piecewise regression analysis and wavelet analysis have established the exact dates of occurrence at different observatories without any *a priori* assumptions. Using wavelets, the 1969 and 1978 jerks have been shown to have different arrival times at the Earth's surface, with the northern hemisphere leading the southern hemisphere by about 2 years (Alexandrescu *et al.*, 1996).

Implications for studies of the Earth's deep interior

The shortest variations in the Earth's internal field that can be seen at the Earth's surface are determined by the rate of change of the magnetic field at the core-mantle boundary (CMB) and by the electrical conductivity of the mantle through which the magnetic signal from the core must pass. From analyses of surface data we know that the duration of individual jerks, as indicated by the width of impulses in the third-time derivative, is no more than about 2 years. However, it

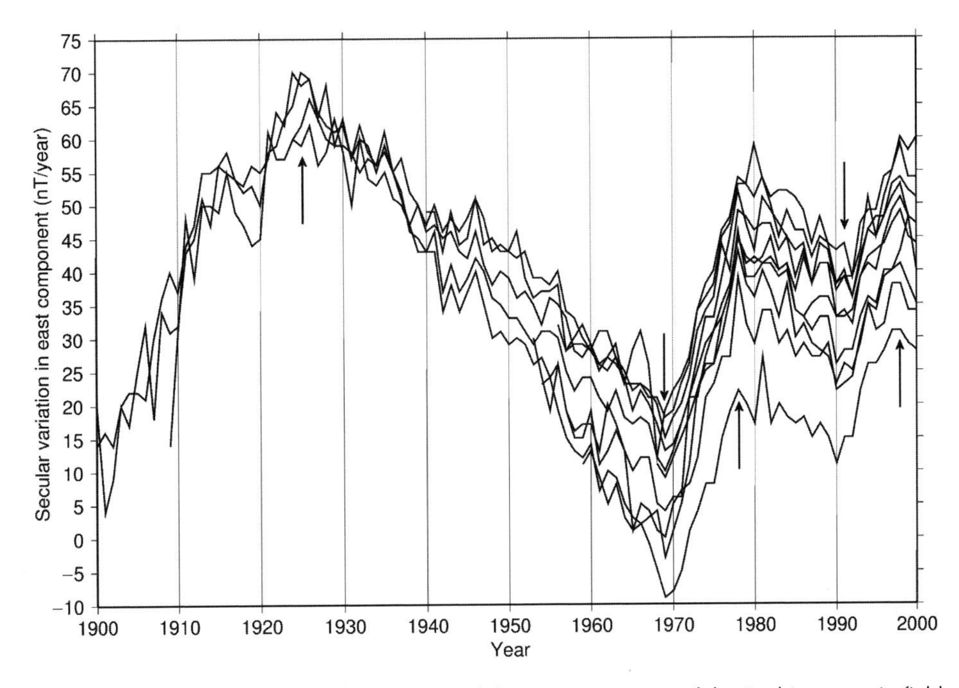

Figure G21 Geomagnetic jerks as seen in the secular variation of the east component of the Earth's magnetic field observed at European observatories. Time of jerks are shown by arrows.

is difficult to make any estimate of the rate of change of the magnetic field at the CMB, or the electrical conductivity of the mantle, from geomagnetic field observations at the Earth's surface. As in most geophysical problems, assumptions have to be made or constraints have to be applied using information from other data sources. In the case of geomagnetic jerks one useful independent dataset is length of day (LOD) observations. Correlation of the occurrences of jerks with marked changes in the deficit LOD (the former appearing to lead the latter by a few years) indicates some form of coupling between the core and the mantle. It has recently been shown that inflexion in the time derivative of splines fitted to filtered LOD data from 1960 onwards coincide remarkably well with geomagnetic jerks, including the late arrivals reported for the southern hemisphere (Holme and de Viron, 2005).

Jerks are most likely to result from dynamic processes inside the Earth that produce changes in the magnetic field near the top of the fluid core. These include the changes caused by advection and shear within the flow. Diffusion is not likely to be an important factor because of the short timescales involved. Core flows can be estimated directly from observatory secular variation data or indirectly from spherical harmonic models of secular variation. Again, assumptions have to be made or constraints applied so that a unique solution is found. One common assumption applied when determining core flows from magnetic data over time intervals of a few decades is the frozen-flux hypothesis with tangentially geostrophic time-varying flows. With this assumption the fine detail of the secular variation, including jerks, can be well fitted (Jackson, 1997) but the models offer little physical explanation of the flows in terms of core processes. Another assumption that also gives good fits to data is steady flow in a core reference frame that is allowed to rotate about the Earth's rotation axis with respect to the mantle (Holme and Whaler, 2001). This, not surprisingly, provides results which are consistent with LOD data.

A time-dependent flow model that comprises only torsional oscillations, recovers jerks very well while not producing spurious jerks where none are observed (Bloxham *et al.*, 2002). This suggests that the origin of jerks is related to the action of torsional oscillations on the local magnetic field at the CMB. If this is correct, then the torsional oscillations provide a physical representation for the origin of jerks, for they are not only a part of the mathematical representation of core flows but, more importantly, are an actual component of flow predicted to occur in the geodynamo. They also provide the angular momentum changes required to explain the LOD changes. However, the specific details of jerk generation in the core are still to be resolved, in particular the mechanism driving torsional oscillations.

Susan Macmillan

Bibliography

Alexandrescu, M., Gibert, D., Hulot, G., Le Mouël, J.-L., and Saracco, G., 1996. Worldwide wavelet analysis of geomagnetic jerks. *Journal of Geophysical Research*, **101**(B10): 21975–21994.

Bloxham, J., Zatman, S., and Dumberry, M., 2002. The origin of geomagnetic jerks. *Nature*, **420**: 65–68.

Courtillot, V., Ducruix, J., and Le Mouël, J.-L., 1978. Sure une accéleration récente de la variation seculaire du champ magnétique terrestre. *Comptes rendus des séances de l' Académie des Sciences*, **287**, D: 1095–1098.

Courtillot, V., and Le Mouël, J.-L., 1984. Geomagnetic secular variation impulses. *Nature*, **311**: 709–716.

Holme, R., and de Viron, O., 2005. Geomagnetic jerks and a high-resolution length-of-day profile for core studies. *Geophysical Journal International*, **160**: 413–413. doi: 10.1111/j.1365-246X.2005.02547.x.

Holme, R., and Whaler, K.A., 2001. Steady core flow in an azimuthally drifting reference frame. *Geophysical Journal International*, **145**: 560–569.

Jackson, A., 1997. Time-dependency of tangentially geostrophic core surface motions. *Physics of the Earth and Planetary Interiors*, **103**: 293–312.

Macmillan, S., 1996. A geomagnetic jerk for the early 1990's. *Earth and Planetary Science Letters*, **137**: 189–192.

Malin, S.R.C., and Hodder, B.M., 1982. Was the 1970 geomagnetic jerk of internal or external origin? *Nature*, **296**: 726–728.

Mandea, M., Bellanger, E., and Le Mouël, J.-L., 2000. A geomagnetic jerk for the end of the 20th century? *Earth and Planetary Science Letters*, **183**: 369–373.

McLeod, M.G., 1985. On the geomagnetic jerk of 1969. *Journal of Geophysical Research*, **90**(B6): 4597–4610.

Cross-references

Decade Variations in LOD
Main Field Modelling
Spherical Harmonics
Time-dependent Models of the Main Magnetic Field
Torsional Oscillations

GEOMAGNETIC POLARITY REVERSALS

Early records of reversals

Natural magnetization of rocks opposed to that of the present day field was observed by David and Brunhes about one century ago. Brunhes concluded that "at a certain moment of the Miocene epoch in the neighborhood of Saint-Flour the north pole was directed upward: it was the south pole which was the closest to central France" (see Laj *et al.*, 2002). Then, Matuyama observed in 1929 other reverse polarities in the magnetic directions of lava flows from Japan and Manchuria. However, early studies did not provide definite evidence that the geomagnetic field had reversed its polarity, in particular because self-reversal in some rocks can produce thermoremanent magnetization antiparallel to the geomagnetic field at cooling time. Existence of polarity reversals was definitely established when K-Ar dating demonstrated that all lavas from the same age have a similar polarity, whatever their geographical position at the Earth surface. First Geomagnetic Polarity Timescales (GTPS) were constructed in the 1960s (Cox, 1969). Then, discovery of marine magnetic anomalies (Figure G22) confirmed seafloor spreading (Vine and Matthews, 1963), and the GTPS was extended to older times (Vine, 1966; Heirtzler *et al.*, 1968; Lowrie and Kent, 1981). Since then, succession of polarity intervals has been extensively studied and used to construct magnetostratigraphic timescales linking biostratigraphies, isotope stratigraphies, and absolute ages (see Opdyke and Channell, 1996, for a review).

Reversal frequency

The geomagnetic polarity timescale constructed from seafloor magnetic anomalies (Figure G23) revealed that polarity reversals were common in Earth's history. Their occurrence, however, is not constant through time. Reversal frequency has increased during the past 85 Ma, since the long Cretaceous interval of normal polarity. Before this superchron, reversal frequency decreased since the Late Jurassic.

Changes in reversal frequency occur with time constant of tens of millions of years or more, of the same order of magnitude of those involved in mantle overturning. It was therefore suggested that reversal frequency is linked to the structure of the mantle, an idea that was reinforced by 3D computations of the geodynamo (Glatzmaier *et al.*, 1999). The question of the existence of a 15 Ma periodicity in reversal frequency (Mazaud *et al.*, 1983) has been largely debated (see Jacobs, 1994).

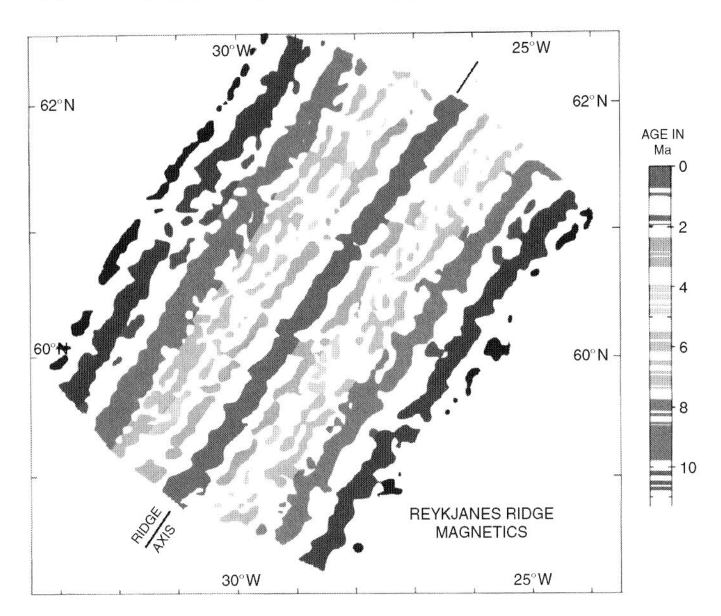

Figure G22 Magnetic structure of seafloor near the Reyjkanes Ridges (after Phinney, 1968).

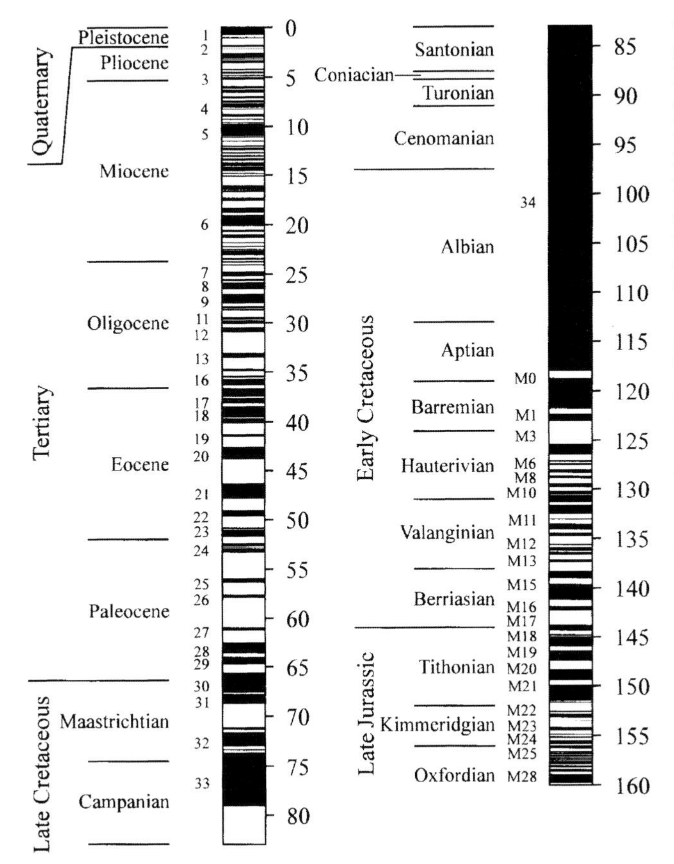

Figure G23 Geomagnetic polarity timescale with magnetic anomaly numbers for the past 160 Ma. (Figure from Merrill *et al.*, 1996).

Geometry and dynamics of the transitional field

Polarity transitions occur so quickly on a geological timescale that it is difficult to find rocks that have preserved in detail variations of the transitional field. Also, sediment and lava may sometimes provide different views of the phenomenon. Lava flows acquire magnetic memory during their cooling, and provide instantaneous pictures of the Earth magnetic field. Polarity transitions are recorded only if eruptions occurred at the time of the transition, and if no alteration or secondary heating affected the recording rocks. Lava may provide absolute paleointensities, by using Thellier, Shaw, or Thellier and Coe determination methods. In contrast, sediments provide continuous records of geomagnetic field changes, but acquisition process may smooth out fastest field changes. Only relative changes of past geomagnetic intensity are obtained, and only with sediment exhibiting limited downcore variations in magnetic mineralogy and magnetic particle grain size.

Despite difficulties to obtain paleomagnetic records of polarity transitions, several characteristics have emerged in the last few decades. The geomagnetic intensity decreases a few thousands of years before directional changes. Then, the magnetic vector exhibit directional changes while intensity remains low. When the direction reaches opposite polarity, the geomagnetic intensity rises to normal values. The total process takes place in few thousand years, and may vary for different reversals. It was also suggested that the geomagnetic field intensity progressively decreases during periods of stable polarity and recovers high values immediately after a transition (Valet and Meynadier, 1993).

The standing field hypothesis and the flooding models

The geometry and the dynamics of the geomagnetic field during polarity transitions have been subject to large debate. Hillhouse and Cox (1976) suggested that if the usual nondipole drift (secular variation) was unchanged during a polarity reversal, which is theoretically envisaged (Le Mouël, 1984), then one should observe large longitudinal swings in reversal records. Their absence in the first records obtained from sediments led these authors to suggest the standing field hypothesis, in which an invariant component dominates the transitional field when the axial dipole component has vanished. The standing field hypothesis, however, was not confirmed. A classical tool used for investigating the morphology of the transitional field is the virtual geomagnetic pole (VGP), defined as the pole of the dipolar field that gives the observed direction of magnetization at studied site. If VGPs obtained at several sites for a given instant in time coincide, then the field has a dipolar structure. If different VGPs are obtained at different sites, then the field was not dipolar (the word "virtual" indicates that VGPs can be calculated for any field, not necessarily dipolar).

Sedimentary records also indicated different VGP paths for the same transition studied at different sites at the Earth's surface, with transitional VGPs moving progressively along longitudinal great circles. This led Hoffman and Fuller (1978) to develop flooding models, in which reversals originate in a localized region of the core and then progressively propagate into other regions.

Stop and go behavior

Higher resolution records were obtained from lava and sedimentary series. They suggested some repetition over successive reversals (Valet and Laj, 1984), and also stop-and-go behavior, with alternating phases of rapid change and stationarity during transitions. This is seen in the Miocene reversal record obtained at Steens Mountain, Oregon (Figure G24) (Prévot *et al.*, 1985), and also in Miocene sedimentary records from Northwest Greece (Laj *et al.*, 1988).

The two preferred bands

An intriguing point was issued by Laj and colleagues (1991), examining the longitudinal distribution of the VGP paths of sedimentary records of reversals over the past 10 Ma. They found that VGP paths tend to follow paths located into two preferred bands, over Americas and eastern Asia. These two preferred bands coincide with regions of fast seismic wave propagation in the lower mantle, in the downward continuation of subducted slabs in the mantle (Figure G25). These two bands also coincide with the main patches of radial magnetic flux at the core-mantle boundary in the present-day field and the historical field (Bloxham and Gubbins, 1985), and in paleofield reconstructions (Constable, 1992; Gubbins and Kelly, 1993). Thus, Laj and colleagues (1991) argued that the persistence of these two preferred bands over time of the order of magnitude of mantle convection was experimental evidence that the geodynamo was constrained by temperature patterns at the core-mantle boundary, as suggested by present day

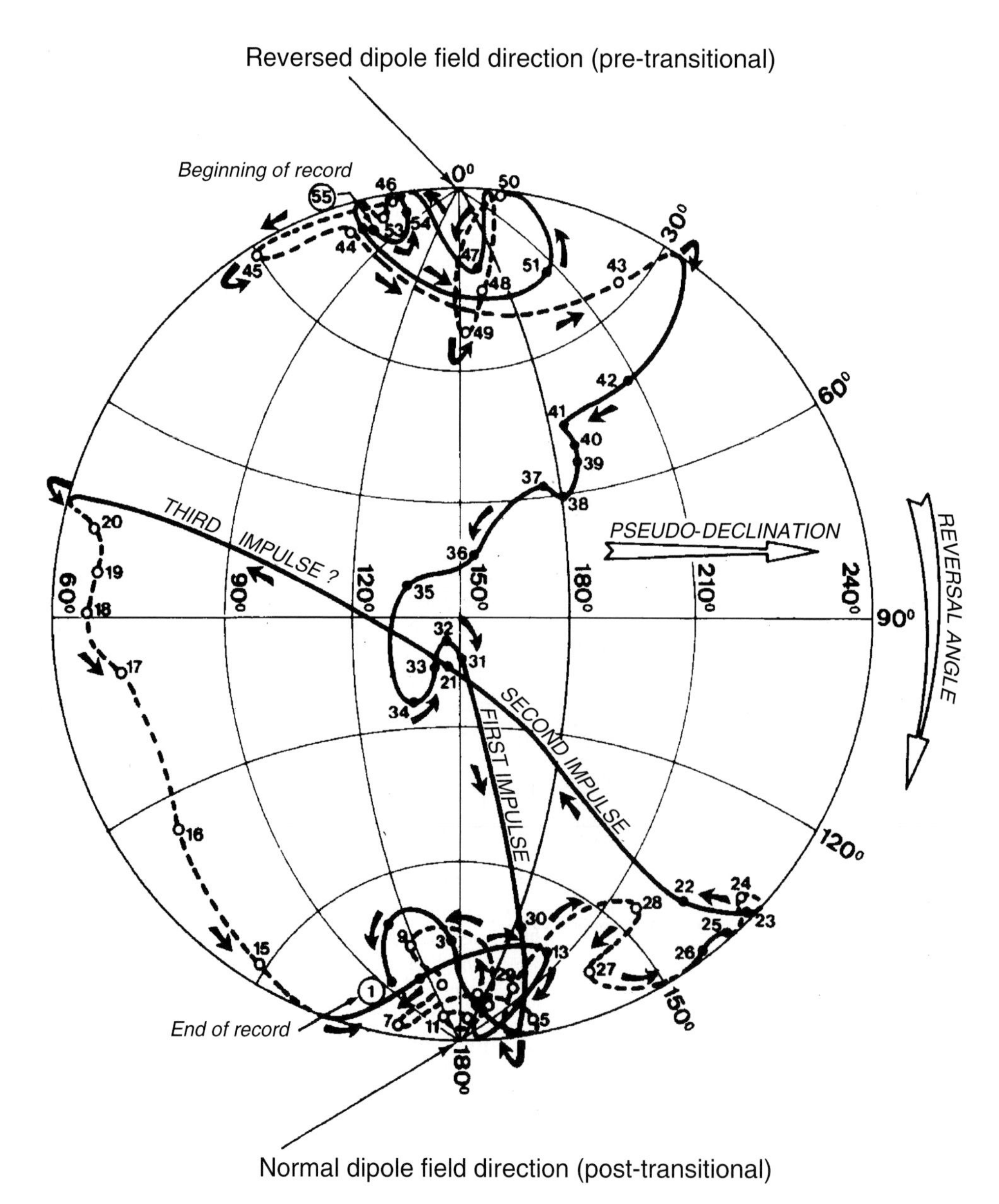

Figure G24 The Steens Mountain directional record. Stereographic projection of field directions after rotation about the east-west horizontal axis, so dipole field directions coincide with the pole of the projection sphere (from Prévot *et al.*, 1985).

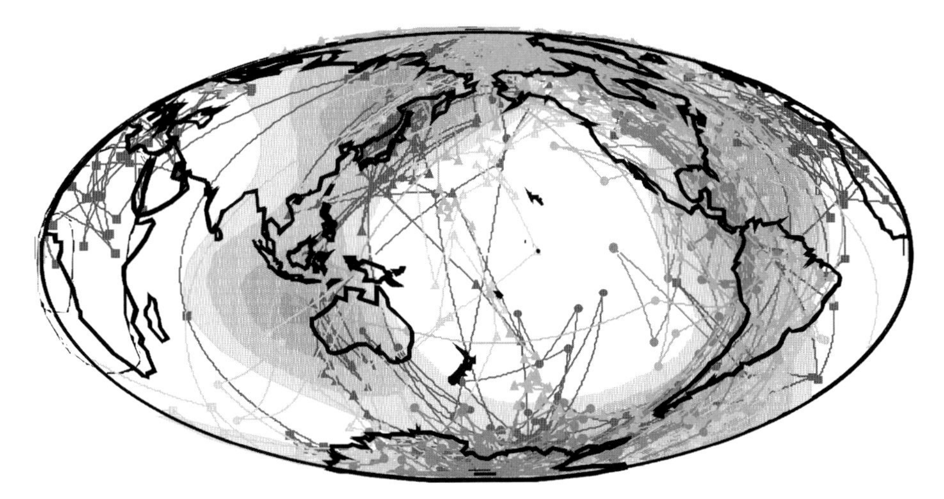

Figure G25 The two preferred bands for VGP paths of sedimentary records of reversals (from Laj *et al.*, 1991). Zones of fast seismic velocity in the lower mantle are indicated in dark (blue in online version of encyclopedy).

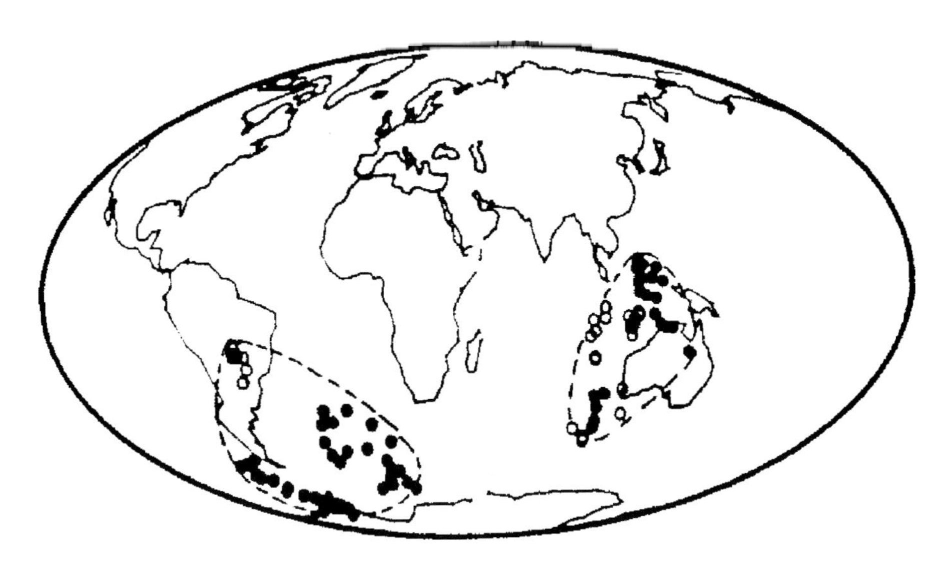

Figure G26 Clusters of VGPs (from Hoffman, 1992).

and historical magnetic observations. The existence of the two preferred bands was and is still debated, in particular because they were not seen in the volcanic compilation of Prévot and Camps (1993). However, another analysis of volcanic databases suggests that the two bands also exist in compilations of volcanic reversals records (Love, 1998). Influence of variable conditions at the core-mantle boundary is suggested in 3D computations of the geodynamo (Glatzmaier *et al.*, 1999).

Toward a consistent picture of polarity reversals?

Progressively, a consistent picture emerges from high-resolution sediment and lava records. In 1992, Hoffman observed long-lived transitional states of the geomagnetic field, with VGPs of volcanic reversal records clustering in the two preferred bands (Figure G26).

More recently, several high-resolution sedimentary records were obtained that exhibited complex field behavior, with VGP loops and clusters reminiscent of volcanic records. Some clusters, but not all, lie in the two preferred bands (Figure G27). Transitional precursors are sometimes observed.

Conclusion

The situation is still complex, because different reversals may document different behavior and the question of the two preferred longitudinal bands is still open. The relation between polarity reversals and excursions is under investigation. Geomagnetic excursions are seen as aborted reversals in which the field may reverse in the liquid outer core, which has timescales of 500 years or less, but not in the solid inner core, where field must change by diffusion with a timescale of $\approx$3 ka. This disparity of dynamical timescales between the inner and outer core is consistent with the presence of several excursions between full reversals (Gubbins, 1999). Overall, the mechanisms that trigger reversals and excursions have to be better understood. Whether or not the present-day axial dipole field decrease corresponds to a reversal or excursion onset, or to a field fluctuation during stable polarity, is not yet known.

Combination of supercomputer 3D simulations (see for instance Glatzmaier and Olson, 2005, for a review) and new high-resolution records from both sediments and lava flows should lead to a better understanding of the mechanisms involved in field generation and

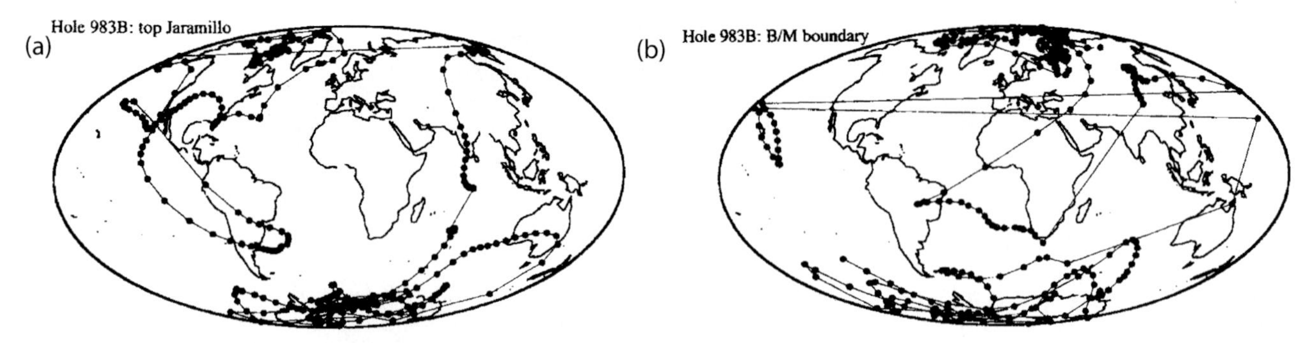

Figure G27 Examples of VGP paths for the Bunhes-Matuyama (a) and upper Jaramillo (b) transitions obtained from high deposition rate sediments in the North Atlantic (after Channell and Lehman, 1997).

polarity transitions, and ultimately of the global dynamics of the Earth's interior. LSCE contribution n° 2407.

Alain Mazaud

Bibliography

Bloxham, J., and Gubbins, D., 1985. The secular variation of the Earth's magnetic field. *Nature*, **317**: 777–781.

Channell, J.E.T., and Lehman, B., 1997. The last two geomagnetic polarity reversals recorded in high-deposition-rate sediment drifts. *Nature*, **389**: 712–715.

Constable, C., 1992. Link between geomagnetic reversal paths and secular variation of the field over the past 5 Ma. *Nature*, **358**: 230–233.

Cox, A., 1969. Geomagnetic reversals. *Science*, **263**: 237–245.

Glatzmaier, G.A., and Olson, P., 2005. Probing the geodynamo. *Scientific American*, **292**: 51–57.

Glatzmaier, G.A., Coe, R., Hongre, L., and Roberts, P.H., 1999. The role of the Earth's mantle in controlling the frequency of geomagnetic reversals. *Nature*, **401**: 885–890.

Gubbins, D., 1999. The distinction between geomagnetic excursions and reversals. *Geophysical Journal International*, **137**: F1–F3.

Gubbins, D., and Kelly, P., 1993. Persistent patterns in the geomagnetic field over the past 2.5 Ma. *Nature*, **365**: 829–832.

Heirtzler, J.R., Dickson, G.O., Herron, E.M., Pitman W.C. II, and Le Pichon, X., 1968. Marine magnetic anomalies, geomagnetic field reversals, and motions of the ocean floor and continents. *Journal of Geophysical Research*, **73**: 2119–2136.

Hillhouse, J.W., and Cox, A., 1976. Brunhes-Matuyama polarity transition. *Earth and Planetary Science Letters*, **29**: 51–64.

Hoffman, K.A., and Fuller, M., 1978. Polarity transition records and the geomagnetic dynamo. *Nature*, **273**: 715–718.

Hoffman, K.A., 1992. Dipolar reversal states of the geomagnetic field and core mantle dynamics. *Nature*, **359**: 789–794.

Jacobs, J.A., 1994. *Reversals of the Earth's Magnetic Field*, 2nd edn. Cambridge: Cambridge University Press, 339 pp.

Laj, C., Guitton, S., Kissel, C., and Mazaud, A., 1988. Complex behavior of the geomagnetic field during three successive polarity reversals, 11–12 Ma BP. *Journal of Geophysical Research*, **93**: 11655–11666.

Laj, C., Mazaud, A., Weeks, R., Fuller, M., and Herrero-Bervera, H., 1991. Geomagnetic reversals paths. *Nature*, **351**: 447.

Laj, C., Kissel, C., and Guillou, H., 2002. Brunhes research revisited: magnetization of volcanic flows and backed clays. *EOS Transactions: American Geophysical Union*, **83**(35): 381, 386–387.

le Mouël, J.L., 1984. Outer core geostrophic flow and secular variation of Earth's magnetic field. *Nature*, **311**: 734–735.

Love, J.J., 1998. Paleomagnetic volcanic date and geometric regularity of reversals and excursions. *Journal of Geophysical Research*, **103**: 12435–12452.

Lowrie, W., and Kent, D.V., 1981. One hundred million years of geomagnetic polarity history. *Geology*, **9**: 392–397.

Mazaud, A., Laj, C., De Seze, L., and Verosub, K.L., 1983. 15 Ma periodicity in the frequency of geomagnetic reversals since 100 Ma Reply to McFadden. *Nature*, **304**: 328–330.

Merrill, R.T., McElhinny, M.W., and McFadden, P.L., 1996. *The Magnetic Field of the Earth.* International Geophysics Series 63. San Diego, CA: Academic Press, 527 pp.

Opdyke, N., and Channell, J.E.T, 1996. *Magnetic Stratigraphy.* International Geophysics Series 64. San Deigo, CA: Academic Press, 346 pp.

Phinney, R.A., 1968. *The History of the Earth's Crust.* Princeton, NJ: Princeton University Press, 244 pp.

Prévot, M., and Camps, P., 1993. Absence of preferred longitudinal sectors from pole from volcanic records of geomagnetic reversals. *Nature*, **366**: 53–57.

Prévot, M., Mankinen, E.A., Grommé, C.S., and Coe, R., 1985. How the geomagnetic field vector reverses polarity. *Nature*, **316**: 230–234.

Valet, J.P., and Laj, C., 1984. Invariant and changing transitional field in a sequence of geomagnetic reversals. *Nature*, **311**: 552–555.

Valet, J.P., and Meynadier, L., 1993. Geomagnetic intensity and reversals during the past four million years. *Nature*, **366**: 234–238.

Vine, F.J., 1966. Spreading of the ocean floor: new evidence. *Science*, **154**: 1405.

Vine, F.J., and Matthews, D.H., 1963. Magnetic anomalies over oceanic ridges. *Nature*, **199**: 947–949.

Cross-references

Paleointensity, Absolute, Determination
Polarity Transition, Paleomagnetic Record

GEOMAGNETIC POLARITY REVERSALS, OBSERVATIONS

Earth's magnetic field exhibits the remarkable property of undergoing a 180° change in directions at geologically frequent, but irregular, intervals. As a result, a compass needle that had been pointing to one geographic pole (such as North), would point to the opposite geographic pole. Why the field reverses polarity remains a mystery and solving this mystery is important for understanding the processes in Earth's fluid outer core responsible for generating the geomagnetic field.

The best record of how frequently the geomagnetic field reverses polarity comes from marine magnetic anomalies (Cande and Kent, 1995). Neither the marine anomaly record nor long stratigraphic sequences of polarity history, often record field directions that fall in between the two stable polarity directions. Given the seafloor spreading and sedimentation rates, this indicates that polarity reversals are rapid events, at least on geological timescales. For example the marine magnetic anomaly record limits the time for a reversal to occur to less than 20 ka.

The first step in attempting to solve the mystery of why Earth's magnetic field reverses is to determine what happens as the field reverses. Paleomagnetic records of polarity transitions are the only source of information about how the geomagnetic field behaves during a reversal. Because polarity reversals are very short events, polarity transition records are difficult to obtain. Not only is it necessary to find a paleomagnetic recorder that provides great enough time resolution to catch the field in the act of reversing, but also it is necessary for that recorder to provide a high fidelity recording of the field, even when the field was weak.

Two very different types of paleomagnetic recorders have repeatedly yielded polarity transition records. Volcanic sequences created by rapid extrusion periods that coincide with a polarity reversal, record the changing field as successive lava flows cool and become permanently magnetized in the transitional field. Sequences of sedimentary rocks or unconsolidated sediments have also proven useful in transition studies. If sedimentation rates are rapid and continuous enough, a record of a polarity transition may result as layers of sediment become magnetized as they accumulate during the reversal.

These two types of paleomagnetic recorders document transitional fields in fundamentally different ways. Most importantly, the way each type of recorder becomes magnetized is very different, and these differences must be considered carefully when interpreting each type of record. A second important difference is the two kinds of recorders provide different types of temporal information about reversing fields.

The processes by which lavas become magnetized are well founded in the theory of thermoremanent magnetization (TRM). This theory explains how grains of magnetic minerals become permanently magnetized as they cool through their magnetic blocking temperatures in the presence of an external field. Because the cooling times for typical lava flows are rapid, it may be assumed that the lava flows provide a spot reading in time of the magnetic field direction and intensity at the time of cooling. TRM acquisition theory also provides a basis for obtaining measures of the strength or intensity of the field, known as absolute paleointensities.

On the other hand, the processes by which sediments become magnetized are less well understood. Although it is clear that many types of sediments can record the geomagnetic field accurately, as evidenced by recording the full polarity directions predicted for their site location, it is not clear how these magnetizations are acquired. A number of studies indicate that a postdepositional remanent magnetization process that locks-in the magnetization at some depth beneath the seafloor magnetizes marine sediments. Of particular interest in polarity transition studies is not so much the depth offset of remanence acquisition, but rather over what depth range the acquisition process occurs simultaneously. If the remanence becomes fixed over a relatively thick depth interval, the sediment in that layer will average out the changes in the field that occur. The resulting magnetization therefore would be an integration of the field during that time, and would provide an average of those changes in the field. As the thickness of sediment over which the remanence becomes blocked in becomes thinner, the magnetization approaches more of a spot reading of the changing field. Unfortunately, the remanence lock-in thicknesses in sediments are not known, and this presents an important difficulty in interpreting sediment records of polarity transitions.

Unlike the theory of TRM, theories of how sediments become magnetized do not provide a basis for determining absolute field intensities. But, if conditions are favorable, it is possible to obtain records of relative changes in field strength (Tauxe, 1993). In other words, sediments are capable of providing information on the increasing or decreasing field strength, but as of yet, it is not possible to calibrate those relative changes to absolute magnitudes of field strength.

In addition to the different ways these two types of recorders become magnetized, there is also a fundamental difference in the temporal resolution provided by each type of recorder. On the timescales over which polarity reversals occur, it is not possible to determine the rates at which lavas are extruded from a volcano, or how much time has elapsed between successive lava flows. This means, that while lavas may provide high fidelity recordings of the field, the timing of the sequence of magnetizations of the flows is unknown. Recent advances in radiometric dating techniques using $^{40}Ar/^{39}Ar$ and K/Ar methods (Singer and Pringle, 1996) are approaching the precision needed to provide bounding limits on the durations of the youngest reversals but cannot yet provide information of timing with reversals.

Sediment records on the other hand, generally provide a much more continuous recording of the field. Correlation of $\partial^{18}O$ records with the Marine Isotopic Stages can provide high-resolution age control through a reversal. Such records provide important information regarding the age, timing, and duration of polarity transitions.

Paleomagnetic studies of polarity transitions result in a sequence of magnetizations that were acquired as a reversal occurred. These results are generally presented as magnetization declination, inclination, and intensity or virtual geomagnetic pole (VGP) latitude as a function of stratigraphic position (either as positions in a sequence of sediments or lava flows). In order to compare records from different geographic locations it is necessary to take into account the fact that even during intervals of full polarity, different field directions are expected at different sites. For full polarity intervals, paleomagnetists do this by using the Geocentric Axial Dipole hypothesis to calculate the equivalent geomagnetic pole for the given magnetization directions (where the site results would place Earth's magnetic north pole). If the magnetization is thought to represent a spot reading (nearly instantaneous) of the field in time, the pole is called a VGP.

Polarity transition data are often presented as VGP positions as a function of stratigraphic position (Figure G28). This method of viewing the data is a convenient way to show the polarity reversal by illustrating how the apparent north magnetic pole (VGP), calculated from the observed directions at a site, moves from one high latitude position to the other.

An additional way of presenting a transition record is to plot the path of the VGP on a world map as it moved from one geographic pole to the other (Figure G29). This method makes it possible to compare records from distant sites and to test the hypothesis that the transitional field was dipolar.

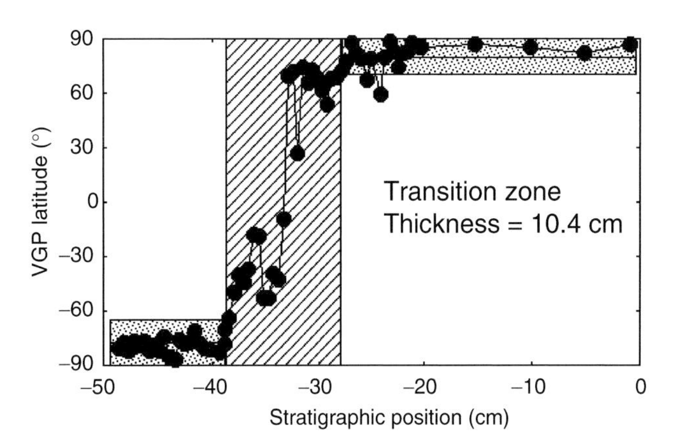

Figure G28 A record of the Lower Jaramillo polarity transition (Gee *et al.*, 1991) plotted as virtual geomagnetic pole position versus stratigraphic position. The full polarity intervals exhibit VGPs, which cluster close to the geographic poles. The dotted intervals represent a measure of the scatter of the VGPs about the full polarity average direction. The progression of VGP positions from one geographic pole to the other defines the polarity transition (shown by dashed interval). In this case, the transitional interval is defined as the zone in which the VGPs fall a statistically significantly distance from the geographic poles.

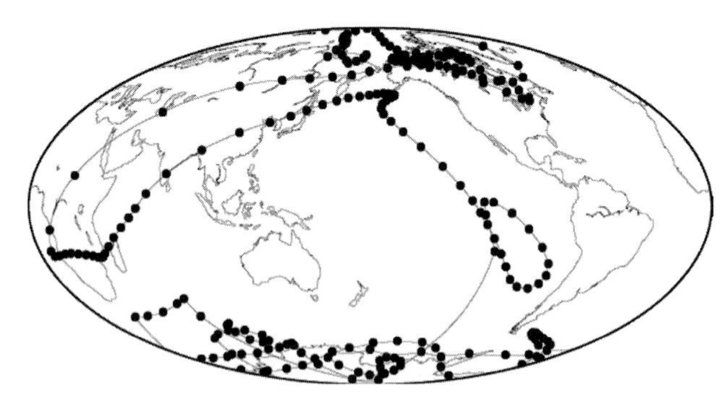

Figure G29 The Matuyama-Brunhes polarity transition obtained from North Atlantic deep-sea sediments (Channell and Lehman, 1997) shown as the path the VGP positions track along as the field gradually changes from reverse polarity (VGPs near the south geographic pole) to normal polarity (VGPs near the north geographic pole).

If the transitional fields were not dipolar, then it is not strictly appropriate to calculate VGPs from the transitional magnetizations, because this calculation assumes a dipolar field. But then some other method is needed for comparing records from distant sites. One method that has been proposed is to plot the changing directions on an equal-area stereographic projection; a method commonly used by paleomagnetists. However, before plotting the directions, the vectors are rotated so that the plot is constructed looking down the full polarity direction (Hoffman, 1984). This method provides a way of comparing the directional behavior observed from different locations without invoking the assumption of dipolar field geometries. This method, however does not illustrate absolute or relative paleointensities.

Although the details of transitional field behavior remain uncertain, the number of paleomagnetic records of polarity transitions now available makes it possible to address the major features of reversals with greater certainty. It is generally agreed that during a reversal, the strength of the geomagnetic field drops to low levels, close to that of the strength of the present-day nondipole field (Merrill and McFadden, 1999). This represents about an 80% drop in intensity. Both volcanic and sedimentary transition records record intensity lows.

The timing of the directional change relative to the intensity change is not tightly constrained; however, the majority of sediment records and many volcanic records show that the directional change occurs while the field is weak. In other words, the directional change begins after the field has weakened and finishes before the field strength increases to full polarity values.

Simple geometrical models of reversals generally suggest that the directional change occurs at slightly different times during the low intensity depending on the site location. For example, the directional change may occur earlier at low latitudes than at high latitudes. Unfortunately, the records available to date do not have enough time control to rigorously test for this result.

The time it takes for a reversal to occur provides a major constraint on the geodynamo process. The reversal durations obtained from the available sediment records indicate that the time it takes for the directional change to occur is $\approx 7 \pm 1$ ka. The intensity change takes longer, perhaps up to 11 ka. These duration estimates are less than estimates of the free-decay time of the geodynamo (20 ka), indicating that reversals do not result from a passive decay of the field (Clement, 2004).

The reversal durations are also dependent on site latitude, with the directional change occurring faster at low latitude than at high latitude (Clement, 2004). Again, this observation is expected based on simple geometrical models of reversals and provides constraints on three-dimensional numerical simulations of the dynamo.

While the features described above are generally agreed upon, there remain several controversial interpretations of transitional field behavior, most of which involve interpreting the transitional field geometries (based on the directional data).

Some of the earliest comparisons of records of the most recent reversal showed that VGP paths from widely separated sites do not coincide (Hillhouse and Cox, 1976). This means that the field geometry was not dipolar, but was more complex. In other words, the reversal did not occur by a simple rotation of a dipole from one geographic pole to the other. This interpretation has held for the majority of reversals for which multiple records are available.

Based on this result, it has traditionally been held that the simplest hypothesis is that, because the field is weak during a reversal, the field geometry would likely be very complex: much like that of the present-day, time-varying, nondipole field. This interpretation is supported by dynamo theory and what is known about the properties of the core. This hypothesis predicts that there should be no systematic variation in intermediate polarity field directions observed at distant sites.

Perhaps the most controversial issue regarding polarity transitions is that several polarity reversals do, in fact, exhibit systematic variations in directions. For the most recent and several older reversals, it has been shown that the distribution of transitional VGPs from multiple sites fall into two, nearly antipodal, preferred longitudinal bands, one passing through the Americas and the other through eastern Asia (Clement, 1991; Laj *et al.*, 1991). The observation that multiple records of the same reversal exhibit VGPs in both longitudinal bands means that the transitional fields were not dipolar, but instead, some other, relatively simple transitional field geometry gave rise to the VGP distribution.

Because there is no known intrinsic property of Earth's outer core or the geodynamo that should give rise to this distribution of VGPs or the recurrence of the pattern in multiple reversals, it has been suggested that lateral variations in the lowermost mantle affect the dynamo and influence the geometry of the transitional fields. However, this interpretation remains controversial because the grouping of longitudinal bands of transitional VGPs comes primarily from sediment records, and because the geographic distribution of available transition records is not wide enough to rigorously demonstrate the grouping.

Some volcanic transition records have been obtained that exhibit clusters of VGPs that fall within one or the other of the preferred longitudinal bands, suggesting that this distribution may not be an artifact of the remanence acquisition process in sediments (Love and Mazaud, 1997). Because clusters of VGPs from multiple lava flows occur, it has been suggested that the transitional field may get temporarily locked into a geometry that produces these VGPs (Hoffman, 1992). If so, this could explain the longitudinal grouping of VGPs from the sediment records by assuming that the sediment records have averaged the dominant field geometry over the reversal. This process would produce a great circle VGP path over the cluster of VGPs from the lava records.

An additional controversy centers over the observation of recurring VGP positions that occur within individual transition records. Several records, some volcanic and some sedimentary, exhibit VGP paths remarkable in that the VGPs return to a position that had occurred previously during the reversal. This observation suggests that the reversal process possesses a memory, at least over the timescales of a single reversal. A few lava records have also been interpreted as exhibiting VGPs that return to similar positions during different reversals, suggesting a memory in the dynamo process that exists over much longer timescales. In both cases, lateral variations in the lowermost mantle are the likeliest candidate for providing such a memory (Hoffman, 1991).

This interpretation has been questioned by suggesting that the recurrent VGP positions may be an artifact of the remanence acquisition processes. If such an artifact is present or a magnetic overprint was acquired at a later age, it is possible that the observed sequence of transitional directions does not correspond to the actual temporal sequence that occurred during the reversal. So far, however, only one example

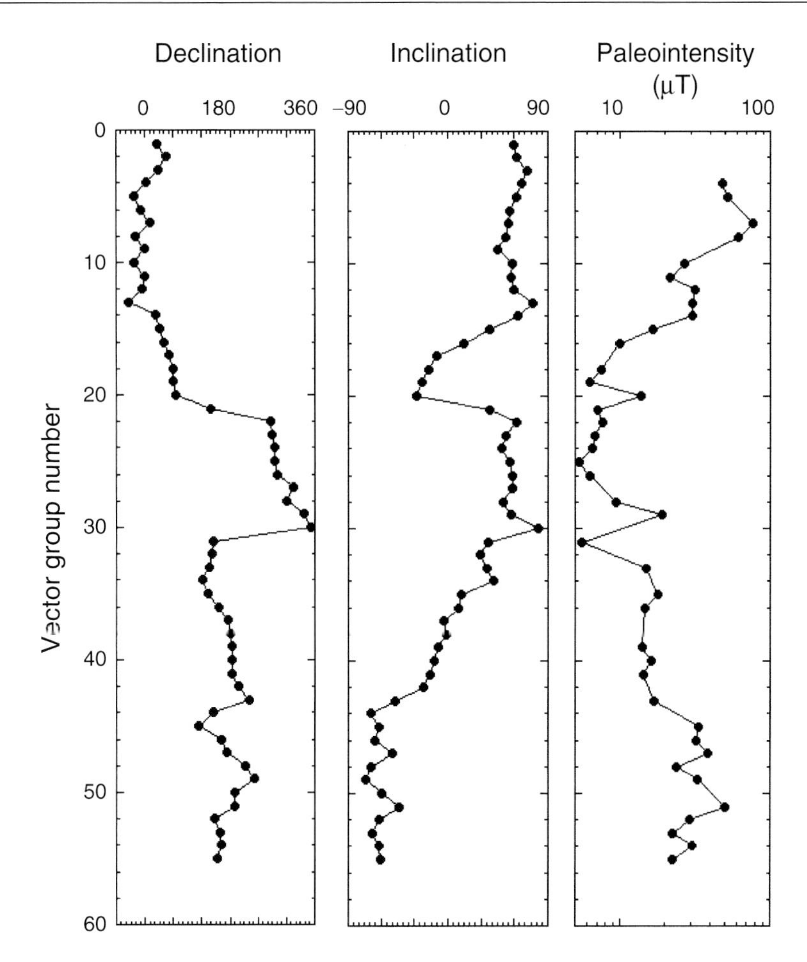

Figure G30 A record of a polarity reversal recorded in lava flows at Steens Mountain, Oregon (Mankinen *et al.*, 1985; Prevot *et al.*, 1985). The changes in magnetization through the reversal are shown here plotted as declination, inclination, and paleointensity in stratigraphic sequence of similar magnetization vectors. The data are plotted versus vector group number because it is assumed that successive flows exhibiting the same magnetizations were extruded rapidly and represent multiple records of the same field and do not necessarily represent times when the field was not changing.

has been found for a remagnetization of intermediate directions in a sediment record of an excursion (Coe and Liddicoat, 1994).

Yet another controversy regards how fast the magnetic field can change during a reversal. The reversal recorded by lava flows at Steens Mountain, Oregon provides evidence that the transitional field may change as fast as degrees per day (Coe and Prevot, 1989; Coe *et al.*, 1995). This rate is extremely fast and in fact is thought to be too fast (Figure G30). This is because the electrical conductivity of Earth's mantle filters field changes produced by the dynamo and should limit just how fast the dynamo-produced field can change at Earth's surface.

The evidence for the rapid changes comes from magnetizations recorded within a single lava flow. The magnetizations differ with position in the flow. Using cooling rate estimates for the flow, the rate at which the field was changing can be estimated. Despite an intensive effort, no evidence has been found to suggest that the different magnetizations result from differences in the magnetic minerals that record the field.

These controversies will likely be resolved as additional polarity transition records are obtained from a greater geographic distribution and as records of older reversals are obtained. These records will help solve the mystery of what happens during a polarity reversal, and that knowledge will in turn help us understand what it is about Earth that gives rise to this fascinating feature of our magnetic field.

Bradford M. Clement

Bibliography

Cande, S.C., and Kent, D.V., 1995. Revised calibration of the geomagnetic polarity timescale for the Late Cretaceous and Cenozoic. *Journal of Geophysical Research*, **100**: 6093–6095.

Channell, J.E.T., and Lehman, B., 1997. The last two geomagnetic polarity reversals recorded in high-deposition-rate sediment drifts. *Nature*, **389**: 712–715.

Clement, B.M., 1991. Geographical distribution of transitional VGPs: evidence for non-zonal equatorial symmetry during the Matuyama-Brunhes geomagnetic reversal. *Earth and Planetary Science Letters*, **104**: 48–58.

Clement, B.M., 2004. Dependence of the duration of geomagnetic polarity reversal on site latitude. *Nature*, **428**(6983): 608–609.

Coe, R.S., and Liddicoat, J.C., 1994. Overprinting of natural magnetic remanence in lake sediments by a subsequent high intensity field. *Nature*, **367**: 57–59.

Coe, R.S., and Prevot, M., 1989. Evidence suggesting extremely rapid field variation during a geomagnetic reversal. *Earth and Planetary Science Letters*, **92**: 292–298.

Coe, R.S., Prevot, M., and Camps, P., 1995. New evidence for extraordinarily rapid change of the geomagnetic field during a reversal. *Nature*, **374**: 687–692.

Gee, J.S., Tauxe, L., Barge, E., Peirce, J.W., Weissel, J.K., Taylor, E., Dehn, J., Driscoll, N.W., Farrell, J.W., Fourtanier, E., Frey, F.A.,

Gamson, P.D., Gibson, I.L., Janecek, T.R., Klootwijk, C.T., Lawrence, J.R., Littke, R., Newman, J.S., Nomura, R., Owen, R.M., Pospichal, J.J., Rea, D.K., Resiwati, P., Saunders, A.D., Smit, J., Smith, G.M., Tamaki, K., Weis, D., and Wilkinson, C., 1991. Lower Jaramillo polarity transition records from the equatorial Atlantic and Indian oceans: *Proceedings of the Ocean Drilling Program, Scientific Results*, **121**: 377–391.

Hillhouse, J.A., and Cox, A., 1976. Brunhes-Matuyama polarity transition. *Earth and Planetary Science Letters*, **29**: 51–64.

Hoffman, K.A., 1984. A method for the display and analysis of transitional paleomagnetic data. *Journal of Geophysical Research*, **137**: 6285–6292.

Hoffman, K.A., 1991. Long-lived transitional states of the geomagnetic field and the two dynamo families. *Nature*, **354**: 273–277.

Hoffman, K.A., 1992. Dipolar reversal states of the geomagnetic field and core-mantle dynamics. *Nature*, **359**: 789–794.

Laj, C., Mazaud, A., Weeks, R., Fuller, M., and Herrero-Bervera, E., 1991. Geomagnetic reversal paths. *Nature*, **351**: 447.

Love, J.J., and Mazaud, A., 1997. A database for the Matuyama-Brunhes magnetic reversal. *Physics of the Earth and Planetary Interiors*, **103**: 207–245.

Mankinen, E.A., Prevot, M., Gromme, C.S., and Coe, R.S., 1985. The Steens Mountain (Oregon) geomagnetic polarity transition. I. Directional history, duration of episodes, and rock magnetism. *Journal of Geophysical Research*, **90**: 10393–10417.

Merrill, R.T., and McFadden, P.L., 1999. Geomagnetic polarity transitions. *Reviews of Geophysics*, **37**: 201–226.

Prevot, M., Mankinen, E.A., Coe, R.S., and Gromme, C.S., 1985. The Steens Mountain (Oregon) geomagnetic polarity transition. 2. Field intensity variations and discussion of reversal models. *Journal of Geophysical Research*, **90**: 10417–10448.

Singer, B.S., and Pringle, M.S., 1996. Age and duration of the Matuyama-Brunhes geomagnetic polarity reversal for $^{40}Ar/^{39}Ar$ incremental heating analyses of lavas. *Earth and Planetary Science Letters*, **139**: 47–61.

Tauxe, L., 1993. Sedimentary records of relative paleointensity of the geomagnetic field: theory and practice. *Reviews of Geophysics*, **31**: 319–354.

Cross-references

Core-Mantle Boundary Topography, Implications for Dynamics
Core-Mantle Boundary Topography, Seismology
Core-Mantle Boundary, Heat Flow Across Geodynamo
Geodynamo, Dimensional Analysis and Timescales
Geodynamo, Numerical Simulations
Magnetic Anomalies, Marine
Nondipole Field
Reversals, Theory

GEOMAGNETIC POLARITY TIMESCALES

Marine magnetic anomaly record

It is well established that Earth's magnetic field has alternated frequently but irregularly between two opposing polarity states for at least most of the Phanerozoic. Older rocks of Proterozoic age that display both polarity states are also known, but whether they correspond to reversals of a dipole field is less firmly established. The time interval during which geomagnetic polarity remains constant is called a polarity chron; long episodes of continuous reversal behavior are called superchrons. The polarity equivalent to the present state is referred to as "normal" and the opposite state as "reversed."

The history of geomagnetic polarity is derived from two sources: the interpretation of lineated marine magnetic anomalies, and magnetic polarity stratigraphy in continuous sedimentary sequences and radiometrically dated igneous rocks. Since the Late Jurassic, when the current phase of seafloor spreading began, these records support and confirm each other. The geomagnetic polarity timescale (GPTS) is most reliable for this time. It is divided into two sequences of alternating polarity, the younger covering the Late Cretaceous and Cenozoic, and the older corresponding to Early Cretaceous and Late Jurassic time. The reversal sequences are referred to as the C-sequence and M-sequence, respectively. They are separated in the oceanic record by the Cretaceous Quiet Zone in which lineated magnetic anomalies are absent. It appears that Earth's magnetic field did not reverse polarity during this time interval, which is referred to as the Cretaceous normal polarity superchron (CNPS). Prior to the Late Jurassic only the polarity record preserved in rocks is available. Knowledge of older geomagnetic polarity history is patchy and, despite some excellent magnetostratigraphic results, largely unconfirmed.

Construction and identification of a polarity chron sequence

Although the pioneering studies of geomagnetic polarity were carried out on radiometrically dated lavas, the marine magnetic anomaly record provides the most extensive, detailed and continuous record of reversal history and forms the basis of all GPTS covering the last 160 Ma of the Earth history. The appearance of lineated large amplitude, long wavelength marine magnetic anomalies (Figure G31) depends on the latitude and direction of the corresponding measurement profile. The first step in constructing a GPTS thus consists of interpreting the anomalies as a block model of alternating magnetization of the oceanic crust responsible for the anomalies. The polarity pattern is used to correlate coeval segments of different profiles and to obtain an optimized model that minimizes local variations in spreading rate on any given profile. The ensuing composite block model constitutes a polarity sequence in which the distances between the block boundaries are proportional to the relative lengths of the polarity chrons. No consideration is usually given to the finite duration of a polarity transition, which is thought to last about 4–6 ka (Clement and Kent, 1984). This time is included in the lengths of polarity chrons, which are measured between the midpoints of polarity transitions. This simplification is generally acceptable but it may be problematic for accurately describing short polarity chrons for which the transitional time may be an appreciable fraction of the duration of the chron.

The anomalies are identified by numbering them in increasing order away from the spreading axis. In the C-sequence the positive magnetic anomalies, corresponding to normal polarity, are numbered and preceded by the letter "C," as anomalies C18, C29, etc. The M-sequence oceanic crust in the North Pacific Ocean was magnetized south of the equator but plate motion has brought it into the northern hemisphere. The positive anomalies, identified by the letter "M" as anomalies M0, M13, etc., correspond to reversely magnetized crust. To reconcile the two numbering schemes, adjacent normally and reversely polarity chrons are paired, so that the younger member has normal polarity. Thus marine magnetic anomaly C15 is associated with polarity chrons C15n and C15r.

To obtain an optimized global model of polarity, which then becomes the polarity sequence for a GPTS, block models for different spreading centers must be compared, stretched, and squeezed differentially. The matching of block models may be visual or more sophisticated. To form an optimized record of the C-sequence of polarity Cande and Kent (1992a) used nine rotation poles to stack polarity records on contemporaneous profiles and determine relative widths of crustal blocks for the modeled reversal sequence. Before a timescale can be generated, the optimized polarity sequence must be dated. Biostratigraphic stage boundaries or other dated levels are correlated by magnetostratigraphy to the polarity

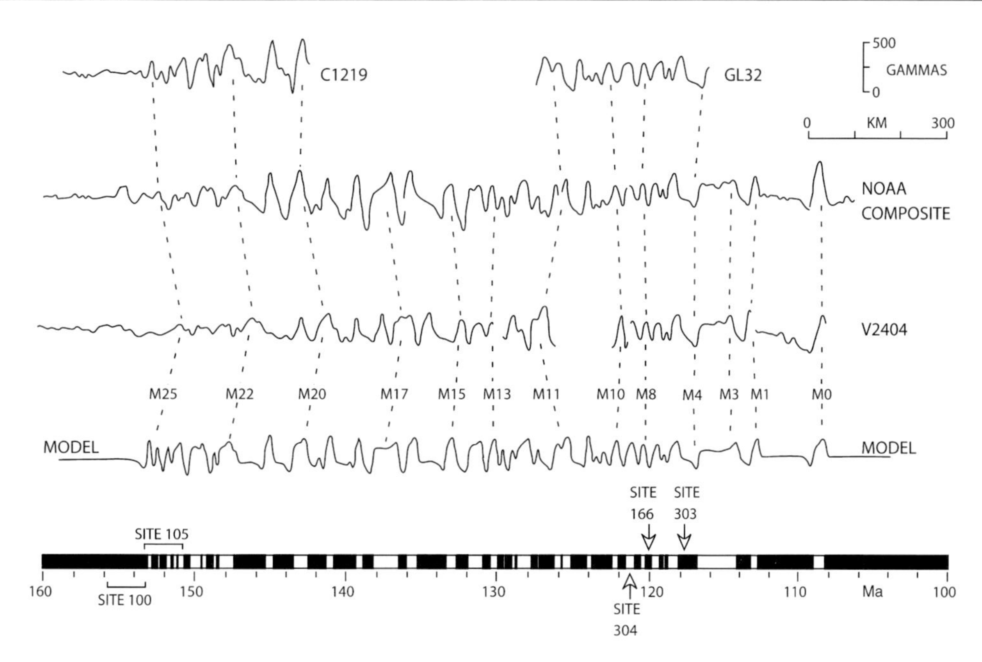

Figure G31 Marine magnetic profiles, correlation tie-lines, and block model of the polarity of oceanic crustal magnetization (black, normal; white, reverse) for M-sequence anomalies in the North Pacific (after Larson and Hilde, 1975).

sequence. Intervening reversal boundaries between the calibration levels are dated by linear interpolation and older boundaries by extrapolation. The largest sources of error in this dating are the "absolute" ages associated with the tie-levels. The relative lengths of the polarity chrons, derived from a crustal block model that assumes constant spreading rate, are known more accurately.

Cenozoic and Late Cretaceous GPTS

The pioneering efforts to determine a GPTS were carried out on marine magnetic anomalies at actively spreading oceanic ridge systems. Heirtzler *et al.* (1968), hereafter referred to as HDHPL68, derived a GPTS from the present time to the Late Cretaceous by matching profiles in different oceans to a reference profile in the South Atlantic, which was judged to be the most likely to have a constant spreading rate. The spreading rate at the South Atlantic Ridge was estimated by correlating distances to young reversals with radiometrically dated reversals from lava sequences. By assuming this spreading rate to be constant, the distance of a polarity reversal from the ridge could be converted to age. However, it was later found that the magnetic anomalies composing anomaly C14 in HDHPL68 were not reproducible in other oceans and thus did not correspond to polarity intervals; moreover, the HDHPL68 timescale did not cover the full range of older anomalies following the Cretaceous Quiet Interval. Improved, extended, and modified, the HDHPL68 timescale served as the basis for several subsequent versions of the GPTS for this time interval (Figure G32). Magnetostratigraphic correlation of the Cretaceous–Tertiary boundary to the upper part of chron C29r in a continental exposure of marine limestones (Alvarez *et al.*, 1977) provided a way of associating age with the older end of the reversal sequence and resulted in improved calibration of the C-sequence GPTS. The magnetostratigraphy also confirmed that the C-sequence began at C33r, as termination of the CNPS. Together with more detailed analysis of the oceanic record, this led to a more complete and accurate GPTS (LaBrecque *et al.*, 1977), hereafter LKC77.

The number of stage boundaries now tied to the polarity record led Lowrie and Alvarez (1981) to propose a modification of the GPTS. They assumed the polarity sequence in LKC77 and best estimates of the "absolute" ages for 11 tie-levels that correlated the stratigraphic and marine magnetic polarity records. Disregarding the effects on seafloor spreading rates they stretched and squeezed the polarity record between the tie-points and obtained a new GPTS. The ensuing timescale resulted in a history of seafloor spreading characterized by sudden large changes in spreading rate. To avoid this, Harland *et al.* (1982) modified the ages of tie-points and obtained a GPTS that gave a smoother seafloor spreading record. One of the most important uses of a GPTS is now recognized to be the ability to attach numeric ages to faunal appearances and extinctions. Berggren *et al.* (1985) and Harland *et al.* (1990) produced GPTS versions in which the biostratigraphy was well tied to the magnetic polarity record, which was derived from that of LKC77.

Cande and Kent (1992a) carried out a detailed reevaluation of the C-sequence marine magnetic anomalies and improved the definitions of the corresponding oceanic block models, resulting in adjustments of the relative lengths of polarity chrons. They associated ages with the improved polarity sequence by fitting a cubic spline curve to nine dated correlation points. In so doing they used a nonstandard age for the Cretaceous-Tertiary boundary. An updated version of their timescale (Cande and Kent, 1995), hereafter referred to as CK95, is the current reference GPTS for the Late Cretaceous and Cenozoic. An archive of Cenozoic-Late Cretaceous timescales has been assembled by Mead (1996).

The Cretaceous normal polarity superchron

The C-sequence and M-sequence polarity chrons are separated by the CNPS. Using the nomenclature for labeling chrons, this might be called C34n or M0n, but it is customary to designate it separately from the adjacent sequences. In the marine magnetic anomaly record the CNPS corresponds to regions of oceanic crust referred to as the Cretaceous Quiet Zone, in which magnetic anomalies, although present, are not lineated. It is believed to represent an interval of time, from about 121 to 83 Ma ago, in which the geomagnetic field had a consistent normal polarity. Despite magnetostratigraphic and deep-tow investigations to find polarity reversals within this 38 Ma interval, there is no incontrovertible evidence of them. The absence of polarity changes

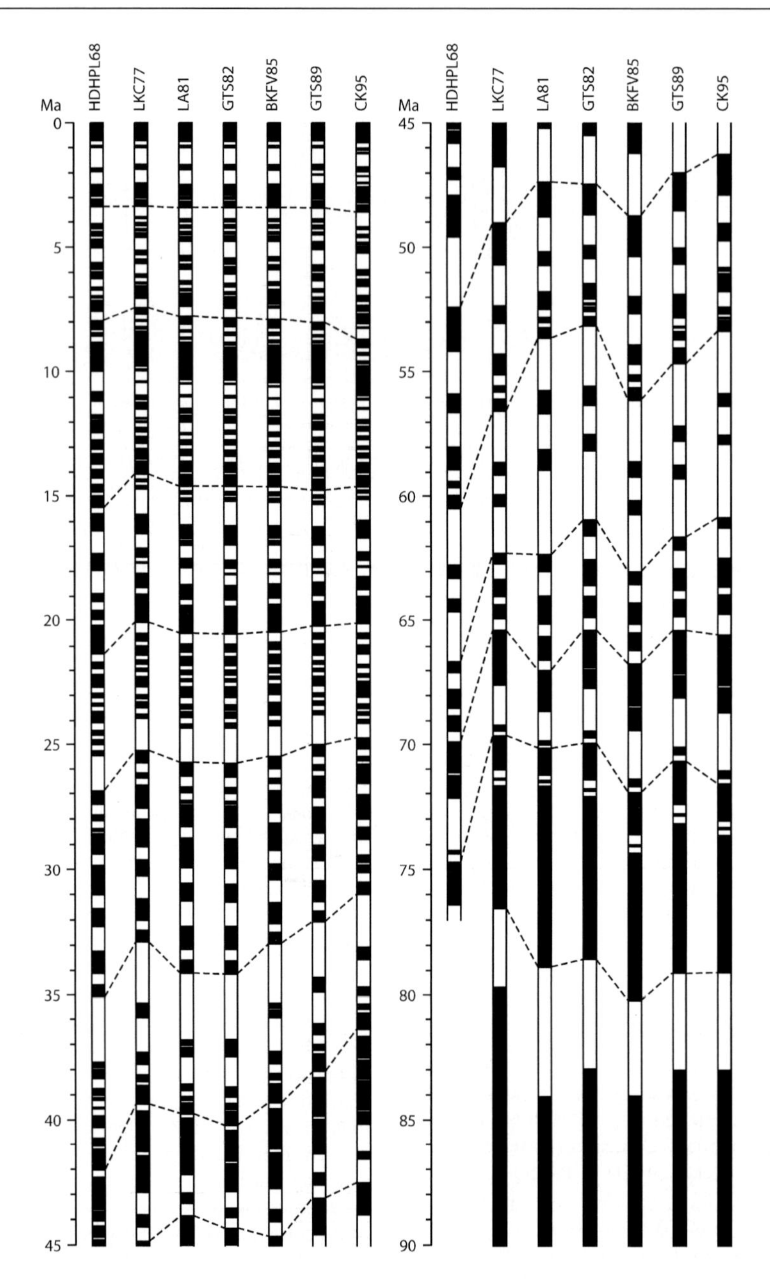

Figure G32 Evolution of the geomagnetic polarity timescale for the Cenozoic and Upper Cretaceous. The authors of individual C-sequence models are as follows: HDHPL68, Heirtzler *et al.* (1968); LKC77, LaBreque *et al.* (1977); LA81, Lowrie and Alvarez (1981); GTS82, Harland *et al.* (1982); BKFV85, Berggren *et al.* (1985); GTS89, Harland *et al.* (1990); CK95, Cande and Kent (1995).

in the CNPS may imply a special behavior of the geodynamo. A few very long polarity chrons, lasting several million years each, are found adjacent to the CNPS.

Early Cretaceous and Late Jurassic GPTS

The oldest regions of oceanic crust characterized by lineated magnetic anomalies were formed during the Late Jurassic. The Phoenix, Japanese, and Hawaiian lineations are found in the Western Pacific, and the Keathley lineations in the western North Atlantic. Larson and Pitman (1972) identified and correlated these sets of anomalies, numbering them from M1 to M22 in order of increasing age. They realized that the three sets of Pacific lineations were inverted with respect to the Keathley sequence. The North Pacific oceanic crust had been magnetized south of the magnetic equator, so that positive anomalies are now found over reversely magnetized crust. Larson and Hilde (1975) refined the reversal record for the Hawaiian lineations, resolving additional anomalies, adding a younger anomaly M0, and extending the older anomaly record to M25 (Figure G31). They dated their timescale (LH75) by estimating the age of magnetic basement at drillholes of the Deep Sea Drilling Project from the paleontological ages of the oldest calcareous fossils found in the holes. The magnetic reversal block model was derived for the Hawaiian lineations, but the ages were determined for sites on other lineation sets and correlated by the magnetic polarity pattern to the Hawaiian set.

The LH75 polarity sequence has formed the basis for subsequent modifications and improvements to the M-sequence GPTS (Figure G33). Later versions (Kent and Gradstein, 1985; Harland *et al.*, 1990) are somewhat better dated, but still rely on bottom ages in groups of drillholes near the ends of the sequence. A new Hawaiian block model with an optimized approximation to a constant spreading rate was derived by Channell *et al.* (1995) after critical comparison of block models for the Hawaiian,

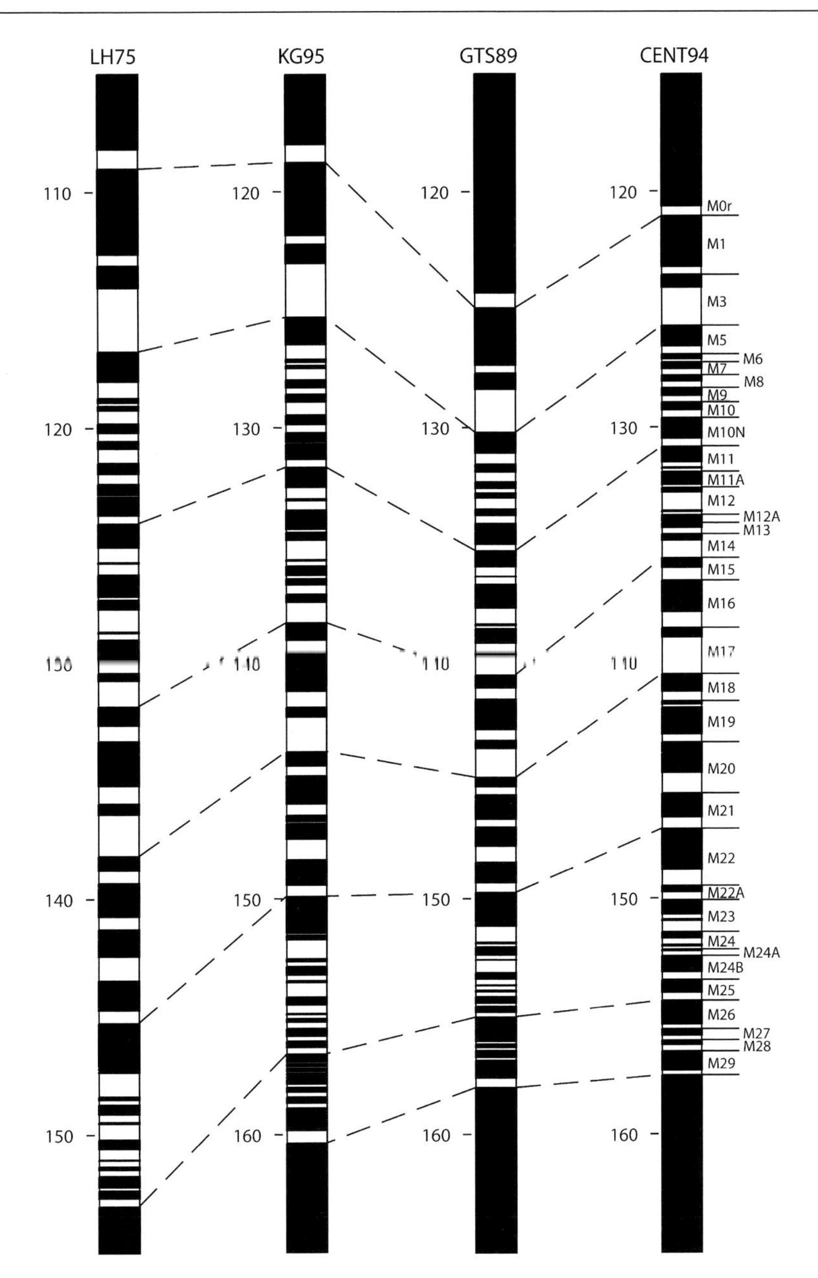

Figure G33 Evolution of the geomagnetic polarity timescale for the Early Cretaceous and Late Jurassic. The authors of individual M-sequence models are as follows: LH75, Larson and Hilde (1975); KG85, Kent and Gradstein (1985); GTS89, Harland *et al.* (1990); CENT94, Channell *et al.* (1995).

Japanese, Phoenix, and Keathley lineations. This model (CENT94), covering magnetic polarity chrons CM0r to CM29r, is probably the optimum current GPTS for the M-sequence anomalies.

Oceanic crust older than chron CM25r was thought to be free of lineated magnetic anomalies and was labeled the Jurassic Quiet Zone by analogy to the Cretaceous equivalent. However, magnetic lineations with low amplitude were subsequently identified in the youngest part of this Quiet Zone, extending the polarity sequence to CM29r (Cande *et al.*, 1978). These weaker old anomalies are related to oceanic crust whose magnetization decreases with increasing age. Even older anomalies, with short wavelengths and low amplitudes, have been detected within the Jurassic Quiet Zone, both from aeromagnetic profiles (Handschumacher *et al.*, 1988) and from deep-towed magnetometer surveys (Sager *et al.*, 1998). Their origin is as yet uncertain. They may have been formed during an episode of high reversal frequency, in which case they would extend the M-sequence from CM29r to CM41r. This would imply that the Jurassic Quiet Zone is different in origin from the Cretaceous Quiet Zone, in which no reversals are thought to have occurred. However, the polarity pattern corresponding to these anomalies has not been established definitively and the reversal sequence has not been confirmed independently by magnetostratigraphy. It is possible that the anomalies may be due to fluctuations of paleomagnetic field intensity, as suggested to explain low amplitude, short wavelength anomalies in the Cenozoic (Cande and Kent, 1992b).

Early Jurassic and Triassic GPTS

The record of geomagnetic polarity prior to the onset of seafloor spreading is much less well known. In the absence of a marine

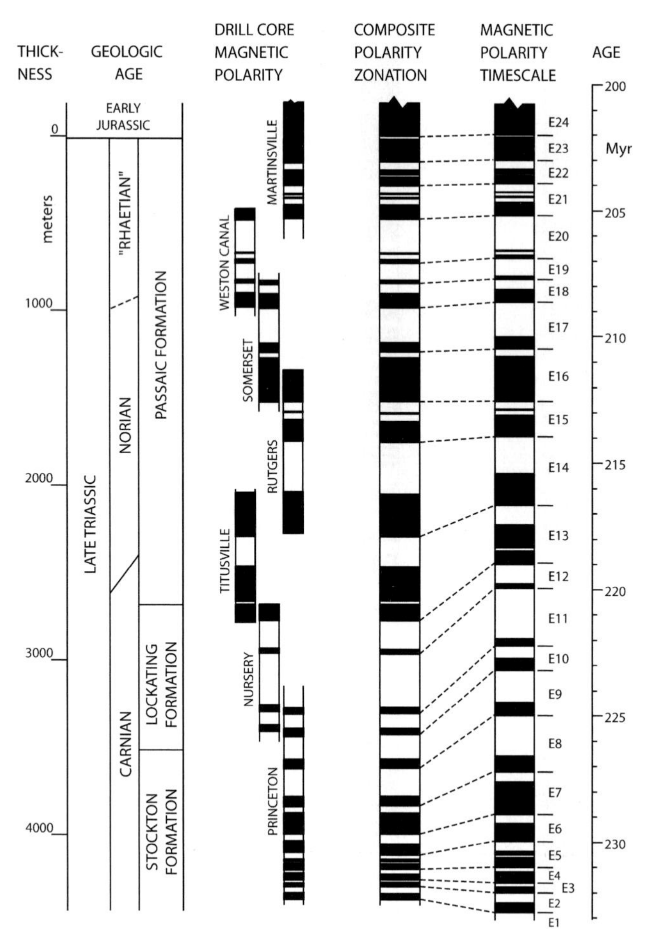

Figure G34 Construction of a GPTS for the Late Triassic based on overlapping magnetostratigraphies from drillholes in the Newark Basin, USA (after Kent *et al.*, 1995). The GPTS is that of Kent and Olsen (1999), calibrated by astrochronological dating of the magnetozones.

magnetic anomaly record, the GPTS must be pieced together from magnetostratigraphic results. This is a large undertaking. Changes of sedimentation rate within a given depositional basin or from one basin to another can strongly modify the "fingerprint" pattern of reversals that is essential for correlation. Investigations in marine limestones of Middle and Early Jurassic age have identified magnetozones, but these have only rarely been confirmed by other magnetostratigraphic results.

Kent *et al.* (1995) determined detailed magnetostratigraphies in overlapping drillholes through continental redbeds of Late Triassic age in the Newark basin. They matched the individual records stratigraphically to produce a composite geomagnetic polarity record (Figure G34). The sediments displayed lithological variations related to cyclical changes in Earth's orbital parameters, of which the 400 ka fluctuation of orbital eccentricity was prominent. Kent and Olsen (1999) used this Milankovitch cycle to convert the polarity sequence to a GPTS (Figure G34), assuming an "absolute" age of 202 Ma for the Jurassic-Triassic boundary. The Newark section currently serves as the standard of reference for the history of geomagnetic polarity in the Late Triassic.

William Lowrie

Bibliography

Alvarez, W., Arthur, M.A., Fischer, A.G., Lowrie, W., Napoleone, G., Premoli Silva, I., and Roggenthen, W.M., 1977. Upper Cretaceous-Paleocene magnetic stratigraphy at Gubbio, Italy. V. Type section for the Late Cretaceous-Paleocene geomagnetic reversal time scale. *Geological Society of America Bulletin*, **88**: 383–389.

Berggren, W.A., Kent, D.V., Flynn, J.J., and Van Couvering, J.A., 1985. Cenozoic geochronology. *Geological Society of America Bulletin*, **96**: 1407–1418.

Cande, S.C., and Kent, D.V., 1992a. A new geomagnetic polarity time scale for the Late Cretaceous and Cenozoic. *Journal of Geophysical Research*, **97**: 13917–13951.

Cande, S.C., and Kent, D.V., 1992b. Ultra-high resolution marine magnetic anomaly-profiles: a record of continuous paleointensity variations? *Journal of Geophysical Research*, **97**: 15075–15083.

Cande, S.C., and Kent, D.V., 1995. Revised calibration of the geomagnetic polarity timescale for the Late Cretaceous and Cenozoic. *Journal of Geophysical Research*, **100**: 6093–6095.

Cande, S., Larson, R.L., and LaBrecque, J.L., 1978. Magnetic lineations in the Pacific Jurassic quiet zone. *Earth and Planetary Science Letters*, **41**: 434–440.

Channell, J.E.T., Erba, E., Nakanishi, M., and Tamaki, K., 1995. Late Jurassic-Early Cretaceous time scales and oceanic magnetic anomaly block models. In Berggren, W.A., Kent D.V., Aubry, M., and Hardenbol, J. (eds.), *Geochronology, Timescales, and Global Stratigraphic Correlation*. Tulsa, Oklahoma: SEPM Special Publication, pp. 51–64.

Clement, B.M., and Kent, D.V., 1984. Latitudinal dependency of geomagnetic polarity transition durations. *Nature*, **310**: 488–491.

Handschumacher, D.W., Sager, W.W., Hilde, T.W.C., and Bracey, D.R., 1988. Pre-Cretaceous evolution of the Pacific plate and extension of the geomagnetic polarity reversal time scale with implications for the origin of the Jurassic "Quiet Zone". *Tectonophysics*, **155**: 365–380.

Harland, W.B., Cox, A.V., Llewellyn, P.G., Pickton, C.A.G., Smith, A.G., and Walters, R., 1982. *A Geologic Time Scale*. Cambridge: Cambridge University Press, 131 pp.

Harland, W.B., Armstrong, R.L., Cox, A.V., Craig, L.E., Smith, A.G., and Smith, D.G., 1990. *A Geologic Time Scale 1989*. Cambridge: Cambridge University Press, 263 pp.

Heirtzler, J.R., Dickson, G.O., Herron, E.M., Pitman, W.C. III, and Le Pichon, X., 1968. Marine magnetic anomalies, geomagnetic field reversals and motions of the ocean floor and continents. *Journal of Geophysical Research*, **73**: 2119–2136.

Kent, D.V., and Gradstein, F.M., 1985. A Cretaceous and Jurassic geochronology. *Geological Society of America Bulletin*, **96**: 1419–1427.

Kent, D.C., and Olsen, P.E., 1999. Astronomically tuned geomagnetic polarity timescale for the Late Triassic. *Journal of Geophysical Research*, **104**: 12,831–12,841.

Kent, D.V., Olsen, P.E., and Witte, W.K., 1995. Late Triassic-Earliest Jurassic geomagnetic polarity sequence and paleolatitudes from drill cores in the Newark rift basin, eastern North America. *Journal of Geophysical Research*, **100**: 14965–14998.

LaBrecque, J.L., Kent, D.V., and Cande, S.C., 1977. Revised magnetic polarity timescale for Late Cretaceous and Cenozoic time. *Geology*, **5**: 330–335.

Larson, R.L., and Hilde, T.W.C., 1975. A revised time scale of magnetic reversals for the Early Cretaceous and Late Jurassic. *Journal of Geophysical Research*, **80**: 2586–2594.

Larson, R.L., and Pitman, W.C. III, 1972. World-wide correlation of Mesozoic magnetic anomalies, and its implications. *Geological Society of America Bulletin*, **83**: 3645–3662.

Lowrie, W., and Alvarez, W., 1981. One hundred million years of geomagnetic polarity history. *Geology*, **9**: 392–397.

Mead, G.A., 1996. Correlation of Cenozoic-Late Cretaceous geomagnetic polarity timescales: an Internet archive. *Journal of Geophysical Research*, **101**: 8107–8109.

Sager, W.W., Weiss, C.J., Tivey, M.A., and Johnson, H.P., 1998. Geomagnetic polarity reversal model of deep-tow profiles from the Pacific Jurassic Quiet Zone. *Journal of Geophysical Research*, **103**: 5269–5286.

Cross-references

Crustal Magnetic Field
Geomagnetic Polarity Reversals
Geomagnetic Polarity Timescales
Magnetic Anomalies, Marine
Magnetic Surveys, Marine
Magnetostratigraphy
Paleomagnetism
Reversals, Theory

GEOMAGNETIC PULSATIONS

Geomagnetic pulsations or micropulsations are ultralow frequency (ULF) plasma waves in the Earth's magnetosphere. These waves have frequencies in the range 1 mHz to greater than 10 Hz and appear as more or less regular oscillations in records of the geomagnetic field (see Figure G35). Geomagnetic oscillations, or ULF pulsations as they are also called, can also be identified in electric field measurements in the ionosphere as well as observations of the electromagnetic field of the magnetosphere as made onboard spacecraft.

The lower frequency pulsations have wavelengths comparable to typical scale lengths of the entire magnetosphere. They may also be interpreted as eigenoscillations of or standing waves in the magnetospheric systems. The higher frequency waves are usually identifiable as proton ion-cyclotron waves in the magnetospheric plasma. The amplitudes of the lower frequency pulsations can reach several tens up to hundreds of nanotesla in the auroral zone region while the higher frequency waves reach amplitudes of the order of a few nanotesla.

The first observation of a geomagnetic pulsation was published in 1861 by the Scottish scientist Balfour Stewart who identified quasisinusoidal variations of the geomagnetic field in records of the Kew observatory after the great magnetic storm that occurred in 1859. The International Geophysical Year (1958–1959) with its large

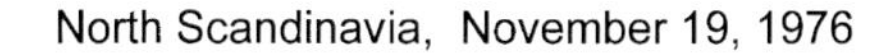

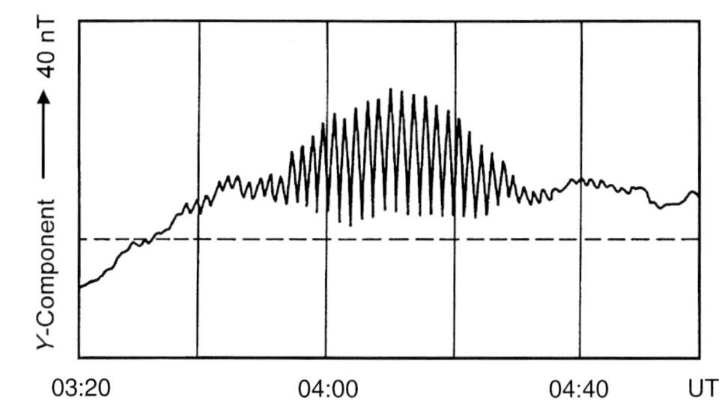

Figure G35 Geomagnetic pulsation of the Pc4 type, recorded at a magnetic observatory in North Scandinavia. The *y*- or east-west component of the geomagnetic field is displayed relative to a quiet day record.

number of coordinated geomagnetic field observations made available a multitude of studies on ULF pulsations, stimulated the interest in this type of geomagnetic field variation and created an active area of research.

The classification scheme (Table G7) of the International Association of Geomagnetism and Aeronomy (IAGA) distinguishes seven different types of geomagnetic pulsations based on their oscillation period and appearance in magnetograms as almost continuous and more irregular pulsations. The two classes, continuous pulsations (Pc) and irregular pulsations (Pi), are usually divided into subclasses. A more detailed overview of this morphological classification was given by Jacobs (1970). Though this classification is still widely in use it is somewhat outdated as the increased understanding of the physical nature and properties of geomagnetic pulsations allows a more in-depth classification based on physical processes and generating mechanisms. The long-period pulsations are at present interpreted as magnetohydrodynamic waves while the short-period pulsations are related to ion-cyclotron waves propagating in the magnetosphere. As for many magnetospheric phenomena the solar wind provides the energy for geomagnetic pulsations, partly directly, partly indirectly.

A direct energy source is plasma waves generated in the solar wind and penetrating the magnetopause. A major source of these waves are plasma instabilities in the upstream region of the near-Earth solar wind where, for example, protons reflected at the magnetospheric bow shock constitute an unstable particle distribution, generating a variety of upstream waves. These waves are convected downstream toward the magnetopause and couple through it into the magnetosphere. Detailed investigations, however, indicate that the transmission of hydromagnetic waves through the magnetopause is a rather inefficient process. Only a small percentage of the energy of the upstream solar wind waves couples to oscillations of the magnetosphere.

A more efficient, solar wind-driven process is the impulsive excitation of plasma waves by sudden impulses from the solar wind. The magnetosphere constitutes a kind of body capable of eigenoscillations. It can be excited much like a bell. The British scientist Jim Dungey in 1954 was the first to analyze such eigenoscillations in more detail. Assuming that the magnetospheric magnetic field is of dipole nature only, he studied the magnetohydrodynamic equations of such a dipole-magnetosphere filled with an ionized gas of spatially depending mass density. The boundaries of this magnetosphere are the magnetopause and the ionosphere, where the geomagnetic field lines are anchored much as strings of a violin are anchored between the pegs and the tailpiece.

He derived the so-called Dungey equations, a set of partial differential equations describing the coupling between toroidal oscillations of the fluid velocity field and the toroidal (poloidal) component of the electric (magnetic) field oscillations in this dipole-magnetosphere. If the excitation of the dipole-magnetosphere is axisymmetric, then the toroidal components of these two fields are decoupled. This in turn

Table G7 The IAGA classification of geomagnetic pulsations

Name	Period range (s)
Continuous	
Pc1	0.2–5
Pc2	5–10
Pc3	10–45
Pc4	45–150
Pc5	150–600
Irregular	
Pi1	1–40
Pi2	40–150

implies that individual field line shells are oscillating independent from each other, much as different violin strings oscillate independently. The oscillation period depends on Alfvén wave velocity along the considered field line. Therefore geomagnetic pulsations can also be used as a diagnostic tool for magnetospheric physics.

This is very much in accord with the observational finding that the periods of geomagnetic pulsations in the Pc4–5 range decrease with decreasing geomagnetic latitude. At lower latitudes the length of field lines anchored between the northern and southern ionosphere is much shorter than at higher latitudes, which causes a smaller oscillation period at low-latitudes. Geomagnetic oscillations can thus be viewed at as standing field line oscillations with fundamental and higher harmonic waves being generated (Figure G36) .

Besides an impulsive excitation of geomagnetic pulsations the so-called Kelvin-Helmholtz instability can drive magnetospheric magnetohydrodynamic waves. At the interface between the solar wind plasma and the magnetosphere plasma, the so-called magnetopause, strong shear flows exists. The solar wind plasma has to flow around the magnetosphere along the magnetopause. Much as atmospheric wind flow over a water surface can cause water waves the velocity shear at the magnetopause destabilizes this boundary and causes surface waves which are coupled into the magnetosphere. The energy for these waves is drained out of the solar wind flow and maximum instability of the magnetopause is expected at the flanks of the magnetosphere, which is in the dawn and dusk hours. Furthermore, the unstable waves should propagate tailward that is they propagate east at dusk and west at dawn. This is indeed observed in ground and satellite observations of geomagnetic pulsations.

As the plasma density distribution is not uniform wave propagation in the magnetosphere may lead to a very interesting physical effect, field line resonance or resonant mode coupling. A Kelvin-Helmholtz instability generated wave is a compressional magnetohydrodynamic wave, which couples to a transverse oscillation, an Alfvén wave, at a region in the magnetosphere where the surface wave's period equals the local eigenperiod of the toroidal field oscillation or Alfvén wave. At this resonance point the oscillation magnitude maximizes and the wave phase changes by 180° when crossing the resonant field line in the radial direction. The physics of this resonant mode coupling between a surface wave and a local eigenmodes is actually a tunneling process (Southwood and Hughes, 1983).

Internal to the magnetosphere a variety of wave sources exist. Most important is the ring current region with its energetic protons. Proton distributions are usually nonthermal and tend to thermalize via interaction with electromagnetic waves. These waves in turn can be generated by plasma instabilities such as the so-called bounce-resonance or drift-mirror instabilities (Samson, 1991; Walker, 2004).

The ring current region is also the source region of many of the Pc1-Pc2 geomagnetic pulsations. During the expansive phase of magnetospheric substorms large numbers of energetic ions are injected from the magnetotail into the inner magnetosphere where they drift westward to create the substorm-enhanced ring current. These energetic particle populations are highly unstable against the ion-cyclotron instability and cause the generation of short-period geomagnetic pulsations (Kangas *et al.*, 1998).

Once geomagnetic pulsations have been generated in the magnetosphere their energy must be dissipated somewhere. Most of this dissipation occurs in the ionosphere where the pulsation-associated electric fields cause current flow, which in turn leads to significant Joule heating of the ionosphere. Local kinetic temperature increases of several thousand Kelvin have been observed. Part of the wave energy is also used to accelerate magnetospheric particles. Such high-energy particles may subsequently hit the atmosphere where they can cause aurora.

Much as the terrestrial magnetosphere also magnetospheres of other planets, in particular those of Mercury, Jupiter, and Saturn exhibit magnetic field oscillations comparable to geomagnetic pulsations (Glassmeier *et al.*, 1999). Magnetic pulsations of these magnetospheric systems have different properties than those at Earth due to, for example, different spatial scales of the oscillating system.

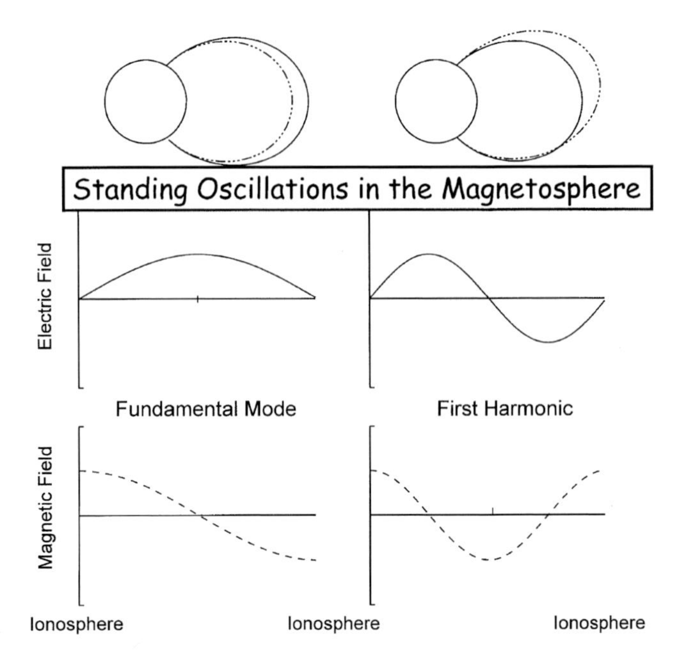

Figure G36 Schematic representation of standing geomagnetic field line oscillations. The perturbed field line is the dashed-dotted line. The electric field always has a node in the ionosphere as large electrical conductivity there shortcuts all electric potential differences.

Karl-Heinz Glaßmeier

Bibliography

Glassmeier, K.H., Othmer, C., Cramm, R., Stellmacher, M., and Engebretson, M., 1999. Magnetospheric field line resonances: a comparative planetology approach. *Surveys in Geophysics*, **20**: 61–109.

Jacobs, J.A., 1970. *Geomagnetic Micropulsations.* Berlin: Springer-Verlag.

Kangas, J., Guglielmi, A., and Pokhotelov, O., 1998. Morphology and physics of short-period magnetic pulsations. *Space Science Reviews*, **83**: 435–512.

Samson, J.C., 1991. Geomagnetic pulsations and plasma waves in the Earth's magnetosphere. In Jacobs, J.A. (ed.), *Geomagnetism*, Vol. 4. London: Academic Press, pp. 481–592.

Southwood, D.J., and Hughes, W.J., 1983. Theory of hydromagnetic waves in the magnetosphere. *Space Science Reviews*, **35**: 301–366.

Walker, A.D.M. (ed.), 2004. *Magnetohydrodynamic Waves in Geospace: The Theory of ULF Waves and Their Interaction with Energetic Particles in the Solar-Terrestrial Environment.* Philadelphia: Institute of Physics Publishing.

Cross-references

Alfvén Waves
Ionosphere
Magnetohydrodynamic Waves
Magnetosphere of the Earth
Ring Current

GEOMAGNETIC REVERSAL SEQUENCE, STATISTICAL STRUCTURE

Introduction

It is now well recognized that the geomagnetic field is produced by dynamo action within the molten iron in the outer core. This is a dynamic process intimately linked to cooling of the core and the rapid spin of the Earth. The equations governing the geodynamo, which have to be solved jointly to obtain a full model of the process, are Maxwell's equations, Ohm's law, the Navier-Stokes' equation, the continuity equation, Poisson's equation, the generalized heat equation, and the equation of state for the material in the outer core. This is a complex, nonlinear set of equations, making it extraordinarily difficult to obtain a full solution. However, the equations are even in **H**, the magnetic field. That is, the equations are insensitive to the sign of **H**, and so if **H** is a solution then so also is –**H**. We know from present-day observations that the geomagnetic field can exist in a relatively stable state in which the field at the Earth's surface is approximately that of a dipole with its axis almost parallel to the Earth's spin axis. Consequently we should expect that there is a similar relatively stable solution, with the same statistical properties as the field we observe today, that simply has the opposite polarity, that is, the north and south poles are swapped. Hence, if there is a mechanism for the field to move from one solution to the other, we should expect to see reversals of the geomagnetic field. The solar magnetic dynamo reverses regularly with a full period of about 22 years (see *Magnetic field of Sun*), so we know that reversal is possible in other self-sustaining dynamos. This then leaves open the question of how we can tell whether the field has reversed polarity in geological time, well before humans started to observe the field and its behavior.

By at least the late 18th century it was recognized that deviation of magnetic compasses could occur because of nearby strongly magnetized rocks. The first observations that the magnetization in certain rocks was actually parallel to the Earth's magnetic field were made independently by Delesse and Melloni. Folgerhaiter extended their work but also studied the magnetization of bricks and pottery. He argued that when a brick or pot was fired in the kiln then the remanent magnetization it acquired on cooling provided a record of the direction of the Earth's magnetic field. With the wisdom of hindsight it is fairly obvious that this would be the case. Volcanic rocks are heated well above the Curie point so the magnetization is free to align with the external magnetic field and becomes locked in as the rock cools. This is known as a *thermoremanent magnetization* (TRM) (*q.v.*) and, in extrusive volcanics, provides a record of the direction of the magnetic field at that locality at a specific point in time.

There are several processes by which a rock will lock in a fossil record of the ancient (or paleo) magnetic field. The fossil magnetism naturally present is termed the *natural remanent magnetization* (NRM) (*q.v.*) and its existence provides us with an opportunity to discover the direction of the geomagnetic field over geological time. David in 1904 and Brunhes in 1906 reported the first discovery of NRM that was roughly opposite in direction to that of the present field and this led to the speculation that the Earth's magnetic field had reversed its polarity in the past. At that time it was not recognized that the field was generated by a self-sustaining dynamo in the outer core, so the possibility of reversal was more exciting than it may seem from our perspective today. In 1926, Mercanton pointed out that if the Earth's field had in fact reversed itself in the past then reversely magnetized rocks should be found in all parts of the world. He demonstrated that this was indeed the case for Quaternary-aged rocks around the world. The speculation gained further support when Matuyama in 1929 observed reversely magnetized lava flows from the past 1 or 2 Ma in Japan and Manchuria. However, doubts about the validity of the field reversal hypothesis surfaced during the 1950s after Néel presented theory that showed it was possible for samples to acquire a magnetization antiparallel to the external field during cooling, a process referred to as self-reversal. Shortly thereafter Nagata and Uyeda found the first laboratory-reproducible self-reversing rock, the Haruna dacite. Subsequently, it was recognized that self-reversal is relatively rare and by the early 1960s it was accepted that the Earth's magnetic field has indeed reversed and that the phenomenon of field reversal has occurred many times. An excellent history of this subject was given by Glen (1982).

A critical component in our understanding of reversals was development in the early 1960s of the K-Ar dating method, which made it possible to date young volcanic rocks with reasonable precision. Consequently, it was possible to undertake systematic studies attempting to define the geomagnetic polarity timescale (GPTS) using joint magnetic polarity and K-Ar age determinations on young lavas. As data rapidly became available, it was established that rocks of the same age had the same polarity of magnetization, helping to confirm that the observed reversals of magnetization were indeed due to reversals of the geomagnetic field itself. A few of the earliest compilations, covering the years 1959–1966, of the GPTS for the past 4 Ma are shown in Figure G37. Not surprisingly, when the field has the polarity that we observe today, it is referred to as normal polarity, and the opposite polarity is referred to as reverse polarity.

As already noted, the solar magnetic field has a periodic reversal, and the first timescale put forward by Cox *et al.* (1963) appeared to be consistent with geomagnetic reversals having a periodicity of about 1 Ma intervals. However, as new data appeared it rapidly became apparent that there was no simple periodicity; some of the observed polarity intervals were nearly a million years in length and some were as short as 0.1 Ma. Furthermore, there did not appear to be any regular pattern to these different lengths. This led to the suggestion that there is a random component to the reversal process and, therefore, to interest in the statistical structure of the geomagnetic reversal sequence.

It is extremely difficult to solve the geodynamo problem, that is, to determine just how the geodynamo operates. Determination and understanding of the statistical structure of the geomagnetic reversal sequence provides insight into the long-term dynamic process and in so doing can provide powerful constraints on geodynamo models. For example, the observation that the solar dynamo reverses with a clear periodicity but the geodynamo reversal process appears random, is probably a reflection of the different boundary conditions on the two dynamos.

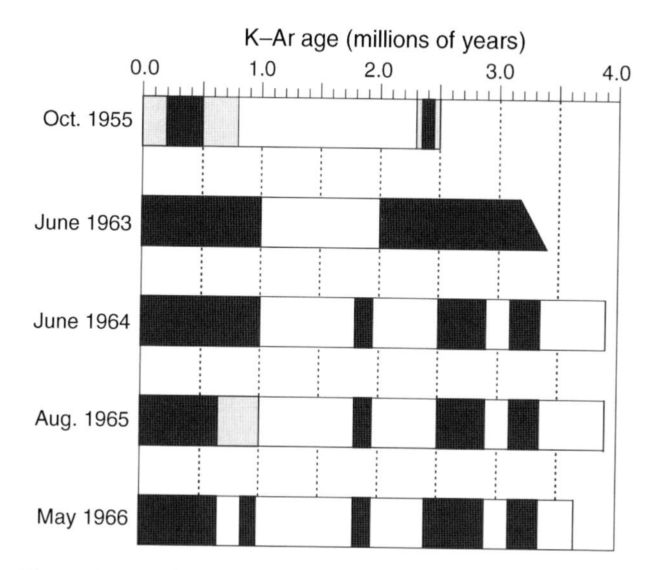

Figure G37 Early compilations of the GPTS. Black represents normal polarity, white represents reversed polarity, and grey indicates uncertain polarity. Abstracted with permission from Cox (1969). © American Association for the Advancement of Science.

When examining a time sequence such as the GPTS it is crucial that the observed events be ordered correctly. This, together with other difficulties, has led to severe problems with the land-based polarity sequence. As one goes further back in time, the absolute error in dating a rock soon becomes larger than the length of the shorter polarity intervals. Because the land-based information comes from combined observations of magnetic polarity from rocks in widely spaced localities worldwide and not from a single continuous sequence, the ordering relies on the accuracy of the K-Ar dates. Hence, the ordering of events is effectively indeterminate for events closer together than the dating error. Consequently the length of the reliable land-based polarity timescale has been too short for reliable estimation of the statistical structure of the GPTS. Recently however, Kent and coworkers (e.g., Kent and Olson, 1999) have shown the potential for GPTS extension in thick, complete continental sedimentary sections.

The most useful data source for development of the GPTS has been marine magnetic anomalies, and the reversal chronology is now quite well determined for the past 165 Ma. The absolute error in the age assigned to individual reversals will still often be greater than the length of the shorter intervals, but the continuous nature of the data source means that the ordering of events will be correct. Furthermore, because of the way ages are assigned in the marine magnetic anomaly timescales, the lengths of the intervals between reversals will typically be about right.

Naturally though, there are difficulties and problems with the marine magnetic anomaly timescales. As we go further back in time the record is more degraded, so the timescale from 120 Ma back to about 165 Ma is less reliable than that from the present back to 80 Ma. Furthermore, because of subduction of the seafloor, any record prior to about 170 Ma has been destroyed.

A major problem with marine magnetic anomalies is that it is fairly difficult to identify all of the very short intervals, particularly if a short interval appears between two relatively long intervals of the opposite polarity. This had led to significant problems in interpreting the GPTS.

Despite the problems in the marine magnetic anomaly timescales, these provide us with our best GPTS and so analyses of the statistical structure of the geomagnetic reversal sequence have typically been performed on these timescales.

Relevant probability distributions

It has been established that individual reversals take only a few thousand years to complete, which is a short time relative to the average interval between reversals. Thus it is a reasonable approximation to assume that the reversals themselves are instantaneous relative to the time constants of interest. It has already been noted that there appears to be a random component in the reversal process, and so it is sensible to look for a probabilistic description. Consider an interval of time Δx short enough that the probability of having two reversals in Δx is negligible, but with probability of $\lambda\Delta x$ of having a single reversal in Δx. This is then a Poisson process, and it is a simple matter to show that the probability density p(x) of interval lengths x between reversals is given by

$$p(x)\mathrm{d}x = \lambda \mathrm{e}^{-\lambda x}\mathrm{d}x, \qquad \text{(Eq. 1)}$$

where λ is the rate of the process and $\mu = 1/\lambda$ is the mean interval length.

It is possible to test if the observed distribution of interval lengths is compatible with the Poisson distribution. Initially it was concluded that there was indeed compatibility but after further testing it was suggested that there were too few short intervals in the observed record. Naidu (1971) showed that a gamma distribution provided a good fit to the then observed intervals of the Cenozoic timescale and Phillips (1977) confirmed this in an extensive study of geomagnetic reversal sequences. For a gamma process the probability density $p(x)$ of interval lengths x is given by

$$\begin{aligned} p(x)\mathrm{d}x &= \frac{1}{\Gamma(k)}(k\Lambda)^k x^{(k-1)}\mathrm{e}^{-k\Lambda x}\mathrm{d}x \\ &= \frac{1}{\Gamma(k)}\lambda^k x^{(k-1)}\mathrm{e}^{-\lambda x}\mathrm{d}x; \quad \lambda = k\Lambda \end{aligned} \qquad \text{(Eq. 2)}$$

where the mean interval length is now given by $\mu = 1/\Lambda = k/\lambda$. $\Gamma(k)$ is the gamma function of k, given by

$$\Gamma(k) = \int_0^\infty z^{k-1}\mathrm{e}^{-z}\mathrm{d}z \qquad \text{(Eq. 3)}$$

If k is an integer, then this is just the factorial function of k–1, i.e., $\Gamma(k) = (k-1)!$.

From Eqs. (1) and (2) it is apparent that the gamma process leads to a family of distributions depending on the value of k, and that the Poisson process is simply the special case of a gamma process with $k = 1$. Figure G38 shows appropriately scaled probability densities for this family of distributions. It is immediately apparent that a gamma process with $k > 1$ has far fewer very short intervals than a Poisson process.

Statistical tools for analyzing the reversal sequence in terms of a renewal process are given by McFadden (1984a) and McFadden and Merrill (1986, 1993).

Implications of a gamma or Poisson process

Consider the general probability of a reversal occurring at an interval Δx. Let x be the interval of time since the most recent reversal (i.e., no reversals have occurred in the interval 0 to x), and let $f(x)\Delta x$ be the probability that a reversal will then occur in the interval x to $(x + \Delta x)$. Hence the function $f(x)$ describes how the instantaneous probability for the occurrence of another reversal varies with the time since the last reversal and is given by

$$f(x) = \frac{p(x)}{1 - \int_0^x p(t)\mathrm{d}t} \qquad \text{(Eq. 4)}$$

Figure G39 shows $f(x)/\lambda$ plotted as a function of $(x\lambda)$. For a Poisson process the occurrence of a reversal has no impact on the probability of a future reversal. That is, the system has no memory. For $k > 1$, the probability for a future reversal drops to 0 as soon as a reversal

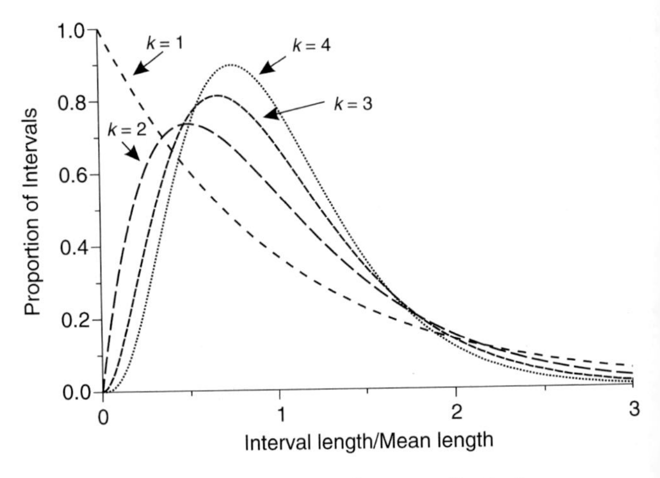

Figure G38 Probability densities of gamma distributions plotted against the interval length scaled to the mean length. The $k = 1$ curve represents a Poisson distribution, which has a relatively large number of shorter intervals compared with the other distributions.

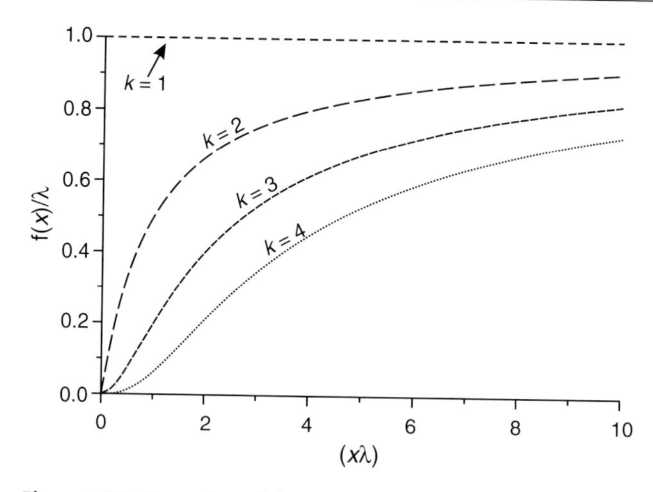

Figure G39 Illustration of the memory (inhibition) in a gamma process and the absence of a memory in the special case of a Poisson process ($k = 1$). After Merrill *et al.* (1996).

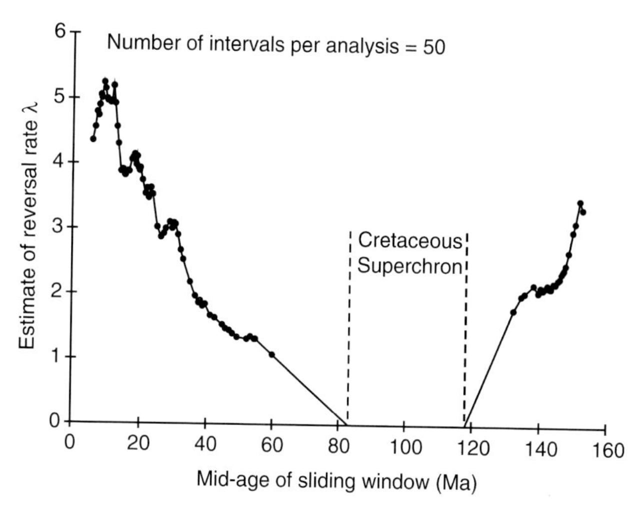

Figure G40 Estimated reversal rate λ for the past 160 Ma. Constructed from the timescales of Kent and Gradstein (1986) and Cande and Kent (1995), following the methods of McFadden (1984a). From Merrill *et al.* (1996).

occurs and then gradually rebuilds to its undisturbed value. That is, the system has a memory of the previous reversal and this memory causes inhibition of future events by depressing their probability of occurrence.

Is the process gamma or Poisson?

As noted above, there may be too few very short intervals in the observed timescale for a Poisson process, but a gamma distribution provides a good fit. Thus it may initially appear as a clear case in favor of a general gamma process, and indeed this was felt to be so for some time. Analysis of early timescales showed that k was quite different for the normal and reverse polarity sequences: at times it was about 4 for the normal polarity sequence while it was close to unity for the reverse polarity sequence (from about 40 to about 25 Ma); and in recent times (the past 15 Ma) it was about 2½ for the reverse sequence while close to unity for the normal sequence. This suggested a substantial asymmetry between normal and reverse polarity, which was surprising when theoretical considerations strongly implied that there should be no asymmetry (see above and *Geomagnetic field, asymmetries*).

However, McFadden (1984a) showed that these estimates of k were not robust and that the actual reversal process is much more likely to be nearly Poisson. Obtaining a reliable GPTS from the marine magnetic anomaly record is not a trivial matter: the major problems relate to accurate dating of the individual events and reliable recognition of the shorter intervals. A small error in the dating of an individual reversal has small consequences. By contrast, there are significant consequences when a short polarity interval is actually missed. When a short interval is not identified it is, in effect, combined with the preceding and succeeding intervals of the opposite polarity. This means that the short interval is missed from its own polarity sequence thereby tending to increase the apparent value of k for that polarity sequence. At the same time, it incorrectly produces a long interval of the opposite polarity that is the sum of the interval preceding the missed short interval, the short interval itself, and the succeeding interval.

McFadden (1984a) has shown that if n intervals from a gamma process with index k are joined together then the resulting interval appears to have been drawn from a gamma process with index nk. Therefore, if the process is Poisson and a short interval is missed, the resulting long interval of the opposite polarity appears to be an observation drawn from a gamma process with $k = 3$. Hence the parameter k can be a fairly sensitive indicator of polarity intervals that have been missed.

McFadden and Merrill (1984) concluded that k for the observed sequence was about 1.25 for the period from about 80 Ma to the present. This suggested that several short intervals had not been identified in the GPTS. McFadden and Merrill (1993) discussed the possible alternative that the reversal process is actually gamma and showed that this would imply that following a reversal the probability for another reversal would be depressed for about 50 ka.

The observed sequence is statistically similar to a sequence that would be obtained by taking a Poisson process with the appropriate rate λ and filtering the sequence so that any interval less than 30 ka in duration is incorporated into the surrounding intervals of opposite polarity. This is consistent with the conclusion of Parker (1997) that the maximum resolution of the GPTS from marine magnetic anomalies is about 36 ka.

Overall it is probably safe to conclude that the reversal process is either Poisson or nearly Poisson, that there is no asymmetry between the normal and reverse sequences, and that the polarity sequence can be considered as a single sequence without regard to the actual polarity of the individual intervals.

Nonstationarity in the reversal process

Recently the reversal rate has been about 4.5 Ma–1, around 40 Ma it was about 2 Ma–1, and from about 83 Ma back to about 119 Ma there were no reversals, the field having normal polarity. Clearly the rate at which the reversal process occurs has not been constant through time. This is of central geophysical interest because the existence of nonstationarity implies a change in the properties of the origin of the process, typically in the boundary conditions.

Estimates of the reversal rate λ, using the reversal chronology of Cande and Kent (1995) for the interval 0 to 118 Ma and that of Kent and Gradstein (1986) for the interval 118–160 Ma, are shown in Figure G40. These estimates have been obtained using a sliding window containing 50 polarity intervals. The Cretaceous Superchron, an interval of about 36 Ma from about 118 to 82 Ma when the polarity was normal and there were no reversals, is very obvious in the sequence and requires explanation.

The characteristic time of these changes is the same as that associated with changes in boundary conditions imposed by the mantle on the outer core (e.g., Jones, 1977; McFadden and Merrill, 1984, 1993, 1995, 1997; Courtillot and Besse, 1987). Suggestions that spatial variations in the core-mantle boundary conditions can affect the reversal rate are now supported by some dynamo theory (e.g., Glatzmaier *et al.*, 1999). The variation in λ shown in Figure G40 led to the interpretation first articulated by McFadden and Merrill (1984), that changes in the core-mantle boundary conditions gradually slowed the reversal process until eventually the process ceased, creating the superchron without any reversals. Gradually the boundary conditions became more favorable for reversals until eventually the reversal process restarted and gradually sped up to its current rate.

Gallet and Hulot (1997), using the same timescales, proposed an alternative nonstationarity. They suggested that the reversal rate had been essentially constant from about 158 to 130 Ma and from about 25 Ma to the present, with an intermediate nonstationary segment including the Cretateous Superchron. However, McFadden and Merrill (2000) developed a statistical test that showed that the data demanded the trends identified in Figure G40. Subsequently, using the methodology developed by McFadden and Merrill (2000), Hulot and Gallet (2003) analyzed the Gradstein *et al.* (1994) and Channell *et al.* (1995) timescales. These timescales update the pre-superchron information. Hulot and Gallet (2003) concluded that these newer timescales show no long-term trend in reversal rate leading into the superchron. Indeed, they suggest that the only precursor to the superchron was perhaps a single unusually long interval (CM1n) just before the superchron. If this is confirmed by future timescales then the interpretation of the causes of the superchron will naturally change.

There is some fine structure observable in Figure G40, and the question arises as to whether this structure is meaningful. Several authors (Mazaud *et al.*, 1983; Negi and Tiwari, 1983; Raup, 1985; Stothers, 1986; Marzocchi and Mulargia, 1990; Mazaud and Laj, 1991; Rampino and Caldeira, 1993) have suggested either a 15- or a 30 Ma periodicity in the reversal chronology record. Hulot and Gallet (2003) have recently suggested a 20 Ma periodicity. Clearly the perceived periodicity is not robust. McFadden (1984b) showed that similar apparent periodicities are produced when using fixed-length sliding windows to analyze a Poisson process with a linear trend in the reversal rate, but that such periodicities are not observed when using sliding windows with a fixed number of intervals (McFadden and Merrill, 1984). Lutz (1985), Stigler (1987), McFadden (1987), and Lutz and Watson (1988) all showed that the perceived periodicities are more likely an artifact of the methods of analysis than real geophysical phenomena.

Acknowledgment

This paper is published with the permission of the Chief Executive Officer, Geoscience Australia.

Phillip L. McFadden

Bibliography

Cande, S., and Kent, D.V., 1995. Revised calibration of the geomagnetic polarity timescale for the Late Cretaceous and Cenozoic. *Journal of Geophysical Research*, **100**: 6093–6095.

Channell, J., Erba, E., Nakanishi, M., and Tamaki, K., 1995. Late Jurassic-Early Cretaceous time scales and oceanic magnetic anomaly block models. In Berggren, W., Kent, D., Aubry, M., and Hardenbol, J. (eds.), *Geochronology, Timescales and Global Stratigraphic Correlation*. Society of Economic Paleontologists and Mineralogists Special Publications, **54**: 51–63.

Courtillot, V., and Besse, J., 1987. Magnetic field reversals, polar wander, and core-mantle coupling. *Science*, **237**: 1140–1147.

Cox, A., 1969. Geomagnetic reversals. *Science*, **163**: 237–245.

Cox, A., Doell, R.R., and Dalrymple, G.B., 1963. Geomagnetic polarity epochs and Pleistocene geochronometry. *Nature*, **198**: 1049–1051.

Gallet, Y., and Hulot, G., 1997. Stationary and nonstationary behavior within the geomagnetic polarity timescale. *Geophysical Research Letters*, **24**: 1875–1878.

Glatzmaier, G.A., Coe, R.S., Hongre, L., and Roberts, P.H., 1999. The role of the Earth's mantle in controlling the frequency of geomagnetic reversals. *Nature*, **401**: 885–890.

Glen, W. 1982. *The Road to Jaramillo. Critical Years of the Revolution in Earth Science.* Stanford: Stanford University Press.

Gradstein, F.M., Agterberg, F.P., Ogg, J.G., Hardenbol, J., van Veen, P., Thierry, J., and Huang, Z., 1994. A Mesozoic timescale. *Journal of Geophysical Research*, **99**: 24051–24074.

Hulot, G., and Gallet, Y., 2003. Do superchrons occur without any palaeomagnetic warning? *Earth and the Planetary Science Letters*, **210**: 191–201.

Jones, G.M., 1977. Thermal interaction of the core and the mantle and long term behaviour of the geomagnetic field. *Journal of Geophysical Research*, **82**: 1703–1709.

Kent, D.V., and Gradstein, F.M., 1986. A Jurassic to recent chronology. In Vogt, P.R., and Tucholke, B.E. (eds.), *The Geology of North America,* Vol. M, The Western North Atlantic Region. Boulder: Geological Society of America.

Kent, D.V., and Olsen, P.E., 1999. Astronomically tuned geomagnetic polarity timescale for the Late Triassic. *Journal of Geophysical Research*, **104**: 12831–12842.

Lutz, T.M., 1985. The magnetic reversal record is not periodic. *Nature*, **317**: 404–407.

Lutz, T.M., and Watson, G.S., 1988. Effects of long-term variation on the frequency spectrum of the geomagnetic reversal record. *Nature*, **334**: 240–242.

Marzocchi, W., and Mulargia, F., 1990. Statistical analysis of the geomagnetic reversal sequences. *Physics of the Earth and Planetary Interiors*, **61**: 149–164.

Mazaud A., and Laj, C., 1991. The 15 Ma geomagnetic reversal periodicity: a quantitative test. *Earth and the Planetary Science Letters*, **107**: 689–696.

Mazaud A., Laj, C., de Seze, L., and Verosub, K.L., 1983. 15 Ma periodicity in the reversal frequency of geomagnetic reversals since 100 Ma. *Nature*, **304**: 328–330.

McFadden, P.L., 1984a. Statistical tools for the analysis of geomagnetic reversal sequences. *Journal of Geophysical Research*, **89**: 3363–3372.

McFadden, P.L., 1984b. 15 Ma periodicity in the frequency of geomagnetic reversals since 100 Ma. *Nature*, **311**: 396.

McFadden, P.L., 1987. Comment on "A periodicity of magnetic reversals?" *Nature*, **330**: 27.

McFadden, P.L., and Merrill, R.T., 1984. Lower mantle convection and geomagnetism. *Journal of Geophysical Research*, **89**: 3354–3362.

McFadden, P.L., and Merrill, R.T., 1986. Geodynamo energy source constraints from paleomagnetic data. *Physics of the Earth and Planetary Interiors*, **43**: 22–33.

McFadden, P.L., and Merrill, R.T., 1993. Inhibition and geomagnetic field reversals. *Journal of Geophysical Research*, **98**: 6189–6199.

McFadden, P.L., and Merrill, R.T., 1995. History of Earth's magnetic field and possible connections to core-mantle boundary processes. *Journal of Geophysical Research*, **100**: 317–316.

McFadden, P.L., and Merrill, R.T., 1997. Asymmetry in the reversal rate before and after the Cretaceous normal polarity superchron. *Earth and the Planetary Science Letters*, **149**: 43–47.

McFadden, P.L., and Merrill, R.T., 2000. Evolution of the geomagnetic reversal rate since 160 Ma: Is the process continuous? *Journal of Geophysical Research*, **105**: 28455–28460.

Merrill, R.T., McElhinny, M.W., and McFadden, P.L., 1996. *The Magnetic Field of the Earth: Paleomagnetism, the Core, and the Deep Mantle.* San Diego, CA: Academic Press.

Naidu, P.S., 1971. Statistical structure of geomagnetic field reversals. *Journal of Geophysical Research*, **76**: 2649–2662.

Negi, J.G., and Tiwari, R.K., 1983. Matching long-term periodicities of geomagnetic reversals and galactic motions of the solar system. *Geophysical Research Letters*, **10**: 713–716.

Parker, R.L., 1997. Coherence of signals from magnetometers on parallel paths. *Journal of Geophysical Research*, **102**: 5111–5117.

Phillips, J.D., 1977. Time-variation and asymmetry in the statistics of geomagnetic reversal sequences. *Journal of Geophysical Research*, **82**: 835–843.
Rampino, M.R., and Caldeira, K., 1993. Major episodes of geologic change: correlations, time structure and possible causes. *Earth and the Planetary Science Letters*, **114**: 215–227.
Raup, D.M., 1985. Magnetic reversals and mass extinctions. *Nature*, **314**: 341–343.
Stigler, S., 1987. A periodicity of magnetic reversals? *Nature*, **330**: 26–27.
Stothers, R.B., 1986. Periodicity of the Earth's magnetic reversals. *Nature*, **332**: 444–446.

Cross-references

Core-Mantle Boundary
Core-Mantle Boundary Topography, Implications for Dynamics
Geodynamo, Numerical Simulations
Geomagnetic Field, Asymmetries
Geomagnetic Polarity Timescales
Magnetic Field of Sun
Reversals, Theory
Superchrons, Changes in Reversal Frequency

GEOMAGNETIC REVERSALS, ARCHIVES

It is now widely accepted that the Earth's magnetic field is generated by electric currents in the iron-rich liquid outer core. A dynamo process converts the energy associated with fluid convection within the Earth's core into magnetic energy. At the surface of the Earth, the field varies on timescales that range over more than 18 orders of magnitude, from less than a millisecond to more than 100 Ma. The most dramatic field variations are reversals. Almost exactly one century ago, Bernard Brunhes (1903, 1905) and his colleague David measured the magnetization of a lava flow, which was magnetized in the opposite direction to the present field. After investigating several possibilities, they convinced themselves that this resulted from a magnetic field with its magnetic north pole close to the south geographic pole. In other words the field would have been flipped in the opposite configuration to the present field with its south magnetic pole (in contrast to the current belief the pole referred as the north magnetic pole is actually a south pole which attracts the northern edge of the magnet) close to the north geographic pole. It took about 50 years before the existence of the geomagnetic reversals was established. Between 1925 and 1935 the discovery of new reversely magnetized rocks from different continents convinced the Japanese Matuyama (1926, 1929) and the French scientist Mercanton (1931) of their existence. However Louis Néel (1955) and his colleagues reported that reversely magnetized rocks could result from specific arrangements of the atoms lattices and the Japanese Uyeda (1958) observed this mechanism in a dacite from Mount Haruna. Further work demonstrated that this process was actually associated with specific chemical configurations and thus very rarely met in natural rocks. In the meantime, dating using new radiometric techniques indicated that rocks of the same age had the same polarity. Hence, it was obvious that the presence of reversed magnetization was a worldwide phenomenon, not caused by self-reversing processes. The discovery of the geomagnetic reversals allowed also understanding why the basaltic seafloor magnetization was organized in alternances of positive and negative giant anomalies parallel to the ridge axis. This pattern actually reflects the succession of the two polarities of the geomagnetic field, which were initially recorded at the ridge axis and then pushed away by the spreading seafloor.

Geomagnetic polarity timescale and the duration of reversals

The first reversals were dated using the ages derived from land sequences. This yielded the construction of the first geomagnetic polarity timescale (GPTS), which covered the past 5 Ma (Cox *et al.*, 1963). Using the ages of these individual reversals, it was then possible to interpolate or to extrapolate on the basis of the width of the marine magnetic anomalies, which led to extend the timescale over the past 160 Ma (Figure G41). Beyond this period, we are faced to the absence of magnetic anomalies. During the past 20 years much activity has been devoted at studying long sequences of exposed sediments or lava flows in order to build up the polarity timescale for the older periods.

The succession of polarity intervals during the past 160 Ma shows the existence of periods of high reversal frequency, which alternate with periods of low frequency. The mean reversal frequency is of the order of 1 reversal per million years and the maximum value does not seem to exceed 6 reversals per million years. During the period extending from 118 to 83 Ma, the field remained in a virtually uninterrupted normal state. These 35 Ma in the upper Cretaceous are known as the "Cretaceous normal superchron (CNS)" (between 120 and 83 Ma). Before the Kiaman reversed superchron stretches undisturbed for 50 Ma from the late Carboniferous to the middle Permian (between 310 and 260 Ma). The period between these two superchrons looks very much like the magnetic record that has followed the CNS. A third superchron, although apparently shorter in duration (20 ka), has been proposed from magnetostratigraphic studies (Pavlov and Gallet, 1998). Thus, in the present state of knowledge superchrons can be seen as a constant characteristic of the polarity timescale. It was frequently claimed that the slow decrease of reversal frequency before the Cretaceous superchron (McFadden and Merrill, 1995) would led to the superchron, thus resulting from a long-term influence of the mantle dynamics on core convection. Recent analysis (Hulot and Gallet, 2003) has shown that no long-term behavior over the 40 Ma preceding the Cretaceous superchron can be seen in the reversal rate that could be invoked as announcing its occurrence 120 Ma ago.

Superchrons might correspond to times when the process responsible for geomagnetic reversals passed below a certain critical threshold, and reversals could start again when that threshold was exceeded again. Other analyses suggest that the superchrons indicate the existence of a nonreversing state of the geodynamo in contrast to its reversing state. Alternatively, changes between the two processes may depend on factors directly connected to the boundary conditions, such as the distribution of heterogeneities within the "D" layer (just above the core-mantle boundary) that affect the flow pattern within the core and therefore may not be related to the intrinsic time constants of the geodynamo.

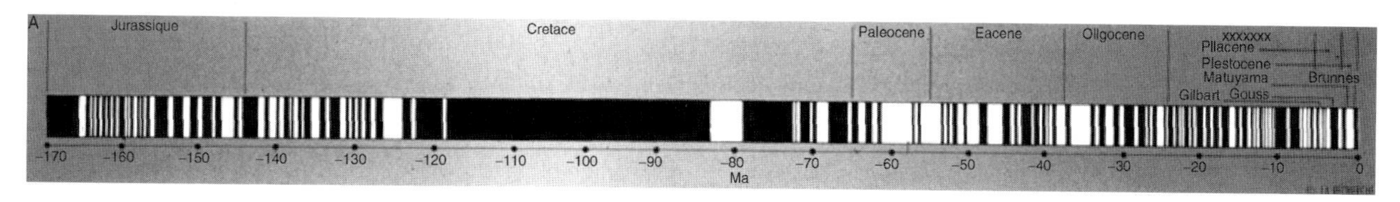

Figure G41 Geomagnetic polarity timescale for the past 160 Ma. Black (white) bars indicate periods of normal (reverse) polarity. Note the existence of the long Cretaceous period without any reversal.

The time span between two successive reversals appears to be random. Many statistical analyses were aimed at detecting periodicities or recurrent features, which would be hidden behind the succession of reversals. The succession of polarity intervals is closely approximated by a Poissonian distribution or at least by a gamma process if one assumes that short events have been missed in the reversal sequence. One can always consider that subsequent refinement of the reversal sequence can change this view. A critical aspect is the presence of field excursions which large deviations from the geocentric axial dipole during periods of very-low-field intensity. Excursions and reversals share common characteristics, which suggest that they could be treated as manifestations of similar processes within the core. In fact many authors proposed that excursions must be seen as aborted reversals whereas others regard them as intrinsic secular variation in presence of a weak dipolar field. In the first case, they should be treated at the same level as reversals in the statistics, which would provide a very different picture of the field. A critical point to answer these problems would be to determine whether excursions existed during the superchrons.

Dating reversals and estimating their duration

The age of the successive reversals is critical for analyzing their succession. As described earlier most reversals were dated by extrapolating the ages of the recent reversals after combining the spreading rates. Very significant progress was accomplished during the past 20 years with the discovery that carbonated sediments revealed the succession of the orbital cyclicities of the Earth (23 ka). The ages of the Pliocene and Pleistocene reversals were refined by using this independent method (Shackleton *et al.*, 1990; Tauxe *et al.*, 1992, 1996; Channel and Kleiven, 2000) which relies on correlating climate proxies (such as oxygen isotope ratios, susceptibility or density variations ... or simply counting the astronomical cycles recorded in continuous sedimentary sequences) with calculated variations of the Earth's orbit. Using seafloor spreading rates for five plate pairs, Wilson (1993) has shown that the errors in the astronomical calibration are not greater than 0.02 Ma (which corresponds to a precessional cycle of the Earth's orbit) and also that spreading rate can remain constant for several million years.

A critical question that comes to mind is how long it takes for the field to reverse from one polarity to the other. Before dealing with this aspect, it is important to determine when a reversal starts and when it ends. There are many observations showing successions of large oscillations (Hartl and Tauxe, 1996; Dormy *et al.*, 2000) preceding or following polarity changes, and it is not clear whether or not they should be incorporated in the reversal process. Such oscillations can be linked to enhanced secular variation in presence of low dipole field. Alternatively, they can be considered as successive attempts by the field to reverse. Important also is the definition of a transitional direction, which must exceed the normal range of secular variation (i.e., the range of the field variations during periods of polarity). Limits on the virtual geomagnetic pole (VGP) latitudes have been mostly used since it is rare that VGPs reach latitudes lower than 60°, although a strict definition should be restrained to positions lower than 45° (episodes of large secular variation can occasionally reach these latitudes).

The sharp transition between magnetic anomalies of opposite polarities, the very narrow thickness of the intervals recording sedimentary columns as well as the small number of lava flows with directions being unambiguously identified as "intermediate" between the two polarities were rapidly convincing and strong evidences that reversals are short phenomena on a geological timescale. The data recorded from the best continuous and documented sedimentary sequences indicate that the jump between the two polarities is shorter than one precessional cycle (23 ka). If we refer to radiometric dating (e.g., K-Ar or Ar-Ar techniques) of lava flows, the problem is delicate because of the sporadic succession of the flows characterized by pulses of eruptions occurring over a short period and long intervals without any magmatic event. A direct consequence is that sequences of lavas provide only very partial records of one or several phases of the reversal but not the entire process. Finally and even more critical is the fact that uncertainties on the ages cannot be ruled out as this is clearly illustrated by the large number of studies performed on the last reversal (Brunhes-Matuyama). The compilation of K-Ar and/or Ar-Ar dating (Quidelleur *et al.*, 2003) for 23 volcanic records indicates an age of 789 ± 8 ka (total error). The tuning of the $\delta^{18}O$ records from sedimentary sequences to orbital forcing models gives an age of 779 ± 2 ka (Tauxe *et al.*, 1996). Recently, Singer *et al.* (2005) mentioned that the astronomical determination is close to the age of the lavas from Maui (Hawaii) (776 ± 2 ka), and therefore that the other volcanic records are related to the onset of the transitional process. Consequently they deduce that a field reversal would require a significant period. One can oppose that there is no reason to consider that the onset of the reversal was initiated at the same time everywhere but above all if we consider the total uncertainties on the ages, there is no significant difference. Actually this uncertainty leaves doubts as to the possibility of constraining the duration of the transition with precision. There is also no reason to consider that all reversals have the same duration. Most studies converge to estimates between 5 and 10 ka but durations as short as 1 ka or as long as 20–30 ka have been proposed also. Clement (2004) recently analyzed the four most recent reversals recorded in sediments with various deposition rates and found a mean average duration of 7 ka for the directional changes (Figure G42). An interesting characteristic is that the mean duration seems to vary with latitude, as expected from simple geometrical models in which the nondipole fields are allowed to persist while the axial dipole decays through zero and then builds in the opposite direction.

The reversing field

Recorded by sediments

There has been a great deal of speculations concerning the processes governing field reversals. In order to decipher the mechanisms it was important to focus on the morphology of the field during the transition from one polarity state to the other. The first major step was to determine whether this "transitional" field would keep its dipolar dominant character. To achieve this goal, the paleomagnetists took advantage of

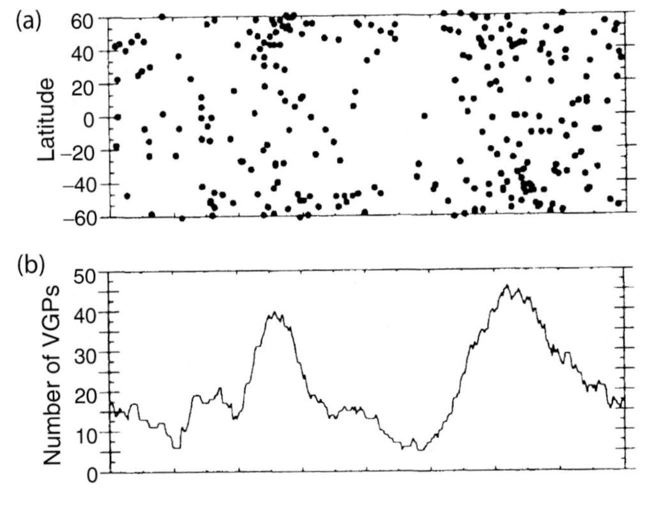

Figure G42 (a) Geographical distribution of Matuyama-Brunhes transitional VGPs derived from a selection of sedimentary records of the last reversal (from Clement, 1991). (b) Longitudinal distribution of transitional VGPs plotted as the number of transitional VGPs in a sliding 30° wide longitude window. Note the presence of two peaks over the American and Asian continents.

the concept of the VGP, which defines the position of the pole assuming that the field is dominantly dipolar. The idea was very simple. If the field remained dipolar, then any vector at the surface of the Earth would point toward the same pole following the rotation of the dipole with its north (south) pole passing to the (south), while in the opposite situation the poles would be different at any site. Using the first records from sediments, it was rapidly shown that the field was not dipolar during reversals (Dagley and Lawley, 1974; Hillhouse and Cox, 1976). In the meantime the first records of paleointensity established that the field intensity was systematically very low during the transitions. This decrease in dipole intensity is necessary before directions depart significantly from dipolar ones (Mary and Courtillot, 1993) and therefore consistent with the concept of a nondipolar field. The nature of this nondipolar field is one of the major questions to elucidate.

Two objectives were pursued in order to provide some answers to this question—the first one relied on the acquisition of multiple records of the same transition. It was proposed that because the Earth's rotation keeps a major role in field regeneration the transitions would be dominated by axisymmetrical components (quadropolar or octupolar). This suggestion could be tested at least at the first order by referring again to the useful concept of the VGPs. Indeed to satisfy the axial symmetry, the VGPs always follow the great circle passing through the observation site (or its antipode), which implies to study the same reversals at different sites. Records from sediments were appropriate because the transitions can be identified without ambiguity and thus correlated from widely separated sites. In the meantime the development of cryogenic magnetometers was very helpful as they provided the possibility of measuring weakly magnetized sediments. The results established that the VGP trajectories were effectively constrained in longitude but in many cases 90° away from the site meridian rather than centered above it.

The second objective was to investigate whether successive reversals were characterized by some recurrent or persistent features. This required studying sequences of reversals at the same site. Again long sequences of marine sediments were appropriate for this kind of study as well as the use of VGP paths, despite the dominance of nondipolar components. The most appealing observation emerged from a selection of sedimentary records (Clement, 1991; Tric *et al.*, 1991), which were all showing VGP paths within two preferred longitudinal bands, over the Americas and eastern Asia. Going one step further Laj *et al.* (1992) noted that these areas coincide with the cold circum Pacific regions in the lower mantle (outlined by seismic tomography), thereby suggesting that density or temperature conditions in the lower mantle could control the geometry of the reversing field. However, this observation was controversial because it relied on a selection of records. Another intriguing characteristic (Valet *et al.*, 1992; McFadden *et al.*, 1993) was the fact that these VGP paths were also found 90° away from the longitude of their sites despite their relatively wide geographic distribution (Figure G43), which could suggest some artifacts in the recording processes. Several studies effectively questioned the fidelity of sediments as recorders of the field variations, particularly during periods of low field intensity. Many factors (compaction, alignment of the elongated particles) reduce the inclination of the magnetization, a process that moves the VGPs paths far away from the longitude of the observation sites (Rochette, 1990; Langereis *et al.*, 1992; Quidelleur and Valet, 1994, 1995; Barton and McFadden, 1996,). There is also some indication that for some sediments the magnetic torque generated by a weak field is too low to provide accurate orientation of the magnetic grains, leaving in this case a prominent role to the hydrodynamic forces. These problems suggest that sediments could not be as appropriate as it was originally thought to study reversals.

Recorded by volcanic lava flows

One must thus turn toward volcanic records keeping in mind that in this case we are faced to the very discontinuous character of the

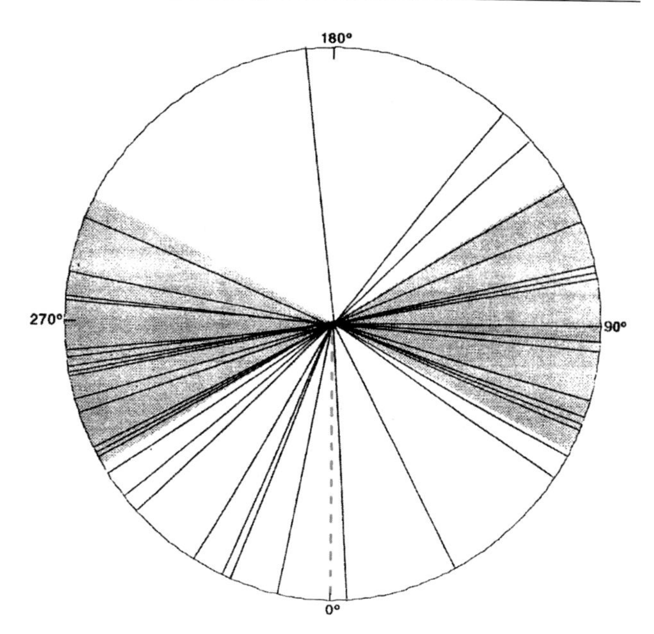

Figure G43 Distance between the mean longitudes of the virtual geomagnetic poles (VGPs) and their site meridian. Note that most lie 90° away from their site longitude.

eruption rates. Following the presentation of preferred longitudinal bands, Prévot and Camps (1993) compiled all volcanic records with VGPs latitudes lower than 60°, considering only one position for poles that were identical or very close to each other. They observed that no preferred longitudinal band emerged from this database. Love (1998) questioned this interpretation, arguing that similar directions from successive flows should not be averaged, because they do not necessarily result from a very rapid succession of eruptions. Instead he treated each individual flow as a single time event (implying no correlation between successive flows (Figure G44), or identically that the duration between flows was larger than the typical correlation times of secular variation). Because VGPs obtained from volcanics can reach latitudes as low as 45–50° during episodes of large secular changes (and of course excursions), it is also important to restrain the analysis to the most transitional directions, i.e., those with VGP latitudes less than 45°. Unfortunately in this case the number of points becomes too small to perform any robust analysis. This illustrates the difficulties of finding detailed records of reversals but also implies and confirms that indeed transitions occur very rapidly.

Using another selection of records, Hoffman (1991, 1992, 1996) pointed out the existence of clusters of VGPs in the vicinity of South America and above western Australia. Because of the apparent longevity of these directions they were interpreted as indicating the existence of a persistent inclined dipolar field configuration during the reversal process. This is an attractive suggestion, which would establish some link between the sedimentary and the volcanic records but limited to a selection of data.

It is striking that these interpretations depend on the chronology of the lava flows. Volcanism is mostly governed by short periods of intense eruptions alternating with quiet intervals. It is usually admitted that the active periods can be very short with respect to the intervals of quiescence. An indirect "magnetostratigraphic" indication has been given by three parallel sections of Hawaii (Herrero-Bervera and Valet, 1999, 2005), which are not distant, by more than a few kilometers. They all recorded the same reversal but do not show the same successions of transitional directions. Clusters of similar directions can be present in one section without being recorded 2 km away. Similarly apparent rapid changes can be observed in detail in one case and be absent 2 km away. In the first case, the field remained in the same position during a period short enough for not being recorded in the nearby section. In the second case, the field variations occurred either over a short time interval or the field moved very rapidly and was thus recorded only at a single location. Thus we cannot rule out that the apparent concentrations of VGPs close to South America and western Australia can be purely coincidental so that it cannot be considered yet as certain that long-lived transitional states represent an actual characteristic of the reversing field.

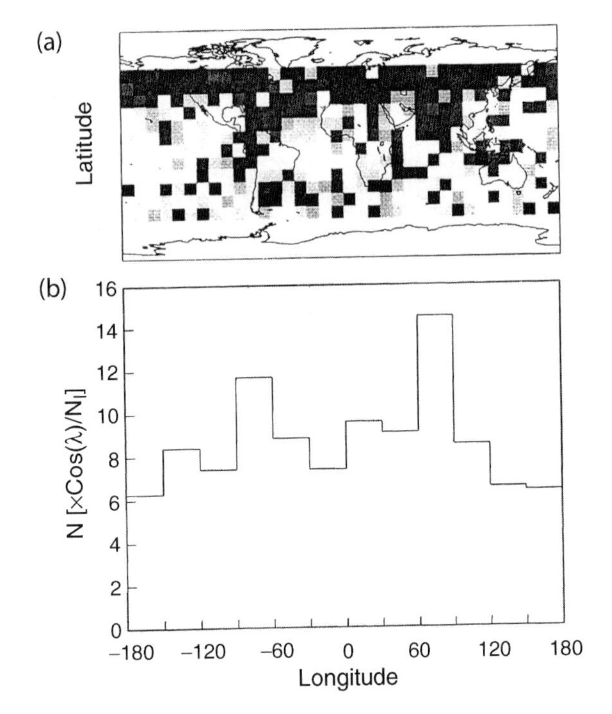

Figure G44 (a) Map of gray-scale histogram of the VGPs from volcanic lava flows with latitudes that are low enough for being considered as transitional (from Love *et al.*, 1998). (b) Histogram showing their longitudinal distribution. All lava flows were considered and thus not necessarily incorporated within a sequence of overlying flows. The similar directions that could be linked to a phase of rapid volcanic eruptions were considered as a result of a random process and thus were all taken in consideration. The database is strongly dominated by Icelandic lava flows.

Another approach is to rely exclusively on volcanic records with well-defined pre- and posttransitional directions and a sufficient number of intermediate directions. These records are more significant in terms of transitional field characteristics because they are not associated with uncertainties about the origin of the directions and their stratigraphic relation. The distribution of poles extracted from this limited database does not display any preferred location, nor does it show any evidence for systematics in the reversal process (Figure G45). Note also the presence of clusters at various longitudes and latitudes.

Despite the difficulties inherent to the interpretation of sedimentary and volcanic transitional directions, the two kinds of data may share some common characteristics. We already noted a possible link between the volcanic clusters and the preferred VGP paths in sediments but outlined some difficulties in reconciling these two aspects. Another important issue is that the two kinds of records are associated with very different timing in the acquisition of their magnetization, the almost instantaneous cooling of the lavas being almost opposite to the slow processes governing the lock-in of magnetization in sediments. It is thus more justified to attempt a comparison by considering sediments with very high deposition rates in order to reach a better resolution (for magnetization acquisition) thus closer to the volcanic characteristics. A few detailed sedimentary records with deposition rates exceeding 5 cm per 1 ka have been published (Valet and Laj, 1984; Clement and Kent, 1991; Channell and Lehman, 1997). A dominant and common characteristic is that they display a complex structure with large directional variations preceding and/or following the transition which reminds the features seen in the volcanics (Mankinen *et al.*, 1985; Chauvin *et al.*, 1990; Herrero-Bervera and Valet, 1999). These large loops share similarities with the secular variation of the present field. This observation reinforces the simplest model initially suggested by Dagley and Lawley (1974) of a rather complex transitional field which would be dominated by nondipole components following the large drop of the dipole field (Valet *et al.*, 1989; Courtillot *et al.*, 1992).

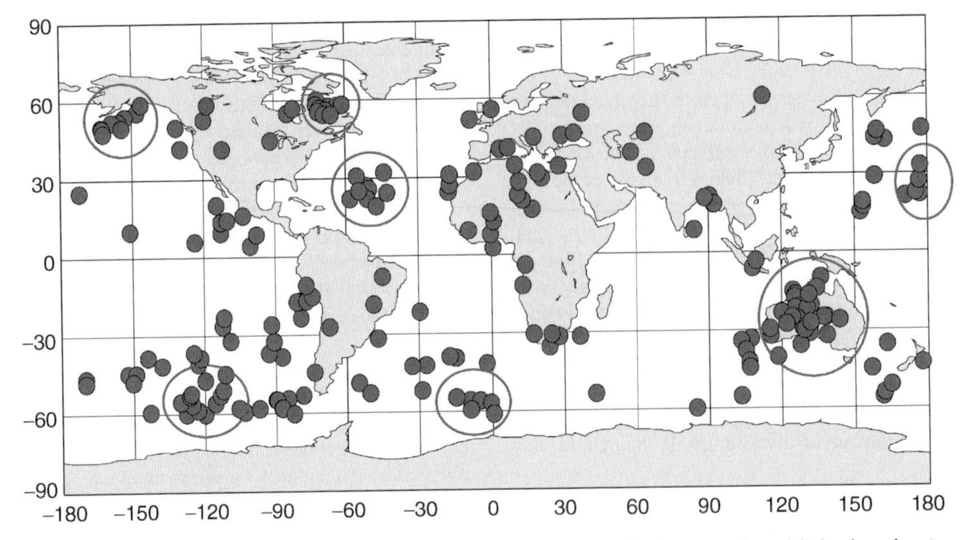

Figure G45 Positions of the VGPs derived from the most detailed volcanic records of reversals published so far. In contrast with Figure G44 no Icelandic sequence of superimposed flows met the selection criteria. Note also the large number of clusters (surrounded by circles) at various locations of the globe.

Fast impulses during reversals?

The existence of fast impulses during the 16 Ma old reversal recorded at the Steens Mountain in Oregon (Mankinen *et al.*, 1985) was suggested from a puzzling progressive evolution of the paleomagnetic directions in the interiors of two transitional lava flows (Coe and Prévot, 1989; Camps *et al.*, 1999). Each lava unit recorded a complete sequence of directions going all the way from that of the underlying flow to the direction of the overlying flow. In the absence of any clear evidence for anomalous rock magnetic properties, these features have been interpreted in terms of very fast geomagnetic changes, which would have reached 10° and 1000 nT per day. For comparison values typical of the present-day secular variation of the field (of internal origin) are of the order of 0.1° and 50 nT per year, i.e., some 10^4 times slower. In this specific case the timing of these fast changes can be constrained by estimates of the cooling times of individual flows.

However, such rapid changes do not seem to be compatible with accepted values of mantle conductivity (Ultré *et al.*, 1995). As a consequence this interpretation of the magnetization generated exciting controversy. Additional detailed investigations have been conducted in order to see whether this situation could not have arisen because of remagnetization of the flows. Remagnetizations of lava flows, yielding complex or unusual directions, have been detected at several locations where reversals have been recorded. A first interesting example was given by Hoffman (1984) from Oligocene basaltic rocks. Valet *et al.* (1998) observed the coexistence of both polarities (with similar characteristics as for the directions recorded at Steens Mountain) within flows marking the last reversal boundary (0.78 Ma) at the Canary Islands, and also in a lava flow associated with the onset of the upper Réunion subchron (2.13 Ma) in Ethiopia. In these cases, a purely geomagnetic interpretation would imply that a full reversal took place in only a few days. Similarly to the Steens Mountain, there is no striking difference between the rock magnetic properties of these units and the rest of the sequence, but a scenario involving thermochemical remagnetization is not incompatible with the results. Thermochemical magnetic overprinting can be particularly serious when it affects a flow emplaced at a time of very low field intensity, sandwiched between flows emplaced at a time of full (stable polarity) intensity. Recent investigation of additional flows at Steens did not shed more light on this problem but rather casts doubts on a geomagnetic interpretation of the paleomagnetic directions (Camps *et al.*, 1999). Therefore, the existence of very large and rapid changes that have been documented from a single site by a unique team remains controversial. In the meantime no alternative explanation has been completely accepted yet.

Field intensity variations across reversals

Variations in field intensity accompanying the reversals provide important and unique information concerning the transition itself but also the evolution of the dipole prior and after the reversals. We mentioned earlier that a significant decrease of the dipole was reported since the earliest studies. It is well established, based on all sedimentary as well as volcanic records that field intensity drops significantly and in most cases these changes last longer than directional changes (see e.g., Lin *et al.*, 1994; Merrill and Mc Fadden, 1999). Note that similar drops have been mentioned in all records of excursions with one exception (Leonhardt *et al.*, 2000). Initially, the records were restrained to the transitional interval or to a few thousand years preceding and following the reversal. Several long records of relative paleointensity have now been obtained using sequences of deep-sea sediments, which make possible to observe the evolution of the field over a long period. These independent records from sediment cores in different areas of the world can be stacked together to extract the evolution of the geomagnetic dipole moment (Guyodo and Valet, 1996, 1999; Laj *et al.*, 2000). There is a remarkable consistency between the first stacks published for the past 200 and 800 ka and a similar approach performed from field measurements immediately above bottom seafloor magnetic anomalies (Gee *et al.*, 2000). There is also a good agreement with the cosmogenic isotopic records (Frank *et al.*, 1997; Baumgartner *et al.*, 1998; Carcaillet *et al.*, 2003, 2004; Thouveny *et al.*, 2004). Apart many other interesting observations a dominant feature is the existence of a large drop of intensity about every 100 ka, which coincide with excursions reported from various sequences in the world.

The most recent curve of relative paleointensity was extended to the past 2 Ma (Valet *et al.*, 2005) and is in good agreement with the absolute dipole moments derived from volcanic lavas, which were used for calibration. It shows that the time-averaged field was higher during periods without reversals but the amplitude of the short-term oscillations remained the same. As a consequence, few intervals of very low intensity and thus less instability are expected during periods with a strong average dipole moment, whereas more excursions and reversals are produced during periods of weak field intensity. Prior to reversals, the axial dipole decays during 60 to 80 ka, but rebuilds itself in the opposite direction in a few thousand years at most (Figure G46). The most complete volcanic records confirm that recovery following a transition is short and culminates to very high values. The detailed volcanic records including determinations of absolute paleointensity provide support for such an asymmetry. Strong posttransitional field values have been reported in a Pliocene reversal recorded at Kauai (Bogue and Paul, 1993) and in the upper Jaramillo (0.99 Ma) subchron recorded from Tahiti (Chauvin *et al.*, 1990) as well as for the lower Mammoth reversal (3.33 Ma) from Hawaii (Herrero-Bervera and Valet, 2005). The same characteristics emerge also from the 60 Ma oldest record obtained so far in Greenland (Riisager and Abrahamsen, 2000), from the 15 Ma old Steens Mountain reversal (Prévot *et al.*, 1985) and from the last reversal (0.78 Ma) recorded from La Palma in the Canary Islands (Valet *et al.*, 1999).

Conclusion and perspectives

Several hundreds reversals have been documented from geological records and it is not unlikely, yet not demonstrated, that reversals always accompanied the existence of the geomagnetic field. Their internal origin neither makes any doubt. Many numerical models for the Earth's dynamo were produced during the last decade, including three-dimensional self-consistent dynamos that exhibit magnetic reversals. However, mostly for computation difficulties the parameters used remain far away from the Earth. This demonstrates the importance of

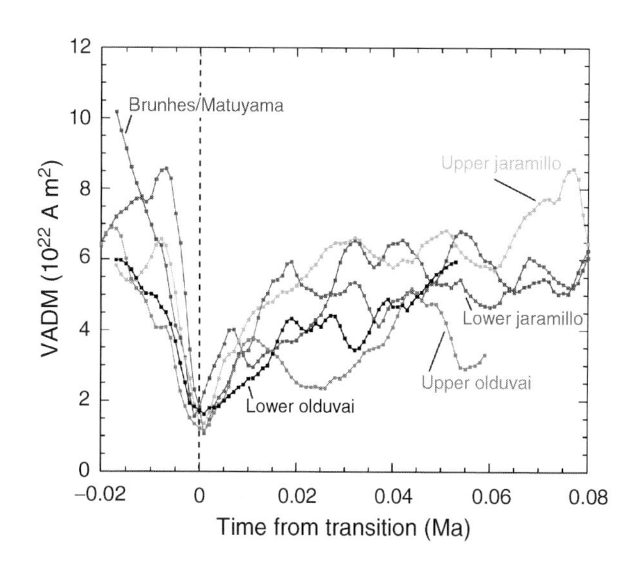

Figure G46 Field intensity variations across the five reversals occurring during the past 2 Ma. In this figure we superimposed the changes in dipole moment during the 80 and 20 ka time intervals, respectively preceding and following each reversal. Note the 60–80 ka long decrease preceding the reversals, and the rapid recovery following the transitions.

accumulating data. For about 30 years the paleomagnetists attempted to acquire as many detailed records as possible using the magnetic memory of sediments and lava flows. One of the first objectives was to determine whether the field keeps its dipolar character when reversing. The complexity of the directional changes shown by the detailed records and the large decrease of the field intensity indicates that the dipolar component strongly decreases by at least 80%, if not vanishing completely. One of the major constraints is the rapidity of the reversal process. There is no clear estimate for reversal duration, which may vary. Indeed it seems easier to isolate transitional directions for some reversals than for some others. There is no estimate for a lower limit of the duration of a transition which could be as short as a few hundreds years, if not less. It is reasonable to consider that the upper limit does not exceed 20 ka. After many years the suitability of sedimentary records (which have the advantage of preserving continuous information on field evolution) has been heavily questioned because their direction of magnetization can be affected by other factors (climate, alignment of the magnetic grains, postdepositional reorientations), particularly in presence of low field intensity. It is thus wise to turn also toward volcanic records despite their intrinsic limits in terms of resolution and dating. If we refer to the existing volcanic database, different views are presently defended regarding the field configuration during the short transitional period. Some claim that there is a dominance of the pole positions within preferred longitudinal bands, particularly within the American and Australian sectors, while others oppose rock magnetic artifacts and defend that the distribution of the transitional directions is typical of a nondipole field that would be similar to the present one. These two views have different implications and impose different constrains. The first one assumes that the lower mantle exerts some control on the reversal processes while the alternative interpretation defends that the transitional field would result from intrinsic processes linked to the dynamic of the core fluid. Another aspect is the existence of precursory events. The complexity of the field evolution prior to reversals depicts some "excursions" of the directions that are interpreted as precursory events. This observation can be linked to the long-term decay of the dipole component prior to the reversal, which is responsible for the complexity of the directional changes observed at the surface. Under this scenario the "precursory" excursions simply reflect the dominance of the nondipole part of the field, which will then prevail during the transition. Finally, a fast and strong recovery takes place immediately after the transition. The amplitude of this restoration phase is certainly critical as recent observations suggest that the dipole field strength could be a dominant factor controlling the frequency of reversals.

Jean-Pierre Valet and Emilio Herrero-Bervera

Bibliography

Barton, C.E., and McFadden, P.L., 1996. Inclination shallowing and preferred transitional VGP paths. *Earth and the Planetary Science Letters*, **140**: 147–157.

Baumgartner, S., Beer, J., Masarik, J., Wagner, G., Meynadier, L., and Synal, H.-A., 1998. Geomagnetic modulation of the ^{36}Cl flux in the Grip ice core, Greenland. *Science*, **279**: 1330–1332.

Bogue, S.W., and Paul, H.A., 1993. Distinctive field behaviour following geomagnetic reversals. *Geophysical Research Letters*, **20**: 2399–2402.

Brunhes, B., 1903. Sur la direction de l'aimantation permanente dans diversees roches volcaniques. *Comptes Rendus De l Academie Des Sciences Paris*, **137**: 975–977.

Brunhes, B., 1905. Sur la direction de l'aimantation dans une argile de Pontfarein. *Comptes Rendus De l Academie Des Sciences Paris*, **141**: 567–568.

Camps, P., Coe, R.S., and Prévot, M., 1999. Transitional geomagnetic impulses hypothesis: geomagnetic fact or rock magnetic artifact? *Journal of Geophysical Research*, **104**(B8): 17747–17758.

Carcaillet, J., Thouveny, N., and Bourlès, D., 2003. Geomagnetic moment instability between 0.6 and 1.3 Ma from cosmonuclide evidence. *Geophysical Research Letters*, **30**(15): 1792.

Carcaillet, J., Bourlès, D.L., Thouveny, N., and Arnold, M., 2004. A high resolution authigenic ^{10}Be/^{9}Be record of geomagnetic moment variations over the last 300 ka from sedimentary cores of the Portuguese margin. *Earth and Planetary Science Letters*, **219**: 397–412.

Channell, J.E., and Kleiven, H.F., 2000. Geomagnetic paleointensity and astronological ages for the Matuyama-Brunhes boundary and the boundaries of the Jaramillo subchron: paleomagnetic and oxygen isotope records from ODP site 983. *Philosophical Transactions of the Royal Society of London A*, **358**: 1027–1047.

Channell, J.E.T., and Lehman, B., 1997. The last two geomagnetic polarity reversals recorded in high deposition-rate sediments drifts. *Nature*, **389**: 712–715.

Chauvin, A., Roperch, P., and Duncan, R.A., 1990. Records of geomagnetic reversals from volcanic islands of French Polynesia. *Journal of Geophysical Research*, **95**: 2727–2752.

Clement, B.M., 1991. Geographical distribution of transitional VGPs: evidence for non zonal equatorial symmetry during the Matuyama-Brunhes geomagnetic reversal. *Earth and Planetary Science Letters*, **104**: 48–58.

Clement, B.M., 2004. Dependence of the duration of geomagnetic polarity reversal. *Nature*, **428**: 637–640.

Coe, R.S., and Prévot, M., 1989. Evidence suggesting extremely rapid field variation during a geomagnetic reversal. *Earth and Planetary Science Letters*, **92**: 292–298.

Courtillot, V., Valet, J.-P., Hulot, G., and Le Mouël, J.-L., 1992. The Earth's magnetic field: which geometry? *EOS Transactions, American Geophysical Union*, **73**(32): 337.

Cox, A., Doell, D.R., and Dalrymple, G.B., 1963. Geomagnetic polarity epochs and Pleistocene geochronometry. *Nature*, **198**: 1049.

Dagley, P., and Lawley, E., 1974. Paleomagnetic evidence for the transitional behaviour of the geomagnetic field. *Geophysical Journal of the Royal Astronomical Society*, **36**: 577–598.

Dormy, E., Valet, J.-P., and Courtillot, V., 2000. Observational constraints and numerical dynamos. *Geochemistry, Geophysics and Geosystems*, **1**: Voll, N°10, doi:10,1029/2000GCD00062.

Frank, M., 1997. A 200 ka record of cosmogenic radionuclide production rate and geomagnetic field intensity from ^{10}Be in globally stacked deep-sea sediments. *Earth and Planetary Science Letters*, **149**: 121–129.

Gee, J., Cande, S.C., Hildebrand, J.A., Donnelly, J.A., and Parker, R.L., 2000. Geomagnetic intensity variations over the past 780 ka obtained from near-seafloor magnetic anomalies. *Nature*, **408**: 827–832.

Guyodo, Y., and Valet, J.P., 1996. Relative variations in geomagnetic intensity from sedimentary records: the past 200 thousand years. *Earth and the Planetary Science Letters*, **143**: 23–26.

Guyodo, Y., and Valet, J.P., 1999. Global changes in intensity of the Earth's magnetic field during the past 800 ka. *Nature*, **399**: 249–252.

Hartl, P., and Tauxe, L., 1996. A precursor to the Matuyama/Brunhes transition-field instability as recorded in pelagic sediments. *Earth and the Planetary Science Letters*, **138**: 121–135.

Herrero-Bervera, E., and Valet, J.-P., 1999. Paleosecular variation during sequential geomagnetic reversals from Hawaii. *Earth and the Planetary Science Letters*, **171**: 139–148.

Herrero-Berrvera, E., and Valet, J.P., 2005. Absolute paleointensity from the Waianae volcanics (Oahu, Hawaii) between the Gilbert-Gauss and the upper Mammoth reversals. *Earth and the Planetary Science Letters*, **234**: 279–296.

Hillhouse, J., and Cox, A., 1976. Brunhes-Matuyama polarity transition. *Earth and the Planetary Science Letters*, **29**: 51–64.

Hoffman, K.A., 1984. Late acquisition of "primary" remanence in some fresh basalts: a case of spurious paleomagnetic results. *Geophysical Research Letters*, **11**: 681–684.

Hoffman, K.A., 1991. Long-lived transitional states of the geomagnetic field and the two dynamo families. *Nature*, **354**: 273–277.

Hoffman, K.A., 1992. Dipolar reversal states of the geomagnetic field and core-mantle dynamics. *Nature*, **359**: 789–794.

Hoffman, K.A., 1996. Transitional paleomagnetic field behavior: preferred paths or patches? *Surveys in Geophysics*, **17**: 207–211.

Hulot, G., and Gallet, Y., 2003. Do superchrons occur without any paleomagnetic warming? *Earth and the Planetary Science Letters*, **210**: 191–201.

Laj, C., Mazaud, A., Weeks, R., Fuller, M., and Herrero-Bervera, E., 1992. Geomagnetic reversal paths. *Nature*, **351**: 447.

Laj, C., Kissel, C., Mazaud, M., Channell, J.E.T., and Beer, J., 2000. North Atlantic paleointensity stack since 75 ka (NAPIS 75) and the duration of the Laschamp event. *Philosophical Transactions of the Royal Society of London A*, **358**: 1009–1025.

Langereis, C.G., van Hoof, A.A.M., and Rochette, P., 1992. Longitudinal confinement of geomagnetic reversal paths. Sedimentary artefact or true field behavior? *Nature*, **358**: 226–230.

Love, J.J., 1998. Paleomagnetic volcanic data and geometric regularity of reversals and excursions. *Journal of Geophysical Research*, **103**(B6): 12435–12452.

Leonhardt, R., Hufenbecher, F., Heider, F., and Soffel, H.C., 2000. High absolute paleointensity during a mid Miocene excursion of the Earth's magnetic field. *Earth and the Planetary Science Letters*, **184**(1): 141–154.

Lin, J.L., Verosub, K.L., and Roberts, P.A., 1994. Decay of the virtual dipole moment during polarity transitions and geomagnetic excursions. *Geophysical Research Letters*, **21**: 525–528.

Mankinen, E.A., Prévot, M., Grommé, C.S., and Coe, R.S., 1985. The Steens Mountain (Oregon) geomagnetic polarity transition. 1. Directional history, duration of episodes and rock magnetism. *Journal of Geophysical Research*, **90**: 10393–10416.

Mary, C., and Courtillot, V., 1993. A three-dimensional representation of geomagnetic field reversal records. *Journal of Geophysical Research*, **98**: 22461–22475.

Matuyama, M., 1929. On the direction of magnetisation of basalt in Japan, Tyosen and Manchuria. *Proceedings of the Imperial Academy of Japan*, **5**: 203.

McFadden, P., and Merrill, R., 1995. Fundamental transitions in the geodynamo as suggested by paleomagnetic data. *Physics of the Earth and Planetary Interiors*, **91**: 253–260.

McFadden, P.L., Barton, C.E., and Merrill, R.T., 1993. Do virtual geomagnetic poles follow preferred paths during geomagnetic reversals? *Nature*, **361**: 342–344.

Mercanton, P.L., 1926. Inversion de l'inclinaison magnétique aux âges géologiques. Nouvelles constattations. *Comptes Rendus De l Academie Des Sciences. Paris*, **192**: 978–980.

Merrill, R.T., and McFadden, P.L., 1999. Geomagnetic polarity transitions. *Reviews of Geophysics*, **37**: 201–226.

Néel, L., 1955. Some theoretical aspects of rock magnetism. *Advances in Physics*, **4**: 191.

Pavlov, V., and Gallet, Y., 1998. Upper Cambrian to Middle Ordovician magnetostratigraphy from the Kulumbe river section (northwestern Siberia). *Physics of the Earth and Planetary Interiors*, **108**(1): 49–59.

Prévot, M., and Camps, P., 1993. Absence of longitudinal confinement of poles in volcanic records of geomagnetic reversals. *Nature*, **366**: 53–57.

Prévot, M., Mankinen, E.A., Coe, R.S., and Grommé, C.S., 1985. The Steens Mountain (Oregon) geomagnetic polarity transition. Field intensity variations and discussion of reversal models. *Journal of Geophysical Research*, **90**: 10417–10448.

Quidelleur, X., and Valet, J.-P., 1994. Paleomagnetic records of excursions and reversals: possible biases caused by magnetization artefacts? *Physics of the Earth and Planetary Interiors*, **82**: 27–48.

Quidelleur, X., Valet, J.-P., Le Goff, M., and Bouldoire, X., 1995. Field dependence on magnetization of laboratory redeposited deep-sea sediments: first results. *Earth and the Planetary Science Letters*, **133**: 311–325.

Quidelleur, X., Carlut, J., Soler, V., Valet, J.-P., and Gillot, P.Y., 2003. The age and duration of the Matuyama-Brunhes transition from new K-Ar data from LaPalma (Canary Islands) and revisited $^{40}Ar/^{39}Ar$ ages. *Earth and the Planetary Science Letters*, **208**: 149–163.

Riisager, P., and Abrahamsen, N., 2000. Paleointensity of west Greenland Paleocene basalts: asymmetric intensity around the C57n-C26r transition. *Physics of the Earth and Planetary Interiors*, **118**: 53–64.

Rochette, P., 1990. Rationale of geomagnetic reversal versus remanence recording process in rocks: a critical review. *Earth and the Planetary Science Letters*, **98**: 33–39.

Shackleton, N.J., Berger, A., and Peltier, W.R., 1990. An alternative astronomical calibration of the lower Pleistocene timescale based on ODP Site 677. *Transactions of the Royal Society of Edinburg Earth Science*, **81**: 251–261.

Singer, B.S., Hoffman, K.A., Coe, R.S., Brown, L.L., Jicha, B.R., Pringle, M.S., and Chauvin, A., 2005. Structural and temporal requirements for geomagnetic reversal deduced from lava flows. *Nature*, **434**: 633–636.

Tauxe, L., Deino, A.D., Behrensmeyer, A.K., and Potts, R., 1992. Pinning down the Brunhes/Matuyama and Upper Jaramillo boundaries: a reconciliation of orbital and isotopic timescales. *Earth and the Planetary Science Letters*, **109**: 561–572.

Tauxe, L., Herbert, T., Shackleton, N.J., and Kok, Y.S., 1996. Astronomical calibration of the Matuyama Brunhes Boundary: consequences for magnetic remanence acquisition in marine carbonates and the Asian loess sequences. *Earth and the Planetary Science Letters*, **140**: 133–146.

Thouveny, N., Carcaillet, J., Moreno, E., Leduc, G., and Nérini, D., 2004. Geomagnetic moment variation and paleomagnetic excursions since 400 ka BP: a stacked record of sedimentary sequences of the Portuguese margin. *Earth and Planetary Science Letters*, **219**: 377–396.

Tric, E., Laj, C., Jehanno, C., Valet, J.-P., Mazaud, A., Kissel, C., and Iaccarino, S., 1991. A detailed record of the upper Olduvai polarity transition from high sedimentation rate marine deposits of the Po valley. *Physics of the Earth and Planetary Interiors*, **65**: 319–336.

Ultré-Guérard, P., and Achache, J., 1995. Core flow instabilities and geomagnetic storms during reversals: the Steens mountain impulsive field variation revisited. *Earth and the Planetary Science Letters*, **139**: 91–99.

Uyeda, S., 1958. Thermoremanent magnetism as a medium of paleomagnetism with special references to reverse thermoremanent magnetism. *Japanese Journal of Geophysics*, **2**: 1–123.

Valet, J.-P., and Laj, C., 1984. Invariant and changing transitional field configurations in a sequence of geomagnetic reversals. *Nature*, **311**(5986): 552–555.

Valet, J.-P., Tauxe, L., and Clement, B.M., 1989. Equatorial and mid-latitudes records of the last geomagnetic reversal from the Atlantic Ocean. *Earth and the Planetary Science Letters*, **94**: 371–384.

Valet, J.P., Tucholka, P., Courtillot, V., and Meynadier, L., 1992. Paleomagnetic constraints on the geometry of the geomagnetic field during reversals. *Nature*, **356**: 400–407.

Valet, J.-P., Kidane, T., Soler, V., Brassart, J., Courtillot, V., and Meynadier, L., 1998. Remagnetization in lava flows recording pretransitional directions. *Journal of Geophysical Research*, **103**(B5): 9755–9775.

Valet, J.P., Meynadier, L., and Guyodo, Y., 2005. Geomagnetic field strength and reversal rate over the past 2 Million years. *Nature*, **435**: 802–805.

Wilson, D., 1993. Confirmation of the astronomical calibration of the magnetic polarity timescale from seafloor spreading rates. *Nature*, **364**: 788–790.

Cross-references

Core motions
Geomagnetic dipole field
Geomagnetic excursion
Geomagnetic hazards
Magnetization, chemical remanent (CRM)
Magnetization, depositional remanent (DRM)
Magnetization, natural remanent (NRM)
Non-dipole field
Paleomagnetic secular variation
Paleomagnetism, deep-sea sediments
Polarity transitions: Radioisotopic Dating

GEOMAGNETIC SECULAR VARIATION

Introduction

The term *secular* comes from the Latin *Seculum* which means the duration of the influence of a powerful family becoming steadily less. The Romans accounted 100 years for that, why it also was synonymous for century. In context of geomagnetism it implies a long-term variation of the Earth's magnetic field.

The observed temporal variations of the Earth's magnetic field cover timescales from milliseconds to a few million years and originate in two distinct (external and internal) source regions with respect to the Earth's surface. With this respect the fluctuations of the external field ranges from milliseconds to a few decades, where the longer periods are related to variations of the solar magnetic field, e.g., the turnover of solar magnetic field (about 22 years). Changes of the internal field are of the order of a few years to millions of years. This variation results from the effect of magnetic induction in the fluid outer core and from effects of magnetic diffusion in the core and the mantle. Here, we distinguish geomagnetic secular variation and paleomagnetic secular variation, where the latter includes temporal changes longer than several hundreds of years, such as reversals (see *Reversals, theory*). The overlap of periods of internal and external sources in the range of a few years to decades can be separated in internal and external contributions applying spherical harmonic analysis (see *Internal external field separation*).

The geomagnetic secular variation was first noticed by Gunter and "Gellibrand" in 1635, who collected measurements of magnetic declination made at Limehouse near London between 1580 and 1634. This component gradually changed over the period of 350 years from 11° E to 24° W in 1820, before turning eastward again. Figure G47 shows the declination measured in London and Danzig (Gdansk) for about 350 years. Whereas the main field is dominated by its dipolar nature, the secular variation is clearly nondipolar, which is reflected in regions of different magnitudes of secular variation. For example, in the pacific region the secular variation appears to be crestless. However, the (geomagnetic) secular variation has been observed to be the feature of the main field and not of the local field.

One other prominent feature of the secular variation is the tendency of isoporic foci (areas of maximum secular variation) to drift westward. Analyzes by Vestine *et al.* (1947) and Bloxham *et al.* (1989) found an averaged drift rate of about $0.3° y^{-1}$.

In addition to the slowly varying secular variation event-like features appear, the so-called geomagnetic jerks. Such jerks show up as a change of sign in the slope of the secular variation, a discontinuity in the second time derivative of the field, most clearly seen in the east (Y) component of the geomagnetic field. For the last 100 years at least seven jerks have been reported (1912, 1925, 1969, 1978, 1983, 1991, and 1999), some of them of global extent. The 1969 event (first described by Courtillot *et al.*, 1978) was widely investigated; on the basis of observatory records. Courtillot *et al.* (1978) and Malin and Hodder (1982) showed its global extent, although it was not evident in all field components. This fact and the coincidental occurrence of jerks and sunspot maxima set off a lively discussion between Alldredge and McLeod in the 1980s (Alldredge, 1984; McLeod, 1985; Backus *et al.*, 1987) about the causative processes of jerks. The general understanding is that jerks are of internal origin, mainly because of two reasons: First, the potential of the solar cycle has approximately the form of zonal spherical harmonics, therefore any contribution to the East component would be small. Jerks are most clearly visible in the East component of European observatories. The second argument is based on a comparison of the strength of the solar maxima adjacent to the 1969 jerk, 1958 and 1980. At both epochs the solar maxima were more pronounced than in 1969, so we would expect two jerks at those epochs, but nothing obvious happened in 1958 and the jerks around 1978 and 1983 do not fit well to the solar maximum in 1980 (see also entry on *Geomagnetic jerks*).

Determination of secular variation models

Before considering the generation of secular variation and its link to other observables of processes in the Earth's core, we shall first describe a modeling approach to map the secular variation at the source region, the core-mantle boundary (CMB).

In principle, the magnetic field and its variation can be separated in parts due to external and internal sources by spherical harmonic analysis (Gauss, 1839). The geomagnetic field is then represented by the so called Gauss coefficients (see entries on *Internal external field separation and Main field modeling*). The variations of the internal field can be modeled by expanding the internal Gauss coefficients in a Taylor series in time about some epoch, t_e, e.g.,

$$g_l^m(t) = g_l^m(t_e) + \dot{g}_l^m(t_e)(t - t_e) + \ddot{g}_l^m \frac{(t - t_e)^2}{2!} + \cdots \qquad \text{(Eq. 1)}$$

where the first time derivative $\dot{g}_l^m$ is the secular variation and the second time derivative $\ddot{g}_l^m$ the secular acceleration. The determinations usually have been truncated after some derivatives. The GSFC(12/66) model (Cain *et al.*, 1967) included second time derivatives and the GSFC(9/80) model of the magnetic field between 1960 and 1980 by Langel *et al.* (1982) included third time derivatives. These

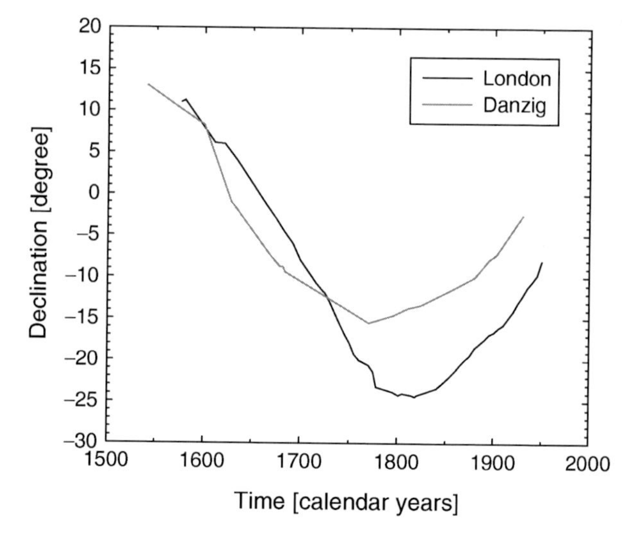

Figure G47 The declination at London and Danzig (Gdansk).

representations are only adequate to represent the temporal variation over short periods, e.g., the GSFC(9/80) is only sufficient for the period 1955–1980. Outside this period the misfit increases drastically.

An alternative attempt to model the secular variation was put forward by Langel *et al.* (1986). The methodology is to model the secular variation directly from first time derivatives of observatory annual means. For example, using the differences between annual means for different years, i.e., $\dot{X} = \Delta X/\Delta t$. Then performing with those $\dot{X}, \dot{Y}, \dot{Z}$ spherical harmonic analysis. Langel *et al.* (1986) achieved a continuous representation of the secular variation model for 1903 to 1982 by fitting each coefficient with cubic B-splines.

In a series of publications Bloxham (1987); Bloxham and Jackson (1989); Bloxham and Jackson (1992) developed a method which gives the most favorable description of the secular variation. This method bases on the simultaneous construction of a time-dependent model of the secular variation and main field. Their description of the time-dependent geomagnetic potential at the CMB follows:

$$V(r,\theta,\phi) = \sum_{l=1}^{L}\sum_{m=0}^{l}\sum_{n=1}^{N}\left(\frac{a}{c}\right)^{l+2}(l+1)\times(g_l^{mn}\cos(m\phi) + h_l^{mn}\sin(m\phi))P_l^m(\cos\theta)M_n(t), \qquad \text{(Eq. 2)}$$

where a is the Earth' radius, c is the radius of the Earth's core, $M_n(t)$ are the temporal basis functions, i.e. cubic B splines. The expansion (2) involves coefficients $\{g_l^{mn}, h_l^{mn}\}$ which are related to the standard Gauss coefficients $\{g_l^m, h_l^m\}$ by

$$g_l^m = \sum_{n=1}^{N} g_l^{mn} M_n(t) \qquad \text{(Eq. 3)}$$

and for the h_l^m likewise. The model is derived to meet some constraints, first it should be spatially smooth. Spatial smoothness, for instance could be controlled by the minimum ohmic dissipation based on the ohmic heating bound of Gubbins (1975). The second constraint to meet is the temporal smoothness of the model, which can be controlled by minimizing the second time derivative of the radial component of the field at the CMB. Further constraints, invoking satellite field models for certain epochs are conceivable as shown by Wardinski (2005) (see entry on *Time-dependent models of the geomagnetic field*).

Periodicities in the geomagnetic secular variation and related phenomena

An analysis of the geomagnetic activity and the solar activity reveals periods which can be attributed to the effect of the sunspot cycle. These periodicities are linked to external field variations, such as fluctuation of the strength and position of the ring current and variations of the strengths of current systems in the magnetosphere. Therefore, geomagnetic field variations exhibit nearly the same periods. If we assume that the effect of sunspot cycle and its harmonics is independent of observatory longitudes, then the effect can be approximated by a series of zonal harmonics. We would expect that these contributions dominantly map into the first degree Gauss coefficients of the internal field. Langel *et al.* (1986) found evidence that some of the periods also exist in higher degree secular variation coefficients ($\dot{g}_l^m, \dot{h}_l^m$). It is most likely, that this is due to induction in the mantle.

Beside the periods related to solar variations there exists a bundle of periods which are supposed to be inherent features of the geomagnetic secular variation. One prominent long period is the 23 (22.9) years period. It is found in the data of geomagnetic observatories (Alldredge, 1977b) as well as in secular variation coefficients (Langel *et al.*, 1986). Although the closeness of this period with the double solar cycle period may refer to a common cause, the origin of the 23 years period seems to be internal. Alldredge (1977b) showed that it does not appear in all analyzed observatories and is not in phase at those observatories at which it is present. Further, Langel *et al.* (1986) argued for an internal origin, because it is only present in a few secular variation coefficients, namely $\dot{g}_1^2, \dot{g}_2^2, \dot{h}_3^3$.

The picture is less clear when considering longer periods such as the 60 years period. Slaucitajis and Winch (1965) provided evidence of this period by an analysis of five observatory data sets. They found an averaged period of 61 ± 6 years. These results were confirmed by Jin and Jin (1989) with a slightly different period of 60 ± 12 years, where the uncertainty is due to an average of the results from an analysis of inclination and declination data. But its true nature is not clear since the match with a period of about 60 years in geomagnetic activity and sunspot numbers suggests an external origin which may cause an inductive 60 years period. On the other hand, there also may be an internal origin of this signal related to the variation of the westward drift and torsional oscillations of the fluid outer core. A comparison of the variation of the "westward drift" of the eccentric dipole (which is part of the secular variation) with the variation of the "length of day" reveals a significant correlation between both variations at about 60 years (Vestine, 1953; Vestine and Kahle, 1968). Braginsky (1970) developed a theory which is capable of characterizing both observables as two aspects of one phenomenon related to torsional magnetohydrodynamic oscillations in the Earth's core. These oscillations can have periods of about 60, 30, and 20 years, and via coupling processes between mantle and core they should produce fluctuation of the Earth's rotation. (see *Core mantle coupling*, *Length of day variations* and *Torsional oscillations* (*q.v.*)). Bloxham *et al.* (2002) suggested that the occurrence of geomagnetic jerks should be linked to "torsional oscillation," therefore also to the length of day variations. Indeed, recent analyzes by Holme and de Viron (2005) and Shirai *et al.* (2005) reveal that "geomagnetic jerks" are directly linked to the processes responsible for changes in the core angular momentum.

Further long periods may exist, such as 18.6 and 9.3 years resulting from the periodic gravitational action of celestial bodies on the equatorial bulge of the Earth. However, their spectral peaks are fairly close to those of the solar activity. Another was predicted by Yukutake (1972) to be the free modes of the electromagnetically coupled core-mantle system, which has a length of about 6.7 years. Currie (1973) may have found this period. For other periods, e.g., 70, 55, 32, 29, and 13 years (Langel *et al.*, 1986) a theory is only vaguely outlined and needs further investigation.

Generation of secular variation

Secular variation results from the effect of magnetic induction in the fluid outer core and from effects of magnetic diffusion in the core and the mantle. These processes are constituted by the induction equation, which follows from Faraday's induction law

$$\frac{\partial \mathbf{B}}{\partial t} = -\nabla \times \mathbf{E}, \qquad \text{(Eq. 4)}$$

where **E** is the electric field, showing that a spatially varying electric field can induce a magnetic field. And further, Ampere's law

$$\frac{1}{\sigma}\mathbf{j} = \mathbf{E} + \mathbf{u} \times \mathbf{B} = \frac{1}{\sigma\mu}\nabla \times \mathbf{B}, \qquad \text{(Eq. 5)}$$

where σ is the electric conductivity, **j** the current density, and μ the magnetic permeability. Combining both and after some algebra we get

$$\frac{\partial \mathbf{B}}{\partial t} = \nabla \times (\mathbf{u} \times \mathbf{B}) + \eta\nabla^2\mathbf{B}. \qquad \text{(Eq. 6)}$$

This is the induction equation, where the first term on the right-hand side displays the advection of the magnetic field due to the fluid motion in the liquid outer core, the second represents the action of magnetic diffusion.

It is generally assumed that the diffusive timescale is factor 300 to 500 to the advective timescale. This implies that on short timescales, i.e., less than 100 years, the secular variation is entirely caused by the rigidly coupled movement of the magnetic field lines with the fluid motion in the liquid outer core and therefore diffusion can be neglected.

$$\frac{\partial \mathbf{B}}{\partial t} = \nabla \times (\mathbf{u} \times \mathbf{B}) \qquad \text{(Eq. 7)}$$

That is the so-called frozen flux hypothesis (see *Alfvén's Theorem and the frozen flux approximation* (*q.v.*)).

The "frozen flux assumption" simplifies the inversion for the fluid motion $\mathbf{u}$ of the liquid outer core (see *Core motions*), but it imparts an incomplete description of the mechanism of the secular variation generation, for two reasons. First, it reduces the algebraic order of the induction equation by one degree, therefore the solution for the fluid flow may not be expected to be complete or correct (Love, 1999). Second, there is evidence that the frozen flux assumption is violated in the last 40 or 50 years. Bloxham and Gubbins (1985); Bloxham (1986) evaluated the magnetic flux through individual flux patches at the CMB. A necessary condition for the "frozen flux approximation" to apply is that the flux F through a patch S on the core surface bounded by a contour of zero radial field must be constant with respect to time

$$F = \int_S B_r \mathrm{d}S = \text{const.} \qquad \text{(Eq. 8)}$$

Patches located in the southern hemisphere show a significant change of the flux during the period 1960–1980. On a global scale the unsigned flux integral

$$|F| = \oint_{\text{CMB}} |B_r| \mathrm{d}S = \text{const.}, \qquad \text{(Eq. 9)}$$

must also be constant with respect to time. Recent analyzes by Holme and Olsen (2005) and Wardinski (2005) suggest that this is not the case for the period 1980–2000. Figure G48 shows the differences of this integral for successive years averaged over the period 1980 to 2000. The given error bars are the sample variance of the integral

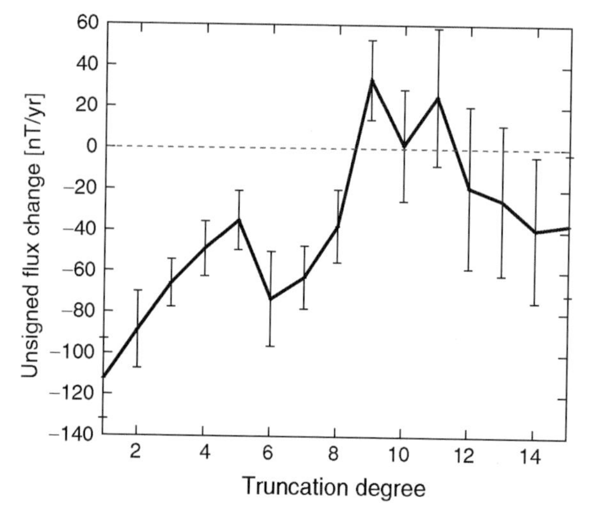

Figure G48 The mean change of the unsigned flux integral for 1980–2000.

for a specific truncation degree. For truncation degrees less than 9 a conservation of unsigned flux is not achieved, for higher degree this seems to be achieved within error margins, but it should be mentioned that the higher degrees of the model more and more reflect the *a priori* beliefs, i.e., spatial and temporal damping applied in the computation of the secular variation model.

In a related discussion about the origin of the "westward drift" it has been suggested that waves rather than a bulk advective motion account for the "westward drift" and most of the secular variation. The possible set of wave types are: inner core oscillation, which can have very short periods of about a year (Gubbins, 1981; Glatzmaier and Roberts, 1996). And, "torsional oscillations" of coaxial cylindrical shells oscillating in a solid-body rotation, for which the force balance is between Lorentz force and inertia. The periods are of the order of decades (Braginsky, 1970; Zatman and Bloxham, 1997, 1999). On longer timescales ($\geq$300 years) magnetohydrodynamic waves which are dependent on Magnetic, Archimedean (buoyancy), and Coriolis force could play a role. The difference between torsional oscillation and MAC-waves is the additional acting of buoyancy in MAC-waves. An interaction of these waves, core-surface flows and the morphology of the magnetic field on a timescale shorter than 300 years is likely. However, it should be mentioned that our distance to core surface and the mantle low conductivity attenuate any details in time and space.

The influence of the mantle on the secular variation

The secular variation as observed at the Earth's surface undergoes two principal processes when it permeates from the core through the mantle. First a geometrical attenuation determined by the factor

$$\left(\frac{a}{c}\right)^{l+2},$$

where $a/c \approx 1.8$. The higher the harmonic number l, the stronger is the geometrical attenuation. The second effect concerns the magnetic diffusion. Here the temporal change of the field is entirely given by the diffusion of the magnetic field through the mantle implicitly assuming that there is no magnetic field generation in the mantle.

The shortest temporal variation of the core's magnetic field that can be seen at the Earth's surface is determined by the conductivity of the mantle and in particular boundary between core and mantle. This region of compositional variation, partial melting and heterogeneous high conductivity, known as D″, may have a thickness of about 200 km and should have a significant impact to the permeability of the magnetic field (see D″).

However, our understanding of the electrical conductivity of the mantle and its lower part is very poor. All method, which have been employed to deduce the mantle conductivity from the observed secular variation are based on the assumption that the mantle conductivity σ varies as a power law of the radius r

$$\sigma = \sigma_0 \left(\frac{c}{r}\right)^{\alpha}. \qquad \text{(Eq. 10)}$$

This model permits high conductivity in the boundary layer as well as low values of conductivity in the upper mantle, well in agreement with "magnetotellurics." The values of σ_0 ranges from 223 S m^{-1} (McDonald, 1957) and 100 000 S m^{-1} (Alldredge, 1977a) depending on the method of analysis (see *Electrical conductivity of the mantle*). A value between 600 and 3000 S m^{-1} (the later preferred by Backus, 1983) would be in good agreement with length of day variation and electromagnetic core-mantle coupling (Stix and Roberts, 1984; Paulus and Stix, 1989; Stewart *et al.*, 1995).

Our knowledge of the mantle conductivity is far from being complete and a rigorous analysis, along theoretical lines as set by Backus (1983) remains to be done. This requires a consolidated knowledge about the secular variation. In order to achieve this the observation of the Earth's magnetic field have to be improved, either in terms of more geomagnetic observatories or satellite missions such as MAGSAT, ØRSTED, and CHAMP.

Ingo Wardinski

Bibliography

Alldredge, L.R., 1977a. Deep mantle conductivity. *Journal of Geophysical Research*, **82**: 5427–5431.

Alldredge, L.R., 1977b. Geomagnetic variations with periods from 13 to 30 years. *Journal of Geomagnetism and Geoelectricity*, **29**: 123–135.

Alldredge, L.R., 1984. A discussion of impulses and jerks in the geomagnetic field. *Journal of Geophysical Research*, **89**: 4403–4412.

Backus, G.E., 1983. Application of mantle filter theory to the magnetic jerk of 1969. *Geophysical Journal of the Royal Astronomical Society*, **74**: 713–746.

Backus, G.E., Estes, R.H., Chinn, D., and Langel, R.A., 1987. Comparing the jerk with other global models of the geomagnetic field from 1960 to 1978. *Journal of Geophysical Research*, **92**: 3615–3622.

Bloxham, J., 1986. The expulsion of magnetic flux from the Earth's core. *Geophysical Journal of the Royal Astronomical Society*, **87**: 669–678.

Bloxham, J., 1987. Simultaneous stochastic inversion for geomagnetic main field and secular variation. I. A large-scale inverse problem. *Journal of Geophysical Research*, **92**: 11597–11608.

Bloxham, J., and Gubbins, D., 1985. The secular variation of earth's magnetic field. *Nature*, **317**: 777–781.

Bloxham, J., and Jackson, A., 1989. Simultaneous stochastic inversion for geomagnetic main field and secular variation II: 1820–1980. *Journal of Geophysical Research*, **94**: 15753–15769.

Bloxham, J., and Jackson, A., 1992. Time-dependent mapping of the magnetic field at the core-mantle boundary. *Journal of Geophysical Research*, **97**: 19537–19563.

Bloxham, J., Gubbins, D., and Jackson, A., 1989. Geomagnetic secular variation. *Philosophical Transactions of the Royal Society of London A*, **329**: 415–502.

Bloxham, J., Dumberry, M., and Zatman, S., 2002. The origin of geomagnetic jerks. *Nature*, **420**: 65–68.

Braginsky, S.I., 1970. Torsional magnetohydrodynamic vibrations in the Earth's core and variations in day length. *Geomagnetism and Aeronomy (English translation)*, **10**: 1–8.

Cain, J.C., Hendricks, S.J., Langel, R.A., and Hudson, W.V., 1967. A proposed model for the International Geomagnetic Reference Field—1965. *Journal of Geomagnetism and Geoelectricity*, **19**: 335–355.

Courtillot, V., Ducruix, J., and Le Mouël, J.-L., 1978. Sur une accélération récente de la variation séculaire du champ magnétique terrestre. *Comptes Rendus de l' Academie des Sciences Paris-Series D*, **287**: 1095–1098.

Currie, R.G., 1973. Geomagnetic line spectra—2 to 70 years. *Astrophysics and Space Science*, **21**: 425–438.

Gauss, C.F., 1839. Allgemeine Theorie des Erdmagnetismus. In Gauss, C.F., and Weber, W. (eds.), *Resultate aus den Beobachtungen des magnetischen Vereins im Jahre 1838*. Leipzig, pp. 1–57.

Glatzmaier, G.A., and Roberts, P.H., 1996. On the magnetic sounding of planetary interiors. *Physics of the Earth and Planetary Interiors*, **98**: 207–220.

Gubbins, D., 1975. Can the Earth's magnetic field be sustained by core oscillations? *Geophysical Research Letters*, **2**: 409–412.

Gubbins, D., 1981. Rotation of the inner core. *Journal of Geophysical Research*, **86**: 11695–11699.

Holme, R., and de Viron, O., 2005. Geomagnetic jerks and a high resolution length-of-day profile for core studies. *Geophysical Journal International*, **160**: 435–440.

Holme, R., and Olsen, N., 2005. Core-surface flow modelling from high resolution secular variation. *Geophysical Journal International*, submitted.

Jin, R.S., and Jin, S., 1989. The approximately 60-year power spectral peak of the magnetic variations around London and the earth's rotation rate fluctuations. *Journal of Geophysical Research*, **94**: 13673–13679.

Langel, R.A., Estes, R.H., and Mead, G.D., 1982. Some new methods in geomagnetic field modeling applied to the 1960–1980 epoch. *Journal of Geomagnetism and Geoelectricity*, **34**: 327–349.

Langel, R.A., Kerridge, D.J., Barraclough, D.R., and Malin, S.R.C., 1986. Geomagnetic temporal change: 1903–1982—a spline representation. *Journal of Geomagnetism and Geoelectricity*, **38**: 573–597.

Love, J.J., 1999. A critique of frozen-flux inverse modelling of a nearly steady geodynamo. *Geophysical Journal International*, **138**: 353–365.

Malin, S.R.C., and Hodder, B.M., 1982. Was the 1970 geomagnetic jerk of internal or external origin. *Nature*, **296**: 726–728.

McDonald, K.L., 1957. Penetration of the geomagnetic secular field through a mantle with variable conductivity. *Journal of Geophysical Research*, **62**: 117–141.

McLeod, M.G., 1985. On the geomagnetic jerk of 1969. *Journal of Geophysical Research*, **90**: 4597–4610.

Paulus, M., and Stix, M., 1989. Electromagnetic core-mantle coupling: the Fourier method for the solution of the induction equation. *Geophysical Astrophysical Fluid Dynamics*, **47**: 237–249.

Shirai, T., Fukushima, T., and Malkin, Z., 2005. Detection of phase disturbances of free core nutation of the Earth and their concurrence with geomagnetic jerks. *Earth, Planets and Space*, **57**: 151–155.

Slaucitajis, L., and Winch, D.E., 1965. Some morphological aspects of geomagnetic secular variation. *Planetary and Space Science*, **13**: 1097–1110.

Stewart, D.N., Busse, F.H., Whaler, K.A., and Gubbins, D., 1995. Geomagnetism, Earth rotation and the electrical conductivity of the lower mantle. *Physics of the Earth and Planetary Interiors*, **92**: 199–214.

Stix, M., and Roberts, P.H., 1984. Time-dependent electromagnetic core-mantle coupling. *Physics of the Earth and Planetary Interiors*, **36**: 49–60.

Vestine, E.H., 1953. On the variations of the geomagnetic field, fluid motions, and the rate of the Earth's rotation. *Journal of Geophysical Research*, **58**: 127–145.

Vestine, E.H., and Kahle, A.B., 1968. The westward drift and geomagnetic secular change. *Geophysical Journal of the Royal Astronomical Society*, **15**: 29–37.

Vestine, E.H., Laporte, L., Lange, I., and Scott, W.E., 1947. The geomagnetic field, its description and analysis, Technical Report Publication 580, Carnegie Institution of Washington.

Wardinski, I., 2005. Core surface flow models from decadal and subdecadal secular variation of the main geomagnetic field, PhD thesis, Freie Universität Berlin (http://www.gfz-potsdam.de/bib/pub/str0507/0507.htm).

Yukutake, T., 1972. The effect of change in the geomagnetic dipole moment on the rate of the Earth's rotation. *Journal of Geomagnetism and Geoelectricity*, **24**: 19–48.

Zatman, S., and Bloxham, J., 1997. Torsional oscillations and the magnetic field within the Earth's core. *Nature*, **388**: 760–763.

Zatman, S., and Bloxham, J., 1999. On the dynamical implications of models of Bs in the Earth's core. *Geophysical Journal International*, **138**: 679–686.

Cross-references

Alfvén's Theorem and the Frozen Flux Approximation
CHAMP
Core-Mantle Coupling, Electromagnetic
Core Motions
Core-Mantle Coupling, Electromagnetic
Core-Mantle Coupling, Thermal
Core-Mantle Coupling, Topographic
D″ and F Layers of the Earth
D″ as a Boundary Layer
D″, Anisotropy
D″, Composition
D″, Seismic Properties
Gauss, Carl Friedrich (1777–1855)
Gellibrand, Henry (1559–1636)
Geomagnetic Jerks
Internal External Field Separation
Length of Day Variations, Decadal
Length of Day Variations, Long Term
Magnetotellurics
Main Field Modeling
Mantle, Electrical Conductivity, Mineralogy
Oscillations, Torsional
Reversals, Theory
Time-dependent Models of the Geomagnetic Field
Westward Drift

GEOMAGNETIC SPECTRUM, SPATIAL

Introduction

Just as for a function defined round a circle it is useful to separate the contributions of different wavelengths/spatial frequencies, so for a function defined on the surface of a sphere it is useful to separate the contributions of different (two-dimensional) wavelengths/spatial frequencies. In the context of geomagnetism it is useful to apply this concept of a spatial spectrum to the main magnetic field (coming from the core) and its time variation (the secular variation), and to the field from the magnetization of the crust.

In the one-dimensional case we call this process Fourier analysis; if we have a variable F, which is known round a circle (and which is therefore a periodic function round the circle), we can express it in the form

$$F(\lambda) = \Sigma_m F_m(\lambda) = \sum\nolimits_n (a_m \cos m\lambda + b_m \sin m\lambda), \qquad \text{(Eq. 1)}$$

where λ is the angle round the circle, measured from some origin. a_m and b_m are numerical coefficients; the integers m, which start from zero, can be thought of as spatial frequencies, or wave numbers. Each term $\cos m\lambda$ or $\sin m\lambda$ is called a harmonic, though this name is sometimes also applied to the numerical values of $a_m \cos m\lambda$ and $b_m \sin m\lambda$. The concepts of frequency and harmonic come from the analogous situation of a variable which is a periodic function in time; this function of time can be separated into harmonics specified by frequencies (in time) which are multiples of the fundamental frequency.

These harmonics are orthogonal; in the sense that the average value round the circle of the product of any two *different* harmonics is zero:

$$\langle F_m F_\mu \rangle = 0 \qquad \text{(Eq. 2)}$$

unless $m = \mu$ (and both harmonics use only either $\cos m\lambda$ or $\sin m\lambda$), where $\langle x \rangle$ represents the average of x round the circle. Therefore, if we expand an arbitrary F in terms of its harmonic components F_m, and then square it, all the cross-terms vanish when we take the average round the circle. So the mean-square value of F round the circle, $\langle F^2 \rangle$, becomes

$$\langle F^2 \rangle = \Sigma_m \langle F_m^2 \rangle = \sum\nolimits_m \langle a_m^2 \cos^2 m\lambda + b_m^2 \sin^2 m\lambda \rangle = \sum\nolimits_m (a_m^2 + b_m^2)/2; \qquad \text{(Eq. 3)}$$

the factor of (1/2) is the mean-square (ms) value of $\cos m\lambda$ or $\sin m\lambda$ round the circle. (For $m = 0$ the mean-square value is in fact 1, but I will ignore this minor complication.) We see that the two harmonics of a given spatial frequency m contribute to the overall mean-square value independently of all the other harmonics.

We can write $F_m(\lambda) = (a_m \cos m\lambda + b_m \sin m\lambda)$ alternatively as $c_m \cos(m\lambda - \varepsilon)$; it is a sinusoidal variation of amplitude $c_m = (a_m^2 + b_m^2)^{1/2}$, and phase ε. Our choice of the origin of the angle λ is arbitrary; while the numerical values of a_m and b_m, and the phase ε, will depend on this choice of origin, the variation of F_m round the real circle, and its amplitude c_m, is independent of this choice.

A plot of c_m^2 against m is called the (power) spectrum of F. (The notation arises because for a wave-motion periodic in time, c_m^2 is proportional to the average power (over the cycle) carried by the wave at that frequency.)

Application to the geomagnetic field

The case of the geomagnetic field is more complicated for two reasons. The first complication is that the field is a vector (at each point in space it has direction as well as magnitude) rather than a scalar (which has only magnitude). However, in regions where there is no significant electric current (in practice this means throughout the upper mantle and the lower atmosphere) we can represent the vector magnetic field $\boldsymbol{B}$ as the gradient of a scalar potential, V,

$$\boldsymbol{B} = -\text{grad}\, V, \qquad \text{(Eq. 4)}$$

and can then work in terms of this scalar variable V.

The second complication is that the relevant geometry is three-dimensional space rather than a one-dimensional circle. As we are concerned with a roughly spherical Earth, it is convenient to use a spherical polar coordinate system (r, θ, λ) based at the center of the Earth, with $\theta = 0$ being the direction of the north geographic pole, and $\lambda = 0$ the direction of the Greenwich meridian. (The angle λ is the conventional longitude, but because the Earth is a slightly oblate ellipsoid the (geocentric) θ is not quite the same as the (geodetic) colatitude.)

Provided that there is also no significant magnetization in the region of interest, it is convenient to approximate the scalar potential V as a finite sum of "spherical harmonics" (see *Harmonics, spherical*). (In practice, although the rocks of the crust are magnetized, even in the crust they produce a field which is small compared with that of the electric currents in the core, and we can often ignore this minor complication. See the last section below for the relative magnitude of the fields.)

We then have

$$V(r,\theta,\lambda) = \sum\nolimits_{n,m} V_n^m(r,\theta,\lambda) = a \sum\nolimits_{n,m} P_n^m(\cos\theta)(g_n^m \cos m\lambda + h_n^m \sin m\lambda)(a/r)^{n+1} \qquad \text{(Eq. 5)}$$

for that part of the field having sources internal to the region of interest. The $P_n^m(\cos\theta)$ are (scaled versions of) the mathematical functions of θ called Legendre polynomials. The different harmonics are labeled by two integers, the degree n and the order m, with $m \leq n$. (Note that some authors use the symbol l rather than n for the degree.) The factor a, the mean radius of the Earth, is incorporated so that the numerical coefficients g_n^m and h_n^m have the dimensions of magnetic field, conventionally expressed in units of nanotesla.

The particular variation with radius in Eq. (5) is valid only for internal sources, and only in a region free from sources of magnetic field. But on any given spherical surface, whatever the source distribution, a scalar such as V can be represented uniquely as a function of (θ, λ) given by the sum of a set of *surface* harmonics, where any radial variation is incorporated numerically into the coefficients. For simplicity, considering a field of internal origin on the surface $r = a$ (approximately the surface of the Earth), we have

$$V(\theta,\lambda) = \sum_{n,m} V_n^m(\theta,\lambda) = \sum_n V_n(\theta,\lambda) \qquad \text{(Eq. 6)}$$

where

$$V_n^m(\theta,\lambda) = aP_n^m(\cos\theta)(g_n^m \cos m\lambda + h_n^m \sin m\lambda). \qquad \text{(Eq. 7)}$$

We can think of these "surface harmonics," $P_n^m(\cos\theta)\cos m\lambda$ and $P_n^m(\cos\theta)\sin m\lambda$, as being the two-dimensional analogs of the one-dimensional Fourier harmonics on a circle.

For a given physical field, the numerical values of the coefficients, g_n^m and h_n^m will depend on the origin of our coordinate system. If, for example, we used the meridian of Paris, rather than that of Greenwich, to define $\lambda = 0$, we would find that the coefficients would be different. However, for a given pair of g_n^m and h_n^m we find that the sum of their squares, $(g_n^m)^2 + (h_n^m)^2$ is a constant, independent of our choice of the origin $\lambda = 0$ (exactly as in one-dimensional Fourier analysis). More generally, even if we also move the $\theta = 0$ axis to another point on the sphere, it turns out that we have

$$C_n^2 = \sum_m [(g_n^m)^2 + (h_n^m)^2] = \text{constant}, \qquad \text{(Eq. 8)}$$

where the sum is over $m = 0,1,2,\ldots,n$. (This particularly simple result is true provided that we use the Schmidt seminormalized definitions for the Legendre polynomials P_n^m, as is conventional in geomagnetism. It is also true if we use fully normalized P_n^m (as do workers in gravity), but *not* if we use the basic mathematical definitions (sometimes denoted by P_{nm}) which give unnormalized functions, the mean-square of which varies with m as well as with n.) In fact any one of the V_n of Eq. (6) is physically the same, whatever our choice of coordinate system; the processes producing the fields do not "know" of our arbitrary choice of coordinate system.

We can think of all the contributions of a given degree n as having essentially the same minimum (surface) wavelength on the sphere, of value (approximately) the circumference divided by degree n. *Mathematically*, the three terms for $n = 1$ correspond to the potentials given by three dipoles at the center, the $n = 2$ terms to five quadrupoles at the center, and so on, but of course *physically* the sources are distributed current systems.

Just as in Fourier analysis the various harmonics are orthogonal round the circle, these surface harmonics are orthogonal over the spherical surface; the average value over the sphere of the product of two *different* surface harmonics is zero. Now using $<\ldots>$ for the average over the sphere $r = a$, the mean-square value of V over the surface can be expressed as the sum

$$\begin{aligned}\langle V^2\rangle &= \sum_n \langle V_n^2\rangle \text{ where } \langle V_n^2\rangle \\ &= a^2 \sum_m [(g_n^m)^2 + (h_n^m)^2]/(2n+1),\end{aligned} \qquad \text{(Eq. 9)}$$

the factor $1/(2n+1)$ is the mean-square value of each harmonic over the surface for our Schmidt seminormalized P_n^m.

However we are using the scalar potential V only as a convenient mathematical simplification. What we are really interested in is the magnetic field $\boldsymbol{B}$ itself. It turns out that if we represent each harmonic $P_n^m(\cos\theta)\cos m\lambda$ or $P_n^m(\cos\theta)\sin m\lambda$ by (say) W_n^m, and put

$$\boldsymbol{B}_n^m = -\text{grad}W_n^m, \qquad \text{(Eq. 10)}$$

then the vector fields $\boldsymbol{B}_n^m$ are themselves also orthogonal over the sphere: using the scalar product for the multiplication, we have

$$\langle \boldsymbol{B}_n^m \cdot \boldsymbol{B}_\nu^\mu \rangle = 0$$

unless $n = \nu$, $m = \mu$ (and both harmonics use only either $\cos m\lambda$ or $\sin m\lambda$);

$$\langle \boldsymbol{B}_n^m \cdot \boldsymbol{B}_n^m \rangle = (n+1). \qquad \text{(Eq. 11)}$$

The factor $(n + 1)$ is the mean-square value over the sphere of the vector fields corresponding to each of the individual harmonics of the degree n part of the scalar potential V (Lowes, 1966). (This factor $(n + 1)$ is valid only if the sources of the field are internal to the sphere. For external sources, e.g., ionospheric or magnetospheric currents, the factor is n.)

So if $\boldsymbol{B}$ is the observed field (X, Y, Z), where X, Y, Z are Cartesian components of the vector, the mean-square value of the field vector over the sphere $r = a$ becomes

$$< \boldsymbol{B} \cdot \boldsymbol{B} >= |\boldsymbol{B}|^2 = \langle X^2 + Y^2 + Z^2 \rangle = \sum_n R_n \qquad \text{(Eq. 12)}$$

Where

$$R_n = (n+1)\sum_m \left[(g_n^m)^2 + (h_n^m)^2\right]. \qquad \text{(Eq. 13)}$$

If we plot R_n as a function of n, we have the geomagnetic spatial spectrum.

An alternative interpretation of Eq. (13) is that if it is multiplied by $1/2\mu_0$ (and the coefficients are expressed in units of tesla) it gives the stored magnetic energy density $\boldsymbol{B \cdot H}/2$ (the energy stored per unit volume in the magnetic field, in J m^{-3}) averaged over the surface; so this spatial spectrum (of the vector field) is sometimes called the geomagnetic power spectrum, or the geomagnetic energy spectrum. (R_n is roughly analogous to the "degree variance" used in gravity. But note that while R_n is zero for $n = 0$ in geomagnetism, as there are no magnetic monopoles, the $n = 0$ term completely dominates in gravity.)

Results for the geomagnetic field

If we apply Eq. (13) to the results of a spherical harmonic analysis of the geomagnetic field of internal origin, as observed globally by satellites such as Ørsted and CHAMP, and plot R_n on a logarithmic scale against n, we obtain the spectrum of Figure G49. Formally, given observation of the field only at and above the Earth's surface, we cannot tell whereabouts inside the surface are the currents producing the field. But if we look at the plot, we see that to a first approximation it can be represented as the sum of two straight lines (two power-law functions on a linear scale). Although there is argument about the exact shape of the best-fitting curves, it is now accepted that (almost all of) the field from harmonics up to about degree $n = 14$ comes from electric currents in the Earth's core, while (almost all of) the field of higher degrees comes from the magnetization of crustal rocks.

As is well known, at the Earth's surface the geomagnetic field is dominated by its dipole component, and indeed the $n = 1$, dipole, point lies well above the first line. But the line is a good fit for values of n from 2 to about 12. If we go back to the definition of Eq. (5), and remember that taking the gradient in effect introduces another power of (a/r), we see that to draw the equivalent plot for a radius r different from a, we have to multiply each R_n by the factor $(a/r)^{2n+4}$. This has more effect on the points for larger n, so that for $r < a$ the line becomes less steep. In fact, if we go down to the radius of the core-mantle boundary (CMB) the line becomes nearly horizontal, and if we continued the extrapolation a little further we would find that just below the CMB the line becomes horizontal; each degree would make about the same contribution to the total ms field. Strictly, because there will be electric currents in this region, this further extrapolation is not allowable, but this result is a convincing argument that these fields do have their origin in the core.

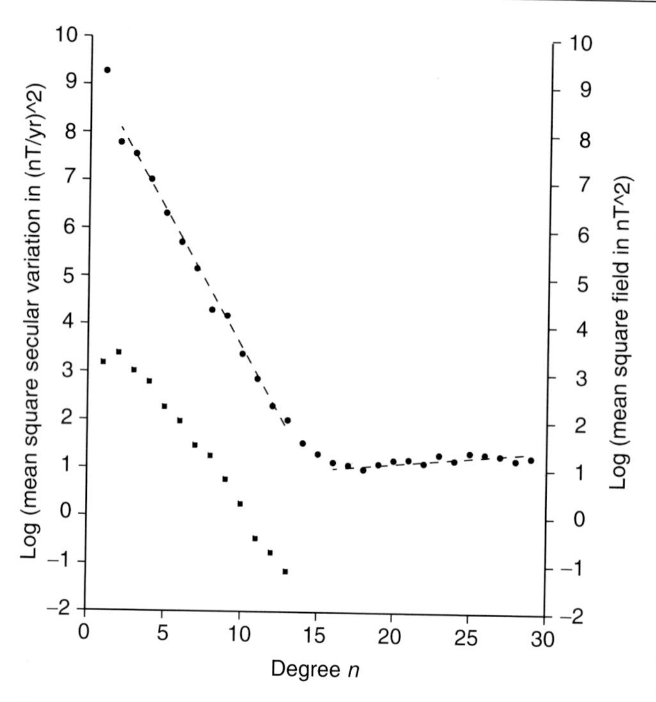

Figure G49 Spatial spectrum of the internal part of the geomagnetic field at 2000.0. The circles are the values of R_n, the mean-square vector field produced by the harmonics of degree n over the surface $r = a$, plotted on a logarithmic scale against n; right-hand scale. The squares are the corresponding values for the secular variation; left-hand scale. The coefficients used are those of CM4, Sabaka *et al.* (2005).

The other line, for higher spatial frequencies/shorter wavelengths, is already sloping upward, so we cannot use such a simple argument to indicate the source depth. The results for even higher frequencies (see below) show that the curve does in fact eventually turn downward, as it must for the crust to contribute only a finite field, and the overall shape is consistent with reasonable models of crustal magnetization.

So there is no doubt that this separation into core and crustal origin must be largely correct; the division is usually put at about $n = 14$. But note that the spectrum of the core field will continue above this value, and that of the crustal field will continue below this value. We do not know the detailed physics that lies behind each of the two spectra, so we can only guess at how they are to be extrapolated. But if we simply extrapolate the straight lines, then we find that at the Earth's surface the core field (of total root-mean-square (rms) magnitude about 45000 nT) would contribute about 10 nT rms from degrees beyond 13, while the crust (total rms about 300 nT) would contribute about 10 nT from degrees below $n = 14$.

While the spectrum of the main field itself is now quite well determined (by measurements from polar-orbiting satellites such as Magsat, Ørsted, and CHAMP), the spectrum of the crustal field (see *Crustal magnetic field*) is much more uncertain. All the points shown on Figure G49 are obtained from satellite measurements. However at satellite altitudes the crustal field is small, and for individual measurements is often not much larger than the instrument noise. Probably more importantly, there are also comparable or larger time-varying fields of external origin, and the separation of the spatial-variation and time-variation of the field seen by a moving satellite is very difficult. In fact, as techniques improve there has been a tendency for the derived crustal spectrum to become lower on a graph such as that of Figure G49! For degrees above about 30–40 it is at present difficult to do a formal spherical harmonic analysis of the satellite data, though it is sometimes possible to do a Fourier, one-dimensional, analysis along individual tracks. Similarly we can use the long tracks of the Project MAGNET airborne vector magnetometer (see *Project MAGNET*), and a much larger number of tracks from oceanographic vessels towing a scalar magnetometer; these approaches can extend the data out to about $n = 1000$. However these are only one-dimensional cross sections of the field variation over the two-dimensional sphere, so some assumptions have to be made (see, e.g., Korte *et al*., 2002); also, the near-surface tracks are almost all over the oceans, where the crust is systematically thinner than under the continents. Such results however do suggest that the mean-square field per degree does in fact start to decrease at higher degrees. In theory, for the very high degrees, we could analyze data from the large number of detailed aeromagnetic surveys which have been carried out (though mostly over land), but there are many problems, and this has not yet been attempted.

This concept of a spatial spectrum can obviously be extended to the secular variation, the time variation of the main field. A complication is that there are many external sources of field variations, of period of a year or less, the effects of which are most easily removed by taking averages over one year. This averaging is most easily done with the long-term surface observations at magnetic observatories and repeat stations; unfortunately these have only a very poor global coverage. Satellites have much better global coverage, and are now also giving continuous coverage in time, though unfortunately their measurements are subject to more interfering effects. However, at present it looks as though we can estimate the secular variation over the last year or so, and hence its spectrum, up to about $n = 10$ (see, e.g., Langlais *et al.*, 2003). The spectrum for 2000.0 is shown on Figure G49; it is much flatter than that of the main field, so that (in this range) at the CMB the shorter wavelengths are varying more rapidly than the longer wavelengths.

History

Lowes (1966) was the first to introduce to the English-speaking community the expression for mean-square field, Eq. (11) of this article, in the context of geomagnetism; he also produced the first power spectrum, analogous to Figure G49, in Lowes (1974). However, essentially the same expression had been introduced earlier in Germany, by Lucke in a colloquium in 1955, referred to by Fanselau and Lucke (1956). Lucke worked in terms of the energy per unit volume stored in a magnetic field, which (in modern notation) is

$$E = (1/2)\boldsymbol{B} \cdot \boldsymbol{H} = (1/2\mu_0)\boldsymbol{B} \cdot \boldsymbol{B}. \qquad \text{(Eq. 14)}$$

So, when he averaged over the sphere $r = a$, he obtained

$$\begin{aligned}\langle E\rangle &= (1/2\mu_0)\langle \boldsymbol{B}.\boldsymbol{B}\rangle = (1/2\mu_0)\sum\nolimits_n (n+1)\left[\left(g_n^m\right)^2+\left(h_n^m\right)^2\right]\\ &= (1/2\pi\mu_0)\sum\nolimits_n R_n,\end{aligned} \qquad \text{(Eq. 15)}$$

or

$$\langle E\rangle = (1/2\mu_0)\sum\nolimits_n R_n, \qquad \text{(Eq. 16)}$$

in the notation of this article.

Lucke's lecture was also referred to by Mauersberger (1956), who produced the same expression for the mean energy density, but using a different derivation. Hence graphs such as those of Figure G49 are sometimes referred to as a "Lowes spectrum," "Lowes-Mauersberger spectrum," or "Mauersberger-Lowes spectrum."

Frank Lowes

Bibliography

Fanselau, von G., and Lucke, O., 1956. Über die Veränderlichkeit des erdmagnetischen Hauptfeldes und seine Theorien. *Zeitshrift für Geophysik*, **22**: 121–216.

Korte, K., Constable, C.G., and Parker, R.L., 2002. Revised magnetic power spectrum of the oceanic crust. *Journal of Geophysical Research*, **107**(B9): doi:10.1029/2001JB1389.

Langlais, B., Mandea, M., and Ultrée-Guérard, P., 2003. High-resolution magnetic field modeling: application to MAGSAT and Ørsted data. *Physics of the Earth and Planetary Interiors*, **135**: 77–79.

Lowes, F.J., 1966. Mean-square values on sphere of spherical harmonic vector fields. *Journal of Geophysical Research*, **71**: 2179.

Lowes, F.J., 1974. Spatial power spectrum of the main geomagnetic field, and extrapolation to the core. *Geophysical Journal of the Royal Astronomical Society*, **36**: 717–730.

Mauersberger, P., 1956. Das Mittel der Energiedichte des geomagnetischen Hauptfeldes an der Erdoberfläche und seine säkulare Änderung, *Gerlands Beitrage Geophysik*, **65**: 207–215.

Sabaka, T.J., Olsen, N., and Purucker, M., 2005. Extending comprehensive models of the Earth's magnetic field with Ørsted and CHAMP data. *Geophysical Journal International*, **159**: 521–547.

Cross references

Crustal Magnetic Field
Harmonics, Spherical
IGRF, International Geomagnetic Reference Field
Main Field Modeling
Nondipole Field
Project Magnet

GEOMAGNETIC SPECTRUM, TEMPORAL

The geomagnetic field varies on a huge range of timescales, and one way to study these variations is by analyzing how changes in the geomagnetic field are distributed as a function of frequency. This can be done by estimating the spectrum of geomagnetic variations. The power spectral density $S(f)$ is a measure of the power in geomagnetic field variations at frequency f. When integrated over all frequencies it measures the total variance in the geomagnetic field.

Figure G50/Plate 2 shows a schematic of the various processes that contribute to the geomagnetic field, and these can be roughly divided according to the frequency range in which they operate. The bulk of Earth's magnetic field is generated in the liquid outer core, where fluid flow is influenced by Earth rotation and the geometry of the inner core. Core flow produces a secular variation in the magnetic field, which propagates upward through the relatively electrically insulating mantle and crust. Short-term changes in core field are attenuated by their passage through the mantle so that at periods less than a few months most of the changes are of external origin. The crust makes a small static contribution to the overall field, which only changes detectably on geological timescales making an insignificant contribution to the long-period spectrum. Above the insulating atmosphere the electrically conductive *ionosphere* (*q.v.*) supports Sq currents with a diurnal variation as a result of dayside solar heating. Lightning generates high-frequency Schumann resonances in the Earth/ionosphere cavity. Outside the solid Earth the *magnetosphere* (*q.v.*), the manifestation of the core dynamo, is deformed and modulated by the solar wind, compressed on the dayside and elongated on the nightside. At a distance of about 3 earth radii, the magnetospheric ring current acts to oppose the main field and is also modulated by solar activity. Although changes in solar activity probably occur on all timescales the associated magnetic variations are much smaller than the changes in the core field at long periods, and only make a very minor contribution to the power spectrum.

With an adequate physical theory to describe each of the above processes one could predict the power in geomagnetic field variations as a function of frequency. The inverse problem is to use direct observations or paleomagnetic measurements of the geomagnetic field to estimate the power spectrum. Power spectral estimation is usually carried out using variants of one of the following well-known techniques: (1) direct spectral estimation based on the fast Fourier transform, and using extensions and improvements to the time-honored periodogram method introduced by Schuster in 1898 to search for hidden periodicities in meteorological data; (2) the autocovariance method which exploits the Fourier transform relationship between the time-domain autocovariance and frequency-dependent power spectral density of a process; (3) parametric modeling schemes like the maximum entropy method based on a discrete autoregressive process. The relative merits of these techniques have been widely discussed (see Constable and Johnson, 2005: or Barton, 1983, for the geomagnetic context), while all have been used in analyzes of geomagnetic intensity and directional variations. It is generally acknowledged (e.g., Percival and Walden, 1993) that direct spectral estimation combined with tapering and averaging of nearly independent spectral estimates can provide high-resolution estimates and be used to optimize the unavoidable trade-offs between variance and bias.

The parameter often chosen to represent the geomagnetic spectrum is the field strength at midlatitudes, or a proxy form for times where it is not possible to obtain a direct measurement. This is the case for paleomagnetic time series derived from lacustrine or marine sediments which provide only directional information and/or relative intensity variations. Barton (1982) combined spectra from lake sediment directional paleomagnetic records with those from full vector data recorded at magnetic observatories and used periodogram analysis in the first attempt to provide an integrated power spectrum for periods ranging from less than a year to 10^5 years. Courtillot and Le Mouël (1988) merged Barton's result with other spectral estimates at longer and shorter periods extending the timescales from seconds to millions of years. They debated whether the result was compatible with an overall $1/f^2$ spectrum, and concluded that it was too early to make such an inference.

A recent version of such a composite spectrum (Constable and Constable, 2004) uses spectral estimates from relative paleointensity variations (Constable *et al.*, 1998) at long periods and is shown in Figure G51 (note that this is an amplitude rather than a power spectrum, that is the square root of power spectral density). Between 10^{-10} and 1 Hz, the spectrum is from Filloux (1987). Above 1 Hz, the results are those of Nichols *et al.* (1988). Internal variations reflecting motions of the fluid core dominate at periods longer than a few months, and the spectrum generally rises toward longer period (low frequency) with *reversals* (*q.v.*) of the dipolar part of the field the dominant influence on 10^5 to 10^6 year timescales. The 11-year sunspot cycle, solar rotation, and Earth's orbit modulate the distortions of the field associated with geomagnetic storms, which themselves have energy in the several hour to several second band. Energy at the daily variation and harmonics comes from diurnal heating of the ionosphere. Lightning creates high frequency energy in the Earth/ionosphere cavity, which resonates at 7–8 s and associated harmonics. At the highest frequencies there is presumed to be a continued fall-off in the natural spectrum. The upturn seen in Figure G51 reflects the dominant influence of human-made sources.

The spectrum in Figure G51 lacks information at very long periods where the spectrum is dominated by the intensity variations associated with changing geomagnetic reversal rate, and between about 10 ka and 10-year periods, where the timescales for processes of internal and external origin overlap. Figure G52/Plate 4c provides a range of estimates at centennial to 50 Ma periods for the spectrum of the geomagnetic *dipole moment* (*q.v.*). The longest period power spectrum is estimated from reversal times given by the magnetostratigraphic

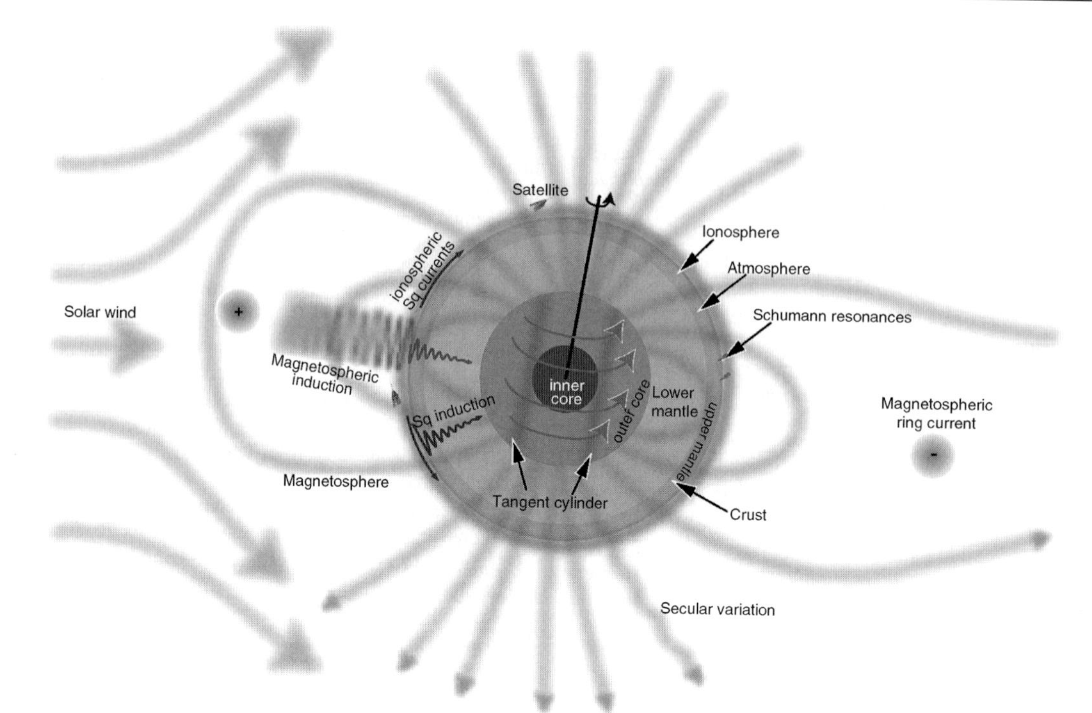

Figure G50/Plate 2 Schematic illustration of the physical processes that contribute to the geomagnetic field, from Plate 1 of Constable and Constable (2004), Copyright American Geophysical Union and reproduced with their and authors' permission.

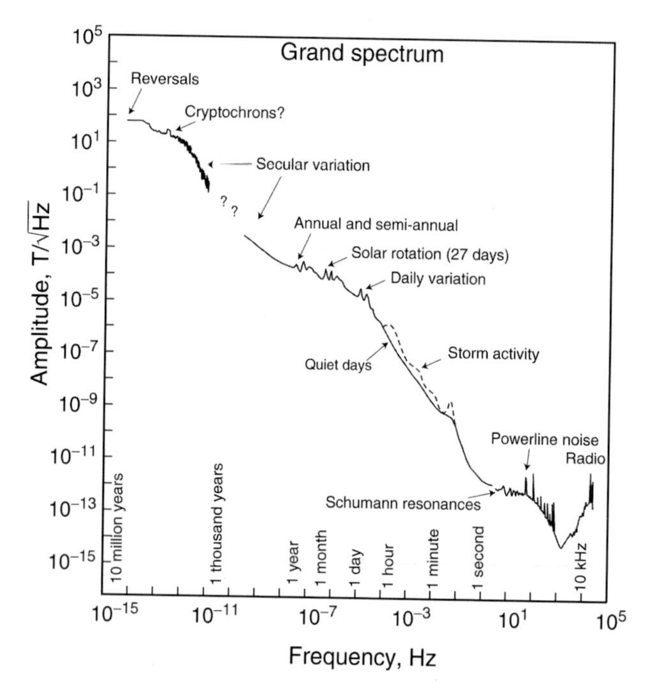

Figure G51 Composite amplitude spectrum of geomagnetic variations as a function of frequency (Constable and Constable, 2004): annotations indicate the predominant physical processes at the various timescales. Copyright American Geophysical Union, 2004, reproduced with their and the authors' permission.

timescale and shown in black. Further information comes from various sedimentary relative paleointensity records with varying accumulation rates and the dipole moment estimate of a time varying global paleomagnetic field model for the past 7 ka.

In constructing a paleomagnetic power spectrum like that in Figure G52/Plate 4c there are a number of challenges. A basic requirement for spectral analysis is a time series of observations, but there is no single record that covers the time span of interest. The magnetostratigraphic record appears to be nonstationary with long-term changes in reversal rate, and provides no information about intensity variation on long timescales. The relative paleointensity records from sediments not only lack an absolute scale, but are usually unevenly sampled in time so that some stable calibration and interpolation scheme is required before using the standard analysis techniques. It is likely that some sediments record a smoothed version of the geomagnetic signal because of low sedimentation rates, while in others it may be necessary to consider the possibility of aliasing. Nongeomagnetic signals may be inadvertently interpreted as arising from geomagnetic variations with time.

The choice of dipole moment (as opposed to some other geomagnetic field parameter) is motivated in large part by the dominance of the geocentric axial dipole when the field is averaged over long time intervals: the strength of the axial dipole is representative of the global field, and may be related to the amount of energy required by the geodynamo or to the geomagnetic reversal rate. Although it is possible that other properties of the field (such as nondipole field contributions) directly reflect particular physical processes controlling the secular variation, resolving such variations in paleomagnetic time series remains controversial. Overall, it is likely that the power in geomagnetic field variations is underestimated.

There is substantial scope for improving the spectra in both Figures G51 and G52/Plate 4c. Relative intensity records from sediments are steadily improving, and new modeling techniques may extend time-varying geomagnetic field models providing better dipole moment estimates on million year timescales. This may give new insight into what controls very long-period secular variation. Many newer observations could be used for direct spectral estimates to replace the current schematic spectrum at periods from decades down to 1 s. Although the general form of the spectrum is quite well understood in this region, detailed analyzes could provide further insight

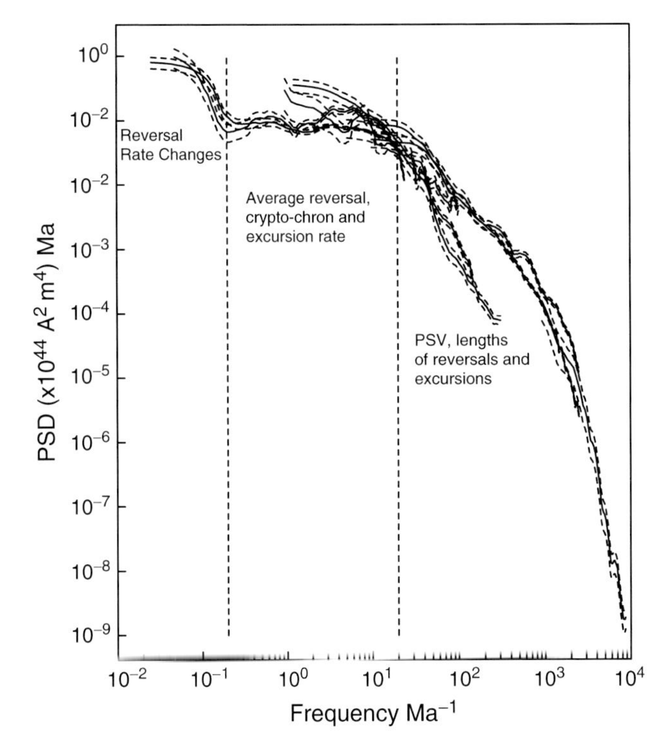

Figure G52/Plate 4c Composite spectrum for the geomagnetic dipole moment constructed from the magnetostratigraphic reversal record with and without cryptochrons (frequencies 10^{-2}–20 Ma^{-1}), various sedimentary records of relative paleointensity (10^0–10^3 Ma^{-1}), and from the dipole moment of a 0–7 ka paleomagnetic field model (10^3–10^4 Ma^{-1}). Figure redrawn from Constable and Johnson (2005).

into the underlying physical processes. Techniques such as wavelet analysis, that can take account of nonstationarity in the underlying geomagnetic processes, have yet to be fully exploited for geomagnetic data (Guyodo *et al.*, 2000) and may prove useful. However, despite the relatively crude nature of existing spectral estimates it is apparent that the geomagnetic power spectrum does not follow a simple power law fall-off with increasing frequency. The form is instead influenced by the characteristic timescales that reflect the distinct physical processes contributing to the geomagnetic field.

Catherine Constable

Bibliography

Barton, C.E., 1982. Spectral analysis of palaeomagnetic time series and the geomagnetic spectrum. *Philosophical Transactions of the Royal Society of London A*, **306**: 203–209.

Barton, C.E., 1983. Analysis of paleomagnetic time series-techniques and applications. *Geophysical Survey*, **5**: 335–368.

Constable, C.G., and Constable, S.C., 2004. Satellite magnetic field measurements: applications in studying the deep earth. In Sparks, R.S.J., and Hawkesworth, C.J., (eds.), *The State of the Planet: Frontiers and Challenges in Geophysics*. Washington, DC: American Geophysical Union, doi: 10.1029/150GM13, pp. 147–160.

Constable, C.G., and Johnson, C.L., 2005. A paleomagnetic power spectrum. *Physics of the Earth and Planetary Interiors*, **153**: 61–63.

Constable, C.G., Tauxe, L., and Parker, R.L., 1998. Analysis of 11 Ma of geomagnetic intensity variation. *Journal of Geophysical Research*, **103**: 17735–17748.

Courtillot, V., and Le Mouël, J.-L., 1988. Time variations of the Earth's magnetic field: from daily to secular. *Annual Review of Earth and Planetary Science*, **16**: 389–476.

Filloux, J.H., 1987. Instrumentation and experimental methods for oceanic studies. In Jacobs, J.A. (ed.), *Geomagnetism*. London: Academic Press, pp. 143–248.

Guyodo, Y., Gaillot, P., and Channell, J.E.T., 2000. Wavelet analysis of relative geomagnetic paleointensity at ODP Site 983. *Earth and Planetary Science Letters*, **184**: 109–123.

Nichols, E.A., Morrison, H.F., and Clarke, J., 1988. Signals and noise in measurements of low-frequency geomagnetic fields. *Journal of Geophysical Research*, **93**: 13743–13754.

Percival, D.B., and Walden, A.T., 1993. *Spectral Analysis for Physical Applications: Multitaper and Conventional Univariate Techniques*. Cambridge: Cambridge University Press.

Cross-references

Dipole Moment Variation
Harmonics, Spherical
Ionosphere
Magnetosphere of the Earth
Mantle
Nondipole Field
Reversals, Theory
Secular Variation Model

GEOMAGNETISM, HISTORY OF

In its present form, the geophysical discipline of geomagnetism is of relatively recent origin (the term "geomagnetism" was coined in 1938 by *Sydney Chapman*, (*q.v.*)), yet interpretations of the Earth's magnetic field have been propounded by scholars from the Middle Ages onward, in the broader context of cosmologies, natural philosophies, and oceanic navigation. In this historical overview, geomagnetism will be considered as the scientific study of the Earth's internal field and its secular change, encompassing both descriptive and causal hypotheses; paleomagnetism and the external field are treated elsewhere. Furthermore, since most of this volume is dedicated to current issues, stress will here be placed on earlier times.

Main developmental stages

The history of geomagnetism can be subdivided into three main periods: firstly, a proto-scientific stage (up to the 16th century), during which awareness slowly grew of the existence of a global magnetic property worthy of investigation. At this time, causal hypotheses almost exclusively identified the heavens as the seat of magnetic attraction. Secondly, an early-modern stage took place (16th to early 19th century), during which directional data (magnetic declination and inclination) were increasingly measured, compiled, and mapped. These efforts led to the discovery of secular variation and causal models involving one to three crustal or nuclear dipoles (terrestrial polar attraction). Thirdly, a modern stage emerged (from the 1830s), characterized by measurement of the full magnetic vector (direction and intensity) in dedicated scientific surveys, observatories, and satellites; by the description of the field as a whole, both at the surface and at the top of the source region; and by the mid-20th century introduction of *geodynamo theory* (*q.v.*), leading to a multitude of numerical and laboratory magnetohydrodynamic simulations.

These developments are summarized per century in Table G8, which pays separate attention to empirical (data and mapping) and theoretical aspects (hypotheses). As is apparent, the main watershed between the early-modern and modern stage pervades both. The most influential scientist in bringing about this change was *Carl Friedrich Gauss*

Table G8 Empirical and theoretical stages in geomagnetism

Century	Data	Mapping	Hypotheses
16th	Maritime	D; interpolation	Polar attraction (celestial, crustal)
17th	Maritime; site series	D; interpolation	Polar attraction (crustal, core)
18th	Maritime; site series; surveys	D, I; interpolation	Polar attraction (core)
19th	Surveys; observatories	D, I, H, F; spherical harmonics (surface)	Geomagnetic field
20th	Surveys, repeat stations, Observatories, satellites	All components; spherical harmonics (surface, CMB)	Geodynamo (fluid outer core)

Notes: D, declination; I, inclination; H, horizontal intensity; F, total intensity; CMB, core-mantle boundary.

(1777–1855) (*q.v.*), who not only invented the first absolute magnetometer (1832) but, moreover, developed for geomagnetic application the mathematical techniques of spherical harmonic expansion (1839) and least-squares analysis (independently discovered by Adrien-Marie Legendre, 1752–1833). Furthermore, around the same time occurred a shift in the realm of data acquisition, up till then performed mostly by navigational practitioners and hydrographers, thereafter predominantly by professional astronomers and physicists. This article will therefore apply the same temporal caesura in discussing geomagnetic data processing and theory formation before, and after Gauss's contributions.

The era of polar attraction

Although the phenomenon of magnetic attraction had been known since antiquity, it was in the Middle Ages that the notions of polarity and orientation were first discovered, and several cultures recorded its potential (or actual) utility. The Chinese record goes back furthest in this respect, with magnetized objects being employed in geomancy from at least the A.D. 6th century, and a wire-suspended magnetic needle first described in 1096. Moreover, in the next few decades appeared the first Chinese references to mariners relying for direction on needles made to float in a bowl of water (Needham, 1962), a technique independently discovered in Europe. By the 13th century, the magnetic needle was commonly relied upon in both Asian and European navigation at sea, in particular when overcast skies made an astronomical fix impossible. In subsequent centuries, the dry-pivoted *compass* (*q.v.*) was moreover increasingly relied upon on land, in surveying and to meridionally align portable instruments and some newly planned official buildings (churches, temples, and palaces). The difference between magnetic and true north was as of yet unappreciated; in keeping with Aristotelian cosmology, the needle was still thought to respect the imagined immutable perfection of the supralunary spheres (Smith, 1992).

This view is epitomized in the "Epistle on the Magnet" (1269) in which engineer Pierre de Maricourt (Petrus Peregrinus) related his experiments with a spherical lodestone. This *magnes rotundus* represented the firmament, its two magnetic poles coinciding with the celestial poles (the only fixed points in a diurnally rotating field of stars). Given this arrangement, magnetic and true meridian would always coincide (declination was zero everywhere), and inclination would change as a function of geographical latitude. The concept of a celestial axial dipole was reiterated over the next two centuries by many medieval and Renaissance authors. Some, however, attributed the magnetic force to the nearby Polestar, relying on the Classical concept of sympathy to establish an occult bond between compass and star. These tenets would be challenged when magnetic declination was finally acknowledged as a real phenomenon, rather than being ascribed to an error in instrument or measurement.

The first tacit evidence thereof can be found on German portable sundials from the second quarter of the 15th century. Craftsmen in Augsburg and Nuremberg then started to mark local declination in the base of the instrument, adjacent to a small inset compass, used to enable proper orientation in the geographical meridian. Around that time, similar markings also started to appear next to the wind rose on some German maps (e.g., by cartographers Etzlaub, Waldseemüller, Ziegler, and Murer). But it would take another century before the spatial variability of the Earth's magnetic field became overwhelmingly clear, and with it the need for global descriptive and causal geomagnetic hypotheses.

The 16th century

The advent of oceanic navigation, without the benefit of land sightings, soundings, or a practical method with which to measure longitude, meant that mariners had to rely on compass and astronomical observations to a hitherto unequalled degree. Off the continental shelves, steering, dead reckoning, log keeping, charting, and sailing instructions consequently came to incorporate true compass directions. Any local magnetic declination had to be measured regularly, and compensated for in courses and calculations. Fortunately, establishing this so-called compass allowance was a fairly simple calculation, based on sighting the Sun or other stars at rise and/or set. As Portuguese and Spanish explorers traversed increasing parts of the Atlantic and Indian Oceans, the peculiar spatial distribution of needle behavior was slowly recorded, collated, and compared, both by navigators at sea and hydrographers at home. In large parts of the Atlantic, the compass northeasted, whereas in the Indian Ocean, variable northwesting was the rule; near-zero declination was furthermore associated with the mid-Atlantic, South Africa, Southeast Asia, and Middle America.

These early data sets seemed to suggest that the Earth's dipole was tilted relative to the Earth's rotation axis, since local declination was assumed to have a direct bearing on the position of the dipole. By applying spherical trigonometry, two observations of declination at places far apart in longitude could provide a cross bearing of two great circles, each through one of the points and at the given angle to true north. Where the two circles intersected, the geomagnetic poles were supposed to lie. Once the position of the dipole was known, a similar calculation at sea would yield the longitude of a place where declination and latitude were measured.

In other words, given a tilted dipole at a fixed position, an observer traveling around the globe at the same latitude would see his compass needle diverge from true north as a function of longitude. At the meridian of the dipole's longitude and on the antimeridian (180° E) the great circle through the place of the observer and the dipole would cut the geographical pole as well, resulting in zero degrees declination; deflection from true north would reach a maximum near 90° longitude east and west from either meridian. Actual measurements of zero declination near the Azores led to the mistaken belief that such a meridional agonic line was found, forming a "natural" indicator of a prime meridian to reckon longitude from. Geomagnetic considerations have thereby greatly affected early-modern cartography. The combination of reigning northeasting in Western Europe and northwesting on the American east coast furthermore led to the conclusion that the dipole was situated on longitude 180° E (of the Azorean prime meridian).

The notion of the tilted dipole neatly concurred with a more ancient conjecture, that of the "magnetic mountain." This lodestone rock, mountain, or island, often located in Arctic regions, was not only supposed to affect compass needles all over the globe, but, according to some legends, could even draw, capture, or destroy iron-bearing vessels that sailed too close by (Balmer, 1956). Several scholars have repeatedly tried to calculate its exact coordinates, among them cartographer Gerard Mercator (1512–1594) and astronomer Johannes Kepler (1571–1631). Others instead interpreted the postulated tilted dipole as evidence that the points of magnetic attraction were identical with the poles of the ecliptic. Nevertheless, by this time, the majority opinion tended to favor magnetic poles on the Earth's surface.

This shift from celestial to terrestrial magnetism was reinforced by the discovery of magnetic inclination. In a qualitative sense, it was made by German mathematician and instrument maker Georg Hartmann (reported in a letter to his patron in 1544). It would take until about 1580 before London compass maker Robert Norman constructed the first inclinometer (a magnetized needle rotating in the vertical plane on a horizontal axis), with which the first quantified readings were taken. Inclination too was briefly considered as a navigational aid; in the 1590s, English mathematician Henry Briggs (1561–1630) assumed an axial dipole and produced a table of dip for each degree of latitude. Its lasting impact, however, was to strengthen the growing conviction that the origin of the Earth's field was located deep inside the planet, instead of on, or above its surface.

The consideration of additional sites of observed zero declination meanwhile led Iberian and Dutch cosmographers to postulate a more complex arrangement of two tilted dipoles, resulting in two perpendicular great circles that quartered the world into alternating sections of northeasting and northwesting. In the Dutch Republic, cartographer Petrus Plancius (1552–1622) turned such a hypothesis into a practical method to find longitude, which was used at sea for about three decades. His compatriot Simon Stevin (1548–1620) instead put forward the first sextupole hypothesis. Other schemes developed by European navigators proposed a locally valid ratio between distance traveled along a parallel of latitude and change in needle stance.

The striking diversity of explanations formulated during the 16th century bears evidence, not just of the numerical paucity of geomagnetic data, but also of poor charts (longitudinal uncertainty) and lack of standardization in instruments and measurement, generating error margins large enough to tentatively confirm the witnessed variety of (one or multiple) tilted dipole arrangements and other explanations.

The 17th century

With the arrival of overseas trading companies, missionary networks, and better instruments, both the quantity and quality of geomagnetic data substantially increased in the 17th century. Moreover, this period witnessed several official efforts to process and publish these measurements, such as by the newly founded scientific societies (Royal Society, Académie Royale des Sciences), and the hydrographic office of some East India Companies. Observations of declination and inclination also appeared in printed compilations, sometimes ordered by latitude or longitude, both for mariners' and scholars' benefit. In addition, the first manuscript chart (now lost) depicting curved isogonics (lines connecting all points with equal declination) was produced in the 1620s by the Italian Jesuit Christoforo Borro (or Bruno), a teacher of navigation in Portugal. Although isogonic charts would never attain prominence as a tool in navigational practice, they did serve to underline that geomagnetism implied more than a collection of isolated data points, enabling global visualization of both *a priori* and empirically derived patterns. The continuing stream of readings from places near and far furthermore stressed the inadequacy and simplification of imposing a tilted dipole model on observed reality.

The nascent geomagnetic discipline therefore avidly sought more appropriate explanations, while being increasingly influenced by competing paradigms of natural philosophy. The stage was set with the publication of *De Magnete* in 1600, by Elizabethan court physician *William Gilbert* (1544–1603) (*q.v.*). This work contained the famous conclusion that "the Earth itself is a great magnet." Gilbert posited an axial dipole field distorted by the attraction of iron-rich continental landmasses (and more localized sources); a single Arctic magnetic pole was thus supplemented by global crustal heterogeneity, dispensing with any perceived regularity. Regarding cosmology, the Englishman maintained that the geomagnetic force was likewise responsible for the Earth's orientation and diurnal rotation, and the orbit of other planets, a view that brought him into conflict with the religious order of the *Jesuits* (*q.v.*). To the likes of Nicolo Cabeo (1629), Athanasius Kircher (1639), and Jacques Grandamy (1645), who supported geocentrism, Gilbert's universal "magnetic philosophy" was anathema; they claimed instead that it was geomagnetism which held the Earth firmly fixed at the center of creation. However, upon closer examination, Jesuit interpretations of the Earth's (seemingly irregular) field proved far from consistent themselves, attributing its source to quasiorganic magnetic "fibers," mines of iron ore, subterranean heat, chemical processes, and the Earth's heterogeneous crustal constitution (Daujat, 1945). Outside of religious circles, the composition of the deep Earth also evoked much speculation and debate; Galileo Galilei postulated intense pressures at the core, while others imagined an infernal furnace there, driving hot sulfurous gases through a huge system of caverns. The notion of a hollow Earth would become even more important after the discovery of secular variation.

The first sustained series of measurements at a single site, which gave rise to the realization that the geomagnetic field was subject to time-dependent change, was compiled in east London (foremost by naval commander William Borough, instrument maker John Marr, and Gresham astronomers Edmund Gunter and Henry Gellibrand. It was *Gellibrand* (*q.v.*) who eventually published all findings in 1635, concluding "a sensible diminution" at London of about seven degrees westerly over the period 1580–1634 (Chapman and Bartels, 1940). It was one of the earliest scientific conclusions based on averaging sets of measurements, a procedure through which observational error was substantially reduced.

The notion of geomagnetic change over time was accepted by scholars across Europe within two decades, forcing a reassessment of all earlier work. Compiled undated measurements became useless, as did all time-invariant geomagnetic hypotheses put forward so far. The acceptance of secular variation thus inaugurated the third phase of geomagnetic models, which introduced a temporal parameter. Nonetheless, this variable was not necessarily always quantified. Kircher, for example, interpreted time-variance as resulting from the slow "generation" and "deterioration" of iron mines in the crust, a view shared by the philosopher René Descartes (1596–1650), who additionally blamed atmospheric circulation as affecting the vortices of tiny magnetic particles he postulated as the essence of the magnetic force (1644). A more atomistic (but equally qualitative) micromechanistic hypothesis was formulated by his compatriot Pierre Gassendi (1592–1655).

Once (geo)magnetic corpuscularism reached English soil, research into magnetic particle circulation was initially promoted by the Royal Society, but by the 1680s, "magnetic philosophy" lost much of its former identity, becoming subsumed in broader theories involving the ether and effluvial emanations of matter in general. Although these speculations would eventually lead to the formulation of the laws of electromagnetism, none of these concepts made much impact on English geomagnetic hypotheses throughout the 17th century. The path followed here led scholars instead to reinstate the notion of polar attraction in dynamic terms, through one or more dipoles precessing around the Earth's rotation axis in hundreds of years. Such solutions were launched by Henry Bond (1639), Henry Phillippes (1659), Robert Hooke (1674), Peter Perkins (1680), *Edmond Halley* (1683 and 1692) (*q.v.*), and Edward Harrison (1696). In addition to an array of parameters to define these geomagnetic models (precessional period and direction, dipole colatitude and longitude at a given year), a physical rationale was added by some, for example, Halley's magnetic core rotating relative to the crust, separated from the latter by a fluid or gaseous intermediary layer. Even more parameters were required in the 18th century multipole variant (several hollow spheres nested around a common kernel), which allowed dipoles at different depths to precess with distinct speeds. This greatly expanded the scope for accommodating the observed asymmetrical global distribution of surface declination, as visualized on the first printed isogonic charts. These were based on the first scientific naval surveys, by Edmond Halley in 1698–1700.

The 18th century

Halley's declination charts were revised twice, by mathematics teachers William Mountaine and James Dodson (1744, 1756). Isogonics were likewise drawn on near-global-scale by engineer Frezier (1717), physicist Van Musschenbroeck (1729, 1744), Captain Nicolaas van Ewyk (1752), cartographer Samuel Dunn (1775), natural philosopher Johann Heinrich Lambert (1777), and astronomer Le Monnier (1778). Physicist Johann Karl Wilcke published the first world chart depicting isoclinics (lines of equal dip) in 1768, based on inclinations observed on board Swedish Eastindiamen. In the last decade of the century, surveyor John Churchman moreover published several editions of his "Magnetic Atlas," which delineated both isogonics and isoclinics worldwide.

These efforts were based on unprecedented data compilations, mostly derived from maritime sources. Mountaine and Dodson's, for instance, used over 50000 geomagnetic observations for their 1756 world chart. More accurate compasses with enlarged sights, a greater volume of world shipping than ever before, and improved infrastructures for processing and disseminating magnetic data, allowed increased data accuracy at higher resolution over larger parts of the globe. Strenuous regimes of daily or hourly observations at fixed locations furthermore offered workers the chance to explore diurnal variation, secular acceleration, and the link between erratic needle behavior and the *aurora polaris*. In this century, major cities such as London and Paris maintained almost uninterrupted series of annual observations at astronomical observatories.

Eighteenth century geomagnetic hypotheses displayed the full gamut of existing explanations, with proponents of static and dynamic tilted dipoles and multipoles in numerous countries. Nevertheless, data and theory still continued to disagree to an appreciable extent. In 1732, Royal Society Fellow Servington Savery was the first to propose that irregular surface topography of the magnetic core might be to blame. Mathematical work by Leonhard Euler (1757) furthermore laid the groundwork for the abandonment of the assumption of pairs of poles in diametrical opposition (Fleury Mottelay, 1922). Implicitly, some earlier multipole solutions had featured nonantipodal configurations, but only after Euler did the fourth phase of geomagnetic hypothesis receive explicit attention, with detailed discussion of each pole's position and movements inside the Earth. In the most extreme cases (from the 1790s), the two disjointed poles of a dipole were assigned different coordinates, directions, and velocities. And like in earlier phases, when a single dipole was no longer deemed sufficient to account for the observations, another one was added. The disjointed quadrupole solution (1819) developed by Norwegian mathematician Christopher Hansteen (1784–1873) is a case in point, incorporating compass readings from 74 sea voyages. But eventually, even four independent magnetic poles proved insufficient, heralding the end of the belief that magnetized needles were solely governed by the attraction of a few, distant, all-powerful geomagnetic poles. The four main phases of geomagnetic hypotheses up till then are illustrated in Figure G53.

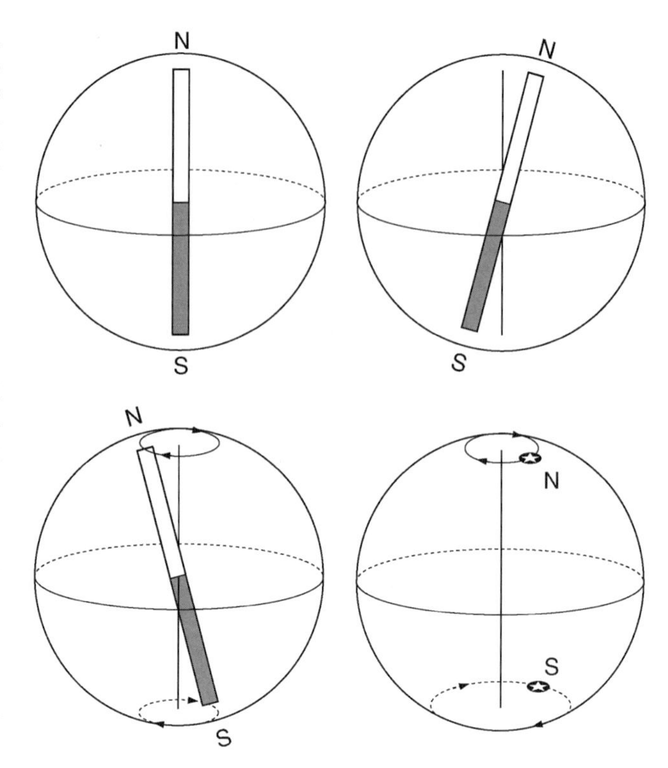

Figure G53 The four phases of geomagnetic hypotheses based on polar attraction. *Top left,* axial dipole (static); *top right,* tilted dipole (static); *bottom left,* precessing dipole (dynamic, tilted, antipodal); *bottom right,* disjointed dipole (dynamic, tilted, nonantipodal). (From Jonkers, A.R.T., *Earth's Magnetism in the Age of Sail,* p. 36, Fig. 2.2. © 2003 Johns Hopkins University Press. Reprinted with permission of The Johns Hopkins University Press.)

The era of the global field

The 19th century

Relative geomagnetic intensity data began to be compiled in earnest from the 1790s, by comparing the time it took a magnetized needle (horizontally or vertically) displaced by a standard distance from its preferred orientation to return to it, or by timing a given number of such oscillations. Admiral de Rossel, on d'Entrecasteaux's voyage in search of the lost expedition of La Pérouse, obtained such readings (1791–1794) relative to Brest. Naturalist Alexander von Humboldt (1768–1859), on his south American explorations (1799–1804), instead chose the number of dip oscillations in 10 min on the magnetic equator (where vertical geomagnetic intensity is lowest) at Micuipampa, Peru as his reference value. Other readings obtained at intermediate latitudes led him to postulate the "law" of decreasing (vertical) magnetic force from the magnetic poles to the magnetic equator. After Humboldt's return to Paris, this initial standard of relative intensity soon became eclipsed by the Paris unit, whereas similar local units based on swing times at London, Christiania (Oslo), and St. Petersburg found more limited application elsewhere. Global isodynamics (lines of equal intensity) were drawn from 1825.

Two periods in the 19th century stand out as of particular significance for geomagnetism: the first of these was 1820–1840. After Hans Christian Ørsted's (1777–1851) discovery in 1820 of electromagnetism, and André Marie Ampère's (1775–1836) "Theorie Mathématique des Phénomènes Électro-dynamiques" (1826), it was Michael Faraday (1791–1867) who discovered electromagnetic induction in 1831, and who applied it to build the first dynamo (a copper disk rotating in a strong magnetic field, with an electrical circuit connecting rim and center). The year thereafter, Carl Friedrich Gauss built the first absolute magnetometer in his Göttingen laboratory (21 May 1832). Units of absolute intensity soon replaced all relative scales, and apart from an increasing number of surveys on land, networks of fixed observatories were set up in numerous countries. Geomagnetism gradually became separated from traditional ferromagnetic studies, championed instead by professional astronomers as one of a range of interconnected "telluric" forces (including heat, volcanism, and atmospheric electricity), in a framework of terrestrial or cosmic physics.

Such phenomena obviously warranted globally coordinated data collection. Already from 1828, von Humboldt had tirelessly advocated international cooperation in geomagnetism, personally establishing a string of observatories in Europe and Asia that made simultaneous measurements at specific intervals. In 1834, Gauss and his colleague Wilhelm Weber (1804–1891) joined this scheme, expanding it to about 50 stations, known as the "Göttingen Magnetic Union" (1836–1841). At the time, the three recorded elements were declination, inclination, and intensity. When these had all been mapped on a global scale for (approximately) the same era (declination 1833, inclination 1836, intensity 1837), it became possible to interpolate the complete magnetic vector for any point on Earth. This is what Gauss set out to do in 1839. He promulgated the idea of representing the magnetic field by the gradient of a potential function, to be written as a linear combination of a series of spherical harmonic coefficients (or Gauss coefficients), to be derived by least-squares analysis. Assuming a spherical Earth, and converting the charted data into *X*, *Y*, and *Z* components at fixed intervals of colatitude and longitude, Gauss obtained the first 24 coefficients (equivalent to an expansion up to degree and order 4). The results, laid down in his 1839 "Allgemeine Theorie des Erdmagnetismus" showed that the strength of the internal field overwhelmingly dominated any possible external sources affecting the surface field.

This conclusion contrasted starkly with the "cosmological" interpretation, which emphasized the importance of the external field. While many workers employed, expanded, and refined Gauss's method in subsequent decades (over 300 such field models have been computed since, Jacobs, 1987), army officer *Edward Sabine* (1788–1883) (*q.v.*), rejected Gauss's theory, reverting instead to Hansteen's disjointed quadrupole as a working hypothesis, the verification of which culminated in the British movement known as the "Magnetic Crusade" of the 1840s. This consisted mainly of two Antarctic exploration voyages (1839–1842) under James Clark Ross (1800–1862), and a network of colonial observatories (1840–1848) run by the British Admiralty, the War Office, and the English East India Company, supplemented by over twenty, mostly European, Russian, and American stations (Cawood, 1979).

At a new geomagnetic observatory at Kew (est. 1841), Sabine and his staff processed the network data, plus those from ship's logbooks, land surveys, and earlier compilations, to be published in 15 "Contributions to Terrestrial Magnetism" (1840–1877) in the Royal Society's "Philosophical Transactions." The most notable results were Sabine's distinction of diurnal variation of internal and external origin, and his 1851 discovery that the periodicity of magnetic storms was correlated with the 11-year sunspot cycle. Nevertheless, some observatories proved less permanent than originally envisaged, and geomagnetic land surveys in the second half of the century often had to rely on more established disciplines for funding, as part of grand empirical programmes.

Whereas France and Russia were the main sponsors of oceanic circumnavigations up to the midcentury (acquiring substantially better coverage in geomagnetic field measurements, not least in the Pacific Ocean), a plethora of nations (Germany, Sweden, Austria-Hungary, Britain, Norway, the United States, Denmark, and others) partook in the Arctic exploration frenzy characteristic of the second peak period of geomagnetism in the 19th century, from about 1870 to 1885. These ventures, partly stimulated by the First International Polar Year (1882–1883), followed the policy adopted on naval surveys of carrying geomagnetic instruments as standard equipment. Other significant empirical contributions were made on the British "Challenger" (1872–1875), the German "Gazelle" (1874–1876), and the Swedish "Vanadis" (1883–1885) expedition.

On land, networks of repeat stations rose to prominence where observatory coverage was sparse (eventually reaching a total of about 3000). These carefully chosen, fixed and marked locations were to be revisited for a few days at regular intervals (1, 2, 5, or 10 years) to study secular and diurnal variation. Time-dependent change on the historical timescale was meanwhile studied in the 1890s by physicist *Willem van Bemmelen* (1868–1941) (*q.v.*), who reconstructed 16- and 17th century isogonics from nautical data in 165 historical ship's logbooks (based in part on original manuscripts). Arctic and observatory data were also reexamined time and again, forming the basis for new spherical harmonic expansions. By this time, geomagnetism (and related geoelectricity) had become almost exclusively the field-theoretical domain of physicists.

The 20th century

Empirical geomagnetism continued to prosper in the 20th century. Besides the European colonial powers, the United States, Russia, and Japan likewise established new observatories and organized new surveys to track the field's appearance and change. Moreover, Antarctica became a novel focal point of investigation, for example, on Robert Falcon Scott's (1868–1912) first Antarctic expedition (1901–1904), and later during the Second International Polar Year (1932–1933). In 1904, *Louis Agricola Bauer* (1865–1932) (*q.v.*), at the *Carnegie Institution in Washington* (*q.v.*), established the "Department of International Research in Terrestrial Magnetism," which employed the freighter "Galilee" (1905–1908) and the nonmagnetic *vessel Carnegie* (1909–1929) (*q.v.*), for oceanic geomagnetic surveys. A similar initiative, with the "Zarya," was launched by the USSR in 1956, just before the International Geophysical Year (1957–1958) renewed interest in geomagnetic observations through its "World Magnetic Survey." From 1953, novel magnetometers were towed at distance behind a ship to avoid deviation effects, a practice still routinely carried out on recent ocean-bound scientific missions.

With the advent of new technology, land observatories no longer recorded the local vector in terms of angles and total intensity, but adopted the three orthogonal vector components (*X*, *Y*, and *Z*). Airborne magnetic surveys (in particular to study the crustal field) became increasingly common in the second half of the century, for example, in the "World Magnetic Modelling and Charting" programme of the US. Defence Mapping Agency. Satellite measurements of the Earth's magnetic field started with the Soviet "Sputnik 3" (1958), maintained in subsequent decades by their "Vostoks" and the American "OGO" series (see *POGO*). In 1979 the United States launched *Magsat* (*q.v.*), the first geomagnetic vector satellite, followed by the European "*Ørsted*" (1999) (*q.v.*), and "*CHAMP*" (2000) (*q.v.*). Once workers were able to compare the near-global, uniform coverage provided by satellites with earlier field maps and historical datasets, the continuity of some static and dynamic features provided theoreticians new food for thought.

Important strides forward were also made in the theoretical domain before that time. At the turn of the 20th century, a great number of *nondynamo theories* (*q.v.*) were launched, based on thermoelectricity and (gravitational or geochemical) charge separation. Another hypothesis explored by several workers is the one often associated with the work of *Patrick M.S. Blackett* (1897–1974) (*q.v.*), which posited that the Earth's magnetic field could be due to its rotation (Good, 1998). Two of Blackett's former students explored this idea: *Edward C. Bullard* (1907–1980) (*q.v.*) suggested that such a distributed source would cause the field to be weaker at greater depth, whereas actual measurements made in deep mine shafts by *S. Keith Runcorn* (1922–1995) (*q.v.*), ultimately concurred with Blackett's own laboratory research, in supporting a negative conclusion.

A different hypothesis, more in keeping with a growing body of seismological evidence that delineated a fluid outer core and solid inner core, originated with the supposition of Joseph Larmor in 1919 that the Sun's magnetic field might operate as a self-exciting dynamo.

Building on the mathematical foundations of electromagnetism laid down by James Clark Maxwell (1831–1879) in 1861–1864, and armed with new methods of analysis, workers in the 1940s and 1950s developed the theory of the fluid geodynamo, which still forms the basis of current understanding regarding geomagnetism. In 1939, Walter Elsasser published "On the Origin of the Earth's Magnetic Field," in which turbulent convective motions were deemed to create inhomogeneities in the liquid metallic interior of the Earth's core, giving rise to thermoelectric currents. After World War II, Elsasser focused on kinematic instabilities, positing in 1946 that a toroidal core field could play a vital role in generating and sustaining a self-exciting dynamo process. In the late 1940s and 1950s, the theory was elaborated further by *Bullard* and *George E. Backus* (*q.v.*), among others. Despite initial skepticism, first after publication of *Cowling's theorem* (1934, *q.v.*), and later, after the demise of the Bullard-Gellman numerical solution, dynamo theory has since gained, in broad outline, near universal acceptance. Nevertheless, important areas of controversy remain (for example, regarding the magnetohydrodynamic wave and frozen flux hypotheses), and to date, no causal geomagnetic model exists that simultaneously and accurately mimics physics and behavior of the Earth's field on all relevant timescales.

By the end of the 20th century, the discipline of geomagnetism, which in the 1930s had split into the three subdisciplines of paleomagnetism, internal field modeling, and investigations of the external field, had initiated a move toward reintegration, combining data, theories, and constraints from all. In addition, field maps no longer only captured the surface field, but likewise displayed geomagnetic features at the top of the source region, at the *Core-Mantle Boundary* (*q.v.*), and increasing attention was being devoted to possible coupling mechanisms at work across this interface. Advances in mathematics, instruments, and computing power likewise opened up vast new horizons to explore. Together, these developments promise geomagnetists in the 21st century many exciting, more comprehensive insights into the myriad fascinating aspects of their chosen field of study.

Art R.T. Jonkers

Bibliography

Balmer, H., 1956. *Beiträge zur Geschichte der Erkenntnis des Erdmagnetismus.* Veröffentlichungen der Schweizerischen Gesellschaft für die Geschichte der Medizin und Naturwissenschaft 20. Aarau: H. R. Sauerländer.

Bullard, E.C., 1954. Homogeneous dynamos and terrestrial magnetism. *Philosophical Transactions of the Royal Society of London, A*, **247**: 213–278.

Cawood, J., 1979. The magnetic crusade: science and politics in Early Victorian Britain. *Isis*, **70**(254): 493–518.

Chapman, S., and Bartels, J., 1940. *Geomagnetism*, 2 vols. Oxford: Clarendon Press.

Daujat, J., 1945. *Origines et Formation de la Théorie des Phénomènes Électriques et Magnétiques.* Exposés d'Histoire et Philosophie des Sciences vol. 989–991. Paris: Hermann.

Fleury Mottelay, P., 1922. *Bibliographical History of Electricity and Magnetism Chronologically Arranged.* London: Charles Griffin.

Good, G.A. (ed.), 1998. *Sciences of the Earth: An Encyclopedia of Events, People, and Phenomena.* New York, London: Garland.

Jacobs, J.A. (ed.), 1987. *Geomagnetism*, 4 vols. London: Academic Press.

Jonkers, A.R.T., 2003. *Earth's Magnetism in the Age of Sail.* Baltimore, MD: Johns Hopkins University Press.

Needham, J., 1962. *Science and Civilisation in China*, vol. 4. Cambridge: Cambridge University Press.

Smith, J.A., 1992. Precursors to Peregrinus: the early history of magnetism and the Mariner's compass in Europe. *Journal of Medieval History*, **18**: 21–74.

Still, A., 1946. *Soul of Lodestone: The Background of Magnetical Science.* Toronto: Murray Hill.

Cross-references

Bauer, Louis Agricola (1865–1932)
Bemmelen, Willem van (1868–1941)
Blackett, Patrick Maynard Stuart, Baron of Chelsea (1897–1974)
Bullard, Edward Crisp (1907–1980)
Carnegie Institution of Washington, Department of Terrestrial Magnetism
Carnegie, Research Vessel
CHAMP
Chapman, Sydney (1888–1970)
Compass
Core-Mantle Boundary
Cowling's Theorem
Elsasser, Walter M. (1904–1991)
Gauss, Carl Friedrich (1777–1855)
Gellibrand, Henry (1597–1636)
Geodynamo
Gilbert William (1544–1603)
Halley, Edmond (1656–1742)
Harmonics, Spherical
Humboldt, Alexander von (1759–1859)
Instrumentation, History of
Jesuits, Role in Geomagnetism
Kircher, Athanasius (1602–1680)
Larmor, Joseph (1857–1942)
Magsat
Nondynamo Theories
Norman, Robert (1560–1585)
Ørsted
POGO (OGO-2, -4, and -6 Spacecraft)
Runcorn, S. Keith (1922–1995)
Sabine, Edward (1788–1883)
Voyages Making Geomagnetic Measurements

GILBERT, WILLIAM (1544–1603)

William Gilbert (sometimes spelled "Gilberd") was born in Colchester, England, in 1544, the son of Jerome, a successful Burgess and Recorder of the town, and his wife Elizabeth. From St John's College, Cambridge, where he graduated M.D. in 1569, and of which foundation he became a Senior Fellow, serving the College in a variety of offices, William Gilbert began to practice medicine in London. Sometime before 1581, he was admitted into the prestigious Royal College of Physicians, and over the years occupied several of its senior offices. His professional career was crowned in 1600 when he was appointed Physician to Queen Elizabeth I, though as that monarch wisely avoided medical attention, one suspects that this distinguished appointment was, in practical terms, a sinecure. When King James I ascended to the throne in March 1603, Gilbert continued as Royal doctor, only to die later that year on November 30, 1603. Gilbert never married.

Gilbert's enduring fame, however, lies not in his medical achievements, which were conventional, but in his pioneering and privately conducted researches into magnetism, geomagnetism, and electrostatics, which he published in *De Magnete, magneticisque corporibus, et de magno magnete tellure; physiologia nova, plurimis et argumentis, et experimentis demonstrata* (London, 1600). For as this long title tells us, Gilbert's book was far more than just a book on magnets, but developed an argument, based on abundant experimental evidences, that magnetism was a property of the terrestrial globe itself. In short, it was the first coherent treatise on geomagnetism, and in this respect, is a foundation text of modern experimental physics, having a profound impact on all subsequent researches, and influencing figures such as Kepler, Galileo, Hooke, Newton, Halley, and the early Fellows of the Royal Society.

At the heart of Gilbert's scientific rationale was his skepticism about aspects of the philosophy of Aristotle, which was enshrined in university curricula across Europe, and which saw the earth as essentially passive and made of a fundamentally different stuff from the dynamic and shining heavens. Magnetism, however, tended to challenge such an attitude, suggesting that the earth possessed a lively dynamic force of its own: a force, moreover, that was amenable to experimental investigation. For it is very clear that Gilbert was part of a tradition of emerging experimental practice in Elizabethan England. A practice sometimes seen as closer in principle to the mechanic practice of artisans than to the refined philosophy of university men, and which Gilbert's younger contemporary, Sir Francis Bacon, would develop into a coherent model of how to investigate nature by means of experiments in his *Novum Organon* (1620) (*q.v.*).

Central to the argument of *De Magnete* is that the earth itself is a spherical magnet, with a north and south pole. Moving between these poles in what we might call a curved field was the invisible magnetic force deriving from the very being, or "soul," of the earth itself. This force was actively present in all magnetite stone and ferrous metals, and Gilbert knew, from writers going back to Petrus Peregrinus in 1269 and before, that suspended iron needles would orientate themselves along it. He also knew from his Elizabethan contemporary, the compass-maker and artisan-scientist, Robert Norman, in 1581, that this force not only acted in a lateral plane, but also in a vertical one, to produce the magnetic dip.

Central to Gilbert's geomagnetic ideas, and probably picked up initially from Peter Peregrinus' *Epistola*, were his experiments conducted upon what he called "terrellae," or little earths. These were spheres of magnetic material, probably lumps of spherically chiseled magnetite, all of which were found to have magnetic poles, equators, and contours, just like the earth. These characteristics were discovered with an instrument, which Gilbert sometimes called his magnetized "Versorium" (presumably from Latin *versare*, "to turn around"), which was a delicately poised magnetic needle, capable of moving both horizontally and vertically when held near the terrella. Over the terrella's magnetic equator, for instance, the needle tended to position itself north-south horizontally, or in a tangent to the equator. But as it moved forever closer to a pole, it dipped in its angle, and stood vertically upon reaching one or other pole. This suggested a continuous force field emanating from and connecting the poles in a series of invisible arches, which were at their flattest over the equator. Gilbert also found that the slight irregularities in the terrella produced in turn local force fields irregularities, which he saw as analogous to local variations in the magnetic field of the earth itself.

The major aspect of Gilbert's genius as an experimental physicist was his realization that one could study and model phenomena with the terrella in the laboratory that were physically identical to the phenomena exhibited by the globe of the earth itself as it hung in space. A concept, indeed, so familiar to modern scientific practice as to be taken for granted, but outrageous for the 16th century, for Aristotle taught not only that the heavens were fundamentally different from the earth, but that beneath the fire, air, and water that surrounded our planet, there was a primary element of Earth. Yet how could this Earth be homogeneous or at one with itself, if parts of it were magnetic and other parts were not?

Gilbert's terrella experiments enabled him to develop a coherent and verifiable model for the Earth's magnetic field, explaining the north-finding properties of compass needles, local irregularities, and the dip. It was arguably the first experimentally based comprehensive theory in the history of physics, and it is hardly surprising that several subsequent generations of scientists found inspiration in his work.

From his laboratory and terrestrial studies, Gilbert then took the portentous step of developing a magnetic cosmology in Book VI of *De Magnete*. For one thing, he argued that the Earth's magnetic field suggested that our planet rotated on its axis, *contra* Aristotle, Ptolemy, and the classical philosophers, who said that it was stationary, with the universe rotating around us. And while he never formally proclaimed himself a Copernican, all of Gilbert's cosmological arguments presumed the Sun, and not the Earth, to be at the center of the solar system. He also argued that, instead of being made of a unique cosmological fifth element, as the ancients had thought, the planets themselves were probably made of magnetic material, and moved through space under the influence of magnetic force fields. Very important in this context, moreover, was his abandonment of Aristotle's cosmological divide of the Moon's orbit, which was believed to separate the terrestrial from the celestial realms. Substantiated in part from Tycho Brahe's recent astronomical discoveries, And in his posthumously published *De Mundo* (Amsterdam, 1651), Gilbert suggested that space was, in effect, homogeneous and empty: without qualitative divisions, yet traversed by magnetic forces which were themselves the sources of all motion. He also speculated that the diffuse light of the Milky Way might be occasioned by masses of very distant stars, no individual star of which we could see from Earth, though here, in some respects, Gilbert was in keeping with earlier medieval writers such as Jean Buridan, Nicolas Oresme, and Simon Tunsted, to name but a few.

There is no evidence to suggest that Gilbert's failure to openly embrace the Copernican theory derived from a fear of religious persecution. Copernicanism only became a contentious issue for the Roman Catholic Church after Galileo used it for his own highly adversarial purposes after 1612, while the Church of England never had any official policies on scientific issues one way or the other. One suspects that his reluctance comes from covering his back professionally, as an eminent physician. Academic medicine was a deeply conservative art in Gilbert's time, and learned physicians risked professional suicide if they openly proclaimed novel ideas, which cast doubt on the time-honored wisdom of the ancients. After all, Gilbert's Royal doctor and Physicians' College colleague of the next generation, William Harvey, found that his published discovery of the circulation of the blood in 1628 badly damaged his practice as a society doctor. Writing a book about magnetism was one thing, but openly espousing a theory, so contradictory to common sense as Copernicanism then seemed in 1600 was risking being branded an unsound man. Not a good trait for a Royal doctor to have attributed to him, indeed!

In addition to his experimental and speculative cosmological work, *De Magnete* also aspired to present a history of and devise a taxonomy for magnetic phenomena. And very significantly, in Chapter 2 of Book II (out of the six books into which *De Magnete* is divided) Gilbert set out his researches into the properties of amber, jet, and other substances, which displayed what he called *Electric* characteristics. (The term, which introduced the words *electric* and *electrical* into the modern world, was derived by Gilbert from the Greek word for amber: electrum.) From his experiments, he differentiated between magnetic phenomena proper, which he saw as innate and permanent properties of the stuff from which God had created the world, and friction-generated *electric* phenomena, which were a short-lasting product of the residual moisture or effluvium of once-fluid substances, such as those resins which solidified into amber, and which could be temporarily excited, and draw things to themselves by rubbing. Though by modern standards Gilbert's explanations were wrong, his recognition that magnetism and electrical phenomena were two quite different forms of attraction was correct.

Gilbert's *De Magnete* was one of those milestone books in the history of science which turned a hitherto vague and confused collection of observations that was magnetics into a coherent discipline, the phenomena of which could be tested at leisure by its readers and applied to new situations and monitored with refined instruments. Its taxonomy of phenomena, moreover, introduced new terms, such as *magnetic polarity* and *electric* into general usage. And very portentously, it began that transformation away from the Aristotelian doctrine of motion to something much more dynamic, and, by its exploration through a series of tests and hypotheses, laid the foundation for modern experimental physics.

Allan Chapman

Bibliography

Stephen, P., 'William Gilbert', 2004. *New Dictionary of National Biography.* Oxford: Oxford University Press.

Taylor, E.G.R., 1954, 1968. *The Mathematical Practitioners of Tudor and Stuart England 1485–1714.* Cambridge: Cambridge University Press, p. 174, no. 31.

William Gilbert, 1600. *De Magnete.* London: Chiswick press. See P. Fleury Mottelay, *William Gilbert of Colchester...on the Great Magnet of the Earth* (Ann Arbor, 1893). Re-issued in facsimile as *De Magnete*, New York: Dover, 1958.

GRAVITATIONAL TORQUE

The expression can be given a wide definition. We will restrict the presentation to the torques mutually exerted one on the other by the three envelopes of the internal Earth (inner core, outer core, and mantle) through the perturbative gravitational force field superimposed on the gravitational field of an equilibrium state—hydrostatic in the fluid core.

The equilibrium state

For the sake of simplicity we consider the inner core and mantle to be rigid. In fact the mantle is convecting, and density anomalies exist, as well as bumps at the core-mantle boundary (CMB). A disturbed gravitational potential results in the mantle, liquid outer core (Wahr and de Vries, 1989), and inner core. Our hydrostatic state is the state in which the mantle, outer core and inner core are at rest with respect to one another under self-gravitation potential $\Psi_0(\hat{r})$ ($\hat{r}$ is the current point), the whole Earth rotating with uniform angular velocity $\overrightarrow{\Omega_0}$ around its polar axis of inertia fixed in space (we ignore precession, although an external gravitational force is responsible for driving it). Let $\rho_0(\hat{r})$ and $p_0(\hat{r})$ be the corresponding density and pressure fields in the liquid outer core. We have:

$$\rho_0\vec{\nabla}(\Psi_0 + U_0) = \vec{\nabla}p_0, \vec{\nabla}\Psi_0 = -\overrightarrow{\Omega_0} \wedge (\overrightarrow{\Omega_0} \wedge \vec{r})$$

$$\nabla^2 U_0 = -4\pi G\rho_0$$

G is the gravitational constant, Ψ_0 the rotational potential. Level surfaces of $U_0 + \Psi_0$, ρ_0, p_0 are parallel. It is essential in the following that U_0 be nonaxisymmetric. In fact ρ_0 changes with time, but we consider time constants much shorter than the time constant of the convection in the mantle. This (pseudo) static equilibrium state corresponds to a minimum gravitational energy, and no mutual torque exists.

We will now consider perturbations from this state and the resulting gravitational torque. We separate different problems for the sake of clarity.

The mantle—inner core gravitational coupling and l.o.d.

Let us ignore in a first step the convection in the core (density heterogeneities responsible for this convection are of the order of $10^{-8}\overline{\rho_0}$, $\overline{\rho_0}$ being the mean core density, whereas those generated by mass anomalies in the mantle are of the order of $10^{-4}\,\overline{\rho_0}$.

Consider that $\rho_0(\hat{r})$ is not axisymmetric, for example

$$\rho_0(\hat{r}) = \rho_0(r)\,(1 + \Delta_2^2(r)\,P_2^2(\cos\theta)\cos 2(\varphi - \varphi_0)),$$

θ and φ being colatitude and longitude; (this quadrupole term is evidenced by seismic studies) and that a difference (A_S–B_S) results between the equatorial moments of inertia of the inner core. Let the mantle and inner core be rotated by angles φ_m and φ_s, respectively, from their equilibrium positions. A restoring gravitational torque is exerted by the mantle on the inner core (and vice versa), proportional to the relative rotation (Buffet, 1996a):

$$\Gamma_i = \bar{\Gamma}(\varphi_m - \varphi_s) \qquad \text{(Eq. 1)}$$

(the action of the outer core is just to contribute to the value of $\bar{\Gamma}$). That means that the inner core is gravitationally locked to—aligned with—the mantle in the equilibrium figure. $\bar{\Gamma}$ can be computed from the gravitational energy, which is minimum in the equilibrium state.

Small oscillations of the inner core and mantle around the rotation axis $\overrightarrow{\Omega_0}$—i.e., oscillations in the length of the day—arise whose period is proportional to the polar momentum of inertia of the inner core (approximately) and inversely proportional to $\bar{\Gamma}$. Periods of few years are obtained with current models of density anomalies in the mantle.

The liquid outer core does not play an important role in the former mechanism, but it can be the seat of the so-called torsional oscillations (Braginsky, 1970; Zatman and Bloxham, 1998; Hide *et al.*, 2000; Dumberry and Bloxham, 2003; Jault and Legaut, 2005) made of rotations $\omega(s)$ of the geostrophic cylinders C(s) of radius s (whose axis is the rotation axis $\overrightarrow{\Omega_0}$). The adjacent cylinders are coupled through electromagnetic forces due to the radial cylindrical component of the main field, and the inner ones are coupled with the solid inner core through magnetic friction. The inner core and mantle are as before coupled by gravitational forces. Again, if the inner core and mantle happen to be moved with respect one to the other, restoring gravitational torques will generate small oscillations around the equilibrium configuration. Their periods are increased up to several decades due to the involvement of geostrophic cylinders (Buffet, 1996b). As for the excitation of these necessarily damped oscillations, it could be due to the dynamo process itself.

The mantle—inner core gravitational coupling and nutations

The system made of the rotating mantle, outer core and inner core displays small motions of the rotation axes of the three envelopes, called nutations. Forced nutations are generated by the lunisolar tidal potential. Free nutations are rotational eigenmodes of the system. Two modes are nearly diurnal in a frame corotating with the mantle ($\overrightarrow{\Omega_0}$), the free core rotation (FCN) and the free inner core nutation (FICN). Two modes have periods much longer than a day in this same frame, the Chandler wobble and the inner core wobble (Mathews *et al.*, 1991).

The gravitational torque exerted by the mantle and outer core on the ellipsoidal inner core has a profound effect on the inner core wobble and FICN. This torque results from the tilt of the inner core (due to the equatorial component of its rotation $\overrightarrow{\omega_s}$) or misalignment of the figure axes of the inner core and the mantle (Xu and Zseto, 1996). In other words, as before in the axial problem, the restoring gravitational torque tends to lock the inner core with respect to the mantle (here the flattening of an axisymmetric earth is to be invoked). If the inner core is rigid, or purely elastic, the locking is strong. If it is viscous, it yields to the gravitational forces and tends to cancel the torque. In the limit of a liquid inner core, the inner core wobble disappears. As for the period of the FICN (in an absolute frame), it may vary, in the absence of magnetic field, from 75 days (liquid inner core) to 485 days (elastic inner core) (Greff-Lefftz *et al.*, 2000).

Gravitational forces due to the convective density heterogeneity in the core

Consider now the independent of (ρ_0) density heterogeneity ρ', thermal or compositional, generating the convective flow $\vec{u}$ driving the dynamo. Extra gravitational and rotational potentials appear:

$$\rho_0 \to \rho_0 + \rho',\ U_0 \to U_0 + U',\ \overrightarrow{\Omega_0} \to \overrightarrow{\Omega_0} + \overrightarrow{\Omega'},\ \Psi_0 \to \Psi_0 + \Psi'$$

A gravitational torque is exerted on the mantle by this density distribution (in this section we ignore the inner core):

$$\overrightarrow{\Gamma_g} = \iiint_M \rho_0 \left(\vec{r} \wedge \vec{\nabla} U' \right) \mathrm{d}v \qquad \text{(Eq. 2)}$$

(we ignore the effect of Ψ' for the sake of simplicity); of course an equal and opposite torque is applied on the core by the mantle.

(a) The axial problem

$\hat{k}$ being the unit vector of $\vec{\Omega}$ (or $\overrightarrow{\Omega_0}$)

$$\Gamma_{gz} = \overrightarrow{\Gamma_g} \cdot \vec{k} = -\iiint_M \rho_0 \frac{\partial U'}{\partial \varphi} \mathrm{d}v = \iiint_M U' \frac{\partial \rho_0}{\partial \varphi} \mathrm{d}v$$

that makes it clear again that only the non axisymmetric part of ρ_0 intervenes in the axial torque. If I_M is the polar moment of inertia of the mantle and ω_z the component of $\overrightarrow{\Omega'}$ along $\overrightarrow{\Omega_0}$, we have

$$I_M \frac{\mathrm{d}\omega'_z}{\mathrm{d}t} = \Gamma_{gz}$$

if $\overrightarrow{\Gamma_g}$ is only acting.

The effect of the gravitational forces on the fluid core itself is more subtle since it does not react like a rigid body. Let us again consider the cylinders $C(s)$ of radius s and thickness ds. The axial gravitational torque acting on $C(s)$ is (Jault and le Mouel, 1989):

$$\Gamma_{gz}(s) = \mathrm{d}s \int \left(\rho' \frac{\partial U_0}{\partial \varphi} - U' \frac{\partial \rho_0}{\partial \varphi} \right) \mathrm{d}z \, \mathrm{d}\varphi$$

s, φ, z being the cylindrical coordinates. An accelerated rotation of $C(s)$ results, with linear velocity $\vec{t_1}(s,t) = t_1(s,t)\,\hat{\varphi}$, governed by the equations

$$I(s) \left(\frac{\partial t_1(s,t)}{\partial t} - \frac{\mathrm{d}\omega'_z}{\mathrm{d}t} \right) = \Gamma_g(s) \qquad \text{(Eq. 3)}$$

$$I_M \frac{\mathrm{d}\omega'_z}{\mathrm{d}t} = -\int_0^a \Gamma_g(s)\, \mathrm{d}s = \Gamma_{gz} \qquad \text{(Eq. 4)}$$

$I(s)$ is the moment of inertia of cylinder $C(s)$, a is the radius of the core. The reason why the geotrophic flow $t_1(s,t)$ is accelerated is that it is the only one for which the growing Coriolis force can be balanced by a growing pressure torque. Equations (3) and (4) would govern changes in ω'_z, i.e., changes in l.o.d generated by gravitational forces linked to ρ', should $\overrightarrow{\Gamma_g}$ be only acting. In fact it is rather artificial to separate here the gravitational torque from the topographic torque (action of the pressure field linked to flow $\vec{u}$ on the CMB bumps). Furthermore the electromagnetic forces which couple the adjacent cylinders (and the outer core with the inner core and the poorly conducting mantle) provide the dissipative mechanism, which prevents rotation of the core relative to the mantle to grow indefinitely.

Unfortunately the evaluation of the gravitational torque, which requires the knowledge of both the anomalous (nonaxisymmetric) part of ρ_0 and the convective density heterogeneity ρ‘, is far from being easy. It could generate significant variations in l.o.d.

(b) The gravitational torque $\overrightarrow{\Gamma_g}$ and the pole motion

Let us consider the equatorial component of the gravitational torque Eq. (2) and the Euler equations for the mantle (i.e., the equations of the equatorial components of the angular momentum of the mantle in a frame rotating with angular velocity $\overrightarrow{\Omega_0}$):

$$A_M \frac{\mathrm{d}\Omega'}{\mathrm{d}t} - \mathrm{i}\,(C_E - A_E)\,\Omega_0\,\Omega' + i\,\Omega_0^2\, c = \Gamma_g$$

with the usual notations: A_M is the equatorial moment of inertia of the mantle, C_E and A_E the polar and equatorial moments of inertia of the whole Earth (taken axisymmetric), $\Omega' = \omega'_1 + i\omega'_2$, $\Gamma_g = \Gamma_{g1} + \mathrm{i}\,\Gamma_{g2}$, ω'_1 and ω'_2 being the two equatorial orthogonal components of $\overrightarrow{\Omega'}$, Γ_{g1} and Γ_{g2} the same components of $\overrightarrow{\Gamma_g}$, $\mathrm{i} = \sqrt{-1}$. And $c = c'_{E1} + \mathrm{i}\, c'_{E2}$ is the change in the Earth equatorial moment of inertia due to the perturbations ρ'.

Again, it is artificial to separate here the gravitational torque from the pressure torque (Hulot and Le Mouel, 1996). The gravitational torque linked to a time varying ρ'—i.e., a flow—might be invoked to excite the Chandler wobble; but short time constants ($\ll$435 days) are requested; or to generate the so-called Markovitz wobble; but balances with other torques have to be considered (Hulot and Le Mouel, 1996).

Note that dynamo modeling could now provide estimates of the gravitational torque due to the corresponding core flow.

Jean-Louis Le Mouel

Bibliography

Braginski, S.I., 1970. Torsional magnetohydrodynamic vibrations in the Earth's core and variations in day length. *Geomagnetism and Aeronomy, English Translation*, **10**: 1–8.

Buffet, B., 1996a. Gravitational oscillations in the length of day. *Geophysical Research Letters*, **23**: 2279–2282.

Buffet, B., 1996b. A mechanism for decade fluctuations in the length of day. *Geophysical Research Letters*, **25**: 3803–3806.

Dumberry, M., and Bloxham, J., 2003. Torque balance, Taylor's constraint and torsional oscillations in a numerical model of the geodynamo. *Physics of the Earth and Planetary Interiors*, **140**: 29–51.

Greff-Lefftz, M., Legros H., and Dehant, V., 2000. Influence on the inner core viscosity on the rotational modes of the Earth. *Physics of the Earth and Planetary Interiors*, **122**: 187–204.

Hide, R., Boggs, D.H., and Dickey, J.O., 2000. Angular momentum fluctuations within the Earth's liquid core and torsional oscillations of the core-mantle system. *Geophysical Journal International*, **143**: 777–786.

Hulot, G., Le Huy, M., and Le Mouel, J.L., 1996. Influence of core flows on the decade variations of the polar motion. *Geophysical and Astrophysical Fluid dynamics*, **82**: 35–67.

Jault, D., and Le Mouel, J.-L., 1989. The topographic torque associated with a tangentially geostrophic motion at the core surface and inferences on the flow inside the core. *Geophysical and Astrophysical Fluid Dynamics*, **48**: 273–296.

Jault, D., and Legaut, G., 2005. Fluid dynamics and dynamos in a strophysics and geophysics. In Soward, A.M., Jones, C.A., Hughes, D.W., and Weiss, N.O., (eds.) *The Fluid Mechanics of Astrophysics and Geophysics Series,* vol. 12, chap. 9. London: Commodity Resource Corporation Press, pp. 277–293.

Mathews, P.M., Buffett, B.A., Herring, T.A., and Shapiro, I.L., 1991. Forced nutations of the Earth: influence of inner core dynamics. I. Theory. *Journal of Geophysical Research*, **96**(B5): 8219–8242.

Wahr, J., and de Vries, D., 1989. The possibility of lateral structure inside the core and its implications for rotation and Earth tide observations. *Geophysical Journal International*, **99**: 511–519.

Xu, S., and Szeto, A.M.K., 1996. Gravitational coupling within the Earth: computation and reconciliation. *Physics of the Earth and Planetary Interiors*, **97**: 95–107.

Zatman, S., and Bloxham, J., 1998. A one-dimensional map of B_s from torsional oscillations of the Earth's core. In Gurnis, M., Wysession, M.E., Knittle, E., and Buffett, B.A., (eds.), *The Core-mantle Boundary Region.* Geophysical Monograph, 28, Geodynamics series. Washington, DC: American Geophysical Union, pp. 183–196.

GRAVITY-INERTIO WAVES AND INERTIAL OSCILLATIONS

Earth's outer core is a predominantly iron, rotating liquid body, bounded by a silicate mantle, that is quasispherical. Thus an understanding of the global dynamics of the core can be had by considering it to be an incompressible fluid contained by an approximately spherical, rigid boundary. If the core were rotating at a constant rate as though it were a solid, individual parcels of fluid would be at rest relative to Earth's surface. A parcel of fluid displaced from equilibrium would experience a buoyancy force if the density were not uniform, and a coriolis force due to Earth's rotation, resulting in what are called *core undertones*, their periods being large compared to Earth's free (elastic) oscillations. The combined result of these two restoring forces would produce gravity-inertio waves or gravito-inertial waves as they are sometimes termed. In Earth's atmosphere they are usually termed inertio-gravity waves while in the Sun they are called *g*-modes; more generally they are described as internal gravity waves modified by rotation or simply gravity waves. If there were no density effects, just inertial waves result. The local motion of a fluid parcel in this latter case is an inertial oscillation. A member of the triply infinite set of inertial modes can be excited in a contained rotating fluid like the core if there is a global source of excitation at an appropriate frequency and will decay by viscous dissipation in the boundary layers and interior. The physics of these modes is discussed below where the range of frequencies for each type of response is found heuristically from basic physical principles.

Inertial oscillations

Starting with the simplest case, assume that there are no density differences from place to place in the core and consider a uniform core rotating at the diurnal rate. A ring of fluid encircling the axis of rotation is held in equilibrium by the balance between centrifugal and radial pressure gradient forces on the ring. Expanding the ring slows it down to conserve angular momentum so that the centrifugal force in the expanded position is now less than the local, axially directed radial pressure gradient. Accordingly, the ring is forced back to its starting position just as it would be if the ring had been initially contracted toward the axis of rotation. Thus the ring of fluid is stable with respect to radial perturbations so it oscillates.

Frequency of oscillations

A ring of fluid displaced in a direction perpendicular to the axis of rotation will oscillate at frequency $\omega = 2\Omega$ where Ω is the assumed steady rotation rate of Earth. If instead of a radial displacement of the fluid ring, it were moved axially with its motion parallel to the rotation axis, no restoring force would exist so that $\omega = 0$. Displacement of the ring in any other direction would produce a component of restoring force only in the direction perpendicular to the rotation axis giving

$$\omega = 2\Omega\cos\theta \qquad \text{(Eq. 1)}$$

where θ is the colatitude of the ring as shown in Figure G54. Thus any frequency of oscillation ω in the range $0 < \omega < 2\Omega$ is allowed so that for the idealized constant density core considered here (see also *adiabatic gradient in the core*), the shortest period of oscillation would be approximately 12 h.

Individual particle motions follow elliptical paths that are the same in every meridional plane. These are axisymmetric inertial oscillations (Aldridge and Toomre, 1969; Greenspan, 1969) that have particle motions predominantly parallel to the rotation axis when $\omega \to 0$ and perpendicular to this axis when $\omega \to 2\Omega$. In a bounded rotating fluid like the core, a doubly infinite set of axially symmetric inertial oscillations can exist, usually called inertial modes. Surfaces on which disturbance pressure vanishes form nodal lines which are ellipses in meridional planes. Some examples of these lines which divide the fluid into cells are given in Figure G55 for a unit sphere. The small ticks at the surface show what is known as the critical colatitude for the mode where θ satisfies Eq. (1). Tangents to the surface at this location define a direction within the fluid, called a *characteristic*, along which small disturbance will propagate as illustrated in Figure G54.

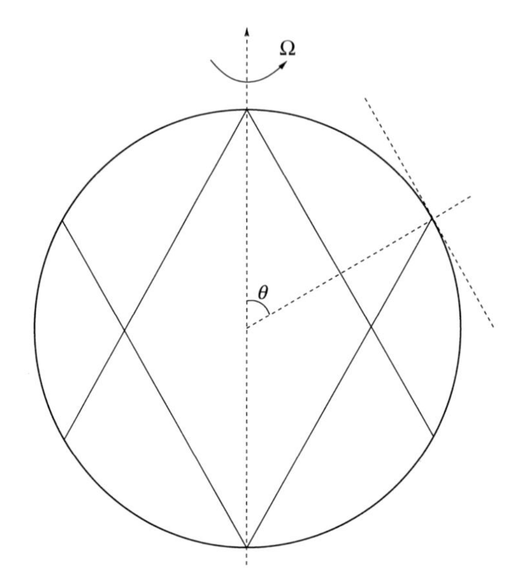

Figure G54 Characteristic lines for an inertial wave of frequency $\omega = \Omega$ propagating in a sphere. The critical colatitude $\theta = 60°$ is shown corresponding to the point where a tangent to the sphere parallels a characteristic line.

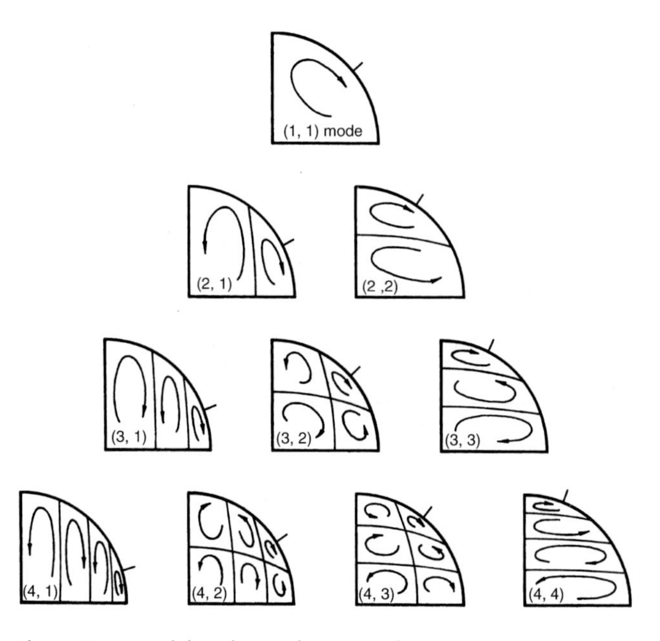

Figure G55 Nodal surfaces of pressure for some axisymmetric inertial oscillations with velocity shown by sketched curves. Index (n, m) corresponds to the mth root of the Legendre polynomial $P_{2n+2}(\omega/2\Omega) = 0$. (From Aldridge and Toomre, 1969. With permission.)

The source of these characteristics is that the differential equation describing the gravity waves is hyperbolic in space and it must satisfy a no-slip boundary condition at the surface of the core-mantle boundary (CMB). A boundary value problem which is hyperbolic (rather than elliptic) is said to be ill-posed. No continuous solutions are guaranteed for such problems so that approximate and laboratory methods must be used to solve them. A notable example is the limiting case $\omega = 0$ (steady motion) where $\theta = 90°$ so the tangent is parallel to the rotation axis; this is a Taylor column, the axial column seen in a rotating fluid when an object is towed steadily through the fluid. These disturbances are *inertial waves* and are discussed below.

Amplitude of oscillations

Inertial oscillations could be excited in Earth's core through coupling of the interior fluid at the boundary by viscous forces. If the boundary motion is sufficiently well-coupled spatially to a mode—e.g., the semidiurnal tide has very similar spatial structure to a mode with azimuthal dependence $\cos 2\phi$ where ϕ is geographic longitude. There are several modes with periods near the semidiurnal tide (Aldridge *et al.*, 1988). Modes coupled to the CMB by viscosity will be excited on a timescale of spin-up or $E^{-1/2}\Omega^{-1}$ where E is the Ekman number for the core. For laminar flow this turns out to be about 5 ka. This is also the timescale for free decay, which is mostly caused by boundary layer dissipation (see also *Boundary layers in the core*) with additional internal dissipation which increases dramatically for modes of smaller spatial structure. Theorems have been developed (Zhang *et al.*, 2001) that imply no dissipation for inertial modes in a sphere; though elegant they are not relevant for real fluids since they only apply to ideal (inviscid) fluids.

Inertial waves

At frequencies $\omega \rightarrow 2\Omega$ the waves travel in the direction of the axis of rotation while at intervening frequencies they travel in directions given by θ in Eq. (1). The particle motion of the waves is perpendicular to the propagation direction since it is the Coriolis force that is acting on the particle. Thus the inertial waves are transverse. The waves are reflected at boundaries such that they maintain equal angles of reflection and incidence with the direction of the rotation axis rather than a normal to the reflecting boundary. Thus these waves will have multiple reflections in a closed container. In a spherical shell of fluid, multiple reflections lead to repeated paths (Hollerbach and Kerswell, 1995; Tilgner, 1999) which are known as *attractors* and have been studied extensively (Rieutord *et al.*, 2001).

The assumption of axial symmetry can be relaxed by introducing azimuthal dependence and this reveals that there is a triple infinity of inertial modes. The critical directions become cones that define the propagation of waves in the fluid.

Inertial waves have been excited in laboratory experiments (Noir *et al.*, 2001) on precession of a spheroidal shell of fluid (see also *Fluid dynamic experiments, Precession and core dynamics*). Since an observer in the rotating frame of reference with the container's spin will see the precession as a diurnal disturbance, $\theta = \arccos(1/2) = 60°$ which corresponds to the characteristics illustrated earlier in Figure G54. It is estimated that the precession will excite inertial waves of amplitude 6×10^{-6} m s^{-1} in a band 20 km wide in Earth's core, thus setting a level for their detectability.

At semidiurnal periods, inertial modes with azimuthal dependence $\cos 2\phi$ that are viscously coupled to the CMB will produce a semidiurnal wobble as angular momentum of Earth must be conserved. In principle this wobble should be detectable using both VLBI and supergravimetry as changes in latitude of observatories but so far only limited success has been achieved in the search for evidence of these modes in the core (Aldridge and Cannon, 1993).

Gravity-inertio waves

The previously ignored buoyancy force can be returned by assuming the fluid is stratified. Stratification is characterized by

$$N^2 = -\frac{g}{\rho}\frac{d\rho}{dr}$$

where N is the Brunt-Väisälä frequency, g is acceleration due to gravity, and $\rho(r)$ is the fluid density assumed to be a function of spherical radius r. (The $-$ sign is included to ensure $N^2 > 0$ as $\rho(r)$ is a decreasing function of r.)

With rotation both coriolis and buoyancy forces act on a fluid parcel. If buoyancy dominates coriolis in the sense that $N > 2\Omega$ then wave gravity wave frequencies are such that

$$2\Omega \leq \omega \leq N$$

while if stratification is very weak so that $N < 2\Omega$, then

$$2\Omega \geq \omega \geq N.$$

In the presence of stratification, the characteristics shown in Figure G54 become curved (Dintrans *et al.*, 1999).

In principle gravity waves should be detectable at Earth's surface since there must be a redistribution of mass when fluid parcels oscillate (Crossley *et al.*, 1991). Changes in gravity are due to this redistribution of mass directly as well as the movement of Earth's surface through the gradient of the gravity field. Amplitudes are extremely small, estimated at 10^{-11} m s^{-2}, and are only likely to be detectable by using a global array of superconducting gravity meters as organized under the Global Geodynamics Project (GGP). Successful detection of gravity waves in Earth's core would constrain the *core's density* (*q.v.*) distribution as has already been demonstrated for the Sun (see also *Helioseismology*).

Instability of inertial modes

The stable inertial modes described above can be destabilized by straining of the fluid streamlines (Kerswell, 1993, 2002). This fact has significant implications for Earth's core as it provides for a mechanism to pump energy parametrically into the core at a diurnal rate and cause a disturbance to grow at a rate determined by the strain. For Earth, this strain is very small leading timescales of instability growth of several thousand years (Aldridge and Baker, 2003) (see also *Turbulence in the core*). The strain comes from two sources, Earth's *precession* (*q.v.*) that introduces a predominately shearing of the core fluid's streamlines and Earth's semidiurnal tide which deforms the streamlines into ellipses. The former is small because of the relatively long period of the precession compared to the diurnal rotation while the latter owes its small size to the amplitude of tidal deformation in the core.

At present there is a small amount of evidence for the existence of parametric instability in Earth's fluid core. Based on laboratory observations of rotating parametric instabilities (Aldridge, 2003), a search has been made for a signature of these instabilities in records of relative *paleointensity* (*q.v.*) obtained from seafloor sediments. Initial results of this work (Aldridge and Baker, 2003) confirmed that a 400 ka record of relative paleointensities yielded a sequence of geophysically plausible growths and decays of magnetic field intensity over several thousand years that is consistent with what would be expected for rotating parametric instabilities. On longer timescales, tidal deformation and precessional forcing have been identified in paleomagnetic records. For example, variation in relative paleomagnetic intensity corresponding to Earth's obliquity has been reported (Kent and Opdyke, 1977). Although the origin of these variations has been often considered to be climatic, recent paleomagnetic records have revealed robust evidence of external forcing that is independent of climatological variations. Spectral analysis of the relative paleointensity record from ocean sediments (Channell *et al.*, 1997) show modulation of intensity corresponding to orbital obliquity and eccentricity periods. The inclination error found in a 2 Ma long record from the west Caroline basin (Yamazaki and Oda, 2002) correlates to orbital eccentricity.

Although gravitational energy released through compositional convection is considered to drive the core's geodynamo that maintains the geomagnetic field, other phenomena like rotational parametric instabilities may prove to play a significant role in maintaining the geodynamo.

Keith Aldridge

Bibliography

Aldridge, K.D., 2003. Dynamics of the core at short periods: theory, experiments and observations. In Jones, C.A., Soward, A.M., and Zhang, K. (eds.), *Earth's Core and Lower Mantle, The Fluid Mechanics of Astrophysics and Geophysics.* London: Taylor & Francis.

Aldridge, K.D., and Baker, R.E., 2003. Paleomagnetic intensity data: a window on the dynamics of Earth's fluid core? *Physics of the Earth and Planetary Interiors*, **140**: 91–100.

Aldridge, K.D., and Cannon, W.H., 1993. A search for evidence of short period polar motion in VLBI and supergravimetry observations. *Proceedings of the IUGG XX Assembly, Symposium U6: Dynamics of the Earth's Deep Interior and Earth Rotation*, American Geophysical Union Geophysical Monograph 72, pp. 17–24.

Aldridge, K.D., and Toomre, A., 1969. Axisymmetric inertial oscillations of a fluid in a rotating spherical container. *Journal of Fluid Mechanics*, **37**: 307–323.

Aldridge, K.D., Lumb, L.I., and Henderson, G., 1988. Inertial modes in the Earth's fluid outer core. *Proceedings of the International Union of Geodesy and Geophysics XIX Assembly, Symposium U2: Instability within the Earth and core dynamics*, American Geophysical Union Monograph 46, pp. 13–21.

Channell, J.E.T., Hodell, D.A., and Lehman, B., 1997. Relative geomagnetic paleointensity and $\delta^{18}O$ at ODP Site 983 (Garder Drift, North Atlantic) since 350 ka. *Earth and Planetary Science Letters*, **153**: 103–118.

Crossley, D., Hinderer, J., and Legros, H., 1991. On the excitation, detection and damping of core modes. *Physics of the Earth and Planetary Interiors*, **116**: 68–97.

Dintrans, B., Rieutord, M., and Valdettaro, L., 1999. Gravito-inertial waves in a rotating stratified spherical shell. *Journal of Fluid Mechanics*, **398**: 271–297.

Greenspan, H., 1969. *The Theory of Rotating Fluids.* Cambridge: Cambridge University Press.

Hollerbach, R., and Kerswell, R.R., 1995. Oscillatory internal shear layers in rotating and precessing flows. *Journal of Fluid Mechanics*, **298**: 327–339.

Kent, D.V., and Opdyke, N., 1977. Paleomagnetic field intensity variations recorded in a Brunhes epoch deep-sea sediment core. *Nature*, **266**: 156–159.

Kerswell, R.R., 1993. The instability of precessing flow. *Geophysical and Astrophysical Fluid Dynamics*, **72**: 107–114.

Kerswell, R.R., 2002. Elliptical instability. *Annual Review of Fluid Mechanics*, **34**: 83–113.

Noir, J., Brito, D., Aldridge, K., and Cardin, P., 2001. Experimental evidence of inertial waves in a precessing spheroidal cavity. *Geophysical Research Letters*, **19**: 3785–3788.

Rieutord, M., Georgeot, B., and Valdettaro, L., 2001. Inertial waves in a rotating spherical shell: attractors and asymptotic spectrum. *Journal of Fluid Mechanics*, **435**: 103–144.

Tilgner, A., 1999. Driven inertial oscillations in spherical shells. *Physical Review E*, **59**: 1789–1794.

Yamazaki, T., and Oda, H., 2002. Orbital influence on Earth's magnetic field: 100000 year periodicity in inclination. *Science*, **295**: 2435–2438.

Zhang, K., Earnshaw, P., Liao, X., and Busse, F.H., 2001. On inertial waves in a rotating fluid sphere. *Journal of Fluid Mechanics*, **437**: 103–119.

Cross-references

Core Density
Core Turbulence
Core, Adiabatic Gradient
Core, Boundary Layers
Fluid Dynamics Experiments
Helioseismology
Paleointensity, Relative, in Sediments
Precession and Core Dynamics

GRÜNEISEN'S PARAMETER FOR IRON AND EARTH'S CORE

Introduction

The Grüneisen parameter γ is a necessary tool for assessing Earth's thermal properties. Here we emphasize aspects of γ that are pertinent to Earth's core. γ has various definitions, many of which are derivable one from another by thermodynamic identities. The following equation for γ quantifies the relationship between the thermal and elastic properties of a solid. The parameter γ can be considered as a measure of the change in pressure P resulting from an increase in energy density at constant volume V. It is dimensionless, since pressure and $\Delta\mathcal{U}/V$ have the same units

$$\gamma = V\left(\frac{\partial P}{\partial \mathcal{U}}\right)_V \quad \text{(Eq. 1)}$$

where $\mathcal{U}$ is the internal energy. From Eq. (1) it is seen that γ connects pressure and energy, and therefore, pressure and temperature. Elasticity properties of Earth's core (especially pressure) come from seismological data. Gamma is useful to obtain energy and temperature from these seismological data. Another example of the importance of γ is in the expression used for thermal pressure at high temperature

$$P_{\mathrm{TH}} = \frac{\gamma}{V} E_{\mathrm{TH}} \quad \text{(Eq. 2)}$$

the so-called Mie-Grüneisen relationship (E_{TH} is the thermal energy and P_{TH} is the thermal pressure). This equation is often used with an isothermal equation of state, $P(V,T_0)$. Equation (1) is the general statement of the relationship between pressure and energy of which Eq. (2) is a special case. Equation (2) can be considered as the historical presentation of γ, attributed to Grüneisen (1926), who derived the equation for pressure as a function of volume and temperature in an early version of lattice dynamics.

For many physical properties of the Earth's interior, we need to know the value of γ of a solid at the pressure and temperature of the Earth's interior, not just at ambient conditions. We need to know how γ changes with T, especially at high T, if at all, and how γ changes with pressure (or rather, volume), especially at high compression.

Thermodynamic derivations of gamma

In order to evaluate Eq. (1) in parameters representing measurable physical properties, start with the following equation

$$\left(\frac{\partial P}{\partial \mathcal{U}}\right)_V = \left(\frac{\partial P}{\partial T}\right)_V \Big/ \left(\frac{\partial \mathcal{U}}{\partial T}\right)_V \qquad \text{(Eq. 3)}$$

The definition of specific heat is $(\partial \mathcal{U}/\partial T)_V = C_V$ at unit mass (all quantities are per unit mass). The numerator on the right side of Eq. (3) is found from calculus: $(\partial P/\partial T)_V = -(\partial V/\partial T)_P/(\partial V/\partial P)_T = \alpha K_T$, where α is the volume thermal expansivity, and K_T is the isothermal bulk modulus. Equation (3) is therefore equivalent to

$$\left(\frac{\partial P}{\partial \mathcal{U}}\right)_V = \frac{\alpha K_T}{C_V} \qquad \text{(Eq. 4)}$$

Using Eq. (4) in Eq. (1), the most useful definition of γ is found:

$$\gamma = V\left(\frac{\partial P}{\partial \mathcal{U}}\right)_V = \frac{\alpha K_T V}{C_V} \qquad \text{(Eq. 5)}$$

γ as given by Eq. (5) is composed of individual measurable physical properties, each of which varies significantly with temperature. The ratio of these properties as given by Eq. (5), however, does not vary greatly with temperature, and often not at all. There are many approximations to Eq. (5) for γ. The few approximations we will use here will be given special subscripts and names.

Another method of finding γ involves adiabatic compression. Start with one of Maxwell's relationships:

$$\left(\frac{\partial T}{\partial V}\right)_S = -\left(\frac{\partial P}{\partial \mathcal{S}}\right)_V \qquad \text{(Eq. 6)}$$

Expand the right side of Eq. (6): $(\partial P/\partial \mathcal{S})_V = (\partial P/\partial T)_V\ (\partial T/\partial \mathcal{S})_V$. By using $(\partial P/\partial T)_P = \alpha K_T$ and $(\partial \mathcal{S}/\partial T)_V = C_V$, the right side of this equation becomes $-T\alpha K_T/C_V$. Using Eq. (5), the right side becomes $T\gamma/V$, so that Eq. (6) can also be written as

$$\gamma = -\left(\frac{\partial \ln T}{\partial \ln V}\right)_S = \left(\frac{\partial \ln T}{\partial \ln \rho}\right)_S \qquad \text{(Eq. 7)}$$

Equation (7) is the thermodynamic basis for finding the adiabatic thermal gradient in the core and mantle.

Equations (1), (2), (5), and (7) are thermodynamically equivalent definitions of γ. The choice of the equation to use depends on the parameters at hand and the result desired.

Lattice dynamic derivations of gamma

The lattice dynamic view of a solid is that of statistical mechanics: a solid is composed of N atoms (where N is Avogadro's number); each atom is an oscillator having three degrees of freedom and connected to neighboring atoms by a spring. The solid's thermodynamic properties are found from the dynamics of $3N$ vibrations with modal frequencies $\nu_1, \nu_2, \nu_3, \ldots, \nu_{3N}$. Each modal vibration arises from a simple harmonic oscillation, the energy of which is given by an Einstein function, and the frequency of which is classically related to the atomic mass and the spring constant. For a monatomic solid, such as iron, all the masses are equal. A standard treatment of this subject shows that the Helmholtz free energy for a monatomic solid of N degrees of freedom is (Slater, 1939)

$$\mathcal{F} = \mathcal{U}_0 + \sum_j^{3N} kT \ln\left(1 - e^{-\hbar\nu_i/\kappa T}\right) \qquad \text{(Eq. 8)}$$

where κ is Boltzmann's constant and $\hbar$ is $h/2\pi$, where h itself is Planck's constant, and where the quantity under the summation sign is the Einstein function.

The isothermal equation of state $P(V)$ is found by differentiating Eq. (8) with respect to V at constant T, giving

$$P = -\left(\frac{\partial \mathcal{U}_0}{\partial V}\right)_T + \frac{1}{V}\sum_j^{3N} \gamma_j \frac{\hbar\nu_j}{\left(e^{\hbar\nu_j/\kappa T} - 1\right)} \qquad \text{(Eq. 9)}$$

Attention is directed to γ_j, called the mode gamma, which arises from the V derivative

$$\gamma_j = -\left(\frac{\partial \ln \nu_j}{\partial \ln V}\right)_T \qquad \text{(Eq. 10)}$$

Although γ_j is dimensionless, its value is influenced by the rate of change of mode frequency with volume. Thus, in order for γ_i to be nonzero, it must change with volume, and for it to be positive, it must decrease with volume. It is customary to make an assumption (called the quasiharmonic assumption) that the mode gammas depend on volume but are independent of temperature.

For core physics, where the temperature is high, the high-temperature limit of the above equations is needed. The Debye temperature Θ_D marks the division between the high temperature region and the quantum state region of a solid. For a monatomic solid, Θ_D is (Anderson, 1995)

$$\Theta_D = 251.2\left(\frac{\rho}{\mu}\right)^{1/3} \nu_m\, k \qquad \text{(Eq. 11)}$$

where ρ is density (in g cm^{-3}), μ is the atomic mass number (55.85 for iron), and ν_m is the mean sound velocity (in km s^{-1}), given in terms of the longitudinal and shear velocities as

$$\frac{3}{\nu_m^3} = \frac{1}{\nu_s^3}\left[2 + \left(\frac{\nu_p}{\nu_s}\right)^3\right] \qquad \text{(Eq. 12)}$$

The numerical factor in Eq. (11) is composed of the atomic constants $\hbar$, k, and N.

Body-centered cubic (bcc) iron is the phase of iron that exists at ambient conditions. There is much data on this well-known phase. We use the properties of bcc iron to explain principles such as the evaluation of Eq. (11), but our chief interest lies in hexagonal close-packed (hcp) iron, which is the most likely pure iron phase at core conditions. In the evaluation of Eq. (11), properties of ambient bcc iron are ρ, 7.87 g cm^{-3}; ν_p, 5.9 km s^{-1}; ν_s, 3.25 km s^{-1}; ν_m, 3.62 km s^{-1} (note that ν_m is only slightly larger than ν_s). This gives $\Theta_D(T = 300) = 415$ K at ambient conditions. Thus, above 415 K, bcc iron is a classical solid, and below, it is a quantum solid. Since the core is in the 5000–6000 K temperature range, its properties are in the high T regime. Since the value of Θ_D/T at core temperatures is of the order of 0.1, the exponential term in Eq. (8) is of the order of 10^{-5}, insignificant compared to unity.

To find the lattice dynamical equation for γ, divide Eq. (5) into two factors as follows:

$$\gamma = \frac{(\alpha K_T)}{C_V/V} \qquad \text{(Eq. 13)}$$

and evaluate each of the two parts separately from Eq. (8). From calculus, $\alpha K_T = (\partial P/\partial T)_V$. $C_V = (\partial \mathcal{U}/\partial T)_V$, where $\mathcal{U}$ is the internal energy. Using Eq. (9),

$$\left(\frac{\partial P}{\partial T}\right)_V = \frac{\hbar}{V}\sum y_j^2\left\{\gamma_j\frac{e^{y_i}}{(e^{y_i}-1)^2}\right\} \quad \text{(Eq. 14)}$$

where $y_j = \hbar v_j/kT$.

Since $\mathcal{U} = \mathcal{F} - (\partial\mathcal{F}/\partial T)_V$, then

$$\mathcal{U} = E_0 + kT\sum_{j=1}^{3N}\frac{y_j}{(e^{y_j}-1)} \quad \text{(Eq. 15)}$$

and

$$C_V = k\sum_{j=1}^{3N}\frac{y_j^2 e^{y_j}}{(e^{y_j}-1)^2} = \sum_{j=1}^{3N} C_{V_j} \quad \text{(Eq. 16)}$$

Following Eq. (13), there are major cancellations, leaving

$$\overline{\gamma} = \frac{\sum_j^{3N}\gamma_j C_{V_j}}{\sum_j^{3N} C_{V_j}} \quad \text{(Eq. 17)}$$

where the $\overline{\gamma}$ means that this Grüneisen parameter may be affected by the approximation of the quasiharmonic assumption. Note that all factors containing $y = \hbar v_j/kT$ have cancelled out, leaving the temperature dependence alone in C_V, but since there is input of C_V in both the numerator and the denominator, there is very little temperature dependence in either the low- or high-temperature regimes. In the very high-temperature regime (such as found in the core), all C_{V_j} are equal to k, so the denominator in Eq. (17) is equal to $3Nk$, while the numerator is equal to $k\sum^{3N}\gamma_j$. Thus, the high-temperature limit of the lattice dynamical Grüneisen parameter is the arithmetic average of all mode gammas (Barron, 1957).

$$\overline{\gamma} = \frac{1}{3N}\sum^{3N}\gamma_j \quad \text{(Eq. 18)}$$

Simplifications made by assuming a Debye solid

Evaluation of Eq. (18) is complicated by the fact that N (Avogadro's number) is quite large. We wish to reduce the summation in this equation from a limit of $3N$ to a lower value. We accomplish this by invoking the assumptions of a Debye solid (Debye, 1912), which are: the solid is monatomic and isotropic; its frequency spectrum (modal frequency versus wave number k) is quadratic with a sharp cutoff at the maximum frequency called the Debye frequency, v_D; the slope of the longitudinal and shear v_j versus k curves is constant; and there are no optic modes. Barron (1957) showed that invoking the properties of a Debye solid reduces the sum over $3N$ modes in Eq. (18) to a sum of 3 modes.

$$\gamma_{ac} = \frac{1}{3}(\gamma_p + 2\gamma_s) \quad \text{(Eq. 19)}$$

where γ_p is related to the longitudinal sound velocity and γ_s to the shear sound velocity. In Eq. (19), γ_{ac} has replaced $\overline{\gamma}$ from Eq. (18) because acoustic information is the sole input to gamma.

The mode gammas can be expressed in terms of velocity derivatives (Anderson, 1995),

$$\gamma_p = \frac{1}{3} - \frac{V}{v_p}\left(\frac{\partial v_p}{\partial V}\right)_T \quad \text{(Eq. 20)}$$

$$\gamma_s = \frac{1}{3} - \frac{V}{v_s}\left(\frac{\partial v_s}{\partial V}\right)_T \quad \text{(Eq. 21)}$$

Sumino and Anderson (1984) evaluated γ for bcc iron using Eq. (5) and experimental values of α, K_T, and C_P. They found that $\gamma = 1.81$. They also evaluated γ, using the acoustic approximation to γ (Eq. (19)) and the measured elastic constant data under pressure and found $\gamma = 1.81$. Since the derivation of Eq. (19) starts with Eq. (5), it is concluded that the quasiharmonic assumption and the assumptions of the Debye solid do not modify the value of γ for bcc iron. This is a way of demonstrating that bcc iron is a Debye solid. Debye's theory of a solid assumes that the whole vibrational spectrum can be represented by the long wave limit where frequency is proportional to the wave number ($v = vk$). All modes are acoustic, each with the same average velocity. Equation (19) can also be written in terms of the derivatives of the bulk modulus and the shear modulus, G

$$\gamma_{ac} = \frac{K_T}{6}\frac{(\partial K_S/\partial P)_T + (4/3)(\partial G/\partial P)_T}{K_S + (4/3)G} + \frac{K_T}{3}\frac{(\partial G/\partial P)_T}{G} - \frac{1}{6} \quad \text{(Eq. 22)}$$

(Stacey and Davis, 2004).

Stacey and Davis (2004) stated, "The acoustic formula (for γ) withstands critical scrutiny well. For application to the lower mantle it has no serious rival."

Thus, for the Debye solid, the $3N$ modal frequencies are replaced by one frequency, v_D, and one gamma,

$$\gamma_D = -\left(\frac{\partial \ln v_D}{\partial \ln V}\right) \quad \text{(Eq. 23)}$$

Since $\hbar v_0 = k\Theta_D$, the above can be replaced by

$$\gamma_D = -\left(\frac{\partial \ln \Theta_D}{\partial \ln V}\right)_T \quad \text{(Eq. 24)}$$

A variant of Eq. (24) is

$$\left(\frac{\Theta_D}{\Theta_{D_0}}\right) = \left(\frac{\rho}{\rho_0}\right)^{\gamma_D} \quad \text{(Eq. 25)}$$

which is true only for a Debye solid.

Equation (24) is used as follows. Measure Θ_D as a function of V, and thus determine γ as a function of V. Equation (25) is used to find a shift in Θ_D corresponding to a shift in density.

A stringent test of the applicability of the theory of Debye solid to properties of bcc iron is that the frequency spectra of the real solid (called the phonon density of states, PDOS) has a sharp cutoff at a frequency corresponding to the Debye temperature.

Proving epsilon (hcp) iron is a Debye solid

In the last section, bcc iron was proven to be a Debye solid. That is only one of many proofs that have been made for this solid. Early in the 20th century, a popular proof was to measure the entropy versus T data for temperatures below the Debye temperature and show that there was good agreement with the entropy calculated from the Debye theory. Unquestionably, bcc iron is taken as a Debye solid. It is not evident however, that hcp iron should be a Debye solid because this phase of iron has hexagonal symmetry, which requires two atoms in the lattice dynamic cell. Consequently, there are vibrational modes in which both atoms move in the cell, but the center of mass remains fixed. Such a vibration is an optical mode, and thus hcp iron has an optical branch in its frequency spectrum, which usually implies that there are modes with frequencies much higher than v_D.

For hcp iron, however, the two masses in the cell have equal values, and lattice dynamic theory shows that in this case the frequencies found in the optic branch are contained within a narrow band anchored close to the maximum acoustic frequency. This means that within the frequency spectrum $f(v)$, called the phonon density of states (PDOS), there is a clustering of modes near the maximum acoustic value of $f(v)$.

A criterion for the validity of the Debye solid is that the high-frequency edge of the PDOS is the same (or nearly the same) as ν_D, which is determined by Θ_D. Debye (1965) said, "it is not too important to know the details of the frequency spectrum because at high temperature one only has to know the number of degrees of freedom [see Eq. (18)] and at low temperature the high frequency modes carry less and less energy." It is the maximum frequency of the PDOS that is significant. The crucial test for hcp iron is to show that the PDOS has a high-frequency, sharp cutoff close in value to ν_D.

A recent paper by Giefers *et al.* (2002) reported the measured PDOS of hcp and bcc iron by nuclear inelastic scattering. Since hcp iron is a high-pressure phase and does not exist at room pressure, the experiment was done at $P = 28$ GPa. An experiment was also done for bcc iron at $P = 0$ GPa (resulting data for both shown in Figure G56). Their plots of the density of states, $G(E)$, are reported in units of energy (meV) instead of the usual units of frequency, but they report the cutoff in terms of temperature, so the comparison with the Debye temperature is eased. Note that for both phases of iron there is a sharp cutoff of the spectrum with only a very small percentage of modes seen above the cutoff. This is proof that both phases are Debye solids. Further proofs come from the value of Θ_D.

The authors report that the value of the temperature corresponding to the cutoff is 511 K for hcp iron (at 29 GPa) and 417 K for bcc iron (at $P = 0$). The value for hcp iron is to be compared with the measurement of Θ_D reported by Anderson *et al.* (2001), 521 K at 29 GPa and discussed in the next section. These values are sufficiently close to conclude that hcp iron is a Debye solid. The value of $\Theta_D = 417$ K for bcc iron reported by Giefers *et al.* (2002) is to be compared with $\Theta_D = 415$ K found from Eq. (11).

Anderson *et al.* (2001) also report that for hcp iron, the value of Θ_D at $P = 0$ is 446 K. This value can also be used to predict the cutoff of bcc iron by the bcc-hcp phase change using Eq. (25), and the ratio of the uncompressed density of bcc iron to hcp iron is (7.87/8.28) with $\gamma = 1.7$. The predicted Θ_D is 410 K, to be compared with 417 K from the cutoff of the PDOS for bcc iron.

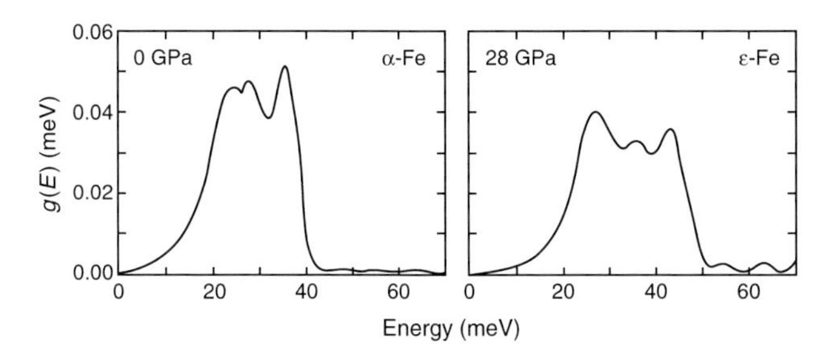

Figure G56 Experimental phonon density of states $g(E)$ of bcc iron and hcp iron by Giefers *et al.* (2002). For hcp iron, the pressure is 28 GPa. The cutoff of bcc iron at $P = 0$ is 40 meV (with a corresponding temperature of 417 K). For hcp iron, the cutoff is 45 meV (with a corresponding temperature of 511 K). The Debye temperature of bcc iron is 415 K, according to Eq. (11), and the Debye temperature of hcp iron is 520 K according to Anderson *et al.* (2001) and as seen in Figure G57.

Experimental determination of lattice γ(V) for iron from Eq. (24)

The state of the art in diamond anvil pressure cells at high pressure has improved so that by using a controlled intense x-ray beam from a synchrotron radiation facility, determination of high quality powder x-ray diffraction data is possible. (The intensity of the diffracted x-ray beam is a measurable function of pressure.) For a Debye solid, the intensity of the beam is given by the mean-square amplitude of atomic displacements $< u^2 >$. Classical x-ray theory relates $< u^2 >$ to the Debye temperature, which Gilvarry (1956) gives as

$$< u^2 > = \frac{c\hbar^2 T}{\mu k \Theta_D^2} \qquad \text{(Eq. 26)}$$

where c is a numerical constant, and μ is the atomic mass. Equation (26) with $c = 3$ is found in Willis and Pryor (1975). Using this equation and the measured values of $< u^2 >$ versus V, Anderson *et al.* (2001) measured the experimental values of Θ versus V for hcp iron from ambient pressure to $P = 300$ GPa; these values are plotted in Figure G57.

The value of γ versus V was then found from Eq. (24). The data are plotted in $\gamma - P$ space as the lower curve in Figure G58. It is seen that at 330 GPa, γ_{vib} decreases with pressure from the zero pressure value of 1.7 to 1.2, the normal trend for nonmetallic solids. But metals have an additional contribution to gamma not found in nonmetallic solids arising from conduction electrons (often called free electrons). This contribution to γ is especially important at high temperature. Since this electron contribution adds to the Helmholtz energy, γ as defined by Eq. (1) will be sensitive to the electronic contribution. This means that for metals, the gamma found in this section is only a part of that defined by Eq. (1) at core temperatures. We therefore give the gamma shown in the lower curve of Figure G58 a special name, γ_{vib}, indicating vibrational energy, or that arising from the PDOS.

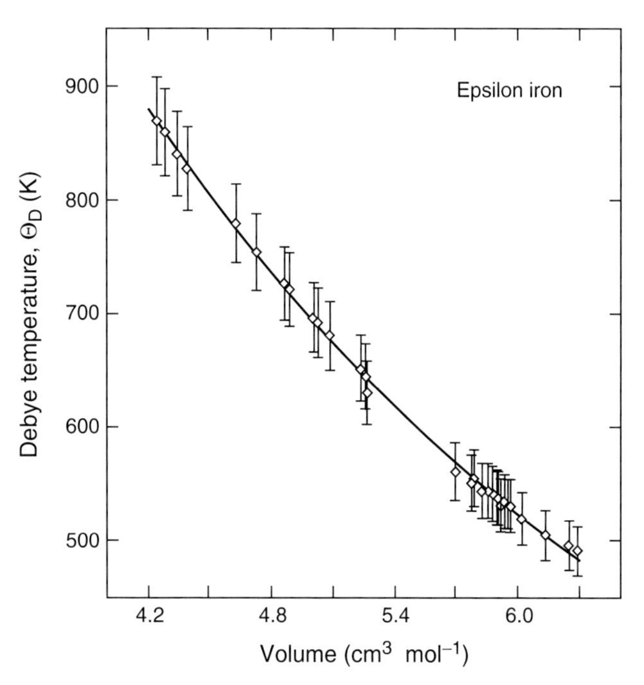

Figure G57 The experimental variation of Debye temperature Θ_D for hcp iron (Anderson *et al.*, 2001). For $P = 28$ GPa, $V = 5.95$ cm^3 mol^{-1} with a corresponding $\Theta_D = 519$ K.

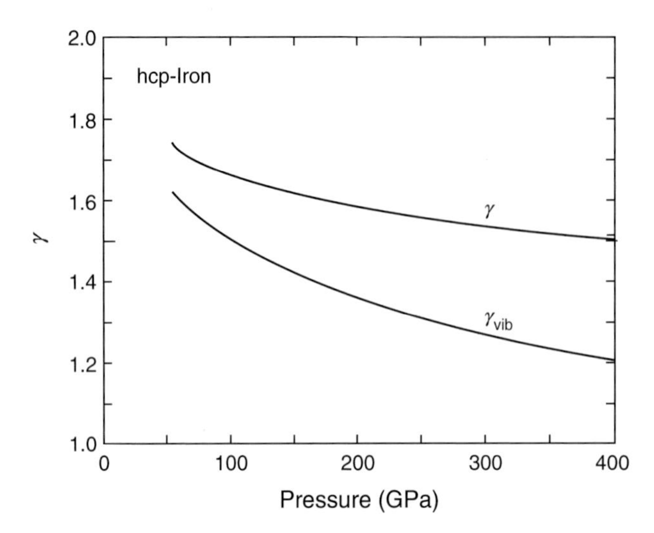

Figure G58 Variation of γ_{vib} and γ for pure hcp iron with pressure (Anderson *et al.*, 2003). For the core range of pressure, γ varies between 1.62 at 135 GPa and 1.51 at 330 GPa. The calculations were made with temperatures from the hcp solidus, so the values of γ represent the solid state edge of melting.

Accounting for the electronic contribution to gamma for iron

The electronic contribution to the specific heat in a metal has a dominant effect at low T and sometimes at high T. The electronic contribution to specific heat, for example, is given by Kittel (1956) as

$$C_{V_e} = \frac{1}{3}\pi^2 D(\varepsilon_F) k^2 T \qquad \text{(Eq. 27)}$$

where $D(\varepsilon_F)$ is the electronic density of states at the Fermi energy level.[1] The important point is that C_{V_e} increases steadily with T at high T, whereas the lattice specific heat levels out for $T > \Theta_D$ and remains independent of T (neglecting anharmonic terms). At 6000 K, the specific heat of iron is 37% greater than the classical value of $3k$ per atom, due to Eq. (27). From Eq. (27), the electronic Grüneisen parameter is

$$\gamma_e = \frac{\partial \ln D(\varepsilon_F)}{\partial \ln \rho} \qquad \text{(Eq. 28)}$$

Numerical evaluation of Eq. (28) (Bukowinski, 1977) gives $\gamma_e = 1.5$. Thus, for iron the thermodynamic gamma of Sections and consists of γ_{vib}, suitably modified by γ_e. The formalism for finding γ for metals is found in Bukowinski (1977); it requires the electronic specific heat found from the electronic density of states (EDOS). Anderson (2002a) found γ_e and γ from γ_{vib} by using the PDOS and C_{V_e}, as presented by Stixrude *et al.* (1997). The theory of Bukowinski (1977) results in the useful equation,

$$\gamma = \frac{C_{V_{vib}}}{C_V}\gamma_{vib} + \frac{C_{V_e}}{C_V}\gamma_e \qquad \text{(Eq. 29)}$$

where

$$C_V = C_{V_{vib}} + C_{V_e} \qquad \text{(Eq. 30)}$$

[1] For further information on the Fermi energy level for metals, see Kittel, Introduction to Solid State Physics, 2nd edn. New York: John Wiley & Sons, 1956.

Some of the detailed calculations for hcp iron are given in Table G9. Values in Table G9 were found for T increasing with P along the solidus of hcp iron, as calculated by Anderson *et al.* (2003). The value of γ descends from 1.62 at $P = 135$ GPa to 1.53 at 330 GPa, a gradual decrease. The values of γ in Table G9 are for the solid state edge of melting, called the solidus. It is seen that γ for the ICB pressure (330 GPa) is 1.53, and for the CMB (135 GPa) it is 1.62. Figure G58 shows γ (the top curve), as well as γ_{vib}, versus P.

First principles calculations of gamma for liquid iron

The value of γ is likely to be different for the liquid state than for the solid state. Verhoogen (1980) first emphasized that the liquid state γ_l should be used for calculations of properties of Earth's core. The difference between the values of the liquid gamma (γ_l) and the solid gamma (γ_s) for iron is quite large at low pressure but small at high pressure. The value of γ_l is 2.44 at $P = 0$ (Stevenson, 1981), but that of γ_s is 1.66 at $P = 0$ (Boehler and Ramakrishnan, 1980) for bcc iron. The value of γ_l decreases with pressure, becoming 1.63 at 30 GPa (Chen and Ahrens, 1997), and then changes much more slowly with further pressure, as we shall see. Measurements and/or calculations of γ_l at core pressures have long been needed. Important progress has been made in the theory of thermodynamic properties of condensed matter by calculations of parameter-free *ab initio* techniques using quantum mechanics. The use of *ab initio* techniques for several decades by many physicists has resulted in considerable advances in understanding of solid state properties. Recently, *ab initio* methods have advanced to the point that successful calculations of properties of the liquid state at high pressure and high temperature have been made.

Using *ab initio* methods, Alfè *et al.* (2002a) calculated 1.51 for γ_l at $P = 330$ GPa for iron along with the liquidus, giving $T_m = 6350$ K. This is quite close to the value of γ_s (1.53) at $P = 330$ GPa for hcp iron reported by Anderson *et al.* (2003), who found $T_m = 6050$ K for the solidus (see Table G9). Thus, $\Delta\gamma$ at the ICB pressure is small, substantially smaller than at the CMB pressure, where $\gamma_l = 1.52$ (Alfè *et al.*, 2002a), while $\gamma_s = 1.62$ (Anderson *et al.*, 2003).

For the solid state, Stixrude *et al.* (1997) reported $\gamma_s = 1.5$ for hcp iron at ICB pressure from their *ab initio* calculations. Alfè *et al.* (2001) reported $\gamma_s = 1.52$ for hcp iron at $P = 330$ GPa, where the solidus is found to be 6250 K, as compared to 1.53, as obtained by Anderson *et al.* (2003). Thus, three laboratories have found very close values of γ_s at 330 GPa, $1.5 < \gamma_s < 1.53$.

The various values for gamma are plotted on the phase diagram of iron in Figure G59 (in the diagram, T is versus V with isobars as solid lines).

We see that as the pressure increases, at core conditions, the difference between liquidus and solidus temperature values decreases and $\Delta\gamma$ decreases, as well. This is interpreted to mean that as pressure increases, the volume of crystallization diminishes. Indeed, Vocadlo *et al.* (2003) report $\Delta V_m = 0.77$ cm mol^{-1} at 330 GPa, while Anderson (2002b) report $\Delta V_m = 0.55$ also at 330 GPa from an analysis of shock wave data. The values of γ_l and γ_s approach each other at deep core pressures.

Core's gamma: Consideration of impurities in iron

Structural changes

One approach for finding the properties of the core is to learn all that is possible of pure iron at core pressure and then insert a certain quantity of light elements into the iron structure and determine the resulting changes. A number of problems arise, one of which is that the crystallographic structure is reported to change from hcp when impurities are added. The consensus is that hcp is the stable phase of pure iron, but there is strong evidence that for the light impurities in iron, light silicon, sulfur, and oxygen, the high-pressure stable structure is bcc.

Table G9 Heat capacity, γ_{vib} and γ along the calculated solidus of hcp iron

V (cm^3 mol^{-1})	T_m (K)[a]	P (GPa)	C_{V_e}[a]	γ_{vib}[a]	γ[a]
6.060	2790	55.0	1.331	1.62	1.74
6.000	2840	58.1	1.336	1.61	1.73
5.800	3040	71.6	1.360	1.57	1.70
5.600	3270	88.0	1.390	1.53	1.68
5.500	3400	97.5	1.407	1.51	1.67
5.300	3690	119.5	1.445	1.47	1.64
5.181	3880	135.0	1.472	1.44	1.62
5.100	4030	146.5	1.491	1.43	1.62
5.000	4230	162.2	1.517	1.41	1.61
4.900	4440	179.7	1.545	1.35	1.60
4.800	4660	199.2	1.575	1.37	1.59
4.700	4910	220.8	1.607	1.35	1.58
4.620	5120	240.0	1.633	1.32	1.56
4.600	4170	244.9	1.640	1.32	1.56
4.500	5460	271.7	1.675	1.30	1.55
4.400	5770	301.7	1.712	1.28	1.54
4.316	6050	330.0	1.743	1.25	1.53
4.300	6100	335.3	1.749	1.25	1.53
4.200	6450	372.8	1.788	1.23	1.52
4.100	6830	414.9	1.827	1.20	1.50

[a]Anderson *et al.* (2003).

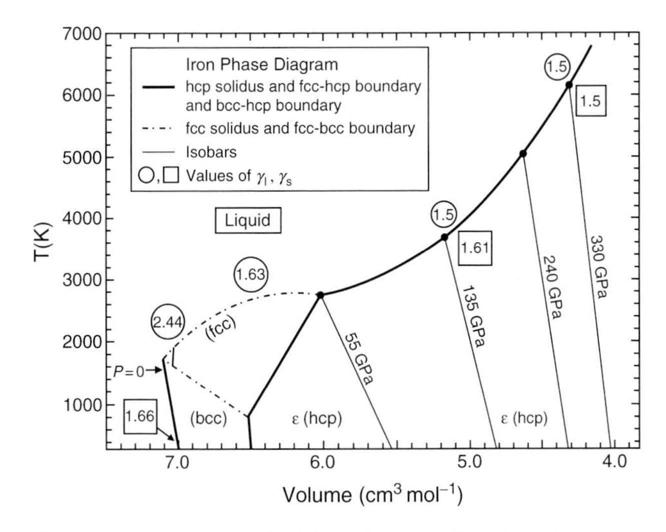

Figure G59 Values of γ_s (solid state) and γ_ℓ (liquid state) plotted on the phase diagram of iron (T versus V with isobars as shown).

Vocadlo *et al.* (2003) suggest that a mole of 4% silicon or sulfur, either in combination or separately, may change the structure of hcp iron to the bcc phase above 3000 K at high pressure.

When light impurities are placed in iron, its density decreases. The percentage change in density, called the "core density deficit" (cdd), is reported to be 4% to 7% for the outer core. The cdd is used to find the ΔT drop at the ICB, giving the value of the liquidus temperature at $P = 330$ GPa for the core. This ΔT is called the "freezing point depression."

The effects of Si, S, and O impurities on the value of γ_l

Alfè *et al.* (2002a) found from *ab initio* calculations that the core density deficit due to O, Si, and S, is 6.6%. The concentrations of these light elements were found in a series of papers in which chemical equilibrium between coexisting solid and liquid was treated. *Ab initio* calculations by Alfè *et al.* (2002b) incorporating as a boundary condition the seismically determined density jump at the inner core-outer core boundary found that substantial oxygen was required in the outer core because silicon and sulfur could not account for the size of the jump. From their conclusions of the amount and type of impurities in the core, Alfè *et al.* (2002c) found the freezing point depression at $\Delta T = -700$ K $\pm$ 100 K. In a subsequent paper, Alfè *et al.* (2002d) reported the freezing point depression to be $\Delta T = -800$ K, a value they have used in subsequent work. They estimated the core density deficit, the isobaric density of pure hcp iron, less the corresponding density of the core, to be 6.6% (Alfè *et al.*, 2002c). The core temperature at the ICB was found by subtracting 800 K from the value 6350 K (the *ab initio* calculation of T_m(330 GPa) for hcp iron), giving 5550 K.

As seen in the previous paragraph, it is concluded from *ab initio* studies that

1. The difference between γ_l for pure iron and γ_l for the core at the ICB is only ~0.02 despite the addition of 9% oxygen and 9% Si/Si to the hcp structure. This difference is trivial.
2. The difference between γ_l and γ_S at the ICB for both the core and hcp iron is ~0.01; again, it is trivial.
3. The value of γ_l at the ICB of the core found by Alfè *et al.* (2002b) is so close to 1.5 that it has been taken to be 1.5 in the paper by Gubbins *et al.* (2003) in which D. Alfè and G.D. Price were coauthors.
4. The value of $\gamma_l = 1.5$ at the ICB is associated with a cdd of 6.6% and a freezing point depression of $-$ 800 K.
5. The value of γ_l is virtually unchanged from the CMB pressure to the ICB pressure.

The effect of Ni in the core on the value of γ_c

The relative abundances of Fe and Ni in the core are probably about 10^{-1} (McDonough and Sun, 1995). Thus, we consider the effect of $Fe_{0.90}$-$Ni_{0.10}$ in place of Fe. Mao *et al.* (1990) found that ρ for $Fe_{0.80}$-$Ni_{0.20}$ at 330 GPa and 300 K is about 2% higher than for pure Fe. Therefore, the alloy $Fe_{0.90}$-$Ni_{0.10}$ at the ICB is 1% denser than hcp iron.

Table G10 Values of γ, T_m, and α for pure hcp iron and Earth's core

Parameters	*Ab initio* calculations[a]	Thermal physics measurements and theory[b]	K-primed EoS and seismic data[c]
Pure hcp iron, solidus, 330 GPa			
γ_s	1.50	1.53	
T_m (K)	6250	6050	
$\alpha(10^{-5}\ K^{-1})$	1.07	0.7	
Pure hcp iron, solidus 135 GPa			
γ_s	1.519	1.62	
T_m (K)	4734	4062	
$\alpha(10^{-5}\ K^{-1})$	1.78	2.5	
Pure hcp iron, liquidus 330 GPa			
γ_l	1.518		
T_m (K)	6350		
$\alpha(10^{-5}\ K^{-1})$	1.072		
Pure hcp iron, liquidus 135 GPa			
γ_ℓ	1.520		
T_m (K)	4734		
$\alpha(10^{-5}\ K^{-1})$	1.78		
Core, liquidus, 330 GPa			
γ_l	1.50		1.391
T_m (K)	5550	5100	5001
$\alpha(10^{-5}\ K^{-1})$	0.99		0.971
Core, liquidus, 135 GPa			
γ_l	1.53		1.443
T_m (K)	4111	4100	3739
$\alpha(10^{-5}\ K^{-1})$	1.70		1.804

[a]Alfè *et al.* (2001, 2002a-d), Vocadlo *et al.* (2003).
[b]Anderson (2002b); Anderson *et al.* (2003), Isaak and Anderson (2003).
[c]Stacey and Davis (2004).

A 1% increase in density would have virtually no effect on γ_l or γ_s because γ is nearly constant with pressure in the vicinity of the ICB (Figures G58 and G59). From the data plotted in both figures, a 1% change in density changes γ only slightly in the third significant figure. Thus, the value of γ_l is insensitive to the addition of nickel to iron.

In the *ab initio* approach, all physical properties are found from derivatives of the Helmholtz free energy, *F*, which is made to account for all atoms (Fe and impurities, if any) in their respective lattice sites. The physical properties of the core are found from the core's free energy. These properties are different from those found from the free energy of pure iron, except for the value of γ_l, which is virtually the same for both cases. How can this be explained? The value of γ_l changes with volume, but in the case of iron, the value of gamma changes very slowly with volume at core pressures, as shown in Figure G58. Consider the drop in temperature called the freezing point depression, $\Delta T_m = -800$ K. The relative change in volume due to freezing point depression is $\Delta V/V = -\Delta\rho/\rho = \alpha\Delta T$ at the ICB. The value of α varies from 1.07×10^{-5} at 6350 K and 330 GPa (for pure iron) to 0.99×10^{-5} and 330 GPa for the core (D. Alfè, personal communication). If we know α and ΔT, then we can calculate $\Delta\rho/\rho$, the percent density change. The average $\alpha \times \Delta T$ gives $\Delta\rho/\rho = -0.85\%$, less than but opposite to the 1% increase arising from placing 10% Ni in the Fe-Ni alloy.

Finding core properties independent of pure iron properties

Stacey (2000) introduced a new equation of state called the K-primed approach because he found all EoSs used in geophysical treatment of Earth's mantle to be in error with regard to the higher derivatives, especially the derivative of K_S with respect to *V*. Stacey successfully applied his K-primed equation of state to the Earth's mantle and core. In Stacey and Davis (2004), thermal core properties were extracted directly from seismic data of Earth without the intermediate step of using properties of pure iron. The values of gamma for the core they report, to three significant figures, are 1.39 at the CMB pressure and 1.44 at the ICB pressure. They also report that the values of T_m for the liquidus are 5000 K and 3739 K, for the ICB and CMB, respectively.

Summary and conclusions

Our prime focus is on the value of γ at pressures of 135 GPa (CMB) and 330 GPa (ICB) for the liquidus and the solidus of hcp iron and for the liquidus of the outer core and the solidus of the inner core. The values of γ and associated properties of thermal expansivity α and melting temperature T_m described in the previous sections are summarized in Table G10.

It is noted that the values of γ_s at the ICB from the two approaches (*ab initio* and thermal physics) agree quite well. There is good agreement on the value of γ_l, but the value reported from the K-primed EoS approach is about 0.1 less than that reported from the *ab initio* approach.

The *ab initio* approach yields the largest range of results, giving values for both γ_l and γ_s and for both hcp iron and the core, whereas the thermal physics approach is limited to γ_s for hcp iron and the inner core. The K-primed EoS approach is limited to γ_l for the outer core and γ_s for the inner core. An advantage of the K-primed EoS approach is that information about the chemistry—in particular, the concentrations of the various impurity elements—is not required to obtain γ_l of the core. However, this is also a disadvantage because no information can be given to geochemists about impurities. In contrast, the *ab initio* approach can give valuable results about core impurities: it was shown that sulfur and silicon as impurities in iron cannot account for the seismic density jump at the ICB, but about 8 mol% oxygen in the outer core is consistent with the seismic jump.

From the results of the *ab initio* calculations, the value of γ_l is virtually constant from the ICB pressure to the CMB pressure, varying by 1.3%. The variation using the K-primed EoS approach is somewhat larger, 5.7%. If γ_l is assumed to be constant over this range, the relationship between T_m and density ρ given by Eq. (7) is simplified,

$$\left(\frac{T_{m_1}}{T_{m_0}}\right) = \left(\frac{\rho_1}{\rho_0}\right)^{\gamma} \qquad \text{(Eq. 31)}$$

Using the ratio of the seismic density of the core at the CMB to that at the ICB (12166.34/9903.49), Eq. (31) becomes

$$\left(\frac{T_{m_1}}{T_{m_0}}\right) = 0.7344 \quad \text{for} \quad \gamma_l = 1.5$$

$$= 0.7466 \quad \text{for} \quad \gamma_l = 1.42$$

Thus, if Eq. (31) is used to find T_m (CMB) from T_m (ICB), the answer will differ by only 1.2% if one uses the *ab initio* gamma in comparison with the K-primed EoS gamma (starting out with the same value of T_{m_0}). If one uses the *ab initio* data or the thermal physics data and is satisfied that the value of $T_m(P)$ is rounded to three significant figures, then the value of γ_l can be rounded to three significant figures for the entire outer core region. The value of $\gamma_l = 1.5$ was used by Gubbins *et al.* (2003) throughout the core in their analysis of the geodynamo. The value of γ_s for the inner core decreases very slowly with the depth from its value at the ICB pressure. Table G9 shows that γ_s (listed in the table as gamma) drops from 1.53 at the ICB (330 GPa) to 1.52 at a pressure of 372.8 GPa; by interpolation gamma is 1.525 at the Earth's center, which has a pressure of 363.9 GPa. One may as well use $\gamma_s = 1.53$ for calculation of T_m at the Earth's center. We need to know the temperature of the core; this requires ΔT_m, the freezing point depression from pure iron. Alfè *et al.* (2002c) report $\Delta T_m = -800$ K. Thus, for the temperature at the ICB pressure, one finds $T_m(\text{core}) = 5\,550$ K from the solidus at ICB (6250 K) for the *ab initio* approach, and $T_m(\text{core}) = 5\,250$ K from the thermal properties approach. The K-primed EoS approach leads to $T_m(\text{core}) = 5000$ K. This calculation using Eq. (31) is $T_m(\text{E center})/T_m(\text{ICB}) = 1.1213$, giving, for Earth's center, 6111 K for the *ab initio* ICB value (5450 K) and 5456 K for the thermal physics ICB value (5250 K). The K-primed EoS approach gives 5030 K for Earth's center.

Acknowledgments

The author gratefully acknowledges the data on γ, $T_m(K)$, and α (listed in Table G10), from Dario Alfè's work (some unpublished), which were sent to him on request.

Orson L. Anderson

Bibliography

Alfè, D., Price, G.D., and Gillan, M.J., 2001. Thermodynamics of hexagonal close packed iron under Earth's core conditions. *Physical Review B*, **64**: 1–16.04123,

Alfè, D., Price, G.D., and Gillan, M.J., 2002a. Iron under Earth's core conditions: thermodynamics and high pressure melting from ab-initio calculations. *Physical Review B*, **65**(118): 1–11.

Alfè, D., Gillan, M.J., and Price, G.D., 2002b. Composition and Earth's core constrained by combining ab initio calculations and seismic data. *Earth and Planetary Science Letters*, **195**: 91–98.

Alfè, D., Gillan, M.J., and Price, G.D., 2002c. Ab-initio chemical potentials of solid and liquid solutions and the chemistry of the Earth's core. *Journal of Chemical Physics*, **116**: 7127–7136.

Alfè, D., Gillan, M.J., and Price, G.D., 2002d. Complementary approach to ab-initio calculations of melting properties. *Journal of Chemical Physics*, **116**: 6170–6177.

Anderson, O.L., 1995. *Equations of State for Geophysics and Ceramic Science*. New York: Oxford University Press, 405 pp.

Anderson, O.L., 2002a. The power balance at the core-mantle boundary. *Physics of the Earth and Planetary Interiors*, **131**: 1–17.

Anderson, O.L., 2002b. The three dimensional phase diagram of iron. In Karato, S., Dehant, V., and Zatman, S. (eds.), *Core Structure and Rotation*. Washington, DC: American Geophysical Union.

Anderson, O.L., Dubrovinsky, L., Saxena, S.K., and Le Bihan, T., 2001. Experimental vibrational Grüneisen ratio value for ε-iron up to 330 GPa at 3000 K. *Geophysical Research Letters*, **28**: 399–402.

Anderson, O.L., Isaak, D.G., and Nelson, V.E., 2003. The high-pressure melting temperature of hexagonal close-packed iron determined from thermal physics. *Journal of Physics and Chemistry of Solids*, **64**: 2125–2131.

Barron, T.H.K., 1957. Grüneisen parameters for the equation of state of solids. *Annals of Physics*, **1**: 77–89.

Boehler, R., and Ramakrishnan, J., 1980. Experimental results on the pressure dependence of the Grüneisen parameter: A review. *Journal of Geophysical Research*, **85**: 6996–7002.

Bukowinski, M.S.T., 1977. A theoretical equation of state for the inner core. *Physics of the Earth and Planetary Interiors*, **14**: 333–339.

Chen, G.Q., and Ahrens, T.J., 1977. Sound velocities of liquid γ and liquid iron under dynamic compression (abstract). *EOS Transactions of the American Geophysical Union*, **78**: P757.

Debye, P., 1912. Theorie der spezifischen wärmen. *Annals of Physics* (Berlin), **39**: 789–839.

Debye, P., 1965. The early days of lattice dynamics. In Wallis, R.I. (ed.) *Lattice Dynamics: Proceedings of an International Conference*. Oxford, UK: Pergamon Press, pp. 9–17.

Giefers, H., Lubbers, P., Rupprecht, K., Workman, G., Alfè, D., and Chumakov, A.I., 2002. Phonon spectroscopy of oriented hcp iron. *High Pressure Research*, **22**: 501–506.

Gilvarry, J.J., 1956. The Lindemann and Grüneisen laws. *Physical Review*, **102**: 308–316.

Grüneisen, E., 1926. The state of a solid body. In *Handbuch der Physik*, vol. 10, Berlin: Springer-Verlag, pp. 1–52, (English translation, NASA RE 2-18-59W, 1959).

Gubbins, D., Alfè, D., Masters, G., Price, G.D., and Gillan, M.J., 2003. Can the Earth's dynamo run on heat alone? *Geophysical Journal International*, **155**(2): 609–622, doi: 10.1046/j.1365-246X.2003.02064.x.

Gubbins, D., Alfè, D., Masters, G., Price, G.D., and Gillan, M., 2004. Gross thermodynamics of 2-component core convection. *Geophysical Journal International*, **157**: 1407–1414.

Isaak, D.G., and Anderson, O.L., 2003. Thermal expansivity of hcp iron at very high pressure and temperature. *Physica B*, **328**: 345–354.

Kittel, C., 1956. *Introduction to Solid State Physics*, 2nd edn. New York: Wiley, 617 pp.

Mao, H.K., Wu, Y., Chen, C.C., Shu, J.F., and Jephcoat, A.P., 1990. Static compression of iron to 300 GPa and $Fe_{0.8}Ni_{0.2}$ alloy to 200 GPa. *Journal of Geophysical Research*, **95**: 21691–21693.

McDonough, W.F., and Sun, S., 1995. The composition of the Earth. *Chemical Geology*, **120**: 228–253.

Slater, J.C., 1939. *Introduction to Chemical Physics*, 1st edn. New York: McGraw-Hill.

Stacey, F.D., 2000. The K-primed approach to high pressure equations of state. *Geophysical Journal International*, **128**: 179–193.

Stacey, F.D., and Davis, P.M., 2004. High pressure equations of state with applications to the lower mantle and core. *Physics of the Earth and Planetary Interiors*, **142**: 137–184.

Stevenson, D.J., 1981. Models of the Earth's core. *Science*, **214**: 611–619.
Stixrude, L., Wasserman, E., and Cohen, R.E., 1997. Composition and temperature of the Earth's inner core. *Journal of Geophysical Research*, **102**: 24729–24739.
Sumino, Y., and Anderson, O.L., 1984. Elastic constants of minerals. In Carmichael, R.S. (ed.) *CRC Handbook of Physical Properties of Rocks*. Boca Raton, FL: CRC Press, pp. 39–138.
Verhoogen, J., 1980. *Energetics of the Earth*. Washington, DC: National Academy of Sciences, 139 pp.
Vocadlo, L., Alfè, D., Gillan, M.J., Wood, I.G., Brodholt, J.P., and Price, G.D., 2003. Possible thermal and chemical stabilization of body-centred-cubic iron in the Earth's core. *Nature*, **424**: 536–538.
Willis, B.T.M., and Pryor, A.W., 1975. *Thermal Vibrations in Crystallography*, 1st edn. London, UK: Cambridge University Press.

Cross-references

Core Composition
Core Density
Core Properties, Theoretical Determination
Core, Adiabatic Gradient
Core-Mantle Boundary, Heat Flow Across
Higgins-Kennedy Paradox
Shock Wave Experiments

H

HALLEY, EDMOND (1656–1742)

Edmond Halley (Figure H1) was born in London in 1656 and educated at St. Paul's School in London and the Queen's College, Oxford, which he left in 1676 without a degree to catalog southern stars and observe a transit of Mercury on the island of St. Helena. He also measured magnetic inclination on this voyage, in the Cape Verdes and on St Helena. On his return the Royal Society elected him a fellow. He visited Johannes Hevelius in Danzig and G.D. Cassini in Paris and observed with them. He became clerk to the Royal Society in 1686. He prompted Newton to write the *Philosophiae Naturalis Principia Mathematica* and published it himself. He was often at sea, surveying and diving for salvage until he was elected Savilian Professor of Geometry in Oxford in 1704. He became the second Astronomer Royal in 1720.

The Westward Drift (q.v.)

In 1680, it was known that the major deviations of the Earth's field from that of a uniformly magnetized sphere appeared to move westward. Halley selected representative observations of the variation (magnetic declination) from 1587 to 1680, and from New Zealand (170° E) to Baffin Bay (80° W). They came from mariners' reports, and perhaps from collections not to be found now. He found that the westward drift was not the same everywhere (Halley, 1683). He argued that the extensive anomalies had much deeper sources than iron near the surface that Descartes (Descartes, 1644) and *Gilbert* (*q.v.*) proposed. He suggested that the Earth had a shell with one pair of poles and a core with another pair, the two in relative rotation (Halley, 1692). Would there be life in the space between the core and the shell? Other worlds in other places and at other times were then seriously contemplated.

The Atlantic cruises

In October 1698, Halley sailed as a captain in the Royal Navy in a ship built for him by the government, to cruise around the Atlantic and observe the magnetic variation for navigation, the first-ever scientific naval surveys (Thrower, 1981). He measured the angle between compass north and geographic north as established by the Sun's position at rise or set. He found his latitude from the elevation of the Sun at noon. Longitude at sea was much more difficult, and sometimes had gross errors. Halley had delays, bad weather, and difficult officers, and was too late in the season to go into the South Atlantic. His second cruise from September 1699 to September 1700 accomplished much more, and he went further into the Antarctic ice than anyone before.

Halley's chart of isogonic lines over the Atlantic hung in the Royal Society building for many years but is now lost. Many printed copies were published, the first isarhythmic chart of any variable to have been published. There may already have been private manuscript charts and *Athanasius Kircher* (*q.v.*) had written of one in his *Magnes*. (Kircher, 1643) Halley claimed, justly, that his chart was the first with isogonic lines to have been printed and published: He was a founder of modern cartography. His lines were known as Halleyan lines, and were soon used for other data such as temperature and depth of water. About two hundred years later *Christopher Hansteen* (*q.v.*) and *W. van Bemmelen* (*q.v.*), having access to the logs of Dutch ships, compiled worldwide isogonic charts at regular epochs. His chart for 1700 is generally close to Halley's where they overlap.

Figure H1 Edmond Halley.

Professor Sir Alan Cook died in August 2004 after completing this article. Final minor changes were made by the editors, who take full responsibility for any omissions or errors.

Aurorae

In 1716 Halley observed an intense auroral display over London (Halley, 1716). He collected reports from distant places and plotted the forms of the auroral arcs. They followed the lines of the Earth's magnetic field and were most intense around the magnetic, not the geographic, pole. Halley argued that matter circulating around the field lines produced the aurorae. He thought the matter leaked out of hollow spaces in the Earth, perhaps between the core and the mantle. He did not explain the colors. As there were then no instruments to detect the daily variation of the geomagnetic field or magnetic storms, Halley could not relate aurorae to events on the Sun.

Halley established entirely new lines of enquiry in his three principal contributions to geomagnetism. He was buried with his wife in the churchyard of St Margaret's Lee, not far from the Royal Observatory where a wall now carries his gravestone with its memorial inscription. There is a tablet in the cloisters of Westminster Abbey, and the observatory at Halley Bay on Antarctica recalls his magnetic and auroral observations and the cruise that went so close to the Antarctic continent. A recent biography is in Cook (1998).

Sir Alan Cook

Bibliography

Cook, A., 1998. *Edmond Halley: Charting the heavens and the seas.* Oxford: Oxford University Press.

Descartes, René, 1644. *Principia Philosophiae.* Amsterdam.

Halley, E., 1683. A theory of the variation of the magnetical compass. *Philosophical Transactions of the Royal Society*, **13**, 208–221.

Halley, E., 1692. An account of the cause of the change of the variation of the magnetic needle, with an hypothesis of the structure of the internal parts of the Earth. *Philosophical Transactions of the Royal Society*, **17**, 563–578.

Halley, E., 1716. An account of the late surprising appearance of lights seen in the air, on the sixth of March last, with an attempt to explain the principal phenomena thereof, as it was laid before the Royal Society by Edmund Halley, J.V.D., Savilian Professor of Geometry, Oxon, and Reg. Soc. Secr. *Philosophical Transactions of the Royal Society*, **29**, 406–428.

Kircher, A., 1643. *Magnes, sive de arte magnetica, opus tripartium.* Rome.

Thrower, N.J.W. (ed.), 1981. *The three voyages of Edmond Halley in the "Paramour", 1698–1701.*, 2nd series, vol. 156, 157. London: Hakluyt Society Publications.

Cross-references

Auroral Oval
Bemmelen, Willem van (1868–1941)
Geomagnetism, History of
Gilbert, William (1544–1603)
Hansteen, Christopher (1784–1873)
Humboldt, Alexander von (1759–1859)
Humboldt, Alexander von and magnetic storms
Jesuits, Role in Geomagnetism
Kircher, Athanasius (1602–1680)
Storms and Substorms, Magnetic
Voyages Making Geomagnetic Measurements
Westward Drift

HANSTEEN, CHRISTOPHER (1784–1873)

Christopher Hansteen (1784–1873) was born in Christiania (now Oslo), Norway. In the years 1816–1861 he was professor of applied mathematics and astronomy at the institution today denoted as the University of Oslo. His pioneering achievements in terrestrial magnetism and northern light research are today widely appreciated (Brekke, 1984; Josefowicz, 2002), though it has not always been the case earlier. Hansteen was selected as university teacher because of his successful participation in a prize competition, answering the question posed by the Royal Danish Academy of Sciences in 1811: "Can one explain all the magnetic peculiarities of the Earth from one single magnetic axis or is one forced to assume several?" The prize, a gold medal, was won by Hansteen. In his treatise he meant to demonstrate the necessity to assume that the Earth possesses two magnetic axes, implying our globe to be a magnetic quadrupole. The terrestrial magnetism became his main scientific interest through the rest of his life. For economic reasons Hansteen's one-volume treatise was not published until 1819, but then in a considerably extended form. The book has the title *Untersuchungen über den Magnetismus der Erde* (Hansteen, 1819). It is still quoted in the literature. This work appeared in print only one year ahead of the discovery of the connection between electricity and magnetism by his Danish friend and colleague H.C. Ørsted. With his well-formulated treatise, Hansteen thus obtained a central position in the development of the geophysical sciences taking place in that period (see Figure H2).

A Norwegian expedition under the leadership of Hansteen operated in Siberia in the years 1828–1830, traveling to the Baikal Sea and crossing the border into China (Hansteen, 1859). One measured the numerical values of the total magnetic field strength, the inclination, and the declination. Hansteen found no evidence of any additional magnetic pole in Siberia. To him this was an enormous disappointment. Nevertheless, in a letter to H.C. Ørsted of June 21, 1841, he proudly relates the written statement of Gauss, that to a large extent, it was the measurements of Hansteen, that had made Gauss devote himself to the study of magnetism.

Furthermore, Hansteen also made contributions to the investigation of northern light phenomena (Hansteen, 1825, 1827) as indicated in Figure H3. Related to his extensive studies of the Earth's magnetic

Figure H2 Christopher Hansteen, drawn by C.W. Eckersberg ca. 1828. Shown is also an instrument constructed by Hansteen for the determination of the magnetic intensity. It exploits the fact that a magnetic needle suspended in a magnetic field, when set in motion; its movement in time will among other factors depend on the field strength. This device received international acclaim and was used during the so-called Magnetic Crusade (Cawood, 1979).

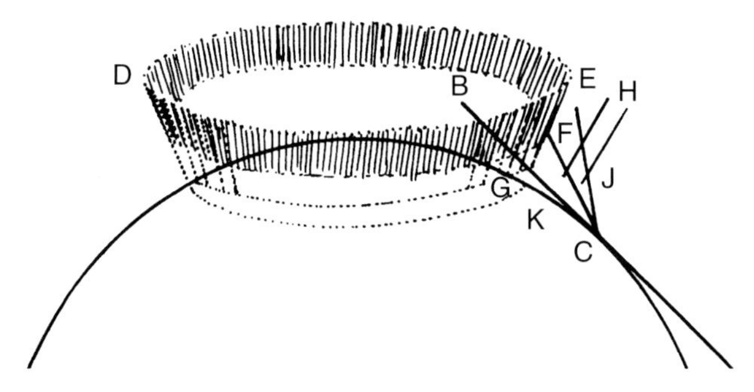

Figure H3 Picture from work of Hansteen (1827) on the concept of the aurora, showing an auroral ring encircling the polar cap. The figure is presumably one of the first drawings of the auroral ring. The illustration also demonstrates how to measure the height of the aurora from one point only.

field he claimed that the center of the auroral ring to be situated somewhere north of Hudson's bay (Brekke, 1984).

At the meeting of the Scandinavian natural scientists in Stockholm in 1842, Christopher Hansteen gave a talk on the development of the theory of terrestrial magnetism (Hansteen, 1842). He points to the discovery of Ørsted of the connection between magnetism and electricity as a possible way to explain the interplay between the processes in Earth's interior and the terrestrial magnetic phenomena. The modern dynamo theory is a development of this highly intuitive idea.

The scientific results from the Siberian expedition were published as late as 1863 (Hansteen and Due, 1863). It represents a dignified finish to Christopher Hansteen's contributions to this part of geoscience.

It deserves attention that Christopher Hansteen was also active as a nation builder in the first 50 years after 1814 of Norwegian independence. Important subjects of his activities were the mapping of the Norwegian costal area together with inland triangulation, exact time determination, and many years of almanac edition, to mention just a few of his undertakings.

Johannes M. Hansteen

Bibliography

Brekke, A., 1984. On the evolution in history of the concept of the auroral oval. *Eos, Transactions of the American Geophysical Union*, **65**: 705–707.

Cawood, J., 1979. The magnetic crusade: science and politics in the early Victorian Britain. *Isis*, **70**: pp. 493–518.

Hansteen, C., 1819. *Untersuchungen über den Magnetismus der Erde. Erster Teil*, Christiania: J. Lehman and Chr. Gröndal.

Hansteen, C., 1825. Forsøg til et magnetisk Holdningskart. *Mag. Naturvidensk*, **2**: pp. 203–212.

Hansteen, C., 1827. On the polar light, or aurora borealis and australis. *Philos. Mag. Ann. Philos*, New Ser., **2**: 333–334.

Hansteen, C., 1842. Historisk Fremstilling af hvad den fra det forloebne Seculums Begyndelse til vor Tid er udrettet for Jordmagnetismens Theorie. In *Proceedings of the Third Meeting of the Scandinavian Natural Scientists,* Stockholm, pp. 68–80.

Hansteen, C., 1859. *Reiseerindringer*, Chr. Tønsbergs Forlag, Christiania.

Hansteen, C., and Due, C. 1863. *Resultate Magnetischer, Astronomischer und Meteorologischer Beobachtungen auf einer Reise nach dem Östlichen Sibirien in den Jahren 1828–1830.* Christiania Academy of Sciences.

Josefowicz, J.G., 2002. *Mitteilungen der Gauss-Gesellschaft* **39**: pp. 73–86.

HARMONICS, SPHERICAL

Introduction

Spherical harmonics are solutions of Laplace's equation

$$\frac{\partial^2 V}{\partial x^2} + \frac{\partial^2 V}{\partial y^2} + \frac{\partial^2 V}{\partial z^2} = 0 \qquad \text{(Eq. 1)}$$

in three dimensions, and they are collected together as homogeneous polynomials of degree l, with $2l + 1$ in each group. The simplest spherical harmonics are the three Cartesian coordinates x, y, z, which are three homogeneous polynomials of degree 1. Inverse distance $1/r$ satisfies Laplace's equation everywhere, except at the point $r = 0$, forming a spherical harmonic of degree -1, and the Cartesian derivatives of inverse distance generate spherical harmonics of higher degree (and order). Because inverse distance is the potential function for gravitation, electric, and magnetic fields of force, the theory of spherical harmonics has many important applications.

It is a simple matter to observe that on taking the gradient of Laplace's equation (1), that if V is a solution, then so also is the vector ∇V, and also the tensor $\nabla(\nabla V)$, showing clearly the need to consider vector and tensor spherical harmonics. As a simple example, the gradients of the Cartesian coordinates x, y, z, lead to $\mathbf{i}, \mathbf{j}, \mathbf{k}$, as vector spherical harmonics, and gradients of $x\mathbf{i}, x\mathbf{j}, x\mathbf{k}, y\mathbf{i}, \ldots$, lead to Cartesian tensors $\mathbf{ii}, \mathbf{ij}, \mathbf{ik}, \mathbf{ji}, \ldots$ as tensor spherical harmonics.

With increasing interest in atomic structure and electron spin, the study of Laplace's equation in four dimensions,

$$\frac{\partial^2 V}{\partial p^2} + \frac{\partial^2 V}{\partial q^2} + \frac{\partial^2 V}{\partial r^2} + \frac{\partial^2 V}{\partial s^2} = 0, \qquad \text{(Eq. 2)}$$

which is satisfied by $1/(p^2 + q^2 + r^2 + s^2)$ and its derivatives, is important. The solution uses surface spherical harmonics from Eq. (1), and "spin weighted" associated Legendre functions.

Spherical polar coordinates

In three dimensions, the spherical polar coordinates of a point P on the surface of the sphere are r, θ, ϕ, where r is the radius of the sphere, θ is the colatitude of the point measured from the point chosen as the north pole of the coordinate system, and ϕ is the east longitude of the point measured from the meridian of longitude chosen as the prime meridian,

$$\begin{aligned} x &= r \sin\theta \cos\phi, \\ y &= r \sin\theta \sin\phi, \\ z &= r\cos\theta. \end{aligned} \qquad \text{(Eq. 3)}$$

If a spherical harmonic of degree l is denoted V_l, then in spherical polars, by virtue of being a homogeneous polynomial of degree l, we may write

$$V_l(r, \theta, \phi) = r^l S_l(\theta, \phi) \qquad \text{(Eq. 4)}$$

and the function $S_l(\theta, \phi)$ is called a surface spherical harmonic of degree l. From the theory of homogeneous functions,

$$(\mathbf{r} \cdot \nabla) V_l = r\frac{\partial V_l}{\partial r} = x\frac{\partial V_l}{\partial x} + y\frac{\partial V_l}{\partial y} + z\frac{\partial V_l}{\partial z} = l\, V_l. \qquad \text{(Eq. 5)}$$

Conversely, it can be shown that if a function V_l satisfies Eq. (5), then it is a homogeneous function of degree l in x, y, z.

The four-dimensional case arises when considering rotations through and angle w, about an axis of rotation with which is in the direction that has colatitude u and east longitude v. For rotations, the hypersphere has unit radius, for we include a value ρ here, and the four-dimensional Cartesian coordinates $p, q, r, s,$ are

$$p = \rho \sin u \cos v \sin \frac{1}{2} w,$$
$$q = \rho \sin u \sin v \sin \frac{1}{2} w,$$
$$r = \rho \cos u \sin \frac{1}{2} w,$$
$$s = \rho \cos \frac{1}{2} w. \qquad \text{(Eq. 6)}$$

With $\rho = 1$, the coordinates $p, q, r, s,$ are better known as quaternions.

Separation of variables

In spherical polar coordinates, the Laplacian $\nabla^2 V$ is

$$\nabla^2 V = \frac{1}{r^2}\left[\frac{\partial}{\partial r}\left(r^2 \frac{\partial V}{\partial r}\right) + \frac{1}{\sin\theta}\frac{\partial}{\partial\theta}\left(\sin\theta\frac{\partial V}{\partial\theta}\right) + \frac{1}{\sin^2\theta}\frac{\partial^2 V}{\partial\phi^2}\right]. \qquad \text{(Eq. 7)}$$

The method of separation of variables is used to solve Laplace's equation in spherical polars. With $V(r, \theta, \phi) = R(r)S_l(\theta, \phi)$, and with $l(l+1)$ as a constant of separation, then

$$\frac{\mathrm{d}}{\mathrm{d}r}\left(r^2 \frac{\mathrm{d}R}{\mathrm{d}r}\right) - l(l+1)R = 0, \qquad \text{(Eq. 8)}$$

and

$$\frac{1}{\sin\theta}\frac{\partial}{\partial\theta}\left(\sin\theta\frac{\partial S_l}{\partial\theta}\right) + \frac{1}{\sin^2\theta}\frac{\partial^2 S_l}{\partial\phi^2} + l(l+1)S_l = 0. \qquad \text{(Eq. 9)}$$

Note that the constant of separation $l(l+1)$ remains unchanged if l is replaced by $-l-1$.

Radial dependence of the spherical harmonics is given by (8),

$$R(r) = A_l r^l + \frac{B}{r^{l+1}}. \qquad \text{(Eq. 10)}$$

Equation 9 can be solved by further separation of variables, using $S_l(\theta, \phi) = \Theta(\theta)\Phi(\phi)$, with m^2 as the constant of separation,

$$\frac{\mathrm{d}^2\Phi}{\mathrm{d}\phi^2} + m^2\Phi = 0, \qquad \text{(Eq. 11)}$$

and

$$\frac{1}{\sin\theta}\frac{\mathrm{d}}{\mathrm{d}\theta}\left(\sin\theta\frac{\mathrm{d}\Theta}{\mathrm{d}\theta}\right) + \left[l(l+1) - \frac{m^2}{\sin^2\theta}\right]\Theta = 0. \qquad \text{(Eq. 12)}$$

From Eqs. (7) and (9), the Laplacian of the surface spherical harmonic $S_l(\theta, \phi)$ is

$$\nabla^2 S_l(\theta, \phi) = -\frac{l(l+1)}{r^2} S_l(\theta, \phi) \qquad \text{(Eq. 13)}$$

Solutions of Eq. (11) are the trigonometric functions $\cos V$ and $\sin V$, and the complex exponential functions $\mathrm{e}^{\mathrm{i}V}$ and $\mathrm{e}^{-\mathrm{i}V}$. The concept of "irreducibility" favors the use of the complex exponential expressions, because, with the transformation $V \to V + \alpha$, the trigonometric functions become mixed,

$$\cos V \to \cos(V+\alpha) = \cos V \cos\alpha - \sin V \sin\alpha,$$
$$\sin V \to \sin(V+\alpha) = \sin V \cos\alpha + \cos V \sin\alpha, \qquad \text{(Eq. 14)}$$

and are "reducible", whereas the complex exponential functions do not become mixed, and are therefore "irreducible."

$$\mathrm{e}^{\mathrm{i}V} \to \mathrm{e}^{\mathrm{i}(V+\alpha)} = \mathrm{e}^{\mathrm{i}V}\mathrm{e}^{\mathrm{i}\alpha},$$
$$\mathrm{e}^{-\mathrm{i}V} \to \mathrm{e}^{-\mathrm{i}(V+\alpha)} = \mathrm{e}^{-\mathrm{i}V}\mathrm{e}^{-\mathrm{i}\alpha}, \qquad \text{(Eq. 15)}$$

Orthogonality of surface spherical harmonics

A number of properties of spherical harmonics can be established using only the definition, namely, that they are homogeneous functions of degree l that satisfy Laplace's equation, and specific mathematical expressions are not required.

Let $S_l(x, y, z)$ and $S_L(x, y, z)$ be two surface spherical harmonics of degree l and L respectively. A vector $\mathbf{F}$ is defined by

$$\mathbf{F} = (\nabla S_l)\overline{S_L} - S_l(\nabla\overline{S_L}), \qquad \text{(Eq. 16)}$$

when, by Eq. (13), the divergence of $\mathbf{F}$ is

$$\nabla\cdot\mathbf{F} = \frac{1}{r^2}\left[-l(l+1)S_l\overline{S_L} + L(L+1)S_l\overline{S_L}\right],$$
$$= \frac{1}{r^2}(L-l)(L+l+1)S_l\overline{S_L}. \qquad \text{(Eq. 17)}$$

Gauss's theorem, applied to a spherical surface and the enclosed spherical volume, is

$$\iiint\limits_{\substack{\text{spherical}\\ \text{volume}}} \nabla\cdot\mathbf{F}\,\mathrm{d}v = \iint\limits_{\substack{\text{spherical}\\ \text{surface}}} \mathbf{F}\cdot\mathrm{d}\boldsymbol{S}, \qquad \text{(Eq. 18)}$$

where, in this case, $\mathrm{d}\boldsymbol{S} = r^2\mathbf{e}_r \sin\theta\,\mathrm{d}\theta\,\mathrm{d}\phi$, and $\mathbf{e}_r$ is the unit radial vector. The vector $\mathbf{F}$ has no radial component, and therefore the spherical surface integral is zero, and the volume integral, after integrating with respect to radius, reduces to

$$\int_0^r\int_0^{2\pi}\int_0^{\pi}\nabla\cdot\mathbf{F}\,\mathrm{d}v = (L-l)(L+l+1)$$
$$\int_0^r \mathrm{d}r\int_0^{2\pi}\int_0^{\pi} S_l(\theta,\phi)S_L(\theta,\phi)\sin\theta\mathrm{d}\theta\mathrm{d}\phi, = 0,$$

and therefore, when $l \neq L$ or when $l \neq -L-1$, then

$$\int_0^{2\pi}\int_0^{\pi} S_l(\theta,\phi)S_L(\theta,\phi)\sin\theta\,\mathrm{d}\theta\,\mathrm{d}\phi = 0. \qquad \text{(Eq. 19)}$$

Legendre polynomials

In the case $m = 0$, the spherical harmonics are independent of longitude ϕ and are said to be "zonal". The differential equation is obtained from Eq. (12) with $m = 0$, namely

$$(1-\mu^2)\frac{\mathrm{d}^2\Theta}{\mathrm{d}\mu^2} - 2\mu\frac{\mathrm{d}\Theta}{\mathrm{d}\mu} + l(l+1)\Theta = 0, \qquad \text{(Eq. 20)}$$

The factor $(1-\mu^2)$ of the second derivative shows that the solution has singularities at $\mu = \pm 1$, corresponding to colatitudes $\theta = 0$ and $\theta = \pi$, at the north and south poles respectively of the chosen reference frame. The two independent solutions of Eq. (20) are

Legendre functions of the first and second kind, $P_l(\mu)$ and $Q_l(\mu)$, respectively, where

$$P_l(\mu) = \frac{1}{2^l l!}\left(\frac{d}{d\mu}\right)^l (\mu^2 - 1)^l, \quad \text{(Eq. 21)}$$

$$Q_l(\mu) = P_l(\mu)\ln\sqrt{\frac{1+\mu}{1-\mu}} - \frac{2l-1}{1l}P_{l-1}(\mu) - \frac{2l-5}{3(l-1)}P_{l-3}(\mu) - \frac{2l-9}{5(l-2)}P_{l-5}(\mu) + \ldots, |\mu| < 1. \quad \text{(Eq. 22)}$$

Legendre polynomials are orthogonal over the range $-1 \le \mu \le 1$, and the normalization of $P_l(\mu)$ has been chosen so that

$$\frac{1}{2}\int_{-1}^{1} P_l(\mu)P_L(\mu)d\mu = \frac{1}{2l+1}\delta_l^L, \quad \text{(Eq. 23)}$$

where δ_l^L is the Kronecker delta.

The descending power series expansion for the Legendre polynomials is

$$P_l(\mu) = \frac{(2l)!}{2^l l! l!}\left[\mu^l - \frac{l(l-1)}{2(2l-1)}\mu^{l-2} + \frac{l(l-1)(l-2)(l-3)}{2\cdot 4(2l-1)(2l-3)}\mu^{l-4} - \ldots\right], \quad \text{(Eq. 24)}$$

and the first few are

$$P_0 = 1, \quad P_1 = \mu, \quad P_2 = \frac{1}{2}(3\mu^2 - 1),$$
$$P_3 = \frac{1}{2}(5\mu^3 - 3\mu), \quad P_4 = \frac{1}{8}(35\mu^4 - 30\mu^2 + 3). \quad \text{(Eq. 25)}$$

Note that $P_l(1) = 1$ and that $P_l(-1) = (-1)^l$ for all values of l.

A series expansion for Legendre functions of the second kind, $Q_l(\mu)$ is

$$Q_l(\mu) = \frac{2^l l! l!}{(2l+1)!}\left[\frac{1}{\mu^{l+1}} + \frac{(l+1)(l+2)}{2(2l+3)}\frac{1}{\mu^{l+3}} + \frac{(l+1)(l+2)(l+3)(l+4)}{2\cdot 4(2l+3)(2l+5)}\frac{1}{\mu^{l+5}} + \ldots\right].$$

Neumann's formula

Neumann's formula, given by

$$\sum_{l=0}^{\infty}(2l+1)P_l(\zeta)Q_l(\mu) = \frac{1}{\mu - \zeta}, \quad |\mu| > |\zeta|, \quad \text{(Eq. 26)}$$

can be used to show that the recurrence relations for $Q_l(\mu)$ are the same as those for the Legendre polynomials $P_l(\mu)$. Therefore, rounding errors, regarded as proportional to $Q_l(\mu)$, in the generation of Legendre polynomials using recurrence relations, are likely to become large at or near the poles. Neumann's formula (26) can also be written in the form

$$Q_l(\mu) = \frac{1}{2}\int_{-1}^{1}\frac{P_l(\zeta)}{\mu - \zeta}d\zeta, \quad \text{(Eq. 27)}$$

which can be used to derive the Christoffel formula (22) for $Q_l(\mu)$.

Generating function

The function

$$V(r,\mu) = \frac{1}{\sqrt{1 - 2\mu h + h^2}} = \sum_{l=0}^{\infty} h^l P_l(\cos\theta) \quad \text{(Eq. 28)}$$

is the generating function for the Legendre polynomials.

If a gravitating particle of mass m is moved from the origin a distance d along the z-axis, regarded as the pole of a coordinate system, then, using the cosine rule of trigonometry, the gravitational potential of the particle (in a region free of gravitating material) is

$$V(r,\theta) = \frac{Gm}{\sqrt{r^2 - 2rd\cos\theta + d^2}},$$
$$V(r,\theta) = \frac{Gm}{d}\sum_{l=0}^{\infty}\left(\frac{d}{r}\right)^{l+1} P_l(\cos\theta), \quad \text{when } r > d,$$
$$V(r,\theta) = \frac{Gm}{d}\sum_{l=0}^{\infty}\left(\frac{r}{d}\right)^{l} P_l(\cos\theta), \quad \text{when } r < d. \quad \text{(Eq. 29)}$$

If the potential of a distribution of matter is independent of azimuth or east longitude, it is said to be zonal. In the case where the potential of the distribution in a source-free region along the z-axis is known to be

$$V(z) = \frac{G}{d}\sum_{l=0}^{\infty}\alpha_l\left(\frac{d}{z}\right)^{l+1} \quad \text{for } z > d,$$

and

$$V(z) = \frac{G}{d}\sum_{l=0}^{\infty}\alpha_l\left(\frac{z}{d}\right)^{l} \quad \text{for } z < d, \quad \text{(Eq. 30)}$$

then, from the continuity of the potential, the potential in regions away from the z-axis is

$$V(r,\theta) = \frac{G}{d}\sum_{l=0}^{\infty}\alpha_l\left(\frac{d}{r}\right)^{l+1} P_l(\cos\theta) \quad \text{for } r > d,$$

and

$$V(r,\theta) = \frac{G}{d}\sum_{l=0}^{\infty}\alpha_l\left(\frac{r}{d}\right)^{l} P_l(\cos\theta) \quad \text{for } r < d. \quad \text{(Eq. 31)}$$

Recurrence relations

Differentiation of the generating function (28) and the Rodrigues formula (21) can be used to derive the recurrence relations for Legendre polynomials. The more important ones are

Bonnet's formula $(l+1)P_{l+1} - (2l+1)\mu P_l + lP_{l-1} = 0,$ (Eq. 32)

$$\frac{dP_{l+1}}{d\mu} = \mu\frac{dP_l}{d\mu} + (l+1)P_l, \quad \text{(Eq. 33)}$$

and

$$(2l+1)P_l = \frac{dP_{l+1}}{d\mu} - \frac{dP_{l-1}}{d\mu}. \quad \text{(Eq. 34)}$$

Multiple derivatives of inverse distance

Parameters ξ and η are required, and they are defined by

$$\xi = -\frac{1}{\sqrt{2}}(x + \mathrm{i}y) \quad = -\frac{1}{\sqrt{2}} r \sin\theta \mathrm{e}^{\mathrm{i}\phi},$$
$$\eta = \frac{1}{\sqrt{2}}(x - \mathrm{i}y) \quad = \frac{1}{\sqrt{2}} r \sin\theta \mathrm{e}^{-\mathrm{i}\phi}. \qquad \text{(Eq. 35)}$$

Partial derivatives with respect to ξ and η are

$$\frac{\partial f}{\partial \xi} = -\frac{1}{\sqrt{2}}\left(\frac{\partial f}{\partial x} - \mathrm{i}\frac{\partial f}{\partial y}\right), \quad \frac{\partial f}{\partial \eta} = \frac{1}{\sqrt{2}}\left(\frac{\partial f}{\partial x} + \mathrm{i}\frac{\partial f}{\partial y}\right). \qquad \text{(Eq. 36)}$$

The solid spherical harmonic, inverse distance, is

$$\frac{1}{r} = \frac{1}{\sqrt{x^2 + y^2 + z^2}} = \frac{1}{\sqrt{z^2 - 2\xi\eta}}, \qquad \text{(Eq. 37)}$$

with multiple derivatives

$$\left(\frac{\partial}{\partial \xi}\right)^m \frac{1}{r} = 1 \cdot 3 \cdot 5 \ldots (2m-1) \frac{\eta^m}{r^{2m+1}},$$
$$\left(\frac{\partial}{\partial \eta}\right)^m \frac{1}{r} = 1 \cdot 3 \cdot 5 \ldots (2m-1) \frac{\xi^m}{r^{2m+1}}. \qquad \text{(Eq. 38)}$$

The $(l-m)^{\text{th}}$ derivative of $1/r^{2m+1}$ with respect to z is

$$(-1)^{l-m}\left(\frac{\partial}{\partial z}\right)^{l-m} \frac{1}{r^{2m+1}} = (2m+1)(2m+3)\ldots(2l-1)$$
$$\left[\frac{z^{l-m}}{r^{2l+1}} - \frac{(l-m)(l-m-1)}{2(2l-1)}\frac{z^{l-m-2}}{r^{2l-1}} + \frac{(l-m)(l-m-1)(l-m-2)(l-m-3)}{2 \cdot 4(2l-1)(2l-3)}\frac{z^{l-m-4}}{r^{2l-3}} - \ldots\right] \qquad \text{(Eq. 39)}$$

The required $2l+1$ independent spherical harmonics of degree l are obtained by differentiating (38) partially with respect to z some $l-m$ times, for $m = -l, -l+1, \ldots -1, 0, 1, \ldots, l-1, l$, when, with the substitutions $z = r\cos\theta$ and $\mu = \cos\theta$,

$$(-1)^{l-m}\left(\frac{\partial}{\partial \eta}\right)^m \left(\frac{\partial}{\partial z}\right)^{l-m} \frac{1}{r} = \frac{1}{r^{l+m+1}} \frac{(l-m)!}{2^l l!} \xi^m$$
$$\left[\frac{(2l)!}{(l-m)!}\mu^{l-m} - \frac{l(2l-2)!}{(l-m-2)!}\mu^{l-m-2} + \frac{l(l-1)(2l-4)!}{2(l-m-4)!}\mu^{l-m-4} - \ldots\right]. \qquad \text{(Eq. 40)}$$

The series in (40) is the $(l+m)^{\text{th}}$ derivative of a series which can be summed by the binomial theorem, and hence

$$(-1)^{l-m}\left(\frac{\partial}{\partial x} + \mathrm{i}\frac{\partial}{\partial y}\right)^m \left(\frac{\partial}{\partial z}\right)^{l-m} \frac{1}{r}$$
$$= (-1)^m \frac{\mathrm{e}^{\mathrm{i}m\phi}}{r^{l+1}} \frac{(l-m)!}{2^l l!}(1-\mu^2)^{m/2}\left(\frac{\mathrm{d}}{\mathrm{d}\mu}\right)^{l+m}(\mu^2-1)^l. \qquad \text{(Eq. 41)}$$

Ferrers normalized functions

Associated Legendre functions are solutions of Eq. (12), and are given in the first instance as the Ferrers normalized functions $P_{l,m}(\mu)$, (Ferrers, 1897), defined by

$$P_{l,m}(\mu) = (1-\mu^2)^{m/2}\left(\frac{\mathrm{d}}{\mathrm{d}\mu}\right)^m P_l(\mu),$$
$$= \frac{1}{2^l l!}(1-\mu^2)^{m/2}\left(\frac{\mathrm{d}}{\mathrm{d}\mu}\right)^{l+m}(\mu^2-1)^l, \quad \text{for } l \geq |m| \qquad \text{(Eq. 42)}$$

See Table H1 for a list of these functions.

Applying Leibnitz's theorem for multiple derivatives of $\left[(\mu-1)^l(\mu+1)^l\right]$ in the function $P_l^{-m}(\mu)$ and then re-arranging terms, gives the expression in terms of $P_l^m(\mu)$,

$$P_{l,-m}(\mu) = \frac{1}{2^l l!}(1-\mu^2)^{-m/2}\left(\frac{\mathrm{d}}{\mathrm{d}\mu}\right)^{l-m}(\mu^2-1)^l$$
$$= (-1)^m \frac{(l-m)!}{(l+m)!} P_{l,m}(\mu). \qquad \text{(Eq. 43)}$$

Both Eqs. (42) and (43) can be used with positive or negative values of m. Caution is needed with definitions based on $|m|$. Equations (42) and (43) can be used to derive the normalization integral for associated Legendre functions with Ferrers normalization,

$$\frac{1}{2}\int_{-1}^{1} P_{l,m}(\mu) P_{L,m}(\mu)\mathrm{d}\mu = \frac{1}{2l+1}\frac{(l+m)!}{(l-m)!}\delta_l^L. \qquad \text{(Eq. 44)}$$

Surface spherical harmonics

Surface spherical harmonics $Y_l^m(\theta,\phi)$ are defined in terms of the Ferrers normalized functions $P_{l,m}(\cos\theta)$

$$Y_l^m(\theta,\phi) = (-1)^m \mathrm{e}^{\mathrm{i}m\phi}\sqrt{(2l+1)\frac{(l-m)!}{(l+m)!}} P_{l,m}(\cos\theta),$$
$$= (-1)^m \mathrm{e}^{\mathrm{i}m\phi}\sqrt{(2l+1)\frac{(l-m)!}{(l+m)!}}\frac{1}{2^l l!}(1-\mu^2)^{m/2}\left(\frac{\mathrm{d}}{\mathrm{d}\mu}\right)^{l+m}(\mu^2-1)^l. \qquad \text{(Eq. 45)}$$

The initial factor $(-1)^m$ is now used following the influential work of Condon and Shortley (1935). See Table H1 for a list of the functions $Y_l^m(\theta,\phi)$, and Figure H4 for tesseral, sectorial, and zonal harmonics.

Therefore, in terms of multiple derivatives of inverse distance, from Eqs. (41) and (45),

$$\frac{1}{r^{l+1}} Y_l^m(\theta,\phi)$$
$$= (-1)^{l-m}\sqrt{\frac{2l+1}{(l-m)!(l+m)!}}\left(\frac{\partial}{\partial x} + \mathrm{i}\frac{\partial}{\partial y}\right)^m \left(\frac{\partial}{\partial z}\right)^{l-m}\frac{1}{r} \qquad \text{(Eq. 46)}$$

The series expression for surface spherical harmonics is

$$Y_l^m(\theta,\phi) = \frac{(2l)!}{2^l l!}\mathrm{e}^{\mathrm{i}m\phi}\sqrt{\frac{2l+1}{(l-m)!(l+m)!}}(1-\mu^2)^{m/2}$$
$$\left[\mu^{l-m} - \frac{(l-m)(l-m-1)}{2(2l-1)}\mu^{l-m-2} + \frac{(l-m)(l-m-1)(l-m-2)(l-m-3)}{2 \cdot 4(2l-1)(2l-3)}\mu^{n-m-4} - \ldots\right]. \qquad \text{(Eq. 47)}$$

Table H1 List of associated Legendre functions $P_l^m(\cos\theta)$ in Ferrers normalization and Schmidt normalization to degree and order six, and the corresponding surface spherical harmonics $Y_l^m(\theta,\phi)$, normalized after Condon and Shortley

l	m	Ferrers $P_{l,m}(\cos\theta)$	Schmidt $P_l^m(\cos\theta)$	$Y_l^m(\theta,\phi)$
0	0	1	1	1
1	0	c	c	$\sqrt{3}\,c$
1	1	s	s	$-\frac{\sqrt{6}}{2}s\mathrm{e}^{\mathrm{i}\phi}$
2	0	$\frac{1}{2}(3c^2-1)$	$\frac{1}{2}(3c^2-1)$	$\frac{\sqrt{5}}{2}(3c^2-1)$
2	1	$3\,sc$	$\sqrt{3}\,sc$	$-\frac{\sqrt{30}}{2}sc\mathrm{e}^{\mathrm{i}\phi}$
2	2	$3s^2$	$\frac{\sqrt{3}}{2}s^2$	$\frac{\sqrt{30}}{4}s^2\mathrm{e}^{2\mathrm{i}\phi}$
3	0	$\frac{1}{2}(5c^3-3c)$	$\frac{1}{2}(5c^3-3c)$	$\frac{\sqrt{7}}{2}(5c^3-3c)$
3	1	$\frac{3}{2}s(5c^2-1)$	$\frac{\sqrt{6}}{4}s(5c^2-1)$	$-\frac{\sqrt{21}}{4}s(5c^2-1)\mathrm{e}^{\mathrm{i}\phi}$
3	2	$15\,s^2c$	$\frac{\sqrt{15}}{2}s^2c$	$\frac{\sqrt{210}}{4}s^2c\,\mathrm{e}^{2\mathrm{i}\phi}$
3	3	$15\,s^3$	$\frac{\sqrt{10}}{4}s^3$	$-\frac{\sqrt{35}}{4}s^3\mathrm{e}^{3\mathrm{i}\phi}$
4	0	$\frac{1}{8}(35c^4-30c^2+3)$	$\frac{1}{8}(35c^4-30c^2+3)$	$\frac{3}{8}(35c^4-30c^2+3)$
4	1	$\frac{5}{2}s(7c^3-3c)$	$\frac{\sqrt{10}}{4}s(7c^3-3c)$	$-\frac{3\sqrt{5}}{4}s(7c^3-3c)\mathrm{e}^{\mathrm{i}\phi}$
4	2	$\frac{15}{2}s^2(7c^2-1)$	$\frac{\sqrt{5}}{4}s^2(7c^2-1)$	$\frac{3\sqrt{10}}{8}s^2(7c^2-1)\mathrm{e}^{2\mathrm{i}\phi}$
4	3	$105\,s^3c$	$\frac{\sqrt{70}}{4}s^3c$	$\frac{3\sqrt{35}}{4}s^3c\mathrm{e}^{3\mathrm{i}\phi}$
4	4	$105\,s^4$	$\frac{\sqrt{35}}{8}s^4$	$\frac{3\sqrt{70}}{16}s^4\mathrm{e}^{4\mathrm{i}\phi}$
5	0	$\frac{1}{8}(63c^5-70c^3+15c)$	$\frac{1}{8}(63c^5-70c^3+15c)$	$\frac{\sqrt{11}}{8}(63c^5-70c^3+15c)$
5	1	$\frac{15}{8}s(21c^4-14c^2+1)$	$\frac{\sqrt{15}}{8}s(21c^4-14c^2+1)$	$-\frac{\sqrt{330}}{16}s(21c^4-14c^2+1)\mathrm{e}^{\mathrm{i}\phi}$
5	2	$\frac{105}{2}s^2(3c^3-c)$	$\frac{\sqrt{105}}{4}s^2(3c^3-c)$	$\frac{\sqrt{2310}}{8}s^2(3c^3-c)\mathrm{e}^{2\mathrm{i}\phi}$
5	3	$\frac{105}{2}s^3(9c^2-1)$	$\frac{\sqrt{70}}{16}s^3(9c^2-1)$	$-\frac{\sqrt{385}}{16}s^3(9c^2-1)\mathrm{e}^{3\mathrm{i}\phi}$
5	4	$945\,s^4c$	$\frac{3\sqrt{35}}{8}s^4c$	$\frac{3\sqrt{770}}{16}s^4c\mathrm{e}^{4\mathrm{i}\phi}$
5	5	$945\,s^5$	$\frac{3\sqrt{14}}{16}s^5$	$-\frac{3\sqrt{77}}{16}s^5\mathrm{e}^{5\mathrm{i}\phi}$
6	0	$\frac{1}{16}(231c^6-315c^4$ $+105c^2-5)$	$\frac{1}{16}(231c^6-315c^4$ $+105c^2-5)$	$\frac{\sqrt{13}}{16}(231c^6-315c^4$ $+105c^2-5)$
6	1	$\frac{21}{8}s(33c^5-30c^3+5c)$	$\frac{\sqrt{21}}{8}s(33c^5-30c^3+5c)$	$-\frac{\sqrt{546}}{16}s(33c^5-30c^3+5c)\mathrm{e}^{\mathrm{i}\phi}$
6	2	$\frac{105}{8}s^2(33c^4-18c^2+1)$	$\frac{\sqrt{210}}{32}s^2(33c^4-18c^2+1)$	$\frac{\sqrt{1365}}{32}s^2(33c^4-18c^2+1)\mathrm{e}^{2\mathrm{i}\phi}$
6	3	$\frac{315}{2}s^3(11c^3-3c)$	$\frac{\sqrt{210}}{16}s^3(11c^3-3c)$	$-\frac{\sqrt{1365}}{16}s^3(11c^3-3c)\mathrm{e}^{3\mathrm{i}\phi}$
6	4	$\frac{945}{2}s^4(11c^2-1)$	$\frac{3\sqrt{7}}{16}s^4(11c^2-1)$	$\frac{3\sqrt{182}}{32}s^4(11c^2-1)\mathrm{e}^{4\mathrm{i}\phi}$
6	5	$10395\,s^5c$	$\frac{3\sqrt{154}}{16}s^5c$	$-\frac{3\sqrt{1001}}{16}s^5c\mathrm{e}^{5\mathrm{i}\phi}$
6	6	$10395\,s^6$	$\frac{\sqrt{462}}{32}s^6$	$\frac{\sqrt{3003}}{32}s^6\mathrm{e}^{6\mathrm{i}\phi}$

In the special case $m=0$, from Eq. (45),

$$Y_l^0(\theta,\phi)=\sqrt{(2l+1)}\frac{1}{2^l l!}\left(\frac{\mathrm{d}}{\mathrm{d}\mu}\right)^l(\mu^2-1)^l=\sqrt{(2l+1)}P_l(\mu), \tag{Eq. 48}$$

Changing the sign of m in Eq. (45) and making use of Eq. (43), gives

$$\begin{aligned}Y_l^{-m}(\theta,\phi)&=(-1)^m\mathrm{e}^{-\mathrm{i}m\phi}\sqrt{(2l+1)\frac{(l+m)!}{(l-m)!}}P_{l,-m}(\cos\theta),\\&=\mathrm{e}^{-\mathrm{i}m\phi}\sqrt{(2l+1)\frac{(l-m)!}{(l+m)!}}P_{l,m}(\cos\theta),\\&=(-1)^m\overline{Y_l^m(\theta,\phi)}.\end{aligned} \tag{Eq. 49}$$

Note that ξ, z, η are related to the surface spherical harmonics of the first degree by

$$\begin{aligned}Y_1^1(\theta,\phi)&=-\frac{\sqrt{6}}{2}\sin\theta\mathrm{e}^{\mathrm{i}\phi}&&=\frac{\sqrt{3}}{r}\xi,\\Y_1^0(\theta,\phi)&=\sqrt{3}\cos\theta&&=\frac{\sqrt{3}}{r}z,\\Y_1^{-1}(\theta,\phi)&=\frac{\sqrt{6}}{2}\sin\theta\mathrm{e}^{-\mathrm{i}\phi}&&=\frac{\sqrt{3}}{r}\eta.\end{aligned} \tag{Eq. 50}$$

The normalization and orthogonality integral is

$$\frac{1}{4\pi}\int_0^{2\pi}\int_0^{\pi}Y_l^m(\theta,\phi)\overline{Y_L^M(\theta,\phi)}\sin\theta\mathrm{d}\theta\mathrm{d}\phi=\delta_l^L\delta_m^M. \tag{Eq. 51}$$

In theoretical physics texts, it is common practice to replace the initial factor $1/4\pi$ in Eq. (51) by applying a factor $\sqrt{1/4\pi}$ to the surface spherical harmonic $Y_l^m(\theta,\phi)$. This factor is sometimes broken up into a factor $\sqrt{1/2\pi}$ applied to the associated Legendre function and a factor $\sqrt{1/2}$ applied to the complex exponential part.

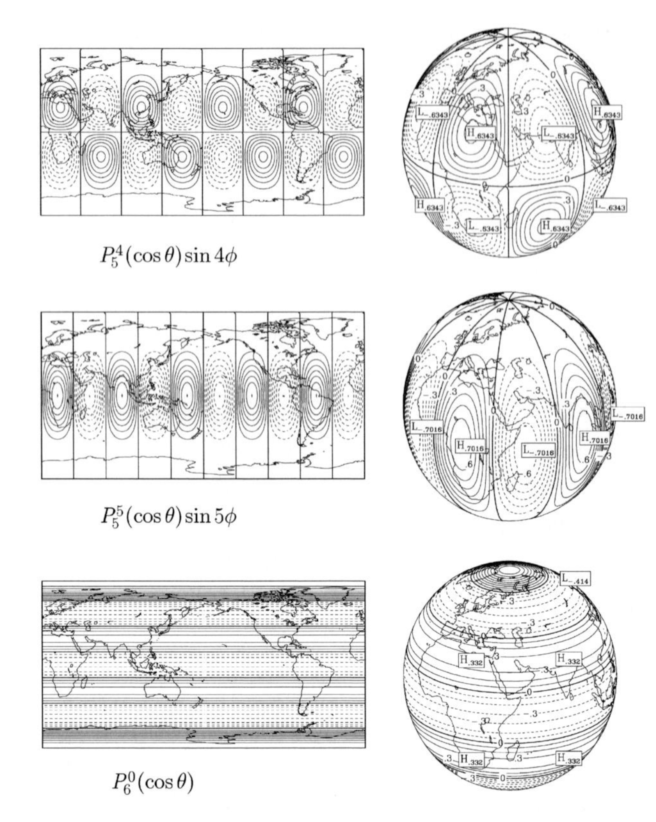

Figure H4 Zonal surface spherical harmonics are of the form $P_l^0(\cos\theta)$; sectorial surface spherical harmonics are of the form $P_l^l(\cos\theta)\cos l\phi$ and $P_l^l(\cos\theta)\sin l\phi$. Tesseral surface spherical harmonics are those that are neither zonal nor sectorial.

The $(2l+1)$ independent solutions of degree l of Laplace's equation, with no logarithmic singularity at the poles, are the $(2l+1)$ solid spherical harmonics,

$$r^l Y_l^m(\theta,\phi), \quad \text{where } m=-l,-l+1,\ldots l-1,\, l. \qquad \text{(Eq. 52)}$$

Given a sufficiently differentiable function $f(\theta,\phi)$ over the surface of a sphere, it can be written as a finite linear combination of surface spherical harmonics

$$f(\theta,\phi)=\sum_{l=0}^{L}\sum_{m=-n}^{n} f_{lm}Y_l^m(\theta,\phi), \qquad \text{(Eq. 53)}$$

where the complex constant coefficients f_{lm} are determined using

$$f_{lm}=\frac{1}{4\pi}\int_0^{2\pi}\int_0^{\pi} f(\theta,\phi)\overline{Y_l^m(\theta,\phi)}\sin\theta\mathrm{d}\theta\mathrm{d}\phi. \qquad \text{(Eq. 54)}$$

Vector spherical harmonics are required for the representation of vector fields over the surface of a sphere.

Schmidt normalized functions

The Schmidt normalized associated Legendre functions $P_l^m(\cos\theta)$ are defined by

$$P_l^0(\cos\theta)=P_{l,0}(\cos\theta),$$
$$P_l^m(\cos\theta)=\sqrt{2\frac{(l-m)!}{(l+m)!}}P_{l,m}(\cos\theta), \quad m\neq 0. \qquad \text{(Eq. 55)}$$

They are normalized to have the same value as the Legendre polynomials and they are widely used in geomagnetism in accordance with a resolution of the Association of Terrestrial Magnetism and Electricity of the International Union of Geodesy and Geophysics (Goldie and Joyce, 1940). Following Schmidt (1899), it is convenient to write the two formulae of Eq. (55) in the form

$$P_l^m(\cos\theta)=\sqrt{\varepsilon_m\frac{(l-m)!}{(l+m)!}}P_{l,m}(\cos\theta) \qquad \text{(Eq. 56)}$$

in which the parameter ε_m is defined by

$$\varepsilon_0=1, \quad \varepsilon_1=\varepsilon_2=\varepsilon_3=\ldots=2, \qquad \text{(Eq. 57)}$$

or alternatively, in terms of the Kronecker delta function

$$\varepsilon_m=2-\delta_m^0. \qquad \text{(Eq. 58)}$$

For expressions $f(\theta,\phi)$ involving real variables only, we may write it as a linear combination of Schmidt normalized functions,

$$f(\theta,\phi)=\sum_{l=0}^{N}\left(g_l^m\cos m\phi+h_l^m\sin m\phi\right)P_l^m(\cos\theta) \qquad \text{(Eq. 59)}$$

where the constant coefficient where the constants g_n^m and h_n^m are determined using

$$g_l^m=(2l+1)\frac{1}{4\pi}\int_0^{2\pi}\int_0^{\pi} f(\theta,\phi)P_l^m(\cos\theta)\cos m\phi\sin\theta\mathrm{d}\theta\mathrm{d}\phi$$
$$h_l^m=(2l+1)\frac{1}{4\pi}\int_0^{2\pi}\int_0^{\pi} f(\theta,\phi)P_l^m(\cos\theta)\sin m\phi\sin\theta\mathrm{d}\theta\mathrm{d}\phi. \qquad \text{(Eq. 60)}$$

Recurrence relations

Recurrence relations for spherical harmonics are derived using derivatives of Eq. (46) and derivatives of the recurrence relations for Legendre polynomials. The spherical polar components for the derivatives used in Eq. (46) defining $Y_n^m(\theta,\phi)$, are

$$\frac{\partial f}{\partial x}+\mathrm{i}\frac{\partial f}{\partial y}=\left(\frac{\partial f}{\partial r}\sin\theta+\frac{1}{r}\frac{\partial f}{\partial\theta}\cos\theta+\frac{\mathrm{i}}{r\sin\theta}\frac{\partial f}{\partial\phi}\right)\mathrm{e}^{\mathrm{i}\phi}, \qquad \text{(Eq. 61)}$$

$$\frac{\partial f}{\partial z}=\frac{\partial f}{\partial r}\cos\theta-\frac{1}{r}\frac{\partial f}{\partial\theta}\sin\theta, \qquad \text{(Eq. 62)}$$

$$\frac{\partial f}{\partial x}-\mathrm{i}\frac{\partial f}{\partial y}=\left(\frac{\partial f}{\partial r}\sin\theta+\frac{1}{r}\frac{\partial f}{\partial\theta}\cos\theta-\frac{\mathrm{i}}{r\sin\theta}\frac{\partial f}{\partial\phi}\right)\mathrm{e}^{-\mathrm{i}\phi}. \qquad \text{(Eq. 63)}$$

The basic set of recurrence relations in which the degree n and order m on the left hand side are changed by -1, 0, or $+1$ on the right-hand side are

$$-(l+1)\sin\theta Y_l^m+\cos\theta\frac{\partial Y_l^m}{\partial\theta}-\frac{m}{\sin\theta}Y_l^m$$
$$=\mathrm{e}^{-\mathrm{i}\phi}\sqrt{\frac{2l+1}{2l+3}(l+m+1)(l+m+2)}Y_{l+1}^{m+1}, \qquad \text{(Eq. 64)}$$

$$-(l+1)\cos\theta Y_l^m-\sin\theta\frac{\partial Y_l^m}{\partial\theta}$$
$$=-\sqrt{\frac{2l+1}{2l+3}(l-m+1)(l+m+1)}Y_{l+1}^m, \qquad \text{(Eq. 65)}$$

$$-(l+1)\sin\theta Y_l^m + \cos\theta\frac{\partial Y_l^m}{\partial\theta} + \frac{m}{\sin\theta}Y_l^m = -e^{i\phi}\sqrt{\frac{2l+1}{2l+3}}(l-m+1)(l-m+2)Y_{l+1}^{m-1}, \qquad \text{(Eq. 66)}$$

$$\frac{\partial Y_l^m}{\partial\theta} - m\cot\theta Y_l^m = e^{-i\phi}\sqrt{(l-m)(l+m+1)}Y_l^{m+1}, \qquad \text{(Eq. 67)}$$

$$\frac{\partial Y_l^m}{\partial\theta} + m\cot\theta Y_l^m = -e^{i\phi}\sqrt{(l+m)(l-m+1)}Y_l^{m-1}, \qquad \text{(Eq. 68)}$$

$$l\sin\theta Y_l^m + \cos\theta\frac{\partial Y_l^m}{\partial\theta} - \frac{m}{\sin\theta}Y_l^m = e^{-i\phi}\sqrt{\frac{2l+1}{2l-1}}(l-m)(l-m-1)Y_{l-1}^{m+1}, \qquad \text{(Eq. 69)}$$

$$l\cos\theta Y_l^m - \sin\theta\frac{\partial Y_l^m}{\partial\theta} = \sqrt{\frac{2l+1}{2l-1}(l+m)(l-m)}Y_{l-1}^m, \qquad \text{(Eq. 70)}$$

$$l\sin\theta Y_l^m + \cos\theta\frac{\partial Y_l^m}{\partial\theta} + \frac{m}{\sin\theta}Y_l^m = -e^{i\phi}\sqrt{\frac{2l+1}{2l-1}}(l+m)(l+m-1)Y_{l-1}^{m-1}. \qquad \text{(Eq. 71)}$$

Most recurrence relations can be derived from this basic set. Schuster (1903) gives a list of recurrence relations useful in ionospheric dynamo theory. Chapman and Bartels (1940) also derive some recurrence relations.

Transformation of spherical harmonics

On using the spinor forms of derivatives (see Rotations; Eq. (45))

$$\frac{\partial}{\partial\xi} = \lambda_1^2, \qquad \frac{1}{\sqrt{2}}\frac{\partial}{\partial z} = \lambda_1\lambda_2, \qquad \frac{\partial f}{\partial\eta} = \lambda_2^2,$$

it follows that solid spherical harmonics can also be written using spinors, in the more symmetric form

$$\frac{1}{r^{l+1}}Y_l^m(\theta,\phi) = (-1)^{l-m}\sqrt{\frac{2l+1}{(l-m)!(l+m)!}}\lambda_1^{l-m}\lambda_2^{l+m}\frac{1}{r}. \qquad \text{(Eq. 72)}$$

The transformation law for spherical harmonics under rotation of the reference frame follows from Eq. (72), using the transformation law (see Rotations; (190))

$$\Lambda_1 = u\lambda_1 + v\lambda_2,$$
$$\Lambda_2 = -\bar{v}\lambda_1 + \bar{u}\lambda_2. \qquad \text{(Eq. 73)}$$

where the Cayley-Klein parameters u and v are given in terms of Euler angles (α,β,γ), by

$$u = \cos\frac{1}{2}\beta e^{i(\gamma+\alpha)/2},$$
$$v = \sin\frac{1}{2}\beta e^{i(\gamma-\alpha)/2}. \qquad \text{(Eq. 74)}$$

If a point P has spherical polar coordinates, colatitude and east longitude (θ,ϕ), and if after a rotation through Euler angles (α,β,γ) the coordinates are (Θ,Φ), then from Eq. (72)

$$\frac{1}{r^{l+1}}Y_l^m(\Theta,\Phi) = (-1)^{l-m}\sqrt{\frac{2l+1}{(l-m)!(l+m)!}}\Lambda_1^{l-m}\Lambda_2^{l+m}\frac{1}{r}. \qquad \text{(Eq. 75)}$$

From the transformation formula for spinors Eq. (73), it follows that

$$\Lambda_1^{l-m}\Lambda_2^{l+m} = (u\lambda_1 + v\lambda_2)^{l-m}(-\bar{v}\lambda_1 + \bar{u}\lambda_2)^{l+m} \qquad \text{(Eq. 76)}$$

and after expanding by the binomial theorem

$$\Lambda_1^{l-m}\Lambda_2^{l+m} = \sum_{j=0}^{l+m}\sum_{k=0}^{l-m}\frac{(l-m)!(l+m)!}{(l-m-k)!k!(l+m-j)!j!} \times u^k\bar{u}^{l+m-j}v^{l-m-k}\bar{v}^{2l-j-k}. \qquad \text{(Eq. 77)}$$

Replace the summing index k by M, where $k = l - M - j$, and with the substitutions

$$u = (\cos\frac{1}{2}\beta)e^{i(\gamma+\alpha)/2}, \qquad v = (\sin\frac{1}{2}\beta)e^{i(\gamma-\alpha)/2}, \qquad \text{(Eq. 78)}$$

we find that

$$\Lambda_1^{l-m}\Lambda_2^{l+m} = \sum_M\sum_j\frac{(-1)^j}{(M-m+j)!(l-M-j)!(l+m-j)!j!} \times (\cos\frac{1}{2}\beta)^{2l-M+m-2j}(\sin\frac{1}{2}\beta)^{M-m+2j}e^{-i(M\alpha+m\gamma)}\lambda_1^{l-M}\lambda_2^{l+M} \qquad \text{(Eq. 79)}$$

The range of summing indices M and j, is such that none of the factorial expressions or powers of trigonometrical functions become negative.

Rotation matrix elements

It follows from Eq. (79) that

$$(-1)^{l-m}\frac{\Lambda_1^{l-m}\Lambda_2^{l+m}}{\sqrt{(l-m)!(l+m)!}} = \sum_{M=-l}^{l}D_{Mm}^l(\alpha,\beta,\gamma)\left[(-1)^{l-M}\frac{\lambda_1^{l-M}\lambda_2^{l+m}}{\sqrt{(L-m)!(L+m)!}}\right],$$

where the functions $D_{Mm}^{\lambda}(\alpha,\beta,\gamma)$ are called "rotation matrix elements" and are written

$$D_{Mm}^l(\alpha,\beta,\gamma) = d_{Mm}^l(\beta)e^{-i(M\alpha+m\gamma)}, \qquad \text{(Eq. 80)}$$

and the purely real function $d_{Mm}^l(\beta)$ is

$$d_{Mm}^l(\beta) = (-1)^{M-m}\sum_j(-1)^j\frac{\sqrt{(l-M)!(l+M)!(l-m)!(l+m)!}}{(M-m+j)!(l-M-j)!(l+m-j)!\,j!} \times (\cos\frac{1}{2}\beta)^{2l-M+m-2j}(\sin\frac{1}{2}\beta)^{M-m+2j}. \qquad \text{(Eq. 81)}$$

See Table H2 for a list of these functions. The form often given for the functions $d_{Mm}^l(\beta)$ is obtained with the substitution $j = m - M + t$, when

$$d_{Mm}^l(\beta) = \sum_t(-1)^t\frac{\sqrt{(l-M)!(l+M)!(l-m)!(l+m)!}}{(m-M+t)!(l+M-t)!(l-m-t)!t!} \times (\cos\frac{1}{2}\beta)^{2l-m+M-2t}(\sin\frac{1}{2}\beta)^{m-M+2t},$$

$$= \sqrt{\frac{(l-M)!(l+M)!}{(l-m)!(l+m)!}}\sum_t(-1)^t\binom{l+m}{l+M-t}\binom{l-m}{t} \times (\cos\frac{1}{2}\beta)^{2l-m+M-2t}(\sin\frac{1}{2}\beta)^{m-M+2t}, \qquad \text{(Eq. 82)}$$

Table H2 List of rotation matrix elements to degree and order 3. Symmetry properties are required to complete the $(2l+1)\times(2l+1)$ table of rotation matrix elements for degree l

l	M	m	Rotation matrix element $d^l_{M,m}(\beta)$	l	M	m	Rotation matrix element $d^l_{M,m}(\beta)$
0	0	0	1	3	1	−1	$\frac{1}{4}\sin^2\frac{\beta}{2}(15\cos^2\beta+10\cos\beta-1)$
1	0	0	$\cos\beta$	3	1	0	$\frac{\sqrt{3}}{2}\cos\frac{\beta}{2}\sin\frac{\beta}{2}(1-5\cos^2\beta)$
1	1	−1	$\sin^2\frac{\beta}{2}$	3	1	1	$\frac{1}{4}\cos^2\frac{\beta}{2}(15\cos^2\beta-10\cos\beta-1)$
1	1	0	$-\sqrt{2}\cos\frac{\beta}{2}\sin\frac{\beta}{2}$	3	2	−2	$\sin^4\frac{\beta}{2}(3\cos\beta+2)$
1	1	1	$\cos^2\frac{\beta}{2}$	3	2	−1	$-\frac{\sqrt{10}}{2}\cos\frac{\beta}{2}\sin^3\frac{\beta}{2}(1+3\cos\beta)$
2	0	0	$\frac{1}{2}(3\cos^2\beta-1)$	3	2	0	$\sqrt{30}\cos^2\frac{\beta}{2}\sin^2\frac{\beta}{2}\cos\beta$
2	1	−1	$\sin^2\frac{\beta}{2}(2\cos\beta+1)$	3	2	1	$\frac{\sqrt{10}}{2}\cos^3\frac{\beta}{2}\sin\frac{\beta}{2}(1-3\cos\beta)$
2	1	0	$-\sqrt{6}\cos\frac{\beta}{2}\sin\frac{\beta}{2}\cos\beta$	3	2	2	$\cos^4\frac{\beta}{2}(3\cos\beta-2)$
2	1	1	$\cos^2\frac{\beta}{2}(2\cos\beta-1)$	3	3	−3	$\sin^6\frac{\beta}{2}$
2	2	−2	$\sin^4\frac{\beta}{2}$	3	3	−2	$-\sqrt{6}\cos\frac{\beta}{2}\sin^5\frac{\beta}{2}$
2	2	−1	$-2\cos\frac{\beta}{2}\sin^3\frac{\beta}{2}$	3	3	−1	$\sqrt{15}\cos^2\frac{\beta}{2}\sin^4\frac{\beta}{2}$
2	2	0	$\sqrt{6}\cos^2\frac{\beta}{2}\sin^2\frac{\beta}{2}$	3	3	0	$-2\sqrt{5}\cos^3\frac{\beta}{2}\sin^3\frac{\beta}{2}$
2	2	1	$-2\cos^3\frac{\beta}{2}\sin\frac{\beta}{2}$	3	3	1	$\sqrt{15}\cos^4\frac{\beta}{2}\sin^2\frac{\beta}{2}$
2	2	2	$\cos^4\frac{\beta}{2}$	3	3	2	$-\sqrt{6}\cos^5\frac{\beta}{2}\sin\frac{\beta}{2}$
3	0	0	$\frac{1}{2}(5\cos^3\beta-3\cos\beta)$	3	3	3	$\cos^6\frac{\beta}{2}$

The transformation law for spherical harmonics is therefore

$$Y_l^m(\Theta,\Phi)=\sum_{M=-l}^{l}D^l_{Mm}(\alpha,\beta,\gamma)Y_l^M(\theta,\phi). \quad \text{(Eq. 83)}$$

Equation (82) is valid for half-odd integer values of the parameters l, M, and m. In particular,

$$D^{\frac{1}{2}}_{-\frac{1}{2},-\frac{1}{2}}(\alpha,\beta,\gamma)=\cos\frac{1}{2}\beta e^{i(\gamma+\alpha)/2}=u,$$

$$D^{\frac{1}{2}}_{\frac{1}{2},-\frac{1}{2}}(\alpha,\beta,\gamma)=\sin\frac{1}{2}\beta e^{i(\gamma-\alpha)/2}=v. \quad \text{(Eq. 84)}$$

The transformation law for spinors, Eq. (73), becomes

$$\begin{pmatrix}\Lambda_1\\ \Lambda_2\end{pmatrix}=\begin{pmatrix}D^{\frac{1}{2}}_{-\frac{1}{2},-\frac{1}{2}}(\alpha,\beta,\gamma) & D^{\frac{1}{2}}_{\frac{1}{2},-\frac{1}{2}}(\alpha,\beta,\gamma)\\ -D^{\frac{1}{2}}_{-\frac{1}{2}\frac{1}{2}}(\alpha,\beta,\gamma) & D^{\frac{1}{2}}_{-\frac{1}{2},-\frac{1}{2}}(\alpha,\beta,\gamma)\end{pmatrix}\begin{pmatrix}\lambda_1\\ \lambda_2\end{pmatrix}. \quad \text{(Eq. 85)}$$

The transformation law for surface spherical harmonics of degree one, is

$$\begin{pmatrix}Y_1^1(\Theta,\Phi)\\ Y_1^0(\Theta,\Phi)\\ Y_1^{-1}(\Theta,\Phi)\end{pmatrix}=\begin{pmatrix}\cos^2\frac{1}{2}\beta e^{i(\alpha+\gamma)} & \sqrt{2}\sin\frac{1}{2}\beta\cos\frac{1}{2}\beta e^{i\gamma} & \sin^2\frac{1}{2}\beta e^{i(-\alpha+\gamma)}\\ -\sqrt{2}\sin\frac{1}{2}\beta\cos\frac{1}{2}\beta e^{i\alpha} & \cos\beta & \sqrt{2}\sin\frac{1}{2}\beta\cos\frac{1}{2}\beta e^{-i\alpha}\\ \sin^2\frac{1}{2}\beta e^{i(\alpha-\gamma)} & -\sqrt{2}\sin\frac{1}{2}\beta\cos\frac{1}{2}\beta e^{-i\gamma} & \cos^2\frac{1}{2}\beta e^{-i(\alpha+\gamma)}\end{pmatrix}\begin{pmatrix}Y_1^1(\theta,\phi)\\ Y_1^0(\theta,\phi)\\ Y_1^{-1}(\theta,\phi)\end{pmatrix} \quad \text{(Eq. 86)}$$

Closure

When a rotation through Euler angles $(\alpha_1,\beta_1,\gamma_1)$ is followed by a second rotation through Euler angles $(\alpha_2,\beta_2,\alpha_2)$, then

$$Y_l^{m'}(\Theta,\Phi)=\sum_{M=-l}^{l}D^l_{Mm'}(\alpha_1,\beta_1,\gamma_1)\,Y_l^M(\theta,\phi),$$

$$Y_l^M(\Theta',\Phi')=\sum_{m'=-l}^{l}D^l_{m'm}(\alpha_2,\beta_2,\gamma_2)\,Y_l^{m'}(\Theta,\Phi),$$

and therefore,

$$Y_l^m(\Theta',\Phi')=\sum_{M=-l}^{l}\left[\sum_{m'=-l}^{l}D^l_{Mm'}(\alpha_1,\beta_1,\gamma_1)D^l_{m'm}(\alpha_2,\beta_2,\gamma_2)\right]Y_l^M(\theta,\phi).$$

However, this is equivalent to a rotation through Euler angles (α,β,γ) where

$$Y_l^m(\Theta',\Phi')=\sum_{M=-l}^{l}D^l_{Mm}(\alpha,\beta,\gamma)\,Y_l^M(\theta,\phi),$$

from which it follows that

$$D^l_{Mm}(\alpha,\beta,\gamma)=\sum_{m'=-l}^{l}D^l_{Mm'}(\alpha_1,\beta_1,\gamma_1)D^l_{m'm}(\alpha_2,\beta_2,\gamma_2). \quad \text{(Eq. 87)}$$

Equation (87) is an important result, with special cases giving the sum rule, the addition theorems of trigonometry, and formulae of spherical trigonometry, including the analogies of Napier and Delambre, as well as the various haversine formulae. For example, in the special case that $\alpha_1=\alpha_2=\alpha_3=0$ and $\gamma_1=\gamma_2=\gamma_3=0$, then $\beta_3=\beta_1+\beta_2$, and

$$d^l_{Mm}(\beta_1+\beta_2)=\sum_{m'=-l}^{l}d^l_{Mm'}(\beta_1)d^l_{m'm}(\beta_2), \quad \text{(Eq. 88)}$$

which is a generalization of all of the sum formulae of trigonometry.

Rodrigues formula

The rotation matrix elements $D^l_{Mm}(\beta)$ have two types of orthogonality. Firstly, they are orthogonal under integration over the range of the Euler angles, $0 \leq \alpha < 2\pi, 0 \leq \beta \leq \pi$, and $0 \leq \gamma < 2\pi$, and secondly, for a fixed value of l, regarded as a $(2l+1) \times (2l+1)$ matrix array, they have matrix orthogonality properties. These properties are most easily derived using the Rodrigues formula, substituting $z = \cos\beta$, when

$$d^l_{Mm}(z) = \frac{(-1)^{l+M}}{2^l}\sqrt{\frac{(l-M)!}{(l+m)!(l-m)!(l+M)!}} \times (1-z)^{(M-m)/2}(1+z)^{(M+m)/2} \left(\frac{\mathrm{d}}{\mathrm{d}z}\right)^{l+M}[(1-z)^{l+m}(1+z)^{l-m}]. \quad \text{(Eq. 89)}$$

The "symmetry", $d^l_{M,m}(\beta) = d^l_{-m,-M}(\beta)$, indicates that there are two other equivalent formulae (Schendel, 1877).

Equation (89), in the case $m = 0$, using Eq. (43), gives

$$d^l_{M,0}(z) = \frac{(-1)^M}{2^l l!}\sqrt{\frac{(l-M)!}{(l+M)!}}(1-z^2)^{M/2}\left(\frac{\mathrm{d}}{\mathrm{d}z}\right)^{l+M}(z^2-1)^l,$$
$$= (-1)^M\sqrt{\frac{(l-M)!}{(l+M)!}}P_{l,M}(z),$$
$$= \sqrt{\frac{(l+M)!}{(l-M)!}}P_{l,-M}(z). \quad \text{(Eq. 90)}$$

A full description of the properties of the rotation matrix elements, including contour integral formulae is given in Vilenkin (1968).

Orthogonality

The D-functions form unitary matrices, with orthonormal rows and columns for each fixed degree l. They also have orthogonality properties under integration over all three Eulerian angles. The derivation makes use of a second equivalent form for $d^l_{Mm}(z)$, namely

$$d^l_{Mm}(z) = \frac{(-1)^{l-m}}{2^l}\sqrt{\frac{(L+M)!}{(l+m)!(l-m)!(L-M)!}} \times (1-z)^{(-M+m)/2}(1+z)^{(-M-m)/2} \left(\frac{\mathrm{d}}{\mathrm{d}z}\right)^{l-M}\left[(1-z)^{l-m}(1+z)^{l+m}\right]. \quad \text{(Eq. 91)}$$

The required property is that

$$\frac{1}{8\pi^2}\int_0^{2\pi}\int_0^{\pi}\int_0^{2\pi} D^l_{Mm}(\alpha,\beta,\gamma)\overline{D^l_{M'm'}(\alpha,\beta,\gamma)}\sin\beta \mathrm{d}\alpha\mathrm{d}\beta\mathrm{d}\gamma = \frac{1}{2l+1}\delta^{l'}_l\delta^{M'}_M\delta^{m'}_m. \quad \text{(Eq. 92)}$$

Special values

At the particular values $0, \pi, 2\pi$ of β, the rotation matrix elements are

$$d^l_{M,m}(0) = \delta^m_M,$$
$$d^l_{M,m}(\pi) = (-1)^{l+M}\delta^{-m}_M = (-1)^{l-m}\delta^{-m}_M,$$
$$d^l_{M,m}(2\pi) = (-1)^{2l}\delta^m_M, \quad \text{(Eq. 93)}$$

with symmetries for general values of β

$$d^l_{M,m}(\beta) = d^l_{-m,-M}(\beta),$$
$$d^l_{m,M}(\beta) = (-1)^{m-M}d^l_{-m,-M}(\beta), \quad \text{(Eq. 94)}$$

and

$$d^l_{M,m}(-\beta) = (-1)^{m-M}d^l_{M,m}(\beta) = d^l_{m,M}(\beta),$$
$$d^l_{M,m}(\pi+\beta) = (-1)^{l-m}d^l_{M,-m}(\beta),$$
$$d^l_{M,m}(\pi-\beta) = (-1)^{l-M}d^\beta_{M,-m}(\beta) = (-1)^{l+m}d^l_{-M,m}(\beta). \quad \text{(Eq. 95)}$$

The sum rule

An important special case of Eq. (89) is that for which $m = 0$, when, from Eq. (75) defining $Y^m_l(\theta,\phi)$, and with $\mu = \cos\beta$, we obtain

$$D^l_{M,0}(\alpha,\beta,\gamma) = d^l_{M,0}(\beta)\mathrm{e}^{-\mathrm{i}M\alpha},$$
$$= (-1)^M\sqrt{\frac{(l-M)!}{(l+M)!}}P_{l,M}(\mu)\mathrm{e}^{-\mathrm{i}M\alpha},$$
$$= \frac{1}{\sqrt{2l+1}}\overline{Y^M_l(\beta,\alpha)}. \quad \text{(Eq. 96)}$$

Therefore, in the special case $m = 0$, the transformation formula (83) becomes

$$Y^0_l(\Theta,\Phi) = \sum_{M=-l}^{l} D^l_{M0}(\alpha,\beta,\gamma)\, Y^M_l(\theta,\phi)\,. \quad \text{(Eq. 97)}$$

From Eqs. (48) and (96), Eq. (97) becomes

$$P_l(\cos\Theta) = \frac{1}{2l+1}\sum_{M=-l}^{l}\overline{Y^M_l(\beta,\alpha)}\, Y^M_l(\theta,\phi)\,. \quad \text{(Eq. 98)}$$

and the sum rule in the form Eq. (98) becomes

$$P_l(\cos\Theta) = \sum_{M=-l}^{l}\frac{(l-M)!}{(l+M)!}P_{l,M}(\cos\beta)P_{l,M}(\cos\theta)\mathrm{e}^{\mathrm{i}M(\phi-\alpha)}. \quad \text{(Eq. 99)}$$

In terms of Schmidt normalized functions, the sum rule takes the well-known form,

$$P_l(\cos\Theta) = \sum_{M=0}^{l}P^M_l(\cos\beta)P^M_l(\cos\theta)\cos m(\phi-\alpha). \quad \text{(Eq. 100)}$$

The case $l = 1$ of Eq. (100) is the cosine rule of spherical trigonometry,

$$\cos\Theta = \cos\beta\cos\theta + \sin\beta\sin\theta\cos(\phi-\alpha), \quad \text{(Eq. 101)}$$

and the case $l = 2$, is the basic rule in the theory of tides, with the longitudinal terms showing the dependence on $M = 0$ long-period terms, $M = 1$ lunar diurnal terms and $M = 2$ lunar semidiurnal terms,

$$P_2(\cos\theta) = P_2(\cos\beta)P_2(\cos\theta) + P^1_2(\cos\beta)P^1_2(\cos\theta)\cos(\phi-\alpha) + P^2_2(\cos\beta)P^2_2(\cos\theta)\cos 2(\phi-\alpha). \quad \text{(Eq. 102)}$$

In the case that $\Theta = 0$, and the spherical triangle collapses to a straight line so that $\beta = \theta$ and $\phi = \alpha$, then since $P_l(1) = 1$ for all values of l, then Eq. (98) becomes

$$\sum_{M=-l}^{l} \overline{Y_l^M(\theta,\phi)}\, Y_l^M(\theta,\phi) = 2l + 1\,. \qquad \text{(Eq. 103)}$$

and Eq. (100) in terms of Schmidt normalized polynomials gives

$$\sum_{M=0}^{l} \left[P_l^M(\cos\theta)\right]^2 = 1, \qquad \text{(Eq. 104)}$$

which can be used to provide a useful check on numerical work.

The result of Eq. (103) is equivalent to the conservation of "lengths" under rotation, and therefore requires that the $(2l+1)\times(2l+1)$ rotation matrices $D^l_{Mm}(\alpha,\beta,\gamma)$ be unitary, i.e.,

$$\sum_{M=-l}^{l} \overline{D^l_{M'M}(\alpha,\beta,\gamma)}\, D^l_{M''M}(\alpha,\beta,\gamma) = \delta^{M''}_{M'}\,. \qquad \text{(Eq. 105)}$$

Schmidt's analysis

Schmidt (1899) derived the transformation formula (83) for spherical harmonics well before the later derivations in the context of the theory of atomic spectra. Schmidt gave the result in terms Schmidt normalized functions, using real variables only, as

$$\begin{aligned} &P_l^m(\cos\Theta)\cos m[\pi - (\Phi+\gamma)] \\ &= \frac{1}{2}\sum_{M=0}^{l} \sqrt{\varepsilon_m\varepsilon_M}\left[(-1)^M d^l_{M,m}(\beta) + d^l_{-M,m}(\beta)\right] \\ &\times P_l^M(\cos\theta)\cos M(\phi-\alpha). \end{aligned} \qquad \text{(Eq. 106)}$$

Because m and $M = 1,2,3,\ldots$ only in the expression for imaginary component, therefore, in this case, $\varepsilon_m = 2$ and $\varepsilon_M = 2$, and

$$\begin{aligned} &P_l^m(\cos\Theta)\sin m[\pi - (\Phi+\gamma)] \\ &= \sum_{M=1}^{l}\left[(-1)^M d^l_{M,m}(\beta) - d^l_{-M,m}(\beta)\right] P_l^M(\cos\theta)\sin M(\phi-\alpha) \end{aligned} \qquad \text{(Eq. 107)}$$

The expression Schmidt developed for the rotation matrix elements was given in a slightly different, but nevertheless equivalent form to that of Eq. (89) where $z = \cos\beta$. In Schmidt's result, $c = \cos\beta$ and $s = \sin\beta$, and

$$\begin{aligned} d^l_{Mm}(\beta) &= (-1)^M\sqrt{\frac{(l-m)!(l-M)!}{(l+m)!(l+M)!}}(1+c)^m s^{M-m} \\ &\left(\frac{\mathrm{d}}{\mathrm{d}c}\right)^M\left[(1-c)^m\left(\frac{\mathrm{d}}{\mathrm{d}c}\right)^m P_l(c)\right]. \end{aligned} \qquad \text{(Eq. 108)}$$

Leibnitz's theorem is used to show the equivalence of Eqs. (108) and (89), where $z = \cos\beta$.

Wigner 3-*j* coefficients

The Wigner 3-*j* coefficients (Wigner, 1931) are just generalizations of the factors that appear in recurrence relations for associated Legendre polynomials, and rotation matrix elements. They have an interesting structure, and are best defined as a sum of products of three combinatorial coefficients. Thus

$$\begin{aligned} \begin{pmatrix} a & b & c \\ \alpha & \beta & \gamma \end{pmatrix} &= (-1)^{a-b-\gamma}\sqrt{\frac{(a-\alpha)!(a+\alpha)!(b-\beta)!(b+\beta)!(c-\gamma)!(c+\gamma)!}{(-a+b+c)!(a-b+c)!(a+b-c)!}} \\ &\times \frac{1}{\sqrt{(a+b+c+1)!}}\sum_t (-1)^t \begin{pmatrix} a-b+c \\ a-\alpha-t \end{pmatrix}\begin{pmatrix} -a+b+c \\ b+\beta-t \end{pmatrix}\begin{pmatrix} a+b-c \\ t \end{pmatrix} \end{aligned} \qquad \text{(Eq. 109)}$$

Conditions on the values of the parameters require that the nine elements of the following 3×3 array, called a Racah square, are all positive,

$$\boxed{\begin{array}{ccc} -a+b+c & a-b+c & a+b-c \\ a-\alpha & b-\beta & c-\gamma \\ a+\alpha & b+\beta & c+\gamma \end{array}} \qquad \text{(Eq. 110)}$$

and that $\alpha+\beta+\gamma = 0$. The following special cases of the 3-*j* coefficients are determined from the series definition of Eq. (109),

$$\left.\begin{aligned} \begin{pmatrix} l & l+1 & 1 \\ m & -m-1 & 1 \end{pmatrix} &= (-1)^{l-m}\sqrt{\frac{(l+m+1)(l+m+2)}{(2l+1)(2l+2)(2l+3)}}, \\ \begin{pmatrix} l & l+1 & 1 \\ m & -m & 0 \end{pmatrix} &= (-1)^{l-m-1}\sqrt{\frac{2(l-m+1)(l+m+1)}{(2l+1)(2l+2)(2l+3)}}, \\ \begin{pmatrix} l & l+1 & 1 \\ m & -m+1 & -1 \end{pmatrix} &= (-1)^{l-m}\sqrt{\frac{(l-m+1)(l-m+2)}{(2l+1)(2l+2)(2l+3)}}, \end{aligned}\right\} \qquad \text{(Eq. 111)}$$

$$\left.\begin{aligned} \begin{pmatrix} l & l & 1 \\ m & -m-1 & 1 \end{pmatrix} &= (-1)^{l-m}\sqrt{\frac{2(l-m)(l+m+1)}{2(2l+1)(2l+2)}}, \\ \begin{pmatrix} l & l & 1 \\ m & -m & 0 \end{pmatrix} &= (-1)^{l-m}\frac{2m}{\sqrt{2l(2l+1)(2l+2)}}, \\ \begin{pmatrix} l & l & 1 \\ m & -m+1 & -1 \end{pmatrix} &= (-1)^{l+m-1}\sqrt{\frac{2(l+m)(l-m+1)}{2l(2l+1)(l+2)}}, \end{aligned}\right\} \qquad \text{(Eq. 112)}$$

$$\left.\begin{aligned} \begin{pmatrix} l & l-1 & 1 \\ m & -m-1 & 1 \end{pmatrix} &= (-1)^{l-m}\sqrt{\frac{(l-m-1)(l-m)}{(2l-1)2l(2l+1)}}, \\ \begin{pmatrix} l & l-1 & 1 \\ m & -m & 0 \end{pmatrix} &= (-1)^{l-m}\sqrt{\frac{2(l-m)(l+m)}{(2l-1)2l(2l+1)}}, \\ \begin{pmatrix} l & l-1 & 1 \\ m & -m+1 & -1 \end{pmatrix} &= (-1)^{l-m}\sqrt{\frac{(l+m-1)(l+m)}{(2l-1)2l(2l+1)}} \end{aligned}\right\}. \qquad \text{(Eq. 113)}$$

An integration formula

The following integration formula can be used to derive the properties of 3-*j* coefficients directly from their definitions,

$$\begin{aligned} &\int_{-1}^{1}(\mu-1)^{l_2-m_2}(\mu+1)^{l_2+m_2}\left(\frac{\mathrm{d}}{\mathrm{d}\mu}\right)^{l_2+l_2-l_3}\left[(\mu-1)^{l_1-m_1}(\mu+1)^{l_1+m_1}\right]\mathrm{d}\mu \\ &= (-1)^{2l_1}(l_1+l_2-l_3)!\begin{pmatrix} l_1 & l_2 & l_3 \\ m_1 & m_2 & m_3 \end{pmatrix} \\ &\times\left[\frac{(l_1-m_1)!(l_1+m_1)!(l_2-m_2)!(l_2+m_2)!(l_3-m_3)!(l_3+m_3)!}{(l_1+l_2-l_3)!(l_1-l_2+l_3)!(-l_1+l_2+l_3)!(l_1+l_2+l_3)!}\right]. \end{aligned} \qquad \text{(Eq. 114)}$$

The 3-j coefficients have two important orthogonality properties

$$\sum_{\substack{m_2,m_3 \\ m_1+m_2+m_3=0}} \begin{pmatrix} l_1 & l_2 & l_3 \\ m_1 & m_2 & m_3 \end{pmatrix} \begin{pmatrix} l_1' & l_2 & l_3 \\ m_1 & m_2 & m_3 \end{pmatrix} = \frac{1}{2l_1+1}\delta_{l_1}^{l_1'}, \quad \text{(Eq. 115)}$$

and Eq. (119) given below.

Products of rotation matrix elements

The integral of a product of three rotation matrix elements is

$$\frac{1}{8\pi^2}\int_0^{2\pi}\int_0^{\pi}\int_0^{2\pi} D^{l_1}_{M_1m_1} D^{l_2}_{M_2m_2} D^{l_3}_{M_3m_3} \sin\beta \, d\alpha d\beta d\gamma = \begin{pmatrix} l_1 & l_2 & l_3 \\ M_1 & M_2 & M_3 \end{pmatrix} \begin{pmatrix} l_1 & l_2 & l_3 \\ m_1 & m_2 & m_3 \end{pmatrix} \quad \text{(Eq. 116)}$$

and therefore, from the orthogonality of rotation matrix elements, it follows that the product of two rotation matrix elements can be written as a sum of rotation matrix elements,

$$D^{l_1}_{M_1m_1}(\alpha,\beta,\gamma) D^{l_2}_{M_2m_2}(\alpha,\beta,\gamma) = \sum_{l_3}(2l_3+1)\begin{pmatrix} l_1 & l_2 & l_3 \\ M_1 & M_2 & M_3 \end{pmatrix} \begin{pmatrix} l_1 & l_2 & l_3 \\ m_1 & m_2 & m_3 \end{pmatrix} \overline{D^{l_3}_{M_3m_3}(\alpha,\beta,\gamma)}. \quad \text{(Eq. 117)}$$

Equation (117) is an important equation because it effectively contains all the recurrence relations for rotation matrix elements, and therefore for surface spherical harmonics. The first orthogonality property of 3-j coefficients applied to the product formula (117) gives

$$\sum_{M_1,M_2} \begin{pmatrix} l_1 & l_2 & l_3 \\ M_1 & M_2 & M_3 \end{pmatrix} D^{l_1}_{M_1,m_1} D^{l_2}_{M_2,m_2} = \begin{pmatrix} l_1 & l_2 & l_3 \\ m_1 & m_2 & m_3 \end{pmatrix} \overline{D^{l_3}_{M_3,m_3}}. \quad \text{(Eq. 118)}$$

Setting $\alpha = \beta = \gamma = 0$ in Eq. (117) gives the second orthogonality property of the 3-j coefficients,

$$\sum_{l_3}(2l_3+1)\begin{pmatrix} l_1 & l_2 & l_3 \\ M_1 & M_2 & M_3 \end{pmatrix} \begin{pmatrix} l_1 & l_2 & l_3 \\ m_1 & m_2 & m_3 \end{pmatrix} = \delta^{m_1}_{M_1}\delta^{m_2}_{M_2}. \quad \text{(Eq. 119)}$$

Vector coupling coefficients

In applications, it is more convenient to use a coupling coefficient, (Varshalovich *et al.*, 1988),

$$C^{l_3,-m_3}_{l_1m_1l_2m_2} = (-1)^{l_1-l_2-m_3}\sqrt{2l_3+1}\begin{pmatrix} l_1 & l_2 & l_3 \\ m_1 & m_2 & m_3 \end{pmatrix}, \quad \text{(Eq. 120)}$$

when the orthogonality properties, Eqs. (115) and (119) are

$$\sum_{m_1+m_2=m_3} C^{l_3m_3}_{l_1m_1l_2m_2} C^{l_3'm_3}_{l_1m_1l_2m_2} = \delta^{l_3'}_{l_3}, \quad \text{with } m_1+m_2=m_3, \quad \text{(Eq. 121)}$$

$$\sum_{l_3} C^{l_3m_3}_{l_1m_1l_2m_2} C^{l_3m_3}_{l_1m_1'l_2m_2'} = \delta^{m_1'}_{m_1}\delta^{m_2'}_{m_2}, \quad \text{with } m_1+m_2=m_3, m_1'+m_2'=m_3. \quad \text{(Eq. 122)}$$

The product formulae (117) and (118), for rotation matrix elements become

$$D^{l_1}_{M_1,m_1} D^{l_2}_{M_2,m_2} = \sum_{l_3} C^{l_3M_3}_{l_1M_1L_2M_2} C^{l_3m_3}_{l_1m_1l_2m_2} D^{l_3}_{M_3,m_3}, \quad \text{(Eq. 123)}$$

$$\sum_{m_1,m_2} C^{l_3m_3}_{l_1m_1l_2m_2} D^{l_1}_{M_1,m_1} D^{l_2}_{M_2,m_2} = C^{l_3M_3}_{l_1M_1l_2M_2} D^{l_3}_{M_3,m_3}, \quad \text{(Eq. 124)}$$

respectively, and in both of which

$$M_1 + M_2 = M_3, \quad \text{and} \quad m_1 + m_2 = m_3$$

Note that the complex conjugate expressions required on the right-hand sides of Eqs. (117) and (118) are not required in the corresponding vector coupling coefficient formulae. This occurs because of the result that

$$D^l_{-M,-m}(\alpha,\beta,\gamma) = (-1)^{M-m}\overline{D^l_{M,m}(\alpha,\beta,\gamma)}. \quad \text{(Eq. 125)}$$

We can now use vector-coupling coefficient to couple together two quantities that transform like spherical harmonics. Consider

$$Y^{m_1}_{l_1}(\Theta_1,\Phi_1) = \sum_{M_1=-l_1}^{l_1} D^{l_1}_{M_1,m_1}(\alpha,\beta,\gamma) Y^{M_1}_{l_1}(\theta_1,\phi_1),$$
$$Y^{m_2}_{l_2}(\Theta_2,\Phi_2) = \sum_{M_1=-l_1}^{l_1} D^{l_2}_{M_2,m_2}(\alpha,\beta,\gamma) Y^{M_2}_{l_2}(\theta_2,\phi_2). \quad \text{(Eq. 126)}$$

When the coupled spherical harmonic expression is formed

$$\sum_{m_1,m_2} C^{l_3m_3}_{l_1m_1l_2m_2} Y^{m_1}_{l_1}(\Theta_1,\Phi_1) Y^{m_2}_{l_2}(\Theta_2,\Phi_2) = \sum_{M_1=-l_1}^{l_1}\sum_{M_2=-l_2}^{l_2}\left[\sum_{m_1,m_2} C^{l_3m_3}_{l_1m_1l_2m_2} D^{l_1}_{M_1m_1} D^{l_2}_{M_2m_2}\right] \times Y^{m_1}_{l_1}(\theta_1,\phi_1) Y^{m_2}_{l_2}(\theta_2,\phi_2). \quad \text{(Eq. 127)}$$

The sums on the right of Eq. (127) can be rearranged and the expressions in square brackets replaced by means of Eq. (124), giving the results that

$$\sum_{m_1,m_2} C^{l_3m_3}_{l_1m_1l_2m_2} Y^{m_1}_{l_1}(\Theta_1,\Phi_1) Y^{m_2}_{l_2}(\Theta_2,\Phi_2) = \sum_{M_3} D^{l_3}_{M_3m_3}\left[\sum_{M_1,M_2} C^{l_3M_3}_{l_1M_1l_2M_2} Y^{M_1}_{l_1}(\theta_1,\phi_1) Y^{M_2}_{l_2}(\theta_2,\phi_2)\right], \quad \text{(Eq. 128)}$$

showing that the coupled expression in square brackets in Eq. (128) obeys the transformation law for spherical harmonics $Y^{M_3}_{l_3}(\Theta,\Phi)$under rotation of the reference frame; that is

$$Y^{M_3}_{l_3}(\Theta,\Phi) = \sum_{M_1,M_2} C^{l_3M_3}_{l_1M_1l_2M_2} Y^{M_1}_{l_1}(\theta_1,\phi_1) Y^{M_2}_{l_2}(\theta_2,\phi_2), \quad \text{(Eq. 129)}$$

so that coupled spherical harmonics on the right of Eq. (129) will transform under rotation of the reference frame like a spherical harmonics.

Vector spherical harmonics

If the unit vectors in the x-, y- and z-directions are denoted $\mathbf{e}_x, \mathbf{e}_y, \mathbf{e}_z$, respectively, the complex reference vectors $\mathbf{e}_1, \mathbf{e}_0, \mathbf{e}_{-1}$, are defined by

$$\mathbf{e}_1 = -\frac{1}{\sqrt{2}}(\mathbf{e}_x + \mathrm{i}\mathbf{e}_y) = \nabla\xi,$$
$$\mathbf{e}_0 = \mathbf{e}_z = \nabla z,$$
$$\mathbf{e}_{-1} = \frac{1}{\sqrt{2}}(\mathbf{e}_x - \mathrm{i}\mathbf{e}_y) = \nabla\eta. \quad \text{(Eq. 130)}$$

If $\bar{\mathbf{e}}_\mu$ denotes the complex conjugate of $\mathbf{e}_\mu$, $\mu = -1, 0, 1$, then the complex reference vectors satisfy

$$\mathbf{e}_\mu \cdot \bar{\mathbf{e}}_\nu = \delta^\nu_\mu, \text{ and } \mathbf{e}_\mu \cdot \mathbf{e}_\nu = (-1)^{\mu-\nu}\delta^{-\nu}_\mu, \quad \text{for } \mu, \nu = 1, 0, -1. \quad \text{(Eq. 131)}$$

Therefore, a vector $\mathbf{B}$, with Cartesian components B_x, B_y, B_z, will have complex reference components B_{-1}, B_0, B_1, such that $\mathbf{B} = -B_1\mathbf{e}_{-1} + B_0\mathbf{e}_0 - B_{-1}\mathbf{e}_1$, and

$$B_1 = -\frac{1}{\sqrt{2}}(B_x + \mathrm{i}B_y),$$
$$B_0 = B_z,$$
$$B_{-1} = \frac{1}{\sqrt{2}}(B_x - \mathrm{i}B_y). \quad \text{(Eq. 132)}$$

Complex reference components of the gradient operator ∇ and the angular momentum operator $\mathbf{L} = -i\,\mathbf{r} \times \nabla$, are obtained, and can be expressed in terms of spherical polar coordinates, r, θ, ϕ, as follows:

$$\nabla_1 = -\frac{1}{\sqrt{2}}\mathrm{e}^{\mathrm{i}\phi}\left(\sin\theta\frac{\partial}{\partial r} + \frac{\cos\theta}{r}\frac{\partial}{\partial\theta} + \frac{\mathrm{i}}{r\sin\theta}\frac{\partial}{\partial\phi}\right),$$
$$\nabla_0 = \cos\theta\frac{\partial}{\partial r} - \frac{\sin\theta}{r}\frac{\partial}{\partial\theta},$$
$$\nabla_{-1} = \frac{1}{\sqrt{2}}\mathrm{e}^{-\mathrm{i}\phi}\left(\sin\theta\frac{\partial}{\partial r} + \frac{\cos\theta}{r}\frac{\partial}{\partial\theta} - \frac{\mathrm{i}}{r\sin\theta}\frac{\partial}{\partial\phi}\right), \quad \text{(Eq. 133)}$$

and

$$L_1 = -\frac{1}{\sqrt{2}}\mathrm{e}^{\mathrm{i}\phi}\left(\frac{\partial}{\partial\theta} + \mathrm{i}\cot\theta\frac{\partial}{\partial\phi}\right),$$
$$L_0 = -\mathrm{i}\frac{\partial}{\partial\phi},$$
$$L_{-1} = -\frac{1}{\sqrt{2}}\mathrm{e}^{-\mathrm{i}\phi}\left(\frac{\partial}{\partial\theta} - \mathrm{i}\cot\theta\frac{\partial}{\partial\phi}\right). \quad \text{(Eq. 134)}$$

We now define three vector spherical harmonics $\mathbf{Y}^m_{l,l+1}(\theta,\phi), \mathbf{Y}^m_{l,l}(\theta,\phi), \mathbf{Y}^m_{l,l-1}(\theta,\phi)$, in terms of complex reference vectors, and also present them in the better known and widely used forms in terms of spherical polars. Thus

$$\frac{1}{r^{l+2}}\sqrt{(l+1)(2l+1)}\mathbf{Y}^m_{l,l+1}(\theta,\phi) = \nabla\left[\frac{1}{r^{l+1}}Y^m_l(\theta,\phi)\right],$$
$$\sqrt{l(l+1)}\mathbf{Y}^m_{l,l}(\theta,\phi) = \mathbf{L}Y^m_l(\theta,\phi),$$
$$r^{l-1}\sqrt{l(2l+1)}\mathbf{Y}^m_{l,l-1}(\theta,\phi) = \nabla\left[r^l Y^m_l(\theta,\phi)\right]. \quad \text{(Eq. 135)}$$

The spherical polar components of the vector spherical harmonics are

$$\sqrt{(l+1)(2l+1)}\mathbf{Y}^m_{l,l+1}(\theta,\phi) = -(l+1)Y^m_l\mathbf{e}_r + \frac{\partial Y^m_n}{\partial\theta}\mathbf{e}_\theta + \frac{1}{\sin\theta}\frac{\partial Y^m_n}{\partial\phi}\mathbf{e}_\phi,$$
$$\sqrt{l(l+1)}\mathbf{Y}^m_{l,l}(\theta,\phi) = \frac{\mathrm{i}}{\sin\theta}\frac{\partial Y^m_n}{\partial\phi}\mathbf{e}_\theta - \mathrm{i}\frac{\partial Y^m_n}{\partial\theta}\mathbf{e}_\phi,$$
$$\sqrt{l(2l+1)}\mathbf{Y}^m_{l,l-1}(\theta,\phi) = l\,Y^m_l\mathbf{e}_r + \frac{\partial Y^m_n}{\partial\theta}\mathbf{e}_\theta + \frac{1}{\sin\theta}\frac{\partial Y^m_n}{\partial\phi}\mathbf{e}_\phi, \quad \text{(Eq. 136)}$$

where $\mathbf{e}_r, \mathbf{e}_\theta, \mathbf{e}_\phi$, are unit vectors in the direction of r, θ, ϕ, increasing, respectively.

However, from the recurrence relations given in Eqs. (64)–(71), and from the spherical polar forms of the vector operators given in Eqs. (133) and (134), it follows immediately that

$$\sqrt{(l+1)(2l+1)}\mathbf{Y}^m_{l,l+1} = \sqrt{\frac{2l+1}{2l+3}} \times \left(\frac{1}{\sqrt{2}}\sqrt{(l+m+1)(l+m+2)}Y^{m+1}_{l+1}\mathbf{e}_{-1} - \sqrt{(l-m+1)(l+m+1)}Y^m_{l+1}\mathbf{e}_0 + \frac{1}{\sqrt{2}}\sqrt{(l-m+1)(l-m+2)}Y^m_{l+1}\mathbf{e}_1\right), \quad \text{(Eq. 137)}$$

$$\sqrt{l(l+1)}\mathbf{Y}^m_{l,l} = \frac{1}{\sqrt{2}}\sqrt{(l+m+1)(l-m)}Y^{m+1}_l\mathbf{e}_{-1} + mY^m_l\mathbf{e}_0 - \frac{1}{\sqrt{2}}\sqrt{(l+m)(l-m+1)}Y^{m-1}_l\mathbf{e}_1, \quad \text{(Eq. 138)}$$

$$\sqrt{l(2l+1)}\mathbf{Y}^m_{l,l-1} = \sqrt{\frac{2l+1}{2l-1}}\left(\frac{1}{\sqrt{2}}\sqrt{(l-m-1)(l-m)}Y^{m+1}_{l-1}\mathbf{e}_{-1} + \sqrt{(l+m)(l-m)}Y^m_{l-1}\mathbf{e}_0 + \frac{1}{\sqrt{2}}\sqrt{(l+m-1)(l+m)}Y^{m-1}_{l-1}\mathbf{e}_1\right). \quad \text{(Eq. 139)}$$

Using the 3-j coefficients listed in Eqs. (111)–(113), it follows that all three Eqs. (137)–(139) can be written as a single expression

$$\mathbf{Y}^m_{l,l+\nu}(\theta,\phi) = (-1)^{l+\nu-1+m}\sqrt{2l+1}\sum_{\mu=-1}^{1}\begin{pmatrix} l+\nu & 1 & l \\ m-\mu & \mu & -m \end{pmatrix}Y^{m-\mu}_{l+\nu}\mathbf{e}_\mu. \quad \text{(Eq. 140)}$$

In terms of the coupling coefficient of Eq. (120)

$$\mathbf{Y}^m_{l,l+\nu}(\theta,\phi) = \sum_{\mu=-1}^{1} C^{l,m}_{l+\nu,m-\mu,1,\mu}Y^{m-\mu}_{l+\nu}\mathbf{e}_\mu. \quad \text{(Eq. 141)}$$

Because of the orthogonality properties of complex reference vectors and surface spherical harmonics, it follows that the vector spherical harmonics are orthogonal under integration over the surface of a sphere:

$$\frac{1}{4\pi}\int_0^{2\pi}\int_0^{\pi}\mathbf{Y}^{m_1}_{l_1,l_1+\mu}(\theta,\phi)\cdot\mathbf{Y}^{m_2}_{l_2,l_2+\nu}(\theta,\phi)\sin\theta\,\mathrm{d}\theta\,\mathrm{d}\phi = \delta^{l_2}_{l_1}\delta^{m_2}_{m_1}\delta^{\mu}_{\nu}. \quad \text{(Eq. 142)}$$

Appendix

Rotations

Introduction

The word "rotation" has related, but nevertheless, different meanings in different disciplines, such as agriculture, medicine, psychology, and mechanics. This article is intended to deal with the theory of rotation about an axis, and to bring together the relationships between the different methods used to describe rotations, including the angle and axis of rotation, the Euler-Rodrigues parameters, quaternions, the Cayley-Klein parameters, spinors, and Euler angles.

Rotation about an axis

Consider a positive rotation of a system of particles without deformation, through an angle V about an axis $\mathbf{n}$, in a reference frame that remains fixed. An origin O is chosen along the axis of rotation, and $\boldsymbol{r}$ is the position vector of a general particle that becomes the vector $\mathbf{R}$. From Figures H5 and H6, it can be seen that

$$\begin{aligned}\mathbf{R} &= \overrightarrow{OQ} = \overrightarrow{OP} + \overrightarrow{PN} + \overrightarrow{NQ},\\ &= \mathbf{r} + [(\mathbf{n}\times\mathbf{r})\times\mathbf{n}](\cos V - 1) + (\mathbf{n}\times\mathbf{r})\sin V. \end{aligned} \quad \text{(Eq. 143)}$$

where N is a point along PM, such that PN is perpendicular to NQ. Therefore,

$$\mathbf{R} = \mathbf{r}\cos V + \mathbf{n}(\mathbf{n}\cdot\mathbf{r})(1-\cos V) + (\mathbf{n}\times\mathbf{r})\sin V. \quad \text{(Eq. 144)}$$

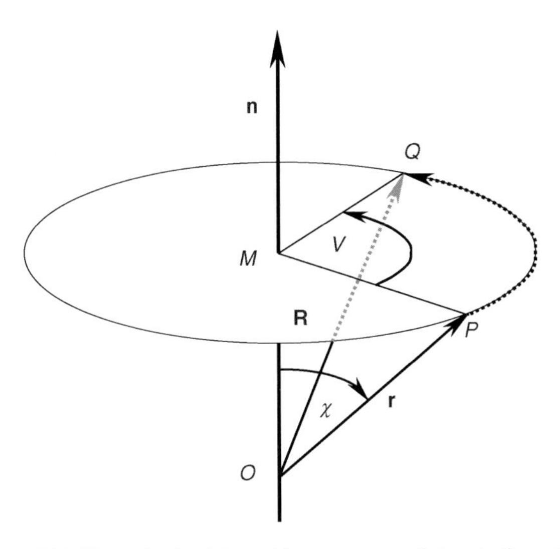

Figure H5 The point P with position vector $\mathbf{r}$ relative to the origin O, is carried to the point Q, with the position vector $\mathbf{R}$, by rotation through an angle V about the axis $\mathbf{n}$.

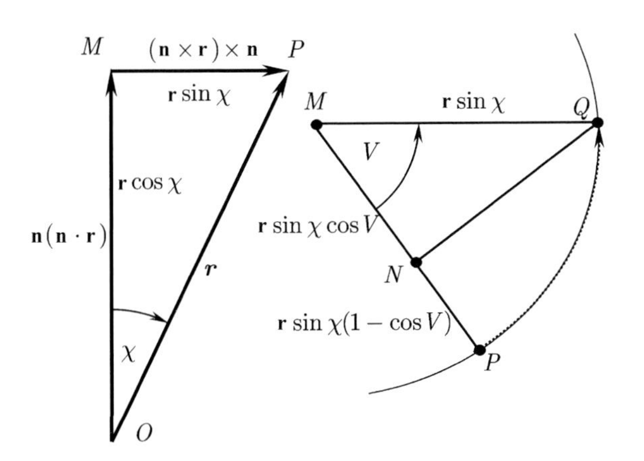

Figure H6 Showing the vector geometry of the rotation through an angle V about an axis of rotation $\mathbf{n}$.

Rotation of the reference frame

The equation for the transformation of the coordinates of P, under rotation of the reference frame, follows directly from Eq. (144) by changing the sign of the angle V. Thus

$$\mathbf{R} = \mathbf{r}\cos V + \mathbf{n}(\mathbf{n}\cdot\mathbf{r})(1-\cos V) - (\mathbf{n}\times\mathbf{r})\sin V. \quad \text{(Eq. 145)}$$

Writing out the Cartesian components of Eq. (145) gives a matrix form

$$\mathbf{R} = \mathbf{A}\cdot\mathbf{r}, \quad \text{where } \mathbf{A} = \mathbf{A}(\mathbf{n}, V), \text{and} \quad \text{(Eq. 146)}$$

$$\mathbf{A}(\mathbf{n}, V) = \begin{pmatrix} \cos V + a^2(1-\cos V) & ab(1-\cos V) + c\sin V & ac(1-\cos V) - b\sin V \\ ab(1-\cos V) - c\sin V & \cos V + b^2(1-\cos V) & bc(1-\cos V) + a\sin V \\ ac(1-\cos V) + b\sin V & bc(1-\cos V) - a\sin V & \cos V + c^2(1-\cos V) \end{pmatrix}. \quad \text{(Eq. 147)}$$

The angle and axis of rotation

The angle of rotation and the axis of rotation can be determined from the trace and the antisymmetric components of the transformation matrix $\mathbf{A}$. The trace can be expressed in a number of different ways,

$$\begin{aligned}\text{trace}\mathbf{A}(\mathbf{n}, V) &= 3\cos V + (a^2+b^2+c^2)(1-\cos V)\\ &= 1 + 2\cos V = 4\cos^2\frac{1}{2}V - 1 = \frac{\sin\frac{3}{2}V}{\sin\frac{1}{2}V}.\end{aligned} \quad \text{(Eq. 148)}$$

The trace of $\mathbf{A}(\mathbf{n}, V)$ is a single valued function of V in the range $0 < V < \pi$, and for rotations V in the range $\pi < V < 2\pi$, the axis of rotation $\boldsymbol{n}$ is reversed. In the case of no rotation when $V = 0$, and $\cos V = 1$, the rotation matrix reduces to $\mathbf{A}(\mathbf{n}, 0) = \mathbf{1}$, and there is no axis of rotation. Having determined the angle of rotation, the Cartesian components (a, b, c) of the axis of rotation are given by

$$\begin{aligned} 2a\sin V &= \mathbf{A}_{23} - \mathbf{A}_{32} = 4a\sin\frac{1}{2}V\cos\frac{1}{2}V,\\ 2b\sin V &= \mathbf{A}_{31} - \mathbf{A}_{13} = 4b\sin\frac{1}{2}V\cos\frac{1}{2}V,\\ 2c\sin V &= \mathbf{A}_{12} - \mathbf{A}_{21} = 4c\sin\frac{1}{2}V\cos\frac{1}{2}V. \end{aligned} \quad \text{(Eq. 149)}$$

The eigenvalues of the rotation matrix $\mathbf{A}(\mathbf{n}, V)$ depend only on the angle of rotation V,

$$\lambda_1 = e^{iV}, \quad \lambda_2 = e^{-iV}, \quad \lambda_3 = 1, \quad \text{(Eq. 150)}$$

and their sum is equal to the trace of the rotation matrix, $1 + 2\cos V$.

Infinitesimal rotations

When the angle of rotation V about the axis $\mathbf{n}$, is written as a differential, or infinitesimal $\mathrm{d}V$, then Eq. (144) for the new position vector $\mathbf{R}$ of a particle with original position vector $\mathbf{dr}$, of a system of particles, rotating as a rigid body, reduces to

$$\mathbf{R} = \mathbf{r} + (\mathbf{n}\times\mathbf{r})\mathrm{d}V. \quad \text{(Eq. 151)}$$

The actual displacement during the rotation is $\mathbf{dr} = \mathbf{R} - \mathbf{r}$, and therefore,

$$\mathbf{dr} = (\mathbf{n} \times \mathbf{r})\mathrm{d}V. \quad \text{(Eq. 152)}$$

If the change takes place over an infinitesimal interval of time, $\mathrm{d}t$, then

$$\begin{aligned}\frac{\mathbf{dr}}{\mathrm{d}t} &= (\mathbf{n} \times \mathbf{r})\frac{\mathrm{d}V}{\mathrm{d}t},\\ &= \mathbf{\Omega} \times \mathbf{r},\end{aligned} \quad \text{(Eq. 153)}$$

where $\mathbf{\Omega}$ is said to be the angular velocity of the particle.

The time rate of change of a scalar function of position $f(\mathbf{r})$ relative to a position vector $\mathbf{r}$ rotating with angle velocity $\mathbf{\Omega}$ is therefore

$$\begin{aligned}\frac{\mathrm{d}f}{\mathrm{d}t} &= \frac{\partial f}{\partial x}\frac{\mathrm{d}x}{\mathrm{d}t} + \frac{\partial f}{\partial y}\frac{\mathrm{d}y}{\mathrm{d}t} + \frac{\partial f}{\partial z}\frac{\mathrm{d}z}{\mathrm{d}t},\\ &= \nabla f \cdot \frac{\mathbf{dr}}{\mathrm{d}t},\\ &= \nabla f \cdot (\mathbf{\Omega} \times \mathbf{r}),\end{aligned} \quad \text{(Eq. 154)}$$

where it is assumed that there is no "local" time derivative, usually denoted $\partial f/\partial t$. By the rules for scalar triple products, we may write

$$\frac{\mathrm{d}f}{\mathrm{d}t} = \Omega \cdot (\mathbf{r} \times \nabla f). \quad \text{(Eq. 155)}$$

At this point it is convenient to introduce the operator $\mathbf{L}$, known as the angular momentum operator in quantum physics, and

$$\mathbf{L}f = -\mathrm{i}\mathbf{r} \times \nabla f = \mathrm{i}\nabla \times (\mathbf{r}f), \quad \text{(Eq. 156)}$$

when the expression Eq. (154) for the time rate of change of $f(\mathbf{r})$ becomes

$$\frac{\mathrm{d}f}{\mathrm{d}t} = \mathrm{i}\Omega \cdot \mathbf{L}f, \text{ and } \left(\frac{\mathrm{d}}{\mathrm{d}t}\right)^n f = (\mathrm{i}\Omega \cdot \nabla\mathbf{L})^n f. \quad \text{(Eq. 157)}$$

The Cartesian components of the angular momentum operator are

$$\begin{aligned}L_x f &= \mathrm{i}\left(z\frac{\partial f}{\partial y} - y\frac{\partial f}{\partial z}\right),\\ L_y f &= \mathrm{i}\left(x\frac{\partial f}{\partial z} - z\frac{\partial f}{\partial x}\right),\\ L_z f &= \mathrm{i}\left(y\frac{\partial f}{\partial x} - x\frac{\partial f}{\partial y}\right).\end{aligned} \quad \text{(Eq. 158)}$$

In spherical polar coordinates, the Cartesian derivatives are

$$\begin{aligned}\frac{\partial f}{\partial x} &= \sin\theta\cos\phi\frac{\partial f}{\partial r} + \frac{\cos\theta\cos\phi}{r}\frac{\partial f}{\partial\theta} - \frac{\sin\phi}{r\sin\theta}\frac{\partial f}{\partial\phi},\\ \frac{\partial f}{\partial y} &= \sin\theta\sin\phi\frac{\partial f}{\partial r} + \frac{\cos\theta\sin\phi}{r}\frac{\partial f}{\partial\theta} + \frac{\cos\phi}{r\sin\theta}\frac{\partial f}{\partial\phi},\\ \frac{\partial f}{\partial z} &= \cos\theta\frac{\partial f}{\partial r} - \frac{\sin\theta}{r}\frac{\partial f}{\partial\theta}.\end{aligned} \quad \text{(Eq. 159)}$$

The spherical polar forms of the angular momentum operators are therefore

$$\begin{aligned}L_x &= \mathrm{i}\left(\sin\phi\frac{\partial}{\partial\theta} + \cot\theta\cos\phi\frac{\partial}{\partial\phi}\right),\\ L_y &= \mathrm{i}\left(-\cos\phi\frac{\partial}{\partial\theta} + \cot\theta\sin\phi\frac{\partial}{\partial\phi}\right),\\ L_z &= -\mathrm{i}\frac{\partial}{\partial\phi}.\end{aligned} \quad \text{(Eq. 160)}$$

The Eq. (160) are used in determining recurrence relations for surface spherical harmonics. Note also that a finite rotation can be done by a succession of a large number of small rotations, which can be used to show that the knowledge of the infinitesimal transformation amounts implicitly to a knowledge of the entire transformation.

Euler Rodrigues parameters

The Euler-Rodrigues parameters $p,q,r,s,$ are defined by

$$\begin{aligned}p &= a\sin\frac{1}{2}V,\\ q &= b\sin\frac{1}{2}V,\\ r &= c\sin\frac{1}{2}V,\\ s &= \cos\frac{1}{2}V,\end{aligned} \quad \text{(Eq. 161)}$$

and are spherical polar coordinates in a four-dimensional space. Greek symbols $\xi, \eta, \zeta, \chi,$ are also used (Whittaker, 1904), as well as λ, μ, v, ρ (Kendall and Moran, 1962).

Note that (a, b, c), regarded as (x, y, z), are spherical harmonics of the first degree, and those of higher degree being generated by multiple derivatives inverse distance. The Euler-Rodrigues parameters (p, q, r, s) are spin-weighted spherical harmonics, with terms of higher degree generated by derivatives of the inverse square of distance.

Resultant of two rotations

When a reference frame is rotated through angle V_1 about an axis $\mathbf{n}_1$, and then rotated through an angle V_2 about an axis $\mathbf{n}_2$, the resulting configuration is equivalent to a single rotation through an angle V_3 about an axis $\mathbf{n}_3$. In terms of rotation matrices

$$\mathbf{A}(\mathbf{n}_3, V_3) = \mathbf{A}(\mathbf{n}_2, V_2)\mathbf{A}(\mathbf{n}_1, V_1). \quad \text{(Eq. 162)}$$

The trace of the product matrix is found to have the form

$$\begin{aligned}1 + 2\cos V_3 = &(\mathbf{n}_1 \cdot \mathbf{n}_2)^2(1 - \cos V_2)(1 - \cos V_2)\\ &- 2(\mathbf{n}_1 \cdot \mathbf{n}_2)\sin V_2 \sin V_1\\ &+ (1 + \cos V_2)(1 + \cos V_1)\end{aligned}$$

leads to a perfect square for $\cos\frac{1}{2}V_3$, the positive square root of which gives

$$\cos\frac{1}{2}V_3 = \cos\frac{1}{2}V_2\cos\frac{1}{2}V_1 - (\mathbf{n}_1 \cdot \mathbf{n}_2)\sin\frac{1}{2}V_2\sin\frac{1}{2}V_1. \quad \text{(Eq. 163)}$$

Similarly, the antisymmetric parts of the product matrix, after some algebra, lead to

$$\begin{aligned}a_3\sin\frac{1}{2}V_3 = &(b_1c_2 - c_1b_2)\sin\frac{1}{2}V_2\sin\frac{1}{2}V_1 + a_1\cos\frac{1}{2}V_2\sin\frac{1}{2}V_1\\ &+ a_2\sin\frac{1}{2}V_2\cos\frac{1}{2}V_1,\\ b_3\sin\frac{1}{2}V_3 = &(c_1a_2 - a_1c_2)\sin\frac{1}{2}V_2\sin\frac{1}{2}V_1 + b_1\cos\frac{1}{2}V_2\sin\frac{1}{2}V_1\\ &+ b_2\sin\frac{1}{2}V_2\cos\frac{1}{2}V_1,\\ c_3\sin\frac{1}{2}V_3 = &(a_1b_2 - b_1a_2)\sin\frac{1}{2}V_2\sin\frac{1}{2}V_1 + c_1\cos\frac{1}{2}V_2\sin\frac{1}{2}V_1\\ &+ c_2\sin\frac{1}{2}V_2\cos\frac{1}{2}V_1.\end{aligned} \quad \text{(Eq. 164)}$$

Using $p,q,r,s,$ as defined in Eq. (161), with suitable subscripts, Eq. (164) leads to the following formulae for the combination of rotations of the reference frame,

$$\begin{aligned} p_3 &= s_2p_1 + r_2q_1 - q_2r_1 + p_2s_1 \\ q_3 &= -r_2p_1 + s_2q_1 + p_2r_1 + q_2s_1 \\ r_3 &= q_2p_1 - p_2q_1 + s_2r_1 + r_2s_1, \\ s_3 &= -p_2p_1 - q_2q_1 - r_2r_1 + s_2s_1. \end{aligned} \qquad \text{(Eq. 165)}$$

The beginnings of vector algebra

The product formula of Eq. (165) contains within it as a special case, the vector product of the axes of rotation, namely $\mathbf{n}_1$ and $\mathbf{n}_2$, and also their scalar product. By choosing the rotation angles V_1 and V_2 to be π radians, equivalent to 180°, then $s_1 = s_2 = \cos\frac{1}{2}\pi = 0$, and the remaining Euler-Rodrigues parameters reduce to the Cartesian components of the rotation axes,

$$(p_1, q_1, r_1)|_{V_1=\pi} = (a_1, b_1, c_1) = \mathbf{n}_1$$

and

$$(p_2, q_2, r_2)|_{V_2=\pi} = (a_2, b_2, c_2) = \mathbf{n}_2 \qquad \text{(Eq. 166)}$$

The product formula (165) reduces to

$$\left.\begin{aligned} a_3 &= c_2b_1 - b_2c_1 \\ b_3 &= a_2c_1 - c_2a_1 \\ c_3 &= b_2a_1 - a_2b_1 \end{aligned}\right\} \text{so that } (a_3, b_3, c_3) \equiv \mathbf{n}_1 \times \mathbf{n}_2, \qquad \text{(Eq. 167)}$$

while the expression for s_3 reduces to the scalar product times -1,

$$s_3|_{V_1,V_2=\pi} = -a_1a_2 - b_1b_2 - c_1c_2 \text{ so that } s_3|_{V_1,V_2=\pi} = -\mathbf{n}_1 \cdot \mathbf{n}_2. \qquad \text{(Eq. 168)}$$

The results of Eqs. (167) and (168) would be written $\mathrm{V}\mathbf{n}_1\mathbf{n}_2$ and $\mathrm{S}\mathbf{n}_1\mathbf{n}_2$ respectively, where V and S refer to "vector" and "scalar," respectively. Maxwell's equations were given in Cartesian form (Maxwell, 1881), and also as equivalent "quaternion expressions" using V and S as in Eqs. (167) and (168). It was the fashion, when using Hamilton's quaternions, to write vectors, such as the electromagnetic momentum ***A*** and the magnetic induction ***B*** in Gothic upper case script, $\mathfrak{A}$ and $\mathfrak{B}$, respectively.

Hyperspherical trigonometry

If the rotation axes $\mathbf{n}_1$ and $\mathbf{n}_2$ subtend an angle Θ between them, and

$$\begin{aligned} \mathbf{n}_1 &= (a_1, b_1, c_1) = (\sin\theta_1\cos\phi_1, \sin\theta_1\sin\phi_1, \cos\theta_1), \\ \mathbf{n}_2 &= (a_2, b_2, c_2) = (\sin\theta_2\cos\phi_2, \sin\theta_2\sin\phi_2, \cos\theta_2), \end{aligned} \qquad \text{(Eq. 169)}$$

then $\mathbf{n}_1 \cdot \mathbf{n}_2 = \cos\Theta$, and hence Eq. (163) for s_3 becomes

$$\cos\frac{1}{2}V_3 = \cos\frac{1}{2}V_1\cos\frac{1}{2}V_2 - \sin\frac{1}{2}V_1\sin\frac{1}{2}V_2\cos\Theta, \qquad \text{(Eq. 170)}$$

where $\cos\Theta$ is the scalar product of $\mathbf{n}_1$ and $\mathbf{n}_2$,

$$\cos\Theta = \cos\theta_1\cos\theta_2 + \sin\theta_1\sin\theta_2\cos(\phi_1 - \phi_2). \qquad \text{(Eq. 171)}$$

Thus Eq. (170) uses half-angles of rotation, and Eq. (171) uses polar coordinates of rotation axes.

Without using half-angles

From Eqs. (163) and (164), the trace and the antisymmetric part of the resultant rotation matrix $\mathbf{A}(\mathbf{n}_3, V_3)$ can be expressed as the product of the Euler Rodrigues parameters,

$$\begin{aligned} 2a_3\sin V_3 &= 4a_3\sin\frac{1}{2}V_3\cos\frac{1}{2}V_3 = 4p_3s_3, \\ 2b_3\sin V_3 &= 4b_3\sin\frac{1}{2}V_3\cos\frac{1}{2}V_3 = 4q_3s_3, \\ 2c_3\sin V_3 &= 4c_3\sin\frac{1}{2}V_3\cos\frac{1}{2}V_3 = 4r_3s_3, \\ 2 + 2\cos V_3 &= 4s_3^2. \end{aligned}$$

Thus, starting with

$$s_3 = \frac{1}{2}\sqrt{\mathrm{trace}[\mathbf{A}(\mathbf{n}_3, V_3)] + 1},$$

then

$$\begin{aligned} p_3 &= [\mathbf{A}(\mathbf{n}_3, V_3)_{23} - \mathbf{A}(\mathbf{n}_3, V_3)_{32}]/2s_3, \\ q_3 &= [\mathbf{A}(\mathbf{n}_3, V_3)_{31} - \mathbf{A}(\mathbf{n}_3, V_3)_{13}]/2s_3, \\ r_3 &= [\mathbf{A}(\mathbf{n}_3, V_3)_{12} - \mathbf{A}(\mathbf{n}_3, V_3)_{21}]/2s_3. \end{aligned}$$

Rigid body rotations

The formulae for rigid body rotations are obtained by replacing (V_1, V_2, V_3) by $(-V_1, -V_2, -V_3)$, when the required formulae are

$$\begin{aligned} p_3 &= s_2p_1 - r_2q_1 + q_2r_1 + p_2s_1, \\ q_3 &= r_2p_1 + s_2q_1 - p_2r_1 + q_2s_1, \\ r_3 &= -q_2p_1 + p_2q_1 + s_2r_1 + r_2s_1, \\ s_3 &= -p_2p_1 - q_2q_1 - r_2r_1 + s_2s_1. \end{aligned} \qquad \text{(Eq. 172)}$$

It is important to distinguish carefully between Eq. (165) for rotation of the reference frame and Eq. (172) for a rigid-body rotation.

Quaternions

The product formula Eq. (172) for combining rigid-body rotations, corresponds exactly to the law for the product $\mathbf{Q}_2\mathbf{Q}_1$ of quaternions $\mathbf{Q}_1$ and $\mathbf{Q}_2$, where

$$\begin{aligned} \mathbf{Q}_1 &= 1s_1 + ip_1 + jq_1 + kr_1, \\ \mathbf{Q}_2 &= 1s_2 + ip_2 + jq_2 + kr_2, \end{aligned} \qquad \text{(Eq. 173)}$$

subject to the rules that 1 is the identity and

$$\begin{aligned} & i^2 = j^2 = k^2 = -1, \\ & jk = i, \quad kj = -i, \\ & ki = j, \quad ik = -j, \\ & ij = k, \quad ji = -k, \\ & ijk = -1. \end{aligned} \qquad \text{(Eq. 174)}$$

It is difficult to see why such a fuss is made over the "rule" that $ijk = -1$, when it is an immediate consequence of the rules $ij = k$ and $k^2 = -1$. Quaternions are useful for the occasional calculation, but if any quantity of such rigid-body rotation calculations has to be done, then clearly the matrix form of Eq. (172) is easier to deal with.

Cayley-Klein parameters

The Cayley-Klein parameters u and v are defined by

$$u = s + ir = \cos\frac{1}{2}V + ic\sin\frac{1}{2}V,$$
$$v = q + ip = (b + ia)\sin\frac{1}{2}V. \qquad \text{(Eq. 175)}$$

With this substitution, the quaternion product formula (165) for the combining of reference frame rotations can be written

$$s_3 + ir_3 = (s_2 + ir_2)(s_1 + ir_1) - (q_2 + ip_2)(q_1 - ip_1),$$
$$-q_3 + ip_3 = -(q_2 - ip_2)(s_1 + ir_1) - (s_2 - ir_2)(q_1 - ip_1), \qquad \text{(Eq. 176)}$$

which becomes

$$u_3 = u_2u_1 - v_2\bar{v}_1,$$
$$-\bar{v}_3 = -\bar{v}_2u_1 - \bar{u}_2\bar{v}_1, \qquad \text{(Eq. 177)}$$

and, in matrix form, Eq. (177) becomes

$$\begin{pmatrix} u_3 \\ -\bar{v}_3 \end{pmatrix} = \begin{pmatrix} u_2 & v_2 \\ -\bar{v}_2 & \bar{u}_2 \end{pmatrix}\begin{pmatrix} u_1 \\ -\bar{v}_1 \end{pmatrix}. \qquad \text{(Eq. 178)}$$

Equation (178) can be regarded as the transformation law for the Cayley-Klein parameters u and v, and, more importantly, can be used to derive the transformation law for homogeneous polynomials of u and v of degree n.

Spinors

Using the elementary results that

$$ax + by = \frac{1}{2}[(a + ib)(x - iy) + (a - ib)(x + iy)],$$
$$\cos V + ic\sin V + \frac{1}{2}(a^2 + b^2)(1 - \cos V) = (\cos\frac{1}{2}V + ic\sin\frac{1}{2}V)^2, \qquad \text{(Eq. 179)}$$

and the vector relationships

$$\begin{pmatrix} X + iY \\ Z \\ X - iY \end{pmatrix} = \begin{pmatrix} 1 & i & 0 \\ 0 & 0 & 1 \\ 1 & -i & 0 \end{pmatrix}\begin{pmatrix} X \\ Y \\ Z \end{pmatrix}, \text{ and}$$
$$\begin{pmatrix} x \\ y \\ z \end{pmatrix} = \begin{pmatrix} \frac{1}{2} & 0 & \frac{1}{2} \\ -\frac{1}{2}i & 0 & \frac{1}{2}i \\ 0 & 1 & 0 \end{pmatrix}\begin{pmatrix} x + iy \\ z \\ x - iy \end{pmatrix}, \qquad \text{(Eq. 180)}$$

one obtains from the basic Eq. (147) for reference frame rotation, that

$$X + iY = (x + iy)(\cos\frac{1}{2}V - ic\sin\frac{1}{2}V)^2$$
$$- 2z(b - ia)\sin\frac{1}{2}V(\cos\frac{1}{2}V - ic\sin\frac{1}{2}V)$$
$$- (x - iy)\left[(b - ia)\sin\frac{1}{2}V\right]^2,$$
$$Z = (x + iy)(b + ia)\sin\frac{1}{2}V(\cos\frac{1}{2}V - ic\sin\frac{1}{2}V)$$
$$+ z\left[\cos^2\frac{1}{2}V + c^2\sin^2\frac{1}{2}V - (1 - c^2)\sin^2\frac{1}{2}V\right]$$
$$+ (x - iy)(b - ia)\sin\frac{1}{2}V(\cos\frac{1}{2}V + ic\sin\frac{1}{2}V),$$
$$X - iY = (x + iy)\left[(a - ib)\sin\frac{1}{2}V\right]^2$$
$$- 2z(b + ia)\sin\frac{1}{2}V(\cos\frac{1}{2}V + ic\sin\frac{1}{2}V)$$
$$+ (x - iy)(\cos\frac{1}{2}V + ic\sin\frac{1}{2}V)^2. \qquad \text{(Eq. 181)}$$

Using the Cayley-Klein parameters u and v defined in Eq. (175), the Eq. (181) become

$$-(X + iY) = -(x + iy)\bar{u}^2 + 2z\bar{u}\,\bar{v} + (x - iy)\bar{v}^2,$$
$$Z = (x + iy)\bar{u}v + z(u\bar{u} - v\bar{v}) + (x - iy)u\bar{v},$$
$$(X - iY) = -(x + iy)v^2 - 2zuv + (x - iy)u^2. \qquad \text{(Eq. 182)}$$

Introducing the parameters ξ and η, defined by

$$\xi = -\frac{1}{\sqrt{2}}(x + iy), \quad \xi' = -\frac{1}{\sqrt{2}}(X + iY),$$
$$\eta = \frac{1}{\sqrt{2}}(x - iy), \quad \eta' = \frac{1}{\sqrt{2}}(X - iY), \qquad \text{(Eq. 183)}$$

the Eq. (182) become

$$\xi' = \xi\,\bar{u}^2 + \sqrt{2}\,z\bar{u}\,\bar{v} + \eta\,\bar{v}^2,$$
$$Z = -\sqrt{2}\,\xi\,\bar{u}v + z(u\bar{u} - v\bar{v}) + \sqrt{2}\,\eta\,u\bar{v},$$
$$\eta' = \xi\,v^2 - \sqrt{2}\,zuv + \eta\,u^2 \qquad \text{(Eq. 184)}$$

These equations are easily solved with $V \rightarrow -V$, when by Eq. (175), $u \rightarrow \bar{u}$ and $v \rightarrow -v$ giving

$$\xi = \xi' u^2 - \sqrt{2}\,Z\,u\bar{v} + \eta'\bar{v}^2,$$
$$z = \sqrt{2}\,\xi' uv + Z(u\bar{u} - v\bar{v}) - \sqrt{2}\,\eta'\bar{u}\,\bar{v},$$
$$\eta = \xi' v^2 + \sqrt{2}\,Z\,\bar{u}v + \eta'\,\bar{u}^2. \qquad \text{(Eq. 185)}$$

The required partial derivatives then follow from Eq. (185) and the chain rule of partial differentiation,

$$\begin{pmatrix} \dfrac{\partial f}{\partial \xi'} \\ \dfrac{1}{\sqrt{2}}\dfrac{\partial f}{\partial Z} \\ \dfrac{\partial f}{\partial \eta'} \end{pmatrix} = \begin{pmatrix} u^2 & 2uv & v^2 \\ -u\,\bar{v} & u\,\bar{u} - v\,\bar{v} & \bar{u}\,v \\ \bar{v}^2 & -2\bar{u}\,\bar{v} & \bar{u}^2 \end{pmatrix}\begin{pmatrix} \dfrac{\partial f}{\partial \xi} \\ \dfrac{1}{\sqrt{2}}\dfrac{\partial f}{\partial z} \\ \dfrac{\partial f}{\partial \eta} \end{pmatrix}. \qquad \text{(Eq. 186)}$$

The substitutions

$$\frac{\partial f}{\partial \xi'} = \Lambda_1^2, \qquad \frac{\partial f}{\partial \xi} = \lambda_1^2,$$
$$\frac{1}{\sqrt{2}}\frac{\partial f}{\partial Z} = \Lambda_1\Lambda_2, \qquad \frac{1}{\sqrt{2}}\frac{\partial f}{\partial z} = \lambda_1\lambda_2,$$
$$\frac{\partial f}{\partial \eta'} = \Lambda_2^2, \qquad \frac{\partial f}{\partial \eta} = \lambda_2^2, \qquad \text{(Eq. 187)}$$

are valid only for spherical harmonic functions, since

$$\nabla^2 f = \frac{\partial^2 f}{\partial x^2} + \frac{\partial^2 f}{\partial y^2} + \frac{\partial^2 f}{\partial z^2},$$
$$= -2\frac{\partial}{\partial \xi}\frac{\partial f}{\partial \eta} + \frac{\partial^2 f}{\partial z^2},$$
$$= -2\lambda_1^2\lambda_2^2 + 2\lambda_1^2\lambda_2^2,$$
$$= 0. \qquad \text{(Eq. 188)}$$

With the substitutions of Eqs. (187), the Eq. (186) becomes

$$\begin{aligned}\Lambda_1^2 &= (u\lambda_1 + v\lambda_2)^2,\\ \Lambda_1\Lambda_2 &= (u\lambda_1 + v\lambda_2)(-\bar{v}\lambda_1 + \bar{u}\lambda_2),\\ \Lambda_2^2 &= (-\bar{v}\lambda_1 + \bar{u}\lambda_2)^2,\end{aligned} \qquad \text{(Eq. 189)}$$

and are equivalent to the 2 × 2 form

$$\begin{aligned}\Lambda_1 &= u\lambda_1 + v\lambda_2,\\ \Lambda_2 &= -\bar{v}\lambda_1 + \bar{u}\lambda_2,\end{aligned} \qquad \text{(Eq. 190)}$$

where the positive square roots are required. In the case when there is no rotation and the angle $V = 0$, for which $u = 1$ and $v = 0$, the Eq. (190) reduce correctly to $\Lambda_1 = \lambda_1$ and $\Lambda_2 = \lambda_2$. In matrix form, Eq. (190) can be written

$$\begin{pmatrix}\Lambda_1\\ \Lambda_2\end{pmatrix} = \begin{pmatrix}u & v\\ -\bar{v} & \bar{u}\end{pmatrix}\begin{pmatrix}\lambda_1\\ \lambda_2\end{pmatrix}. \qquad \text{(Eq. 191)}$$

The parameters Λ_1, Λ_2, and λ_1, λ_2 are called spinors, and although from their basic definition given in Eq. (187), they appear to be the square roots of differential operators, they are in fact, apart from the transformation formula Eq. (191) (and representations of "spin weighted" spherical harmonics), used only in combinations of integer powers of differential operators.

The 2 × 2 matrix in Eq. (191) has already been derived in Eq. (178) from the product formula for the Euler-Rodrigues parameters *p,q,r,s,* based on the complex forms (the Cayley-Klein parameters) *u* and *v*, arising solely from the need for a formula for the resultant of two successive rotations.

Euler angles (α, β, γ)

In this system the (x, y, z) reference frame is rotated through an angle α about the z–axis to form the (x', y', z') reference frame, which is rotated though an angle β about the y'-axis to form the (x'', y'', z'') reference frame, which is rotated through an angle γ about the z''–axis to form the final (X, Y, Z) frame (see Figure H7). The point with coordinates of $\mathbf{r}$ becomes a point with coordinates $\mathbf{R}$ in the rotated reference frame, and the relationship can be written

$$\mathbf{R} = \mathbf{A}(\alpha, \beta, \gamma)\mathbf{r}, \qquad \text{(Eq. 192)}$$

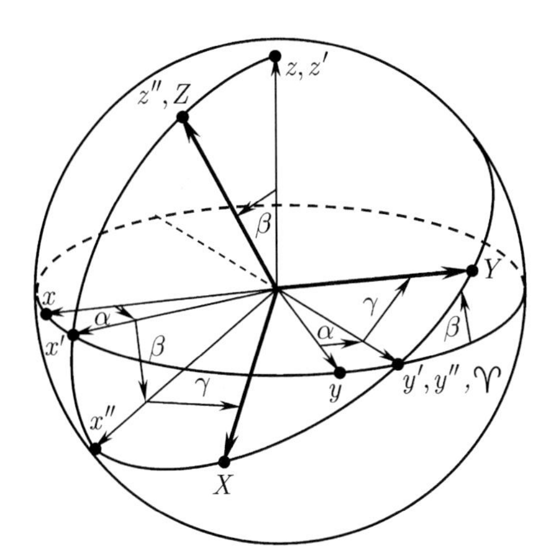

Figure H7 The *Oxyz* reference frame is rotated through Euler angles (α, β, γ) to become the *OXYZ* reference frame.

where the matrix $\mathbf{A}(\alpha, \beta, \gamma)$ has the form

$$\mathbf{A}(\alpha, \beta, \gamma) = \begin{pmatrix} \cos\alpha\cos\beta\cos\gamma - \sin\alpha\sin\gamma & \sin\alpha\cos\beta\cos\gamma + \cos\alpha\sin\gamma & -\sin\beta\cos\gamma\\ -\cos\alpha\cos\beta\sin\gamma - \sin\alpha\cos\gamma & -\sin\alpha\cos\beta\sin\gamma + \cos\alpha\cos\gamma & \sin\beta\sin\gamma\\ \cos\alpha\sin\beta & \sin\alpha\sin\beta & \cos\beta \end{pmatrix}. \qquad \text{(Eq. 193)}$$

For the reverse rotation, the Euler angles (α, β, γ) are replaced by $(-\gamma, -\beta, -\gamma)$,

$$\mathbf{r} = \mathbf{A}(-\gamma, -\beta, -\alpha)\mathbf{R}. \qquad \text{(Eq. 194)}$$

The matrix can be written in terms of its factors, which derive from the three rotations:

$$\begin{pmatrix}X\\ Y\\ Z\end{pmatrix} = \begin{pmatrix}\cos\gamma & \sin\gamma & 0\\ -\sin\gamma & \cos\gamma & 0\\ 0 & 0 & 1\end{pmatrix}\begin{pmatrix}\cos\beta & 0 & -\sin\beta\\ 0 & 1 & 0\\ \sin\beta & 0 & \cos\beta\end{pmatrix} \times \begin{pmatrix}\cos\alpha & \sin\alpha & 0\\ -\sin\alpha & \cos\alpha & 0\\ 0 & 0 & 1\end{pmatrix}\begin{pmatrix}x\\ y\\ z\end{pmatrix}. \qquad \text{(Eq. 195)}$$

Comparing the off-diagonal terms of $\mathbf{A}(\mathbf{n}, V)$ of Eq. (147), and $\mathbf{A}(\alpha, \beta, \gamma)$ of Eq. (193), gives

$$\begin{aligned}\sin\alpha\sin\beta &= cb(1-\cos V) - a\sin V,\\ \sin\beta\sin\gamma &= cb(1-\cos V) + a\sin V,\\ -\sin\beta\cos\gamma &= ca(1-\cos V) - b\sin V,\\ \cos\alpha\sin\beta &= ac(1-\cos V) + b\sin V,\\ -\cos\alpha\cos\beta\sin\gamma - \sin\alpha\cos\gamma &= ab(1-\cos V) - c\sin V,\\ \sin\alpha\cos\beta\cos\gamma + \cos\alpha\sin\gamma &= ba(1-\cos V) + c\sin V,\end{aligned} \qquad \text{(Eq. 196)}$$

and from Eq. (196) it follows that

$$\begin{aligned}a\sin V &= \frac{1}{2}\sin\beta(\sin\gamma - \sin\alpha),\\ b\sin V &= \frac{1}{2}\sin\beta(\cos\gamma + \cos\alpha),\\ c\sin V &= \frac{1}{2}(1+\cos\beta)\sin(\gamma+\alpha).\end{aligned} \qquad \text{(Eq. 197)}$$

The right-hand sides of Eq. (197) can be written as a product of two terms,

$$\begin{aligned}a\sin V &= 2\sin\frac{1}{2}\beta\sin\frac{1}{2}(\gamma-\alpha)\cdot\cos\frac{1}{2}\beta\cos\frac{1}{2}(\gamma+\alpha),\\ b\sin V &= 2\sin\frac{1}{2}\beta\cos\frac{1}{2}(\gamma-\alpha)\cdot\cos\frac{1}{2}\beta\cos\frac{1}{2}(\gamma+\alpha),\\ c\sin V &= 2\cos\frac{1}{2}\beta\sin\frac{1}{2}(\gamma+\alpha)\cdot\cos\frac{1}{2}\beta\cos\frac{1}{2}(\gamma+\alpha),\end{aligned} \qquad \text{(Eq. 198)}$$

from which we obtain

$$(a^2+b^2)\sin^2 V = 4\sin^2\frac{1}{2}\beta\cos^2\frac{1}{2}\beta\cos^2\frac{1}{2}(\gamma+\alpha). \qquad \text{(Eq. 199)}$$

Comparing the diagonal terms of $\mathbf{A}(\mathbf{n}, V)$ of Eq. (147), and $\mathbf{A}(\alpha, \beta, \gamma)$ of Eq. (193), gives

$$\begin{aligned}\cos\alpha\cos\beta\cos\gamma - \sin\alpha\sin\gamma &= \cos V + a^2(1-\cos V),\\ -\sin\alpha\cos\beta\sin\gamma + \cos\alpha\cos\gamma &= \cos V + b^2(1-\cos V),\\ \cos\beta &= \cos V + c^2(1-\cos V),\end{aligned} \quad \text{(Eq. 200)}$$

from the third of Eq. (200) we obtain

$$\sin^2\frac{1}{2}\beta = \sin^2\frac{1}{2}V - c^2\sin^2\frac{1}{2}V = (a^2+b^2)\sin^2\frac{1}{2}V \quad \text{(Eq. 201)}$$

From Eqs. (199) and (201)

$$\cos^2\frac{1}{2}V = \cos^2\frac{1}{2}\beta\cos^2\frac{1}{2}(\gamma+\alpha),$$

and the positive square root is required, because in the case in which $\alpha = \gamma = 0$, the angle of rotation $V = \beta$. Therefore

$$\cos\frac{1}{2}V = \cos\frac{1}{2}\beta\cos\frac{1}{2}(\alpha+\gamma). \quad \text{(Eq. 202)}$$

Euler-Rodrigues parameters and Euler angles

Combining the results of Eq. (161) for rotation about **n** axis, with Eqs. (198) and (202),

$$\begin{aligned}p &= a\sin\frac{1}{2}V = \sin\frac{1}{2}\beta\sin\frac{1}{2}(\gamma-\alpha),\\ q &= b\sin\frac{1}{2}V = \sin\frac{1}{2}\beta\cos\frac{1}{2}(\gamma-\alpha),\\ r &= c\sin\frac{1}{2}V = \cos\frac{1}{2}\beta\sin\frac{1}{2}(\gamma+\alpha),\\ s &= \cos\frac{1}{2}V = \cos\frac{1}{2}\beta\cos\frac{1}{2}(\gamma+\alpha),\end{aligned} \quad \text{(Eq. 203)}$$

and these important equations relate the axis and angle of rotation with the Euler angle formulation of the same rotation.

Cayley-Klein parameters and Euler angles

Using the Cayley-Klein parameters, u, v, defined in Eq. (175), and the results of Eq. (203),

$$\begin{aligned}u &= s + \mathrm{i}r = \cos\frac{1}{2}V + \mathrm{i}c\sin\frac{1}{2}V = \cos\frac{1}{2}\beta e^{\mathrm{i}(\gamma+\alpha)/2},\\ v &= q + \mathrm{i}p = (b+\mathrm{i}a)\sin\frac{1}{2}V = \sin\frac{1}{2}\beta e^{\mathrm{i}(\gamma-\alpha)/2}.\end{aligned} \quad \text{(Eq. 204)}$$

In terms of the rotation matrix elements $D^l_{Mm}(\alpha,\beta,\gamma)$, (see *Spherical harmonics*), and the generalized Legendre functions $d^l_{M,m}(\beta)$, the parameters u and v are

$$\begin{aligned}u &= D^{\frac{1}{2}}_{-\frac{1}{2},-\frac{1}{2}}(\alpha,\beta,\gamma) = d^{\frac{1}{2}}_{-\frac{1}{2},-\frac{1}{2}}(\beta)e^{\mathrm{i}(\alpha+\gamma)/2},\\ v &= D^{\frac{1}{2}}_{\frac{1}{2},-\frac{1}{2}}(\alpha,\beta,\gamma) = d^{\frac{1}{2}}_{\frac{1}{2},-\frac{1}{2}}(\beta)e^{\mathrm{i}(-\alpha+\gamma)/2}.\end{aligned} \quad \text{(Eq. 205)}$$

Spherical harmonics

Briefly, spherical harmonics are multiple derivatives of inverse distance,

$$(-1)^{l-m}\left(\frac{\partial}{\partial\eta}\right)^m\left(\frac{\partial}{\partial z}\right)^{l-m}\frac{1}{r} = \frac{1}{r^{l+1}}\sqrt{\frac{(l+m)!(l-m)!}{2l+1}}Y_l^m(\theta,\phi). \quad \text{(Eq. 206)}$$

and on using the spinor forms of derivatives given in Eqs. (187)–(206), it follows that solid spherical harmonics can also be written in a more symmetric, spinor form,

$$\frac{1}{r^{l+1}}Y_l^m(\theta,\phi) = (-1)^{l-m}\sqrt{\frac{2l+1}{(l-m)!(l+m)!}}\lambda_1^{l-m}\lambda_2^{l+m}\frac{1}{r}. \quad \text{(Eq. 207)}$$

Using the transformation law (191) for spinors, under rotation of the reference frame, the transformation law for spherical harmonics is easily derived. This leads, for example, to formulae in spherical (and hyperspherical) trigonometry, and to the identification of vector and tensor quantities, which are formed into "irreducible" parts, with profound physical implications, on account of their identification as spherical harmonics in their own right.

Denis Winch

Bibliography

Chapman, S., and Bartels, J., 1940. *Geomagnetism.* London: Oxford Clarendon Press.

Condon, E.U., and Shortley, G.H., 1935. *The Theory of Atomic Spectra.* Cambridge: Cambridge University Press. [7th printing 1967.]

Ferrers, Rev. N.M., 1897. *An Elementary Treatise on Spherical Harmonics and Subjects Connected with them.* London: Macmillan and Co.

Goldie, A.H.R., and Joyce, J.W., editors, 1940. Proceedings of the 1939. Washington Assembly of the Association of Terrestrial Magnetism and Electricity of the International Union of Geodesy and Geophysics. International Union of Geodesy and Geophysics (IUGG), Edinburgh: Neill & Co., Bulletin 11, part 6, 550.

Kendall, M.E., and Moran, P.A.P., 1962. *Geometrical Probability.* London: Griffin.

Maxwell, J.C., 1881. *A Treatise on Electricity and Magnetism.* 2nd edition. London: Oxford Clarendon Press.

Schendel, L., (1877). Zusatz zu der Abhandlung über Kugelfunktionen S. 86 des 80. Bandes. *Crelle's Journal*, **82**: 158–164.

Schmidt, A., 1899. Formeln zur Transformation der Kugelfunktionen bei linearer Änderung des Koordinatensystems. *Zeitschrift für Mathematik*, **44**: 327–338.

Schuster, A., 1903. On some definite integrals, and a new method of reducing a function of spherical coordinates to a series of spherical harmonics. *Philosophical Transactions of the Royal Society of London*, **200**: 181–223.

Varshalovich, D.A., Moskalev, A.N., and Khersonskii, V.K., 1988. *Quantum Theory of Angular Momentum. Irreducible Tensors, Spherical Harmonics, Vector Coupling Coefficients, 3nj Symbols.* Singapore: World Scientific Publishing Co.

Vilenkin, N.J., 1968. *Special Functions and the Theory of Group Representations*, American Mathematical Society, Second printing 1978. Translated from 1965 Russian original by V.N. Singh.

Wigner, E., 1931. *Gruppentheorie und ihre Anwendungen auf die Quantenmechanik und Atomsspektren*, Braunschweig. Translation: Griffin, J.J. (ed.) (1959). *Group Theory and its Application to the Quantum Mechanics of Atomic Spectra.* New York and London: Academic Press.

Whittaker, E.T., 1904. *Treatise on the Analytical Dynamics of Particles and Rigid Bodies.* Cambridge: Cambridge University Press.

HARMONICS, SPHERICAL CAP

Spherical cap harmonics are used to model data over a small region of the Earth, either because data are only available over this region or because interest is confined to this region. The three-dimensional region considered in this article is that of a spherical cap (Figure H8), although one can also consider only the two-dimensional surface of a spherical cap. A three-dimensional model is used where the data are vector data that represent a field with zero curl and divergence, whereas a two-dimensional model is used for general fields with no such constraints.

In a source free region of the Earth, the vector magnetic field **B** can be expressed as the gradient of a scalar harmonic potential V. (This scalar potential is "harmonic" since its Laplacian is zero.) The potential, and therefore the field, can then be expressed mathematically as a series of basis functions, each term of the series being harmonic by design. This harmonic solution is usually identified by the coordinate system used. For example, when the coordinates chosen are rectangular coordinates, we say the field is expressed in terms of rectangular harmonics. These would be useful when dealing with a local or very small portion of the Earth's surface, such as in mineral exploration. When the coordinate system is the whole sphere, we speak of spherical harmonics (see *Harmonics, spherical*). These are used for investigating global features of the field. This article will discuss the harmonic solution for the case of a spherical cap coordinate system, which involves wavelengths intermediate between the local and the global solutions.

Basis functions

The basis functions for the series expansion of the potential over a spherical cap region are found in the usual way by separating the variables in the given differential equations (that the curl and divergence of the field are zero) and solving the individual eigenvalue problems subject to the appropriate boundary conditions (e.g., Smythe, 1950; Sections 5.12 and 5.14). The boundary conditions include continuity in longitude, regularity at the spherical cap pole, and the appropriate Sturm-Liouville conditions on the basis functions and their derivatives at the boundary of the cap (e.g., Davis, 1963, Section 2.4). The details have been given by Haines (1985a), and computer programs in Fortran by Haines (1988).

Let r denote the radius, θ the colatitude, and λ the east longitude of a given spherical cap coordinate. This coordinate system is, of course, identical to the usual spherical or polar coordinate system, except that the colatitude θ must be less than θ_0, the half angle of the spherical cap (Figure H8). Also, the spherical cap pole is not usually the geographic North Pole, in which case the geographic coordinate system is rotated to the new spherical cap pole giving new spherical cap colatitudes and longitudes. In both spherical and spherical cap systems, the radius r must lie between the outer radius of any current sources within the Earth and the inner radius of any current sources within the ionosphere, and the longitude λ takes on the full 360° range. The harmonic solution of the potential $V(r,\theta,\lambda)$ applicable to this three-dimensional spherical cap region is then given by:

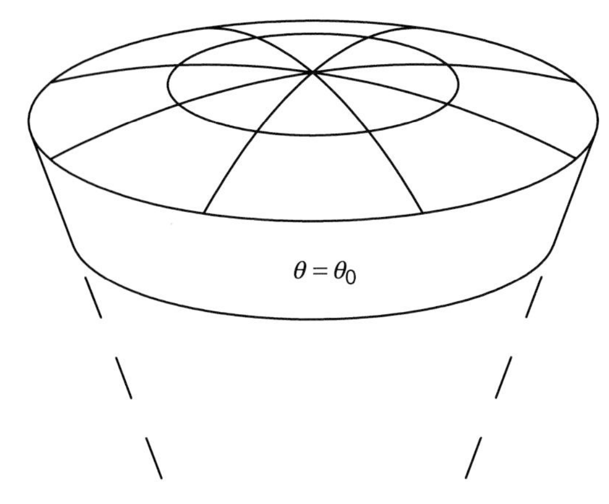

Figure H8 Three-dimensional spherical cap region: colatitude $\theta \le \theta_0$. Thickness of cap indicates radial coverage of data.

$$\begin{aligned} V(r,\theta,\lambda) &= a\sum_{k=0}^{K_i}\sum_{m=0}^{k}(a/r)^{n_k(m)+1}P^m_{n_k(m)}(\cos\theta) \\ &\times [g_k^{m,i}\cos(m\lambda)+h_k^{m,i}\sin(m\lambda)] \\ &+ a\sum_{k=1}^{K_e}\sum_{m=0}^{k}(r/a)^{n_k(m)}P^m_{n_k(m)}(\cos\theta) \\ &\times [g_k^{m,e}\cos(m\lambda)+h_k^{m,e}\sin(m\lambda)] \end{aligned} \qquad \text{(Eq. 1)}$$

where a is some reference radius, usually taken as the radius of the Earth, and $P^m_{n_k(m)}(\cos\theta)$ is the associated Legendre function of the first kind. It is usual in geomagnetism for the Legendre functions P to be Schmidt-normalized (Chapman and Bartels, 1940; Sections 17.3 and 17.4). The subscript of P is known as the degree of the Legendre function and the superscript is known as the order; k is referred to as the index and simply orders the real (usually nonintegral) degrees $n_k(m)$. The g_k^m and h_k^m are the coefficients, which are each further identified with the superscript i or e to denote internal or external sources, respectively. The internal source terms involve powers of (a/r) while the external source terms involve powers of (r/a). If one intended to fit only internal (or external) sources, only the internal (or external) terms of the expansion would be used. The truncation indices K_i and K_e are the maximum indices for the internal and external series, respectively.

The potential can easily be transformed into a function of time t as well as space, $V(r,\theta,\lambda,t)$, by simply making the coefficients functions of time, $g_k^m(t)$ and $h_k^m(t)$, and expanding these coefficients in terms of some temporal basis functions, such as cosine functions or Fourier functions or whatever is appropriate.

The degree $n_k(m)$ is chosen so that

$$\frac{dP^m_{n_k(m)}(\cos\theta_0)}{d\theta} = 0 \quad \text{when } k-m=\text{even} \qquad \text{(Eq. 2)}$$

and

$$P^m_{n_k(m)}(\cos\theta_0) = 0 \quad \text{when } k-m=\text{odd} \qquad \text{(Eq. 3)}$$

where the Legendre functions $P_n^m(\cos\theta_0)$ are here considered to be functions of n, given the order m and the cap half-angle θ_0. The index k simply starts at m and is incremented by 1 each time a root is found to one of the Eqs. (2) or (3). This choice of k is analogous to the case of ordinary spherical harmonics ($\theta_0 = 180°$), and in fact for that case the degree $n_k(m)$ is simply the integer k. Table H3 gives the roots $n_k(m)$ for $\theta_0 = 30°$, up to $k = 8$. Figure H9 shows how $P^m_{n_k(m)}(\cos\theta)$ varies over a 30° cap, up to index 4, both for $k-m=$ even and for $k-m=$ odd.

We can see how the $P^m_{n_k(m)}(\cos\theta)$, $0 \le \theta \le \theta_0$, for $k-m=$ even, are analogous to the cosine functions $\cos(n\theta)$, $0 \le \theta \le \pi$, in that each has zero slope at the upper boundary (θ_0 or π, respectively). In fact, we can think of $P^m_{n_k(m)}$ as being defined on the interval $[-\theta_0, +\theta_0]$, just as $\cos(n\theta)$ is defined on $[-\pi,+\pi]$, so that an expansion over the spherical cap using $P^m_{n_k(m)}(\cos\theta)$, $k-m=$ even, as basis functions is analogous to an expansion in Cartesian coordinates over $[-\pi,+\pi]$ using $\cos(n\theta)$ as basis functions. The extension of $P^m_{n_k(m)}$ to the $[-\theta_0, 0]$ interval of course, takes place on the meridian 180° away from the meridian on which the $[0,+\theta_0]$ interval lies, which is why the $P^m_{n_k(m)}(\cos\theta)$ must be zero at $\theta = 0$ when $m \ne 0$ (Figure H9a).

Table H3 $n_k\,(m)$ for $\theta_0 = 30°$

k	m: 0	1	2	3	4	5	6	7	8
0	0.00								
1	4.08	3.12							
2	6.84	6.84	5.49						
3	10.04	9.71	9.37	7.75					
4	12.91	12.91	12.37	11.81	9.96				
5	16.02	15.82	15.62	14.92	14.18	12.13			
6	18.94	18.94	18.58	18.22	17.39	16.50	14.29		
7	22.02	21.87	21.72	21.25	20.76	19.81	18.80	16.42	
8	24.95	24.95	24.69	24.42	23.84	23.24	22.19	21.07	18.55

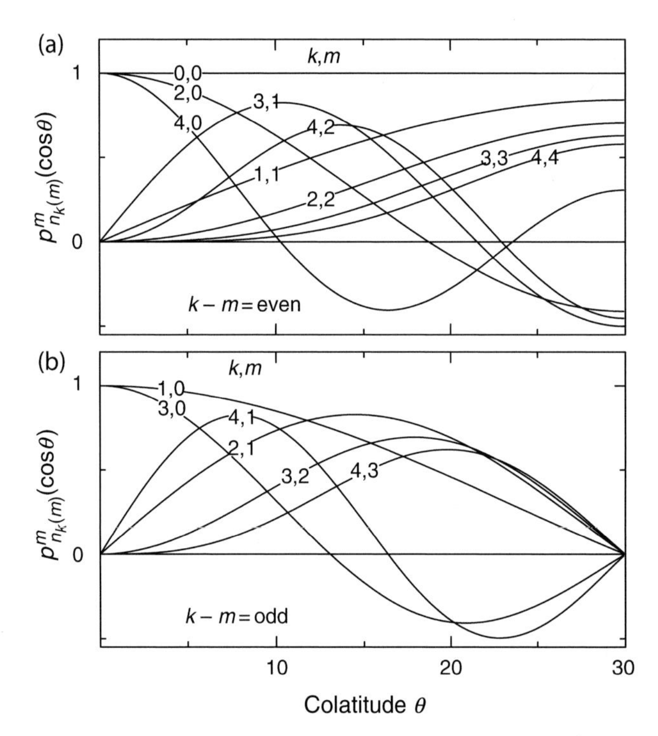

Figure H9 Legendre functions $P^m_{n_k(m)}(\cos\theta)$ of nonintegral degree $n_k(m)$ (given in Table H3) and integral order m, up to index $k = 4$: (a) those that have zero slope at the cap boundary $\theta_0 = 30°$, and (b) those that are zero at the cap boundary.

Similarly, the $P^m_{n_k(m)}(\cos\theta)$, for $k - m =$ odd, are analogous to the sine functions $sin(n\theta)$, $0 \leq \theta \leq \pi$, in that each is zero at the upper boundary. Here, of course, $P^m_{n_k(m)}(\cos\theta)$ must be nonzero at $\theta = 0$ when $m = 0$ (Figure H9b). So an expansion over the spherical cap using $P^m_{n_k(m)}(\cos\theta)$, $k - m =$ odd, as basis functions is analogous to an expansion in Cartesian coordinates using $\sin(n\theta)$ as basis functions.

Finding the Legendre functions with zero slope (Eq. (2)) or zero value (Eq. (3)) at the boundary of the expansion region is therefore analogous to finding the Fourier cosine and sine basis functions (on the interval $[0, \pi]$) applicable to a given expansion region in Cartesian coordinates. Those with zero slope at the boundary are able to fit nonzero functions there, while those that are zero at the boundary are only able to fit functions that go to zero there. Of course, we are considering Eq. (1) to be uniformly convergent, not merely convergent in mean square (or "in the mean"). So to fit an arbitrary function over the region, one only needs to include the basis functions with zero slope at the boundary (like the cosine functions). On the other hand, if one wishes to fit the derivative of the arbitrary function, these basis functions would not do (unless that arbitrary function has zero slope at the boundary). However, the basis functions that are zero (and whose derivatives are therefore not zero) at the boundary, will be able to fit uniformly the derivative of that arbitrary function. Including basis functions both with zero slope and zero function at the boundary thus permits the uniform expansion of functions with both arbitrary values and arbitrary derivatives at the boundary of the expansion region.

The Legendre functions thus defined comprise two sets of separately orthogonal functions, with respect to the weight function $\sin\theta$. That is, $P^m_{n_j(m)}(\cos\theta)$ and $P^m_{n_k(m)}(\cos\theta)$ are orthogonal when $j \neq k$ and when $j - m$ and $k - m$ are either both odd or both even. However, they are not orthogonal when $j - m =$ even and $k - m =$ odd (or *vice versa*).

The vector magnetic field B

The vector magnetic field **B** is given by the negative gradient of Eq. (1). That is, we compute the vertical and the horizontal spherical cap north and east components of **B** by differentiating Eq. (1) with respect to the spherical cap coordinates r, θ, and λ, respectively (with appropriate signs and of course including the metrical coefficients 1, r, and $r\sin\theta$). Note that only when the spherical cap pole is the geographic North Pole, are these components the usual geographic components.

We can see now why we need the spherical cap basis functions with $k - m =$ odd. It is because the north component of the field involves a differentiation with respect to θ. If the north component of the data happened to be nonzero at $\theta = \theta_0$, the north component of the model would not be uniformly convergent if only basis functions with $dP(\cos\theta_0)/d\theta = 0$ were included in the model. Of course, if the north component is not being modeled, the second set of basis functions are not strictly required (the vertical anomaly field from Magsat data, e.g., was modeled by Haines, 1985b). In this case, the constraint on the derivative of P results in the field having zero slope, with respect to θ, at the cap boundary.

Even if one did not need to fit a derivative, it can still be an advantage to include basis functions that allow for a nonzero derivative at the boundary of the analysis region. That advantage is in faster convergence (a smaller number of coefficients for a given truncation error), as discussed by Haines (1990).

Mean square and cross product values for the internal and external harmonics of the vector field **B** have been derived by Lowes (1999), in terms of the harmonics of the potential, as well as for their horizontal and radial components. Power spectra have been defined and expressed by Haines (1991), and the relationship of spherical cap harmonics to ordinary spherical harmonics has been described by De Santis *et al.* (1999).

Subtraction of a reference field

The price paid for having uniform convergence over a spherical cap by including both sets of basis functions is that some of these functions are mutually nonorthogonal. Although most are mutually orthogonal (those within each of the two sets and all those of different order by virtue of the orthogonality of the longitude functions), even this sparse nonorthogonality can have an adverse effect on the coefficient solution, particularly at high truncation levels. This is because the nonorthogonality results in an ill-conditioned least-squares matrix, which will have some nondiagonal terms. The smaller the computer word size and the larger the field values being fitted, the more serious is this problem. It is therefore advantageous to subtract a spherical harmonic reference field from the data, do the spherical cap harmonic fit on the resulting residuals, and simply add the reference field back on when the full field is required. Which particular harmonic reference field is used for this purpose is unimportant; the idea is simply to have smaller numbers in the modeling process so as not to lose too much numerical accuracy for the given computer word size.

A second difficulty with spherical cap and other regional models, is the accumulation of errors in upward or downward continuation of the field (see *Upward and downward continuation*). The field at the continuation point, of course depends on field values outside the cap, as well as inside, and so unless the effect of these outside values could somehow be compensated for by more complicated boundary conditions, there will be an error in continuing a model based only on field values inside the cap. Here again, we can get considerable relief from subtraction of global reference fields, which are models of data from the whole Earth. Haines (1985a, Figures 3–6) shows the kinds of errors to expect in upward continuing fields to 300 and 600 km, and the lower errors when subtracting an *International Geomagnetic Reference Field (IGRF)* (*q.v.*) prior to analysis.

General fields over a cap surface

Previously, we have discussed fitting a harmonic function in a three-dimensional region of space over a spherical cap, as portrayed in Figure H8. However, the spherical cap basis functions can also be used to expand a general function over a (two-dimensional) spherical cap surface ($\theta \leq \theta_0, r = r_0$). That is, there is no constraint for this surface field to be harmonic since the radial field is not being defined; it can represent, on the surface, fields that are very complex in three-dimensional space. This is analogous to the expansion theorem for the whole sphere (e.g., Courant and Hilbert, 1953, Chapter VII, Section 5.3).

The vertical field is able to play the role of the general function by considering only the internal field (i.e., putting $K_e = 0$), and setting $r = a$. The factor $-[n_k(m)+1]$ arising from the differentiation with respect to r can also be set to unity. Similarly, the north and east components can be made to play the role of north and east derivatives of a general function by similar modifications (see Haines, 1988, p. 422 for details). This latter aspect can be used to model electric fields (horizontal derivatives of an electric potential) over a surface in the ionosphere. Although an electric potential satisfies Poisson's equation in three-dimensional space, it can be treated as a general function on the spherical surface.

This surface expansion is in fact the solution of the two-dimensional eigenvalue problem in the two variables (colatitude and longitude) whose functions (Legendre and trigonometric) were orthogonalized in the solution of the three-dimensional differential equation. This is a common technique in expansion methods. In Cartesian coordinates, for example, the solution to Laplace's equation in three-dimensional space ("Rectangular Harmonics") gives rise to a surface expansion (two-dimensional "Fourier Series") simply by finding the eigensolutions in the coordinates whose (trigonometric) functions were orthogonalized there. Again, these Fourier series allow the expansion of a very large class of functions, certainly functions that are in no way constrained as are the rectangular harmonic functions of three-dimensional space. Of course, we can go down another dimension and expand one-dimensional functions in either of the surface variables. This gives an expansion, on the spherical cap, in trigonometric functions for longitude and in Legendre functions for colatitude, analogously to the one-dimensional trigonometric Fourier series in Cartesian coordinates.

Example application

Regional magnetic field models over Canada have been produced every 5 years since 1985 using spherical cap harmonics. For each model, the IGRF at an appropriate epoch is subtracted from the data, and a spherical cap harmonic model of the residuals is then determined. (This also provides an estimate of the "error" or wavelength limitations of the IGRF that was subtracted.) The final model, obtained by adding the IGRF back on to the spherical cap model, is referred to as the Canadian Geomagnetic Reference Field (CGRF). Details of the data and processing methods used for the 1995 CGRF have been given by Haines and Newitt (1997).

G.V. Haines

Bibliography

Chapman, S., and Bartels, J., 1940. *Geomagnetism*, Vol. II. New York: Oxford University Press.

Courant, R. and Hilbert, D., 1953. *Methods of Mathematical Physics*, translated and revised from the German original, Vol. I. New York: Interscience Publishers.

Davis, H.F., 1963. *Fourier Series and Orthogonal Functions*. Boston: Allyn and Bacon.

De Santis, A., Torta, J.M., and Lowes, F.J., 1999. Spherical cap harmonics revisited and their relationship to ordinary spherical harmonics. *Physics and Chemistry of the Earth (A)*, **24**: 935–941.

Haines, G.V., 1985a. Spherical cap harmonic analysis. *Journal of Geophysical Research*, **90**: 2583–2591.

Haines, G.V., 1985b. Magsat vertical field anomalies above 40° N from spherical cap harmonic analysis. *Journal of Geophysical Research*, **90**: 2593–2598.

Haines, G.V., 1988. Computer programs for spherical cap harmonic analysis of potential and general fields. *Computers and Geosciences*, **14**: 413–447.

Haines, G.V., 1990. Modelling by series expansions: a discussion. *Journal of Geomagnetism and Geoelectricity*, **42**: 1037–1049.

Haines, G.V., 1991. Power spectra of subperiodic functions. *Physics of the Earth and Planetary Interiors*, **65**: 231–247.

Haines, G.V., and Newitt, L.R., 1997. The Canadian Geomagnetic Reference Field 1995. *Journal of Geomagnetism and Geoelectricity*, **49**: 317–336.

Lowes, F.J., 1999. Orthogonality and mean squares of the vector fields given by spherical cap harmonic potentials. *Geophysical Journal International*, **136**: 781–783.

Smythe, W.R., 1950. *Static and Dynamic Electricity*, 2nd ed. New York: McGraw-Hill.

Cross-references

Harmonics, Spherical
IGRF, International Geomagnetic Reference Field
Upward and Downward Continuation

HARTMANN, GEORG (1489–1564)

Hartmann was born on February 9, 1489, at Eggolsheim, Germany. After studying theology and mathematics at Cologne around 1510, he spent some time in Rome, where he ranked Andreas, the brother of Nicholas Copernicus, amongst his friends. Hartmann belonged to that class of Renaissance scholar who, while the recipient of a learned education, also had a fascination with mechanics, horology, instrumentation, and natural phenomena. His principal claim to fame as a student of magnetism lay in his discovery that in addition to the compass needle pointing north, it also had a dip, or inclination, out of the horizontal. He claims to have noticed, for instance, that in Rome, the needle of a magnetic compass dipped by 6° towards the north. Although his numerical quantification of this phenomenon was wrong, the compass does, indeed, dip in Rome, as it does in most nonequatorial locations.

Although Hartmann never published his discovery, he did communicate it in a letter of 4 March 1544 to Duke Albert of Prussia. Unfortunately, this remained unknown to the wider world for almost the next three centuries, until it was finally printed in 1831. Hence, Hartmann's discovery had no influence on other early magnetical researchers, and credit for the discovery of the "dip" went to the Englishman *Robert*

Norman (*q.v.*), who published his own independent discovery in *A Newe Attractive* (London, 1581).

Hartmann was a priest by vocation, and held several important benefices. He settled in Nuremberg in 1518, where he was no doubt in his element, as that city was one of Europe's great centers for the manufacture of clocks, watches, instruments, ingenious firearms, and other precision metal objects. He collected astronomical and scientific manuscripts and was associated with the Nuremberg observational astronomer Johann Schöner, as well as Joachim Rheticus, who saw Copernicus' *De Revolutionibus* through the press at Nuremberg in 1543. Hartmann died in Nuremberg on April 9, 1564.

Allan Chapman

Bibliography

Heger, K., 1924. Georg Hartmann von Eggolsheim. *Der Fränkische Schatzgräber* **2**: 25–29.

Hellmann, G. (ed.), *Rara Magnetica 1269–1599*, Neudrucke von Schriften und Karten über Meteorologie und Erdmagnetismus no. 10 (Berlin, 1898). Reprints Hartmann's letter to Duke Albert of Prussia. Letter first noticed in Prussian Royal Archives, Königsberg; see J. Voigt, in Raumer's *Historisches Taschenbuch* II (Leipzig, 1831), 253–366.

Ritvo, L.B., 'Georg Hartmann', *Dictionary of Scientific Biography* (Scribner's, New York, 1972–).

Zinner, Ernst, *Deutsche und niederländische astronomische Instrumente des 11–18 Jahrhunderts* (Munich, 1956).

HELIOSEISMOLOGY

Helioseismology is the study of the interior of the Sun using observations of waves on the Sun's surface. It was discovered in the 1960s by Leighton, Noyes, and Simon that patches of the Sun's surface were oscillating with a period of about 5 min. Initially these were thought to be a manifestation of convective motions, but by the 1970s it was understood from theoretical works by Ulrich and Leibacher and Stein and from further observational work by Deubner that the observed motions are the superposition of many global resonant modes of oscillation of the Sun (e.g., Deubner and Gough, 1984, Christensen-Dalsgaard, 2002). The Sun is a gaseous sphere, generating heat through nuclear fusion reactions in its central region (the core) and held together by self-gravity. It can support various wave motions, notably acoustic waves and gravitational waves; these set up global resonant modes. The modes in which the Sun is observed to oscillate are predominantly acoustic modes, though the acoustic wave propagation is modified by gravity and the Sun's internal stratification, and by bulk motions and magnetic fields. The observed properties of the oscillations, especially their frequencies, can be used to make inferences about the physical state of the solar interior. The excitation mechanism is generally believed to be turbulent convective motions in subsurface layers of the Sun, which generate acoustic noise. This is a broadband source, but only those waves that satisfy the appropriate resonance conditions constructively interfere to give rise to resonant modes.

There are many reasons to study the Sun. It is our closest star and the only one that can be observed in great detail, so it provides an important input to our understanding of stellar structure and evolution. The Sun greatly influences the Earth and near-Earth environment, particularly through its outputs of radiation and particles, so studying it is important for understanding solar-terrestrial relations. And the Sun also provides a unique laboratory for studying some fundamental physical processes in conditions that cannot be realized on Earth.

From the 1980s up to the present day, much progress has been made in helioseismology through analysis of the Sun's global modes of oscillation. Some of the results, on the Sun's internal structure and its rotation, are given below. Though not discussed here, similar studies are now also beginning to be made in more limited fashion for some other stars, in a field known as asteroseismology. In the past decade, global mode studies of the Sun have been complemented by so-called local helioseismology, techniques such as tomography which have been used to study flows in the near-surface layers and flows and structures under sunspots and active regions of sunspot complexes. Local helioseismology is also discussed below.

Solar oscillations

The Sun's oscillations are observed in line-of-sight Doppler velocity measurements of the visible solar disk, and in measurements of variations of the continuum intensity of radiation from the surface. The latter are caused by compression of the radiating gas by the waves. Spatially resolved measurements are obtained by observing separately different portions of the visible solar disk; but the motions with the largest horizontal scales are also detectable in observations of the Sun as a star, in which light from the whole disk is collected and analyzed as a single time series. To measure the frequencies very precisely, long, uninterrupted series of observations are desirable. Hence, observations are made from networks of dedicated small solar telescopes distributed in longitude around the Earth (networks such as the Global Oscillation Network Group [GONG] making 1024×1024-pixel resolved observations and the Birmingham Solar Oscillation Network [BiSON] making Sun-as-a-star observations) or from space (for instance, the Solar and Heliospheric Orbiter [SOHO] satellite has three dedicated helioseismology instruments onboard, in decreasing order of resolution MDI, VIRGO and GOLF). In Doppler velocity, the amplitudes of individual modes are of order $10\,\mathrm{cm\,s^{-1}}$ or smaller, their superposition giving a total oscillatory signal of the order of a few kilometers per second. The highest amplitude modes have frequencies ν of around 3 mHz.

The outer 30% of the Sun comprises a convectively unstable region called the convection zone (e.g., Christensen-Dalsgaard *et al.*, 1996). The solar oscillations are stochastically excited by turbulent convection in the upper part of the convection zone. The modes are both excited and damped by their interactions with the convection. Although large excitation events such as solar flares may occasionally contribute, the dominant excitation is probably much smaller-scale and probably takes place in downward plumes where material previously brought to the surface by convection has cooled and flows back into the solar interior. These form a very frequent and widespread set of small-scale excitation sources.

The fluid dynamics of the solar interior is described by the fluid dynamical momentum equation and continuity equation, an energy equation and Poisson's equation for the gravitational potential. On the timescales of the solar oscillations of interest here, the bulk of the solar interior can be considered to be in thermal equilibrium and providing a static large-scale equilibrium background state in which the waves propagate. Treating the departures from the time-independent equilibrium state as small perturbations with harmonic time dependence $\exp(\mathrm{i}\omega t)$, where $\omega \equiv 2\pi\nu$ is an (as yet) unknown resonant angular frequency, t is time and $\mathrm{i} = \sqrt{-1}$, the above mentioned equations can be linearized in perturbation quantities and constitute a coupled set of linear equations describing the waves. The timescales of the solar oscillations are sufficiently short that the energy equation can be replaced by the condition that the perturbations of pressure (p) and density (ρ) are related by an adiabatic relation. At high frequencies, pressure forces provide the dominant restoring force for the perturbed motions, giving rise to acoustic waves. A key quantity for such acoustic waves is the adiabatic sound speed c, which varies with position inside the Sun: $c^2 = \Gamma_1 p/\rho$, where Γ_1 is the first adiabatic exponent ($\Gamma_1 = 5/3$ for a perfect monatomic gas). To an excellent approximation, $p \propto \rho T/\mu$, where T is temperature and μ is the mean molecular weight of the gas, so

$c \propto T^{1/2}$ inside the Sun. Except for wave propagating exactly vertically, inward-propagating waves get refracted back to the surface at some depth, because the temperature and hence sound speed increase with depth. Near the surface, outward-propagating waves also get deflected, by the sharply changing stratification in the near-surface layers. Hence the acoustic waves get trapped in a resonant cavity and form modes, with discrete frequencies ω. To a good approximation, the Sun's structure is spherically symmetric, hence, the horizontal structure of the eigenfunctions of the modes is given by spherical harmonics Y_l^m, where integers $l (l \geq 0)$ and $m (-l \leq m \leq l)$ are respectively called the *degree* and *azimuthal order* of the mode. The structure of the eigenfunction in the radial direction into the Sun is described by a third quantum number n, called the *order* of the mode: the absolute value of n is essentially the number of nodes in the (say) pressure perturbation eigenfunction between the center and the surface of the Sun. Hence, each resonant frequency can be labeled with three quantum numbers thus: ω_{nlm}. The labeling is such that at fixed l and m, the frequency ω_{nlm} is a monotonic increasing function of n.

In a wholly spherically symmetrical situation the frequencies would be independent of m. However, the Sun's rotation, as well as any other large-scale motions, thermal asphericities, and magnetic fields break this degeneracy and introduce a dependence on m in the eigenfrequencies. The modes of different m but with the same values of n and l are called a multiplet: it is convenient also to introduce the mean multiplet frequency ω_{nl}.

High-frequency modes, with positive values of the order n, are essentially acoustic modes (p modes): for them, the dominant restoring force is pressure. Low-frequency modes, with negative values of n, are essentially gravity modes (g modes) set up by gravity waves, which can propagate where there is a stable stratification. There is an intermediate mode with $n = 0$: this is the so-called fundamental or f mode. For large values of l, the f mode has the physical character of a surface gravity mode. The observed global modes of the Sun are p modes and f modes. Modes of low degree (mostly $l = 0, 1, 2, 3$) are detectable in observations of the Sun as a star: low-degree p modes are sensitive to conditions throughout the Sun, including the energy-generating core. Spatially resolved observations have detected p and f modes up to degrees of several thousand: such modes are no longer global in character, but nonetheless measuring their properties conveys information about the Sun's outer subsurface layers. Internal g modes have not unambiguously been observed to date, and they are expected to have small amplitudes at the surface: this is because their region of propagation is the stably stratified radiative interior and they are evanescent through the intervening convection zone.

The computed mean multiplet frequencies for modes of a current model of the Sun are shown in Figure H10. Points on each branch of the dispersion relation (each corresponding to a single value of n) have been joined with continuous curves.

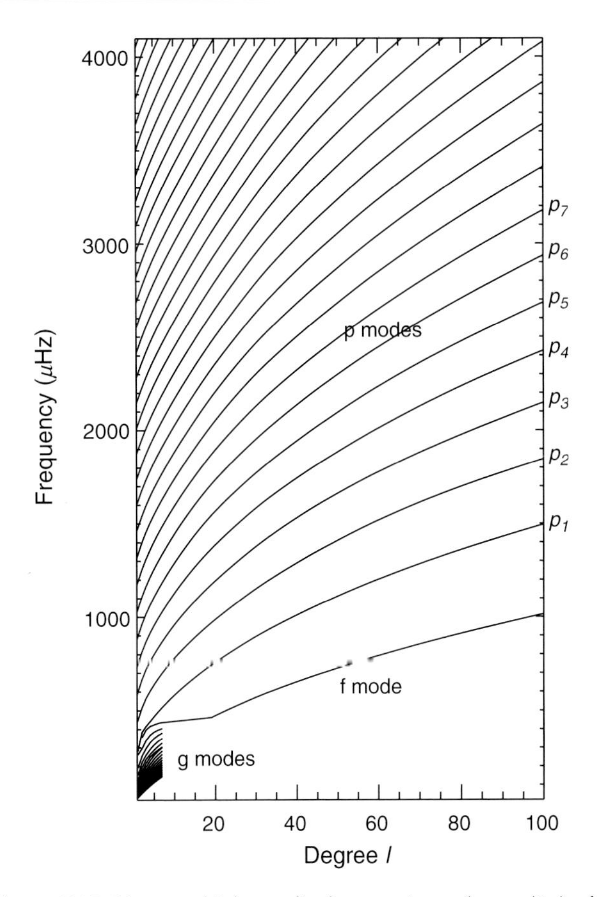

Figure H10 Mean multiplet cyclic frequencies $\nu_{nl} (= \omega_{nl}/2\pi)$ of p, f, and g modes of a solar model, as a function of the mode degree l. The low-order p modes are labelled according to the value of the mode order n. The g modes are only illustrated up to $l = 7$ and for $|n| \leq 20$.

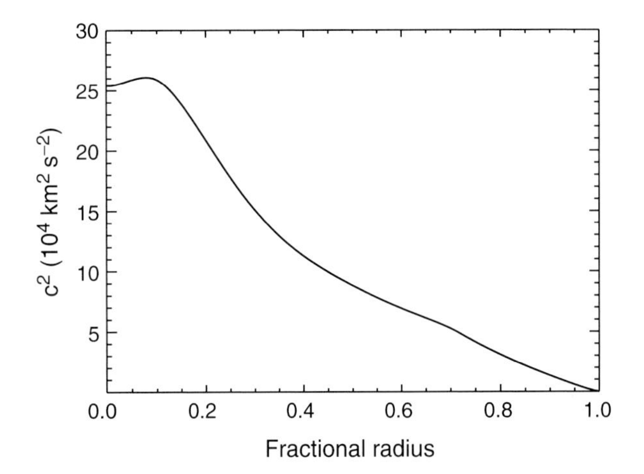

Figure H11 Square of the adiabatic sound speed c inside a model of the present Sun, as a function of fractional radius (center at 0, photospheric surface at 1.0).

Results from global-mode helioseismology

The frequencies of the Sun's global modes are estimated by making suitable spatial projections and temporal Fourier transformations of the observed Doppler or intensity fluctuations. From these, using a variety of fitting or inversion techniques (similar to those used in other areas of, e.g., geophysics and astrophysics), properties of the solar interior can be deduced. Indeed, helioseismology has borrowed and adapted various approaches and ideas on inversion from geophysics, notably those of G. Backus and F. Gilbert (Backus and Gilbert, 1968, 1970). One of the principal deductions from helioseismology has been the sound speed as a function of position in the Sun (e.g., Gough *et al.*, 1996). The value of this deduction lies not in the precise value of the sound speed but in the inferences that follow concerning the physics that determines the sound speed. The sound speed in a solar model that is closely in agreement with the helioseismic data is illustrated in Figure H11. As discussed above, the adiabatic sound speed c is related to temperature T, mean molecular weight μ, and adiabatic exponent Γ_1 by $c^2 \propto \Gamma_1 T/\mu$, where the constant of proportionality is the gas constant. The general increase in sound speed with depth reflects the increase in temperature from the surface to the center of the

Sun. The gradient of the temperature, and hence of the sound speed, is related to the physics by which heat is transported from the center to the surface. In the bulk of the Sun, that transport is by radiation; but in the outer envelope the transport is by convection. A break in the second derivative of the sound speed near a fractional radius of 0.7 indicates the transition between the two regimes and has enabled helioseismology to determine the location of the base of the convection zone: this is important because it delineates the region in which chemical elements observed at the surface are mixed and homogenized by convective motions. Beneath the convection zone, the gradients of temperature and sound speed are influenced by the opacity of the material to radiation: it has thus been possible to use the seismically determined sound speed to find errors in the theoretical estimates of the opacity, which is one of the main microphysical inputs for modeling the interiors of stars. Not readily apparent on the scale of Figure H11 is the spatial variation of Γ_1 and hence of sound speed in the regions of partial ionization of helium and hydrogen in the outer 2% of the Sun. This variation depends on the equation of state of the material and on the abundances of the elements: it has been used to determine that the fractional helium abundance by mass in the convection zone, which is poorly determined from surface spectroscopic observations, is about 0.25. This is significantly lower than the value of 0.28 believed from stellar evolutionary models to have been the initial helium abundance of the Sun when it formed. The deficiency of helium in the present Sun's convection zone is now understood to arise from gravitational settling of helium and heavier elements out of the convection zone and into the radiative interior over the 4.7 Ba that the Sun has existed as a star. This inference is confirmed by an associated slight modification to the sound speed profile beneath the convection zone.

The dip in the sound speed at the center of the Sun is the signature of the fusion of hydrogen to helium that has taken place over the Sun's lifetime: the presence of helium increases the mean molecular weight μ, decreasing the sound speed. Thus the amount of build-up of helium in the core is an indicator of the Sun's age. Apart from inversions, a sensitive indicator of the sound speed in the core is the difference $\nu_{nl} - \nu_{n-1,l+2}$ between frequencies of p modes of low degree $(l = 0, 1, 2, 3)$ differing by two in their degree. Such pairs of modes penetrate into the core and have similar frequencies and hence very similar sensitivity to radial structure in the outer part of the Sun, but have different sensitivities in the core.

Another major deduction of global-mode helioseismology has been the rotation as a function of position through much of the solar interior (Figure H12) (e.g., Thompson *et al.*, 1996, 2003). In the convection zone, it has been discovered that the rotation varies principally with latitude and rather little with depth: at low solar latitudes the rotation is fastest, with a rotation period of about 25 days; while at high latitudes the rotation periods in the convection zone are in excess of 30 days. These rates are consistent with deductions of the surface rotation from spectroscopic observations and from measurements of motions of magnetic features such as sunspots. The finding was at variance with some theoretical expectations that the rotation in the convection zone would be constant on Taylor columns. At low- and mid-latitudes there is a near-surface layer of rotational shear, which may account for the different rotational speeds at which small and large magnetic features are observed to move. Near the base of the convection zone, the latitudinally differential rotation makes a transition to latitudinally independent rotation. This gives rise to a layer of rotational shear at low and high latitudes, which is called the tachocline. It is widely believed that the tachocline is where the Sun's large-scale magnetic field is generated by dynamo action, leading to the 11-year solar cycle of sunspots and the large-scale dipole field. Deeper still the rotation appears to be consistent with solid-body rotation, presumably caused by the presence of a magnetic field. In the core, there is some hint of a slower rotation, but the uncertainties on the deductions are quite large: nonetheless, some earlier theoretical predictions that the core would rotate much faster than the surface, a relic of the Sun's faster rotation as a young star, are strongly ruled out by the seismic observations. Superimposed on the rotation of the convection zone are weak but coherent migrating bands of faster and slower rotation (of amplitude only a few meters per second, compared with the surface equatorial rotation rate of about 2 km s^{-1}), which have been called torsional oscillations: the causal connection between the migrating zonal flows and the sunspot active latitudes, which also migrate during the solar cycle, is as yet uncertain. There have also been reported weak variations, with periodicities around 1.3 years, in the rotation rate in the deep convection zone and in the vicinity of the tachocline. To date there has been no direct helioseismic detection of magnetic field in the region of the tachocline: indeed, because the pressure increases rapidly with depth, a field there would have to have a strength of order 10^6 Gauss to have a significant direct influence on the mode frequencies. Thus detecting

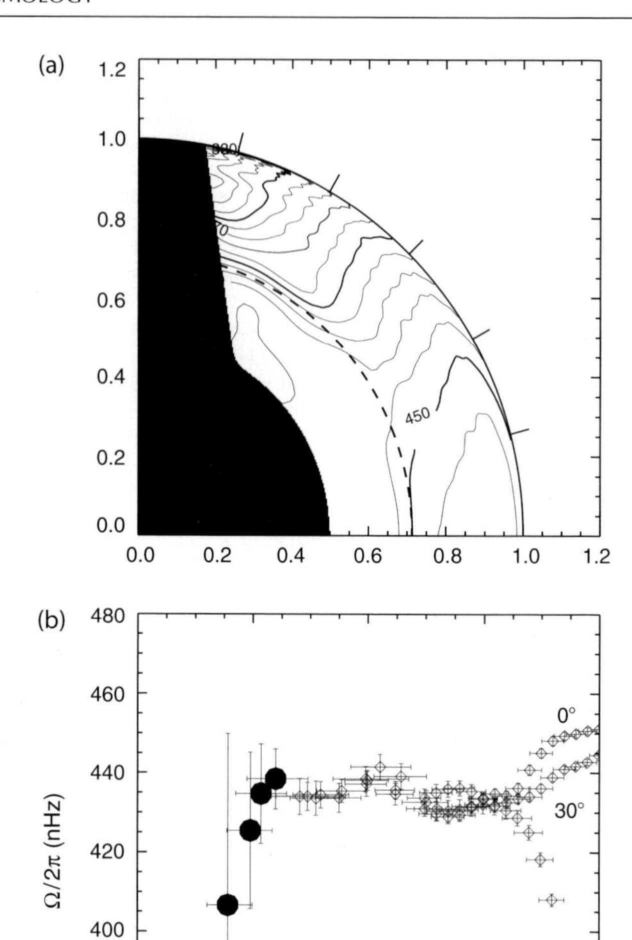

Figure H12 (a) Contour plot of the rotation rate inside the Sun inferred from MDI data. The rotation axis is up the *y*-axis, the solar equator is along the *x*-axis. Contour spacings are 10 nHz; contours at 450, 400, and 350 nHz are thicker. The shaded region indicates where a localized solution has not been possible with these data. (b) The rotation rate deeper in the interior, on three radial cuts at solar latitudes $0°, 30°$, and $60°$, using data from Sun-as-a-star observations by BiSON and spatially resolved observations by the LOWL instrument.

the effect of a temporally varying magnetic field on the angular momentum may be the most likely way to infer seismically the presence of such a field in the deep interior. On the other hand, near-surface magnetic fields can influence mode frequencies in a detectable way, because the pressure is much lower there; and there is strong evidence that mode frequencies vary over the solar cycle in a manner that is highly correlated with the temporal and spatial variation of photospheric magnetic field.

Local helioseismology

Analysis of the Sun's global mode frequencies has provided an unprecedented look at the interior of a star, but such an approach has limitations. In particular, the frequencies sense only a longitudinal average of the internal structure. To make more localized inferences, the complementary approaches of local helioseismology are used. One such technique is to analyze the power spectrum of oscillations as a function of frequency and the two horizontal components of the wavenumber in localized patches. To obtain good wavenumber resolution, the patches are usually quite large: square tiles of up to about 2×10^5 km on the side. The technique is known as ring analysis, because at fixed frequency the p-mode power lies on near-circular rings in the horizontal wavenumber plane. By performing inversions for the depth dependence under each tile, maps with horizontal resolution similar to the size of the tiles can be obtained of structures and particularly of flows in the outer few per cent of the solar interior (e.g., Toomre, 2003). As well as the zonal flows, the meridional (i.e., northward and southward) flow components have also been measured. Beneath the surface, these are generally found to be poleward in both hemispheres down to the depth at which the ring analyzes lose resolution: this depth is of order 10^4 km. However, the analyses indicate that the flow patterns in the northern but not the southern hemispheres changed markedly in 1998–2000 as the Sun approached the maximum of its 11-year magnetic activity cycle.

Other local techniques include time-distance helioseismology and acoustic holography. So for example, in time-distance helioseismology, the travel-time of waves between different points on the surface of the Sun are used to infer wavespeed and flows under the surface (e. g., Kosovichev, 2003). This can be achieved on much smaller horizontal scales than have yet been achieved by ring analysis, down to just a few thousand kilometers. Since individual excitation events are rarely if ever seen, the travel-times are isolated from the measured Doppler velocities across the solar disk by cross-correlating pairs of points, or sets of points. Using high-resolution observations such as those from the MDI instrument onboard SOHO, travel-times can be measured between many different locations on the solar surface, and with different spatial separations which in turn provide different depth sensitivities to subsurface conditions. By comparing the measured travel-times with those of a solar model, inversion techniques similar to those used in global mode helioseismology have been used to map conditions in the convection zone using travel-time data. In particular, wavespeed anomalies and flows under sunspots and active regions of strong magnetic fields on the surface have been mapped. An unexpected finding has been that the wavespeed beneath sunspots (which in such regions is not just the sound speed but is modified by the magnetic field) is increased relative to the spot's surroundings, *except* in a shallow layer of up to 5000km depth where the wavespeed is decreased. Such studies are providing much-needed constraints on models of sunspot structure and theories of their origin, and on models of the emergence of magnetic flux from the solar interior.

Michael J. Thompson

Bibliography

Backus, G. and Gilbert, F., 1968. The resolving power of gross Earth data. *Geophysical Journal*, **16**: 169–205.

Backus, G. and Gilbert, F., 1970. Uniqueness in the inversion of inaccurate gross Earth data. *Philosophical Transactions of the Royal Society of London, Series A*, **266**: 123–192.

Christensen-Dalsgaard, J., 2002. Helioseismology. *Reviews of Modern Physics*, **74**: 1073–1129.

Christensen-Dalsgaard, J., *et al.*,1996. The current state of solar modeling. *Science*, **272**: 1286–1292.

Deubner, F.-L., and Gough, D.O., 1984. Helioseismology: oscillations as a diagnostic of the solar interior. *Annual Review of Astronomy and Astrophysics*, **22**: 593–619.

Gough, D.O., *et al.*,1996. The seismic structure of the Sun. *Science*, **272**: 1296–1300.

Kosovichev, A., 2003. Telechronohelioseismology. In Thompson, M.J., and Christensen-Dalsgaard, J. (eds.), *Stellar Astrophysical Fluid Dynamics*. Cambridge: Cambridge University Press, pp. 279–296.

Thompson, M.J., Christensen-Dalsgaard, J., Miesch, M.S., and Toomre, J., 2003. The internal rotation of the Sun. *Annual Review on Astronomy and Astrophysics*, **41**: 599–643.

Thompson, M.J., *et al.*,1996. Differential rotation and dynamics of the solar interior. *Science*, **272**: 1300–1305.

Toomre, J., 2003. Bridges between helioseismology and models of convection zone dynamics. In Thompson, M.J., and Christensen-Dalsgaard, J., (eds.), *Stellar Astrophysical Fluid Dynamics*. Cambridge: Cambridge University Press, pp. 299–314.

Cross-references

Dynamo, Solar
Harmonics, Spherical
Magnetic Field of Sun
Proudman-Taylor Theorem

HIGGINS-KENNEDY PARADOX

For the student learning about properties of the Earth's core it may be surprising that the core is solid at its center where its temperature is highest, while it becomes liquid at a distance of about 1220 km from the center where the temperature is lower. This property is caused by the dependence of the melting temperature T_m on the pressure p. That the melting temperature T_m of nearly all materials increases with pressure has been a well-known property for a long time and finds its most simple expression in Lindemann's law

$$\frac{1}{T_m}\frac{dT_m}{dp} = 2\left(\gamma - \frac{1}{3}\right)/k \qquad \text{(Eq. 1)}$$

where k is the compressibility along the melting curve $T_m(p)$ and γ is the Grüneisen parameter which, in thermodynamics, is defined by

$$\gamma = \frac{\alpha k_T}{\rho c_v} = \frac{\alpha k_S}{\rho c_p}. \qquad \text{(Eq. 2)}$$

Here, ρ denotes the density, α is the coefficient of thermal expansion, and k_T and k_S are the isothermal and adiabatic compressibilities, while c_v and c_p are the specific heats at constant volume and constant pressure, respectively. The Grüneisen parameter γ plays a prominent role in studies of the thermal state of planetary interiors since it assumes a value of the order unity for all condensed materials and does not vary much with pressure or temperature in contrast to the material properties on the right hand sides of Eq. (2). Using thermodynamic relationships one finds (see *Grüneisen's parameter for iron and Earth's core*)

$$\gamma = -\frac{T}{V}\frac{\partial V}{\partial T}\bigg|_S = -\frac{\partial \ln V}{\partial \ln T}\bigg|_S \qquad \text{(Eq. 3)}$$

which indicates that γ describes the negative slope in logarithmic plots of volume versus temperature under adiabatic compression. The γ used in relationship (1) is not necessarily the thermodynamic one, since the melting temperature may be influenced by electronic contributions and other effects which are not taken into account in Eqs. (2) and (3). The simple approximate relationship (1) called Lindemann's law is far from a rigorous law in any mathematical sense and is mentioned here only as an example of many similar "laws" that have been derived in the literature. For a review and a recent paper, see Jacobs (1987) and Anderson *et al.* (2003).

The adiabatic temperature gradient is defined by (see *Core, adiabatic gradient*)

$$\left.\frac{\partial T}{\partial p}\right|_S = \frac{\alpha T}{\rho c_p} = \frac{\gamma T}{k_S} \qquad \text{(Eq. 4)}$$

Here, the isentropic compressibility k_S is well determined in the Earth's core by seismic data, since the relationship $k_S = \rho(V_p^2 - \frac{4}{3}V_s^2)$ holds where V_p and V_s are the velocities of propagation of p (compressional)- and s (shear)-waves. The adiabatic temperature gradient defines a state where the exchange of fluid parcels from different radii can be accomplished without gain or loss of energy when dissipative effects such as those connected with viscous friction can be neglected. When the increase of temperature with pressure is less than the adiabatic gradient, a stably stratified state is obtained which requires an input of energy for the exchange of fluid parcels from different radii. When the temperature increases more strongly by a finite amount ε than the adiabatic gradient, the static state becomes unstable and convection sets in. In fact, for dimensions as large as those of the core, the amount ε is minute, such that a state of convection corresponds essentially to an adiabatic temperature distribution. Since convection flows of sufficient strength are needed to generate the geomagnetic field (Busse, 1975) it has always been assumed that the temperature field of the liquid outer core must be close to an adiabatic one except, perhaps, close to the core-mantle boundary.

In their paper of 1971, Higgins and Kennedy shattered this confidence in the adiabatic temperature distribution. On the basis of the extrapolation of numerous experimental results measured at relatively low pressures, they suggested that the melting temperature T_m in the core depends only weakly on the pressure and that the adiabatic temperature would exhibit a much steeper dependence. Accordingly they claimed that the latter temperature distribution could only be compatible with a frozen outer core in contradiction to all seismic evidence, while any temperature distribution at the melting temperature or above would imply a stably stratified core in contradiction to the dynamo hypothesis of the origin of geomagnetism. This is the Higgins-Kennedy core paradox.

The publication of the paper by Higgins and Kennedy (1971) stimulated numerous attempts to circumvent the paradoxial situation. It turns out that the melting temperature $T_m(p)$ can coincide with an isentropic temperature distribution if a suspension is assumed of small solid particles; the melting and freezing of which contributes the correct energies for a neutral exchange of fluid parcels from different radii (Busse, 1972; Malkus, 1973). Stacey (1972) pointed out that the adiabat corresponding to the temperature at the inner core-outer core boundary may well lie above the melting temperature of the outer core since the latter corresponds to that of iron alloyed with a significant amount of light elements. It is thus considerably lower than the temperature at the boundary of the inner core, which corresponds to the melting of nearly pure iron or an iron-nickel alloy. In later years rather convincing arguments have been put forward which cast doubts on the validity of the extrapolation of experimental data carried out by Higgins and Kennedy. Their proposed melting temperature together with their assumed value of γ is far removed from Lindemann's law (1). On the other hand, modern theoretical studies support relationships similar to Eq. (1). Stevenson (1980) and others argued that a weak pressure dependence of $T_m(p)$ is compatible only with an unphysically low value of the Grüneisen parameter γ. Indeed, comparing Eqs. (1) and (4) one finds that

$$\frac{dT_m}{dp} < \left.\frac{\partial T}{\partial p}\right|_S \qquad \text{(Eq. 5)}$$

can be satisfied only for $\gamma < \frac{2}{3}$, which contrasts with the value of about 1.5 of γ found for most liquid metals at high pressures. For details on the various theoretical arguments we refer to the comprehensive review given by Jacobs (1987). For a recent review of experimental measurements see Boehler (2000). Although it is unlikely today that a situation as imagined by Higgins and Kennedy exists in the Earth's core, this possibility cannot be excluded entirely. In view of our ignorance about properties of the core, it seems advisable to keep the core paradox in the back of one's mind in thinking about problems of planetary interiors (see *Dynamos, planetary and satellites*).

Friedrich Busse

Bibliography

Anderson, O.L., Isaak, D.G., and Nelson, V.E., 2003. The high-pressure melting temperature of hexagonal close-packed iron determined from thermal physics. *Journal of Physics and Chemistry of Solids*, **64**: 2125–2131.

Boehler, R., 2000. High-pressure experiments and the phase diagram of lower mantle and core materials. *Reviews of Geophysics*, **38**: 221–245.

Busse, F.H., 1972. Comments on paper by G. Higgins and G.C. Kennedy, The adiabatic gradient and the melting point gradient in the core of the Earth, *Journal of Geophysical Research*, **77**: 1589–1590.

Busse, F.H., 1975. A necessary condition for the geodynamo. *Journal of Geophysical Research*, **80**: 278–280.

Higgins, G. and Kennedy, G.C., 1971. The adiabatic gradient and the melting point gradient in the core of the Earth. *Journal of Geophysical Research*, **76**: 1870–1878.

Jacobs, J.A., 1987. *The Earth's Core*, 2nd edn, London: Academic Press.

Malkus, W.V.R., 1973. Convection at the melting point: a thermal history of the Earth's core. *Geophysical Fluid Dynamics*, **4**: 267–278.

Stacey, F.D., 1972. Physical Properties of the Earth's Core. *Geophysical Surveys*, **1**: 99–119.

Stevenson, D.J., 1980. Applications of liquid state physics to the Earth's core. *Physics of the Earth and Planetary Interiors*, **22**: 42–52.

Cross-references

Core, Adiabatic Gradient
Dynamos, Planetary and Satellite
Grüneisen's Parameter for Iron and Earth's Core

HUMBOLDT, ALEXANDER VON (1759–1859)

Alexander von Humboldt (Figure H13) was born in Berlin in 1769 as the son of a nobleman and former officer of the Prussian Army. He studied at the universities of Frankfurt/Oder and Göttingen, at the trade academy of Hamburg and at the mining academy of Freiberg in Saxonia. In his studies he was interested in all aspects of nature from botany and zoology to geography and astronomy. His friendship with Georg Forster who had participated in Cook's second voyage and whom he met in Göttingen had a strong influence on him and

Figure H13 Alexander von Humboldt in 1832. (Lithography of F.S. Delpech after a drawing of François Gérard.)

motivated him to see his vocation as a scientific explorer. First he entered, however, the career as a mining inspector in the service of the Prussian state. But when his mother died in 1796—he had lost his father already when he was only nine years old—and he became the heir of a considerable fortune, he declined the offer of the directorship of the Silesian mines and started the realization of his long held plan for a scientific expedition. In 1798, he embarked on his six years voyage to South and North America from which he returned in 1804 with huge collections of scientific data and materials. It was in the preparation for this expedition that he was instructed by the French mathematician and nautical officer Jean-Charles de Borda in the measurements of the components of the Earth's magnetic field and throughout the years of his voyage he collected magnetic data in addition to geodetic, meteorological, and other ones. One of his findings was the decrease of magnetic intensity with latitude which, apparently, had not been clearly recognized until that time since measurements had been focused on the direction of the field.

A. von Humboldt continued his magnetic measurements on his voyage with his friend Gay-Lussac to Italy in 1805 and back to Berlin. Here he was offered a wooden cabin, which allowed him to take readings of the declination each night for several months, a task he shared with the astronomer Oltmann. In December 1805 he was lucky in observing strong fluctuations of the magnetic field while an aurora borealis occurred. Humboldt coined the term "Magnetischer Sturm" for the period of strong oscillations of the magnetometer needle. Nowadays the term "magnetic storm" is generally accepted for this phenomenon. Because of the defeat of Prussia by Napoleon's army von Humboldt had to stay longer in Berlin than he had planned and returned to Paris only in late 1807. Here he was occupied with the edition of the scientific results of his American voyage until 1827. It is worth noting that during a shorter journey to Berlin in 1826, he stopped in Göttingen to meet Carl Friedrich Gauss (see *Gauss, Carl Friedrich*) for the first time in person.

A. von Humboldt and Gauss were the most prominent German scientists of their time and they had been in contact by letters for quite a while. They had high regards for each other even though they were opposites in their styles of research. A. von Humboldt was one of the last universally educated scientists interested in the descriptive comprehension of all natural phenomena, while the mathematician and physicist Gauss used primarily deductive analysis in his research. The study of the Earth's magnetic field was one of their common interests. Their interaction intensified after von Humboldt had invited Gauss to participate in the 7th Assembly of the Society of German Scientists and Physicians, which took place in the fall of 1828 in Berlin. During that time Gauss stayed at von Humboldt's house. The latter had left Paris reluctantly in 1827 to follow a call from the Prussian King Friedrich Wilhelm III to assume the position of a chamberlain at the Berlin court. In his free time, von Humboldt continued his scientific work among which magnetic measurements had a high priority. For this purpose his friend A. Mendelson-Bartholdy (father of the famous composer) had provided a place in his garden where von Humboldt built an iron free wooden cabin for his magnetometers.

Gauss pursued his studies of the Earth's magnetic field quite independently from those of von Humboldt, which caused the latter some irritation. Von Humboldt had thought that he had motivated Gauss to do magnetic measurements when the latter visited him in Berlin in 1828. But Gauss' interests in geomagnetism went back to a time 40 years earlier, as he mentions in a letter of 1833, and in 1806 he had already contemplated a description of the field in terms of spherical harmonics. At the Berlin Assembly, Gauss had met the promising physicist Wilhelm Weber and had arranged that this young man got a professorship in Göttingen in 1831. In the following six years, an intense and highly productive collaboration between Gauss and Weber ensued on all kinds of electromagnetic problems. In the course of this research Gauss developed his method of the absolute determination of magnetic intensity (see *Gauss' determination of absolute intensity*) and built together with Weber appropriate instruments. It now became possible to calibrate instruments locally independently from any other. For the purpose of absolute measurements of the magnetic field and its variations in time at different places on the Earth, Gauss and Weber initiated the "Göttinger Magnetischer Verein."

Alexander von Humboldt was also interested in simultaneous measurements of the geomagnetic field at different geographic locations in order to determine, for instance, whether magnetic storms are of terrestrial origin or depended on the position of the Sun. Already in 1828 he had arranged for coincident measurements in Paris and inside a mine in Freiberg (Saxonia). When he received a glorious reception at the Russian court at St. Petersburg in 1829 at the end of his expedition to Siberia, von Humboldt used the opportunity to suggest the creation of a network of stations throughout the Russian empire for the collection of magnetic and meteorological data. In the following years such stations were indeed installed including one in Sitka, Alaska, which at that time belonged to Russia. Von Humboldt realized that he had to persuade British authorities in order to achieve his goal of a nearly worldwide distribution of stations. In 1836 he wrote to the Duke of Sussex whom he had gotten to know during his student days at the University of Göttingen and who was now the president of the Royal Society. For an English translation of this letter see the paper by Malin and Barraclough (1991). Von Humboldt's recommendations for the establishment of permanent magnetic observatories in Canada, St. Helena, Cape of Good Hope, Ceylon, Jamaica, and Australia were well received. Besides realizing these proposals, the British Government went a step further and organized an expedition to Antarctica under the direction of Sir James Clark Ross with magnetic measurements as one of its main tasks. Based on the observatory data, Sir Edward Sabine (see *Sabine, Edward*) who supervised the network of stations could later establish a connection between magnetic storms and sunspots and demonstrate in particular the correlation between the 11-year sunspot cycle and a corresponding periodicity of magnetic storms.

In the course of his continuing research on geomagnetism, von Humboldt realized the superiority of Gauss' method of measurement and he also often expressed his admiration for Gauss' theoretical work on the representation of the geomagnetic field in terms of potentials and the separation of internal and external sources in particular. From today's point of view it is obvious that Gauss made the more fundamental contributions to the field of geomagnetism. But Alexander von Humboldt's lasting legacy has been the organization of a worldwide cooperation in the gathering of geophysical data with simultaneous measurements at prearranged dates; the influence of which can still be felt today. A. von Humboldt died in 1859 at the age of nearly 90 years. For further details on von Humboldt's scientific endeavours the reader is referred to the book by Botting (1973).

Friedrich Busse

Bibliography

Botting, D., 1973. *Humboldt and the Cosmos*, George Rainbird Ltd., London.

Malin, S.R.C. and Barraclough, D.R., 1991. Humboldt and the Earth's magnetic field *Quarterly Journal of the Royal Astronomical Society* **32**: 279–293.

Cross-references

Gauss, Carl Friedrich (1777–1855)
Gauss' Determination of Absolute Intensity
Sabine, Edward (1788–1883)

HUMBOLDT, ALEXANDER VON AND MAGNETIC STORMS

Baron Alexander von Humboldt (September 14, 1769–May 6, 1859) was a great German naturalist and explorer, universal genius and cosmopolitan, scientist, and patron. Application of his experience and knowledge gained through his travels and experiments transformed contemporary western science. He is widely acknowledged as the founder of modern geography, climatology, ecology, and oceanography. This article focuses on Alexander von Humboldt's contribution to the science of geomagnetism (Busse, 2004), which is not so widely known.

Historical background on geomagnetism

With the publication of *De Magnete* by William Gilbert in A.D. 1600, the Earth itself was seen as a great magnet, and a new branch of physics, geomagnetism, was born. The first map of magnetic field declination was made by Edmund Halley in the beginning of 18th century.

Alexander von Humboldt prepared a chart of "isodynamic zones" in 1804, based on his measurements made during his voyages through the Americas (1799–1805). He swung a dip needle in the magnetic meridian plane, and observed the number of oscillations during 10 min. He noticed that the number of oscillations were at maximum near the magnetic equator, and decreased to the north and to the south. This indicated a regular decrease of the total magnetic intensity from the poles to the equator. The publication of this result stimulated further investigations on the Earth's magnetic field (Chapman and Bartels, 1940). Von Humboldt, together with Guy Lussac, improvised the method to measure the horizontal intensity by observing the time of oscillation of a compass needle in the horizontal plane. They took the first measurements of the relative horizontal intensity of the Earth's magnetic field on a journey to Italy in 1807.

From May 1806 to June 1807 in Berlin, Humboldt and a colleague observed the local magnetic declination every half hour from midnight to morning. On December 21, 1806, for six consecutive hours, von Humboldt observed strong magnetic deflections and noted the presence of correlated northern lights (aurora) overhead. When the aurora disappeared at dawn, the magnetic perturbations disappeared as well. Von Humboldt concluded that the magnetic disturbances on the ground and the auroras in the polar sky were two manifestation of the same phenomenon (Schröder, 1997). He gave this phenomenon involving large-scale magnetic disturbances (possibly already observed by George Graham) the name "Magnetische Ungewitter," or magnetic storms (von Humboldt, 1808). The worldwide net of magnetic observatories later confirmed that such "storms" were indeed worldwide phenomena.

Alexander von Humboldt organized the first simultaneous observations of the geomagnetic field at various locations throughout the world after his return from his South American journey. Wilhelm Weber, Karl F. Gauss, and he organized the Göttingen Magnetische Verein (Magnetic Union); and from 1836–1841, simultaneous observations of the Earth's magnetic field were made in nonmagnetic huts at up to 50 different locations, marking the beginning of the magnetic observatory system. Through his diplomatic contacts, Alexander von Humboldt was instrumental in the establishment of a number of magnetic observatories around the world, especially in Britain, Russia, and in countries then under the British and Russian rule, e.g., cities such as Toronto, Sitka, and Bombay, etc. He constructed his own iron-free magnetic observatory in Berlin in 1832. These observatories have played an important role in the development of the science of geomagnetism. With the beginning of the space age, our knowledge about the near-Earth space environment, the magnetosphere-ionosphere system in general, and geomagnetic storms in particular, has improved dramatically.

Intense geomagnetic storms and their causes

A geomagnetic storm is characterized by a main phase during which the horizontal component of the Earth's low-latitude magnetic fields are significantly depressed over a time span of one to a few hours. This is followed by a recovery phase, which may extend for $\sim$10 h or more (Rostoker, 1997). The intensity of a geomagnetic storm is measured in terms of the disturbance storm-time index (Dst). Magnetic storms with Dst < -100 nT are called intense and those with Dst < -500 nT are called superintense. Geomagnetic storms occur when solar wind-magnetosphere coupling becomes intensified during the arrival of fast moving solar *ejecta* like interplanetary coronal mass ejections (ICMEs) and fast streams from the coronal holes (Gonzalez *et al.*, 1994) accompanied by long intervals of southward interplanetary magnetic field (IMF) as in a "magnetic cloud" (Klein and Burlaga, 1982). It is now well established that the major mechanism of energy transfer from the solar wind to the Earth's magnetosphere is magnetic reconnection (Dungey, 1961). The efficiency of the reconnection process is considerably enhanced during southward IMF intervals (Tsurutani and Gonzalez, 1997), leading to strong plasma injection from the magnetotail towards the inner magnetosphere causing intense auroras at high-latitude nightside regions. Further, as the magnetotail plasma gets injected into the nightside magnetosphere, the energetic protons drift to the west and electrons to the east, forming a ring of current around the Earth. This current, called the "ring current," causes a diamagnetic decrease in the Earth's magnetic field measured at near-equatorial magnetic stations (see Figure H14 for the results from one such station). The decrease in the equatorial magnetic field strength, measured by the Dst index, is directly related to the total kinetic energy of the ring current particles (Dessler and Parker, 1959; Sckopke, 1966); thus the Dst index is a good measure of the energetics of the magnetic storm. The Dst index itself is influenced by the interplanetary parameters (Burton *et al.*, 1975).

One obviously cannot directly determine the solar/interplanetary causes of storm events prior to the space age. However, based on the recently gained knowledge on solar, interplanetary, and magnetospheric physics, one can make these determinations by a process of elimination. For example, by examining the profile of magnetic storms using ground magnetic field data, storm generation mechanisms can be identified (Tsurutani *et al.*, 1999). From the ground magnetometer data shown in Figure H14, reports on the related solar flare event (Carrington, 1859; Hodgson, 1859), and reports on the concomitant aurora (taken from newspapers and private correspondence, Kimball,

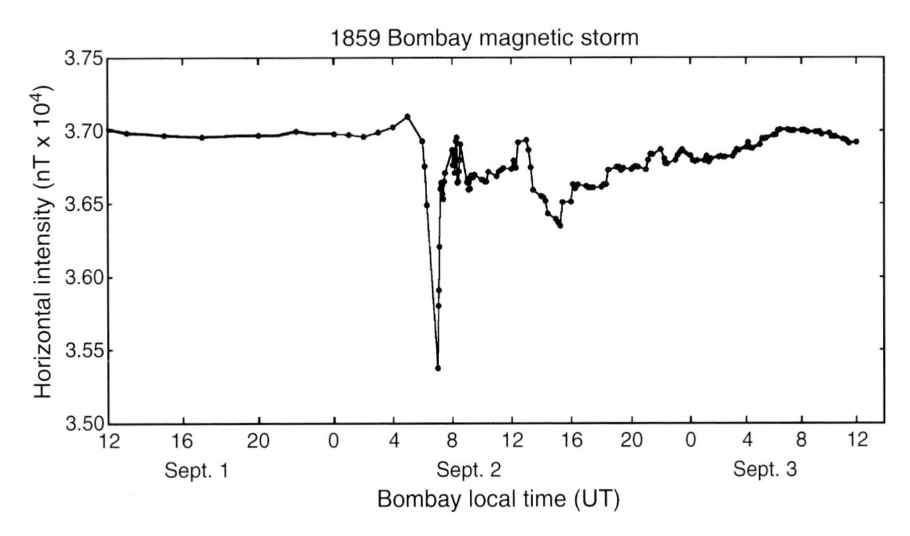

Figure H14 The Colaba (Bombay) magnetogram for the September 1–2, 1859 geomagnetic storm. The peak near 0400 UT September 2, is due to the storm sudden commencement (SSC) caused probably by the shock ahead of the magnetic cloud. This was followed by the storm "main phase" which lasted for about one hour and a half. Taken from Tsurutani *et al.* (2003).

1960), Tsurutani *et al.* (2003) were able to deduce that an exceptionally fast (and intense) magnetic cloud was the interplanetary cause of the superintense geomagnetic storm of September 1–2, 1859 with a Dst $\sim$–1760 nT. This large value of Dst is consistent with the decrease of $\Delta H = 1600 \pm 10$ nT recorded at Colaba (Bombay), India (Figure H14). The supposition that the intense southward IMF was due to a magnetic cloud was surmised by the simplicity and short duration of the storm in the ground magnetic field data. Main phase compound events or "double storms" (Kamide *et al.*, 1998) can be ruled out by the (simple) storm profile. Compound stream events (Burlaga *et al.*, 1987) can also be eliminated by the storm profile. The only other possibility that might be the cause of the storm is sheath fields associated with an ICME. This can be ruled out because the compression factor of magnetic fields following fast shocks is only approximately four times (Kennel *et al.*, 1985). Thus with quiet interplanetary fields being typically $\sim$3 to 10 nT, the compressed fields would be too low to generate the inferred interplanetary and magnetospheric electric fields for the storm. Thus by a process of elimination the interplanetary fields that caused this superintense storm have been determined to be part of a fast, intense magnetic cloud.

Geomagnetic storms produce severe disturbances in Earth's magnetosphere and ionosphere, creating so-called adverse space weather conditions. They pose major threats to space- and ground-based technological systems on which modern society is becoming increasingly dependent. The intense magnetic storms during October 29–31, 2003, the so-called "Halloween storms", associated with the largest X-ray solar flare of the solar cycle 23, caused damage to 28 satellites, ending the operational life of two, disturbed flight routes of some airlines, telecommunications problems, and power outage in Sweden. These Halloween storms with Dst $\sim$–400 nT were about four times less intense than the superintense storm of 1859. One can imagine the loss to society if a magnetic storm similar to the superintense storm of 1859 were to occur today!

Acknowledgments

GSL would like to thank Prof. Y. Kamide for the kind hospitality during his stay at STEL, Nagoya University, Japan. Portions of the research for this work were performed at the Jet Propulsion Laboratory, California Institute of Technology, under contract with the National Aeronautics and Space Administration.

G.S. Lakhina, B.T. Tsurutani, W.D. Gonzalez, and S. Alex

Bibliography

Burlaga, L.F., Behannon, K.W., and Klein, L.W. 1987. Compound streams, magnetic clouds, and major geomagnetic storms. *Journal of Geophysical Research*, **92**: 5725.

Burton, R.K., McPherron, R.L., and Russell, C.T. 1975. An empirical relationship between interplanetary conditions and Dst. *Journal of Geophysical Research*, **80**: 4204.

Busse, F., 2004. Alexander von Humboldt, this volume.

Carrington, R.C., 1859. Description of a singular appearance seen in the Sun on September 1, 1859. *Monthly Notices of the Royal Astronomical Society,* XX, **13**.

Chapman, S. and Bartels, J., 1940. *Geomagnetism*, vol. II, Oxford University Press, New York, pp. 913–933.

Dessler, A.J., and Parker, E.N., 1959. Hydromagnetic theory of magnetic storms. *Journal of Geophysical Research*, **64**: 2239.

Dungey, J.W., 1961. Interplanetary magnetic field and the auroral zones. *Physical Research Letters*, **6**: 47.

Gonzalez, W.D., Joselyn, J.A., Kamide, Y., Kroehl, H.W., Rostoker, G., Tsurutani, B.T., and Vasyliunas, V.M., 1994. What is a geomagnetic storm? *Journal of Geophysical Research*, **99**: 5771.

Hodgson, R.,1859. On a curious appearance seen in the Sun. *Monthly Notices of the Royal Astronomical Society London,* XX, **15**.

Kamide, Y., Yokoyama, N., Gonzalez, W., Tsurutani, B.T., Daglis, I.A., Brekke, A., and Masuda, S., 1998. Two-step development of geomagnetic storms. *Journal of Geophysical Research*, **103**: 6917.

Kennel, C.F., Edmiston, J.P., and Hada, T. 1985. A quarter century of collisionless shock research. In Stone, R.G. and Tsurutani, B.T. (eds.), *Collisionless Shocks in the Heliosphere: A Tutorial Review.* Washington, DC: American. Geophysical Union, Vol. 34, p. 1.

Kimball, D.S., 1960. A study of the aurora of 1859. *Sci. Rpt. 6, UAG-R109*, University of Alaska.

Klein, L.W. and Burlaga, L.F., 1982. Magnetic clouds at 1 AU. *Journal of Geophysical Research*, **87**: 613.

Rostoker, G., 1997. Physics of magnetic storms. In Tsurutani, B.T., Gonzalez, W.D., Kamide, Y., and Arballo, J.K. (eds.), *Magnetic Storms.* Geophysical Monograph 98, AGU, Washington DC, p. 149.

Schröder, W., 1997. Some aspectss of the earlier history of solar-terrestrial physics. *Planetary and Space Science*, **45**: 395.

Sckopke, N., 1966. A general relation between the energy of trapped particles and the disturbance field near the Earth. *Journal of Geophysical Research*, **71**: 3125.

Tsurutani, B.T. and Gonzalez, W.D., 1997. Interplanetary causes of magnetic storms: a review. In Tsurutani, B.T., Gonzalez, W.D.,

Kamide, Y., and Arballo, J.K. (eds.), *Magnetic Storms.* Geophysical Monograph 98, AGU, Washington DC, p. 77.
Tsurutani, B.T., Kamide, Y., Arballo, J.K., Gonzalez, W.D., and Lepping, R.P., 1999. Interplanetary causes of great and superintense magnetic storms, *Physics and Chemistry of the Earth*, **24**: 101.
Tsurutani, B.T., Gonzalez, W.D., Lakhina, G.S., and Alex, S., 2003. The extreme magnetic storm of September 1–2, 1859. Journal of Geophysical Research, 108, A7, 1268, doi:10.1029/2002JA009504.
von Humboldt, A., 1808. *Annalen der Physik*, **29**: 425.

Cross-references

Gilbert, William (1544–1603)
Halley, Edmond (1656–1742)

I

IAGA, INTERNATIONAL ASSOCIATION OF GEOMAGNETISM AND AERONOMY

History

The origin of the International Association of Geomagnetism and Aeronomy (IAGA) can be traced to the Commission for Terrestrial Magnetism and Atmospheric Electricity, part of the International Meteorological Organization, which was established in 1873. The Commission planned the geomagnetic observation campaign for the First International Polar Year, 1882–1883. Following the end of World War I, the International Research Council (IRC) was established to promote science through international cooperation. At a meeting of the IRC in July 1919, the International Geodetic and Geophysical Union was formed, with Terrestrial Magnetism and Electricity as Section D, and with its leadership provided by the International Meteorological Organization. In 1930, the International Geodetic and Geophysical Union agreed to cooperate with the International Meteorological Organization to organize a Second International Polar Year, 50 years after the first. The Union changed its name to the International Union of Geodesy and Geophysics (IUGG), its Sections were renamed Associations, and the International Association of Terrestrial Magnetism and Electricity (IATME) came into existence.

At the Brussels IUGG General Assembly in 1951, upper atmosphere scientists lobbied to have their interests recognized within IATME. Following the Brussels Assembly, *Sydney Chapman* (*q.v.*), who had previously suggested replacing the term "Terrestrial Magnetism" by "Geomagnetism" coined the term "Aeronomy" and suggested adoption of the name International Association of Geomagnetism and Aeronomy. The expansion of the scope of IATME was ratified at the General Assembly in Rome in 1954, the new title IAGA was agreed, and aeronomy was defined as "the science of the upper atmospheric regions where dissociation and ionization are important."

In 1950, Lloyd Berkner proposed a Third International Polar Year to provide motivation to re-equip geophysical observatories, many of which had been damaged or destroyed during World War II. The International Council of Scientific Unions adopted and broadened the scope of the idea, and designated July 1957 to December 1958 the International Geophysical Year (IGY). Observations were made in a number of IUGG discipline areas, including geomagnetism. The IGY resulted in a leap forward in geophysics through coordinated observational campaigns and through the establishment of new observatories and the World Data Center system. It was the beginning of the space age, and marked the start of the modern era for IAGA science, which has expanded to include solar-terrestrial interactions and studies of the Sun and the planets.

Present structure and organization

In 2005, IAGA is one of seven scientific associations of the IUGG. It is an international nongovernmental organization deriving the majority of its funding from the IUGG member nations. An Executive Committee, elected by member countries, runs the Association, and the scientific work is organized through a structure defined in the Association's Statutes and By-Laws:

Division I: Internal Magnetic Fields;
Division II: Aeronomic Phenomena;
Division III: Magnetospheric Phenomena;
Division IV: Solar Wind and Interplanetary Field;
Division V: Geomagnetic Observatories, Surveys, and Analyses.
Interdivisional Commission on History
Interdivisional Commission on Developing Countries

Several of the Divisions have Working Groups in specialist topic areas and establish Task Groups to deal with specific issues.

IAGA's purpose and linkages

IAGA's "mission" is defined by the Association's first Statute, which states that IAGA should promote scientific studies of international interest and facilitate international coordination and discussion of research. An important defining characteristic of IAGA is, therefore, to encourage inclusiveness in the scientific community, making excellent science accessible to scientists worldwide. There is a particular commitment to the less developed countries, and to the free exchange of scientific data and information.

One way in which IAGA achieves its objectives is through the organization of meetings. IAGA Scientific Assemblies are held every 2 years, and in conjunction with IUGG General Assemblies every 4 years. Smaller scale meetings and specialized workshops make IAGA science accessible to a wider audience and help younger scientists and scientists from developing countries to accelerate their learning. IAGA sponsors several such meetings each year. They are held in all parts of the world, an important factor in enabling attendance by scientists with limited resources.

Meetings enable scientific results to be presented and debated, new ideas generated, and collaborations to be established. The major Assemblies are also used to conduct divisional and working group business meetings, and matters of interest to IAGA are presented at open meetings convened for discussions between the IAGA Executive, the official delegates from member countries, and individual scientists. Resolutions are adopted as a formal means for IAGA to express views on scientific matters. Support is often given to scientific initiatives under consideration by national or international agencies that will benefit IAGA science. For example, a series of IAGA Resolutions supported the initiation of and extensions to the International Equatorial Electrojet Year (IEEY) project (1991–1994). The project not only produced good science, but also resulted in investment in observational facilities at low-latitudes, benefiting scientists in the less developed world. Although the IEEY has finished, the impetus it gave to research in the equatorial regions continues. Similarly, IAGA has supported the Decade of Geopotential Research, which appears likely to achieve its goal of securing uninterrupted geomagnetic field satellite survey measurements spanning a decade (including the *Ørsted* and *CHAMP* missions).

IAGA science is also promoted through collaboration with other bodies with similar interests. There are links with the other IUGG Associations and with Inter-Association bodies including *Studies of the Earth's Deep Interior* (SEDI) (*q.v.*) and the Working Group on Electric and Magnetic Studies on Earthquakes and Volcanoes (EMSEV). There are formal contacts for liaison with the International Lithosphere Program (ILP), the Scientific Committee on Antarctic Research (SCAR), the Committee on Space Research (COSPAR), and the Scientific Committee on Solar Terrestrial Physics (SCOSTEP).

IAGA cooperates with the World Data Center system on the definition of geomagnetic data exchange formats and management and preservation of analog and digital databases. The Association has provided strong support to INTERMAGNET (*q.v.*), the international program promoting the modernization of magnetic observatory practice and the distribution of data in near real time. IAGA advises bodies such as the International Organization for Standardization (ISO).

IAGA science in the 21st century

IAGA science, because of the pervasiveness of the geomagnetic field and its interactions with charged particles and electrically conducting materials, is useful for studies of properties and processes in practically all parts of the solid Earth, the atmosphere, and the surrounding space environment. As well as covering a vast range of length scales, IAGA science covers timescales from seconds to billions of years. Modern-day observatories record rapid variations during magnetic storms caused by the interaction of the solar wind with the magnetosphere; the imprinting of the paleomagnetic field in rocks provides records of geodynamic changes on geological timescales.

While IAGA provides an international focus for fundamental research resulting in advances in understanding in specialist areas, national and international funding for research often focuses on issues of societal concern. IAGA science is providing answers to many important questions, and through its links to other bodies and projects the Association is able to foster the building of the interdisciplinary teams required to address complex problems of interest to society.

For instance, a natural goal for IAGA scientists is to be able to understand and model the whole Sun-Earth system including the complex interactions and feedbacks controlling the transfer of energy momentum and matter between parts of the system. Research in this area is proving relevant to the problem of how to disentangle natural from anthropogenic causes of climate change. This area of science also underpins the understanding of how "space weather" conditions affect the risk to technological systems and human activities on the ground and in space. For example, during magnetic storms, electrical power distribution grids, radio communications, GPS accuracy, and satellite operations can be adversely affected. Also in the geohazards area, EMSEV is charged with establishing firm scientific understanding of the generation mechanisms of any signals that may help to mitigate the effects of earthquakes and volcanoes.

IAGA is responsible for the production of the International Geomagnetic Reference Field, used in a variety of scientific and "real world" applications, including navigation, and hydrocarbons exploration and production. The Association is responsible for the definition of the most widely used magnetic activity indices, and works closely with the International Service for Geomagnetic Indices, the body responsible for their production and distribution.

IAGA has a long history, and its science remains vigorous and relevant. Rapid advances in scientific understanding are resulting from improved instruments, better observations and data analysis techniques, the wealth of satellite data now routinely available, and the power of modern computer technology. As the 50th anniversary of the IGY approaches, IAGA is promoting the concept of an "Electronic Geophysical Year" (eGY), for 2007–2008, taking advantage of the modern capability to link distributed computing resources to multiple remote sources of data and modeling codes to address scientific problems. This initiative is in line with the Association's mission to promote international scientific cooperation and collaboration, and has the potential to advance the ability of scientists in developing countries to participate in leading-edge research.

(The principal point of contact with the Association is the IAGA Secretary General, and IAGA communicates with its members and the public through its Web site and through issues of *IAGA News*.)

David Kerridge

Bibliography

Naoshi Fukushima, 1995. History of the International Association of Geomagnetism and Aeronomy (IAGA). *IUGG Chronicle*, **226**: 73–87.

The IAGA Web site: http://www.iugg.org/IAGA/

Cross-references

CHAMP
Chapman, Sydney (1888–1970)
Ørsted
SEDI

IDEAL SOLUTION THEORY

Consider two different substances; mix them together and in general they will form a *solution*, like sugar and coffee, for example. We call *solvent* the substance present in the largest quantity (coffee), and *solute* the other (sugar). In general solutions may have more than one solute, and/or more than one solvent, but for simplicity we will focus here only on binary mixtures.

The *behavior* of solutions can be understood in terms of the *chemical potential* μ_i, which represents the constant of proportionality between the energy of the system and the amount of the specie i (Wannier, 1966):

$$\mu_i = \left(\frac{\partial E}{\partial N_i}\right)_{S,V} \qquad \text{(Eq. 1)}$$

where E is the internal energy of the system, S is the entropy, V is the volume, and N_i is the number of particles of the specie i. Alternative equivalent definitions of the chemical potential are (Wannier, 1966; Mandl, 1997):

$$\mu_i = \left(\frac{\partial F}{\partial N_i}\right)_{T,V} = \left(\frac{\partial G}{\partial N_i}\right)_{T,p} = -T\left(\frac{\partial S}{\partial N_i}\right)_{E,V} \qquad \text{(Eq. 2)}$$

where F and G are the Helmholtz and Gibbs free energies of the system, T is the temperature, and p is the pressure.

We recall the statistical mechanics definition of the Helmholtz free energy for a classical system (Frenkel, 1996):

$$F = -k_B T \ln \frac{1}{\Lambda^{3N} N!}\int_V \mathrm{d}\mathbf{R}_1 \ldots \int_V \mathrm{d}\mathbf{R}_N \mathrm{e}^{-U(\mathbf{R}_1,\ldots,\mathbf{R}_N;T)/k_B T}, \qquad \text{(Eq. 3)}$$

where $U(\mathbf{R}_1, \ldots, \mathbf{R}_N; T)$ is the potential energy which depends on the positions $(\mathbf{R}_1, \ldots, \mathbf{R}_N)$ of all the particles in the system and on T, k_B is the Boltzmann constant and $\Lambda = h/(2\pi M k_B T)^{1/2}$ is the thermal wavelength, with M being the nuclear mass and h the Plank's constant.

Consider now a solution with N_A particles of solvent A and N_X particles of solute X, with $N = N_A + N_X$. The Helmholtz free energy of this system is:

$$F = -k_B T \ln \frac{1}{\Lambda_A^{3N_A} \Lambda_X^{3N_X} \mu_A! N_X!} \int_V d\mathbf{R}_1 \ldots \int_V d\mathbf{R}_N e^{-U(\mathbf{R}_1,\ldots,\mathbf{R}_N;T)/k_B T}. \quad \text{(Eq. 4)}$$

According to (Eq. 2), we have:

$$\mu_X = \left(\frac{\partial F}{\partial N_X}\right)_{T,V} = F(N_A, N_X + 1) - F(N_A, N_X), \quad \text{(Eq. 5)}$$

which can be evaluated using (Eq. 4)

$$\mu_X = -k_B T \ln \frac{1}{\Lambda_X^3 (N_X + 1)} \frac{\int_V d\mathbf{R}_1 \ldots \int_V d\mathbf{R}_N \int_V d\mathbf{R}_{N+1} e^{-U(\mathbf{R}_1,\ldots,\mathbf{R}_N,\mathbf{R}_{N+1};T)/k_B T}}{\int_V d\mathbf{R}_1 \ldots \int_V d\mathbf{R}_N e^{-U(\mathbf{R}_1,\ldots,\mathbf{R}_N;T)/k_B T}}. \quad \text{(Eq. 6)}$$

The ratio of the two integrals is an extensive quantity, but μ_X is an intensive quantity; therefore it is useful to rewrite the expression as follows:

$$\mu_X = -k_B T \ln \frac{N}{(N_X + 1)} \left\{ \frac{1}{N \Lambda_X^3} \frac{\int_V d\mathbf{R}_1 \ldots \int_V d\mathbf{R}_N \int_V d\mathbf{R}_{N+1} e^{-U(\mathbf{R}_1,\ldots,\mathbf{R}_N,\mathbf{R}_{N+1};T)/k_B T}}{\int_V d\mathbf{R}_1 \ldots \int_V d\mathbf{R}_N e^{-U(\mathbf{R}_1,\ldots,\mathbf{R}_N;T)/k_B T}} \right\} \quad \text{(Eq. 7)}$$

so that the value in curly brackets is now independent on system size. By setting $c_X = N_X/N$(which in the limit of large N and N_X is the same as$(N_X + 1)/N)$) and

$$\tilde{\mu}_X = -k_B T \ln \left\{ \frac{1}{N \Lambda_E^3} \frac{\int_V d\mathbf{R}_1 \ldots \int_V d\mathbf{R}_N \int_V d\mathbf{R}_{N+1} e^{-U(\mathbf{R}_1,\ldots,\mathbf{R}_N,\mathbf{R}_{N+1};T)/k_B T}}{\int_V d\mathbf{R}_1 \ldots \int_V d\mathbf{R}_N e^{-U(\mathbf{R}_1,\ldots,\mathbf{R}_N;T)/k_B T}} \right\} \quad \text{(Eq. 8)}$$

we can rewrite the chemical potential in our final expression:

$$\mu_X(p, T, c_X) = k_B T \ln c_X + \tilde{\mu}_X(p, T, c_X). \quad \text{(Eq. 9)}$$

The first term of (Eq. 9) depends only on the number of particles of solute present in the solution, while the second term is also responsible for all possible chemical interactions. For small concentration of solute we can make a Taylor expansion of $\tilde{\mu}_X$:

$$\tilde{\mu}_X = \mu_X^0 + \lambda c_X + o(c_X^2), \quad \text{(Eq. 10)}$$

where $\lambda = (\partial \tilde{\mu}_X / \partial c_X)_{p,T}$. If the solution is so dilute that the particles of the solute do not interact with each other, we can stop the expansion to the first term:

$$\mu_X(p, T, c_X) = k_B T \ln c_X + \mu_X^0(p, T). \quad \text{(Eq. 11)}$$

We define *ideal solution* a system in which (Eq. 11) is strictly satisfied.

To find an expression for the chemical potential of the solvent we employ the Gibbs-Duhem equation, which for a system at constant pressure and constant temperature reads (Wannier, 1966):

$$\sum_i N_i d\mu_i = 0. \quad \text{(Eq. 12)}$$

In particular, in our two-components system the Gibbs-Duhem equation implies:

$$c_A d\mu_A + c_X d\mu_X = 0, \quad \text{(Eq. 13)}$$

which gives (Alfè *et al.*, 2002a):

$$\mu_A(p, T, c_X) = \mu_A^0(p, T) + (k_B T + \lambda_X(p, T)) \ln(1 - c_X) + \lambda_X(p, T) c_X + O(c_X^2), \quad \text{(Eq. 14)}$$

where μ_A^0 is the chemical potential of the pure solvent. To linear order in c_X, this gives:

$$\mu_A(p, T, c_X) = \mu_A^0(p, T) - k_B T c_X + O(c_X^2). \quad \text{(Eq. 15)}$$

Notice that this expression for μ_A is not restricted to *ideal solutions*. Though μ_A^0 is the chemical potential of the pure solvent, observe that μ_X^0 is *not* the chemical potential of the pure solute, unless the validity of (Eq. 11) extends all the way up to $c_X = 1$.

Volume of mixing

It is often interesting to study the change of volume of a solution as a function of the concentration of the solute. To this end, it is useful to express the volume of the system as the partial derivative of the Gibbs free energy with respect to pressure, taken at constant temperature and number of particles:

$$V = (\partial G / \partial p)_{T, N_X, N_A}. \quad \text{(Eq. 16)}$$

If we now add to the system one particle of solvent at constant pressure, the total volume changes by v_A, and becomes $V + v_A$. We call v_A the *partial molar volume* of the solvent. The total Gibbs free energy is $G + \mu_A$, so that according to (Eq. 16) $v_A = (\partial \mu_A / \partial p)_{T, N_X, N_A} = (\partial \mu_A / \partial p)_{T, c_X}$, where the last equality stems from the fact that μ_A only depends on N_X and N_A through the molar fraction c_X (we assume here that c_X does not change when we add one particle of solvent to the system, this is obviously true if the number of atoms of solvent N_A is already very large). The partial volume in general depends on c_X, p, and T, but under the assumption of ideality $v_A = (\partial \mu_A^0 / \partial p)_{T, c_X}$, and it depends only on p and T. In an ideal solution v_A is the same as in the pure solvent.

Similarly, the partial molar volume of the solute is: $v_X = (\partial \mu_X / \partial p)_{T, c_X}$, which becomes independent on c_X under the *assumption* of ideality. Notice that this is *not* in general equal to the partial volume of the pure solute.

As an illustration of the applicability of the *ideal solution* approximation, I mention the recent first-principles calculations of the density of the Earth's liquid outer core as a function of the concentration of light impurities like sulfur, silicon, and oxygen in liquid iron. As reported in (Gubbins *et al.*, 2004), explicit first-principles calculations of the density of the core were not able to resolve any departure from the prediction of *ideal solution theory*. However, we shall see below that for other properties *ideal solution theory* is not necessarily a good working hypothesis.

Solid-liquid equilibrium

We want to study now the conditions that determine the equilibrium between solid and liquid, and in particular how the solute partitions between the two phases. Thermodynamic equilibrium is reached when the Gibbs free energy of the system is at its minimum (Wannier, 1966; Mandl, 1997), and therefore, $0 = dG = d(G^l + G^s)$, where superscripts s and l indicate quantities in the solid and in the liquid, respectively. In a multicomponent system, the Gibbs free energy can be expressed in terms of the chemical potentials of the species present in the system (Wannier, 1966; Mandl, 1997):

$$G = \sum_i N_i \mu_i. \quad \text{(Eq. 17)}$$

Using (Eq. 17) and the Gibbs-Duhem equation (12), we obtain:

$$dG = \sum_i \mu_i dN_i. \quad \text{(Eq. 18)}$$

If the system is isolated, particles can only flow between the solid and the liquid, and we have $dN_i^s = -dN_i^l$, which implies:

$$dG = \sum_i dN_i(\mu_i^l - \mu_i^s). \quad \text{(Eq. 19)}$$

If $\mu_i^l < \mu_i^s$ there will be a flow of particles from the solid to the liquid region ($dN_i > 0$), so that the Gibbs free energy of the system is lowered. The opposite will happen if $\mu_i^l > \mu_i^s$. The flow stops at equilibrium, which is therefore reached when $\mu_i^l = \mu_i^s$. In particular, in our two-components system, the equilibrium between solid and liquid implies that the chemical potentials of both solvent and solute are equal in the solid and liquid phases:

$$\mu_X^s(p, T_m, c_X^s) = \mu_X^l(p, T_m, c_X^l);$$
$$\mu_A^s(p, T_m, c_X^s) = \mu_A^l(p, T_m, c_X^l); \quad \text{(Eq. 20)}$$

where T_m is the melting temperature of the solution at pressure p. Using (Eq. 9), we can rewrite the first of the two equations above as:

$$\tilde{\mu}_X^s(p, T_m) + kT_m \ln c_X^s = \tilde{\mu}_X^l(p, T_m) + kT_m \ln c_X^l; \quad \text{(Eq. 21)}$$

from which we obtain an expression for the ratio of concentrations of solute between the solid and the liquid:

$$c_X^s / c_X^l = \exp\{[\tilde{\mu}_X^l(p, T_m) - \tilde{\mu}_X^s(p, T_m)]/kT_m\}. \quad \text{(Eq. 22)}$$

In general $\tilde{\mu}_X^l < \tilde{\mu}_X^s$, because the greater mobility of the liquid can usually better accommodate particles of solute, and therefore their energy (chemical potential) is lower. This means that the concentration of the solute is usually smaller in the solid.

Equation (22) was used by Alfè *et al.* (2000, 2002a,b) to put constraints on the composition of the Earth's core. The constraints came from a comparison of the calculated density contrast at inner core boundary, and that obtained from seismology, which is between $4.5\% \pm 0.5\%$ (Shearer and Masters, 1990) and $6.7\% \pm 1.5\%$ (Masters and Gubbins, 2003). This density contrast is significantly higher than that due to the crystallization of pure iron, and therefore must be due to the partitioning of light elements between solid and liquid. This partitioning for some candidate impurities can be obtained by calculating $\tilde{\mu}_X^l$ and $\tilde{\mu}_X^s$. Alfè *et al.* (2000, 2002a,b) considered sulfur, silicon, and oxygen as possible impurities, and using first-principles simulations, in which the interactions between particles were treated using quantum mechanics, obtained the partitions for each impurity. The calculations showed that for both sulfur and silicon $\tilde{\mu}_X^l$ and $\tilde{\mu}_X^s$ are very similar, which means that c_X^s and c_X^l are also very similar, according to (Eq. 22). As a result, the density contrast of a Fe/S or a Fe/Si system is not much different from that of pure Fe, and still too low when compared with the seismological data. By contrast, for oxygen $\tilde{\mu}_e^l$ and $\tilde{\mu}_X^s$ are very different, and the partitioning between solid and liquid is very large. This results in a much too large density contrast, which also does not agree with the *seismological* data. The conclusion from these calculations was that none of these binary mixtures can be viable for the core. The density contrast can of course be explained by ternary or quaternary mixtures. Assuming no cross-correlated effects between the chemical potentials of different impurities, Alfè *et al.* (2002a,b) proposed an inner core containing about 8.5% of sulfur and/or silicon and almost no oxygen, and an outer core containing about 10% of sulfur and/or Si, and an additional 8% of oxygen.

It is worth noting that these calculations (Alfè *et al.*, 2000, 2002a,b) also showed that the dependence of $\tilde{\mu}_X^s$ and $\tilde{\mu}_X^l$ on c_X^s and c_X^l was significant, and therefore departure from *ideal solution* behavior. If nonideal effects were ignored, the calculations would result in an outer core containing about 12% of sulfur and/or silicon, and about 6% of oxygen.

Shift of freezing point

The partitioning of the solute between the solid and the liquid is generally responsible for a change in the melting temperature of the mixture with respect to that of the pure solvent. To evaluate this, we expand the chemical potential of the solvent around the melting temperature of the pure system, T_m^0:

$$\mu_A(p, T_m, c_X) = \mu_A(p, T_m^0, c_X) - s_A^0 \delta T + \cdots \quad \text{(Eq. 23)}$$

where $\delta T = (T_m - T_m^0)$, and $s_A^0 = -(\partial \mu_A / \partial T)_{T=T_m^0}$ is the entropy of the pure solvent at T_m^0. We now impose continuity across the solid-liquid boundary:

$$\mu_A^{0s}(p, T_m^0) - s_A^{0s}\delta T - kT_m c_X^s = \mu_A^{0l}(p, T_m^0) - s_A^{0l}\delta T - kT_m c_X^l, \quad \text{(Eq. 24)}$$

where we have considered only the linear dependence of μ_A on c_X (See Eq. 15). Noting that $\mu_A^{0s}(p, T_m^0) = \mu_A^{0l}(p, T_m^0)$ we have

$$\delta T = \frac{kT_m}{s_A^{0l} - s_A^{0s}}(c_X^s - c_X^l). \quad \text{(Eq. 25)}$$

Since usually $c_X^s < c_X^l$, there is generally a *depression* of the freezing point of the solution.

Using (Eq. 25), Alfè *et al.* (2002a,b) estimated a depression of about 600–700 K of the melting temperature of the core mixture with respect to the melting temperature of pure Fe, and they suggested an inner core boundary temperature of about 5600 K.

Dario Alfè

Bibliography

Alfè, D., Gillan, M.J., and Price, G.D., 2000. Constraints on the composition of the Earth's core from *ab initio* calculations. *Nature*, **405**: 172–175.

Alfè, D., Gillan, M.J., and Price, G.D., 2002a. *Ab initio* chemical potentials of solid and liquid solutions and the chemistry of the Earth's core. *Journal of Chemical Physics*, **116**: 7127–7136.

Alfè, D., Gillan, M.J., and Price, G.D., 2002b. Composition and temperature of the Earth's core constrained by combining *ab initio* calculations and seismic data. *Earth and Planetary Science Letters*, **195**: 91–98.

Frenkel, D., and Smit, B., 1996. *Understanding Molecular Simulation*. San Diego, CA: Academic Press.

Gubbins, D. *et al.*, 2004. Gross thermodynamics of two-component core convection. *Geophysical Journal International*, **157**: 1407–1414.

Mandl, F., 1997. *Statistical Physics*, 2nd edn. New York: John Wiley & Sons.

Masters, T.G., and Gubbins, D., 2003. On the resolution of the density within the Earth. *Physics of the Earth and Planetary Interiors*, **140**: 159–167.
Shearer, P., and Masters, T.G., 1990. The density and shear velocity contrast at the inner core boundary. *Geophysical Journal International*, **102**: 491–498.
Wannier, G.H., 1966. *Statistical Physics*. New York: Dover Publications.

IGRF, INTERNATIONAL GEOMAGNETIC REFERENCE FIELD

History of the IGRF

The IGRF is an internationally agreed global spherical harmonic model of the Earth's magnetic field whose sources are in the Earth's core (see *Harmonics, spherical* and *Main field modeling*). It is revised every 5 years under the auspices of the International Association of Geomagnetism and Aeronomy (see *IAGA, International Association of Geomagnetism and Aeronomy*).

The concept of an IGRF grew out of discussions concerning the presentation of the results of the World Magnetic Survey (WMS). The WMS was a deferred element in the program of the International Geophysical Year, which, during 1957–1969, encouraged magnetic surveys on land, at sea, in the air, and from satellites and organized the collection and analysis of the results. At a meeting in 1960, the Committee on World Magnetic Survey and Magnetic Charts of IAGA recommended that, as part of the WMS program, a spherical harmonic analysis be made using the results of the WMS, and this proposal was accepted. Another 8 years of argument and discussion followed this decision and a summary of this, together with a detailed description of the WMS program, is given by Zmuda (1971). The first IGRF was ratified by IAGA in 1969.

The original idea of an IGRF had come from global modelers, including those who produced such models in association with the production of navigational charts. However, the IGRF as it was first formulated was not considered to be accurate or detailed enough for navigational purposes.

The majority of potential users of the IGRF at this time consisted of geophysicists interested in the geological interpretation of regional magnetic surveys. An initial stage in such work is the removal of a background field from the observations that approximates the field whose sources are in the Earth's core. With different background fields being used for different surveys, difficulties arose when adjacent surveys had to be combined. An internationally agreed global model, accurately representing the field from the core, eased this problem considerably.

Another group of researchers who were becoming increasingly interested in descriptions of the geomagnetic field at this time were those studying the ionosphere and magnetosphere and behavior of cosmic rays in the vicinity of the Earth. This remains an important user community today.

Development of the IGRF

The IGRF has been revised and updated many times since 1969 and a summary of the revision history is given in Table I1 (see also Barton, 1997, and references therein).

Each generation of the IGRF comprises several constituent models at 5-year intervals, some of which are designated definitive. Once a constituent model is designated definitive it is called a Definitive Geomagnetic Reference Field (DGRF) and it is not revised in subsequent generations of the IGRF.

New constituent models are carefully produced and widely documented. The IAGA Working Group charged with the production of the IGRF invites submissions of candidate models several months in advance of decision dates. Detailed evaluations are then made of all submitted models, and the final decision is usually made at an IAGA Assembly if it occurs in the appropriate year, otherwise by the IAGA Working Group. The evaluations are also widely documented. The coefficients of the new constituent models are derived by taking means (sometimes weighted) of the coefficients of selected candidate models. This method of combining several candidate models has been used in almost all generations as, not only are different selections of available data made by the teams submitting models, there are many different methods for dealing with the fields which are not modeled by the IGRF, for example the ionospheric and magnetospheric fields and crustal fields. The constituent main field models of the most recent generation of the IGRF (IAGA, 2005) extend to spherical harmonic degree 10 up to and including epoch 1995.0, thereafter they extend to degree 13 to take advantage of the excellent coverage and quality of satellite data provided by Ørsted and CHAMP (see *Ørsted* and *CHAMP*). The predictive secular variation model extends to degree 8.

Future of the IGRF

Firstly, no model of the geomagnetic field can be better than the data on which it is based. An assured supply of high-quality data distributed evenly over the Earth's surface is therefore a fundamental prerequisite for a continuing and acceptably accurate IGRF. Data from magnetic observatories (see *Observatories, overview*) continue to be the most important source of information about time-varying fields. However their spatial distribution is poor and although data from other sources such as repeat stations (see *Repeat stations*), Project MAGNET (see *Project MAGNET*), and marine magnetic surveys (see *Magnetic surveys, marine*) have all helped to fill in the gaps,

Table I1 Summary of IGRF history

Full name	Short name	Valid for	Definitive for
IGRF 10th generation (revised 2004)	IGRF-10	1900.0–2010.0	1945.0–2000.0
IGRF 9th generation (revised 2003)	IGRF-9	1900.0–2005.0	1945.0–2000.0
IGRF 8th generation (revised 1999)	IGRF-8	1900.0–2005.0	1945.0–1990.0
IGRF 7th generation (revised 1995)	IGRF-7	1900.0–2000.0	1945.0–1990.0
IGRF 6th generation (revised 1991)	IGRF-6	1945.0–1995.0	1945.0–1985.0
IGRF 5th generation (revised 1987)	IGRF-5	1945.0–1990.0	1945.0–1980.0
IGRF 4th generation (revised 1985)	IGRF-4	1945.0–1990.0	1965.0–1980.0
IGRF 3rd generation (revised 1981)	IGRF-3	1965.0–1985.0	1965.0–1975.0
IGRF 2nd generation (revised 1975)	IGRF-2	1955.0–1980.0	–
IGRF 1st generation (revised 1969)	IGRF-1	1955.0–1975.0	–

the best spatial coverage is provided by near-polar satellites. Measurements made by the POGO satellites (1965–1971) (see *POGO (OGO-2, -4, and -6 spacecraft)*), *Magsat* (1979–1980), POGS (1990–1993), *Ørsted* (1999–), and *CHAMP* (2000–) have all been utilized in the production of the IGRF.

Secondly, the future of the IGRF depends on the continuing ability of the groups who have contributed candidate models to the IGRF revision process to produce global geomagnetic field models. This ability is dependent on the willingness of the relevant funding authorities to continue to support this type of work.

Thirdly, the continued interest of IAGA is a necessary requirement for the future of the IGRF. This is assured as long as there is, as at present, a large and diverse group of IGRF-users worldwide. One reason why the IGRF has gained the reputation it has is because it is endorsed and recommended by IAGA, the recognized international organization for geomagnetism.

A topic still under discussion is how best to extend the IGRF backward in time. The current generation includes non-definitive models at 5-year intervals covering the interval 1900.0–1940.0, and, more importantly, some of the earlier DGRFs are of questionable quality. Constructing an internationally acceptable model that describes the time variation better than the IGRF using splines is one way forward (see *Time-dependent models of the geomagnetic field*).

Susan Macmillan

Bibliography

Barton, C.E., 1997. International Geomagnetic Reference Field: the seventh generation. *Journal Geomagnetism and Geoelectricity*, **49**: 123–148.

International Association of Geomagnetism and Aeronomy (IAGA), Division V, Working Group VMOD: Geomagnetic Field Modeling, 2005. The 10th-Generation International Geomagnetic Reference Field. *Geophysical Journal International*, **161**, 561–565.

Zmuda, A.J., 1971. The International Geomagnetic Reference Field: Introduction, *Bulletin International Association of Geomagnetism and Aeronomy*, **28**: 148–152.

Cross-references

CHAMP
Harmonics, Spherical
IAGA, International Association of Geomagnetism and Aeronomy
Magnetic Surveys, Marine
Magsat
Main Field Modeling
Observatories, Overview
Ørsted
POGO (OGO-2, -4 and -6 spacecraft)
Project MAGNET
Repeat Stations
Time-dependent Models of the Geomagnetic Field

INDUCTION ARROWS

Introduction

The history of continuous magnetic field observations in Germany dates back to the 19th century (see *Observatories in Germany*). Modern observatory work began in 1930, with the foundation of the geomagnetic observatory in Niemegk, south of Potsdam. When a second observatory started operation in the late 1940s in Wingst, northern Germany, scientists soon noticed reversed signal amplitudes in the vertical field components between the two locations.

This observation was unexpected because magnetic field disturbances were known to be generated by large-scale processes in the Earth's atmosphere, far away from the surface of the Earth (see *Natural sources for EM induction studies*). Hence, for external sources, the magnetic field variations should have been very similar, given the relatively short distance between the two observatories. However, it was soon speculated that this peculiar behavior may be caused by electromagnetic induction of currents into an electrically conducting region deep in the earth's interior. When portable magnetometers became available, this phenomenon was investigated more systematically. With measurements following a N-S oriented profile at selected field sites, Schmucker (1959) could not only confirm the observatory results but could also confine the width of this North German anomaly to approximately 100 km.

Northern Germany was not the only place where magnetic field variations were studied; similar conductivity anomalies were reported from many other parts of the world. Over the years, some of the largest conductivity anomalies on Earth were identified with studies relating vertical magnetic fields to horizontal magnetic fields (see *Geomagnetic deep sounding*).

The induction arrow

Wiese (1962) and Parkinson (1962) independently developed graphical methods to locate the observed "induction effect," thereby inventing the induction arrow (IA). Both original definitions of the IA are based on a time-domain approach which is not in use any more. However, we still draw our IAs in either the Wiese or the *Parkinson (q.v.)* convention according to their original designs (see Hobbs, 1992).

Functions that interrelate vertical or horizontal magnetic fields are known as transfer functions (see *Transfer functions*). If lateral conductivity variations exist within the subsurface on length scales equal to or greater than the penetration depth of the induced horizontal magnetic fields, a vertical magnetic field component is generated.

The frequency-domain, geomagnetic transfer functions $T_x(\omega), T_y(\omega)$ are defined as (dependence on frequency assumed):

$$B_z = T_x B_x + T_y B_y \qquad \text{(Eq. 1)}$$

x, y, and z denote north, east, and vertical directions, respectively. B is the magnetic field component in [T]. The IA is a graphical representation of the complex vertical magnetic field transfer function **T**. The real and imaginary parts of T_x and T_y can be combined to form $\pm(T_{xr}, T_{yr})$ and (T_{xi}, T_{yi}). The visual display of the IAs in maps is a comprehensive presentation of changes in vertical magnetic field anomalies, both, as a function of frequency and location. In Parkinson convention the $(-T_{xr}, -T_{yr})$ is plotted, which tends to point toward regions of higher conductivity. The IA in the Wiese convention (T_{xr}, T_{yr}) tends to point away from areas of higher conductivity. Table I2 gives the definitions of the IAs in the Wiese and Parkinson conventions.

Figure I1 shows an example. The conductivity contrast associated with a fault zone can be either due to a juxtaposition of different lithologies (upper panel of Figure I1) or the damaged rocks within a fault zone can become conductive to form a conductive channel (*bottom panel* of Figure I1). If the conductivity contrast is assumed to extend indefinitely, both cases resemble a two-dimensional (2D) problem.

Table I2 The definition of induction arrows in Parkinson and Wiese conventions

	Real arrow		Imaginary arrow	
Convention	Length	Angle	Length	Angle
Parkinson	$\sqrt{T_{xr}^2 + T_{yr}^2}$	$\arctan\left(\frac{-T_{xr}}{-T_{yr}}\right)$	$\sqrt{T_{xi}^2 + T_{yi}^2}$	$\arctan\left(\frac{T_{yi}}{T_{xi}}\right)$
Wiese		$\arctan\left(\frac{T_{xr}}{T_{yr}}\right)$		

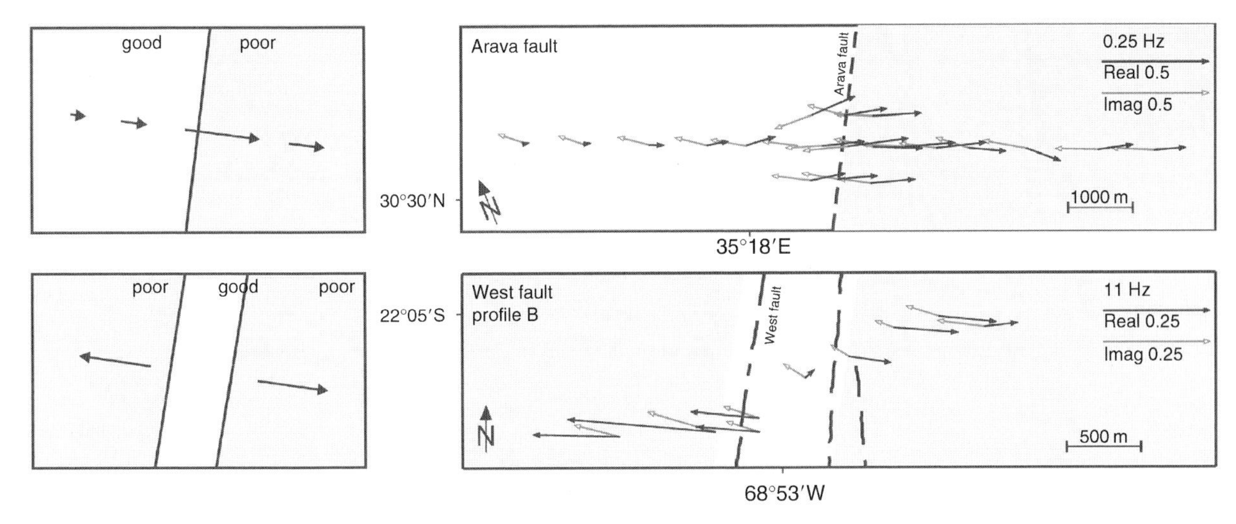

Figure I1 Expected and observed behavior of induction arrows (IA) for two different fault structures. In Wiese convention real (black) arrows point away from good conductors; imaginary (gray) arrows are parallel or antiparallel to the real arrows, given 2D Earth structure. Across a conductivity contrast (*top panel*) real IAs point away from the more conductive side and are largest over the less conductive side. Measurements from the Dead Sea Transform, locally known as the Arava Fault, are representative of this type of fault structure. A conductive fault zone sandwiched between more resistive units is characterized by IAs pointing away from the fault on both sides and small IAs over the fault zone conductor. The West Fault in Chile is representative of this type of fault structure. (Figure after Ritter *et al.*, 2005.)

The expected behavior for the real IA is indicated on the left-hand side of the figure. In 2D, the real and imaginary IAs point parallel or antiparallel to each other, and perpendicular to the strike of the structure. The largest real IAs are observed close to a conductivity contrast, they become smaller with increasing distance from that boundary. A larger conductivity contrast will also cause a longer arrow. Although vertical magnetic field response functions do not contain direct information about the underlying conductivity structure, they are extremely valuable in detecting the lateral extension and the strike of conductivity anomalies.

In the absence of lateral resistivity gradients IAs vanish. Real and imaginary IAs pointing obliquely indicate the influence of off-profile features of a more complicated three-dimensional Earth. The observed IAs on the right-hand side of Figure I1 show this influence to some extend but overall, they are still in good agreement with the simple 2D scenarios on the left. If IAs show great variation, both between sites and between real and imaginary arrows then a simple interpretation in terms of geoelectric strike direction is not possible.

Many scientists, including the author, prefer the term induction vector over IA because we construct and treat IAs like vectors, i.e., their length and the direction have a physical meaning. They are however, not vectors in a strict mathematical sense. IAs caused by a superposition of two different lateral conductivity contrasts which are linked by inductive coupling will generally not be the same as the vector constructed by adding the individual IAs. Furthermore, IAs are very sensitive to *electrical anisotropy* (*q.v.*) (Pek and Verner, 1997).

Oliver Ritter

Bibliography

Hobbs, B.A., 1992. Terminology and symbols for use in studies of electromagnetic induction in the Earth. *Surveys of Geophysics*, **13**: 489–515.

Parkinson, W., 1962. The influence of continents and oceans on geomagnetic variations. *Geophysical Journal*, **2**: 441–449.

Pek, J., and Verner, T., 1997. Finite-difference modelling of magnetotelluric fields in two-dimensional anisotropic media. *Geophysical Journal International*, **132**: 535–548.

Ritter, O., Hoffmann-Rothe, A., Bedrosian, P.A., Weckmann, U., and Haak, V., 2005. Electrical conductivity images of active and fossil fault zones. In Bruhn, D., and Burlini, L., (eds.), *High-Strain Zones: Structure and Physical Properties,* Vol. 245. London: Geological Society of London Special Publications, pp. 165–186.

Schmucker, U., 1959. Erdmagnetische Tiefensondierung in Deutschland 1957/59. Magnetogramme und erste Auswertung. Abh. Akad. Wiss.Götingen. Math.-Phys. Klasse: Beiträge zum Int. geophys. Jahr 5.

Wiese, H., 1962. Geomagnetische Tiefentellurik Teil II: die Streichrichtung der Untergrundstrukturen des elektrischen Widerstandes, erschlossen aus geomagnetischen Variationen. *Geofisica Pura e Applicata*, **52**: 83–103.

Cross-references

Anisotropy, Electrical
Geomagnetic Deep Sounding
Natural Sources for EM Induction Studies
Observatories in Germany
Parkinson, Wilfred Dudley (1919–2001)
Transfer Functions

INDUCTION FROM SATELLITE DATA

In addition to sources of internal origin, magnetic satellites record the variations in Earth's external magnetic field caused by interaction of the ionosphere and magnetosphere with the solar wind and radiation, along with the secondary magnetic fields induced in Earth. The ratio of the internal to external field may be used to probe the electrical conductivity structure of the planet. Satellite induction studies date from Didwall's (1984) early work with the POGO satellite data. Olsen (1999a) and Constable and Constable (2004a) provide reviews of satellite induction methods. Table I3 presents a list of past, present, and future magnetic satellite missions that are relevant to induction studies (*see MAGSAT, Ørsted, POGO*).

Table I3 Satellite missions relevant to magnetic induction studies, as of July, 2005

Satellite	Type	Launch date	Data to	Inclination	Altitude (km)
POGO 2	Scalar	October 14, 1965	October 2, 1967	87.3°	410–1510
POGO 4	Scalar	July 28, 1967	January 19, 1969	86.0°	410–910
POGO 6	Scalar	June 5, 1969	April 26, 1971	82.0°	400–1100
MAGSAT	Vector	October 30, 1979	May 6, 1980	96.8	352–561
Ørsted	Vector	February 23, 1999	Present	96.6°	500–850
SAC-C	Vector	November 18, 2000	Present	98.2°	702
CHAMP	Vector	July 15, 2000	Present	87.3°	200–454
SWARM	Three vector sats	Proposed 2008			450–550

Longer period EM variations penetrate deeper into Earth, because EM fields decay exponentially with a characteristic length given by the skin depth, or

$$z_s \approx 500\sqrt{T/\sigma}\ \mathrm{m}$$

where σ is electrical conductivity (S m^{-1}) and T is the period of the field variations (s). Traditionally, studies of deep Earth conductivity have relied on either the *magnetotelluric* (MT) method (*q.v.*) or the *geomagnetic depth sounding* (GDS) method (*q.v.*). For MT studies, time series measurements of horizontal magnetic and electric fields are transformed into frequency domain Earth impedance (see *Transfer functions for EM*). In GDS, observatory records of three-component magnetic fields (or data from portable arrays of instruments), along with an assumption about the geometry of the inducing field, are transformed into a similar impedance. MT studies have the advantage that they are largely independent of the geometry of the external fields, but because induced electric fields decay with increasing period, MT studies are limited to periods of less than a few days. GDS studies, on the other hand, have been extended out to the period of the 11-year sunspot cycle, the longest wherein field variations are external to Earth's core.

The usual assumption about the geometry of the external field variations is that it is dominated by the fundamental *spherical harmonic* (*q.v.*) aligned with Earth's internal dipole field called P_1^0. This is based both on observations of field geometry (e.g., Banks, 1969) and the morphology of the equatorial *ring current* (*q.v.*), which is composed mainly of oxygen ions circulating at a distance of 2–9 Earth radii (a review of the ring current is given by Daglis *et al.*, 1999). Within the satellite orbit, the magnetic field **B** can be expressed as the gradient of a scalar potential Φ

$$\mathbf{B} = -\mu_0 \nabla \Phi \qquad \text{(Eq. 1)}$$

(μ_0 is the magnetic permeability of space) where for P_1^0 geometry Φ is described by a spherical harmonic representation in radial distance from Earth's center r, Earth radius a_0, and geomagnetic colatitude θ:

$$\Phi_1^0(r,\theta) = a_0 \left\{ i_1^0(t)\left(\frac{a_0}{r}\right)^2 + e_1^0(t)\left(\frac{r}{a_0}\right) \right\} P_1^0(\cos\theta). \qquad \text{(Eq. 2)}$$

In GDS studies, knowledge of the field geometry allows observatory time series measurements of vertical and horizontal components of the magnetic field to be separated into the internal (i_1^0) and external (e_1^0) coefficients, which may be Fourier transformed into the frequency domain to derive a complex geomagnetic response function of period:

$$Q_1^0(T) = i_1^0(T)/e_1^0(T). \qquad \text{(Eq. 3)}$$

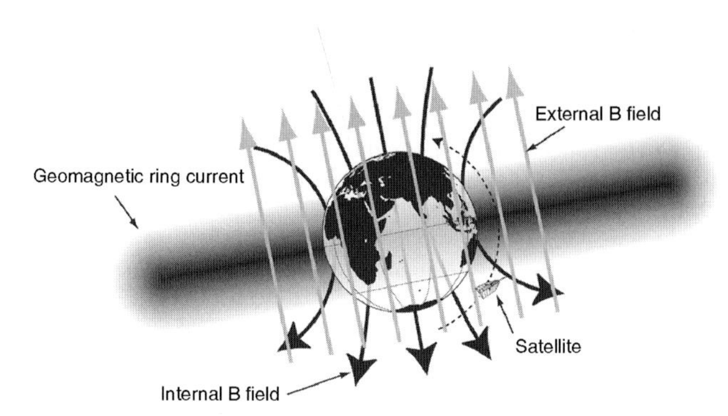

Figure I2 The equatorial ring current generates an external magnetic field of uniform P_1^0 geometry (grey arrows). The external field variations induce an internal field (black arrows) which is also described by P_1^0 geometry but that decays with altitude. The different altitude dependence and different directional behavior with latitude allows a separation of the internal and external fields using satellite measurements made during pole to pole passes.

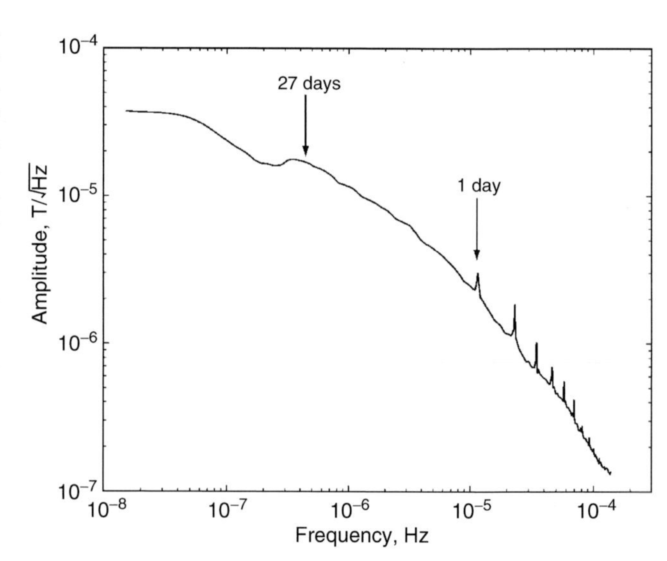

Figure I3 Amplitude spectrum taken from 13 years of the hourly Dst geomagnetic index. A broad peak in energy is seen around the 27-day solar rotation period, and narrow peaks at periods of 1 day and harmonics. The daily variations are generated mainly in the ionosphere, and are not P_1^0 geometry. Power at frequencies lower than 10^{-7} Hz is probably dominated by secular variation in the internal magnetic field.

The more familiar MT apparent resistivity and phase can be derived from the real and imaginary components of $Q_1^0(T)$, as can the inductive scale length C-response (see *transfer functions for EM*).

Magnetic satellites, collecting either scalar (total field magnitude only) or vector (three component) data, offer an opportunity to take GDS induction studies beyond the global observatory network. There are numerous advantages to using satellites for these types of studies:

1. Global coverage can be achieved using satellites, while the magnetic observatory network is sparse, nonuniform, and restricted to land.
2. During each mission magnetic measurements are made using a single, high-quality instrument having a single calibration and which is removed from the distorting effects of crustal magnetic anomalies. Satellite magnetic sensors include scalar proton precession magnetometers, which are often augmented by vector fluxgate magnetometers oriented by star cameras (see *Magnetometers*).
3. Extensive spatial coverage means that the geometry of the inducing magnetic fields can be estimated, instead of being assumed.

Point (3) is illustrated in Figure I2. As a satellite makes a pass across the equator from pole to pole, the geometry of the magnetic field variations is different for the internal and external components. Variations in satellite altitude provide additional discrimination to fit the separate geometries of the internal and external fields.

Induction is powered by variations in the ring current intensity caused by interactions with the solar wind. During magnetic *storms* (*q.v.*), these may amount to a few hundred nanoteslas. A proxy for ring current intensity is provided by the Dst (disturbed storm time) magnetic index, a weighted average of horizontal fields at four low latitude magnetic observatories. Ring current variations are observed at periods of a few hours to many months or years, but are strongest at about 1 month because the solar wind is modulated by solar rotation. Figure I3 shows an amplitude spectrum of the Dst index; power at the daily harmonics is caused by contamination from the daily variation in the magnetic field, generated in the ionosphere and by the Chapman-Ferraro currents outside the magnetosphere, and is more complicated than simple P_1^0 geometry. Power at periods greater than 1 month is contaminated by secular variation of the main *magnetic field* (*q.v.*), of internal origin and unsuitable for induction studies. For GDS observatory studies, it is difficult to reject these contaminations and these periods have to be avoided, but satellite data have some power to reject non-P_1^0 geometries.

The above discussion ignores the contributions to the magnetic field that come from the main (internal) field and the *crustal field* (*q.v.*). These parts of the field must be removed before an induction analysis can be carried out, and therein lies one of the great difficulties of satellite induction. Without additional information, it is impossible to

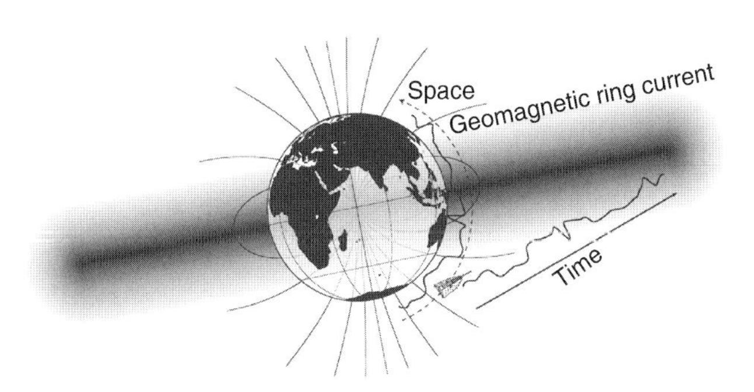

Figure I4 Without additional information, it is impossible to distinguish between temporal variations in the external magnetic field and spatial variations in the internal magnetic field sampled as the magnetic satellite flies though its orbit.

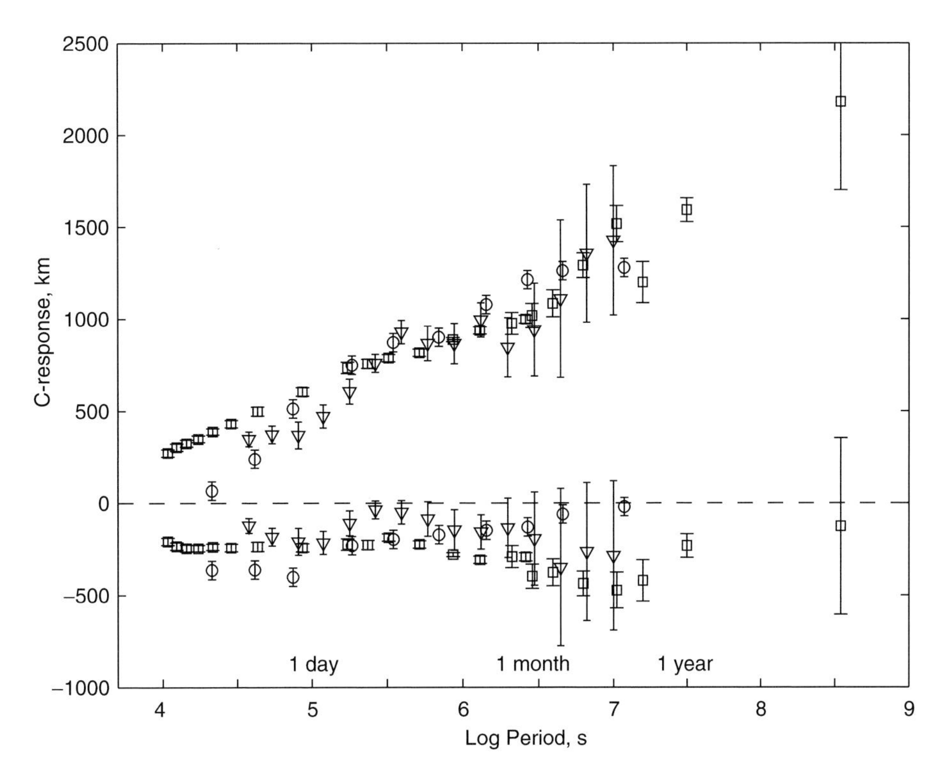

Figure I5 Satellite responses from the MAGSAT (circles, from Constable and Constable, 2004b) and Ørsted (triangles, from Olsen *et al.*, 2003) missions, compared with Olsen's (1999b) long period observatory response for Europe (squares). The real components of C are positive, the imaginary are negative; error bars are one standard deviation.

distinguish between time variations in the ring current source field and spatial variations in the main and crustal fields sampled as the spacecraft flies over Earth. This is illustrated in Figure I4. A model of the main field that includes secular variation, along with a model of the crustal fields and possibly the daily variation, electrojets, and field-aligned currents must be used to remove all these elements before the induction analysis can be carried out. A convenient tool for removing these contributions is the Comprehensive Field Model of Sabaka *et al.* (2004).

Figure I5 shows GDS response functions from the MAGSAT and Ørsted satellite missions, compared with a response function generated using magnetic observatories. The agreement is generally good, and radial conductivity models generated from observatory data also fit the satellite responses fairly well. However, differences do exist. At short periods, the satellite response is influenced by oceans (Constable and Constable, 2004b). At long periods, there may be contamination of the source-field geometry, and some differences are undoubtedly due to the various analysis techniques. The subject of magnetic satellite induction is relatively young and many issues remain to be addressed. For example, there are complications to the ring current geometry that have largely been ignored (e.g., Balasis *et al.*, 2004). The daily variation in the magnetic field has variously been considered a source of noise (Constable and Constable, 2004b) and a signal for 3D induction studies (Velimsk and Everett, 2005). Ultimately, the goal of satellite induction is to move beyond global response functions and radial conductivity models toward true 3D conductivity structure.

Steven Constable

Bibliography

Balasis, G., Egbert, G.D., and Maus, S., 2004. Local time effects in satellite estimates of electromagnetic induction transfer functions. *Geophysical Research Letters*, **31**: L16610, doi:10.1029/2004GL020147.

Banks, R.J., 1969. Geomagnetic variations and the conductivity of the upper mantle. *Geophysical Journal of the Royal Astronomical Society*, **17**: 457–487.

Constable, C.G., and Constable, S.C., 2004a. Satellite magnetic field measurements: applications in studying the deep Earth. In Sparks, R.S.J., and Hawkesworth, C.T. (eds.) *The State of the Planet: Frontiers and Challenges in Geophysics,* Geophysical Monograph 150. Washington, DC: American Geophysical Union, pp. 147–159.

Constable, S., and Constable, C., 2004b. Observing geomagnetic induction in magnetic satellite measurements and associated implications for mantle conductivity. *Geochemistry Geophysics Geosystems*, **5**: Q01006, doi:10.1029/2003GC000634.

Daglis, I.A., Thorne, R.M., Baumjohann, W., and Orsini, S., 1999. The terrestrial ring current: origin, formation, and decay. *Reviews of Geophysics*, **37**: 407–438.

Didwall, E.M., 1984. The electrical conductivity of the upper mantle as estimated from satellite magnetic field data. *Journal of Geophysical Research*, **89**: 537–542.

Olsen, N., 1999a. Induction studies with satellite data. *Surveys in Geophysics*, **20**: 309–340.

Olsen, N., 1999b. Long-period (30 days–1 year) electromagnetic sounding and the electrical conductivity of the lower mantle beneath Europe. *Geophysical Journal International*, **138**: 179–187.

Olsen, N., Vennerstrøm, S., and Friis-Christensen, E., 2003. Monitoring magnetospheric contributions using ground-based and satellite magnetic data. In Reigber, Ch., Luehr, H., and Schwintzer, P. (eds.), *First CHAMP Mission Results for Gravity, Magnetic and Atmospheric Studies.* Berlin: Springer-Verlag, pp. 245–250.

Sabaka, T.J., Olsen, N., and Purucker, M.E., 2004. Extending comprehensive models of the Earth's magnetic field with Orsted and CHAMP data. *Geophysical Journal International*, **159**: 521–547.

Velimsk, J., and Everett, M.E., 2005. Electromagnetic induction by *Sq* ionospheric currents in heterogeneous Earth: modeling using ground-based and satellite measurements. In Reigber, Ch., Luehr, H., Schwintzer, P., and Wickert, J. (eds.) *Earth Observation with CHAMP Results from Three Years in Orbit.* Berlin: Springer-Verlag, pp. 341–346.

Cross-references

CHAMP
Crustal Magnetic Field
EM, Marine Controlled Source
Geomagnetic Deep Sounding
Geomagnetic Spectrum, Temporal
Harmonics, Spherical
Magnetometers, Laboratory
Magnetosphere of the Earth
Magnetotellurics
Magsat
Ørsted
POGO (OGO-2, -4, and -6 spacecraft)
Ring Current
Secular Variation Model
Storms and Substorms, Magnetic
Transfer Functions

INHOMOGENEOUS BOUNDARY CONDITIONS AND THE DYNAMO

Thermal core-mantle coupling and geodynamo

Modern observations indicate that the Earth's magnetic field takes the form of an approximate geocentric axial dipole, i.e., a dipole positioned at the Earth's center and aligned with the axis of rotation. Paleomagnetic measurements indicate that, when it is averaged over a sufficiently long time, this has been the case for the observable past. This suggests that the observed Earth's magnetic field is primarily the result of a geodynamo consistent with turbulent flows and rapid rotation (Moffatt, 1978). In consequence, the geocentric axial dipole is widely employed to provide a reference state for the Earth's magnetic field. Any persistent departure from the axial dipole is indicative of the external influence such as inhomogeneous boundary conditions on the geodynamo imposed by the overlying mantle (Hide, 1967; Bloxham and Gubbins, 1987; Bloxham, 2000).

Convection is occurring in both the Earth's outer core and mantle. However, there exists a huge difference between the timescales of mantle and core convection. Core convection has an overturn time of several hundred years while that of mantle convection is a few hundred million years. As far as the Earth's mantle is concerned, the core-mantle boundary is a fixed temperature interface. As far as the Earth's fluid core is concerned, the core-mantle boundary is rigid and imposes a nearly unchanging heat-flux heterogeneity on the core convection and dynamo.

The structure of the inhomogeneous boundary condition can be inferred from the seismic tomography of the Earth's lower mantle (for example, Masters *et al.*, 1996; van der Hilst *et al.*, 1997). By assuming that the hotter regions of mantle correspond to the lower seismic velocity while the cold regions are related to the anomalously high seismic velocity, the inhomogeneous boundary condition at the core-mantle boundary for the core convection and dynamo can be determined. The temperature variations above the core-mantle boundary impose a nearly unchanging heat-flux heterogeneity on the core dynamo.

Paleomagnetic and historical magnetic field measurements suggest persistent distinct patterns of variation of the geomagnetic field taking place in different regions of the Earth. For example, secular variation under the Pacific region is much slower than that under the Atlantic region (Bloxham and Gubbins, 1985; McElhinny *et al.*, 1996). This longitudinal asymmetry of the geomagnetic field, the so-called Pacific dipole window, may be explained by thermal heterogeneity at the lower mantle. Bloxham and Gubbins (1987) proposed that thermal interaction whereby lateral variations in heat flux across the core-mantle boundary drive core flows directly and influence the preexisting deep convection. Furthermore, comparison of the geomagnetic field at the core surface with lower mantle temperature or seismic velocity suggests a close connection between them (Gubbins and Richards, 1986; Gubbins and Bloxham, 1987) and order-of-magnitude estimates also suggest the variations are large enough to drive the core flow required for secular variation of the Earth's magnetic field (Bloxham and Gubbins, 1987).

The inhomogeneous boundary condition in heat flux or temperature at the core-mantle boundary, together with the strong effect of rotation, drive fluid motions directly in the outer core which influence the generation of geomagnetic field in the core. Moreover, an inhomogeneous boundary condition may also influence the core convection being driven from below by locking it to the nonuniform boundary condition (Zhang and Gubbins, 1993; Sarson *et al.*, 1997).

Inhomogeneous boundary conditions: models

With a uniform thermal boundary, convection takes place when and only when a sufficient radial temperature gradient, measured by the size of the Rayleigh number, exists to drive it. With a nonuniform thermal boundary, there are two driving parameters: one measures the radial buoyancy and the other the strength of the lateral heating. It follows that how an inhomogeneous boundary condition affects the core convection and dynamo can be illustrated by a simple nonlinear amplitude equation in the form

$$\dot{A} = (\mu + i)A - A|A|^2 + \epsilon, \qquad \text{(Eq. 1)}$$

where the flow amplitude A is a complex variable, $i = \sqrt{-1}, \epsilon$ is a positive parameter measuring the strength of the imposed inhomogeneous boundary condition (thermal or nonthermal) at the base of mantle, and μ is a positive parameter like the Rayleigh number in connection with the deep core convection. The amplitude equation is simple but provides a helpful mathematical framework in the understanding of complicated nonlinear dynamics affected by an imposed inhomogeneous boundary condition.

Without lateral heterogeneity of the lower mantle, $\epsilon = 0$, convection is always in the form of an azimuthally traveling wave, such as convection rolls aligned with the axis of rotation. With the influence of the lateral heterogeneity $\epsilon \neq 0$, a steady equilibrium solution A_0 becomes possible and can be obtained by setting $\dot{A}_0 = 0$. Denoting $A_0 = X_0 + iY_0$ and $Z = |A_0|^2 = X_0^2 + Y_0^2$, we obtain the cubic equation

$$Z^3 - 2\mu Z^2 + Z(\mu^2 + 1) - \epsilon^2 = 0 \qquad \text{(Eq. 2)}$$

for steady solutions of convection, which can be solved analytically using a standard formula. The stability of a steady solution A_0 can be investigated by linearizing the nonlinear equation by letting

$$A = (X_0 + x) + i(Y_0 + y), \qquad \text{(Eq. 3)}$$

where x and y are small perturbations of X_0 and Y_0. Stability of the nonlinear equilibrium A_0 is then related to the two linear equations

$$\dot{x} = (\mu - 3X_0^2 - Y_0^2)x - (1 + 2X_0Y_0)y, \qquad \text{(Eq. 4)}$$

$$\dot{y} = (1 - 2X_0Y_0)x + (\mu - 3Y_0^2 - X_0^2)y. \qquad \text{(Eq. 5)}$$

The corresponding growth rate σ of the perturbations, or stability of the boundary-driven steady solution, is then described by

$$\sigma = (\mu - 2|A_0|^2) \pm (|A_0|^4 - 1)^{1/2}. \qquad \text{(Eq. 6)}$$

Two important features are evident from the stability of the boundary-driven steady solution: (i) a Hopf bifurcation (traveling wave convection) cannot occur if the amplitude is sufficiently large, $|A_0| > 1$; (ii) a necessary condition for the occurrence of a Hopf bifurcation is $\epsilon < \sqrt{2}$. In other words, the core convection and dynamo in the presence of an inhomogeneous core-mantle boundary can be locked into the boundary provided that the effect is sufficiently strong, $\epsilon > \sqrt{2}$. It follows that the Pacific dipole window may reflect the thermal structure of mantle convection under the Pacific region.

Complex three-dimensional numerical simulations in rotating spherical geometry show a similar behavior revealed by the above simple model, demonstrating that an inhomogeneous boundary condition imposed by the lower mantle can alter the properties of core convection and dynamo in a fundamental way (Zhang and Gubbins, 1992, 1993; Sarson *et al.*, 1997; Olson and Christesen, 2002). The mantle convection imposes a different length scale upon the system of the core convection and dynamo. When a geodynamo model uses the inhomogeneous boundary condition derived from seismic tomography, it can produce an anomalous magnetic field and westward fluid velocity which are largely consistent with the observed features of the Earth's magnetic field (Olson and Christensen, 2002).

Keke Zhang

Bibliography

Bloxham, J., 2000. Sensitivity of the geomagnetic axial dipole to thermal core-mantle interactions. *Nature*, **405**(6782): 63–65.

Bloxham, J., and Gubbins, D., 1985. The secular variation of the Earth's magnetic field. *Nature*, **317**: 777–781.

Bloxham, J., and Gubbins, D., 1987. Thermal core-mantle interactions. *Nature*, **325**: 511–513.

Gubbins, D., and Bloxham, J., 1987. Morphology of the geomagnetic field and implications for the geodynamo. *Nature*, **325**: 509–511.

Gubbins, D., and Richards, M., 1986. Coupling of the core dynamo and mantle: thermal or topographic? *Geophysical Research Letters*, **13**: 1521–1524.

Hide, R., 1967. Motions of the earth's core and mantle, and variations of the main geomagnetic field. *Science*, **157**: 55–56.

Masters, G., Johnson S., Laske, G., and Bolton, H., 1996. A shear-velocity of the mantle. *Philosophical Transactions of the Royal Society of London, Series A*, **354**: 1385–1411.

McElhinny, M.W., McFadden, P.L., and Merrill, P.T., 1996. The myth of the Pacific dipole window. *Earth and Planetary Science Letters*, **143**: 13–22.

Moffatt, H.K., 1978. *Magnetic Field Generation in Electrically Conducting Fluids.* Cambridge, England: Cambridge University Press.

Olson, P., and Christensen U.R., 2002. The time-averaged magnetic field in numerical dynamos with nonuniform boundary heat flow. *Geophysical Journal International*, **151**(3): 809–823.

Sarson, G.R., Jones, C.A., and Longbottom, A.W., 1997. The influence of boundary region heterogeneities on the geodynamo. *Physics of the Earth and Planetary Interiors*, **104**: 13–32.

van der Hilst, R.D., Widiyantoro S., and Engdahl E.R., 1997. Evidence for deep mantle circulation from global tomography. *Nature*, **386**: 578–584.

Zhang, K., and Gubbins, D., 1992. On convection in the Earth's core forced by lateral temperature variations in the lower mantle. *Geophysical Journal International*, **108**: 247–255.

Zhang, K., and Gubbins, D., 1993. Convection in a rotating spherical fluid shell with an inhomogeneous temperature boundary condition at infinite Prandtl number. *Journal of Fluid Mechanics*, **250**: 209–232.

Cross-references

Core, Magnetic Instabilities
Core Motions
Magnetoconvection
Magnetohydrodynamic Waves
Proudman-Taylor Theorem

INNER CORE ANISOTROPY

Evidence for anisotropy of inner core

Anisotropy is the general term used to describe a medium whose properties (i.e., elastic properties in the case of seismic anisotropy) depend on orientation. Seismic waves in an anisotropic medium travel with different speeds depending on both their particle motion and propagation directions. The Earth's solid inner core was found to possess significant anisotropy in seismic velocity in the 1980s and early 1990s. Before that time, the inner core was known to us as a featureless small solid ball (at a radius of 1220 km) of iron-nickel alloy with some light elements.

The evidence for the anisotropy of the inner core came from two different kinds of observations: directional variations of travel-times of seismic body waves that go through the inner core and anomalous splitting, splitting not explainable simply by Coriolis and ellipticity effects, of the Earth's normal modes that are sensitive to the structure of the Earth's core. Anomalies in inner core arrival times and in normal-mode splitting were observed in early 1980s (Masters and Gilbert, 1981; Poupinet *et al.*, 1983) and the hypothesis that the inner core is anisotropic was first proposed in 1986 (Morelli *et al.*, 1986).

Subsequent studies using travel-time data (e.g., Shearer *et al.*, 1988) generally favored the existence of inner core anisotropy. However, the interpretation that anomalous splitting of core-sensitive modes is primarily caused by the anisotropy of the inner core was controversial (e.g., Widmer *et al.*, 1992). Strong support for the inner core anisotropy was provided in the early 1990s when new sets of information started to emerge. Measurements of high quality differential travel-times of PKP waves (waves that pass through the Earth's core) show large travel-time anomalies (2–6 s) for rays traveling through the inner core nearly parallel to the Earth's spin axis (polar paths) (Creager, 1992; Song and Helmberger, 1993; Vinnik *et al.*, 1994). Reanalysis of arrival time data also confirmed large travel-time anomalies (Shearer, 1994; Su and Dziewonski, 1995). At the same time, Tromp (1993) demonstrated that a simple transversely isotropic inner core explained the anomalously split core-sensitive modes and the larger, newly observed PKP differential travel-time anomalies reasonably well.

An example of the evidence for the inner core anisotropy from travel-time observations is shown in Figure I6. In analyzing travel-time data, residuals of travel-times that seismic waves (phases) take from an earthquake to arrive at recording stations are often formed by subtracting the times predicted for a standard Earth model from measured travel-times; similarly, residuals of differential travel-times are formed

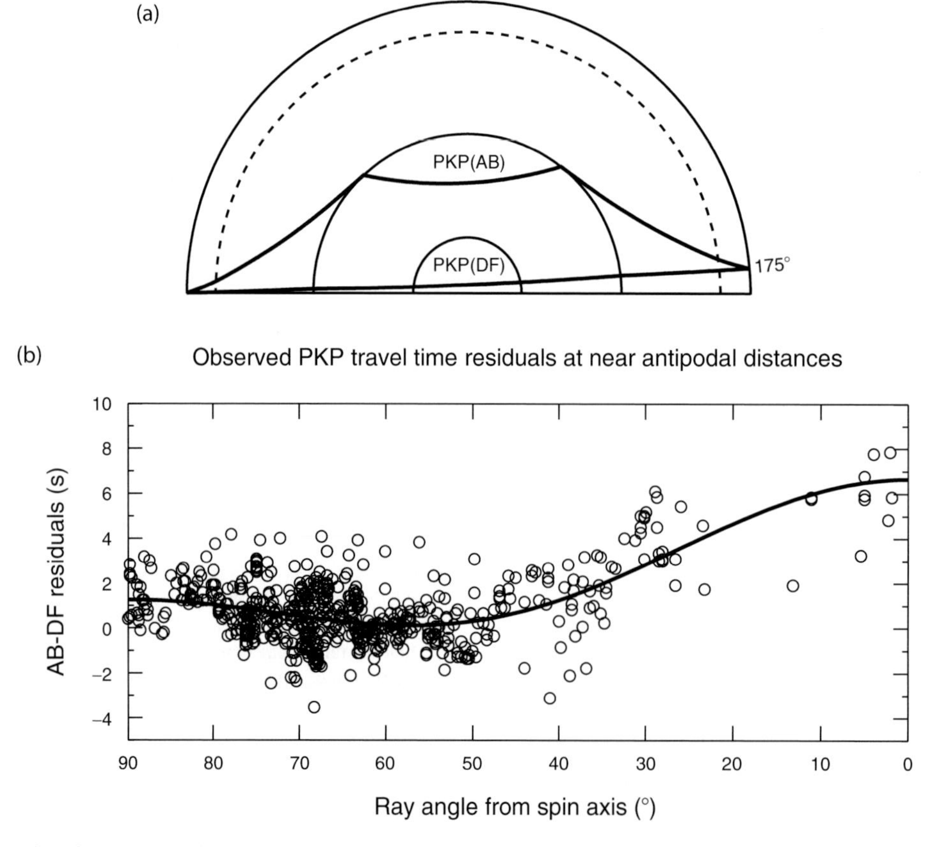

Figure I6 (a) Ray paths of PKP(AB) and PKP(DF) phases at near antipodal distances. The PKP(AB) phase turns at mid-outer core and the PKP(DF) phase passes through the inner core, sampling the bulk of the inner core at near antipodal distances as shown in this example. (b) Residuals of differential PKP(AB)-PKP(DF) travel-times for near antipodal paths (modified from Sun and Song, 2002). Assuming uniform cylindrical anisotropy in the inner core, the curve is the best fit to the data.

by subtracting the time differences between two phases concerned predicted for a standard Earth model from measured differential travel-times. Figure I6b shows the residuals of differential between AB and DF branches of PKP waves (Figure I6a). Here the PKP(AB) is used as a reference phase to form the differential travel times, which eliminates bias from uncertainties in earthquake origin times and reduces the biases from earthquake mislocations and heterogeneous upper mantle; heterogeneity in the outer core is assumed to be small because of efficient mixing in the fluid core (Stevenson, 1987). The residuals for the polar paths are systematically larger than those of the equatorial paths, suggesting faster wave speed along the NS direction through the inner core than along the equatorial plane. Assuming cylindrical anisotropy with fast axis parallel to the spin axis, this data set indicates an average anisotropy of about 2.5% in P-velocity in the inner core.

Three-dimensional structure of inner core anisotropy and heterogeneity

The presence of significant anisotropy in the inner core is now well accepted. The anisotropy appears to be dominantly cylindrical, with the axis of symmetry aligned approximately with the NS spin axis of the Earth. The detailed structure of the inner core anisotropy is still being mapped out. Some studies suggest the inner core has a simple constant anisotropy except for the innermost inner core (Ishii and Dziewonski, 2002). Other studies, however, suggest that the inner core anisotropy is more complex, varying both laterally and in depth, although some of the complexity may be due to contamination from the lowermost mantle heterogeneity (Breger *et al.*, 2000; Tromp, 2001).

The outermost part of the inner core appears nearly isotropic (Shearer, 1994). The thickness of the weak anisotropy layer varies from upper 100–250 km in the western inner core to upper 400 km or more in the eastern inner core (Creager, 2000; Souriau *et al.*, 2003). The transition from isotropy in the upper inner core to strong anisotropy in lower inner core under Central America appears to be sharp enough to cause multipathing of seismic waves (Song and Helmberger, 1998).

There is also growing evidence for strong lateral variation in inner core structure. The inner core appears to vary on all scales, from the scale of half a hemisphere to the scale of a few kilometers (Vidale and Earle, 2000; Cormier and Li, 2002). At the uppermost 100 km of the inner core, the P-velocity in the quasi-eastern hemisphere (40° E–180° E) is isotropically faster than the quasi-western hemisphere (180° W–40° E) by about 0.8% (Niu and Wen, 2001). At intermediate depth, 100–400 km below the inner core boundary (ICB), the inner core possesses a hemispherical pattern of a different form. The quasi-western hemisphere is strongly anisotropic but the quasi-eastern hemisphere is nearly isotropic (Tanaka and Hamaguchi, 1997). Deeper into the inner core, significant anisotropy (~3%) seems to exist in both hemispheres (Creager, 2000; Souriau *et al.*, 2003); however, in the innermost 300–400 km of inner core, it is suggested that the form of the anisotropy is distinctly different (Ishii and Dziewonski, 2002; Beghein and Trampert, 2003).

Sources of inner core anisotropy

Apparent seismic velocity anisotropy can, in general, arise from preferred orientation of anisotropic crystals or from lamination of a solid. The properties of the inner core are widely believed to be consistent with those of anisotropic iron crystals in the hexagonal close-packed (hcp) phase (e.g., Brown and McQueen, 1986). Thus, the anisotropy is believed to be due to a preferred orientation of the hcp iron (e.g., Stixrude and Cohen, 1995). However, the mechanisms responsible for creating such a preferred alignment are under debate.

One category of the proposed models involves texturing established during the solidification of iron crystals at the surface of the inner core. Possible mechanisms of solidification texturing include dendritic growth of iron crystals as they solidify (Bergman, 1997) and the development of anisotropic polycrystalline solid due to the presence of thermal gradient and initial nucleation in the melt (Brito *et al.*, 2002).

However, solidification texturing alone cannot explain the observed depth dependence of the anisotropy. The other category of the proposed models involves the alignment of iron crystals from plastic deformation in the inner core after solidification. Proposed mechanisms of plastic deformation in the inner core include solid state thermal convection driven by internal heating (Jeanloz and Wenk, 1988); flow induced by Maxwell stress (stress due to magnetic field) (Karato, 1999; Buffett and Wenk, 2001); and flow induced by a differential stress field created by preferential growth of the inner core in the equatorial belt, which results from more efficient heat transport in the equator than near the polar regions (Yoshida *et al.*, 1996).

Implications of inner core anisotropy

The inner core anisotropy has important implications for improving our understanding of the structure, composition, and dynamics of the Earth's deep interior. It has been suggested that the inner core is rotating relative to the mantle (Song and Richards, 1996). Detailed mapping of the lateral variations of the inner core anisotropic structure is crucial for quantification of the rotation rate (Creager, 1997), which would contribute to our understanding of the geodynamo (e.g., Glatzmaier and Roberts, 1995; Kuang and Bloxham, 1997) and could influence our interpretation of decadal variations in the length of day (Buffett and Glatzmaier, 2000). The understanding of the source of the inner core anisotropy is expected to improve our understanding of the inner core evolution and its interactions with the outer core (e.g., Romanowicz *et al.*, 1996; Buffett, 2000). Viable models for the origin of anisotropy must be compatible with the estimates of symmetry, depth-dependence and lateral variations. The basic cylindrical symmetry of the inner core anisotropy probably reflects the strong influence of rotation on the dynamics of the fluid core. The absence of anisotropy from the uppermost region of the inner core suggests that any texture acquired during solidification is weak. However, a weak initial texture could be subsequently amplified by grain growth and plastic deformation. In this case, the rate of increase in anisotropy with depth is closely related to the relative rates of inner core growth, grain growth, and strain accumulation. The existence of sharp increase with depth and strong lateral variations of inner core anisotropy poses serious restrictions of the mechanisms responsible for the anisotropy. Improved models of the anisotropic structure of the inner core from seismic imaging will be vital in addressing these issues.

Xiaodong Song

Bibliography

Beghein, C., and Trampert, J., 2003. Robust normal mode constraints on inner-core anisotropy from model space search. *Science*, **299**: 552–555.

Bergman, M.I., 1997. Measurements of elastic anisotropy due to solidification texturing and the implications for the Earth's inner core. *Nature*, **389**: 60–63.

Breger, L., Tkalcic, H., and Romanowicz, B., 2000. The effect of D" on PKP(AB-DF) travel time residuals and possible implications for inner core structure. *Earth and Planetary Science Letters*, **175**: 133–143.

Brito, D., Elbert, D., and Olson, P., 2002. Experimental crystallization of gallium: ultrasonic measurement of elastic anisotropy and implications for the inner core. *Physics of the Earth and Planetary Interiors*, **129**: 325–346.

Brown, J.M., and McQueen, R.G., 1986. Phase-transitions, Gruneisen-parameter, and elasticity for shocked iron between 77-GPa and 400-GPa. *Geophysical Journal of the Royal Astronomical Society*, **91**: 7485–7494.

Buffett, B.A., 2000. Dynamics of the Earth's core. In Karato, S., Forte, A.M., Liebermann, R.C., Masters, G., and Stixrude, L. (eds.),

Earth's Deep Interior: Mineral Physics and Tomography from the Atomic to the Global Scale, American Geophysical Union Monograph 117. Washington, DC: American Geophysical Union, pp. 37–62.

Buffett, B.A., and Glatzmaier, G.A., 2000. Gravitational braking of inner-core rotation in geodynamo simulations. *Geophysical Research Letters*, **27**: 3125–3128.

Buffett, B.A., and Wenk, H.R., 2001. Texturing of the inner core by Maxwell stresses. *Nature*, **413**: 60–63.

Cormier, V.F, and Li, X., 2002. Frequency-dependent seismic attenuation in the inner core. 2. A scattering and fabric interpretation. *Journal of Geophysical Research*, **107**(B12): doi:10.1029/2002 JB001796.

Creager, K.C., 1992. Anisotropy of the inner core from differential travel times of the phases PKP and PKIKP. *Nature*, **356**: 309–314.

Creager, K.C., 1997. Inner core rotation rate from small-scale heterogeneity and time-varying travel times. *Science*, **278**: 1284–1288.

Creager, K.C., 2000. Inner core anisotropy and rotation. In Karato, S., Forte, A.M., Liebermann, R.C., Masters, G., and Stixrude, L. (eds.), *Earth's Deep Interior: Mineral Physics and Tomography from the Atomic to the Global Scale*, American Geophysical Union Monograph 117. Washington, DC: American Geophysical Union, pp. 89–114.

Glatzmaier, G.A., and Roberts, P.H., 1995. A three-dimensional convective dynamo solution with rotating and finitely conducting inner core and mantle. *Physics of the Earth and Planetary Interiors*, **91**: 63–75.

Ishii, M., and Dziewonski, A.M., 2002. The innermost inner core of the earth: evidence for a change in anisotropic behavior at the radius of about 300 km. *PNAS*, **99**(22): 14026–14030.

Jeanloz, R., and Wenk, H.R., 1988. Convection and anisotropy of the inner core. *Geophysical Research Letters*, **15**: 72–75.

Karato, S., 1999. Seismic anisotropy of the Earth's inner core resulting from flow induced by Maxwell stresses. *Nature*, **402**: 871–873.

Kuang, W.J., and. Bloxham, J., 1997. An earth-like numerical dynamo model. *Nature*, **389**: 371–374.

Masters, G., and Gilbert, F., 1981. Structure of the inner core inferred from observations of its spheroidal shear modes. *Geophysical Research Letters*, **8**: 569–571.

Morelli, A., Dziewonski, A.M., and Woodhouse, J.H., 1986. Anisotropy of the inner core inferred from PKIKP travel-times. *Geophysical Research Letters*, **13**: 1545–1548.

Niu, F.L., and Wen, L.X., 2001. Hemispherical variations in seismic velocity at the top of the Earth's inner core. *Nature*, **410**: 1081–1084.

Poupinet, G., Pillet, R., and. Souriau, A., 1983. Possible heterogeneity of the Earth's core deduced from PKIKP travel-times. *Nature*, **305**: 204–206.

Romanowicz, B., Li, X.D., and Durek, J., 1996. Anisotropy in the inner core: could it be due to low-order convection? *Science*, **274**(5289): 963–966.

Shearer, P.M., 1994. Constraints on inner core anisotropy from PKP(DF) travel-times. *Journal of Geophysical Research*, **99**: 19,647–19,659.

Shearer, P.M., Toy, K.M., and Orcutt, J.A., 1988. Axi-symmetric Earth models and inner-core anisotropy. *Nature*, **333**: 228–232.

Song, X.D., and Helmberger, D.V., 1988. Seismic evidence for an inner core transition zone. *Science*, **282**: 924–927.

Song, X.D., and Helmberger, D.V., 1993. Anisotropy of Earth's inner core. *Geophysical Research Letters*, **20**: 2591–2594.

Song, X.D., and Richards, P.G., 1996. Observational evidence for differential rotation of the Earth's inner core. *Nature*, **382**: 221–224.

Souriau, A., Garcia, R., and Poupinet, G., 2003. The seismological picture of the inner core: structure and rotation. *Comptes Rendus Geoscience*, **335**(1): 51–63.

Su, W.J., and Dziewonski, A.M., 1995. Inner core anisotropy in three dimensions. *Journal of Geophysical Research*, **100**: 9831–9852.

Sun, X.L., and Song, X.D., 2002. PKP travel times at near antipodal distances: implications for inner core anisotropy and lowermost mantle structure. *Earth and Planetary Science Letters*, **199**: 429–445.

Stevenson, D.J., 1987. Limits on lateral density and velocity variations in the Earth's outer core. *Geophysical Journal of the Royal Astronomical Society*, **88**: 311–319.

Stixrude, L., and Cohen, R.E., 1995. High-pressure elasticity of iron and anisotropy of Earth's inner core. *Science*, **267**: 1972–1975.

Tanaka, S., and Hamaguchi, H., 1997. Degree one heterogeneity and hemispherical variation of anisotropy in the inner core from PKP (BC)-*PKP*(DF) times. *Journal of Geophysical Research*, **102**: 2925–2938.

Tromp, J., 1993. Support for anisotropy of the Earth's inner core from free oscillations. *Nature*, **366**: 678–681.

Tromp, J., 2001. Inner-core anisotropy and rotation. *Annual Review of Earth and Planetary Sciences*, **29**: 47–69.

Vidale, J.E., and Earle, P.S., 2000. Fine-scale heterogeneity in the Earth's inner core. *Nature*, **404**(6775): 273–275.

Vinnik, L., Romanowicz, B., and Breger, L., 1994. Anisotropy in the center of the inner-core. *Geophysical Research Letters*, **21**(16): 1671–1674.

Widmer, R.W., Masters, G., and Gilbert, F., 1992. Observably split multiplets—data analysis and interpretation in terms of large-scale aspherical structure. *Geophysical Journal International*, **111**: 559–576.

Yoshida, S., Sumita, I., and. Kumazawa, M., 1996. Growth-model of the inner core coupled with the outer core dynamics and the resulting elastic anisotropy. *Journal of Geophysical Research*, **101**: 28085–28103.

Cross-references

Core Convection
Core Properties, Theoretical Determination
Geodynamo, Energy Sources
Geodynamo, Numerical Simulations
Grüneisen's Parameter for Iron and Earth's Core
Inner Core Composition
Inner Core Oscillation
Inner Core Rotation
Inner Core Rotational Dynamics
Inner Core Seismic Velocities
Lehmann, Inge (1888–1993)

INNER CORE COMPOSITION

Why is the inner core important?

The solid inner core is complex and not yet fully understood. This is hardly surprising given that the temperature of the Earth's *core* (*q.v.*) is in the range 5000–6000 K and inner core pressures are ~330–360 GPa. Knowledge of the exact composition and structure of the Earth's inner core would enable a better understanding of the internal structure and dynamics of the Earth as a whole; in particular, better constraints on *core composition* (*q.v.*) would not only have fundamental implications for models of the formation, differentiation, and evolution of the Earth, but would also enable successful interpretation of seismic observations which have revealed *inner core anisotropy* (*q.v.*), layering and heterogeneity (e.g., Creager, 1992; Song, 1997; Beghein and Trampert, 2003; Cao and Romanowicz, 2004; Koper *et al.*, 2004). The elastic anisotropy of the inner core is well established: *seismic velocities of the inner core* (*q.v.*) show P-wave velocities ~3% faster along the polar axis than in the equatorial plane (e.g., Creager, 1992). However, more recent seismic observations suggest further complexity. The evidence is for a seismically isotropic upper layer, with lateral variations in thickness of ~100–400 km, overlaying an irregular, nonspherical transition region to an anisotropic lower layer (Song and Helmberger, 1998; Ouzounis and Creager, 2001; Song and Xu, 2002). The existence of an isotropic upper layer implies that the magnitude of the seismic anisotropy in the lower inner core must be significantly greater than previously thought, possibly as much as 5%–10%. The observed layering also implies that the upper and lower inner core are compositionally or structurally different. Small-scale heterogeneities in the Earth's inner core have also been observed, possibly associated with phase changes, a "mushy layer," melt inclusions and/or compositional differences (Cao and Romanowicz, 2004; Koper *et al.*, 2004). The question now arises as to the mechanisms by which such complexity can occur, all of which could be highly compositionally dependent. Therefore, we cannot test any hypotheses for the observed features without knowing the inner core composition.

The inner core is not just made of iron

The exact composition of the Earth's inner core is not very well known. On the basis of cosmochemical and geochemical arguments, it has been suggested that the core is an iron alloy with possibly as much as ~5 wt% Ni and very small amounts (only fractions of a wt% to trace) of other siderophile elements such as Cr, Mn, P, and Co (McDonough and Sun, 1995). On the basis of materials density/sound-wave velocity systematics, Birch (1964) further concluded that the core is composed of iron that is alloyed with a small fraction of lighter elements. The light alloying elements most commonly suggested include S, O, Si, H, and C, although minor amounts of other elements, such K, could also be present (e.g., Poirier, 1994; Gessmann and Wood, 2002). From seismology it is known that the density jump across the inner core boundary is between ~4.5% and 6.7% (Shearer and Masters, 1990; Masters and Gubbins, 2003) indicating that there is more lighter element alloying in the outer core. The evidence thus suggests that the outer core contains ~5%–10% light elements, while the inner core has ~2%–3% light elements. Our present understanding is that the Earth's solid inner core is crystallizing from the outer core as the Earth slowly cools and the partitioning of the light elements between the solid and liquid is therefore crucial to understanding the evolution and dynamics of the core.

Which phase of iron exists in the inner core?

Before tackling the more detailed problem of inner *core composition* (*q.v.*) in terms of alloying elements, it is essential to know what crystalline structure(s) are present and also their seismic properties, as these will greatly affect both interpretations made from seismic data and also the possible alloying mechanism. Many experimentalists have put an enormous effort over the last 10–15 years into obtaining a phase diagram of pure iron under core conditions, but above relatively modest pressures and temperatures there is still much uncertainty. Experimental techniques have evolved rapidly in recent years, and today, using diamond anvil cells or *shock wave experiments* (*q.v.*), the study of minerals at pressures up to ~200 GPa and temperatures of a few thousand Kelvin are possible. These studies, however, are still far from routine and results from different groups are often in conflict. At low pressures and temperatures the phase diagram of iron is well understood: the body-centered-cubic (bcc) phase is stable at ambient conditions, transforming to the hexagonal close-packed phase (hcp) at high pressure (>10–15 GPa) and to the face-centered-cubic phase at high temperature (>1200 K). Both experiments and also theoretical calculations of the static, zero-Kelvin solid (Stixrude *et al.*, 1997; Vočadlo *et al.*, 2000), have suggested that the hcp phase has a wide stability field at high pressures *and* temperatures right to the conditions of the Earth's inner core. However, this assumption that iron must have the hcp structure at core conditions has recently been challenged (Brown, 2001; Beghein and Trampert, 2003), especially if the presence of lighter elements is included (Lin *et al.*, 2003). It now seems very possible that a bcc phase might be formed (Vočadlo *et al.*, 2003). Previously, the bcc phase of iron was considered an unlikely candidate for an inner core-forming phase because athermal calculations show it to be elastically unstable at high pressures, with an enthalpy considerably higher than that of hcp iron (Stixrude and Cohen, 1995; Söderlind *et al.*, 1996; Vočadlo *et al.*, 2000). However, more recent *ab initio* molecular dynamics calculations at core pressures and temperatures have suggested that the bcc phase of iron becomes entropically stabilized at core temperatures. Although for pure iron the thermodynamically most stable phase is still the hcp phase, the free energy difference is so very slight that a small amount of light element impurity could stabilize the bcc phase at the expense of the hcp phase (Vočadlo *et al.*, 2003).

Alloying elements

It is generally assumed that the small amount of nickel alloyed to iron in the inner core is unlikely to have any significant affect on *core properties* (*q.v.*) as nickel and iron have sufficiently similar densities to be seismically indistinguishable, and addition of small amounts of nickel are unlikely to appreciably change the physical properties of iron. However, the presence of light elements in the core does have an affect on core properties. The light element impurities most often suggested are sulfur, oxygen, and silicon. Although these alloying systems have been experimentally studied up to pressures of around 100 GPa (e.g., Li and Agee, 2001; Lin *et al.*, 2003; Rubie *et al.*, 2004) there seems little prospect of obtaining experimental data for iron alloys at the highly elevated pressures and temperatures of the Earth's inner core.

An alternative approach to understanding inner *core composition* (*q.v.*) is to simulate the behavior of these iron alloys with *ab initio* calculations which are readily able to access the pressures and temperatures of the inner core. Alfè *et al.* (2000, 2002) calculated the chemical potentials of iron alloyed with sulfur, oxygen, and silicon. They developed a strategy for constraining both the impurity fractions and the temperature at the inner core boundary (ICB) based on the supposition that the solid inner core and liquid outer core are in thermodynamic equilibrium at the ICB. For thermodynamic equilibrium the chemical potentials of each species must be equal both sides of the ICB, which fixes the ratio of the concentrations of the elements in the liquid and in the solid, which in turn fixes the densities. If the core consisted of pure iron, equality of the chemical potential (the Gibbs free energy in this case) would tell us only that the temperature at the ICB is equal to the *melting temperature of iron* (*q.v.*) at the ICB pressure of 330 GPa. With impurities present, the *ab initio* results reveal a major qualitative difference between oxygen and the other two impurities: oxygen partitions strongly into the liquid, but sulfur and silicon, both partition equally in the solid and liquid. Having established the partitioning coefficients, Alfè *et al.* (2002) then investigated whether the known densities of the outer and inner core, estimated from seismology, could be matched by one of their calculated binary systems. For sulfur and silicon, their ICB density discontinuities were considerably smaller than the known seismological value at that time of 4.5% ± 0.5% (Shearer and Masters, 1990); for oxygen, the discontinuity was markedly greater than that from seismology. Therefore none of these binary systems were plausible, i.e., the core *cannot* be made solely of Fe/S, Fe/Si, or Fe/O. However, the seismic data can clearly be matched by a ternary/quaternary system of iron and oxygen together with sulfur and/or silicon. A system consistent with seismic data could contain 8 mol% oxygen and 10 mol% sulfur and/or silicon in the outer core, and 0.2 mol% oxygen and 8.5 mol% sulfur and/or silicon in the inner core (Alfè *et al.*, 2002). However, it should be remembered that it is likely that several other light elements could exist in the inner core and would therefore have to be considered before a true description of inner core composition could be claimed.

Lidunka Vočadlo

Bibliography

Alfè, D., Gillan, M.J., and Price, G.P., 2000. Constraints on the composition of the Earth's core from ab initio calculations. *Nature*, **405**: 172–175.

Alfè, D., Gillan, M.J., and Price, G.P., 2002. *Ab initio* chemical potentials of solid and liquid solutions and the chemistry of the Earth's core. *Journal of Chemical Physics*, **116**: 7127–7136.

Beghein, C., and Trampert, J., 2003. Robust normal mode constraints on inner-core anisotropy from model space search. *Science*, **299**: 552–555.

Birch, F., 1964. Density and composition of the mantle and core. *Journal of Geophysical Research*, **69**: 4377–4388.

Brown, J.M., 2001. The equation of state of iron to 450 GPa: another high pressure solid phase? *Geophysical Research Letters*, **28**: 4339–4342.

Cao, A., and Romanowicz, B., 2004. Hemispherical transition of seismic attenuation at the top of the Earth's inner core. *Earth and Planetary Science Letters*, **228**: 243–253.

Creager, K.C., 1992. Anisotropy of the inner core from differential travel-times of the phases PKP and PKIKP. *Nature*, **356**: 309–314.

Gessmann, C.K., and Wood, B.J., 2002. Potassium in the Earth's core? *Earth and Planetary Science Letters*, **200**: 63–78.

Koper, K.D., Franks, J.M., and Dombrovskaya, M., 2004. Evidence for small-scale heterogeneity in Earth's inner core from a global study of PkiKP coda waves. *Earth and Planetary Science Letters*, **228**: 227–241.

Li, J., and Agee, C.B., 2001. Element partitioning constraints on the light element composition of the Earth's core. *Geophysical Research Letters*, **28**: 81–84.

Lin, J-F., *et al.*, 2003. Sound velocities of iron-nickel and iron-silicon alloys at high pressures. *Geophysical Research Letters*, **30**: doi:10.1029/2003GL018405.

Masters, G., and Gubbins, D., 2003. On the resolution of density within the Earth. *Physics of the Earth and Planetary Interiors*, **140**: 159–167.

McDonough, W.F., and Sun, S-S., 1995. The composition of the Earth. *Chemical Geology*, **120**: 223–253.

Ouzounis, A., and Creager, K.C., 2001. Isotropy overlying and isotropy at the top of the inner core. *Geophysical Research Letters*, **28**: 4331–4334.

Poirier, J-P., 1994. Light elements in the Earth's outer core: a critical review. *Physics of the Earth and Planetary Interiors*, **85**: 319–337.

Rubie, D.C., Gessmann, C.K., and Frost, D.J., 2004. Partitioning of oxygen during core formation of the Earth and Mars. *Nature*, **429**: 58–61.

Shearer, P., and Masters, G., 1990. The density and shear velocity contrast at the inner core boundary. *Geophysical Journal International*, **102**: 491–498.

Söderlind, P., Moriarty, J.A., and Wills, J.M., 1996. First-principles theory of iron up to Earth's core pressures: structural, vibrational and elastic properties. *Physical Review B*, **53**: 14,063–14,072.

Song, X., 1997. Anisotropy of the Earth's inner core. *Reviews in Geophysics*, **35**(3): 297–313.

Song X., and Helmberger, D.V., 1998. Seismic evidence for an inner core transition zone. *Science*, **282**: 924–927.

Song, X., and Xu, X., 2002. Inner core transition zone and anomalous PKP(DF) waveforms from polar paths. *Geophysical Research Letters*, **29**: 1–4.

Stixrude, L., and Cohen, R.E., 1995. Constraints on the crystalline structure of the inner core—mechanical instability of bcc iron at high pressure. *Geophysical Research Letters*, **22**: 125–128.

Stixrude, L., Wasserman, E., and Cohen, R.E., 1997. Composition and temperature of the Earth's inner core. *Journal of Geophysical Research*, **102**: 24729–24739.

Vočadlo. L., *et al.,* 2000. Ab initio free energy calculations on the polymorphs of iron at core conditions. *Physics of the Earth and Planetary Interiors*, **117**: 123–127.

Vočadlo. L., *et al.,* 2003. Possible thermal and chemical stabilisation of body-centred-cubic iron in the Earth's core. *Nature*, **424**: 536–539.

Cross-references

Core Composition
Core Properties, Physical
Core Properties, Theoretical Determination
Core Temperature
Inner Core Anisotropy
Inner Core Seismic Velocities
Melting Temperature of Iron in the Core, Experimental
Melting Temperature of Iron in the Core, Theory
Shock Wave Experiments

INNER CORE OSCILLATION

Earth's inner core, identified in relatively recent time (*Lehmann,* 1936) (*q.v.*), can be considered to be an elastic solid on timescales of seismic waves. With a radius of 1220 km and hence containing less than 1% of Earth's volume, the inner core's importance to our understanding of Earth structure far exceeds both its relative size and our knowledge of its properties.

Decadal changes in Earth's magnetic field, linked to *torsional oscillations* (*q.v.*) of the fluid outer core, can produce corresponding oscillations of the inner core through electromagnetic coupling. Since the inner core is well coupled gravitationally to the mantle, both changes in *LOD* (*q.v.*) and polar motion of the mantle on a decade timescale could be used to constrain torsional oscillations (Dumbery and Bloxham, 2003) (see also *Inner core rotation*).

Gravitational force is responsible for holding the slightly denser inner core at Earth's center in the surrounding fluid core. Thus the inner core will return to this central equilibrium position if displaced axially and subsequently oscillate at periods of a few hours. This fact was initially recognized by Slichter (1961) who also realized that Earth's rotation would lead to two additional circular modes, one prograde and one retrograde if the inner core were displaced in a direction perpendicular to the rotation axis.

Theoretical values of three modes of oscillation, often referred to as the *Slichter triplet*, depend on the physical properties of the core described by the Earth model chosen, as well as the method used to calculate their periods. In the case of the Earth model 1066A (Gilbert and Dziewonski, 1975), for example, the buoyancy oscillation period without Earth rotation or degenerate period, is reported as 4.599 h by Rogister (2003), 4.309 h by Rieutord (2002), and 4.45471 h by Smylie *et al.* (2001). The central period of the triplet has been calculated using normal mode theory (Rogister, 2003), yielding estimates near 5.309 h for the Earth model PREM (Dziewonski and Anderson, 1981) and 4.529 h for 1066A which closely agree with those from perturbation methods (e.g., Dahlen and Sailor, 1979; Crossley *et al.*, 1992) that ignored ellipticity and centrifugal effects. A central period of 4.255 h for 1066A has been found by Rieutord (2002) who reminded us that the buoyancy of the inner core is determined by the difference between the mean density of the inner core and its surrounding fluid rather than the density jump at the inner core-outer core boundary. An even shorter central period of 3.7926 h is predicted (Smylie *et al.*, 2001) for the Cal8 Earth model (Bullen and Bolt, 1985) with a higher density inner core that experiences a larger restoring force when displaced.

The Slichter triplet has been the subject of search in gravimetric records since a movement of the inner core will produce a small change in Earth's gravitational field. Following the Chilean earthquake of May 22, 1960, a peak in the gravimetric record near 86 min was tentatively identified as an inner core oscillation, implying a much denser inner core than would be consistent with existing Earth models. Significant attempts had been made to detect the Slichter triplet by locating gravimeters near the South Pole where the diurnal and semidiurnal tidal effects are small (Rydelek and Knopoff, 1984) but there was no identification of the triplet.

In recent years, stable, low noise, superconducting gravimeters have allowed for the measurement of changes in gravity down to $\pm 2 \times 10^{-11}$ m s^{-2} in Earth's gravity field of about 9.8 m s^{-2}. Thus, long records can be analyzed in the frequency domain for three spectral peaks separated by precisely the amount predicted by the Earth model chosen. Even with the above sensitivity, however, the signal-to-noise ratios are so small that no one has yet observed the Slichter triplet in a single gravimetric record. By combining several superconducting gravimetric records with a wide geographical distribution, through a method known as product spectral analysis (Smylie *et al.*, 1993), a claim has been made (Smylie *et al.*, 2001) to have observed the Slichter triplet at periods 3.5822, 3.7656, and 4.0150 h with a statistical error of 0.04%. These observed values are reported to correspond to periods of 3.5168, 3.7926, and 4.1118 h predicted from the Cal8 model assuming the fluid core is inviscid. This identification and the method of analysis reported by Smylie *et al.* (2001) are yet to be widely accepted. Other groups have either failed to find the Slichter triplet (Hinderer *et al.*, 1995) or, more recently, have reported several possible candidate signals for the Slichter triplet depending on the Earth model chosen (Rosat *et al.*, 2004). Still others have cautioned (Florsch *et al.*, 1995) that there is a high likelihood in finding significant fits to a triplet of spectral peaks even if one chooses the frequencies at random.

The rotational splitting of the Slichter triplet should include a contribution from viscosity of the liquid outer core as noted by Smylie (1999). Using boundary layer theory, he calculated the viscosity of the outer core fluid in a layer next to the inner core boundary, based on the small discrepancy between the observed and predicted triplet periods. He found a very high kinematic viscosity, close to 10^7m^2 s^{-1}, that was interpreted as confirming the semifluid nature of the layer at the base of the fluid core as the source of energy for the geodynamo. Such a large viscosity, however, brings into question the validity of results obtained by using only the leading term of the asymptotic series, as pointed out by Rieutord (2002). Future models that include dissipation due to turbulence (see *Core turbulence*) and partial solidification in the boundary layer around the inner core would decrease estimates of viscosity of the fluid core.

Periods of the Slichter triplet are extremely sensitive to the mean density of the *inner core* (*q.v.*), about 3 h for each gm cm^{-3} for the axial mode (Smylie *et al.*, 2001), and thus would provide an independent estimate of that important quantity. Confirmation of the measured periods of the Slichter triplet will provide the basis for an estimation of the density jump at the inner core-outer core boundary, that is central to estimation of the inner core's age and its role in a geodynamo driven by compositional convection. Further improvement in signal-to-noise ratio of gravity measurements and a larger excitation of the inner core, whose source remains unknown, are needed to validate the detection of the Slichter triplet.

Keith Aldridge

Bibliography

Bullen, K.E., and Bolt, B.A., 1985. *An Introduction to the Theory of Seismology.* Cambridge: Cambridge University Press.

Crossley, D., Rochester, M., and Peng, Z., 1992. Slichter modes and Love numbers. *Geophysical Research Letters*, **19**: 1679–1682.

Dahlen, F.A., and Sailor, R.V., 1979. Rotational and elliptical splitting of the free oscillations of the Earth. *Geophysical Journal of the Royal Astronomical Society*, **58**: 609–623.

Dumbery, M., and Bloxham, J., 2003. Torque balance, Taylor's constraint and torsional oscillations in a numerical model of the geodynamo. *Physics of the Earth and Planetary Interiors*, **140**: 29–51.

Dziewonski, A.M., and Anderson, D.L., 1981. Preliminary Reference Earth Model (PREM). *Physics of the Earth and Planetary Interiors*, **25**: 297–356.

Florsch, N., Legros, H., and Hinderer, J., 1995. The search for weak harmonic signals in a spectrum with applications to gravity data. *Physics of the Earth and Planetary Interiors*, **90**: 197–210.

Gilbert, F., and Dziewonski, A.M., 1975. An application of normal mode theory to the retrieval of structural parameters and source mechanisms from seismic spectra. *Philosophical Transactions of the Royal Society of London, Series A*, **278**: 187–269.

Hinderer, J., Crossley, D., and Jensen, O., 1995. A search for the Slichter triplet in superconducting gravimeter data. *Physics of the Earth and Planetary Interiors*, **90**: 221–241.

Lehmann, I., 1936. *P′. Publication Bureau Central Seismologique International, Series A*, **14**: 87–115.

Rieutord, M., 2002. Slichter modes of the Earth revisited. *Physics of the Earth and Planetary Interiors*, **131**: 269–278.

Rogister, Y., 2003. Splitting of seismic-free oscillations and of the Slichter triplet using the normal mode theory of a rotating, ellipsoidal earth. *Physics of the Earth and Planetary Interiors*, **140**: 169–182.

Rosat, S., Hinderer, J., Crossley, D., and Boy, J.P., 2004. Performance of superconducting gravimeters from long-period seismology to tides. *Journal of Geodynamics*, **38**: 461–476.

Rydelek, P., and Knopoff, L., 1984. Spectral analysis of gapped data: search for the mode ${}_1S_1$ at the South Pole. *Journal of Geophysical Research*, **89**: 1899–1911.

Slichter, L.B., 1961. The fundamental free mode of the Earth's inner core. *Proceeding of the National Academy of Sciences USA*, **47**: 186–190.

Smylie, D.E., 1999. Viscosity near earth's solid inner core. *Science*, **284**: 461–463.

Smylie D.E., Hinderer, J., Richter, B., and Ducarme, B., 1993. The product spectra of gravity and barometric pressure in Europe. *Physics of the Earth and Planetary Interiors*, **80**: 135–157.

Smylie, D.E., Francis, O., and Merriam, J.B., 2001. Beyond tides—determination of core properties from superconducting gravimeter observations. *Journal of the Geodetic Society of Japan*, **47**: 364–372.

Cross-references

Core Density
Core Turbulence
Inner Core Rotation
Lehmann, Inge (1888–1993)
Length of Day Variations, Decadal
Oscillations, Torsional

INNER CORE ROTATION

Observational evidence of inner core rotation is based on changes in the travel-time of seismic waves. The rotation rate appears to be a few tenths of a degree per year eastward with respect to the mantle.

Early speculations

The inner core with radius of about 1215 km resides concentrically within the much larger fluid outer core, which has a low viscosity (e.g., Poirier, 1998). Patterns of convection within the fluid core associated with the geodynamo are presumed to undergo temporal variations because the Earth's magnetic field is changing on timescales ranging over several orders of magnitude. It is therefore reasonable to speculate (Gubbins, 1981; Anderson, 1983) that the inner core might have a rotation rate somewhat different from that of the rest of the solid Earth, which is dominated by the daily rotation. If such relative rotation could be detected, it would provide information on the energy of convection patterns that maintain the geomagnetic field.

The study of inner core motion relative to the mantle and crust is difficult, because of the remoteness of the inner core (more than 5150 km from the Earth's surface where the nearest observations can be made) and its small size (about 0.7% of the Earth's volume). Also there are intrinsic difficulties in telling whether a spherical object is rotating, unless a marker on or within the object can be identified and tracked if it moves.

First claims, based on seismological observations and implications

The first published claim of observational evidence for inner core rotation, relative to the mantle and crust, was given by Song and Richards (1996). They noted that seismic waves originating in the South Sandwich Islands (in the southernmost South Atlantic) and recorded in Alaska, which had traveled through the inner core, appeared to have traveled systematically faster for earthquakes in the later years during the period from 1967 to 1995, compared to observations of seismic waves from earlier earthquakes in this time period. The rate of travel-time decrease amounted to about 0.01 s y^{-1}— a value that was close to the precision of measurement.

More detailed studies by Creager (1997), Song (2000), and Li and Richards (2003), have provided additional support for travel-time change of about this same value, for seismic P-waves crossing the inner core. Figure I7 shows a clear example of PKIKP waves (which pass through the inner core) showing a faster arrival for the later earthquake, when the seismograms are aligned on PKP waves (which avoid the inner core).

Song and Richards (1996) interpreted the travel-time change as an effect of anisotropy, which causes P-waves to travel with speeds that depend on direction relative to a crystalline axis that rotates with the inner core. Creager (1997) persuasively argued for a more important marker of inner core rotation, namely a lateral gradient of the P wavespeed (increasing from east to west) in the part of the inner core traversed by PKIKP waves. He and Song (2000), Richards (2000), and Li and Richards (2003) concluded the observed travel-time changes indicate an eastward rotation of the inner core amounting to a few tenths of a degree per year.

The concept of an inner core rotating fast enough to be detected on a human timescale has attracted numerous investigators since 1996. Dehant *et al.* (2003) describe inner core research in mineral physics, seismology, geomagnetism, and geodesy. The rate of inner core rotation is an indication of the vigor of convection in the outer core, associated with the geodynamo. A nonzero rotation rate can be used to place limits on the outer core's viscosity (Buffett, 1997).

Counterclaims, and additional methods and evidence

Several papers since 1996 have argued that the seismological evidence for inner core rotation is equivocal. Thus, Souriau and Poupinet (in Dehant *et al.*, 2003) claim the reported travel-time changes of PKIKP waves on the path between the South Sandwich Islands and Alaska are an artifact of mislocated earthquakes. Early reports of purported changes in the absolute arrival times (not differential times) of PKIKP waves were later dismissed as based on inadequate evidence.

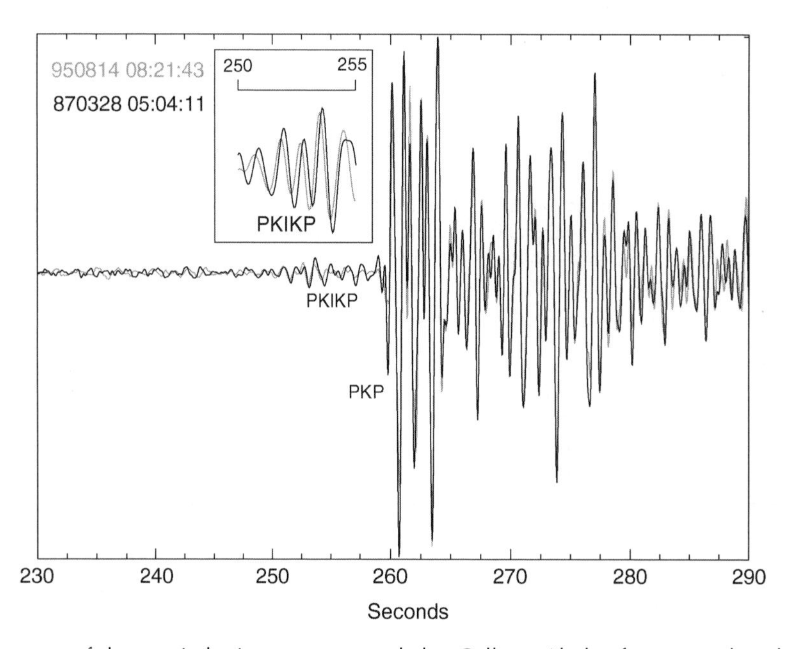

Figure I7 One minute segments of short-period seismograms recorded at College, Alaska, for two earthquakes in the South Sandwich Islands (March 28, 1987 and August 14, 1995). For 30 s following the PKP arrival, and for an additional 3 min (not shown here), these seismograms (passed in the band from 0.6 to 3 Hz) show excellent waveform agreement for signals that have traversed the Earth but not via the inner core. For the PKIKP waveform, from time 250 to 255 s, an insert shows an expanded view of the narrowband filtered version of the two arrivals (in the passband from 0.8 to 1.5 Hz) that have traversed the inner core. With the two seismograms aligned on the PKP phase, it is seen that the PKIKP phase of the later event (shown in gray) traveled slightly faster. An explanation is that the inner core rotated during the 8-year period between the earthquakes, in a manner that provided a faster path for the later PKIKP signal through the inner core.

Laske and Masters have used normal mode data to study inner core inhomogeneities. Some modes appear to indicate eastward rotation, others westward, and their paper in Dehant *et al.* (2003) concludes the rate is only marginally indicative of a small eastward rotation, about $0.15° \, y^{-1}$ (but alternatively estimated as $0.34 \pm 0.13° \, y^{-1}$ if the normal modes likely to be most contaminated by upper mantle structure are excluded). Vidale and Earle (2000) used backscatter from within the inner core, following PKIKP waves, to find an eastward rotation of the inner core amounting to a few tenths of a degree per year.

It appears that the strongest claims of evidence for inner core rotation derive from differential traveltimes, for earthquakes separated by several years and which occur at essentially the same location, generating very similar waveforms. The evidence for inner core rotation is still under debate, and consensus on inner core rotation will likely depend on whether examples such as that given in Figure I7 can be accumulated, since waveform doublets avoid artifacts of event mislocation. A report on 18 high-quality doublets with time separation of up to 35 years in the South Sandwich Islands region, observed at up to 58 stations in and near Alaska, provides such an accumulation (see Zhang *et al.*, 2005).

Paul G. Richards and Anyi Li

Bibliography

Anderson, D.L., 1983. A new look at the inner core of the Earth. *Nature*, **302**: 660.

Buffett, B.A., 1997. Geodynamic estimates of the viscosity of the Earth's inner core. *Nature*, **388**: 571–573.

Creager, K.C., 1997. Inner core rotation rate from small-scale heterogeneity and time-varying travel times. *Science*, **278**: 1284–1288.

Dehant, V., Creager, K.C., Karato, S., and Zatman, S. (eds.), 2003. *Earth's Core: Dynamics, Structure, Rotation*, Geodynamics series 31. Washington, DC: American Geophysical Union, p. 279.

Gubbins, D., 1981. Rotation of the inner core. *Journal of Geophysical Research*, **86**: 11695–11699.

Li, A., and Richards, P.G., 2003. Using doublets to study inner core rotation and seismicity catalog precision. *G-Cubed*, **4**: 1072, doi:10.1029/2002GC000379.

Poirier, J.P., 1998. Transport properties of liquid metals and viscosity of the Earth's core. *Geophysical Journal of the Royal Astronomical Society*, **92**: 99–105.

Richards, P.G., 2000. Earth's inner core—discoveries and conjectures. *Astronomy and Geophysics*, **41**: 20–24.

Song, X., 2000. Joint inversion for inner core rotation, inner core anisotropy, and mantle heterogeneity. *Journal of Geophysical Research*, **105**: 7931–7943.

Song, X., and Richards, P.G., 1996. Seismological evidence for differential rotation of the Earth's inner core. *Nature*, **382**: 221–224.

Vidale, J.E., and Earle, P.S., 2000. Slow differential rotation of the Earth's inner core indicated by temporal changes in scattering. *Nature*, **405**: 445–448.

Zhang, J., Song, X., Li, Y., Richards, P.G., Sun, X., and Waldhauser, F., 2005. Inner core differential motion confirmed by earthquake waveform doublets. *Science*, **310**(5752): 1279.

Cross-references

Core Motions
Geodynamo
Geodynamo, Energy Sources
Geomagnetic Spectrum, Temporal
Inner Core Anisotropy
Inner Core Composition
Inner Core Seismic Velocities
Lehmann, Inge (1888–1993)
Length of Day Variations, Long Term

INNER CORE ROTATIONAL DYNAMICS

For several decades after its discovery in 1936, the Earth's solid inner core (IC) played only a passive role in geodynamics. Now the ICs material properties, internal structure, and rotational dynamics are lively concerns of geomagnetism, seismology, and geodesy. Suppose the mantle spins about its figure axis $\mathbf{e}_3$ at rate Ω relative to the stars, i.e., $2\pi/\Omega = 1$ sidereal day (sd). In principle the IC can exhibit:

a. differential rotation, i.e., spin relative to the mantle at angular rate $\Delta\omega$ about the IC figure axis $\mathbf{i}_3$ (where $\mathbf{i}_3$ is possibly inclined at angle ϵ to $\mathbf{e}_3$) and
b. small periodic changes in orientation of $\mathbf{i}_3$ relative to $\mathbf{e}_3$ (wobble) and relative to the stars (nutation).

The spinning IC can be regarded as a gyroscope bathed in the surrounding liquid outer core (OC), and will orient itself relative to the mantle in response to the torques exerted on it. Changes in OC flow will cause *fluid pressure torques* at the inner core boundary (ICB). The dominantly iron IC (see *Inner core composition*) is expected to have an electrical conductivity comparable to that of the OC, allowing it to be penetrated by changing magnetic fields generated by dynamo action in the OC, and experience *electromagnetic torques* as a result of Lenz's law. The OC viscosity is probably so small near the ICB that viscous torques are negligible (see *Core viscosity*). The vital importance of *gravitational torques*, exerted on the body of the IC not only by the OC but also by the mantle, has been realized only in the past two decades.

Braginsky (1964) first noted that one boundary condition on the geomagnetic dynamo is a (possibly nonzero) value of $\Delta\omega$ fixed by the requirement of zero net electromagnetic torque on the IC. Gubbins (1981) pointed out that any equatorial component of $\Delta\omega$ would be shielded by the electrically conducting lower mantle, and (by modeling the OC as a solid spherical shell rotating with the mantle) showed that a steady axial component could result from the balance of two electromagnetic torques: one exerted by the dynamo-generated field in the OC, and the other induced by the nonzero $\Delta\omega$ itself. Such a torque balance was incorporated in their 2-D numerical hydromagnetic dynamo model by Hollerbach and Jones (1993), who found that the IC (by virtue of its finite electrical conductivity and solidity) played a crucial role in stabilizing dynamo operation. This was confirmed by the numerical 3-D model of a self-sustaining geodynamo created by Glatzmaier and Roberts (1995). This model predicted that, under the balance of both electromagnetic and viscous torques (assuming an artificially large value for OC viscosity for computational reasons), the IC rotates dominantly eastward relative to the mantle at a few degrees per year, changing on a timescale of 500 years. However Kuang and Bloxham (1997) found that a much smaller, more realistic, value of OC viscosity leads to an alternative dynamo in which $\Delta\omega$ fluctuates, on a timescale of several thousand years, between eastward and westward values. (See *Geodynamo, numerical simulations* for details.) Buffett and Glatzmaier (2000) showed that dynamo models incorporating gravitational torques between the mantle and IC yield much smaller values of $\Delta\omega$, on average about 0.17 or $0.02° \, y^{-1}$ eastward according as viscous torques do or do not act on the ICB.

Seismological evidence that the IC departed from isotropy and radial stratification began to emerge nearly two decades ago (see *Inner core anisotropy* and *Inner core composition* for details). Using such nonuniformities to directly determine IC rotation became an exciting new branch of seismology when Song and Richards (1996) realized that any tilt of the anisotropy axis away from the rotation axis would cause a systematic change with time in the travel-times of P-waves through the IC from the same source location to the same seismograph. They used 30 years of data to estimate a tilt of about 10° and $\Delta\omega \cong 1° \, y^{-1}$ eastward. Su *et al.* (1996), analyzing a different but equally long data set, found $\Delta\omega \cong 3° \, y^{-1}$. However most subsequent studies, involving the passage of a particular IC feature (anisotropy or lateral heterogeneity) across a particular ray path, led to smaller values of $\Delta\omega$, of order $0.1° \ y^{-1}$ (e.g., Creager, 1997; Poupinet *et al*., 2000; Vidale *et al.*, 2000; Isse and Nakanishi, 2002). A very different seismological approach (Laske and Masters, 1999) examined the effects of such nonuniformities on the spectra of nine free oscillations sampling the IC, and yielded $\Delta\omega = 0.01 \pm 0.21° \ y^{-1}$. Because such vibrations are comparatively unaffected by errors in locating seismic events and by local IC structure, and are independent of the earthquake source mechanism, this result seems particularly robust and provides strong evidence that $\Delta\omega$ is likely to be small, possibly not significantly different from zero.

The latter result is consistent with the inference that the IC is likely to be gravitationally locked to the mantle, because the gravitational torque exerted on the IC by mantle heterogeneities, when the IC and mantle are not in their equilibrium orientation, greatly exceeds any other (e.g., electromagnetic) torque that might act to maintain such a misalignment (Buffett, 1996). A more extensive analysis of gravitational coupling between IC and mantle by Xu *et al.* (2000) showed that IC superrotation cannot be supported by such torques alone, and that unless $\epsilon < 3°$ the corresponding values of $\Delta\omega$ far exceed those inferred from seismology. However, Buffett (1996) pointed out that a nonzero $\Delta\omega$ could be sustained if the IC viscosity is low enough ($<5 \times 10^{16}$ Pa s) to allow its shape to deform as the IC moves out of alignment with the mantle, but not so low as to quickly annul any misalignment.

Any seismological indications of a value for $\Delta\omega$, whether zero or nonzero, constitute only a recent snapshot of its possible behavior on the longer timescales of the geodynamo. Hollerbach (1998) concluded that a value for $\Delta\omega$ will indicate the average angular velocity of the OC just above the ICB, but is unlikely to suggest a reliable estimate of magnetic field strength at the ICB unless seismology can detect fluctuations in $\Delta\omega$ induced by torsional oscillations in the OC. At present, seismological data is not inconsistent with, but can hardly be regarded as constraining, recent dynamo models.

Periodic changes in orientation of $\mathbf{i}_3$ relative to $\mathbf{e}_3$ are a particularly interesting subset of the Earth's free oscillations, involving a redistribution of angular momentum within the rotating planet. The presence of the IC potentially adds two free modes to the Earth's polar motion spectrum: an ultralong period prograde inner core wobble (ICW) and a nearly diurnal retrograde wobble accompanied by a much larger, and prograde, long period free inner core nutation (FICN) The first discussions of the ICW (Busse, 1970; Kakuta *et al.,* 1975) neglected gravitational coupling of the IC to the mantle. Models of IC rotational dynamics, taking into account gravitation, fluid pressure at the ICB, and elastic deformation using Earth properties from PREM, predict periods for the FICN of 475 sd and for the ICW about 2400 sd (Mathews *et al*., 1991). However, the first reported detection of the FICN, in an analysis of data from very long baseline radio interferometry (VLBI) by Mathews *et al.* (2002), indicated a period in the range 930–1150 sd. Buffett *et al.* (2002) attributed the lengthening over previous predicted periods to IC-OC electromagnetic coupling, due to penetration of the IC by a magnetic field of 7 mT at the ICB. So far the ICW remains below the level of detection in the data from VLBI and gravimetry (Guo *et al*., 2005). For the case of a rigid IC and mantle coupled only by gravitation and fluid pressure, Guo and Ning (2002) found that nonzero $\Delta\omega$ has no effect on the FICN period but slightly shortens the ICW period, and inclination ϵ slightly lengthens both periods. Dumberry and Bloxham (2002) showed that electromagnetic torques applied to the ICB by a radial magnetic field of a few microteslas can drive the Markowitz wobble, a somewhat enigmatic feature of the Earth's polar motion in which the rotation pole traces out (relative to the crust) a highly elliptical path with period about 30 years.

No longer treated with benign neglect, the rotational dynamics of the IC now provides essential content in geodynamo models, an interpretive challenge to seismologists probing the deepest interior of the Earth, and intriguing new features for geodesists exploring the small changes in the length of day and the motion of the pole.

Michael G. Rochester

Bibliography

Braginsky, S.I., 1964. Kinematic models of the Earth's hydromagnetic dynamo. *Geomagnetizm i Aeronomiya*, **4**: 572–583.

Buffett, B.A., 1996. A mechanism for decade fluctuations in the length of day. *Geophysical Research Letters*, **23**: 3803–3806.

Buffett, B.A., and Glatzmaier, G.A., 2000. Gravitational braking of inner-core rotation in geodynamo simulations. *Geophysical Research Letters*, **27**: 3125–3128.

Buffett, B.A., Mathews, P.M., and Herring, T.A., 2002. Modelling of nutation and precession: effects of electromagnetic coupling. *Journal of Geophysical Research*, **107**(B4): ETG 5.

Busse, F.H., 1970. The dynamical coupling between inner core and mantle of the Earth and the 24-year libration of the pole. In Mansinha, L., Smylie, D.E., and Beck, A.E. (eds.), *Earthquake Displacement Fields and the Rotation of the Earth.* Dordrecht: D. Reidel, pp. 88–98.

Creager, K.C., 1997. Inner core rotation rates from small-scale heterogeneity and time-varying travel times. *Science*, **278**: 1284–1288.

Dumberry, M., and Bloxham, J., 2002. Inner core tilt and polar motion. *Geophysical Journal International*, **151**: 377–392.

Glatzmaier, G.A., and Roberts, P.H., 1995. A three-dimensional convective dynamo solution with rotating and finitely conducting inner core and mantle. *Physics of the Earth and Planetary Interiors*, **91**: 63–75.

Gubbins, D., 1981. Rotation of the inner core. *Journal of Geophysical Research*, **86**: 11695–11699.

Guo, J.Y., Greiner-Mai, H., and Ballani, L., 2005. Spectral search for the inner core wobble in Earth's polar motion. *Journal of Geophysical Research,* **110**(B10): 10.1029/2004JB003377.

Guo, J.Y., and Ning, J.S., 2002. Influence of inner core rotation and obliquity on the inner core wobble and the free inner core nutation. *Geophysical Research Letters*, **29**(8): 45.

Hollerbach, R., 1998. What can the observed rotation of the Earth's inner core reveal about the state of the outer core? *Geophysical Journal International*, **135**: 564–572.

Hollerbach, R., and Jones, C.A., 1993. A geodynamo model incorporating a finitely conducting inner core. *Physics of the Earth and Planetary Interiors*, **75**: 317–327.

Isse, T., and Nakanishi, I., 2002. Inner core anisotropy beneath Australia and differential rotation. *Geophysical Journal International*, **151**: 255–263.

Kakuta, C., Okamoto, I., and Sasao, T., 1975. Is the nutation of the solid inner core responsible for the 24-year libration of the pole? *Publications of the Astronomical Society of Japan*, **27**: 357–365.

Kuang, W., and Bloxham, J., 1997. An Earth-like numerical dynamo model. *Nature*, **389**: 371–374.

Laske, G., and Masters, G., 1999. Limits on differential rotation of the inner core from an analysis of the Earth's free oscillations. *Nature*, **402**: 66–69.

Mathews, P.M., Buffett, B.A., Herring, T.A., and Shapiro, I.I., 1991. Forced nutations of the Earth: influence of inner core dynamics. *Journal of Geophysical Research*, **96B**: 8219–8257.

Mathews, P.M., Herring, T.A., and Buffett, B.A., 2002. Modelling of nutation and precession: new nutation series for nonrigid Earth and insights into the Earth's interior. *Journal of Geophysical Research*, **107**(B4): ETG 3.

Poupinet, G., Souriau, A., and Coutant, O., 2000. The existence of an inner core super-rotation questioned by teleseismic doublets. *Physics of the Earth and Planetary Interiors*, **118**: 77–88.

Song, X., and Richards, P., 1996. Seismological evidence for differential rotation of the Earth's inner core. *Nature*, **382**: 221–224.

Su, W.J., Dziewonski, A.M., and Jeanloz, R., 1996. Planet within a planet: rotation of the inner core of the Earth. *Science*, **274**: 1883–1887.
Vidale, J.E., Dodge, D.A., and Earle, P.S., 2000. Slow differential rotation of the Earth's inner core indicated by temporal changes in scattering. *Nature*, **405**: 445–448.
Xu, S., Crossley, D.J., and Szeto, A.M.K., 2000. Variations in length of day and inner core differential rotation from gravitational coupling. *Physics of the Earth and Planetary Interiors*, **117**: 95–110.

Cross-references

Core Viscosity
Geodynamo, Numerical Simulations
Inner Core Anisotropy
Inner Core Composition

INNER CORE SEISMIC VELOCITIES

The inner core was discovered in 1936 by the Danish seismologist Inge Lehmann (1888–1993) from the observation of short-period P-wave arrivals in the shadow zone of the core, at a distance where no P-wave would arrive if the core was totally fluid. Since her discovery, major advances concerning the structure of the inner core have been made thanks to seismological studies, combined with observational and computational results in mineral physics. A good knowledge of inner core structure is important to specify the chemical and mineralogical nature of the iron alloy which constitutes the center of the Earth, and to constrain the Earth's differentiation process. It also has important outcomes for the thermodynamical and geodynamical properties of the Earth.

Investigation tools: seismic waves and normal modes

The structure and properties of the inner core are mostly constrained by two body waves: PKIKP, a P-wave transmitted through the inner core, and PKiKP, a P-wave reflected at the inner core boundary (ICB). They give information on the P-wave velocity and attenuation inside the inner core, on the topography and properties of ICB, and on the S-velocity and density contrasts at the ICB, as these parameters influence the P-wave reflection and transmission coefficients. In theory, the S-velocity is constrained by the phase PKJKP, which propagates as an S-wave in the inner core, but its observation is still controversial. The normal modes sensitive to inner core give information on the mean P- and S-velocities and mean density in the inner core. However, most of the modes sensitive to inner core structure have stronger energy in the mantle and in the liquid core.

The phase PKIKP, also called PKPdf, is often compared to core phases having their turning points inside the liquid core, namely PKPbc (which turns at the base of the liquid core with a path in the mantle very close to that of PKPdf), and PKPab (which turns in the middle of the liquid core) (See Figure I8, and Kennett's article on seismic phases). The differential travel-times between PKPbc and PKPdf, or between PKPab and PKPdf, are less sensitive than absolute travel-times to hypocenter mislocations, origin time errors, and upper mantle heterogeneities. PKPab, however, is strongly affected by the heterogeneities in the D″ layer at the base of the mantle.

Mean radial model

The P-velocity profile inside the inner core is rather well constrained. The P-velocity increases from about 11.02 ± 0.02 km s^{-1} at the radius of ICB (1219 ± 2 km) to 11.26 km s^{-1} at the center of the Earth (Dziewonski and Anderson, 1981; Kennett *et al.*, 1995). The P-velocity jump at the ICB is about 0.6–0.7 km s^{-1}.

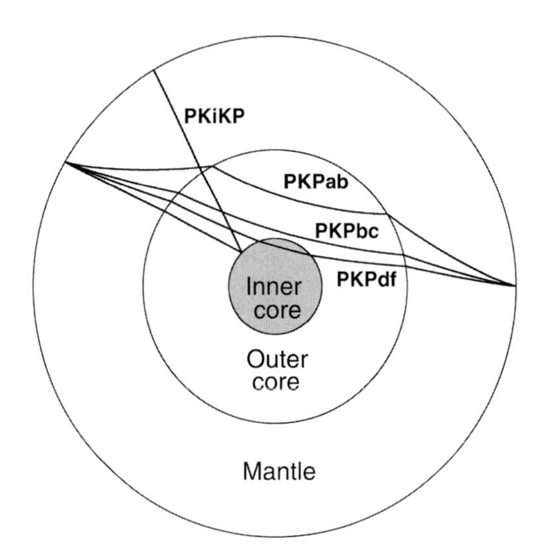

Figure I8 Main seismological phases sampling the inner core, and reference phases sampling the liquid outer core which are used for comparison (see *Seismic phases*).

The S-velocity is mostly constrained from the periods of normal modes, which first gave evidence of the rigidity of the inner core (Dziewonski and Gilbert, 1971), and from the modeling of amplitudes of core phases with synthetic seismograms (Müller, 1973). It increases from 3.50 kms^{-1} at the ICB to 3.67 kms^{-1} at Earth's center. This leads to a high Poisson's ratio $\sigma \approx 0.44$ inside the inner core (compared to 0.28–0.30 in the mantle), which is characteristic of solid metals.

The density is poorly constrained by the seismic data. The density jump at the ICB is deduced from the P-wave reflection coefficient at vertical incidence, and from the normal modes. A very large range of possible values have been obtained, the most recent ones being in the range 0.6–1.0 g cm^{-3} (Shearer and Masters, 1990; Cao and Romanowicz, 2004; Masters and Gubbins, 2004).

The velocity and density profiles are smooth, suggesting the absence of chemical stratification inside the inner core. Assuming the inner core results from the solidification of the liquid core, and taking into account the pressure at inner core condition ($P = 330$ GPa at the surface to 360 GPa at Earth's center), the velocity and density values favor an inner core composition depleted in light elements compared to the liquid core. Estimation of the temperature in the inner core, 5500–6500 K (see *Core temperature*), is strongly dependent on the presence and nature of these light elements, as well as the power released by compositional buoyancy, which is available to maintain the geodynamo.

The attenuation γ inside the inner core (or the quality factor $Q = 1/\gamma$) is estimated from the broadening of the peaks of the core-sensitive eigenmodes, and from the amplitude variations of PKPdf. Normal modes generally lead to much higher Q-values than body waves. This discrepancy is possibly due to the different contributions in bulk and shear dissipation; it is however, not fully understood. For P-waves, the comparison of the amplitudes of PKPdf and PKPbc leads to a Q-value of the order of 200 in the uppermost 100 km of the inner core, increasing to about 400 between 150 and 300 km below ICB. The presence of liquid inclusions is proposed, at least in the external part of the inner core, to account for the low value of the quality factor (see Romanowicz and Durek, 2000, for a review).

Anisotropy

Since the first observation of large-scale travel-time anomalies of PKPdf (Poupinet *et al.*, 1983), there have been many studies revealing the existence of a strong anisotropy in the inner core. Paths parallel to the Earth's rotation axis ("polar paths") are about 3.5% faster than those parallel to the equatorial plane. This anisotropy is well established from absolute travel-times of PKPdf (e.g., Morelli *et al.*, 1986; Shearer, 1994; Su and Dziewonski, 1995), and from differential travel-times of core phases (e.g., Creager, 1992) (Figure I9). It takes about 1210 s for a P-wave to cross the Earth in the equatorial plane, 7 s less to cross it along the pole axis, once the influence of Earth ellipticity has been corrected. The anomalous splitting of inner core-sensitive modes is another argument in favor of P-anisotropy (Woodhouse *et al.*, 1986; Tromp, 1993; Romanowicz *et al.*, 1996; Ishii *et al.*, 2002); it may also give insight into S-anisotropy (Beghein and Trampert, 2003).

The question has arisen whether the anomalous propagation times and mode splittings could originate outside the inner core. A structure in the liquid core, in particular heterogeneity or anisotropy inside the inner core tangent cylinder or beneath the polar caps, which is compatible with normal modes (Romanowicz and Bréger, 2000), is ruled out by the propagation times of core phases (Souriau *et al.*, 2003), and seems in any case hardly acceptable from a geodynamical point of view. On the other hand, an anomalous ellipticity of the inner core is ruled out by the PKiKP travel-times (Souriau and Souriau, 1989).

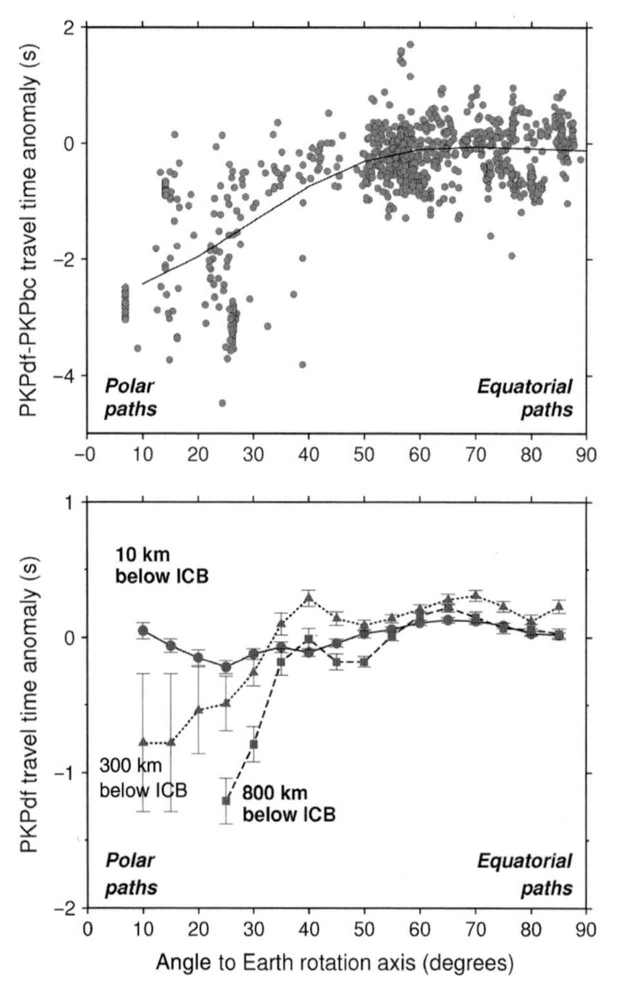

Figure I9 Evidence for anisotropy in the inner core. *Top*: Travel-time anomalies (with respect to a mean model) of the inner core seismic phase PKPdf, compared to the liquid core phase PKPbc, as a function of the angle of the ray with respect to the Earth's rotation axis. Data sample the uppermost 500 km of the inner core. A faster propagation (low residuals) is observed for polar paths. The curve corresponds to a uniform structure with hexagonally symmetric anisotropy. *Bottom*: Same for absolute PKPdf travel-times extracted from bulletin data, for rays turning at different depths below inner core boundary (ICB). Note the absence of anisotropy at the top of the inner core. Data have been binned to avoid the predominance of some oversampled regions.

The favored explanation for anisotropy is the preferred orientation of anisotropic iron crystals. The hexagonal closed-packed form, Fe-ϵ, is often proposed (Stixrude and Cohen, 1995), but other structures of iron are also possible. The orientation of elliptical fluid inclusions provides another possible explanation. The mechanism able to generate a preferred orientation is however debated; processes such as convection in the inner core, solidification in the presence of a magnetic field, solidification texturing at the ICB, or growth in a stress field, have been proposed (see reviews and discussions in Buffett, 2000; Bergman, 2003).

The depth variation and lateral variation of anisotropy are important for understanding its physical nature and origin. Early studies proposed an anisotropy with cylindrical symmetry and possibly a slightly tilted symmetry axis (Creager, 1992; Su and Dziewonski, 1995). This tilt is however not strongly required by the data (Souriau *et al.*, 1997). The variation with depth of anisotropy reveals that the uppermost part of the inner core is isotropic (Song and Helmberger, 1995). The thickness of the isotropic layer seems to vary with longitude, from 100 km beneath the western hemisphere to about 400 km beneath the eastern hemisphere (Creager, 1999; Garcia and Souriau, 2000) (Figure I10). This difference explains the apparent hemispherical pattern of the anisotropy (Tanaka and Hamaguchi, 1997). However, an artifact due to strong heterogeneities in the lowermost mantle along some particular polar paths (in particular the very anomalous path from South Sandwich Islands to Alaska) cannot completely be ruled out.

At greater depth, there are some presumptions that the anisotropy may be different in the central 300–400 km of the inner core. Ishii and Dziewonski (2002) observe an orientation of the slow axis at 45° from the pole axis for waves penetrating deep in the inner core. From a normal mode inversion, Beghein and Trampert (2003) find a fast axis perpendicular to the Earth's rotation axis in the lowermost 400 km, which could suggest the presence of a different phase of iron.

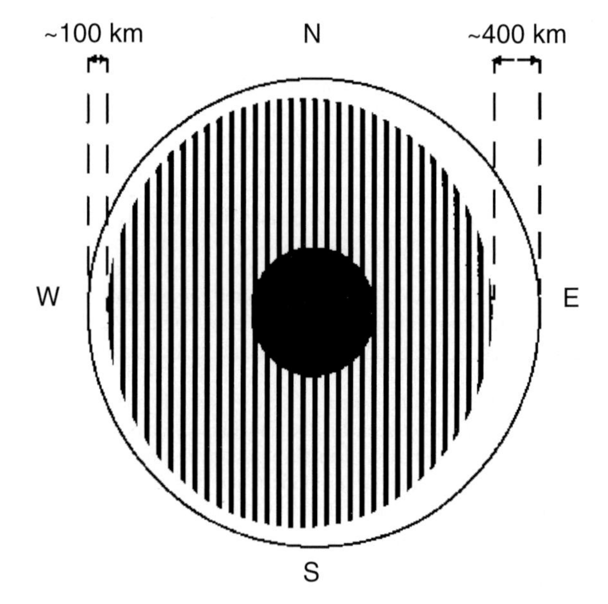

Figure I10 A possible model of inner core anisotropy explaining body wave data. Meridian cross-section. Structure with 3% anisotropy and fast axis parallel to Earth rotation axis surrounded by an isotropic layer of variable thickness. A different form of anisotropy may be present in the central 300–400 km of the inner core.

Heterogeneities

Even though most of the propagation time anomalies are well explained by anisotropy, it is important to estimate whether heterogeneities could also contribute to these anomalies. Voigt's isotropic average of P-wave velocity seems almost invariant between the western and eastern hemispheres (Creager, 1999), indicating the same chemical composition of the two hemispheres. However, in the uppermost 50 km, a hemispherical P-heterogeneity has been detected by modeling the PKP waveforms sampling the ICB region (e.g., Niu and Wen, 2001). The absence of heterogeneity in the range of wavelengths 200–1000 km is evidenced from stochastic methods: the heterogeneity level is less than 0.3% at any depth (Garcia and Souriau, 2000). In the range 50–200 km, many studies have reported the possible existence of heterogeneities. However, it is not possible to prove that the heterogeneities effectively take place in the inner core, and not in the mantle.

On the other hand, the splitting functions of inner core-sensitive normal modes exhibit large-scale variations, a result which seems in contradiction with the previous ones. Note, however, that modes are mostly sensitive to S-heterogeneities, and that single modes give access only to the even part of the structure of the inner core (Laske and Masters, 2003).

Heterogeneities at very short wavelength have been detected in the uppermost 300 km of the inner core from the energy present in the coda of the PKiKP waves (Vidale and Earle, 2000). The observations are explained by scatterers of size $\sim$2 km, with velocity contrast of 1.2%. The modeling of the inner core attenuation also suggests the presence of scatterers: small heterogeneities of length scale $\sim$10 km with velocity perturbations of about 8%, due to the boundaries between single or ordered groups of crystals with various orientations, could explain the mean level of attenuation, as well as the apparent anisotropy in velocity and attenuation (Cormier and Li, 2002).

Conclusion

Our knowledge of inner core structure has been considerably improved during the last two decades, thanks to an increasing number of digital records with worldwide distribution, and to the availability of about 40 years of bulletin data. Major observational difficulties come from the perturbing effect of the D" layer, at the base of the mantle and from the uneven sampling of the Earth, in particular the scarcity of polar paths, that are crucial for investigating the anisotropy. The deployment of permanent ocean bottom observatories will help to overcome this difficulty in a near future.

There is general agreement about a 3% to 3.5% anisotropic internal structure with fast axis parallel to the Earth's rotation axis, surrounded by an isotropic layer. Other features, such as the hemispherical variation of the thickness of the isotropic layer, or the possibility of a different anisotropy in the deeper inner core, are still debated. Numerous questions remain unanswered, concerning in particular the S-velocity model, the P-anisotropy near the center of the Earth, and the physical cause of the anisotropic P-wave velocity.

An accurate knowledge of the inner core heterogeneities appears also necessary for being able to quantify a possible superrotation of inner core with respect to the mantle. Such analyses rely on the possibility to track inner core heterogeneities over several years or several decades. Although the results are still controversial, with most recent rotation rates in the range -0.2 to $+0.4^{\circ}\mathrm{yr}^{-1}$ (Souriau and Poupinet, 2003), they will certainly improve in the future with the availability of longer and denser series of data.

Annie Souriau

Bibliography

Beghein, C., and Trampert, J., 2003. Robust normal mode constraints on inner-core anisotropy from model space search. *Science*, **299**: 552–555.

Bergman, M., 2003. Solidification of the Earth's core. In Dehant, V. *et al.*, (eds.), *Earth's core, Dynamics, structure, Rotation,* Geodynamics Series 31. Washington, DC: American Geophysical Union, pp. 105–128.

Buffett, B.A., 2000. Dynamics of the Earth's core. In Karato, S. *et al.*, (eds.), *Earth's Deep Interior,* Geophysics Monograph 117. Washington, DC: American Geophysical Union, 37–62.

Cao, A., and Romanowicz, B., 2004. Constraints on density and shear velocity contrasts at the inner core boundary. *Geophysical Journal International*, **157**: 1–6.

Cormier, V.F., and Li, X., 2002. Frequency dependent seismic attenuation in the inner core. Part II. A scattering and fabric interpretation. *Journal of Geophysical Research*, **107**: 2362, doi:10.1029/2002JB001796.

Creager, K.C., 1992. Anisotropy in the inner core from differential travel times of the phases PKP and PKIKP. *Nature*, **356**: 309–314.

Creager, K.C., 1999. Large-scale variations in inner core anisotropy. *Journal of Geophysical Research*, **104**: 23127–23139.

Dziewonski, A.M., and Anderson, D.L., 1981. Preliminary reference Earth model. *Physics of the Earth and Planetary Interiors*, **25**: 297–356.

Dziewonski, A.M., and Gilbert, F., 1971. Solidity of the inner core of the Earth inferred from normal mode observations. *Nature*, **234**: 465–466.

Garcia, R., and Souriau, A., 2000. Inner core anisotropy and heterogeneity level. *Geophysical Research Letters*, **27**: 3121–3124; and correction *Geophysical Research Letters*, **28**: 85–86, 2001.

Ishii, M., and Dziewonski, A.M., 2002. The innermost inner core of the earth: evidence for a change in anisotropic behavior at the radius of about 300 km. *PNSA*, **99**: 14026–14030.

Ishii, M., Tromp, J., Dziewonski, A.M., and Ekström, G., 2002. Joint inversion of normal mode and body wave data for inner core anisotropy. 1. Laterally homogeneous anisotropy. *Journal of Geophysical Research*, **107**: 2379, doi:10.1029/2001JB000712.

Kennett, B.L.N., Engdahl, E.R., and Buland, R., 1995. Constraints on seismic velocities in the Earth from travel times. *Geophysical Journal International*, **122**: 108–124.

Laske, G., and Masters, G., 2003. The Earth's free oscillations and the differential rotation of the inner core. In Dehant, V. *et al.*, (eds.), *Earth's Core, Dynamics, Structure, Rotation,* Geodynamics Series 31. Washington, DC: American Geophysical Union, pp. 5–22.

Masters, G., and Gubbins, D., 2003. On the resolution of density within the Earth. *Physics of the Earth and Planetary Interiors*, **140**: 159–167.

Morelli, A., Dziewonski, A.M., and Woodhouse, J.H., 1986. Anisotropy of the inner core inferred from PKIKP travel-times. *Geophysical Research Letters*, **13**: 1545–1548.

Müller, G., 1973. Amplitude studies of core phases. *Journal of Geophysical Research*, **78**: 3469–3490; and correction, 1977, *Journal of Geophysical Research*, **82**: 2541–2542.

Niu, F., and Wen, L., 2001. Hemispherical variations in seismic velocity at the top of the Earth's inner core. *Nature*, **410**: 1081–1084.

Poupinet, G., Pillet, R., and Souriau, A., 1983. Possible heterogeneity in the Earth's core deduced from PKIKP travel times. *Nature*, **305**: 204–206.

Romanowicz, B., and Bréger, L., 2000. Anomalous splitting of free oscillations: a reevaluation of possible interpretations. *Journal of Geophysical Research*, **105**: 21559–21578.

Romanowicz, B., and Durek, J.J., 2000. Seismological constraints on attenuation in the Earth: a review. In Karato, S. *et al.*, (eds.), *Earth's Deep Interior,* Geophysics Monograph 117. Washington, DC: American Geophysical Union, pp. 161–179.

Romanowicz, B., Li, X.-D., and Durek, J., 1996. Anisotropy in the inner core: Could it be due to low-order convection? *Science*, **274**: 963–966.

Shearer, P.M., 1994. Constraints on inner core anisotropy from PKP(DF) travel times. *Journal of Geophysical Research*, **99**: 19647–19659.

Shearer, P.M., and Masters, G., 1990. The density and shear velocity contrast at the inner core boundary. *Geophysical Journal International*, **102**: 491–498.
Song, X., and Helmberger, D.V., 1995. Depth dependence of anisotropy of Earth's inner core. *Journal of Geophysical Research*, **100**: 9805–9816.
Souriau, A., and Poupinet, G., 2003. Inner core rotation: a critical appraisal. In Dehant, V. *et al.*, (eds.), *Earth's Core, Dynamics, Structure, Rotation,* Geodynamics Series 31. Washington, DC: American Geophysical Union, pp. 65–82.
Souriau, A., and Souriau, M., 1989. Ellipticity and density at the inner core boundary from subcritical PKiKP and PcP data. *Geophysical Journal International*, **98**: 39–54.
Souriau, A., Moynot, B., and Roudil, P., 1997. Inner core rotation: facts and artefacts. *Geophysical Research Letters*, **24**: 2103–2106.
Souriau, A., Teste, A., and Chevrot, S., 2003. Is there any structure inside the liquid core? *Geophysical Research Letters*, **30**: 1567, doi:10.1029/2003GL017008.
Stixrude, L., and Cohen, R., 1995. High pressure elasticity of iron and anisotropy of Earth's inner core. *Journal of Geophysical Research*, **102**: 24729–24739.
Su, W., and Dziewonski, A.M., 1995. Anisotropy in three dimensions. *Journal of Geophysical Research*, **100**: 9831–9852.
Tanaka, S., and Hamaguchi, H., 1997. Degree one heterogeneity and hemispherical variation of anisotropy in the inner core from PKP(BC)-PKP(DF) times. *Journal of Geophysical Research*, **102**: 2925–2938.
Tromp, J., 1993. Support for anisotropy of the Earth's inner core from free oscillations. *Nature*, **366**: 678–681.
Vidale, J.E., and Earle, P.S., 2000. Fine-scale heterogeneity in the Earth's inner core. *Nature*, **404**: 273–275.
Woodhouse, J.H., Giardini, D., and Li, X., 1986. Evidence for inner core anisotropy from free oscillations. *Geophysical Research Letters*, **13**: 1549–1552.

Cross-references

Core Composition
Core Density
Core Temperature
D″, Seismic Properties
Inner Core Anisotropy
Inner Core, PKJKP
Inner Core Rotation
Inner Core Tangent Cylinder
Seismic Phases

INNER CORE TANGENT CYLINDER

The Earth's core consists of a solid inner core of radius $r_i = 1\,220$ km, and a fluid outer core of radius $r_o = 3\,480$ km. The obvious geometry to study it would therefore appear to be spherical (or perhaps spheroidal, if one insists on taking into account the slight flattening induced by the rotation). However, one could also view it in terms of cylindrical geometry, in which case a new feature emerges, namely the inner core tangent cylinder, the cylinder just touching the inner core and parallel to the axis of rotation. Figure I11 shows this tangent cylinder C, as well as how the outer core is then divided into regions inside and outside C, with the region inside C further subdivided into two parts above and below the inner core. The purpose of this article is to discuss why it might be sensible to consider the Earth's core in terms of cylindrical geometry, and what the significance of this tangent cylinder therefore is.

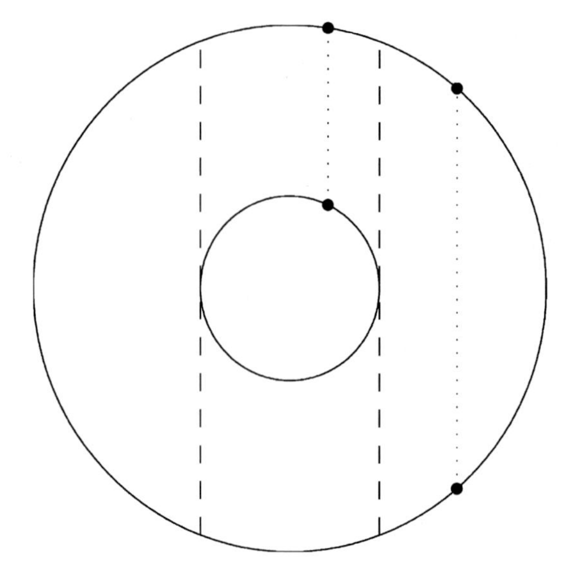

Figure I11 The circles represent the inner and outer core boundaries; the dashed lines then show the location of the inner core tangent cylinder. The dotted lines are the paths over which (Eq. 4) is to be integrated, with the dots at the ends indicating where the no normal flow boundary conditions are to be imposed. Note how these boundary conditions change abruptly across the tangent cylinder. See also Hollerbach and Proctor (1993) for the details of how this constant of integration $\mathbf{U}_0$ is determined (or not).

Cylindrical coordinates

To see why cylindrical coordinates in fact arise quite naturally, we begin by considering the Navier-Stokes equation governing the fluid flow in the outer core. In its simplest, Boussinesq form, this takes the form

$$Ro\left[\frac{\partial}{\partial t} + \mathbf{U}\cdot\nabla\right]\mathbf{U} + 2\hat{\mathbf{e}}_z \times \mathbf{U} = -\nabla p + E\nabla^2\mathbf{U} + \mathbf{F}_\mathrm{M} + \mathbf{F}_\mathrm{B}, \quad \text{(Eq. 1)}$$

where $\mathbf{F}_\mathrm{M}$ and $\mathbf{F}_\mathrm{B}$ are the magnetic and buoyancy forces, respectively, the Rossby number $Ro = \eta/\Omega r_o^2$ measures the importance of inertia, and the Ekman number $E = \nu/\Omega r_o^2$ the importance of viscosity, both relative to the Coriolis force $2\hat{\mathbf{e}}_z \times \mathbf{U}$. Here $\mathbf{\Omega}$ is the Earth's rotation rate $(7.27 \times 10^{-5}\ \mathrm{rad\,s^{-1}}), \eta \approx 2\ \mathrm{m^2\,s^{-1}}$ and $\nu \approx 10^{-6}\ \mathrm{m^2\,s^{-1}}$ the magnetic diffusivity and viscosity, respectively. Inserting the numbers, we therefore find that $Ro = O(10^{-9})$ and $E = O(10^{-15})$, indicating the extreme dominance of rotation over inertia and viscosity.

If the dominant balance in the Navier-Stokes equation is thus

$$2\hat{\mathbf{e}}_z \times \mathbf{U} \approx -\nabla p + \mathbf{F}_\mathrm{M} + \mathbf{F}_\mathrm{B}, \quad \text{(Eq. 2)}$$

then taking the curl (and using $\nabla\cdot\mathbf{U} = 0$) one obtains

$$-2\frac{\partial}{\partial z}\mathbf{U} \approx \nabla \times (\mathbf{F}_\mathrm{M} + \mathbf{F}_\mathrm{B}). \quad \text{(Eq. 3)}$$

So we see that even though the geometry of the core is most obviously defined in terms of spherical coordinates (r, θ, ϕ), the dynamics of the outer core naturally, indeed almost inevitably, lead to cylindrical coordinates (z, s, ϕ). And as we saw above, once one acknowledges the relevance of cylindrical coordinates, one invariably must consider the tangent cylinder as well.

Shear layers on C?

One might suppose that the solution of (Eq. 3) is then simply

$$\mathbf{U} = -\frac{1}{2}\int^{z} \nabla \times (\mathbf{F}_{\mathrm{M}} + \mathbf{F}_{\mathrm{B}})\mathrm{d}z' + \mathbf{U}_0(s,\phi), \qquad \text{(Eq. 4)}$$

where the constant of integration $\mathbf{U}_0(s,\phi)$ is to be determined by the appropriate boundary conditions (no normal flow at $r = r_i$ and r_o). There are in fact severe difficulties with this approach (see *Taylor's condition*). However, here we simply wish to consider what happens as one crosses C. In particular, the boundary conditions to be imposed change discontinuously as one crosses C, as indicated in Figure I11. If the boundary conditions change abruptly though, one must expect that the constant of integration $\mathbf{U}_0(s,\phi)$ will also change, so according to (Eq. 4) the flow itself will be discontinuous everywhere along the tangent cylinder.

Any finite viscosity will of course smooth these discontinuities out into shear layers of small but finite thickness. Figure I12 shows two examples of such layers, at $E = 10^{-5}$. The panel on the left shows contours of the angular velocity in the so-called Stewartson layer problem, in which one imposes a slight differential rotation on the inner core (such as may indeed exist in the Earth—see *inner core rotation*).

In the axisymmetric Stewartson layer problem, the only component of the flow that would be discontinuous in the $E \to 0$ limit is the angular velocity U_ϕ/s, that is, a component of the flow tangential to C. If one considers the more general nonaxisymmetric problem, one finds that even the normal component U_s may be discontinuous. The right panel in Figure I12 shows such an example, and sure enough, we note that the layer that resolves the discontinuity at finite E is more severe than it was in the Stewartson layer problem.

Both the axisymmetric and nonaxisymmetric layers shown in Figure I12 are purely nonmagnetic. If one includes a magnetic field, it seems likely that the tension in the field lines would tend to suppress such strong shear layers. At least in the nonaxisymmetric case, however, there is also a price to pay: Hollerbach and Proctor (1993) showed that a shear layer can only be avoided if the forcing (that is, $\mathbf{F}_{\mathrm{M}} + \mathbf{F}_{\mathrm{B}}$) satisfies a certain integral constraint on C, and Hollerbach (1994b) showed further that the magnetic field does indeed evolve so as to satisfy this constraint. It would thus appear that there probably are no shear layers on C (the axisymmetric Stewartson layer is similarly suppressed by a magnetic field), but that instead one has certain integral constraints on C. The dynamical significance—and indeed the physical interpretation—of these constraints is not fully understood though.

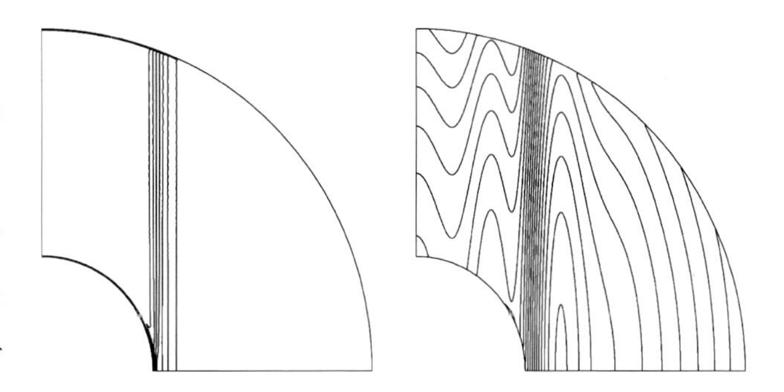

Figure I12 The panel on the left is from Hollerbach (1994a), and shows contours of the angular velocity U_ϕ/s in the axisymmetric Stewartson layer problem. The panel on the right is from Hollerbach and Proctor (1993), and shows contours of U_s in a nonaxisymmetric shear layer problem. Both layers are at $E = 10^{-5}$.

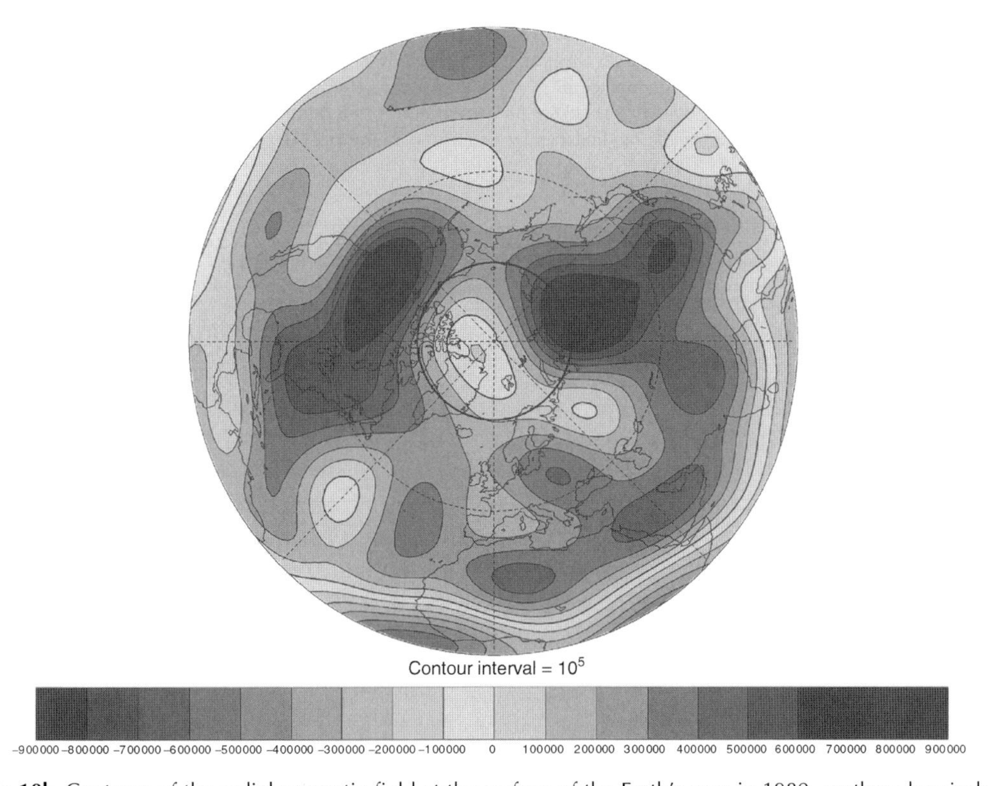

Figure I13/Plate 10b Contours of the radial magnetic field at the surface of the Earth's core in 1980, northern hemisphere, in equal area projection. The circle denotes the end of the tangent cylinder. Blue indicates a region where field lines enter the core, yellow to red where they exit, and the zero contour is a "magnetic equator" where the field is horizontal. Note the concentration of flux into two lobes in the northern hemisphere just outside the tangent cylinder, and the lack of flux inside the tangent cylinder.

Inside versus outside C

There are also a number of issues where it is not so much C itself, but rather whether one is inside or outside that matters. For example, the fact that the region inside C is further subdivided into parts above and below the inner core means that inside C Taylor's condition consists of two distinct integral constraints (at any given cylindrical radius s), whereas outside C there is only one integral over the entire extent of the cylindrical shell. What effect this has on the geodynamo is not known for certain, but given the general importance of Taylor's condition, it could be significant.

Another difference between inside versus outside C is how the height of fluid columns parallel to the rotation axis varies with s. Inside C, dH/ds is positive, whereas outside it is negative. The reason this may be important is that so-called Rossby waves (e.g., Salmon, 1998) propagate in the direction $\nabla H \times \mathbf{\Omega}$, that is, westward inside C, eastward outside. However, as with the Stewartson layer, classical Rossby waves are a nonmagnetic phenomenon more usually associated with oceanography or meteorology, so it is not entirely clear how relevant they are to the Earth's core. Similar effects have also been observed though in convection-driven hydromagnetic waves (Zhang and Gubbins, 2002), so it seems that this distinction between inside and outside C does indeed persist into the magnetic regime as well.

Speaking of convection, numerous simulations have shown that the overall pattern of convection appears to be quite different inside versus outside C. (See *Core convection* or Kono and Roberts, 2002 for a summary of some of these results.) The reasons for this are not always apparent, but one effect that is likely to play a role is simply that inside C gravity and the rotation vector are largely parallel, whereas outside C they are predominantly perpendicular. Given the powerful influence that both of these effects have on convection (with gravity driving the whole process, of course), it seems likely that their relative orientation will also be important.

The inner core may also impart some stability to the geodynamo, because the magnetic field can only change inside it by diffusion, whereas in the outer core it can change by the much more rapid process of advection (Hollerbach and Jones, 1993). This longer timescale of the inner core has been used to explain the preponderance of *geomagnetic excursions* (*q.v.*), or aborted reversals, over full reversals of the geomagnetic field (Gubbins, 1999). The tangent cylinder may enhance this stabilizing effect by providing a larger, cylindrical volume of "dead" space where the flow does not affect the magnetic field to any great extent. In particular, we note that the inner core by itself constitutes only 4% of the whole core volume, whereas the total region inside the tangent cylinder constitutes 18%. If the fluid inside C is relatively quiescent, it would therefore considerably boost this stabilizing influence.

Observed effects on the Earth's field

Surprisingly, the effect of the inner core tangent cylinder is spectacularly visible in the Earth's magnetic field when projected down to the core surface. Figure I13/Plate 10b shows the vertical component

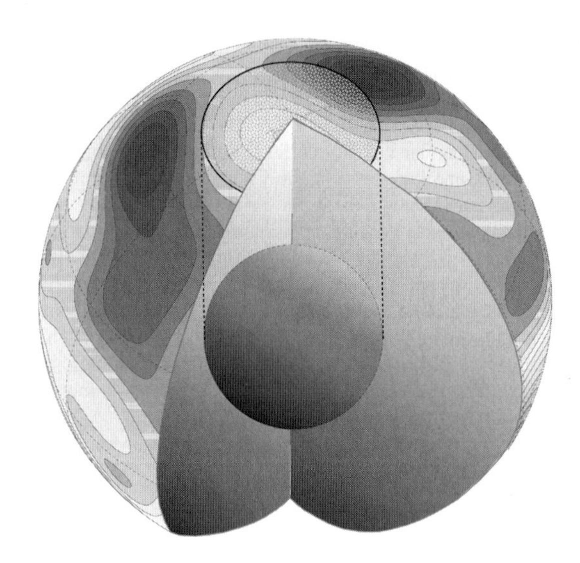

Figure I14/Plate 10a Contours of the same magnetic field as in Figure I13. The inner core is shown in red and the tangent cylinder is indicated, along with the circle where it intersects the surface in the northern hemisphere.

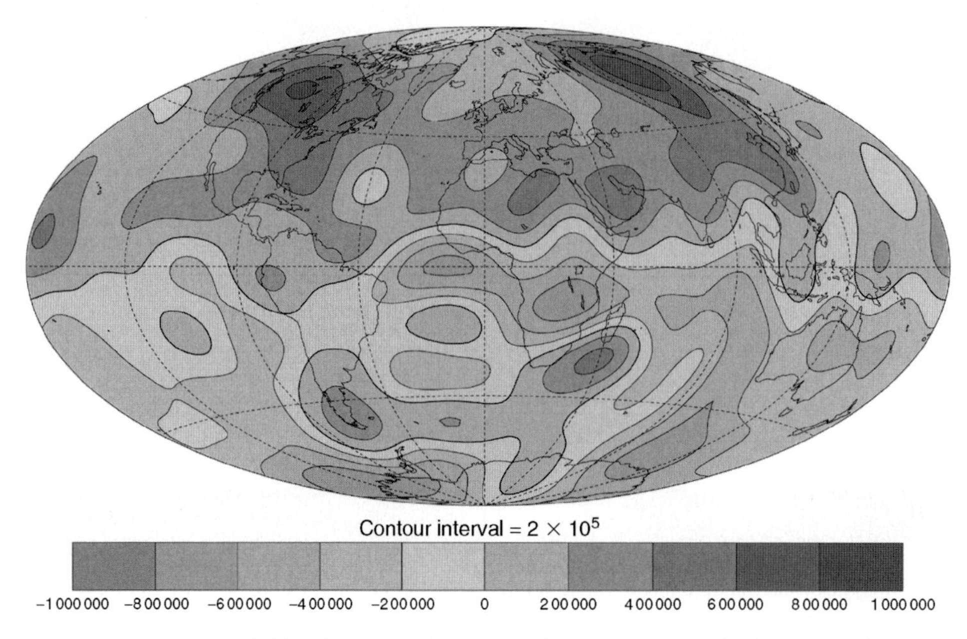

Figure I15/Plate 10c The same magnetic field at the core surface in Aitoff projection. Note the four main lobes, symmetrical about the equator and distributed outside the tangent cylinder.

of the magnetic field at the surface of the core in the northern hemisphere. The dark circle is the top of the tangent cylinder, with radius 20°. Figure I14/Plate 10a shows the same field with a cutaway liquid core to illustrate the tangent cylinder. The vertical field is strongest not at the pole, which would be the case for an axial dipole, but in two lobes just outside the tangent cylinder. Similar lobes are present in the southern hemisphere: these are the four main lobes of the geomagnetic field responsible for the predominance of its dipole character; they are less mobile than the features at lower latitude associated with the *westward drift (q.v.)*.

The pattern of lobes is symmetrical about the equator, Figure I15/ Plate 10c, suggestive of straight convection rolls touching the inner core (see *Convection, nonmagnetic rotating*). Numerical simulations and kinematic dynamo calculations show magnetic flux being concentrated at the surface by shear around the tangent cylinder and by downwelling associated with the convection rolls (see *Geodynamo, numerical simulations* and Gubbins *et al.*, 1999). It appears that the geomagnetic field is not generated strongly inside the tangent cylinder, and that most of the generation occurs just outside, as in some of the models. Note that the radial field is almost zero within the tangent cylinder in the northern hemisphere, and in Figure I13/Plate 10b there is a similar null patch in the southern hemisphere.

This lack of field near the poles probably explains why virtual paleomagnetic poles (VGPs), those determined from a single site measurement of the magnetic field direction assuming a *geocentric axial dipole (q.v.)*, usually lie some distance (typically 10°) from the geographic pole. The same is true of the present-day field. Only when a time average is done to produce a paleomagnetic pole, by averaging many samples from a rock unit that spans many thousands of years, is a reliable estimate obtained for the geographic pole (see *Pole, paleomagnetic*), and even then the match is not perfect. The departure of the VGP from the geographic pole is caused by the nonaxisymmetric nature of the field near the poles; this largely averages out with time as the main lobes drift. The departure of the resulting paleomagnetic pole is usually attributed to a nondipolar time average. For further discussion of the time average see Gubbins (1998).

Rainer Hollerbach and David Gubbins

Bibliography

Gubbins, D., 1998. Interpreting the paleomagnetic field. In Gurnis, M., Buffett, B., Wysession, M., and Knittle, E. (eds.), *The Core-Mantle Boundary Region.* AGU Geophysical Monograph, Geodynamics Series. Washington, DC: American Geophysical Union, pp. 167–182.

Gubbins, D., 1999. The distinction between geomagnetic excursions and reversals. *Geophysical Journal International*, **137**, F1–F3.

Gubbins, D., Barber, C.N., Gibbons, S., and Love, J.J., 1999. Kinematic dynamo action in a sphere. II. Symmetry selection. *Proceedings of the Royal Society of London, Series A*, **456**: 1669–1683.

Hollerbach, R., 1994a. Magnetohydrodynamic Ekman and Stewartson layers in a rotating spherical shell. *Proceedings of the Royal Society of London, Series A*, **444**: 333–346.

Hollerbach, R., 1994b. Imposing a magnetic field across a nonaxisymmetric shear layer in a rotating spherical shell. *Physics of Fluids*, **6**: 2540–2544.

Hollerbach, R., and Jones, C.A., 1993. Influence of the Earth's inner core on geomagnetic fluctuations and reversals. *Nature*, **365**: 541–543.

Hollerbach, R., and Proctor, M.R.E., 1993. Nonaxisymmetric shear layers in a rotating spherical shell. In Proctor, M.R.E., Matthews, P.C., and Rucklidge, A.M. (eds.), *Solar and Planetary Dynamos.* Cambridge: Cambridge University Press, pp. 145–152.

Kono, M., and Roberts, P.H., 2002. Recent geodynamo simulations and observations of the geomagnetic field. *Reviews of Geophysics*, **40**(4): 1013, doi:10.1029/2000RG000102.

Salmon, R., 1998. *Lectures on Geophysical Fluid Dynamics.* Oxford: Oxford University Press.

Zhang, K., and Gubbins, D., 2002. Convection-driven hydromagnetic waves in planetary fluid cores. *Mathematical and Computer Modelling*, **36**: 389–401.

Cross-references

Anelastic and Boussinesq Approximations
Convection, Nonmagnetic Rotating
Core Convection
Dynamos, Kinematic
Earth Structure, Major Divisions
Geocentric Axial Dipole Hypothesis
Geodynamo, Numerical Simulations
Geomagnetic Excursion
Inner Core Rotation
Pole, Paleomagnetic
Proudman-Taylor Theorem
Taylor's Condition
Westward Drift

INNER CORE, PKJKP

General background

The inner core was detected by Lehmann in 1936 (Lehmann, 1936) based on observations of P-waves through the inner core (later called PKIKP phases) that were unexplained by the Earth models of that time. In her noteworthy paper Lehmann concluded: "The question of the existence of the inner core cannot however be regarded solely from a seismological point of view, but must be considered also in its other geophysical aspects." Indeed, in 1940 Birch hypothesized that the inner core may be the result of a liquid-solid phase change in iron (Birch, 1940). A solid inner core implies that shear waves can be locally present, and the phase PKJKP with a shear wave leg in the inner core should exist (Figure I16). The letter J in this phase notation denotes the shear wave path in the inner core, whereas I represents an inner core P-wave path. Similarly, the letter K denotes a P-wave passage through the outer core, and P and S stand for mantle P- and S-wave legs, respectively.

Thus, to confirm the inner core's solidity, the search for observations of PKJKP started in the 1940s. In his article on the history of inner core research, Bolt (1987) recalls Bullen's attempts to find PKJKP (see also Bullen, 1951), and numerous other seismologists have tried to identify PKJKP as well. The detection of PKJKP, however, has proven to be extremely difficult. The solidity of the inner core was eventually established in 1971 based on normal mode observations (Dziewonski and Gilbert, 1971), but identification of PKJKP remained a challenge to many seismologists.

PKJKP identifications

PKJKP is difficult to detect because of its small amplitude. This is a consequence of its large spreading (see Figure I16), inefficient P-to-S and S-to-P conversions at the inner core boundary, and significant inner core attenuation (Doornbos, 1974). Furthermore, interference with other larger amplitude phases severely hampers a proper interpretation. PKJKP cannot be identified on individual seismograms, and therefore stacking of large number of seismograms is required. If different stations are used in the stack, slowness information can be obtained, which is essential for an unambiguous identification.

Julian *et al.* (1972) were the first to report observations of PKJKP. They stacked data of five earthquakes with magnitudes ranging from 6.0–6.3 recorded by the Large Aperture Seismic Array (LASA) in

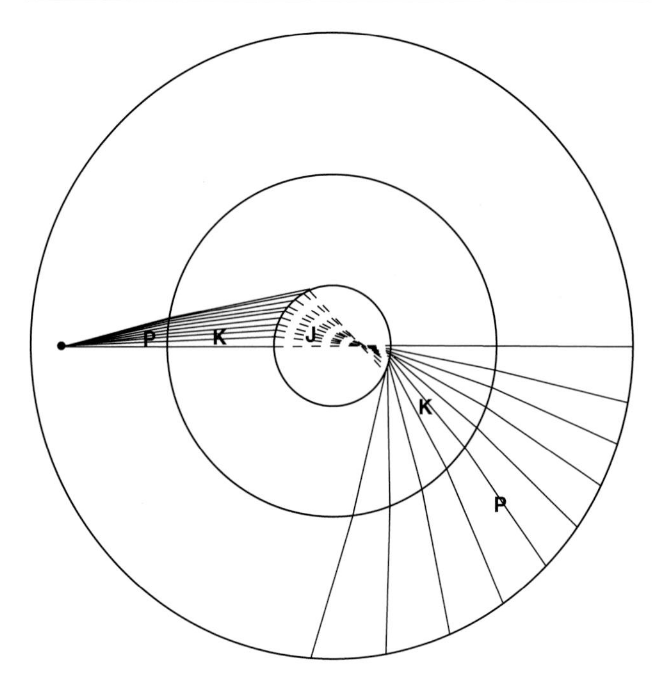

Figure I16 Schematic ray paths of PKJKP through mantle (P), outer core (K), and inner core (J).

Montana (USA). This first PKJKP claim was, however, controversial because the inferred inner core shear velocity is at odds with the shear velocity determined from the normal mode data (3.5–3.6 km s^{-1}). It has been speculated that the observed signal was in fact the SKJKP phase instead of PKJKP, but the data were not reanalyzed.

It took more than 60 years after the detection of the inner core before two new reports of PKJKP were made. Both papers rely on advanced signal detection techniques combined with data from large magnitude (7.9–8.2) deep earthquakes. Okal and Cansi (1998) applied a multichannel correlation method to data of the 1996 Flores Sea event for eight stations of the French short-period seismic network. They identified PKJKP and the surface-reflected pPKJKP phase. Deuss *et al.* (2000) used 47 globally recorded broadband seismograms for the same Flores Sea event and employed a nonlinear stacking technique based on phase coherency. They compared the data stack with theoretical stacks computed from normal mode seismograms: (1) for a model with a solid inner core, and (2) for a model with a fluid inner core. Evidently, the difference between these two synthetic stacks is due to the presence of J-phases. Deuss *et al.* (2000) found that, due to interference, PKJKP cannot be identified unambiguously in the data of the Flores Sea event, but the combination of pPKJKP and SKJKP can be identified in the data.

Although recent claims have now been made, it remains difficult to observe the J-phases. For instance, Deuss *et al.* (2000) showed that the global broadband data do not allow identification of the J-phases for the magnitude 8.2 Bolivia earthquake of 1994 due to interference effects of other signals. For the same reason they dispute the PKJKP identification of Okal and Cansi (1998) in their data set.

Implications of PKJKP observations

PKJKP and other J-phases may now be detected under very favorable conditions: a very big earthquake ($M \approx 8$), a station distribution with a large number of stations in the appropriate distance range, a favorable source depth and radiation pattern, and use of a sophisticated stacking technique. The detection itself, however, is not crucial anymore as the shear velocity is determined from normal mode data. Yet, observations of J-phases can still be important because they contain information about attenuation and anisotropy, constraining the physical state of the inner core. The issue of inner core attenuation is briefly addressed by Deuss *et al.* (2000). They find that the shear-wave attenuation as given by reference model PREM (Dziewonski and Anderson, 1981) is in agreement with their data. It corresponds to a relatively high level of attenuation possibly due to partial melting (Loper and Fearn, 1983). Inner core shear wave anisotropy may be investigated by directional dependence of the PKJKP observations (e.g., parallel and perpendicular to the Earth's rotation axis). After the detection, this could be the next challenge in PKJKP research. However, normal mode measurements proved to be more efficient to determine these properties (Beghein and Trampert, 2003).

All in all, PKJKP is a phase that does not easily reveal the information it contains.

Hanneke Paulssen

Bibliography

Beghein, C., and Trampert, J., 2003. Robust normal mode constraints on inner core anisotropy from model space search. *Science*, **299**: 552–555.

Birch, F., 1940. The alpha-gamma transformation of iron at high pressures, and the problem of the Earth's magnetism. *American Journal of Science*, **238**: 192–211.

Bolt, B.A., 1987. 50 years of studies on the inner core. *EOS, Transactions of the American Geophysical Union*, **68**: 73–81.

Bullen, K.E., 1951. Theoretical amplitudes of the seismic phase PKJKP. *Monthly Notices of the Royal Astronomical Society, Geophysics Supplement*, **6**: 163–167.

Deuss, A., Woodhouse, J.H., Paulssen, H., and Trampert, J., 2000. The observation of inner core shear waves. *Geophysical Journal International*, **142**: 67–73.

Doornbos, D.J., 1974. The anelasticity of the inner core. *Geophysical Journal of the Royal Astronomical Society*, **38**: 397–415.

Dziewonski, A.M., and Anderson, D.L., 1981. Preliminary reference Earth model. *Physics of the Earth and Planetary Interiors*, **25**: 297–356.

Dziewonski, A.M., and Gilbert, F., 1971. Solidity of the inner core of the Earth inferred from normal mode observations. *Nature*, **234**: 465–466.

Julian, B.R., Davies, D., and Sheppard, R.M., 1972. PKJKP. *Nature*, **235**: 317–318.

Lehmann, I., 1936. P', *Bur. Centr. Seismol. Int. A., Trav. Sci.*, **14**: 87–115.

Loper, D.E., and Fearn, D.R., 1983. A seismic model of a partially molten inner core. *Journal of Geophysical Research*, **88**: 1235–1242.

Okal, E.A., and Cansi, Y., 1998. Detection of PKJKP at intermediate periods by progressive multichannel correlation. *Earth and Planetary Science Letters*, **164**: 23–30.

Cross-references

Core Composition
Inner Core Anisotropy
Inner Core Composition
Inner Core Seismic Velocities
Lehmann, Inge (1888–1993)

INSTRUMENTATION, HISTORY OF

Geomagnetic instruments before 1800: instruments in early navigation and science

Scientists and various practitioners have used a bewildering array of magnetic instruments over the last thousand years or so: the mariner's compass (many types), circumferentors, declinometers, variation compasses, dip circles or inclinometers, theodolite magnetometers, variometers, bifilar magnetometers, Lloyd's and Schmidt's balances, earth inductors, and more recently, nuclear magnetic resonance magnetometers, saturable-core magnetometers, induction magnetometers, and more. This rich history, however, is characterized by short bursts of inventive activity followed by longer periods of diffusion of instrumental designs. Because this volume includes articles on the more common contemporary magnetic instruments (see *Archaeology, magnetic methods*; *Compass*; *Magnetometers, laboratory*; *Magnetic surveys, marine*; and *Observatories, instrumentation*), this article emphasizes the earlier history of magnetic instruments and their use. It sets some current instruments against this backdrop (see also *History of geomagnetism*).

Although magnetism as a property of the lodestone was known since antiquity in both China and Greece, the first magnetic instruments unequivocally appear in the historical record in the 12th century CE (see *Compass*). Strong evidence now shows that the magnetic compass was in use in Europe around 1150 and in China around 1100 (Smith, 1992, pp. 25 and 33–38), with evidence the Chinese knew something of the magnetic needle's directionality for centuries. The first accounts that can be dated are the *Meng Chhi Pi Than*, a chronicle written by Shen Kua in 1088, and Alexander Neckam's *De Nominibus Utensilium*, written between 1175 and 1183. Shen Kua not only mentioned a "wet" compass (a needle floating on water) and balancing a needle on a horizontal edge, but also magnetic declination. Neckam's was the first European reference to a compass needle on a pivot (a "dry" compass), but it is possible he had never used one and was only reporting an increasingly common practice on ships at sea. The best known magnetic investigator of the 13th century, Petrus Peregrinus (Pierre de Maricourt), wrote his famous *Epistola . . . de Magnete* in 1269. Peregrinus was a consummate experimenter and careful thinker. Familiar with the compasses used and discussed for about a century, he developed improved wet and dry compasses, with a scale of 360° attached (Smith, 1992, pp. 70–72). His significance lies in the clarity and completeness of his summation of magnetic knowledge.

From 1200 until around 1600, use of the compass and knowledge of its phenomena were restricted mainly to navigators and cartographers, although poets and philosophers discussed it frequently from the mid-13th century. Early 15th century German makers of sundials had a new use of the compass, the orientation of portable sundials. The mid- to late 16th century was a period of instrumental innovation and investigation of the nature of environmental magnetism, indeed, the beginning of geomagnetism. The Portuguese Pedro Nunes and João de Castro and the Englishmen *Robert Norman*, William Borough, and *William Gilbert* (*q.v.*) stirred interest in Earth's magnetism partly because of phenomena revealed by instrumental improvements.

These magnetic workers represent a mixing of the practical goals of navigators and more esoteric goals of improving the instruments and studying the phenomena revealed by them. Nunes and de Castro developed an improved "variation compass" and began paying closer attention to magnetic declination in the 1530s (Fara, 1998, p. 136). Norman focused on the dipping needle, conceived the dip circle (first described in print in 1581), and concluded that the seat of magnetism is within the Earth. Borough concentrated on improving instruments for measuring magnetic declination and Gilbert assembled the most complete résumé of magnetic experiments, a *summa* of magnetic knowledge to the time. No fundamental changes in instrument design occurred until near 1800. Nevertheless, two fundamental discoveries were made using improved instruments: the discovery by *Henry Gellibrand* (*q.v.*) in 1634 that declination varies over time and by George Graham in 1722 that declination varies daily according to a set pattern. During the 18th century artisans modified the variation compass with lighter needles, a box to enclose the needle, and the addition of reading microscopes and vernier scales. Another 18th century innovation was the tall azimuth visor, which allowed mariners to sight celestial objects high above the horizon.

Navigators and surveyors continued into the 18th and 19th centuries to be among the primary users of compasses. Their compasses "compensated" for declination by rotating the compass card so the needle pointed east or west of north (and by other means) and for inclination by weighting one end of the needle. In the 18th century, John England (of London) introduced steel needles in 1705, as did the Dutch in 1731 and P. Le Maire and d'Après de Mannevillette in 1745. Gowin Knight and John Smeaton, among others, continued to experiment with new designs (Fara, 1998, p. 136). More improvements were aimed at ease of use and dependability. Innovations were often encouraged by the British Board of Longitude, the French Academy of Sciences, and other institutions.

Ever since Robert Norman first discussed it in the late 16th century, the dip circle had been the least reliable magnetic instrument. The vertical needle was supported by a metal axis with pointed ends, inserted in glass or crystal cups. Friction often stopped the needle short of the direction of dip. Instrument makers tried many methods to minimize this friction, from lining the axle points and cups with gold or bell metal to adding "friction" wheels (McConnell, 1980, pp. 16–17; Good, 1998a, pp. 175–177).

Natural philosophers meanwhile, fascinated by the detail and complexity of the phenomena, worked with artisans to modify instruments. In 1777, Charles Coulomb proposed using a thread to suspend the needle and he derived the mathematical theory of torsional suspensions (see *Oscillations, torsional*). Jean Jacques Cassini soon used this type instrument at the Paris Observatory; Tiberius Cavallo, *Christopher Hansteen* (*q.v.*), and ultimately *Carl Friedrich Gauss* (*q.v.*) in the 1830s, elaborated on and perfected fiber-suspension instruments (Multhauf and Good, 1987, pp. 5–20; Good, 1998b). The instrument makers responsible for these devices included Étienne Lenoir, Thomas Jones, and George Dolland.

Instrument makers produced a variety of instruments for both navigation (and surveying) and for scientific investigation. The former instruments stressed durability and ease of use, the latter precision. Scientific use, however, included both fixed observatories—where size of the instrument was no concern—and expeditions by land or sea, in which size mattered. From the 1780s on, expeditions included geomagnetic research more often, though not until the late 19th century were dedicated geomagnetic surveys more commonplace.

The first completely new line of instruments to appear after the compass and dip circle started out simply as a different way to use these two instruments. George Graham observed the oscillating needle of a dip circle in 1723 but did not draw any conclusion regarding the intensity of magnetic force (McConnell, 1980, p. 26). Frederick Mallet used an oscillating compass needle in 1769 to measure the horizontal magnetic force. And in 1776 Jean Charles Borda discussed oscillations of a dipping needle as a measure of total magnetic force. An oscillating needle indicated magnetic intensity in the same way that a pendulum measured gravity: the shorter the period, the stronger the force. Until the 1830s, however, the results of different instruments could not be compared and were called "relative" measurements.

Soon many investigators were improving intensity instruments, developing methods to counteract problems like friction and the asymmetry of needles. Soon, too, scientists carried out magnetic intensity measurements on several expeditions: Jean-Françoise La Pérouse made measurements lost at sea in the 1780s. E.P.E. De Rossel, with the D'Entrecasteaux expedition, counted dip circle oscillations in 1792 and 1793 during a circumnavigation. *Alexander von Humboldt* (*q.v.*) continued and popularized the study of magnetic intensity during his expedition to South America, and Humboldt and J.L. Gay-Lussac continued this around Europe (Parkinson, 1998, p. 448). Humboldt used both instruments but began a trend among researchers to prefer to measure horizontal intensity with an oscillating, suspended needle. This led to a famous and popular instrument made by Christopher Hansteen in 1824, a needle suspended by a fiber. But the type was already widely established and used by others, too, including Coulomb

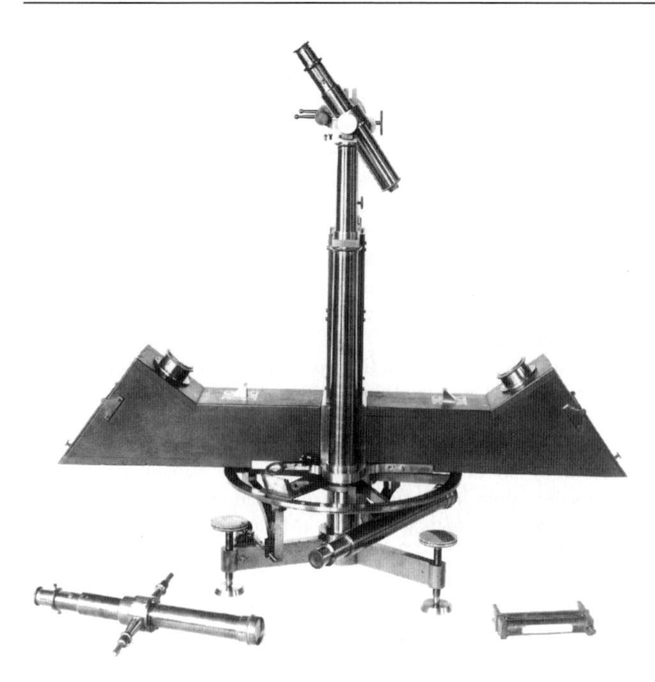

Figure I17 Variation transit by H.P. Gambey, Paris. This instrument was used by A.D. Bache in his magnetic survey of Pennsylvania in the 1830s. (Photo courtesy of Smithsonian Institution.)

(McConnell, 1980, pp. 26–29). Hansteen used it for indicating both declination and horizontal intensity. Other magnetic researchers across Europe used and designed fiber-suspension instruments, including François Arago and C.F. Gauss (Multhauf and Good, 1987, pp. 15–16). Around 1830, the word magnetometer was applied to these instruments that measured magnetic intensity.

By 1800, all of the forms of magnetic instruments based on magnetic needles, for both application and research, were available, but the most dramatic innovations were still to come. The efforts by Arago and Humboldt to improve the quality of magnetic observations and to popularize them established a solid trade in magnetic instruments. The favorite maker of magnetic instruments of Arago and Humboldt, H.P. Gambey of Paris, supplied variation and dip instruments around the globe during the 1820s–1840s (see Figure I17). Soon, Gauss and Weber would change how the instruments were used and how the data obtained were analyzed, bringing on a new period of instrumental innovation in the 1830s.

Classical geomagnetic instruments, from Gauss and Weber to the 1920s

When Carl Friedrich Gauss and Wilhelm Weber turned to geomagnetism in the 1830s, they developed several new geomagnetic instruments and procedures. Gauss and Weber developed a much larger fiber-suspension instrument than usual for determining declination. Their magnetic needle weighed over 2 kg (and later ones 10 kg or more), in contrast with, e.g., Hansteen's needles measured in grams, to mitigate the effects of air currents. They used a much longer suspension, about 2 m. Hansteen's suspensions were under 20 cm. Most importantly, Gauss and Weber borrowed an innovation from J.C. Poggendorff. He had fixed a mirror to the end of the needle and observed it from about 5 m away using an observing telescope, arranged so that he could see in the mirror a scale fixed to the base of the tripod of the telescope. Gauss wrote that this new instrument, the Gauss-Weber magnetometer, let them measure declination with a dramatically higher precision, similar to that known in astronomy. Indeed, he wrote that this instrument made possible the scientific study of declination and horizontal intensity (Gauss and Weber, 1837, p. 11).

Gauss and Weber also changed the fiber-suspension, horizontal magnetometer fundamentally by changing the way it was used. Gauss announced a two-step procedure in 1832 for calculating Earth's horizontal magnetic intensity in mechanical terms, i.e., length, mass, and time. He called this an "absolute" method (see *Gauss' determination of absolute intensity*). Gauss and Weber's magnetometer described above was so large it filled the building. It became the standard observatory instrument for the Magnetische Verein, an association of magnetic observatories, which collaborated with Gauss and Weber in the 1830s and 1840s. They worked with instrument makers Moritz Meyerstein of Göttingen, F.W. Breithaupt of Kassel, and others to supply more than a dozen of these instruments for observatories from the United States to Russia. They also developed the first bifilar magnetometer for direct indication of changes in the horizontal magnetic force and several portable instruments. In the bifilar instrument, the two-filament suspension was twisted so that the magnetic needle was held perpendicular to the magnetic meridian. As the Earth's magnetic force changed, the needle moved (Multhauf and Good, 1987, pp. 16–19). This instrument type was later termed a "variometer" because it indicated directly the variation of a magnetic variable. Hence, while they did not use the word, Gauss and Weber's bifilar magnetometer was the first variometer.

Besides Gauss and Weber, *Humphrey Lloyd* (*q.v.*) most carefully analyzed magnetic instruments and their use in the 1830s. Like Gauss and Weber, Lloyd investigated sources of error and ways to optimize instruments. He developed "Lloyd's balance" to indicate changes in the vertical magnetic force (hence the second variometer) and an indirect method of measuring the vertical force through induction, in addition to his own designs for unifilar and bifilar magnetometers, both for observatories and for field use. He, like Weber, recognized that no portable magnetic instrument could match the precision of an observatory magnetometer. Also like Weber, he designed portable instruments of the greatest possible precision. Lloyd's instruments supplanted Gauss and Weber's for the British "Magnetic Crusade" of the 1840s and 1850s. C.J.B. Riddell illustrated and explained the selection of magnetic instruments used by British observers during the Magnetic Crusade, including their shortcomings, in his *Magnetical Instructions* (1844). He even included a catalog of the instruments, available from three British instrument-making shops.

Clearly, the mid-19th century witnessed a remarkable burst in instrumental innovation. The best illustration of this goes far beyond any instruments seen before: photographically self-registering instruments. Introduced in 1846, they transformed magnetic research. Variants came and passed quickly at first, but by the 1850s the market settled to just a few models, especially those by Charles Brooks and Patrick Adie, both of London. Self-registering instruments used a tightly focused light beam, directed by mirrors attached to the instruments' magnets, to trace a graph of varying declination, intensity, etc., on a moving film or paper. These instruments removed the drudgery of oft-repeated visual measures. The data cascaded and overwhelmed magnetic workers, requiring greater staffing and more computational power.

National or regional magnetic surveys in France, England, the United States, and India, for example, in the late 19th century organized and routinized magnetic measurements. This spurred agencies in several countries to produce lightweight, robust field instruments. The "Kew-pattern" dip circle and magnetometer became standard in Britain, its colonies, and in the United States (Figures I18 and I19, Multhauf and Good, 1987, Figure 63). For the French magnetic survey, elegant, gem-like magnetometers and dip circles were fashioned of brass (Figure I20; Multhauf and Good, 1987, Figures. 38, 61, 64). The Prussian survey adopted a design by J.F.A.M. Eschenhagen, which other agencies decided was more appropriate for observatories because of its complexity (Multhauf and Good, 1987, Figure 41).

From the 1890s until about 1915, the US Coast and Geodetic Survey and then the *Carnegie Institution of Washington's Department of Terrestrial Magnetism* (DTM) (*q.v.*) systematically extended the traditional instruments based on magnetic needles and induction almost as far as they could go. The culmination was the DTM "Universal Magnetometer," (Figures I21 and I22) which packaged nearly everything needed for a complete magnetic survey in Tibet or deepest Amazonia into a wooden box barely 20 cm square and weighing 17 kg.

Schmidt's vertical balance, designed by Adolf Schmidt in 1915 at the Prussian Meteorological Institute, improved the measurement of vertical intensity. Instruments by Askania-Werke AG were very popular in prospecting in the mid-20th century. The Askania firm resulted from the combination of Werkstätten für Praezisions-Mechanik und Optik of Berlin with Centralwerkstatte Dessau in 1921.

Figure I18 Kew pattern magnetometer by John Dover of London (Photo courtesy of Smithsonian Institution.)

Figure I20 Magnetometer and dip circle by Brunner-Chasselon, Paris. These instruments were developed for the French magnetic survey and were purchased by the US Coast and Geodetic Survey in the 1880s. (Photo courtesy of Smithsonian Institution.)

Figure I19 Kew-pattern dip circle by John Dover. (Photo courtesy of Smithsonian Institution.)

Figure I21 DTM universal magnetometer, open to show needles and suspension. (Photo courtesy of Carnegie Institution of Washington.)

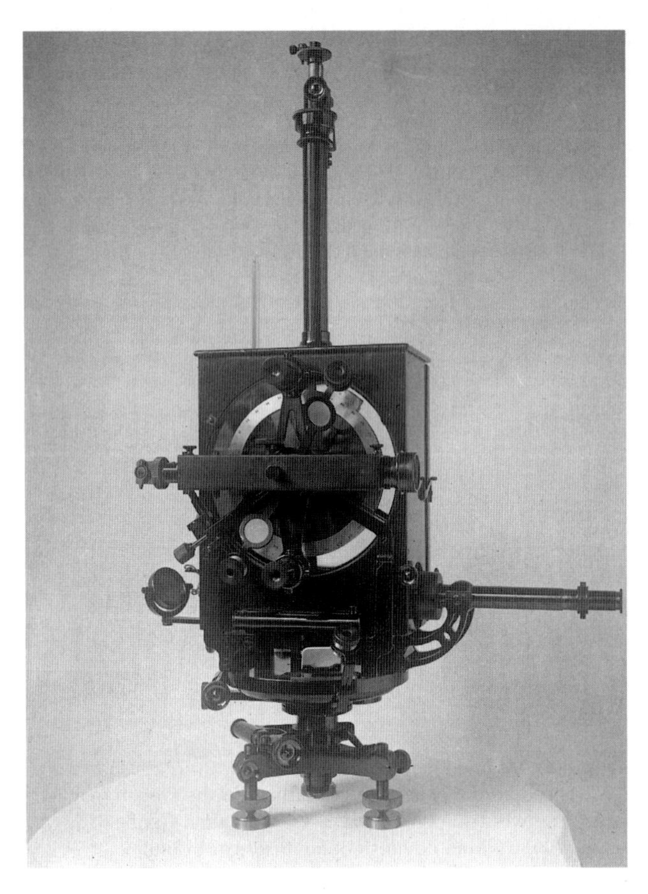

Figure I22 DTM universal magnetometer, closed. Set up for intensity measurements. (Photo courtesy of Carnegie Institution of Washington.)

The ultimate instruments along traditional lines were devised by Daniel La Cour of Denmark for the Second International Polar Year (1932–1933). His quartz horizontal magnetometer (QHM) and balance magnétique zero (BMZ) were "semiabsolute" instruments, which only required calibration yearly (Parkinson, 1998, p. 450; Chapman and Bartels, 1940, 1: 89–93). The QHM's high standard of accuracy within 1γ or nT kept the La Cour instruments in use even into the 1980s (Forbes, 1987). Nevertheless, even in the 1920s and 1930s instruments were beginning to change dramatically again.

Geomagnetic instruments since 1920: a second generation

As Parkinson has written, advances in magnetic materials and improvements in electronics ushered in the "second generation" of magnetic instruments around World War II(Parkinson, 1998, p. 450). Efforts to develop electrical methods of measuring geomagnetic variables started on simpler lines, shortly after 1900, under Arthur Schuster, L.A. Bauer, Adolf Schmidt, and others. These first instruments of the second generation used accurately determined electrical currents and tight engineering standards to measure magnetic variables. The Carnegie's DTM developed the first practical field earth inductor in 1913. Earth inductors improved the accuracy of inclination measurements to 0.1 min and eliminated time-consuming procedures with dip circles. In 1921, S.J. Barnett also developed a sine galvanometer at the DTM for absolute measurement of horizontal intensity.

Frank Smith, at Britain's National Physical Laboratory (NPL) in Teddington, developed one of the first standard instruments to use Helmholtz-Gaugain coils to measure a magnetic variable, in this case horizontal force. He did this in 1920, following an idea that Schuster suggested in 1914. According to one estimate, the Schuster-Smith magnetometer required only 8 min to produce and reduce an absolute measure of H, whereas the Gaussian oscillation/deflection method required 2.5 h. This magnetometer had an accuracy for horizontal intensity of 1γ or nT. In 1928, D.W. Dye developed a similar instrument at the NPL for vertical intensity, accurate to between 2 and 4γ.

The US Naval Research Laboratory and Gulf Research and Development perfected absolute magnetometers based on Helmholtz-Gaugain coils in the 1930s and 1940s, with one eye especially on improving measurement of vertical intensity (always problematic) and the other eye on greater magnetic stability and greater accuracy in measuring horizontal intensity and declination as well. Victor Vacquier started developing a vertical-force instrument in 1940 in cooperation with the US Coast and Geodetic Survey. Vacquier continued during the War, and announced the Gulf Absolute Magnetometer in 1945. Commercialized forms of Helmholtz-Gaugain coil instruments became standards at observatories for decades.

The biggest changes came, however, when researchers started applying totally new materials and ideas to magnetic measurements from 1940 onward. Vacquier developed the saturable-core magnetometer (or fluxgate magnetometer) about 1940 for the Gulf Research and Development Corporation (see *Magnetometer*). He based it on the first saturable-core compass, patented in 1931 by H.P. Thomas, and a simple, saturable-core magnetometer by H. Aschenbrenner and G. Goubau of 1936. The core of the sensing element has a high magnetic permeability and is surrounded by an excitation coil. This is a vector instrument, which measures magnetic intensity in the direction of the sensor core.

World War II spurred on development of a fluxgate magnetometer in the United States for the detection of German submarines. Vacquier elaborated his "fluxgate" instrument and method. In 1941 Vacquier and a team developed a device called the magnetic airborne detector (MAD). The challenge was to align the magnetometer along Earth's total magnetic vector, a goal achieved by a stabilizing system driven by input from two mutually perpendicular "orienting" fluxgates in a plane perpendicular to the sensing fluxgate. One system used servo motors to move the instrument in a gimbal and a second system used an air-driven gyroscope by Sperry Gyroscope to maintain orientation. After the War, investigators in many countries developed numerous variants of these airborne magnetometers. Some were towed behind naval ships or airplanes and contributed to military purposes, but these instruments also led ultimately to the discovery of mid-ocean, seafloor spreading and to the mapping of mineral deposits. Other versions of fluxgates gradually replaced classic variometers (Forbes, 1987, p. 111) in magnetic observatories, or took their place in satellites (see *Magsat and Ørsted*) and planetary probes.

Other new types of magnetometers were introduced in the late 20th century, including proton precession, Superconducting Quantum Interference Devices (SQUIDs), optically pumped sensors, and others (see *Magnetometers*). The use of these newer instruments is discussed in other articles in this volume (see *Aeromagnetic surveying*; *Archaeology, magnetic methods*; *Magnetic surveys, marine*; and *Repeat stations*). The most important change in instrumentation in the last half-century, however, may have been the automation of observation and the direct connection of sensors to data storage and computational facilities (see *Observatories, automation*; *Observatories, instrumentation*). But perhaps it is too early to have historical perspective on these events. (See also Campbell, 2003, Chapter 4 for a current discussion of "Measurement methods," pp. 189–227.)

Gregory A. Good

Bibliography

Campbell, W.H., 2003. Measurement methods. In Campbell, W.H. (ed.), *Introduction to Geomagnetic Fields*, 2nd edn. Cambridge, UK: Cambridge University Press, pp. 189–227.

Chapman, S., and Bartels, J., 1940. *Geomagnetism*, 2 vols. Oxford: Clarendon Press, especially Vol. 1, pp. 29–95.
Fara, P., 1998. Compass, variation. In Bud, R., and Warner, D.J. (eds.), *Instruments of Science: An Historical Encyclopedia*. New York: Garland Publishing, pp. 136–138, 709 pp.
Forbes, A.J., 1987. General instrumentation. In Jacobs, J.A. (ed.), *Geomagnetism*, 2 vols. London: Academic Press, pp. 51–142.
Good, G.A., 1998a. Dip circle. In Bud, R. *et al.*, (eds.), Instruments of Science: An Historical Encyclopedia. New York and London: Garland Publishing, pp. 175–177.
Good, G.A., 1998b. Magnetometer. In Bud, R., *et al.*, (eds.), Instruments of Science: An Historical Encyclopedia. New York and London: Garland Publishing, pp. 368–371.
Gauss, C.F., and Weber, W., 1837. *Resultate aus den Beobachtungen des magnetischen Vereins im Jahre 1836*. Göttingen: Dieterischsche Buchhandlung.
McConnell, A., 1980. *Geomagnetic Instruments before 1900: An Illustrated Account of their Construction and Use*. London: Harriet Wynter Ltd., 75 pp.
Multhauf, R.P., and Good, G., 1987. A brief history of geomagnetism and a catalog of the collections of the National Museum of American History. In *Smithsonian Studies in History and Technology No. 48*. Washington, DC: Smithsonian Institution Press, 87 pp.
Parkinson, W.D., 1998. Instruments, geomagnetic. In Good, G.A. (ed.), *Sciences of the Earth: An Encyclopedia of Events, People, and Phenomena*. New York and London: Garland Publishing, pp. 446–453.
Smith, J., 1992. Precursors to Peregrinus: the early history of magnetism and the mariner's compass in Europe. *Journal of Medieval History*, **18**: 21–74.

Cross-references

Aeromagnetic Surveying
Archaeology, Magnetic Methods
Carnegie Institution of Washington, Department of Terrestrial Magnetism
Compass
Gauss, Carl Friedrich (1777–1855)
Gellibrand, Henry (1597–1636)
Gilbert, William (1544–1603)
Hansteen, Christopher (1784–1873)
Humboldt, Alexander von (1759–1859)
Lloyd, Humphrey (1808–1881)
Magnetic Surveys, Marine
Magnetometers, Laboratory
Norman, Robert (1560–1585)
Observatories, Automation
Observatories, Instrumentation
Oscillations, Torsional
Repeat Stations

INTERIORS OF PLANETS AND SATELLITES

Introduction and background

The planets and satellites (or moons) of our solar system are of several types. The terrestrial planets and moons are bodies like the Earth, which are made mainly of rocks and metals like iron. The inner planets, Mercury, Venus, Earth, and Mars are terrestrial planets. The Moon and some of the satellites of the planets in the outer solar system are terrestrial type bodies. The outer solar system planets are the gaseous giants Jupiter and Saturn and the icy giants Uranus and Neptune. The gaseous giants are made mainly of hydrogen while the icy giants are built mainly from things like water and methane. The substances in the gaseous and icy giants are under enormous pressures and temperatures, so their physical states are very different from those that we encounter on Earth. At the very centers of Jupiter and Saturn there may be cores that resemble the terrestrial planets of the inner solar system. The satellites of the solar system reside mostly in orbits around the outer planets. Only Earth's moon and the small moons of Mars, Phobos and Deimos, circle the inner terrestrial planets. The outer solar system satellites are either terrestrial type bodies, like Jupiter's moon Io, or icy moons like Jupiter's moons Ganymede and Callisto, Saturn's moon Titan, and Neptune's moon Triton. As the name implies, the icy satellites are made mainly of a mixture of water ice, rock, and metal, with the ice making up typically around 50% of the moon by mass. Pluto and its moon Charon are icy bodies much like the icy satellites of the large outer planets. The interiors of many of the planets and moons of the solar system are depicted in Figure I23/Plate 12. The internal structures of all these bodies are based on theoretical models strongly constrained by observations, as discussed in the remainder of this article.

Density and the interior of a planet

How do we know what the planets and moons are made of? There are clues from what we can identify either on the surfaces of these bodies or in their gaseous envelopes (elemental atmospheric compositions,

Figure I23/Plate 12 Cutaway views of what lies beneath the surfaces of many of the planets and moons of the solar system. These models of the planetary interiors are based on a variety of geophysical observations as discussed in this article. Starting at the top and proceeding down to the right and then to the left are Neptune, Uranus, Saturn, Jupiter, Mars, Earth, Venus, and Mercury. Jupiter's four Galilean moons are arranged in line to the left of the planet (in order from right to left are Io, Europa, Ganymede, and Callisto). Earth's moon lies to its lower right. The bodies are not shown to proper relative scale. The figure was prepared by Calvin J. Hamilton and included here with permission.

chemical compositions and mineralogy of surface materials, surface geology), but one of the main indicators of overall composition is the density of the body, the ratio of its mass to volume. The sizes and masses of almost all planets and moons have been known for some time from astronomical observations, but the most current and accurate values are based on measurements by spacecraft that have landed on, orbited, or flown close by the bodies. The orbits of spacecraft are determined by the gravitational fields of the planets and moons, especially the fields of those bodies that are close to the spacecraft. The gravitational field of a body is the result of its mass and the way the mass is distributed inside the body. Therefore, tracking a spacecraft using its radio communication signal determines its orbit or trajectory from which the gravitational field and mass of a body can be inferred. The size and shape of a planet or moon can be obtained from images of the body acquired telescopically or by cameras onboard spacecraft. Detailed topographic maps of several bodies, e.g., Venus, Moon, and Mars, have been made by radar and laser altimeters on orbiting spacecraft. The volumes of planetary bodies follow directly from these types of measurements. Planets and moons are nearly spherical in shape and often the specification of a single average radius suffices to characterize the size of the body. Table I4 lists the masses, radii, and densities of the planets and major moons of the solar system.

The density of a planet or moon immediately tells us something about its interior because the materials from which planets are formed have characteristic densities. Rocks typically have densities about 3000 kg m^{-3}, metals have densities higher than about 5000 kg m^{-3}, and ices have densities of about 1000 kg m^{-3}. The density of any material depends on pressure and temperature, variables that take on extreme values deep inside large planets. Accordingly, the significance of a planet's density also depends on its size. The inner terrestrial planets and the Moon have densities between about 3300 and 5500 kg m^{-3} (Table I4). These bodies are mostly made of metal (iron) and rock. The Moon is relatively small, and its density reveals that it must be predominantly rock. Mercury's radius is only about 50% larger than the Moon's radius, but Mercury's much higher density requires that it be mostly metal compared to the mostly rocky Moon. The moons Ganymede and Titan are somewhat larger than the planet Mercury (Some moons are larger than some planets!), but they have densities of only about 2000 kg m^{-3}. Ice must be a major component of these moons to explain such a low density. Triton, Pluto, and Charon are similar icy bodies.

Two-layer model of planetary interiors

The inference of interior composition and structure from density can be made more quantitative by constructing a model of a planet or moon. One of the simplest models of a planet is the so-called two-layer model in which the planet is represented by a spherical core of radius r_c and density ρ_c surrounded by a spherical shell of outer radius R and density ρ_s. The average density $\overline{\rho}$ of such a model planet is

$$\overline{\rho} = \rho_c \left(\frac{r_c}{R}\right)^3 + \rho_s \left(1 - \frac{r_c^3}{R^3}\right) \qquad \text{(Eq. 1)}$$

The average density $\overline{\rho}$ is an observed quantity and is known. The quantities on the right-hand side of the equation ρ_c, ρ_s, and r_c/R are model parameters and are unknown and to be determined. There are three unknowns in this simple model, core and shell densities and fractional core radius (r_c/R) and only one equation to constrain their values, so it is not possible to obtain a unique model of the planet. However, if, for example, reasonable values of the model densities can be assumed, then the fractional core radius can be determined. Of course, the model can be made more deterministic by adding additional observational constraints (equations) as we discuss later.

The simple two-layer model is in fact a reasonable first-order representation of the interiors of many planets and satellites. This is because many planets and moons are differentiated, i.e., the rock and metal have separated into a metallic core surrounded by a rocky shell or mantle. This likely happens during the early evolution of a planet when temperatures in its interior are high enough to melt the heavy metal (iron) which subsequently sinks and accumulates into a core at the center of the planet. We know this has happened for the Earth because seismological observations of the Earth's interior have detected the Earth's metallic core and have measured its density and radius. The Earth occupies a special position in the discussion of planetary interiors because we have observations of the Earth's interior, mainly from seismology, that are unavailable for other planets and moons. The one exception to this is the Moon; we also have seismic observations of the lunar interior from seismic stations placed on the Moon's surface during the Apollo exploration era. However, these data are not as informative about the Moon's interior as are the seismic data sets collected on Earth. For example, the lunar seismic data are unable to confirm the presence or absence of a small metallic core in the Moon. Seismic instruments placed on the surface of Mars by the

Table I4 Properties of planets and satellites

Body	Mass (kg)	Radius (km)	Density (kg m^{-3})	C/MR^2
Mercury	3302.2×10^{20}	2439.7	5430	–
Venus	4869×10^{21}	6051.8	5240	–
Earth	5973.7×10^{21}	6371.0	5514.8	0.33078
Moon	7347.7×10^{19}	1737.1	3341	0.3935
Mars	641.91×10^{21}	3397	3940	0.365
Jupiter	1898.7×10^{24}	71492	1330	0.25
Io	8.932×10^{22}	1821.6	3527.5	0.378
Europa	4.800×10^{22}	1565.0	2989	0.346
Ganymede	1.482×10^{23}	2631.2	1942	0.3115
Callisto	1.076×10^{23}	2410.3	1834	0.3549
Saturn	568.51×10^{24}	60268	700	0.22
Titan	1.345×10^{23}	2575	1880	–
Uranus	86.849×10^{24}	25559	1300	0.23
Neptune	102.39×10^{24}	24764	1640	0.24
Triton	2139.8×10^{19}	1352.6	2065	–
Pluto	$\approx 1250 \times 10^{19}$	≤ 1200	$\approx$1750–2150	–
Charon	$\approx 180 \times 10^{19}$	≈ 600	$\approx$1300–2200	–

Viking spacecraft did not provide data that could be used to determine the planet's internal structure. For Earth, seismology has not only revealed the metallic core and the rocky mantle around it, but it has also shown that the core consists of a solid inner part and a liquid outer shell. Thus we talk about the inner core and the outer core of the Earth (Figure I23/Plate 12).

Two-layer model of Mercury

Mercury, Venus, Mars, and the Galilean satellite Io are all believed to have metallic cores made mostly of iron surrounded by rocky shells or mantles (Figure I23/Plate 12). The first-order interiors of these bodies can be represented by a two-layer model. In the case of Mercury, the existence of a metallic core is based almost entirely on its density. Mercury's high density (Table I4) requires that it consist of more than 50% iron by mass, and it is likely that the iron has concentrated into a central core. If we assume that the density of Mercury's core is 8000 kg m^{-3} (a reasonable value for a mostly iron core) and that the density of its mantle is 3000 kg m^{-3} (a representative value for rock) then according to (Eq. 1), the radius of Mercury's core would be nearly 80% of the planet's radius. Mercury can be thought of as a ball of iron with a relatively thin coating of rock (Figure I23/Plate 12). This result can also be seen in Figure I24, a graphical representation of (Eq. 1). Figure I24 plots $\rho_c/\bar{\rho}$ against r_c/R for different values of $\rho_s/\bar{\rho}$. With the above values of ρ_c, ρ_s, and $\bar{\rho}$, $\rho_c/\bar{\rho}$ is about 1.5, $\rho_s/\bar{\rho}$ is about 0.55, and the position of Mercury on the plot would correspond to r_c/R almost equal to 0.8. For the Moon, on the other hand, $\rho_c/\bar{\rho}$ would be in excess of 2, $\rho_s/\bar{\rho}$ would be nearly 1 (Table I4), and the position of the Moon on the plot would correspond to small values of r_c/R.

Mercury is also known to have a magnetic field that is believed to be generated by a dynamo in its metallic core. The existence of Mercury's magnetic field supports the existence of its metallic core if the magnetic field is indeed generated in the way the Earth's magnetic field is produced, by convective motions of the highly electrically conducting core metallic fluid, i.e., by a core dynamo. Dynamo action requires a core to be at least partially fluid, so if Mercury's magnetic field is produced by a dynamo, then not only must Mercury have a metallic core, but its core must also have a substantial liquid portion. Earth's metallic core is partially solidified, a result of the cooling of the core over geologic time. Dynamo action in the Earth occurs in its molten outer core. Mercury's core could also be solidifying from the inside out, similar to Earth's core, but the present operation of a dynamo in Mercury means that the planet, like Earth, has a liquid outer core sufficiently thick for a dynamo to operate. We will learn much more about the interior of Mercury with the successful completion of the Messenger mission. The spacecraft is presently on its way to orbit the planet; it is planned to make detailed measurements of Mercury's gravitational and magnetic fields and Mercury's rotational state, observations that can reveal a great deal about the interior of the body.

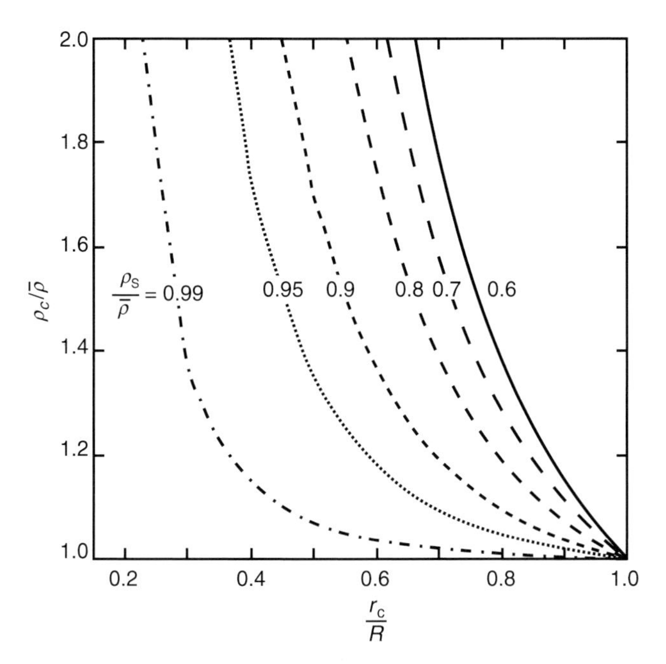

Figure I24 Normalized core density $\rho_c/\bar{\rho}$ versus normalized core radius r_c/R for different values of normalized mantle density $\rho_s/\bar{\rho}$ in a two-layer model of a planet.

Two-layer model of Venus

Like Mercury, the main constraint on the interior of Venus is its density. Venus is slightly smaller than Earth and its density is about 5% less than the density of the Earth (Table I4). Based on these similar properties, it is generally believed that the interior of Venus is like the interior of Earth, i.e., that Venus has a metallic, mainly iron core with a radius about half the planet's radius surrounded by a rocky mantle (Figure I23/Plate 12). Many spacecraft have orbited and landed on Venus, and we are quite certain that the planet does not have a magnetic field. So, unlike Mercury, we cannot use the existence of a magnetic field to support the existence of a metallic core inside Venus. Instead, we must explain why, if Venus is like Earth in its interior, Venus does not have a magnetic field like Earth does. There are several possible explanations for this, including the absence of a solid inner core in Venus. Because Venus is somewhat smaller than Earth and the pressure at its center is less than at the center of the Earth, Venus' core, though cooling over geologic time, may not have yet cooled sufficiently to freeze out an inner core. Inner core solidification might be necessary to drive a dynamo in the outer liquid core of a planet because the freezing of the inner core releases energy to drive the dynamo in the form of the latent heat of solidification and gravitational potential energy. Solidification of the Earth's inner core might play an essential role in driving the geodynamo. It might also be that Venus' core is simply not cooling fast enough to have a thermally driven dynamo because the planet's mantle cannot convect efficiently enough to remove a lot of heat from the core. Venus has no plate tectonics, an indication that its mantle convects sluggishly. Only when we can determine the extent of solidification of Venus' core through seismic exploration of the planet will we be able to identify the reason for the planet's lack of a magnetic field.

Moment of inertia and interior of a planet, Mars and the Moon

When we come to discuss Mars and the Moon we are fortunate to have a crucial piece of information in addition to density that tells us a great deal about the distribution of mass inside a body. This is the moment of inertia C of the body about its rotation axis. The moment of inertia of a point mass about a rotation axis is the product of the mass and the square of the perpendicular distance of the mass to the axis. Planets and moons are not point masses, so their moments of inertia represent a summation or integral of contributions to the total moment of inertia from all the elements of mass within the body. It is easy to calculate the moments of inertia of idealized bodies. For example, the moment of inertia of a constant density sphere of radius R and total mass M is $0.4MR^2$. Since the actual moment of inertia of a planetary body is some huge number, it is convenient to describe a planet's moment of inertia by scaling it to MR^2. The moment of inertia factor of a constant density sphere is then $C/MR^2 = 0.4$. The moment of inertia factors of planets and moons are less than 0.4 because these bodies are denser at depth than they are near the surface. The more a planet's mass is concentrated near its center the smaller is its moment of inertia. Planets and moons with metallic cores are obviously denser at their centers since iron is much denser than rock. Density also increases with depth in a planet because of the increase in pressure with depth. Extraordinary pressures experienced deep within planets compress rock and make it denser. The extreme pressures can even rearrange the minerals of the rock to create denser phases of the minerals. The tendency of pressure to increase density at depth in a planet occurs despite the effect of enhanced temperature deep within a planet. Most materials including rocks decrease in density with increasing temperature. Pressure overwhelms temperature as far as density is concerned. The moment of inertia factors of Earth, Moon, and Mars are given in Table I4. These are the only planets in the solar system for which we have direct measurements of the moment of inertia.

Moment of inertia and two-layer models, Mars

We will discuss how the moment of inertia of a planet can be determined, but let us first see how this property can be used to learn more about a planet's internal distribution of mass than could be deduced from density alone. The simple two-layer model of a planet provides an illustrative tool for this purpose. As discussed above, the two-layer model has three unknown quantities we would like to determine, ρ_c, ρ_s, and r_c/R. Equation (1) and the known density provide one constraint on the three unknowns. The equation for the moment of inertia of a two-layer spherical body and the value of the moment of inertia provide a second constraint. The equation is

$$\frac{C}{MR^2} = 0.4\left[\frac{\rho_s}{\bar{\rho}} + \frac{(\rho_c - \rho_s)}{\bar{\rho}}\left(\frac{r_c}{R}\right)^5\right] \quad \text{(Eq. 2)}$$

and the left-hand side is known. There are now two equations for three unknowns, and we can construct a suite of models for a planet by considering, for example, a range of plausible values of r_c/R. Figure I25 shows the results of using (Eqs. 1 and 2) together with the known values of density and moment of inertia factor for the planet Mars (Table I4). The figure plots core density against r_c/R and represents the equation

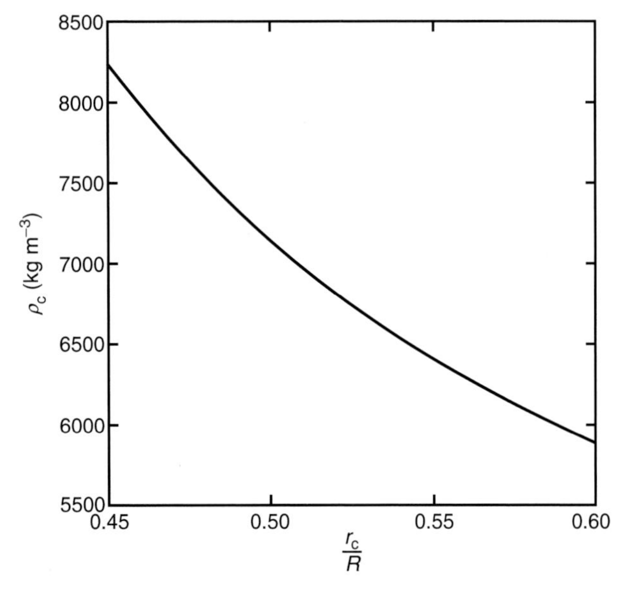

Figure I25 Core density as a function of normalized core radius r_c/R for a two-layer model of Mars constrained by the values of average density and axial moment of inertia.

$$\rho_c\ (\mathrm{kg\,m^{-3}}) = 3\,935.0\left[\frac{\left(\frac{0.365}{0.4}\right)\left(1-\left(\frac{r_c}{R}\right)^3\right) - 1 + \left(\frac{r_c}{R}\right)^5}{\left(\frac{r_c}{R}\right)^5 - \left(\frac{r_c}{R}\right)^3}\right] \quad \text{(Eq. 3)}$$

For plausible core densities corresponding to densities of iron or iron mixed with lighter elements like sulfur, values of r_c/R range from about 0.45 to 0.60. If the normalized core radius is about 0.45, then the density of the core is about 8250 kg m^{-3} typical of the density of a pure Fe core. If the normalized core radius is 0.60, then the density of the core is about 5900 kg m^{-3} typical of the density of a core containing both iron and sulfur as FeS. While the cores of terrestrial planets are composed mainly of metals like iron and nickel, the cores can also contain other lighter elements such as S or O. Indeed, based on the seismically determined density of the Earth's core and the measured density of iron at high temperature and pressure, it is certain that light elements are in the Earth's core in addition to Fe and Ni. The density and moment of inertia of Mars leave no doubt that Mars has a metallic core and rocky mantle (Figure I23/Plate 12). For plausible core densities the core radius is between about 1530 (0.45R) and 2040 km (0.6R) depending on core composition; the more iron-rich the core, the smaller it is. The mantle density is about 3400–3500 kg m^{-3} depending on core size; smaller mantle densities correspond to larger cores. The results we find for Mars show that even simple two-layer models can provide strong constraints on certain global properties of planets and moons.

Measuring moment of inertia from a planet's rotation

How do we determine the moment of inertia of a planet like Mars? Basically, we must be able to observe how a planet's state of rotation responds to external torques exerted on the planet. The rotation of a body is determined by an equation analogous to Newton's second law of motion for the linear movement of a body. Newton's second law states that the force applied to a body equals the product of its mass and acceleration. The analogous principle for angular motion states that the torque applied to a body equals the product of its moment of inertia and angular acceleration. The torque about an axis is the product of a force with the perpendicular distance between the force and the axis. Torques on planets arise from the gravitational pull of the Sun or the gravitational attraction of a planet's satellite. The torque applied to the Earth by the Moon significantly affects the Earth's rotation. Mercury and Venus have no satellites to influence their rotational states. Phobos and Deimos exert torques on Mars, but their combined effect is much less than the torque exerted on Mars by the Sun. The rotation or spin axes of Earth and Mars are tilted about axes normal to their orbital planes (the obliquities of Earth and Mars are 23.45° and 25.19°, respectively), and the gravitational torques on them cause their spin axes to precess about the orbit normals. Measurement of the precession rate together with knowledge of the forcing gravitational torque allows the determination of the axial moment of inertia from the form of Newton's law for angular motion. Astronomical observations have revealed the precession rate of the Earth's spin axis and in turn the Earth's moment of inertia. Radio tracking of the Mars Pathfinder Lander has yielded similar information for Mars. Laser ranging observations of the Moon made possible by reflectors placed on the lunar surface during the Apollo program have determined the Moon's rotational state and its moment of inertia. Moments of inertia of planets and satellites are summarized in Table I4. The table includes the direct measurements and values inferred indirectly by other means as discussed below.

Moment of inertia and interior of the Moon

The moment of inertia factor of the Moon ($C/MR^2 = 0.3935$, Table I4) is sufficiently close to 0.4 that the interior could be rock with an almost constant density. Density would increase slightly with depth in such a body due to the increase of pressure with depth and the small increase of density could account for the small decrease in the moment of inertia factor below 0.4. Alternatively, the moment of inertia factor of the Moon is also consistent with it having a small iron core a few hundred kilometers in radius at its center (Figure I23/Plate 12). Though we do have seismic data for the Moon, these data do not constrain the possible existence of a metallic core. However, magnetic data for the Moon indicate the likelihood that it has an iron core. Magnetometers were placed on the lunar surface and in orbit around the Moon during the Apollo exploration program, and these instruments showed that while the Moon has no large-scale magnetic field at present, the lunar crust is magnetized. It is generally believed that the crust was magnetized at a time in the past when the Moon had an active dynamo, and the crustal rocks cooled through the Curie temperature in the presence of a lunar magnetic field. If the remanent magnetization of the Moon's crust was created in this way, then the Moon must have a metallic core that supported a dynamo in the past. The lack of a magnetic field at present could be due to the cooling of the Moon over geologic time, which has left it with a nearly solidified core in which dynamo action can no longer occur. Other explanations are also possible, but the cooling of a relatively small body like the Moon is expected to have substantially solidified its core. The lunar core may not be entirely solidified, or even partially frozen, according to an interpretation of the lunar rotational data from the laser ranging observations.

Mars crustal magnetization

A magnetometer on the Mars Global Surveyor spacecraft discovered that the southern highlands crust of Mars is also magnetized, and, in certain regions of Mars, even more strongly magnetized than Earth's most magnetic rocks (see *Magnetic field of Mars*). The southern highlands crust of Mars dates from the earliest phase of Martian evolution, the so-called Noachian, and its magnetization suggests that Mars had a magnetic field generated by an internal core dynamo in the distant past. The existence of a metallic core in Mars, as deduced from its density and moment of inertia, is consistent with the inference of a core from the magnetization of the Martian crust. The absence of a Martian dynamo at present could be explained in ways similar to those that offer explanations for the lack of present magnetic fields on Venus and the Moon. These include solidification of the core to the point where a dynamo cannot operate in the remaining liquid outer core or the inability of the mantle to cool the core at a rate large enough to drive a dynamo through thermal convection in the core. The detection of the solar gravitational tide on Mars through radio tracking of the orbiting Mars Global Surveyor has led to the conclusion that at least some outer portion of the Martian core is not solidified.

Gravitational field and a planet's interior

Much can be learned about a planet's interior from measurements of its external gravitational field since the gravitational field of a body arises directly from the mass inside it. Each element of mass in a planet or satellite contributes to the external gravitational field according to Newton's law of gravitation, which states that the gravitational force exerted by an element of mass is directly proportional to the element's mass and inversely proportional to the square of the distance between the mass element and the point at which the force acts. The net gravitational force at a point outside a planet is obtained by summing the forces exerted by each mass element over the entire distribution of mass inside the body, taking proper account of the directional nature of the gravitational force. When a planet is in hydrostatic equilibrium (see below), its moment of inertia and internal structure can be deduced directly from the gravitational field measurements. The Earth is reasonably close to this state, but Mars is not, due to the uncompensated portion of the Tharsis uplift. The Moon is not in hydrostatic equilibrium, and it is not known to what extent Mercury and Venus might depart from this state. Thus, though we know the gravitational field of Venus quite well, we cannot use this knowledge to determine its moment of inertia and deep internal structure and, in particular, to verify that it has a metallic core like Earth. There are other bodies in the solar system that we think are in hydrostatic equilibrium. These include the outer planets with fluidlike interiors and the Galilean moons of Jupiter, Io, Europa, Ganymede, and Callisto. Measurements of the gravitational fields of these moons by the Galileo spacecraft have been used to deduce their internal structures.

The gravitational field of any planet is generally represented mathematically by a function known as the gravitational potential V; the gravitational force on a unit mass is the gradient or directional derivative of V. The potential is, in turn, given by a series of terms known as a spherical harmonic expansion. *Spherical harmonics* (*q.v.*) are functions of latitude and longitude that provide a basis for the representation of potential functions like the gravitational field external to a distribution of mass. These potential fields are solutions of Laplace's equation. The terms in the spherical harmonic expansion of the external gravitational field depend not only on latitude and longitude, but they also include a dependence on the radial distance r to the center of mass of the planet; the terms are inversely proportional to powers of r. The leading term in the spherical harmonic expansion of the external gravitational field is GM/r (where G is the universal gravitational constant and M is the mass of the planet). This is the gravitational field that would exist if all the mass of the planet were concentrated in a point at its center of mass. The next terms, there are five of them, are all proportional to r^{-3}. These terms can be thought of as arising from the ellipsoidal distortion of the planet's mass distribution. One of these terms, $(GM/r)(R/r)^2 C_{20} (3 \sin^2 \phi - 1/2)$, is of particular importance because the coefficient C_{20} (sometimes denoted by $-J_2$) can be used, under certain circumstances, to determine the planet's axial moment of inertia. There is no term in the external gravitational potential proportional to r^{-2} because the origin of the coordinate system is fixed at the planet's center of mass (the center of mass is the point inside a mass distribution about which the mass distribution has no net moment).

Hydrostatic equilibrium: determining C from the gravitational field

The gravitational coefficient C_{20} can be used to determine C/MR^2 when a planet is in a state of hydrostatic equilibrium. A planet in hydrostatic equilibrium behaves like a fluid in response to forces exerted on the planet. Over geologically long times, rocks can behave like fluids and flow in response to forces exerted on them. The motions are very slow, but over geologic time the rocks in a planet can deform and move over large distances. This is the process that enables plate tectonics and mantle convection on Earth and allows the Earth, for example, to adjust to changes in the mass loading of its surface by postglacial rebound.. Fluidlike deformation of rock is facilitated by high temperature so that subsolidus creep generally occurs beneath a relatively cold and strong outer shell of a planet called the lithosphere. If the lithosphere is thick enough and strong enough to support the stresses associated with surface or subsurface mass anomalies, then the planet will not be in hydrostatic equilibrium. This is probably why the Moon is not in hydrostatic equilibrium, and a similar explanation might apply to the small planet Mercury if we find that it is also not in hydrostatic equilibrium. We are not safe in using C_{20} to determine C/MR^2 for Mercury or Venus, but the procedure can be applied to the Galilean satellites of Jupiter.

For a body in hydrostatic equilibrium that is distorted by rotation and tides, it can be shown that C/MR^2 is related to C_{20} by the Radau equation

$$\frac{C}{MR^2} = \frac{2}{3} - \frac{4}{15}\left[\frac{4 - k_f}{1 + k_f}\right]^{1/2} \qquad \text{(Eq. 4)}$$

where k_f is the fluid Love number of the planet, which depends on its internal mass distribution and is given in terms of C_{20} by

$$k_f = \frac{6}{5q_r}(-C_{20}) \qquad \text{(Eq. 5)}$$

for a tidally and rotationally distorted satellite in synchronous rotation about a planet (e.g., one of the Galilean moons orbiting Jupiter), or by

$$k_f = \frac{3}{q_r}(-C_{20}) \qquad \text{(Eq. 6)}$$

for a planet distorted by rotation only (e.g., Earth or Mars), where

$$q_r = \frac{\omega^2 R^3}{GM} \qquad \text{(Eq. 7)}$$

and ω is the rotational angular velocity of the body.

For Earth and Mars it is unnecessary to try to estimate C/MR^2 from C_{20} using the Radau equation because C/MR^2 has been determined directly from the observed rates of precession of the planets' rotation axes. However, it is useful to carry out the exercise to see how the approach works for these planets and to assess how far these planets are out of hydrostatic equilibrium. For bodies like the Galilean satellites of Jupiter, the only way we can estimate C/MR^2 is to deduce it from the gravitational fields (C_{20}) measured by radio tracking the Galileo spacecraft as it made multiple flybys of the moons. For Earth, $q_r = 0.0034498$ and $C_{20} = -0.0010826$, and the Radau equation yields $C/MR^2 = 0.332$, very close to the value of $C/MR^2 = 0.3308$ deduced from precession of the Earth's rotation axis. For Mars, $q_r = 0.0045702$ and $C_{20} = -0.001964$, and the Radau equation yields $C/MR^2 = 0.376$, not close enough to the value of $C/MR^2 = 0.365$ deduced from observations of the precession rate of the rotation axis of Mars (by radio tracking of the Mars Pathfinder Lander and Mars Global Surveyor orbiter) to yield reliable models of the Martian interior. The reason, as noted above, is that Mars is not in hydrostatic equilibrium because the Tharsis highland is a major mass load that is partially supported by the strength of the Martian lithosphere.

Galilean moons of Jupiter

A strong case can be made that the Galilean moons of Jupiter, Io, Europa, Ganymede, and Callisto are sufficiently close to hydrostatic equilibrium for their axial moments of inertia and internal structures to be inferred from their gravitational field coefficients C_{20}. These satellites are distorted by the tidal forces from the massive Jupiter and by the centrifugal forces of their rotations. The Galilean moons are in synchronous rotation about Jupiter (they keep the same face to Jupiter as they rotate about it) with periods of days (Table I5); the satellites are reasonably rapid rotators. Because they do not have the strength to resist the large tidal and rotational forces over geological time, their internal masses flow and adopt an equilibrium ellipsoidal fluid shape. Rotation flattens the moons along their rotational axes and produces equatorial bulges perpendicular to the axes of rotation. Each moon is also elongated by tidal forces along an axis that passes from its center of mass to the center of mass of Jupiter. The redistribution of mass into an ellipsoidal shape is the source of the contributions to the gravitational field represented by coefficients like C_{20}.

Table I5 Orbital parameters of moons

Body	Average distance from planet (km)	Orbital period (days)
Io	4.22×10^5	1.77
Europa	6.71×10^5	3.55
Ganymede	10.70×10^5	7.15
Callisto	18.83×10^5	16.7
Titan	1.222×10^6	15.94
Triton	3.55×10^5	5.88

The gravitational fields of all the Galilean satellites have been measured by radio tracking the trajectory of the Galileo spacecraft as it repeatedly orbited Jupiter and flew by the moons. The subtle deflection of the spacecraft trajectory as it passes close to a moon and feels the pull of the moon's gravitational field provides a direct measure of the field. The Galileo spacecraft arrived at Jupiter in December of 1995 and explored the Jovian system until it was targeted to crash into Jupiter on September 21, 2003. The many flybys of the Galilean moons by the spacecraft over this nearly 8-year period provided all the gravitational data on which present models of the satellite interiors are based. Multiple flybys of Io were particularly useful in establishing that it is in hydrostatic equilibrium. A polar flyby of Io at an altitude of 300 km and equatorial flybys at altitudes between 100 and 200 km enabled independent determination of rotational and tidal contributions to the gravitational field. These separately determined gravitational field components are in a ratio consistent with the equilibrium ellipsoidal distortion of Io under Jovian tidal forces and rotation of Io. Io's actual shape, as determined from spacecraft images of the satellite, is also consistent with the equilibrium ellipsoid. Io is certainly in hydrostatic equilibrium, a condition expected for a body so hot in its interior that its mantle might be substantially partially molten. Even without melting, the rocks should be so hot that they should easily flow over geologic time. Io's high-temperature internal state is revealed by its active volcanism; Io is the most volcanically active body in the solar system. The energy for all this volcanic activity comes from the dissipation of tidally forced distortions of the body. The time-varying tidal distortions giving rise to heating are deformations in addition to the steady tidal distortion that gives rise to Io's quadrupole gravitational field. These additional tidal deformations are caused by the eccentricity of Io's orbit around Jupiter. There are no similar data to more rigorously establish the hydrostaticity of the other Galilean satellites.

Io

Table I4 summarizes the physical characteristics of the Galilean satellites of Jupiter including, in particular, the densities and moments of inertia that are used to construct models of the moons' interiors. These properties are based mainly on the data acquired by the Galileo spacecraft. Two- and three-layer models of the interiors of the moons, based on these properties and the modeling approach discussed above, are shown in Figure I23/Plate 12. Io, about the size of the Moon, is differentiated into a metallic core and rocky mantle. If Io's core is pure iron with a density of about 8 090 kg m^{-3} then its radius is 650 km or about 36% of Io's outer radius. The mass of this Fe core is about 10% of Io's total mass. If the core consists of a eutectic Fe-FeS alloy with a density of about 5150 kg m^{-3}, then its radius is about 950 km or about 36% of Io's outer radius. The mass of this Fe-FeS core is about 20% of Io's total mass. The magnetometer on the Galileo spacecraft did not detect an Ionian magnetic field on its flybys of the satellite. The lack of an operative dynamo in Io's metallic core might be explained by the inability of the hot mantle to cool the core. If the heat flow from the core is too small, the core would not convect and there would be no core motions to drive a dynamo. It seems unlikely, given the volcanic activity of the satellite and the large amount of heat it radiates, that Io's core could have solidified to the point where dynamo action would not be possible. It is more likely that Io's core is completely molten but not convecting.

Europa

Europa is somewhat smaller than Io and the Moon. Like Io, Europa has a metallic core surrounded by a rocky mantle (Figure I23/Plate 12).

In addition, it has a layer of water about 100 km thick that covers the rock mantle. At the surface, the water is frozen, but beneath the surface the water may be liquid, i.e., Europa might have an internal ocean. The ocean could be extremely deep and extend all the way down to the silicate mantle. The ocean could also be a relatively thin layer of liquid water within a largely ice shell. The average density and the moment of inertia of a planet or satellite provide information on the internal density distribution. We cannot infer from these data whether an iron core is solid or liquid or whether a water shell is ice or liquid. This is because the density changes associated with freezing of water or molten iron are too small to be distinguished in simple models of planetary interiors based on only mean density and moment of inertia. Europa is not volcanically active like Io, but its icy surface is disrupted by fractures and ridges indicative of internal activity in the moon. In fact, the appearance of the surface, with some regions so broken apart that they resemble the tops of floating icebergs, is one of the reasons, it is believed, that Europa has a liquid water ocean beneath its surface. A second reason to suspect the existence of an internal liquid water ocean in Europa is the detection of electrical currents inside Europa by the magnetometer on the Galileo spacecraft. The currents are driven by magnetic field changes felt in the moon as it orbits Jupiter and senses the planet's magnetic field. The currents in turn produce magnetic fields that are detected by the Galileo spacecraft. Rocks and ice are not good enough conductors of electricity to support the currents, but salty water, like the Earth's ocean, is conductive enough. So, the evidence for a subsurface ocean in Europa is all indirect, but the possibility that the ocean really exists is so important (an internal ocean would provide an environment conducive to life) that Europa will undoubtedly be a focus of future planetary exploration. The Galileo magnetometer did not detect an Europan magnetic field (a steady global magnetic field produced by a dynamo), so like Io, there is no dynamo operating in Europa's core. The reason for lack of dynamo action in Europa's core is probably different from the explanation for Io because Europa is not intensely heated by tidal dissipation as is Io. Europa may not have a dynamo for the same reason the Moon lacks a dynamo at present; these small moons may have cooled sufficiently that their cores are largely solidified and unable to support a dynamo. Other explanations are also possible. Europa's core could vary in radius from about 13% to 45% of Europa's full radius, depending on the composition of the core.

Ganymede

Ganymede is the largest moon in the solar system. It is even larger than the planet Mercury (Tables I4 and I5). It is also the only moon in the solar system known to have a magnetic field. Its magnetic field is larger than Mercury's magnetic field. Ganymede's moment of inertia is smaller than that of any of the terrestrial planets or other Galilean satellites (Tables I4 and I5). Its density (about 2000 $\mathrm{kg\,m^{-3}}$) is intermediate between that of water (about 1000 $\mathrm{kg\,m^{-3}}$) and rock (about 3000 $\mathrm{kg\,m^{-3}}$), and accordingly it consists of about half rock (+ metal) and half water by mass. Ganymede is therefore known as an icy satellite. Like Europa, Ganymede consists of a metallic core surrounded by a rock mantle and a water-ice shell (Figure I23/Plate 12). However, unlike Europa, the outer ice shell is enormously thick (as thick as about 800 km). The small moment of inertia ($C/MR^2 = 0.3115$) requires a large density difference between the core and the outer shell; this is provided by the density difference between water and iron. The detection of induced magnetic fields by the Galileo magnetometer, similar to the induced fields found around Europa, supports the possibility that Ganymede, like Europa, has an internal liquid water ocean embedded in its largely frozen outer shell of ice. The internal ocean on Ganymede would be buried deeper below the surface than it is on Europa. Though Ganymede's surface has been modified by internal dynamical processes (endogenic modification of the surface), it does not display the features found on Europa that suggest an ocean just below the surface. The core is the site of the dynamo that creates Ganymede's magnetic field; the existence of the magnetic field is indirect proof of the existence of Ganymede's metallic core. Moreover, we know that Ganymede's core is either entirely molten or liquid in an outer shell thick enough to sustain a convectively-driven dynamo. The possibility that Ganymede's magnetic field is produced by the global magnetization of a shell of rock in its interior cannot be completely ruled out, but it is considered less likely than a core dynamo. Even in this case, Ganymede would have to have had a dynamo operating in a metallic core in the past to create the magnetic field in which the crust was magnetized. Curiously, perhaps, the interior of Ganymede looks very much like Io with the addition of a thick ice shell. However, tidal dissipation does not presently heat the interior of Ganymede.

Callisto

Callisto is another icy satellite, just slightly smaller and slightly less dense than Ganymede. However, both its interior and surface are very different from Ganymede. Though Callisto is also about half rock and half water by mass, its C/MR^2 is only slightly smaller than 0.4 (Table I4), implying that its interior is not too different from a uniform mixture of ice and rock with nearly constant density. The reduction of C/MR^2 from 0.4 is due in part to the densification of ice with depth inside Callisto (ice transforms to higher density phases at sufficiently high pressures) and the partial separation of rock from ice in the relatively shallow subsurface layers of Callisto. However, in the bulk of Callisto at depth, ice and rock are still intimately mixed as they are in the material which accreted to form Callisto (Figure I23/Plate 12). In particular, Callisto does not have a metallic core as do the other Galilean satellites. The lack of a magnetic field associated with Callisto (based on observations by the Galileo spacecraft) is consistent with the absence of a core. Callisto is said to be only partially differentiated, where in this context, differentiation is the process that separates ice from rock and usually would involve the melting of the ice of an ice-rock mixture. Ganymede is differentiated not only in the sense that its ice and rock have separated, but its metal and rock have also separated through the melting of the primordial rock component early in the formation of the satellite. The term differentiation, as traditionally applied to terrestrial planets, usually refers to the melting of rock and the separation of molten iron into a central core. Why Ganymede got hot enough inside to fully differentiate and separate ice from rock and also form a metal core while similarly sized Callisto failed to get hot enough to even melt most of the ice in the primordial ice-rock mixture from which it formed is still an unsolved problem. The answer might lie in the orbital-dynamical history of the satellites; Ganymede might have been intensely tidally heated in the past, while Callisto would have escaped such a fate since Callisto is much farther from Jupiter and not involved in orbital resonances with the other Galilean satellites. Electromagnetic induction signals have been detected in several flybys of Callisto by the Galileo spacecraft, indicating that Callisto, like Europa and Ganymede, has an internal liquid water ocean buried deep beneath its surface. How such a global ocean could exist in a largely undifferentiated satellite is not understood.

Titan

Saturn's large satellite Titan is an icy satellite with a density of about 1880 $\mathrm{kg\,m^{-3}}$ (Table I4). It is in synchronous rotation about Saturn at an average orbital distance of about 1.2 million km (Table I5). There are no rotational or gravitational data to constrain interior models of Titan, so present ideas about the moon are based on its density. It is generally thought that ice and rock are separated in Titan as they are in Ganymede. However, unlike Ganymede, Titan does not have a magnetic field, a fact confirmed by magnetometer data from a recent flyby of Titan by the Cassini spacecraft. The absence of a magnetic field is only a weak constraint on whether or not Titan has a metallic core; either Titan does not have an iron core, or it has a core, but not an operative dynamo in the core. Future flybys of Titan by the Cassini spacecraft should provide information about Titan's gravitational field that will help us infer its internal structure similar to the inferences about the interiors of the icy Galilean satellites made from measurements of their gravitational fields by the Galileo spacecraft. Titan is distinguished by its massive atmosphere of nitrogen (95%) and methane (5%). The surface pressure on Titan is about 1.5 times the pressure at Earth's surface. Titan is the only moon in the solar system with such a substantial atmosphere whose existence holds clues to the formation and evolution of the satellite and the nature of its interior.

Pluto, Charon, and Triton

The small minor planet Pluto and its comparably sized moon Charon are additional examples of icy solar system bodies. Their densities are somewhat uncertain; Pluto's density lies between about 1750 and 2150 $kg\,m^{-3}$, and Charon's density lies in the range 1300–2200 $kg\,m^{-3}$ (Table I4). We have little other information about these bodies, so it is impossible to conclude one way or another about the separation of ice from rock in their interiors. In addition to ice and rock, Pluto and Charon may contain organic material in their interiors. Pluto and Charon are likely to be typical of many other icy objects orbiting at the edge of the solar system in the so-called Kuiper belt. Pluto is very similar in size and density and surface composition to Neptune's icy moon Triton (Table I4). Nitrogen and methane ices (in addition to water ice) have been detected on the surfaces of Pluto and Triton, and both bodies have thin predominantly nitrogen atmospheres. Triton is in an unusual orbit around Neptune; the orbit is highly inclined and is retrograde (the inclination of the orbit is about 157°). Triton's orbit suggests that it formed elsewhere in the outer solar system and was captured from heliocentric orbit by Neptune. Triton was likely strongly heated during this capture event, and it is therefore expected to have differentiated into a rock core surrounded by an ice mantle. Theoretical considerations suggest that Pluto might also be differentiated, though not by heating during capture.

Giant planets in the outer solar system

Jupiter

The outer planets are large, fluidlike bodies with low densities compared to the densities of the terrestrial planets (Table I4). Jupiter, the largest planet in the solar system, is composed largely of hydrogen and helium (Figure I23/Plate 12). Hydrogen and helium are gases in the outer layers of Jupiter (H_2 and He are 86.4% and 13.6%, respectively, by number, in Jupiter's atmosphere), but pressures are so large in the deep interior of Jupiter that its hydrogen and helium are more like dense liquids than gases at large depths. Hydrogen transforms to a metallic, highly electrically conducting phase at a depth perhaps as shallow as 20% of Jupiter's radius. The dynamo that produces Jupiter's magnetic field could reside in this metallic hydrogen region of the planet's interior. The molecular hydrogen envelope could be electrically conducting enough near the base of the layer to support dynamo action. There may be a relatively small core at the center of Jupiter as large as several Earth masses (up to about 10 Earth masses) and made of rock, metal, and ices (water, ammonia, and methane). Jupiter is so massive that it is contracting under its own gravitation, a process that releases gravitational potential energy in the form of heat. This heat is radiated from Jupiter's atmosphere; it has been measured and has been found to be comparable to the amount of heat Jupiter receives from the Sun and reradiates to space. Models of Jupiter's interior are based on theoretical considerations and are constrained by its density and spacecraft observations of the planet's gravitational field described by gravitational coefficients such as $C_{20} = -J_2$ (and higher degree coefficients like J_4 and J_6).

Saturn

Saturn is similar to Jupiter in its interior (Figure I23/Plate 12). It is a smaller planet (Table I4) and contains relatively more helium. The molecular to metallic phase transition in Saturn occurs deeper within the planet (at a depth of about 0.5 Saturn radius) as compared to Jupiter. This is because the pressure at which the phase change occurs is found deeper within Saturn as a consequence of the planet's smaller size and gravity. Like Jupiter, Saturn also has a magnetic field likely generated by a dynamo in the metallic hydrogen part of its interior. Saturn might also have a rock, metal, ice core at its center.

Uranus and Neptune

Uranus and Neptune are known as icy giants (Figure I23/Plate 12). They are similar to each other in mass and size but smaller than Jupiter and Saturn (Table I4). Their densities are about the same as that of Jupiter but about twice as large as that of Saturn. They are believed to have outer envelopes of hydrogen and helium, but deep interiors consisting mostly of ices such as water, methane, and ammonia. They might also have rock, metal, ice cores of about an Earth mass at their centers. Uranus and Neptune have magnetic fields that must be generated in highly electrically conducting regions of their icy interiors.

Crusts of terrestrial planets

The discussion above has emphasized the deep interiors of planets and moons. The outer layers of the terrestrial planets all have rocky crusts, and the outer regions of the giant planets all have gaseous atmospheres. Of course, some terrestrial planets also have atmospheres. The rocky crusts of Mercury, Venus, Earth, Moon, and Mars are compositionally distinct from the rocks of the underlying mantles. The crusts of the terrestrial planets formed by differentiation from the primordial mantle rocks, i.e., they are formed by melting of the original mantle rocks and separation of the melt from the unmelted residuum (the present mantle rocks). Separation was achieved by the gradual upward migration of the lighter magma that later solidified to form the crust. This process is going on today at the mid-ocean ridges on Earth, the sites of seafloor spreading. We know a great deal about the Earth's crust because we have the rocks themselves (at least the rocks of the upper crust) and data from seismic, magnetic, and gravity studies that reveal values of crust density and thickness. We also have rock samples and seismic, magnetic, and gravity data that reveal the nature of the lunar crust. Meteorites, believed to be from Mars, the so-called SNC meteorites (Shergotty-Nakhla-Chassigny), provide information about the composition and mineralogy of the planet. For Venus, Mars, and the Moon there are considerable data about their crusts from gravity, topography, and geochemical observations by orbiting spacecraft. Landers on Venus and Mars have provided *in situ* analyses of the surface rocks. The crusts of the terrestrial planets are essentially basaltic in character, similar to the rocks that form the crust beneath the Earth's oceans. On Earth there are actually two distinct types of crust, oceanic and continental. The rocks of the continents are compositionally distinct from those of the ocean floor. Continental rocks are richer in silica and formed from multiple stages of melting and differentiation in the presence of water. The oceanic crust is about 5 km thick on average while the continental crust is much thicker, about 40 km thick. Because of the concentration of the crust in the northern hemisphere beneath Eurasia, there is a hemispheric dichotomy in the Earth's crustal thickness.

It is not known if there is crust on the other terrestrial planets similar to the continental crust of Earth. There are two crustal types on the Moon, the anorthositic crust of the lunar highlands (the light crystallization product of the lunar magma ocean) and the basaltic crust of the lunar maria (the infill of the lunar impact basins). There is a hemispheric dichotomy in crustal thickness on the Moon with the farside of the Moon having a substantially greater thickness than the nearside. The lunar crust has also been thinned beneath the South Pole Aitken basin. There may be two crustal types on Venus, one type forming the high-standing plateaus like Ishtar Terra, and the other forming the volcanic rises like Beta Regio. The *in situ* analyses by the Venera

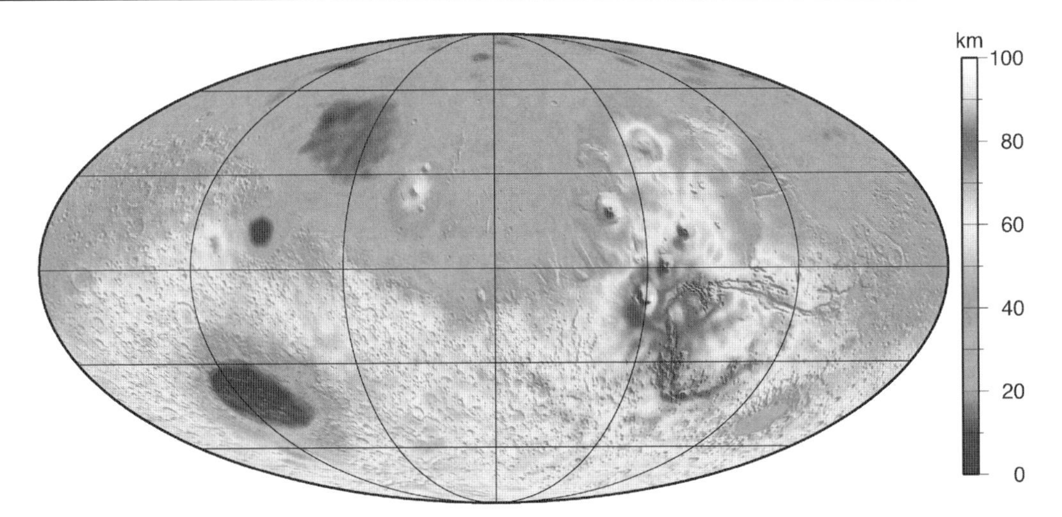

Figure I26/Plate 13b Map of crust thickness of Mars based largely on gravitational and topographic data from the Mars Global Surveyor spacecraft (Neumann *et al.*, 2004).

landers have revealed basaltic compositions in the Venusian lowland plains. Mars also has a crustal dichotomy. The primordial crust of the southern hemisphere stands higher and is thicker than the primordial crust of the northern hemisphere which has been mantled with a thin covering of younger and compositionally distinct crust. (The average thickness of the crust of Mars is between about 50 and 100 km.) The crustal dichotomies of the terrestrial planets were likely formed by internal dynamical processes, although major impacts, particularly on the Moon (e.g., the large South Pole Aitken basin) could also have played a role. The crustal dichotomies are probably the cause of the offsets between the centers of mass and the centers of figure of the terrestrial planets.

Less is known about the crust of Mercury. Remote sensing observations indicate that Mercury has a basaltic crust. If the Messenger spacecraft successfully orbits Mercury and returns anticipated data, we will know as much about Mercury' s crust as we do about the crusts of the other terrestrial planets.

The main source of information about the thicknesses of the crusts of the terrestrial planets (other than Earth) derives from the topography and gravity data obtained by orbiting spacecraft. These data have enabled the construction of global crust thickness maps of the Moon and Mars. The observed gravity variations are first modified to account for the effects of topography using the measured topography. The extra mass associated with elevated regions and the mass deficit associated with low areas contribute to the measured variations in gravity. The resulting gravity variations are known as Bouguer gravity anomalies. While the Bouger gravity map accounts for topography effects, it does not take account of the fact that variations in topography are generally isostatically compensated by variations in crust thickness. According to a geophysical principle known as Airy compensation, the masses of material columns above a certain depth of compensation are balanced; the crust is thicker beneath elevated topography compared to average crust so that the extra mass of the topography is balanced by an equal mass deficit associated with the replacement of heavy mantle rock by lighter thickened crust. Similarly, the crust is thinned beneath topographic depressions, so that the mass deficit of negative topography is balanced by an equal mass excess associated with the replacement of thinned crust by heavier mantle rock. By attributing the Bouguer gravity anomalies to crustal thickness variations, those variations can be determined. However, the depth of compensation or average crustal thickness must be assumed or constrained by some other observations. In this way, the crustal thickness map of Mars shown in Figure I26/Plate 13b has been derived from topography and gravity measurements obtained by the Mars Global Surveyor spacecraft. The map assumes that the average crust thickness on Mars is 50 km. The crustal dichotomy between the northern lowlands and southern highlands is immediately apparent. The strongly magnetized region of the Martian crust lies in the southern highlands (see *Magnetic field of Mars*, chapter and figure), but not all of the southern hemisphere crust is magnetized. The principle of Airy compensation is well established for Earth where it can be tested against seismic measurements of crustal thickness. The thickest crust on Earth lies below its most mountainous regions.

Concluding remarks

Seismic measurements directly probe a planet's interior, determining structure and providing constraints on composition and mineralogy. We have such observations for Earth, but the only other planet that has been seismically studied is the Moon. Nevertheless, we have learned a good deal about the interiors of other planets and moons in our solar system by observing how some of them respond dynamically to tidal torques and by measuring their densities and external gravitational and magnetic fields. Still other observations of the geology and composition of their surfaces provide additional clues about what lies inside them. Though we only have observed most of the bodies shown in Figure I23/Plate 12 from the Earth, or from orbiting and flyby spacecraft, it is rather remarkable that we can draw the cutaway views shown in the figure to reveal what these planets and moons are like inside.

Sources

The data and models on which this article is based are discussed in the following papers: Mercury (Schubert *et al.*, 1988; Spohn *et al.*, 2001; van Hoolst and Jacobs, 2003; Solomon, 2003), Venus (Schubert *et al.*, 1997), Earth (Schubert and Walterscheid, 2000), Moon (Hood, 1986; Khan *et al.*, 2004), Mars (Schubert *et al.*, 1992; Spohn *et al.*, 1998; Yoder *et al.*, 2003; Neumann *et al.*, 2004), Jupiter (Guillot *et al.*, 2004), Galilean satellites (Schubert *et al.*, 2004), Saturn (Hubbard and Stevenson, 1984; Guillot, 1999), Titan (Hunten *et al.*, 1984; Grasset *et al.*, 2000; Stevenson, 1992; Sohl *et al.*, 2003). Uranus (Hubbard and Marley, 1989; Hubbard *et al.*, 1991; Podolak *et al.*, 1991), Neptune (Hubbard *et al.*, 1995), Triton (McKinnon *et al.*, 1995), Pluto and Charon (Stern, 1992; McKinnon *et al.*, 1997).

Gerald Schubert

Bibliography

Grasset, O., Sotin, C., and Deschamps, F., 2000. On the internal structure and dynamics of Titan. *Planetary and Space Science*, **48**: 617–636.

Guillot, T., 1999. A comparison of the interiors of Jupiter and Saturn. *Planetary and Space Science*, **47**: 1183–1200.

Guillot, T., Stevenson, D.J., Hubbard, W.B., and Saumon, D., 2004. The interior of Jupiter. In Bagenal, F., Dowling, T.E., and McKinnon, W.B. (eds.), *Jupiter: The Planet, Satellites, and Magnetosphere.* Cambridge: Cambridge University Press, pp. 35–57.

Hood, L.L., 1986. Geophysical constraints on the lunar interior. In Hartmann, W.K.R., Phillips, J., and Taylor, G.J. (eds.), *Origin of the Moon.* Houston, TX: Lunar and Planetary Institute, pp. 361–410.

Hubbard, W.B., and Marley, M.S., 1989. Optimized Jupiter, Saturn, and Uranus interior models. *Icarus*, **78**: 102–118.

Hubbard, W.B., and Stevenson, D.J., 1984. Interior structure of Saturn. In Gehrels, T., and Matthews, M.S. (eds.), *Saturn.* Tucson, AZ: The University of Arizona Press, pp. 47–87.

Hubbard, W.B., Nellis, W.J., Mitchell, A.C., Holmes, N.C., Limaye, S.S., and McCandless, P.C., 1991. Interior structure of Neptune: comparison with Uranus. *Science*, **253**: 648–651.

Hubbard, W.B., Podolak, M., and Stevenson, D.J., 1995. The interior of Neptune. In Cruikshank, D.P., (ed.), *Neptune and Triton.* Tucson, AZ: The University of Arizona Press, pp. 109–138.

Hunten, D.M., Tomasko, M.G., Flasar, F.M., Samuelson, R.E., Strobel, D.F., and Stevenson, D.J., 1984. Titan. In Gehrels, T., and Matthews, M.S., (eds.), *Saturn.* Tucson, AZ: The University of Arizona Press, pp. 671–759.

Khan, A., Mosegaard, K., Williams, J.G., and Lognonné, P., 2004. Does the Moon possess a molten core? Probing the deep lunar interior using results from LLR and Lunar Prospector. *Journal of Geophysical Research*, **109**: E09,007, doi:10.1029/2004JE002,294.

McKinnon, W.B., Lunine, J.I., and Banfield, D., 1995. Origin and evolution of Triton, In Cruikshank, D.P., (ed.), *Neptune and Triton.* Tucson, AZ: The University of Arizona Press, pp. 807–877.

McKinnon, W.B., Simonelli, D.P., and Schubert, G., 1997. Composition, internal structure, and thermal evolution of Pluto and Charon. In Stern, S.A., and Tholen, D.J., (eds.), *Pluto and Charon.* Tucson, AZ: The University of Arizona Press, pp. 295–343.

Neumann, G.A., Zuber, M.T., Wieczorek, M.A., McGovern, P.J., Lemoine, F.G., and Smith, D.E., 2004. Crustal structure of Mars from gravity and topography. *Journal of Geophysical Research*, **109**: E08,002, doi:10.1029/2004JE002,262.

Podolak, M., Hubbard, W.B., and Stevenson, D.J., 1991. Model of Uranus' interior and magnetic field. In Bergstralh, J.T., Miner, E.D., and Matthews, M.S., (eds.), *Uranus.* Tucson, AZ: The University of Arizona Press, pp. 29–61.

Schubert, G., and Walterscheid, R.L., 2000. Earth. In Cox, A.N. (ed.), *Allen's Astrophysical Quantities,* 4th edn. New York: Springer-Verlag, pp. 239–292.

Schubert, G., Ross M.N., Stevenson D.J., and Spohn, T., 1988. Mercury's thermal history and the generation of its magnetic field. In Vilas, F., Chapman, C.R., and Matthews, M.S. (eds.), *Mercury.* Tucson, AZ: The University of Arizona Press, pp. 429–460.

Schubert, G., Solomon, S.C., Turcotte, D.L., Drake, M.J., and Sleep, N.H., 1992. Origin and thermal evolution of Mars. In Kieffer, H.H., Jakosky, B.M., Snyder, C.W., and Matthews, M.S. (eds.), *Mars.* Tucson, AZ: The University of Arizona Press, pp. 147–183.

Schubert, G., Solomatov, V.S., Tackley, P.J., and Turcotte, D.L., 1997. Mantle convection and the thermal evolution of Venus. In Bougher, S.W., Hunten, D.M., and Phillips, R.J. (eds.), *Venus II: Geology, Geophysics, Atmosphere, and Solar Wind Environment.* Tucson, AZ: The University of Arizona Press, pp. 1245–1287.

Schubert, G., Anderson, J.D., Spohn, T., and McKinnon, W.B., 2004. Interior composition, structure and dynamics of the Galilean satellites. In Bagenal, F., Dowling, T.E., and McKinnon, W.B., (eds.), *Jupiter: The Planet, Satellites, and Magnetosphere.* Cambridge, UK: Cambridge University Press, pp. 281–306.

Sohl, F., Hussmann, H., Schwentker, B., and Spohn, T., 2003. Interior structure models and tidal Love numbers of Titan. *Journal of Geophysical Research*, **108**: 5130, doi:10.1029/2003JE002,044.

Solomon, S.C., 2003. Mercury: the enigmatic innermost planet. *Earth and Planetary Science Letters*, **216**: 441–455.

Spohn, T., Sohl, F., and Breuer, D., 1998. Mars. *Astronomy and Astrophysics Review*, **8**: 181–236.

Spohn, T., Sohl, F., Wieczerkowski, K., and Conzelmann, V., 2001. The interior structure of Mercury: what we know, what we expect from BepiColombo. *Planetary and Space Science*, **49**: 1561–1570.

Stern, S.A., 1992. The Pluto-Charon system. *Annual Review on Astronomy and Astrophysics*, **30**: 185–233.

Stevenson, D.J., 1992. Interior of Titan. In *ESA SP-338, Proceedings Symposium on Titan,* Noordwijk, Netherlands: European Space Agency, pp. 29–33.

Van Hoolst, T., and Jacobs, C., 2003. Mercury's tides and interior structure. *Journal of Geophysical Research*, **108**: 7–(1–16), doi:10.1029/2003JE002,126.

Yoder, C.F., Konopliv, A.S., Yuan, D.N., Standish, E.M., and Folkner, W.M., 2003. Fluid core size of Mars from detection of the solar tide. *Science*, **300**: 299–303.

Cross-references

Harmonics, Spherical
Magnetic Field of Mars

INTERNAL EXTERNAL FIELD SEPARATION

Introduction

The process of separation of the magnetic field can be applied to the main magnetic field or to the daily variation fields of solar and lunar origin, or to disturbance fields or disturbance daily variation fields. Each application has a different purpose. The results of the separation show that the Earth's main magnetic field is dominantly of internal origin, while the daily variation and disturbance fields have external fields that are greater than the internal fields and therefore originate above the Earth.

The results of the analyses provide the information needed for studies of dynamo processes in the Earth's interior, in the ionosphere, in the deep oceans, and for distributions of electrical conductivity within those regions. In studies of magnetic daily variations of solar and lunar origin, the amplitude ratio and phase angle differences of the internal and external fields are used as the electrical response function of the Earth.

The internal and external fields correspond to two independent vector spherical harmonics that are orthogonal under integration over the surface of a sphere. However, there is a third, independent vector spherical harmonic, orthogonal to the internal and external fields, referred to as the "nonpotential" field, which can be used to resolve the residual differences between horizontal components of the field that cannot be represented by "potential" fields, meaning internal and external fields. Coefficients of the nonpotential field determined from observatory data are not statistically significant, indicating that the electrical current system associated with the maintenance of the Earth's electrostatic charge, called the "global electrical circuit," cannot be determined from magnetic observatory data. With the availability of satellite magnetic data, and the possibility of field-aligned currents at low-Earth orbit altitudes, it is quite likely that satellite magnetic data will provide significant nonpotential field coefficients.

Field separations can be done globally or regionally. Global analyses are usually done using spherical harmonics, but some analyses have been done using surface integral methods, (Price and Wilkins, 1963). A variety of methods are available for regional analyses, including spherical cap harmonic analysis (see *Harmonics, spherical cap*) and spectral methods. Global analyses using spherical harmonics and the method of least-squares are the simplest and most easily understood. Spherical cap harmonic analyses are useful for interpolation of field values, but do not separate the field accurately. Other methods have problems of implementation, as well as problems with aliasing or poor choice of spectral cut-off. In the study of magnetic daily variations, an estimate of the local direction and strength of the overhead current system is easily obtained by rotating the horizontal magnetic field variation components clockwise through 90°.

Spherical harmonic analysis

The first spherical harmonic analysis on a global scale was that of the Earth's main magnetic field by Gauss (1838). He assumed that the Earth was a sphere, and that all measurements of the field were made on the surface of the sphere. There is an implied assumption that there are no surface electrical currents on the surface of the sphere and that there are no electrical currents flowing with a radial component, e.g., Earth-air, or field-aligned. Thus, the electrical current density will be zero, $\mathbf{J} = 0$. With all these assumptions, together with the absence of displacement currents, Ampere's law for the magnetic field strength $\mathbf{H}$ reduces to $\nabla \times \mathbf{H} = 0$. In regions of constant magnetic susceptibility, the equation for the magnetic flux density $\mathbf{B}$ is therefore $\nabla \times \mathbf{B} = 0$, so that the magnetic flux density is the gradient of a scalar potential, that is, $\mathbf{B} = -\nabla V$, where the scalar potential function V is a function of the radius r of the sphere, the colatitude θ, and the east longitude ϕ.

The magnetic flux density has no divergence, because there are no free magnetic poles, quite in contrast to the existence of free electric charges. This condition is written $\nabla \cdot \mathbf{B} = 0$, and therefore, the scalar potential V satisfies Laplace's equation, $\nabla^2 V = 0$.

Solutions of Laplace's equations are called spherical harmonics (see *Harmonics, spherical*), and the solutions $V(r, \theta, \phi)$ are called solid spherical harmonics, to distinguish them from surface spherical harmonics $P_n^m(\cos\theta)\cos m\phi$ and $P_n^m(\cos\theta)\sin m\phi$, which do not satisfy Laplace's equation.

The functions $V_{\text{int}}(r, \theta, \phi)$ and $V_{\text{ext}}(r, \theta, \phi)$, defined by

$$\begin{aligned} V_{\text{int}}(r,\theta,\phi) &= a\sum_{n=1}^{N}\frac{a^{n+1}}{r^{n+1}}\sum_{m=0}^{n}\left(g_{ni}^m\cos m\phi + h_{ni}^m\sin m\phi\right) \\ &\quad\times P_n^m(\cos\theta), \quad r > a, \\ V_{\text{ext}}(r,\theta,\phi) &= a\sum_{n=1}^{N}\frac{r^n}{a^n}\sum_{m=0}^{n}\left(g_{ne}^m\cos m\phi + h_{ne}^m\sin m\phi\right) \\ &\quad\times P_n^m(\cos\theta), \quad r < a, \end{aligned} \tag{Eq. 1}$$

are the potential functions for sources internal to the reference sphere $r = a$ and for sources external to the reference sphere, respectively. Subscripts i and e have been applied to coefficients g_n^m and h_n^m to indicate internal or external fields. The initial factor a in (Eq. 1) has been included so that the g and h coefficients will have the dimensions of magnetic flux density. Gauss's first analysis of the main field to degree and order $N = 4$ used only the term $V_{\text{int}}(r, \theta, \phi)$, on the assumption that the geomagnetic main field is a term of internal origin only. Analysis of the field gives no indication as to the location of the internal and external sources. Therefore, the potential of the main field, or its variations, is represented as a linear combination of the two types given in (Eq. 1), and applied to field measurements made in the region between the ground and the ionosphere, or, in the case of satellite magnetic data, between the ionosphere and the magnetopause.

The current function representation

The field of internal origin can be represented by a system of surface currents flowing over the surface of a sphere, concentric with and within the minimum sphere over which observations are made. A stream function $\Psi(\theta, \phi)$, with units of amperes, called the current function, gives the surface current density $\mathbf{K}(\theta, \phi)$ (in amperes per meter) in the form

$$\mathbf{K}(\theta,\phi) = \frac{1}{R\sin\theta}\frac{\partial\Psi}{\partial\phi}\mathbf{e}_\theta - \frac{1}{R}\frac{\partial\Psi}{\partial\theta}\mathbf{e}_\phi = -\mathbf{e}_r \times \nabla\Psi = \nabla \times (\mathbf{e}_r\Psi) \tag{Eq. 2}$$

The vector $\nabla\Psi$ is in the direction of $\Psi(\theta, \phi)$ increasing, that is, directed toward maxima, and from (Eq. 2) the electrical current flow is counterclockwise around maxima when seen from above. Note however, that the maxima of $\Psi(\theta, \phi)$ can be negative and the minima positive.

The current function is denoted $\Psi(\theta, \phi)$, with units of amperes, and has a spherical harmonic representation

$$\Psi(\theta,\phi) = \sum_{n=1}^{N}\sum_{m=0}^{n}\left(J_{nc}^m\cos m\phi + J_{ns}^m\sin m\phi\right)P_n^m(\cos\theta), \tag{Eq. 3}$$

Subscripts c and s have been added to the coefficients J_n^m to indicate application to $\cos m\phi$ and $\sin m\phi$ respectively. For a current function in the ionosphere, of radius R, to represent the magnetic daily variations, whose coefficients have been determined relative to a sphere of radius a, the current function coefficients in amperes are given by

$$J_{nc}^m = -(a)_{\text{km}}\frac{10}{4\pi}\frac{2n+1}{n+1}\frac{R^n}{a^n}(g_{ne}^m)_{\text{nT}}, \quad J_{ns}^m = -(a)_{\text{km}}\frac{10}{4\pi}\frac{2n+1}{n+1}\frac{R^n}{a^n}(h_{ne}^m)_{\text{nT}}. \tag{Eq. 4}$$

For a current function within the Earth to represent main field coefficients that have been determined relative to a sphere of radius a, the current function coefficients in amperes are given by

$$J_{nc}^m = (a)_{\text{km}}\frac{10}{4\pi}\frac{2n+1}{n}\frac{a^{n+1}}{R^{n+1}}(g_{ni}^m)_{\text{nT}}, \quad J_{ns}^m = (a)_{\text{km}}\frac{10}{4\pi}\frac{2n+1}{n}\frac{a^{n+1}}{R^{n+1}}(h_{ni}^m)_{\text{nT}}. \tag{Eq. 5}$$

Global magnetic field separation

In studies of the geomagnetic daily variations, current systems in the ionosphere are the basic source, and eddy currents are induced in the Earth by the ionospheric currents, with only a very small contribution from the magnetospheric current system. Therefore, for the study of the geomagnetic daily variations, the potential function must include terms of both internal and external origin, with the expectation that the terms of external origin will be the greater of the two, being the basic source of the phenomenon. The two types of potential can be separated by numerical calculations as follows, the process being the title of this article, "internal-external field separation."

Magnetic flux density field components

The three components of the geomagnetic field are denoted X, Y, and Z. The northward component is X, the eastward component Y, and the vertically downward component Z. The components, X, Y, and Z in that order, form a right-handed orthogonal system.

The mathematical expressions for the field components over the reference sphere, $r = a$, are

$$X = \frac{1}{a}\frac{\partial V}{\partial\theta}\bigg|_{r=a}, \quad Y = -\frac{1}{a\sin\theta}\frac{\partial V}{\partial\phi}\bigg|_{r=a}, \quad Z = \frac{\partial V}{\partial r}\bigg|_{r=a}, \tag{Eq. 6}$$

and therefore in a region free of magnetic sources and electric currents, the ambient magnetic field can be derived from fields of internal and external origin, and the components will have the mathematical form

$$X(a,\theta,\phi)=\sum_{n=1}^{N}\sum_{m=0}^{n}\left[(g_{ni}^{m}+g_{ne}^{m})\cos m\phi+(h_{ni}^{m}+h_{ne}^{m})\sin m\phi\right]\frac{dP_n^m(\cos\theta)}{d\theta},$$

$$Y(a,\theta,\phi)=\sum_{n=1}^{N}\sum_{m=0}^{n}\left[(g_{ni}^{m}+g_{ne}^{m})\sin m\phi-(h_{ni}^{m}+h_{ne}^{m})\cos m\phi\right]\frac{m}{\sin\theta}P_n^m(\cos\theta),$$

$$Z(a,\theta,\phi)=\sum_{n=1}^{N}\sum_{m=0}^{n}\left\{\left[-(n+1)g_{ni}^{m}+n g_{ne}^{m}\right]\cos m\phi+\left[-(n+1)h_{ni}^{m}+n h_{ne}^{m}\right]\sin m\phi\right\}P_n^m(\cos\theta). \quad \text{(Eq. 7)}$$

Because the radial dependence of the internal and external fields is known, (Eq. 7) can be easily adapted for calculations using magnetic field measurements which are not made over the surface of a sphere, and where the coordinates of the point at which each measurement is made is given in terms of geocentric coordinates, r, θ, ϕ. From the equations for the horizontal components of the field, namely X and Y, it will be seen that the coefficients involve only $(g_{ni}^m + g_{ne}^m)$ and $(h_{ni}^m + h_{ne}^m)$, so that from the X component only, it is possible to determine the Y components. However, from these two elements only, it is not possible to separate the coefficients into internal and external parts, meaning it is not possible to determine g_{ni}^m and g_{ne}^m separately. However, the coefficients $\left[-(n+1)g_{ni}^m + ng_{ne}^m\right]$ and $\left[-(n+1)h_{ni}^m + n\,h_{ne}^m\right]$ as determined from the radial or vertical component Z, will allow the separation of the coefficients and therefore allow separation of the field into parts of internal and external origin, respectively.

The nonpotential field

It will be seen from (Eq. 7), that the coefficients from the northward component of the field X, obtained in the form $(g_{ni}^m + g_{ne}^m)$ and $(h_{ni}^m + h_{ne}^m)$, can be used to estimate the values of the eastward component Y. Should the estimated values of Y not be in agreement with the observed values of Y, then to resolve the situation, a nonpotential field is introduced, as shown below by coefficients g_{nv}^m and h_{nv}^m. The radial dependence required for the internal and external fields is quite definite, but the radial dependence of the nonpotential field depends upon the hypothesis chosen for the origin of the fields; for example, Earth-air currents, or field-aligned currents.

The nonpotential field $\mathbf{B}_v(r,\theta,\phi)$ is

$$\begin{aligned}\mathbf{B}_v(r,\theta,\phi) &= \frac{1}{\sin\theta}\frac{\partial V_v}{\partial\phi}\mathbf{e}_\theta - \frac{\partial V_v}{\partial\theta}\mathbf{e}_\phi \\ &= -\mathbf{r}\times\nabla V_v(r,\theta,\phi), \\ &= \nabla\times[\mathbf{r}V_v(r,\theta,\phi)], \end{aligned} \quad \text{(Eq. 8)}$$

and has no radial component. The scalar function $V_v(r,\theta,\phi)$ (with units of magnetic flux density) is given by

$$V_v(r,\theta,\phi) = \sum_{n=1}^{N}\sum_{m=0}^{n}\frac{q_n^m(r)}{q_n^m(a)}(g_{nv}^m\cos m\phi + h_{nv}^m\sin m\phi)P_n^m(\cos\theta). \quad \text{(Eq. 9)}$$

The function $q_n^m(r)$ cannot be determined from analysis of theory or data alone, but requires a hypothesis to be made on the form of the electrical current system.

The field components of the magnetic field over the reference sphere, $r = a$, are

$$X(a,\theta,\phi) = \sum_{n=1}^{N}\sum_{m=0}^{n}[(g_{ni}^m+g_{ne}^m)\cos m\phi + (h_{ni}^m+h_{ne}^m)\sin m\phi]\frac{dP_n^m}{d\theta} + (g_{nv}^m\sin m\phi - h_{nv}^m\cos m\phi)\frac{m}{\sin\theta}P_n^m, \quad \text{(Eq. 10)}$$

$$Y(a,\theta,\phi) = \sum_{n=1}^{N}\sum_{m=0}^{n}[(g_{ni}^m+g_{ne}^m)\sin m\phi - (h_{ni}^m+h_{ne}^m)\cos m\phi] \times \frac{m}{\sin\theta}P_n^m - (g_{nv}^m\cos m\phi + h_{nv}^m\sin m\phi)\frac{dP_n^m}{d\theta}, \quad \text{(Eq. 11)}$$

$$Z(a,\theta,\phi) = \sum_{n=1}^{N}\sum_{m=0}^{n}\left\{\left[-(n+1)g_{ni}^m + ng_{ne}^m\right]\cos m\phi + \left[-(n+1)h_{ni}^m + nh_{ne}^m\right]\sin m\phi\right\}P_n^m(\cos\theta). \quad \text{(Eq. 12)}$$

The electrical current system comes only from the nonpotential field as

$$\begin{aligned}\mathbf{J}_v(r,\theta,\phi) &= \frac{1}{\mu_0}\nabla\times\mathbf{B}_v(r,\theta,\phi) \\ &= \frac{1}{\mu_0 r}\sum_{n=0}^{N}\sum_{m=0}^{n}\frac{1}{q_n^m(a)}\left\{q_n^m(r)n(n+1)\,\mathbf{e}_r + \frac{d}{dr}[rq_n^m(r)]\right. \\ &\quad \times\left.\left(\mathbf{e}_\theta\frac{\partial}{\partial\theta} + \frac{\mathbf{e}_\phi}{\sin\theta}\frac{\partial}{\partial\phi}\right)(g_{nv}^m\cos m\phi + h_{nv}^m\sin m\phi)P_n^m(\cos\theta)\right\}\end{aligned} \quad \text{(Eq. 13)}$$

The hypothesis of Earth-air currents requires $q_n^m(r) = 1/r$ when the current system of (Eq. 13) reduces to a radial term only. An expression for a field of internal origin, along whose field lines the electrical current system travels, can also be found (Winch *et al.*, 2005).

The representation of (Eqs. 10–12) can be interpreted as a representation of the magnetic field over a sphere in terms of the three independent and orthogonal vector spherical harmonics, usually given in complex form as $\mathbf{Y}_{n,n+1}^m(\theta,\phi)$, $\mathbf{Y}_{n,n}^m(\theta,\phi)$, $\mathbf{Y}_{n,n-1}^m(\theta,\phi)$.

Magnetic daily variations

The magnetic daily variations of solar and lunar type originate in the ionosphere by a dynamo process, so that in contrast to the main magnetic field in which the internal coefficients are almost the only significant terms, the magnetic daily variations are dominantly of external origin. The extra complication of dependence on time adds an extra dimension to the complexity of the analysis. A very basic assumption of local time t^* dependence on time, where $t^* = t + \phi$, and t is universal time, can reduce the complexity, but ignores significant nonlocal time terms that arise because of the inclination of the Earth's geomagnetic and geographic axes. A less restrictive assumption is that of westward movement only, since the solar and lunar magnetic variations are driven by the Sun, one can expect that eastward moving variations will be mostly insignificant.

The analysis begins with Fourier analysis of the elements X, Y, Z, at each observatory. Thus

$$\begin{aligned}X(a,\theta,\phi,t) &= \sum_{M=1}^{4}[A_{XM}(a,\theta,\phi)\cos Mt + B_{XM}(a,\theta,\phi)\sin Mt], \\ Y(a,\theta,\phi,t) &= \sum_{M=1}^{4}[A_{YM}(a,\theta,\phi)\cos Mt + B_{YM}(a,\theta,\phi)\sin Mt], \\ Z(a,\theta,\phi,t) &= \sum_{M=1}^{4}[A_{ZM}(a,\theta,\phi)\cos Mt + B_{ZM}(a,\theta,\phi)\sin Mt].\end{aligned} \quad \text{(Eq. 14)}$$

It is the standard practice to do separate spherical harmonic analyses of the A and B coefficients, to obtain internal and external coefficients

$$g_{nAi}^{mM},\ h_{nAi}^{mM} \text{ and } g_{nAe}^{mM},\ h_{nAe}^{mM},\quad g_{nBi}^{mM},\ h_{nBi}^{mM} \text{ and } g_{nBe}^{mM},\ h_{nBe}^{mM}.$$

of the potential functions $V_A^M(r,\theta,\phi)$ and $V_B^M(r,\theta,\phi)$, where

$$V_A^M(r,\theta,\phi) = a\sum_{n=1}^{N}\sum_{m=0}^{n}\left[\left(\frac{a}{r}\right)^{n+1}(g_{nAi}^{mM}\cos m\phi + h_{nAi}^{mM}\sin m\phi) + \left(\frac{r}{a}\right)^{n}(g_{nAe}^{mM}\cos m\phi + h_{nAe}^{mM}\sin m\phi)\right]P_n^m(\cos\theta), \quad \text{(Eq. 15)}$$

$$V_B^M(r,\theta,\phi) = a\sum_{m=1}^{N}\sum_{n=0}^{m}\left[\left(\frac{a}{r}\right)^{n+1}(g_{nBi}^{mM}\cos m\phi + h_{nBi}^{mM}\sin m\phi) + \left(\frac{r}{a}\right)^{n}(g_{nBe}^{mM}\cos m\phi + h_{nBe}^{mM}\sin m\phi)\right]P_n^m(\cos\theta), \quad \text{(Eq. 16)}$$

where only a limited range of m is used, for example, $m = M-1, M$, and $M+1$.

However, if one were to include the following expression for $V_B^M(r,\theta,\phi)$, which contains only the g and h coefficients appearing in $V_A^M(r,\theta,\phi)$, along with the least-squares analysis of $V_A^M(r,\theta,\phi)$,

$$V_B^M(r,\theta,\phi) = a\sum_{n}\sum_{n}\left[\left(\frac{a}{r}\right)^{n+1}(h_{nAi}^{mM}\cos m\phi - g_{nAi}^{mM}\sin m\phi) + \left(\frac{r}{a}\right)^{n}(h_{nAe}^{mM}\cos m\phi - g_{nAe}^{mM}\sin m\phi)\right]P_n^m(\cos\theta), \quad \text{(Eq. 17)}$$

then the combination

$$\begin{aligned} &V_A^M(r,\theta,\phi)\cos Mt + V_B^M(r,\theta,\phi)\sin Mt \\ &= a\sum_{n=1}^{N}\sum_{m=0}^{n}\left(\frac{a}{r}\right)^{n+1}\left[g_{nAi}^{mM}\cos(m\phi+Mt) + h_{nAi}^{mM}\sin(m\phi+Mt)\right]P_n^m(\cos\theta) \\ &+ a\sum_{n=1}^{N}\sum_{m=0}^{n}\left(\frac{r}{a}\right)^{n}\left[g_{nAe}^{mM}\cos(m\phi+Mt) + h_{nAe}^{mM}\sin(m\phi+Mt)\right]P_n^m(\cos\theta). \end{aligned} \quad \text{(Eq. 18)}$$

consists only of terms with argument $m\phi + Mt$, which are westward moving. In the special case that $m = M$, the terms with argument $m\phi + Mt$ reduce to local time t^*, and $mt^* = m(t+\phi)$.

The use of (Eqs. 15 and 17) together doubles the number of equations of condition for the westward moving terms. Because the magnetic variation "follows the Sun" and moves westward, the eastward moving terms are therefore much smaller, and in the first instance, can be regarded as random noise. Analysis of residuals after removing the westward moving part can be done by using an expression for $-V_B^M(r,\theta,\phi)$ in place of (Eq. 17) for $V_B^M(r,\theta,\phi)$.

Separation using Cartesian components

For a Cartesian coordinate system at the center of the Earth, the x-axis directed toward the intersection of the equator and the Greenwich meridian, the y-axis at the intersection of the equator and the 90°E meridian, and the z-axis directed toward the North Pole. If Π, Γ, Ξ, denote field components parallel to the x-, y- and z- axes, then

$$\begin{aligned} \Pi &= -Z\sin\theta\cos\phi - X\cos\theta\cos\phi - Y\sin\phi, \\ \Gamma &= -Z\sin\theta\sin\phi - X\cos\theta\sin\phi + Y\cos\phi, \\ \Xi &= -Z\cos\theta + X\sin\theta. \end{aligned} \quad \text{(Eq. 19)}$$

The method requires the *surface* spherical harmonic analysis of the Cartesian field elements, Π, Γ, Ξ. Coefficients $a_n^m, b_n^m, c_n^m, d_n^m, e_n^m, f_n^m$, are determined using

$$\begin{aligned} \Pi &= \sum_{n=0}^{N+1}\sum_{m=0}^{n}(a_n^m\cos m\phi + b_n^m\sin m\phi)P_n^m(\cos\theta), \\ \Gamma &= \sum_{n=0}^{N+1}\sum_{m=0}^{n}(c_n^m\cos m\phi + d_n^m\sin m\phi)P_n^m(\cos\theta), \\ \Xi &= \sum_{n=0}^{N+1}\sum_{m=0}^{n}(e_n^m\cos m\phi + f_n^m\sin m\phi)P_n^m(\cos\theta). \end{aligned} \quad \text{(Eq. 20)}$$

By forming specified orthogonal combinations of the computed coefficients, the internal, external, and nonpotential coefficients are determined independently, so that, for example, if one decided to have only internal terms in the model, there is no need to repeat the calculation (see Winch, 1968 for the necessary details).

Surface integral method

Kahle and Vestine (1963) put forward a method of surface integrals for separating a potential function $V(\theta,\phi)$, calculated as a set of numerical values over the globe, into its terms of internal and external origin. The method first requires the determination of the numerical values of $V(\theta,\phi)$ by integration of the given X values with respect to colatitude θ and integration of Y values with respect to east longitude ϕ. The reconciliation of values of potential from the two different elements is discussed by Price and Wilkins (1963), and the method applied to the analysis of Sq variations for the International Polar Year.

Using the known potential function V, the internal and external components of V are determined from a form of Green's third identity,

$$V_{\text{ext}} - V_{\text{int}} = \frac{1}{2\pi}\int_S (V + 2aZ)\frac{dS}{r}. \quad \text{(Eq. 21)}$$

The separation into internal and external parts is done using the numerical arrays determined by (Eq. 21) and the known array

$$V(\theta,\phi) = V_{\text{ext}}(\theta,\phi) + V_{\text{int}}(\theta,\phi). \quad \text{(Eq. 22)}$$

Weaver (1964) describes a method for separating a local geomagnetic field into its external and internal parts using two-dimensional Fourier analyses of the X, Y, and Z components. The formulae for separating the X field are

$$X_e = \frac{1}{2}(X + M_1 Z), \quad X_i = \frac{1}{2}(X - M_1 Z),$$

where

$$M_1 Z = \frac{1}{2\pi}\int_{-\infty}^{\infty}\int_{-\infty}^{\infty} Z(u,v)\frac{x-u}{\left[(x-u)^2 + (y-v)^2\right]^{3/2}}\,du\,dv.$$

A simple method

In order to determine the direction and strength of the external field in magnetic daily variation studies, without the need for any analysis at all, the horizontal components can be rotated clockwise through 90°. For the Sq field, it can be shown that the method is accurate to within 10% of the maximum overhead current circulation, with largest errors occurring in middle and equatorial latitudes, e.g., Winch (1966). The method is useful for checking results of other more complicated forms of analysis and interpolation.

Denis Winch

Bibliography

Gauss, C.F., 1838. Allgemeine Theorie des Erdmagnetismus, Resultate aus den Beobachtungen des magnetischen Vereins im Jahre 1838.

(Reprinted in Werke, Bd. 5, 121, see also the translation by Mrs. Sabine and Sir John Herschel, Bart, Taylor, Richard, 1841. *Scientific Memoirs Selected from the Transactions of Foreign Academies of Science and Learned Societies and from Foreign Journals*, **2**: 184–251.)
Kahle, A.B., and Vestine, E.H., 1963. Analysis of surface magnetic fields by integrals. *Journal of Geophysical Research*, **68**: 5505–5515.
Price, A.T., and Wilkins, G.A., 1963. New methods for the analysis of geomagnetic fields and their application to the Sq field of 1932–1933. *Philosophical Transactions of the Royal Society of London, Series A*, **256**: 31–98.
Weaver, J.T., 1964. On the separation of local geomagnetic fields into external and internal parts. *Zeitschrift für Geophysik*, **30**: 29–36.
Winch, D.E., 1966. The Sq overhead current system approximation. *Planetary and Space Science*, **14**: 163–172.
Winch, D.E., 1968. Analysis of the geomagnetic field by means of Cartesian components. *Physics of the Earth and Planetary Interiors*, **1**: 347–360.
Winch, D.E., Ivers, D.J., Turner, J.P.R., and Stening, R.J., 2005. Geomagnetism and Schmidt quasi-normalization. *Geophysical Journal International*, **160**: 487–504.

Cross-references

Harmonics, Spherical
Harmonics, Spherical Cap

IONOSPHERE

The ionosphere is the weakly ionized region of the upper atmosphere above 60 km altitude where free electrons and ions form a plasma that influences radio wave propagation and conducts electrical currents. The plasma is created by solar extreme-ultraviolet and X-ray radiation impinging on the atmosphere and at high latitudes in the northern and southern *auroral ovals* (*q.v.*), by energetic particles precipitating from the *magnetosphere* (*q.v.*). Plasma is destroyed through chemical reactions that lead to electron-ion recombination. The ionizing solar radiation varies by about a factor of two over the 11-year solar sunspot cycle, which results in significant solar-cycle variations in the plasma density. Figure I27 shows typical altitude profiles of the mid-latitude electron density at day and night for years of minimum and maximum solar activity. For historical reasons relating to radio sounding of the ionosphere, the region of maximum density above 150 km is called the *F*-region (or layer), the region around the secondary maximum density at about 110 km is called the *E* region, and the region below about 90 km is called the *D* region.

The ionosphere refracts, reflects, retards, scatters, and absorbs radio waves in a manner that depends on the radio wave frequency. Around-the-Earth communications are possible by utilizing ionospheric and ground reflections of waves at frequencies below about 3–30 MHz (100 to 10 m wavelength), depending on the peak electron density. However, frequent collisions between electrons and air molecules in the *D* region remove energy from the radio waves, leading to partial or even complete absorption, depending on the electron density. At higher frequencies, radio waves penetrate entirely through the ionosphere, allowing radio astronomy and communications with spacecraft. Nevertheless, such signals can still be degraded by refraction and scattering off of small-scale density irregularities. In the case of geolocation signals like those of the Global Positioning System (GPS), variable and uncertain ranging errors can be introduced by ionospheric signal retardation. The retardation is proportional to the total electron content (TEC), or height-integrated electron density. A typical global pattern of TEC is shown in Figure I28. The TEC tends to be larger in winter than in summer, because of slower chemical loss in winter.

The ionospheric electrons and ions are strongly influenced by the geomagnetic field, and tend to move in spirals along geomagnetic-field lines. In the *F*-region, where collisions with air molecules are infrequent, the plasma diffuses rapidly along the magnetic field, but not across it. Winds in the upper atmosphere can also push the ions along the field lines. Plasma motion across magnetic field lines, on the other hand, is produced by electric fields. In the presence of an electric field **E** and a magnetic field **B**, electrons and ions drift perpendicular to both fields at the velocity $\mathbf{E} \times \mathbf{B}/B^2$. In the upper *F*-region, where ion lifetimes are long, the ions and electrons can be transported vertically and horizontally over significant distances between the times they are produced and lost, giving rise to a pattern of plasma density that depends on the geomagnetic-field configuration. As seen in Figure I28, a relative minimum in TEC tends to form along the magnetic equator in the afternoon and evening. It is created by an upward $\mathbf{E} \times \mathbf{B}/B^2$ plasma drift during the day, with subsequent rapid plasma diffusion along the magnetic field, driven downward and away from the magnetic equator by gravity and plasma pressure gradients. Maxima in

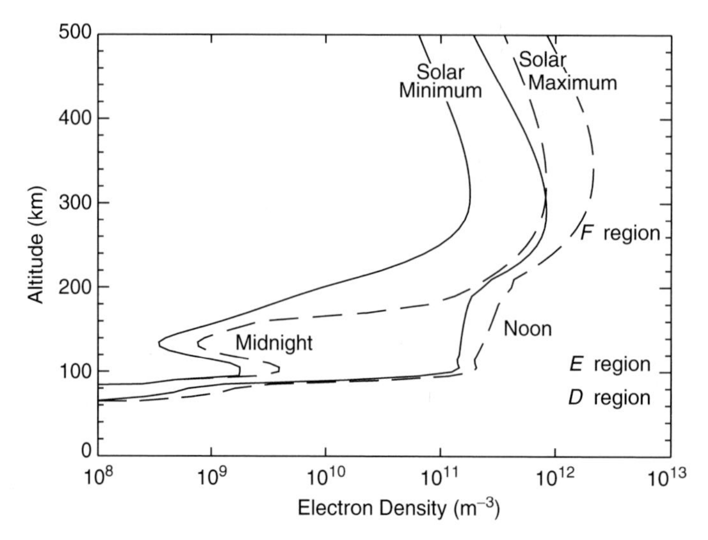

Figure I27 Altitude profiles of electron density at 18°N, 67°W, September equinox, representative of noon and midnight, solar minimum (solid lines) and solar maximum (dashed lines), with the *F*, *E*, and *D* regions indicated.

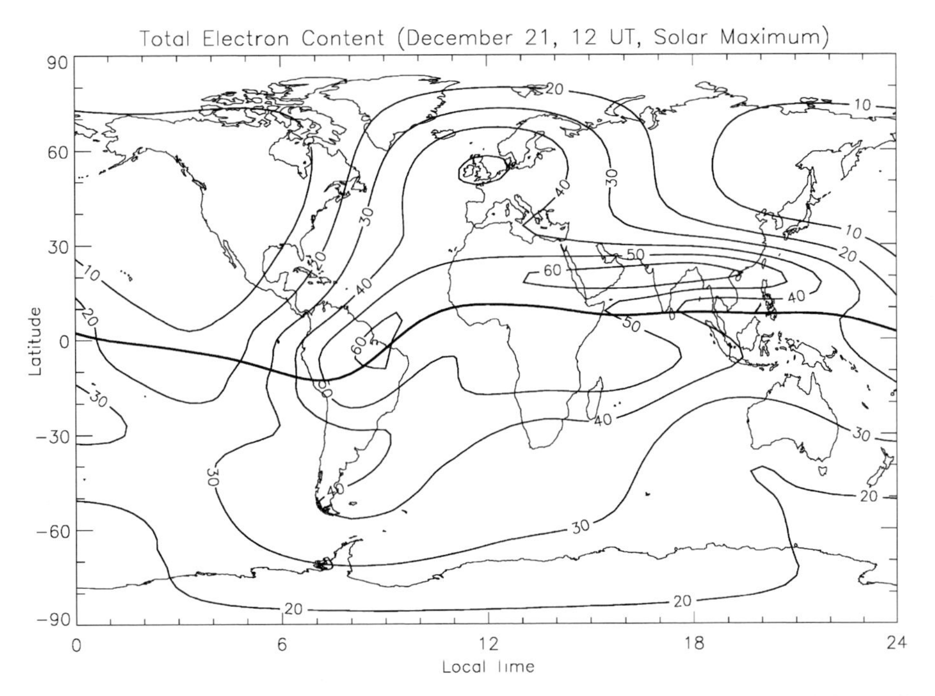

Figure I28 Global map of TEC at 12 UT, December solstice, solar maximum. Local time increases with longitude as shown on the bottom scale. Contours are spaced at intervals of 10×10^{16} electrons per square metre. The thick solid line is the magnetic equator.

the *F*-region ion density build up about 1500 km on either side of the magnetic equator.

The geomagnetic influence on ion and electron mobility makes the electrical conductivity of the ionosphere highly anisotropic. The relatively free motion of electrons along the magnetic-field direction gives rise to large conductivity in that direction. In fact, the large parallel conductivity extends far into space, out to many Earth radii, and permits electric currents to flow relatively easily along geomagnetic-field lines from one hemisphere of the Earth to the opposite hemisphere.

In the direction perpendicular to the geomagnetic field, current can flow in response to an electric field only when the electrons and positive ions move at different velocities. Their common drift at the velocity $\mathbf{E} \times \mathbf{B}/B^2$ in the upper *F*-region does not produce current. However, at lower altitudes, ion-neutral collisions deflect the ion motion away from the velocity $\mathbf{E} \times \mathbf{B}/B^2$, while the electrons continue to drift at that velocity so that current flows perpendicular to the magnetic field. Part of this current, called Pedersen current, is in the direction of $\mathbf{E}$, and is associated with a Pedersen conductivity. Another part of the current, called Hall current, flows in the $\mathbf{B} \times \mathbf{E}$ direction, opposite to the electron velocity, and is associated with a Hall conductivity. Figure I29 shows typical daytime altitude profiles of the parallel, Hall, and Pedersen conductivities. In the *E*-region the conductivities vary by about 40% over the solar cycle, but in the *F*-region the percentage variation is much greater. Unlike the TEC, which represents primarily *F*-region plasma, the height-integrated Pedersen and Hall conductivities vary mainly with the *E*-region plasma densities, which have short lifetimes and therefore depend directly on the solar and auroral ionization. They maximize during daylight, and at night in the auroral zone.

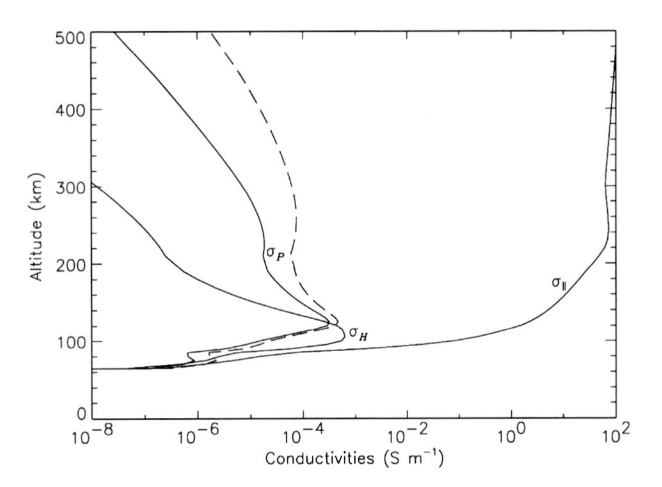

Figure I29 Solid lines: altitude profiles of the noon-time parallel ($\sigma_{\parallel}$), Pedersen (σ_P), and Hall (σ_H) conductivities at 18°N, 67°W, September equinox, solar minimum. Dashed line: Pedersen conductivity at solar maximum.

The electric field that drives current in the conductive medium is the field present in the reference frame moving with the air, $\mathbf{E} + v \times \mathbf{B}$, where v is the wind velocity. Upper atmospheric winds drive electric current by moving the conductive medium through the geomagnetic field, thereby driving an ionospheric dynamo. Convergent or divergent wind-driven currents create space charge and an electrostatic field $\mathbf{E}$ that drives offsetting divergent or convergent current. The ionospheric dynamo is the main source of quiet-day ionospheric electric fields and currents, and of their associated geomagnetic daily variations, at middle and low latitudes (see *Periodic external fields*). At high magnetic latitudes, electric fields and currents connect along geomagnetic-field lines with sources in the outer magnetosphere. They intensify greatly during magnetic storms (see *Storms and substorms*), and can produce strong heating and dynamical changes in the upper atmosphere and ionosphere.

For further information about the ionosphere and ionospheric processes, see Kelley (1989) and Schunk and Nagy (2000).

Arthur D. Richmond

Bibliography

Kelley, M.C., 1989. *The Earth's Ionosphere.* San Diego: Academic Press.

Schunk, R.W., and Nagy, A.F., 2000. *Ionospheres: Physics, Plasma Physics and Chemistry,* Cambridge University Press.

Cross-references

Auroral Oval
Magnetosphere
Periodic External Fields
Storms and Substorms, Magnetic

IRON SULFIDES

General

Iron sulfides are generally quoted as minor magnetic minerals and the interest of paleomagnetists for this family of minerals progressively developed only during the last 10–15 years. This was due partly to the fact that their occurrence was originally believed to be restricted to peculiar geological environments (i.e., sulfidic ores, anoxic sulfate-reducing sedimentary environments) and partly to their metastability with respect to pyrite (FeS_2), which is paramagnetic. Magnetic iron sulfides were therefore not expected to carry a stable remanent magnetization and to survive over long periods of geological time in sedimentary environments. However, their occurrence as main carriers of a remanent magnetization stable through geological times has been increasingly reported in recent years from a large variety of rock types, primarily as a result of more frequent application of magnetic methods to characterize the magnetic mineralogy in paleomagnetic samples. The recognition of the widespread occurrence of magnetic iron sulfides as stable carriers of natural remanent magnetizations (NRMs) in rocks propelled specific researches on their fundamental magnetic properties.

Pyrrhotite: magnetic properties

Among magnetic iron sulfides, the pyrrhotite solid solution series ($Fe_{1-x}S$; with x varying between 0 and 0.13) exhibits a wide range of magnetic behaviors. Stochiometric pyrrhotite FeS (troilite) shows a hexagonal crystal structure based upon the NiAs structure, with alternating c-planes of Fe and S (Figure I30). In each Fe-layer the Fe^{2+} ions are ferromagnetically coupled, while neighboring Fe-layers are coupled antiferromagnetically to each other via intervening S^{2-} ions. The nonstoichiometric pyrrhotites are cation deficient. The increase of Fe deficiency affects both the crystallographic and magnetic structures: ordering of Fe vacancies leads to an alternation of partially and fully filled Fe layers, the hexagonal structure distorts to monoclinic and the magnetic ordering turns from antiferromagnetic to ferrimagnetic. Pyrrhotite shows strong magnetocrystalline anisotropy, with the easy directions of magnetization confined in the crystallographic basal plane (the magnetic crystalline anisotropy constant K_1 is positive and estimated at ca. 10^4 J m^{-3}, see Dunlop and Özdemir, 1997). The crystallographic c-axis perpendicular to the basal plane is the axis of very hard magnetization.

The more Fe-rich, hexagonal, pyrrhotites (F_9S_{10}, and possible $F_{10}S_{11}$, $F_{11}S_{12}$) are antiferromagnetic at room temperature and are characterized by a λ transition at temperatures between ca. 180°C and 220°C, depending on composition, above which they exhibit ferrimagnetism. The λ transition represents a change in the vacancy-ordering pattern of the crystal structure and is distinctive and diagnostic of hexagonal (Fe-rich) pyrrhotites. The Curie temperature for the different compositions of hexagonal pyrrhotites varies between 210°C and 270°C.

The most Fe-deficient pyrrhotite (F_7S_8) is monoclinic and ferrimagnetic at room temperature, with a Curie temperature of ca. 325°C. It has no λ transition, but it shows a diagnostic low-temperature transition in remanence and coercivity at 30–35 K (Rochette *et al.*, 1990).

Natural pyrrhotite usually occurs as a mixture of superstructures of monoclinic and hexagonal types, which results in an intermediate overall composition. Magnetic properties of pyrrhotite and their dependence from grain size and temperature have been investigated in detail in a number of specific studies (e.g., Clark, 1984; Dekkers, 1988, 1989, 1990; Menyeh and O'Reilly, 1991, 1995, 1996; Worm *et al.*, 1993; O'Reilly *et al.*, 2000). Monoclinic ferrimagnetic pyrrhotite (F_7S_8) has a saturation magnetization (J_s) of 18 A m^2 kg^{-1} at room temperature. The magnetic susceptibility of pyrrhotites is field independent in grains smaller than 30 μm, however for larger grains the magnetic susceptibility (χ) and its field dependence increase with increasing grain size (1×10^{-5}–7×10^{-5} m^3 kg^{-1}). Conversely, the coercivity parameters all increase with decreasing grain size. The room temperature coercive force (H_c) ranges from 10 to 70 kA m^{-1} (i.e., 12–88 mT for coercive force expressed as magnetic induction B_c)

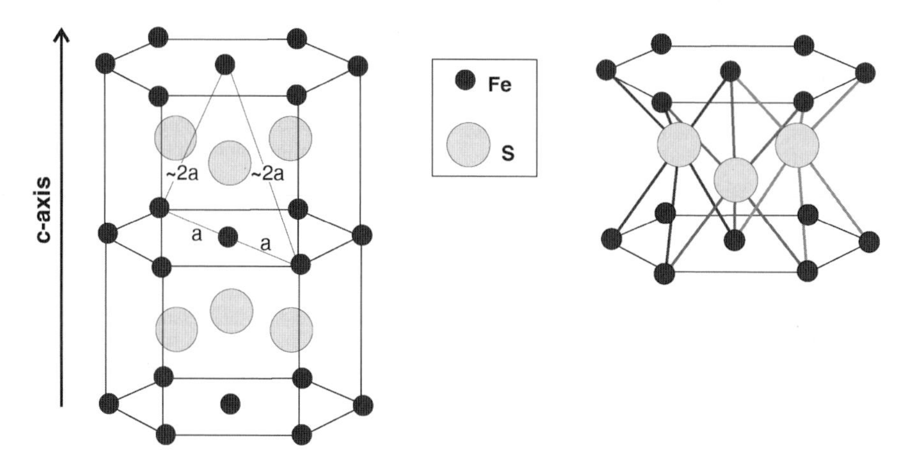

Figure I30 Sketch of the crystalline structure of stoichiometric troilite (FeS). The structure is basically hexagonal, with alternating layers of Fe^{2+} and S^{2-} ions. The c-axis of hexagonal symmetry is the axis of very hard magnetization and is perpendicular to the basal planes. Elemental magnetic moments are parallel within a particular cation basal plane. The alternating Fe^{2+} layers define the two magnetic sublattices with oppositely directed magnetic moments. In nonstoichiometric monoclinic pyrrhotite Fe_7S_8, the cation vacancies are preferentially located on one of the two magnetic sublattices, giving rise to ferrimagnetism.

and for synthetic powders with grain size in the range 1–24 μm, it could be fitted by a power-law dependence of the form $H_c \propto L^n$, where L is the particle size and $n = -0.38$ (O'Reilly *et al.*, 2000). During hysteresis measurements and isothermal remanent magnetization (IRM) acquisition experiments pyrrhotite powders approach complete saturation only in fields greater than 1 T, with the magnetic hardness increasing with decreasing grain sizes (O'Reilly *et al.*, 2000).

In synthetic monoclinic pyrrhotites the critical size for the transition from the single-domain (SD) to the multidomain (MD) state, with typical lamellar domains normal to the *c*-axis, is estimated at a mean value of ca. 1 μm. The micromagnetic structures of hexagonal synthetic pyrrhotites are far more complex than that of monoclinic pyrrhotites, with typical wavy walls and a significantly smaller size of the individual magnetic domains (O'Reilly *et al.*, 2000).

The remanent magnetization carried by pyrrhotites can be of prime importance for paleomagnetic studies, but various factors may complicate the paleomagnetic interpretation of the data obtained from the classical demagnetization treatments. During alternating field (AF) demagnetization pyrrhotite-bearing samples were reported to acquire significant gyromagnetic and rotational remanent magnetization (RRM) in fields higher than 20 mT (Thomson, 1990) and upon heating at temperatures greater than 500°C during thermal demagnetization pyrrhotite transforms irreversibly to magnetite and, via subsequent further oxidation and heating, to hematite (Dekkers, 1990), producing new magnetic phases that may acquire new remanent magnetizations in the laboratory. Furthermore, self-reversal phenomena of pyrrhotite have been earlier reported under various laboratory experiments (Everitt, 1962; Bhimasankaram, 1964; Bhimasankaram and Lewis, 1966). Recent studies linked such phenomena to crystal twinning (Zapletal, 1992) and/or to a close coexistence of pyrrhotite and magnetite crystals, the latter being nucleated from direct oxidation of the pyrrhotite grains during heating (Bina and Daly, 1994).

Greigite: magnetic properties

The iron sulfide greigite (Fe_3S_4) is the ferrimagnetic inverse thiospinel of iron and its crystalline structure is comparable to that of magnetite (Fe_3O_4) in which sulfur replace oxygen atoms. In the cubic close-packed crystal structure of greigite, the tetrahedral A-sites are filled by Fe^{3+} ions and the octahedral B-sites are filled half by Fe^{3+} ions and half by Fe^{2+} ions (Figure I31).

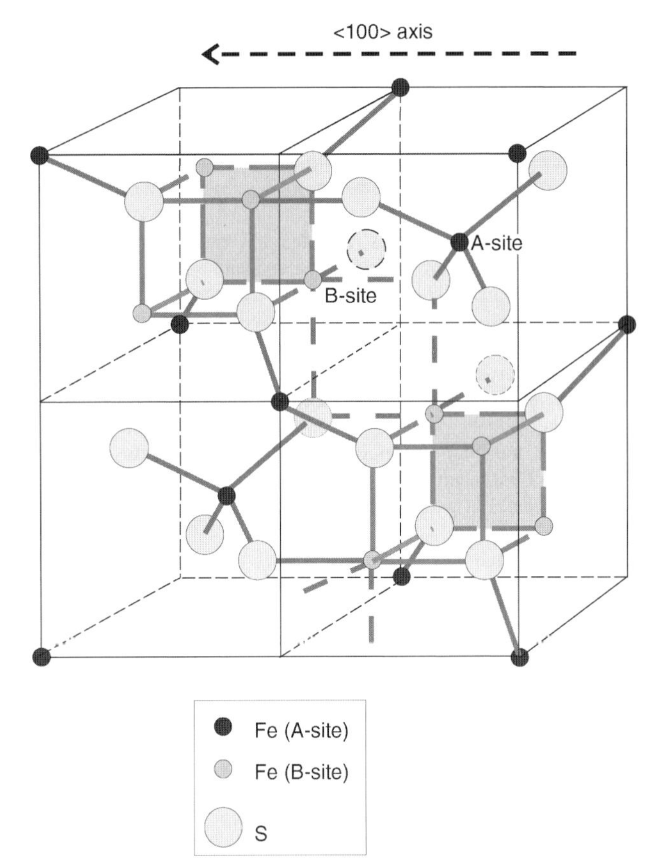

Figure I31 Sketch of the ½ of a unit cell of greigite (Fe_3S_4). The crystalline structure of greigite is basically the same inverse spinel structure typical for magnetite (Fe_3O_4) in which S^{2-} ions substitute O^{2-} ions. Cations are both in tetrahedral (A-site) and octahedral (B-site) coordination with S^{2-} ions. In greigite the easy axis of magnetization is the <100> crystallographic axis.

Greigite is characterized by a high magnetocrystalline anisotropy (with a positive magnetic crystalline anisotropy constant K_1 estimated at ca. 10^3 J m^{-3}, see Diaz Ricci and Kirschvink, 1992; Dunlop and Özdemir, 1997) and its magnetic easy axis is aligned along the <100> crystallographic direction. Basic magnetic properties at room temperature in synthetic greigite powders were systematically investigated by Dekkers and Schoonen (1996), indicating a minimum lower bound for the saturation magnetization (J_s) of 29 A m^2 kg^{-1}, a magnetic susceptibility (χ) between ca. 5×10^{-5} and 20×10^{-5} m^3 kg^{-1} and intermediate coercivities (i.e., IRM acquisition curves saturate after application of fields $\geq$0.7–1 T).

Greigite is unstable during heating to temperature higher than ca. 200°C. As a consequence of such instability the magnetic susceptibility and the magnetization in greigite-bearing sediments both undergo significant changes during heating (see Krs *et al.*, 1990, 1992; Reynolds *et al.*, 1994; Roberts, 1995; Horng *et al.*, 1998; Sagnotti and Winkler, 1999; Dekkers *et al.*, 2000): a major drop is generally observed between ca. 250°C and 350°C, reflecting decomposition of greigite in nonmagnetic sulfur, pyrite, and marcasite, followed by a dramatic increase above 350–380°C and a subsequent decrease above 400–450°C, indicating progressive production of new magnetic phases (pyrrhotite, then magnetite/maghemite, and finally hematite). Thermal decomposition of greigite precludes direct determination of its Curie temperature.

Greigite does not show any low temperature (5–300 K) phase transition (Roberts, 1995; Torii *et al.*, 1996; Dekkers *et al.*, 2000), though a broad J_s maximum peak value was observed at 10 K during cooling from 300 to 4 K (Dekkers *et al.*, 2000).

Typical for greigite are: low to intermediate coercivities, with spectra that partly overlap those of magnetite and pyrrhotite (i.e., for natural greigite-bearing sediments and synthetics greigite the published values for the coercivity of remanence (B_{cr}) ranges from 20 to 100 mT, while for the coercivity (B_c) varies from 13 to 67 mT), maximum unblocking temperatures in the range 270–380°C, the lack of low-temperature phase transitions and the presence of distinct stable SD properties (i.e., Roberts, 1995). With regards to these latter properties, in particular, greigite-bearing sediments show: (a) SD-like hysteresis ratios, with M_{rs}/M_s often exceeding 0.5 (where M_{rs} is the saturation remanent magnetization and M_s is the saturation magnetization) and B_{cr}/B_c often lower than 1.5, (b) high values of the SIRM/k ratio (where SIRM is the saturation IRM and k the low-field magnetic susceptibility), (c) a sensitivity to field impressed anisotropy and (d) a marked tendency for acquisition of gyromagnetic remanent magnetization (GRM) and RRM. GRM and RRM acquisition in greigite-bearing samples has been investigated in the detail in several specific studies (Snowball, 1997a,b; Hu *et al.*, 1998; Sagnotti and Winkler, 1999; Stephenson and Snowball, 2001; Hu *et al.*, 2002) that have shown that greigite has the highest effective gyrofield (B_g) reported so far for all magnetic minerals (of the order of several hundred microteslas for a peak AF of 80 mT) and that gyromagnetic effects are powerful indicators for the presence of greigite in sediments.

GRM and RRM are effects due to the application of an AF on SD grains. They are produced whenever there is an asymmetry in the number of magnetic moments that flip in a particular sense during the AF treatment. In other words, such remanences appear in any system where a particular sense of flip predominates (Stephenson, 1980a,b). RRM was explained in terms of a gyromagnetic effect associated with the irreversible flip of SD particles during rotation of the sample in an AF (Stephenson, 1985).

The typical SD properties of natural greigite grains may be explained by its intrinsic magnetic and crystalline structure, with the magnetocrystalline anisotropy dominating the magnetization process. The theoretically estimated size range of stability for prismatic greigite SD grains extends well beyond that of magnetite, suggesting that elongated greigite crystals may be in a SD state even for very large sizes (up to several micrometers) (Diaz Ricci and Kirschvink, 1992). Moreover, direct magnetic optical observations indicated that the single- to two-domain transition in greigite may occur for grain sizes of 0.7–0.8 μm (Hoffmann, 1992) and that greigite usually occurs in framboidal aggregates of grains individually smaller than 1 μm (i.e., Jiang *et al.*, 2001).

The maximum value for the M_{rs}/M_s ratio in SD grains with shape anisotropy is 0.5. Conversely, if magnetocrystalline anisotropy controls the hysteresis behavior and the <100> axis is the easy axis of magnetization, it would be expected that M_{rs}/M_s would approach a value of 0.832 as B_{cr}/B_c approach unity. Under the same circumstances if <111> is the easy axis of magnetization M_{rs}/M_s is predicted to approach 0.866 (O'Reilly, 1984; Dunlop and Özdemir, 1997).

A magnetic method for discriminating between greigite and pyrrhotite in paleomagnetic samples has been proposed by Torii *et al.* (1996), based on the thermal demagnetization of a composite IRM and relying upon the instability and alteration of greigite during thermal heating at temperatures above 200°C.

Notwithstanding its metastable properties and magnetic instability, the importance of greigite for paleomagnetism and magnetostratigraphy has been particularly stressed in recent years since it has been widely recognized as a carrier of stable chemical remanent magnetization (CRM) in lacustrine and marine sediments, with ages ranging from the Cretaceous to the Present (e.g., Snowball and Thomson, 1990; Snowball, 1991; Hoffmann, 1992; Krs *et al.*, 1990, 1992; Roberts and Turner, 1993; Hallam and Maher, 1994; Reynolds *et al.*, 1994; Roberts, 1995; Roberts *et al.*, 1996; Sagnotti and Winkler, 1999), as well as in soils (Fassbinder and Stanjek, 1994).

Fe-Ni sulfides

Smythite $(Fe, Ni)_9S_{11}$, pentlandite $(Fe, Ni)_9S_8$, and other complex Fe-Ni sulfides were occasionally reported in sediments (i.e., Krs *et al.*, 1992; Van Velzen *et al.*, 1993) in association with pure Fe sulfides, but their magnetic properties have not been studied in the detail so far.

Occurrences: formation, preservation, problems, and potential for use in paleomagnetism

Troilite is common in meteorites and lunar rocks, but not on Earth.

Pyrrhotites are ordinary magnetic carriers in magmatic, hydrothermal, and metamorphic rocks. Pyrrhotite in metamorphic rocks has been shown to acquire postmetamorphic partial thermomagnetic remanent magnetization (pTRM) during the uplift of mountain belts and has been used as a thermometer for postmetamorphic cooling (thermopaleomagnetism) and for dating and evaluation of exhumation rates through various reversals of the Earth magnetic field (i.e., Crouzet *et al.*, 1999, 2001a,b).

Pyrrhotite has also been recognized as the main remanence carrier in Shergotty-Nakhla-Cassigny type (SNC) Martian meteorites, and inferred as a main magnetic phase in the crust of Mars with significant implication for a proper evaluation of magnetic anomalies on such planet (Rochette *et al.*, 2001).

Pyrrhotite and greigite occur also as authigenic phases in geologically young or recent sediments deposited under anoxic, sulfate-reducing conditions (Figure I32).

The presence of magnetic iron sulfides in sediments is of special interest, since they have important implication for magnetostratigraphy and environmental magnetism. Reduction diagenesis and authigenesis can significantly affect the magnetic mineralogy of sediments (Karlin and Levi, 1983; Karlin, 1990a,b; Leslie *et al.*, 1990a,b). In particular, magnetic iron sulfides can be produced through a number of different processes: (1) bacterially mediated synthesis of single-domain greigite, in the form of magnetosomes produced by magnetotactic bacteria (Mann *et al.*, 1990; Bazylinski *et al.*, 1993), and/or (2) precipitation from pore waters, and/or (3) dissolution of detrital iron oxides and subsequent precipitation of iron sulfides.

Authigenic growth of iron sulfides is a common and relatively well-understood process in anoxic sedimentary environments with relatively high organic carbon contents (e.g., Berner, 1984; Leslie *et al.*, 1990a,b; Roberts and Turner, 1993). Authigenic magnetic iron sulfides

(a)

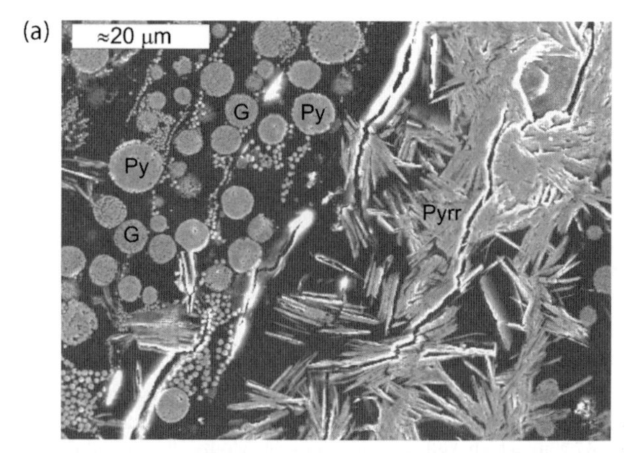

(b)

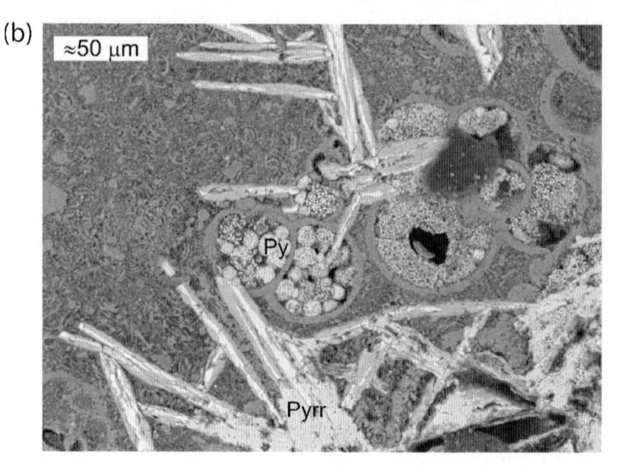

Figure I32 (a) Backscattered electron micrograph of an iron sulfide nodule from the Valle Ricca section, Italy (see Florindo and Sagnotti, 1995). The nodule shows evidence for multiphase sulfidization and contains framboidal pyrite (Py) and greigite (G), which always occurs with finer crystal sizes than the co-occurring pyrite, and intergrown plates of hexagonal pyrrhotite (Pyrr). (Courtesy of Andrew P. Roberts and Wei-Teh Jiang.) (b) Backscattered electron micrograph of a Trubi marl sample from Sicily (see Dinarès-Turell and Dekkers, 1999) showing framboidal pyrite (Py) filling foraminifera shells and pyrrhotite (Pyrr) intergrown plates dispersed in the rock matrix. (Courtesy of Jaume Dinarès-Turell.)

(pyrrhotite and greigite) are intermediate mineral phases within the chemical iron reduction series that eventually forms stable paramagnetic pyrite (FeS_2; Berner, 1969, 1970, 1984; Canfield and Berner, 1987). The process can be mediated by bacterial activity that reduces pore-water sulfate (SO_4^{2+}) to sulfide (S^-) associated with consumption of organic carbon during burial. Hydrogen sulfide (H_2S) then progressively reacts with detrital iron minerals to ultimately produce pyrite. Intermediate magnetic iron sulfides are metastable with respect to pyrite in the presence of excess H_2S. The major factors controlling pyrite formation and the preservation of the (magnetic) intermediate phases in marine sediments are the amounts of dissolved sulfate, reactive iron detrital minerals, and decomposable organic matter (Berner, 1970, 1984). Intermediate magnetic iron sulfides may be preserved in all cases when the limited availability of sulfide, reactive iron and/or organic matter prevents completion of the processes that result in formation of pyrite. In normal marine environments, sulfate and reactive detrital iron minerals are practically unlimited and the major controlling factor on the pyritization process appears to be the availability of detrital organic matter (Kao *et al.*, 2004). In this case, the small amounts of sulfide produced will rapidly react with dissolved iron; the rapid consumption of sulfide means that formation of intermediate greigite is favored over pyrite. However, in some settings, organic carbon may have different sources and greigite has also been reported to form during later burial as a consequence of the diffusion of hydrocarbons and gas hydrates through permeable strata (Reynolds *et al.*, 1994; Thompson and Cameron, 1995; Housen and Mousgrave, 1996).

Greigite is the more common magnetic iron sulfide in fine-grained sediments; pyrrhotite, however, has also been reported in various fine-grained sediments, often alongside greigite (Linssen, 1988; Mary *et al.*, 1993; Roberts and Turner, 1993; Dinarès-Turell and Dekkers, 1999; Horng *et al.*, 1998; Sagnotti *et al.*, 2001; Weaver *et al.*, 2002) (Figure I32). The relative abundance of reactive iron versus organic matter appears to be the controlling factor for the transformation pathway of initial amorphous FeS into greigite or into pyrrhotite (Kao *et al.*, 2004). Compared to greigite, pyrrhotite is favored by more reducing environments (i.e., lower Eh) and higher concentration of H_2S, both implying a higher consumption of organic carbon (Kao *et al.*, 2004). Even under appropriate diagenetic conditions, however, monoclinic pyrrhotite (Fe_7S_8) formation will be extremely slow below $\sim 180°C$, which makes it a highly unlikely carrier of early diagenetic remanences in sediments (Horng and Roberts, 2006). The abundance of monoclinic pyrrhotite in regional metamorphic belts makes it a likely detrital rather than authigenic magnetic mineral in marginal basins in such settings (Horng and Roberts, 2006).

During nucleation and crystal growth, authigenic greigite and pyrrhotite acquire a CRM, which contributes to the total NRM of the sediments. Authigenic formation of magnetic iron sulfides is generally believed to occur at the very first stages of diagenesis and NRM carried by magnetic iron sulfides were used for detailed paleomagnetic studies, assuming they reflect primary components of magnetization (e.g., Tric *et al.*, 1991). However, many studies documented a significant delay between the deposition of the sediment and the formation of magnetic iron sulfides, implicating late diagenetic magnetizations. In particular, sediments bearing magnetic iron sulfides were often reported to carry CRM that are antiparallel either to those carried by coexisting detrital magnetic minerals and of opposite polarity with respect to the polarity expected for the age of the rock unit (e.g., Florindo and Sagnotti, 1995; Horng *et al.*, 1998; Dinarès-Turell and Dekkers, 1999; Jiang *et al.*, 2001; Weaver *et al.*, 2002; Oms *et al.*, 2003; Roberts and Weaver, 2005; Sagnotti *et al.*, 2005). Such studies demonstrate that magnetic iron sulfides can carry stable magnetizations with a wide range of ages and that a syndepositional age should not be automatically assumed.

Leonardo Sagnotti

Bibliography

Bazylinski, D.A., Heywood, B.R., Mann, S., and Frankel, R.B., 1993. Fe_3O_4 and Fe_3S_4 in a bacterium. *Nature*, **366**: 218.

Berner, R.A., 1969. Migration of iron and sulfur within anaerobic sediments during early diagenesis. *American Journal of Science*, **267**: 19–42.

Berner, R.A., 1970. Sedimentary pyrite formation. *American Journal of Science*, **268**: 1–23.

Berner, R.A., 1984. Sedimentary pyrite formation: an update. *Geochimica et Cosmochimica Acta*, **48**: 605–615.

Bhimasankaram, V.L.S., 1964. Partial magnetic self-reversal of pyrrhotite. *Nature*, **202**: 478–480.

Bhimasankaram, V.L.S., and Lewis, M., 1966. Magnetic reversal phenomena in pyrrhotite. *Geophysical Journal of the Royal Astronomical Society*, **11**: 485–497.

Bina, M., and Daly, L., 1994. Mineralogical change and self-reversal magnetizations in pyrrhotite resulting from partial oxidation; geophysical implications. *Physics of the Earth and Planetary Interiors*, **85**: 83–99.

Canfield, D.E., and Berner, R.A., 1987. Dissolution and pyritization of magnetite in anoxic marine sediments. *Geochimica et Cosmochimica Acta*, **51**: 645–659.

Clark, D.A., 1984. Hysteresis properties of sized dispersed monoclinic pyrrhotite grains. *Geophysical Research Letters*, **11**: 173–176.

Crouzet, C., Ménard, G., and Rochette, P., 1999. High-precision three-dimensional paleothermometry derived from paleomagnetic data in an alpine metamorphic unit. *Geology*, **27**: 503–506.

Crouzet, C., Rochette, P., and Ménard, G., 2001a. Experimental evaluation of thermal recording of successive polarities during metasediments uplift. *Geophysical Journal International*, **145**: 771–785.

Crouzet, C, Stang, H., Appel, E., Schill, E., and Gautam, P., 2001b. Detailed analysis of successive pTRMs carried by pyrrhotite in Himalayan metacarbonates: an example from Hidden Valley, Central Nepal. *Geophysical Journal International*, **146**: 607–618.

Dekkers, M.J., 1988. Magnetic properties of natural pyrrhotite. I. Behavior of initial susceptibility and saturation-magnetisation-related rock-magnetic parameters in a grain-size dependent framework. *Physics of the Earth and Planetary Interiors*, **52**: 376–394.

Dekkers, M.J., 1989. Magnetic properties of natural pyrrhotite. II. High- and low-temperature behavior of Jrs and TRM as a function of grain size. *Physics of the Earth and Planetary Interiors*, **57**: 266–283.

Dekkers, M.J., 1990. Magnetic monitoring of pyrrhotite alteration during thermal demagnetization. *Geophysical Research Letters*, **17**(6): 779–782.

Dekkers, M.J., and Schoonen, M.A.A., 1996. Magnetic properties of hydrothermally synthesized greigite (Fe_3S_4). I. Rock magnetic parameters at room temperature. *Geophysical Journal International*, **126**: 360–368.

Dekkers, M.J., Passier, H.F., and Schoonen, M.A.A., 2000. Magnetic properties of hydrothermally synthesized greigite (Fe_3S_4). II. High- and low-temperature characteristics. *Geophysical Journal International*, **141**: 809–819.

Diaz Ricci, J. C., and Kirschvink, J.L., 1992. Magnetic domain state and coercivity prediction for biogenic greigite (Fe_3S_4): a comparison of theory with magnetosome observations. *Journal of Geophysical Research*, **97**: 17309–17315.

Dinarès-Turell, J., and Dekkers, M.J., 1999. Diagenesis and remanence acquisition in the Lower Pliocene Trubi marls at Punta di Maiata (southern Sicily): palaeomagnetic and rock magnetic observations. In Tarling, D.H., and Turner, P. (eds.), *Palaeomagnetism and Diagenesis in Sediments*. Geological Society of London, Special Publication, **151**: 53–69.

Dunlop, D.J., and Özdemir, Ö., 1997. *Rock Magnetism: Fundamentals and Frontiers*. New York: Cambridge University Press.

Everitt, C.W.F., 1962. Self-reversal of magnetization in a shale containing pyrrhotite. *Philosophical Magazine Letters*, **7**: 831–842.

Fassbinder, J.W.E., and Stanjek, H., 1994. Magnetic properties of biogenic soil greigite (Fe_3S_4). *Geophysical Research Letters*, **21**: 2349–2352.

Florindo, F., and Sagnotti, L., 1995. Palaeomagnetism and rock magnetism in the upper Pliocene Valle Ricca (Rome, Italy) section. *Geophysical Journal International*, **123**: 340–354.

Hallam, D.F., and Maher, B.A., 1994. A record of reversed polarity carried by the iron sulphide greigite in British early Pleistocene sediments. *Earth and Planetary Science Letters*, **121**: 71–80.

Hoffmann, V., 1992. Greigite (Fe_3S_4): magnetic properties and first domain observations. *Physics of the Earth and Planetary Interiors*, **70**: 288–301.

Horng, C.-S., and Roberts, A.P., 2006. Authigenic or detrital origin of pyrrhotite in sediments? Resolving a paleomagnetic conundrum. *Earth and Planetary Science Letters*, **241**: 750–762.

Horng, C.-S., Torii, M., Shea, K.-S., and Kao, S.-J., 1998. Inconsistent magnetic polarities between greigite- and pyrrhotite/magnetite-bearing marine sediments from the Tsailiao-chi section, southwestern Taiwan. *Earth and Planetary Science Letters*, **164**: 467–481.

Housen, B.A., and Musgrave, R.J., 1996. Rock-magnetic signature of gas hydrates in accretionary prism sediments. *Earth and Planetary Science Letters*, **139**: 509–519.

Hu, S., Appel, E., Hoffmann, V., Schmahl, W.W., and Wang, S., 1998. Gyromagnetic remanence acquired by greigite (Fe_3S_4) during static three-axis alternating field demagnetization. *Geophysical Journal International*, **134**: 831–842.

Hu, S., Stephenson, A., and Appel, E., 2002. A study of gyroremanent magnetisation (GRM) and rotational remanent magnetisation (RRM) carried by greigite from lake sediments. *Geophysical Journal International*, **151**: 469–474.

Jiang, W.T., Horng, C.S., Roberts, A.P., and Peacor, D.R., 2001. Contradictory magnetic polarities in sediments and variable timing of neoformation of authigenic greigite. *Earth and Planetary Science Letters*, **193**: 1–12.

Kao, S.J., Horng, C.S., Liu, K.K., and Roberts, A.P., 2004. Carbon-sulfur-iron relationships in sedimentary rocks from southwestern Taiwan: influence of geochemical environment on greigite and pyrrhotite formation. *Chemical Geology*, **203**: 153–168.

Karlin, R., 1990a. Magnetic mineral diagenesis in suboxic sediments at Bettis site WN, NE Pacific Ocean. *Journal of Geophysical Research*, **95**: 4421–4436.

Karlin, R., 1990b. Magnetite diagenesis in marine sediments from the Oregon continental margin. *Journal of Geophysical Research*, **95**: 4405–4419.

Karlin, R., and Levi, S., 1983. Diagenesis of magnetic minerals in recent hemipelagic sediments. *Nature*, **303**: 327–330.

Krs, M., Krsova, M., Pruner, P., Zeman, A., Novak, F., and Jansa, J., 1990. A petromagnetic study of Miocene rocks bearing microorganic material and the magnetic mineral greigite (Solokov and Cheb basins, Czechoslovakia). *Physics of the Earth and Planetary Interiors*, **63**: 98–112.

Krs, M., Novak, F., Krsova, M., Pruner, P., Kouklíková, L., and Jansa, J., 1992. Magnetic properties and metastability of greigite-smythite mineralization in brown-coal basins of the Krusné hory Piedmont, Bohemia. *Physics of the Earth and Planetary Interiors*, **70**: 273–287.

Leslie, B.W., Hammond, D.E., Berelson, W.M., and Lund, S.P., 1990a. Diagenesis in anoxic sediments from the California continental borderland and its influence on iron, sulfur, and magnetite behavior. *Journal of Geophysical Research*, **95**: 4453–4470.

Leslie, B.W., Lund, S.P., and Hammond, D.E., 1990b. Rock magnetic evidence for the dissolution and authigenic growth of magnetic minerals within anoxic marine sediments of the California continental borderland. *Journal of Geophysical Research*, **95**: 4437–4452.

Linssen, J.H., 1988. Preliminary results of a study of four successive sedimentary geomagnetic reversal records from the Mediterranean (Upper Thvera, Lower and Upper Sidufjall, and Lower Nunivak). *Physics of the Earth and Planetary Interiors*, **52**: 207–231.

Mann, S., Sparks, N.H.C., Frankel, R.B., Bazylinski, B.A., and Jannasch, H.W., 1990. Biomineralization of ferrimagnetic greigite (Fe_3S_4) and iron pyrite (FeS_2) in a magnetotactic bacterium. *Nature*, **343**: 258–261.

Mary, C., Iaccarino, S., Courtillot, V., Besse, J., and Aissaoui, D.M., 1993. Magnetostratigraphy of Pliocene sediments from the Stirone River (Po Valley). *Geophysical Journal International*, **112**: 359–380.

Menyeh, A., and O'Reilly W., 1991. The magnetization process in monoclinic pyrrhotite (Fe_7S_8) particles containing few domains. *Geophysical Journal International*, **104**: 387–399.

Menyeh, A., and O'Reilly, W., 1995. The coercive force of fine particles of monoclinic pyrrhotite (Fe_7S_8) studied at elevated temperature. *Physics of the Earth and Planetary Interiors*, **89**: 51–62.

Menyeh, A., and O'Reilly, W., 1996. Thermoremanent magnetization in monodomain pyrrhotite Fe_7S_8. *Journal of Geophysical Research*, **101**: 25045–25052.

Oms, O., Dinarès-Turell, J., and Remacha, E., 2003. Paleomagnetic results on clastic turbidite systems in compressional settings: example from the Eocene Hecho Group (Southern Pyrenees, Spain). *Studia Geophysica et Geodaetica*, **47**: 275–288.

O'Reilly, W., 1984. *Rock and Mineral Magnetism.* Glasgow: Blackie.

O'Reilly, W., Hoffmann, V., Chouker, A.C., Soffel, H.C., and Menyeh, A., 2000. Magnetic properties of synthetic analogues of pyrrhotite ore in the grain size range 1–24 μm. *Geophysical Journal International*, **142**: 669–683.

Reynolds, R.L., Tuttle, M.L., Rice, C.A., Fishman, N.S., Karachewski, J.A., and Sherman, D.M., 1994. Magnetization and geochemistry of greigite-bearing Cretaceous strata, North Slope Basin, Alaska. *American Journal of Science*, **294**: 485–528.

Roberts, A.P., 1995. Magnetic properties of sedimentary greigite (Fe_3S_4). *Earth and Planetary Science Letters*, **134**: 227–236.

Roberts, A.P., and Turner, G.M., 1993. Diagenetic formation of ferrimagnetic iron sulphide minerals in rapidly deposited marine sediments, South Island, New Zealand. *Earth and Planetary Science Letters*, **115**: 257–273.

Roberts, A.P., and Weaver, R., 2005. Multiple mechanisms of remagnetization involving sedimentary greigite (Fe_3S_4). *Earth and Planetary Science Letters*, **231**: 263–277.

Roberts, A.P., Reynolds, R.L., Verosub, K.L., and Adam, D.P., 1996. Environmental magnetic implications of greigite (Fe_3S_4) formation in a 3 Ma lake sediment record from Butte Valley, northern California. *Geophysical Research Letters*, **23**: 2859–2862.

Rochette, P., Fillion, G., Matteí, J.L., and Dekkers M.J., 1990. Magnetic transition at 30–34 K in Fe_7S_8: insight into a widespread occurrence of pyrrhotite in rocks. *Earth and Planetary Science Letters*, **98**: 319–328.

Rochette, P., Lorand, J.P., Fillion, G., and Sautter, V., 2001. Pyrrhotite and the remanent magnetization of SNC meteorites: a changing perspective on Martian magnetism. *Earth and Planetary Science Letters*, **190**: 1–12.

Sagnotti, L., and Winkler, A., 1999. Rock magnetism and palaeomagnetism of greigite-bearing mudstones in the Italian peninsula. *Earth and Planetary Science Letters*, **165**: 67–80.

Sagnotti, L., Macrí, P., Camerlenghi, A., and Rebesco, M., 2001. Environmental magnetism of Antarctic Late Pleistocene sediments and interhemispheric correlation of climatic events. *Earth and Planetary Science Letters*, **192**: 65–80.

Sagnotti, L., Roberts, A.P., Weaver, R., Verosub, K.L., Florindo, F., Pike, C.R., Clayton, T., and Wilson, G.S., 2005. Apparent magnetic polarity reversals due to remagnetization resulting from late diagenetic growth of greigite from siderite. *Geophysical Journal International*, **160**: 89–100.

Snowball, I.F., 1991. Magnetic hysteresis properties of greigite (Fe_3S_4) and a new occurrence in Holocene sediments from Swedish Lapland. *Physics of the Earth and Planetary Interiors*, **68**: 32–40.
Snowball, I.F., 1997a. Gyroremanent magnetization and the magnetic properties of greigite-bearing clays in southern Sweden. *Geophysical Journal International*, **129**: 624–636.
Snowball, I.F., 1997b. The detection of single-domain greigite (Fe_3S_4) using rotational remanent magnetization (RRM) and the effective gyro field (Bg): mineral magnetic and palaeomagnetic applications. *Geophysical Journal International*, **130**: 704–716.
Snowball, I.F., and Thompson, R., 1990. A stable chemical remanence in Holocene sediments. *Journal of Geophysical Research*, **95**: 4471–4479.
Stephenson, A., 1980a. Gyromagnetism and the remanence acquired by a rotating rock in an alternating field. *Nature*, **284**: 48–49.
Stephenson, A., 1980b. A gyroremanent magnetization in anisotropic magnetic material. *Nature*, **284**: 49–51.
Stephenson, A., 1985. The angular dependence of rotational and anhysteretic remanent magnetization in rotating rock samples. *Geophysical Journal of the Royal Astronomical Society*, **83**: 787–796.
Stephenson, A., and Snowball, I.F., 2001. A large gyromagnetic effect in greigite. *Geophysical Journal International*, **145**: 570–575.
Thomson, G.F., 1990. The anomalous demagnetization of pyrrhotite. *Geophysical Journal International*, **103**. 425–430.
Thompson, R., and Cameron, T.D.J., 1995. Palaeomagnetic study of Cenozoic sediments in North Sea boreholes: an example of a magnetostratigraphic conundrum in a hydrocarbon producing area. In Turner P., and Turner, A. (eds.), Palaeomagnetic Applications in Hydrocarbon Exploration and Production. Geological Society of London, Special Publication, **98**: 223–236.
Torii, M., Fukuma, K., Horng, C.S., and Lee, T.Q., 1996. Magnetic discrimination of pyrrhotite- and greigite-bearing sediment samples. *Geophysical Research Letters*, **23**: 1813–1816.
Tric, E., Laj, C., Jehanno, C., Valet, J.P., Kissel, C., Mazaud, A., and Iaccarino, S., 1991. High-resolution record of the upper Olduvai polarity transition from Po Valley (Italy) sediments: support for dipolar transition geometry? *Physics of the Earth and Planetary Interiors*, **65**: 319–336.
Van Velzen, A.J., Dekkers, M.J., and Zijderveld, J.D.A., 1993. Magnetic iron-nickel sulphides in the Pliocene and Pleistocene marine marls from the Vrica section (Calabria, Italy). *Earth and Planetary Science Letters*, **115**: 43–55.
Weaver, R., Roberts, A.P., and Barker, A.J., 2002. A late diagenetic (synfolding) magnetization carried by pyrrhotite: implications for paleomagnetic studies from magnetic iron sulphide-bearing sediments. *Earth and Planetary Science Letters*, **200**: 371–386.
Zapletal, K., 1992. Self-reversal of isothermal remanent magnetization in a pyrrhotite (Fe_7S_8) crystal. *Physics of the Earth and Planetary Interiors*, **70**: 302–311.

Cross-references

Demagnetization
Magnetization, Chemical Remanent (CRM)

J

JESUITS, ROLE IN GEOMAGNETISM

The Jesuits are members of a religious order of the Catholic Church, the Society of Jesus, founded in 1540 by Ignatius of Loyola. From 1548, when Jesuits established their first college, their educational work expanded rapidly and in the 18th century, in Europe alone, there were 645 colleges and universities and others in Asia and America. As an innovation in these colleges, special attention was given to teaching of mathematics, astronomy, and the natural sciences. This tradition has been continued in modern times in the many Jesuit colleges and universities and this tradition thus spread throughout the world. Jesuits' interest in geomagnetism derived from teaching in these colleges and universities. In many of these colleges and universities observatories were established, where astronomical and geophysical observations were made (Udías, 2003). Missionary work in Asia, Africa, and America, where scientific observations were also made and some observatories established, was another factor in the role of Jesuits in geomagnetism. In this work we must distinguish two periods. The first, from 1540 to 1773, ended with the suppression of the Jesuit order. The second began in 1814 with its restoration and lasts to our times.

Early work on magnetism

Terrestrial magnetism attracted Jesuits' attention from very early times (Vregille, 1905). José de Acosta, a missionary in America, in 1590 described the variation of the magnetic declination across the Atlantic and the places where its value was zero, one of them at the Azores islands. His work is quoted by *William Gilbert* (*q.v.*). In 1629, Nicolò Cabbeo (1586–1650) was the first Jesuit to write a book dedicated to magnetism, *Philosophia Magnetica*, where he collected all that was known in his time, together with his own experiments and observations. Some of his work is based on that of a previous Jesuit Leonardo Garzoni (1567–1592), professor at the college of Venice, who could well be the first Jesuit interested in such matters. Cabbeo was opposed to Gilbert on the origin of terrestrial magnetism. The best-known early work on magnetism by a Jesuit is that of *Athanasius Kircher* (1601–1680) (*q.v.*), *Magnes sive de arte magnetica*, published in 1641. Martin Martini (1614–1661), who sent many magnetic observations from China to Kircher, proposed to him in 1640 to draw a map with lines for the magnetic declination. Had his suggestion been followed this would have been the first magnetic chart (before that of *Edmund Halley* (*q.v.*)).

A curious work is that of Jacques Grandami (1588–1672), *Nova demonstratio inmobilitatis terrae petita ex virtute magnetica* (1645), in which, in order to defend the geocentric system, he tried to show that the Earth does not rotate because of its magnetic field. Among the best observations made in China are those of Antoine Gaubil (1689–1759), who mentioned that the line of zero declination has with time a movement from east to west. His observations and those of other Jesuits in China were published in France in three volumes between 1729 and 1732. In 1727, Nicolas Sarrabat (1698–1739) published *Nouvelle hypothèse sur les variations de l'aiguille aimantée*, which was given an award by the Académie des Sciences of Paris. In 1769, Maximilian Hell (1720–1792), director of the observatory in Vienna, made observations of the magnetic declination during his journey to the island of Vardö in Lapland, at a latitude of 70° N, where he observed the transit of Venus over the solar disk.

Magnetic observatories

The Jesuit order was suppressed in 1773 and restored in 1814. From this time work on geomagnetism was taken up again at the new Jesuit observatories (Udías, 2000, 2003). A total of 72 observatories were founded throughout the world. Magnetic stations were installed in 15 of them: five in Europe, one in North America, four in Central and South America, and five in Asia, Africa, and the Middle East. Some of those magnetic stations in Europe were among the first to be in operation. Observatories in Central and South America, Asia, and Africa provided for some time the only magnetic observations in those regions. Details of these observatories can be found in Udías (2003). At present, most of them have been either closed or transferred to other administration.

The first of these observatories was established in 1824 at the Collegio Romano (Rome). There, in 1858, Angelo Secchi (1818–1878) began magnetic observations, using a set of magnetometers, a declinometer, and an inclinometer. He studied the characteristics of the periodic variations of the different components of the magnetic field and tried to relate magnetic variations with solar activity, considering the Sun to be a giant magnet at a great distance. Relations between geomagnetism and solar activity were to become a favorite subject among Jesuits. In the same year, 1858, magnetic observations begun at Stonyhurst College Observatory (Great Britain). Stephen J. Perry (1833–1889, Figure J1) began his work on geomagnetism, carrying out three magnetic surveys: two in France in 1868 and 1869 and the third in Belgium in 1871 (Cortie, 1890). In each of these surveys, at each station, careful measurements were made of the horizontal component of the magnetic field, magnetic declination, and inclination or dip (Perry, 1870). In Belgium, Perry found large magnetic anomalies related to coal mines. In order to study the relation

Figure J1 Stephen J. Perry (1833–1889), Director of Stonyhurst Observatory.

Figure J2 Ebro Observatory, Roquetas, Tarragona, Spain.

between solar and terrestrial magnetic activity, which was still a controversial subject, Perry at Stonyhurst began a series of observations of sunspots, faculae, and prominences in 1881. For this purpose he installed direct-vision spectroscopes and photographic-grating spectrometers and made large drawings of the solar disk (27 cm diameter). Perry collaborated with *Edward Sabine* (*q.v.*) in this work. Perry participated in several scientific expeditions. The most important was to Kerguelen Island in 1874 to observe the transit of Venus; there he carried out a very comprehensive program of magnetic observations. His project of collecting and comparing all his magnetic and solar observations was never completed due to his untimely death during a scientific expedition to the Lesser Antilles to observe a solar eclipse. In 1874 he was elected a Fellow of the Royal Society for his work in terrestrial magnetism. Perry's successor Walter Sidgreaves (1837–1919) completed the work and showed the correlation between magnetic storms and the maxima of sunspots (Sidgreaves, 1899–1901). The continuous magnetic observations from 1858 to 1974 at Stonyhurst may be one of the longest series at the same site.

Solar-terrestrial relations were also the main subject of Haynald Observatory, founded in 1879 by Jesuits in Kalocsa (Hungary). Between 1885 and 1917, Gyula Fényi (1845–1927) carried out a long program of magnetic and solar observations. He concentrated his efforts on the study of sunspots and prominences, making detailed observations (about 40000 of them) and proposing some interesting, for that time, new ideas about their nature. Magnetic observations began in 1879.

The Manila Observatory (Philippines) was founded by Spanish Jesuits in 1865. Martín Juan (1850–1888) was trained in geomagnetism by Perry in Stonyhurst. Juan brought new magnetic instruments to Manila where he took charge of the magnetic section in 1887. In 1888, he carried out a magnetic field survey on various islands of the archipelago. His death did not allow him to finish the work; this was done in 1893 by Ricardo Cirera (1864–1932), who extended the survey to the coasts of China and Japan (Cirera, 1893).

In the observatory of Zikawei, founded in 1872 near Shanghai, magnetic instruments for absolute measurements and variations were installed in 1877; instruments were moved to two nearby sites, first Lukiapang in 1908 and then Zose in 1933. Results were published in nine volumes between 1908 and 1932 (*Études sur le magnetisme terrestre*, pp. 1–39). Joseph de Moidrey (1858–1936) carried out some early work on the secular variation in the Far East region (de Moidrey and Lou, 1932). These observations continued until 1950 when Jesuits were expelled from China by the Communist government.

After returning to Spain, Cirera founded in 1904 the Observatorio del Ebro, Roquetas (Tarragona; Figure J2), dedicated specifically to the study of the relations of solar activity and terrestrial magnetism and electricity. The observatory was equipped from the beginning with the most up-to-date instrumentation for this purpose and has kept updating its instrumentation since then. Luis Rodés (1881–1939) studied the influence of various forms of solar activity, mainly sunspots and prominences on the terrestrial magnetic and electric fields (Rodés, 1927). After the Spanish civil war, work was resumed by Antonio Romañá (1900–1981) and José O. Cardús (1914–). Since 1958, Ebro has been the base of the International Association of Geomagnetism and Aeronomy *(IAGA)* (*q.v.*) Commission for rapid magnetic variations. Romañá and Cardús held office in IAGA for many years.

Maintaining magnetic observatories in Africa was a difficult task, which Jesuits undertook early. One of the earliest observatories was established in 1889 in Antananarive, Madagascar. Continuous magnetic observations were made until 1967 when the observatory was transferred to the University of Madagascar. A magnetic station was established in the observatory founded in 1903 in Bulawayo, Zimbabwe. Edmund Goetz (1865–1933), its director for 23 years, in 1909 and 1914 carried out two magnetic field surveys, the first with a profile from Broken Hill (Kabwe, Zambia) to the "Star of the Congo Mine" (Congo) and the second in Barotseland (Zambia) covering a distance of more than 200 miles from Kazungula to Lealui (Goetz, 1920). The third observatory was a state observatory entrusted to and directed by Jesuits in Ethiopia. This observatory arose from a recommendation of the Scientific Committee of the International Geophysical Year in 1955 for a magnetic station near the magnetic equator. The station was established in Addis Ababa in 1958 and directed by Pierre Gouin (1917–) for 20 years. Gouin with the collaboration of Pierre Noel Mayaud (1923–), a Jesuit working at the Institute de Physique du Globe in Paris, studied temporal magnetic variations and magnetic storms. Mayaud participated in French expeditions to the Antarctica, studied the magnetic activity in the polar regions and worked mainly on the nature and practical use of geomagnetic indices (Mayaud, 1980). Early magnetic observations were also made by Jesuits in Central and South America. These observations began in 1862 in Belen Observatory (Havana, Cuba) and continued to about 1920. In Puebla (Mexico), magnetic observations were made between 1877 and 1914.

Summary

As an important part of their scientific tradition Jesuits dedicated much of their efforts to the study of terrestrial magnetism. Jesuits' dedication to science can be explained by their peculiar spirituality, which unites prayer and work and finds no activity too secular be turned into prayer. Teaching mathematics and observing the magnetic field of the Earth or of the Sun are activities, as has been shown, that a Jesuit finds perfectly compatible with his religious vocation and through which he can find God in his life. Early work on magnetism was carried out in the 17th and 18th centuries. Jesuit missionaries during those centuries carried out magnetic observations, which were analyzed and published in Europe. Modern work was done through the establishment of observatories. In at least 15 of these observatories magnetic observations were made. Jesuits contributed to the earliest magnetic observations in Africa, Asia, and Central and South America.

Agustín Udías

Bibliography

Cirera, R., 1893. *El magnetismo terrestre en Filipinas.* Manila: Observatorio de Manila, 160 pp.

Cortie, A.L., 1890. *Father Perry, F.R.S.: The Jesuit Astronomer.* London: Catholic Truth Society, 113 pp.

Goetz, E., 1920. Magnetic observations in Rhodesia. *Transactions of the Royal Society of South Africa*, **8**(4): 297–302.

Mayaud, P.N., 1980. *Derivation, Meaning, and Use of Geomagnetic Indices.* Geophysical Monograph 22. Washington, D.C.: American Geophysical Union.

Moidrey, J. de and Lou, F. 1932. Variation séculaire des éléments magnétiques en extreme orient. *Études sur le magnetisme terrestre 39, Observatoire de Zikawei*, **9**: 1–21.

Perry, S., 1870. Magnetic survey of the west of France in 1868. *Philosophical Transactions Royal Society of London*, **160**: 33–50.

Rodés, L., 1927. Some new remarks on the cause and propagation of magnetic storms. *Terrestrial Magnetism and Atmospheric Electricity*, **32**: 127–131.

Sidgreaves, W., 1899–1901. On the connection between solar spots and Earth magnetic storms. *Memoirs of the Royal Astronomical Society*, **54**: 85–96.

Udías, A., 2000. Observatories of the Society of Jesus, 1814–1998. *Archivum Historicum Societatis Iesu*, **69**: 151–178.

Udías, A., 2003. *Searching the Heavens and the Earth: The History of Jesuit Observatories.* Dordrecht: Kluwer Academic, 369 pp.

Vregille, P. de, 1905. Les jésuites et l'étude du magnétisme terrestre. *Études*, **104**: 492–511.

Cross-references

Geomagnetism, History of
Gilbert, William (1544–1603)
Halley, Edmond (1656–1742)
IAGA, International Association of Geomagnetism and Aeronomy
Kircher, Athanasius (1602–1680)
Observatories, Overview
Sabine, Edward (1788–1883)

K

KIRCHER, ATHANASIUS (1602–1680)

Athanasius Kircher was born in Geisa near Fulda (Germany) on May 2, 1601 (or 1602—there is a disagreement in his biographies). After his primary education at the Jesuit school in Fulda, he joined (October 1618) the Jesuit order in Paderborn, where he began to study Latin and Greek. As a consequence of the Thirty Years War he went to Münster and Cologne, where he studied philosophy. He taught for a short time at the secondary school in Coblenz and afterward in Heiligenstadt. Later he went to Mainz to study theology, where he was ordained priest in 1628 and at the end of his Jesuit training was appointed professor of ethics and mathematics at the Jesuit University in Würzburg. He again migrated as a result of the war and went to Avignon (France) and was finally appointed professor of mathematics and oriental languages at the Roman College, which is now the well-known Gregorian University in Rome, where he died on November 27, 1680.

One cannot use modern standards of scientific evaluation to understand his real value, as Kircher's thinking and his methods are typical for the 17th century. First of all, he was a polymath interested in almost every kind of knowledge. His books (over 30 printed volumes) range from oriental languages—Arabic grammar; translation of old Coptic books and an extensive research to decipher Egyptian hieroglyphics: *Lingua Egyptiaca restituta* (1643), *Oedypus Egyptiacus* (1652), *Obeliscus Pamphilius* (1650)—to physics—optics: *Ars Magna Lucis et Umbra* (1646); magnetism: *Ars Magnesia* (1631), *Magnes, sive de Arte Magnetica* (1641, 1643, 1654), *Magneticum Naturae Rerum* (1667); music: *Musurgia Universalis sive Ars Magna consoni and disoni* (1650), *Phonurgia Nova* (1673); geophysics: *Mundus Subterraneus* (1664); and also medicine: *Scrutinium physico-medicum contagiosae luis quae pestis dicitur* (1657). He also wrote on biblical subjects such as *Arca Noe* (1675) and *Turris Baabel* (1679), in which he combines science and archeological imagination. This imagination together with his knowledge of many different subjects inclined him to encyclopedism as in his works *Itinerarium Extaticum* (1656) and *Iter Extaticum II* (1657), in which he discusses theories of the solar system and beyond while telling the fictitious story of an imaginary journey through Earth and celestial bodies.

Kircher tries to find some sort of unity among the many different subjects with which he dealt, not only in written language—*Polygraphia nova universalis* (1663)—but also in the much more universal concept of knowledge itself as in *Ars Magna Sciendi* (1669). Although mathematical combinations in Kircher's book are far removed from the cabalistic usage, which was so frequent in his time, there is a misleading use of the word "magnetism" in his writings: for him magnetism with its two polarities is a metaphorical symbol of the dualism in things and forces acting in nature: attraction and repulsion; love and hatred; and light and darkness. This use allows him to talk about the magnetism of celestial bodies, music, medicine, love, and to print, perhaps for the first time in a scientific book, the word electromagnetism (Kircher, 1654, p. 451), although this word has nothing to do with the concept introduced many years afterward by Maxwell.

He was certainly dominated by Baroque culture, redundant in many aspects of his expression, overlengthy in his writings, and not easy to read in his Latin, which is far removed from the classical language. That Kircher was accepted during his lifetime as one of the best scholars in Europe is proven by the fact that the Imperial Court in Vienna invited Kircher to be professor of astronomy in the University of Vienna as successor of Kepler, who died in 1632. It was only by the influence of Cardinal Barberini and of Pope Urban VIII that he was prevented from accepting the post, and was called to Rome. They accepted the request of the French gentleman Nicolaus Claude de Peiresc, who wanted Kircher to continue his work on the interpretation of Egyptian hieroglyphics. Since his death importance of Kircher as a scientist has been much debated although now his value seems to be accepted again.

Of the three books that Kircher wrote on magnetism the first *Ars Magnesia* (his first printed book) is a short booklet of 48 pages and seems to be the printed edition of a talk he gave some years earlier when he was teaching at Heiligenstadt. That can be inferred from the explanation that follows the title where he says he will deal with the nature, force, and effects of the magnet together with several applications of it. In the third book, *Magneticum Naturae Rerum*, the word "magneticum" is taken in the metaphorical sense that includes all forces that act on all things in nature.

The second book *Magnes sive de Arte Magnetica Opus Tripartitum* (Kircher, 1654) (a copy of its 3rd edition is in the library of the Ebro Observatory) contains the scientific contribution of Kircher to geomagnetism. It is divided into three "books:"

1. *On the Nature and Properties of the Magnet*
2. *The Use of the Magnet*
3. *The Magnetic World*

In the first book Kircher tries to investigate what magnetism is, using as a starting point what was known in his time: the natural existence of loadstones, the fact that iron may become magnetic with two poles acting in opposite senses on other magnetic bodies. He also accepts the hypothesis of the Earth being a magnet after repeating Gilbert's experiments with the *terrella*, but it is easy to see that the data is too poor to explain what magnetism is.

In the second book, Kircher shows with the help of several experiments how magnetic force is distributed in magnetic bodies of different shapes and how to find what he calls the barycentre. He tries to show that in a spherical body magnetism propagates linearly to the poles in its interior and spherically to the poles in the exterior. He finds that both poles have equal force and explains how to make magnetic needles to measure inclination and declination accurately. For him it is a well-established fact that "in the Boreal Hemisphere there is a distribution of inclination in accordance with latitude, whether it exists also past the equator I do not know, as to my knowledge nobody has studied these concordances" (Kircher, 1654, p. 295). At this point he accepts the idea already given by Gilbert and prints a table of magnetic inclinations for each degree of northern latitudes. He includes his own calculations and some of Gilbert's values (Kircher 1654, p. 306).

Declination is treated quite differently in Kircher's works: he accepts that its value is related with geographical position but he does not believe that it depends only on longitude because it deviates from the local meridian either eastward or westward and also because it changes values at different latitudes on the same meridian. The solution of this point is of greatest importance in geography and navigation. He also realized the errors of many of these values: "there is no doubt about declination, the only doubt is in its quantity (=value)" and a little afterward he adds "the only way to get out of this problem is to make all around the world as many reliable observations as possible" (Kircher, 1654, p. 313) and for Kircher it was easy to see that his Jesuit order, with many colleges in Europe and missionaries traveling to America and Asia, could be a reliable instrument to perform this task.

After talks with his colleagues in Rome and with the approval of his superior general, he prepared a *Geographical Consilium* (an instruction manual) about declination and inclination, what kind of instruments were needed, how to use them, and how to report the results. On the occasion of one of the regular meetings of Jesuits from around the world in Romc in 1639, Kircher collected names of Jesuits who could do the work and wrote letters to all of them.

Three long tables are the result: the first one (Table K1) "Magnetic declination on the oceans observed by Jesuits" is a long list of nearly 200 geographical points with their declination and latitude. The latitudes extend as far as from 81°0′N to 54°27′S, and the declinations from 25°0′E to 35°30′W. Table K2 "Longitudes and latitudes of several places with their declination" seems to be arranged in increasing longitude from 0°17′ to 359°40′, but longitudes from 347°25′ to 359°40′ are included between 41°10′ and 42°7′. Declinations go from 0°0' to 33°. There is no indication whether they are to the east or to the west, but the fact that longitudes from 42° to 346° are listed at the end of the table may suggest some elaboration of the data, which are not explained by Kircher. Table K3 is the list of declinations observed in Europe following Kircher's request. The names of some 57 authors and of 66 cities in Europe with their declination are mentioned in it. Values in the list extend from Lisbon to Constantinople and from Vilna to Malta. Kircher adds six other cities: Aleppo (Syria), Alexandria (Egypt), Goa and Narfinga (India), and Canton and Macau (China). Behind this third table Kircher adds the copy of a letter from Francis Brescianus, who observed in his travel westward to Canada the following declination changes: from Europe to a place near the Azores the declination was around 2° E or 3° E, at the Azores it was about 0°, from there for some 300 leagues it increased regularly up to around 21° or 22° and from there it decreased again but more quickly as one approached America. In Quebec it was 16° and further west, in the land of the Huron Indians, some 200 leagues from Quebec, it was only 13°. Kircher adds also a letter from his pupil in Rome, P. Martin s.I. with 13 declinations but their geographical positions are too vague as "in the meridian of river *Laurentij Marques*" (Kircher, 1654, p. 348). In this letter Martin suggested a method for drawing a world chart of declination. This suggestion was not followed by Kircher and the first isogonic chart of the Atlantic was not published until 1701 by Edmund Halley (Chapman and Bartels, 1940).

Unfortunately, the final result of all this observational work was never completed. As Kircher says: "When I was keeping the work, composed with no small effort, in my Museum and waiting for the right moment to publish it for the good of the Republic of Letters, it was secretly removed by one of those people who came to me almost every day from all around the world to see my Museum" (Kircher, 1654, p. 294).

This combined and centralized effort is probably the first one of that kind in magnetism and must be considered as the great contribution of Kircher to the advancement of geomagnetism. It is certainly not a milestone on the long way of progress in science but it is an important step in it.

Oriol Cardus

Bibliography

Chapman, S., and Bartels J., 1940. *Geomagnetism.* Oxford: The Clarendon Press.

Diccionario Histórico de la Compañía de Jesús, 2001. *Comillas* (4 volumes).

Kircher, A., 1654. *Magnes sive de Arte Magnetica.* 3rd edn. Rome.

Virgille P. de, 1905. Les Jesuites et l'étude du magnetisme terrestre. *Etudes*, 492–511.

Cross-references

Gilbert, William (1544–1603)
Halley, Edmond (1656–1742)
Observatories, Overview

L

LANGEL, ROBERT A. (1937–2000)

A NASA (UnitedStates) scientist who served as the project scientist for the Magsat mission. Trained as an applied mathematician and physicist, Langel was born in Pittsburgh, Pennsylvania. He was a strong advocate of high-accuracy vector magnetic measurements from satellite. These measurements were used to demonstrate the break in the power spectra at about spherical harmonic degree 13 that represents the transition from core-dominated processes to lithospheric-dominated processes (Figure L1; Langel and Estes, 1982). Tantalizing suggestions of such a break had previously been reported from total field-intensity measurements acquired on an around-the-world magnetic profile (Alldredge *et al.*, 1963). Langel also made use of the high-accuracy vector measurements, and magnetic-observatory measurements, for the coestimation of internal (core and lithosphere), external (ionosphere and magnetosphere), and induced magnetic fields. These comprehensive models of the geomagnetic field were developed in collaboration with T. Sabaka (NASA) and N. Olsen (Denmark). Earlier workers had relied on a sequential approach, beginning with the largest (core) fields. Langel also pioneered mathematical techniques for the merging of *a priori* long-wavelength lithospheric magnetic field information with shorter wavelength observations (Figure L1; Purucker *et al.*, 1998) using inverse techniques. This is necessary because the longest-wavelength lithospheric magnetic fields are inaccessible to direct observation as they overlap with the much-larger magnetic fields originating in the Earth's outer core. Significant earlier developments with forward models came from Cohen (1989) and Hahn *et al.* (1984). Langel authored or coauthored in excess of 94 peer-reviewed publications and helped train a generation of geomagnetists by virtue of his work on Magsat and his teaching at Purdue University (UnitedStates) and Copenhagen University (Denmark).

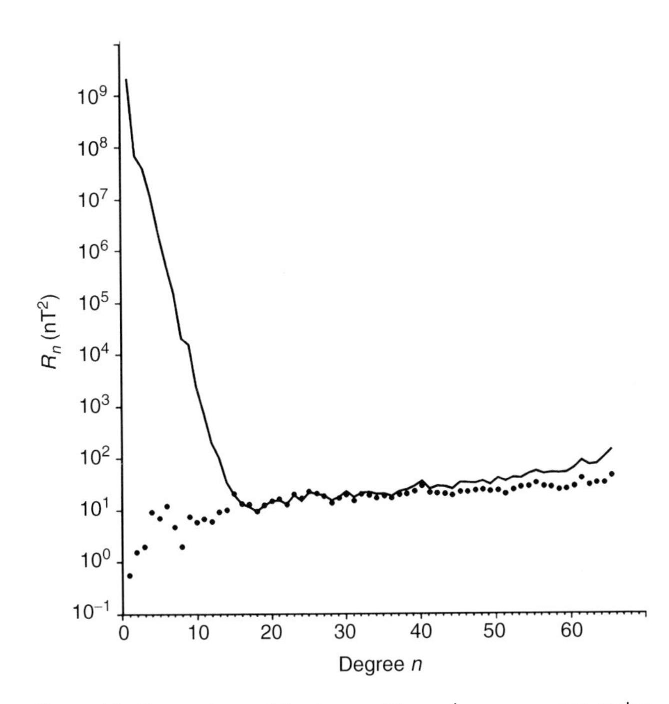

Figure L1 Comparison of the Lowes-Mauersberger spectra at the Earth's surface for a recent comprehensive model (line) of the core and lithospheric fields by Sabaka *et al.* (2004) and a lithospheric field model (symbols) by Fox Maule *et al.* (2005). R_n is the mean square amplitude of the magnetic field over a sphere produced by harmonics of degree n.

Bob was devoted to his family and church. His religiosity was well known among his colleagues, and he undertook missionary work, lead Bible study groups, and worked with teenagers.

Michael E. Purucker

Bibliography

Alldredge, L.R., Vanvoorhis, G.D., and Davis, T.M., 1963. A magnetic profile around the world. *Journal of Geophysical Research*, **68**: 3679–3692.

Cohen, Y., 1989. Traitements et interpretations de donnees spatiales in geomagnetisme: etude des variations laterales d'aimantation de la lithosphere terrestre. *Docteur es sciences physiques these,* Paris: Institut de Physique du Globe de Paris, 95 pp.

Fox Maule, C., Purucker, M., Olsen, N., and Mosegaard, K., 2005. Heat flux anomalies in Antarctica revealed by satellite magnetic data. *Science*, 10.1126/science.1106888.

Hahn, A., Ahrendt, H., Meyer, J., and Hufen, J.H., 1984. A model of magnetic sources within the Earth's crust compatible with the field measured by the satellite Magsat. *Geologischer Jahrbuch A*, **75**: 125–156.

Langel, R.A., 1987. The main geomagnetic field. In Jacobs, J. (ed.), *Geomagnetism.* New York: Academic Press, pp. 249–512.

Langel, R.A., and Estes, R.H., 1982. A geomagnetic field spectrum. *Geophysical Research Letters*, **9**: 250–253.
Langel, R.A., and Hinze, W.J., 1998. *The Magnetic Field of the Earth's Lithosphere.* Cambridge: Cambridge University Press.
Lowes, F.J., 1974. Spatial power spectrum of the main geomagnetic field, and extrapolation to the core. *Geophysical Journal of Royal Astronomical Society*, **36**: 717–730.
Purucker, M., Langel, R., Rajaram, M., and Raymond, C., 1998. Global magnetization models with a priori information. *Journal of Geophysical Research*, **103**: 2563–2584.
Sabaka, T.J., Olsen, N., and Langel, R.A., 2002. A comprehensive model of the quiet-time, near-Earth magnetic field: phase 3. *Geophysical Journal International*, **151**: 32–68.
Sabaka, T.J., Olsen, N., and Purucker, M., 2004, Extending comprehensive models of the Earth's magnetic field with Oersted and CHAMP data. *Geophysical Journal International*, **159**: 521–547.
Taylor, P., and Purucker, M., 2000. Robert A. Langel III (1937–2000). *EOS. Transactions of the American Geophysical Union*, **81**(15): 159.

Cross-references

Harmonics, Spherical
IAGA, International Association of Geomagnetism and Aeronomy
Magsat

LAPLACE'S EQUATION, UNIQUENESS OF SOLUTIONS

Potential theory lies at the heart of geomagnetic field analysis. In regions where no currents flow the magnetic field is the gradient of a potential that satisfies Laplace's equation:

$$\mathbf{B} = -\nabla V \qquad \text{(Eq. 1)}$$

$$\nabla^2 V = 0 \qquad \text{(Eq. 2)}$$

We can find the geomagnetic potential V everywhere in the source-free region by solving Eq. (2) subject to appropriate boundary conditions. Equation (1) means that we are dealing with a single scalar, the geomagnetic potential, and Eq. (2) means that we only have to know that scalar on surfaces rather than throughout the entire region. This reduces the number of measurements enormously. I shall discuss here two idealized situations appropriate to geomagnetism and paleomagnetism: measurements on a plane in Cartesian coordinates (x, y, z) with z down, continued periodically in x and y, appropriate for a small survey, and measurements on a spherical surface $r = a$ in spherical coordinates (r, θ, ϕ), where a is Earth's radius. Simple generalizations are possible to more complex surfaces.

It is well known that solutions of Laplace's equation within a volume Ω are unique when V is prescribed on the bounding surface $\partial\Omega$, the so-called *Dirichlet* conditions. This is of no use to us because we cannot measure the geomagnetic potential directly. We can, however, measure the component of magnetic field normal to the surface, which also guarantees uniqueness (the so-called *Neumann* conditions). Thus measurement of a single component of magnetic field on a surface can be used to reconstruct the magnetic field throughout the current-free region, provided the sources are entirely inside or outside the measurement surface. When both internal and external sources are present two components need to be measured on a single surface, when it is also possible to separate the parts (see *Internal external field separation*). Here I shall focus on the case of purely internal sources.

Components other than the vertical also provide uniqueness. As illustration consider the general solution of Laplace's equation in Cartesian coordinates:

$$V(x,y,z) = \sum_{k,l,m} C_m \exp(mz) \cos(kx + \ell y + \alpha_{k,\ell}), \qquad \text{(Eq. 3)}$$

where $m = \sqrt{k^2 + \ell^2}$ is positive to ensure a finite solution as $z \to -\infty$ and the $\{\alpha_{k,\ell}\}$ are phase factors. If $B_z = \partial V/\partial z$ is known on $z = 0$ we can differentiate Eq. (3) with respect to z, set $z = 0$, and determine the arbitrary coefficients $\{C_{k,\ell}, \alpha_{k,\ell}\}$ from the usual rules for Fourier series. This is the case of Neumann boundary conditions. Of course, solutions of Eq. (3) are periodic in x and y with periods $2\pi/k$ and $2\pi/l$, respectively. B_x and B_y, however, do not give unique solutions. Differentiating Eq. (3) with respect to x and putting $z = 0$, for example, allows determination of all the Fourier coefficients except those with $k = 0$. The ambiguity in the solution takes the form $V_0(y,z) = \sum_l C_l \exp(-lz) \cos(ly + \alpha_l)$. Similarly, measurement only of B_y leaves an ambiguity independent of y. Knowing B_x and B_y overdetermines the problem.

The spherical case is slightly different. B_r ($-Z$ in conventional geomagnetic notation) provides Neumann boundary conditions and so will give a unique solution. B_ϕ (east) allows an axisymmetric ambiguity similar to that for the cartesian B_x and B_y cases. Surprisingly, $-B_\theta$ (north) gives a unique solution; the condition that V be finite at the poles provides the necessary constraint to remove the arbitrary parts of the solution. Proof involves the general solution of Laplace's equation as a spherical harmonic expansion (see *Main field modeling*) in similar fashion to the use of the Fourier series solution in the cartesian case (see *Magnetic anomalies, modeling*). More elegant formal proofs of these standard results making use of Green's theorem are to be found in texts on potential theory (e.g., Kellogg (1953)).

Formal mathematical results dictate the type of measurements we must make in order to obtain meaningful estimates of the geomagnetic field in the region around the observation surface. Clearly in the real situation other complications arise from finite resolution and errors, but with a uniqueness theorem we can at least be confident that the derived field will converge toward the correct field as the data are improved; for this to happen we must measure the right components of magnetic field, but this is not always possible. Instrumentation has changed over the years, dictating changes in the type of measurement available. Table L1 gives a summary of the bias in available datasets.

Backus (1970) addressed the problem of uniqueness when only F is known on the surface. He showed that, for certain magnetic field configurations, there exist pairs of magnetic fields $\mathbf{B}_1$ and $\mathbf{B}_2$, which satisfy $\nabla \cdot \mathbf{B} = 0$ and $\nabla \times \mathbf{B} = 0$ outside $\partial\Omega$ and such that $|\mathbf{B}_1| = |\mathbf{B}_2|$ everywhere on $\partial\Omega$, but $|\mathbf{B}_1| \neq |\mathbf{B}_2|$ elsewhere. This result led to the decision to make vector measurements on the MAGSAT satellite. In practice $\mathbf{B}$ obtained from intensity measurements alone can differ by 1000 nT from the correct vector measurements. The difference takes the form of a spherical harmonic series with terms dominated by $l = m$, now called the Backus ambiguity or series. Additional information does provide uniqueness (e.g., locating the dip equator (Khokhlov *et al.*, 1997)).

Backus (1968, 1970) gives details of the proof in both two and three dimensions. The 2D proof uses complex variable theory and is particularly simple and illuminating. We use the complex potential $w(z)$, where $z = x + iy$. The field is given by the real and imaginary parts of dw/dz, and $|dw/dz|$ is known on the bounding circle. Focus attention on

$$\zeta(z) = i \log dw/dz = \xi + i\eta = -\arg|dw/dz| + i \log|dw/dz|. \qquad \text{(Eq. 4)}$$

Both ξ and η satisfy Laplace's equation as the real and imaginary parts of an analytic function, except where $dw/dz = 0$. When such singularities exist, taking a circuit enclosing the source region gives an increase of 2π in the argument of the logarithm. This defines the ambiguity. The number of singularities determines the number of different solutions; singularities exist if the field has more than two dip poles on the surface.

Paleomagnetic and older geomagnetic data sets are dominated by directional measurements. Early models made the reasonable but

Table L1 Datasets are dominated by the components shown for each epoch

Date	Dominant component	Comment
–1700	D	Compass only, I from 1586 restricted to Europe
1700–1840	D, I	No absolute intensity until *Gauss'* method (*q.v.*)
1840–1955	D, I, F	Magnetic surveys generally include all components
1955–1980	F	Proton magnetometer gives absolute F, satellites not orientated
1980–	X, Y, Z	MAGSAT starts global vector measurements
Paleomagnetism	D, I	Paleointensity is a difficult and inaccurate measurement
Archeomagnetism	F	Sample orientation usually not available
Borehole samples	I, F	Declination usually not logged in boreholes

Note: Relative horizontal intensities were measured by *von Humboldt* (*q.v.*) before 1840, but in a global model these give no more information than does direction.

Table L2 Summary of uniqueness results

Component	Ambiguity	Comment
Z (down)	Unique	Neumann boundary condition
X (north)	Unique	
Y (east)	Any axisymmetric solution	$Y = 0$ for any axisymmetric field
F (total intensity)	Usually nonunique	Backus effect
F + location of dip equator	Unique	
D, I (direction)	Arbitrary multiplicative constant	n constants in general
D + partial H (horizontal)	Unique	H needed on line joining dip poles

Note: $2n$ is the number of dip poles on the measurement surface.

unproven assumption that direction determined the magnetic field up to a single multiplicative factor. Proctor and Gubbins (1990) found a result in 2D using Backus' method above, focussing instead on the potential

$$\zeta(z) = \log dw/dz = \log|dw/dz| + i\arg|dw/dz|. \quad \text{(Eq. 5)}$$

The above argument then leads to a general solution with one arbitrary multiplicative constant plus $n - 1$ additional arbitrary constants, where $2n$ is the number of dip poles.

As with the intensity problem, extension of the 2D proof into 3D is not straightforward. Proctor and Gubbins (1990) conjectured a similar result in 3D and provided an axisymmetric example. They also gave a practical recipe for determining the ambiguity using standard methods of inverse theory. The problem is homogeneous in the sense that, if $\mathbf{B}_1$ and $\mathbf{B}_2$ are fields derived from harmonic potentials and are parallel on $\partial\Omega$, then any linear combination of the form $a_1\mathbf{B}_1 + a_2\mathbf{B}_2$ is also a solution. Thus, having found one solution, $\mathbf{B}_1$ say, a linear error analysis around $\mathbf{B}_1$ will yield a direction in solution space along $\mathbf{B}_1 - \mathbf{B}_2$ that defines an annihilator for the problem. Homogeneity then guarantees that this linearized analysis will be valid globally. Application of this procedure to the axisymmetric example did yield both solutions, but only when a very large number of spherical harmonics were used in their representation, confirming Backus' view that uniqueness demonstrations on solutions restricted to finite spherical harmonic series could be misleading.

Hulot *et al.* (1997) finally found a formal proof in 3D using the homogeneity of the problem. The key is to show that the solution space is linear; the proof follows using relatively recent general results from potential theory. They show that, if n is the number of loci where the field is known to be either zero or normal to the surface then the dimension of the solution space (number of independent arbitrary constants) is $n - 1$. This result holds whether sources are outside or inside the surface. For the related problem of gravity, which allows monopoles, the dimension is n.

It is rather easy to prove that directional measurements throughout a volume provide uniqueness to within a single arbitrary constant (Bloxham, 1985). Given one solution $\mathbf{B}$ others must have the form $\alpha\mathbf{B}$ and must be solenoidal and curl-free:

$$\nabla \cdot (\alpha\mathbf{B}) = \mathbf{B} \cdot \nabla\alpha = 0 \quad \text{(Eq. 6)}$$

$$\nabla \times (\alpha\mathbf{B}) = -\mathbf{B} \times \nabla\alpha = 0. \quad \text{(Eq. 7)}$$

Thus $\nabla\alpha = 0$ and α must be a constant. Knowing the field everywhere inside the volume of measurement enables us to construct the field component normal to the boundary and solve Laplace's equation outside the volume of measurement, all within a single-multiplicative constant.

The earliest historical data sets are dominated by declination. In this case we know the direction of the horizontal field and location of the dip poles. Suppose the potential field $B_r + \alpha\mathbf{B}_h$ fits the data; $\mathbf{B}_h$ is known but B_r and α are unknown. The radial component of $\nabla \times \mathbf{B}$ does not involve B_r or $\partial/\partial r$, so $\nabla_h \times (\alpha\mathbf{B}_h) = 0$ and

$$\nabla\alpha \times \mathbf{B}_h = 0. \quad \text{(Eq. 8)}$$

The unknown α is therefore invariant along contours of $\mathbf{B}_h$; measurement of horizontal intensity along any line joining the dip poles is therefore sufficient to guarantee uniqueness of $\mathbf{B}_h$, which in turn would allow us to determine the potential and $\mathbf{B}$. The coverage of additional data required is therefore reduced enormously from the whole surface to a line. Unfortunately horizontal intensity is not available for the early historical epoch, and paleoinclinations are too inaccurate to be of any use (Hutcheson and Gubbins, 1990), but inclination did become available from the early 18th century on voyages from Europe round Cape Horn and through the Magellan Straits into the Pacific, which represents a fair approximation to the ideal case of horizontal intensity between the north and south dip poles.

Uniqueness results are summarized in Table L2. Practical aspects of the problem are discussed further by Lowes *et al.* (1995).

David Gubbins

Bibliography

Backus, G.E., 1968. Application of a nonlinear boundary value problem for Laplace's equation to gravity and geomagnetic intensity surveys. *Quarterly Journal of Mechanics and Applied Mathematics*, **21**: 195–221.

Backus, G.E., 1970. Nonuniqueness of the external geomagnetic field determined by surface intensity measurements. *Journal of Geophysical Research*, **75**: 6339–6341.

Bloxham, J., 1985. Geomagnetic secular variation. PhD thesis, Cambridge University, Cambridge.

Hulot, G., Khokhlov, A., and Mouël, J.L.L., 1997. Uniqueness of mainly dipolar magnetic fields recovered from directional data. *Geophysical Journal International*, **129**: 347–354.

Hutcheson, K., and Gubbins, D., 1990. A model of the geomagnetic field for the 17th century. *Journal of Geophysical Research*, **95**: 10,769–10,781.

Kellogg, O.D., 1953. *Foundations of Potential Theory.* New York: Dover.

Khokhlov, A., Hulot, G., and Mouël, J.L.L., 1997. On the Backus effect—I. *Geophysical Journal International*, **130**: 701–703.

Lowes, F.J., Santis, A.D., and Duka, B., 1995. A discussion of the uniqueness of a Laplacian potential when given only partial field information on a sphere. *Geophysical Journal International*, **121**: 579–584.

Proctor, M.R.E., and Gubbins, D., 1990. Analysis of geomagnetic directional data. *Geophysical Journal International*, **100**: 69–77.

Cross-references

Gauss, Carl Friedrich (1777–1855)
Internal External Field Separation
Magnetic Anomalies, Modeling
Magsat
Main Field Modeling

LARMOR, JOSEPH (1857–1942)

Joseph Larmor is known in geomagnetism for the precession frequency of the proton that now bears his name and his seminal paper of 1919 proposing the dynamo theory for the first time. His main purpose was to explain the solar magnetic field, but he also had the geomagnetic field in mind.

Born in 1857 in Magherhall, County Antrim, and educated at the Royal Belfast Academical Institution and Queen's College, Belfast, Larmor showed an early ability in mathematics and classics. He was senior wrangler in the Mathematics Tripos at Cambridge, beating J.J. Thomson to second place. He succeeded George Stokes as Lucasian Professor of Mathematics in 1903, and was knighted in 1909.

Larmor worked at the dawn of modern physics and was concerned with the aether and with unifying electricity and matter. His greatest work, *Aether and Matter*, was revolutionary and inspiring. His contributions are often compared with those of Lorentz. His discovery of what we now call *Larmor precession* was incidental: he showed that in a magnetic field, an electron orbit precesses with angular velocity proportional to the magnetic field strength while remaining unchanged in form and inclination to the magnetic field. The same principle applies to a particle, such as a proton, that possesses a magnetic moment, and the frequency of precession of a proton in an applied magnetic field is called the *Larmor frequency*. The proton magnetometer, developed 30 years after his death, determines the absolute intensity of a magnetic field by measuring the precession frequency.

Later in his career he contributed to many of the geophysical problems of the day, publishing on the Earth's precession and irregular axial motion, and on the effects of viscosity on free precession, which Harold Jeffreys later used in his argument against mantle convection. Larmor published on the origin of sunspots, which G.E. Hale had shown were associated with strong magnetic fields, and became a leading authority on geomagnetism. This was a time when theories of the origin of the magnetic fields of the Earth and of the Sun abounded. His 1919 paper, presented at the British Association for the Advancement of Science meeting in Bournemouth (Larmor, 1919) is only two pages long; he dismisses permanent magnetization and convection of electric charge as candidates for both bodies. For the Sun he considers three theories: the dynamo theory, electric polarization by gravity or centrifugal force, and intrinsic polarization of crystals. He dismisses the last two in the case of the Earth because of the secular variation, arguing that they predict proportionality between the magnetic field and spin rate in contrast to the observation of a changing dipole moment and constant spin rate. His version of the dynamo theory is quite explicit and requires fluid motion in meridian planes. It was this very specific theory that was tested by Cowling (1934) in establishing his famous antidynamo theorem (see *Cowling's theorem*).

According to Eddington (1942–1944) he was a retiring and complicated man, refusing a celebration in his honor organized by Cambridge University. His lectures were "ill-ordered and obscure but... even the examination-obsessed student could perceive that here he was coming to an advanced post of thought, which made all his previous teaching seem behind the times." He was one of only three lecturers *E. C. Bullard* (*q.v.*) bothered with during his undergraduate days. Darrigol (2000) says "Larmor's physics was freer and broader than conceptual rigor and practical efficiency commanded." In similar vein, Bullard once described him to me as a lazy man, which I took to refer to his rather casual outline of the dynamo theory and failure to follow it up with any mathematical analysis, of which he was eminently capable, leaving Cowling to pick it up a decade later. But how right he was! How much better to take the credit for a correct theory and avoid the disappointments of the next half-century, from the antidynamo theorems to the numerical failure of Bullard and Gellman's attempt at solution, and the graft of the subsequent 30 years of mathematical and computational effort it has taken to put the dynamo theory onto a secure footing!

David Gubbins

Bibliography

Cowling, T.G., 1934. The magnetic field of sunspots. *Monthly Notices of Royal Astronomical Society*, **94**: 39–48.

Darrigol, O., 2000. *Electrodynamics from Ampère to Einstein.* Oxford: Oxford University Press.

Eddington, A.S., 1942–1944. Joseph Larmor. *Obituary Notices of Fellows of the Royal Society*, **4**: 197–207.

Larmor, J., 1919. How could a rotating body such as the Sun become a magnet?. *Reports of the British Association*, **87**: 159–160.

Cross-references

Bullard, Edward Crisp (1907–1980)
Cowling, Thomas George (1906–1990)
Cowling's Theorem
Dynamo, Bullard-Gellman
Magnetometers, Laboratory
Nondynamo Theories

LEHMANN, INGE (1888–1993)

In 1936, Miss Inge Lehmann (Figure L2) proposed a new seismic shell of radius 1400 km at the center of the Earth, which she called the "inner core." It was already known that low seismic velocities in the outer core created a shadow at angular distances between 105° and 142° and that seismic waves diffracted some considerable distance into the shadow. Wiechert also attributed seismic arrivals before 142° to diffraction, but Gutenberg and Richter (1934) had already found their

Figure L2 Inge Lehmann (Reprinted with permission from the Royal Society).

amplitude too large and frequency too high to fit with the diffraction theory. Measurements of the vertical component of motion clinched the interpretation as waves reflected or refracted through a large angle. Lehmann's paper remains a classic example of lateral thinking, careful interpretation, and cautious conclusions. The proposed radius was close to the currently accepted one of 1215 km.

The inner core was rapidly accepted into the new Earth models being constructed by Gutenberg and Richter (1938) in the United States and Jeffreys (1939) in the U.K., although the scattered nature of the arrivals in the shadow zone demanded a thick transition zone at the bottom of the outer core, where the seismic velocity increased gradually. In the first edition of his book on seismology, Bullen (1947) labeled the Earth's layers A-G, with A as the crust and G the inner core, with F denoting the inner core transition. The need for this layer was largely removed when seismic arrays showed that much of the scattered energy came not from the inner-core boundary but from the core-mantle boundary (Haddon and Cleary, 1974). The inner core is vital in the theory of geomagnetism because its slow accretion, a consequence of the Earth's secular cooling, powers the geodynamo and influences core convection (see *Inner core tangent cylinder*). The F-layer lives on in name at least, despite the removal of seismological evidence, in the form of the required boundary layer between the inner and outer cores.

Although the inner core was probably the most spectacular of Lehmann's discoveries, she enjoyed a long and distinguished career in seismology, remaining active throughout her long retirement. Born in Copenhagen into a distinguished family, she benefited from an exceptionally enlightened early education and read mathematics at the University of Copenhagen from 1908. A spell in Cambridge in 1910 came as a shock because of the restrictions placed on women there (Cambridge did not grant women degrees until much later).

Lehmann's career as a seismologist began in 1925 when she was appointed assistant to Professor N.E. Norland, who was engaged in establishing a network of seismographs in Denmark. She met Beno Gutenberg at this time. In 1928 she was appointed chief of the seismological department of the Royal Danish Geodetic Institute, a post she held until retirement in 1953. In retirement she enjoyed frequent visits to Lamont and other institutions in the United States and Canada. It seems her international reputation was strong before her work was recognized at home in Denmark. She was elected Foreign Member of the Royal Society in 1969, was awarded the Bowie Medal (the American Geophysical Union highest honor) in 1971, and now has an AGU medal named in her honor.

Further details of her life may be found in Bolt (1997).

David Gubbins

Bibliography

Bolt, B.A., 1997. Inge Lehmann. *Biographical Memoirs of Fellows of the Royal Society*, **43**: 28285–28301.

Bullen, K.E., 1947. *An Introduction to the Theory of Seismology.* London: Cambridge University Press.

Gutenberg, B., and Richter, C.F., 1938. P' and the Earth's core. *Monthly Notices of Royal Astronomical Society, Geophysical Supplement*, **4**: 363–372.

Haddon, R.A.W., and Cleary, J.R., 1974. Evidence for scattering of seismic PKP waves near the mantle-core boundary. *Physics of the Earth and Planetary Interiors*, **8**: 211–234.

Jeffreys, H., 1939. The times of the core waves. *Monthly Notices of Royal Astronomical Society, Geophysical Supplement*, **4**: 548–561.

Lehmann, I., 1936. *Publication's Bureau Centrale Seismologique Internationale Series A*, **14**: 87–115.

Cross-reference

Inner Core Tangent Cylinder

LENGTH OF DAY VARIATIONS, DECADAL

It is a truism to say that the Earth rotates once per day. However, the length of each day is not constant, but varies over timescales from everyday to millions of years. Perhaps the best-known variation is the gradual slowing of the rotation rate due to the tidal interaction between the Earth and the moon (see *Length of day variations, long-term*). Historical observations are well fit by an increase in length of day (LOD) of about 1.4 ms per century, most clearly seen in the timing of eclipses recorded by ancient civilizations. (Note that there are 86400 seconds in a day, so these millisecond changes are of order 1 part in 10^8 of the basic signal.) Geological evidence also supports increase in LOD over Earth history (Williams, 2000). At the other end of the scale, modern measurements from the global positioning system (GPS) or very-long baseline interferometry (VLBI) demonstrate a signal on yearly and sub-yearly timescales, again of a few milliseconds in magnitude. These variations are dominated by angular momentum exchange between the atmosphere and the solid Earth; put crudely, winds blowing against mountain topography (particularly north-south ranges such as the Rockies in America) push the solid Earth, speeding up or slowing down the rate of rotation. Atmospheric angular momentum calculated from models of the global circulation can explain almost all the short-period signals. The small remaining residual (of amplitude ~0.1 ms) shows strong coherence with estimates of oceanic angular momentum (Marcus *et al.*, 1998), and the hydrological cycle is also probably important, by causing small variations in the Earth's moment of inertia. However, surface observations are unable to explain a decadal fluctuation, again of a few milliseconds in LOD. This signal was first recognized in the early 1950s, and because the core is the only fluid reservoir in the Earth capable of taking up the requisite amount of angular momentum and due to its correlation with various geomagnetic data (in particular records of magnetic intensity and declination at certain magnetic observatories), it was assumed to result from exchange of angular momentum between the fluid core and solid mantle. More robust evidence was provided by Vestine (1953), who demonstrated a good correlation between the change in length of day (ΔLOD) signal and core angular momentum (CAM) estimated from observations of *westward drift* (*q.v.*).

The connection between core processes and ΔLOD was put on a rigorous footing by Jault *et al.* (1988). They realized that models of surface core flow reflect motions throughout the core and can be used to estimate changes in angular momentum of the core, matching changes in angular momentum of the solid Earth required to explain the observed decadal ΔLOD. Most processes in the core have characteristic timescales which are either of centuries or longer (for example the planetary waves that have been proposed as an explanation for *westward drift* (*q.v.*)), or which are short, of order days (*Alfvén waves* (*q.v.*) or inertial waves). *Torsional oscillations* (*q.v.*), however, are thought to have characteristic periods of decades. These are solid-body rotations of fluid cylinders concentric with the Earth's spin axis, about that axis. Such motions carry angular momentum about the spin axis, and extend to the core surface. Therefore, changes in CAM can be calculated using surface flow models. Such a calculation ignores the inner core, and assumes uniform density for the liquid core; however, the value calculated is not sensitive to these assumptions. Most commonly, a surface flow is used that is assumed to be tangentially geostrophic (the force balance in the horizontal direction is assumed to be between pressure gradients and Coriolis force). The core surface flow is expanded in a poloidal-toroidal decomposition in spherical coordinates (r, θ, φ) (r is radius, θ is colatitude, and φ is longitude), such that

$$\mathbf{v} = \sum_{l=1}^{\infty}\sum_{m=0}^{l} \nabla_{\mathrm{H}}\left(r s_l^m Y_l^m\right) + \nabla_{\mathrm{H}} \times \left(\mathbf{r} t_l^m Y_l^m\right) \quad \text{(Eq. 1)}$$

where ∇_{H} is the horizontal part of the gradient operator, and s_l^m and t_l^m are poloidal and toroidal scalars, respectively, corresponding to surface *spherical harmonics* (*q.v.*) $Y_l^m(\theta, \varphi)$ of degree l and order m. Then, core angular momentum depends only on two large-scale toroidal, zonally symmetric, surface flow harmonics. Using standard parameters for the Earth, the relation

$$\delta T = 1.138\left(\delta t_1^0 + \frac{12}{7}\delta t_3^0\right) \quad \text{(Eq. 2)}$$

is obtained, where δT is the change in LOD in milliseconds and the changes in the toroidal flow coefficients are measured in km y^{-1}. The t_1^0 harmonic corresponds to a uniform rotation of the core surface about the Earth's rotation axis. Figure L3 shows a calculation of Jackson *et al.* (1993) using the ufm *time-dependent model of the main field* (*q.v.*) to calculate the predicted LOD variations, compared with observations of McCarthy and Babcock (1986). Despite the highly idealized theory, the agreement is striking, especially over the last century.

The correlation post-1900 between LOD and calculated core-angular momentum is remarkably robust, being evident for many different strengths of flow (explaining more or less of the secular variation) and for different physical assumptions about the flow (for example, assuming that the flow is toroidal, or steady but within a drifting reference frame). By 1900 the network of *magnetic observatories* (*q.v.*) was globally distributed; these stations provide by far the most sensitive and accurate measurements of the global magnetic secular variation from which the flows are calculated. Before 1900, as more observatory data become available, the time variation of the field models reflects changes in data quality as well as any physical processes; these changes map into the flow models, giving unphysical results and poorer correlation with observed ΔLOD. There is some evidence that this problem can be overcome by calculating simpler flow models; a notably better correlation is provided by the original analysis (Vestine, 1953) of westward drift!

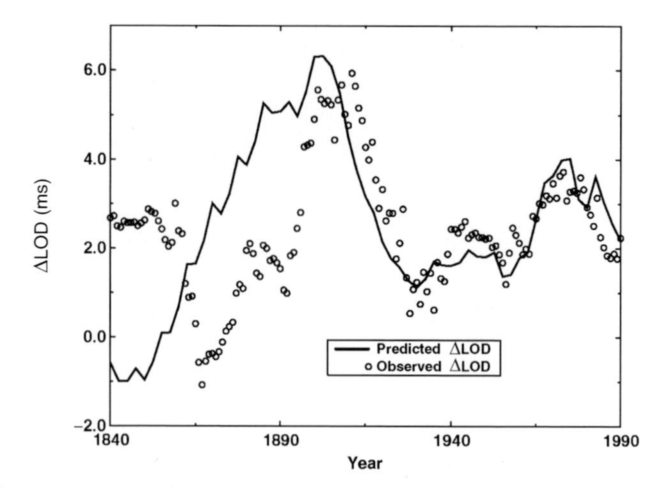

Figure L3 Comparison of observed decadal variations in length of day LOD (with a long-term trend of 1.4 ms/century removed) with the prediction from a model of surface core flow (following Jackson *et al.*, 1993).

More recent work has considered whether the core could have a role to play in angular momentum exchange at shorter periods. It has been argued that the core may well change the phase of the observed LOD signal compared with forcing from the atmosphere and oceans, but there is no strong evidence in subdecadal ΔLOD of a signal from the core.

The study of LOD variations as described here is essentially phenomenological—there is no need to assume any particular mechanism for the exchange of angular momentum between the core and the mantle. How the exchange of angular momentum may be effected, and the influence of the dynamics of the *torsional oscillations* (*q.v.*), are discussed elsewhere.

Richard Holme

Bibliography

Jackson, A., Bloxham, J., and Gubbins, D., 1993. Time-dependent flow at the core surface and conservation of angular momentum in the coupled core-mantle system. In Le Mouël, J.-L., Smylie, D.E., and Herring T. (eds.), *Dynamics of the Earth's Deep Interior and Earth Rotation*, pp. 97–107, AGU/IUGG.

Jault, D., Gire, C., and Le Mouël, J.L., 1988. Westward drift, core motions, and exchanges of angular momentum between core and mantle. *Nature*, **333**: 353–356.

Marcus, S.L., Chao, Y., Dickey, J.O., and Gregout, P., 1998. Detection and modeling of nontidal oceanic effects on Earth's rotation rate. *Science*, **281**: 1656–1659.

Vestine, E.H., 1953. On variations of the geomagnetic field, fluid motions and the rate of the Earths rotation. *Journal of Geophysical Research*, **38**: 37–59.

Williams, G.E., 2000. Geological constraints on the Precambrian history of earth's rotation and the moon's orbit. *Reviews of Geophysics*, **38**: 37–59.

Cross-references

Alfvén Waves
Core-Mantle Coupling, Electromagnetic
Core-Mantle Coupling, Topographic
Harmonics, Spherical
Length of Day Variations, Long-Term
Observatories, Overview
Oscillations, Torsional
Time-Dependent Models of the Geomagnetic Field
Westward Drift

LENGTH OF DAY VARIATIONS, LONG-TERM

Tidal friction

The Moon raises tides in the oceans and solid body of the Earth. The Sun does likewise, but to a lesser extent. Let us first consider the action of the Moon. Due to the anelastic response of the Earth's tides, the tidal bulges can be thought of as being carried ahead of the sublunar point by the Earth's rotation. This misalignment of the tidal bulges produces a torque on the Moon, which increases its orbital angular momentum and drives it away from the Earth, while reducing the Earth's rotational angular momentum by the opposite amount. The rate of rotation of the Earth is reduced through the action of tidal friction, which occurs very largely in the oceans. By analogy, the solar tides make a smaller contribution to slowing down the Earth, but the reciprocal effect on the Sun is negligible. As the Earth slows down the length of the day (LOD) increases. The change in the rate of rotation of the Earth under this mechanism is usually termed its tidal acceleration. Beside this long-term change, there are relative changes between the rotation of the mantle and the core on shorter timescales. The transfer of angular momentum, which produces these changes, is caused by core-mantle coupling and the astronomical results discussed here shed light on the timescale of the geomagnetic processes involved.

The tidal acceleration of the Earth has been measured reliably in the following way. Analysis of the perturbations of near-Earth satellites produced by lunar and solar tides, together with the requirement that angular momentum be conserved in the Earth-Moon system, leads to an empirical relation between the retardation of the Earth's spin and the observed tidal acceleration of the Moon (see, for example, Christodoulidis *et al.* (1988)). Lunar laser ranging gives an accurate value for the Moon's tidal acceleration and inserting this in the relation gives the result $-6.1 \pm 0.4 \times 10^{-22}$ rad s^{-2} for the total tidal acceleration of the Earth (Stephenson and Morrison, 1995). This result is equivalent to a rate of increase in the LOD of $+2.3 \pm 0.1$ ms per century (ms/100 y). For the interconversion of units see Table 2 of Stephenson and Morrison (1984). These satellite and lunar laser ranging measurements were obtained from data collected over the past 30 years or so. However, they can be applied to the past few millennia because the mechanism of tidal friction has not changed significantly during this period (see, for example, Lambeck, 1980, Section 10.5).

While the tidal component of the Earth's acceleration can be derived from recent high-precision observations, the actual long-term acceleration, which is the sum of the tidal and nontidal components, cannot be measured directly because it is masked by the relatively large decade fluctuations. Instead, observations from the historical past, albeit crude by modern standards, have to be used. By far the most accurate data for measuring the Earth's rotation before the advent of telescopic observations (ca. A.D. 1620) are records of eclipses. Useful records of eclipses extend back to about 700 B.C.

Eclipses and the Earth's rotation

Early eclipse observations fall into two main independent categories: untimed reports of total solar eclipses; and timed measurements of solar and lunar eclipse contacts. The tracks of total solar eclipses on the Earth's surface are narrow and distinct. The retrospective calculation of their occurrence is made by running back the Sun and Moon in time along their apparent orbits around the Earth, the Moon's motion having been corrected for the lunar tidal interaction with the Earth. Unless special provision is made, this computation presupposes the uniform progression of time and hence the uniform rotation of the Earth on its axis. For each observation, the displacement in longitude between the computed position of the track of totality and the actual observed track measures the cumulative correction to the Earth's rotational phase due to variations in its rate of rotation over the intervening period. This displacement in degrees, divided by 15, gives the correction in hours to the Earth's "clock," usually designated by ΔT.

Stephenson and Morrison (1995) analyzed all the available reliable records of solar and lunar eclipses during the period 700 B.C. to A.D. 1600 from ancient Babylon, China, Arab, and European sources in order to measure ΔT at epochs in the past. For full translations of the various historical records see Stephenson (1997). The untimed data consist of unambiguous descriptions of total solar eclipses from known places on particular dates, but without timing. The measurement of time is not necessary because of the narrowness of the belt of totality (a few minutes of time) relative to the quantity being measured (ΔT), which amounts to several hours in the era B.C. The timed data, on the other hand, include a timing of the occurrence of one or more of the contacts during solar and lunar eclipses.

Their results for ΔT from the untimed data are shown in Figure L4. Some less-critical data have been omitted here from the figure for the sake of clarity. A single parabola does not adequately represent all the data, so Stephenson and Morrison fitted a smooth curve to the data using cubic splines, with economical use of knots in order to follow the degree of smoothness in the observed record after A.D. 1600. The details of the fitting procedure can be found in their 1995 paper. Their results from the independent timed data are very close to the untimed data, and are not reproduced here.

Long-term accelerations in rotation

The parabola in Figure L4 corresponds physically to an acceleration of $-4.8 \pm 0.2 \times 10^{-22}$ rad s^{-2} in the rotation of the Earth. This is significantly less algebraic than the acceleration expected on the basis of tidal friction alone, $-6.1 \pm 0.4 \times 10^{-22}$ rad s^{-2}. Some other mechanism, or mechanisms, must account for the accelerative component of $+1.3 \pm 0.4 \times 10^{-22}$ rad s^{-2}. This nontidal acceleration may be associated with the rate of change in the Earth's oblateness attributed to viscous rebound of the solid Earth from the decrease in load on the polar caps following the last deglaciation about 10000 years ago (Peltier and Wu, 1983; Pirazzoli, 1991). From an analysis of the acceleration of the motion of the node of near-Earth satellites, a present-day fractional

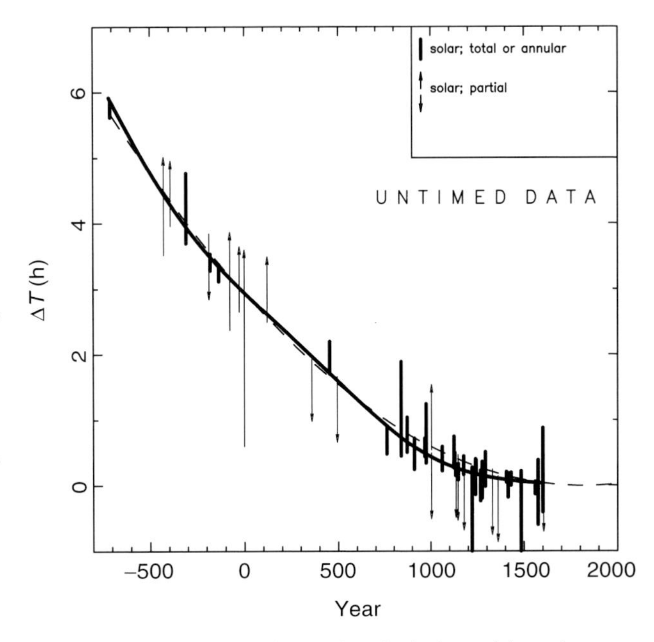

Figure L4 Corrections to the Earth's clock derived from the differences between the observed and computed paths of (untimed) solar eclipses. The vertical lines are not error bars, but solution space, anywhere in which the actual value is equally likely to lie. Arrowheads denote that the solution space extends several hours in that direction. The solid line has been fitted by cubic splines. The dashed line is the best-fitting parabola.

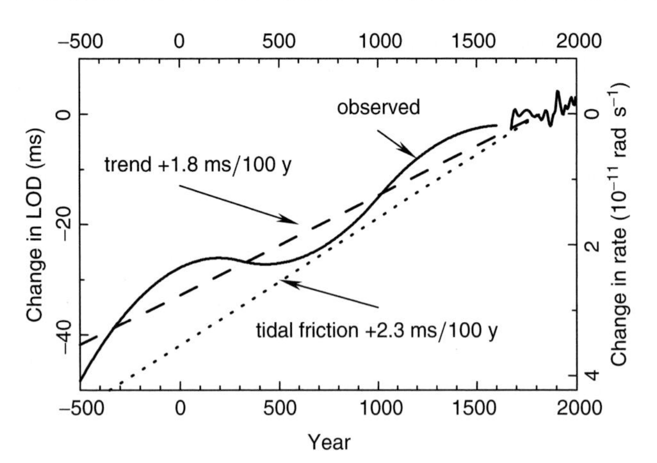

Figure L5 Changes in LOD –500 to +2000 obtained by taking the first time-derivative along the curves in Figure L4. The change due to tidal friction is +2.3 ms/100 y. The high-frequency changes after +1700 are taken from Stephenson and Morrison (1984).

rate of change of the Earth's second zonal harmonic J_2 of $-2.5 \pm 0.3 \times 10^{-11}$ per year has been derived, which implies an acceleration in the Earth's rotation of $+1.2 \pm 0.1 \times 10^{-22}$ rad s^{-2}. This is consistent to within the errors of measurement with the result from eclipses, assuming an exponential rate of decay of J_2 with a relaxation time of not less than 4000 years. Other causes may include long-term core-mantle coupling and small variations in sea level associated with climate changes.

Long-term changes in the length of the day

The first time-derivative along the curves in Figure L4 gives the change in the rate of rotation of the Earth from the adopted standard. This standard is equivalent to a LOD of 86400 SI s. The change in the LOD is plotted in Figure L5. The wavy line is the derivative along the cubic splines in the period 500 B.C. to A.D. 1600. After A.D. 1600, the higher resolution afforded by telescopic observations reveals the decade fluctuations (see *Length of day variations, decadal*), which are usually attributed to core-mantle coupling (see *Core-mantle coupling*) (see Ponsar *et al.*, 2003). Similar fluctuations are undoubtedly present throughout the whole historical record, but the accuracy of the data is not capable of resolving them. All that can be resolved is a long-term oscillation with a period of about 1500 years and amplitude comparable to the decade fluctuations. The observed acceleration of $-4.8 \pm 0.2 \times 10^{-22}$ rad s^{-2} in Figure L4 is equivalent to a trend of $+1.8 \pm 0.1$ ms/100 y in the LOD. In our considered opinion, no reliable eclipse data suitable for LOD determinations are available before ca. 700 B.C.

Paleorotation

The rotation of the Earth about 400 Ma ago can, in principle, be measured from the seasonal variations in growth patterns of fossilized corals and bivalves. This leads to estimates of the number of days in the year at that time, and hence the LOD (the year changes very little). Lambeck (1980) reviews the data and arrives at the value $-5.2 \pm 0.2 \times 10^{-22}$ rad s^{-2} (= $+1.9 \pm 0.1$ ms/100 y in LOD) for the tidal acceleration over the past 4×10^8 years, which implies that tidal friction was less in the past.

L.V. Morrison and F.R. Stephenson

Bibliography

Christodoulidis, D.C., Smith, D.E., Williamson, R.G., and Klosko, S.M., 1988. Observed tidal braking in the Earth/Moon/Sun system. *Journal of Geophysical Research*, **93**: 6216–6236.

Lambeck, K., 1980. *The Earth's Variable Rotation.* Cambridge UK: Cambridge University Press.

Peltier, W.R., and Wu, P., 1983. History of the Earth's rotation. *Geophysical Research Letters*, **10**: 181–184.

Pirazzoli, P.A., 1991. In Sabadini, R., Lambeck, K., and Boschi, E. (eds.), Glacial Isostacy, Sea-level and Mantle Rheology. Dordrecht: Kluwer, pp. 259–270.

Ponsar, S., Dehant, V., Holme, R., Jault, D., Pais, A., and Van Hoolst, T., 2003. The core and fluctuations in the Earth's Rotation. In Dehant, V., Creager, K.C., Karato, S., and Zatman, S. (eds.), *Earth's Core: Dynamics, Structure, Rotation* Vol 31. Washington, DC: American Geophysical Union Geodynamics Series, pp. 251–261.

Stephenson, F.R., 1997. *Historical Eclipses and Earth's Rotation.* Cambridge UK: Cambridge University Press.

Stephenson, F.R., and Morrison, L.V., 1984. Long-term changes in the rotation of the Earth: 700 B.C. to A.D. 1980. In Hide, R. (ed.), *Rotation in the Solar System.* The Royal Society, London. Reprinted in *Philosophical Transactions of the Royal Society of London, Series A*, **313**: 47–70.

Stephenson, F.R., and Morrison, L.V., 1995. Long-term fluctuations in the Earth's rotation: 700 B.C. to A.D 1990. *Philosophical Transactions of the Royal Society of London, Series A*, **351**: 165–202.

Cross-references

Core-Mantle Coupling-Electromagnetic
Geomagnetic Spectrum, Temporal
Halley, Edmond (1656–1742)
Length of Day Variations, Decadal
Mantle, Electrical Conductivity, Mineralogy
Westward Drift

LLOYD, HUMPHREY (1808–1881)

Humphrey Lloyd (Figure L6) was a distinguished and influential experimental physicist whose entire career was pursued at Trinity College, Dublin (O'Hara, 2003).

Figure L6 Humphrey Lloyd.

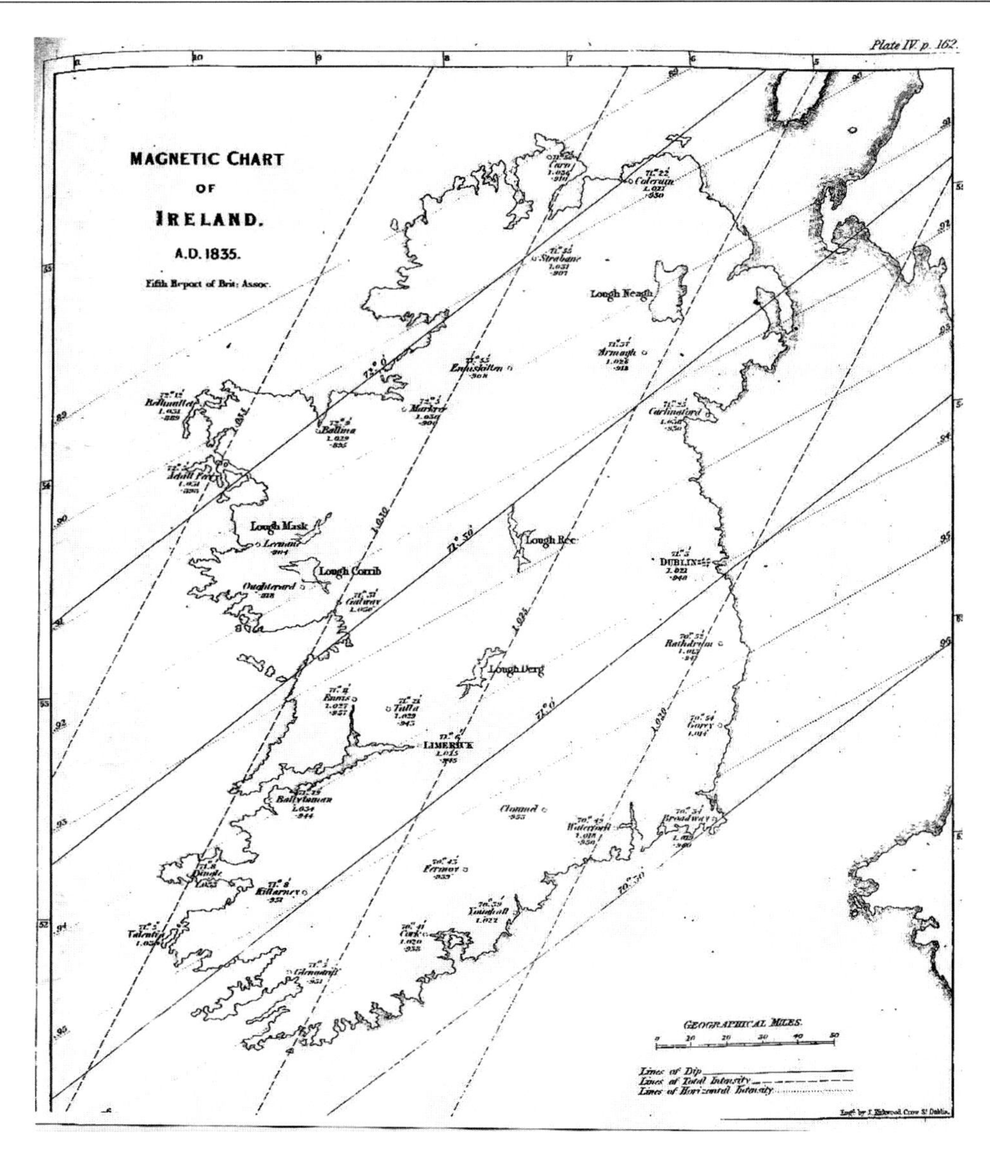

Figure L7 Lloyd's magnetic survey of Ireland, undertaken with Ross and Sabine.

In optics, he is remembered for his experimental confirmation of Hamilton's prediction of conical refraction, his demonstration of interference fringes with a pair of mirrors, and his authoritative report on the wave theory of light for the British Association. He had a lifelong interest in terrestrial magnetism and played an important role in the establishment of magnetic observatories, beginning in the 1830s.

His first venture into geomagnetism was a survey of Ireland (Figure L7) (Lloyd *et al.*, 1836) undertaken, at the instigation of the British Association, with Ross and *Sabine* (*q.v.*) in 1834 and 1835. In addition to dip and the horizontal component of the Earth's field, measured following the oscillation method of *Hansteen* (*q.v.*), simultaneous measurements of dip and intensity were made by a static balance method of his own devising.

Following the lead of *von Humboldt* (*q.v.*) and *Gauss* (*q.v.*) and with the detailed advice of the latter, Lloyd proceeded to have his Magnetical Observatory built in the grounds of the college. This charming edifice can still be viewed, but in the grounds of University College Dublin, some miles distant from its original home.

He conceived and designed many of the instruments that were to be used in the observatory: most were made by Thomas Grubb of Dublin. These included his balance magnetometer and a bifilar magnetometer, which he had developed to measure the horizontal component.

Lloyd participated enthusiastically in Gauss's Europe-wide Magnetic Union, and when such observatories came to be established throughout the British colonies, at the urging of *Sabine* (*q.v.*) Lloyd was heavily involved. The Dublin facility served as a model for those far-flung stations and new observers were trained in the college. The association of European, Russian, and British observatories represented what Lloyd called "a spirit unparalleled in the history of science." They constituted the first global network devoted to a scientific project. (see *Observatories, overview*). The specific idea of a network for magnetic observations lives on today in Intermagnet.

Even his honeymoon in 1840 was subsumed in the busy program of overseas observatory visits that followed. Meanwhile his inventiveness continued to show itself in new and refined designs for instrumentation: for example an induction magnetometer for the vertical component (1841).

The current notion that terrestrial magnetism was linked to meteorological phenomena led him to study the latter, with a characteristically systematic approach. He also explored the conjecture that internal electric currents were responsible for the Earth's field, tried to relate this to the diurnal variation, and proposed new experiments to test this theory.

Lloyd's father Bartholomew had been an inspiring leader of the college as a reforming mathematician and provost. Humphrey followed his example in becoming provost, and served as president of the Royal Irish Academy and the British Association. He was recognized with such awards as the German "Pour le Mérite" and Fellowship of the Royal Society. At home, he was awarded the Royal Irish Academy's Cunningham Medal.

He wrote 8 books and 64 papers and reports that record the achievements and insights of a remarkable experimentalist, organizer, and educator. A bust and a number of portraits survive in his college. There his father had created a new impetus in mathematics; the son accomplished the same for experimental science.

On his death, the Erasmus Smith Chair of Experimental and Natural Philosophy passed to George Francis Fitzgerald, together with much of Lloyd's energy and ideals. He had "struck an almost twentieth century note in his emphasis on the importance to a university of the cultivation of scientific research" (McDowell and Webb, 1982).

Deanis Weaire and J.M.D. Coey

Bibliography

Lloyd, H., Sabine, E., and Ross, J.C., 1836. Observations on the Direction and Intensity of the Terrestrial Magnetic Force in Ireland. In the Fifth Report of the British Association for the Advancement of Science. London: John Murray, pp. 44–51.

McDowell, R.B., and Webb, D.A., 1982. *Trinity College Dublin, 1592–1952: An Academic History.* London: Cambridge University Press.

O'Hara, J.G., 2003. Humphrey Lloyd 1800–1881. In McCartney, M., and Whitaker, A. *Physicists of Ireland.* Bristol: IOP Publishing, pp. 44–51.

Cross-references

Gauss, Carl Friedrich (1777–1855)
Hansteen, Christopher (1784–1873)
Humboldt, Alexander von (1759–1859)
Observatories, Overview
Sabine, Edward (1788–1883)

M

MAGNETIC ANISOTROPY, SEDIMENTARY ROCKS AND STRAIN ALTERATION

The relationship between anisotropy of magnetic susceptibility (AMS) fabrics (the shape and orientation of the AMS ellipsoid) and rock fabrics has been studied for over 40 years. Many studies have employed magnetic fabrics as an independent data set to compare with remanent magnetization directional data. Magnetic fabrics can also be very sensitive indicators of low-intensity strain, though deformation exceeding 10% to 20% often modifies or destroys the fabric to the point of making a functional interpretation impossible. This is in part due to diagenetic processes; for example, reduction, pressure solution, recrystallization, fluid migrations, and growth of new mineral constituents occurring with greater degrees of strain. Thus, AMS techniques are most often successful in investigating tectonic strains on weakly deformed sedimentary rocks.

While AMS fabrics are most often easily interpreted in sedimentary rocks, the method is nonetheless complicated by the great range of primary sedimentary fabrics requiring correct interpretation before any modification to that fabric can be understood. It is important to note that AMS ellipsoids and grain-shape often do not correspond. Recent studies have pointed out the strong dependence of the AMS ellipsoid shape on mineral composition, for example, which can have a much larger effect than strain modification. Slight variations in ferromagnetic (or ferrimagnetic) trace minerals, for example, can strongly influence the shape and magnitude of the AMS ellipsoid, to the extent that composition-dependent variation may completely obscure the effects of strain. Finally, modeling and experimental studies have demonstrated that the relationship between incremental strain and the change in degree of anisotropy is highly complex. Some studies have determined power-law relationships, but with a great deal of variability reflecting differences in deformational style, intensity, and minerals that carry the anisotropy. Despite such complications, AMS methods as fabric indicators in sedimentary rocks have the advantages of being rapid, nondestructive, and result from the averaging of a susceptibility tensor from a very large population of magnetic grains. Weak deformation distinguished by AMS techniques may often be undetectable by nonmagnetic methods.

Depositional (primary) sedimentary fabrics and processes

AMS studies have been directed at analyzing deep-sea current flow, bioturbation, fluvial and lacustrine fabrics, wind-deposited sand and loess fabrics, to name a few, as interpreted from AMS ellipsoids. Since sedimentary rocks should nearly always have a primary magnetic fabric, any fabric acquired as a result of later deformation will be superimposed upon a preexisting depositional fabric. Therefore, interpretation of altered AMS fabrics must begin with interpretation of the earlier (primary) fabric, which controls any subsequent alteration.

Studies of depositional AMS fabrics in sediments show that the shape and orientation of the susceptibility ellipsoid can be affected by a great variety of sedimentological processes predominantly formed during the deposition of particles from suspension in water or air (for a review, see Ellwood, 1980). Initial sedimentary fabrics are generally determined by gravity and hydrodynamics. Thus, to a first approximation, the shape and orientation of the AMS ellipsoid (which is the sum of the susceptibilities of a large number of grains present in the measured sample) is determined by both the characteristics of the grains themselves, and by the velocity and direction of transport of the medium through which they are carried. In general, sedimentary depositional fabrics are comparatively low in both susceptibility ($<10^{-4}$ SI units) and in anisotropy ($<5\%$). Typical primary depositional fabrics tend to be characterized by vertical or subvertical minimum susceptibility axes (K_3) (that is, the magnetic foliation is near the bedding plane) consistent with a conceptual model of grains settling flat on a surface. The maximum susceptibility axis (K_1) is often related to the transport or postdepositional current direction. However, this simple conceptual model can be complicated by factors such as the velocity of grains during deposition, the angle of slope on which grains settle, bulk mineralogy, compaction, fluid migration, and chemical changes such as growth of authigenic minerals or oxidation of iron-containing minerals, which may modify or completely destroy the original depositional fabric.

Many studies of deep-sea sediments have been aimed at describing variables such as current velocity or transport mechanisms through interpretation of the orientation of AMS ellipsoids, particularly the orientation of the maximum susceptibility axis (K_1) or the lineation (K_1/K_2). A number of studies have successfully used AMS to orient sediment cores used for studies of paleomagnetic remanence, and many investigations have described and identified the degree to which sampling techniques used for deep-sea sediments can distort the samples, thus affecting the measured AMS fabric.

In deep-sea sediments, both current-parallel and current-perpendicular alignments of the maximum susceptibility (K_1) axes are reported in a variety of depositional environments. One might imagine that the long axis of an elongate grain would align with current direction like a weather vane. In rock fabric studies, however, there are many reported instances of "rolling fabrics," in which elongated axes of grains are oriented perpendicular to the current direction as a result of rolling along the surface of the depositional plane in the direction of current flow.

In general, AMS studies in deep-sea sediments call attention to the need for a better understanding of the relationship between the shape and orientation of AMS ellipsoids and current flow, such that transport mechanisms and sedimentation processes in deep-sea sediments (for example, turbidites, contourites, and channel deposits), can be identified using AMS data. Recent studies indicate bioturbation is a widespread and continuous process, which disturbs deep-sea sediments. Flood *et al.* (1985) evaluated the effect of bioturbation on AMS fabrics in deep-sea sediments and found characteristic alteration from weakly oblate, to secondary prolate fabrics in the uppermost centimeters of sediment. More study is needed to assess the degree to which deep-sea sediment fabrics are generally affected by bioturbation.

Grain orientations determined by microscopic measurement and AMS measurements in experimentally produced fluvial and wind-deposited (eolian) sands were compared by Taira (1989). The study demonstrates the usefulness of AMS fabric in identification of sediment depositional processes. From analysis of experimentally produced fabrics, five distinct modes of deposition (gravity, flat bed and sloping bed, current, grain collision, and viscous flow) were distinguished, each with characteristic fabric parameters. In general, the maximum susceptibility was found to be oriented parallel to the current (shear stress) direction, whereas the minimum susceptibility was found to be oriented near normal, but with a distinct angle to the plane of shear stress (the imbrication angle). The five depositional types recreated in experimentally deposited sediments were found to fall into distinct groupings when two AMS measurement parameters are plotted against one another. The magnetic foliation/lineation (q) was plotted against the angle of imbrication of the minimum susceptibility axis (K_3) relative to the bedding plane (β).

Many useful measurement parameters, such as q and β, which describe the magnitude, shape, or orientation of the AMS ellipsoid, have been defined. See *Magnetic susceptibility, anisotropy* (*q.v.*) for a complete treatment.

Strain alteration of sedimentary fabrics

Many studies have successfully correlated the orientation of the principal axes of the AMS ellipsoid with mesoscopic rock fabrics. More recent efforts have concentrated on correlating the AMS ellipsoid and the finite strain ellipsoid (for a review, see Tarling and Hrouda, 1993). While these studies have, in the main, found that the finite strain and susceptibility ellipsoids are coaxial in orientation, the compelling goal of such research is to quantitatively relate the AMS fabric of a strained rock with its measured strain (using passive markers for example). Such studies have been complicated by variations in magnetic mineralogy, for example, and attempts to correlate the magnitudes of the finite strain ellipsoid and the AMS ellipsoid continue to be a challenge (for a review, see Borradaile, 1988).

Some features of strain-induced modification to sedimentary AMS fabrics are well documented. A basic idea, common to interpretation of AMS in strain studies, is the concept that as primary principal axes are deformed, they will be deflected away from their original direction. As the intensity of strain increases, magnetic foliations tend to tilt away from the bedding plane, forming a "girdle" of poles parallel to the direction of shortening. Thus, fabrics can often show girdling of principal directions as the axes progressively reorient from originally clustered principal directions. This has allowed some workers to use AMS to reverse the strain from a rock in order to correct the remanent magnetization directions, resulting in improved clustering of the remanence directions (for a review, see Cogne and Perroud, 1987).

Tectonic lineations

Secondary magnetic lineations acquired through transitional tectonic strains are reported (often termed a "pencil structure," Graham, 1978; Kligfield *et al.*, 1981). A pencil structure can be thought of as analogous to the intersection fabric between cleavage and bedding commonly used in structural analysis. In a true pencil structure, individual grains are mechanically and progressively rotated toward a direction perpendicular to the direction of principal stress, producing a lineation normal to the shortening direction. While fabrics of this type are often reported, clearly, magnetic susceptibility axes and grain-shape do not correspond in all of these fabrics. For instance, many AMS fabrics found in weakly strained sediments in young compressional mountain belts show K_1 in the bedding plane and perpendicular to the direction of tectonic transport, much like a pencil structure. These AMS maximum axes are often well-grouped perpendicular to the direction of principal stress, and are usually interpreted as having a tectonic origin. Such well-grouped magnetic lineations, however, are often reported in visually undeformed sediments, in which an analogous rigid deformation of the rock fabric seems very unlikely with such weak deformation.

Such fabrics can be explained by observing that, when principal axes have nearly equal magnetic susceptibilities, exchanging of axes can be instantaneous with very small increments of strain. A reorientation of 90° in the maximum susceptibility axis in very weakly strained rocks, with no girdling of principal axes, is the result. Consider an assemblage of magnetic grains with an initial sedimentary fabric, having a foliation in the bedding plane and K_1 oriented in a downstream direction (for example, perpendicular to a compressional mountain range). Any shortening parallel to the bedding plane will have the effect of tilting the grains away from the bedding plane, thereby weakening the K_1 of the bulk fabric, but leaving K_2 unchanged. At some point, the susceptibility of K_1 and K_2 will become equal, and with the next increment of strain their orientations will abruptly exchange. In this way, a "pencil structure" might seem to develop with very weak strains as the susceptibility of initially well-clustered K_1 axes are diluted by small changes in the grain orientations. This is most likely to occur with K_1 and K_2 very nearly equal in magnitude. Exchanging susceptibility axes, as in this example, can result in the "inverted fabrics" reported by some studies; grains will not physically reorient with very low strain, but the bulk AMS ellipsoid may.

An evolution of syndeformational magnetic fabrics was suggested in a study of sediments from a Neogene accretionary complex by Kanamatsu *et al.* (1996). In that study, sedimentary AMS fabrics were found to follow a path of decreasing shape parameter (T) and decreasing degree of corrected anisotropy (P') with the evolution from sedimentary through transitional to "tectonic" fabrics. That is, AMS ellipsoids were found to vary during weak progressive strain from oblate to prolate in shape, while anisotropy decreased (the shape parameter (T) and corrected anisotropy (P') have been adopted as standard parameters to evaluate AMS ellipsoids, and are discussed in Jelinek, 1981).

Strain-induced changes in mineral composition

Since AMS is determined from the combined susceptibility of all paramagnetic, diamagnetic, and ferromagnetic minerals in a sample, increased intensities of strain produces an important caveat for AMS studies. The growth of minerals through dissolution and recrystallization processes accompanying increased strain hampers interpretation of AMS ellipsoids. Housen and van der Pluijm (1990) demonstrated control of paramagnetic chlorite on AMS fabrics during development of slaty cleavage. Since magnetic anisotropy increases with increasing degree of crystallographic preferred orientation, the AMS ellipsoid can be determined by the crystallographic principal axes. Such studies point to the need for a thorough understanding of the paramagnetic and diamagnetic matrix as well as ferromagnetic constituents, especially under conditions of progressive strain. Fortunately, rock magnetic methods are providing increasingly sophisticated means to separate ferromagnetic and paramagnetic contributions to the susceptibility tensor.

For example, the anisotropy of anhysteretic remanent magnetization (AARM) (see *Magnetization, anhysteretic remanent (ARM), q.v.*) is a method unaffected by paramagnetic constituents. In many cases, a reasonable approximation of single mineral anisotropies can increasingly be obtained. These methods will continue to provide exciting opportunities to advance the state of research relating magnetic fabrics and rock fabrics.

Peter D. Weiler

Bibliography

Borradaile, G.J., 1988. Magnetic susceptibility, petrofabrics and strain. *Tectonophysics*, **156**: 1–20.

Cogne, J.P., and Perroud, H., 1987. Unstraining paleomagnetic vectors: the current state of debate. *EOS Transactions of American Geophysical Union*, **68**(34): 705–712.

Ellwood, B.B., 1980. Induced and remanent properties of marine sediments as indicators in depositional processes. *Marine Geology*, **38**: 233–244.

Flood, R.D., Kent, D.V., Shor, A.N., and Hall, F.R., 1985. The magnetic fabric of surficial deep-sea sediments in the HEBBLE area (Nova Scotian Continental Rise). *Marine Geology*, **66**: 149–167.

Graham, R.H., 1978. Quantitative deformation studies in the Permian rocks of the Alpes-Maritimes. *Memoire de Bureau de Recherches Geologiques et Minieres*, **91**: 219–238.

Housen, B.A., and van der Pluijm, B.A., 1990. Chlorite control of correlations between strain and anisotropy of magnetic susceptibility. *Physics of the Earth and Planetary Interiors*, **61**: 315–323.

Jelinek, V., 1981. Characterization of the magnetic fabric of rocks. *Tectonophysics*, **79**: 63–67.

Kanamatsu, T., Herrero-Bervera, E., Taira, A., Ashi, J., and Furomoto, A.S., 1996. Magnetic fabric development in the Tertiary accretionary complex in the Boso and Miura Peninsulas of Central Japan. *Geophysical Research Letters*, **23**: 471–474.

Kligfield, R., Owens, W.H., and Lowrie, W., 1981. Magnetic susceptibility anisotropy, strain, and progressive deformation in Permian sediments from the Maritime Alps (France). *Earth and Planetary Science Letters*, **55**: 181–189.

Taira, A., 1989. Magnetic fabrics and depositional processes. In Taira, A., and Masuda, F. (eds.), *Sedimentary Facies in the Active Plate Margins*. Tokyo: Terra Science, pp. 43–77.

Tarling, D.H., and Hrouda, F., 1993. *The Magnetic Anisotropy of Rocks*. New York: Chapman & Hall.

Cross-references

Magnetic Susceptibility, Anisotropy
Magnetization, Anhysteretic Remanent (ARM)

MAGNETIC ANOMALIES FOR GEOLOGY AND RESOURCES

Magnetic anomalies and geological mapping

Knowledge of the geology of a region is the scientific basis for resource exploration (petroleum, solid minerals, groundwater) the world over. Among the variety of rock types to be found in the Earth's crust, many exhibit magnetic properties, whether a magnetization induced by the present-day geomagnetic field, or a remanent magnetization acquired at some time in the geological past, or a combination of both. Mapping the patterns of magnetic anomalies attributable to rock magnetism has proved to be a very effective way of reconnoitering large areas of geology at low cost per unit area. The fact that most sedimentary rocks and surface-cover formations (including water) are effectively nonmagnetic means that the observed anomalies are attributable to the underlying igneous and metamorphic rocks (the so-called "magnetic basement"), even where they are concealed from direct observation at the surface. Anomalies arising from the magnetic basement are only diminished in amplitude and extended in wavelength through the extra vertical distance between source and magnetometer imposed by the nonmagnetic sediment layers. Thus *aeromagnetic surveys* (*q.v.*) are able to indicate the distribution of bedrock lithologies and structures virtually everywhere. Interpretation of magnetic anomaly patterns can then lead to maps of (hidden) geology that give direction to the exploration process (Figure M1/Plate 7b). In igneous and metamorphic ("hard rock") terranes, outlines of local areas promising for the occurrence of ore bodies can be delineated for closer follow-up studies. In the case of petroleum exploration, interpretation of the structure of the underlying basement can help understanding of basin development and help locate areas for (costly) seismic studies and drilling. Similar economies in the exploration process can be made through exploitation of aeromagnetic surveys in the systematic mapping of potential groundwater resources, of particular importance in the arid and semiarid areas of the world.

Mineral exploration

The amplitude, shape, and internal texture of magnetic anomalies may be used to indicate the likelihood of finding of certain ore types. Initially, aeromagnetic measurements were used in direct prospecting for magnetic iron ores and in the indirect detection of certain classes of magmatic-hosted Ni-deposits, kimberlite pipes for diamonds, and so forth. Hydrothermally altered areas in magnetic environments are often detectable as low magnetic zones and may be prospective for Au and Pb. The largest concentrated source of magnetic anomaly in terms of magnetic moment is the Kursk low-grade magnetic iron-quartzite ore formation in Ukraine that can be measured even at satellite altitudes. Most of the major satellite anomalies are, however, attributable to entire provinces of relatively magnetic rocks, such as Northern Fennoscandia, of which the Kiruna magmatic iron ore is but a tiny part.

Most ore deposits within the crystalline basement are, either in themselves or through their host rocks, accompanied by magnetic

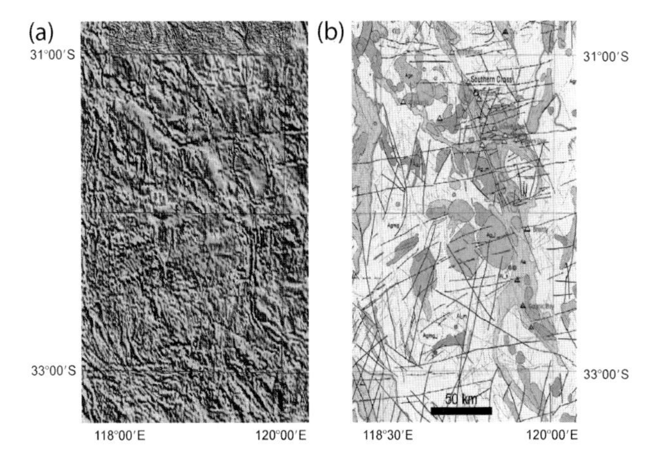

Figure M1/Plate 7b (a) Magnetic anomaly patterns over part of Western Australia recorded in various aeromagnetic surveys. (b) Geological interpretation of the data shown in (a). (Courtesy of Geoscience Australia and the Geological Survey of Western Australia).

anomalies. These anomalies are often used further in the closer evaluation of the extent and geometry of a deposit and in assessing the mineral potential of other comparable geological formations. At the reconnaissance stage of mineral assessment, area selection and prospecting—particularly of soil-covered areas of crystalline basement—geological mapping can be driven in large part by interpretation of detailed aeromagnetic anomaly maps. These provide, in a cost-effective and environmentally friendly way, a reliable picture of the underlying subsurface, including the location and extent of geological units and their lithology, structure, and deformation.

Continental and oceanic anomaly mapping

Given that most countries have a national program of mineral resource management, the foundation of which is the geological mapping program at an appropriate scale (say 1:1000000), ambitious national programs of aeromagnetic anomaly mapping have been instigated generally to supplement and accelerate geological mapping (for example Canada, Australia, former Soviet Union, the Nordic countries). National aeromagnetic anomaly maps, together with gravity anomaly maps, have therefore become preeminent in the geophysical support of geological reconnaissance. Given that gamma-ray spectrometer surveys are now usually flown simultaneously with aeromagnetic surveys at little additional cost, they form a third common type of geophysical support for geological mapping.

Over 70% of the Earth is covered by oceans. The geology here is totally obscured, even over the continental shelves. Until the advent of systematic ocean exploration in the second half of the 20th century, little was known of the geological evolution of the deep oceans. The patterns of magnetic anomaly stripes discovered, paralleling the mid-ocean ridges (Figure M2/Plate 9e from Verhoef *et al.*, 1996), became one of the leading lines of evidence in support of continental drift and, eventually, global tectonics (see *Vine-Matthews-Morley hypothesis*). This revelation must count as one of the most profound advances in our understanding of the Earth's history and its mode of development.

Being of igneous origin, the rocks of the oceanic crust solidify and cool at the mid-ocean ridge while inheriting a magnetic field direction from the erstwhile geomagnetic field. Repeated reversals of that field are recorded symmetrically either side of the ocean-spreading axis as new crust is relentlessly added, a few centimeters per year, at the axis itself. While people's interest in mineral resources of the deep ocean is still limited, understanding the previous locations of the continents, particularly during the past 200 Ma for which ocean floor can still be found, is central to our understanding of geological processes in this period. The sedimentary rocks laid down on the passive continental margins now separated by "new" oceans host a great deal of the world's hydrocarbon reserves.

Ocean crust is eventually recycled into the Earth's mantle via subduction zones with the result that ocean crust *in situ* older than about 200 Ma is no longer to be found. Evidence of the earlier 95% of geological history is therefore confined to the continents. Here a great deal of geological complexity is revealed (Figure M3/Plate 7d; Zonenshain *et al.*, 1991), reflecting repeated orogenesis (mountain building) and metamorphism since the time of the oldest known rocks, dating from the Archean. The patterns of repeated continental collisions and separations evident from more recent geology can be extrapolated into this past. However, poor rock exposure in most of the oldest, worn-down areas of the world (Precambrian shields) hampers their geological exploration. Aeromagnetic surveys assist markedly here, though understanding at the scale of whole continents often necessitates maps extending across many national frontiers, as well as across oceans where present continents were formally juxtaposed.

What holds for investigations into geology and resources over regions or areas quite locally (say, scale 1:250000, a scale typical for geological reconnaissance) also holds for national compilations of larger countries (say scale 1:1000000) and at continental scale (1:5 million or 1:10 million). Magnetic anomaly mapping is arguably even more useful at such scales since it represents unequivocal physical data coverage, part of a nation's (or a continent's) geoscience data-infrastructure. The view it offers to the geological foundations of continents is therefore of prime importance to improved understanding of global geology. Repeated cycles of continental collision, coalescence, and rifting-apart have led to the present-day arrangement of the igneous and metamorphic rocks of the continental crust, as revealed in continental scale images of magnetic anomalies, such as Australia (Figure M4/Plate 7a; Milligan and Tarlowski, 1999).

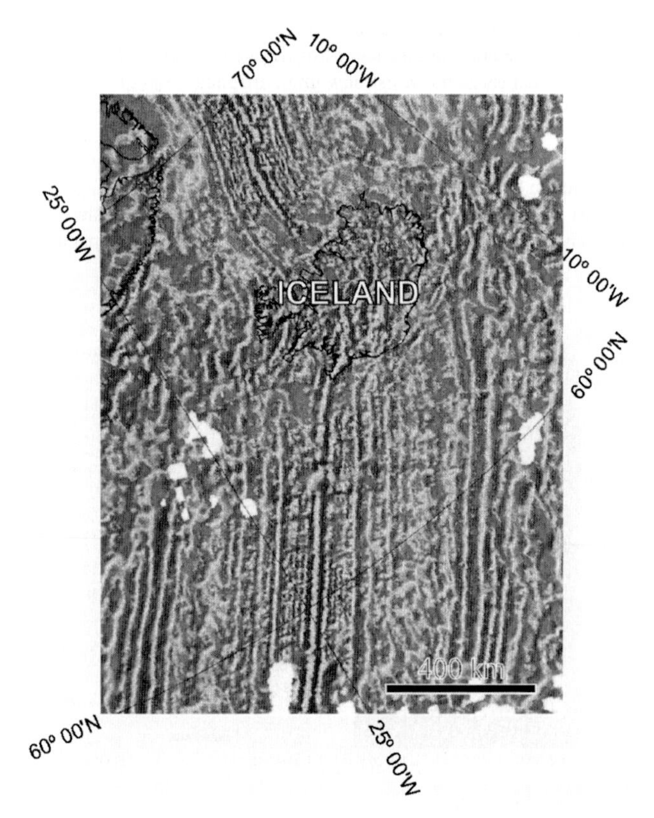

Figure M2/Plate 9e Magnetic anomaly patterns in the North Atlantic Ocean showing the symmetry of the anomalies either side of the mid-Atlantic Ridge in the vicinity of Iceland. (Detail from Verhoef *et al.*, 1996, courtesy of the Geological Survey of Canada).

Continental and global compilations of magnetic anomalies

(Aero-)magnetic surveys usually record only the total strength of the magnetic field at any given point, thus avoiding the need for any precision orientation of the magnetometer. After suitable corrections have been applied for temporal field variations during the weeks or months of a survey, subtraction of the long wavelength (more than several hundred kilometers) components of the field leaves local anomalous values that should be comparable from one survey to another. Thus it is possible—though, in practice, challenging—to link many hundreds of surveys together to make national or continental scale coverages (Tarlowski *et al.*, 1996; Fairhead *et al.*, 1997).

The long wavelength components of magnetic anomalies must be defined in an internationally coherent way, for which purpose the *IGRF* (*q.v.*) was designed and is periodically updated by IAGA. Anomaly definitions still vary greatly between surveys for various

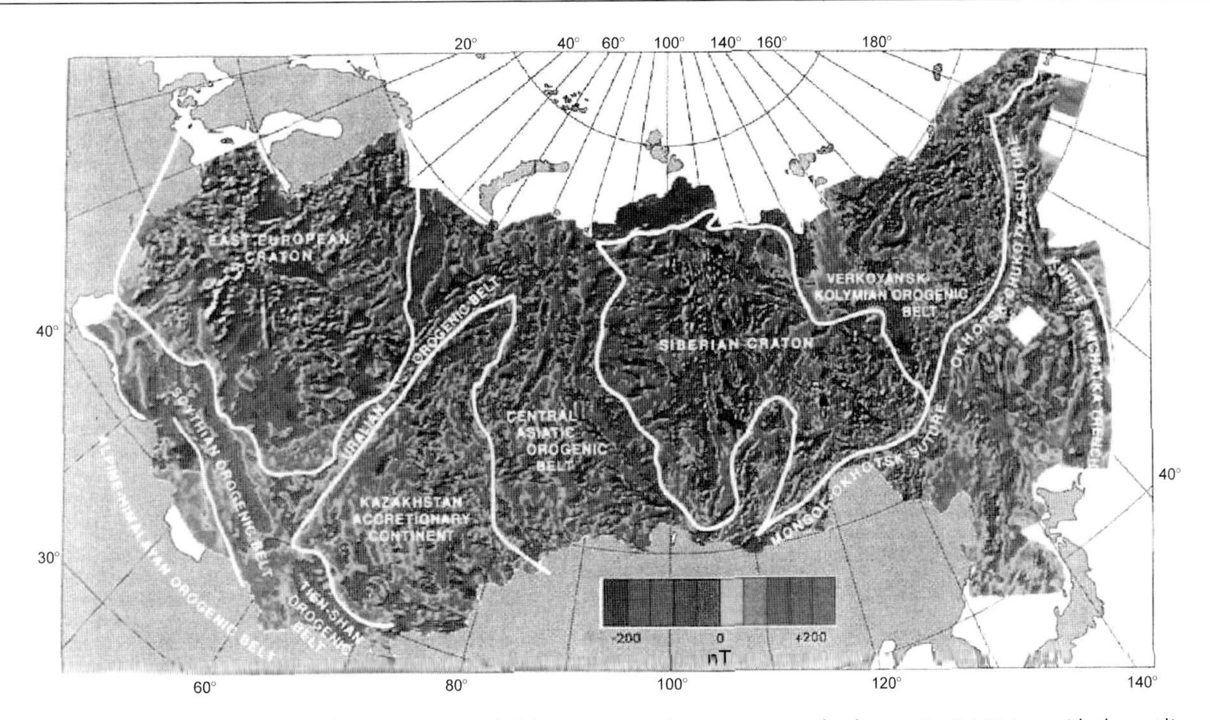

Figure M3/Plate 7d Magnetic anomaly patterns recorded in aeromagnetic surveys over the former Soviet Union with the outlines of some major tectonic domains added. Russia. (From Zonenshain *et al.*, 1991, courtesy of the American Geophysical Union).

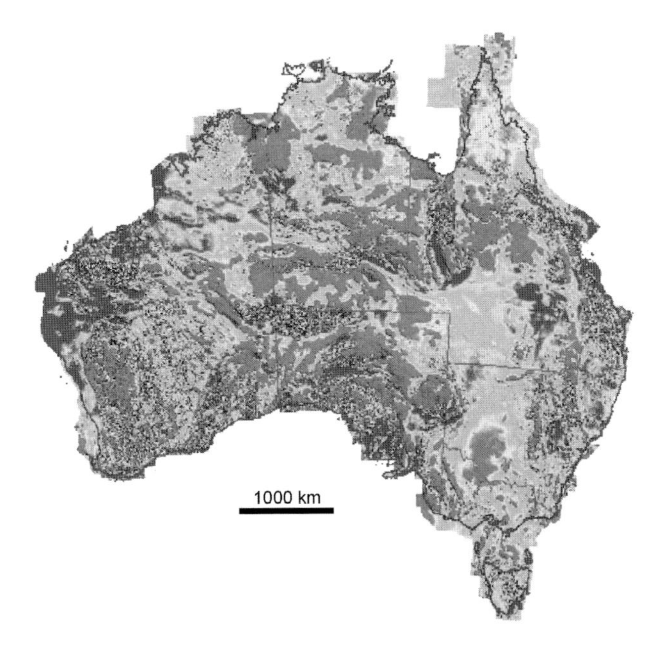

Figure M4/Plate 7a *Magnetic Anomaly Map of Australia*, 3rd edn (Milligan and Tarlowski, 1999). (Courtesy of Geoscience Australia).

reasons and, in addition to consistent use of the IGRF, national and international cooperation is required to link and level together the variety of magnetic surveys to a common level. Magnetic anomaly data exist over most of the Earth's surface, mostly a patchwork of airborne surveys on land and marine traverses at sea (Reeves *et al.*, 1998). In 2003, IAGA appointed a Task Group to oversee the compilation of such a global magnetic anomaly map (www.ngdc.noaa.gov/IAGA/vmod/TaskGroupWDMAM-04July12s.pdf) that endeavors to complete its work in time for the 2007 IUGG General Assembly.

Magnetic mineralogy

The physical link between geological formations and their magnetic anomalies is the magnetic properties of rocks (Clark and Emerson, 1999). These are often measured, for example, in connection with Ocean Drilling Program (ODP) analyses, paleomagnetic studies, geological mapping, and mineral prospecting. ODP data provide local information at widespread oceanic locations giving a global coverage. International paleomagnetic databases represent the remanent magnetic properties acquired in the geological past (see *Paleomagnetism*). Magnetomineralogical studies reveal that by far the most common magnetic source mineral of Precambrian shield areas is magnetite. So far, the largest national campaign of magnetic property mapping was that carried out in the former Soviet Union. The results were presented as analog maps. Most of the world's resource of digital petrophysical data for the continents was collected in the Fennoscandian Shield by the Nordic countries (Korhonen *et al.*, 2002a,b). Even so, these data represent only a small part of the crystalline basement of NW Europe. More, similar data sets are required if we are to understand how well this information represents crustal rocks more globally.

The results from Fennoscandia show that, when plotted on a diagram of induced magnetization against density (Figure M5a) the samples form two populations, A and B. Population A represents the paramagnetic range of susceptibilities defined by Curie's law. Compositional variation of Fe- and Mn-oxides correlates with density, the denser, more basic (mafic) rock lithologies being more magnetic than acid (silicic) ones by up to an order of magnitude. This population is only capable of causing anomalies less than about 25 nT, however. A second population of rocks (B), mostly acid in chemistry, represents the ferrimagnetic range of susceptibilities, mainly due to variations in the abundance and grain size of magnetite. This population is two

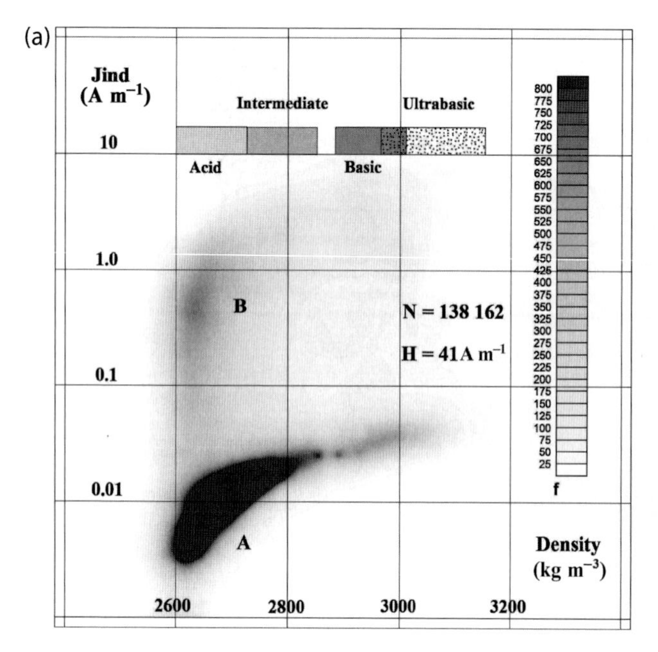

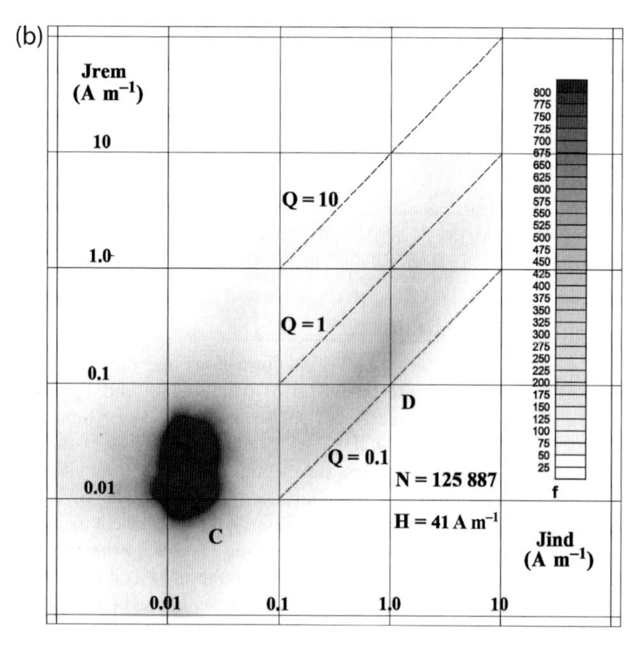

Figure M5 (a) Bulk density and induced magnetization in the Fennoscandian Shield (redrawn from Korhonen *et al.* 2002a). The lower population (A) contains the majority of rock samples, and represents the paramagnetic range of susceptibilities defined by Curie's law. A second population of rocks (B), mostly acid in chemistry, represents the ferrimagnetic range of susceptibilities, due mainly to variations in the abundance and grain size of magnetite. This population is two orders of magnitude more magnetic than the average of the first population (A). (b) Induced and remanent magnetization in the Fennoscandian Shield (redrawn from Korhonen *et al.*, 2002b). For the relatively few rock samples (D) that depart from a low level of magnetization (C), the Q-ratio is mostly less than 1.0, indicating the predominance of induced magnetization over remanent. A few exceptions are, however, highly magnetic. (Courtesy of Geological Surveys of Finland, Norway, and Sweden and the Ministry of Natural Resources of the Russian Federation).

orders of magnitude more magnetic than the average of the first population (A). Population B rocks represent most sources of local, induced magnetic anomalies. Average susceptibilities vary typically from 0.04 to 0.02 SI units for these ferrimagnetic geological formations, but much variation is found from one formation to another.

Another important parameter is the relative proportion of ferrimagnetic (population B) rocks to the effectively "nonmagnetic" paramagnetic (population A) rocks in any given area. For example, it is only a few percent in the magnetic "low" of central Fennoscandia but almost 100% in the northern Fennoscandian "high." Overall in Fennoscandia the average value is about 25%. In oceanic areas, by contrast, it approaches 100%.

Spatial contrasts in magnetic rock properties that give rise to the local magnetic anomalies encountered in mineral exploration are attributable to such factors as (a) the aforementioned bimodal nature of magnetic mineralogy (populations A and B), (b) the effects of magnetic mineralogy and grain size, (c) the history of magnetization and demagnetization, and (d) the variation between induced and remanent magnetization. These are related in turn to geological causes such as initial rock lithology, chemical composition, oxygen fugacity, and metamorphic history.

The ratio of remanent to induced magnetization varies typically from 0.1 to 20, corresponding to rocks containing coarse-grained fresh magnetite (most susceptible to induced magnetization) via altered and fine-grained magnetites to pyrrhotite (with a very stable remanent magnetization). Figure M5b shows the results for induced and remanent magnetization from the Fennoscandian Shield. For the relatively few rock samples (population D) that depart from a low level of magnetization (population C), the *Q*-ratio is mostly less than 1.0, indicating the predominance of induced magnetization over remanent as a source of magnetic anomalies. A few exceptions are, however, highly magnetic (above 1.0 A m^{-1}). For increasingly large source bodies, variations in the direction of all local remanent magnetizations cause the net remanent magnetization to sum up more slowly than the consistently oriented induced magnetization. Hence the effects of remanent magnetization are relatively more important in magnetic anomalies measured close to source bodies (such as on the ground) than farther away (from an aircraft or satellite). This effect is even more noticeable at magnetizations above 1 A m^{-1}, where *Q*-values tend to approach or even exceed 1.0 (Figure M5b).

Colin Reeves and Juha V. Korhonen

Bibliography

Clark, D.A., and Emerson, D.W., 1991. Notes on rock magnetisation characteristics in applied geophysical studies. *Exploration Geophysics*, **22**: 547–555.

Fairhead, J.D., Misener, D.J., Green, C.M., Bainbridge, G., and Reford, S.W., 1997. Large scale compilation of magnetic, gravity, radiometric and electromagnetic data: the new exploration strategy for the 90s (Paper 103). In Gubbins, A.G. (ed.), *Proceedings of Exploration'97: Fourth Decenniel International Conference on Mineral Exploration*, Toronto, September 15–18, GEO/FX, 1068pp.

Korhonen, J.V., Aaro, S., All, T., Elo, S., Haller, L.Å., Kääriäinen, J., Kulinich, A., Skilbrei, J.R., Solheim, D., Säävuori, H., Vaher, R., Zhdanova, L., and Koistinen, T., 2002a. Bouguer Anomaly Map of the Fennoscandian Shield 1:2000000. Geological Surveys of Finland, Norway and Sweden and Ministry of Natural Resources of Russian Federation. ISBN 951-960-818-7.

Korhonen, J.V., Aaro. S., All, T., Nevanlinna, H., Skilbrei, J.R., Säävuori, H., Vaher, R., Zhdanova, L., and Koistinen, T., 2002b. Magnetic Anomaly Map of the Fennoscandian Shield 1:2000000. Geological Surveys of Finland, Norway and Sweden and Ministry of Natural Resources of Russian Federation. ISBN 951-690-817-9.

Milligan, P.R., and Tarlowski, C., 1999. *Magnetic Anomaly Map of Australia*, 3rd edn, 1:5000000. Geoscience Australia, Canberra.

Reeves, C., Macnab, R., and Maschenkov, S., 1998. Compiling all the world's magnetic anomalies. *EOS, Transactions of the American Geophysical Union*, **79**: 338.

Tarlowski, C., McEwin, A.J., Reeves, C.V., and Barton, C.E., 1996. Dewarping the composite aeromagnetic anomaly map of Australia using control traverses and base stations. *Geophysics*, **61**: 696–705.

Verhoef, J., Macnab, R., Roest, W. *et al.*, 1996. Magnetic anomalies, Arctic and North Atlantic Oceans and adjacent land areas. *Geological Survey of Canada Open File*, 3282c.

Zonenshain, L.P., Verhoef, J., Macnab, R., and Meyers, R., 1991. Magnetic imprints of continental accretion in the USSR. *EOS, Transactions of the American Geophysical Union*, **72**: 305, 310.

Cross-references

Aeromagnetic Surveying
IGRF, International Geomagnetic Reference Field
Paleomagnetism
Vine-Matthews-Morley Hypothesis

MAGNETIC ANOMALIES, LONG WAVELENGTH

Long-wavelength anomalies are static or slowly varying features of the geomagnetic field, and originate largely within the lithosphere. These anomalies stand in contrast to the rapidly time varying features characteristic of even longer wavelengths, which originate within the outer core. An inflection point, or change of slope, in the geomagnetic power spectrum (Figure M6) can be seen at degree 13 and is a manifestation of the relatively sharp transition from core-dominated processes to lithospheric-dominated processes. Long-wavelength anomalies (Figure M7a/Plate 5c) are most easily recognized from near-Earth satellites at altitudes of 350–750 km, and these altitudes define the shortest wavelengths traditionally associated with such geomagnetic features. The lithospheric origin of these features was firmly established by comparison with the marine magnetic record of seafloor spreading in the North Atlantic (LaBrecque and Raymond, 1985). Virtually identical features have now been recognized in satellite magnetic field records from POGO (1967–1971), Magsat (1979–1980), Ørsted (1999–), and CHAMP (2000–). Long-wavelength anomalies were first recognized by Cain and coworkers in about 1970 on the basis of total field residuals of POGO data.

Although electrical conductivity contrasts (Grammatica and Tarits, 2002) and motional induction of oceanic currents (Vivier *et al.*, 2004) can produce quasistatic long-wavelength anomalies, the largest contributors to long-wavelength anomalies are induced (M_i) and remanent (M_r) magnetization in the Earth's crust. Contributions from the uppermost mantle may also be of importance, at depths where temperatures do not exceed the Curie temperature (T_c) of the relevant magnetic mineral. The earth's main field (H) is the inducing magnetic field responsible for induced magnetization of lithospheric materials. $M_i = kH$ expresses the linear relationship between the inducing field and induced magnetization, true for small changes in the inducing field. k is the volume magnetic susceptibility, treated here as a dimensionless scalar quantity, and reflects the ease with which a material is magnetized. If M does not return to zero in the absence of H, the resulting magnetic field is said to be remanent or permanent. Thus $M = M_r + M_i$, and the relative strength of the two contributions is referred to as the Koenigsberger ratio or $Q = M_r/M_i$.

Inversions of long-wavelength anomaly observations into lithospheric source functions, for example magnetic crustal thickness

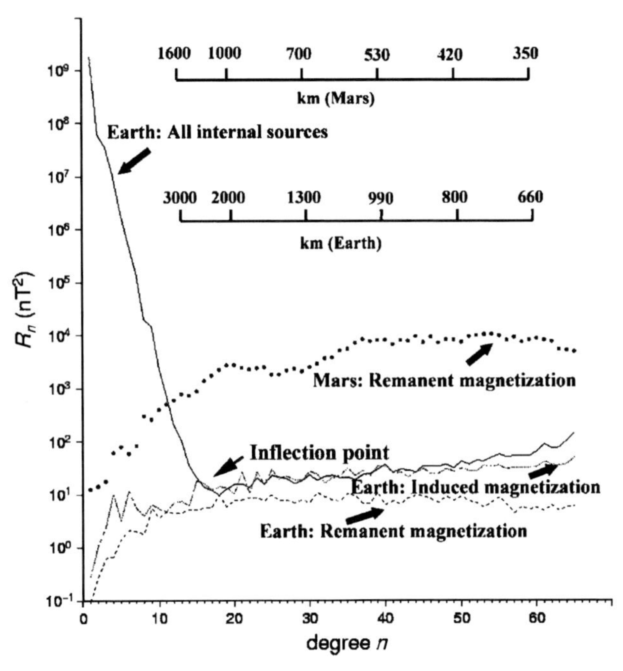

Figure M6 Comparison of the Lowes-Mauersberger (R_n) spectra at the surface of the Earth and Mars for a variety of internal fields. The inflection point in the terrestrial power spectra represents the sharp transition from core processes at low n to lithospheric processes at higher n. R_n is the mean square amplitude of the magnetic field over a sphere produced by harmonics of degree n. The terrestrial spectrum of all internal sources comes from Sabaka *et al.* (2004), the Martian remanent spectrum is derived from Langlais *et al.* (2004), the terrestrial induced spectrum is derived from Fox Maule *et al.* (2005), and the terrestrial remanent magnetization spectrum (of the oceans, and hence a minimum value) was derived from Dyment and Arkani-Hamed (1998).

(Figure M7b/Plate 5d) are subject to many caveats. Simple solutions are preferred, which also agree with other independently determined lithospheric properties. Specific caveats with respect to inversions of these magnetic field observations are that (1) direct inversion, in the absence of priors, can uniquely determine only an integrated magnetization contrast, (2) a remarkably diverse assemblage of magnetic annihilators (Maus and Haak, 2003) exist, which produce vanishingly small magnetic fields above the surface, and (3) the longest wavelength lithospheric magnetic signals are obscured by overlap with the core and it is formally impossible to separate them.

Unresolved research questions include (1) the continuing difficulty of signal separation, especially with respect to external fields (Sabaka *et al.*, 2004), and the particular problem of resolving north-south features from polar-orbiting satellites, (2) the relative importance of magnetic crustal thickness variations and magnetic susceptibility variations in producing long-wavelength anomalies, (3) the relative proportions of induced and remanent magnetization in the continents and oceans, (4) the mismatch between the observed long-wavelength fields at satellite altitude, and surface fields upward continued to satellite altitude, (5) the separation of long-wavelength anomalies caused by motional induction of large-scale ocean currents, (6) the isolation and relative importance of shorter wavelength anomalies (between 660 and 100 km wavelength), and finally (7) the origin of the order of magnitude difference between the observed lithospheric magnetic fields of the Earth and Mars.

Michael E. Purucker

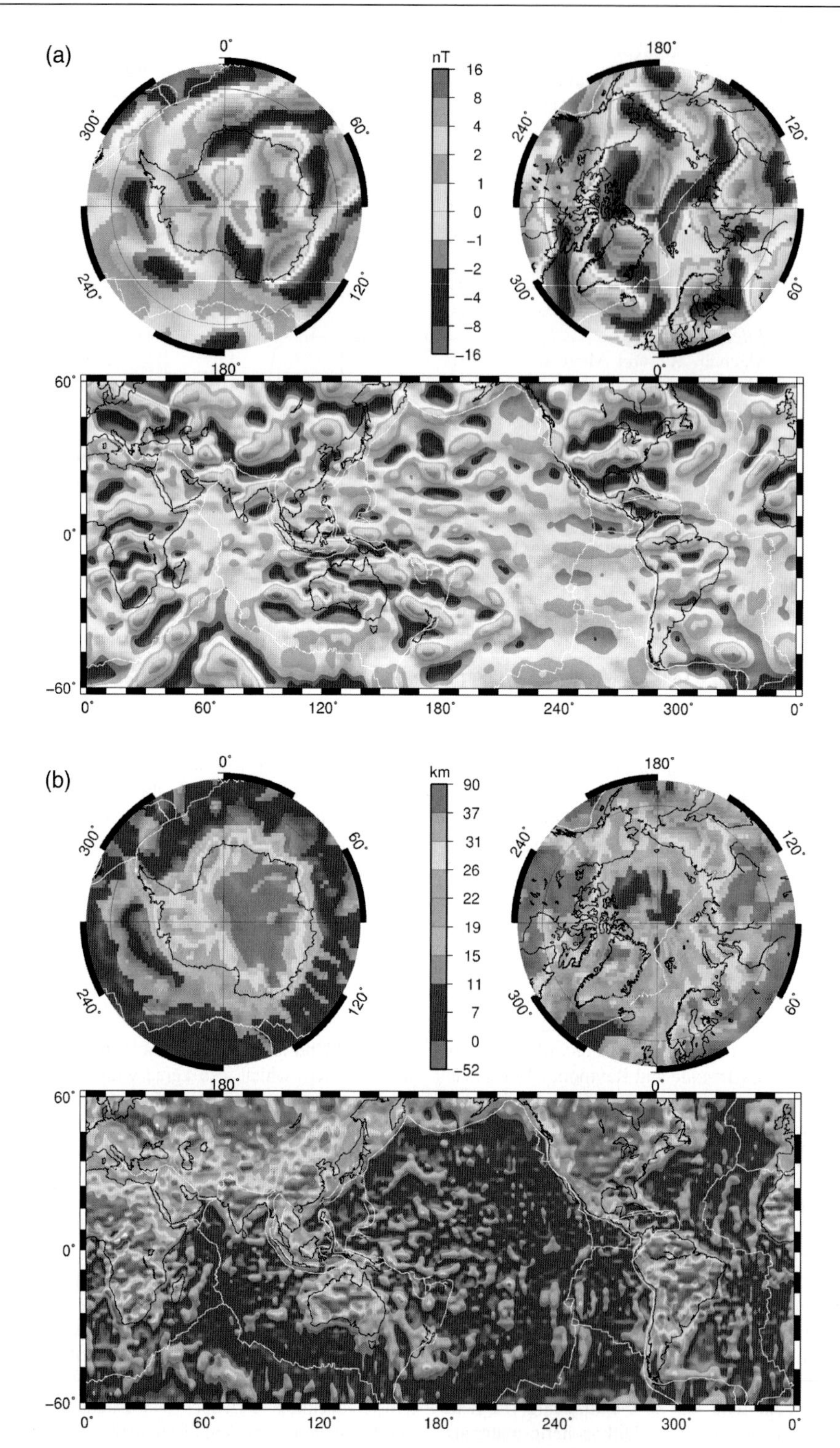

Figure M7/Plate 5c,d Long-wavelength anomalies (a) in the total field as seen by the CHAMP magnetic field satellite at an altitude of 400 km (MF-4 model of GFZ-Potsdam available at www.gfz-potsdam.de/pb2/pb23/SatMag/model.html). Spherical harmonic degrees between 16 and 90 are included within this map. The magnetic crustal thickness (b), derived as described in Fox Maule *et al.* (2005), explains the observations in (a). The map uses as a starting model the 3SMAC (Nataf and Ricard, 1996) compositional and thermal model of the crust and mantle. 3SMAC is modified in an iterative fashion with the satellite data, after first removing a model of the oceanic remanent magnetization (Dyment and Arkani-Hamed, 1998), until the magnetic field predicted by the model matches the observed magnetic field. A unique solution is obtained by assuming that induced magnetizations dominate in continental crust, and that vertical thickness variations dominate over lateral susceptibility variations (Purucker *et al.*, 2002). A starting model such as provided by 3SMAC is necessary to constrain wavelengths longer than about 2600 km. Longer wavelengths are obscured by overlap with the core field. The white lines delineate plate boundaries, transform faults, and mid-ocean ridges. Illumination on these shaded relief maps is from the east.

Bibliography

Dyment, J., and Arkani-Hamed, J., 1998. Contribution of lithospheric remanent magnetization to satellite magnetic anomalies over the world's oceans. *Journal of Geophysical Research*, **103**: 15423–15441.

Fox Maule, C., Purucker, M., Olsen, N., and Mosegaard, K., 2005. Heat flux anomalies in Antarctica revealed by satellite magnetic data. *Science*, 10.1126/science.1106888.

Grammatica, N., and Tarits, P., 2002. Contribution at satellite altitude of electromagnetically induced anomalies arising from a three-dimensional heterogeneously conducting Earth, using SQ as an inducing source field. *Geophysical Journal International*, **151**: 913–923.

LaBrecque, J.L., and Raymond, C.A., 1985. Seafloor spreading anomalies in the MAGSAT field of the North Atlantic. *Journal of Geophysical Research*, **90**: 2565–2575.

Langel, R.A., and Hinze, W.J., 1998. *The Magnetic Field of the Earth's Lithosphere.* Cambridge: Cambridge University Press.

Langlais, B., Purucker, M., and Mandea, M., 2004. Crustal magnetic field of Mars. *Journal of Geophysical Research*, **109**: E02008, doi:10.1029/2003JE002048.

Maus, S., and Haak, V., 2003. Magnetic field annihilators: invisible magnetisation at the magnetic equator. *Geophysical Journal International*, **155**: 309–313.

Nataf, H., and Ricard, Y., 1996. 3SMAC: an a priori tomographic model of the upper mantle based on geophysical modeling. *Physics of the Earth and Planetary Interiors*, **95**: 101–122.

Purucker, M., Langlais, B., Olsen, N., Hulot, G., and Mandea, M., 2002. The southern edge of cratonic North America: evidence from new satellite magnetometer observations. *Geophysical Research Letters*, **29**(15): 8000, doi:10.1029/2001GL013645.

Sabaka, T.J., Olsen, N., and Purucker, M., 2004. Extending comprehensive models of the Earth's magnetic field with Ørsted and CHAMP data. *Geophysical Journal International*, **159**: 521–547.

Vivier, F., Maier-Reimer, E., and Tyler, R.H., 2004. Simulations of magnetic fields generated by the Antarctic Circumpolar Current at satellite altitude: can geomagnetic measurements be used to monitor the flow? *Geophysical Research Letters*, **31**: L10306, doi:10.1029/2004GL019804.

Cross-references

CHAMP
Harmonics, Spherical
Magnetic Anomalies, Marine
Magnetic Field of Mars
Magnetic Surveys, Marine
Magsat
Ørsted
POGO (OGO-2, -4, and -6 spacecraft)

MAGNETIC ANOMALIES, MARINE

Magnetic anomalies are perturbations of the geomagnetic field due to Earth structure. They are seen when one subtracts a regional magnetic field from a series of observed readings. They have a different character over surface of the deep oceans than on they do on land. The unique nature of marine magnetic anomalies became apparent in the early 1950s when submarine detecting magnetometers were used for geophysical exploration by research ships. Over the oceans important marine anomalies may have an amplitude of a few tens to a few hundreds nanotesla and a wavelength of about 100 km, depending upon the depth of water. Previously some geomagnetic field measurements had been made on nonmagnetic ships but these readings were taken so far apart that the short anomalies were not defined. A few aircraft measurements had been made but they were high above the ocean surface reducing the size of the anomaly or were made over the continental shelf where anomalies have a different character from those in the deep ocean.

Early observations of marine anomalies

When research ships started taking measurements the magnetometer was towed behind the ship so the magnetic materials of which the ship was constructed did not influence the readings. Furthermore the intensity of the field, rather than its directional components, was measured thereby eliminating the need to know the orientation of the magnetometer. The first generation of marine magnetometers was of the fluxgate type, which had a small oriented special metal core. This core was mechanically oriented in the direction of the Earth's field by nulling the field measured by secondary cores, which were at right angles to the main core. This arrangement, while awkward, was used at sea for several years. The first systematic survey at sea was made by Scripps Institute of Oceanography using a fluxgate magnetometer towed by a Coast and Geodetic Survey ship in the Pacific Ocean off the western United States. When a regional magnetic field was subtracted from the observed field to get the anomalies, these anomalies were seen to be linear patterns, oriented nearly north-south, but with occasional sharp east-west offsets in the linear pattern (Vacquier, 1959). The source of these linear anomalies was unknown because there was no obvious seafloor feature, which could cause them.

At about the same time as the Pacific measurements were made miscellaneous cruise measurements off the US east coast, near the continental shelf showed major marine magnetic anomalies which were subparallel to the coast. They turned onto land and split into several parts over the states of Georgia and Florida.

Early shipboard measurements had shown a large magnetic anomaly over the center of the mid-Atlantic Ridge in several places. To investigate this in a systematic way, a low-level aeromagnetic survey was made over the part of the Reykjanes Ridge south of Iceland (Figure M8) (the Reykjanes, using a US Naval Oceanographic Office aircraft flying out of southern Iceland; Heirtzler *et al.*, 1966). This survey also showed linear anomalies, paralleling the ridge axis and symmetrically located about the axis of the Reykjanes Ridge. This suggested that the origin of these linear anomalies was related to the structure of the ridge.

The origin of the anomalies

In 1963, while the Reykjanes Ridge was being flown, an important paper was published by Fred Vine and Drummond Matthews (Vine and Matthews, 1963) pointing out that the linear anomalies in the ocean might be caused by normally and reversely magnetized rocks of the ocean floor. These authors suggested, furthermore, that these rocks can be represented by magnetized blocks and they may have acquired their magnetization when they were extruded at the axis of the mid-ocean ridge in times past. Presumably the blocks moved out from both sides of the ridge. This is the basis of the seafloor spreading theory. The magnetism was acquired at the axis, while the upwelling magma cooled. It is called remanent magnetization. As the material moves away from the axis it is in the ambient geomagnetic field and so secondary magnetism is induced in the rocks. That is called induced magnetization. Thus, the magnetization has two components, although the remanent component is usually stronger.

Applying Vine-Matthews theory to the Reykjanes Ridge data and knowing the times of the most recent few reversals of the field it was possible to learn that the seafloor of the Reykjanes Ridge is spreading at about 1.5 cm y^{-1} away from the ridge axis. Reykjanes Ridge anomalies which were further from the axis allowed one, for the first time, to know the ages of the various reversals as far back as 10 Ma, assuming that the seafloor spreading rate has been constant.

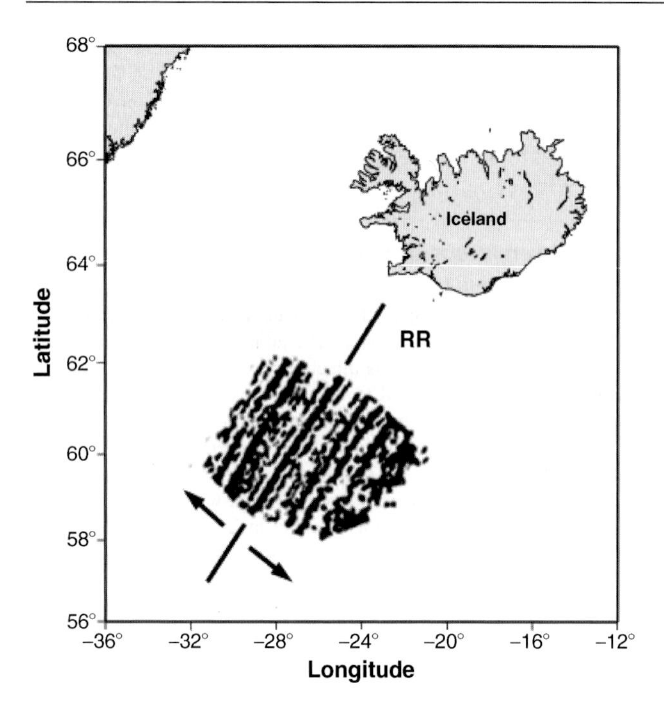

Figure M8 Marine magnetic anomalies over the Reykjanes Ridge (RR) south of Iceland. Positive anomalies are shaded black (after Heirtzler *et al.*, 1966).

This knowledge of the relationship between marine magnetic anomalies, reversals of the geomagnetic field, and seafloor spreading rates created a flurry of activity in marine geophysics in the 1960s. A new simplified (proton precession) magnetometer was developed, analytical methods were undertaken to relate the oceanic rocks to the anomaly observed. Cruises were undertaken in many parts of the world to extend the geomagnetic reversal timescale and to study seafloor spreading and the tectonics of the ocean floor. (See Figure M9/Plate 8: World map of isochrons of the ocean floor).

The rocks of the ocean floor are usually covered with a thick layer of sediments. The sediments are weakly magnetized so they do not interfere with the magnetic anomalies caused by the rocks beneath. However, if a sediment core is taken and measured in the laboratory it will also show reversals of past geomagnetic field directions just like anomalies measured across the ocean surface. Thus one can date sediments using the timescale of geomagnetic field reversals.

Some geomagnetic field anomalies can be observed by magnetic field satellites in near-Earth orbit, although anomalies over the oceans are small and difficult to analyze. Magnetic measurements have also been made by magnetometers towed near the deep seafloor. These have proved especially useful in studying the transition between normally and reversely magnetized rocks.

Using magnetic anomalies to date the ocean floor

The age of the seafloor is related to the distance of the magnetized blocks from the ridge axis, not the distance of the magnetic anomaly, from the ridge axis. Thus it is necessary to see how the position of the anomalies is related to the position of the source rocks. A computer

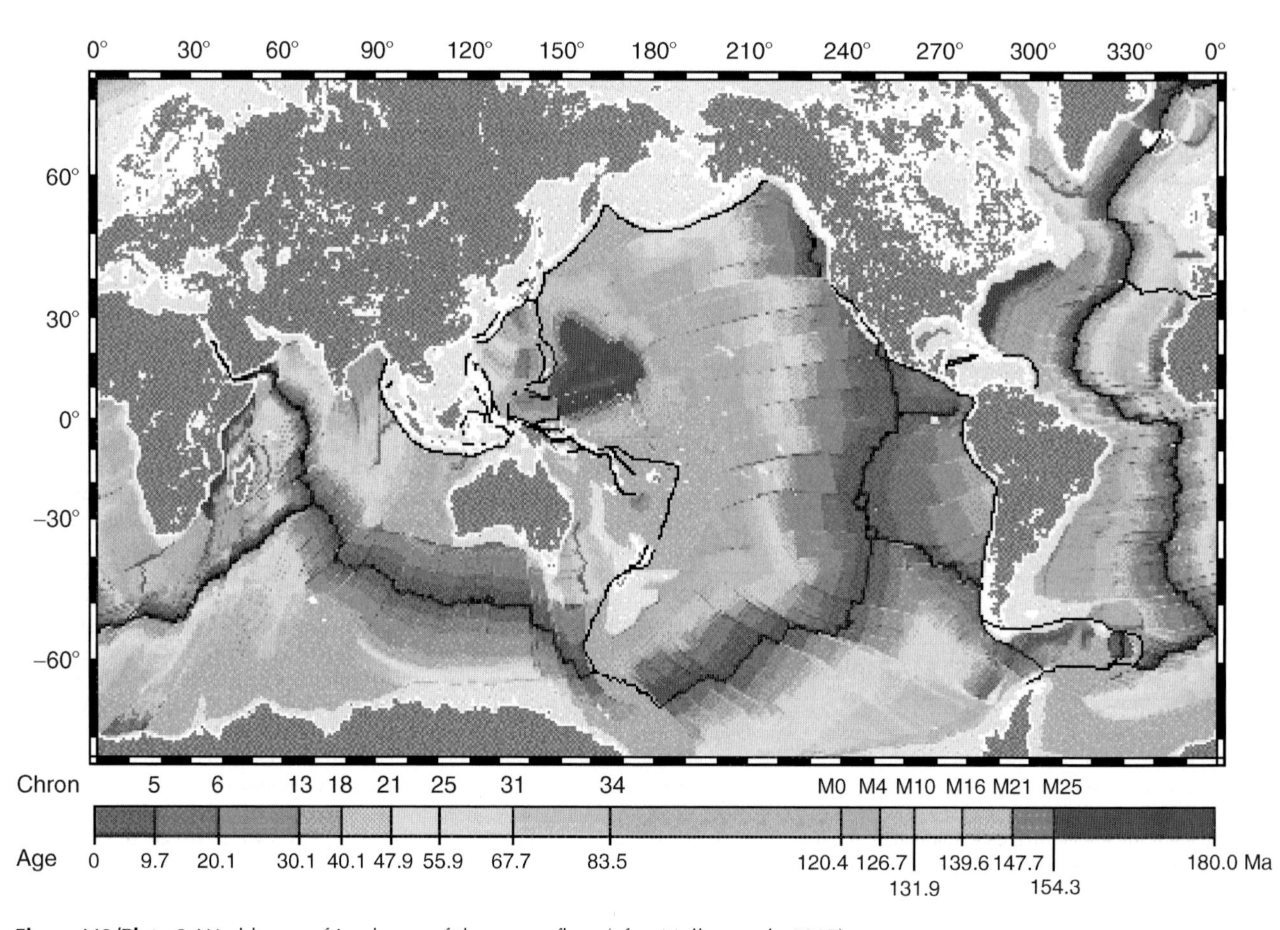

Figure M9/Plate 8 World map of isochrons of the ocean floor (after Muller *et al.*, 2005).

program is commonly used to calculate the anomaly over a block. These programs assume the block is infinitely long in some direction (2D) rather than having a finite three-dimensional shape (3D), making the calculation simpler. The shape of the source block is changed until the calculated anomaly matches the observed anomaly. This technique is called a forward anomaly calculation. It is inherently nonunique in that several different shapes may cause the same anomaly. Other types of programs assign one dimension, for example the thickness, and then use the observed anomalies to locate the source body. Other programs convert the observed spatial series to a frequency spectrum and filter the spectrum to locate the bodies. It is also possible to mathematically convert the observed anomaly series to show how it would look at the magnetic pole. This transformed anomaly will be directly over its causative body.

Normal and reverse intervals of the geomagnetic field occur at irregular time intervals creating a pattern of magnetic anomalies with irregular spacing. Since this same irregular pattern is found in the different oceans of the world it aids in dating the anomalies in one ocean if they were already dated in another ocean. To refer to anomalies along an anomaly profile, the anomalies have been given numbers 1 (at the axis) to 34 (about 83 Ma). Intermediate intervals are assigned a "chron" notation (see *Geomagnetic polarity timescale*, *q.v.*).

Older marine magnetic anomalies

Long shipboard magnetic profiles extending across the oceans showed a similar sequence of magnetic anomalies, although the shape varies a bit with latitude and the azimuth of the magnetic lineation. This suggested that seafloor spreading had been constant in each major ocean, although the spreading rate may be different from one ocean to another. In the South Atlantic anomaly 34 is close to the continental margins of South America and South Africa, indicating that the South Atlantic Ocean opened about 83 Ma ago. In the North Atlantic anomaly 34 does not reach the continental margins, because the North Atlantic Ocean is older than the South Atlantic Ocean.

In the Northwest Pacific the anomalies are linear but not subparallel to existing spreading centers. The spreading center, which created them, has since disappeared and the seafloor spreading has changed direction. Many are older than anomaly 34 and are assigned a number prefixed by the letter M to show that they be part of another (Mesozoic) series. Near the edges of the Pacific Ocean there are subduction zones where the seafloor turns under the continents. The magnetic anomaly pattern is extinguished at these subduction zones, such as off Alaska, off the Aleutians, off Japan, and at the Kermadac Trench north of New Zealand. It was found that, in the South Pacific Ocean, anomalies are symmetric about the ridge axis, even when the anomalies are separated by more than 1000 km. The Indian Ocean and the Arctic Ocean south of Australia abound with dated magnetic anomalies, although the oldest are not as old as the oldest in the Northwest Pacific.

By using a long profile in the South Atlantic Ocean the sequence of magnetic reversals (geomagnetic timescale) were originally determined for the last 80 Ma, where there were 171 reversals. By comparisons with dated rocks from the seafloor the original geomagnetic timescale has been slightly revised (Cande and Kent, 1995).

James R. Heirtzler

Bibliography

Cande, S.C., and Kent, D.V., 1995. Revised calibration of the geomagnetic polarity timescale for the late Cretaceous and Cenozoic. *Journal of Geophysical Research*, **100**: 6093–6095.

Heirtzler, J.R., LePichon, and Xavier Gregory, B.J., 1966. Magnetic anomalies over the Reykjanes Ridge. *Deep-Sea Research*, **13**: 427–443.

Muller, R.D., Roest, W.R., Royer, Y.-Y., Gahagan, L.M., and Sclater, J.G., 2005. Digital isochrons of the ocean floor. Available at: htto://gdcinfo.agg.nrcan.gc.ca/app/images/agemap.gif

Vacquier, V.V., 1959. Measurement of horizontal displacements along faults in the ocean floor. *Nature*, **183**: 452–453.

Vine, F.J., and Matthews, D.H., Magnetic anomalies over oceanic ridges. *Nature*, **199**: 947–949.

Cross-references

Geomagnetic Polarity Timescale
Geomagnetic Polarity Timescales
Magnetization, Oceanic Crust
Magnetization, Remanent, Application
Seamount Magnetism

MAGNETIC ANOMALIES, MODELING

Introduction

The magnetic anomalies of planetary lithosphere reflect the lateral variations of magnetization in the upper most part of the lithosphere that is colder than the Curie temperature of its magnetic minerals. The variations may arise from several processes, such as intrusive bodies with different magnetization than the country rocks, the juxtaposition of crustal blocks of different magnetization through plate tectonics, and the creation of the oceanic crust during different core field polarity periods.

A magnetic anomaly is the projection of the magnetic field of a body in the direction of the local geomagnetic field. The measured magnetic data have contributions from the core field, the lithospheric field, the external field, the field arising from the induced currents in the mantle because of the time variations of the external field, and the measurement errors. The core and the lithospheric fields are time-invariant within a few years, while the external field is strongly time-dependent. A time-varying external field with a period of τ is translated to a special variation of wavelength $\lambda = \tau v$ when the magnetic data is acquired by a vehicle (airplane, ship, or satellite) moving at speed v, and contaminates the lithospheric field of the same wavelength. To avoid the contamination, the magnetic data acquired during the quiet times of the external field are generally used. The core field, with amplitudes ranging from 30000 to 70000 nT, dominates the lithospheric field, which has usually an amplitude of a few hundred nanotesla near the surface and less than 20 nT at satellite altitudes of $\sim$400 km. The instrument errors are nowadays negligible, but the measurement error can be appreciable especially for the satellite magnetic data. An attitude error of only 10″ in the orientation of the coordinate axes fixed to a satellite can produce an error on the order of 2–4 nT, depending on the satellite's latitude, which is comparable to a vector component of the lithospheric magnetic field at the satellite elevations. One possible way of removing the error arising from the attitude is to use the magnetic intensity data, which is independent of the orientation of the coordinate system. Let $\vec{F}$ denote the measured magnetic field and assume that it consists of the core field $\vec{B}$ and the lithospheric field $\vec{f}$, ignoring the external field. Then

$$|\vec{F}| = |\vec{B} + \vec{f}| \quad \text{(Eq. 1)}$$

The magnetic anomaly T is the difference between the intensity of the observed field and that of the local core field,

$$T = |\vec{F}| - |\vec{B}| \quad \text{(Eq. 2)}$$

which to the first approximations is reduced to

$$T = \vec{f} \cdot \hat{B} \tag{Eq. 3}$$

where $\hat{B}$ is the unit vector in the core field direction. Almost all aeromagnetic measurements on Earth result in magnetic anomalies as defined in Eq. (3). In case of a planet with no core field, such as Mars, a magnetic anomaly simply refers to the magnetic field of the lithosphere.

This article addresses modeling of the magnetic anomalies of the planetary lithosphere. The first section describes the forward modeling, where the shape and magnetization of a magnetic body is given and the magnetic anomaly of the body is to be determined. The second section considers the inverse problem, where the magnetic anomalies are measured and the shapes and magnetization of the magnetic source bodies are to be determined.

Forward modeling

The magnetic potential V of a magnetic body is

$$V(\vec{r}) = -\frac{\mu_0}{4\pi}\iiint \vec{M}(\vec{r}_0) \cdot \nabla \frac{1}{|\vec{r} - \vec{r}_0|} \mathrm{d}v_0 \tag{Eq. 4}$$

where μ_0 is the magnetic permeability of free space ($4\pi \times 10^{-7}$ H m^{-1}), $\vec{M}(\vec{r}_0)$ is the magnetic dipole moment per unit volume, called magnetization, and the integration is over the entire volume of the body. Equation (4) is quite complicated especially when the body is irregular in shape and has heterogeneous magnetization. In forward modeling it is usually assumed that the magnetic body is uniformly magnetized in a given direction. The size and magnetization of the source body is determined by fitting the model anomaly to the observed one. To better fit the model field to the observation, it is often required to divide the body into several parts with different magnetization, but each part has still a uniform magnetization. The constant magnetization assumption reduces Eq. (4) to

$$V(\vec{r}) = -\frac{\mu_0}{4\pi}\iint \frac{\vec{M}}{|\vec{r} - \vec{r}_0|} \cdot \vec{\mathrm{d}s}_0 \tag{Eq. 5}$$

upon the application of the divergence theorem. $\vec{\mathrm{d}s}_0$ is a surface element and the integration is over the entire surface of the body. This equation shows that the magnetic field outside a uniformly magnetized body is equivalent to the magnetic field of magnetic poles (north or south) distributed on the surface of the body. A uniformly magnetized body acts by its surface rather than its entire volume. For example, to determine the magnetic field of a vertically upward magnetized vertical prism it is sufficient to assume a north pole distribution on the upper surface and a south pole distribution on the bottom surface. An interesting case is a uniformly magnetized infinite horizontal layer of constant thickness, which produces no magnetic field outside regardless of its magnetization intensity and direction. Such a layer is used as an annihilator in the inverse problems (e.g., Parker and Huestis, 1974).

Equation (4) can also be written as

$$V(r) = -\left(\frac{\mu_0}{4\pi\rho G}\right)\vec{M} \cdot \nabla \iiint \frac{G\rho}{|\vec{r} - \vec{r}_0|} \mathrm{d}v_0 \tag{Eq. 6}$$

for a uniformly magnetized body, where ρ denotes an arbitrary constant density of the body and G is the gravitational constant. This equation is further reduced to

$$V(r) = \left(\frac{\mu_0 M}{4\pi\rho G}\right)\hat{M} \cdot \nabla U(\vec{r}) \tag{Eq. 7}$$

where $\hat{M}$ is the unit vector of the magnetization and $U(\vec{r})$ denotes the gravitational potential of the body with the density ρ. Equation (7) emphasis that the magnetic field outside a uniformly magnetized body is just the gradient of the gravitational potential of the body in the direction of its magnetization vector, with a multiplying factor $(\mu_0 M/4\pi\rho G)$ called the Poisson factor. This tremendously simplifies the calculations of the magnetic field of a body, from integration of an integrand involving vector functions to that includes only a scalar function. A simple example is the magnetic potential of a uniformly magnetized sphere of radius R and magnetization $\vec{m}_0$ centered at $\vec{r}_0$, which according to Eq. (7) is

$$V(\vec{r}) = -\left(\frac{\mu_0}{4\pi}\right)\frac{4\pi}{3}R^3\vec{m}_0 \cdot \nabla \frac{1}{|\vec{r} - \vec{r}_0|} \tag{Eq. 8}$$

which is equivalent to the magnetic potential of a magnetic dipole located at the center of the sphere and has a dipole moment of $\vec{P}(\vec{r}_0) = (4\pi/3)R^3\vec{m}_0$.

Several techniques developed in the space domain, to determine the magnetic field of an arbitrary-shaped body, have been used for forward modeling of an isolated magnetic anomaly (see Chapter 9 of Blakely, 1996 for details, also Telford *et al.*, 1990). Many of the methods use the above-mentioned characteristics of the magnetic field of a uniformly magnetized body. Here I briefly explain two methods; one uses the above characteristics and the other does not. The first (e.g., Talwani, 1965) subdivides a magnetic body into thin horizontal slabs, thin enough to reliably assume the walls of the slab are vertical, thus reducing the integral in the z-direction of Eq. (4) to summation over the slabs. The method uses the divergence theorem, a two-dimensional version of Eq. (5), to reduce the remaining surface integrals in Eq. (4) into a single line integral, the line describes the circumference of the slab, which is approximated by a polygon for easy integration. This method has been extensively used in aeromagnetic and satellite magnetic anomaly modeling (e.g., Plouff, 1976). A simple example of this method is the vertical prism discussed above. Because the magnetic body is divided into many thin slabs, the method can also handle a magnetic body with depth-dependent magnetization. The other method is the dipole array method where a magnetic body is subdivided into small volume elements, small enough to approximate the magnetization inside a volume element by a constant vector and the magnetic field of the volume element by that of a magnetic dipole located at the center of the volume element. This method does not require a uniform magnetic body, because it is always possible to subdivide the body into very many small volume elements of constant magnetization. Although the formulation in the rectangular coordinate system is straightforward, there has been some incorrect formulation in spherical coordinate system (see Dyment and Arkani-Hamed, 1998 for details). The dipole array method has been used in modeling satellite magnetic anomalies of Earth and Mars, where dipole arrays are placed on the surface of the planets (e.g., Mayhew, 1979; Purucker *et al.*, 2000; Langlais *et al.*, 2004). The space-domain methods, including the two explained, generally require a huge amount of mathematical manipulations and computer time. But the main problem with these methods is that they do not provide insight to the general relationship between the magnetic anomalies and the magnetization of the crust, that the magnetic anomalies arise from lateral variations of the vertically integrated magnetization. This is an important issue for understanding the resolving power of magnetic data. The advantage of the space-domain methods is the sharp boundaries of the model magnetic bodies. This enables us to integrate magnetic data with other geophysical data, such as seismic, and obtain a better model for a given magnetic anomaly. The Fourier domain methods, on the contrary, yield lateral variations of the magnetization where boundaries of the magnetic bodies are not sharply defined.

In the following I explain the Fourier domain technique used in forward modeling. The Fourier domain formulation better presents the relationship between the magnetization of the body and its magnetic field.

Consider a magnetic layer of magnetization $\vec{M}(\vec{r}_0)$ with an upper surface S_2 and a lower surface S_1. The surfaces are piecewise continuous. In a rectangular coordinate system with the z-axis upward, Eq. (4) is written as

$$V(\vec{r}) = -\frac{\mu_0}{4\pi}\int_{-\infty}^{\infty}\int_{-\infty}^{\infty}\left[\int_{S_1}^{S_2}\vec{M}(\vec{r}_0)\cdot\nabla\frac{1}{|\vec{r}-\vec{r}_0|}\mathrm{d}z_0\right]\mathrm{d}x_0\mathrm{d}y_0 \quad \text{(Eq. 9)}$$

Taking two-dimensional Fourier transform of this equation yields (Arkani-Hamed and Strangway, 1986)

$$V_{\xi\eta} = -\frac{\mu_0}{2}\frac{\mathrm{e}^{-Kz}}{K}\vec{\beta}\cdot\vec{\alpha}_{\xi\eta} \quad \text{(Eq. 10)}$$

where

$$\vec{\beta} = (i\xi, i\eta, -K) \quad \text{(Eq. 11)}$$

and

$$K = (\xi^2 + \eta^2)^{1/2} \quad \text{(Eq. 12)}$$

is the two-dimensional wave number and ξ and η are wave numbers is the x and y directions. $\vec{\alpha}_{\xi\eta}$ is the Fourier transform of

$$\vec{\alpha} = \int_{S_1}^{S_2}\vec{M}\mathrm{e}^{Kz_0}\mathrm{d}z_0. \quad \text{(Eq. 13)}$$

Equations (10) and (13) show that the magnetic anomaly arises from the lateral variations of the bulk magnetization (the vertically integrated magnetization) of the source body, rather than detailed variations of magnetization with depth. In a local region where the Fourier domain formulation is applicable the core field has almost a constant direction $\hat{B}_0$. The Fourier transform of the magnetic anomaly is then reduced to

$$T_{\xi\eta} = -(\hat{B}_0\cdot\vec{\beta})V_{\xi\eta} \quad \text{(Eq. 14)}$$

Therefore, the main task is to determine the magnetic potential of a body.

Figure M10 shows the magnetic anomaly of a rectangular body with a square horizontal cross section of 1×1 km and a thickness of 100 m. The top of the prism is at the surface and its sides are in the east and north directions. The body is located at the geomagnetic north pole (column 1), at 45° N geomagnetic latitude (column 2), and at the geomagnetic equator (column 3). It is magnetized in the direction of the core field with an intensity of 1 A m^{-1}. The magnetic anomalies are calculated at 100 m (row 1), 500 m (row 2), and 1 km (row 3) altitudes. The figure illustrates the complex characteristics of the magnetic anomaly that change drastically with the body location and the observation elevation. At the 100 m elevation the anomaly delineates the edges of the body, all the four edges at the north, only the north and south edges at the equator, and mainly the north and south edges at 45° N but slightly shifted. As the elevation increases the anomalies over the edges migrate toward the center and finally merge to give rise to a single positive anomaly at the north, a single negative anomaly at the equator, but a skewed anomaly at 45° N. The anomaly is directly over the body at the north and at the equator, but is significantly shifted toward the equator at 45° N. Moreover, the amplitude of the anomaly changes as the body moves from the pole toward the equator. This is due to the changes in the direction of the geomagnetic field, and thus the magnetization. In practice the magnetization intensity, which is assumed constant in these examples also decreases at lower latitudes, because of the decrease in the intensity of the magnetizing geomagnetic field.

The difficulty in relating the magnetic anomalies to their source bodies (Figure M10) is usually alleviated by applying a reduction-to-pole operator that reduces an anomaly to that as if the body is located at the north pole and is magnetized by the geomagnetic field there (e.g., Baranov, 1957; Bhattacharyya, 1965). That is, the anomalies seen in column 2 are reduced to those in column 1. On a regional scale where the core field direction and intensity change appreciably over the entire map a differential reduction-to-pole operator (Arkani-Hamed, 1988) is required to take into account these changes. However, all reduction-to-pole operators are singular at the geomagnetic equator and cannot be used to reduce the anomalies in column 3 to those in column 1.

Inverse modeling

The inverse modeling of a local magnetic anomaly map is simplified because the core field is constant over the entire map. Blakely (1996) discusses this issue in a great detail. However, the inverse modeling of a global map is quite complicated. This is largely due to the fact that the direction and intensity of the core field drastically change over the globe (see Figure M10). Unlike the gravity anomaly of a body that attains the same maximum value directly over the body regardless of its location, the magnetic anomaly has quite complex characteristics, which is a source of great confusion among those who treat the magnetic anomaly maps similar to the gravity anomaly maps.

Here I briefly discuss major points of a generalized inversion technique developed by Arkani-Hamed and Dyment (1996) under the assumption that the anomalies arise from the induced magnetization, which is largely the case in the continents (e.g., Schnetzler and Allenby, 1983). It takes into account the global variations of the core field and transforms a global magnetic anomaly map into a global magnetic susceptibility contrast map of the lithosphere. Unlike the magnetic anomaly map, the magnetic susceptibility contrast map delineates the very nature of the magnetic properties of the lithosphere. Even a magnetization map does not delineate the nature of the magnetic lithosphere, because it is resulted from the multiplication of the magnetic susceptibility and the core field intensity which changes by more than a factor of 2 over the globe. The geomagnetic coordinate system is adopted throughout this inversion technique and the technique is based on spherical harmonic representations of the magnetic anomalies and the magnetic susceptibility distribution, which are the natural presentations on a global scale.

Let V be the magnetic potential of the lithosphere at an observation point $\vec{r}$, and $\hat{B}$ denote the unit vector in the core field direction at that point. The magnetic anomaly T is

$$T = -\hat{B}\cdot\nabla V \quad \text{(Eq. 15)}$$

Let $\hat{B}_1$ be the unit vector along the dipole component of the core field,

$$\hat{B}_1 = -\frac{(2\cos\theta\hat{r} + \sin\theta\hat{\theta})}{(1 + 3\cos^2\theta)^{1/2}} \quad \text{(Eq. 16)}$$

and write

$$\hat{B} = \hat{B}_1 + \vec{\delta}B \quad \text{(Eq. 17)}$$

where $\hat{r}$ and $\hat{\theta}$ are the unit vectors in the r and the colatitude θ directions, respectively, and $\vec{\delta}B$ denotes the contribution to $\hat{B}$ from the non-dipole part of the core field. Also, let

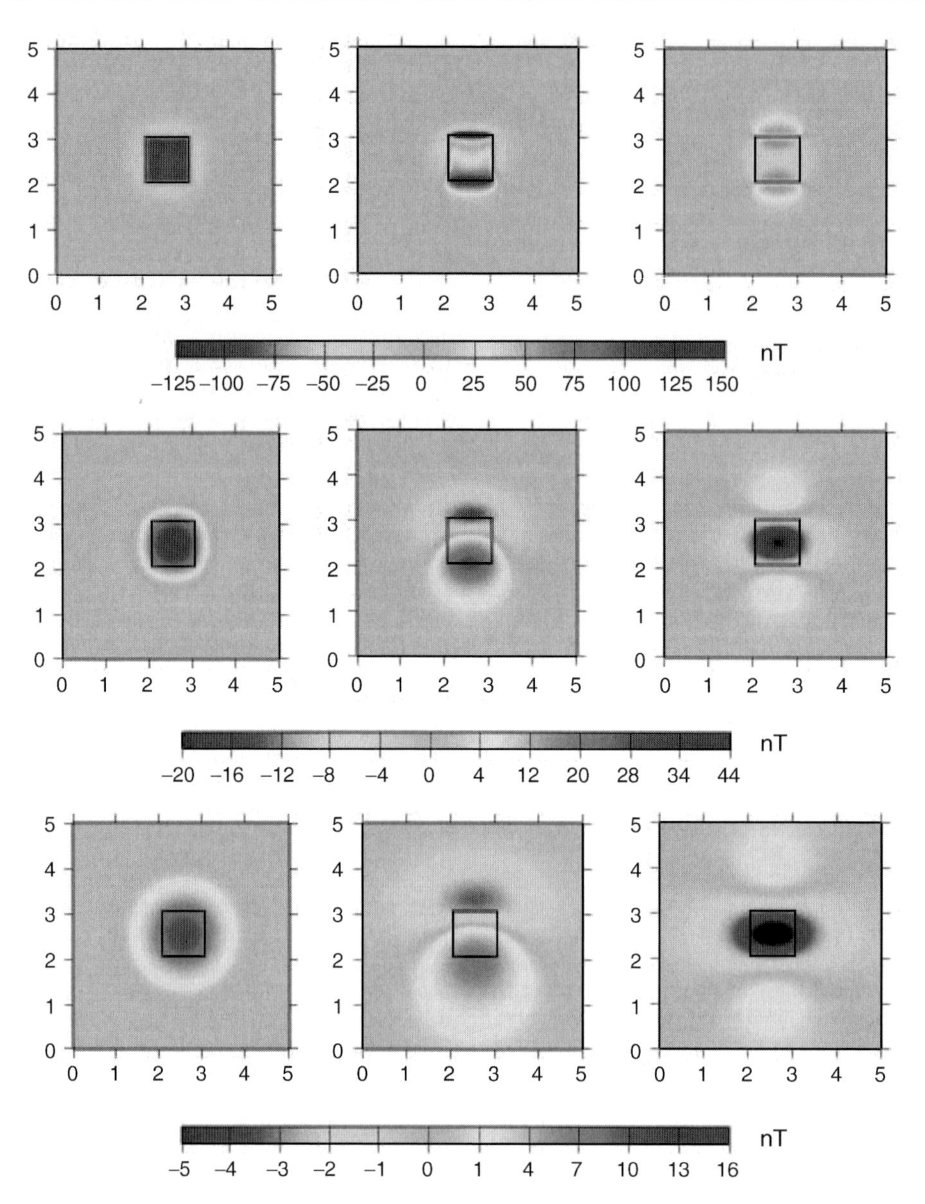

Figure M10 Magnetic anomaly of a vertical prism of 1 × 1 km square cross section and 100 m thickness, having 1 A m^{-1} magnetization. The first column shows the anomalies at the north pole, the second column shows the anomalies at 45° N latitude, and the third column shows the anomalies at the equator. The first row shows the anomalies at 100 m, the second row at 500 m, and the third row at 1000 m elevations.

$$\tau = (1 + 3\cos^2\theta)^{1/2}T \qquad \text{(Eq. 18)}$$

and

$$A = (1 + 3\cos^2\theta)^{1/2}\vec{\delta}B \cdot \nabla V \qquad \text{(Eq. 19)}$$

A is a measure of the contribution to the magnetic anomalies arising from the coupling of the nondipole part of the core field with the magnetic potential of the lithosphere. Now expanding V, τ, and A in terms of the spherical harmonics, and putting Eqs. (16)–(19) into Eq. (15) yields coupled equations that are solved iteratively to determine the spherical harmonic coefficients of the magnetic potential using those of the observed magnetic anomalies. It is worth noting, however, that the inversion from a global magnetic anomaly map to a magnetic potential map is unstable if the core field is dipolar (Backus, 1970). Although the entire core field is used in the calculations, the strong dominance of the dipole component of the core field still tends to enhance the sectorial harmonics of the potential, though slightly.

The next step is to calculate the magnetic susceptibility contrasts in the lithosphere from the magnetic potential thus obtained. Let $\sigma(\vec{r}_0)$ denote the magnetic susceptibility of a volume element dv_0 located at the point $\vec{r}_0$ in the lithosphere, then the induced magnetization of the volume element $\vec{M}(\vec{r}_0)$ is

$$\vec{M}(\vec{r}_0) = \frac{1}{\mu_0}\sigma(\vec{r}_0)\vec{B}'(\vec{r}_0)dv_0 \qquad \text{(Eq. 20)}$$

in which $\vec{B}'(r_0)$ is the core field at that point. As mentioned earlier, the most one can achieve is to determine the vertically (radially) averaged magnetization of the lithosphere regardless of the technique used. This is the fundamental nonuniqueness of the inversion of magnetic anomalies. Therefore, we model the magnetic part of the lithosphere by a thin magnetic spherical shell, and seek for a vertically averaged magnetic susceptibility, i.e., we let

$$\sigma(\vec{r}_0) = \sigma_0(\theta_0, \phi_0) \quad \text{(Eq. 21)}$$

Similarly, the core field is assumed to be constant with depth within the shell, i.e.,

$$\vec{B}'(\vec{r}_0) = \vec{B}'_0(\theta_0, \phi_0) \quad \text{(Eq. 22)}$$

The core field intensity changes by less than 3% in the upper 50 km of the Earth and its direction hardly changes with depth in this region. Now decomposing $\vec{B}'_0$ into a dipole part, $\vec{B}'_1$, and a nondipole part, $\vec{\delta B}'$,

$$\vec{B}'_0 = \vec{B}'_1 + \vec{\delta B}' \quad \text{(Eq. 23)}$$

and letting W denote the magnetic potential of the shell arising from the magnetization induced by the nondipole part of the core field, we obtain

$$W = \frac{1}{4\pi}\int \sigma_0(\theta_0, \phi_0)\vec{\delta B}'(\theta_0, \phi_0) \cdot \nabla \frac{1}{|\vec{r} - \vec{r}_0|}\,\mathrm{d}v_0 \quad \text{(Eq. 24)}$$

Expanding σ and W in spherical harmonic and putting Eqs. (20)–(24) into Eq. (4) results again in coupled equations that are solved iteratively for the spherical harmonic coefficients of the magnetic susceptibility contrast from those of the magnetic potential.

The coupled equations that are iteratively solved for the magnetic potential using the magnetic anomaly, and for the magnetic susceptibility contrast using the magnetic potential, show that the spherical harmonic coefficients of different degrees of the magnetic susceptibility contrasts couple to those of the core field to produce a given spherical harmonic coefficient of the observed magnetic anomalies. This is a fundamental characteristic of a nonlinear inverse problem.

The above inversion technique is applied to the global magnetic anomaly map of the Earth (Figure M11/Plate 6a) derived from POGO and Magsat data that includes the spherical harmonics of degree 15–60 (Arkani-Hamed *et al.*, 1994). The harmonic coefficients of the magnetic anomalies with degree lower than 15 are dominated by the core field, and those higher than 60 are largely of nonlithospheric origin. The magnetic lithosphere is modeled by a spherical shell of thickness 40 km. Figure M12/Plate 6b shows the resulting vertically averaged magnetic susceptibility contrasts. Detailed interpretation of the features seen in the figure is beyond the scope of this article. Here some basic characteristics of the susceptibility contrast map that illustrate the effects of the inversion process are briefly explained.

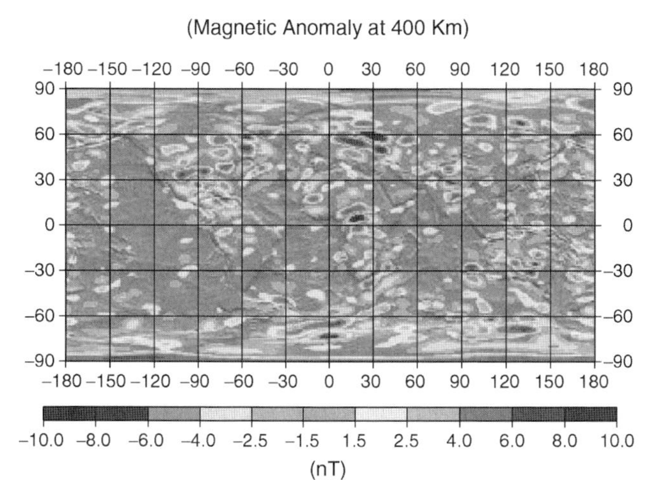

Figure M11/Plate 6a Magnetic anomaly of the Earth's lithosphere at 400 km altitude derived from POGO and Magsat data (Arkani-Hamed *et al.*, 1994).

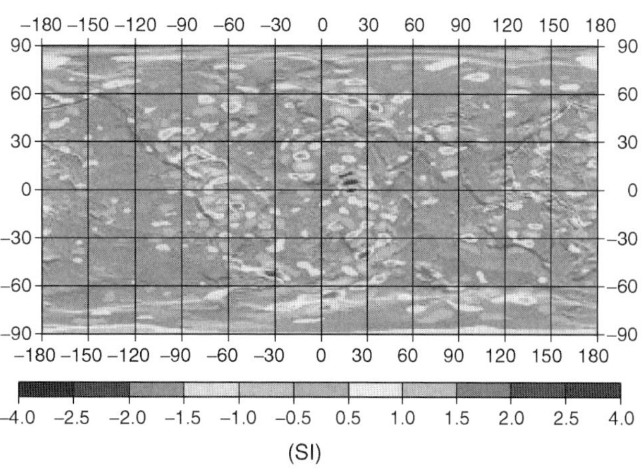

Figure M12/Plate 6b Magnetic susceptibility contrasts of a 40 km thick spherical shell in the upper part of the lithosphere required to give rise to the anomalies seen in Figure M11/Plate 6a.

Comparison of Figures M11/Plate 6a and M12/Plate 6b shows the major effects of the inversion. The anomalies near the polar regions have retained their sign, whereas the anomalies in the equatorial region have changed their sign. The magnetic susceptibility contrasts associated with the magnetic anomalies in the midlatitudes are shifted poleward with respect to the magnetic anomalies. Moreover, the amplitude of the magnetic anomalies near the poles is significantly greater than those in the equatorial region, whereas the magnetic susceptibilities of their source bodies are comparable. Also, the inversion procedure in effect includes downward continuation. The higher degree harmonics of the resulting magnetic susceptibility contrast map have enhanced compared to the corresponding harmonics of the magnetic anomaly map at satellite altitudes. Consequently, some of the broad features in the anomaly map are divided into two or more small size features in the susceptibility contrast map.

It is worth reminding that Figure M12/Plate 6b shows the vertically averaged magnetic susceptibility contrasts within a spherical shell of 40 km thickness that can give rise to the magnetic anomalies seen in Figure M11/Plate 6a. Figure M12/Plate 6b shows the contrasts rather than the absolute magnetic susceptibility. The areas with positive and negative susceptibility contrasts should be regarded as the high- and low-magnetic areas. A constant magnetic susceptibility is usually added to the resulting susceptibility contrast map to render all values positive, without affecting the observed magnetic anomalies. This is due to the fact that a spherical shell of uniform magnetic susceptibility produces no magnetic anomaly when magnetized by an internal magnetic field such as the core field (Runcorn, 1975). The nonuniqueness of the inverse problem in magnetic anomaly modeling, which is discussed in detail by Arkani-Hamed and Dyment (1996, see their Appendix B), concludes a large range of susceptibility distribution that give rise to no magnetic anomalies outside the lithosphere. The Runcorn model is one of the simple cases. These susceptibility distributions can be added to a resulting susceptibility map without affecting the resulting magnetic anomaly map.

Conclusions

This article presents both forward and inverse modeling of magnetic anomalies. In the forward modeling the basic characteristics of the anomalies are discussed in some detail, and a Fourier domain technique is presented in detail. The magnetic anomaly of a vertical prism of square cross section is shown at three elevations and at three latitude locations, to illustrate its complex behavior. The inversion of a global magnetic anomaly map into a global magnetic susceptibility contrasts map consists of two nonlinear operations. The first inverts a magnetic anomaly map into a magnetic potential map, while the second transforms the potential map into a magnetic susceptibility contrast map. The nonlinearity arises from the variations in the direction and intensity of the core field over the globe. The inversion technique removes the adverse effects of these variations, and results in the magnetic susceptibility contrasts that represent the very nature of the vertically averaged magnetic properties of the lithosphere.

Acknowledgment

This research was supported by the Natural Sciences and Engineering Research Council (NSERC) of Canada.

Jafar Arkani-Hamed

Bibliography

Arkani-Hamed, J., 1988. Differential reduction to the pole of regional magnetic anomalies. *Geophysics*, **53**: 1592–1600.

Arkani-Hamed, J., and Dyment, J., 1996. Magnetic potential and magnetization contrast of Earth's lithosphere. *Journal of Geophysical Research*, **101**: 11,401–11,425.

Arkani-Hamed, J., and Strangway, D.W., 1986. Magnetic susceptibility anomalies of lithosphere beneath eastern Europe and the Middle East. *Geophysics*, **51**: 1711–1724.

Arkani-Hamed, J., Langel, A.L., and Purucker, M., 1994. Scalar magnetic anomaly maps of Earth derived from POGO and Magsat data. *Journal of Geophysical Research*, **99**: 24075–24090.

Backus, G.E., 1970. Non-uniqueness of the external geomagnetic field determined by surface intensity measurements. *Journal of Geophysical Research*, **75**: 6339–6341.

Baranov, V., 1957. A new method for interpretation of aeromagnetic maps: pseudo-gravimetric anomalies. *Geophysics*, **22**: 359–383.

Bhattacharyya, B.K., 1965. Two-dimensional harmonic analysis as a tool for magnetic interpretation. *Geophysics*, **30**: 829–857.

Blakely, R., 1996. *Potential Theory in Gravity and Magnetic Applications.* Cambridge: Cambridge University Press.

Dyment, J., and Arkani-Hamed, J., 1998. Equivalent source magnetic dipoles revisited. *Geophysical Research Letters*, **25**: 2003–2006.

Langlais, B., Purucker, M.E., and Mandea, M., 2004. Crustal magnetic field of Mars. *Journal of Geophysical Research*, **109**: E02008, doi:10.1029/2003JE002048.

Mayhew, M.A., 1979. Inversion of satellite magnetic anomaly data. *Journal of Geophysical Research*, **45**: 119–128.

Parker, R.L., and Huestis, S.P., 1974. The inversion of magnetic anomalies in the presence of topography. *Journal of Geophysical Research*, **79**: 1587–1593.

Plouff, D., 1976. Gravity and magnetic fields of polygonal prisms and application to magnetic terrain corrections. *Geophysics*, **41**: 727–741.

Purucker, M.E., Ravat, D., Frey, H., Voorhies, C., Sabaka, T., and Acuna, M., 2000. An altitude normalized magnetic map of Mars and its interpretation. *Geophysical Research Letters*, **27**: 2449–2452.

Runcorn, S.K., 1975. On the interpretations of lunar magnetism. *Physics of the Earth and Planetary Interiors*, **10**: 327–335.

Schnetzler, C.C., and Allenby, R.J., 1983. Estimation of lower crustal magnetization from satellite derived anomaly field. *Tectonophysics*, **93**: 33–45.

Talwani, M., 1965. Computation with the help of a digital computer of magnetic anomalies caused by bodies of arbitrary shape. *Geophysics*, **30**: 797–817.

Telford, W.M., Geldard, L.P., and Sheriff, R.E., 1990. *Applied Geophysics*, 2nd edn. New York: Cambridge University Press.

MAGNETIC DOMAINS

Introduction

The existence of a paleomagnetic record testifies to the ability of magnetic minerals in rocks to retain their natural remanent magnetizations (NRMs) over geologic time. In the early days of paleomagnetism, it was thought that the stable components of NRM mainly resided in extremely small magnetic mineral grains, which occupied the single-domain (SD) state. However, it is now recognized that, due to their very small size and scarcity, SD particles may not be the major carriers of NRM in many rocks. Instead, it is more likely that much of the NRM is carried by grains which, by virtue of their larger sizes, are subdivided into two or more magnetic domains (Figure M13).

For a component of NRM to survive over geologic time, a particle's domain structure must remain stable in two ways. First, the initial domain state, whether it consists of a single domain or several domains, must resist being reset by the various physical and chemical forces which can affect a rock after emplacement, such as moderately elevated temperature, moderate pressure, slight chemical alteration, and shifts in the direction and magnitude of the Earth's field. In short, the fundamental domain structure—that is, the number of domains and the overall geometric style in which they are arranged in a crystal—should not change greatly over time. Second, in particles consisting of two or more domains, the domain walls (the transition regions between adjacent domains) present initially must remain locked in their positions, despite the aforementioned perturbations. Likewise, coercivities in a population of SD particles must remain sufficiently high over geologic time, so that paleomagnetic directions are not lost. Thus, magnetic domain structure lies at the heart of paleomagnetic stability.

In this article, we first discuss some of the fundamental principles and theoretical results of both classical magnetic domain theory and more recent micromagnetic models. Next, we discuss results of experiments to investigate how domain structure depends on grain size, applied field, and temperature in natural magnetic minerals and their synthetic analogs. Here, we focus on particles that contain two or more domains, such particles being prevalent in most rocks. For more detailed treatments of these subjects, the reader is referred to standard texts such as Chikazumi (1964), Cullity (1972), Stacey and Banerjee (1974), Dunlop and Özdemir (1997), and other references listed at the end of this article.

Classical theory of magnetic domain structure

Theories of magnetic domain structure are based on three energies: exchange energy, anisotropy energy, and magnetostatic energy. In most

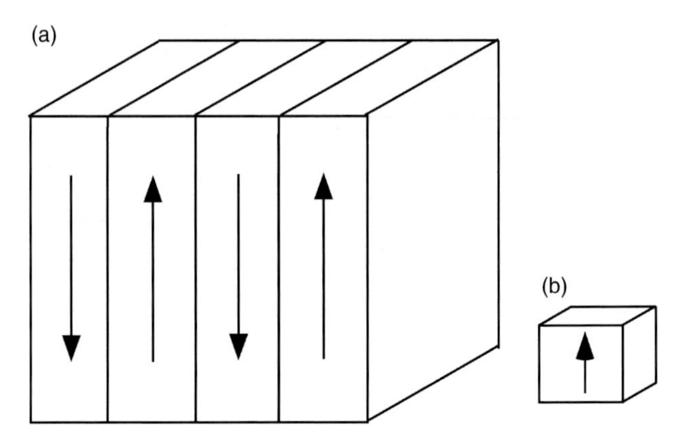

Figure M13 Illustrations of (a) a multidomain cube containing four domains and (b) a single-domain cube of uniform magnetization.

classical theories, the first two energies produce the surface energy of the domain wall. Magnetostatic energy, the third energy, is the potential energy required to assemble an array of "magnetic free poles" where magnetic spins terminate at the domains' surfaces. Magnetostatic energy is the fundamental reason why particles subdivide into two or more domains because, in so doing, magnetostatic energy is reduced. However, as discussed below, this subdivision carries an energy "price": the energy of domain walls. Consequently, below a certain transition size particles occupy the SD state. Above this size, particles normally contain two or more domains, the number of which generally increases with the size of the particle. In this section, we discuss how traditional theories predict and how domain structure in a given material depends on grain size and temperature.

Exchange energy

The origin of permanent magnetism was originally postulated by Weiss (1907) in terms of a "molecular field" that promotes the alignment of atomic spins in a ferromagnetic material. Subsequently, the molecular field was shown to be the phenomenological expression of a quantum-mechanical phenomenon, called the exchange interaction. Seminal calculations by Heisenberg (1928) demonstrated that the exchange energy between two adjacent spin vectors $\boldsymbol{S}_1$ and $\boldsymbol{S}_2$ is given by

$$E_{ex} = -2J_{ex}\boldsymbol{S}_1 \cdot \boldsymbol{S}_2 = -2J_{ex}\boldsymbol{S}_1\boldsymbol{S}_2 \cos\theta_{12}$$

Here, J_{ex} is the exchange integral of the specific material and θ_{12} is the angle between the two adjacent spins. In the vast majority of natural and synthetic elements and compounds, the exchange integral is negative; in this case, adjacent spins which are antiparallel yield the configuration of lowest energy. Such materials are incapable of permanent magnetism. In those few materials whose exchange integral is positive, the configuration of lowest energy is achieved when neighboring spins are aligned. This gives rise to the rare phenomenon of spontaneous magnetization (M_s) and remanence. Such materials possess a Curie temperature, at which spontaneous magnetism vanishes, because (a) thermal agitation disrupts the alignment of adjacent spins, and (b) thermal expansion of the lattice increases interatomic (or intermolecular) distances and thus reduces the effective strength of E_{ex}.

Magnetocrystalline anisotropy energy

Magnetic anisotropy causes magnetic properties to depend on the direction in which they are measured. For example, due to magnetocrystalline anisotropy, it requires a lower field to saturate a magnetite crystal along the <111> directions than along <100>. Magnetocrystalline anisotropy arises from the coupling between spins and orbits of the electrons. Thus, work is required to rotate spins out of the directions of lowest energy determined by spin-orbit coupling. These crystallographic directions of lowest energy are often referred to as "easy" directions of magnetization, whereas the directions of highest energy are the "hard" directions. Easy and hard directions depend on the specific material and its crystal structure. In hexagonal crystals, for example, the work per unit volume required to rotate the spontaneous magnetization vector M_s away from the c-axis through an angle θ is:

$$E = K_0 + K_1 \sin^2\theta + K_2 \sin^4\theta + \cdots \text{higher order terms.}$$

Here, K_0, K_1, K_2 are the material's magnetocrystalline anisotropy constants (units of erg cm^{-3} in cgs). When $K_1 > 0$ and $K_2 > -K_1$, the c-axis is the easy direction of magnetization, as in cobalt.

In a cubic material, the magnetocrystalline anisotropy energy density is given by:

$$E = K_0 + K_1(\alpha_1^2\alpha_2^2 + \alpha_2^2\alpha_3^2 + \alpha_1^2\alpha_3^2) + K_2(\alpha_1^2\alpha_2^2\alpha_3^2) + \cdots \text{higher order terms.}$$

Here, α_1, α_2, and α_3 are the direction cosines of the angles between the spontaneous magnetization vector M_s and the crystal axes [100], [010], and [001], respectively. Often, K_1 is much larger than K_2, so that higher powers in the expression above can be neglected. In materials such as magnetite, both K_1 and K_2 are negative at and above room temperature; this results in <111> being easy directions. When $K_1 > 0$ and $K_2 = 0$, then <100> are easy directions, as in iron (e.g., see Chikazumi, 1964; Cullity, 1972).

Stress anisotropy

Also due to spin-orbit coupling, the lattice of a ferromagnetic material will spontaneously strain below the Curie temperature. Conversely, application of external stress or accumulation of internal strain can rotate the spontaneous magnetization away from the easy direction given by crystalline anisotropy. In a cubic material the energy resulting from an applied stress σ is

$$E = -3/2\lambda_{100}\sigma(\alpha_1^2\gamma_1^2 + \alpha_2^2\gamma_2^2 + \alpha_3^2\gamma_3^2) - 3\lambda_{111}\sigma(\alpha_1\alpha_2\gamma_1\gamma_2 + \alpha_2\alpha_3\gamma_2\gamma_3 + \alpha_3\alpha_1\gamma_3\gamma_1)$$

where α_1, α_2, and α_3 are the direction cosines of M_s with respect to [100], [010], and [001], respectively. The γ_i are the equivalent direction cosines of σ; λ_{100} and λ_{111} are the material's magnetostriction constants. Tension is assumed to be positive. When $\lambda_{100} = \lambda_{111}$, as in the case of isotropic magnetostriction, then the magnetoelastic energy reduces to the simple form

$$E = 3/2\lambda\sigma \sin^2\theta$$

where θ is the angle between M_s and σ and where $\lambda = \lambda_{100} = \lambda_{111}$. In this simple case, energy is minimum when (a) M_s is parallel to σ and $\lambda\sigma > 0$ or (b) M_s is perpendicular to σ and $\lambda\sigma < 0$. Thus in case (a) the application of stress will cause M_s to rotate toward the stress axis, whereas in case (b) M_s will rotate toward the direction perpendicular to the axis of stress.

Magnetostatic energy

The work, per unit volume, required to assemble a population of magnetic "free poles" into a particular configuration is called the magnetostatic energy. Magnetostatic calculations even for the simplest domain configurations are complex and usually require numerical methods. One of the simplest examples was addressed by Kittel (1949), who analyzed a semi-infinite plate of thickness L that contained lamellar domains of uniform width D, with spontaneous magnetizations normal to the plate surface (Figure M14). Walls were

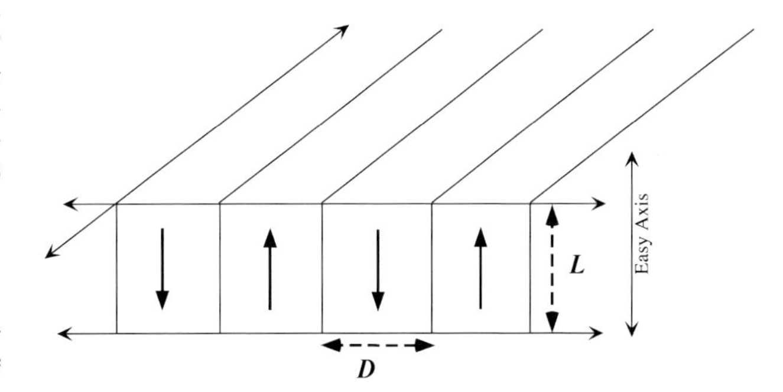

Figure M14 Illustration of the semi-infinite, magnetized plate of thickness L, whose magnetostatic energy was calculated by Kittel (1949). The plate contains domains of identical width D, with magnetizations perpendicular to the plate's surface. Walls are assumed to be infinitely thin.

assumed to be of negligible thickness with respect to the domains' widths and, therefore, the magnetostatic energies due to the walls' magnetic moments were not taken into consideration.

Using the magnetic potential and a Fourier series approach, Kittel obtained $E_m = 1.705\ M_s^2 D$, where E_m is the magnetostatic energy, per unit area of plate surface (erg cm^{-2}, in cgs).

Note, however, that the expression above is an approximation, because Kittel did not account for magnetostatic interactions among all combinations of polarized slabs on the top and bottom surfaces of the plate. Subsequently, these extra energies were included in calculations by Rhodes and Rowlands (1954). Also, they relaxed the assumption of semi-infinite geometry and addressed finite, rectangular grains. Their numerical calculations yielded "Rhodes and Rowlands" functions, with which one may calculate the total magnetostatic energy of rectangular grains of specified relative dimensions.

Energy and width of the domain wall

In the classical sense, a domain wall is the transition region where the spontaneous magnetization changes direction from one domain to the next. According to classical theories, each domain is spontaneously magnetized in one direction and is clearly distinct from the wall. (Micromagnetic theories relax this assumption and will be discussed in a later section.)

The two most important energies that affect the domain wall's energy and width are (1) exchange energy and (2) anisotropy energy, the latter being due either to magnetocrystalline anisotropy or stress. The magnetostatic energy of the wall itself also plays a role; this energy was first analyzed by Amar (1958) and grows especially important when the wall width approaches that of the particle.

Were magnetocrystalline energy acting alone, the spin vectors would change directions abruptly from one domain to the next. For example, if a substance possessed very strong, uniaxial crystalline anisotropy and exchange energy was very much weaker, then the lowest energy "transition" between two adjacent domains would virtually consist of two adjacent spins pointing 180° apart.

This kind of abrupt transition usually involves a large amount of exchange energy, however, because exchange energy is minimized when adjacent spins are parallel. Because minimization of anisotropy energy alone would produce an infinitely thin wall, while minimization of exchange energy alone would produce an infinitely broad wall, the spins in a wall rotate gradually from one domain to the next (Figure M15). This produces a wall with finite width.

By this model, spins in a wall reach an equilibrium configuration when magnetocrystalline and exchange energies are balanced. For the 180° Bloch wall shown in Figure M15 in which the spins rotate through 180°, the wall energy per unit area of wall surface is

$$E_w = 2(J_{ex}S^2\pi^2 K/a)^{1/2} = 2\pi(AK)^{1/2}$$

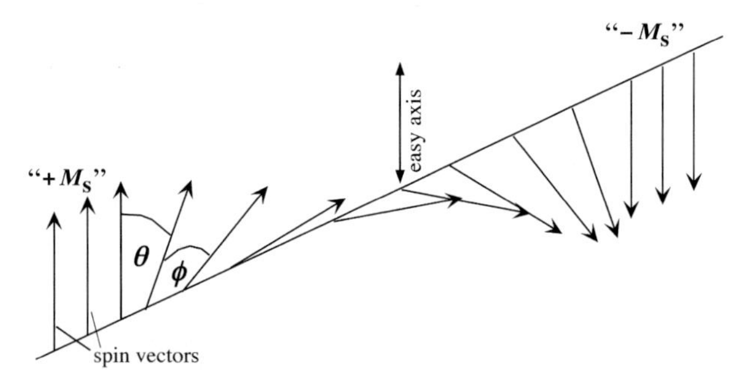

Figure M15 Illustration of spins in a 180° Bloch wall. θ is the angle between a spin and the easy axis of magnetization; ϕ is the angle between adjacent spins.

where J_{ex} is exchange integral, S is spin, K is anisotropy constant due to magnetocrystalline anisotropy energy and/or stress energy, A is exchange constant (equal to $J_{ex}S^2/a$), and a is lattice constant. The width of a 180° wall is

$$\delta_w = (JS^2\pi^2/Ka)^{1/2} = \pi(A/K)^{1/2}$$

In magnetite, values predicted for E_w and δ_w are approximately 0.9 erg cm^{-2} and 0.3 μm, respectively (e.g., see Dunlop and Özdemir, 1997).

Domain width versus grain size

For the simple, planar domain structure illustrated in Figure M14, domain width D can be calculated for the lowest energy state by minimizing the sum of magnetostatic and wall energies per unit volume of material. One obtains the familiar half-power law derived originally by Kittel (1949):

$$D = (1/M_s)(E_w L/1.705)^{1/2}$$

By this model, domain width increases with the half-power of crystal thickness, so long as the crystal occupies the state of absolute minimum energy. Experimental determinations of domain width versus grain thickness will be discussed in a subsequent section.

Rocks, however, rarely contain magnetic minerals in the shape of thin platelets that might be fair approximations to the semi-infinite plate of Kittel's model. Moskowitz and Halgedahl (1987) calculated the number of domains versus grain size in rectangular particles of $x = 0.6$ titanomagnetite ("TM60": $Fe_{2.4}Ti_6O_4$) containing planar domains separated by 180° walls. Magnetostatic energy, including that due to the walls' moments, was determined with the method developed by Amar (1958), based on Rhodes and Rowlands' method (1954). It was assumed that grains occupied states of absolute minimum energy. The effects of two dominant anisotropies were investigated: magnetocrystalline anisotropy (zero stress) and a uniaxial stress (σ = 100 MPa) which was strong enough to completely outweigh the crystalline term. Their calculations yielded two principal results: (1) particles encompassing a wide range of grain sizes can contain the same number of domains, and (2) a plot of N (number of domains) versus L (grain thickness) is fitted well by a power law $N \propto L^{1/2}$. It follows that domain width D also follows a half-power law in L. Thus, the general form of functional dependence of D on L is the same for finite, rectangular grains, and semi-infinite plates.

Single-domain/two-domain transition size

Grains containing only two or three domains can rival the remanences and coercivities of SD grains, and such particles are referred to as being "pseudosingle-domain," or PSD. PSD grains can be common in many rocks and, in terms of interpreting rock magnetic behavior, it is important to know their size ranges in different magnetic minerals. The onset of PSD behavior begins at the single-domain–two-domain transition size, d_0.

In general, a particle will favor a two-domain over a SD state because the magnetostatic energy associated with two domains is much lower than that of a saturated particle. However, below d_0 the energy price of adding a domain wall is too great to produce a state of minimum energy. At d_0 the total energy of the two-domain state (E_{2D}) and the SD state are equal; above d_0, $E_{2D} < E_{SD}$, while $E_{2D} > E_{SD}$ below it.

In zero applied field this transition size depends on the material's magnetic properties, its state of stress, the particle shape, and temperature. Moskowitz and Banerjee (1979) calculated total energies of SD, two-domain, and three-domain cubes of stress-free magnetite containing lamellar domains at room temperature. Magnetostatic energy of the walls was included in their calculations (Amar, 1958). They obtained $d_0 \approx 0.08$ μm.

Similar calculations for TM60 by Moskowitz and Halgedahl (1987) yielded $d_0 \approx 0.5$ μm for unstressed particles at room temperature. Raising the stress level to 100 MPa shifted d_0 upward to 1 μm.

Domains and domain walls at crystal surfaces

In relatively thick crystals, domains and domain walls may change their geometric styles near and at crystal surfaces, in order to lower the total magnetostatic energy with respect to that of the "open" structure shown in Figure M14. The particular style depends largely on the dominant kind of anisotropy, as well as on the relative strengths of magnetostatic and anisotropy energies. When $2\pi M_s^2/K \gg 1$ in a uniaxial material, prism-shaped closure domains bounded by 90° walls, in which spins rotate through 90° from one domain to the next, may completely close off magnetic flux at the crystal surface. Lamellar domains separated by 180° walls may fill the crystal's volume (Figure M16a). In this case, the two main energies originate from the walls and from magnetoelastic energy due to magnetostrictive strain where closure domains abut body domains. Closure domain structures such as these can subdivide further into elaborate arrays of smaller closure and nested spike domains, if the crystal is sufficiently thick. Because closure domains can either greatly reduce or completely eliminate magnetostatic energy, the body domains can be several times broader than predicted for the Kittel-like, "open" structure.

When $2\pi M_s^2/K \ll 1$ in a uniaxial substance like barium ferrite, a large amount of anisotropy energy results when M_s is perpendicular to the "easy" axis. Therefore, the style of surface closure shown in Figure M16a is energetically unfavorable. Instead, a very different style of surface domain structure may evolve in thick crystals. Walls which are planar within the body of the crystal can become wavy at the surface (Figure M16b). In extremely thick crystals, wavy walls can alternate with rows of reverse spikes. These elaborate surface domain structures lower magnetostatic energy by achieving a closer mixture of "positive" and "negative" free magnetic poles (e.g., see Szymczak, 1968).

Large crystals governed by cubic magnetocrystalline anisotropy reduce surface flux through prism-shaped closure domains at the surface. When <100> are easy directions, as in iron, closure domains are bounded by 90° walls (Figure M16a). When <111> are easy directions, as in magnetite, closure domains are bounded by 71° and 109° walls and, within the closure domains, M_s is canted with respect to the crystal surface (Figure M16c).

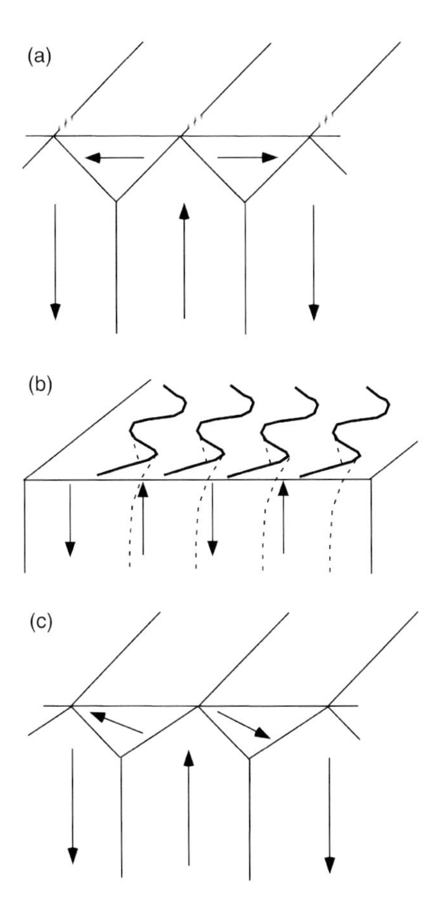

Figure M16 Three styles in which domains and domain walls may terminate at a crystal surface. (a) Illustration of prism-shaped surface closure domains at the surface of a material which is either uniaxial, with $2\pi M_s^2/K \gg 1.0$, or cubic, such as iron, whose easy axes are along <100>. Here, 90° walls separate closure domains from the principal "body" domains that fill most of the crystal. Arrows indicate the sense of spontaneous magnetization within the domains. (b) Illustration of wavy walls at the surface of a uniaxial material with $2\pi M_s^2/K < 1.0$. Waviness dies out with increasing distance from the surface. (c) Prism-shaped closure domains at the surface of a cubic material, such as magnetite, whose easy directions of magnetization are along <111>. Closure domains and body domains are separated by 71° and 109° walls.

Temperature dependence of domain structure

Understanding how domain structure evolves during both heating to and cooling from the Curie point is crucial to understanding the acquisition and thermal stability of thermal remanent magnetization (TRM). If the number of domains changes significantly during cooling from the Curie point, it is reasonable to hypothesize that TRM will not become blocked until the overall domain structure reaches a stable configuration.

According to Kittel's original model (Figure M14), grains will nucleate (add) domain walls and domains with heating in zero field, if the wall energy drops more rapidly with increasing temperature than does the magnetostatic term. Energywise, in this first case a particle can "afford" to add domains with heating. Conversely, during cooling from the Curie point a grain will denucleate (lose) domains and domain walls if wall energy rises more quickly than does M_s^2 with decreasing temperature. If wall energy drops less rapidly with increasing temperature than does M_s^2, then the opposite scenarios apply. Of course, such behavior relies on the assumptions that the particle is able to maintain a global energy minimum (GEM) domain state at all temperatures and that the total magnetostatic energy of the walls themselves can be ignored.

Using Amar's (1958) model, Moskowitz and Halgedahl (1987) calculated the number of domains between room temperature and the Curie point in parallelepipeds of TM60. As discussed earlier, they investigated two cases: dominant crystalline anisotropy (zero stress) and high stress ($\sigma = 100$ MPa). Magnetostatic energy from wall moments was included in the calculations. At all temperatures they assumed that particles occupied GEM domain states. Because the temperature dependences of the material constants A (exchange constant) and λ (magnetostriction constant) for TM60 had not been constrained well by experiments, they ran six models to bracket the least rapid and most rapid drops in wall energy with heating.

In TM60 grains larger than a few micrometers, most of their results gave an increase in the number of domains with heating. Exceptions to this overall pattern were cases in which walls broadened so dramatically with increasing temperature that they nearly filled the particle and rendered nucleation unfavorable. During cooling from the Curie point in zero field, the domain "blocking temperature"—that is, the temperature below which the number of domains remained constant—increased both with decreasing grain size and with internal stress.

Micromagnetic models

In contrast to classical models of magnetic domain structure, micromagnetic models do not assume the presence of discrete domains—that is, relatively large volumes in a crystal where the spontaneous magnetization points in a single direction. Instead, micromagnetic models allow the orientations of M_s-vectors to vary among extremely small subvolumes, into which a grain is divided. A stable configuration is obtained numerically when the sum of exchange, anisotropy, and magnetostatic energies is minimum. To reduce computation time, all micromagnetic models for magnetite run to date assume unstressed, defect-free crystals.

Moon and Merrill (1984, 1985) were the first in rock magnetism to construct one-dimensional (1D) micromagnetic models for defect-free magnetite cubes. To simplify calculations, they assumed that magnetite was uniaxial. In their models, they subdivided a cube into K thin, rectangular lamellae. Within a Kth lamella, the spin vectors were parallel to the lamella's largest surface but oriented at angle θ_k with respect to the easy axis of anisotropy. The distribution of angles was varied throughout the cube until a state of minimum energy was obtained. Because the θ_k's varied continuously across the grain, there was no sharp demarcation between domains and domain walls. For practical purposes, a wall's effective width was defined as the region in which the moments rotated most rapidly. Nucleation was modeled by "marching" a fresh wall from a particle's edge into the interior. A nucleation barrier was determined by calculating the maximum rise of total energy as the wall moved to its final location and as preexisting walls adjusted their positions to accommodate the new wall. The denucleation barrier was determined by reversing the nucleation process.

Moon and Merill's major breakthrough was the discovery that a particle can occupy a range of local energy minimum (LEM) domain states. Each LEM state is characterized by a unique number of domains and is separated from adjacent states by energy barriers. As illustrated in Figure M17, the GEM state is the configuration of lowest energy but, owing to the energy barriers between states, a LEM state can be quite stable as well. Any given LEM can be stable over a broad range of grain sizes.

Several authors have extended micromagnetic calculations for magnetite to two- and three-dimensions (e.g., Williams and Dunlop, 1989, 1990; Newell *et al.*, 1993; Xu *et al.*, 1994; Fabian *et al.*, 1996; Fukuma and Dunlop, 1998; Williams and Wright, 1998). In 2D models, grains are subdivided into rods. In 3D models, the crystal is subdivided into a multiplicity of extremely small, cubic cells (e.g., ≈ 0.01 μm on a side), within each of which M_s represents the average magnetization over several hundred atomic dipole moments. Each cell's M_s-vector is oriented at angle θ with respect to the easy axis, and the θ's are varied independently with respect to their neighbors until an energy minimum is achieved. Owing to the extremely large number of cells and the even larger number of computations, energy calculations begin with an initial guess of how the final, minimum energy structure might appear. To date, the largest magnetite cubes addressed by 3D models are only a few micrometers in size, due to limitations of computing time (e.g., Williams and Wright, 1998).

2D and 3D micromagnetic models yield a variety of exotic, nonuniform configurations of magnetization, such as "flower" and "vortex" states. Analogous to Moon and Merrill's results, these models yield both LEM and GEM states, although very different in their M_s-structures from those of 1D models. For example, Figure M18 illustrates a "flower" state in a cube magnetized parallel to the z-axis. The flower state is reminiscent of a classical SD state of uniform magnetization, except that the M_s-vectors are canted at and near the crystal surface. Increasing the cube size makes other nonuniform states energetically favorable (e.g., Fabian *et al.*, 1996). For example, according to 3D models of magnetite cubes between 0.01 and 1.0 μm in size, the flower state is the lowest energy state between about 0.05 and 0.07 μm (Williams and Wright, 1998). Further increase of cube size yields a vortex state, in which the M_s-vectors circulate around a closed loop within the grain. By Williams and Wright's calculations, both flower and vortex states are stable between about 0.07 and 0.22 μm, although the vortex state has the lower energy. Raising the cube size to the 0.22–1.0 μm range results in the flower state becoming unstable, so that the vortex state is the only possible state. In relatively large cubes of magnetite—e.g., 4 or 5 μm—2D and 3D models yield M_s-configurations closely approaching those of

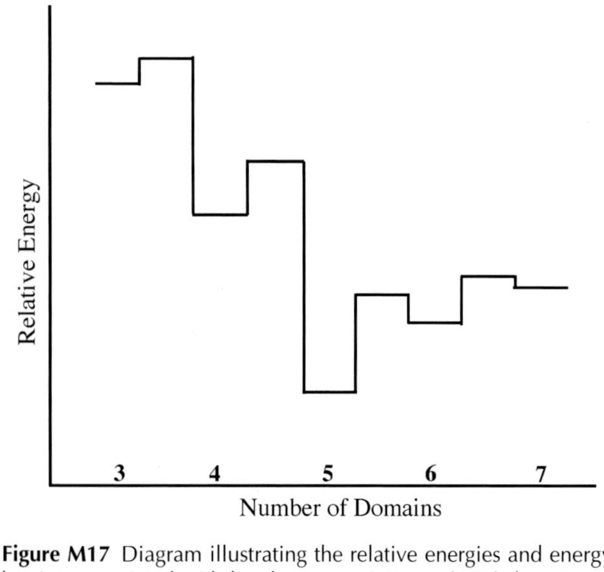

Relative Energy
3 4 5 6 7
Number of Domains

Figure M17 Diagram illustrating the relative energies and energy barriers associated with local energy minimum (LEM) domain states. Each LEM state is characterized by a unique number of domains and is separated from adjacent states by nucleation and denucleation energy barriers. In this diagram, the global energy minimum, or GEM, domain state has five domains.

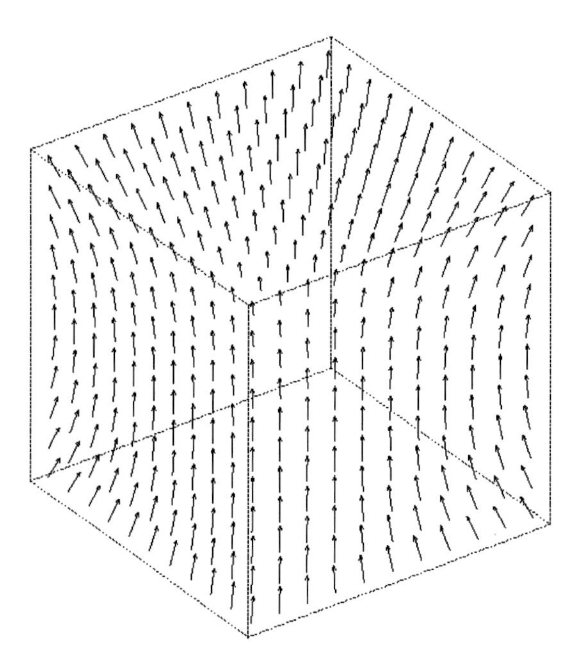

Figure M18 Illustration of a "flower" state obtained through three-dimensional micromagnetic modeling of a cube largely magnetized along the cube's z-axis.

classical domain structures expected for magnetite; some models predict "body" domains separated by domain walls, with closure domains at the surface (e.g., Xu *et al.*, 1994; Williams and Wright, 1998).

According to several micromagnetic models of hysteresis in submicron magnetite, magnetization reversal can occur through LEM-LEM transitions (e.g., flower to vortex state). Reversal can take place through almost independent reversals of the particle's core and outer shell (e.g., Williams and Dunlop, 1995).

Dunlop *et al.* (1994) used 1D micromagnetic models to investigate transdomain TRM. Transdomain TRM—that is, acquisition of TRM through LEM-LEM transitions—had been proposed earlier by Halgedahl (1991), who observed denucleation of walls and domains in titanomagnetite during cooling (see below for a discussion of these results). In particular, Halgedahl's observations strongly suggested that denucleation could give rise to SD-like TRMs in grains that, after other magnetic treatments, contained domain walls. Furthermore, Halgedahl (1991) found that a grain could "arrive" in a range of LEM states after replicate TRM acquisitions.

To determine whether transdomain TRM could be acquired by stress-free, submicron magnetite particles free of defects, Dunlop *et al.* (1994) calculated the energy barriers for all combinations among single domain-two-domain-three-domain transitions with decreasing temperatures from the Curie point of magnetite in a weak external field. They assumed that LEM-LEM transitions were driven by thermal fluctuations across LEM-LEM energy barriers and that thermal-equilibrium populations of LEM states were governed by Boltzmann statistics. According to their results, after acquiring TRM most populations would be overwhelmingly biased toward GEM domain states, and an individual particle should not exhibit a range of LEM states after several identical TRM runs.

Using renormalization group theory, Ye and Merrill (1995) arrived at a very different conclusion. According to their calculations, short-range ordering of spins just below the Curie point could give rise to a variety of LEM states in the same particle after replicate coolings.

These conflicting theoretical results are discussed below in the context of experiments.

Domain observations

Methods of imaging domains and domain walls

Rock magnetists have mainly used three methods to image domains and domain walls: the Bitter pattern method, the magneto-optical Kerr effect (MOKE), and magnetic force microscopy (MFM). Two other methods—transmission electron microscopy (TEM) and off-axis electron holography—have been used in a very limited number of studies on magnetite.

The Bitter method images domain walls through application of a magnetic colloid to a smooth, polished surface carefully prepared to eliminate residual strain from grinding (e.g., see details in Halgedahl and Fuller, 1983). Colloid particles are attracted by the magnetic field gradients around walls, giving rise to Bitter patterns. When viewed under reflected light, patterns appear as dark lines against a grain's bright, polished surface. With liquid colloid, this method can resolve details as small as about 1 μm. If colloid is dried on the sample surface and the resultant pattern is viewed in a scanning electron microscope (SEM), features as small as a few tenths of one micrometer can be resolved (Moskowitz *et al.*, 1988; Soffel *et al.*, 1990).

The MOKE is based on the rotation of the polarization plane of incident light by the magnetization at a grain's surface. Unlike the Bitter method, the MOKE images domains themselves, rather than domain walls. Domains appear as areas of dark and light, the result of their contrasting magnetic polarities (Hoffmann *et al.*, 1987; Worm *et al.*, 1991; Heider and Hoffmann, 1992; Ambatiello *et al.*, 1999). The MOKE has the same resolving power as the Bitter method.

In the MFM, a magnetized, needle-shaped tip is vibrated above the highly polished surface of a magnetic sample. Voltage is induced in the tip by the magnetic force gradients resulting from domains and domain walls. The effects of any surface topography are removed by scanning the surface with the nonmagnetic tip of an atomic force microscope. The MFM can resolve magnetic features as small as 0.01 μm (Williams *et al.*, 1992; Proksch *et al.*, 1994; Moloni *et al.*, 1996; Pokhil and Moskowitz, 1996, 1997; Frandson *et al.*, 2004).

Styles of domains observed in magnetic minerals of paleomagnetic significance

In rock magnetism, the majority of domain observation studies have focused on four magnetic minerals, all important to paleomagnetism: pyrrhotite (Fe_7S_8), titanomagnetite of roughly intermediate composition (near $Fe_{2.4}Ti_6O_4$, or TM60), magnetite (Fe_3O_4), and hematite (Fe_2O_3).

Owing to its high magnetocrystalline anisotropy constant and relatively weak magnetostriction constant, pyrrhotite behaves magnetically as a uniaxial material. When studied with the Bitter method, pyrrhotite often exhibits fairly simple domain patterns, which suggest lamellar domains separated by 180° walls (Figure M19a) (Soffel, 1977; Halgedahl and Fuller, 1983).

Despite being cubic, intermediate titanomagnetites rarely, if ever, exhibit the arrays of closure domains, 71°, and 109° walls that one would predict. Instead, these minerals usually display very complex patterns of densely spaced, curved walls. Possibly, these complex structures result from varying amounts of strain within the particle (Appel and Soffel, 1984, 1985). Occasionally, grains exhibit simple arrays of parallel walls, such as that shown in Figure M20 (e.g., Halgedahl and Fuller, 1980, 1981), but it is not unusual to observe wavy walls alternating with rows of reverse spikes (Halgedahl, 1987; Moskowitz *et al.*, 1988). Both simple and wavy patterns suggest a dominant, internal stress that yields a uniaxial anisotropy, although the origin of this stress is still unclear. On one hand, it could originate from the mechanical polishing required to prepare samples for domain studies. Alternatively, in igneous rock samples stress could be generated during cooling, due to differences in the coefficients of thermal expansion among the various minerals.

Small magnetite grains randomly dispersed in a rock or a synthetic rock-like matrix generally display simple arrays of straight domain walls, if the domains' magnetizations are sufficiently close to being parallel to the surface of view (e.g., Worm *et al.*, 1991; Geiß *et al.*, 1996) (Figure M21). In such samples, however, there appears to be a paucity of closure domains, perhaps the result of observation surfaces being other than {100} planes.

In an early study by Bogdanov and Vlasov (1965), Bitter patterns of 71° and 109° walls were observed on cleavage planes of a few small magnetite particles obtained by crushing a natural crystal. Similarly, from Bitter patterns Boyd *et al.* (1984) found both closure domains and networks of 180°, 71°, and 109° walls in several magnetite particles in a granodiorite. Body domains accompanied by closure domains were also reported by Smith (1980), who applied the TEM method to a

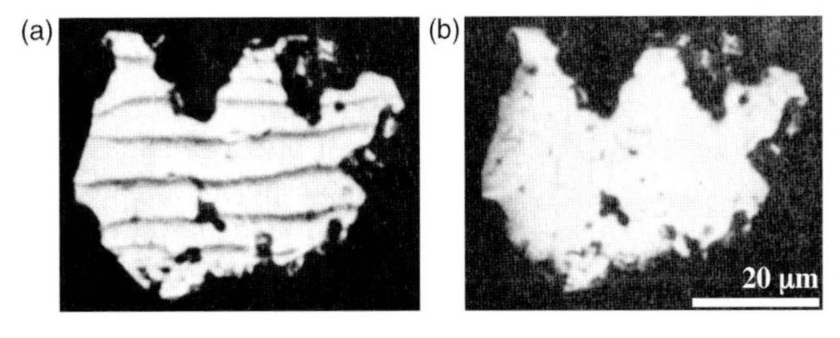

Figure M19 Bitter patterns on a particle of natural pyrrhotite (a) after demagnetization in an alternating field of 1000 Oe and (b) in an apparently SD-like state after acquiring saturation remanence in 15 kOe.

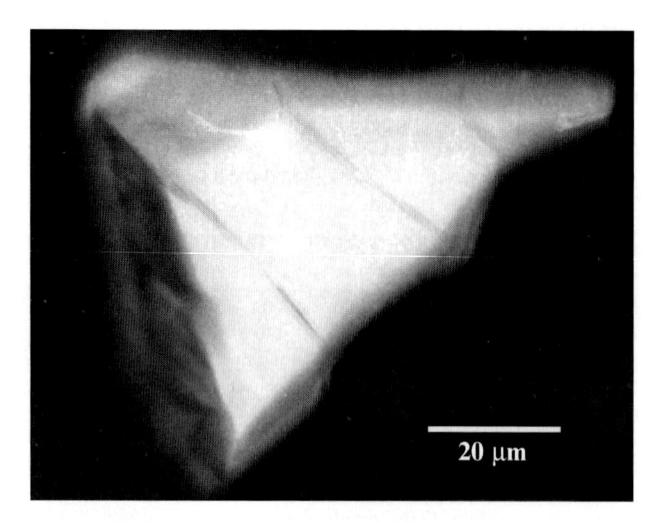

Figure M20 Bitter pattern on a grain of intermediate titanomagnetite ($x \approx 0.6$) in oceanic basalt drilled near the Mid-Atlantic Ridge.

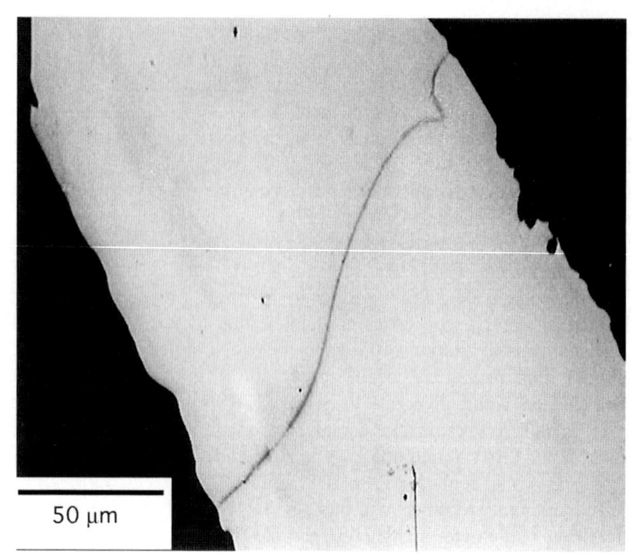

Figure M22 Bitter pattern of a wall on a very large (approximately 150 μm width, 1 mm diameter) platelet of natural hematite from Elba, Italy. At most this platelet exhibits only 1–2 principal walls, although small edge domains are often observed. Apparently, the wall is bowing around a defect near the upper right-hand edge of the photograph. Only a small part of the crystal surface is shown.

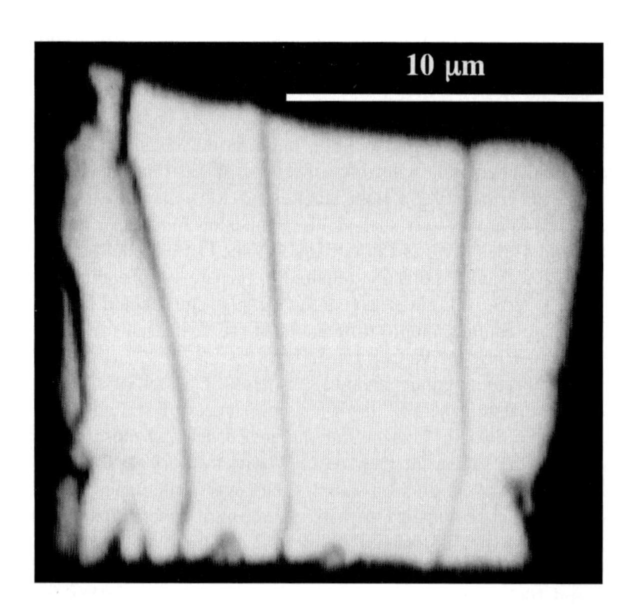

Figure M21 Bitter pattern on a grain of magnetite synthesized with the glass-ceramic method. In this particular state of magnetization, the grain contains four walls, whose lengths are commensurate with the particle's length. Note that one wall is pinned very near the particle's extreme left-hand edge. The small triangular patterns at the lower edge of the particle represent walls which enclose small reverse spike domains.

few small magnetite crystals in rock. Note that this method images domains through deflection of the electron beam by the magnetizations within domains. Consequently, the TEM method reveals volume domains, not just their surface manifestations. Similarly, large magnetite crystals cut and polished on {100} planes almost invariably exhibit the closure domains bounded by 71° and 109° walls expected for truly multidomain magnetite (e.g., Bogdanov and Vlasov, 1966; Özdemir and Dunlop, 1993, 1997; Özdemir *et al.*, 1995).

To date, domain studies on hematite have been limited to large (e.g., 100 μm–1 mm) platelets. Even large crystals such as these contain very few walls and a fairly simple domain structure, owing to hematite's weak spontaneous magnetization (about 2 emu cm^{-3}) and low magnetostatic energy (Figure M22) (Halgedahl, 1995, 1998).

Observed number of domains versus grain size

Both Kittel's original model of domains in a semi-infinite platelet and calculations for finite grains by (e.g., Moskowitz and Halgedahl, 1987) lead to the prediction that domain width $D \propto L^{1/2}$, where L is plate or particle thickness. These predictions are supported by domain studies of natural magnetic minerals, which generally yield a power-law dependence of D on L, although the power may differ somewhat from 0.5.

In rock magnetism, Soffel (1971) was the first to study the grain-size dependence of the number of domains in a paleomagnetically important magnetic mineral. From Bitter patterns on grains of natural, intermediate ($x \approx 0.55$) titanomagnetite in a basalt, Soffel determined that, on average, N (number of domains) $\propto L^{1/2}$, where L is the average particle size. This relation translates to D (domain width) $\propto L^{1/2}$. Extrapolation of these data to $N = 1$ (or, equivalently, to $D = L$) yielded a single domain-two domain transition size of 0.6 μm for this composition. By assuming that Kittel's model could be applied to roughly equidimensional grains without serious errors and that $M_s = 100$ emu cm^{-3} for $x \approx 0.55$, Soffel obtained a wall energy density of about 1 erg cm^{-2}.

Bitter patterns on natural pyrrhotite in the Bad Berneck diabase from Bavaria were studied both by Soffel (1977) and by Halgedahl and Fuller (1983). Halgedahl and Fuller determined the dependence of domain width on grain size for three states of magnetization: NRM, after alternating field demagnetization (AFD) in a peak field of 1000 Oe, and saturation remanence imparted in a field of 15 kOe. Particles that contained walls in these three different states yielded similar power-law dependences of D on L: $D \propto L^{0.43}$ (NRM), $D \propto L^{0.40}$ (AFD), and $D \propto L^{0.45}$ (saturation remanence). The single domain-two-domain boundary sizes estimated for the three states also were similar, falling between 1.5 and 2 μm. These results were consistent with those of Soffel (1977).

Likewise, Geiß *et al.* (1996) obtained $D \propto L^{0.45}$ from Bitter patterns on synthetic magnetite particles grown in a glass-ceramic matrix (Worm and Markert, 1987). They obtained a single domain-two-domain transition size for magnetite at approximately 0.25 μm.

In contrast to the results discussed above for titanomagnetite, pyrrhotite, and glass-ceramic magnetite, one of Dunlop and Özdemir's (1997) data compilations for magnetite samples from several very different provenances yielded $D \propto L^{0.25}$. The cause of the discrepancy between this result and those obtained from other samples is not understood.

Domain wall widths

When studied in the SEM, patterns of dried magnetic colloid afford high-resolution views of domain walls—or, more precisely, the colloid accumulations around walls. Moskowitz *et al.* (1988) applied this method to polycrystalline pellets of synthetic titanomagnetite substituted with aluminum and magnesium ($Fe_{2.2}Al_{0.1}Mg_{0.1}Ti_{0.6}O_4$). Henceforth, this composition is referred to as "AMTM60." Dried Bitter patterns on unpolished surfaces were virtually identical in style to those common to materials controlled by strong uniaxial anisotropy, such as barium ferrite. Patterns indicated closely spaced stripe domains, sinusoidally wavy walls, and elaborate wavy walls, which alternated with nested arrays of reverse spikes. Direct measurements from SEM photographs yielded wall widths between 0.170 and 0.400 μm.

In the same study, Moskowitz *et al.* (1988) also measured the wavelengths, amplitudes, and domain widths of the sinusoidally wavy patterns. Using these parameters and the experimental value of spontaneous magnetization for AMTM60, they applied Szymczak's (1968) model to estimate the exchange constant and the uniaxial anisotropy constant, the latter presumably due to residual stress generated, while the pellet cooled from the sintering temperature.

Very high-resolution images of domains and domain walls have been obtained with MFM on magnetite. The first images were reported by Williams *et al.* (1992) from a {110} surface on a large crystal of natural magnetite. They recorded a magnetic force profile across a 180° wall. This record indicated that the spins within the wall reversed their polarity of rotation along the length of the wall, demonstrating that walls in real materials can be much more complex than those portrayed by simple models.

In a study of glass-ceramic magnetite particles with MFM, Pokhil and Moskowitz (1996, 1997) found that, along its length, an individual wall can be subdivided into several segments of opposite polarity. The segments are separated by Bloch lines, the transition regions where polarity changes sense. Rather than being linear, subdivided walls zigzag across a grain. Analogous to wavy walls, the zigzags help to reduce a grain's total magnetostatic energy. The number of Bloch lines within any specific wall was found to vary with repeated AF demagnetization treatments. Thus walls, like particles, can occupy LEM states.

Owing to its high-resolution capabilities, the MFM can provide estimates of wall width. Proksch *et al.* (1994) obtained MFM profiles across a 180° wall on a {110} surface of natural magnetite. The deconvolved signal yielded a half-width, assumed to be commensurate with wall width, of about 0.21 μm.

Experimental evidence for local energy minimum domain states

In their study of Bitter patterns on natural pyrrhotite, Halgedahl and Fuller (1983) noted that grains of virtually identical size could contain very different numbers of domains, despite these same particles having undergone the same magnetic treatments. Moreover, they found that an individual particle could arrive in very different domain states—i.e., with different numbers of walls—after different cycles of minor hysteresis. Clearly, such particles did not always occupy a domain state of absolute minimum energy.

These observations led Halgedahl and Fuller (1983) to the conclusion that a particle could occupy domain states other than the ground state. Shortly thereafter, Moon and Merrill (1984, 1985) dubbed these states as LEM states, on the basis on their one-dimensional micromagnetic results for magnetite. Subsequent domain observation studies by Geiß *et al.* (1996) provided strong evidence for LEM states in magnetite as well.

LEM states and thermomagnetic treatments: hysteresis

A particularly unexpected type of LEM state in pyrrhotite and intermediate titanomagnetite is a SD-like state that certain particles can occupy after being saturated in a strong external field, even though these same particles readily accommodate walls in other states of magnetization. Bitter patterns on intermediate ($x \approx 0.6$) titanomagnetite in oceanic basalt and on natural pyrrhotite in diabase were studied during hysteresis by Halgedahl and Fuller (1980, 1983). An electromagnet was built around the microscope stage for these experiments, permitting one to track the evolution of walls continuously, while varying the applied field. In states of saturation remanence, most particles in a large population contained one or more walls, as expected on the basis of previous theories. However, a significant percentage—nearly 40%—of the finer (5–15 μm) particles observed appeared saturated after the maximum field was shut off (Figure M19a and b). Within a large population of such grains it was found that the number of domains was described well by a Poisson distribution. Halgedahl and Fuller (1980, 1983) proposed that grains, which failed to nucleate walls, could make a substantial contribution to saturation remanence. Furthermore, such grains might explain much of the known decrease in the ratio of saturation remanent magnetization to saturation magnetization with increasing particle size. This hypothesis was extended to the case of weak-field TRM.

Particles in SD-like states remained saturated, until nucleation was accomplished by applying a back-field of sufficient magnitude. In some cases, the nucleating field was strong enough to sweep a freshly nucleated wall across the particle and thereby result in a state of saturation (or near-saturation) in the opposite sense. In other cases, the new wall moved until it was stopped by a pinning site. In the smaller grains studied (e.g., smaller than 20 μm) the nucleation field dropped off with increasing grain size L according to a power law in $L^{-1/2}$.

Boyd *et al.* (1984) reported Bitter patterns on natural magnetite grains carrying saturation remanence, and these patterns suggested SD-like states. These results were surprising, in view of magnetite's strong tendency to self-demagnetize.

Subsequent domain observations and hysteresis studies of hematite platelets from Elba, Italy demonstrated that, after exposure to strong fields, even large crystals could arrive in states suggesting near-saturation. Often, however, small, residual domains clung to the platelet's surface in these states. Back-fields were necessary to nucleate principal domains, and these nucleation fields also followed a power-law in $L^{-1/2}$ (Halgedahl, 1995, 1998).

Results from titanomagnetite and pyrrhotite were interpreted by Halgedahl and Fuller (1980, 1983) in light of previous theoretical and experimental work on high-anisotropy, industrial materials synthesized to produce magnetically hard, permanent magnets. In such materials, nucleation of walls is an energetically difficult process (Brown, 1963; Becker, 1969, 1971a,b, 1976). According to Brown's original theory, the nucleation of a fresh wall requires the following condition to be fulfilled:

$$2K/M_s < |\vec{H}_{appl} = \vec{H}_d|$$

Here, $2K/M_s$ is the local anisotropy field at the nucleation site; H_d is the local demagnetizing field at the site, assumed to be opposite in sense to M_s; and H_{appl} is the field applied to the sample. In samples where back-fields are necessary to accomplish nucleation, H_{appl} is opposite to the sample's original saturation magnetization. Wall nucleation may occur either at crystal defects where $2K/M_s$ is anomalously low or where the local demagnetizing field H_d is anomalously high, as at shape defects on a crystal's surface. Note that Brown's theory of nucleation differs from the kind of "global" nucleation resulting from micromagnetic models. Brown-type nucleation may depend on the presence of defects, which fulfill the above condition. By contrast, micromagnetic models assume that particles are defect-free.

However, SD-like states also can result from submicroscopic walls being trapped by surface and volume defects (Becker, 1969, 1971a,b, 1976). There are two cases. First, when a saturating field is shut off, a submicroscopic, virgin wall can nucleate but remain pinned at a local defect. In the second case, a strong applied field may be insufficient to completely saturate a particle. When this occurs, preexisting domain walls may be driven into surface or volume defects and become trapped. In both cases, the particle appears to be saturated at the top of a hysteresis loop. Experimentally, the saturation remanent state appears to be SD-like. A back-field, opposite in sense to the maximum field, is required to break free these residual walls and the submicroscopic domains which they enclose.

These two phenomena can be distinguished by observing both the position where a reverse nucleus first appears in a grain and the field required for its appearance. If complete saturation of a grain, followed by nucleation of a fresh wall, has occurred, then the nucleus will always appear at the same location and in the same back-field, regardless of the maximum field's polarity. Of course, the submicroscopic nucleus could have been trapped by a defect just after it was nucleated initially. If the grain was not saturated completely, then both the location where the nucleus appears and the field required to expand this nucleus and make it visible depend on the direction in which a wall was driven in the first place and on the back-field's magnitude—that is, on the specific trapping site (Becker, 1969, 1971a,b, 1976; Halgedahl and Fuller, 1983).

The style of domain structure and the range of LEM states that a particle can occupy can depend on thermomagnetic history. In a study of Bitter patterns on natural, polycrystalline pyrrhotite, Halgedahl and Fuller (1981) discovered that crystallites displayed arrays of undulating walls on their surfaces after acquiring TRM in a weak external field. By contrast, after AF demagnetization in a strong peak field the same crystallites exhibited planar walls. Evidently, cooling through the TRM blocking temperature locked in a high-temperature configuration of walls, whose curved shapes promoted lower magnetostatic energy than did a planar geometry.

As indicated by Moon and Merrill's theoretical results for magnetite, a particular LEM state can be stable across a broad range of grain size. This prediction is born out by experiments and calculations by Halgedahl and Ye (2000) and Ye and Halgedahl (2000), who investigated the effects of mechanical thinning on domain states in natural pyrrhotite particles. In their experiments, several individual pyrrhotite grains in a diabase were mechanically thinned and their Bitter patterns observed after each thinning step. Despite some grains being thinned to about one-fourth of their initial diameter, the widths of surviving domains and positions of surviving walls remained unaffected (Figure M23). Neither nucleations nor denucleations were observed, although calculations indicated that thinning would cause significant changes in GEM domain states.

Bitter patterns on crystallites of polycrystalline AMTM60 were studied after many replicate TRM acquisition and AF demagnetization experiments (Halgedahl, 1991). In each particle, the number of domains varied from one experiment to the next, describing a distribution of TRM states. For states of weak-field TRM, this distribution could be broad and could include SD-like states (Figure M24). After replicate AF demagnetizations, however, typical distributions were narrow and clustered about a most probable state, possibly the GEM state.

(a)

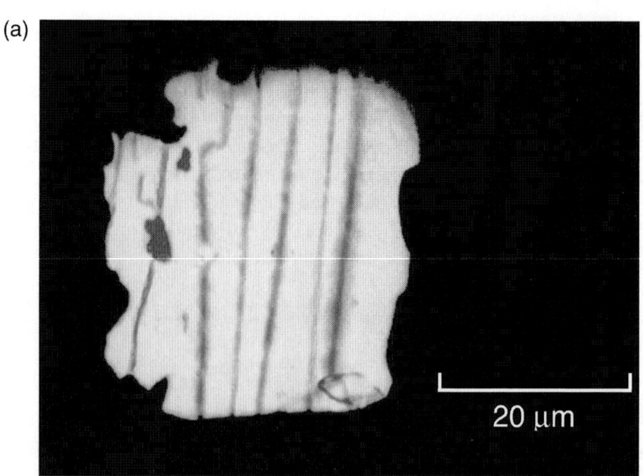

(b)

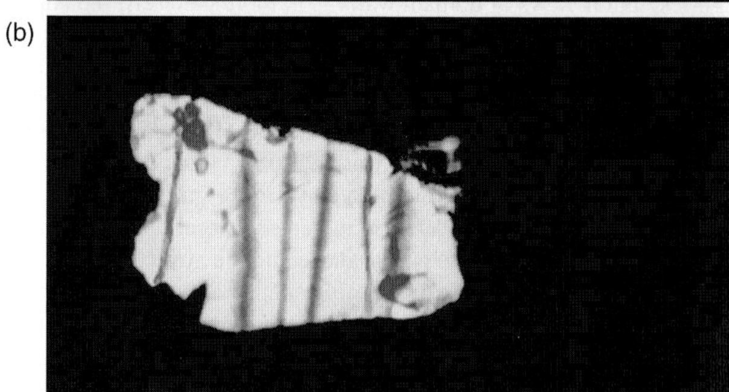

(c)

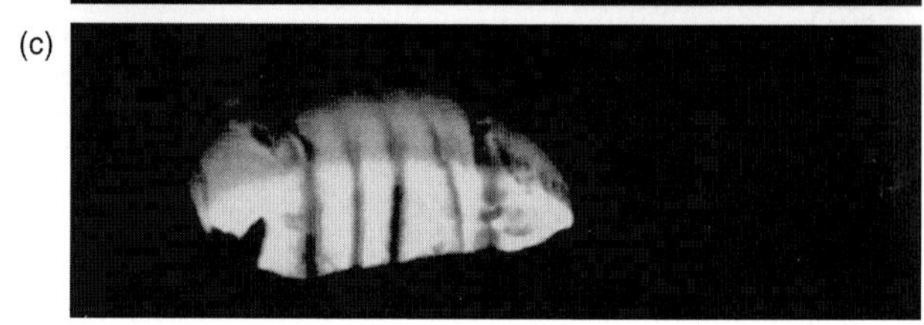

Figure M23 Bitter patterns on a grain of natural pyrrhotite in a diabase before and after mechanical thinning. (a) Initial state, before thinning, (b) after thinning the particle to about one-half of its original length along a direction parallel to the trends of the Bitter lines, and (c) final state, after the particle has been thinned to about one-fourth of its original length.

Thermal evolution of magnetic domain structures observed at elevated temperatures

The manner in which domain structure evolves with temperature carries clear implications for TRM. In very early models of TRM, such changes were either ignored or were assumed not to occur. Instead, some models assumed that the number of domains remained constant from the Curie point to room temperature in a weak external field, and that the blocking of TRM was controlled solely by the growth of energy barriers associated with defects which, eventually, locked walls in place.

Yet changes in the number of domains with temperature could profoundly affect how pseudosingle-domain and multidomain particles acquire TRM. Temperature-induced nucleations or denucleations of walls should trigger sudden changes of the internal demagnetizing field. As a result, preexisting walls, which survive a domain transition, could be dislodged from imperfections where they were pinned initially at higher temperatures.

In rock magnetism, domain observation experiments to study the thermal response of domain structure have focused on two magnetic minerals: (a) magnetite, owing to its great importance to the paleomagnetic

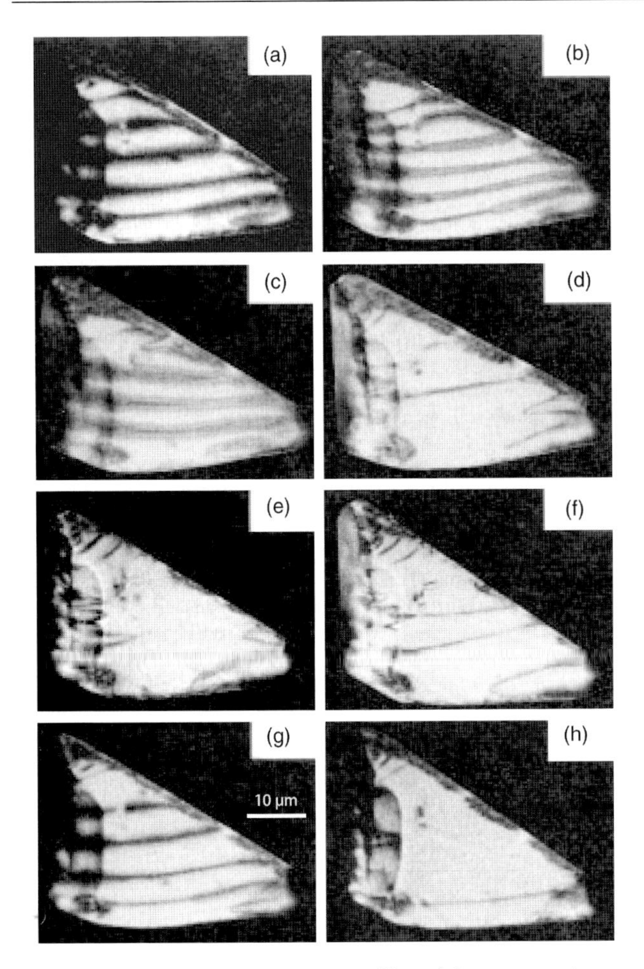

Figure M24 Bitter patterns on a crystallite of titanomagnetite ("AMTM60": see text) after each of eight replicate TRM acquisition runs in the Earth's field.

signal carried by many rocks, and (b) intermediate titanomagnetite, owing to its significance to marine magnetic anomalies and to its moderately low-Curie temperature, which renders study of Bitter patterns feasible at moderate temperatures. High-temperature experiments have yielded highly variable results.

The first domain observations on magnetite above room temperature in the Earth's field were made by Heider *et al.* (1988), using particles of hydrothermally recrystallized magnetite embedded in an epoxy matrix. Depending on the particle, Bitter patterns could be followed to approximately 200°C, above which temperature the patterns grew too faint to distinguish against a grain's bright background. Surprisingly, heating to very moderate temperatures drove certain walls across much of a particle; in some cases, denucleation occurred. Upon cooling, walls reassembled in a similar, though not identical, arrangement to that observed initially at room temperature. In some cases repeated thermal cycling between room temperature and about 200°C produced different numbers of domains in the same particle.

Ambatiello *et al.* (1995) used the MOKE to study domain widths versus temperature in several large (>5 mm) crystals of natural magnetite cut and polished on {110} planes. At room temperature, these planes were dominated by broad (40–90 μm), lamellar domains separated by 180° walls, terminating in closure domains at crystal edges. On {111} planes they found complex, nested arrays of very small closure domains, which finely subdivided the main closure structure.

Unlike the Bitter technique, the MOKE is applicable to the Curie point of magnetite (580°C), at least in theory. In practice, the amount of Kerr rotation in magnetite is very small even at room temperature; this rotation decreases progressively as M_s drops with heating. As the Curie point is approached, the contrast between adjacent domains grows exceedingly faint. For this reason, Ambatiello and colleagues successfully imaged domains to 555°C, but no higher.

According to their observations, domain widths generally increased with heating, and such changes were thermally reversible. However, the temperature that triggered changes depended on the provenance of the sample. In some samples, domain widths remained nearly constant until significant broadening of domains occurred above 400°C. In other samples, domains began to broaden at far lower temperatures.

To explain the overall character of their results, Ambatiello *et al.* (1995) calculated the thermal dependence of domain widths on the basis of the closed, "Landau-Lifshitz"-type structure shown in Figure M16c. Their calculations included the magnetoelastic energy associated with closure domains, as well as the "μ^*" correction to magnetostatic energy, originating from small deviations of spins away from the easy axis near a crystal surface. There were several discrepancies between experimental and model results. One mechanism which they put forth to explain these discrepancies was that heating promoted an even finer subdivision into small domains at the crystal surface. This would cause a greater collapse of magnetostatic energy than predicted from their model and thus a more pronounced broadening of body domains than expected.

Bitter patterns observed at elevated temperatures on natural, intermediate titanomagnetites have been reported by Soffel (1977), Metcalf and Fuller (1987a,b, 1988), and Halgedahl (1987). Because the magnetic force gradients around a wall weaken rapidly with heating, walls attract increasingly less magnetic colloid as the Curie point is approached. Consequently, these authors could follow patterns from room temperature to about 10° to 20° below their samples' Curie points (Soffel, 1977: T_c = 105°C; Metcalf and Fuller, 1987a,b, 1988, Halgedahl, 1987: T_c approximately 150°C). During heating, patterns gradually faded to obscurity, with few significant changes in domain structure.

Remarkably, some titanomagnetite particles studied by Metcalf and Fuller (1987a,b, 1988) displayed no Bitter lines after cooling from the Curie point in the Earth's field. This observation provided supporting evidence for Halgedahl and Fuller's (1983) proposal that weak-field TRM acquired by populations of pseudosingle-domain grains could, in part, be attributable to particles which failed to nucleate walls during cooling.

The possible importance of LEM states to TRM acquisition was raised by work on synthetic, polycrystalline AMTM60 (Halgedahl, 1991). Observations focused on grains that displayed simple "Kittel-like" patterns suggesting lamellar domains (Figure M14). The sample was cycled repeatedly between room temperature and the Curie point of 75°C in the Earth's field, although patterns lost definition near 70°C. Bitter patterns were observed continuously throughout heatings and coolings.

As discussed above and shown in Figure M24, replicate TRM experiments often produced different numbers of domains in the same particle. Some of these domain states suggested that the particle was entirely saturated, with no visible Bitter lines, or nearly saturated, with only small spike domains at grain boundaries.

During heating, few changes in domain patterns were observed, except for the usual fading as the Curie temperature was approached. SD-like states were quite thermally stable, exhibiting no changes during thermal cycling until the sample was heated nearly to, or above, the Curie point. As a result of heating above a certain critical temperature, followed by cooling, a SD-like state could be transformed to a LEM state with several domain walls.

Continuous observation of Bitter patterns on AMTM60 during cooling revealed that denucleation was the mechanism by which a particle arrived in a final LEM state. Walls which initially nucleated just below T_c grew visible after the sample cooled slightly through just a few degrees. With further cooling, however, the number of domains often decreased either through contraction of large, preexisting spike domains, or by straight walls moving together and coalescing into spikes. In both cases, the spikes often would collapse altogether. Nucleations were never observed during cooling, once Bitter patterns grew visible. In some cases, denucleation left behind large volumes which, apparently, contained no walls, although walls were present elsewhere in the grain. In other cases, denucleation left behind a grain that appeared nearly saturated, but for small edge domains. In both cases denucleation appeared to produce anomalously large moments in particles that contained several walls after other TRM runs (Halgedahl, 1991).

Results of the experiments described above on AMTM60 lend further support to the hypothesis that particles in metastable SD states can make significant contributions to stable TRM acquired in weak fields. Furthermore, particles which do contain domain walls can carry anomalously large TRM moments, if denucleation leaves behind large volumes in which walls are absent.

These experimental results are at odds with theoretical results by Dunlop *et al.* (1994) for LEM-LEM transitions in magnetite at high temperatures. At present this disagreement has not been explained fully. However, it is important to note that the model of Dunlop *et al.* (1994) assumes that (1) such transitions are driven by thermal fluctuations, and that (2) occupation frequencies of LEM states follow Boltzmann statistics for thermal equilibrium, before domains are blocked in. It is reasonable to expect that thermally activated jumps among LEM states occur very rapidly and discontinuously during times on the order of 1 s or less; rapid transitions such as these are impossible to track under a microscope. By contrast, many of the denucleations observed by Halgedahl (1991) occurred via continuous wall motions easily tracked over laboratory time scales of several seconds. It is questionable, therefore, whether the changes exhibited by AMTM60 were driven by the thermal fluctuation mechanism modeled by Dunlop *et al.* (1994). Perhaps the denucleations observed are not stochastic processes. Additional analyses and experiments are needed to determine the controlling mechanisms.

Observed temperature dependence of magnetic domain structure: low temperatures

Magnetite undergoes two types of transitions at low temperatures, which can profoundly affect domain structure and remanence (e.g., see detailed discussions in Stacey and Banerjee, 1974; Dunlop and Özdemir, 1997). First, at the isotropic point (approximately 130 K) the first magnetocrystalline anisotropy constant, K_1, passes through zero as it changes sign from negative at temperatures above the transition to positive below it. Second, at the Verwey transition, T_v (approximately 120 K), magnetite undergoes a crystallographic transition from cubic to monoclinic.

At the isotropic point domain walls should broaden dramatically, because wall width is proportional to $K^{-1/2}$, K being the crystalline anisotropy constant. It follows that, by cooling through the isotropic point, walls may break free of narrow defects, which pinned them at higher temperatures. At the Verwey transition the easy axis of magnetization changes direction. In multidomain and pseudosingle-domain magnetite, this thermal passage should cause domain structure and domain magnetizations to reorganize completely, so that much of an initial remanence acquired at room temperature would be demagnetized by cooling below T_v. Low-temperature demagnetization has proved useful in removing certain spurious components of NRM, which are surprisingly resistant to thermal demagnetization above room temperature. Thus, it is important to determine how these transitions affect domain structure.

Using an MFM specially adapted to operate at low temperature, Moloni *et al.* (1996) were the first to image domains in magnetite at temperatures near the two transitions. Their sample was a synthetic magnetite crystal cut and polished on a {110} surface. At room temperature, the crystal displayed 180°, 71°, and 109° walls on {110}, as expected for multidomain magnetite. At 77 K—that is, below the Verwey transition—they observed both straight, 180° walls separating broad, lamellar domains, and wavy walls accompanied by reverse spikes. Both types of domain structure were consistent with a uniaxial anisotropy arising from monoclinic crystal structure and for which $2\pi M_s^2/K < 1$. They interpreted the broad, "body" domains as being magnetized along <100> easy axes within {110} planes. The wavy walls were interpreted to indicate domains with magnetizations directed along an easy axis that lay out-of-plane. This mixture of domain styles was thought to reflect lateral variations in the easy axis, due to *c*-axis twinning below T_v. As the crystal was warmed to a few degrees below T_v the domain structure disappeared entirely below the instrument's noise level, evidently undergoing a complete reorganization near the crystallographic transition.

Ultrahigh-resolution imaging of micromagnetic structures

Very recently, off-axis electron holography in the TEM has been used to image magnetic microstructures in submicron magnetite intergrown with ulvospinel (Harrison *et al.*, 2002). This method enables one to image the magnetization vectors within very small, single particles, at a resolution approaching nanometers. Harrison *et al.* (2002) discovered that clusters of magnetically interacting blocks of magnetite could assume both vortex and multidomain-like states. This promising method should prove highly fruitful in future experimental tests of micromagnetic models.

Susan L. Halgedahl

Bibliography

Amar, H., 1858. Magnetization mechanism and domain structure of multidomain particles. *Physical Review*, **111**: 149–153.

Ambatiello, A., Fabian, K., and Hoffmann, V., 1999. Magnetic domain structure of multidomain magnetite as a function of temperature: observation by Kerr microscopy. *Physics of the Earth and Planetary Interiors*, **112**: 55–80.

Appel, E., and Soffel, H.C., 1984. Model for the domain state of Ti-rich titanomagnetites. *Geophysical Research Letters*, **11**: 189–192.

Appel, E., and Soffel, H.C., 1985. Domain state of Ti-rich titanomagnetites deduced from domain structure observations and susceptibility measurements. *Journal of Geophysics*, **56**: 121–132.

Becker, J.J., 1969. Observations of magnetization reversal in cobalt-rare-earth particles. *IEEE Transactions on Magnetics*, **MAG-5**: 211–214.

Becker, J.J., 1971a. Magnetization discontinuities in cobalt-rare-earth particles. *Journal of Applied Physics*, **42**: 1537–1538.

Becker, J.J., 1971b. Interpretation of hysteresis loops of cobalt-rare-earths. *IEEE Transactions on Magnetics*, **MAG-7**: 644–647.

Becker, J.J., 1976. Reversal mechanism in copper-modified cobalt-rare-earths. *IEEE Transactions on Magnetics*, **MAG-12**: 965–967.

Bogdanov, A.K., and Ya. Vlasov, A., 1965. Domain structure in a single crystal of magnetite. (English trans.). *Izvestiya Akademii Nauk, SSSR, Earth Physics*, series no. 1., 28–32.

Bogdanov, A.A., and Ya Vlasov, A., 1966. The domain structure of magnetite particles. (English trans.), *Izvestiya Akademii Nauk, SSSR, Physics, Solid Earth*, **9**: 577–581.

Boyd, J.R., Fuller, M., and Halgedahl, S., 1984. Domain wall nucleation as a controlling factor in the behaviour of fine magnetic particles in rocks. *Geophysical Research Letters*, **11**: 193–196.

Brown, W.F., 1963. *Micromagnetics.* New York: John Wiley, 143 pp.

Chikazumi, S., 1964. *Physics of Magnetism.* New York: John Wiley, 664 pp.

Cullity, B.D., 1972. *Introduction to Magnetic Materials*. Addison-Wesley, Reading, MA, 666 pp.

Dunlop, D.J., and Özdemir, O., 1997. *Rock Magnetism: Fundamentals and Frontiers*. Cambridge: Cambridge University Press, UK, 573 pp.

Dunlop, D.J., Newell, A.J., and Enkin, R.J., 1994. Transdomain thermoremanent magnetization. *Journal of Geophysical Research*, **99**: 19,741–19,755.

Fabian, K., Kirchner, A.,Williams, W., Heider, F., and Leibl, T., Three-dimensional micromagnetic calculations for magnetite using FFT. *Geophysical Journal International*, **124**: 89–104.

Foss, S., Moskowitz, B., and Walsh, B., 1996. Localized micromagnetic perturbation of domain walls in magnetite using a magnetic force microscope. *Applied Physics Letters*, **69**: 3426–3428.

Foss, S., Moskowitz, B.M., Proksch, R., and Dahlberg, E.D., Domain wall structures in single-crystal magnetite investigated by magnetic force microscopy. *Journal of Geophysical Research*, **103**: 30,551–30,560.

Frandson, C., Stipp, S.L.S., McEnroe, S.A., Madsen, M.B., and Knudsen, J.M., 2004. Magnetic domain structures and stray fields of individual elongated magnetite grains revealed by magnetic force microscopy (MFM). *Physics of the Earth and Planetary Interiors*, **141**: 121–129.

Fukuma, K., and Dunlop, D.J., Grain-size dependence of two-dimensional micromagnetic structures for pseudo-single-domain magnetite (0.2–2.5 μm). *Geophysical Journal International*, **134**: 843–848.

Geiß, C.E., Heider, F., and Soffel, H.C., Magnetic domain observations on magnetite and titanomaghemite grains (0.5–10 μm). *Geophysical Journal International*, **124**: 75–88.

Halgedahl, S.L., 1987. Domain pattern observations in rock magnetism: progress and problems. *Physics of the Earth and Planetary Interiors*, **46**: 127–163.

Halgedahl, S.L., 1991. Magnetic domain patterns observed on synthetic Ti-rich titanomagnetite as a function of temperature and in states of thermoremanent magnetization. *Journal of Geophysical Research*, **96**: 3943–3972.

Halgedahl, S.L., 1995. Bitter patterns versus hysteresis behavior in small single particles of hematite. *Journal of Geophysical Research*, **100**: 353–364.

Halgedahl, S.L., 1998. Barkhausen jumps in larger versus small platelets of natural hematite. *Journal of Geophysical Research*, **103**: 30,575–30,589.

Halgedahl, S., and Fuller, M., 1980. Magnetic domain observations of nucleation processes in fine particles of intermediate titanomagnetite. *Nature*, **288**: 70–72.

Halgedahl, S.L., and Fuller, M., 1981. The dependence of magnetic domain structure upon magnetization state in polycrystalline pyrrhotite. *Physics of the Earth and Planetary Interiors*, **26**: 93–97.

Halgedahl, S., and Fuller, M., 1983. The dependence of magnetic domain structure upon magnetization state with emphasis upon nucleation as a mechanism for pseudo-single domain behavior. *Journal of Geophysical Research*, **88**: 6505–6522.

Halgedahl, S.L., and Ye, J., 2000. Observed effects of mechanical grain-size reduction on the domain structure of pyrrhotite. *Earth and Planetary Science Letters*, **176**(3): 457–467.

Harrison, T.J., Dunin-Borkowski, R.E., and Putnis, A., 2002. Direct imaging of nanoscale magnetic interactions in minerals. *Proceedings of the National Academic Sciences*, **99**: 16,556–16,561.

Heider, F., 1990. Temperature dependence of domain structure in natural magnetite and its significance for multi-domain TRM models. *Physics of the Earth and Planetary Interiors*, **65**: 54–61.

Heider, F., and Hoffmann, V., 1992. Magneto-optical Kerr effect on magnetite crystals with externally applied fields. *Earth and Planetary Science Letters*, **108**: 131–138.

Heider, F., Halgedahl, S.L., and Dunlop, D.J., 1988. Temperature dependence of magnetic domains in magnetite crystals. *Geophysical Research Letters*, **15**: 499–502.

Heisenberg, W., 1928. Zur Theorie des Ferromagnetismus. *Zeitschrift fuer Physik*, **49**: 619–636.

Hoffmann, V., Schafer, R., Appel, E., Hubert, A., and Soffel, H., 1987. First domain observations with the magneto-optical Kerr effect on Ti-ferrites in rocks and their synthetic equivalents. *Journal of Magnetism and Magnetic Materials*, **71**: 90–94.

Kittel, C., 1949. Physical theory of ferromagnetic domains. *Reviews of Modern Physics*, **21**: 541–583.

Metcalf, M., and Fuller, M., 1987a. Magnetic remanence measurements of single particles and the nature of domain patterns in titanomagnetites. *Geophysical Research Letters*, **14**: 1207–1210.

Metcalf, M., and Fuller, M., 1987b. Domain observations of titanomagnetites during hysteresis at elevated temperatures and thermal cycling. *Physics of the Earth and Planetary Interiors*, **46**: 120–126.

Metcalf, M., and Fuller, M., 1988. A synthetic TRM induction curve for fine particles generated from domain observations. *Geophysical Research Letters*, **15**: 503–506.

Moloni, K., Moskowitz, B.M., and Dahlberg, E.D., 1996. Domain structures in single crystal magnetite below the Verwey transition as observed with a low-temperature magnetic force microscope. *Geophysical Research Letters*, **23**: 2851–2854.

Moon, T., and Merrill, R.T., 1984. The magnetic moments of non-uniformly magnetized grains. *Physics of the Earth and Planetary Interiors*, **34**: 186–194.

Moon, T.S., and Merrill, R.T., 1985. Nucleation theory and domain states in multidomain magnetic material. *Physics of the Earth and Planetary Interiors*, **37**: 214–222.

Moskowitz, B.M., and Banerjee, S.K., 1979. Grain size limits for pseudosingle domain behavior in magnetite: implications for paleomagnetism. *IEEE Transactions on Magnetics*, **MAG-15**: 1241–1246.

Moskowitz, B.M., and Halgedahl, S.L., Theoretical temperature and grain-size dependence of domain state in $x = 0.6$ titanomagnetite. *Journal of Geophysical Research*, **92**: 10,667–10,682.

Moskowitz, B.M., Halgedahl, S.L., and Lawson, C.A., 1988. Magnetic domains on unpolished and polished surfaces of titanium-rich titanomagnetite. *Journal of Geophysical Research*, **93**: 3372–3386.

Muxworthy, A.R., and Williams, W., Micromagnetic calculations of hysteresis as a function of temperature in pseudo-single domain magnetite. *Geophysical Research Letters*, **26**: 1065–1068.

Newell, A.J., Dunlop, D.J., and Williams, W., 1993. A two-dimensional micromagnetic model of magnetization and fields in magnetite. *Journal of Geophysical Research*, **98**: 9533–9549.

Özdemir, O., and Dunlop, D.J., 1993. Magnetic domain structures on a natural single crystal of magnetite. *Geophysical Research Letters*, **20**: 1835–1838.

Özdemir, O., and Dunlop, D.J., 1997. Effect of crystal defects and internal stress on the domain structure and magnetic properties of magnetite. *Journal of Geophysical Research*, **102**: 20, 211–20,224.

Özdemir, O., Xu, S., and Dunlop, D.J., 1995. Closure domains in magnetite. *Journal of Geophysical Research*, **100**: 2193–2209.

Pokhil, T.G., and Moskowitz, B.M., 1996. Magnetic force microscope study of domain wall structures in magnetite. *Journal of Applied Physics*, **79**: 6064–6066.

Pokhil, T.G., and Moskowitz, B.M., 1997. Magnetic domains and domain walls in pseudo-single-domain magnetite studied with magnetic force microscopy. *Journal of Geophysical Research*, **102**: 22,681–22,694.

Proksch, R.B., Foss, S., and Dahlberg, E.D., 1994. High resolution magnetic force microscopy of domain wall fine structures. *IEEE Transactions on Magnetics*, **30**: 4467–4472.

Rhodes, P., and Rowlands, G., 1954. Demagnetizing energies of uniformly magnetised rectangular blocks. *Proceedings of the Leeds Philosophical and Literary Society, Science Section*, **6**: 191–210.
Smith, P.P.K., 1980. The application of Lorentz electron microscopy to the study of rock magnetism. *Institute of Physics Conference Series*, **52**: 125–128.
Soffel, H., 1971. The single-domain-multidomain transition in natural intermediate titanomagnetites. *Zeitschrift fuer Geophysik*, **37**: 451–470.
Soffel, H.C., 1977. Domain structure of titanomagnetites and its variation with temperature. *Journal of Geomagnetism and Geoelectricity*, **29**: 277–284.
Soffel, H., 1977. Pseudo-single-domain effects and single-domain multidomain transition in natural pyrrhotite deduced from domain structure observations. *Journal of Geophysics*, **42**: 351–359.
Soffel, H.C., Aumuller, C., Hoffmann, V., and Appel, E., 1990. Three-dimensional domain observations of magnetite and titanomagnetites using the dried colloid SEM method. *Physics of the Earth and Planetary Interiors*, **65**: 43–53.
Stacey, F.D., and Banerjee, S.K., 1974. *The Physical Principles of Rock Magnetism.* Amsterdam: Elsevier, 195 pp.
Szymczak, R., 1968. The magnetic structure of ferromagnetic materials of uniaxial structure. *Electronics Technology*, **1**: 5–43.
Weiss, P., 1907. L'hypothèse du champ moléculaire et la propriété ferromagnétique. *Journal of Physique*, **6**: 661–690.
Williams, W., and Dunlop, D.J., 1989. Three-dimensional micromagnetic modelling of ferromagnetic domain structure. *Nature*, **337**: 634–637.
Williams, W., and Dunlop, D.J., 1990. Some effects of grain shape and varying external magnetic fields on the magnetic structure of small grains of magnetite. *Physics of the Earth and Planetary Interiors*, **65**: 1–14.
Williams, W., and Dunlop, D.J., 1995. Simulation of magnetic hysteresis in pseudo-single-domain grains of magnetite. *Journal of Geophysical Research*, **100**: 3859–3871.
Williams, W., and Wright, T.M., 1998. High-resolution micromagnetic models of fine grains of magnetite. *Journal of Geophysical Research*, **103**: 30,537–30,550.
Williams, W., Hoffmann, V., Heider, F., Goddenhenreich, T., and Heiden, C., 1992. Magnetic force microscopy imaging of domain walls in magnetite. *Geophysical Journal International*, **111**: 417–423.
Worm, H.-U., and Markert, H., 1987. The preparation of dispersed titanomagnetite particles by the glass-ceramic method. *Physics of the Earth and Planetary Interiors*, **46**: 263–270.
Worm, H.-U., Ryan, P.J., and Banerjee, S.K., 1991. Domain size, closure domains, and the importance of magnetostriction in magnetite. *Earth and Planetary Science Letters*, **102**: 71–78.
Xu, S., Dunlop, D.J., and Newell, A.J., 1994. Micromagnetic modelling of two-dimensional domain structures in magnetite. *Journal of Geophysical Research*, **99**: 9035–9044.
Ye, J., and Halgedahl, S.L., 2000. Theoretical effects of mechanical grain-size reduction on GEM domain states in pyrrhotite. *Earth and Planetary Science Letters*, **178**: 73–85.
Ye, J., and Merrill, R.T., 1995. The use of renormalization group theory to explain the large variation of domain states observed in titanomagnetites and implications for paleomagnetism. *Journal of Geophysical Research*, **100**: 17,899–17,907.

Cross-references

Magnetic Properties, Low-Temperature
Magnetization, Isothermal Remanent (IRM)
Magnetization, Thermal Remanent (TRM)
Paleomagnetism

MAGNETIC FIELD OF MARS

Introduction

Strong magnetic anomalies have been detected over the south hemisphere of Mars. The Mars missions prior to Mars Global Surveyor (MGS) detected no appreciable magnetic field around the planet, which led to the conclusion that the Martian core field is weaker than Earth's by more than an order of magnitude. However, orbiting at elevations as low as 100–200 km during the science phase and aerobreaking phase, MGS detected a very strong crustal magnetic field, as strong as 200 nT, over the ancient southern highlands (Acuna *et al.*, 1999), indicating that the Martian crust is more magnetic than Earth's by more than an order of magnitude.

There is evidence from the Martian meteorites that a magnetic field as strong as $\sim$3000 nT has existed on the surface of Mars. The oldest Martian meteorite (ALH84001) formed before 4 Ga (e.g., Collinson, 1997; Kirschvink *et al.*, 1997; Weiss, *et al.*, 2002; Antretter *et al.*, 2003) and the young Martian meteorites that have crystallization ages ranging from 0.16 to 1.3 Ga (Nyquist *et al.*, 2001) are magnetized in a weak field of less than 3000 nT (e.g., Cisowski, 1986; Collinson, 1997; McSween and Treiman, 1998). Whether the meteorites were magnetized by a weak core field or by the local crustal field is still debated. There is good evidence that the strong anomalies of the crust have existed for the last $\sim$4 Ga (Arkani-Hamed, 2004a).

The magnetometer on board of MGS has provided immense amount of magnetic data since it resumed its mapping phase when its highly elliptical orbit was reduced to an almost circular polar orbit at a mean elevation of $\sim$400 km. Besides the magnetometer, MGS carried an electron reflectometer, which has also indirectly provided data on the magnetic field of Mars (e.g., Mitchell *et al.*, 2001). This article is concerned with the magnetometer data. The first section presents a spherical harmonic model of the magnetic field of Mars that is derived from the mapping-phase magnetic data. The second section is concerned with the first-order interpretation of the anomalies. The final section summarizes the major points of the article.

The spherical harmonic model of the magnetic field

The first magnetic anomaly map of Mars was derived by Acuna *et al.* (1999) from the low-altitude MGS data acquired at 100–200 km elevations during the science phase and aerobreaking phase. It was presented without altitude corrections. Subsequently, Purucker *et al.* (2000) reduced the radial component anomalies to a constant elevation of 200 km, and Arkani-Hamed (2001a) derived a 50° spherical harmonic model of the magnetic potential at 120 km altitude using all three vector components of the magnetic field. The later magnetic field models of Mars by Arkani-Hamed (2002), Cain *et al.* (2003), and Langlais *et al.* (2004) were obtained using all three components of the magnetic field acquired at the low- and high-altitude phases.

A highly repeatable and reliable magnetic anomaly map of Mars is essential for the investigation of the relationship between the tectonic features and the magnetization of the Martian crust. MGS has provided a huge amount of high-altitude magnetic data acquired within 360–420 km altitudes. Although the high altitude of the spacecraft limits the resolution of the data (e.g., Connerney *et al.*, 2001; Arkani-Hamed, 2002), the huge amount of the data provides an opportunity to derive a highly repeatable and accurate magnetic anomaly map at that altitude. For this purpose Arkani-Hamed (2004b) used the nighttime radial component of the high-altitude data, which is least contaminated by noncrustal sources, and selected the most common features on the basis of covariance analysis. The vast amount of the data allowed him to divide the entire data into two almost equal sets, acquired during two different periods separated by more than a year, and to derive two separate spherical harmonics models of the magnetic field. The spherical harmonic model of the radial component of the magnetic field R is expressed as

$$R(r,q,j) = \sum_{n=1}^{N} (n+1)(a/r)^{n+2} \sum_{m=0}^{n} (C_{nm}\cos m\varphi + S_{nm}\sin m\varphi)P_m^n(\cos\theta)$$

where a is the mean radius of Mars (3393 km); r, θ, and φ are distance from the center, colatitude, and east longitude (the coordinate system is centered at the center of mass of Mars); $P_m^n(\cos\theta)$ is the Schmidt-normalized associated Legendre function of degree n and order m; C_{nm} and S_{nm} are the Gauss coefficients; and N denotes the highest degree harmonics retained in the model. The covariance analysis of the two models showed that the covarying coefficients of the models are almost identical over harmonics of degree up to 50, with correlation coefficients greater than 0.95. These harmonics are highly repeatable and hence reliable. Figure M25/Plate 13a shows the radial component of the magnetic field of Mars at 100 km altitude, derived by averaging the covarying harmonic coefficients of the two models up to degree 50 and downward continuing from 400 to 100 km altitude. The color bar is saturated to better illustrate the weak anomalies.

Interpretation of the magnetic anomalies

Figure M25/Plate 13a shows strong magnetic anomalies arising from the remanent magnetization of the crust. This section provides first order interpretation of the anomalies.

The magnetic anomalies in the northern lowlands are very weak compared to those in the south. Many buried impact craters in the lowlands suggest similar ages for the crust in the north and south (Frey *et al.*, 2001). There is no striking evidence that the composition of the Martian crust had a distinct north-south dichotomy before the formation of the lowlands that would have prevented the formation of strong magnetic bodies in the north. The weak anomalies in the lowlands are most likely the relicts of stronger ones that existed before the formation of the lowlands. Zuber *et al.* (2000) disputed the impact origin of the lowlands and suggested crustal thinning due to a giant mantle plume. According to this hypothesis, the magnetic source bodies must have been partially demagnetized by the thermal effects of the plume, and to a lesser extent by near-surface low-temperature hydration (e.g., Arkani-Hamed, 2004a).

Figure M25/Plate 13a shows that the giant impact basins Hellas (41° S, 70° E), Argyre (50° S, 316° E), and Isidis (13° N, 88° E) are almost devoid of magnetic anomalies. These basins were formed by large impacts ~4 Ga, which resulted in strong shock pressures and high temperatures, capable of demagnetizing the crust beneath the basins (Hood *et al.*, 2003; Mohit and Arkani-Hamed, 2004; Kletetschka *et al.*, 2004). The crust beneath these basins is completely demagnetized to a distance of ~0.8 basin radius, where the shock pressure exceeded ~3 GPa, and partially demagnetized to ~1.4 radius where the pressure exceeded 2 GPa. This observation together with the fact that many intermediate-size impacts that created craters of 200–500 km diameter have little demagnetization effect led Mohit and Arkani-Hamed (2004) to suggest that the magnetic carriers of Martian crust have high coercivity. The resulting demagnetization depends on the coercivity of the magnetic minerals, magnetite, hematite, and pyrrhotite, suggested for the Martian crust (e.g., Kletetschka *et al.*, 2000, 2004; Hargraves *et al.*, 2001). Cisowski and Fuller (1978) found that remanence with a coercivity of 70 mT was ~20% demagnetized after a shock of 1 GPa and 70% after a shock of 4 GPa. Under a shock of 1 GPa, the remanence of the multidomain hematite and magnetite are reduced by 20% and 68%, respectively (Kletetschka *et al.*, 2002, 2004). Shock experiments on high coercivity (300 mT) single-domain pyrrhotite samples showed that a shock of 1 GPa removed 50% of the magnetization at room temperature, and they were completely demagnetized by shocks exceeding 2.75 GPa, undergoing a transition to a paramagnetic phase (Rochette *et al.*, 2003). The high-pressure magnetic measurements in a diamond anvil cell showed that magnetite behaved as single domain with high coercivity at high pressures (Gilder *et al.*, 2002). The survival of the magnetic anomalies beneath intermediate-size craters can be partly due to the increase of the coercivity of already high-coercive magnetic carriers of the Martian crust during the passage of the shock wave.

The magnetic anomalies over Tharsis bulge are much weaker than those in the south. The bulge formed through major volcanic activities in Noachian and Early Hesperian, although minor volcanism likely continued to the recent past (e.g., Hartmann and Neukum, 2001).

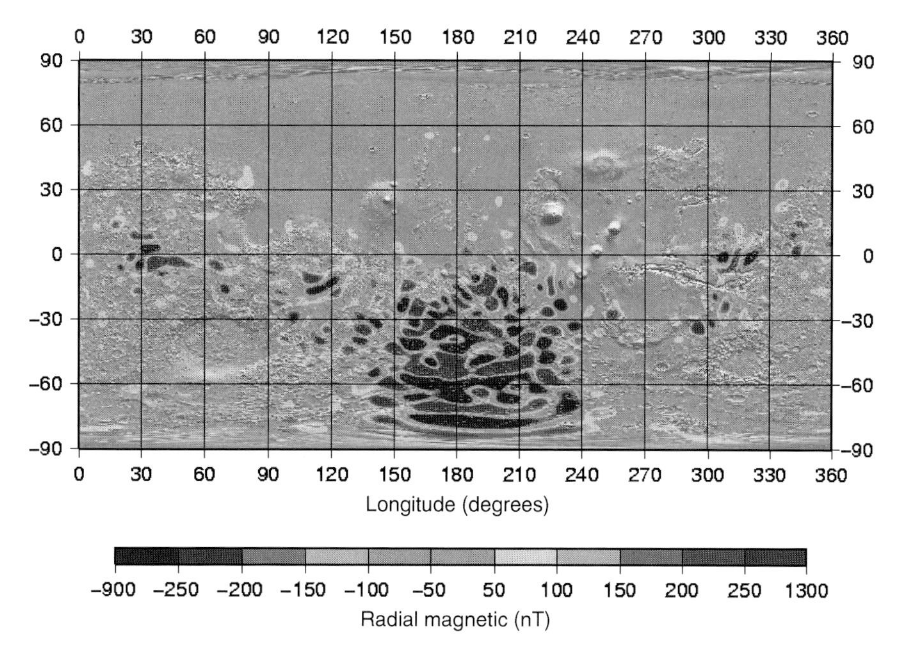

Figure M25/Plate 13a The radial component magnetic map of Mars at 100 km altitude derived from the high-altitude MGS data using the spherical harmonic of degree up to 50. The color bar is saturated to better illustrate the weak anomalies. Units are in nanotesla.

There are many lines of evidence that the volcanic layers of Tharsis are not significantly magnetized, they are formed either in the absence of a core dynamo or in the waning period of the dynamo (Arkani-Hamed, 2004b). Many places of Tharsis have been punctured by tectonic processes, which could have created detectable magnetic anomalies if the Tharsis plains were appreciably magnetized. Here two major features, Valles Marineris and shield volcanoes, are discussed in some detail.

Figure M25/Plate 13a shows no distinct magnetic edge effects associated with Valles Marineris, a canyon of over 5 km mean depth, 100–400 km width, and ~3500 km length. The models proposed for the formation of this giant canyon (e.g., Schultz, 1997; Tanaka, 1997; Wilkins and Schultz, 2003) indicate a vast amount of mass wasting. Many of the proposed models imply that upper strata of the canyon floor are porous. The porous floor and possible long-existing water make the rocks more susceptible to demagnetization by low-temperature hydration. The magnetization of mid-ocean ridge basalt on Earth declines from ~20 to ~5 A m^{-1} in the first ~30 Ma, likely a result of low-temperature hydration by circulating pore water (e.g., Bliel and Petersen, 1983). Also a positive Bouguer anomaly of ~150 mGal is associated with the central part of the canyon, implying crustal thinning and mantle uplift of ~15 km. If the Tharsis crust were strongly magnetized before the formation of the canyon, it is to be expected that the crust in the canyon would be partly demagnetized, giving rise to detectable magnetic edge effects. The absence of the expected edge effects indicates that the preexisting Tharsis plains were not significantly magnetized. Figure M25/Plate 13a shows that the formation of the canyon has ruptured the magnetic anomalies in the eastern part of the canyon between 290° E and 320° E. The anomalies must have existed before the formation of Valles Marineris that occurred during Late Noachian to Early Hesperian (e.g., Anderson *et al.*, 2001). Figure M25/Plate 13a also shows no magnetic signature associated with the shield volcanoes Olympus, Arsia, Pavonis, and Ascreaus, suggesting that the underlying preexisting Tharsis plains had not been appreciably magnetized. Otherwise, the thermal effects of the shield volcanoes could have demagnetized part of them and created low-magnetic patches in the magnetized plains, giving rise to detectable magnetic anomalies.

There was no active core dynamo during the formation of the shield volcanoes, from Early Hesperian to Late Amazonian. Hood and Hartdegen (1997) estimated the magnetic anomaly that would have been produced by the entire volcanic structure if were magnetized by a core field an order of magnitude weaker than the Earth's core field. The lack of such a magnetic anomaly indicates that no core dynamo existed during the formation of shield volcanoes to magnetize the volcanic flows.

Many lines of evidence suggest that the core dynamo decayed during Early Noachian. The lack of magnetic anomalies associated with giant impact basins, the absence of magnetic edge effect of Valles Marineris, and the fact that the huge amount of volcanic flows of Syria Planum, occurred from Noachian to Early Amazonian (e.g., Anderson *et al.*, 2001) and those of Olympus and Tharsis mounts, occurred from Early Hesperian to Late Amazonian, show no magnetic signatures strongly argue against an active core dynamo since Early Noachian (Arkani-Hamed, 2004b).

Attempts have been made to estimate the paleomagnetic pole position of Mars. Arkani-Hamed (2001b) modeled 10 widely distributed relatively isolated small magnetic anomalies. Assuming that the source bodies were magnetized by a dipole core field, he found that most of the pole positions were clustered within a 30° circle centered at 25° N, 230° E. Moreover, both north and south magnetic poles were found in the cluster, suggesting reversal of the core field polarity. Using the two weak magnetic anomalies near the geographic north pole, Hood and Zacharian (2001) found paleomagnetic poles in general agreement with Arkani-Hamed (2001b). Arkani-Hamed and Boutin (2004) used magnetic profiles from the low- and high-altitude data to model nine anomalies. The magnetic pole positions determined from their models showed clustering of the poles at the same general region, but not as tightly as those of Arkani-Hamed (2001b).

The magnetic field of Mars provides information about the rotational dynamics of the planet. None of the paleomagnetic poles are close to the present rotation pole. If the axis of dipole core field were close to the axis of rotation at the time the magnetic sources acquired magnetization, which is the case for the terrestrial planets at present, the paleomagnetic pole positions suggest appreciable, 20°–60°, polar wander since the magnetic bodies were magnetized. Polar wander of 20° to 120° were proposed on the basis of geological features (Murray and Malin, 1973; Schultz and Lutz-Garihan, 1982; Schultz and Lutz-Garihan, 1988). Approximating the topographic mass of Tharsis bulge by a surface mass, Melosh (1980) predicted polar wander of up to 25° from the effects of the bulge on the moment of inertia of the planet. Willeman (1984) suggested that compensation of Tharsis bulge would limit the amount of polar wander to less than 10°. The theoretical studies of polar wander by Spada *et al.* (1996) showed that Mars could have undergone polar wander by as much as 70°, in response to surface loads such as Olympus and Tharsis mounts.

Summary

Figure M25/Plate 13a presents the most recent and highly accurate spherical harmonic model of the radial component of the magnetic field of Mars derived from the high-altitude mapping-phase magnetic data from MGS, using harmonics of degree up to 50. It shows that the northern lowland formation and the large impacts that created the giant basins Hellas, Isidis, and Argyre have significantly demagnetized the crust and there was no active core dynamo to remagnetize the crust. The absence of magnetic edge effects associated with Valles Marineris, and the lack of magnetic signatures of the large shield volcanoes Olympus and Tharsis mounts imply that core dynamo did not exist from Late Noachian to Late Amazonian, and likely to the present. The fact that intermediate-size craters, with diameters 200–500 km, show no sign of demagnetization of the underlying crust indicates that the magnetic carriers of the Martian crust have high coercivity. The clustering of the paleomagnetic poles far from the present rotation axis implies polar wander of Mars by about 20°–50° since the magnetic source bodies were magnetized. Finally, the presence of both north and south paleomagnetic poles in the cluster suggests that the core field of Mars had polarity reversals.

Acknowledgment

This research was supported by the Natural Sciences and Engineering Research Council (NSERC) of Canada.

Jafar Arkani-Hamed

Bibliography

Acuna, M.H. *et al.*, 1999. Global distribution of crustal magnetization discovered by the Mars Global Surveyor MAG/ER experiment. *Science*, **284**: 790–793.

Anderson, R.C. *et al.*, Primary centers and secondary concentrations of tectonic activity through time in the western hemisphere of Mars. *Journal of Geophysical Research*, **106**: 20,563–20,585.

Antretter, M., Fuller, M., Scott, E., Jackson, M., Moskowitz, B., and Soleid, P., 2003. Paleomagnetic record of Martian meteorite ALH84001. *Journal of Geophysical Research*, **108**(E6): 5049, doi:10, 1029/2002JE001979.

Arkani-Hamed J., 2001a. A 50 degree spherical harmonic model of the magnetic field of Mars. *Journal of Geophysical Research*, **106**: 23,197–23,208.

Arkani-Hamed, J., 2001b. Paleomagnetic pole positions and poles reversals of Mars. *Geophysical Research Letters*, **28**: 3409–3412.

Arkani-Hamed, J., 2002. An improved 50-degree spherical harmonic model of the magnetic field of Mars, derived from both high-altitude and low-altitude observations. *Journal of Geophysical Research*, **107**: 10.1029/2001JE001835.

Arkani-Hamed, J., 2004a. Timing of the Martian core dynamo. *Journal of Geophysical Research*, **109**(E3): E03006, doi:10.1029/2003JE002195.

Arkani-Hamed J., 2004b. A coherent model of the crustal magnetic field of Mars. *Journal Of Geophysical Research*, **109**: E09005, doi:10.1029/2004JE002265.

Arkani-Hamed, J., and Boutin, D., 2004. Paleomagnetic poles of Mars: Revisited. *Journal of Geophysical Research*, **109**: doi:10.1029/2003JE0029.

Bliel, U., and Petersen, N., 1983. Variations in magnetization intensity and low-temperature titano-magnetite oxidation of ocean floor basalts. *Nature*, **301**: 384–388.

Cain, J.C., Ferguson, B., and Mozzoni, D., 2003. An $n = 90$ model of the Martian magnetic field. *Journal of Geophysical Research*, **108**: 10.1029/2000JE001487.

Cisowski, S.M., 1986. Magnetic studies on Shergotty and other SNC meteorites. *Geochemica Cosmochemica Acta*, **50**: 1043–1048.

Cisowski, S., and Fuller, M., 1978. The effect of shock on the magnetism of terrestrial rocks. *Journal of Geophysical Research*, **83**: 3441–3458.

Collinson, D.W., 1997. Magnetic properties of Martian meteorites: implications for an ancient Martian magnetic field. *Planetary Science*, **32**: 803–811.

Connerney, J.E.P., Acuna, M.H., Wasilewski, P.J., Kletetschka, G., Ness, N.F., Remer, H., Lin, R.P., and Mitchell, D.L., 2001. The global magnetic field of Mars and implications for crustal evolution. *Geophysical Research Letters*, **28**: 4015–4018.

Frey, H., Shockey, K.M., Frey, E.L., Roark, J. H., and Sakimoto, S.E.H., 2001. A very large population of likely buried impact basins in the northern lowlands of Mars revealed by MOLA data. *Lunar and Planetary Science Conference XXXII*, Abstr. 1680.

Gilder, S.A., Le Goff, M., Peyronneau, J., and Chervin, J., 2002. Novel high pressure magnetic measurements with application to magnetite. *Geophysical Research Letters*, **29**: 10,1029/2001GL014227, 2002.

Hargraves, R.B., Knudsen, J.M., Madsen, M.B., and Bertelsen, P., 2001. Finding the right rocks on Mars. *EOS: Transactions, American Geophysical Union*, **82**: 292–293.

Hartmann, W.K., and. Neukum, G., 2001. Cratering chronology and the evolution of Mars. *Space Science Reviews*, **96**: 165–194.

Hood, L.L., and Hartdegen, K., 1997. A crustal magnetization model for the magnetic field of Mars: a preliminary study of the Tharsis region. *Geophysical Research Letters*, **24**: 727–730.

Hood, L.L., and Zacharian, A., 2001. Mapping and modeling of magnetic anomalies in the northern polar region of Mars. *Journal of Geophysical Research*, **106**: 14601–14619.

Hood, L.L., Richmond, N.C., Pierazzo, E., and Rochette, P., Distribution of crustal magnetic fields on Mars: Shock effects of basin-forming impacts. *Geophysical Research Letters*, **30**(6): 1281, doi:10.1029/2002GL016657.

Kletetschka, G., Wasilewski, P.J., and Taylor, P.T., 2000. Mineralogy of the sources for magnetic anomalies on Mars. *Meteoritics and Planetary Science*, **35**: 895–899.

Kletetschka, G., Wasilewski, P.J., and Taylor, P.T., 2002. The role of hematite-ilmenite solid solution in the production of magnetic anomalies in ground- and satellite-based data. *Tectonophysics*, **347**: 167–177.

Kletetschka, G., Connerney, J.E.P., Ness, N.F., and Acuna, M.H., 2004. Pressure effects on Martian crustal magnetization near large impact basins. *Meteoritics and Planetary Science*, **39**: 1839–1848.

Kirschvink, J.L., Maine, A.T., and Vali, H., 1997. Paleomagnetic evidence of a low-temperature origin of carbonate in the Martian meteorite ALH84001. *Science*, **275**: 1629–1633.

Langlais, B., Purucker, M.E., and Mandea, M., 2004. Crustal magnetic field of Mars. *Journal of Geophysical Research*, **109**: E002008, doi:10.1029/2003JE002048.

McSween, H.Y., and Treiman, A.H., 1998. Martian meteorites, Chapter 6 in Planetary materials. *Reviews in Mineralogy*, **36**: 53.

Melosh, H.J., 1980. Tectonic patterns on a reoriented planet: Mars. *Icarus*, **44**: 745–751.

Mitchell, D.L. *et al.*, 2001. Probing Mars' crustal magnetic field and ionosphere with the MGS electron reflectometer. *Journal of Geophysical Research*, **106**: 23,419–23,427.

Mohit, P.S., and Arkani-Hamed, J., 2004. Impact demagnetization of the Martian crust. *Icarus*, **168**: 305–317.

Murray, B.C., and Malin, M.C., 1973. Polar wandering on Mars. *Science*, **179**: 997–1000.

Nyquist, L.E. 2001. Ages and geological history of Martian meteorites. In Kallenbach, R., Geiss, J., and Hartmann, W.K. (eds.), *Chronology and Evolution of Mars*. Dordrecht, the Netherlands: Kluwer Academic Publishers.

Purucker, M., Ravat, D., Frey, H., Voorhies, C., Sabaka, T., and Acuna, M., 2000. An altitude-normalized magnetic map of Mars and its interpretation. *Geophysical Research Letters*, **27**: 2449–2452.

Rochette, P., Fillion, G., Ballou, R., Brunet, F., Ouladdiaf, B., and Hood, L., 2003. High pressure magnetic transition in pyrrhotite and impact demagnetization on Mars. *Geophysical Research Letters*, **30**(13): 1683, doi:10.1029/2003GL017359.

Schultz, P.H., and Lutz-Garihan, A.B., 1982. Grazing impacts on Mars: a record of lost satellites. *Journal Geophysical Research*, [illegible]

Schultz, P.H., and Lutz-Garihan, A.B., 1988. Polar wandering of Mars. *Icarus*, **73**: 91–141.

Schultz, R.A., 1997. Dual-process genesis for Valles Marineris and troughs on Mars, presented at the XXVIII Lunar and Planetary Science Conference, Houston, Texas.

Spada, G. 1996. Long-term rotation and mantle dynamics of the Earth, Mars and Venus. *Journal of Geophysical Research*, **101**: 2253–2266.

Tanaka, K.L., 1997. Origin of Valles Marineris and Noctis Labyrinthus, Mars, by structurally controlled collapses and erosion of crustal materials, presented at the XXXVIII Lunar and Planetary Science Conference, Houston, Texas.

Weiss, B.P., Vali, H., Baudenbacher, F.J., Kirschvink, J.L., Stewart, S.T., and Schuster, D.L., 2002. Records of an ancient Martian field in ALH84001. *Earth and Planetary Science Letters*, **201**: 449–463.

Wilkins, S.J., and Schultz, R.A., 2003. Cross faults in extensional settings: stress triggering, displacement localization, and implications for the origin of blunt troughs at Valles Marineris, Mars. *Journal of Geophysical Research*, **108**: E6, 5056, doi:1029/2002JE001968.

Willeman, R.J., 1984. Reorientation of planets with elastic lithospheres. *Icarus*, **60**: 701–709.

Zuber, M.T. 2000. Internal structure and early thermal evolution of Mars from Mars Global Surveyor topography and gravity. *Science*, **287**: 1788–1793.

MAGNETIC FIELD OF SUN

The Sun has been observed to exhibit a breathtaking variety of magnetic phenomena on a vast range of spatial and temporal scales. These vary in spatial scale from the solar radius down to the limit of present resolution of the most powerful satellite instrumentation, and with durations varying from minutes to hundreds of years, encompassing the famous 11-year sunspot cycle. This dynamic and active field, which is visible in extreme ultraviolet wavelengths as shown in Figure M26/Plate 14a, is responsible for all solar magnetic phenomena, such as sunspots, solar flares, coronal mass ejections and the solar wind, and also heats the solar corona to extremely high temperatures. These have important terrestrial consequences, causing severe magnetic storms and major disruption to satellites, as well as having a possible impact on

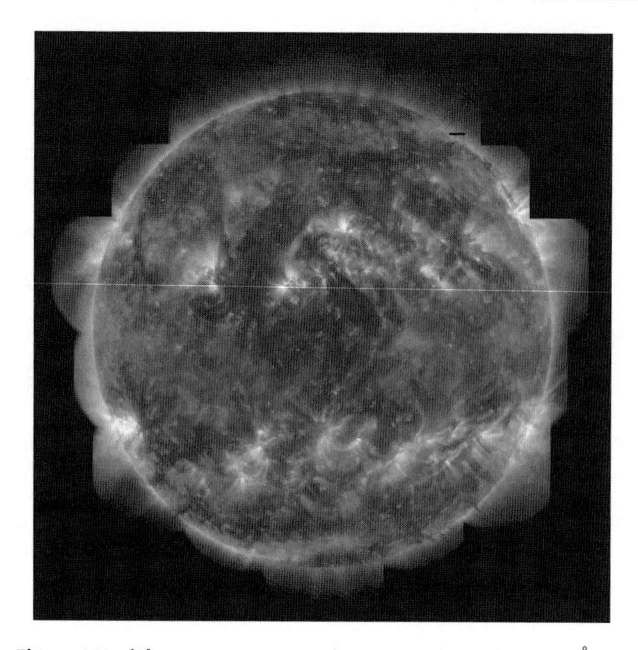

Figure M26/Plate 14a Compound extreme ultraviolet (171 Å) full-disk image of the Sun taken by the Transition Region and Coronal Explorer (TRACE) satellite. The two bands on either side of the Sun's equator contain bright features indicating magnetic activity.

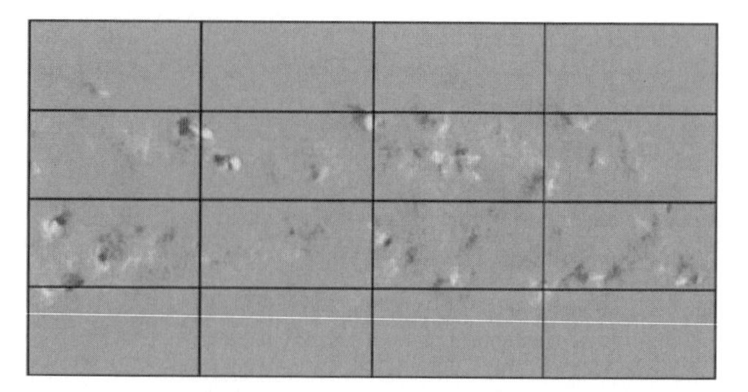

Figure M27 Hale's polarity law: Magnetogram of the solar surface as a function of longitude and latitude. White indicates strong radial field pointing out of the solar surface, while black indicates strong radial field pointing into the solar surface. Notice that the sense of the active regions is opposite in the northern and southern hemispheres.

the terrestrial climate. The magnetic field also plays a major role in the evolution of the Sun as a star, with magnetic braking drastically reducing the angular momentum of the star as it evolves over its lifetime (Mestel, 1999).

This article will review the observations of the solar magnetic field and briefly indicate the features that need to be understood in order to formulate a theory for the generation and evolution of such a field.

The large-scale solar field

Although magnetic fields in the Sun are observed on a whole spectrum of length-scales, it is usual to classify magnetic phenomena either as large- or small-scale events. Here the large-scale magnetic field is defined to be that which is observed on scales comparable to the solar radius, with systematic properties, spatial organization, and temporal coherence.

The most obvious manifestation of the large-scale solar magnetic field is the appearance of sunspots at the solar surface. Sunspots have been systematically observed in the West since the early 17th century when Galileo utilized the newly invented telescope to document the passage of dark spots across the solar surface. Although the 11-year period for the cycle of sunspot activity was discovered by Schwabe in 1843, the connection between sunspots and the magnetic field of the Sun was only determined when Hale in 1908 used the Zeeman splitting of spectral lines to establish the presence of a strong magnetic field in sunspots (Hale, 1908). By then it could clearly be seen that sunspots appear in bipolar pairs, with a leading and trailing spot nearly aligned with the solar equator. Magnetic observations of sunspots indicate that they have systematic properties—often described as Hale's polarity laws. These state that the magnetic polarities of the leading and trailing spots are opposite, with the sense of the polarity being opposite in each hemisphere of the Sun (see Figure M27). Moreover, the polarities of the bipolar sunspot pairs that are observed at the solar surface reverse after each minimum in sunspot activity. Finally, Joy's law states that there is a systematic tilt (of the order of 4°) in the alignment of sunspot pairs with respect to the solar equator, with the leading spot being closer to the equator. Taken together, these indicate that sunspots are the surface manifestation of an underlying magnetic field with an opposite sense in each hemisphere. This magnetic field is cyclic, with a mean period of 22 years.

Sunspot pairs are interpreted as an indicator for an azimuthal (toroidal) B_ϕ field. A radial (poloidal) component of the magnetic field, B_r, may also be detected. The radial field is particularly visible at the solar poles with a distinct and opposite polarity at each pole. This polar field reverses at the *maximum* of sunspot activity and the polarity is such that between solar maximum and solar minimum the radial field has the same polarity as the trailing spots of the sunspot pairs in the same hemisphere. Hence B_ϕ and B_r fields are nearly in antiphase with $B_\phi B_r < 0$ for most of the time (Stix, 1976). Also particularly visible at high latitudes are ephemeral active regions. These are small regions that do not systematically obey Hale's polarity laws (although they do show a preference for the same orientation as sunspots) and have a lifetime of a few days. It is now believed that these ephemeral active regions are either reprocessed flux from active regions or the result of local regeneration in the solar convection zone.

An important measure of the topology of the large-scale solar magnetic field is given by the current helicity of the magnetic field $\mathcal{H}_j = \langle \mathbf{j} \cdot \mathbf{B} \rangle$, where $\mathbf{j} = 1/\mu_0 \nabla \times \mathbf{B}$ is the electric current. This is a difficult quantity to measure, although some information about the radial contribution to the current helicity $\langle j_r B_r \rangle$ can be estimated from vector magnetograms of active regions. The measurements are consistent with the premise that magnetic structures possess a preferred orientation corresponding to a left-hand screw in the northern hemisphere (with an opposite sense in the southern hemisphere). Another crucial topological measure of the magnetic field is the magnetic helicity $\mathcal{H} = \langle \mathbf{A} \cdot \mathbf{B} \rangle$ (where $\mathbf{B} = \nabla \times \mathbf{A}$). Although some estimates have been made of the magnetic helicity from vector magnetograms, it is almost impossible to measure this important property of the solar magnetic field with any great certainty.

Spatiotemporal variability

In addition to the systematic properties of the solar magnetic field indicated by Hale's and Joy's laws, these large-scale features follow a systematic spatiotemporal pattern. At the beginning, of a cycle spots appear on the Sun at latitudes of around ±27°; as the cycle progresses the location of activity drifts toward the equator and the spots then die out (at a latitude of around ±8°) in the next minimum. It is important to note that it is the location of magnetic activity rather than individual sunspot pairs—which have a lifetime of at most a few months—that migrates toward the solar equator. This migration of activity produces the familiar butterfly diagram first exhibited by Maunder (1913). An up-to-date version showing the incidence of sunspots as a function of colatitude is in Figure M28a. It is clear from this figure that the level of sunspot activity has been approximately the same in each hemisphere, and that the solar magnetic field has a high degree of symmetry. The above observations indicate that the solar field is primarily axisymmetric and dipolar, although a small nonaxisymmetric component of the solar field has been observed.

In addition to the basic solar cycle, sunspot activity undergoes further significant temporal variability. By using the (arbitrarily defined) sunspot number (e.g., Eddy, 1976), the variation in magnetic activity over the past 400 years may be obtained. An up-to-date record of sunspot numbers is shown in Figure 28b. This figure shows that an average cycle is temporally asymmetric, with a sharp rise from minimum to maximum of duration 3–6 years being followed by a gradual decay lasting between 5 and 8 years. The amplitude of the cycle has been shown to be anticorrelated with the length of the rise time and that of the solar cycle itself. The solar cycle appears to be chaotic with an amplitude that is modulated aperiodically. Despite the shortness of the time sequence, the existence of a characteristic timescale for the modulation of approximately 90 years (the Gleissberg cycle) has been postulated—although this is open to debate. What is clear, however, is the presence in the record of an episode with a dearth of sunspots, which lasted for about 70 years at the end of the 17th century. This interruption is known as the Maunder minimum. There is no doubt that this is a real phenomenon, and this is not due to the lack of [illegible] on the part of the observers (see Ribes and Nesme-Ribes, 1993). It is also interesting to note that at the end of the Maunder minimum large-scale solar magnetic activity started again primarily in the southern hemisphere of the Sun and hence the field was not dipolar for a cycle. Although small asymmetries between sunspot activity in the northern and southern hemispheres have been observed since the Maunder minimum, when the field is strong the field is largely dipolar, with only a small quadrupole component as noted above. It is only when the field is weak and the dipole component is small that a comparable quadrupolar contribution can be seen.

More useful information about the temporal variability of the solar magnetic activity can be gained by the use of proxy data (see e.g., Beer, 2000). Magnetic fields in the solar wind modulate the cosmic ray flux entering the Earth's atmosphere. These high-energy particles lead to the production of radioisotopes, including ^{10}Be and ^{14}C, and so the abundance of these isotopes can act as a measure of solar magnetic activity—the abundance is anticorrelated with magnetic activity. The isotope ^{10}Be is preserved in polar ice-cores and its production rate, together with the sunspot number, is shown in Figure M29. The figure clearly shows the presence of the Maunder minimum and that, although sunspot activity is largely shut off during this period, cyclic magnetic activity continued with a period of approximately 9 years. The ^{10}Be record extends back over 50 ka and analysis of this record clearly shows both continued presence of the 11-year solar cycle. Moreover analysis indicates that the Maunder minimum is not an isolated event, with regularly spaced minima (termed grand minima) interrupting the record of activity with a significant recurrent timescale of 205 years (Wagner *et al*., 2001). The variations in ^{14}C production confirm this pattern of recurrent grand minima with a timescale of about 200 years. Moreover, both of these radioisotope records show significant power at a frequency that corresponds to roughly 2100 years. It appears as though grand minima occur in bursts, with this as a well-defined period of recurrence (J. Beer, private communication).

Taken together, these data indicate that the solar magnetic field undergoes a considerable amount of temporal variability on scales varying from days to hundreds of years.

The small-scale solar magnetic field

In addition to the large-scale magnetic field with systematic properties described above, magnetic field is also observed on scales comparable with the solar convective scales and smaller. The magnetic network is seen to coincide largely with the network forming the boundaries of supergranular convection. The network has a different form at solar maximum and minimum. At solar minimum it consists of mixed polarity, vertically oriented bundles of magnetic flux, while at solar maximum the network is dominated by the presence of large unipolar regions in close proximity to the bipolar active regions. The magnetic network is therefore only weakly linked to the solar magnetic cycle. The flux bundles in the magnetic network are dynamic and are observed to move randomly across the solar surface, with a motion that is consistent with two-dimensional diffusion.

The magnetic network is also interspersed with weak stochastic magnetic fields of random orientations. This intranetwork field is

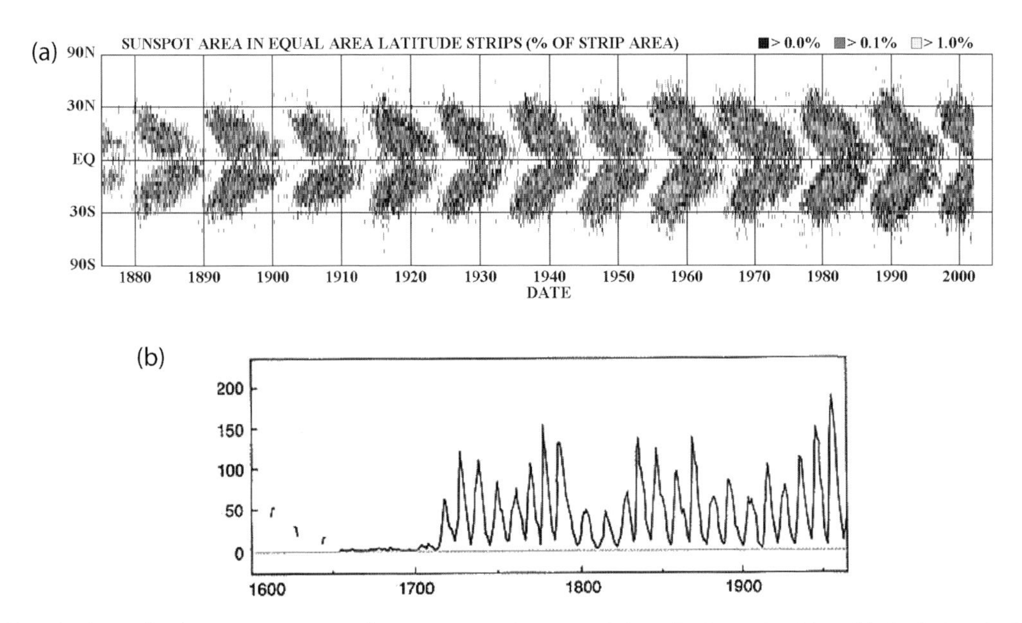

Figure M28 (a) Solar butterfly diagram: Location of sunspots as a function of time (horizontal axis) and latitude (vertical axis). As the cycle progresses sunspots migrate from midlatitudes to the equator. (Courtesy of D.N. Hathaway). (b) Time series for yearly sunspot group numbers. Notice the Maunder minimum at the end of the 17th century.

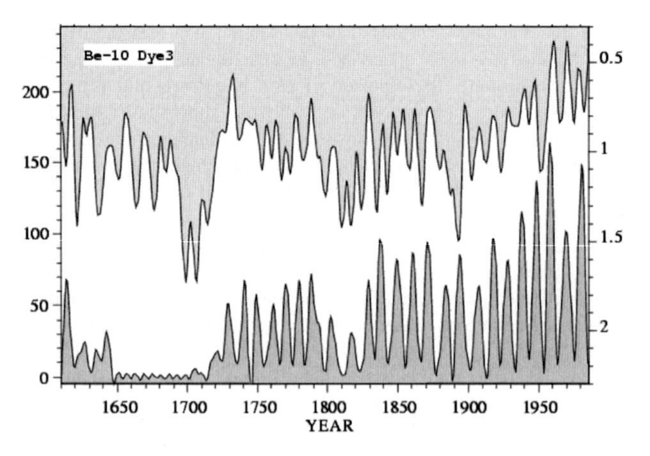

Figure M29 Comparison of the proxy ^{10}Be data from the dye 3 ice core (measured in 10000 atoms g^{-1}), filtered using a low pass (6 year) filter, with the filtered (6 year) sunspot group number as determined by Hoyt and Schatten (1998). Note that the ^{10}Be is anticorrelated with sunspot activity.

weaker than the network field and has a significantly shorter dynamical timescale associated with it (on the scale of days as opposed to months). There appears to be little or no correlation of this field with the solar cycle and the behavior of this field is consistent with it being generated by the action of a local dynamo (see *Dynamo, solar*) situated in the upper reaches of the solar convection zone. All these small-scale magnetic features visible at the solar surface migrate slowly toward the solar pole with a velocity of between 1 and 10 m s^{-1}. This migration is due to the presence of a systematic large-scale meridional flow at the solar surface. The meridional flow has also been detected by helioseismic inversions of the solar interior and is believed to continue poleward to a depth of $0.9R_\odot$, where $R_\odot$ is the solar radius.

Solar cycle indices—irradiance, flows and the shape of the corona

The solar magnetic cycle can not only be detected in the large-scale magnetic features such as sunspots. Many other solar indices follow the 11-year activity cycle. Of particular interest terrestrially is the variation in the solar luminosity (the so-called solar constant) that occurs over the solar cycle. Total solar irradiance does vary systematically in phase with the solar cycle. In the visible spectrum the modulation varies only weakly, with the effect of bright magnetic faculae nearly exactly canceling out (but just overcoming) that of dark sunspots. However in the far-ultraviolet and X-ray part of the spectrum the modulation is large and systematic (see e.g., Frölich, 2000). The shape of the solar corona and strength of the solar wind is also correlated with the solar cycle, and these in turn interact with the geomagnetic field and are measurable in the AA and AP geomagnetic indices and visible in the record of aurorae.

In addition, the presence of the magnetic field is also responsible for driving flows that are correlated with the solar activity cycle. This pattern of parallel belts of faster slower rotation was initially discovered by Doppler measurements of the solar surface and dubbed torsional oscillations. It is now known from helioseismic inversions for the internal solar rotation (Vorontsov *et al.*, 2002) that these oscillations have an amplitude of approximately 5–10 m s^{-1} and that they extend deep into the solar interior—reaching at least halfway into the solar convective zone (see Figure M30/Plate 14b). These bands of zonal flows, which have an equatorward and poleward propagating branch, have a period of 11 years, which is consistent with them being driven by the Lorentz force associated with the magnetic field.

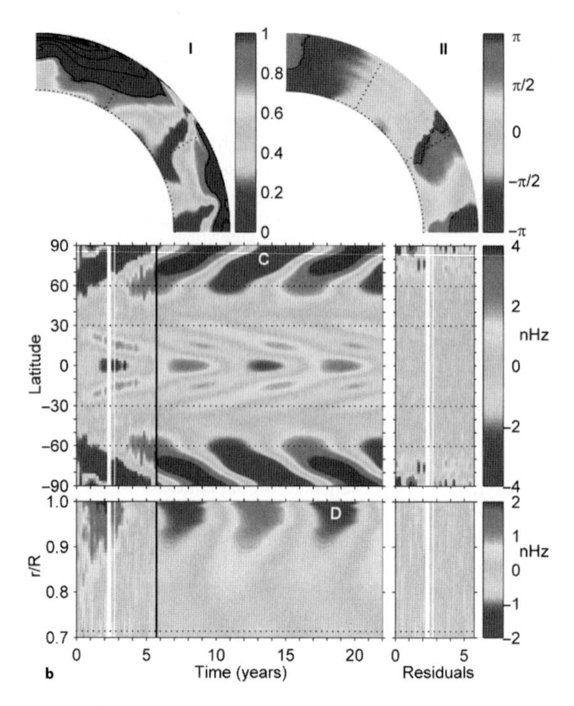

Figure M30/Plate 14b *Top panel*: The rotational variation as a function of time and latitude at radius $r = 0.98R_\odot$ are shown for the first nearly 6 years, and thereafter the time series is continued by exhibiting the 11-year harmonic fit. Shown to the right are the residuals from the fit, on the same color scale. *Bottom panel*: As for above, but showing the rotation variation as a function of depth instead of latitude, at latitude 20°. (From Vorontsov *et al.*, 2002).

Stellar magnetic fields

The Sun is just one example of a moderately rotating star whose magnetic properties can be observed in detail. Much can be learned about the magnetic behavior of the Sun, by examining the properties of nearby magnetically active stars that have similar properties (e.g., age and rotation rate). The most important results indicating magnetic activity in stars are obtained by measuring the level of Ca^+, H, and K emission. The rate of Ca^+ emission in the Sun is well-known to be correlated with solar magnetic fields. Results from the Mount Wilson Survey indicate the level of magnetic activity in a relatively large sample of slow and moderate rotators. It is now clear that, for stars of a fixed spectral type, there is a range of activity, but that this scatter is a function of age. It can be shown that the level of activity in a star can be represented as a function of the inverse Rossby (or Coriolis) number $\sigma = \Omega\tau_c$, where Ω is the angular velocity of the star and τ_c is a suitable convective timescale (Brandenburg *et al.*, 1998).

For slow rotators, cyclic magnetic activity may be deduced from variations in the Ca^+ emission and it is clear that the cycle period of activity decreases with increasing rotation rate. Moreover as the rotation rate is increased the magnetic activity of the stars becomes more disordered, with a transition between cyclic, doubly periodic and chaotic activity occurring as the rotation rate is increased. On studying this relationship in greater detail it becomes apparent that the stars fall into two distinct groups, an active and an inactive branch (Brandenburg *et al.*, 1998). This is enough to discourage the extension of the properties of slowly rotating stars to cover those with a greater angular momentum.

Direct measurement of magnetic fields in stars is also possible using a variety of techniques, including broadband polarimetry and photometry (see e.g., Rosner, 2000). The luminosity of a highly active star may vary by up to 30%, whereas the known variation in solar luminosity is only 0.2%. Whether an increase in magnetic activity in a star is positively or negatively correlated with an increase in luminosity depends upon the absolute level of activity in the star. It is also found that the level of magnetic activity is proportional to the inverse Rossby number of the star—a result that is consistent with the emission data. Clearly, as there is a small sample of stars for which we have a detailed record of magnetic activity and this record is comparatively short, there is a need to continue these observations as they give us great understanding of the magnetic properties of our nearest star—the Sun.

S. Tobias

Bibliography

Beer, J., 2000. Long-term indirect indices of solar variability. *Space Science Reviews*, **94**: 53–66.

Brandenburg, A., Saar, S.H., and Turpin, C.R., 1998. Time evolution of the magnetic activity cycle period. *The Astrophysical Journal*, **498**: L51–L54.

Eddy, J.A., 1976. The Maunder minimum. *Science*, **192**: 1189–1202.

Fröhlich, C., 2000. Observations of irradiance variations. *Space Science Reviews*, **94**: 15–24.

Hale, G.E., 1908. On the probable existence of a magnetic field in sunspots. *Astrophysics Journal*, **28**: 315–343.

Hoyt, D.V., and Schatten, K.H., 1998. *The Role of the Sun in Climate Change*. New York: Oxford University Press.

Maunder, E.W., 1913. Note on the distribution of sunspots in heliographic latitude. *Monthly Notices of the Royal Astronomical Society*, **64**: 747–761.

Mestel, L., 1999. *Stellar Magnetism*. Oxford: Clarendon Press.

Ribes, J.C., and Nesme-Ribes, E., 1993. The solar sunspot cycle in the Maunder minimum AD 1645–AD 1715. *Astronomy and Astrophysics*, **276**: 549–563.

Rosner, R., 2000. Magnetic fields of stars: using stars as tools for understanding the origins of cosmic magnetic fields. *Philosophical Transactions of the Royal Society of London, Series. A*, **358**: 689–708.

Stix, M., 1976. Differential rotation and the solar dynamo. *Astronomy and Astrophysics*, **47**: 243–254.

Wagner, G. *et al.*, 2001. Presence of the solar de vres cycle (205 years) during the last ice age. *Geophysical Research Letters*, **28**: 303.

Vorontsov, S.V., Christensen-Dalsgaard, J., Schou, J., Strakhov, V.N., and Thompson, M.J., 2002. Helioseismic measurement of solar torsional oscillations. *Science*, **296**: 101–103.

Cross-references

Dynamo, Solar

MAGNETIC INDICES

Magnetic indices are simple measures of magnetic activity that occurs, typically, over periods of time of less than a few hours and which is recorded by magnetometers at ground-based observatories (Mayaud, 1980; Rangarajan, 1989; McPherron, 1995). The variations that indices measure have their origin in the Earth's ionosphere and magnetosphere. Some indices having been designed specifically to quantify idealized physical processes, while others function as more generic measures of magnetic activity. Indices are routinely used across the many subdisciplines in geomagnetism, including direct studies of the physics of the upper atmosphere and space, for induction studies of the Earth's crust and mantle, and for removal of disturbed-time magnetic data in studies of the Earth's deep interior and core. Here we summarize the most commonly used magnetic indices, using data from a worldwide distribution of observatories, those shown in Figure M31 and whose sponsoring agencies are given in Table M1.

Range indices K and K_p

The 3-h K integer index was introduced by Bartels (1938) as a measure of the range of irregular and rapid, storm-time magnetic activity. It is designed to be insensitive to the longer term components of magnetic variation, including those associated with the overall evolution of a magnetic storm, the normal quiet-time diurnal variation, and the very much longer term geomagnetic secular variation arising from core convection. The K index is calculated separately for each observatory, and, therefore, with an ensemble of K indices from different observatory sites, the geography of rapid, ground level magnetic activity can be quantified.

When it was first implemented, the calculation of K relied on the direct measurement of an analog trace on a photographic record. Today, in order to preserve continuity with historical records, computer programs using digital data mimic the original procedure. First, the diurnal and secular variations are removed by fitting a smooth curve to 1-min horizontal component (H) observatory data. The range of the remaining data occurring over a 3-h period is measured. This is then converted to a quasilogarithmic K integer, $0, 1, 2, \ldots, 9$, according to a scale that is specific to each observatory and which is designed to normalize the occurrence frequency of individual K values among the many observatories and over many years.

A qualitative understanding of the K index and its calculation can be obtained from Figure M32. There we show a trace, Figure M32a, of the horizontal intensity at the Fredericksburg observatory recording magnetically quiet conditions during days 299–301 of 2003, followed by the sudden commencement in day 302 and the subsequent development of the main and recovery phases of the so-called great Halloween Storm. In Figure M32b we show, on a logarithmic scale, the range of the Fredericksburg data over discrete 3-h intervals, and in Figure M32c we show the K index values themselves. Note the close correspondence between the magnetogram, the log of the range and the K index. This storm is one of the 10 largest in the past 70 years since continuous measurements of storm size have been routinely undertaken. For more information on this particular storm, see the special issue of the *Journal of Geophysical Research*, A9, **110**, 2005.

Planetary-scale magnetic activity is measured by the K_p index (Menvielle and Berthelier, 1991). This is derived from the average of fractional K indices at 13 subauroral observatories (Table M1) in such a way as to compensate for diurnal and seasonal differences between the individual observatory K values. The final K_p index has values $0, 0.3, 0.7, 1.0, 1.3, \ldots$ etc. For illustration, in Figure M33 we show magnetograms from the 13 observatories contributing to K_p, recording the Halloween Storm of 2003, along with the K_p index itself. The distribution of observatories is far from uniform, with a predominant representation from North America and Europe, and very little representation from the southern hemisphere. In fact, in Figure M33, it is easy to see differences during the storm in the magnetograms among the different regional groupings of observatories. Although geographic bias is an obvious concern for any index intended as a measure of planetary-scale magnetic activity, K_p has proven to be very useful for scientific study (e.g., Thomsen, 2004). And, since it has been continuously calculated since 1932, K_p lends itself to studies of magnetic disturbances occurring over many solar cycles.

There are several other indices related to the K and K_p. A_k and A_p are linear versions of K and K_p. K_n, A_n, K_s, and A_s are similar to K_p and A_p except that they use, respectively, northern and southern hemisphere observatories; their global averages are K_m and A_m. The *aa* index is like the K_p except that it utilizes only two, roughly antipodal, observatories, one in the northern hemisphere and one in the southern hemisphere. *aa* has been continuously calculated since 1868, making it one of the longest historical time series in geophysics.

Auroral electrojet indices AU, AL, AE, AO

During magnetic storms, particularly during substorms, magnetospheric electric currents are often diverted along field lines, with

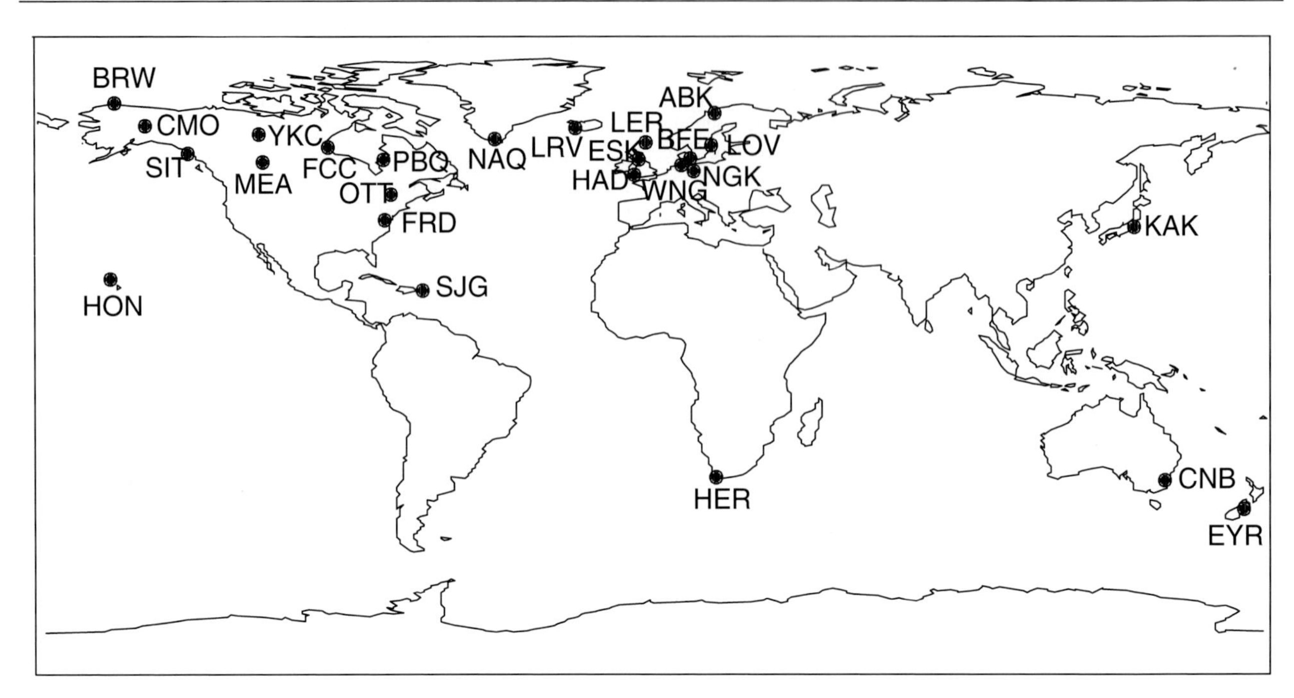

Figure M31 Map showing geographic distribution of magnetic index observatories.

Table M1 Summary of index observatories used here

Agency	Country	Observatory	Observatory	Index
Geoscience Australia	Australia	Canberra	CNB	K_p
Geological Survey of Canada	Canada	Fort Churchill	FCC	AE
Geological Survey of Canada	Canada	Meanook	MEA	K_p
Geological Survey of Canada	Canada	Ottawa	OTT	K_p
Geological Survey of Canada	Canada	Poste-de-la-Baleine	PBQ	AE
Geological Survey of Canada	Canada	Yellowknife	YKC	AE
Danish Meteorological Institute	Denmark	Brorfelde	BFE	K_p
Danish Meteorological Institute	Denmark	Narsarsuaq	NAQ	AE
GeoForschungsZentrum Potsdam	Germany	Niemegk	NGK	K_p
GeoForschungsZentrum Potsdam	Germany	Wingst	WNG	K_p
University of Iceland	Iceland	Leirvogur	LRV	AE
Japan Meteorological Agency	Japan	Kakioka	KAK	D_{st}
Geological and Nuclear Science	New Zealand	Eyerewell	EYR	K_p
National Research Foundation	South Africa	Hermanus	HER	D_{st}
Swedish Geological Survey	Sweden	Abisko	ABK	AE
Swedish Geological Survey	Sweden	Lovoe	LOV	K_p
British Geological Survey	United Kingdom	Eskdalemuir	ESK	K_p
British Geological Survey	United Kingdom	Hartland	HAD	K_p
British Geological Survey	United Kingdom	Lerwick	LER	K_p
US Geological Survey	United States	Barrow	BRW	AE
US Geological Survey	United States	College	CMO	AE
US Geological Survey	United States	Fredericksburg	FRD	K_p
US Geological Survey	United States	Honolulu	HON	D_{st}
US Geological Survey	United States	San Juan	SJG	D_{st}
US Geological Survey	United States	Sitka	SIT	K_p

current closure through the ionosphere. To measure the auroral zone component of this circuit, Davis and Sugiura (1966) defined the auroral electrojet index AE. Ideally, the index would be derived from data collected from an equally spaced set of observatories forming a necklace situated underneath the northern and southern auroral ovals. Unfortunately, the southern hemispheric distribution of observatories is far too sparse for reasonable utility in calculating AE, and the northern hemispheric observatories only form a partial necklace, due to the present shortage of reliable observatory operations in northern Russia. Progress is continuing, of course, to remedy this shortcoming, but for

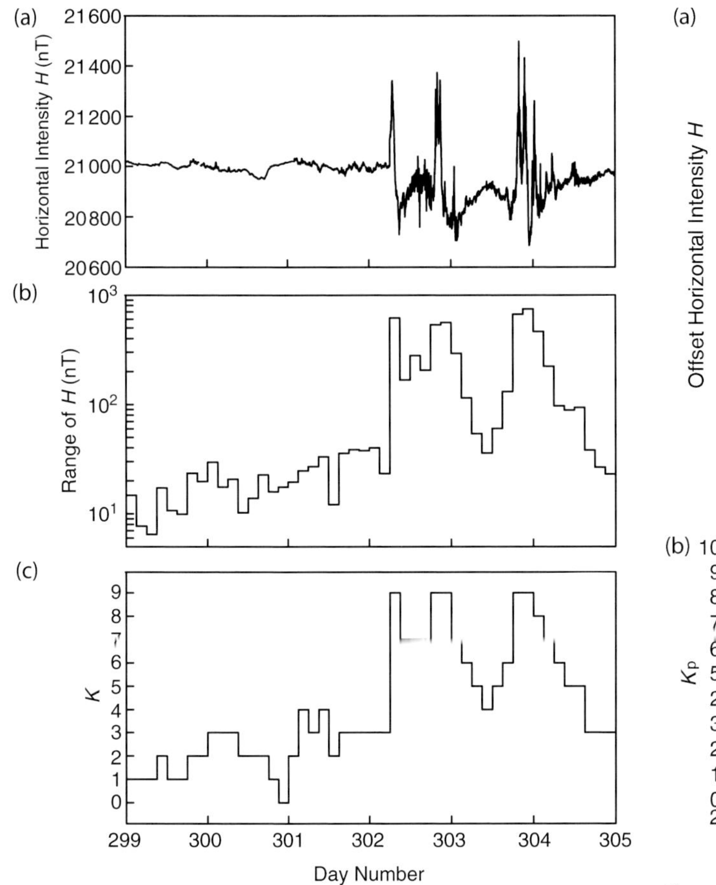

Figure M32 Example of (a) magnetometer data, horizontal intensity (H) from the Fredericksburg observatory recording the Halloween Storm of 2003, (b) the maximum range of H during discrete 3 h intervals, and (c) the K index for Fredericksburg.

now the partial necklace of northern hemisphere observatories is used to calculate an approximate AE.

The calculation of AE is relatively straightforward. One-min resolution data from auroral observatories are used, and the average horizontal intensity during the five magnetically quietest days is subtracted. The total range of the data from among the various AE observatories for each minute is measured, with AU being the highest value and AL being the lowest value. The difference is defined as AE = AU − AL, and for completeness the average is also defined as AO = 1/2(AU − AL). For illustration, in Figure M34, we show magnetograms from the eight auroral observatories contributing to AE, during the Halloween Storm of 2003, along with the AE and its attendant relatives.

Equatorial storm indices D_{st} and A_{sym}

One of the most systematic effects seen in ground-based magnetometer data is a general depression of the horizontal magnetic field as recorded at near-equatorial observatories (Moos, 1910). This is often interpreted as an enhancement of a westward magnetospheric equatorial ring current, whose magnetic field at the Earth's surface partially cancels the predominantly northerly component of the main field. The storm-time disturbance index D_{st} (Sugiura, 1964) is designed to measure this phenomenon. D_{st} is one of the most widely used indices in academic research on the magnetosphere, in part because it is well motivated by a specific physical theory.

The calculation of D_{st} is generally similar to that of AE, but it is more refined, since the magnetic signal of interest is quite a bit smaller.

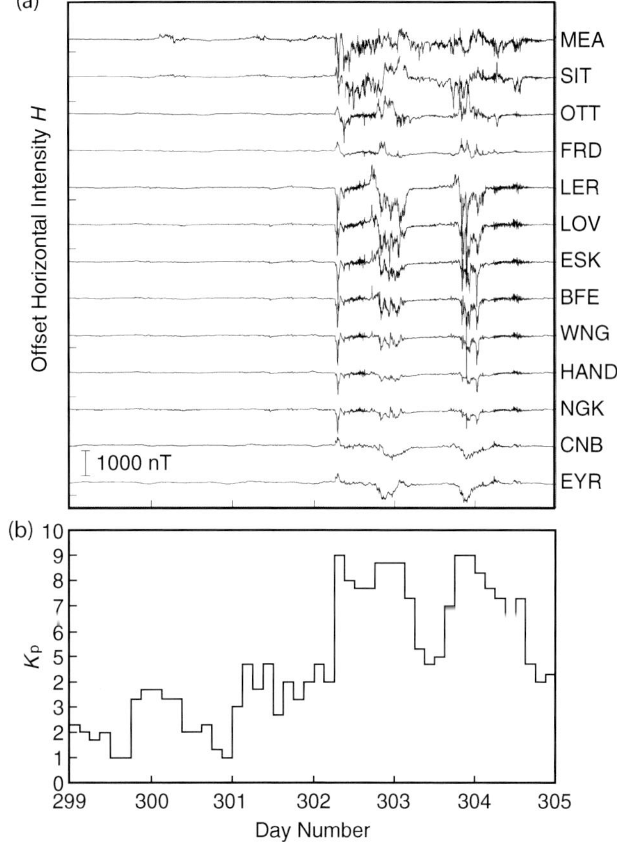

Figure M33 Example of (a) magnetometer data, horizontal intensity (H), from the observatories used in the calculation of the K_p index, together with (b) the corresponding K_p index. The observatories have been grouped into North American, European, and southern hemisphere regions in order to highlight similarities of the data within each region and differences in the data across the globe.

One-min resolution horizontal intensity data from low-latitude observatories are used, and diurnal and secular variation baselines are subtracted. A geometric adjustment is made to the resulting data from each observatory so that they are all normalized to the magnetic equator. The average, then, is the D_{st} index. It is worth noting that, unlike the other indices summarized here, D_{st} is not a range index. Its relative A_{sym} is a range index, however, determined by the difference between the largest and smallest disturbance field among the four contributing observatories.

In Figure M35 we show magnetograms from the four observatories contributing to D_{st} and A_{sym}, for the Halloween storm of 2003, along with the indices themselves. The commencement of the storm is easily identified, and although the magnetic field is very disturbed during the first hour or so of the storm, the disturbance shows pronounced longitudinal difference, and hence a dramatically enhanced A_{sym}. With the subsequent worldwide depression of H through to the beginning of day 303 the storm is at its main phase of development. During this time D_{st} becomes increasingly negative. It is of interest to note that it is during this main phase that AE is also rapidly variable, signally the occurrence of substorms with the closure of magnetospheric electric currents through the ionosphere. AE diminishes during the recovery period of the storm as D_{st} also pulls back for its most negative values and A_{sym} is diminished. Toward the end of day 303 the second act of this complicated storm begins, with a repeat of the observed relationships of the various indices.

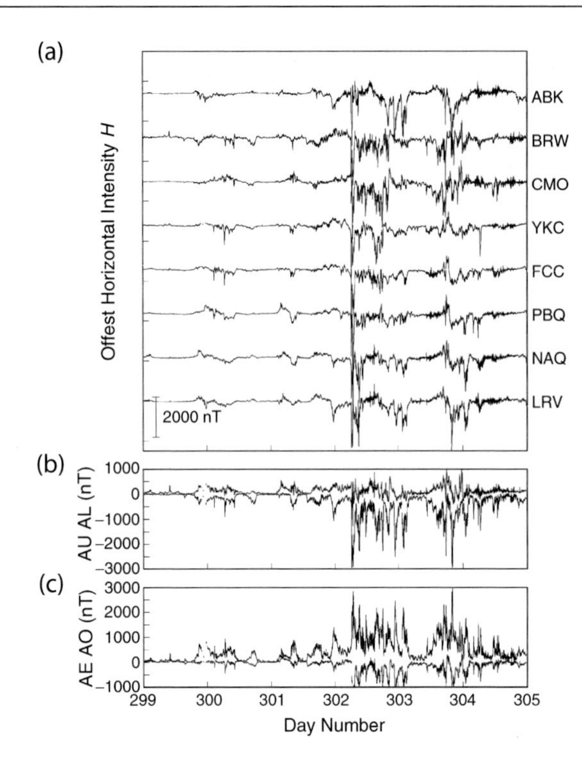

Figure M34 Example of (a) magnetometer data, horizontal intensity (H), from the observatories used in the calculation of the AE indices, together with (b) the corresponding AU and AL indices and the (c) AE and AO indices.

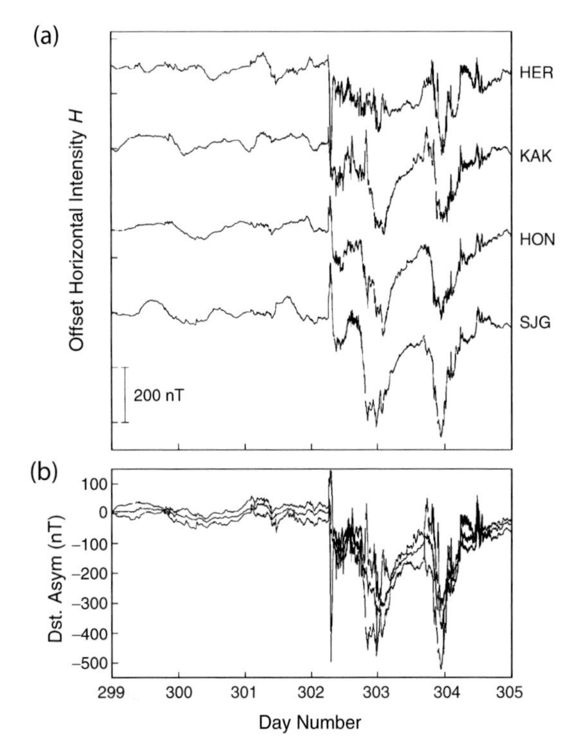

Figure M35 Example of (a) magnetometer data, horizontal intensity (H), from the observatories used in the calculation of the D_{st} indices, together with (b) the corresponding D_{st} plotted as the center trace and the maximum and minimum disturbance values, the difference of which is A_{sym}.

Availability

Magnetic indices are routinely calculated by a number of different agencies. Intermagnet agencies routinely calculate K indices for their observatories (www.intermagnet.org). The GeoForschungsZentrum in Potsdam calculates K_p (www.gfz-potsdam.de). The Kyoto World Data Center calculates AE and D_{st} (swdcwww.kugi.kyoto-u.ac.jp). Other agencies supporting the archiving and distribution of the indices include the World Data Centers in Copenhagen (web.dmi.dk/fsweb/projects/wdcc1) and Boulder (www.ngdc.noaa.gov), as well as the International Service of Geomagnetic Indices in Paris (www.cetp.ipsl.fr).

Jeffrey J. Love and K.J. Remick

Bibliography

Bartels, J., 1938. Potsdamer erdmagnetische Kennziffern, 1 Mitteilung. *Zeitschrift für Geophysik*, **14**: 68–78, 699–718.

Davis, T.N., and Sugiura, M., 1966. Auroral electrojet activity index AE and its universal time variations. *Journal of Geophysical Research*, **71**: 785–801.

Mayaud, P.N., 1980. *Derivation, Meaning, and Use of Geomagnetic Indices, Geophysical Monograph 22.* Washington, DC: American Geophysical Union.

McPherron, R.L., 1995. Standard indices of geomagnetic activity. In Kivelson, M.G., and Russell, C.T. (eds.), *Introduction to Space Physics.* Cambridge, UK: Cambridge University Press, pp. 451–458.

Menvielle, M., and Berthelier, A., 1991. The K-derived planetary indices—description and availability. *Reviews of Geophysics*, **29**: 415–432.

Moos, N.A.F., 1910. *Colaba Magnetic Data, 1846 to 1905. 2. The Phenomenon and its Discussion.* Bombay, India: Central Government Press.

Rangarajan, G.K., 1989. Indices of geomagnetic activity. In Jacobs, J.A. (ed.), *Geomagnetism* , Vol. 2. London, UK: Academic Press, pp. 323–384.

Sugiura, M., 1964. Hourly values of equatorial D_{st} for the IGY. *Annals of the International Geophysical Year*, **35**: 945–948.

Thomsen, M.F., 2004. Why K_p is such a good measure of magnetospheric convection. *Space Weather*, **2**: S11004, doi:10.1029/2004SW000089.

Cross-references

IAGA, International Association of Geomagnetism and Aeronomy
Ionosphere
Magnetosphere of the Earth

MAGNETIC MINERALOGY, CHANGES DUE TO HEATING

Mineralogical alterations occur very often in rocks subjected to thermal treatment. Laboratory heating may cause, in many cases, not only magnetic phase transformations, but also changes in the effective magnetic grain sizes, the internal stress, and the oxidation state. The presence or absence of such alterations is crucial to the validity and success of numerous magnetic studies.

The basic assumption in paleointensity determinations in the measurement of anisotropy of thermoremanent magnetization is that the rock is not modified during the different successively applied heating treatments. For simple thermal demagnetization, the occurrence of mineralogical alteration can introduce errors in the determination of the magnetic carrier if the latter had undergone transformation a at lower temperature than its unblocking temperature. Acquisition of parasitic chemical remanent magnetization is also possible if the heating was not made in a perfectly zero magnetic field. Formation of magnetic grains with very short relaxation times can lead to a large variation in the magnetic viscosity, making the stable components of the remanent magnetization sometimes immeasurable during thermal demagnetization. The interpretation of the results obtained by the Lowrie's method (1990) of identification of magnetic minerals by their coercivity and unblocking temperatures can be biased by mineralogical alteration, because rock can undergo mineralogical alteration instead of thermal demagnetization.

Analysis of the Curie curve (the measurement of the susceptibility in the low- or high magnetic field as a function of the temperature) is a key method for the identification of magnetic minerals, because each mineral has its Curie temperature (corresponding to a change from a ferrimagnetic to a paramagnetic state). On a Curie curve with increasing temperature, the Curie temperature is shown as a strong decrease in the susceptibility, sometimes preceded by an increase in susceptibility (a Hopkinson peak). However, mineralogical alteration can also increase or decrease susceptibility so it is not possible to discriminate the Curie temperature or the Hopkinson peak from mineralogical alteration from the shape of a simple curve directly until the highest temperature is reached, except if the heating and cooling curves are identical.

Example of mineralogical transformations due to heating

The increase of susceptibility during heating is mainly due to the formation of iron oxides. A decrease in susceptibility is often related to transformation of these oxides, such as in the oxidation of magnetite to hematite. Magnetite growth between 500 and 725°C (susceptibility increase) and the hematization of the magnetite at a higher temperature (susceptibility decrease) have been pointed out in a study of biotite granites (Trindade *et al.*, 2001). Furthermore, magnetic oxides can be formed from iron sulfides (pyrite, pyrrhotite, greigite, troilite), carbonates (siderite, ankerite), silicates, other iron oxides or hydroxides (e.g., Schwartz and Vaughan, 1972; Dekkers, 1990a,b). During heating, hexagonal pyrrhotite (antiferromagnetic) can be transformed first, to monoclinic pyrrhotite (ferrimagnetic) by partial oxidation to magnetite (Bina *et al.*, 1991). Pyrrhotite oxidizes mostly to magnetite at higher temperatures. Siderite oxidizes to magnetite or maghemite when exposed to air, even at room temperature but the oxidation is faster during heating (Ellwood *et al.*, 1986; Hirt and Gehring, 1991). Maghemite converts to hematite in the temperature range from 250 to 750°C, depending on the grain size, degree of oxidation, and the presence of defects in the crystallographic lattice (Verwey, 1935; Özdemir, 1990). In the range 250–370°C, goethite is transformed generally to very fine grains of hematite (Gehring and Heller, 1989; Dekkers, 1990b). Lepidocrocite starts to transform into superparamagnetic (SP) maghemite at ~175°C, with further conversion of this maghemite into hematite at around 300°C (Gehring and Hofmeister, 1994). In an oxidative atmosphere, ferrihydrite transforms into hematite (Weidler and Stanjek, 1998). But, ferrihydrite heated in the presence of an organic reductant forms magnetite and/or maghemite, or magnetite-maghemite intermediates (Campbell *et al.*, 1997). Exsolved magnetite has been found in plagioclases, pyroxene, and micas. With increasing temperature phyllosilicates undergo different reactions (Murad and Wagner, 1998). Thus, new phases developed at high temperatures depend on the composition of the original clay minerals, the firing temperature, and the atmosphere around the samples (Osipov, 1978).

During thermal treatment, ferrimagnetic minerals can also be affected by changes in the magnetic grain size (alteration, grain growth, or grain breaking due to fast heating and cooling). Therefore, grain size dependence of magnetic susceptibility is weak for pseudosingle-domain (PSD) and multidomain (MD) magnetite grains (Hartstra, 1982), but becomes important if the grain size variation is a transformation into SP—Single-Domain (SD) or a SD—PSD transformation. The susceptibility of the SD grains is lower than for the other grain sizes (Stacey and Banerjee, 1974).

Homogenization of the distribution of crystal defects can also have significant effects, leading to an increase in the susceptibility of large grains and in the redistribution of the domain wall. Inversion by filling up by new cations of the vacancy sites within the crystal lattice (Bina and Henry, 1990) can give significant susceptibility variations. This last mechanism and grain breaking could be at the origin of susceptibility variation since the heating takes place at relatively low temperatures.

Methods of analysis

A change of color of the sample is an indication of mineralogical alteration, but not all mineralogical transformations of magnetic minerals do lead to such color changes. Moreover, such change can be limited to a very fine surface layer and concern only to paramagnetic or diamagnetic minerals. Different investigations are therefore necessary to determine the magnetic mineralogy characteristics.

To investigate the mineralogical alteration of magnetic minerals due to heating, as a first possibility the simple observation of a thin section or conduct a microprobe analysis. In practice, such an approach is often difficult since alteration can have affected a limited part of the magnetic minerals, which can be moreover of a very small size. It is therefore easy to miss the modified minerals. However, when it works, this method yields precise indications about the affected minerals.

Transformation of a nonferrimagnetic mineral to another nonferrimagnetic mineral does not introduce a significant variation in susceptibility. Since the mineralogical change of the ferrimagnetic phase mostly corresponds on the contrary to the susceptibility change, measurement of susceptibility is generally a simple, fast, and efficient method to point out the alteration of the ferrimagnetic part. The least time consuming and common approach uses the monitoring of the variations of bulk magnetic susceptibility measured at room temperature and then after subsequent thermal steps, for example during thermal demagnetization.

The different parts of Curie curve measured during heating and cooling respectively, are similar if no alterations have occurred. The nonreversibility of the curve, on the contrary, indicates mineralogical changes. Hrouda *et al.* (2003) proposed to make several Curie curves for the same sample, with increasing maximum temperature, until obtaining a nonreversible curve. A faster approach is to apply cooling as soon as a significant susceptibility change occurs during heating. If the cooling curve is similar to the heating curve, the sample can be then heated again until the next significant susceptibility variation, etc. (Figure M36). Comparison of the heating and cooling curves can yield key information about the affected and altered minerals. If a Curie point occurs only on the heating curve, the corresponding mineral not disappeared. On the other hand, if this Curie point appears on the heating curve at a higher temperature than a transformation, it is

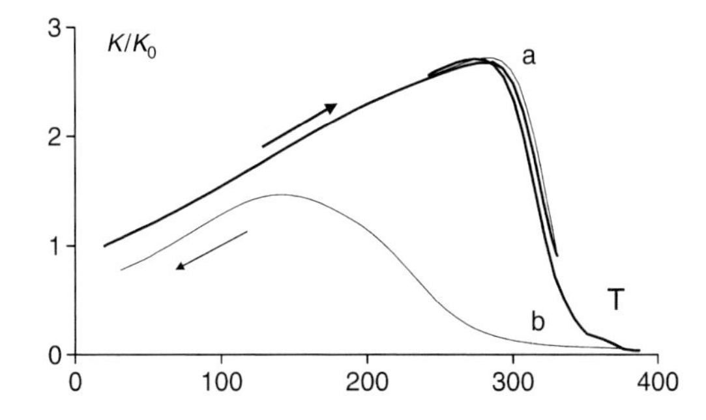

Figure M36 Thermomagnetic curve (low field normalized susceptibility K/K_0 as a function of the temperature T in °C) of a dolerite sample during heating (thick line) and cooling (thin line). Partial loop "a" corresponds to an absence of mineralogical alteration (reversible curve) contrary to the final part "b" of the loop (irreversible curve).

often difficult or even impossible to determine if this Curie temperature was associated with the preexisting or the newly formed mineral. A Curie temperature occuring only on the cooling curve is related to a newly formed mineral.

Van Velzen and Zijderveld (1992) proposed a complementary method for monitoring mineralogical alteration by using not only the susceptibility but several remanent magnetization characteristics. This time-consuming method allows a much more precise study of the effect of heating. However, if several different alterations had occurred during the same thermal treatment, they cannot be distinguished using these approaches.

Hysteresis loop analysis is another classical method in rock magnetism, giving, in particular, data on grain size and coercivities of the magnetic minerals. The sample is first subjected to a strong magnetic field to obtain the saturation of the magnetization. The total magnetization is then measured during progressive decrease of the field followed by the application of an increasing field in the opposite direction until obtaining the saturation of the magnetization in this opposite direction ("descending" curve d, Figure M37). The "ascending" curve a is then obtained by measurement of the total magnetization during a decrease of the field in the opposite direction followed by an increase in the field in the direction of the initial field. To further investigate the mineralogical alteration, hysteresis loops at room temperature are measured before and after heating (Figure M37a). Difference hysteresis loops (J_{b-a}) are obtained by subtraction for a same field (H) value of the measured magnetization value before (J_b) and after (J_a) heating in curves d and a, respectively (Figure M37b). They are sometimes relatively complicated curves, but using the separation method proposed by Von Dobeneck (1996) in two curves (half-difference and half-sum), the interpretation becomes much easier (Figure M37c). In particular, it is possible to discriminate characteristics (coercivity, magnetic field for saturation) of disappearing and occurring magnetic components, even during the same thermal treatment (Henry *et al.*, 2005).

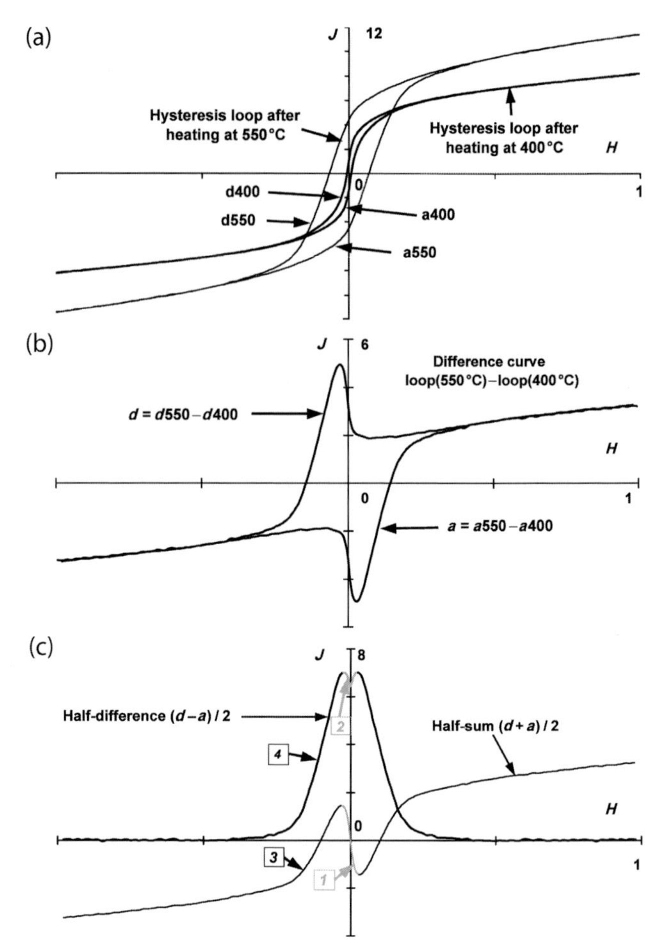

Figure M37 Example of hysteresis loops (ascending—a and descending—d curves) measured at room temperature on dolerite sample after heating at 400 and at 550°C (a) and corresponding difference loop (b). From curves (c) obtained from half-sum and half-difference of the curves b, transformation of a magnetic component with low coercivity and saturation field to another magnetic phase with higher coercivity and saturation field can be inferred (Henry *et al.*, 2004): Part of the curves pointing out disappearance (in grey) of component with low coercivity (1) and moderate saturation field (2), and occurrence (in black) of component with higher coercivity (3) and high saturation field (4). Field H in T, Magnetization in A m^2 kg^{-1}.

Applications

If heating at a temperature T applied in laboratory introduces a mineralogical change, previous heating at the same temperature T should have given the same transformation. The studied rocks in their present state were therefore not subjected to this temperature before being sampled. The minimum temperature to develop mineralogical change, therefore, corresponds to the maximum temperature possibly undergone by the rock in the past (Hrouda *et al.*, 2003). This is therefore the maximum paleotemperature indicator. However, that means obviously, the maximum temperature since the rock has the present mineralogical composition, i.e. sometimes relatively recently in case of weathering and low-temperature oxidation.

Modification of the magnetic fabric as a result of heating has been applied to different types of rocks (see *Magnetic susceptibility, anisotropy, effects of heating*). The outcome was sometimes a simple enhancement of the fabric, but significant changes of the fabric have also been obtained. It is moreover possible to determine the anisotropy of magnetic susceptibility of the ferrimagnetic minerals formed or that have disappeared during successive heatings. To this aim, the tensor resulting from the subtraction of the tensors measured before and after heating is used (Henry *et al.*, 2003). When a same magnetic fabric is obtained from several following thermal steps, it cannot be related to the randomly oriented ferrimagnetic minerals. Instead, the newly formed fabric must be related to characteristics of the preexisting rock. By comparing this ferrimagnetic mineral fabric with the initial whole rock fabric, it is possible to distinguish cases where heating simply enhances of preexisting fabric from those where thermal treatment induces a different fabric. Relative to the preheated fabric, this different fabric may simply be an inverse fabric or one whose principal susceptibility axes are oriented in a different direction, relative to petrostructural elements other than those defining the initial magnetic fabric.

Important applications concerning all the methods, assumes that no transformation occurred during heating. It is fundamentally important to verify this assumption when paleointensity or anisotropy of thermoremanent magnetization studies are carried out. It is important to point out that mineralogical alteration can also have important implications for all the studies using heating, such as Lowrie's (1990) method of identification of magnetic minerals (using corercivity and unblocking temperatures) or even simple thermal demagnetization.

Bernard Henry

Bibliography

Bina, M., Corpel, J., Daly, L., and Debeglia, N., 1991. Transformation de la pyrrhotite en magnetite sous l'effet de la temperature: une source potentielle d'anomalies magnétiques. *Comptes Rendus de l'Académie des Sciences, Paris.* **313**(II): 487–494.

Bina, M., and Henry, B., 1990. Magnetic properties, opaque mineralogy and magnetic anisotropies of serpentinized peridotites from ODP hole 670A near Mid-Atlantic ridge. *Physics of the Earth and Planetary Interiors*, **65**: 88–103.

Campbell, A.S., Schwertmann, U., and Campbell, P.A., 1997. Formation of cubic phases on heating ferrihydrite. *Clay Minerals*, **32**: 615–622.

Dekkers, M.J., 1990a. Magnetic monitoring of pyrrhotite alteration during thermal demagnetisation. *Geophysical Research Letters*, **17**: 779–782.

Dekkers, M.J., 1990b. Magnetic properties of natural goethite—III. Magnetic behaviour and properties of minerals originating from goethite dehydration during thermal demagnetisation. *Geophysical Journal International*, **103**: 233–250.

Ellwood, B.B., Balsam, W., Burkart, B., Long, G.J., and Buhl, M.L., 1986. Anomalous magnetic properties in rocks containing the mineral siderite: paleomagnetic implications. *Journal of Geophysical Research*, **91**: 12779–12790.

Gehring, A.U., Heller, F., and 1989. Timing of natural remanent magnetization in ferriferous limestones from the Swiss Jura Mountains. *Earth and Planetary Science Letters*, **93**: 261–272.

Gehring, A.U., and Hofmeister, A.M., 1994. The transformation of lepidocrocite during heating: a magnetic and spectroscopic study. *Clays and Clay Minerals*, **42**(4): 409–415.

Hartstra, R.L., 1982, Grain size dependence of initial susceptibility and saturation magnetization-related parameters of four natural magnetites in the PSD-MD range. *Geophysical Journal of the Royal Astronomical Society*, **71**: 477–495.

Henry, B., Jordanova, D., Jordanova, N., Souque, C., and Robion, P., 2003. Anisotropy of magnetic susceptibility of heated rocks. *Tectonophysics*, **366**: 241–258.

Henry, B., Jordanova, D., Jordanova, N., and Le Goff, M., 2005. Transformations of magnetic mineralogy in rocks revealed by difference of hysteresis loops measured after stepwise heating: Theory and cases study. *Geophysical journal International*, **162**: 64–78.

Hirt, A.M., and Gehring, A.U., 1991. Thermal alteration of the magnetic mineralogy in ferruginous rocks. *Journal of Geophysical Research*, **96**: 9947–9953.

Hrouda, F., Müller, P., and Hanak, J., 2003. Repeated progressive heating in susceptibility vs. temperature investigation: a new palaeotemperature indicator? *Physics and Chemistry of the Earth*, **28**: 653–657.

Lowrie, W., 1990. Identification of ferromagnetic minerals in a rock by coercivity and unblocking temperature properties. *Geophysical Research Letters*, **17**: 159–162.

Murad, E., and Wagner, U., 1998. Clays and clay minerals: The firing process. *Hyperfine Interactions*, **117**: 337–356.

Osipov, J., 1978. *Magnetism of Clay Soils (in Russian)*. Moscow: Nedra.

Özdemir, Ö., 1990. High temperature hysteresis and thermoremanence of single-domain maghemite. *Physics of the Earth and Planetary Interiors*, **65**: 125–136.

Schwartz, E.J., and Vaughan, D.J., 1972. Magnetic phase relations of pyrrhotite. *Journal of Geomagnetism and Geoelectricity*, **24**: 441–458.

Stacey, F.D., and Banerjee, S.K., 1974. *The Physical Principles in Rock Magnetism*. Amsterdam: Elsevier, 195 pp.

Trindade, R.I.F., Mintsa Mi Nguema, T., and Bouchez, J.L., 2001. Thermally enhanced mimetic fabric of magnetite in a biotite granite. *Geophysical Research Letters*, **28**: 2687–2690.

Van Velzen, A.J., and Zijderveld, J.D.A., 1992. A method to study alterations of magnetic minerals during thermal demagnetization applied to a fine-grained marine marl (Trubi formation, Sicily). *Geophysical Journal International*, **110**: 79–90.

Verwey, 1935. The crystal structure of γFe_2O_3 and γAl_2O_3. *Z. Krist., Zeitschrift für Kristallographie* **91**: 65–69.

Von Dobeneck, T., 1996. A systematic analysis of natural magnetic mineral assemblages based on modeling hysteresis loops with coercivity-related hyperbolic basis functions. *Geophysical Journal International*, **124**: 675–694.

Weidler, P., and Stanjek, H., 1998. The effect of dry heating of synthetic 2-line and 6-line ferrihydrite. II. Surface area, porosity and fractal dimension. *Clay Minerals*, **33**: 277–284.

Cross-references

Chemical Remanent Magnetization
Depth to Curie Temperature
Magnetic Remanence, Anisotropy
Magnetic Susceptibility
Magnetic Susceptibility, Anisotropy
Magnetic Susceptibility, Anisotropy, Effects of Heating
Paleointensity: Absolute Determination Using Single Plagioclase Creptals

MAGNETIC PROPERTIES, LOW-TEMPERATURE

Introduction

Use of magnetic measurements at cryogenic temperatures for characterizing magnetic mineralogy of rocks was initiated in the early 1960s, when it was realized that several minerals capable to carry natural remanent magnetization (NRM), e.g., magnetite and hematite, show distinctive magnetic phase transitions below room temperature. In the last decade, low-temperature magnetometry of rocks and minerals has seen a new boost due to increasing availability of commercial systems capable to carry out magnetic measurements down to and below 4.2 K. Low-temperature magnetometry has the potential to complement conventional high-temperature methods of magnetic mineralogy while offering an advantage of avoiding chemical alteration due to heating. This is especially important in the case of sedimentary rocks, which alter much more readily upon heating. However, additional complications may arise because of a possible presence in a rock of mineral phases showing ferrimagnetic or antiferromagnetic ordering below room temperature. On the other hand, these minerals are often of a diagnostic value by themselves, being the signature of various rock-forming processes. In all, low-temperature magnetometry is a valuable new tool in rock and environmental magnetism.

Several factors control low-temperature behavior of remanent magnetization and low-field susceptibility of minerals and rocks. Phase transitions, which may occur below room temperature, have the most profound effect. Also of importance is the temperature variation of the intrinsic material properties such as magnetocrystalline anisotropy and magnetostriction. Low-temperature magnetic properties of minerals are also affected by their stoichiometry and degree of crystallinity. Last but not least, low-temperature variation of remanence and magnetic susceptibility is generally grain-size dependent. In particular, ultrafine (say, $<$20 nm) grains show the distinct behavior called superparamagnetism, which manifests itself in a nearly exponential decrease of remanence with increasing temperature due to a progressive unblocking of magnetization by thermal activation. On the other hand, all the above factors have relatively little effect on saturation magnetization, which is primarily determined by chemical composition of the material.

Measurements that can be used to characterize low-temperature magnetic properties of minerals and rocks include the following. Thermal demagnetization of a saturation isothermal remanent magnetization (SIRM) given at a low temperature, typically 10 K, is a relatively rapid experiment and the most frequently used. Low-temperature magnetic phase transitions generally manifest themselves in SIRM vs temperature curves and are most useful as diagnostic features. However, in some cases signature of a magnetic-phase transition can be confused with that of a ferrimagnetic to paramagnetic transition if the transition temperatures are in the same range. An example is the 34-K phase transition in mineral pyrrhotite, which is fairly close to Néel temperatures of minerals siderite and rhodochrosite (38 and 32 K, respectively). It is thus advised to use SIRM demagnetization only for a reconnaissance study of mineral phases present. A useful method to distinguish between the two above cases is to measure a SIRM acquired at room temperature during a zero-field cycle to a low temperature. Magnetic phase transitions will be recorded in this experiment, while Néel temperatures will not.

For some materials low-temperature magnetic properties measured on warming may depend on the mode of preceding cooling, i.e., on whether a sample has been cooled in a zero (zero-field cooling, ZFC) or in a strong magnetic field (field cooling, FC). In particular, this is the case for magnetically hard minerals, in which thermoremanent magnetization (TRM) acquired during cooling in a strong field to a given low temperature is much higher than a SIRM, which can be induced by a single application of the same field at the same temperature. Demagnetization curves of the two remanences will then be strongly discrepant but converge near room temperature. Another variant of ZFC vs FC measurements, which must not be confused with the former one, consists in measuring magnetization during warming in a relatively weak, usually 2–20 mT, DC field, after sample being cooled once in a zero and second time in the DC field equal to the measurement field. This method can be used to characterize the superparamagnetic behavior. The two magnetization vs temperature curves would diverge if the blocking of superparamagnetic moments occurs below room temperature.

Measurements of magnetic hysteresis loops as a function of temperature potentially would yield the most detailed information on the magnetic mineralogy of the sample. However, this is a rather time-consuming experiment, and relatively little has been reported so far on low-temperature hysteresis properties of minerals and rocks.

Variation of AC susceptibility at low temperatures is also a useful characteristics of the magnetic mineralogy. An important advantage of susceptibility measurements is that they allow to detect minerals that show antiferromagnetic ordering below Néel temperatures. On the other hand, low-temperature susceptibility signal due to ferrimagnetic and antiferromagnetic phases is progressively masked by a contribution from paramagnetic minerals which increases proportionally to $1/T$.

The remainder of this article presents an overview of low-temperature magnetic properties of most common iron minerals found in terrestrial rocks.

Magnetite and maghemite

Magnetite (Fe_3O_4), iron ferrite, is the longest known and most ubiquitous natural magnetic material and probably also the most extensively studied. It occurs, in a varying concentration, in most volcanic, sedimentary, and metamorphic rocks, and forms economically significant iron ore deposits. Furthermore, magnetite can be formed biogenically by a wide variety of organisms, from prokaryotes to humans (cf. Kirschvink *et al.*, 1985).

Low-temperature anomalies in the magnetic susceptibility and magnetization of magnetite were discovered in 1910–1920s (Weiss and Renger, 1914; Weiss and Forrer, 1929). However, only much later it was proposed (Verwey, 1939; Verwey and Haayman, 1941) that these anomalies, together with those in electrical properties, heat capacity, etc., are due to the phase transition, which is now called the Verwey transition. The transition is accompanied by a slight monoclinic distortion of the cubic spinel crystalline lattice of magnetite stable at ambient temperature, which results in a large, by over an order of magnitude, increase of the magnetocrystalline anisotropy. The transition temperature (T_V) is 125 K in stoichiometric crystals with low internal stress (Walz, 2002). Substitution of iron with vacancies (Aragón *et al.*, 1985; Aragón, 1992; Özdemir *et al.*, 1993) and foreign metal cations (Miyahara, 1972; Kozłowski *et al.*, 1996a,b; Brabers *et al.*, 1998) are known to lower the T_V and eventually suppress the transition. Stresses have the same effect (Samara, 1968; Rozenberg *et al.*, 1996). The low-temperature behavior of the magnetic properties, and particularly of a remanent magnetization during cooling from room temperature in zero magnetic field, is also strongly affected by the isotropic point at about 130 K, the temperature where the first constant of magnetocrystalline anisotropy K_1 momentarily turns zero (Bickford, 1950; Syono, 1965). The isotropic point temperature appears much less sensitive to nonstoichiometry than the T_V (Kąkol and Honig, 1989; Kąkol *et al.*, 1994). A review of low-temperature magnetic properties of magnetite from the point of view of rock magnetism has been recently given by Muxworthy and McClelland (2000a).

The Verwey transition is easily recognizable in the demagnetization curves of a SIRM acquired at a low temperature (e.g., 10 K), as a sharp drop of magnetization (Figure M38). The temperature where the magnetization decrease rate is maximal is usually taken as a transition temperature. This has a justification in studies on large nearly perfect synthetic magnetite crystals (Aragón, 1992), where the magnetization drop occurs within 1 K. However, in fine grains magnetization decreases more gradually, particularly when magnetite is slightly oxidized (Figure M39, Özdemir *et al.*, 1993), and such an assignment of the transition temperature may be somewhat arbitrary. Perhaps for this reason the exact values of T_V in rocks are seldom reported, and the observation of the transition is taken as only a qualitative test for the presence of magnetite. Meanwhile, actual values of transition temperatures can be of interest because of their sensitivity to minute variations of oxygen stoichiometry and metal substitution, thus providing information inaccessible with other methods. Recent examples that illustrate use of the low-temperature magnetometry of magnetite to solve magnetomineralogical problems include: (i) a method to detect the presence of biogenic magnetite produced by magnetotactic bacteria, showing a distinctive difference between the demagnetization curves of SIRMs given below the Verwey transition after ZFC and FC, respectively (Moskowitz *et al.*, 1993); (ii) a model explaining the magnetic susceptibility signal in Chinese loess by the effect of natural grass fires (Kletetschka and Banerjee, 1995); and (iii) tracing of chemical alteration of goethite (α-FeOOH) during heating (Özdemir and Dunlop, 2000).

Magnetite and magnetite-bearing rocks can be further characterized by a variation of a remanence acquired at room temperature during a zero-field cycle crossing the Verwey temperature. On initial cooling,

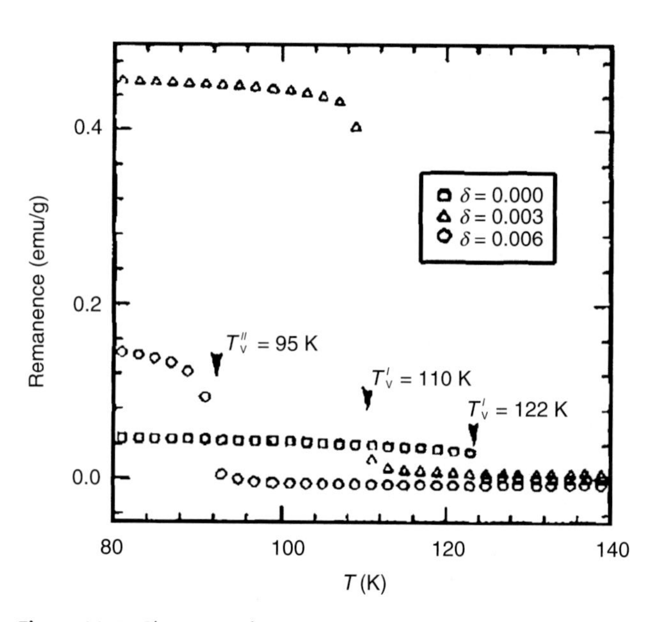

Figure M38 Change in the remanent moment obtained by cooling in a magnetic field of 1 T to 4.5 K during warming in a zero field, for samples of $Fe_{3-\delta}O_4$ with $\delta = 0.000$ (squares), 0.003 (triangles), and 0.006 (circles). (After Aragón (1992) © American Physical Society, with the permission of the publisher.)

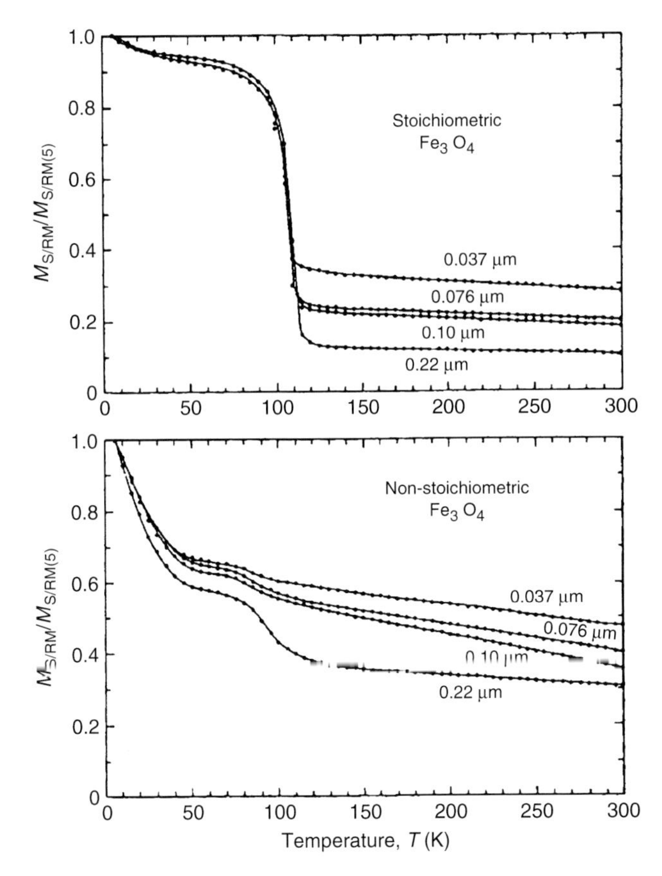

Figure M39 Normalized saturation isothermal remanence (SIRM) curves of submicron magnetites during warming from 5 to 300 K in zero field. *Top*: reduced (apparently stoichiometric) magnetites; *bottom*: surface-oxidized magnetites. The grain size is indicated on the curves. (After Özdemir *et al.* (1993) © American Geophysical Union, with the permission of the author and the publisher.)

magnetization is lost between ca. 200 K and the T_V; demagnetization is grain-size dependent, being by far the most effective in large multidomain grains, where it is essentially complete at the isotropic point (Dunlop, 2003). Below the transition temperature, magnetization is nearly constant until on warming the temperature reaches the vicinity of T_V. The behavior on crossing the T_V from below depends primarily on stress: in low-stress samples remanence tends to be further lost (Muxworthy and McClelland, 2000b), while highly stressed ones show a rebound (Heider *et al.*, 1992; King and Williams, 2000). In the paleomagnetic practice, cycling the rock sample through the isotropic point and Verwey temperature is known as the low-temperature cleaning (LTC, Ozima *et al.*, 1964). LTC should at least partly remove the unwanted secondary components of NRM carried by large multidomain grains of magnetite. However, perhaps because physical mechanisms of remanence demagnetization during LTC remained elusive for so long (Muxworthy and McClelland, 2000b; Dunlop, 2003), it has been seldom used in paleomagnetic studies.

Magnetic properties of the low-temperature monoclinic phase of magnetite, although of no direct importance in paleomagnetism, proved to be an interesting subject of study itself, receiving much recent interest (Schmidbauer and Keller, 1996; Muxworthy, 1999; Özdemir, 2000; Kosterov, 2001, 2002; Özdemir *et al.*, 2002; Smirnov and Tarduno, 2002; Kosterov, 2003). In particular, it appears that below the Verwey transition temperature dependences of the magnetic hysteresis parameters are qualitatively different in truly multidomain and in smaller, pseudosingle-domain grains, respectively: in multidomain grains both M_{rs} and the coercive force are much higher after a ZFC while a more complex relationship between ZFC and FC hysteresis parameters is observed in pseudosingle-domain grains (Kosterov, 2001, 2002). Magnetic properties of the low-temperature phase also strongly depend on magnetite stoichiometry (Özdemir *et al.*, 1993; Kosterov, 2002, 2003), which is a result of the variation of the magnetocrystalline anisotropy with the degree of oxidation (Kąkol and Honig, 1989). Memorizing the value of a small to moderate magnetic field (<100 mT) applied during cooling through the Verwey transition has been demonstrated (Smirnov and Tarduno, 2002).

Maghemite (γ-Fe_2O_3) is a fully oxidized analog of magnetite, in which 1/9 of iron cations in the spinel lattice are replaced with vacancies. It is a common constituent of soils (Taylor and Schwertmann, 1974) and also has been found in deep-sea sediments (Harrison and Peterson, 1965). Unlike magnetite, maghemite does not show phase transitions below room temperature, and remanence given at a given low temperature decreases monotonously on warming (Morrish and Watt, 1958; Moskowitz *et al.*, 1993). SIRM given at room temperature changes reversibly on ZFC to a low temperature (de Boer and Dekkers, 1996).

Titanomagnetites

Titanomagnetites are spinel iron-titanium oxides of the composition $Fe_{3-x}Ti_xO_4$. Stable single-phase oxides exist for x between 0 (magnetite) and 0.96 (Kąkol *et al.*, 1991a). The end member with nominal composition Fe_2TiO_4 is known as mineral ulvöspinel. A titanomagnetite with $x \sim 0.6$ (TM60) is the dominant magnetic mineral in fresh submarine basalts; in subaerial basalts, unoxidized titanomagnetites typically have compositions TM60-TM70 (Dunlop and Özdemir, 1997, Section14.4). Ti-rich titanomagnetites have only a local significance in sediments, where the eroded material from volcanic rocks is readily available. Titanomagnetites of intermediate compositions are found relatively less frequently in nature than those close to magnetite and TM60, respectively.

Titanomagnetites with compositions richer in Ti than TM04 do not exhibit the Verwey transition (Kozłowski *et al.*, 1996a). However, the variation of the magnetocrystalline anisotropy remains similar to that in magnetite for the wide range of compositions. At room temperature, K_1 is negative; with the decrease of temperature its absolute value first increases and then starts to decrease approaching the isotropic point. With the increase of x the isotropic point temperature first shifts toward lower temperatures, being below 77 K for the material in the TM18-TM36 range (Syono, 1965; Kąkol *et al.*, 1991b), and then starts to increase again, so that the material with the nominal composition TM41 has the isotropic point of about 125 K, very close to that of magnetite (Kąkol *et al.*, 1991b). For even more Ti-rich compositions, K_1 is positive at all temperatures below 300 K (Syono, 1965; Kąkol *et al.*, 1991b) and increases sharply to very high values below 200 K.

Low-temperature dependences of SIRM and low-field susceptibility for millimeter-size single crystals of titanomagnetite in the composition range from TM0 to TM60 have been reported by Moskowitz *et al.* (1998, Figure M40). Titanomagnetites from TM05 to TM28 show a sharp decrease of SIRM between 40 and 80 K. Low-field susceptibility of these titanomagnetites increases rapidly in this temperature range. For TM05 it reaches a maximum at about 90 K and then slightly decreases; for TM19 and TM28 the susceptibility increase just becomes considerably slower above ~110 K. Since the material under study is multidomain, this behavior is likely controlled by the temperature variation of magnetocrystalline anisotropy. Compositions around TM40 show a somewhat more complex behavior: SIRM first drops rapidly in about the same range as for more Ti-poor compositions, and then shows further, less pronounced decrease between 120 and 150 K. The latter should be due to the isotropic point at about 125 K (Kąkol *et al.*, 1991b). In even more Ti-rich titanomagnetites TM55-TM60 SIRM decreases rapidly between 50 and 70 K, and more

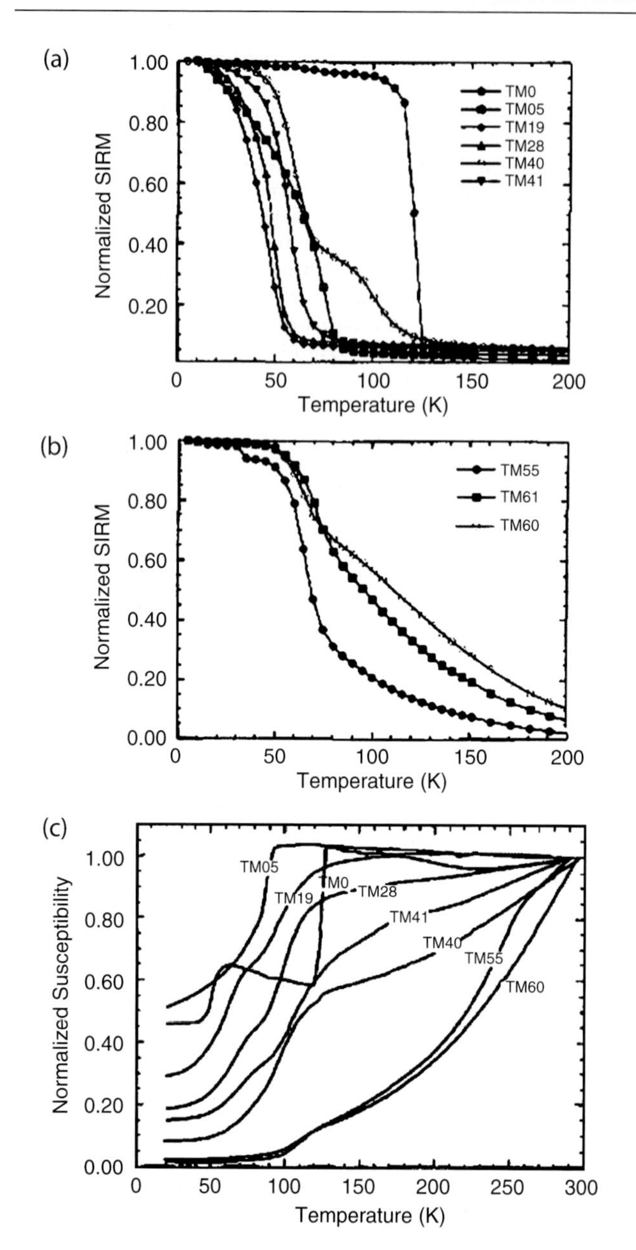

Figure M40 Normalized saturation isothermal remanence (SIRM) curves of synthetic titanomagnetites during zero-field warming from 5 to 300 K: (a) TM0-TM41 and (b) TM55-TM61. SIRM was given in a 2.5 T field at 5 K. Single-crystal samples are denoted by solid symbols and polycrystalline samples are denoted by open symbols. (c) Normalized in-phase susceptibility curves of synthetic titanomagnetites during warming from 15 to 300 K. Single-crystal samples are TM0, TM05, TM19, TM28, TM41, TM55, and polycrystalline samples are TM40 and TM60. The in-phase susceptibility was measured at 1000 Hz. (After Moskowitz *et al.* (1998), with the permission of the author and the publisher, Elsevier.)

gradually above this temperature, bearing some resemblance to the behavior of ultrafine superparamagnetic grains.

Magnetic hysteresis of intermediate titanomagnetites as a function of low temperature has been studied only on millimeter-sized single-crystal samples, for a rather limited number of compositions from TM40 to TM80, and in the magnetic fields not exceeding 1.5 T, which is likely to be too low to fully saturate the samples, especially more Ti-rich ones (Tucker, 1981; Schmidbauer and Readman, 1982). The most distinctive feature of these compounds is a large increase of the coercive force with decreasing temperature, which is due to the sharp rise of the magnetocrystalline anisotropy. For compositions from TM50 to TM80 the rise of the H_c starts at about 150 K and for TM40 at 70 K (Schmidbauer and Readman, 1982). In addition, both M_{rs} and the coercive force appear to depend on the mode of preceding cooling, the latter being much higher and the former considerably lower after ZFC.

It must be borne in mind that the above experiments were done on large crystals, necessarily multidomain. To the best of author's knowledge, no low-temperature magnetic studies have been carried out on synthetic intermediate titanomagnetites of micron to submicron grain size. The low-temperature data on rocks, which contain such grains, are also scarce. Moskowitz *et al.* (1998) have reported contrasting low-temperature SIRM(T) curves for basaltic samples from the center and a chilled margin of an Icelandic dike, respectively. Samples from the center of the dike yielded curves closely resembling those for synthetic Ti-rich ($x > 0.4$) titanomagnetites, while samples from the dike margin showed curves with a break-in-slope around 50 K. It may be suggested that rapid cooling of the dike margin had produced a titanomagnetite phase with $x \sim 0.2$, and the discrepancy between SIRM(T) curves of these samples and those of synthetic titanomagnetite of similar composition is due to a difference in grain size. Similar SIRM(T) curves have been observed in late Pleistocene to Holocene sediments from Zacapu basin in central Mexico, where the abundant titanomagnetite of composition ranging from TM20 to TM55 comes from nearby volcanic activity (Ortega *et al.*, 2002). Further complications are likely to arise if titanomagnetites are oxidized to form titanomaghemites, as is the case for most submarine basalts older than several millions of years. In such rocks, Matzka *et al.* (2003) have observed even a self-reversal of SIRM given at room temperature during a low-temperature cycle in zero field. Magnetization crossed to negative values around 70 K during the cooling leg but returned reversibly to initial values upon the subsequent warming. In summary, much has yet to be done to fully understand the low-temperature behavior of naturally occurring intermediate titanomagnetites.

Titanohematites

The composition of minerals belonging to the titanohematite series can be described by the chemical formula $Fe_{2-y}Ti_yO_3$, the end members being minerals hematite (α-Fe_2O_3) and ilmenite ($FeTiO_3$), respectively. These minerals crystallize in the rhombohedral system; the space group describing crystal symmetry is $R\bar{3}C$ for $y < 0.5$ and $R\bar{3}$ for $y > 0.5$. Compositions close to both hematite and ilmenite are quite common in igneous rocks, while truly single-phase titanohematites of an intermediate composition ($0.2 < y < 0.8$) can be obtained only in the laboratory by quenching from high temperatures. Natural intermediate titanohematites of nominal composition $0.5 \leq y \leq 0.7$, found in rapidly chilled dacitic pyroclastic rocks, received much interest in rock magnetism due to the ability to acquire a self-reversed TRM (Uyeda, 1958; Hoffman, 1992). Magnetic properties of titanohematites are determined by their composition: (i) for $0 \leq y < 0.5$ an overall ordering is antiferromagnetic with a weak parasitic ferromagnetism due to a spin canting (Dzyaloshinsky, 1958; Moriya, 1960); (ii) compositions with $0.5 < y < 0.7$ are ferrimagnetic; (iii) titanohematites with $0.7 < y \leq 0.93$ show rather complex magnetic structures, which will be discussed in more detail below; (iv) compositions with y greater than about 0.93 show antiferromagnetism. Curie temperatures of titanohematites decrease approximately linearly with increasing ilmenite content up to $y \sim 0.93$, reaching the values below room temperature at $y \sim 0.75$ (Nagata, 1961).

Hematite is the most stable iron oxide in oxidizing conditions and, as such, is found in many sedimentary and in some highly oxidized volcanic rocks. Its small net saturation moment of about 0.4 A m^2 kg^{-1} (Morrish, 1994) results primarily from a spin canting. Another mechanism that can produce a magnetic moment is a presence of defects or impurity atoms in the hematite crystalline lattice (Bucur, 1978). Since hematite is much less strongly magnetic than magnetite or maghemite, the two latter minerals dominate the NRM in the case when concentrations of these minerals are comparable. However, in an important class of sedimentary rocks called red beds, NRM is almost entirely carried by hematite.

Below room temperature, bulk hematite exhibits the so-called Morin transition (Morin, 1950). At the Morin transition spin directions change from the basal plane to the rhombohedral *c*-axis, and the spin canting, and the magnetic moment associated with it, seems to vanish. The defect moment, however, remains largely intact. This explains, at least qualitatively, the peculiar behavior of room temperature remanence being cycled in a zero field through the Morin transition (Haigh, 1957; Gallon, 1968; Bucur, 1978): magnetization lost on cooling is largely, but not completely, restored on warming (Figure M41). Still unresolved is an existence of the transition temperature hysteresis during magnetization cycling in a zero field. Transition temperatures determined from cooling curves are often lower than those determined from warming curves, sometimes by as much as 20°.

Alike the Verwey transition in magnetite, the Morin transition is sensitive to impurities. Titanium is known to suppress the transition in concentrations as low as less than 1% (Kaye, 1962), while the transition exists until 10% of Al substitution (da Costa *et al.*, 2002). The transition becomes more gradual for highly aluminous hematites, in the sense that low- and high-temperature phases appear to coexist over a considerable range of temperatures, and the effective transition temperature is lowered to about 180 K for hematite with ~10% Al. The transition temperature is lowered and the transition itself appears considerably suppressed in fine grains (Bando *et al.*, 1965; Schroeer and Nininger, 1967; Nininger and Schroeer, 1978; Amin and Arajs, 1987). This may be the reason why the Morin transition is rarely found in hematite-bearing rocks (Dekkers and Linssen, 1989).

Titanohematites with y up to about 0.7 do not show any distinctive low-temperature magnetic properties (Ishikawa and Akimoto, 1957).

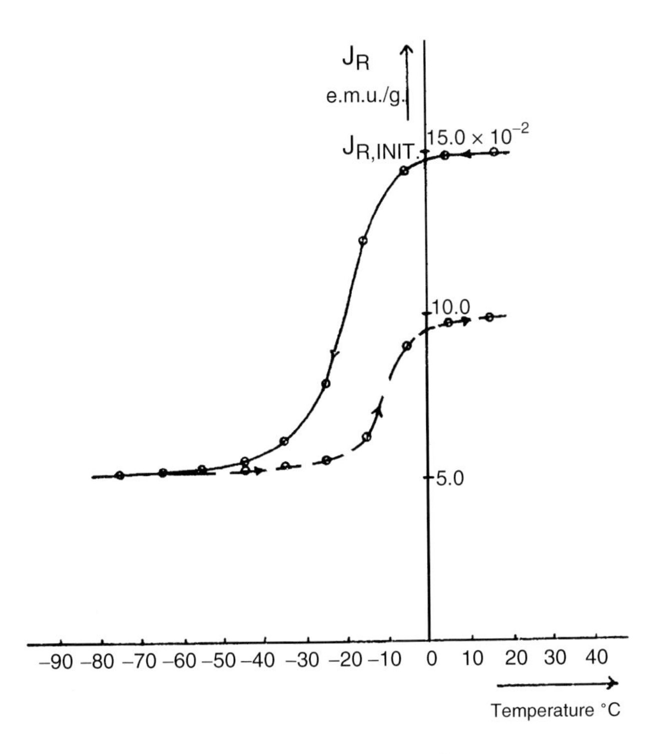

Figure M41 Variation of IRM (0.6 T), acquired at room temperature in a dispersed hematite powder, during a zero-field cycle to 193 K (−80°C). (Modified from Haigh (1957), with the permission of the publisher, Taylor & Francis Ltd., http://www.tandf.co.uk/journals.)

On the contrary, more Ti-rich compositions show fairly complex behavior (Figure M42). At very low temperatures, a spin glass structure seems to arise (Arai and Ishikawa, 1985; Arai *et al.*, 1985a,b; Ishikawa *et al.*, 1985), the spin glass transition temperatures T_{sg} being considerably lower than the respective Curie temperatures. Between these two transition points compositions with $y \sim 0.8$–0.9 show superparamagnetic behavior (Ishikawa *et al.*, 1985). The spin glass phase is capable to carry a relatively strong remanence, which disappears at the T_{sg}. The coercive force also increases sharply below this temperature reaching 300–600 mT at 4.2 K. It must be noted, however, that the above studies have been carried out using the synthetic materials. Whether their natural analogs would show similar magnetic properties remains largely unknown.

The second end member of the titanohematite series, ilmenite, orders antiferromagnetically below its Néel temperature of 57±2 K (Senftle *et al.*, 1975). Antiferromagnetism in the low-temperature phase persists up to about 6.6% of Fe^{3+} substitution, while Néel temperatures shift to below 50 K (Thorpe *et al.*, 1977). Ilmenite therefore contributes only to induced magnetization. A susceptibility peak at 45 K, which may be due to ilmenite, has been reported for a basalt sample from the Hawaiian deep drill hole SOH-1 (Moskowitz *et al.*, 1998).

Iron hydroxides

Hydrous iron oxides are produced commonly by weathering of iron-bearing rocks in ambient conditions. They are also ubiquitous in

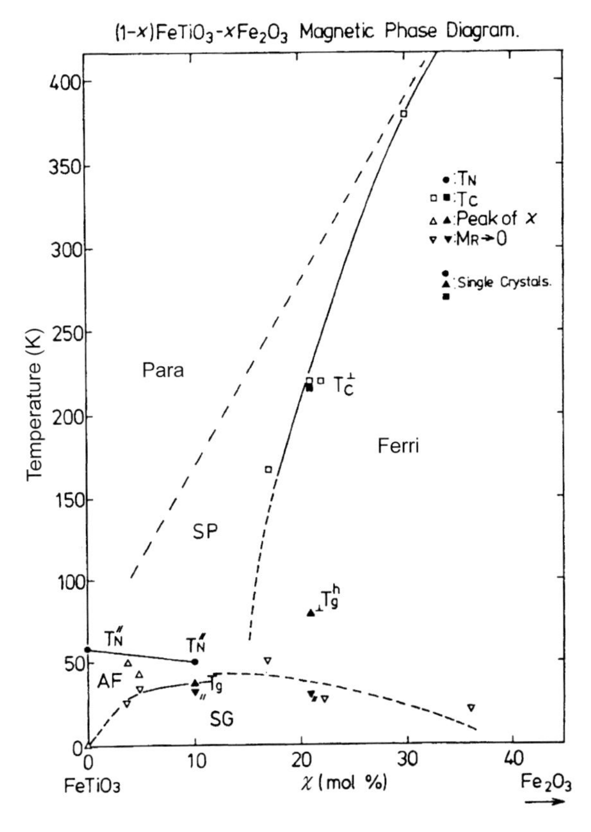

Figure M42 Low-temperature magnetic-phase diagram of ilmenite-rich part of the $FeTiO_3$-Fe_2O_3 system. Down triangles indicate the temperature where TRM disappears, up triangles the temperature of the susceptibility peak. Closed symbols refer to single crystals, open symbols to polycrystalline samples. Note that $x \equiv 1-y$. (After Ishikawa *et al.* (1985), with the permission of the publisher, the Physical Society of Japan.)

marine sediments (Murray, 1979). Of various iron hydroxides, the orthorhombic form, goethite (α-FeOOH), is the most stable. Goethite is an antiferromagnet but, like hematite, often carries a small magnetic moment believed to be of defect origin (Strangway *et al.*, 1968; Hedley, 1971). Below its Néel temperature of 120°C (in well-crystalline samples) goethite is extremely magnetically hard, requiring fields in excess of 10 T to approach the magnetic saturation. An intense TRM, nearly equal to the spontaneous magnetization J_s, can be acquired by cooling from T_N in a strong field. Upon warming in a zero field it is demagnetized almost linearly with temperature until fully vanishing somewhat below T_N, but on the consequent cooling in a zero field a considerable part of TRM is restored (Figure M43a, Rochette and Fillion, 1989). TRM is far greater than an IRM that could be acquired by application of the same field at a low temperature. Therefore, FC and ZFC warming curves of goethite are very strongly different for both bulk and nanocrystalline goethite (Figure M43b, Guyodo *et al.*, 2003). Both features can be used as a discriminatory criterion for this mineral.

Other iron hydroxides often found in sediments and soils include lepidocrocite (γ-FeOOH) and ferrihydrite ($5Fe_2O_3{\cdot}9H_2O$). These minerals are paramagnetic at room temperature and do not contribute to NRM; however, both can transform to ferrimagnetic phases (maghemite) on moderate heating. Recent studies of the low-temperature magnetic properties (Zergenyi *et al.*, 2000; Hirt *et al.*, 2002) have revealed that below their ordering temperatures (~100 K for ferrihydrite and 50–60 K for lepidocrocite) these minerals can acquire a considerable, on the order of several tenths of A m^2 kg^{-1}, remanence which decays nearly exponentially on heating in zero field, mimicking the superparamagnetic behavior of ultrafine particles of strongly magnetic minerals, e.g., magnetite or maghemite (Figure M44).

Iron sulfides

Of a variety of iron sulfides occurring in nature, monoclinic 4C pyrrhotite (Fe_7S_8) is the most extensively studied. The magnetic transition at 30–34 K, initially discovered in early 1960s (Besnus and Meyer, 1964), has been since found in a number of natural and synthetic pyrrhotites (Rochette, 1988; Dekkers, 1989; Dekkers *et al.*, 1989; Rochette *et al.*, 1990). Phenomenologically, the behavior of the SIRM and the coercive force on crossing the transition has much in common with the much more extensively studied Verwey transition in magnetite. Part of a SIRM given at some temperature below the transition is demagnetized at the latter. A SIRM acquired above the transition, when cycled through the transition in zero magnetic field is also partly demagnetized, and a rebound of magnetization at the transition is observed during warming (Figure M45). The latter feature is grain-size dependent, becoming progressively less pronounced in finer grains. Pyrrhotite grains several tens of microns in size show an irreversible loss of remanence even during a zero-field cycle to 77 K (Dekkers, 1989). The M_{rs}/M_s ratio and the coercive force increase sharply below the transition, indicating that the low-temperature phase is in the single-domain magnetic state. These properties can serve as discriminatory criteria for monoclinic pyrrhotite in the case when heating the sample above room temperature is impossible or undesirable.

Thiospinel of iron, mineral greigite (Fe_3S_4) once thought to be rare, is being identified in growing number of environments, contributing to NRM of sediments and soils (see e.g., Snowball and Torii, 1999 and references therein). Its low-temperature magnetic properties have been the subject of several recent studies (Roberts, 1995; Snowball and Torii, 1999; Dekkers *et al.*, 2000); however, no distinctive features between 4.2 K and room temperature have been observed. Generally, SIRM given at a low temperature decreases monotonously on warming in zero-field albeit at different rates (Roberts, 1995), while room-temperature SIRM hardly show any changes on cooling to 4 K (Dekkers *et al.*, 2000, Figure M46). The latter study also reported on the magnetic hysteresis measurements on synthetic greigite. A peak

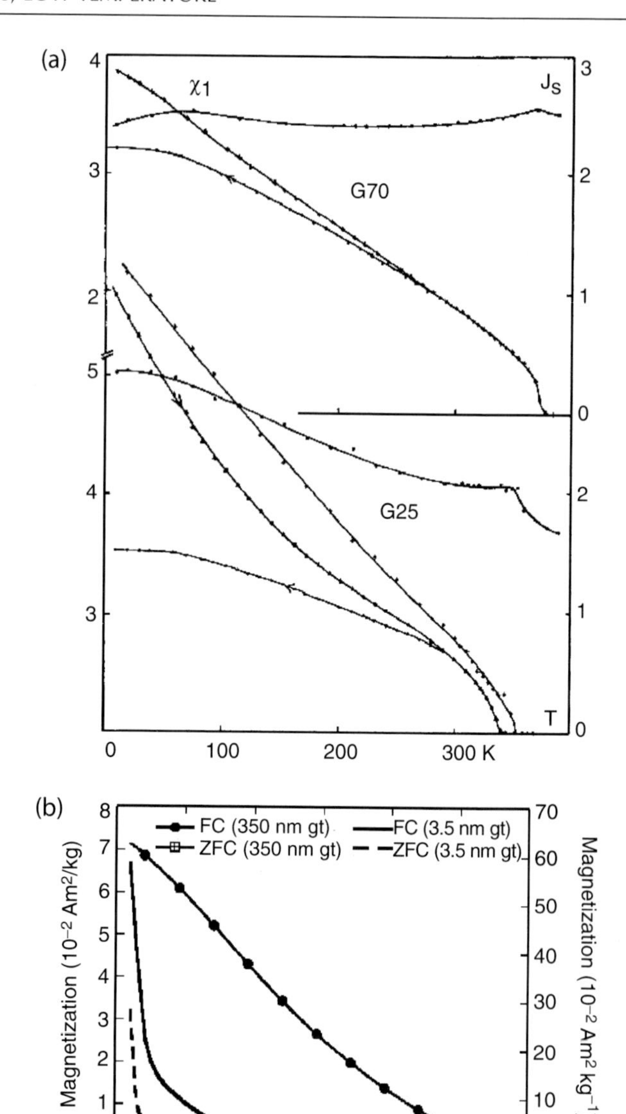

Figure M43 (a) Saturation magnetization J_s (crosses) in 10^{-1} A m^2 kg^{-1} and high-field susceptibility χ_l (open circles) in 10^{-9} m^3 kg^{-1} vs temperature for two goethite samples. J_s and saturation remanence J_{rs} are equal for the G70 sample. The upper curve of J_{rs} (triangles) is obtained by heating peak TRM at 4.2 K and the lower one (dots) by cooling peak TRM from 290 K. (b) Field-cooled (FC) and zero-field-cooled (ZFC) remanence curves for 350-nm (left vertical axis) and 3.5-nm goethite (right vertical axis) samples. (After Rochette and Fillion (1989) and Guyodo *et al.* (2003) © American Geophysical Union, with the permission of the authors and the publisher.)

in M_{rs} was found at 10 K, and both the coercive force and coercivity of remanence increased significantly below at about 30 K. Reasons for this behavior are not clear.

Pyrite (FeS_2) is often found in sediments affected by diagenesis. Theoretical calculations show that pyrite and its polymorph marcasite are low-spin paramagnets whose magnetic susceptibility is temperature independent (Hobbs and Hafner, 1999). However, in practice an increase of susceptibility is often observed in both compounds below

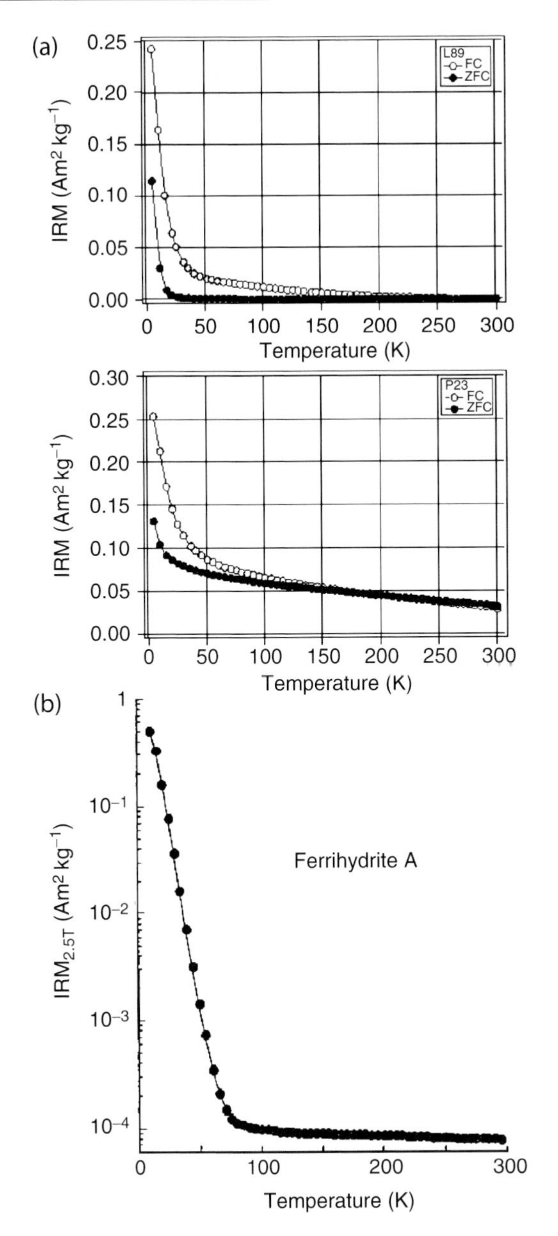

Figure M44 (a) Thermal demagnetization of IRM acquired at 5 K in a 2.5 T field during warming to 300 K after cooling the sample in zero-field cooling (ZFC) or in a 2.5 T field (FC) for two lepidocrocite samples with different crystallite size. (b) Thermal demagnetization of IRM acquired at 10 K in a 2.5 T field during warming to 300 K for a ferrihydrite sample. (After Hirt *et al.* (2002) and Zergenyi *et al.* (2000) © American Geophysical Union, with the permission of the authors and the publisher.)

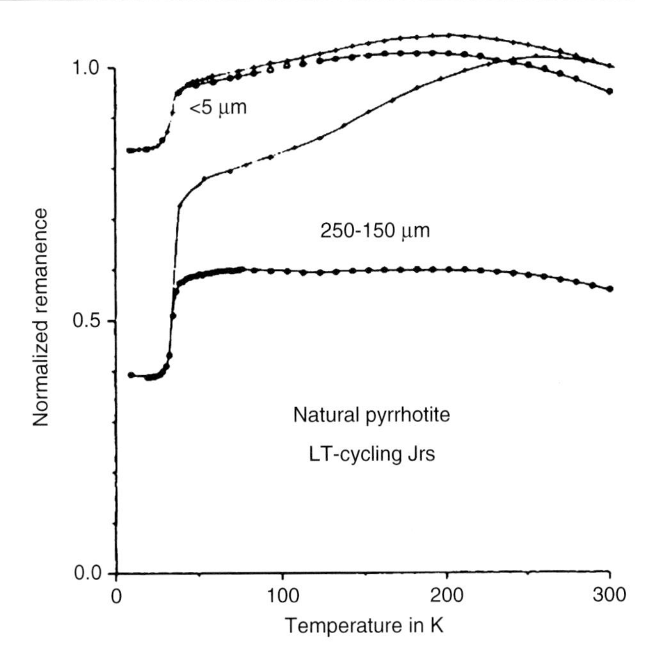

Figure M45 Examples of the 34-K transition in samples of pyrrhotite with contrasting grain size. Crosses correspond to cooling and circles to rewarming. (Modified from Dekkers *et al.* (1989) © American Geophysical Union, with the permission of the author and the publisher.)

10 K (Burgardt and Seehra, 1977; Seehra and Jagadeesh, 1979). Most likely it is due to otherwise undetectable impurities showing usual Curie-Weiss paramagnetism.

Pyrrhotites richer in iron than Fe_7S_8 are antiferromagnetic at room temperature, but transform to a ferrimagnetically ordered phase at ~200°C, and eventually to monoclinic pyrrhotite on further heating. Other Fe-S minerals (mackinawite, smythite) have been often considered metastable in ambient conditions. These minerals received so far a limited interest in rock magnetism, and their magnetic properties at low temperatures are largely unknown.

Siderite and rhodochrosite

Siderite (iron carbonate, $FeCO_3$) and rhodochrosite (manganese carbonate, $MnCO_3$) usually precipitate jointly in anoxic sedimentary environments when iron and manganese are present in soluble form. The two compounds may form solid solutions spanning the entire range of intermediate compositions. Although paramagnetic at room temperature, siderite has some importance in paleomagnetism, since secondary maghemite is produced by weathering siderite-bearing rocks, and also because siderite easily converts to magnetite at heating to ~400°C, thereby putting an effective limit to thermal demagnetization (Ellwood *et al.*, 1986, 1989). Siderite orders antiferromagnetically below the Néel temperature of 37–40 K (natural samples of varying purity, cf. Jacobs, 1963; Robie *et al.*, 1984). $MnCO_3$ is a canted antiferromagnet with slightly lower Néel temperature of 32 K (synthetic crystal, Borovik-Romanov *et al.*, 1981). Both compounds can acquire remanence below T_N, which is lost quite sharply on warming in zero field. These minerals could be thus confused with pyrrhotite, which shows the magnetic transition in the same temperature range. However, siderite and rhodochrosite can still be distinguished from pyrrhotite since both minerals show a large (over one order of magnitude for siderite) difference between SIRMs acquired after FC and ZFC, respectively (Housen *et al.*, 1996; Frederichs *et al.*, 2003), not observed for pyrrhotite (Figure M47). It might be expected that cycling of room-temperature SIRM in siderite- and pyrrhotite-bearing samples will also distinguish between the two minerals.

Vivianite

Vivianite, iron hydrophosphate ($Fe_3(PO_4)_2{\cdot}8H_2O$), usually precipitates in moderately to highly productive, iron-rich sedimentary environments. At low temperatures, vivianite shows antiferromagnetic ordering in each of two sublattices formed by Fe^{2+} ions occupying nonequivalent positions (van der Lugt and Poulis, 1961). The two sublattices are canted by approximately 42° (Forsyth *et al.*, 1970). However, conflicting results have been reported on changes of the antiferromagnetic order with temperature. Forstat *et al.* (1965) have

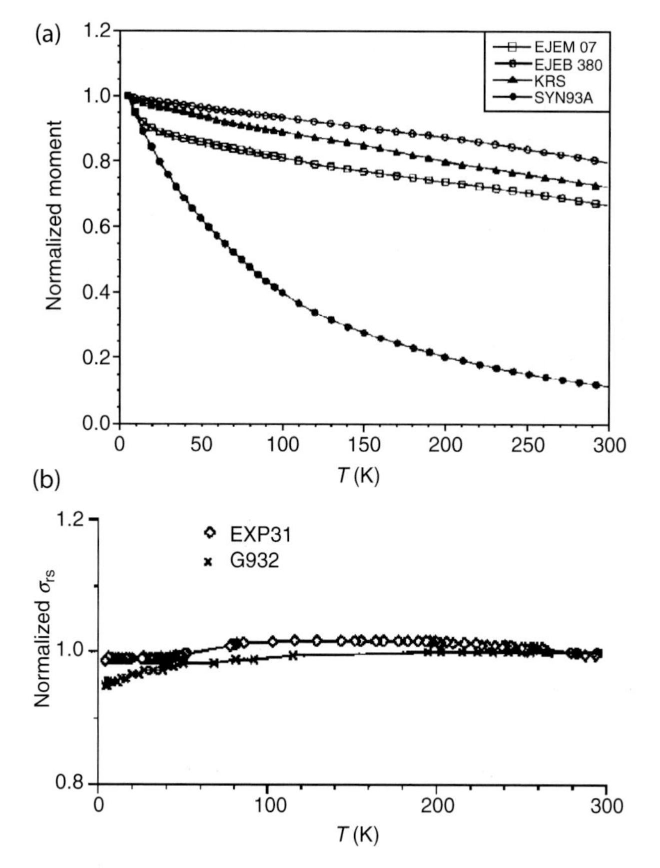

Figure M46 (a) Thermal demagnetization of IRM acquired at 5 K in a 2.5 T field for four greigite samples. (b) Behavior of IRM (1.2 T) acquired at room temperature for synthetic greigite during zero-field cooling to 4 K. Sample EXP31 was stored for ~22 months before the measurements, whereas sample G932 was a "fresh" sample. (After Roberts (1995) and Dekkers *et al.* (2000), with the permission of the authors.)

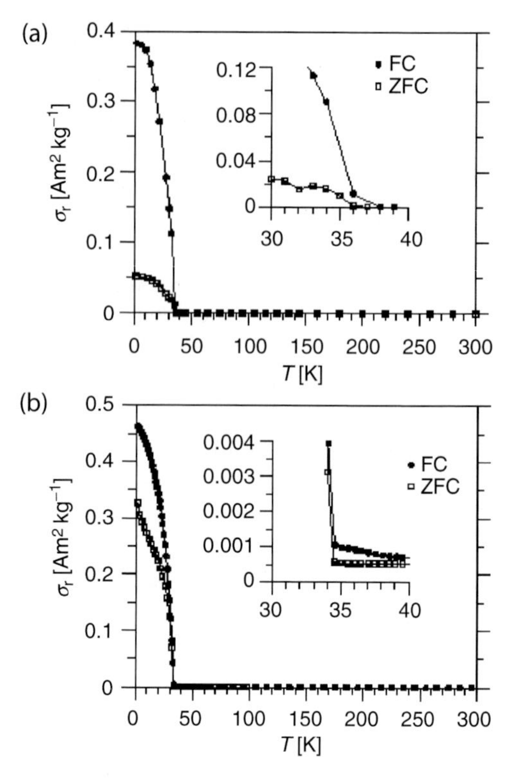

Figure M47 Warming curves of IRM acquired at 2 K in a 5 T magnetic field after cooling in zero field (open symbols) and in a 5 T field (solid symbols) for well-crystalline siderite (a) and rhodochrosite (b). (After Frederichs *et al.* (2003), with the permission of the author and the publisher, Elsevier.)

found two sharp anomalies in the specific heat at 9.60 and 12.40 K, respectively, which were attributed to two distinct antiferromagnetic-paramagnetic transitions. On the other hand, in a neutron diffraction experiment Forsyth *et al.* (1970) found only one antiferromagnetic-paramagnetic transition at 8.84 ± 0.05 K. A recent rock magnetic study (Frederichs *et al.*, 2003) has confirmed previous data on the magnetic susceptibility, and found that below 10 K vivianite can acquire only a small remanence on the order of 10^{-3} A m^2 kg^{-1}.

Iron silicates

Silicates are the most abundant minerals in the Earth's crust. They often contain iron in either ferrous (Fe^{2+}) or ferric (Fe^{3+}) form. At low temperatures, iron-rich silicates typically order antiferromagnetically, and the observed Néel temperatures range from <10 K to about 100 K (Coey *et al.*, 1982; Coey and Ghose, 1988). Examples are fayalite (Fe_2SiO_4) and ferrosilite ($FeSiO_3$), iron-rich end members of olivine and pyroxene series, respectively. For the former Néel temperature is 65 K (Santoro *et al.*, 1966), and for the latter 43 K (Sawaoka *et al.*, 1968). However, iron-rich silicates are rare in nature and in most cases iron cations are diluted with nonmagnetic cations such as Mg^{2+} and Al^{3+}. With increasing nonmagnetic dilution, Néel temperatures are progressively lowered and the magnetic ordering eventually suppressed. In the olivine series $(Fe_xMg_{1-x})_2SiO_4$ the antiferromagnetic transition disappears for $x \sim 0.3$ (Hoye and O'Reilly, 1972). Some sheet silicates, while having the antiferromagnetic ground state, can exhibit the magnetic hysteresis and acquire remanent magnetization at low temperatures. An example is mineral minnesotaite, or ferrous talc, of chemical formula $Fe_3Si_4O_{10}(OH)_2$, which acquires an IRM(2 T) of about 40 A m^2 kg^{-1} at 4.2 K (Ballet *et al.*, 1985). The underlying mechanism of this behavior is that in minnesotaite the metamagnetic transition from an antiferromagnetic to a ferromagnetic state occurs in an anomalously low field of about 0.5 T. After once being subjected to fields higher than the metamagnetic threshold, the sample follow hysteresis loop never to return to the antiferromagnetic state (Coey and Ghose, 1988). In summary, silicate minerals may have certain potential to contribute to the low-temperature magnetic signal of rocks, particularly to the magnetic susceptibility; whether and how often this is really the case remains to be seen.

Andrei Kosterov

Bibliography

Amin, N., and Arajs, S., 1987. Morin temperature of annealed submicronic α-Fe_2O_3 particles. *Physical Review B*, **35**: 4810–4811.

Aragón, R., 1992. Magnetization and exchange in nonstoichiometric magnetite. *Physical Review B*, **46**: 5328–5333.

Aragón, R., Buttrey, D.J., Shepherd, J.P., and Honig, J.M., 1985. Influence of nonstoichiometry on the Verwey transition. *Physical Review B*, **31**: 430–436.

Arai, M., and Ishikawa, Y., 1985. A new oxide spin glass system of $(1-x)FeTiO_3$-xFe_2O_3. III. Neutron scattering studies of magnetization processes in a cluster type spin glass of $90FeTiO_3$-$10Fe_2O_3$. *Journal of the Physical Society of Japan*, **54**: 795–802.

Arai, M., Ishikawa, Y., Saito, N., and Takei, H., 1985a. A new oxide spin glass system of $(1-x)FeTiO_3$-xFe_2O_3. II. Neutron scattering studies of a cluster type spin glass of 90$FeTiO_3$-10Fe_2O_3. *Journal of the Physical Society of Japan*, **54**: 781–794.

Arai, M., Ishikawa, Y., and Takei, H., 1985b. A new oxide spin glass system of $(1-x)FeTiO_3$-xFe_2O_3. IV. Neutron scattering studies on a reentrant spin glass of 79$FeTiO_3$-21Fe_2O_3 single crystal. *Journal of the Physical Society of Japan*, **54**: 2279–2286.

Ballet, O., Coey, J.M.D., Mangin, P., and Townsend, M.G., 1985. Ferrous talk—a planar antiferromagnet. *Solid State Communications*, **55**: 787–790.

Bando, Y., Kiyama, M., Yamamoto, N., Takada, T., Shinjo, T., and Takaki, H., 1965. The magnetic properties of α-Fe_2O_3 fine particles. *Journal of the Physical Society of Japan*, **20**: 2086.

Besnus, M.J., and Meyer, A.J.P., 1964. *Nouvelles données expérimentals sur le magnétisme de la pyrrhotine naturelle*. In *Proceedings of the International Conference on Magnetism*. Nottingham: Institute of Physics, pp. 507–511.

Bickford, L.R. Jr., 1950. Ferromagnetic resonance absorption in magnetite single crystals. *Physical Review*, **78**: 449–457.

Borovik-Romanov, A.S., Egorov, V.M., Ikornikova, N.Y., Kreines, N.M., Ozhogin, V.I., and Prozorova, L.A., 1981. Antiferromagnetic carbonates of the 3d elements. *Soviet Physics—Crystallography*, **26**: 623–630.

Brabers, V.A., Walz, M.F., and Kronmüller, H., 1998. Impurity effects upon the Verwey transition in magnetite. *Physical Review B*, **58**: 14163–14166.

Bucur, I., 1978. Experimental study of the origin and properties of the defect moment in single domain haematite. *Geophysical Journal of the Royal Astronomical Society*, **55**: 589–604.

Burgardt, P., and Seehra, M.S., 1977. Magnetic susceptibility of iron pyrite (FeS_2) between 4.2 and 620 K. *Solid State Communications*, **22**: 153–156.

Coey, J.M.D., and Ghose, S., 1988. Magnetic phase transitions in silicate minerals. In Ghose, S., Coey, J.M.D., and Salje, E. (eds.), *Structural and Magnetic Phase Transitions in Minerals*. New York: Springer-Verlag, pp. 162–184.

Coey, J.M.D., Moukarika, A., and Ballet, O., 1982. Magnetic order in silicate minerals. *Journal of Applied Physics*, **53**: 8320–8325.

da Costa, G.M., Van San, E., De Grave, E., Vandenberghe, R.E., Barrón, V., and Datas, L., 2002. Al hematites prepared by homogeneous precipitation of oxinates: material characterization and determination of the Morin transition. *Physics and Chemistry of Minerals*, **29**: 122–131.

de Boer, C.B., and Dekkers, M.J., 1996. Grain-size dependence of the rock magnetic properties for a natural maghemite. *Geophysical Research Letters*, **23**: 2815–2818.

Dekkers, M.J., 1989. Magnetic properties of natural pyrrhotite. II. High- and low-temperature behaviour of J_{rs} and TRM as a function of grain size. *Physics of the Earth and Planetary Interiors*, **57**: 266–283.

Dekkers, M.J., and Linssen, J.H., 1989. Rock magnetic properties of fine-grained natural low-temperature haematite with reference to remanence acquisition mechanisms in red beds. *Geophysical Journal International*, **99,** 1–18.

Dekkers, M.J., Mattéi, J.L., Fillion, G., and Rochette, P., 1989. Grain-size dependence of the magnetic behavior of pyrrhotite during its low-temperature transition at 34 K. *Geophysics Research Letters*, **16**: 855–858.

Dekkers, M.J., Passier, H.F., and Schoonen, M.A.A., 2000. Magnetic properties of hydrothermally synthesized greigite (Fe_3S_4). II. High- and low-temperature characteristics. *Geophysical Journal International*, **141**: 809–819.

Dunlop, D.J., 2003. Stepwise and continuous low-temperature demagnetization. *Geophysical Research Letters*, **30**: 1582, doi: 10.1029/2003GL017268.

Dunlop, D.J., and Özdemir, Ö., 1997. *Rock Magnetism: Fundamentals and Frontiers*. Cambridge: Cambridge University Press.

Dzyaloshinsky, I., 1958. A thermodynamic theory of "weak" ferromagnetism of antiferromagnetics. *Journal of Physics and Chemistry of Solids*, **4**: 241–255.

Ellwood, B.B., Balsam, W., Burkart, B., Long,. G.J., and Buhl, M.L., 1986. Anomalous magnetic properties in rocks containing the mineral siderite: paleomagnetic implications. *Journal of Geophysical Research*, **91**: 12779–12790.

Ellwood, B.B., Burkart, B., Rajeshwar, K., Darwin, R.L., Neeley, R.A., McCall, A.B., Long, G.J., Buhl, M.L., and Hickcox, C.W., 1989. Are the iron carbonate minerals, ankerite and ferroan dolomite, like siderite, important in paleomagnetism? *Journal of Geophysical Research*, **94**: 7321–7331.

Forstat, H., Love, N.D., and McElearney, J., 1965. Specific heat of $Fe_3(PO_4)_2{\cdot}8H_2O$. *Physical Review*, **139**: A1246–1248.

Forsyth, J.B., Johnson, C.E., and Wilkinson, C., 1970. The magnetic structure of vivianite, $Fe_3(PO_4)_2{\cdot}8H_2O$. *Journal of Physics C: Solid State Physics*, **3**: 1127–1139.

Frederichs, T., von Dobeneck, T., Bleil, U., and Dekkers, M.J., 2003. Towards the identification of siderite, rhodochrosite, and vivianite in sediments by their low-temperature magnetic properties. *Physics and Chemistry of Earth*, **28**: 669–679.

Gallon, T.E., 1968. The remanent magnetization of haematite single crystals. *Proceedings of the Royal Society of London A*, **303**: 511–524.

Guyodo, Y., Mostrom, A., Lee Penn, R., and Banerjee, S.K., 2003. From nanodots to nanorods: Oriented aggregation and magnetic evolution of nanocrystalline goethite. *Geophysical Research Letters*, **30**: 1512, doi: 10.1029/2003GL017021.

Haigh, G., 1957. Observations on the magnetic transition in hematite at $-15°C$. *Philosophical Magazine*, **2**: 877–890.

Harrison, C.G.A., and Peterson, M.N.A., 1965. A magnetic mineral from the Indian Ocean. *American Mineralogist*, **50**: 704–712.

Hedley, I.G., 1971. The weal ferromagnetism of goethite (α-FeOOH). *Zeitschrift für Geophysik*, **37**: 409–420.

Heider, F., Dunlop, D.J., and Soffel, H.C., 1992. Low-temperature and alternating field demagnetization of saturation remanence and thermoremanence in magnetite grains (0.037 to 5 mm). *Journal of Geophysical Research*, **97**: 9371–9381.

Hirt, A.M., Lanci, L., Dobson, J., Weidler, P., and Gehring, A.U., 2002. Low-temperature magnetic properties of lepidocrocite. *Journal of Geophysical Research*, **107**: 10.1029/2001JB000242.

Hobbs, D., and Hafner, J., 1999. Magnetism and magneto-structural effects in transition-metal sulphides. *Journal of Physics: Condensed Matter*, **11**: 8197–8222.

Hoffman, K.A., 1992. Self-reversal of thermoremanent magnetization in the ilmenite-hematite system: Order-disorder, symmetry, and spin alignment. *Journal of Geophysical Research*, **97**: 10883–10895.

Housen, B., Banerjee, S.K., and Moskowitz, B.M., 1996. Low-temperature magnetic properties of siderite and magnetite in marine sediments. *Geophysical Research Letters*, **23**: 2843–2846.

Hoye, G.S., and O'Reilly, W., 1972. A magnetic study of the ferro-magnesian olivines $(Fe_xMg_{1-x})_2SiO_4$, $0 < x < 1$. *Journal of Physics and Chemistry of Solids*, **33**: 1827–1834.

Ishikawa, Y., and Akimoto, S., 1957. Magnetic properties of the $FeTiO_3$-Fe_2O_3 solid solution series. *Journal of the Physical Society of Japan*, **12**: 1083–1098.

Ishikawa, Y., Saito, N., Arai, M., Watanabe, Y., and Takei, H., 1985. A new oxide spin glass system of $(1-x)FeTiO_3$-xFe_2O_3. I. Magnetic properties. *Journal of the Physical Society of Japan*, **54**: 312–325.

Jacobs, I.S., 1963. Metamagnetism of siderite ($FeCO_3$). *Journal of Applied Physics*, **34**: 1106–1107.

Kąkol, Z., and Honig, J.M., 1989. Influence of deviation from ideal stoichiometry on the anisotropy parameters of magnetite $Fe_{3(1-\delta)}O_4$. *Physical Review B*, **40**: 9090–9097.

Kąkol, Z., Sabol, J., and Honig, J.M., 1991a. Cation distribution and magnetic properties of titanomagnetites $Fe_{3-x}Ti_xO_4$ $(0 \leq x < 1)$. *Physical Review B*, **43**: 649–654.

Kąkol, Z., Sabol, J., and Honig, J.M., 1991b. Magnetic anisotropy of titanomagnetites $Fe_{3-x}Ti_xO_4$, $0 \leq x \leq 0.55$. *Physical Review B*, **44**: 2198–2204.

Kąkol, Z., Sabol, J., Stickler, J., Kozłowski, A., and Honig, J.M., 1994. Influence of titanium doping on the magnetocrystalline anisotropy of magnetite. *Physical Review B*, **49**: 12767–12772.

Kaye, G., 1962. The effect of titanium on the low temperature transition in natural crystals of haematite. *Proceedings of the Physical Society*, **80**: 238–243.

King, J.G., and Williams, W., 2000. Low-temperature magnetic properties of magnetite. *Journal of Geophysical Research*, **105**: 16427–16436.

Kirschvink, J.L., Jones, D.S., and Mac Fadden, B.J., (eds.), 1985. *Magnetite Biomineralization and Magnetoreception in Organisms: A New Biomagnetism.* New York: Plenum Press.

Kletetschka, G., and Banerjee, S.K., 1995. Magnetic stratigraphy of Chinese loess as a record of natural fires. *Geophysical Research Letters*, **22**: 1341–1343.

Kosterov, A.A., 2001. Magnetic hysteresis of pseudo-single-domain and multidomain magnetite below the Verwey transition. *Earth and Planetary Science Letters*, **186**: 245–253.

Kosterov, A.A., 2002. Low-temperature magnetic hysteresis properties of partially oxidized magnetite. *Geophysical Journal International*, **149**: 796–804.

Kosterov, A.A., 2003. Low-temperature magnetization and AC susceptibility of magnetite: effect of thermomagnetic history. *Geophysical Journal International*, **154**: 58–71.

Kozłowski, A., Kąkol, Z., Kim, D., Zaleski, R., and Honig, J.M., 1996a. Heat capacity of $Fe_{3-\alpha}M_\alpha O_4$ (M=Zn, Ti, $0 \leq \alpha \leq 0.04$). *Physical Review B*, **54**: 12093–12098.

Kozłowski, A., Metcalf, P., Kąkol, Z., and Honig, J.M., 1996b. Electrical and magnetic properties of $Fe_{3-z}Al_zO_4$ ($z < 0.06$). *Physical Review B*, **53**: 15113–15118.

Matzka, J., Krása, D., Kunzmann, T., Schult, A., and Petersen, N., 2003. Magnetic state of 10–40 Ma old ocean basalts and its implications for natural remanent magnetization. *Earth and Planetary Science Letters*, **206**: 541–553.

Miyahara, Y., 1972. Impurity effects on the transition temperature of magnetite. *Journal of the Physical Society of Japan*, **32**: 629–634.

Morin, F.J., 1950. Magnetic susceptibility of αFe_2O_3 and αFe_2O_3 with added titanium. *Physical Review*, **78**: 819–820.

Moriya, T., 1960. Anisotropic superexchange interaction and weak ferromagnetism. *Physical Review*, **120**: 91–98.

Morrish, A.H., 1994. *Canted Antiferromagnetism: Hematite.* Singapore; River Edge, NJ: World Scientific.

Morrish, A.H., and Watt, L.A.K., 1958. Coercive force of iron oxide micropowders at low temperatures. *Journal of Applied Physics*, **29**: 1029–1033.

Moskowitz, B.M., Frankel, R.B., and Bazylinski, D.A., 1993. Rock magnetic criteria for the detection of biogenic magnetite. *Earth and Planetary Science Letters*, **120**: 283–300.

Moskowitz, B.M., Jackson, M., and Kissel, C., 1998. Low-temperature magnetic behavior of titanomagnetites. *Earth and Planetary Science Letters*, **157**: 141–149.

Murray, J.W., 1979. Iron oxides. In Burns, R.G., (ed.), *Marine Minerals.* Washington, DC: Mineralogical Society of America, pp. 47–98.

Muxworthy, A.R., 1999. Low-temperature susceptibility and hysteresis of magnetite. *Earth and Planetary Science Letters*, **169**: 51–58.

Muxworthy, A.R., and McClelland, E., 2000a. Review of the low-temperature magnetic properties of magnetite from a rock magnetic perspective. *Geophysical Journal International*, **140**: 101–114.

Muxworthy, A.R., and McClelland, E., 2000b. The causes of low-temperature demagnetization of remanence in multidomain magnetite. *Geophysical Journal International*, **140**: 115–131.

Nagata, T., 1961. *Rock Magnetism.* Tokyo: Maruzen Co.

Nininger, R.C. Jr., and Schroeer, D., 1978. Mössbauer studies of the Morin transition in bulk and microcrystalline α-Fe_2O_3. *Journal of Physics and Chemistry of Solids*, **39**: 137–144.

Ortega, B., Caballero, C., Lozano, S., Israde, I., and Vilaclara, G., 2002. 52000 years of environmental history in Zacapu basin, Michoacan, Mexico: the magnetic record. *Earth and Planetary Science Letters*, **202**: 663–675.

Özdemir, Ö., 2000. Coercive force of single crystals of magnetite at low temperatures. *Geophysical Journal International*, **141**: 351–356.

Özdemir, Ö., and Dunlop, D.J., 2000. Intermediate magnetite formation during dehydration of goethite. *Earth and Planetary Science Letters*, **177**: 59–67.

Özdemir, Ö., Dunlop, D.J., and Moskowitz, B.M., 1993. The effect of oxidation on the Verwey transition in magnetite. *Geophysical Research Letters*, **20**: 1671–1674.

Özdemir, Ö., Dunlop, D.J., and Moskowitz, B.M., 2002. Changes in remanence, coercivity and domain state at low temperature in magnetite. *Earth and Planetary Science Letters*, **194**: 343–358.

Ozima, M., Ozima, M., and Nagata, T., 1964. Low temperature treatment as an effective means of "magnetic cleaning" of natural remanent magnetization. *Journal of Geomagnetism and Geoelectricity*, **16**: 37–40.

Roberts, A.P., 1995. Magnetic properties of sedimentary greigite (Fe_3S_4). *Earth and Planetary Science Letters*, **134**: 227–236.

Robie, R.A., Haselton, H.T. Jr., and Hemingway, B.S., 1984. Heat capacities and entropies of rhodochrosite ($MnCO_3$) and siderite ($FeCO_3$) between 5 and 600 K. *American Mineralogist*, **69**: 349–357.

Rochette, P.,1988. *La Susceptibilité Anisotrope des Roches Faiblement Magnétiques: Origines et Applications*, Thèse d'Etat, Université de Grenoble.

Rochette, P., and Fillion, G., 1989. Field and temperature behavior of remanence in synthetic goethite: paleomagnetic implications. *Geophysical Research Letters*, **16**: 851–854.

Rochette, P., Fillion, G., Mattéi, J.L., and Dekkers, M.J., 1990. Magnetic transition at 30–34 Kelvin in pyrrhotite: insight into a widespread occurrence of this mineral in rocks. *Earth and Planetary Science Letters*, **98**: 319–328.

Rozenberg, G.Kh., Hearne, G.R., Pasternak, M.P., Metcalf, P.A., and Honig, J.M., 1996. Nature of the Verwey transition in magnetite (Fe_3O_4) to pressures of 16 GPa. *Physical Review B*, **53**: 6482–6487.

Samara, G.A., 1968. Effect of pressure on the metal-nonmetal transition and conductivity of Fe_3O_4. *Physical Review Letters*, **21**: 795–797.

Santoro, R.P., Newham, R.E., and Nomura, S., 1966. Magnetic properties of Mn_2SiO_4 and Fe_2SiO_4. *Journal of Physics and Chemistry of Solids*, **27**: 655–666.

Sawaoka, A., Miyahara, S., and Akimoto, S., 1968. Magnetic properties of several metasilicates and metagermanates with pyroxene structure. *Journal of the Physical Society of Japan*, **25**: 1253–1257.

Schmidbauer, E., and Keller, R., 1996. Magnetic properties and rotational hysteresis of Fe_3O_4 and γ-Fe_2O_3 particles ~250 nm in diameter. *Journal of Magnetism and Magnetic Materials*, **152**: 99–108.

Schmidbauer, E., and Readman, P.W., 1982. Low temperature magnetic properties of Ti-rich Fe-Ti spinels. *Journal of Magnetism and Magnetic Materials*, **27**: 114–118.

Schroeer, D., and Nininger, R.C. Jr., 1967. Morin transition in α-Fe_2O_3 microcrystals. *Physical Review Letters*, **19**: 632–634.

Seehra, M.S., and Jagadeesh, M.S., 1979. Temperature-dependent magnetic susceptibility of marcasite (FeS_2). *Physical Review B*, **20**: 3897–3899.

Senftle, F.E., Thorpe, A.N., Briggs, C., Alexander, C., Minkin, J., and Griscom, D.L., 1975. The Néel transition and magnetic properties of terrestrial, synthetic, and lunar ilmenites. *Earth and Planetary Science Letters*, **26**: 377–386.

Smirnov, A.V., and Tarduno, J.A., 2002. Magnetic field control of the low-temperature magnetic properties of stoichiometric and cation-deficient magnetite. *Earth and Planetary Science Letters*, **194**: 359–368.

Snowball, I., and Torii, M., 1999. Incidence and significance of magnetic iron sulphides in Quaternary sediments and soils. In Maher, B.A., and Thompson, R. (eds.), *Quaternary Climates, Environments and Magnetism*. Cambridge: Cambridge University Press, pp. 199–230.

Strangway, D.W., Honea, R.M., McMahon, B.E., and Larson, E.E., 1968. The magnetic properties of naturally occurring goethite. *Geophysical Journal of the Royal Astronomical Society*, **15**: 345–359.

Syono, Y., 1965. Magnetocrystalline anisotropy and magnetostriction of Fe_3O_4-Fe_2TiO_4 series—with special application to rocks magnetism. *Japanese Journal of Geophysics*, **4**: 71–143.

Taylor, R.M., and Schwertmann, U., 1974. Maghemite in soils and its origin. I. Properties and observations on soil maghemites. *Clay Minerals*, **10**: 289–298.

Thorpe, A.N., Minkin, J.A., Senftle, F.E., Alexander, C., Briggs, C., Evans, H.T. Jr., and Nord, G.L. Jr., 1977. Cell dimensions and antiferromagnetism of lunar and terrestrial ilmenite single crystals. *Journal of Physics and Chemistry of Solids*, **38**: 115–123.

Tucker, P., 1981. Low-temperature magnetic hysteresis properties of multidomain single-crystal titanomagnetite. *Earth and Planetary Science Letters*, **54**: 167–172.

Uyeda, S., 1958. Thermo-remanent magnetism as a medium of palaeomagnetism, with special reference to reverse thermo-remanent magnetism. *Japanese Journal of Geophysics*, **2**: 1–123.

van der Lugt, W., and Poulis, N.J., 1961. The splitting of the nuclear magnetic resonance lines in vivianite. *Physica*, **27**: 733–750.

Verwey, E.J.W., 1939. Electronic conduction of magnetite (Fe_3O_4) and its transition point at low temperatures. *Nature*, **44**: 327–328.

Verwey, E.J.W., and Haayman, P.W., 1941. Electronic conductivity and transition point of magnetite ("Fe_3O_4"). *Physica*, **8**: 979–987.

Walz, F., 2002. The Verwey transition—a topical review. *Journal of Physics: Condensed Matter*, **14**: R285–R340.

Weiss, P., and Forrer, R., 1929. La saturation absolue des ferromagnétiques et les lois d'approche en fonction du champ et de la température. *Annales de Physique*, 10ᵉ série, **12**: 279–374.

Weiss, P., and Renger, K., 1914. Die anfängliche Permeabilität von Eisen und Magnetit in Funktion der Temperatur und die Abhängigkeit der Umwandlungspunkte von der Feldstärke. *Archiv für Electrotechnik*, **2**: 406–418.

Zergenyi, R.S., Hirt, A.M., Zimmermann, S., Dobson, J.B., and Lowrie, W., 2000. Low-temperature magnetic behavior of ferrihydrite. *Journal Geophysical Research*, **105**: 8297–8303.

Cross-references

Iron Sulfides
Magnetic Mineralogy, Changes due to Heating
Magnetic Proxy Parameters
Rock Magnetism
Rock Magnetism, Hysteresis Measurements

MAGNETIC PROXY PARAMETERS

Introduction

As a rule, a natural system is complex and the result of the interplay of numerous constituting parameters. To fully describe such systems, ideally each (independent) parameter must be known. However, their mere number would imply a substantial analytical operation and in practice one should look for parameters that describe the essential parts of the system. Because these parameters can only describe the system in an approximate fashion they are termed "proxy parameters" or "proxies." There exists a myriad of proxies including the chemical analysis of inorganic or organic compounds, the number and types of living and fossil biota, and the physical properties of a sample, such as grain-size distribution, porosity, magnetic properties, etc. Basically, each physical, chemical, or biological parameter that is measurable reasonably quickly at reasonable cost may serve as proxy to describe a complete system. One should realize that up to over several thousands of data points are rapidly gathered in a case study. Combinations of two or more of those parameters may yield valuable proxies as well. Another important condition for a parameter to serve as proxy is that its meaning must be understood. This pertains to the analytical point of view as well as to its functioning in the natural system. In the following, we outline the merit and potential of the so-called mineral-magnetic proxies.

From the analytical viewpoint, the assets of mineral-magnetic or rock-magnetic proxy parameters are manifold. They include sensitivity, rapidity, ease of sample preparation, comparatively modestly priced instrumentation for data acquisition, nondestructiveness, and in particular the ability to sense grain-size variation in the ultrafine grain-size range. Magnetic properties refer to bulk sample properties in contrast to (electron) microscopical techniques where often only a small fraction of a sample can be analyzed. Therefore, mineral-magnetic parameters are well suited as proxy parameters. As drawbacks should be mentioned that some of the parameters are nonunique, and that their scales are most often utilized in a relative sense. Often ratios of several parameters are used to reduce nonuniqueness. The meaning of a ratio may change depending on the absolute values of numerator and denominator; hence, may vary among case studies. Proper interpretation of a mineral-magnetic data set therefore requires expert training and a feel for potential variability in the data cloud.

Magnetism

Mineral-magnetic parameters refer to the concentration and properties of iron oxides (and some iron sulfides), minerals that occur in trace amounts in any rock, soil, or even organic tissue. In quantum mechanics, electrons are assigned orbital and spin magnetic moments. For magnetism, the spin moment is most important: each electron can be regarded as a microscopic magnet. In most compounds electrons act independently and consequently a very low-magnetic moment is measured in applied magnetic fields and no permanent or remanent magnetic moment can exist. This is the case in diamagnetic and paramagnetic substances. Diamagnetism—occurring in compounds with no unpaired electron spins—is characterized by a small but negative magnetic moment in applied fields. It is a property of all matter because all paired electrons yield a diamagnetic moment. Paramagnetism—characteristic of compounds with unpaired electron spins—yields a positive magnetic moment that outweighs diamagnetic moments very quickly.

By virtue of exchange interaction, a property described also by quantum mechanics, the outermost electrons of transition elements may act collectively in some compounds, provided that their orbitals sufficiently overlap. For the first row transition elements, this situation occurs in metal oxides. In nature, iron is the most important element and the collective electron behavior is termed ferromagnetism (*sensu lato*). Several classes of ferromagnetism are distinguished: ferromagnetism, ferrimagnetism, and antiferromagnetism (Figure M48).

Ferromagnetic and ferrimagnetic materials have very large magnetic moments in applied fields, several orders of magnitude larger than a typical paramagnet. Antiferromagnetic materials ideally would have a net magnetic moment of zero but in reality usually a weak moment persists. Collective spin behavior gives rise to the existence of permanent or remanent magnetization, i.e., magnetic moments that remain after removal of an external magnetic field. The magnetic properties of ferromagnets (*sensu lato*) are dependent on mineral type, temperature, grain size, and strain state of the grain. Subtle variations in magnetic properties can thus be interpreted in terms of changing concentration, grain size, or strain state of the magnetic materials. In a geoscientific context, in turn, these can be interpreted along the lines of changes in catchment area, in climate, in diagenesis, or in oxidation degree when dealing with a series of lava flows. This is exploited when utilizing magnetic properties as proxy parameters.

Magnetic minerals

In nature, magnetic minerals are largely restricted to iron-titanium oxides that may contain minor amounts of substituted aluminum, chromium, manganese, and magnesium. The most important terrestrial magnetic mineral is magnetite (Fe_3O_4), one of the endmembers of the titanomagnetite solid-solution series (the other endmember is ulvöspinel, Fe_2TiO_4). Hematite (α-Fe_2O_3) is another important mineral; it is one of the endmembers of the titanohematite solid-solution series (the other endmember is ilmenite, $FeTiO_3$). Goethite (α-FeOOH), pyrrhotite (Fe_7S_8), and greigite (Fe_3S_4) are often referred to as minor magnetic minerals. Some copper minerals show magnetic properties as well but their occurrence is restricted to certain iron types. Metallic iron is very rare in terrestrial environments but occurs prominently in meteorites. Magnetic minerals can be distinguished from each other by their Curie point temperature, the temperature above which collective behavior is lost and a paramagnetic substance remains. When dealing with antiferromagnetic minerals the temperature at which collective behavior is lost, is referred to as Néel point temperature. In practice Curie and Néel temperature are used interchangeably. When cooling below the Curie or Néel temperature, collective spin coupling sets in again. For most magnetic minerals the Curie (Néel) temperature is several hundreds of degree Celsius, i.e., far above room temperature (see Table M2). If we would extend the measurement range down to a few Kelvin, many silicate minerals will order magnetically as well. However, this is not routinely done because it requires specific instrumentation that is not widely available.

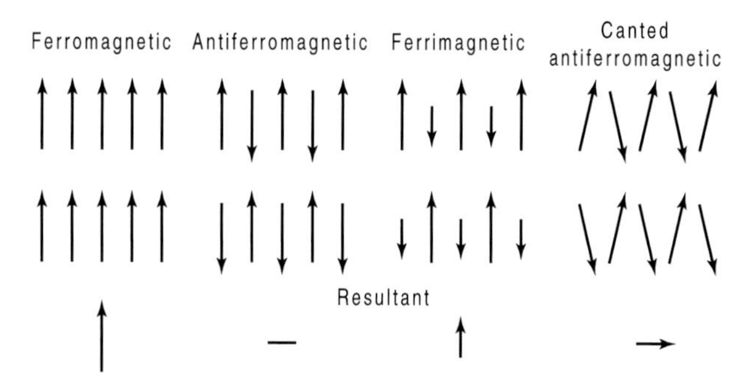

Figure M48 Classes of magnetism. The resultant magnetic moment is indicated below.

Grain-size indicators

Small grains have a different number of magnetic domains—zones with a uniform magnetization direction within a grain—than large grains of the same magnetic mineral. Hence, magnetic properties that reflect somehow the number of domains, are grain-size indicative. From the smallest grain size to the largest, four main domain structure types are distinguished: superparamagnetic, single-domain, pseudosingle-domain, and multidomain (Figures M49 and M50). The boundaries between these domain types are gradual, but the change in domain-state dependent properties essentially forms a continuum. Superparamagnetic grains are that small that they cannot support a stable domain configuration: upon a changing external field the spin configuration

Table M2 Some natural magnetic minerals, Curie (T_C) or Néel temperature (T_N) and magnetic structure below T_C or T_N. More detailed descriptions of the magnetic properties of most of these phases are given by Hunt *et al.* (1995), Dunlop and Özdemir (1997), Maher and Thompson (1999), or Evans and Heller (2003)

Mineral name	Composition	T_C or T_N	Magnetic structure
Magnetite-ulvöspinel	Fe_3O_4-Fe_2TiO_4	578°C to −155°C	Ferrimagnetic
Magnetite-hercynite	Fe_3O_4-$FeAl_2O_4$	578°C to 339°C	Ferrimagnetic
Magnetite-jacobsite	Fe_3O_4-Fe_2MnO_4	578°C to 350°C	Ferrimagnetic
Magnetite-chromite	Fe_3O_4-Fe_2CrO_4	578°C to 30°C	Ferrimagnetic
Hematite-ilmenite	α-Fe_2O_3-$FeTiO_3$	675°C to −170°C	Canted antiferromagnetic-ferrimagnetic[a]
Maghemite	γ-Fe_2O_3	645°C[b]	Ferrimagnetic
Goethite	α-FeOOH	120°C	Antiferromagnetic
Akagenéite	β-FeOOH	26°C	Antiferromagnetic
Bernallite	$Fe(OH)_3$	154°C	Canted(?) antiferromagnetic[c]
Feroxyhite	δ'-FeOOH	177°C	Ferrimagnetic
Pyrrhotite	Fe_7S_8-$Fe_{11}S_{12}$	325°C	Ferrimagnetic-antiferromagnetic[d]
Greigite	Fe_3S_4	350°C ?[e]	Ferrimagnetic
Iron	Fe	770°C	Ferromagnetic
Lepidocrocite	γ-FeOOH	−196°C	Antiferromagnetic[f]
Siderite	$FeCO_3$	−238°C	Antiferromagnetic[f]
Ferrihydrite	$Fe_5HO_8 \cdot 4H_2O$	−158°C to −248°C[f]	Speromagnetic[f,g]

[a]Above the Morin transition, hematite has a canted antiferromagnetic structure which results in a macroscopic magnetic moment, its structure becomes ferrimagnetic with increasing Ti-content. As in the titanomagnetite solid-solution series, high Ti-contents yield Curie temperatures below room temperature.
[b]Most maghemites invert to hematite before their Curie temperature is reached.
[c]The structure of bernallite is not known with certainty.
[d]Only monoclinic pyrrhotite (Fe_7S_8) is ferrimagnetic at room temperature, the other pyrrhotite structures become ferrimagnetic on heating above ~200°C.
[e]On heating, greigite decomposes before it reaches its Curie temperature. Therefore, the temperature listed should be regarded with great caution.
[f]The last three minerals listed are paramagnetic at room temperature but they order magnetically at low temperatures, so their presence can be demonstrated with magnetic methods. Furthermore, on heating a sample above room temperature these paramagnetic minerals chemically alter into magnetic minerals.
[g]Speromagnetism is characterized by short-range antiferromagnetic spin coupling. On longer ranges the magnetic ordering is random. The range of ordering temperatures of ferrihydrite is related to its crystallinity.

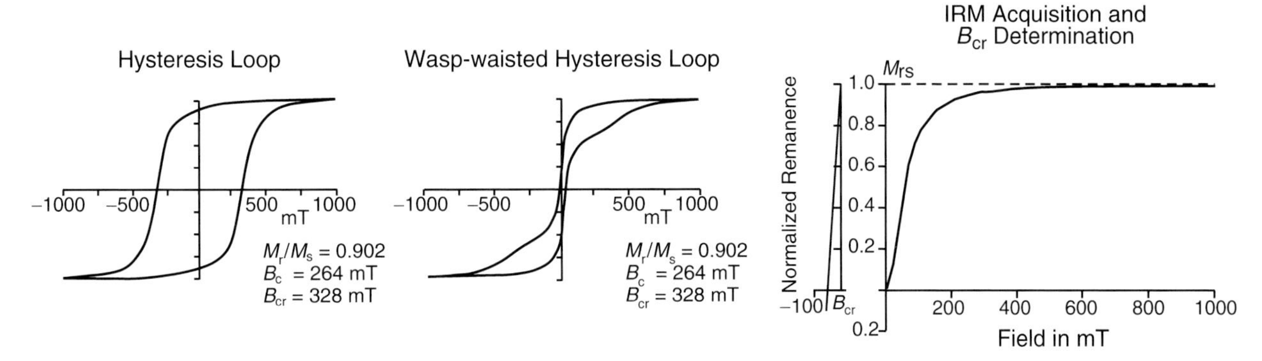

Figure M49 Examples of magnetic hysteresis loops (left, central). (*Left*) Hysteresis loop of an ensemble of single-domain grains. Single-domain grains are characterized by square hysteresis loops; loops of pseudosingle-domain and multidomain grains are increasingly slender and have inclined slopes. (*Central*) So-called "wasp-waisted" hysteresis loop. The central section is smaller than the outer parts. Wasp-waisted loops are typically of mixed ensembles with contrasting coercivities: either a mixture of two magnetic minerals (magnetite and hematite in the present case) or a mixture of superparamagnetic and larger grains of the same mineral. (*Right*) Determination of an IRM acquisition curve and the remanent coercive force.

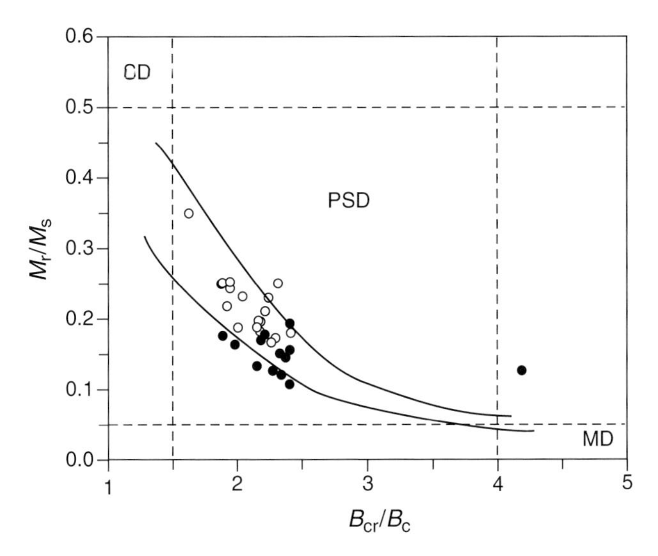

Figure M50 An example of a so-called Day plot. The solid lines represent the experimental calibration by Day *et al.* (1977). SD, single domain; PSD, pseudosingle-domain; MD, multidomain. Data points are from Icelandic lava with stable (open circles) and unstable (closed circles) behavior of the natural remanent magnetization. Dunlop (2002) showed that SP grains plot upward and to the right of the calibrated band.

conforms to the new situation rapidly (= on laboratory timescale). At room temperature, magnetite grains smaller than 20–25 nm in size are superparamagnetic. Single-domain grains contain one magnetic domain and are paleomagnetically most stable (25–80 nm). The stability can be expressed in terms of relaxation time. Pseudosingle-domain grains (80 nm—~10–15 μm) contain "a few" domains (up to ~10); they are paleomagnetically very stable. Multidomain grains (>10–15 μm) contain many domains (>~10); they are paleomagnetically less stable.

The number of domains is a function of the grain's size and shape, and of the saturation magnetization of the magnetic material. In essence, a grain of a material with a high-saturation magnetization like magnetite will contain (much) more domains than a grain with the same size but of material with a low-saturation magnetization like hematite (see Table M3). Therefore we need to know the magnetic mineralogy before we can make a proper grain-size estimate with magnetic methods. Macroscopically, the domain state is reflected in the so-called hysteresis loop (Figure M49 left and central panels), the response of a sample to a cyclic applied magnetic field. If remanent magnetization occurs, the loops are open (the surface area corresponds to the energy contained in the magnetic system). The maximum value of the magnetization is the saturation magnetization (M_s or σ_s), the corresponding remanent magnetization (when the external field is reduced to zero) is termed the remanent saturation magnetization (SIRM or saturation isothermal remanent magnetization, also M_{rs} or σ_{rs}). The field values where M is zero are termed coercive force (B_c or H_c). The field value where the remanence corresponds to zero is labeled remanent coercive force (B_{cr} or H_{cr}, cf. Figure M49 right panel). B_c and B_{cr} go up with decreasing grain size. The M_{rs}/M_s and B_{cr}/B_c ratios are grain-size indicative, their plot against each other is referred to as "Day" plot (Figure M50). Ferro- and ferrimagnetic materials have lower values for their coercive forces than antiferromagnetic materials (see Table M3). Minerals with low coercivity values are termed magnetically "soft" while those with high values are "hard."

It is important to note that coercive force values are concentration independent when magnetic interaction can be neglected. To a first-order description, this is warranted in many natural situations. Under this condition, values of magnetizations and various remanences scale linear with the concentration of the magnetic mineral. By dividing two concentration-dependent parameters, a concentration-independent ratio is obtained that contains information on grain size or the oxidation degree of the magnetic mineral. This idea of correcting for concentration forms the basis of many magnetic proxies.

Magnetic proxy parameters

The concept of using magnetic properties as proxy parameters and correlation tool in a geoscientific context was put forward in the late 1970s mostly by Oldfield and his coworkers; the article in *Science* (Thompson *et al.*, 1980) is often considered as the formal definition of "environmental magnetism" as a subdiscipline. Before briefly highlighting some magnetic proxies, we need to introduce two other important mineral-magnetic parameters, the low-field or initial susceptibility (χ_{in} or κ_{in}) and the anhysteretic remanent magnetization (ARM).

The initial susceptibility is the magnetization of a sample in a small applied field (up to a few times the intensity of the geomagnetic field, 30–60 μT) divided by that field. Measurement takes a few seconds and requires no specific sample preparation; instrumentation is sensitive. This makes initial susceptibility an attractive proxy parameter, especially for correlation purposes. All materials have a magnetic susceptibility, also paramagnets and diamagnets. The initial susceptibility

Table M3 Saturation magnetization, typical coercive forces, remanent coercive forces and field required to reach saturation of some magnetic minerals (at room temperature)

Mineral	Saturation magnetization (A m^2 kg^{-1})	Coercive force[a] (mT)	Remanent coercive force[a] (mT)	Saturation field (T)
Magnetite	92[b]	5–80	15–100	<0.3[c]
Maghemite	~65–74[c]	5–80	15–100	<0.3[d]
Hematite	0.1–0.4	100–500	200–800	~1–~5
Iron	217	<1–10	~5–15	<0.1[d]
Goethite	0.01–1	>1000	>2000	~10–>20
Pyrrhotite	18	8–100	9–115	up to 1[d]
Greigite	~30[e]	15–40	35–70	<0.3[d]

[a]In the indicative ranges given the smallest number refers to multidomain particles and the largest to single-domain particles. Needle-shaped particles of strong magnetic material can have higher values for the coercivity than quoted in the ranges above.
[b]With increasing substitution of Ti, the saturation magnetization decreases.
[c]The ideal maghemite composition has a saturation magnetization of 74 A m^2 kg^{-1}, in practice lower values are often determined.
[d]Oxidized coatings and differential composition often increase coercivity and fields in which saturation is reached.
[e]The purity of greigite samples is problematic, the saturation magnetization value should be considered a minimum.

Table M4 Some important magnetic proxy parameters

Parameter/ratio	Name	Indicative of
χ_{in}	Low-field susceptibility	Combination of ferromagnetism, paramagnetism, and diamagnetism; ferromagnetism (s.l.) when concentration of ferrimagnets is >1%
χ_{hifi}	High-field susceptibility	Paramagnetism and diamagnetism
χ_{ferri}	Ferrimagnetic susceptibility	Ferro-/ferrimagnetism ($=\chi_{in}-\chi_{hifi}$)
χ_{ARM}	Anhysteretic susceptibility	Single-domain material, particularly magnetite and maghemite
HIRM	Hard IRM	Antiferromagnetic minerals: hematite and goethite
S	S-ratio	"Soft" IRM/"hard" IRM; "magnetite"/ "hematite" ratio in terms of magnetic remanence
ARM/SIRM		Grain size
SIRM/χ_{in}		Grain size, for low SIRM values concentration influence as well. Magnetic sulfides have high values
ARM/χ_{in}		Grain size
M_{rs}/M_s		Grain size
B_{cr}/B_c		Grain size
M_s/χ_{ferri}		Superparamagnetic material
χ_{FD}%	Frequency dependence of susceptibility $((\chi_{lowfreq}-\chi_{highfreq})/\chi_{lowfreq})*100\%$	Senses window in the superparamagnetic grain-size range, instrumentally dictated

of a geologic sample is therefore a composite of the diamagnetic, paramagnetic, and ferromagnetic (*sensu lato*) contributions, something which should be kept in mind when interpreting susceptibility profiles. Ferro- and ferrimagnetic materials have very high initial susceptibilities so that for concentrations larger than ~1% the measured susceptibility may be equated to the ferromagnetic susceptibility. For lower concentrations—the rule in sediment and soil samples—the paramagnetic and diamagnetic contributions can be substantial. Superparamagnetic grains have a high value for the initial susceptibility, much higher than the bulk value of the same material. The variation of initial susceptibility with grain size is small for single domain and larger grains. The superparamagnetic threshold size is dependent on the measurement frequency. By measuring at different frequencies the frequency dependence of the initial susceptibility can be calculated. This is a proxy for the relative amount of superparamagnetic grains.

The ARM is induced by an asymmetric alternating field, which is reduced from a peak value to zero. Often equipment for alternating field demagnetization is used, the asymmetry is created by a small direct current bias field. Typical values for the peak alternating field are 100–300 mT, whereas the bias field can vary between 30 and 100 μT. The ARM intensity is a function of the bias field, for low bias fields (say <100 μT) it scales linearly. Hence, by dividing the ARM intensity by the bias field value, a field-independent parameter is created, referred to as ARM susceptibility (χ_{ARM}) through its analogy with the low-field susceptibility. Some instruments allow to have the bias field switched on in desired windows of the decreasing alternating field to sense particular coercivity windows, such ARMs are referred to as partial ARM, labeled pARM. ARM is considered to be a good analog of thermoremanent magnetization created by cooling through the Curie temperature in a small magnetic field. Single-domain grains are sensed in particular by ARM.

Most proxy parameters are based on combinations of low-field susceptibility and/or various remanent magnetizations because they are most quickly measured (measurement of a complete hysteresis loop still takes 3 min on dedicated instrumentation, operator attendance is, however, required). Instrumentation that can measure ARMs and IRMs in a fully automated fashion has meanwhile developed, a distinct advantage (Table M4).

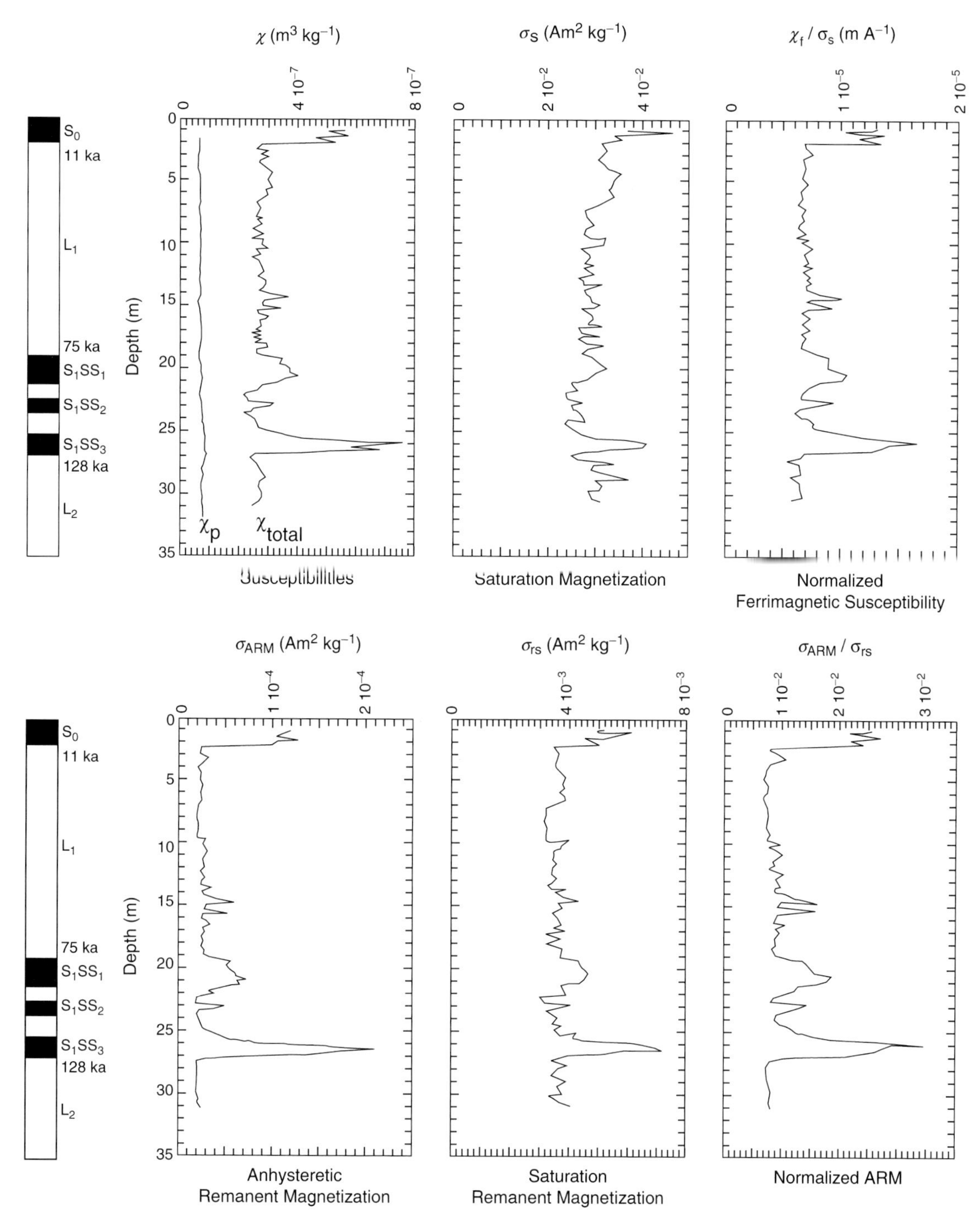

Figure M51 Magnetic properties from the last 140 ka of a loess-paleosol sequence at Xining, western Loess Plateau, China (simplified after Hunt *et al.* 1995). (a) (*Outermost left*): Lithological column with ages for the paleosols. *S*, soil; *L*, loess, numbering starts from the Holocene soil S_0. In the S_1 complex three subunits are distinguished: SS_1, SS_2, and SS_3. (*Left-hand panel*): Susceptibility. χ_p, the susceptibility of paramagnetic minerals (measured in a high-magnetic field), appears to be relatively constant in this profile. χ_{total} refers to the low-field susceptibility. The ferrimagnetic contribution to the low-field susceptibility equals $\chi_{total}-\chi_p$. (*Central panel*): Saturation magnetization which is used as a normalizer to remove concentration effects. (*Right-hand panel*): The ratio χ_f/σ_s, indicative of SP grains, increases in the paleosols. It is also slightly increased in a zone in L_1, which probably indicates a warmer period in the glacial period in which L_1 was deposited. Note that no soil is visually evident in this zone. (b) (*Outermost left*): see Figure M51 (*Left-hand panel*): ARM intensity expressed on a mass-specific basis (σ_{ARM}). ARM values are higher in soils than in loess. (*Central panel*): SIRM intensity expressed on a mass-specific basis (σ_{rs}). Note that the pattern is noisier than that of the ARM and that is differs from that of σ_s (see a) above. (*Right-hand panel*): Higher ratios of σ_{ARM}/σ_r indicate greater abundances of SD and PSD particles in the soils. Not only the SP fraction, but also the SD and the PSD fractions increase in soils. Subtle paleoclimatic variations can also be traced within and between lithological units, e.g., paleosol S_1SS_3 is much more developed than S_1SS_1 and, in particular, than S_1SS_2. The climate change which led to the formation of the present-day soil, S_0, was apparently sharp because no intermediate values (such as for S_1SS_3) are present.

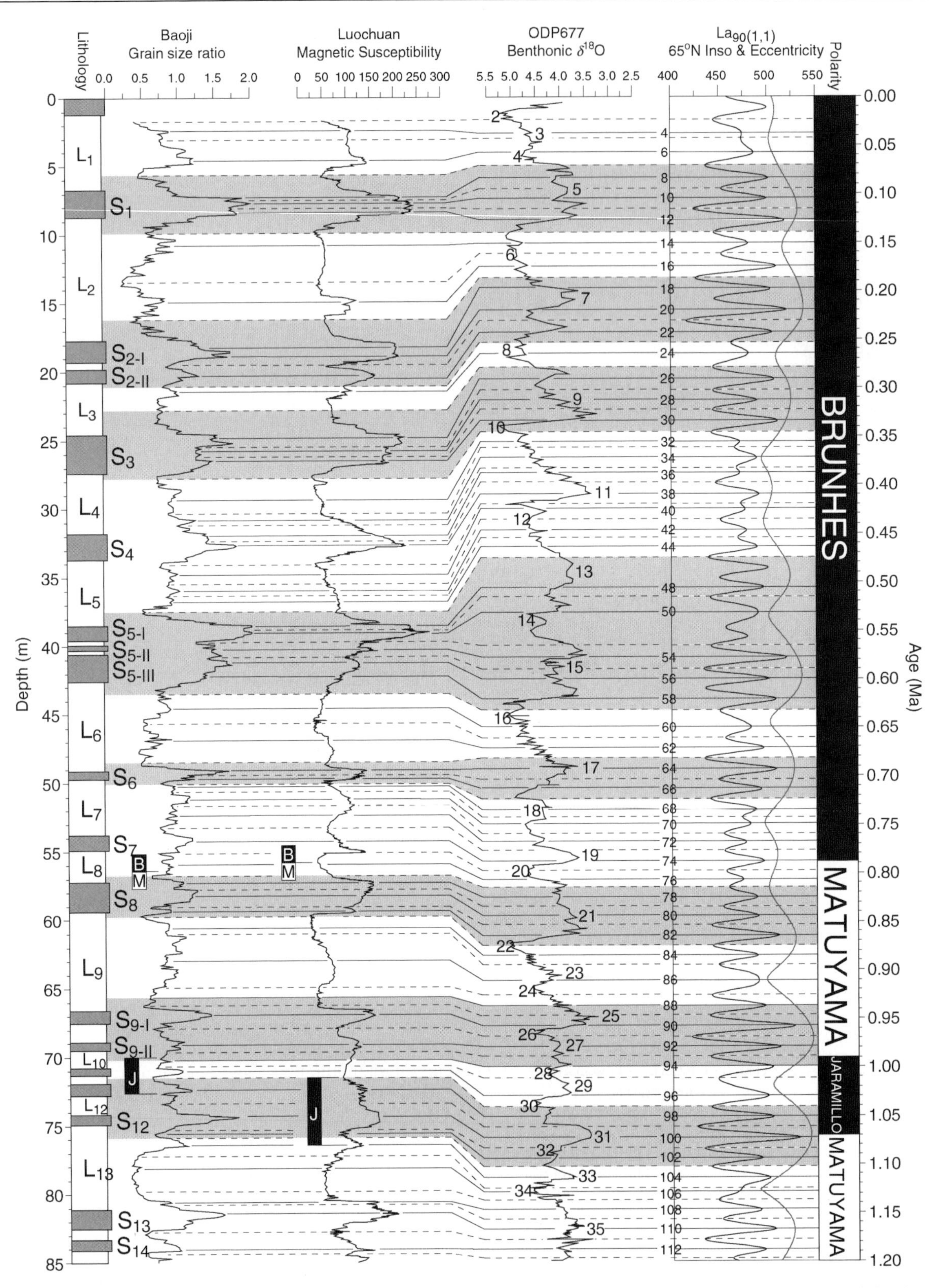

Figure M52 Correlation of the Baoji grain-size ratio (<2 μm (%)/>10 μm (%)) and Luochan low-field magnetic susceptibility (χ_{in}; fitted to the Baoji depth scale) to the 65° N insolation curve of Laskar (1990) (i-cycle numbers shown on the left) and the ODP677 benthonic $\delta^{18}O$ record (Marine isotope stages adjacent to curve). The eccentricity component from the Laskar (1990) solution is displayed for comparison with the longer term trends observed within the loess record. Correlation points linking warm features are shown with full lines while those for cold features are shown with dashed lines.

Some examples

The loess-paleosol climate archive was studied numerous times since Heller and Liu (1982) reported reliable magnetostratigraphic data from the Chinese Loess Plateau (CLP). The climatic conditions on the CLP are broadly determined by the relative intensities of the summer and winter monsoons. The summer monsoon delivers moisture toward the CLP, so strong summer monsoons indicate a wetter (and warmer) climate whereas strong winter monsoons (more and coarser dust) herald dry and cold climate conditions. Hunt *et al.* (1995b) utilized the magnetic proxy approach to highlight subtle differences within loess-paleosol profiles on the CLP (see Figure M51). Mainly qualitative observations in the loess-soil profile were confirmed whereas very subtle differences within the loess horizons were detected. Soils that are indicative for a wetter climate are characterized by higher values for χ_{in}, in line with earlier observations. By using χ_{in} as proxy for the intensity of the summer monsoon and the grain-size ratio as proxy for the winter monsoon, Heslop *et al.* (2000) were able to tune the entire "yellow" loess record (2.6 Ma) to the astronomical solution of Laskar (1990) providing a high-resolution timescale for the loess-paleosol deposits (Figure M52). With this high-resolution timescale, climatic and environmental changes can be put into a regional and also global perspective.

Mineral-magnetic proxy studies from the marine realm are numerous as well. Many studies report on the sensitivity of magnetic proxies to climate change since the mid-1980s. Robinson (1986) showed for North Atlantic sediments fairly close to the Mid-Atlantic spreading ridge west from Spain that concentration-dependent and -independent magnetic parameters and ratios reflect changes in climate caused by changes in provenance of the sediment flux. He argued for an eolian source and found indications for ice-rafted debris. Later work among others by Stoner and colleagues (1996) documented the sensitivity of North Atlantic sediments, in particular from higher latitudes, to record Heinrich events, short-duration rapid climate variations. The minor importance of reductive diagenetic processes in most of these sediments and distinct magnetic properties of the source areas allow interpretation of subtle variations in the record in terms of climate change. Reductive diagenesis obscures the magnetic fingerprint of source areas (e.g., Bloemendal *et al.*, 1992). The presence of pyrite (FeS_2) and magnetic sulfides testifies anoxic conditions. Joint mineral-magnetic and geochemical analysis can be used to evaluate the extent of diagenesis (e.g., Dekkers *et al.*, 1994; Passier *et al.*, 2001). Oxidative diagenesis has a typical magnetic signature (Passier *et al.*, 2001), which was utilized by Larrasoaña *et al.* (2003a) to provide a proxy for bottom-water ventilation in the Mediterranean Sea during sapropel formation (Figure M53). They distinguish three types of sapropel environment: (i) sapropels without oxidation fronts with high organic matter contents inferred to be deposited under severely anoxic conditions; (ii) sapropels with both dissolution and oxidation fronts, the mostly occurring situation referred to as "anoxic"; and (3) sapropels without dissolution front but with a well-developed oxidation front, referred to as "less anoxic." The third group implies no excess sulfide being generated within the sapropel for a dissolution front formed below it. These sapropels are most prone to be reoxidized completely and therefore easily escape visual detection. They can, however, be traced with magnetic (and other) proxies.

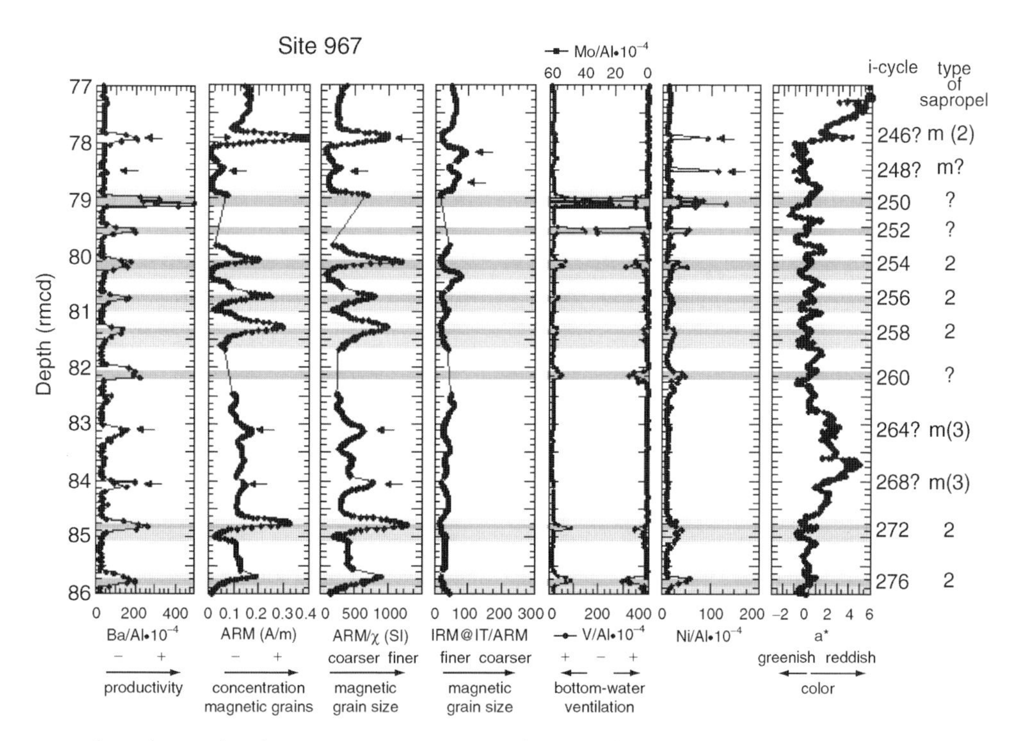

Figure M53 A series of geochemical and magnetic proxies measured on samples from ODP Site 967 (Eastern Mediterranean) with their meaning underneath the abscissas (after Larrasoaña *et al.*, 2003a). Sediments constitute an alternation of marly limestones and marls with sapropels intercalated (sapropels are sediments with more than 2% organic matter and therefore dark-brown to black colored). Sapropeletic sediments are indicated by darkest gray shading, dissolution zones with intermediate gray, and reoxidized zones with lightest gray. The insolation cycle (i-cycle) codification implies an age range from ~2.9 to ~2.5 Ma. The sapropels class into groups (2) and (3). Arrows indicate the position of completely reoxidized (missing, m) sapropels on the basis of magnetic and geochemical proxies. In cases where assignment to a sapropel type is not certain because of lack of magnetic data due to unavailability of the core segment for rock magnetic analysis, a question mark is put.

Mixed magnetic mineralogy, magnetic interaction, and thermal relaxation

Often magnetite-like minerals dominate the magnetic signal. However, more detailed analysis has shown that the occurrence of mixed magnetic mineralogy is the rule rather than the exception. The absolute values of HIRM ("hard" IRM) as an indicator for the amount of hematite have been used as a proxy for dust input by Larrasoaña *et al.* (2003b). They gave the samples at 0.9 T IRM which was demagnetized at 120 mT peak AF to cope with the limited dynamic range of the magnetometer.

Also the *S*-ratio is an example of a parameter to distinguish between "magnetite" and "hematite" ("soft" IRM and "hard" IRM; with respect to a saturating forward IRM and various back field IRMs). In particular, S_{300} is being used as indicator for the magnetite-hematite ratio. S_{300} equals 1 (or -1 depending on the definition, see Bloemendal *et al.*, 1992 and Kruiver and Passier, 2001) would represent magnetite only and that lower values would indicate the presence of hematite. Magnetite fully saturates in a 300 mT field and hematite is (much) harder, yielding lower *S*-ratios. Up to now, the ratio has been used as a qualitative indicator only. The interpretation of the S_{300} in terms magnetite and hematite has been criticized by Kruiver and Passier (2001) who modelled that for oxidized magnetite the forward S_{300} would yield values lower than one and that when more than two magnetic minerals are present the *S*-ratio can be seriously flawed. So, the meaning of lower *S*-ratios needs to be assessed on a case-by-case basis.

Another way to quantify mixed magnetic mineralogy is the so-called IRM component analysis, the decomposition of a measured IRM acquisition curve into several components with the use of model functions. Initially, cumulative log-Gaussian curves were used based on the experimental observation by Robertson and France (1994) that such curves closely conform to measured IRM acquisition curves. Kruiver *et al.* (2001) proposed the use of the *F*-test (and *t*-test where appropriate) to judge upon the number of coercivity fractions required for an optimal fit from a statistical viewpoint. The cumulative

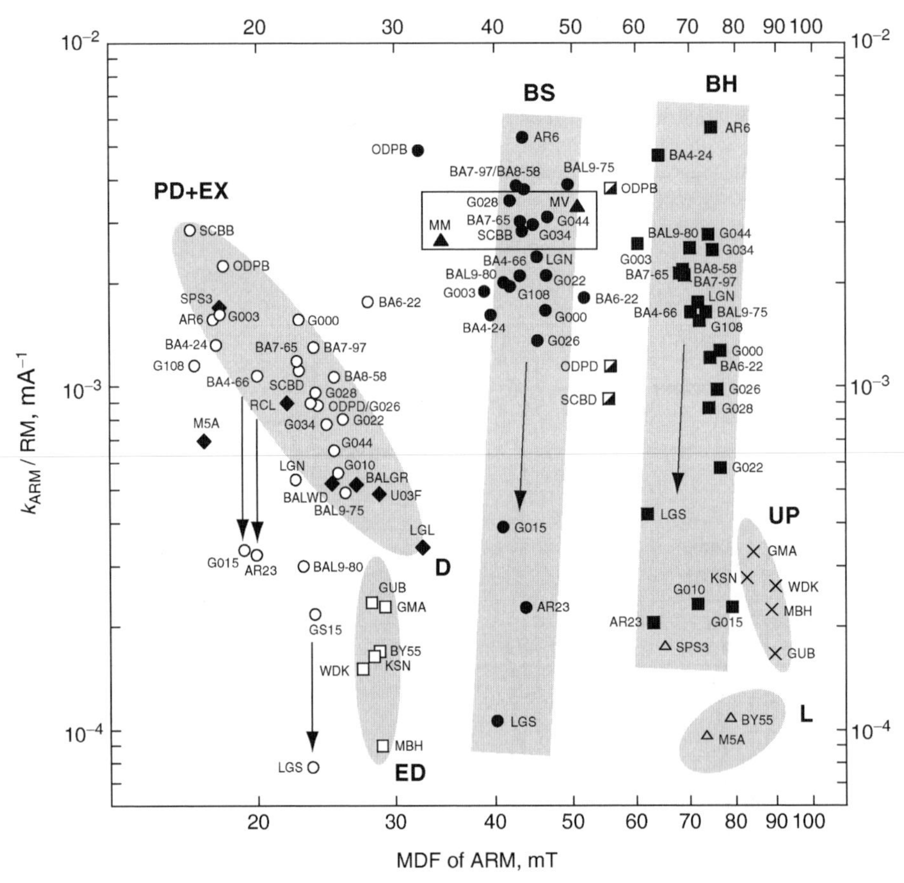

Figure M54 Magnetic fingerprint diagram of MDF_{ARM} vs k_{ARM}/IRM (after Egli, 2004). Summary of the magnetic properties of ARM (300 mT peak AF, 0.1 mT bias field) and IRM (300 mT acquisition field to avoid magnetization that cannot be AF demagnetized) for all iron spinel components identified by Egli (2004). The magnetic components group into different clusters, indicated by gray ellipses and rectangles. White letters classify all components into low-coercivity magnetosomes (biogenic soft, BS: dots), high-coercivity magnetosomes (biogenic hard, BH: squares), ultrafine extracellular magnetite (EX: circles), pedogenic magnetite (PD: diamonds), detrital particles transported in water systems (D: diamonds), windblown particles (eolian dust, ED: open squares), atmospheric particulate matter produced by urban pollution (UP: crosses), and a maghemite component in loess (L: open triangles). Other components, as BM (dots) and BI (half-filled squares), have been measured in a few samples only and are not labeled. The open rectangle indicates the range of values measured in samples of cultured magnetotectic bacteria (triangles). GS15 labels the measurement of extracellular magnetite particles produced by a cultured dissimilatory iron-reducing microorganism. Its properties are influenced by the strong magnetostatic interactions within clumps of these particles which form spontaneously during the sample preparation. Smaller values of MDF_{ARM} and larger values of k_{ARM}/IRM are expected for these particles when they are dispersed in natural sediments: in this case, GS15 would probably fall into the cluster labeled with PD+EX. Arrows indicate the decrease of k_{ARM}/IRM observed during anoxic conditions in lake sediments.

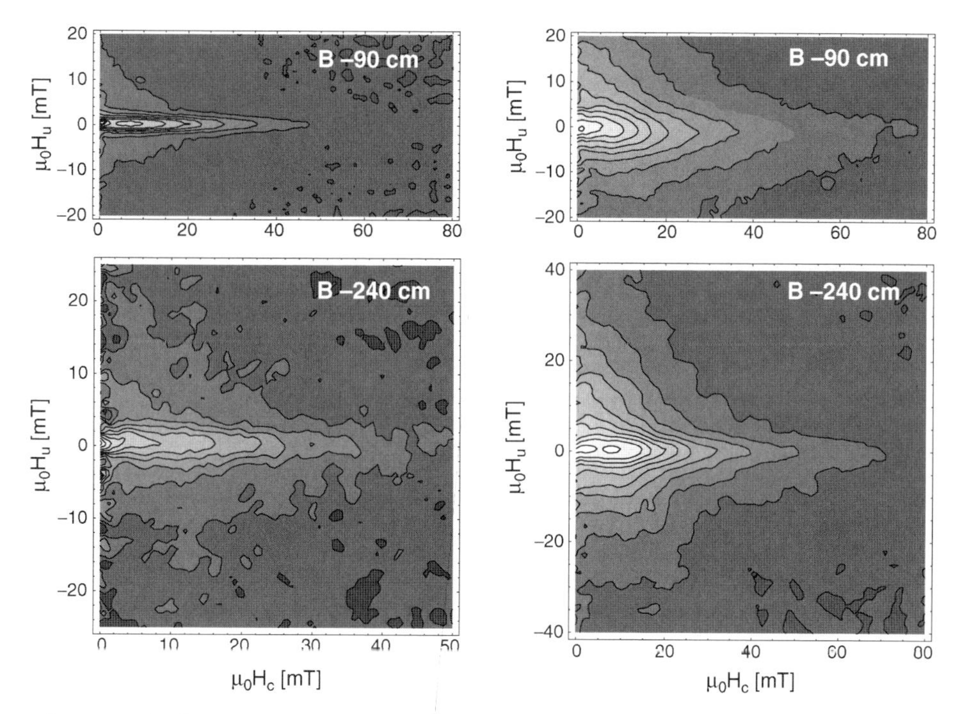

Figure M55 FORC diagrams of paleosols (*top panels*) and loesses (*bottom panels*) from Moravia (Czech Republic) before (*left panels*) and after three times ferrous iron acid-ammonium-oxalate extraction (*right panels*) to dissolve the finest magnetite particles (after Van Oorschot *et al.*, 2002). After extraction, coercivity maxima shift toward the right, indicating dissolution of the finest particles.

log-Gaussian model function would be valid in the absence of magnetic interaction. Egli (2003) extended the use of the model functions to a set of skewed generalized Gaussian functions where skewness and kurtosis are additional free parameters. He showed that many natural samples would require skewed model functions for optimal fits (kurtosis appeared to be not significant). Heslop *et al.* (2004) utilized Preisach modeling to document that thermal relaxation and magnetic interaction would yield negatively skewed distributions already for low levels of interaction and thermal relaxation supporting the observations of Egli (2003, 2004). In Figure M54, the summary diagram of Egli (2004) is reproduced. He processed a series of lake sediments, marine sediments, soils, loesses, polluted samples, and magnetotactic and extracellular bacteria with the procedure described by Egli (2003). He concluded that a plot of the median destructive field (MDF, the peak alternating field strength at which 50% of a remanent magnetization is demagnetized) of the ARM vs the ARM susceptibility divided by the IRM discriminates the grouping most instructively. In many natural samples combinations of these groups are present and observed variability could be described meaningfully by these "endmember groups." It should be kept in mind that the low IRM peak acquisition field (300 mT) used in the approach excludes hematite, goethite, and other antiferromagnets from the analysis.

A nice way to illustrate the occurrence of magnetic interaction is by the so-called first-order reversal curve (FORC) plots (Figure M55), an approach introduced into environmental magnetism by Pike *et al.* (1999) and Roberts *et al.* (2000). On the abscissa coercivity is plotted, whereas on the ordinate magnetic interaction is expressed. Zero magnetic interaction occurs when $\mu_0 H_u$ equals zero, most contour density in this area in the samples before sequential extraction implies minor interaction which was taken by Van Oorschot *et al.* (2002) that IRM component analysis was warranted. The plots are normalized.

Proxy for anthropogenic pollution

A number of studies have produced reasonable correlations between magnetic proxy parameters, in particular low-field susceptibility, and a series of contaminants, either of organic or of inorganic chemical origin. The underlying reason for the magnetic methods to yield fairly consistent results under certain conditions is the sorption capacity of fine-grained iron oxides and sulfides. For airborne particulate matter—for example from industrial fly ash, general industrial activity, or urban traffic—magnetic parameters may provide an estimate of the pollution load (e.g., Dekkers and Pietersen, 1992; Morris *et al.*, 1995; Matzka and Maher, 1999; Xie *et al.*, 2001; Muxworthy *et al.*, 2002). The suitability of magnetic parameters for monitoring of pollution loads in some central European countries has been investigated by the MAGPROX team. In a number of settings magnetic susceptibility is a good indicator of atmospherically derived particulate matter. A recent example of the magnetic expression of the dust load in tree leaves includes the work by Hanesch *et al.* (2003), the impact of an industrial zone could be evaluated straightforwardly. In the same area, a province in Austria, heavy metal concentrations in soils could be traced magnetically by low-field susceptibility patterns (Hanesch and Scholger, 2002).

Summary

Subtle changes in magnetic properties reflect changes in magnetic mineralogy, grain size, oxidation degree, stoichiometry, strain state, etc. These can be understood and explained in terms of changing provenance areas, climatic conditions, diagenetic regimes, and—under certain conditions—anthropogenic pollution. Therefore, combined with analytical assets: sensitivity, rapidity, nondestructiveness, and comparative ease of sample preparation, they are an attractive set of proxy parameters. The interpretive value can be enhanced by utilizing several magnetic proxies together with a few geochemical proxies, for example, elemental analysis. Increased use of magnetic proxies is foreseen as a consequence of methodological advances in unraveling mixed magnetic mineralogy and further establishment of more quantitatively based parameters.

M.J. Dekkers

Bibliography

Bloemendal, J., King, J.W., Hall, F.R., and Doh, S.-J., 1992. Rock magnetism of Late Neogene and Pleistocene deep-sea sediments: relationship to sediment source, diagenetic processes, and sediment lithology. *Journal of Geophysical Research*, **97**: 4361–4375.

Day, R., Fuller, M., and Schmidt, V.A., 1977. Hysteresis properties of titanomagnetites: grain-size and compositional dependence. *Physics of the Earth and Planetary Interiors*, **13**: 260–267.

Dekkers, M.J., Langereis, C.G., Vriend, S.P., van Santvoort, P.J.M., and de Lange, G.J., 1994. Fuzzy *c*-means cluster analysis of early diagenetic effects on natural remanent magnetisation acquisition in a 1.1 Myr piston core from the Central Mediterranean. *Physics of the Earth and Planetary Interiors*, **85**: 155–171.

Dekkers, M.J., and Pietersen, H.S., 1992. Magnetic properties of low-Ca fly ash: a rapid tool for Fe-assessment and a proxy for environmental hazard. In Glasser, F.P. *et al.* (eds.), *Advanced Cementitious Systems: Mechanisms and Properties, Material Research Society Symposium Proceedings*, **245**: 37–47

Dunlop, D.J., 2002. Theory and application of the Day plot (M_{rs}/M_s versus H_{cr}/H_c) 1. Theoretical curves and tests using titanomagnetite data. *Journal of Geophysical Research*, **107**: doi: 10.1029/2001JB000486.

Dunlop, D.J., and Özdemir, Ö., 1997. *Rock Magnetism: Fundamentals and Frontiers.* Cambridge:Cambridge University Press, 573 pp.

Egli, R., 2003. Analysis of the field dependence of remanent magnetisation curves. *Journal of Geophysical Research*, **108**: doi:10.1029/ 2002JB002023.

Egli, R., 2004. Characterization of individual rock magnetic components by analysis of remanence curves. 1. Unmixing natural sediments. *Studia Geophysica et Geodaetica*, **48**: 391–446.

Evans, M.E., and Heller, F., 2003. *Environmental Magnetism—Principles and Applications of Enviromagnetics.* San Diego: Academic Press, 299 pp.

Hanesch, M., and Scholger, R., 2002. Mapping of heavy metal loadings in soils by means of magnetic susceptibility measurements. *Environmental Geology*, **42**: 857–870.

Hanesch, M., Scholger, R., and Rey, D., 2003. Mapping dust distribution around an industrial site by measuring magnetic parameters of tree leaves. *Atmospheric Environment*, **37**: 5125–5133.

Heller F., and T.-S., Liu 1982. Magnetostratigraphical dating of loess deposits in China. *Nature*, **300**: 431–433.

Heslop, D., Langereis, C.G., and Dekkers, M.J., 2000. A new astronomical timescale for the loess deposits of Northern China. *Earth and Planetary Science Letters*, **184**: 125–139.

Heslop, D., McIntosh, G., and Dekkers, M.J., 2004. Using time- and temperature-dependent Preisach models to investigate the limitations of modelling isothermal remanent magnetization acquisition curves with cumulative log Gaussian functions. *Geophysical Journal International*, **157**: 55–63.

Hunt, C.P., Moskowitz, B.M., and Banerjee, S.K., 1995. Magnetic properties of rock and minerals. In Ahrens, T.J. (ed.), *Rock Physics and Phase Relations: A Handbook of Physical Constants*, Vol. 3. American Geophysical Union, pp. 189–204.

Hunt, C.P., Banerjee, S.K., Han, J., Solheid, P.A., Oches, R., Sun, W., and Liu, T., 1995b. Rock-magnetic properties of climate change in the loess-paleosol sequences of the western Loess Plateau of China. *Geophysical Journal International*, **123**: 232–244.

Kruiver, P.P., Dekkers, M.J., and Heslop, D., 2001. Quantification of magnetic coercivity components by the analysis of acquisition curves of isothermal remanent magnetisation. *Earth and Planetary Science Letters*, **189**: 269–276.

Kruiver, P.P., and Passier, H.F., 2001. Coercivity analysis of magnetic phases in sapropel S1 related to variations in redox conditions, including an investigation of the *S*-ratio. *Geochemistry Geophysics Geosystems*, **2**(12): doi:10.1029/2001GC000181.

Larrasoaña, J.C., Roberts, A.P., Stoner, J.S., Richter, C., and Wehausen, R., 2003a. A new proxy for bottom-water ventilation in the eastern Mediterranean based on diagenetically controlled magnetic properties of sapropel-bearing sediments. *Palaeogeography Palaeoclimatology Palaeoecology*, **190**: 221–242.

Larrasoaña, J.C., Roberts, A.P., Rohling, E.J., Winklhofer, M., and Wehausen, R., 2003b. Three million years of monsoon variability over the Northern Sahara. *Climate Dynamics*, **21**: 689–698.

Laskar, J., 1990. The chaotic motion in the solar system: a numerical estimate of the size of the chaotic zones. *Icarus*, **88**: 266–291.

MAGPROX network. http://www.uni-tuebingen.de/geo/gpi/ag-appel/projekte/envmag/magprox/index.html

Maher, B.A., and Thompson, R. (eds.), 1999. *Quaternary Climates, Environments and Magnetism.* Cambridge: Cambridge University Press, 390 pp.

Matzka, J., and Maher, B.A., 1999. Magnetic biomonitoring of roadside tree leaves: identification of spatial and temporal variations in vehicle-derived particulates. *Atmospheric Environment*, **33**: 4565–4569.

Morris, W.A., Versteeg, J.K., Bryant, D.W., Legzdins, A.E., McCarry, B.E., and Marvin, C.H., 1995. Preliminary comparisons between mutagenicity and magnetic susceptibility of respirable airborne particulate. *Atmospheric Environment*, **29**: 3441–3450.

Muxworthy, A.R., Schmidbauer, E., and Petersen, N., 2002. Magnetic properties and Mössbauer spectra of urban atmospheric particulate matter: a case study from Munich, Germany. *Geophysical Journal International*, **150**: 558–570.

Passier, H.F., de Lange, G.J., and Dekkers, M.J., 2001. Rock-magnetic properties and geochemistry of the active oxidation front and the youngest sapropel in the Mediterranean. *Geophysical Journal International*, **145**: 604–614.

Pike, C.R., Roberts, A.P., and Verosub, K.L., 1999. Characterizing interactions in fine magnetic particle systems using first order reversal curves. *Journal of Applied Physics*, **85**: 6660–6667.

Roberts, A.P., Pike, C.R., and Verosub, K.L., 2000. FORC diagrams: a new tool for characterizing the magnetic properties of natural samples. *Journal of Geophysical Research*, **105**: 28461–28475.

Robertson, D.J., and France, D.E., 1994. Discrimination of remanence-carrying minerals in mixtures, using isothermal remanent magnetisation acquisition curves. *Physics of the Earth and Planetary Interiors*, **82**: 223–234.

Robinson, S., 1986. The late Pleistocene palaeoclimatic record of North Atlantic deep-sea sediments revealed by mineral-magnetic measurements. *Physics of the Earth and Planetary Interiors*, **42**: 22–47.

Stoner, J.S., Channell, J.E.T., and Hillaire-Marcel, C., 1996. The magnetic signature of rapidly deposited detrital layers from the deep Labrador Sea: relationship to North Atlantic Heinrich Layers, *Paleoceanography*, **11**: 309–325.

Thompson, R., Bloemendal, J., Dearing, J.A., Oldfield, F., Rummery, T.A., Stober, J.C., and Turner, G.M., 1980. Environmental applications of magnetic measurements. *Science*, **207**: 481–486.

Van Oorschot, I.H.M., Dekkers, M.J., and Havlicek, P., 2002. Selective dissolution of magnetic iron oxides with the acid-ammonium-oxalate/ferrous-iron extraction technique-II. Natural loess and palaeosol samples. *Geophysical Journal International*, **149**: 106–117.

Xie, S., Dearing, J.A., Boyle, J.F., Bloemendal, J., and Morse, A.P., 2001. Association between magnetic properties and element concentrations of Liverpool street dust and its implications. *Journal of Applied Geophysics*, **48**: 83–92.

Cross-references

Archeology, Magnetic Methods
Biomagnetism
Environmental Magnetism
Paleointensity, Relative, in Sediments
Paleomagnetism
Rock Magnetism

MAGNETIC REMANENCE, ANISOTROPY

It has been known since the early 1950s that rocks and sediments display a magnetic anisotropy when constituent mineral grains have a preferred orientation (Graham, 1954; Hargraves, 1959). Magnetic fabric is usually described by the anisotropy of magnetic susceptibility (AMS), measured in weak applied fields. With this method all minerals in a rock or sediment contribute to the susceptibility. Therefore, the observed anisotropy is the sum of the individual mineral components, their specific susc\eptibility anisotropy and their preferred alignment. Sincc nonfcrromagnctic matrix mincrals oftcn makc up thc bulk composition of a rock or sediments, the low-field susceptibility of these minerals can dominate the measured signal. The anisotropy of magnetic remanence (AMR) is only dependent on the ferromagnetic grains (s.l.) in a rock. Since the number of different ferromagnetic phases is more limited, the source of the AMR is easier to distinguish, and the degree of anisotropy is less sensitive to mineral variation (Jackson, 1991).

In early studies several authors noted that certain rocks that display AMS also have natural remanent magnetizations (NRMs) that are shallowly inclined toward the plane of high susceptibility (Howell *et al.*, 1958; Hargraves, 1959; Fuller, 1960). This observation promoted the first studies examining the anisotropy of remanent magnetization. Since then numerous investigations have used the anisotropy of remanent magnetization to determine if the ferromagnetic minerals are preferentially aligned, which may in turn affect the characteristic remanent magnetization (ChRM) of a rock.

Theoretical background

The physical theory of magnetization and the magnetic behavior of minerals is given elsewhere in this volume. Detailed coverage on the theory of AMS and AMR is available in several textbooks and review papers (Jackson, 1991; Rochette *et al.*, 1992; Tarling and Hrouda, 1993; Borradaile and Henry, 1997; Tauxe, 2002). A cursory description necessary to understand the basis of anisotropy of remanent magnetization follows.

The magnetization $\mathbf{M}$ of a material is a vector sum of two components: the induced magnetization $\mathbf{M}_i$ and remanent magnetization $\mathbf{M}_r$,

$$\mathbf{M} = \mathbf{M}_i + \mathbf{M}_r$$

The induced magnetization can be expressed as

$$\mathbf{M}_i = \chi\mathbf{H},$$

where χ is the susceptibility, a material constant that can be described by a tensor of second order, and $\mathbf{H}$ is the inducing field. The remanent magnetization is the magnetization that remains after the material is removed from an applied magnetic field and is carried by ferrimagnetic minerals.

A mathematical description of a weak-field magnetic remanence is analogous to AMS. This has been described in Stacey and Banerjee (1974) as

$$\mathbf{M}_r = k_r\mathbf{H},$$

where k_r is the remanent susceptibility, which is also a material constant that can be described by a tensor of second order for linearly anisotropic magnetizations.

$$k_r = \begin{bmatrix} k_{r11} & k_{r12} & k_{r13} \\ k_{r21} & k_{r22} & k_{r23} \\ k_{r31} & k_{r32} & k_{r33} \end{bmatrix}$$

It can be represented geometrically by an ellipsoid with principal axes parallel to the eigenvectors of k_r and lengths equal to the corresponding eigenvalues, where $k_{r1} \geq k_{r2} \geq k_{r2}$. It must be noted that $\mathbf{M}_r$ should be both linearly dependent on the intensity of the inducing field and reversible.

In a rock or sediments, we are interested in the total contribution of all ferrimagnetic minerals to the observed signal. If a material is isotropic, i.e., the minerals show no preferred orientation, then $\mathbf{M}_i$ will be along the direction of the applied field and $\mathbf{M}_r$ will be in the direction of the field that produced the remanence. If, however, the minerals in a rock or sediment have a preferred orientation then the acquired magnetization will lie between the applied field direction and the direction of the easy axis of magnetization. The easy axis of magnetization may be related to the shape of the grain, as is the case for ferrimagnetic spinel. Minerals that display shape anisotropy have a high spontaneous magnetization and a slight departure from equidimensional habit leads to the magnetization being preferentially aligned along the long axis of a grain. Crystallography may also control the direction of magnetization, due to the geometric distribution of atoms within the crystallographic lattice. The ferrimagnetic minerals, hematite and pyrrhotite, are controlled by magnetocrystalline anisotropy. Compared to the value along the *c*-axis, the susceptibility in the basal plane is two orders of magnitude higher in hematite and three orders of magnitude higher in pyrrhotite. Stress within a mineral may also lead to an anisotropic acquisition of magnetization, known as magnetostriction. It arises from the strain dependence of the anisotropy constants and has been observed in titanomagnetite.

Types of remanent magnetization

A rock or sediment can acquire different types of remanent magnetizations. Some types of remanence are acquired by natural processes, whereas other types are imparted under laboratory conditions. When rocks form from molten lava or sediments are deposited, the magnetic moments of ferromagnetic minerals align statistically with the prevailing Earth's magnetic field and receive a remanent magnetization, known as the NRM.

The NRM that is found in rocks that are formed from molten lava is called a thermal remanent magnetization (TRM). When individual ferrimagnetic minerals cool through their Curie temperature or Néel temperature, the ferrimagnetic grains take on the magnetization of the ambient magnetic field. TRM is very stable over time. Sediments acquire their remanent magnetization as they are deposited. The ferrimagnetic grains align statistically in the ambient field as they settle and the acquired magnetization is called a detrital remanent magnetization (DRM). As the grains are cemented in the matrix of the sediment, the resulting magnetization is a postdetrital remanent magnetization (pDRM).

It is also possible to give a material a TRM in the laboratory by heating the material above the Curie or Néel temperature of the constituent ferrimagnetic phases and cooling it in a field of known direction and intensity. Other laboratory-acquired remanent magnetizations include anhysteretic remanent magnetization (ARM) and isothermal remanent magnetization (IRM). An ARM is acquired when an alternating field (H_{AC}) is superimposed on a weak DC bias field, which serves to impart a magnetization in a known direction. All ferrimagnetic minerals with a coercivity less than or equal to H_{AC} will be remagnetized. Anhysteretic remanence shows grain-size dependence in magnetite (Figure M56a), such that single-domain grains (0.03–0.05 μm) have a higher remanence than larger multidomain grains. Grains of superparamagnetic size effectively carry no remanent magnetization at temperatures where they are superparamagnetic.

An IRM is acquired in a strong DC field (H_{IRM}, generally >10 mT) of known intensity and direction. All ferrimagnetic grains with a coercivity less than or equal to H_{IRM} will be remagnetized. Since IRM is a strong-field remanence the imparted magnetization is not linearly related to H_{IRM}. If the applied field is below saturation and if only

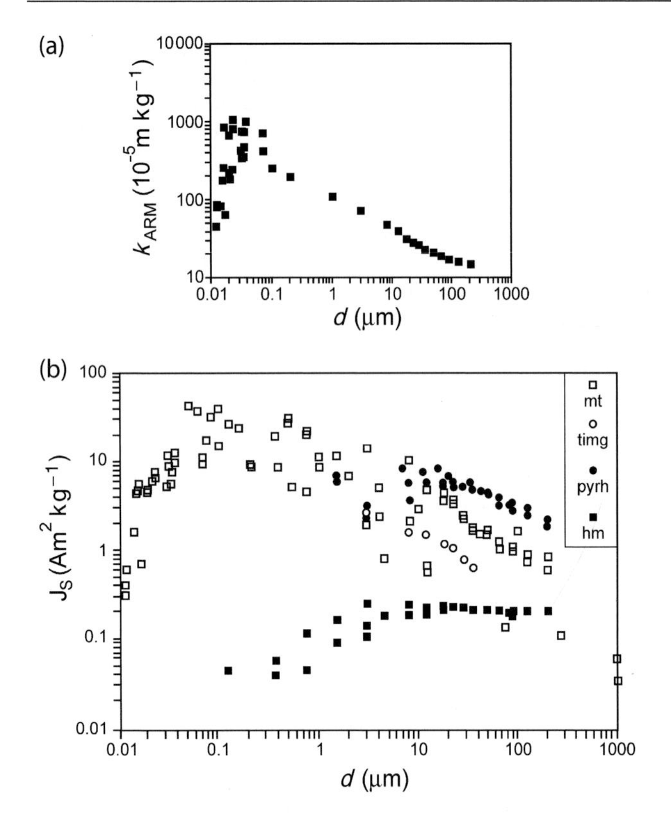

Figure M56 (a) Variation in the intensity of anhysteretic remanent susceptibility (k_a, intensity of ARM normalized for the applied DC field) as a function of grain size in magnetite. (b) Variation in intensity of saturation IRM in different magnetic minerals as a function of grain size (after Jackson, 1991).

the first-order terms are considered, a symmetric matrix can be assumed for constant H_{IRM} (Cox and Doell, 1967; Stephenson *et al.*, 1986; Jackson, 1991). Saturation IRM also shows a grain-size dependence for (titano-)magnetite, hematite, and pyrrhotite (Figure M56b).

Measurement of the anisotropy of magnetic remanence

All methods that measure the AMR require a laboratory magnetization to be imparted along a set of known directions. The most commonly applied laboratory magnetization is an ARM, although several studies have used other types of magnetization, such as IRM and TRM. Since the remanent anisotropy is a symmetric tensor of second rank, at least six independent measurements must be made to define the tensor. Many measurement schemes for the AMS have been proposed over the years (see Borradaile and Henry, 1997), based on early suggestions of Nye (1957), who suggested 13 positions (Figure M57, excluding position 5 and 9), and Girdler (1961), who suggested nine positions (positions 1–4, 6–8, 10, 11). Cox and Doell (1967) first described a method to measure the anisotropy of IRM (AIRM), using 15 independent positions (1–11, with positions 1, 3, 5, and 7 measured twice). McCabe *et al.* (1985) proposed using ARM to define the remanent anisotropy; they used the same positions as used by Girdler (1961) to measure AMS.

Independent of the type of remanence that is applied and the number of positions that are used, a general procedure is followed. First a sample is demagnetized and its remanent magnetization is measured as a basis for the "demagnetized" or natural remanent state. A laboratory magnetization is then applied along the first axis, the imparted remanent magnetization is measured, and the component due to the imparted magnetization is found by vector subtraction of the natural remanence. The sample is again demagnetized to remove the imparted magnetization and the demagnetized state is measured as a control. This procedure is repeated along all the selected field positions, and the measurements are used to define the matrix k_r. Differences between the best-fit ellipsoid and the individual measurements give the residuals, and these can be minimized by using a least-squares computation. Jackson (1991) proposed that a goodness-of-fit (GoF) can be determined from

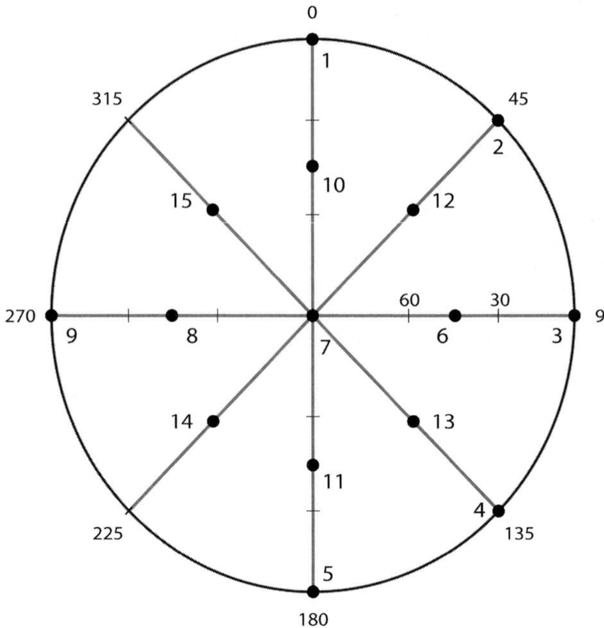

Figure M57 Lower hemisphere, equal area plot that shows the directions of magnetic field used to investigate the acquisition of anisotropic remanent magnetization.

$$\text{GoF} = \text{RMS}/(k_{r1} - k_{r3}) \times 100\%$$

where RMS is the root mean squared residuals normalized to k_{mean}. The expression is an estimate of how well the anisotropy ellipsoid is resolved. The ellipsoid can be considered well resolved for GoF values under 10%; typically values are on the order of a few percent.

Stephenson *et al.* (1986) suggested an alternative method to obtain the ellipsoid of anisotropy of remanence. They measured the individual diagonal and off-diagonal matrix elements directly. Here a laboratory remanent magnetization is applied along the sample *x*-axis and the three components of the magnetization are measured. The sample is demagnetized and the same field is next applied along the sample *y*-axis and all the three components of the magnetization are measured. After demagnetization the field is now applied along the sample *z*-axis and the magnetization is measured. This yields nine values that overdefine the tensor and yield a measure of the precision of fit. This method is probably reliable for IRM, since the imparted remanence is relatively large, allowing for accurate determination of the magnetization components oriented normal to the applied field direction. It is less reliable for ARM and TRM where imparted remanences are relatively weak.

Anisotropy of anhysteretic remanent magnetization (AARM)

The anisotropy of ARM has also been called the anisotropy of anhysteretic susceptibility (AAS), since normalizing the magnetization by the field strength converts it to a pseudosusceptibility. An ARM is acquired from the DC bias field, often on the order of 0.05–0.1 mT, superimposed on the AC field. The strength of the bias field is on the order of one to two times that of the Earth's magnetic field. The intensity of the AC field can be chosen as desired, but is generally less than 300 mT, due to limits of commercially available AF demagnetizing units. This limitation means that an ARM can only be imparted to minerals with low coercivity, such as (titano-)magnetite or pyrrhotite. A detailed description of the measurement procedure is outlined above and has been given in McCabe *et al.* (1985).

Since the coercivity of ferromagnetic minerals, particularly magnetite and titanomagnetite, is dependent on grain size, it is possible to impart an ARM to a particular grain-size fraction during demagnetization in the DC bias field. This is accomplished by switching on the DC bias field in a specific AC field range or coercivity window during demagnetization. The imparted magnetization is known as a partial ARM (pARM). Jackson *et al.* (1988) demonstrated that the method can be used to distinguish if a specific grain-size fraction is responsible for the observed AARM. For example, a high coercivity window (e.g., 80–150 mT) will only magnetize single-domain magnetite, whereas a low coercivity window (e.g., 0–30 mT) would magnetize larger multidomain grains (see "Geological Applications" sections for a practical example).

Anisotropy of isothermal remanent magnetization (AIRM)

The advantage of AIRM is that a relatively strong DC field is applied to the sample, such that higher coercivity phases may also be magnetized. As stated above, IRM is not linearly related to the applied field; however, a symmetric matrix can be assumed if only the first-order terms are considered. Jelinek (1996) proposed a nonlinear AIRM model that more precisely describes the magnetization phenomena, in which the anisotropy is also described by a symmetric second-order tensor. The typical strength of the applied field is between 5 and 60 mT (Tarling and Hrouda, 1993). Cox and Doell (1967), who proposed the method originally, used an applied field of 700 mT, which is above the saturation of the ferromagnetic phases. The measurement procedure is the same as outlined above. The main difficulty with their method is the complete demagnetization between each imparted magnetization step. If lower fields are used, it should be possible to demagnetize with alternating field (AF) demagnetization. Cox and Doell (1967) rotated their samples in the 700 mT field of an electromagnet to randomize the magnetization, since complete randomization of magnetic domains is reversible.

As with AARM, Cox and Doell (1967) suggested that it is possible to examine a partial IRM. In their case, they proceeded as follows. After measuring the sample magnetized in a 700 mT field, the sample can be subjected to AF demagnetization in a significantly smaller field, e.g., 100 mT. The remaining magnetization is then remeasured. The sample can then be further demagnetized in a stronger field, e.g., 200 mT, and then remeasured. In this manner a set of deviation vectors is produced,

$$D = J_r - J_m H/H$$

where J_r is the measured remanent magnetization, which is produced in the field H, and J_m is the mean amplitude of a set of J_r vectors averaged over all the results from the different directions of the applied field. This gives a set of $D(H')$ for each value of the alternating field demagnetization (H').

Several cautionary notes should be made concerning the method and the magnetization process in AIRM. The first caution is with regard to the accuracy with which a field strength can be applied. A field strength between 5 and 60 mT is in a range where the magnetization of a sample is quickly acquired for (titano-)magnetite and iron sulfides (Figure M58). Slight variations in field strength may lead to apparent anisotropy that is not related to the preferred orientation of ferromagnetic grains. Figure M58 shows a typical IRM acquisition curve for a pelagic limestone. In this example a variation in IRM intensity on the order of 2%, comparable to the required resolution in the study of AMR, would occur if the field strength varied by only ±2 mT.

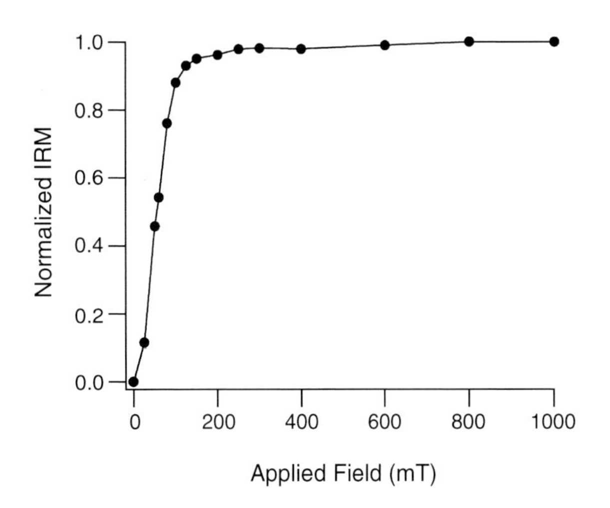

Figure M58 Example of an IRM acquisition curve for a magnetite-bearing, pelagic limestone.

Caution must also be exercised when measuring the AIRM in pyrrhotite-bearing rocks. A study by de Wall and Worm (1993) demonstrated that the shape of the ellipsoid and degree of anisotropy are dependent on the applied field in rocks containing pyrrhotite. Although the orientations of the principal axes are not affected, they showed that the degree of anisotropy in a sample of isometric pyrrhotite ore varied between 3.8 and 1.05 for applied field strengths between 1.5 and 100 mT, respectively. The authors attributed this effect to the control of the observed anisotropy by magnetocrystalline anisotropy in low applied fields, and by magnetostatic anisotropy, which arises from the cylindrical sample shape, in strong fields. Jackson and Borradaile (1991) reported a similar dependence of AIRM and field strength in hematite-bearing slates, but attributed the effect to different coercivity fractions in the rocks.

It has been noted by Tauxe *et al.* (1990) that following repeated application of an IRM, the coercivity of the specimen increased after exposure to the first high field, so that in spite of demagnetization to remove the previous magnetization, some grains affected by the first field application were no longer affected by subsequent field applications. The authors attributed this effect to hematite grains with metastable domains, which change domain state during the AIRM experiment.

Anisotropy of thermal remanent magnetization (ATRM)

Heating a sample above the Curie or Néel temperature of its ferromagnetic phases and cooling it in an applied field will produce a TRM. As outlined in the general procedure, the sample must be first demagnetized. This can be achieved either with AF demagnetization or by heating the rock above its Curie temperature (T_C) or Néel temperature (T_N), and cooling in zero field. The rock is then magnetized in the first field direction by reheating above T_C or T_N and cooling in a field of fixed intensity and rate of cooling. It is important that the field intensity and rate of cooling is the same in all heating runs. After measurement of the imparted TRM, the sample is again demagnetized. At least six independent heating steps are necessary to define the tensor, 12 if the sample cannot be demagnetized in alternating fields and must be reheated above the blocking temperature. It is important to control that no chemical alteration has occurred, which could produce new ferromagnetic minerals and affect the acquired TRM. Often a thermomagnetic curve is made on a small piece of the sample to check that the ferromagnetic mineralogy is thermally stable.

As for a pARM, a partial TRM (pTRM) can be acquired if the field applied during cooling is turned on only when cooling through a particular temperature range. This requires accurate monitoring of the temperature during cooling.

Geological applications

Several examples are presented below to illustrate how AMR can differ from AMS and the type of geologic information that can be obtained from remanent fabrics. The most common method used in examining AMR is AARM. The experimental method is relatively simple and this magnetization method avoids many problems that have been found in applying IRM or TRM. The method is suitable for magnetite- and possibly pyrrhotite-bearing rocks, but not for high coercivity phases, such as hematite. In this case, AIRM and ATRM are more suitable methods. An example is given for AMR in hematite-bearing rocks. In the examples below, the magnitudes of the principal axes of the anisotropy ellipsoids are defined as $k_1 \geq k_2 \geq k_3$.

AMS and AARM

If the same processes leading to grain orientation affect the ferromagnetic and paramagnetic minerals in a rock, one would expect that the magnetic fabric of both components should be similar. The orientation of the AARM ellipsoid is close to the AMS ellipsoid, and the main difference between the two fabrics is found in terms of the degree of anisotropy, lineation, and foliation. If the two mineral components formed at separate times in the history of the rock or if deformation mechanisms acting on the different mineral components varied, then it is possible that the susceptibility and remanence magnetic fabrics could be different.

The first example illustrates how rock rheology can lead to two different magnetic fabrics. The Lower Paleozoic stratigraphic sequence from the Central Appalachian Fold and Thrust Belt includes the Lower Ordovician Coburn limestone, which is overlain by the Reedsville/ Martinsburg shale in central Pennsylvania, USA. Both lithologies underwent the same deformation during the Alleghenian orogeny. The AMS was measured with an Agico KLY-2 susceptibility bridge, and the susceptibility fabric in the limestones can be characterized by a prolate ellipsoid with k_1 subparallel to pencil structure in the overlying shales, which is also the structural trend of the major fold structures, and k_2 and k_3 distributed in a girdle normal to the k_1 axes (Figure M59). The prolate fabric is the result of a horizontal tectonic compaction superimposed on the vertical bedding compaction. The shales, on the other hand, have oblate AMS ellipsoids with k_3 subparallel to the pole to bedding and k_1 aligned within the bedding plane, subparallel to the pencil structure or fold axis. The bedding compaction controls the susceptibility fabric in the shales.

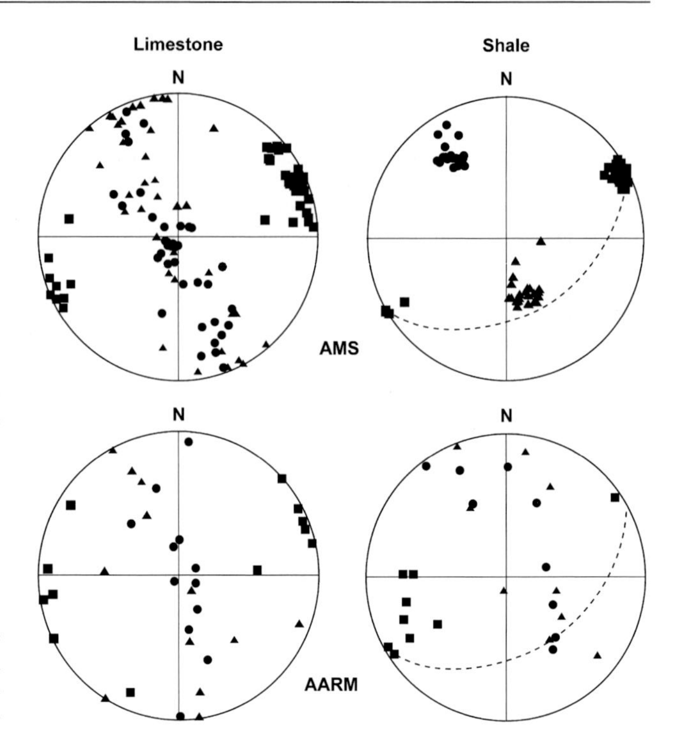

Figure M59 Lower hemisphere, equal area plot showing the orientation of the principal axes of the AMS (upper plots) and AARM (lower plots) for the Coburn limestone (*left*) and Reedsville/ Martinsburg shale (*right*). The principal axes k_1 are squares, k_2 are triangles, and k_3 are circles in this and subsequent figures. Bedding plane in shales is shown with the dashed line.

Magnetite is the major ferromagnetic phase in both lithologies. Anisotropy of ARM was investigated using a 0.1 mT DC bias field with a 150 mT alternating field, and the measurement procedure of McCabe *et al.* (1985). The AARM of the limestones is similar to the AMS fabric both in the orientation of the principal axes and the shape of the ellipsoid (Figure M59). This suggests that the susceptibility and remanent fabrics in the limestone are both controlled by magnetite. The AARM in the shales, however, is characterized by prolate ellipsoids, in which k_1 is well-grouped and subparallel to the pencil directions and k_2 and k_3 are distributed in a girdle in a plane with k_1 as its pole (Figure M59). This remanent fabric shows that the magnetite grains have been more strongly affected by the horizontal compaction. The difference between the AMS and AARM fabrics can be explained by the fact that paramagnetic clays and phyllosilicates dominate the AMS fabric, which is largely controlled by bedding compaction. The magnetite grains acted as rigid particles in a passive matrix during deformation, and were therefore quicker to respond to the horizontal compression.

In the second example the AARM was determined in the Ordovician Martinsburg Formation from Lehigh Gap in eastern Pennsylvania using ARMs applied in two different coercivity ranges. Samples were taken from the part of the outcrop in which pencil structures are well developed and a bedding and incipient cleavage can be identified. Magnetite and pyrrhotite are the main ferromagnetic phases in the rocks, as determined from IRM acquisition and thermal demagnetization of a multiple component IRM. The AARM was first determined using a 0.035 mT DC bias field with an alternating field of 30 mT, and then using a 0.1 mT DC bias field with an alternating field of 100 mT, in which the DC bias field was turned on between 60 and 100 mT. The first application would only affect coarser ferromagnetic grains, whereas the higher coercivity range would selectively magnetize finer grain sizes (Figure M56a). Figure M60a shows that the low

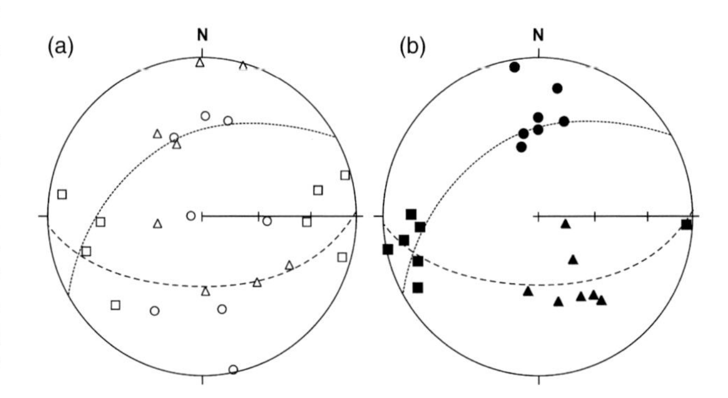

Figure M60 Lower hemisphere, equal area plots of the AARM for (a) low-coercivity range (0.035 mT bias field, 0–30 mT AF), and (b) high-coercivity range (0.1 mT bias field, 60–100 mT AF). The bedding plane is shown with the dotted line and the cleavage plane with the dashed line.

coercivity AARM fabric is not well defined, but the anisotropy is loosely related to tectonic compaction. This suggests that the coarser grain were less efficient in orienting in the strain field. The high coercivity fabric is well-defined and characterized by a triaxial ellipsoid, in which the k_3 axes are subparallel to the pole to cleavage, and k_1 axes mirror the pencil lineation.

AIRM

Tan and Kodama (2002) investigated paleomagnetic directions in the Mississippian Mauch Chunk Formation in eastern Pennsylvania. They found that the inclination of the ChRM was shallower than the expected direction and used AMR to see if the ferromagnetic grains showed preferential flattening in the bedding plane. Hematite is the sole ferromagnetic carrier in the red beds, and the ChRM was isolated in thermal demagnetization above 670°C. IRM acquisition curves indicate that 85%–90% of the saturation IRM is activated in an applied field of 1.2 T. The AMS in the rocks has a degree of foliation on the order of 1.01 to 1.04, which would not result in a significant inclination flattening. An IRM was applied along nine directions following the method of McCabe *et al.* (1985) using a 1.2 T impulse field, and the tensor solution was found using a method outlined in Tauxe (2002). Samples were heated to 670°C and cooled in zero field, so that only the ChRM-bearing hematite retains an IRM. In this manner the magnetic fabric due to the grains that are responsible for the ChRM can be isolated. The AIRM shows that the k_3 axes are well-grouped along the pole to bedding in both the undemagnetized and thermally demagnetized samples (Figure M61). The k_1 and k_2 axes lie in the bedding plane and are slightly better grouped in the undemagnetized samples. Both sets of data show the same degree of lineation, but the demagnetized samples show a slightly higher degree of foliation (Figure M61). In both AIRM cases the degree of flattening is almost an order of magnitude higher than in the AMS. Tan and Kodama (2002) used the AIRM and AMS data to obtain an individual particle anisotropy factor, which was then used to correct the paleomagnetic inclination. They could show that, after correction, the expected inclination was in good agreement with a European igneous paleopole of the same age.

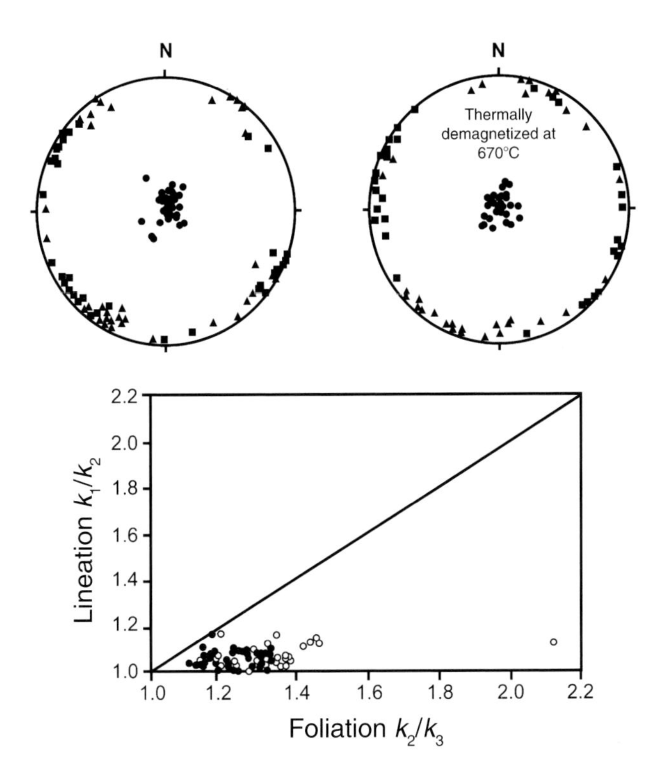

Figure M61 Lower hemisphere, equal area plots of the AIRM for undemagnetized samples (upper left) and thermally demagnetized samples (upper right) from red beds of the Mauch Chunk Formation. Principal axes ratios (lower plot) for the undemagnetized (solid circles) and demagnetized (open circles) samples (after Tan and Kodama, 2002).

AMS, AARM, and ATRM

Selkin *et al.* (2000) made a combined study of AMS, ATRM, and AARM in an anorthositic layer from the Upper Banded Series of the Precambrian Stillwater complex. The susceptibility fabric, which was measured with a KLY-2 susceptibility bridge, is compatible with the preferred orientation of plagioclase crystals in the samples, where k_3 is normal to the foliation plane of the rock and k_1 is close to the *c*-axis orientation of plagioclase (Figure M62a). Thermal demagnetization data indicates that titanium-poor magnetite is the main ferromagnetic phase in the samples. The AARM was measured in a 0.05 mT DC bias field and 180 mT alternating field using six different positions. The ATRM was determined by heating the samples to 600°C and cooling in a 25 μT field in air; six positions were used to define the tensor. The two different remanent fabrics are similar, but distinctly different from the susceptibility fabric (Figure M62b,c). The AARM and ATRM ellipsoids are prolate with k_1 well-grouped, coinciding with the pole to the foliation plane defined by the plagioclase crystals and k_3 constrained in the direction of the plagioclase *c*-axes. The authors interpret these results as showing that single-domain magnetite grains are responsible for the remanent anisotropy. Needles of single-domain magnetite have been identified by TEM in plagioclase in various intrusions, including

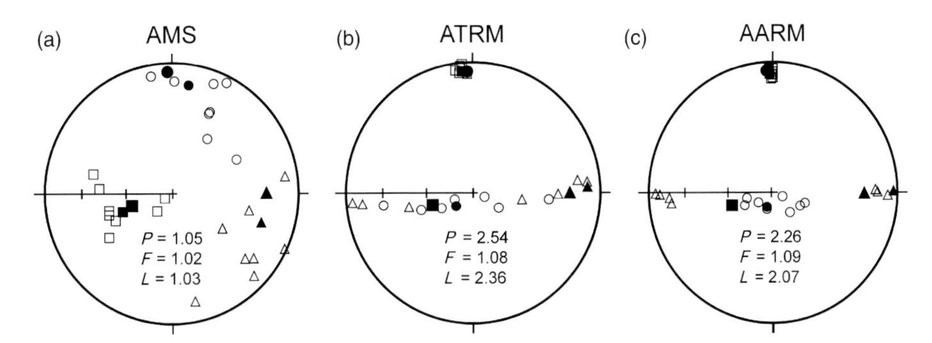

Figure M62 Lower hemisphere, equal area plot for the magnetic fabric of an anorthosite sample, measured by (a) AMS, (b) ATRM, and (c) AARM. Large black symbols show the principal axes of the plagioclase fabric, and small black symbols are the site mean averages of the individual sample measurements (open symbols). P is the degree of anisotropy (k_1/k_3), L is the degree of lineation (k_1/k_2), and F is the degree of foliation (k_2/k_3) (after Selkin *et al.*, 2000).

the Stillwater complex, and often the needles are perpendicular to the plagioclase *c*-axes. Further support for this interpretation was obtained from hysteresis loops made along the direction of the principal axes of the remanence ellipsoids. Single-domain magnetite could also be responsible for an inverted magnetic fabric in susceptibility anisotropy. However, the same fabric would be obtained if paramagnetic minerals are responsible for the AMS. The AMR was used to correct estimates of paleointensity, which were affected by the preferential alignment of the magnetite grains.

Ann M. Hirt

Bibliography

Borradaile, G.J., and Henry, B., 1997. Tectonic applications of magnetic susceptibility and its anisotropy. *Earth-Science Reviews*, **42**: 49–93.

Cox, A., and Doell, R.R., 1967. Measurement of high-coercivity magnetic anisotropy. In Collinson, D.W., Creer, K.M., and Runcorn, S.K. (eds.), *Methods in Paleomagnetism.* Amsterdam: Elsevier, pp. 477–482.

de Wall, H., and Worm, H.-U., 1993. Field dependence of magnetic anisotropy in pyrrhotite: effects of texture and grain shape. *Physics of the Earth and Planetary Interiors*, **76**: 137–149.

Fuller, M.D., 1960. Anisotropy of susceptibility and the natural remanent magnetization of some Welsh slates. *Nature*, **186**: 790–792.

Girdler, R.W., 1961. The measurement and computation of anisotropy of magnetic susceptibility of rocks. *Geophysical Journal of the Royal Astronomical Society*, **5**: 34–45.

Graham, J.W., 1954. Magnetic susceptibility anisotropy, an unexploited petrofabric element. *Bulletin of the Geological Society of America*, **65**: 1257–1258.

Hargraves, R.B., 1959. Magnetic anisotropy and remanent magnetization in hemo-ilmenite ore deposits at Allard Lake, Quebec. *Journal of Geophysical Research*, **64**: 1565–1578.

Howell, L.G., Martinez, J.D., and Statham, E.H., 1958. Some observations on rock magnetism. *Geophysics*, **23**: 285–298.

Jackson, M., 1991. Anisotropy of magnetic remanence: a brief review of mineralogical sources, physical origins, and geological applications, and comparison with susceptibility anisotropy. *Pure and Applied Geophysics*, **136**: 1–28.

Jackson, M.J., and Borradaile, G., 1991. On the origin of the magnetic fabric in purple Cambrian slates of north Wales. *Tectonophysics*, **194**: 49–58.

Jackson, M., Gruber, W., Marvin, J., and Banerjee, S.K., 1988. Partial anhysteretic remanence and its anisotropy: Applications and grain size-dependence. *Geophysical Research Letters*, **15**: 440–443.

Jelinek, V., 1996. Theory and measurement of the anisotropy of isothermal remanent magnetization of rocks. *Travaux Géophysique*, **37**: 124–134.

McCabe, C., Jackson, M., and Ellwood, B.B., 1985. Magnetic anisotropy in the Trenton limestone: results of a new technique, anisotropy of anhysteretic susceptibility. *Geophysical Research Letters*, **12**: 333–336.

Nye, J.F., 1957. *Physical Properties of Crystals.* London: Oxford University Press.

Rochette, P., Jackson, M.J., and Aubourg, C., 1992. Rock magnetism and the interpretation of anisotropy of magnetic susceptibility. *Reviews of Geophysics*, **30**: 209–226.

Selkin, P.A., Gee, J.S., Tauxe, L., Meurer, W.P., and Newell, A.J., 2000. The effect of remanence anisotropy on paleointensity estimates: a case study from the Archean Stillwater Complex. *Earth and Planetary Science Letters*, **183**: 403–416.

Stacey, F.D., and Banerjee, S.K., 1974. *The Physical Principles of Rock Magnetism*. Amsterdam: Elsevier Scientific Publishing Company.

Stephenson, A., Sadikun, S., and Potter, D.K., 1986. A theoretical and experimental comparison of the anisotropies of magnetic susceptibility and remanence in rocks and minerals. *Geophysical Journal of the Royal Astronomical Society*, **84**: 185–200.

Tauxe, L., Constable, C., Stokking, L., and Badgley, C., 1990. Use of anisotropy to determine the origin of characteristic remanent magnetization in the Siwalik red beds of northern Pakistan. *Journal of Geophysical Research—Solid Earth and Planets,* **95**: 4391–4404.

Tan, X.D., and Kodama, K.P., 2002. Magnetic anisotropy and paleomagnetic inclination shallowing in red beds: evidence from the Mississippian Mauch Chunk Formation, Pennsylvania. *Journal of Geophysical Research—Solid Earth*, **107**B11):2311, doi: 10.1029/2001JB001636.

Tarling, D.H., and Hrouda, F., 1993. *The Magnetic Anisotropy of Rocks.* London: Chapman & Hall.

Tauxe, L., 2002. *Paleomagnetic Principles and Practice.* Dordrecht: Kluwer Academic Press.

Cross-references

Magnetic Susceptibility, Anisotropy
Magnetization, Anhysteretic Remanent (ARM)
Magnetization, Isothermal Remanent (IRM)
Magnetization, Natural Remanent (NRM)
Magnetization, Thermoremanent (TRM)

MAGNETIC SHIELDING

There are numerous uses for magnetic shields in research and consumer electronics. In geophysical research, magnetic shields are used in paleomagnetic laboratories to protect samples and improve the performance of sample magnetometers. Research shields range in size from 10 cm to 15 m.

The irony in magnetic shielding theory is that a shield actually adds another magnetic field. This new field is specifically designed to be equal and opposite to the existing field, resulting in a nullification of, not a removal of the ambient magnetic fields. This is analogous to gravitational forces at Lagrangian points, or the balance of centripetal and gravitational forces on orbiting bodies, or noise cancellation by antiphase headphones.

Materials used

Diverse materials and techniques are used for magnetic shielding. The choice depends on the kind of fields to be opposed (e.g., AC vs DC) and the specific application. In general, shielding can be accomplished by using one of two physical phenomenon, either electrical conductivity or strong magnetic responses. Electrical conductors carry currents that produce a magnetic field perpendicular to the current direction. To generate these currents in a conductive sheet, there must be an original, varying magnetic field. Within the conductor, these currents produce a new magnetic field in the opposite direction to this exciting field (Lenz's law). By surrounding a site with a good conductor, the volume will be shielded from varying fields coming from all directions. The important parameters for efficient shielding by conductors are: conductivity and thickness of the shielding material, frequency of the magnetic field, and the completeness of enclosure. Of course, electrical wires can be powered externally to produce a magnetic field of almost any configuration, including the nullification case of shielding. This will be considered in detail later.

Materials with strong magnetic properties, usually metals containing Fe, Ni, or Co, can react to changing magnetic fields by generating an oppositely directed field (magnetic permeability) or can be tuned to oppose a steady field (magnetic remanence). Efficient shielding by magnetic materials depends on the material, its treatment history, as well as the ambient field characteristics of frequency, strength, gradient, and source direction.

Sources of magnetic fields

The goal of magnetic shielding can be either to nullify ambient magnetic fields (external source shielding, described in detail later) or to nullify local, instrument-generated magnetic fields (internal source shielding). Many consumer electronics and medical diagnostic instruments generate large local magnetic fields. Here the goal of shielding is to protect other circuits and conductors within the instruments from unwanted induced currents, and to allow the normal functioning of other, nearby electronic equipment. The larger the source field, the more important shielding becomes. Significant care and expense is taken to shield the immediate environment from monitors and TVs (cathode-ray tube types), and nuclear magnetic resonance (NMR) instruments. Since these internal sources generate rapidly varying magnetic fields, the shielding material must have either high magnetic permeability or high conductivity.

External source shielding has the challenge of encountering a different magnetic environment (differences in frequency, strength, and direction) in each location. In theory, the entire range of the electromagnetic spectrum could be shielded against, but an equally broad range of shielding materials would need to be used simultaneously. Great design simplifications can be realized by restricting the shielding to a specific frequency or a narrow range of frequencies.

Shielding in paleomagnetic research

Paleomagnetic research uses shields to protect samples during remanence measurements and during demagnetization experiments. In general, shielding design focuses on the reduction of DC fields (stable in the time domain). Most laboratory sites have large field gradients owing to the steel in pipes, ducts, building framework, furniture, and research equipment. Where space permits, some laboratories have fabricated specific, isolated buildings without steel components, to create a low-gradient environment. Isolation from varying field sources is another critical design feature. These sources can be: vehicles (moving or parked), elevators (especially piston types), steel doors, steel containers, mass spectrometers, magnetic mineral separators, circuit panel boxes, and electric-powered mass transit.

Three different shielding techniques are used in paleomagnetic laboratories: coils, high-permeability magnetic sheets, and remanent magnetic sheets. Current flow in coils of wire can be adjusted to produce any magnetic field desired, including a field that matches, but is in the opposite direction to, a component of the ambient field. Sets of coils can be configured to control a component of the field over a specified volume, e.g., a pair of coils, spaced apart a distance equal to their radius (the Helmholtz arrangement). For a review of other multicoil arrangements see Kirschvink (1992). The advantages of coil systems are: relatively low cost, open arrangement, flexible positioning, and adjustable field strength (net fields other than zero). Disadvantages are: small-shielded volume relative to apparatus size (shielded volume to coil volume about 1:100), the need for a large DC power supply, continuous cost for electrical power, and difficult access to the shielded center of a three-component set.

High-permeability magnetic materials are extensively used for research shielding. Nickel-rich metals are common, along with newly developed materials such as glass and amorphous metals. Rolled into cylinders, these shields can be arranged concentrically to provide excellent shielding. Design considerations for these cylindrical shields are: orientation (optimum direction of cylinder axis is perpendicular to the field), open or closed ends (open-ended cylinders have higher fields at least 2 diameters inside), and treatment history (both metallurgical and magnetic degaussing). Small room-sized shields have been made from high-permeability sheet material (Patton and Fitch, 1962). There are hybrid designs for these larger shields that utilize low-cost remanent materials for the bulk of the shielding. The high-permeability magnetic materials are used only as an inner shield, taking advantage of their ability to react to (shield against) weak and variable magnetic fields. The advantages of high-permeability magnetic materials are: effective shielding over a wide range of frequency and field strengths, including DC fields. Disadvantages are: cost, annealing and processing restrictions on size and shape, and permeability reduction if deformed.

Remanent magnetic materials have been extensively used (since 1980) for large volume, room-sized shields (Scott and Frohlich, 1985). Originally developed for the transformer industry, silicon steel alloys are used as magnetostatic shields in many paleomagnetic laboratories. Magnetostatic or DC shielding incorporates a series of design simplifications that allow the fabrication of large (15 m) and user-friendly enclosures (open doorways and windows). Therefore, an entire paleomagnetic laboratory along with thousands of specimens can be contained inside a shielded enclosure. Remanent shielding materials must be magnetically tuned (using anhysteretic remanence) to generate the balanced opposing (or shielding) field. The advantages of remanent magnetic materials are: low cost, wide range of size, and high magnetic field strength. Disadvantages are: postfabrication tuning (remagnetization) is required, and ineffective shielding against variable magnetic fields.

One distinguishing feature of shields that use magnetic materials is the concentric layers of shielding material, referred to as shielding stages. While the typical two-stage shield provides a sufficient field reduction for many applications, shields with up to six stages have been fabricated for specific research applications. Each shielding stage can only respond to the magnetic field where it is located, so that interior stage shields respond to much smaller fields than the initial (outermost) shield. Therefore, interior shields are less effective in both absolute strength and percent reduction in residual fields. This is caused by the reduced magnetic alignment in weaker magnetic fields, called the Langevin alignment function. This applies to both high-permeability materials (permeability is much smaller in lower fields) and in remanence shields (the smaller biasing field lowers the strength of anhysteretic remanence). Using a three-stage shield as an example, we start with the outer stage in the Earth's field (50000 nT), which can reduce the interior field by 95% (down to 2500 nT). The middle stage is in the 2500 nT field, and can reduce the interior field by 90% (down to 250 nT). The inner stage is in the 250 nT field, and can reduce the interior field by 80% (down to 50 nT). Most shields are assembled starting with the outer shield, progressively adding stages to the interior. The residual interior field provided by a magnetic shield is proportional to both the initial strength of the ambient magnetic field, and to the magnitude of pre-existing field gradients.

The spacing between shielding stages is approximately 10% of the shield diameter. Why so far apart? And why not fill the interval between the stages with shielding material? Wills (1899) examined these questions in detail for both the two- and three-stage cases. Counterintuitively, his numerical (2D) models showed that for two-stage shields the residual field was five times higher if the area between the shielding stages was solid. Theoretically, this decreased shielding with increased mass is analogous to the decreased field from a single-domain particle as it increases in volume to a multidomain state. Thick magnetic material will magnetically interact with itself, in a process of self-cancellation that produces a small external field. Likewise, the spacing between stages allows each stage to be magnetically independent, instead of being linked by dipolar interaction into a lower energy state. Wide spacing between stages is also an important design feature when large field gradients are present.

If the goal is to shield against high-frequency fields (>100 Hz), then thick electrically conductive sheets are used. Thickness is an advantage in this case, where the conductivity phenomenon of skin-depth is important. Induced electric currents should be able to flow uninterrupted in any direction, thus generating the opposing magnetic field. With conductive materials, multiple shielding stages are not as useful as concentrating the material in a single, thick layer. At low frequencies (<100 Hz), the rate of change of the magnetic field will not be able to generate sufficient current flow to create the opposing magnetic field. In the case of biological research, the shielding must respond to a wide range of frequencies, therefore dual material shields are usually fabricated. One of the design challenges is that these two materials react orthogonally. Conductive material react to magnetic fields that arrive perpendicular to the sheet, as compared to magnetic material that react to magnetic fields that arrive parallel to the sheet.

Gary R. Scott

Bibliography

Kirschvink, J.L., 1992. Uniform magnetic fields and double-wrapped coil systems: improved techniques for the design of bioelectromagnetic experiments. *Bioelectromagnetics*, **13**: 401–411.

Patton, B.J., and Fitch, J.L., 1962. Design of a room-size magnetic shield. *Journal of Geophysical Research*, **67**: 1117–1121.

Scott, G.R., and Frohlich, C., 1985. Large-volume magnetically shielded room, a new design and material. In Kirschvink, J.L., Jones, D.S., and Macfadden, B.J. (eds.), *Magnetite Biomineralization and Magnetoreception in Organisms.* New York: Plenum Press, pp. 197–220.

Wills, A.P., 1899. On the magnetic shielding effect of trilamellar spherical and cylindrical shells. *Physical Review*, **9**: 193–213.

MAGNETIC SURVEYS, MARINE

Definition

A marine magnetic survey is the measurement of Earth's magnetic field intensity or its components (such as vertical component) along a series of profiles over an area of interest with the objective of measuring the magnetism of the ocean floor. The seafloor is known to be highly magnetized in a pattern of magnetic stripes that record the history of Earth's magnetic field behavior, which in turn, allows marine scientists to determine the age and history of seafloor spreading in the ocean basins (see *Vine-Matthews-Morley hypothesis*). Magnetic surveys are also used to locate a concentration or absence of magnetic minerals in the search for economic ore deposits (Sheriff, 1981). Variations in the observed magnetic field from the predicted regional field are called magnetic anomalies and are attributed to variations in the distribution of materials having different magnetic polarity, remanent magnetization, and/or susceptibility. The SI units of measurement for Earth's magnetic field intensity are nanoteslas (nT) or in the older cgs system gammas, where 1 gamma is equal to 1 nT. Earth's magnetic field intensity varies between 30000 and 60000 nT and magnetic anomalies measured at sea are typically in the range of 1–1000 nT.

History of marine magnetic surveying

Magnetic field measurements have been made at sea since the early days of marine exploration, although these consisted primarily of directional measurements for declination and inclination (see *History of geomagnetism, Voyages making geomagnetic measurements, Halley, Carnegie Institute of Terrestrial Magnetism*). One of the first observations that magnetic fields at sea may be caused by seafloor geologic sources was made by Cole (1908) who compiled a declination chart of the western Atlantic Ocean around Bermuda in which "sources of considerable disturbance" could be discerned. The research vessel (R/V) Carnegie (see *Carnegie research vessel*) carried out the first global marine magnetic surveys for the purpose of mapping Earth's magnetic field. During World War II, the rise of submarine warfare precipitated the development of magnetic airborne detectors (MAD); a self-orienting fluxgate magnetometer flown by an aircraft above the ocean to detect submerged submarines (Rumbaugh and Alldredge, 1949). This development allowed for continuous magnetic field intensity measurements to be made at sea. After the war, one of these magnetometers was modified into a towed marine magnetometer (Heezen *et al.*, 1953) and in 1948 the first shipboard magnetic anomaly profile was collected across the Atlantic Ocean from Dakar to Barbados with R/V Atlantis. The profile crossed the then unknown Mid-Atlantic Ridge and recorded a strong anomaly over the rift valley that is now known to be the Brunhes central anomaly that marks the locus of the active mid-ocean ridge spreading center where new crust is being created at the seafloor. Strong magnetic lows were also found over the flanks of the ridge, which are now known to be the reversed crust of the Matuyama Chron and again part of the seafloor spreading magnetic stripes (see *Vine-Matthews-Morley hypothesis*).

In 1954, Arthur Raff at Scripps Institution of Oceanography began using a fluxgate magnetometer based on the Lamont design. Raff worked with Russell Varian and Martin Packard to calibrate the ship-towed magnetometer against their newly developed proton precession magnetometer that provided an absolute value of Earth's magnetic field intensity (Glen, 1982). Proton precession magnetometers do not suffer from drift problems or require calibration like the fluxgate sensors and is the standard tool of today's magnetic surveying practice. These early surveys proved successful and Arthur Raff teamed up with geophysicist Ronald Mason visiting from Imperial College in London to tow their magnetometer as part of a large bathymetric survey off the west coast of North America using the US Coast and Geodetic Survey (USCGS) ship Pioneer. This survey was published by Raff and Mason (1961) and became one of the key data sets leading to the acceptance of the ideas about how the marine magnetic stripe pattern was recorded by the ocean crust and subsequently preserved by the seafloor spreading process (Dietz, 1961; Hess, 1962).

Today, modern marine magnetometers have a precision of between 1 and 0.1 nT using proton precession sensors, optically pumped cesium magnetometers or more recently, the Overhauser nuclear precession magnetometer (see *Observatory instrumentation*). While magnetometers are still towed behind ships for conventional surveys, magnetic measurements are also made from a variety of seagoing platforms including deep-diving submersibles (Macdonald *et al.*, 1983), deep-towed sleds (Macdonald, 1977), remotely operated vehicles (ROVs) (Johnson *et al.*, 2002), and autonomous underwater vehicles (AUVs) (Tivey *et al.*, 1997). Airborne magnetic surveys are also an important method of surveying particularly over the sediment-dominated continental shelf areas in the hunt for natural resources of economic value such as oil and gas. The US Naval Oceanographic Office also ran an airborne magnetic survey program called Project MAGNET for many years with the aim of mapping Earth's magnetic field intensity on a global scale (see *Project MAGNET*).

Methodology, analysis, and processing

Sea surface marine magnetic surveys are typically obtained by towing a magnetometer sensor behind a ship. The sensor is towed at the end of a cable with electrical conductors that enable the sensor to be powered and communicated to from the ship (Figure M63). The sensor is towed between 3 and 6 ship lengths behind the ship in order to reduce the magnetic effects of the ship on the measurement. The optimal lateral line spacing for a survey is directly related to the water depth because this is the filtering distance over which the magnetic field has propagated. The typical depth of the ocean ranges between 2 and 6 km so that a line spacing of 2 to 10 km is usually sufficient for most surveys (Figure M64). Magnetic surveying is often done in conjunction with other geophysical techniques such as swath bathymetric surveying, gravity and seismic profiling, any of which may determine the ultimate line spacing required. The best approach to magnetic surveying is to collect profiles perpendicular to the magnetic "grain" or striping that characterizes the ocean basins.

The first step in processing is the removal of noise spikes that may have been produced from external sources, such as motors and generators. At equatorial latitudes it is also usually necessary to remove the diurnal variation in Earth's magnetic field due to the combined effects

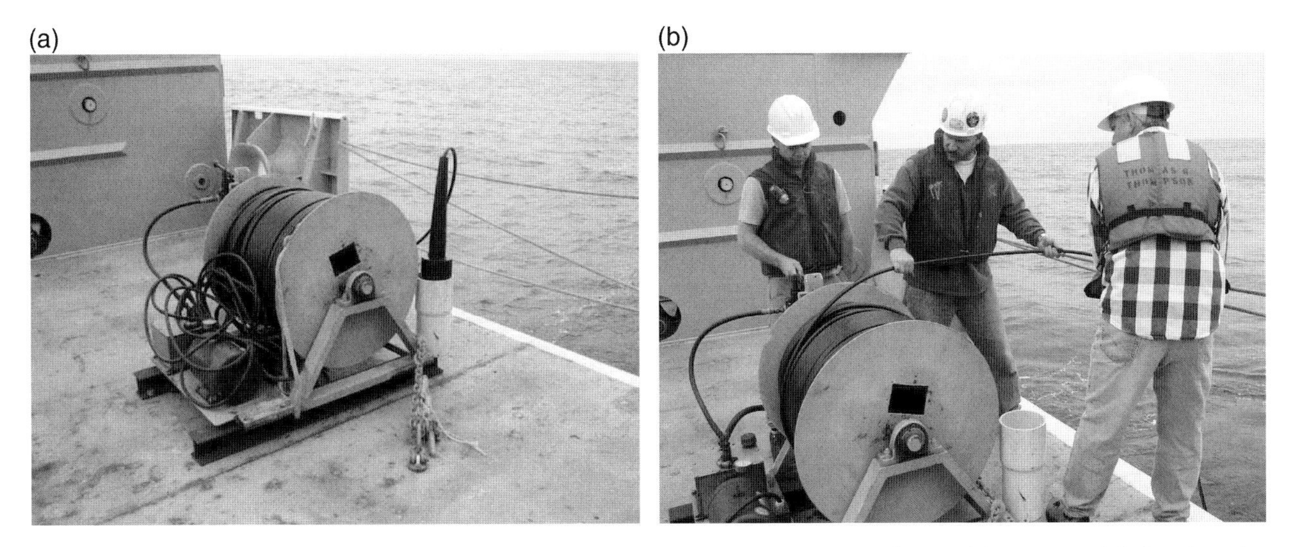

Figure M63/Plate 9a (a) Photograph showing a marine proton precession magnetometer and tow cable on a winch before launch. (Photo taken by Maurice Tivey, Woods Hole Oceanographic Institution.) (b) Photograph showing the deployment of a marine proton precession magnetometer from the stern of a research vessel. (Photo taken by Maurice Tivey, Woods Hole Oceanographic Institution.)

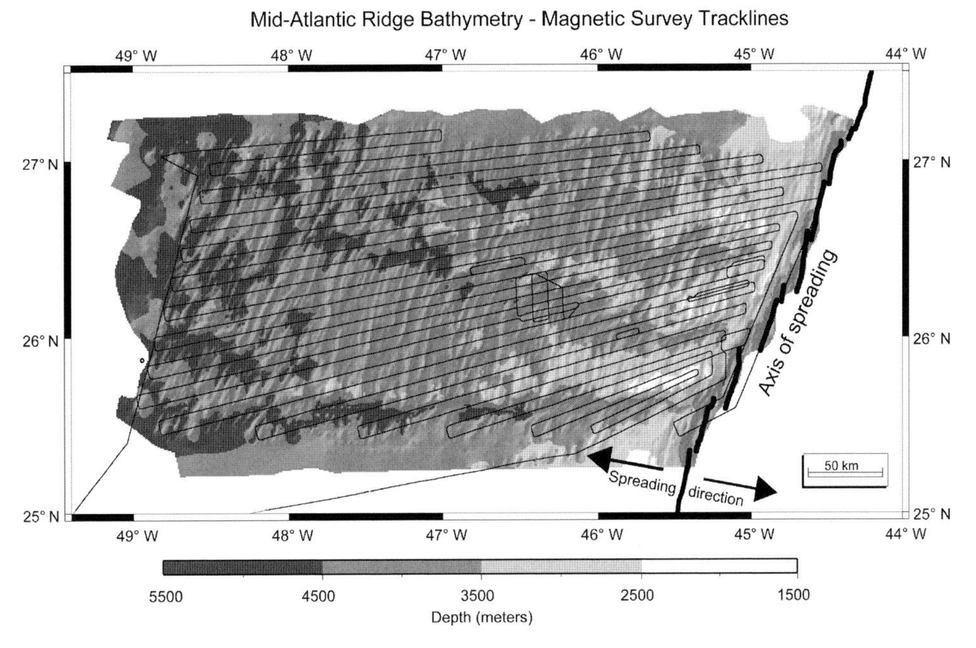

Figure M64 Map showing the seafloor depth for a portion of the western flank of the Mid–Atlantic Ridge near 26° N. Bold black lines indicate the location of the mid–ocean ridge spreading axis and arrows indicate direction of seafloor spreading. Data from Tivey and Tucholke (1998). Thin lines indicate shiptracks.

of the equatorial electrojet and the weak equatorial geomagnetic field. Diurnal records can be obtained from nearby geomagnetic observatories on land or with a temporarily installed base station on the seafloor. The next step is to merge the magnetic time series data with the ship's position. In the modern era this is easily accomplished with Global Positioning Satellite (GPS) data, but in the past this was perhaps the largest source of error, especially in remote areas of the ocean. Once the position of the magnetic field measurement is known the regional magnetic field is removed to reveal the residual "magnetic anomaly" field. Typically, the predicted regional field is obtained from the global International Geomagnetic Reference Field (see *IGRF, DGRF*) spherical harmonic model for the appropriate date and geographical location of the survey. The suitably processed data can then be plotted as wiggle plots (Figure M65) or depending on the data density gridded into a contour map (Figure M66).

Magnetic anomaly highs represent places where the local magnetic field exceeds the regional field and magnetic lows are where the local magnetic field is less than the regional field. For marine magnetic surveys these high- and low-magnetic anomalies correspond to magnetized zones of normal and reverse polarity ocean crust, respectively. The source of marine magnetic anomalies is primarily the volcanic rocks that comprise ocean crust and typically not the overlying

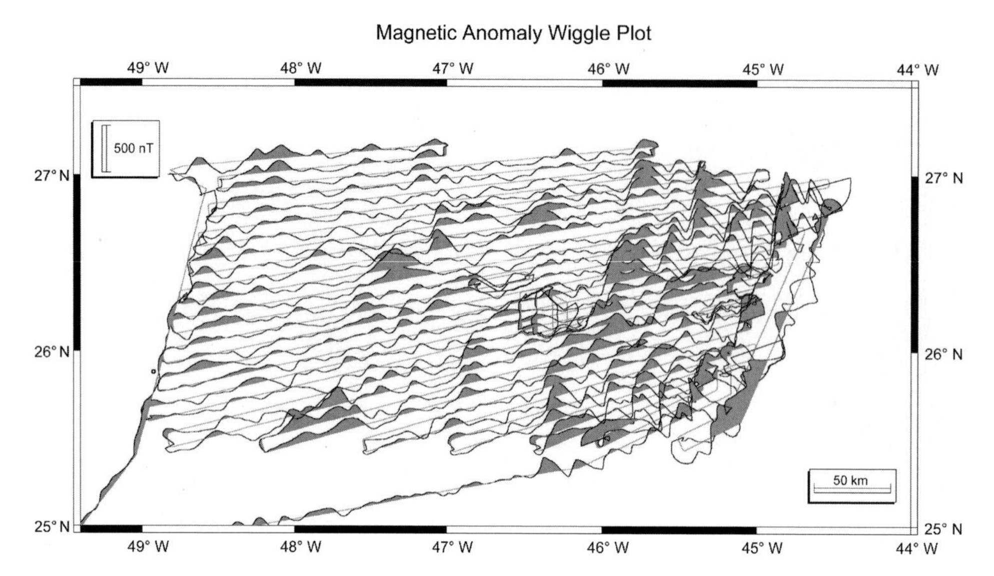

Figure M65 Map showing the magnetic anomaly wiggles along the tracklines of a marine magnetic survey on the western flank of the Mid–Atlantic Ridge near 26° N. Positive anomalies are shaded gray and represent a first–order normal polarity crust. Negative anomalies are unshaded and represent the reverse polarity crust. Data from Tivey and Tucholke (1998).

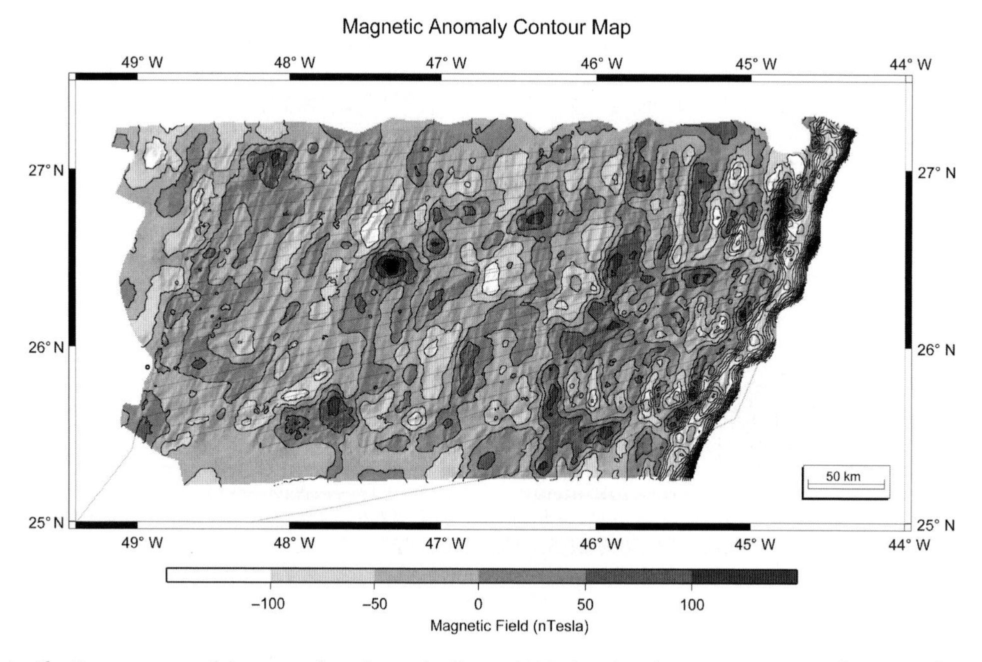

Figure M66/Plate 9b Contour map of the same data shown in Figure M65 showing the magnetic anomaly stripes that characterize the seafloor over most of the ocean basins. Contour interval is 50 nT. Thin black lines indicate the ship's tracklines. Map based on data from Tivey and Tucholke (1998).

sediments. The magnetic anomalies can be correlated and identified with the predicted anomalies from the geomagnetic polarity timescale (GPTS) and thereby allow for an isochron map of the ocean crust to be generated (Figure M67). This allows for a history of seafloor spreading to be documented (see Figure M68) and for spreading rates to be calculated.

Magnetic anomalies vary in their morphology according to their latitude of formation. Magnetic anomalies arise not only from the magnitude of the magnetization but also the direction of the magnetization. Unlike gravity, which is directed toward Earth's center, magnetic field lines originate on south poles and end on north poles. Thus, magnetic anomalies are centered over the magnetized body at the poles, but offset and phase-shifted at midlatitudes with magnetic lows and highs marking the north and south ends of a normal polarity magnetized body. At the equator a normal polarized body produces a magnetic low. These phase shifts can be compensated for by either reducing the anomaly to the pole (see *Reduction to pole*) or by computing the crustal magnetization (Figure M67).

Magnetic surveys are also collected with deep-towed sensors or other moving platforms such as submersibles or AUVs (see for example Tivey *et al.*, 1997, 2003). Such surveys are usually detailed studies encompassing small areas of the seafloor (<10 km). Additional

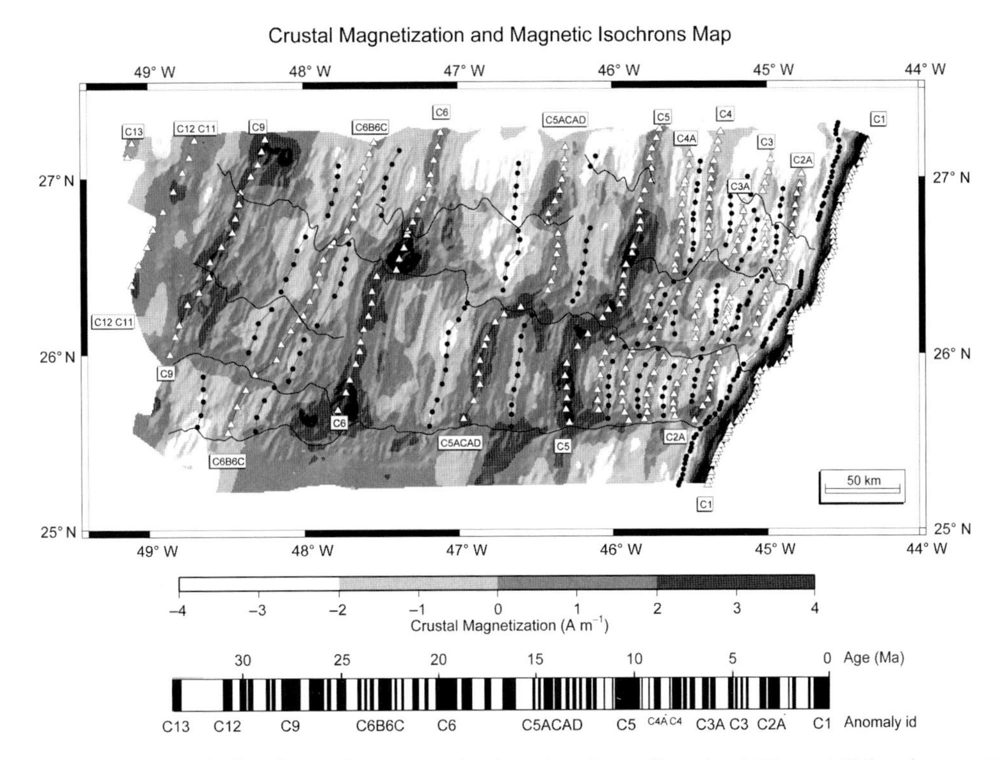

Figure M67/Plate 9c Map of the calculated crustal magnetization based on the seafloor depth (Figure M64) and magnetic anomaly data (Figure M66). Black dots and adjoining lines indicate the reverse polarity isochrons, whereas the white triangle symbols and adjoining lines indicate normal polarity chrons. Selected major chrons are identified and can be converted to age using the geomagnetic polarity timescale shown at the bottom of the figure. Map based on data from Tivey and Tucholke (1998).

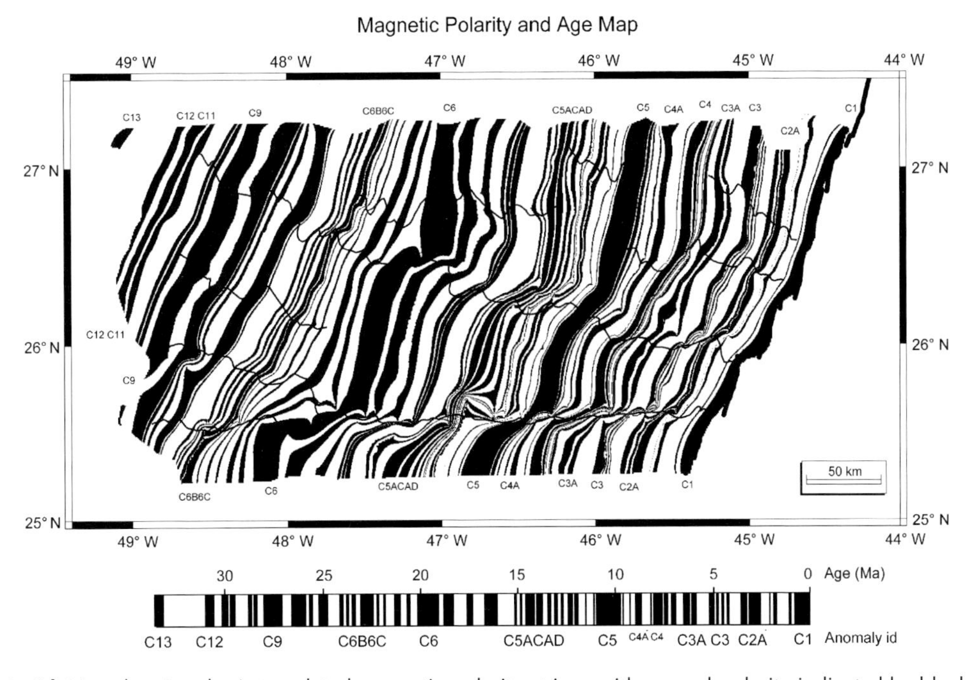

Figure M68/Plate 9d Map showing the interpolated magnetic polarity stripes with normal polarity indicated by black and reverse polarity in white. Thin black lines indicate the boundaries of individual spreading ridge segments through time. The geomagnetic polarity timescale is shown at the bottom to correlate with age.

processing steps are required in the analysis of these kinds of data. The depth and height above the seafloor of the vehicle or platform must be known with some precision. Various algorithms exist to continue the data from an uneven observation surface onto a level observation plane above the topography (see Tivey *et al.*, 2003). If the type of magnetic sensor used in the survey is a vector magnetometer rather than an absolute sensor then the orientation of the vehicle (i.e., its attitude) must also be known. Finally, navigation requires specialized acoustic ranging to determine the vehicle's position underwater. Once these steps have been completed then the data can be treated like sea surface data by removing the regional field to obtain an anomaly and then calculating a source crustal magnetization.

Maurice A. Tivey

Bibliography

Cole, J.F., 1908. Magnetic declination and latitude observations in the Bermudas. *Terrestrial Magnetism and Atmospheric Electricity*, **13**: 49–56.

Dietz, R.S., 1961. Continent and ocean basin evolution by spreading of the seafloor. *Nature*, **190**: 854–857.

Glen, W., 1982. *The Road to Jaramillo. Critical Years of the Revolution in Earth Science.* Stanford, CA: Stanford University Press, 459 pp.

Heezen, B.C., Ewing, M., and Miller, E.T., 1953. Trans-Atlantic profile of total magnetic intensity and topography, Dakar to Barbados: *Deep Sea Research*, **1**: 25–33.

Hess, H.H., 1962. History of ocean basins. In Engel, A.E.J., James, H.L., and Leonard, B.F. (eds.), *Petrologic Studies, the Burlington Volume.* Boulder, CO: Geological Society of America, pp. 599–620.

Johnson, H.P., Hautala, S.L., Tivey, M.A., Jones, C.D., Voight, J., Pruis, M., Garcia-Berdeal, I., Gilbert, L.A., Bjorklund, T., Fredericks, W., Howland, J., Tsurumi, M., Kurakawa, T., Nakamura, K., O'Connell, K., Thomas, L., Bolton, S.,Turner, J., 2002. Survey studies hydrothermal circulation on the northern Juan de Fuca ridge. *EOS: Transactions of the American Geophysical Union*, **83**: 73–79.

Macdonald, K.C., 1977. Near-bottom magnetic anomalies, asymmetric spreading, oblique spreading, and tectonics of the Mid-Atlantic Ridge near lat 37°N. *Geological Society of America Bulletin*, **88**: 541–555.

Macdonald, K.C., Miller, S.P., Luyendyk, B.P., Atwater, T.M., and Shure, L., 1983. Investigation of a Vine-Matthews magnetic lineation from a submersible: the source and character of marine magnetic anomalies. *Journal of Geophysical Research*, **88**: 3403–3418.

Raff, A.D., and Mason, R.G., 1961. A magnetic survey off the west coast of North America, 40-N to 52-N. *Geological Society of America Bulletin*, **72**: 1267–1270.

Rumbaugh, L.H., and Alldredge, L.R., 1949. Airborne equipment for geomagnetic measurements. *Transactions of American Geophysical Union*, **30**: 836–848.

Sheriff, R.E., 1981. *Encyclopedic Dictionary of Exploration Geophysics*, 6th edn. Tulsa, OK: Society of Exploration Geophysicists, 266 pp.

Tivey, M.A., Bradley, A., Yoerger, D., Catanach, R., Duester, A., Liberatore, S., and Singh, H., 1997. Autonomous underwater vehicle maps seafloor. *EOS*, **78**: 229–230.

Tivey, M.A., Schouten, H., and Kleinrock, M.C., 2003. A near-bottom magnetic survey of the Mid-Atlantic Ridge axis at 26°N: Implications for the tectonic evolution of the TAG segment. *Journal of Geophysical Research*, **108**: 2277, 10.1029/2002JB001967.

Tivey, M.A., and Tucholke, B.E., 1998. Magnetization of 0-29 Ma ocean crust on the Mid-Atlantic Ridge, 25°30′ to 27°10′N. *Journal of Geophysical Research*, **103**: 17807–17826.

Cross-references

Carnegie Institution of Washington,Department of Terrestrial Magnetism
Carnegie, Research Vessel
Geomagnetism, History of
Halley, Edmond (1656–1742)
IGRF, International Geomagnetic Reference Field
Instrumentation, History of
Observatories, Instrumentation
Project MAGNET
Reduction to Pole
Vine-Matthews-Morley Hypothesis
Voyages Making Geomagnetic Measurements

MAGNETIC SUSCEPTIBILITY, ANISOTROPY

Introduction

The preferred orientation of minerals is typical of almost all rock types. In some rocks, for example, metamorphic mica-schist, it is very strong and visible to the naked eye, while in others, like basalt and massive granite, it is very weak and detectable only by sensitive instruments. It develops during various geological processes, such as by water flow in sediments, by lava or magma flow in volcanic and plutonic rocks, or by ductile deformation in metamorphic rocks, and in turn, these processes can be assessed from it. The preferred orientation of rock-forming minerals has been measured in thin sections using microscope and universal stage analysis since the beginning of the 20th century, while today, more sophisticated techniques have been developed (e.g., X-ray pole figure goniometry, neutron pole figure goniometry, and electron backscatter diffractography).

Magnetic minerals, mostly occurring in rocks in accessory amounts, show preferred orientation. This orientation, called magnetic fabric, can be advantageously investigated by means of magnetic anisotropy, a technique based on the directional variability in magnetic properties. For example, the magnetic susceptibility of hematite or pyrrhotite crystal is an order of magnitude higher within the basal plane than along the *c*-axis, thus constituting the magnetocrystalline anisotropy. In magnetite grains, the susceptibility is the highest along the longest dimension of the grain, and lowest along the shortest dimension regardless of the orientation of crystal lattice. This anisotropy is called shape anisotropy. If the magnetic grains are oriented preferentially in a rock, the rock shows magnetic anisotropy. Modern instruments are sensitive enough to be able to measure anisotropy in almost all rock types with sufficient accuracy. In addition, this technique is so sensitive that in rocks with a very weak-preferred orientation of minerals (e.g., some volcanic rocks), it is the only method that gives reasonable results. This technique is also extremely fast, an order of magnitude faster than classical methods of structural analysis. For these reasons, anisotropy has experienced broad use in many branches of geology and geophysics and has become one of the most important techniques of modern structural geology.

Principles of anisotropy of magnetic susceptibility

The magnetization (M) of a magnetically isotropic substance is dependent on the intensity of the magnetizing field (H) as follows:

$$M = kH$$

where k is the magnetic susceptibility, independent of the field in paramagnetic and diamagnetic substances and being a complex function of the field in ferromagnetic *sensu lato* (in a broad sense) substances. In weak fields, in which the so-called initial susceptibility is measured, the susceptibility is also constant in ferromagnetic *sensu lato* substances.

If in magnetically anisotropic rocks, the magnetization is no longer parallel to the magnetizing field, then, the relationship between the magnetization and the intensity of the weak field is described in a more complex way by a set of three equations:

$$\begin{aligned} M_1 &= k_{11}H_1 + k_{12}H_2 + k_{13}H_3 \\ M_2 &= k_{21}H_1 + k_{22}H_2 + k_{23}H_3 \\ M_3 &= k_{31}H_1 + k_{32}H_2 + k_{33}H_3 \end{aligned}$$

where M_i ($i = 1, 2, 3$) are the components of the magnetization vector, H_j ($j = 1, 2, 3$) are the components of the vector of the intensity of magnetic field, and k_{ij} ($k_{ij} = k_{ji}$) are the components of the symmetric second-rank susceptibility tensor that reflects the directional variation of susceptibility in its shape. The susceptibility is the highest in one direction and the lowest in a perpendicular direction. In these two directions and in a direction perpendicular to them, the magnetization is parallel to the field. In the Cartesian coordinate system defined by these directions, the above equations simplify to:

$$M_1 = k_{11}H_1$$
$$M_2 = k_{22}H_2$$
$$M_3 = k_{33}H_3$$

The components $k_{11} \geq k_{22} \geq k_{33}$, often denoted as $k_1 \geq k_2 \geq k_3$, or $k_{max} \geq k_{int} \geq k_{min}$, are called the principal susceptibilities (the maximum, intermediate, and minimum susceptibilities) and their directions are the principal directions. The susceptibility tensor, even though mathematically rigorous, is simple and elegant and is nonillustrative from a geological point of view. Fortunately, it can be geometrically represented by the susceptibility ellipsoid or AMS ellipsoid (for details see Nye, 1957) that are quite illustrative and frequently used in geological interpretations of AMS; their definitions are clear from Figure M69. The axes of the AMS ellipsoid are parallel to the principal directions and the semiaxes lengths equal the principal susceptibilities.

AMS can be measured with various instruments (e.g., low-field torque meter, AC induction bridge, anisotropy spinner magnetometer), their principles being described in another entry of this volume. Nevertheless, the principle of the AMS determination lies in measuring the susceptibility of a specimen in at least six independent directions, and subsequently fitting the susceptibility tensor to these data using the least-squares method (Figure M70). Measuring susceptibility in more than six directions enables the error in fitting the tensor as well as the errors in determining the principal susceptibilities and principal directions to be evaluated. For each principal direction, two error angles are determined that delimit a region within which the true principal direction lies with a probability of 95% (Figure M71; for details see Jelínek, 1977).

The theory of AMS in rocks works with field-independent susceptibility that results from the linear relationship between magnetization and the magnetizing field. This is valid in diamagnetic and

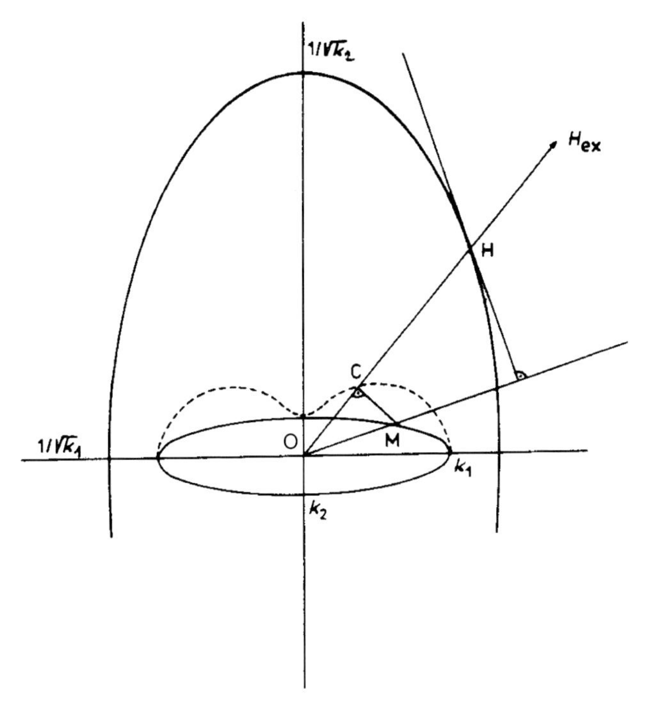

Figure M69 Definition of the susceptibility ellipsoid and AMS ellipsoid. Legend: OH, direction of external magnetic field H_{ex}; OM, anisotropic magnetization induced by H_{ex}; OC, component of magnetization parallel to H_{ex} or the quantity measured by usual methods. If \vert H_{ex} \vert = 1, OC is the susceptibility parallel to H_{ex} (directional susceptibility). Loci of H, where $OH = 1/\sqrt{OC}$, represents the susceptibility ellipsoid (heavy line) and the loci of M (thin line) is the AMS ellipsoid. The dashed line represents the directional behavior of the directional susceptibility; note that it is not represented by an ellipsoid. (Adapted from Nagata, 1961.)

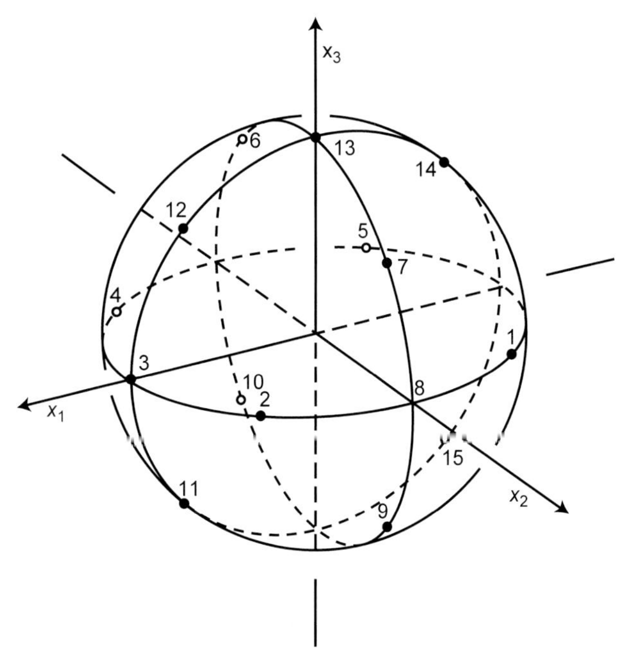

Figure M70 Rotatable design (independent of the coordinate system) of measuring directions used in the Jelínek (1977) method for the AMS determination. Besides, some instruments measure a slowly rotating specimen; consequently, a much larger set of directional data is obtained. (After Jelínek, 1977.)

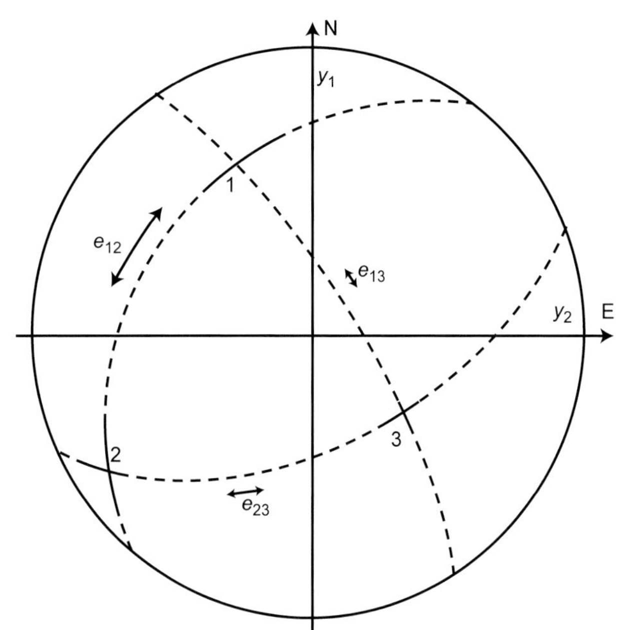

Figure M71 Confidence angles (e_{12}, e_{23}, e_{13}) that define the regions where the true principal directions lie with probability of 95%. Specimen of granodiorite. (After Jelínek, 1977.)

paramagnetic minerals by definition, as well as in ferromagnetic minerals *sensu lato*, provided that the fields used in common AMS meters are weak enough so that the initial susceptibility is measured. Recently it has been shown that this is true for magnetite, while the susceptibility of pyrrhotite, hematite, and titanomagnetite may be field-dependent. This may theoretically result in the invalidity of the AMS theory in rocks whose AMS is carried by these minerals. Fortunately, the recent investigations have also shown that this effect does not manifest in the orientations of the principal susceptibilities and the AMS ellipsoid shapes. It affects only the degree of AMS, that can be corrected by relatively simple means (de Wall, 2000; Hrouda, 2002). Consequently, the simple and elegant linear AMS theory may be still in use.

The causes of the magnetic anisotropy of rocks are usually summarized as follows: (1) the alignment of nonequidimensional grains, (2) the lattice alignment of crystals with magnetocrystalline anisotropy, (3) the alignment of magnetic domains, (4) the stringing together of magnetic grains (this is special case of (1)), (5) stress-induced anisotropy, and (6) exchange anisotropy (Banerjee and Stacey, 1967). Extensive investigations of magnetic anisotropy have shown that the dominant factors are (1) and (2) (Tarling and Hrouda, 1993).

The eccentricity and shape of the AMS ellipsoid can be characterized by parameters derived from the relationship between the principal susceptibilities. More than 30 such parameters have been introduced (see Tarling and Hrouda, 1993), despite two are sufficient to do so. The most commonly used are:

$P = k_1/k_3$—degree of AMS

$P_j = \exp\sqrt{\{2[(\eta_1-\eta)^2 + (\eta_2-\eta)^2 + (\eta_3-\eta)^2]\}}$—corrected degree of AMS

$L = k_1/k_2$—magnetic lineation

$F = k_2/k_3$—magnetic foliation

$T = (2\eta_2-\eta_1-\eta_3)/(\eta_1-\eta_3)$—shape parameter

$q = (k_1 - k_2)/[(k_1 + k_2)/2 - k_3]$—shape parameter

where $\eta_1 = \ln k_1$, $\eta_2 = \ln k_2$, $\eta_3 = \ln k_3$, $\eta = (\eta_1 + \eta_2 + \eta_3)/3$. The parameters P and P_j, being interrelated as $P_j = P^a (a = \sqrt{T^2/3})$, indicate the intensity of the preferred orientation of the magnetic minerals in a rock. The parameters L and F characterize the intensity of the linear-parallel and planar-parallel orientations, respectively. The direction of magnetic lineation is parallel to the maximum susceptibility, while the magnetic foliation pole is parallel to the minimum susceptibility (Figure M72). The parameters T and q characterize the shape of the AMS ellipsoid. If $0 < T < +1$ or $0 < q < 0.7$ the AMS ellipsoid is oblate (the magnetic fabric is planar); $T = +1$ or $q = 0$ means that the AMS ellipsoid is rotationally oblate. If $-1 < T < 0$ or $0.7 < q < 2$ the AMS ellipsoid is prolate (the magnetic fabric is linear); $T = -1$ or $q = 2$ means that the AMS ellipsoid is rotationally prolate.

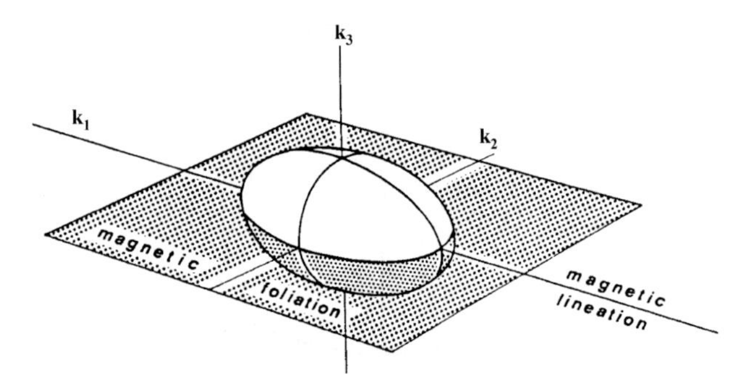

Figure M72 Definition of magnetic foliation and magnetic lineation with respect to the AMS ellipsoid. (Adapted from Siegesmund *et al.*, 1995.)

The eccentricity and shape of the AMS ellipsoid can be represented graphically using two types of AMS plots (Figure M73a,b). In the first type (Figure M73a), the magnetic foliation (F) is plotted on the abscissa and the magnetic lineation (L) on the ordinate, analogous to the Flinn plot used by structural geologists. In the second type of plot (Figure M73b), the degree of AMS (P or P_j) is as abscissa and the T parameter is the ordinate. The first plot type is convenient for the AMS of deformed rocks because it formally resembles the deformation plot used by structural geologists. The second plot type is convenient in cases when the degree of AMS is of primary interest.

The orientations of the directions of the principal susceptibilities are usually presented on the lower hemisphere of an equal-area projection. The general convention is to use square symbols for k_1, triangles for k_2, and circles for k_3.

AMS of minerals

Whereas in paleomagnetism only ferromagnetic minerals *sensu lato* are important, most frequently represented by magnetite, hematite, and pyrrhotite, since they carry a remanent magnetization, in the AMS are important not only these minerals, but also some paramagnetic ferromagnesian silicates, as for example olivine, pyroxene, augite, hornblende, mica, and diamagnetic calcite. The AMS of these minerals are presented in Table M5, in terms of the degree of AMS, the shape parameter, and the AMS type. The AMS of hematite and pyrrhotite is magnetocrystalline in origin and extremely strong. In ferromagnesian silicates and in calcite, the AMS is also magnetocrystalline but much weaker.

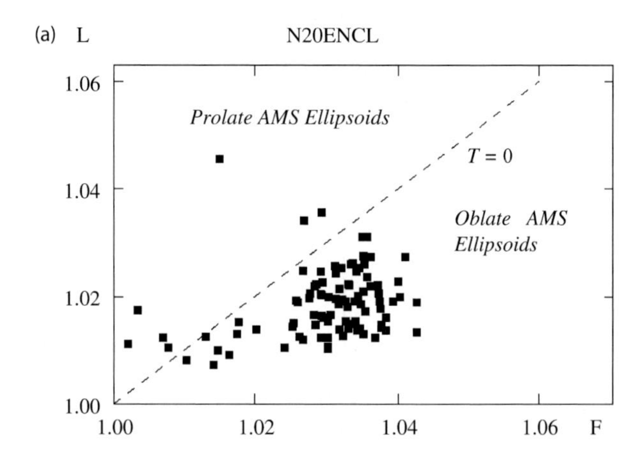

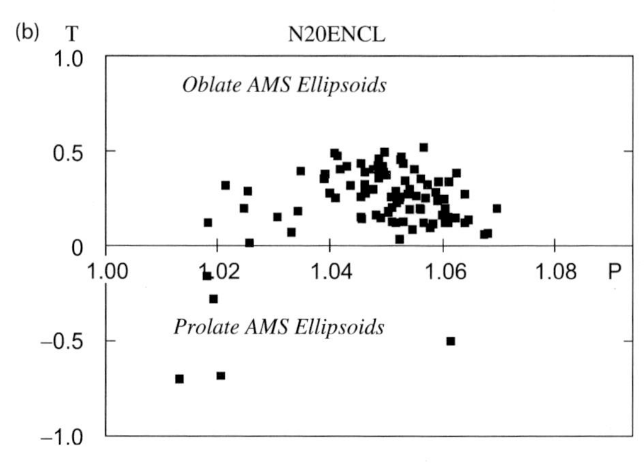

Figure M73 Magnetic anisotropy plots. (a) *L-F* (Flinn type) plot and (b) *P-T* (Jelínek type) plot.

Table M5 Grain AMS of various minerals (adapted from Hrouda (1993)

Mineral	P	T	AMS type
Magnetite	1.1–3.0	Variable	Shape
Hematite	>100	~1	Magnetocrystalline
Pyrrhotite	100–10000	~1	Magnetocrystalline
Actinolite	1.1–1.2	−0.4–0.4	Magnetocrystalline
Hornblende	1.665	−0.51	Magnetocrystalline
Crocidolite	1.098	−0.25	Magnetocrystalline
Glaucophane	1.205	0.10	Magnetocrystalline
Chlorite	1.2–1.7	~1	Magnetocrystalline
Biotite	1.2–1.6	~1	Magnetocrystalline
Phlogopite	1.3	0.95	Magnetocrystalline
Muscovite	1.4	0.44	Magnetocrystalline
Quartz	1.01	–	Magnetocrystalline
Calcite	1.11	1	Magnetocrystalline
Aragonite	1.15	0.80	Magnetocrystalline

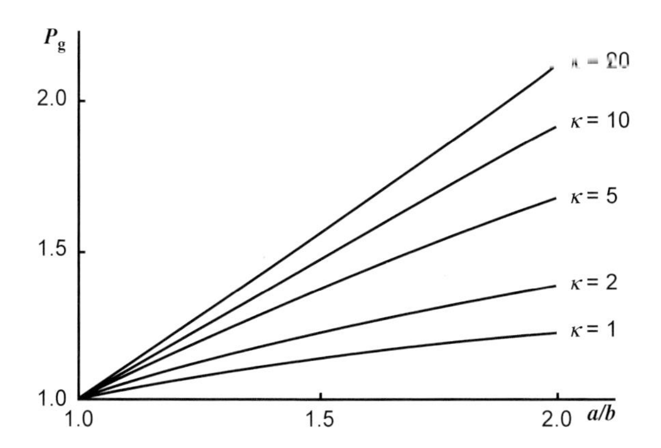

Figure M74 Relation between the degree of grain AMS (P_g) and the dimension ratio *a/b* of prolate spheroid (titano-)magnetite grains for various values of intrinsic susceptibility (SI). (Adapted from Hrouda and Tarling, 1993.)

Magnetite displays shape anisotropy depending on the eccentricities of the grains and on the intrinsic susceptibility of the mineral. The degree of grain AMS is then (Uyeda *et al.*, 1963)

$$P = (1 + \kappa N_c)/(1 + \kappa N_a),$$

where κ is the intrinsic susceptibility, which is isotropic and dependent only on the chemical composition of the mineral, and N_a and N_c which are the demagnetizing factors along the longest and shortest grain axes, respectively, solely depending on the grain shape. The quantitative relationship, though straightforward, is relatively complex, described by formulas containing elliptic integrals. Figure M74 shows this relationship quantitatively. It is obvious that grains with a lower intrinsic susceptibility show a lower degree of AMS than the grains with a higher intrinsic susceptibility having exactly the same shapes. Consequently, the AMS data for magnetite presented in Table M5 are only informative.

Rock susceptibility, in general, is controlled by all minerals present in a rock. Figure M75 shows model contributions of individual mineral groups. In strongly magnetic rocks, with a bulk susceptibility higher than 5×10^{-3} (SI units), the effect of mafic silicates is negligible and the rock susceptibility is effectively controlled by magnetite only. In weakly magnetic rocks, with bulk susceptibility less than 5×10^{-4} (SI units), the content of ferromagnetic minerals is often so low that the susceptibility is effectively controlled by mafic silicates. In rocks with a bulk susceptibility between 5×10^{-4} and 5×10^{-3} (SI units), the AMS, in general, is controlled by both ferromagnetic and paramagnetic minerals. If the bulk susceptibility is less than 5×10^{-5} (SI units), neither the effect of diamagnetic fraction (the calcite in marble) is negligible.

Quantitative relationship between AMS and mineral preferred orientation

The AMS of rocks is controlled not only by the intensity of the preferred orientation of magnetic minerals, but also by their grain AMS. The effect of these factors is obvious from a simple model in which the degree of AMS of the rock (P_s) is plotted against the degree of AMS of the magnetic grains (P_g) for different intensities of the preferred orientation of the principal axes (characterized by parameter C in Figure M76). If the grain anisotropy is small (similar to that of magnetite and phyllosilicates), the rock AMS is controlled by both the grain AMS and the intensity of the preferred orientation. If the grain AMS is high (as in hematite and pyrrhotite), the rock AMS is controlled solely by the intensity of the preferred orientation of the magnetic axes.

The preferred orientation of minerals in rocks is traditionally represented by density diagrams presented on equal-area projections. The density that maps the preferred orientation of a specific crystal axis can be described mathematically by the orientation tensor, defined as:

$$\mathbf{E} = \frac{1}{N}\begin{vmatrix} \Sigma l_i^2 & \Sigma l_i m_i & \Sigma l_i n_i \\ \Sigma m_i l_i & \Sigma m_i^2 & \Sigma m_i n_i \\ \Sigma n_i l_i & \Sigma m_i n_i & \Sigma n_i^2 \end{vmatrix}$$

where l_i, m_i, n_i are the direction cosines of the ith linear element represented by unit vector and N is the number of the linear elements considered (Scheidegger, 1965). The principal values of this tensor ($E_1 \geq E_2 \geq E_3$) have the following property:

$$E_1 + E_2 + E_3 = 1.$$

Consequently, $E_1 > E_2 = E_3$ represents a cluster type of distribution, whereas $E_1 = E_2 > E_3$ corresponds to a girdle-type pattern (Figure M77). In a perfect cluster, in which all linear elements are parallel, $E_1 = 1$, $E_2 = 0$, $E_3 = 0$. In a perfect girdle, $E_1 = 0.5$, $E_2 = 0.5$, $E_3 = 0$.

A straightforward relationship exists between the orientation tensor and the susceptibility tensor (for a summary see Ježek and Hrouda, 2000). For a rock, whose AMS is carried by a single magnetic mineral, the relationship is:

$$\mathbf{k} = K\mathbf{I} + \Delta\mathbf{E}$$

where $\mathbf{k}$ is the rock susceptibility tensor, $\mathbf{I}$ is the identity matrix. For magnetically rotational oblate grains, $K = K_2 = K_3$, $\Delta = K_1–K$ (K_1, K_2, K_3 are the grain principal susceptibilities) and $\mathbf{E}$ is the orientation tensor of the grain minimum susceptibility axes. For magnetically rotational oblate grains, $K = K_1 = K_2$, $\Delta = K–K_3$, and $\mathbf{E}$ is the orientation tensor of the grain maximum susceptibility axes. For the grains displaying so-called perfectly triaxial AMS ($K_1–K_2 = K_2–K_3$), $K = K_2$, $\Delta = K_1–K_2 = K_2–K_3$ and $\mathbf{E} = \mathbf{E}_x–\mathbf{E}_z$ is the Lisle orientation tensor, where $\mathbf{E}_x$ and $\mathbf{E}_z$ are the Scheidegger orientation tensors of the maximum and minimum axes, respectively.

Since the rock susceptibility is affected by all minerals present in the rock, the theoretical rock susceptibility can be calculated if the susceptibility tensors of all minerals and their relative contributions are known. Unfortunately, this is never known in practice and studies of the quantitative relationship between the AMS and preferred orientation are made only in such rocks in which the contribution of one mineral to the rock susceptibility is dominant. In this case, the model rock susceptibility is:

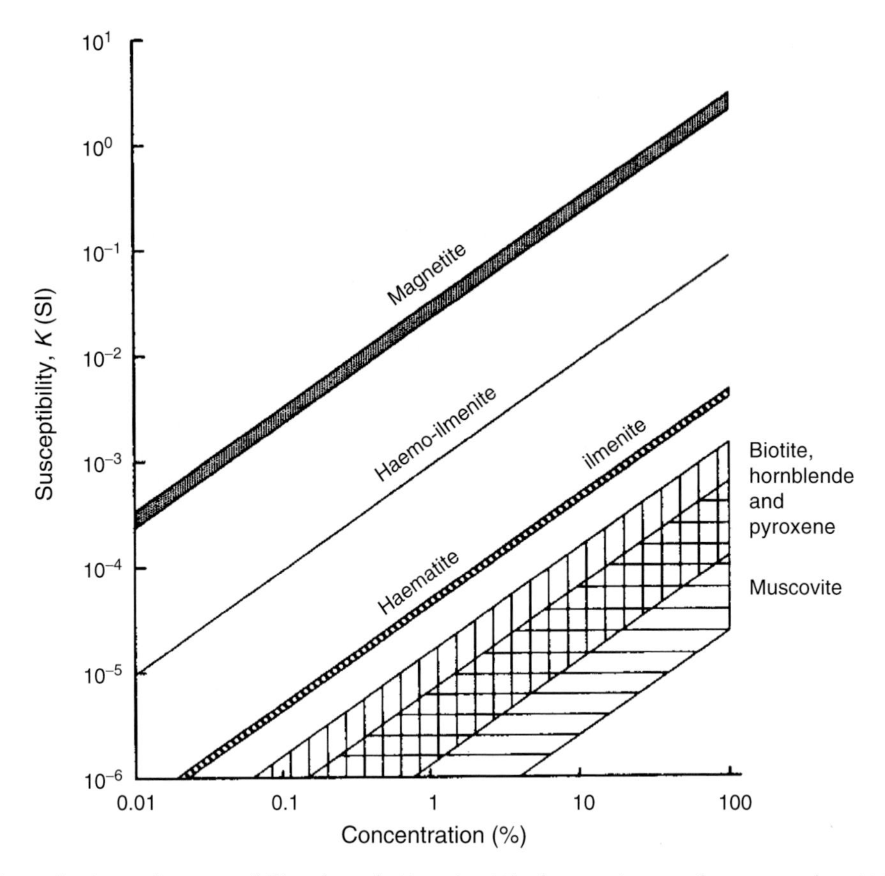

Figure M75 Mineral contribution to the susceptibility of a rock. Note that 1% of magnetite contributes more than 100% of phyllosilicates. (Adapted from Hrouda and Kahan, 1991.)

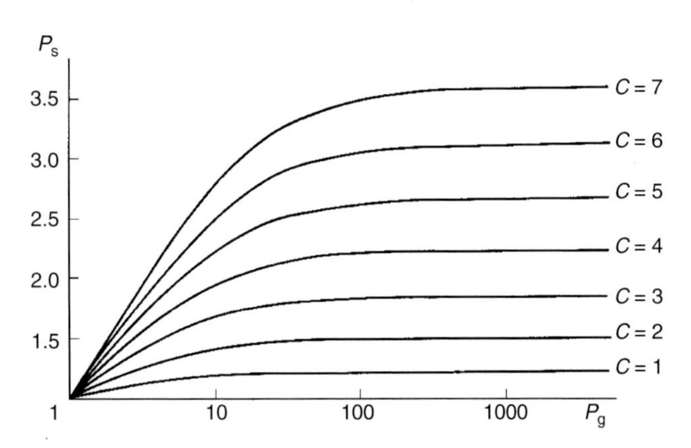

Figure M76 The dependence of the degree of AMS of a model rock (P_s) on the grain degree of AMS (P_g) and on the parameter C characterizing the intensity of the preferred orientation of the axes of magnetically uniaxial grains. (Adapted from Hrouda, 1980.)

$$\mathbf{k} = p\Sigma\mathbf{K} = p\Sigma(\mathbf{O}\mathbf{K}_d\mathbf{O}'),$$

where **k** is the rock susceptibility tensor, **K** is the magnetic grain susceptibility tensor, $\mathbf{K}_d$ the diagonal from the grain susceptibility tensor, **O** the matrix specifying the mineral orientation ($\mathbf{O}'$ is the transposed matrix of **O**), and p the percentage of a magnetic mineral in a rock.

The orientation of magnetic minerals can be investigated either through direct measurement of individual grains (microscope and universal stage methods), through determining the orientation tensor using X-ray pole figure goniometry, or electron backscatter diffraction with orientation contrast. Studies of this type were made for hematite (deformed hematite ore, Minas Gerais, Brazil, Hrouda *et al*., 1985), biotite (Bíteš orthogneiss, Hrouda and Schulmann, 1990), chlorites (Lüneburg *et al*., 1999; Chadima *et al*., 2004), and calcite (Carrara marble, Owens and Rutter, 1978). Results from these studies show that the principal susceptibilities of the whole rock measured are oriented in the same way as those calculated from the crystallographic axis pattern. The degree of AMS and the shape parameter are not always in agreement with that calculated from the crystallographic axis pattern. The differences may reflect the situation that the AMS and X-ray measurements are often not executed on exactly the same specimens, and the respective specimen volumes differ substantially (1 cm^3 vs thin section). In addition, minor effect of other magnetic minerals cannot be excluded. Nevertheless, the agreement between the theoretical and measured AMS can be considered to be satisfactory. Some of these results are illustrated in Figure M78.

Resolution of AMS into paramagnetic and ferromagnetic components

In rocks with the bulk susceptibility between 5×10^{-4} and 5×10^{-3}, the AMS is, in general, controlled by both ferromagnetic and paramagnetic minerals. As these mineral groups may behave differently in various geological situations, it is desirable to resolve the rock AMS into its ferromagnetic and paramagnetic components. This resolution is usually made through measuring the AMS in various strong magnetic fields of the order of tesla in which the ferromagnetic and paramagnetic minerals behave in a different way (described in an another part

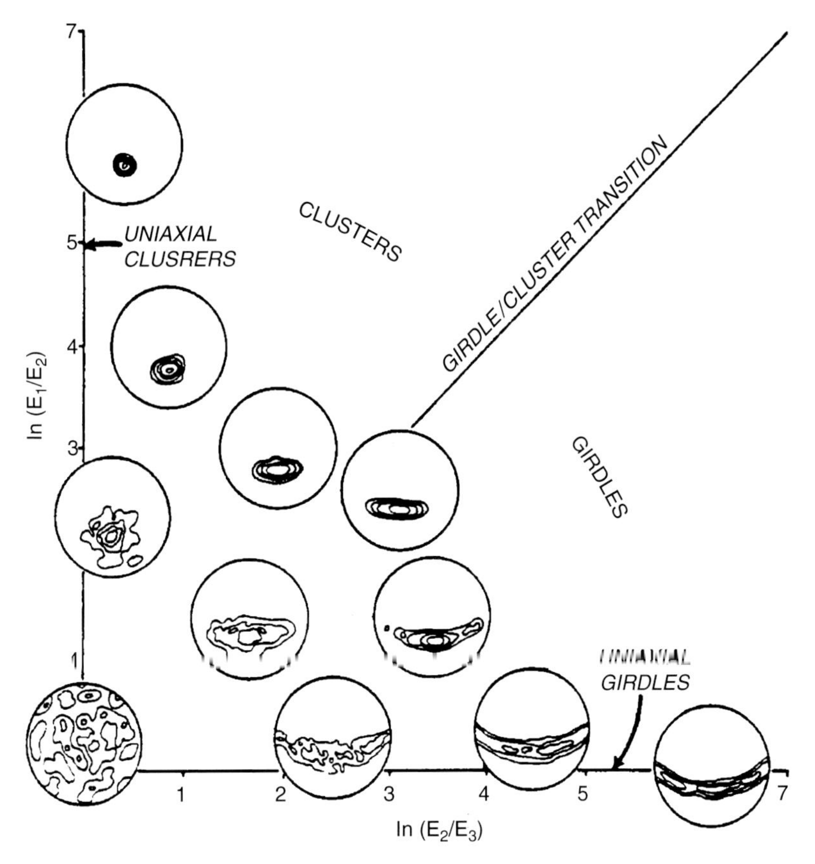

Figure M77 Relationship between preferred orientation of a linear element and principal values of the orientation tensor. (Adapted from Woodcock, 1977.)

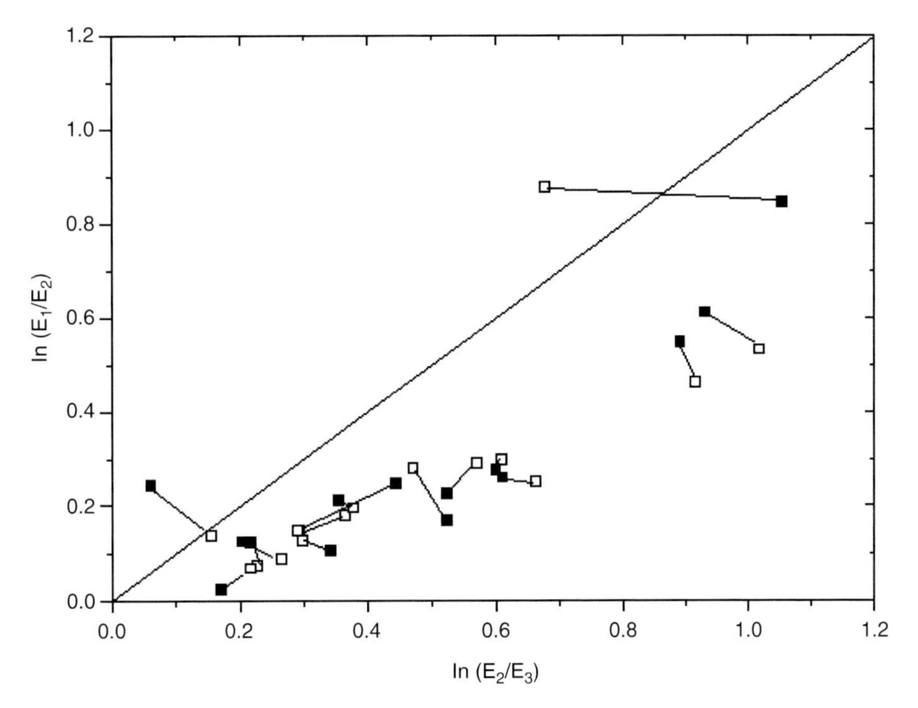

Figure M78 Relationship between orientation tensor calculated from mineral preferred orientation (*c*-axes of chlorite) and from AMS. Constructed from Chadima *et al.* (2004) data.

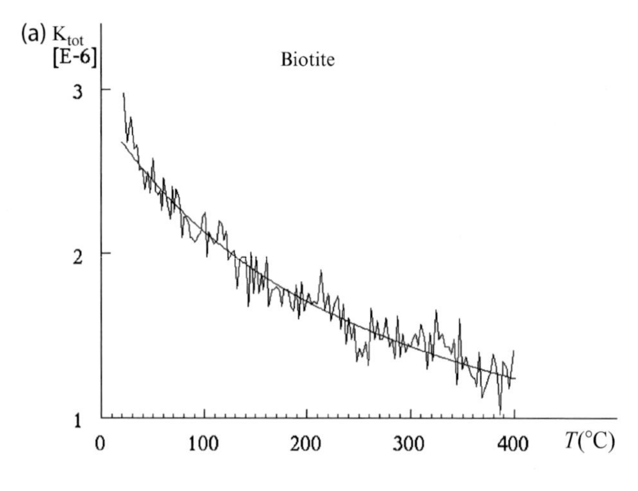

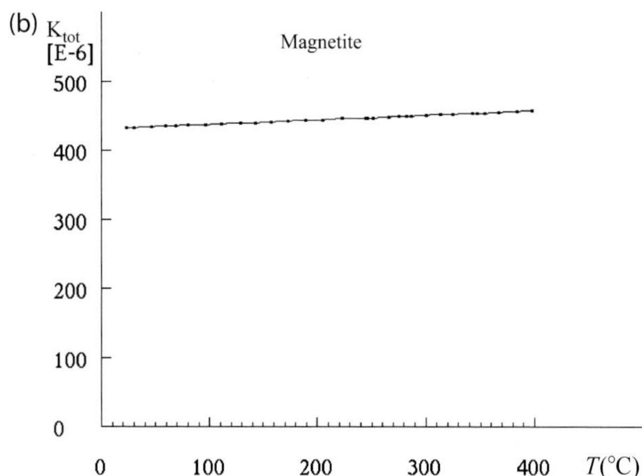

Figure M79 Susceptibility vs temperature relationship in the temperature interval between room temperature and 200°C for biotite (a) and magnetite (b).

of this volume) or at variable temperatures in which the ferromagnetic and paramagnetic minerals also behave differently.

Variation of the susceptibility of paramagnetic minerals with temperature is represented by a hyperbola, while in ferromagnetic minerals it is a complex curve in case of magnetite characterized by acute susceptibility decreases at about $-155°C$, the Verwey transition, and at about $580°C$, the Curie temperature. Fortunately, in some temperature intervals, typically between room temperature and $200°C$, the ferromagnetic susceptibility is either constant or follows only a mildly sloped straight line (Figure M79).

The susceptibility vs temperature curve of a rock containing both ferromagnetic and paramagnetic minerals for the case in which the ferromagnetic susceptibility is constant within the considered temperature interval, is:

$$k = p_p C/T + p_f k_f,$$

and for the case in which the ferromagnetic susceptibility is represented by only a mildly sloped straight line, it is:

$$k = p_p C/T + p_f(bT + a),$$

where k is the rock susceptibility, k_f the ferromagnetic susceptibility, p_p and p_f the percentages of paramagnetic and ferromagnetic fractions, respectively, C the paramagnetic constant and T the absolute temperature, a and b are constants.

The first equation is that of a hyperbola offset along the susceptibility axis, and the second is a combination of a hyperbola and a straight line. By fitting the hyperbola and straight line either horizontal (first equation), or sloped (second equation) to k vs T curve using the least-squares method, one obtains the paramagnetic susceptibility contribution ($p_p C/T$) and the ferromagnetic susceptibility contribution ($p_f k_f$ or $p_f(bT + a)$). These methods were developed by Hrouda (1994) and Hrouda *et al.* (1997), respectively, and are broadly used in resolving the mean bulk susceptibility measured on powder specimens into paramagnetic and ferromagnetic components.

Richter and van der Pluijm (1994) developed a technique for the AMS resolution into the paramagnetic and ferromagnetic components based on susceptibility measurements at low temperatures (between the temperature of liquid nitrogen and room temperature). At these temperatures, the contribution of paramagnetic fraction to the rock susceptibility is amplified because of the hyperbolic course. By plotting the reciprocal susceptibility against temperature one can easily visualize the linear paramagnetic susceptibility contribution and separate it from the ferromagnetic contribution. If this measurement is made in six directions, one can obtains six directional paramagnetic and ferromagnetic susceptibilities from which the paramagnetic and ferromagnetic susceptibility tensor can be calculated in a standard way (Figure M80).

Thermal enhancement of the magnetic fabric

The AMS is often measured using the instrumentation of a paleomagnetic laboratory. Paleomagnetic techniques, such as AF demagnetization and thermal demagnetization, have then been logically tested as to whether they are also applicable to the AMS (Urrutia-Fucugauchi, 1981). It has been shown that the AF demagnetization has only weak effect on the AMS and its use ceased. On the other hand, the effect of thermal demagnetization on the AMS can be strong. However, thermal demagnetization mechanisms are different in AMS than in remanent magnetization. Thermal demagnetization affects the induced magnetization very weakly, probably negligibly, but may induce the creation of new magnetic phases or even new minerals. If this creation is made in a zero magnetic field, it would have virtually no effect on the remanent magnetization, but would have a very strong effect on the AMS. Often, the newly created phases are more magnetic than the original phases and may be coaxial to the old phases. Thermal demagnetization, in fact, enhances the original magnetic fabric. Newly created phases can also be noncoaxial, and their creation may obscure the original magnetic fabric. If the new creation is confined to a phase of different origin than the original phase, thermal treatment may "visualize" the cryptic fabric (Henry *et al.*, 2003). Even though positive results may be obtained in this way in some cases, this technique cannot be recommended to be used in a routine way because the process of the creation of new magnetic phases due to specimen heating is very complex, and our knowledge about this is far from complete.

Statistical evaluation of the AMS data

The AMS technique is so fast that several specimens are usually measured in a site. The site data or the data from geological bodies should be evaluated statistically. The simplest approach is to consider the individual principal directions as vectors and to process them using a vector statistics, for example the Fisher statistics. Unfortunately, the principal directions do not have a Fisherian distribution, being orthogonal and bipolar, so various modifications of the Fisher statistics have been proposed (Ellwood and Whitney, 1980; Park *et al.*, 1988).

To avoid these problems Jelínek (1978) has developed a statistical method that is based on averaging out the individual components of the susceptibility tensors in one coordinate system (e.g., geographical) after they have been normalized against their mean susceptibility. In this way the mean tensor (and the orientations of its principal directions) as well as the elliptical confidence regions (in terms of confidence angles) about each principal direction are computed from the

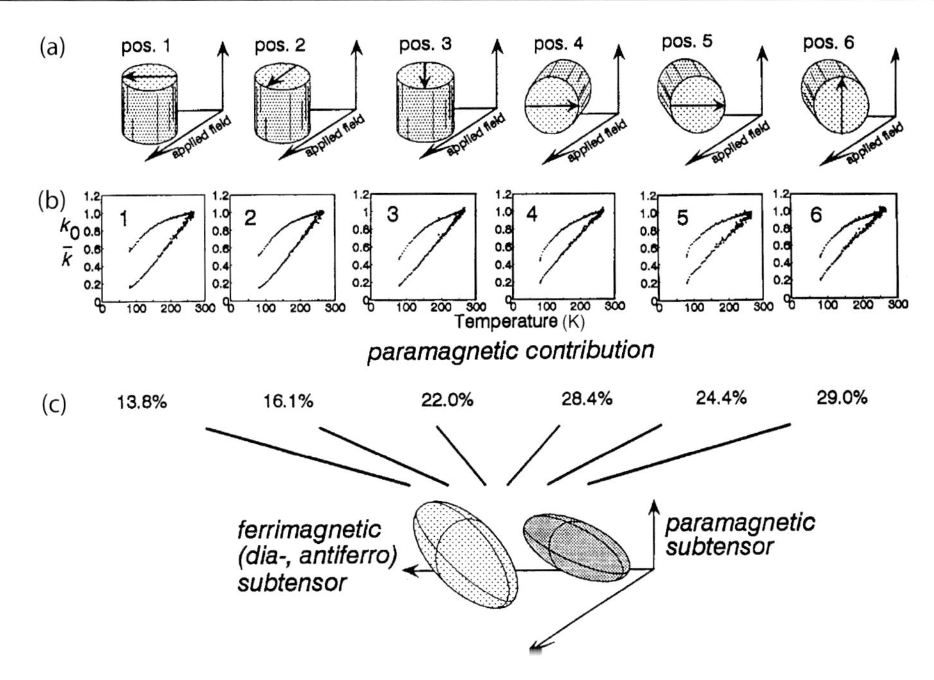

Figure M80 Principle of the resolution of AMS into paramagnetic and ferromagnetic components using measurement of low-temperature susceptibility. (a) Six orientations of a specimen used to construct the susceptibility tensor, (b) the relative amount of paramagnetic and ferromagnetic susceptibilities is determined in six orientations from heating curves, the upper curve being the original heating curve and the lower curve being the paramagnetic remnant, and (c) the resolved paramagnetic and ferromagnetic subtensors. (Adapted from Richter and van der Pluijm, 1994.)

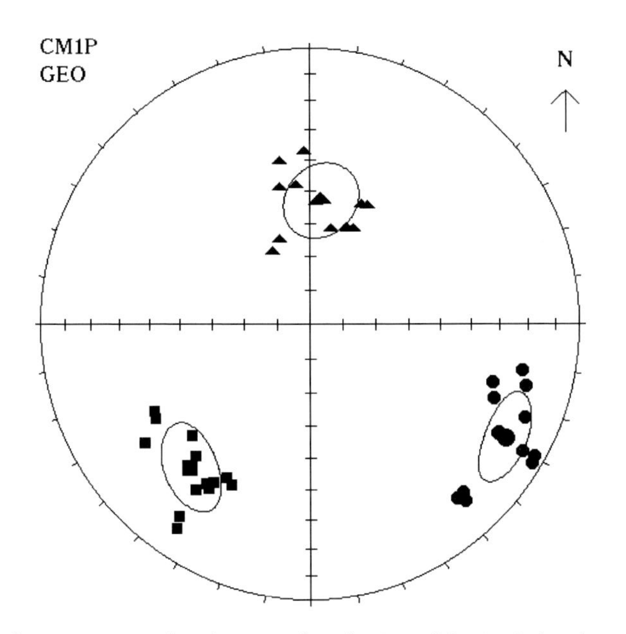

Figure M81 Results of statistical evaluation of the AMS data from a locality of the Čistá granodiorite made by the Jelínek (1978) method. Small symbols indicate individual specimens, and large symbols indicate the mean tensor. The confidence regions around the mean principal directions are elliptic on a sphere, being slightly deformed due to the equal area projection.

variance-covariance matrix (Figure M81). This technique seems to be the most frequently used at present.

In data sets with widely scattered principal directions, neither analytical method provides an adequate visual representation of the observed data. To overcome these disadvantages, Constable and Tauxe (1990) developed a bootstrap method which delineates the confidence regions using Monte Carlo probability modeling.

Statistical evaluation of the AMS data is a complex problem and further investigations are required to evaluate the effectiveness of the individual methods (for discussion see Ernst and Pearce, 1989; Tarling and Hrouda, 1993; Borradaile, 2001).

Geological applications of AMS

Magnetic fabric in sedimentary rocks

The AMS in sedimentary rocks provides information on the deposition and compaction processes (for a summary, see Hamilton and Rees, 1970; Rees and Woodall, 1975; Rees, 1983; Taira, 1989; Tarling and Hrouda, 1993; Hrouda and Ježek, 1999). In natural sedimentary rocks unaffected by later deformation, the relationship that is primarily investigated is the agreement between the magnetic foliation and lineation and the sedimentary external structures (sole markings, flute casts, groove casts) and internal structures (cross-bedding, current-ripple lamination, symmetric and asymmetric ripples). It has been found that the magnetic foliation is always oriented near the bedding/compaction plane, while the magnetic lineation is mostly roughly parallel to the near-bottom water current directions determined using sedimentological techniques (see Figure M82). Less frequently, the magnetic lineation may be perpendicular to the current direction, which is typical of the flysch sediments of the lowermost *A* member of the Bouma sequence. The degree of AMS is relatively low and the AMS ellipsoid is, in general, oblate.

In artificial sediments deposited through grain by grain (or from thin suspension) deposition from still or running water onto a flat or sloping bottom under controlled conditions in tanks and flumes in laboratories, the AMS ellipsoid is clearly oblate; the magnetic foliation dips less than 15° from the bedding toward the origin of flow and the magnetic lineation is parallel to the direction of flow, slightly plunging toward the origin of flow.

During deposition from medium-concentrated suspensions, the AMS ellipsoids can be clearly triaxial to slightly prolate, in which the

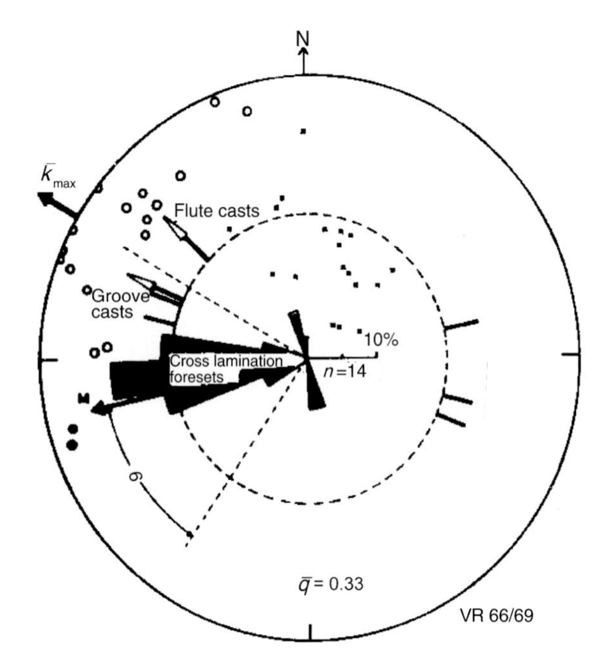

Figure M82 Magnetic and sedimentary fabrics in sediments of the Rosario Formation, La Jolla, California (crosses, magnetic foliation poles; open circles, magnetic lineation). Equal-area projection on lower hemisphere (projection plane = bedding plane). (Adapted from von Rad, 1971.)

magnetic foliation dips less than 15° toward the origin of flow, and the magnetic lineation is perpendicular to the direction of flow and to the dip of magnetic foliation. During deposition from very concentrated grain dispersions onto a sloping bottom, the AMS ellipsoids are triaxial. The magnetic foliation dips 25°–30° toward the origin of the flow and the magnetic lineation is parallel both to the flow and to the dip of magnetic foliation, and plunges slightly toward the origin of flow.

During the process of diagenesis, the originally sedimentary magnetic fabric may be slightly modified due to ductile deformation accompanying this process. If the ductile deformation is represented by vertical shortening due to the loading by the weight of the overlying strata, the degree of AMS and the oblateness of the AMS ellipsoid increase, whereas the magnetic foliation and lineation retain their orientations (e.g., Lowrie and Hirt, 1987). If the ductile deformation is represented by the bedding parallel shortening or by the bedding parallel simple shears or by both, the degree of AMS initially decreases and only later increases when the deformation is strong enough to overcome the initial compaction. The magnetic fabric initially becomes more planar, and only later it does become more triaxial or even linear. The magnetic lineation deviates gradually from the direction of flow toward that of maximum strain, often creating a bimodal pattern. The magnetic foliation remains initially near the bedding. After a stronger strain, it deviates from it, creating a girdle in the magnetic foliation poles that are perpendicular to the magnetic lineation.

Magnetic fabric in volcanic rocks

The AMS of volcanic rocks is, in general, very weak, reflecting a very poor dimensional orientation of the magnetic minerals (mostly titanomagnetites) in these rocks. In spite of this, their AMS can, in general, be measured precisely because the volcanic rocks are often strongly magnetic, and the AMS seems to be the quickest method to be able to reliably investigate the weakly preferred orientation of the minerals.

Since the first investigations of the AMS of volcanic rocks it has been clear that it reflects the dimensional orientation of magnetic minerals created during a lava flow. The magnetic foliation is often found to be near or even parallel to the flow plane in lava flows, sills, and dikes (Ernst and Baragar, 1992; Cañon-Tapia *et al.*, 1994; Raposo and Ernesto, 1995; Hrouda *et al.*, 2002a). The magnetic lineation is mostly parallel to the lava flow directions (Figure M83), even though perpendicular or oblique relationships can also be rarely found.

AMS enables not only the lava flow process to be investigated, but also the rheomorphic flow in ignimbrites and the motions of tuffs and tuffites (MacDonald and Palmer, 1990).

Magnetic fabric in plutonic rocks

In plutonic rocks, the AMS is one of the most powerful tools of structural analysis, because it can efficiently measure the magnetic fabric even in massive granites that are isotropic at the first sight. The AMS of plutonic rocks varies from very weak (typical of volcanic rocks) to extraordinarily strong (typical of metamorphic and/or strongly deformed rocks).

Granites, the most important representatives of plutonic rocks, usually display a bimodal distribution of their magnetic susceptibility. One mode corresponds to susceptibilities in the order of 10^{-5} to 10^{-4} SI units and are due to weakly magnetic (Dortman, 1984) or paramagnetic (Bouchez, 2000) granites. These are equated with the S (sedimental) type granites, such as the ilmenite-bearing granites. The other mode has susceptibilities in the order of 10^{-3}–10^{-2} SI units, and this is due to magnetic or ferromagnetic granites. These are equated with the I (igneous) type granites such as magnetite-bearing granites. In magnetic granites, the AMS investigates the preferred orientation of magnetite by grain shape. In weakly magnetic granites, the AMS reflects the preferred orientation of mafic silicates (mainly biotite, and less frequently amphibole) by crystal lattice.

The plutonic rocks that have suffered no postintrusive deformation show the magnetic fabric created during the process of magma emplacement. The characteristic features of this magnetic fabric are as follows: the degree of AMS is relatively low, indicating only weak preferred orientation of magnetic minerals created during the liquid flow of magma; the magnetic fabric ranges from oblate to prolate according to the local character of the magma flow; the magnetic foliation is parallel to the flow plane and the magnetic lineation is parallel to the flow direction; the magnetic foliations are steep in stocks and upright sheetlike bodies in which the magma flowed vertically (Figure M84). On the other hand, it is oblique or horizontal in the bodies where the magma could not ascend vertically and moved in a more complex way. The magnetic lineation can be vertical, horizontal, or oblique according to the local direction of the magma flow. The magnetic fabric elements usually show a close relationship to the shapes of magmatic bodies and to magmatic structural elements, if observable. Many examples are shown in the book by Bouchez *et al.* (1997).

Some plutonic rocks exhibit a very high degree of AMS and a very high mean magnetic susceptibility. The magnetic fabric in these rocks usually indicates a preferred orientation of magnetite by grain shape that is younger than the fabric of the surrounding silicates and takes its shape from highly anisometric intergranular (interfoliation) spaces. This magnetic fabric is usually termed as the mimetic magnetic fabric (e.g., Hrouda *et al.*, 1971).

Some granites have suffered tectonic ductile deformation after their emplacement. In this process, the originally intrusive magnetic fabric is overprinted or even obliterated by the deformational magnetic fabric. The degree of AMS of such rocks is often much higher, because the ductile deformation is a relatively efficient mechanism for the reorientation of magnetic minerals. The magnetic foliations and lineations deviate from the directions of the intrusive fabric elements toward the directions of the principal strains. A good example includes granites of the West Carpathians whose magnetic fabrics are coaxial with those of surrounding metamorphic rocks and covering sedimentary rocks as result of ductile deformation associated with retrogressive deformation acting during creation and motion of the West Carpathian nappes (Hrouda *et al.*, 2002b).

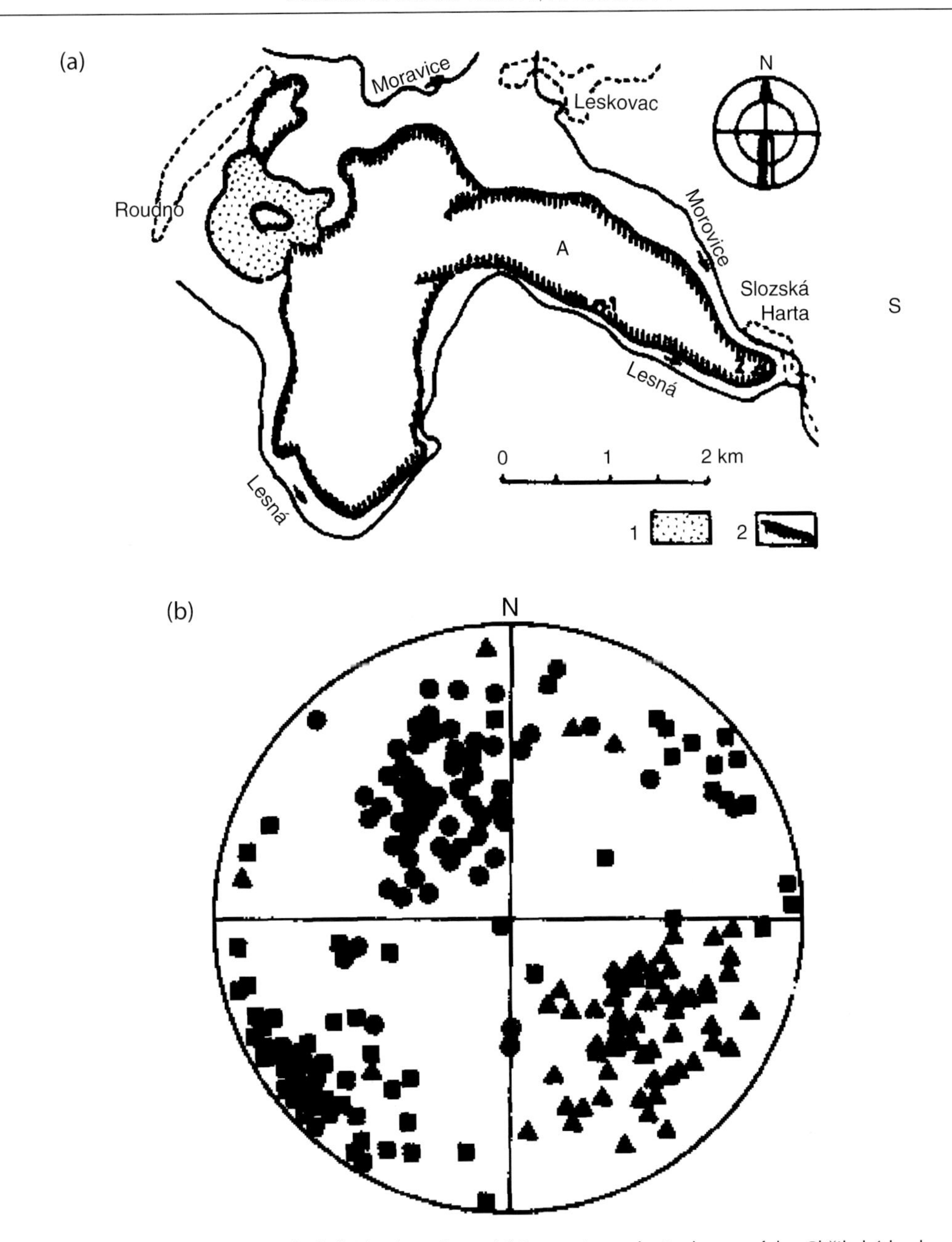

Figure M83 Magnetic fabric in basalt of the Chřibský les lava flow. (a) Synoptic geological map of the Chřibský les lava flow (A) and its environs (Legend: 1, basalt tuff; 2, boundaries of lava bodies), (b) orientations of principal susceptibilities in the quarry 1 (Legend: triangle, magnetic lineation; square, intermediate susceptibility; circle, magnetic foliation pole). Equal-area projection on lower hemisphere. (Adapted from Kolofíkova, 1976.)

AMS of metamorphic and deformed rocks

The AMS of metamorphic rocks in general and even of low-grade metamorphic rocks is considerably higher than the AMS in undeformed sedimentary and volcanic rocks. Hence, the mechanism orienting magnetic grains during low-grade metamorphism is very effective. It probably involves ductile deformation or recrystallization in an anisotropic stress field operating during metamorphism. The AMS therefore enables these processes to be studied.

AMS is an extremely powerful tool in investigating the progressive modification of the sedimentary magnetic fabric by ductile deformation that takes place in accretionary prisms (Figure M85). After sediment deposition, the AMS ellipsoid is usually oblate with the magnetic foliation near the bedding, and if a magnetic lineation is present, it is near the current direction. In initial stages of deformation, represented by shortening along the bedding, the AMS ellipsoid changes from oblate to triaxial; the degree of AMS may slightly decrease, the magnetic foliation remains parallel to bedding, but the magnetic lineation reorients perpendicular to the shortening direction. In progressing deformation, spaced cleavage develops and the magnetic lineation becomes parallel to the bedding/cleavage intersection line. The AMS ellipsoid becomes prolate. If the deformation continues, giving rise to the development of slaty cleavage, the AMS ellipsoid becomes oblate again, the degree of AMS considerably increases, the magnetic foliation pole reorientates parallel to the cleavage and the magnetic lineation remains parallel to the bedding/cleavage intersection line. If the deformation is very strong, the magnetic lineation may be dip-parallel in the cleavage plane.

AMS can be used in revealing the origin of folds (Hrouda, 1978; Hrouda *et al.*, 2000). Various types of folds differing in terms of the

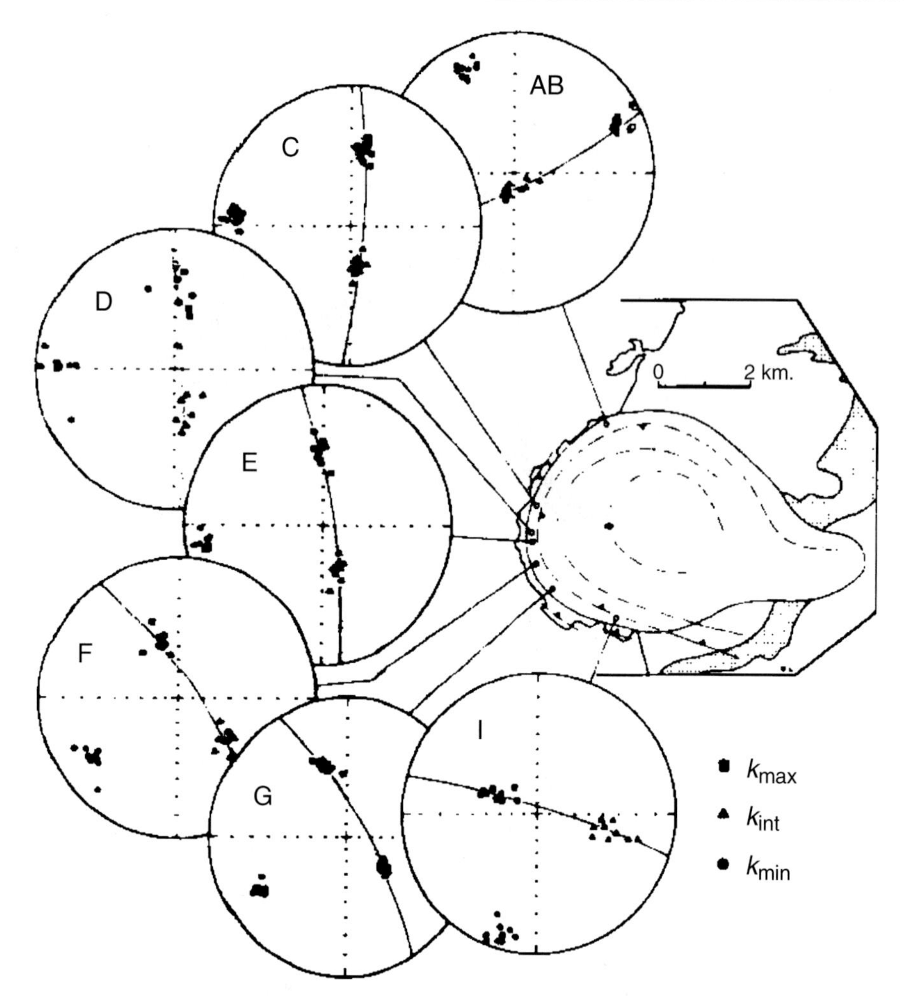

Figure M84 Magnetic fabric in the Flamanville granite, Normandy, France. Note steep magnetic foliations conforming the magmatic body shape. (Adapted from Cogné and Perroud, 1988.)

relation of the magnetic fabric to the fold curve were found, and simple techniques for the recognition of unfoldable and homogeneous folds were elaborated.

AMS can also be used in deciphering the sense of a shear movement (sinistral or dextral) (Rathore and Becke, 1980). This method is based on the assumption that the minimum susceptibility direction rotates during a simple shear movement in a manner similar to that of the minimum strain direction.

AMS can also be used as a strain indicator of deformed igneous rocks. During the process of deformation, the degree of AMS, in general, increases and the principal directions reorientate into the directions parallel to the strain directions. Studies of this type were made in dikes, volcanic bodies, and granitic rocks (e.g., Henry, 1977).

Strain analysis is one of the most laborious techniques of structural analysis and it is confined to rocks containing convenient strain indicators (oolites, concretions, reduction spots, lapilli, fossils). For this reason, many attempts have been made to use the AMS as a strain indicator. After revealing a close correlation between the directions of principal susceptibilities and principal strains (for examples, see Tarling and Hrouda, 1993), the quantitative relationship between the AMS and strain has been investigated theoretically through mathematical modeling (for review, see Hrouda, 1993), empirically through examining natural rocks with known strain (for review, see Borradaile, 1991), and experimentally through deforming rocks and rock analogs in the laboratory (for review, see Borradaile and Henry, 1997).

The first set of formulae describing a quantitative relationship between the AMS and strain was obtained through empirical studies of the AMS and strain (determined through investigation of natural strain markers—reduction spots) in Welsh slates (Figure M86) (Kneen, 1976; Wood *et al.*, 1976):

$$L = (k_1/k_2) = (s_1/s_2)^{a}$$
$$F = (k_2/k_3) = (s_2/s_3)^{a}$$
$$P = (k_1/k_3) = (s_1/s_3)^{a}$$

or in terms of natural strain:

$$\ln L = a(e_1 - e_2)$$
$$\ln F = a(e_2 - e_3)$$
$$\ln P = a(e_1 - e_3)$$

where $k_1 \geq k_2 \geq k_3$ are the principal susceptibilities, $s_1 \geq s_2 \geq s_3$ are the principal strains, and $\varepsilon_1 = \ln s_1$, $\varepsilon_2 = \ln s_2$, $\varepsilon_3 = \ln s_3$, are the principal natural strains. In later empirical studies, several other formulae were suggested, but as shown by Hrouda (1982), all of them can be converted, within an error of a few percent, into the above formulae.

In theoretical studies pioneered by Owens (1974), no simple formula for the AMS-to-strain relationship has been found. However, the curves

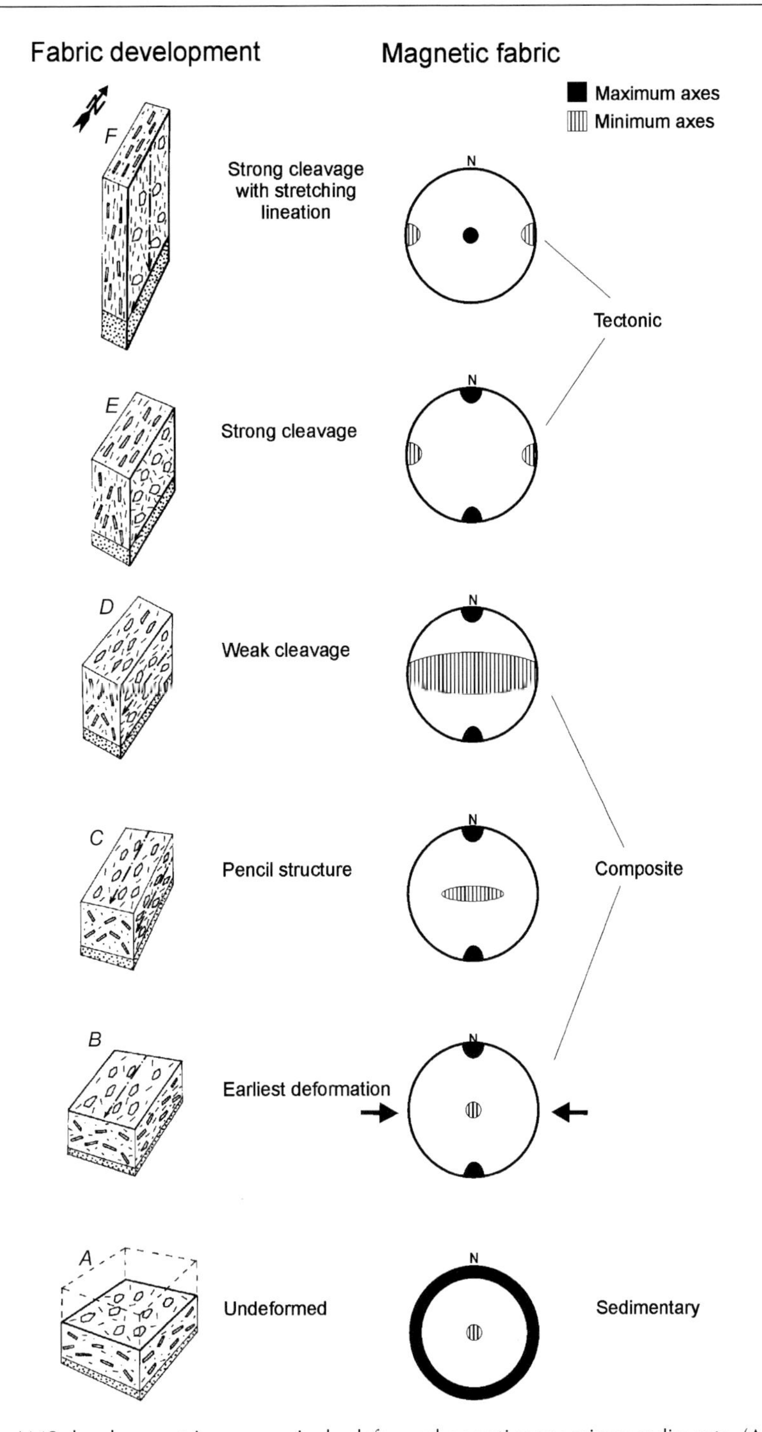

Figure M85 Scheme of the AMS development in progressively deformed accretionary prisms sediments. (Adapted from Pares *et al.*, 1999.)

representing the $\ln P$ vs $(\varepsilon_1 - \varepsilon_3)$ relationship (for summary, see Hrouda, 1993), considering various AMS carriers (magnetite, hematite, pyrrhotite, and paramagnetic mafic silicates), are represented by monotonously rising and only gently curved lines (Figure M87). In small strain ranges, mainly in their initial parts, the lines do not differ very much from straight lines. Consequently, using a reasonable simplification, the theoretical AMS-to-strain relationship can be described in the same way as the empirical AMS-to-strain relationship.

In experimental studies, the AMS of rock analogs deformed in laboratory was investigated (Borradaile and Alford, 1987, 1988). The rock analogs were relatively strongly magnetically anisotropic before deformation, and to describe the AMS to strain relationship the following formula respecting the predeformational AMS was used:

$$P' - P'_0 = B(\varepsilon_1 - \varepsilon_3),$$

where P'_0 and P' are the corrected degrees of AMS before and after deformation, respectively, and B is a constant. Analysis by Hrouda (1991) shows that this relationship can also be converted into the relationship presented for empirical data.

It is obvious from the above analysis that the AMS-to-strain relationship obtained through empirical, theoretical, and experimental studies

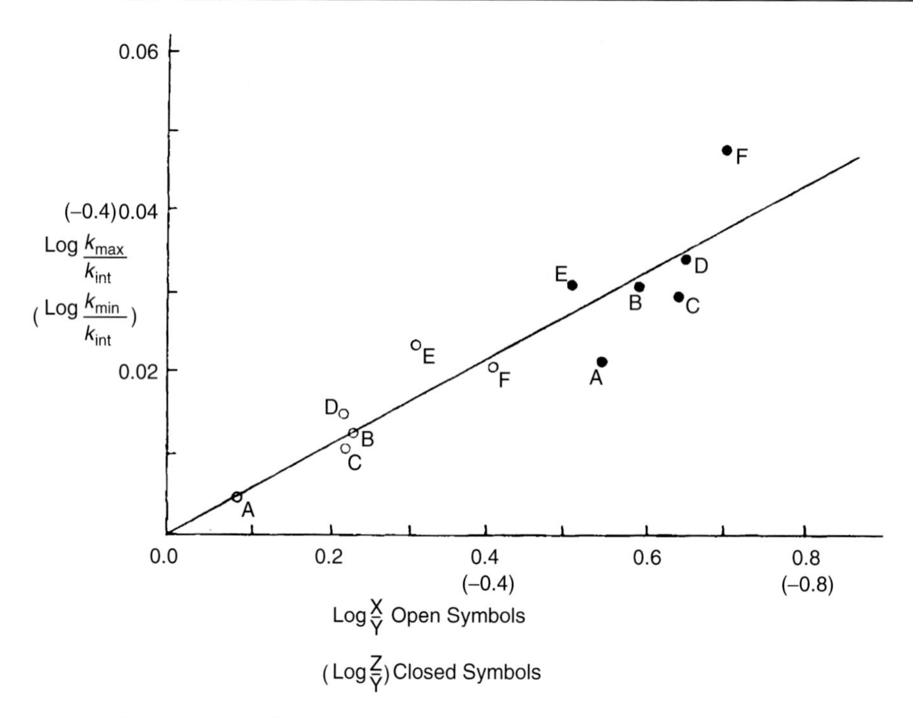

Figure M86 AMS to strain relationship in Welsh slates. (Adapted from Kneen, 1976.)

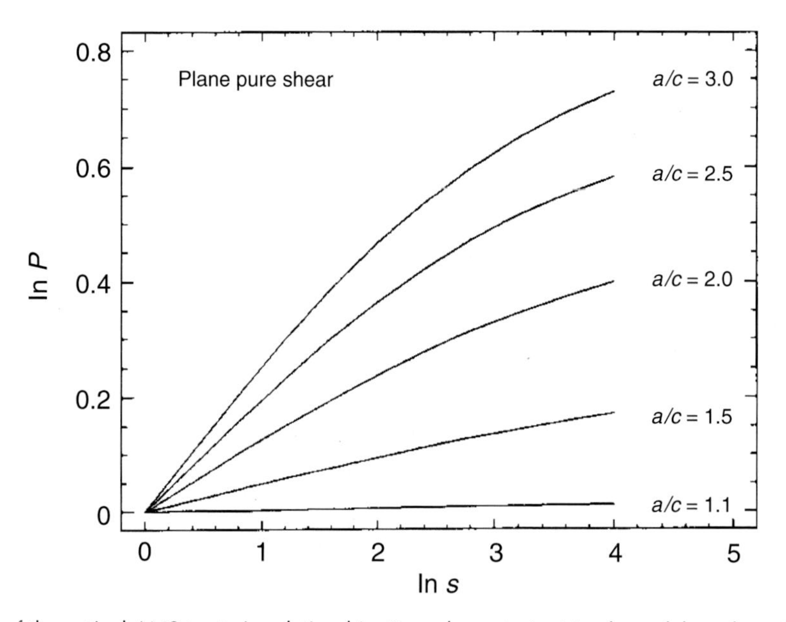

Figure M87 An example of theoretical AMS to strain relationship. Pure shear strain, March model, prolate ellipsoids magnetite grains with variable aspect ratios (*a/c*). (Adapted from Hrouda, 1993.)

can, with a reasonable simplification, be represented by the formulae first derived for the empirical relationship. Unfortunately, the above studies have also shown that the a values are not the same for all rock types, but vary according to the mineral carrying the AMS (Tarling and Hrouda, 1993) and also probably by the lithology of the rock investigated.

Even though a great majority of the empirical studies had revealed a relatively close correlation between AMS and strain, some studies had obtained results suggesting no correlation or even inverse correlation between the AMS and strain (Borradaile and Mothersill, 1984; Borradaile and Tarling, 1984). This implies that some rocks may have had complex deformation histories in which various rock components responded in different ways to overall strain and such rocks should not be used in the strain determination via AMS. Nevertheless, it is believed that in homogeneously deformed rocks, the AMS can serve as a quantitative strain indicator, which is evident from the great majority of empirical and experimental studies.

František Hrouda

Bibliography

Banerjee, S.K., and Stacey, F.D., 1967. The high-field torque-meter method of measuring magnetic anisotropy of rocks. In Collinson, D.W., Creer, K.M., and Runcorn, S.K. (eds.) *Methods in Paleomagnetism*. Amsterdam: Elsevier, pp. 470–476.

Borradaile, G.J., 1991. Correlation of strain with anisotropy of magnetic susceptibility (AMS). *PAGEOPH*, **135**: 15–29.

Borradaile, G.J., 2001. Magnetic fabrics and petrofabrics: their orientation distribution and anisotropies. *Journal of Structural Geology*, **23**: 1581–1596.

Borradaile, G., and Alford, C., 1987. Relationship between magnetic susceptibility and strain in laboratory experiments. *Tectonophysics*, **133**: 121–135.

Borradaile, G.J., and Alford, C., 1988. Experimental shear zones and magnetic fabrics. *Journal of Structural Geology*, **10**: 895–904.

Borradaile, G.J., and Henry, B., 1997. Tectonic applications of magnetic susceptibility and its anisotropy. *Earth-Science Reviews*, **42**: 49–93.

Borradaile, G.J., and Mothersill, J.S., 1984. Coaxial deformed and magnetic fabrics without simply correlated magnitudes of principal values. *Physics of the Earth and Planetary Interiors*, **35**: 294–300.

Borradaile, G.J., and Tarling, D., 1984. Strain partitioning and magnetic fabrics in particulate flow. *Canadian Journal of Earth Sciences*, **21**: 694–697.

Bouchez, J.-L., 2000. Anisotropie de susceptibilité magnétique et fabrique des granites. *Comptes Rendus Académie des Sciences Paris*, Sciences de la Terre et des planétes, **330**: 1–14.

Bouchez, J.-L., Hutton, D.W.H., and Stephens, W.E. (eds.), 1997. *Granite: From Segregation of Melt to Emplacement Fabric*. Dordrecht: Kluwer Academic Publishers, 358 pp.

Canon-Tapia, E., Walker, G.P.L., and Herrero-Bervera, E., 1994. Magnetic fabric and flow direction in basaltic Pahoehoe lava of Xitle Volcano, Mexico. *Journal of Volcanology and Geothermal Research*, **65**: 249–263.

Chadima, M., Hansen, A., Hirt, A.M., Hrouda, F., and Siemes, H., 2004. Phyllosilicate preferred orientation as a control of magnetic fabric: evidence from neutron texture goniometry and low and high-field magnetic anisotropy (SE Rhenohercynian Zone of Bohemian Massif). In: Martín-Hernández, F., Lüneburg, C.M., Aubourg, C., and Jackson, M. (eds.). Magnetic Fabric: Methods and Applications. *Geological Society*, London, Special Publications, **238**: 361–380.

Constable, C., and Tauxe, L., 1990. The bootstrap for magnetic susceptibility tensors. *Journal of Geophysical Research*, **95**: 8383–8395.

de Wall, H., 2000. The field dependence of AC susceptibility in titanomagnetites: implications for the anisotropy of magnetic susceptibility. *Geophysical Research Letters*, **27**: 2409–2411.

Dortman, N.B. (ed.), 1984. *Physical Properties of Rocks and Mineral Deposits* (in Russian). Moscow: Nedra, 455 pp.

Ellwood, B.B., and Whitney, J.A., 1980. Magnetic fabric of the Elberton granite, Northeast Georgia. *Journal of Geophysical Research*, **85**: 1481–1486.

Ernst, R.E., and Baragar, W.R.A., 1992. Evidence from magnetic fabric for the flow pattern of magma in the Mackenzie giant radiating dyke swarm. *Nature*, **356**: 511–513.

Ernst, R.E., and Pearce, G.W., 1989. Averaging of anisotropy of magnetic susceptibility data. In Agterberg, F.P., and Bonham-Carter, G. F. (eds.), *Statistical Applications in the Earth Sciences*, Geological Survey of Canada Paper 89-9, pp. 297–305.

Hamilton, N., and Rees, A.I., 1970. The use of magnetic fabric in palaeocurrent estimation. In Runcorn, S.K. (ed.) *Palaeogeophysics*. London: Academic Press, pp. 445–463.

Henry, B., 1977. Relations entre deformations et propriétés magnétiques dans des roches volcaniques des Alpes francaises. *Mémoires du B.R.G.M.* **91**: 79–86.

Henry, B., Jordanova, D., Jordanova, N., Souque, C., and Robion, P., 2003. Anisotropy of magnetic susceptibility of heated rocks. *Tectonophysics*, **366**: 241–258.

Hrouda, F., 1978. The magnetic fabric in some folds. *Physics of the Earth and Planetary Interiors*, **17**: 89–97.

Hrouda, F., 1980. Magnetocrystalline anisotropy of rocks and massive ores: a mathematical model study and its fabric implications. *Journal of Structural Geology*, **2**: 459–462.

Hrouda, F., 1982. Magnetic anisotropy of rocks and its application in geology and geophysics. *Geophysical Surveys*, **5**: 37–82.

Hrouda, F., 1991. Models of magnetic anisotropy variation in sedimentary sheets. *Tectonophysics*, **186**: 203–210.

Hrouda, F., 1993. Theoretical models of magnetic anisotropy to strain relationship revisited. *Physics of the Earth and Planetary Interiors*, **77**: 237–249.

Hrouda, F., 1994. A technique for the measurement of thermal changes of magnetic susceptibility of weakly magnetic rocks by the CS-2 apparatus and KLY-2 Kappabridge. *Geophysical Journal International*, **118**: 604–612.

Hrouda, F., 2002. Low-field variation of magnetic susceptibility and its effect on the anisotropy of magnetic susceptibility of rocks. *Geophysical Journal International*, **150**: 715–723.

Hrouda, F., and Ježek, J., 1999. Magnetic anisotropy indications of deformations associated with diagenesis. In Tarling, D.H., and Turner, P. (eds.), *Palaeomagnetism and Diagenesis in Sediments*. London: Geological Society, Special Publications 151, pp. 127–137.

Hrouda, F., and Kahan, S., 1991. The magnetic fabric relationship between sedimentary and basement nappes in the High Tatra Mts. (N Slovakia). *Journal of Structural Geology*, **13**: 431–442.

Hrouda, F., and Schulmann, K., 1990. Conversion of magnetic susceptibility tensor into orientation tensor in some rocks. *Physics of the Earth and Planetary Interiors*, **63**: 71–77.

Hrouda, F., Chlupacova, M., and Rejl, L., 1971. The mimetic fabric of magnetite in some foliated granodiorites, as indicated by magnetic anisotropy. *Earth Science and Planetary Interiors*, **11**: 381–384.

Hrouda, F., Siemes, H., Herres, N., and Hennig-Michaeli, C., 1985. The relation between the magnetic anisotropy and the *c*-axis fabric in a massive hematite ore. *Journal of Geophysics*, **56**: 174–182.

Hrouda, F., Jelínek, V., and Zapletal, K., 1997. Refined technique for susceptibility resolution into ferromagnetic and paramagnetic components based on susceptibility temperature-variation measurement. *Geophysical Journal International*, **129**: 715–719.

Hrouda, F., Krejčí, O., and Otava, J., 2000. Magnetic fabric in folds of the Eastern Rheno-Hercynian Zone. *Physics and Chemistry of the Earth* (A), **25**: 505–510.

Hrouda, F., Chlupáčová, M., and Novák, J.K., 2002a. Variations in magnetic anisotropy and opaque mineralogy along a kilometer deep profile within a vertical dyke of the syenogranite porphyry at Cínovec (Czech Republic). *Journal of Volcanology and Geothermal Research*, **113**: 37–47.

Hrouda, F., Putiš, M., and Madarás, J., 2002b. The Alpine overprints of the magnetic fabrics in the basement and cover rocks of the Veporic Unit (Western Carpathians, Slovakia). *Tectonophysics*, **359**: 271–288.

Jelínek, V., 1977. The statistical theory of measuring anisotropy of magnetic susceptibility of rocks and its application. *Geofyzika*, n.p. Brno, 88 pp.

Jelinek, V., 1978. Statistical processing of anisotropy of magnetic susceptibility measured on groups of specimens. *Studia Geophysica et Geodaetica*, **22**: 50–62.

Ježek, J., and Hrouda, F., 2000. The relationship between the Lisle orientation tensor and the susceptibility tensor. *Physics and Chemistry of the Earth* (A), **25**: 469–474.

Kolofikova, O., 1976. Geological interpretation of measurement of magnetic properties of basalts on example of the Chribsky les lava flow of the Velky Roudny volcano (Nizky Jesenik Mts.) (in Czech). *Časopis pro mineralogii a geologii*, **21**: 387–396.

Kneen, S.J., 1976. The relationship between the magnetic and strain fabrics of some haematite-bearing Welsh slates. *Earth and Planetary Science Letters*, **31**: 413–416.

Lowrie, W., and Hirt, A.M., 1987. Anisotropy of magnetic susceptibility in the Scaglia Rossa pelagic limestone. *Earth and Planetary Science Letters*, **82**: 349–356.
Lüneburg, C.M., Lampert, S.A., Lebit, H.D., Hirt, A.M., Casey, M., and Lowrie, W., 1999. Magnetic anisotropy, rock fabrics and finite strain in deformed sediments of SW Sardinia (Italy). *Tectonophysics*, **307**: 51–74.
MacDonald, W.D., and Palmer, H.C., 1990. Flow directions in ash-flow tuffs: a comparison of geological and magnetic susceptibility measurements, Tshirege member (upper Bandelier Tuff), Valles caldera, New Mexico, USA. *Bulletin of Volcanology*, **53**: 45–59.
Nagata, T., 1961. *Rock Magnetism*. Tokyo: Maruzen.
Nye, J.F., 1957. *Physical Properties of Crystals*. Oxford: Clarendon Press.
Owens, W.H., 1974. Mathematical model studies on factors affecting the magnetic anisotropy of deformed rocks. *Tectonophysics*, **24**: 115–131.
Owens, W.H., and Rutter, E.H., 1978. The development of magnetic susceptibility anisotropy through crystallographic preferred orientation in a calcite rock. *Physics of the Earth and Planetary Interiors*, **16**: 215–222.
Pares, J.M., van der Pluijm, B.A., and Dinares-Turell, J., 1999. Evolution of magnetic fabrics during incipient deformation of mudrock (Pyrenees, northern Spain). *Tectonophysics*, **307**: 1–14.
Park, J.K., Tanczyk, E.I., and Desbarats, A., 1988. Magnetic fabric and its significance in the 1400 Ma Mealy diabase dykes of Labrador, Canada. *Journal of Geophysical Research*, **93**: 13689–13704.
von Rad, U., 1971. Comparison between "magnetic" and sedimentary fabric in graded and cross-laminated sand layers, Southern California. *Geologische Rundschau*, **60**: 331–354.
Raposo, M.I.B., and Ernesto, M., 1995. Anisotropy of magnetic susceptibility in the Ponta Grossa dyke swarm (Brazil) and its relationship with magma flow direction. *Physics of the Earth and Planetary Interiors*, **87**: 183–196.
Rathore, J.S., and Becke, M., 1980. Magnetic fabric analyses in the Gail Valley (Carinthia, Austria) for the determination of the sense of movements along this region of the Periadriatic Line. *Tectonophysics*, **69**: 349–368.
Rees, A.I., 1983. Experiments on the production of transverse grain alignment in a sheared dispersion. *Sedimentology*, **30**: 437–448.
Rees, A.I., and Woodall, W.A., 1975. The magnetic fabric of some laboratory-deposited sediments. *Earth and Planetary Science Letters*, **25**: 121–130.
Richter, C., and van der Pluijm, B.A., 1994. Separation of paramagnetic and ferrimagnetic susceptibilities using low temperature magnetic susceptibilities and comparison with high field methods. *Physics of the Earth and Planetary Interiors*, **82**: 111–121.
Scheidegger, A.E., 1965. On the statistics of the orientation of bedding planes, grain axes, and similar sedimentological data. *US Geological Survey Professional Paper*, **525-C**: 164–167.
Siegesmund, S., Ullemeyer, K., and Dahms, M., 1995. Control of magnetic rock fabrics by mica preferred orientation: a quantitative approach. *Journal of Structural Geology*, **17**: 1601–1613.
Taira, A., 1989. Magnetic fabrics and depositional processes. In Taira, A., and Masuda, F., (eds.), *Sedimentary Facies in the Active Plate Margin*. Tokyo: Terra Publications, pp. 43–77.
Tarling, D.H., and Hrouda, F., 1993. *The magnetic anisotropy of rocks*. London: Chapman & Hall, 217 pp.
Urrutia-Fucugauchi, J., 1981. Preliminary results on the effects of heating on the magnetic susceptibility anisotropy of rocks. *Journal of Geomagnetism and Geoelectricity*, **33**: 411–419.
Uyeda, S., Fuller, M.D., Belshe, J.C., and Girdler, R.W., 1963. Anisotropy of magnetic susceptibility of rocks and minerals. *Journal of Geophysical Research*, **68**: 279–292.
Wood, D.S., Oertel, G., Singh, J., and Bennet, H.G., 1976. Strain and anisotropy in rocks. *Philosophical Transactions of the Royal Society of London, Series A*, **283**: 27–42.
Woodcock, N.H., 1977. Specification of fabric shapes using an eigenvalue method. *Geological Society of America Bulletin*, **88**: 1231–1236.

Cross-references

Fisher Statistics
Magnetic Remanence, Anisotropy
Magnetic Susceptibility

MAGNETIC SUSCEPTIBILITY, ANISOTROPY, EFFECTS OF HEATING

Study of the anisotropic properties of magnetic susceptibility and remanent magnetizations is an active field of research in paleomagnetism. These studies have important applications in petrofabrics, structural geology, metamorphism, rock magnetism, volcanology, and tectonics. Here we concentrate on the anisotropy of magnetic susceptibility (AMS) of rocks measured at low magnetic fields, and in particular, on the effects of laboratory heating on the AMS of rocks. Mineralogical changes resulting from heating samples in the laboratory have long been recognized and studied (see Table M6). The potential use of temperature-induced effects to investigate composite magnetic fabrics needs further development.

Introduction

The magnetic susceptibility of rocks given by the ratio between magnetization (M) and applied magnetic field (H) exhibits anisotropic properties. These properties have been used to investigate the characteristics of rock fabric (Tarling and Hrouda, 1993).

$$M_i = k_{i,j} H_j$$

The susceptibility $k_{i,j}$ of anisotropic samples is a second-order tensor (Nye, 1957), which is usually represented in terms of the principal susceptibility axes (maximum k_1, intermediate k_2, and minimum k_3). The

Table M6 Laboratory data on heating-induced thermochemical reactions

From	To	°C
Igneous		
Impure titanomagnetites	Magnetite	>300
Magnetite	Maghemite	150–250
Olivines	Magnetite	>300
Pyrite	Magnetite	350–500
Maghemite	Hematite	350–450
Magnetite	Hematite	>500
Pyroxenes	Magnetite	>600
Sediments		
Siderite	Magnetite	>200
Lepidocrocite	Maghemite	220–270
Goethite	Hematite	110–120
Maghemite	Hematite	350–450
Pyrite	Magnetite	350–500
Magnetite	Hematite	>500
Hematite	Magnetite	>550

Note: The actual temperatures are strongly dependent on grain size and shape, redox conditions, rate of heating, etc., and are only preliminary and indicative. Similarly, the magnetic product resulting from heating is dependent on the specific redox conditions at that temperature and it is likely that many of these reactions take place slowly at lower temperatures under natural conditions—for example, goethite may change to magnetite at 100°C–120°C under low oxidation conditions (after Tarling, 1983; Tarling and Hrouda, 1993).

AMS is generally measured at low magnetic fields, and several instrumental systems have been developed and commercially produced in the last decades (e.g., Kappabridge, Saphire, Bartington, Digico, Molspin, and PAR-SM2). AMS is also determined from measurements in cryogenic magnetometers.

AMS is usually analyzed in terms of the orientation of the principal susceptibility axes and the AMS parameters. The parameters based on the magnitudes of principal susceptibilities have been proposed to quantify the anisotropy degree, lineation, foliation, and shape of susceptibility ellipsoid (Tarling and Hrouda, 1993). The magnitude and shape of susceptibility ellipsoid have been determined from various plots, which include the Flinn diagram of lineation as a function of foliation and the plot of shape parameter as a function of anisotropy degree (Jelinek, 1981). They permit distinction of susceptibility ellipsoid, with prolate, oblate, and triaxial (neutral) ellipsoids.

In paleomagnetic and rock magnetic studies, several methods have been developed to investigate the temperature dependence of magnetic properties and remanent magnetizations (Tarling, 1983; Dunlop and Özdemir, 1997). These studies use laboratory heating or thermal treatment on samples under controlled conditions (e.g., thermal demagnetization, thermomagnetic measurements, temperature-dependent susceptibility, and paleointensity experiments). Understanding the effects of heating on the magnetic properties and magnetic mineralogy is then an active research field. This note is concerned with the effects of laboratory heating on the AMS of rocks. Thermal treatment has been proposed to enhance AMS in rocks, and used to investigate on temperature-induced fabric changes.

The low-field magnetic susceptibility changes as a result of growth/destruction/transformation of iron oxides, sulfides and carbonates, and hydroxides. Oxidation and reduction processes occur at different temperature ranges during thermal treatment. Magnetic oxides can result from transformation of iron sulfides (pyrite, pyrrhotite). Several methods have been implemented to study mineral alteration during laboratory heating (e.g., Dekkers, 1990; Van Velzen and Zijderveld, 1992), and to promote, reduce, or avoid oxidation/reduction of magnetic minerals (for instance, use of inert or reducing atmospheres in heating/cooling chambers of ovens).

Susceptibility is a bulk property, resulting from relative contributions of the diamagnetic, paramagnetic, and ferromagnetic minerals. It depends on its mineralogy, grain size and shape, mineral distribution, crystallographic preferred orientation, layering, etc. Interpretation of AMS in terms of rock fabric can often be complex, particularly for composite or multicomponent fabrics. Thermal treatment has been proposed to enhance and quantify given components of AMS fabric (Urrutia-Fucugauchi, 1979, 1981), where heating produces new magnetic minerals from the phyllosilicates and ferromagnetic minerals, which mimic the crystallographic structure in micas and clays.

Temperature-dependent mineral transformations

Mineralogical changes resulting from heating samples in the laboratory have long been identified and studied (e.g., Abouzakhm and Tarling, 1975; Urrutia-Fucugauchi, 1979, 1981). Susceptibility increases and decreases during heating results mainly from formation of iron-titanium oxides (titanomagnetites and magnetite) and transformation of oxides (hematization of magnetite), respectively. Magnetic oxides can be produced from a range of minerals and processes, such as iron sulfides, carbonates and silicate minerals, hydroxides, and other iron oxides (e.g., Tarling, 1983; Borradaile *et al.*, 1991; Tarling and Hrouda, 1993; Henry *et al.*, 2003). Lithologies in which iron oxides are not present, other minerals play an important role and constitute special cases that require further investigation.

Pyrrhotite-bearing rocks can display changes due to transformation of antiferromagnetic hexagonal pyrrhotite to ferrimagnetic monoclinic pyrrhotite (Bina *et al.*, 1991). Most often, complex series of mineralogical changes occur at different temperature ranges. Oxidation of pyrrhotite at high temperature results in further susceptibility increase (Dekkers, 1990). Magnetite formation between 500 and 725°C result in susceptibility increase, which can be followed by decrease at higher temperature as a result of hematization (biotite granites; Trinidade *et al.*, 2001). Mineral transformations depend not only on mineral type, but also on grain size, mineral impurities, time, atmosphere conditions (inert, oxidizing, or reducing conditions), and water content. Temperature-induced changes in the bulk susceptibility and AMS have often been related to the formation of magnetite. Perarnau and Tarling (1985) proposed magnetite formation from vermiculite at 450°C and from illite at >850°C.

Thermal enhancement of AMS

Early studies of AMS thermal enhancement used samples from Late Precambrian tillites, Late Jurassic red beds, and Late Tertiary volcanics (Urrutia-Fucugauchi, 1981). Thermal treatment resulted in enhancing given components of the AMS fabric in the first two cases, and no significant changes in the AMS axis pattern for the volcanic rocks. Thermal enhancement was proposed as a method to study composite fabrics (Urrutia-Fucugauchi, 1979). Subsequent studies applying thermal treatment to glacial deposits, slates, sandstones, and red beds supported that heating enhanced the primary AMS fabric and pointed out the complexity of heating-induced effects on magnetic mineralogy (e.g., Urrutia-Fucugauchi and Tarling, 1983; Perarnau and Tarling, 1985; Borradaile *et al.*, 1991; Borradaile and Lagroix, 2000).

An example of simple thermal enhancement of AMS is illustrated in Figure M88. Samples studied are Cretaceous sandstones sampled near a fault system in Venezuela (Perarnau and Tarling, 1985). The initial fabric is poorly defined with mixed maximum, intermediate, and maximum axes (Figure M88a). Maximum and minimum axes are oriented normal to bedding, and maximum, intermediate, and minimum axes conform a girdle distribution on the bedding plane. After heating to 350°C, the AMS axes pattern changes, resulting in minimum axes preferentially aligned normal to bedding, and maximum axes on the bedding plane and oriented to the north (Figure M88b). The thermal enhanced fabric is interpreted as a depositional sedimentary fabric, resulting from new magnetic minerals that mimic the crystallographic structure and shape of micas and clays. Samples from sites closer to the fault show a fault-parallel oblateness effect (Perarnau and Tarling, 1985).

Composite AMS fabrics are in general difficult to interpret, and several methods have been proposed in an attempt to separate the different mineral contributions (e.g., Hrouda and Jelinek, 1990). Other methods using measurements at high and low fields, low-temperature observations, etc. have also been used contributions (e.g., Rochette and Fillion, 1988; Pares and Van der Pluijm, 2002).

An example of magnetic fabric enhancement of a given component of a composite fabric in glacial sediments induced by laboratory heating is summarized in Figures M89 and M90. Samples used come from a Late Precambrian Port Askaig Formation from southwestern Scotland (Urrutia-Fucugauchi, 1981; Urrutia-Fucugauchi and Tarling, 1983). The mean susceptibility showed little change up to 300–400°C and increased markedly after heating at 400 and 500°C (Figure M90). The orientation of principal susceptibility axes before thermal treatment shows a pattern with vertical minimum axes and maximum axes on the bedding plane (with some scatter; Figure M89a). This pattern conforms that expected for a depositional sedimentary fabric in sediments deposited under the influence of dominant currents (indicated by the lineation). With thermal treatment up to 500°C, there are marked changes in bulk susceptibility and AMS axes pattern (Figure M89b). The maximum axes are grouped close to paleovertical and minimum axes lie in the paleohorizontal plane. Thus, heating-induced changes result in an inverse fabric. The scatter in AMS axes is reduced, which gives a well-defined NW-SE lineation (Figure M89a,b). This new fabric was interpreted in terms of compressional stresses acting parallel to bedding on strata free to extend in directions normal to bedding and compression. Deformation results on lineation increase eventually exceeding foliation, with predominantly prolate-shaped ellipsoids.

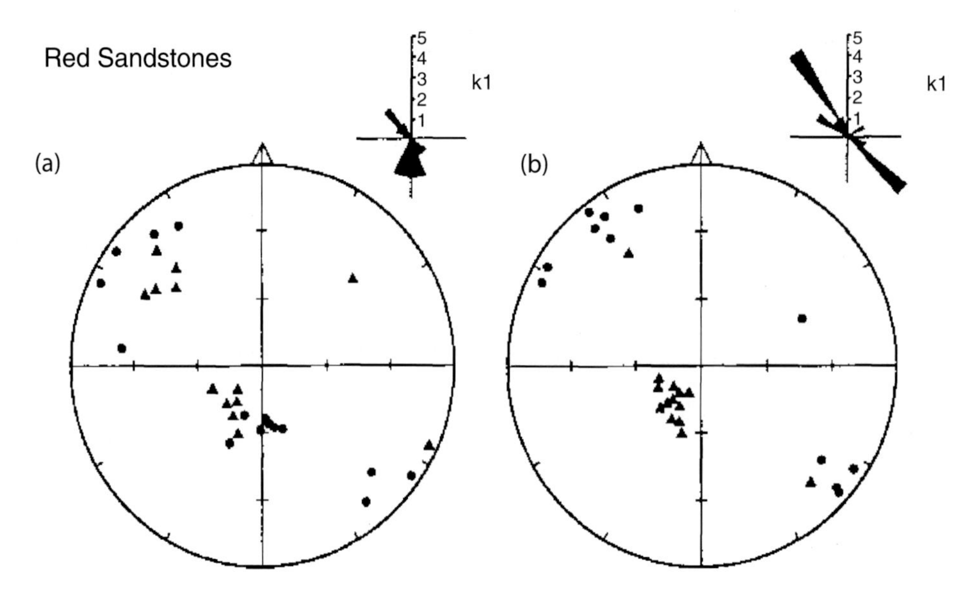

Figure M88 Example of thermal enhancement of AMS (for explanation see text).

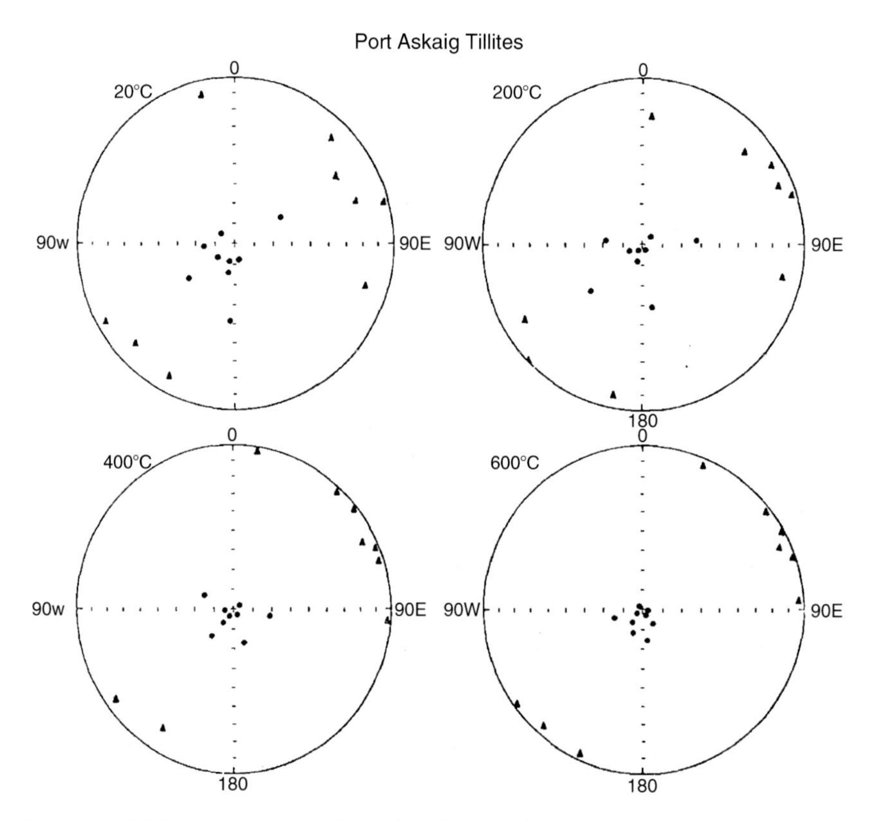

Figure M89 Example of magnetic fabric enhancement induced by laboratory heating. Orientation of principal susceptibility axes before and after thermal treatment (for explanation see text).

The initial fabric can be totally obliterated, resulting in oblate ellipsoids but with different orientation to bedding. Maximum axes align normal to bedding and minimum axes parallel to compression (Graham, 1966). The AMS after heating to 500 °C then suggests a magnetic fabric that is strain-controlled with compressional stresses locally NW-SE oriented, in agreement with geologic and petrofabric studies which indicate strong deformation of this region during the Caledonian orogeny (Borradaile, 1979). AF and thermal demagnetization of remanence suggests that magnetic carriers are slightly oxidized titanomagnetites. Thermal demagnetization to 300 °C–400 °C produces further oxidation with some hematization accompanied by increase of coercive force and Curie temperature. Heating to higher temperatures results in the formation of new magnetite from paramagnetic minerals, which accounts for the marked increase of mean susceptibility (Figure M90). Formation of new magnetic minerals occurs preferentially controlled by existing tectonic fabrics.

Borradaile and Lagroix (2000) presented data for high-grade gneisses from the Kapuskasing granulite zone. They used measurements of AMS

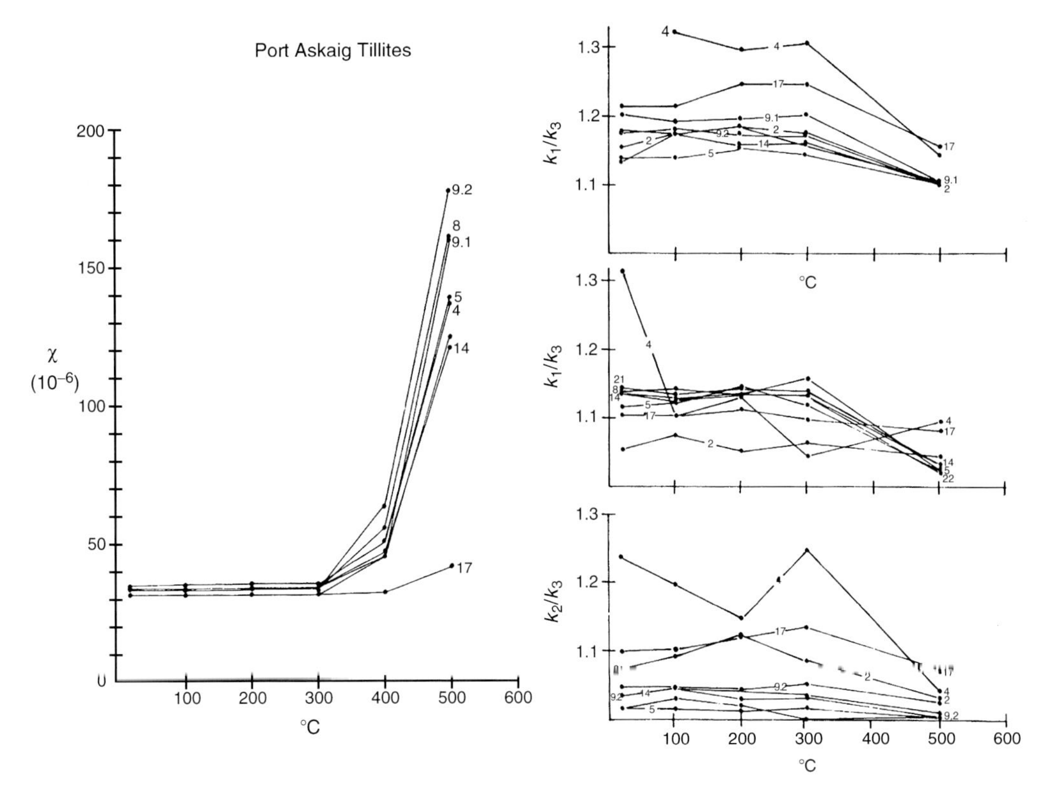

Figure M90 Example of magnetic fabric enhancement induced by laboratory heating: mean susceptibility (for explanation see text).

and anisotropy of anhysteretic remanent magnetization (AARM) to define the magnetic fabric. AMS was mainly due to the contributions of magnetite and silicate minerals. Thermal treatment does not result in increase of mean susceptibility, but definition of AMS and AARM axial orientation is improved. Heating results in enhancement of the magnetite subfabric, which can then be quantified. This magnetite subfabric reveled by the thermal treatment is also shown in the heated AARM fabric, which is also preferentially enhanced by the laboratory heating.

Other examples of successful thermal enhancement of component of composite fabrics have been observed in granites, particularly in S-type granites in which fabrics arise from biotite and tourmaline minerals (Trinidade *et al.*, 2001; Mintsa *et al.*, 2002). The AMS of paramagnetic phyllosilicates in granites generally correlate well with the crystallographic fabric (e.g., Borradaile and Werner, 1994). Mintsa *et al.* (2002) presented results for the Carnmenellis granite, where thermal treatment above 500 °C results in enhancement of the biotite fabric overcoming the tourmaline contribution. Secondary magnetite produced by laboratory heating within biotite crystals mimics the phyllosilicate cleavage, which in turn corresponds to magmatic (linear) flow structures in the granitic body. Comparison of the AMS fabric lineations before and after the laboratory thermal treatment shows clear differences between the composite initial fabric produced by the biotite fabric and the (inverse) tourmaline fabric (Mintsa *et al.*, 2002).

Discussion

Measurement of magnetic susceptibility has long been used to monitor changes in magnetic minerals produced by heating of rocks in the laboratory. The effects of heating on the AMS and the enhancement of components of AMS fabric open new exciting possibilities in petrofabric studies. In the simple case, heating produces new magnetic minerals, which mimic the crystallographic structure in micas and clays (Urrutia-Fucugauchi, 1979, 1981).

The initial studies showed, however, that heating-induced changes in AMS are more complex than simple enhancement of the magnetic fabric. Recent studies investigating different lithologies (e.g., granites, gabbros, diorites, ignimbrites, volcanoclastic sediments, loess, and limestones) confirm that heating does not always result in simple AMS fabric enhancement (Henry *et al.*, 2003). Henry *et al.* (2003) distinguish three different cases when comparing the heating-induced fabric with the preexisting fabric. First case is simple enhancement of preexisting fabric or masked poorly defined initial fabric. Second case corresponds to formation of an inverse fabric. Third case is the formation of a different fabric, which shows principal susceptibility axes related to petrostructural elements, but differing from the initial fabric.

In the cases where thermal treatment results in increasing the bulk susceptibility and anisotropy degree, heating-induced changes facilitate the measurement of the susceptibility tensor and principal susceptibility axes. Susceptibility increase during heating mainly results from the formation of iron-titanium oxides, principally titanomagnetites and magnetite. Susceptibility decrease results from transformation of oxides, like hematization of magnetite. Iron-titanium oxides could be produced from a range of minerals and processes; most often including iron sulfides, iron carbonates, iron silicate minerals, hydroxides, and other iron oxides (Tarling, 1983; Henry *et al.*, 2003). Lithologies in which iron oxides do not play a dominant role present special cases that are still not fully investigated for magnetic properties.

Study of AMS changes induced by progressive thermal treatment appears as a useful method to investigate the fabric of rocks. Laboratory stepwise heating has potential for enhancement of AMS and revealing masked or cryptic fabrics. Further investigation of heating-induced effects in mineralogy, grain size, and textural changes, applied systematically to different lithologies, is clearly required. The methodology permits studying composite fabrics and may give further insight on the relationship of AMS to the different petrofabrics arising from stratification, depositional conditions, current-induced features, strain effects, cleavage, microfracturing, etc.

Acknowledgments

Partial support for the magnetic fabric studies has been provided by DGAPA projects IN-116201. Thanks are due to Editor E. Herrero-Bervera for the invitation to contribute, encouragement, and useful comments.

Jaime Urrutia-Fucugauchi

Bibliography

Abouzakhm, A.G., and Tarling, D.H., 1975. Magnetic anisotropy and susceptibility from northwestern Scotland. *Journal of the Geological Society London*, **131**: 983–994.

Bina, M., Corpel, J., Daly, L., and Debeglia, N., 1991. Transformation de la pyrrhotite en magnetite sous l'effet de la temperature: une source potentielle d'anomalies magnetiques. *Comptes Rendus de l'Académie des Sciences de Paris*, **313**: 487–494.

Borradaile, G.J., 1979. Strain study of the Caledonides in the Islay region, S.W. Scotland: implications for strain histories and deformation mechanisms in greenschists. *Journal of the Geological Society London*, **136**: 77–88.

Borradaile, G.J., and Henry, B., 1997. Tectonic applications of magnetic susceptibility and its anisotropy. *Earth-Science Reviews*, **42**: 49–93.

Borradaile, G.J., and Lagroix, F., 2000. Thermal enhancement of magnetic fabrics in high grade gneisses. *Geophysical Research Letters*, **27**: 2413–2416.

Borradaile, G.J., and Werner, T., 1994. Magnetic anisotropy of some phyllosilicates. *Tectonophysics*, **235**: 223–248.

Borradaile, G.J., Mac Kenzie, A., and Jensen, E., 1991. A study of colour changes in purple-green slate by petrological and rock-magnetic methods. *Tectonophysics*, **200**: 157–172.

Dekkers, M.J., 1990. Magnetic monitoring of pyrrhotite alteration during thermal demagnetization. *Geophysical Research Letters*, **17**: 779–782.

Dunlop, D.J., and Özdemir, O., 1997. *Rock Magnetism: Fundamentals and Frontiers.* Cambridge: Cambridge University Press, 573 pp.

Graham, J.W., 1966. Significance of magnetic anisotropy in Appalachian sedimentary rocks. In Steinhart, J.S., and Smith, T.J. (eds.), *The Earth Beneath the Continents.* Geophysical Monograph Series 10. Washington, DC: American Geophysical Union, pp. 627–648.

Jelinek, V., 1981. Characterization of the magnetic fabric of rocks. *Tectonophysics*, **79**: 63–67.

Henry, B., Jordanova, D., Jordanova, N., Souque, C., and Robion, P., 2003. Anisotropy of magnetic susceptibility of heated rocks. *Tectonophysics*, **366**: 241–258.

Hrouda, F., and Jelinek, V., 1990. Resolution of ferrimagnetic and paramagnetic anisotropies in rocks, using combined low-field and high-field measurements. *Geophysical Journal International*, **103**: 75–84.

Mintsa Mi Nguema, T., Trinidade, R.I.F., Bouchez, J.L., and Launeau, P., 2002. Selective thermal enhancement of magnetic fabrics from the Carnmenellis granite (British Cornwall). *Physics and Chemistry of Earth*, **27**: 1281–1287.

Nye, J.F., 1957. *Physical Properties of Crystals.* London: Oxford University Press, 322 pp.

Pares, J.M., and Van der Pluijm, B.A., 2002. Phyllosilicate fabric characterization by low-temperature anisotropy of magnetic susceptibility (LT-AMS). *Geophysical Research Letters*, **29**: 2215, doi:10.1029/2002GL015459.

Perarnau, A., and Tarling, D.H., 1985. Thermal enhancement of magnetic fabric in Cretaceous sandstone. *Journal of Geological Society London*, **142**: 1029–1034.

Rochette, P., and Fillion, C., 1988. Identification of multicomponent anisotropies in rocks using various field and temperature values in a cryogenic magnetometer. *Physics of the Earth and Planetary Interiors*, **51**: 379–386.

Tarling, D.H., 1983. *Palaeomagnetism.* London: Chapman & Hall, 379 pp.

Tarling, D.H., and Hrouda, F., 1993. *The Magnetic Anisotropy of Rocks.* London: Chapman & Hall, London, 217 pp.

Trinidade, R.I.F., Mintsa Mi Nguema, T., and Bouchez, J.L., 2001. Thermally enhanced mimetic fabric of magnetite in a biotite granite. *Geophysical Research Letters*, **28**: 2687–2690.

Urrutia-Fucugauchi, J., 1979. Variation of magnetic susceptibility anisotropy versus temperature. Thermal cleaning for magnetic anisotropy studies? European Geophysical Union Meeting, Vienna, Austria.

Urrutia-Fucugauchi, J., 1981. Preliminary results on the effects of heating on the magnetic susceptibility anisotropy of rocks. *Journal of Geomagnetism and Geoelectricity*, **33**: 411–419.

Urrutia-Fucugauchi, J., and Tarling, D.H., 1983. Palaeomagnetic properties of Eaocambrian sediments in northwestern Scotland: implications for world-wide glaciation in the Late Precambiran. *Paleogeography Paleoclimatology Paleoecology*, **1**: 325–344.

Van Velzen, A.J., and Zijderveld, J.D.A., 1992. A method to study alterations of magnetic minerals during thermal demagnetization applied to a fine-grained marine marl (Trubi formation, Sicily). *Geophysics Journal International*, **110**: 79–90.

MAGNETIC SUSCEPTIBILITY, ANISOTROPY, ROCK FABRIC

Although the anisotropy of magnetic susceptibility (AMS) of rocks first called the attention of scientists because of its possible influence on the direction of the remanent magnetization, it was relatively soon realized that the AMS of a rock is in general too weak to exert a noticeable influence in the paleomagnetic record (e.g., Uyeda *et al.*, 1963). On the other hand, due to the close relationship that exists between the crystalline structure of minerals and their AMS (Nye, 1960), it was well established that the AMS should bear a very close relationship with the mineral fabric of the rock. It is therefore not surprising that the modern interest in the study of the AMS of rocks resides mainly in its value as a petrofabric indicator rather than as a paleomagnetic tool.

There are several sources for the AMS of rock. The most common of these sources is the anisotropy arising from the crystalline structure of the minerals (Nye, 1960). Almost all minerals will have an AMS that is controlled by these sources, although their bulk susceptibility will be generally small. A second source of AMS is found only in some minerals with a strong permeability. These minerals will show an anisotropy that arises from the overall shape of the grain rather than from its crystalline structure. For this reason, this source of AMS is called shape anisotropy. Finally, when two or more ferromagnetic minerals are in close proximity to each other, they can interact magnetically and such interaction becomes the source of a magnetic anisotropy. If the distance between the grains is too small (<2 grain radius) the interactive effects can be so strong that they will overcome the shape anisotropy of individual minerals (Cañón-Tapia, 1995). All of these sources of anisotropy contribute in different amounts to the total AMS measured in the laboratory. In general, the contributions are not simple to express because they depend on the relative proportion of minerals with a given value of bulk susceptibility and degree of anisotropy (Cañón-Tapia, 2001). Traditionally, the presence of a mineral phase with a high value of bulk susceptibility has been assumed to indicate an AMS dominated by such phase, but recent evidence suggests that the main source of anisotropy will reside in the most anisotropic phase even if such phase has a lower value of bulk susceptibility (Borradaile and Gauthier, 2003). Therefore, some attention has to be given to the mineral composition of the rock type under examination to assess the probable sources of its AMS, and hence to make a meaningful interpretation.

The variables responsible for the acquisition of a rock fabric also depend on the type of rock that is considered. Although a mineral fabric is acquired as a response to deformation forces, the total amount of deformation differs greatly from rock to rock, as do the actual mechanism of deformation experienced by the rock. Consequently, the AMS of a rock specimen needs to be examined to the light of the lithology under consideration to reach a meaningful interpretation. Notably, this lithological dependence of the AMS has been disregarded through most of the history of the AMS method probably because advances in the identification of the forces responsible for fabric acquisition of different lithologies have been uneven in time.

Undoubtedly, the study of fabric structures in sedimentary and deformed rocks has a longer history. Therefore it is not surprising that the earlier works establishing a link between sedimentary features (bedding) and AMS were made on these lithologies (Ising, 1942; Granar, 1957). Among the first attempts to explain theoretically the magnetic fabric acquisition of sedimentary rocks is the work by Rees (1961), followed by experiments in which different sedimentation conditions were explored (Rees, 1966, 1968; Hamilton *et al.*, 1968). These works showed that the influence of the magnetic moment induced in a ferromagnetic particle that is about to be deposited will only become important for very small particles, because the magnetic moment induced in the particle is proportional to the cube of the particle dimension whereas the influence of either gravitational or hydrodynamic forces is proportional to its fourth power. Also, the formation of ripples and cross-lamination in sediments deposited under relatively strong currents was shown to affect the measured AMS because each sample might contain different sets of cross-laminations each with a different orientation. Nevertheless, it was concluded that in general the AMS measurements of sediments reflects the gravitational and hydrodynamic forces acting during deposition, and therefore this method yields results equivalent to those concerning the preferred orientation of particles as obtained from optical observation.

Departures from the AMS obtained under experimental conditions (or so-called "primary fabrics") were considered to be "anomalous" by Hamilton and Rees (1970). The causes of the anomalous fabrics were associated with deformation (whether penecontemporaneous with sediment deposition as for example by gravitational slump or by tectonic-related deformation), mineral alteration, or with the activity of plants and burrowing animals. In all of these cases, the magnetic fabric was predicted to yield a larger site-scatter of directions of the principal susceptibilities of individual specimens than it would be found if the sediments in the site had formed by normal depositional processes. Consequently, the scatter of AMS axes became to be synonymous of fabric alteration processes.

In contrast to sediments, prior to the 1970s the interest in the fabric of igneous rocks was almost null (Smith, 2002). Therefore, it is not surprising that most assumptions concerning the interpretation of AMS in sediments were adopted indiscriminately when the first studies of the AMS of igneous rocks were made. In particular, the idea of a mineral fabric as the result of the preferred orientation of minerals already present during viscous flow of melted rock became easily adopted as equivalent to the processes of fabric acquisition studied in sediments (Khan, 1962). As time passed, however, it was realized that the AMS of igneous rocks presents much more variability than the AMS of sediments, leading to a series of confusions and contradictions concerning the utility of AMS in this type of rocks (for a recent historical recount focusing on this type of rocks see Cañón-Tapia, 2004). Although some discrepancies can exists concerning the exact mechanism responsible for the fabric acquisition of a particular rock, most workers will nowadays acknowledge that the mechanisms of deformation of igneous rocks are varied and complex, and have a different character than the (deformation) mechanisms responsible for the acquisition of a mineral fabric in sedimentary rocks. Consequently, different assumptions should be made when interpreting AMS measurements in either rock type.

Another consequence of the advances made in the understanding of emplacement mechanisms of igneous rocks and magmatic deformation, is that the AMS of this type of rocks is not unique and it does not remain constant in the same unit, as previously assumed. These characteristics of the AMS are common to all mineral fabrics that are acquired during magmatic deformation, provided the density of suspended minerals is not very large. In these cases, the movement of the particles of almost any shape can be approximated by the equations of movement of ellipsoidal particles derived by Jeffery (1922). The movement of these particles will be periodic, and therefore a cyclic behavior of the mineral fabric (passing from well-defined mineral orientations to random distributions of mineral grains) will be found as a function of total shear. Consequently, systematic variations of AMS are predicted as the result of the velocity gradients commonly found in igneous rocks, and therefore, a nonunique AMS signal is to be expected. Nevertheless, as the variations of the fabric are directly related to the magmatic deformation, it is possible to devise sampling strategies (e.g., Cañón-Tapia and Chávez-Álvarez, 2004) that provide enough evidence allowing us to extract useful information concerning the late aspects of the emplacement of these type of rocks (including but not limited to) the local flow direction.

Although the same basic equations could be used in principle to model the mineral fabric of a sedimentary rock, it should be noted that in these cases, other factors like the rolling-over effect of elongated grains over the deposition bed or the mechanical interaction of a much larger population of grains, will result in conditions that differ from those found in igneous rocks. Consequently, the AMS of both rock types might reflect completely different processes.

In summary, although the AMS of all rock types bears a close relation to their mineral fabric, the basic assumptions required for the correct interpretation of this type of results are lithology-dependent. Nevertheless, the AMS method remains a very useful petrofabric tool, provided that the interpretation of results is made with a clear understanding of the physical conditions relevant for each particular case.

Edgardo Cañón-Tapia

Bibliography

Borradaile, G.J., and Gauthier, D., 2003. Interpreting anomalous magnetic fabrics in ophiolite dikes. *Journal of Structural Geology*, **25**: 171–182.

Cañón-Tapia, E., 1995. Single-grain versus distribution anisotropy: a simple three-dimensional model. *Physics of the Earth and Planetary Interiors*, **94**: 149–158.

Cañón-Tapia, E., 2001. Factors affecting the relative importance of shape and distribution anisotropy in rocks: theory and experiments. *Tectonophysics*, **340**: 117–131.

Cañón-Tapia, E., 2004. Anisotropy of magnetic susceptibility of lava flows and dykes: an historical recount. In Martín-Hernández, F., Lüneburg, C., Aubourg, C., and Jackson, M., (eds.), *Magnetic Fabric Methods and Applications*. London: Geological Society, Special Publication, 238, pp. 205–225.

Cañón-Tapia, E., and Chávez-Álvarez, M.J., 2004. Theoretical aspects of particle movement in flowing magma: implication for the anisotropy of magnetic susceptibility of dykes. In Martín-Hernández, F., Lüneburg, C., Aubourg, C., and Jackson, M., (eds.), *Magnetic Fabric Methods and Applications*. London: Geological Society, Special Publication, 238, pp. 227–249.

Granar, L., 1957. Magnetic measurements on Swedish varved sediments. *Arkiv för Geofysik*, **3**: 1–40.

Hamilton, N., and Rees, A.I., 1970. The use of magnetic fabric in paleocurrent estimation. In Runcorn, S.K. (ed.) *Paleogeophysics*. London: Academic Press, pp. 445–464.

Hamilton, N., Owens, W.H., and Rees, A.I., 1968. Laboratory experiments on the production of grain orientation in shearing sand. *Journal of Geology*, **76**: 465–472.

Ising, E., 1942. On the magnetic properties of varved clays. *Arkiv för Matematik, Astronomi och fysik*, **29A**: 1–37.

Jeffery, G.B., 1922. The motion of ellipsoidal particles immersed in a viscous fluid. *Proceedings of the Royal Society of London*, **102**: 161–179.

Khan, M.A., 1962. The anisotropy of magnetic susceptibility of some igneous and metamorphic rocks. *Journal of Geophysical Research*, **67**: 2873–2885.
Nye, J.F., 1960. *Physical Properties of Crystals.* Oxford: Oxford University Press, 323 pp.
Rees, A.I., 1961. The effect of water currents on the magnetic remanence and anisotropy of susceptibility of some sediments. *Geophysical Journal*, **5**: 235–251.
Rees, A.I., 1966. The effect of depositional slopes on the anisotropy of magnetic susceptibility of laboratory deposited sands. *Journal of Geology*, **74**: 856–867.
Rees, A.I., 1968. The production of preferred orientation in a concentrated dispersion of elongated and flattened grains. *Journal of Geology*, **76**: 457–465.
Smith, J.V., 2002. Structural analysis of flow-related textures in lavas. *Earth Science Reviews*, **57**: 279–297.
Uyeda, S., Fuller, M.D., Belshé, J.C., and Girdler, R.W., 1963. Anisotropy of magnetic susceptibility of rocks and minerals. *Journal of Geophysical Research*, **68**: 279–291.

Cross-references

Magnetic Anisotropy, Sedimentary Rocks and Strain Alteration
Magnetic Susceptibility, Anisotropy
Magnetic Susceptibility, Anisotropy, Effects of Heating
Magnetic Susceptibility, Anisotropy, Rock Fabric
Magnetic Susceptibility, Low-Field
Susceptibility
Susceptibility, Measurements of Solids
Susceptibility, Parameters, Anisotropy

MAGNETIC SUSCEPTIBILITY (MS), LOW-FIELD

All mineral grains are "susceptible" to become magnetized in the presence of a magnetic field, and MS is an indicator of the strength of this transient magnetism within a material sample. MS is very different from remanent magnetism (RM), the magnetization that accounts for the magnetic polarity of materials. MS in sediments is generally considered to be an indicator of iron, ferromagnesian or clay mineral concentration, and can be quickly and easily measured on small samples. In the very-low-inducing magnetic fields that are generally applied, MS is largely a function of the composition, concentration, grain size, and morphology of the magnetizable material in a sample. It is also somewhat variable in direction (MS has anisotropy) according to magnetocrystalline anisotropy, mineral distributions, and grain morphology. MS has the advantage of being quickly and easily measured on small, friable, unoriented samples using commercially available devices such as balanced coil induction systems (susceptibility bridges). Most of these instruments generate alternating magnetic fields at from 500 Hz to 5 kHz. Sensitivity is better at higher frequencies so most of the commercially available bridges operate at around 5 kHz.

Magnetizable materials in sediments include not only the *ferrimagnetic* and *antiferromagnetic* minerals, such as the iron oxide minerals magnetite and maghemite, and iron sulfide and sulfate minerals, including pyrrhotite and greigite, that may acquire an RM (required for reversal magnetostratigraphy), but also any other less magnetic compounds, *paramagnetic* compounds. The important paramagnetic minerals in sediments include the clays, particularly chlorite, smectite, and illite, ferromagnesian silicates such as biotite, pyroxene, and amphiboles, iron sulfides including pyrite and marcasite, iron carbonates such as siderite and ankerite, and other iron- and magnesium-bearing minerals.

In addition to the ferrimagnetic and paramagnetic grains in sediments, calcite and/or quartz may also be abundant, as are organic compounds. These compounds typically acquire a very weak negative MS when placed in inducing magnetic fields, that is, their acquired MS is opposed to the low magnetic field that is applied. The presence of these *diamagnetic* minerals reduces the MS in a sample. Therefore factors such as changes in biological productivity or organic carbon accumulation rates may cause some variability in MS values.

Low-field MS, as used in most reported studies, is defined as the ratio of the induced moment ($\mathbf{M}_i$ or $\mathbf{J}_i$) to the strength of an applied, very low-intensity magnetic field ($\mathbf{H}_j$), where

$$\mathbf{J}_i = \chi_{ij}\mathbf{H}_j \text{ (density-specific)} \quad \text{(Eq. 1)}$$

or

$$\mathbf{M}_i = k_{ij}\mathbf{H}_j \text{ (volume-specific)} \quad \text{(Eq. 2)}$$

In these expressions, MS in SI units is parameterized as k, indicating that the measurement is relative to a 1 m^3 volume and therefore is dimensionless; MS parameterized as χ indicates measurement relative to a mass of 1 kg, and is given in units of $m^3\ kg^{-1}$. Both k and χ have anisotropy.

Bulk (initial) measurements of MS are often performed without consideration of the anisotropy of samples, and for most samples the bulk MS error that is due to this anisotropy is usually small. The anisotropy effect is further reduced when samples are crushed before the MS is measured. Unlike the remanence of a sample, MS is much less susceptible to remagnetization than is the remanent magnetization in rocks. MS can be measured on small, irregular lithic fragments, and on the highly friable material that is difficult to sample for RM measurement.

Applications and interpretations of bulk (initial) low-field magnetic susceptibility

MS variability from sample to sample results primarily from changes in the amount of its iron, clay, and ferromagnesian constituents. This variability in sedimentary sequences is primary, produced during deposition, and secondary, resulting from alteration of the deposited sequence. The primary factors include weathering, erosion and deposition rates, and biological productivity. These factors are tied to climate, tectonic and volcanic processes, availability of nutrients, and local (relative) or global sea-level changes. Secondary processes include pedogenesis, diagenesis, and within-sediment redox effects driven by sulfate-reducing bacteria.

Studies of marine sedimentary sequences

Marine sediments are composed of terrigenous, detrital and eolian components, and a biogenic component of the carbonate and/or siliceous tests of marine organisms. There is also a component of sediment derived from marine weathering processes, but this is generally minor. The MS signature mainly represents contributions from the terrigenous and biogenic components. Because the biogenic component is only weakly diamagnetic, in general, the terrigenous component dominates the MS. Therefore, the processes controlling the influx into the marine environment of terrigenous components will usually account for the MS variations observed. Thus, climate cycles that cause erosion and transport of terrigenous material will result in a cyclic signal in the MS recorded in the deposited sediments. There are many factors that can modify this signature, but while the expected variations may be reduced in magnitude, usually the character of the cycles will remain and can be extracted. Superimposed on these cyclic variations are contributions from unique events, such as volcanic eruptions or meteorite impacts, which may be of local, regional, or global significance.

Continuous and discrete measurements on unlithified sedimentary cores; marine and lacustrine

Routine MS measurements were not possible on sediments until the 1960s when published instructions for building instruments became available and when the first commercially produced instruments entered the market. Early work on sediments was difficult because of the low sensitivity of these instruments. Besides instrumental problems, there were ambiguities in units reported, making interpretations difficult. In the 1970s, publications of MS of discrete samples from marine piston cores began to appear in the literature. In the 1980s, better instrumental sensitivity made it possible to build instruments for continuous MS measurement on cores and these were applied to marine and lacustrine sequences. Much of this work was directed toward correlating cores. In unlithified sediments an inverse correlation between percent carbonate and MS variability, observed as early as the middle 1970s, was interpreted to result from variations in marine organic carbonate productivity (Ellwood and Ledbetter, 1977). More recent work looking at the spectral character of these MS variations has shown that the cycles observed are correlated to climate cyclicities.

Ultimately, the MS in most marine and lacustrine sediments can be used as a proxy for physical processes responsible for delivering the detrital and eolian components to the system. These processes are both regional and global and include climate, volcanism and tectonic processes. In marine sediments they also include relative sea-level changes and eustacy. (See Evans and Heller (2003) for a general discussion of unlithified marine and lacustrine sediments.)

Surface sediment measurements from pilot and trigger core-top samples

MS has proved useful in some cases. For example, the MS has been measured on a large number of cores from throughout the Argentine Basin in the South Atlantic Ocean and the variability is characterized (Sachs and Ellwood, 1988). They showed that the detrital components responsible for the MS originated from two primary sources. These components are brought into and along the western margin of the basin from Antarctica by Antarctic Bottom Water (AABW), and into the basin from rivers along the eastern margin of South America. Both sources exhibit relatively high MS values at their sources. These values exhibit a rapid down-current MS reduction away from the source. AABW and other bottom currents then disperse detrital components throughout the basin, so that the majority of MS values from the basin show similar values.

Results from lithified marine sedimentary rock sequences

During diagenesis the MS of marine sediment decreases. This occurs because ferrimagnetic grains that dominate in unlithified sediments are generally converted to paramagnetic phases during burial and diagenesis, mainly by the action of sulfate-reducing bacterial organisms. This process transforms much of the iron to paramagnetic phases such as iron sulfide and carbonate minerals. In general, however, this process does not remove the iron from the system, so that total iron content is conserved. These new phases then contribute, along with the other paramagnetic detrital components present, such as clays, and the remaining authigenic ferrimagnetic constituents, to the MS in lithified sediments. Measured MS values for most marine limestones, marls, shales, siltstones, and sandstones range from 1×10^{-9} to 1×10^{-7} m^3 kg^{-1}. Shales and some marls have the greatest anisotropy, but for most of these samples it is less than 2%. Limestones, marls, siltstones, and sandstones are usually less than 1%. Shale and marl samples are friable and measurement is done generally on small broken fragments, further reducing the anisotropy effect. Therefore, the error introduced by the anisotropy of MS is minimal. A test of the MS variability in single limestone beds has demonstrated the nonvarying nature of the MS over distances up to 25 km (Ellwood *et al.*, 1999). Diagenesis also modifies some of the paramagnetic constituents that make up the terrigenous component of marine sediments. In many cases clays dominate much of the detrital/eolian component, and these are generally not destroyed but may be altered. The iron is still conserved in the secondary clays, as is the paramagnetic behavior of these materials. Other common detrital components observed in marine rocks are the ferromagnesian minerals including biotite, tourmaline, and other compounds (see Ellwood *et al.*, 2000 for discussion).

Ultimately, following diagenesis, the MS observed for most marine sedimentary rocks can be used as a proxy for physical processes responsible for delivering the detrital/eolian component of these sediments into sedimentary basins. As observed for unlithified sediments, these processes in lithified sediments are both regional and global and include climate, volcanism, tectonic processes, relative sea-level changes, and eustacy.

Presentation of MS data from lithified sequences

For presentation purposes and inter data set comparisons, the bar log format, similar to that previously established for magnetic polarity data presentations, is recommended. These bar logs should be accompanied by both raw and smoothed MS data sets. Here, raw MS data (dashed line in Figure M91) are smoothed using splines (solid line with circles indicating data point locations shown in Figure M91). The following bar log plotting convention is used; if the MS cyclic trends increase or decrease by a factor of two or more, and if the change is represented by two or more data points, then this change is assumed to be significant [illegible] by filled (high MS values) or open (low MS values) bar logs (Figure M91). This method is best employed when high-resolution data sets are being analyzed (large numbers of closely spaced samples) and helps resolve variations associated with anomalous samples. Such variations may be due to weathering effects, secondary alteration and metamorphism, longer term trends due to factors such as plate-driven eustacy, as opposed to shorter term climate cycles, or event sequences such as impacts (Ellwood *et al.*, 2003), and to other factors. In addition, variations in detrital input between localities, or a change in detrital sediment source is resolved by developing and comparing bar logs between different localities. It is also recommended that MS data be represented in log plots due to the range of MS values often encountered. However, MS data variations in log plots for some ranges ($1–3 \times 10^{-x}$ vs $4–9 \times 10^{-x}$) may appear to exhibit large variations that are artifacts of the presentation style. Therefore, bar logs help to reduce these ambiguities.

Climate cycles and other MS variations

MS results for almost all studies of marine sedimentary rocks show many levels of cycles. Some of these cycles are very long-term and are interpreted as resulting from transgressive-regressive variations (T-R cycles) due to sea-level rise and fall associated with eustacy. Other cycles are much shorter and are interpreted to result from climatically driven processes (Ellwood *et al.*, 2001a,b). Figure M91 is an example of cyclic MS data from limestone—marl couplets from immediately below the Danian-Selandian (Paleocene) stage boundary at Zumaia, Spain. These couplets have been interpreted to result from climate variability, shown to represent climate precession (19–23 ka; Dinares-Turell *et al.*, 2003).

In addition to these cycles, there are distinct MS peaks that are correlated to maximum flooding surfaces (MFS; see discussion in section "Climate Cycles and Other MS Variations"), and to sudden influxes of detrital material into sedimentary basins by turbidity currents or other sediment suspensions. There are also large, relatively sudden shifts or offsets in MS magnitudes within sequences. The Danian-Selandian boundary illustrated in Figure M91 may be such a case. If "knife edge" in character, these shifts have been interpreted as possible unconformities; several tests have demonstrated such an effect. However, not all such abrupt changes will be "unconformities," but may instead be "pseudounconformities" representing missing core sections, unrecognized faults, or other problems. Work by da Silva and Boulvain (2002) for Upper Devonian rocks in Belgium has shown a direct, high

Figure M91 (a) Uppermost Danian limestone (resistant)—marl couplets at Zumaia, Spain. (b) MS for the interval shown. High MS values are associated with the marls. Note the character change in the basal Selandian shales (Ellwood *et al.*, in preparation).

correlation between both third and fourth order T-R cycles and MS variations determined for those rocks.

Large magnitude and rapid shifts in MS (represented by multiple data points) are interpreted to result from geological processes that reflect fundamental changes in the sediments being delivered to marine basins. For example, the sharp change from low MS in the Permian to high MS in the Triassic (Hanson *et al.*, 1999) is argued to result from a sudden dramatic influx into ocean basins of detrital material that helped to cause the mass extinctions in the marine fauna at this time (Thoa *et al.*, 2004). A similar pulse in MS is associated with the K-T boundary and is attributed to erosion following the meteorite impact at that time (Ellwood *et al.*, 2003).

Correlations

MS studies of Phanerozoic marine rock sequences have concentrated on the geological boundaries known as Global Boundary Stratotype Sections and Points (GSSPs) that are currently being defined by the International Geological Congress's (IGC) Sub-commission on Stratigraphy. A GSSP represents a "point" or level in a stratigraphic sequence that defines the beginning of a new geological stage in chronostratigraphic time. GSSPs are becoming the basis for the modern Geological Timescale. Today, about half of the more than 90 GSSPs for the Phanerozoic have been formally defined. IGC intends to have all GSSPs formally defined by the year 2008. All of these boundaries are well studied and much of the biostratigraphy has been published. Most stage boundaries are defined based on the first occurrence of a new fossil species, while a few use other criteria. For example, the Paleocene-Eocene GSSP in Upper Egypt near Luxor is defined by a negative carbon isotope shift in the sequence exposed there and in most other Paleocene-Eocene sequences. The biostratigraphy for this section is also well known and the MS signature is distinctive (Figure M92).

The MS of several of these GSSPs has been studied and a number of results have been published (for examples see Hanson *et al.*, 1998 and Thoa *et al.*, 2004 for MS results from the Permian-Triassic boundary, and Crick *et al.*, 2001 for the MS of the Silurian-Devonian boundary. The Silurian-Devonian boundary represents the first GSSP, established in 1972). The purpose of this work has been to use MS as a viable tool for correlating geological sequences, especially for the Paleozoic where polarities are not well defined and where remagnetization is a significant problem. Published and unpublished work on GSSPs and well-studied associated sections has demonstrated that MS variations can be correlated between sections (Crick *et al.*, 1997).

It has been argued that MS tracks processes that control the influx of detrital and eolian grains into the marine environment. For example, sequence stratigraphic studies have demonstrated that at times when sea level is at a maximum, producing an MFS, at the end of a transgressive cycle there is a large influx of detrital material into the marine environment. There is a corresponding MS high associated with these MFSs. However, as a general rule, during transgressive cycles, especially in distal marine sequences, MS is observed to decrease. This happens because detrital sediments are usually trapped near shore during transgressions. During regressions, when base level is lowered due to falling sea level, erosion flushes detrital sediment into ocean basins and MS increases in basinal sediments during these times. Thus T-R cycles play an important role in creating some cyclic MS variations.

Results of the MS work in these rocks have shown that the MS is mineral-dependent but may be independent of the macrolithology. Reworking of sediments by ocean currents, and the variable deposition of clays may produce large MS changes in both limestones and shales, even though in general the lithology may appear to be relatively non-varying. In addition, while local subsidence may cause a lithologic change from shale to limestone or limestone to shale, the MS may remain relatively constant (Figure M93).

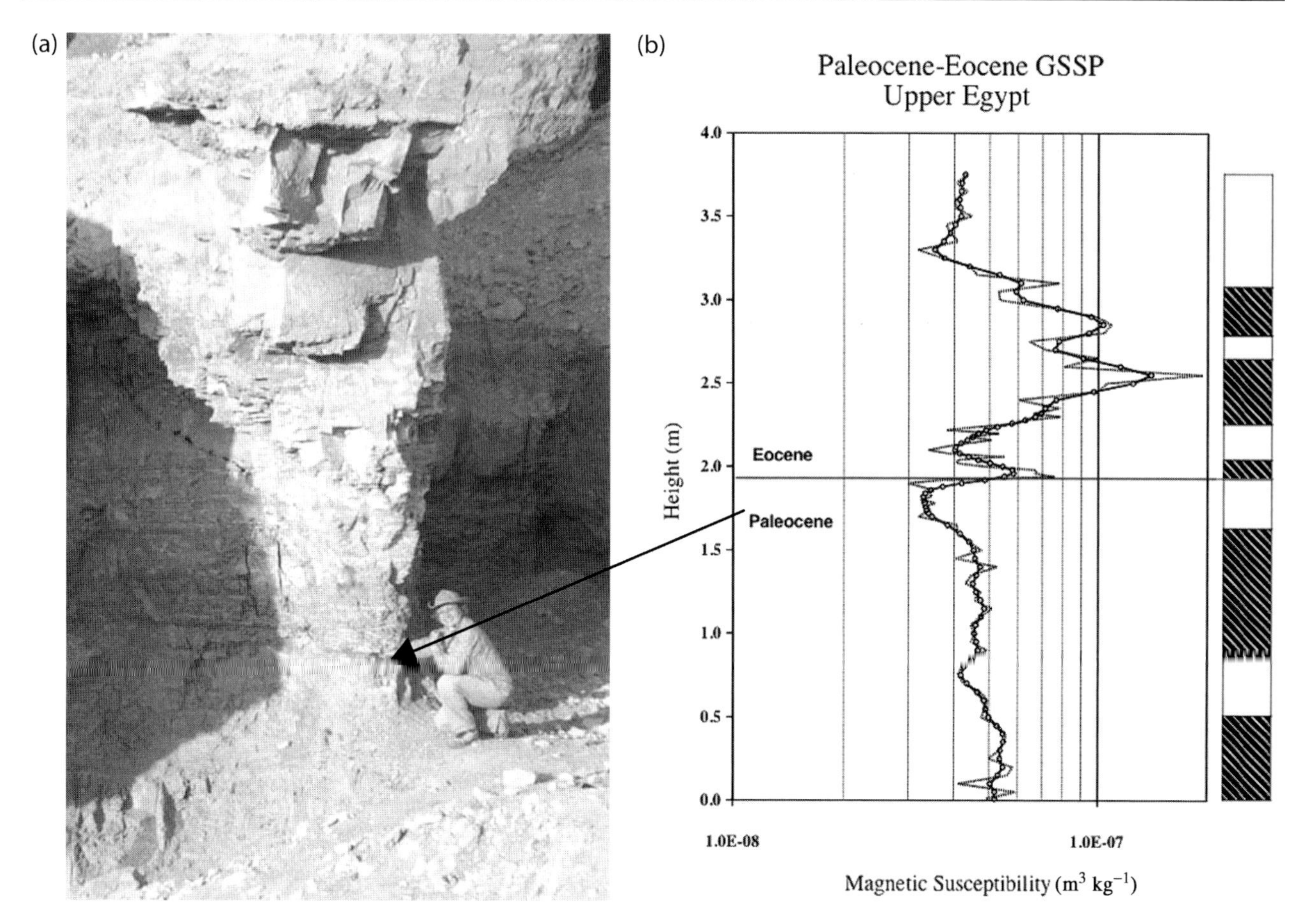

Figure M92 (a) The Paleocene-Eocene GSSP section exposed near Luxor, Egypt. (b) MS variation for the section composed mainly of shales (Ellwood *et al.*, 2007). Bar log convention as discussed in "Presentation of MS Data from Lithified Sequences" section.

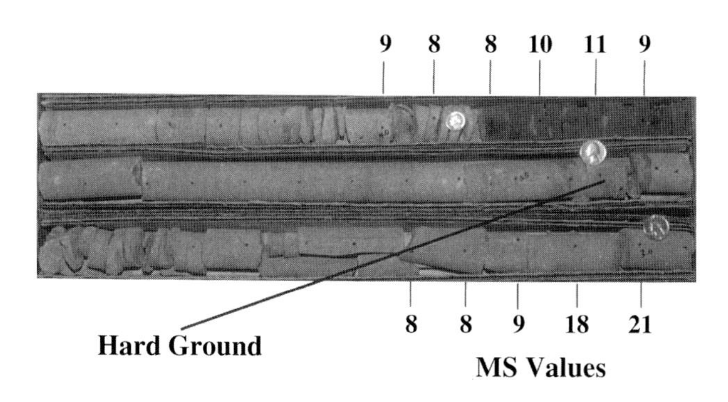

Figure M93 MS variation in a limestone (light) to shale (dark) sequence measured with a portable susceptibility bridge. Note the MS during the transition from limestone to shale is essentially nonvarying (*top*), whereas MS variation within the limestone (*bottom*) shows a large range.

Studies of sedimentary sequences exposed during archeological excavations within limestone caves and deep rock shelters

Archaeological artifacts found during excavation are usually encased in sediments. In most cases, the microenvironment within caves and deep rock shelters protects deposited sediments from the large-scale alteration effects produced by pedogenesis. Artifacts, lithic fragments, and other materials are generally interpreted to have been deposited as part of a stratigraphic sequence and are isochronous with the encasing sediments. Careful microstratigraphic excavations expose stratigraphic sequences that can be sampled for many purposes, including the study of faunal and floral materials, geochemical studies, dating, and characterization of MS variability. For this purpose, sediment is usually collected as a set of continuous samples that allows evaluation of a reasonably complete archaeological sequence. Such a sequence is illustrated in Figure M94a.

The source of these sediments can be fairly complex. In most cases, pedogenesis of the limestone-surrounding caves is the source of eroded soil that can be blown, washed, or tracked in from outside the cave. Sediments may be deposited from streams flowing through or within caves or derived as residues from dissolution of the containing limestones. Because the sedimentary sequences in caves have been shown to accumulate relatively rapidly, limestone residues are considered to be only a minor source of the observed sediments, leaving fluvial and soil materials as the dominant sedimentary sources. Some of the material transported into caves is loess and the MS usually reflects the glacial-interglacial variations observed in these loess sequences (loess MS studies are discussed below.)

The MS of accumulating sediments derived from limestone pedogenesis in archaeological stratigraphic sequences varies with changing climate. In general, periods of warm climate exhibit increases in pedogenesis that produce extremely fine-grained ferrimagnetic iron oxide minerals, primarily magnetite and maghemite. During cold periods, pedogenic activity is reduced, thus reducing the production of ferrimagnetic components. As surface soils are blown, tracked or washed into caves and rock shelters, the MS reflects the climate at the time

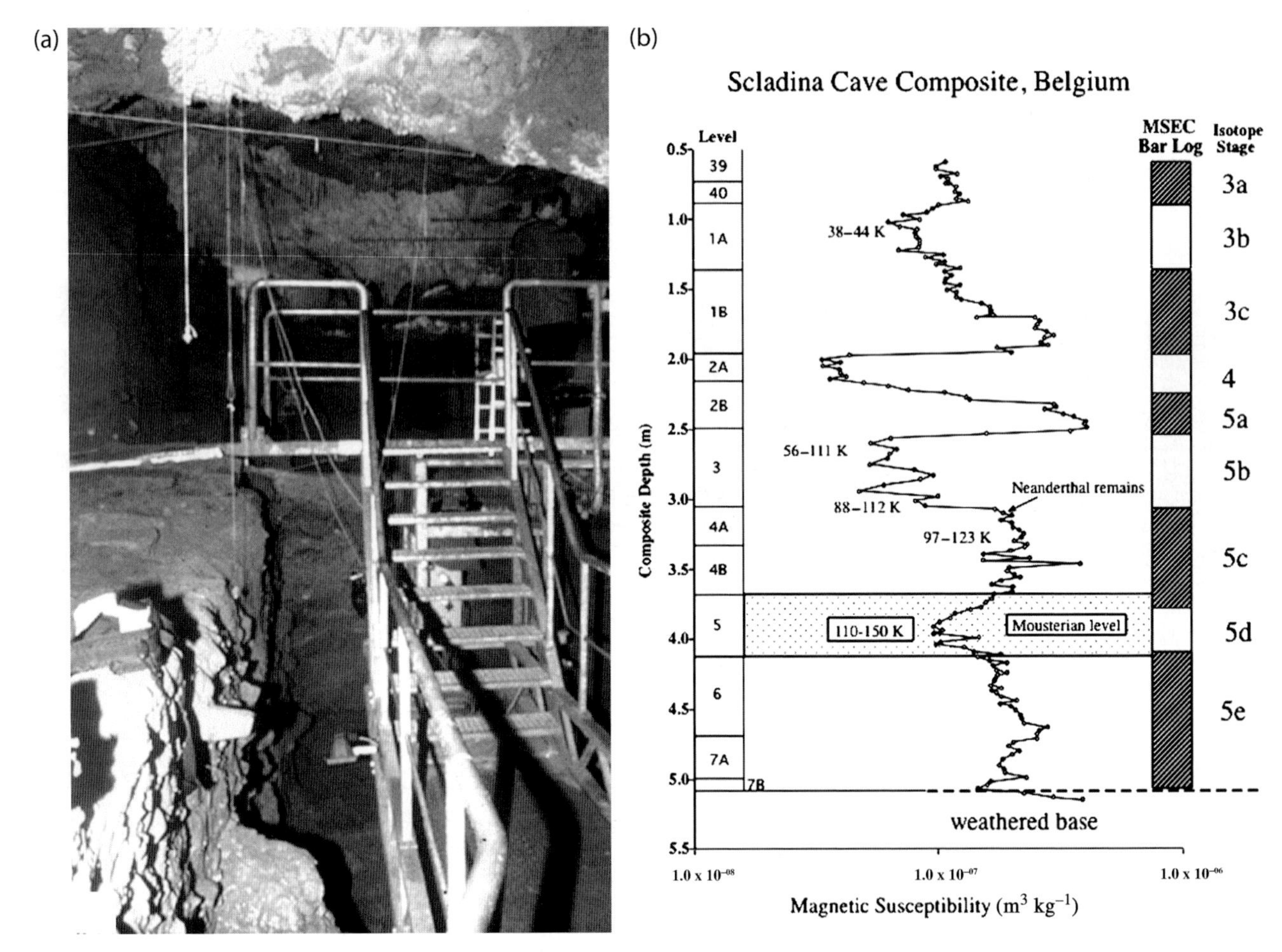

Figure M94 (a) Interior excavations in Scladina Cave, Belgium. (b) MS of Scladina Cave sediments. The data points correspond to samples from several overlapping sections within the cave that were correlated with the archaeological levels shown on the left. MS measurements have then been used to construct the MS bar log which in turn is compared with the marine oxygen isotope results for the last ~130 ka (Imbrie *et al.*, 1984; Martinson *et al.*, 1987). For comparison with the MS results, the $\delta^{18}O$, as a function of time, equivalent to a bar log that represents the times of relatively warm (hatched) and relatively cool (white) climates. The $\delta^{18}O$ isotope stage number assignments are based on the dates obtained by Otte *et al.* (1993, 1998). The ages for the samples from the cave, in thousands of years, and their locations within the sequence are indicated, as are the lower Mousterian level (lithic artifacts produced by Neanderthals) and the location of Neanderthal skeletal remains. The assigned ages for the top and the bottom of the MS warm zones have been based on the assumption that the top and bottom of each level are equivalent in age to the $\delta^{18}O$ ages of the isotope substages (from Ellwood *et al.*, 2004).

the material is accumulating. This has been tested in a number of cases against faunal, floral, and lithic data sets, by relating MS data to dated chronologies, and by studies of the concentration and character of the dominant magnetic components in these sediments. All are consistent in most cases with the climate interpretation given here.

Fluvial sediments within caves are usually deposited as quickly accumulating, thinly laminated sequences of clays, silts, and sands, and thus easily identified as fluvial. In MS studies these sequences usually do not show much variation. Areas of sedimentary mixing or disturbance are also distinctive because the mixing process radically reduces the MS variability. In addition, in some archaeological contexts there are alteration effects that produce distinctive, much reduced MS magnitudes that are essentially nonvarying through sequences that are expected to show climate variations. In many cases the sediments from the top-down to midlevel in a sequence show good cyclicity, whereas below that is observed a distinctive offset to much lower MS values. Rock magnetic analyses of these materials show that the primary magnetite or maghemite has been altered and either replaced by paramagnetic iron oxyhydroxides like goethite, or removed by dissolution effects.

Results from archaeological excavations within limestone caves and deep rock shelters

Correlations

Excellent correlations have been shown among archaeological sites in Europe using MS (Ellwood *et al.*, 2001b). These sites range in age from ~3000 B.P. to ~50000 B.P. and are based on a modified version of the graphic correlation method (Shaw, 1964) and available ^{14}C dates from each site for which data are reported. MS high (warm) and low (cool) values from individual sequences are differentiated diagrammatically by developing bar logs (an example is given in Figure M94b). To these bar log diagrams are added archaeological industries, levels, and available dates. The archaeological lithic industries and dates constrain the MS bar logs to a chronology that allows comparison to a previously developed (and continuously evolving) standard sequence (composite standard, CS). The boundaries between warm and cool MS climate proxies are assigned ages by comparison to the CS making it possible to compare these boundaries to other cave sequences.

Climate cycles

Results from MS data show cycles, which correlate with climate cycles. Both short- and long-term cycles have been resolved from cave sequences. Neoglacial cycles, with a ~2500 year cyclicity, well known from Quaternary climate studies and also reported from work on the Greenland GISP2 Ice Core, have been observed in a number of caves in Europe (Ellwood *et al.*, 1996). Longer term cyclicities (Figure M94b, from Ellwood *et al.*, 2004) have been correlated to ~20 ka fluctuations observed in the marine oxygen isotopic record (Imbrie *et al.*, 1984).

Relative dating

Relative dating can be accomplished by comparing known dates with MS bar log age assignments, either by using dates assigned by comparison with the MS standard sequence (CS) or with the marine isotope curve. Results from this method have been used to date Neanderthal remains in Scladina Cave in Belgium (Ellwood *et al.*, 2004). This result was further tested using high-precision radiometric dating and found to be consistent within the errors assigned by these methods.

Studies of loess

MS work on Chinese loess sequences in the early 1980s showed a strong correlation with the marine oxygen isotopic record and MS variability. Much of this work has been summarized by Evans and Heller (2003). It was observed that during interglacial periods MS values were high, while during glacial times MS values were low. This was attributed to differences in pedogenic rates, with the production of more ferrimagnetic minerals during interglacial times. In the main, at least for the Chinese sequences, the MS variations in loess appear to provide a proxy for climate fluctuations. However, there are a number of problems with such a simple interpretation. The original protolith from which the loess was derived has an important effect on final MS variability, and in some cases has been shown to produce an inverse MS relative to the Chinese loess sequences, with interglacials showing lower MS values than during glacial times. This is due to erosion of coarse magnetite from an original, high-magnetite-content protolith, and this magnetite is accumulated in loess sequences during glacial times. However, pedogenesis during interglacial times oxidizes the original magnetite, producing a suite of magnetic minerals with a significantly weaker MS. In addition, during interglacials, deep pedogenesis can affect underlying loess sequences, causing ambiguities in the observed MS results.

Grain size estimates based on frequency dependence of MS

There is a grain-size dependence to MS that results from measurement at varying frequencies. This effect has been loosely employed in some rock magnetic studies using balanced coil instruments that can measure MS at varying frequencies. There is only one such instrument currently commercially available and it allows measurements at two frequencies, ~500 Hz and ~5 kHz. Measurements of the same sample at both frequencies have been compared and results interpreted to represent grain size differences between samples.

There are at least three significant problems with this method.

1. An instrument that only operates at two frequencies is not capable of measurement at intermediate frequencies where responses in varying size ranges may be important.
2. In general, it is assumed that while coarse, multidomain ferrimagnetic grains respond quickly to inducing magnetic fields at both low and high frequencies, the more stable, single-domain or pseudosingle-domain grains do not respond as quickly to high-frequency-inducing fields. This results in a lag in higher frequency fields that is reflected in lower MS values for samples containing these grains. However, there is a third size range of ferrimagnetic particles, the ultrafine-grained superparamagnetic grains produced during pedogenesis. While these are generally unimportant in remanence studies, due to Brownian motion effects, they respond rapidly to both low- and high-frequency-inducing magnetic fields, in much the same way as do multidomain grains, and thus are not distinguishable from larger grains.
3. MS originates mainly from a large range of paramagnetic and ferrimagnetic grain compositions and grain sizes. Because only two frequencies are used in frequency-dependence studies, the true behavior of grains at higher, intermediate, and lower frequencies is poorly known or unknown in most cases. Clearly, much more work needs to be done before reliable interpretations of MS data from variable frequency measurements are useful.

Brooks B. Ellwood

Bibliography

Crick, R.E., Ellwood, B.B., El Hassani, A., Feist, R., and Hladil, J., 1997. Magnetosusceptibility event and cyclostratigraphy (MSEC) of the Eifelian-Givetian GSSP and associated boundary sequences in North Africa and Europe. *Episodes*, **20**: 167–175.

Crick, R.E., Ellwood, B.B., El Hassani, A., Hladil, J., Hrouda, F., and Chlupac, I., 2001. Magnetostratigraphy susceptibility of the Pridoli-Lochkovian (Silurian-Devonian) GSSP (Klonk, Czech Republic) and a Coeval sequence in Anti-Atlas Morocco. *Palaeogeography Palaeoclimatology Palaeoecology*, **167**: 73–100.

da Silva, A.-C., and Boulvain, F., 2002. Sedimentology, magnetic susceptibility and isotopes of a Middle Frasnian carbonate platform: Tailfer section, Belgium. *Facies*, **46**: 89–101.

Dinares-Turell, J., Baceta, J.I., Pujalte, V., Orue-Etxebarria, X., Bernaola, G., and Lorito, S., 2003. Untangling the Palaeocene climatic rhythm: as astronomically calibrated Early Palaeocene magnetostratigraphy and biostratigraphy at Zumaia (Basque basin, northern Spain). *Earth and Planetary Science Letters*, **216**: 483–500.

Ellwood, B.B., and Ledbetter, M.T., 1977. Antarctic bottom water fluctuation in the Vema Channel: effects of velocity changes on particle alignment and size. *Earth and Planetary Science Letters*, **35**: 189–198.

Ellwood, B.B., Petruso, K.M., and Harrold, F.B., 1996. The utility of magnetic susceptibility for detecting paleoclimatic trends and as a stratigraphic correlation tool: an example from Konispol cave sediments, SW Albania. *Journal of Field Archaeology*, **23**: 263–271.

Ellwood, B.B., Crick, R.E., and El Hassani, A., 1999. The magneto-susceptibility event and cyclostratigraphy (MSEC) method used in geological correlation of Devonian rocks from Anti-Atlas Morocco. *AAPG Bulletin*, **83**: 1119–1134.

Ellwood, B.B., Crick, R.E., El Hassani, A., Benoist, S., and Young, R., 2000. Magnetosusceptibility event and cyclostratigraphy (MSEC) in marine rocks and the question of detrital input versus carbonate productivity. *Geology*, **28**: 1135–1138.

Ellwood, B.B., Crick, R.E., Garcia-Alcalde Fernandez, J.L., Soto, F.M., Truyols-Massoni, M., El Hassani, A., and Kovas, E.J., 2001a. Global correlation using magnetic susceptibility data from Lower Devonian rocks. *Geology*, **29**: 583–586.

Ellwood, B.B., Harrold, F.B., Benoist, S.L., Straus, L.G., Gonzalez-Morales, M., Petruso, K., Bicho, N.F., Zilhão, Z., and Soler, N., 2001b. Paleoclimate and intersite correlations from Late Pleistocene/Holocene cave sites: results from Southern Europe. *Geoarchaeology*, **16**: 433–463.

Ellwood, B.B., MacDonald, W.D., Wheeler, C., and Benoist, S.L., 2003. The K-T Boundary in Oman: Identified Using Magnetic Susceptibility Field Measurements with Geochemical Confirmation. *Earth Planetary Science Letters*, **206**: 529–540.

Ellwood, B.B., Harrold, F.B., Benoist, S.L., Thacker, P., Otte, M., Bonjean, D., Long, G.L., Shahin, A.M., Hermann, R.P., and Grandjean, F., 2004. Magnetic susceptibility applied as an age-depth-climate relative dating technique using sediments from Scladina Cave, a Late Pleistocene cave site in Belgium. *Journal of Archaeological Science*, **31**: 283–293.

Ellwood, B.B., Kafafy, A., Kassab, A., Tomkin, J.H., Abdeldayem, A., Obaidalla, N., Willson, K., and Thompson, D.E., 2007. Magnetostratigraphy susceptibility used for high resolution correlation among Paleocene/Eocene boundary sequences in Egypt, Spain and the USA. SEPM Special Publication, in press.

Evans, M.E., and Heller, F., 2003. *Environmental Magnetism: Principles and Applications of Enviromagnetics.* San Diego, CA: Academic Press, p. 299.

Hansen, H.J., Lojen, S., Toft, P., Dolenec, T., Yong, J., Michaelsen, P., and Sarkar, A., 1999. Magnetic susceptibility of sediments across some marine and terrestrial Permo-Triassic boundaries. In *Proceedings of the International Conference on Pangea and the Paleozoic-Mesozoic Transition, March 9–11, 1999.* China University of Geosciences, Hubei, China, pp. 114–115.

Imbrie, J., Hays, J.D., Martinson, D.G., McIntyre, A., Mix, A.C., Morley, J.J., Pisias, N.G., Prell, W.L., and Shackleton, N.J., 1984. The orbital theory of Pleistocene climate: support from a revised chronology of the Marine Delta ^{18}O record. In Berger, A.L., Imbrie, J., Hays, J., Kukla, G., and Saltzman, B. (eds.), *Milankovitch and Climate, Part I.* Boston, MA: Reidel, pp. 269–305.

Martinson, D.G., Pisias, N.G., Hays, J.D., Imbrie, T.C., and Shackleton, N.J., 1987. Age dating and the orbital theory of the ice ages: development of a high-resolution 0 to 300 000-year chronostratigraphy. *Quaternary Research*, **27**: 1–29.

Otte, M., Toussaint, M., and D., Bonjean 1993. Découverte de restes humains immatures dans les niveaux moustériens de la grotte Scladina à Andenne (Belgique). *Bulletin Mémoires de la Société d'Anthropologie de Paris*, **5**: 327–332.

Otte, M., Patou-Mathis, M., and Bonjean D. (eds.), 1998. *Recherches aux grottes de Sclayn, Volume 2, L'Archéologie.* Liège, ERAUL 79, Service de Préhistoire, Université de Siége.

Sachs, S.D., and Ellwood, B.B., 1988. Controls on magnetic grain-size variations and concentrations in the Argentine Basin, South Atlantic Ocean. *Deep Sea Research*, **35**: 929–942.

Shaw, A.B., 1964. *Time in Stratigraphy.* New York: McGraw-Hill, 365 pp.

Thoa, N.T.K., Huyen, D.T., Ellwood, B.B., Lan, L.T.P., and Truong, D.N., 2004. Determination of Permian-Triassic boundary in limestone formations from Northeast of Vietnam by paleontological and MSEC methods. *Journal of Sciences of the Earth*, **26**: 222–232.

Cross-references

Anisotropy of Magnetic Susceptibility (AMS)
Archeomagnetism
Iron Sulfides
Magnetic Anisotropy, Sedimentary Rock and Strain Alteration
Magnetic Susceptibility
Magnetization, Natural Remanent (NRM)
Magnetostratigraphy
Polarity

MAGNETIZATION, ANHYSTERETIC REMANENT

Anhysteretic remanent magnetization (ARM) is a magnetization an assemblage of magnetic particles acquires when it is subjected to an alternating field (AF) of gradually decreasing amplitude (H_{AF}) with a constant decrement (ΔH_{AF}/cycle) simultaneously with a steady, unidirectional DC field (H_{DC}). The ARM is measured when both AF and DC fields are zero. Typically, the DC field is maintained while the AF is slowly ramped down to zero, and then reduced to zero. For isotropic samples, the direction of the ARM is parallel to H_{DC} but the intensity will depend on the amplitudes and relative orientations of the AF and DC fields. The maximum ARM intensity occurs when H_{AF} and H_{DC} are parallel. For a standard laboratory experiment, $H_{DC} \ll H_{AF}$ with $H_{AF} = 100$–200 mT and $H_{DC} = 0.05$–0.1 mT. H_{DC} is also referred to as a bias field because it produces a statistical preference or biases the direction of ARM along H_{DC}. A special case of ARM is a partial ARM (pARM) produced by a DC field applied only for a limited time between two AF field values ($0 \leq H_1, H_2 \leq H_{max}$). A related parameter is the ARM susceptibility, $\chi_{arm} = dM_{arm}/dH_{DC}$ ($H_{DC} \to 0$) where M_{arm} is the ARM intensity. Typically, χ_{arm} is determined from a single measurement using one DC field value, $\chi_{arm} = M_{arm}/H_{DC}$. For mass-normalized ARM (in A m^2 kg^{-1}) per unit DC field (in A m^{-1}), the SI unit for ARM susceptibility is kg m^{-3}. For volume-normalized ARM (in A m^{-1}), the SI unit for ARM susceptibility is dimensionless.

With the possible exception of lightning-induced remanence (see *Magnetization, isothermal remanent (IRM)*), ARM is not a naturally occurring remanence like TRM, DRM, or CRM. However, laboratory-produced ARM is used in a variety of applications in paleomagnetism. For example, ARM-based methods have been developed to (1) estimate absolute paleointensity from igneous rocks and relative paleointensity from lake and marine sediments; (2) characterize magnetic carriers and determine domain state and grain size; (3) detect magnetic fabrics in rocks and sediments; and (4) study the fundamental aspects of magnetism. Anhysteretic magnetization also forms the basis for AC-bias magnetic recording.

The combined action of an AF and DC field ($H_{AF} \gg H_{DC}$) is much more efficient in producing a remanence that is more resistant to demagnetization than an isothermal remanent magnetization (IRM), which results from the application of a DC field alone. Two fundamental characteristics of ARM distinguish it from an IRM given in the same DC field: (1) $M_{arm} \gg M_{irm}$, and (2) in order to demagnetize the ARM, AF fields approximately equal to the coercivity ($H_c \gg H_{DC}$) must be applied, whereas AF or DC fields equal to H_{DC} are needed to demagnetize the IRM. In a classic study, Rimbert (1959) was the first to demonstrate analogous properties between ARM and TRM (see *Magnetization, thermoremanent (TRM)*). Therefore, ARM could be used as a surrogate for TRM in laboratory experiments (Rimbert, 1959; Gillingham and Stacey, 1971; Levi and Merrill, 1976; Dunlop and Argyle, 1997), thus avoiding the need to heat the sample to high temperatures and the possibility of mineralogical alteration. The AF field for ARM acquisition acts as the randomizing agent in analogous fashion as temperature for TRM acquisition. Consider a single-domain (SD) grain with uniaxial anisotropy and coercivity, h_c. The following illustration is applicable whether we are dealing with moment rotation between easy axes in SD particles or wall displacement in multidomain particles. In this case, the magnetic moment is pinned in the positive direction along the anisotropy axis and will undergo an irreversible rotation into the opposite orientation when a field greater than h_c is applied along the negative direction. Starting from a large AF field such that H_{AF} is greater than h_c, the AF field overcomes the anisotropy barrier and the magnetization will follow the oscillating field, i.e., the large AF field initially saturates the sample during each cycle. However, once $H_{AF} < h_c$, the particle moment is blocked into a stable position along the anisotropy axis closest to the DC field direction. As H_{AF} is reduced to zero, little or no change in magnetization occurs. For an ensemble of SD particles with randomly oriented anisotropy axes, a statistical bias is produced resulting in a net remanence along H_{DC}. If $H_{DC} = 0$ during the application of the AF field, the net remanence will be zero and the process is simply AF demagnetization (see *Demagnetization*).

ARM can be an unwanted spurious magnetization produced during AF demagnetization of natural remanent magnetization (NRM). This type of ARM can be generated by (1) insufficient magnetic shielding of external DC fields or (2) distorted or asymmetric AF waveforms that produce higher-order even harmonics from the fundamental AF drive frequency (Collinson, 1983). In either case, a small net DC bias is present during the AF demagnetization process that can produce an ARM of similar or even greater magnitude, than the initial or partially demagnetized NRM signal. Therefore, it is important to provide adequate shielding around the AF coils to attenuate external DC fields, such as the Earth's magnetic field or laboratory fields. This is usually achieved with multilayer, mu-metal shielding. Asymmetries in the AF waveform can be attenuated by tuning the AF coil to the fundamental frequency. Spurious ARM can also be significantly reduced by doing a double-demagnetization at each demagnetization step. In this technique, the sample is (1) demagnetized along three orthogonal axes ($+x$, $+y$, $+z$), (2) NRM measured, (3) demagnetized along $-x$, $-y$, and $-z$, (4) NRM remeasured, and (5) the vector sum of steps (2) and (4) is determined. Collinson (1983) provides detailed reviews of AF and ARM instrumentation.

Experimental properties of ARM

The grain size dependences for weak-field ARM ($H_{DC} < 0.2$ mT) and its median destructive field (MDF) in magnetite from various sources are shown in Figure M95. The intensity of ARM peaks within the single-domain size range ($d < 0.1$ μm) and then drops off approximately as d^{-n} ($n = 0.5$–1) for pseudosingle-domain (PSD) and multidomain grain sizes. Below 0.05 μm, ARM intensity also falls with decreasing grain size as particles become superparamagnetic and their

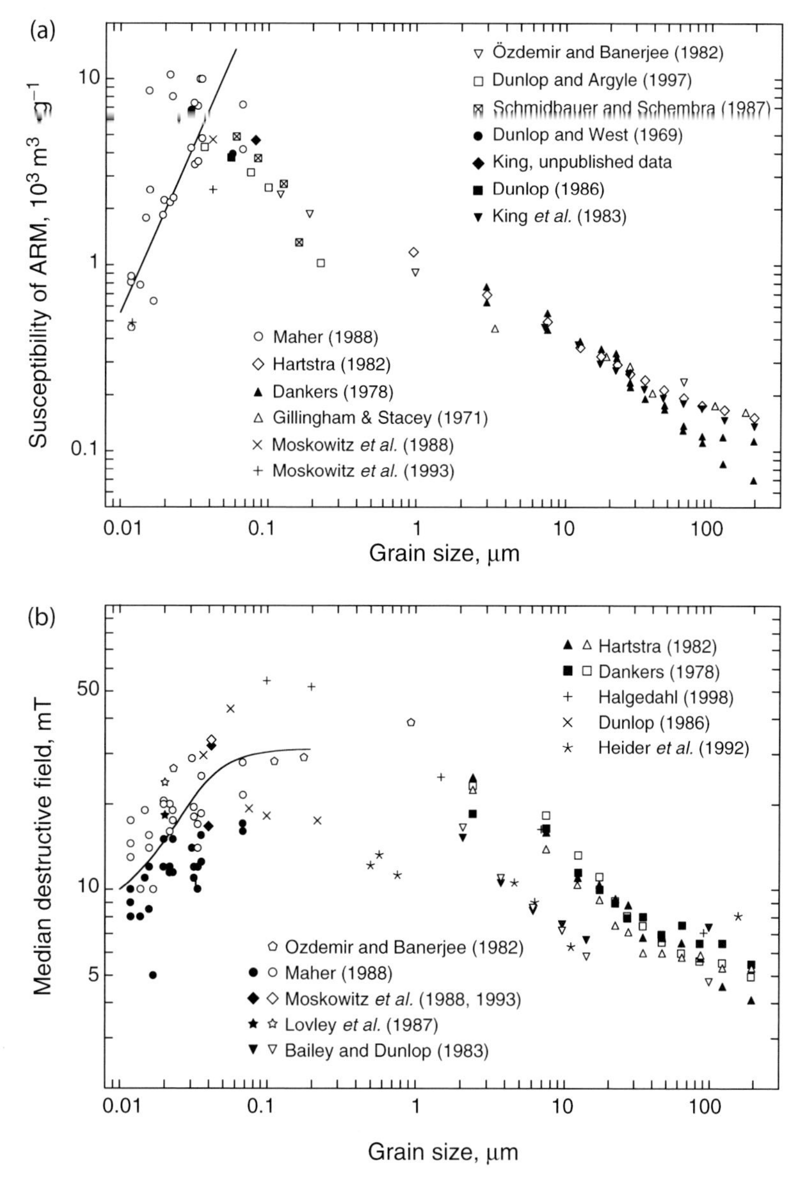

Figure M95 (a) ARM susceptibility as a function of magnetite grain size. (b) Median destructive fields of SIRM (symbols in left legend) and ARM (symbols in right legend). The experimental data are synthetic and biogenic magnetites from various sources. (Reprinted from Egli and Lowrie, 2002, with kind permission of the authors).

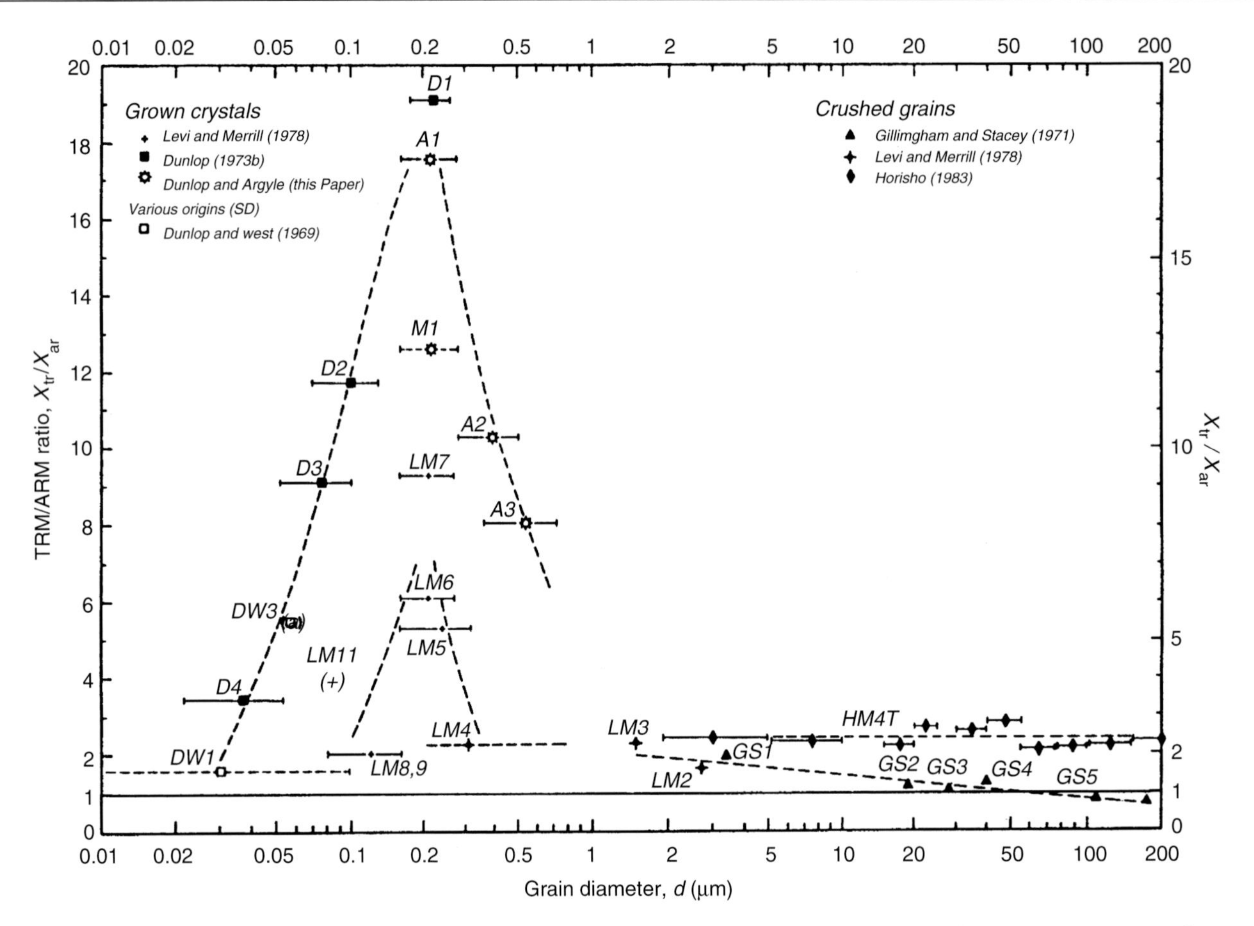

Figure M96 The ratio of weak-field TRM and ARM, and susceptibility ratio (χ_{trm}/χ_{arm}) as a function of magnetite grain size. The experimental data are synthetic magnetites from various sources. (Reprinted from Dunlop and Argyle, 1997, with kind permission of the authors).

ability to acquire remanence is lost. This grain size trend is similar in form to other grain size-dependent trends in magnetic properties (e.g., TRM, SIRM, H_c) and is a hallmark of the change in micromagnetic states from superparamagnetic to stable single- to multidomain configurations with increasing grain size. For single-domain particles, ARM and TRM have many similar experimental characteristics (Dunlop and Özdemir, 1997) including: (1) comparable resistance to AF demagnetization (Levi and Merrill, 1976), (2) linear dependence on H_{DC} for weak fields, $H_{DC} < 0.2$ mT (Dunlop and Argyle, 1997), and (3) pARMs that obey laws analogous to the three Thellier laws for partial TRM: additivity, reciprocity, and independence (Yu *et al.*, 2002a,b, 2003; see *Magnetization, thermoremanent* (*TRM*)). These experimental observations form the basis for ARM methods of paleointensity determinations (e.g., Shaw, 1974; Kono, 1978; Tauxe *et al.*, 1995), which avoid multiple heating to high temperatures that are necessary for the conventional Thellier-Thellier method for paleointensity (Dunlop and Özdemir, 1997).

Whereas qualitative similarities exist between ARM and TRM as exemplified by properties (1)–(3) in the previous paragraph, TRM is still a more efficient remanence acquisition process than ARM for a given DC field. This is demonstrated by the grain size dependence of the ratio of intensities for weak-field TRM and ARM in magnetite (Figure M96), which show TRM/ARM ratios greater than 1 for all grain sizes (Levi and Merrill, 1976; Dunlop and Argyle, 1997). While the TRM/ARM ratio is between 1 and 2 and practically independent of grain size above 1 μm, ratios exceeding 10 occur within a narrow size range between 0.1 and 0.5 μm. This contrast in behavior for grain sizes above and below $d = 1$ μm is possibly related to different micromagnetic remanent states produced by field-cooling from above the Curie temperature for TRM (remanence acquired at elevated temperature) or field-cycling at room temperature for ARM (remanence acquired at room temperature). For instance, micromagnetic models suggest that for the 0.1–0.5 μm size range, high-moment metastable single- or two-domain states are blocked in during TRM acquisition, but only low-moment vortex ground states result from the AF cycling during ARM acquisition (Dunlop and Argyle, 1997). In contrast, for $d > 1$ μm magnetites, TRM/ARM ratios between 1 and 2 suggest that similar micromagnetic states, probably with many domains, are produced during either field-cooling or field-cycling.

ARM theories in single- and multidomain particles

For an assembly of noninteracting SD particles undergoing only field-assisted switching (i.e., no thermal fluctuations or time-dependent switching), the intensity of ARM is theoretically independent of H_{DC} (and will equal the saturation remanence) and the initial ARM susceptibility is infinite (Wohlfarth, 1964). However, these theoretical predictions are not observed in synthetic or natural materials. For instance, Figure M97 clearly demonstrates that the shape of the ARM acquisition curve is strongly affected by particle concentration, where high concentrations of particles result in stronger dipolar (magnetostatic) interactions between magnetic particles. The effect of increasing interactions is to lower the efficiency of ARM acquisition. Dunlop and West (1969) used the Néel-Preisach model for interactions to explain weak-field ARM and TRM behavior in several synthetic and natural samples containing

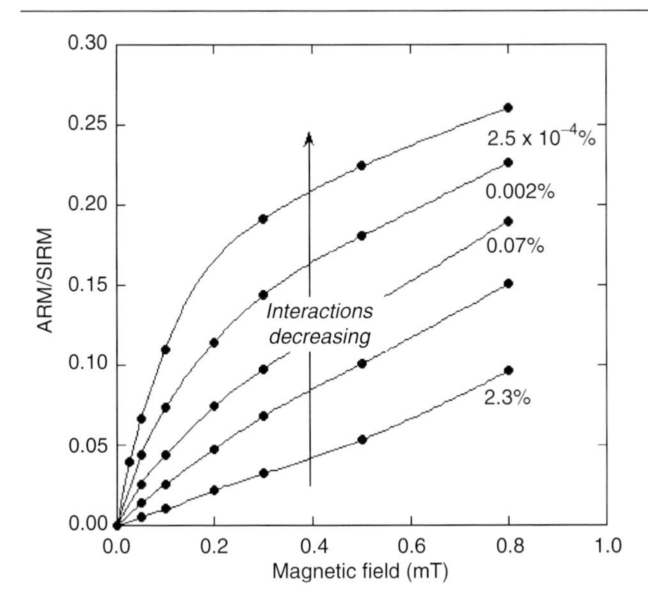

Figure M97 ARM acquisition curves for samples containing dispersed grains of magnetite in different concentrations. The curves are normalized by saturation remanence and the magnetic field is the DC bias field applied during AF demagnetization. The magnetite sample contains a mixture of SD and PSD grains. (Redrawn from Sugiura, 1979).

SD particles of iron oxides. In the Néel-Preisach model, interactions are accounted for in particle assemblies by assuming each particle has an asymmetric rectangular hysteresis loop, with the amount of asymmetry being proportional to the strength of the interaction field. A distribution of interaction fields, determined by separate experiments, is then used to model the ARM acquisition curves. A shortcoming to this type of model is that moment switching occurs instantaneously, while it is known that time-dependent switching via thermal fluctuations is important in SD and small PSD particles according to the Néel theory of thermally activated magnetization (Dunlop and Özdemir, 1997). ARM theories for SD particles incorporating interparticle, dipolar interactions, and thermal fluctuations have been proposed by Jeap (1971), Walton (1990), and Egli and Lowrie (2002). The following equation was derived by Jeap (1971) for ARM for interacting SD grains:

$$p_{\mathrm{arm}} = \tanh(\mu_0 V M_s / kT (B H_0 - \lambda p_{\mathrm{arm}})$$
$$B \approx (M_{sb}/M_{s0})(T_0/T_b)^{1/2}$$

where $p_{\mathrm{arm}} = M_{\mathrm{arm}}/M_{\mathrm{rs}}$, M_s is the saturation magnetization, M_{rs} is the saturation remanence, k is Boltzmann's constant, V is grain volume, T is absolute temperature, λ is the mean interaction field, and the subscripts 0 and b refer to room temperature (T_0) and the blocking temperature (T_b). A field-blocking condition, derived from Néel's relaxation theory, is assumed whereby the alternating field necessary to reverse the magnetization in a given time interval corresponds to a relaxation time that makes $\ln(f_0 t) = 25$, where f_0 is the frequency factor usually taken as 10^9 Hz. Note, that when $B = 1$, the equation reduces to the SD equation for TRM with interactions.

In the most recent theoretical work, Egli and Lowrie (2002) show that even in noninteracting systems, ARM can be described solely by thermal fluctuation effects. Therefore, in dilute systems, such as in many rocks and sediments containing low concentrations of magnetic carriers, ARM properties are controlled by intrinsic parameters (coercivity and grain volume) rather than by interactions. The SD ARM equation derived by Egli and Lowrie (2002) is

$$p_{\mathrm{arm}} = \tanh(\chi_{\mathrm{arm}} H_{\mathrm{DC}} / M_{\mathrm{rs}})$$
$$\chi_{\mathrm{arm}} = 1.797 \mu_0 M_{\mathrm{rs}} \left(\frac{V M_s}{kT\sqrt{\mu_0 H_K}} \right)^{3/2} \ln^{1/3}\left[\frac{0.35 F_0}{f_{\mathrm{ac}} \Delta H_{\mathrm{AF}} \sqrt{\mu_0 H_K}} \left(\frac{kT}{V M_s} \right)^{3/2} \right]$$

where H_k is the anisotropy field, f_{ac} is the AF frequency, and ΔH_{AF} is the decay rate of AF. This model predicts an increase in ARM intensity proportional to d^2 within the SD size range ($d < 60$ nm). ARM intensity is also predicted to be weakly dependent on the characteristics of the AF. Experimental results on natural and synthetic samples gave results compatible with theoretical predictions.

For non-single-domain grains, Gillingham and Stacey (1971) proposed a multidomain theory of ARM incorporating domain wall translation under the influence of the self-demagnetizing field. The self-demagnetizing field itself can be thought of as an internal interaction field. Theory predicts that for DC fields in which domain wall translation occurs ($H_{\mathrm{DC}} < H_c$), $M_{\mathrm{arm}} \propto H_{\mathrm{DC}}$. Reasonable agreement has been obtained between theory and experiment for magnetite grains with $d > 40$ μm, which most likely contain true multidomain states. However, experimental ARM intensities were significantly larger than predicted for $d < 40$ μm or for particle sizes within the pseudosingle domain size range (Gillingham and Stacey, 1971). Dunlop and Argyle (1997) developed a PSD theory for ARM by combining theories incorporating a wall displacement term linearly dependent on H_{DC} (an MD process) and a wall moment term proportional to the hyperbolic tangent function (an SD process). Reasonable fits to experimental ARM acquisition data for magnetite particles in the 0.2–0.5 μm size range were obtained with this theory.

Applications

Absolute paleointensity

Shaw (1974) proposed an alternative paleointensity method (now called the Shaw method) to the conventional Thellier-Thellier method, employing ARM and a single-step laboratory TRM heating. First, the NRM of the sample is stepwise AF demagnetized. Then an ARM (designated ARM1) is imparted and subsequently AF is demagnetized. Next, the sample is given a total TRM by heating the sample once above its Curie temperature and cooling in the same DC field as used in the ARM experiment. The TRM is then AF demagnetized using the same step as the one used for ARM. Finally, a second ARM (designated ARM2) is given and AF is demagnetized. If the sample does not experience any chemical alteration during the TRM step then a plot of ARM1 vs ARM2 using the peak AF as the common parameter should be linear with a slope of 1.0. In practice, a portion of the coercivity spectrum is selected that gives the best fit to a line of slope 1.0. Within this coercivity window it is assumed that the TRM is likewise unchanged and a plot of NRM vs TRM using the peak AF as the common parameter yields an estimate of the paleofield. Idealized Shaw diagrams are shown in Figure M98. This method is considerably quicker than the Thellier-Thellier method because less heating is required. Furthermore, samples are heated only once above the Curie temperature, which reduces the potential for chemical alteration that can occur during the multiple heatings used in the Thellier-Thellier method. Kono (1978) showed experimentally that the Shaw method could be as reliable as the Thellier-Thellier method for samples that undergo little or no chemical change during heating. However, a disadvantage of the Shaw method is that the entire experiment is unusable if alteration takes place during the single-step heating to the Curie temperature. Thermochemical alteration of the sample can lead to changes in its TRM capacity and produce erroneous paleointensity results. Modifications of the original Shaw method have been proposed to correct for limited amounts of thermochemical alteration

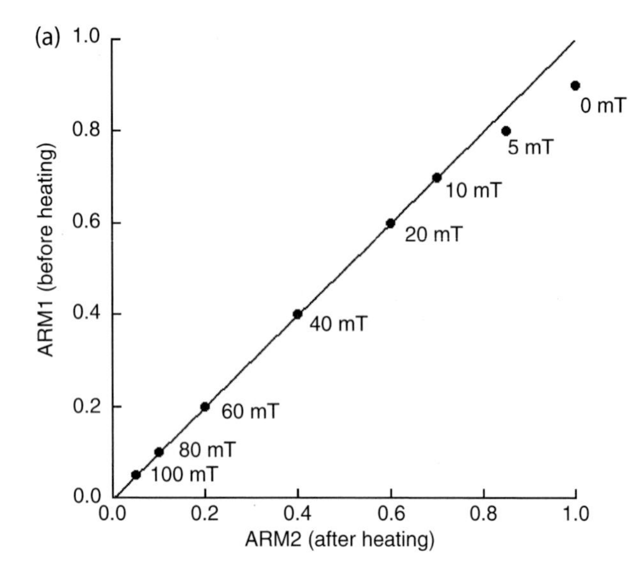

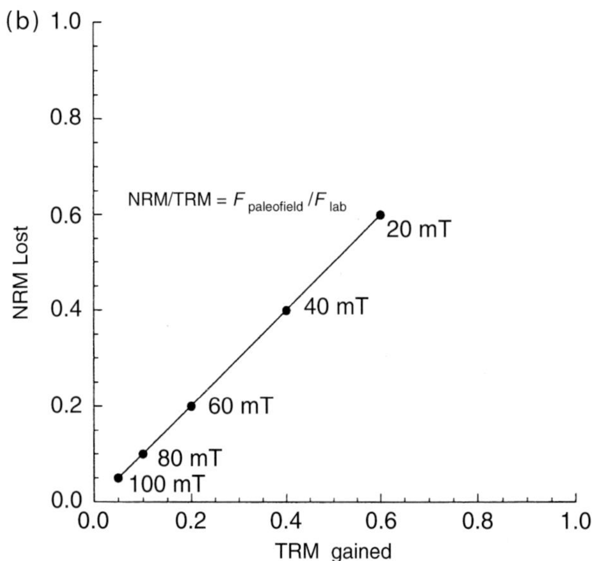

Figure M98 Idealized Shaw diagrams for absolute paleointensity determination. (a) ARM1/ARM2 diagram is used to test samples for thermochemical alteration during heating. ARM1 (before heating) and ARM2 (after heating) are AF-demagnetized in identical steps (5, 10, 20,...,100 mT). The solid line represents the slope where ARM1/ARM2 = 1.0. Points that fall off the line (0, 5, 10 mT) may indicate part of the coercivity spectrum that was altered during heating. (b) NRM/TRM diagram is used to estimate the paleointensity. The line with slope = 1.0 from part (a) between 20 and 100 mT delimits the range of identical AF demagnetization characteristics of NRM and TRM within which the TRM and NRM may be compared with each other for estimating paleointensity.

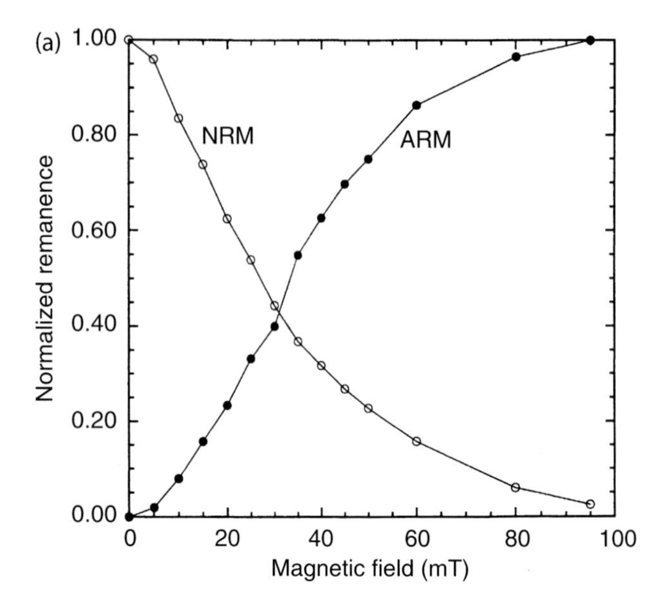

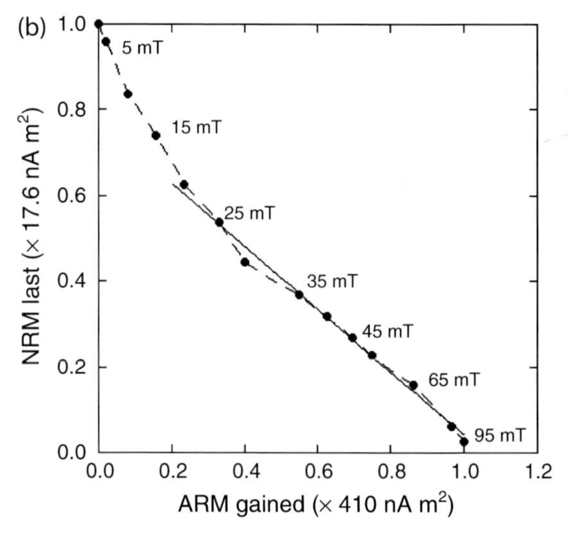

Figure M99 Pseudo-Thellier method for relative paleointensity in sediments. (a) Normalized remanence curves for stepwise AF-demagnetization of NRM and the subsequent acquisition of partial ARM in the identical steps as the NRM demagnetization. (b) NRM/ARM plot (also called an Arai plot) for the data in (a). The Arai plot shows ARM gained during magnetization in specified AF fields vs NRM left after demagnetization in the same AF fields. The solid line represents the best-fit line through the data and the slop gives the relative paleointensity. (Data taken from Tauxe *et al.*, 1995).

during the TRM heating step (Kono, 1978; Rolph and Shaw, 1985). In these methods, a "slope correction" is (1) applied to the NRM/TRM slope using slopes of linear segments of the ARM1/ARM2 plot over "unaltered" coercivity windows or (2) applied individually to TRM data points using ARM1/ARM2 ratios measured at the same AF level as the corresponding TRM point.

Besides the Shaw method and its modified forms, several ARM techniques were developed earlier, which made use of the theoretical relationships between the intensities of ARM and TRM (Banerjee and Mellama, 1974; Collinson, 1983). Here, the laboratory heating step was eliminated and replaced entirely by room-temperature ARM. However, the main shortcoming of such an approach was the difficulty in determining the correct calibration factor, R = ARM/TRM, for particular samples. Here R is known to be a function of grain size, interactions, and applied field (Figures M96 and M97) and would vary from sample to sample. These methods were used primarily to estimate paleointensities from lunar samples containing iron or iron-nickel grains, which are highly susceptible to chemical alteration even when heated or slightly heated above room temperature (see review by Collinson, 1983).

Relative paleointensity

For marine and lacustrine sediments in which NRM is a detrital remanent magnetization (DRM), ARM is used as a normalizing parameter to remove the effects on the NRM intensity signal due to depth variations in concentration, grain size, and composition of the magnetic carriers of remanence (e.g., Tauxe, 1993; Valet, 2003 for comprehensive reviews). According to this hypothesis, down-core variations in NRM/ARM values should theoretically reflect relative variations in paleointensity of the Earth's field rather than variations in physical properties (grain size, concentration) of the magnetic particles in the sediment. The idea was originally proposed by Johnson *et al.* (1975a,b) and Levi and Banerjee (1976). These authors argued that the magnetic particles responsible for the stable portion of the NRM signal, namely, SD- and PSD-sized grains, will be the same grain size fraction that is most efficient in acquiring ARM (see Figure M95). King *et al.* (1983) propose three uniformity criteria for sediments under which NRM/ARM normalization provides a reliable measure of relative paleointensity: (1) magnetite is the predominate remanence carrier (uniformity in composition); (2) grain size varies between 1 and 15 μm (uniformity in grain size); and (3) variability in magnetite concentration is less than 20–30 times the minimum concentration (uniformity in concentration).

Another approach, which uses ARM to determine relative paleointensity in sediments, is the pseudo-Thellier method of Tauxe *et al.* (1995). In this procedure, NRM lost in successive AF demagnetization steps is compared to partial ARMs gained in matching AF steps, similar to the pNRM-pTRM steps in the Thellier-Thellier method. When NRM lost is plotted against the pARM gained for each matching AF step, the slope of the resulting line is proportional to the relative paleointensity (Figure M99). This method has two advantages over the conventional NRM/ARM normalization. (1) It allows identification of possible overprints due to viscous remanent magnetization (VRM), particularly at low AF demagnetization steps (<15 mT), which can bias the NRM/ARM estimate. (2) The pNRM-pARM data can be statistically analyzed in the same way as conventional Thellier-Thellier data and, most importantly, provide error estimates on the relative paleointensity determination.

Magnetic granulometry

Magnetic granulometry is the determination of grain size of magnetic materials based on the discrimination between magnetic relaxation effects (superparamagnetic and SD particles) and domain processes (MD grains). An approximate method to distinguish fine grains (<20 μm) from coarse grains (>20 μm), known as the Lowrie-Fuller test and modified by Johnson *et al.* (1975a,b), compares the AF demagnetization spectra of a weak-field ARM and a strong-field SIRM. If the ARM is less stable or "softer" to AF demagnetization than the corresponding SIRM (where both parameters are normalized to their initial values before demagnetization) then the remanence is assumed to be carried by grains in multidomain states. If the opposite trend is observed, then the remanence must be carried by grains in SD or PSD states (e.g., Dunlop and Özdemir, 1997). Originally, this experimental method was developed as a simple way to screen samples to determine if the NRM in magnetite-bearing igneous or sedimentary rocks resided in SD or small PSD grains, which are likely to carry the stable, primary component of magnetization, or in MD grains, which are likely to be susceptible to secondary remagnetization. Subsequent experimental and theoretical modeling have shown that other factors such as particle volume and coercivity distributions and crystal defect density can complicate the original Lowrie-Fuller test by producing "false positives" (i.e., SD-type behavior for MD grains and vice versa). Dunlop and Özdemir (1997) provide a recent review of the theoretical and experimental work related to the Lowrie-Fuller test.

Banerjee *et al.* (1981) introduced a magnetic granulometry technique that uses the ratio of anhysteretic susceptibility (χ_{arm}) to initial magnetic susceptibility (χ_0). The method was designed as a rapid way for determining the relative size variations of magnetite in marine and lake sediment cores and has found wide use as a grain size proxy in environmental magnetism studies because it can provide information about past variations in sediment source regions, transport paths, and postdepositional alteration. The basis of the method is that ARM is enhanced in the fine-grained SD fraction (less than 1.0 μm, see Figure M95), whereas χ_0 is relatively independent for the coarse-grained PSD and MD fraction. The ratio χ_{arm}/χ_0, therefore, should vary inversely with grain size, assuming that by taking the ratio, any effect due to variations in the concentration of magnetic carriers is removed (King *et al.*, 1983). Data is presented either as bivariate plots of χ_{arm} vs χ_0 or a plot of χ_{arm}/χ_0 as a function of core depth. Figure M100 shows the experimental efficacy of the method using data from synthetic samples of magnetite with known grain sizes. However, it is important to note that the effects of particle interactions can significantly modify grain size estimates of natural magnetic mineral assemblages using such an approach. In addition, any ultrafine-grained (<0.05 μm), superparamagnetic particles present will contribute high susceptibility but very low ARM, yielding χ_{arm}/χ_0 values that could be misinterpreted as a coarse-grained MD signal. Other authors have proposed variations of this method using χ_{arm}/SIRM or χ_{arm}/χ_{FD} (where χ_{FD} is the difference between χ_0 measured at low and high frequency) to discriminate SP from MD particle (Maher, 1988; Oldfield, 1994). An example of the χ_{arm}/χ_0 method is shown in Figure M101 for lake sediments from Minnesota (USA) revealing relative changes in magnetic grain sizes due to paleoclimatic and anthropogenic events over the last 1000 years.

Anisotropy of anhysteretic remanent magnetization

The directional dependence of ARM in a sample, commonly termed as the anisotropy of anhysteretic remanent magnetization (AARM), is used as a tool for quantitative analysis of petrofabrics in rocks and sediments and can provide information on depositional, emplacement,

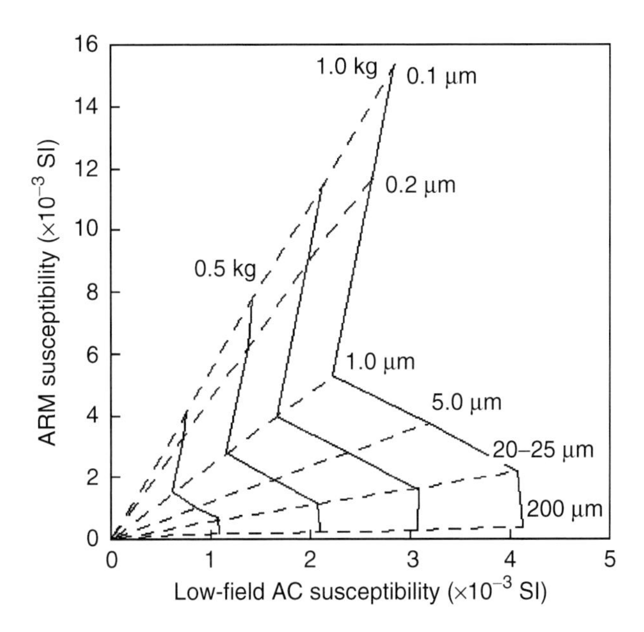

Figure M100 Model of ARM susceptibility vs low-field AC susceptibility for detecting magnetite grain size variations in natural samples. The model is based on data for sized magnetites. The slope of the line connecting the origin to the individual sample point can determine relative grain size variations of magnetite in natural samples. High slopes indicate finer grain sizes. Concentration of magnetite can be estimated from the distance from the origin to an individual data point. (Data taken from King *et al.*, 1983).

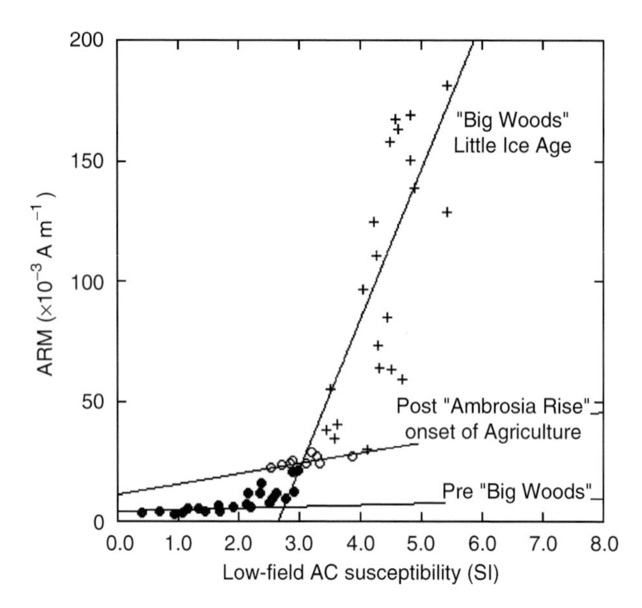

Figure M101 An example of an ARM vs low-field AC susceptibility diagram for recent lake sediments from Long Lake, MN (USA). The solid lines with different slopes are correlated with major paleoclimatic and anthropogenic events within the drainage basin during the Holocene. (Redrawn from Banerjee *et al.*, 1981).

or tectonic histories of igneous, metamorphic, and sedimentary rocks (see review by Jackson, 1991). Similar to the traditional method of anisotropy of magnetic susceptibility (AMS, see *Magnetic susceptibility, anisotropy*), AARM is used to determine preferred orientations of magnetic mineral grains in a rock. However, because AARM fabrics reside exclusively in the ferrimagnetic, remanence-carrying fraction (e.g., magnetite, titanomagnetite, pyrrhotite), it may provide information about complementary or different geological events or processes from those responsible for AMS fabrics, which reflects an average of contributions from all nonmagnetic (paramagnetic and diamagnetic) and ferrimagnetic mineral phases present. The use of AARM in magnetic fabric studies was pioneered by the work of McCabe *et al.* (1985) and Jackson *et al.* (1988).

In the weak-field limit ($H < 0.2$ mT), where ARM acquisition is approximately linear in applied field, the ARM anisotropy can be described by a second-rank tensor and represented by an ellipsoid, mathematically equivalent to the AMS tensor. Mathematical and statistical treatment of AARM tensor data follows procedures developed for AMS data (e.g., Tauxe, 1998; Tarling and Hrouda, 1993). AARM is analyzed in terms of anisotropy of remanence tensor calculated by subjecting a sample to repeated ARMs along different orientations and measuring the intensity of ARM acquired along the applied field direction for each orientation. At least six independent orientations are needed to determine the tensor. Usually ARM is imparted along 9 to 15 different orientations and the tensor is determined by least-squares analysis (McCabe *et al.*, 1985; Jackson, 1991). Jackson *et al.* (1988) developed a method employing pARMs over selective coercivity windows instead of total ARMs for AARM analysis. Unlike the

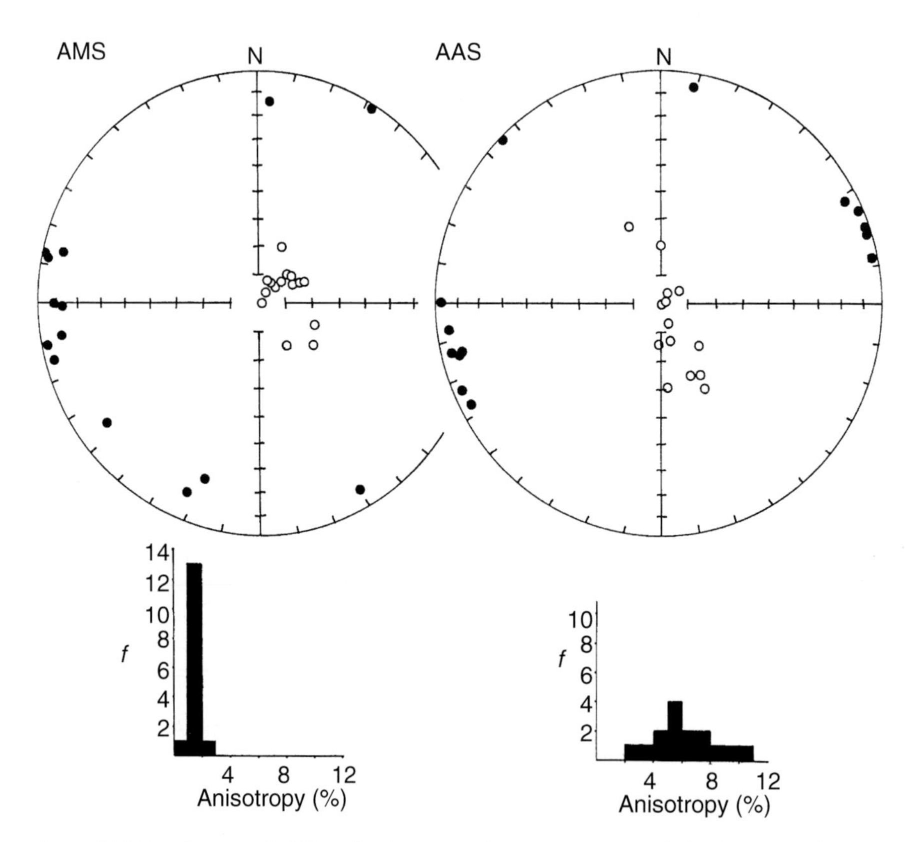

Figure M102 Comparison of ARM anisotropy (AAS) and anisotropy of magnetic susceptibility (AMS) in the Trenton limestone. The directional data are plotted on equal area projections. The degree of ARM anisotropy is greater and its principal axes are more tightly grouped than the corresponding AMS data. (Reprinted from McCabe *et al.*, 1985, with kind permission of the authors.)

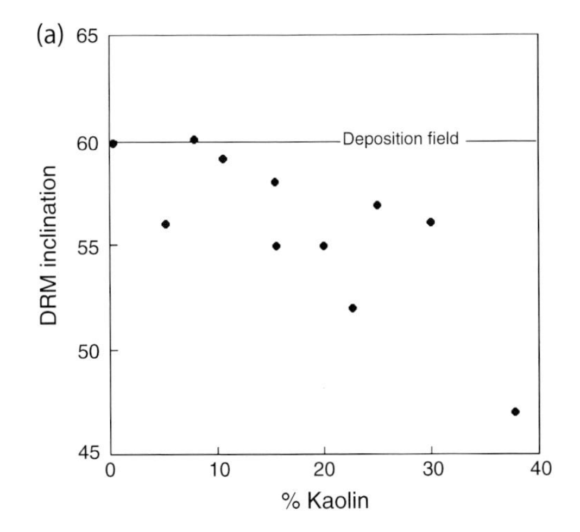

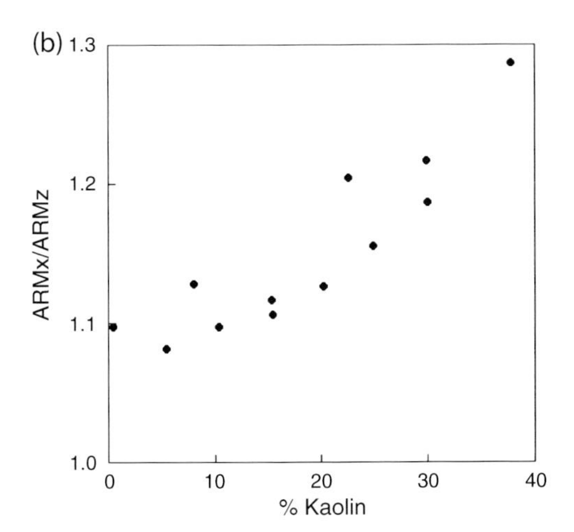

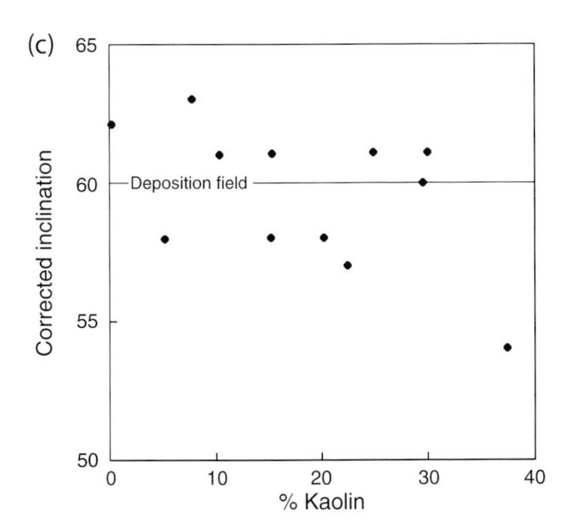

Figure M103 (a) Magnetic inclinations of experimental DRM as a function of kaolin content. The synthetic sediments were deposited under controlled conditions in an applied field with an inclination of 60°. Inclination error increases with % kaolin due to clay-magnetite electrostatic interactions. (b) ARM anisotropy as a function of kaolin content. (c) Corrected inclination values using the ARM anisotropy model of Jackson *et al.* (1991). (Redrawn from Jackson, 1991).

standard ARM method, the use of pARMs has the advantage of being able to isolate magnetic fabrics from discrete coercivity, or grain size, fractions. For instance, the different fabrics carried by SD and MD grain fractions can be distinguished from one another.

AARM and other remanence anisotropy methods (e.g., IRM) have several advantages over AMS; most notably, remanence-based methods (1) provide higher signal-to-noise ratios, particularly in weakly magnetic or weakly anisotropic rocks, and (2) can provide information related to the fidelity of the orientation of the NRM vector (i.e., whether NRM reflects magnetic alignment by the paleofield or alignment by a petrofabric) because both are carried by the same remanence-carrying phases. Figure M102 shows a comparison of AMS and AARM data for samples of the Trenton limestone (USA), taken from the work of McCabe *et al.* (1985). The AARM data shows a much more clearly defined west-southwesterly lineation with a higher degree of anisotropy than the companion AMS results. In this case, the lower degree of anisotropy determined from the AMS data resulted from a dilution effect on the susceptibility due to the diamagnetism ($\chi < 0$) of the calcite matrix and, therefore, was not as sensitive a measure as the AARM for detecting the petrofabrics. Other examples of fabrics determined by a combination of AMS- and AARM-derived data are given in a review by Jackson (1991).

Another example of the usefulness of AARM is its application in the field of sedimentary paleomagnetism. Jackson *et al.* (1991) developed a quantitative model for correcting inclination shallowing carried out by magnetite grains in sediments by measuring the ARM tensor, where ARM is assumed to activate the same coercivity spectrum responsible for DRM acquisition. The model assumes that whatever processes lead to inclination shallowing (e.g., compaction) result in a horizontally flattened distribution of long axis orientations of magnetic particles. In general terms, the **DRM** vector is related to the applied field vector (**H**) through the following equation:

$$\mathbf{DRM} = (k_d)\mathbf{H}$$

where k_d is the anisotropic detrital remanence tensor. According to the theory, the AARM tensor (k_a) can be used as an approximation for k_d and the deposition field vector can be obtained from

$$\mathbf{H} = (k_d)^{-1}\mathbf{DRM} \cong (k_a)^{-1}\mathbf{DRM}$$

Figure M103 shows results from controlled deposition experiments using synthetic sediments composed of magnetite, silica, and kaolin (Jackson *et al.*, 1991). Both inclination error (Figure M103a) and ARM anisotropy (Figure M103b) are direct functions of kaolin concentration, presumably because of electrostatic interactions between clay and magnetite particles. Correcting the inclination values by multiplying the measured DRM vector by the inverse of k_a according to equation YY yields inclination values (Figure M103c) closer to the actual depositional field orientation and provides confirmation of the theoretical model proposed by Jackson *et al.* (1991). Therefore, this method can provide more accurate determinations of the direction of the ancient paleofield in magnetite-bearing sediments and sedimentary rocks.

Bruce M. Moskowitz

Bibliography

Banerjee, S.K., and Mellama, J.P., 1974. A new method for the determination of paleointensity from ARM properties of rocks. *Earth and Planetary Science Letters*, **23**: 177–184.

Banerjee, S.K., King, J.W., and Marvin, J.M., 1981. A rapid method for magnetic granulometry with applications to environmental studies. *Geophysical Research Letters*, **8**: 333–336.

Collinson, D.W., 1983. *Methods in Rock Magnetism and Paleomagnetism: Techniques and Instrumentation.* New York: Chapman & Hall.
Dunlop, D.J., and West, G., 1969. An experimental evaluation of single-domain theories. *Reviews of Geophysics*, **7**: 709–757.
Dunlop, D.J., and Argyle, K.S., 1997. Thermoremanence, anhysteretic remanence and susceptibility of submicron magnetites: nonlinear field dependence and variation with grain size. *Journal of Geophysical Research*, **102**: 20 199–20 210.
Dunlop, D.J., and Özdemir, Ö., 1997. *Rock Magnetism: Fundamentals and Frontiers.* New York: Cambridge University Press.
Egli, R., and Lowrie, W., 2002. Anhysteretic remanent magnetization of fine magnetic particles. *Journal of Geophysical Research*, **107** (B10): 2209, doi:10.1029/2001JB0000671.
Gillingham, E.W., and Stacey, F.D., 1971. Anhysteretic remanent magnetization (ARM) in magnetite Grains. *Pure and Applied Geophysics*, **91**: 160–165.
Jackson, M.J., 1991. Anisotropy of magnetic remanence: a brief review of mineralogical sources, physical, origins, and applications and comparisons with susceptibility anisotropy. *Pure and Applied Geophysics*, **136**: 1–28.
Jackson, M.J., Gruber, W., Marvin, J., and Banerjee, S.K., 1988. Partial anhysteretic remanence and its anisotropy: applications and grain size dependence. *Geophysical Research Letters*, **15**: 440–443.
Jackson, M.J., Banerjee, S.K., Marvin, J.A., Lu, R., and Gruber, W., 1991. Detrital remanence inclination errors and anhysteretic remanence anisotropy: quantitative model and experimental results. *Geophysical Journal International*, **104**: 95–193.
Jeap, W.F., 1971. Role of interactions in magnetic tapes. *Journal of Applied Physics*, **42**: 2790–2794.
Johnson, H.P., Lowrie, W., and Kent, D., 1975a. Stability of anhysteretic remanent magnetization in fine and coarse magnetite and maghemite particles. *Geophysical Journal of the Royal Astronomical Society*, **41**: 1–10.
Johnson, H.P., Kinoshita, H., and Merrill, 1975b. Rock magnetism and paleomagnetism of some North-Pacific deep sea sediments. *Geological Society of America Bulletin*, **86**: 412–420.
King, J.W., Banerjee, S.K., and Marvin, J., 1983. A new rock-magnetic approach to selecting samples for geomagnetic paleointensity studies: applications to paleointensity for the last 4000 years. *Journal of Geophysical Research*, **88**: 5911–5921.
Kono, M., 1978. Reliability of paleointensity methods using alternating field demagnetization and anhysteretic remanence. *Geophysical Journal of the Royal Astronomical Society*, **54**: 241.
Levi, S., and Banerjee, S.K., 1976. On the possibility of obtaining relative paleointensities from lake sediments. *Earth and Planetary Science Letters*, **29**: 219–226.
Levi, S., and Merrill, R.T., 1976. A comparison of ARM and TRM in magnetite. *Earth and Planetary Science Letters*, **32**: 171–184.
Maher, B.A., 1988. Magnetic properties of some synthetic submicron magnetites. *Geophysical Journal of the Royal Astronomical Society*, **94**: 83–96.
McCabe, C., Jackson, M.J., and Ellwood, B.B., 1985. Magnetic anisotropy in the Trenton limestone: results of a new technique, anisotropy of anhysteretic susceptibility. *Geophysical Research Letters*, **12**: 333–336.
Oldfield, F., 1994. Toward the discrimination of fine grained ferrimagnets by magnetic measurements in lake and near-shore marine sediments. *Journal of Geophysical Research*, **99**: 9045–9050.
Rimbert, F., 1959. Contribution À L'étude De L'action De Champs Alternatifs Sur Les Aimantations Rémanentes Des Roches: Applications Géophysiques. *Revue de l'Institut Francais du Petrole*, **14**: 17–54.
Rolph, T.C., and Shaw, J., 1985. A new method of paleofield magnitude correction for the thermally altered samples and its applications to Lower Carboniferous lavas. *Geophysical Journal of the Royal Astronomical Society*, **80**: 773–781.
Shaw, J., 1974. A new method of determining the magnitude of the palaeomagnetic field: application to five historic lavas and five archaeological samples. *Geophysical Journal of the Royal Astronomical Society*, **39**: 133–141.
Sugiura, N., 1979. ARM, TRM and magnetic interactions: concentration dependence. *Earth and Planetary Science Letters*, **42**: 451–455.
Tarling, D.H., and Hrouda, F., 1993. The Magnetic Anisotropy of Rocks, London: Chapman and Hall.
Tauxe, L., 1993. Sedimentary records of relative paleointensity of the geomagnetic field: theory and practice. *Reviews of Geophysics*, **31**: 319–354.
Tauxe, L., 1998. *Paleomagnetic Principles and Practice.* Dordrecht: Kluwer Academic Publishers.
Tauxe, L., Pick, T., and Kok, Y.S., 1995. Relative paleointensity in sediments: a pseudo Thellier approach. *Geophysical Research Letters*, **22**: 2885–2888.
Valet, J-P., 2003. Time variations in geomagnetic intensity. *Reviews of Geophysics*, **41**(1): 1004, doi:10.1029/2001RG000104.
Walton, D., 1990. A theory of anhysteretic remanent magnetization of single domain grains. *Journal of Magnetism and Magnetic Materials*, **87**: 369–374.
Wohlfarth, E.P., 1964. A review of the problem of fine-particle interactions with special reference to magnetic recording. *Journal of Applied Physics*, **35**: 783–790.
Yu, Y., Dunlop, D.J., and Özdemir, Ö., 2002a. Partial anhysteretic remanent magnetization in magnetite: 1. Additivity. *Journal of Geophysical Research,* 107, 10.1029/2001JB001249.
Yu, Y., Dunlop, D.J., and Özdemir, Ö., 2002b. Partial anhysteretic remanent magnetization in magnetite: 2. Reciprocity. Journal of Geophysical Research, 107, 10.1029/2001JB001269.
Yu, Y., Dunlop, D.J., and Özdemir, Ö., 2003. Testing the independence law of partial ARMs: implications for paleointensity determination. *Earth and Planetary Science Letters*, **208**: 27–39.

Cross-references

Demagnetization
Magnetic Susceptibility, Anisotropy
Magnetization, Isothermal Remanent (IRM)
Magnetization, Thermoremanent (TRM)

MAGNETIZATION, CHEMICAL REMANENT (CRM)

Chemical remanent magnetism (CRM) is imparted to ferro- and ferrimagnetic minerals by chemical processes, at temperatures below their Curie points, in the presence of an effective magnetic field. Here, chemical processes are considered broadly to include but not be limited to modifications in oxidation state, phase changes and crystal growth. The effective magnetic field is the resultant vector field acting on the chemically-altered material, including the external and various interaction fields.

Nearly a century and half ago, Beetz (1860) discovered CRM during laboratory electrolytic depositions of iron; these observations supported Weber's hypothesis that some atoms possess intrinsic magnetization. These results were confirmed by Maurain (1901 and 1902) with electrolytic depositions of iron and nickel in the presence of external fields. Koenigsberger (1938, part 1, p. 122 & part 2, p. 319) noted the presence of coherent remanence in some sedimentary rocks, which he called *crystallization remanence,* and he advanced the hypothesis that it was "impressed by the Earth's field at temperatures between about 100°C and 500°C during the time of lattice changes in magnetite which result very probably from unmixing of Fe_2O_3."

With the rapid growth of paleomagnetism after World War II, it became apparent to many rock magnetists that the remanence of many sediments, especially red beds, was at least partly controlled by magnetic minerals chemically precipitated subsequent to deposition (e.g., Blackett, 1956).

In rocks, CRM is usually a secondary remanence; this is an important reason that in geophysics CRM studies lag behind investigations of TRM (Thermoremanent magnetization) and DRM (Depositional remanence), which are usually responsible for the primary remanence. In most paleomagnetic studies, it is advantageous to select samples that retain their primary TRM or DRM; secondary CRM is a nuisance to be avoided, if possible. However, in nature, CRM can rarely be entirely neglected. If a secondary CRM is superimposed on an extant primary TRM or DRM, one of the first objectives would be to remove (demagnetize) the CRM in order to expose the primary remanence. In many cases, especially for older, chemically-altered sediments, CRM is the dominant characteristic remanence, the primary DRM having been obliterated by processes similar to those that are responsible for the CRM.

Theory for CRM

Many CRM properties can be explained with Néel's (1949, 1955) thermal fluctuations theory for super-paramagnetic (SP) particles, which has been highly successful at explaining TRM and many other properties of remanence. We consider the an assemblage of non-interacting, uniformly magnetized SP particles with identical volumes, v, which have uniaxial anisotropy and aligned easy axes. For thermal equilibrium and in the presence of an external magnetic field, H, applied parallel to the easy axes, the volume magnetization, M, is given by the equation

$$M = nM_S(T)\tanh[M_S(T)vH/kT] \quad \text{(Eq. 1)}$$

The argument $[M_S(T)vH/kT]$ is the alignment factor, which vanishes for $H = 0$. $M_S(T)$ is the spontaneous saturation magnetization of the grains at temperature T; n is the number of magnetic particles per unit volume; k is Boltzmann's constant; T, the absolute temperature (°K).

When the field is removed, M decays, seeking the new state of thermal equilibrium, $M = 0$, at a rate determined by the relaxation time, τ. M is given by the equation

$$M = M_0 \exp[-t/\tau] \quad \text{(Eq. 2)}$$

where M_0 represents the initial magnetization; t is the elapsed time since the field was removed. For the particle assemblage described above, Néel derived an equation for τ, and in zero external magnetic field,

$$\tau = C^{-1} \exp[Kv/kT] \quad \text{(Eq. 3)}$$

where K is the uniaxial anisotropy energy per unit volume; C is the frequency factor whose value is on the order of $10^9\ s^{-1}$. C is a function of T and T-dependent material properties; however, C's variation with temperature is significantly less than the exponential factor, and, in comparison, C is usually treated as constant. By contrast, τ, which is also a measure of the remanence stability, varies orders of magnitude in response to modest changes of the argument (Kv/kT).

We now consider isothermal CRM production at temperature T_A, caused by crystal growth of magnetic particles. The precipitating ferro- or ferrimagnetic particles grow from atomic/molecular paramagnetic nuclei to larger exchange-coupled SP grains with spontaneous magnetization, M_S. In an external field and thermal equilibrium, the particles are aligned according to Eq. (1). When the field is removed, and as long as the thermal fluctuations (kT) can easily overcome the anisotropy energy barriers (Kv), $\tau \ll \tau_L$, M decays quickly following Eq. (2), and there is no remanence. (τ_L is a characteristic laboratory time on the order of minutes.) As the particle volumes increase, it becomes more difficult for thermal fluctuations to overcome the growing barriers to domain rotations. Because of the exponential dependence of τ on v in Eq. (3), the magnetization changes to a stable CRM over a very narrow range of volumes, Δv. At volume v_{AB}, $\tau \gg \tau_L$, and CRM is said to be blocked. v_{AB} is the critical blocking volume at T_A. That is, at temperature T_A, for $v < v_{AB}$, the particle assemblage is super-paramagnetic, while for $v > v_{AB}$, the magnetization is blocked as stable CRM in single domain (SD) particles. This is analogous to TRM production, where v is considered constant, and τ increases as T cools below the Curie point, T_C. Above the blocking temperature, T_B, the SP magnetic moments are aligned according to Eq. (1) with $\tau \ll \tau_L$, so that when the external field is removed, the magnetization quickly decays following Eq. (2). At lower temperatures $T < T_B$, $\tau \gg \tau_L$, and the magnetization is blocked as stable TRM.

CRM is usually difficult to distinguish from other remanences such as TRM and VRM (viscous remanence), because of overlapping stabilities and possible associations between CRM, partial-TRM (PTRM) and high-temperature VRM. For example, grain growth CRM at $T_A > T_R$, where T_R denotes room temperature, is in particles that have grown beyond the blocking volume, v_{AB}; chemically precipitated particles with volumes $v < v_{AB}$ do not contribute to the CRM. On cooling from T_A to T_R, a PTRM will be produced in particles with blocking temperatures $T_A > T_B > T_R$. Also, ubiquitous time effects might contribute significant VRM superimposed on the CRM. In addition, the resultant remanence is likely to grow on cooling from T_A to T_R, due to the increase in M_S on cooling for most magnetic minerals. Such complexities make it difficult to uniquely isolate CRM from other remanences, hence it is probable that CRM occurrences in the paleomagnetic record are more numerous than is usually recognized.

As long as the magnetic ensemble consists of non-interacting, homogeneously magnetized SD particles, τ and the magnetic stability increase exponentially with particle volume. For multi-domain (MD) particles, τ decreases with increasing volume, principally because it becomes progressively easier to alter the remanence by domain wall movements, as opposed to rotating the magnetic moments of SD particles. These considerations of remanence stability and relaxation times were demonstrated experimentally by coercivity measurements of dispersed fine particles of ferromagnetic metals (Fe, Ni, Co), as the grain sizes increased due to progressive heat treatments (e.g., Meiklejohn, 1953; Becker, 1957). Zero initial coercivity in the SP region is followed by increasing coercivity with particle sizes, presumably in the SD size range; the coercivity then decreases for larger presumably MD particles (Figure M104). Haigh (1958) first applied Néel's theory and the grain size dependence of the coercivity to rock magnetism to explain CRM properties in growing particles of magnetite obtained during laboratory reduction of hematite. Kobayashi (1959) also examined CRM in magnetite obtained from hematite reduction and showed that CRM stability with respect to both alternating fields and thermal demagnetization was much greater than for IRM (isothermal remanence) and very similar to the stability of TRM (Figure M105). Kobayashi (1961) produced CRM in cobalt grains precipitated from Cu-Co alloy and showed that the specific CRM intensity had a similar bell-shaped grain size dependence as the coercivity, increasing from zero for SP particles, attaining a maximum value and then decreasing for inhomogeneously magnetized MD particles (Figure M106). Grain growth CRM can explain many examples in paleomagnetism, where the natural remanence (NRM) is predominantly a secondary remanence in chemically-altered rock and where new particles of a chemically-nucleated magnetic phase have grown beyond the SP size range.

CRM in Igneous Rocks

General Considerations

In igneous rocks, CRM can be produced by phase transformations or nucleation and growth of new magnetic minerals at temperatures

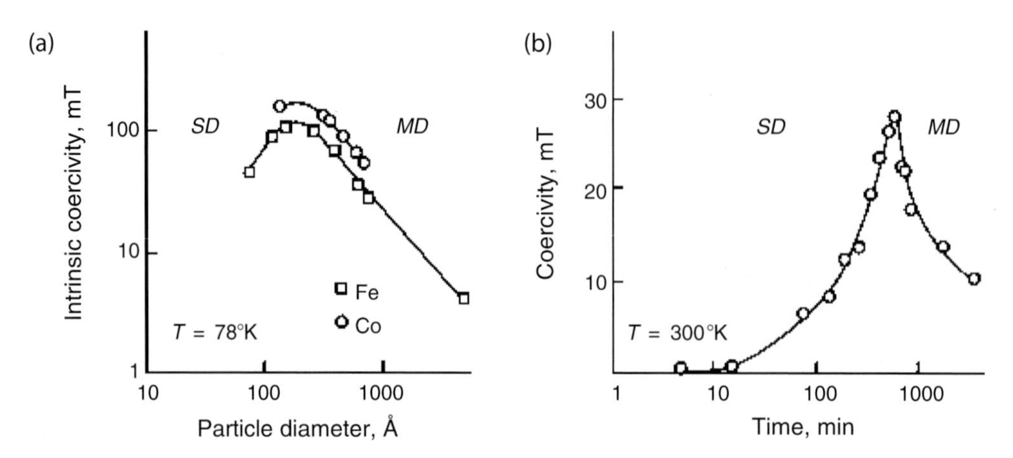

Figure M104 (a) Intrinsic coercive force of iron and cobalt as a function of particle size at liquid nitrogen temperature (modified from Meiklejohn, 1953). (b) Change in coercive force of Cu(98%)Co(2%) alloy as a function of annealing time (particle size) at 700°C, measured at 300 K (modified from Becker, 1957).

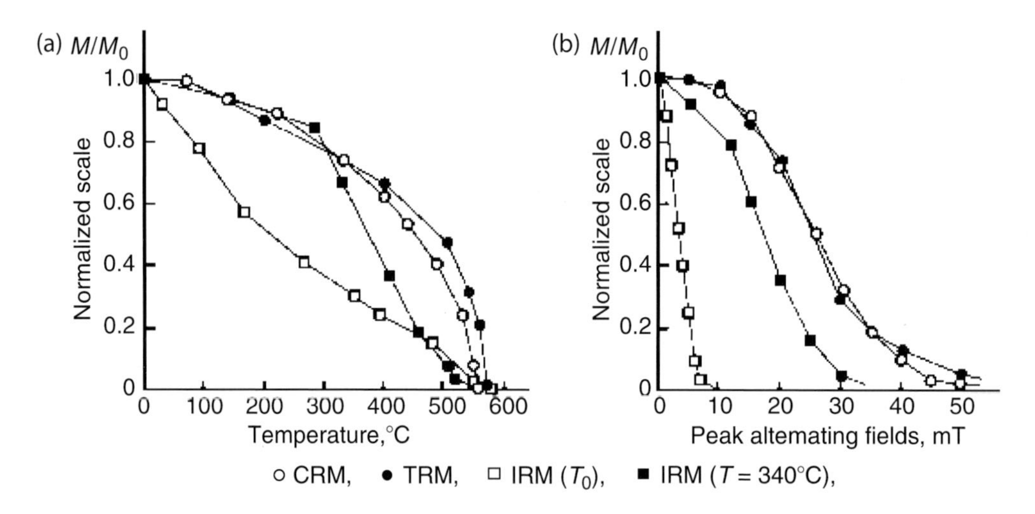

Figure M105 (a) Thermal, (b) alternating fields (AF) demagnetization of remanence in magnetite. All remanence measurements were made at room temperature. CRM was produced at 340°C with external fields of (a) 0.3 mT, (b) 1 mT. Total TRM was produced in (a) 0.3 mT, (b) 0.05 mT. IRM(T_0) at room temperature was produced in fields of (a) 20 mT, (b) 3 mT. IRM(T) was produced at 340°C in fields of (a) 2 mT, (b) 1 mT. CRM (T = 340°C) and IRM (T = 340°C) were cooled to T_0 in zero field prior to being demagnetized (modified from Kobayashi, 1959).

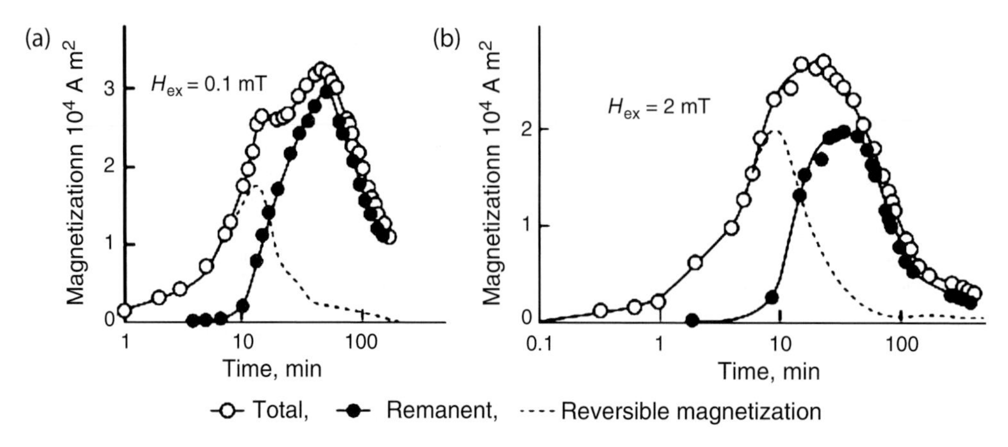

Figure M106 Total, remanent, and reversible magnetization of Cu-Co alloy as a function of annealing time (particle size) at 750°C, measured at 750°C. The reversible magnetization is the difference between the total magnetization (in presence of an applied field) and the remanence (modified from Kobayashi, 1961).

below the Curie point of the new magnetic species. The effective magnetic field at the site of the new magnetic particles determines the direction and intensity of the CRM. In igneous rocks, magnetic interactions among multiple phases, such as during exsolution of the iron-titanium solid solution series, might significantly modify the external field through magnetostatic or exchange interactions with phase(s) having higher blocking temperatures. Occasionally, such interactions might produce CRM with oblique directions or opposite polarity to the external magnetic field and the existing magnetic phase (negative magnetic interactions), and in rare cases a self-reversal may result (Néel, 1955).

CRM Origin of Marine Magnetic Anomalies

The first remanence of freshly extruded submarine basalts is TRM in stoichiometric titanomagnetites, $xFe_2TiO_4(1\text{-}x)Fe_3O_4$, with $x \sim 0.6 \pm 0.1$, that is, about 60% molar ulvöspinel, with Curie points between 100°–150°C (Readman and O'Reilly, 1972). The initial magnetic phase is transformed at the sea floor by topotactic low temperature oxidation to cation deficient titanomaghemites. The oxidation is thought to proceed by a net cation migration out of the crystal lattice to accommodate a higher Fe^{3+}/Fe^{2+} ratio and accompanying changing proportions of Fe:Ti. The resulting cation deficient phases retain the cubic crystal structure with smaller lattice dimensions, higher Curie points, approaching 500°C, and lower saturation magnetizations, responsible for the rapid diminution of the amplitudes of the marine magnetic anomalies away from spreading centers (e.g., Klitgord, 1976). Therefore, marine magnetic anomalies over the world's oceans can be considered to be preserved predominantly as CRM in oxidized Fe-Ti oxides.

The secondary CRM in submarine basalts usually retains the same polarity as the initial TRM recorded upon extrusion. This is indicated by the agreement, for overlapping time intervals, between the polarity time scales from marine magnetic anomalies, continental lavas and marine sediments, as well as the magnetic polarity of oriented dredged and drilled submarine basalts. The agreement of CRM and TRM directions is further supported by laboratory low temperature oxidation experiments of predominantly SD titanomagnetites (e.g., Marshall and Cox, 1971; Johnson and Merrill, 1974; Özdemir and Dunlop, 1985). The superexchange interactions vary with changes in cation distribution and lattice dimensions, as indicated by the reduced saturation magnetization and higher Curie points for the more oxidized titanomaghemites. However, it is possible that the orientations of the sub-lattice magnetic moments remain intact during low temperature oxidation of the titanomagnetite minerals, so that the ensuing CRM retains or inherits the original TRM direction.

Evidence from magnetic anomalies and magnetic properties of drilled oceanic basalts suggests an increase of the magnetization of extrusive submarine basalts of oceanic crust older than about 40 Ma (e.g., Johnson and Pariso, 1993). At present, the data are too sparse for one to be confident of the generality of this phenomenon or to select from several mechanisms that might be responsible.

CRM Influence on Paleointensity Studies

At present, only TRM can be used for obtaining absolute paleointensities of the Earth's magnetic field (see *Paleointensity from TRM*). Hence, all paleointensity methods require that the NRM be essentially pure TRM, or that the TRM can be readily isolated from the NRM. The paleointensity methods compare the NRM of each specimen to a new laboratory TRM produced in a known laboratory field, H_L. Because chemical and mineralogical alterations of specimens during laboratory heatings are common and often preclude reliable paleointensity determinations, the different paleointensity methods apply various pre- and post-heating tests to assess the extent of chemical changes on the remanence and paleointensity experiments. Several paleointensity methods use a single heating to above the specimen's highest Curie temperature to produce a total laboratory TRM. However, because reaction rates increase with temperature, such a procedure tends to maximize chemical alterations.

The Thelliers' double heating method (Thellier and Thellier, 1959) was developed to diminish this problem by gradually heating the samples in steps from room temperature, T_R, to the highest blocking temperature. At each temperature step, $T_I > T_R$, the samples are heated twice; to determine both the thermally demagnetized partial-NRM (PNRM) and the acquired partial-TRM (PTRM) between T_I and T_R. The Thellier method depends on the additivity and independence of PTRMs acquired in different temperature intervals; that is, total-TRM = ΣPTRM. Also, it is assumed that the unknown paleointensity, H_U, was constant throughout the remanence acquisition process and that H_U and H_L are sufficiently small that both TRMs are linearly proportional to the imposed field. When these conditions are satisfied and in the absence of chemical changes on heating, H_U can be calculated from the ratio $PNRM(T_I,T_R)/PTRM(T_I,T_R) = H_U/H_L$ and should be the same for each temperature interval. Every temperature step gives an independent paleointensity value. These consistency checks, provided by several independent paleointensity estimates at the different temperatures, are the primary asset of the Thellier method. The data can be displayed on a PNRM versus PTRM plot (an Arai diagram), with data corresponding to the different temperature steps (Figure M107). Ideal behavior in the Thellier sense implies linear data with a slope equal to $-H_U/H_L$ (line A, Figure M107). The Thelliers' procedure is well suited to detect the onset of chemical alterations, which are more common at higher temperatures, and are often expressed as deviations from the ideal straight line. Another feature of the Thelliers' procedure is the PTRM check, where a PTRM is repeated at a lower temperature $T_P < T_I$. The PTRM check provides information about changes in the PTRM capacity of magnetic particles with $T_B \leq T_P$. For these reasons, the Thelliers' procedure is usually considered to be the most reliable paleointensity method.

High-temperature chemical alterations during the Thelliers' paleointensity procedure is one of the more common causes for failed or abbreviated paleointensity experiments. If the CRM is expressed as a

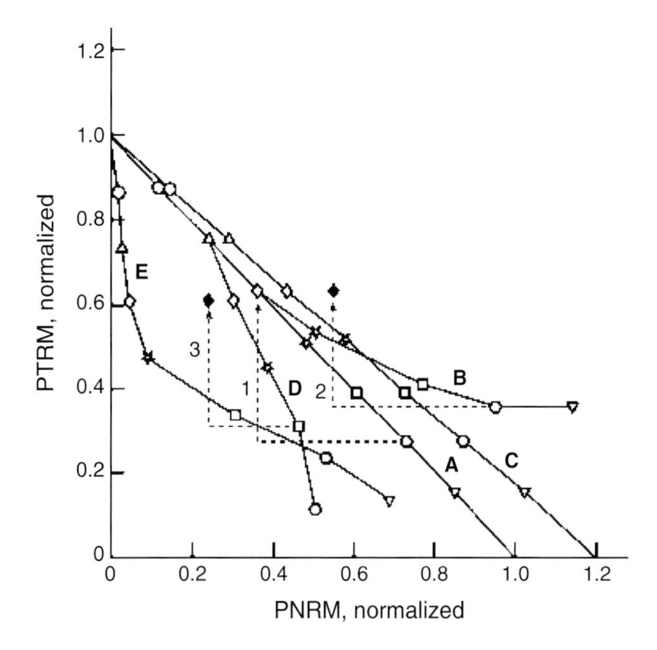

Figure M107 PNRM-PTRM diagrams for five hypothetical Thellier paleointensity experiments, A-E, discussed in the text. Like symbols indicate identical temperatures. Dashed lines 1, 2, and 3 indicate PTRM checks between the designated temperatures for experiments A, B, and D, respectively. Solid diamonds refer to unsuccessful PTRM checks.

greater PTRM capacity, which increases with temperature, the data will form a concave-up PNRM-PTRM plot (Figure M107, curve B). Provided the lower temperature data are linear and the PTRM checks show no evidence of alteration, then the lower temperature data can be used to calculate a paleointensity (Figure M107 curve B, points T1-T3). Recent studies suggest that at least 50% of the NRM should be used to obtain reliable results (e.g., Chauvin *et al.*, 2005).

CRM production that increases the PTRM capacity may result from precipitation of new magnetic particles or from unmixing of titanomagnetite grains to a more Fe-rich phase with higher saturation magnetization. Chemical modifications, which lead to higher PTRM, usually cause the PNRM-PTRM points to lie above the ideal line, and the calculated paleointensity will be lower than the actual paleofield. Alternatively, if the chemical alterations decrease the PTRM potential by destroying magnetic particles or by transforming them to a phase with lower intrinsic magnetic moments, then the PNRM-PTRM points will plot below the ideal line, with higher apparent paleointensities than the actual values (e.g., Figure M107, curve D and PTRM check 3). For a special case, where CRM acquisition grows linearly with temperature, the PNRM-PTRM plot might be linear (Figure M107, line C); however, these data plot above the ideal line (Figure M107, line A), and the calculated paleointensity would be lower than its actual value. This result emphasizes that linear PNRM-PTRM data are *a necessary but not sufficient* condition for obtaining reliable paleointensities.

The adverse effects of chemical alterations and CRM on paleointensity studies do not always arise from heatings in the laboratory. It is also possible that the NRM is not a pure TRM but contains a significant CRM component. Yamamoto *et al.* (2003) suggested that high temperature CRM contributes to the NRM of the Hawaiian 1960 lava, which results in higher than expected paleointensities. Alternatively, low-temperature hydrothermal alteration might produce CRM in new magnetic particles that contribute to the NRM. If these particles have not grown significantly beyond their blocking volumes, v_{AB}, they would be demagnetized at $T \geq T_A$, leading to rapid decrease of the NRM. This scenario might explain the precipitous diminution of the NRM observed for some basalts, with decreases on the order of 20% to more than 50% in the first few temperature steps of the Thellier experiment. When this decrease in NRM cannot be attributed to viscous remanence, it is possible that the NRM is augmented by CRM. It is no longer pure TRM.

CRM in Sedimentary Rocks

Oxidized Red Sediments

Red beds are a broad and loosely defined category of highly oxidized sediments with colors ranging from brown to purple, usually resulting from secondary fine particles of hematite, maghemite and/or ferric oxyhydroxide. The color is a complex function of the mineralogy, chemical composition and particle sizes of the iron oxides, as well as the impurity cations and their concentrations; however, for paleomagnetism, color is unimportant. Red sediments have been used extensively for paleomagnetism since the late 1940s, because they are widely distributed geographically and with respect to geologic time. In addition, the remanence of red sediments is often stable and sufficiently intense for paleomagnetic measurements, even with early-generation magnetometers. Already in the 1950s, it was deduced that low temperature oxidation was responsible for transforming the original magnetite to fine particles of hematite, maghemite, and/or goethite, which provide the pigment and CRM of red beds (Blackett, 1956).

Larson and Walker (1975) studied CRM development during early stages of red bed formation in late Cenozoic sediments; they showed that in their samples CRM occurred in several authigenic phases including hematite and goethite. The CRM, which obscured the original DRM, had formed over multiple polarity intervals, as indicated by different polarities in several generations of authigenic minerals. Complex multi-generation patterns of CRM, with several polarities within single specimens, have also been observed in Paleozoic and Mesozoic red beds. The influence of the secondary CRM on the primary remanence depends on the relative stability and intensity of the CRM carriers as compared with the primary DRM. In many cases the DRM may have been entirely obliterated by diagenetic processes, and the CRM is the dominant characteristic remanence. However, there are examples of red beds, where the primary DRM in specularite hematite remains the characteristic remanence with respect to CRM (e.g., Collinson, 1974). Sometimes distinct CRM components can be isolated by thermal demagnetization, selective leaching in acids (e.g., Collinson, 1967) and removal of altered phases of sediment by selective destructive demagnetization (Larson, 1981).

During the past more than five decades, paleomagnetic studies of red sediments have contributed significantly to magnetostratigraphy, plate tectonics and rock magnetism. Many data of apparent polar-wander paths are from red beds, where it is assumed that the CRM was produced soon after deposition, so that the paleomagnetic pole accurately represents the depositional age of the sediments. Moreover, CRM is not subject to inclination shallowing, which often affects the primary DRM. The utility of red sediments for high resolution studies of the geomagnetic field and paleosecular variation is limited and depends on how pervasive the CRM is as compared to the primary DRM, the time lag between the CRM and initial DRM, and the duration of CRM production.

Non-red Sediments

Here we discuss CRM in a subset of non-red mostly carbonate sediments, whose remanence is usually much weaker than for red beds. The low remanence intensity of these sediments was a key reason that they were generally excluded from paleomagnetic investigations until the introduction in the early 1970s of cryogenic magnetometers, capable of measuring minute signals, down to the 10^{-9}–10^{-10} Gauss range. Since then, there has been an explosion of studies of weakly magnetized non-red sediments. For example, it has been shown that the characteristic remanence of some early Paleozoic carbonate sequences was acquired in the late Paleozoic, hundreds of millions of years younger than their biostratigraphic ages (e.g., McCabe *et al.*, 1983). Magnetic extracts from these diagenetically altered sediments contained essentially pure magnetite particles, whose botryoidal and spheroidal forms have been used to infer their secondary origin, and they are thought to be responsible for the secondary characteristic remanence of these sediments. This conclusion has been buttressed by electron microscope observations (Figure M108) of in-situ authigenic magnetites in Paleozoic limestones (Suk *et al.*, 1993). In the absence of evidence of significant heating of these sediments, their remanence has been attributed to low temperature CRM in secondary magnetite, which might have been produced by "diagenetic alteration of preexisting iron sulfides (e.g., framboidal pyrites)" (McCabe *et al.*, 1983). For Miocene dolomites and limestones of the Monterey Formation, Hornafius (1984) concluded that the secondary remanence, presumably a low temperature CRM, resides in diagenetic magnetite produced by partial oxidation of pyrite upon the introduction of oxygenated meteoric groundwaters to the formation. CRM in some Paleozoic carbonates resides in hematite particles (e.g., Elmore *et al.*, 1985) produced by diagenetic dedolomitization, where oxidizing fluids with high calcium contents cause calcite replacing dolomite (McCabe and Elmore, 1989).

The presence of magnetite (and siderite, $FeCO_3$) in oil impregnated sediments was discovered by Bagin and Malumyan (1976), and Donovan *et al.* (1979) reported correspondence of near surface magnetic anomalies over an oil field with a higher concentration of magnetic minerals in the sediments. Paleomagnetic studies of remagnetized hydrocarbon-impregnated Paleozoic sediments (McCabe *et al.*, 1987; Benthien and Elmore, 1987) indicate a relationship between hydrocarbon migration and the precipitation of authigenic magnetite particles, carrying the secondary CRM. This scenario is supported by extracted magnetite spherules up to several tens of microns in diameter and

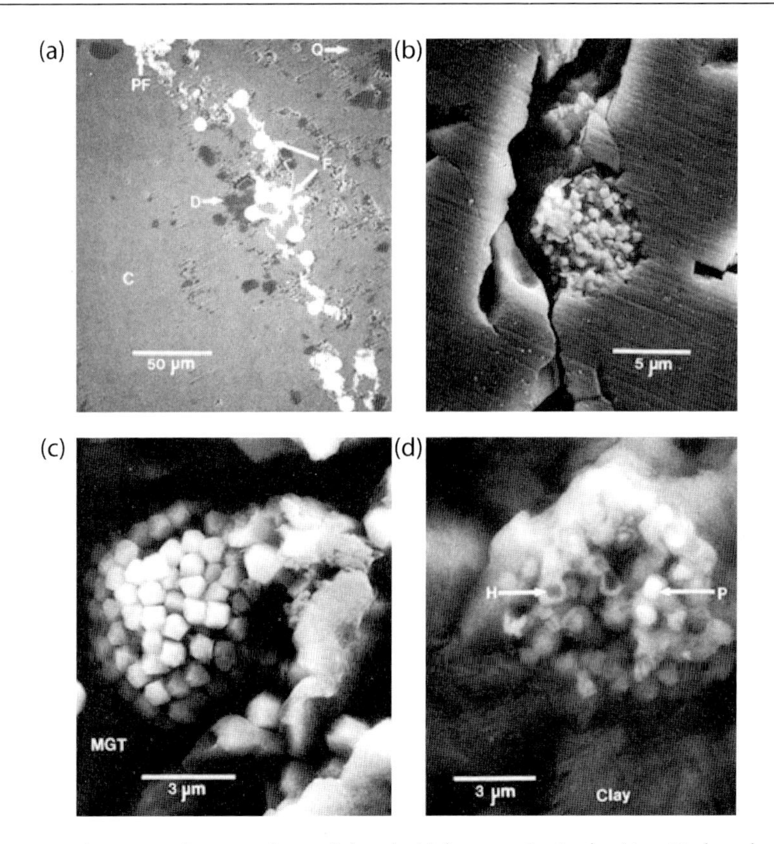

Figure M108 Scanning electron microscope images of pseudoframboidal magnetite in the New York carbonates. Symbols are MGT, magnetite; PF, pseudoframboid; F, framboid; P, pyrite; C, calcite; D, dolomite; Q, quartz; and H, hole. (a) Densely distributed framboids and pseudoframboids in a calcite matrix with occasional occurrence of dolomite and quartz; backscattered electron image (BEI). (b) Cross section of a pseudoframboid in a microcrack showing individual octahedral/cubo-octahedral crystals; secondary electron image (SEI). (c) A pseudoframboid in a microcrack showing almost perfect spherical shape (SEI). (d) An imperfectly spherical magnetite pseudoframboid in a void showing pyrite cores or voids within originally homogeneous pyrite crystals. Layered iron-rich clay minerals surround the grain (SEI) (from Suk *et al.*, 1993).

the presence of other authigenic textures in the sediments. All these studies suggest a net reduction of ferric ions in oxides, hydroxides and silicates, caused by the biodegradation (oxidation) of the hydrocarbons and the precipitation of more reduced iron oxide phases such as magnetite (Fe_3O_4), siderite ($FeCO_3$) and wustite (FeO). Of these, only magnetite is ferrimagnetic; hence it is responsible for the CRM and for being preferentially extracted during magnetic separations.

Rapidly deposited marine and lacustrine sediments are increasingly being used to study high-resolution behavior of the Earth's magnetic field, including secular variation and relative paleointensities. However, to accurately interpret the sedimentary record, geochemical processes that influence the magnetic signal must be understood. In anoxic and suboxic environments, bacterial sulfate reduction produces H_2S, which reacts with the detrital iron oxides to precipitate sulfide minerals (Berner, 1970, 1984). An abundance of sulfate favors reactions that produce relatively more stable pyrite (FeS_2), which does not carry remanence. When the sulfate supply is more limited, the formation of ferrimagnetic pyrrhotite (Fe_7S_8) and/or greigite (Fe_3S_4) is preferred. Pyrrhotite formation is less common in sediments because it is thought to require pH > 11 (Garrels and Christ, 1965), which is outside the range of values measured in sedimentary pore waters. However, for extremely low sulfur activity, it is possible for pyrrhotite to form in sediments.

For anoxic sediments from the Gulf of California and suboxic hemipelagic muds from the Oregon continental slope with sedimentation rates exceeding 1m/kyr, Karlin and Levi (1983, 1985) documented very rapid, dramatic decreases with depth of the intensity of NRM and artificial remanences, paralleled by downcore decrease in porewater sulfate and systematic growth in solid sulfur, mainly as pyrite (Figure M109). In both environments, the remanence resides in fine-grain nearly pure magnetite. These data suggest that early oxidative decomposition of organic matter leads to chemical reduction of the ferrimagnetic minerals and other iron oxides, which are subsequently sulfidized and pyritized with depth. Changes in the remanence intensity and stability are consistent with selective dissolution of the smaller particles, causing downcore coarsening of the magnetic fraction. In these environments, there was no evidence for the formation of authigenic magnetitic minerals. In this example, there is no CRM formation; rather, the sediments experience chemical demagnetization via dissolution. The chemical processes cause substantial reduction of the remanence intensity, while the directions appear to be unaffected.

In other suboxic marine environments, characterized by lower sedimentation rates, on the order of centimeters/kyrs, CRM in authigenic magnetite particles accompanies oxidative decomposition of organic matter immediately above the Fe-reducing zone (Karlin *et al.*, 1987; Karlin, 1990). Some of the smaller authigenic magnetite particles are subsequently dissolved downcore on entering the zone of Fe-reduction (Figure M110). In the past approximately fifteen years, an increasing number of paleomagnetic studies have identified CRM in ferrimagnetic iron sulfides, pyrrhotite and greigite, in a variety of marine and lacustrine settings (e.g., Roberts and Turner, 1993; Reynolds *et al.*, 1999; Weaver *et al.*, 2002; Sagnotti *et al.*, 2005). While in many cases the Fe-sulfides are formed during early diagenesis upon initial burial, they can also result from later diagenesis, deeper in the sections.

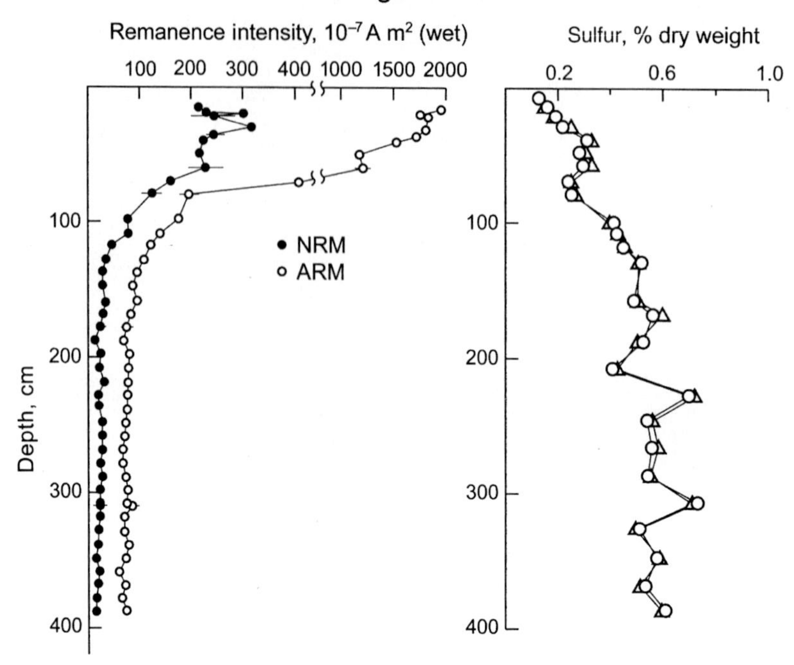

Figure M109 Downcore profiles of magnetic intensities and solid sulfur for Kasten core W7710-28, Oregon continental slope. The magnetization intensities were partially AF demagnetized at 15 mT. Solid sulfur concentrations of total (circles) and acid-insoluble (triangles) fractions on a sulfate-free basis, measured by X-ray fluorescence (modified from Karlin and Levi, 1983).

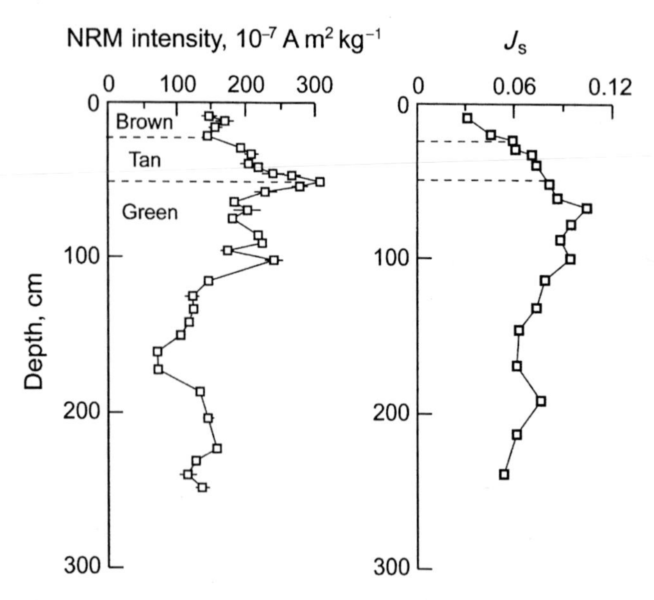

Figure M110 Downcore profiles of the NRM intensity, partially AF demagnetized at 20 mT, and the saturation magnetization for core TT11 from NE Pacific Ocean (modified from Karlin, 1990).

Recently, Roberts and Weaver (2005) described multiple mechanisms for CRM involving sedimentary greigite. The resolution of paleomagnetic time series would be compromised in sediments where CRM occurs in authigenic magnetic minerals, formed below the remanence lock-in depths. In light of this analysis t, it is possible that sediments discussed earlier in this section experienced multiple episodes of CRM. The first might arise from diagenesis of magnetite and other iron oxides to iron sulfides during initial burial. A later CRM might be produced in magnetite due to diagenesis of iron sulfides or, possibly, in hematite due to dedolomitization in an oxidizing environment, depending on the sediment composition and prevailing geochemistry.

Concluding Remarks

This report on CRM is not exhaustive and reflects the interests, biases and limitations of the author. It is an update of a similar article written over fifteen years ago (Levi, 1989). In the future, as paleomagnetists address more difficult tectonic and geomagnetic questions, requiring data from structurally more complex, metamorphosed, and older formations, it will be increasingly likely that CRM will contribute to the NRM. Paleomagnetists have become more adept at isolating different remanence components, using detailed and varied demagnetization procedures. It is usually assumed that the most resistant remanence, whether with respect to increasing temperatures, alternating fields, or a particular leaching agent is also the primary component. However, CRM stabilities are highly variable, and this assumption is unlikely to be satisfied universally. During the past fifteen years, there has been progress in understanding several aspects of CRM, including (a) the recognition that even for some very young subaerial lavas the NRM may comprise a low-temperature CRM component, and (b) that in some active sedimentary environments, diagenesis leads to CRM in ferrimagnetic iron sulfides. A more comprehensive understanding of CRM is needed to assist paleomagnetists to interpret complex, often multicomponent, NRMs with probable CRM overprints. This goal would be advanced by conducting controlled field and laboratory CRM experiments to (1) recognize the varied geochemical environments that produce different magnetic minerals and their associated CRMs; (2) determine the ranges of magnetic and mineralogical stabilities with respect to different demagnetization procedures and for isolating different CRM components; and (3) develop procedures for identifying the timing and sequencing of multi-component CRMs.

Shaul Levi

Bibliography

Bagin, V.I., and Malumyan, L.M., 1976. Iron containing minerals in oil-impregnated sedimentary rocks from a producing rock mass of Azerbaidzhan. *Izvestiya, Academy of Sciences, USSR. Physics of the solid earth* (English translation) **12**: 273–277.

Becker, J.J., 1957. Magnetic method for the measurement of precipitate particle size in a Cu-Co alloy. *Journal of Metals*, **9**: 59–63.

Beetz, W., 1860. Über die inneren Vorgänge, welche die Magnetisierung bedingen. *Annalen der Physik und Chemie*, **111**: 107–121.

Benthien, R.H., and Elmore, R.D., 1987. Origin of magnetization in the phosphoria formation at Sheep Mountain, Wyoming: a possible relationship with hydrocarbons. *Geophysical Research Letters*, **14**: 323–326.

Berner, R.A., 1970. Sedimentary pyrite formation. *American Journal of Science*, **268**: 1–23.

Berner, R.A., 1984. Sedimentary pyrite formation: an update. *Geochimica et Cosmochimica Acta*, **48**: 605–615.

Blackett, P.M.S., 1956. *Lectures on Rock Magnetism*. Jerusalem: The Weizmann Science Press of Israel, 131 pp.

Chauvin, A., Roperch, P., and Levi, S., 2005. Reliability of geomagnetic paleointensity data: the effects of the NRM fraction and concave-up behavior on paleointensity determinations by the Thellier method. *Physics of the Earth and Planetary Interiors*, **150**(4): 265–286, doi:10.1016/j.pepi.2004.11.008.

Collinson, D.W., 1967. Chemical demagnetization. In Creer, K.M., and Runcorn, S.K. (eds.) *Methods in Paleomagnetism*. Amsterdam: Elsevier, pp. 306–310.

Collinson, D.W., 1974. The role of pigment and specularite in the remanent magnetism of red sandstone. *Journal of the Royal Astronomical Society*, **38**: 253–264.

Donovan, T.J., Forgey, R.L., and Roberts, A.A., 1979. Aeromagnetic detection of diagenetic magnetite over oil fields. *American Association of Petroleum Geologists Bulletin*, **63**: 245–248.

Elmore, R.D., Dunn, W., and Peck, C., 1985. Absolute dating of a diagenetic event using paleomagnetic analysis. *Geology*, **13**: 558–561.

Garrels, R.M., and Christ, C.L., 1965. *Solutions, Minerals and Equilibria*. New York: Harper & Row, 450 pp.

Haigh, G., 1958. The process of magnetization by chemical change. *Philosophical Magazine*, **3**: 267–286.

Hornafius, J.S., 1984. Origin of remanent magnetization in dolomite from the Monterey Formation. In Garrison, R.E., Kastner, M., and Zenger, K.H. (eds.), *Dolomites of the Monterey Formation and Other Organic-Rich Units*. Society of Economic Paleontologists and Mineralogists pp. 195–212. Pacific Section Publication No. 41.

Johnson, H.P., and Merrill, R.T., 1973. Low-temperature oxidation of a titanomagnetite and the implication for paleomagnetism. *Journal of Geophysical Research*, **78**: 4938–4949.

Johnson, H.P., and Pariso, J.E., 1993. Variations in oceanic crustal magnetization: systematic changes in the last 160 million years. *Journal of Geophysical Research*, **98**: 435–445.

Karlin, R., 1990. Magnetic mineral diagenesis in suboxic sediments at Bettis site W-N, NE Pacific Ocean. *Journal of Geophysical Research*, **95**: 4421–4436.

Karlin, R., and Levi, S., 1983. Diagenesis of magnetic minerals in recent hemipelagic sediments. *Nature*, **303**: 327–330.

Karlin, R., and Levi, S., 1985. Geochemical and sedimentological control of the magnetic properties of hemipelagic sediments. *Journal of Geophysical Research*, **90**: 10373–10392.

Karlin, R., Lyle, M., and Heath, G.R., 1987. Authigenic magnetite formation in suboxic marine sediments. *Nature*, **326**: 490–493.

Klitgord, K.D., 1976. Sea-floor spreading: the central anomaly magnetization high. *Earth and Planetary Science Letters*, **29**: 201–209.

Kobayashi, K., 1959. Chemical remanent magnetization of ferromagnetic minerals and its application to rock magnetism. *Journal of Geomagnetism and Geoelectricity*, **10**: 99–117.

Kobayashi, K., 1961. An experimental demonstration of the production of chemical remanent magnetization with Cu-Co alloy. *Journal of Geomagnetism and Geoelectricity*, **12**: 148–164.

Koenigsberger, J.G., 1938. Natural residual magnetism of eruptive rocks. *Terrestrial Magnetism and Atmospheric Electricity*, **43**, 119–130, part 1: part 2: 299–320.

Larson, E.E., 1981. Selective destructive demagnetization, another microanalytic technique in rock magnetism. *Geology*, **9**: 350–355.

Larson, E.E., and Walker, T.R., 1975. Development of CRM during early stages of red bed formation in late Cenozoic sediments, Baja, California, *Geological Society of America Bulletin*, **86**: 639–650.

Levi, S., 1989. Chemical remanent magnetization. In James, D.E. (ed.) *The Encyclopedia of Solid Earth Geophysics*. London, UK: Van Nostrand Reinhold Ltd., pp. 49–58.

Marshall, M., and Cox, A., 1971. Effect of oxidation on the natural remanent magnetization of titanomagnetite in suboceanic basalt. *Nature*, **230**: 28–31.

Maurain, Ch. 1901. Propriétés des dépots électolytiques de fer obtenus dans un champ magnétique. *Journal of Physique*, **3**(10): 123–135.

Maurain, Ch. 1902. Sur les propriétés magnétiques de lames trés minces de fer et de nickel. *Journal of Physique*, **4**(1): 90–151.

McCabe, C., and Elmore, R.D., 1989. The occurrence and origin of late Paleozoic remagnetization in the sedimentary rocks of North America. *Reviews of Geophysics*, **27**: 471–494.

McCabe, C., Van der Voo, R., Peacor, D.R., Soctese, R., and Freeman, R., 1983. Diagenetic magnetite carries ancient yet secondary remanence in some Paleozoic sedimentary carbonates. *Geology*, **11**: 221–223.

McCabe, C., Sassen, R., and Saffer, B., 1987. Occurrence of secondary magnetite within biodegraded oil. *Geology*, **15**: 7–10.

Meiklejohn, W.H., 1953. Experimental study of coercive force of fine particles. *Reviews of Modern Physics*, **25**: 302–306.

Néel, L., 1949. Théorie du traînage magnétique des ferromagnétiques en grains fins avec applications aux terres cuites. *Annales de Géophysique*, **5**: 99–136.

Néel, L., 1955. Some theoretical aspects of rock magnetism. *Advances in Physics*, **4**: 191–243.

Özdemir, Ö., and Dunlop, D.J., 1985. An experimental study of chemical remanent magnetizations of synthetic monodomain titanomaghemites with initial thermoremanent magnetizations. *Journal of Geophysical Research*, **90**: 11513–11523.

Readman, P.W., and O'Reilly, W., 1972. Magnetic properties of oxidized (cation deficient) titanomagnetites. *Journal of Geomagnetism and Geoelectricity*, **24**: 69–90.

Reynolds, R.L., Rosenbaum, J.G., van Metre, P., Tuttle, M., Callender, E., and Goldin Alan 1999. Greigite (Fe_3S_4) as an indicator of drought—the 1912–1994 sediment magnetic record from White Rock Lake, Dallas, Texas, USA. *Journal of Paleolimnology*, **21**: 193–206.

Roberts, A.P., and Turner, G.M., 1993. Diagenetic formation of ferromagnetic iron sulphide minerals in rapidly deposited marine sediments, South Island, New Zealand. *Earth and Planetary Science Letters*, **115**: 257–273.

Roberts, A.P., and Weaver, R., 2005. Multiple mechanisms of remagnetization involving sedimentary greigite Fe_3S_4. *Earth and Planetary Science Letters*, **231**(3–4): 263–277, doi:10.1016/j.epsl.2004.11.024.

Sagnotti, S., Roberts, A.P., Weaver, R., Verosub, K.L., Florindo, F., Pike, C.R., Clayton, T., and Wilson, G.S., 2005. Apparent magnetic polarity reversals due to remagnetization resulting from late diagenetic growth of greigite from siderite. *Geophysical Journal International*, **160**: 89–100.

Suk, D., Van der Voo, R., and Peacor, D.R., 1993. Origin of magnetite responsible for remagnetization of early Paleozoic limestones of New York State. *Journal of Geophysical Research*, **98**: 419–434.

Thellier, E., and Thellier, O., 1959. Sur l'intensité du champ magnétique terrestre dans le passé historique et géologique. *Annales de Géophysique*, **15**: 285–376.

Weaver, R., Roberts, A.P., and Barker, A.J., 2002. A late diagenetic (syn-folding) magnetization carried by pyrrhotite: implications for paleomagnetic studies from magnetic iron sulphide-bearing sediments. *Earth and Planetary Science Letters*, **200**: 371–386.

Yamamoto, Y, Tsunakawa, H., and Shibuya, H., 2003. Palaeointensity study of the Hawaiian 1960 lava: implications for possible causes of erroneously high intensities. *Geophysical Journal International*, **153**: 263–276.

MAGNETIZATION, DEPOSITIONAL REMANENT

The magnetization acquisition processes in unconsolidated sediments have been long studied (e.g., Johnson *et al.*, 1948; King, 1955; Granar, 1958). The early studies showed that magnetic minerals in the sediments align along the ambient magnetic field during deposition through the water column. The magnetization resulting from the sedimentation process has been referred as depositional or detrital remanent magnetization (DRM). The magnetization acquisition process is still not well understood, and the role of the complex interplay of processes occurring during deposition, water-sediment interface processes, burial, and compaction, etc., require further analyses. The characteristics and stability of the remanent magnetization of unconsolidated sediments are determined by the composition, grain size, and shape of individual grains. During deposition in aqueous media, the magnetic particles are subject to the aligning force of the ambient magnetic field, plus the gravitational and dynamic forces. In tranquil conditions alignment of magnetic grains is relatively effective in depositional timescales, which are affected by Brownian forces. Nagata (1961) showed that equilibrium with the ambient magnetic field is attained in a scale of 1 s. Therefore, saturation magnetization should be expected in natural depositional systems, with the DRM intensity independent of the ambient magnetic field intensity. However, this is not the case, and the magnetization intensity is related to the intensity of the ambient magnetic field (Johnson *et al.*, 1948). The DRM intensities lower than saturation values have been related to misalignment effects of Brownian motion of submicron ferrimagnetic grains (Collinson, 1965; Stacey, 1972). Near the water-sediment interface, flow conditions may become relatively stable and simple by having laminar flow (Granar, 1958) but still the deposition process is complex. The interplay and characteristics of the bottom sediments result in a variety of fabrics in the deposited sediments.

In general, reorientation of magnetic minerals occurring after deposition and before consolidation of the sediment is referred as postdepositional DRM. Perhaps, the most notable distinction between depositional DRM and postdepositional DRM is the occurrence of the so-called inclination error present in depositional DRM (Johnson *et al.*, 1948; King, 1955; Granar, 1958). If I_H is the inclination of the ambient Earth's magnetic field at the time of sediment deposition, then the inclination of magnetization I_S can be expressed in terms of

$$I_S = \arctan(f \tan I_H)$$

where f is a factor that is determined experimentally. The inclination error has been ascribed to deposition of elongated grains with along-axis magnetizations tending to lie parallel to the sediment interface and deflecting the magnetization toward the horizontal plane. Laboratory experiments have been conducted to evaluate effects of the intensity of the ambient magnetic field, size and shapes of the magnetic grains, deposition on horizontal, inclined, and irregular surfaces, bottom currents, etc. (e.g., Johnson *et al.*, 1948; King, 1955; Rees, 1961). The inclination error has been observed in laboratory experiments, where the angular difference can be as high as 20° (King, 1955), but it is smaller (5–10°) or absent in natural sediments.

Barton *et al.* (1980) studied the change with time in the DRM acquisition process of laboratory-deposited sediments and found that in less than 2 days, there was no appreciable inclination error. In natural conditions, postdepositional DRM presents no significant inclination error. One of the major differences between laboratory experiments and natural conditions is the deposition rate. The time taken for realignment has been estimated in a few years and is apparently related to the water content of the sediments. Verosub *et al.* (1979) experimentally reexamined the role of water content in acquisition of postdepositional DRM and suggested that small-scale shear-induced liquefaction is the main magnetization process. There are also several additional factors involved; for instance bottom water currents, changing water levels, presence of organic matter, biological activity (bioturbation), particle flocculation, floccule disaggregation, dewatering, etc.

In general, it appears that rapidly deposited sediments show inclination and bedding inclination errors, similar to those observed in laboratory experiments. Slowly deposited or high-porosity sediments show small or no inclination error. Postdepositional DRM will realign the magnetization direction; this process may occur in short timescales of days or months, but may occur in periods of years or decades following deposition (Tarling, 1983).

In addition to studies of secular variation and magnetostratigraphy in sedimentary sequences, there has also been much interest in determining relative paleointensities from sedimentary records (Tauxe, 1993). Long records of relative paleointensities have been derived from marine and lake sedimentary sequences, and results have been compared with volcanic records and other records. There have been also several attempts to examine the effects of depositional factors in the DRM intensity, including for instance the effects of clay mineralogy, electrical conductivity of sediments, pH and salinity (Lu *et al.*, 1990; Van Vreumingen, 1993; Katari and Tauxe, 2000). Katari and Bloxham (2001) examined the effects of sediment aggregate sizes on the DRM intensities, and proposed that intensity is related to viscous drag that produces misalignment of magnetic particle aggregates. They argue that interparticle attractions arising from clectrostatic or van der Waals forces and/or biologically mediate flocculation results in formation of aggregates (which present a log-normal size distribution) preventing settling of individual smaller grains.

The depositional and postdepositional DRM in laboratory experiments and naturally deposited sediments have been intensively studied; nevertheless, further work is required to understand the complex interplay of processes and then develop magnetization acquisition models (e.g., Verosub, 1977; Tarling, 1983; Tauxe, 1993; Katari and Bloxham, 2001).

Jaime Urrutia-Fucugauchi

Bibliography

Barton, C., McElhinny, M., and Edwards, D., 1980. Laboratory studies of depositional DRM. *Geophysical Journal of the Royal Astronomical Society*, **61**: 355–377.

Collinson, D., 1965. Depositional remanent magnetization in sediments. *Journal of Geophysical Research*, **70**: 4663–4668.

Granar, 1958. Magnetic measurements on Swedish varved sediments. *Arkiv for Geofysik*, **3**: 1–40.

Irving, E., 1964. *Paleomagnetism and its Application to Geological and Geophysical Problems.* New York: Wiley.

Johnson, E., Murphy, T., and Wilson, O., 1948. Pre-history of the Earth's magnetic field. *Terrestrial Magnetism and Atmospheric Electricity*, **53**: 349.

Katari, K., and Bloxham, J., 2001. Effects of sediment aggregate size on DRM intensity: a new theory. *Earth and Planetary Science Letters*, **186**: 113–122.

Katari, K., and Tauxe, L., 2000. Effects of pH and salinity on the intensity of magnetization in redeposited sediments. *Earth and Planetary Science Letters*, **181**: 489–496.
King, R.F., 1955. The remanent magnetism of artificially deposited sediments. *Monthly Notices of the Royal Astronomical Society*, **7**: 115–134.
Lu, R., Banerjee, S., and Marvin, J., 1990. Effects of clay mineralogy and the electrical conductivity of water in the acquisition of depositional remanent magnetism in sediments. *Journal of Geophysical Research*, **B95**: 4531–4538.
McElhinny, M.W., 1973. *Palaeomagnetism and Plate Tectonics.* Cambridge Earth Science Series. Cambridge: Cambridge University Press, 358 pp.
Nagata, T., 1961. *Rock Magnetism.* Tokyo, Japan: Mazuren, 350 pp.
Rees, A.I., 1961. The effect of water currents on the magnetic remanence and anisotropy of susceptibility of some sediments. *Geophysical Journal of the Royal Astronomical Society*, **5**: 235–251.
Stacey, F., 1972. On the role of Brownian motion in the control of detrital remanent magnetization of sediments. *Pure and Applied Geophysics*, **98**: 139–145.
Tarling, D.H., 1983. *Palaeomagnetism, Principles and Applications in Geology, Geophysics and Archaeology.* London: Chapman & Hall, 379 pp.
Tauxe, L., 1993. Sedimentary records of relative paleointensity of the geomagnetic field: theory and practice. *Reviews of Geophysics*, **31**: 319–354.
Van Vreumingen, M., 1993. The influence of salinity and flocculation upon the acquisition of remanent magnetization in some artificial sediments. *Geophysical Journal International*, **114**: 607–614.
Verosub, K.L., 1977. Depositional and postdepositional processes in the magnetization of sediments. *Reviews of Geophysics and Space Physics*, **15**: 129–143.
Verosub, K.L., Ensley, R.A., and Ulrick, J.S., 1979. The role of water content in the magnetization of sediments. *Geophysical Research Letters*, **6**: 226–228.

MAGNETIZATION, ISOTHERMAL REMANENT

Introduction

As indicated by the name, isothermal remanent magnetization (IRM) is a *remanent magnetization* (RM) acquired without the aid of changes in temperature. There are actually many processes that can produce RMs isothermally (see, e.g., *CRM*, *VRM*, *DRM*, and *ARM*), but by convention we restrict the use of the term IRM to denote a remanence resulting from the application and subsequent removal of an applied DC field. IRMs are thus intimately related to *hysteresis*. Conventional nomenclature uses M_R for the IRM determined by measurement of a hysteresis loop (and M_{RS} if the cycle reaches saturation), whereas IRM and SIRM are typically used for remanence measurements made with a zero-field magnetometer after magnetization on a separate instrument (electromagnet, pulse magnetizer, etc.).

Weak fields, such as the geomagnetic field, are generally very ineffective in producing IRMs, and therefore IRMs are rarely significant components of NRM. (A notable exception is found in rocks struck by lightning, as discussed below.) IRMs are therefore primarily of interest as laboratory-produced remanences, and are used in sample characterization, in studying the mechanisms affecting stable remanence, and as a normalizing parameter for estimation of relative *paleointensity*. The attributes of principal interest are: the *intensity* (dipole moment per unit mass or volume); the *coercivity* (range of fields over which significant acquisition occurs) or *stability* (resistance to demagnetization by thermal, alternating-field, and/or low-temperature methods); and the *anisotropy* (dependence on orientation of the applied field).

Grain-scale processes and remanent states of individual particles

In a sufficiently strong applied field, the magnetic structure inside a ferromagnetic particle is simple: uniform magnetization parallel to the field, i.e., saturation magnetization (M_S). With decreasing applied field intensity, interesting complexities develop in response to the internal demagnetizing field, magnetocrystalline anisotropy and stress: magnetization may rotate away from the applied field direction toward an easy axis determined by grain shape, crystal orientation or stress; the magnetization may become nonuniform, with spins arranged in "flower" or "vortex" structures; domains may nucleate and grow, with magnetizations oriented at high angles to those of neighboring domains. Upon complete removal of the saturating applied field, the particle reaches its saturated remanent state, with a net intensity $M_{RS} \leq M_S$, oriented in general with a positive component parallel to the applied field direction, i.e., $M_{RS//} > 0$ (although self-reversal of IRM has been reported in rare cases, e.g., Zapletal, 1992). The ratio M_{RS}/M_S for an individual particle thus ranges from 0 to 1, depending on the saturated remanent domain state, and $M_{RS//}/M_S$ depends additionally on easy-axis orientation(s) with respect to the applied field direction.

For an ideal *single-domain* (uniformly magnetized) particle, M_{RS} is equal in magnitude to M_S, and the ratio $M_{RS//}/M_S$ is thus determined by the particle anisotropy (number and symmetry of easy axes) and orientation (Gans, 1932; Wohlfarth and Tonge, 1957). For an ideal SD particle with *uniaxial* (e.g., shape) anisotropy, oriented at an arbitrary angle ϕ to the applied field, $M_{RS//}/M_S = \cos(\phi)$ and $M_{RS\perp}/M_S = \sin(\phi)$; $M_{RS//}/M_S$ thus ranges from 0 for an applied field perpendicular to the easy axis (i.e., the moment rotates 90° as the field is removed) to 1 for the parallel case. Magnetocrystalline anisotropy often involves multiple symmetrically equivalent easy axes (e.g., the cubic body diagonals in magnetite, and the cube edges in Ti-rich titanomagnetites). For a *multiaxial* SD particle, the maximum angle (ϕ_{max}) that may separate the applied field from the nearest easy axis is reduced (Figure M111), and the minimum ratio $M_{R//}/M_S$ is accordingly increased. For an individual cubic SD magnetite grain, $\phi_{max} = 54.7°$ and thus $0.577 \leq M_{RS//}/M_S$.

Above a threshold grain size, uniformly magnetized remanent states become energetically unfavorable, and vortex (Williams and Wright, 1998) or *multidomain* structures consequently develop, lowering the total energy by reducing the net remanence of the particle. In a two-domain particle with antiparallel domain magnetizations, the net remanence is primarily due to domain imbalance; in asymmetrically shaped grains the minimum-energy state may yield a significant net (imbalance) moment (Fabian and Hubert, 1999). Even when domain moments exactly cancel in two-domain grains, the domain-wall moment ("psark," Dunlop, 1977) provides a small but stable net particle moment. Increasing domain multiplicity generally results in more perfect mutual cancellation of both the domain moments and the wall moments (which have been observed to alternate in polarity across magnetite grains with several domains, Pokhil and Moskowitz, 1997), and there is consequently a strong reduction in $M_{RS//}/M_S$ with increasing particle size.

Remanent states: ensembles of particles

Except in very coarse-grained rocks, even a centimeter-sized cubic or cylindrical specimen contains a very large number of ferrimagnetic particles. The net IRM of a population of grains is simply the vector sum of the individual grain remanent moments, and thus its magnitude depends on the number of remanence-carrying particles, their size (domain-multiplicity) distribution, particle-level anisotropy, and orientation distribution (i.e., population-level anisotropy) relative to the applied-field direction. In interacting assemblages, each particle is influenced not only by the externally applied field but also by the fields of neighboring particles, so the spatial distribution of grains is an additional important factor (Muxworthy *et al.*, 2003).

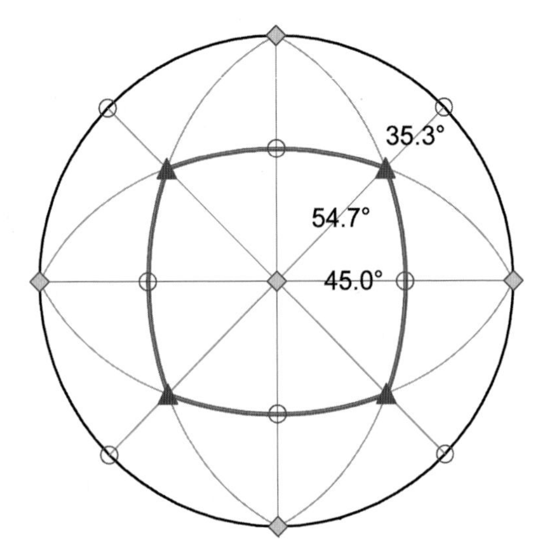

Figure M111 Cube edge <100> directions (diamonds) are the magnetocrystalline easy axes for high-Ti titanomagnetites. Due to the cubic symmetry, an applied field can be oriented no more than $\theta_{max} = 54.7°$ from the nearest easy axis of an individual grain. For a randomly oriented assemblage of ideal SD grains with <100> easy axes, $M_{RS}/M_S = 0.832$ (Gans, 1932). This is most easily calculated in the particle coordinate system, by noting that an IRM parallel to the polar easy axis results from application of a sufficiently strong field oriented anywhere within the "capture area" comprised of the eight adjacent congruent spherical triangles (outlined in bold). For each particle, $M_{R//}/M_S = \cos(\phi)$, and for the assemblage M_{RS}/M_S is obtained by integrating $\cos(\phi)$ over the "capture area." Body diagonal <111> directions (triangles) are the magnetocrystalline easy axes for magnetite at temperatures above the 130 K isotropic point; the "capture area" for these consists of six congruent spherical triangles, and integration of $\cos(\phi)$ over this area yields $M_R/M_S = 0.866$ for a randomly oriented assemblage of ideal SD grains with <111> easy axes.

Orientation distribution and bulk anisotropy

The simplest case to consider is that of a perfectly aligned population of identical single-domain particles, for which the ratio $M_{RS//}/M_S$ of the assemblage equals that of each individual particle: $M_{RS//}/M_S = \cos(\phi)$, where ϕ is the angle between the applied field and the nearest easy-axis direction in the aligned assemblage. Note that in this case, as for an individual particle, the net IRM generally has both parallel and transverse components.

A perfectly aligned population represents an extreme not often approached in natural materials, most of which are considerably closer to the opposite limiting case: isotropic (random or quasiuniform) orientation distribution (OD). For an ensemble of SD particles with an isotropic OD, $M_{RS//}/M_S$ is determined entirely by the particle-level anisotropy (Gans, 1932; Stoner and Wohlfarth, 1948; Wohlfarth and Tonge, 1957), and the net transverse component is effectively zero. For uniaxial anisotropy, grain moments rotate on average 60° to the nearest easy axis when the applied field is removed, yielding $M_{RS//}/M_S = M_{RS}/M_S = \cos(60°) = 0.5$. (This result is obtained by integrating parallel components over the assemblage, equivalent to integrating $\cos(\phi)$ over the hemisphere surrounding the applied field orientation, within which the proximal easy axes are distributed.) With multiaxial anisotropies the average rotation angle is significantly reduced and M_{RS}/M_S is concomitantly increased. For an ensemble of cubic SSD particles with <100> easy axes (e.g., TM60) and a uniform OD, the expected $M_{RS//}/M_S$ ratio is calculated by integrating $\cos(\phi)$ over the "capture area" surrounding a cube edge orientation (Figure M111), yielding $M_{RS//}/M_S = M_{RS}/M_S = 0.832$. For <111> easy axes (e.g., magnetite), integration over the region surrounding a body diagonal orientation yields $M_{RS//}/M_S = M_{RS}/M_S = 0.866$ (Gans, 1932; Stoner and Wohlfarth, 1948).

For the imperfect antiferromagnets hematite and goethite, as well as the hard ferrimagnet pyrrhotite, the applied fields typically used in rock magnetism (≤1 T produced by electromagnets; ≤5 T produced by typical superconducting magnets) are insufficient to achieve saturation. When these minerals occur with fine-grain sizes and/or impure compositions, applied fields exceeding 60 T may be required to saturate the remanence (Mathé *et al.*, 2005). A hard magnetocrystalline anisotropy largely confines the remanence to the basal planes of hematite and pyrrhotite, and to the *c*-axis for goethite. Hematite's basal-plane anisotropy has both a weak triaxial magnetocrystalline component and a uniaxial magnetoelastic component (Banerjee, 1963; Dunlop, 1971). When the uniaxial anisotropy dominates $M_{RS}/M_S = 0.637$. In the absence of strong uniaxial anisotropy, the hexagonal symmetry results in much higher ratios, up to 0.955 (Wohlfarth and Tonge, 1957; Dunlop, 1971).

When particle moments are diminished by the development of nonuniform remanent states, the ratio M_{RS}/M_S of the population decreases proportionally. For magnetite there is a more or less continuous decrease in the remanence ratio with increasing grain size. Data compiled by Dunlop (1995) for the size range 40 nm $\leq d \leq$ 400 μm show that M_{RS}/M_S varies in proportion to d^{-n} ($0.5 \leq n \leq 0.65$), with two separate size-dependent trends: stress-free hydrothermally grown grains have significantly lower remanence ratios than similarly sized crushed grains or grains grown by the glass ceramic method.

Bulk IRM anisotropy is controlled both by particle-level anisotropy and by the orientation distribution of the population. Like other bulk magnetic anisotropies, the anisotropy of IRM (AIRM) has been used to characterize rock fabric for geological applications and for evaluating directional fidelity of NRM (Fuller, 1963; Cox and Doell, 1967; Daly and Zinsser, 1973; Stephenson *et al.*, 1986; Jackson, 1991; Kodama, 1995). IRM differs from weak-field induced and remanent magnetizations (e.g., susceptibility, ARM, and TRM) in that it is generally a nonlinear function of applied field, and consequently the tensor mathematics used to describe linear anisotropic magnetizations are not valid for AIRM, except when very weak magnetizing fields (≤ a few millitesla) are used. Nevertheless it is a relatively rapid and very sensitive method of characterizing magnetic fabric (Potter, 2004).

"Partial IRMs" and coercivity spectrum analysis

Partial TRMs (pTRMs) and partial ARMs (pARMs) are generated when a weak but steady bias field is applied during some part, but not all, of the cooling or AF decay. Clearly no direct analog can exist for IRM, but "pIRM" equivalents can be produced physically (by partial demagnetization of an IRM), or calculated mathematically (by subtraction of IRMs acquired in different fields). Partial IRMs and their ratios are very widely used for sediment characterization in environmental magnetism (Thompson and Oldfield, 1986; King and Channell, 1991; Oldfield, 1991; Verosub and Roberts, 1995; Dekkers, 1997; Maher and Thompson, 1999). For example the hard fraction, HIRM, is determined by subtracting the IRM acquired in 300 mT (IRM_{300}) from SIRM, to estimate the contribution of antiferromagnetic minerals (e.g., hematite and goethite) to the saturation remanence. HIRM, as conventionally defined, is an absolute, concentration-dependent parameter. The soft fraction is more often quantified in relative terms through the so-called *S*-ratios, calculated from measurements of SIRM and of the IRM subsequently acquired in an oppositely directed field of 100 mT (IRM_{-100}) or 300 mT (IRM_{-300}): $S_{100} = -IRM_{-100}/SIRM$, and $S_{300} = -IRM_{-300}/SIRM$. These ratios range from −1 (for samples containing only hard antiferromagnets) to +1 (for samples dominated by soft ferrimagnets).

A more thorough method of deconstructing an SIRM uses application of two successive orthogonal applied fields of decreasing strength to reset the remanence of the intermediate- and low-coercivity fractions;

the resultant "triaxial" IRM is then thermally demagnetized to determine the unblocking temperature distribution associated with each coercivity fraction (Lowrie, 1990). The threshold fields are selected for isolation of different mineralogical and grain-size fractions; Lowrie (1990) used 5 T for the initial SIRM (sufficient to magnetize hematite and goethite at least partially); 0.4 T for the first orthogonal overprint (to reorient the remanence of the ferrimagnets magnetite and pyrrhotite); and 0.12 T for the second overprint (to realign the remanence of soft MD carriers).

For many geological applications, recognition and quantification of a few discrete IRM-carrying fractions is adequate, but finer subdivision of SIRM into a quasicontinuous coercivity distribution can be made by differentiation of the IRM acquisition curve (Dunlop, 1972). Starting from a demagnetized state, application of successively stronger fields results in growth of IRM as domain structures change and single-domain moments are reoriented, each at a critical field determined by particle anisotropy and orientation. The resulting *coercivity spectrum* can be analytically decomposed if it assumed that the distributions each follow some prescribed form, typically lognormal (Figure M112) (Robertson and France, 1994; Stockhausen, 1998; Heslop *et al.*, 2002; Egli, 2003, 2004 a,b).

Starting from the SIRM state, application of successively larger opposite-polarity "backfields" causes the net remanence first to decrease and then to grow in the reverse direction, eventually reaching the negative SIRM state; this process is often termed "DC demagnetization" but it is perhaps more useful to think of it as a part of the isothermal remanent hysteresis cycle. The *remanent coercivity* H_{CR} is defined as the field-axis intercept of the remanent hysteresis loop, exactly as the bulk coercivity H_C is defined for the in-field loop.

Interaction effects

In the absence of interparticle interactions, the IRM acquisition curve for SD ensembles has a simple and direct relationship to the "DC demagnetization" curve (Wohlfarth, 1958). In each case, a particular applied field activates the same set of particles; in the former case randomly oriented moments (in the demagnetized ensemble) are replaced with a net IRM, whereas in the backfield case a net IRM is replaced with its opposite-polarity equivalent. The change in the latter case is exactly twice that in the former, and thus the relationship between the initial acquisition curve $M_{IRA}(H)$ and the backfield curve $M_{IRB}(-H)$ is: $M_{IRB}(-H) = \mathrm{SIRM}-2M_{IRA}(H)$. For AF demagnetization, each peak AF erases the remanence imprinted by the corresponding DC applied field, so $M_{IR}(\tilde{H}) = \mathrm{SIRM}-M_{IRA}(H_{DC})$, i.e., there is a mirror symmetry of the IRM acquisition and AF demagnetization curves for noninteracting SD populations.

There are two commonly used graphical illustrations of these relationships. The *Henkel plot* graphs $M_{IRB}(-H)$ as a function of $M_{IRA}(H)$ (Henkel, 1964). For noninteracting SD ensembles, the relationship is linear, with a slope of -2 (Figure M113a). The *Cisowski plot* (Cisowski, 1981) or *crossover plot* (Symons and Cioppa, 2000) simultaneously graphs the acquisition curve $M_{IRA}(H_{DC})$ together with the AF-demagnetization curve $M_{IR}(\tilde{H})$ and/or the appropriately rescaled DC-backfield curve $M'_{IRB}(H) = 0.5(\mathrm{SIRM} + M_{IRB}(-H))$, as functions of the AC and DC fields (Figure M113b). On this plot, noninteracting SD populations are indicated by mirror symmetry, with an intersection at the point whose field coordinate is equal to both the median destructive field (MDF; also referred to as $\tilde{H}_{1/2}$, Dankers, 1981) and the median acquisition field (MAF; also referred to as H'_{CR} (Dankers, 1981)), and whose magnetization coordinate is equal to 0.5*SIRM.

A *crossover ratio* $R < 0.5$ (equivalent to a trajectory on the Henkel plot that "sags" below the line of slope −2) indicates that MAF > MDF, in other words that the ensemble is harder to magnetize than to demagnetize. Such behavior is a hallmark of either a negatively interacting SD population or MD carriers (for which self-demagnetization produces the equivalent effect); these two possibilities often cannot be definitively distinguished on the basis of IRM data alone (Dunlop and Özdemir, 1997). Crossover ratios $R > 0.5$ are virtually never observed, in part due to the extreme rarity of MD-free magnetic assemblages in natural materials, but further suggesting that interactions in SD assemblages, when they occur, are primarily negative.

Theoretical treatment of interactions in multidomain assemblages has been hindered by the analytical and numerical intractability of modeling interacting nonuniformly magnetized grains. However, recent results obtained at the resolution limit of micromagnetic modeling (Muxworthy *et al.*, 2003) shows that in general, interaction begins

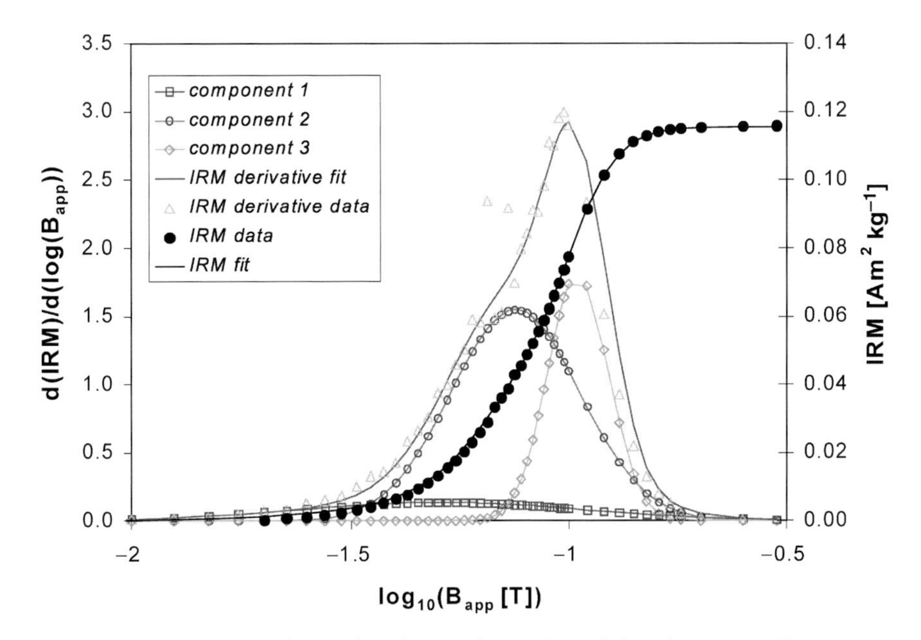

Figure M112 Coercivity spectrum analysis (methods of Heslop *et al.*, 2002), modeling the measured IRM acquisition (solid circles) as the sum of log-Gaussian distributions, in this case with three components (open symbols). Integrating and summing yields the model IRM curve (solid black). The individual components can be interpreted as different mineral and/or grain-size populations.

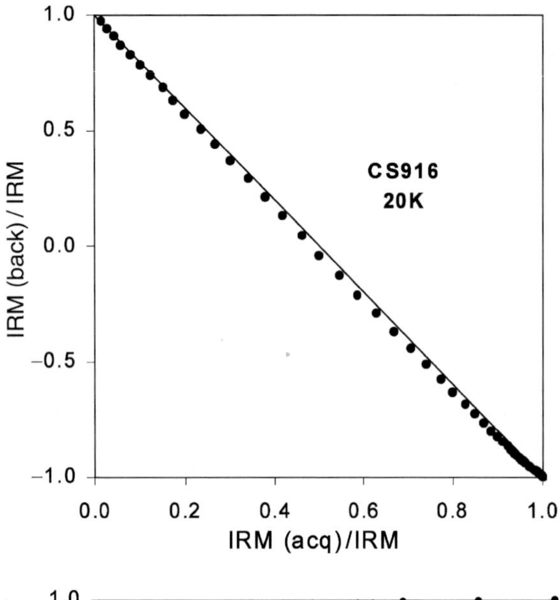

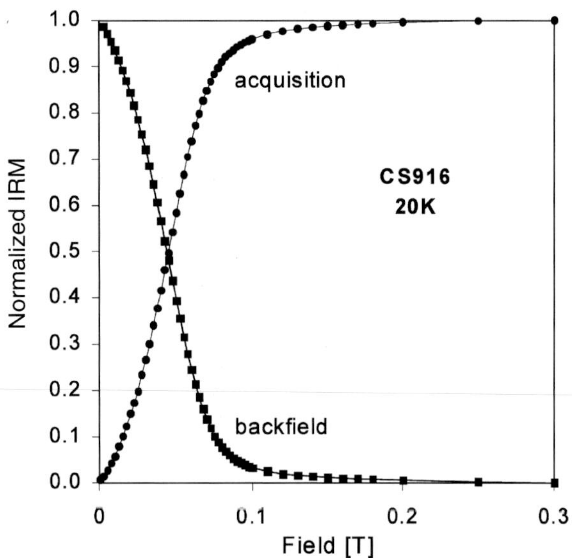

Figure M113 *Top*: Henkel (1964) plot for a sample with ideal noninteracting SD behavior. *Bottom*: Cisowski (1981) plot of the same data (backfield data rescaled and reflected). After Worm and Jackson (1999), with permission of the authors and publisher.

to have significant effects for SD and PSD grains (30–250 nm) when the separation between particles is less than or equal to twice the particle diameter; this is comparable to the threshold for susceptibility interaction effects (Hargraves *et al.*, 1991; Stephenson, 1994; Cañon-Tapia, 1996). Interestingly, the model results show that IRM interactions are primarily negative (decreasing M_{RS}/M_S) for single-domain ensembles, but positive (increasing M_{RS}/M_S) for grains >100 nm.

Kinetic effects: time-dependent IRM (TDIRM)

Néel (1949a,b, 1955) theory provides the basis for understanding phenomena such as magnetic *viscosity*, *thermoremanence*, and *frequency-dependent susceptibility*, by specifying how magnetization kinetics depend on factors including temperature, applied field, and particle size. At room temperature, and for typical experimental timescales of milliseconds to hours, thermal fluctuations are most significant for nanometric particles; for magnetite the transition from superparamagnetic to stable SD behavior occurs under these conditions at approximately 20–30 nm. A highly sensitive method of probing the SP-SSD threshold uses IRM acquisition with varying exposure times (Worm, 1999).

At any fixed absolute temperature T, the initial magnetization M_0 of an assemblage of single-domain grains grows or decays in time (t), approaching an equilibrium value in a constant applied field H:

$$M(t) = M_0 e^{-t/\tau} + M_{eq}(1 - e^{-t/\tau}) \quad \text{(Eq. 1)}$$

where the equilibrium magnetization M_{eq} is:

$$M_{eq}(T,H) = M_S \tanh\left(\frac{\mu_0 V M_S H}{kT}\right) \quad \text{(Eq. 2)}$$

and the relaxation time τ is:

$$\tau(T,H) = \tau_0 \exp\left(\frac{V M_S(T) H_K(T)(1 - H/H_K(T))^2}{2kT}\right) \quad \text{(Eq. 3)}$$

In these equations V is grain volume, H_K is microscopic coercivity, k is Boltzmann's constant (1.38×10^{-23} J K^{-1}), μ_0 is the permeability of free space ($4\pi \times 10^{-7}$ H m^{-1}), and τ_0 is approximately 10^{-9} s. Very fine (small V) and/or very soft (low H_K) grains have shorter relaxation times, and equilibrate more quickly, than do larger and/or magnetically harder SD grains.

Note that relaxation time drops very sharply as the applied field H approaches H_K. IRM is thus acquired quickly, and then becomes stabilized (τ increasing sharply) when the applied field is removed. Time-dependent IRM is significant in populations of SD particles with narrow distributions of V and H_K, i.e., with a narrow distribution of relaxation times in the applied field H. Exposure times less than the mean relaxation time are too short for significant acquisition of remanence, but exposure times one or two orders of magnitude larger may allow substantial equilibration with the applied field. The time-dependence of IRM in such assemblages is much stronger than the frequency-dependence of susceptibility (χ_{fd}). For example, ash-flow tuffs from Yucca Mountain Nevada have a χ_{fd} of up to 35% per decade, the highest known for any geological material; in contrast their IRM intensities increase by as much as 800% with a tenfold increase in exposure time (Worm, 1999). The time-dependence is effectively amplified for IRM by the strong asymmetry between acquisition and decay rates.

Lightning strikes and IRM in nature

Lightning has long been recognized as a significant mechanism for overprinting natural remanence (Matsuzaki *et al.*, 1954; Cox, 1961; Graham, 1961), due primarily to the strong magnetic fields locally generated by the intense currents involved in lightning strikes. These currents, typically 10^4–10^5 A, are dominantly upward-directed (i.e., electrons flow down from the cloud to the ground along the ionized flow path), and most lightning flashes include three or four separate upward current pulses ("return strokes"), spaced about 50 ms apart (e. g., Krider and Roble, 1986).

The magnetic field B generated by a straight filamentary current I of effectively infinite length, at a point located at a perpendicular distance r, has a magnitude $\mu_0 I/(2\pi r)$. For example, a 30 kA current produces a field of 60 mT at $r = 10$ cm, sufficient to impart a significant IRM in most materials. The orientation of the field is perpendicular to both I and r, i.e., the field lines are circles, centered on and perpendicular to the filamentary current. Both of these characteristics, concentric directional geometry and $1/r$ field dependence, have been observed by Verrier and Rochette (2002) in IRM overprints in samples collected around known lightning strikes (Figure M114).

However in most cases the patterns are not so clear. Although the current in a lightning bolt flows along an effectively infinite (typically

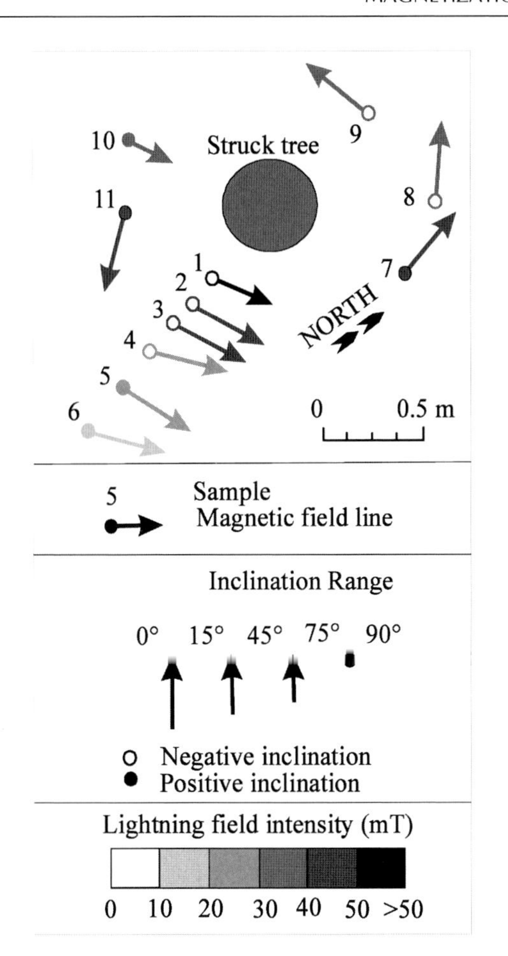

Figure M114 Orientation and intensity of lighting-induced IRM in samples collected around a tree struck by lightning. Modified from Verrier and Rochette (2002), with kind permission of the authors and publisher.

5 km) and relatively straight vertical path between cloud and ground, the path along and below the ground surface is much shorter and more irregular, as shown by the geometry of fulgurites, which are formed when rocks or sediments are melted by lightning. (Temperatures in the ionized channel typically reach 30000 K, five times the temperature of the sun's surface). This irregularity may be expected to disrupt the simple cylindrical symmetry of the induced magnetic field. In the absence of simple directional symmetry, the diagnostic feature for identification of lightning-produced IRM is a combination of high NRM intensity (or more specifically, high NRM/SIRM ratios) and moderate-to-low resistance to AF demagnetization (Cox, 1961; Graham, 1961); thermal demagnetization is generally much less effective at separating lightning overprints (e.g., Tauxe *et al.*, 2003).

Quantitative estimates of lightning currents (Wasilewski and Kletetschka, 1999; Verrier and Rochette, 2002; Tauxe *et al.*, 2003) require both a paleointensity determination and an estimate of the separation between sample location and lightning strike. Wasilewski and Kletetschka (1999) and Tauxe *et al.* (2003) estimate lightning field intensities by finding the lab field that produces an IRM of the same intensity as the lightning overprint. Because lightning remanences may decay with time, this approach yields a minimum estimate of the magnetizing field associated with a strike. Verrier and Rochette (2002) use an alternative approach based not on the overprint intensity but on the AF amplitude required to erase it.

Mike Jackson

Bibliography

Banerjee, S.K., 1963. An attempt to observe the basal plane anisotropy of hematite. *Philosophical Magazine*, **8**: 2119–2120.

Cañon-Tapia, E., 1996. Single-grain versus distribution anisotropy: a simple three-dimensional model. *Physics of the Earth and Planetary Interiors*, **94**: 149–158.

Cisowski, S., 1981. Interacting vs. non-interacting single-domain behavior in natural and synthetic samples. *Physics of the Earth and Planetary Interiors*, **26**: 77–83.

Cox, A., 1961. Anomalous remanent magnetization of basalt. *US Geological Survey Bulletin*, **1083-E**: 131–160.

Cox, A., and Doell, R.R., 1967. Measurement of high-coercivity anisotropy. In Collinson, D.W., Creer, K.M., and Runcorn, S.K. (eds.) *Methods in Palaeomagnetism.* Amsterdam: Elsevier, pp. 477–482.

Daly, L., and Zinsser, H., 1973. Étude comparative des anisotropies de susceptibilité et d'aimantation rémanente isotherme: conséquences pour l'analyse structurale et le paléomagnétisme. *Annales de Géophysique*, **29**: 189–200.

Dankers, P.H.M., 1981. Relationship between median destructive field and coercive forces for dispersed natural magnetite, titanomagnetite, and hematite. *Geophysical Journal of the Royal Astronomical Society*, **64**: 447–461.

Dekkers, M.J., 1997. Environmental magnetism: an introduction. *Geologie en Mijnbouw*, **76**: 163–182.

Dunlop, D.J., 1971. Magnetic properties of fine-particle hematite. *Annales de Géophysique*, **27**: 269–293.

Dunlop, D.J., 1972. Magnetic mineralogy of unheated and heated red sediments by coercivity spectrum analysis. *Geophysical Journal of the Royal Astronomical Society*, **27**: 37–55.

Dunlop, D.J., 1977. The hunting of the 'psark'. *Journal of Geomagnetism and Geoelectricity*, **29**: 293–318.

Dunlop, D.J., 1995. Magnetism in rocks. *Journal of Geophysical Research B: Solid Earth*, **100**: 2161–2174.

Dunlop, D.J., and Özdemir, Ö., 1997. *Rock Magnetism: Fundamentals and Frontiers.* Cambridge: Cambridge University Press, 573 pp.

Egli, R., 2003. Analysis of the field dependence of remanent magnetization curves. *Journal of Geophysical Research—Solid Earth*, **108**(B2): 2081, doi:10.1029/2002JB002023.

Egli, R., 2004a. Characterization of individual rock magnetic components by analysis of remanence curves. 2. Fundamental properties of coercivity distributions. *Physics and Chemistry of the Earth*, **29**: 851–867.

Egli, R., 2004b. Characterization of individual rock magnetic components by analysis of remanence curves. 3. Bacterial magnetite and natural processes in lakes. *Physics and Chemistry of the Earth*, **29**: 869–884.

Fabian, K., and Hubert, A., 1999. Shape-induced pseudo-single-domain remanence. *Geophysical Journal International*, **138**: 717–726.

Fuller, M., 1963. Magnetic anisotropy and paleomagnetism. *Journal of Geophysical Research*, **68**: 293–309.

Gans, R., 1932. Über das magnetische Verhalten isotroper Ferromagnetika. *Annalen der Physik*, **15**: 28–44.

Graham, K.W.T., 1961. The remagnetization of a surface outcrop by lightning currents. *Geophysical Journal of the Royal Astronomical Society*, **6**: 85–102.

Hargraves, R.B., Johnson, D., and Chan, C.Y., 1991. Distribution anisotropy: the cause of AMS in igneous rocks?. *Geophysical Research Letters*, **18**: 2193–2196.

Henkel, O., 1964. Remanenzverhalten und Wechselwirkungen in hartmagnetischen Teilchenkollektiven. *Physica Status Solidi*, **7**: 919–924.

Heslop, D., Dekkers, M.J., Kruiver, P.P., and van Oorschot, I.H.M., 2002. Analysis of isothermal remanent magnetization acquisition curves using the expectation-maximization algorithm. *Geophysical Journal International*, **148**: 58–64.

Jackson, M.J., 1991. Anisotropy of magnetic remanence: a brief review of mineralogical sources, physical origins, and geological

applications, and comparison with susceptibility anisotropy. *Pure and Applied Geophysics*, **136**: 1–28.
King, J.W., and Channell, J.E.T., 1991. Sedimentary magnetism, environmental magnetism, and magnetostratigraphy. *Reviews of Geophysics* Suppl. (IUGG Report—Contributions in Geomagnetism and Paleomagnetism), 358–370.
Kodama, K.P., 1995. Remanence anisotropy as a correction for inclination shallowing: a case study of the Nacimiento Formation. *Eos, Transactions of the American Geophysical Union*, **76**: F160–F161.
Krider, E., and Roble, R.E., 1986. *The Earth's Electrical Environment.* Washington, DC: National Academy Press.
Lowrie, W., 1990. Identification of ferromagnetic minerals in a rock by coercivity and unblocking temperature properties. *Geophysical Research Letters*, **17**: 159–162.
Maher, B.A., and Thompson, R., 1999. *Quaternary Climates, Environments and Magnetism.* Cambridge: Cambridge University Press, 390 pp.
Matsuzaki, H., Kobayashi, K., and Momose, K., 1954. On the anomalously strong natural remanent magnetism of the lava of Mount Utsukushi-ga-hara. *Journal of Geomagnetism and Geoelectricity*, **6**: 53–56.
Muxworthy, A.R., Williams, W., and Virdee, D., 2003. Effect of magnetostatic interactions on the hysteresis parameters of single-domain and pseudo-single-domain grains. *Journal of Geophysical Research*, **108**: 2517.
Néel, L., 1949a. Influence des fluctuations thermiques sur l'aimantation de grains ferromagnétiques très fins. *Comptes rendus hebdomadaires des séances de l'Académie des Sciences (Paris), Série B*, **228**: 664–666.
Néel, L., 1949b. Théorie du traînage magnétique des ferromagnétiques en grains fins avec applications aux terres cuites. *Annales de Géophysique*, **5**: 99–136.
Néel, L., 1955. Some theoretical aspects of rock magnetism. *Advances in Physics*, **4**: 191–243.
Oldfield, F., 1991. Environmental magnetism: a personal perspective. *Quaternary Science Reviews*, **10**: 73–85.
Pokhil, T.G., and Moskowitz, B.M., 1997. Magnetic domains and domain walls in pseudo-single-domain magnetite studied with magnetic force microscopy. *Journal of Geophysical Research B: Solid Earth*, **102**: 22681–22694.
Potter, D.K., 2004. A comparison of anisotropy of magnetic remanence methods—a user's guide for application to palaeomagnetism and magnetic fabric studies. In Martin-Hernández, F., Lüneburg, C.M., Aubourg, C., and Jackson, M. (eds.), *Magnetic Fabric: Methods and Applications,* Vol. 238. London: The Geological Society of London. Geological Society Special Publications, pp. 21–36.
Robertson, D.J., and France, D.E., 1994. Discrimination of remanence-carrying minerals in mixtures, using isothermal remanent magnetization acquisition curves. *Physics of the Earth and Planetary Interiors*, **82**: 223–234.
Rochette, P., Mathé, P.-E., Esteban, L., Rakoto, H., Bouchez, J.-L., Liu, Q., and Torrent, J., 2005. Non-saturation of the defect moment of goethite and fine-grained hematite up to 57 Teslas, *Geophysical Research Letters*, 32, doi:10.1029/2005GL024196.
Stephenson, A., 1994. Distribution anisotropy: two simple models for magnetic lineation and foliation. *Physics of the Earth and Planetary Interiors*, **82**: 49–53.
Stephenson, A., Sadikun, S., and Potter, D.K., 1986. A theoretical and experimental comparison of the anisotropies of magnetic susceptibility and remanence in rocks and minerals. *Geophysical Journal of the Royal Astronomical Society*, **84**: 185–200.
Stockhausen, H., 1998. Some new aspects for the modelling of isothermal remanent magnetization acquisition curves by cumulative log Gaussian functions. *Geophysical Research Letters*, **25**: 2217–2220.
Stoner, E.C., and Wohlfarth, E.P., 1948. A mechanism of magnetic hysteresis in heterogeneous alloys. *Philosophical Transactions of the Royal Society of London, Series A*, **240**: 599–602.
Symons, D.T.A., and Cioppa, M.T., 2000. Crossover plots: a useful method for plotting SIRM data in paleomagnetism. *Geophysical Research Letters*, **27**: 1779–1782.
Tauxe, L., Constable, C., Johnson, C.L., Koppers, A.A.P., Miller, W.R., and Staudigel, H., 2003. Paleomagnetism of the southwestern USA recorded by 0–5 Ma igneous rocks. *Geochemistry, Geophysics, Geosystems*, **4**(4): 8802, doi:10.1029/2002GC000343.
Thompson, R., and Oldfield, F., 1986. *Environmental Magnetism.* London: Allen & Unwin, 227 pp.
Verosub, K.L., and Roberts, A.P., 1995. Environmental magnetism: past, present, and future. *Journal of Geophysical Research B: Solid Earth*, **100**: 2175–2192.
Verrier, V., and Rochette, P., 2002. Estimating peak currents at ground lightning impacts using remanent magnetization. *Geophysical Research Letters*, **29**: 1867, doi:10.1029/2002GL015207.
Wasilewski, P., and Kletetschka, G., 1999. Lodestone: nature's only permanent magnet—what it is and how it gets charged. *Geophysical Research Letters*, **26**: 2275–2278.
Williams, W., and Wright, T.M., 1998. High-resolution micromagnetic models of fine grains of magnetite. *Journal of Geophysical Research*, **103**: 30537–30550.
Wohlfarth, E.P., 1958. Relations between different modes of acquisition of the remanent magnetization of ferromagnetic particles. *Journal of Applied Physics*, **29**: 595–596.
Wohlfarth, E.P., and Tonge, D.G., 1957. The remanent magnetization of single-domain ferromagnetic particles. *Philosophical Magazine*, **2**: 1333–1344.
Worm, H.U., 1999. Time-dependent IRM: a new technique for magnetic granulometry. *Geophysical Research Letters*, **26**: 2557–2560.
Worm, H.-U., and Jackson, M., 1999. The superparamagnetism of Yucca Mountain Tuff. *Journal of Geophysical Research B: Solid Earth*, **104**: 25,415–25,425.
Zapletal, K., 1992. Self-reversal of isothermal remanent magnetization in a pyrrhotite (Fe_7S_8) crystal. *Physics of the Earth and Planetary Interiors*, **70**: 302–311.

Cross-references

Magnetic Anisotropy, Sedimentary Rocks and Strain Alteration
Magnetic Domain
Magnetization, Anhysteretic Remanent (ARM)
Magnetization, Chemical Remanent (CRM)
Magnetization, Depositional Remanent (DRM)
Magnetization, Natural Remanent (NRM)
Magnetization, Thermoremanent (TRM)
Magnetization, Viscous Remanent (VRM)

MAGNETIZATION, NATURAL REMANENT (NRM)

Natural remanent magnetization (NRM) is remanent magnetization that has been acquired naturally (i.e., not artificially acquired in a laboratory). It is the remanent magnetization of a sample (such as rock or baked archaeological material) that is present before any laboratory experiments are carried out. Magnetization is usually measured normalized to either the sample volume (units of A m^{-1}) or sample mass (units of A m^2 kg^{-1}). NRM can consist of one or more types of magnetization depending on the history of the sample. Rock samples that have been exposed to the Earth's magnetic field for many millions of years may have experienced several different processes of magnetization during that time, producing multiple components of magnetization. Components acquired at the time of rock formation are termed primary and later components are termed secondary. The NRM is then the vector sum of all the naturally acquired components of magnetization:

$$\mathrm{NRM} = \mathrm{Primarycomponent} + \sum_i \mathrm{Secondarycomponents}_i$$

$i = 0, 1, 2, \ldots, n$, where n is the number of secondary components. Both primary and secondary components in a rock can record geological events (for example, the time of formation, metamorphism, or an impact event) and are of interest in unraveling the geological history of the rock. Generally, however, it is the primary remanence acquired on formation that is of most interest to the paleomagnetist. Depending on the intensity and duration of the events responsible for secondary magnetizations (also known as overprints), the latter may partially or completely replace the primary magnetization.

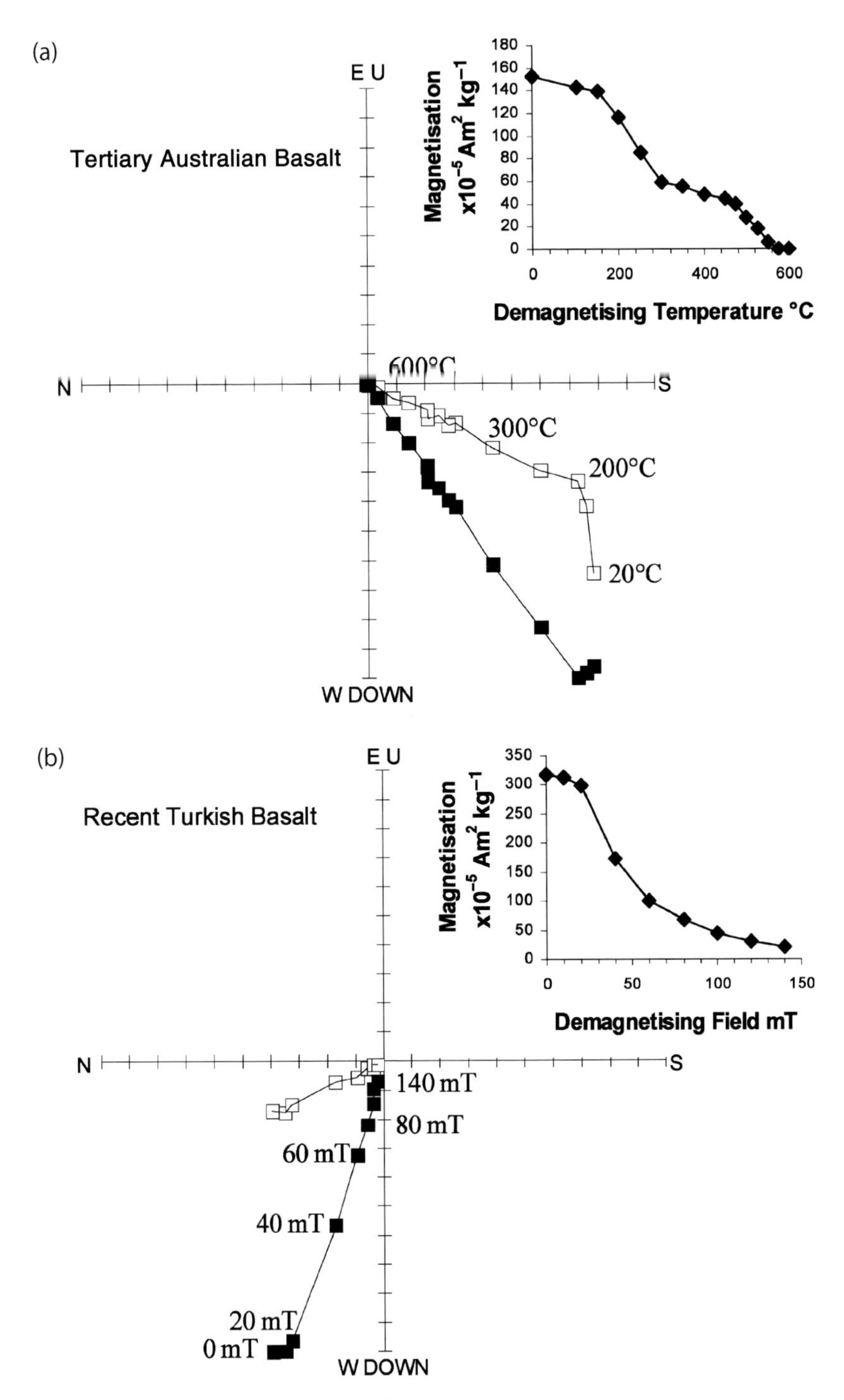

Figure M115 Example of thermal and AF demagnetization. Solid (open) symbols on the orthogonal vector plot (OVP) represent vertical (horizontal) components of magnetization. (a) Thermal demagnetization of Australian Tertiary basalt. There is a viscous secondary component that is removed by 200°C to leave the primary component of remanence. (b) AF demagnetization of recent Turkish basalt. There is a small viscous secondary component that is removed by 20 mT to leave the primary component of remanence.

Primary and secondary remanence

When igneous rocks and baked archaeological materials cool on formation, they acquire a primary thermoremanent magnetization (TRM) (*q.v.*) in the ambient geomagnetic field as the temperature falls below the Curie temperature of the magnetic minerals. A later thermal event below the Curie point will unblock part of the TRM and replace it with a later partial thermal remanent magnetization (pTRM) secondary remanence. If magnetic minerals grow in the presence of a magnetic field, a *chemical remanent magnetization (CRM)* (*q.v.*) is formed when a critical blocking volume is attained. The remanence in sedimentary rocks is often a CRM. CRMs can also be produced in igneous and metamorphic rocks. In sediments, the primary remanence is usually a *depositional remanent magnetization (DRM)* (*q.v.*) acquired when ferromagnetic particles are aligned in the geomagnetic field during deposition. In sedimentary rocks, dewatering during the prolonged process of lithification will usually have enhanced the alignment of these grains with the geomagnetic field to produce a postdepositional detrital magnetization (PDRM).

Samples may acquire a magnetization at ambient temperatures over prolonged lengths of time in a weak magnetic field such as the Earth's. This is called *viscous remanent magnetization (VRM)* (*q.v.*) and its importance is a function of time spent in the field, the strength of the field, and the grain size and composition of the material. VRM is the most common form of secondary remanence and is present, to varying degrees, in all palaeomagnetic samples. Lightning reaching the ground is capable of inducing magnetization in rocks in a very strong, essentially instantaneous magnetic field. This is called an *isothermal remanent magnetization (IRM)* (*q.v.*) and can often completely obscure the primary remanence. Meteorites are extraterrestrial samples that can acquire a secondary shock remanent magnetization (SRM), both in the event that excavated the material from the parent body and during impact at arrival on Earth.

Provided the rock has isotropic properties, each of these types of magnetization is in the direction of the magnetic field at the time of formation. The intensity of magnetization is dependent on many things (for example, the number, composition, and grain size of magnetic minerals) and also depends on the strength of the magnetic field. It is only possible to determine the absolute intensity of the field from samples containing a TRM (see *Palaeointensity*).

Isolation of NRM components

To isolate the different components of NRM, partial demagnetization techniques are used, as the different components will normally have different stabilities to the demagnetization procedures. The techniques used to resolve the components are thermal and alternating field (AF) demagnetization. Components acquired by differing mechanisms will usually have contrasting blocking temperature and coercivity spectra. During thermal demagnetization, samples are heated to successively higher temperatures in a magnetic field-free space so that magnetization below the treatment temperature is unblocked and randomized. As the temperature increases, more and more of the remanence will be removed until the Curie temperature is reached and the sample is completely demagnetized. With AF demagnetization, the peak AF field is increased until magnetic minerals with the highest coercivity are demagnetized (or maximum laboratory peak AF field is reached). Figure M115 shows an example of thermal and AF demagnetization. The characteristic remanent magnetization (ChRM) is the earliest acquired component of magnetization that can be isolated. So for both examples in Figure M115 this is the primary TRM that the basalt acquired on cooling.

Mimi J. Hill

Bibliography

Butler, R.F., 1992. *Paleomagnetism: Magnetic Domains to Geologic Terranes*. Boston, MA: Blackwell Science.

Tarling, D.H., 1983. *Palaeomagnetism Principles and Applications in Geology, Geophysics and Archaeology*. London: Chapman & Hall.

Cross-references

Magnetization, Chemical Remanent (CRM)
Magnetization, Depositional Remanent (DRM)
Magnetization, Isothermal Remanent (IRM)
Magnetization, Thermoremanent (TRM)
Magnetization, Viscous Remanent (VRM)
Paleointensity

MAGNETIZATION, OCEANIC CRUST

In the late 1950s, the first magnetometers adapted to sea surface measurements became available for the scientific community, leading to the discovery of magnetic lineations on the oceanic crust. Since this discovery, first described by Fred Vine and Drummond Matthews in 1963 (the work of Morley and La Rochelle at the same time should also be mentioned), the nature and thickness of the magnetized crust has long been a subject of debate. The depth of the Curie point isotherm for magnetite (580°C) is on the order of a few hundred meters to tens of kilometers below the seafloor (depending on crustal age) and represents a depth limit for remanent magnetization. Below this isotherm, all rocks from the oceanic crust could give rise to a remanent magnetic anomaly signal. However, several decades of magnetic studies on the oceanic crust showed that magnetization is complex and varies by several orders of magnitude depending on the structure of oceanic crust, types of oceanic rocks, and alteration processes.

Magnetic properties of the oceanic crust can be studied by the analysis of magnetic surveys made using sensing instruments mounted on submersibles, towed by ships, or incorporated in satellites. Depending on the distance between the survey and the crustal sources, the focus of remote magnetic analysis can be at very different scales. Submersible surveys can be used to map, for example, the meter scale contrast of magnetization above hydrothermal structures or between flows along ridge axes. By contrast, satellite surveys will give information on the properties of wide oceanic areas and the deepest magnetic sources. In between these extreme scales, sea surface magnetic surveys adapted to the kilometer scale are the reference. Local information can also be collected using downhole magnetometer tools in holes drilled within the crust. When available, the magnetic characterization of drilled or dredged oceanic rocks is a way to directly access magnetization of different rock types and is complementary to remote sensing measurements. Ophiolites, ancient ocean-crust fragments on continents, offer a way to easily access properties of deeper lithologic units. However, the obduction process probably affects mineralogical properties of rocks in such material, and it has been observed that less intense magnetizations are found compared to drilled or dredged oceanic samples. Altogether the fragmented information collected since the early 60's allows a global pattern of magnetization of the oceanic crust to be retreived.

In the following, we present the global structure of the oceanic crust and its variability, then we review the magnetic properties of the extrusive and intrusive basalts, the gabbroic sections, and peridotites.

Global structure of the oceanic crust

In 1972, a model for "typical" oceanic crust structure was derived from ophiolite observation and seismic velocity analysis. Often referred to as "Penrose structure," following the name of the international conference on ophiolites where it was established, this layered oceanic crust is as follows: 500 m of basalts, 1500 m of dikes, and 5000 m of gabbros, which together give an average thickness of 7 km of oceanic crust upon unaltered mantle. Since the establishment of this basic model, drilling and dredging show that deep materials, consisting of altered mantle rocks and gabbros, can be seen at outcrops. This more complex three-dimensional structure of the crust is due to spatial and temporal variations of the magmatic and tectonic processes and is more likely to be

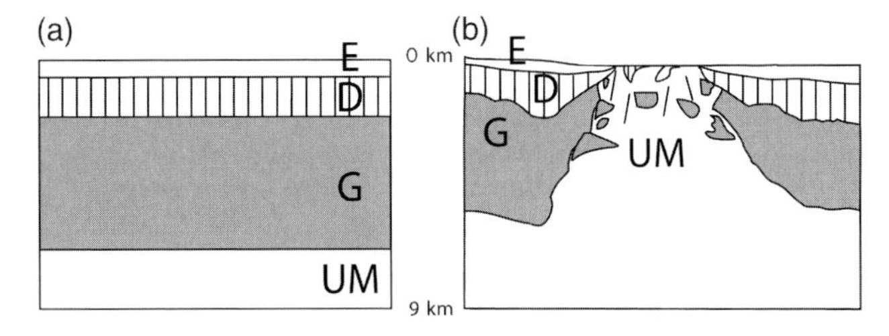

Figure M116 Sketch illustrating (a) the uniform oceanic crust following the model of a Penrose layered crust and (b) an example of idealized portion of crust with a discontinuity in its center allowing exposure of ultramafic outcrops (modified after Cannat *et al.*, 1995). UM, ultramafic rocks; G, Gabbros, D, Dikes, E, Extrusives.

adapted to slow spreading ridges. Ultramafic rocks such as serpentinized peridotites, also called serpentinites, are notably reported in these new crustal models (Cannat *et al.*, 1995). Two "idealized" views of ocean crust are represented in Figure M116, showing a Penrose "layer cake" crust and a discontinuous crust with serpentinite emplaced at the surface. Serpentinized peridotites are altered mantle rocks but will be considered as part of the magnetic oceanic crust, which is described below.

Basalt layer

The extrusive basalt layer, also referred to as seismic layer 2A (layer 1 being the sediment cover), consists of pillow and sheeted lava erupted at the ridge axes, pillows being more widely associated with slow spreading ridges and sheet flows with fast spreading ridges. The surface of flows often consists of a few centimeters of rapidly quenched glasses. Magnetic properties of extrusive basalts of the oceanic crust were the subject of numerous studies as a result of the important amount of dredged and drilled samples made available to the scientific community. The magnetic minerals in oceanic basalts are predominantly titanomagnetites associated with various amounts of titanomaghemites, an alteration product of titanomagnetites. The primary natural remanent magnetization (NRM) of basalts is acquired during the extrusion process and is a thermal remanent magnetization. Magnetic properties can show significant variations between flows depending, in particular, on iron and titanium content but also at the within-flow scale depending on local crystallization conditions and alteration progression. On average, NRM of young basalts is in the range of 10–20 A m^{-1}, but can reach extreme values on the order of 80 A m^{-1} or higher for young flows with high iron content. The Koenigsberger parameter, which is defined as a ratio of remanent magnetization to induced magnetization, is almost always greater than one in basalts and can often reach 100 confirming that magnetization of the basalt layer is dominantly of remanent origin. This property favors the magnetic recording of past inversions of the polarity of Earth's magnetic field leading to the pattern of lineated magnetic anomalies. Microscopic observations and Curie temperatures dominantly within 150–250°C for young unoxidized basalts characterize a primary population of titanomagnetic grains with 60% ulvospinel (called TM60). In addition to this dominant population within the bulk part of the lava, fine-grained titanomagnetite with a broader composition (between 0% and 50% ulvospinel) is observed in the rapidly quenched glassy part (Zhou *et al.*, 2000). As a consequence, higher Curie temperatures and lower NRMs are commonly found at the pillow margin. However, magnetic anomalies are caused by basalts rather than the thin glassy part and the high magnetization of young basalts led some authors during the late 1960s to suggest that rocks from only the first few hundred meters belonging to the extrusive basaltic layers could actually be responsible for most, if not all, of the magnetic signal observed by Vine and Mathews (e.g. Klitgord *et al.*, 1975). This led to the first magnetic models of the oceanic crust as a uniformly highly magnetized basalt layer. However, extensive magnetic measurements on oceanic rocks made available, in particular by the Deep Sea Drilling Project (DSDP) program, in the late 1970s and early 1980s showed that magnetization of the basaltic extrusive layers was not homogeneous. This is mostly a consequence of the low temperature alteration of titanomagnetites to titanomaghemite, caused by circulation of seawater in the rocks pores and cracks (O'Reilly, 1984). During the process, Curie temperature progressively increases, whereas NRM decreases typically down to values on the order of a few A m^{-1}. A secondary, less intense remanent magnetization replace part of the original thermal remanent magnetization. The magnetic record of Earth's magnetic field inversions in the upper oceanic crust is nevertheless well preserved, because this secondary magnetization remains parallel to the original direction despite its chemical origin (for a review on this topic see Dunlop and Ozdemir, 1997). Depending on the alteration environment, secondary minerals can also be created at the expense of magnetite, iron sulfides being an example often associated to hydrothermal alteration. The common occurrence of secondary viscous magnetization contributes to lowering of the magnetic signal. These findings corroborate the idea that to account for the magnetic anomaly signal, deeper seated rocks are more highly magnetized than first expected.

Dike layer

Below the extrusive basalt layer, dike complexes are the first intrusive bodies. Oceanic intrusive rocks have been the subject of fewer studies than basalts due to increasing difficulties in sample accessibility. Rocks from escarpments, tectonized areas, and ophiolites are thus overrepresented and might not represent the norm. From what we know, the sheeted dike layer (also called layer 2B) shows contrasting magnetizations. The much slower cooling rate of intrusive rocks results in an initial magnetic population of large multidomain titanomagnetite grains. Curie temperatures are reported to be quite homogeneous and close to 580°C indicating magnetite as the main carrier of the magnetization. This is the result of widespread oxidation-exsolution of titanomagnetite grains. This process by which the multidomain grains exsolved to a solid solution with pure magnetite could take place during the slow initial cooling (the so-called deuteric oxidation common in subaerial lavas) but the presence of the characteristic lamellar structures associated with such a process is not always observed in submarine dikes. It is likely that low temperature and hydrothermal alteration make a significant contribution to exsolution in this context. Hydrothermal alteration also plays an important role in the loss of iron from grains, recrystallization of magnetite, and replacement of titanomagnetite crystals by silicate minerals (Smith and Banerjee, 1986). Typical remanent magnetization in dikes is typically 10^{-1}–10^{-2} A m^{-1}, but stronger magnetizations on the order of 1 A m^{-1} are found in the freshest samples. Koenigsberger ratios are lower than in basalts but usually above one (ophiolites dikes have lower Koenigsberger ratio) showing that magnetization of submarine dikes is for its most part of remanent origin. This remanent magnetization is mainly acquired during alteration and its relationship with the geomagnetic field at the time of emplacement is not yet clear.

Gabbros

Gabbros are found below oceanic basalts and are coarse-grained igneous rocks (seismic layer 3). The dominant magnetic mineral in gabbros is magnetite, occurring as a primary mineral produced by the exsolution of large titanomagnetite crystals at high temperature and as a secondary mineral typically as recrystallization along cracks. Curie temperatures in gabbros are consequently dominantly found in the range 550°C–590°C. The intensity of magnetization is quite variable spanning several orders of magnitude and depends on the concentration of magnetic minerals, which is often rather low, giving on average NRM in the range 10^{-1} A m^{-1}. In a few sites, ferrogabbro and gabbros rich in olivine were reported and give rather high NRM values due to their high concentration of magnetite of primary origin (in the case of ferrogabbro) or resulting from serpentinization (in the case of olivine-rich gabbros) (Dick *et al.*, 2000). The Koenigsberger ratio is often close to one, but usually shows that remanent magnetization is dominant over induced magnetization. Remanent magnetization is often reported to be quite resistant to alternating field demagnetization showing good stability, this magnetic behavior, characteristic of fine-grained population, arises from the subdivision of magnetite in lamellae structures within exsolved coarse titanomagnetite crystals, secondary magnetite is often observed as fine-grained population as well. The contribution of secondary magnetization of chemical origin (CRM) is difficult to isolate and seems relatively low. Presumably gabbros can preserve a faithful record of the magnetic field acquired during the cooling of the units shortly after intrusion at the spreading axis. Gabbros could contribute to the magnetic anomalies but their low NRM and depth usually make them marginal contributors. Gabbros show important chemical variability and the effect of alteration on magnetic properties is difficult to single out; studies made on sets of metamorphosed gabbros show NRMs and Curie temperatures in the same range as unaltered gabbros (Kent *et al.*, 1978).

Contribution from deeper material: Peridotites

In "typical" Penrose crust, the lower crust is restricted to the gabbroic layer and no contribution from deeper material is reported. However, as the occurrence of serpentinized peridotites (also called serpentinites) within the crust was more and more observed (see Figure M116), their contribution to magnetic signal became obvious. Unaltered peridotites in the mantle have a paramagnetic behavior. Magnetization of peridotites is acquired during the serpentinization process, which is the reaction of the silicate minerals of peridotites (mostly olivine) with water producing serpentine and pure magnetite. The amount of magnetite produced during the serpentinization process is not proportional to the serpentinization progression resulting in complex iron remobilization history (Oufi *et al.*, 2002). Even in fully serpentinized peridotites, a very different behaviors are described with, in particular, Koenigsberger ratios well below or above unity. Curie temperatures are clustering around 580°C as expected for magnetite. Remanent magnetization is variable in the range of 10^{-1} –10 A m^{-1}, peak values of 25 A m^{-1} have been reported, which makes these serpentinite samples as magnetized as young basalts (Oufi *et al.*, 2002). From microscopic observation, magnetite grains can be large and are concentrated along the veins of serpentines, which is believed to indicate the preferential direction of fluid circulation. In this case, large grains contribute to a low NRM and Konigsberger ratio and a strong induced magnetization. Magnetite is also found as small single-domain grains disseminated in the serpentinized peridotite matrix, higher NRM and larger Koenigsberger ratios are then observed. Serpentinites with such a magnetic mineralogy could be important contributors to magnetic anomalies. Nevertheless no clear evidence of serpentinized peridotites recording a reverse magnetization have yet been established.

Conclusion

The typical range in two of the most used magnetic parameters, NRM and Curie temperature, are synthesized on Figure M117 for different rocks within the crust. The effect of serpentinization and low temperature alteration is represented. As shown and discussed earlier, postemplacement alteration has a significant effect on magnetic properties of rocks and contributes to the magnetic heterogeneity of the crust. Due to their shallow depth, strong magnetic intensity and remanent behavior oceanic basalts are thought to be the most important contributors to marine magnetic anomalies allowing the establishment of the sequence of reversals of the Earth's magnetic field. In addition,

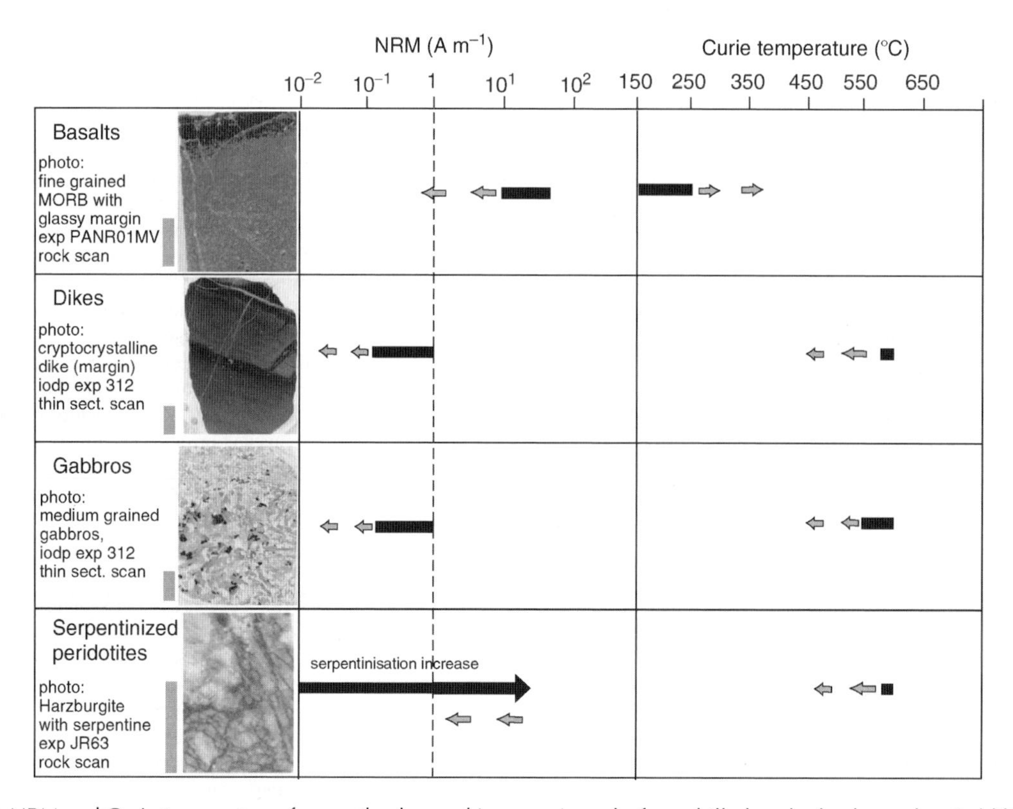

Figure M117 NRM and Curie temperatures frequently observed in oceanic rocks from drilled or dredged samples. Bold lines represent the average values for "fresh" rocks, the black arrow represents the evolution with serpentinization and small grey arrows show the evolution of these magnetic properties during low temperature alteration process. Scale bar for photo is 5 mm.

although this is still a matter of debate, oceanic basalts could provide a record of the magnetic field intensity variations within polarity intervals, as suggested by the worldwide occurrence of small scale magnetic variations (called "tiny wiggles") within the main magnetic anomaly pattern (Cande and Kent, 1992). In any case, it should be mentioned again that the local crustal architecture is probably complex (Karson *et al.*, 2002) and is a major factor to take into account in the analysis of marine magnetic signals.

Julie Carlut and Helene Horen

Bibliography

Cannat, M.C. *et al.*, 1995. Thin crust, ultramafic exposures, and rugged faulting patterns at the Mid-Atlantic Ridge (22° N–24° N). *Geology*, **23**: 149–152.

Cande, S.C., and Kent, D.V., 1992. Ultrahigh resolution marine magnetic anomaly profiles—a record of continuous paleointensity variations. *Journal of Geophysical Research*, **97**(B11): 15075–15083.

Dick, H.J.B. *et al.*, 2000. A long in situ section of the lower ocean crust: results of ODP Leg 176 drilling at the Southwest Indian Ridge. *Earth and Planetary Science Letters*, **179**: 31–51.

Dunlop, D., and Ozdemir, O., 1997. *Rock Magnetism*. Cambridge: Cambridge University Press, 573 pp.

Kent, D.V. *et al.*, 1978. Magnetic properties of dredged oceanic gabbros and the source of marine magnetic anomalies. *Geophysical Journal of the Royal Astronomical Society*, **55**: 513–537.

Karson, J.A. *et al.*, 2002. Structure of the uppermost fast spread oceanic crust exposed at the Hess Deep Rift: implications for subaxial processes at the East Pacific Rise. *Geochemistry, Geophysics, Geosystems*, **3**: DOI:10.1029/2001GC000155.

Klitgord, K.D. *et al.*, 1975. An analysis of near bottom magnetic anomalies: seafloor spreading and the magnetized layer. *Geophysical Journal of the Royal Astronomical Society*, **43**: 387–424.

Morley, L., and La Rochelle, A., 1963. Paleomagnetism and the dating of geological events. *Transactions of the Royal Society of Canada*, **1**: App 31.

O'Reilly, W., 1984. *Rock and Mineral Magnetism*. London: Blackie, 220 pp.

Oufi, O. *et al.*, 2002. Magnetic properties of variably serpentinized abyssal peridotites. *Journal of Geophysical Research*, **107**(B5): 10.1029/2001JB000549.

Smith, G.M., and Banerjee, S.K., 1986. Magnetic structure of the upper kilometer of the marine crust at deep sea drilling project hole 504B, Eastern Pacific Ocean. *Journal of Geophysical Research*, **91**(B10): 10337–10354.

Vine, F.J., and Matthews, D.H., 1963. Magnetic anomalies over oceanic ridges. *Nature*, **199**: 947–949.

Zhou, W. *et al.*, 2000. Variable Ti-content and grain size of titanomagnetite as a function of cooling rate in very young MORB. *Earth and Planetary Science Letters*, **179**: 9–20.

MAGNETIZATION, PIEZOREMANENCE AND STRESS DEMAGNETIZATION

With the advent of paleomagnetic studies in the 1940s, scientists began to wonder how the effect of stress, from burial, folding, etc., would influence the magnetic remanence of rocks. By the mid-1950s, an intense research effort was underway aimed at understanding piezoremanence, which is the remanent magnetization produced by stress, as well as the opposite effect of stress demagnetization. The laboratory experiments suggested that the magnetic signals of rocks were sensitive to stresses typically found around fault zones; and it was calculated that such stress-induced changes would in turn modify the local magnetic field. Scientists thought that earthquakes could be predicted by monitoring magnetic-field variations.

How pressure (stress) affects magnetic remanence

The origin of a spontaneous magnetic remanence is commonly associated with electron exchange between iron (or other transition metals such as Cr, Ni, etc.) atoms. For the iron oxide magnetite (Fe_3O_4), the most abundant magnetic mineral in the Earth's crust, the exchange is indirect, passing through an oxygen atom. The spontaneous magnetization of magnetite is generated in magnetic lattices established by the crystallographic arrangement of the atoms. The sum of the magnetic moments from the lattices comprises the net spontaneous magnetization of the mineral. The bond lengths and bond angles between iron and oxygen atoms of a particular lattice control the magnetic intensities and directions of that lattice. Thus, when a stress acts on a material, the bond lengths and angles change, which will in turn modify the lattices' magnetic moments. This is why the nature of the stress, be it uniaxial, hydrostatic (equal on all sides), compressive, or tensile, can have important consequences on the magnetic properties of materials. If a material compresses isotropically under a pure hydrostatic stress, bond lengths will decrease, yet bond angles will remain unchanged; whereas, uniaxial or shear stress will change the bond angles as well as the bond lengths. Such changes depend on the volumetric compressibility of the magnetic mineral. For example, the volume of magnetite decreases by about 2% per gigapascal (GPa) (note that 1 GPa = 10 kilobars (kbar); for the Earth, 1 kbar corresponds to 3.5 km depth). This is why one needs very large stresses to significantly modify the electronic configuration and the magnetic lattice networks.

Size and shape are other factors influencing the magnetic properties of materials. In extremely small grains, electron exchange exists, but the spontaneous moment does not remain in a fixed direction within the crystal due to thermal fluctuations, even at room temperature. These grains are called superparamagnetic because they have no spontaneous moment in the absence of an external magnetic field, yet in the presence of even a small external field, the randomizing thermal energy is overcome, and the grains display magnet-like behavior. At a critical size, about 0.05 μm for needle-shaped magnetite grains, the spontaneous magnetic moment becomes fixed within the crystal, thus resembling a dipolar magnet with a positive and a negative pole. Such grains are called single domain. At even greater volumes, the magnetic energy of the grain becomes too great and new magnetic domains grow such that the magnetic vectors of the neighboring domains are oriented in directions that diminish the overall magnetization of the crystal. These are called multidomain grains. Multidomain grains often have a net moment because the sum of the magnetic vectors of all the domains does not exactly equal zero.

The magnetizations of single- and multidomain grains react differently to an imposed stress, just as they do with temperature or applied fields. Hydrostatic stress may cause the net spontaneous moment in a single-domain grain to reorient to a new position depending on the grain's shape, presence of crystal defects, etc. Under subhydrostatic stress, the magnetization in a single-domain grain will rotate and change in intensity because different crystallographic axes will experience different stresses, which will deform the exchange network. Under any stress condition, the individual domains in a multidomain grain may grow or shrink to compensate for the deformation, which will modify the magnetic intensity and direction of the grain. Uniaxial stresses are more efficient than hydrostatic stresses at causing domain reorganization.

In sum, two processes contribute to the piezoremanent behavior of materials, both of which can act independently or simultaneously at a given stress condition. The first is a microscopic effect that modifies the electron exchange couple, and the second is more of a macroscopic effect where the magnetization becomes reoriented within a grain. Each type of magnetic mineral and the domain states of each mineral have their own particular piezoremanent and stress demagnetization characteristics. Because rocks are usually composed of a spectrum of magnetic grain sizes, and sometimes more than one type of magnetic mineral, in order to understand how and why the magnetic properties of a rock change in response to an imposed stress, one must know the composition of the magnetic phases and their size distributions. Understanding how stress affects a rock must be viewed in a statistical sense by summing the changes in magnetization of each grain. The type of imposed stress must also be specified.

Details surrounding magnetic measurements at high pressure

Magnetic measurements at high pressures are difficult for several reasons. For one, the pressure vessel apparatus must be non-, or only slightly, magnetic. This is a tall order because most materials strong enough to transmit high pressures are steel alloys—most of which are magnetic or interfere with the electronics of the detector. Second, the sample's magnetic strength should be detectable by the measuring system and be significantly higher than the pressure vessel. In order to maximize the magnetic signal, one wants as many magnetic grains as possible in the sample. It follows that larger samples are more magnetic, and more easily measured than smaller ones—not because the density of magnetic grains is greater, but because of a greater number (or mass) of magnetic grains. As pressure equals force divided by area, a trade-off exists between pressure and sample size, which explains why most experimental work on this problem has focused on large, rock samples. The majority of experiments have been performed using uniaxial stresses and pressures of less than 1 GPa. Indirect measurements of the magnetic states of minerals at much higher pressure cells have been employed since the 1960s using Mössbauer spectroscopy, which measures magnetic interactions between electrons and the nucleus. Although Mössbauer data yields important information about the existence and type of magnetic network in materials, it does not quantify the strength or direction of the moment or their stress dependencies and thus is beyond the scope of this summary. Direct measurements of the full magnetic vector to significantly higher pressures (to 40 GPa) are now becoming possible with the development of diamond anvil cell technology. This new method also allows one to experiment on magnetic minerals whose composition and domain state are well characterized.

An important observation of the high-pressure experiments preformed under uniaxial stress is that magnetic properties change in relation to the maximum stress axis direction. For stress demagnetization (i.e., stress applied in the absence of an applied magnetic field), the magnetization intensity decreases faster parallel to the maximum stress direction rather than perpendicular to it. When pressure is applied in the presence of a magnetic field, magnetic vectors in the grains rotate toward the direction perpendicular to the maximum stress direction. Both magnetic susceptibility and magnetization increase perpendicular to the maximum stress direction and decrease parallel to the maximum stress direction (Figure M118). This has important implications because several experimental apparatuses can only measure magnetic parameters in one direction. Obviously this will lead to a bias of the interpretations of the data.

A special case of stress demagnetization and piezoremanent magnetization is called shock demagnetization and shock magnetization. The term shock is added when the application and removal of a stress occurs in a brief instant in time, such as during a meteorite impact or by firing a projectile at a magnetic sample. To date no study has established that either stress demagnetization or piezoremanence is a time-dependent process, so the difference in terminology should be considered semantic.

Stress demagnetization

Despite the experimental restrictions, several things are known about the magnetic properties of rocks and minerals under stress. The effect of stress in the absence of an external magnetic field has a permanent demagnetizing effect on a rock composed of several magnetic grains. In this case, permanent means that once the pressure is released, the magnetic moment does not grow back to its original, precompressed value because the magnetic vector directions of the grains have been randomized, thus lowering the net magnetization of the rock. It does not mean that the individual magnetic grains have lost their capacity to be magnets. This makes stress demagnetization similar to alternating field demagnetization. Place a rock in a decaying alternating field in a null or very weak external field and the rock will demagnetize. However, the magnetic intensity of a rock placed in a decaying alternating field in the presence of an external field will often increase to levels

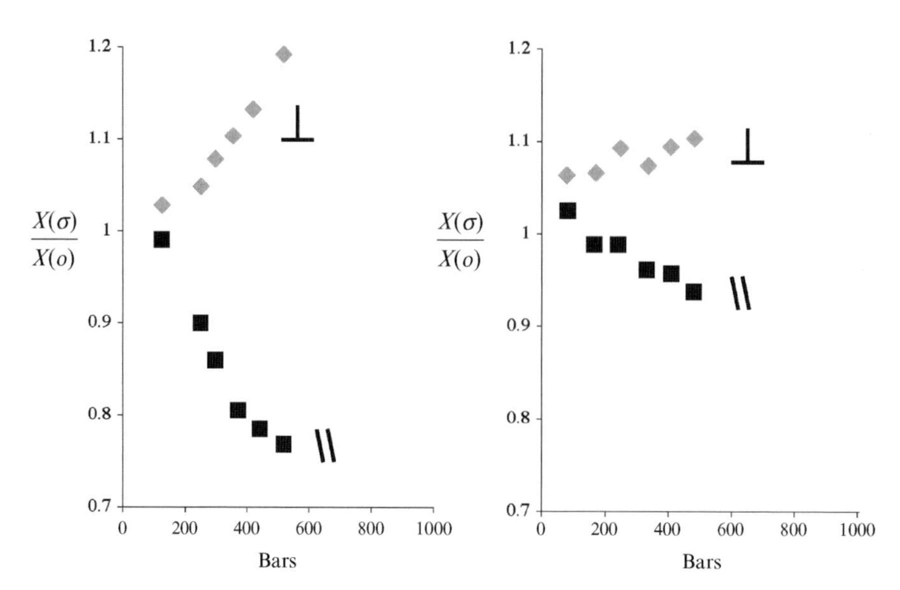

Figure M118 Relative change in magnetic susceptibility (χ) measured perpendicular and parallel to the maximum compression direction in two titanomagnetite samples (Curie temperatures of 250°C and 200°C, respectively) (after Kean *et al.*, 1976).

higher than its starting intensity. The same is true for pressure (stress) demagnetization and its converse, piezoremanent magnetization.

Figure M119 shows a good example of the differences between hydrostatic and uniaxial stress demagnetization of a diabase rock containing large (20–300 μm) titanomagnetite (Curie temperature = 535°C) grains. The pressures employed in these experiments are relevant to the Earth's crust, as 100 MPa (which equals 1 kbar or about 3.5 km depth) is well within the seismogenic zone, i.e., the upper region of the Earth's crust where earthquakes are generated. One quickly sees that uniaxial stress demagnetizes this material about two times faster than hydrostatic stress; e.g., at 100 MPa, hydrostatic and uniaxial pressures demagnetize roughly 20% and 40% of the original, noncompressed magnetization intensity, respectively. This is true whether the material is under pressure, or if it has been decompressed. Upon decompression, a demagnetization effect also exists, which is greater under uniaxial than hydrostatic stress. It should also be noted that during pressure cycling, once a rock has been compressed and then decompressed from a given pressure, it resists further stress demagnetization up until the maximum pressure that the rock has previously experienced. Once that pressure is exceeded, the rock continues demagnetizing in a manner coherent with the previously defined path. The pressure history dependence can be likened to a work or magnetic hardening effect owing to the strain state

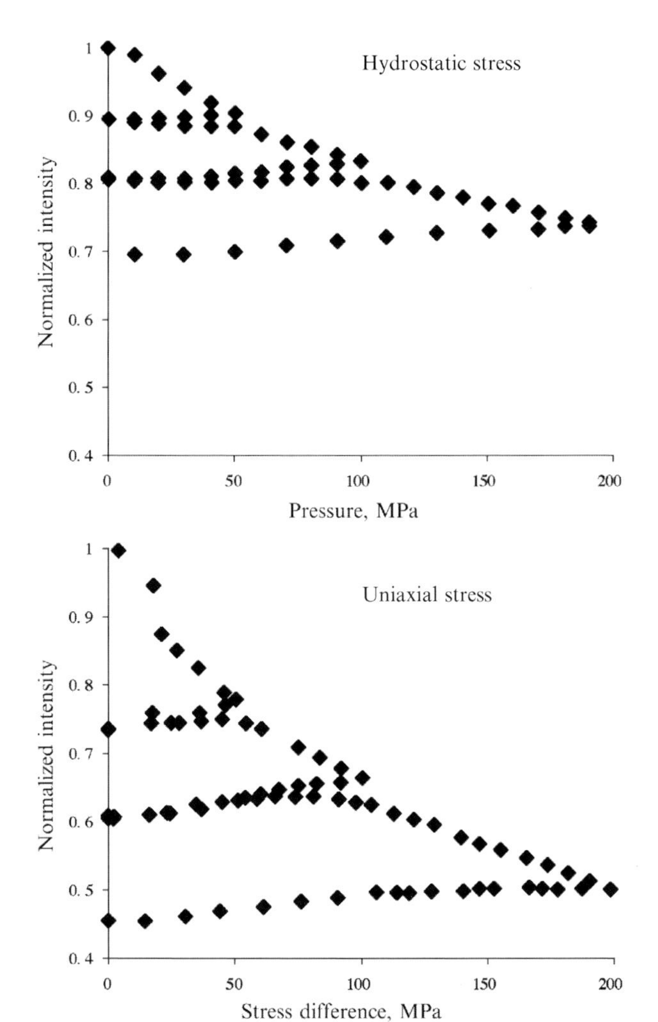

Figure M119 An example of the differences between hydrostatic and uniaxial stress demagnetization for a diabase sample containing multidomain titanomagnetite (after Martin and Noel, 1988).

of the grains. Multidomain grains are more sensitive to stress than single-domain grains because the magnetic remanence of multidomain grains is controlled by domain wall migration, which is highly sensitive to dislocations and other imperfections in the crystal.

Piezoremanent magnetization

Imagine a rock that has acquired its magnetization by cooling in the presence of the Earth's magnetic field. The magnetic intensity of the rock will be proportional to the strength of the field in which it cooled, the number and type of magnetic grains, their sizes and shapes, and the way the grains are geometrically distributed in space. For a rock whose grains are randomly distributed (no preferred orientation), the magnetic moment of each nonspherical grain will tend to lie parallel to its long-axis, favoring the long-axis direction pointing toward the magnetic field direction. If a relatively low magnetic field is then applied, say three times that of the present Earth's field, the rock's magnetization will slightly increase, proportional to the amount of magnetic material in the rock (as defined by its magnetic susceptibility). After the field is removed, the magnetization will decay reversibly back to its original value. When relatively high fields, say 100 times the Earth's field, are applied, the magnetization directions of the grains rotate parallel to the field, leading to a greater net magnetization of the rock. Upon removal of the field, the rock's magnetic intensity will decrease, but will still remain above the value prior to the application of the strong field. This is due to an irreversible reorganization of the individual moments in each grain. As successively higher fields are applied and then removed, the magnetic moments of the grains become more and more aligned parallel to one another, again due to irreversible rotation, leading to greater and greater magnetic intensities. At some applied field value, the moments of all the grains become aligned to the extent that successively higher fields can no longer cause irreversible rotations and the rock's magnetic intensity remains constant. This point is called the saturation remanence of isothermal remanent magnetization (SIRM) and the plot of magnetic intensity as a function of applied field is an isothermal remanent magnetization (IRM) curve (Figure M120).

When a magnetic material is subject to stress in the presence of a magnetic field, the material will acquire a remanence proportional to the strength of the field and the level and type of stress. Figure M120 shows four IRM curves obtained from the same titanium-free, multidomain magnetite sample (a) in the stress-free state (had never been subjected to any pressure), (b) when the sample was compressed and held at a hydrostatic pressure of 3.13 GPa, (c) when the sample was further compressed and held at a hydrostatic pressure of 5.96 GPa, and (d) after decompression ($P = 0$) from $P = 5.96$ GPa. Piezoremanence refers to the phenomenon where the application of stress in the presence of a magnetic field increases the material's magnetization above the level that it would have acquired if the same field had been applied in the absence of stress; e.g., by definition, a piezoremanent magnetization lies anywhere in the gray region above the IRM curve (a) in Figure M120. For this multidomain magnetite example, one observes that the magnetization gained at any applied field increases with increasing pressure (Figure M120, curves b and c). This means that pressure (stress) enhances irreversible rotations of the grains' moments, by domain reorganization, more than an applied field does in the absence of stress. After pressure is released, the domain structures of the grains are even more disposed to irreversible rotation than before (Figure M120, curve d).

The sequence that stress and magnetic fields are applied in will modify a rock's magnetism in different ways and is again dependent on the species of magnetic mineral and its domain state. For the case in Figure M120, one observes that the net magnetization of a multidomain magnetite sample having been (1) brought up in pressure, (2) subject to an applied field, and then (3) removed from the field (curves b and c) is less than having been (1) brought up in pressure, (2) removed from pressure, (3) subject to an applied field, and then

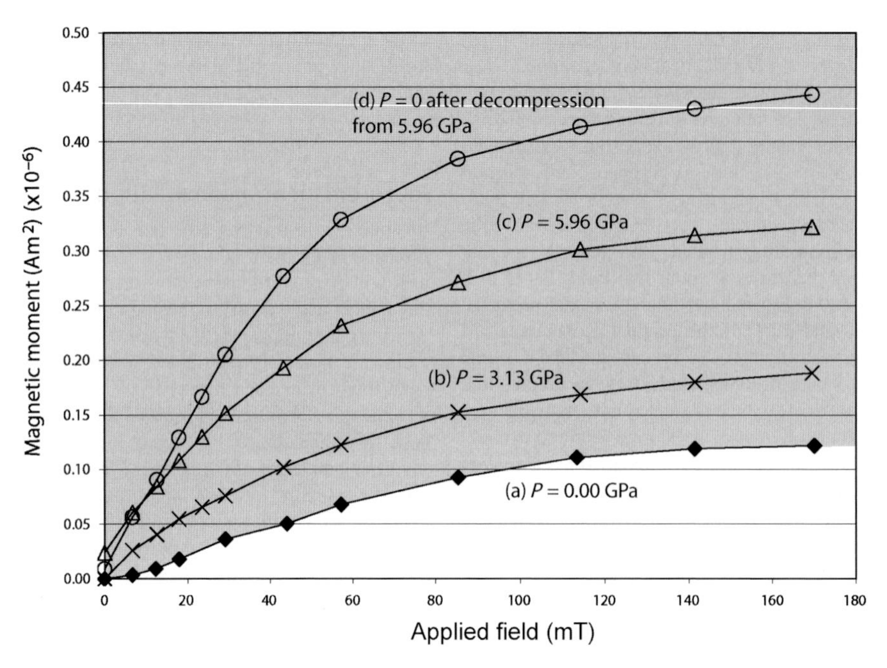

Figure M120 Isothermal remanent magnetization (IRM) curves for the same sample of titanium-free, multidomain magnetite at (a) initial (zero pressure), (b) and (c) under hydrostatic pressures, and (d) at zero pressure after decompressing from 5.96 GPa. Any point lying in the gray area is a piezoremanent magnetization. (S. Gilder, M. LeGoff, J.C. Chervin and J. Peyronneau, unpublished data).

(4) removed from the field (curve d). The opposite is generally true for single-domain magnetite grains. The reason for this is partly due to the way domains facilitate irreversible rotation. Other contributions to irreversibility may come from grain-size reduction (e.g., breaking multidomain grains into single-domain grains) or a permanent modification of the exchange couple. Note that the magnetization of a sample (1) brought up in pressure, (2) subject to an applied field, (3) removed from the field and then (4) removed from pressure may or may not exhibit piezoremanence. This depends on whether compressive stress produces more irreversible rotation than the demagnetization effect that accompanies decompression.

Piezoremanence and stress demagnetization in nature

On Earth, piezoremanent magnetization and stress demagnetization are relevant anywhere transient or permanent stresses provoke a measurable change either in a short-term sense (e.g., during an earthquake) or in a long-term sense (e.g., during the build up or relaxation of stress in the crust due to tectonic motion). Mathematical models of fault motion during earthquakes estimate local deviations in the magnetic field intensity and may reach 1–10 nT (the present Earth's field intensity ranges from 30000 to 60000 nT), which is well within the measurable limit of about 0.1 nT. The size of the seismomagnetic effect depends on several parameters including: magnetic mineralogy, domain state, magnetization direction and intensity of the country rock, direction and intensity of the Earth's field, strike and dip of the fault, sense and amount of motion on the fault, and earthquake magnitude and depth. An important unknown is the stress history of the rock. As shown above in Figure M119, rocks undergo a magnetic hardening during stress demagnetization. So, for a significant change in the rock's magnetization to occur, the rock must be stressed to higher levels than it has previously seen. This could be why piezoremanent effects during an earthquake have never been witnessed. On the other hand, it is difficult to set up and maintain an array of magnetometers over very long time-periods. As earthquakes cannot be predicted, the chances of being at the right place at the right time are extremely low.

Long-term changes in the local magnetic field have been documented around faults and are thought to coincide with nonseismic stress loading due to plate motion. Moreover, long-term changes in the local magnetic environment were observed when large reservoirs were filled with water. The changes were attributed to ground loading and the piezomagnetic effect. Magnetic field variations around volcanoes have been observed during or preceding volcanic eruptions. Although the changes were originally attributed to stress-field changes during the charge or withdraw of magma, closer examination of the data suggests that the magnetic variations were induced by fluid circulation. A similar debate exists for certain subduction zones that exhibit long wavelength magnetic anomalies. Subduction of cold lithosphere depresses the Curie isotherm to depths down to 100 km. Thus, magnetic minerals can be displaced to, or form at, great depths and still be above the Curie isotherm. Note that the Curie temperature of titanomagnetite increases by 2° $kbar^{-1}$. Some workers attribute the anomalies to piezoremanence while others believe it is related to fluid flow in the subduction zone.

Meteorite craters are among the most likely place to find the effects of piezoremanence or stress demagnetization. The presence of shatter cones, pseudotachylite, microscopic textures such as planar deformation features, and large volumes of melted rocks, all suggest that pressures during impact range from about 5 to more than 30 GPa, for reasonably sized (>20 km diameter) craters. However, finding clear-cut evidence that the magnetism of the country rocks has been affected by shock is not so easy because most craters are old (>1 Ga) and have been severely eroded, deeply buried, or the thermal effects of the impact have not been eroded deeply enough to expose the shocked rocks. Moreover, after several hundreds of millions of years, the original magnetic mineralogy of the shocked rocks may become altered by oxidation, or new magnetic minerals may grow. However, the fact that aeromagnetic anomalies are characteristic features of meteorite impacts, there must be some long-term memory of the country rock's magnetization. Some scientists have found that the shocked rocks are remagnetized in the direction of the Earth's field at the time of impact. If true, one can perform an impact test by collecting samples along a radial transect away from the center. The remagnetization direction should be most prevalent near the center and eventually die out with distance.

An intriguing case for shock demagnetization has been proposed for Mars. The southern hemisphere of the Martian crust is highly magnetic except for two very large (>500 km diameter) circular regions that have significantly lower magnetic intensities. Some scientists believe these areas of low magnetic intensity are due to stress demagnetization that occurred during meteorite impacts. Although the magnetic mineral in Martian rocks is presently thought to be magnetite, little is known about the piezomagnetic and stress demagnetization behavior of this substance at the cold (−9°C to −128°C) temperatures of the red planet's surface. Despite some 50 years of research, our knowledge of piezoremanence and stress demagnetization remains limited. A renaissance in this research field, partially due to recent space exploration, will surely lead to exciting discoveries in the future.

Stuart Alan Gilder

Bibliography

Kean, W., Day, R., Fuller M., and Schmidt, V., 1976. The effect of uniaxial compression on the initial susceptibility of rocks as a function of grain size and composition of their constituent ferromagnetic minerals. *Journal of Geophysical Research*, **81**: 861–872.

Martin, R., and Noel, J., 1988. The influence of stress path on thermoremanent magnetization. *Geophysical Research Letters*, **15**: 507–510.

Cross-references

Magnetic Remanence, Anisotropy
Magnetization, Thermoremanent (TRM)

MAGNETIZATION, REMANENT, AMBIENT TEMPERATURE AND BURIAL DEPTH FROM DYKE CONTACT ZONES

Introduction

Nagata (1961, p. 320) proposed using the directions of remanent magnetization in the contact zone of an igneous intrusion as a geothermometer. Carslaw and Jaeger (1959) and Jaeger (1964) applied heat conduction theory to calculate the variation in maximum reheating temperature with distance from a cooling intrusion of simple geometry. Schwarz (1976, 1977) carried out the first study in which the paleomagnetism across a dike contact and theoretical maximum temperature curves were used to estimate the ambient temperature of the host rocks and the depth of burial of the present erosion surface at the time of dike intrusion.

In principle, the depth of burial method can be applied to the contacts of a variety of intrusions with different shapes. However, calculating the maximum reheating temperature in the vicinity of an intrusion with complicated geometry is difficult. To date, depth of burial studies that use remanent magnetization have usually been confined to dike contacts. The case of a dike contact is used to illustrate the method in this article.

Magnetic overprinting in the vicinity of a dike

The depth of burial method is based on the thermal overprinting of remanent magnetization in host rocks near dike contacts. It has been reviewed by Buchan and Schwarz (1987).

When a dike intrudes, temperatures in adjacent host rocks rise above the highest magnetic unblocking temperatures (T_{ub}). During subsequent cooling in the Earth's magnetic field, the dike and this adjacent

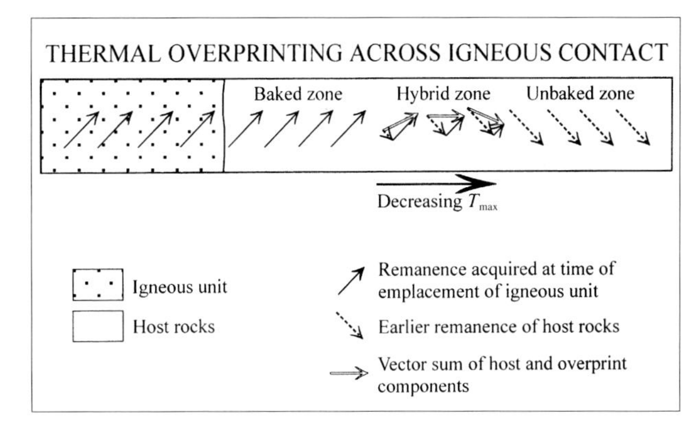

Figure M121 Directions of remanent magnetization with distance from the contact of an igneous intrusion. It is assumed that the magnetic mineralogy of the host rock is similar throughout the profile and that overprinting in the baked and hybrid zones is a thermoremanent magnetization.

"baked" zone acquire magnetic remanences with similar directions (Figure M121). Far from the dike contact, in the "unbaked" zone, peak temperatures are insufficient to thermally reset even the lowest blocking temperature component (see *Baked contact test*). Between the baked and unbaked zones lies a region of hybrid magnetization where reheating is only sufficient to result in thermal resetting of the lower blocking temperature portion of the host rock magnetization. If, in this "hybrid" zone,[1] the older and younger components of magnetization can be separated on the basis of their blocking temperature spectra, then the maximum temperature (T_{max}) reached following dike intrusion can be determined as a function of distance from the contact (Schwarz, 1977). This maximum temperature may be corrected for the effects of magnetic viscosity during prolonged heating in the field. Using heat conduction theory (Carslaw and Jaeger, 1959; Jaeger, 1964), the ambient temperature of the host rock just prior to dike intrusion can then be calculated. Depth of burial of the present erosion surface is determined by dividing the ambient temperature of the host rock by the estimated geothermal gradient at the time of intrusion.

It should be noted that the study of the magnetic hybrid zone according to the procedure of Schwarz (1977) also yields the most rigorous type of baked contact test (the baked contact profile test), a field test for establishing the primary nature of the remanent magnetization of an intrusion (see *Baked contact test*).

Assumptions

Several assumptions are made in the following discussion of the depth of burial method applied to the remanent magnetization across a dike contact (Buchan and Schwarz, 1987).

1. The dike can be approximated by an infinite sheet.
2. The dike was formed by a single magma pulse of short duration.
3. Heat transfer occurs solely by conduction.
4. Dike intrusion results in a (partial) thermal resetting of the remanence of the host rock, without chemical changes.
5. Magnetic mineralogy in the hybrid zone is relatively simple, so that the effects of magnetic viscosity can be estimated from published time-temperature curves.

Complicating factors such as prolonged duration of magma flow in the dike, multiple magma pulses, groundwater, volatiles escaping from the dike, and chemical resetting of remanence in the hybrid zone are

[1] In some publications the baked, hybrid, and unbaked zones are referred to as the contact, hybrid, and host (magnetization) zones, respectively.

usually difficult to take into account (see Buchan *et al.*, 1980; Delaney, 1982, 1987; Delaney and Pollard, 1982; Schwarz and Buchan, 1989). They are not addressed in this article.

Sampling

Oriented paleomagnetic samples are collected along a continuous profile that includes the dike, and the baked, hybrid, and unbaked zones of the host rocks (e.g., Figure M122). The horizontal distance of each sample from the contact is recorded, along with the horizontal width of the dike. If the dike is tilted from vertical, the perpendicular distance of each sample from the contact and the thickness of the dike should be calculated.

Determining maximum reheating temperature (T_{max}) in the hybrid zone

Each hybrid sample is thermally demagnetized in a stepwise fashion. In general, purely viscous components are eliminated at low temperatures to reveal magnetization that is the sum of a host component and a component acquired at the time of emplacement of the igneous unit (Figure M121). Upon thermal demagnetization to higher temperatures, the remanence direction of each hybrid sample moves progressively along a great circle path to the host direction (e.g., Figure M123). This reflects the fact that the T_{ub} spectra of the overprint and host magnetization components are essentially discrete, with the overprint T_{ub} spectrum occupying a range immediately below that of the host component.

T_{max}, the maximum temperature attained at a given locality in the hybrid zone following dike intrusion is given by the highest T_{ub} of the overprint component. It can most easily be obtained using an orthogonal component (or Zijderveld) plot of the horizontal and vertical projection of the thermal demagnetization data (e.g., Figure M124). T_{max} is given by the temperature at which the straight-line segments through the overprint and host components intersect (Dunlop, 1979; McClelland Brown, 1982; Schwarz and Buchan, 1989; Dunlop and Özdemir, 1997). In the example that is shown in Figure M124 the intersections are fairly sharp so that T_{max} values are readily determined. Note that the interpretation of more complicated orthogonal component plots involving multiple magnetic minerals and chemical overprinting are discussed in Schwarz and Buchan (1989).

The T_{max} values determined from the orthogonal component plots of Figure M124 decrease progressively across the hybrid zone with increasing distance from the dike contact as expected for thermal overprinting due to the dike emplacement.

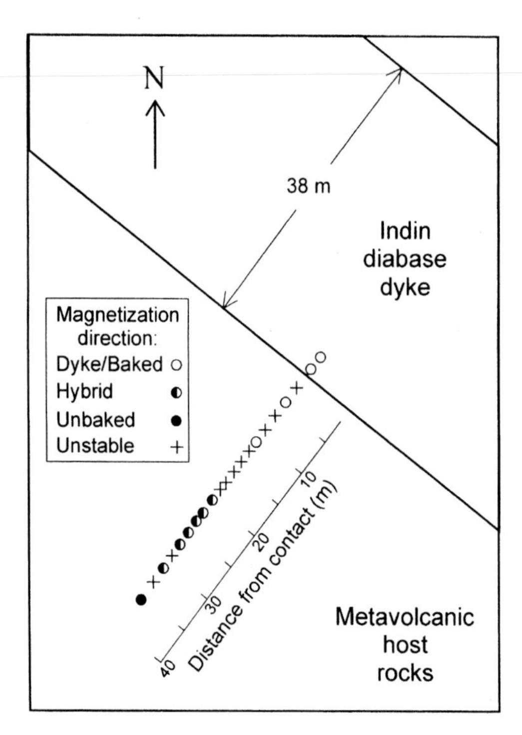

Figure M122 Sampling profile perpendicular to contact of a ca. 2200–2100 Ma Indin diabase dike of the Slave Province of the Canadian Shield (modified after Schwarz *et al.*, 1985). The dike intrudes Archean metavolcanic rocks.

Correcting T_{max} for magnetic viscosity

The maximum blocking temperatures (T_{max}) that are reset in a given thermal event are dependent upon the length of time over which the elevated temperatures are maintained (Dodson, 1973; Pullaiah *et al.*, 1975). Therefore, T_{max} must be corrected using published time-temperature curves for the appropriate magnetic mineral (e.g., Pullaiah *et al.*, 1975).

Determining ambient temperature of host rock at time of dike emplacement (T_{amb})

The maximum temperature (T_{max}) attained at a particular locality in the host rock following dike intrusion equals the sum of the maximum temperature increase (ΔT_{max}) due to heat from the dike and the ambient temperature (T_{amb}) of the present surface just before the dike was emplaced.

Therefore, in the hybrid zone, where T_{max} falls within the overall unblocking spectrum, T_{amb} is calculated from the equation:

$$T_{amb} = T_{max} - \Delta T_{max} \qquad \text{(Eq. 1)}$$

Different heat conduction models can be utilized to determine the maximum increase in temperature (ΔT_{max}) in the contact zone of a dike (Carslaw and Jaeger, 1959; Jaeger, 1964; Delaney, 1987).

The simplest model that accounts for the latent heat of crystallization of the dike magma is one in which the heat released during solidification is added to the initial magma temperature as an equivalent temperature increase (Jaeger, 1964). This model gives a reasonable approximation to more rigorous models for distances greater than a quarter of a dike width from the contact, and has been employed in a number of studies of dike contacts where the hybrid zone is relatively far from the contact (e.g., Schwarz, 1977; Buchan and Schwarz, 1981; Schwarz and Buchan, 1982; Schwarz *et al.*, 1985; Hyodo *et al.*, 1993; Oveisy, 1998).

Using this simple model, the maximum temperature increase at a distance x from the center of a dike of thickness $2d$ is calculated using the following equations from Jaeger (1964), modified slightly to allow for nonzero T_{amb}:

$$\begin{aligned}\Delta T_{max} &= [V + (L/c) - (T_{max} - \Delta T_{max})] \\ &\times \frac{1}{2}[\mathrm{erf}\{(x+d)/2d\tau^{1/2}\} - \mathrm{erf}\{(x-d)/2d\tau^{1/2}\}]\end{aligned} \qquad \text{(Eq. 2)}$$

where $\tau = \kappa t/d^2$, $\kappa = k/(\rho c)$, and erf (u) is the error function. V and L are respectively the intrusion temperature and latent heat of crystallization of the dike magma, κ, k, ρ, and c are respectively the diffusivity, conductivity, density, and specific heat of the host rock, t is the time elapsed since intrusion of the dike, and τ is dimensionless time.

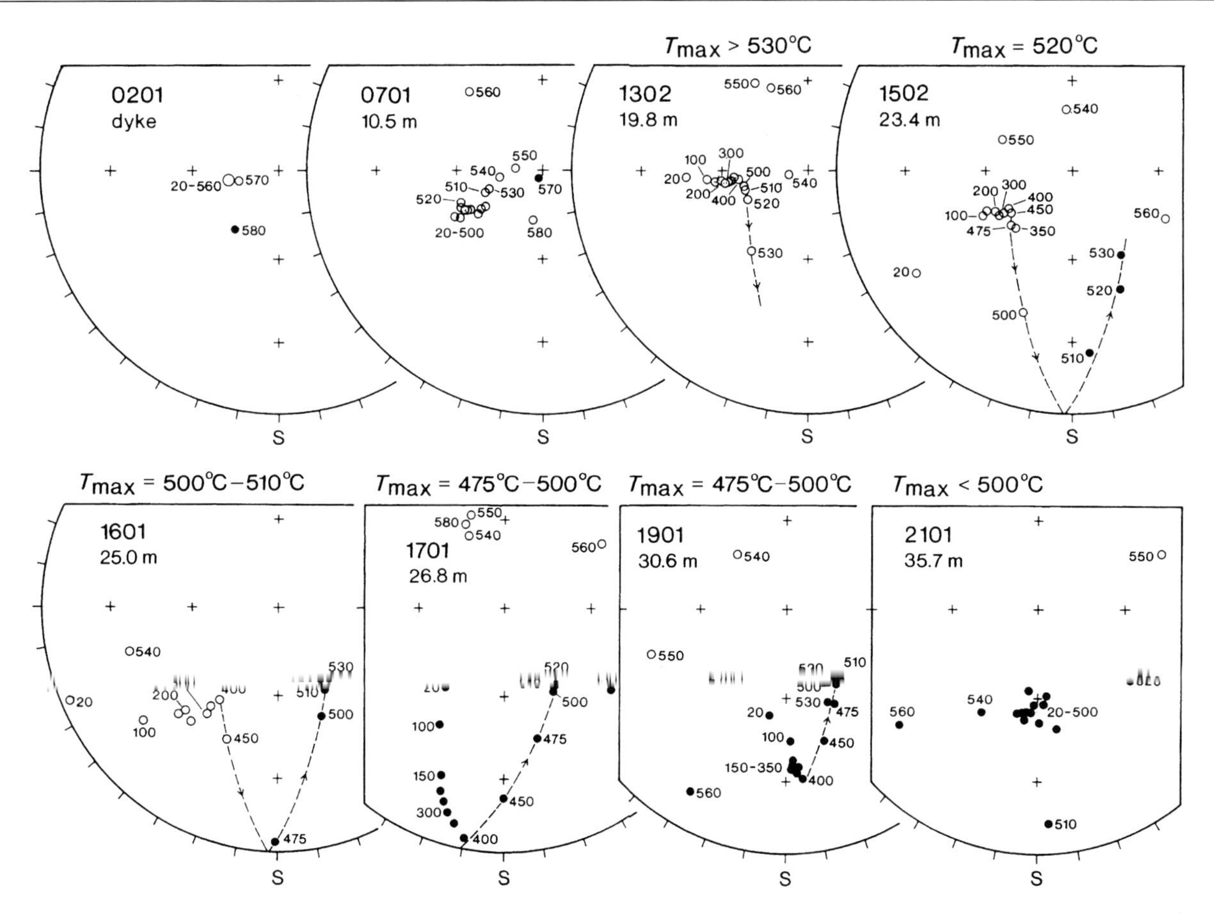

Figure M123 Thermal demagnetization characteristics of selected samples from the Indin dike profile of Figure M122 (after Schwarz *et al.*, 1985). Distance to the contact is given for each sample. Directions are plotted on equal area nets with closed (open) symbols indicating positive (negative) inclinations. Heat-treatment temperatures are indicated in degree Celsius. Data points that represent the great circle between dike and host magnetization directions are connected by a dashed line. T_{max} values, estimated from the experimental data before correction for magnetic viscosity, are given for hybrid zone and host zone samples.

τ_{max} is determined from the equation:

$$\tau_{max} = (x/d)/\ln[(x + d)/(x - d)] \qquad \text{(Eq. 3)}$$

which is satisfied when the temperature reaches its maximum value, T_{max}.

In this model, the required thermal parameters are assumed to be constant. Typical values are: $V = 1150°C$ and $L = 3.77 \times 10^5$ J kg^{-1} for mafic dike rocks; $c = 1.26 \times 10^3$ J kg^{-1}°C^{-1} for mafic host rocks; and $c = 1.09 \times 10^3$ J kg^{-1}°C^{-1} for granitic host rocks (e.g., Touloukian *et al.*, 1981).

The simple heat conduction model described above is not valid within a quarter of a dike width of the contact (Jaeger, 1964), because it does not adequately account for the release of the latent heat of crystallization during cooling of the dike or the temperature dependence of thermal conductivity and diffusivity (Delaney, 1987). In some magnetic studies, hybrid samples have been obtained close to the dike (Buchan *et al.*, 1980; Symons *et al.*, 1980; McClelland Brown, 1981), so that the simple model is inappropriate. The authors in these studies applied more rigorous analytical or numerical models described by Carslaw and Jaeger (1959).

The most sophisticated analysis of heat conduction models for a dike cooling by conduction has been described by Delaney (1987, 1988). He demonstrated the advantage of numerical solutions over analytical solutions, especially at and close to the dike contact. Delaney (1988) published computer programs to calculate the temperature in a dike and host rocks as they cool conductively, in which the temperature dependence of thermal properties and the latent heat of crystallization are taken into account. These programs have been applied to the case of remanent magnetization in a dike contact by Adam (1990).

Finally it should be noted that T_{amb} may be determined in one of two ways. It can be calculated from individual T_{max} values (e.g., Schwarz and Buchan, 1982). Alternatively, when there is a sufficiently wide hybrid zone, T_{amb} can be determined by comparing the whole experimentally determined profile of T_{max} values with the theoretical T_{max} curves of maximum temperature based on conductive heating of the dike (McClelland Brown, 1981; Buchan and Schwarz, 1987).

Determining depth of burial of the present erosion surface at the time of dike emplacement

The depth of burial of the present erosion surface can be determined by dividing T_{amb} (determined from Eq. (1)) by the paleogeothermal gradient for the sampling area at the time of dike intrusion. To estimate the paleogeothermal gradient, the present-day geothermal gradient must be corrected for the decrease in heat generation resulting from the decay of radiogenic isotopes since the time of dike intrusion (e.g., Jessop and Lewis, 1978), and in some areas for the effects of

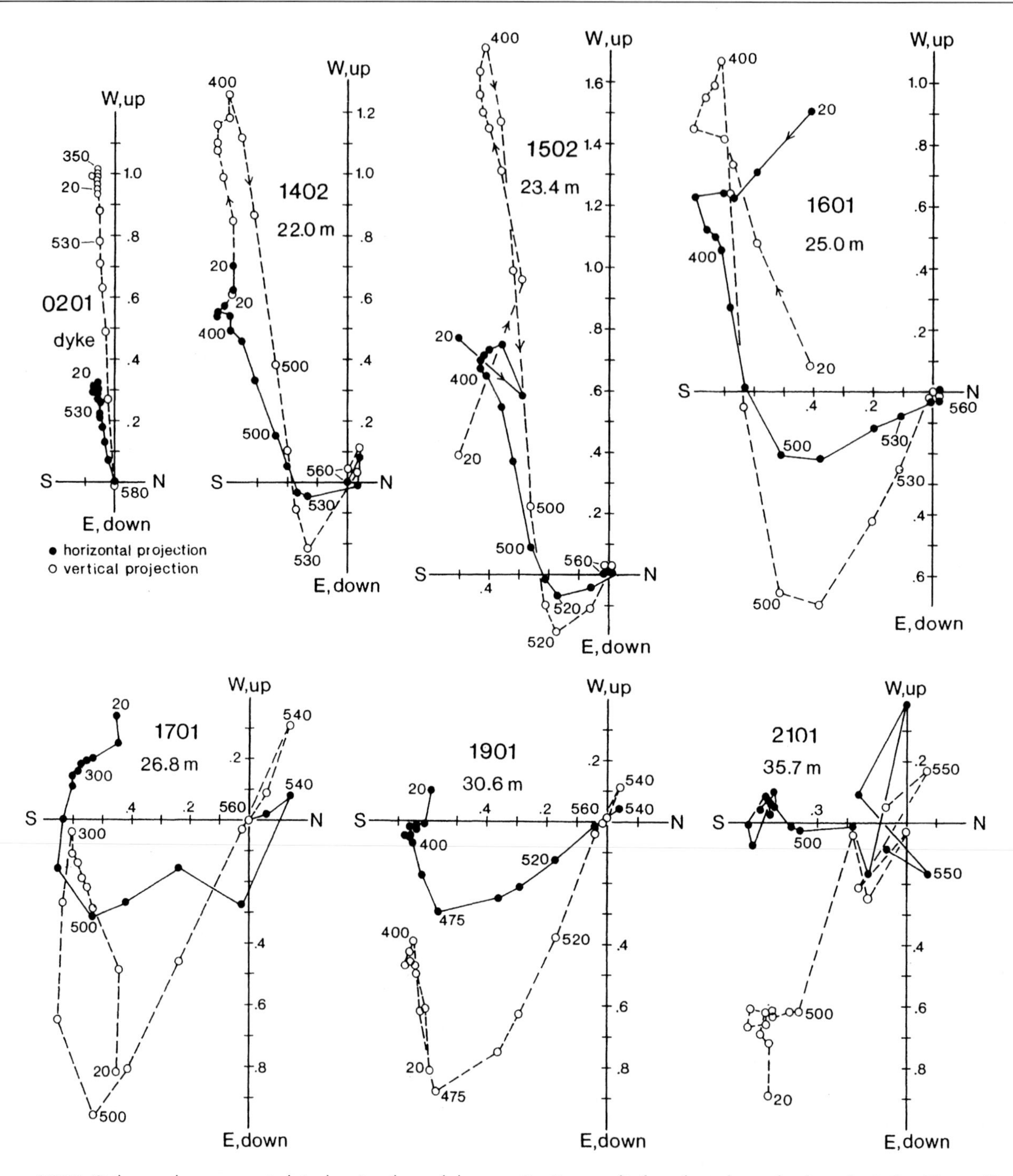

Figure M124 Orthogonal component plots showing thermal demagnetization results for selected samples from the Indin dike profile of Figure M122 (after Schwarz *et al.*, 1985). Peak temperatures are indicated in degree Celsius. Closed and open symbols indicate projections onto the horizontal and N-S vertical planes, respectively.

glacial perturbation (e.g., Jessop, 1971). The uncertainties involved in estimating the paleogeothermal gradient are significant and can lead to relatively large uncertainties in the depth of burial, even if T_{amb} is well constrained (see Schwarz *et al.*, 1985; Hyodo *et al.*, 1993).

Kenneth L. Buchan

Bibliography

Adam, E., 1990. Temperature ambiante de la roche hôte deduite de l'histoire thermique d'un dyke: apport du paleomagnetisme. B.Eng. thesis, École Polytechnique, Montreal: Université de Montréal, 62 pp.

Buchan, K.L., and Schwarz, E.J., 1981. Uplift estimated from remanent magnetization: Munro area of Superior Province since 2150 Ma. *Canadian Journal of Earth Sciences*, **18**: 1164–1173.

Buchan, K.L., and Schwarz, E.J., 1987. Determination of the maximum temperature profile across dyke contacts using remanent magnetization and its application. In Halls, H.C., and Fahrig, W.F., (eds.), *Mafic Dyke Swarms*. Geological Association of Canada Special Paper 34, pp. 221–227.

Buchan, K.L., Schwarz, E.J., Symons, D.T.A., and Stupavsky, M., 1980. Remanent magnetization in the contact zone between Columbia Plateau flows and feeder dykes: evidence for groundwater layer at time of intrusion. *Journal of Geophysical Research*, **85**: 1888–1898.

Carslaw, H.S., and Jaeger, J.C., 1959. *Conduction of Heat in Solids*, 2nd edn. New York: Oxford University Press, 510 pp.

Delaney, P.T., 1982. Rapid intrusion of magma into wet rock: groundwater flow due to pore pressure increases. *Journal of Geophysical Research*, **87**: 7739–7756.

Delaney, P.T., 1987. Heat transfer during emplacement and cooling of mafic dykes. In Halls, H.C., and Fahrig, W.F. (eds.), *Mafic Dyke Swarms*. Geological Association of Canada Special Paper 34, pp. 31–46.

Delaney, P.T., 1988. Fortran 77 programs for conductive cooling of dikes with temperature dependent thermal properties and heat of crystallization. *Computers and Geosciences*, **14**: 181–212.

Delaney, P.T., and Pollard, D.D., 1982. Solidification of basaltic magma during flow in a dike. *American Journal of Science*, **282**: 856–885.

Dodson, M.H., 1973. Closure temperature in cooling geochronological and petrological systems. *Contributions to Mineralogy and Petrology*, **40**: 259–274.

Dunlop, D.J., 1979. On the use of Zijderveld vector diagrams in multicomponent paleomagnetic studies. *Physics of the Earth and Planetary Interiors*, **20**: 12–24.

Dunlop, D.J., and Özdemir, Ö., 1997. *Rock magnetism: fundamentals and frontiers*. Cambridge: Cambridge University Press, 573 pp.

Halls, H.C., 1986. Paleomagnetism, structure and longitudinal correlation of Middle Precambrian dykes from northwest Ontario and Minnesota. *Canadian Journal of Earth Sciences*, **23**: 142–157.

Hyodo, H., York, D., and Dunlop, D., 1993. Tectonothermal history in the Mattawa area, Ontario, Canada, deduced from paleomagnetism and $^{40}Ar/^{39}Ar$ dating of a Grenville dike. *Journal of Geophysical Research*, **98**: 18001–18010.

Jaeger, J.C., 1964. Thermal effects of intrusions. *Reviews of Geophysics*, **2**(3): 711–716.

Jessop, A.M., 1971. The distribution of glacial perturbation of heat flow. *Canadian Journal of Earth Sciences*, **8**: 162–166.

Jessop, A.M., and Lewis, T., 1978. Heat flow and heat generation in the Superior Province of the Canadian Shield. *Tectonophysics*, **50**: 55–77.

McClelland Brown, E., 1981. Paleomagnetic estimates of temperatures reached in contact metamorphism. *Geology*, **9**: 112–116.

McClelland Brown, E., 1982. Discrimination of TRM and CRM by blocking-temperature spectrum analysis. *Physics of the Earth and Planetary Interiors*, **30**: 405–411.

Nagata, T., 1961. *Rock Magnetism*. Tokyo, Japan: Maruzen Company Ltd., 350 pp.

Oveisy, M.M., 1998. Rapakivi granite and basic dykes in the Fennoscandian Shield: a palaeomagnetic analysis, PhD thesis, Luleå, Sweden: Luleå University of Technology.

Pullaiah, G., Irving, E., Buchan, K.L., and Dunlop, D.J., 1975. Magnetization changes caused by burial and uplift. *Earth and Planetary Science Letters*, **28**: 133–143.

Schwarz, E.J., 1976. Vertical motion of the Precambrian Shield from magnetic overprinting. *Bulletin of the Canadian Association of Physicists*, **32**: 3.

Schwarz, E.J., 1977. Depth of burial from remanent magnetization: the Sudbury Irruptive at the time of diabase intrusion (1250 Ma). *Canadian Journal of Earth Sciences*, **14**: 82–88.

Schwarz, E.J., and Buchan, K.L., 1982. Uplift deduced from remanent magnetization: Sudbury area since 1250 Ma ago. *Earth and Planetary Science Letters*, **58**: 65–74.

Schwarz, E.J., and Buchan, K.L., 1989. Identifying types of remanent magnetization in igneous contact zones. *Physics of the Earth and Planetary Interiors*, **68**: 155–162.

Schwarz, E.J., Buchan, K.L., and Cazavant, A., 1985. Post-Aphebian uplift deduced from remanent magnetization, Yellowknife area of Slave Province. *Canadian Journal of Earth Sciences*, **22**: 1793–1802.

Symons, D.T.A., Hutcheson, H.I., and Stupavsky, M., 1980. Positive test of the paleomagnetic method for estimating burial depth using a dike contact. *Canadian Journal of Earth Sciences*, **17**: 690–697.

Touloukian, Y.S., Judd, W.R., and Roy, R.F., (eds.), 1981. *Physical Properties of Rocks and Minerals*. New York: McGraw-Hill, 548 pp.

Cross-reference

Baked Contact Test

MAGNETIZATION, REMANENT, FOLD TEST

Establishing the age of the remanence acquisition with respect to the origin of the rock unit or age of structural deformation events is critical in the interpretation of paleomagnetic data. It is also important to establish that the magnetic minerals carry a stable magnetization over geological timescales. These two related questions remain difficult problems, which often affect the application of paleomagnetism to tectonic, stratigraphic, and paleogeographic problems (Cox and Doell, 1960; Irving, 1964). It is therefore not surprising that early in the development of the paleomagnetic method, several field and laboratory tests were developed and applied to a range of geological contexts. Among them, an elegant simple answer was put forward to constrain the age of remanence acquisition by using field information by Graham (1949). He realized that in folded rocks a simple test based on rotation of the magnetization directions about the local strike could determine if the remanence was acquired before the deformation event —if magnetization directions were dispersed in present-day coordinates and clustered upon tilt rotation (Figure M125a). Magnetization directions that cluster in present-day coordinates and disperse upon tilt rotation indicate that remanence acquisition occurred after the deformation event (Figure M125b). At the time Graham (1949) developed this test, the multivectorial nature of natural remanent magnetization (NRM) had not been properly recognized, and the demagnetization techniques for isolating remanent components had not been developed. The fold test nevertheless worked well as applied to the deformed sedimentary strata of the Rose Hill Formation from Maryland, mainly because the NRMs are univectorial and the strata are tilted to large angles without major structural complexities (French and Van der Voo, 1979). The Graham fold test can be applied to study single deformed structures, including a fold within an undeformed sequence, or over wide deformed areas with sampling at sites with varying bedding tilts.

The impact of paleomagnetic data in the study of regional tectonic processes, particularly in the establishment of the continental drift theory, and the increasing use of demagnetization and statistical techniques in investigating multi-vectorial remanences made the use of field tests fundamental tools for paleomagnetists (Irving, 1964; McElhinny, 1973). Field tests to investigate the magnetization stability and the timing and mode of remanent magnetization acquisition include

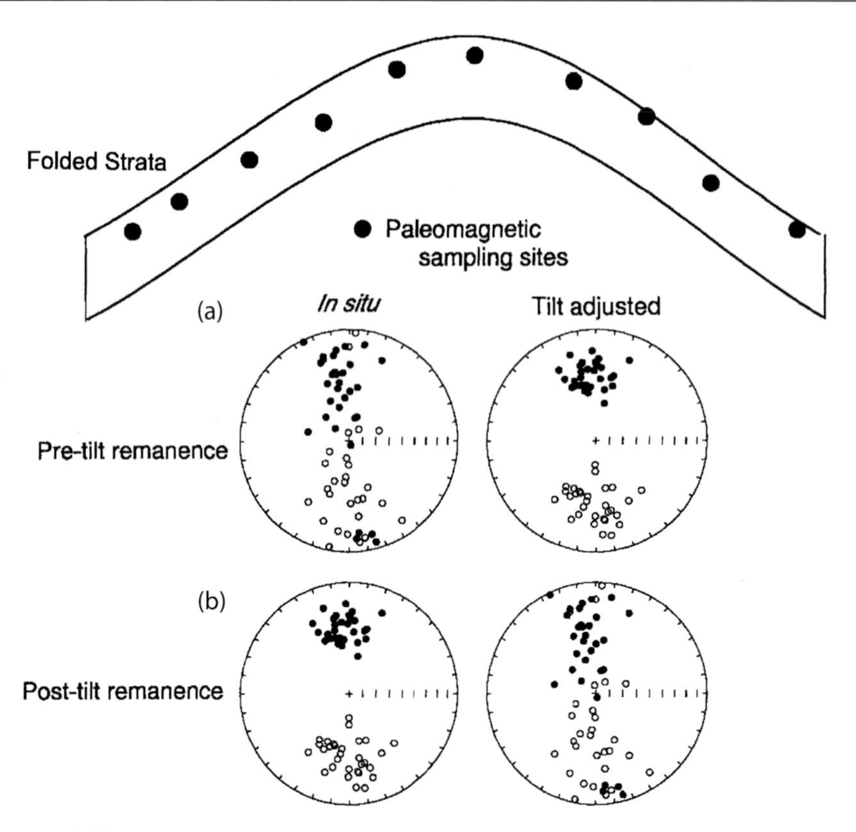

Figure M125 Paleomagnetic fold test.

the (a) fold test, (b) conglomerate test, (c) baked contact test, (d) reversal test, and (e) consistency test. Looking at the many applications developed over the years can easily assess the importance of the field tests in paleomagnetic research.

Development of the fold test in paleomagnetism has involved incorporation of statistical criteria to evaluate the significance of the test, advances in separating multicomponent magnetizations, advances in paleomagnetic instrumentation providing more sensitive instrumentation and methods to measure weak magnetizations, and application to a wide range of deformed rock units as well as local and regional tectonic structures.

Statistical criteria

Application of statistical criteria in evaluating the test significance was first attempted by McElhinny (1964), by using the concentration parameter k of the Fisher (1953) statistics to distinguish between pre- and postfolding magnetization data sets. The criterion is an application of the statistical test for comparing the precisions using the Fisherian concentration parameter of two separate groups of sample directions with the precision of unfolded directions (k_a) compared with the precision for *in situ* directions (k_b). The ratio k_a/k_b is referred to F-ratio tables at given confidence limits and with equal degrees of freedom ($2N - 1$), where N is the number of sample directions studied.

The application of statistical criteria permitted the use of the fold test in a wide range of deformed structures, and McElhinny (1964) fold test is still used in paleomagnetic studies. The test is applied in spite of the problem noticed by McFadden and Jones (1981) that the use of Fisher statistics for directional data sets, which are not Fisher distributed, is not valid. This occurs either way if magnetizations are pre- or postfolding. For instance, if directions are Fisher distributed in *in situ* coordinates, then they are not tilt-corrected coordinates. McFadden and Jones (1981) proposed a test based on comparison of Fisher distributed data sets derived from opposite limbs of a fold in *in situ* and tilt-corrected coordinates. The new test again implies no internal distortion of magnetization and more intensive field sampling, but it can be applied under less stringent conditions than the McElhinny (1964) test.

Two further statistical significance tests based on comparison of distributions of magnetization directions, which require less demanding sampling strategies than the test by McFadden and Jones (1981), have been proposed by McFadden (1990) and Bazhenov and Shipunov (1991). The test of Bazhenov and Shipunov (1991) considers different structural attitudes by making division of bedding plane data into groups, which gives greater flexibility in the sampling strategy. Nevertheless, the tests have been little used in paleomagnetic studies mainly because their statistical significance is relatively weak (e.g., Weil and Van der Voo, 2002).

Evaluation of fold test data sets has also been examined in terms of a parameter estimation problem, by applying simple bootstrap or Monte Carlo approaches. In these tests, large numbers of sample data are generated to determine confidence limits and significance of the maximum clustered directional distribution (Fisher and Hall, 1990; Tauxe *et al.*, 1991; Watson and Enkin, 1993; Tauxe and Watson, 1994). Tauxe and Watson (1994) developed a test using eigen analysis of directional data sets that estimates the degree to which directions from different structural attitude sites are parallel. In this way, directional data sets do not need to conform to a Fisher distribution, and confidence limits are derived from bootstrap and parametric bootstrap techniques according to data set size.

McFadden (1998) discussed the underlying assumption of parametric estimation fold tests that the magnetization directions show maximum clustering with the strata in the position magnetization was acquired. This is not the case when folding does not conform to simple cylindrical horizontal axis geometry and additional factors are involved like plunging folds, unaccounted overprints, etc.

Jaime Urrutia-Fucugauchi

Bibliography

Bazhenov, M.L., and Shipunov, S.V., 1991. Fold test in paleomagnetism: new approaches and reappraisal of data. *Earth and Planetary Science Letters*, **104**: 16–24.

Chan, L.S., 1988. Apparent tectonic rotations, declination anomaly equations, and declination anomaly charts. *Journal of Geophysical Research*, **93**: 12151–12158.

Cogné, J.P., and Perroud, H., 1985. Strain removal applied to paleomagnetic directions in an orogenic belt; the Permian red slates of the Alpes Maritimes, France. *Earth and Planetary Science Letters*, **72**: 125–140.

Cox, A., and Doell, R.R., 1960. Review of paleomagnetism. *Bulletin of the Geological Society of America*, **71**: 647–768.

Facer, R.A., 1983. Folding, Graham's fold tests in paleomagnetic investigations. *Geophysical Journal of the Royal Astronomical Society*, **72**: 165–171.

Fisher, N.I., and Hall, P., 1990. New statistical methods for directional data—I. Bootstrap comparison of mean directions and the fold test in paleomagnetism. *Geophysical Journal International*, **101**: 305–313.

Fisher, R.A., 1953. Dispersion on a sphere. *Proceedings of the Royal Society of London Series A*, **217**: 295–305.

French, A.N., and Van der Voo, R., 1979. The magnetization of the Rose Hill Formation at the classic site of Graham's fold test. *Journal of Geophysical Research*, **48**: 7688–7696.

Graham, J.W., 1949. The stability and significance of magnetism in sedimentary rocks. *Journal of Geophysical Research*, **54**: 131–167.

Irving, E., 1964. *Paleomagnetism and Its Application to Geological and Geophysical Problems.* New York: Wiley.

Kodama, K.P., 1988. Remanence rotation due to rock strain during folding and the stepwise application of the fold test. *Journal of Geophysical Research*, **93**: 3357–3371.

MacDonald, W.D., 1980. Net rotation, apparent tectonic rotation, and the structural tilt correction in paleomagnetic studies. *Journal of Geophysical Research*, **85**: 3659–3669.

McElhinny, M.W., 1964. Statistical significance of the fold test in palaeomagnetism. *Geophysical Journal of the Royal Astronomical Society*, **8**: 338–340.

McElhinny, M.W., 1973. *Paleomagnetism and Plate Tectonics.* Cambridge Earth Science Series, Cambridge University Press, 358 pp.

McFadden, P.L., 1990. A new fold test for paleomagnetic studies. *Geophysical Journal International*, **103**: 163–169.

McFadden, P.L., 1998. The fold test as an analytical tool. *Geophysical Journal International*, **135**: 329–338.

McFadden, P.L., and Jones, D.L., 1981. The fold test in paleomagnetism. *Geophysical Journal of the Royal Astronomical Society*, **67**: 53–58.

Stamatakos, J., and Kodama, K.P., 1991a. Flexural flow folding and the paleomagnetic fold test: an example of strain reorientation of remanence in the Mauch Chunk Formation. *Tectonics*, **10**: 807–819.

Stamatakos, J., and Kodama, K.P., 1991b. The effects of grain-scale deformation on the Bloomsburg Formation pole. *Journal of Geophysical Research*, **96**: 17919–17933.

Stewart, S., 1995. Paleomagnetic analysis of plunging fold structures: errors and a simple fold test. *Earth and Planetary Science Letters*, **130**: 57–67.

Tauxe, L., and Watson, G.S., 1994. The fold test: an eigen analysis approach. *Earth and Planetary Science Letters*, **122**: 331–341.

Tauxe, L., Kylstra, N., and Constable, C., 1991. Bootstrap statistics for paleomagnetic data. *Journal of Geophysical Research*, **96**: 11723–11740.

Van der Plujim, B.A., 1987. Grain-scale deformation and fold test-evaluation of syn-folding remagnetization. *Geophysical Research Letters*, **14**: 155–157.

Watson, G.S., and Enkin, R.J., 1993. The fold test in paleomagnetism as a parameter estimation problem. *Geophysical Research Letters*, **20**: 2135–2137.

Weil, A.B., and Van der Voo, R., 2002. The evolution of the paleomagnetic fold test as applied to complex geologic situations, illustrated by a case study from northern Spain. *Physics and Chemistry of the Earth*, **27**: 1223–1235.

MAGNETIZATION, THERMOREMANENT

Introduction

Thermoremanent magnetization (TRM) is acquired when magnetic minerals cool in a weak magnetic field ***H*** from above their Curie temperatures. TRM is the most important remanent magnetization used in paleomagnetism. It is almost always close to parallel to the field which produced it, and its intensity is proportional to the strength of the field for weak fields like the Earth's. The TRM of rocks is therefore a vast [illegible]
Earth's lithospheric plates and the history of the geomagnetic field.

The primary natural remanent magnetization of an igneous rock or a high-grade metamorphic rock is a TRM. Newly erupted seafloor lavas at mid-ocean ridges acquire an intense TRM on cooling below the Curie temperature T_C. Rapid cooling also results in fine grain size, which makes the TRM highly stable, so that oceanic basalts are excellent recorders of the paleomagnetic field. However, the TRM is largely replaced within at most a million years of formation by chemical remanent magnetization (CRM) of reduced intensity.

The magnetic signal recorded in archeological materials such as pottery, bricks, and the walls and floors of the ovens in which they were fired at high temperature is also a pure TRM residing in single-domain (SD) and pseudosingle-domain (PSD) grains. These materials should be ideal for determinations of the paleointensity of the Earth's field at the time of firing because they are strongly magnetized and the minerals have been stabilized physically and chemically well above their Curie temperatures before their initial cooling.

TRM results from thermally excited changes in magnetization. In the case of SD grains, it is a frozen high-temperature equilibrium distribution between two microstates in which the spins of all atoms are either parallel or antiparallel to an applied field ***H***. At temperatures below T_C, there are large perturbations of a crystal's spin structure, leading to transitions between structures of different types. With cooling, transitions become more difficult, as evidenced by a rapidly increasing relaxation time τ. Eventually, decreasing thermal energy and increasing energy barriers E_B between states prevent further transitions and TRM is frozen in very abruptly at a blocking temperature T_B.

The TRM of multidomain (MD) grains results from the blocking of domain walls at positions determined by the externally applied field ***H***, the internal demagnetizing field H_D of the grain, and the pinning effect of lattice defects.

Theories of TRM and reviews of experimental data have been given by Néel (1949, 1955), Everitt (1961, 1962), Stacey (1958), Schmidt (1973), Day (1977), and Dunlop and Özdemir (1997).

Relaxation time and single-domain TRM

Néel's (1949, 1955) theory of relaxation time deals with noninteracting uniaxial SD grains with saturation magnetization $\boldsymbol{M}_S$ and volume V. $\boldsymbol{M}_S$ results from the parallel exchange coupling of atomic spins and is a strong function of temperature T near T_C, but is weakly dependent on T at low temperatures. Néel's SD theory makes use of

Stoner and Wohlfarth's (1948) results for coherent rotation in spheroidal SD grains aligned with a weak field $\boldsymbol{H}$. The magnetic moment $V\boldsymbol{M}_S$ has a choice of orientations controlled by the easy axes of anisotropy. Uniaxial shape anisotropy is dominant in minerals such as magnetite with high M_S. Shape anisotropy produces two energy minima, corresponding to spins in one or the other direction along the longest axis of the grain. These two minima define the two SD microstates. At 0 K, spins would be exchange coupled exactly parallel; but at ordinary temperatures, thermal excitations perturb the spin lattice, resulting in a steady decrease in $M_S(T)$. At high temperatures, reversals of grain moments are also excited as $M_S(T)$ drops rapidly toward zero at T_C.

At ordinary temperatures, spontaneous reversals of an SD grain are unlikely because the energy barrier between microstates due to shape anisotropy is much larger than the available thermal energy ($\approx 25kT$ for experimental times of a few minutes, where k is Boltzmann's constant). Close to T_C, where E_B is small, the energy barrier becomes comparable to $25kT$ and SD moments can reverse with the help of thermal excitations. This thermally excited condition is called superparamagnetism. The transition from a superparamagnetic state to a stable magnetic state is quite sharp and defines the blocking temperature T_B.

At any time, SD particles are either in state 1, in which moments are aligned with $\boldsymbol{H}$, or state 2, in which moments are antiparallel to $\boldsymbol{H}$. The transition probability set by the energy barrier ΔE_B between states leads to a relaxation time for exponential magnetization decay

$$1/\tau = 1/\tau_0 \exp(-E_B/kT) = 1/\tau_0 \exp[-(\mu_0 V M_S H_K/2kT)(1-H/H_K)^2] \quad \text{(Eq. 1)}$$

where $\tau_0 = 10^{-9}$–10^{-10} s is the atomic reorganization time for transitions between microstates and microcoercivity $H_K = (N_b - N_a)M_S$ for shape anisotropy. N_a and N_b are the demagnetizing factors when $\boldsymbol{M}_S$ is directed parallel or perpendicular, respectively, to the long axis. The thermally activated transitions between states 1 and 2 cause the average moment of a grain ensemble to relax to a thermal equilibrium value

$$M(t) = M(0)\exp(-t/\tau) + M(\infty)[1 - \exp(-t/\tau)]. \quad \text{(Eq. 2)}$$

The thermal equilibrium magnetization above and at the blocking temperature T_B is

$$M_{eq} = M(\infty) = M_S \tanh(\mu_0 V M_S H/kT). \quad \text{(Eq. 3)}$$

The dependence of τ on temperature by Eq. (1) is very significant. As T decreases, the energy barriers increase and τ grows exponentially, changing from a few seconds to millions of years over a narrow temperature interval. This rapid change leads to the concept of the blocking temperature at which $\tau = t$ (experimental time).

$$T_B = [\mu_0 V M_{SB} H_{KB}/2k \ln(t/\tau_0)](1 - H/H_{KB})^2, \quad \text{(Eq. 4)}$$

where M_{SB} and H_{KB} are the values of M_S and H_K at the blocking temperature.

Equation (1) also predicts a strong dependence of τ on grain size. SD grains only slightly above superparamagnetic size have relaxation times greater than the age of the earth, but relaxation times of only a few minutes at T_B of a few hundred degrees Celsius.

Thus, on cooling below T_B in an applied field, all of the orientations of the SD moments remain fixed as their M_S increases and TRM is frozen in:

$$M_{TRM} = M_{RS}(T) \tanh(\mu_0 V M_{SB} H/kT_B), \quad \text{(Eq. 5)}$$

where M_{RS} is the saturation value of TRM. Single-domain M_{TRM} should be relatively intense because $\tanh(\alpha)$ saturates rapidly as $\alpha = \mu_0 V M_{SB} H/kT_B$ increases and TRM then approaches the SD saturation remanence M_{RS}.

The high thermal stability of TRM is also accounted for, because M_{TRM} can only be unfrozen and reset to zero by reheating to the unblocking temperature T_{UB} in $H = 0$. According to Eq. (4), if $H \ll H_K$, then $T_{UB} \approx T_B$. TRM can only be demagnetized by reheating to its original high blocking temperature.

Experimental single-domain TRM

According to Eq. (5), the intensity of weak-field TRM is proportional to the applied field strength H, since $\tanh(\alpha) \propto \alpha$ for small α. The proportionality between M_{TRM} and H in weak fields (Figure M126) has been verified by measuring TRM acquisition curves in the weak-field region for synthetic oxidized SD titanomagnetites $Fe_{2.3}Al_{0.1}Ti_{0.6}O_4$ over a range of oxidation parameters $0.15 < z < 0.41$ (Özdemir and O'Reilly, 1982). Figure M126 also shows a decrease in the intensity of TRM with increasing degree of maghemitization. This fall in intensity is about the same as that reported to take place in submarine basalts with increasing distance from mid-ocean ridges.

Although Eq. (5) gives a reasonable match to observed absolute TRM intensity in the weak-field region, the Néel SD theory is less successful in explaining TRM acquisition for higher fields, $H \gtrsim 1$ Oe or 0.1 mT. The TRM intensity predicted by Eq. (5) reaches saturation at quite low fields. Experimentally, SD TRM usually saturates in much larger fields, on the order of 100 Oe or 10 mT (Figure M127). This discrepancy has been attributed to either angular dispersion of particle axes (Stacey and Banerjee, 1974) or particle interactions. Dunlop and West (1969) modified Eq. (5) by introducing an interaction field distribution that was estimated from experimental Preisach diagrams.

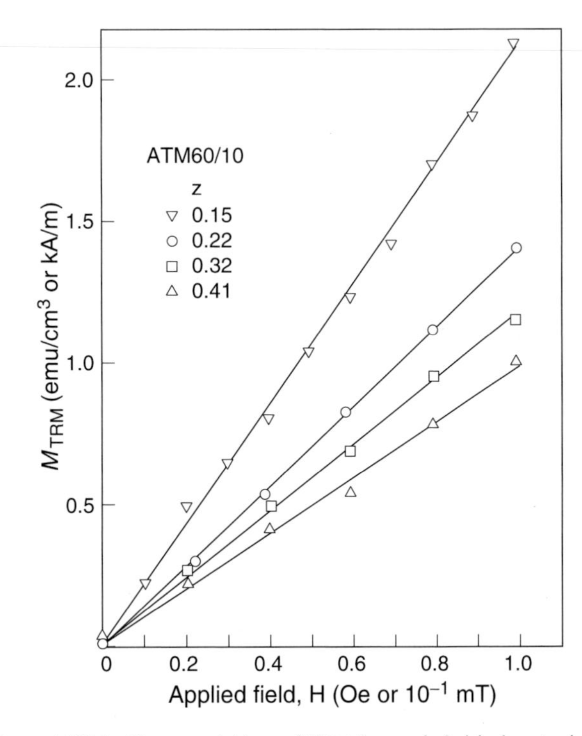

Figure M126 The acquisition of TRM in weak fields for single domain titanomaghemites with composition $Fe_{2.5}Al_{0.1}Ti_{0.4}O_4$ (ATM/60). TRMs are linear with H and decrease with increasing oxidation parameter z (after Özdemir and O'Reilly, 1982).

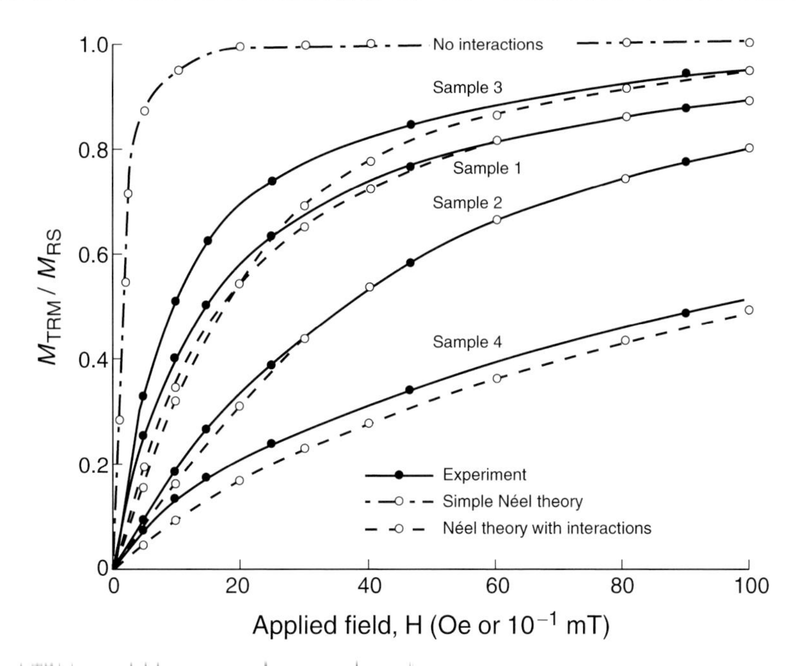

Figure M127 Experimental TRM acquisition curves for a number of synthetic magnetites with mean grain sizes from 0.04 to 0.22 μm compared to the prediction of Néel's (1949) theory of noninteracting SD grains. The theory cannot explain the rapid initial rise of the experimental curves combined with their gradual approach to saturation (after Dunlop and West, 1969).

TRM intensity varies strongly with grain size d. Eq. (5) predicts an increase in M_{TRM} in proportion to increasing grain volume V. Figure M128 shows the grain-size dependence of TRM intensity for SD hematites. M_{TRM} increases in almost exact proportion to grain size d, not d^3. A similar dependence is observed for magnetite grains smaller than the critical SD size $d_0 \approx 0.1$ μm, the threshold between SD and non-SD states (Figure M129). Above d_0, M_{TRM} decreases with increasing grain size, as d^{-1} below 1 μm and as $d^{-0.55}$ above 1 μm. Thus, the Néel SD theory does not explain experimental TRM data for magnetite over most of its natural grain-size range. However, Néel's SD theory is more relevant to TRM in SD hematites because hematite has a small M_S (≈ 2 kA m^{-1}) and a large d_0.

One of the important properties of SD TRM is the equality of the blocking temperature T_B during field cooling and the unblocking temperature T_{UB} during zero-field reheating. Experimentally, as in Figure M126, and theoretically according to Eq. (5), the intensity of a weak-field TRM, and the intensities of all the partial TRMs with different T_B values that compose it, are proportional to the applied field strength. The equality of T_B and T_{UB} for weak fields and the proportionality of TRM and pTRM intensities to field H form the basis for determining paleofield intensity by the Thellier and Thellier (1959) method.

Field dependence of blocking temperature has been examined by Sugiura (1980) and Clauter and Schmidt (1981). $T_B(H)$ as predicted by Eq. (4) has been verified directly by measuring pTRMs covering the entire blocking-temperature spectrum for SD magnetites. Sugiura (1980) obtained a close match of theoretical and experimental T_B spectra by assuming a single value of V and subdividing the H_K spectrum into 11 fractions (Figure M130). The blocking-temperature spectrum shifts to lower temperature as H increases. Unblocking temperatures during zero-field heating are therefore always higher than blocking temperatures observed during in-field cooling (Dunlop and West, 1969).

Equation (4) predicts that blocking and unblocking temperatures vary with grain volume V. Experimentally determined blocking temperatures (Dunlop, 1973) agree well with the theoretical blocking temperatures calculated with a 3D micromagnetic model (Winklhofer *et al.*, 1997). Small SD grains have lower blocking and unblocking temperatures than large SD grains. The predicted quadratic dependence of T_B on the applied field is confirmed experimentally.

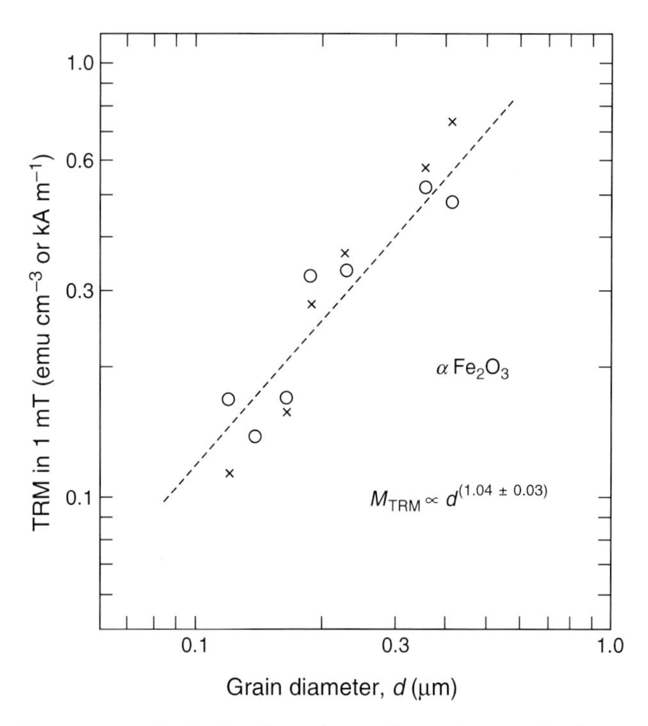

Figure M128 Grain-size dependence of TRM intensity for the single-domain hematites. The dashed line has slope 1.04 showing that M_{TRM} is very nearly proportional to d. Open circles: experimental data points; crosses: theoretical TRM values from Néel (1949) SD theory (after Özdemir and Dunlop, 2002).

The dependence of TRM intensity on the rate of cooling is of practical interest to paleomagnetists studying slowly cooled orogens.

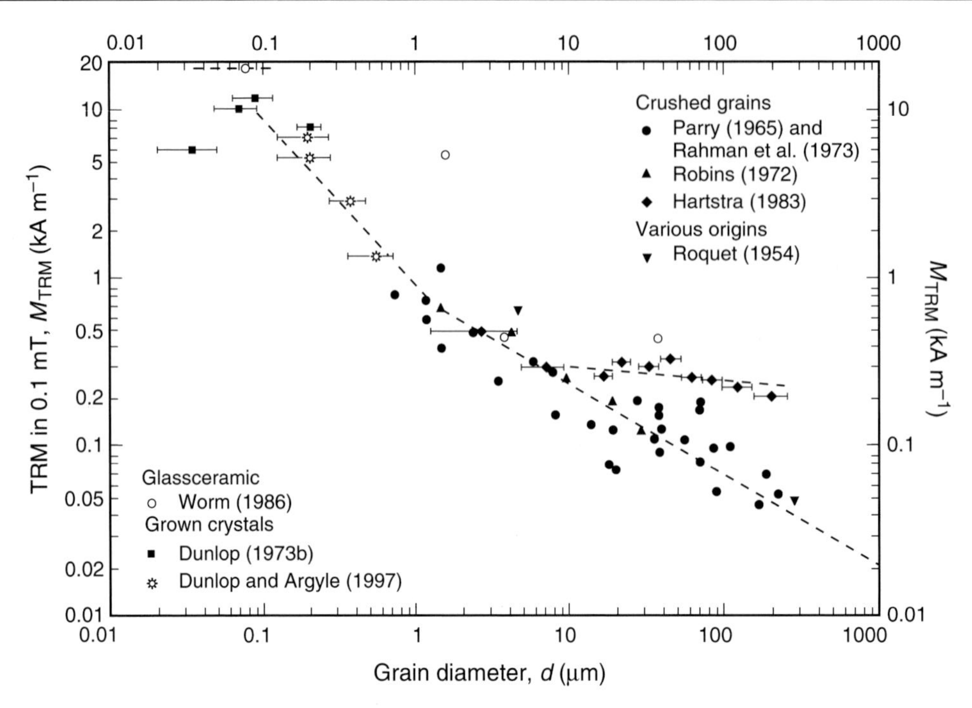

Figure M129 Weak-field TRM as a function of magnetite grain size d. M_{TRM} decreases approximately as d^{-1} between 0.1 and ≈1 μm and less strongly as $d^{-0.55}$ above 1 μm (after Dunlop and Argyle, 1997).

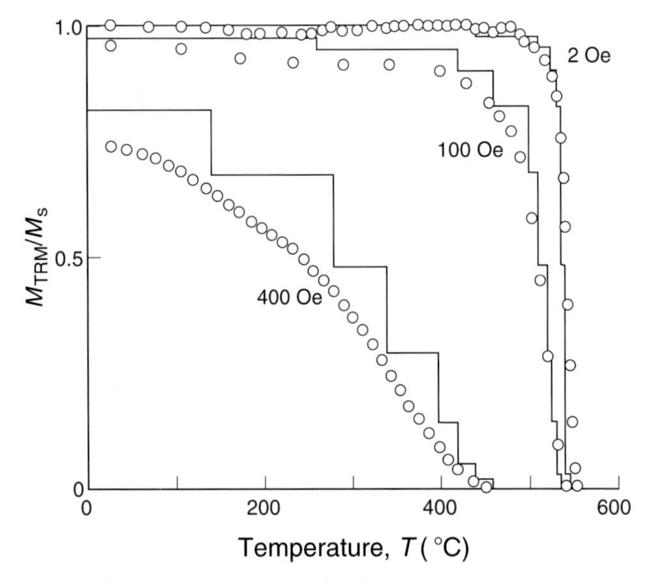

Figure M130 Continuous cumulative TRM spectra measured during cooling of single domain magnetite (points and solid curves) compared with theoretical stepwise spectra for $H = 2$, 100, and 400 Oe (after Sugiura, 1980).

Dodson and McClelland-Brown (1980) and Halgedahl *et al.* (1980) predicted that the intensity of TRM in SD grains should increase for longer cooling times, possibly by as much as 40% between laboratory and geological settings. Experiments by Fox and Aitken (1980) for baked clay samples containing SD magnetites showed that there was about a 7% decrease in the intensity of TRM when the cooling time changed from 2.5 h to 3 min.

Partial TRM and paleofield intensity determination

According to Eq. (4), T_B depends on grain volume V and microscopic coercivity H_K. Any real rock has a distribution $f(V, H_K)$ of both V and H_K, and as a result will have a spectrum of blocking temperatures. Therefore, the total TRM of a rock will consist of a spectrum of partial TRMs, each carried by grains with similar (V, H_K). When a rock cools from T_1 to T_2 in a field $\boldsymbol{H}$, SD ensembles with $T_2 \leq T_B \leq T_1$ pass from the unblocked to the blocked condition, acquiring a partial TRM, $\boldsymbol{M}_{PTRM}$ $(T_1, T_2, \boldsymbol{H})$.

Thellier (1938) showed that SD partial TRMs follow experimental laws of additivity, reciprocity, and independence. *Independence*: Partial TRMs acquired in different temperature intervals are mutually independent in direction and intensity. Each partial TRM disappears over its own blocking-temperature interval. *Additivity*: Partial TRMs produced by the same $\boldsymbol{H}$ have intensities that are additive. This is expected theoretically because the blocking-temperature spectrum can be decomposed into nonoverlapping fractions, each associated with one of the partial TRMs. The total TRM is the sum of partial TRMs covering the entire blocking-temperature interval from T_C to room temperature because each partial TRM contains a unique part of the T_B spectrum of the total TRM. *Reciprocity*: The partial TRM acquired between T_1 and T_2 during cooling in $\boldsymbol{H}$ is thermally demagnetized over the interval (T_2, T_1) when heated in zero field. In other words, the blocking and unblocking temperatures are identical for weak fields.

Since partial TRM acquired at T_B during cooling is erased at T_B during heating, one can replace the natural remanent magnetization (NRM) M_{NRM} of a rock in a stepwise fashion with a laboratory TRM. This is done in practice by a series of double heatings, the first in zero field and the second in a known laboratory field H_{LAB}. If the NRM of a rock was acquired as a TRM in an ancient geomagnetic field H_A, M_{NRM} should be proportional to H_A. The intensity of the ancient geomagnetic field is then given by

$$H_A = M_{NRM}H_{LAB}/M_{TRM},$$

each pair of heating steps giving an independent estimate of H_A. This is the Thellier and Thellier (1959) method of paleointensity determination as modified by Coe (1967).

Multidomain TRM

In large grains containing many domain walls, TRM acquisition results from several processes: domain wall pinning, domain nucleation or denucleation, and nucleation failure. A multidomain grain contains crystal imperfections such as inclusions, voids, and line defects such as dislocations, all of which create surrounding stress fields. These defects tend to pin domain walls. Voids and inclusions reduce the volume of a wall and create magnetic poles, affecting both the wall energy and demagnetizing energy. Dislocations pin walls by the magnetoelastic interaction between their stress fields and the spins in the wall, which are rotated out of easy axes (Özdemir and Dunlop, 1997).

Pinned domain walls can be thermally activated just as SD moments can. Wall displacements in response to a weak field applied at high temperatures, like reversals of SD grains, can be frozen in by cooling to room temperature, resulting in TRM. The Néel (1955), Everitt (1962), and Schmidt (1973) multidomain TRM theories consider the wall-defect interaction and the field dependence of the blocking temperature. These theories follow Néel (1949) SD theory fairly closely except that V becomes the volume of one Barkhausen jump of a domain wall and H_K becomes the critical field for such a Barkhausen jump.

McClelland and Sugiura (1987) observed that during cooling below the blocking range, after H had been zeroed, multidomain partial TRM did not remain constant but decreased in real terms, i.e., $M(T)/M_S(T)$. This observation compromises the concept of the blocking temperature, because evidently, magnetization can partially relax during cooling below T_B. Later Shcherbakov *et al.* (1993) developed a kinetic model that invokes nucleation or denucleation of domain walls as a way of explaining changes in $M(T)$ below T_B.

The only direct observations of domain nucleation during zero-field warming or cooling are by Heider *et al.* (1988) and Ambatiello *et al.* (1999) for magnetite and by Metcalf and Fuller (1987) and Halgedahl and Fuller (1983) for titanomagnetite. In each case, any changes in the number or widths of domains occurred either close to T_C or in a range of T where anisotropy changes rapidly, just above room temperature in magnetite. In intermediate T ranges, where most observations of partial TRM relaxation have been made, there is no evidence for nucleation.

A related phenomenon is nucleation failure. Titanomagnetite grains field-cooled from T_C sometimes nucleate fewer than the equilibrium number of domain walls, and occasionally none at all (Halgedahl and Fuller, 1980). The latter structure is of particular interest because single-domain TRM is so intense that a small fraction of such grains in metastable SD states could account for multidomain TRM. In particular, the size dependence of TRM in grains larger than SD size is then explicable.

There are fundamental differences between SD and most MD TRM models. The internal demagnetizing field $\boldsymbol{H}_D = -N\boldsymbol{M}$ favors a configuration of domain walls with minimum net moment. Loosely pinned domain walls may undergo a series of Barkhausen jumps during heating and cooling. Therefore, there is no single blocking temperature for a given domain wall or for a grain containing many walls since a jump by one wall alters the internal field at the location of every other wall. This is an alternative way of explaining TRM relaxation below T_B.

Néel and Schmidt two-domain TRM theories

Néel's (1955) 2D theory is based on the temperature dependence of the wall-defect interaction and the wall displacement driven by the internal demagnetizing field $\boldsymbol{H}_D = -N\,\boldsymbol{M}$ where $\boldsymbol{M}$ is the local magnetization vector. A domain wall, which has been displaced by H, will reequilibrate its position at each new temperature during cooling under the influences of $\boldsymbol{H}_D$ and of local potential wells created by interaction

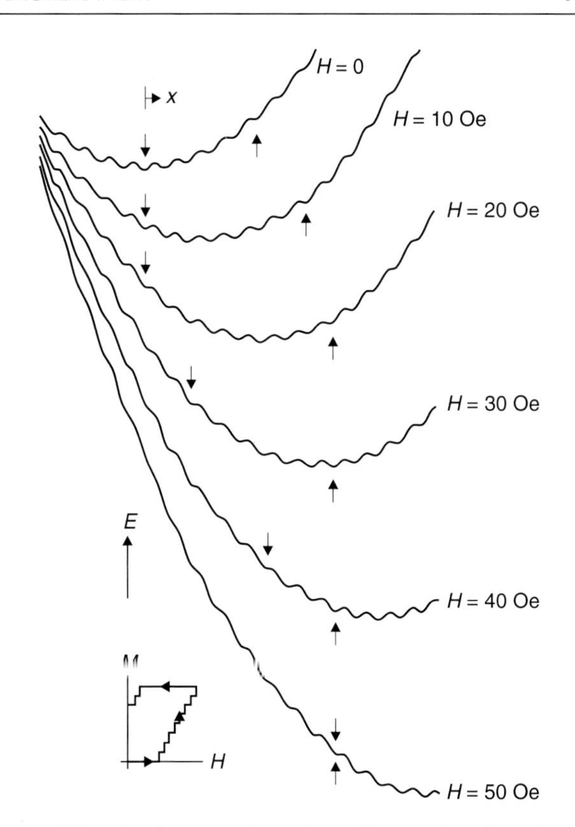

Figure M131 Total energy, $E_W + E_D + E_H$, as a function of wall displacement x for various fields (after Schmidt, 1973). Downward arrows show successive local energy minima in which the wall is trapped as H increases from 0 to 50 Oe. Upward arrows indicate minima where the wall is pinned as H decreases from 50 to 0 Oe. The wall occupies different minima in increasing and decreasing fields of the same strength, giving rise to magnetic hysteresis and remanence.

between the domain walls and lattice defects (Dunlop and Özdemir, 1997). The pinning strength is measured by the microcoercivity H_C and determined by the barrier height between adjacent local energy minimums (LEMs). Figure M131 shows the total energy in a two-domain grain with many identical wall energy (E_W) barriers, each with the same coercivity H_C for positive and negative jumps. The central position of the wall is favored by the parabolic energy well due to the demagnetizing energy E_D. As the applied $\boldsymbol{H}$ increases, the parabolas are tilted to the right by the field energy E_H and the wall jumps from one LEM to another in a series of Barkhausen jumps. Repeated jumps of the wall generate the ascending part of the hysteresis loop, which has a slope $1/N$. The barriers to wall motion are not symmetric in increasing and decreasing $\boldsymbol{H}$. As $\boldsymbol{H}$ decreases, the wall is pinned in a different set of LEMs and generates a descending loop with the same slope $1/N$. When $\boldsymbol{H}$ becomes zero, the grain is left with a displaced wall and a net moment, which is an isothermal remanent magnetization.

As the temperature decreases, the demagnetizing field and the barriers between wells grow. At the blocking temperature T_B, the potential barriers E_W begin to grow more rapidly than the demagnetizing field H_D pushing the wall back toward the demagnetized state and the wall is trapped. TRM is frozen in. The basic result of Néel's 2D theory is that

$$M_{TRM} = 2H^{1/2}H_C^{1/2}/N. \qquad \text{(Eq. 6)}$$

Equation (6) predicts absolute TRM intensity at room temperature. The TRM is blocked by the growth of energy barriers to wall motion and thermal fluctuations are ignored. This is referred to as field-blocked TRM.

Néel (1955) pointed out that field blocking of TRM occurs very close to T_C if H is small. At such high temperatures, thermal fluctuations can unpin a domain wall below its blocking temperature T_B. Néel introduced a thermal fluctuation field H_F such that a wall cannot be blocked until

$$H_C(T_B) = H_F, \quad \text{(Eq. 7)}$$

where $H_C(T_B)$ is the thermal fluctuation blocking temperature. TRM intensity in the presence of thermal fluctuations is given by

$$M_{TRM} = H_F^{1/2} H_C^{1/2} H/N. \quad \text{(Eq. 8)}$$

Thermally blocked TRM should be proportional to H as observed experimentally for small fields.

Schmidt's (1973) MD model is based on the temperature variation of total energy in a two-domain grain with a single wall (Figure M131). The model has the same physical basis as Néel's 2D theory, but the coercivity appears only as a derived quantity. The total energy has contributions due to the interactions between the domain wall and the applied field, the demagnetizing field, and lattice defects. The temperature dependence of wall-defect interaction is contained in a term m^p, where m is the reduced domain magnetization $m(T) = M/M_S(T)$. The index p takes a value between 2 and 10 depending on the relative importance of magnetostriction or magnetocrystalline anisotropy in the wall pinning. The basic results of the Schmidt model are that

$$H_C(T) \propto m(T)^{p-1}, \quad \text{(Eq. 9)}$$

$$M_{TRM} \propto H^{1-1/(p-1)}, \quad \text{(Eq. 10)}$$

$$M_{TRM} \propto H_C^{1/(p-1)}. \quad \text{(Eq. 11)}$$

These are identical to Néel's predictions (Eq. (6)) when $p = 3$.

Stacey multidomain TRM theory

Stacey (1958) considered the blocking temperature T_B to be independent of the applied field and neglected the interaction between the domain wall and lattice defects. At high temperatures, $T_B < T < T_C$, the walls are highly mobile and take up positions in which the internal field $H_I = H - NM$ is zero. Above T_B, the magnetization is given by $M = H/N$. During cooling below T_B, the walls' displacements are frozen, so that M changes only by the reversible increase in $M_S(T)$. At room temperature, the TRM is given by

$$M_{TRM} = H\, M_S/[NM_{SB}(1 + N\chi_0)], \quad \text{(Eq. 12)}$$

where χ_0 is the initial volume susceptibility. Equation (12) seemingly predicts the linear dependence of TRM on the applied field observed experimentally for weak fields. In reality, as Néel (1955) showed, blocking temperature T_B is itself dependent on H. TRM as given by Eq. (12) actually varies approximately as $H^{1/2}$, a dependence observed experimentally only when $H > 1$ mT.

Fabian phenomenological TRM model

Fabian (2000) has proposed a phenomenological approach in which a sample is described in terms of a distribution function $\chi(T_B, T_{UB})$. This formulation establishes no direct link between (T_B, T_{UB}) and physical properties like grain size and coercivity, but recognizes that for MD grains $T_B \neq T_{UB}$ in general, i.e., Thellier reciprocity does not hold. With this simple approach, it is possible to explain many observed properties of multidomain partial TRMs and to justify the kinematic equation for partial TRM relaxation during cooling or heating (Shcherbakov *et al.*, 1993). Although not a theory of TRM acquisition in the fundamental sense, the Fabian model is useful in describing and connecting a body of experimental data and making predictions about the TRM properties of a particular sample based on its measured $\chi(T_B, T_{UB})$.

Experimental multidomain TRM

There are relatively few testable predictions of multidomain TRM theories. The prediction that TRM intensity should increase with increasing H_C (Eqs. (6), (8), (11)) was verified in a general way in Néel's original paper and the test has not been improved upon since. The predicted dependence on applied field H has received more attention. Tucker and O'Reilly (1980) showed that in the low-field region ($H \leq 1$ kA m^{-1}, where 80 A m^{-1} $\approx$ 1 Oe = 0.1 mT), the TRM of large titanomagnetite crystals, as predicted by Eq. (8), was more or less proportional to H (Figure M132). For intermediate field strengths, the field dependence was as a power of $H < 1$ (Eqs. (6) or (10)), and saturation occurred around 10 kA m^{-1}. Similar results for magnetite and magnetite-bearing rocks were reported by Dunlop and Waddington (1975) and for pyrrhotite by Menyeh and O'Reilly (1998).

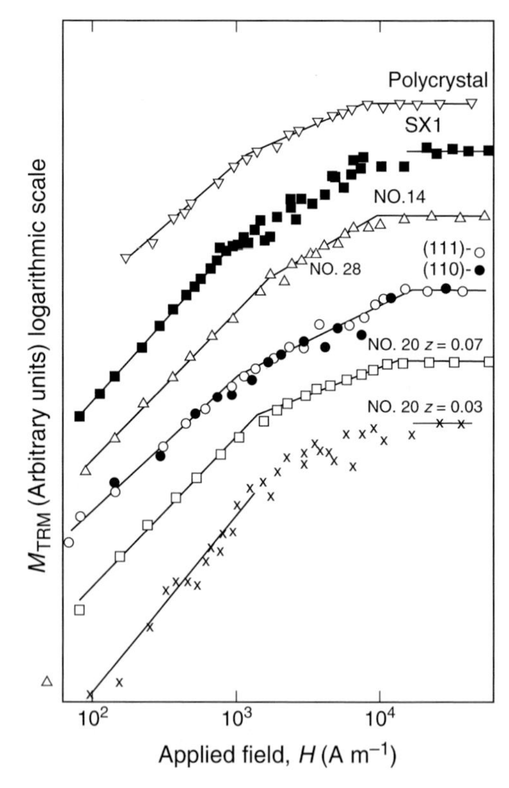

Figure M132 TRM acquisition curves for millimeter-size single crystals of titanomagnetite with composition $Fe_{2.4}Ti_{0.6}O_4$ (TM60), plotted bilogarithmically (after Tucker and O'Reilly, 1980). The experimental data have been fitted by a series of linear segments corresponding to power law dependences of TRM on applied field. TRM intensity is proportional to H for fields below about 2 mT, then increases approximately as $H^{1/2}$ until it saturates at the value $M_{RS} = H/N$, in agreement with Néel's (1955) theory of TRM due to domain wall displacement limited by self-demagnetization.

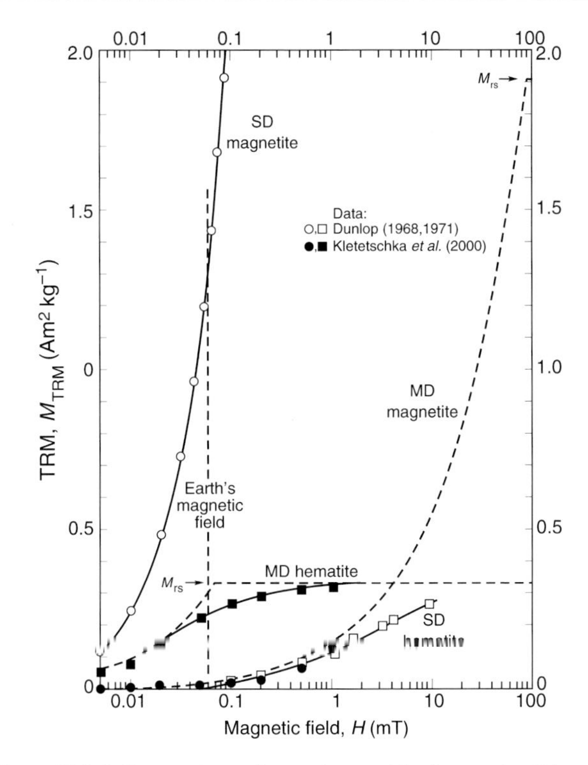

Figure M133 Comparison of experimental (points and solid curves) and theoretical (dashed curves) TRM values for SD and MD hematites and magnetites (Dunlop and Kletetschka, 2001). The contrast in TRM intensities for two minerals is due to the internal demagnetizing field $H_D = -NM$. At saturation, $H_D \approx 200$ mT for magnetite, but for hematite $H_D \approx 1$ mT, making it much easier for a small field like the Earth's to push the magnetic domain walls to their limiting positions.

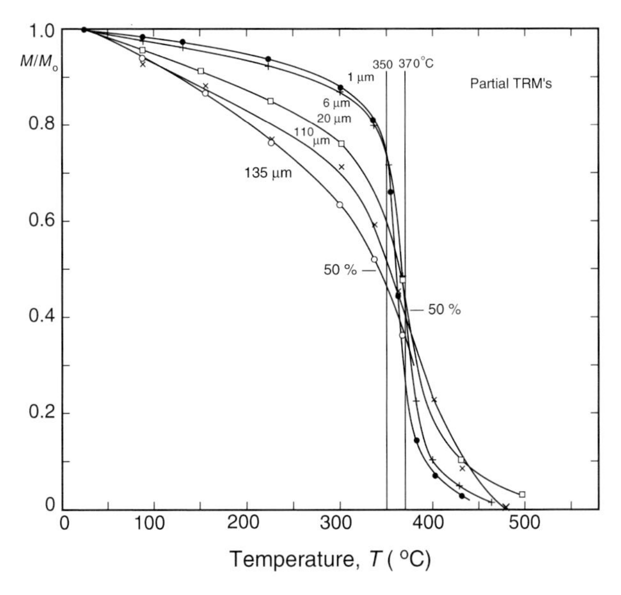

Figure M134 Stepwise thermal demagnetization of partial TRMs produced by a small field applied during cooling from 370 to 350°C in magnetites ranging from small PSD to MD size. About 50%–90% of the remanence unblocks at temperatures below or above the PTRM blocking temperature range. Tailing of thermal demagnetization curves is not confined to large MD grains but is significant even for fairly small PSD grains (after Dunlop and Özdemir, 2001).

An interesting sidelight on multidomain TRM behavior is the fact that for H of the order of the Earth's magnetic field, weakly magnetic hematite ($M_S \approx 2$ kA/m) in a multidomain state has a much stronger TRM than MD magnetite ($M_S = 480$ kA m^{-1}) or SD hematite (Figure M133). The reason is that the weak M_S of hematite results in a negligible demagnetizing field H_D, so that even a small H is sufficient to drive walls to their limiting positions and saturate the TRM (Dunlop and Kletetschka, 2001). MD magnetite does not approach saturation TRM except in fields 3 orders of magnitude larger because of its correspondingly greater H_D.

The model of Fabian (2000) was a response to recent experimental work on multidomain partial TRMs. The thermal demagnetization data of Figure M134 are one example. Partial TRMs produced by applying H over a narrow blocking interval T_B = 370–350°C are demagnetized over increasingly broad intervals of unblocking temperature T_{UB} as the grain size increases. While partial TRMs of 1 and 6 μm magnetites demagnetized mainly over the original T_B interval, partial TRMs of 110 and 135 μm grains demagnetized almost entirely outside the 370°C–350°C interval. This almost total violation of reciprocity led Fabian (2000) to treat T_B and T_{UB} as independent variables.

Özden Özdemir

Bibliography

Ambatiello, A., Fabian, K., and Hoffmann, V., 1999. Magnetic domain structure of multidomain magnetite as a function of temperature: observations by Kerr microscopy. *Physics of the Earth and Planetary Interiors*, **112**: 55–80.

Coe, R.S., 1967. The determination of paleointensities of the earth's magnetic field with emphasis on mechanisms which could cause non-ideal behaviour in Thellier's method. *Journal of Geomagnetism and Geoelectricity*, **19**: 157–179.

Clauter, D.A., and Schmidt, V.A., 1981. Shifts in blocking temperature spectra for magnetite powders as a function of grain size and applied magnetic field. *Physics of the Earth and Planetary Interiors*, **26**: 81–92.

Day, D., 1977. TRM and its variation with grain size. *Journal of Geomagnetism and Geoelectricity*, **29**: 233–265.

Dodson, M.H., and McClelland-Brown, E., 1980. Magnetic blocking temperatures of single-domain grains during slow cooling. *Journal of Geophysical Research*, **85**: 2625–2637.

Dunlop, D.J., 1973. Thermoremanent magnetization in submicroscopic magnetite. *Journal of Geophysical Research*, **78**: 7602–7613.

Dunlop, D.J., and Argyle, K.S., 1997. Thermoremanence, anhysteretic remanence and susceptibility of submicron magnetites: nonlinear field dependence and variation with grain size. *Journal of Geophysical Research*, **102**: 20199–20210.

Dunlop, D.J., and Kletetschka, G., 2001. Multidomain hematite: a source of planetary magnetic anomalies? *Geophysical Research Letters*, **28**: 3345–3348.

Dunlop, D.J., and Özdemir, Ö., 1997. *Rock Magnetism: Fundamentals and Frontiers*. Cambridge and New York: Cambridge University Press.

Dunlop, D.J., and Özdemir, Ö., 2001. Beyond Néel theories: thermal demagnetization of narrow-band partial thermoremanent magnetizations. *Physics of the Earth and Planetary Interiors*, **126**: 43–57.

Dunlop, D.J., and Waddington, E.D., 1975. The field dependence of thermoremanent magnetization in igneous rocks. *Earth and Planetary Science Letters*, **25**: 11–25.

Dunlop, D.J., and West, G.F., 1969. An experimental evaluation of single domain theories. *Reviews of Geophysics*, **7**: 709–757.

Everitt, C.W.F., 1961. Thermoremanent magnetization. I. Experiments on single domain grains. *Philosophical Magazine*, **6**: 713–726.

Everitt, C.W.F., 1962. Thermoremanent magnetization. III. Theory of multidomain grains. *Philosophical Magazine*, **7**: 599–616.

Fabian, K., 2000. Acquisition of thermoremanent magnetization in weak magnetic fields. *Geophysical Journal International*, **142**: 478–486.
Fox, J.M.W., and Aitken, M.J., 1980. Cooling-rate dependence of thermoremanent magnetisation. *Nature*, **283**: 462–463.
Halgedahl, S.L., and Fuller, M., 1980. Magnetic domain observations of nucleation processes in fine particles of intermediate titanomagnetite. *Nature*, **288**: 70–72.
Halgedahl, S.L., and Fuller, M., 1983. The dependence of magnetic domain structure upon magnetization state with emphasis on nucleation as a mechanisms for pseudo-single-domain behaviour. *Journal of Geophysical Research*, **88**: 6506–6522.
Halgedahl, S.L., Day, R., and Fuller, M., 1980. The effect of cooling rate on the intensity of weak-field TRM in single domain magnetite. *Journal of Geophysical Research*, **85**: 3690–3698.
Heider, F., Halgedahl, S.L., and Dunlop, D.J., 1988. Temperature dependence of magnetic domains in magnetite crystals. *Geophysical Research Letters*, **15**: 499–502.
McClelland, E., and Sugiura, N., 1987. A kinematic model of TRM acquisition in multidomain magnetite. *Physics of the Earth and Planetary Interiors*, **46**: 9–23.
Menyeh, A., and O'Reilly, W., 1998. Thermoremanence in monoclinic pyrrhotite particles containing few domains. *Geophysical Research Letters*, **25**(18): 3461–3464.
Metcalf, M., and Fuller, M., 1987. Domain observations of titanomagnetites during hysteresis at elevated temperatures and thermal cycling. *Physics of the Earth and Planetary Interiors*, **46**: 120–126.
Néel, L., 1949. Théorie du traînage magnétique des ferromagnétiques en grains fins avec applications aux terres cuites. *Annales de Geophysique*, **5**: 99–136.
Néel, L., 1955. Some theoretical aspects of rock magnetism. *Advances in Physics*, **4**: 191–243.
Özdemir, Ö., and Dunlop, D.J., 1997. Effect of crystal defects and internal stress on the domain structure and magnetic properties of magnetite. *Journal of Geophysical Research*, **102**: 20211–20224.
Özdemir, Ö., and Dunlop, D.J., 2002. Thermoremanence and stable memory of single-domain hematite. *Geophysical Research Letters*, **29**:doi:10.1029/2002GL015597.
Özdemir, Ö., and O'Reilly, W., 1982. An experimental study of thermoremanent magnetization acquired by synthetic monodomain titanomaghemites. *Journal of Geomagnetism and Geoelectricity*, **34**: 467–478.
Schmidt, V.A., 1973. A multidomain model of thermoremanence. *Earth and Planetary Science Letters*, **20**: 440–446.
Shcherbakov, V.P., McClelland, E., and Shcherbakova, V.V., 1993. A model of multidomain thermoremanent magnetization incorporating temperature-variable domain structure. *Journal of Geophysical Research*, **98**: 6201–6216.
Stacey, F., 1958. Thermoremanent magnetization (TRM) of multidomain grains in igneous rocks. *Philosophical Magazine*, **3**: 1391–1401.
Stacey, F., and Banerjee, S.K., 1974. *The Physical Principles of Rock Magnetism*. Elsevier, Amsterdam.
Stoner, E.C., and Wohlfarth, E.P., 1948. A mechanism of magnetic hysteresis in heterogeneous alloys. *Philosophical Transactions of the Royal Society of London*, **A240**: 599–642.
Sugiura, N., 1980. Field dependence of blocking temperature of single-domain magnetite. *Earth and Planetary Science Letters*, **46**: 438–442.
Thellier, E., 1938. Sur l'aimantation des terres cuites et ses applications géophysiques. *Annales de. l'Institut de Physique du Globe, Université de Paris*, **16**: 157–302.
Thellier, E., and Thellier, O., 1959. Sur l'intensité du champ magnétique terrestre dans le passé historique et géologique. *Annales de Géophysique*, **15**: 285–376.
Tucker, P., and O'Reilly, W., 1980. The acquisition of thermoremanent magnetization by multidomain single-crystal titanomagnetite. *Geophysical Journal of the Royal Astronomical Society*, **60**: 21–36.
Winklhofer, M., Fabian, K., and Heider, F., 1997. Magnetic blocking temperatures of magnetite calculated with a three-dimensional micromagnetic model. *Journal of Geophysical Research*, **102**: 22695–22709.

Cross-references

Archeomagnetism
Magnetic Anisotropy, Sedimentary Rocks and Strain Alteration
Magnetic Domain
Magnetic Susceptibility
Magnetization, Chemical Remanent (CRM)
Magnetization, Isothermal Remanent (IRM)
Magnetization, Natural Remanent (NRM)
Magnetization, Viscous Remanent (VRM)

MAGNETIZATION, THERMOREMANENT, IN MINERALS

Thermoremanent magnetization

Perhaps the best understood of the primary magnetizations, of natural rocks and specimens, is thermal remanent magnetization (TRM). Most of the natural rocks are magnetized primarily by the geomagnetic field ($\sim$30000 nT) and acquire natural remanent magnetization (NRM). Magnetic minerals acquire TRM when they are contained within the rock that is cooled in an external magnetic field from temperatures above the minerals' blocking temperatures. Blocking of remanent magnetization at a specific temperature results in locking of a specific direction and intensity of magnetization as it becomes stable on the timescale of the TRM acquisition.

The generally recognized first-order theory of TRM can be applied only to small uniformly magnetized grains (Néel, 1949) and it provides a reasonable explanation for the intensity of TRM vs inducing field. The theory explains the changes of stability of TRM with temperature and with the inducing field, explaining how a rock can maintain a TRM record for billions of years. This theory can also provide an explanation on how the secondary component of magnetization of a rock can be removed from the primary one.

However, besides being useful only for small single-domain (SD) grains it is also less successful in explaining TRM acquisition for more intense ambient fields ($\sim$1 mT). According to Neel's theory, the TRM reaches saturation, in lower fields than actually measured on specimens ($\sim$10 mT). This effect was assigned to particle interaction and long axis dispersions (Stacey and Banerjee, 1974).

In general, the TRM is not carried only by a small single-domain (SD) fraction of magnetic mineral grains. When the volume of the grain is larger ($>$0.5 μm for magnetite mineral) the demagnetizing field (caused by magnetic sources distributed on the surface of the grain) has a slightly different geometry than the pattern of the uniform magnetization (Figure M135). This geometry causes inhomogeneity of magnetization of the larger grains that is replaced by more energetically favorable state containing domain walls bounding volume with reversed magnetic moments.

Neel first attempted to construct the theory for multidomain materials (Néel, 1955). Stacey (1963) and Everitt (1962) applied a concept of Barkhausen discontinuities, small jumps of the domain walls into the new position, believed to be due to crystal imperfections. The resistance of the domain wall motion was thought to be responsible for remanent magnetism in MD grains. As the smaller MD grains were

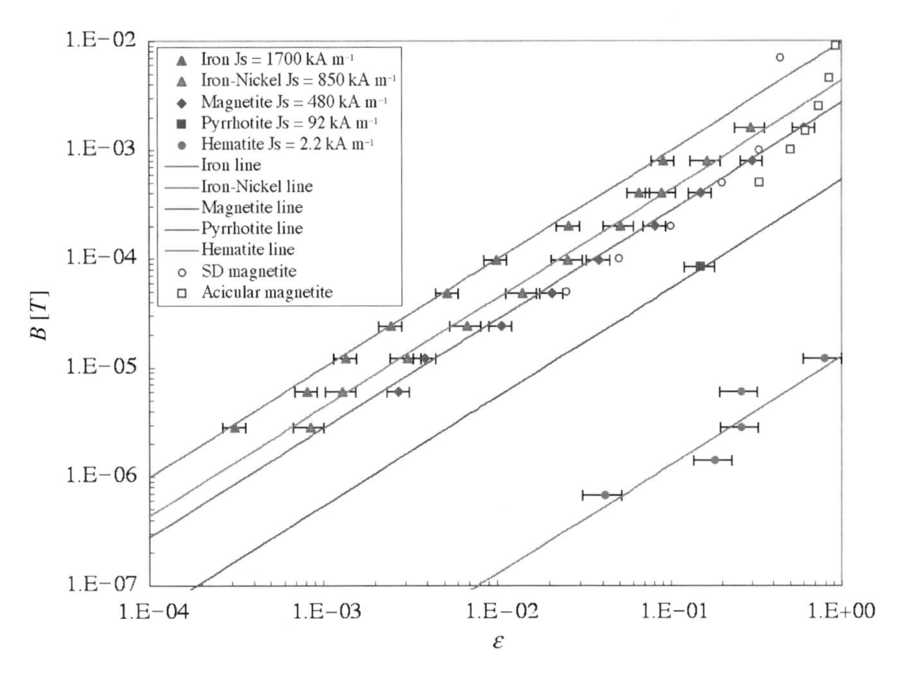

Figure M135 Intensities of the ambient magnetic field B (tesla) against thermoremanent efficiencies ε of hematite (Fe_2O_3), magnetite (Fe_3O_4), iron-nickel alloy (FeNi) and iron (Fe). The pyrrhotite data (Fe_7S_8) are from Dekkers (1989). Straight lines are drawn according to $B = aJ\varepsilon$, where a is a dimensionless constant equal to 0.0046 (see Figure M137a) at 300 K, and $J = \mu_0 J_s$ where μ_0 is permeability of vacuum and J_s is saturation magnetization at 300 K. Magnetization efficiency ε is defined as $\varepsilon = M_{tr}/J_{sr}$, the ratio of thermoremanence to saturation remanence. Single-domain (SD) and acicular (elongated crystal parallel to the applied field) magnetite data are redrawn from Dunlop and Argyle (1997) and Dunlop and West (1969), respectively.

more stable than larger MD grains, Stacey invented the term pseudo-single-domain (PSD) grains for grains slightly larger than SD grains (Stacey, 1963).

Because, in general, small grains of magnetite have 2–3 orders of magnitude larger TRM intensity than larger MD grains. For some time it was thought that the TRM of MD grains is negligible (Hargraves and Young, 1969; Hoye and Evans, 1975). More small grains are present in rocks than apparently visible, resulting from the formation of the iron oxides through oxidation inside the silicates.

The intensity of the remanent magnetization acquired by rocks is determined by an unknown strength of the ambient magnetic field, an unknown magnetic mineral composition, and an unknown temperature history of the sample. Stacey pointed out in his theory of multidomain TRM (Stacey, 1958) that because the demagnetizing energy falls off more slowly with temperature than any other, the condition under which TRM is first acquired is simply the minimization of the internal field. This guarantees that at least at this temperature the TRM is related only to the magnetostatic energy and the demagnetizing energy.

Néel's theory of MD TRM is incomplete since it fails to describe many aspects of pTRM (partial TRM) behavior (Néel, 1955; Shcherbakova *et al.*, 2000). There the blocking occurs at temperature T_b when magnetic coercivity increases high enough to pin domain walls against the demagnetizing field. For TRM of SD grains at room temperature, $M_{tr}(T_r)$, magnetic remanence was frozen in high temperature equilibrium distribution achieved by thermally excited transitions among the different magnetic states. Transitions cease below the T_b, because in the course of cooling, the energy barriers between different magnetization states grow larger than the available thermal energy. For both SD and MD states the resulting magnetization, composed of many magnetic moments, is in the direction of and for mineral specific field range (Kletetschka *et al.*, 2004) proportional to the applied magnetic field B. Efficiency $\varepsilon(T_r)$ of $M_{tr}(T_r)$ of SD grains of saturation remanence $J_{sr}(T_r)$, volume V, and saturation magnetization $J_s(T_b)$ is (Néel, 1949):

$$\epsilon(T_r) = \frac{M_{tr}(T_r)}{J_{sr}(T_r)} = \tanh\left(\frac{\mu V J_s(T_b)B}{kT_b}\right), \qquad \text{(Eq. 1)}$$

with $\mu_0 = 4\pi \times 10^{-7}$, and $k = 1.38 \times 10^{-23}$ J K^{-1}.

On the timescale 50–100 s:

$$\left(\frac{\mu_0 V J_s(T_b) B_c(T_b)}{kT_b}\right) = 2\ln(f_0 t) \approx 50,$$

where $B_c(T_b)$ is a critical field for moment rotations in the absence of thermal energy (microcoercivity), and frequency of moment fluctuation $f_0 \approx 10^9$ s^{-1}. Therefore, from Eq. (1) one can derive for small fields ($\epsilon \ll 1$):

$$\epsilon(\mathrm{T_r}) = \frac{50B}{B_c(T_b)} \qquad \text{(Eq. 2)}$$

In most fine-grained magnetic material, a typical efficiency $\epsilon(T_r)$ of thermoremanent magnetization $M_{tr}(T_r)$ acquired in the geomagnetic field is about 1% (Wasilewski, 1977, 1981; Cisowski and Fuller, 1986; Kletetschka *et al.*, 2000a). This small efficiency is consistent with the $M_{tr}(T_r)$ acquisition curves for magnetite (Dunlop and Waddington, 1975; Tucker and O'Reilly, 1980; Özdemir and O'Reilly, 1982) with grain sizes covering the range from the single-domain (SD) to multidomain (MD) magnetic states. However, $M_{tr}(T_r)$ experiments with hematite (Kletetschka *et al.*, 2000b,c; Dunlop and Kletetschka, 2001; Kletetschka *et al.*, 2002) showed ε >10%.

Experimental TRM constrains for theory

The following part is largely reiteration of the results from Kletetschka *et al.* (2004). There in an attempt to reconcile the contrast between TRM acquisition of hematite and magnetite series of magnetic M_{tr}

acquisitions performed using distinct magnetic materials (Kletetschka *et al.*, 2004): iron (Fe), iron-nickel (FeNi), magnetite (Fe_3O_4), hematite (α-Fe_2O_3), and resistance wires MWS-294R and ALLOY52. Figure M135 shows the field B required to reach efficiency $\epsilon(T_r)$ of the $M_{tr}(T_r)$ acquisition for equidimensional samples and a literature sample of acicular magnetite (Dunlop and Argyle, 1997) in which the crystals are highly elongated parallel to the applied field. Data near and at saturation, where the simple power law breaks down (see Figure M136), were excluded from the data set. The data set includes literature data for pyrrhotite (Dekkers, 1989). Remarkably, each mineral is restricted to its own line with the unit slope in the log$B$-logϵ space. Another important feature reported in Kletetschka *et al.* (2004) is that the larger the $J_s(T_r)$ (see legend) the larger the field B required to achieve a predefined efficiency level ϵ.

An increase of the minerals' $J_s(T_r)$ is equivalent to a similar increase in opposing demagnetizing field $H_d(T_r)$ (Dunlop and Özdemir, 1997; Kletetschka *et al.*, 2000b, c; Dunlop and Kletetschka, 2001) as well as critical fields B requiring material to reach the saturation magnetization at T_b. The demagnetizing field at saturation relates to $B_c(T_b)$, above which the energy minima become unstable causing the magnetic moment to irreversibly rotate in an absence of thermal fluctuations. Low $J_s(T_r)$ value is associated with low value of both $B_c(T_b)$ and H_d and leads to a large critical SD size and large magnetic domain wall spacing while large $J_s(T_r)$ implies large $B_c(T_b)$ (or H_d) causing a fine scale of the individual domains. Other effects, like magnetostriction, anisotropy, and exchange constants, may also cause changes in the overall magnetic domain size.

Data in Figure M135 are representative of multidomain (MD) magnetic materials (1 mm grain size). However, the $M_{tr}(T_r)$ varies with the grain size according to the domain type (Kletetschka *et al.*, 2004). For example, MD magnetite has $M_{tr}(T_r)$ that increases with decreasing grain size (Dunlop, 1990). Similar grain size dependence has been observed at saturation for $J_{sr}(T_r)$ (Dunlop, 1990) ($\epsilon = 1$). Thus, efficiency $\epsilon(T_r)$ of $M_{tr}(T_r)$ rather than just $M_{tr}(T_r)$ reduces the grain size dependence to a minimum, and a line separation in the log$B$-logϵ plot can be used to identify the magnetic mineralogy in ideal circumstances.

The insensitivity of the log$B$-log$\epsilon(T_r)$ plot to various grain sizes is illustrated in Figure M135, where the $M_{tr}(T_r)$ efficiency for SD magnetite (literature data; Dunlop and Argyle, 1997) correlates with MD magnetites; this breaks down when mineral size becomes so small that it is near or in the superparamagnetic size range (Dunlop and Argyle, 1997). The grain size independence is resolved in Figure M136 where we have literature data of various $M_{tr}(T_r)$ acquisitions of titanomagnetite with disparate magnetic domain states identified by the specific grain sizes (Dunlop and Waddington, 1975; Tucker and O'Reilly, 1980; Özdemir and O'Reilly, 1982). Despite much stronger $M_{tr}(T_r)$ of fine vs large magnetic grains, all sizes appear to have identical acquisitions when normalized by saturation remanence $J_{sr}(T_r)$.

When neglecting effects near saturation ($\epsilon < 0.3$), this linear dependence predicts approximate maximum values of the magnetic field that can be recorded by a specific material (Figure M135). Near the saturation, the magnetization is not linear with the applied field (Figure M136) owing to Eq. (1) when $\epsilon \approx 1$. Thus, the magnetic fields at which $\epsilon \approx 1$ should be the magnetic fields that define the values of intrinsic $B_c(T_b)$ for the specific mineral. Knowledge of $B_c(T_b)$ fields in principle can be used for dating of the magnetization according to known magnetization viscous decay curves (Heller and Markert, 1973; Borradaile, 1996). $B_c \propto M_s$ for shape anisotropy, $B_c \propto \lambda/M_s^n$ for magnetoelastic anisotropy (Syono and Ishikawa, 1963b; Moskowitz, 1993) (for $n > 2$) and $B_c \propto K/M_s^n$ for crystalline anisotropy (Syono and Ishikawa, 1963a; Fletcher and O'Reilly, 1974) (for $n > 8$) where n is the experimentally determined exponent. Both K and λ go to zero much faster than M_s when approaching Curie temperature T_c. Energy minimum, related to magnetic ordering, becomes shallower when approaching T_b due to thermal fluctuations (Stacey, 1958). Thus, for the purpose of magnetic remanence blocking near T_c we may consider only the shape anisotropy: $B_k \propto M_s$. The distribution of demagnetization field vectors (tensors in general; Dunlop and Özdemir, 1997) relates to $B_c(T_b)$, the nature of the resulting $M_{tr}(T_r)$ and the ϵ (T_r) dependencies. The $B_c(T_b)$ fields are small in minerals with low $J_s(T_r)$ causing them to reach saturation ($\epsilon(T_r) = 1$) in much lower applied fields B

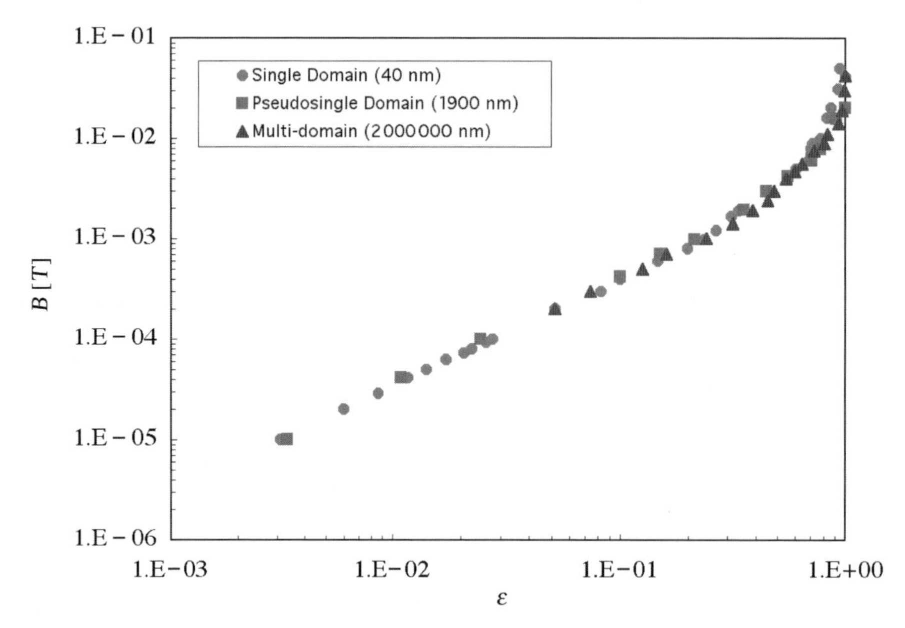

Figure M136 Acquisition fields are plotted against thermoremanent efficiency for contrasting domain states of titanomagnetite. Data for multidomain and pseudosingle-domain mineral are from Tucker and O'Reilly (1980). Data for single-domain minerals are from Özdemir and O'Reilly (1982).

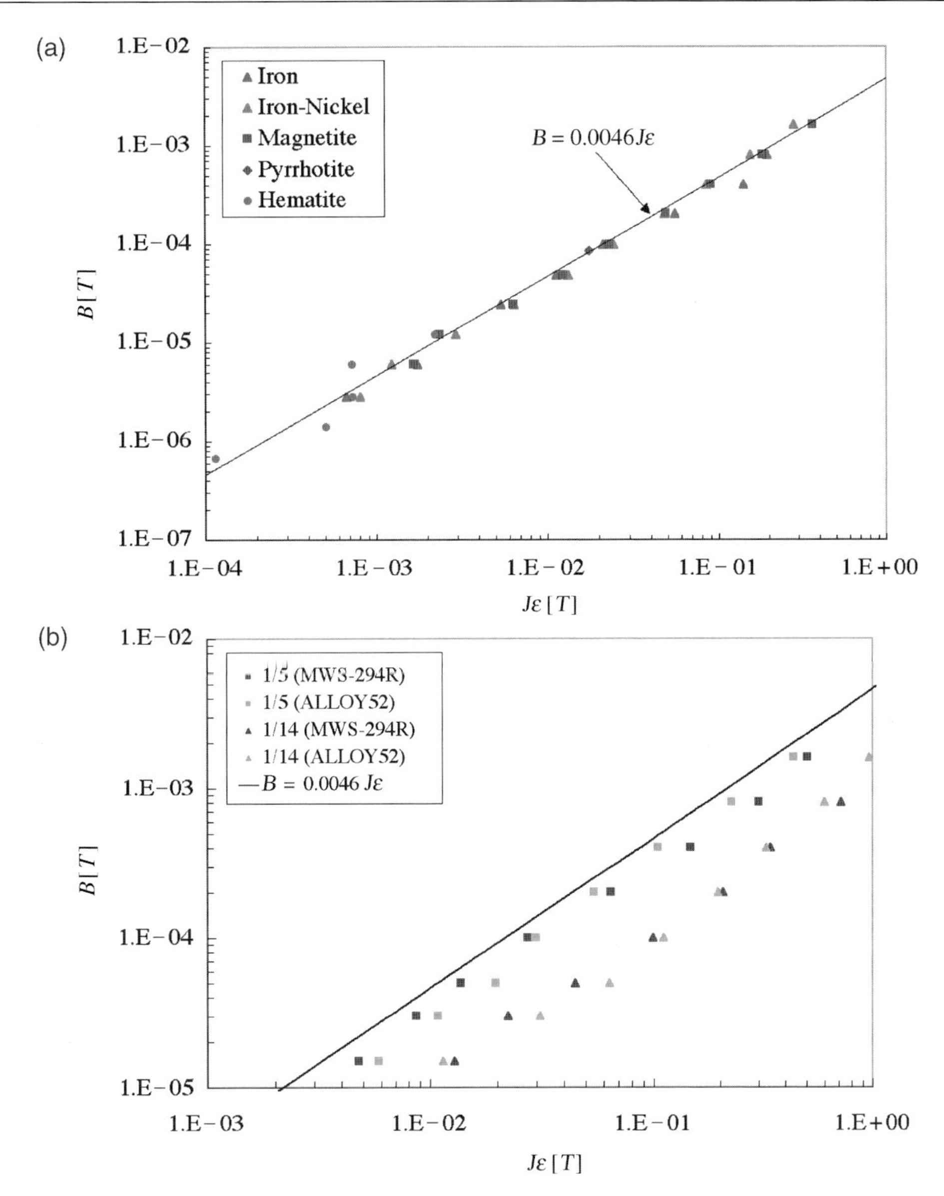

Figure M137 Magnetic acquisition fields are plotted against $J\epsilon$, which is the saturation magnetization $J = \mu_0 J_s$ multiplied by efficiency ϵ of various materials at 300 K. (a) Equidimensional grains of iron, iron-nickel, magnetite, pyrrhotite, and hematite define a straight line that is a result of a linear fit to all of the data. For J_s at 300 K this fit has the form of $B = (4.6 \pm 0.3) \times 10^{-3} J\epsilon$. The linear regression coefficient is $R = 0.97$. (b) Effect of shape (nonequidimensional crystals) on thermoremanent acquisition fields for wire materials (MWS-294R and ALLOY52) with small (1/5) and large (1/14) predefined length to diameter ratios compared with the predicted acquisition (solid line) for equidimensional materials. Measurements were made with wires aligned parallel to the applied field.

(Kletetschka *et al.*, 2000c; Dunlop and Kletetschka, 2001). Larger $B_c(T_b)$ in minerals with large $J_s(T_r)$ creates larger resistance against acquisition of $M_{tr}(T_r)$ and requires larger magnetizing fields to achieve the saturation ($\epsilon(T_r) = 1$). For example, because magnetoelastic and crystalline constants go to zero much faster than M_s close to T_C, hematite has low $B_c(T_b)$ (due to shape anisotropy) at the point at which TRM is acquired in contrast to its high B_c at room temperature caused by high magnetoelastic anisotropy.

The fundamental role of $J_s(T_r)$ in mineral specific $M_{tr}(T_r)$ acquisition can be crystal clear by taking the data from Figure M135, multiplying the magnetic efficiency $\epsilon(T_r)$ by $J(T_r) = \mu_0 J_s(T_r)$ (Figure M137a). Remarkably the resulting data set completely eliminates the effect of the demagnetizing field during the $M_{tr}(T_r)$ acquisition. Figure M137a suggests that mineral $M_{tr}(T_r)$ acquisitions can be in general approximated (linear regression coefficient $R = 0.97$) by following a linear fit:

$$B = a(T)J(T)\epsilon(T_r), \qquad \text{(Eq. 3)}$$

where $a = (4.6 \pm 0.3) \times 10^{-3}$ with 95% confidence level for $T = T_r = 300$ K. The product $J_s(T_r)\epsilon(T_r)$ is essentially the $M_{tr}(T_r)$ normalized by the squareness ratio $J_{sr}(T_r)/J_s(T_r)$ of the hysteresis loop. This linear behavior (Figure M137a) indicates that all magnetic minerals should contribute to a planetary thermoremanent magnetic anomaly (e.g., intense magnetic anomalies detected on Mars (Acuña *et al.*, 1999) with the same, squareness $J_{sr}(T_r)/J_s(T_r)$-normalized, $M_{tr}(T_r)$ intensity. Because $J_s(T_r)$

eliminates the mineral dependence observed in Figure M135 produced byvariation in $B_c(T_b)$ in different minerals, $B_c(T_b) = 0.23J_s(T_r)$ in Eq. (2) leads to the empirically observed relationship equation (3). Equation (3) breaks down if the Curie temperature of the magnetic material gets near or below 300 K. With decreasing temperature, J_s increases and reaches a maximum at absolute zero temperature unless the material undergoes a phase transition (e.g. Verwey transition for magnetite). Extending the trend of the published $J_s(T)$ curves (Dunlop and Özdemir, 1997) into 0 K (ignoring any phase transitions) results in increase of J_s by a factor of 1.05 for iron, 1.00 for hematite, 1.10 for magnetite, and 1.25 for pyrrhotite. This change of J_s values has a negligible effect on Eq. (3) and still results in a near perfect linear relationship where $a = (4.2 \pm 0.3) \times 10^{-3}$ with 95% confidence level for $T{=}T_0{=}0$ K and $B_c(T_b) = 0.21J(T_0)$. Using magnetic constants at absolute zero temperature, the problem of Curie temperature is eliminated and Eq. (3) can be applied for any magnetic material.

As the microcoercivity $B_c(T_b)$ modifies the $M_{tr}(T_r)$ acquisition, the shape, magnetostriction, and crystalline anisotropy of the carriers should have significant influence on the $M_{tr}(T_r)$ acquisition curves. For example, the length vs diameter ratio of the carrier should reduce or increase the effect of $B_c(T_b)$ or the demagnetizing field (Dunlop and Özdemir, 1997) for sample lengths parallel or perpendicular to the field and thus shift the $M_{tr}(T_r)$ acquisitions into lower or higher field intensities, respectively. In Figure M135, $M_{tr}(T_r)$ acquisition for acicular (elongated crystals parallel to the applied field) magnetite (Dunlop and West, 1969) with diameter to length ratio 1:7 violates the equidimensionality assumption. Although the data are for magnetic fields near saturation of the magnetite, the demagnetizing field due to elongation causes these grains to acquire magnetization at lower fields than equidimensional magnetite grains. This shape effect was verified experimentally (Kletetschka *et al.*, 2004) by measuring $M_{tr}(T_r)$ acquisition in industrial wires (MWS-294R and ALLOY52) with length to diameter ratios 1:5 and 1:14 (Figure M137b), where the longer wires, parallel to the field, required lower fields to acquire the predicted intensity of magnetization.

The effect of placing wires perpendicular to the applied field should be equal and opposite to placing them parallel. Consequently for a large number of randomly oriented, elongated grains (such as is frequently the case in igneous rocks) ϵ should follow the same relationships as for the single equidimensional grains used in this study. This makes the TRM relationship far more applicable to those who study natural materials. However, this information would have to be accompanied by the caveat that, for the relationship to hold for multiple grains, the grains would have to be identical in size and composition (equal J_s, J_{sr}, and M_{tr}).

It is important to emphasize that a substitution of ϵ to Eq. (3)

$$B = a(T_0)J_s(T_0)\frac{M_{tr}(T)}{J_{sr}(T)} \qquad \text{(Eq. 4)}$$

represents the first ever means to obtain a paleointensity determination using measurable quantities that does not involve the comparison of a TRM imparted in the lab with that acquired in nature. Practical considerations may pose a serious hindrance to it ever being used as such because the above equation would not be satisfied by bulk values and natural grains, capable of retaining a remanence over geological time, would be too small to be measured individually. Possible solutions involve isolating and amassing grains with sufficiently similar properties, decomposition of bulk values using FORC diagrams, and so on to satisfy the requirements of Eq. (4).

Gunther Kletetschka

Bibliography

Acuña, M.H., Connerney, J.E.P., Ness, N.F., Lin, R.P., Mitchell, D., Carlson, C.W., McFadden, J., Anderson, K.A., Rème, H., Mazelle, C., Vignes, D., Wasilewski, P., and Cloutier, P., 1999. Global distribution of crustal magnetization discovered by the Mars global surveyor MAG/ER experiment. *Science*, **284**: 790–793.

Borradaile, G.J., 1996. An 1800-year archeological experiment in remagnetization. *Geophysical Research Letters*, **23**(13): 1585–1588.

Cisowski, S., and Fuller, M., 1986. Lunar paleointensities via the IRMs normalization method and the early magnetic history of the Moon. In Hartmann, W.K., Phillips, R.J., and Taylor, G.J. (eds.), *Origin of the Moon*. Houston: Lunar and Planetary Institute, pp. 411–424.

Dekkers, M.J., 1989. Magnetic properties of natural pyrrhotite. II. High- and low-temperature behavior of J_{rs} and TRM as a function of grain size. *Physics of the Earth and Planetary Interiors*, **57**: 266–283.

Dunlop, D.J., 1990. Developments in rock magnetism. *Reports on Progress in Physics*, **53**: 707–792.

Dunlop, D.J., and Argyle, K.S., 1997. Thermoremanence, anhysteretic remanence and susceptibility of submicron magnetites: nonlinear field dependence and variation with grain size. *Journal of Geophysical Research-Solid Earth*, **102**(B9): 20,199–20,210.

Dunlop, D.J., and Kletetschka, G., 2001. Multidomain hematite: a source of planetary magnetic anomalies? *Geophysical Research Letters*, **28**(17): 3345–3348.

Dunlop, D.J., and Özdemir, Ö., 1997. Rock magnetism: fundamentals and frontiers. In Edwards, D. (ed.), *Cambridge Studies in Magnetism*, Vol. 3. Cambridge: Cambridge University Press, 573 pp.

Dunlop, D.J., and Waddington, E.D., 1975. Field-dependence of thermoremanent magnetization in igneous rocks. *Earth and Planetary Science Letters*, **25**(1): 11–25.

Dunlop, D., and West, G., 1969. An experimental evaluation of single-domain theories. *Reviews of Geophysics*, **7**: 709–757.

Everitt, C.W.F., 1962. Thermoremanent magnetization II: experiments on multidomain grains. *Philosophical Magazine*, **7**: 583–597.

Fletcher, E.J., and O'Reilly, W., 1974. Contribution of Fe^{2+} ions to the magnetocrystalline anisotropy constant K_1 of $Fe_{3-x}Ti_xO_4$ ($0 < x < 0.1$). *Journal of Physics C*, **7**: 171–178.

Hargraves, R.B., and Young, W.M., 1969. Source of stable remanent magnetism in Lambertville diabase. *American Journal of Science*, **267**: 1161–1177.

Heller, F., and Markert, H., 1973. Age of viscous remanent magnetization of Hadrians wall (Northern-England). *Geophysical Journal of the Royal Astronomical Society*, **31**(4): 395–406.

Hoye, G.S., and Evans, M.E., Remanent magnetizations in oxidized olivine. *Geophysical Journal of the Royal Astronomical Society*, **41**: 139–151.

Kletetschka, G., Taylor, P.T., Wasilewski, P.J., and Hill, H.G.M., 2000a. The magnetic properties of aggregate polycrystalline diamond: implication for carbonado petrogenesis. *Earth and Planetary Science Letters*, **181**(3): 279–290.

Kletetschka, G., Wasilewski, P.J., and Taylor, P.T., 2000b. Hematite vs. magnetite as the signature for planetary magnetic anomalies? *Physics of the Earth and Planetary Interiors*, **119**(3–4): 259–267.

Kletetschka, G., Wasilewski, P.J., and Taylor, P.T., 2000c. Unique thermoremanent magnetization of multidomain sized hematite: implications for magnetic anomalies. *Earth and Planetary Science Letters*, **176**(3–4): 469–479.

Kletetschka, G., Wasilewski, P.J., and Taylor, P.T., 2002. The role of hematite-ilmenite solid solution in the production of magnetic anomalies in ground and satellite based data. *Tectonophysics*, **347** (1–3): 166–177.

Kletetschka, G., Acuna, M.H., Kohout, T., Wasilewski, P.J., and Connerney, J.E.P., 2004. An empirical scaling law for acquisition of thermoremanent magnetization. *Earth and Planetary Science Letters*, **226**(3–4): 521–528.

Moskowitz, B.M., 1993. High-temperature magnetostriction of magnetite and titanomagnetites. *Journal of Geophysical Research*, **98**: 359–371.

Néel, L., 1949. Théorie du traînage magnétique des ferromagnétiques en grains fins avec applications aux terres cuites. *Annales de Géophysique*, **5**: 99–136.
Néel, L., 1955. Some theoretical aspects of rock magnetism. *Advances in Physics*, **4**: 191–243.
Özdemir, Ö., and O'Reilly, W., 1982. An experimental study of the intensity and stability of thermoremanent magnetization acquired by synthetic monodomain titanomagnetite substituted by aluminium. *Geophysical Journal of the Royal Astronomical Society*, **70**: 141–154.
Shcherbakova, V.V., Shcherbakov, V.P., and Heider, F., 2000. Properties of partial thermoremanent magnetization in pseudosingle domain and multidomain magnetite grains. *Journal of Geophysical Research-Solid Earth*, **105**(B1): 767–781.
Stacey, F.D., 1958. Thermoremanent magnetization (TRM) of multidomain grains in igneous rocks. *Philosophical Magazine*, **3**: 1391–1401.
Stacey, F.D., 1963. The physical theory of rock magnetism. *Advances in Physics*, **12**: 45–133.
Stacey, F.D., and Banerjee, S.K., 1974. *The Physical Principles of Rock Magnetism*. Amsterdam: Elsevier, 195 pp.
Syono, Y., and Ishikawa, Y., 1963a. Magnetocrystalline anisotropy of $x\mathrm{Fe_2TiO_4}.(1-x)\mathrm{Fe_3O_4}$. *Journal of the Physical Society of Japan*, **18**: 1230–1231.
Syono, Y., and Ishikawa, Y., 1963b. Magnetostriction constants of $x\mathrm{Fe_2TiO_4}.(1-x)\mathrm{Fe_3O_4}$. *Journal of the Physical Society of Japan*, **18**: 1231–1232.
Tucker, P., and O'Reilly, W., 1980. The acquisition of thermoremanent magnetization by multidomain single-crystal titanomagnetite. *Geophysical Journal of the Royal Astronomical Society*, **63**: 21–36.
Wasilewski, P.J., 1977. Magnetic and microstructural properties of some lodestones. *Physics of the Earth and Planetary Interiors*, **15**: 349–362.
Wasilewski, P.J., 1981. Magnetization of small iron-nickel spheres. *Physics of the Earth and Planetary Interiors*, **26**: 149–161.

Cross-references

Blocking Temperature
Crystalline Anisotropy
Curie Temperature
Demagnetization Field
Demagnetizing Energy
Empirical Law
Hematite
Iron
Iron-Nickel
Magnetic Ordering
Magnetite
Magnetoelastic Anisotropy
Microcoercivity
Multidomains (MD)
Mysterisis Loop
Natural Remanent Magnetization (NRM)
Néel Theory
Phase Transition
Pseudo Single Domain (PSD)
Pyrrhotite
Remanent Efficiency of Magnetization (REM)
Saturation Magnetization
Saturation Remanence
Shape Anisotropy
Single Domains (SD)
Squareness Ratio
Super Paramagnetic
Thermal Remanent Magnetization (TRM)
Titanomagnetite
Verwey Transition
Viscous Decay

MAGNETIZATION, VISCOUS REMANENT (VRM)

Introduction

Viscous magnetization is the gradual change of magnetization with time in an applied magnetic field $\boldsymbol{H}$. Brief exposure of a ferromagnetic material to a field results in isothermal remanent magnetization (IRM). The additional remanence produced by a longer field exposure is *viscous remanent magnetization* (VRM). The longer the exposure time t, the stronger is the VRM.

Viscous remagnetization is the time-dependent change of VRM or other remanences, such as thermoremanent magnetization (TRM), depositional remanent magnetization (DRM), or chemical remanent magnetization (CRM), in response to a change in the direction or strength of $\boldsymbol{H}$. In nature, such field changes are due to secular variation, excursions, polarity transitions, or plate motion. In laboratory experiments, but never in nature, samples may be exposed to zero field and the *viscous decay* of their magnetization measured.

The natural remanent magnetization (NRM) of rocks, sediments, and soils usually includes a VRM produced at ambient temperature by exposure to the weak ($\lesssim$100 μT) Earth's magnetic field during the Brunhes normal polarity epoch. This VRM obscures the useful paleomagnetic information residing in older components of the NRM. Removing the VRM is the purpose of standard "cleaning" procedures, such as alternating field (AF) and thermal demagnetization.

Brunhes-epoch VRM is distinctive because it is roughly parallel to the present-day local geomagnetic field. It is usually weaker than TRM and DRM and also "softer" or more easily cleaned than these older NRMs. However, magnetically "hard" or high-coercivity minerals like hematite, goethite, pyrrhotite, and some compositions of titanomagnetites and titanohematites have hard VRM that is not easy to AF demagnetize.

Rocks that have been exposed to the Earth's field at elevated temperatures for long times acquire *thermoviscous magnetization* (TVRM), which may replace part or all of their primary NRM. Usually these rocks are slowly cooling plutons, or have been deeply buried in sedimentary basins, volcanic piles, or mountain belts. TVRM blurs the distinction between thermal processes like TRM and partial TRMs and time-dependent processes like VRM and viscous remagnetization.

In reality, even ambient temperature is sufficiently high compared to 0 K that time t and temperature T are interwoven in VRM and viscous overprinting. The fundamental mechanism of all viscous processes is the slow, continuous approach of the magnetization $\boldsymbol{M}$ to its thermal equilibrium value in field $\boldsymbol{H}$ at T, aided by thermal fluctuations of the magnetization in each crystal. Fortunately, the approach to equilibrium is usually extremely sluggish (hence *viscous*) at ordinary temperatures. If this were not so, there would be no surviving NRM of ancient origin, and no paleomagnetism or seafloor magnetic anomalies to track plate tectonic movements.

Although most viscous magnetization changes are driven by thermal fluctuations, another possible source of viscous effects is slow diffusion of lattice defects that pin domain walls in multidomain crystals. This *diffusion after-effect* is most significant in titanomagnetites over geologically short times (Moskowitz, 1985).

Theory of single-domain VRM

The Néel (1949, 1955) theory of time- and temperature-dependent magnetization deals for simplicity with ensembles of N identical uniaxial single-domain grains, all mutually aligned and having volume V, anisotropy constant $K(T)$, and spontaneous magnetization $M_s(T)$. At time $t = 0$, magnetic field $\boldsymbol{H}$ is applied parallel to the anisotropy axis. At time t, n^+ grains have their moments $+VM_s$ in the direction of $\boldsymbol{H}$ and n^- grains have their moments $-VM_s$ opposite to $\boldsymbol{H}$.

The distribution function $n(t)/N = (n^+ - n^-)/N$, which equals $M(t)/M_s$, obeys the kinetic equation

$$dn/dt = -[n(t) - n_{eq}]/\tau, \quad \text{(Eq. 1)}$$

where the equilibrium value $n_{eq}/N = \tanh(VM_sH/kT)$ and the *relaxation time* τ, due to thermal activation of moments over the energy barrier VK due to anisotropy, is

$$1/\tau = 2f_0 \exp[-VK(T)/kT], \quad \text{(Eq. 2)}$$

with $f_0 \approx 10^{-10}\ s^{-1}$.

At constant temperature, the solution to Eq. (1) is

$$n(t) - n_{eq} = [n(0) - n_{eq}]\exp(-t/\tau), \quad \text{(Eq. 3)}$$

an exponential decay of the out-of-equilibrium distribution function and magnetization. Néel introduced an approximation that replaces the real exponential decay by a step function,

$$n(t) = n(0) \quad \text{if } t \le \tau$$
$$n(t) = n_{eq} \quad \text{if } t > \tau. \quad \text{(Eq. 4)}$$

In the case of TRM, where τ changes rapidly with small changes in T, Néel's approximation leads to the definition of the *blocking temperature* T_B, at which the ensemble rapidly passes from a thermal equilibrium state to an out-of-equilibrium state during cooling, or the reverse during heating, e.g., in thermal demagnetization. The blocking concept is less useful in isothermal viscous magnetization because τ remains constant and all changes are a result of changing t.

Any real sample contains many ensembles with different values of V and K. The grain distribution $f(V, K)dV\,dK$ is generally poorly determined, but the formal solution of Eq. (3) for acquisition of magnetic moment $m(t)$ by all ensembles, starting from a demagnetized state, is

$$m(t) = \int\int VM_s n_{eq}[1 - \exp(-t/\tau)]\,f(V,K)dVdK. \quad \text{(Eq. 5)}$$

Numerical solutions can be found for specified forms of $f(V, K)$.

Time and temperature dependence of VRM

If $f(V, K)$ is constant or nearly constant over a range of (V, K), the convolution of the exponential time dependence for a single ensemble (Eq. (3)) with $f(V, K)$, as expressed by Eq. (5), leads to a logarithmic dependence of $m(t)$ on t (see Dunlop, 1973, or Dunlop and Özdemir, 1997, Chapter 10 for details). Experimentally most ferromagnetic materials on a laboratory timescale exhibit viscous changes that are indeed approximately proportional to $\log t$. Figure M138 illustrates the measured time dependence of VRM for oxidized and unoxidized pyroclastics on a t scale from 15 min to 28 days (about 4 decades of t). Only a limited range of (V, K) is activated in a few decades of t and the VRM increases more or less linearly on a $\log t$ scale.

According to Eq. (2), for T constant, VK is proportional to $\log 2f_0\tau$. Using Néel's blocking approximation (4), it is then possible to transform $f(V, K)$ directly into the corresponding distribution $g(\log t)$. In this way, one can predict the *viscosity coefficient* $S = \partial M/\partial \log t$ for a sample whose grain distribution $f(V, K)$ is known.

Another approach (Walton, 1980) is to calculate S directly from Eq. (5) without explicitly invoking a blocking approximation. If we replace K by a suitable average value and approximate n_{eq}/N by VM_sH/kT, since H is small, we find

$$S = \int\int (M_s^2H/kT)(t/\tau)[1 - \exp(-t/\tau)]V^2Nf(V)dV. \quad \text{(Eq. 6)}$$

The integrand is of the form $x(1-e^{-x})$, with x itself exponential. It peaks sharply at $x = 1$ or $t = \tau$, which is in fact the blocking condition. All nonexponential factors can be assigned their blocking values, e.g., $V \to V_B = (kT/K)\log 2f_0t$, and taken outside the integral, giving

$$S = Ik^2N(M_s^2T^2/K^3)H(\log 2f_0t)^2 f[kT/K)\log 2f_0t], \quad \text{(Eq. 7)}$$

the integral I being approximately 1.

Assuming that the anisotropy is due to grain shape, so that $K(T) \propto M_s^2(T)$, and that the grain distribution is uniform over the range of ensembles affected in time t, we obtain

$$S = CH(T/M_s^2)^2(\log 2f_0t)^2, \quad \text{(Eq. 8)}$$

whereas if $f(V) \propto 1/V$, as for the larger grains in a lognormal distribution,

$$S = C'H(T/M_s^2)^2 \log 2f_0t, \quad \text{(Eq. 9)}$$

C, C' being constants. Either Eq. (8) or (9) predicts a $\log t$ dependence for VRM intensity over times short enough that $\log t \ll \log 2f_0 \approx 23.7$, i.e., for a few decades of t, as observed. Deviations from $\log t$ dependence will become obvious even at short times for very viscous samples, which scan larger fractions of the grain distribution, and at longer times for all samples. Deviations will appear earlier if Eq. (8) applies rather than Eq. (9).

VRM of multidomain and interacting grains

Viscous magnetization due to thermal activation of domain walls has been treated theoretically in the weak-field limit by Néel and in the

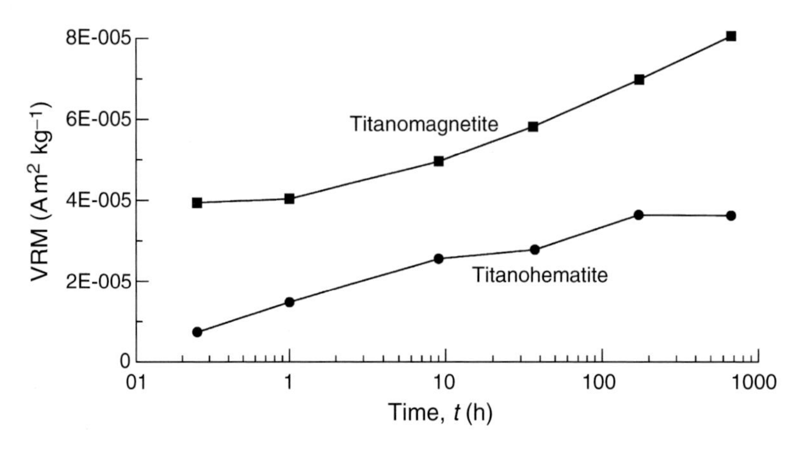

Figure M138 Viscous remanent magnetization measured after exposure times of 15 min to 4 weeks to a 200 μT field (data: Saito *et al.*, 2003). Upper curve, unoxidized pyroclastic sample containing large grains of titanomagnetite. Lower curve, oxidized pyroclastic containing hematite and titanohematite.

presence of self-demagnetizing fields by Stacey (see Dunlop, 1973, for a review). All theories predict a $\log t$ dependence of VRM for geologically short times. The viscosity coefficient S varies with temperature as T/M_s or $(T/M_s)^{1/2}$. The theories resemble single-domain formulations with V in Eq. (2) replaced by the volume V_{act} activated in a single Barkhausen jump. The frequency constant f_0 is $\approx 10^{-10}$ s^{-1}, as for single-domain grains. Viscous magnetization due to activation of entire walls should be negligible because $V_{act} \gg V$ of a single-domain grain. It must be that segments of walls are activated past single pinning sites.

In nature, titanomagnetite and pyrrhotite crystals sometimes fail to renucleate equilibrium domain structures following saturation, resulting from heating to the Curie temperature. A possible mechanism for multidomain viscous magnetization is thermally activated nucleation of domains in parts of the crystal where the internal field is close to the critical nucleation field. This process has been termed *transdomain VRM*. Nucleation events have been observed in magnetite during small changes of T (Heider *et al.*, 1988) but are not yet documented for changing t at constant T.

Interacting single-domain grains have been approached in two different ways. One method is to treat them collectively by increasing V from the single-grain value. This approach preserves the experimentally observed $\log t$ behavior but it is not known at what level of interaction the picture breaks down. Another approach is to deal with individual particles under the influence of a randomly varying interaction field. Walton and Dunlop (1985) predicted deviation from $\log t$ behavior when the interactions are strong. Their theory gave a good fit to viscous magnetization data for an interacting single-domain assemblage with known $f(V)$.

Experimental results

Grain size dependence of viscosity coefficients

Room-temperature viscosity coefficients S have their highest values in single-domain grains with volumes V just above the critical superparamagnetic volume V_B. Figure M139 illustrates the data for magnetite. Data sets for other minerals are even smaller. Between the magnetite critical superparamagnetic size (0.025–0.03 μm) and the maximum size for single-domain behavior (0.07–0.08 μm), S decreases by at least a factor 4. The decrease may actually be larger and more precipitous than shown. Even in carefully sized samples, there is always a fraction of very fine grains. This fine fraction probably controls the short-term viscous behavior, because, according to Eqs. (2) and (4), only grains with very small V will be activated in the short times used in laboratory experiments. Similar "contaminating" ultrafine grains may explain the constant baseline value of S in grains larger than single-domain size.

The twofold increase in S over the upper pseudosingle-domain region (2–15 μm) is probably real and due to the increasing ease with which segments of walls can escape from their pins as the scale of the domains grows. On the other hand, large multidomain grains 80–100 μm in mean size have significantly lower S values.

In summary, strong viscous magnetization is found mainly in the finest single-domain grains. However, small multidomain (so-called pseudosingle-domain) grains are also significantly viscous.

Observations on lunar rocks and soils (see Dunlop, 1973 and Dunlop and Özdemir, 1997, Chapter 17 for summaries) confirm this trend. Lunar soils and loosely welded soil breccias containing nearly superparamagnetic iron particles are among the most viscous materials known, whereas crystalline rocks and mature breccias containing multidomain iron exhibit a consistent but small viscous magnetization. The viscous nature of lunar samples greatly complicated the study of their other magnetic properties during the Apollo missions.

Temperature dependence of viscous magnetization

Viscous magnetization data for a single-domain magnetite sample (mean grain size 0.037 μm) at and above room temperature appear in Figure M140. Zeroing the field H for each measurement would have changed the initial state for subsequent measurements. Therefore viscous induced magnetization in the presence of H was measured.

Over the short time spans examined (2–4 decades of t), M_{VP} is linear in $\log t$, as predicted by Eqs. (2), (8), and (9). The viscosity coefficient S, the slope of each data run, increases steadily with increasing T. This too is as predicted by theory. However, the inherent temperature dependence of S, in Eqs. (8) and (9) for example, tends to be obscured by the variability of the grain distribution $f(V, K)$, which is scanned with a narrow thermal "window," as expressed by the factor $f\,(kT/K)\log 2f_0t]$ in Eq. (7).

Very different temperature dependences of viscous magnetization have been reported for magnetite and titanomagnetites of various grain sizes by different authors. In many cases, the higher-T results are distinctly nonlinear in $\log t$. However, a good theoretical match is evident in Figure M140 between the single-domain magnetite data (Dunlop, 1983) and the theory of equations (6)–(9), using the measured $f(V)$ of this sample (Walton, 1983).

Field dependence of viscous magnetization

Experiments confirm the theoretical prediction (e.g., Eqs. (6)–(9)) that VRM intensity is proportional to field strength H for weak fields. The most detailed studies are those of Creer (1957) on single-domain hematite and Le Borgne (1960) on soils containing single-domain magnetite and maghemite. Linear behavior was observed for 0.05 mT $\leq H \leq$ 1 mT.

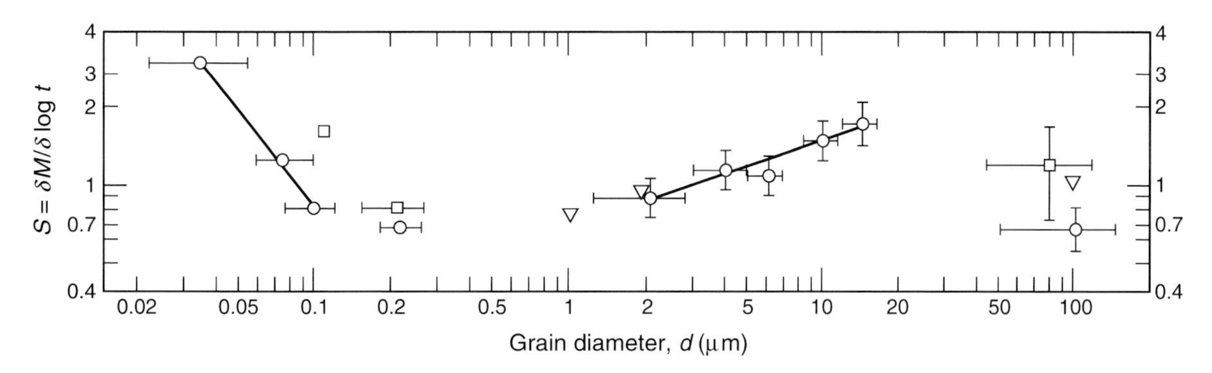

Figure M139 Viscosity coefficients S for sized grains of magnetite. Data: triangles, Shimizu (1960); circles, Dunlop (1983); squares, Tivey and Johnson (1984).

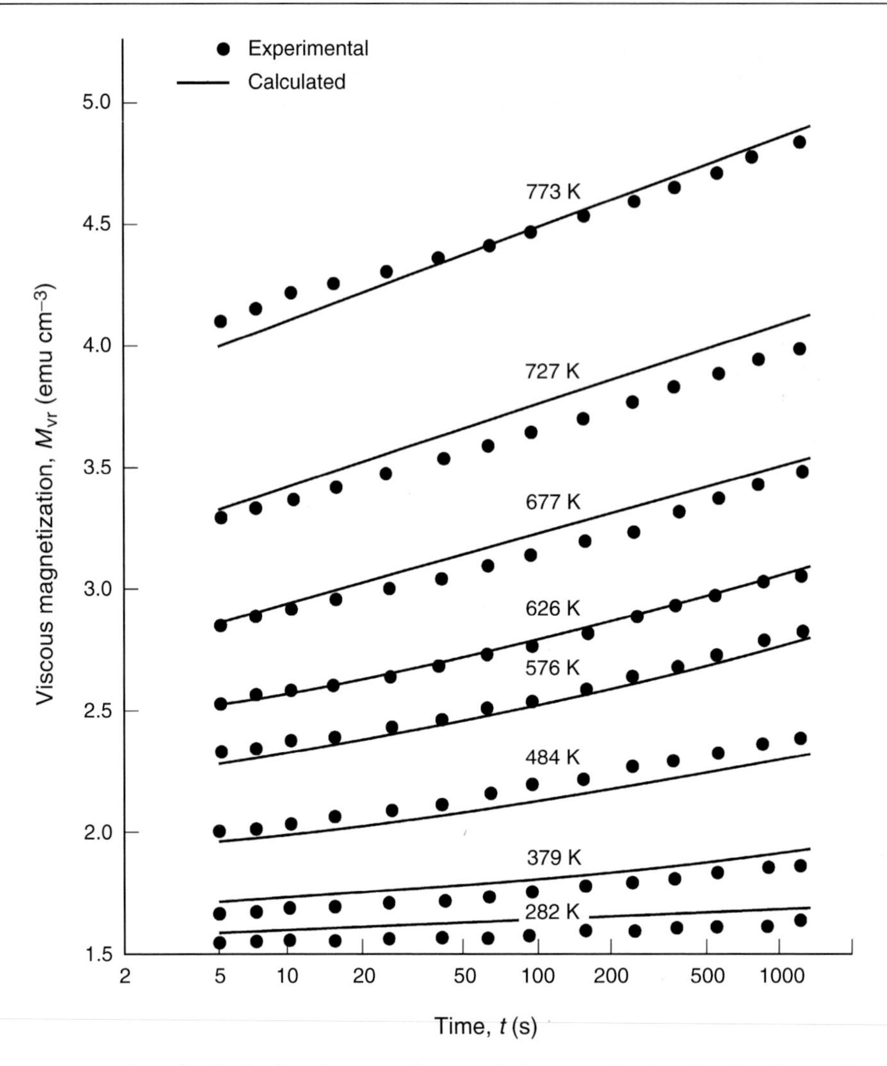

Figure M140 Viscous magnetization of a single-domain magnetite sample (mean grain size 0.037 μm) at various temperatures. Dots: measured results, Dunlop (1983); lines: theoretical fits, Walton (1983).

Viscous magnetization at short times: frequency dependent and quadrature susceptibility

Viscous changes in magnetization can be detected over very short times by measuring initial susceptibility χ (short-term induced M, normalized by H) as a function of the frequency f of an alternating field (AF) $\tilde{H}$. In effect, each cycle of the AF restarts the viscous magnetization experiment. A second method is to measure the quadrature (90° out-of-phase) component of susceptibility χ_q, which represents a time-delayed response to $\tilde{H}$.

In the classic experiments of Mullins and Tite (1973) on soils containing single-domain size magnetite and maghemite, $\partial\chi/\partial\log f$ and χ_q were constant over the tested range of f (66–900 Hz, 1.4 decades of t). No frequency dependence or quadrature susceptibility was detected for multidomain magnetite.

A pair of measurements of χ at two standard frequencies is routinely used in environmental magnetic studies as a test for the presence of nearly superparamagnetic magnetite or maghemite. Because the frequencies are preset, there is no attempt to tune the viscosity measurement to other minerals or to domain states other than single-domain. Much more granulometric information could be obtained with a little extra effort by scanning a wider range of f and by extending the measurements to temperatures other than room temperature (e.g., Jackson and Worm, 2001).

Demagnetizing VRM

AF demagnetization

The AFs $\tilde{H}$ used in frequency-dependent susceptibility measurements have small amplitudes, usually ~0.1 mT, and do not seriously modify preexisting remanences. If larger AFs are applied, they entrain single-domain moments or domain walls, causing them to oscillate and destroy VRM (or any other remanence, for sufficiently large $\tilde{H}$). To demagnetize a previously magnetized sample, $\tilde{H}$ is slowly decreased from its maximum value at a rate $\ll f$, leaving about equal numbers of moments or domains in one or the other polarity. AF demagnetization is carried out in a stepwise fashion. The maximum field is increased in steps, starting from small values, so as to gradually erase remanence components of increasing AF coercivity.

VRM is often the NRM component of lowest coercivity and therefore erased first. Indeed the strategy of AF cleaning in paleomagnetism is firmly rooted in this assumption.

Strong fields lower the energy barrier VK to rotation of single-domain moments or the corresponding "barrier" (pinning energy) for domain wall motion. In the single-domain case, Eq. (2) is modified to

$$1/\tau_H = f_0 \exp[-(VK(T)/kT)(1 - M_s H/2K)^2]. \qquad \text{(Eq. 10)}$$

One consequence is hard VRM in minerals with high K/M_s, like hematite, goethite, pyrrhotite, elongated single-domain iron, and many compositions of titanomagnetite, titanomaghemite, and titanohematite. VRM is produced by a weak field and Eq. (2) applies, but it is demagnetized by a strong field $\tilde{H}$ for which Eq. (10) is

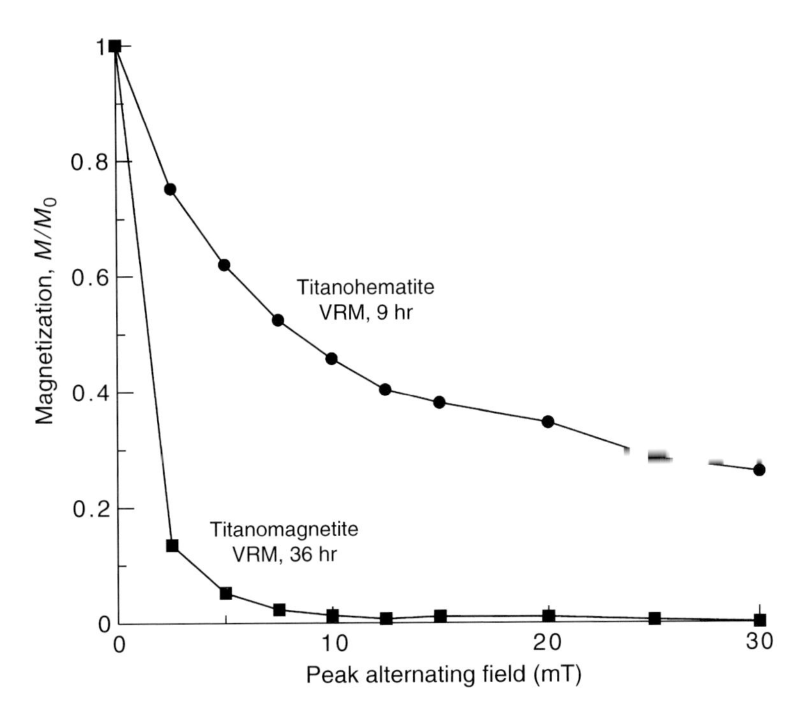

Figure M141 AF demagnetization of some of the VRMs of Figure M138 (data: Saito *et al.*, 2003). VRM of the unoxidized titanomagnetite is erased by ≈2.5 mT. VRM of the oxidized sample containing titanohematite is much more resistant.

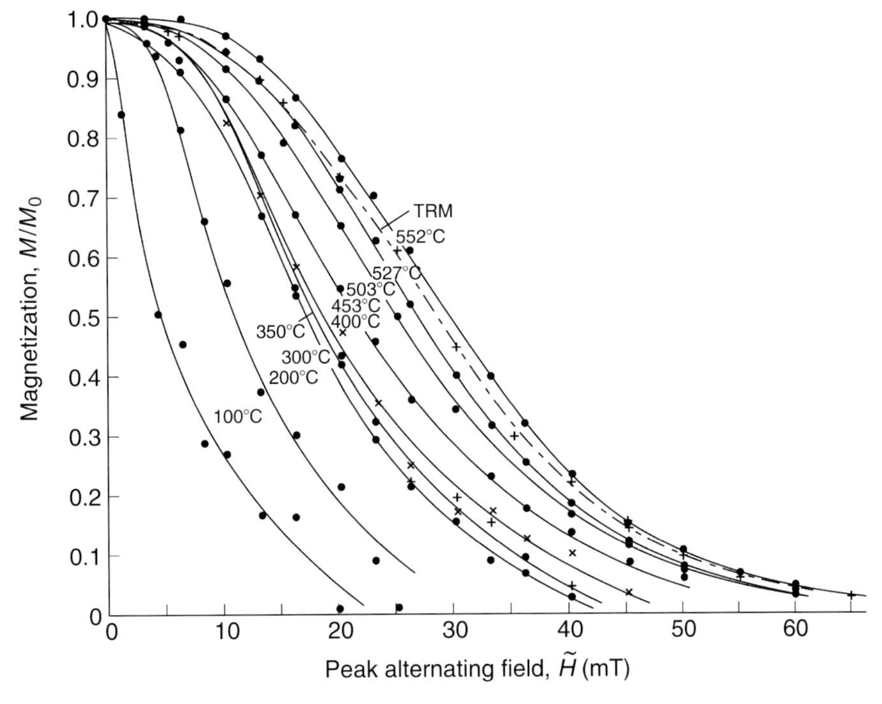

Figure M142 AF demagnetization of VRMs produced by 2.5 h exposure to a 50 μT field at the temperatures shown (after Dunlop and Özdemir, 1990). VRMs produced at higher temperatures are increasingly resistant to AF cleaning.

appropriate. Using the blocking approximations $\tau = t$ for the last ensemble to acquire VRM and $\tau_H = 1/f$ when $H = \tilde{H}$ for AF cleaning of this same ensemble, we have by combining (2) and (10),

$$\tilde{H} = (2K/M_s)\,[1 - (\log f_0/f)^{1/2}/(\log 2f_0 t)^{1/2}]. \qquad \text{(Eq. 11)}$$

The resistance of VRM to AF cleaning is thus proportional to K/M_s, the other factor being independent of mineral properties.

Figure M141 confirms this prediction. One sample is an unoxidized pyroclastic containing iron-rich titanomagnetite with low coercivity K/M_s. VRMs produced over all ranges of t are very soft, demagnetizing almost completely for $\tilde{H} = 2.5$ mT. The other sample is oxidized and its VRM is carried by titanohematite of much higher coercivity. VRMs for this sample are more resistant to AF demagnetization, surviving to $\tilde{H} = 10$–20 mT. This is only a moderately hard VRM, because in fact both these samples contain multidomain-size grains, not single-domain material to which Eq. (11) strictly applies. However, practically any rock whose NRM is due mainly to hematite has VRM that cannot be entirely removed by $\tilde{H} = 100$ mT (Biquand and Prévot, 1971).

VRM produced in nature at above-ambient temperature and subsequently AF cleaned at room temperature is also relatively hard. Figure M142 shows an example for a single-domain magnetite sample that was given VRMs in 2.5 h runs at temperatures as high as 552°C (the Curie point is 580°C). The effect of T is to shift the window of $f(V, K)$ affected viscously in $t = 2.5$ h to grains with increasingly high K (see Dunlop and Özdemir, 1997, Chapter 10).

Ambient-temperature VRM in magnetically hard minerals and elevated-temperature VRM in both hard and soft minerals are both difficult to remove completely or cleanly (without removing part of the primary NRM as well) by AF demagnetization. A more satisfactory method is thermal demagnetization.

Thermal demagnetization

Because VRM is a thermoviscous relaxation phenomenon, thermal demagnetization should be the most efficient way of erasing VRM. Randomizing single-domain moments or domain wall positions is accomplished by random thermal excitations, resulting in a truly demagnetized state, not the polarized condition that follows AF cleaning. From Eq. (2), the laboratory temperature T_L to which a single-domain sample must be heated (in zero field) on a timescale t_L (typically an hour or less) to erase VRM produced over a long time t_N in nature at temperature T_N is given by

$$T_L \log 2f_0 t_L / M_s^2(T_L) = T_N \log 2f_0 t_N / M_s^2(T_N), \qquad \text{(Eq. 12)}$$

(Pullaiah *et al.*, 1975), assuming $K(T) \propto M_s^2(T)$ as for shape anisotropy.

Substituting $t_N = 0.78$ Ma for VRM acquired over the Brunhes epoch at surface temperatures, one obtains for magnetite $T_L \approx 250$°C. This is indeed the typical reheating temperature needed to completely erase magnetite VRM, although AF cleaning often serves very well in this case. For pyrrhotite, with a different Curie temperature (320°C) and $M_s(T)$ variation, the corresponding T_L is ≈ 130°C.

Figure M143 illustrates how complete erasure of VRM is detected. The NRM vector of this sample (Milton Monzonite, SE Australia), plotted in vector projection, reveals four components whose removal requires successively higher heatings. The VRM, produced by the Earth's field over $\approx$100 ka at $\approx$ 165°C (Dunlop *et al.*, 2000), and the primary TRM are carried by single-domain pyrrhotite, the two CRMs by magnetite and hematite. The junction between the VRM and TRM vectors is sharp because single-domain thermoviscous remanences separate cleanly in heating. The vector direction changes abruptly at the junction temperature because the TRM was produced by a reverse-polarity field and the VRM by a normal-polarity field. The junction temperature T_L is >130°C because T_N was above ambient while the rock was buried and VRM was being produced. Using

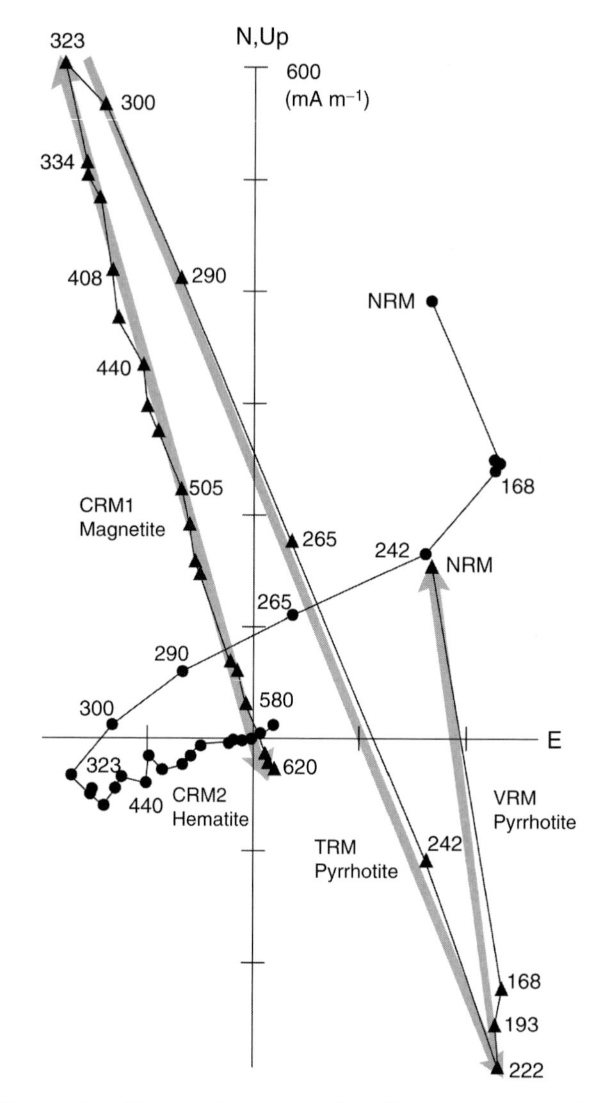

Figure M143 Thermal demagnetization of NRM of a sample of the Milton Monzonite (after Dunlop *et al.*, 1997). The pyrrhotite primary TRM and VRM separate cleanly in thermal demagnetization to 222°C. Triangles and circles: vertical- and horizontal-plane vector projections, respectively.

$t_N = 100$ ka, $T_N = 165$°C, and $M_s(T)$ for pyrrhotite in Eq. (12) yields $T_L = 222$°C, as observed.

Contours of constant (V, K) are often plotted on a time-temperature diagram (Pullaiah *et al.*, 1975), which can then be used as a nomogram for locating matching (T, t) pairs. Figure M144 is such a diagram, with contours for magnetite based on Eq. (12). It is clear from the data plotted, some of which are from the Milton Monzonite, that single-domain magnetite samples (0.04 μm, A) obey Eq. (12) but samples with pseudosingle-domain (20 μm, B) and multidomain (135 μm, C) magnetites follow contours with smaller slopes. That is, these latter grains require more heating to remove their VRMs than the single-domain theory predicts.

Observations of "anomalously high" unblocking temperatures in thermal demagnetization of VRM are well documented in the literature. Initially it seemed possible to reconcile the data by using the Walton (1980) single-domain theory outlined in an earlier section. From Eq. (9),

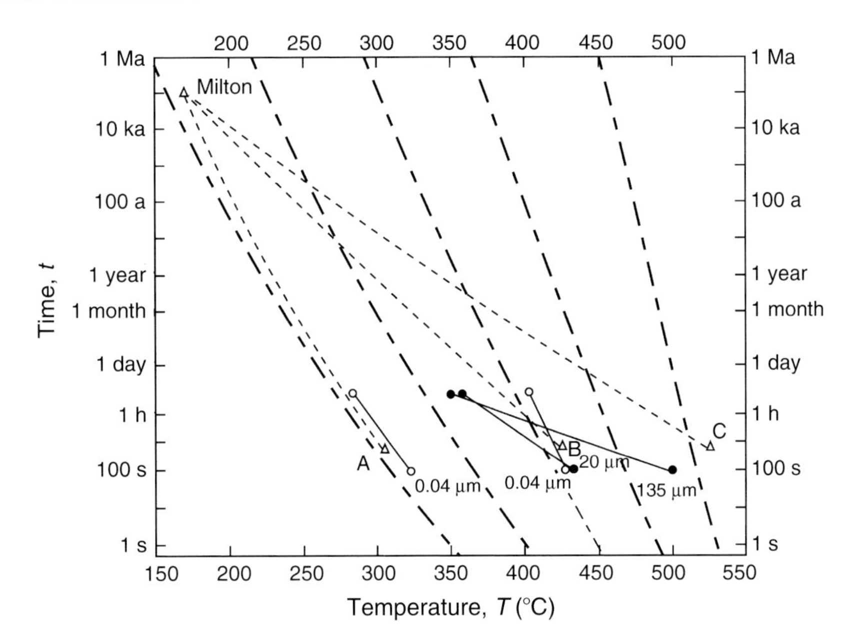

Figure M144 Theoretical time-temperature contours based on Eq. (12) compared to data from thermal demagnetization of VRMs produced at elevated temperatures for synthetic magnetites and samples of the Milton Monzonite (after Dunlop and Özdemir, 2000). Single-domain samples agree well with the contours. Larger pseudosingle-domain and multidomain samples are more difficult to demagnetize than predicted.

$$T_L(\log 2f_0t_L)^2/M_s^2(T_L) = T_N(\log 2f_0t_N)^2/M_s^2(T_N), \qquad \text{(Eq. 13)}$$

which produces shallower contours resembling the trend of some of the deviant data (Middleton and Schmidt, 1982).

However, on reflection it became clear that Eqs. (12) and (13) actually answer different questions (Enkin and Dunlop, 1988). Equation (13) tells us what exposure time t_L to field $\boldsymbol{H}$ at temperature T_L will produce the *same VRM intensity* as the original exposure to $\boldsymbol{H}$ for time t_N at T_N. The question that is relevant to erasure of VRM by thermal demagnetization, and is answered by Eq. (12), is what exposure t_L to $\boldsymbol{H}$ at T_L will reactivate the *same ensembles* (V, K) as the exposure in nature to $\boldsymbol{H}$ for time t_N at T_N. Because VRM intensity increases with rising T even if $f(V, K)$ is constant (e.g., Eq. (7)), the two answers will always be different.

This has now been shown beyond doubt by the experiments of Jackson and Worm (2001), shown in Figure M145. They carried out experiments on the magnetite-bearing Trenton Limestone below room temperature at fixed, very short times t and varied T to either achieve unblocking of the same (V, K) ensembles (upper graph) or produce the same viscous magnetization intensity (lower). The former data sets agree well with the Pullaiah *et al.* (1975) contours from Eq. (12), while the latter are close to the predicted Walton-Middleton-Schmidt contours from Eq. (13).

Discrepancies between the predictions of Eq. (12) and thermal demagnetization results for rocks are mainly attributable to VRMs of grains larger than single-domain size. The larger the grain size, the greater the discrepancy, as illustrated by Figure M144. It is well known that the Thellier (1938) law of independence of partial TRMs is increasingly violated as grain size increases. One manifestation of this violation is that partial TRMs produced over narrow T intervals do not demagnetize over the same intervals, so that a clean separation of partial TRMs is no longer possible using thermal demagnetization.

VRMs behave analogously. The *average* demagnetization temperature of both VRMs and partial TRMs is close to that predicted by Eq. (12) (Dunlop and Özdemir, 2000) but domain walls continue to move toward a demagnetized state at higher T, producing a "tail" of unblocking temperatures. Unfortunately only the ultimate highest unblocking temperature is easily detected experimentally, and it is not predictable by any current theory.

To further complicate matters, the initial state of a multidomain grain can strongly affect its magnetic behavior, including thermal demagnetization. Figure M146 demonstrates that VRMs produced in an AF demagnetized sample or in one thermally demagnetized and then heated in a zero field to the VRM production temperature are subsequently erased by further heating of $\leq 100°C$. But a sample cooled in a zero field directly from the Curie point to the VRM temperature acquires a VRM that requires $\approx 250°C$ of further heating to remove. These experiments use continuous thermal demagnetization, which differs from standard stepwise demagnetization by not revisiting room temperature for each measurement. Although these results may not be directly applicable to standard paleomagnetic practice, they do make the point that any multidomain sample retains a memory of its entire past history and this may influence its later behavior in as yet unpredictable ways.

Applications of viscous magnetization

Magnetic granulometry

Viscous magnetization is a sensitive probe of narrow (V, K) bands of the grain distribution. If K has a narrow spread compared with V, inversion of Eq. (5) by Laplace transforms yields $f(\tau)$, i.e., $f(V)$. A cruder estimate, which is adequate in most cases, is to use a blocking approximation as in Eq. (7) to obtain $f(V_B) = f(kT\log 2f_0t/K)$. Small V_B can be accessed by using the frequency dependence of χ. Large V_B can be activated in VRM experiments above room temperature.

Estimating Brunhes-chron VRM

In very viscous rocks, particularly unoriented or partially oriented seafloor samples, one would like to know if a large part of the NRM is

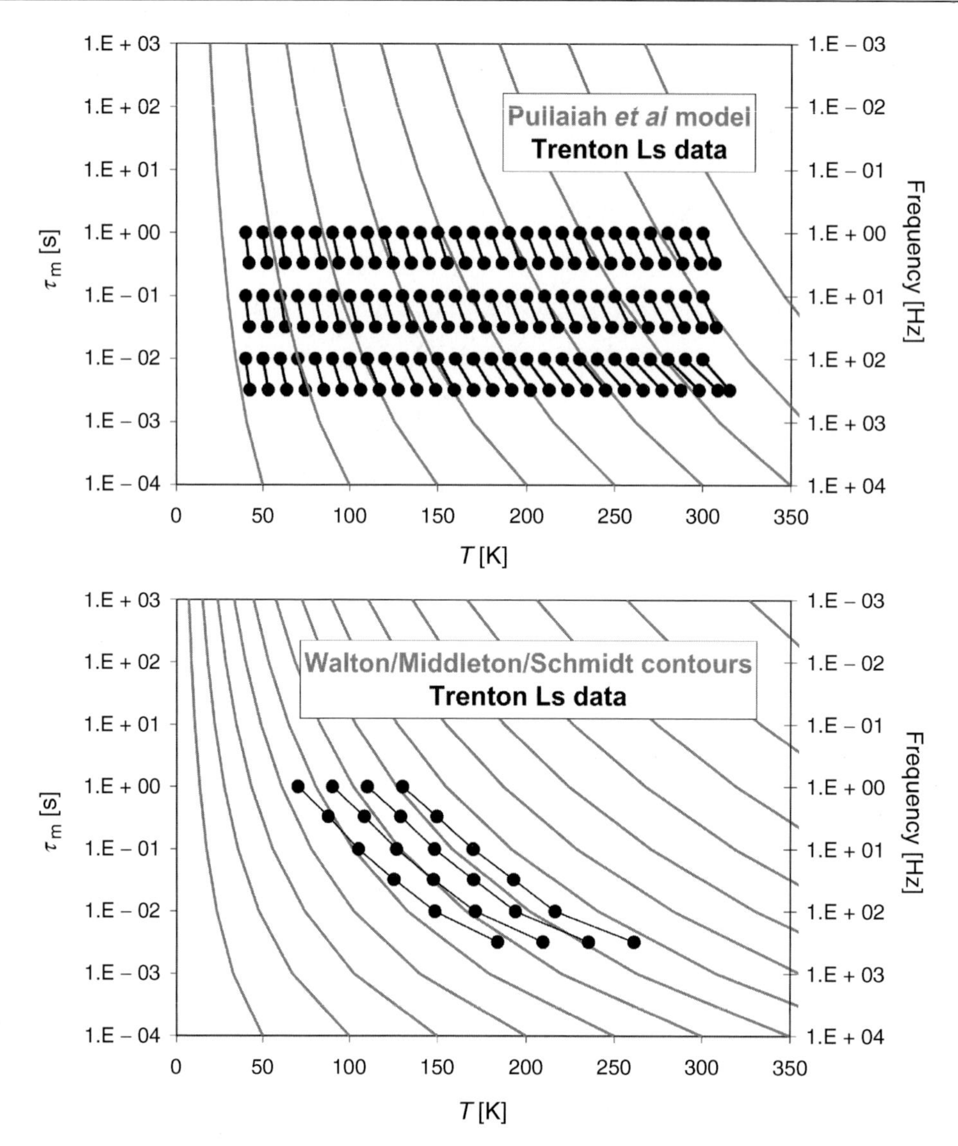

Figure M145 Comparison of equivalent (τ, T) sets determined from frequency-dependent susceptibility data at low temperatures for the magnetite-bearing Trenton Limestone (after Jackson and Worm, 2001). The upper data compare reactivation of equivalent grain ensembles (V, K) and are in good agreement with Eq. (12). The lower data compare equal viscosity coefficients at different T and agree reasonably well with Eq. (13).

VRM produced during the Brunhes chron (the past 0.78 Ma). For samples magnetized during an earlier normal-polarity chron, there is not much difference between the directions of primary TRM and recent VRM. No sharp junctions like those of Figure M143 appear in thermal demagnetization trajectories.

Since relaxation times of $\sim$1 Ma are inaccessible in laboratory experiments, two different methods have been used. First is simple extrapolation of laboratory data, assuming S remains constant over many decades of t beyond the laboratory scale. This is certainly unjustified with the more viscous shallow oceanic rocks, e.g., doleritic flows (Lowrie and Kent, 1978). With less viscous rocks, like those of Figure M138, simple extrapolation implies that a substantial part of the NRM could be VRM, but a precise quantitative estimate is not possible (Saito *et al.*, 2003).

The second method is to activate long relaxation times in the laboratory in mild heatings above room temperature. For titanomagnetite of mid-ocean ridge compositions, the heatings must remain below the Curie point (150–200°C). Another criterion is that chemical alteration must be prevented. Thermal demagnetization of the NRM, using Eq. (12) or Figure M144 as a guide to the (T_L, t_L) combination needed, is a better approach than production of a new VRM. In the latter experiment it is not possible to reactivate at T_L all the ensembles (V, K) that carried the room-temperature VRM.

Paleothermometry

If the approximate residence time t_N over which VRM was acquired can be estimated (within about an order of magnitude, since log t_N is involved), but the burial temperature during VRM production is unknown, Eq. (12) or Figure M144 can again be used to analyze laboratory thermal demagnetization data and estimate T_N. This method is useful for estimating depth of burial in a sedimentary basin as a guide to hydrocarbon potential. It of course only works well if the VRM has single-domain carriers. A recent variant, which is capable

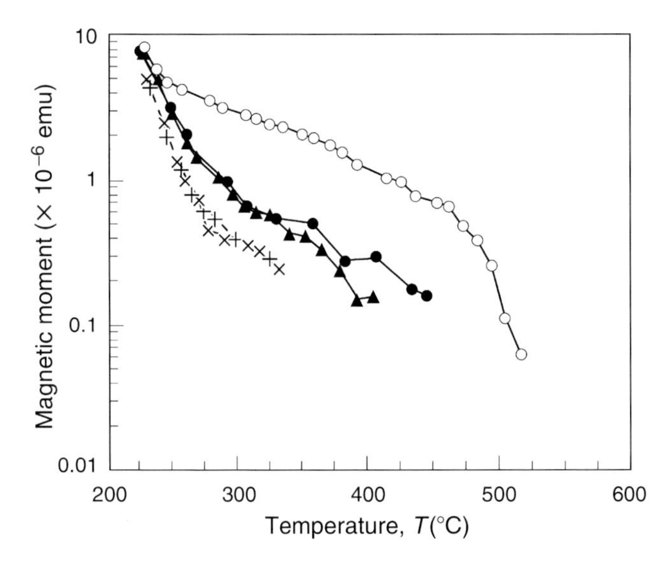

Figure M146 Continuous thermal demagnetization of VRMs produced at 225°C in a synthetic multidomain magnetite sample with different initial states: crosses, AF demagnetized; solid circles and triangles, thermally demagnetized [illegible]; open circles, zero-field cooled from the Curie point to 225°C (after Halgedahl, 1993).

in principle of estimating both t_N and T_N independently, is to analyze thermal demagnetization data for VRMs carried by *two* minerals, e.g., magnetite and pyrrhotite (Dunlop *et al.*, 2000).

Cooling-rate dependence of TRM

Another thermoviscous effect is the variation of TRM intensity depending on cooling rate. For single-domain grains, about 5% increase in intensity is predicted for the slowest versus the fastest practical laboratory cooling rates. The theory is not simple because T and t vary simultaneously but the few experimental data that exist agree quite well with predictions (Fox and Aitken, 1980). There is some evidence that for multidomain grains, TRM intensity may *decrease* for slower cooling (Perrin, 1998). The enormous range of cooling rates present in nature when metamorphic terrains are uplifted over millions or tens of millions of years by erosional unroofing may lead to much larger changes than can be measured in the laboratory. So far no method of simulating or estimating changes over these very long timescales has been proposed.

VRM as an archeological dating tool

When blocks of stone were reoriented in building historical or more ancient structures, viscous remagnetization of their NRMs began. This "clock" is less than ideal because the changes in M are roughly in proportion to $\log t$, so that the resolution $dM/dt = t^{-1}\, dM/d\log t = S/t$ decreases with increasing age. The classic study by Heller and Markert (1973) found reasonable values, 1.6–1.8 ka, for the age of Hadrian's Wall (Roman, northern England) from the viscous behavior of two of three blocks tested. This success is somewhat unanticipated because it implies that S is constant for times of historical length, contrary to most laboratory observations. Multidomain grains, with their lower S values, may perhaps succeed where single-domain grains would fail.

More recently, Borradaile has successfully used ad hoc relations between t_N and T_N for specific limestones used in the construction of buildings and monuments in England and Israel to interpret VRM thermal demagnetization data of samples from other structures of unknown age (see Borradaile, 2003 and references therein). The success of this approach depends on constructing a calibration (t_N, T_N) curve using samples from buildings of known age. It is specific to a certain area and a particular building stone, and requires substantial labor in establishing the calibration curve, but is undeniably useful to archeologists thereafter.

VRM in rocks

Lunar rocks owe their magnetic properties to metallic iron. Lunar soils and low-grade breccias contain ultrafine (≤0.02 μm) single-domain grains which are potently viscous. High-grade breccias, basalts, and anorthosites have weaker viscous magnetizations originating in multidomain iron. For a full discussion, see Dunlop (1973) or Dunlop and Özdemir (1997, Chapter 17).

Terrestrial submarine basalts are rather weakly viscous. Coarse-grained massive flows, on the other hand, can be extremely viscous (Lowrie and Kent, 1978). The VRM is likely due to large homogeneous titanomagnetite grains with easily moved walls. VRM of titanomaghemites is very dependent on the degree of oxidation (see Dunlop and Özdemir, 1997, Chapter10).

For Tertiary and early Quaternary subaerial basalts, Brunhes epoch VRM averages about one-quarter of total NRM intensity (Prévot, 1981). Soft VRM is due to homogeneous multidomain titanomagnetite grains (cf. Figure M141). Relatively hard VRM is carried by single-domain size magnetite-ilmenite intergrowths formed by oxyexsolution. Thermal cleaning to ≤300°C erases all Brunhes epoch VRM, as Eq. (12) suggests.

Red sediments of all types have relatively strong and hard VRMs. Creer's (1957) study remains the classic. Among magnetite-bearing sedimentary rocks, limestones have been the most thoroughly studied (e.g., Borradaile, 2003).

Plutonic rocks, especially intermediate and felsic ones, tend to be quite viscous. The VRM is generally due to multidomain magnetite in grains ≈100 μm in size and is easily AF demagnetized ($\tilde{H} < 10$ mT). The primary TRM sometimes resides in elongated single-domain magnetite grains within silicate host minerals, and is physically as well as thermoviscously distinct from VRM.

Soils and baked clays, e.g., pottery and bricks, owe their viscous magnetization to fine single-domain magnetite and maghemite. The classic study is Le Borgne (1960).

David J. Dunlop

Bibliography

Biquand, D., and Prévot, M., 1971. AF demagnetization of viscous remanent magnetization in rocks. *Zeitschrift für Geophysik*, **37**: 471–485.

Borradaile, G.J., 2003. Viscous magnetization, archaeology and Bayesian statistics of small samples from Israel and England. *Geophysical Research Letters*, **30**(10): 1528, doi:10.1029/2003GL016977.

Creer, K.M., 1957. The remanent magnetization of unstable Keuper marls. *Philosophical Transactions of the Royal Society of London*, **A250**: 130–143.

Dunlop, D.J., 1973. Theory of the magnetic viscosity of lunar and terrestrial rocks. *Reviews of Geophysics and Space Physics*, **11**: 855–901.

Dunlop, D.J., 1983. Viscous magnetization of 0.04–100 μm magnetites. *Geophysical Journal of the Royal Astronomical Society*, **74**: 667–687.

Dunlop, D.J., and Özdemir, Ö., 1990. Alternating field stability of high-temperature viscous remanent magnetization. *Physics of the Earth and Planetary Interiors*, **65**: 188–196.

Dunlop, D.J., and Özdemir, Ö., 1997. *Rock Magnetism: Fundamentals and Frontiers.* Cambridge: Cambridge University Press.

Dunlop, D.J., and Özdemir, Ö., 2000. Effect of grain size and domain state on thermal demagnetization tails. *Geophysical Research Letters*, **27**: 1311–1314.

Dunlop, D.J., Schmidt, P.W., Özdemir, Ö., and Clark, D.A., 1997. Paleomagnetism and paleothermometry of the Sydney Basin. 1. Thermoviscous and chemical overprinting of the Milton Monzonite. *Journal of Geophysical Research*, **102**: 27271–27283.
Dunlop, D.J., Özdemir, Ö., Clark, D.A., and Schmidt, P.W., 2000. Time-temperature relations for the remagnetization of pyrrhotite (Fe_7S_8) and their use in estimating paleotemperatures. *Earth and Planetary Science Letters*, **176**: 107–116.
Enkin, R.J., and Dunlop, D.J., 1988. The demagnetization temperature necessary to remove viscous remanent magnetization. *Geophysical Research Letters*, **15**: 514–517.
Fox, J.M.W., and Aitken, M.J., 1980. Cooling-rate dependence of thermoremanent magnetisation. *Nature*, **283**: 462–463.
Halgedahl, S.L., 1993. Experiments to investigate the origin of anomalously elevated unblocking temperatures. *Journal of Geophysical Research*, **98**: 22443–22460.
Heider, F., Halgdahl, S.L., and Dunlop, D.J., 1988. Temperature dependence of magnetic domains in magnetite crystals. *Geophysical Research Letters*, **15**: 499–502.
Heller, F., and Markert, H., 1973. The age of viscous remanent magnetization of Hadrian's Wall (northern England). *Geophysical Journal of the Royal Astronomical Society*, **31**: 395–406.
Jackson, M., and Worm, H.-U., 2001. Anomalous unblocking temperatures, viscosity and frequency-dependent susceptibility in the chemically-remagnetized Trenton limestone. *Physics of the Earth and Planetary Interiors*, **126**: 27–42.
Le Borgne, E., 1960. Étude expérimentale du traînage magnétique dans le cas d'un ensemble de grains magnétiques très fins dispersés dans une substance non magnétique. *Annales de Géophysique*, **16**: 445–494.
Lowrie, W., and Kent, D.V., 1978. Characteristics of VRM in oceanic basalts. *Journal of Geophysics*, **44**: 297–315.
Middleton, M.F., and Schmidt, P.W., 1982. Paleothermometry of the Sydney Basin. *Journal of Geophysical Research*, **87**: 5351–5359.
Moskowitz, B.M., 1985. Magnetic viscosity, diffusion after-effect, and disaccommodation in natural and synthetic samples. *Geophysical Journal of the Royal Astronomical Society*, **82**: 143–161.
Mullins, C.E., and Tite, M.S., 1973. Magnetic viscosity, quadrature susceptibility, and frequency dependence of susceptibility in single-domain assemblies of magnetite and maghemite. *Journal of Geophysical Research*, **78**: 804–809.
Néel, L., 1949. Théorie du traînage magnétique des ferromagnétiques en grain fins avec applications aux terres cuites. *Annales de Géophysique*, **5**: 99–136.
Néel, L., 1955. Some theoretical aspects of rock magnetism. *Advances in Physics*, **4**: 191–243.
Perrin, M., 1998. Paleointensity determination, magnetic domain structure, and selection criteria. *Journal of Geophysical Research*, **103**: 30591–30600.
Prévot, M., 1981. Some aspects of magnetic viscosity in subaerial and submarine volcanic rocks. *Geophysical Journal of the Royal Astronomical Society*, **66**: 169–192.
Pullaiah, G., Irving, E., Buchan, K.L., and Dunlop, D.J., 1975. Magnetization changes caused by burial and uplift. *Earth and Planetary Science Letters*, **28**: 133–143.
Saito, T., Ishikawa, N., and Kamata, H., 2003. Identification of magnetic minerals carrying NRM in pyroclastic-flow deposits. *Journal of Volcanology and Geothermal Research*, **126**: 127–142.
Shimizu, Y., 1960. Magnetic viscosity of magnetite. *Journal of Geomagnetism and Geoelectricity*, **11**: 125–138.
Thellier, E., 1938. Sur l'aimantation des terres cuites et ses applications géophysiques. *Annales de l'Institut de Physique Globe de l'Université de Paris*, **16**: 157–302
Tivey, M., and Johnson, H.P., 1984. The characterization of viscous remanent magnetization in large and small magnetite particles. *Journal of Geophysical Research*, **89**: 543–552.
Walton, D., 1980. Time-temperature relations in the magnetization of assemblies of single-domain grains. *Nature*, **286**: 245–247.
Walton, D., 1983. Viscous magnetization. *Nature*, **305**: 616–619.
Walton, D., and Dunlop, D.J., 1985. The magnetization of a random assembly of interacting moments. *Solid State Communications*, **53**: 359–362.

Cross-references

Archeomagnetism
Iron Sulfides
Magnetic Domain
Magnetic Susceptibility
Magnetization, Natural Remanent (NRM)
Magnetization, Thermoremanent (TRM)

MAGNETOCONVECTION

Magnetoconvection refers to the thermal driving of flow within an electrically conducting fluid in the presence of an imposed magnetic field. In rapidly rotating systems like the Earth's fluid outer core, magnetoconvection offers a key insight into the intricate interaction between the effects of rotation and magnetic field. The simplest problem of magnetoconvection is given by a rotating plane horizontal layer of an electrically conducting Boussinesq fluid across which a uniform vertical magnetic field is imposed (Chandrasekhar, 1961; Roberts, 1978; Soward, 1979; Aurnou and Olson, 2001). The mathematical simplicity of the problem provides an essential understanding of the fundamental magnetohydrodynamic processes taking place in the Earth's fluid core. We shall use this example to illustrate these primary features of general magnetoconvection. For convenience and to avoid unnecessary complication, we assume that the fluid has constant viscosity ν, thermal diffusivity κ, and magnetic diffusivity λ. The thickness of the horizontal layer is taken to be d. We further suppose that magnetoconvection is stationary and that the amplitude of the convection is sufficiently small so that the nonlinear effects can be safely ignored. The lower boundary of the layer is heated to maintain a constant negative vertical temperature gradient β. When β is small, magnetoconvection cannot occur and the static equilibrium is described by

$$\mathbf{u}_0 = 0, \quad \mathbf{B} = \mathbf{k}B_0, \quad \nabla T_0 = -\beta\mathbf{k}, \quad \nabla p_0 + \mathbf{k}g\rho = 0 \quad \text{(Eq. 1)}$$

where $\mathbf{u}_0$ is the fluid velocity, $\mathbf{k}$ is the unit upward vector, p_0 is the total pressure (kinematic plus magnetic), ρ is the density of the fluid, g is the acceleration due to gravity, $\mathbf{B}$ is the imposed vertical uniform magnetic field, and T_0 is the conduction temperature. When β is sufficiently large, magnetoconvection takes place and modifies the basic static state. The problem of magnetoconvection is characterized by the Rayleigh number R, the Taylor number T_a, and the Chandrasekhar number Q, which are defined as

$$R = \frac{\alpha\beta g d^4}{\nu\kappa}, \quad T_a = \left(\frac{2\Omega d^2}{\nu}\right)^2, \quad Q = \frac{B_0^2 d^2}{\mu\nu\rho\lambda} \quad \text{(Eq. 2)}$$

where μ is the magnetic permeability and Ω is the angular velocity of the Earth. The Rayleigh number R provides a measure of the vertical buoyancy force, the Chandrasekhar number Q represents the ratio of the Lorentz force to the viscous force, and the Taylor number T_a is the squared ratio of the Coriolis force to the viscous force. The main characteristics of magnetoconvection can be elucidated in terms of these three physical parameters.

By assuming that the fluid layer has an infinitely horizontal extent, solutions of stationary magnetoconvection at onset can be expressed in the form

$$\mathbf{u}(x, y, z) \rightarrow \mathbf{u}(z) \exp[i(a_x x + a_y y)], \tag{Eq. 3}$$

where $\mathbf{u}$ is the velocity of the magnetoconvection, and a_x and a_y are the horizontal wave numbers with the total horizontal wave number a defined as

$$a^2 = a_x^2 + a_y^2. \tag{Eq. 4}$$

When the temperatures of the upper and lower boundaries are held constant, the flow velocity is stress-free at the boundaries and the magnetic field satisfies the stress-free-type condition discussed by Chandrasekhar (1961), the Rayleigh number R at the onset of magnetoconvection is given by

$$R = \frac{\xi^2}{a^2[\xi^4 + Q\pi^2]} \left[(\xi^2 + Q\pi^2)\xi^4 + T_a \pi^2 \xi^2 + Q(\xi^4 + Q\pi^2)\pi^2\right], \tag{Eq. 5}$$

where $\chi^2 \equiv a^2 + \pi^2$. The physically realizable form of magnetoconvection is characterized by the smallest value of the Rayleigh number, which is referred to as the critical Rayleigh number R_c. The subtle physics contained in Eq. (5) can be clearly unfolded by considering a number of extreme cases.

Magnetoconvection with strong field, weak rotation

When the system is slowly rotating in the absence of an imposed magnetic field, Chandrasekhar (1961) showed that the critical Rayleigh number R_c and the corresponding critical wave number of magnetoconvection a_c are given by

$$R_c \approx \frac{27\pi^4}{4}, \quad a_c \approx \frac{\pi}{\sqrt{2}}, \quad \text{for } Q = 0 \tag{Eq. 6}$$

However, both the critical Rayleigh number R_c and wave number a_c increase dramatically when a strong magnetic field is present

$$R_c \approx Q\pi^2, \quad a_c \approx \left(\frac{\pi^4}{2}\right)^{1/6} Q^{1/6}, \quad \text{for } Q \gg 1 \tag{Eq. 7}$$

Comparison between Eqs. (6) and (7) indicates that the role of the magnetic force is strongly stabilizing: the presence of an intense magnetic field requires a much larger Rayleigh number to excite magnetoconvection.

Magnetoconvection with weak field, strong rotation

The Proudman-Taylor theorem states that infinitesimal, nonmagnetic steady flows in a rotating inviscid fluid are two-dimensional with respect to the direction of the rotation axis. It follows that the Proudman-Taylor constraint must be broken in order that convection can occur in rapidly rotating systems. In consequence, a large viscous force in association with small-scale convection cells is usually required. When the system rotates rapidly and the imposed magnetic field is weak, we obtain from Eq. (5)

$$R_c \approx \frac{3}{2^{2/3}} (\pi^2 T_a)^{2/3}, \quad a_c \approx \left(\frac{\pi^2}{6}\right)^{1/6} T_a^{1/6}, \quad \text{for } T_a \gg 1 \tag{Eq. 8}$$

Comparison of Eqs. (6) and (8) indicates that the role of the Coriolis force is also strongly stabilizing: the effect of rapid rotation makes convection difficult.

Magnetoconvection with magnetic and Coriolis forces in balance

When the rotational effect is predominant ($T_a \gg 1$ and $Q \ll 1$), magnetoconvection is ineffective, invoking large viscous effects in connection with the short horizontal length of the flow; when the effect of the magnetic field is dominant ($Q \gg 1$ and $T_a \ll 1$), magnetoconvection has similar features. However, when the two inhibiting effects, rotation and magnetic field, act simultaneously and have a comparable strength ($Q \approx T_a^{1/2}$), magnetoconvection operates in the optimum state: both the critical Rayleigh number R_c and the critical wave number a_c reach an overall minimum. In other words, convection can be most readily excited when Lorentz and Coriolis forces are of comparable amplitude.

Several examples calculated from Eq. (5) are shown in Figure M147 for $T_a^{1/2} = 0, 10^4, 10^6$. The overall minimum for R_c and a_c can be explained by the fact that the effect of the magnetic field relaxes the

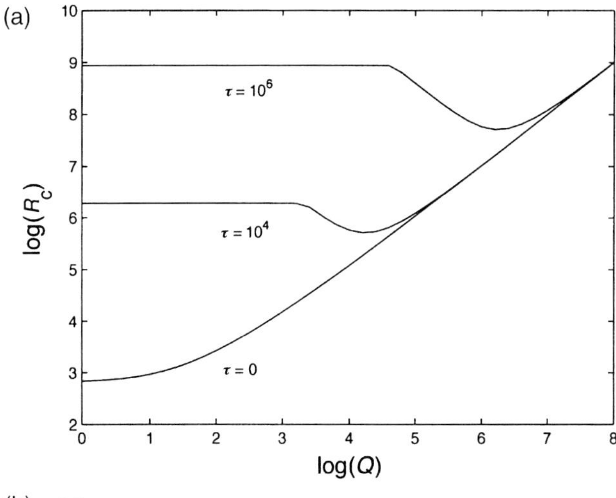

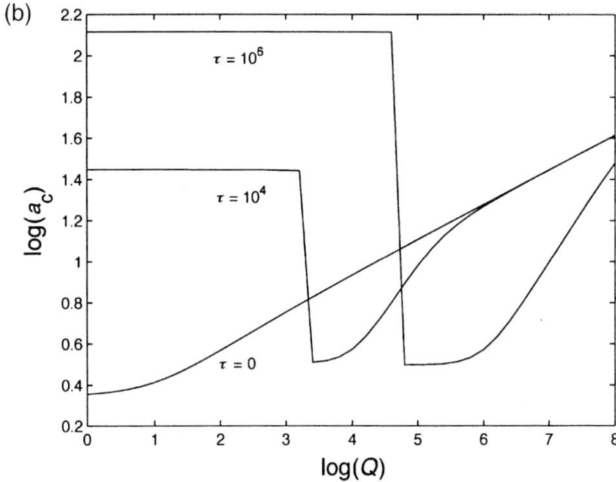

Figure M147 The critical Rayleigh number R_c (*top panel*) and the corresponding wave number a_c (*lower panel*), calculated using Eq. (5), are shown as functions of the Chandrasekhar number Q for different values of τ, where $\tau = T_a^{1/2}$.

Proudman-Taylor constraint, increases the length scale of the magnetoconvection cell and, hence, makes the system convect more readily and efficiently. This characteristic of magnetoconvection has led to an important suggestion that the dynamo of the Earth's core operates in the regime $T_a^{1/2} \approx Q$ where the geodynamo as a thermal engine is most effective.

Magnetoconvection in spherical geometry

In rotating spherical geometry, magnetoconvection in the presence of an imposed azimuthal field whose strength is proportional to distance from the rotation axis has been extensively studied (for example, Fearn, 1979, 1998; Proctor, 1994). Spherical magnetoconvection exhibits similar features in a plane layer: the critical Rayleigh number R_c reaches an overall minimum as the magnetic field strength increases to $Q = O(T_a^{1/2})$ at which the Lorentz and Coriolis forces are of comparable size. For larger values of Q, the effects of the magnetic field inhibit convection and thus R_c increases with growing Q; for smaller values of Q, the small scale of the convection cells resulting from the rotational constraint leads to extremely large R_c.

It should be pointed out, however, that spherical magnetoconvection in the presence of a more realistic magnetic field that satisfies electrically insulating boundary conditions shows a quite different behavior (Fearn and Proctor, 1983; Zhang, 1995). It was found that the two-dimensionality of purely thermal convection survives under the influence of a strong Lorentz force and that there exist no optimum values of Q that can give rise to an overall minimum of the critical Rayleigh number (Zhang, 1995). The value of R_c is a monotonically, smoothly decreasing function of Q. This is because the more realistic magnetic field can become unstable when Q is sufficiently large (Zhang and Fearn, 1993).

Keke Zhang and Xinhao Liao

Bibliography

Aurnou, J.M., and Olson, P.L., 2001. Experiments on Rayleigh-Bénard convection, magnetoconvection and rotating magnetoconvection in liquid gallium. *Journal of Fluid Mechanics*, **430**: 283–307.

Chandrasekhar, S., 1961. *Hydrodynamic and Hydromagnetic Stability.* Oxford: Clarendon Press.

Fearn, D.R., 1979. Thermal and magnetic instabilities in a rapidly rotating sphere. *Geophysics, Astrophysics, and Fluid Dynamics*, **14**: 103–126.

Fearn, D.R., 1998. Hydromagnetic flow in planetary cores. *Reports on Progress in Physics*, **61**: 175–235.

Fearn, D.R., and Proctor, M.R.E., 1983 Hydromagnetic waves in a differentially rotating sphere. *Journal of Fluid Mechanics*, **128**: 1–20.

Proctor, M.R.E., 1994. Convection and magnetoconvection in a rapidly rotating sphere. In Proctor, M.R.E., and Gilbert, A.D. (eds.), *Lectures on Stellar and Planetary Dynamos.* Cambridge: Cambridge University Press, pp. 75–115.

Roberts, P.H., 1978. Magneto-convection in a rapidly rotating fluid. In Roberts, P.H., and Soward, A.M. (eds.), *Rotating Fluids in Geophysics.* London: Academic Press, pp. 421–435.

Soward, A.M., 1979. Thermal and magnetically driven convection in a rapidly rotating fluid layer. *Journal of Fluid Mechanics*, **90**: 669–684.

Zhang, K., 1995. Spherical shell rotating convection in the presence of toroidal magnetic field. *Proceedings of the Royal Society of London Service A*, **448**: 245–268.

Zhang, K., and Fearn, D., 1993. How strong is the invisible component of the magnetic field in the Earth's core? *Geophysical Research Letters*, **20**: 2083–2088.

Cross-references

Core Convection
Core Motions
Geodynamo
Magnetohydrodynamic Waves
Proudman-Taylor Theorem

MAGNETOHYDRODYNAMIC WAVES

Introduction

Magnetohydrodynamic waves are propagating disturbances found in electrically conducting fluids permeated by magnetic fields where magnetic tension provides a restoring force on fluid parcels moving across field lines. The role played by magnetohydrodynamic waves, transporting disturbances in the flow and magnetic field and connecting disparate regions of the fluid, is crucial to our understanding of hydromagnetic systems. Magnetohydrodynamic waves in the Earth's liquid iron outer core have been proposed as the origin of changes of the Earth's magnetic field taking place on timescales of decades to centuries, and are thus of interest to both geomagnetists and paleomagnetists.

In the Earth's outer core, in addition to the magnetic forces acting on the electrically conducting fluid, we must also consider Coriolis forces resulting from planetary rotation, buoyancy forces due to gravity acting on density gradients and the constraints placed on flow by spherical shell geometry. Magnetohydrodynamic waves could be excited by convection-driven instabilities (Braginsky, 1964), topographically as flow is forced over bumps at the core-mantle boundary (Hide, 1966), by instabilities of the background magnetic field (Acheson, 1972), or even tidally due to deviations of rotating core geometry from exact sphericity (Kerswell, 1994).

This article focuses on the likely properties of magnetohydrodynamic waves in the Earth's outer core and provides a review of attempts to observe them. After a brief account of the history of investigations into magnetohydrodynamic waves in the section "Historical Review," the physics underpinning their existence will be described in the section "Force Balance and Waves in Rapidly Rotating Hydromagnetic Fluids." In particular, attention will focus on the emergence of a new characteristic timescale associated with such waves in rotating magnetohydrodynamic systems. Dispersion relations for magnetohydrodynamic waves when magnetic, buoyancy (Archimedes), and Coriolis forces are of equal importance (MAC waves) will be derived and interesting properties are noted in the section "Dispersion Relations for MC/MAC Waves in the Absence of Diffusion." The influence of diffusion, spherical geometry, and nonlinear effects on the waves will be discussed in the sections "Effects of Diffusion on MC/MAC Waves," "Influence of Spherical Geometry on MC/MAC Waves," and "Nonlinear Magnetohydrodynamic Waves," respectively. In the section "Magnetohydrodynamic Waves in a Stratified Ocean at the Core Surface," the suggestion that a stratified layer may exist at the top of the Earth's outer core is described and the type of waves that could be present there will be discussed. Finally, in the section "Magnetohydrodynamic Waves as a Mechanism for Geomagnetic Secular Variation," attempts to identify the presence of magnetohydrodynamic waves in the Earth's outer core through observations of the Earth's magnetic field will be reviewed and suggestions made as to how the wave hypothesis of geomagnetic secular variation could be tested using a combination of dedicated modeling and rapidly improving high-resolution observations.

For further details, the interested reader should consult the overviews by Hide and Stewartson (1972) and Braginsky (1989) or look in the textbooks by Moffatt (1978) or Davidson (2001). More technical reviews of the subject include Roberts and Soward (1972), Acheson and Hide (1973), Eltayeb (1981), Proctor (1994), and Zhang and Schubert (2000).

Historical review

Study of magnetohydrodynamic waves, especially with a focus on geophysical applications has a rich history and has captured the attention of some of the finest applied mathematicians and theoretical geophysicists over the past 50 years. Alfvén (1942) initiated the study of magnetohydrodynamic waves, investigating the simplest possible scenario where a balance of magnetic tension and inertia gives rise to waves, which became known as Alfvén waves in his honor (see *Alfvén Hannes* and *Alfvén waves*). Lehnert (1954) deduced that rapid rotation of the fluid system would lead to the splitting of plane Alfvén waves into two circularly polarized, transverse waves, one with period similar to inertial waves (a consequence of the intrinsic stability endowed to fluids by rotation) and a second with a much longer period. The latter represents a new, fundamental, timescale for rotating hydromagnetic systems which we shall refer to as the magnetic-Coriolis (MC) timescale. Chandrasekhar (1961) studied the effects of buoyancy on rotating magnetic systems, focusing primarily on axisymmetric motions invariant about the rotation axis. Braginsky (1964, 1967) realized the importance of nonaxisymmetric disturbances and showed that if magnetic, buoyancy, and Coriolis forces were equally important, fast inertial modes and slower magnetic modes would again result, but with periods also dependent on the strength of stratification. He christened these waves dependent on magnetic, buoyancy (Archimedes), and Coriolis forces as "MAC waves."

Hide (1966) was the first to consider the influence of spherical geometry on magnetohydromagnetic waves in a rotating fluid, studying the effects of the variation of Coriolis force with latitude. He showed that the resulting MC waves (commonly called MC Rossby waves) had the correct timescale to account for some parts of the geomagnetic secular variation, particularly its westward drift (see *Westward drift*). Malkus (1967) studied MC waves in a full sphere considering the special case when the background field increased in strength with distance from the rotation axis.

Elteyab (1972), Roberts and Stewartson (1974), Busse (1976), Roberts and Loper (1979), and Soward (1979) have demonstrated the importance of including magnetic and thermal diffusion in models of MAC waves, showing that the most unstable MAC waves in plane layer and annulus systems often occur on diffusive timescales. Elteyab and Kumar (1977) and Fearn (1979) carried out the first numerical studies of magnetohydrodynamic waves to include the effects of both buoyancy and diffusion in a rotating, spherical geometry. Fearn and Proctor (1983) went on to consider the effect of more geophysical physically plausible background magnetic fields and nonzero mean azimuthal flows. Most recently, Zhang and Gubbins (2002) have discussed the properties of convection-driven MAC and MC waves in a spherical shell geometry, studying a variety of background field configurations.

Force balance and waves in rapidly rotating hydromagnetic fluids

A physical understanding of MC waves can be achieved through consideration of the force balance in a rotating, electrically conducting, inviscid fluid that involves inertia, magnetic tension resisting flow across field lines (see *Alfvén waves* and *Magnetohydrodynamics*), and Coriolis forces acting normal to flows and to the axis of rotation. Coriolis forces are well known for causing circulating eddies in the atmosphere (e.g., hurricanes) and arise because, in a rotating reference frame, inertial motions follow curved trajectories rather than straight lines. It is useful to think about rotation imparting vorticity to a fluid, in the same way that magnetic fields impart tension perpendicular to magnetic field lines; vorticity imparts tension perpendicular to vortex lines (which lie parallel to the rotation axis) leading to a restoring force when fluid flows across them.

In this system, four possible force balances are conceivable. The first three require rapid fluid motions while the final is only possible for slow fluid motions. They are as follows:

1. When magnetic forces are much stronger than Coriolis forces; magnetic tension alone balances inertia and disturbances are communicated by Alfvén waves (see *Alfvén waves*).
2. When Coriolis forces are much stronger than magnetic forces; vortex tension balances inertia and disturbances are communicated by inertial waves.
3. When magnetic and Coriolis forces are of similar strength; a combination of magnetic field tension and vortex tension balances inertia and disturbances are communicated by inertial magnetic Coriolis (inertial-MC) waves.
4. When fluid motions are slow so that inertia is unimportant in the leading order force balance but magnetic and Coriolis forces are of similar strength; in this scenario, magnetic and vortex tension are in balance and disturbances are communicated by MC waves.

Balance (4) thus permits the existence of a new class of slow wave in rotating hydromagnetic systems, which is absent in nonmagnetic and nonrotating systems. Time dependence in this case arises only through changes in the magnetic field that, via the Lorentz force, produces changes in the fluid flow.

An estimate of the MC timescale can be obtained by performing a scale analysis of the important terms in the equations for conservation of momentum and magnetic induction. In the momentum equation, Coriolis and magnetic (Lorentz) forces are in balance, so $2\Omega U = B^2/\rho L_{MC}$, where Ω is the angular rotation rate, U is a typical velocity scale, B is a typical magnetic field strength, L_{MC} is a typical length scale over which changes associated with MC waves occur, ρ is the density of the fluid, and is the fluid's magnetic permeability. We also know that for a highly conducting fluid, changes in the magnetic field come primarily from advection, so scale analysis of the induction equation ignoring diffusion yields $B/T_{MC} = UB/L_{MC}$, where T_{MC} is a typical timescale over which changes associated with MC waves occur. Substituting this expression for U into the force balance leads to the relation $T_{MC} = 2\Omega L_{MC}^2 \rho / B^2$. The quantity $v_A = B/(\rho)^{1/2}$ has units of velocity and is the phase speed of Alfvén waves (see *Alfvén waves*). Estimates for these quantities in the Earth's core are $\boldsymbol{\Omega} = 7.3 \times 10^{-5}\,\mathrm{s}^{-1}$, $\rho = 1 \times 10^4\,\mathrm{kg}$, $= 4\pi \times 10^{-7} T^2\,\mathrm{m\,kg^{-1}\,s^2}$, so that $T_{MC} = 10^{-6} L_{MC}^2 / B^2$. Neither the magnetic field strength nor the length scale associated with its variation in the Earth's core is well known. Taking $B = 5 \times 10^{-4}\,\mathrm{T}$ as suggested by observations at the core surface and $L_{MC} = 3.5 \times 10^6\,\mathrm{m}$, the core radius, yields $T_{MC} \approx 1.5 \times 10^6$ y. Equally plausibly, if we are consider a wave with azimuthal wave number 8, and assume that the field inside the outer core is 10 times the observed core surface field strength we find $T_{MC} \approx 235$ years. The coincidence between the latter MC wave timescale and that of geomagnetic secular variation motivates attempts to link the two phenomena.

In the Earth's core, it is likely that buoyancy forces (either thermal or compositional) could also be important in the primary force balance (see *Core convection*). In the remainder of this article we shall therefore generalize our discussion to include buoyancy, which modifies the MC timescale to a MAC timescale because Archimedes forces are now present. In the next section we present an outline of the derivation of the dispersion relation for MAC waves.

Dispersion relations for MC/MAC waves in the absence of diffusion

To focus the discussion, while keeping mathematics to a minimum, we shall consider a rather basic model of a rapidly rotating, electrically conducting, incompressible fluid in an infinite three-dimensional domain, where there are no dissipative processes (viscous, magnetic, or thermal diffusion) operating. Buoyancy forces are included via the

Boussinesq model, with the degree of stratification depending on the magnitude of the background temperature gradient.

We shall work in Cartesian coordinates $(\hat{x}, \hat{y}, \hat{z})$ with the axis of rotation along $\hat{z}$, a uniform background field $\boldsymbol{B}_0 = (B_{0x}\hat{x} + B_{0y}\hat{y} + B_{0z}\hat{z})$ and a uniform background temperature gradient of $-\beta\hat{z}$. If α is the thermal expansivity of the fluid, then density is determined by the relation $\rho = \rho_0(1 + \alpha\Theta)$ where Θ is the perturbation from the background temperature field, so that in a gravity field of $-g\hat{z}$ there will be a buoyancy force of magnitude $g\alpha\Theta$ in the $\hat{z}$ direction.

We shall consider small perturbations $(\boldsymbol{u}, \boldsymbol{b}, \Theta)$ about a state of no motion (so the background velocity field is zero) and consider only slow motions so that inertial terms can be neglected and attention can focus on the MC force balance. The linearized equations governing the evolution of small perturbations are then the momentum equation

$$\underbrace{2\boldsymbol{\Omega}\times\boldsymbol{u}}_{\substack{\text{Coriolis}\\\text{acceleration}\\\text{due to rotation}}} = \underbrace{-\frac{1}{\rho}\nabla p}_{\substack{\text{acceleration}\\\text{due to pressure}\\\text{gradient}}} + \underbrace{\frac{1}{\mu\rho}(\boldsymbol{B}_0\cdot\nabla)\boldsymbol{b}}_{\substack{\text{acceleration}\\\text{due to}\\\text{field tension}}} + \underbrace{g\alpha\Theta\hat{z}}_{\substack{\text{buoyant}\\\text{acceleration}}}, \qquad \text{(Eq. 1)}$$

the induction equation

$$\underbrace{\frac{\partial \boldsymbol{b}}{\partial t}}_{\substack{\text{Change in the}\\\text{magnetic field}}} = \underbrace{(\boldsymbol{B}_0\cdot\nabla)\boldsymbol{u}}_{\substack{\text{Stretching of magnetic}\\\text{field by fluid motion}}}, \qquad \text{(Eq. 2)}$$

and temperature equation

$$\underbrace{\frac{\partial \Theta}{\partial t}}_{\substack{\text{Change in the}\\\text{temperature field}}} = \underbrace{\beta(\hat{z}\cdot\boldsymbol{u})}_{\substack{\text{advection of temperature}\\\text{field by fluid motion}}}. \qquad \text{(Eq. 3)}$$

Taking $\partial/\partial t(\nabla\times)$ the momentum equation (1) to eliminate pressure gives

$$2(\boldsymbol{\Omega}\cdot\nabla)\frac{\partial \boldsymbol{u}}{\partial t} = \frac{(\boldsymbol{B}_0\cdot\nabla)}{\rho}\frac{\partial}{\partial t}(\nabla\times\boldsymbol{b}) + g\alpha\frac{\partial}{\partial t}(\nabla\times\Theta\hat{z}), \qquad \text{(Eq. 4)}$$

while taking the curl $(\nabla\times)$ of the induction equation (2) we find,

$$\frac{\partial}{\partial t}(\nabla\times\boldsymbol{b}) = (\boldsymbol{B}_0\cdot\nabla)(\nabla\times\boldsymbol{u}). \qquad \text{(Eq. 5)}$$

Substituting from Eq. (5) into Eq. (4) for $\partial/\partial t(\nabla\times\boldsymbol{b})$ gives a vorticity equation quantifying the MAC balance with terms arising from Coriolis forces on the left-hand side and terms arising from the magnetic and buoyancy forces on the right-hand side

$$2(\boldsymbol{\Omega}\cdot\nabla)\frac{\partial \boldsymbol{u}}{\partial t} = \frac{(\boldsymbol{B}_0\cdot\nabla)^2}{\rho}(\nabla\times\boldsymbol{u}) + g\alpha\frac{\partial}{\partial t}(\nabla\times\Theta\hat{z}). \qquad \text{(Eq. 6)}$$

Operating on Eq. (6) with $((B_0\cdot\nabla)^2\nabla\times)/\rho$, we can then eliminate $((B_0\cdot\nabla)^2\nabla\times\boldsymbol{u})/\rho$ from the term on the left-hand side by using Eq. (6) once again. By utilizing the well-known relation for incompressible fluids that $\nabla\times\nabla\times\boldsymbol{u} = -\nabla^2\boldsymbol{u}$, and then taking the dot product with $\hat{z}$ while noting that $\hat{z}\cdot(\nabla\times\Theta\hat{z}) = 0$ and $\hat{z}\cdot(\nabla\times\nabla\times\Theta\hat{z}) = -(\partial^2/\partial x^2 + \partial^2/\partial y^2)\Theta = -\nabla_H^2\Theta$ leaves

$$4(\boldsymbol{\Omega}\cdot\nabla)^2\frac{\partial^2}{\partial t^2}u_z = -\left[\frac{(\boldsymbol{B}_0\cdot\nabla)^2}{\rho}\right]^2\nabla^2 u_z - g\alpha\frac{(\boldsymbol{B}_0\cdot\nabla)^2}{\rho}\nabla_{\mathrm{H}}^2\frac{\partial\Theta}{\partial t}.$$

Finally, we make use of the temperature Eq. (3) to eliminate $\partial\Theta/\partial t$ and obtain a sixth-order equation in u_z, which we shall refer to as the diffusionless MAC wave equation

$$\left(4(\boldsymbol{\Omega}\cdot\nabla)^2\frac{\partial^2}{\partial t^2} + \left[\frac{(\boldsymbol{B}_0\cdot\nabla)^2}{\rho}\right]^2\nabla^2 - g\alpha\beta\frac{(\boldsymbol{B}_0\cdot\nabla)^2}{\rho}\nabla_{\mathrm{H}}^2\right)u_z = 0.$$

Properties of diffusionless MAC waves can now be deduced by substitution of plane traveling wave solutions of the form $u_z = \mathbf{Re}\{\hat{u}_z \mathrm{e}^{i(\boldsymbol{k}\cdot\boldsymbol{r}-\omega t)}\}$, where $\boldsymbol{k}$ is the wavevector and ω is the angular frequency

$$4(\boldsymbol{\Omega}\cdot\boldsymbol{k})^2\omega^2 - \left[\frac{(\boldsymbol{B}_0\cdot\boldsymbol{k})^2}{\rho}\right]^2 k^2 - \frac{(\boldsymbol{B}_0\cdot\boldsymbol{k})^2}{\rho}g\alpha\beta(k_x^2 + k_y^2) = 0. \qquad \text{(Eq. 7)}$$

This expression can be written more concisely by observing that terms in it correspond to characteristic natural frequencies for magnetic-inertial (Alfvén) waves, gravity waves, and inertial waves in rotating fluids, respectively

$$\omega_{\mathrm{M}}^2 = \frac{(\boldsymbol{B}_0\cdot\boldsymbol{k})^2}{\rho}, \quad \omega_{\mathrm{A}}^2 = \frac{g\alpha\beta(k_x^2 + k_y^2)}{k^2}, \quad \omega_{\mathrm{C}}^2 = \frac{4(\boldsymbol{\Omega}\cdot\boldsymbol{k})^2}{k^2}, \qquad \text{(Eq. 8)}$$

so Eq. (7) simplifies to

$$\omega_{\mathrm{C}}^2\omega^2 - \omega_{\mathrm{M}}^4 - \omega_{\mathrm{M}}^2\omega_{\mathrm{A}}^2 = 0. \qquad \text{(Eq. 9)}$$

Solving for ω gives the necessary condition (or dispersion relation) that must be satisfied by the angular frequency and wavevectors of plane MAC waves,

$$\omega = \pm\frac{\omega_{\mathrm{M}}^2}{\omega_{\mathrm{C}}}\left(1 + \frac{\omega_{\mathrm{A}}^2}{\omega_{\mathrm{M}}^2}\right)^{1/2} = \pm\frac{k(\boldsymbol{B}_0\cdot\boldsymbol{k})^2}{2\rho(\boldsymbol{\Omega}\cdot\boldsymbol{k})}\left(1 + \frac{g\alpha\beta\rho(k_x^2 + k_y^2)}{k^2(\boldsymbol{B}_0\cdot\boldsymbol{k})^2}\right)^{1/2}. \qquad \text{(Eq. 10)}$$

Note that this is singular if $\boldsymbol{B}_0\cdot\boldsymbol{k} = 0$ or if $\boldsymbol{\Omega}\cdot\boldsymbol{k} = 0$, so diffusionless MAC waves cannot propagate normal to magnetic field lines or the rotation axis. Their frequency depends strongly on their wavelength (i.e., they are highly dispersive) and on their direction (i.e., they are anisotropic). In the special case when the background magnetic field and the direction of the rotation axis are parallel to the direction of wave propagation, and when buoyancy forces are absent ($\alpha = 0$), the dispersion relation simplifies to $\omega = B_0^2k^2/2\Omega\rho$ or $T_{\mathrm{MC}} = 2\Omega\rho L_{\mathrm{MC}}/B_0^2$ as was deduced from scaling arguments in the previous section. The phase speed of the waves is then $c = \omega/k = B_0^2k/2\Omega\rho$ and it is seen that waves with shorter wavelengths travel faster.

Effects of diffusion on MC/MAC waves

So far we have neglected the influence of any source of dissipation (viscous, magnetic, or thermal diffusion) on the system in order to simplify both the mathematical analysis and the physical picture. It is now necessary to consider their effects. Naively, we might expect the presence of dissipation should merely damp disturbances and irreversibly transform energy to an unusable form. Although such processes undoubtedly occur, they are not the only effects of the presence of diffusion. Perhaps more importantly, diffusion adds extra degrees of freedom to the system and facilitates the destabilization of waves that are stable in the absence of diffusion (see Roberts and Loper, 1979). This rather counterintuitive effect means that instability of MAC/MC waves can occur for smaller unstable density or magnetic field gradients than if no diffusion were present. In fact, such diffusive instability turns out to be possible, even in the presence of a stable density gradient.

The diffusive instability mechanism works most effectively when the oscillation frequency matches the rate of diffusion, so the timescale of the most unstable MC/MAC waves will be that of the diffusion process that is facilitating the instability. Diffusion thus introduces new preferred timescales into the MC/MAC wave problem.

To include diffusion in the mathematical description of MAC waves, we must replace the operator $\partial/\partial t$ by $(\partial/\partial t - \nu\nabla^2)$ in the momentum equation, by $(\partial/\partial t - \eta\nabla^2)$ in the induction equation and by $(\partial/\partial t - \kappa\nabla^2)$ in the heat equation. Retaining the acceleration term from the momentum equation and including the Laplacian (diffusion) terms before the substitution of plane wave solutions results in a more complicated dispersion relation for diffusive MAC waves

$$\left(\omega_C^2(\omega + i\eta k^2)^2 - \left[(\omega + i\nu k^2)(\omega + i\eta k^2) - \omega_M^2\right]^2\right)(\omega + i\kappa k^2) + \omega_A^2(\omega + i\eta k^2)\left[(\omega + i\nu k^2)(\omega + i\eta k^2) - \omega_M^2\right] = 0 \qquad \text{(Eq. 11)}$$

Restricting ourselves to the conditions present in the Earth's outer core, where we expect ohmic diffusion to dominate viscous and thermal diffusion ($\eta \gg \nu, \kappa$) and where we can again neglect inertial accelerations when considering slow oscillations, this expression simplifies to

$$\left(\omega_C^2(\omega + i\eta k^2)^2 - \omega_M^4\right)\omega - \omega_M^2\omega_A^2(\omega + i\eta k^2) = 0. \qquad \text{(Eq. 12)}$$

The link to diffusionless MC waves becomes apparent if buoyancy forces are negligible ($\omega_A = 0$) when Eq. (12) reduces to,

$$\omega = \frac{\omega_M^2}{\omega_C} - i\eta k^2. \qquad \text{(Eq. 13)}$$

Here the classical damping role of magnetic diffusion is obvious, causing MC waves with shorter wavelengths to decay in amplitude more quickly than MC waves with longer wavelengths. More detail on diffusive MC waves and their consequences for on geodynamo simulations can be found in Walker *et al.* (1998).

Influence of spherical geometry on MC/MAC waves

The Earth's outer core is not an infinite plane layer, but a thick spherical shell with an inner radius approximately one-third of its outer radius. How does spherical shell geometry influence propagation properties, stability, and the planform of magnetohydrodynamic waves? It appears that when the magnetic field is strong enough, and the Lorentz force dominates the force balance in the momentum equation, then the spherical boundaries play a secondary role. On the other hand, when the magnetic field is weak, the influence of the Coriolis force and its latitudinal variations caused by the spherical geometry are crucial. In the absence of any certain knowledge of the strength of the magnetic field in the Earth's core it is unclear if spherical shell geometry has a controlling influence on magnetohydrodynamic waves there, so the safest course is to use spherical geometry and study a variety of magnetic field strengths.

Hide (1966) was the first to appreciate the importance of the latitudinal dependence of the Coriolis force for waves in the Earth's core. He developed a simple analytical model of MC waves retaining only the linear variation of Coriolis force with latitude (this is known to meteorologists and oceanographers as a β plane model and is the necessary ingredient for the restoring force responsible for Rossby waves). We shall refer to Hide's waves as magnetic-Coriolis (MC) Rossby waves. He showed they would propagate westward in a thick spherical shell and could have a timescale similar to that of geomagnetic secular variation.

Eltayeb and Kumar (1977) and Fearn (1979) included both thermal buoyancy and diffusion and worked in spherical geometry. They were confronted by a rather complex scene, with several different mechanisms giving rise to different types of magnetohydrodynamic waves, any of which could potentially be important in the Earth's core. They identified four distinct regimes where different waves were favored. Only a brief overview of the four possible regimes is presented here.

Type I: Magnetically modified, buoyancy-driven Rossby waves

When the magnetic field is very weak, wave motion is essentially that produced by convection in a rapidly rotating sphere (i.e., thermal Rossby waves or Busse rolls—see *Core convection*). The flow consists of columnar rolls parallel to the rotation axis, arranged on a cylindrical shell that intersects the outer boundary at midlatitudes. At the onset of convection and in the absence of any mean azimuthal flow these waves drift eastward on a thermal diffusion timescale. The planform of the waves is columnar because the Coriolis force promotes invariance along the rotation axis (see *Proudmann-Taylor theorem*). The magnetic field acts only as a small perturbation and actually stabilizes the system, increasing the critical Rayleigh number compared to the nonmagnetic systems. Instability is driven by the component of buoyancy perpendicular to the rotation axis and is balanced primarily by the Coriolis force (varying with latitude due to the spherical geometry) and viscous diffusion.

Type II: Buoyancy-driven magneto-Rossby waves

With stronger magnetic fields, buoyancy is balanced by the magnetic (Lorentz) force as well as the Coriolis force. The planform of Type II waves is similar to those of Type I waves and they too propagate on the thermal diffusion timescale. Both westward and eastward propagation of these waves is possible, depending on the relative magnitudes of magnetic and thermal diffusion. It should be noted that the magnetic field now plays a destabilizing roll, catalyzing the onset of convection. The dominant role of the uniform imposed magnetic field causes an increase in the length scale so that the number of waves fitting around a cylindrical shell decreases, while the latitude of the rolls moves toward where the magnetic field is strongest. Figure M148 shows the form of the radial magnetic field disturbance produced by a magneto-Rossby wave in spherical geometry when the imposed magnetic field increases linearly with distance from the rotation axis (the force-free field of Malkus, 1967). This figure is the result of an eigenvalue calculation (using the code of Jones *et al.*, 2003) used to determine the most unstable wave in a regime when magnetic and Coriolis forces are approximately of equal magnitude.

Type III: Buoyancy-driven MAC waves

When magnetic forces are much stronger than the Coriolis forces, boundary curvature associated with spherical geometry plays a less important role and the most unstable wave is of diffusive MAC-type, again propagating on the thermal diffusion timescale. The planform of the waves is no longer that of columnar rolls because the strong magnetic fields permit departures from z independence. Both westward- and eastward-propagating waves are possible in this regime.

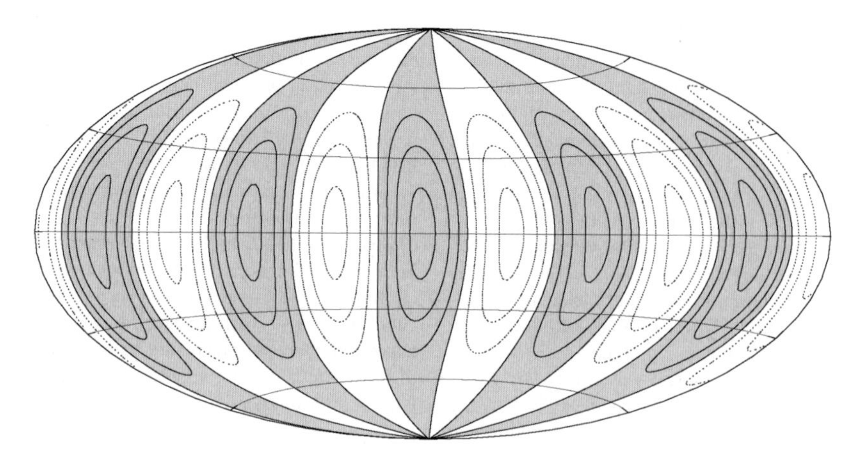

Figure M148 Anomalies in the radial magnetic field (B_r) at $r = 0.95r_0$ produced by a marginally critical, $m = 5$, buoyancy-driven magneto-Rossby wave. The imposed magnetic field is purely toroidal and increases linearly in magnitude with distance from the rotation axis. Units are arbitrary because no nonlinear saturation mechanisms are included in this model. Gray regions with solid contours indicate negative field anomalies, and white regions with dotted contours indicate positive field anomalies.

Type IV: Magnetically driven MAC waves

When the magnetic field becomes sufficiently strong or complex, then MC/MAC waves can be produced by either diffusive (resistive) or ideal instability of the background magnetic field. The resulting waves are of diffusive MC/MAC or diffusionless MC/MAC type and propagate on either the magnetic diffusion or on the MC/MAC timescale. They do not require the presence of buoyancy for their existence, and can even occur when the background density field is stable.

Fearn and Proctor (1983) have considered the additional effect of the presence of a background azimuthal flow (including shear) and found that this tends to stabilize diffusive MAC waves. They observed that such waves are localised at the extrema of the shear, moving with an azimuthal speed equal to the fluid velocity at that point. This indicates MC/MAC waves could perhaps be preferentially excited in zonal jets and would drift by advection rather than propagation, which could perhaps be of relevance at low latitudes in the Earth's outer core (see the section "Magnetohydrodynamic Waves as a Mechanism for Geomagnetic Secular Variation").

Nonlinear magnetohydrodynamic waves

All the magnetohydrodynamic waves discussed up to now have been linear in nature. This implies that (i) waves can simply be superposed without considering any mutual interaction and (ii) there is no feedback between the waves and the rest of the system. This scenario is unphysical because unstable waves can grow without limit, but is nonetheless useful for determining the types of waves most easily excited in a particular regime of interest. Early studies by Braginsky (1967) and Roberts and Soward (1972), though deriving linear equations for MAC waves riding on general background states, emphasized the importance of understanding nonlinear feedback processes. They noted that waves are determined by the background state, but the background state is itself altered by the waves. In seeking to interpret geophysical observations indicative of hydromagnetic waves in the Earth's core, we should remember that linear analysis is only formally valid for small perturbations to an artificial, steady background state and cannot tell us how waves will evolve, saturate, interact with each other, or what flow structures might result from nonlinear bifurcations of the waves. Attempts to understand such processes deserve a concerted theoretical and numerical modeling effort in the future.

Some progress in understanding nonlinear MAC waves has already been made. El Sawi and Eltayeb (1981) have derived higher order equations for MAC waves in a plane layer in the presence of a slowly varying background mean flow. Their equations describe the evolution of diffusionless MAC wave amplitude via the conservation of wave action (wave energy divided by wave frequency per unit volume). This conservation law tells us that wave energy increases, at the expense of the energy of the background state, whenever a wave moves into a region where its frequency is higher. More recently Ewen and Soward (1994) have derived equations describing the evolution of the amplitude of diffusive MAC waves in the limit of a weak magnetic field. They find that an azimuthal (geostrophic) mean flow is driven by magnetic forces resulting from the MAC waves and is linearly damped by viscous diffusion at the boundary.

A start has also been made at numerically investigating the nonlinear evolution of magnetohydrodynamic waves in rapidly rotating, convecting spherical shells. As the unstable density gradient is increased, it is observed that the system undergoes bifurcations from steadily traveling magnetohydrodynamic waves to vacillating wave motions for which both temporal and spatially symmetries have been broken (see *Magnetoconvection*).

Study of nonlinear MC and MAC waves is still in its infancy and may yet yield important and exciting insights that will help us to better understand how magnetohydrodynamic waves might manifest themselves in the Earth's core.

Magnetohydrodynamic waves in a stratified ocean at the core surface?

There has been some debate over the possibility of a stratified layer or "inner ocean" at the top of the Earth's outer core and the MAC waves that would be supported there (Braginsky, 1999). This stratified layer has yet to be observed seismically, though its existence seems plausible on thermodynamic grounds with light fluid released during the solidification of the inner core expected to pond below the core-mantle boundary. Oscillations of such a layer would be of shorter period than the MC/MAC waves expected in the body of the outer core due to the presence of an additional restoring force due to density stradification.

The dynamics of a stably stratified layer would be dominated by its thin spherical shell geometry. There would undoubtedly be many similarities with the water ocean on the Earth's surface, but with the additional complications caused by the presence of magnetic forces. In particular, MC Rossby waves, which rely on the change in the Coriolis force with latitude for their existence, are likely to be present within such an ocean. Braginsky has developed models of both axisymmetric and nonaxisymmetric disturbances of such a stably stratified layer and has suggested they could be responsible for short period geomagnetic secular variation.

Unfortunately, MAC waves and MC Rossby waves in a hidden ocean at the top of the outer core are not the only possible source of short period geomagnetic secular variation-torsional oscillations within the body of the core (see *Oscillations, torsional*) are an equally plausible explanation. Until the existence of the hidden ocean of the core can be confirmed, study of magnetohydrodynamic waves that may exist there will remain of primarily theoretical interest.

Magnetohydrodynamic waves as a mechanism for geomagnetic secular variation

It has been common knowledge since the time of Halley that the Earth's magnetic field changes significantly over decades to centuries. Perhaps the most striking aspect of this geomagnetic secular variation is the westward motion of field features (see Westward drift). Several explanations have been proposed for this Westward drift, but today there are two widely accepted candidate mechanisms. The first involves bulk fluid motion at the surface of the outer core that advects magnetic field features. Bullard *et al.* (1950) originally envisaged this involving westward flow of all the fluid close to the core surface, but modern core flow inversions (see *Core motions*) have refined this suggestion—it now appears that a westward equatorial jet under the Atlantic hemisphere is sufficient to explain much of the westward drift of the geomagnetic field observed at the surface. The source of this proposed equatorial jet is still debated, but geodynamo models indicate that it could be produced by nonlinear inertial forces that are a by-product of columnar convection in a sphere, or by thermal winds due to an inhomogeneous heat flux into the mantle (see *Inhomogeneous boundary conditions and the dynamo*). The second possible mechanism is that motivating the inclusion of this article in an *Encyclopedia of Geomagnetism and Paleomagnetism*—propagation of magnetohydrodynamic waves in the Earth's outer core. We shall henceforth refer to this mechanism as the wave hypothesis.

Hide (1966) and Braginsky (1967) proposed the wave hypothesis on theoretical grounds and each attempted to test it through the consideration of available records of the geomagnetic secular variation. It is informative to review these pioneering attempts before considering other possible ways to search for the presence of magnetohydrodynamic waves in the Earth's outer core.

Hide (1966) identified three major factors in favor of the wave hypothesis: (i) MC waves had periods comparable with the timescale of geomagnetic secular variation; (ii) MC waves had dispersion times comparable with the timescale of geomagnetic secular variation; and (iii) shorter wavelength MC waves had larger phase velocities and similarly higher order spherical harmonic components of the geomagnetic field drifted faster. His observational analysis was based on the mean westward drift rates of spherical harmonics up to degree 4, from seven previous publications, spanning 135 years from 1830 to 1965. Despite the failure of detailed comparisons between the predicted and observed drift rates for individual spherical harmonics, this study was instrumental in persuading many geophysicists that a wave origin for geomagnetic secular variation was worth serious consideration.

Braginsky (1967, 1972, 1974) sought to confirm the wave hypothesis by comparing his own theoretical predictions of the spectrum of diffusionless MAC waves to an observationally inferred spectrum of geomagnetic secular variation. Despite the poor quality of data available in the early

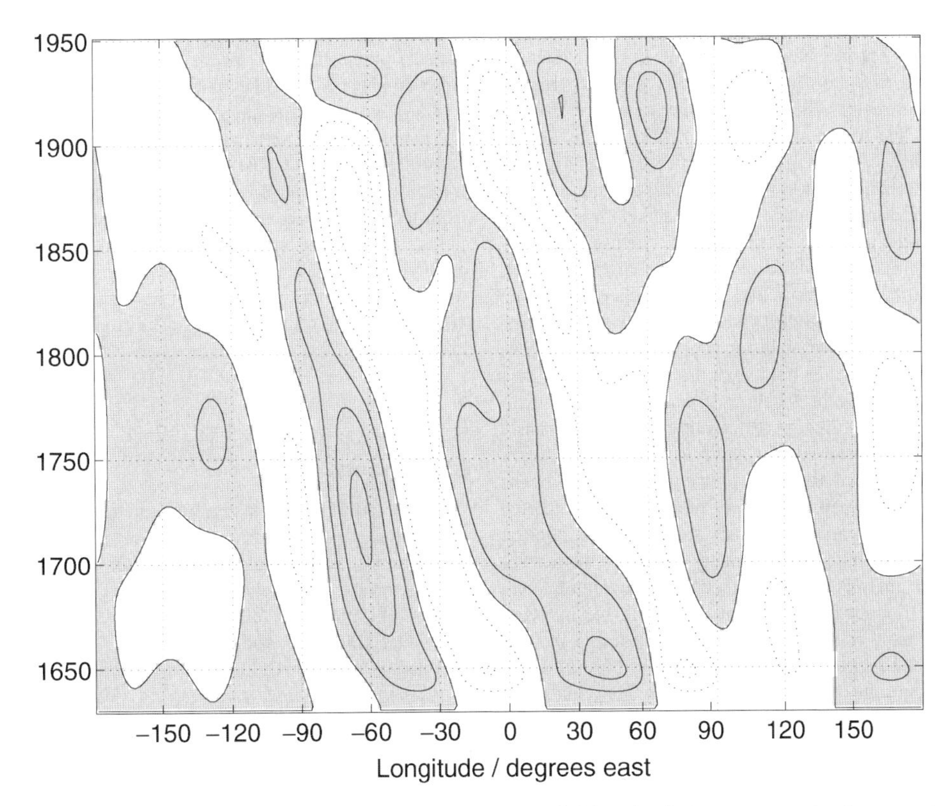

Figure M149 Equatorial time-longitude plot of processed radial magnetic field B_r (with time-averaged axisymmetric field and field variations with timescales longer than 400 years removed) from the historical field model *gufm1*, (Jackson *et al.*, 2000). Field evolution in the azimuthal direction at the equator is shown and consists of spatially and temporally coherent wavelike anomalies, with dominant azimuthal wave number $m = 5$ and moving consistently westward. The contour lines are at intervals of 5×10^4 nT. Gray regions with solid contours indicate negative field anomalies, and white regions with dotted contours indicate positive field anomalies.

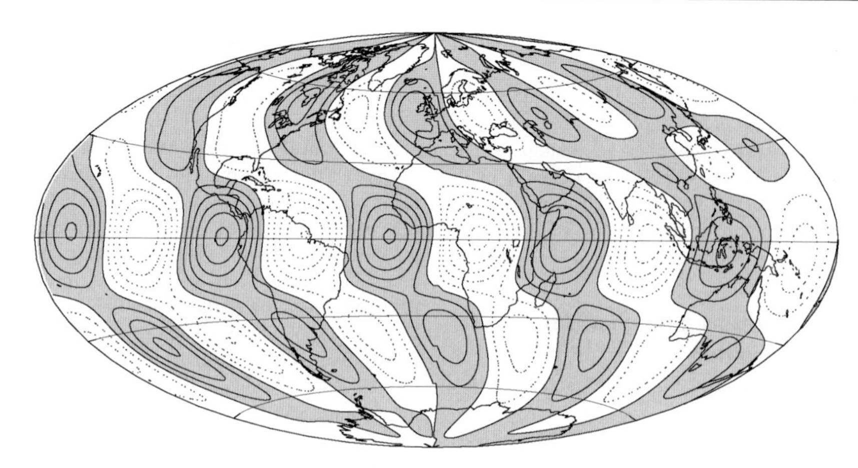

Figure M150 Snapshot from 1830 of the $m = 5$ radial magnetic field, core surface signal responsible for the wavelike pattern of field evolution observed at low latitudes in the historical geomagnetic field model *gufm1*. This snapshot was obtained by restricting the period of field variations to between 125 and 333 years and the wave number of field variations to $m = 5$ (i.e., FK filtering). The contour lines are at intervals of 4×10^3 nT. Gray regions with solid contours indicate negative field anomalies, and white regions with dotted contours indicate positive field anomalies.

1970s, Braginsky's efforts were important in demonstrating that the wave hypothesis was at least compatible with observations.

In the last 20 years there have been significant advances in our observational knowledge of the Earth's magnetic field and its evolution. Using over 365000 historical observations from maritime records, observatories, surveys, and satellites, time-dependent models of the global magnetic field have now been constructed covering the past 400 years (see *Time-dependent models of the geomagnetic field*). Use of regularised inversion methods has enabled the construction of images of the magnetic field at the core surface allowing us to map its evolution at the edge of the source region. This technique has shown that the origin of the westward drift of magnetic field is not due to the whole field drifting westward but rather is due to the drift of small patches of intense magnetic field (see *Westward drift*), particularly at low latitudes in the Atlantic hemisphere. Advances have also been made in our understanding of field evolution over longer time intervals, notably through the use of archeomagnetic and lake sediment records to construct global field models for the past 3 ka (Korte and Constable, 2005). Although lacking the resolution of the historical models, these models offer the first glimpse of the long-term behavior of flux patches at the core surface and with the inclusion of more data hold great promise for the future.

Can recent observational advances help us in our attempts to evaluate the wave hypothesis by facilitating the identification of previously obscured signatures of magnetohydrodynamic waves? The existence of images of the evolution of the field at the core surface opens up fresh possibilities for comparing not only timescales of predicted wave motions, but also geographical features such as the latitude at which waves occur, their equatorial symmetry and their dispersive properties. Space-time analysis of geomagnetic field evolution, and particularly the use of time-longitude diagrams (as employed by oceanographers to study Rossby waves), can be carried out during the historical epoch (Finlay and Jackson, 2003). As shown in Figure M149, after the removal of the time-averaged axisymmetric field and those components of the field varying on timescales longer than 400 years, spatially and temporally coherent (wavelike) evolution of the radial magnetic field is observed at the equator. The clearest signal in the time-longitude plot has an azimuthal wave number of $m = 5$, a period of around 250 years and travels at $\sim$17 km yr^{-1} westwards. Figure M150 shows the result of frequency-wave number filtering to recover the spatial structure of this wave at the core surface. The domination of aspects of geomagnetic secular variation by a single wave number disturbance suggests that a magnetohydrodynamic wave (perhaps driven by an instability) might currently be present at low latitudes in Earth's outer core.

The major challenge for theoreticians is to keep pace with improving observations and construct models accurate enough to predict the structure of observable space-time features caused by magnetohydrodynamic waves in the core. Modeling of convection-driven magnetohydrodynamic waves (see *Magnetoconvection*) suggests these might propagate too slowly to account for observed azimuthal field motions. However, thermal and magnetic winds in the core (see *Thermal wind*) could advect wave patterns at the required speeds and produce spatially and temporally coherent patterns of field evolution consistent with observations.

The hypothesis that magnetohydrodynamic waves produce directly observable changes in the geomagnetic field cannot yet be conclusively confirmed or rejected. The difficulty in rigorously testing the hypothesis is twofold. Firstly, one requires robust predictions of observable signatures from theoretical wave models that take account of realistic background fields, diffusion, spherical geometry, and nonlinearity. Secondly, one requires self-consistent, high-fidelity observations over a long period of time to assess model predictions in a statistically significant way. Almost 40 years after the wave hypothesis was proposed this remains a tall order, but we are approaching the stage where such tests will be feasible.

Christopher Finlay

Bibliography

Acheson, D.J., 1972. On hydromagnetic stability of a rotating fluid annulus. *Journal of Fluid Mechanics*, **52**: 529–541.

Acheson, D.J., and Hide, R., 1973. Hydromagnetics of rotating fluids. *Reports on Progress in Physics*, **36**: 159–221.

Alfvén, H., 1942. Existence of electromagnetic-hydrodynamic waves. *Nature*, **150**: 405–406.

Braginsky, S.I., 1964. Magnetohydrodynamics of the Earth's core. *Geomagnetism and Aeronomy*, **4**: 898–916 (English translation, 698–712).

Braginsky, S.I., 1967. Magnetic waves in the Earth's core. *Geomagnetism and Aeronomy*, **7**: 1050–1060 (English translation, 851–859).

Braginsky, S.I., 1972. Analytical description of the geomagnetic field of past epochs and determination of the spectrum of magnetic waves in the core of the Earth. *Geomagnetism and Aeronomy*, **12**: 1092–1105 (English translation, 947–957).

Braginsky, S.I., 1974. Analytical description of the geomagnetic field of past epochs and determination of the spectrum of magnetic waves in the core of the Earth. II. *Geomagnetism and Aeronomy*, **14**: 522–529 (English translation, 441–447).

Braginsky, S.I., 1989. Magnetohydrodynamic waves within the Earth. In James, D. E. (eds.), *Encyclopedia of Solid Earth Geophysics*. Kluwer Academic Publishers.

Braginsky, S.I., 1999. Dynamics of the stably stratified ocean at the top of the core. *Physics of the Earth and Planetary Interiors*, **111**: 21–34.

Bullard, E.C., Freedman, C., Gellman, H., and Nixon, J., 1950. The westward drift of the Earth's magnetic field. *Philosophical Transactions of the Royal Society of London*, **243**: 67–92.

Busse, F.H., 1976. Generation of planetary magnetism by convection. *Physics of the Earth and Planetary Interiors*, **12**: 350–358.

Chandrasekhar, S., 1961. *Hydrodynamic and Hydromagnetic Stability*. Oxford: Oxford University Press.

Davidson, P.A., 2001. *An Introduction to Magnetohydrodynamics*. Cambridge: Cambridge University Press.

El Sawi, M., and Eltayeb, I.A., 1981. Wave action and critical surfaces for hydromagnetic-inertial-gravity waves. *Quarterly Journal of Mechanics and Applied Mathematics*, **34**: 187–202.

Eltayeb, I.A., 1972. Hydromagnetic convection in a rapidly rotating fluid layer. *Proceedings of the Royal Society of London Series A*, **326**: 229–254.

Eltayeb, I.A., 1981. Propagation and stability of wave motions in rotating magnetic systems. *Physics of the Earth and Planetary Interiors*, **24**: 259–271.

Eltayeb, I.A., and Kumar S., 1977. Hydromagnetic convective instability of a rotating self-gravitating fluid sphere containing a uniform distribution of heat sources. *Proceedings of the Royal Society of London Series A*, **353**: 145–162.

Ewen, S.A., and Soward, A.M., 1994. Phase mixed rotating magnetoconvection and Taylor's condition. I. Amplitude equations. *Geophysical and Astrophysical Fluid Dynamics*, **77**: 209–230.

Fearn, D.R., 1979. Thermal and magnetic instabilities in a rapidly rotating sphere. *Geophysical and Astrophysical Fluid Dynamics*, **14**: 102–126.

Fearn, D.R., and Proctor M.R.E., 1983. Hydromagnetic waves in a differentially rotating sphere. *Journal of Fluid Mechanics*, **128**: 1–20.

Finlay, C.C., and Jackson A., 2003. Equatorially dominated magnetic field change at the surface of Earth's core. *Science*, **300**: 2084–2086.

Hide, R., 1966. Free hydromagnetic oscillations of the Earth's core and the theory of geomagnetic secular variation. *Philosophical Transactions of the Royal Society of London Series A*, **259**: 615–647.

Hide R., and Stewartson K., 1972. Hydromagnetic oscillations in the Earth's core. *Reviews of Geophysics and Space Physics*, **10**: 579–598.

Jackson, A., Jonkers, A.R.T., and Walker, M.R., 2000. Four centuries of geomagnetic secular variation from historical records. *Philosophical Transaction of the Royal Society of London*, **358**: 957–990.

Jones, C.A., Mussa A.I., and Worland S.J., 2003. Magnetoconvection in a rapidly rotating sphere: the weak field case. *Proceedings of the Royal Society of London Series A*, **459**: 773–797.

Kerswell, R.R., 1994. Tidal excitation of hydromagnetic waves and their damping in the Earth. *Journal of Fluid Mechanics*, **274**: 219–241.

Korte, M., and Constable, C., 2005. Continuous global geomagnetic field models for the past 7 millennia II: CALS7K.1. *Geochemistry, Geophysics, Geosystems*, **6**(1): doi:10.1029/2004GC00801

Lehnert B., 1954. Magnetohydrodynamic waves under the action of the Coriolis force. *Astrophysical Journal*, **119**: 647–654.

Malkus, W.V.R., 1967. Hydromagnetic planetary waves. *Journal of Fluid Mechanics*, **90**: 641–668.

Moffatt H.K., 1978. *Magnetic Field Generation in Electrically Conducting Fluids*. Cambridge: Cambridge University Press.

Proctor, M.R.E., 1994. Convection and magnetoconvection. In Proctor, M.R.E., and Gilbert, A.D. (eds.), *Lectures on Solar and Planetary Dynamos*. Cambridge: Cambridge University Press, 97–115.

Roberts, P.H., and Loper, D.E., 1979. On the diffusive instability of some simple steady magnetohydrodynamic flows. *Journal of Fluid Mechanics*, **90**: 641–668.

Roberts, P.H., and Soward, A.M., 1972. Magnetohydrodynamics of the Earth's core. *Annual Review of Fluid Mechanics*, **4**: 117–153.

Roberts, P.H., and Stewartson, K., 1974. On finite amplitude convection in a rotating magnetic system. *Philosophical Transactions of the Royal Society of London*, **277**: 287–315.

Soward, A.M., 1979. Convection-driven dynamos. *Physics of the Earth and Planetary Interiors*, **20**: 281–301.

Walker, M.R., Barenghi, C.F., and Jones C.A., 1998. A note on dynamo action at asymptotically small Ekman number. *Geophysical and Astrophysical Fluid Dynamics*, **88**: 261–275.

Zhang, K., and Gubbins, D., 2002. Convection-driven hydromagnetic waves in planetary fluid cores. *Mathematical and Computer Modelling*, **36**: 389–401.

Zhang, K., and Schubert, G., 2000. Magnetohydrodynamics in rapidly rotating spherical systems. *Annual Review of Fluid Mechanics*, **32**: 409–443.

Cross-references

Alfvén Waves
Alfvén, Hannes Olof Gösta
Core Convection
Core Motions
Inhomogeneous Boundary Conditions and the Dynamo
Magnetoconvection
Magnetohydrodynamics
Oscillations, Torsional
Proudman-Taylor Theorem
Thermal Wind
Time-Dependent Models of the Geomagnetic Field
Westward Drift

MAGNETOHYDRODYNAMICS

Introduction

Magnetohydrodynamics (MHD) is the study of the flow of electrically conducting fluids in the presence of magnetic fields. It has significant applications in technology and in the study of planets, stars, and galaxies. Here the main focus will be on its role in explaining the origin and properties of the geomagnetic field.

The interaction between fluid flow and magnetic field defines the subject of MHD and explains much of its fascination (and complexity). The magnetic field $\mathbf{B}$ influences the fluid motion $\mathbf{V}$ through the Lorentz force, $\mathbf{J} \times \mathbf{B}$. The electric current density $\mathbf{J}$ is affected by the fluid motion through the electromotive force (emf), $\mathbf{V} \times \mathbf{B}$. The most famous offspring of this marriage of hydrodynamics to electromagnetism are the Alfvén waves, a phenomenon absent from the two subjects separately (see *Alfvén waves*). In fact, many consider the discovery of this wave by Hannes Alfvén in 1942 to mark the birth of MHD (see *Alfvén, Hannes*). Initially MHD was often known as hydromagnetics, but this term has largely fallen into disuse. Like MHD, it conveys the unfortunate impression that the working fluid is water. In reality, the electrical conductivity of water is so small that MHD effects are essentially absent. Moreover, many fluids used in MHD experiments are antipathetical to water. Even as fluid mechanics is now more widely employed than hydrodynamics, the terms magnetofluid mechanics or magnetofluid dynamics, which are already sometimes employed, may ultimately displace MHD.

Since electric and magnetic fields are on an equal footing in electromagnetism (EM), it may seem strange that the acronym EMHD is not preferred over MHD. In many systems, however, including the Earth's core, the application of EM theory in its full, unapproximated form would add complexity without compensating enlightenment. It suffices to apply the nonrelativistic version of EM theory that existed in the 19th century before Maxwell, by introducing displacement currents, cast the theory into its present-day form. In this pre-Maxwell theory, electric and magnetic fields are not on an equal footing; the magnetic field is the master and the electric field the slave. Consequently MHD is an appropriate acronym whereas EMHD is not. We exclude situations, such as those characteristic of electrohydrodynamics (EHD), in which an electric field, **E**, is externally applied that is large compared with the typical electric fields in MHD, which are created entirely by the emf $\mathbf{V} \times \mathbf{B}$.

The pre-Maxwell theory is described in the Section on Electrodynamics. It requires that the characteristic flow speed $\mathcal{V}$ is small compared with the speed of light c, and that the characteristic timescale $\mathcal{T}$ is large compared with $\mathcal{L}/c$, where $\mathcal{L}$ is the characteristic length scale. The theory to be described also demands that $\mathcal{L}$ is large compared with the Debye length, which is the distance over which the number densities of electrons and ions can differ substantially from one another.

To simplify the discussion below, some verbal abbreviations will usually be employed. In particular, velocity will mean fluid velocity, density will stand for the mass density, field will mean magnetic field, current will be short for electric current density, conductor will mean conductor of electricity, potential will signify electric potential, and core will refer to the Earth's core. Script letters $\mathcal{L}$, $\mathcal{T}$, $\mathcal{V}$, $\mathcal{B}$, $\mathcal{J}$, $\mathcal{E}$, etc., will indicate typical magnitudes of length, time, velocity, field, current, electric field, etc. In the geophysical application, FOC means fluid outer core; SIC, solid inner core; CMB, core-mantle boundary; and ICB, inner core boundary.

Electrodynamics

Nonrelativistic electromagnetism

The pointwise form of the pre-Maxwell EM equations are, in SI units,

$$\nabla \times \mathbf{B} = \mathbf{J}, \quad \text{(Eq. 1)}$$

$$\nabla \times \mathbf{E} = -\partial_t \mathbf{B}, \quad \text{(Eq. 2)}$$

$$\nabla \cdot \mathbf{B} = 0, \quad \text{(Eq. 3)}$$

$$\nabla \cdot \mathbf{D} = \vartheta, \quad \text{(Eq. 4)}$$

where **D** is the electric displacement, **H** is the magnetizing force and ϑ is the electric charge density; t is time and $\partial_t = \partial/\partial t$. Equations (1) and (2) are, respectively, Ampère's law and Faraday's law. The sources on the right-hand sides of Eqs. (1) and (4) must satisfy charge conservation and, in pre-Maxwell theory, this requires

$$\nabla \cdot \mathbf{J} = 0, \quad \text{(Eq. 5)}$$

which is consistent with Eq. (1). Equation (2) shows that, if Eq. (3) holds for any t, it holds for all t.

Equations (1)–(5) are invariant under the Galilean transformation:

$$\mathbf{x}' = \mathbf{x} - \mathbf{U}t, \quad \text{(Eq. 6)}$$

$$t' = t, \quad \text{(Eq. 7)}$$

where **x** is the position vector and **U** is constant. The fields and sources transform as

$$\mathbf{B}' = \mathbf{B}, \quad \text{(Eq. 8)}$$

$$\mathbf{E}' = \mathbf{E} + \mathbf{U} \times \mathbf{B}, \quad \text{(Eq. 9)}$$

$$\mathbf{H}' = \mathbf{H}, \quad \text{(Eq. 10)}$$

$$\mathbf{D}' = \mathbf{D} + \mathbf{U} \times \mathbf{H}/c^2, \quad \text{(Eq. 11)}$$

$$\mathbf{J}' = \mathbf{J}, \quad \text{(Eq. 12)}$$

$$\vartheta' = \vartheta + \nabla \cdot (\mathbf{U} \times \mathbf{H})/c^2. \quad \text{(Eq. 13)}$$

According to (6) and (7), it follows that $\partial'_t = \partial_t + \mathbf{U} \cdot \nabla$, from which it is readily verified that the primed variables also obey Eqs. (1)–(5). The presence of the terms involving $1/c^2$ in Eqs. (11) and (13) may seem anomalous in a theory based on $c \gg \mathcal{L}/\mathcal{T}$. However, since $\mathbf{H}/\mathbf{D} \sim c^2\mathbf{B}/\mathbf{E}$ (see the Section on Constitutive relations), it follows that all terms in Eq. (11) are of the same order, as are those in Eq. (13). Equation (13) has the curious consequence that the free charge density is frame-dependent.

We shall be interested in systems in which, on some surface S, the properties of the medium change abruptly. We exclude cases in which S separates the same material in different physical states and through which the material passes, changing its state as it does so, as for a shock discontinuity. The same integral laws that led to Eqs. (1)–(5) imply that

$$[\![\mathbf{n} \times \mathbf{H}]\!] = \mathbf{C}, \quad \text{on } S, \quad \text{(Eq. 14)}$$

$$[\![\mathbf{n} \times (\mathbf{E} + \mathbf{U} \times \mathbf{B})]\!] = \mathbf{0}, \quad \text{on } S, \quad \text{(Eq. 15)}$$

$$[\![\mathbf{n} \cdot \mathbf{B}]\!] = 0, \quad \text{on } S, \quad \text{(Eq. 16)}$$

$$[\![\mathbf{n} \cdot \mathbf{D}]\!] = \Sigma, \quad \text{on } S, \quad \text{(Eq. 17)}$$

$$[\![\mathbf{n} \cdot \mathbf{J}]\!] = -\nabla_S \cdot \mathbf{C}, \quad \text{on } S, \quad \text{(Eq. 18)}$$

where **C** and Σ are the surface current and the surface charge density on S; ∇_S is the two-dimensional surface divergence on S; $[\![Q]\!] = Q_1 - Q_2$ is the difference in the limiting values of any quantity Q at a point P of S from sides 1 and 2; the unit vector **n** to S is directed out of side 2 and into side 1; **U** is now the velocity of P although only the component of **U** along **n** is actively involved. The Galilean transformation, when supplemented by

$$\Sigma' = \Sigma + \mathbf{n} \cdot [\![\mathbf{U} \times \mathbf{H}]\!]/c^2, \quad \text{(Eq. 19)}$$

preserves Eqs. (14)–(18). Equations (13) and (19) may also be written as

$$\vartheta' = \vartheta - \mathbf{U} \cdot \mathbf{J}/c^2, \quad \text{(Eq. 20)}$$

$$\Sigma' = \Sigma - \mathbf{U} \cdot \mathbf{C}/c^2. \quad \text{(Eq. 21)}$$

Constitutive relations

The pre-Maxwell equations must be supplemented by relations that define the physical nature of the medium in which they are applied. We shall suppose that the medium is isotropic and homogeneous so that, at a point P within it and in a reference frame moving with the velocity $\mathbf{U} = \mathbf{V}(\mathbf{x}_P)$ of P,

$$\mathbf{H}' = \mathbf{B}'/\mu, \quad \text{(Eq. 22)}$$

$$\mathbf{D}' = \epsilon\mathbf{E}', \quad \text{(Eq. 23)}$$

$$\mathbf{J}' = \sigma\mathbf{E}', \quad \text{(Eq. 24)}$$

where the proportionality variables are the permeability μ, the permittivity ϵ, and the electrical conductivity σ. Except in the crust, the temperature within the Earth everywhere exceeds the Curie point, at which permanent magnetism ceases to exist. We therefore assume that μ is the permeability of free space μ_0; the permittivity of free space is ϵ_0 and $c^2 = 1/\mu_0\epsilon_0$. When translated back to the laboratory frame, (22)–(24) are, according to Eqs. (8), (10), and (12),

$$\mathbf{H} = \mathbf{B}/\mu_0, \quad \text{(Eq. 25)}$$

$$\mathbf{D} = \epsilon\mathbf{E} + (\epsilon - \epsilon_0)\mathbf{V} \times \mathbf{B}, \quad \text{(Eq. 26)}$$

$$\mathbf{J} = \sigma(\mathbf{E} + \mathbf{V} \times \mathbf{B}). \quad \text{(Eq. 27)}$$

Wherever **H** occurs, we shall now use Eq. (25) to remove it, in favor of **B**.

Equation (27) is the generalization of *Ohm's law* to a moving conductor, and is centrally important. It may be written equivalently as

$$\mathbf{E} = -\mathbf{V} \times \mathbf{B} + \eta\nabla \times \mathbf{B}, \quad \text{(Eq. 28)}$$

where $\eta = 1/\sigma\mu_0$ is the *magnetic diffusivity*. This form provides a convenient way of removing **E** in favor of **B**; as mentioned in the Section Introduction, **E** plays a subsidiary role in MHD. Nevertheless, it is worth observing that, by Eqs. (3)–(5), (26), and (28),

$$\vartheta = -\epsilon_0\nabla \cdot (\mathbf{V} \times \mathbf{B}). \quad \text{(Eq. 29)}$$

Unlike the case of a stationary conductor for which an initial ϑ flows to the boundaries in a time of order η/c^2 (i.e., instantaneously according to pre-Maxwell theory), the free charge density in a moving conductor is generally nonzero, and takes the value (29).

EM energy conservation, ohmic dissipation, Lorentz force

It follows from Eqs. (1) and (2) that

$$\partial u_t^B + \nabla \cdot \mathbf{I}^B = -\mathbf{E} \cdot \mathbf{J}, \quad \text{(Eq. 30)}$$

where u^B is the EM energy per unit volume and $\mathbf{I}^B$ is the EM energy flux (often called "the Poynting flux"):

$$u^B = B^2/2\mu_0, \quad \text{(Eq. 31)}$$

$$\mathbf{I}^B = \mathbf{E} \times \mathbf{B}/\mu_0. \quad \text{(Eq. 32)}$$

When integrated over a small volume v surrounding a point P in the medium, Eq. (30) states that the rate of increase of magnetic energy in v is diminished by the outward flux of magnetic energy across its surface s and by the rate at which the EM field transfers its energy to the material contents of v, as determined from Eq. (28) as

$$-\mathbf{E} \cdot \mathbf{J} = -\mathbf{V} \cdot \mathbf{L} - Q^J, \quad \text{(Eq. 33)}$$

where

$$\mathbf{L} = \mathbf{J} \times \mathbf{B}, \quad \text{(Eq. 34)}$$

$$Q^J = J^2/\sigma \geq 0. \quad \text{(Eq. 35)}$$

Here **L**, the Lorentz force, is the force per unit volume exerted by the EM field on the medium, so that the net force and torque (about the origin of **x**) on the contents of a volume v are

$$\boldsymbol{f} = \int_v \mathbf{L}dv = \int_v \mathbf{J} \times \mathbf{B}dv, \quad \text{(Eq. 36)}$$

$$\boldsymbol{\Gamma} = \int_v \mathbf{x} \times \mathbf{L}dv = \int_v \mathbf{x} \times (\mathbf{J} \times \mathbf{B})dv. \quad \text{(Eq. 37)}$$

The work done by the Lorentz force diminishes u^B at the rate $\mathbf{V} \cdot \mathbf{L}$. When negative, this represents a transfer of kinetic energy to EM energy. The second term on the right-hand side of Eq. (33), which is necessarily nonnegative, gives the rate at which the EM field loses energy through ohmic heating, also called Joule losses.

By using Eqs. (1), (3), and (34) it is possible, and often useful, to replace the Lorentz force by an equivalent EM stress tensor:

$$L_i = \nabla_j \pi_{ij}^B, \quad \text{(Eq. 38)}$$

$$\pi_{ij}^B = \frac{B_i B_j}{\mu_0} - \frac{B^2}{2\mu_0}\delta_{ij}. \quad \text{(Eq. 39)}$$

Through Eq. (38), **f** and **Γ** may be replaced by surface integrals:

$$f_i = \oint_s \pi_{ij}^B ds_j, \quad \text{(Eq. 40)}$$

$$\Gamma_i = \oint_s \epsilon_{ijk} x_j \pi_{km}^B ds_m, \quad \text{(Eq. 41)}$$

where, to obtain Eq. (41) the symmetry of π^B has been invoked. Alternate forms of Eqs. (40) and (41) are

$$\mathbf{f} = \oint_s \left[\frac{\mathbf{B}(\mathbf{B} \cdot d\mathbf{s})}{\mu_0} - \frac{B^2}{2\mu_0}d\mathbf{s}\right], \quad \text{(Eq. 42)}$$

$$\boldsymbol{\Gamma} = \oint_s \left[\frac{\mathbf{x} \times \mathbf{B}}{\mu_0}(\mathbf{B} \cdot d\mathbf{s}) - \frac{B^2}{2\mu_0}\mathbf{x} \times d\mathbf{s}\right]. \quad \text{(Eq. 43)}$$

The first term on the right-hand side of Eq. (42) represents a tension in the field lines; the second term contributes an isotropic magnetic pressure, $P_m = B^2/2\mu_0$, which is about 4 atm for a field strength of 1 T. These interpretations will be useful in the Section on the perfect conductor below.

It is worth emphasizing that the magnetic field alone contributes to the EM force on the conductor, to the EM stresses, and to the EM energy. According to pre-Maxwell theory, the corresponding contributions made by the electric field are smaller by a factor of order $(\mathcal{V}/c)^2$.

The induction equation; the magnetic Reynolds number

On eliminating **E** between Eqs. (2) and (28), it is found that **B** obeys the *induction equation*,

$$\partial_t\mathbf{B} = \nabla \times (\mathbf{V} \times \mathbf{B}) - \nabla \times (\eta\nabla \times \mathbf{B}). \quad \text{(Eq. 44)}$$

This equation is sometimes called "the dynamo equation." This is a misnomer, since Eq. (44) is applicable in many situations totally unrelated to dynamo theory.

When η is constant (the usual assumption), Eq. (3) shows that Eq. (44) can be written as

$$\partial_t \mathbf{B} = \nabla \times (\mathbf{V} \times \mathbf{B}) + \eta \nabla^2 \mathbf{B}. \quad \text{(Eq. 45)}$$

This clearly exposes the evolution of $\mathbf{B}$ as a competition between EM induction (through $\nabla \times (\mathbf{V} \times \mathbf{B})$) and ohmic diffusion (through $\eta \nabla^2 \mathbf{B}$), and it is convenient to introduce a dimensionless measure of the relative importance of these effects:

$$R_{\mathrm{m}} = \mathcal{VL}/\eta = \mu_0 \sigma \mathcal{VL}. \quad \text{(Eq. 46)}$$

In analogy with the familiar (kinetic) Reynolds number $R_{\mathrm{k}} = \mathcal{VL}/\nu$, used in fluid mechanics to quantify the effects of viscosity, R_{m} is called the *magnetic Reynolds number.* (Here ν is kinematic viscosity.) An illuminating way of writing Eq. (45) is

$$d_t \mathbf{B} = \mathbf{B} \cdot \nabla \mathbf{V} - \mathbf{B} \nabla \cdot \mathbf{V} + \eta \nabla^2 \mathbf{B}, \quad \text{(Eq. 47)}$$

where $d_t = \partial_t + \mathbf{V} \cdot \nabla$ is the motional derivative, the derivative following the fluid motion. This shows that, unless there is velocity shear ($\nabla \mathbf{V} \neq \mathbf{0}$), the magnetic field will simply be advected by the motion ($d_t \mathbf{B}$) while being continually diffused by electrical resistance (through $\eta \nabla^2 \mathbf{B}$).

MHD processes in rotating bodies such as the core are usually most easily studied using a reference frame that rotates with the body (see the Section on Classical and Coriolis MHD). Even though $\mathbf{E}$ is then different from the electric field in the inertial frame, the induction equation is unchanged, provided ∂_t, d_t, and $\mathbf{V}$ are defined relative to the rotating frame.

Solutions to the induction equation are subject to the boundary conditions (14)–(18). The first question to ask concerns the role of the surface current $\mathbf{C}$. From a physical standpoint, it is clear that, when $\eta \neq 0$, an infinitely thin concentration of current, even if it could be set up initially, would be instantly diffused into a layer of large, but finite, $\mathbf{J}$. This shows that $\mathbf{C} = 0$ when $\eta \neq 0$. Nevertheless, when $R_{\mathrm{m}} \gg 1$, it is found that, where material properties of the medium change abruptly, a thin "boundary layer" may arise, which, because $\mathbf{J}$ is so large, carries a finite total current $\mathbf{C}$. When focusing on phenomena on scales $\mathcal{L}$ large compared with the thickness, δ_{m}, of the boundary layer, it may be convenient (even though $\eta \neq 0$) to pretend that the boundary layer is infinitely thin, i.e., is a surface current $\mathbf{C}$ (see *Core, boundary layers*).

A good example is provided by a rotating solid sphere of radius a across which a uniform field, $\mathbf{B}_0$, is applied; the sphere rotates about an axis perpendicular to $\mathbf{B}_0$ with angular velocity $\mathbf{\Omega}$ (see also *Dynamo, Herzenberg*). The appropriate magnetic Reynolds number is $R_{\mathrm{m}} = \Omega a^2/\eta$. To an observer on the surface S of the sphere and rotating with it, the applied field will seem to be oscillatory and, if $R_{\mathrm{m}} \gg 1$, it will therefore penetrate only a short distance, of order $\delta_{\mathrm{m}} = R_{\mathrm{m}}^{-1/2} a = (\eta/\Omega)^{1/2}$, into the sphere. This phenomenon is well known in the EM of solid conductors and is called the "skin effect," because the induced currents are confined to a thin "skin" on S. In the limit $R_{\mathrm{m}} \to \infty$, the skin becomes infinitely thin, i.e., a surface current $\mathbf{C}$. This creates a dipolar magnetic field, $\mathbf{b}$ that completely excludes $\mathbf{B}_0$ from the interior of the sphere: $\mathbf{n} \cdot (\mathbf{b} + \mathbf{B}_0) = 0$ on S; Swirling motions in a fluid conductor can similarly expel flux from within them.

For the remainder of this subsection, we suppose that $R_{\mathrm{m}} = O(1)$ and $\mathbf{C} = \mathbf{0}$. Then, of Eqs. (14)–(18), only Eqs. (14)–(16) are significant:

$$[\![\mathbf{B}]\!] = 0, \quad \text{on } S, \quad \text{(Eq. 48)}$$

$$[\![\mathbf{n} \times \mathbf{E}]\!] = 0, \quad \text{on } S. \quad \text{(Eq. 49)}$$

By Eq. (28), condition (49) can also be written as $[\![\mathbf{n} \times (\mathbf{V} \times \mathbf{B} - \eta \nabla \times \mathbf{B})]\!] = 0$, hence removing all reference to $\mathbf{E}$ when solving Eq. (44) subject to Eqs. (48) and (49). This again highlights the unimportance of $\mathbf{E}$ relative to $\mathbf{B}$. If there is any interest in finding $\mathbf{E}$, ϑ, and Σ, these can be determined after Eq. (44) has been solved, by applying Eqs. (28), (29), and (17).

Condition (18) is superfluous since, according to Eq. (1), it is satisfied when $[\![\mathbf{n} \times \mathbf{B}]\!] = 0$ is. Similarly, when Eq. (49) is obeyed, so is $[\![\mathbf{n} \cdot \mathbf{B}]\!] = 0$, by Eq. (2). Thus, only four of the five scalar conditions (48) and (49) are independent. According to Eqs. (3) and (48), $[\![\mathbf{n} \cdot \nabla(\mathbf{n}.\mathbf{B})]\!] = 0$, but, because $[\![\eta]\!]$ is generally nonzero, $[\![\mathbf{n} \times \mathbf{J}]\!] \neq 0$, so that $[\![\mathbf{n}.\nabla(\mathbf{n} \times \mathbf{B})]\!] \neq 0$. Conditions (48) and (49) apply at the ICB but usually the SIC is modeled as a solid of the same conductivity as the fluid core ($[\![\eta]\!] = 0$), and moving with it at the ICB ($[\![\mathbf{V}]\!] = 0$). Then, by Ohm's law $[\![\mathbf{J}]\!] = 0$ and $[\![\nabla \mathbf{B}]\!] = 0$.

Geophysically, a more significant application of Eqs. (48) and (49) is when S is the CMB. Because the mantle is a relatively poor conductor compared with the core, it is often assumed that the entire region $\widehat{\mathcal{V}}$ above the CMB is electrically insulating. Then, distinguishing variables in $\widehat{\mathcal{V}}$ by a circumflex and setting $\widehat{\mathbf{J}} \equiv 0$ in Eq. (1), we have

$$\nabla \times \widehat{\mathbf{B}} = 0, \quad \text{(Eq. 50)}$$

$$\nabla \cdot \widehat{\mathbf{B}} = 0. \quad \text{(Eq. 51)}$$

(This also ignores all permanent magnetism in the crust.) Equations (50) and (51) then show that $\widehat{\mathbf{B}}$ is a potential field:

$$\widehat{\mathbf{B}} = -\nabla \widehat{\mathbf{\Phi}}, \quad \text{(Eq. 52)}$$

where

$$\nabla^2 \widehat{\mathbf{\Phi}} = 0. \quad \text{(Eq. 53)}$$

This is the basis for the Gauss representation of the observed geomagnetic field. The general solution of Eq. (53) is a linear combination of spherical harmonics, $r^n S_n(\theta, \phi)$ and $r^{-n-1} S_n(\theta, \phi)$, the sources of which are respectively outside the sphere of radius r and within it (see *Harmonics, spherical*). Here (r, θ, ϕ) are spherical coordinates, with origin at the geocenter O and $\theta = 0$ as north polar axis; the function $S_n(\theta, \phi)$ is a surface harmonic, $(g_n^m \cos m\phi + h_n^m \sin m\phi) P_n^m(\theta)$, where $P_n^m(\theta)$ is the Legendre function. For the main geomagnetic field, the $r^n S_n(\theta, \phi)$ terms are found to be negligibly small at the Earth's surface and, when we assume that the mantle contains no sources of field, this is also true at the core surface, $r = R_1$. The representation of $\widehat{\mathbf{B}}$ into a sum of $r^{-n-1} S_n(\theta, \phi)$ terms then holds throughout $r \geq R_1$, and (since magnetic monopoles do not exist), it is dominated at large r by the first interior harmonic, $r^{-2} S_1(\theta, \phi)$, which corresponds to a magnetic dipole. This leads to the important condition,

$$\widehat{\mathbf{\Phi}} = O(r^{-2}), \quad \text{as} \quad r \to \infty, \quad \text{(Eq. 54)}$$

which expresses, in a succinct mathematical way, the fact that the source of the main geomagnetic field lies entirely beneath the core surface.

Conditions (48) and (49) hold on the CMB, but Eq. (49) does not restrict solutions of Eq. (44). To see this, let us represent $\mathbf{B}$ and $\mathbf{E}$ in a well-known way:

$$\mathbf{B} = \nabla \times \mathbf{A}, \quad \text{(Eq. 55)}$$

$$\mathbf{E} = -\partial_t \mathbf{A} - \nabla \Psi. \quad \text{(Eq. 56)}$$

The Coulomb gauge is convenient in pre-Maxwell theory, so that

$$\nabla \cdot \mathbf{A} = 0, \quad \text{(Eq. 57)}$$

$$\nabla^2 \Psi = -\vartheta/\epsilon, \quad \text{(Eq. 58)}$$

the second of which follows from Eqs. (4) and (23). Continuity of $\mathbf{n}\cdot\mathbf{B}$ and $\mathbf{n}\times\mathbf{E}$ imply that $[\![\mathbf{n}\times\mathbf{A}]\!] = 0$ and $[\![\Psi]\!] = 0$ on S. Free charge may be exist in $\widehat{\mathrm{V}}$, but it is reasonable to suppose that $\widehat{\vartheta}$, and therefore $\widehat{\Psi}$ also, vanish as $r \to \infty$. The Poisson equation (58) then determines $\widehat{\Psi}$. The resulting $[\![\mathbf{n}\cdot\nabla\Psi]\!]$ is generally nonzero, and defines Σ on S (see Eq. (17)).

In short, Eq. (49) does no more than determine $\widehat{\mathbf{E}}$ in $\widehat{\mathrm{V}}$ and Σ on S; it places no restriction on $\mathbf{B}$. The essential difference between Eqs. (48) and (49) is that surface layers of "magnetic charge," analogous to surface layers of electric charge, do not exist. This means that, while a potential field $\widehat{\mathbf{\Phi}}$ can always be found that satisfies Eq. (54) together with either $[\![\mathbf{n}\cdot\mathbf{B}]\!] = 0$ or $[\![\mathbf{n}\times\mathbf{B}]\!] = 0$, to demand that it satisfies both these conditions places a restriction on $\mathbf{B}$ on S. This restriction expresses the fact that all sources of $\mathbf{B}$ are currents flowing beneath S. It is sometimes called "the dynamo condition."

The kinematic dynamo

A self-excited dynamo is a device that maintains the magnetic field indefinitely. Here "indefinitely" means that $\mathbf{B} \not\to 0$ as $t \to \infty$; "maintains" means that the only sources of $\mathbf{B}$ are currents flowing in the conductor itself, i.e., there are no external current-carrying coils even at infinity. The magnetic field is a by-product of the current flow (see Eq. (1)), and in a self-excited dynamo, the motion creates from that field the electromotive force that produces the required current flow. There is no conflict with energy conservation. As seen in the Section on Energy conservation, ohmic dissipation, Lorentz force, the field creates a Lorentz force and this in general opposes the motion and attempts to bring the fluid to rest, an example of Lenz's law. Some agency has to maintain the motion, and provide the energy necessary to make good the ohmic losses Q^J. The self-excited dynamo is, in principle, no different from generators of electricity in a power station.

Theorists distinguish two types of self-excited dynamo: the kinematic dynamo and the MHD dynamo. In a kinematic dynamo, the flow $\mathbf{V}$ maintaining $\mathbf{B}$ is specified; in an MHD dynamo the energy sources driving the flow are specified and $\mathbf{B}$ and $\mathbf{V}$ are determined simultaneously by solving the full MHD equations. These equations are nonlinear and therefore much harder to solve than the induction equation, which is linear in $\mathbf{B}$ and is all that is needed (together with the matching to a source-free external field $\widehat{\mathbf{B}}$) in studying the kinematic dynamo. Not surprisingly, much is now known about kinematic dynamos, and many solutions exist (see *Dynamos, kinematic*). In some kinematic models, the conductor fills all space (see *Dynamo, Herzenberg* and *Dynamo, Ponomarenko*). Because of applications of the dynamo theory to the Earth and other heavenly bodies, it is more usual to assume that the conductor occupies a simple, simply-connected body, V, such as a sphere, the exterior $\widehat{\mathrm{V}}$ of which is electrically insulating. (See the Section on The induction equation; the magnetic Reynolds number and see *Dynamo, Backus; Dynamo, Braginsky; Dynamo, Bullard-Gellman;* and *Dynamo, Gailitis*. See also *Dynamos, fast* and *Dynamos, periodic*.)

Maintenance of $\mathbf{B}$ obviously implies maintenance of u^B. By assumption, external sources do not bring magnetic energy into a system from the outside so that, by Eqs. (30)–(33), the integral of $\mathbf{V}\cdot\mathbf{L}$ over V must be negative to offset the ohmic losses Q^J, which is nonnegative at all points in V. A necessary condition for dynamo action is therefore that $\mathbf{V}$ is "sufficiently large," i.e., R_{m} is "big enough" (see *Antidynamo and bounding theorems*). For convenience, we shall later say that $R_{\mathrm{m}} \gtrsim 100$ in a functioning dynamo, but it has to be clearly understood that a condition such as that is far from sufficient to ensure dynamo action. There are a number of antidynamo theorems that define circumstances in which a dynamo will not work, no matter how large R_{m} is (see *Antidynamo and bounding theorems*). The most significant of these is a result due to Cowling (1933) (see *Cowling's theorem*). This states that

> *An axisymmetric magnetic field cannot be maintained by dynamo action.*

In what follows, axisymmetric quantities will be distinguished by an overbar. More generally, $\overline{Q}(r,\theta)$ will be the ϕ—average of a scalar quantity $Q(r,\theta,\phi)$; the remaining asymmetric part, Q', of Q will be denoted by a prime: $Q' = Q - \overline{Q}$. An axisymmetric field consists of a zonal magnetic field, $\overline{\mathbf{B}}_\phi$, in the direction of increasing ϕ and a meridional field, $\overline{\mathbf{B}}_{\mathrm{M}}$, that has only r and θ components $\overline{B}_r$ and $\overline{B}_\theta$. The essence of Cowling's theorem is that, although motions can induce $\overline{B}_\phi$ from $\overline{\mathbf{B}}_{\mathrm{M}}$ (see Section on the perfect conductor), they cannot do the reverse; asymmetric field and flow are required to produce an emf $(\overline{\mathbf{V}' \times \mathbf{B}'})_\phi$ that creates the current, $\overline{J}_\phi$, necessary to maintain $\overline{\mathbf{B}}_{\mathrm{M}}$. This means that "dynamo fields are necessarily three-dimensional," i.e., their components must depend on r, θ, and ϕ. (Axisymmetric $\mathbf{V}$ are not, however, excluded by the theorem; e.g., see *Dynamo, Ponomarenko*.)

Despite the antidynamo theorems, self-excited kinematic models have been constructed and have been modeled in the laboratory (see *Dynamos, experimental*). These successes and earlier work going back to Parker (1955) have highlighted the importance of symmetry-breaking; dynamos flourish if the motions lack mirror-symmetry (see *Geodynamo, symmetry properties*). A flow is mirror-symmetric if a mirror ($z = 0$, say) exists such that

$$V_x(x,y,-z) = V_x(x,y,z),\ V_y(x,y,-z) = V_y(x,y,z),$$
$$V_z(x,y,-z) = -V_z(x,y,z). \qquad \text{(Eq. 59)}$$

The simplest manifestations of such symmetry breaking are helical motions. Helicity is defined as $\mathbf{V}\cdot\boldsymbol{\omega}$, where $\boldsymbol{\omega} = \nabla\times\mathbf{V}$ is the vorticity of the flow. It is a measure of how mirror-symmetric the flow is; like a common carpenter's screw, a helical flow does not look the same when viewed in a mirror. Helicity is created naturally in convecting systems that are rotating; the rising and falling motions created by buoyancy acquire vertical vorticity through the action of the Coriolis force. The word helicity is a term in general use to describe any quantity that is the scalar product of a vector with its curl, such as the magnetic helicity, $\mathbf{A}\cdot\mathbf{B}(=\mathbf{A}\cdot\nabla\times\mathbf{A})$, and the current helicity, $\mathbf{B}\cdot\mathbf{J}(=\mathbf{B}\cdot\nabla\times\mathbf{B}/\mu_0)$. Helical flows tend to produce helical fields, as measured by the magnetic and current helicities.

Although examples exist of kinematic dynamos that function even though the motions are nonhelical, there is little doubt that helical dynamos are more efficient, i.e., they function at smaller values of R_{m}. The simplest manifestation of this is the *alpha-effect*, in which helical motions produce a mean emf $\overline{\mathbf{V}'\times\mathbf{B}'}$ that is proportional to $\overline{\mathbf{B}}$ and is usually written as $\alpha\overline{\mathbf{B}}$. This includes an emf $\alpha\overline{B}_\phi$ in the ϕ-direction that drives the ϕ-component of $\overline{\mathbf{J}}$ that creates the axisymmetric meridional part, $\overline{\mathbf{B}}_{\mathrm{M}}$, of $\mathbf{B}$ needed to defeat Cowling's theorem (see above). This simple fact largely explains the popularity of mean field dynamos, where only the axisymmetric part of $\mathbf{B}$ is sought, induction by the asymmetric parts being parameterized by an ansatz, such as the simple choice $\overline{\mathbf{V}'\times\mathbf{B}'} = \alpha\overline{\mathbf{B}}$ just made (see *Dynamos, mean field*).

The perfect conductor

A perfect conductor is one that has infinite electrical conductivity ($\eta = 0$). Since $\mathbf{J}$ must be finite even though $\sigma = \infty$, Ohm's law (27) implies that

$$\mathbf{E} = -\mathbf{V}\times\mathbf{B}. \qquad \text{(Eq. 60)}$$

The induction equation becomes

$$\partial_t\mathbf{B} = \nabla\times(\mathbf{V}\times\mathbf{B}) \quad \text{when} \quad \eta = 0. \qquad \text{(Eq. 61)}$$

In a perfect conductor, the magnetic Reynolds number, R_{m}, is infinite; see Eq. (46). Although $R_{\mathrm{m}} \gg 1$ in many applications of MHD (including some considered here), $R_{\mathrm{m}} = \infty$ is an idealization that is never achieved. It is however an abstraction that is often useful. This is a consequence of Eq. (61) from which it follows that (see *Alfvén's theorem and the frozen flux approximation*).

Magnetic flux tubes move with a perfect conductor as though "frozen" to it.

Frozen flux provides a useful way of picturing MHD processes when R_m is large, one that employs some of the concepts of stress and energy of the Section on EM energy conservation, ohmic dissipation, Lorentz force. Consider the way in which fluid motions can exchange energy with the field. Imagine a straight flux tube, initially of cross-sectional area A_0, lying in a compressible fluid. Suppose a motion compresses the tube uniformly, so that, while remaining straight, its cross-sectional area is reduced to A $(< A_0)$. The density of the fluid it contains increases from its initial value of ρ_0 to $\rho = \rho_0(A_0/A)$. If the compression is performed fast enough, heat conduction is negligible. In such an adiabatic process, the internal energy of the tube increases, so that its temperature rises. According to Alfvén's theorem, the flux of the magnetic field in the tube cannot change, so that the field B_0 initially within the tube increases in strength to $B = B_0A_0/A$, i.e., by the same factor as the density; thus B/ρ remains at its initial value, B_0/ρ_0. The magnetic energy per unit length contained in the tube increases, however, from $A_0(B_0^2/2\mu_0)$ to $A(B^2/2\mu_0) = [A_0(B_0^2/2\mu_0)](A_0/A)$. In short, the kinetic energy of the compressing motion has been transformed into internal energy and magnetic energy. The reverse happens if $A > A_0$; the flux trapped in an intense tube tends to expand into surroundings where the field is weaker. For these reasons, sound traveling across a field moves faster than it would in the absence of the field; the magnetic pressure created by the field intensifies the restoring force in the compressions and rarefactions of the wave.

The transformation of kinetic energy into magnetic energy also occurs in an incompressible fluid when motions stretch field lines against their tension. The process may be likened to the storing of elastic energy in a stretched rubber band or to the energy transmitted to a violin string by plucking it. If the flux tube of cross-sectional area A_0 containing field B_0 is lengthened from L_0 to L, its cross-section will decrease in the same proportion $(A = A_0L_0/L)$ and the field within it will increase by the same factor $(B = B_0L/L_0)$. The magnetic energy it contains, which is proportional to B^2, is enhanced by a factor of $(L/L_0)^2$ from $(B_0^2/2\mu_0)L_0A_0$ to $(B^2/2\mu_0)LA = (B^2/2\mu_0)L_0A_0 = [(B_0^2/2\mu_0)L_0A_0](L/L_0)^2$. If $L = L_0 + \delta$ where $\delta \ll L_0$, the increase in magnetic energy is $(B_0^2/\mu_0)A_0\delta$. This is the work done by the applied force in stretching the tube by δ in opposition to the magnetic tension $(B_0^2/\mu_0)A_0$ of the field lines.

It was mentioned in the Section on the induction equation; the magnetic Reynolds number that fluid motions can create $\overline{B}_\phi$ from $\overline{\mathbf{B}}_\mathrm{M}$. Alfvén's theorem provides a picture of the process: a zonal shearing motion $\overline{V}_\phi$ drags the lines of force of $\mathbf{B}_\mathrm{M}$ out of their meridional planes, i.e., it gives them a $\overline{B}_\phi$ component. Dating from a time when the zonal shear, $\overline{V}_\phi/r\sin\theta$, was frequently denoted by ω, this is often called *the omega effect*. Taking Alfvén's theorem literally, $\overline{V}_\phi$ will continue to wind the field lines around the symmetry axis for as long as the motion is maintained, and $\overline{B}_\phi$ and u^B will increase monotonically. The magnetic stresses grow as the field lines become increasingly stretched in the ϕ-direction until at last the agency that creates $\overline{V}_\phi$ can no longer maintain it.

Alfvén's theorem leads to a new phenomenon, the *Alfvén wave*. Consider again the straight flux tube, of cross-sectional area A in an incompressible fluid, and imagine that a transverse displacement bends the tube, carrying its contents with it, in obedience to the theorem. The tension $\tau = AB^2/\mu_0$ of the tube acts, as in a stretched string, to shorten the tube. A wave results, which moves in each direction along the tube with speed $V_\mathrm{A} = \sqrt{(\tau/M)}$, where $M = \rho A$ in the mass per unit length of the string. The wave velocity is therefore

$$\mathbf{V}_\mathrm{A} = \mathbf{B}/(\mu_0\rho)^{1/2}, \quad \text{(Eq. 62)}$$

which is usually called the *Alfvén velocity*. It is the speed with which energy can be transmitted along the field lines. On taking $\mathcal{B} = 0.01$ T and $\rho = 10^4$ kg m^{-3} as typical of the core, it is found that $\mathcal{V}_\mathrm{A} \sim 0.1$ m s^{-1}. The MHD or *Alfvénic timescale*,

$$\tau_\mathrm{A} = \mathcal{L}/\mathcal{V}_\mathrm{A}, \quad \text{(Eq. 63)}$$

is often significant. It provides an estimate of how quickly an MHD system responds to changes in its state. Alfvén waves are the simplest example of a wider class of waves; see the Section on the CMHD for ideal fluids and *Magnetohydrodynamic waves*.

It was seen in the Section on the induction equation; the magnetic Reynolds number that $\mathbf{C}$ may be nonzero at a surface S of discontinuity when $\eta = 0$ on one side of S. This surface current is subjected to a Lorentz force that is computed best through the EM stress tensor (39). The discontinuity in magnetic stress is then

$$[\![\pi_{ij}^B n_j]\!] = \frac{1}{\mu_0}(\mathbf{n}\cdot\mathbf{B})(\mathbf{C}\times\mathbf{n}) - [\![P_\mathrm{m}]\!]\mathbf{n}. \quad \text{(Eq. 64)}$$

This can be written, in analogy with $\mathbf{J}\times\mathbf{B}$, as $\mathbf{C}\times\frac{1}{2}(\mathbf{B}_1+\mathbf{B}_2)$. Other stresses must act to compensate for this discontinuity. For example, if $\mathbf{n}\cdot\mathbf{B} = 0$, the discontinuity in $P_\mathrm{m} = B^2/2\mu_0$ can be compensated by a discontinuity in the kinetic pressure P that makes the total pressure, $P_\mathrm{total} = P + P_\mathrm{m}$, continuous. Even if a discontinuity in P_total could be set up initially, it would immediately be radiated away from S sonically. If $\mathbf{n}\cdot\mathbf{B} \neq 0$, the situation is more complicated but it is helpful to recall (See the Section on the induction equation; the magnetic Reynolds number) that the surface current is an idealization for $R_\mathrm{m} \to \infty$ of a thin boundary layer in which other processes may be significant. For example, the surface force $(\mathbf{n}\cdot\mathbf{B})(\mathbf{C}\times\mathbf{n})/\mu_0$ in Eq. (64) represents the limit of a volumetric force that would create a large velocity shear in the boundary layer; associated with this is a large viscous stress that may be able to balance the magnetic stress, as in the Hartmann layer (See the Section on the Classical MHD). In the absence of such effects, one must assume that $\mathbf{C} = 0$ when $\mathbf{n}\cdot\mathbf{B} \neq 0$. Even if a surface current could be set up initially, it would immediately be Alfvénically radiated away from S.

The imperfect conductor, reconnection

Some caution has to be exercised in applying Alfvén's theorem. According to the theorem, the topology of field lines cannot change. Fluid lying on a field line always remains on that field line; reconnection of field lines is impossible. This makes the dynamo problem meaningless; the magnetic flux trapped in the conductor is never lost, and one cannot even find out how the conductor acquired that flux originally. Magnetic field is gained, retained, or lost only by its diffusion relative to the conductor, and diffusion happens only when σ and R_m are finite. Nevertheless, Alfvén's theorem is useful in picturing what happens when R_m is large. As the conductor moves, it loses flux on the *EM diffusion timescale*, τ_η, but this may be replenished by a shearing processes (such as those considered above) on the *advective timescale* τ_V; here

$$\tau_\eta = \mathcal{L}^2/\eta, \quad \text{(Eq. 65)}$$

$$\tau_V = \mathcal{L}/\mathcal{V}. \quad \text{(Eq. 66)}$$

For a dynamo to be successful, renewal of flux must balance or exceed its destruction, i.e., τ_V must be shorter than τ_η, implying that $R_\mathrm{m} = \tau_\eta/\tau_V \gtrsim O(1)$. The westward drift velocity on the CMB of recognizable features of the geomagnetic field is of order 3×10^{-4} m s^{-1}. It is likely however that part of this motion is due to diffusion of the features relative to the conductor. A magnetic feature of scale $\mathcal{L}$ can diffuse relative to the conductor at a drift speed c_d of order $\eta/\mathcal{L}$. We shall therefore assume that the actual fluid velocity is less than 3×10^{-4} m s^{-1} and estimate that $\mathcal{V} = 10^{-4}$ m s^{-1}. Taking $\mathcal{L} \sim 2 \times 10^6$ m and $\eta = 2$ m^{2}s^{-1}, this gives $R_\mathrm{m} \sim 100$.

Because of the small scale of laboratory apparatus, it is hard to achieve $O(1)$ values of R_m. For the same reason, it is hard to demonstrate dynamo action using liquid metals (but see *Dynamos, experimental*). The frozen flux picture is of limited usefulness in describing MHD experiments with liquid metals. When $R_\mathrm{m} \ll 1$, EM diffusion is rapid ($\tau_\eta \ll \tau_\mathcal{V}$) and, in the limit, instantaneous. The left-hand side of Eq. (44) is then negligible compared with $\nabla \times (\eta\nabla \times \mathbf{B})$, and, on uncurling this equation, we obtain the limiting form that is the antithesis of Eq. (60):

$$\mathbf{J} = \sigma(-\nabla\Psi + \mathbf{V} \times \mathbf{B}). \qquad \text{(Eq. 67)}$$

This is equivalent to the statement that $\mathbf{E}$ is a potential field ($\mathbf{E} = -\nabla\Psi$), the potential of which can be obtained by solving the Poisson equation that follows from Eq. (67):

$$\nabla^2\Psi = \nabla \cdot (\mathbf{V} \times \mathbf{B}). \qquad \text{(Eq. 68)}$$

The first term on the right-hand side of Eq. (67) is *nonlocal*, i.e., it depends on the form of $\mathbf{V}$ and $\mathbf{B}$ everywhere. The second term creates a current $\sigma\mathbf{V}\times\mathbf{B}$ and a *local* Lorentz force $\sigma(\mathbf{V}\times\mathbf{B})\times\mathbf{B}$, from the prevailing $\mathbf{V}$ and $\mathbf{B}$. The latter may be written as $-\mathbf{V}_\perp/\tau_\mathrm{m}$, where $\mathbf{V}_\perp = \mathbf{V} - (\mathbf{V}\cdot\mathbf{B})\mathbf{B}/B^2$ is the part of $\mathbf{V}$ that is perpendicular to $\mathbf{B}$, and τ_m is the *magnetic damping time*:

$$\tau_\mathrm{m} = \rho/\sigma\mathcal{B}^2 = \eta/\mathcal{V}_\mathrm{A}^2. \qquad \text{(Eq. 69)}$$

This shows that, when $R_\mathrm{m} \ll 1$, the primary effect of a field is to damp-out motions across magnetic field lines, rather like an anisotropic friction (Davidson, 2001).

Because of the small scale of laboratory apparatus, it is also hard to demonstrate Alfvén waves in the laboratory. After the wave has traveled a distance $\mathcal{V}_\mathrm{A}\tau_\eta$, the comoving induced currents have essentially diffused away. The condition that a wave can cross the system before disappearing is $\mathcal{V}_\mathrm{A}\tau_\eta \gtrsim \mathcal{L}$ or $Lu \gtrsim 1$, where Lu is the *Lundquist number*:

$$Lu = \mathcal{V}_\mathrm{A}\mathcal{L}/\eta = \sqrt{(\tau_\eta/\tau_\mathrm{m})}. \qquad \text{(Eq. 70)}$$

If $\mathcal{B} = 0.1$ T is the strength of a field applied across a liquid sodium system of scale $\mathcal{L} = 0.1$ m, we have $\mathcal{V}_\mathrm{A} \sim 3$ m s^{-1}, $\eta \sim 1$ m^2 s^{-1}, $\tau_\mathrm{m} \sim 0.1$ s, $\tau_\eta \sim 0.01$ s, and $Lu \sim 0.3$.

Some effects of turbulence

Many fluid systems occurring in nature are turbulent. The fluid flow within them exists on length scales that range from $\mathcal{L}_\mathrm{macro}$, comparable with the dimensions $\mathcal{L}$ of the system, to $\mathcal{L}_\mathrm{micro} \ll \mathcal{L}_\mathrm{macro}$, where by macro is meant large scales and by micro small scales (though still much larger than intermolecular distances). The corresponding magnetic Reynolds numbers range from $R_\mathrm{m,macro} = \mathcal{V}_\mathrm{macro}\mathcal{L}_\mathrm{macro}/\eta \gtrsim O(1)$ to $R_\mathrm{m,micro} = \mathcal{V}_\mathrm{micro}\mathcal{L}_\mathrm{micro}/\eta \ll R_\mathrm{m,macro}$. In a violently turbulent environment such as the solar convection zone, even $R_\mathrm{m,micro} \gtrsim 1$ so that these scales of motion can generate a small-scale field by themselves. From seeds planted by an epoch-making paper by Parker (1955), a new subject grew called *mean field electrodynamics* in which the concepts of mirror-symmetry and α-effect were central (see, for example, Krause and Rädler, 1980). It was shown that the microscale field and motion not only generate a macroscale field by the α-effect but also destroy it by an enhanced *turbulent diffusivity*, η_T, vastly greater than the molecular (or radiative) value, η. This was advanced as an explanation of why the solar field reverses its polarity on a timescale, $\tau_{\eta,\mathrm{T}} = \mathcal{L}^2/\eta_\mathrm{T}$, of 11 years even though $\tau_\eta = \mathcal{L}^2/\eta$ is greater than the age of the Sun. During the 22-year cycle, sunspots are seen as markers for a large-scale field that drift equatorward relative to the fluid with a velocity of order $c_\mathrm{d,T} = \eta_\mathrm{T}/\mathcal{L}$.

If the Earth's core were as turbulent as the Sun, the timescales of the geomagnetic field would be similarly shortened. The paleomagnetic evidence indicates however that polarity reversals are completed in a time comparable with $\tau_\eta = \mathcal{L}^2/\eta$ which, taking $\mathcal{L} \sim 10^6$ m and $\eta = 2$ m^2 s^{-1}, is of order 10^4 years. This strongly suggests not only that turbulence in the core is much less violent than in the convection zone of the Sun (which is hardly surprising) but also that $R_\mathrm{m,micro}$ in the core is too small for EM induction by small eddies to affect the large-scale field significantly. If this is the case, the concepts of mean field electrodynamics are not applicable to the core, the turbulent α-effect is negligibly small, and η_T does not dominate η. We shall therefore ignore η_T in what follow. According to these arguments, the geodynamo operates on the macroscale, for which the concepts of mirror-symmetry, helicity, and α-effect apply only in a global nonlocal sense, as for the waves to be mentioned in the Section CMHD for ideal fluids below.

The microscales are, nevertheless, significant in another way; they considerably enhance the energy requirements Q^J of the geodynamo. The current densities associated with the macro- and microscales are related to their respective field strengths, $\mathcal{B}_\mathrm{macro}$ and $\mathcal{B}_\mathrm{micro}$ by Eq. (1):

$$\mathcal{J}_\mathrm{macro} \sim \mathcal{B}_\mathrm{macro}/\mu_0\mathcal{L}_\mathrm{macro}, \qquad \mathcal{J}_\mathrm{micro} \sim \mathcal{B}_\mathrm{micro}/\mu_0\mathcal{L}_\mathrm{micro}. \qquad \text{(Eq. 71)}$$

Although it may be true that $\mathcal{B}_\mathrm{micro} \ll \mathcal{B}_\mathrm{macro}$, it is also true that $\mathcal{L}_\mathrm{micro} \ll \mathcal{L}_\mathrm{macro}$. It follows that $\mathcal{J}_\mathrm{micro}$ may be comparable with $\mathcal{J}_\mathrm{macro}$, and therefore its contribution to $Q^J = J^2/\sigma$ may be of a similar size. As shown in the Section on The imperfect conductor, reconnection, the microscale fields for which $R_\mathrm{m,micro} \ll 1$ create a Lorentz force, of order $-\rho\mathbf{V}_\perp/\tau_\mathrm{m}$, and this acts like an anisotropic friction. The energy expended working against this friction is dissipated ohmically (see also *Core turbulence*).

Fluid dynamics

Energy sources

It is clear that an energy source is needed to maintain fluid flow in the face of viscous friction; in addition, an MHD system must expend further energy ohmically. In laboratory systems, the required energy is frequently supplied by pumps that force fluid flow through pipes or channels, and the magnetic field is usually generated externally. In contrast, many naturally occurring fluid systems are forced into motion by buoyancy forces and, because of their vast dimensions $\mathcal{L}$, can easily satisfy the condition $R_\mathrm{m} \gtrsim 100$ necessary for dynamo action at quite modest flow speeds $\mathcal{V}$. Moreover, these systems are often rotating, and the convective motions may then lack the mirror-symmetry that is detrimental to dynamo action; in other words, they possess helicity (see the Section on The kinematic dynamo). Not surprisingly, therefore, they are often able to equip themselves with the magnetic field that transforms them into MHD systems.

Buoyancy is not the only way of creating fluid flow; laboratory motions are sometimes produced by moving the walls of the container; and, in a similar way, flows in planetary cores can be forced at their boundaries. For example, motions in the core are affected by the lunisolar precession. This causes the Earth's axis of rotation to describe a cone of semiangle $23\frac{1}{2}^\circ$ about an axis perpendicular to the ecliptic in a period of about 26 ka. The associated motion of the CMB affects the fluid flow in the core and may even be the main mechanism driving it (see Tilgner, 2005, and *Precession and core dynamics*). So far, a definitive assessment of the geophysical importance of precession is lacking.

Because MHD turbulence is poorly understood, it is hard to estimate the total energy requirements of the motions and fields in the Earth's core. Because the core is a liquid metal, its kinematic viscosity $\nu(\sim 10^{-6}$ m^2 s$^{-1})$ is substantially less than its magnetic diffusivity $\eta(\sim 2$ m^2 s$^{-1})$, i.e., its *magnetic Prandtl number*,

$$P_m = \nu/\eta, \quad \text{(Eq. 72)}$$

is very small ($\sim 10^{-6}$). Assuming that the dissipation length scale for the viscous energy dissipation Q^ν is not much smaller than that for the magnetic energy dissipation Q^J, it follows that $Q^\nu \ll Q^J$ (except possibly in boundary layers). The total ohmic dissipation from the macroscales has been estimated to lie between 0.1 and 2 TW (1 TW $= 10^{12}$ W) with the microscales possibly contributing just as much (see Section on Some effects of turbulence). Although a definitive figure is so far unobtainable, the fact that the Earth has possessed a magnetic field for most, and perhaps all, of its existence proves that the necessary power requirements have been met throughout that time, though the source of that power is not completely certain (see *Geodynamo, energy sources*). This article will focus mainly on core convection, but only in a simplified way. Even though it is often argued that compositional buoyancy is as (or even more) important as thermal buoyancy in driving core motions, only the latter will be included here, as this will suffice to illustrate the salient points. For a geophysically more faithful treatment, see Braginsky and Roberts (1995, 2003) and *Convection, chemical* and *Core convection* (*q.v.*). For MHD models of the geodynamo, see *Geodynamo* and *Geodynamo, numerical simulations* (*q.v.*).

Basic equations

The aim in much of the remainder of this article is to explore the roles of the Coriolis, Lorentz, and buoyancy forces. An attempt will be made to raise the level of complexity gradually and to start from a model that includes these forces as simply as possible by using the Boussinesq model (see *Anelastic and Boussinesq approximations*). Then

$$\nabla \cdot \mathbf{V} = 0, \quad \text{(Eq. 73)}$$

$$\partial_t \mathbf{V} + \mathbf{V} \cdot \nabla \mathbf{V} + 2\boldsymbol{\Omega} \times \mathbf{V} = -\nabla \Pi - \alpha T \mathbf{g} + \mathbf{J} \times \mathbf{B}/\rho + \nu \nabla^2 \mathbf{V}. \quad \text{(Eq. 74)}$$

In this approximation, the density ρ of the fluid is uniform, so that mass conservation reduces to the condition (73) of incompressibility. Momentum conservation (74) is expressed in the reference frame that corotates with the mantle, at a constant angular velocity $\boldsymbol{\Omega}$. The resulting Coriolis acceleration, $2\boldsymbol{\Omega}\times\mathbf{V}$, appears on the left-hand side of Eq. (74). The centrifugal acceleration, $\boldsymbol{\Omega} \times (\boldsymbol{\Omega} \times \mathbf{x})$, has been combined with the kinetic pressure, P, to form the reduced pressure, $\Pi = P/\rho - \frac{1}{2}(\boldsymbol{\Omega} \times \mathbf{x})^2$. The last two terms in Eq. (74) represent the Lorentz and viscous forces per unit mass. The buoyancy force per unit mass is $-\alpha T \mathbf{g}$, where T is the temperature, α is the thermal expansion coefficient, and $\mathbf{g}$ is the gravitational acceleration, assumed constant in time.

The temperature is governed by an energy equation of the form

$$d_t T = \kappa \nabla^2 T + Q/\rho C_p, \quad \text{(Eq. 75)}$$

where κ is the thermal diffusivity and Q comprises the various sources of heat: $Q = Q^\nu + Q^J + Q^R$, where Q^R arises from internal sources, if any, such as dissolved radioactivity; C_p is the specific heat at constant pressure. We shall initially consider nonmagnetic systems, but, when later we generalize to MHD situations, Eqs. (73)–(75) must be solved in conjunction with Eq. (3) and the induction equation (45).

The boundary conditions for $\mathbf{B}$ developed in Section Electrodynamics must be supplemented by boundary conditions on $\mathbf{V}$ and T, such as

$$\mathbf{V} = 0, \quad \text{on the CMB}, \quad \text{(Eq. 76)}$$

$$T = T_{\text{CMB}}, \quad \text{on the CMB}, \quad \text{(Eq. 77)}$$

where T_{CMB} is constant. Equation (76) expresses the no-slip condition that, at their common interface, the fluid moves with the solid, which has zero velocity in the chosen reference frame. The SIC contains only about 5% of the total mass of the core, and its low inertia makes its rotation rate susceptible to the torques (mainly gravitational and magnetic) to which it is subjected (see Eq. (43)). These feature in an equation of motion for its angular velocity $\boldsymbol{\Omega}_{\text{SIC}}$ (see *Inner core rotational dynamics*). For simplicity, we shall assume that

$$\mathbf{V} = 0, \quad \text{on the ICB}, \quad \text{(Eq. 78)}$$

$$T = T_{\text{ICB}}, \quad \text{on the ICB}, \quad \text{(Eq. 79)}$$

where T_{ICB} is constant. When viscous forces are ignored, the differential order of Eq. (74) is reduced, and it is possible to satisfy only one condition on $\mathbf{V}$ at the boundaries; the first of Eqs. (76) and (78) are then replaced by

$$\mathbf{n} \cdot \mathbf{V} = 0, \quad \text{on the CMB and ICB}, \quad \text{for } \nu = 0. \quad \text{(Eq. 80)}$$

The relative importance of the various terms of Eqs. (73)–(75) is quantified by several dimensionless groupings, of which the (kinetic) Reynolds number $R_k = \mathcal{VL}/\nu$ has already been defined. It quantifies the relative sizes of the inertial and viscous forces. Among the dimensionless parameters describing convection, special mention should be made of the *Rayleigh number*, Ra and the usual (thermal) *Prandtl number*, P_t:

$$Ra = g\alpha\beta\mathcal{L}^4/\nu\kappa, \quad \text{(Eq. 81)}$$

$$P_t = \nu/\kappa, \quad \text{(Eq. 82)}$$

where β is, in the Boussinesq approximation, a typical measure of the temperature gradient, e.g., the difference ΔT between the temperatures of the ICB and CMB divided by the depth of the fluid core. (In a layer as thick as the FOC, however, the compressibility of the fluid is significant. Convection is better described by the anelastic approximation and a better choice of β is then the difference between the actual temperature gradient and the adiabatic temperature gradient [see *Core, adiabatic gradient; Anelastic and Boussinesq approximations*].) The Rayleigh number is a measure of how effective the buoyancy force is in overcoming the diffusive processes that oppose convection.

A popular way to gain insight into convective processes, popular because the theory is relatively tractable, is to study the onset of convection. A motionless conductive state is defined in which heat is carried across the system by thermal conduction; the linear stability of this state is then analyzed. It is found that when the control parameter, Ra, reaches some critical or marginal value, Ra_c, the conduction solution becomes convectively unstable. The eigenfunction corresponding to Ra_c gives the structure of the marginal state of motion. In the absence of Coriolis and Lorentz forces, Ra_c is typically of order 10^3 and the marginal mode is a pattern of overturning convection cells, having about the same horizontal scale as the depth $\mathcal{L}$ of the convecting layer.

Although studies of this type provide some insight into convective flows, they clearly have limited value. They predict that, when $Ra > Ra_c$, the convective motions increase without limit, though in reality feedback from nonlinearities, such as the inertial term $\mathbf{V} \cdot \nabla\mathbf{V}$ in Eq. (74), prevents this. The enhanced amplitudes and modified structure of the convective motions as Ra is increased from Ra_c are topics that are beyond the competence of linear theory and are also beyond the scope of this article. Here we shall use the results of linear stability analyzes to obtain clues about how Coriolis and Lorentz forces affect thermal convection.

Classical theory of rotating fluids

In this section we shall suppose that $\mathbf{B} \equiv 0$; initially, buoyancy is also excluded ($\mathbf{g} \equiv 0$). We give a rudimentary account of relevant concepts in the theory of rotating fluids. For a more complete treatment, see the classic text of Greenspan (1968).

Two dimensionless numbers quantify the importance of viscosity and inertia relative to the Coriolis force, the *Ekman number*, E, and the *Rossby number*, Ro:

$$E = \nu/\Omega\mathcal{L}^2, \qquad \text{(Eq. 83)}$$

$$Ro = \mathcal{V}/\Omega\mathcal{L}. \qquad \text{(Eq. 84)}$$

The Ekman number may be called a *structural parameter*, since, like P_m and P_t, its value is determined by the rotation and physical properties of the Earth, none of which relate to MHD processes in the core. In contrast, the Ro may be called a *response parameter* since its value, like R_m and R_k, depends on the response of the system to the physical state in which it finds itself. The molecular viscosity, ν_M, is uncertain but is commonly estimated to be about 10^{-6} m^2 s^{-1}. Using this for ν in Eq. (83) and taking $\mathcal{L} = 2 \times 10^6$ m as before, we obtain $E \sim 10^{-15}$. Such a small value suggests that large-scale momentum is transported more effectively by small turbulent eddies than by molecular processes, and that E should be estimated using a larger, turbulent viscosity, ν_T. A popular paradigm (see, for example, Braginsky and Meytlis, 1990) assumes that $\nu_T \sim \eta$, but even then E does not exceed 10^{-9}. Taking, as before, $\mathcal{V} = 2 \times 10^{[illegible]}$ m s^{-1}, we have $Ro \sim 10^{[illegible]}$.

From the estimates just made, we see that the FOC is a *rapidly rotating fluid*, defined as one for which

$$E \ll 1, \qquad \text{(Eq. 85)}$$

$$Ro \ll 1. \qquad \text{(Eq. 86)}$$

The ratio $Ro/E = \mathcal{VL}/\nu$ is the (kinetic) Reynolds number R_k and is large. For the geodynamo to exist, $R_m = \mathcal{VL}/\eta$ must be of order 10^2 and, since $P_m \ll 1$, Eq. (85) follows from Eq. (86). The smallness of Ro means that we may discard one of the inertial terms, $\mathbf{V} \cdot \nabla\mathbf{V}$, from Eq. (74). Looking ahead to the Section on the Classical and Coriolis MHD, we shall recognize that the nonlinear feedback equilibrating the geodynamo is the Lorentz force rather than the inertial force.

The remaining inertial term, $\partial_t\mathbf{V}$, in Eq. (74) is responsible for *inertial waves*. These are most easily studied by abbreviating Eq. (74) to

$$\partial_t\mathbf{V} + 2\boldsymbol{\Omega}\times\mathbf{V} = -\nabla\Pi, \qquad \text{(Eq. 87)}$$

and by solving this equation and Eq. (73) subject to Eq. (80). The waves are dispersive and are obtained by assuming that $\mathbf{V}$ is proportional to $\exp(\mathrm{i}\omega t)$. The spectrum of possible ω is discrete, with infinitely many possible eigenvalues ω, all in the range $|\omega| \leq 2\Omega$ (see Greenspan, 1968). The closer $|\omega|$ is to zero, the denser the packing of eigenvalues and the more two-dimensional the eigenfunctions are with respect to $\boldsymbol{\Omega}$, which we shall take to be in the z-direction. The extreme case, $\omega = 0$, gives a single, infinitely degenerate eigenfunction called the *geostrophic mode*. For this mode, $\partial_t\mathbf{V} = \mathbf{0}$ and Eqs. (73) and (87) give

$$2\boldsymbol{\Omega} \cdot \nabla\mathbf{V} = \mathbf{0}. \qquad \text{(Eq. 88)}$$

Stated in words (see *Proudman-Taylor theorem*):

The slow steady inviscid motion of a rotating inviscid fluid is two-dimensional with respect to the rotation axis.

Taking $\boldsymbol{\Omega} = \Omega\mathbf{1}_z$, Eq. (88) gives

$$\mathbf{V} = \mathbf{V}(x, y). \qquad \text{(Eq. 89)}$$

(Here and elsewhere $\mathbf{1}_q$ is the unit vector in the direction of increasing coordinate q.) Since the CMB and ICB have been assumed to be spherical (though the following result is equally valid for an axisymmetric boundary), Eqs. (80) and (89) imply that

$$\mathbf{V} = \overline{V}_G(s)\mathbf{1}_\phi, \qquad \text{(Eq. 90)}$$

where (s, ϕ, z) are cylindrical coordinates. The function $\overline{V}_G(s)$ is arbitrary; this is the infinite degeneracy referred to above. The geostrophic flow is axisymmetric, zonal, and constant on *geostrophic cylinders*, $\mathcal{C}(s)$. These are cylinders of constant radius s. Figure M151 shows a typical geostrophic cylinder and also a particularly significant one, $\mathcal{C}_2 = \mathcal{C}(R_2)$, which touches the SIC on its equator and is therefore called the tangent cylinder, abbreviated here to "TC" (see *Inner core tangent cylinder*).

For further insight into the geostrophic mode, consider the effect of a buoyancy force produced by an axisymmetric temperature distribution $\overline{T}$ that creates a density stratification frozen into the fluid ($\kappa = 0$). Then Eq. (74) gives

$$2\boldsymbol{\Omega} \times \overline{\mathbf{V}} = -\nabla\Pi - \alpha\overline{T}\mathbf{g}. \qquad \text{(Eq. 91)}$$

Since $\nabla \times \mathbf{g} = \mathbf{0}$, this implies, instead of Eq. (88), that

$$2\boldsymbol{\Omega} \cdot \nabla\overline{\mathbf{V}} = -\alpha\mathbf{g} \times \nabla\overline{T}. \qquad \text{(Eq. 92)}$$

Since $\mathbf{g}(= -g\mathbf{1}_r)$ is radial, this reduces to

$$\partial_z\overline{V}_\phi = (g\alpha/2\Omega r)\partial_\theta\overline{T}, \qquad \text{(Eq. 93)}$$

which when integrated gives

$$\overline{\mathbf{V}} = [\overline{V}_T(s, z) + \overline{V}_G(s)]\mathbf{1}_\phi. \qquad \text{(Eq. 94)}$$

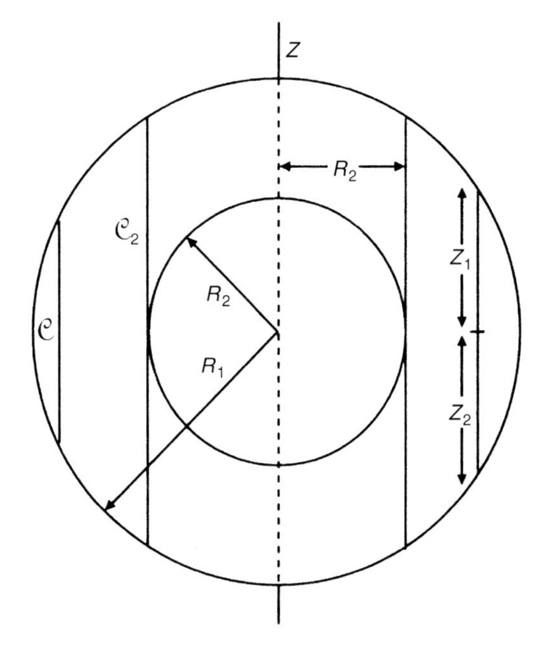

Figure M151 A typical geostrophic cylinder $\mathcal{C}$ and a particularly significant geostrophic cylinder $\mathcal{C}_2$ called the "tangent cylinder" that touches the inner core $r = R_2$ at its equator. The cylinder of radius s meets the surface of the outer core, $r = R_1$, on the latitude circles $z = z_1 \equiv \sqrt{(R_1^2 - s^2)}$ and $z = z_2 \equiv -\sqrt{(R_1^2 - s^2)}$. If $s < R_2$, there is a northern geostrophic cylinder meeting the inner core at $z = \sqrt{(R_2^2 - s^2)}$ and a southern geostrophic cylinder meeting it at $z = -\sqrt{(R_2^2 - s^2)}$.

The flow $\overline{V}_\mathrm{T}$ is called the "thermal wind" (see *Thermal wind*). Its magnitude is $O(g\alpha\Delta T/\Omega)$ and, if this is comparable with our assumed characteristic velocity $\mathcal{V} = 2 \times 10^{-4}$ m s^{-1}, the pole-equator temperature difference ΔT is, for $\alpha \sim 10^{-5}$ K^{-1}, of the order 10^{-4} K. This may be regarded as typical of the temperature differences between rising and falling fluid throughout the core though, more precisely, when the compressibility of the core is properly allowed for, it is typical of the excess or deficit of the temperature relative to the adiabat. The state (94) may be subject to asymmetric *baroclinic instabilities*. These are studied by, for example, Gill (1982).

There is clearly some arbitrariness in Eq. (94), since all or any part of $\overline{V}_\mathrm{G}$ can be absorbed into $\overline{V}_\mathrm{T}$. A convenient way of removing this arbitrariness is to introduce a *geostrophic average*. The geostrophic average of a scalar field $Q(\mathbf{x}, t)$ is denoted by angle brackets and is defined for $s > R_2$ by:

$$\langle Q\rangle(s,t) = \frac{1}{A(s)}\int_{\mathcal{C}(s)} Q(\mathbf{x},t)\mathrm{d}S = \frac{1}{2z_1}\int_{-z_1}^{z_1} \overline{Q}(s,z,t)\mathrm{d}z. \quad \text{(Eq. 95)}$$

Here $\pm z_1(s) = \pm\sqrt{(R_1^2 - s^2)}$ gives the z-coordinates of the latitude circles where the geostrophic cylinder C(s) meets the CMB (see Figure M151); $A = 4\pi s z_1(s)$ is the area of the cylinder and $\mathrm{d}S = s\mathrm{d}\phi\mathrm{d}z$. The *ageostrophic part* of $Q(\mathbf{x}, t)$ is what is left over after the geostrophic part has been subtracted: $\widetilde{Q}(\mathbf{x},t) = Q(\mathbf{x},t) - \langle Q\rangle(s,t)$. The geostrophic part of a vector field $\mathbf{Q}(\mathbf{x}, t)$ is defined by

$$\langle \mathbf{Q}\rangle(s,t) = \langle Q_\phi\rangle(s,t)\mathbf{1}_\phi. \quad \text{(Eq. 96)}$$

It is particularly significant because all the angular momentum about Oz contained between the shells $\mathcal{C}(s)$ and $\mathcal{C}(s + \mathrm{d}s)$ is accounted for by $V_\mathrm{G}(s)$. The separation (94) is made unique by defining $\overline{\mathbf{V}}_\mathrm{G} = \langle\overline{\mathbf{V}}\rangle$.

The definitions of $\langle Q\rangle$ and $\langle\mathbf{Q}\rangle$ just given apply only when $s > R_2$. Inside the TC, there is a geostrophic average for the fluid to the north of the inner core and another for the fluid to the south.

The thermal wind is illustrative of a more general situation: the response of a rotating fluid to forcing. We consider

$$\partial_t\mathbf{V} + 2\mathbf{\Omega}\times\mathbf{V} = -\nabla\Pi + \mathbf{F}, \quad \text{(Eq. 97)}$$

where $\mathbf{F}(\mathbf{x}, t)$ is assigned and the response $\mathbf{V}$ is sought. There are interesting cases in which the forcing has a high frequency. For example, in the rotating frame, the precessionally driven flow considered earlier has the same timescale as the inertial waves. Also, diurnal frequencies may arise from core turbulence. But, on the timescale $\mathcal{T}$ of large scale convection, the modified Rossby number, $1/\Omega\mathcal{T}$, is small. The response to $\mathbf{F}$ is then a combination of free inertial waves (the solution to the homogeneous problem (87) defined by Eq. (97)) and a particular solution that varies slowly, on the same timescale as $\mathbf{F}$. This is the part of $\mathbf{V}$ of greatest interest; the free inertial waves are of lesser significance and can be filtered out by discarding the $\partial_t\mathbf{V}$ term in Eq. (97), so dispensing with the inertial force in its entirety.

Although discarding $\partial_t\mathbf{V}$ is tempting, some caution should be exercised. The geostrophic average of Eq. (97) is

$$\partial_t\overline{V}_\mathrm{G} + 2\Omega\langle\overline{V}_\mathrm{s}\rangle = \langle F_\phi\rangle. \quad \text{(Eq. 98)}$$

The second term is proportional to the mass flux out of $\mathcal{C}$, and is by Eqs. (73) and (80):

$$\langle\overline{V}_\mathrm{s}\rangle \equiv \frac{1}{A(s)}\int_{\mathcal{C}} \mathbf{V}\cdot\mathrm{d}\mathbf{S} = 0, \quad \text{(Eq. 99)}$$

so that Eq. (98) becomes

$$\partial_t\overline{V}_\mathrm{G} = \langle F_\phi\rangle. \quad \text{(Eq. 100)}$$

If its left-hand side is omitted, there is no solution to Eq. (100) unless $\langle F_\phi\rangle = 0$. When $\mathbf{F}$ is the buoyancy force $-\alpha T\mathbf{g}$ alone, $\langle F_\phi\rangle = 0$ because $\mathbf{g}$ has no ϕ-component. If $\mathbf{F}$ contains the Lorentz force, however, it is not necessarily true that $\langle F_\phi\rangle = 0$, a point to which we return in the Section on Classical and Coriolis MHD. Meanwhile we see that one way of evading the difficulty that arises when $\langle F_\phi\rangle \neq 0$, while at the same time removing the free inertial waves, is to retain only the geostrophic part of the inertial force, replacing Eq. (74) by

$$\partial_t\overline{\mathbf{V}}_\mathrm{G} + 2\mathbf{\Omega}\times\mathbf{V} = -\nabla\Pi - \alpha T\mathbf{g} + \mathbf{J}\times\mathbf{B}/\rho + \nu\nabla^2\mathbf{V}. \quad \text{(Eq. 101)}$$

This is often employed in preference to Eq. (74).

We have so far ignored the effects of viscosity. The smallness (85) of E encourages an asymptotic approach to determining core flow. Conceptually, the core is divided into boundary layers, in which viscosity is significant, and the remaining mainstream in which viscosity does not act at leading order (see Eq. (97)). The most important boundary layers in rotating fluids are the Ekman layers, the thicknesses of which are of order $\delta_\nu = (\nu/\Omega)^{1/2} = E^{1/2}\mathcal{L} \sim 0.1$ m (see Greenspan, 1968 and *Core, boundary layers*). The main task of the Ekman layers is to reconcile mainstream flows with the no-slip conditions (76) and (78). For example, the mainstream $\overline{\mathbf{V}}$ obtained from Eq. (97) will not in general obey Eq. (76) at the latitude circle (s, z_1), and an Ekman layer forms that smoothly joins $\overline{\mathbf{V}}(s, z_1)$ to the angular velocity of the CMB (which is zero by choice of reference frame). Associated with the Ekman layer is a significant process called Ekman pumping. Although the solution of Eq. (97) had to satisfy Eq. (80) to leading order, mass conservation in the Ekman layer requires that, at the next order, a meridional mainstream flow $\overline{\mathbf{V}}_M(s,z)$ exists of order $E^{1/2}\overline{\mathbf{V}}(s, z_1)$.

This provides an alternative way of overcoming the difficulty encountered when $\langle F_\phi\rangle \neq 0$. The outward mass flux $\langle V_\mathrm{s}\rangle$ from $\mathcal{C}(s)$, represented by the second term in Eq. (98), can be balanced by an inward radial flux from the Ekman layers on the northern and southern radial caps, $\mathcal{N}(s)$ and $\mathcal{S}(s)$, that form the ends of $\mathcal{C}(s)$; see Figure M152 where, for simplicity, complications from the rotation of the

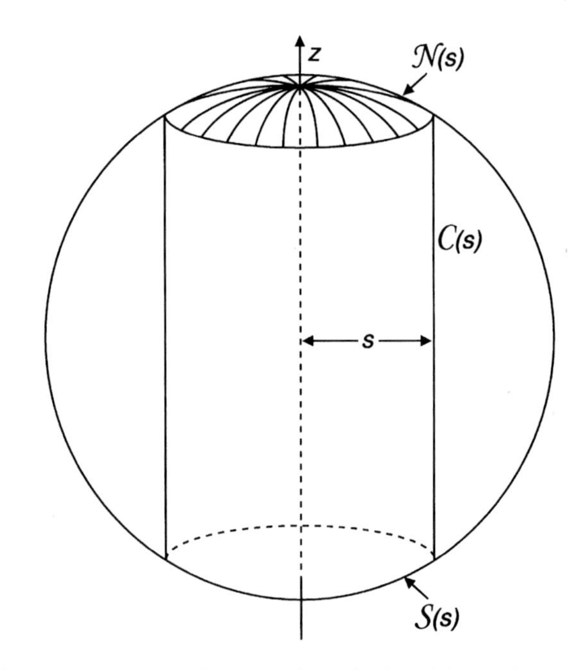

Figure M152 A typical geostrophic cylinder $\mathcal{C}(s)$ of radius s and the two spherical caps, $\mathcal{N}(s)$ and $\mathcal{S}(s)$, at its ends. The flux of fluid out of $\mathcal{C}$ through the curved surface of radius s is equal to the pumping from the Ekman layers on the spherical caps (see text).

SIC have been evaded by ignoring its existence. Pumping through the caps vitiates Eq. (99) and leads to a modification of Eq. (100) in which a frictional term of order $E^{1/2}\Omega\overline{V}_G$ is added to the left-hand side. This determines the spin-up timescale, $\tau_{\text{spin up}}$, on which changes in the angular velocity of the mantle are transmitted to the fluid core. By comparing $\partial_t\overline{V}_G$ with $E^{1/2}\Omega\overline{V}_G$, we see that $\tau_{\text{spin up}} \sim E^{-1/2}\Omega^{-1} \sim E^{1/2}\tau_v$, or about 10^5 y. Here τ_v is the *viscous decay time*:

$$\tau_v = \mathcal{L}^2/\nu. \qquad \text{(Eq. 102)}$$

For $\mathcal{L} = 2 \times 10^6$ m and $\nu_M = 10^{-6}$ m^2 s^{-1}, this is over 10^{11} y; for $\nu_T \sim \eta$ it is $O(\tau_\eta)$.

The effects of viscosity are also felt in complicated shear layers that surround the TC whenever the SIC does not move with the mantle. For a discussion of these "Stewartson layers," see *Core, boundary layers*.

Viscosity also plays a crucial role in thermal convection. We shall consider only the case (85) of small E; for simplicity, we suppose that $P_t > 1$. We ignore the SIC and suppose that heat sources are distributed uniformly throughout the core. If these sources are weak, they create only a spherically symmetric temperature distribution $\overline{T}_c(r)$ that carries heat out of the core by thermal conduction. Although the associated density distribution is top-heavy, the diffusion of heat and momentum prevents convective instability; but, if the heat sources are gradually increased, weak convection occurs as soon as Ra exceeds a critical value, Ra_c. For $Ra = Ra_c$, the flow is asymmetric in structure, not axisymmetric. It takes the form of a "cartridge belt" of two-dimensional cells, often called *Taylor cells*, regularly spaced round the axis of rotation and drifting in longitude about that axis (Roberts, 1968; Busse, 1970; Jones *et al.*, 2000). Adjacent cells spin around their axes in opposite directions in a sequence of cyclonic and anticyclonic vortices, with vorticity respectively parallel and antiparallel to $\mathbf{\Omega}$. (See *Convection, nonmagnetic rotating*.) Although, when Ra exceeds Ra_c, nonlinear interactions create other flow components (including an easily excited geostrophic motion), the cartridge belt structure is maintained, provided Ra is not too large.

The name Taylor cell is a useful reminder of the Proudman-Taylor theorem, which the flow is trying to obey by being as two-dimensional as possible, consistent with allowing convection to occur at all. We have already seen that, if $\nu = 0$, small amplitude motion must be geostrophic, since the thermal forcing $T_c(r)$ is independent of latitude. But geostrophic motions have no radial components to carry heat outward. Convection can occur only if the buoyancy force is large enough to break the rotational constraint of the theorem. Thus, although Ra_c would be $O(1)$ if E were $O(1)$, the critical Rayleigh number is large when $E \ll 1$; in fact $Ra_c = O(E^{-4/3}) \sim 10^{20}$, and the number of cells in the cartridge belt is then of order $E^{-1/3} \sim 10^5$, i.e., the scale $\mathcal{L}_\perp$ of the motions perpendicular to $\mathbf{\Omega}$ is $O(E^{1/3}\mathcal{L}) \sim 30$ m. Convective heat transport is mainly in the s-direction, i.e., away from the rotation axis (Busse, 1970). The instability criterion $Ra_c = O(E^{-4/3})$ may also be written as $N_c = (\nu\kappa/\mathcal{L}^4)^{1/2}E^{-2/3}$, where $N = \sqrt{(g\alpha\beta)}$ is the buoyancy frequency that, if imaginary, would usually be called the *Brunt-Väisälä frequency* (Gill, 1982). It is worth pointing out that, in the absence of diffusive effects, the density stratification would be unstable when $N_c = O(\Omega)$. This value of N_c is greater than $(\nu\kappa/\mathcal{L}^4)^{1/2}E^{-2/3}$ by a large factor, of order $E^{-1/3}$. In other words, the effects of diffusion are destabilizing. There is also a basic difference between the convective instability that onsets when $Ra = O(E^{-4/3})$ and the *Rayleigh-Taylor instability* of the diffusionless system. The latter releases gravitational energy in a single spasm that transforms a top-heavy mass distribution into a bottom-heavy configuration. The heat conduction necessary to renew constantly the top-heavy distribution is (by assumption) absent, and therefore the continuously overturning motions characteristic of thermal convection do not occur. We return to this topic in the Section on CMHD for imperfect fluids.

Classical and Coriolis MHD

Classical MHD

In this section we shall suppose that $\mathbf{\Omega} \equiv 0$ and initially also exclude buoyancy ($\mathbf{g} \equiv 0$). We give a rudimentary account of relevant concepts in the MHD in nonrotating systems. For a more complete treatment, see Davidson (2001).

Much of classical MHD theory was originally developed to explain laboratory flows of liquid metals, and two dimensionless numbers were introduced to quantify the importance of the Lorentz force relative to the viscous and inertial forces, the *Hartmann number*, Ha, and the *interaction parameter*, N. These can be written in several equivalent ways:

$$Ha = \mathcal{B}\mathcal{L}\left(\frac{\sigma}{\rho\nu}\right)^{1/2} = \frac{\mathcal{B}\mathcal{L}}{\sqrt{(\mu_0\rho\nu\eta)}} = \frac{\mathcal{V}_A\mathcal{L}}{\sqrt{(\nu\eta)}} = \left(\frac{\tau_v}{\tau_m}\right)^{1/2} = \frac{Lu}{\sqrt{P_m}}, \qquad \text{(Eq. 103)}$$

$$N = \frac{\sigma\mathcal{B}^2\mathcal{L}}{\rho\mathcal{V}} = \frac{\mathcal{B}^2\mathcal{L}}{\mu_0\rho\eta\mathcal{V}} = \frac{\mathcal{V}_A^2\mathcal{L}}{\eta\mathcal{V}} = \frac{\tau_V}{\tau_m} = \frac{Lu}{Al} = \frac{Ha^2}{R_k}. \qquad \text{(Eq. 104)}$$

The current $\mathcal{J}$ created by motions $\mathcal{V}$ in a field $\mathcal{B}$ is $\mathcal{J} \sim \sigma\mathcal{V}\mathcal{B}$ so that the Lorentz force per unit volume is approximately $\sigma\mathcal{B}^2\mathcal{V}$. This is greater than the inertial acceleration $\rho\mathbf{V}\cdot\nabla\mathbf{V} \sim \rho\mathcal{V}^2/\mathcal{L}$ by a factor of N and exceeds the viscous force per unit volume, which is of order $\rho\nu\mathcal{V}/\mathcal{L}^2$, by a factor of Ha^2 which is often called the *Chandrasekhar number*, particularly in the literature dealing with magnetoconvection (see *Magnetoconvection*). Even though R_m and Lu may be very small in laboratory systems (see the Section on the imperfect conductor, reconnection), Ha may be large because of the smallness of P_m for liquid metals; the Lorentz force can therefore profoundly affect the motion. In large systems for which $R_m \gtrsim 1$, a more appropriate measure of $\mathcal{J}$ may be $\mathcal{B}/\mu_0\mathcal{L}$, with a corresponding Lorentz force of order $\mathcal{B}^2/\mu_0\mathcal{L}$ per unit volume. This is greater than the viscous force by order N/P_m and exceeds the inertial force by order $1/Al^2$, where Al is the *Alfvén number* (also sometimes called the magnetic Mach number):

$$Al = \mathcal{V}/\mathcal{V}_A. \qquad \text{(Eq. 105)}$$

It is frequently argued that an MHD system capable of generating its own magnetic field will favor a dynamical balance in which inertial and Lorentz forces are comparable ($Al = O(1)$) so that the magnetic and kinetic energies are comparable also: $u^B/u^K = Al^{-2} = O(1)$. In a highly rotating system however, this is not necessarily true (see the Section on the CMHD for imperfect fluids).

MHD boasts an analogue of the Proudman-Taylor theorem. This has already been foreshadowed in the Section on the perfect conductor, where an anisotropic friction was identified for $R_m \ll 1$ that acts to damp out motions perpendicular to $\mathbf{B}$ in a time of order τ_m. If $\tau_m \ll \tau_v$ and τ_V, i.e., if Ha and N are so large that viscous and inertial forces are negligible, the steady state form of Eq. (74) becomes

$$0 = \nabla(\rho\Pi) + \mathbf{J} \times \mathbf{B}, \qquad \text{(Eq. 106)}$$

which is the equation for magnetostatic equilibrium. Nevertheless, because $Ha \gg 1$, small motions create large enough currents to make the Lorentz force significant in Eq. (106). Let $\mathbf{B} = \mathbf{B}_0 + \mathbf{b}$, where $\mathbf{B}_0$ is a uniform field applied across the system and $\mathbf{b}(\ll B_0)$ is the small response to the flow. Then $\mathbf{J} = \mathbf{j} = \nabla \times \mathbf{b}/\mu_0$ and Eq. (106) give

$$\mathbf{B}_0 \cdot \nabla\mathbf{J} = 0. \qquad \text{(Eq. 107)}$$

If it is also true that $R_m \gg 1$ so that magnetic diffusion is negligible, Eq. (73) and the steady induction equation (47) gives

$$\mathbf{B}_0 \cdot \nabla \mathbf{V} = \mathbf{0}. \quad \text{(Eq. 108)}$$

The MHD two-dimensional theorem follows from Eqs. (107) and (108):

In a steady dissipationless MHD system across which a uniform magnetic field is applied, the flow and current are two dimensional with respect to the direction of the applied field.

Taking $\mathbf{B}_0 = B_0 \mathbf{1}_z$, we have

$$\mathbf{V} = \mathbf{V}(x, y), \quad \text{(Eq. 109)}$$

$$\mathbf{J} = \mathbf{J}(x, y). \quad \text{(Eq. 110)}$$

This result may be compared with Eq. (89).

Solutions such as Eqs. (109) and (110), obtained on the assumption that R_k and R_m are large, are subject to boundary conditions such as

$$[\![\mathbf{n} \cdot \mathbf{B}]\!] = 0, \quad \text{on the boundaries;} \quad \text{(Eq. 111)}$$

$$\mathbf{n} \cdot \mathbf{V} = 0, \quad \text{on the boundaries;} \quad \text{(Eq. 112)}$$

cf. Eq. (80). Boundary layers are required to make the solutions satisfy the full set of wall conditions; these are known as Hartmann layers. As for the Ekman layer, the Hartmann layers are of a type sometimes called "active" or "controlling," since mainstream solutions involving unknown functions, like Eqs. (109) and (110), cannot be completely determined until certain "Stewartson conditions" that are demanded by the Hartmann layer are applied:

$$P_m^{1/2} [\![\mathbf{n} \times \mathbf{V}]\!]_{bl} = \mathrm{sgn}(\mathbf{n} \cdot \mathbf{B}) [\![\mathbf{n} \times \mathbf{B}]\!]_{bl} / \mu_0 \rho. \quad \text{(Eq. 113)}$$

Here the unit normal $\mathbf{n}$ is directed from the boundary layer into the fluid and the suffix bl signifies that the discontinuity is right across the Hartmann layer, i.e., is the difference between mainstream and boundary values; Eq. (113) is valid only where $\mathbf{n} \cdot \mathbf{B}$ is nonzero.

In the limit of zero dissipation ($\nu \to 0$, $\eta \to 0$), the Hartmann layer becomes a combined surface current $\mathbf{C}$ and surface vorticity $\mathbf{Z}$, defined in analogy to Eq. (14) as $[\![\mathbf{n} \times \mathbf{V}]\!] = \mathbf{Z}$. The origins of Eq. (113) can be qualitatively understood as in the Section on the perfect conductor by demanding that the viscous force associated with the shear generated by the integrated Lorentz force, $(\mathbf{n} \cdot \mathbf{B})(\mathbf{n} \times \mathbf{C})/\mu_0$, balances that force, and by arguing that the emf created by $\mathbf{Z}$ is $(\mathbf{n} \cdot \mathbf{B})\mathbf{Z}$ and generates $\mathbf{C}$. This also shows that the thickness of the Hartmann layer is of order $\delta_M = \mathcal{L}/Ha = (\nu\eta)^{1/2}/\mathcal{V}_A$.

What is perhaps most remarkable is that Eq. (113) applies even in the absence of diffusion ($\nu = \eta = 0$) but that nevertheless it depends on the ratio P_m of the two diffusivities! When $\mathbf{B}$ is nonuniform and the boundaries are curved, $\nabla_S \cdot \mathbf{C}$ is generally nonzero, so that the Hartmann layer pumps current into or out of the mainstream (see Eq. (18)).

The MHD two-dimensional theorem has implications for thermal convection resembling those that the Proudman-Taylor theorem has for rotating convection (see the Section on the Classical theory of rotating fluids). Consider a Bénard layer, i.e., a uniform horizontal layer of fluid, of depth $\mathcal{L}$, heated from below and cooled from above. Suppose a large uniform vertical field of strength $\mathcal{B}$ is applied ($Ha \gg 1$). If $\nu = 0$, weak motions must satisfy Eq. (109); by Eq. (112), they are horizontal everywhere and therefore have no z-component to carry heat across the layer. Convection can occur only if the buoyancy force is large enough to break the magnetic constraint of the two-dimensional theorem. Thus, although Ra_c would be $O(1)$ if Ha were $O(1)$, the critical Rayleigh number is large when $Ha \gg 1$; in fact $Ra_c = O(Ha^2)$, and the (horizontal) scale $\mathcal{L}_\perp$ of the motions perpendicular to $\mathbf{B}_0$ is $O(Ha^{-1/2}\mathcal{L})$. If the applied field is horizontal instead of vertical, it has no effect on the marginal mode, in which the pattern of convection consists of horizontal rolls parallel to $\mathbf{B}_0$, for which the two dimensional theorem applies, though now the two-dimensional motions have a vertical component that transports heat across the layer. (See *Magnetoconvection*.)

CMHD for ideal fluids

In a way reminiscent of MHD itself (see the Introduction), the combination of the classical theories of MHD and rotating fluids gives rise to new phenomena that are absent from the two subjects in isolation. The subject becomes so different that it deserves its own name and acronym. The obvious choice, RMHD, has already been appropriated by both "relativistic MHD" and "reduced MHD," so it is called here "Coriolis magnetohydrodynamics" or "CMHD." Phenomena in which diffusion plays a central role will be described in the Section on the CMHD for imperfect fluids. Here we focus on ideal flows and, to illustrate as dramatically as possible how different CMHD is from MHD, we shall suppose that the *magnetic Rossby number*,

$$Ro_m = \mathcal{V}_A / \Omega \mathcal{L}, \quad \text{(Eq. 114)}$$

is small:

$$Ro_m \ll 1. \quad \text{(Eq. 115)}$$

In the Earth's core, $Ro_m = O(10^{-3})$.

One of the ways in which the magnetic field transforms the classical theory of rotating fluids is by removing the geostrophic degeneracy of the inertial waves discussed in the Section on the Classical theory of rotating fluids. These are replaced by a discrete set of torsional waves (see *Oscillations, torsional*). These arise because the geostrophic cylinders sketched in Figure M151 are coupled together by the s-component of $\mathbf{B}$, i.e., the component that threads the cylinders together. It was seen in the Section on the Classical theory of rotating fluids that the Coriolis force associated with geostrophic motions can be absorbed into the pressure gradient. Since this otherwise dominating force is then essentially absent, the remaining forces become influential, including the inertial term $\rho \partial_t \mathbf{V}$. The torsional waves therefore resemble Alfvén waves and have the same timescale, $\tau_A = \mathcal{L}/\mathcal{V}_A$, where $\mathcal{V}_A = \sqrt{\langle V_A^2 \rangle}$ is now based on the rms strength of B_s on the geostrophic cylinder $\mathcal{C}(s)$:

$$\mu_0 \rho \langle V_A^2 \rangle (s, t) = \langle B_s^2 \rangle (s, t) = \frac{1}{A(s)} \int_{\mathcal{C}(s)} B_s^2(s, \phi, z, t) \mathrm{d}S. \quad \text{(Eq. 116)}$$

Even if $\sqrt{\langle B_s^2 \rangle}$ were as small as 10^{-3} T, the Alfvénic timescale τ_A is less than a decade. This may be compared with the timescale of the ageostrophic waves which (see below) is of order 10^3 y. It is also short in comparison with the timescales, $\tau_{\text{spin up}}$ and τ_η, of the diffusive processes. The high frequency character of the torsional waves means that in describing them we may ignore the time-dependence of $\langle B_s \rangle$ and all diffusive effects.

The dissipationless geostrophic component of $\mathbf{V}$ obeys the inviscid form of Eq. (98), now written as

$$\rho \partial_t \overline{V}_G = \langle (\mathbf{J} \times \mathbf{B})_\phi \rangle, \quad \text{(Eq. 117)}$$

and by the induction equation (61) for a perfect conductor. (The term $2\Omega \langle V_s \rangle$ has been omitted from Eq. (117) because of mass conservation; see Eq. (99).) In a steady state, $\langle (\mathbf{J} \times \mathbf{B})_\phi \rangle = 0$, i.e.,

$$\int_{\mathcal{C}(s)} (\mathbf{J} \times \mathbf{B})_\phi \mathrm{d}S = 0. \quad \text{(Eq. 118)}$$

This important result, due to Taylor (1963) defines what are called *Taylor states* (see *Taylor's condition*).

If Eq. (118) does not hold initially, a torsional wave is launched in which the geostrophic cylinders oscillate about a Taylor state, $\mathbf{V}_G^T(s)$. Denoting by $v(s,t) = \overline{V}_G(s,t) - \overline{V}_G^T(s)$ the departure of the geostrophic state from this Taylor state, Braginsky (1970) showed that the associated shear $\zeta = v/s$ satisfies the *torsional wave equation*

$$\frac{\partial^2 \zeta}{\partial t^2} = \frac{1}{s^2 A}\frac{\partial}{\partial s}\left[s^2 A \langle V_A^2 \rangle \frac{\partial \zeta}{\partial s}\right]. \qquad \text{(Eq. 119)}$$

The waves transport the z-component, $\rho s \overline{V}_\phi$, of the angular momentum density to and fro across the core, but do not change the integrated angular momentum of core flow:

$$M = \int_{\text{SIC}} \rho s V_\phi \mathrm{d}V = \rho_0 \int_0^{R_1} s\overline{V}_G A \mathrm{d}s. \qquad \text{(Eq. 120)}$$

In the simplest application of Eq. (119), the existence of the SIC is ignored. The ordinary differential equation obtained by substituting $\zeta(s,t) = Z(s)\exp(\mathrm{i}\omega t)$ into Eq. (119) has regular singularities at both $s = 0$ and $s = R_1$ where $A = 0$. The implicit requirement that Z be bounded at both these points transforms Eq. (119) into an eigenvalue problem for the torsional wave frequencies, ω.

The picture is complicated by the SIC. This has a composition similar to the FOC but, being a solid, its conductivity is even greater and the justification for ignoring its resistivity is even stronger. Since ohmic and viscous diffusion are absent, the boundary layer at the ICB is of Hartmann type (see the Section on Classical MHD) so that Eq. (113) applies and P_m is involved. Since however $P_m \ll 1$, the condition (113) reduces to $[\![\mathbf{B}]\!] = 0$. In consequence,

$$\zeta = \Omega_{\text{SIC}}, \quad \text{for } s < R_2. \qquad \text{(Eq. 121)}$$

In other words, in torsional wave motion, the entire contents of the tangent cylinder $\mathcal{C}_2$, both solid and liquid, move together as a solid body. (For an alternative point of view, see Jault and Légaut, 2005.) As Buffett (1996) showed, this part of the torsional motion is mainly determined by the gravitational couple exerted by the mantle on the SIC and by the EM torque on $\mathcal{C}_2$, which can be computed from Eq. (43). As in the Section on the Basic equations, the simplified condition $\Omega_{\text{SIC}} = 0$ is adopted here. The torsional waves therefore exist only outside the tangent cylinder and the appropriate condition of wave reflection at $\mathcal{C}_2$ is $\zeta(R_2) = 0$.

If viscous and ohmic dissipation is allowed for, it is plausible that the torsional waves will die out, so returning the system to the Taylor state, $\overline{V}_G^T$. Roberts and Soward (1972) demonstrated that viscosity would quench the waves on the spin-up timescale, but over such a long period the evolution of $\mathbf{B}$, and therefore of $\sqrt{\langle V_A^2 \rangle}$, would be significant and the coefficients in Eq. (119) would change secularly. It is plausible that ohmic dissipation in the fluid and in the mantle (assumed weakly conducting) will quench the waves even more rapidly on the τ_η timescale of the core, or perhaps (since it differs less from τ_A) on the even shorter τ_η timescale of the mantle.

Electromagnetic coupling, either directly across the CMB, or indirectly via the inner core with the gravitational coupling between the mantle and inner core acting as catalyst (Buffett, 1996), provides one of three ways in which angular momentum can be exchanged between the core and the mantle, the other two being topographic coupling, which depends on the roughness of the CMB, and viscous coupling, which is generally thought to be negligibly small. Since the total angular momentum of the Earth is unchanged by such internal processes, a correlation should exist between variations in the length of the day and changes in the angular momentum of the core. Jault *et al.* (1988) were the first to establish such a connection. From an analysis of the geomagnetic record they could trace the evolution of the core's angular momentum and could show that it was in antiphase with the mantle's angular momentum.

The excitation of torsional oscillations is inexplicable according to the ideas presented here but it seems clear that it must be due to some other higher frequency processes in the core, such as turbulence. For further analysis see Jault (2003), *Oscillations, torsional*; *Core-mantle coupling, electromagnetic*; *Core-mantle coupling, gravitational;* and *Core-mantle coupling, topographic*. Bloxham *et al.* (2002) have associated torsional waves with the observed secular variation "impulses" (see *Geomagnetic jerks*).

Consider next the ageostrophic waves. When $Ro_m = O(1)$, the waves are an inextricable mixture of inertial and Alfvén waves, but when Eq. (115) holds, little remains of the Alfvén wave! This illustrates again how Coriolis forces can transform classic MHD. When $Ro_m \ll 1$, the inextricable mixture neatly separates into two families: inertial waves and *slow waves*, also known as "Lehnert waves." The former differ little from the inertial waves discussed in the Section on the Classical theory of rotating fluids; they are "fast," since their timescale is diurnal. The latter are very unlike Alfvén waves. They are called "slow" because their timescale, $\tau_s = \Omega\mathcal{L}^2/\mathcal{V}_A^2$, is of order 10^3 y. This is comparable with the timescale of the secular variation of the main geomagnetic field (see *Geomagnetic secular variation*). The corresponding velocity, $c_s = \mathcal{L}/\tau_s = \mathcal{V}_A^2/\Omega\mathcal{L}$, is of order 10^{-4} m s^{-1}, which the speed at which discernable magnetic features at the CMB drift westward. The time dependence of the slow waves is governed by the left-hand side, $\partial_t \mathbf{B}$, of the induction equation (61); the inertial forces in Eq. (74) play essentially no role. For this reason the waves are sometimes called MC *waves*, to emphasize that the essential ingredients governing their propagation are the magnetic and Coriolis forces alone (and of course the pressure gradient). The relative unimportance of the inertial waves suggests that MC waves can be analyzed in a clearer way by discarding the inertial forces in Eq. (74). If this is done, however, the torsional waves are filtered out too. By retaining the key part of the inertial force as in Eq. (101), both the torsional and MC waves are retained but the inertial waves are removed (see also *Magnetohydrodynamic waves*).

MC waves, like inertial waves, are dispersive but possess abundant helicity, making them potentially efficient components of a dynamo. Although they have axisymmetric parts, they are characteristically large-scale planetary waves, i.e., they are asymmetric and, if they have finite amplitude, they create an emf $\overline{(\mathbf{V}' \times \mathbf{B}')}_\phi$ that drives a current $\overline{J}_\phi$ that potentially can defeat Cowling's theorem (see the Section on the kinematic dynamo). They are, however, obliterated by ohmic diffusion in traveling a distance of order $c_s\tau_\eta = (\mathcal{V}_A^2/\Omega\eta)\mathcal{L}$. Buoyancy forces can prevent their demise. This provides a strong motivation for studying MAC *waves*, which are MC waves modified by the insertion of Archimedean (buoyancy) forces. As in the example discussed in the Section on the Classical theory of rotating fluids, it may be expected that, if buoyancy is strong enough, instability occurs. More specifically, instability occurs when the buoyancy frequency N is of order $\mathcal{V}_A/\mathcal{L}$, i.e., when the *magnetic Rayleigh number*, Ra_m is $O(1)$, where

$$Ra_m = g\alpha\beta\mathcal{L}^2/\mathcal{V}_A^2. \qquad \text{(Eq. 122)}$$

If we accept the estimate $\mathcal{V} = O(g\sigma\Delta T/\Omega)$ made in the Section on the Classical theory of rotating fluids, we may rewrite Eq. (122) as

$$Ra_m = \Omega\mathcal{V}\mathcal{L}/\mathcal{V}_A^2 = \mathcal{V}/c_s. \qquad \text{(Eq. 123)}$$

From $\Omega = 7\times10^{-4}$ s^{-1}, $\mathcal{V} = 2\times10^{-4}$ m s^{-1}, $\mathcal{L} = 10^6$ m, and $\mathcal{V}_A = 0.1$ m s^{-1}, we obtain $Ra_m = 1$. This suggests that MAC wave instability is relevant to core dynamics.

The simplest demonstrations of MAC waves and their instabilities (Braginsky, 1964) assume that they infinitesimally perturb a uniform applied field $\mathbf{B}_0$ in a plane layer of fluid. Braginsky (1967) also studied small amplitude MAC waves and instabilities in a spherical system in which the imposed field $\mathbf{B}_0$ is zonal. In this case the waves travel in the $\pm\phi$-directions with speeds of order c_s. Braginsky (1964) suggested that the geomagnetic secular variation and westward drift are manifestations of MAC waves. Since c_s and $\mathcal{V}$ are approximately equal when $Ra_m = 1$, it appears that our method (see Section on The imperfect conductor, reconnection) of estimating $\mathcal{V}$ from the speed of westward drift speed c_s is defensible. MAC wave instability is, as in the Section on the Classical theory of rotating fluids, a single spasm of overturning. To maintain the waves continuously it is necessary to invoke diffusion, and this (as will be seen in the Section CMHD for imperfect fluids) is destabilizing.

CMHD for imperfect fluids

Although the core is rapidly rotating in the classical sense ($E \ll 1$, $Ro \ll 1$), it is not rapidly rotating according to the criterion $\Lambda \ll 1$ of CMHD, where Λ is the *Elsasser number*, a dimensionless parameter that can be written in several equivalent ways:

$$\Lambda = \frac{\mathcal{V}_A^2}{\Omega\eta} = \frac{R_m Ro}{Al^2} = Ha^2 E = \frac{\tau_\eta}{\tau_s} = \frac{\tau_\Omega}{\tau_m} = \left(\frac{\delta_E}{\delta_M}\right)^2 = Ro_m Lu. \quad \text{(Eq. 124)}$$

Far from being small, the value of Λ in the core is better described by

$$\Lambda = O(1). \quad \text{(Eq. 125)}$$

Equation (125) defines the scale for *strong fields*:

$$\mathcal{B} = \sqrt{(\rho\Omega/\sigma)} = \sqrt{(\mu_0\rho\Omega\eta)}, \quad \text{in strong field regime.} \quad \text{(Eq. 126)}$$

This, like the Elsasser number, is independent of the length scale $\mathcal{L}$. According to Eq. (126), $\mathcal{J} = O(\sigma\mathcal{V}\mathcal{B}) = O(\mathcal{V}\sqrt{(\sigma\rho\Omega)})$, so that the Lorentz force, $\mathbf{J} \times \mathbf{B}$, is of order $\rho\Omega\mathcal{V}$, i.e., the Lorentz and Coriolis force, $2\mathbf{\Omega} \times \mathbf{V}$, are about equal in magnitude. This suggests that the primary force balance in the core is between $2\mathbf{\Omega} \times \mathbf{V}$ and $\mathbf{J} \times \mathbf{B}$. It is often said that the fact that the magnetic compass needle points approximately north proves that the Coriolis force dominates core dynamics. This however is an oversimplification. The Coriolis force has a preferred direction, that of $\mathbf{\Omega}$. To counter the rotational constraint (see below), the magnetic field configures itself so that it and the Lorentz force share the same preferred direction.

In the same way that $R_m = O(1)$ numerically underestimates the magnetic Reynolds number required for dynamo action (see Section on the kinematic dynamo), it appears from the Section on Classical MHD that Eq. (125) numerically underestimates Λ. Writing Eq. (122) as

$$Ra_m = R_m/\Lambda, \quad \text{(Eq. 127)}$$

we see that marginal MAC wave instability ($Ra_m \sim 1$) requires that $\Lambda = O(100)$. Numerical estimates bear this out: if we take $\mathcal{V}_A = 0.1 \text{ m s}^{-1}$ and $\eta = 2 \text{ m}^2 \text{ s}^{-1}$ as before, we obtain $\Lambda \sim 100$ from Eq. (124), and $\mathbf{B} \sim 0.01$ T, rather than the 10^{-3} T given by Eq. (126). Far from being rapidly rotating, the core is, according to this value of Λ, slowly rotating. Equation (124) shows that the magnetic energy density u^B exceeds the kinetic energy density u^K in the rotating frame by a factor of order $Al^{-2} = \Lambda/R_m Ro \sim 10^5$, assuming as before that $\Lambda \sim 100$, $R_m \sim 100$ and $Ro \sim 10^{-5}$.

A striking way in which CMHD differs from the classical theories of rotating and MHD flows is that the two-dimensional theorems of those subjects are no longer so constraining. If $Ha \gtrsim E^{-1/2}$, the Lorentz force weakens the Proudman-Taylor constraint; if $E \lesssim Ha^{-2}$, the Coriolis force weakens the MHD constraint. When $\Lambda(= Ha^2 E)$ is $O(1)$, both constraints have diminished force. This has very significant consequences for CMHD convection. Suppose that the Bénard layer of the Section on Classical MHD rotates about a vertical axis. It is found that $Ra_c = O(E^{-1})$ for $E \to 0$, instead of the $Ra_c = O(E^{-4/3})$ for the same rotating system when nonmagnetic (see the Section on Classical theory of rotating fluids). The horizontal scale of motions is comparable in the marginal state with the vertical scale $\mathcal{L}$, in contrast to the Section on Classical theory of rotating fluids where the length scale $\mathcal{L}_\perp$ in directions perpendicular to $\mathbf{\Omega}$ was found to be small, $O(E^{1/3}\mathcal{L})$. The result $Ra_c = O(E^{-1})$ may be written as $Ra_{mod,c} = O(1)$, where

$$Ra_{mod} = g\alpha\beta\mathcal{L}^2/\Omega\kappa = RaE \quad \text{(Eq. 128)}$$

is the *modified Rayleigh number*, which is often used in preference to Ra in CMHD convection studies. The fact that this, like Λ and $\mathcal{L}_\perp$, is independent of ν serves to emphasize how Lorentz forces completely take over from viscous forces in breaking the Proudman-Taylor constraint when $\Lambda = O(1)$.

The Bénard example is the simplest to show how convection adapts to changes in field strength. The convection pattern arranges itself to provide the Lorentz force necessary to break the rotational constraint. For example, if $\mathbf{B}_0$ is horizontal and Λ is small, the convection pattern in the marginal state consists of horizontal rolls that are *perpendicular* to $\mathbf{B}_0$ and not parallel as in the Section Classical MHD. For larger Λ, the rolls counter the Coriolis force by becoming inclined to $\mathbf{B}_0$ at an angle that diminishes as Λ increases.

This approach to rotating magnetoconvection is so simple that it borders on the simplistic. It ignores processes that may be significant in a naturally occurring CMHD system: the effects of nonlinearity, boundary curvature, velocity shear, self-generation of field, etc. In particular, it assumes that the magnetic field in which convection occurs is *given*. One might wonder whether a dynamo, which is required to generate its own magnetic field, would expend more energy in doing so than it saved by recouping its convective energy losses. This qualm can be most easily allayed by investigating finite amplitude convection in the Bénard layer, in the absence of the applied field $\mathbf{B}_0$.

Suppose Ra is progressively increased from zero. Convection does not occur until Ra reaches its critical value, $Ra_c = O(E^{-4/3})$, for which convective motions of horizontal scale $\mathcal{L}_\perp = O(E^{1/3}\mathcal{L})$ become possible. As Ra is enhanced, $\mathcal{L}_\perp$ increases, $\mathbf{V}$ grows in amplitude and, at another critical value Ra_B of Ra, $\mathcal{L}_\perp$ becomes $O(E^{1/6}\mathcal{L})$ and $\mathcal{V}$ becomes $O(E^{-1/6}\eta/\mathcal{L})$ so that dynamo action occurs ($R_m = \mathcal{L}_\perp\mathcal{V}/\eta = O(1)$), and supports a field of infinitesimal amplitude. As Ra increases further, $\mathcal{L}_\perp$ becomes larger as do $\mathbf{V}$ and $\mathbf{B}$, but the system remains in a *weak field* state in which the Hartmann number is $O(1)$, i.e., by Eq. (103),

$$\mathcal{B} = \sqrt{(\mu_0\rho\nu\eta)}/\mathcal{L}, \quad \text{in weak field regime.} \quad \text{(Eq. 129)}$$

Unaided by viscosity, the Lorentz force is still too weak to counter the rotational constraint, but this becomes less and less true as the imposed Ra is made larger still; $\mathcal{L}_\perp$ increases toward $\mathcal{L}$ and $\mathcal{B}$ approaches an asymptote at $Ra = Ra_A$ (see Figure M153). The weak field regime exists only for $Ra_B < Ra < Ra_A$; it terminates at the asymptote. As Ra increases to Ra_B there is *runaway field growth* from $Ha = O(1)$ to the strong field regime where $Ha = O(E^{-1/2})$.

The scenario just described, apart from the final runaway process, has been demonstrated by Soward (1974). His analysis showed that the asymptote $Ra = Ra_A$ exists. Strong field solutions were obtained numerically by Jones and Roberts (2000). Once the strong field dynamo is established, hysteresis occurs: the dynamo can continue to function even if Ra is reduced below Ra_B (Rotvig and Jones, 2002). The dynamo is then in a "subcritical state." It may even be that it can regenerate the field when $Ra < Ra_c$, but this has not been demonstrated; it is clear however that the strong field branch is bounded below by some value Ra_{min} of Ra of order E^{-1} or larger, i.e., the dynamo will fail if the buoyancy forces cannot make good the ohmic losses.

It may be wondered whether the spherical convecting system considered in the Section on Classical theory of rotating fluids would behave in a fundamentally different way from the plane layer. This qualm is significantly harder to allay because the analytical and numerical

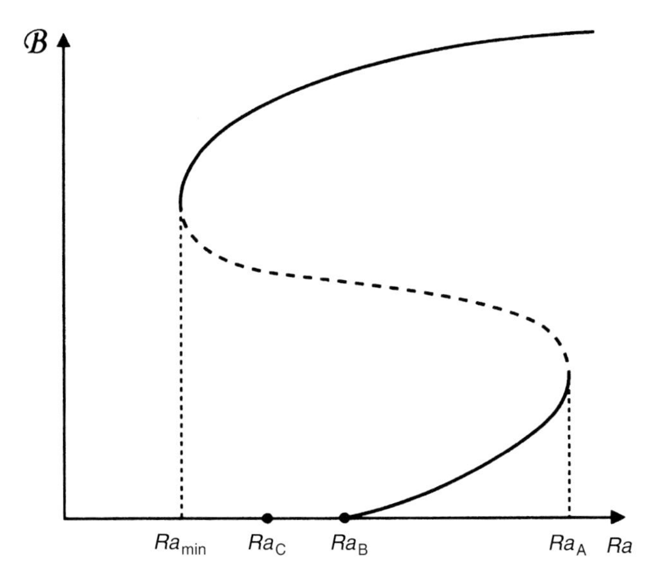

Figure M153 A cartoon showing the weak and strong field branches of the plane layer dynamo (full lines). The dashed curve displays the unstable branch connecting them (see text for full explanation).

approaches are much more difficult to implement. There seems, however, to be no reason to doubt that, by opposing the rotational constraint, a magnetic field will generally allow convection to occur on large scales, as planetary waves, with correspondingly less ohmic and viscous energy losses, and that a strong field branch of dynamo solutions will result. Numerical simulations of convective dynamos have demonstrated that the field adapts itself to oppose the Coriolis force. They have also confirmed the increase in the scale, $\mathcal{L}_\perp$, created by the magnetic field; far fewer Taylor cells lie on the cartridge belt of the Section on Classical theory of rotating fluids. Through the analysis of magnetic features on the CMB, Gubbins and Bloxham (1987) have inferred flux concentrations that may be associated with the ends of four, but no more than four, Taylor cells (see also Kahle *et al.*, 1967).

In the parameter range just considered ($E \ll 1$, $P_t = O(1)$, $P_m = O(1)$), the criterion $Ra_m = O(1)$ of the Section on CMHD for ideal fluids for MAC wave instability is essentially the same (for $\Lambda = O(1)$) as the criterion $Ra_{mod} = O(1)$ for thermal instability. This range is however irrelevant to the core, where $E \ll 1$, $P_t = O(1)$, $P_m \ll 1$. It is hard to make progress in this difficult parameter range, and the theory is still somewhat speculative. Since $Ra_m/Ra_{mod} = (P_m/P_t)\Lambda^{-1} = O(\kappa/\eta) \sim 10^{-5}$, thermal convection can occur in the range $1 < Ra_{mod} < Ra_m$ in which MAC wave instability is impossible. Since, by Eq. (127), $R_m < \Lambda \sim 100$ in this range of Ra_{mod}, convective motions are too weak to maintain a dynamo. Only when the convection is strongly supercritical and the MAC wave bifurcation is reached ($Ra_{mod} \sim 10^6$, $Ra_m \sim 1$) can magnetic field be generated. (For analogous behavior in a simpler system, see Jones *et al.*, 1976.)

The observed geomagnetic field shows a tendency to drift in longitude (see *Westward drift*). Three mechanisms may contribute to this motion: field diffusion, MAC wave propagation and actual fluid motions. Field lines can diffuse with velocity $c_d = \eta/\mathcal{L}$ relative to the moving conductor (see the Section on The imperfect conductor, reconnection), and this velocity is comparable for $R_m = O(1)$ with $\mathbf{V}$ and c_s. It is therefore hard to decide quantitatively how much of the observed drift to attribute to each of the mechanisms. A CMHD model of Roberts and Stewartson (1975) demonstrated a clear preference for westward flux diffusion (see also Acheson, 1972). The recent geodynamo simulations of Glatzmaier and Roberts (1997) have shown a striking propensity for flux diffusion to the west despite zonal fluid motion predominantly to the east.

As shown in the Sections on the Classical theory of rotating fluids and Classical MHD, boundary layers arise when $E \ll 1$ and also when $Ha \gg 1$. Another characteristic of CMHD is that, when $\Lambda(= Ha^2E)$ is $O(1)$, neither Ekman nor Hartmann layers exist; they are replaced by a composite *Ekman-Hartmann layer*. Its thickness, δ_{EH}, is of the same order as δ_E and δ_M. It pumps both fluid and current into or out of the mainstream beyond the boundary layer. As for the Ekman and Hartmann layers, the Ekman-Hartmann layer provides a link through which angular momentum can be exchanged between the fluid and surrounding solid walls. This is accomplished by a viscous stress and, if the wall is electrically conducting, by a magnetic stress (see Eq. (39)). The role of boundary layers in geomagnetic simulations has been controversial (see *Core, boundary layers*).

Paul H. Roberts

Bibliography

Acheson, D.J., 1972. On the hydromagnetic stability of a rotating fluid annulus. *Journal of Fluid Mechanics*, **52**: 529–541.

Bloxham, J., Zatman, S., and Dumberry, M., 2002. The origin of geomagnetic jerks. *Nature*, **420**: 65–68.

Braginsky, S.I., 1964. Magnetohydrodynamics of Earth's core. *Geomagnetism and Aeronomy*, **4**: 698–712.

Braginsky, S.I., 1967. Magnetic waves in the Earth's core. *Geomagnetism and Aeronomy*, **7**: 851–859.

Braginsky, S.I., 1970. Torsional magnetohydrodynamic vibrations in the Earth's core and variations in day length. *Geomagnetism and Aeronomy*, **10**: 1–8.

Braginsky, S.I., and Meytlis, V.P., 1990. Local turbulence in the Earth's core. *Geophysical and Astrophysical Fluid Dynamics*, **55**: 71–87.

Braginsky, S.I., and Roberts, P.H., 1995. Equations governing convection in Earth's core and the Geodynamo. *Geophysical and Astrophysical Fluid Dynamics*, **79**: 1–97.

Braginsky, S.I., and Roberts, P.H., 2003. On the theory of convection in the Earth's core. In Ferriz-Mas, A., and Núñez, M. (eds.), *Advances in Nonlinear Dynamos*. London: Taylor & Francis, pp. 60–82.

Buffett, B.A., 1996. A mechanism for decade fluctuations in the length of day. *Geophysical Research Letters*, **23**: 3803–3806.

Busse, F.H., 1970. Thermal instabilities in rapidly rotating systems. *Journal of Fluid Mechanics*, **44**: 441–460.

Cowling, T.G., 1933. The magnetic field of sunspots. *Monthly Notices of the Royal Astronomical Society*, **140**: 39–48.

Davidson, P.A., 2001. *An Introduction to Magnetohydrodynamics*. Cambridge, UK: Cambridge University Press.

Gill, A.E., 1982. *Atmosphere-Ocean Dynamics*. New York: Academic Press.

Glatzmaier, G.A., and Roberts, P.H., 1997. Simulating the geodynamo. *Contemporary Physics*, **38**: 269–288.

Greenspan, H.P., 1968. *The Theory of Rotating Fluids*. Cambridge, UK: Cambridge University Press.

Gubbins, D., and Bloxham, J., 1987. Morphology of the geomagnetic field and implications for the geodynamo. *Nature*, **325**: 509–511.

Jault, D., 2003. Electromagnetic and topographic coupling, and LOD variations. In Jones, C.A., Soward, A.M., and Zhang, K. (eds.), *Earth's Core and Lower Mantle*. London: Taylor & Francis, pp. 56–76.

Jault, D., and Légaut, G., 2005. Alfvén waves within the core. In Soward, A.M., Jones, C.A., Hughes, D.W., and Weiss, N.O. (eds.), *Fluid Dynamics and Dynamos in Astrophysics and Geophysics*. Boca Raton, FL: CRC Press, pp. 277–293.

Jault, D., Gire, G., and Le Mouël, J.-L., 1988. Westward drift, core motions and exchanges of angular momentum between core and mantle. *Nature*, **333**: 353–356.

Jones, C.A., and Roberts, P.H., 2000. Convection-driven dynamos in a rotating plane layer. *Journal of Fluid Mechanics*, **404**: 311–343.

Jones, C.A., Moore, D.R., and Weiss, N.O., 1976. Axisymmetric convection in a cylinder. *Journal of Fluid Mechanics*, **73**: 353–388.

Jones, C.A., Soward, A.M., and Mussa, A.I., 2000. The onset of thermal convection in a rapidly rotating sphere. *Journal of Fluid Mechanics*, **405**: 157–179.
Kahle, A.B., Ball, R.H., and Vestine, E.H., 1967. Comparison of estimates of surface fluid motions of the Earth's core at various epochs. *Journal of Geophysical Research*, **72**: 4917–4925.
Krause, F., and Rädler, K.-H., 1980. *Mean-field Magnetohydrodynamics and Dynamo Theory.* Oxford, UK: Pergamon Press.
Lehnert, B., 1954. Magnetohydrodynamic waves under the action of the Coriolis force, Part I. *Astrophysics Journal*, **119**: 647–654.
Parker, E.N., 1968. Hydromagnetic dynamo models. *Astrophysics Journal*, **122**: 293–314.
Roberts, P.H., 1968. On the thermal instability of a rotating-fluid sphere containing heat sources. *Philosophical Transactions of the Royal Society of London Series A*, **263**: 93–117.
Roberts, P.H., and Soward, A.M., 1972. Magnetohydrodynamics of the Earth's core. *Annual Review of Fluid Mechanics*, **4**: 117–152.
Roberts, P.H., and Stewartson, K., 1975. On double-roll convection in a rotating magnetic system. *Journal of Fluid Mechanics*, **68**: 447–466.
Rotvig, J., and Jones, C.A., 2002. Rotating convection-driven dynamos at low Ekman number. *Physical Review E*, **66**: 056308.
Soward, A.M., 1974. A convection-driven dynamo. I. Weak-field case. *Philosophical Transactions of the Royal Society of London Series A*, **275**: 611–651.
Taylor, J.B., 1963. The magnetohydrodynamics of a rotating fluid and the Earth's dynamo problem. *Proceedings of the Royal Society of London Series A*, **274**: 274–283.
Tilgner, A., 2005. Precession driven dynamos. *Physics of Fluids*, **17**: 034104.

Cross-references

Alfvén Waves
Alfvén, Hannes Olof Gösta
Alfvén's Theorem and the Frozen Flux Approximation
Anelastic and Boussinesq Approximations
Antidynamo and Bounding Theorems
Convection, Nonmagnetic Rotating
Core Turbulence
Core, Adiabatic Gradient
Core, Boundary Layers
Cowling's Theorem
Dynamo, Backus
Dynamo, Braginsky
Dynamo, Bullard-Gellman
Dynamo, Gailitis
Dynamo, Herzenberg
Dynamo, Ponomarenko
Dynamos, Experimental
Dynamos, Fast
Dynamos, Kinematic
Dynamos, Mean Field
Dynamos, Periodic
Geodynamo
Geodynamo, Energy Sources
Geodynamo, Numerical Simulations
Geodynamo, Symmetry Properties
Geomagnetic Jerks
Geomagnetic Secular Variation
Harmonics, Spherical
Inner Core Rotational Dynamics
Inner Core Tangent Cylinder
Magnetohydrodynamic Waves
Oscillations, Torsional
Proudman-Taylor Theorem
Taylor's Condition
Thermal Wind
Westward Drift

MAGNETOMETERS, LABORATORY

Magnetometers in paleomagnetic laboratories are used for the determination of the geomagnetic field directions and intensities that have been recorded in rocks, and also for the discrimination of magnetic minerals and their magnetic properties. For paleomagnetism, the main consideration is to measure that magnetization vector in a rock, whereas for environmental magnetism the requirement is to identify a sample's magnetic mineralogy and grain size distribution. In either case the magnetometers employed can operate by one of two fundamentally different processes. Either the magnetization of the sample can be used to induce a current in a sensing coil, or the magnetization can create a force by interacting with an instrument-generated magnetic field gradient, and the force measured by a variety of methods.

Magnetometers used for determination of paleomagnetic vectors

Standard paleomagnetic samples are 2.5 cm diameter cores, approximately 2.5 cm in length, and most paleomagnetic magnetometers will be designed for this sample size. The major exception of this will be whole core sections commonly collected from lake and deep-sea sediments and used in environmental studies. These cores are typically 10 cm in diameter and can be several meters in length.

Astatic magnetometers

Astatic magnetometers were the most common type of instrument in paleomagnetic laboratories in the 1950s and 1960s. They are a force-type magnetometer, where the magnetization of a sample is used to exert a force on a pair of identical suspended magnets. The bar magnets are set with their axis horizontal and are suspended vertically above each other via a torsional wire (Figure M154). Although in

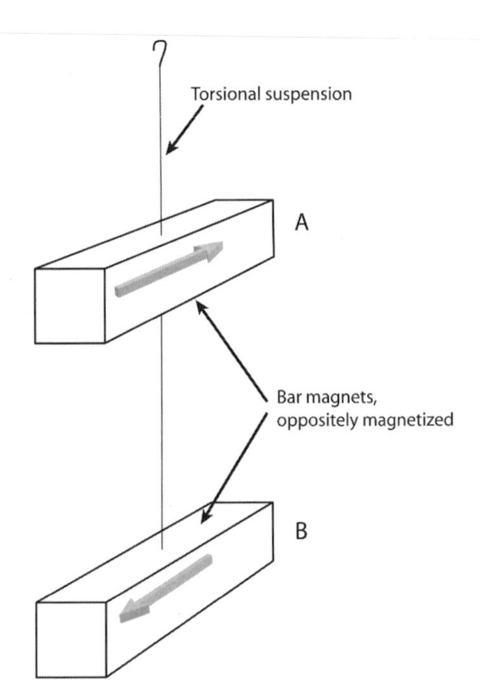

Figure M154 The arrangement of magnets in an astatic magnetometer. If sample is either placed in the horizontal plane containing magnet B, or vertically below B, then only the magnetization component normal to that of B is measured. If the sample is placed midway between A and B, then the vertical component of its magnetization is measured.

principle, just one magnet could be used, the vertical pairing of the magnets increases accuracy by making the deflection insensitive to uniform field gradients. Addition of a third magnet (parastatic magnetometers) reduces sensitivity to changes in vertical field gradients. The accuracy of astatic magnetometers largely depends on the magnets having a large magnetization and small moment of inertia, but typically sensitivities are less than 10^{-8} A m^2. One of the first such instruments was used for paleomagnetism by Blackett (1952) and similar instruments were in common usage in paleomagnetic laboratories into until the mid-1970s.

Spinner magnetometers

These are a range of magnetometers that induce a signal in a static sensor by spinning the sample about its own axis. The spinning sample will produce a time-varying magnetic flux with respect to one or more stationary sensors. The sensor determines the magnetization component in the plane normal to the sample rotation axis. Within this plane, the magnetization vector is found with reference to a fixed direction by comparing the sensor output to a signal linked to the rotation mechanism (Collinson, 1983). The overall sensitivity of the spinner magnetometer will depend on the speed at which the sample is rotated and the sensitivity of the sensors themselves, and so a wide range of sensitivities are possible depending on the type of sensor employed. The simplest sensor is an induction coil placed around the sample chamber, and in this case the signal output will increase with sample size (limited by the induction coil size) and with the rotation frequency. The highest sensitivities for such instruments approach 10^{-10} A m^2. Fluxgate sensors (see *Observatories, instrumentation*) may be used instead of induction coils or increased sensitivity can be obtained using a superconducting quantum interference device (SQUID) detector (see *Cryogenic magnetometers* below).

Cryogenic magnetometers (see also *Rock magnetometer, superconducting*)

Cryogenic magnetometers are currently the most sensitive device for measuring magnetizations, commonly achieving sensitivities of 10^{-12} A m^2. They consist of a superconducting coil that surrounds the sample to be measured. The magnetization of the sample changes the magnetic flux that passes through the sensing coil, which then induces a current to flow through the sensor coil. The current is detected by a SQUID, and these magnetometers are sometimes referred to as SQUID magnetometers. There are two types of SQUIDs, direct current (DC) and radio frequency (RF), depending on the exact method used to detect the current induced in the sensing coil. DC SQUIDs are more expensive but much more sensitive, and it is this type that is usually used in modern paleomagnetic magnetometers.

SQUID magnetometers can only measure the change in magnetic flux through its sensing coils, and there will be an output signal even when no sample is present. Thus, in order to determine a sample's magnetization, it must be cycled in and out of the sensing coil and the background reading subtracted.

The main disadvantage of cryogenic magnetometers is the requirement of having to keep the sensing coil superconductive, since this normally requires immersion in liquid helium (recently 4.2 K cryocoolers have been used without the need of liquid helium). High-temperature SQUIDs are able to operate at liquid nitrogen temperatures but the increased thermal noise prevents then from attaining the sensitivities available for conventional SQUIDs.

Magnetometers used for discrimination of magnetic minerals

Alternating gradient magnetometers (AGM)

Electromagnets are used to produce an alternating magnetic field gradient across a region in which a sample is placed. The sample is mounted

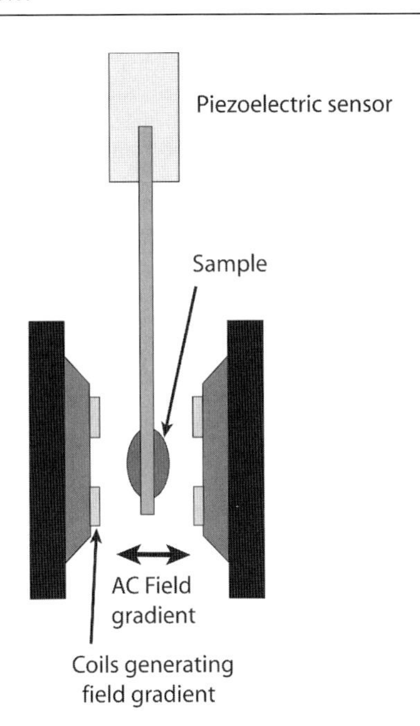

Figure M155 A schematic of an alternating gradient magnetometer. The force on the sample is generated by a set of field coils, and the motion of the sample is proportional to its magnetization.

at one end of a vertical cantilever, the other end of which is attached to the bottom of a piezoelectric transducer (see Figure M155). The alternating field gradient exerts an oscillatory force on the sample, and the bending of the transducer produces a voltage proportional to the amplitude of the oscillation.

The amplitude of the displacement of the rod at the sample is then given by $A = M \cdot \nabla H\,(Q/k)$, where M is the sample magnetization, ∇H is the field gradient generated by the instrument, Q is the mechanical quality factor of the cantilever (its resonance frequency divided by its resonance width), and k is the effective spring constant of the rod. The AGM needs to be operated at the resonance frequency of the cantilever system, and this will change with the weight of the sample. The sensitivity of the instrument is limited by thermal noise in the mechanical oscillator, and typical sensitivities are comparable to cryogenic magnetometers ($\sim 10^{-12}$ A m^2).

A DC field may also be applied during measurements, allowing the magnetization as a function of field to be determined. Often a heater or cryostat is also fitted to such instruments allowing both field and temperature variation of the magnetization to be determined (O'Grady *et al.*, 1993).

AGMs are not well suited to measuring paleomagnetic samples since they only measure the magnetization component along the applied field gradient, and the sample size is relatively small (usually a few millimeters in size) compared with the standard paleomagnetic sample size of 2.5 cm diameter cores. AGMs are good for characterization of magnetic minerals and determining the magnetization dependence as a function of applied field or temperature.

Vibrating sample magnetometers

These are similar to AGMs, but rather than using a field gradient, the sample is mechanically vibrated at a fixed frequency (see Figure M156). The sensing coils will have an emf induced within them due to the changing flux produced by the vibrating sample. For a sample magnetization M, the induced voltage V in the sensing coils is given by $V = \mu_0 GM\omega A \sin \omega t$, where ω is the angular frequency of the vibration, and A is the vibration amplitude. G is a factor relating to the geometry of the sensing coils. A uniform magnetic field in the region of the sample is provided by an electromagnet, or superconducting solenoid if fields greater than 3 T are required. A heater or cryostat can be fitted for variable temperature measurements.

VSMs are usually at least an order of magnitude less sensitive than AGMs but have the advantage of being able to handle much larger sample sizes (Lindemuth *et al.*, 2000), and are thus more suitable for standard paleomagnetic samples. The VSM, like the AGM, is generally used for high-field measurements, such as hysteresis properties. The large metal poles of the DC electromagnets used to generate the uniform field will have a residual field even when no current flows through the coils (the residual remanence of the pole pieces is avoided by using a field-controlled (with Hall sensor) power supply). The difficulty in reducing the ambient field in the region of the sample to a sufficiently low value makes these instruments unsuitable for low field (remanence) measurements.

Translation balance

A translation balance in its simplest form consists of a rod suspended by wires so it is free to make small movements along its axis (see Figure M157). The sample is placed at one end of the rod, and a field is created perpendicular to the rod axis, by means of a large electromagnet. The field produced by the electromagnet will have a gradient along the rod axis, and the force on the magnetic sample will therefore be given by $F = M\, dH_y/dx$, where M is the sample magnetization (see Figure M157). This force causes a translation of the rod, which is calibrated against the samples magnetization. Such instruments will normally have an oven surrounding the sample so that the magnetization can be measured as a function of temperature, allowing the Curie point to be determined. Sensitivities of 10^{-9} A m^2 can be achieved with these instruments.

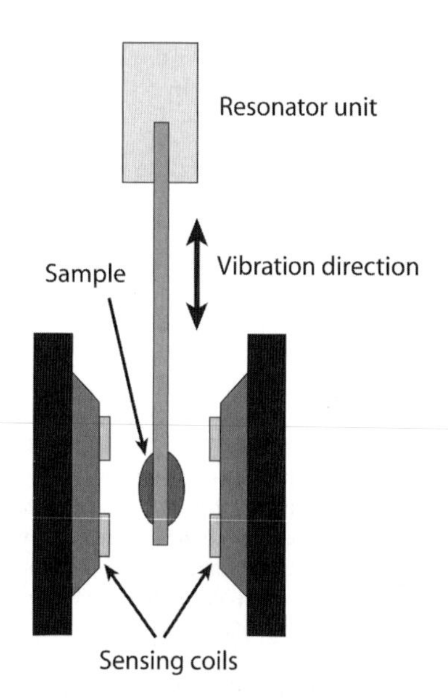

Figure M156 A schematic of a vibrating sample magnetometer. The sample is vibrated mechanically, and the induced signal in the sensing coils will be proportional to its magnetization.

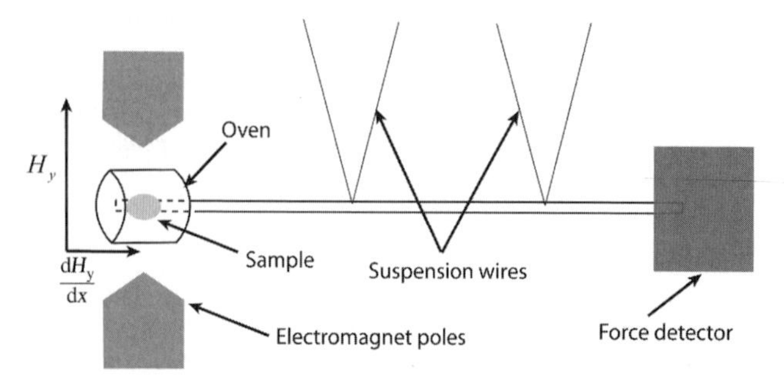

Figure M157 The principle elements of a horizontal translation balance. An electromagnet produces a large magnetizing field H_y along the y-axis and a field gradient along the x-axis that generates a force on the sample proportional to its magnetization.

Cryogenic magnetometers

SQUID magnetometers (described above) are also used in magnetic mineralogy. Here, the requirement for liquid helium to keep the sensing coils superconducting can also be used to cool the sample, and provide large fields (greater than 5 T) in superconducting solenoids. It is therefore possible to measure over a wide range of temperatures (down to less than 2 K) and fields. However SQUID sensors do not allow continuous measurement of magnetization versus field strength. To change the field within a superconducting coil it must be switched out of its superconducting state. This requirement makes field dependency observations much more time consuming compared to similar measurements on a VSM or AGM.

Wyn Williams

Bibliography

Blackett, P.M.S., 1952. A negative experiment relating to magnetism and the Earth's rotation. *Philosophical Transactions of the Royal Society of London, Series A*, **245**: 309–370.
Collinson D.W., 1983. Methods in rock magnetism and palaeomagnetism. *Techniques and Instrumentation.* London: Chapman & Hall.
O'Grady, K., Lewis V.G., and Dickson D.P.E., 1993. Alternating gradient force magnetometry: applications and extension to low temperatures. *Journal of Applied Physics*, **73**: 5608–5613.
Lindemuth, J., Krause J., and Dodrill B., 2000. Finite sample size effects on the calibration of vibrating sample magnetometer. *IEEE Transactions on Magnetics*, **37**: 2752–2754.

Cross-references

Compass
Gauss' Determination of Absolute Intensity
Instrumentation, History of
Observatories, Automation
Observatories, Instrumentation
Rock Magnetometer, Superconducting

MAGNETOSPHERE OF THE EARTH

The Earth's magnetosphere is the region surrounding the planet, above the outer atmosphere and *ionosphere* (*q.v.*), which contains and is controlled by the Earth's magnetic field. It extends from an altitude of ~500 km above the Earth's surface to an outer boundary which is formed by the interaction of the planetary magnetic field with the solar wind, the plasma (charged particle) gas that streams continuously outward from the Sun. On the dayside, the compressive effect of the solar wind confines the field to a region extending ~10 Earth radii from the Earth, while on the nightside the field is stretched into a long comet-like tail which extends typically in excess of ~1000 Earth radii. The Earth's radius, R_E ~6400 km, is thus a convenient measure of magnetospheric spatial scales. Electric currents flowing in these regions, whose magnetic effects can be observed at the Earth's surface, include those flowing at the solar wind-magnetosphere boundary, those produced by the drift of energetic charged particles inside the magnetosphere, and those associated with the coupling between the magnetosphere and the ionosphere Ohtani *et al.* (2000). Magnetic perturbations observed at the Earth's surface due to these three sources peak typically at a few tens, a few hundreds, and a few thousands of nanoteslas, respectively, and vary on a range of timescales from minutes to hours and days. In this article we first concentrate on the overall structure and basic physics of the coupled solar wind-magnetosphere-ionosphere system, and then consider the dynamic processes which occur, leading to strong temporal variations.

Structure of the magnetosphere

In Figure M158 we show a cross-section through the Earth's magnetosphere in the noon-midnight meridian plane, where the arrowed solid lines indicate magnetic field lines. The small symbols indicate the main plasma populations, while the arrows show their principal flows. The small dashes show the solar wind plasma, together with the magnetospheric "boundary layer" plasma populations derived directly from it. The direction of the Sun is toward the left, so that the solar wind flows from left to right as indicated. The solar wind derives from hydrogen gas in the solar atmosphere, which is heated sufficiently strongly in the solar corona, to temperatures in excess of a million degrees, that the atoms are fully broken up by collisions to form a plasma of electrons and protons (plus a few percent α-particles from helium), which then streams continuously out into the solar system at speeds $\sim$500 km s^{-1}. At the orbit of the Earth, the proton and electron number densities are $\sim$10 cm^{-3}, sufficiently low that the particles in the gas are collision-free, meaning that the mean free path for particle collisions is comparable with or larger than the size of the system. A second source of plasma in the magnetosphere derives from the Earth's upper atmosphere, which is relatively cool ($\sim$1000 K) but is partially ionized by solar far-UV and X-rays at altitudes above $\sim$100 km, forming the ionosphere. These charged particles, consisting of protons and heavy ions, principally singly-charged oxygen, together with corresponding electrons, can also stream out of the topside ionosphere into the magnetosphere, as indicated by the small crosses in the figure. In the ionosphere, collisions between ions and neutral atoms are significant at altitudes up to $\sim$500 km. Above this altitude, however, the ionospheric ions also become collision-free due to falling densities, $\sim$500 km thus being chosen (rather arbitrarily) above as the "base" of the magnetosphere, where it interfaces with the ionosphere. Like the solar wind, therefore, the plasma of charged particles in the Earth's magnetosphere is also collision-free. Although the Earth's neutral atmosphere is detectable throughout the magnetosphere to distances of at least $\sim$10 R_E (forming a hydrogen "geocorona" originating from the photodissociation of water vapor lower in the atmosphere), the behavior of neutral and charged particle populations within it are largely decoupled due to the lack of collisions, apart from the occasional charge-exchange reaction between them.

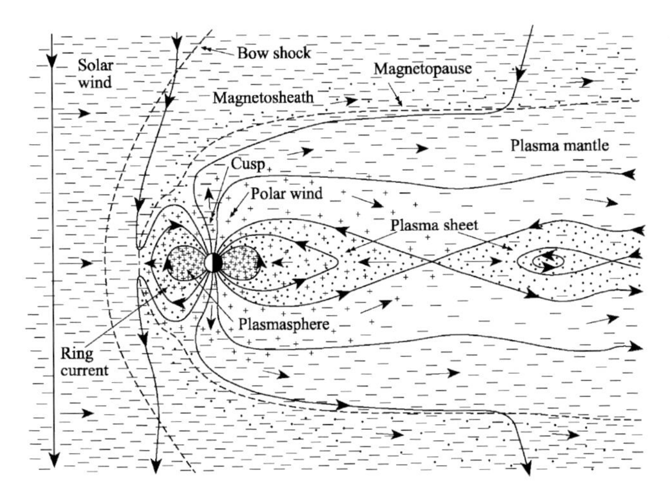

Figure M158 Sketch of a cross section through the Earth's magnetosphere in the noon-midnight meridian plane. The direction of the Sun is toward the left, such that the solar wind blows from left to right. The arrowed solid lines show magnetic field lines, while the heavy dashed lines show locations of principal field and plasma discontinuities at the bow shock and magnetopause. The small symbols indicate the principal plasma populations present, while the arrows indicate their direction of flow. The dashes indicate the solar wind, and the plasma populations derived directly from it in the magnetosheath and magnetosphere (cusp and plasma mantle). The crosses indicate plasma originating from the Earth's ionosphere (plasmasphere and polar wind), while the dots indicate the heated plasma of mixed origin that originates in the tail (plasma sheet and ring current).

When collisions can be ignored, the motion of ions and electrons in the plasma is governed by the forces of the large-scale fields prevailing, and for both solar wind and magnetosphere the most important forces by far are electromagnetic. Solar and planetary gravity is only important in the regions close to the bodies concerned, i.e., in the solar corona near the Sun, and in the ionosphere near the Earth. The motion of charged particles in a large-scale electromagnetic field consists of a number of components, which take place on generally widely separated timescales. The first is a rapid gyration of the particles around the magnetic field lines in nearly circular orbits whose radius is generally tiny compared with the scale size of the systems. The second is a uniform motion along the magnetic field lines, which is subject to the magnetic mirror force which repels particles from regions of increasing field strength along the field lines (i.e., is directed away from regions where field lines converge). The mirror force is produced by the action of the field strength gradient on the magnetic dipole formed by the gyrating particle. In the quasidipolar field of a planetary magnetosphere, for example, the field strength along a given field line is weakest at the equator and increases greatly toward the planet (see Figure M158). This configuration can therefore trap charged particles almost indefinitely, the mirror force resulting in a bounce motion between mirror points in the northern and southern hemisphere. Only those particles that are directed almost along the field lines at the equator, which thus have small magnetic moments, can reach down to the upper layers of the atmosphere. The third motion consists of large-scale plasma flows transverse to the magnetic field lines, such as that shown for the solar wind in Figure M158. Such flows are associated with an electric field $\boldsymbol{E}$ directed transverse to the magnetic field $\boldsymbol{B}$, related to the plasma velocity $\boldsymbol{V}$ by the vector relation $\boldsymbol{E} = -\boldsymbol{V} \times \boldsymbol{B}$. (The effect of additional cross-field drifts associated with magnetic field inhomogeneity will be discussed further below.) Inspection of the flows indicated in Figure M158 shows that the electric field is directed everywhere out of the plane of the diagram, though it is not everywhere of equal strength. In such "crossed" electric and magnetic fields, charged particles drift perpendicular to the magnetic field with velocity $\boldsymbol{V} = (\boldsymbol{E} \times \boldsymbol{B})/\boldsymbol{B}^2$ independent of their charge or mass, this expression being exactly equivalent to the vector relation for $\boldsymbol{E}$ given above.

This "$\boldsymbol{E} \times \boldsymbol{B}$ drift" has a very special property, first discovered by Alfvén in 1942 (see *Alfvén's theorem*), that particles whose centers of gyration lie initially on the same field line as each other, remain on the same field line as each other for all time, thus showing why plasma structures are strongly organized by magnetic field lines, as indicated by various features in Figure M158. We may picture the collective behavior as one in which the magnetic field and the plasma are "frozen" together. However, we can think either of the plasma as moving and carrying the field lines, or equivalently of the field lines as moving and carrying the plasma. Which of these pictures is the most appropriate for a given system depends upon the relative energies in the plasma and magnetic field. Thus, for example, since the plasma energy exceeds the magnetic energy in the solar wind, the solar magnetic field is carried outward "frozen" into the expanding plasma flow, forming a large-scale interplanetary magnetic field (IMF) that pervades the entire solar system (see Figure M158). The strength of the IMF at the Earth's orbit is typically ~5–10 nT, directed, on the average, in the ecliptic plane at an angle of ~45° to the Earth-Sun line. The latter tilt is due to the rotation of the Sun, which winds the IMF into a spiral form as the field lines are carried out into the solar system by the plasma flow. On the other hand, in the quasidipolar field regions of the magnetosphere and ionosphere, where the magnetic energy dominates, we instead think of the field lines as moving, transporting the plasma.

We may apply this "frozen-in" concept to the interaction between the solar wind and the planetary magnetic field. Since the solar wind and IMF are frozen together, as well as the planetary field and planetary plasma (e.g., from the ionosphere), then when these two media interact they will not mix. Instead, the solar wind confines the planetary field to a cavity surrounding the planet, around which it flows, as first deduced by Chapman and Ferraro in 1931 (see *Chapman, Sydney*). This magnetic cavity is the planet's magnetosphere, whose outer boundary, shown by the dashed line in Figure M158, is called the "magnetopause." A "bow shock" forms ahead of the cavity, also shown by a dashed line in Figure M158, due to the fact that the magnetosphere represents a blunt obstacle in the supersonic solar wind flow. Across the shock the solar wind is slowed, compressed, and heated, forming the turbulent "magnetosheath" layer located between the shock and the magnetopause boundary.

The size of the magnetospheric cavity is set by the condition of pressure balance at the boundary. A simple estimate of the distance of the equatorial boundary at noon (the minimum distance in the direction facing the Sun) can be made by equating the ram pressure of the solar wind on one side of the boundary, with the magnetic pressure ($B^2/2\mu_0$) of the compressed planetary field on the other. With typical solar wind values the radial distance of the boundary is estimated to be ~10 R_E on this basis, as observed, a position which may vary by factors of up to two in either direction under extreme solar wind conditions. The strength of the compressed planetary field just inside the boundary is typically ~60 nT, representing a planetary dipole field of ~30 nT enhanced by a factor of ~2 by the electric current flowing in the magnetopause boundary. This latter current is termed the "Chapman-Ferraro" current, and inspection of Figure M158 with Ampere's law in mind shows that it flows out of the plane of the diagram in the equatorial region on the dayside, closing over the magnetopause into the plane of the diagram over the polar regions and on the nightside. These rings of magnetopause current produce a perturbation magnetic field in the near-Earth magnetosphere whose strength is typically a few tens of nanoteslas directed northward (i.e., upward in Figure M158), hence enhancing the horizontal field at low and middle latitudes at the Earth's surface.

Frozen-in behavior of the field and plasma associated with the $\boldsymbol{E} \times \boldsymbol{B}$ drift is not, however, a universally valid description. In a collision-free medium it is broken in particular by the presence of additional particle drifts, the most significant of which cause ions and electrons to drift in opposite directions across the field (thus producing a current) due to the presence of gradients in the strength and direction of the magnetic field. These "field inhomogeneity" drifts are proportional to the field gradients, and also to the particle energy, thus being more important for particles of higher energy. When the field gradients are weak, these additional drifts are important only for particles in the high-energy tail of the energy distribution, and frozen-in motion then represents a useful organizing concept for the bulk of the plasma population. This limit applies essentially throughout the solar wind, and through most of the Earth's magnetosphere. However, when the field gradients are very strong, such that the motion of the bulk of the plasma particles is affected, then the frozen-in picture breaks down. One such place where this happens is the magnetopause boundary, where the field strength and direction in general switch rapidly from magnetosheath to magnetosphere values across the magnetopause current sheet. The simplest theoretical description of the consequence of frozen-flux breakdown is that the magnetic field diffuses through the plasma, locally, in the region of the strong gradient. As first pointed out by Dungey in 1961, this then allows magnetic field lines to become joined across the boundary, producing "open" magnetic field lines which pass from the solar wind at one end, through the magnetopause, to the Earth's polar regions at the other. This process is called "magnetic reconnection." Two newly reconnected open field lines are shown in Figure M158 passing through the dayside magnetopause shortly after reconnection has taken place near the equator. Sharply bent magnetic field lines exert a tension force on the plasma like the force of rubber bands (the force per unit volume being $\boldsymbol{j} \times \boldsymbol{B}$, where $\boldsymbol{j}$ is the current density in the plasma), in this case accelerating the boundary plasma poleward away from the equator, such that the field lines also contract poleward, releasing energy to the plasma and allowing further reconnection to proceed at the equator. Subsequently, the open field lines are carried downstream frozen into the magnetosheath flow, and are stretched out into a long cylindrical comet-like tail. This tail consists of two lobes, D-shaped in cross-section, one connected to the northern polar region at Earth, the other to the southern, as indicated in Figure M158. Observations show that the tail lobe field lines remain open typically for a few hours, such that with a downstream speed of ~500 km s^{-1}, the tail is typically ~1000 R_E long.

The open field lines form magnetic pathways along which the magnetosheath plasma may enter the magnetosphere. Such plasma thus flows along newly opened field lines to form a boundary layer adjacent to the dayside magnetopause, and the "cusp" population as it then moves down toward the Earth (see the magnetospheric dashed regions in Figure M158). The majority of the particles, however, are repelled by the magnetic mirror force as the field strength increases near the Earth, and hence move back out again toward the outer magnetosphere. Due to the antisunward motion of the open field lines, however, the cusp plasma flows back out into the lobes of the tail, in the region adjacent to the magnetopause, forming the "plasma mantle" population. As the open field lines are carried down the tail, so the field lines and mantle plasma sink in toward the center plane of the tail, followed by further entry of antisunward flowing magnetosheath plasma at the tail magnetopause, such that the mantle grows wider and with increasing density at larger distances. Plasma from the Earth's polar ionosphere also flows into the lobes (cross symbols in Figure M158), but because of its low velocity along the field lines (~10 km s^{-1}), it does not reach far down the tail on the few-hour timescale that the lobe field lines remain open. Overall, the plasma density in the inner part of the tail lobes is very low, ~0.01–0.1 cm^{-3}, and with temperatures typically of order a few tens to a few hundred electron-volts, most of the system energy resides in the lobe magnetic field. (Note that while temperatures in the solar and terrestrial ionized atmospheres are generally quoted in Kelvin, as above, the temperatures of magnetospheric plasmas are usually indicated by the mean or typical energy of the particles W, in eV. To convert between them, we note that for a near-Maxwellian velocity distribution $T(\mathrm{K}) \approx 10^4\ W(\mathrm{eV})$. Thus, for example, typical hot magnetospheric plasma of ~1 keV mean energy corresponds to a temperature of ~10^7 K.)

The residency of the oppositely directed open field lines in the two tail lobes is terminated when they sink into the center plane of the tail and reconnect in the equatorial current sheet that separates them, as shown by the X-shaped field configuration on the right side of Figure M158. On the tailward side of the tail reconnection site the lobe field lines become disconnected from the Earth, and the $\boldsymbol{j} \times \boldsymbol{B}$ (rubber band) tension force accelerates the tail plasma rapidly away from the Earth, where it eventually rejoins the solar wind. On the Earthward side, the process forms new "closed" field lines, connected to the Earth at both ends, which similarly contract rapidly earthward, compressing and heating the lobe plasma as they do so. This hot plasma, containing both solar wind and ionospheric contributions, is termed the "plasma sheet" population, and is shown by the dotted region in Figure M158. In the near-Earth tail it is characterized by densities ~0.1–1 cm^{-3}, and electron and ion temperatures of ~0.5 and ~5 keV, respectively, but both density and temperatures increase as the plasma flows earthward into the quasidipolar inner magnetosphere and is further compressed and heated. Throughout these regions the hot plasma sheet population carries a current, due to the field-inhomogeneity effects mentioned above, which is directed westward, i.e., out of the plane of Figure M158 on the nightside of the Earth, and into the plane of the diagram on the dayside. On the nightside within the tail, the current paths reach the tail boundary, and close over the magnetopause to form two D-shaped solenoids lying back-to-back, as required by the lobe magnetic field structure. Nearer the Earth, the current paths may close wholly round the Earth to form a plasma "*ring current*" (*q.v.*). Overall, these currents produce a perturbation field at the Earth, which is directed southward (i.e., downward in Figure M158), and is usually a few tens of nanoteslas in amplitude. However, as we discuss below, these currents become greatly enhanced during *magnetic storms* (*q.v.*), producing perturbation fields up to an order of magnitude larger.

The closed field lines and hot plasma produced in the tail subsequently flow sunward around the Earth in the quasidipolar magnetosphere, eventually reaching the dayside magnetopause where they become open again, allowing the hot magnetospheric plasma to escape into the magnetosheath, as indicated by the dotted region outside the magnetopause. Overall, therefore, reconnection at the magnetopause and in the tail results in the formation of a large-scale cyclical flow of field lines and plasma through the Earth's magnetosphere, first discussed by Dungey in 1961. The overall cycle time is ~12 h, of which the field lines spend ~4 h open in the tail lobes, and ~8 h closed flowing sunward through the central magnetosphere. The solar wind is not, however, the only source of momentum that drives magnetospheric flow. A second source resides in the rotation of the Earth, and the coupled rotation of the Earth's upper atmosphere. Collisions between ions and neutral atmospheric particles in the ionosphere produce a frictional force on the feet of the magnetospheric field lines which tends to spin them up toward rotation with the planet, a flow called "corotation." In general, the overall magnetospheric flow results from a summation of the effects of the planetary and solar wind driving forces. The corotation effect tends to produce a flow in the equatorial plane increasing linearly with the distance from the planet (i.e., a flow with the planetary angular velocity), while in the simplest picture of a dipole magnetic field and uniform cross-system electric field, the equatorial Dungey-cycle flow increases as the cube of the distance (i.e., as the inverse of the field strength). Reflection on these relative dependencies then shows that corotation will dominate close-in, and Dungey-cycle convection at larger distances. For Earth, the boundary between these flow regimes is located typically at ~5 R_E in the equatorial plane, about half-way to the dayside magnetopause. The hot plasma flowing in from the tail is therefore excluded from this inner core of corotating field lines, as indicated in Figure M158, which is instead filled to relatively high densities (~100–1000 cm^{-3}) by cold (~1–10 eV) hydrogen and oxygen plasma from the ionosphere. This population is termed the plasmasphere, and is indicated by the near-Earth region of dense crosses in Figure M158.

Magnetosphere-ionosphere coupling

The magnetospheric flows discussed above produce corresponding motions of the field and plasma in the polar ionosphere, resulting in large-scale current systems flowing between these regions. Figure M159 shows a view looking down on the North Polar Region, with noon (the direction toward the Sun) at the top of the diagram, dusk to the left, dawn to the right, and midnight at the bottom. The central circular dashed line shows the outer boundary of the "open" field region (located typically at magnetic latitudes of ~75° at noon and ~70° at midnight), while the arrowed solid lines show plasma streamlines. The open field lines within the dashed line boundary,

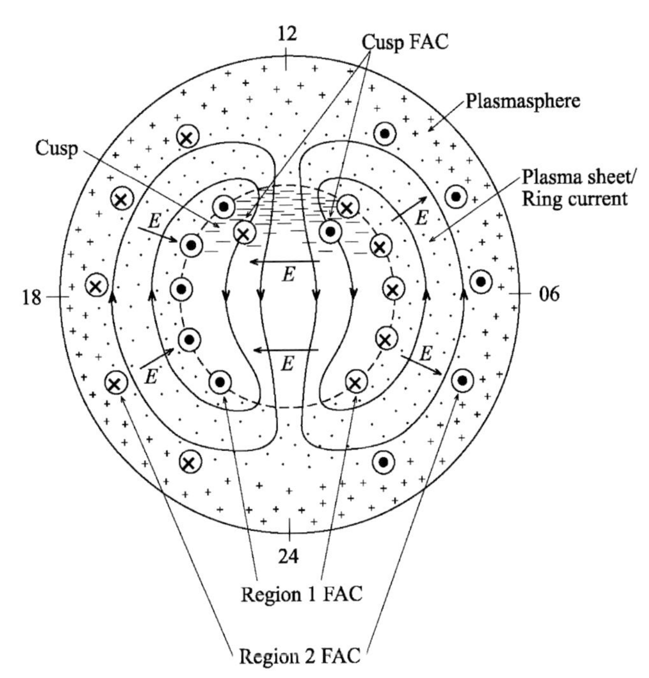

Figure M159 Sketch looking down onto the northern polar ionosphere, with noon (the direction toward the Sun) at the top of the diagram, dusk to the left, dawn to the right, and midnight at the bottom. The dashed line shows the boundary between open and closed magnetic field lines, while the arrowed solid lines show the streamlines of the Dungey-cycle plasma flow driven by solar wind-magnetosphere interaction. The short arrows marked $\boldsymbol{E}$ indicate the directions of the electric field associated with the flow, $\boldsymbol{E} = -\boldsymbol{V} \times \boldsymbol{B}$ (the magnetic field points into the northern ionosphere), while the circled dots and crosses indicate regions of upward and downward field-aligned current (FAC) flow, respectively. The small symbols indicate the principal plasma regimes into which the field lines map in the magnetosphere, with the same format as Figure M158.

mapping to the northern tail lobe, flow antisunward from noon to midnight at speeds typically of a few 100 m s^{-1}, while the closed field lines at lower latitudes return sunward. This twin-vortex flow is the signature of the Dungey-cycle in the ionosphere. The $-\boldsymbol{V} \times \boldsymbol{B}$ electric field associated with the flow is directed from dawn to dusk across the open field line region as indicated, reversing in sense to poleward at dusk and equatorward at dawn. The small symbols also show the magnetospheric plasma populations into which the ionospheric field lines map, the symbol code corresponding to that in Figure M158. Ions and electrons from the cusp (dashes) and plasma sheet (dots) precipitate into the upper atmosphere, excite the atoms (typically at altitudes ~100–300 km), and cause them to emit photons, giving rise to one component of auroral light emissions (see *Auroral oval*) Paschmann *et al.* (2002) and Sandholt *et al.* (2001). They also ionize the neutral atoms and increase the plasma density of the ionosphere, a process that is especially important at night when the solar production mechanism is inoperative.

The Dungey-cycle flow drives two components of electric current in the lower ionosphere, both resulting from the effects of collisions between ionospheric ions and atmospheric neutral particles. At altitudes below ~120 km, these collisions become sufficiently frequent that the ions do not move with the field lines at all, but are essentially tied to the generally smaller motions of the upper atmosphere (i.e., the winds in the neutral thermosphere). However, the electron motion remains almost unaffected by collisions, frozen to the field line motion throughout the ionosphere, to altitudes below ~100 km. Consequently, in the layer between ~100 and 120 km, ionospheric electrons flow around the streamlines shown in Figure M159, carrying a current in the opposite direction, while the ions in the layer remain almost motionless, tied to the neutral atmosphere. This current, directed transverse to both the electric and magnetic fields is thus termed a Hall current, and with a height-integrated ionospheric conductivity of ~10 mho and typical ionospheric flows of several hundred meters per second, it produces equal and opposite magnetic perturbations on either side of the ionospheric Hall layer which are typically ~100–200 nT in strength. At the Earth's surface in the northern hemisphere these perturbation fields are roughly in the direction of the overhead ionospheric electric field (opposite in the southern hemisphere). These currents are nondissipative (i.e., $\boldsymbol{j} \cdot \boldsymbol{E} = 0$), and in principle can close wholly within the ionosphere (for uniform conductivity), flowing round the plasma flow streamlines.

The second current component flows at a slightly higher altitude of ~120–150 km, where the ion motion is affected by collisions with neutrals, but is not yet wholly dominated by them. In this layer collisions provide the ions with some mobility in the direction of the electric field, while continuing to move approximately with the electrons along the plasma streamlines. An ion current thus flows in the direction of $\boldsymbol{E}$, termed the Pedersen current, and its magnitude is such that the $\boldsymbol{j} \times \boldsymbol{B}$ force of the current, directed along the plasma streamlines, just balances the frictional drag on the ionospheric plasma due to ion-neutral collisions in the ionosphere. A key fact about the Pedersen current is that it cannot, in principle, close within the ionosphere itself, but is instead part of a large-scale current system which flows between the magnetosphere and ionosphere, which imposes the flow of the former onto the latter in the presence of ionospheric ion-neutral drag. It can be seen in Figure M159, for example, that the Pedersen currents, directed along $\boldsymbol{E}$, flow away from the open-closed field line boundary on both sides at dawn, while flowing toward the boundary from both sides at dusk. Current continuity then requires the presence of field-aligned currents (FACs), which flow down the field lines from the magnetosphere to the ionosphere at the dawn boundary, and up the field lines from the ionosphere to the magnetosphere at the dusk boundary. These are called the "region 1" FACs, whose direction is indicated in Figure M159 by the circled dots and crosses for upward and downward currents, respectively. These FACs close across the field lines at large distances in the plasma which is the source of the momentum producing the flow, i.e., in the magnetosheath plasma at the tail magnetopause. The magnetic perturbations produced by the overall current circuit bends the magnetic field lines in such a way that the $\boldsymbol{j} \times \boldsymbol{B}$ force slows the magnetosheath plasma, while transferring the momentum to the ionosphere to maintain the flow against collisional frictional drag. Overall, this current system is solenoidal in nature, such that the main magnetic effects are contained within the effective current solenoid between the ionosphere and magnetopause. Thus although the ionospheric height-integrated Pedersen conductivity is generally comparable with the height-integrated Hall conductivity, such that the two height-integrated current components are of similar intensity in the ionosphere, the combined contribution of the ionospheric Pedersen current and the associated FACs approximately cancels underneath the ionosphere, and produces a much weaker magnetic perturbation on the ground than that of the Hall current. Above the ionospheric Pedersen layer, however, the perturbation field produced by the Pedersen-FAC system is directed approximately opposite to the plasma streamlines in the northern hemisphere, and along the streamlines in the southern hemisphere. Typical amplitudes in the region just above the Pedersen layer are again a few 100 nT. The total current flowing in the "region 1" FAC system is typically ~2–3 MA.

We also note from Figure M159 that Pedersen current continuity at the equatorward boundary of the Dungey-cycle flow (typically located at magnetic latitudes of ~70° at noon and ~65° at midnight), also requires the presence of upward FACs at dawn and downward FACs at dusk. These are called the "region 2" FAC system, in which the upward currents at dawn close in the downward currents at dusk via field-perpendicular currents flowing westward in the inner edge of the equatorial plasma sheet population in the nightside magnetosphere (and then via ionospheric Pedersen currents, the "region 1" FAC, and the tail magnetopause). The total current flowing in the "region 2" FACs is a little less than that in the "region 1" FACs, due to the fact that the latter currents are fed by Pedersen currents from both open and closed field line regions. We also note that regions of upward-directed FAC are often associated with bright "discrete" auroras (e.g., structured curtain-like forms), thus forming another component of the auroral emissions. This is because upward currents are carried primarily by warm plasma sheet electrons, which flow down the field lines into the ionosphere. In order to produce a downward electron flux which is sufficient to carry the upward current, the electrons may be accelerated downward by an electric field directed upward along the field lines (this being another scenario in which the frozen-in picture breaks down). Typically the required field-aligned voltages are a few kilovolts, such that the precipitating electrons carry a sufficient energy flux to produce a bright aurora.

Magnetosphere dynamics

The above discussion has centered on the plasma physics principles which govern the structure and properties of the coupled magnetosphere-ionosphere system, focusing particularly on the nature of the current systems that flow within it. For clarity of exposition we have viewed the system as being steadily driven, principally by the solar wind, but also by planetary rotation. However, it is now important to emphasize that the Earth's plasma environment is almost never in such a steady state, but generally varies strongly with time Cowley *et al.* (2003). There are two reasons for this. The first is that the interplanetary medium which impinges on the Earth's field itself varies strongly with time, on timescales from a few minutes to the ~27-day rotation period of the Sun. There are also variations over the 11-year period of the solar cycle. The second factor is that even when the solar wind is relatively steady, the processes in the key interaction regions at the magnetopause and in the tail show a propensity for pulsed behavior. We will now outline these behaviors, beginning here with those which are driven essentially directly by variations in the interplanetary medium.

Although the density and velocity of the solar wind determine the size of the magnetospheric cavity and the speed with which open flux tubes are transported to the tail, by far the most important upstream parameter which determines the overall nature of the magnetospheric interaction with the solar wind, and hence the dynamics of the Earth's plasma-field environment, is the direction and strength of the IMF. This determines how much flux is reconnected at the magnetopause per unit time, where on the magnetopause it is reconnected, and how the newly formed open tubes move into the tail. Although the IMF on average lies in the ecliptic plane at the ~45° spiral angle mentioned above, pointing either toward or away from the Sun along this direction, variable north-south fields are also produced by a variety of effects, such as waves and other disturbances propagating in the wind outward from the Sun. Open flux production is largest, and the Dungey-cycle flow and related currents strongest, when the IMF is directed southward opposite to the Earth's equatorial field (i.e., points downward in Figure M158, as drawn in the figure), and is weak when the IMF is directed northward. From Faraday's law the overall magnetic flux transfer in the Dungey-cycle, in Wb s^{-1}, is equal to the voltage of the cross-system electric field associated with the flow, in volts. This cannot be measured directly in the magnetosphere, but can routinely be measured at ionospheric heights using a network of ground-based radars, which determine the flow in the polar ionosphere. These observations show that when the IMF, of strength ~5–10 nT, points south, the voltage associated with the flow is ~100 kV, i.e., the flux throughput in the Dungey-cycle is ~100 k Wb s^{-1}. This compares with voltages in the solar wind across the magnetospheric diameter of about five times this value. Reconnection is thus ~20% efficient during such southward field intervals, in the sense that ~20% of the interplanetary magnetic flux which is directed toward the magnetosphere in the solar wind becomes reconnected with the terrestrial field. At the same time, of course, ~80% of the magnetic flux is deflected around the sides of the magnetosphere, such that most of the impinging solar wind flow is so deflected, as envisaged in Chapman and Ferraro's early considerations. Nevertheless, it is the breakdown of perfect deflection at the ~20% level which is critical to magnetospheric dynamics. As the IMF direction rotates away from southward, however, reconnection becomes less efficient. For example, when the IMF lies in its average direction in the ecliptic plane, the voltage values fall to ~50 kV, while when it rotates to northward they fall to "background" values of ~20 kV or less. The Dungey-cycle flow in the magnetosphere-ionosphere system is therefore strongly modulated by the sense and magnitude of the north-south component of the IMF, and is found to respond rapidly to changes in this component on timescales down to a few minutes.

The form of the flow is also found to change not only in response to the north-south field component of the IMF, usually referred to as the z component (with B_z positive northward), but also to the component which points into or out of the plane of the diagram in Figure M158, referred to as the y component (with B_y positive out of the plane of the diagram). When dayside reconnection takes place in the presence of IMF B_y, the $\boldsymbol{j} \times \boldsymbol{B}$ force on newly opened flux tubes pulls the open tubes "sideways" (i.e., east-west) as well as poleward, oppositely in the two hemispheres. This is illustrated schematically in Figure M160, which shows a view of the dayside magnetosphere looking from the direction of the Sun shortly after reconnection has taken place with an IMF having a positive B_y component (pointing from left to right in the figure). The $\boldsymbol{j} \times \boldsymbol{B}$ (rubber band) tension force pulls the newly opened field lines westward (to the left) in the dayside cusp in the northern hemisphere, and simultaneously eastward (to the right) in the dayside cusp in the southern hemisphere. The open field lines are correspondingly carried asymmetrically into the tail lobes, preferentially into the dawn side of the northern lobe, and into the dusk side of the southern lobe, an effect that tends to twist the tail structure on the nightside. The corresponding ionospheric flows in the northern hemisphere are shown in the upper row of diagrams in Figure M161, for near-zero or negative IMF B_z and various IMF B_y as indicated. Again noon is at the top of each diagram and dusk to the left. In this case the region of open field lines is relatively expanded and the twin cell Dungey-cycle flows well-developed due to the negative IMF B_z, while the IMF B_y component produces dawn-dusk asymmetries in the flow with newly opened cusp field lines flowing west for B_y positive (as indicated in Figure M160), and east for B_y negative. The simultaneous east-west flows are opposite in the southern hemisphere. These variations of the flow in the cusp region (and beyond) produce corresponding variations in the Hall and Pedersen-FAC current system that can be sensed both on the ground and above the ionosphere according to the general principles outlined above. The perturbations observed at magnetic observatories lying under the cusp (at ~75° magnetic latitude near noon) can be used in particular to reconstruct the large-scale structure of the IMF over long intervals from historic magnetic records.

Observations indicate that reconnection at the dayside magnetopause tends to occur preferentially in regions where the magnetospheric and magnetosheath field adjacent to the magnetopause are antiparallel, such that the "magnetic gradients" across the boundary are largest. When the IMF turns from southward to northward, these regions migrate from low to high latitudes, such that reconnection can then take place between northward IMF and open field lines in the tail lobes that were produced by earlier intervals of southward field. Such "lobe reconnection" does not change the amount of open flux in the tail, but causes it to circulate within the region of open field lines, as the newly reconnected field lines flow around the dayside magnetopause and back into the lobe. The patterns of lobe circulation produced by this effect are shown in the lower row of diagrams in Figure M161, which also indicate that the amount of open flux present is usually smaller under these conditions. Again, these patterns of ionospheric flow are associated with corresponding patterns of magnetic perturbations above and below the ionosphere, produced by the corresponding Hall and Pedersen-FAC currents that are driven.

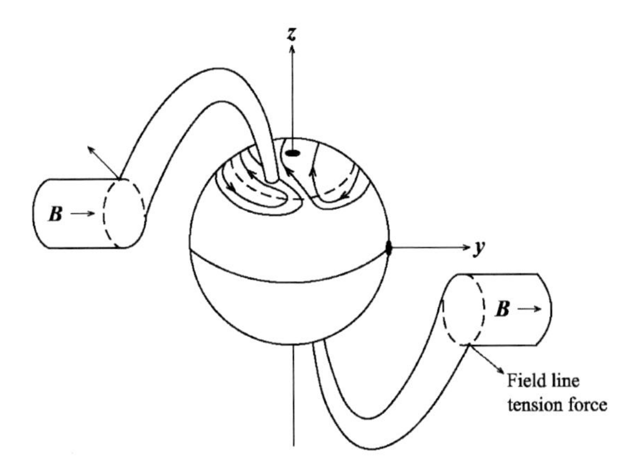

Figure M160 Schematic of newly opened field lines on the dayside of the magnetosphere, following low-latitude reconnection in the presence of an IMF with a positive B_y component (directed toward the right). The view is from the direction of the Sun. Under these conditions the $\boldsymbol{j} \times \boldsymbol{B}$ (rubber band) tension force pulls open field lines westward toward dawn (*left*) in the northern hemisphere, and simultaneously eastward toward dusk (*right*) in the southern hemisphere, as shown. This leads to dawn-dusk asymmetries in the dayside ionospheric flow on open field lines as indicated schematically, which are oppositely directed in opposite hemispheres.

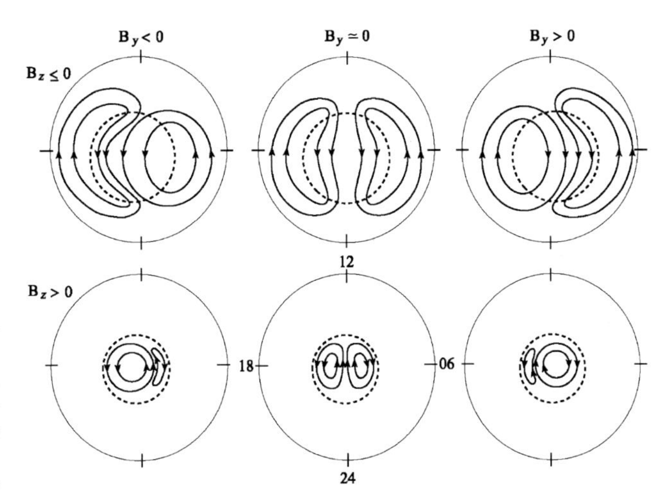

Figure M161 Sketches of the ionospheric flow in the northern hemisphere ordered according to the direction of the IMF, in a format similar to Figure M159. The dashed line shows the open-closed field line boundary, while the arrowed solid lines show the plasma streamlines. The upper row show flows for IMF B_z near zero and negative, while the lower row shows flows for IMF B_z significantly positive. The three columns show flows for IMF B_y negative (*left*), near zero (*center*), and positive (*right*). Dawn-dusk asymmetries are simultaneously oppositely directed in the southern hemisphere.

Magnetospheric substorms

The above sections have discussed the flows and currents that are driven in the coupled magnetosphere-ionosphere system by the Dungey-cycle, viewed as a quasisteady process that is strongly modulated by variations in the strength and direction of the IMF. Observations show, however, that even when interplanetary conditions are relatively steady, the driving processes in the key interaction regions at the magnetopause and in the geomagnetic tail are not. Reconnection on the dayside, in particular, often occurs as waves that propagate east-west over the magnetopause from the dayside toward the tail, which recur on timescales of ~5–10 min. These "flux transfer events" (FTEs) give rise to pulsed magnetic and plasma signatures at the magnetopause, and pulsed flow, current, plasma injection, and auroras in the cusp ionosphere on the dayside. Due to the typically overlapping nature of the effect of these pulses, however, the overall flow and current modulations are generally modest.

The pulsed behavior in the tail takes place on longer timescales of around 1–2 h, however, and produces major perturbations of the flow and current in the nightside magnetosphere-ionosphere system, termed magnetospheric substorms (see *Storms and substorms*). Substorms are initiated by intervals of southward-directed IMF, typically southward fields of a few nanoteslas lasting for several tens of minutes. Such intervals, which happen rather frequently (often several per day), are sufficient to produce enhanced open flux production at the dayside magnetopause which in turn enhances the radius and field strength of the lobes of the tail. The current carried by the plasma sheet at the center of the tail is also correspondingly intensified, while the thickness of this current layer is also found to decrease, within the near-Earth tail, to only a few thousand kilometers. After ~30–40 min of such tail development, termed the "growth phase" of the substorm, a disruption is observed to occur within the plasma sheet. The origin and development of the disruption remain controversial, but it involves the formation within a minute or two of a new reconnection region within the plasma sheet in the central tail at down-tail distances of 20–30 R_E. Reconnection at this site first pinches off the tailward portion of the preexisting plasma sheet, forming a closed-loop plasmoid field structure, illustrated schematically on the right-hand side of Figure M158, which is expelled tailward into the solar wind at speeds in excess ~500 km s^{-1}. Subsequently, open field lines in the tail lobes are also reconnected, reducing the amount of open flux in the system. Earthward of the new reconnection region, new closed field lines collapse toward the Earth, "dipolarizing" the extended tail field lines, and heating and compressing the plasma sheet plasma. Precipitation from this hot plasma into the nightside ionosphere produces an expanding patch of bright active auroras, called the "substorm auroral bulge," and also considerably enhances the density of the nightside ionosphere. Consequently the height-integrated Hall and Pedersen conductivities in this "bulge" are also strongly enhanced, from ~1 mho or less in the presubstorm nightside ionosphere, to values of ~10–100 mho. This has the effect of suppressing the flow and electric field within the perturbed region, but even so intense currents flow within the region, consisting principally of a westward-directed Hall current or "substorm electrojet." The total current flow is typically ~0.5–1 MA, yielding magnetic perturbations on the ground underneath the electrojet of several hundred nanoteslas. The disturbance field is directed equatorward, weakening the horizontal component of the planetary dipole field in both hemispheres. Current continuity is maintained by FACs which flow from the tail plasma sheet into the ionospheric electrojet at its eastern end, and return from the ionosphere to the plasma sheet at its western end, as shown in Figure M162, forming the "substorm current wedge". The FACs then close in the cross-tail current on either side of the dipolarized region, and from thence, over the tail lobe magnetopause as shown. It can therefore be seen that the current wedge is associated with a reduction in the cross-tail current within the dipolarized region, the FACs being associated with the shears in the tail field that occur in the interface between the dipolarized field within the wedge, and the (as yet) undipolarized taillike field outside.

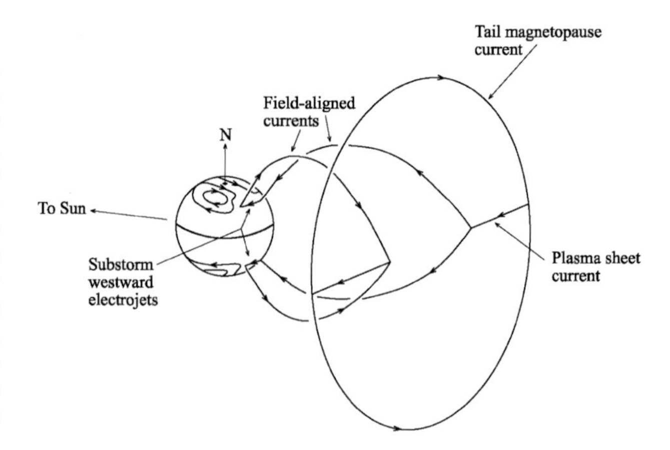

Figure M162 Sketch showing the form of the nightside "current wedge" current system that is excited during the expansion phase of magnetospheric substorms. A portion of the plasma sheet current in the center of the tail is diverted along the field lines north and south toward the Earth, and closes in the ionospheric substorm westward electrojets. The arrowed vortices drawn on the sphere representing the Earth's ionosphere indicate the concurrent Dungey-cycle flow, similar to Figures M159 and M161.

As indicated above, substorm tail reconnection and field dipolarization begin in a localized region near the center of the tail (generally displaced to the dusk side of midnight), but then spreads in both directions across the tail. The bright auroral and electrojet current region in the ionosphere correspondingly starts as a small oval region in the premidnight sector, located typically at ~65° magnetic latitude, and then spreads toward dusk and dawn, as well as poleward as flux continues to be reconnected and plasma compressed and heated in the nightside tail. This is called the "expansion phase" of the substorm, and was first described by Akasofu in 1964. Sometimes the auroral bulge expands to cover much of the nightside polar ionosphere, though it does not generally reach to the magnetic pole itself. After 20–30 min the auroral bulge reaches its maximum size, and the auroral intensity and the currents then decline, signaling the start of the substorm "recovery phase" which typically lasts a further 30–40 min. In the tail, "recovery" is associated with the down-tail propagation of the new substorm reconnection region to large distances from Earth, such that the plasma sheet re-forms in the near-Earth tail. On the ground underneath the bulge, where the magnetic effects are dominated by the overhead electrojet current, the magnetic depression in the horizontal field typically grows rapidly and impulsively during the expansion phase, and then declines somewhat more gradually during recovery, giving rise to a signature which is called a "magnetic bay" in high-latitude magnetic records. Substorm signatures are also observed in nightside midlatitude magnetic fields, but in this case both the ionospheric currents and the FACs of the current wedge contribute to the form of the overall disturbance.

Several substorm cycles often occur each day, driven by individual episodes of modest southward field in the IMF, producing magnetic disturbances in the polar region under the electrojet of ~100–1000 nT which grow and decay over intervals of ~1 h. During rarer extended intervals of strong southward IMF, however, impulsive electrojet activity is essentially continuously present in the high-latitude nightside ionosphere, with variable magnetic disturbances on the ground underneath peaking at ~1000–2000 nT. These high-latitude perturbations occur simultaneously with worldwide field depressions of typically ~50–250 nT which are produced under the same conditions. These are termed geomagnetic storms and will be discussed in the next section. The largest high-latitude disturbance observed since systematic records began in 1957 was of ~3000 nT, corresponding to ~5% of the planetary field, this being the largest magnetic effect produced at the Earth's surface due to external currents. At other times, however, the IMF may remain small and northward-pointing for extended intervals, leading to prolonged intervals of "magnetic quiet" on the ground, when only the variations due to the effects of the Sq-current system are present, driven by the daily thermal-tidal motions of the upper atmosphere.

Geomagnetic storms

Usually the north-south component of the IMF is a few nanoteslas in magnitude and fluctuates in sense on timescales of minutes to a few tens of minutes, as indicated above. This drives variable Dungey-cycle flows and substorms in the magnetosphere, as discussed in the previous sections. However, under some rather rare but specific circumstances the IMF can become very strong (e.g., several tens of nanoteslas), and can remain southward-pointing for extended intervals of several hours. Under these circumstances a geomagnetic storm is produced at the Earth in which the horizontal (northward) field is depressed globally over intervals of hours and days, the depression typically maximizing at a few tens to a few hundreds of nanoteslas. These effects are masked at high-latitudes, however, by the stronger and more variable magnetic perturbations associated with magnetosphere-ionosphere coupling discussed in the previous section.

For reasons that will be outlined below, the occurrence of magnetic storms tends to follow the 11-year solar sunspot cycle, as first noted by Sabine in 1852, but on average there are roughly 10 storms each year whose midlatitude field depression exceeds 50 nT, and one that exceeds 250 nT. The largest depression observed since systematic records began in 1957 was of ~600 nT during a storm in March 1989, which occurred near solar maximum. After a variable "initial phase" to be discussed below, the magnetic depression typically grows fairly gradually over an interval of several hours (~2–12 h), termed the "main phase," and then decays even more gradually over the following several days (~1–5 days), termed the "recovery phase." These effects result from a prolonged enhancement of the Dungey-cycle flow, which is produced by the prolonged interval of strong southward IMF. During such intervals the boundary between corotating and Dungey-cycle flow in the inner equatorial magnetosphere shrinks significantly in size, such that the outer layers of the preexisting plasmasphere are stripped off and flow out to the dayside magnetopause. Correspondingly, the hot plasma produced in the tail flows inward to replace it, significantly enhancing the westward-directed "ring current" produced by the differential drift of energetic ions and electrons in the inhomogeneous quasidipolar planetary magnetic field, as outlined above. The perturbation field produced by the hot inner plasma is a major contributor to the "main phase" field depression at Earth. However, the picture is complicated by the fact that the inflowing hot plasma is usually asymmetric in its distribution around the Earth, leading to partial current "rings" that close in the ionosphere via "region 2"-type FACs. These FACs also contribute to the overall magnetic disturbance, as do the combined "fringing" fields of the enhanced plasma sheet and tail currents on the nightside. During the recovery phase, however, Dungey-cycle convection is reduced, so that these current systems also decline. Hot "ring current" plasma now marooned within the newly expanded corotating region slowly decays due to wave-scattering followed by precipitation into the midlatitude atmosphere, and due to charge-exchange with neutral atoms of the geocorona. The outer newly corotating flux tubes also refill with cold plasma from the ionosphere on timescales of a few days, thus re-forming a more extended plasmasphere.

As indicated at the beginning of this section, the extended intervals of strong southward IMF which excite geomagnetic storms, are associated with specific phenomena in the interplanetary medium which link storms with processes on the Sun and the solar cycle. Two principal mechanisms are responsible. In the first, a large loop-like field structure located in the solar corona suddenly becomes unstable and is ejected away from the Sun, forming a "coronal mass ejection" (CME). The ejection speed is sometimes very rapid, exceeding ~1000 km s^{-1}, such that the CME plasma and frozen-in field plow into the slower solar wind ahead of it. This creates a giant shock wave, which propagates through the solar system compressing and heating the ambient solar wind medium ahead of the CME "driver gas". Behind the shock the IMF is also compressed to high field strengths. When such a shock impinges on the Earth, the magnetosphere is impulsively compressed, such that the horizontal field at the Earth's surface is suddenly increased, typically by a few tens of nanoteslas, corresponding to the effect of the increased Chapman-Ferraro current at the magnetopause. What happens after this depends on the direction of the enhanced IMF in the compressed solar wind and CME driver gas. These fields may have any orientation depending on the individual circumstances of the event, and may also fluctuate in direction during the event. A magnetic storm "main phase" occurs only if the enhanced IMF points southward during some extended interval after the shock wave has passed. If this occurs (typically in about one in six events), the impulsive field enhancement at the Earth's surface produced by the shock is called the "storm sudden commencement" (SSC), while the variable interval between the SSC and the onset of the main phase (when the upstream IMF happens to turn southward) is called the "initial phase" of the storm. If such an enduring interval of southward field does not occur, and with it no "main phase" field depression, then the shock-related compression event is simply termed a "sudden impulse". The connection between these events and the solar cycle results from the fact that ~3 CMEs occur each day near solar cycle maximum (only a small number of which impinge on Earth), reducing to one every several days during solar cycle minimum. The connection with solar flares, first noted by Carrington in 1860, results from the fact that flares often occur near the sites in the solar corona where CMEs have formed.

During the declining phase of the solar cycle, extended intervals of strong southward IMF can also be produced by another interplanetary mechanism, which thus can also generate geomagnetic storms. The Sun typically produces two types of quasisteady solar wind outflow (as opposed to CMEs), a "slow" variable wind of ~300–500 km s^{-1} from regions surrounding closed field regions of the solar corona, and a "fast" steady wind of ~800 km s^{-1} from open-field "coronal holes." During solar maximum, the solar field is typically very disordered, and with it the solar wind outflow. During the declining phase of the cycle, however, large-scale off-axis coronal holes tend to form, which migrate slowly toward the poles of the Sun as activity decreases toward solar minimum. Initially, however, these coronal holes may extend down to the solar equator, such that during each solar rotation the Earth experiences a pattern of fast and slow solar wind output that may persist in form for many solar rotations. The corotating variable solar wind source regions on the Sun thus give rise to "corotating interaction regions" (CIRs) in the interplanetary medium, in which fast plasma outflow from an equatorial coronal hole "runs into" slower plasma that was emitted earlier into the same direction from a noncoronal hole source region on the rotating Sun. This compresses the plasma and may enhance the frozen-in field by factors of ~5–10 above the usual values. As the solar wind velocity subsequently drops, however, after the passage of the material from the coronal hole, a rarefaction region is created in the solar wind, and with it regions of weak IMF occur. Geomagnetic storms can be excited by the enhanced fields of the compression phase of CIRs, during intervals in which the solar wind speed is increasing locally at the Earth, if the enhanced field happens to point southward for a significant interval. Such storms, however, are typically not as intense as those generated by CMEs, because the field directions tend to fluctuate more during these events, as opposed to the more ordered field structures produced by CMEs. In addition, these events are not associated with interplanetary shocks, such that no SSC and "initial phase" occurs prior to the onset of the main phase.

Summary

Overall, it can be seen from this discussion that a variety of interlinked and rather complex plasma physical processes affect the magnetic field observed at the Earth's surface, and throughout the magnetospheric region containing the Earth's magnetic field. The current systems involved are those concerned with the magnetopause boundary of the magnetosphere, the hot plasma that flows inside it, and the current systems that link the magnetosphere and ionosphere and are associated with the auroras. These produce peak perturbation fields at the Earth's surface of typically a few tens, a few hundreds, and a few thousands of nanoteslas, respectively, varying on a range of timescales from minutes during substorms to hours and days during worldwide geomagnetic storms. These current systems are linked to the Sun and solar activity through the solar wind outflow, which forms the Earth's magnetosphere and conditions its dynamics. The variable output of the Sun similarly produces effects on a wide range of timescales, from the minute scales associated with interplanetary shocks, up to the ~11 year timescales of the solar activity cycle.

Stanley W.H. Cowley

Bibliography

Cowley, S.W.H., Davies, J.A., Grocott, A., Khan, H., Lester, M., McWilliams, K.A., Milan, S.E., Provan, G., Sandholt, P.E., Wild, J.A., and Yeoman, T.K., 2003. Solar wind-magnetosphere-ionosphere interactions in the Earth's plasma environment. *Philosophical Transactions A*, **361**: 113–126.

Ohtani, S.-I., Fujii, R., Hesse, M., and Lysak, R.L. 2000. *Magnetospheric Current Systems*, Geophysical Monograph 118. Washington, DC: American Geophysical Union.

Paschmann, G., Haaland, S., and Treumann, R. (eds.), 2002. *Auroral Plasma Physics, Space Science Review*, Vol. 103.

Sandholt, P.E., Carlson, H.C., and Egeland, A., 2001. *Dayside and Polar Cap Aurora.* Dordrecht: Kluwer Academic Publisher.

Cross-references

Alfvén's Theorem and the Frozen Flux Approximation
Auroral Oval
Chapman, Sydney (1888–1970)
Ionosphere
Ring Current
Storms and Substorms

MAGNETOSTRATIGRAPHY

Magnetostratigraphy refers to the description, correlation, and dating of rock sequences by means of magnetic parameters. A rock interval in which a magnetic parameter has a constant value is called a magnetozone. Most commonly, the parameter is the polarity of the Earth's magnetic field during acquisition of a primary magnetization that is contemporaneous with the rock formation. Magnetozones of alternating polarity yield a magnetic polarity stratigraphy. This form of magnetostratigraphy plays an important role in the construction of geomagnetic polarity timescales. However, in principle, any rock magnetic parameter can serve as a magnetostratigraphic indicator.

Although the most widespread and successful use of magnetostratigraphy has been in sediments and sedimentary rocks, the method can be applied to any succession of layered rocks. In igneous rocks and high-deposition rate sediments, magnetostratigraphy provides fine details of geomagnetic field behavior. For example, the analysis of paleomagnetic vectors in radiometrically dated lava flows from Steen's Mountain (Oregon, USA) showed details of the behavior of magnetic field direction and intensity during a Miocene polarity reversal (Prévot *et al.*, 1985). The study gave an estimated duration of about 4500 years for the polarity transition. Other estimates range from 3500 to 10000 years. The time intervals (chrons), during which the geomagnetic field maintains constant normal or reverse polarity, are many times longer than the transitional interval, and may last hundreds of thousands to millions of years; they are interspersed with shorter subchrons lasting only some tens of thousands of years. The reversals apparently occur at random separations, so that a sequence of reversals defines polarity chrons and subchrons of very variable lengths. Each polarity reversal is a global feature, which is very useful for magnetic stratigraphy because a short sequence of 10–20 reversals forms a distinctive pattern, like a fingerprint, which can be used for geological correlations on a worldwide basis.

Marine magnetic anomaly sequences

Seafloor spreading since the Middle Jurassic has created sequences of lineated marine magnetic anomalies in the ocean basins, which have yielded the most complete and detailed record of geomagnetic polarity for this time interval. An episode of alternating polarity began about 157 Ma ago in the Late Jurassic and lasted until about 120 Ma ago in the Early Cretaceous; it gave rise to the M-sequence of marine magnetic anomalies. The field apparently rested in a state of normal polarity for the next 37 Ma. A further episode of reversals began 83 Ma ago in the Late Cretaceous and has lasted throughout the Cenozoic until the present; it produced a sequence of magnetic anomalies that is referred to here as the C-sequence. It is possible that the known C-sequence is incomplete. Short wavelength, low amplitude magnetic anomalies (so-called "tiny wiggles" or "cryptochrons") that could correspond to very short-polarity chrons have been observed in the marine magnetic anomaly record that defines the C-sequence, especially in the Oligocene and Paleocene. The origin of these magnetic anomalies is uncertain because they can be modeled equally well as short polarity intervals and intensity fluctuations of the paleomagnetic field (Cande and LaBrecque, 1974; Cande and Kent, 1992b).

Major magnetic anomalies in each sequence are numbered in order of increasing age. The corresponding polarity chron is identified with the prefix "C" in the C-sequence and "CM" in the M-sequence; the suffix "n" or "r" is appended to designate normal or reverse polarity, respectively. Thus, the C-sequence of polarity chrons extends from C1n to C33r, and the M-sequence extends from CM0r to CM29r. The two anomaly sequences are separated in the oceanic record by the Cretaceous Quiet Zone (equivalent to C34n) in which lineated magnetic anomalies are not found. It corresponds to a time interval, the Cretaceous Normal Polarity Superchron (CNPS), in which the Earth's magnetic field did not reverse polarity for more than 37 Ma. Regions of the oceanic crust older than CM29r form a Jurassic Quiet Zone, which, like the Cretaceous Quiet Zone, is characterized by the absence of correlatable magnetic anomalies at the ocean surface. Deep-tow magnetometer studies have revealed a tentatively correlated sequence of short wavelength, low amplitude magnetic anomalies numbered M29-M41 (Sager *et al.*, 1998). These anomalies suggest that the Jurassic Quiet Zone may result from a high reversal frequency, in contrast to the origin of the CNPS.

Geomagnetic polarity timescales

The association of radiometric ages with key biostratigraphic stage boundaries, which have been correlated by magnetostratigraphy to the marine polarity record, yields a dated geomagnetic polarity timescale (GPTS). Several GPTS have been proposed for each reversal sequence, accompanying refinements in defining the fundamental polarity record and improvements in confirming, correlating, and dating the polarity chrons. A comparison of the different GPTS that have been proposed for the Late Cretaceous and Cenozoic is given by Opdyke and Channell (1996). A databank of the C-sequence timescales has been compiled and stored electronically on-line (Mead, 1996). The optimum GPTS for this time interval is the CK95 timescale (Cande and Kent, 1995), based upon a reevaluation of the C-sequence marine magnetic anomalies and the corresponding block models (Cande and Kent, 1992a).

The M-sequence of marine anomalies is not as well dated as the C-sequence and it also may still be incomplete. The old end of the

M-sequence merges into the Jurassic Quiet Zone, whose origin is less well understood than its Cretaceous counterpart. A comparison of different GPTS for the M-sequence shows that appreciable discrepancies still exist. The optimum GPTS for this part of the record is currently the CENT94 timescale (Channell *et al.*, 1994), which is based on revised block models for the origin of the marine magnetic anomalies and covers magnetic polarity chrons CM0 to CM29. Because of their still uncertain origin, anomalies M29-M41 are not yet accepted as polarity chrons that can be included in the GPTS.

For each polarity sequence, tie-levels are established by magnetostratigraphic correlation of radiometrically dated biostratigraphic stage boundaries or other datum-levels to the C- and M-sequence polarity record. The rest of each timescale is dated by conversion of polarity boundary locations to numeric ages by interpolation between the tie-levels. There are several inherent sources of error in this process, including errors of absolute dating of the stage boundary or other tie-levels, stratigraphic errors in correlation, etc. Consequently, the "absolute ages" in a GPTS are known less exactly than the relative lengths of the polarity chrons.

Magnetic polarity stratigraphy

Methodology

Shallow, unconsolidated sediments from lakes and seas have been sampled with gravity-driven piston cores, whereas older sediments from deep ocean basins have been sampled with drilled cores in the international Ocean Drilling Project (ODP). Indurated sedimentary rocks can be sampled by ordinary paleomagnetic techniques. Long, continuous, continental outcrops or drillholes are necessary for successful magnetostratigraphy. Marine and nonmarine sedimentary rocks that have not been adversely disturbed by local tectonic conditions have given good results. Marine rocks can be dated paleontologically, whereas cyclostratigraphy has enabled the dating of some continental deposits.

The sample and data processing methods are the same as in a customary paleomagnetic study, and strict criteria have been proposed to define the quality of a magnetostratigraphy (Opdyke and Channell, 1996). It is possible to measure and partially demagnetize whole cores of sediments virtually continuously with specially designed pass-through magnetometers. More commonly, the cores are subsampled, either using long, so-called u-channels or by taking conventional 1 in. paleomagnetic samples at discrete intervals. Whatever their source, the prepared samples are treated using standard paleomagnetic methods such as progressive stepwise demagnetization. In soft wet sediments this treatment is limited to alternating field demagnetization, which in many cases is quite adequate, as the carrier of remanent magnetization is often primary magnetite. Secondary magnetic phases cannot be excluded. In soft marine sediments obtained from piston cores and ODP they are usually unimportant, but in lake sediments the possible presence and effects of secondary magnetizations carried by minerals such as hematite and greigite must be taken into account. In indurated sedimentary rocks such as limestones and redbeds, the primary magnetic mineral is usually magnetite, but the magnetic mineralogy may be complicated by postdepositional growth of secondary ferromagnetic minerals. Of these, hematite is the most common with serious consequences for paleomagnetic interpretation. In this case, progressive partial thermal demagnetization is the most important laboratory technique for establishing the direction of the primary magnetization on which the magnetostratigraphy is based.

The stable magnetization directions can be used to calculate the latitude of the virtual geomagnetic pole (VGP) at the time of formation of a sample. The stratigraphic plot of the VGP latitude, or in some cases the inclination of the magnetization, defines magnetozones of common polarity (Figure M163). Normal polarity magnetozones are customarily shaded black and reverse polarity zones are left unshaded. Paleontological analyses on samples from the same section based on ammonites, foraminifera, nannofossils, or other fossil types, define a biostratigraphy that is tied to the magnetostratigraphy. In this way the locations of major biostratigraphic stage boundaries are correlated to the magnetic reversal sequence.

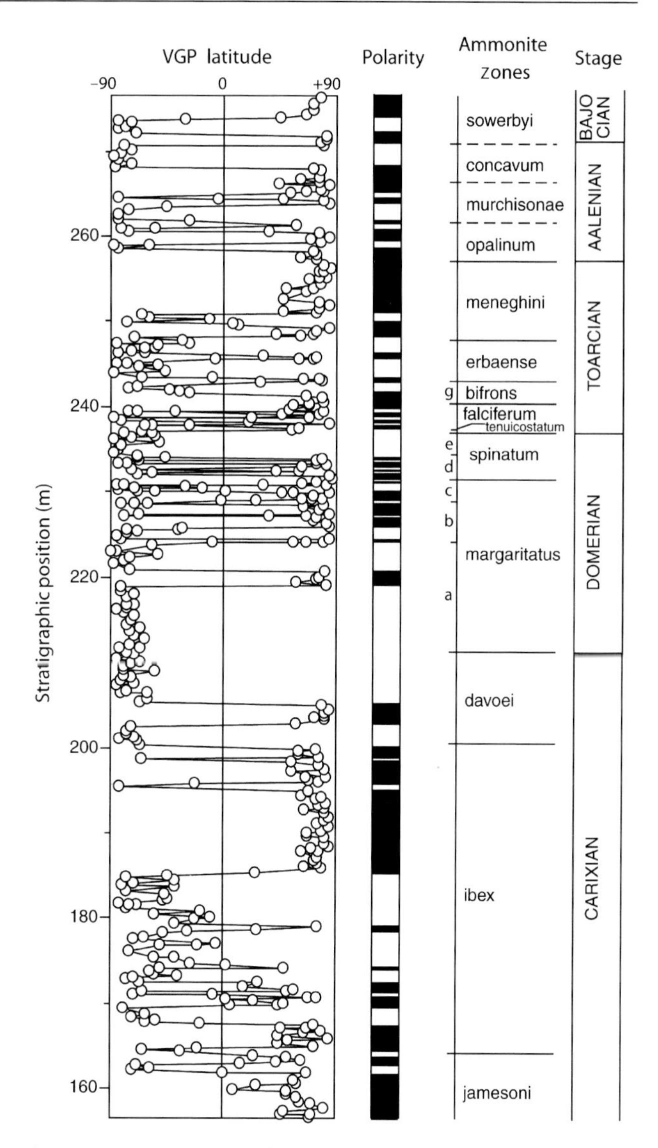

Figure M163 Lower Middle Jurassic magnetic polarity stratigraphy from the Breggia Gorge, southern Switzerland (after Horner and Heller, 1983).

Magnetostratigraphic results often show individual samples with polarity opposite to adjacent samples. This is often due to misorientation of the sample, or inadequate magnetic cleaning, but it may be that some represent very short-lasting polarity chrons.

Magnetostratigraphic verification of the C- and M-sequences

Magnetic polarity stratigraphy is most successful when a reference record is available, such as that provided by marine magnetic anomalies since the Late Jurassic. Numerous investigations of magnetic polarity stratigraphy have been carried out in marine limestones and marls to verify, correlate, and date the history of geomagnetic polarity for this time interval (Opdyke and Channell, 1996). Studies in continental exposures of marine limestones have confirmed the main features of the C- and M-sequences of polarity reversals. Improvements in drilling technology and the use of pass-through cryogenic magnetometers have yielded excellent magnetostratigraphies in several ODP studies.

Coordinated biostratigraphic and magnetostratigraphic investigations have correlated most important paleontological stage and substage boundaries from the Late Jurassic to the present with the corresponding geomagnetic polarity sequences.

The reversal history of the Pleistocene and Upper Pliocene was established in the 1960s jointly from radiometrically dated lavas and modern deep-sea sediments (Opdyke, 1972). During the 1970s and 1980s, magnetostratigraphic research confirmed the marine magnetic record in Paleogene and older rocks. The verification of Miocene polarity chrons was delayed by high reversal rate, weak magnetization, and unstable remanent magnetizations in many sections. However, coordinated magnetostratigraphy, biostratigraphy, and cyclostratigraphy in marly sections in Sicily and Crete confirmed and dated the Early Pliocene and Upper Miocene polarity sequence (Krijgsman *et al.*, 1995). Recent results from ODP sites have given excellent "fingerprint" correlation of Early Miocene polarity chrons (Parés and Lanci, 2004).

Magnetostratigraphic studies in limestone sections in the Umbrian Apennines and Southern Alps confirmed and dated geomagnetic polarity history from the Jurassic-Cretaceous boundary to the Oligocene-Miocene boundary (Channell and Erba, 1992; Channell *et al.*, 1979, 2000; Lowrie *et al.*, 1980, 1982; Channell and Medizza, 1981; Lowrie and Alvarez, 1981, 1984). The Cretaceous-Tertiary boundary was correlated within polarity chron C29r and the Santonian-Campanian stage boundary was located just older than polarity chron C33r (Alvarez *et al.*, 1977; Lowrie and Alvarez, 1977) (Figure M164).

Occasional reports of reversed polarity zones in magnetostratigraphic sections have given rise to speculation about the continuity of the 37-Ma long CNPS and the origin of the Cretaceous Quiet Zone. However, in each instance the magnetostratigraphic results have been associated with a rock magnetic complication, such as remagnetization in redbeds, lack of independent verification, or possible sample inversion. The paleomagnetic evidence supports the interpretation of the CNPS as an uninterrupted interval of constant normal polarity. It may represent a distinct behavioral regime of the geomagnetic field, different in its nature from the preceding and following episodes of reversing polarity.

The youngest reverse polarity chron CM0r of the M-sequence is located just above the Barremian-Aptian boundary (Channell and Erba, 1992). Correlation of the Jurassic-Cretaceous boundary with the M-sequence anomalies was found to be slightly dependent on the paleontological dating scheme used to define the boundary (Lowrie and Channell, 1984; Ogg, 1984; Lowrie and Ogg, 1986). Magnetozones in the Kimmeridgian and Tithonian stages of the Upper Jurassic were correlated in magnetostratigraphic sections in southern Spain with M-sequence anomalies M18-M25 (Ogg *et al.*, 1984) (Figure M165). Many details of these correlations and polarity sequences have been further confirmed in ODP drillcores.

In addition to the well-documented magnetic anomalies of the M-sequence and C-sequence, the marine magnetic record contains many short wavelength, low-amplitude magnetic anomalies (tiny wiggles). The origin of these features is uncertain, and they have been designated as "cryptochrons." They can be modeled equally well as very short polarity chrons with typical durations of 10 ka or less, or as intensity fluctuations of the paleomagnetic field (Cande and LaBrecque, 1974; Cande and Kent, 1992b). They have generally not been identified in detailed magnetostratigraphic investigations. Although some may indeed represent polarity reversals, most are probably due to paleointensity variations and are not part of the reversal sequence.

Magnetostratigraphy older than the M-sequence

Magnetostratigraphy in older rocks suffers from the absence of a marine magnetic anomaly record. Geomagnetic polarity history must be pieced together from magnetic stratigraphy alone. This requires overlapping sections to confirm the polarity sequence. Until now, the pre-M29 magnetic anomalies have not been securely and independently established in the oceanic record. Their appearance resembles the controversial tiny wiggles found in the Cenozoic record (Cande and Kent, 1992b) and they may be due to paleointensity fluctuations rather than polarity reversals. Deep-tow studies are in progress to verify the anomaly correlations. The assumed polarity sequence must then be confirmed and dated independently by magnetostratigraphy. This is a difficult task. Middle Jurassic limestone sections have given good magnetostratigraphic data but the magnetozones correlate only tenuously with the marine polarity record corresponding to anomalies M29-M41 (Steiner *et al.*, 1985, 1987).

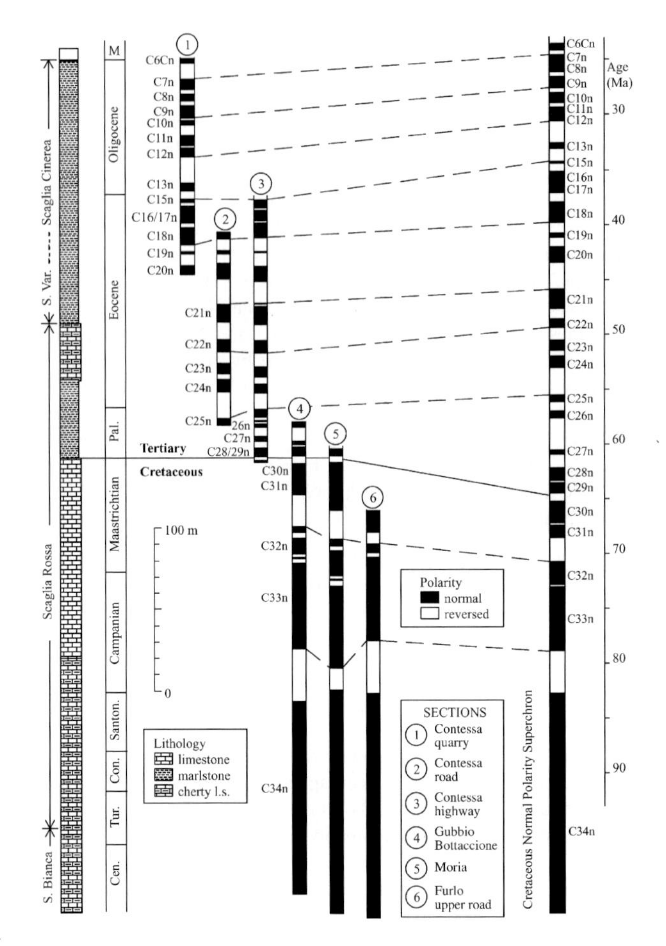

Figure M164 Correlation of Cenozoic and Upper Cretaceous limestone sections in the Umbrian Apennines, Italy, with the oceanic polarity record on the basis of magnetic polarity stratigraphy (after Lowrie and Alvarez, 1981).

Lower Jurassic results from the Breggia Gorge in Switzerland (Horner and Heller, 1983) showed a very well-defined magnetostratigraphy for the Carixian (early Pliensbachian) to Bajocian stages of the Early to Middle Jurassic (Figure M163). This record stands as the reference magnetic polarity stratigraphy for this part of the timescale. Because of the rareness of comparable sites with suitable magnetic properties for magnetic polarity analysis, it has only been verified in part. Moreau (2002) recently established the magnetostratigraphy of a drillcore from the Paris Basin and correlated it successfully to the Pliensbachian part of the Breggia Gorge record.

It is a great challenge for current and future magnetostratigraphic research to piece together the history of geomagnetic polarity reversals for earlier periods in the Mesozoic and Paleozoic. The magnetic polarity stratigraphy of the Late Triassic is a focus of current activity. Excellent results have been obtained in overlapping drillcores from continental redbeds from the Newark Basin in New Jersey, USA (Kent *et al.*, 1995). This study required subtle analysis of

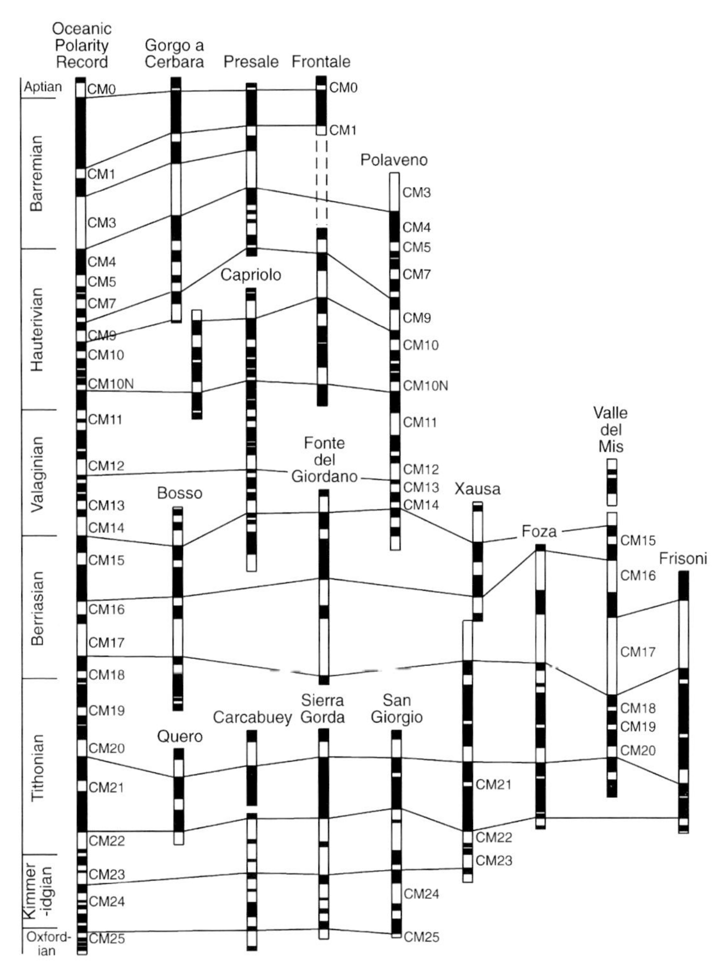

Figure M165 Correlation of Lower Cretaceous and Upper Jurassic limestone sections with the polarity record derived from the M-sequence of oceanic magnetic anomalies (after Lowrie, 1989).

thermal demagnetization data; the polarity stratigraphy was dated by a combination of paleontology and Milankovich-related cyclostratigraphy. As the Breggia Gorge is for the Lower Jurassic, the magnetostratigraphy of the Newark Basin is the standard for comparison for the Late Triassic. Confirmation by correlation with other sections, especially of marine origin, is in progress.

Rock magnetic stratigraphy

Susceptibility stratigraphy

Although polarity stratigraphy is the best-known magnetic method for establishing the stratigraphic variation in a section and for correlating between stratigraphic sections, other magnetic methods are also possible. For example, magnetic susceptibility, remanent magnetizations, and coercivity parameters have all been invoked in magnetic studies of young sediments to show small variations in mineralogy or grain size that might have an environmental or paleoclimatic cause.

Magnetic susceptibility is a measure of the ease with which a magnetization can be induced in a sample. It is determined by all the mineral grains in a sample; that is, by the type, concentrations, and grain sizes of any ferromagnetic minerals, as well as by the paramagnetic and diamagnetic (i.e., nonferromagnetic) minerals in the sample. If magnetite is present, its high intrinsic susceptibility may dominate the susceptibility of a sample, and fluctuations in susceptibility within a sedimentary sequence may be related to variations in magnetite content.

An example is shown in Figure M166. The loess deposits of Central China form great thicknesses of sediment in which loess and paleosol layers alternate with each other. The loess are windblown sediments that form in cold, dry conditions, whereas the paleosols are thought to form under warmer, moist conditions. The natural remanent magnetization of the deposits gives a well-defined magnetic polarity stratigraphy that correlates the oldest layers with the end of the Gauss Epoch (top of Chron C2An.1n) about 2.6 Ma ago (Heller and Liu, 1984; Heller and Evans, 1995). The magnetostratigraphy of samples from a borehole at Luochuan in China showed that the loess deposits were much older than previously thought. The magnetic susceptibility correlated with the lithology, showing higher values in the paleosols than in the loess layers. The magnetic mineral in both lithologies is dominantly magnetite. The magnetite enhancement in the paleosols may be related to *in situ* formation of a new phase of magnetite as a result of the change to a warmer, moister climate. This interpretation has been carried a stage further in an attempt to estimate the record of paleoprecipitation. The existence and interpretation of magnetite enhancement in the paleosols is still a controversial subject. Nevertheless, the example demonstrates the potential of magnetic measurements in an environmental context.

Coercivity stratigraphy

In contrast to magnetic susceptibility, the remanent magnetization of a sample, whether naturally acquired or artificially produced by an applied magnetic field in the laboratory, is carried only by ferromagnetic minerals.

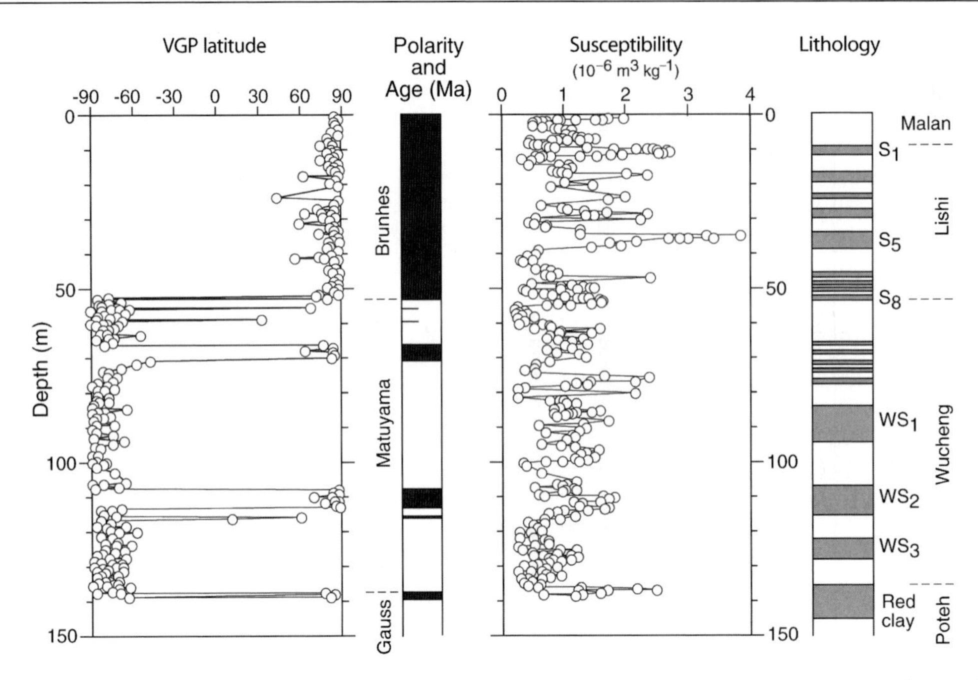

Figure M166 Magnetic polarity stratigraphy and susceptibility stratigraphy of loess and paleosol deposits in a borehole at Luochuan, China (after Heller and Evans, 1995; courtesy of F. Heller).

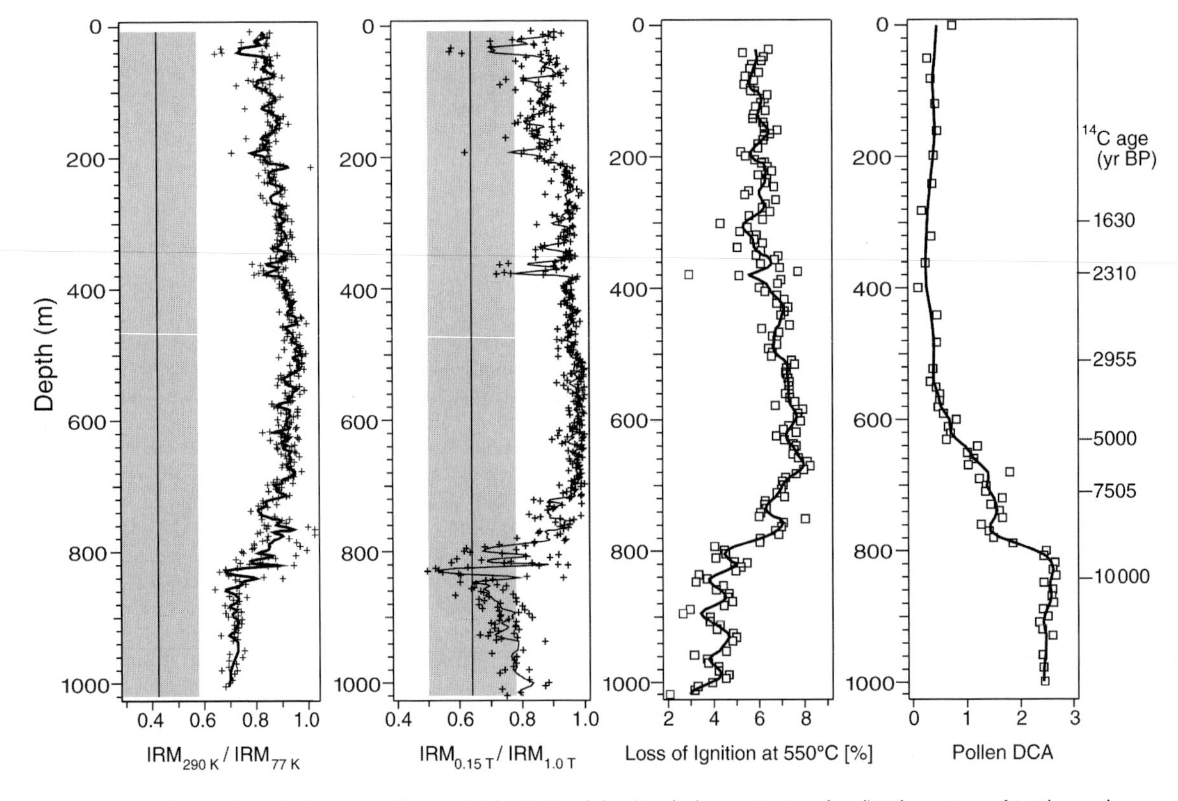

Figure M167 Magnetic stratigraphy in a core from a high Alpine lake (Bachalpsee, Switzerland). The ratios of isothermal remanent magnetizations (IRM) acquired at 290 and 77 K, and in fields of 0.15 and 1 T at room temperature, are determined by variations in coercivity of ferromagnetic minerals. They are compared to climatic proxies that reflect organic content and vegetation changes (after Lanci *et al.*, 1999).

The most common are magnetite, hematite, maghemite, and pyrrhotite. The ability to retain a remanent magnetization is a coercivity parameter, and is dependent on grain size, temperature, and the applied magnetic field. A commonly used remanent magnetization is produced in the laboratory by applying a magnetic field to a sample and then removing it. The residual magnetization is called the isothermal remanent magnetization (IRM). Different grain-size fractions are activated by different applied fields, and at different temperatures. For example, ultrafine-grained magnetite that is superparamagnetic (unable to hold a remanent magnetization) at room temperature becomes stable at low temperature. The IRM of a sample containing superparamagnetic grains of magnetite is thus larger at low temperature than at room temperature. The ratio of IRM acquired at room temperature (290 K) and after immersion in liquid nitrogen at a temperature of 77 K (written $IRM_{290\ K}/IRM_{77\ K}$) is an indication of the relative amounts of superparamagnetic magnetite in a sediment. Similarly, the ratio of IRM acquired in fields of 0.15 and 1.0 T (written $IRM_{0.15\ T}/IRM_{1.0\ T}$) is an indication of the relative magnitude of low-coercivity magnetite and higher coercivity grains in a sample. For example, in a magnetostratigraphic section of Upper Cretaceous Scaglia Rossa limestone at Gubbio, Italy, the variation of this parameter showed the relative amounts of magnetite and hematite in the section (Lowrie and Alvarez, 1977). In a sediment core from Bachalpsee (Figure M167), a high-alpine lake in the Swiss Alps, this ratio has a value around 0.8 at the bottom of the core. About 80% of the IRM is carried by low-coercivity magnetite at this depth, which corresponds to the Younger Dryas biozone. The remainder may be due to high-coercivity magnetite or even harder magnetic minerals. Higher in the core, where the ratio is almost 1.0, low-coercivity magnetite is virtually the only ferromagnetic mineral present. The ratio decreases again upward in the core. The parallel behavior of the $IRM_{290\ K}/IRM_{77\ K}$ ratio shows that grain-size variation of the magnetite is probably the cause of both curves.

The loss of ignition at 550°C, an indicator of organic carbon content, parallels the magnetic profiles. The ^{14}C ages on samples from the core show that all the major changes that took place began 10 ka ago at the end of the last ice age. Analysis of pollens showed a reduction in coniferous types as more deciduous types were transported into the lake accompanying the climatic warming in the early Holocene. These changes are reflected in the magnetite content of the lake sediments. The coercivity parameters provide magnetic stratigraphies that reflect directly the paleoclimatic changes.

William Lowrie

Bibliography

Alvarez, W., Arthur, M.A., Fischer, A.G., Lowrie, W., Napoleone, G., Premoli Silva, I., and Roggenthen, W.M., 1977. Upper Cretaceous-Paleocene magnetic stratigraphy at Gubbio, Italy. V. Type section for the late Cretaceous-Paleocene geomagnetic reversal timescale. *Geological Society of America Bulletin*, **88**: 383–389.

Cande, S.C., and Kent, D.V., 1992a. A new geomagnetic polarity timescale for the Late Cretaceous and Cenozoic. *Journal of Geophysical Research*, **97**: 13917–13951.

Cande, S.C., and Kent, D.V., 1992b. Ultra-high resolution marine magnetic anomaly-profiles: a record of continuous paleointensity variations? *Journal of Geophysical Research*, **97**: 15075–15083.

Cande, S.C., and Kent, D.V., 1995. Revised calibration of the geomagnetic polarity timescale for the Late Cretaceous and Cenozoic. *Journal of Geophysical Research*, **100**: 6093–6095.

Cande, S.C., and LaBrecque, J.L., 1974. Behaviour of the Earth's palaeomagnetic field from small scale marine magnetic anomalies. *Nature*, **247**: 26–28.

Channell, J.E.T., and Erba, E., 1992. Early Cretaceous polarity chrons CM0 to CM11 recorded in northern Italian land sections near Brescia. *Earth and Planetary Science Letters*, **108**: 161–179.

Channell, J.E.T., Erba, E., Nakanishi, M., and Tamaki, K., 1995. Late Jurassic-Early Cretaceous timescales and oceanic magnetic anomaly block models. In Berggren, W.A., Kent, D.V., Aubry, M., and Hardenbol, J. (eds.), *Geochronology, Timescales, and Global Stratigraphic Correlation*. SEPM Special Publication, pp. 51–64.

Channell, J.E.T., Erba, E., Muttoni, G., and Tremolada, F., 2000. Early Cretaceous magnetic stratigraphy in the APTICORE drill core and adjacent outcrop at Cismon (Southern Alps, Italy), and correlation to the proposed Barremian-Aptian boundary stratotype. *Geological Society of America Bulletin*, **112**: 1430–1443.

Channell, J.E.T., Lowrie, W., and Medizza, F., 1979. Middle and Early Cretaceous magnetic stratigraphy from the Cismon section, northern Italy. *Earth and Planetary Science Letters*, **42**: 153–166.

Channell, J.E.T., and Medizza, F., 1981. Upper Cretaceous and Paleogene magnetic stratigraphy and biostratigraphy from the Venetian (Southern) Alps. *Earth and Planetary Science Letters*, **55**: 419–432.

Heller, F., and Evans, M.E., 1995. Loess magnetism. *Reviews of Geophysics*, **33**: 210–240.

Heller, F., and Liu, T.S., 1984. Magnetism of Chinese loess deposits. *Geophysical Journal of the Royal Astronomical Society*, **77**: 125–141.

Horner, F., and Heller, F., 1983. Lower Jurassic magnetostratigraphy at the Breggia Gorge (Ticino, Switzerland) and Alpe Turati (Como, Italy). *Geophysical Journal of the Royal Astronomical Society*, **73**: 705–718.

Kent, D.V., Olsen, P.E., and Witte, W.K., 1995. Late Triassic-earliest Jurassic geomagnetic polarity sequence and paleolatitudes from drill cores in the Newark rift basin, eastern North America. *Journal of Geophysical Research*, **100**: 14965–14998.

Krijgsman, W., Hilgen, F.J., Langereis, C.G., Santarelli, A., and Zachariasse, W.J., 1995. Late Miocene magnetostratigraphy, biostratigraphy and cyclostratigraphy in the Mediterranean. *Earth and Planetary Science Letters*, **136**. 475–494.

Lanci, L., Hirt, A.M., Lowrie, W., Lotter, A.F., Lemcke, G., and Sturm, M., 1999. Mineral magnetic record of Late Quaternary climatic changes in a high Alpine lake. *Earth and Planetary Science Letters*, **170**: 49–59.

Lowrie, W., 1988. Magnetostratigraphy and the geomagnetic polarity record. *Cuàdernos de Geologia Ibérica*, **12**: 95–120.

Lowrie, W., and Alvarez, W., 1977. Late Cretaceous geomagnetic polarity sequence: detailed rock- and palaeomagnetic studies of the Scaglia Rossa limestone at Gubbio, Italy. *Geophysical Journal of the Royal Astronomical Society*, **51**: 561–581.

Lowrie, W., and Alvarez, W., 1981. One hundred million years of geomagnetic polarity history. *Geology*, **9**: 392–397.

Lowrie, W., and Alvarez, W., 1984. Lower Cretaceous magnetic stratigraphy in Umbrian pelagic limestone sections. *Earth and Planetary Science Letters*, **71**: 315–328.

Lowrie, W., Alvarez, W., Premoli Silva, I., and Monechi, S., 1980. Lower Cretaceous magnetic stratigraphy in Umbrian pelagic carbonate rocks. *Geophysical Journal of the Royal Astronomical Society*, **60**: 263–281.

Lowrie, W., Alvarez, W., Napoleone, G., Perch-Nielsen, K., Premoli Silva, I., and Toumarkine, M., 1982. Paleogene magnetic stratigraphy in Umbrian pelagic carbonate rocks: the Contessa sections, Gubbio. *Geological Society of America Bulletin*, **93**: 414–432.

Lowrie, W., and Channell, J.E.T., 1984. Magnetostratigraphy of the Jurassic-Cretaceous boundary in the Maiolica limestone (Umbria, Italy). *Geology*, **12**: 44–47.

Lowrie, W., and Ogg, J.G., 1986. A magnetic polarity timescale for Early Cretaceous and Late Jurassic. *Earth and Planetary Science Letters*, **76**: 341–349.

Mead, G.A., 1996. Correlation of Cenozoic-Late Cretaceous geomagnetic polarity timescales: an Internet archive. *Journal of Geophysical Research*, **101**: 8107–8109.

Moreau, M.-G., Bucher, H., Bodergat, A.-M., and Guex, J., 2002. Pliensbachian magnetostratigraphy: new data from Paris Basin (France). *Earth and Planetary Science Letters*, **203**: 755–767.

Ogg, J.G., 1984. Comment on: Magnetostratigraphy of the Jurassic-Cretaceous boundary in the Maiolica limestone (Umbria, Italy). *Geology*, **12**: 701.

Ogg, J.G., Steiner, M.B., Oloriz, F., and Tavera, J.M., 1984. Jurassic magnetostratigraphy. 1. Kimmeridgian-Tithonian of Sierra Gorda and Carcabuey, southern Spain. *Earth and Planetary Science Letters*, **71**: 147–162.

Opdyke, N.D., 1972. Paleomagnetism of deep-sea cores. *Reviews of Geophysics and Space Physics*, **10**: 213–249.
Opdyke, N.D., and Channell, J.E.T., 1996. *Magnetic Stratigraphy.* San Diego, CA: Academic Press, 346 pp.
Parés, J.M., and Lanci, L., 2004. A Middle Eocene-Early Miocene magnetic polarity stratigraphy in Equatorial Pacific sediments (ODP Site 1220). In Channell, J.E.T., Kent, D.V., Lowrie, W., and Meert, J. (eds.), *Timescales of the Paleomagnetic Field.* American Geophysical Union, Geophysical Monograph Series, 145, American Geophysical Union, Washington, DC, pp. 130–140.
Prévot, M., Mankinen, E.A., Grommé, C.S., and Coe, R.S., 1985. How the geomagnetic field vector reverses polarity. *Nature*, **316**: 230–234.
Sager, W.W., Weiss, C.J., Tivey, M.A., and Johnson, H.P., 1998. Geomagnetic polarity reversal model of deep-tow profiles from the Pacific Jurassic Quiet Zone. *Journal of Geophysical Research*, **103**: 5269–5286.
Steiner, M.B., Ogg, J.G., Melendez, G., and Sequeiros, L., 1985. Jurassic magnetostratigraphy. 2. Middle-Late Oxfordian of Aguilon, Iberian Cordillera, northern Spain. *Earth and Planetary Science Letters*, **76**: 151–166.
Steiner, M., Ogg, J., and Sandoval, J., 1987. Jurassic magnetostratigraphy. 3. Bathonian-Bajocian of Carcabuey, Sierra Harana and Campillo de Arenas (Subbetic Cordillera, southern Spain). *Earth and Planetary Science Letters*, **82**: 357–372.

Cross-references

Geomagnetic Polarity Reversals
Geomagnetic Polarity Timescales
Magnetic Susceptibility
Magnetization, Isothermal Remanent (IRM)
Magnetization, Remanent
Superchrons, Changes in Reversal Frequency

MAGNETOTELLURICS

Introduction

Magnetotellurics (MT) is the use of natural electromagnetic signals to image subsurface electrical conductivity structure through electromagnetic induction. The physical basis of the magnetotelluric method was independently discovered by Tikhonov (1950) and Cagniard (1953). After a debate over the length scale of the incident waves the technique became established as an effective exploration tool (Vozoff, 1991; Simpson and Bahr, 2005).

Basic method of magnetotellurics

Electromagnetic waves are generated in the Earth's atmosphere and magnetosphere by a range of physical processes (Vozoff, 1991). Below a frequency of 1 Hz, most of the signals originate in the magnetosphere as *periodic external fields* including *magnetic storms and substorms* and *micropulsations*. These signals are normally incident on the Earth's surface. Above a frequency of 1 Hz, the majority of electromagnetic signals originate in worldwide lightning activity. These signals travel through the resistive atmosphere as waves and when they strike the surface of the Earth most of the signal is reflected. However, a small fraction is transmitted into the Earth and is refracted vertically downward, owing to the decrease in propagation velocity (Figure M168). The oscillating magnetic field of the wave generates electric currents in the Earth through electromagnetic induction, and the signal propagation becomes diffusive, resulting in signal attenuation with depth. The signals diffuse a distance into the Earth that is defined as the skin depth, δ, in meters by

$$\delta = \frac{1}{\sqrt{\pi\mu\sigma f}} \approx \frac{503}{\sqrt{\sigma f}}$$

where σ is the conductivity (S m^{-1}), f is the frequency (Hz), and μ is the magnetic permeability. The skin depth is inversely related to the frequency and thus low frequencies will penetrate deeper into the Earth. The impedance of the Earth is defined as

$$Z_{xy} = \frac{E_x}{H_y}$$

where E_x is the horizontal electric field and H_y is the orthogonal, horizontal, magnetic field. From this impedance, apparent resistivity can be defined as

$$\rho_{xy} = \frac{1}{2\pi f \mu}|Z_{xy}|^2$$

and the electric and magnetic field components will have a phase difference

$$\Phi_{xy} = \arg(Z_{xy}).$$

In general, the impedance is written as a tensor,

$$\begin{bmatrix} E_x \\ E_y \end{bmatrix} = \begin{bmatrix} Z_{xx} & Z_{xy} \\ Z_{yx} & Z_{yy} \end{bmatrix} \begin{bmatrix} H_x \\ H_y \end{bmatrix},$$

which relates the horizontal components of the electric and magnetic fields. The impedance relates the applied magnetic fields to the resulting electric fields and can be considered a *transfer function*. Note that the apparent resistivity depends on the ratio of the electric and magnetic field components. This makes the MT method simple to apply by combining values of E_x and H_y recorded at different times. The other useful characteristic is that the direction of the incident wave does not affect the value of apparent resistivity. The apparent resistivity can be considered an average value of the Earth's resistivity over a hemisphere of radius δ. Thus, by computing apparent resistivity as

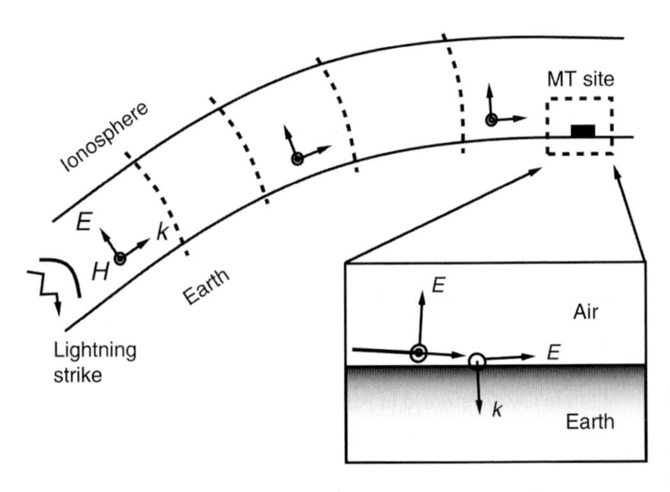

Figure M168 Propagation of electromagnetic waves from a distant lightning strike to the location where MT data is recorded. The resistive atmosphere forms a waveguide between the conductive Earth and ionosphere. The electric field (E) and magnetic field (H) are both orthogonal to the direction of propagation (k). Note that the electromagnetic energy travels as a wave in the atmosphere, but diffuses in the Earth. This type of signal propagation occurs above 1 Hz.

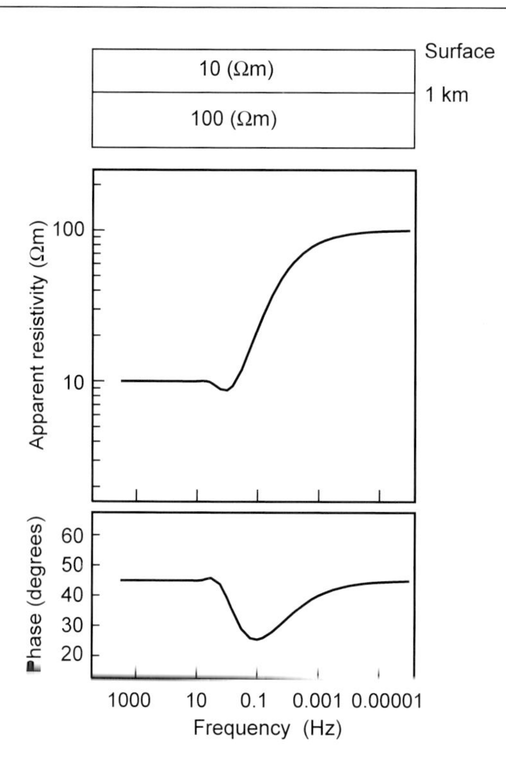

Figure M169 Variation of apparent resistivity and phase that would be measured at the surface of a two-layer Earth model. Note that the depth sounding of resistivity is achieved by varying the frequency of the signal. The dip in apparent resistivity below 10 Ωm at 1 Hz is a resonance phenomenon.

a function of frequency, the variation of resistivity with depth can be determined. This is illustrated in Figure M169. At high frequency (1000–10 Hz) the apparent resistivity is equal to the true resistivity of the upper layer. As the frequency decreases, the skin depth increases and the MT signal penetrates further into the Earth and the apparent resistivity rises. The MT phase (Φ_{xy}) is the phase delay between the electric and magnetic fields at the Earth's surface. The apparent resistivity and phase are related through

$$\Phi_{xy} = 45\left(1 + \frac{d(\log_{10} \rho_{xy})}{d(\log_{10} f)}\right)$$

where Φ_{xy} is in degrees. Thus, when the apparent resistivity increases with decreasing frequency, the phase will be less than 45°. Similarly, a decrease in resistivity will correspond to a phase greater than 45°. At the lowest frequency, the apparent resistivity asymptotically approaches the true resistivity of the lower layer, and the phase returns to 45°. Note that the phase is sensitive to changes in subsurface resistivity with depth. For a multilayer model, MT data can reliably determine the conductance of a layer. Conductance is the vertically integrated conductivity, and for a uniform layer the conductance is the product of conductivity and thickness. A consequence of the *inverse problem of electrical conductivity* is that MT data cannot individually determine the conductivity and thickness of a layer. Thus layers with differing values of conductivity and thickness, but the same overall conductance cannot be distinguished with MT.

Early studies analyzed MT data in terms of a one-dimensional (1D) conductivity model. In this class of model, conductivity only varies with depth. This approach is sometimes valid in locations where the geoelectric structure does not change rapidly in the horizontal direction. However, it is usually necessary to consider at least a two-dimensional (2D) Earth model. In this case, the apparent resistivity

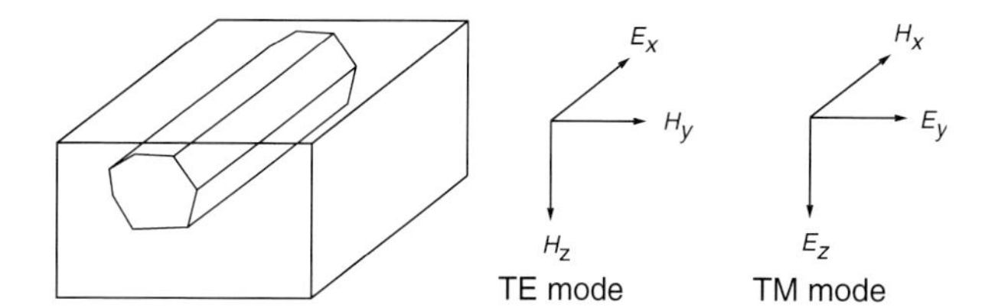

Figure M170 Geometry of electromagnetic field components over a two-dimensional Earth. The transverse electric (TE) mode is also called the E-polarization. Similarly, the transverse magnetic (TM) mode is also called the B-polarization.

computed from E_x and H_y will differ from that derived from E_y and H_x and the application of a 1D MT analysis can give misleading results. For a 2D Earth, E_x is dependent only on H_y and H_z, and these three field components comprise the transverse electric (TE) mode with the impedance (Z_{xy}) computed from E_x and H_y. The transverse magnetic (TM) mode comprises the H_x, E_y, and E_z field components, with the impedance (Z_{yx}) computed from E_y and H_x (Figure M170). In a 2D Earth with the x-axis parallel to the geoelectric strike direction, the impedance tensor can be written as:

$$\begin{bmatrix} E_x \\ E_y \end{bmatrix} = \begin{bmatrix} 0 & Z_{xy} \\ Z_{yx} & 0 \end{bmatrix} \begin{bmatrix} H_x \\ H_y \end{bmatrix}$$

The TE mode is most sensitive to along-strike conductors. In the TM mode the electric current flows across the boundaries between regions of differing resistivities, which causes electric charges to build up on interfaces. Thus the TM mode is more effective than the TE mode at locating interfaces between regions of differing resistivity.

If the subsurface structure is three-dimensional (3D) then all four elements of the impedance tensor are nonzero. Progress has been made in the last decade in 3D MT modeling and inversion. However, if a single profile of MT stations is available, and 3D effects can be shown to be small, then a 2D analysis can be used. If the subsurface conductivity structure exhibits *electrical anisotropy*, this will influence the measured impedance tensor. However, it can be difficult to convincingly distinguish heterogeneity from anisotropy in MT data. Small-scale, near-surface bodies can generate electric charges on their boundaries. If the body is small, then insignificant electromagnetic induction occurs and the only effect is *galvanic distortion*. This changes the magnitude of the electric field at the surface and can cause a static shift, which is a frequency-independent offset in the apparent resistivity curve (Jones, 1988). The phase curve is not affected. Static shifts are an example of spatial aliasing. A range of techniques is used to remove static shifts, and include external measurements of surface resistivity and estimation of the static shift coefficient in modeling and inversion.

Magnetotelluric data collection and time series processing

MT data are recorded in the time-domain, with the electric fields measured using dipoles 50–200 m in length that are connected to the ground with nonpolarizing electrodes. Audiomagnetotelluric (AMT) data (10000–1 Hz) typically sample the upper 1–2 km and are often used in mineral exploration (see *EM, industrial uses*). Magnetic fields are measured with induction coils, and in noisy environments the natural signals are supplemented with a transmitter. This modified technique is termed controlled-source audio magnetotellurics (CSAMT). Broadband MT data (1000–0.001 Hz) are used for sounding to mid-crustal depths. Induction coils are generally used and a recording time of one day is required. In the presence of excessive cultural noise additional recording may be needed. Noise can originate in a wide range of

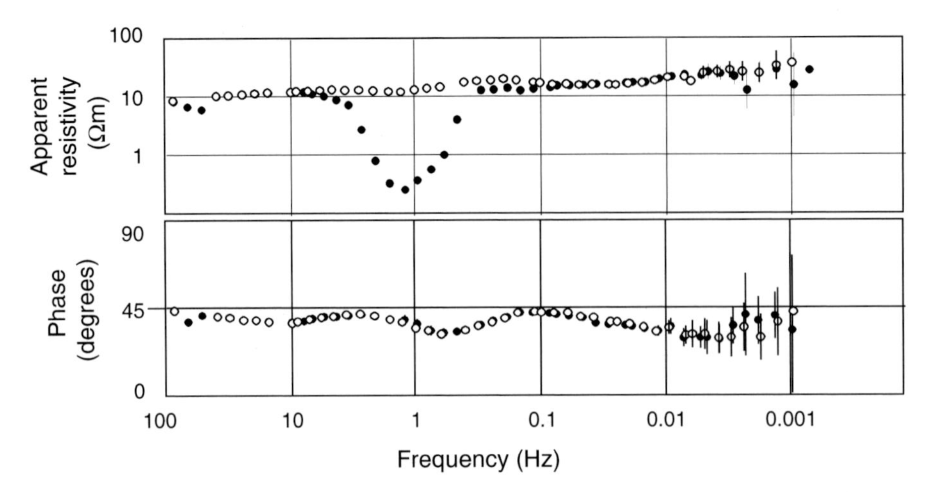

Figure M171 Estimates of apparent resistivity and phase at an MT station in central California in 1999. The open circles were derived from remote reference processing, while the black circles were derived from local data only. The MT data are contaminated by magnetic noise due to ocean wave-induced ground motion. Note the downward bias in the apparent resistivity in the frequency band 3–0.3 Hz when local MT data processing is used.

sources, including power lines, cathodically protected pipelines, railways, water pumps, and electric fences. Long-period magnetotelluric (LMT) data measure very low frequencies (1–0.0001 Hz) and are used for imaging the lower crust and upper mantle. A specialized LMT instrument is used with a *fluxgate magnetometer*, solar panels, and low power electronics. MT data can also be collected on the seafloor.

MT time series data are processed to yield frequency-domain estimates of apparent resistivity and phase. Modern processing schemes compute fast Fourier transforms of subsections of the time series and then utilize robust statistical techniques to average the multiple estimates of the impedance. The application of robust statistics has dramatically improved the quality of responses and allowed many types of noise to be effectively suppressed (Jones *et al.* 1989; Egbert, 1997). In MT data collection, time series data should be recorded simultaneously at several locations to allow for the removal of noise at the measurement location through the remote-reference method (Gamble *et al.* 1979). This is important even in locations with minimal cultural noise (Figure M171). In this example, ground motion from ocean waves caused oscillations of the magnetic sensors and resulted in the apparent resistivity being artificially low in the band 3–0.3 Hz. When the data were processed with a remote reference, the bias was removed.

Magnetotelluric data interpretation

Before modeling or inverting MT data, it is vital to understand the dimensionality of MT data. Tensor decomposition is a common approach and several techniques exist (Groom and Bailey, 1989; Bahr, 1991). Each method determines how well the measured MT impedance data can be fit to a 2D geoelectric model and gives an estimate of the geoelectric strike direction. It is common for a well-defined, consistent geoelectric strike to only be defined in a subset of an MT dataset. Decomposition can also determine if shallow conductivity structures are causing *galvanic distortion* of the surface electric fields. *Galvanic distortion* is a more general case of the process responsible for static shifts, and changes both the magnitude and direction of the electric fields. In extreme cases, near-surface structure can also cause distortion of the magnetic field through intense current channeling (Lezaeta and Haak, 2003).

Once the dimensionality has been understood, and distortion addressed, MT can be *forward modeled* or *inverted* in 1D, 2D, or 3D to recover a model of subsurface electrical conductivity. The *inverse problem of electrical conductivity* is nonunique (Berdichevsky and Dmitriev, 2002), which implies that a finite set of MT data containing noise can be reproduced by an infinite number of geoelectric models. To overcome this nonuniqueness and select a preferred model, additional constraints must be imposed on the solution. One of the most successful methods is to require that the geoelectric model derived from the inversion satisfies both the MT data and some additional requirements (regularization). In the absence of any other geoelectric information, the most common requirement is that the resistivity model should be as spatially smooth as possible in the horizontal and vertical directions (Constable *et al.*, 1987). Widely used inversion algorithms for MT data include those of Rodi and Mackie (2001) and Siripunvaraporn and Egbert (2000). The fit of the model is usually measured in term of the root-mean-square (rms) misfit of the predicted model response to the measured data. An rms misfit significantly greater than one indicates that the inversion is incapable of fitting the MT data, and usually indicates excessive noise in the data, or 3D effects that cannot be physically reproduced by a 2D inversion algorithm. A misfit significantly less than one indicates that either the error bars were too large or that the data is being over fit. In this second scenario, the resistivity model usually appears spatially rough, with the appearance of a checkerboard.

The MT method is now routinely used in both commercial exploration and in research. Commercial applications include exploration for minerals, hydrocarbons, and geothermal resources (see *EM, industrial uses*). Researchers use MT to study the structure of the continents and the dynamics of plate boundaries (Brown, 1994) and also in *EM, regional studies*. MT measurements are also made on the seafloor for both commercial and academic investigations (see *EM, marine controlled source*).

Martyn Unsworth

Bibliography

Bahr, K., 1991. Geological noise in magnetotelluric data: a classification of distortion types. *Physics of the Earth and Planetary Interiors*, **66**: 24–38.

Berdichevsky, M.N., and Dmitriev, V.I., 2002. *Magnetotellurics in the Context of the Theory of Ill-posed Problems.* Tulsa, OK: Society of Exploration Geophysicists.

Brown, C., 1994. Tectonic interpretation of regional conductivity anomalies. *Surveys in Geophysics*, **15**: 123–157.

Cagniard, L., 1953. Basic theory of the magnetotelluric method of geophysical prospecting. *Geophysics*, **18**: 605–635.

Constable, S.C., Parker, R.L., and Constable, C.G., 1987. Occam's inversion: a practical algorithm for generating smooth models from electromagnetic sounding data. *Geophysics*, **52**: 289–300.

Egbert, G.D., 1997. Robust multiple-station magnetotelluric data processing. *Geophysical Journal International*, **130**: 475–496.

Gamble, T.B., Goubau, W.M., and Clarke, J., 1979. Magnetotellurics with a remote reference. *Geophysics*, **44**: 53–68.

Groom, R.W., and Bailey, R.C., 1989. Decomposition of magnetotelluric impedance tensors in the presence of local three-dimensional galvanic distortion. *Journal of Geophysical Research*, **94**: 1913–1925.

Jones, A.G., 1988. Static shift of MT data and its removal in a sedimentary basin environment. *Geophysics*, **53**: 967–978.

Jones, A.G., Chave, A.D., Egbert, G.D., Auld, D., and Bahr, K., 1989. A comparison of techniques for magnetotelluric response function estimates. *Journal of Geophysical Research*, **94**: 14201–14213.

Lezaeta, P., and Haak, V., 2003. Beyond magnetotelluric decomposition: induction, current channeling, and magnetotelluric phases over 90°. *Journal of Geophysical Research*, **108**, doi:10.1029/2001JB000990.

Rodi, W., and Mackie, R.L., 2001, Nonlinear conjugate gradients algorithm for 2-D magnetotelluric inversion. *Geophysics*, **66**: 174.

Simpson, F., and Bahr, K., 2005. *Practical Magnetotellurics.* Cambridge: Cambridge University Press, p. 270.

Siripunvaraporn, W., and Egbert, G.D., 2000. An efficient data subspace inversion for two-dimensional magnetotelluric data. *Geophysics*, **65**: 791–803.

Tikhonov, A.N., 1950. Determination of the electrical characteristics of the deep strata of the Earth's crust. *Doklady Akademii Nauk, SSSR*, **73**(2): 295–297.

Vozoff, K., 1991. The Magnetotelluric method. In Nabighian, M.N. (ed.), *Electromagnetic Methods in Applied Geophysics,* Vol. 2, Chapter 8. Tulsa, OK: Society of Exploration Geophysicists.

Cross-references

Anisotropy, Electrical
EM Modeling, Forward
EM Modeling, Inverse
EM, Industrial Uses
EM, Marine Controlled Source
EM, Regional Studies
Galvanic Distortion
Magnetometers, Laboratory
Periodic External Fields
Storms and Substorms, Magnetic
Transfer Functions

MAGSAT

NASA (US) scientific satellite (1979–1980) made the first precise, globally distributed measurements of the vector magnetic field near Earth. The satellite flew at an altitude of 300–550 km, in a near-polar inclination. Researchers modeling the Magsat data have developed representations of the internal field extending from *spherical harmonic* (*q.v.*) 1–36 for the vector field (Lowe *et al.*, 2001), with additional resolution to 65 for the scalar field (Sabaka *et al.*, 2004). Magsat also contributed significantly to the understanding of the ionospheric magnetic field, and the meridional current systems through which the satellite flew. The most demanding aspect of making vector magnetic measurements from space is in precise alignment, which must be at the arc-second level. Magnetic field measurements have been made from near-Earth space since 1958, when Sputnik 3 (USSR) collected fluxgate magnetometer measurements of the field with accuracies of about 100 nT. Magsat's measurements of the near-Earth vector field were not improved upon, or even repeated, until the launchings of *Ørsted* (*q.v.*) (1999) and *CHAMP* (*q.v.*) (2000), led by Denmark and Germany, respectively. Follow-on magnetic field missions by NASA focused on planetary magnetic fields (e.g., Mercury, Venus, Mars, and the outer planets), on the Earth's magnetosphere, and the dynamics of the high-latitude ionosphere.

The Magsat project scientist was *Robert A. Langel* (*q.v.*), the project manager was Gilbert Ousley, and the spacecraft design, development, and testing was done by the Applied Physics Laboratory (APL) of John Hopkins University (Potemra *et al.*, 1981). The vector magnetometer was designed and built at Goddard Space Flight Center's Laboratory for Extraterrestrial Physics by Mario Acuña (instrument scientist) and his group. Scientists and scientific organizations in the United States, Europe, and Japan formed the largest group of users of Magsat observations. Magsat weighed 182 kg, and cost 19.2 million US dollars from inception through production of the finished data products. Magsat was modeled on the Small Astronomical Satellite (SAS-3), built by APL in 1975. SAS-3 utilized a Doppler tracking system for position determination to within 60 m vertically and 200 m horizontally, and had two star trackers that could provide attitude information to 10″. However, the magnetic fields associated with these star trackers and the spacecraft itself would have introduced unacceptably large errors into the magnetic field determinations. Therefore, a deployable scissors boom and an infrared attitude transfer system (ATS) were developed for Magsat that put the scalar and vector magnetometers six meters away from the star trackers, and the body of the spacecraft (Figure M172). The deployable boom was designed, developed, and tested by APL, and the ATS was based on one used for submarines and adapted for Magsat by the Barnes Engineering Company. The sun-synchronous orbit of Magsat, resulting in a sampling of the magnetic field only at dawn and dusk local time, was a compromise dictated by the carrying capacity of the Scout launch vehicle. The chosen orbit, and mission lifetime (October 1979 through June 1980) allowed for a maximum exposure of the fixed solar array to the sun, a near-constant thermal environment, and a fixed flight attitude that allowed the star trackers to always face away from the earth.

The orientation of the vector magnetometer was determined, at 0.25 s intervals, using the two star cameras. If only one star camera was tracking an identified star, a precision sun sensor immediately adjacent to the vector magnetometer was used instead. Jumps in the vector data of 10–15 nT can occur at locations where the method of determining the orientation of the vector magnetometer changes. The system for transferring the attitude determined by the star camera on the spacecraft body to the vector magnetometer on the end of the boom involved two optical systems, one for pitch and yaw (utilizing a plane mirror), and a second for roll (using a dihedral mirror). As predicted, the system for measuring roll had the largest errors, estimated at about eight times the pitch and yaw error.

Magsat carried both scalar and vector magnetometers. The scalar magnetometer, a two-sensor, cesium 133 vapor optically pumped magnetometer, was used to calibrate the vector magnetometer. The two-sensor arrangement was chosen to minimize null zones, where the magnetometers are incapable of sensing the ambient field. In practice, intermittent noise in the cesium lamp excitation circuits caused the tracking filters to lose lock, and reduced the amount of data returned to well below the nominal 8 Hz rate. System errors in the determination of the scalar field were 0.5–1.5 nT. The vector instrument was a triaxial fluxgate magnetometer with a dynamic range of ±2000 nT, and digitally controlled current sources to increase its range to ±64000 nT. The vector instrument sampled the ambient field 16 times per second with a resolution of ±0.5 nT. Each of the three orthogonal sensors was wound with platinum wire on a highly permeable toroidal magnetic core. The sensors were mounted on a stable ceramic block and temperature controlled to minimize alignment shifts with respect to the reference mirrors of the ATS.

Each observation has associated with it an attitude uncertainty, and details on the sensors that were used to calculate the attitude (Langel

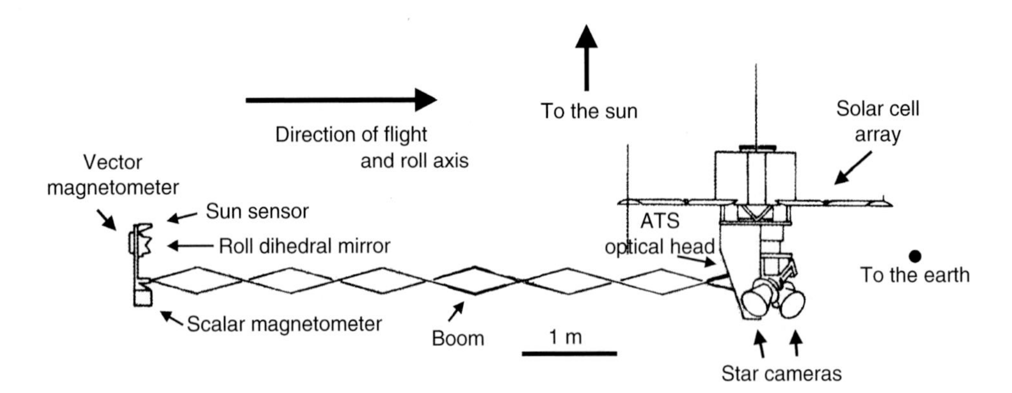

Figure M172 Line drawing of Magsat, its orientation, and critical systems. The roll axis is along the orbital velocity vector, the yaw axis is vertical, and the pitch axis completes the orthogonal coordinate system.

et al., 1981). The mission requirements specified that the three axes of the vector magnetometer be determined within 60″, although in practice it was often known within 20″, corresponding to an attitude error of 5 nT. A detailed error budget can be found in Langel and Hinze (1998) in Table 3.4. The observations can be accessed (as daily files) at the FTP server (ftp.spacecenter.dk/data/magnetic-satellites/Magsat) of the Danish National Space Center.

Michael E. Purucker

Bibliography

Langel, R.A., Berbert, J., Jennings, T., and Horner, R., 1981. *Magsat Data Processing: A report for investigators*, NASA Technical Memorandum 82160.

Langel, R.A., and Hinze, W.J., 1998. *The Magnetic Field of the Earth's Lithosphere: The Satellite Perspective.* Cambridge: Cambridge University Press.

Lowe, D.A.J., Parker, R.L., Purucker, M.E., and Constable, C.G., 2001. Estimating the crustal power spectrum from vector Magsat data. *Journal of Geophysical Research*, **106**: 8589–8598.

Potemra, T.A. *et al.*, 1980. *John Hopkins APL Technical Digest.* The Magsat issue, July-September, 1980. 12 articles, pp. 162–232.

Sabaka, T.J., Olsen, N., and Purucker, M., 2004. Extending comprehensive models of the Earth's magnetic field with Oersted and CHAMP data. *Geophysical Journal International*, **159**: 521–547.

Cross-references

CHAMP
Harmonics, Spherical
Langel, Robert A. (1937–2000)
Laplace's Equation, Uniqueness of Solutions
Ørsted
POGO (OGO-2, -4, and -6 Spacecraft)

MAIN FIELD MAPS

Historically, in navigating long distances, the compass was as important an instrument for indicating direction as the sandglass was for marking time. However, until the 17th century, everyone believed that magnetic north coincided with true north, and mariners were unaware of this variation (see *Geomagnetism, history of*); consequently, as they sailed west they found their position did not correspond with their location according to charts. Having learned that true north differed from magnetic north, instrument makers in some of the northern countries produced compasses in which the compass card was mounted on the magnetic needle in alignment with the amount of magnetic variation. Thus, while the needle pointed to magnetic north, the *fleur-de-lis* on the compass card indicated true north. By this arrangement the compass had a built-in correction for the magnetic variation. This was fine, as long as voyages were limited to regions where the amount of variation did not appreciably change.

The earliest magnetic charts—before 1700

As the geomagnetic poles lie at some distance from the geographical poles, the deviation of the compass needle (known as the "magnetic variation" by mariners and "magnetic declination" by geophysicists) can vary considerably over the Earth's surface and can be directed either to the East (positive) or to the West (negative).

What led people to include magnetic information on charts? The first practical reason was navigation, and the earliest map that included magnetic information was a road map, around 1500, indicating the pilgrim's route from Germany to Rome (*Das ist der Rom Weg*). On this map, Erhard Etzlaub represented a pictorial pocket sundial (known at that time as a compass), on which he clearly showed an easterly declination (Hellmann, 1909).

During the 16th century, the number of voyages increased, and nautical charts were being drawn. In 1576, William Borough produced a chart for Martin Frobisher's first voyage in search of a North-West passage. On this manuscript, numerical values for declination were indicated for several locations situated along the likely trip (Barraclough, 2000). A global map indicating magnetic poles, magnetic meridians, and the magnetic equator was published by Guillaume de Nautonnier (1602–1604). In Edward Wright's book (1610), a chart with values for declination is included. Between 1625 and 1634, Jean Guérard, a French pilot and hydrographer, produced seven world maps; one of them includes declination values, marked in small circles at the places of measurements (Ultré-Guérard and Mandea, 2000).

If knowledge of the declination was acquired during the 15th century, the variation of declination with geographical position became more widely known around the end of 15th century. One very interesting application was to use the declination to find the longitude at sea. A sailor can find the latitude by the length of day or by the height of the Sun or other known star above the horizon. To measure longitude one needs to know, at the same moment, the time at the departure place, and the time aboard ship. Then, measuring longitude means measuring the time. However, until John Harrison's (1760) very precise ship-chronometer, it was not obvious how to measure time with enough accuracy. One proposed method was to use the magnetic compass, and thus declination, to find the longitude. According to Hellman (1909) Columbus was the first to have the idea, but, this has since been discredited (Mitchell, 1937). Mitchell indicated that in 1544, Sebastian Cabot (Pilot Major of Spain) produced a planisphere on which the contour of zero declination was indicated. Another attempt to solve the longitude problem using declination was made by Guillaume de Nautonnier (1602–1604); for more details about Nautonnier's map and work see Mandea (2000). In another book, published in 1609 by Antony Linton (strongly influenced by Nautonnier), it was suggested that the declination resulting from a model with different positions for the magnetic and geographic poles could be represented on a sphere by "*helique* and *tortuos* lines".

The earliest magnetic charts—1700 onward

Feeling the need for more accurate magnetic charts of the Atlantic Ocean, the lordships of the British Admiralty lent Edmund Halley a small sailing ship, *Paramore*, and instructed him to carry out a magnetic survey of the Atlantic Ocean and its bordering lands. From Halley's survey measurements the first magnetic map of the Atlantic was compiled (Figure M173/Plate 1). This earliest surveying, magnetic contour

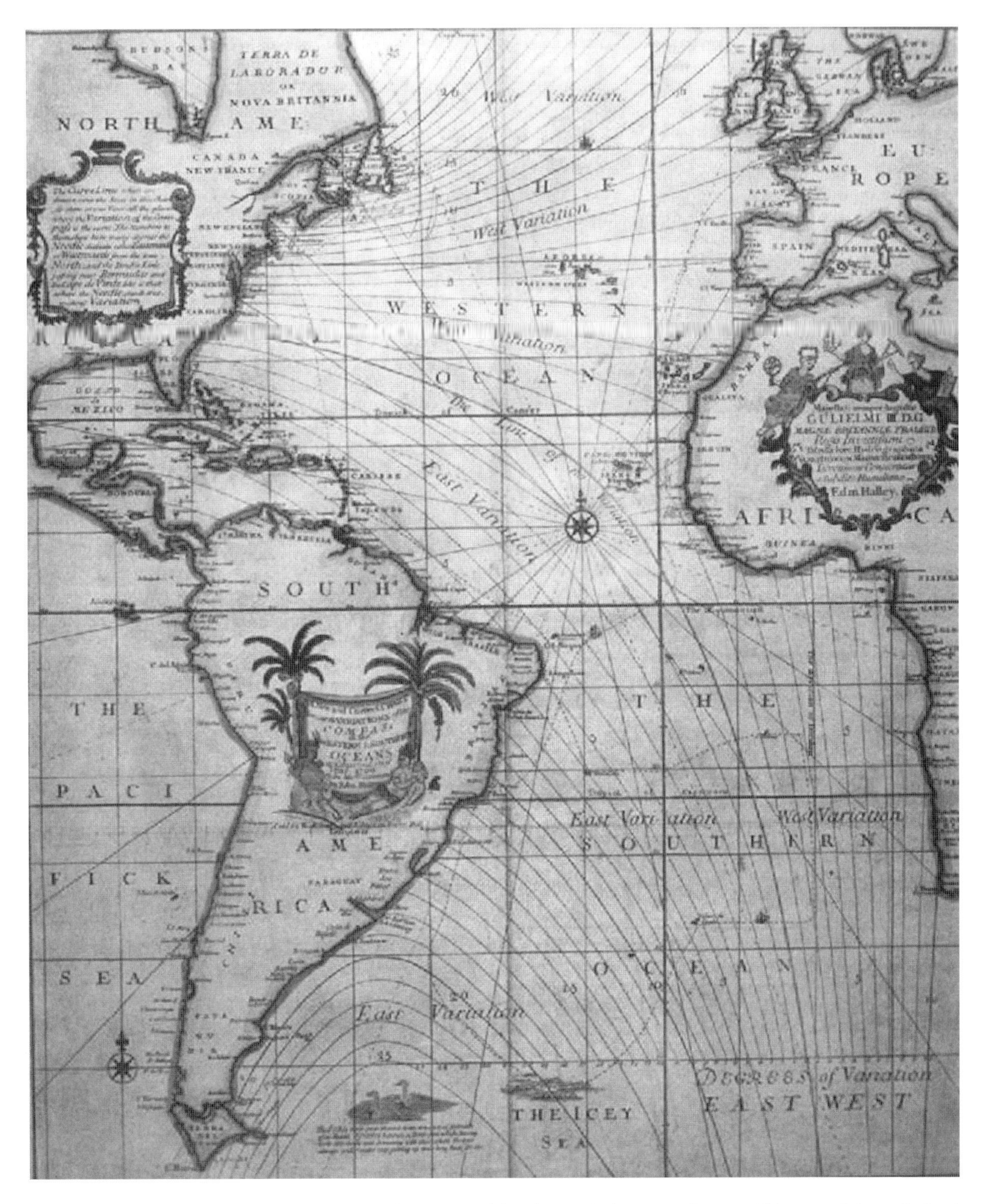

Figure M173/Plate 1 Halley's Atlantic chart. This is the version of the chart with the cartouche over Africa containing a dedication to King William III.

chart was published by Halley in 1700, and was entitled "*A New and Correct CHART Shewing the Variations of the Compass in the Western and Southern Oceans as Observed in ye Year 1700 by his Maties Command* by Edmund Halley." This map was the first published chart to include isolines of declinations (called for a while "Halleyan lines," until Hanssen used the term "isogone" for a contour line of equal declination). This map became widely used as it allowed navigators to estimate declination variation when they could not get sight of an astronomical body (see Clark (2000) for more details).

William Whiston produced the first known charts showing the direction of magnetic inclination for south-eastern England, in 1719 and 1720. The contours are straight lines and are thus a very smooth representation of the distribution of inclination (see Howarth (2003) for more details). In 1768, Wilcke published the first approximately global inclination map, based on observations made during sea voyages.

The first chart of magnetic field intensity, covering the Northern part of South America, was produced by Alexander von Humboldt (Humboldt and Biot, 1804). On this map the geographic and magnetic equators are indicated, and five "isodynamic zones" are defined by the average number of oscillations per minute of the magnetized needle used during Humboldt's expeditions (Malin and Barraclough, 1991). To quote Hellman (1909), the first "true" magnetic field intensity map was produced by Hansteen (1824) for a region of the Northern part of Europe. The intensity values shown on this chart are in relative units ("Humboldt units"), as it was produced before Gauss' demonstration on how to measure the magnetic field in absolute units.

Official magnetic maps

The naval authorities from different countries have published world magnetic charts for different epochs. The first official chart was of declination for the epoch 1858, produced by F.J. Evans, under the superintendence of the hydrographer of the Royal Navy. The first German chart was published in 1880, with the United States following in 1882. Other countries, such as France and Spain, have published copies of the British charts with titles and text translated into the appropriate language. Hellmann (1909) gives details of the earlier examples of these charts.

Present-day maps are produced from different geomagnetic field models for the main field and its secular variation (see *Main field modeling*; *Geomagnetic secular variation*). Geomagnetic models such as the World Magnetic Model (WMM) and the International Geomagnetic Reference Field (IGRF) only predict the values of that portion of the field originating in the fluid outer core. In this respect, they are accurate to within 1° for 5 years into the future, after which they need to be updated. Maps produced from these models are used widely in civilian navigation systems.

World Magnetic Model (WMM)

The WMM is a product of the United States National Geospatial-Intelligence Agency (NGA). The US National Geophysical Data Center (NGDC) and the British Geological Survey (BGS) produce the WMM as the standard model for navigation and attitude/heading referencing systems for the US Department of Defense, the UK Ministry of Defence, the North Atlantic Treaty Organization (NATO), and the World Hydrographic Office (WHO). The model, associated software, and documentation are distributed by NGDC on behalf of NGA and used widely in civilian navigation systems. The model is produced at 5-year intervals, with the current model, WMM-2005, expiring on December 31, 2009 (McLean *et al.*, 2004).

International Geomagnetic Reference Field (IGRF)

Production of the IGRF is an international collaborative effort relying on cooperation between magnetic field modelers, institutes, and agencies responsible for collecting and publishing geomagnetic field data (see *IGRF, International Geomagnetic Reference Field*). The first model, designated IGRF 1965, was of spherical harmonic degree and order 8, in both main field and secular variation (Langel, 1987). The latest version of the model—the 10th generation—includes models of the core field at 5-year intervals from 1900 to 2005, and a secular variation model for 2005–2010. Magnetic charts based on different generations of IGRF are available on several Web sites, including http://swdcwww.kugi.kyoto-u.ac.jp/igrf/; http://www.geomag.bgs.ac.uk/navigation.html.

Maps accuracy

The WMM and IGRF/DGRF geomagnetic models, and the charts produced from these models characterize only that portion of the Earth's magnetic field that is generated in the Earth's fluid outer core (the main magnetic field). The portions of the geomagnetic field generated by the Earth's upper mantle and crust, and by the ionosphere and magnetosphere, are not represented (see *Internal external field separation*). Consequently, a magnetic sensor such as a compass or magnetometer may observe spatial and temporal magnetic anomalies when referenced to these models. In particular, certain local, regional, and temporal magnetic declination anomalies can exceed 10°. Anomalies of this magnitude are not common but they do exist. Local anomalies distort the WMM or IGRF predictions, and frequently induce an error of 3–4° (ferromagnetic ore deposits; geological features, particularly of volcanic origin, such as faults and lava beds; topographical features such as ridges, trenches, seamounts, and mountains; ground that has been hit by lightning and possibly harboring fulgurites; cultural features such as power lines, pipes, rails, and buildings; personal items such as crampons, ice axe, stove, steel watch, hematite ring, or even a belt buckle).

Main field and secular variation maps at the Earth's surface

To create an accurate main field model, data with good global coverage and as low a noise level as possible are needed. The Danish Ørsted and German CHAMP satellite data sets satisfy these requirements. Both satellites provide high-quality vector and scalar data at all latitudes and longitudes, but not during all time periods needed for modeling. These satellite data were augmented with ground observatory data, which have been available almost continuously over the period of interest, although with poorer spatial coverage. Used together, satellite and observatory data provide an exceptional quality data set for modeling the behavior of the main magnetic field in space and time, and for producing main field magnetic maps for a given epoch. Such maps at the Earth's surface field were drawn using a main field model based on data from the Ørsted magnetic field satellite (Neubert *et al.*, 2001). For the main field and its secular variation representation some other models are available, derived from Ørsted, CHAMP and Ørsted-2/SAC-C measurements, such as IDEMM, CO2, OSVM and OIFM (see http://www.dsri.dk/Oersted/Field_models/ for more details).

Among the multitude of geomagnetic models based on satellite data, POMME, derived from Ørsted and CHAMP data, is used here to draw maps for three field components. This model provides a representation of the strongest contributions to the geomagnetic field: the time-varying core field, the ring current field modulated by the Dst index, a time-averaged magnetospheric field, the penetration of the horizontal part of the Interplanetary Magnetic Field (IMF), and the fields induced by Earth's rotation in the external fields (Maus *et al.*, 2004). From POMME coefficients, synthetic data for the main field components are computed for the whole Earth's surface, on a grid of $30' \times 30'$. Each set of synthetic data is then interpolated and plotted as maps, generally on global scale. In Figures M174-M176 maps of declination, inclination, and total intensity for the epoch 2002.5, and their secular variations, derived from the POMME model are shown. The isolines on the declination map (isogones) show the distribution

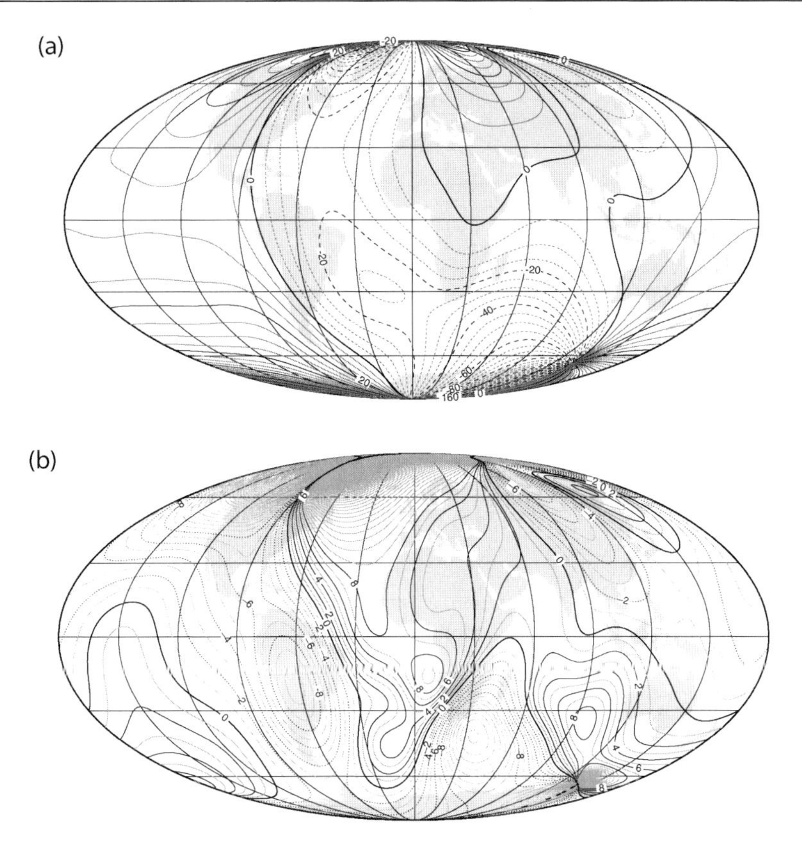

Figure M174 Contour maps of declination (degrees) at 2002.5 and secular variation of declination (arc-min y^{-1}), at the Earth's surface, computed from POMME model (Maus *et al.*, 2004).

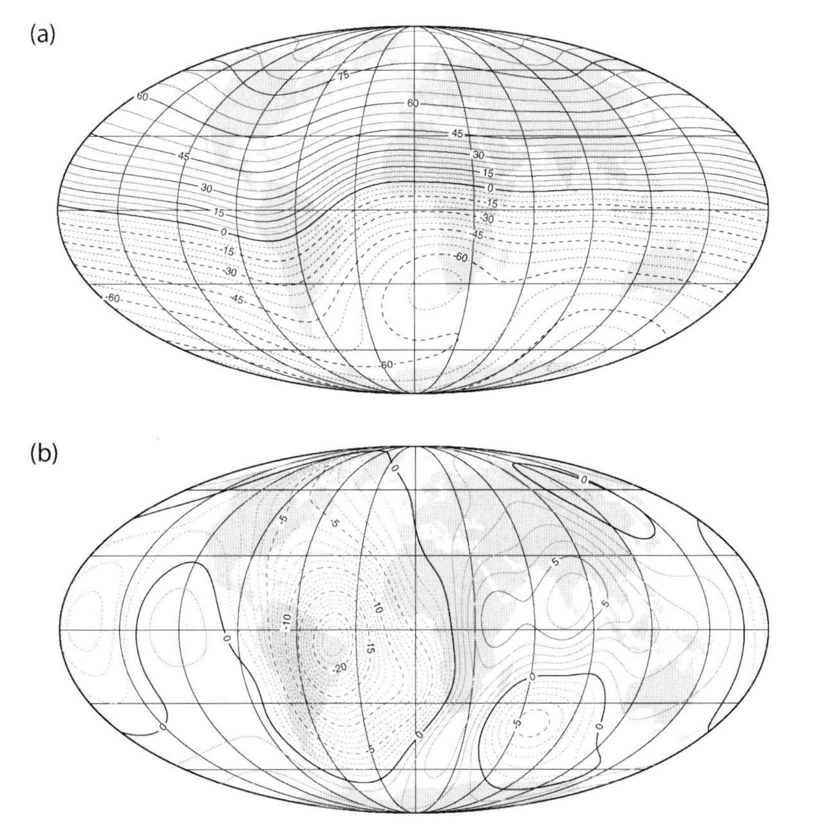

Figure M175 Contour maps of inclination (degrees) at 2002.5 and secular variation of inclination (arc-min y^{-1}), at the Earth's surface, computed from POMME model (Maus *et al.*, 2004).

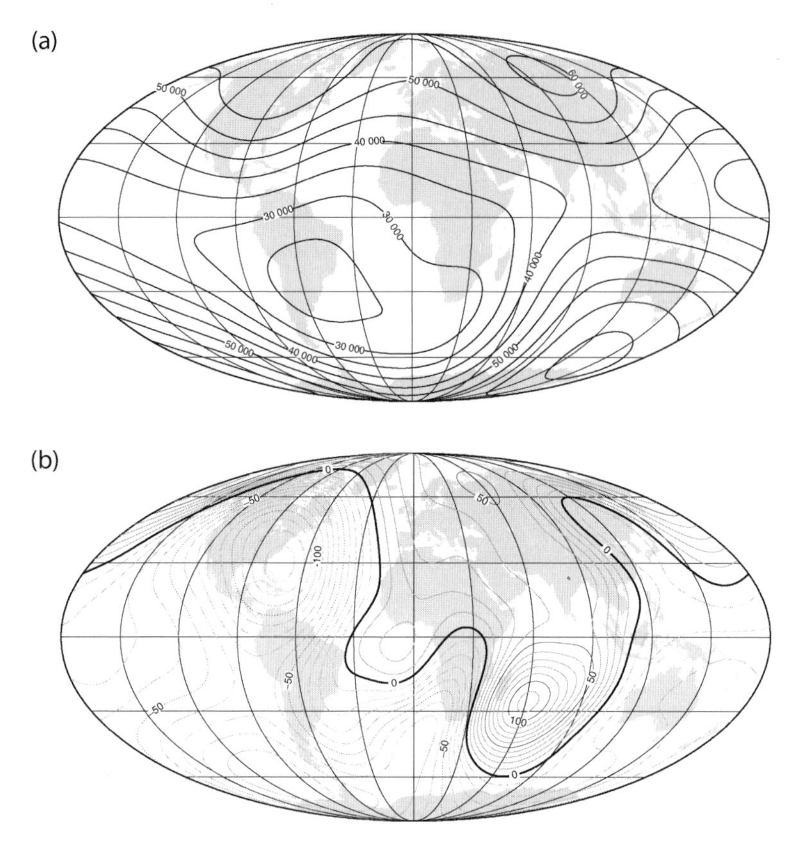

Figure M176 Contour maps of total intensity (nT) at 2002.5 and secular variation of total intensity (nT y^{-1}), at the Earth's surface, computed from POMME model (Maus *et al.*, 2004).

of the declination (or of the compass needle eastward or westward from the true north), the isolines on the inclination map (isoclines) show the distribution of the inclination (angle between the geomagnetic field vector and its horizontal projection), and the isolines on the total field intensity map (isodynamics) show the distribution of the total geomagnetic field (the scalar value of the geomagnetic field vector). The lines of equal secular variation (isopods) are used for dates outside the epoch for which the main field charts have been drawn.

Unfortunately, the same potential theory that allows us to generate these maps also tells us that, formally, little more about the origin of the field can be said: the field certainly originates within the Earth, but where in the Earth cannot be distinguished. However, by looking at the structure of the field—and making assumptions about its sources—further inferences can be made. The spectrum of the field has clearly two parts, first a rapidly declining part down to wavelengths of approximately 3000 km, with a much more gentle decline at shorter wavelengths (see *Geomagnetic spectrum, spatial*). The presented maps reflect only the first part of the spectrum corresponding to the core field. The lithospheric contributions could be included in extended models, expanded to high degrees. The crustal field is almost constant in time, which can be inferred from all available marine, aeromagnetic, and high-resolution Ørsted, CHAMP, and future satellite data, measured at all times. However, these extended models and maps would differ significantly in format and form from the current main field models and maps.

Mioara Mandea

Bibliography

Barraclough, D.R., 2000. Four hundred years in geomagnetic field charting and modelling. In Schröder, W. (ed.), *Geomagnetism. Research Past and Present.* Bremen-Roennebeck: IAGA, pp. 93–111.

Clark, T.D.G., 2000. Edmond Halley's voyages in the Paramore and the first isogonic chart of the Earth's magnetic field. In Schröder, W. (ed.), *Geomagnetism. Research Past and Present.* Bremen-Roennebeck: IAGA, pp. 61–71.

Hansteen, C., 1824. Magnetiske Intensitets-Iagttagelser, Anstillede paa forskjellige Reiser i den nordlige Deel af Europa. *Magazin for Naturvidenskaberne*, **4**: 268–316.

Hellmann, G., 1909. Magnetische Kartographie in historisch-kritischer Darstellung. *Abhandlungen des Königlichen Preussischen meteorologischen Instituts* **3**: 61.

Howarth, R.J., 2003. Fitting geomagnetic fields before the invention of least squares: II. William Whiston's isoclinic maps of Southern England (1719 and 1721). *Annals of Science*, **60**: 63–84.

Humboldt, A., von and Biot, J.-B., 1804. Sur les variations du magnétisme terrestre à différentes latitudes. *Lu par M.Biot à la clsse des sciences mathématiques et physiques de l'Institut National, le 26 frimaire an 13*, pp. 24.

Langel, R.A., 1987. Main field. In Jacobs, J.A. (ed.), *Geomagnetism.* London: Academic Press, pp. 249–512.

Mandea, M., 2000. French magnetic observation and the theory at the time of DE MAGNETE. In Schröder, W. (ed.), *Geomagnetism. Research Past and Present.* Bremen-Roennebeck: IAGA, pp. 73–80.

Maus, S., Lühr, H., Balasis, G., Rother, M., and Mandea, M., 2004. Introducing POMME, the POtsdam magnetic model of the Earth in CHAMP. In *Earth Observation with CHAMP, Results from Three Years in Orbit.* Berlin-Heidelberg: Springer, pp. 293–298.

McLean, S., Macmillan, S., Maus, S., Lesur, V., Thomson, A., and Dater, D., 2004. The US/UK World Magnetic Model for 2005–2010, *NOAA Technical Report NESDIS-NGDC-1.*

Mitchell, A.C., 1937. Chapters in the history of terrestrial magnetism. II. The discovery of the magnetic declination. *Terrestrial Magnetism and Atmospheric Electricity*, **42**: 241–280.

Nautonnier, G., 1602–1604. Mecometrie de leymant, c'est a dire La maniere de mesvrer les longitudes par le moyen de l'eymant. *Venes: ches l'autheur*, 343 pp.

Neubert, T., Mandea, M., Hulot, G., von Frese, R., Primdahl, F., Jorgenson, J.L., Friis-Christensen, E., Stauning, P., Olsen, N., and Risbo, T., 2001. High-precision geomagnetic field data from the Ørsted satellite. *EOS*, **82**: 81–87.

Ultré-Guérard P., and Mandea, M., 2000. Declination and longitude in France in the early 17th century. In Schröder, W. (ed.), *Geomagnetism. Research Past and Present.* Bremen-Roennebeck: IAGA, pp. 81–92.

Wright, E., 1610. Certaine errors in Navigation, detected and corrected, Printed by Feelix Kingstrõ. London, 354 pp.

Cross-references

Geomagnetic Secular Variation
Geomagnetic Spectrum, Spatial
Geomagnetism, History of
IGRF, International Geomagnetic Reference Field
Internal External Field Separation
Main Field Modeling

MAIN FIELD MODELING

The history of modeling the geomagnetic field stretches back four centuries, to William Gilbert's *De Magnete*, the first book on geomagnetism. His assertion, that the Earth is a great magnet, can be regarded as the first model of the geomagnetic field. The development of geomagnetic field models is explained by the need to draw magnetic maps. The practical reasons were mainly connected with navigation: the use of the magnetic compass and portable sundials that included a compass needle, the problem of determining longitude at sea, and also the construction of road maps on land. These practical applications in turn provided the motivation to seek a more fundamental understanding of the geomagnetic field.

Before modeling the geomagnetic field, some fundamental properties of the field had to be discovered, like the existence of magnetic variation or declination, the spatial variation of declination, the existence of magnetic inclination or dip, and the secular variation. The history and definitions of these fundamental discoveries are given elsewhere. Here, some different ways of modeling the geomagnetic field since William Gilbert's first attempt in 1600 are given. A milestone in main field modeling was achieved in 1839, when Gauss published in his famous description of the geomagnetic field using spherical harmonics.

Empirical models: dipoles and magnetized spheres

The first geomagnetic field model was a physical one, the terrella built out of lodestone and used by Gilbert in his experiments. He investigated the behavior of very small magnetized iron needles on the surface of such a sphere, and compared the results with what was known about the magnetic field from direct measurements made with compass needles and dip needles. Gilbert believed that the positions of the magnetic poles of the Earth were the same as those of the rotation poles. He also believed that declination, which with his assumption would have been zero everywhere, was caused by the effects of the continents, whose magnetic rocks attracted the needles toward them. In his experiments Gilbert modeled this effect by using an irregularly shaped sphere, and he concluded that "the variation in any one place is constant."

At about the same time that Gilbert wrote his famous book, a French cartographer, Nautonnier de Castelfranc came to a similar conclusion concerning the terrestrial source of the geomagnetic field (Mandea, 2000). However, Nautonnier believed that the positions of the magnetic and geographic poles were not coincident, and this property explained the phenomenon of declination. Nautonnier produced the first known chart showing magnetic poles, magnetic meridians, and the magnetic equator.

The first attempt to model inclination was published by Edward Wright in 1610, who gave an empirical expression, obtained from a geometrical construction, apparently devised to give 0° at the equator, 90° at the pole, and an acceptable value of dip for the position of London (note that his tables of inclination were not very accurate for locations other than London). Wright's model did not take into account the secular variation. This was first addressed some years later, when Henry Bond attempted to take into account the secular variation by using a tilted dipole model, with the magnetic poles moving around the geographical poles (Barraclough, 2000).

Halley's theory of the Earth's magnetic field has been lauded as a great achievement. In his first paper, Halley used observational data to examine previous theories (Barraclough, 2000). Finding the theories lacking in certain respects, Halley proposed a new model, which explained the observations qualitatively. Having presented his data, and discussed the shortcomings of previous theories, Halley turned to his own hypothesis. He suggested that the pattern of declination could be explained by four magnetic poles, roughly situated at: N1 (75°N, 129°W); N2 (83°N, 6°W); S1 (70°S, 120°E); S2 (74°S, 95°W). He had no means to quantitatively describe the effects that these poles would have, but he demonstrated that they did qualitatively explain the pattern of variation seen in his data. Nine years later he further developed these ideas by proposing a geomagnetic field model which could incorporate the secular variation (Halley, 1692). He suggested that the Earth consisted of an outer shell with poles corresponding to N1 and S1, and an inner shell with poles corresponding to N2 and S2. The inner shell rotated with respect to the outer one, and consequently these poles would shift with time and result in a secular change in declination. At least in part to provide an observational test of this model, he undertook an extensive survey (accomplished in his three voyages in the Atlantic Ocean between 1698 and 1701) to remedy the lack of data in ocean areas. Other priorities evidently prevented him from using these new data to refine his model, but he used them to produce magnetic charts, far more useful and important than his four-pole model.

In 1757, Euler introduced a new model, in which the two magnetic poles were not antipodal (what would today be called *an eccentric dipole model*). Using a method of successive approximations he found the positions of these poles for 1757. During the 19th century, Halley's four-pole model was not forgotten. Hansteen collected as many measurements of declination and inclination as he could find and used them to derive a model with four poles, including their changes in time (Barraclough, 2000). Although the concept of modeling the geomagnetic field using two magnetic axes and four magnetic poles is today only of historical interest, the data collected and published by Hansteen have proved to be very valuable in studying how the field has evolved with time (see *Geomagnetism, history of*).

In 1832, Carl Friedrich Gauss and Wilhelm Weber began investigating the theory of terrestrial magnetism after Alexander von Humboldt attempted to obtain Gauss's assistance in making magnetic observation on a grid of points around the Earth. Gauss was excited by this prospect and by 1840 he had written three important papers on the subject, all of which dealt with the current theories on terrestrial magnetism, including Poisson's equation, the potential theory equation (in 1812 Poisson discovered that Laplace's equation is valid only outside of a solid), and the absolute measure for magnetic force. In *Allgemeine Theorie des Erdmagnetismus* (1839) Gauss showed that there can only be two poles in the globe and used the Laplace equation to aid him with his magnetic field calculations (see *Gauss, Carl Friedrich*).

Mathematical models: spherical harmonic analysis

The Earth's magnetic field is continuously changing, in such a way that it is impossible to predict accurately what the field will be at any point in the very distant future (for example, declination is accurately described by geomagnetic models to within 1° for 5 years). By measuring the magnetic field, its spatial variation and changes over a period of years, can be observed. This structure and variation can most conveniently be expressed in terms of spherical harmonics (see *Harmonics, spherical*). A review of the technique's development can be also found in Langel (1987) or Merrill *et al.* (1996).

The two Maxwell equations relating to the magnetic field are:

$$\nabla \times \vec{H} = \vec{J} + \frac{\partial \vec{D}}{\partial t} \quad \text{(Eq. 1)}$$

$$\nabla \cdot \vec{B} = 0 \quad \text{(Eq. 2)}$$

where $\vec{H}$ is the magnetic field, $\vec{B}$ is the magnetic induction, $\vec{J}$ is the current density, and $\partial\vec{D}/\partial t$ is the electric displacement current density. In a region that does not contain sources of magnetic field (from the Earth's surface up to about 50 km), it is reasonable to assume that $\vec{J} = 0$ and $\partial\vec{D}/\partial t = 0$, so $\nabla \times \vec{H} = 0$, meaning that the vector field is conservative in the region of interest, and the magnetic field $\vec{H}$ can be expressed as $\vec{H} = -\vec{\nabla} V$, where V is a scalar potential. Because $\vec{B} = \mu_0 \vec{H}$ above the Earth's surface (where $\mu_0 = 4\pi \times 10^{-7}\ \mathrm{H\,m^{-1}}$), it follows that $\nabla \cdot \vec{H} = 0$ and that V has to satisfy Laplace's equation:

$$\nabla^2 V = 0 \quad \text{(Eq. 3)}$$

In spherical coordinates (r, θ, ϕ), which are the radial distance, the geocentric colatitude, and the longitude, Laplace's equation takes the form:

$$\frac{1}{r}\frac{\partial^2}{\partial r^2}(rV) + \frac{1}{r^2 \sin\theta}\frac{\partial}{\partial\theta}\left(\sin\theta\frac{\partial V}{\partial\theta}\right) + \frac{1}{r^2\sin^2\theta}\frac{\partial^2 V}{\partial\phi^2} = 0 \quad \text{(Eq. 4)}$$

Equation (4) has, as a solution, a product of three expressions (the first is to be only a function of r, the second only a function of θ, and the third only a function of ϕ) and can be solved by separation of variables. A completely general solution is provided by spherical harmonics, and can be written as:

$$V(r,\theta,\phi) = a\sum_{n=1}^{\infty}\sum_{m=0}^{n}\left[\left(A_{n,m}\cos m\phi + B_{n,m}\sin m\phi\right)\left(\frac{a}{r}\right)^{n+1} + \left(C_{n,m}\cos m\phi + D_{n,m}\sin m\phi\right)\left(\frac{r}{a}\right)^{n}\right]P_{n,m}(\cos\theta) \quad \text{(Eq. 5)}$$

where $A_{n,m}$, $B_{n,m}$, $C_{n,m}$, $D_{n,m}$ are called spherical harmonic coefficients, and $P_{n,m}(\cos\theta)$ are the associated Legendre polynomials. In geomagnetism the partially normalized Schmidt functions $P_n^m(\cos\theta)$ are used.

The $P_n^m(\cos\theta)$ are functions of colatitude only. They are quasisinusoidal oscillations having $n - m + 1$ nodes. Figure M177 illustrates the variations of some of these functions.

The surface harmonics $P_n^m(\cos\theta)\sin m\phi$ and $P_n^m(\cos\theta)\cos m\phi$ divide the surface of the sphere into regions defined by the intersections of latitudinal zones and longitudinal sectors. The three situations are: (i) when $m = 0$, the surface harmonics are described by the Legendre polynomials and they are referred to as *zonal harmonics*; (ii) when $n = m$, the surface harmonics are referred to as *sectorial harmonics*, and (iii) when $0 < m < n$, the surface is divided in $2m(n - m + 1)$ regions and the surface harmonics are referred to as *tesseral harmonics*. Some examples are given in Figure M178.

Application of spherical harmonics to the Earth's magnetic field involves writing the magnetic scalar potential as the sum of two contributions:

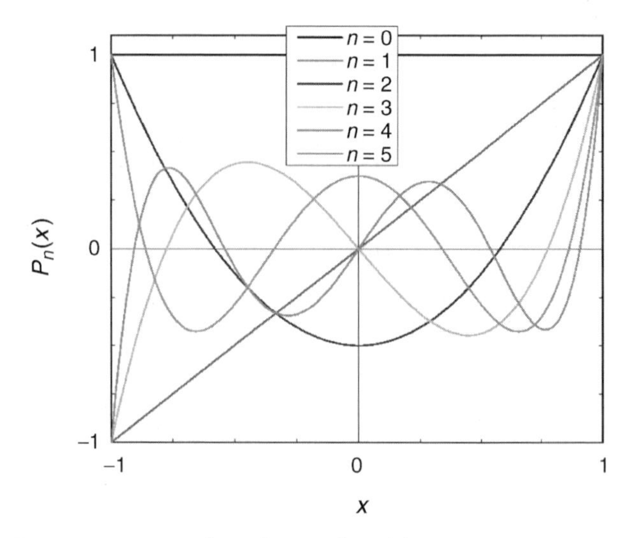

Figure M177 Legendre polynomials $P_n(x)$ up to $n = 5$.

$$V = V_{\text{int}} + V_{\text{ext}} \quad \text{(Eq. 6)}$$

where V_{int} and V_{ext} are the internal and external scalar potential, respectively (see *Internal external field separation*). These two potentials can be represented by spherical harmonic expansions:

$$V_{\text{int}} = a\sum_{n=1}^{N_i^{\max}}\left(\frac{a}{r}\right)^{n+1}\sum_{m=0}^{n}\left(g_n^m\cos(m\phi) + h_n^m\sin(m\phi)\right)P_n^m(\cos(\theta)) \quad \text{(Eq. 7)}$$

$$V_{\text{ext}} = a\sum_{n=1}^{N_e^{\max}}\left(\frac{r}{a}\right)^{n}\sum_{m=0}^{n}\left(q_n^m\cos(m\phi) + s_n^m\sin(m\phi)\right)P_n^m(\cos(\theta)) \quad \text{(Eq. 8)}$$

where the reference radius is $a = 6\,371.2$ km. Measurements of the magnetic field are used to estimate the so-called Gauss spherical harmonic coefficients ((g_n^m, h_n^m) for internal sources, and (q_n^m, s_n^m) for external sources), which in principle uniquely describe the geomagnetic field outside source regions. The summations in Eqs. (7) and (8) are truncated at maximum degree $N_i^{\max}$ and $N_e^{\max}$, respectively.

In practice $\vec{B}$ is not measured everywhere on the Earth's surface. The coefficients (g_n^m, h_n^m) and (q_n^m, s_n^m) are obtained by a least-squares fit to the data. Gauss originated this method, and with the small number of observatory measurements he had at that time, he took $N_i^{\max} = 4$ and $N_e^{\max} = 0$. Even today, there are fewer than 200 magnetic observatories, and they are very unevenly distributed. A problem for main field modeling is that magnetic observatories, as well as being poorly distributed over the globe, may be situated in areas with large local crustal magnetic anomalies. In contrast, low-altitude satellites in near-polar orbits rapidly provide excellent global data coverage and are much less affected by crustal magnetic fields, which are considerably attenuated at satellite altitude. A problem with magnetic satellite data is that the full vector measurements need very accurate orientation of the magnetometer. Originally it was hoped (in the POGO and OGO missions) that one could measure the field strength $|\vec{B}|$, which is much easier, and a very high density of measurements would then amount to knowing the vector field. Unfortunately, it was shown that even if $|\vec{B}|$ is known exactly everywhere, a unique field model cannot be obtained (Khokhlov *et al.*, 1999). There exist pairs of magnetic fields $\vec{B}_a$ and $\vec{B}_b$ satisfying $\nabla \times \vec{H} = 0$ and $\nabla \cdot \vec{H} = 0$, outside the considered sphere and such that $|\vec{B}_a| = |\vec{B}_b|$ everywhere on the sphere, but $\vec{B}_a \neq \vec{B}_b$. These two

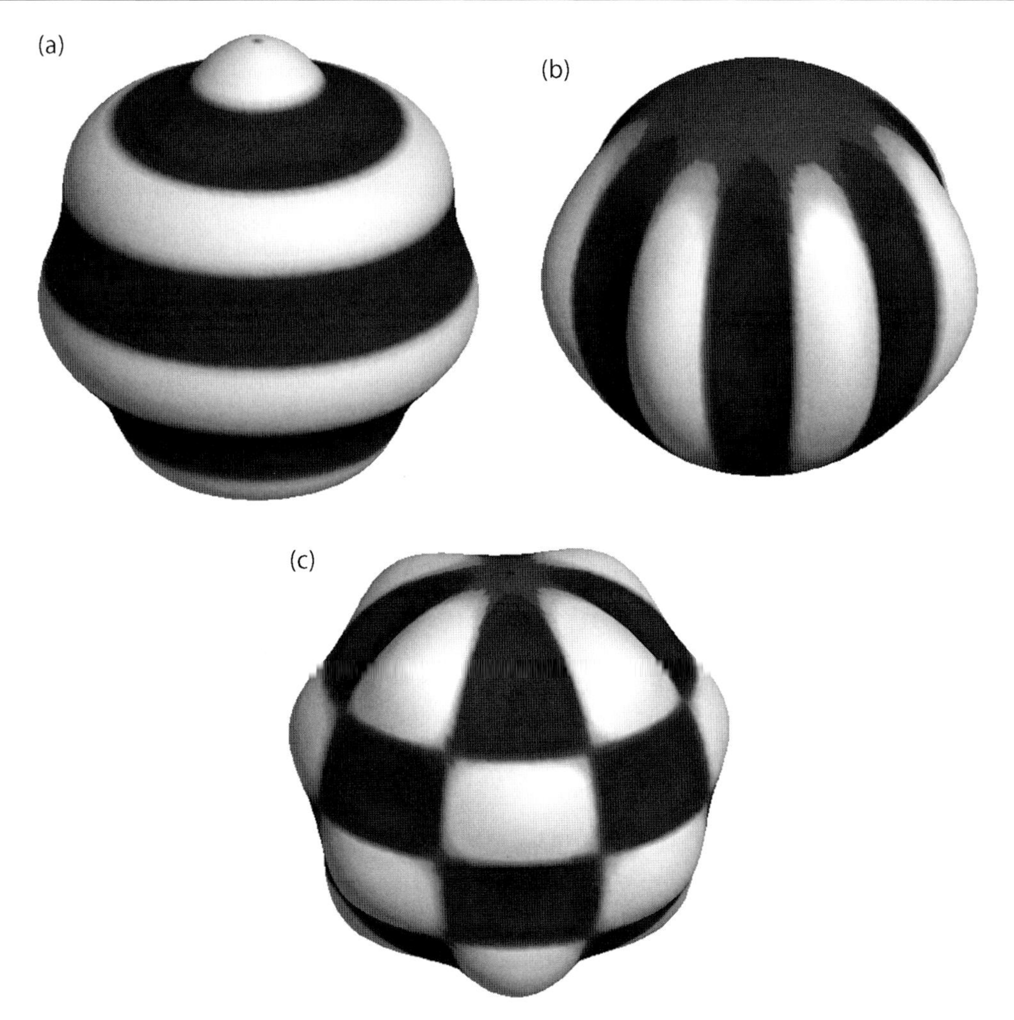

Figure M178 Surface harmonics showing three cases, with $n = 7$ and (a) $m = 0$ (*zonal harmonics*); (b) $n = m$ (*sectorial harmonics*), and (c) for $0 < m < n$, when the surface is divided in $2m(n - m + 1)$ regions (*tesseral harmonics*), with $m = 4$ in this example.

fields would give exactly the same values of total intensity, but two different models. The details of nonuniqueness of fields based on total field observations are given by Backus (1970). Empirically it has been observed that fitting MAGSAT intensity data produces models whose $\vec{B}$ can differ from the vector field components measured by the satellite by some *thousands* of nanoteslas near the equator.

Spherical harmonic analysis is an appropriate way of modeling the geomagnetic field. Considering only the part of the geomagnetic field whose sources lie beneath the Earth's surface, Eq. (7) is used. However, this equation describes both the main field (with sources in the core) and the crustal field (with sources in the Earth's crust).

The Gauss coefficients can be interpreted in terms of "sources" at the Earth's center. The first term, g_1^0, is associated with the geocentric dipole oriented along the vertical axis, with dipole moment $g_1^0 4\pi a^3/\mu_0$ (see *Geomagnetic dipole field*). The next two terms, g_1^1, h_1^1, correspond to geocentric dipoles oriented along the two horizontal axes. The magnitude of the *geocentric dipole* is given by $M = (4\pi a^3/\mu_0)((g_1^0)^2 + (g_1^1)^2 + (h_1^1)^2)^{1/2}$, and it is tilted at roughly 11° from the rotation axis. The potential for a dipole falls off as r^{-2}, and the strength of the field components as r^{-3}. Similarly, for $n = 2$ the terms represent the *geocentric quadrupole* (potential falls off as r^{-3}, and the strength as r^{-4}), for $n = 3$ the terms represent the *geocentric octupole* (potential falls off as r^{-4}, and the strength as r^{-5}), and so forth.

The first three Gauss coefficients (g_0^1, g_1^1, h_1^1) represent the *dipole* field, while the remaining terms represent the *nondipole field*. The lowest degree terms correspond to the largest wavelength features of the field, as can be appreciated by considering the zonal harmonics. The first Gauss coefficient has a $\cos\theta$ dependence, so at distance r the "wavelength" of the associated magnetic feature is $2\pi r$. Generally speaking the term g_n^m has a wavelength of $2\pi r/n$.

The internal field at the Earth's surface contains clearly defined components from the core at least up to degree $N_i = 13$, beyond which they begin to become dominated by those from the crust. The maximum associated wavelengths for the crustal components are of the order of a few thousand kilometers. The relative importance of the two internal contributions is clearly illustrated by using the normalized power in spherical harmonics of the internal geomagnetic field. The spectrum of a model is constructed by plotting the contribution of each degree n of the spherical harmonic expansion to the mean square field over the Earth's surface (see *Geomagnetic spectrum, spatial*). The power at degree n is computed from the equation (Lowes, 1974):

$$W_n = (n+1)\sum_m \left((g_n^m)^2 + (h_n^m)^2\right). \quad \text{(Eq. 9)}$$

High-quality data sets, such as those provided by the magnetic satellite surveys MAGSAT, Ørsted, and CHAMP (Langel and Estes, 1985,

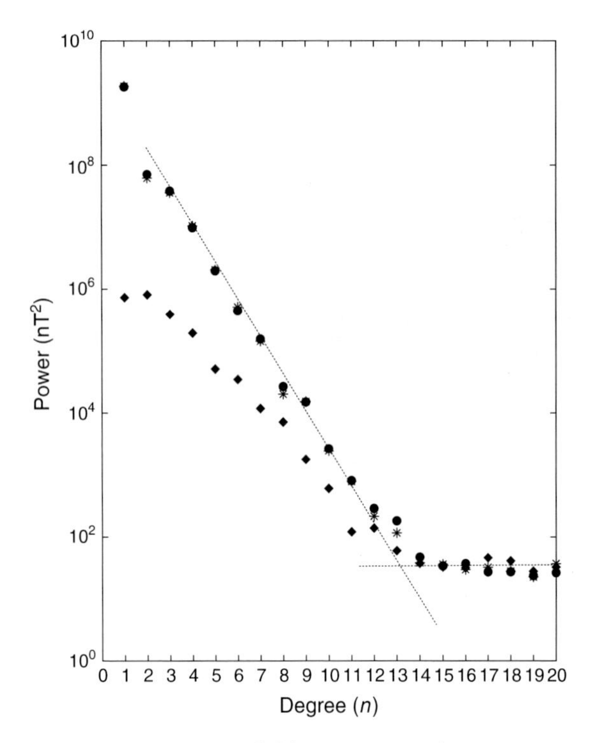

Figure M179 Geomagnetic field spectra up to degree $n = 20$ at the Earth's surface, from two internal models, computed from MAGSAT (stars) and Ørsted (filled circle) data. The difference between these two spectra is also shown (filled diamond).

Neubert *et al.*, 2001, Reigber *et al.*, 2002) have allowed the computation of spherical harmonic models up to high degree and order. Figure M179 shows the power spectra from models obtained from MAGSAT and Ørsted data, and also the difference between them (Langlais *et al.*, 2003). The dipole term obviously stands alone, the main field dominates the signal for degrees less than 13, while the crustal field dominates for degrees larger than 15.

The Gauss coefficients are also time-dependent. Indeed, the temporal variations of the geomagnetic field have an extremely wide spectrum, ranging over more than 20 orders of magnitude (Langel, 1987). In order to resolve time variations of internal origin, the secular variation, about 1 year of continuous observations is needed. If the secular variation is assumed to be constant over short timescales, it can be included in Eq. (7) by adding a secular-variation potential, V_{sv}, truncated to N_{sv}^{max}:

$$V_{sv} = a\sum_{n=1}^{N_{sv}^{max}}\left(\frac{a}{r}\right)^{n+1}\sum_{m=0}^{n}(t-T_0)\left(\dot{g}_n^m\cos(m\phi)+\dot{h}_n^m\sin(m\phi)\right)P_n^m(\cos(\theta)) \quad \text{(Eq. 10)}$$

where $\dot{g}_n^m$ and $\dot{h}_n^m$ are the time derivatives of the internal Gauss coefficients. T_0 denotes the reference time (i.e., the epoch of the main field model), and t is the considered epoch.

Since the geomagnetic field changes in space and time, magnetic observations must be made continually and models have to be generated to accurately represent the magnetic field. For example, the International Geomagnetic Reference Field (IGRF) models are adopted every 5 years (see *IGRF, International Geomagnetic Reference Field*) by the International Association of Geomagnetism and Aeronomy (Mandea and Macmillan, 2000). These models are simply truncated series (up to degree and order 10 or, more recently, 13). An extreme cutoff like this is, of course, unrealistic, and downward continuation of the models to the core-mantle boundary is unstable. Other methods, including stochastic inversion, have been used to produce reasonable models of the field at the core-mantle boundary (see *Time-dependent models of the main magnetic field*).

Mathematical models: other methods

The mathematical properties of the spherical harmonic representation make it a convenient method for describing the magnetic field. However, this representation does not have direct physical significance; some important limitations are linked to the precision in observations, their distribution and to the use of truncated spherical harmonic expansions. Indeed, the truncation of the spherical harmonic series followed by straightforward least squares inversion leads to solutions that are strongly dependent on the chosen truncation level. Changing the truncation level modifies all the coefficients, due to spatial aliasing of the higher-order harmonics (Whaler, 1986). Some alternative modeling methods exist, such as harmonic spline modeling (Shure *et al.*, 1982). This approach produces models smoothed according to some objective criteria (e.g., minimizing the complexity of some well-defined parameters of the field, such as the energy outside the core, the mean-square radial field component at the core-mantle boundary, the surface gradient of the radial field, or imposing a lower bound on Ohmic losses in the core) which are nevertheless, consistent with the data. Short-wavelength fields originating in both the core and the crust are suppressed. More details about this method can be found elsewhere (e.g., Parker and Shure, 1982; Shure *et al.*, 1982; Gubbins, 1983).

To represent both the global and the regional pattern of the magnetic field and to overcome the limitations of spherical harmonic analysis, a new method to model the Earth's magnetic field is desirable. Wavelet techniques may play a role in the solution of this challenging problem. Recently, a new representation of the magnetic field on the sphere has been sought, the chosen approach making a direct relation between spherical harmonics and wavelets. In Holschneider *et al.* (2003) the main result is a theoretical description of the wavelets on the sphere in order to use them in field modeling. The first comparisons between the spherical harmonic and wavelet bases show that wavelets are able to reproduce the spherical harmonics well.

Any kind of geomagnetic field modeling can be described as seeking a solution to optimize the cost function "*error + smoothness*". In general, the smoothness can be associated with some (physically plausible) norms. It is in general implemented through some *a priori* damping of the basis functions in the inversion. As it stands, it does not depend on the family of basis functions in which the actual numerical computations are carried out. Therefore, in the limit of infinitely many functions and without any numerical errors, it is possible to get the same models, whether spherical harmonics, wavelets, or any other complete family of functions are used. However, given finite resources it does matter, since for a fixed number of basis functions, wavelets have better approximation properties than spherical harmonics. Moreover, in modeling strongly heterogeneous fields, such as those due to external currents, wavelets may play an additional role, since they allow a more geometric description of the fields.

Mioara Mandea

Bibliography

Backus, G.E., 1970. Nonuniqueness of the external geomagnetic field determined by surface intensity measurements. *Journal of Geophysical Research Space Physics*, **75**: 6339–6341.

Barraclough, D.R., 2000. Four hundred years in geomagnetic field charting and modelling. In Schröder, W. (ed.) *Geomagnetism. Research Past and Present.* Bremen-Roennebeck: IAGA, pp. 93–111.

Gauss, C.F., 1839. Allgemeine Theorie des Erdmagnetismus, Resultate aus den Beobachtungen des magnetischen Vereins im Jahre 1838. Printed by Wilhelm Weber, Leipzig.

Gubbins, D., 1983. Geomagnetic field analysis I—stochastic inversion. *Geophysical Journal of the Royal Astronomical Society*, **73**: 641–652.
Halley, E., 1692. An account of the causes of the change of the variation of the magnetical needle; with an hypothesis of the structure of the internal parts of the Earth. *Philosophical Transactions of the Royal Society of London*, **195**: 208–221.
Holschneider, M., Chambodut, A., and Mandea M., 2003. From global to regional analysis of the magnetic field on the sphere using wavelet frames. *Physics of the Earth and Planetary Interiors*, **135**: 107–124.
Khokhlov, A., Hulot G., and Le Mouël, J.-L., 1999. On the Backus effect–II. *Geophysical Journal International*, **137**: 816–820.
Langel, R.A., 1987. Main field. In Jacobs, J.A. (ed.) *Geomagnetism*. London: Academic Press, pp. 249–512.
Langel, R.A., and Estes, R.H., 1985. The near-Earth magnetic field at 1980 determined from MAGSAT data. *Journal of Geophysical Research*, **90**: 2495–2509.
Langlais, B., Mandea, M. and Ultré-Guérard, P., 2003. High-resolution magnetic field modeling: application to MAGSAT and Ørsted data. *Physics of the Earth and Planetary Interiors*, **135**: 77–91.
Lowes, F.J., 1974. Spatial power spectrum of the main geomagnetic field and extrapolation to the core. *Geophysical Journal of the Royal Astronomical Society*, **36**: 717–730.
Mandea, M., 2000. French magnetic observation and the theory at the time of DE MAGNETE. In Schröder, W. (ed.), *Geomagnetism:* [illegible]
Mandea, M., and Macmillan, S., 2000. International geomagnetic reference field—the eighth generation. *Earth Planets Space*, **52**: 1119–1124.
Merrill, R.T., McElhinny, M.W., and McFadden P.L., 1996. *The Magnetic Field of the Earth*. London: Academic Press, p. 531.
Neubert, T., Mandea, M., Hulot, G., von Frese, R., Primdahl, F., Jorgenson, J.L., Friis-Christensen, E., Stauning, P., Olsen, N. and Risbo, T., 2001. High-precision geomagnetic field data from the Ørsted satellite. *EOS*, **82**: 81–87.
Parker, T.L., and Shure, L., 1982. Efficient modelling of the Earth's magnetic field with harmonic splines. *Geophysical Research Letters*, **9**: 812–815.
Reigber, Ch., Luehr, H., and Schwintzer, P., 2002. CHAMP mission status. *Advances in Space Research*, **30**: 129–134.
Shure, L., Parker, R.L., and Backus, G.E., 1982. Harmonic splines for geomagnetic modelling. *Physics of the Earth and Planetary Interiors*, **28**: 215–229.
Whaler, K.A., 1986. Geomagnetic evidence for fluid upwelling at the core-mantle boundary. *Geophysical Journal of the Royal Astronomical Society*, **86,** 563–586.

Cross-references

Gauss, Carl Friedrich (1777–1855)
Geomagnetic Dipole Field
Geomagnetic Spectrum, Spatial
Geomagnetism, History of
Harmonics, Spherical
IGRF, International Geomagnetic Reference Field
Internal External Field Separation
Time-dependent Models of the Geomagnetic Field

MAIN FIELD, ELLIPTICITY CORRECTION

The first spherical harmonic analyses of the geomagnetic main field ignored the ellipticity of the Earth. This was not a serious omission, since the radial departures of the Earth's surface from a sphere never amount to more than 0.17%. However, as the quality of data improved, it was thought that they might justify inclusion of a correction for ellipticity.

The Earth's surface is closely approximated by an oblate spheroid with eccentricity 0.082. This affects both the position of an observation and the observation itself, since the horizontal component, for example, is measured tangential to the spheroid rather than perpendicular to the line joining the site to the Earth's center. Early analyses that took account of this ellipticity (Schmidt, 1898; Adams, 1900; Jones and Melotte, 1953) first determined coefficients for a spherical Earth, then adjusted these for ellipticity. A better method (Cain *et al.*, 1963) was based directly on the original observations.

Adjustment from geodetic to geocentric coordinates

A closer approximation to the Earth's surface than a sphere is an oblate spheroid. That adopted by the International Astronomical Union and subsequently used with the International Geomagnetic Reference Field (see *IGRF, International Geomagnetic Reference Field*) has flattening $f = 1/298$, where $f = 1 - b/a$. Here a denotes the equatorial radius (6378.388 km) and b the polar radius (6356.912 km). This corresponds to a mean radius $a_0 = (a^2b)^{1/3} = 6371.2$ km and ellipticity $e = (1 - b^2/a^2)^{1/2} = 0.082$. Departures from this spheroid are well-known and determined, but are too small to be of concern in the context of geomagnetism.

Observations made at or near the surface of the Earth are referred to geodetic coordinates, with north and east tangential to the surface and vertical perpendicular to it. (Satellite data are usually provided in geocentric coordinates.) Geodetic latitude, ϕ', is the angle between the equatorial plane and the vertical. Height, h, is measured vertically from the surface (see Figure M180). In geocentric coordinates, position is defined by r, ϕ, λ, where r denotes the radial distance from the center of the Earth, ϕ denotes latitude—the angle between the equatorial plane and a line drawn from the center of the Earth, and λ denotes longitude measured east from the Greenwich meridian. In geocentric coordinates, north and east are perpendicular to the radial direction and vertical is along it. The direction of east and λ are the same in both coordinate systems.

Two operations are needed to convert the observations from geodetic to geocentric coordinates. (1) The site must be converted from h, ϕ', λ, to r, ϕ, λ. This may be done using the relations

$$r^2 = h^2 + 2hA + (a^4 \cos^2 \phi' + b^4 \sin^2 \phi')/A^2,$$

$$\cos \phi = \cos \phi' \cos \delta + \sin \phi' \sin \delta$$

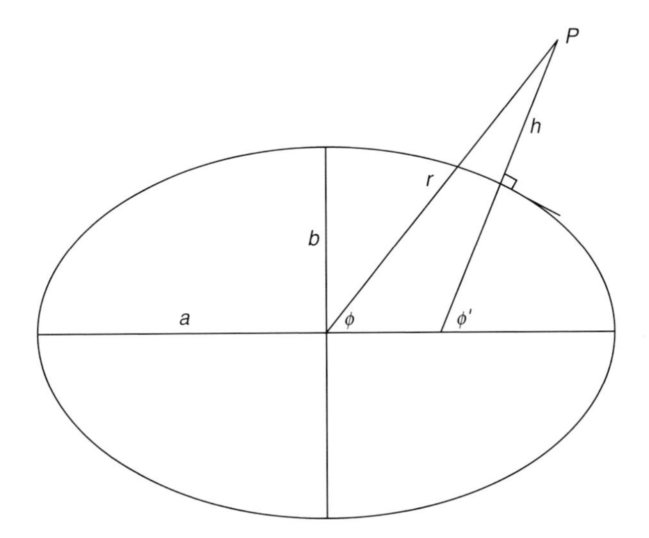

Figure M180 Geodetic (ϕ', h) and geocentric (ϕ, r) coordinates of point P.

and

$$\sin\phi = \sin\phi' \cos\delta - \cos\phi' \sin\delta,$$

where $\delta = \phi' - \phi$,

$$\cos\delta = (h + A)/r,$$

$$\sin\delta = [(a^2 - b^2)\cos\phi' \sin\phi']/Ar$$

and

$$A = (a^2 \cos^2\phi' + b^2 \sin^2\phi')^{1/2}.$$

(2) The magnetic elements X, Y, Z, the geodetic north, east and vertically downward geomagnetic components, respectively, must be rotated to X', Y, Z', the geocentric equivalents. The appropriate relations are

$$X' = X\cos\delta - Z\sin\delta$$

and

$$Z' = Z\cos\delta + X\sin\delta.$$

The treatment of nonorthogonal elements may be deduced from that described in *main field modeling* (*q.v.*)

Once the data are in geocentric coordinates, a standard spherical harmonic analysis may be performed. Magnetic values synthesized from spherical harmonic coefficients may, if required, be converted back into geodetic coordinates.

Discussion

Since the advent of computers, it is just as easy to take account of the Earth's ellipticity as to ignore it, so the adjustment is now universally incorporated as outlined above. Whether or not it is justified is debatable. Precomputer analysts (Adams, 1900; Jones and Melotte, 1953) concluded that the additional labor was not justified. Before satellite data were available it was thought that there might be an advantage when upward extrapolation was involved (Kahle *et al.*, 1964; Cain *et al.*, 1965) or when external as well as internal coefficients were being determined (Winch, 1967). For modeling the main magnetic field of internal origin, the correction for ellipticity is harmless but probably unnecessary (Malin and Pocock, 1969).

A related point, which is probably only of academic interest, concerns the separation of the field into internal and external parts (see *Internal external field separation*). The boundary between internal and external should be the surface of the Earth, but, using the above method, the boundary is the reference sphere, with radius a_0, which passes some 14 km above the poles and is about 7 km below the surface at the equator. Any sources within the region between the ellipsoid and the reference sphere would be wrongly classified. For example, crustal sources near the equator would be deemed to be external. It would not help to increase the radius of the reference sphere to a, since any magnetic fields due to currents in the lower polar atmosphere would then be considered to be of internal origin.

Stuart R.C. Malin

Bibliography

Adams, J.C., 1900. *Terrestrial Magnetism. Scientific Papers*, Vol. 2. Cambridge: Cambridge University Press, pp. 243–640.

Cain, J.C., Daniels, W.E., Hendricks, S.J., and Jensen, D.C., 1965. An evaluation of the main geomagnetic field, 1940–1962. *Journal of Geophysical Research*, **70**: 3647–3674.

Jones, H.S., and Melotte, P.J., 1953. The harmonic analysis of the Earth's magnetic field for epoch 1942. *Monthly Notices of the Royal Astronomical Society, Geophysical Supplement*, **6**: 409–430.

Kahle, A.B., Kern, J.W., and Vestine, E.H., 1964. Spherical harmonic analyses for the spheroidal Earth. *Journal of Geomagnetism and Geoelectricity*, **16**: 229–237.

Malin, S.R.C., and Pocock, S.B., 1969. Geomagnetic spherical harmonic analysis. *Pure and Applied Geophysics*, **75**: 117–132.

Schmidt, A., 1898. Der magnetische Zustand der Erde zur Epoch 1885.0. *Archiv der deutschen Seewarte, Hamburg*, **21**(2): 76 pp.

Winch, D.E., 1967. An application of oblate spheroidal harmonic functions to the determination of geomagnetic potential. *Journal of Geomagnetism and Geoelectricity*, **19**: 49–61.

Cross-references

Harmonics, Spherical
IGRF, International Geomagnetic Reference Field
Internal External Field Separation
Main Field Modeling

MANTLE, ELECTRICAL CONDUCTIVITY, MINERALOGY

The electrical conductivity profile in the mantle can be obtained by *magnetotelluric* (*q.v.*) and *geomagnetic deep sounding* (*q.v.*) methods (Figure M181). Although the profiles obtained can have significant differences from each other, the following features can be seen in the most conductivity models. Conductivity in the upper mantle is relatively low, i.e., 10^{-4}–10^{-2} S m^{-1}. It increases with increasing depth to the top of the lower mantle. At the top of the lower mantle, conductivity is 10^{0} S m^{-1}. Conductivity does not increase significantly in the lower mantle. Some studies have shown that the electrical conductivity of the uppermost mantle to 100 km depth is about 10^{-2}~10^{-1} S m^{-1}, known as the mantle high conductive layer (HCL). Local variations of electrical conductivity are large at shallow depths, and become smaller with increasing depth.

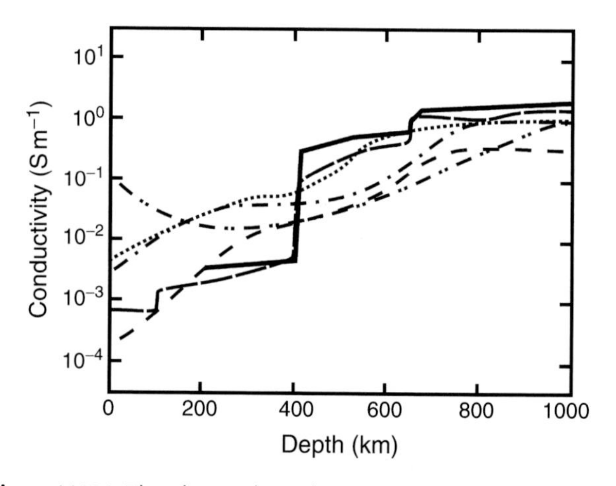

Figure M181 The electrical conductivity of the upper part of the mantle. Long dashed curve: north Pacific region, dotted curve: northeastern China, short dashed curve: Canadian shield, one-dotted dashed curve: southwestern United States, two-dotted dashed curve: Hawaii. Thick solid curve: laboratory electrical conductivity model (Xu *et al.*, 2000).

Geochemical and petrological studies suggest that the major constituent of the uppermost mantle is olivine, with an approximate composition of $(Mg_{0.9}Fe_{0.1})_2SiO_4$. With an increasing pressure, olivine transforms to its high-pressure polymorph, wadsleyite at about 410 km depth (olivine-wadsleyite transition). Wadsleyite transforms to a further high-pressure polymorph, ringwoodite, at about 520 km depth (wadsleyite-ringwoodite transition). Ringwoodite dissociates to $(Mg,Fe)SiO_3$ perovskite and (Mg,Fe)O ferropericlase at 660 km depth (postspinel transition). These minerals form a network in each region and therefore, the electrical conductivity of these minerals would be primarily responsible for that of the mantle itself. Additionally, some minor phases such as silicate melt that have high conductivity may also have important roles.

Electrical conduction mechanisms

In metals, free electrons transfer electric charges. However, the mantle minerals are essentially ionic solids in which most electrons are bound to ions. Therefore, carriers other than free electrons play an important role in transferring electric charges. For example, if an electron is released from an ion, an electric hole is created in the ion, and the hole itself can migrate to carry an electric charge.

For the migration of carriers other than a free electron, defects play important roles in crystalline ionic solids. Here, defects in ferromagnesian silicates are briefly described using the Kröger-Vink notation in which the base character refers to the defect species, that is, ion, electron, or hole. An electron or an electron hole is expressed by element symbols, "e" and "h," respectively. The vacancy of an ion is expressed by "V". The subscript refers to the site, that is, Mg, Si, O, or an interstitial site that is expressed as "I". The dots or slashes as superscripts refer to the number of positive or negative excess charges compared to the site in a normal lattice. "x" as a superscript refers to no excess charge. In this notation, a released electron or a hole is expressed as "e'" or "$h^{\cdot}$", respectively. The normal Mg^{2+}, Si^{4+}, O^{2+} ions in their own sites are expressed as $Mg_{Mg}{}^{X}$, $Si_{Si}{}^{X}$, and $O_O{}^{X}$ respectively. Fe ions usually substitute the Mg site, which is expressed as $Fe_{Mg}{}^{X}$. The Fe ion can have 3+ charges. The Fe^{3+} ion in the Mg site has an extra plus charge and is expressed as $Fe_{Mg}{}^{\cdot}$. The vacancy has a charge with an opposite sign to that of the ion that should occupy the site. The vacancies for Mg, Si, and O sites are expressed as $V_{Mg}{}''$ $V_{Si}{}''''$, and $V_O{}''$, respectively. A small cation can exist interstitially in the crystalline lattice. A typical interstitial ion is H^+, which is expressed as $H_I{}^{\cdot}$.

The major mantle minerals contain significant amounts of Fe and, therefore, electron hopping from Fe^{2+} to Fe^{3+} ions could be the dominant electron conduction mechanism (Figure M182). An electron has to jump across an energy barrier, ΔE_h and, therefore the electrical conductivity, σ, is usually expressed as

$$\sigma = \sigma_0 \exp(-\Delta E_h/kT) \qquad \text{(Eq. 1)}$$

where k is the Boltzmann constant and T is absolute temperature. The ΔE_h is called the activation energy. The preexponential term σ_0 could be a function of temperature; however, it is often regarded as a constant with a limited temperature range. For electron hopping, the energy barrier should be low.

Electron hopping can be regarded as the migration of $Fe_{Mg}{}^{\cdot}$ (Figure M182). The surrounding oxygen ions and cations, respectively, are attracted and repulsed by the excess positive charge, which causes a distortion of the lattice in the neighborhood (Figure M183). The migration of $Fe_{Mg}{}^{\cdot}$ is associated with this lattice distortion. The unit of $Fe_{Mg}{}^{\cdot}$ and its induced lattice distortion is called a small polaron. Migration of a small polaron is a thermally activated process:

$$\sigma = \sigma_0 \exp(-\Delta E_p/kT) \qquad \text{(Eq. 2)}$$

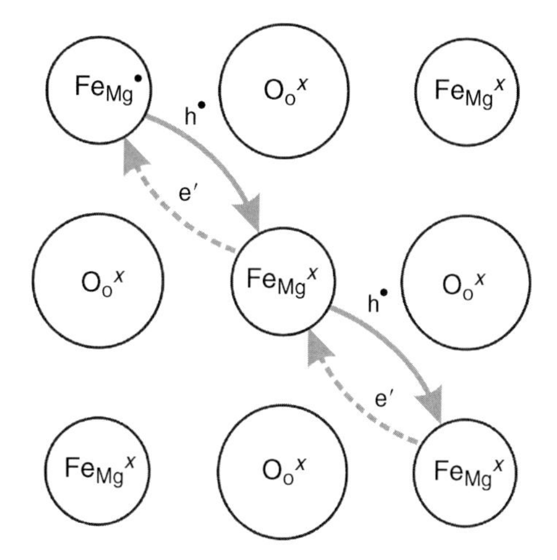

Figure M182 Electrical conduction by electron hopping between $Fe_{Mg}{}^{X}$ and $Fe_{Mg}{}^{\cdot}$. It can be regarded as migration of the hole.

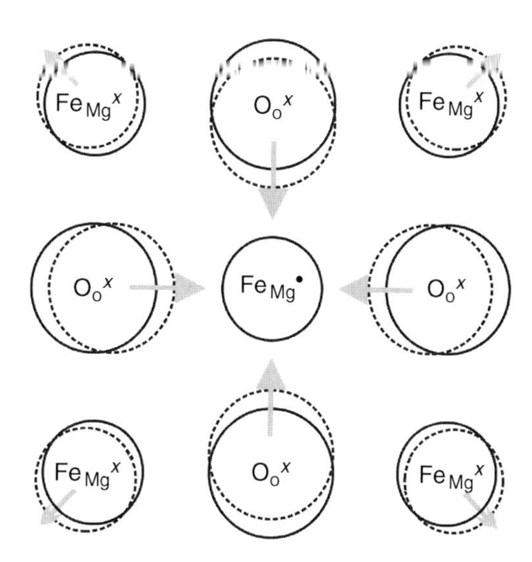

Figure M183 Small polaron model. The $Mg_{Mg}{}^{X}$ (and $Si_{Si}{}^{X}$) and $O_O{}^{X}$ ions are repulsed and attracted by the excess plus charge of $Fe_{Mg}{}^{\cdot}$, which induces distortion of the lattice.

where ΔE_p is the activation energy for the small polaron. A small polaron is less mobile than an electron itself, resulting in $\Delta E_p \gg \Delta E_h$. ΔE_p ranges from 0.5 to 2 eV for mantle minerals.

The charges are carried by $Fe_{Mg}{}^{\cdot}$ and, therefore, the electrical conductivity should be a function of the populations of $Fe_{Mg}{}^{\cdot}$ and $Fe_{Mg}{}^{X}$. Electrical conductivity thus increases with increasing total Fe content. With increasing oxygen partial pressure, the population of $Fe_{Mg}{}^{\cdot}$ increases; therefore, the electrical conductivity of the mantle minerals increases with increasing oxygen partial pressure even with constant Fe content.

The bodily motions of ions transfer electric charges, which is called ionic conduction. The best known material for ionic conduction is zirconia (ZrO_2), in which the oxygen ion is highly mobile and carries minus charges whereas the zirconium ion is very immobile and retains the crystalline lattice. In the case of ferromagnesian silicate or oxide, the migration of $Mg_{Mg}{}^{X}$ via $V_{Mg}{}''$ can be an effective conduction mechanism and is regarded as the migration of $V_{Mg}{}''$ (Figure M184).

Si_{Si}^X and O_O^X are much less mobile than is Mg_{Mg}^X and the migrations of these two ions do not work as effective conduction mechanisms.

Ionic conduction is also a thermally activated process, and can be written as:

$$\sigma = \sigma_0 \exp(-\Delta E_i/kT) \quad \text{(Eq. 3)}$$

where ΔE_i is the activation energy for ionic conduction. The bodily motion of ions causes a very strong distortion of the lattice and so the activation energy for ionic conduction would be much larger than that of a small polaron ($\Delta E_i \gg \Delta E_p$). This indicates that the temperature dependence of ionic conduction is much larger than that of small polaron conduction. Ionic conduction can dominate small polaron conduction at higher temperatures.

The population of V_{Mg}'' increases with that of $Fe_{Mg}^{\cdot}$ so that the charge balance is maintained. The increase in oxygen partial pressure increases $Fe_{Mg}^{\cdot}$ and, therefore increases V_{Mg}''. Hence, the ionic conductivity of V_{Mg}'' increases with increasing oxygen partial pressure.

Ionic conductivity is linked with the diffusion coefficient by the Nernst-Einstein relation:

$$\sigma = q^2 nD/kT \quad \text{(Eq. 4)}$$

where q is the charge of the carrier, n is the concentration of the carrier, and D is the diffusion coefficient. The diffusion is a thermally activated process:

$$D = D_0 \exp(-\Delta H_d/kT) \quad \text{(Eq. 5)}$$

From Eqs. (4) and (5), we have

$$\sigma = (q^2 n D_0/kT) \exp(-\Delta H_d/kT) \quad \text{(Eq. 6)}$$

Therefore, the preexponential term for ionic conduction is inversely proportional to the absolute temperature. The migration of a small polaron is also a diffusion process and its preexponential term should also be inversely proportional to the absolute temperature.

A certain amount of water could be transported into the deep mantle by the subduction process. Although water should form hydrous minerals, olivine and its high-pressure polymorphs can contain a certain amount of water. The rate of diffusion of a proton is very high and so transportation of an electric carrier by H^+ (proton conduction) could have an important role in electrical conduction in the deep mantle. Proton conduction is a form of ionic conduction and so its temperature dependence should be also expressed by Eq. (3).

Though it is usually difficult to specify the electrical conduction mechanism, an indication may come from the sign of the dominant charge carrier that can be inferred from the thermopower. The thermopower, Q, is the ratio of the voltage, ΔV, generated across a sample with a temperature difference, ΔT, to ΔT:

$$Q = -\Delta V_T/\Delta T. \quad \text{(Eq. 7)}$$

The voltage, ΔV_T, is generated because the charge carrier has a larger energy at a higher temperature. Hence, the sign of the charge carrier is the same as that of the thermopower, Q.

At high pressures, compression can change the energy barrier. We introduce the activation enthalpy, ΔH, as follows:

$$\Delta H = \Delta E + P\Delta V \quad \text{(Eq. 8)}$$

where P is pressure and ΔV is the activation volume.

In small polaron conduction, the energy barrier for migration of a hole decreases with increasing pressure because the distance between ions shortens. Hence, it is expected that small polaron conductivity should increase with an increasing pressure ($\Delta V_a < 0$). In ionic conduction, the corridor for the bodily migration of ions narrows with increasing pressure, resulting in an increase in the energy barrier. Hence ionic conduction is expected to have negative pressure dependence ($\Delta V_a > 0$).

Electrical conductivity of mantle minerals

In this section, the electrical conductivities of the major mantle minerals are explained, which are summarized in Figure M185.

Olivine

The electrical conductivity of olivine has been extensively studied. The measurement of thermoelectric effects has shown that the charge of the dominant carrier of olivine is positive at temperatures below 1600 K, whereas it becomes negative above 1700 K. This suggests that there should be two electrical conduction mechanisms. The mechanism dominant at lower temperatures is considered to be the small polaron conduction by $Fe_{Mg}^{\cdot}$. The mechanism dominant at higher temperatures is considered to be the migration of V_{Mg}'' (ionic conduction).

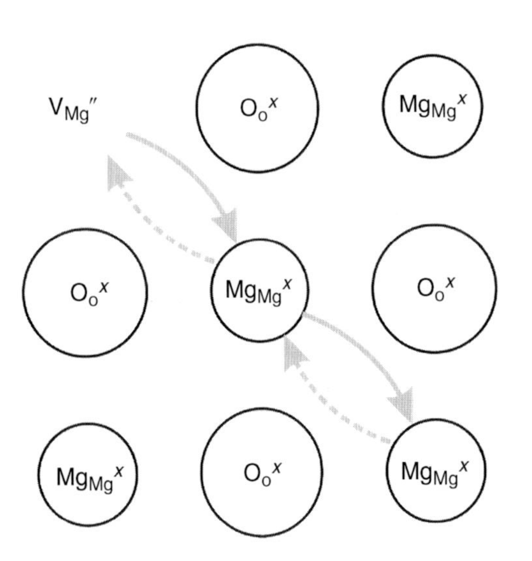

Figure M184 Ionic conduction. Mg_{Mg}^X diffuses through V_{Mg}'', which causes the migration of V_{Mg}''.

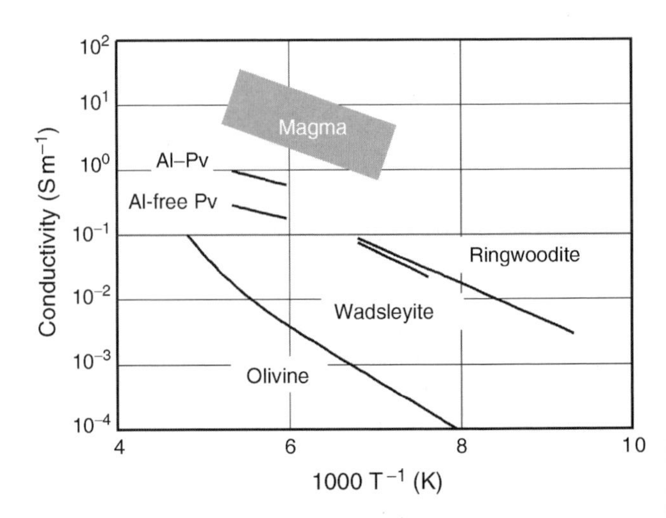

Figure M185 Electrical conductivity of olivine, wadsleyite, ringwoodite, Al-free perovskite, and Al-bearing perovskite against reciprocal temperature. The shaded area denotes the conductivity of various kinds of silicate magmas.

It is reported that the electrical conductivity of natural olivine with Mg/(Mg + Fe) = 0.91 at temperatures of 1000–1770 K under an oxygen partial pressure of 10^{-4} Pa can be expressed as the sum of two exponential functions:

$$\sigma = 10^{2.4}(\mathrm{S\ m^{-1}})\exp[-1.6(\mathrm{eV})/kT] + 10^{9.2}(\mathrm{S\ m^{-1}})\exp[-4.2(\mathrm{eV})/kT]. \quad \text{(Eq. 9)}$$

The activation energy of ionic conduction (4.2 eV) is much larger than that of small polaron conductivity (1.6 eV). The conductivity increases by a power of 1.8 of the Fe/(Mg + Fe) ratio. Conductivity also increases by one-seventh power of the oxygen partial pressure. The pressure dependence is small. It is reported that the activation volume is +0.6 cm^3 mol^{-1}. The plus sign of the activation volume is not consistent with the small polaron model.

Olivine has an orthorhombic crystal structure, resulting in anisotropy of the electrical conductivity. Conductivity in the [001] crystallographic direction is about twice those in the [100] and [010] directions. Anisotropy of electrical conductivity in the upper mantle could be observed in the future.

Olivine has an M_2SiO_4 stoichiometry, where M = Mg, Fe, Ni, Ca…, and does not contain H^+ as a first approximation. However, a small amount of H^+ up to the order 1000 ppm can be contained in olivine at high pressures. Diffusion of H^+ in olivine is very fast so that even a small amount of H^+ cause high-electrical conductivity.

Wadsleyite and Ringwoodite

Ringwoodite has a spinel structure, in which structural units A and B are arranged alternatively. Unit A consists of a SiO_4 tetrahedron, and unit B consists of MgO_6 octahedra. Wadsleyite has a modified spinel structure. The structural units of wadsleyite are the doubles of those of ringwoodite. Unit AA is a Si_2O_7 unit, and unit BB consists of MgO_6 octahedra. Thus, although wadsleyite has a more complex structure than ringwoodite, the structures of these two minerals are fairly similar.

The electrical conductivities of these two minerals also have similar values, which may reflect their structural similarity. The conductivities of these minerals are about two orders of magnitude higher than olivine. The conduction mechanisms of these minerals have not been clarified. They could be a small polaron of $Fe_{Mg}^{\cdot}$, because the carrier charge is plus. Their activation enthalpies are about 1 eV, which are considerably smaller than that of olivine (1.7 eV). That is, the temperature dependences of the electrical conductivities of these two minerals are much smaller than that of olivine. The conductivities of these two minerals were measured over only a reasonably small temperature range, that is, 1173–1473 K. The preexponential terms are 100–1000 S m^{-1}, which are indistinguishable from that of olivine.

The electrical conductivity of wadsleyite is about two orders of magnitude higher than that of olivine. Hence the olivine-wadsleyite transition causes a significant conductivity jump. In contrast, because conductivities of wadsleyite and ringwoodite are quite similar, the wadsleyite-ringwoodite transition does not cause a conductivity jump.

A certain amount of water (up to 3%) can be contained in wadsleyite and ringwoodite through substituting silicon. However, the contribution of proton conduction to the electrical conductivity of wadsleyite and ringwoodite has not yet been clarified.

Perovskite

The composition of mantle perovskite is primarily $(Mg,Fe)SiO_3$. In addition to these components, perovskite can also contain up to 25% Al_2O_3. In the lower mantle, perovskite would contain up to 6% Al_2O_3. The Mg site is probably too large for Al. Hence, Al should predominantly substitute the Si site (Al_{Si}'). To keep the charge valance, the Fe in the Mg site tends to be trivalent ($Fe_{Mg}^{\cdot}$) if perovskite contains an Al_2O_3 component.

The electrical conductivity of Al-free $(Mg_{0.9}Fe_{0.1})SiO_3$ perovskite is $10^{-0.8}$–$10^{-0.5}$ S m^{-1} at 1700–1900 K. The conductivity of $(Mg_{0.9}Fe_{0.1})SiO_3$ perovskite with 3% Al_2O_3 is about half an order of magnitude higher than that of Al-free perovskite; that is, $10^{-0.3}$–10^0 S m^{-1} in the same temperature range. The higher electrical conductivity of the Al-bearing perovskite implies that the dominant conduction mechanism is the small polaron of $Fe_{Mg}^{\cdot}$. The activation enthalpies of Al-free and Al-bearing perovskites have similar values, 0.6–1.0 eV, even smaller than those of wadsleyite and ringwoodite.

The electrical conductivity of perovskite at the top of the lower mantle is 10^0 S m^{-1}, and is about half order of magnitude higher than that of ringwoodite. Therefore, the postspinel transition should cause a conductivity jump by half an order of magnitude.

The electrical conductivity of perovskite has very little pressure dependence. Activation volume is about −0.1 cm^3 $mole^{-1}$. The negative sign of the activation volume (i.e., conductivity increases with increasing pressure) is consistent with the small polaron model.

Activation energy is relatively small, and the temperature gradient in the mantle is small (0.2–0.5 K km^{-1}). Therefore, the electrical conductivity of perovskite should be nearly constant and of the order of 10^0 S m^{-1} throughout the lower mantle.

Magma

Silicate melts (magma) generally have higher conductivity than solid minerals. Their conductivity is 10^0–10^2 S m^{-1} at temperatures above 1500 K. The high conductivity of silicate melts is probably because of the high mobility of atoms in silicate melts. Hence the dominant conduction mechanism is considered to be ionic conduction.

Under the conditions found in the mantle, the wetting angle of silicate melts is significantly small. Hence, silicate melts easily form a network structure in mantle rocks even though their fraction is very low. The network of the melts works as a conduction path and therefore the presence of a small amount of silicate melts should drastically increase electrical conductivity. The conductivity anomalies observed in the mantle could be attributable to a certain degree of melting of mantle rocks.

Explanation of the electrical conductivity of the mantle

A laboratory electrical conductivity model is shown with observations in Figure M181. Generally, mantle minerals with a higher-pressure stability field have higher electrical conductivity. This fact is primarily responsible for the increase of the electrical conductivity in the upper mantle. The activation enthalpy of mantle minerals with a higher-pressure stability field is lower, which means they have smaller temperature dependence. This is one of the reason why the uniform electrical conductivity is more in the deeper mantle.

Conductivity increases by two orders of magnitude with the olivine-wadsleyite transition, whereas it increases by half an order of magnitude with the postspinel transition. The seismic velocity jumps associated with the 410- and 660-km zone are both significant. By contrast, the conductivity jump associated with the olivine-wadsleyite transition is four times larger than that of the postspinel transition. Therefore, the conductivity jump associated with the 410-km discontinuity should be much larger than that of the 660-km discontinuity, if indeed the latter is observed at all.

The conductivity of the HCL at the top of the upper mantle is too high to be explained by the conductivity of normal mantle olivine. Some additional contribution is required. One possibility is that partial melting occurs in the region and the melt forms an electrically conductive path in the mantle rock. Another possibility is that olivine in the region contains considerable amounts of hydrogen and hence the electrical conductivity is significantly raised by proton conduction. For now, the lack of sufficient experimental evidence means we cannot posit an explanation for this situation.

Tomoo Katsura

Bibliography

Poirier, J.-P., 2000. *Introduction to the physics of the Earth's interior, 2nd edn.* Cambridge: Cambridge University Press.

Utada, H., Koyama, T., Shimizu, H., and Chave, A.D., 2003. A semi-global reference model for electrical conductivity in the mid-mantle beneath the north Pacific region. *Geophysical Research Letters*, **30**: 1194, doi:10.1029/2002GL016092.

Xu, Y., Shankland, T.J., and Poe, B.T., 2000. Laboratory-based electrical conductivity in the Earth's mantle. *Journal of Geophysical Research*, **105**: 27865–27875.

Cross-references

Earth Structure, Major Divisions
Electrical Conductivity of the Core
Electromagnetic Induction
Geomagnetic Deep Soundings
Magnetotellurics

MANTLE, THERMAL CONDUCTIVITY

The thermal conductivities of familiar igneous rocks, such as basalt, at laboratory pressure, are typically $\kappa \approx 2.5$ W m^{-1} K^{-1}. Taking a high-temperature specific heat, $C_P = 1200$ J K^{-1} kg^{-1} and $\rho = 2800$ kg m^{-3} as the basalt density, the corresponding thermal diffusivity is $\eta = \kappa/\rho C_P \approx 7.4 \times 10^{-7}$m^2 s^{-1}. For mantle minerals we find $\kappa \approx 4.0$ W m^{-1} K^{-1}, $\eta \approx 1.0 \times 10^{-6}$ m^2 s^{-1}. Thermal conductivity of the crust and uppermost mantle (the lithosphere) can differ little from the conductivities of laboratory samples, but for the deep mantle, the effects of high pressure as well as high temperature must be considered. If we nevertheless assume that there is no dramatic variation in diffusivity, then thermal diffusion is too slow to influence the deep structure or thermal history. Heat transport in the mantle is by convection and conduction is important only in boundary layers with steep temperature gradients. In the context of geomagnetism this means especially the D″ layer at the bottom of the mantle, adjacent to the core.

Lattice conductivity and the effect of pressure

We have only a crude theory of thermal conductivity for complex materials, such as rocks. While we know that lattice conductivity is due to phonons, the quanta of thermal vibration, a calculation of conductivity requires an estimate of the mean free path of phonons, the average distance traveled between the scattering events that randomize them, and there are several scattering mechanisms, some of which are difficult to quantify. In a hypothetical, perfect single crystal, scattering would arise only from interactions between phonons. This is the best understood of the scattering processes and so has received most attention in discussions of mantle conductivity. It is a consequence of anharmonicity, or nonlinearity in atomic bond forces. If these forces were perfectly linear, that is, if the forces on displaced atoms were simply proportional to their displacements, then their oscillations would be harmonic or sinusoidal, with no interaction between superimposed oscillations (phonons would not scatter one another). The anharmonicity of real materials is apparent as positive thermal expansion coefficients and is described theoretically in terms of the *Grüneisen parameter*, γ (*q.v.*). The fundamental physics of the process is discussed by Kittel (1971); Poirier (1991) summarized its application to geophysics, emphasizing an expression, due originally to A.W. Lawson, representing the variation of conductivity with temperature and pressure.

Phonon-phonon scattering, which leads to Lawson's formula, is observable only under highly idealized conditions. For common rocks Lawson's formula overestimates conductivity by a factor of order 10. In these materials, phonon scattering is due to their interactions with crystal imperfections and the random distribution of different elements, even including different isotopes of the same element. The temperature and pressure dependences of "impurity scattering" are not represented by the theory of phonon-phonon scattering. The extreme case of an "imperfect crystal" is a glass, in which the atomic arrangement resembles that of a liquid. Glasses have much lower thermal conductivities than the corresponding crystalline solids and both the temperature and pressure dependences are opposite to those of pure, simple solids. The result is a very confused picture of phonon scattering in chemically and crystallographically complicated materials, such as rocks, making it impossible to extrapolate reliably to the base of the mantle. It appears safest to assume that there is no dramatic variation in conductivity with depth in the mantle and to verify that this leads to no conflict with what we understand about the properties of D″ (see below).

Radiative heat transfer

Materials that are transparent at infrared wavelengths may transmit heat radiatively, giving an additional component of thermal conductivity at high temperature. This is an important heat transport mechanism in the interiors of stars and was introduced to geophysics by S.P. Clark. In the simple case of a "gray body," a generalization of the ideal black body of radiation laws, in which the absorption of radiation with depth of penetration is the same for all wavelengths, the radiative conductivity is

$$\kappa_R = \frac{16}{3}\frac{n^2}{\epsilon}\sigma T^3$$

where n is refractive index, $\sigma = 5.67 \times 10^{-8}$ W m^{-2} K^{-4} is the Stefan-Boltzmann constant and ϵ is the opacity, the reciprocal of the average distance traveled by a photon before absorption. For an idea of how significant this may be we can calculate the value of ϵ required to give a radiative conductivity equal to a plausible lattice conductivity at a temperature of 2500 K (near to the base of the mantle). We find $\varepsilon \approx 2500$ m^{-1}, corresponding to a penetration depth of 0.4 mm. Thus, a quite modest transparency would suffice to make this an important consideration.

The opacities of mantle minerals are controlled primarily by their iron contents, especially if the iron occurs as both Fe^{2+} and Fe^{3+} ions, providing an absorption mechanism by charge transfer. Electron energy bands are broadened by pressure, making this process easier by spreading the wavelength range of absorption. Lower mantle minerals are believed to be primarily $(Mg,Fe)SiO_3$ perovskite and (Mg,Fe)O magnesiowustite, both with substantial iron contents, so that, although the question is not conclusively resolved, it appears unlikely that radiative transfer is important in the lower mantle.

Conduction of core heat into the D″ layer

The structure of D″ is complicated by compositional heterogeneity, making a direct interpretation of seismological observations in terms of temperature difficult, at best. However, we have a reasonable idea of the thickness of the D″ zone and if this is accepted as representative of the thickness of the thermal boundary layer (see D″ *as a boundary layer*) then we can use it to estimate the diffusivity and hence conductivity. D″ is modelled as a 150 km thick layer in PREM (Dziewonski and Anderson, 1981) and this value is assumed here. An uncertainty of 50 km or so is of no consequence as we can only make a rough calculation.

The temperature at the core-mantle boundary, estimated as 3740 K from core properties (see *Core, thermal conduction*), is almost 1000 K higher than the lower mantle temperature estimated by extrapolating adiabatically from the 670 km deep phase transition zone (2760 K). We must, therefore, consider a 1000 K temperature increment across the boundary layer. This means that the mantle material in contact with the core has a viscosity that is lower by a factor of at least 10^4 than that of the mantle a few hundred kilometers higher, concentrating convective flow in a very thin layer at the bottom. A theory of the mechanism by which this flow feeds narrow hot plumes is given by Stacey and Loper (1983), but a simple analog represents the essential features.

As the hot material is skimmed off into narrow buoyant plumes, there is a general collapse of the mantle on to the core to replace it. The analog is the ablation of materials from the surfaces of meteorites and spacecraft entering the atmosphere. A steady, exponential profile of temperature, T, is established, moving inward as surface material is removed, so that

$$T - T_0 = (T_S - T_0)\exp(-h/H)$$

where h is the distance from the heated surface, at temperature T_S, and T_0 is the temperature well removed from the ablating surface. Our interest is in the scale height, H, which we identify with the thickness of D″

$$H = \eta/v$$

η being thermal diffusivity and v is the speed of the collapse of the mantle on to the core. If the core to mantle heat flux is $\dot{Q}/A$ (per square meter) then

$$\dot{Q}/A = v\rho C_P(T_S - T_0)$$

and if we estimate this as 0.033 W m^{-2} (5×10^{12} W over the surface of the core-mantle boundary) then these relationships give $\eta = 7.6 \times 10^{-7}$ m^2 s^{-1}, in fortuitously close coincidence with the value for basalt at zero pressure.

Although this calculation is simplified and approximate, it encourages the view that thermal diffusivity does not vary greatly through the mantle. With the higher density at the core-mantle boundary the corresponding conductivity is $\kappa \approx 5$W m^{-1} K^{-1}, higher than for laboratory samples, but not dramatically so.

A general conclusion

Thermal conductivity of the deep mantle is poorly constrained by either observation or theory, in spite of repeated efforts to understand it, but we have no evidence that it differs greatly from the conductivity of the uppermost mantle or laboratory samples of basic or ultrabasic rock. In this circumstance core heat is conducted into a boundary layer at the base of the mantle, 100 to 200 km thick, softening it and allowing for its removal in buoyant plumes. Convection, not conduction, is responsible for heat transport up the several thousand kilometer depth of the mantle.

Frank D. Stacey

Bibliography

Dziewonski, A.M., and Anderson, D.L., 1981. Preliminary reference Earth model. *Physics of the Earth and Planetary Interiors*, **25**: 297–356.

Kittel, C., 1971. *Introduction to Solid State Physics*, 4th edn. New York: Wiley.

Poirier, J.-P., 1991. *Introduction to the Physics of the Earth's Interior.* Cambridge: Cambridge University Press.

Stacey, F.D., and Loper, D.E., 1983. The thermal boundary layer interpretation of D″ and its role as a plume source. *Physics of the Earth and Planetary Interiors*, **33**: 45–55.

Cross-references

Core, Thermal Conduction
D″ as a Boundary Layer
Grüneisen's Parameter for Iron and Earth's Core

MATUYAMA, MOTONORI (1884–1958)

Born in Oita Prefecture, Japan, as a son of a Buddhist priest, he graduated from the physics department, Kyoto University, 1911. He became lecturer in 1913, associate professor in 1916, and then professor in the Department of Geology, Kyoto University, from 1922 to 1946. He studied at the University of Chicago between 1933 and 1935. After retirement from Kyoto University, he served as president of the Yamaguchi University, from 1949 to his death. For his contribution to paleomagnetism and gravity, Matuyama was elected to Member of Japan Academy in 1950.

In the 1920s, Matuyama conducted a systematic survey of the natural remanent magnetization (NRM) of Quaternary and late Tertiary volcanic rocks in Japan, Korea, and Northeast China (Manchuria at that time). The determination of the dipole moment was done by the method of Gauss (spherical harmonic analysis). In the results, he noted that the youngest rocks are always magnetized to directions similar to the present day field, while many of the early Quaternary and Tertiary rocks had NRMs almost oppositely directed. In all cases, there were not many rocks with the NRMs in the intermediate directions. On the basis of these observations, Matuyama concluded that the Earth's magnetic field was in the reversed state in the late Tertiary (Miocene) and again in the early Quaternary (Matuyama, 1929). This early report was later picked up by Allan Cox and Richard Doell, who named the reversed interval just before the current normal period (Brunhes epoch) after him (Matuyama epoch). "Matuyama" thus became one of the most often referred personal names of Japanese origin. "Matuyama chron" is often quoted in the studies such as magnetostratigraphy, seafloor spreading, sedimentology, geomagnetic reversal timescale, and so on.

Earlier in his career, Matuyama performed gravity survey of a coral reef in the Marshall Islands using an Eötvös deviatoric gravimeter. This was one of the earliest application of gravity methods to the study of the crustal structure, especially in the coral islands. Gravity continued to be one of his main research subjects and, inspired by the work of Vening Meinesz in the Indonesian Archipelago, he also measured the gravity near the Japan Trench in a submarine of the Japanese Navy (Matuyama, 1934). This led to the discovery of the westward offset of the center of the negative gravity belt from the trench axis, which later became an important factor for considering the origin of the great earthquakes in the Sanriku region. Other subjects he studied include the rheological properties of the ice, which was undertaken in collaboration with T.C. Chamberlin of the University of Chicago when he stayed in the United States

Masaru Kono

Bibliography

Matuyama, M., 1929. On the direction of magnetization of basalt in Japan, Tyosen, and Manchuria. *Proceedings of the Imperial Academy (Tokyo)*, **5**: 203–205.

Matuyama, M., 1934. Measurement of gravity over the Nippon Trench on board the I.J. submarine Ro–57, Preliminary report. *Proceedings of the Imperial Academy (Tokyo)*, **10**: 626–628.

MELTING TEMPERATURE OF IRON IN THE CORE, EXPERIMENTAL

Introduction

There are two major techniques commonly used for ultrahigh pressure melting studies: shock wave experiments and the internally heated diamond anvil cell (DAC). In shock compression experiments, the sample is subjected to high pressures and high temperatures by dynamic processes (see *Shock wave experiments*). In DAC experiments, pressure is generated by pressing two opposing diamond anvils, while heating is applied resistively and/or using laser heating. Both techniques have been extensively applied to study high pressure melting of iron. However, accurate determination of melting is exceedingly difficult at extremely high pressure and temperature conditions. The associated weaknesses—the short timescale in shock compression and the small sample size in DAC—are reflected by the large uncertainties and the discrepancy among literature values on melting temperature of iron in the core.

In this section, we begin with a brief description of the generation of simultaneous high pressures and temperatures in the DAC, followed by a discussion on the observation of the onset of the melting transition and the determination of pressure-temperature conditions associated with the observation. A brief review is given on DAC experiments in studying the melting temperature of iron at high pressures. The effect of light elements on the melting temperature is discussed in the end.

For more details on this topic, the following review articles are suggested. See *Melting temperature of iron in the core, theory* for theoretical approaches in this area.

1. Jephcoat, A.P., and Besedin, S.P., 1996. Temperature measurement and melting determination in the laser-heated diamond-anvil cell. *Philosophical Transactions of the Royal Society of London, Series A*, **354**(1711): 1333–1360.
2. Shen, G., and Heinz, D.L., 1998. High pressure melting of deep mantle and core materials. In Hemley, R.J. (ed.), *Ultrahigh Pressure Mineralogy: Physics and Chemistry of the Earth's Deep Interior.* Washington, DC: Mineralogical Society of America, pp. 369–396
3. Boehler, R., 2000. High pressure experiments and the phase diagram of lower mantle and core materials. *Reviews of Geophysics*, **38**: 221–245.

Diamond cell technique

Diamond as anvil and window

Diamond is a premier anvil material because it is the hardest material known, and it is transparent from infrared to hard X-rays, thus providing a window for probing samples by various methods and for heating samples by lasers. With the DAC technique, pressures beyond that at the center of the Earth have been reported (Xu *et al.*, 1986). The principal components of a DAC are two opposing diamond anvils with a gasket in between. A drilled hole in the center of the gasket serves as sample chamber. For melting experiments at very high pressures, typical dimensions of a sample chamber are 10–20 μm thick and ~50 μm in diameter.

Heating diamond anvil cells

An easy way to provide heating is to heat the entire DAC externally. However, there is a temperature limit of ~1500 K with this method. To generate extreme high temperatures to melt iron at high pressures, internal heating methods are generally employed. One is the resistive wire heating, in which iron is used as a conductor wire and heating is applied electrically (Boehler, 1986; Mao *et al.*, 1987). With this method, temperatures can go as high as its melting point, but pressures are limited because of deformation of the electrical leads as pressure increases. The other is the laser-heating technique, in which near- or mid-infrared lasers (Nd:YAG, Nd:YLF, CO_2) are used for heating samples in DAC. There are, in principle, no pressure and temperature limitations associated with this technique. The challenge is to control pressure and temperature conditions, to measure them accurately, and to characterize the sample at defined conditions. A typical sample configuration is shown in Figure M186.

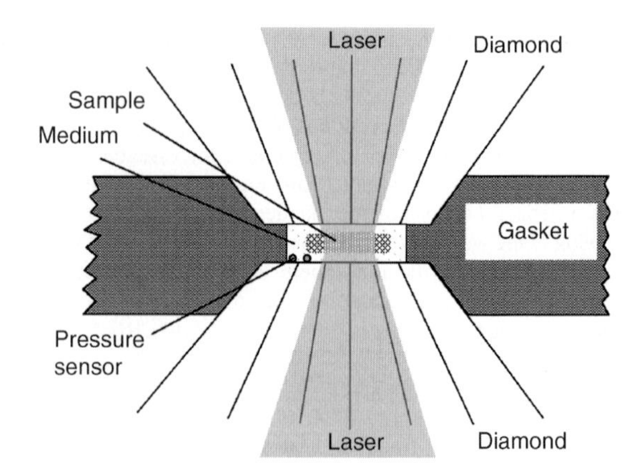

Figure M186 A typical sample configuration in laser-heated diamond anvil cell experiment.

Pressure measurement

Pressures in the laser-heated DAC are often measured by the ruby fluorescence method (Mao *et al.*, 1986) before and after laser heating or during heating from unheated ruby chips away from the heating area. When the laser-heating experiment is coupled with synchrotron radiation, pressures are estimated by measuring unit cell volumes of samples whose pressure-temperature-volume equations of state are known. For pressures measured before and after laser heating, thermal pressure (caused by the material's thermal expansion) has to be considered in estimating pressures at high temperature. Thermal pressure may be minimized by choosing a hydrostatic pressure medium and a small sample-to-medium ratio, and/or characterized with the use of highly collimated synchrotron radiation.

Temperature measurement

Temperature measurement is based on the collected radiation from thermal emission through the Planck radiation law: $I_\lambda = c_1\varepsilon(\lambda)\lambda^{-5}/[\exp(c_2/\lambda T)-1]$, where I_λ is spectral intensity, λ is wavelength, T is temperature, $c_1 = 2\pi hc^2 = 3.7418 \times 10^{-16}$ Wm^2, $c_2 = hc/k = 0.014388$ mK, and $\varepsilon(\lambda)$ is emissivity with $\varepsilon(\lambda) = 1$ for a black body. This method covering a range of wavelengths is often called *spectral radiometry*, as compared to the general pyrometry in which thermal radiation at only one wavelength is measured. One major advantage of spectral radiometry is that, assuming wavelength-independent emissivity, temperatures can be measured independently from the absolute emissivity. However, emissivity is often found to be wavelength dependent. Use of the available data on the wavelength-dependent emissivity at ambient pressure (de Vos, 1954) gives about a 5% temperature correction downward at 3000 K and about 12% downward at 5000 K when compared with the results from the assumption of wavelength-independent emissivity. Unfortunately, such a correction cannot be accurately made because the wavelength dependence of emissivity is not currently known at high pressures and high temperatures, and it can be strongly dependent on the surface finish of a DAC sample. This is the main limitation in the absolute accuracy of temperature measurement by the spectral radiometry method.

Melting criteria

Melting is thermodynamically defined as equilibrium between a solid and a liquid. When materials melt, their physical properties, such as density, viscosity, absorption properties, and electrical resistance, undergo a sudden change. Such property changes are characteristic for a first-order phase transition and are often used for recognition of melting. Unlike other first-order phase transitions, melting is characterized by the loss of long-range order and resistance to shear. To definitively identify melting, both of these two characteristics should be documented. Visual optical observation is a common way to determine whether melting has taken place. It is obvious that fluid flow observation is a good measure of the loss of resistance to shear. Therefore it has been widely used by almost all groups in the world (Saxena *et al.*, 1994; Jeanloz and Kavner, 1996; Jephcoat and Besedin, 1996; Boehler, 2000). However, visual observation (fluid flow) is less obvious as pressure increases, making it difficult to unambiguously define the onset of melting. Synchrotron X-ray diffraction has been combined with laser-heated DAC and used for melting studies to document the loss of long-range order upon melting. Melting at high pressure was identified by the appearance of diffuse scattering from the melt with the simultaneous loss of crystalline diffraction signals (Shen *et al.*, 2004).

Synchrotron Mössbauer spectroscopy (SMS) provides another promising method to identify melting. SMS measures the atomic thermal displacement that can be used for documenting rigidity of a material. Upon melting the strong elastic resonance signal diminishes in SMS. The mean-square thermal displacement of atoms (Lamb-Mössbauer factor) can be measured as a function of temperature, providing a plot for determining the onset of melting. Since measurement of SMS takes only a few seconds to minutes, it holds a great potential for high pressure melting studies.

Melting of core material at high pressures

Iron

Iron is the major component of the core. Knowledge of the melting curve of iron can constrain the temperature of the inner core boundary and anchor the Earth's temperature profiles (see *Core temperature*). At pressures below 50 GPa, the phase diagram of iron is reasonably well known (Figure M187). At ambient pressure, the stable phase is α-Fe with a body-centered-cubic (bcc) structure. At high pressure it transforms to ε-Fe with hexagonal-close-packed (hcp) structure. At high temperature there is a large stability field for γ-Fe with face-centered-cubic (fcc) structure. Although a controversial new solid phase (called β-Fe) was reported (Saxena *et al.*, 1995; Andrault *et al.*, 1997; Dubrovinsky *et al.*, 1998), the existence of this phase was not confirmed in later experiments (Shen *et al.*, 1998; Kubo *et al.*, 2003; Ma *et al.*, 2004). For the melting curve, there is a converging consensus at pressures below 60 GPa, reflected by a narrow uncertainty range in Figure M187. As pressure increases, uncertainties in phase boundaries, including the melting curve, become large as shown by wide bands in Figure M187. The width of the bands represents the scatter in literature data from recent years. Factors causing these uncertainties include those in pressure determinations (neglecting thermal pressure, different equations of state, and/or different standard materials), in temperature determinations (large temperature gradient and temporal variation, chromatic aberration in optics), and in sample characterizations (different melting criteria, transition kinetics).

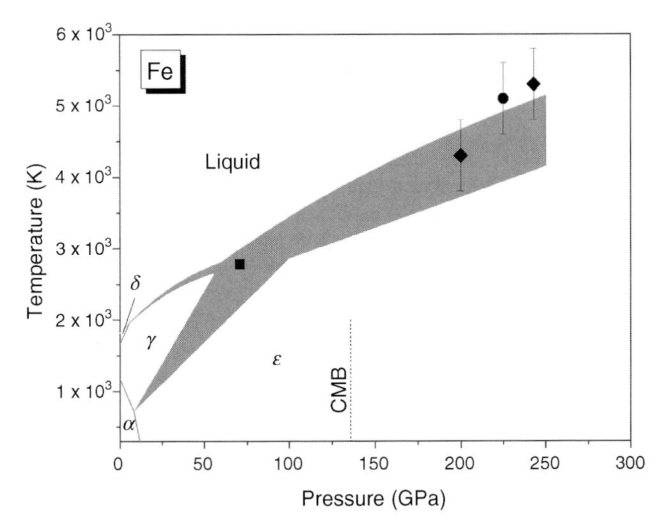

Figure M187 Iron phase diagram. Shaded areas represent the range of literature values in recent years with static diamond anvil cell experiments. Symbols are shock-wave data: square, Ahrens *et al.* (2002); diamonds, Brown and McQueen (1986); circle, Nyuyen and Holmes (2004). Brown and McQueen's point at 200 GPa was interpreted as a solid-solid transition. All other points are referred as melting.

Uncertainties in the melting curve and the γ-ε transition lead to significant variations in the location of γ-ε-*liquid* triple point. Knowledge of its location is important because it is the starting point used for extrapolating the melting curve of ε-Fe to core pressures. As shown in Figure M187, the slope of γ-ε transition ranges from 25 to 40 K GPa^{-1}, placing the triple point between 60 and 100 GPa. Such large uncertainties in the slope of the γ-ε transition mainly arise from the coexisting nature of these two phases in the pressure-temperature range, causing difficulties in identifying the boundary. The uncertainties are also contributed by the pressure-temperature determination associated with the use of different standard materials and/or different equations of state.

The melting data for iron above the triple point are scarce and scattered, reflecting the difficulty level of such experiment. It appears that the early DAC data (Boehler, 1993; Saxena *et al.*, 1994) represent a lower bound of the melting curve in this region. The later experimental data falls at higher melting temperatures (Shen *et al.*, 1998; Ma *et al.*, 2004). The shock wave data (Brown and McQueen, 1986; Nguyen and Holmes, 2004) lie close to the upper bound. Extrapolating the data to the inner core boundary, the melting temperature of iron is between 4800 and 6000 K at 330 GPa (see e.g., *Grüneisen's parameter for iron and Earth's core*).

Effects of light elements

As shown by Birch (1952), the density of the Earth's outer core is about 10% too low to be pure iron, so the core must also contain some light elements. At present, there is no consensus on the dominant light element in the core (see *Core composition* and *Inner core composition*).

At ambient pressure, the addition of a small amount of light elements decreases the melting temperature of iron by up to a few hundred degrees. Data at high pressure are scarce and limited in the Fe-FeO-FeS system (Usselmann, 1975; Urakawa *et al.*, 1987; Ringwood and Hibberson, 1990; Knittle and Jeanloz, 1991; Williams *et al.*, 1991; Boehler, 1992; Fei *et al.*, 1997). Eutectic melting was found in the Fe-FeS system, while a solid solution mechanism was suggested for Fe-FeO system (Knittle and Jeanloz, 1991; Boehler, 1992). Boehler (1996) pointed out that the melting curves in the Fe-FeS-FeO system converge instead of diverge with increasing pressure to that of pure iron.

However, several difficulties need to be addressed and overcome in melting experiments of multicomponent systems using DAC. The laser-heating spot could be on a scale similar to nonuniform samples, which could cause large experimental errors. Large temperature gradients in radial and axial directions could lead to compositional gradients (e.g., Soret diffusion), causing incomplete mixing during heating and melting. Significant developments are thus needed for constraints of light element effects on the melting temperature of iron.

Guoyin Shen

Bibliography

Ahrens, T.J., Holland, K.G., and Chen, G.Q., 2002. Phase diagram of iron, revised-core temperatures. *Geophysical Research Letters*, **29**: doi:10.1029/2001GL014350.

Andrault, D., Fiquet, G., Kunz, M., Visocekas, F., and Hausermann, D., 1997. The orthorhombic structure of iron: an in situ study at high-temperature and high-pressure. *Science*, **278**: 831–834.

Birch, F., 1952. Elasticity and composition of Earth's interior. *Journal of Geophysical Research*, **57**: 227–286.

Boehler, R., 1986. The phase diagram of iron to 430 kbar. *Geophysical Research Letters*, **13**: 1153–1156.

Boehler, R., 1992. Melting of Fe-FeO and Fe-FeS systems at high pressures: constraints on core temperatures. *Earth and Planetaty Science Letters*, **111**: 217–227.

Boehler, R., 1993. Temperatures in the Earth's core from melting-point measurements of iron at high static pressures. *Nature*, **363**: 534–536.

Boehler, R., 1996. Melting of mantle and core materials at very high pressures. *Philosophical Transactions of the Royal Society of London, Series A*, **354**: 1265–1278.

Boehler, R., 2000. High pressure experiments and the phase diagram of lower mantle and core materials. *Reviews of Geophysics*, **38**: 221–245.

Brown, J.M., and McQueen, R.G., 1986. Phase transitions, Grüneisen parameter, and elasticity for shocked iron between 77 GPa and 400 GPa. *Journal of Geophysical Research*, **91**: 7485–7494.

de Vos, J.C., 1954. A new determination of the emissivity of tungsten ribbon. *Physica*, **20**: 690–714.

Dubrovinsky, L.S., Saxena, S.K., and Lazor, P., 1998. Stability of β-phase: a new synchrotron x-ray study of heated iron at high pressure. *European Journal of Mineralogy*, **10**: 43–47.

Fei, Y., Bertka, C.M., and Finger, L.W., 1997. High pressure iron sulfur compound, Fe_3O_2, and melting relations in the Fe-FeS system. *Science*, **275**: 1621–1623.

Jeanloz, R., and Kavner, A., 1996. Melting criteria and imaging spectroradiometry in laser-heated diamond-cell experiments. *Philosophical Transactions of the Royal Society of London, Series A*, **354**: 1279–1305.

Jephcoat, A.P., and Besedin, S.P., 1996. Temperature measurement and melting determination in the laser-heated diamond-anvil cell. *Philosophical Transactions of the Royal Society of London, Series A*, **354**: 1333–1360.

Knittle, E., and Jeanloz, R., 1991. The high pressure phase diagram of $Fe_{0.94}O$: a possible constituent of the Earth's core. *Journal of Geophysical Research*, **96**: 16169–16180.

Kubo, A., Ito, E., Katsura, T., Shinmei, T., Yamada, H., Nishikawa, O., Song, M., and Funakoshi, K., 2003. Phase equilibrium study of iron using sintered diamond (SD) anvils: absence of beta phase. *PEPI*, **30**: 1126.

Ma, Y.Z., Somayazulu, M., Mao, H.K., Shu, J.F., Hemley, R.J., and Shen, G., 2004. In situ x-ray diffraction studies of iron to the Earth core conditions. *Physics of the Earth and Planetary Interiors*, **144**: 455–467.

Mao, H.K., Xu, J., and Bell, P.M., 1986. Calibration of the ruby pressure gauge to 800 kbar under quasi-hydrostatic conditions. *Journal of Geophysical Research*, **91**: 4673–4676.

Mao, H.K., Bell, P.M., and Hadidiacos, C., 1987. Experimental phase relations of iron to 360 kbar and 1400 C, determined in an internally heated diamond anvil apparatus. In Manghnani, M.H., and Syono, Y. (eds.), *High Pressure Research in Mineral Physics*. Tokyo: Terra Scientific Publishing Company/American Geophysical Union, pp. 135–138.

Nguyen, J.H., and Holmes, N.C., 2004. Melting of iron at the physical conditions of the Earth's core. *Nature*, **427**: 339–342.

Ringwood, A.E., and Hibberson, W., 1990. The system Fe-FeO revisited. *Physics and Chemistry of Minerals*, **17**: 313–319.

Saxena, S.K., Shen, G., and Lazor, P., 1994. Temperatures in Earth's core based on melting and phase transformation experiments on iron. *Science*, **264**: 405–407.

Saxena, S.K., Dubrovinsky, L.S., and Haggkvist, P., 1995. X-ray evidence for the new phase β-iron at high pressure and high temperature. *Geophysical Research Letters*, **23**: 2441–2444.

Shen, G., Mao, H.K., Hemley, R.J., Duffy, T.S., and Rivers, M.L., 1998. Melting and crystal structure of iron at high pressures. *Geophysical Research Letters*, **25**: 373–376.

Shen, G., Prakapenka, V.B., Rivers, M.L., and Sutton, S.R., 2004. Structure of liquid iron at pressures up to 58 GPa. *Physical Review Letters*, **92**: 185701.

Urakawa, S., Kato, M., and Kumazawa, M., 1987. Experimental study of the phase relation in the system Fe-Ni-O-S up to 15 GPa. In Manghnani, M.H., and Syono, Y. (eds.), *High Pressure Research in Mineral Physics*. Tokyo: Terra Scientific Publishing Company, pp. 95–111.

Usselmann, T.M., 1975. Experimental approach to the state of the core. I. The liquidus relations of the Fe-rich portion of the Fe-Ni-S system from 30 to 100 kb. *American Journal of Science*, **275**: 278–290.

Williams, Q., Knittle, E., and Jeanloz, R., 1991. The high pressure melting curve of iron: a technical discussion. *Journal of Geophysical Research*, **96**: 2171–2184.

Xu, J., Mao, H.K., and Bell, P.M., 1986. High pressure ruby and diamond fluorescence: observations at 0.21 to 0.55 TPa. *Science*, **232**: 1404–1406.

Cross-references

Core Composition
Core Temperature
Grüneisen's Parameter for Iron and Earth's Core
Inner Core Composition
Melting Temperature of Iron in The Core, Theory
Shock Wave Experiments

MELTING TEMPERATURE OF IRON IN THE CORE, THEORY

An accurate knowledge of the melting properties of Fe is particularly important, as the temperature distribution in the core is relatively uncertain and a reliable estimate of the melting temperature of Fe at the pressure of the inner-core boundary (ICB) puts a constraint on core temperatures. However, there is much controversy over its high pressure melting behavior (e.g., see Shen and Heinz, 1998). Static compression measurements of the melting temperature (T_m) with the diamond anvil cell (DAC) have been made up to ~200 GPa, but even at lower pressures results for T_m disagree by several hundred kelvins. Shock experiments are at present the only available method to determine melting at higher pressures, but their interpretation is not simple, and there is a scatter of at least 2000 K in the reported T_m of Fe at ICB pressures.

Since both quantum mechanics and experiments suggest that Fe melts from the hexagonal close-packed (hcp) structured phase in the pressure range immediately above 60 GPa, this entry will focus on the equilibrium between hcp Fe and liquid phases. Several approaches for determining T_m from theory have been adopted, including quantum mechanically based methods (e.g., Alfè *et al.*, 2002a,b), and more empirical methods such as dislocation melting models (Poirier and Shankland, 1993), or those based on thermal physics (Stacey and Irvine, 1977).

The condition for two phases to be in thermal equilibrium at a given temperature, T, and pressure, P, is that their Gibbs free energies, $G(P, T)$, are equal. To determine T_m at any pressure from theory, it desirable to

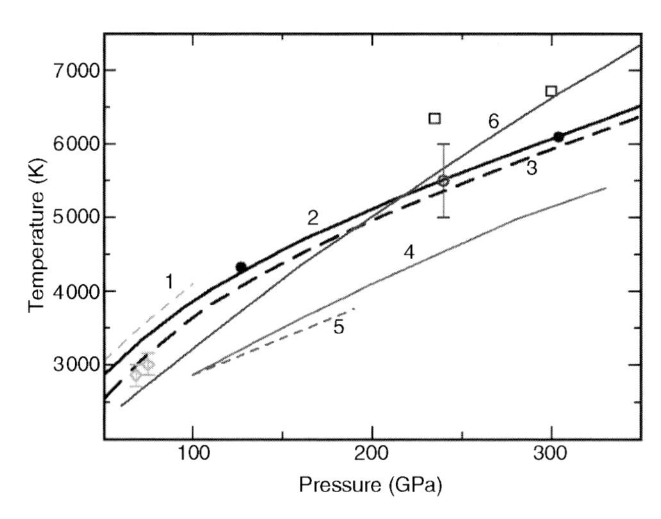

Figure M188 The calculated high pressure melting curve of Fe of Alfè *et al.* (line 2) passes through the shock wave datum (open circle) of Brown and McQueen. Other data shown includes: The melting curve of Alfè *et al.*, corrected for the GGA pressure error (line 3), Belonoshko's melting curve (line 6), Belonoshko's melting data corrected for errors in potential fitting (black dots), Laio *et al.*'s melting curve (line 4), Boehler's DAC curve (line 5), Shen's data (diamonds), Yoo's shock data (open squares), William's melting curve (line 1).

calculate G for the solid and liquid phases as a function of T and determine where they are equal. In fact, it is usual to calculate the Helmholtz free energy, $F(V, T)$, as a function of volume, V, and hence obtain the pressure through the relation $P = -(\partial F/\partial V)_T$ and G through its definition $G = F + PV$.

To obtain melting properties with useful accuracy, free energies must be calculated with high precision, because the free energy curves for liquid and solid cross at a shallow angle. It can readily be shown that to obtain T_m with a technical precision of 100 K, noncanceling errors in G must be reduced below 10 meV. It has been shown that *ab initio* molecular dynamics can be used to determine the free energies of liquid and solid of Al and Cu with sufficient accuracy as to reproduce the known melting temperature to within 100 K. Alfè *et al.* (1999, 2002a) used this method to calculate the melting curve of Fe up to core pressures, and their results, shown in Figure M188, give the melting temperature of Fe at ICB pressures to be ~6200 K, with an error of ±300 K. For pressures $P < 200$ GPa (the range covered by DAC experiments) their curve lies ~900 K above the values of Boehler (1993) and ~200 K above the more recent values of Shen *et al.* (1998). Their curve falls significantly below the shock-based estimates for the T_m of Yoo *et al.* (1993), in whose experiments temperature was deduced by measuring optical emission (however, the difficulties of obtaining temperature by this method in shock experiments are well known), but accords almost exactly with the shock data value of Brown and McQueen (1986) and the new data of Nguyen and Holmes (1999). For melting at ICB pressure, Alfè *et al.* (1999, 2002a) calculate $\Delta V_m/V = 1.8\%$, $\Delta S_m = 1.05R$ J K^{-1}mol^{-1}, and so obtain a latent heat of fusion of 55 kJmol^{-1}.

There are other ways of determining the melting temperature of a system by *ab initio* methods, including performing simulations that model coexisting liquid and crystal phases. The melting temperature of such a system can then be inferred by seeing which of the two phases grows during the course of a series of simulations at different temperatures. This approach has been used by Laio *et al.* (2000) and by Belonosko *et al.* (2000) to study the melting of Fe. In their studies they modeled Fe melting using interatomic potentials fitted to *ab initio* surfaces. Alfè *et al.* (2002b) discovered, however, that these fitted potentials did not simultaneously describe the energy of the liquid and crystalline phases with the same precision, and so these simulations do not represent the true melting behavior of Fe, but rather that of the fitted potential. Alfè *et al.* (2002b) found a way to correct for these shortcomings, by calculating the free energy differences between the model and the *ab initio* system for both the liquid and solid phases. This difference in free energy between liquid and solid can then be transformed into an effective temperature correction. When this was done to Belonosko's data, there was excellent agreement (see Figure M188) with previously determined melting curve for Fe determined by Alfè *et al.* (2002a) using the technique based on direct determination of free energies.

Other more approximate theoretical approaches are also in accord with the inferences from more accurate quantum mechanical studies. Poirier and Shankland (1993) obtained a value of T_m for Fe at 330 GPa of 6100 ± 100 K by using a dislocation melting model, Stacey and Irvine (1977) estimated $T_m = 6050$ K from a Lindemann's law-based analysis, Stixrude *et al.* (1997) estimated T_m to be ~6150 K, and Anderson (2003) found a T_m of 5900 ± 300 K on the basis of thermodynamic analysis. Thus with *ab initio* calculations having been shown to be robust, there now seems to be an emerging consensus on the melting temperature Fe at ICB pressure to be between 6 000 and 6400 K. There is still scope for further work on the difficult problem of the modeling of melting, but for the high pressure melting of Fe at least, it now appears that there may be as many problems with reconciling divergent experimental data than there are in obtaining accurate predictions of T_m from *ab initio* studies.

David Price

Bibliography

Alfè, D., Gillan, M.J., and Price, G.D., 1999. The melting curve of iron at the pressures of the Earth's core from ab initio calculations. *Nature*, **401**: 462–464.

Alfè, D., Price, G.D., and Gillan, M.J., 2002a. Iron under Earth's core conditions: liquid-state thermodynamics and high-pressure melting curve from ab initio calculations. *Physics Review*, **B65**: 165118, 1–11.

Alfè, D., Gillan, M.J., and Price, G.D., 2002b. Complementary approaches to the ab initio calculation of melting properties. *Journal of Chemical Physics*, **116**: 6170–6177.

Anderson, O.L., 2003. The three-dimensional phase diagram of iron. In Dehant, V., Creager, K.C., Karato, S-I., and Zatman, S. (eds.), *Earth's Core*. AGU, Geodynamics Series, **31**: 83–104.

Belonosko, A.B., Ahuja, R., and Johansson, B., 2000. Quasi-ab initio molecular dynamic study of Fe melting. *Physics Review Letters*, **84**: 3638–3641.

Boehler, R., 1993. Temperature in the Earth's core from the melting point measurements of iron at high static pressures. *Nature*, **363**: 534–536.

Brown, J.M., and McQueen, R.G., 1986. Phase transitions, Grüneisen parameter and elasticity for shocked iron between 77 GPa and 400 GPa. *Journal of Geophysical Research*, **91**: 7485–7494.

Laio, A., Bernard, S., Chiarotti, G.L., Scandolo, S., and Tosatti, E., 2000. Physics of iron at Earth's core conditions. *Science*, **287**: 1027–1030.

Nguyen, J.H., and Holmes, N.C., 1999. Iron sound velocities in shock wave experiments. In Furnish, M.D., Chhabildas, L.C., and Hixson, R.S. (eds.) *Shock Compression of Condensed Matter—1999*. Melville, NY: American Institute of Physics, pp. 81–84.

Poirier, J.P., and Shankland, T.J., 1993. Dislocation melting of iron and the temperature of the inner-core boundary, revisited. *Geophysical Journal International*, **115**: 147–151.

Shen, G.Y., and Heinze, D.L., 1998. High pressure melting of deep mantle and core materials. In Hemley, R.J. (ed.), *Ultrahigh-pressure Mineralogy*. Reviews in Mineralogy, **37**: 369–398.

Stacey, F.D., and Irvine, R.D., 1977. Theory of melting: the thermodynamic basis of Lindemann's Law. *Australian Journals of Physics*, **30**: 631–640.

Stixrude, L., Wasserman, E., and Cohen, R.E., 1997. Composition and temperature of the Earth's inner core. *Journal of Geophysical Research*, **102**: 24729–24739.
Williams, Q., Jeanloz, R., Bass, J., Svendsen, B., and Ahrens, T.J., 1987. The melting curve of iron to 250 GPa—a constraint on the temperature at Earth's center. *Science*, **236**: 181–182.
Yoo, C.S., Holmes, N.C., Ross, M., Webb, D.J., and Pike, C., 1993. Shock temperatures and melting of iron at Earth core conditions. *Physics Review Letters*, **70**: 3931–3934.

Cross-references

Core Properties, Physical
Core Properties, Theoretical Determination
Core Temperature
Core, Adiabatic Gradient
Inner Core
Melting Temperature of Iron in the Core, Experimental
Shock Wave Experiments

Magnons (spin waves)

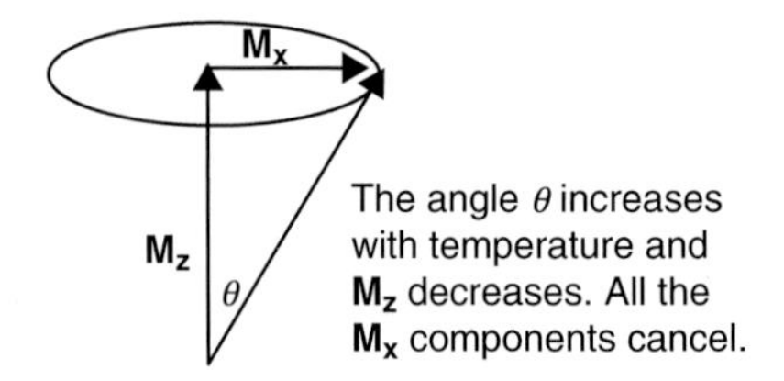

Looking down on the $\mathbf{M_x}$ components

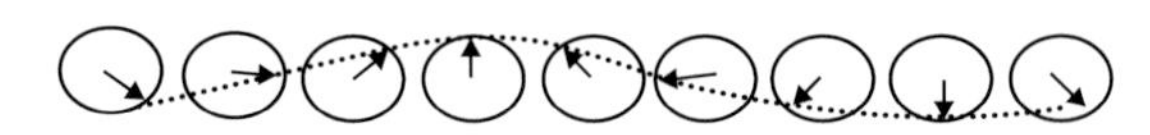

We can see the "spin wave" or magnon that decreases the magnetisation.
More magnons = less magnetisation

Figure M189 Magnons.

MICROWAVE PALEOMAGNETIC TECHNIQUE

The microwave technique is used to determine the strength of the Earth's magnetic field in the past. It is usually applied to igneous rock such as lava and dike or the surrounding material that has been heated by them (baked contact), as well as heated archaeological material (ceramics, fireplaces, etc.). Samples are placed inside a resonant cavity and exposed to microwaves for a few seconds (typically 10 s). The microwave exposure is directly analogous to raising the temperature of the sample in the conventional Thellier technique, the main difference being that when the sample is exposed to microwaves most of the energy is absorbed by the magnetic system whereas in direct heating the entire sample absorbs energy.

By concentrating the energy in the magnetic system the bulk sample is not heated significantly. This reduces the probability of laboratory thermal alteration and provides a more accurate determination of the ancient magnetic field strength.

Microwaves magnons and temperature

A simple model for ferromagnetic minerals is a large number of spinning electrons, one for each ion. Each spin has a magnetic moment $\mathbf{M} = g\mu_b S$, where g is the spectroscopic splitting factor, μ_b is the Bohr magnetron, and S is the spin (Kittel, 1966).

In a ferromagnetic mineral there is a very strong force between neighboring spinning electrons that tries to align all of the spins (exchange coupling). At very low temperatures when the thermal energy is near to zero the spins align and the total magnetization of the mineral is greatest. This total magnetization is expressed as a strong internal field in the mineral. As the temperature increases the energy increases and the individual spin magnetic moments move away from perfect alignment. As the spin magnetic moments move away from perfect alignment they precess around the internal field of the mineral so that only a component ($\mathbf{M}_z$) remains aligned and the total magnetization of the mineral reduces (Figure M189).

All of the spins precess at the same frequency (H where is the gyromagnetic ratio and H is the magnetic field) but each is slightly out of phase with its neighbor. Looking along the field direction ($\mathbf{M}_z$ direction) the $\mathbf{M}_x$ components form a "spin wave" that passes through the mineral (Figure M189). The amplitude of the spin wave increases in quantized jumps equivalent to the complete reversal of one spin magnetic moment and so the total magnetization of the mineral changes by 2**M**. A spin wave that is quantized in this way is called a magnon. For each magnon that is excited the magnetic moment reduces by 2**M**. This is a simplified picture of spin wave generation in ferromagnetic minerals but it helps us to understand how the magnetic moment of a sample is reduced by the generation of magnons.

In the case of conventional thermal demagnetization, ferromagnetic magnetization decreases with increasing temperature until it is completely removed at the Curie temperature. Looking at this process in more detail, the external thermal energy (the high temperature in an oven, say) generates phonons (lattice vibrations) in order to equilibrate the external energy with the internal energy. The phonons in turn generate magnons in the magnetic system in order to equilibrate the energy in the magnetic system. When sufficient magnons are generated the sample is demagnetized and the equivalent external temperature that caused this is called the Curie temperature.

Using microwaves it is possible to excite the magnons directly by ferromagnetic resonance without applying heat to the sample (Walton *et al.*, 1992, 1993). If this can be done quickly the high-energy magnetic system will not have time to pass energy into the sample, generate phonons and raise its temperature. In theory microwaves can demagnetize a sample in the same way as heating a sample but without raising its temperature. In practice it takes a few seconds of microwave exposure to generate sufficient magnons to completely demagnetize a sample, the normal exposure time is 10 s and this means that some phonons are generated. Also, since most samples are good insulators, there is some dielectric heating of the sample in the microwave field. The combination of these two effects is to raise the temperature of the sample slightly, usually less than 150°C (Hill *et al.*, 2000).

Dielectric heating is frequency-dependent, the higher the frequency the more heat is generated. Magnon generation is also frequency-dependent, higher frequency generates more magnons. The sample size is also frequency-dependent, higher frequency means a smaller resonant cavity and therefore smaller samples. Experiments at 8 and 14 GHz show clearly that at 14 GHz samples are more easily demagnetized and are heated less than at 8 GHz. Much above 14 GHz dielectric heating becomes dominant and the samples are heated more. For microwave demagnetization without heating the 14–16 GHz range seems optimal (Figure M190).

Recent experiments demonstrate that high (low) microwave power is equivalent to high (low) temperature (Hill *et al.*, 2002) and so microwave demagnetization may be used instead of thermal

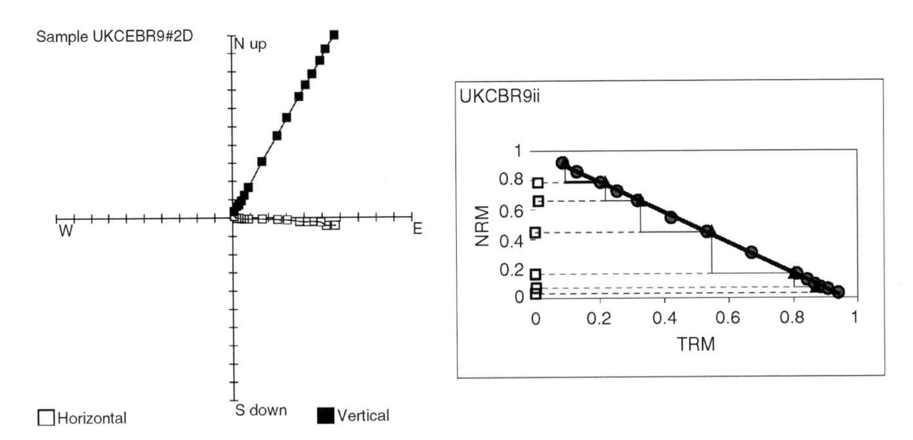

Figure M190 Orthogonal 18 GHz microwave demagnetization plot (*left*) and 18 GHz microwave Thellier paleointensity plot with PTRM checks and multidomain tail checks (*right*). The maximum microwave power used was 40 W for 10 s. $H_{pal} = 51.9 \pm 0.4$ mT, $q = 140.4$.

demagnetization to determine paleomagnetic directions. Microwave demagnetization has the added advantage that it can be used on samples that undergo severe alteration on heating and are unsuitable for thermal analysis (i.e., sulfide-bearing rocks).

John Shaw

Bibliography

Hill, M.J., and Shaw, J., 2000. Magnetic field intensity study of the 1960 Kilauea lava flow, Hawaii, using the microwave palaeointensity technique. *Geophysical Journal International*, **142**: 487–504.

Hill, M.J., Gratton, M.N., and Shaw, J., 2002. A comparison of thermal and microwave palaeomagnetic techniques using lava containing laboratory induced remanence. *Geophysical Journal International*, **151**: 157–163.

Walton, D., Shaw, J., Share, J.A., and Hakes, J., 1992. Microwave demagnetisation. *Journal of Applied Physics*, **71**: 1549–1551.

Walton, D., Share, J.A., Rolph, T.C., and Shaw, J., 1993. Microwave magnetisation. *Geophysical Research Letters*, **20**: 109–111.

Cross-references

Demagnetization
Depth to Curie temperature
Magnetization, Thermoremanent (TRM)
Paleointensity

N

NAGATA, TAKESI (1913–1991)

Born in Aichi Prefecture, Japan; graduated from the Physics Department, University of Tokyo, 1936; research associate, Earthquake Research Institute, University of Tokyo, 1937; associate professor, 1941 and professor in the Department of Geophysics, University of Tokyo, 1952–1974; director of the National Polar Research Institute, 1973–1984; adjunct professor of University of Pittsburgh, 1962–1977; Nagata was quite active in international cooperation and served as bureau member of the International Union of Geodesy and Geophysics (IUGG) for 1960–1963, president of the International Association of Geomagnetism and Aeronomy (IAGA) for 1967–1971, and vice-president of the Scientific Committee on Antarctic Research (SCAR) for 1972–1976.

His doctoral thesis was devoted to a systematic experimental study of the properties of the thermoremanent magnetization (TRM) in volcanic rocks (Nagata, 1943). This included, concurrent with the work of Emile Thellier in France, the discovery of proportionality of the TRM to the magnetizing field and additivity of partial TRM. In 1951, Nagata and his colleague, and students Syun-iti Akimoto and Seiya Uyeda made a great sensation in geophysics community, by the discovery of the self-reversal of the TRM in Haruna dacite. The field reversal vs self-reversal continued to be a topic of hot controversy through 1950s and 1960s until Allan Cox and others established the polarity reversal timescale. Nagata and his colleagues studied not only the TRM, but also various forms of remanent magnetization, which are found in nature as well as are produced in the laboratory. These included viscous remanent magnetization (VRM, with Yoshio Shimuzu), chemical remanent magnetization (CRM, with Kazuo Kobayashi), pressure-induced remanent magnetization (PRM, with Hajimu Kinoshita), and shock remanent magnetization (SRM). The textbook *Rock Magnetism*, which was first published in 1953 and then revised in 1961, summarized all these works in rock magnetism, and was very influential to the community till very recent times.

Nagata also took part in studying the magnetic properties of the lunar rocks collected by the Apollo project and meteorites of various types. Nagata's scientific interest covered all the aspects of geomagnetism and the newly developing space sciences. Besides rock magnetism and paleomagnetism, he studied the main magnetic field, secular variation, tectonomagnetic effects, ionosphere, magnetic storms, and aurorae. Nagata became a very famous scientist in Japan and was influential in Japanese scientific policy on developing the polar exploration, rocket and satellite experiments, and even on the earthquake prediction program. His popularity arose mostly because he was the leader of the Antarctic expeditions for three consecutive years (1956–1958). This project was started as a part of the Japanese contribution to the International Geophysical Year Program (1957). The expedition built the Syowa Base on the Ongul Island in Antarctica and started various scientific observations, which are continued to this day. One of the prominent achievements of these expeditions was the discovery of meteorites on the ice field in 1969. Nagata strongly promoted continued search in the following years, and more than 16000 meteorites were recovered from the Antarctic ice surfaces since then. For the scientific achievement and for public services, he was elected to Member of Japan Academy (1983), and was awarded the Japan Academy Prize (1951), the Order of Culture of Japan (1974), and the gold medal of the Royal Astronomical Society, UK (1984).

Masaru Kono

Bibliography

Nagata, T., 1943. The natural remanent magnetism of volcanic rocks and its relation to geomagnetic phenomena. *Bulletin of Earthquake Research Institute*, **21**: 1–196.

Nagata, T., Akimoto, S., and Uyeda, S., 1951. Reverse thermoremanent magnetism. *Proceedings of the Japan Academy*, **27**: 643–645.

Nagata, T., 1961. *Rock Magnetism*, revised edition. Tokyo: Maruzen, 350 pp.

NATURAL SOURCES FOR ELECTROMAGNETIC INDUCTION STUDIES

Time changes of the magnetic field, regardless whether they are of internal (core) or of external (in the atmosphere, ionosphere, and magnetosphere) origin, produce secondary currents in the Earth's interior. Observations of the superposition of the primary (inducing) and secondary (induced) electromagnetic field allows for a determination of the electrical conductivity of the Earth's interior.

Mantle conductivity can be probed "from below" using signals originating from the outer core. This method requires a precise determination of the field during rapid and isolated events, like *geomagnetic jerks* (*q.v.*), and some *a priori* assumptions about the kinematics of the fluid motion at the top of the core. Conductivity of mantle and crust can also be studied "from above" by the analysis of field changes of external origin, as the induced currents of these fluctuations modify the observed electromagnetic field. Both methods require good knowledge of the time-space structure of the inducing field. The first approach has its strength in estimating the conductivity of the deep mantle, whereas the second approach is suitable for probing the conductivity of the crust and upper mantle down to the depths of 1000 km or so. This article summarizes natural sources of external origin only.

Two approaches are in use: the *Geomagnetic Deep Sounding* (*q.v.*) (GDS) method utilizes time changes of the vertical and horizontal magnetic field components; whereas *Magnetotellurics* (*q.v.*) (MT) uses the horizontal components of magnetic and electric fields. GDS is used for global soundings and with natural sources in the period range between hours and years, whereas MT is mostly used for regional studies and with periods between milliseconds and a few hours due to the decrease of the electric field strength with decreasing frequency.

Natural geomagnetic variations in the period range between a few hours and a couple of days have been used for more than one century to infer the electrical conductivity of the Earth's mantle in the depth range of a few hundred kilometers. The dominant sources for these variations are the magnetospheric ring-current and the ionospheric current system generated by wind systems in the upper atmosphere. The latter produces periodic external field variations (periods of 24 h and harmonics), whereas the magnetospheric ring-current contributes to field variations in the whole period range between minutes and months. These two sources are still the most important ones for inferring mantle conductivity in the depth range of a few hundred kilometers.

Time-space structure of external sources

At the Earth's surface, temporal variations of the magnetic field of external origin, $\mathbf{B}^{\text{ext}} = -\nabla V^{\text{ext}}$, can be described by means of a scalar potential V^{ext}. For modeling its spatial dependency, the flat-Earth approximation may be used to study phenomena that have a spatial scale much smaller than the Earth's radius. Studying phenomena at longer scales requires a spherical approach in which V^{ext} is expanded in a series of *spherical harmonics* (*q.v.*)

$$V^{\text{ext}}(t,r,\theta,\phi) = a\sum_{n=1}^{N}\sum_{m=0}^{n}\left(q_n^m(t)\cos\phi + s_n^m(t)\sin\phi\right)\left(\frac{r}{a}\right)^n P_n^m(\cos\theta)$$
$$= a\mathrm{Re}\left\{\sum_{n=1}^{N}\sum_{m=0}^{n}\varepsilon_n^m(t)\left(\frac{r}{a}\right)^n P_n^m(\cos\theta)\exp \mathrm{i}m\phi\right\} \quad \text{(Eq. 1)}$$

where (r,θ,ϕ) are radius, colatitude, and longitude of a spherical coordinate system, respectively, with a as the Earth's radius, P_n^m as the associated Legendre function of degree n and order m, and $\varepsilon_n^m = q_n^m - is_n^m$ as the corresponding external expansion coefficients. A similar expansion holds for the internal (induced) magnetic field, $\mathbf{B}^{\text{int}} = -\nabla V^{\text{int}}$, with expansion coefficients ι_n^m and radial dependence $(a/r)^{n+1}$ instead of $(r/a)^n$. The horizontal scale-length of a mode of degree n is $\lambda_n = 2\pi a/(n+1) \approx 40000/(n+1)$ km.

If conductivity depends only on depth (1D conductivity), each expansion coefficient of the external field induces only one internal coefficient of same degree n and order m, and the ratio of their Fourier components (time dependence $e^{i\omega t}$), $Q_n(\omega) = \iota_n^m(\omega)/\varepsilon_n^m(\omega)$, depends only on degree n.

Transfer functions (*q.v.*) like Q_n or the C-response

$$C_n = \frac{a}{n+1}\frac{1-[(n+1)/n]Q_n}{1+Q_n}$$

depend on the frequency ω as well as on the spatial scale λ_n (i.e., spherical harmonic degree n) of the source field, and on the electrical conductivity distribution. Dependence of the C-response on spatial scale λ_n and period is shown in Figure N1 for a typical model of mantle conductivity. The horizontal parts of the curves present the range of λ_n, for which source effects (the dependence of C_n on λ_n) are negligible and C_n converge to the zero wave number response C_0. For the given conductivity model, this is true for variations of period $T < 1$ min. In the limit of no induction for which $|C_n| = \lambda_n/2\pi$, the response (and hence the induced field) is determined by the geometry of the source rather than Earth's conductivity. Because $|C_n| \leq \lambda_n/2\pi$, the maximum depth for which conductivity can be determined is given by the spatial scale λ_n of the source; with the chosen conductivity model it is for instance not possible to probe the Earth's conductivity using measurements of the equatorial electrojet (EEJ) for periods >2 h.

GDS requires a precise description of the time-space structure of the potential V^{ext} of the external field, but knowledge of the location of the external sources (currents) is not required—provided they are external to a sphere including the observation point (i.e., external to Earth for ground observations). This means that the determination of the equivalent source is sufficient for induction studies. The following section focuses on the morphology rather than on the physical nature of the various geomagnetic variations that are used for EM induction studies.

Classification of geomagnetic variations

Table N1 presents a classification of magnetic variations in dependence on period and Figure N2 shows a typical spectrum of the north component of the geomagnetic field for a midlatitude site.

Magnetospheric ring-current

The *ring current* (*q.v.*) in the *magnetosphere* (*q.v.*) is probably the source that has been used for most studies of mantle conductivity. To first approximation, its spatial structure can be described by the single spherical harmonic $P_1^0(\cos\theta_d)$, where θ_d is geomagnetic (dipole) colatitude. Hence only the expansion coefficient $q_1^0 = \varepsilon_1^0$ of Eq. (1) is used. Although the spatial structure of individual magnetic storms is often more complicated and requires more coefficients, the validity of this P_1^0-assumption *on average* seems to be remarkably good, at least for periods longer than a few days. For a P_1^0 source, and a 1D conductivity, only one internal coefficient, ι_1^0, is induced, and the sum of external (inducing) and internal (induced) potential is

$$V = V^{\text{ext}} + V^{\text{int}} = \left[\varepsilon_1^0 + \iota_1^0(r/a)^3\right]\cos\theta_d \quad \text{(Eq. 2)}$$

This allows for a determination of the C-response from magnetic data of an individual site from observations of $B_r = \left(-\varepsilon_1^0 + 2\iota_1^0\right)\cos\theta_d$ and $B_\theta = \left(\varepsilon_1^0 + \iota_1^0\right)\sin\theta_d$ at that site ($r = a$) according to

$$C_1(\omega) = -\frac{a\tan\theta_d}{2}\frac{B_r(\omega)}{B_\theta(\omega)} \quad \text{(Eq. 3)}$$

This approach, known as Z/H-method, has been used by most researchers to infer mantle conductivity at a few hundred kilometer depths with periods between few days and several months.

Daily variations

Periodic external field (*q.v.*) variations at nonpolar latitudes (Sq variations) are caused by diurnal wind systems in the upper atmosphere, which produce electric currents in the E-layer of the *ionosphere* (*q.v.*) between 100 and 130 km altitude. Typical peak-to-peak amplitudes at midlatitudes are 20–50 nT; amplitudes during solar maximum are about twice as large as those during solar minimum. Sq variations are restricted to the sunlit (i.e., dayside) hemisphere, and thus depend mainly on local time $T = t + \phi$ (t and ϕ are UT and longitude expressed in radians). The external potential (see Eq. (1)) of such local

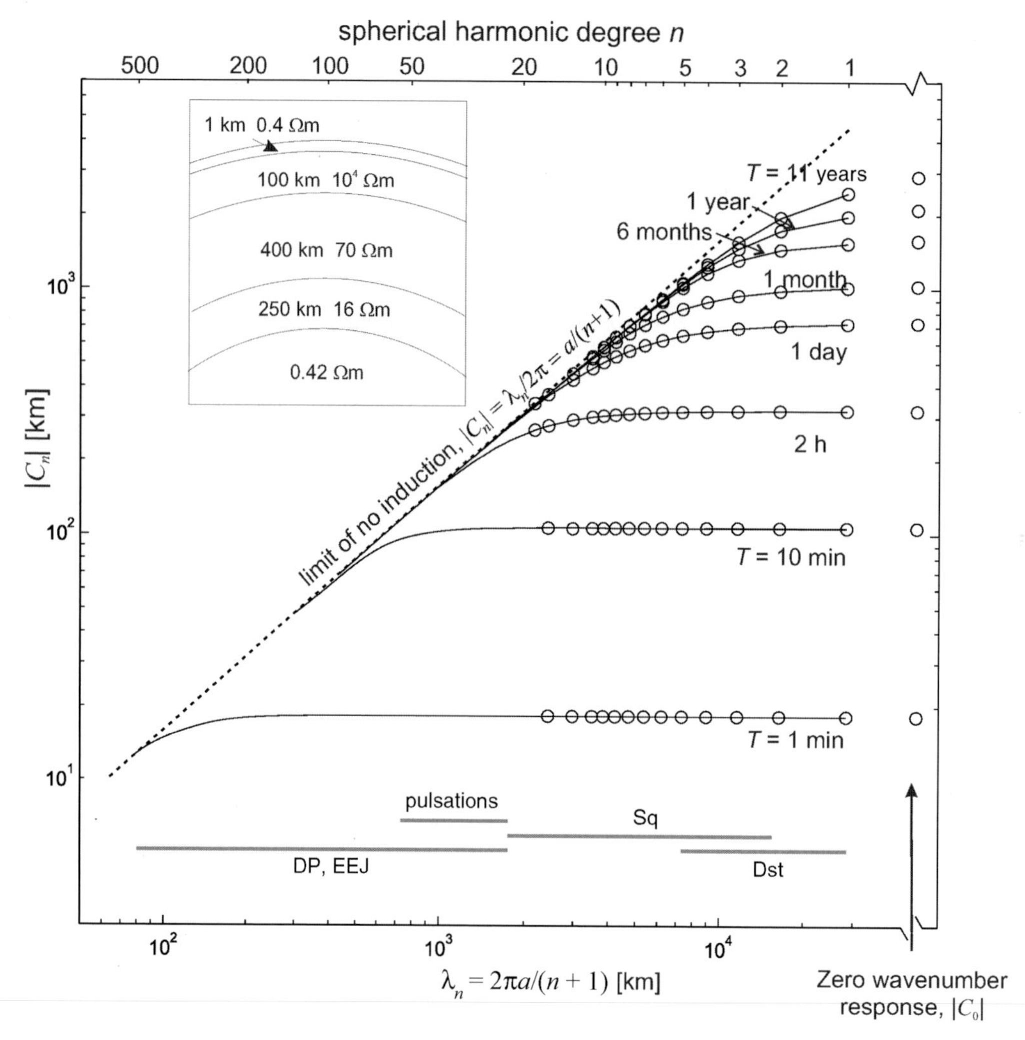

Figure N1 Absolute value of the C-response for a typical conductivity model (shown in the inset) as a function of horizontal scale-length λ_n and period T of the source.

Table N1 Classification of geomagnetic variations (after Schmucker 1985, modified), with typical periods, amplitudes, and penetration depths

Type of variation	Symbol	Typical period	Typical amplitude	Typical penetration depth
Solar cycle variations		11 years	10–20 nT	>2000 km
Annual variation		12 months	5 nT	1500–2000 km
Semiannual variation		6 months	5 nT	
Storm-time variation	Dst	Hours to weeks	50–500 nT	300–1000 km
Regular daily variation				
At midlatitudes	Sq	24 h and harmonics	20–50 nT	300–600 km
At low latitudes	EEJ		50–100 nT	
Substorms	DP	10 min to 2 h	100 nT (1000 nT at p.l.)	100–300 km
Pulsations (=Ultra low frequency waves)	ULF	0.2–600 s		20–100 km
regular	pc	150–600 s (pc5)	10 nT (100 nT at p.l.)	
continuous		45–150 s (pc4)	2 nT	
pulsations		5–45 s (pc2,3)	0.5 nT	
		0.2–5 s (pc1)	0.1 nT	
Irregular transient pulsations	pi	1–150 s	1 nT	
Extreme low-frequency waves	ELF sferics	1/5–1/1000 s	<0.1 nT	Tens of meters-kilometers
Schumann resonance oscillations		1/8 s	<0.1 nT	
Very low-frequency waves, whistlers	VLF	10^{-5}–10^{-3} s		Few meters-tens of meters

Note: If amplitude depends significantly on latitude, values are also given for polar latitudes (p.l., dipole latitude >65°).

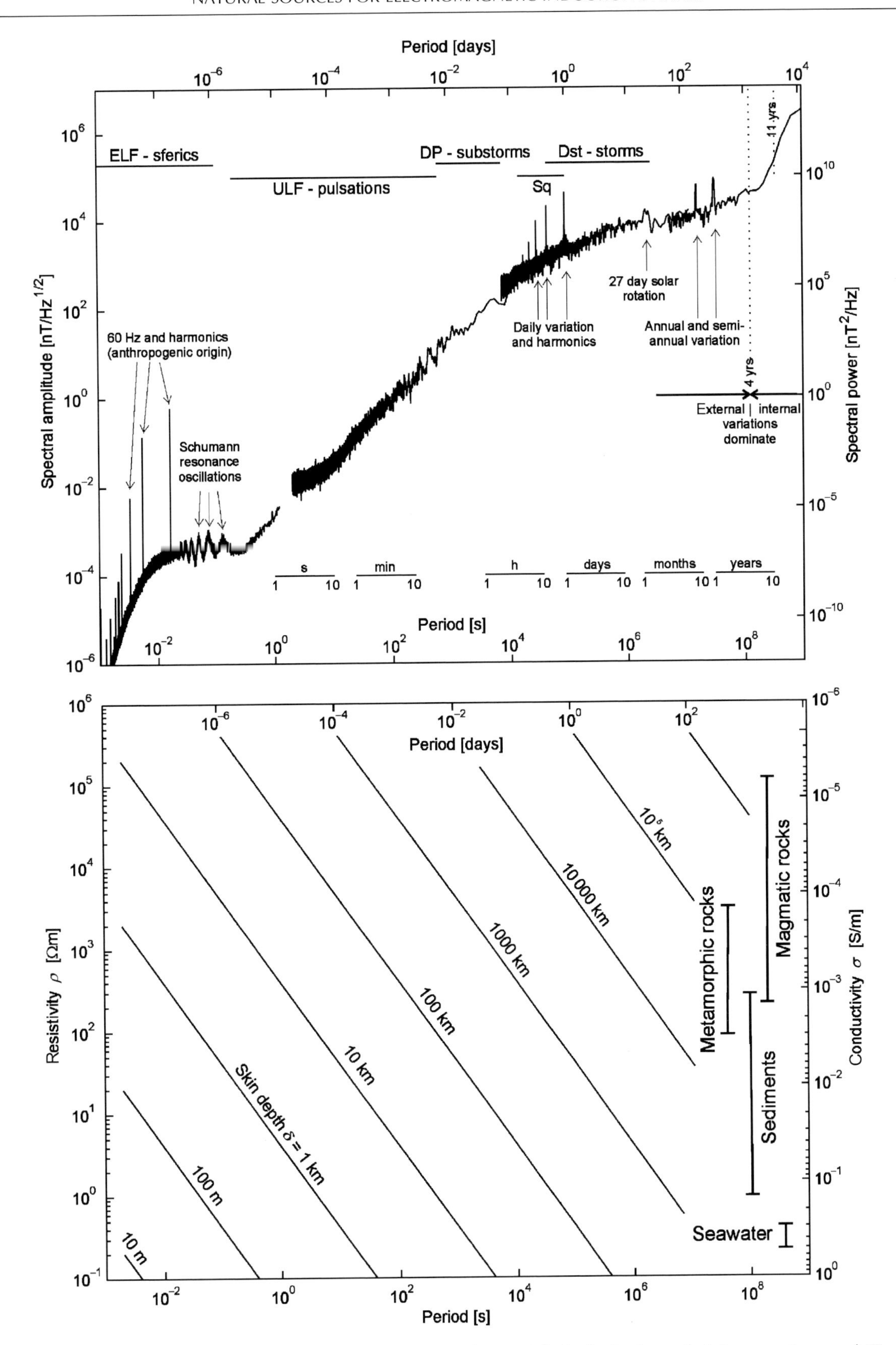

Figure N2 *Top*: Amplitude of magnetic variations (north component) at a midlatitude site for periods between 1 ms and 30 years, estimated using data from the sites Niemegk/Germany (2 h–30 years), Kakioka/Japan (2 s–2 h), and Socorro/USA (1 ms–1 s). *Bottom*: Skin depth $\delta = \sqrt{\rho T/\pi\mu_0}$ for a uniform half-space in dependence on resistivity ρ and period T.

time phenomena consists of elementary modes of the form $\varepsilon_n^m (r/a)^n P_n^m \exp imT$, with periods of 24/m h. Coefficients representing most of the power are those with $n = m + 1$, i.e., $\varepsilon_2^1, \varepsilon_3^2, \ldots$. Typically, four Fourier terms, $m = 1$–4 (periods 24,...,8 h), are used, which results in a somewhat restricted period range of a factor of four; therefore Sq variations provide conductivity estimates only in a rather limited depth range between roughly 300 and 600 km. The earliest attempts (Schuster, 1889; *Chapman* (*q.v.*), 1919) to determine mantle conductivity have used Sq as the inducing source.

There is about a fivefold enhancement of the daily variation near the dip-equator, called EEJ. Its latitudinal width is about 6°–8°, which requires spherical harmonics up to degree $n = 20$–30 or higher.

Seasonal and solar cycle variation

The amplitudes of Sq variations depend on the season and on the 11-year cycle of solar activity, but because of vanishing ionospheric conductivity (due to vanishing solar radiation) ionospheric currents are almost zero at night. This requires ionospheric currents that vary with season and solar cycle (i.e., there exists a true long-period *variation* in addition to a *modulation* of the daily variation). In addition, there is also a pronounced seasonal (mainly semiannual) and solar cycle variation of the magnetospheric ring current. Induction studies using semiannual, annual, and 11-year signals provide information on the conductivity of the lower mantle. However, reliable estimates of transfer functions at such long periods are hampered by contamination by short-term secular variation of the core field. The main spherical coefficients are ε_1^0 (semiannual, solar cycle variation) and ε_2^0 (annual variation).

Magnetospheric substorms, DP variations

Magnetic storms and substorms (*q.v.*) are irregular variations caused by the interaction of the Earth's magnetic field with the solar wind. Ionospheric currents in the auroral regions, called polar electrojets (PEJ), are linked to the magnetosphere by field-aligned currents. Both currents are highly dependent on solar wind conditions. Their magnetic field variation, called disturbed polar (DP) variation, has a typical timescale of a few minutes to hours and a horizontal scale-length of a few hundred kilometers.

Geomagnetic pulsations, ULF and ELF waves, sferics, and whistlers

Geomagnetic pulsations (*q.v.*), i.e., ultralow-frequency (ULF) waves, cover roughly the period range from fractions of a second to a few minutes. On the basis of their waveform, pulsations are classified into *pulsations continuous* (pc) and *pulsations irregular* (pi). Their amplitudes range from 100 to 0.1 nT, with longer periods exhibiting larger amplitudes. Typical amplitudes are 1 nT at midlatitudes and 10 nT at polar latitudes for a period of 30 s. There are several mechanisms that produce pulsations; resonant oscillation of the Earth's main field in response to changes in the solar wind is one of them. The typical spatial size of pulsations at midlatitudes is a thousand kilometers and longer.

Extremely low-frequency (ELF) waves covering periods from milliseconds to seconds originate from thunderstorm lightnings and are called atmospherics of sferics. This part of the spectrum shows, in particular, Schumann resonance oscillations (with peaks at 8 Hz and multiples) caused by standing waves in the waveguide formed by the Earth's surface and the ionosphere. Whistlers are sferics that travel along Earth's magnetic field lines from one hemisphere to the other. As higher frequencies travel faster than the lower ones the lightning wave is dispersed into a whistler tone.

Nils Olsen

Bibliography

Chapman, S., 1919. The solar and lunar diurnal variations of terrestrial magnetism. *Philosophical Transactions of the Royal Society of London, Series A*, **218**: 1–118.

Schmucker, U., 1985. Sources of the geomagnetic field. *Landolt-Börnstein,* Berlin-Heidelberg: Springer-Verlag. New-Series, 5/2b, section 4.1.

Schuster, A., 1889. The diurnal variations of terrestrial magnetism. *Philosophical Transactions of the Royal Society of London, Series A*, **180**: 467–518.

Cross-references

Chapman, Sydney (1888–1970)
Geomagnetic Deep Sounding
Geomagnetic Jerks
Geomagnetic Pulsations
Harmonics, Spherical
Ionosphere
Magnetosphere of the Earth
Magnetotellurics
Periodic External Fields
Ring Current
Storms and Substorms, Magnetic
Transfer Functions

NÉEL, LOUIS (1904–2000)

Louis Néel was born in Lyon, France. He followed the classic French academic path to excellence: starting from the "classes préparatoires Maths Sup" of Lycée du Parc in Lyon, he joined the Ecole Normale Supérieure de Paris and later obtained the "Agrégation de Physique." His early vocation for magnetism led him to defend a thesis under the supervision of Pierre Weiss in 1932 in Strasbourg, where he set up the theoretical basis that grounded his later discovery of antiferromagnetism (Néel, 1936; the term antiferromagnetism was later coined by Bitter). Mainly, his input was to introduce the concepts of quantum mechanics in magnetism (following Heisenberg exchange interaction concept), turning the uniform molecular field of Pierre Weiss into a local field. He became professor in Strasbourg in 1937 and gained

Figure N3 Louis Néel (1904–2000).

international recognition soon after. During World War II, he moved to Grenoble where he spent the rest of his career. His central contribution to modern magnetism was quite lately acknowledged by a Nobel Prize, shared in 1970 with Hannes Halfven. Both of them, top physicists, have grounded a discipline of geophysics: rock magnetism for the former, magnetospheric and space plasma physics for the latter. One can appreciate his place in modern magnetism knowing that his scientific contributions are at the origin or central to the understanding of antiferromagnetism, ferrimagnetism, fine-grained magnetism, thermoremanence, viscosity theory in single- and multidomains, theories of the Rayleigh law and hysteresis cycle, domain walls, exchange and surface anisotropies, coupling interaction and self-reversal, for the most notable. To get a more detailed account of his scientific contributions one can refer to his Nobel lecture (Néel, 1970) as well as to Prévot and Dunlop (2001).

Closer to our concerns, Louis Néel gained interest in paleomagnetism very early mostly through its rapid passage as the director of Clermont-Ferrand Geophysical Observatory in 1931 (30 years after B. Brunhes). At that time the nature of natural remanent magnetization (NRM) was a full mystery and this may have triggered a good part of Louis Néel theoretical breakthroughs. His most prized contribution in rock magnetism concerns the single domain theory of relaxation time, thermoremanence, and viscous remanence (Néel, 1949). This theory turned paleomagnetism from magic to science. In the original paper the theory was validated using the experimental results of Thellier on baked clays, going against the often-observed reluctance of physicists to use natural samples as a valid model. Nicely, this rather exotic contribution for a physicist nowadays is the most cited paper of Louis Néel, mostly in the nongeophysics literature. This is a beautiful example of the input of Earth sciences into physics, balancing the more common backward input. On the other hand, Louis Néel and his coworkers were able to synthesize and measure the highly anisotropic magnetization of the predicted ordered form of Fe-Ni alloy (Néel *et al.*, 1964). As this ordering appears only below 320°C, that is, at a temperature where atomic diffusion is negligible, they used a trick to enhance this diffusion: neutron bombardment. Much later, another trick was found—ultralow cooling rates—only available in meteorite parent bodies; the Fe-Ni ordered phase, tetrataenite, was "rediscovered" (e.g., Wasilewski, 1988).

A great lesson from Louis Néel's career (Néel, 1991) is that while being so sharp in his personal scientific work he was so broad in his interests and inspiration. His hobby was woodwork. It would be a large error to remember him mainly as a theoretician: he demonstrated great skills as an experimentator and as a man of action and organization. When the Nazi troops approached Strasbourg in 1939 he organized practically by himself the evacuation of the most sensitive equipments of the Strasbourg University in a train convoy. Soon after he was responsible for battleship demagnetization to avoid the triggering of magnetic mines moored at the entrance of French ports. It took him less than 6 months to set up original ship-size demagnetization systems in several major ports and efficiently "immunize" more than 500 ships. He was fully involved in practical and engineering magnetism, for example through military research or through the push of improving the magnetic materials used in recording or electromechanical devices. When he arrived in Grenoble, there were a handful of physicists in the university. When he retired from his leading position there 30 years after, thousands of people were involved in various fundamental and applied physics research institutes. As the leading builder of this top European center for solid-state physics, he early recognized that subdisciplines of theoretical and experimental physics have to work in close connection to solve the complex problems facing modern physics. As a result, it is nowhere else than in Grenoble, where it is possible to walk with the same sample from a neutron reactor to a synchrotron x-ray source via world record high-magnetic field facilities, finding all possible types of magnetic, spectroscopic, and other measuring devices in between. The Louis Néel scientific impetus is still visible today in the strong contribution of Grenoble laboratories—first of all, the Louis Néel laboratory: http://lab-neel.grenoble.cnrs.fr—to the advancement of magnetism (du Trémolet de Lacheisserie *et al.*, 2002). These textbooks include a foreword by Louis Néel, acknowledging the fact that he was going on to actively participate in the science community until the term of his tremendously filled life.

Pierre Rochette

Bibliography

Du Trémolet de Lacheisserie, E., Gignoux, D., and Schlenker, M., 2002. *Magnetism, (two volumes) I-Fundamentals, II-Applications.* Dordrecht: Kluwer, 507 and 517 pp.

Néel, L., 1936. Théorie du paramagnétisme constant application au manganese. *Comptes Rendus De l'Académie des Sciences Paris*, **203**: 304–306.

Néel, L., 1949. Théorie du traînage magnétique des ferromagnétiques en grains fins avec applications aux terres cuites. *Annales De Géophysique*, **5**: 99–136.

Néel, L., 1970. Magnetism and the local molecular field, Nobel Prize lecture. Available at: http://www.nobel.se/physics/laureates/1970/Néel-lecture.html.

Néel, L., 1991. Un siècle de Physique. Paris: Odile Jacob, 365 pp.

Néel, L., Paulevé, J., Dautreppe, D., and Laugier, J., 1964. Magnetic properties of an iron-nickel single crystal ordered by neutron bombardment. *Applied Physics*, **35**: 873–876.

Prévot, M., and Dunlop, D., 2001. Louis Néel: 40 years of magnetism. *Physics of the Earth and Planetary Interiors*, **126**: 3–6.

Wasilewski, P., 1988. Magnetic characterization of the new mineral tetrataenite and its contrast with isochemical taenite. *Physics of the Earth and Planetary Interiors*, **52**: 150–158.

NONDIPOLE FIELD

The nondipole (ND) field is that part of the internal geomagnetic field remaining after the major geocentric dipole contribution has been removed. It is distinct from the nonaxial-dipole (NAD) field for which only the component of the geocentric dipole that is parallel to Earth's rotation axis is subtracted. Figure N4a/Plate 4a shows the strength of the total scalar field at Earth's surface, with the spatial variations dominated by the dipole field, while in Figure N4b/Plate 4b the dipole contribution has been subtracted to reveal the substantially more complex nondipole field. Two source regions contribute to the ND field: the dynamo in Earth's core that is also responsible for the dipole part of the geomagnetic field produces the largest part; the other source is Earth's lithosphere (see *Crustal magnetic field*). Nondipole field contributions are significant, but contribute only a small fraction of the average magnetic energy at the surface, as can be seen in Figure N5a, which shows $\langle \vec{B}_l \cdot \vec{B}_l \rangle_{r=a}$, the squared average value of the field strength over the Earth's surface, average radius $r = a$, as a function of spherical harmonic degree, l. This *geomagnetic spatial power spectrum* (*q.v.*) falls off rapidly with increasing l (decreasing wavelength), up to about degree 12, then flattens out and remains roughly constant out to the shortest resolvable wavelengths. The ND field between degrees 2 and 11 is dominated by sources in Earth's core, while above degree 15 the core contribution is overwhelmed by that from lithospheric magnetic anomalies (see *Magnetic anomalies, modeling*; *Magnetic anomalies, long wavelength*; and *Magnetic anomalies, marine*). Between degrees 11 and 15, it is difficult to isolate the primary source, although time variations (Figure N5b) in the core part at these spatial scales will be better characterized with new high-quality satellite data. Temporal variations in the lithospheric field occur on geological timescales, but direct measurements over the past few centuries will only sense changes in inducing fields. These are very small at timescales of the order of a year or longer.

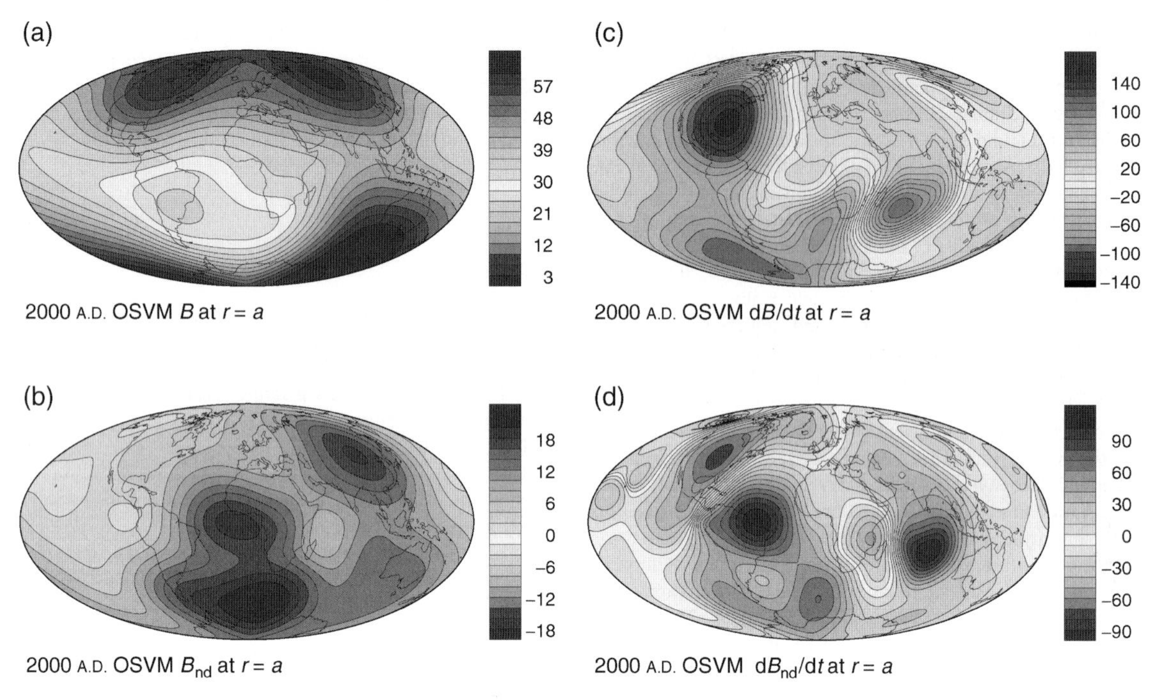

Figure N4/Plate 4a-d (a) Geomagnetic field strength B in μT and (b) its secular variation dB/dt in nT/y, evaluated at Earth's surface ($r = 6371.2$ km) using the geomagnetic field model OSVM for the epoch 2000. Lower panels, (c) and (d) are the nondipole field strength, B_{nd} in μT, and its rate of change, (nT/y). Note different scales for each panel.

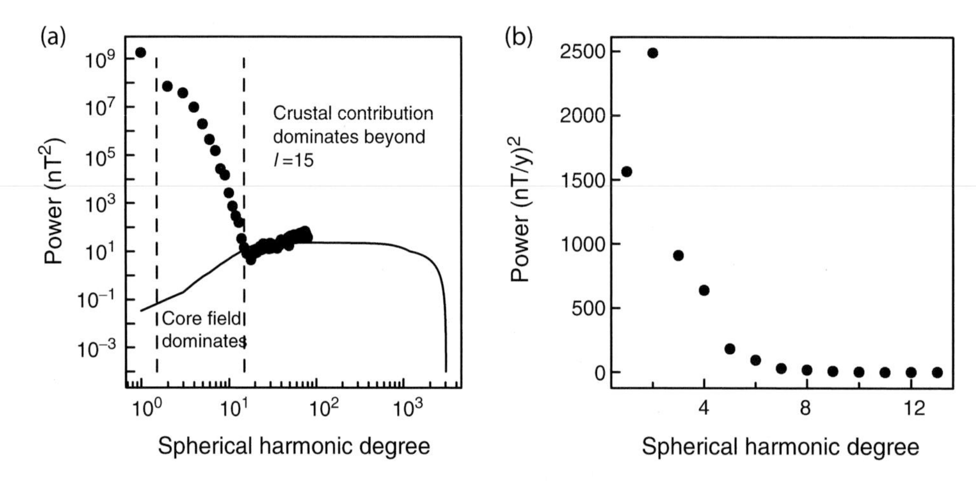

Figure N5 The spatial power spectrum of the geomagnetic field (a) and the spatial power spectrum of the secular variation (b) evaluated at Earth's surface ($r = 6371$ 2 km). In (a) black dots are for a satellite field model for epoch 2000 as a function of degree l, solid line gives the crustal power spectrum derived from *Project MAGNET* (*q.v.*) aeromagnetic data after removal of core field contributions.

Inferences about the historical ND field rely on *time-dependent models of the main magnetic field* (*q.v.*), extending from 1590 to the present (the GUFM model of Jackson *et al.*, 2000). Recently developed paleofield models (Continuous Archeomagnetic and Lake Sediment, CALS7K.2, for 0–7 ka, version 2) model of Korte and Constable, 2005) extend to millennial timescales and are likely to improve as more data become available. The Oersted secular variation model (OSVM) for epoch 2000.0 of Figure N4/Plate 4a-d (Olsen, 2002) is based on recent satellite and observatory data with excellent spatial coverage, and has much higher resolution than historical or paleofield models.

For the year 2000, the ND field is lower in the Pacific than the Atlantic or Asian hemisphere (Figure N4b/Plate 4b). Major ND contributions are in the Central and South Atlantic, where the total field seems anomalously low (Figure N4a/Plate 4a), and beneath Australia and Eastern Asia, where it is on average rather high. The South Atlantic anomaly is of some concern since the relatively low-geomagnetic field results in diminished geomagnetic shielding from cosmic radiation, presenting a hazard to low-Earth-orbiting satellites.

Figure N4c,d show the *geomagnetic secular variation* (*q.v.*), the rate of change with time, in the total and nondipole parts of the scalar field for the year 2000. The ND field is increasing in some regions and decreasing in others, with the largest rates of change in the Atlantic or Asian hemisphere. At present the secular variation in the geomagnetic field is predominantly in the ND part of the field. Figure N5b shows the maximum power is at degree 2. In time-varying historical models, much of the secular variation is manifest by *westward drift* (*q.v.*) of features in the ND field, but the westward drift is confined to the Atlantic or Asian hemisphere, with features modified or dying away before reaching the Pacific region. Paleofield records from distant sites are uncorrelated, supporting the view that the longevity of drifting ND features is insufficient to carry them for a full global circuit. For the second half of the 20th century variations in the ND field are well fit by dynamical models that comprise steady fluid flow in the outer core, and *torsional oscillations* (*q.v.*): such models predict observed *length of day variations* (*q.v.*, see also Jault, 2003), as well as the sudden changes known as *geomagnetic jerks* (*q.v.*). Westward drift of features in the ND field is well documented for the Atlantic or Asian hemisphere (Bloxham *et al.*, 1989), but definitive identification of poleward propagation in the ND field has proved elusive, perhaps because longer timescales are involved and the quality as well as the temporal and spatial distribution of records deteriorates with increasing age.

The longevity of both static and drifting features in the ND field is poorly documented, but both paleofield records and global models derived from them indicate substantial variations on timescales of hundreds to several thousands of years and probably longer. These variations are large in spatial scale, exhibiting substantial coherence over continental size geographic regions.

A fundamental tenet of paleomagnetism, embodied in the *geocentric axial dipole hypothesis* (*q.v.*), is that when the geomagnetic field is averaged over long timescales, nonaxial-dipole contributions can be considered negligible. That this is approximately true can be seen in Figure N6/Plate 4e-h, where the vertical component of the nonaxial-dipole (NAD) field is shown for the year A.D. 2000, along with averages for the most recent 400 years, 7 ka and 5 Ma. The magnitude diminishes with increasing averaging interval, but there remain small (apparently time-varying) NAD field contributions on all timescales. The spatial scale of the residual contributions increases with averaging interval, supporting the idea that short wavelength features are associated with shorter timescales. The short-term averages in Figure N6a,b/Plate 4e,f exhibit a number of similarities to one another, which seem quite distinct from the similarities between the longer term averages in Figure N6c,d/Plate 4g,h.

The similarities in structure for the long-term averages are generally positive NAD radial fields at equatorial and low latitudes and generally negative NAD radial fields at high latitude. This overall latitudinal variation in NAD radial field contributes to the persistent "far-sided effect" in virtual geomagnetic poles derived from paleomagnetic directional observations and detected in early work that identified small departures from the geocentric axial dipole hypothesis. This reflects a (latitudinally varying) negative deviation of the inclination from that predicted by a geocentric axial dipole field (Merrill *et al.*, 1996). For a several million year average such an effect can be largely explained by a persistent axial magnetic quadrupole with a moment of the order of a few percent of that of the axial dipole. The influence of any axial quadrupole on paleomagnetic directions is largest at the equator, generally visible at low- to midlatitudes, and basically so small as to be undetectable at high latitudes. The existence of a small but persistent geocentric axial quadrupole contribution in addition to the axial dipole is the only feature on which all magnetic field models for the interval 0–5 Ma agree: however, the size of the estimated contribution varies by about a factor of three (McElhinny, 2004).

Although it is often supposed that nonzonal (longitudinally varying) contributions to the field will average out on long timescales, this remains controversial. Small contributions persist in the model in Figure N6d/Plate 4h and statistical models of *paleomagnetic secular variation* for the time interval 0–5 Ma invoke significant variations attributed to nonzonal quadrupolar fields. Some researchers attribute all of the nonzonal structure to poor quality and spatial distribution of the data (McElhinny, 2004). Others (including the author) believe that the persistent quadrupole is overemphasized in many models because of the poor geographic coverage (see Gubbins, 1998, 2003; Gubbins and Gibbons, 2004), and that one might expect hemispheric differences arising from thermal *core-mantle coupling* (*q.v.*).

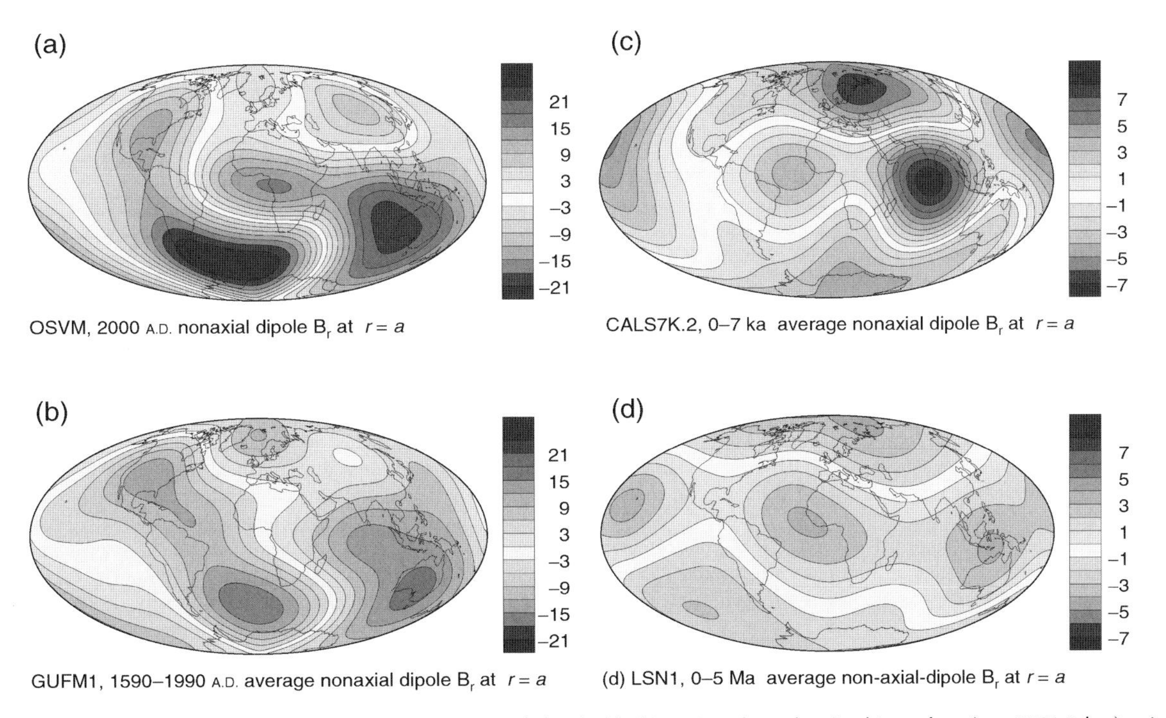

Figure N6/Plate 4e-h (a) Vertical component of the nonaxial dipole filed in μT evaluated at Earth's surface ($r = 6371.2$ km) using the geomagnetic field models OSVM for A.D. 2000, (b) GUFM averaged over 400 years, (c) CALS7K.2 averaged over 7 ka, and (d) LSN1 (Johnson and Constable, 1997), from normal polarity lava and marine sediment directions averaged over 5 Ma. Note scales for (c) and (d) differ by factor of 3 from (a) and (b).

It is notable that the persistent nonzonal structures in Figure N6c,d/Plate 4g,h do have a number of similarities, despite the fact that they are derived from very different kinds of data. It is likely that much of the detail in Figure N6c,d will change as new modeling efforts take advantage of improved data sets. It remains unclear whether field behavior in the Pacific hemisphere is persistently different over thousands or millions of years: paleofield models for 0–7 ka indicate substantial secular variation, but they are not yet good enough to allow a definitive determination of whether the ND field is consistently anomalous there. The same holds true for longer time intervals. More effort needs to be invested in discriminating among the various data sets and viable models.

The temporal evolution of the South Atlantic anomaly from the historical to the present field has given rise to speculation that complex field structures in this region may be a sign of impending geomagnetic reversal. The longevity of this feature is unknown, making it an interesting target for future study. There is no evidence that the nondipole field is substantially increased or diminished during a reversal: most reversal records indicate field strengths around 10%–20% of the prereversal field at their lowest point. It is possible that the NAD field may differ for successive polarity chrons, but this remains at the limit of current data resolution. The possibility of larger nondipole field contributions in the ancient past is widely discussed (van der Voo and Torsvik, 2004; Courtillot and Besse, 2004), and for Precambrian times it cannot be ruled out with currently available data (Dunlop and Yu, 2004).

Catherine Constable

Bibliography

Bloxham, J., Gubbins, D., and Jackson, A., 1989. Geomagnetic secular variation. *Philosophical Transactions of the Royal Society of London, Series A*, **329**: 415–502.

Courtillot, V., and Besse, J., 2004. A long-term octupolar component in the geomagnetic field. In Channell, J.E.T., Kent, D.V., Lowrie, W., and Meert, J.G. *Timescales of the Paleomagnetic Field,* AGU Geophysical Monograph 145. Washington, DC: American Geophysical Union, pp. 59–74.

Dunlop, D.J., and Yu, Y., 2004. Intensity and polarity of the geomagnetic field during Precambrian time. In Channell, J.E.T., Kent, D.V., Lowrie, W., and Meert, J.G. (eds.), *Timescales of the Paleomagnetic Field*, AGU Geophysical Monograph 145. Washington, DC: American Geophysical Union, pp. 85–100.

Gubbins, D.G., 1998. Interpreting the paleomagnetic field. In Gurnis, M. *The Core-Mantle Boundary Region.* Washington, DC: AGU Geodynamics Series 28, pp. 167–182.

Gubbins, D.G., 2003. Thermal core-mantle interactions. In Dehant, V., Creager, K.C., Karato, S., and Zatman, S. (eds.), *Earth's Core: Dynamics, Structure, Rotation*, AGU Geodynamics Series 31. Washington, DC: American Geophysical Union, pp. 163–179.

Gubbins, D.G., and Gibbons, S.J., 2004. Low Pacific secular variation. In Channell, J.E.T., Kent, D.V., Lowrie, W., and Meert, J.G. (eds.), *Timescales of the Paleomagnetic Field*, AGU Geophysical Monograph 145. Washington, DC: American Geophysical Union, pp. 279–286.

Jackson, A., Jonkers, A.R.T., and Walker, M.R., 2000. Four centuries of geomagnetic secular variation from historical records. *Philosophical Transactions of the Royal Society of London, Series A*, **358**: 957–990.

Jault, D., 2003. Electromagnetic and topographic coupling, and LOD variations. In Jones, C., Soward, A., and Zhang, K. (eds.), *Earth's Core and Lower Mantle. The Fluid Mechanics of Astrophysics and Geophysics.* London: Taylor & Francis, pp. 56–76.

Johnson, C.L., and Constable, C.G., 1997. The time-averaged geomagnetic field: global and regional biases for 0–5 Ma. *Geophysical Journal International*, **131**: 643–666.

Korte, M., and Constable, C.G., 2005. Continuous geomagnetic models for the past 7 millennia II: CALS7K. *Geochemistry, Geophysics, Geosystems*, **6**(2): Q02H16 DOI 10.1029/2004GC000801.

McElhinny, M.W., 2004. Geocentric axial dipole hypothesis: a least squares perspective. In Channell, J.E.T., Kent, D.V., Lowrie, W., and Meert, J.G. (eds.), *Timescales of the Paleomagnetic Field*, AGU Geophysical Monograph 145. Washington, DC: American Geophysical Union, pp. 1–12.

Merrill, R.T., McElhinny, M.W., and McFadden, P.L., 1996. *The Magnetic Field of the Earth: Paleomagnetism, the Core and the Deep Mantle.* San Diego, CA: Academic Press.

Olsen, N., 2002. A model of the geomagnetic field and its secular variation for epoch 2000. *Geophysical Journal International*, **149**: 454–462.

Van der Voo, R., and Torsvik, T., 2004. The quality of the European Permo-Triassic paleopoles and its impact in Pangea reconstructions. In Channell, J.E.T., Kent, D.V., Lowrie, W., and Meert, J.G. (eds.), *Timescales of the Paleomagnetic Field*, AGU Geophysical Monograph 145. Washington, DC: American Geophysical Union, pp. 29–42.

Cross-references

Core-Mantle Coupling, Thermal
Crustal Magnetic Field
Geocentric Axial Dipole Hypothesis
Geomagnetic Jerks
Geomagnetic Secular Variation
Geomagnetic Spectrum, Spatial
Length of Day Variations, Decadal
Length of Day Variations, Long-Term
Magnetic Anomalies, Long Wavelength
Magnetic Anomalies, Marine
Magnetic Anomalies, Modeling
Oscillations, Torsional
Project MAGNET
Time-Dependent Models of the Geomagnetic Field
Westward Drift

NONDYNAMO THEORIES

The internal magnetic field of Earth or any planet is normally considered to have three components at most: A large, globally coherent field is attributed to a dynamo operating deep within the planet; a spatially complex, usually much smaller field is attributed to permanent magnetism of near-surface rocks, and a small time-varying magnetotelluric field is induced by the external time-varying field. For Earth at least, the separation of these three is not difficult because of their very different spatial and temporal characteristics. There are some bodies for which the last two fields dominate. For example, the Martian field appears to be dominated by permanent magnetism in the crust, and the change in field near Europe is dominated by an induced field, presumably caused by electrical currents induced in a salty ocean beneath an icy shell. see *Dynamos, Planetary and Satellite* for a discussion of the conventional explanations of observed fields. This section is concerned with other alternatives that do not easily fit into one of three categories of field defined above, with a particular emphasis on the possibility that a large global field might have a nondynamo origin.

There is at present strong acceptance of a dynamo origin for most of Earth's magnetic field. The same acceptance applies to fields of comparable magnitude in other planets or stars Parker (1979). Historically, this acceptance was slow to develop, even though the essential physical ideas (Faraday's law in particular) are over 150 years old and the concept of a dynamo existed long before general acceptance of its validity.

This slow acceptance can be attributed primarily to the difficulty of deriving the desired dynamo behavior from the governing equations. Without the advent of modern computers, beginning in the 1950s, it would still be difficult to convince oneself that a homogeneous dynamo really works. We now know that this difficulty is real, in the sense that many possible fluid flows do not create a dynamo, but that the flows that occur in convecting, rotating planetary cores are indeed suitable for field generation. During the long period of uncertainty about dynamos, many alternatives were considered and advocated. Alternatives are still advocated occasionally. These remain interesting for several reasons. First, they might be relevant for Earth or other planets as an increment to a dynamo field or where the total field is small. Second, the criteria for their presence or their failure might tell us about the conditions within planets. Last (but not least), the dynamo mechanism is still sufficiently poorly understood that one should be open to alternatives.

These nondynamo theories can be categorized in three groups: (1) Theories that invoke well-established physical principles (phenomena that are known to produce magnetic fields in the lab). (2) Theories that may have a physical basis but invoke properties or conditions far removed from those that are compatible with existing laboratory data or material properties. (3) Theories that go beyond established physical principles. We discuss these various ideas in roughly that order and define their specific failings. A summary is then provided for the common failings that these alternatives frequently share. Further details can be found in the bibliography.

Primordial currents and fields

A current inside the core does not immediately disappear when the energy source for its generation is turned off. This raises the possibility that the present magnetic field is the consequence of a freely decaying current distribution of ancient or primordial origin. However, this is not an acceptable explanation because the free decay-time of the field is of order a few tens of thousand years at most (for Earth) and paleomagnetic evidence shows that the field has persisted throughout most of geologic time.

Unexpected or exotic permanent magnetism

In addition to the readily demonstrated permanent magnetism associated with some minerals (e.g., magnetite) in cold, near-surface crustal rocks, one should consider the possibility that some mineral deep within Earth (and perhaps the solid inner core) exhibits permanent magnetism. This cannot be excluded with certainty. Moreover, permanent magnetism (unlike most of the mechanisms discussed here) is in principle capable of providing a large field. However, neither is there any laboratory evidence of high-temperature permanent magnetism in any material of relevance to Earth's interior, nor is there any theoretical support for this idea from *ab initio* quantum mechanical calculations. Solid metallic iron loses its ferromagnetism at modest pressures relative to Earth's deep interior (and the temperatures are in any event high relative to any plausible Curie point). Moreover, any permanent magnet explanation would fail to explain the well-established temporal variation of the field. Rotating permanent magnets can also exhibit a dipole aligned with the rotation axis (the *gyromagnetic effect* Barnett, 1933) arising from the fundamental connection between angular momenta and magnetic moments for nuclei and atoms, but for the Earth's slow rotation rate this fails by ten orders of magnitude to explain Earth's field.

Off-diagonal condensed matter physics effects

Any disequilibrium or gradient in the electronic chemical potential can lead to currents and hence (through Ampere's law) to magnetic fields. Some of these effects are "off-diagonal" in the sense that they involve a relationship among vectors or components of vectors that one would normally treat as independent. The best known and most frequently invoked of these is the *thermoelectric effect*: An emf and electrical current that arises when both compositional and thermal differences are present. The most likely application of this would be that the emf arises at the boundary between a metallic core and a semiconducting mantle when there are temperature variations along that boundary. For reasonable parameter choices, an emf of order millivolts can arise. The associated currents are poloidal and the resulting field is accordingly toroidal. Despite the undisputed reality of this effect, two problems are immediately apparent in its application to explaining a large magnetic field: First, the likely emfs provided are typically two or more orders of magnitude less than dynamo emfs (which are of order 1–10 V typically). Second, the resulting field is toroidal rather than poloidal.

Other off-diagonal effects are even less attractive. The *thermomagnetic effect* (or equivalently the Nernst-Ettinghausen effect) provides a current and hence a field when there is a preexiting field and thermal gradient. In this case, one would hope to use this effect to amplify a seed magnetic field. The magnitude of the effect is too small by orders of magnitude and the proposed amplification is probably not possible, even in principle. The *Hall effect* is well known but offers no method of amplification: It merely changes the conductivity tensor.

Rotating electric fields and charges

According to fundamental electricity and magnetism, magnetic fields can be related to moving charges. In everyday experience, this is usually associated with materials that are very close to electrically neutral on all scales much larger than atomic scale. However, a rotating body with net charge (or charge separation, i.e., macroscopic capacitor) could provide a magnetic field. There is no large net charge associated with Earth (though there can be localized large separation of charge in the atmosphere, e.g., as in thunderstorms) so this mechanism can have no relevance unless the Earth is a large, charged capacitor with the two "plates" at different mean distances from Earth's center. The problem is that one requires very large voltage differences in order to produce modest magnetic fields. In SI units, the electric field E (in V m^{-1}) required to explain a field of strength B (in tesla) is $\sim c^2B/v$, where v is the rotational velocity. This suggests $E \sim 10^{11}$ V m^{-1}, which is physically impossible (it will ionize atoms). This must not be confused with the sustained emf and voltage (only a few volts) that arise through induction during dynamo generation in an electrical conductor. The latter continually drives a current, whereas the former must be associated with a capacitor that never discharges. A "capacitor" (meaning any macroscopic charge separation) can in principle exist even in a conducting region. For example, the thermoelectric effect discussed above causes charge separation. However, the associated electric fields are totally insignificant.

An interesting but equally inadequate origin for an electric field is the differential effect of gravity on electrons and ions. This *compression effect* can create a voltage difference of a few volts in planets (and is actually important in the diffusion and gravitational settling within white dwarf stars) but it drives no sustained current (i.e., does not have an energy source) and is undetectable in Earth by many orders of magnitude.

Magnetic monopoles

Dirac postulated the possible existence of magnetic monopoles. From the perspective of modern elementary particle physics, the idea of a magnetic "charge" is still acceptable. They are predicted in some grand unified theories. However, they are also predicted to have immense masses by conventional elementary particle standards, suggesting that they are exceeding rare at best. There are a few claimed "sightings" of monopoles in laboratory experiments, but none of these has been confirmed. From the perspective of Earth and planets, the telltale signal of a net magnetic charge would be a field that decays as inverse square rather than inverse cube. There is no evidence for a contribution of this

Table N2 Nondynamo theories

Mechanism	Physical basis	Significance and possible magnitude
Primordial	Finite decay time of currents	Decay time is too short compared to age of the Earth
Unexpected (noncrustal) ferromagnetism permanent magnetism		No suitable candidate minerals exists
Gyromagnetism	Dipole associated with a rotating ferromagnet	Too small by at least 10^{10}
Thermoelectric effect	Differences in temperature and composition	Field too small and wrong geometry
Thermomagnetic effect	Currents arising in presence of thermal gradients and magnetic field	Amplification too small; no regeneration of field
Hall effect	Additional current in presence of magnetic field	Small change in existing field
Rotating electric fields	Current arising from rotating charged capacitor	Requires impossibly large charge or electric fields
Compression effect	Gravity acts differently on electrons and massive nuclei	Too small by $\sim 10^{10}$
Magnetic monopoles	Permitted in some fundamental particle physics theories	Not observed
Magnetic "Bodes law" hypothesized fundamental connection between magnetic moment and angular momentum	No observational or theoretical basis	

form. One could in principle "hide" their presence by hypothesizing equal numbers of magnetic charges of each sign (i.e., equal numbers of South and North poles that are spatially separated within Earth) but in the absence of any laboratory evidence, this hypothesis has no support.

New physical theories associated with rotation

Since Earth's dipole is approximately aligned with the rotation axis, it is natural to think of some (unspecified) physical process that creates a magnetic dipole aligned with the angular momentum vector for any rotating body. This is in the sprit of the previously mentioned gyromagnetic effect, except that it would be different physics (i.e., would exist even for a nonmagnetic material) and would be important only for very large accumulations of material. One could hypothesize, for example, that $M \propto J^n$, where M is the magnetic moment, J is the angular momentum, and n is some exponent. This proposed proportionality has become known as the *Magnetic Bode's law* (in very crude analogy with the Titus-Bode geometric progression of planetary orbital spacings), except that the label is now more often used to describe some poorly understood scaling for the dynamo process rather than hypothesized new physics. (Note that the original "Bode's law" is not in fact a physical law.).

The basic idea that there is some universal linear relationship between angular momentum and magnetic moment sufficient to explain Earth's magnetic field was disproved in the laboratory by Blackett during his development of highly sensitive magnetometers in the 1950s. The strictly empirical claimed proportionality observed when one plots M and J for planets on a log-log plot (magnetic Bodes law) deserves to be in ill repute because it covers many orders of magnitude and most of the range is explained by the fact that dipole moments scale as planet radius cubed, while angular momenta scale as planet radius to the fifth power. (The other parameters in this hypothesized relationship vary by smaller amounts.) The magnetic Bodes "law" can be used as a pedagogical example of how to mislead oneself and others by plotting two quantities that have an underlying common factor and thereby falsely concluding a relationship.

Summary

Table N2 summarizes nondynamo theories, their basis (or otherwise) in fundamental physics, and their possible significance and magnitude.

With the exception of permanent magnetism, these proposals are either real effects that are too small (usually by many orders of magnitude) or effects that have no established reality. The bibliography includes reference to two papers on the planet Mercury, which illustrate the extent to which alternatives can continue to have some credibility in an instance where the field is much smaller than Earth and the data are limited.

David J. Stevenson

Bibliography

Aharonson, O., Zuber, M.T., and Solomon, S.C., 2004. Crustal remanence in an internally magnetized nonuniform shell: a possible source for Mercury's magnetic field? *Earth and Planetary Science Letters*, **218**: 261–268.

Barnett, S.J., 1933. Gyromagnetic effects: history, theory, and experiments. *Physica*, **13**: 241.

Elsasser, W.M., 1939. On the origin of the Earth's magnetic field. *Physical Review*, **55**: 489.

Inglis, D.R., 1955. Theories of the Earth's magnetism. *Review of Modern Physics*, **27**: 212.

Merrill, R.T., McElhinney, M.W., and McFadden, P.L., 1996. *The Magnetic Field of the Earth.* New York: Academic Press, 531 pp.

Parker, E.N., 1979. *Cosmical Magnetic Fields: Their Origin and Their Activity.* Oxford: Clarendon Press; New York: Oxford University Press, 841 pp.

Poirier, J-P., 1991. *Introduction to the Physics of the Earth's Interior.* New York: Cambridge University Press, p. 191.

Roberts, P.H., and Glatzmaier, G.A., 2000. Geodynamo theory and simulations. *Review of Modern Physics*, **72**: 1081–1123.

Russell, C.T., 1993. Magnetic fields of the terrestrial Planets. *Journal of Geophysical Research-Planets*, **98**: 18681–18695.

Stevenson, D.J., 1974. Planetary magnetism. *Icarus*, **22**: 403–415.

Stevenson, D.J., 1983. Planetary magnetic fields. *Reports on Progress in Physics*, **46**: 555–620.

Stevenson, D.J., 1987. Mercury's magnetic field—a thermoelectric dynamo? *Earth and Planetary Science Letters*, **82**: 114–120.

Cross-references

Dynamos, Planetary and Satellite

NORMAN, ROBERT (FLOURISHED 1560–1585)

Nothing is known of the birth and death dates, parentage, or precise marital status of Robert Norman. All that we have to go on, from information contained in his two published works and a few other scattered fragments, is that he served at sea for 18 or 20 years before settling in the seafaring district of Ratcliff, on the north bank of the Thames, London, as a maker of navigational instruments, and in particular of marine compasses. It was to be in this latter profession that he won both a contemporary and an enduring fame in the history of geomagnetism. In *The Newe Attractive* (London, 1581), Norman published the first serious study of the magnetic dip. For while the dip had been discovered by the German Georg Hartmann around 1544, Hartmann had only communicated his discovery in a private letter that was not to become known to the wider world until 1831, so that Robert Norman's discovery, announced in print in 1581, can be rightly credited as independent and original.

Although a working instrument maker, Norman clearly possessed all the right instincts for an original research scientist. He made a long-standing study of the behavior of the compass, and found that to make a ship's compass needle lie properly flat on its suspension pin in London, he had to weight the south end of the needle, for it invariably dipped down to the north. A small blob of wax was used [illegible] that, in London, the magnetic needle dips at an angle of 71°50′. This prompted him to next investigate how the needle behaved elsewhere, which in turn brought him to the understanding that the Earth's magnetic field seemed to act in more than just the horizontal plane. Norman realized that knowledge of the magnetic dip, when used in conjunction with the position of the usual horizontal compass needle, could be of great benefit to position-finding at sea. For if navigators and researchers could build up a detailed knowledge of the position of the horizontal and dip needles for different parts of the globe, then a sailor could derive a better knowledge of his position. (A similar logic was to lie behind William Gilbert's practical uses of magnetism and Edmond Halley's magnetic surveys of the Atlantic, over a century later.) Unlike Gilbert, however, Norman never developed a coherent theory about what magnetism was or how it occurred.

Robert Norman belonged to a class of men who were coming to play an increasingly significant role in the wider rise of science, especially in England: intelligent, ingenious craftsmen, sailors, and experienced travelers, who while not "learned" in the sense of having received a Latinate classical education, were to approach the study of natural phenomena from a refreshing and original angle. Unfettered as they were by classical cautions about the unreliability of sense-knowledge, and outside the linguistic and philosophical games of the European universities, they formed that cadre of men whose praises were soon to be sung by Dr. William Gilbert, Sir Francis Bacon, and the early fellows of the Royal Society. It would be wrong, however, to see Norman, and enterprising men like him, as simple tradesmen who were fond of writing. For, although not classically educated, Norman seems to have had a good reading and even a translating knowledge of Dutch, and was encouraged in his magnetical researches by none other than William Borough, himself a student of magnetism and comptroller of the Royal Navy, to whom *The Newe Attractive* was dedicated. It is perhaps more appropriate to see Norman as a successful entrepreneur who financed his scientific researches on the strength of a successful business: a pattern that would be very common in British science down to the late 19th century.

Geomagnetism, moreover, was in many ways a more challenging field of research for such men to become occupied with, for while mathematical-scale graduators and cross-staff makers became involved in astronomy and oceanography, the study of magnetism was more innately mysterious. This meant that it was much more "philosophical," or what we would call research-physics-based, in so far that it—like optics—involved wrestling with and discovering the mathematical behavior of invisible and seemingly intangible natural forces. Indeed, the essentially practical nature of physics research as conducted by men such as Norman, he explicitly contrasted with the approach of "learned" scholars, who preferred to remain in their studies and find answers to all things within their books. What is more, Norman, like contemporary navigation writers such as Edward Wright, was a pioneer of the English language as a medium of scientific and technical discourse. His *The Safegarde of Saylers, or, Great Rutter* (1584, 1590) used contemporary material, which Normon had translated from Dutch vernacular navigation manuals, often called "Rutters" (or Routers), and was, in fact, part of a vernacular technical and navigation book movement that included not only Dutch and English, but also French, Spanish, and Italian materials. Indeed, all the great sea powers of the age produced vernacular authors who wrote instruction books and tables for their non-Latinate sea captains who now faced, [illegible]

And what began as vernacular books for sailors became increasingly the norm for learned scientists by the 17th century, and by the 1660s the Royal Society and the French Académie regarded it as normal to publish the researches in their mother tongue. In this way, therefore, Norman and his magnetic researches played a role in the growth of a new intellectual tradition. Norman's published work also helped to stimulate wider English researches into geomagnetism. Edmund Gunter, Henry Gellibrand, and those early Gresham College, London, professors who began to monitor the westward drift of the compass variation at Limehouse between 1622 and 1634 (in the wake of William Borough's work in the same place 50 years earlier), *learned* men as they were, took at least part of their queue from Norman's example.

Norman's instruments were clearly prized, for Dr. Christopher Merret, an early fellow of the Royal Society, recorded taking one of Norman's "Clinatories" or dip-needles for repair by Henry Bond nearly a century after it had been made.

It is sad that the wider details of the life of a figure of such importance as Robert Norman in the early history of geomagnetism should have been lost.

Allan Chapman

Bibliography

Bennett, J.A., 2004. Robert Norman. *New DNB*. Oxford: Oxford University Press.

Chapman, A., 1998. Gresham College: Scientific instruments and the advancement of useful knowledge in seventeenth-century England. *Bulletin of the Scientific Instrument Society*, **56**: 6–13.

Norman, R., 1581. *The Newe Attractive*. London.

Taylor, E.G.R., (1954, 1968). *The Mathematical Practitioners of Tudor and Stuart England 1485–1714*. London: Cambridge University Press, p. 173 no. 29.

O

OBSERVATORIES, OVERVIEW

Geomagnetic observatories carry out continuous and accurate monitoring of the strength and direction of the Earth's magnetic field over many years, making measurements at least every minute. Observatory data reveal how the field is changing on a wide range of timescales from seconds to centuries, and this is important for understanding processes both inside and outside the Earth. It is estimated that there are approximately 180 observatories currently operating around the world (Figure O1) (for more information on observatories in particular countries or regions of the world see *Observatories in . . .* and Figure O2/Plate 3). The distribution of observatories is largely determined by the location of habitable land and by the availability of local expertise, funds, and energy supply, and as result, it is uneven and a little sparse in some regions. Many observatories have had to move because of encroaching urbanization. However, the continuity of their data series is generally maintained by simultaneous observations at the old site and new site over a period of time to allow site differences to be established.

The distribution of data in time is shown in Figure O3. International scientific campaigns such as the first International Polar Year in 1882/1883, the second International Polar Year in 1932/1933, and the International Geophysical Year in 1957/1958 encouraged the opening of many observatories around the world. Geomagnetism is a cross-disciplinary science, and as a result, observatories are run by a wide variety of institutes whose interests range from geology, mapping, geophysics (including seismology and earthquake prediction), meteorology to solar-terrestrial physics, and astronomy.

There are two main categories of instruments at an observatory. The first category comprises variometers that make continuous measurements of elements of the geomagnetic field vector but in arbitrary units, for example millimeters of photographic paper in the case of photographic systems or electrical voltage in the case of fluxgates (see *Observatories, instrumentation* and *Observatories, automation*). Both analog and digital variometers require temperature-controlled environments and installation on extremely stable platforms (though some modern systems are suspended and therefore compensate for platform tilt). Even with these precautions they can still be subject to drift. They operate with minimal manual intervention but the resulting data are not absolute.

The second category comprises absolute instruments that can make measurements of the magnetic field in terms of absolute physical basic

Figure O1 Locations of currently operating observatories.

Figure O2/Plate 3 Continued

Figure O2/Plate 3 The Danish Meteorological Institute suspended triaxial fluxgate magnetometer being adjusted at Alibag observatory in India (a). This variometer can measure the magnetic field variations along the axes of three orthogonal sensors every second. Another instrument also run in continuous mode at observatories is a proton precession magnetometer. It is an absolute instrument and measures the strength of the magnetic field. In order to convert the variometer data into absolute data, manual observations are necessary. For this a fluxgate theodolite, or DI-flux, is widely used. (b) Shows the fluxgate theodolite at Kakadu observatory in Australia. The fluxgate sensor is fixed to the telescope of the non-magnetic theodolite and is used to indicate when its axis is orthogonal to the magnetic field, and therefore the orientation of the field (see *Observatories, instrumentation* and *Observatories, overview*). (c) Fürstenfeldbruck observatory, Germany; (d) Resolute observatory, Canada; (e) Qeqertarsuaq (Godhavn) observatory, Greenland; (f) Tamanrasset observatory, Algeria; (g) Hartland observatory, UK; (h) Dumont D'Urville observatory, Antarctica; (i) Tihany observatory, Hungary. These buildings, some in the vernacular style, are made of non-magnetic material and can be maintained at stable temperatures if housing variometers.

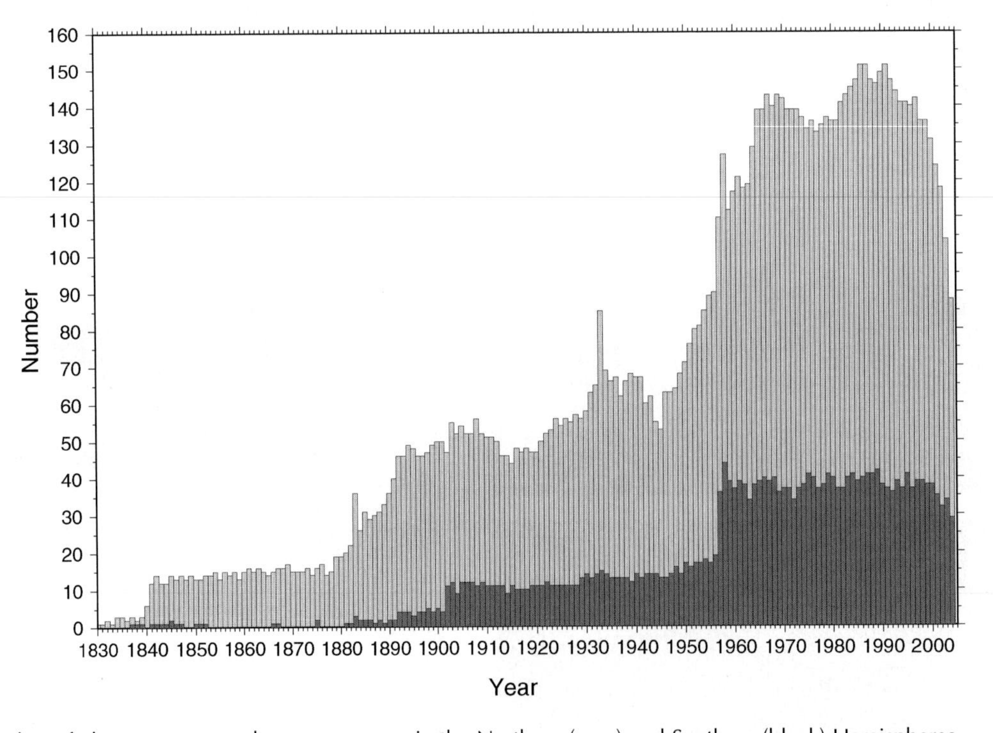

Figure O3 Number of observatory annual means per year in the Northern (grey) and Southern (black) Hemispheres.

units or universal physical constants. The most common types of absolute instrument are the fluxgate theodolite for measuring declination and inclination and the proton precession magnetometer for measuring total intensity. In the former the basic unit is an angle. The fluxgate sensor mounted on the telescope of a non-magnetic theodolite is used to detect when it is perpendicular to the magnetic field vector. True north is determined by reference to a fixed mark of known azimuth. This can be determined astronomically or by using a gyro attachment. In a proton precession magnetometer, the universal physical constant is the gyromagnetic ratio of the proton and the basic unit is time (frequency). Measurements with a fluxgate theodolite can only be made manually while a proton magnetometer can operate automatically.

Final observatory data are produced by adjusting the variometer data by application of a baseline, so that they closely fit the absolute data. In this process any long-term drifts or steps in the variometer data and differences between the sites of the variometers and absolute instruments are effectively removed. Various other instrument effects may also be included in the baselines. These are scaling factors, offsets, sensor alignments, temperature responses, and timing errors. Many of these effects are insignificant with modern digital variometers. If there is a backup variometer, ideally housed in a different building on the observatory site, filling of gaps resulting from human-made noise and data loss may also be possible. Once a continuous clean time series of data reduced to the observatory reference location is obtained, final observatory products are produced and disseminated. These include 1-min means, hourly means, annual means, and *K*-indices.

There have been many efforts to standardize the operation of magnetic observatories and to expand the network. Current international efforts include the IAGA Observatory Workshops and INTERMAGNET but there are also many efforts made by individual institutes (see *Observatories in ...*). The International Association of Geomagnetism and Aeronomy (see *IAGA, International Association of Geomagnetism and Aeronomy*) runs workshops every 2 years where information is exchanged on observatory practice and instrumentation. IAGA has also published a guide to observatory practice (Jankowski and Sucksdorff, 1996). Since the late 1980s INTERMAGNET (*Inter*national *R*eal-time *Mag*netic observatory *Net*work) has contributed to the establishment of global standards in the measurement, recording and dissemination of digital geomagnetic data (see *Observatories, INTERMAGNET*). There are now about 100 observatories operating to INTERMAGNET standards.

The satellite magnetic survey missions of the International Decade of Geopotential Research, which started with the launch of the Ørsted satellite in 1999 (see *Ørsted* and *CHAMP*), along with increased appreciation of the effects of space weather on technological systems during magnetic storms (see *Geomagnetic hazards*) has fueled recent interest in geomagnetic observatory data. The many programs to replace analog systems with digital systems will also increase the utility of observatory data. The future for geomagnetic observatories is hopefully a bright one.

Susan Macmillan

Bibliography

Jankowski, J., and Sucksdorff, C., 1996. *IAGA Guide for Magnetic Measurements and Observatory Practice*. International Association of Geomagnetism and Aeronomy.

Cross-references

CHAMP
Geomagnetic Hazards
IAGA, International Association of Geomagnetism and Aeronomy
Observatories, Automation
Observatories in ...
Observatories, Instrumentation
Observatories, INTERMAGENT
Observatories Program in ...
Ørsted

OBSERVATORIES, INSTRUMENTATION

Introduction

The main task of a geomagnetic observatory is to observe the natural geomagnetic field for an extended period of time (at least 1 year) by performing continuous measurements of the three-dimensional geomagnetic vector. Observatory instrumentation is responsible for this measurement task. As we are dealing with a very wide-frequency spectrum of signals, whose characteristic times extend from centuries to subsecond intervals (see *Geomagnetic spectrum, temporal*), issues of long-term instrument reliability, standardization of the instrumentation, continuity in data formats, as well as fast sampling capability are to be considered carefully. As only the natural geomagnetic field is of interest here, observatory instrumentation should be completely nonmagnetic: it should not modify the direction or amplitude of the geomagnetic vector in a measurable way. If, however, the sensors are magnetic, or produce magnetic fields for the sake of the measurement procedure, it must be ensured that those effects are eliminated by the measurement protocol and that no perturbation on neighboring instruments arises.

For the instrumentation to perform successfully, the observatory must answer to strict magnetic conditions: low magnetic gradient and identity of field changes over the area. Moreover, one is led to consider the whole observatory site as part of the observatory instrumentation; hence the quality of measurement depends on the magnetic hygiene of the surroundings and the buildings, on low magnetic noise conditions, as well as on the stability of the pillars bearing the field orientation sensors. Detailed information about geomagnetic instrumentation in observatories can be found in Jankowski and Sucksdorff, 1996.

Short history

The first magnetic observatory instrument—the magnetic compass—was discovered in China already before the year A.D. 1000. Early observatories—starting in the 16th century—used an oversize compass needle so as to ensure sufficient resolution on the declination angle. The compass was later complemented by the dip circle for measuring the magnetic inclination. Still later, projected field components would be measured by using setups with current-carrying coils and calibrated magnets. A (virtual) visit to the Museum of the History of Science in Florence will allow you to see those early instruments. Continuous observations were performed by an observer making a reading every 2 h or so and writing the values down in a notebook. The photographic paper recorder appeared in the 19th century, allowing uninterrupted graphical registration of the field. The invention of the fluxgate in the early 1900s permitted the development of the triaxial electronic variometer and the DI-flux, which are in use today in most magnetic observatories. The introduction of the proton precession magnetometer (Packard and Varian, 1954) allowed an important operational simplification and improvement in the measurement accuracy of the magnetic induction (see *Instrumentation, history of*).

Geomagnetic metrology

To completely describe the geomagnetic vector, each measurement must sample three independent vector components. The three most popular coordinate systems for describing the geomagnetic vector are:

Cartesian: X (North component), Y (East component), Z (vertical component, positive downward).

Cylindrical: D (magnetic declination angle, positive East), H (horizontal component).

CRLF: Z Spherical: D, I (magnetic inclination angle, positive downward), F (magnetic field induction intensity).

Component magnetometers will measure projections of the vector with a large constant part, like X and Z, while orientation magnetometers will be sensitive to the orientation of the field with respect to the sensor and measure small projections like D or Y. Component and orientation magnetometers are not yet capable of continuously measuring all the field components with the required absolute accuracy. Therefore, an observatory measurement setup uses:

- a variometer to measure the variation of the field components about baseline values in a continuous and unattended way at the required sampling rate, say 1 Hz and
- absolute measurements performed regularly (say 2 per week) by an observer with the adequate instrumentation (DI-flux, proton magnetometer; see Figure O4) to establish the values of the baselines.

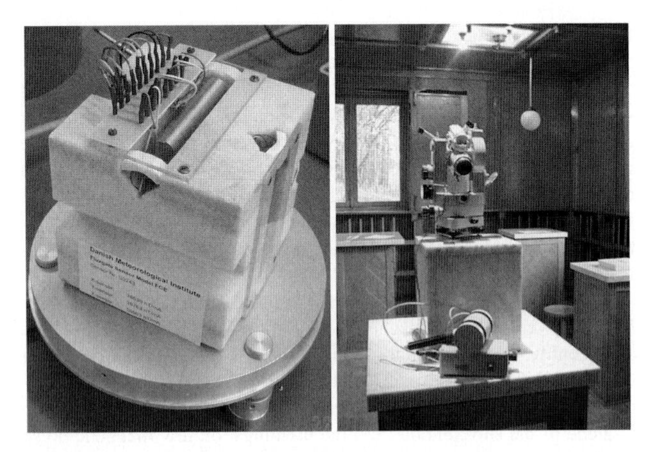

Figure O4 Instrumentation as used in most magnetic observatories of the world. The left panel represents a triaxial fluxgate variometer sensor. Right panel: the DI-flux magnetometer mounted on top of the pillar consists of a nonmagnetic theodolite with a fluxgate sensor on its telescope. In the foreground, an Overhauser proton magnetometer consisting of the cylindrical sensor and the electronic console.

This makes the observatory measurement essentially a two-step process, where the second part is performed manually. Postprocessing merges the two data sets to produce a final data record having the accuracy of the absolute instruments at the variometer's sampling rate. The expression "absolute measurement" means here that the observations must be fully traceable to metrological standards for the magnetic induction and that the orientation of the geomagnetic vector is measured with respect to the local vertical and to geographic north.

It is expected that in the future fully automatic observatories will be available (see *Observatories, automation*) where the absolute measurement part will be also unattended. Observers are striving to make absolute measurements with an accuracy of 0.1 nT for inductions and 1 s of arc for orientation. The INTERMAGNET consortium (see *Observatories, INTERMAGNET*) has laid down observatory instrumentation specifications for its members. This conveniently provides guidelines for observatories wishing to upgrade their instrumentation.

In less-modern observatories, variometers consist of suspended magnets and magnetic balances (Wienert, 1970). The continuous record of the field's components is recorded as curves on a piece of photographic paper, the magnetogram. Elongated axial magnets, suspended so as to rotate freely around the vertical, will orient their magnetic axis along the geomagnetic meridian, and will thus indicate its direction, providing a declination variometer. If an auxiliary stationary magnet creates a bias field so as to position the suspended magnet with its magnetic axis east-west, the suspended magnet's motion will follow the changes in the horizontal component, allowing it to be recorded. Magnetic balances compare the torque exerted by the vertical component on a magnet with the one exerted by gravity on a constant mass using a knife-edge balance. In all cases, the recording of the magnet's motion is ensured by a light beam falling on a mirror at the magnet's extremity. The horizontal beam, deflected by this mirror, will impress a time-dependent curve on a vertically streaming photographic paper. The streaming speed is usually about 20 mm h^{-1}, but for rapid run magnetographs it may go up to 200 mm h^{-1}.

In modern observatories, the most popular magnetometers are based on the proton precession effect and the fluxgate sensor. In former Soviet bloc countries, the Bobrov magnetometer, which consists of a magnet suspended by a taut quartz fiber, is successfully used in its servo version, where the magnet's orientation is kept fixed by the action of a coil. The coil current is easily fed to a digital data acquisition system to produce computer readable files.

The proton precession magnetometer

It is used in observatories for measuring the magnetic induction B in the range of 10–100 μT with accuracy down to 0.1 nT. Here the free precession frequency f of protons is the observable and the formula

$$2\pi f = \gamma'_p B \qquad \text{(Eq. 1)}$$

allows the measurement of the magnetic induction by way of the measurement of a frequency. The standard for the magnetic induction is thus conveniently converted to a frequency standard, which is widely available. The quantity γ'_p is the proton gyromagnetic ratio at low field for a spherical H_2O sample at 25 °C. The recommended value given by CODATA and adopted by IAGA in 1992 is (Cohen and Taylor, 1987)

$$\gamma'_p = 2.67515255 \mathrm{T}^{-1}\,\mathrm{s}^{-1} \qquad \text{(Eq. 2)}$$

The sensor consists of a bottle containing the proton-rich fluid and is surrounded by a coil, serving the dual purpose of applying periodically a polarizing field to the liquid and picking up the signal from the precessing protons after cutting off the polarizing field. An electronic console will amplify the precession signal and perform a frequency measurement of it with the required accuracy. This measurement is then scaled using γ'_p to give the field induction intensity in teslas.

This quite remarkable instrument is one of the few sensors able to measure directly the modulus of a vector without using computation from components or performing a leveling or tedious orientation procedure. This is of course due to the fact that the free protons have only the ambient magnetic field to orient themselves.

Modern proton magnetometers can be quite compact and have high resolution and sampling rate (0.01 nT and 3 Hz are typical). The Overhauser types (Sapunov *et al.*, 2001) in particular are responsible for this progress and have the additional benefit of low power operation. They are therefore a first choice in modern observatories instrumentation, providing automatic and continuous observation of the geomagnetic induction field intensity.

The optically pumped magnetometer

Optically pumped magnetometers (OPM) are based on the splitting of the energy levels of some alkali metal and helium atoms in a magnetic field. The optical pumping scheme allows measuring this energy splitting with very high resolution (Alexandrov and Bonch-Bruevich, 1992).

OPMs have been used at magnetic observatories, both for total field modulus measurement and, when complemented by a set of bias coils, for observing the geomagnetic field direction changes. One of the first digital observatories was based on a rubidium OPM (Alldredge and Saldukas, 1964).

Like proton magnetometers, the OPMs are scalar magnetometers but, to the contrary of the proton magnetometer, deliver a continuous stream of data, in the form of a frequency modulated by the value of the field. This frequency, with name of Zeeman, is linked to the measured magnetic induction by the Breit-Rabi polynomial formula. In a few cases, when the Zeeman peaks corresponding to the different energy-splits in the atoms are resolved, the polynomial coefficients can be calculated from fundamental physical constants, and the OPM has then absolute accuracy. This is the case of the OPMs based on potassium, ^{3}He, and ^{4}He vapor (Alexandrov and Bonch-Bruevich, 1992; Gilles *et al.*, 2001).

OPMs have a high sensitivity with some approaching the 0.1 pT $\mathrm{Hz}^{-1/2}$ noise level. This performance makes them attractive for prospection work, aeromagnetic survey, space-based observations, and field stabilizers. However their high initial and maintenance cost due to the short lifetime of the gas-discharge pumping lamp has limited their use in the observatory. It looks like this deficiency will be overcome in the future in view of quick progress in laser-pumping lamps (Gilles *et al.*, 2001).

Induction coil magnetometer

The induction coil (or search-coil) magnetometer's (ICM) operation principle is based on the Maxwell equation which can be written in integral form as

$$e = -\frac{d\Phi}{dt},$$

where e is the electromotive force (emf), Φ is the magnetic flux, and t is the time. In its simplest form an ICM contains a coil with many turns of copper wire connected to the input of a voltage amplifier.

As it is seen from the equation, an ICM can be used only for the measurement of a time-varying magnetic flux. Its component collinear to the coil's axis intercepts the coil loops and generates an emf at the coil's terminals, which is further amplified to an easily measurable level.

To increase the ICM sensitivity, a high permeability ferromagnetic material is used as a core inside the multiturn winding. Due to their relative simplicity, ICMs are widely used for many applications, mainly in geophysics. In geomagnetic observatories they are used for the study of Earth's magnetic field pulsations.

Their operational frequency band covers from about 10^{-4} till 10^{+7} Hz and the measurement dynamic range covers the range from fractions of femtotesla till tens of tesla. In spite of an apparent simplicity, the creation of a high-class ICM needs complicated calculations to establish an optimal matching of the sensor coil with the amplifier (Korepanov *et al.*, 2001).

The fluxgate magnetometer

This instrument is used for measuring the component of the magnetic field vector along its sensor's axis. The fluxgate principle uses the nonlinear field/induction relationship of an easily saturable ferromagnetic core. The core, usually a rod or a ring, is subjected to both the DC field to be measured, and an auxiliary AC field produced by a coil and an electronic oscillator. This offset sinusoidal excitation will create a distorted AC signal, in a pick-up coil surrounding the core. The detection of its even harmonics provides a DC signal proportional to the field to be measured.

Fluxgate sensors work best in small axial fields, therefore they are often operated within a third current-carrying coil, which cancels off the main part of the DC field. This main part may be calculated by a servo scheme, or be constant. Otherwise, the fluxgate sensor is operated with the field essentially normal to the sensor, resulting in a small field component along the core axis. The fluxgate will then be sensitive to its orientation to the field.

Observatory fluxgate variometers will include three fluxgate sensors arranged orthogonally on a stable support made of marble or quartz. This trihedron is then oriented by the variometer frame according to the three components set one wishes to observe (see Figure O4).

Absolute measurements of declination and inclination

Measuring the orientation of the geomagnetic vector with respect to a coordinate frame pointing to the geographic north and to the local vertical represents a metrological challenge, especially if one wishes to attain 1″ direction accuracy on this fluctuating vector. In fact, several measurements are taken so as to correct for systematic instrumental defects. Therefore the variometer of the observatory is used for taking into account any field changes during the measurement procedure. The state of the instrumentation art is now provided by a device called "DI-flux" or "DIM," which is assembled from a nonmagnetic theodolite and a fluxgate sensor mounted on the telescope. The magnetic axis of the fluxgate should be nearly parallel to the optical axis of the telescope (see Figure O4). With a DI-flux so configured, a measurement protocol suggested by Lauridsen (1985) allows the error-free determination of D and I.

Jean L. Rasson

Bibliography

Alexandrov, E.B., and Bonch-Bruevich, V.A., 1992. Optically pumped magnetometers after three decades. *Optical Engineering*, **31**: 711–717.

Alldredge, L.R., and Saldukas, I., 1964. An automatic standard magnetic observatory. *Journal of Geophysical Research*, **69**: 1963–1970.

Cohen, E.R., and Taylor, B.N., 1987. The 1986. CODATA recommended values of the fundamental physical constants. *Journal of Research of the National Bureau of Standards (U.S.)*, **92**: 85–95.

Gilles, H., Hamel, J., and Chéron, B., 2001. Laser pumped [4]He magnetometer. *Review of Scientific Instruments*, **72**: 2253–2260.

Jankowski, J., and Sucksdorff, C., 1996. *Guide for Magnetic Measurements and Observatory Practice.* Boulder: International Association of Geomagnetism and Aeronomy.

Korepanov, V., Berkman, R., Rakhlin, L., Klymovych, Ye., Pristai, A., Marussenkov, A., and Afanassenko, M., 2001. Advanced field magnetometers comparative study. *Measurement*, **29**: 137–146.

Lauridsen, K.E., 1985. Experiences with the DI-fluxgate magnetometer inclusive theory of the instrument and comparison with other methods. *Danish Meteorological Institute Geophysical Papers*, R-71.

Packard, M., and Varian, R., 1954. Free nuclear induction in the Earth's magnetic field (abstract only). *Physical Review*, **93**: 941.

Sapunov, V., Denisov, A., Denisova, O., and Saveliev, D., 2001. Proton and Overhauser magnetometers metrology. *Contributions to Geophysics & Geodesy*, **31**: 119–124.

Wienert, K.A., 1970. *Notes on Geomagnetic Observatory and Survey Practice.* Brussels: UNESCO.

Cross-references

Compass
Gauss' Determination of Absolute Intensity
Geomagnetic Spectrum, Temporal
Instrumentation, History of
Magnetometers, Laboratory
Observatories, Automation
Observatories, INTERMAGNET

OBSERVATORIES, AUTOMATION

In 1847 magnetic recording by photography was introduced at the Royal Observatory at Greenwich. This early example of observatory automation was prompted by the difficulty and expense of maintaining a schedule of hourly observations coupled with the scientific desire to see a fuller spectrum of magnetic variations. Economics is a major factor behind the present-day move to automate as many observatory operations as possible. If Canada, for example, were to place a full-time observer at each of its 13 observatories, the cost of operating the network would roughly double. Fully automated observatories can be deployed in remote locations where staff is unavailable, including the ocean bottom. Automation is also a natural consequence of the necessity for observatories to meet the present-day scientific requirements for digital data. Digital magnetometer systems naturally lend themselves to automation.

All aspects of observatory operations are subject to full or partial automation: (1) data collection, (2) data telemetry, (3) data processing, (4) data dissemination, (5) error detection, and (6) absolute observations. The natural end point of the automation process is an automatic magnetic observatory (AMO). Not all aspects of observatory operations need to be completely automated for the observatory to be considered an AMO. However, (1), (2), and (6) must be fully automated and (3), (4), and (5) must be at least partially automated. Although several systems previously or currently in use have been called "automatic observatories" none truly deserves the designation.

Observatories may be classified as fully staffed if there is an operator on duty everyday, unstaffed if the observatory is visited less than once per month, or partially staffed if visits to the observatory fall between these two extremes. A fully staffed observatory may be highly automated; an unstaffed observatory must be highly automated. Regardless of the level of staffing or the degree of automation, the quality of the data obtained from the observatory should meet accepted international standards such as those promulgated by INTERMAGNET. In particular, the final data disseminated by the observatory should have an absolute accuracy of better than 5 nT, so that data may be used to study secular variation and other long-period changes in the magnetic field. The requirement for absolute accuracy, as opposed to relative accuracy, distinguishes magnetic observatories from other magnetometer installations. To define absolute accuracy consider the difference between the true value of the magnetic field, $\boldsymbol{B}$, and the reported value, $\boldsymbol{Br}$; given by $\delta = \epsilon + \sigma$ where ϵ denotes a slowly varying or systematic error and σ denotes a random error. To achieve high absolute accuracy, ϵ is treated as a parameter that can be determined through a series of calibration, or "absolute", observations that are used to correct the output values of the magnetometer. If data are only used to study short-period variations, relative accuracy, rather than absolute accuracy, is important. We assume that ϵ remains constant over the time interval t_2–t_1, so that the difference $\boldsymbol{Br}(t_2)$–$\boldsymbol{Br}(t_1)$ will accurately represent $\boldsymbol{B}(t_2)$–$\boldsymbol{B}(t_1)$, provided σ is small, even though ϵ, and therefore $\boldsymbol{B}(t_1)$ and $\boldsymbol{B}(t_2)$ are not accurately known.

Alldredge (1962) designed a highly automated system, ASMO (Automatic Standard Magnetic Observatory) which was intended to produce outputs similar to those obtained by scaling standard magnetograms. However, since 1-min values were recorded to tape, it was obvious that many additional types of analyses could be carried out. ASMO consisted of a rubidium magnetometer located at the center of a two-axis Helmholtz coil system. This was oriented such that the total vector field would be biased to produce sequential changes in the angle of declination and inclination. When installed at Castle Rock, California, the system featured data transmission via telephone line to a receiving and recording center 160 km away (Blesch, 1965). Thus, it fulfilled several, but not all, of the requirements of an AMO.

The fluxgate magnetometer rapidly superceded the vector proton magnetometer as the primary instrument in automated systems. For example, AMOS, deployed in Canada in 1969 (Delaurier *et al.*, 1974), consisted of a triaxial fluxgate sensor mounted inside a square Helmholtz coil system so that the sensor essentially operated in zero field. D, H, Z (and F) were sampled every minute and recorded on tape. By 1974, it was agreed that the system was at least as reliable as the photographic recorder it was intended to replace and could in fact record large, dynamic magnetic storms, which the photographic system was incapable of recording. AMOS featured an innovative telephone verification system (TVS) that permitted remote diagnosis of malfunctions. Given the lack of stability of fluxgate magnetometers of that era, the deployment of AMOS did not eliminate or reduce the need for absolute observations. There was no longer a need for full-time personnel at the observatories, but part-time contractors were still needed to perform absolute observations and other minor duties. Thus, of the six areas of automation, AMOS addressed only the areas (1) and (5).

Other early systems that incorporated fluxgate magnetometers include those installed at Erdmagnetisches Observatorium Wingst in 1980 (Schultz, 1983), at British observatories in 1979 (Forbes and Riddick, 1984) and at French high (southern) latitude observatories in 1972 and at Chambon-la-Forêt (Bitterly *et al.*, 1986).

By the mid-1980s advances in magnetometry, computers, and telemetry increased the possibilities for automation. Although developments took place in many countries, the UK and United States (quickly joined by France and Canada) spearheaded a movement to establish modern standards for digital magnetometers and data. The initiative, which evolved into INTERMAGNET, also promoted rapid telemetry and automatic dissemination of data. Any observatory that adopts INTERMAGNET standards and specifications will possess a system capable of providing digital data that can be collected, processed, and distributed automatically. Such an observatory will typically possess a late generation fluxgate magnetometer system controlled by and recording data to a computer or other microprocessor-controlled data collection platform (DCP). Data will be filtered in the DCP to form 1-min values, which will be transmitted by satellite or phone line to the host institute and/or an INTERMAGNET geomagnetic information node (GIN), where, after some basic error checking, they can be obtained automatically by users. Thus an INTERMAGNET observatory is likely to achieve complete automation of (1), almost complete automation of (2), and at least some automation of (3) and (4).

One of the first such systems was ARGOS (Automatic Remote Geomagnetic Observatory System), which became operational at the three UK observatories in 1987. ARGOS consisted of a three-component fluxgate and a proton precession magnetometer controlled by a computer, which also acted as the data logger. Data were transmitted to a central office by phone line. The US system was similar in concept; however, it used a ring core fluxgate and the (GOES) for transmission of data. The Canadian system, Canadian magnetic observatory system (CANMOS), used the same fluxgate magnetometer as the US system and also used the GOES for data transmission. It also included an improved remote diagnostic system. The French system included an improved magnetometer operating in feedback coils. Similar automated magnetometer systems are now (2005) in operation at approximately 100 observatories.

Subsequent improvements to magnetometers and data collection systems have been concerned with magnetometer stability and sampling rates. Efforts have been made to make the output of the fluxgate magnetometer as stable as possible, thus reducing the need for independent absolute observations. Tilt-reducing sensor suspensions are capable of eliminating the effects of pier movement. Temperature coefficients of fluxgate magnetometers typically range from 0.1 to 1 nT K^{-1} (Rasson, 2001) leading to errors of up to 10 nT. Temperature coefficients can be determined for some systems but for others it is simpler to keep the system at a constant temperature. Most observatories recognize the need of the scientific community for data at a faster sampling rate than 1-min; many are now providing 1-s data.

Many observatories currently rely on GOES, Meteosat, or GMS for telemetering 1-min data to the host institute. However, this method of telemetry is incapable of coping with the increasing need for near-real time 1-Hz data. The Internet, which is increasingly accessible at even remote locations, is becoming the method of choice for many institutes, which operate networks with satellite and phone systems serving as backup.

It is unlikely that data processing will ever be carried out without at least some human intervention. There will always be instances when judgment is necessary to decide on the validity of a datum or series of data. A certain amount of data processing may be carried out in the DCP; the rest is normally carried out after the data are received at the host institute. Tasks that have been automated include removal of large spikes, plotting of magnetograms, production of indices, computation of hourly means, and computation and addition of preliminary baselines. More difficult and perhaps impossible to completely automate are such tasks as detection and removal of small spikes, especially at auroral latitudes; handling of offsets; and calculation of final baselines, including the detection and removal of spurious absolute observations.

Sodankylä and Tromsö were the first observatories to use the Internet to display and download data (Linthe, 2001). Today many observatories have taken advantage of the World Wide Web to provide access to their data and products such as magnetogram plots, minute data (and faster sampled data as well), hourly, monthly, and annual mean values, indices, and observatory and network descriptions.

Automated error detection and diagnosis is extremely important when an institute runs a network of unstaffed or partially staffed observatories. Frequently the part-time operator at the site can be instructed on how to fix a problem once the nature of the problem has been established. In addition, some problems can be corrected remotely without on site human intervention. Tasks that can be performed remotely include rebooting the computer, adjusting the time, changing temperature of an instrument enclosure, adjusting pier differences, removing spikes or periods of bad data, and erasing old files. Parameters that can be monitored include temperature, humidity, pier tilt, and ΔF, the difference between F measured with a proton magnetometer and F calculated from the values of a vector magnetometer.

To be considered a true AMO an observatory must either automate its method of performing absolute observations or, alternatively, construct as system that is so stable that absolute observations need be carried out infrequently. With one notable exception, the vector proton magnetometer (VPM) has been the instrument of choice for either of these approaches. One of the most successful of these has been Kasmmer (Kakioka Automatic Standard Magnetometer), first installed at Kakioka Observatory in 1973. Both the British and US systems have employed VPMs, known as DIDD (delta I-delta D) as a means of providing baseline control for the fluxgate magnetometers. Recently the US Geological Survey and the Eötvös Loránd Geophysical Institute of Hungary have collaborated to produce a highly stable DIDD featuring a compact spherical coil design. Evaluations of the system have proved encouraging (Pankratz, 1998; Csontos *et al.*, 2001; Schott *et al.*, 2001). Nevertheless, the DIDD is still susceptible to pier tilt and other forms of misalignment and can therefore be considered only quasi-absolute. Independent absolute observations are still required to meet the standards mentioned earlier. However, in a region otherwise devoid of observatories, one equipped with a good quasi-absolute DIDD would certainly be better than no observatory at all. Meanwhile, efforts continue to improve VPMs. In France, a ^{4}He pumped magnetometer is being developed which will eliminate the need for a coil system (Gravrand *et al.*, 2001); consideration is also being considered to developing a tilt-compensating suspension for the Hungarian or US system.

A totally different approach to automating absolute observations has been taken by Rasson (1996): the automation of a declination-inclination magnetometer (DIM). The DIM consists of a single-axis fluxgate sensor mounted on a nonmagnetic theodolite; by rotating and tumbling the telescope in a prescribed manner the angle of declination and inclination can be obtained. Rasson has been able to automate this process through the use of piezoelectric motors, digital readout of the circles, accurate recording of tilt and the use of an autocollimator for sighting the reference mark. Development is also underway of an alternative version using a north-seeking gyroscope to eliminate the need for a reference mark; this would potentially make it possible to install the device on an ocean bottom observatory magnetometer, thus fulfilling a long-standing dream of the observatory community.

Lawrence R. Newitt

Bibliography

Alldredge, L.R., 1962. A Proposed Automatic Standard Magnetic Observatory. U.S. Department of Commerce, Coast and Geodetic Survey, Washington.

Bitterly, J., Cantin, J.M., Burdin, J., Schlich, R., Folques, J., and Gilbert, D., 1986. Digital recording of variations in the Earth's magnetic field in French observatories: Description of equipment and results for the period 1972–86. In *Proceedings of the International Workshop on Magnetic Observatory Instruments*, Geological Survey of Canada Paper 88-17: 75–81.

Blesch, J., 1965. An automatic magnetic observatory. *Geophysical Technical Memorandum No.* 21, Varian Associates.

Csontos, A., Hegymegi, L., Heilig, B., and Kõmendi, A., 2001. The test results of the Delta AI@ Delta AD@ (DIDD) measurement system at the Tihany Geomagnetic Observatory of Elgi. *Contributions to Geophysics and Geodesy*, **31**: 83–89.

Delaurier, J.M., Loomer, E.I., Jansen van Beek, J., and Nandi, A., 1974. Editing and evaluating recorded geomagnetic components at Canadian observatories. *Publication of the Earth Physics Branch*, **44(9)**: 235–242.

Forbes, A.J., and Riddick, J.C., 1984. The digital recording system operated at the U.K. magnetic observatories. *Geophysical Surveys*, **6**: 393–405.

Gravrand, O., Khokhlov, A., Le Mouël, J.-L., and Léger, J.M., 2001. On the calibration of a vectorial, ^{4}He pumped magnetometer. *Earth Planets Space*, **53**: 949–958.

Linthe, H.-J., 2001. How geomagnetic observatories present collected data. *Contributions to Geophysics and Geodesy*, **31**: 151–158.

Pankratz, L.W., Test Results on a New Hungarian/US Delta I–Delta D (DIDD) Quasi-absolute Spherical Coil System, 1998. *VIIth IAGA Workshop on Magnetic Observatory Instruments, Data Acquisition and Processing, Scientific Technical Report STR98/21,* 147–149 GeoForschunsZentrum, Potsdam.

Rasson, J.L., 1996. Report on the progress in the design of an automated diflux. In *Proceedings of the VIth Workshop on Geomagnetic Observatory Instruments Data Acquisition and Processing*, Publication scientifique et technique No. 003, Institut Royal Meteorologique-De Belgique, pp. 190–194.

Rasson, J.L., 2001. The status of the world-wide network of magnetic observatories, their location and instrumentation. *Contributions to Geophysics and Geodesy*, **31**: 427–439.

Schott, J.J., Boulard, V., Pérès, A., Cantin, J.M., and Bitterly, J., 2001. Magnetic component measurements with the DIDD. *Contributions to Geophysics and Geodesy*, **31**: 35–42.

Schultz, G., 1983. Experience with a digitally recording magnetometer system at Wingst geomagnetic observatory. *Deutsche Hydrographische Zeitschrift*, **36**: 173–190.

Cross-references

Observatories, INTERMAGNET
Observatories, Overview

OBSERVATORIES, INTERMAGNET

INTERMAGNET (*Inter*national *R*eal-time *Mag*netic Observatory *Net*work) was created in order to establish a worldwide network of cooperating digital magnetic observatories. Those observatories agree to adopt modern standard specifications for measuring and recording equipment and to exchange data in close to real time. Moreover, INTERMAGNET extends technical support for maintaining and upgrading existing magnetic observatories as well as for establishing new ones.

INTERMAGNET defines standards for the measurement and recording of the geomagnetic field, considering the state of the art (see *Observatories, instrumentation*). INTERMAGNET is constituted from existing groups whose task is geomagnetic observatory measurement. The abbreviation IMO is used to indicate an INTERMAGNET Magnetic Observatory (Green *et al.*, 1998).

Short history

The idea of a near real-time observatory network arose independently from discussions inside the Geological Survey of Canada and between US Geological Survey (USGS) and British Geological Survey in 1986 during an IAGA workshop in Ottawa (see *IAGA, International Association of Geomagnetism and Aeronomy*). The idea was expanded to a global concept at the International Union of Geophysics and Geodesy (IUGG) assembly in Vancouver in 1987 as Institut de Physique du Globe de Paris joined. The network was based on magnetic observatories communicating as meteorological data collecting platforms and transmitting

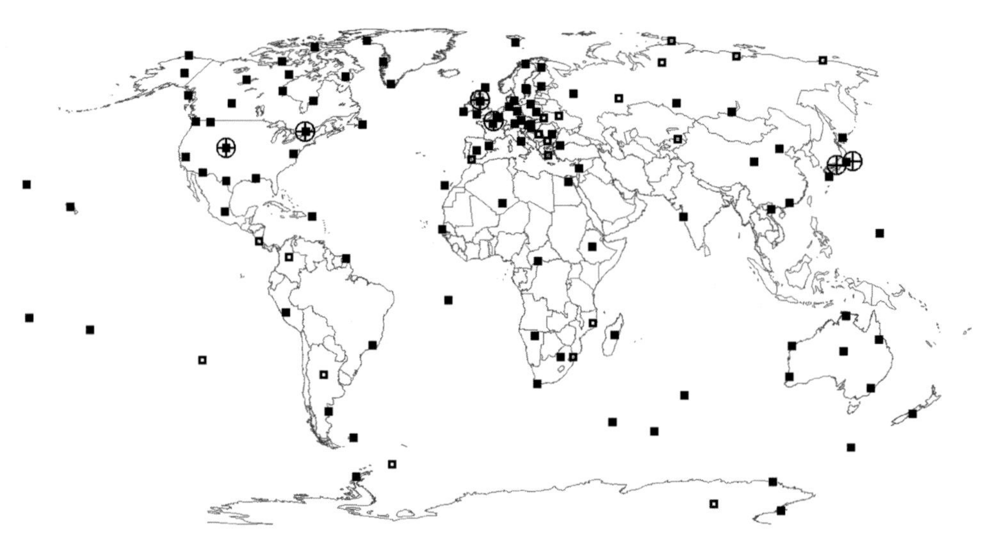

Figure O5 INTERMAGNET magnetic observatories as full squares and Geomagnetic Information Nodes (GIN) as crossed circles representing the global coverage in 2005. Open squares indicate some possible future IMOs.

their data to the meteorological satellites GOES and METEOSAT. Geomagnetic Information Nodes (GINs) in Golden, Ottawa, Hartland, and Paris equipped with downlinks collected the data from the satellites for archiving and distribution. The first transatlantic transmissions took place in late 1988 and by early 1990, four USGS, two Canadian, three French, and one British observatory were operating in INTERMAGNET mode. The first CD-ROM with definitive data (see below) from 41 observatories in 11 countries was produced in 1991.

The network predictably evolved by increasing the amount of IMOs and GINs, reaching 100 and 6, respectively, in 2005. The Geostationary Meteorological Satellite (GMS) (Japan) was added in order to cover the Asian continent. Where possible, data transfers by e-mail from IMO to GIN replaced satellite transmission, for reasons of simplicity and cost.

In 1997 a Web site was created at www.intermagnet.org

INTERMAGNET management

The Executive Council (EXCON totaling four members) establishes policy for INTERMAGNET, deals with questions of international participation and data exchange and communicates with national agencies and international scientific and funding agencies. It is assisted and advised by the 14 members of the Operations Committee (OPSCOM).

The OPSCOM advises the Executive Council on the technical issues that arise within INTERMAGNET. This includes matters relating to instrumentation, data processing, and communications. The committee is also responsible for establishing and maintaining standards and formats for the global exchange of data under the INTERMAGNET program, for designing and publishing the INTERMAGNET Technical Manual and producing the annual CD-ROM.

EXCON and OPSCOM meet about once a year for a few days in order to manage the network and plan for the future.

The data

INTERMAGNET data consists of time series of the geomagnetic vector, sampled at the round minute and carefully filtered to avoid aliasing effects. This data, collected at the IMOs represented on Figure O5 (full squares), is continuously available from the GINs within 72 h. The data come in different accuracies: reported (as recorded—near real-time), adjusted (corrected for artificial spikes and jumps), and definitive (reduced to baseline so that it has absolute accuracy). The latter is made available a few months after the end of each year at the earliest and surely with the production of the yearly CD-ROM.

People needing real-time will use the reported or adjusted data, available from the GINs in daily ASCII files either in INTERMAGNET imfv1.22 or IAGA 2002 format (e-mail request or ftp).

Access to the recent definitive data is through the Web site. Older data is available on the CD-ROMs. Definitive data come in monthly files in binary code. Browsers are available for easy perusal and inspection.

The data is freely available for *bonafide* scientific users. If any commercial aspects are involved, the user should seek a financial arrangement with the IMO directly.

All practical details about the data and for accessing them are available from the Web site where the Technical Manual is also available for download.

The future

A large part of the activities of INTERMAGNET is devoted to improve the global coverage of its network. Therefore future activities will be oriented toward filling gaps in the network, mainly in the former Soviet Union, Latin America, and the oceans (see Figure O5). Many oceanic island sites are still available for installing new IMOs but lack of staff may impose fully automatic IMOs there (see *Observatories, automation*). Open ocean sites are required to obtain the ultimate global coverage (Rasson, 2001) and INTERMAGNET is keen to support and advise groups contemplating the installation of ocean bottom magnetic observatories.

Another development is the introduction of data sampling at 1 Hz, which is in demand from the space weather community. In 2003, INTERMAGNET defined an internal data file format (based on IAGA 2002 format) for interchange of this type of data. In the next years, it is anticipated that 1 Hz data will progressively become the norm in IMOs and that GINs will start disseminating and archiving them.

Jean L. Rasson

Bibliography

Green, A.W., Coles, R.L., Kerridge, D.J., and LeMouël J.-L., 1998. INTERMAGNET, today and tomorrow. In *Proceedings of the VIIth Workshop on Geomagnetic Observatory Instruments, Data*

Acquisition and Processing '96, Scientific Technical Report STR98/21, GFZ-Potsdam, pp. 277–286.
Rasson, J.L., 2001. The status of world-wide network of magnetic observatories, their location and instrumentation. *Contributions to Geophysics and Geodesy*, **31**: 427–453.

Cross-references

Geomagnetic Secular Variation
Harmonics, Spherical
IAGA, International Association of geomagnetism and Aeronomy
Internal External Field Separation
Main Field Maps
Main Field Modeling
Observatories, Automation
Observatories, Instrumentation
Observatories, Program in Australia
Time-Dependent Models of the Geomagnetic Field

OBSERVATORIES, PROGRAM IN AUSTRALIA

The first Australian magnetic observatory was "Rossbank" Observatory at Hobart, established in 1840 by the *British Association* with support of the *Royal Society* during an expedition led by Captain James Clark Ross in command of *HMS Erebus* and *HMS Terror* (Savours and McConnell, 1982). Regular magnetic observations continued there until the end of 1854. By the initiative of Dr. Georg Neumayer an observatory was established by the Government of Victoria in Melbourne in 1858 (Dooley, 1958), beginning one of the longest continuous series of magnetic observations in the world (Figure O6). The original site of the Melbourne Observatory was in the Flagstaff Gardens but it was moved in 1862 to the Botanical Gardens, the geologically more suitable site originally chosen. Subsequent encroachment of electric tramways and railways forced the move in 1919 to Toolangi, where the observatory operated until 1986, at which time it had been replaced by the Canberra Magnetic Observatory. The Carnegie Institute of Washington's Department of Terrestrial Magnetism established Watheroo Magnetic Observatory in 1919 (McGregor, 1979), beginning the continuous recording of geomagnetic variations on the western side of the continent. The Bureau of Mineral Resources, Geology and Geophysics (BMR) was formed in 1946 (renamed the Australian Geological Survey Organization in August 1992; then Geoscience Australia (GA) in August 2001) soon after which both the magnetic observatories operating in Australia at that time were transferred to its geophysical observatories program. A new magnetic observatory at Gnangara replaced the Watheroo Observatory in 1958. The analogue variometer that operated at Watheroo was moved to Gnangara and operated until mid-1990 when it was replaced with digital instrumentation. In recent years there have been numerous burglaries and vandalism on the site and there are plans to relocate to a less vulnerable site in the vicinity.

In the 1950s BMR established observatories in the Australian Antarctic Territory (AAT) and sub-Antarctic islands. Macquarie Island Observatory, established in 1952 is still in operation. Apart from the short-lived observatory at Cape Denison, Commonwealth Bay, the first permanent observatory in AAT (which still operates) is at Mawson, established in 1955. The original instruments and buildings at Mawson came from Heard Island Observatory. Observatories at Wilkes (established by the USA IGY program), Port Moresby in Papua New Guinea, and Darwin were completed in time for the 1957–1958 International Geophysical Year (IGY). Wilkes Observatory was transferred to the Australian Government after the IGY and operated by BMR until its closure in 1967. Observatory operations were are also maintained at Casey and Davis in collaboration with the Australian Antarctic Division (McGregor, 2000).

Absolute control at Davis ceased in 2001. The observatory at Port Moresby was established by BMR and operated from 1957 to mid-1995. It was transferred to the Geological Survey of Papua New Guinea in 1978. Absolute control ceased when the absolute magnetometers were stolen in November 1993.

The observatory at Canberra was established in 1978 after the Observatories Group of the BMR was transferred from Melbourne to Canberra in the early 1970s. As Canberra was eventually to replace Toolangi Observatory, and so contribute to a number of global geomagnetic indices, it was important to operate both observatories for as long as possible to enable the *K*-indices from the observatories to be statistically compared and station differences to be established. Routine absolute observations at Toolangi ceased in 1979 although the variometers continued and *K*-indices were scaled until early 1986.

Before Canberra Observatory was commissioned, all BMR's magnetic observatories made analogue recordings on photographic paper, from which hourly mean values were scaled. Since its beginning in 1978 only digital instrumentation has operated at the Canberra Observatory. All observatories commissioned subsequently were equipped to digitally record variations in three orthogonal components as well as the total field, initially at 1-min intervals, but presently at 1-sec intervals.

In 1983, the first of a new generation of digital observatories in Australia was established at Charters Towers. The variometers are installed deep within a disused mine tunnel in the side of a hill on the site of the Seismograph Station. The Learmonth Magnetic Observatory was installed on the site of the Learmonth Solar Observatory (operated jointly by the Australian Ionospheric Prediction Service and the US Air Force) in late 1986. Small concrete vaults in the ground house the variometer and proton magnetometer sensors. In mid-1992 an observatory was established at Alice Springs on a compound of the CSIRO Arid Zone research. It remains the only Australian observatory further than 120 km from the sea and so free from oceanic induction effects. The variometers are housed in concrete vaults similar to those at Learmonth.

The most recently established observatory is that at Kakadu in the Northern Territory, located on the South Alligator Ranger Station of the Australian Nature Conservation Agency, where recording began in March 1995. Problems with lightning strikes have plagued the observatory, resulting in a number of long breaks in data acquisition and instrumentation changes.

All Australian observatories now have 1-sec digital computer-based data acquisition with data available in near real-time via either dial-up or network connections.

The main basis of the 5-yearly regional field models is the large collection of ground, airborne, marine, and satellite survey data. Observatory data, supplemented by repeat station data provide the main basis for the secular change part of the models. Together with the ongoing repeat station surveys comprising 15 super repeat stations that cover Australia, South-West Pacific Islands, and Papua New Guinea, it has been estimated that, together with the New Zealand and Indonesian observatories, GA's Geomagnetism program monitor variations of the Earth's magnetic field over approximately an eighth of the Earth's surface. Figure O7 shows the current Australian Observatory network and super repeat stations.

Calibration of observatory variometers is achieved by regular absolute observations with magnetometers that have been extensively intercompared. Final observatory data are thus consistent across the continent. Past calibrations of quartz horizontal magnetometers (QHMs) at the Danish Meteorological Institute and recent intercomparisons at IAGA Geomagnetism Workshops have ensured Australian absolute magnetometers are kept to international standards. Since the 1980s, classical absolute magnetometers have gradually been replaced by the declination and inclination magnetometer (DIM) together with the proton precession magnetometer (PPM) as the principal means of calibrating the observatory variometers. The accuracy of data produced by the Australian observatories has been enhanced by the availability of a magnetometer calibration facility at the Canberra Magnetic Observatory, in operation since 2000, providing highly accurate scale-values, relative sensor orientations, and temperature sensitivities of three-component observatory

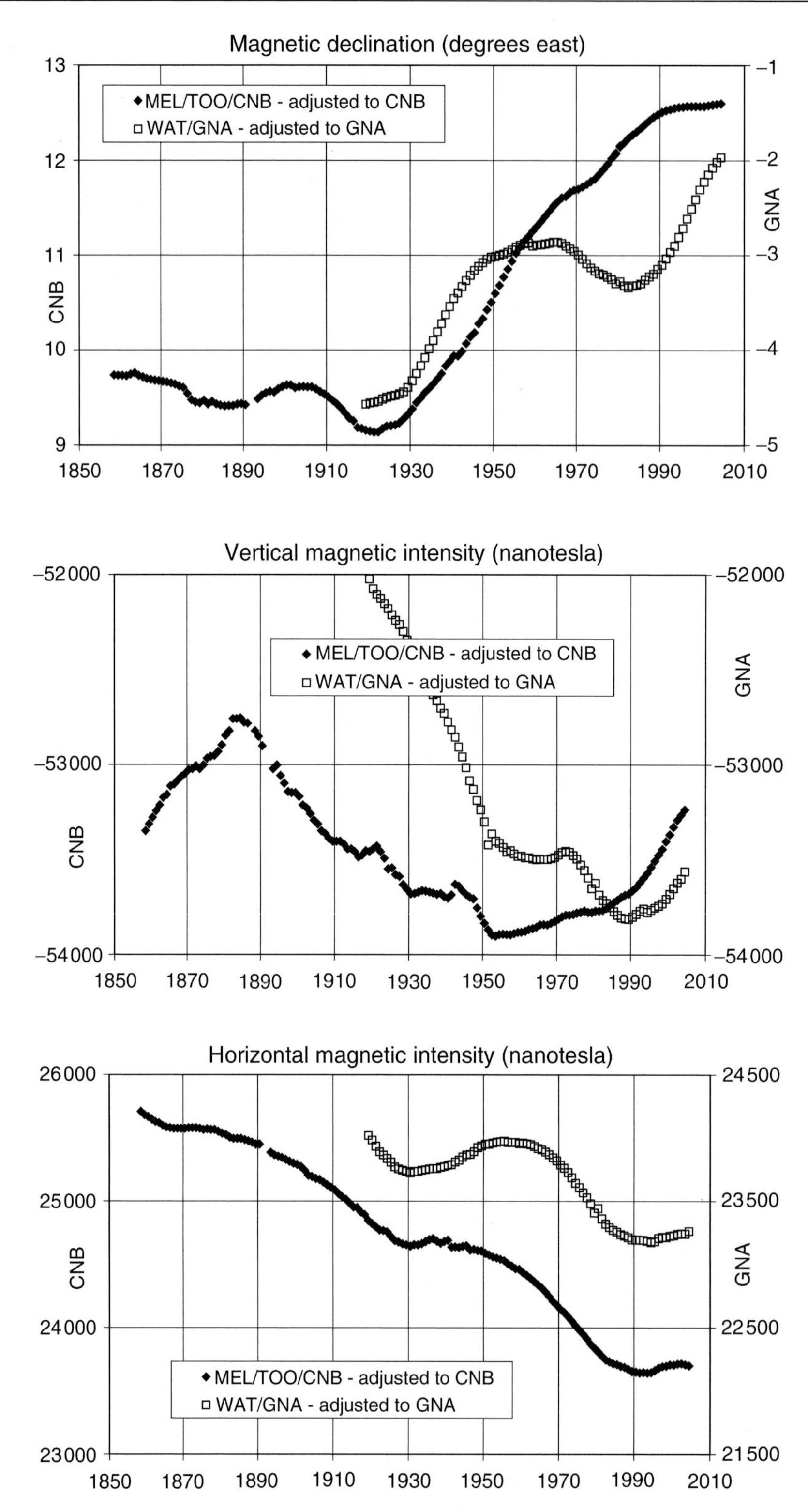

Figure O6 Annual mean values in magnetic declination, vertical intensity, and horizontal intensity from the succession of observatories at Melbourne, Toolangi, and Canberra adjusted to Canberra; and Watheroo and Gnangara adjusted to Gnangara.

variometers. These parameters were previously derived from sequences of absolute observations through baselines.

Although the emphasis of GA's program in geomagnetism is toward solid-earth geophysics, with a time resolution of 1 s and observational accuracy of the order of tenths of a nanotesla, information of relevance to upper-atmospheric, magnetospheric, and solar physics is available. One-minute vector data and hourly mean values are regularly provided to the world data centers in Boulder, USA and Copenhagen, Denmark. Data from most observatories are also transmitted by e-mail daily or more frequently to the Edinburgh GIN of the INTERMAGNET project.

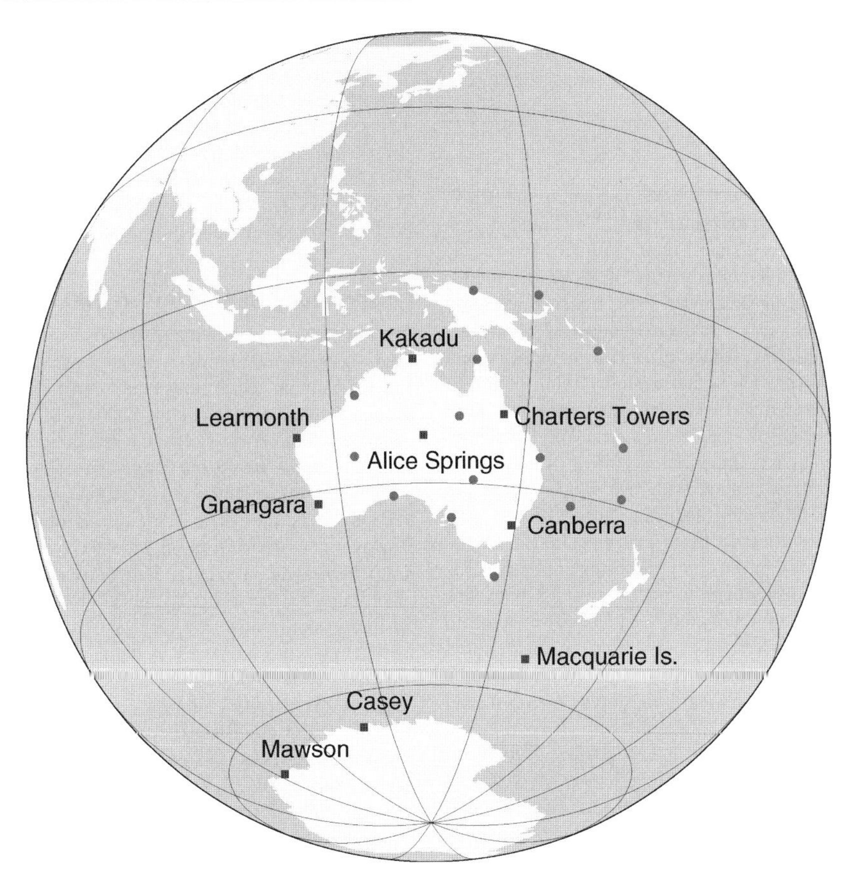

Figure O7 Locations of magnetic observatories (named squares) and super repeat stations (round dots) operated by Geoscience Australia in 2005.

K-indices are is scaled from magnetogram records from the Canberra, Gnangara, and Mawson observatories, applying the classical method of hand-scaling until late 2002. As this index is still in demand, and as no computer algorithms has quite been able to emulate the results of manual scaling, a computer-assisted method of manual scaling has been developed at GA. Using the LRNS (Hattingh *et al.* 1989) algorithm to produce an estimate of the undisturbed daily (regular solar or SR) variation that overlays the daily magnetogram on a computer screen, the SR curve is manipulated on the screen into a form that would be applied if scaling manually. Australian observatories contribute toward the global Kp-index and its derivatives, the aa-index and the am-index.

GA maintains a corporate database that has been gradually populated with data and information from all projects undertaken by the organization. There is a long history of geomagnetic operations at GA as well as holdings of data acquired before the organization existed. Software has been developed that enables geomagnetic data to be loaded into the corporate database as soon as it is transmitted to headquarters. This, together with the steady population with historical data, has allowed access to Australian geomagnetic data via the Internet.

Detailed descriptions of the observatories, their operations, histories, and data summaries can be found in the *Australian Geomagnetism Report* series, especially from 1993 when the contents of the report was expanded beyond monthly summaries. The more recent issues are available on the GA Web site: http://www.ga.gov.au

Peter A. Hopgood

Bibliography

Dooley, J.C., 1958. Centenary of Melbourne-Toolangi magnetic observatory. *Journal of Geophysical Research*, **63**: 731–735.

Hattingh, M., Loubser, L., and Nagtegaal, D., 1989. Computer K-index estimation by a new linear-phase, robust, non-linear smoothing method. *Geophysical Journal International*, **99**: 533–547.

Hopgood, P.A., 1993–2003. *Australian Geomagnetism Report* (series). AGSO & GA.

McGregor, P. M., 1979. Australian magnetic observatories. *BMR Journal of Australian Geology and Geophysics*, **4**: 361–371.

McGregor, P.M., 2000. Observatory geophysics, 1947–1998. *Aurora (ANARE Club Journal)*, **19**(3): 3–21.

Savours, A., and McConnell, A., 1982. The history of the Rossbank observatory, Tasmania. *Annals of Science*, **39**: 527–564.

Cross-references

Coast Effect of Induced Currents
Department of Terrestrial Magnetism, Carnegie Institution of Washington
Geomagnetic Secular Variation
History of Geomagnetism
Intermagnet Observations
Magnetic Indices (contents entry only)
Main Field Modelling
Observatories in Antarctica
Repeat Stations

OBSERVATORIES, PROGRAM IN THE BRITISH ISLES

History

In his survey of magnetic observatories that have operated in the British Isles, Robinson (1982) lists 18 locations, including 11 in England, 4 in Scotland, 2 in the Republic of Ireland, and 1 on the island of Jersey. The earliest two observatories were both constructed in 1838: in the grounds of Trinity College, Dublin; and in Greenwich Park, London, where an astronomical observatory had been established following a warrant issued by King Charles II in 1675. The earliest known measurement of magnetic declination at Greenwich is that made by John Flamsteed in 1680, and declination measurements were made there regularly from 1816, to assist in the calibration of ships' compasses. The initial program of observations at Dublin and Greenwich consisted of two-hourly measurements. Both observatories, along with that at Makerstoun in Scotland, participated in the Göttingen Magnetic Union (1836–1841), which promoted simultaneous measurements at cooperating observatories, and certain days were designated "term days" when observations were made every 5 min. A British artillery officer, Edward Sabine, was a chief protagonist of the efforts by the Göttingen Magnetic Union to establish a global network of observatories, and through this, several observatories were set up by the British in its colonies at the time (see *Gauss, Carl Friedrich*; *Humbolt, Alexander von*; *Sabine, Edward*; and *Geomagnetism, history of*).

The laborious nature of meteorological and magnetic observations stimulated a drive to develop automatic recording devices, and Charles Brooke designed the photographic magnetographs that were brought into operation in Greenwich in 1847 (Brooke, 1847). Francis Ronalds, working at Kew Observatory, which had been built at Richmond, London, for King George III to observe the transit of Venus in 1769, also produced a photographic magnetograph at around the same time as Brooke's. Ronalds' instrument was later redesigned by John Welsh to produce the Kew-pattern magnetograph that was subsequently installed in many observatories around the world (Stewart, 1859). The instrumental developments by Brooke and Ronalds established the standard technique employed for magnetic observatory recording worldwide for more than a century (see *Instrumentation, history of*).

Observations at Dublin Observatory were short-lived, continuing only until 1850. A meteorological observatory was established on Valentia Island, Kerry, in 1867 and regular absolute magnetic observations commenced there in 1888. The observatory was moved to the mainland, close to the town of Cahirciveen, in 1892, but kept the name Valentia Observatory. Continuous recording instruments were first installed in 1953. Met Éireann, the Irish meteorological service, now runs the observatory.

Geomagnetic measurements in London were to become impossible because of electrification of the railway and tramway systems. Kew Observatory suffered disturbances from about 1900 and with compensation from the tramway company responsible, Eskdalemuir Observatory, in the Southern Uplands of Scotland, was built. Eskdalemuir was selected as one of very few places in Great Britain that was more than 10 miles from the then extensive rail network. Construction work at Eskdalemuir started in 1904 and a full program of magnetic observations began in 1908. Later, the geomagnetic work carried out at Greenwich was similarly affected and was transferred to Abinger Observatory, on Leith Hill, London, in 1924. Observations were made there until April 1957 when, once more, disturbances from electrified railways reached intolerable levels. A transfer of operations was planned and a magnetic observatory was constructed close to the village of Hartland, Devon. Hartland Observatory opened in 1957, in time for the observatory to participate in the International Geophysical Year (IGY). Overlapping measurements were made at the times of the Greenwich-Abinger and the Abinger-Hartland moves, establishing site differences and enabling the records from the three observatories to be combined.

Lerwick Observatory in the Shetland Isles was established as a meteorological station in 1919 and geomagnetic measurements began there in 1922. Lerwick, Eskdalemuir, and Hartland are the three magnetic observatories in operation in the UK in 2005. They are run by the British Geological Survey (BGS), a component body of the Natural Environment Research Council. Lerwick and Eskdalemuir continue to have roles as meteorological stations, and seismological equipment is operated at all the observatories.

Observatory operations in 2005

The main objective of a magnetic observatory is to record geomagnetic field variations continuously, over the long term, at a stable location, and to maintain the accuracy required to produce data of the quality needed for studies of the slow changes of the main geomagnetic field generated in the Earth's core, the secular variation. The combination of magnetographs, based on suspended magnet instruments and photographic recording, supported by regular absolute observations, enabled the UK observatories to achieve this objective over many decades. However, the observatory operations were labor-intensive, both in running the magnetographs and in processing the analog photographic recordings. By the 1970s, technical developments in fluxgate and proton precession magnetometers gave an alternative set of observatory instruments, and advances in digital data acquisition and computing presented new opportunities for observatory automation. The BGS began a development program, and automatic digital systems were adopted as the observatory standard recording equipment at the three UK observatories on January 1, 1983 (see *Geomagnetic secular variation*; *Observatories, instrumentation*).

Since 1983, technology has continued to develop, bringing significant improvements to magnetometer performance and data acquisition and communications. The latest generation systems used at the UK observatories were commissioned on January 1, 2003, again based on the combination of fluxgate and proton precession magnetometers. The fluxgate magnetometer output is sampled once per second and the data are transmitted to the BGS offices in Edinburgh within minutes. A fluxgate-theodolite combination is used to make regular (manual) absolute measurements of magnetic declination and inclination, with the proton magnetometer providing absolute values of total field strength.

Instruments similar to those at the UK observatories are used at Valentia Observatory. Efforts to improve the global coverage of observatories continue and to this end observatories were set up by the BGS on Ascension Island and at Port Stanley in the Falklands Islands in 1992 and 1994, respectively.

Uses of the observatory data

In 2005, the combination of data from magnetic survey satellites, such as Ørsted and CHAMP, and observatories worldwide, is providing a rich resource for research into core processes. The data also bring practical benefits through their use in the production of global and national magnetic field models and charts used for navigation. Here, observatory data are of great importance because of the information they give on the secular variation, providing the basis for estimation of future values of the geomagnetic field at a given location (see *Ørsted*; *CHAMP*; *Main field maps*; *Main field modeling*; and *IGRF, International Geomagnetic Reference Field*).

Observatories also provide data on the relatively rapidly varying magnetic fields of ionospheric and magnetospheric origin. These "disturbance fields," indicative of "space weather" conditions, are characterized by various geomagnetic activity indices (Mayaud, 1980). Each of the UK observatories has provided data for the computation of the Kp-index since its creation in 1932. Data from the Greenwich-Abinger-Hartland series, together with data from Australian observatories, have been used to construct the aa-index, which extends from 1868, providing valuable information on long-term changes in solar-terrestrial interactions. (Both the Kp- and the aa-indices are computed for 3-h intervals.) When space weather conditions are such that a major geomagnetic disturbance, a magnetic storm, is in progress, operations in space and on the ground can be at risk. Global geomagnetic activity indices, such as Kp, are used widely as measures of disturbance. Satellite management, assessment of propagation conditions for radio and GPS signals, and geophysical surveying are examples of applications where indices are used (see *Storms and substorms, magnetic* and *Geomagnetic hazards*).

By collecting 1-s samples, the UK observatories are producing data useful for research into external magnetic fields. Data at this time resolution are required for, and are applied to, modeling geomagnetically induced currents, which is of interest to the electricity distribution industry. A further modern-day application, important to hydrocarbons production around the UK, is the use of accurate magnetic reference data to correct wellbore survey measurements acquired using magnetic survey tools. Data from the UK magnetic observatories are used in the analysis of data collected in measurement-while-drilling operations, and this application demands the rapid access to the data that the UK observatory operations provide. This application has resulted in observatories being set up on Sable Island, offshore Nova Scotia, and at Prudhoe Bay in northern Alaska by Halliburton with the assistance of the BGS.

Outlook

Lerwick, Eskdalemuir, Hartland, Valentia, Ascension, and Port Stanley are all INTERMAGNET observatories and hence are members of a coordinated global monitoring network operating to high modern standards (see *Observatories, INTERMAGNET*). This echoes the participation of observatories in the British Isles and Empire in the Göttingen Magnetic Union. These observatories are providing data for scientific research, for global data products, and are also finding local applications for the data they produce. The scientific and "real world" demands for the data are as strong today as at any time in the past.

David Kerridge

Bibliography

Brooke, C., 1847. Description of apparatus for automatic registration of magnetometers and other meteorological instruments by photography. *Philosophical Transactions of the Royal Society of London*, **137**: 69–77.

Mayaud, P.N., 1980. *Derivation, Meaning and Use of Geomagnetic Indices*. American Geophysical Union, Geophysical Monograph 22. Washington, DC: American Geophysical Union.

Robinson, P.R., 1982. Geomagnetic observatories in the British Isles. *Vistas in Astronomy*, **26**: 347–367.

Stewart, B., 1859. An account of the construction of the self-recording magnetographs at present in operation at the Kew Observatory. *British Association Report 1859*, pp. 200–228.

Cross-references

CHAMP
Gauss, Carl Friedrich (1777–1855)
Geomagnetic Hazards
Geomagnetic Secular Variation
Geomagnetism, History of
Humbolt, Alexander von (1759–1859)
IGRF, International Geomagnetic Reference Field
Instrumentation, History of
Magnetometers, Laboratory
Main Field Maps
Main Field Modeling
Observatories, Instrumentation
Observatories, INTERMAGNET
Ørsted
Sabine, Edward (1788–1883)
Storms and Substorms, Magnetic

OBSERVATORY PROGRAM IN FRANCE

The French contribution to the history of measuring the Earth's magnetic field is a long one, which began at least as early as the 16th century, with the first declination measurement performed by Künstler Bellarmatus in 1541. In 1667, the Académie des Sciences decided to build an observatory in Paris and take declination measurements. Before beginning construction, an initial measurement was made on June 21, 1667 using a 5-in. needle. Subsequent measurements were performed from 1667 onward, sometimes continued as a family tradition (La Hire, Cassini, Maraldi). There are fewer measurements of inclination than declination at the Paris Observatory, owing mainly to its late discovery; but also, it was of less interest to navigators, and presented a greater problem for accurate measurements. More details about the observations of declination and inclination in the Paris region are given in Alexandrescu *et al.* (1996).

The study of geomagnetism in France has benefited from the centralized nature of the French governments through the last centuries, as only a few site changes are to be noted, and a long continuous measurement series is now available. Indeed, the Paris series for declination and inclination span more than four and three centuries, respectively. The available annual means of declination (1541–2004) and inclination (1671–2004), adjusted to the Chambon la Forêt Observatory are presented in Figure O8.

Chambon la Forêt—the French National Magnetic Observatory. The year 1883 marks the creation of the French National Magnetic Observatory. This was installed successively at Parc Saint-Maur (1883–1901), Val-Joyeux (1901–1936), and Chambon-la-Forêt (1936 to present day), as a consequence of the encroachment of urbanization in the Paris area and of the resulting electromagnetic pollution. From the beginning of the 1980s, absolute measurements have been made using the same fluxgate theodolite and consistent procedures. Three field components are recorded continuously by three different fluxgate variometer systems, and the total field is measured by two Overhauser proton magnetometers. From these continuous time series, 1-min values are produced. Longer series, for example, hourly means (since 1936) or monthly means (since 1883) are also available.

Bureau Central de Magnétisme Terrestre (in 2004). The Institut de Physique du Globe de Paris (IPGP)[1], the Ecole et Observatoire des Sciences de la Terre de Strasbourg (EOST)[2], and the Institut de Recherche pour le Développement de Bondy (IRD, ex-ORSTOM)[3] are responsible for the magnetic observatories located in French territories or maintained in cooperation with local institutes in foreign countries. The IPGP, EOST, and IRD consolidate their activities in a "Bureau Central de Magnétisme Terrestre" (BCMT),[4] and they contribute to running 16 observatories in total (Table O1). Many of these are in the Southern Hemisphere or in regions of the world with few other observatories.

Some of the main tasks of the BCMT are (1) gathering the observatory data and publishing them according to the IAGA/INTERMAGNET recommendations; (2) controlling installation and function of magnetic observatory equipments; (3) specifying procedures for data acquisition and processing; (4) disseminating observatory data and results by year-books and CDs; (5) maintaining continuous cooperation programs with international associations and local institutes in the various countries where the observatories are located.

Cooperation programs. Cooperation programs between BCMT and different institutes have been established, mainly to run observatories in remote areas where local support is lacking (Algeria, Central African Republic, China, Ethiopia, Lebanon, Madagascar, Russia, Senegal, Vietnam). All French observatories or those run in cooperation have

[1] http://www.ipgp.jussieu.fr/
[2] http://eost.u-strasbg.fr/
[3] http://www.ird.fr/
[4] http://obsmag.ipgp.jussieu.fr/

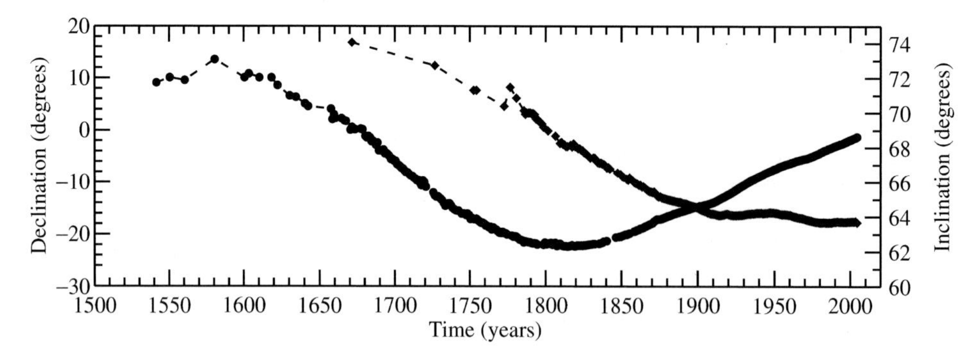

Figure O8 Annual means of declination (full circles, left axis) and inclination (full diamonds, right axis) adjusted to the Chambon la Forêt Observatory.

Table O1 Observatories operating with assistance from BCMT, in 2004

Code	Name	Latitude (°)	Longitude (°)	Altitude (m)	Start-date
AAE	Addis Ababa	9.030	38.765	2442	1958
AMS	Martin de Vivies	−37.833	77.567	48	1981
BNG	Bangui	4.437	18.565	395	1952
BOX	Borok	58.030	38.972	137	1977
CLF	Chambon la Foret	48.024	2.260	145	1936
CZT	Port Alfred	−46.433	51.867	155	1974
DRV	Dumont Durville	−66.665	140.007	30	1957
KOU	Kourou	5.100	307.400	10	1996
LZH	Lanzhou	36.083	103.850	1560	1959
MBO	Mbour	14.392	343.042	10	1952
PAF	Port Aux Francais	−49.350	70.220	15	1957
PHU	Phutuy	21.033	105.967	5	1978
PPT	Papeete	−17.550	210.380	342	1968
QSB	Qsaybeh	35.645	33.870	525	2000
TAM	Tamanrasset	22.800	5.530	1395	1932
TAN	Antananarivo	−18.920	47.550	1375	1890

full absolute control, providing 1-min magnetic field values measured by a vector magnetometer, and an optional scalar magnetometer, all with a resolution of 0.1 nT (see *Observatories, instrumentation*). More details about the instruments used through time are given in Bitterly *et al.* (1999).

All of these observatories are operated according to the principles and conditions necessary and desirable for maintaining a service of rapid magnetic observatory data exchange. The daily data quality control, the weekly absolute measurements, and the definitive data processing are done according to INTERMAGNET standards (see *Observatories, INTERMAGNET*). The observatory data are used for many specific uses, from characterizing the magnetic activity to modeling magnetic field contributions (see *Observatories, overview*).

Mioara Mandea

Bibliography

Alexandrescu, M., Courtillot, V., and Le Mouël, J.-L., 1996. Geomagnetic field direction in Paris since the mid-XVIth century. *Physics of the Earth and Planetary Interiors*, **98**: 321–360.

Bitterly, J., Mandea Alexandrescu, M., Schott, J.-J., and Vassal, J., 1999. Contribution de la France à l'observation cu champ magnétique terrestre. In B *Comité National Français de Géodésie et Géophysique, Rapport Quadriennal 1995–1998.* Toulouse, pp. 145–158.

Cross-references

Observatories, Instrumentation
Observatories, INTERMAGNET
Observatories, Overview

OBSERVATORIES, PROGRAM IN USA

The Geomagnetism Program of the US Geological Survey has, for over a century now, monitored the Earth's magnetic field through a network of magnetic observatories and conducted scientific analysis on the data collected. The program traces its origins to the Reorganization Act of 1843, in which Congress authorized the creation of a coastal survey agency, as part of the Treasury Department, that was responsible for, among other things, geomagnetic surveys. The 19th century saw the establishment of relatively short-lived magnetic stations, as well as the production of declination maps for the United States

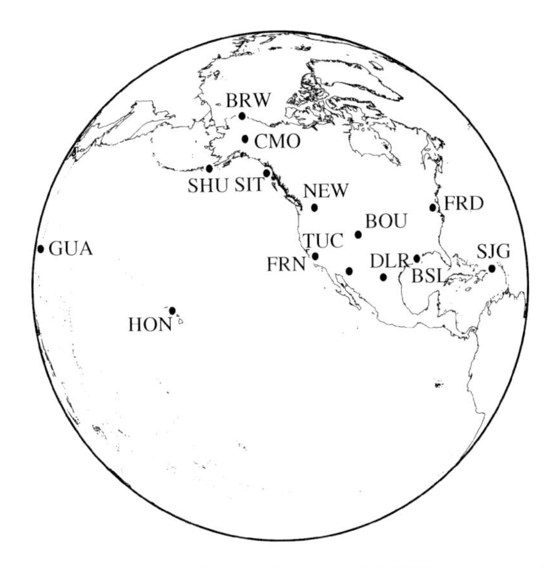

Figure O9 The geographic distribution of USGS geomagnetic observatories, identified by their three-letter IAGA codes.

and territories. With the purchase of Alaska, coastal surveys became an increasingly higher priority, and in 1889 the Coast and Geodetic Survey, with a Division of Terrestrial Magnetism, was established. The first essentially permanent geomagnetic observatories were established under the Division's leadership of Dr. *Louis A. Bauer* and Dr. John A. Fleming: Cheltenham Maryland Observatory was established in 1900, subsequently moved to the Fredericksburg site in 1956; Sitka Alaska Observatory was established in 1901 and that of Honolulu Hawaii in 1902. Soon after these observatories became operational, it was found that the Sitka and Honolulu magnetometers were also sensitive to local earthquakes, and so seismometers were installed at the sites. In part, because of this colocation of instruments, the magnetic and seismological programs in the Coast and Geodetic Survey were united in 1925 under the Division of Geomagnetism and Seismology. Over the years, the Geomagnetism Program has evolved in response to the needs of the United States and in response to changes in the nation's various federal agencies. In 1903 the Coast and Geodetic Survey was transferred to the newly organized Department of Commerce, and in 1970 the survey became part of the National Oceanic and Atmospheric Administration (NOAA). In 1973, the US Geological Survey of the Department of the Interior assumed responsibility for the nation's Geomagnetism and Seismology programs.

Today, Geomagnetism is one of four programs, in addition to the National Earthquake Hazards, the Global Seismic Network, and the Landslides Programs, represented by the USGS Central Region Geohazards Team in Golden, Colorado. Unlike the Earthquake Hazards Program, which supports many different projects, based primarily in Menlo Park and in Golden, the National Geomagnetism Program is a self-contained entity within the USGS and the team. A major part of the program is concerned with operating and maintaining magnetic observatories located in the United States and its territories (see Figure O9). The observatories, which have modern digital acquisition systems, are designed to produce long time series of stable magnetometer data having high accuracy and resolution. The observatory data are collected, transported, and can be disseminated in near-real time. The program has made a considerable investment in computing technology to enable efficient processing and management of data. By necessity, the network, and everything associated with handling the data, is technologically elaborate; it consists of many finely tuned components, each of which need to be operated in careful synchronization. The USGS observatories form an important part of the INTERMAGNET network, within which the USGS has an important leadership role. The USGS Geomagnetism Program, working in cooperation with other Federal Government departments and bureaus, most particularly the NOAAs Space Environment Center and the US Air Force Weather Agency, is an integral part of the Federal Government's National Space Weather Program. Further details about the USGS Geomagnetism Program are available at http://geomag.usgs.gov

Jeffrey J. Love and J.B. Townshend

Cross-references

Bauer, Louis Agricola (1865–1932)
Fleming, John Adam (1877–1956)
IGRF, International Geomagnetic Reference Field
Observatories, Automation
Observatories, INTERMAGNET
Observatories, Overview

OBSERVATORIES IN ANTARCTICA

Some historical landmarks

The history of magnetic measurements in Antarctica and the surrounding oceans can be traced back to the expeditions looking for the South Magnetic Pole (see Fogg, 1992 for a review). Until the turn of the 19th/20th century, all measurements were performed at sea, on ships including *Gauss* (E. Drygalski expedition, 1901–1903) and *Discovery* (first R.F. Scott expedition, 1901–1904), both being equipped with a magnetic observatory (Lüdecke, 2003). The first confirmed landing (although certainly not really the first one), can be credited to C. Borchgrevink's expedition (1893–1895), whose party reached Cape Adare and made some magnetic measurements there (Fogg, 1992). If we define the Antarctic region as that being south of 60° S the longest span of data is that collected at the observatory of Orcadas del Sur (South Orkneys, IAGA code ORC). This observatory was founded by R.C. Mossman in 1903 during the expedition led by W.S. Bruce, 1902–1904, (Moneta, 1951). Many annual means of the magnetic elements are available from this observatory from 1905 up to now. Other early observatories had a rather short or intermittent life, for example, Hut Point (Discovery Bay) set up by the first R.F. Scott expedition (1902–1903), Cape Evans (1911–1912) opened during the tragic second expedition headed by R.F. Scott, Cape Denison (1912–1913) opened by D. Mawson, and Little America (ran intermittently at various locations from 1929 to 1958). Things changed after World War II. It seems that P.N. Mayaud pioneered this new era with the observatory of Port Martin built in 1951 (Mayaud, 1953). Unfortunately, the station burned down 2 years later. Neither the first International Polar Year (1882–1883), which came too early for Antarctica, nor the second IPY (1932–1933) that was launched during an economically depressed time gave to scientific activities in Antarctica. However, the first gave Geophysical Year (1957–1958) boosted the opening of several observatories increasing the total number from around 4 to 14. All of the observatories recorded field variations on analog photographic magnetograms according to common practice in use before the advent of digital records in the 1970s. The first reported digital records were performed at Dumont d'Urville in 1969.

Situation at the beginning of the 21st century

At present, more or less permanent sites may be divided into three types: standard observatories with or without regular absolute measurements, unmanned magnetometer chains devoted to external field studies, and repeat stations, the latter two being beyond the scope of this review. According to J.L. Rasson (2001), there are a total of 44 observatories in the region with data for various time spans.

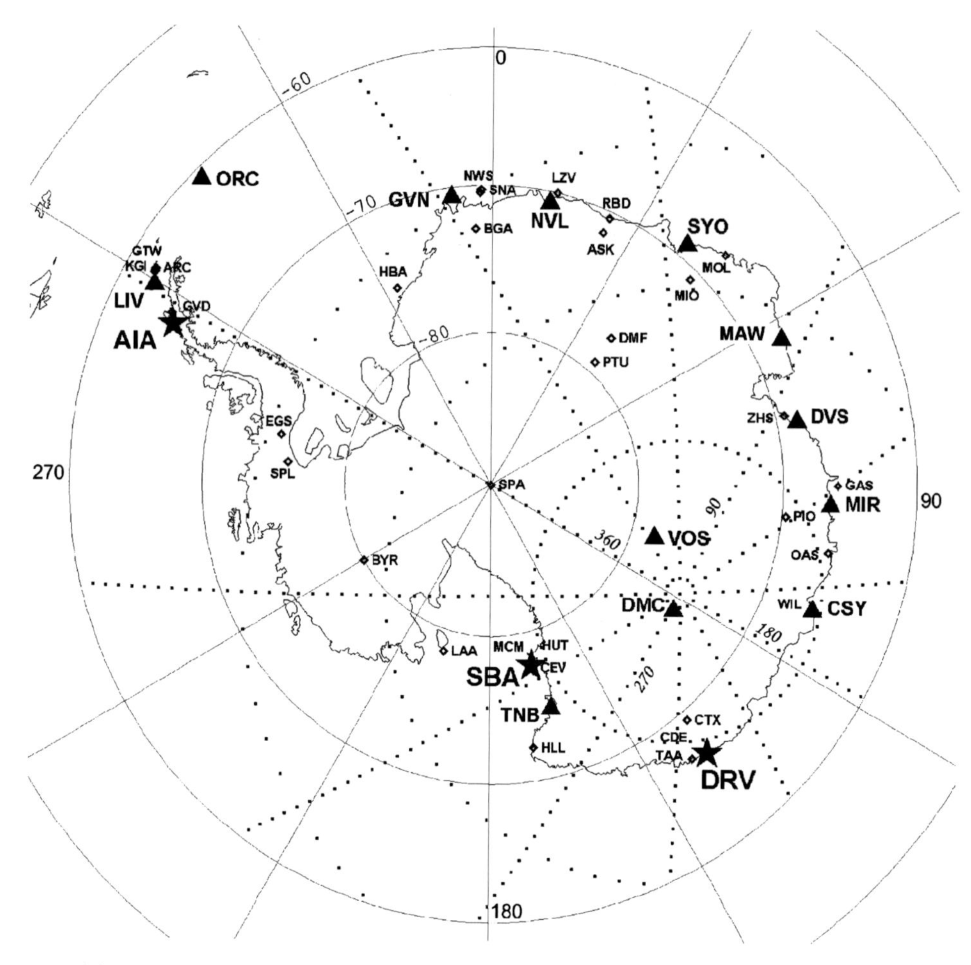

Figure O10 Location of the magnetic observatories in Antarctica. Large full triangles: open; stars: INTERMAGNET observatories; small open squares: closed. Dotted lines: geomagnetic coordinates according to the 2005 IGRF model.

Figure O10 displays the locations of all known observatories with respect to both geographic and geomagnetic coordinates. It shows that only a handful of the observatories were located on the ice cap. Vostok is the only survival of this on-land group, not taking into account the new base currently opening at DomeC (French-Italian Concordia station, DMC). ORC is the only observatory, opened before the first IPY that continues its activity today. The time spans of the observatories set up around the first IPY are highly variable, ranging from 1 year to nearly the full interval 1957–2004. In 2005, there are around a dozen observatories still running. To be termed as observatories it is necessary that they yield relative records complemented by baseline control afforded by absolute measurements. The quality of this control is variable. It ranges from episodical measurements performed during summer campaigns to measurements made regularly throughout the year at the staffed bases hosting qualified observers. There are currently three INTERMAGNET observatories (AIA, DRV, and SBA) in the region (see *Observatories, INTERMAGNET*).

The history of magnetic observatories in Antarctica closely reflects the history of the exploration of this mysterious and barely accessible continent. Their data are particularly useful for southern auroral and polar cap studies as well as tracking changes in the main field in this large region.

Acknowledgments

Updated information were provided by A. Kadokura (NIPR, Japan), V.O. Papitashvili (University of Michigan, Ann Arbor), P. Crosthwaite (Geoscience Australia), and A. Eckstaller (Alfred Wegener Institute, Bremen, Germany).

Jean-Jacques Schott and Jean L. Rasson

Bibliography

Fogg, G.E., 1992. *A history of Antarctic Science.* New York: Cambridge University Press.

Lüdecke, C., 2003. Scientific collaboration in Antarctica (1901–04): a challenge in times of political rivalry. *Polar Record*, **39**: 35–48.

Mayaud, P.N., 1953. Champ magnétique moyen et variation séculaire en Terre Adélie au 1er janvier 1952. *Comptes Rendies Académic des Sciences Paris*, **256**: 954–956.

Moneta, J.M., 1951. *Cuatro Años en las Orcadas del Sur.* Peuser: Buenos Aires.

Rasson, J.L., 2001. The status of the world-wide network of magnetic observatories, their location and instrumentation. *Proceedings IXth IAGA Workshop, 2000, Contributions to Geophysics and Geodesy*, **31**: 427–454.

Cross-references

Geomagnetism, History of
IAGA, International Association of Geomagnetism and Aeronomy
Magnetometers
Observatories, INTERMAGNET
Observatories, Overview

OBSERVATORIES IN BENELUX COUNTRIES

Introduction

Benelux is a small geographical entity, which encompasses the countries Belgium, the Netherlands, and The Grand Duchy of Luxembourg. Although the Benelux countries cover a small area, there has been interest there for magnetic observations since early on. As a result, many observatories have been created by the Benelux countries at home and abroad. Luxembourg has no observatory of its own, only field geomagnetic measurements have ever been made (Flick and Stomp, 2002).

Short history

The Dutch scientist Pieter van Musschenbroek was the first to continuously observe the magnetic declination starting 1728 in Utrecht and later Leiden until 1758. These measurements initiated a long series of observations in the Netherlands, with additional observatory locations in De Bilt (1898) and Witteveen (1938). However, in 1988 the management of the Dutch Meteorological Institute KNMI decided to stop funding magnetic observations in the Netherlands. This effectively ended the 260-year data series in the Netherlands (Schreutelkamp, 2001).

It was also the Dutch who decided in 1826 to build a magnetic observatory in Brussels. This was masterminded by the scientist Adolphe Quetelet and the institution was created almost at the time of the birth of the Belgian state. The location of the observatory changed three times within Brussels, because of the presence of local anomalies. In 1888 the observatory was located in Uccle at the outskirts of the Belgian capital. The century long series in Brussels had to be interrupted completely in 1952 due to cultural noise, while a new observatory was being constructed in the Namur province. This became the Dourbes (DOU) magnetic observatory in use today. It joined the INTERMAGNET consortium in 2002 (see *Observatories, INTERMAGNET*).

The University of Liège recorded the geomagnetic field in Cointe starting 1880, but its first full-size observatory was built in Manhay (MAB) for the International Polar Year (IPY) of 1932–1933 by Marcel Dehalu. DOU staff took over MAB operation in 1990.

Worldwide dissemination

The founding of the Dutch Royal Magnetical and Meteorological Observatory at Batavia, Indonesia, starts the dissemination of worldwide observatories from the Benelux countries, with the first "exotic" magnetic data being recorded in 1884. Magnetic observations are still being made near Jakarta today, although the site of the observatory had to be moved first to Kuyper and finally to Tangerang to avoid cultural magnetic noise (see *Bemmelen, Willem van*).

Binza was installed in Belgian Congo and started delivering data in 1953. Many magnetic observatories were created during the International Geophysical Year of 1957–1958: the observatories of Lwiro and Elizabethville were created by Belgium in Congo as well as the Base Roi Baudouin (RBD) in Antarctica. The Netherlands installed Hollandia and Paramaribo in Indonesia and Dutch Guyana, respectively (see Figure O11).

Present situation

From the 14 observatories indicated on Figure O11, only three remain today: DOU, MAB, both in Belgium, and Tangerang (TNG) in Indonesia. Unfortunately, cultural magnetic noise is on the rise in all of them, so that short period signals smaller than 1 nT, especially in the vertical component, are increasingly difficult to study.

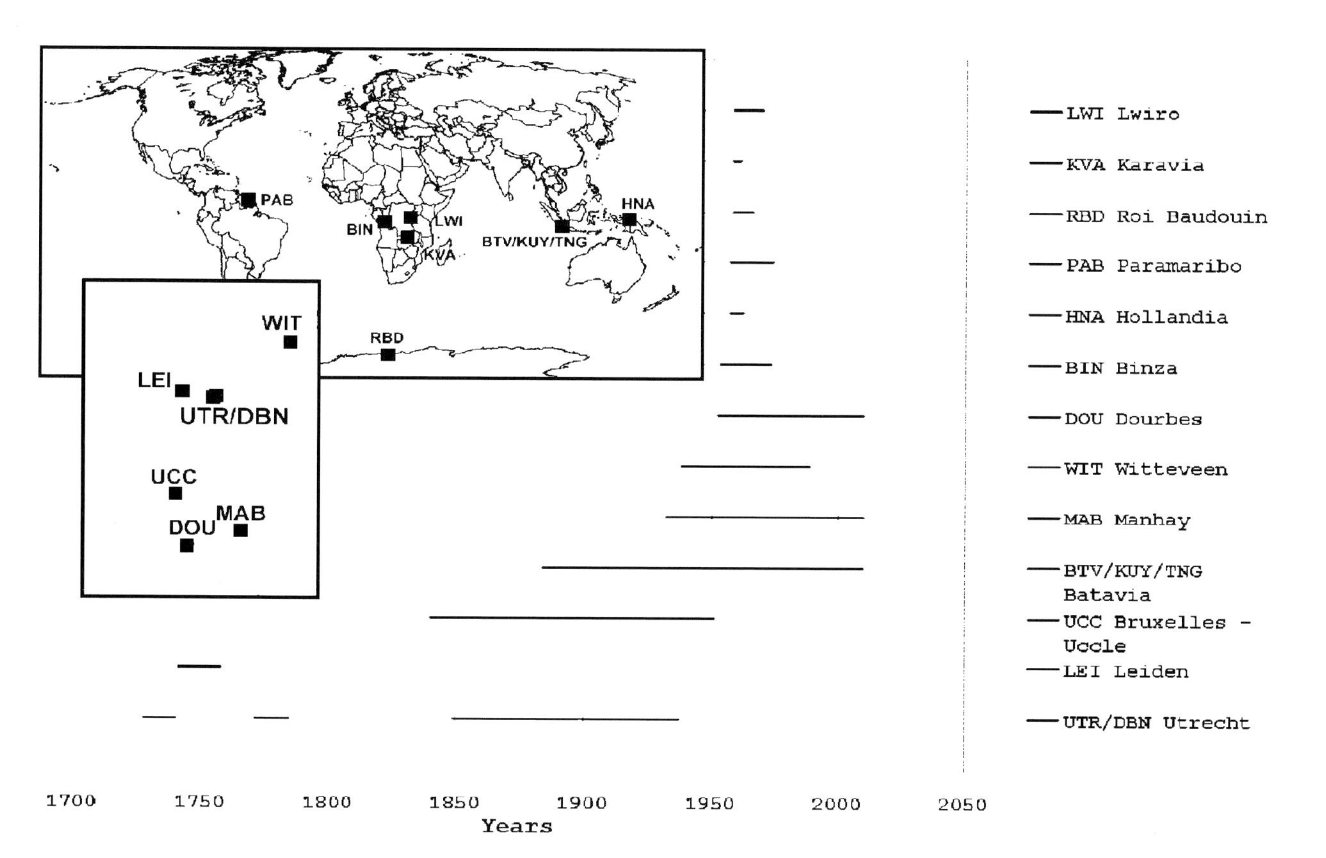

Figure O11 Timelines for the life of the observatories in and installed by the Benelux countries. The insert shows their location.

Dutch customers involved in aeronautical and topographic mapping continued to request declination information over the Netherlands after the closure of the Dutch magnetic service so that from 1999 onward, the team from DOU performs yearly repeat measurements at several Dutch stations, including former Witteveen Observatory, for producing the isogonal maps.

Future perspectives

The Belgian observatories DOU and MAB will remain open for the foreseeable future—especially as there is demand from the socioeconomic sector for magnetic products and services in the Benelux. There is a search going on for a new site in Belgium with reduced noise, especially as space weather research and forecasts needs clean short period signals.

There are no plans in the Netherlands to resume the operation of Witteveen Observatory. The TNG Observatory is in need of a better observation site and improved instrumentation; international collaboration with the Indonesian Meteorological and Geophysical Agency is attempting this now.

There are talks between the Democratic Republic of Congo and Belgium to resume operation in Binza, whose magnetic pavilions are reported to be in good condition. A new Belgian Antarctic base might also host a magnetic observatory 200 km south of RBD as part of a contribution to the IPY in 2007–2008.

Jean L. Rasson

Bibliography

Flick, J.A., and Stomp, N., 2002. *Sciences de la Terre au Luxembourg—Réminiscences.* Luxembourg: Musée National d'Histoire Naturelle et ECGS.

Schreutelkamp, F.H., 2001. Het aardmagnetische veld ontrafeld. *Zenit*, **28**: 136–141, 186–190.

De Vuyst, A., 1962. La variation de la Déclinaison Magnétique en Belgique de 1828 à 1960.5. Contributions de l'Institut Royal Météorologique de Belgique, 68.

Cross-references

Bemmelen, Willem van (1868–1941)
Geomagnetism, History of
Observatories, INTERMAGNET

OBSERVATORIES IN CANADA

The origin of the Canadian Magnetic Observatory Network (CANMON) can be traced back to the founding of the Toronto Magnetic Observatory in 1839. Except for a few temporary magnetic observatories, such as those established during the First International Polar Year (1882–1883), Toronto (later Agincourt) was the only magnetic observatory operating in Canada until 1916, when Meanook was established. The network grew substantially during the years preceding and following the International Geophysical Year (1957–1958) and now consists of 13 observatories, operated by the Geological Survey of Canada, Natural Resources Canada (GSC).

Distribution of observatories

The CANMON extends over 40° of magnetic latitude, from 54.3° N to 86.4° N, and includes three zones of differing magnetic activity: subauroral, auroral, and polar cap. The distribution of observatories relative to the three zones is shown in Figure O12. All observatories, including those that are no longer in operation, are listed in Table O2. Inuvik should be in operation by 2007.

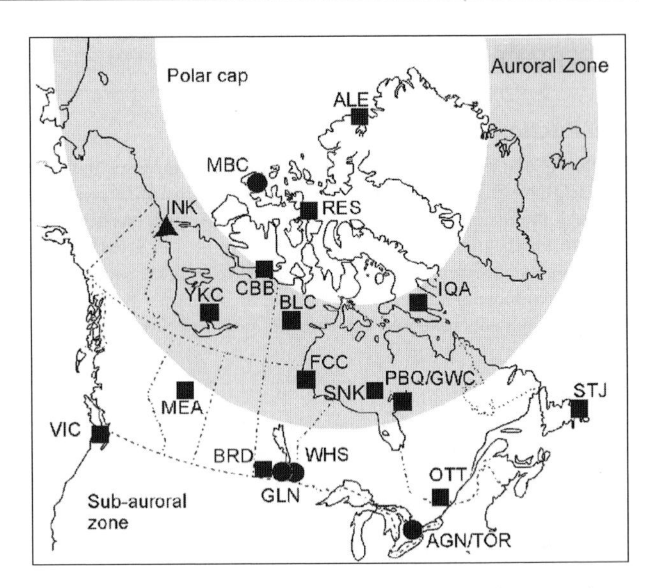

Figure O12 Distribution of Canadian observatories: currently operating (large squares); closed (circles); planned (triangle).

Although the stations shown in Figure O12 are the only true magnetic observatories in Canada, several other magnetic variometer networks have been established to measure the unusual magnetic conditions that result from the presence of the polar cap and auroral zone. These include the CANOPUS (now CARISMA) network, the MACCS network, and the THEMIS network, as well as several other minor networks and single stations. In all, more than 80 permanent or semipermanent magnetometers are currently operating in Canada.

Instrumentation

All magnetic observatories in the CANMON except Alert are part of INTERMAGNET and are operated to INTERMAGNET standards (see *Observatories, INTERMAGNET*). Each observatory is equipped with CANMOS (Canadian Magnetic Observatory System), a modular system whose various components are controlled by a desktop PC, which also serves as the data-acquisition system. The system is equipped with two magnetometers: a three-component ring-core fluxgate for recording variations in the X (north), Y (east), and Z (vertical) components, and an Overhauser magnetometer for recording total intensity. The sensor of the three-component magnetometer is mounted on a novel tilt-compensating suspension (Trigg and Olson, 1990), and both sensor and electronics are housed in a temperature-controlled enclosure to minimize drift. The system transmits 1-min data via the GOES satellite to downlinks located at Ottawa and Golden, Colorado. At some observatories, 1-s data are transmitted in real time via the Internet. The system is also accessible via telephone modem, allowing remote monitoring and diagnosis of any problems.

Each observatory is also equipped with a declination-inclination magnetometer (DIM), consisting of a steel-free theodolite equipped with a single-axis fluxgate sensor used for weekly absolute observations.

Data availability

Both preliminary and definitive data are available, filtered over the following intervals: 1 s, 5 s, and 1 min. Definitive data undergo rigorous quality control and are calibrated using information derived from the DIM and Overhauser magnetometers. Preliminary data are checked for major spikes only. Preliminary 1-min data are available within 24 min. For those observatories that transmit data via the Internet, the delay

Table O2 Canadian magnetic observatories

Name	IAGA code	Geographic		Geomagnetic		First year	Status
Alert	ALE	82.497	297.647	86.89	23.2	1961	Operating
Mould Bay	MBC	76.315	240.638	79.90	275.2	1962	Closed, 1995
Resolute Bay	RES	74.690	265.105	82.90	302.9	1952	Operating
Cambridge Bay	CBB	69.123	254.969	76.53	303.0	1972	Operating
Baker Lake	BLC	64.318	263.988	73.17	322.1	1951	Operating
Iqaluit	IQA	63.753	291.482	73.99	5.2	1994	Operating
Yellowknife	YKC	62.820	245.518	69.25	299.0	1975	Operating
Fort Churchill	FCC	58.759	265.912	67.94	328.4	1957	Operating
Poste-de-la-Baleine (Great Whale River)	PBQ GWC	55.277	282.255	65.46	351.8	1967	Operating, will close 2007 (relocated, 1985)
Sanikiluaq	SNK	56.536	280.769	67.23	350.6	2006	Replaces PBQ
Meanook	MEA	54.616	246.653	61.57	306.2	1916	Operating
Brandon	BRD	49.875	260.025	58.60	324.2	2006	Operating
Glenlea	GLN	49.645	262.880	58.66	327.6	1982	Variometer since 1997; closed 2005
Whiteshell	WHS	49.700	264.75	58.89	330.1	1976	Closed, 1980
Victoria	VIC	48.420	236.580	54.04	297.6	1956	Operating
St. John's	STJ	47.595	307.323	57.15	24.0	1968	Operating
Ottawa	OTT	45.403	284.448	55.64	355.3	1968	Operating
Agincourt (Toronto)	AGN TOR	43.783	280.733	53.94	350.8	1839	Closed, 1968 (relocated 1[illegible])

is at most a few minutes. Minute data may be plotted and downloaded from either the GSC Geomagnetic Laboratory's Web site (www.gsc.nrcan.gc.ca/geomag/) or from the INTERMAGNET site (http://www.intermagnet.org).

One- and 5-s data that are transferred in real time are available almost instantly from the geomagnetic laboratory site. They are also available from the Canadian Geospace Monitoring Program data portal (ssdp.ca). Definitive data are normally available about 3 months after year's end. They replace the preliminary data at all sampling intervals on all the above-mentioned Web sites. In addition, 1-min definitive data are available on the INTERMAGNET CD and from World Data Center Web sites.

Innovation

Magnetic observatories must evolve and adapt to meet the needs of the scientific community that they serve. In addition to this scientific imperative, development of the CANMON has been influenced by the logistical, environmental, and economic challenges brought about by the remoteness of many sites in this vast network. The presence of the auroral zone and the proximity of the north magnetic pole also present challenges in instrument design. These factors have provided impetus for innovation. Many instrumental and procedural developments that are now considered common place were pioneered or quickly adopted for use in CANMON. Most notable were the use of the DIM as an absolute instrument, which began in the mid-1950s (Serson and Hannaford, 1956; Loomer, 1961) and the development of a semiautomated, digital network, which began in 1969 with the installation of the AMOS system at St. John's (Delaurier *et al.*, 1974).

Lawrence R. Newitt and Richard Coles

Bibliography

Delaurier, J.M., Loomer, E.I., Jansen van Beek, J., and Nandi. A., 1974. Editing and evaluating recorded geomagnetic components at Canadian observatories. *Publication of the Earth Physics Branch*, **44**(9): 235–242.

Loomer, E.I., 1961. Record of observations at Resolute Bay Magnetic Observatory 1957–1956 with a summary of earlier observations. *Publications of the Dominion Observatory*, **34**(2): 25–131.

Serson, P.H., and Hannaford, W., 1956. A portable electrical magnetometer. *Canadian Journal of Technology*, **34**: 232–243.

Trigg, D.F., and Olson, D.G., 1990. Pendulously suspended magnetometer sensors. *Review of Scientific Instruments*, **61**: 2632–2636.

Cross-references

Observatories, Automation
Observatories, INTERMAGNET
Observatories, Overview

OBSERVATORIES IN CHINA

At present, there are 43 magnetic observatories in China's mainland. Forty of them are sponsored by the China Earthquake Administration (CEA), formally China Seismological Bureau (CSB) and State Seismological Bureau (SSB). Three others are sponsored by the Institute of Geology and Geophysics, Chinese Academy of Sciences (IGGCAS), formally the Institute of Geophysics, Chinese Academy of Sciences (IGCAS). Among the 40 sponsored by the CEA, 3 new ones are under construction and are expected to operate in 2007. All of the CEA observatories will be equipped with at least two sets of triaxial fluxgate magnetometers, one fluxgate theodolite, and one proton precession magnetometer in 2007. For the most important observatories, an additional fluxgate theodolite and continuous recording of total field by the Overhauser effect magnetometer will also be deployed.

History

There were four major periods for the development of the magnetic observatories in China's mainland.

Before the International Geophysical Year (IGY) 1957–1958

The first magnetic observatory in China was constructed in Beijing in 1870 by Russia. It ceased working in 1882. There had been other six magnetic observatories with different periods of operation but all of them had ceased working before 1944. One exception was Sheshan (SSH) Observatory which was constructed in 1874 by French missionaries and is still in operation now.

During the IGY

During the IGY, construction of seven magnetic observatories including Beijing (BJI), Changchun (CNH), Wuhan (WHN), Guangzhou (GZH), Lanzhou (LZH), Lhasa (LSA) and Urumqi (WMQ) observatories was initiated. Together with SSH, these eight observatories became the backbone of the Chinese magnetic network.

From 1966 to 1979

The 1966 Xingtai Earthquakes triggered the beginning of the Chinese research on earthquake prediction. Various observation methods were utilized including magnetic observation. More than 200 magnetic stations were set up around the country in the following years. These stations were sponsored by different organizations and/or institutions with a variety of observational procedures and quality controls.

After 1979

In 1979, all of the magnetic observatories and stations were put under the administration of the SSB. An organization (now the Geomagnetic Network of China) in the Institute of Geophysics, SSB (IGSSB, now IGCEA) was responsible for the technical support and data management of the magnetic network. The network was readjusted several times taking into account the distribution and the observational environment of the stations. Half of the stations ceased working. Meanwhile, instrumentation at 29 of the stations was improved to make them operate as observatories. Later on, three other observatories were set up in Mohe (1989), Beijing Mingtomb (BMT, 1993), and Sanya (1995) by the IGCAS (now the IGGCAS).

Activities

All of the existing observatories have been producing yearbooks, *K*-indices, and catalogs of magnetic storms. The Geomagnetic Network of China is in charge of the quality control and dissemination of the data from the CEA observatories. These magnetic observations have been widely used in research work and applications related to geomagnetic field.

The relatively short distances among the observatories have had advantages. For example, the detection of errors in data by using the method of interobservatory comparison is justifiable if the observatories are close together. It also helps the scientists in research work on earthquake prediction and regional characteristics of the magnetic field.

Data from more than 15 observatories have been archived in the World Data Center system. Some of the observatories have been participating in international projects such as the Sino-American, Sino-Japanese, and Sino-French projects. Three observatories have participated in INTERMAGNET (see *Observatories, INTERMAGNET*). With the improvement of the instrumentation at the observatories, it is believed that there will be more INTERMAGNET magnetic observatories from China's mainland in the near future.

Dongmei Yang

Cross-reference

Observatories, INTERMAGNET

OBSERVATORIES IN EAST AND CENTRAL EUROPE

The first systematic magnetic observations in Central and Eastern Europe were made in Prague in 1839 by Karl Kreil (1798–1862), who was assistant director, and from 1845 director of the Prague Observatory and professor of astronomy at the Prague University. Measurements of declination started in 1830. Kreil constructed and improved magnetic instruments (using, among others, Gauss's method of measuring the horizontal component) and organized a rational system of magnetic observations. From 1843 he also carried out a magnetic survey of the Austrian Empire. In 1851, he was appointed the first Director of the Central Institute of Meteorology and Geomagnetism (ZAMG) in Vienna. The geomagnetic observatory started operating there in 1852.

The first Hungarian magnetic observatory was founded in Buda in 1871, but had to be closed in 1889 due to increased industrial activity, which rendered the measurements unreliable. The magnetic observations continued in O'Gyalla under the auspices of the Royal Hungarian Meteorological Institute. The geomagnetic observatory at O'Gyalla was officially opened in 1900. The observatory was a witness of the turbulent political development in Central Europe. Whereas many observatories had to change their locations due to industrial noise, O'Gyalla, having kept its original location, changed its name twice (Stará Ďala in 1924 and Hurbanovo in 1948) and the country it belonged to five times. In spite of this, Hurbanovo (IAGA code HRB) is the oldest operating observatory in Central and Eastern Europe.

Another geomagnetic observatory was established in 1880 in Pola (Istria Peninsula) under the auspices of the Hydrographical Institute of the Austro-Hungarian Imperial and Royal Navy. The data were also used for reducing Italian magnetic surveys. The observatory was in operation till 1925.

Before World War I, magnetic observations were made at several other places, but the operation was usually short (less then 10 years) and the only measured element was declination. Five observatories were founded between the wars: Swider near Warsaw (1920, SWI) and Hel (1932, HLP) in Poland, Wien Auhof (1929, WIA) in Austria, Jassy (1931, JSS) in Romania, and Panagyurishte (1937, PAG) in Bulgaria. On the other hand, Prague Observatory, located in the city center since 1839, was finally closed down in 1926.

The present shape of the observatory network was formed after World War II in a climate of increased interest in geosciences, also supported by the International Geophysical Year in 1957/1958. Due to the growth of industrial and urban magnetic noise more attention was paid to the choice of the sites. Of course, it was not easy to foresee future development. As regards the older observatories, only Hel, Hurbanovo, and Panagyurishte were able to keep their original sites.

The Vienna Observatory was moved in 1955 from Auhof to Wien Kobenzl (WIK). As this location is also not perfect, transfer to a quieter place is being considered. The Prague Observatory was replaced in 1946 by Pruhonice near Prague (PRU) but rapid expansion of the city and construction of DC-powered railways resulted in a deterioration of this location. The observatory was moved to Budkov (BDV) in south Bohemia, to a sparsely populated area, in 1967. Two observatories were built in Hungary. The Tihany (THY) Observatory was founded in 1953 in a protected area on the northern shore of Lake Balaton. The tourism boom brought more traffic to this area than had been expected; nevertheless, the conditions are still satisfactory. Another observatory was built in 1956–1957 in Nagycenk (NCK) near Sopron. Although the primary aim of the observatory was for the study of the Earth's electromagnetic field of external origin, it produces standard geomagnetic data. The Swider Observatory was affected by magnetic noise from Warsaw. An appropriate location was found in a rural region about 50 km south of Warsaw in Belsk (BEL) and regular measurements started in 1965. The Institute of Geophysics of the Polish Academy of Sciences has also been operating the Hornsund (HRN) Observatory in Spitsbergen since 1978. The Romanian observatory,

Surlari (SUA), was established in 1943 north-east of Bucharest. The Serbian observatory, Grocka (GCK), began its operations during the International Geophysical Year in 1957.

Most of the observatories joined the INTERMAGNET network (number in brackets shows the first year the data were presented on INTERMAGNET CD-ROM): Tihany (1991), Belsk (1993), Nagycenk (1993), Budkov (1994), Hurbanovo (1997), Hel (1998), Surlari (1999), and Hornsund (2002).

Two research groups are active in geomagnetic instrumentation development. Torsion photoelectric magnetometers (TPM), based on Bobrov-system quartz variometers, are being constructed in Belsk since 1977. In these magnetometers, angular deflections of variometer magnet mirrors are converted, by means of photoelectric convertors, into electric voltage changes in a system with a strong negative feedback. The Tihany Observatory is developing digital data acquisition systems DIMARS and DIMARK. They are designed for geomagnetic observatories to measure and process output signals of vector fluxgate and Overhauser proton precession magnetometers according to INTERMAGNET standards.

Pavel Hejda

Cross-reference

INTERMAGNET

OBSERVATORIES IN GERMANY

Introduction

The tradition of geomagnetic observations in Germany spans back to the 16th century. In the early beginning, only the declination was observed for the purposes of navigation and mining by means of compasses. During the 19th century the theories and measurement methods of geomagnetic intensities were developed in Germany by Gauss and Weber.

The first "real" magnetic observatories in Germany were established by Gauss and Weber in Göttingen; by Alexander von Humboldt in Berlin and by Johann von Lamont in Munich. Today, three geomagnetic observatories are in operation in Germany: Fürstenfeldbruck, Niemegk, and Wingst, which all participate in the worldwide INTERMAGNET network. Table O3 includes essential information about the German observatories.

Development and international contribution

Early time series may be reconstructed from the notations by mining engineers or sailors, inscriptions in old instruments and maps, and by correcting these series with more recent observations. In Knothe (1987), the declination observation series of Krakau (today Krakow, Poland), Danzig (today Gdansk, Poland), Freiburg, Munich, and Clausthal are plotted together with those of Prague, Böckstein (Austria), Copenhagen, Paris, and London. Of course, these series are affected with large uncertainties, but they are the earliest certificates of magnetic observations in Germany.

The German universal scientist Alexander von Humboldt (1769–1859) initiated from 1827 onward with own studies and by inspiring others progress in terrestrial magnetism in Germany. The mathematician Carl Friedrich Gauss (1777–1855), inspired by Humboldt, formulated the general theory of earth magnetism, developed the measurement method for the absolute determination of the intensity and introduced a system of physical units for geomagnetism from 1829 onward. When Wilhelm Weber (1804–1891) came to Göttingen in 1831, one of the most fruitful collaborations devoting to science started. In 1833, they initiated the construction of a nonmagnetic house for geomagnetic measurements in Göttingen. This magnetic observatory soon became the center of the "Göttinger Magnetischer Verein (Göttingen Magnetic Union)." This union, initiated by Humboldt, Gauss, and Weber, was the first worldwide geomagnetic observatory network, which existed from 1836 to 1841 and included 53 observatories, 18 of them non-European ones. All observatories took readings from their magnetometers all 5 min during appointed days. The results were published by Gauss and Weber together with scientific articles.

Table O3 German geomagnetic observatories

Observatory	Latitude (North)	Longitude (East)	Existence
Berlin	52.515°	13.393°	1836–1872
Fürstenfeldbruck	48.165°	11.277°	1939–
Göttingen	51.533°	9.950°	1833–1902
Maisach	48.200°	11.260°	1927–1937
Munich	48.147°	11.608°	1841–1925
Niemegk	52.070°	12.680°	1930–
Potsdam	52.380°	13.060°	1890–1931
Seddin	52.280°	13.010°	1907–1931
Wilhelmshaven	53.530°	8.150°	1878–1919, 1931–1936
Wingst	53.750°	9.070°	1938–

Observatories

Table O3 includes those German observatories that were in operation for long periods, and which provided vector data of the Earth magnetic induction vector. Table O4 shows general information about these observatories, while Table O5 contains the used main instruments.

Summary and outlook

The German scientists Alexander von Humboldt, Carl Friedrich Gauss, and Wilhelm Weber initiated the establishment of the first magnetic observatories in Germany during the 1930s. They created the scientific basements of the measurement and mathematical description of terrestrial magnetism and developed suitable instruments. Further essential theoretical and instrumental contributions were performed by Johann von Lamont (1805–1879) and Adolf Schmidt (1860–1944). The "Göttinger Magnetischer Verein" (Göttingen Magnetic Union), initiated by Humboldt, Gauss, and Weber was the first worldwide geomagnetic observatory network.

Due to several reasons, the first magnetic observatories (Göttingen, Berlin, and Munich) that were established in the frame of the Göttingen Magnetic Union do not exist any more. Industrial and urban disturbances forced the movement of the observatories from their first locations within the cities to places far away from scientific centers. Today there are three observatories Fürstenfeldbruck (successor of Munich and Maisach), Niemegk (successor of Potsdam and Seddin), and Wingst (successor of Wilhelmshaven). The time series span back to 1840 (Munich-Maisach-Fürstenfeldbruck), 1890 (Potsdam-Seddin-Niemegk), and 1878 (Wilhelmshaven-Wingst). The observatories cooperate closely, though they are sponsored by different scientific institutions: Fürstenfeldbruck by the Institute of General and Applied Geophysics of Munich University, Niemegk by the GeoForschungsZentrum Potsdam, and Wingst by the GeoForschungsZentrum Potsdam and the Bundesamt für Seeschifffahrt und Hydrographie Hamburg (German maritime authority).

All three observatories produce minute, hourly, daily, monthly, and annual mean values and *K*-indices. They send reported data in quasi real-time and definitive data to INTERMAGNET. The observations contribute in the frame of the worldwide network of magnetic observatories to the database for scientific research of the Earth interior and the

Table O4 General information about German observatories

Observatory	Previous station	Observation period	Observers in charge	Buildings	Other observations
Berlin	–	1836–1872	J.F. Encke (1836–1865) and J.G. Galle (1836–1872)	One nonmagnetic house at the Berlin Astronomical Observatory	–
Fürstenfeldbruck	Munich, Maisach	Operated continuously since 1937	F. Burmeister (1937–1957), K. Wienert (1957–1978), M. Beblo (1978–2003), and J. Matzka (2003–)	Five nonmagnetic, five others	Seismological, telluric currents, and air electricity
Göttingen	–	1833–1902	C.F. Gauss (1833–1855), W. Weber (1833–1867), E. Schering (1867–1897), and E. Wiechert (1897–1902)	One nonmagnetic house at the Göttingen Astronomical Observatory	Seismological
Maisach	Munich	1927–1937	F. Burmeister (1927–1937)	One nonmagnetic hut and use of brewery basement	Seismological, telluric currents, and air electricity
Munich	–	1836–1925; observing gaps of several years	J. v. Lamont (1836–1879), C. Feldkirchner (1879–1886), F. Schwarz (1898–1902), J.B. Messerschmitt (1902–1912), F. Bidlingmaier (1912–1914), Lutz (1914–1919), and F. Burmeister (1919–1925)	Three nonmagnetic houses at the Munich Astronomical Observatory	Seismological, telluric currents, and air electricity
Adolf Schmidt Observatory Niemegk	Potsdam, Seddin	Operated continuously since 1930, with a short break of nine months 1945/1946	A. Nippoldt (1930–1936), R. Bock (1936–1939), G. Fanselau (1939–1968), H. Schmidt (1969–1979), K. Lengning (1980–1982), A. Best (1983–1998), and H.-J. Linthe (1999–)	Seven nonmagnetic houses and 10 others	Telluric currents
Potsdam	–	1890–1931	M. Eschenhagen (1890–1901), A. Schmidt (1901–1927), and A. Nippoldt (1927–1931)	Two nonmagnetic houses	–
Seddin	Potsdam	1907–1931	A. Schmidt (1901–1927) and A. Nippoldt (1927–1931)	Two nonmagnetic houses, one other	–
Wilhelmshaven	–	1878–1919; 1931–1936	C. Börgen (1878–1909), F. Bidlingmaier (1909–1912), and P. Meier (1931–1936)	One nonmagnetic house	Meteorological and hydrographical observations
Wingst	Wilhelmshaven	1938–	O. Meyer (1938–1954), D. Voppel (1954–1974), G. Schulz (1974–2004), and H.-J. Linthe (2004–)		

Table O5 Instruments used in the German observatories

Observatory	Absolute instruments	Variometers
Berlin	Magnetometer for intensity determination by Meyerstein, Göttingen; inclinometer by Gambey; declinometer by Baumann	–
Fürstenfeldbruck	Magnetic theodolite (D, H) and oscillation box (H) and earth inductor by Schulze, Potsdam; proton vector magnetometer after Nelson (F, H, Z; 1964; Wienert); DI-flux theodolite Zeiss 010B/Bartington MAG010H (D, I; 1992); Overhauser proton magnetometer GSM19 (F; 1996)	Two sets of variometers with photographic recording (H, D, Z) by Schulze, Potsdam, and Edelmann, Munich; three-component fluxgate variometers by FGE and MAGSON (H, D, Z; 1996)
Göttingen	Declinometer (D), and magnetometer for intensity (H) after Gauß by Meyerstein, Göttingen; inclinometers (I) by Robinson (1840) and Meyerstein (1847); earth inductor (I) after Weber by Meyerstein (1868)	–
Maisach	Two magnetic theodolites (D, H) by Bamberg, Berlin and Tesdorpf, Stuttgart; earth inductors by Schulze, Potsdam and Edelmann, Munich	One set of variometers with photographic recording (H, D, Z) by Edelmann
Munich	Meyerstein (Göttingen), declinometer (D), and intensity instrument (H); magnetic theodolite (D, H, Z; 1841) by Lamont; magnetic theodolite (D, H) and inclinometer (I) by Bamberg, Berlin (1899); earth inductor (I) by Edelmann, Munich (1908)	Magnetograph (1847) by Lamont
Adolf Schmidt observatory Niemegk	Magnetic theodolites by Wanschaff, Berlin and after A. Schmidt (D and H); oscillation box by Schulze, Potsdam (H); earth inductor by Schulze, Potsdam (I); self-made proton magnetometers (H. Schmidt, 1968); DI-flux theodolite Zeiss 010B/Bartington MAG01H (D, I; 1996)	Four sets of classical variometers (H, D, Z) by Schulze, Potsdam and Mating and Wiesenberg, Potsdam; three-component fluxgate variometers FGE and MAGSON (H, D, Z; 1996)
Potsdam	Magnetic theodolite by Edelmann, Munich (D, H); inclinometer by Bamberg, Berlin and earth inductor by L. Weber (I); magnetic theodolite by Wanschaff (D, H; 1893); earth inductor by Schulze, Potsdam (I; 1901); oscillation box (H) and theodolite after A. Schmidt (D, H; 1923)	One set of variometers (H, D, Z) by Carpentier, Paris, with photographic recording equipment by Wanschaff, Berlin (main system); one set of variometers (H, D, Z) by Wild-Edelmann, Munich with visual reading (auxiliary system)
Seddin	–	One set of variometers (X, Y, Z) by Schulze, Potsdam
Wilhelmshaven	Declinometer after Neumayer by Bamberg, Berlin (D); magnetic survey theodolite by Lamont, Munich (H); earth inductor after Weber and inclinometer by Dover (I)	One set of variometers (H, D, Z), visually readable, by Lamont, Munich; one set of variometers (H, D, Z) with photographic recording of Kew type by Adie, London
Wingst	Magnetic theodolite (D, H), oscillation box (H), earth inductor (I) by Schulze, Potsdam; proton vector magnetometer after Voppel (F, H, Z; 1969); DI-flux theodolite Zeiss 010B/Bartington MAG01H (D, I; 1992) and Overhauser proton magnetometer GSM19 (F; 1992)	One set of variometers with photographic recording by Schulze, Potsdam (H, D, Z); 2 three-component fluxgate variometers FGE (H, D, Z; 1993)

solar terrestrial relations. The Adolf Schmidt Observatory, Niemegk is further responsible for the repeat station measurements and the construction of magnetic maps of Germany. Niemegk Observatory further carries out the Kp service for the International Service of Geomagnetic Indices of the IAGA. Niemegk and Wingst are Kp stations.

Hans-Joachim Linthe

Bibliography

Knothe, C., 1987. Secular variations of the magnetic declination in middle Europe during the last 500 years, derived mostly from mine surveying. In *Proceedings of the IAGA-Symposium Space-Time-Structure of the Geomagnetic Field.* HHI-Report No. 21 (1987), Berlin, pp. 90–98.

Neunhöfer, H., Börngen, M., Junge, A., and Schweitzer, J., (eds), 1997. *Zur Geschichte der Geophysik in Deutschland Jubiläumsschrift Deutsche Geophysikalische Gesellschaft 1922–1997,* Hamburg: Deutsche Geophysikalische Gesellschaft.

Schroeder, W., and Wiederkehr, K.-H., 2002. Geomagnetic research in the 19th century. *Acta Geodaetica, Geophysica et Hungarica*, **37**(4): 445–466.

OBSERVATORIES IN INDIA

Magnetic observatories have been running in India for more than 180 years. Three of the oldest observatories, namely Madras Observatory (1822–1881), Shimla Observatory (1841–1845), and Trivandrum Observatory (1841–1871) participated in the international collaboration venture involving simultaneous magnetic measurements at 50 observatories all over the globe organized by the Göttingen Magnetic

Union 1836–1841. The Colaba Observatory (Geographic latitude 18°53′36″N; longitude 72°48′54″E) was started in 1823 by the East India Company. Initially it was undertaking mainly meteorological observations to support the function of the Bombay Harbor. The first regular magnetic field observations from Colaba were started in the year 1841. In view of the proposed introduction of electric trams running on direct current in Bombay city in the year 1900, the then Director Dr. Nanabhai A.F. Moos could envisage its deleterious effect on the accuracy of magnetic measurements at Colaba Observatory. He made enormous efforts to choose an alternate site at Alibag (Geographic latitude 18°38′17″N; longitude 72°52′21″E), free from possible electromagnetic noises and in proximity to Colaba. The location of Alibag Observatory is at about 30 km south-south east of Bombay facing the Arabian Sea. The two historic nonmagnetic buildings, built completely with nonmagnetic Porbunder sandstone, brass, or copper fittings at Alibag, are still in good condition. Alibag Magnetic Observatory became operational from April 1904. These records overlapped with the Colaba records for a period of two years (1904–1906) and subsequently recordings at Colaba were stopped. Thus at present, Colaba-Alibag magnetic observatory data form a long series of more than 160 years.

At present there are 13 permanent magnetic observatories in India (Figure O13), and the Indian Institute of Geomagnetism (IIG), Mumbai, operates nine of these. The Alibag Magnetic Observatory joined the INTERMAGNET in 1993 when a Narod Ring Core Fluxgate magnetometer from USGS was installed there. Recently, analog variation magnetometers have been supplemented with digital fluxgate magnetometers manufactured by Dannish Meteorological Institute (DMI) at Tirunelveli, Pondicherry, Nagpur, Vishakhapatnam, and Jaipur observatories operated by IIG. Modern absolute instruments, such as declination inclination magnetometer (DIM), have been installed at most of the observatories, to record and maintain the stability in the baseline, which is a crucial parameter in computing the final absolute values from variation data. A brief description of the magnetic observatories currently operational is given below. Also included is a map showing the locations of the magnetic observatories in India.

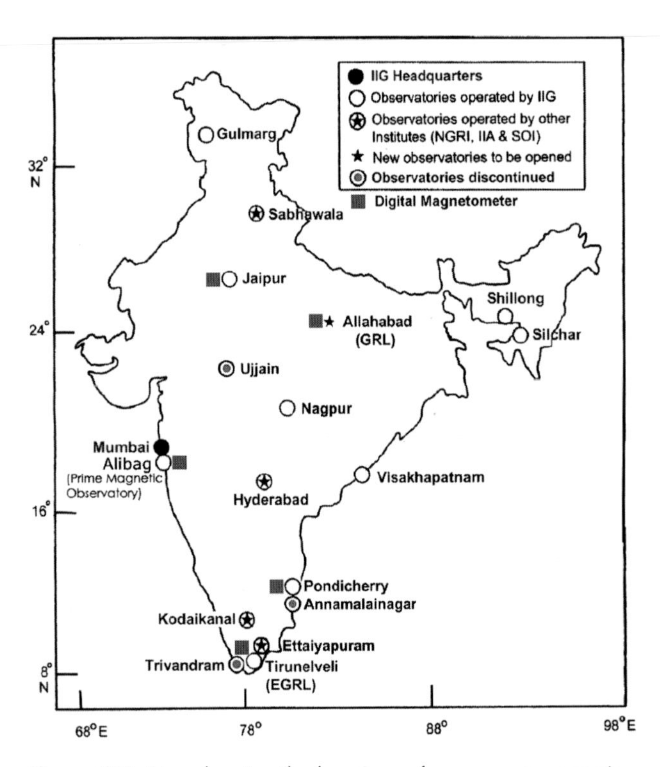

Figure O13 Map showing the locations of permanent magnetic observatories in India.

Alibag Magnetic Observatory: ABG (Operated by IIG, Mumbai), Geographic (18.63° N, 72.87° E), Geomagnetic (at 2000.0 10.03° N, 145.97° E).

This is the prime magnetic observatory of India. It was established in April 1904 as a successor to Colaba Magnetic Observatory, Bombay. The absolute instruments are vector proton precession magnetometer (VPPM), quartz horizontal magnetometer (QHM), zero balance magnetometer (BMZ), Kew No. 7 Magnetometer, and a DIM. The variometers used are Kew magnetometer, Askania magnetometer, La Cour variometer, and Izmiran II variometer.

Ettaiyapuram Magnetic Observatory: ETT (Operated by National Geophysical Research Institute, NGRI, Hyderabad), Geographic (9.00° N, 78.00° E), Geomagnetic (0.04° S, 149.99° E).

The Ettaiyapuram Magnetic Observatory was established in October 1979. The setup includes VPPM, QHM, BMZ, DIM, and a La Cour variometer.

Gulmarg Magnetic Observatory: GUL (Operated by IIG, Mumbai), Geographic (34.25° N, 74.41° E), Geomagnetic (25.56° N, 149.35° E).

Gulmarg Magnetic Observatory started functioning in September 1977. It has QHM, VPPM, DIM, and an Izmiran II variometer.

Hyderabad Magnetic Observatory: HYB (Operated by NGRI, Hyderabad), Geographic (17.42° N, 78.55° E), Geomagnetic (8.29° N, 151.29° E).

Hyderabad Magnetic Observatory was established in December 1964. The setup has VPPM, QHM, BMZ, DIM, and a La Cour variometer.

Jaipur Magnetic Observatory: JAI (Operated by IIG, Mumbai), Geographic (26.91° N, 75.80° E), Geomagnetic (17.98° N, 149.64° E).

The Jaipur Observatory was initially established in the year 1975 in the Malvia Regional Engineering College Campus. The observatory operations were discontinued from the year 1988. This observatory was restarted in September 2002. The absolute instruments are VPPM, QHM, BMZ, and DIM. The variation instrument is an Izmiran IV variometer.

Kodaikanal Magnetic Observatory: KOD (Indian Institute of Astrophysics, IIA, Bangalore), Geographic (10.23° N, 77.47° E), Geomagnetic (1.23° N, 149.58° E).

Kodaikanal Magnetic Observatory was started in January 1949. The setup includes Kew Magnetometer No. 3, galvanometer, Earth inductor, La Cour variometer, and Watson's variometer.

Nagpur Magnetic Observatory: NGP (Operated by IIG, Mumbai), Geographic (21.15° N, 79.08° E), Geomagnetic (11.96° N, 152.14° E).

Nagpur Magnetic Observatory was commissioned in June 1991. The setup has VPPM, DIM, and an Izmiran II variometer.

Pondicherry Magnetic Observatory: PON (Operated by IIG, Mumbai), Geographic (11.92° N, 79.92° E), Geomagnetic (2.70° N, 152.13° E).

Pondicherry Magnetic Observatory was established in January 1993 as a replacement for the Annamalainagar Observatory, Geographic (11.40° N, 79.68° E), which operated from 1957 to 1993. Its location is on the edge of the equatorial electrojet belt, which is a region under the influence of ionospheric current systems associated with the dip equator. The absolute instruments are DIM, PPM, VPPM, QHM, and BMZ. The variometer is an Izmiran II variometer.

Sabhawala Magnetic Observatory: SAB (Operated by Survey of India, SOI, Dehra Dun), Geographic (30.33° N, 77.80° E), Geomagnetic (21.21° N, 151.86° E).

Sabhawala Magnetic Observatory was commissioned in January 1964. The setup includes QHM, BMZ, La Cour variometer, and Askania variometer.

Silchar Magnetic Observatory: SIL (Operated by IIG, Mumbai), Geographic (24.93° N, 92.82° E), Geomagnetic (14.83° N, 165.38° E).

Silchar Magnetic Observatory was commissioned in November 1998. The setup has instruments DIM, PPM, VPPM, QHM, and Izmiran II variometer.

Shillong Magnetic Observatory: SHL (Operated by IIG, New Panvel), Geographic (25.56° N, 91.88° E), Geomagnetic (15.69° N, 164.72° E).

Shillong Magnetic Observatory started functioning from the middle of the year 1975. It operates QHM, BMZ, VPPM, DIM, and Izmiran II variometer.

Tirunelveli Magnetic Observatory: TIR (Operated by IIG, Mumbai), Geographic (8.42° N, 77.48° E), Geomagnetic (0.57° S, 149.42° E).

The Tirunelveli Observatory started in March 1996 as successor to IIGs Trivandrum Magnetic Observatory (Geographic: 8.48° N, 76.95° E), which was in operation from 1957 to 1999. The absolute instruments are DIM, PPM, BMZ, and QHM. The variation instrument is an Izmiran II variometer.

Vishakhapatnam Magnetic Observatory: VSK (Operated by IIG, Mumbai), Geographic (17.68° N, 83.32° E), Geomagnetic (8.17° N, 155.89° E).

Vishakhapatnam Magnetic Observatory started functioning from July 1994. The setup includes DIM, VPPM, QHM, and Izmiran II variometer.

The observatory network in India is ideal for studying equatorial, low- and midlatitude geomagnetic phenomena and space weather. The Indian magnetic data is published by IIG. It is planned to preserve the entire Colaba-Alibag unique data series by scanning and digitizing the old magnetograms. In addition, IIG is operating World Data Center for Geomagnetism, Mumbai. For more details, one can visit the Web site of the Indian Institute of Geomagnetism, Mumbai, at iigs.iigm.res.in.

Gurbax S. Lakhina and S. Alex

Cross-references

Observatories, Instrumentation
Observatories, INTERMAGNET

OBSERVATORIES IN ITALY

Monitoring the Earth's magnetic field is carried out by geomagnetic observatories all over the world. In Italy, the first observatory was founded in 1880, when Pietro Tacchini, the director of the Central Meteorological Institute (*Ufficio Centrale di Meteorologia*), launched an initiative to study the distribution of the Earth's magnetic field over the Italian territory. Prior to this, an initial attempt had been made by Father Francesco Denza (1834–1894). In 1881, Tacchini proposed to build an observatory in Rome, envisaging a nonmagnetic facility on the Palatino hill. However this effort was hampered by administrative difficulties. At the end of the 19th century, electromagnetic noise, generated mainly by the development of electrified transportation, affected Rome as well as other Italian cities. Meanwhile the International Commission for Terrestrial Magnetism was urging the Italian scientific community to establish a magnetic observatory in southern regions to facilitate the southern extension of the European network. The geophysical observatories in Catanzaro and in Messina, both located in Southern Italy, were then proposed. However the strong earthquake of 1905 in Sicily, which had dramatic consequences in Messina, scuppered plans to establish the magnetic observatory in the area.

In the meantime, repeat station measurements have been conducted by Luigi Palazzo in several locations of the Italian territory. These surveys indicated that satisfactory data reduction was achievable using the magnetic data of the observatory of Pola, operating since 1881 in the Istria Peninsula (Latitude 44.87° N, Longitude 13.85° S). At that time, Istria belonged to the Austro-Hungarian Empire. When World War I ended, Istria became part of Italy. Unfortunately, due to the electromagnetic disturbances caused again by the city tramways, geomagnetic recordings of the vertical component of the field were suspended. With a further increase of the noise, in 1923, Pola was closed.

It was then in 1932 that a section of the Italian Navy, the *Istituto Idrografico della Marina*, initiated regular absolute geomagnetic measurements at Castellaccio (near Genova, Latitude 44.83° N, Longitude 8.93° E) along with variational recording using Pola Observatory's equipment. Castellaccio Observatory operated regularly and published annual bulletins until 1962.

In the early fifties, the National Geophysical Institute (*Istituto Nazionale di Geofisica—ING*) undertook the task to identify a suitable location, from a geological and an environmental standpoint, for a national geomagnetic observatory. Accurate surveys conducted in the area of Preturo (Latitude 42.38° N, Longitude 13.32° E) near the city of L'Aquila in central Italy, indicated an appropriate site. Funding was secured in 1958 from Franco Molina. Preturo had been working for two years as a variometer station. In 1960, absolute measurements started to be carried out on a regular basis allowing the calculation of the absolute values of the Earth's magnetic field. This observatory is known internationally as L'Aquila.

Although in different epochs, the three geomagnetic observatories (Pola, Castellaccio, and L'Aquila) operated regularly during the last century for a number of years, allowing the secular variation of declination and the horizontal component in Italy to be determined since 1881.

Another magnetic observatory was then installed by ING in the north of the country, at Castello Tesino (Latitude 46.05° N, Longitude 11.65° E), and fitted with automatic recording equipment. Qualified personnel perform absolute measurements regularly on a monthly basis, guaranteeing the quality of the data.

Both L'Aquila and Castello Tesino are currently operating and recording field values every minute, with regular transmission of data to the data center at Rome.

Under the auspices of the Italian National Antarctic Research Program, ING built and currently operate the Antarctic Italian Geomagnetic Observatory Terra Nova Bay (Latitude 74.69° S, Longitude 164.12° E) located on the coast of the Ross Sea.

Massimo Chiappini

OBSERVATORIES IN JAPAN AND ASIA

Magnetic observations in Japan

Modern geomagnetic observation in Japan began on March 15, 1883, in Tokyo, as an initiative of the Japanese government to join the First Polar Year. In conjunction, a geomagnetic survey over all Japanese islands was conducted. A well-equipped observatory in Tokyo was started in 1897 by the meteorological agency and continued until 1917. Unfortunately, most of the records were burnt during the 1923 Tokyo earthquake. Because of the magnetic noise due to the construction of a DC train system in Tokyo, the observatory was moved. A standard magnetic observatory in Kakioka, near Tokyo, was established in 1913 and the observations are continuing.

There are currently four agencies conducting geomagnetic observations in Japan.

1. The Meteorological Agency of Japan operates three magnetic observatories: Kakioka, Kanoya, and Memambetsu. These observatories provide continuous geomagnetic data with precise absolute measurements. There is also one unmanned observatory, Chichijima. The main purpose of these observatories is for monitoring volcanic or earthquake activity, and it also deploys instruments for temporary observations during a volcanic eruption or strong earthquake. The observations are also used for research.

Table O6 Listing of magnetic observatories, operated by the government in Japan

Station name	Abb.	GGLAT.	Long.	GMLAT.	Elev.	Year	*H*	*D*	*I*
Chichijima	CBI	27.10	142.18	18.47	155	2003	32609	–03.61	37.28
Esashi	ESA	39.24	141.35	30.46	396	2000	29095	–06.87	52.68
Hatizyo	HTY	33.08	139.82	24.21	220	2002	31346	–07.47	46.11
Kakioka	KAK	36.23	140.19	27.37	36	2003	29980	–07.06	49.72
Kanoya	KNY	31.42	130.88	21.89	107	2003	32776	–05.93	44.97
Kanozan	KNZ	35.26	139.96	26.39	342	2000	30366	–06.71	48.57
Memambetsu	MMB	43.91	144.19	35.35	42	2003	26182	–08.78	58.09
Mizusawa	MIZ	39.11	141.21	30.32	130	2000	28646	–07.89	52.92

2. The Geographical Survey Institute of Japan operates three magnetic observatories—Kanozan, Mizusawa, and Esashi—and 11 unmanned continuous observation sites. They conduct magnetic surveys at repeat stations and publish magnetic charts of Japan every 10 years.
3. The Hydrographic and Oceanographic Department has one observatory, Hatizyo. They conduct surveys by ship or airplane over the sea around the Islands of Japan and publish magnetic charts.
4. The National Institute of Advanced Industrial Science and Technology (AIST) conducts Aeromagnetic surveys over the Islands of Japan and publishes magnetic charts.

Other than the observatories mentioned above (see also Table O6), there are many observation sites for scientific research that mainly belong to universities or national institutions.

Magnetic observation in Indonesia

The Indonesian Meteorological and Geophysical Agency (IMGA) operates three geomagnetic observatories: Tondano Geomagnetic Observatory (TND, 1.29° N, 124.95° E) in North Sulawesi, Tangerang Geomagnetic Observatory (TNG, 6.17° S, 106.63° E) in Jakarta, and Tuntungan Geomagnetic Observatory (TTG, 3.50° N, 98.56° E) in North Sumatra. Geomagnetic observations at TNG started in 1964 using the Ruska type photographic paper variograph. TNG was upgraded in 1991 with a digital magnetometers. The observatory Tondano began collecting observations in 1991 with digital fluxgate and proton magnetometer. Tuntungan Geomagnetic Observatory (TTG) has been observing the geomagnetic field since 1980 using a La Cour photographic variograph, and after 2003 was equipped with a digital magnetometer.

Magnetic observation in Vietnam

The Hanoi Institute of Geophysics, National Centre for Natural Science and Technology of Vietnam currently operates four geomagnetic observatories in Vietnam; Cha Pa (CPA, 22.35° N, 103.83° E), Bac Lieu (BCL, 9.3° N, 105.7° E), Phu Thuy (PHU, 21.03° N, 105.95° E), and Dalat (DLT, 11.93° N, 108.48° E). The magnetic observations at CPA and PHU began in 1957 and 1978, respectively.

Toshihiko Iyemori and Heather McCreadie

OBSERVATORIES IN LATIN AMERICA

Introduction

During colonial times, geomagnetism in Latin American countries was restricted to declination measurements for cartography and the definition of borders. A large survey was performed by E. Halley (see *Halley, Edmond*), in 1700, when he prepared his famous magnetic chart of the South Atlantic. Following Halley's work, it was not until 1880 that a large field survey was made by the Dutch scientist Van Rickjervosel in the Brazilian coastal area. That observational work was completed by the publication of the first Brazilian magnetic chart. Since no observatories existed in those times, they could be considered as the prehistoric phase of geomagnetism in Latin America.

Birth and development of Latin American magnetic observatories

In the first half of the 20th century a considerable effort was made by the Carnegie Institution of Washington (see *Carnegie Institution of Washington, Department of Terrestrial Magnetism*) to improve geomagnetism in South America. Observatories were installed in Peru and Argentina. Local efforts were made in various countries, sometimes with the assistance of the Inter-American Geodetic Survey, and the Pan-American Institute of Geography and History (PAIGH), to install other observatories. This phase lasted from 1920 to 1960. Chiripa Magnetic Observatory in Costa Rica is a good example of that international cooperation.

Lately, Peruvian observatories (Huancayo and Ancón) received considerable assistance from Japanese institutions, mainly from Tokyo University. In spite of a good beginning, some observatories lost their initial quality and some of them were closed.

In 1980, PAIGH started a campaign, under the leadership of the National Observatory of Brazil, to reorganize and modernize all observatories in Latin America. The most important part of that campaign was a series of Latin American geomagnetic schools, where the training of technicians and also the installation of modern instruments transformed old and unproductive observatories into operational ones. Important international contribution for these improvements came from the action of two IAGA Divisions, Division V (observatories, instruments, surveys, and analyses) and the Interdivision Commission on Developing Countries (see *IAGA, International Association of Geomagnetism and Aeronomy*). From this, the observatories of Trelew (Argentina) and La Habana (Cuba) received new magnetic instruments, such as digital fluxgate variometers and modern absolute instruments.

Besides the observatories mentioned above and others included in the summary below, there are some variation stations (where no absolute observations are carried out) that are operated for specific research projects. Among these stations the most important are located in Brazil, Peru, and Mexico.

It is interesting to note the importance and responsibility of Latin American magnetic observatories, not only for the great latitude range of the group (66°), but also for the occurrence of two important geomagnetic phenomena in the region—the equatorial electrojet and the South Atlantic magnetic anomaly.

Table O7 gives data for all observatories in the region.

Table O7 Summary

Country	Name of observatory	Geographic coordinates		Year of installation	Actual status
		Latitude	Longitude		
Argentina	Trelew	48.2° S	204.7° W	1957	Normal operation INTERMAGNET station
	Pilar	31.7° S	296.1° W	1905	Irregular operation
	Orcadas	60.7° S	315.6° W	1904	Irregular operation
	La Quiaca	22.1° S	294.4° W	1920	Closed in 1975
	Las Acacias	35.0° S	302.3° W	1957	Not operating since 1997
Bolivia	Patacamaya	17.3° S	292.1° W	1983	Irregular operation
Brazil	Vassouras	22.4° S	316.4° W	1915	Classical instruments up to 1993 INTERMAGNET (1993) Original Instruments operation as backup
	Tatuoca	1.2° S	311.5° W	1957(IGY)	Classical instruments normal operation
	Brasilia	14.9° S	312.1° W		Programmed for Dic.2005
	São Martinho	29.4° S	53.8° W		Programmed for Dic.2005
Chile	Isla De Pascoa	–	–	1957(IGY)	Closed
Costa Rica	Chiripa	10.4° N	275.1° W	1984	Classical. Normal operation
Colombia	Fuquene	5.5° N	280.3° W	1954	Classical. Normal operation
Cuba	La Habana	23.1° N	277.5° W	1964	INTERMAGNET IN 1997
Mexico	Teoloyucan	19.7° N	260.8° W	1914 moved in 1978	Classical instruments INTERMAGNET 2002
Peru	Huancayo	12.0° S	284.7° W	1922 by CIW	Classical instruments INTERMAGNET 2002
	Ancon	11.7° S	282.9° W	1984	Digital variometer and modern absolute instruments
	Arequipa	16.4° S	288.5° W	1957	Classical instruments Normal operation

Acknowledgments

Data for each observatory were furnished by the local person in charge of the station.

Luiz Muniz Barreto

Cross-references

Carnegie Institution of Washington, Department of Terrestrial Magnetism
Halley, Edmond (1656–1742)
IAGA, International Association of Geomagnetism and Aeronomy

OBSERVATORIES IN NEW ZEALAND AND THE SOUTH PACIFIC

New Zealand has been involved in five magnetic observatories over the last century. In New Zealand, there has been a succession of three observatories located at Christ-church, Amberley, and Eyrewell. The others are Apia in Samoa and Scott Base in the Antarctic. The Apia and Scott Base observatories were only intended to operate for limited durations, but Apia has continued for over 100 years and Scott Base for nearly 50 years.

Observatories in New Zealand

Dr. Schmidt had drawn attention to the desirability of an observatory in New Zealand (Schmidt, 1897), and referring to this paper in his presidential address to Section A of the Australasian Association for the Advancement of Science Mr. Barrachi said, "The establishment of a magnetic observatory in New Zealand seems to be a duty that colony owes to the Scientific world" (Farr, 1916). With support from the commander in chief of the Australian Naval Station and the Shipmasters' Association of New Zealand, government funding was obtained for both a survey and the establishment of a magnetic observatory.

The observatory was established in the botanic gardens in Christ-church using Adie magnetographs and a new Kew magnetometer (No. 1C.Sc.I.Co) and Dip Circle No.147 for absolute observations. Christ-church soon became magnetically disturbed by electric trams. A substation, which eventually replaced Christ-church, was established at Amberley 43 km north of Christ-church using Eschenhagen magnetographs. In 1957, a set of La Cour variometers was installed at Amberley (J. A. Lawrie, personal communication).

In the mid-1960s a DC power line was erected to transfer power between the North and South Islands of New Zealand. It consists of two independent conductors (mounted on the same pylons where the line is over land) with a common earth return. Each conductor was capable of independently carrying up to 1200 A (since increased to 1800 A). At the observatory, this produced a vertical field proportional to the net current flowing in the two conductors. The maximum disturbance was 12 nT. A sensor was installed under the lines to measure the net current flowing in the conductors and provide a correction field around the Z variometer.

In 1977, development around Amberley forced the establishment of a new observatory at Eyrewell, about 22 km northwest of the original Christ-church Observatory. Digital recording instruments designed and built by the author were introduced in June 1990 allowing it to become an INTERMAGNET observatory in 1994.

In February 2005, a new Danish Meteorological Institute suspended fluxgate magnetometer and a Gem GSM-90 Overhauser magnetometers were installed. These will become the prime recording instruments after a suitable period of comparison with the existing instruments.

Apia observatory

In November 1899 at a meeting of the council entrusted with the preparation of the German South-Polar Expedition, Adolf Schmidt proposed that a magnetic observatory should be set up in Samoa at the same time when the stations were to be set up by the expedition in Antarctica, to enable simultaneous records. Dr. Otto Tetens sailed to Samoa taking with him two Eschenhagen variometers, a *Z*-balance and a magnetic theodolite for absolute measurements. He arrived in June 1902 (Angenheister, 1974).

When World War I broke out, Samoa was occupied by 1500 soldiers from New Zealand. Dr. Gustav Heinrich Angenheister, who was serving as director, managed to convince the English commander that the observatory was a scientific institution and needed to be continued. He remained at the observatory during and after the war until 1920, working under conditions of extreme hardship and financial difficulties. At the end of the war, the League of Nations mandated control of the German Samoan Islands to New Zealand.

Funding for the observatory was provided by the New Zealand government and the Carnegie Institution, assisted by an annual grant form the British Admiralty (Sommerville, 1923). Eventually, the observatory came under the control of the New Zealand Department of Scientific and Industrial Research (DSIR). New La Cour variometers were installed by Lawrie in 1958.

Control of the observatory remained with the DSIR until a few years after independence in 1962 and was then handed over to the Samoan government. The DSIR still provided some support by way of technical advice and with the processing and publication of the data.

In 1990, Samoa was devastated by tropical cyclone Ofa which removed the roof from the variometer house. This turned out to be a blessing in disguise because repairs to the roof would have cost more than a new digital system. As part of New Zealand's aid program, the New Zealand Department of Foreign Affairs and Trade commissioned the DSIR to supply and install a digital system as used at Eyrewell. The installation of the digital system in December 1991 allowed Apia to become an INTERMAGNET observatory in 1998.

Scott base observatory

Originally Scott Base was to be at Butter Point on the western side of McMurdo Sound, but it was considered unsuitable for the Trans-Antarctic Expedition. The present site on the eastern side of Hut Point Peninsula on Ross Island was chosen instead. It is not a good site for a magnetic observatory as Ross Island is composed of highly magnetic basalt scoria (Cullington, 1961). However there was no alternative.

Two prefabricated buildings were erected in 1957, one each for absolute observations and the La Cour variometer systems. Most buildings have been replaced over the years but these two buildings are still used for their original purpose.

A digital fluxgate magnetometer was installed at Scott Base in December 1990 and it became an INTERMAGNET observatory in 1996. The instruments will be upgraded at the end of 2005 as has been done at Eyrewell.

Lester A. Tomlinson

Bibliography

Angenheister, G.G., 1974. Geschichte des Samoa-Observatoriums von 1902 bis 1921. In *Zur Geschichte der Geophysik.* Berlin: Springer-Verlag, pp. 43–66.

Cullington, A.L., 1961. *Scott Base Observatory Magnetic Results for International Geophysical Year 1957–58.* Wellington: Government Printer.

Farr, C., Coleridge, 1916. *A Magnetic Survey of the Dominion of New Zealand.* Wellington: New Zealand Government Printer.

Schmidt, A., 1897. On the distribution of magnetic observatories. *Terrestrial Magnetism*, XI, p. 30.

Sommerville, D.M.Y., 1923. Editorial note. In *Apia Observatory Report 1921.* Wellington: Government Printer.

OBSERVATORIES IN NORDIC COUNTRIES

The Nordic countries comprise Denmark, the Faeroe Islands, Finland, Greenland, Iceland, Norway, and Sweden. All but the Faeroe Island houses one or more geomagnetic observatories. The observatories in operation as of 2005 are listed in Table O8.

They are all equipped with modern fluxgate magnetometers—the majority with the suspended Danish FGE type—for the continuous recording, and are using standard D/I-flux theodolite along with proton precession magnetometer for absolute calibration (see *Observatories, instrumentation*). Digital recording was introduced during the 1980s and 1990s at all of them.

Denmark and Greenland

The geomagnetic observatories in Denmark and Greenland have been part of the Danish Meteorological Institute in Copenhagen ever since the Copenhagen Observatory was established in 1891. Already in 1907 the Copenhagen Observatory was moved to Rude Skov 20 km to the north due to increasing magnetic disturbance in the city. The observatory remained here until 1984 when relocation to its present site at Brorfelde 40 km west of Copenhagen took place.

The first magnetic observatory at Greenland was established in Godhavn (today Qeqertarsuaq), at Greenland's west coast in 1926, where it is still found. The history of the North Greenland Observatory dates back to the Second International Polar Year (IPY) 1932/1933, when a magnetic observatory was operated in Thule. A permanent observatory was set up there in 1947, but had to be abandoned after only a few years due to the establishment of the US Air Force base at Thule. Finally, operation was resumed in 1956 at its present site of Qaanaaq (previously called Thule) 100 km to the north. The South Greenland Observatory of Narsarsuaq was added in 1967.

Finland

The history of geomagnetic observatories in Finland dates back to Göttingen Magnetic Union (see *Geomagnetism, history of*) when an observatory was set up in Helsinki by the University of Helsinki. Regular operation commenced in 1844 (Nevanlinna, 2004). In 1912 it was relocated within Helsinki, and finally in 1953 it was replaced by the present observatory at Nurmijärvi 40 km north of Helsinki. It is operated by the Finnish Meteorological Institute.

The observatory of northern Finland in Sodanlylä has a precursor in the First IPY station of 1882/1983. In 1914, the Finish Academy of Science and Letters set up a permanent geophysical observatory, a magnetic observatory being part of it (Kataja, 1973). Today the Sodankylä Geophysical Observatory is organized as a branch of the University of Oulu.

Iceland

The magnetic observatory at Iceland started recording in 1957. It is located at Leirvogur 12 km ENE of Reykjavik, and is maintained by the University of Iceland.

Norway

Similar to Finland, the first Norwegian magnetic observatory originated in the activities of the Göttingen Magnetic Union. The astronomical

Table O8 Magnetic observatories in the Nordic countries operating in 2005

Observatory name	IAGA code	Country	Geographic latitude	Geographic longitude	First annual mean
Qaanaaq	THL	Greenland	77 29° N	69 10° W	1956
Hornsund	HRN	Norway	77 00° N	15 33° E	1979
Bear Island	BJN	Norway	74 30° N	19 00° E	1951
Tromsø	TRO	Norway	69 40° N	18 56° E	1930
Qeqertarsuaq	GDH	Greenland	69 15° N	53 32° W	1926
Abisko	ABK	Sweden	68 22° N	18 49° E	1946
Sodankylä	SOD	Finland	67 22° N	26 32° E	1914
Leirvogur	LVR	Iceland	64 11° N	21 42° W	1958
Dombås	DOB	Norway	62 05° N	09 07° E	1952
Narsarsuaq	NAQ	Greenland	61 11° N	45 26° W	1967
Nurmijärvi	NUR	Finland	60 31° N	24 39° E	1953 (1844)
Uppsala	UPS	Sweden	59 54° N	17 21° E	1998 (1928)
Brorfelde	BFE	Denmark	55 38° N	11 40° E	1980 (1891)

Note: NUR, UPS, and BFE are direct successors to earlier observatories. The first year of the total annual mean series is added in parenthesis.

observatory in Oslo was then furnished also as a magnetic observatory and began regular observations in 1842. The University of Oslo kept it running until 1932 (Wasserfall, 1950). In the meantime, a magnetic station had been set up at Dombås 300 km north of Oslo in 1916, but not until 1950 was this station upgraded to a magnetic observatory. It is today run jointly by the universities in Bergen and Tromsø. The magnetic observatory in Tromsø was established in 1930 as part of the newly founded Auroral Observatory. From 1972 it is part of the University of Tromsø.

Bear Island was the site of a Polish research station including a magnetic observatory during the Second IPY 1932/1933. The magnetic observations were continued by the Auroral Observatory in Tromsø, but not until 1951 were the calibration and stability good enough to produce reliable annual means. Today it is run by the University of Tromsø. Hornsund is a Polish research station maintained since the International Geophysical Year in 1957/1958 by the Polish Academy of Science in the southern part of Svalbard. From 1978 the station has been manned all year and includes a magnetic observatory.

Sweden

The observatory of southern Sweden dates back to 1928 when the Lovø Observatory in Stockholm was built. A new observatory near Uppsala is replacing Lovø and is effective from 2005. In northern Sweden, the magnetic station at Abisko was set up by the Swedish Academy of Science in 1921, but before 1945 absolute calibrations were quite scarce. Both are operated today by the Swedish Geological Survey.

Truls Lynne Hansen

Bibliography

Magnetic yearbooks

Descriptions of the observatories, their results, instruments, and methods are found in the so-called magnetic yearbook. Over the years they have been issued under changing forms and titles. Below the most recent issues are listed. Today the distribution of magnetic data by the Internet is important and this is done through INTERMAGNET (see *Observatories, INTERMAGNET*) and the World Data Centers as well as through the institutes' own Web sites.

Leirvogur Magnetic Results 2004, Rauvísindastofnun Háskólans, Reykjavík 2005.

Magnetic Results 2000 Brofelde, Qeqertarsuaq, Qaanaaq and Narsarsuaq Observatories, Danish Meteorological Institute, Data report 02–01, Copenhagen, 2002.

Magnetic Results Nurmijärvi Geophysical Observatory 2002, Finnish Meteorological Institute, Helsinki, 2003.

Magnetic Results Sodankylä 2002, Sodankylä Geophysical Observatory Publication No. 94, Oulu, 2003.

Geomagnetic Observatory Data 2002: Lovö and Abisko, Geofysiska meddelanden Cb31, Sveriges geologiska undersökning, Uppsala, 2004.

Magnetic Observations 1998, The Auroral Observatory—University of Tromsø, Tromsø, 2000.

The Magnetic Station at Dombås, Observations 1998, Institute of Solid Earth Physics, University of Bergen, Bergen, 2000.

Others

Kataja, Eero, 1973. The Sodankylä Geophysical Observatory in 1973. Veröffentlichungen des geophysikalischen Observatoriums der finnischen Akademie der Wissenschaften, No 56/1.

Nevanlinna, H., 2004. Results of the Helsinki magnetic observatory 1844–1912. *Annales de Geophysicae*, **22**: 1691–1704.

Wasserfall, K.F., 1950. A study of the secular variation of magnetic elements based on data for D, *I*, and *H* for Oslo 1820–1948. *Journal of Geophysical Research*, **55**: 275–299.

Cross-references

Geomagnetism, History of
Magnetometers
Observatories, Overview
Observatories, Instrumentation
Observatories, INTERMAGNET

OBSERVATORIES IN RUSSIA

The first regular measurements of the geomagnetic field in Russia were started in St. Petersburg in 1829. Over the course of the next century geomagnetic observatories equipped with the most modern, by contemporary standards, instruments, were established in Ekaterinburg (1836), Tiflis (1844), Pavlovsk (1878), Irkutsk (1886), and Kazan (1909). The first polar observatory in the world was open in 1924 in

Table O9 List of Russian geomagnetic observatories (GO) and stations (S)

	Station (abbr.)	Latitude (φ), Longitude (λ_E)	Year found	Resolution (nT) rate (s)	Available 1-min data	Aff.
GO	Heiss Island (HIS)	80.62; 58.05	1959	0.1 nT; 1 s	1997–1999, 2005	1
GO	Dixon (DIK)	73.54; 80.56	1933	0.1 nT; 1 s	1990, 1993–2005	1R
GO	Cape Chelyuskin (CCS)	77.72; 104.28	1935	0.1 nT; 1 s	1997–2005	1R
GO	Tiksi (TIK)	71.6; 129.0	1944	0.1 nT; 1 s	1995–2005	1R
GO	Pebek (PBK)	70.1; 170.9	2000	0.1 nT; 1 s	2000–2005	1R
S	Vize Island (VIZ)	79.5; 77.0	1982	1 nT; 1 min	1997–2005	1
GO	Amderma (AMD)	69.5; 61.4	1924	0.1 nT; 1 s	1997–1998, 2005	1
GO	Vostok (VOS)	–78.45; 106.87	1958	0.1 nT; 1 s	1995–2002, 2004–05	1R
GO	Mirny (MIR)	–66.55; 93.02	1956	0.1 nT; 1 s	1995–2005	1R
GO	Novolazarevskaya (NVL)	–70.77; 11.83	1960	0.1 nT; 1 s	2002–2005	1R
S	Kotelny (KTN)	76.0; 137.9	1994	0.1 nT; 1 s	1994–2000, 2002–04	2
S	Tiksi (TIX)	71.6; 128.9	1992	0.1 nT; 1 s	1992–2004	2
S	Chokurdakh (CHD)	70.6; 147.9	1992	0.1 nT; 1 s	1992–2004	2
S	Zyryanka (ZYK)	65.8; 150.8	1994	0.1 nT; 1 s	1994–2004	2
S	Zhigansk (ZGN)	66.8; 123.4	1994	0.1 nT; 1 s	2005	2
GO	Yakutsk (YAK)	62.0; 129.6	1932	0.1 nT; 1 s	2005	2
GO	Paratunka (PTK)	52.90; 158.43	1967	0.1 nT; 1 s	1992–2004	3
GO	Stekolny (MGD)	60.12; 151.02	1932	0.1 nT; 1 s	1992–2004	3
GO	Irkutsk (IRT)	52.46; 104.04	1886	0.01 nT, 1 s	1996–2004	4
GO	Norilsk (NOK)	69.40; 88.10	1965	0.1 nT; 1 s	2002–2005	4R
S	Loparskaya (MMK)	68.63; 33.25	1953	1.5 nT; 10 s	1993–2005	5
GO	Lovozero (LOZ)	67.97; 35.02	1957	1.5 nT; 10 s	1996–1999	5R
				0.005 nT; 0.1 s	1999–2005	
GO	Moscow (MOS)	55.48; 37.31	1944	1 nT; 1 min	1986–2005	6R
GO	Borok (BOX)	58.03; 38.97	1957	0.1 nT; 5 s	1998–2005	7R
GO	Novosibirsk (NVS)	54.85; 83.23	1967	0.01 nT, 1 s	2003–2005	8R
GO	Ekaterinburg (ARS)	56.43; 58.56	1836	0.1 nT; 1 s	1997–2005	9
S	Tomsk (TMK)	56.47; 84.93	1958	0.2 nT; 0.05 s	1997–2005	10R
S	Popov Island (PPI)	43.7; 132.0	2000	0.1 nT; 1 s	1997–2004	11

strait Matochkin Shar (close to modern Amderma). The next step, aimed at economic development of the Russian Arctic, was made in connection with the Second Polar Year (1932–1933). The observatories Yakutsk, Srednekan (Magadan), Dixon, Tikhaya Bay, Cape Chelyuskin, and Wellen, were set up in this period. The International Geophysical Year (1957–1958) and the subsequent beginning of the period of space research played a crucial role in building the present network of geomagnetic observations. About 50 magnetic observatories and stations were operated on the territory of former Soviet Union by 1985 (we define an observatory or a station by availability or absence of absolute geomagnetic measurements, respectively). The situation changed dramatically in connection with the disintegration of the Soviet Union and economic problems after 1991.

At present geomagnetic observations in Russia are carried out by about 30 observatories and stations, but quality of observations, determined by the type of instruments, is quite variable. Some observatories, equipped with modern instruments, provide digital data of high resolution, whereas others, using outdated devices, continue to record the magnetic variations in analog format or with insufficiently precise absolute measurements. In Table O9 we list only those geomagnetic observatories (GO) and stations (S), which provide the geomagnetic data in digital form. The table contains the name of observatory, abbreviation, geographical coordinates, the data resolution (in nT) and sampling rate (in s), and years covered by digital data. The observatories and stations are operated by various institutions of the Russian Hydrometeorological Service and Russian Academy of Science. These institutions are listed below and the last column of the table (Aff.) corresponds to the appropriate number in the list of affiliations. Letter R indicates that the data are presented in real time at the institution Web sites:

1. Arctic and Antarctic Research Institute (AARI), St. Petersburg. Web site: http://www.aari.nw.ru/clgmi/geophys/index_ru.htm
2. Yu.G. Shafer Institute of Cosmophysical Research and Aeronomy (IKFIA), Siberian Branch of RAS, Yakutsk; http://ikfia.ysn.ru,http://denji102.geo.kyushu-u.ac.jp/denji/obs/cpmn/cpmn_obs_e.html
3. Institute of Space Research and Radiowave Propagation (IKIR), Petropavlovsk at Kamchatka; ikir.kamchatka.ru
4. Institute of Solar-Terrestrial Physics, Irkutsk; iszf.irk.ru
5. Polar Geophysical Institute (PGI), Murmansk; http://pgi.kolasc.net.ru
6. Institute of Earth Magnetism, Ionosphere, and Radiowave Propagation (IZMIRAN), Moscow; http://www.izmiran.ru/magnetism/mos_data.htm
7. Schmidt's Institute of Earth Physics (IFZ), Moscow; http://geo-data.borok.ru
8. Altay-Sayan Branch of the Geophysical Survey of RAS, Novosibirsk; http://www.gs.nsc.ru/russian/ionka.html
9. Institute of Geophysics, The Ural Branch of RAS, Ekaterinburg
10. Siberian Physics-Technical Institute, Tomsk; http://sosrff.tsu.ru
11. V.I.Il'ichev Pacific Oceanological Institute (POI), Far Eastern Branch of RAS, Vladivostok; http://poi.febras.ru

Observations at stations KTN, TIX, CHD, ZYK, PTK, MGD, and PPI are carried out in the framework of the Australian-Japan-Russian project "Circum-pan Pacific Magnetometer Network" (CPMN). Data from stations PBK, TIK, CCS, NOR, and DIK form the basis for calculation of the AE index, characterizing the intensity of magnetic disturbances in the auroral zone.

Oleg Troshichev

OBSERVATORIES IN SOUTHERN AFRICA

The Early history of 1600 till 1932

This overview of geomagnetism and its historical development in southern Africa covers a period of approximately 400 years till present. During the era of exploration between 1600 and 1700, visiting seamen from Europe have been responsible for the early magnetic observations in South Africa. In fact, the earliest recorded observation of magnetic declination on land was made at Mosselbay in 1595 by Cornelis Houtman (Beattie, 1909), commander of a Dutch fleet on its way to India, who obtained a value of 0°. The first systematic observations in South Africa resulted from the establishment, in 1841, of a worldwide network of observation stations. One of these stations was built in the grounds of the Royal Observatory at the Cape of Good Hope, with the Royal Artillery detachment responsible for all measurements. After the magnetic observatory was destroyed by fire in 1852, a period of 80 years elapsed before a magnetic observatory worthy of name was again established in South Africa. During that period, observations of magnetic declination had been made by surveyors in different parts of the interior of the country. The first magnetic survey of the Union of South Africa, comparable with that of developed countries in the Northern Hemisphere, was carried out by Beattie (1909), assisted by Prof. J.T. Morrison of the University of Stellenbosch between 1898 and 1906.

The Period 1932 till present

The Polar Year of 1932–1933 prompted the establishment of a magnetic observatory at the University of Cape Town in 1932 under the leadership of Prof. Alexander Ogg (Trigonometrical Survey Office, 1939). International organizations provided grants for buildings and the loan of instruments for standard observations. The subsequent electrification of the suburban railway network however created a disturbing influence, to the extent that accurate observations became almost impossible. A new site was identified in Hermanus, because it was sufficiently remote from electric railway disturbances and had been proven by a magnetic survey to be suitable in other respects. The Hermanus Magnetic Observatory officially commenced operation on January 1, 1941. One of the most recent major contributions to geomagnetism in southern Africa was the institution of a long-term magnetic secular variation program in 1938–1939. The program consisted initially of 44 permanent field stations covering, with a fairly uniform distribution, the whole of South Africa, Namibia, and Botswana, but has since been expanded to Zimbabwe as well, with a total of 75 stations. To ensure exact reoccupation during subsequent surveys, concrete beacons were erected at each field station to mark the position of the instruments during observations.

A major undertaking during the period 1960–1975 was the establishment of other magnetic observatories in southern Africa. In 1964, an opportunity arose to establish a magnetic observatory on the premises of the "Forschungsstation Jonathan Zenneck," a permanent ionospheric observation station of the Max Planck Institut für Aeronomie outside Tsumeb in Namibia. Two more recording stations were also established in South Africa—one at Hartebeesthoek in 1972, and the other at Grahamstown in 1974. Problems at Grahamstown eventually led to the closure of the station in 1980. Since then both Hermanus and Hartebeesthoek became INTERMAGNET Observatories, with Tsumeb obtaining the same status in November 2004. The data from Hermanus are used together with those of Honolulu, San Juan, and Kakioka to derive the Dst index, with Hermanus being the only Southern Hemisphere magnetic Observatory of these four stations.

Since the first observatory was founded in Cape Town in 1841, geomagnetism has grown in stature where it is now part of a research program at the Hermanus Magnetic Observatory, exploiting ground observations as well as data from low-earth orbit satellites like Ørsted and CHAMP to investigate the peculiar behavior of the field over southern Africa.

Pieter Kotzé

Bibliography

Beattie, J.C., 1909. *Report of a Magnetic Survey of South Africa.* London: Cambridge University Press.

Trigonometrical Survey Office, 1939. *Results of Observations Made at the Magnetic Observatory.* University of Cape Town.

OBSERVATORIES IN SPAIN

History

The history of magnetic observatories in Spain, as elsewhere in the world, started in the last quarter of the 19th century. Some measurements of the geomagnetic field at earlier times have been recorded, for example the geomagnetic chart of the Iberian Peninsula made by Lamont in 1868, but nothing indicates continuous observations.

San Fernando Observatory (astronomical and geophysical observatory maintained up to the present time by the Spanish Navy, SFS—36°27.7′N, 6°12.3′W) received a set of instruments for absolute measurements around 1842. But the first records of regular measurements at this place were not till 1877. An Adie variometer was received on 1875, but continuous recording did not start till 1891.

The Madrid Astronomical Observatory (today named Observatorio Astronómico Nacional; 40°24.5′N, 3°41.2′W) started the publication of daily measurements of the geomagnetic field in 1879. They were stopped in 1901 due to the installation of electric tramways in Madrid.

At the same time, Jesuits founded magnetic sections in two meteorological observatories in Spanish colonies, Belen (from 1862—23°8.24′N, 82°21.30′W), in Havana (Cuba), and Manila (from 1889—14°34.7′N, 120°58.5′E) in the Philippine Islands. Both continued after the independence of these countries in 1898.

At the beginning of the 20th century, in 1904, the Jesuits founded Ebre (or Ebro) Observatory (EBR-40°49.2′N, 0°29.6′E). The observation program was extensive, including telluric currents, solar activity, seismology, and atmospheric electricity. It was used as the reference observatory for the first Spanish geomagnetic chart completed in 1927.

In 1934, the state founded Toledo Observatory (TOL—39°53.0′N, 4°2.8′W). It was designated as the new central geophysical observatory in Spain. Geomagnetic measurements started in 1934, but the Spanish Civil War delayed the final set-up of the instruments, and regular observations were not undertaken till 1947.

As a contribution to the International Geophysical Year in 1957–1958, Spain decided to install some new geomagnetic observatories. Two were placed in the Iberian Peninsula: Almeria (1955–1989; ALM; 36°51.2′N, 2°27.6′W) and Logroño (1957–1980; LOG; 42°27.5′N, 2°30.3′W); another, Las Mesas (1961–1992; TEN; 28°28.6′N, 16°15.7′W), in Tenerife Island and the last one, Moca (1958–1971; MOC; 3°20.6′N, 8°39.6′E), in Equatorial Guinea, at that time a Spanish colony.

Present status

As elsewhere in the world, magnetic perturbations due to human activity have forced magnetic observatories to move or to shut down. Present status is as follows:

San Fernando observatory (SFS)

Toward the end of the 1970s (1978), the electrification of the Sevilla-Cádiz railway forced the movement of the station 8 km northeast of the original location. Present instrumentation consists of a variometric station with feedback torsion photoelectric sensors, a declinometer/inclinometer (YOM MG2KP), and a Geometrics G-856 proton magnetometer. Since April 1998, to minimize the scatter, observations are restricted to periods when the railway is nonelectric. Two nighttime absolute measurements are made per week.

Although averaged data are not affected, new anthropogenic high frequency (some 2 nT rms) interferences are detected in the records. Thus, since 1997, tests have been made at different sites in order to select a new location. This has been chosen at 40 km distance from the present one. Works have already started. The new instrumentation envisaged for deployment there, consists of a fluxgate triaxial tilt-compensated variometer (model FGE) plus an Overhauser scalar magnetometer. For the absolute instrument the previously mentioned YOM MG2KP declinometer/inclinometer will be used.

Bulletins from this observatory have been published for the period 1891–1978 and from 1991 onward.

San Pablo observatory (SPT)

Located at 50 km southwest of Toledo, in 10 ha grounds free from disturbances, close to San Pablo de los Montes town, at 922 m above sea level (39°32.8′N, 4°20.9′W), it started operations in 1981. This was when Toledo Observatory closed due to the railway electrification. The variometric instrumentation consists of a tilt-compensated FGE station and a Geomag M390, both equipped with Overhauser GSM90 scalar magnetometers. As backup equipment there is also a vector magnetometer equipped with Geometrics G856AX sensors and there is a plan to incorporate a new dIdD sensor recently acquired from Gem Systems. Absolute observations are performed weekly with a couple of declinometer/inclinometer 010B Zeiss theodolites with Bartington fluxgate sensors. Bulletins from Toledo were published from 1947 to 1981 and those from San Pablo from 1982 onward.

Güimar observatory (GUI)

This observatory is located at 27 km south of Santa Cruz de Tenerife (Canary Islands) and at 4 km from Güimar town, at 868 m above sea level (28°19.3′N, 16°26.4′W). It has been operating since February 1993 after the closure of Las Mesas due to contamination from Santa Cruz city, located only 1 km away. Its pavilions are located in a 20 ha area of public mountain, free from human-made disturbances. The volcanic character of the terrain means that there is a large crustal bias for this observatory and that large gradients are present across the site. It is equipped with two tilt-compensated FGE stations with Overhauser GSM90 and GSM19 scalar magnetometers. The absolute instruments and the observation program are the same as at SPT. Published bulletins from Las Mesas cover the period from 1961 to 1992 and those of Güimar from 1992 onward.

Ebre observatory (EBR)

Ebre observatory started operation in 1904 but since 1973 the records have been disturbed because of the electrification of the railway. As a result, the rapid run magnetograph recordings had to be abandoned. Recently, a new location away from the disturbance has been selected and new instrumentation deployed. This comprises a tilt-compensated FGE and an Overhauser GSM90 scalar magnetometer. Continuous data are downloaded regularly via telephone and the site is visited once per week for the absolute measurements. These are done with a declinometer/inclinometer 015B Zeiss theodolite with an ELSEC 810A fluxgate sensor. For backup equipment there is a fluxgate triaxial tilt-compensated Geomag M390 station (also equipped with an Overhauser GSM90 scalar magnetometer) and an ARGO system, developed by the British Geological Survey (BGS), all located at the old site. The ARGO system comprises a dDdI semiabsolute vector magnetometer and a triaxial fluxgate sensor. Absolute measurements are also performed 5 days a week at the old site using another ELSEC 810A DI-flux, but with a 010B Zeiss theodolite.

This observatory has developed a procedure to semiautomatically digitize old classic records on photographic paper. Bulletins are available from the period 1910–1937 and from 1995 onward.

Livingston Island observatory (LIV)

Recently Ebre Observatory started a research program in the Antarctic that included the installation of a new geomagnetic observatory (LIV) at the Spanish Antarctic Station, Juan Carlos I (Livingston Island, South Shetland Islands, 62°39.7′S, 60°23.7′W). The main instrument is again a BGS dDdI vector magnetometer running all year round, although absolute measurements (with 015B Zeiss theodolite and an ELSEC 810A sensor) are only made during the 3 months of the summer survey. Data have been transmitted via the METEOSAT data collection system to the Edinburgh INTERMAGNET GIN, and presently through GOES-E to the Ottawa GIN. Data have been published since December 1996.

Miquel Torta and Josep Batlló

Cross-references

Crustal Magnetic Field
Jesuits, Role in Geomagnetism
Observatories in Antartica

OCEAN, ELECTROMAGNETIC EFFECTS

The oceans play a special role in electromagnetic induction, due to their relatively high conductivity and the dynamo effect of ocean currents. Typical crystalline rocks of the Earth's crust have electrical conductivities in the range of 10^{-6}–10^{-2} S m^{-1}, (Siemen [1/Ω]). In comparison, seawater, with 2.5–6 S m^{-1}, is a very good conductor. Electrical currents are induced in the oceans by two different effects: Induction by time-varying external fields and induction by motion of the seawater through the Earth's main field (see *Main field*).

Ocean conductivity

While fresh water has a very low conductivity, a salt concentration of about 35 g L^{-1} turns the oceans into good conductors. In contrast to the electron and electron-hole conductivity of solid-Earth materials, the carriers of charge in the oceans are the hydrated ions of the dissolved salts. Positive ions are called anions and negative ions are called cations. The conductivity of seawater depends on the number of dissolved ions per volume (i.e., salinity) and the mobility of the ions (i.e., temperature and pressure). Conductivity increases by the same amount with a salinity increase of 1 g L^{-1}, a temperature increase of 1°C, and a depth (i.e., pressure) increase of 2000 m. The change in conductivity is dominated by temperature, with a range of about 2.5 S m^{-1} for cold, deep water to about 6 S m^{-1} for warm surface water.

Motional induction

When the ions are carried by ocean flow through the Earth's magnetic field, they are deflected by the electromagnetic Lorentz force. The direction of the Lorentz force on the anions is given by the right-hand rule: thumb pointing in the direction of ocean flow, index finger in the direction of the magnetic field, then the middle finger indicates the direction of the deflection of the anions. The cations are deflected into the opposite direction. This separation of charge sets up large-scale electric fields. Depending on the conductivity structure of the ocean and solid Earth, these fields drive electric currents, which in turn generate secondary magnetic fields.

Governing equations

Motional induction is governed by the following equations: The local current density $\mathbf{J}$ in a conductor moving with velocity $\mathbf{u}$ is given by

$$\mathbf{J} = \sigma(\mathbf{E} + \mathbf{u} \times \mathbf{B}_{\text{main}}), \qquad \text{(Eq. 1)}$$

where σ is the local conductivity and $\mathbf{B}_{\text{main}}$ is the ambient main magnetic field (see *Main field*). The $\mathbf{u} \times \mathbf{B}_{\text{main}}$ term is the driver of the motionally induced electric current, while the induced electric field $\mathbf{E}$ counteracts this driver and results in the actual current $\mathbf{J}$. This current generates a magnetic field $\mathbf{B}_{\text{ocean}}$ as

$$\nabla \times \mathbf{B}_{\text{ocean}} = \mu_0 \mathbf{J}. \qquad \text{(Eq. 2)}$$

Combining Eqs. (1) and (2), and rearranging, yields an equation for the induced electric field:

$$\mathbf{E} = \frac{1}{\mu_0 \sigma} \nabla \times \mathbf{B}_{\text{ocean}} - \mathbf{u} \times \mathbf{B}_{\text{main}}. \qquad \text{(Eq. 3)}$$

While the irrotational (divergent) part of this electric field is maintained by an induced charge, the rotational part is induced by the temporal variations of the ocean magnetic signal as

$$\partial_t \mathbf{B}_{\text{ocean}} = -\nabla \times \mathbf{E}. \qquad \text{(Eq. 4)}$$

Taking the curl of Eq. (3) and combining with Eq. (4) gives the motional induction equation:

$$\partial_t \mathbf{B}_{\text{ocean}} = \nabla \times \left(\mathbf{u} \times \mathbf{B}_{\text{main}} - \frac{1}{\mu_0 \sigma} \nabla \times \mathbf{B}_{\text{ocean}} \right), \qquad \text{(Eq. 5)}$$

which is valid inside the oceans and the solid Earth. The fields above the surface are then determined by Laplace's equation $\nabla^2 V = 0$ for the scalar potential V of the magnetic field $\mathbf{B}_{\text{ocean}} = -\nabla V$ in the current-free region outside of the Earth.

Electric fields of motional induction

Already in 1832, Faraday predicted a motionally induced electric field for the river Thames but failed to detect it, due to a lack of adequate instrumentation. Such fields were then observed in 1851 by Wollaston in a telegraphic cable across the English Channel. Young demonstrated in 1920 the recording of electric fields in the oceans by ship-towed electrodes. Submarine cables and towed electrodes are still the principal methods of measuring ocean flow by electromagnetic methods. Overviews of the theory and instrumentation are given by Sanford (1971) and by Filloux (1973).

Magnetic fields of motional induction

Motionally induced magnetic fields can be divided into toroidal (strictly horizontal) and poloidal (both horizontal and vertical) parts. The toroidal magnetic field is generated by electric currents closing in vertical planes and is estimated to reach 100 nT in amplitude. It is confined to the oceans and solid Earth and is therefore only observable by subsurface measurements, for example by ship-towed or ocean-bottom magnetometers. Indeed, Lilley *et al.* (2001) showed that a magnetometer lowered by a cable from a ship can be used to determine vertical profiles of the ocean velocity.

The much weaker poloidal field, with amplitudes up to 10 nT, results from electric currents closing horizontally. It has a significant vertical component and reaches remote land and satellite locations. Much attention has been given to the periodic magnetic signals of ocean flow, which is driven by the lunar tides. In interpreting tidal magnetic signals one has to take into account that lunar tidally driven winds in the upper atmosphere induce tidal electric fields and currents in the ionosphere that also contribute to the magnetic signal at these periods. However, the strength of the ionospheric currents depends on the conductivity of the ionosphere, which varies with the time of the day. This additional solar-time modulation of the ionospheric lunar tidal fields can be used to separate them from the oceanic lunar tidal signals (Malin, 1970). The dominant M2 lunar tidal ocean flow magnetic signal has also been identified in CHAMP (see *CHAMP*) satellite data (Tyler *et al.*, 2003). A map of this signal at surface level is displayed in Figure O14.

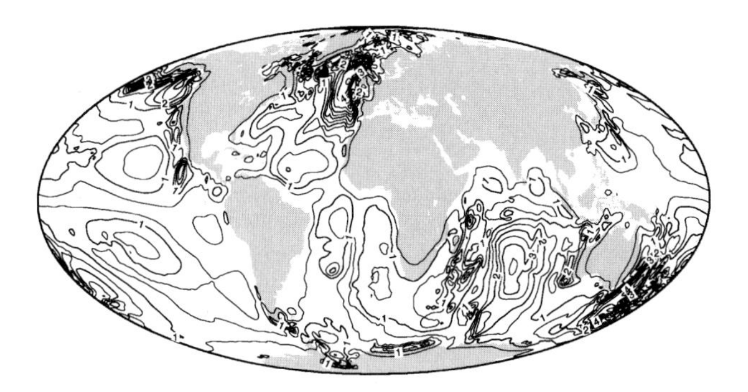

Figure O14 Predicted amplitude of the vertical component of the magnetic field at the Earth's surface generated by ocean flow due to the lunar M2 tide. (Reproduced from Kuvshinov and Olsen, 2005b.)

Using high-resolution ocean circulation models predictions of the magnetic signal of nontidal ocean flow have been made by several authors. Amplitudes at surface level are predicted to be comparable to those of tidal ocean flow signals. However, the magnetic signal of the steady ocean circulation is impossible to distinguish from the crustal magnetic field (see *Crustal magnetic field*). Observations must therefore concentrate on the much weaker time-varying part of the signal. For example, the annual variation in the Indian Ocean is predicted to generate a signal with only about 0.3 nT at satellite altitude. It is a great challenge to separate this weak ocean signal from ionospheric fields, which have similar periodicities and may even share some of the atmospheric driving forces. At this point of time, nontidal ocean currents have not yet been convincingly identified in land observatory or satellite magnetic measurements.

Induction by time-varying magnetic fields

External magnetic fields, generated in the ionosphere (see *Ionosphere*) and the magnetosphere (see *Magnetosphere of the Earth*), vary strongly in time, exhibiting regular daily variations (see *Geomagnetic spectrum, temporal*) and occasional strong magnetic storms (see *Storms and substorms*) caused by coronal mass ejections from the sun. These time-varying magnetic fields induce electric fields and currents in the oceans, generating secondary, induced magnetic fields.

Governing equations

A time-varying magnetic field induces an electric field $\mathbf{E}$ in the oceans, given by

$$\partial_t \mathbf{B} = -\nabla \times \mathbf{E}, \qquad \text{(Eq. 6)}$$

where $\mathbf{B}$ is the sum of the inducing and the secondary induced magnetic field. The induced electric field drives electric currents

$$\mathbf{J} = \sigma \mathbf{E}, \qquad \text{(Eq. 7)}$$

generating a secondary-induced magnetic field:

$$\mu_0 \mathbf{J} = \nabla \times \mathbf{B}_{\text{induced}}. \qquad \text{(Eq. 8)}$$

Combining Eqs. (6), (7), and (8) yields

$$\partial_t \mathbf{B} = -\nabla \times \frac{1}{\mu_0 \sigma} \nabla \times \mathbf{B}_{\text{induced}}, \qquad \text{(Eq. 9)}$$

which is the equation for induction by time-varying magnetic fields inside the oceans. In contrast to the motional induction discussed above, the oceans are assumed here to be stationary in the Earth-fixed reference frame. The field in the source-free region above the Earth's surface and below the external current systems is again given by Laplace's equation.

Skin depth

Low-frequency electromagnetic waves penetrate deeper into an electrical conductor than high-frequency waves. For a periodic external field with frequency ν, Eq. (9) demands that the amplitude in a uniform conductor decays exponentially with depth. The depth at which the amplitude is reduced to $1/e$ is called the skin depth δ, given by

$$\delta = \sqrt{\frac{1}{\pi \mu_0 \sigma \nu}} \approx \frac{270\text{m}}{\sqrt{\nu}} \qquad \text{(Eq. 10)}$$

where the frequency ν is given in hertz and a seawater conductivity of 3.5 S m^{-1} is assumed. Thus, to effectively penetrate through a seawater column of 5 km, magnetic disturbances have to have periods longer than about 6 min. However, even up to periods of several days, induced fields are strongly influenced by the highly conducting oceans.

Ocean induction by magnetic storms

While daily variations during solar quiet conditions generate significant induction in the oceans, a much stronger effect is caused by magnetic storms. Strong magnetic storms generate ocean-induced magnetic fields reaching magnitudes of more than 100 nT. An overview of induction in the oceans by solar quiet and storm time fields is given by Fainberg (1980). The induced currents are concentrated near the coasts, the horizontal boundaries of the conductor. Their magnetic signal is also known as the coast effect (see *Coast effect of induced currents*). The ocean-induced fields can be predicted for a given external storm signature using a 3D conductivity model of the oceans, crust, and mantle. The induction equation (9) is solved either directly in the time domain, or separately for all contributing frequencies with a subsequent inverse Fourier transform to the time domain. The result of a simulation using the latter approach (Kuvshinov and Olsen, 2005a) for a strong magnetic storm is shown in Figure O15.

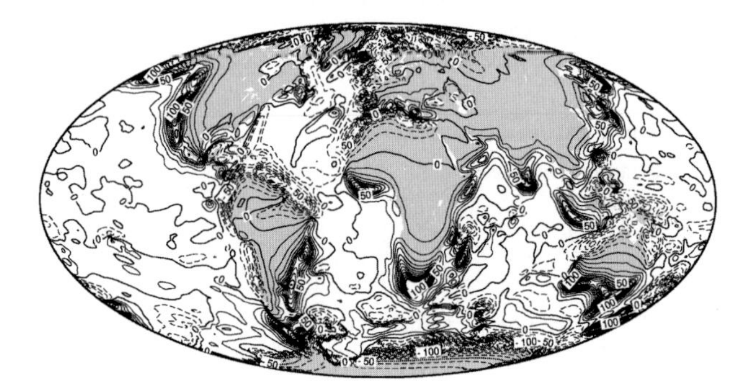

Figure O15 Magnetic field induced in the oceans by the Bastille-day magnetic storm of July 15, 2000. Shown is the vertical component of the induced magnetic field at 21:30 universal time at the Earth's surface. (Reproduced from Kuvshinov and Olsen, 2005a.)

Stefan Maus

Bibliography

Fainberg, E.B., 1980. Electromagnetic induction in the world ocean. *Geophysical Surveys*, **4**: 157–171.

Filloux, J.H., 1973. Techniques and instrumentation for study of natural electromagnetic induction at sea. *Physics of the Earth and Planetary Interiors*, **7**: 323–338.

Kuvshinov, A., and Olsen, N., 2005a. Modelling the ocean effect of geomagnetic storms at ground and satellite altitude. In *Earth Observation with CHAMP.* Berlin: Springer, 353–358.

Kuvshinov, A., and Olsen, N., 2005b. 3-D Modelling of the magnetic fields due to Ocean tidal flow. In *Earth Observation with CHAMP.* Berlin: Springer, 359–365.

Lilley, F.E.M., White, A., and Heinson, G.S., 2001. Earth's magnetic field: ocean current contributions to vertical profiles in deep oceans. *Geophysical Journal International*, **147**: 163–175.

Malin, S.R.C., 1970. Separation of lunar daily geomagnetic variations into parts of ionospheric and oceanic origin. *Geophysical Journal of the Royal Astronomical Society*, **21**: 447–455.

Sanford, T.B., 1971. Motionally induced electric and magnetic fields in the sea. *Journal of Geophysical Research*, **76**: 3476–3492.

Tyler, R., Maus, S., and Lühr, H., 2003. Satellite observations of magnetic fields due to ocean tidal flow. *Science*, **299**: 239–241.

Cross-references

CHAMP
Coast Effect of Induced Currents
Crustal Magnetic Field
Geomagnetic Spectrum, Temporal
Ionosphere
Magnetosphere
Main field
Storms and Substorms

OLDHAM, RICHARD DIXON (1858–1936)

A British geologist and seismologist, born on July 31, 1858 in Dublin and died on July 15, 1936 in Llandrindod Wells. He was the third son of Thomas Oldham (1816–1878), Prof. of Geology at Trinity College, Dublin, and director of the Geological Surveys of Ireland and India. Richard Dixon Oldham was educated in England at first in Rugby, later at the Royal School of Mines. In 1879, he followed his father's footsteps by moving to India and working at the Geological Survey of India. During his time in India he wrote over 40 publications related to geology and tectonics of India, many of them about earthquakes and their observations in India. After moving back to England in 1903, he worked for sometime together with John Milne (1850–1913) on the Isle of Wight. Later, because of his ill health, he spent the winters in Southern France studying the history of the Rhône delta and the summers in Llandrindod Wells working on seismological problems. In 1908, he was honored with the Lyell Medal by the Geological Society and in 1911 he was elected a Fellow of the Royal Society. In geomagnetism, Oldham is known for his establishment of the Earth's core.

Oldham started his career in India with the publication of his father's collection of macroseismical and geological observations in the context of the large 1869 earthquake in Silchar (Assam). With this background, he was predestined for a similar investigation of the huge Assam earthquake of June 12, 1897. The results of his study are published in a famous monograph of 1899 and numerous additional smaller publications. Since the first teleseismic observations of seismic waves by Ernst von Rebeur-Paschwitz (1861–1895), surface and body waves were known not only to exist but also to have different propagation velocities (e.g., Rebeur-Paschwitz, 1895). Oldham's most important result for modern seismology was that he for the first time could identify onsets in seismograms belonging to both body-wave types as proposed by the theory of elastic waves, which propagate with different velocities through the Earth: compressional waves (later named P-waves) are faster than shear waves (later named S-waves) (Oldham, 1899, 1900).

In the last decade of the 19th century, Emil Wiechert (1861–1928) deduced from moment of inertia, ellipticity, and mean density of the Earth the need for dividing the Earth in two principle parts: a less dense mantle of rock material and a much denser core of compressed iron (Wiechert, 1896, 1897; Brush, 1980, 1982). However, the seismological proof could not be given in these early days of seismology. During his first years back in England, Oldham analyzed and compiled teleseismic observations from the collection available at Shide, Milne's Observatory on the Isle of Wright. In 1906, he published a paper in which he claimed to have observed and modelled the Earth's core by delayed P- and S-phase observations due to lower seismic velocities inside the core. Oldham himself was not very convinced by his delayed P observations because of their scatter (Oldham, 1906). More important for his argumentation was the delay of S onsets at distances beyond 120° epicentral distance. From these data, Oldham concluded that the Earth's core has a size of about 0.4 of the Earth's radius and that both P- and S-wave velocities are much smaller than in the Earth's mantle. As Wiechert (1907) pointed out, Oldham's delayed S-phase observations are mostly SS phases, that is, the reflections of S phases at the Earth's surface, an interpretation later accepted also by Oldham (Oldham, 1919).

The correct deciphering of the main structural elements of the Earth's interior was then the work of the next generation of seismologists: Beno Gutenberg (1889–1960) could estimate the depth of the *core-mantle boundary* (*q.v.*) at 2890 km, calculate travel-time curves for core phases, and correctly identify in seismograms the P phases, which had been reflected from, or which had traversed the core (Gutenberg, 1913, 1914). Harold Jeffreys (1891–1989) showed that the Earth's core must have the rheological behavior of a fluid, and that therefore no S phases can pass it (Jeffreys, 1926) and in 1936 Inge Lehmann (1888–1993) discovered the inner core of the Earth (Lehmann, 1936, 1987).

However, Oldham was the first who discovered and published about the shadow effect of the Earth's core due to lower seismic velocities. Together with his work on seismic phase types, Oldham is one of the pioneers for the application of seismic-wave phenomena to investigate the structure of the Earth's interior.

Johannes Schweitzer

Bibliography

Brush, S.G., 1980. Discovery of the Earth's core. *American Journal of Physics*, **48**: 705–724.

Brush, S.G., 1982. Chemical history of the Earth's core. *EOS, Transactions, American Geophysical Union*, **63**: 1185–1186, 1189.

Gutenberg, B., 1913. Über die Konstitution des Erdinnern, erschlossen aus Erdbebenbeobachtungen. *Physikalische Zeitschrift*, **14**: 1217–1218.

Gutenberg, B., 1914. Über Erdbebenwellen. VII A. Beobachtungen an Registrierungen von Fernbeben in Göttingen und Folgerungen über die Konstitution des Erdkörpers. *Nachrichten von der Königlichen Gesellschaft der Wissenschaften zu Göttingen, Mathematisch-physikalische Klasse*, **1914**: 125–177.

Jeffreys, H., 1926. The rigidity of the Earth's central core. *Monthly Notices of the Royal Astronomical Society, Geophysical Supplement*, **1**: 371–383.

Lehmann, I., 1936. P'. *Publications du Bureau Central Séismologique International, Série A, Travaux Scientifique*, **14**: 87–115.

Lehmann, I., 1987. Seismology in the days of old. *EOS, Transactions, American Geophysical Union*, **68**: 33–35.

Oldham, R.D., 1899. Report on the great earthquake of 12th June 1897, Chapter 25, The unfelt earthquake. *Memoirs of the Geological Survey India*, **29**: 226–256.

Oldham, R.D., 1900. On the propagation of earthquake motion to great distances. *Philosophical Transactions of the Royal Society of London*, **194**: 135–174.

Oldham, R.D., 1906. The constitution of the interior of the Earth, as revealed by earthquakes. *The Quarterly Journal of the Geological Society of London*, **62**: 456–473.

Oldham, R.D., 1919. The interior of the Earth. *The Geological Magazine*, New series 6, Decade 6, **56**: 18–27.

Rebeur-Paschwitz, E.V., 1895. Horizontalpendel-Beobachtungen auf der Kaiserlichen Universitäts-Sternwarte zu Strassburg 1892–1894. *Beiträge zur Geophysik*, **2**: 211–536.

Wiechert, E., 1896. Über die Beschaffenheit des Erdinnern. *Schriften der Physikalisch-ökonomischen Gesellschaft zu Königsberg in Preußen, Sitzungsberichte*, **37**: 4–5.

Wiechert, E., 1897. Über die Massenvertheilung im Innern der Erde. *Nachrichten von der Königlichen Gesellschaft der Wissenschaften zu Göttingen, Mathematisch-physikalische Klasse*, **1897**: 221–243.

Wiechert, E., 1907. Über Erdbebenwellen. Theoretisches über die Ausbreitung der Erdbebenwellen. *Nachrichten von der Königlichen Gesellschaft der Wissenschaften zu Göttingen, Mathematisch-physikalische Klasse*, **1907**: 413–529.

Cross-references

Core Composition
Core Density
Core-Mantle Boundary
Core Viscosity
Inner Core Seismic Velocities
Lehmann, Inge (1888–1993)
Seismic Phases

ØRSTED

This geomagnetic research satellite (Figure O16/Plate 5b), named after the Danish scientist Hans Christian Ørsted (1777–1851), is the first satellite mission after *Magsat* (*q.v.*) (1979–1980) for high-precision mapping of the Earth's magnetic field. It was launched with a Delta-II rocket from Vandenberg Air Force Base (California) on February 23, 1999 into a near polar orbit. Being the first satellite of the *International Decade of Geopotential Research* (IAGA, 1997), it has been a model for other missions, like *CHAMP* (*q.v.*) and *Swarm* (Friis-Christensen *et al.*, 2006). Since optimal utilization of magnetic data for improved field modeling requires firm knowledge of all contributing sources, there was an endeavor to bring together experts in internal sources (core and crustal field, electromagnetic induction) and in external sources (current systems in the *ionosphere* (*q.v.*) and *magnetosphere* (*q.v.*)) in the *Ørsted International Science Team* (OIST) very early in the planning phase of the project (Friis-Christensen and Skøtt, 1997). Sixty-four research groups from 14 countries joined the OIST, giving impetus to the mission (Stauning *et al.*, 2003). Ørsted has also had an impact on the collection of ground data, as is evident from the significantly increased number of geomagnetic observatories that deliver data to the World Data Centers in support of the mission. Ørsted external field science is coordinated by the Danish Meteorological Institute (DMI), internal field science is coordinated by the Danish National Space Center (DNSC, formerly Danish Research Institute, DSRI). Satellite control is managed by the industrial company Terma A/S (Birkerød, Denmark).

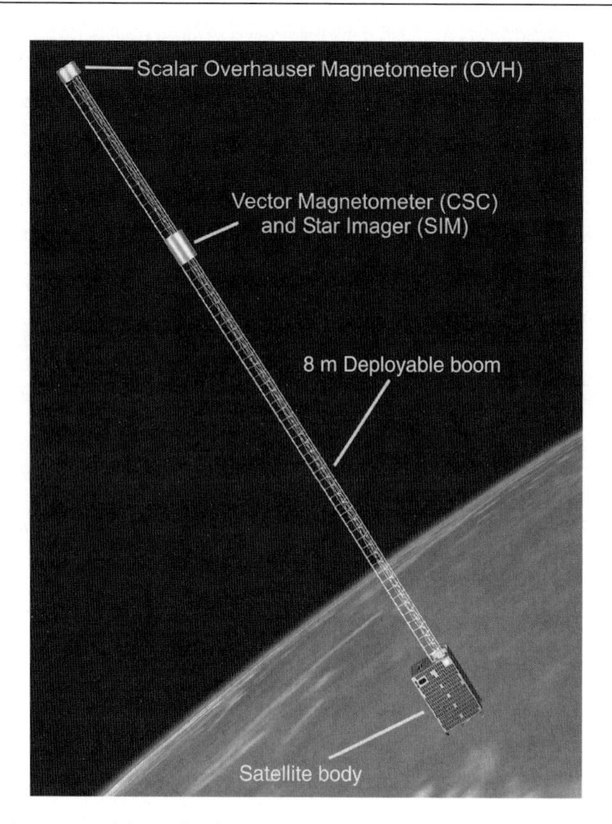

Figure O16/Plate 5b The Ørsted satellite.

Satellite and orbit

Ørsted was built in a joint effort of various Danish research institutions and companies with significant contribution from space agencies in the United States and Europe. The satellite weighs 62 kg, measures 34 × 45 × 72 cm and contains an 8 m long boom, deployed shortly after launch, carrying two *magnetometers* (*q.v.*). The satellite is gravity gradient stabilized; attitude maneuvers are performed using magnetic torquers. The orbit has an inclination of 96.5°, a period of 100.0 min, a perigee at 643 km and an apogee at 881 km (decreased to respectively 99.7 min, 632 km, and 860 km after 7 years in space). The orbit plane is slowly drifting, the local time of the equator crossing decreases by 0.91 min day^{-1}, starting from an initial local time of 1426 on February 23, 1999 for the north-going track, cf. the upper part of Figure O17. Nominal lifetime of the mission was 14 months (2 months commissioning phase + 12 months science phase), but after almost 8 years in space the satellite is still healthy and provides high-precision magnetic data.

Instrumentation

A proton precession Overhauser magnetometer (OVH), measuring the magnetic field intensity with a sampling rate of 1 Hz, is mounted at the top of the deployable 8 m long boom. This instrument, developed by LETI (Grenoble/F), has an absolute accuracy better than 0.5 nT. Two meters away, at a distance of 6 m from the satellite body, is the optical bench with the compact spherical coil (CSC) fluxgate vector magnetometer mounted closely together with the *st*ar *im*ager (SIM); these instruments were developed by the Danish Technical University and DNSC. The CSC samples the magnetic field at 100 Hz (burst mode, at polar latitudes) or 25 Hz (normal mode) with a resolution better than 0.1 nT and is calibrated using the field intensity measured by the OVH (Olsen *et al.*, 2003). After calibration, the agreement between the two magnetometers is better than 0.33 nT rms.

A copy of the Ørsted boom and payload (but with a scalar helium magnetometer provided by NASA/JPL instead of the Overhauser magnetometer), called Ørsted-2, was launched in November 2000 onboard the Argentinean satellite SAC-C. However, due to a broken connection in a coaxial cable, no high-precision attitude data (and hence no reliable vector data) are available.

Data accuracy and availability

Ørsted data are available at the Ørsted Science Data Centre (http://www.dmi.dk/projects/oersted/SDC) and at www.spacecenter.dk/data/magnetic-satellites. Scalar measurements (from the OVH if available, and from the CSC otherwise) are provided at 1 Hz sampling rate in the MAG-F files. The overall accuracy of the scalar data is estimated to be better than 0.5 nT. Vector data (at the sampling rate of 1.135 s of the SIM) are distributed as MAG-L files. The biggest limitation of the vector data is the star imager attitude accuracy. It is highly anisotropic: determination of the SIM bore-sight direction (pointing) is more accurate than determination of the rotation around bore-sight (rotation), resulting in a relatively higher noise of the SIM rotation angle. In addition, the rotation angle is more sensitive to distorting effects like instrument blinding (for instance by the moon). This attitude anisotropy results in correlated errors between the magnetic components, which have to be taken into account when deriving field models, as demonstrated by Holme (2000). Let $\mathbf{n}$ be the unit vector of the SIM bore-sight, and let $\mathbf{B}$ be the observed magnetic field vector. The magnetic residual vector (observation minus model field) $\delta\mathbf{B} = (\delta B_B, \delta B_\perp, \delta B_3)$ can be transformed such that the first component, δB_B, is in the direction of $\mathbf{B}$ (and therefore to first order is equal to the scalar residual), the second component, $\delta B_\perp$, is aligned with $\mathbf{n} \times \mathbf{B}$, and the third component, δB_3, is aligned with $\mathbf{B} \times (\mathbf{n} \times \mathbf{B})$. The last two components are perpendicular to the magnetic field. Attitude noise may result in significant noise in the magnetic vector components: since the nominal bore-sight direction is almost East-West, typical rotational attitude noise will result in noise in B_r of up to 10 nT amplitude at low latitudes. Whenever possible, the user should be aware of this effect and use data in the (δB_B, $\delta B_\perp$, δB_3)-frame rather than (B_r, B_θ, B_ϕ).

The boom experiences occasional jerks associated with thermal expansion/contraction, especially near the terminator. The resulting boom oscillations, mainly around the boom axis, have a frequency close to 0.5 Hz and are seen in the magnetic vector components as occasional damped oscillations with amplitudes of a few nanoteslas. These technically induced signals should not be confused with geomagnetic pulsations of natural origin.

Accuracy of δB_B, $\delta B_\perp$, and δB_3 is estimated to be better than 1, 2, and 8 nT, respectively. Vector data in an Earth-Centered-Earth-Fixed frame (B_r, B_θ, B_ϕ) and the SIM bore-sight direction $\mathbf{n}$ are available together with time and position in the MAG-L files. Data at higher sampling rates are available upon request. The lower part of Figure O17 shows the availability of scalar data (MAG-F, black) and of vector data (MAG-L, light grey). Since attitude data (dark grey) are essential for providing vector data, drop outs of the SIM (for instance due to thermal problems, or due to blinding of the instrument by the Sun, Moon, or Earth) limit the availability of Ørsted vector data. The satellite was fully illuminated by the Sun from July to November 2000, from August 2002 to February 2003, and from January to May 2005 (shaded area in Figure O17 top). Since the satellite was not designed for this situation (nominal lifetime of 14 months ended in April 2000), thermal problems resulted in decreased data availability during these periods, as shown in the bottom panel of Figure O17.

As secondary data products, magnetic field models that have been derived using Ørsted data are provided at www.spacecenter.dk/projects/oersted/models/.

Scientific results

After almost 8 years of operation, about 150 scientific papers using Ørsted data have been published. The satellite and first science results were presented in an overview article in EOS (Neubert *et al.*, 2001).

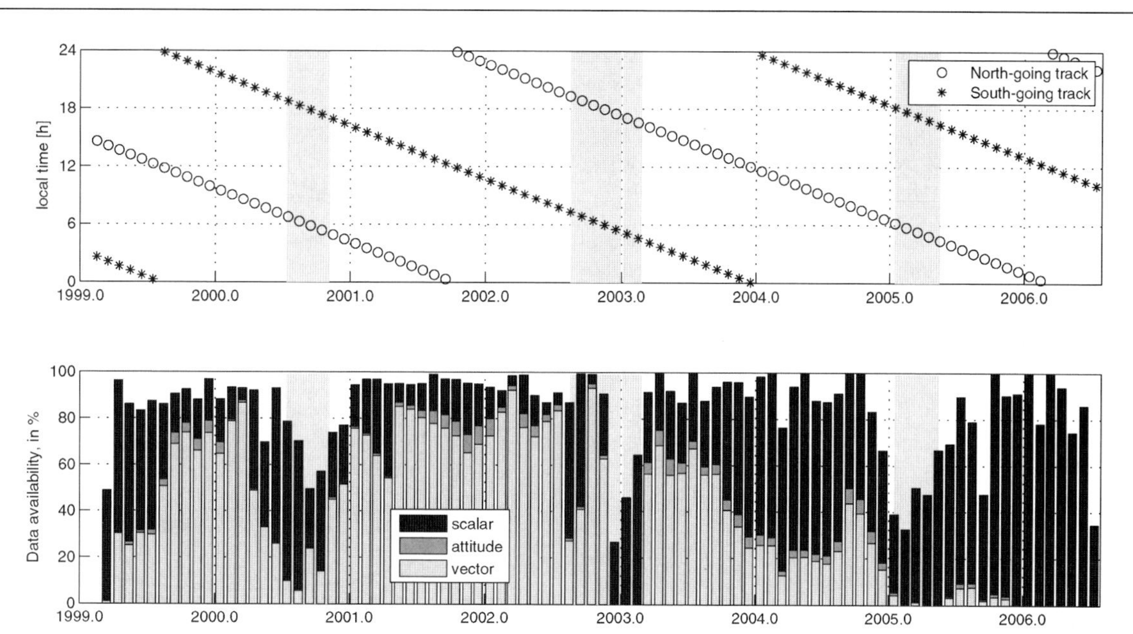

Figure O17 Local time evolution of the Ørsted orbit (*top*) and data availability (*bottom*), as of July 2006.

A parade of *magnetic field* (*q.v.*) models of increasing complexity and accuracy have been derived from Ørsted observations, starting from models that describe a snapshot of the field at a specific epoch, like the *IGRF, International Geomagnetic Reference Field* (*q.v.*) 2000 or the Ørsted Initial Field Model *OIFM* (Olsen *et al.*, 2000) to models that include the temporal changes of the field (Olsen, 2002, Oslen *et al.*, 2006). Spherical harmonic coefficients estimated using more than 6 years of Ørsted observations are robust up to degree $n = 40$ for the static terms, and up to at least $n = 15$ for the linear time terms (see *Secular variation models*). Comparing two models estimated from Ørsted data (epoch 2000) and Magsat data (epoch 1980), respectively, Hulot *et al.* (2002) derived core-flow models from the observed mean secular variation, down to length scales previously inaccessible.

Ørsted's altitude makes an investigation of the *crustal magnetic field* (*q.v.*) difficult, but combination with other satellite (*POGO* (*q.v.*), *Magsat* (*q.v.*), *CHAMP* (*q.v.*)) and ground data allows for an improved separation of internal and external sources, and thereby better crustal field determination. This was shown by Sabaka *et al.* (2004) in their *Comprehensive Model, CM4*, which determined crustal field contributions that are reliable at least up to degree $n = 50$.

Although the main objective of Ørsted is the precise mapping of Earth's internal field, the satellite has contributed to improved understanding of ionospheric and magnetospheric current systems. A much more detailed determination of their variability with season and local time is possible due to the greatly improved data distribution compared to previous missions. By taking the curl of the magnetic field residual (observation minus field model), Christiansen *et al.* (2002) derived mean field-aligned currents (FAC) in dependence on season, local time, and interplanetary magnetic field (solar wind) conditions and found that the entire polar caps are filled with currents even during very quiet solar wind conditions. The influence of these quiet time polar cap currents on core and crustal field models is still an unsolved problem. Embedded in the large-scale average current systems are small-scale (few hundreds of meters) current structures (filaments), especially in the cusp region, as found by Neubert and Christiansen (2003).

Nils Olsen

Bibliography

Christiansen, F., Papitashvili, V.O., and Neubert, T., 2002. Seasonal variations of high latitude field-aligned current systems inferred from Ørsted and MAGSAT observations. *Journal of Geophysical Research*, **107**: doi:10.1029/2001 JA900104.

Friis-Christensen, E., and Skøtt, C., 1997. Ørsted compendium-contributions from the International Science Team. *DMI Scientific Report 97-1*, Copenhagen.

Friis-Christensen, E., Lühr, H., and Hulot, G., 2006. Swarm: a constellation to study the Earth's magnetic field. *Earth, Planets and Space*, **58**: 351–358.

Holme, R., 2000. Modelling of attitude error in vector magnetic data: application to Ørsted data. *Earth, Planets and Space*, **52**: 1187–1197.

Hulot, G., Eymin, C., Langlais, B., Mandea, M., and Olsen, N., 2002. Small-scale structure of the geodynamo inferred from Ørsted and Magsat satellite data. *Nature*, **416**: 620–623.

International Association of Geomagnetism and Aeronomy (IAGA), 1997. Resolution No. 1, 8th IAGA Scientific Assembly, Uppsala, Sweden.

Neubert, T., and Christiansen, F., 2003. Small-scale, field-aligned currents at the top-side ionosphere. *Geophysical Research Letters*, **30**(19): 2010, doi:10.1029/2003GL017808.

Neubert, T., Mandea, M., Hulot, G., von Frese, R., Primdahl, F., Jørgensen, J.L. Friis-Christensen, E., Stauning, P., Olsen, N., and Risbo, T., 2001. Ørsted satellite captures high-precision geomagnetic field data. *EOS, Transactions of the American Geophysical Union*, **82**(7): 81, 87, and 88.

Olsen, N., 2002. A model of the geomagnetic field and its secular variation for epoch 2000. *Geophysical Journal International*, **149**(2): 454–462.

Oslen, N., Luehr, H., Sabaka, T.J., Mandea, M., Rother, M., Tøffner-Clausen, L., and Choi, S., 2006. CHAOS—A Model of Earth's Magnetic Field derived from CHAMP, Ørsted, and SAC-C magnetic satellite data, *Geophys. J. Int.*, **166**: 67–75.

Olsen, N., Holme, R., Hulot, G., Sabaka, T., Neubert, T., Toeffner-Clausen, L., Primdahl, F., Jørgensen, J., Legèr, J.-M., Barraclough, D., Bloxham, J., Cain, J., Constable, C., Golovkov, V., Jackson, A., Kotze, P., Langlais, B., Macmillan, S., Mandea, M., Merayo, J.,

Newitt, L., Purucker, M., Risbo, T., Stampe, M., Thomson, A., and Voorhies, C., 2000. Ørsted initial field model. *Geophysical Research Letters*, **27**(22): 3607–3610.
Olsen, N., Tøffner-Clausen, L., Sabaka, T.J., Brauer, P., Merayo, J.M.G., Jørgensen, J.L., Léger, J.-M., Nielsen, O.V., Primdahl, F., and Risbo, T., 2003. Calibration of the Ørsted vector magnetometer. *Earth, Planets and Space*, **55**: 11–18.
Sabaka T.J., Olsen, N., and Purucker, M.E., 2004. Extending comprehensive models of the Earth's magnetic field with Ørsted and CHAMP. *Geophysical Journal International*, **159**(2): 521–547.
Stauning, P., Lühr, H., Ultré-Guérard, P., LaBreque, J., Purucker, M., Primdahl, F., Jørgensen, J.L., Christiansen, F., Høeg, P., and Lauritsen, K.B., 2003. OIST-4 Proceedings: 4th Oersted International Science Team Conference. *DMI Scientific Report 03-09*, Copenhagen.

Cross-references

CHAMP
Crustal Magnetic Field
Geomagnetic Secular Variation
IGRF, International Geomagnetic Reference Field
Ionosphere
Magnetometers, Laboratory
Magnetosphere of the Earth
Magsat
Main Field Modeling
POGO (OGO-2, -4, and -6 spacecraft)

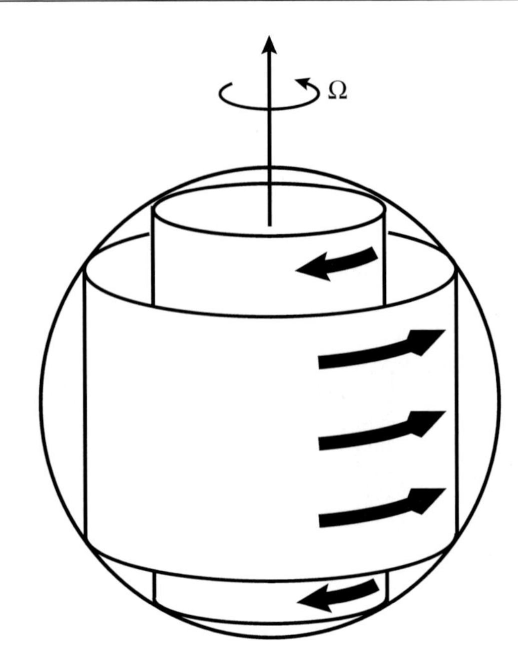

Figure O18 Torsional oscillations in the core. The direction of the azimuthal cylindrical flow indicated by the black arrows oscillates in time. The Earth's axis of rotation points in the direction of Ω.

OSCILLATIONS, TORSIONAL

Torsional oscillations in the Earth's fluid core are the azimuthal oscillations of rigid cylindrical surfaces coaxial with the rotation axis (see Figure O18). The pattern of the time-dependent part of the fluid motion at the surface of the core revealed from geomagnetic data inversion is highly suggestive of the presence of these waves inside the core, with velocity amplitudes on the order of 10 km year^{-1} and typical periods of decades (see Figure O19). The observation of torsional oscillations has important consequences for theoretical models of the geodynamo and provides a window through which we can observe some aspects of core dynamics.

Rigid azimuthal cylindrical flows of this sort, also called geostrophic flows, can be expected in the fluid core because rapid rotation prevents large variations in velocity along the rotation axis (see *Proudman-Taylor theorem*), and because purely azimuthal flows are not affected by the solid spherical boundaries of the mantle and the inner core. However, the Lorentz force prevents the cylindrical surfaces at different radii to rotate freely relative to one another. The magnetic field that permeates the core tends to be "frozen" in the fluid (see *Alfvén's theorem*) and a differential rotation of the cylinders shears the radial component of the field in the azimuthal direction. By Lenz' law, this produces a force that opposes further relative motion between the cylindrical surfaces. The radial component of the magnetic field behaves as if it were elastic strings attached to the cylindrical surfaces, and provides a restoring force for the establishment of waves that propagate in the direction perpendicular to the rotation axis. These are torsional oscillations, or torsional vibrations, as they were originally called in Braginsky's seminal work on the subject (Braginsky, 1970). Since the restoring force is purely magnetic, torsional oscillations are a type of Alfvén wave.

Strictly speaking, torsional oscillations are not a geostrophic flow because they result from an azimuthal force balance between the fluid acceleration and Lorentz forces. However, since the force balance in the direction pointing away from the rotation axis remains geostrophic and the form of the flow is identical to that of geostrophic flows, it is convenient to think of torsional oscillations as time-dependent geostrophic flows.

The importance of geostrophic flows in the core and their intimate connection to the dynamics governing the geodynamo was first established by Taylor (1963). He showed that if one integrates the azimuthal component of the momentum equation over the surface of cylinders coaxial with the rotation axis, the Lorentz force is the only term in the leading order force balance that does not identically vanish. This imposes a morphological constraint on the magnetic field in the core, namely that the axial Lorentz torque must vanish at all times, i.e.,

$$\int_\Sigma ((\nabla \times \mathbf{B}) \times \mathbf{B})_\phi \mathrm{d}\Sigma = 0 \qquad \text{(Eq. 1)}$$

where $\mathrm{d}\Sigma = s\mathrm{d}\phi\mathrm{d}z$ and $(s\ \phi\ z)$ are cylindrical coordinates (see *Taylor's condition*). When this constraint is not satisfied, one possibility to balance the Lorentz torque is by an azimuthal acceleration of the cylindrical surface

$$\frac{\partial \mathcal{V}_\phi(s)}{\partial t} = \frac{1}{\rho\mu_\mathrm{o}} \int_\Sigma ((\nabla \times \mathbf{B}) \times \mathbf{B})_\phi \mathrm{d}\Sigma \qquad \text{(Eq. 2)}$$

where $\mathcal{V}_\phi(s)$ is the geostrophic velocity, ρ is the density, and μ_o is the permeability of free space. The class of motion described by the above equation is therefore not directly influenced by the Coriolis force, but only by the magnetic field. The above system allows oscillatory behavior of geostrophic motion (i.e., torsional oscillations) about an equilibrium position where the Lorentz torque vanishes. The damping of these oscillations by diffusion of the magnetic field naturally brings the system back toward this equilibrium. Torsional oscillations are therefore an essential ingredient of the geodynamo as they always allow the system to relax toward a state where Eq. (1) is satisfied everywhere in the core.

Braginsky (1970) was the first to exploit this theoretical concept in an effort to explain geophysical observations. He sought to explain the decade variations in the length of day (LOD) as exchanges of angular momentum between the mantle and the core, with the angular momentum of the core carried by torsional oscillations. This type of fluid motion, he argued, would also be consistent with a part of the observed geomagnetic secular variation. Braginsky established the wave equation for the torsional oscillations predicted by J.B. Taylor. For $\mathcal{V}_\phi(s)$ proportional to $\exp(-\mathrm{i}\omega t)$, it is given by

$$-\omega^2 \rho s^2 h \mathcal{V}_\phi(s) = \frac{\mathrm{d}}{\mathrm{d}s}\left[\frac{s^3 h}{\mu_o} < (B_s)^2 > \frac{\mathrm{d}}{\mathrm{d}s}\left(\frac{\mathcal{V}_\phi(s)}{s}\right)\right] - \mathrm{i}\omega f(s) \quad \text{(Eq. 3)}$$

where s and h are, respectively, the radius and height of the cylinder, $f(s)$ is the torque at the ends of the cylinders due to surface forces at the fluid-solid boundaries, B_s is the s-component of the field which is assumed to remain steady on the timescale of the oscillations, and $<>$ denotes average over the cylinder surface. Note that the restoring force does not involve the steady azimuthal part of the magnetic field, as this component is not sheared by torsional oscillations.

Solutions of Eq. (3) depend on the knowledge of the radial magnetic field everywhere in the core and on the physical properties near the boundaries that enter $f(s)$. Exact solutions are then not obtainable without solving the dynamo problem as a whole. Nevertheless, estimates of the form and periodicity of the natural modes of oscillation can be obtained for simple magnetic field morphologies and simple models for fluid–solid coupling. An order of magnitude estimate for the period of the fundamental mode is given by

$$\tau_{to} \approx c\left(\frac{\rho\mu_o}{< (B_s)^2 >}\right)^{1/2} \quad \text{(Eq. 4)}$$

where c is the radius of the core. As an example, using a typical value of $B_s = 0.5$ mT, we find $\tau_{to} \sim 25$ years, which corresponds to the timescale of the changes in LOD and of a part of the geomagnetic secular variation.

Braginsky's original idea that torsional oscillations can both explain the LOD changes while being consistent with the secular variations of the magnetic field has since received further support. Jault *et al.* (1988) have reconstructed the changes in geostrophic velocities between 1969 and 1985 from maps of the flow at the top of the core which best explain the secular variation of the magnetic field. The changes in core angular momentum calculated from these motions correspond roughly to the changes in mantle angular momentum required to explain the LOD variations during that period. Jackson *et al.* (1993) subsequently showed that this correlation extends back to 1900. In addition, the variations in time of the rigid cylinder flow suggest large exchanges of angular momentum inside the core and relatively little exchange with the mantle, indicating that the motion is likely to be that of waves of the type consistent with torsional oscillations, a fact that was later verified by Zatman and Bloxham (1997). An example of the time-dependent geostrophic flows in the Earth's core, which possibly represent torsional oscillations, are shown in Figure O19 for the period between 1900 and 1990. Recently, Bloxham *et al.* (2002) showed that rigid cylindrical flows with shorter timescale and wavelength also have wave characteristics suggestive of torsional oscillations. This provides additional support for their presence in the Earth's core and indicates that higher modes are also excited. The same study also showed that a time-dependent flow model solely comprised of torsional oscillations can reproduce the part of the secular variation known as geomagnetic jerks. This suggests that the explanation for geomagnetic jerks is connected to torsional oscillations and further underlines their role in the observable part of the dynamics taking place inside the core.

Torsional oscillations are expected on theoretical grounds and we have good evidences that they occur in the Earth's core. Therefore, they represent one of the most robust links between geophysical observations on historical timescales and the dynamics responsible for maintaining the field against Ohmic decay on geologic timescales. Theoretical models of the dynamical processes involved in the geodynamo must then be consistent with torsional oscillations. Conversely, they can be used to extract a wealth of physical quantities and dynamics in the core that are otherwise not directly observable. For instance, Zatman and Bloxham (1997) have used the observed torsional oscillations to build a model of the steady rms value of B_s on cylinder surfaces inside the core. Similarly, Buffett (1998) used Eq. (3) in order to constrain physical parameters near the core-mantle boundary and the nature of the torque that transfers the angular momentum between the mantle and the core.

While a large part of the secular variation of the magnetic field can be explained by a combination of steady flows and torsional oscillations,

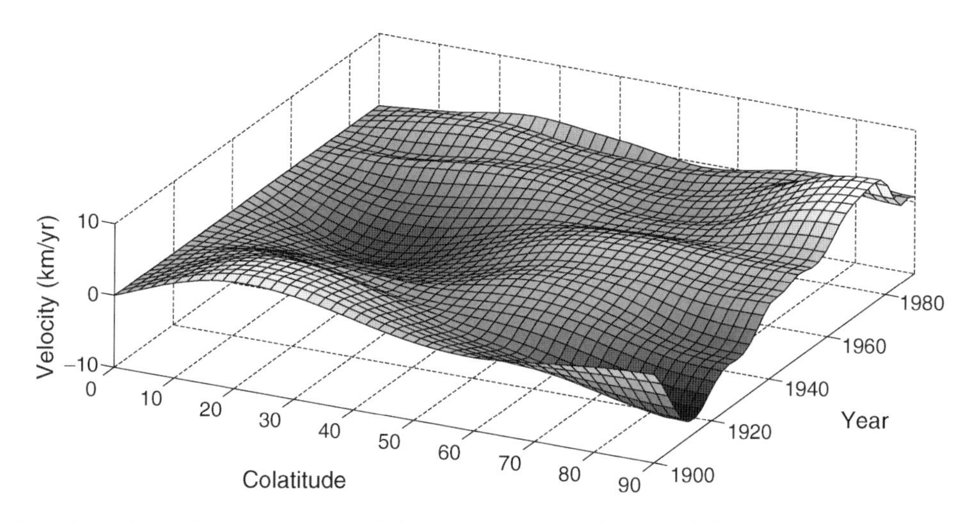

Figure O19 The time-dependent axisymmetric, equatorially symmetric part of the azimuthal velocity at the surface of the core between 1900 and 1990 from inverted *ufm1* field model (Bloxham and Jackson, 1992). Colatitudes (θ) are given in degrees. If this part of the velocity extends rigidly inside the core, then it represents time-dependent velocities of rigid cylindrical surfaces (geostrophic flows) with radius $s = r_c \sin\theta$, where r_c is the (spherical) radius of the core. The undulations in radius and time are suggestive of a propagating wave consistent with torsional oscillations.

a part of the signal remains unaccounted for by this simple model of the dynamics. The unexplained signal may be due to nonaxisymmetric and/or meridional flows associated with the propagation of magnetohydrodynamic waves. The connection between these waves and torsional oscillations remains to be established. Another important issue still unresolved is that of the excitation of torsional oscillations. Estimates of the coupling with the mantle suggest that they should be damped after a few periods, which indicates that an efficient excitation mechanism must exist. The behavior described above, where torsional oscillations occur with respect to a steady equilibrium state, is correct only if the waves are excited by a mechanism that acts suddenly and then plays no further role in the dynamics. A perhaps more likely scenario is the one where a forcing term needs to be added to Eq. (3) and plays an active role in the dynamics at all times. If this latter view is correct, the resulting time-dependent geostrophic flows would consist of a combination of forced and free "torsional oscillations". The nature of this forcing is at present unknown but perhaps it simply consists of the continual changes in the magnetic field, and hence of the Lorentz torque on any cylinder, produced by the convective dynamics in the core. The changes in the Lorentz torque may occur on many different timescales but those that are close to the natural modes of torsional oscillations would provide efficient excitation. In any case, the elucidation of the excitation mechanism will likely lead to an improved understanding of the physical processes involved in the geodynamo.

A promising future avenue to further our understanding of torsional oscillations and of their excitation is through numerical models of the geodynamo (see *Geodynamo, numerical simulations*). This is because the time span of the simulations can far exceed the historical record, and also because the simulations allow direct access to field variables at every location in the core and thus offer a means to examine the details of the dynamics. Since torsional oscillations are one of our most robust observations of core dynamics, realistic numerical models of the geodynamo should include them. Oscillations of geostrophic flows have been observed in a numerical model (Dumberry and Bloxham, 2003) and their excitation is consistent with the above scenario of torques produced by the convective dynamics. However, these oscillations do not follow Eq. (3) because the current limits in computation prevent the use of realistic Earth parameters in the model, and the extrapolation of the results to the real Earth remains uncertain. Nevertheless, the ability of the numerical model to produce geostrophic flows is a step in the right direction. With numerical models becoming increasingly closer to Earth-like conditions, there is hope that they will soon encompass realistic torsional oscillations.

Mathieu Dumberry

Bibliography

Bloxham, J., Zatman, S., and Dumberry, M., 2002. The origin of geomagnetic jerks. *Nature*, **420**: 65–68.

Bloxham, J., and Jackson, A., 1992. Time-dependent mapping of the magnetic field at the core-mantle boundary. *Journal of Geophysical Research*, **97**: 19537–19563.

Braginsky, S.I., 1970. Torsional magnetohydrodynamic vibrations in the Earth's core and variations in day length. *Geomagnetizm I Aeronomiya*, **10**: 1–10

Buffett, B.A., 1998. Free oscillations in the length of day: inferences on physical properties near the core-mantle boundary. In Gurnis, M., Wysession, M.E., Knittle, E., and Buffett, B.A. (eds.) *The Core-Mantle Boundary Region,* AGU Geophysical Monograph, Vol. 28 of Geodynamic series. Washington, DC: American Geophysical Union, pp. 153–165.

Dumberry, M., and Bloxham, J., 2003. Torque balance, Taylor's constraint and torsional oscillations in a numerical model of the geodynamo. *Physics of the Earth and Planetary Interiors*, **140**: 29–51.

Jackson, A., Bloxham, J., and Gubbins, D., 1993. Time-dependent flow at the core surface and conservation of angular momentum in the coupled core-mantle system. In Le Mouël, J.-L., Smylie, D.E., and Herring, T. (eds.) *Dynamics of the Earth's Deep Interior and Earth Rotation,* AGU Geophysical Monograph, Vol. 72. Washington, DC: American Geophysical Union, pp. 97–107.

Jault, D., Gire, C., and Le Mouël, J.L., 1988. Westward drift, core motions and exchanges of angular momentum between core and mantle. *Nature*, **333**: 353–356.

Taylor, J.B., 1963. The magneto-hydrodynamics of a rotating fluid and the Earth's dynamo problem. *Proceedings of the Royal Society of London, Series A*, **274**: 274–283.

Zatman, S., and Bloxham, J., 1997. Torsional oscillations and the magnetic field within the Earth's core. *Nature*, **388**: 760–763.

Cross-references

Alfvén's Theorem and the Frozen Flux Approximation
Alfvén Waves
Core Motions
Geodynamo, Numerical Simulations
Geomagnetic Jerks
Geomagnetic Secular Variation
Length of Day Variations, Decadal
Proudman-Taylor Theorem
Taylor's Condition
Time-Dependent Models of the Geomagnetic Field

P

PALEOINTENSITY: ABSOLUTE DETERMINATIONS USING SINGLE PLAGIOCLASE CRYSTALS

Single plagioclase crystals can contain minute magnetic inclusions, capable of recording and preserving a thermoremanent magnetization. Thellier paleointensity experiments using plagioclase feldspars from a historic lava on Hawaii provide a benchmark for the method. The comparison of rock magnetic and Thellier data from feldspars and whole rocks from older lavas indicate that the feldspars are less susceptible to experimental alteration. This resistance is likely related to the lack of clays and protection of the magnetic minerals by the encasing silicate crystals.

Thellier data sets based on single plagioclase crystals are available from continental flood basalt provinces formed during the Cretaceous Normal Polarity Superchron. These data suggest paleomagnetic dipole moments of approximately 12×10^{22} A m^2, significantly higher than the average paleointensity thought to characterize times of frequent geomagnetic reversals. A correlation between intervals of low-reversal frequency and high-geomagnetic field strength is supported, as seen in some numerical simulations of the geodynamo.

On longer timescales, the magnetization held by plagioclase and other silicate crystals can be used to investigate the Proterozoic and Archean field. Data from plagioclase crystals separated from dikes demonstrate how outstanding questions related to growth of the solid inner core and the magnetic field of the young Earth can be addressed in future investigations.

Whole rock paleointensity analyses

The Thellier double heating method (Thellier and Thellier, 1959) is arguably the most rigorous means of learning about past field strength. But when applied to whole rocks, does this approach insure that the past field strength is accurately recovered? In tests using historic lavas, the Thellier method clearly works well because it retrieves the known field strength. Recent studies of lavas formed during the last 5 Ma have applied rigorous selection criteria to detect experimental alteration and nonideal behavior related to magnetic mineral domain state (e.g., Valet, 2003). But in older rocks, the effects of weathering are important. Clay minerals begin to form a progressively larger portion of a whole rock's matrix. Magnetic mineral phenocrysts undergo low-temperature oxidation.

During the successive heating steps required by the Thellier method, fine-grained magnetic minerals can form from clays (Cottrell and Tarduno, 2000). At relatively high, temperature treatments the alteration can be obvious, it is seen in experimental checks when a partial thermoremanent magnetization (TRM) is imparted at a lower temperature to check for the growth of magnetic minerals. The initial stages of alteration, however, are subtle and can be difficult to detect.

Low-temperature oxidation, or maghemitization, results in a fundamental change in the nature of the magnetization (e.g., Özdemir and Dunlop, 1985). Rather than a TRM, the basis of the Thellier approach, the magnetization can become a partial or complete chemical remanent magnetization (CRM). The accuracy of CRM in preserving the original geomagnetic field strength is unclear (Dunlop and Özdemir, 1997).

There are other reasons for concern about paleointensity estimates derived from older rocks. Many of the virtual dipole moments based on Thellier data from the whole rock and submarine glass samples are similar to values that characterize younger geomagnetic excursions and reversal transitions (Tarduno and Smirnov, 2004). Taken at face value, these data imply that the field has been extraordinarily energetic during the last 10 Ma. But factors that could change the magnetic field energy are generally associated with much longer timescales. Natural and experimental alteration of lavas and submarine basaltic glass tends to lower paleointensity values (Smirnov and Tarduno, 2003), suggesting that the apparent difference in field strength before and after 10 Ma is an artifact.

These issues motivate a search for natural paleomagnetic recorders that are less susceptible to alteration in nature and in the laboratory. Single plagioclase crystals are one such alternative recorder of geomagnetic field history.

Plagioclase feldspar paleointensity analyses

Transmission electron microscopy (TEM) analyses of magnetic extracts from plagioclase separated from basaltic lavas indicate that they can contain equal to slightly elongated magnetic particles 50–350 nm in size (Cottrell and Tarduno, 1999, 2000) (see Figure P1). Studies of the directional dependence of magnetic hysteresis indicate little anisotropy in the plagioclase crystals that could adversely affect Thellier analyses (Cottrell and Tarduno, 1999, 2000).

Unblocking temperatures indicate that the magnetic carriers in plagioclase have compositions that match those of the whole rock. Thermal demagnetization through warming of a saturation isothermal remanent magnetization acquired at low temperature has shown the presence of a blurred Verwey transition (the transition from cubic to monoclinic crystalline symmetry; Verwey, 1939) in plagioclase crystals separated from Cretaceous basalts of the Arctic (Tarduno *et al.*, 2002) (see Figure P1). These data, together with the thermal unblocking

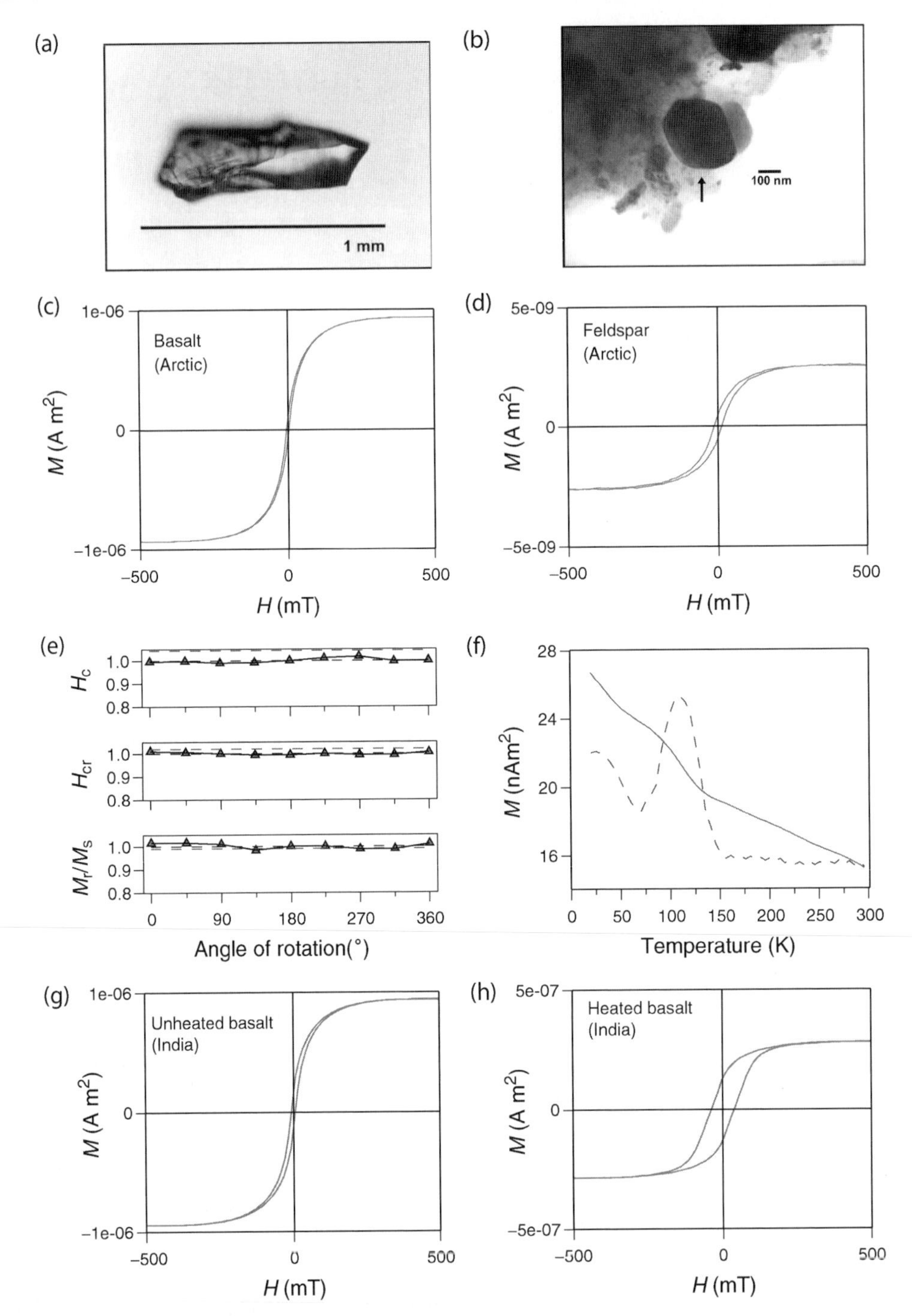

Figure P1 Optical, transmission electron microscope (TEM) and rock magnetic analyses of whole rocks, plagioclase crystals and magnetic separates. (a-f) Examples from the Strand Fiord basalts of the high Arctic after Tarduno *et al.* (2002). (g, h) Examples from the Rajmahal Traps of eastern India after Cottrell and Tarduno (2000). (a) Typical plagioclase crystal used for rock magnetic and paleointensity experiments. (b) TEM image of a magnetic separate from a plagioclase crystal. (c,d) Magnetic hysteresis data (slope corrected) for a whole rock basalt sample and plagioclase feldspar. In general, plagioclase crystals show single domain to pseudo-single domain behavior, while whole rocks display pseudo-single to multidomain characteristics. The former are better suited for Thellier paleointensity analyses. (e) Normalized hysteresis parameters versus rotation angle measured on a plagioclase crystal. The crystal was rotated by 45° increments on the stage of a Princeton Measurements Alternating Gradient Force Magnetometer parallel (solid line and triangles) and perpendicular (dashed line; only mean shown) P1 probes. Abbreviations: M_r/M_s, saturation remanence/saturation magnetization ratio; H_{cr} coercivity of remanence; H_c, coercivity. The lack of systematic variations of these parameters indicates that anisotropy is not a significant concern for Thellier paleointensity experiments. (f) Warming curve of a magnetization (solid line) acquired by a plagioclase crystal at 20 K in a 2.5 T field. Dotted line is the inverse of the derivative. (g,h) Slope corrected magnetic hysteresis curves for unheated and heated whole rock samples. The curve for the heated sample documents the growth of a fine-grained magnetic phase. The heating increments applied were those of a typical Thellier experiment. Magnetic hysteresis curves for plagioclase crystals did not show significant changes after heating (see discussion in Cottrell and Tarduno, 2000).

characteristics and TEM observations, further suggest the plagioclase crystals contain magnetic inclusions of composition similar to that of magnetic grains in the whole rock. In the case of the low-temperature data, the carrier is likely a low Ti titanomagnetite. These observations also suggest that the magnetic particles are inclusions rather than exsolved magnetic particles (Cottrell and Tarduno, 1999, 2000). The latter, which are characteristic of plagioclase in slowly cooled plutonic rocks, typically have end-member mineralogies and distinct shapes, most notably the magnetic "needles", tens of millimeters long, that are sometimes seen oriented along crystallographic axes (e.g., Davis, 1981).

In a test of the method, Thellier analyses of plagioclase crystals separated from a 1955 flow from Kilauea in Hawaii yielded paleointensity estimates that agreed with values reported in detailed comparisons of paleointensity methods based on whole rocks (Coe and Grommé, 1973) and from magnetic observatory data (Cottrell and Tarduno, 1999). The demagnetization of an oriented plagioclase from a Cretaceous lava flow of the Rajmahal Traps demonstrated that the plagioclase recorded the same paleomagnetic record as the whole rock (Cottrell and Tarduno, 2000).

Further comparisons of rock magnetic data from whole rock samples and plagioclase crystals taken from a Rajmahal lava flow demonstrated that the plagioclase crystals were less altered by Thellier heatings (Cottrell and Tarduno, 2000). Specifically, magnetic hysteresis data from heated whole rock samples indicated the formation of a fine-grained magnetic phase; this behavior was not seen in the plagioclase crystals (see Figure P1). This difference in rock magnetic behavior with heating paralleled differences seen in the fidelity and absolute values of paleointensity data. Although whole rock samples meeting paleointensity reliability criteria (e.g., Coe, 1967) yielded paleointensity values that were within error of those derived from single plagioclase crystals, such samples were rare. In general, the whole rocks seldom met reliability criteria and yielded low nominal paleointensity values. The differences can be attributed to the formation of fine-grained magnetite from clays in the groundmass of whole rock samples, resulting in the anomalous acquisition of TRM during Thellier experiments (and a bias toward low values).

Paleointensity during the Cretaceous Normal Polarity Superchron

The experimental alteration, nonideal behavior, and associated high rates of sample reject that typify the Thellier method as applied to whole rocks have generally resulted in a piecemeal approach to the definition of the past geomagnetic field. Directional and paleointensity data are seldom derived from the same rocks. The ability to derive paleointensity values from plagioclase crystals from long lava sequences provides the chance to look at all parts of the time-averaged field. Examining the potential relationships between geomagnetic reversal frequency, secular variation, field morphology and paleointensity is still a daunting task. But if relationships exist, they should be best expressed during superchrons, intervals tens of millions of years long with a few (or no) reversals.

Several continental flood basalt provinces were formed during the Cretaceous Normal Polarity Superchron; these provide an opportunity to gain a more complete view of the past geomagnetic field. For example, Thellier paleointensity analyses of single plagioclase have been derived from distinct lava units of the Rajmahal Traps. The term "lava unit" is used because sometimes it is necessary to combine results from adjacent lava flows that could have been erupted in a very short time. These units were selected from a larger paleomagnetic data set so that they span secular variation. This was gauged in two ways. First, the stratigraphic position of the lava units was considered. Second, the angular dispersion of paleomagnetic directions of the select units (derived from analyses of whole rock samples) simulated that of a complete sampling of the lava sequence. The data allow the calculation of a mid-Cretaceous paleomagnetic dipole moment. The resulting value of $12.5 \pm 1.4 \times 10^{22}$ A m^2 (Tarduno *et al.*, 2001) is higher than the present-day field, or estimates of the long term average field during the last 200 Ma.

A similar study has been conducted on lavas of the high Canadian Arctic, which were also erupted during the Cretaceous Normal Polarity Superchron. Plagioclase feldspars separated from these lavas have also yielded Thellier paleointensity estimate from distinct lava units. As in the case of the Rajmahal Traps, the units are thought to average secular variation because geological indicators show the passage of time, and the angular dispersion of paleomagnetic directions recorded by whole rock samples matches that of a larger data set that spans the entire sequence (Tarduno *et al.*, 2002). The paleomagnetic dipole moment of $12.7 \pm 0.7 \times 10^{22}$ A m^2 suggested by these data agrees with the value from the Rajmahal Traps and indicates that high paleointensity of the latter rocks was not an isolated event within the Cretaceous Normal Polarity Superchron (see Figure P2) . The data from the Rajmahal Traps and Arctic basalts support a correlation between high field strength and low reversal frequency, as suggested in early models (Cox, 1968; see also discussion by Banerjee, 2001) and in more recent numerical simulations of the geodynamo (Glatzmaier *et al.*, 1999).

Precambrian field strength

Mafic dikes of Proterozoic to Archean age are exposed on several continents. These contain feldspars that could carry magnetic inclusions and hence they are potential geomagnetic field recorders. This may provide a means of defining the field during a time interval that may have seen the onset of solid inner core growth. To explore this possibility, Smirnov *et al.* (2003) studied the 2.45 Ga Burakovka intrusion of the Karelian Craton (Russia). Thin mafic border dikes were sampled to minimize the influence of cooling rate. Clear feldspars were selected for study; in other Proterozoic-Archean dike provinces clouded feldspars reflecting exsolution are common (Halls and Zhang, 1998). Magnetic hysteresis parameters indicate multidomain-like behavior of whole rock samples and pseudosingle to single domain (SD) behavior for plagioclase crystals separated from the Karelian dikes. Whole rock samples also show a small anisotropy as recorded by systematic variations in magnetic hysteresis data, presumably recording a flow fabric within the dikes. No significant anisotropy was observed in the plagioclase crystals. The latter observation is important because it indicates that there is not a preferred alignment of elongated particles in the feldspars that could bias TRM acquisition in Thellier experiments.

Transmission electron microscopy analyses suggest that the magnetic inclusions in the plagioclase are equal to slightly elongated and range in size between 50 and 250 nm. Thermal demagnetization data of a saturation remanence imparted on single plagioclase crystals at 10 K are characterized by a well-defined Verwey transition at $\sim$120 K, indicating the presence of stoichiometric magnetite, similar to that reported from whole rock samples.

These rock magnetic data demonstrate the feasibility of the single plagioclase paleointensity approach as applied to select Proterozoic-Archean rocks. Results from only four dikes are available to date from Karelia and it cannot be expected that these adequately record secular variation. Nevertheless, the Thellier paleointensity data are within the range of modern field values. Secular variation and field morphology during this Proterozoic-Archean interval also look amazingly similar to that of the last 5 Ma (Smirnov and Tarduno, 2004).

Discussion and summary

Prévot *et al.* (1990) concluded that geomagnetic reversal rate and paleointensity are decoupled. The contrast between this interpretation and that proposed here arises from differences in data selection. Only a few paleointensity results from rocks older than 10 Ma are based on multiple, independent cooling units that span significant secular variation. Standard paleointensity criteria exclude those lavas with larger multidomain magnetic carriers. A more widespread factor limiting the utility of lavas is weathering. The formation of clays can ultimately

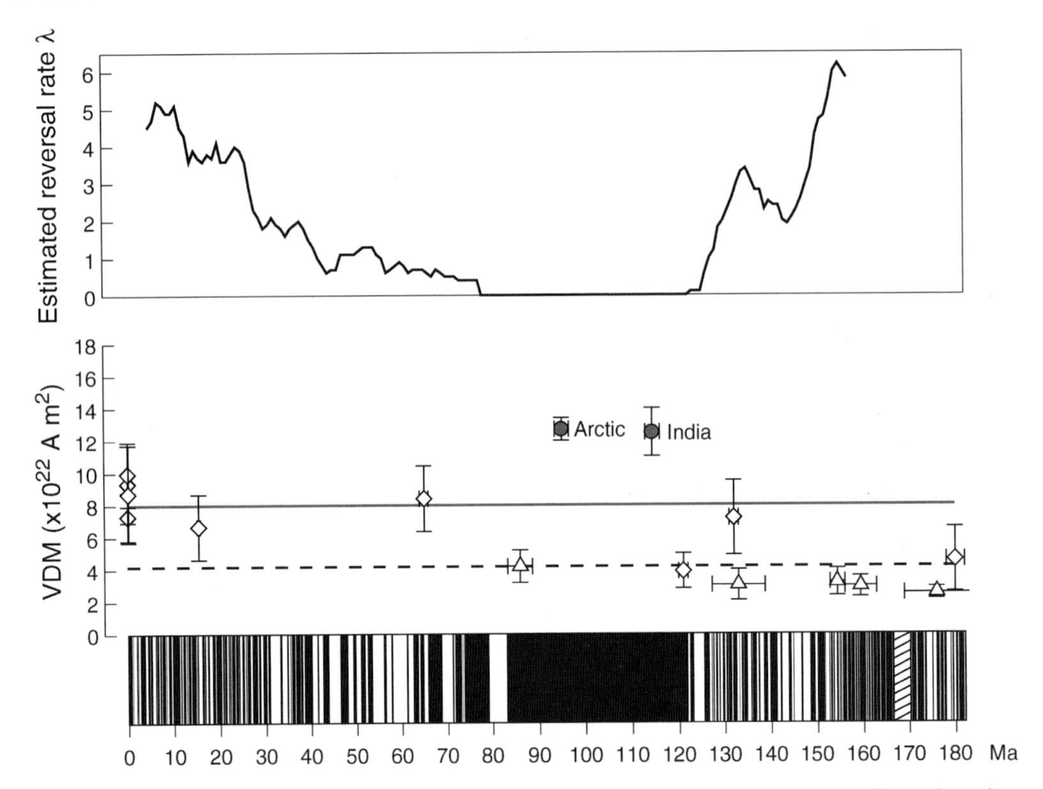

Figure P2 Geomagnetic reversal timescale, select Thellier results (1σ uncertainties shown), and reversal rate (based on a 10-Ma sliding window). Data sources are those selected by Tarduno *et al.* (2001, 2002), with the addition of a new data set from Goguitchaichvili *et al.* (2002). Only studies based on greater than 25 results shown. Data types: triangles, baked contacts; diamonds, basalt whole rocks; circles, plagioclase crystals (Tarduno *et al.*, 2001, 2002). Dashed horizontal line, a proposed mean Cretaceous-Cenozoic value based on Thellier analysis of submarine basaltic glass (Juarez *et al.*, 1998); Solid horizontal line, modern field intensity.

result in the formation of new magnetic minerals during Thellier experiments. In addition, magnetic mineral phenocrysts in the groundmass are transformed by low-temperature oxidation, and CRMs replace TRMs. Some of these effects lead to obvious failures of experimental reliability tests. Others are subtler that have probably led to a bias toward underestimates of field strength.

Because of these limitations, it is premature to apply statistical treatments to the entire basalt virtual dipole moment data set (e.g., Heller *et al.*, 2002) to draw conclusions on the geodynamo. Although there are high-resolution, time-averaged data sets from older lavas (e.g., Kosterov *et al.*, 1997) natural and laboratory alteration will fundamentally limit progress in understanding long-term paleointensity history. This is the prime motivation for a continued pursuit of plagioclase-based Thellier analyses.

Plagioclase feldspar separated from lavas contains minute magnetic inclusions, which appear to have been protected from weathering. Thellier analyses using such crystals have been benchmarked using historic lava. Subsequent tests show that plagioclase crystals are less susceptible to experimental alterations than whole rocks. Because the plagioclase feldspars can be collected from long basalt sequences, they afford a means to obtain joint paleointensity and secular variation data. This further provides the opportunity to investigate relationships between the frequency of geomagnetic reversals and the morphology, secular variation and intensity of Earth's magnetic field. These relationships should be best expressed during superchrons.

Available data from Thellier analyses of plagioclase crystals, coupled with paleomagnetic data from lavas covering various latitude bands, indicate that the time-averaged field during the Cretaceous Normal Polarity Superchron was remarkably strong and stable. A high-field intensity during the Cretaceous Normal Polarity Superchron has also been reported from continued analyses of submarine basaltic glass (Tauxe and Staudigel, 2004). The field was also overwhelmingly dipolar, lacking significant octupole components (Tarduno *et al.*, 2002). These observations suggest that the basic features of the geomagnetic field are intrinsically related. Superchrons may reflect times when the nature of core-mantle boundary heat flux allows the geodynamo to operate at peak efficiency.

Thellier analyses of plagioclase separated from lavas can be used to further examine the reversing field, including times such as the Late Jurassic of peak reversal frequency (McElhinny and Larson, 2003). The approach can also be used to examine the Permian-Carboniferous Kiaman Reversed Polarity Superchron. The Precambrian history of the geomagnetic field bracketing inner core growth is relatively unexplored. Approaches utilizing rigorous Thellier analyses applied to single plagioclase crystals, and other single silicate grains, hold significant promise for revealing this history.

John A. Tarduno, Rory D. Cottrell, and Alexei V. Smirnov

Bibliography

Banerjee, S.K., 2001. When the compass stopped reversing its poles. *Science*, **291**: 1714–1715.

Coe, R.S., 1967. The determination of paleointensities of the Earth's magnetic field with emphasis on mechanisms which could cause nonideal behaviour in Thelliers method. *Journal of Geomagnetism and Geoelectricity*, **19**: 157–179.

Coe, R.S., and Grommé, C.S., 1973. A comparison of three methods of determining geomagnetic paleointensities. *Journal of Geomagnetism and Geoelectricity*, **25**: 415–435.

Cottrell, R.D., and Tarduno, J.A., 1999. Geomagnetic paleointensity derived from single plagioclase crystals. *Earth and Planetary Science Letters*, **169**: 1–5.

Cottrell, R.D., and Tarduno, J.A., 2000. In search of high fidelity geomagnetic paleointensities: A comparison of single crystal and whole rock Thellier-Thellier analyses. *Journal of Geophysical Research*, **105**: 23579–23594.
Cox, A., 1968. Lengths of geomagnetic polarity intervals. *Journal of Geophysical Research*, **73**: 3247–3260.
Davis, K.E., 1981. Magnetite rods in plagioclase as the primary carrier of stable NRM in ocean floor gabbros. *Earth and Planetary Science Letters*, **55**: 190–198.
Dunlop, D.J., and Özdemir, Ö., 1997. *Rock Magnetism, Fundamentals and Frontiers*, Cambridge: Cambridge University Press, 573 pp.
Glatzmaier, G.A., Coe, R.S., Hongre, L., and Roberts, P.H., 1999. The role of the Earth's mantle in controlling the frequency of geomagnetic reversals. *Nature*, **401**: 885–890.
Goguitchaichvili, A., Alva-Valdivia, L.M., Urrutia, J., and Morales, J., 2002. On the reliability of Mesozoic Dipole Low: New absolute paleointensity results from Paraná flood basalts (Brazil). *Geophysical Research Letters*, **29**: 10.1029/2002GL015242.
Halls, H.C., and Zhang, B.X., 1998. Uplift structure of the southern Kapuskasing zone from 2.45 Ga dike swarm displacement. *Geology*, **26**: 67–70.
Heller, R., Merrill, R.T., and McFadden, P.L., 2002. The variation of intensity of earth's magnetic field with time. *Physics of the Earth and Planetary Interiors*, **131**: 237–249.
Juarez, T., Tauxe, L., Gee, J.S., and Pick, T., 1998. The intensity of the Earth's magnetic field over the past 160 million years. *Nature*, **394**: 878–881.
Kosterov, A.A., Prévot, M., and Perrin, M., 1997. Paleointensity of the Earth's magnetic field in the Jurassic: New results from a Thellier study of the Lesotho Basalt, southern Africa. *Journal of Geophysical Research*, **102**: 24859–24872.
McElhinny, M.W., and Larson, R.L., 2003. Jurassic dipole low defined from land and sea data. *Eos, Transactions of the American Geophysical Union*, **84**: 362–366.
Özdemir, Ö., and Dunlop, D.J., 1985. An experimental study of chemical remanent magnetizations of synthetic monodomain titanomaghemites with initial thermoremanent magnetizations. *Journal of Geophysical Research.*, **90**: 11513–11523.
Prévot, M., Derder, M.E., McWilliams, M.O., and Thompson, J., 1990. Intensity of the Earth's magnetic field: Evidence for a Mesozoic dipole low. *Earth and Planetary Science Letters*, **97**: 129–139.
Smirnov, A.V., and Tarduno, J.A., 2003. Magnetic hysteresis monitoring of Cretaceous submarine basaltic glass during Thellier paleointensity experiments: Evidence for alteration and attendant low field bias. *Earth and Planetary Science Letters*, **206**: 571–585.
Smirnov, A.V., Tarduno, J.A., and Pisakin, B.N., 2003. Paleointensity of the early geodynamo (2.45 Ga) as recorded in Karelia: a single crystal approach. *Geology*, **31**: 415–418.
Smirnov, A.V., and Tarduno, J.A., 2004. Secular variation of the Late Archean-Early Proterozoic geodynamo. *Geophysical Research Letters*, **31**, L16607.
Tarduno, J.A., Cottrell, R.D., and Smirnov, A.V., 2001. High geomagnetic field intensity during the mid-Cretaceous from Thellier analyses of single plagioclase crystals. *Science*, **291**: 1779–1783.
Tarduno, J.A., Cottrell, R.D., and Smirnov, A.V., 2002. The Cretaceous Superchron Geodynamo: Observations near the tangent cylinder. *Proceedings of the National Academy of Sciences*. **99**: 14020–14025.
Tarduno, J.A., and Smirnov, A.V., 2004. The paradox of low field values and the long-term history of the geodynamo. In Channell, J.E.T., Kent, D.V., Lowrie, W., and Meert, J.G., (eds.), *Timescales of the Paleomagnetic Field*. American Geophysical Union Geophysical Monograph Series 145. Washington, DC: American Geophysical Union.
Tauxe, L., and Staudigel, H., 2004. Strength of the geomagnetic field in the Cretaceous Normal Superchron: New data from submarine basaltic glass of the Troodos Ophiolite. *Geochemistry, Geophysics and Geosystems*, **5**, Q02H06.
Thellier, E., and Thellier, O., 1959. Sur l'intensité du champ magnétique terrestre dans le passé historique et géologique. *Annales Geophysicae*, **15**: 285–376.
Valet, J.P., 2003. Time variations in geomagnetic intensity. *Reviews of Geophysics*, **41**: 1004.
Verwey, E.J.W., 1939. Electronic conduction of magnetite (Fe_3O_4) and its transition point at low-temperature. *Nature*, **44**: 327–328.

Cross-references

Geodynamo, Numerical Simulations
Geomagnetic Dipole Field
Inner Core
Magnetization, Thermoremanent (TRM)
Paleomagnetism

PALEOINTENSITY, ABSOLUTE, TECHNIQUES

Introduction

The unique possibility to extract the ancient field intensity from magnetic remanence of rocks is to duplicate the acquisition of magnetization by laboratory experiments in presence of a field with a known intensity. This requires a precise knowledge of the mechanisms that govern the magnetization, which is actually the major difficulty inherent to the paleointensity experiments. Acquisition of remanence by strong magnetic fields does not concern natural rocks because the geomagnetic field has a weak intensity (unless they are affected by lightning but this has no interest for paleomagnetism). Another way of magnetizing rocks is by thermal activation, which leads to rotation of some magnetic moments in the direction of the field. The natural magnetization relevant from this process is called a thermoremanent magnetization (TRM). Thermoremanence is acquired by most igneous rocks during cooling from above the Curie temperature (T_c) in the presence of the Earth's magnetic field. Magnetic relaxation, in which remanent magnetization of an assemblage of magnetic grains decays with time, is the most straightforward effect of thermal activation. As temperature decreases through a critical temperature called "blocking temperature" (T_b), an individual magnetic grain experiences a dramatic increase in relaxation time, which will continue down to room temperature. Thus the orientation of the magnetic moment can remain stable at the surface temperature and it will be resistant to effects of magnetic fields after original cooling. Depending on specific parameters inherent to the grain, this magnetic memory can be preserved for billions of years. Rock specimens contain many magnetic grains that can have different characteristics (size, shape etc.) and are thus frequently associated with a distribution of blocking temperatures T_b) from a few tens of degrees up to T_c of the magnetic mineral carrying the magnetization.

Thellier-Thellier technique and derivatives

A very specific characteristic of the total TRM is that it can be separated into the successive magnetizations that were acquired in distinct temperature intervals. For example, TRM of an igneous rock containing magnetite can be broken into portions acquired within successive windows of blocking temperatures from its Curie temperature (T_c) of 580°C down to room temperature. The portion of TRM blocked in any particular T_b window is referred to as "partial TRM," abbreviated pTRM. The total TRM is the vector sum of the pTRMs contributed by all blocking temperature windows. The Individual pTRMs depend only on the magnetic field during cooling through their respective T_b intervals and are not affected by magnetic fields applied during cooling through lower temperature intervals. This is the law of additivity of pTRM, which is essential for paleointensity experiments.

Indeed, as described above, paleointensity experiments rely on a direct comparison between thermal demagnetization of the natural remanent magnetization (NRM, original magnetization that was acquired during cooling of the material) and acquisition of laboratory partial thermoremanent magnetization (pTRM) in presence of a known field. In principle, NRM and TRM of single domain grains of magnetite retain the same proportionality in the applied field over the entire temperature spectrum (Dunlop and Özdemir, 1993). Thus the NRM/TRM ratio must remain constant. In the opposite situation it is impossible to perform suitable experiments of paleointensity because the response of magnetization to the field is not the same in the laboratory than it was during the initial cooling of the rock. In most cases, this is caused by changes in magnetic mineralogy that either altered the natural remanent magnetization (NRM) or which were produced during the laboratory experiments.

The initial and unique reliable technique for absolute paleointensity was proposed and experimented by Thellier and Thellier (1959). Basically the experiment involves a first heating and cooling of the sample in a known field at say 120C followed by measurement of the total magnetization $M_1 = +\text{TRM}\ (0\text{–}120°\text{C}) + \text{NRM}\ (120°\text{C} - T_c)$. The second heating and cooling are performed in presence of a field in the opposite direction and the total magnetization is thus $M_2 = -\text{TRM}\ (0\text{–}120°\text{C}) + \text{NRM}\ (120°\text{C} - T_c)$. The residual NRM is obtained by vectorial subtraction (M_1-M_2) while the TRM is given after substitution of the NRM. This operation is repeated at increasing temperatures up to T_c.

Coe (1967) modified the original Thellier protocol and proposed to demagnetize first the NRM in zero field and then to remagnetize the sample. However, measuring the NRM first does not allow one to detect remagnetization components that could be produced during the subsequent heating at the same temperature in presence of the field and carried by grains with high blocking temperatures. In order to solve this problem (see Figure P3), it is preferable to impart the TRM before heating the sample in zero field (Aitken *et al.*, 1988; Valet *et al.*, 1998; Valet, 2003). The two techniques have their own advantages.

The method of determining the final field value is to calculate the slope of an "Arai" plot (Nagata, 1963), which depicts the amount of NRM lost as a function of pTRM gained for each successive temperature interval (in principle both should keep the same proportion for every temperature interval). The paleointensity is simply obtained after multiplying the slope by the field intensity applied during the experiments. The basic assumption inherent to Thellier experiments is constant proportionality between the NRM lost and the TRM gained after each heating step. Any departure from this behavior cannot be accepted. Unfortunately this ideal situation is not that frequent and linear diagrams over a very wide range of temperatures represent rarely more than 20%–30% of the samples. It is certainly arbitrary to rely on a specific segment of temperature because the quality of the determinations is linked to the amount of NRM involved. We can only be confident if a paleointensity estimate has been derived from a single straight line involving low (say <350°C) and high temperatures (beyond the domain of viscous components). Several indices have been defined (Coe *et al.*, 1978) to evaluate the quality of the determinations. They rely on the percent of initial magnetization involved in the segment selected for calculation, quality of the fit and dispersion of the data points.

Limits imposed by magnetomineralogical changes

Apart from their time-consuming aspect, paleointensity experiments are frequently affected by remagnetization caused by alteration of magnetic minerals during heating. It is thus not only essential to detect these changes during the experiments but also to improve the success rate by selecting appropriate samples. Concerning the first aspect, Thellier and Thellier (1959) introduced the pTRM checks that test for duplication of a TRM at one or several previous temperatures. If the new value is different from the previous one, then the magnetic grains were affected during the previous heating steps at higher temperatures. Unfortunately, the pTRM checks increase the number of heatings, making the full experiment more time-consuming, but they are unavoidable and essential to detect alteration. It is quite justified to consider that the absence of pTRM checks unvalidates the quality of a paleointensity experiment.

In some cases remagnetization can produce grains with blocking temperatures higher than the last heating step. In this case, the NRM itself can be affected and partly lost but the corresponding changes will not necessarily be detected by pTRM checks. There is linear relation between the NRM and the TRM; but the directions of the NRM fail to go through the origin of the demagnetization diagrams when plotted in sample coordinates. This is caused by remagnetization in the field direction of the oven. Kosterov and Prévot (1998) made the point that irreversible changes due to alteration can occur even at moderate temperatures. This was observed by measuring hysteresis loops after different heating steps following a previous suggestion by Haag *et al.* (1995). These changes were not always detected by standard pTRM checks, which again emphasizes the need for multiple and complementary verifications, at least for a limited number of samples per

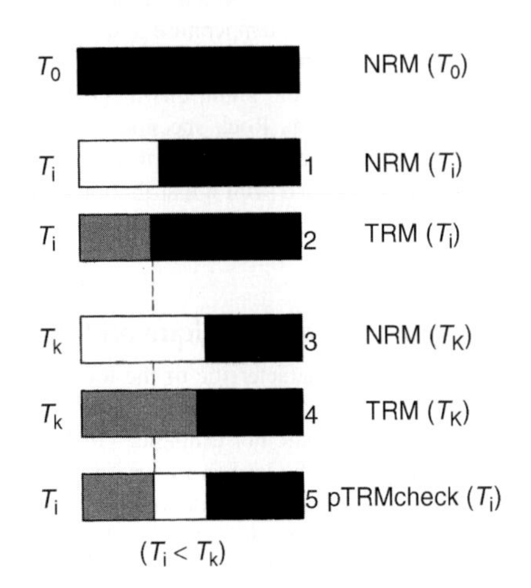

Figure P3 Schematic representation of the procedure for a Thellier-Thellier experiment according to the "modified" Coe protocol. In this case, a TRM is performed first instead of heating in zero field (see text). Black (gray) bars show temperature intervals with remaining NRM (TRM). White bars are for zero magnetization.

lava flow. McCLelland and Briden (1996) and Valet *et al.* (1996) proposed to detect this problem by adding a third heating in zero field, which is particularly appropriate and efficient in the case of the modified Coe version. Indeed if the NRM was affected during the previous heating in presence of field, the repeated experiment will evidently result into a different vector. This approach is possible but more delicate with the standard Thellier-Thellier procedure.

Preselection of the specimens using nontime-consuming techniques rapidly became a critical matter. The problem is that such experiments must necessarily be conducted on adjacent specimens (but from the same core) keeping in mind that there is always a little risk that inhomogeneous magnetic characteristics alter the quality of the selection. Thermomagnetic analyses (measurement of progressive loss of induced susceptibility (K) or saturation remanent magnetization (J_{rs}) while heating up to Curie temperature and its recovery during subsequent cooling) provide useful information about the magnetic minerals and their evolution upon heating. Magnetomineralogical changes have a direct consequence on these two parameters and produce irreversible heating and cooling curves. This test has thus proved very efficient to rule out inappropriate samples, but it does not warranty that those samples which pass the test will necessarily be suitable for paleointensity experiments.

Problems inherent to multidomain grains

Another aspect to consider is the behavior of multidomain (MD) and pseudo-single domain (PSD) grains. Significant progress has been made in this direction. The presence of MD and PSD grains is quite common in lava samples and they can have important consequences. A complete review concerning the theoretical and phenomenological developments on this matter is given by Dunlop and Özdemir (1997). MD and PSD grains do not obey the experimental law of additivity of partial TRMs acquired in nonoverlapping temperature intervals by single domain grains. A direct consequence is the existence of two different slopes in the NRM-TRM plots between low (<350°C) and high (350–580°C) temperatures. Such behavior is caused by the difference between the blocking (T_b) temperature required to randomize the orientation of a portion of magnetic grains and thus to create a pTRM during cooling in presence of field and the unblocking (T_{ub}) temperature needed to randomize the orientation of the magnetic grains with magnetization acquired at T_b. Heating at temperature T_i produce a wide range of T_{ub}'s, both lower and higher than T_b, due to repeated domain wall jumps during heating. The anomalous high T_{ub}'s form a tail extending to the Curie point, which has been widely discussed in the literature (Bol'shakov and Shcherbakova, 1979; Worm *et al.*, 1988; Halgedahl, 1993; McCLelland and Sugiura, 1987; McCLelland and Shcherbakov, 1995; McCLelland *et al.*, 1996). In a recent paper Dunlop and Özdemir (2000) described how low T_{ub} tails can explain the typical concave shape of the Arai diagrams inherent to multidomain grains. Several theoretical and experimental studies (Bol'shakov and Shcherbakova, 1979; Dunlop and Xu, 1994; Xu and Dunlop, 1994) have shown that demagnetization of the NRM can begin at low temperatures, while a significant part of TRM is acquired at high temperatures. Dunlop and Özdemir (2000) suggested that a significant part of the NRM with low T_{ub} can be easily removed whereas pTRM acquisition is delayed by the presence of high T_b. Thus NRM thermal demagnetization will greatly outweigh pTRM acquisition and this effect persists until the NRM is about half demagnetized.

The presence of MD grains can be detected in curves of continuous or thermal demagnetization by their characteristic tails (Bol'shakov and Shcherbakova, 1979). Following experiments by Vinogradov and Markov (1989), Shcherbakova and Shcherbakov (2000) noticed that TRM acquired between room temperature and 300°C after heating to the Curie temperature cannot be removed completely by heating in the zero field in presence of PSD and MD grains while this is not the case for single domain. The remains of the proportion of magnetization is a measurement of the tail and depends on the amount of MD grains. This approach can thus be recommended provided that standard hysteresis parameters confirm the diagnostic. It is frequent that a part of the NRM-TRM spectrum be touched by multidomain grains, which in this case are not clearly reflected by concave up Arai diagrams. McClelland and Briden (1996), Valet *et al.* (1996) and Dunlop and Özdemir (1997) suggested to compare the NRM measured at T_i before and after heating in presence of field. Riisager and Riisager (2001) recently demonstrated the efficiency of the test (that they renamed "pTRM-tail check") on Paleocene lavas from the Faeroe Islands as well as on baked sediments. All these results indicate that concave diagrams must not be used to extract paleointensity information but it is important to perform appropriate tests to detect MD grains.

Variations in the experimental protocol

In an attempt to reduce the number of heatings, Kono and Ueno (1977) proposed to apply a field perpendicular to the NRM direction in order to extract the NRM and the TRM by performing a single heating at each step. In addition to technical difficulties positioning the sample within the oven, the authors noticed that this approach requires exceptionally stable components and has greater experimental errors. Hoffman *et al.* (1989) suggested that one method of reducing alteration is to split the sample into a number of subsamples (usually about 10), each heated to a particular temperature, once in a zero field and again in a known field. The NRM and the pTRM values are then normalized using the initial intensity of the subsample. This approach relies on the assumption that subspecimens from the same sample are characterized by similar magnetic properties, which is not always true. The natural physical properties of a lava often vary over a few centimeters. Sherwood (1991) evaluated the multispecimen approach and found no improvement with respect to the classical Thellier technique and no evidence for reduced alteration.

Several suggestions have been made to increase the number of determinations. Valet *et al.* (1996) proposed a correction technique taking into account the deviations caused by creation or destruction of magnetization. Corrections rely on the hypothesis that the loss (or gain) of TRM induced by mineralogical transformations is indicated by difference between the pTRM check and the TRM that was initially measured at the same step. In a detailed paper aimed at detecting alteration with high-unblocking temperatures, McClelland and Briden (1996) reached basically the same conclusions. Corrections can be applied if (1) the NRM was not affected and (2) the alteration product has unblocking temperatures lower than the temperature of the last heating. These two conditions can be checked by (1) double demagnetization of the NRM (if NRM demagnetization was performed first) and/or (2) scrutinizing the evolution of the NRM in sample coordinates. The suitability of corrections was tested on contemporaneous lava flows. The technique was reasonably successful and gave field determinations within a typical error between 10% and 20% inherent to paleointensity studies. However, it is frequently questioned because there is no clear indication about the reliability of field determinations. For this reason corrections remain speculative particularly in the absence of determinations without corrections for the same flow.

A complete different track was followed by Tanaka *et al.* (1995) who performed continuous direct measurements of the total vector at high temperatures. They defended that this approach can give reliable determinations whereas the standard method does not. The sample was heated in a cryogenic magnetometer with a small furnace installed immediately outside one of the access holes and the measurements were done within the magnetometer. The authors noticed that the drop in temperature did not exceed 1°C at 400°C when moving the sample from the furnace into the magnetometer. The technique involves two successive heatings (with and without field) over successive temperature intervals and does not require cooling the sample to room temperature except at T_{i-1}. The operation is repeated at incremental steps and allows determination of the NRM lost and the TRM gained within each successive temperature interval rather than in a cumulative manner.

A very similar approach has been developed recently by LeGoff and Gallet (2004) with a vibrating magnetometer. The system called "Triaxe" vibrates horizontally along a single direction and comprises three orthogonal sets of pickup coils. Heating is produced by a 2.5 kHz alternating current in a bifilaresistor, avoiding electromagnetic interference in the pickup coils. Heating and cooling rates can be adjusted up to a maximum of 60°C min^{-1}. The sensor and heating-cooling systems are centered inside three perpendicular Helmholtz coils coaxial with the magnetometer axes. Thus a major advantage of this system is that the TRM can be produced in any direction and in particular along the same direction as the initial NRM, avoiding any problems related to anisotropy of the TRM. The same system is used also to produce a zero field. This equipment has been tested with success on archeological material and compared to results obtained with the classical Thellier experiments. Given the gain in precision in rapidity and the versatility provided by this apparatus, there is no doubt that it will be extensively used in future for archeo- and paleointensity studies.

Recent improvements using alternative materials

Pick and Tauxe (1993a) proposed that submarine basaltic glasses would be an ideal material for absolute paleointensity. The magnetic population of submarine volcanic glasses is consistent with very fine Ti-magnetite grains in the superparamagnetic or SD state. There was some concern about the primary origin of the magnetization because of the very fine grain sizes but consistent determinations were obtained from nicely linear Arai plots by Pick and Tauxe (1993a; 1993b), Meija *et al.* (1996), Carlut and Kent (2000), which rules out this possibility. These results must be considered along with recent studies by Zhou *et al.* (1999, 2000) using electron microscope analysis, which indicate that submarine glasses do not seem to be sensitive to alteration on a timescale of a few million years.

Volcanic glasses could thus be ideal candidates for studies of absolute paleointensity. However some aspects still prevent basaltic glasses from several paleomagnetic applications. In most if not all studies published so far, the glass samples were not oriented and the direction of the magnetic field could not be recovered. It is thus impossible to deal with the full vector (which somehow is a paradox after having measured directions without intensity for a very long period). This is even more critical in the case of submarine volcanics since there is a technical problem linked to the difficulty of collecting submarine samples (Cogné *et al.*, 1995) oriented with respect to the geographic coordinates. In this case it is impossible to check the consistency of the magnetic polarity of the glass samples with the underlying crustal magnetic anomaly. This may change in future with the development of new collecting tools. Another limitation is the weak remanent magnetization of the small glassy fragments containing very few crystals. The total moment being on the order of some 10^{-9} A m^2 or less and the measurements are frequently noisy. The irregular size of small chips of glass requires special mounting to allow accurate orientation of the samples for the measurements. Carlut and Kent (2000) proposed a new mounting technique (based on potassium silicate instead of salt encapsulation) to measure magnetic moments as low as 5×10^{-10} A m^2. Below this value, measurements are seldom possible and thus the material cannot be used for paleointensity. Lastly lavas enriched in magnetic oxides along ridge segments can induce high magnetic anomalies and thus add a significant contribution to the main magnetic field. There is no doubt that volcanic glasses are promising to study long-term changes in the average field intensity. It is interesting that the time-averaged field value derived so far from the experiments conducted with this material is lower than the one derived from lava flows.

Cottrell and Tarduno (1999) suggested that the use of plagioclase crystals may provide a viable source of paleointensity data. The major argument is that these crystals are less affected by alteration during heating experiments. The plagioclase grains are picked after crushing the samples, but the direction can be obtained from other specimens taken from the same core. The paleointensity results for 10 plagioclase crystals taken from a sample drilled in the 1955 Kilauea lava flow (Hawaii) indicates a mean paleofield value of 33.8 μT, 5% lower than the expected field of about 36 μT measured at the Honolulu magnetic observatory. Since no field measurement has been performed directly above the flow this could be explained by local anomalies (Coe and Gromme, 1973; Baag *et al.*, 1995; Valet and Soler, 1999). It is important to report that the field determinations have been obtained from low temperature intervals. At higher temperatures, remanence decay was rapid and measurement errors became significant. Such difficulties for measuring extremely low intensities (sometimes $<5 \times 10^{-12}$ A m^2) may be a limiting factor on a wider use of this technique. In a subsequent study, Cottrell and Tarduno (2000) studied 113-Ma-old basalts from the Rajmahal traps using whole rocks and single plagioclase crystals. The paleointensity determinations derived from the whole rock measurements were of lower quality and gave lower values than the plagioclase crystals, which according to the authors was caused by a fine-grained magnetic phase formed during the Thellier-Thellier heating experiments.

Two recent studies indicate that these new materials, plagioclases and volcanic glasses may converge but actually being complementary. Tarduno *et al.* (2001) reported 56 experiments on plagioclase crystals from basalts that formed during the quiet period of the Cretaceous polarity chron and found a strong time-averaged dipole moment, three times greater than during periods with reversals. Similar strong paleointensity was obtained independently by Tauxe and Staudigel (2004) who measured submarine Cretaceous basaltic glasses from the Trodos Ophiolite.

Conclusion

Paleointensity studies remain extremely delicate and time-consuming. However the past two decades of research have seen many developments, which will improve the selection of the samples and the success rate of the determinations. Another important factor is that the conjugation of techniques but also the possibility of using different materials makes now possible a critical evaluation of the data and thus an estimate of their degree of confidence.

Jean-Pierre Valet

Bibliography

Aitken, M.J., Adrian, A.L., Bussell, G.D., and Winter, M.B., 1989. Geomagnetic intensity variation during the last 4000 years. *Physics of the Earth and Planetary Interiors*, **56**: 49–58.

Baag, C., Helsley, C.E., Xu, S.Z., and Lienert, B.R., 1995. Deflection of paleomagnetic directions due to magnetization of the underlying terrain. *Journal of Geophysical Research* **100**(B7): 10013–10027.

Bol'shakov, A.S., and Sherbakova, V.V., 1979. A thermomagnetic criterion for determining the domain structure of ferrimagnetics. *Izvestia Academia of Science, USSR Physics. Solid Earth, Engl. Transl.*, **15**: 111–117.

Carlut, J., and Kent, D.V., 2000. Paleointensity record in zero-age submarine basalt glass: Testing a new dating technique for recent MORBS. *Earth and Planetary Science Letters*, **183**: 389–401.

Coe, R.S., 1967. Paleointensities of the Earth's magnetic field determined from Tertiary and Quaternary rocks. *Journal of Geophysical Research*, **72**: 3247–3262.

Coe, R.S., and Gromme, C.S., 1973. A comparison of three methods of determining geomagnetic paleointensities. *Journal of Geomagnetism and Geoelectricity*, **25**: 415–435.

Coe, R.S., Gromme, S., and Mankinen, E.A., 1978. Geomagnetic intensities from radiocarbon-dated lava flows on Hawaii and the

question of the Pacific nondipole low. *Journal of Geophysical Research*, **83**: 1740–1756.
Cogné, J.P., Francheteau, J., and Courtillot, V., 1995. Large rotation of the easter microplate as evidenced by oriented paleomagnetic samples from the sea-floor. *Earth and Planetary Science Letters*, **136**: 213–222.
Cottrell, R.D., and Tarduno, J.A., 1999. Geomagnetic paleointensity derived from single plagioclase crystals. *Earth and Planetary Science Letters*, **169**: 1–5.
Cottrell, R.D., and Tarduno, J.A., 2000. In search of high-fidelity geomagnetic paleointensities: a comparison of single plagioclase crystals and whole rock Thellier-Thellier analyses. *Journal of Geophysical Research*, **105**: 23579–23594.
Dunlop, D.J., and Xu, S., 1994. Theory of partial thermoremanent magnetization in multidomain grains. 1. Repeated identical barriers to wall motion single microcoercivity. *Journal of Geophysical Research*, **99**: 9005–9023.
Dunlop, D.J., and Özdemir, Ö., 1993. Thermal demagnetization of VRM and pTRM of single domain magnetite: no evidence for anomalously high unblocking tempertaures. *Geophysical Research Letters*, **20**: 1939–1942.
Dunlop, D.J., and Özdemir, Ö., 1997. *Rock Magnetism*. Cambrige: Cambridge University Press, 573 pp.
Dunlop, D.J., and Özdemir, Ö., 2000. Effect of grain size and domain state on thermal demagnetization tails. *Geophysical Research Letters*, **27**: 1311–1314.
Haag, M., Dunn, J.R., and Fuller, M., 1995. A new quality check for absolute paleointensities of the earth magnetic field. *Geophysical Research Letters*, **22**: 3549–3552.
Halgedahl, S.L., 1993. Experiments to investigate the origin of anomalously elevated unblocking temperatures. *Journal of Geophysical Research*, **98**: 22443–22460.
Hoffman, K.A., Constantine, V.L., and Morse, D.L., 1989. Determination of absolute paleointensity using a multi-specimen procedure. *Nature*, **339**: 295–297.
Kono, M., and Ueno, N., 1977. Paleointensity determination by a modified Thellier method. *Physics of the Earth and Planetary Interiors*, **13**: 305–314.
Kosterov, A.A., and Prévot, M., 1998. Possible mechanisms causing failure of paleointensity experiments in some basalts. *Geophysical Journal International*, **134**: 554–572.
Le Goff, M., and Gallet, Y., 2004. A new three-axis vibrating sample magnetometer for continuous high-temperature magnetization measurements: applications to paleo- and archeointensity determinations. *Earth and Planetary Science Letters*, **229**: 31–43.
McClelland, E., and Briden, J.C., 1996. An improved methodology for Thellier-type paleointensity determnation in igneous rocks and its usefulness for verifying primary thermoremanence. *Journal of Geophysical Research*, **101**(B10): 21995–22013.
McClelland, E., Muxworthy, A.R., and Thomas, R.M., 1996. Magnetic properties of the stable fraction of remanence in multidomain (MD) magnetite grains: single-domain or MD? *Geophysical Research Letters*, **23**: 2831–2834.
McClelland, E., and Sugiura, N., 1987. A kinematic model of pTRM acquisition in multidomain magnetite. *Physics of the Earth and Planetary Interiors*, **46**: 9–23.
McClelland, E., and Shcherbakov, V.P., 1995. Metastability of domain state in multi-domain magnetite: consequences for remanence acquisition. *Journal of Geophysical Research*, **100**: 3841–3857.
Mejia, V., Opdyke, N.D., and Perfit, M.R., 1996. Paleomagnetic field intensity recorded in submarine basaltic glass from the East Pacific Rise, the last 69 KA. *Geophysical Research Letters*, **23**: 475–478.
Nagata, T., Arai, Y., and Momose, K., 1963. Secular variation of the geomagnetic total force during the last 5000 years. *Journal of Geophysical Research*, **68**: 5277–5281.
Pick, T., and Tauxe, L., 1993a. Geomagnetic paleointensities during the Cretaceous normal superchron measured using submarine basaltic glass. *Nature*, **36**: 238–242.
Pick, T., and Tauxe, L., 1993b. Holocene paleointensities: Thellier experiments on submarine basaltic glass from the east pacific Rise. *Journal of Geophysical Research*, **98**(B10): 17949–17964.
Sherwood, G.J., 1991. Evaluation of a multispecimen approach to paleointensity determination. *Journal of Geomagnetism and Geoelectricity*, **43**: 341–349.
Tanaka, H., Athanassopoulos, J.D.E., Dunn, J.R., and Fuller, M., 1995. Paleointensity determinations with measurements at high temperature. *Journal of Geomagnetism and Geoelectricity*, **47**: 103–113.
Tarduno, J.A., Cotrell, R.D., and Smirnov, A.V., 2001. High geomagnetic intensity during the mid-Cretaceous fromThellier analyses of single plagiocalse crystals. *Science*, **291**: 1779–1780.
Tauxe, L., and Staudigel, H., 2004. Strength of the geomagnetic field in the Cretaceous Normal Superchron: New data from submarine basaltic glass of the Troodos Ophiolite, *Geochemistry, Geophysics, and Geosystems*, **5**: Q02H06, doi:10.1029/2003GC000635.
Thellier, E., and Thellier, O., 1959. Sur l'intensité du champ magnétique terrestre dans le passé historique et géologique. *Annales Geophysicae*, **15**: 285–376.
Valet, J.P., 2003. Time variations in geomagnetic intensity, *Reviews of Geophysics*, **41**(1): 1004.
Valet, J.P., Brassart, J., Le Meur, I., Soler, V., Quidelleur, X., Tric, E., and Gillot, P.Y., 1996. Absolute paleointensity and magnetomineralogical changes. *Journal of Geophysical Research*, **101**: 25029–25044.
Valet, J.P., Tric, E., Herrero-Bervera, E., Meynadier, L., and Lockwood, J.P., 1998. Absolute paleointensity from Hawaiian lavas younger than 35ka. *Earth and Planetary and Science Letters*, **161**: 19–32.
Valet, J.P., Brassart, J., Le Meur, I., Soler, V., Quidelleur, X., Tric, E., and Gillot, P.Y., 1996. Absolute paleointensity and magnetomineralogical changes. *Journal of Geophysical Research*, **101**: 25029–25044.
Valet, J.P., and Soler, V., 1999. Magnetic anomalies of lava fields in the Canary islands. Possible consequences for paleomagnetic records. *Physics of the Earth and Planetary Interiors*, **115**: 109–118.
Worm, H.U., Jackson, M., Kelso, P., and Banerjee, S.K., 1988. Thermal demagnetization of partial thermoremanent magnetization. *Journal of Geophysical Research*, **96**: 12196–12204.
Xu, S., and Dunlop, D.J., 1994. The theory of partial thermoremanent magnetization in multidomain grains. 2. Effect of microcoercivity distribution and comparison with experiment. *Journal of Geophysical Research*, **99**: 9025–9033.
Zhou, W., Van der Voo, R., and Peacor, D.R., 1999. Preservation of pristine titanomagnetite in older ocean-floor basalts and its significance for paleointensity studies. *Geology*, **27**: 1043–1046.
Zhou, W., Van der Voo, R., Peacor, D., and Zhang, Y., 2000. Variable Ti-content and grain size of titanomagnetite as a function of cooling rate in very young MORB. *Earth and Planetary Science Letters*, **179**: 9–202000.

Cross-references

Archeology, Magnetic Methods
Archeomagnetism
Baked Contact Test
Magnetic Domain
Magnetic Mineralogy, Changes Due to Heating
Magnetization, Natural Remanent (NRM)
Magnetization, Thermoremanent (TRM)
Magnetization, Thermoremanent, in Minerals
Paleointensity, Absolute, Determination

PALEOINTENSITY, RELATIVE, IN SEDIMENTS

What is "relative" paleointensity?

The intensity of depositional remanent magnetization (DRM) in sediments is not absolute, that is, DRM intensity is not solely a function of the ambient field strength at the time of deposition (see Magnetization, depositional remanent, DRM). The intensity of DRM recorded in sediments, when treated with a normalization process, is termed "relative paleointensity." The term "relative" is used because, although the ratio of two intensity values is considered an accurate measure of the percent difference in field strength, a single intensity value is not a stand-alone measure of the Earth's geomagnetic dipole moment (see Figure P4). Sedimentary records must be calibrated against absolute intensity measurements derived from volcanic samples or archeological baked clays (see Magnetization, thermoremanent; Archeomagnetism).

This situation exists because the intensity of DRM is dependent on the amount of ferrimagnetic material present in the sediment sample, the magnetic domain-state of the ferrimagnetic particles, and the mineralogy of the recording assemblage. For example, a package of sediment deposited in the presence of a weak magnetic field may acquire a strong remanence if the sediment contains a large amount of magnetite. In contrast, a package of sediment deposited in a strong field may have a very weak signal if the sediment contains very little magnetite, such as in deep-sea carbonates or biosiliceous ooze (see Figure P5). These nonfield effects must be removed in order to isolate the dependence of DRM on the ambient field intensity. Typically, the intensity of the natural remanent magnetization (see Magnetization, natural remanent, NRM) is divided by a correction factor that is sensitive to the amount of remanence-carrying grains present in the sample. The parameter NRM/(correction factor) yields the intensity of NRM per amount of remanence carrying grains.

The need to correct for nonfield effects has long been recognized. Johnson *et al.* (1948) suggested dividing DRM by isothermal remanent magnetization (IRM) to account for differences in the "magnetization potential" of the sediment (see Magetization, isothermal remanent, IRM). They further suggested that a DRM versus applied field curve derived from redeposition experiments could be used to calibrate NRM intensity in natural sediments. Kent (1973) explored this idea by carrying out a laboratory redeposition experiment in which natural deep-sea sediments were stirred in the presence of an applied field to simulate bioturbation. The resulting deposits were free of inclination error and the intensity of remanence was linearly proportional to the applied field over the range 20–120 μT. This elegant experiment demonstrated that a linear relationship between DRM intensity and ambient field strength existed when an invariant sediment assemblage was repeatedly redeposited. However, invariant sediment assemblages rarely exist in nature. Natural sediment assemblages vary with time as a consequence of environmental processes controlling sediment supply, transport, and deposition. This makes sediments interesting and valuable for their ability to monitor environmental processes and climate change (see Environmental magnetism). This is also responsible for the complexity of the magnetic recording processes in sediment.

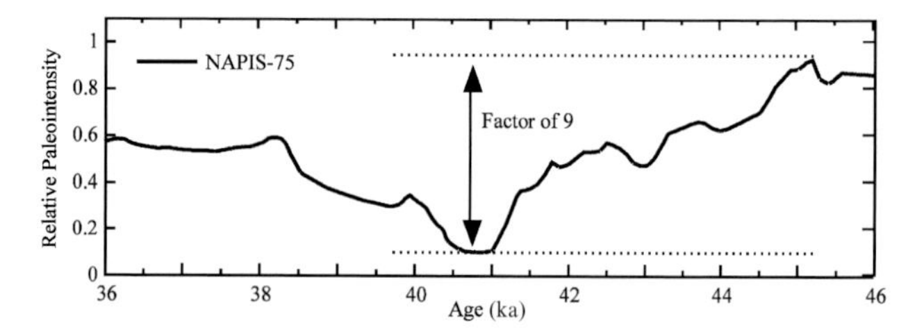

Figure P4 The interval 36–46 ka from the North Atlantic Paleointensity Stack (NAPIS-75) (Laj *et al.*, 2000). The data set has been divided by its mean value, so that the mean is equal to 1. The Laschamp geomagnetic excursion, referred to as the Laschamp Event, is manifested as a 1–2 ka interval of very low intensity at 40–41.5 ka. The Laschamp Event is preceded by an intensity peak at ~45 ka, at which time the geomagnetic field was nine times stronger than during the Laschamp Event. Relative paleointensity records such as NAPIS-75 can reveal the percent change in the geomagnetic field intensity, but not the absolute field strength.

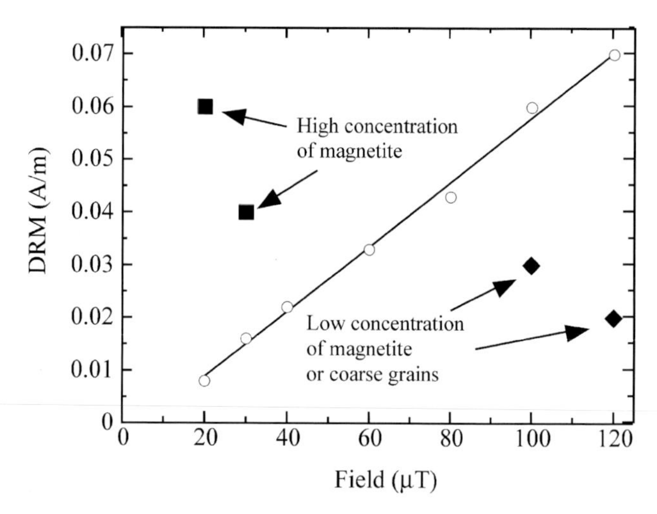

Figure P5 Conceptual model of the intensity of depositional remanent magnetization (DRM) as a function of ambient field strength. Ideally, DRM would be linearly dependent on the amplitude of the ambient field (open circles). In sediments, DRM intensity is dependent on the concentration of remanence carrying grains and the magnetic domain state of the particles. Solid squares represent sediment deposited in the presence of a weak field, but carry a strong DRM due a high concentration of magnetite. Diamonds represent sediment deposited in a strong ambient field, but carry a weak DRM due to low concentration of magnetite, or inefficient recorders (coarse grains, hematite) (After Tauxe, 1993).

Normalization parameters and processes

King *et al.* (1983) proposed a set of guidelines to place limitations on the acceptable amount of sedimentological variability, and limit paleointensity studies to sediments most likely to yield reliable data. The "King Criteria," later modified by Tauxe (1993), attempt to ensure the homogeneity of the magnetic mineral assemblage. If the sedimentary sequence being studied is homogeneous, then one can be confident that any observed features in the geomagnetic field record represent true field behavior, rather than artifacts due to the sedimentology. The uniformity criteria were originally developed to test the suitability of sediments for normalization using anhysteretic remanent magnetization (see Magnetization, anhysteretic remanent, ARM). Hence, the concentration dependence of ARM and the grain-size dependence of ARM observed in magnetite strongly influenced the development of these criteria. These criteria have since become a general set of guidelines, regardless of the normalization method chosen.

First, the sediments should record a strong, stable, single component of remanence whose orientation is appropriate for the site latitude. If the orientation of the remanence vector is inaccurate, then intensity is likely to be inaccurate. Second, the carrier of the NRM should be magnetite. Any mineral with a spontaneous magnetic moment will experience a torque in the presence of an ambient magnetic field.

The stronger the particle's moment, the stronger the torque, and presumably the greater the efficiency of aligning with the field. Magnetite, Fe_3O_4, has the strongest spontaneous moment of the commonly occurring iron-bearing minerals. Further, magnetite and titanomagnetite ($Fe_{3-x}Ti_xO_4$) are the most abundant magnetic minerals in sediments. Therefore magnetite and titanomagnetite are important carriers of DRM in most sediments. Accessory minerals such as hematite, maghemite, greigite and pyrrhotite have been shown to be carriers of a stable DRM in a few rare cases (Kodama, 1982; Tauxe and Kent, 1984; Hallam and Maher, 1994). Generally, hematite has a very low-saturation magnetization and therefore it experiences a weak magnetic torque. Thus its ability to align accurately with the ambient field is questionable. Further, hematite may form in situ long after deposition, such as in redbeds, thus acquiring a chemical remanence (see Magnetization, chemical remanent, CRM) rather than DRM. Maghemite and greigite are more commonly associated with low-temperature oxidation and iron-sulfur diagenesis, respectively (e.g., Roberts and Turner, 1993; Reynolds *et al.*, 1994; Snowball and Torri, 1999), implying the alteration of the primary magnetization and acquisition of a CRM at some unknown time after deposition. Ideally, magnetite should be the only magnetic mineral present in the sediment, to avoid the possibility that the normalization method will activate grains that do not carry DRM.

Third, the magnetite particle size should be 1–10 μm. Superparamagnetic particles (SP, <30 nm for magnetite) are thermally unstable at room temperature and do not carry DRM. Stable single domain (SSD) grains (≈30 nm–1 μm for magnetite) have strong moments and experience a strong torque in an ambient field, in addition to weaker viscous drag due to their smaller size. However, fine SSD grains are susceptible to the influence of Brownian motion (Stacey, 1972). Lu *et al.* (1990) also suggested that fine SSD grains adhere to clay minerals and are rotated out of alignment with the ambient field when the clays are compacted into horizontal orientations. Multi-domain (MD) grains have low magnetization and therefore experience a weaker magnetic torque. Larger grains are more strongly influenced by the gravitational force than by the magnetic force (see Magnetic domain). PSD grains, corresponding to the 1–10 μm size range in magnetite, are assumed to be the most efficient recorders of DRM.

Fourth, the concentration of magnetite should not vary by more than a factor of 10. This limitation on concentration variability is based on the effects of particle interaction during ARM acquisition. The concentration of magnetic material is typically monitored using magnetic susceptibility (k), saturation isothermal remanent magnetization (SIRM), or saturation magnetization (M_S).

The normalization parameters currently in use are k, anhysteretic remanent magnetization (ARM), and (SIRM). Two other methods, the pseudo-Thellier method (Tauxe *et al.,* 1995) and the integral method (Brachfeld *et al.*, 2003), are normalization procedures that have been explored, but not widely used at present. There is not one single normalization parameter that is applicable to all sedimentary records. In selecting the most appropriate normalization parameter for a given sedimentary record, it is necessary to thoroughly characterize the composition, concentration, and grain size of the magnetic recording assemblage, which may be a small subset of the entire magnetic mineral assemblage.

Magnetic susceptibility is logistically and economically the most easily measured magnetic parameter. Susceptibility is a routine measurement conducted on sediment cores collected by ocean research vessels. The ease of measurement is one reason for the continued use of k as a normalization parameter. However, k is measured in the presence of an applied alternating field, whereas DRM is a remanent magnetization. MD and SP grains that do not contribute to DRM contribute strongly to k and lead to "over-correction." The ratio DRM/k is artificially low if these grains are present (Brachfeld and Banerjee, 2000). In addition, several studies have shown strong correspondence between k and $\delta^{18}O$. The presence of Milankovitch periodicities in k profiles from oceanic, lacustrine, and loess records (e.g., Bloemendal and deMenocal, 1989; deMenocal *et al.*, 1991; Peck *et al.*, 1994; Bloemendal *et al.*, 1995), suggests that climatic factors influence the supply of magnetic material to marine, lacustrine, and aeolian sediment. In such cases there is a risk that the climatically induced variability in k, or any potential normalization parameter, could be transferred to the normalized intensity record (e.g., Kok, 1990; Schwartz *et al.*, 1996; 1998; Lund and Schwartz, 1999; Guyodo *et al.*, 2000).

Levi and Banerjee (1976) laid out the arguments in favor of ARM as a normalization parameter. Levi and Banerjee (1976) analyzed a sedimentary record from Lake St. Croix, Minnesota. They compared the alternating field (AF) coercivity spectra of the NRM, ARM, and SIRM. The coercivity spectra of NRM and ARM were nearly identical, while that of a SIRM was much softer (Figure P6). They argued that ARM activated the same grains that carried the NRM, which is the critical factor for selecting a normalization parameter. In addition, since the coercivity spectra of NRM and ARM were identical, the ratio NRM/ARM did not vary with AF demagnetization level, whereas the ratio NRM/IRM (and hence the assumed paleointensity) varied strongly with the demagnetization level.

Complications with ARM normalization arise from strong grain-size and concentration dependencies. Below 1 μm (approximately), small changes in grain-size result in large changes in ARM acquisition (Dunlop and Özdemir, 1997). Above 1 μm, ARM acquisition is less dependent on grain size (Dunlop and Özdemir, 1997). This is the basis for uniformity criterion 3, which states that sediments suitable for paleointensity studies contain magnetite in the 1–10 μm size range. Sediments in which 0.1–1 μm magnetite is the carrier of remanence would be susceptible to miscorrection using ARM. Even relatively small changes in the grain-size of DRM carriers could translate into very large changes in ARM acquired.

ARM acquisition is strongly dependent on the concentration of magnetic material in a sample. Sugiura (1979) imparted ARM to synthetic sediments with varying volume fractions of magnetite. A sample with 2.5-ppm (parts per million) magnetite acquired an ARM that was four times stronger than a sample with 2.3% magnetite. At high magnetite

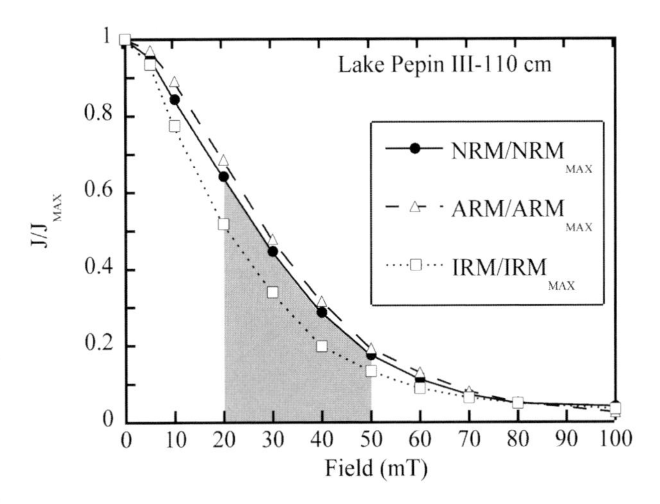

Figure P6 Stepwise alternating field demagnetization of NRM, ARM, and IRM carried by a sediment sample from Lake Pepin, Minnesota (Brachfeld and Banerjee, 2000). Levi and Banerjee (1976) suggested that the most appropriate normalization parameter is the one that activates the same grains that carry the NRM. This is assessed using the alternating field (AF) demagnetization coercivity spectrum. For this sample, the coercivity spectrum of the ARM is nearly identical to the NRM, while the IRM spectrum is softer. ARM is the most appropriate normalization parameter for this sediment sample. The shaded region denotes the area under the NRM demagnetization curve, which is the basis for the integral normalization method (see text).

concentrations, ARM acquisition is less efficient due to particle interactions, which tend to demagnetize the assemblage. The criterion that magnetite concentration vary by no more than a factor of 10 was designed to ensure that the mechanism of ARM acquisition, be it efficient or inefficient, is at least consistent throughout the entire sediment sequence.

Isothermal remanent magnetization (IRM) is imparted by exposing a sample to a steady field for several seconds, then reducing the field nearly instantaneously to zero. By applying stepwise increasing fields and measuring the remanence after each step, one can obtain an IRM acquisition curve. At some field strength the sample acquires its maximum possible signal, at which point the sample has acquired a saturation isothermal remanent magnetization. SIRM and the magnitude of the saturating field vary for different minerals. Magnetite saturates in fields of 100–300 mT. Hematite and goethite requires fields in excess of several tesla to achieve saturation (Dunlop and Özdemir, 1997). SIRM acquisition is less sensitive to grain interactions than ARM acquisition because the applied field which is very strong can overcome the demagnetizing effect of grain interactions. The danger in using SIRM is that—at high fields, nearly all of the magnetic minerals will acquire an IRM or SIRM, even those grains that do not contribute to the DRM. In particular, MD grains that do not contribute to DRM will contribute disproportionately to SIRM.

Tauxe *et al.* (1995) proposed the pseudo-Thellier method of relative paleointensity normalization. The pseudo-Thellier method is an analog of the Thellier. Thellier double heating experiment that is applied to absolute intensity samples. In the pseudo-Thellier method, one plots NRM lost at each AF demagnetization level versus ARM gained at that AF demagnetization level. Relative paleointensity is calculated as the slope of the best-fit line fit to the data (see Figure P7). This method is particularly useful for identifying viscous components of magnetization, which are typically removed after the 5–20 mT demagnetization step. Viscous components are manifested as a line segment with a slightly steeper slope. One can identify these components from the pseudo-Thellier plot, and avoid these intervals when calculating the best-fit line.

Brachfeld *et al.* (2003) proposed an integration method of normalization. This method expands on Levi and Banerjee (1976) who argued that the coercivity spectrum of the normalization parameter should match the coercivity spectrum of the NRM. The integral method involves integrating the area under the NRM/NRM_{MAX} demagnetization curve over an optimal AF range, for example 20–50 mT. The normalization parameter, ARM/ARM_{MAX} is integrated over the same interval. Relative paleointensity is calculated as

$$\frac{\int_{20\,\mathrm{mT}}^{50\,\mathrm{mT}} \mathrm{NRM}(H_{\mathrm{AF}})\mathrm{d}H_{\mathrm{AF}}}{\int_{20\,\mathrm{mT}}^{50\,\mathrm{mT}} \mathrm{ARM}(H_{\mathrm{AF}})\mathrm{d}H_{\mathrm{AF}}}$$

(see Figure P6). The goal of this method is to ensure that the same coercivity spectrum is reflected in both the NRM and the normalization parameter.

There are several ways to assess the quality and accuracy of a normalized intensity record. The simplest and most common method is to apply several normalization parameters. It has traditionally been assumed that if several normalization parameters yield similar profiles, then the sediment is homogeneous and any of the parameters is an adequate normalizer. Normalized intensity records can be judged by comparison with records from other sites, for example other sedimentary records, absolute intensity records, or records of cosmogenic isotopes (^{10}Be, ^{36}Cl), whose production rates are modulated by geomagnetic field intensity (Frank *et al.*, 1997; Frank, 2000; Baumgartner *et al.*, 1998; Wagner *et al.*, 2000a,b).

Normalized intensity records from the same geographic region should yield similar patterns and amplitudes if properly normalized. Unfortunately, the amplitude of normalized intensity features can be quite different depending on the normalization method used (Schwartz *et al.*, 1996, 1998; Lund and Schwartz, 1999; Brachfeld and Banerjee, 2000). Several workers have suggested that the variable amplitudes of crests and troughs as a function of normalization method are the result of subtle grain-size variations that are not removed during normalization (Amerigian, 1977; Sprowl, 1985; Lehman *et al.*, 1994, 1998; Schwartz *et al.*, 1996; Williams *et al.*, 1998; Brachfeld and Banerjee, 2000).

An internal check on the quality of a normalized intensity record is to examine correlations between rock magnetic parameters and normalized intensity. Harrison and Somayajulu (1966) pointed out that the paleointensity estimate should not be correlated with rock magnetic parameters, which they assessed using a linear regression correlation coefficient. Tauxe and Wu (1990) brought this test into the frequency domain by calculating coherence spectra for the normalized paleointensity and the normalization parameters. This permits the identification of frequency bands where paleointensity and rock-magnetic parameters are responding to the same external influence, climatic or otherwise. Tauxe and Wu (1990) proposed that true normalized intensity should not be coherent with its specific normalization parameter. Coherence is typically tested between NRM/*k* and *k*, NRM/ARM and ARM, and NRM/SIRM and SIRM. The best relative paleointensity record is the one that shows no coherence with its normalization parameter. Brachfeld and Banerjee (2000) recommended that normalized intensity also be tested for coherence with grain-size parameters such as the median destructive field of the NRM (NRM_{MDF}) or the coercivity of remanence (H_{CR}).

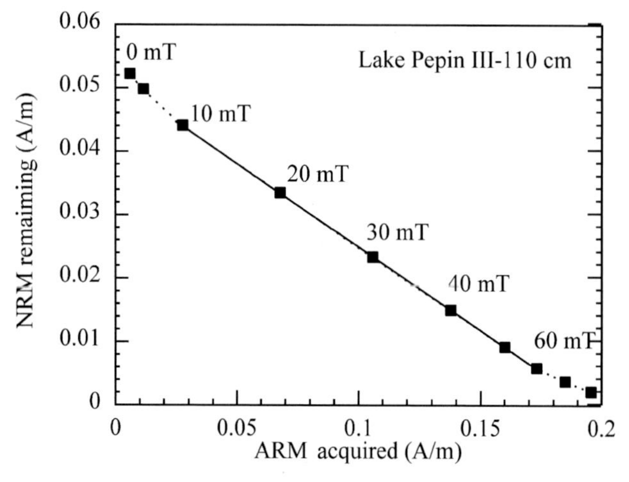

Figure P7 The pseudo-Thellier plot for a sediment sample from Lake Pepin, Minnesota. Tauxe *et al.* (1995) devised this alternate normalization method to simulate the Arai diagram obtained through the Thellier. Thellier absolute intensity analysis. NRM remaining at each demagnetization level is plotted against ARM gained at the same AF level. Viscous components of remanence are identified by the change in slope of the trend line. This sample has a viscous component of remanence that is removed after AF demagnetization at the 10 mT level. Relative paleointensity for this sample is calculated as the slope of the trend line from 20–60 mT.

Paleointensity as a correlation and dating tool

Until the late 1980s, the study of geomagnetic field behavior recorded in sediments was largely undertaken for magnetostratigraphy and to study the geodynamo (see Magnetostratigraphy; Paleomagnetism). Beginning in the late 1980s-early 1990s, several researchers noticed the globally coherent pattern of geomagnetic field paleointensity variations over the last 50 ka–100 ka years (Constable and Tauxe, 1987; Tauxe and Valet, 1989; Tauxe and Wu, 1990; Tric *et al.*, 1992; Meynadier *et al.*, 1992). Soon thereafter, advances in instrumentation enabling the rapid measurement of continuous u-channel subsamples

(Tauxe *et al.*, 1983; Weeks *et al.*, 1993; Nagy and Valet 1993) led to a large increase in the number of long, continuous, sedimentary records analyzed for geomagnetic paleointensity variations. The resulting database of globally distributed records enabled the construction of stacks (weighted averages of many records), which are now used as millennial-scale correlation and dating tools.

The dipole component of the geomagnetic field exhibits temporal variations in its intensity on timescales of hundreds to millions of years. Variations in the geomagnetic dipole are synchronously experienced everywhere on the globe. Therefore, a specific geomagnetic field paleointensity feature represents the same "instant" in time everywhere on Earth. Sedimentary records of geomagnetic field paleointensity from nearly all of Earth's shallow seas, deep ocean basins, and large lakes demonstrate the global coherence of features with wavelengths of 10^4–10^5 years (Figure P8, Tric *et al.*, 1992; Meynadier *et al.*, 1992; Schneider, 1993; Yamazaki *et al.*, 1995; 1999; Guyodo and Valet, 1996, 1999; Laj *et al.*, 1996, 2000, 2004; Lehman *et al.*, 1996; Peck *et al.*, 1996; Verosub *et al.*, 1996; Channell *et al.*, 1997; 2000; Roberts *et al.*, 1997; Schwartz *et al.*, 1998; Stoner *et al.*, 1998, 2000, 2002; Kok and Tauxe, 1999; Channell and Kleiven 2000; Guyodo *et al.*, 2001; Nowaczyk *et al.*, 2001; Sagnotti *et al.*, 2001; Mazaud *et al.*, 2002; Thouveny *et al.*, 2004; Valet *et al.*, 2005). A common characteristic of these records is the presence of several intervals of very low intensity during the past 800 ka. These are interpreted as geomagnetic excursions, which often involve large but short-lived directional changes (>45° deviation of the virtual geomagnetic pole from its time-averaged position) followed by a return of the vector to its previous state (see *Geomagnetic excursion*). Several of the sedimentary records listed above suggest that excursions are associated with a reduction in field strength to less than 50% of its present day value, but are not always associated with a directional change.

These intervals of weak geomagnetic field intensity are partly responsible for peaks in radionuclide production rates. Peaks in ^{10}Be and ^{36}Cl concentration interpreted as the Mono Lake Event and the Laschamp Event have been observed in the Vostok, Dome C, Byrd, Camp Century, and GRIP ice cores (Beer *et al.*, 1984, 1988, 1992; Yiou *et al.*, 1985, 1997; Raisbeck *et al.*, 1987; Baumgartner *et al.*, 1997, 1998; Wagner *et al.*, 2000b). Excursions, whether detected through paleomagnetism or cosmogenic isotopes, are marker horizons that can be used to correlate records from widely separated regions. However, an excursion is one single tie point. A more powerful correlation and dating tool uses time series of geomagnetic paleointensity variations and cosmogenic nuclide production rates for stratigraphic correlation.

Correlation and dating using geomagnetic paleointensity is a "tuning" method. The paleointensity record from the site of interest is compared to a well-dated reference curve. By matching peaks and troughs between the two records, one can import the reference curve ages to the study area. This method has been used to construct chronologies for sites that lacked materials for radiocarbon dating and lacked calcite for oxygen-isotope stratigraphy (Stoner *et al.*, 1998; Sagnotti *et al.*, 2001; Brachfeld *et al.*, 2003).

Any single record of geomagnetic field variability is not suitable for a reference curve. Any single record may contain errors in its chronology or contamination of its remanent magnetization. Guyodo and Valet (1996) proposed using a "stack" to combine many records, thereby enhancing the signal to noise ratio, averaging out any flaws present in the individual records, and extracting the broadscale global features of the geomagnetic field. Guyodo and Valet (1996) showed that a distinctive pattern of geomagnetic field paleointensity was observable in deep-sea sedimentary records from around the globe (see Figure P8). They produced a 200 ka global stack of 19 paleointensity records, named Sint-200, which they suggested could be used as a millennial-scale correlation and dating tool. Each of the 19 constituent records in Sint-200 had its own oxygen-isotope stratigraphy. By correlating paleointensity variations in an undated sediment core with the Sint-200 target curve, one could import the SPECMAP oxygen isotope stratigraphy to the core site.

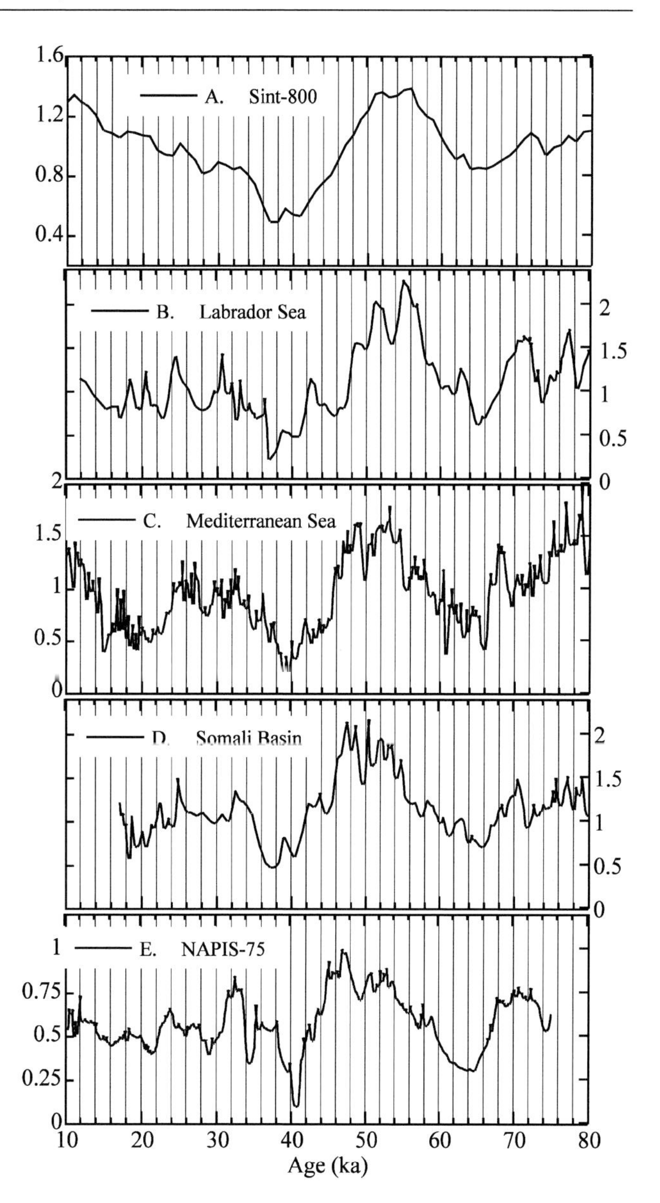

Figure P8 A comparison of globally distributed paleointensity records for the interval 10–80 ka. (A) Sint-800 (Guyodo and Valet, 1999) is a stack comprised of 33 records. Short wavelength features are not preserved in this low-resolution stack. (B) Labrador Sea (Stoner *et al.*, 1998), (C) Mediterranean Sea (Tric *et al.*, 1992), (D) Somali Basin (Meynadier *et al.*, 1992), (E) NAPIS-75 (Laj *et al.*, 2000). Features with wavelengths of 5–10 ka can be recognized and correlated from the Labrador Sea south to the Somali Basin.

The constituent records of Sint-200, and its subsequent extension to the 800 ka Sint-800 (Guyodo and Valet, 1999) were all deep-sea sediment cores with relatively low-sedimentation rates. Therefore, short wavelength features are not preserved in Sint-800. Two newer stacks with higher temporal resolution, the 75 ka North Atlantic paleointensity stack (NAPIS-75) and the South Atlantic paleointensity stack (SAPIS), were constructed from high-sedimentation rate records from sediment drifts in the North Atlantic and the South Atlantic, respectively (Laj *et al.*, 2000; Stoner *et al.*, 2002). NAPIS-75 is a stack of six individual high-resolution records (sedimentation rates = 20 to 30 cm/ka) from cores recovered from the Nordic seas and the North Atlantic. The stacks cover the time interval 10–75 ka, providing overlap with the radiocarbon timescale. The NAPIS-75 chronology is linked to the GISP2 ice core chronology. First, Heinrich layers and a volcanic ash were used as tie points to correlate the six cores. Second, a fine scale correlation was achieved by matching magnetic susceptibility oscillations and ARM oscillations, which put all six cores on a common depth scale (Kissel *et al.*, 1999). The resulting stack was placed on the GISP2 age model of Grootes and Stuiver (1997) by correlating the planktonic foraminifera $\delta^{18}O$ record in one of the sediment cores with the $\delta^{18}O$ record in the GISP2 ice core (Voelker *et al.*, 1998; Kissel *et al.*, 1999).

There are striking similarities between NAPIS-75 and a synthetic geomagnetic field record calculated from ^{36}Cl and ^{10}Be data obtained from the GRIP/GISP2 ice cores (Baumgartner *et al.*, 1997, 1998; Finkel and Nishiizumi, 1997; Yiou *et al.*, 1997; Wagner *et al.*, 2000a,b), which suggests a common geomagnetic origin. Features with a 1000 to 2000-year wavelength can be recognized in both records. The millennial scale features of NAPIS-75, coupled to the precise GISP2 timescale, constitutes a highly efficient tool for correlating and dating cores in different oceans basins around the world (Stoner *et al.*, 2000, 2002), particularly in the Arctic Ocean and southern Ocean, where $\delta^{18}O$ stratigraphy is not available. The extension of NAPIS-75 to NAPIS-300, a 300 ka stack, is now in progress (C. Laj and C. Kissel, personal communication).

SAPIS was constructed from five cores from the South Atlantic (Stoner *et al.*, 2002). The five cores were placed on a common depth scale by correlating magnetic susceptibility, oxygen isotope stratigraphy, geomagnetic inclination, and relative paleointensity. The age model for SAPIS comes largely from one core, Ocean Drilling Program (ODP) Site 1089, whose oxygen isotope stratigraphy was correlated to a nearby core with 14 radiocarbon dates, and by correlating the oxygen isotope stratigraphy to SPECMAP (Stoner *et al.*, 2003).

NAPIS and SAPIS display remarkable similarities over the interval 30–65 ka. The two records display similar peak-to-trough amplitudes, and features with 5-ka wavelengths can be recognized and correlated between the two records. Some differences were present, which the authors ascribed to subtle flaws in chronology—uncorrected environmental influences on the magnetic mineral assemblage, or genuine differences in the nondipole geomagnetic field behavior at these two sites. Nevertheless, the excellent correspondence between the two stacks over key intervals, for example 30–45 ka, illustrates the potential to use geomagnetic field paleointensity as a global, millennial-scale correlation and dating tool.

A high resolution global paleointensity stack (GLOPIS-75, Laj *et al.*, 2004) is a stacked record consisting of 24 paleointensity records distributed world wide, and all with sedimentation rates exceeding 7 cm/ka. The 24 paleointensity records were each individually correlated to NAPIS-75, which was used as the master curve. Each record was resampled every 200 years, normalized to a mean value of 1, and stacked arithmetically (Laj *et al.*, 2004). An iterative process was applied to successively remove discrete intervals in the individual records that may be unreliable, followed by recalculation of the stack using the remaining points. The resulting GLOPIS-75 record covers the interval 10–75 ka.

GLOPIS-75 was calibrated to absolute virtual apparent dipole moment values (VADM) using the archeomagnetic dataset of Yang *et al.* (2000) and globally distributed volcanic data spanning 0–20 ka (Laj *et al.*, 2002). The main trends in this calibrated global stack are very similar to NAPIS-75. There are several noteworthy features. The Mono Lake Event and the Laschamp Event are dated at 34.6 ka and 41 ka, respectively, precise ages that are tied to the GISP2 ice core age model. The duration of these excursions are 1100 years for the Mono Lake Event and 1000–1900 years for the Laschamp Event (Laj *et al.*, 2004). Refinements to GLOPIS-75 can be expected as the number and distribution of paleointensity records increases, particularly with the addition of new records currently under construction from the North Atlantic, Arctic and Antarctic, and with the addition of records spanning 0–20 ka.

Summary

The intensity of DRM recorded by sediments is influenced by the concentration, domain state, and composition of the magnetic mineral assemblage. These nonfield effects on the DRM are removed by normalizing the DRM with a laboratory-derived correction factor such as *k*, ARM, or SIRM. No single normalization parameter is appropriate for all sediments. The magnetic mineral assemblage of each sedimentary record must characterized in order to select the most appropriate normalization parameter. High-quality sedimentary geomagnetic paleointensity records are now being used as global correlation and dating tools. Features with wavelengths of 5 ka, and possibly shorter, can be recognized and correlated between the Northern and Southern hemispheres. By synchronizing geomagnetic intensity records, we may now have the ability to determine millennial-scale leads and lags in Earth's climate systems. Ongoing studies are extending paleointensity stacks further back through time (Valet *et al.*, 2005), and investigating the extent to which finescale features can be recognized and correlated. This includes marine, estuarine, and lacustrine records, with the goal of expanding the temporal and spatial distribution of records, particularly in the Southern Hemisphere (Gogorza *et al.*, 2000, 2002; Stoner *et al.*, 2002, 2003; Brachfeld *et al.*, 2003) and the Arctic (Nowaczyk *et al.*, 2001; St. Onge *et al.*, 2003, 2005; King *et al.*, 2005; Stoner *et al.*, 2005).

Acknowledgments

I thank Carlo Laj, Catherine Kissel, Yohan Guyodo, Joe Stoner, and Laure Meynadier for providing their datasets.

Stefanie Brachfeld

Bibliography

Amerigian, C., 1977. Measurement of the effect of particle size variation on the DRM to ARM ratio in some abyssal sediments. *Earth and Planetary Science Letters*, **36**: 434–442.

Baumgartner, S., Beer, J., Suter, M., Diitrich-Hannen, B., Synal, H.-A., Kubik, P.W., Hammer, C., and Johnsen, S., 1997. Chlorine-36 fallout in the Summit Greenland Ice Core Project ice core. *Journal of Geophysical Research*, **102**: 26659–26662.

Baumgartner, S., Beer, J., Masarik, J., Wagner, G., Meynadier, L., and Synal, H.-A., 1998. Geomagnetic modulation of the ^{36}Cl flux in the GRIP ice core, Greenland. *Science*, **279**: 1330–1332.

Beer, J., Andrée, M., Oeschger, H., Siegenthaler, U., Bonani, G., Hofmann, H., Morenzoni, E., Nessi, M., Suter, M., Wölfli, W., Finkel, R., and Langway, C., Jr. 1984. The Camp Century ^{10}Be record: implications for long-term variations of the geomagnetic dipole moment. *Nuclear Instruments and Methods in Physics Research*, **B5**: 380–384.

Beer, J., Siengenthaler, U., Bonani, G., Finkel, R.C., Oeschger, H., Suter, M., and Wölfi, W., 1988. Information on past solar activity and geomagnetism from ^{10}Be in the Camp Century ice core. *Nature*, **331**: 675–679.

Beer J., Bonani, G.S., Dittrich, B., Heller, F., Kubik, P.W., Tungsheng, L., Chengde, S., and Suter, M., 1992. ^{10}Be and magnetic susceptibility in Chinese loess. *Geophysical Research Letters*, **20**: 57–60.

Bloemendal, J., Liu, X.-M., and Rolph, T.C., 1995. Correlation of the magnetic susceptibility stratigraphy of Chinese loess and the marine oxygen isotope record: Chronological and palaeoclimatic implications. *Earth and Planetary Science Letters*, **131**: 371–380.

Bloemendal, J., and deMenocal, P.B., 1989. Evidence for a change in the periodicity of tropical climate cycles at 2.4 Myr from whole core magnetic susceptibility measurements. *Nature*, **342**: 897–900.

Brachfeld, S.A., Domack, E.W., Kissel, C., Laj, C., Leventer, A., Ishman, S.E., Gilbert, R., Camerlenghi, A., and Eglinton, L.B., 2003. Holocene history of the Larsen Ice Shelf constrained by geomagnetic paleointensity dating. *Geology*, **31**: 749–752.

Brachfeld, S., and Banerjee, S.K., 2000. A new high-resolution geomagnetic relative paleointensity record for the North American Holocene: A comparison of sedimentary and absolute intensity data. *Journal of Geophysical Research*, **105**: 821–834.

Channell, J.E.T., Hodell, D.A., and Lehman, B., 1997. Relative geomagnetic paleointensity and delta ^{18}O at ODP site 983 (Gardar Drift, North Atlantic) since 350 ka. *Earth and Planetary Science Letters*, **153**: 103–118.

Channell, J.E.T., and Kleiven, H.F., 2000. Geomagnetic palaeointensity and astrochronological ages for the Matuyama-Brunhes boundary and boundaries of the Jaramillo subchron: palaeomagnetic and oxygen isotope records from ODP Site 983. *Philosophical Transactions of the Royal Society London A*, **358**: 1027–1047.

Channell, J.E.T., Stoner, J.S., Hodell, D.A., and Charles, C.D., 2000. Geomagnetic paleointensity for the last 100 kyr from the sub-Antarctic South Atlantic: a tool for interhemispheric correlation. *Earth and Planetary Science Letters*, **175**: 145–160.

Constable, C.G., and Tauxe, L., 1987. Paleointensity in the pelagic realm: Marine sediment data compared with archeomagnetic and lake sediment records. *Geophysical Journal of the Royal Astronomical Society*, **90**: 43–59.

deMenocal, P., Bleomendal, J., and King, J.W., 1991. A rock-magnetic record of monsoonal dust deposition to the Arabian Sea: Evidence for a shift in the mode of deposition at 2.4 Ma. *Proceedings of the Ocean Drilling Program- Scientific Results*, **117**: 389–407.

Dunlop, D.J., and Özdemir, Ö., 1997. *Rock Magnetism*, Cambridge: Cambridge University Press.

Finkel, R.C., and Nishiizumi, N., 1997. Beryllium-10 concentrations in the Greenland Ice Sheet Project 2 ice core. *Journal of Geophysical Research*, **102**: 26,699–26,706.

Frank, M., 2000. Comparison of cosmogenic radionuclide production and geomagnetic field intensity over the last 200000 years. *Philosophical Transactions of the Royal Society London A*, **358**: 1089–1107.

Frank, M., Schwarz, B., Baumann, S., Kubik, P.W., Suter, M., and Mangini, A., 1997. A 200 kyr record of cosmogenic radionuclide production rate and geomagnetic field intensity from ^{10}Be in globally stacked deep-sea sediments. *Earth and Planetary Science Letters*, **149**: 121–129.

Gogorza, C.S.G., Sinito, A.M., Lirio, J.M., Nuñez, H., Chaparro, M., and Vilas, J.F., 2002. Paleosecular variations 0–19000 years recorded by sediments from Escondido Lake (Argentina). *Physics of the Earth and Planetary Interiors*, **133**: 35–55.

Gogorza, C.S.G., Sinito, A.M., Tommaso, I.D., Vilas, J.F., Creer, K. M., and Nuñez, H., 2000. Geomagnetic secular variations 0–12 kyr as recorded by sediments from Lake Moreno (southern Argentina). *Journal of South American Earth Sciences*, **13**: 627–645.

Grootes, P.M., and Stuiver, M., 1997. ^{18}O/^{16}O variability in Greenland snow and ice with 10^3 to 10^5 year time resolution. *Journal of Geophysical Research*, **102**: 26455–26470.

Guyodo, Y., Acton, G.D., Brachfeld, S., and Channell, J.E.T., 2001. A sedimentary paleomagnetic record of the Matuyama chron from the western Antarctic margin (ODP Site 1101). *Earth and Planetary Science Letters*, **191**: 61–74.

Guyodo, Y., Gaillot, P., and Channell, J.E.T., 2000. Wavelet analysis of relative geomagnetic paleointensity at ODP Site 983. *Earth and Planetary Science Letters*, **184**: 109–123.

Guyodo, Y., and Valet, J.-P., 1999. Global changes in intensity of the Earth's magnetic field during the past 800 kyr. *Nature*, **399**: 249–252.

Guyodo, Y., and Valet, J.-P., 1996. Relative variations in geomagnetic intensity from sedimentary records: the past 200000 years. *Earth and Planetary Science Letters*, **143**: 23–36.

Hallam, D.F, and Maher, B.A., 1994. A record of reversed polarity carried by the iron sulphide greigite in British Early Pleistocene sediments. *Earth and Planetary Science Letters*, **121**: 71–80.

Harrison, C.G.A., and Somayajulu, B.L.K., 1966. Behavior of the Earth's magnetic field during a reversal. *Nature*, **212**: 1193–1195.

Johnson, E.A., Murphy, T., and Torreson, O.W., 1948. Pre-history of the Earth's magnetic field, *Terrestrial Magnetism and Atmospheric Electricity* **53**: 349–372.

Kent, D.V., 1973. Post depositional remanent magnetization in deep sea sediments. *Nature*, **198**: 32–34.

King, J., Banerjee, S.K., and Marvin, J., 1983. A new rock-magnetic approach for selecting sediments for geomagnetic paleointensity studies: application to paleointensity for the last 4000 years. *Journal of Geophysical Research*, **88**: 5911–5921.

King, J.W., Heil, C., O'Regan, M., Moran, K., Gattacecca, J., Backman, J., Jakobsson, M., and Moore, T., 2005. Paleomagnetic results from the Pleistocene sediments of Lomonosov Ridge, Central Arctic Ocean, IODP Leg 302, *Eos, Transactions, American Geophysical Union* **86**(52), Fall Meet. Suppl. Abstract GP44A-04.

Kissel, C., Laj, C., Labeyrie, L., Dokken, T., Voelker, A., and Blamart, D., 1999. Rapid climatic variations during marine isotopic stage 3: magnetic analysis of sediments from nordic seas and North Atlantic. *Earth and Planetary Science Letters*, **171**: 489–502.

Kok, Y.S., 1999. Climatic influence in NRM and ^{10}Be-derived geomagnetic paleointensity data. *Earth and Planetary Science Letters*, **166**: 105–119.

Kok, Y.S., and Tauxe, L., 1999. A relative geomagnetic paleointensity stack from Ontong-Java plateau sediments for the Matuyama. *Journal of Geophysical Research*, **104**: 25401–25413.

Kodama, K.P., 1982. Magnetic effects of maghemitization of Plio-Pleistocene marine sediments, northern California. *Journal of Geophysical Research*, **87**: 7113–7125.

Laj, C., Kissel, C., Garnier F., and Herrero-Bervera, E., 1996. Relative geomagnetic field intensity and reversals for the last 1.8 My from a central equatorial Pacific core. *Geophysical Research Letters*, **23**: 3393–3396.

Laj, C., Kissel, C., Mazaud, A., Channell, J.E.T., and Beer, J., 2000. North Atlantic Paleointensity Stack since 75 ka (NAPIS-75) and the duration of the Laschamp event. *Philosophical Transactions of the Royal Society London Series A*, **358**: 1009–1025.

Laj, C., Kissel, C., Mazaud, A., Michel, E., Muscheler, R., and Beer, J., 2002. Geomagnetic field intensity, North Atlantic Deep Water circulation and atmospheric Δ^{14}C during the last 50 kyr. *Earth and Planetary Science Letters*, **200**: 179–192.

Laj, C., Kissel, C., and Beer, J., 2004. High resolution paleointensity stack since 75 kyr (GLOPIS-75) calibrated to absolute Values. In Channell, J.E.T., Kent, D.V., Lowrie, W., and Meert, J., (eds.), *Timescales of the Internal Geomagnetic Field*. American Geophysical Union Geophysical Monograph, **145**: 255–265.

Lehman, B., Laj, C., and Kissel, C., 1998. Relative paleointensity determinations from marine sediments: an empirical correction for grain size variations. *Annales Geophysicae, Supplement I*, **16,** 206. European Geophysical Society,

Lehman, B., Laj, C., Kissel, C., Mazaud, A., Paterne, M., and Labeyrie, L., 1996. Relative changes of the geomagnetic field intensity during the last 280 kyear from piston cores in the Açores area. *Physics of the Earth and Planet. Interios*, **93**: 269–284.

Lehman, B., Laj, C., and Kissel, C., 1994. Improvements in selection criteria for geomagnetic paleointensity determinations based on magnetic hysteresis analysis of sediments from the central North Atlantic Ocean. *Eos, Transactions, American Geophysical Union*, **75**: 192.

Levi, S., Banerjee, S.K., and 1976. On the possibility of obtaining relative paleointensities from lake sediments. *Earth and Planetary Science Letters*, **29**: 219–226.

Lu, R., Banerjee, S.K., and Marvin, J., 1990. Effects of clay mineralogy and the electrical conductivity of water on the acquisition of depositional remanent magnetization in sediments. *Journal of Geophysical Research*, **95**: 4531–4538.

Lund, S.P., and Schwartz, M., 1999. Environmental factors affecting geomagnetic field palaeointensity estimates from sediments, In Maher, B.A., and Thompson, R., (eds.), *Quaternary Climates,*

Environments and Magnetism. Cambridge: Cambridge University Press, pp. 323–311.

Mazaud, A., Sicre, M.A., Ezat, U., Pichon, J.J., Duprat, J., Laj, C., Kissel, C., Beaufort, L., Michel, E., and Turon, J.L., 2002. Geomagnetic-assisted stratigraphy and sea surface temperature changes in core MD94-103 (Southern Indian Ocean): possible implications for North-South climatic relationships around H4. *Earth and Planetary Science Letters*, **201**: 159–170.

Nagy, E., and Valet, J.-P., 1993. New advances for paleomagnetic studies of sediment cores using u-channels. *Geophysical Research Letters*, **20**: 671–674.

Nowaczyk, N.R., Frederichs, T.W., Kassens, H., Norgaard-Pedersen, N., Spielhagen, R.F., Stein, R., and Weiel, D., 2001. Sedimentation rates in the Makarov Basin, Central Arctic Ocean—A paleo—and rock-magnetic approach. *Paleoceanography*, **16**: 368–389.

Peck, J.A., King, J.W., Colman, S.M., and Kravchinsky, V.A., 1996. An 84-kyr record from the sediments of Lake Baikal, Siberia. *Journal of Geophysical Research*, **101**: 11365–11385.

Peck, J.A., King, J.W., Colman, S.M., and Kravchinsky, V.A., 1994. A rock-magnetic record from Lake Baikal, Siberia: Evidence for late Quaternary climate change. *Earth and Planetary Science Letters*, **122**: 221–238.

Raisbeck, G.M., Yiou, F., Bourles, D., Lorius, C., Jouzel, J., and Barkov, N.I., 1987. Evidence for two intervals of enhanced ^{10}Be deposition in Antarctic ice during the last glacial period. *Nature*, **326**: 273–277.

Reynolds, R.L., Tuttle, M.L., Rice, C.A., Fishman, N.S., Karachewski, J.A., and Sherman, D.M., 1994. Magnetization and geochemistry of greigite-bearing Cretaceous strata, North Slope Basin, Alaska. *American. Journal of Science*, **294**: 485–528.

Roberts, A.P., Lehman, B., Weeks, R.J., Verosub, K.L., and Laj, C., 1997. Relative paleointensity of the geomagnetic field over the last 200000 years from ODP Sites 883 and 884, North Pacific Ocean. *Earth and Planetary Science Letters*, **152**: 11–23.

Roberts, A.P., and Turner, G.M., 1993. Diagenetic formation of ferrimagnetic iron sulphide minerals in rapidly deposited marine sediments, South Island, New Zealand. *Earth and Planetary Science Letters*, **115**: 257–273.

Sagnotti, L., Macrí, P., Camerlenghi, A., and Rebessco, M., 2001. Antarctic environmental magnetism of Antarctic Late Pleistocene sediments and interhemispheric correlation of climatic events. *Earth and Planetary Science Letters*, **192**: 65–80.

Schwartz, M., Lund, S.P., and Johnson, T.C., 1996. Environmental factors as complicating influences in the recovery of quantitative geomagnetic field paleointensity estimates from sediments. *Geophysical Research Letters*, **23**: 2,693–2,696.

Schwartz, M., Lund, S.P., and Johnson, T.C., 1998. Geomagnetic field intensity from 71 to 12 ka as recorded in deep-sea sediments of the Blake Outer Ridge, North Atlantic Ocean. *Journal of Geophysical Research*, **103**: 30407–30416.

Schneider, D.A., 1993. An estimate of Late Pleistocene geomagnetic intensity variation from Sulu Sea sediments. *Earth and Planetary Science Letters*, **120**: 301–310.

Snowball, I.F., and Torii, M. 1999. Incidence and significance of ferrimagnetic iron sulphides in Quaternary studies, In Maher, B.A., and Thompson, R. (eds.), *Quaternary Climates and Magnetism*. Cambridge: Cambridge University Press, pp. 199–230.

Sprowl, D.R., 1985. The paleomagnetic record from Elk Lake, MN, and its implications, PhD Dissertation thesis, University of Minnesota.

Stacey, F.D., 1972. On the role of Brownian Motion in the control of detrital remanent magnetization of sediments. *Pure and Applied Geophysics.*, **98**: 139–145.

St-Onge, G., Stoner, J.S., and Hillaire-Marcel, C., 2003. Holocene paleomagnetic records from the St. Lawrence Estuary, eastern Canada: centennial- to millennial scale geomagnetic modulation of cosmogenic isotopes. *Earth and Planetary Science Letters*, **209**: 113–130.

St.Onge, G., Stoner, J., Rochon, A., and Channell, J.E., 2005. High resolution Holocene paleomagnetic secular variation records from the Eastern and Western Canadian Arctic. *Eos, Transactions, American Geophysical Union*, **86**(52), Fall Meet. Suppl. Abstract GP44A-03.

Stoner, J.S., Channell, J.E.T., and Hillaire-Marcel, C., 1998. A 200 ka geomagnetic chronostratigraphy for the Labrador Sea: Indirect correlation of the sediment record to SPECMAP. *Earth and Planetary Science Letters*, **159**: 165–181.

Stoner, J.S., Channell, J.E.T., Hillaire-Marcel, C., and Kissel, C., 2000. Geomagnetic paleointensity and environmental record from Labrador Sea core MD95–2024: global marine sediment and ice core chronostratigraphy for the last 110 kyr. *Earth and Planetary Science Letters*, **183**: 161–177.

Stoner, J.S., Laj, C., Channell, J.E.T., and Kissel, C., 2002. South Atlantic and North Atlantic geomagnetic paleointensity stacks (0–80 ka): implications for inter-hemispheric correlation. *Quaternary Science Reviews*, **21**: 1141–1151.

Stoner, J.S., Channell, J.E.T., Hodell, D.A., and Charles, C.D., 2003. A ~580 kyr paleomagnetic record from the sub-Antarctic South Atlantic (Ocean Drilling Program Site 1089). *Journal of Geophysical Research*, **108**, doi:10.1029/2001JB001390.

Stoner, J.S., Francus, P., Bradley, R.S., Patridge, W., Abbott, M.A., Retelle, M.J., Lamoureux, S., and Channell, J.E.T., 2005. Abrupt shifts in the position of the north magnetic pole from Arctic lake sediments: Relationship to archeomagnetic jerks. *Eos, Transactions, American Geophysical Union* **86**(52), Fall Meet. Suppl. Abstract GP44A-02.

Sugiura, N., 1979. ARM, TRM, and magnetic interactions: concentration dependence. *Earth Planetary and Science Letters*, **42**: 451–455.

Tauxe, L., LaBreque, J.L., Dodson, R., Fuller, M., and Dematteo, J., 1983. "U"-channels—a new technique for paleomagnetic analysis of hydraulic piston cores. *Eos, Transactions, American Geophysical Union*, **64**: 219.

Tauxe, L., and Kent, D.V., 1984. Properties of a detrital remanence carried by hematite from study of modern river deposits and laboratory redeposition experiments. *Geophysical Journal of the Royal Astronomical Society*, **77**: 543–561.

Tauxe, L., and Valet, J.-P., 1989. Relative paleointensity of the Earth's magnetic field from marine sedimentary records: A global perspective. *Physics of the Earth and Planetary Interiors*, **56**: 59–68.

Tauxe, L., and Wu, G., 1990. Normalized remanence in sediments of the western equatorial Pacific; Relative paleointensity of the geomagnetic field. *Journal of Geophysical Research*, **95**: 12337–12350.

Tauxe, L, 1993. Sedimentary records of relative paleointensity and the geomagnetic field: theory and practice. *Reviews of Geophysics*, **31**: 319–354.

Tauxe, L., Pick, T., and Kok, Y.S., 1995. Relative paleointensity in sediments: a pseudo-Thellier approach. *Geophysical Research Letters*, **22**: 2885–2888.

Thouveny, N., Carcaillet, J., Moreno, E., Leduc, G., and Nérini, D., 2004. Geomagnetic moment variation and paleomagnetic excursions since 400 kyr B.P: a stacked record from sedimentary sequences of the Portuguese margin. *Earth and Planetary Science Letters*, **219**: 377–396.

Tric, E., Valet, J.-P., Tucholka, P., Paterne, M., La Beyrie, L., Guichard, F., Tauxe, L., and Fontugne, M., 1992. Paleointensity of the geomagnetic field during the last 80000 years, *Journal of Geophysical Research*, **97**: 9,337–9,351.

Valet, J.-P., Meynadier, L., and Guyodo, Y., 2005. Geomagnetic dipole strength and reversal rate over the past two million years. *Nature*, **435**: 802–805.

Verosub, K.L., Herrero-Bervera, E., and Roberts, A.P., 1996. Relative geomagnetic paleointensity across the Jaramillo subchron and the Matuyama/Brunhes boundary. *Geophysical Research Letters*, **23**: 467–470.

Voelker, A., Sarnthein, M., Grootes, P.M., Erlenkeuser, H., Laj, C., Mazaud, A., Nadeeau, M.J., and Schleicher, M., 1998. Correlation of marine ^{14}C ages from the Nordic Sea with GISP2 isotope record: implication for ^{14}C calibration beyond 25 ka B.P. *Radiocarbon*, **40**: 517–534.

Weeks, R., Laj, C., Endignoux, L., Fuller, M., Roberts, A., Manganne, R., Blanchard, E., and Goree, W., 1993. Improvements in long-core measurement techniques: applications in palaeomagnetism and palaeoceanography. *Geophysical Journal International*, **114**: 651–662.

Wagner, G., Masarik, J., Beer, J., Baumgartner, S., Imboden, D., Kubik, P.W., Synal, H.-A., and Suter, M., 2000a. Reconstruction of the geomagnetic field between 20 and 60 kyr B.P from cosmogenic radionuclides in the GRIP ice core. *Nuclear Instruments and Methods In Physics Research Section B*, **172**: 597–604.

Wagner, G., Beer, J., Laj, C., Kissel, C., Masarik, J., Muscheler, R., and Synal, H.-A., 2000b. Chlorine-36 evidence for the Mono Lake event in the Summit GRIP ice core. *Earth and Planetary Science Letters*, **181**: 1–6.

Williams, T., Thouveny, T.N., and Creer, K.M., 1998. A normalised intensity record from Lac du Bouchet: geomagnetic palaeointensity for the last 300 kyr. *Earth and Planetary Science Letters*, **156**: 33–46.

Yamazaki, T., 1999. Relative paleointensity of the geomagnetic field during Brunhes Chron recorded in North Pacific deep-sea sediment cores: orbital influence? *Earth and Planetary Science Letters*, **169**: [illegible]

Yamazaki, T., Ioka, N., and Eguchi, N., 1995. Relative paleointensity of the geomagnetic field during the Brunhes chron. *Earth and Planetary Science Letters*, **136**: 525–540.

Yang, S., Odah, H., and Shaw, J., 2000. Variations in the geomagnetic dipole moment over the last 12000 years. *Geophysical Journal International*, **140**: 158–162.

Yiou, F., Raisbeck, G.M., Bourles, D., Lorius, C., and Barkov, N.I., 1985. ^{10}Be in ice at Vostok Antarctica during the last climatic cycle. *Nature*, **316**: 616–617.

Yiou, F., Raisbeck, G.M., Baumgartner, S., Beer, J., Hammer, C., Johnsen, S., Jouzel, J., Kubik, P.W., Lestringuez, J., Stiévenard, M., Suter, M., and Yiou, P., 1997. Beryllium-10 in the Greenland Ice Core Project ice core at Summit, Greenland. *Journal of Geophysical Research*, **102**: 26,783–26,794.

Cross-references

Archeomagnetism
Environmental Magnetism
Geomagnetic Excursion
Magnetic Domain
Magnetic Susceptibility
Magnetization, Anhysteretic Remanent (ARM)
Magnetization, Chemical Remanent (CRM)
Magnetization, Depositional Remanent (DRM)
Magnetization, Isothermal Remanent (IRM)
Magnetization, Natural Remanent (NRM)
Magnetization, Thermoremanent (TRM)
Magnetostratigraphy
Paleomagnetism
Rock-Magnetism

PALEOMAGNETIC FIELD COLLECTION METHODS

The primary aim of paleomagnetic research is to reconstruct the orientation of the Earth's magnetic field at a given location from the study of a rock unit of known age. For this purpose, it is required to collect samples from the unit to be studied with the purpose of determining their magnetization and magnetic mineralogy in the laboratory. In order to conduct consistency tests of the direction of magnetization of the unit of interest, it is advisable to sample the unit at several localities as widely as possible. This procedure helps to avoid biasing effects introduced by undetected tectonic complications or other, probably more local alteration effects. In return, more effort is required in collecting samples in the field and the adoption of a well-defined hierarchy that needs to be followed during sampling is also needed.

In paleomagnetic parlance, a site refers to the location of each place where a particular sedimentary bed, or cooling igneous unit (i.e., one lava flow or one dyke) is exposed and samples are collected. A sample refers to each oriented piece of rock separately that is retrieved from the outcrop, whereas a specimen refers to smaller pieces of rock, which are well-determined in shape and dimensions obtained from one sample. Commonly, six to eight samples are required within 5–10 m of outcrop at a single site to average out the nonsystematic errors possibly made during sample orientation. At the laboratory, each sample will be split in specimens that can be inserted in the holder of the available instrument that is used for the measurement of magnetization and other laboratory tests. In practice, it is a good idea to have samples large enough to provide multiple specimens, even when only one specimen per sample is all that is required for the measurement of magnetization. This is the case because if more than one specimen per sample is measured at each site, the averaging of the orientation errors will not be unbiased unless the specimens measured for each sample is equal in number. In any case, little is gained in accuracy if more than three specimens from each sample are measured in one site.

Although this hierarchy is easy to follow in general, there are some aspects that require closer examination. For example, if the only access to the rock is through a long drill core (for example those obtained in exploration wells), then following the above definitions we would be limited to only one sample per site (unless more than one core is retrieved at the same location), as little would be gained in terms of accuracy of orientation by measuring more than one specimen per rock-unit traversed by the core. In thick enough units, however, it might prove to be convenient to measure six to eight specimens from the same unit to get a more representative estimation of the magnetization of the unit at that site (core). Certainly, in these conditions the error in the reconstructed magnetization (even when it might only refer to the inclination as the horizontal orientation of the core might not be available) introduced by the error in the orientation of the core will not be averaged out, but at least some internal checks for consistency and homogeneity of the magnetic properties of each rock unit can be made in this form.

As for the actual samples, these can be grouped in three main types depending on the procedure followed for their collection. The most common type of paleomagnetic sample is collected by using a portable drill with a water-cooled diamond bit. The drill is commonly powered by a gasoline engine, but in some occasions it might be necessary to use an electric drill that uses a rechargeable battery. The diameter of the cores drilled in this form is commonly ~2.5 cm and the length depends on the hardness of the rock, the type of drill-bit used and the expertise of the operator. After coring the outcrop to the desired depth and the drill apparatus has been retrieved, an orientation devise is introduced in the space left by the bit. The orientation devise consists of a long tube that has a slot extending from one of its extremes, and a swiveling surface with some type of level attached to the other. The devise is rotated until the surface on the outer extreme becomes horizontal, allowing the correct orientation of the core before retrieving it from the outcrop. The slot in the lower end of the orienting devise (now inserted in the rock) is used to mark the sample, commonly with a brass rod. The orientation is made using a magnetic or a solar compass (or both) attached to the orientation devise, and by reading the angle of the core relative to either the horizontal or the vertical plane. The accuracy of orientation with this method is about ±2°. After the completion of orientation, the orienting devise is retrieved and the core is broken from the outcrop, labeled, and packed.

Another type of paleomagnetic sample is collected by the simpler method of orienting a block of the outcrop that is later broken and labeled. These types of samples are commonly collected when the lithology is not suitable for coring. There are logistic complications for transporting the coring equipment (including water and gasoline) to the outcrop, there are laws forbidding the use of coring equipment (as for example in some national parks or reserve areas), or in desperation when the coring equipment breaks down in the middle of a field season. Although simpler in logistics, this method of sample collection is not as good as the coring method because of the limitations in the accuracy of orientation and on the need of collecting joint blocks that are more likely to have been affected by weathering than the massive portions of the outcrop. It also requires carrying of an excess material that has to be thrown away after specimens have been sliced out of the sample in the laboratory.

The third main type of paleomagnetic sample is collected in the form of core sample obtained from lake- or sea-bottom cores or from exploration wells. These samples are typically ~10 cm diameter and can be from a few meters to some hundreds of meters and even kilometers long. There are two main types of coring techniques in this category: piston coring (in which the bit is introduced forcefully into the sediment) and rotary drills (similar in all respects to the first type of samples discussed above, but with drills that are much larger and not likely to be handled by just one bare-handed operator). Both types of cores are commonly azimuthally unoriented and the direction of the core is assumed to be vertical. Evidently, these types of samples provide less reliability in terms of accuracy of orientation, which often involve the operation of expensive machinery and may not yield recovery of the entire core, but as already mentioned they can be the only access to one particular rock unit.

In some particular situations, paleomagnetic samples can be collected by using very specific (and sometimes very ingenious) methods that often combine some aspects of the three main types just mentioned. For example, loose sediments that are accessible in outcrop cannot be suitable for drilling or for block sampling. In this case, samples can be collected by inserting small plastic boxes in the sediment and orienting them before removing the material around to retrieve the box. Some degrees of impregnation with an epoxic resin might also be used for this purpose, or any other method that allows the operator to retrieve a reasonably well oriented sample.

In summary, the most important aspect to have in mind when collecting samples for any paleomagnetic work is the ability to orient the samples. Although in some circumstances the orientation must be necessarily limited to a distinction between the up and down directions, therefore limiting the utility of such samples.

Edgardo Cañón-Tapia

Cross-references

Paleomagnetism

PALEOMAGNETIC SECULAR VARIATION

The source of the Earth's magnetic field has been the subject of scientific study for more than 400 years (e.g., Gilbert, 1600). At present we believe that most of the field measured at the Earth's surface is of internal origin, generated by hydromagnetic dynamo action in the liquid-iron outer core. Historic measurements of the geomagnetic field have documented its primarily dipolar spatial structure at the Earth's surface and its temporal variability, which is termed as secular variation. One notable characteristic of the Earth's magnetic field and secular variation is its full vector nature, with significant space-time variability in both directions and intensity. Recent historic secular variation (HSV) studies (e.g., Thompson and Barraclough, 1982; Bloxham and Gubbins, 1985. Olson *et al.*, 2002) have characterized the global pattern of short-term secular variation and have related its variability to the core dynamo process.

Paleomagnetic studies make it clear, however, that the Earth's magnetic field has undergone a wider range of temporal and spatial variability than has been seen in historic times. Geomagnetic field polarity reversals have occurred intermittently in time (e.g., Cande and Kent, 1995) and the intervening time intervals of stable dipole polarity contain paleomagnetic secular variation (PSV) larger in amplitude and broader in frequency content than HSV. PSV studies have also documented occasional excursions (Watkins, 1976; Verosub and Banerjee, 1977), which are anomalous PSV fluctuations that may be aborted polarity reversals or represent a fundamentally different multipolar state of the geomagnetic field (Lund *et al.*, 1998, 2001).

PSV is estimated from the paleomagnetic study of archeological materials, unconsolidated sediments, and rocks. The paleomagnetic methods used to recover PSV data are well documented (e.g., Butler, 1992. Tauxe, 1993. Merrill *et al.*, 1998; Dunlop and Özdemir, 2001), and everyone noticed that rather different methods are normally used to recover estimates of paleomagnetic field direction and intensity. Therefore, historically, PSV directional data usually do not have associated paleointensity estimates and vice versa. However, over the last decade that tendency has finally been balanced by the development of numerous high-resolution full-vector PSV records.

This article surveys our current knowledge of PSV; it will provide an overview of PSV data sources, methods of PSV analysis, long-term characteristics of PSV, and models for PSV behavior. The survey will discuss both intensity and directional variability. Special attention will be paid to the relationship between PSV and HSV, the evidence for or against long-term stationarity of PSV, the relationship of PSV to excursions, and the characteristics of PSV that may be useful in dynamo studies.

PSV data

PSV data come from a wide variety of paleomagnetic studies that can be separated into three groups based on the type of sediment or rock measured, the degree of detail in stratigraphic sampling, and the degree of age control for each study. The three resulting PSV data groups are (1) studies of Quaternary-aged sequences of unconsolidated sediments, lava flows, or archeological materials, which can be dated in detail by radiocarbon methods or oxygen isotope stratigraphy, and which are sampled in detail sufficient to resolve the temporal pattern of PSV variability (termed waveform information); (2) studies of older sediment or lava flow sequences that have waveform information but no detailed age control; and (3) studies of any aged rock or sediment sequences that have poor within sequence age control and no waveform information (sequential data are not serially correlated). The first type of PSV study can be used for a full spectrum of time series analyses (waveform, spectral, or statistical analyses); the second type of study can be used for waveform and statistical analyses; the third type of study is only suitable for statistical analysis.

The materials normally used for detailed paleomagnetic studies of PSV are archeological materials (kilns, fire pits, etc.), lava flows, and lake or marine sediment sequences. Each of these materials has inherent advantages and disadvantages for the accurate recording of PSV and the accumulation of paleomagnetic records from all three materials in parallel is the ideal way to establish regional patterns of PSV. Figure P9 illustrates the use of multiple PSV records derived from multiple

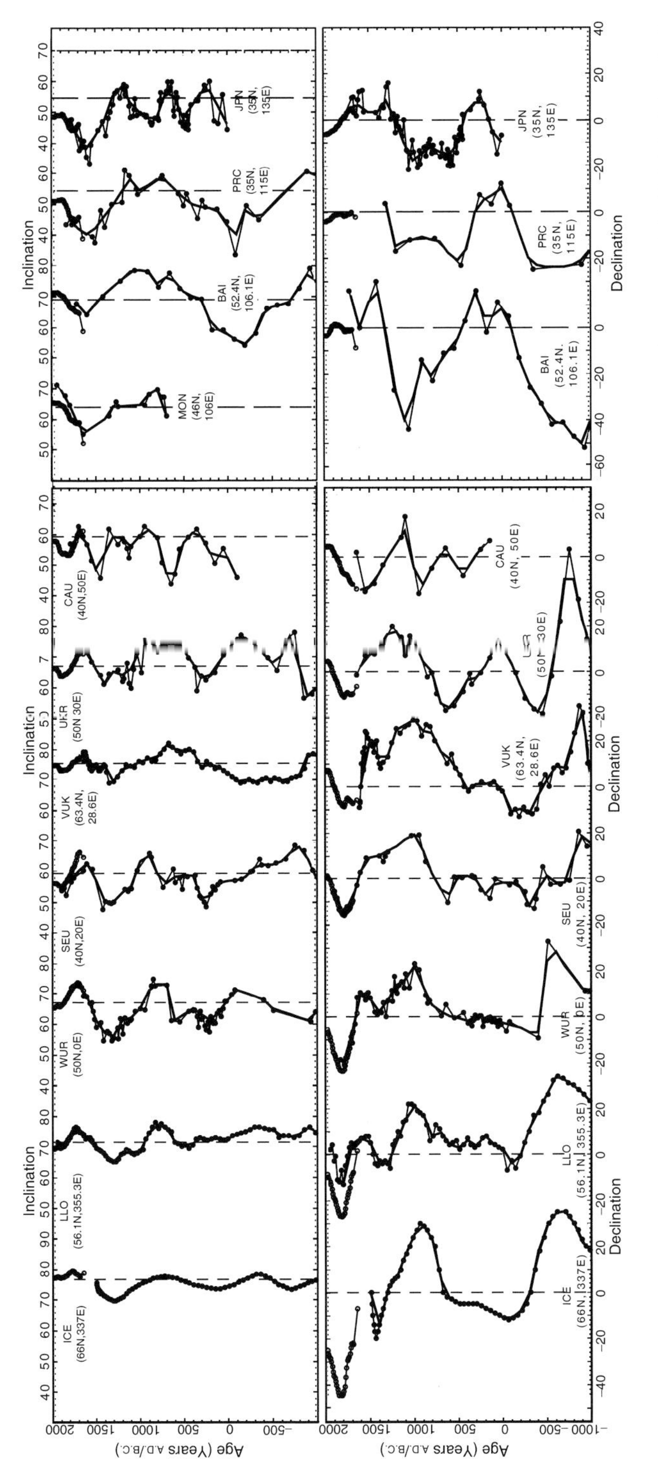

Figure P9 Comparison of paleomagnetic and historic secular variation directional data from Eurasia. Composite PSV records, summarized in Constable *et al.* (2000), extend from Iceland (ICE) to Japan (JPN). Solid dots are original paleomagnetic data, heavy solid lines are interpolated time-series (PSVMOD1.0 available at http://earth.usc.edu/~slund), and open circles in the last 400 years are estimated historical field variations based on spherical harmonic analysis.

data sources (data summarized in Constable *et al.*, 2000) to reconstruct the paleomagnetic directional variability across Eurasia for the last 3 ka. PSV records ICE, Loch Lomand, Scotland (LLO), Lake Vukonjarvi, Finland (VUK), and Lake Baikal, Russia (BAI) are derived from lake sediment sequences, while the other six PSV records are based on archeological and lava-flow measurements. The solid dots in Figure P9 are the actual paleomagnetic measurements, the heavy solid lines are interpolated equi-spaced time series (PSVMOD1.0, available at http://earth.usc.edu/~slund), and open circles in the most recent 400 years are estimated historical secular variation based on spherical harmonic analysis.

Lava flows and archeological materials have the advantage that their natural remanent magnetization (NRM) is normally a thermoremanent magnetization (TRM). A TRM is acquired over time intervals of less than a few minutes to days by heating a material to high temperatures and cooling it in the presence of the geomagnetic field. Therefore, the TRM retains a truly "instantaneous" record of PSV. The primary disadvantage of archeological materials is their scarcity prior to about 2000 years B.P. The primary disadvantage of lava flows is the difficulty in finding sufficient radiocarbon dated flows within a region to develop a long duration, composite PSV record; only a few such studies are currently available in the whole world (e.g., Champion, 1980. Holcomb *et al.*, 1986; Kissel and Laj, 2004).

Sediments acquire their NRM, termed as depositional or post depositional remanent magnetization (DRM/PDRM), due to mechanical alignment of magnetic grains with the geomagnetic field while they are in the water column or in interstitial spaces just below the sediment-water interface. The grains are subsequently locked into that orientation by grain-grain contacts during dewatering, normally within 10–20 cm of the sediment-water interface. The primary advantage of sediment sequences is their potential to provide continuous, high-resolution PSV records far back in time from many sites around the world. The primary drawback to sediments, however, is the lower resolution of the DRM/PDRM recording process due to some degree of inherent smoothing of the PSV signal during remanence acquisition near the sediment-water interface. In most high-resolution sediment records, the smoothing interval can be estimated to be less than 100 years in duration, but further study is necessary to establish the role of smoothing in a better manner in the acquisition process of DRM/PDRM.

An added complexity associated with all PSV data is the limited extent to which they can be compared to HSV data. This difficulty in correlation occurs because (1) PSV data does not have the broad spatial (global) sampling distribution of HSV data, (2) the inherent vector resolution of PSV data (2° to 4° $\alpha 95$ at best) is significantly lower than the resolution of HSV data (typically 1° $\alpha 95$ or better), and (3) the radiometric ages associated with PSV have relatively large errors (ca. ±100 years or greater). PSV records derived from sediments have the added disadvantage, noted previously, of not recovering instantaneous estimates of secular variation due to inherent DRM/PDRM smoothing. Because of these differences, it is difficult to compare the space-time structures of HSV and PSV data in detail, even though site measurements of PSV and HSV may agree within the limit of data resolution (e.g., Figure P9). What we can hope is that analysis of PSV data will yield spatial and/or temporal characteristics that relate, in some way, to observed HSV characteristics.

Three examples of total-vector PSV records are shown in Figures P10-P12. Figure P10 is a Holocene PSV record from wet lake sediments of Lake St. Croix (Lund and Banerjee, 1985; Lund and Schwartz, 1999), Figure P11 is a late-Pleistocene PSV record from uplifted (dry) lake sediments surrounding Mono Lake (Liddicoat and Coe, 1979; Lund *et al.*, 1985), and Figure P12 is a late-Pleistocene PSV record recovered from deep-sea sediments of the western North Atlantic Ocean (Lund *et al.*, 2001, 2005).

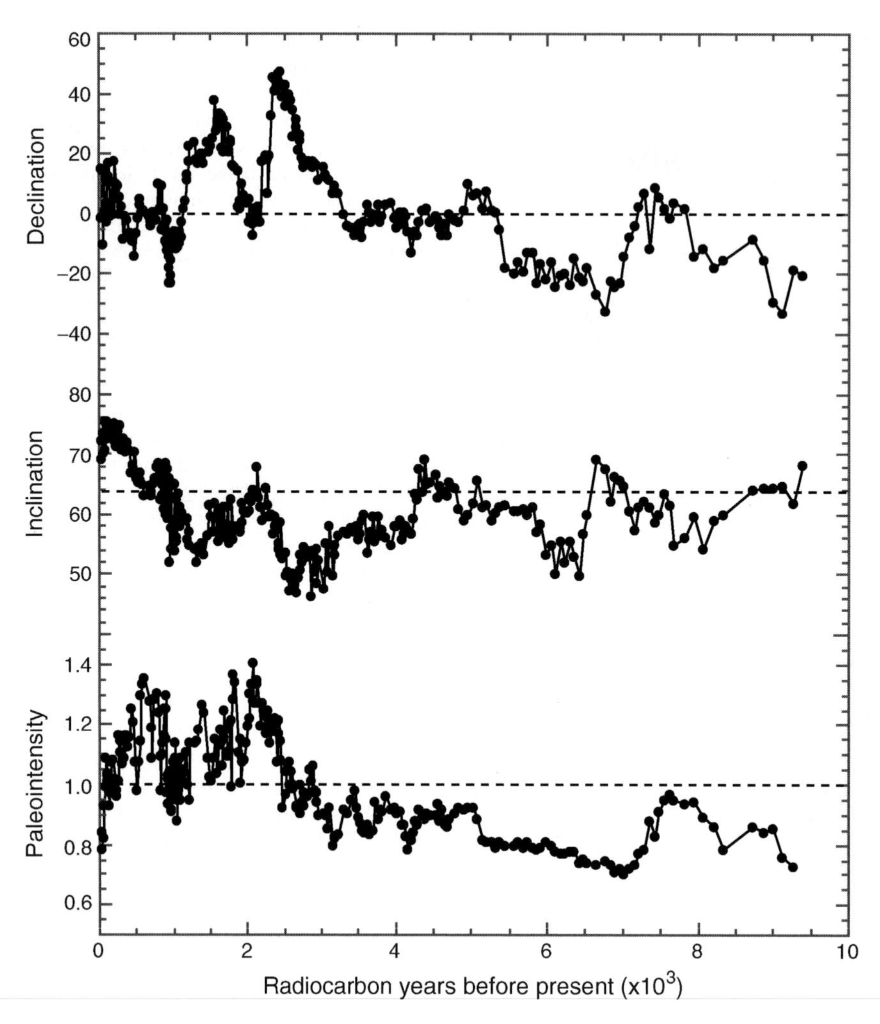

Figure P10 Holocene full-vector paleomagnetic secular variation record recovered from Lake St. Croix, Minnesota (USA) (Lund and Banerjee, 1985; Lund and Schwartz, 1999).

Time-series analysis of PSV

The conceptual framework for analysis of PSV data is based on the fact that HSV records are too short in duration to adequately characterize the totality of temporal field variability, while PSV records are too scattered spatially to routinely describe the prehistoric spatial field variability. We can hope, however, to identify spatial and short duration temporal components of the HSV that may relate to long term temporal components of the PSV. We can also attempt to improve the spatial estimates of PSV by, perhaps, appealing to analogs in HSV. Such a coordinated analysis of PSV and HSV can perhaps address a key to an unresolved question in secular variation studies: What is the mapping function between the observed spatial and temporal variations of the geomagnetic field? Only with a coherent view of the total spatiotemporal variability of the historic and prehistoric geomagnetic field can we properly evaluate models of the core dynamo process, which is the source of the field variability.

Analysis of PSV data uses a variety of time-series and modeling techniques in order to delineate the spatial and temporal characteristics of PSV. Time-series techniques that can be applied to PSV data will be considered in this section; modeling techniques will be considered in the next section. Time-series techniques are classified into three broad categories: waveform analysis, spectral analysis, and statistical analysis. Each of these techniques has unique advantages for characterizing a particular type of PSV data and thereby providing a point of comparison with HSV data.

Waveform analysis

PSV records that display good serial correlation between adjacent data points (e.g., Figures P9-P12) can be used to assess the actual time evolution of geomagnetic field variability. This time evolution can be characterized either by evaluating simple PSV features, such as maxima or minima of inclination, declination, or paleointensity, or more distinctive sets of features termed vector waveforms (Lund, 1996). One type of vector waveform that may have fundamental importance for dynamo studies is the vector-looping pattern termed circularity (Runcorn, 1959. Skiles, 1970) often exhibited by HSV and PSV records (e.g., Bauer, 1895. Creer, 1983. Lund, 1996). Comparisons of simple features or vector waveforms may take place (1) within individual paleomagnetic records, (2) between records of different sites, and (3) between PSV and HSV records (where they overlap in time), and can be used to assess the temporal evolution of secular variation. Comparisons of the amplitudes and phase relationships of PSV features or waveforms with their spatial counterparts(?) in global maps of the present-day field may be used to assess the spatial-temporal mapping of secular variation. Below we discuss three different types of waveform comparison that document distinctive PSV characteristics.

The first type of waveform comparison that should be considered is between PSV records and HSV records from the same sites. For example, Figure P9 shows a transect of PSV records for the last 3 ka across Eurasia. Open circles in the interval A.D. 1600 to A.D. 2000 estimate the historic secular variation for each site based on spherical

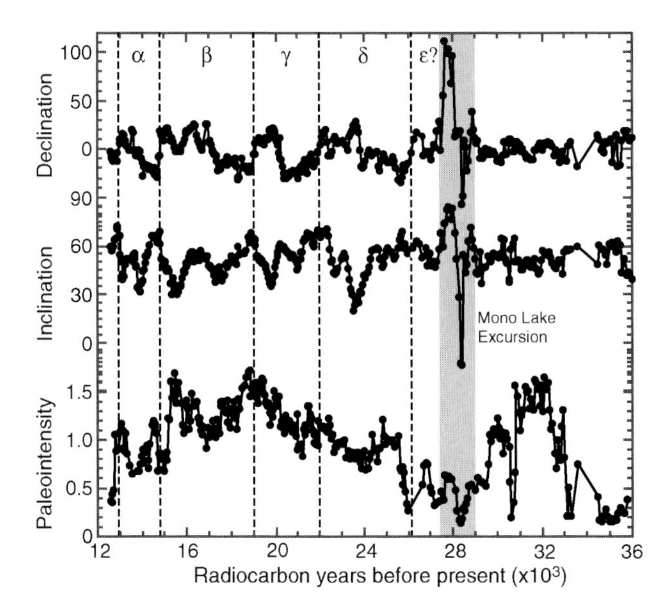

Figure P11 Late Pleistocene full-vector paleomagnetic secular variation record recovered from uplifted lake sediments exposed at Mono Lake, California (USA) (Liddicoat and Coe, 1979; Lund *et al.*, 1988). Note the presence of the Mono Lake excursion near 28000 radiocarbon years B.P. and indication of at least four cycles of a complex repeating vector waveform (α-ε?) associated with the excursion.

harmonic analyses. Such comparisons indicate that only the largest amplitude, longest period (~200–400 years) HSV features can be correlated with the PSV variability (indicated by solid circles in Figure P9). This limited correlation is due to the fact that PSV records only resolve features greater than about 4° in amplitude and a few hundred years in duration. Even then, this correlation shows that high-resolution PSV records can accurately record and extend the long-term temporal variation only hinted at in HSV records.

The second type of comparison, between different high-resolution PSV records from the same region (e.g., Europe or East Asia), can be used to assess the spatial extent of PSV features. For example, Figure P9 shows late Holocene PSV records from seven different sites in Europe (Iceland, ICE to CAU) spread over 70° of longitude (~7000 km) and four different sites in East Asia (MON to Japan, JPN) spread over 30° of longitude (~3000 km). Within each of these regions, it is readily apparent that a large number of directional features can be correlated among the records. (i.e., not to say that the records are identical, but variations in sampling rate and signal to noise ratios or errors in data acquisition and analysis probably can explain most of the differences in single records). One interpretation of these observations is that PSV features with periods longer than a few hundred years must correspond to spatial features that are observable in historic maps of the geomagnetic field and have spatial domains of several 1000 km.

A similar comparison between PSV records from different geographic regions is more problematic. For example, the comparison between European and East Asian PSV records in Figure P9 indicates that there is no simple correlation between directional records from the two regions that preserves phase relationships or long-term trends in the directional data. Similarly, there is no simple vector pattern that

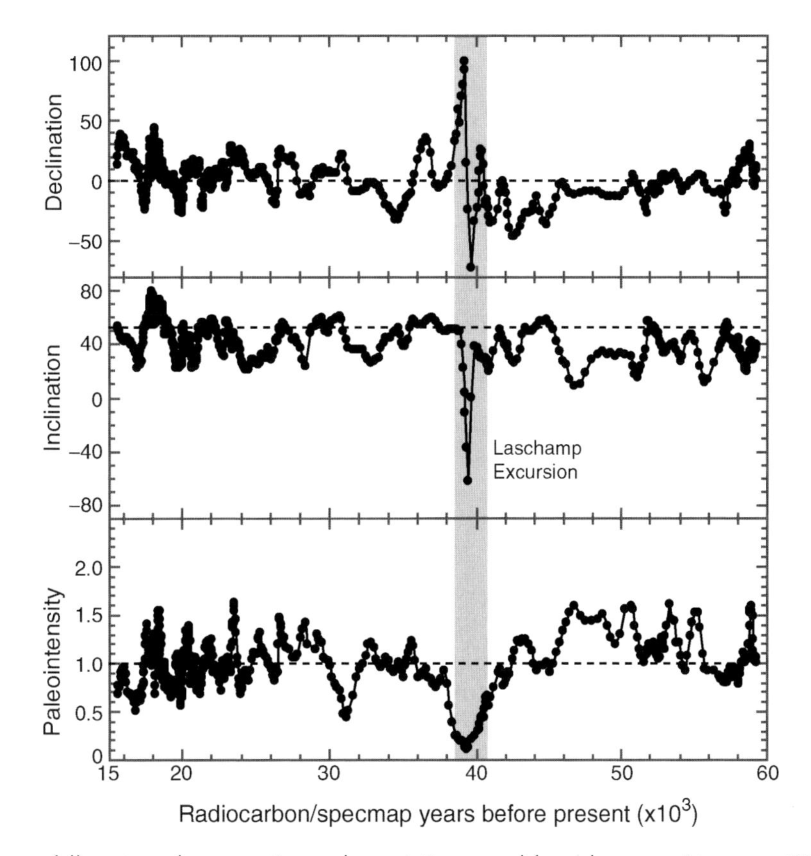

Figure P12 Late Pleistocene full-vector paleomagnetic secular variation record from deep-sea piston core JPC-14, western north Atlantic Ocean (Lund *et al.*, 2001, 2005). Note the presence of the Laschamp excursion near ~40000 years B.P.

can be traced all the way across Eurasia for the last 3000 years. It thus appears that straightforward PSV vector correlations break down beyond perhaps 5000 km. Thompson (1984) noted a similar scale of spatial coherence in the correlation of HSV waveforms. He determined that HSV could be broken down into about nine different regions over the Earth's surface. Within each region, HSV waveforms are broadly correlative, but between regions the patterns are significantly different (and uncorrelated except for an overall vector average approximating the axial dipole expectation). This lack of global-scale correlation appears to be present in directional PSV records as well and indicates that there is no persistent westward (or eastward) drift of distinctive geomagnetic features over thousands of years, as has been suggested most recently by Merrill *et al.* (1998).

Paleointensity, on the other hand, does show clear evidence of global-scale coherence in its variability. Comparisons of paleointensity records from around the world (e.g., Guyodo and Valet, 1999; Laj *et al.*, 2000; Stoner *et al.*, 2003) show clear evidence that most long-term (>1000 years) paleointensity variability is correlatable on a global scale. For example, Figure P13 shows two high-resolution relative paleointensity records recovered from deep-sea sediments on almost opposite sides of the Earth (Stott *et al.*, 2002). The records identify 30 paleointensity features that are synchronous in age (~500 years resolution) based on oxygen isotope age determinations, including a distinctive paleointensity low associated with directional records of the Laschamp Excursion at each site. One future task of PSV studies is to rationalize how significantly different patterns of directional variability can exist around the world in the presence of such globally coherent paleointensity variability.

A third type of waveform comparison can be made between PSV features or vector waveforms within individual paleomagnetic records. Such comparisons have occasionally identified distinctive vector waveforms that seem to recur every 2500 to 4000 years. The PSV record from Mono Lake, shown in Figure P11, illustrates this recurrence pattern. At least four recurrences (α-δ) of basically the same complex waveform can be noted (Lund *et al.*, 1988). With each recurrence, the waveform is slightly altered; however, the general pattern persists for more than 16 ka. It is likely that this distinctive waveform has evolved out of the Mono Lake excursion (waveform ϵ? in Figure P11). Similar recurring patterns have been noted in the Holocene PSV record from Lake St. Croix (Figure P10; Lund and Banerjee, 1985) in a late Pleistocene PSV record from Russia (Ekman *et al.*, 1987), and after the Summer Lake Excursion recorded in lake sediments from Oregon (Negrini *et al.*, 1994).

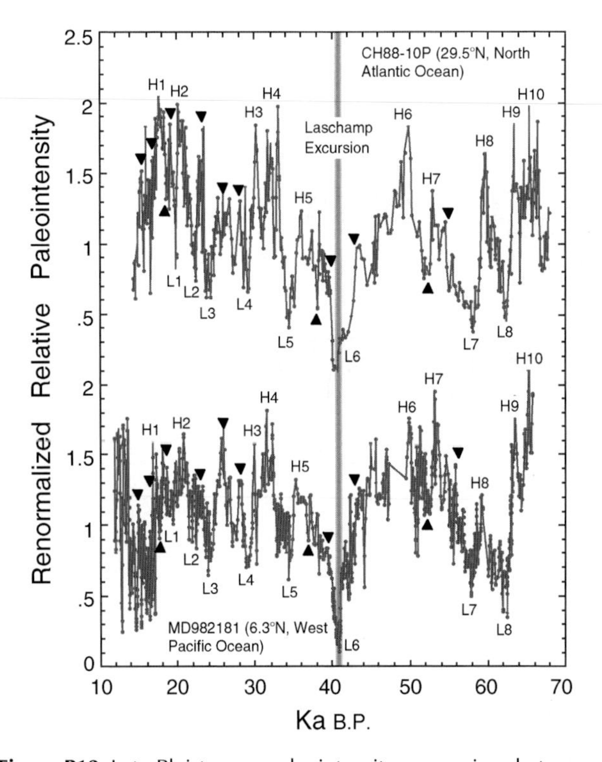

Figure P13 Late-Pleistocene paleointensity comparison between sediment PSV records from the North Atlantic Ocean and western Equatorial Pacific Ocean (Stott *et al.*, 2002). A number of distinctive and correlatable paleointensity features are noted in the two records; independent oxygen isotope stratigraphies suggest that these features are synchronous (~500 year uncertainty).

Recurring vector waveforms can also be identified by their distinctive sequences of vector looping (circularity). Such looping patterns (Runcorn, 1959) are characteristic of both PSV (e.g., Creer, 1983) and HSV (e.g., Bauer, 1895) and have been suggested (Skiles, 1970) to be an indicator of westward or eastward drift of the geomagnetic field (although there is no significant evidence of long-term persistent drift of the field as noted above). However, the correlation between observed looping and drift is not unique (Dodson, 1979). Large amplitude loops, often associated with recurring vector waveforms, last about 1000 to 1400 years; small amplitude loops have also been noted that last about 500–800 years. Looping intervals might be used as an indicator of the simplest temporal scale for PSV coherence at individual sites.

Excursions

One type of vector waveform deserves special note—magnetic field excursions, which are anomalous PSV fluctuations defined by virtual geomagnetic poles (VGPs) located more than 45° away from the geographic pole. It is clear that excursions do occur; it is often not clear, however, what is their waveform morphology or whether some excursions are really artifacts of field-laboratory measurement errors. There is growing evidence (e.g., Lund *et al.*, 2001, 2005) that at least seventeen excursions have occurred in the Brunhes Epoch (0–780000 years B.P.). One distinctive element of all these excursions is that they occur within intervals of anomalously low global-scale paleointensity (see Figures P11-P13). Most excursion records are difficult to correlate around the world because of uncertainties in their age estimates. But, three of the most recent excursions are becoming better understood: the Mono Lake excursion (CA. 28000 years B.P.; Denham and Cox, 1971; Liddicoat and Coe, 1979; Figure P11), The Laschamp excursion (CA. 41000 years B.P.; Bonhommet and Zahringer, 1969; Figures P12 and P13) and the Blake Event (CA. 125000 years B.P.; Smith and Foster, 1969). For each of these excursions, there are now multiple independent records from around the world, which are sufficiently well dated to correlate the records and estimate that individual excursion records are synchronous to ~500-year resolution. There are also a number of high-resolution excursion records (e.g., Liddicoat and Coe, 1979; Lund *et al.*, 2005.rpar; that assess the local waveform patterns of excursions and the surrounding normal PSV. However, there are still not enough high-resolution excursion records to be certain about their global pattern of field variability or relationship to normal PSV or magnetic field reversals.

Spectral analysis

Spectral analysis describes the frequency content of PSV over timescales of 10^2–10^5 years. Traditional ideas suggest that PSV should be a band limited process. That is, that the spectral power of PSV should markedly diminish beyond some cutoff period on the order of 10^4 years. PSV records longer than the cutoff period should then be stationary in a statistical sense and have an average field vector direction that is constant through time for a given site. The axial dipole hypothesis, a cornerstone of plate tectonic reconstructions, assumes that the PSV process is stationary and that each site's average field vector (during intervals of normal polarity) satisfies the formula,

$\tan I = 2\tan\lambda$, $D = 0°$, where λ is the site paleolatitude. One important goal of PSV studies is to test the long-term validity of the axial dipole hypothesis and the notion of a stationary PSV spectrum with limited power at $>10^4$ years.

Spectral analysis also describes the characteristic distribution of continuous spectral power within the overall PSV process (Barton, 1982, 1983) and identifies whether there are distinct frequency bands within the continuous spectrum that have higher than average spectral power. Knowledge of the PSV power spectrum and its potential changes in time and space is critical for better understanding the relationship between PSV and the core dynamo process that generates it.

The primary limitation in recovering detailed estimates of the PSV spectrum is the quality of age estimates associated with the PSV records. Records dated using radiocarbon methods over the last 40 ka or so may have systematic errors with respect to "true" time due to radiocarbon reservoir effects. They may also have random errors on the order of 10^2 years or worse due to random dating errors. Older PSV records can also be dated using oxygen isotope stratigraphies, but such age estimates have random uncertainties on the order of 10^3 years.

Despite these limitations, important and convincing spectral estimates of late-Quaternary PSV have been recovered from several regions around the world: Europe, North America, the Far East, Australia, and South America. Figure P14 shows the stacked PSV spectrum of unit vector (RMS of scalar inclination and declination spectra) and paleointensity results from three deep-sea sediment PSV records in the western North Atlantic Ocean that are >50 ka in duration. These results provide a good overview of the long-term PSV spectrum. The unit-vector and paleointensity spectra are continuous in their power distribution, with the largest amplitude secular variation occurring at periods much greater than 10^3 years (far beyond the range of HSV). Both spectra are red (power-law relationship between frequency and power) for periods <5 ka–12 ka, and white (~constant power as a function of frequency) for longer periods with no indication of a marked decrease in spectral power at the longest observed periods (~50 ka). These characteristics suggest that the geomagnetic field is not stationary (band-limited) over timescales of less than about 10^5 years. The corner frequencies (boundaries between dominantly red and white spectra, indicated by arrows in Figure P14), are about 5 ka for the unit-vector spectrum and about 12000 years for the paleointensity spectrum. One interpretation of the corner frequency is that the red spectrum describes the intrinsic dynamics of field variability associated with the dynamo process, while the white spectrum describes random perturbations of that process or longer-term external forces acting on that process (e.g., Channell *et al.*, 1998). In this scenario, the corner frequency in paleointensity of ~12 ka defines the longest time constant of normal dynamo activity. The fact that the unit-vector corner frequency is only ~5 ka is due to the observation that vector variations only reflect nonaxial dipole field variability, not total field variability. It is probably not coincidental that the longest repeating vector waveforms noted in individual PSV records approach the unit-vector corner frequency of ~5 ka. There is also evidence of selected spectral bands in the red portion of the spectra with larger than average spectral power. That is the pattern noted in most other PSV spectra determined from shorter duration records (e.g., Barton, 1982, 1983). These spectral bands probably define the detailed pattern of dynamo activity in a region at any one time, but none of these components are probably periodic on a long-term spatial or temporal scale and they should tend to smear out when spectra from different time intervals or spatial regions are stacked.

Most of these spectral characteristics have no temporal analog in HSV because of the short time span of historic measurement. One might, however, attempt to relate the long-term PSV spectral characteristics to the present day geomagnetic field spatial spectrum. The spatial field due to core sources has a cutoff near spherical harmonic degree 12 and the spectral power decreases quickly from a maximum at harmonic degree 1 (dipole terms). Only spatial terms of about degree 6 or less have vector amplitudes large enough to be recorded in PSV., One might therefore hypothesize that the long-term PSV spectrum must be

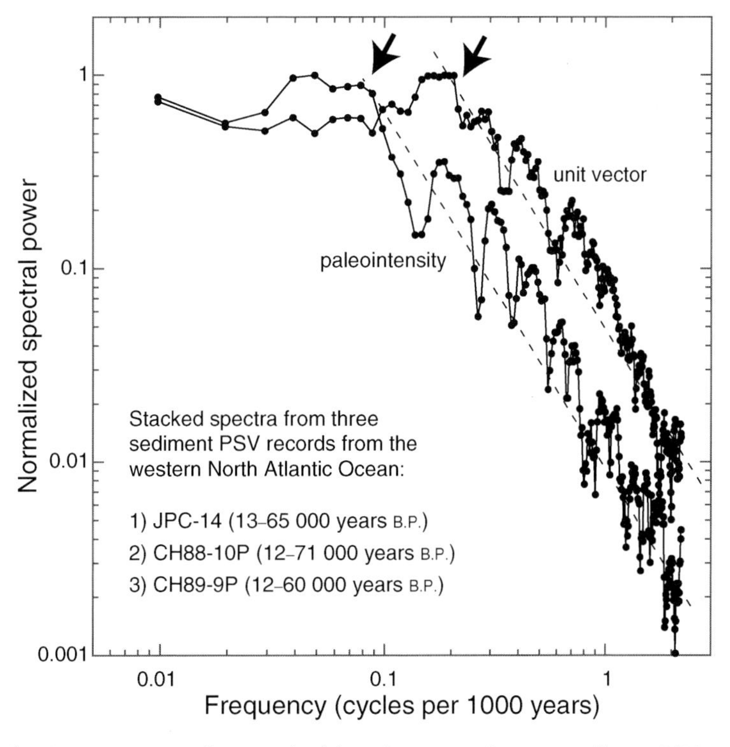

Figure P14 Paleointensity and unit-vector spectra for a stack of three long-term deep-sea sediment PSV records from the western North Atlantic Ocean. One of the records, JPC-14, is shown in Figure P12. Arrows indicate corner frequencies, which separate "white" and "red" portions of the spectra.

due to temporal variations of spatial components of the present-day field with spherical harmonic degree 5–6, under normal conditions.

Statistical analysis

The aspect of PSV that is easiest to analyze is its statistical behavior averaged over some interval of time. For this reason, statistical properties of PSV, within time windows in the order of 10^5 or 10^6 years, were the first PSV characteristics to be compared spatially, and even today they are the only PSV characteristics that can be easily compared on a global scale. Such comparisons provide the strongest evidence relating to possible long-term stationarity of geomagnetic field behavior, the axial dipole hypothesis, and the potential global pattern of selected PSV characteristics. Statistical study of PSV follows two very different paths on the basis of sampling frequency and age control of the paleomagnetic measurements. In the first approach, paleomagnetic field directions in undated rock sequences are measured under the assumption that the age difference of successive rock units is large compared to the longest period of PSV (often assumed to be about 10000 years). Each data point is therefore assumed to be an independent random value picked from an assumed frequency-band-limited PSV process. Data sets from small regions, averaged over 10^5 or 10^6 years, are then statistically analyzed and compared with some global model of the expected statistical behavior. (Even if the underlying assumptions of spectral content are wrong, as they likely are, results from this type of statistical analysis should not be seriously biased if the data are truly randomly spaced in time.) The second approach is to measure well-dated (either by radiocarbon or oxygen isotope methods) paleomagnetic sequences where the sampling interval is less than the shortest period of PSV (about 30 years). It is not often feasible to find sequences with such short time spacing, but useful information can be obtained with sample intervals up to perhaps 250 years (and even longer under special circumstances). Statistics are estimated from equispaced time-series derived from the dated paleomagnetic records. This method permits spatial comparison of statistical parameters averaged over much shorter time intervals, on the order of 10^3–10^5 years. In such studies, it is likely that the statistics does not represent a "stationary" estimate of space-time field variability but rather a "statistical snapshot" of an evolving space-time field structure.

Statistical analyses of the probability distributions of both field vectors and their equivalent VGPs from single sites indicate that neither distribution is typically Fisherian (Fisher, 1953; spherical analog of the Gaussian or normal distribution). For example, Brock's (1971) results from equatorial Africa (Figure P15) show that both field vectors and VGPs tend to have somewhat elliptical distributions. Engebretson and Beck (1978) have summarized the statistical parameters normally used to characterize the shape of the probability distribution. It is probable that shape statistics vary systematically as a function of latitude (and longitude?) and future studies of shape statistics may provide important added characteristics of the Earth's long-term PSV. As a starting point, it is worth noting that the range of vector variability (as a function of latitude) noted in historic maps of the geomagnetic field is comparable to the range of vector variability noted in individual PSV records from the late Quaternary. This suggests that, even though secular variation changes on timescales far beyond the range of historic measurements, the normal range of its spatial variability is completely present in the observed historic field. A similar observation has been made for the long-term pattern of PSV angular dispersion by McFadden *et al.* (1988).

Currently, the two statistical parameters most often measured in PSV studies are the ΔI anomaly, which is the site mean inclination minus the expected axial dipole field inclination and the angular dispersion associated with a site's vector (or equivalent VGP) variability. The global pattern of the ΔI anomaly estimates how well the axial dipole hypothesis, the cornerstone of plate tectonic reconstructions, fits the actual geomagnetic field behavior. The global pattern of angular dispersion should provide some measure of the spatial pattern of intrinsic "energy" or "dynamics" in the core dynamo process. This variability may be due to differing proportions of dipole versus nondipole field variability (see summary in Merrill *et al.*, 1998) or it could be interpreted as the relative importance of primary versus secondary family spherical harmonic components (McFadden *et al.*, 1988), two orthogonal components of the geomagnetic field that may have fundamental relationships to dynamo theory (Roberts and Stix, 1972).

The ΔI anomaly was perhaps first quantified by Wilson in 1970 who noticed that the average paleomagnetic pole positions associated with individual geographic regions (e.g., Australia, Europe, North America) were always farther from the sampled region than the known geographic pole. This offset, termed the farsided effect, is due to paleomagnetic inclinations that are systematically more negative than their axial dipole expectation (negative ΔI anomaly), on average. McElhinny *et al.* (1996) have determined the global ΔI anomaly for the last 5 Ma (see Figure P16), on average, and noted that the ΔI anomaly is mostly negative with a maximum anomaly near the Equator and a latitudinal variation that is symmetric about the Equator and

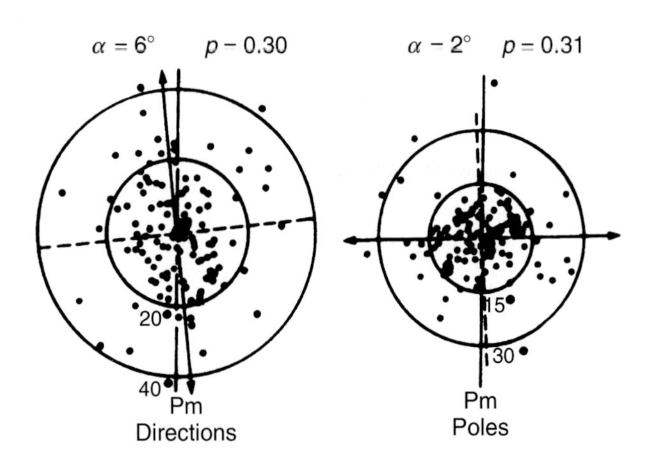

Figure P15 Statistical distribution of PSV directions and virtual geomagnetic poles (VGPs) from Equatorial Africa (Brock, 1971).

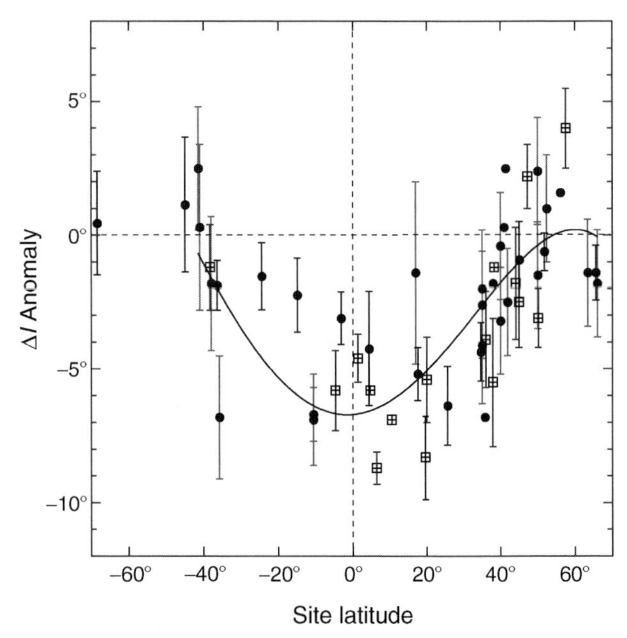

Figure P16 ΔI anomaly as a function of latitude for selected time intervals: solid symbols—last 5 Ma average, open circles last few thousand years average, open squares, last ~10–20 ka average.

zonal (any site along a line of latitude will have the same magnitude of ΔI anomaly). Their analysis also indicated that ΔI anomaly has persisted with the same general pattern and magnitude for the last 5 Ma during both normal and reversed polarity. Other workers (e.g., Coupland and Van der Voo, 1980; Livermore *et al.*, 1984; Schneider and Kent, 1990) have noted that a similar pattern of ΔI anomaly has persisted for at least the last 100 million years, but with slow variations in the maximum anomaly magnitude. Statistical studies of shorter-term equispaced PSV time-series (e.g., Lund, 1985; Lund *et al.*, in review) indicate that the ΔI anomaly averaged over 10^3 years (see open circles in Figure P16) and 10^4 years (see open squares in Figure P16) also show the same general ΔI anomaly pattern noted in the longer-term data. However, Constable and Johnson (2000) have argued that nonzonal components of the ΔI anomaly are significant and have persisted for the last several million years, with some Equatorial regions having more significant ΔI anomalies than other regions and some high-latitude regions actually having positive ΔI anomalies. These observations all suggest that the spatial pattern of the ΔI anomaly is persistent but nonstationary in its detailed pattern over timescales greater than $\sim 10^4$ years. The implication of the ΔI anomaly for plate tectonic studies is that paleolatitude estimates derived from paleomagnetic studies may be in error by as much as 8° depending on time and paleolatitude. Similar analyses of declination have shown no significant deviations in declination values from 0° over the last several million years.

Analysis of PSV angular dispersions (either directions or their equivalent virtual geomagnetic poles, VGPs) has established that this parameter also displays a distinctive zonal pattern of amplitude variation with latitude. The average latitudinal pattern of VGP angular dispersion for the last 5 Ma (McFadden *et al.*, 1988) is shown in Figure P17. Lund (1985) and Lund *et al.* (in review) have noted, however, that VGP angular dispersion averaged over 10^3 and 10^4 year intervals (open circles and open squares in Figure P17) using equispaced PSV time-series is significantly lower than the 5 Ma average, and Merrill and McFadden (1988) have shown that it can vary on much longer (10^7 year) timescales as well. A variety of parametric models, summarized in Merrill *et al.* (1998), have been developed to attempt to explain the observed spatial pattern of angular dispersion in an *ad hoc* manner. More recently, McFadden *et al.* (1988) were able to relate the long-term average pattern of VGP angular dispersion to historical field observations and developed model G, based on the relative importance of primary versus secondary family magnetic field components (Roberts and Stix, 1972), to explain the latitude dependence.

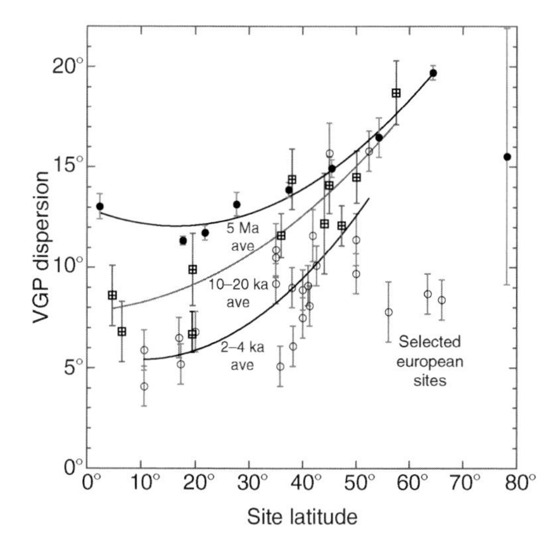

Figure P17 VGP angular dispersion as a function of latitude for selected time intervals: solid symbols—last 5 Ma average, open circles last few thousand years average, open squares, last ~10–20 ka average.

These statistical observations indicate clearly that PSV varies in a nonstationary manner over a variety of timescales, a pattern consistent with the PSV spectrum described above, and may be related to the continuing time evolution of regional PSV waveforms. One discordant note in this discussion of PSV nonstationarity is the qualitative sense that, even though PSV statistics may vary quantitatively in a nonpredictable manner over time, the PSV vector mean for any site is heavily "attracted" to its axial dipole expected direction. This may be due to the symmetry associated with the Earth's rotation, one of the driving forces of the dynamo. But, whatever be the cause, it seems prudent, perhaps, to think of PSV as "quasi-stationary" in its behavior based on the overall sense of "attraction" to such a "fixed point" (the axial dipole expectation) in time and space.

Models of PSV

An alternative method for the analysis of PSV is to develop models for the observed field variability. Such models may be conceptual in nature with their primary purpose being to qualitatively estimate the style of variability that a potential dynamo source might generate, or the models may be more quantitative (essentially mathematical simulation models), with their primary purpose being to replicate observed PSV. The mathematical simulation models may be developed from more conceptual models, with sources that may have some basis in reality, or purely from empirical (unrealistic) inputs. A third group of models that characterize the actual hydromagnetic process that generates the Earth's core field is beyond the scope of this survey.

Conceptual models

Previously, various conceptual models have been proposed to qualitatively explain characteristic features of HSV. Three models that have been discussed most often are (1) dipole wobble, (2) westward (or eastward) drift of the total nondipole field, and (3) standing and drifting nondipole fields. These conceptual models have also been called upon to qualitatively explain specific components of PSV or to justify and physically explain specific mathematical simulation models of PSV.

Dipole wobble has been suggested as one component of HSV and PSV based on the presence of an 11.5° offset in the present day dipole field, its persistence during historic time, and the indication from paleomagnetic data that the field has an average declination of 0° over timescales of 10^4 years or longer. Therefore, the average dipole field direction must have moved prehistorically and perhaps it has "wobbled" irregularly.

The paleomagnetic evidence for dipole wobble is problematic because of non uniqueness. PSV at a single site is really the nonaxial dipole variation (dipole wobble plus true nondipole variation) at that site. The proportion of dipole wobbles versus true nondipole contributions to PSV can only be assessed by analyzing globally distributed paleomagnetic data. Merrill and McElhinny (1983) carried out such an analysis of Northern Hemisphere archeomagnetic data for the last 2000 years and suggested that a significant dipole wobble component does exist. Dipole wobble contributions have also been estimated from the statistical analysis of angular dispersion (e.g., McFadden *et al.*, 1988); however, the proportion of dipole wobble depends upon the proposed model of dipole wobble variability.

Westward drift has been suggested as an important element of HSV based on the observation that temporal changes in nondipole field components at the Earth's surface are due primarily to the westward drift in time of the spatial nondipole field components. The cause of westward drift has been related to differential rotation of the fluid outer core, where the field is generated, versus the overlying lithosphere. The importance of westward drift is complicated by the fact that some areas of the Earth have exhibited eastward drift and other areas have exhibited no drift at all during historic times (e.g., Thompson, 1984). Paleomagnetic evidence for westward drift comes from a variety of

PSV observations, of which a few are unique. The circularity of PSV data has long been associated with westward or eastward drift of the paleomagnetic field, although other nonaxial dipole variations could produce the same effect. The recurring waveforms, noted earlier, may indicate westward (or eastward) drift of a complex nondipole waveform that changes very slowly in time compared to the time it takes for the waveform to drift entirely around the Earth (2500–4000 years). In such a model, similar waveform and spectral characteristics should be noted at all sites along a line of latitude. The Holocene waveform comparisons noted earlier, however, do not appear to be compatible with this model. Regional PSV comparisons within Europe, East Asia, or North America separately display waveform correlations that are consistent with a westward drift model; however, no simple correlation can apparently be made between the three regional data sets. Such a correlation between regional PSV records is necessary if persistent westward (or eastward) drift is a predominant aspect of PSV.

Standing nondipole field components have been proposed to improve the fit of drifting nondipole field components to the total HSV. If truly present, their origin might be related to standing components of fluid flow near the core mantle boundary caused by irregularities in the boundary conditions. The presence of standing nondipole components in the paleomagnetic record, however, is very difficult to evaluate because of problems of nonuniqueness and the uncertainties of spatial PSV behavior. To the extent that standing nondipole components might produce nonzonal components of I and D, their importance must be below the level of noise associated with parametric statistical analyses of long-term PSV (average I, D; AI; angular dispersion), for all of these parameters are apparently zonal in their spatial distribution. However, in the study of late Quaternary PSV waveforms, standing nondipole sources have been suggested as reasonable (but nonunique) alternatives to westward drift to explain the observed waveform variability within individual paleomagnetic records.

Mathematical simulation models

Models that are more quantitative have also been applied to HSV and PSV. Spherical harmonic models separate the field into dipole and nondipole components and may include secular change coefficients for predicting short-term temporal variations. The primary drawback to spherical harmonic models is their general lack of relevance to the underlying physical causes of the Earth's internal magnetic field. (The exception to this may be the separation of spherical harmonic coefficients into primary and secondary family field components.) Models based on the variability of multiple localized sources in the outer core have occasionally been used as alternatives to spherical harmonic models. These models, which may use a distribution of dipoles, current loops, or wave patterns in their formulation, are more appealing in that those sources may mimic that part of the core process associated with observed nondipole foci observed at the Earth's surface (e.g., Thompson, 1984).

Spherical harmonic models are hard to apply to PSV because of the inherent timing uncertainties associated with PSV data and because of the poor spatial distribution of most PSV data. Even so, several recent summaries of PSV for the last few thousand years (e.g., Hongre *et al.*, 1998; Constable *et al.*, 2000; Korte and Constable, in press), based on worldwide (but poorly distributed) sites, have begun to give us a low-degree spherical harmonic view of PSV. Figure P18, for example, shows the geomagnetic field radial flux (Br) and non-axial dipole radial flux (Br-anomaly) at the core-mantle boundary for two prehistoric epochs based on SHA of PSV time-series (Constable *et al.*, 2000). The advantage of PSV derived SHA data sets is that they can be tied to HSV derived SHA data sets and used to extend our view of the true global-space-time pattern of secular variation back thousands of years beyond the range of HSV. In this way, we can finally begin to address the global space-time pattern of secular variation on scales appropriate to the dynamics of PSV and the geodynamo.

Historically, time-averaged PSV statistical parameters, such as the ΔI anomaly and vector dispersion, have been more amenable to time-averaged spherical harmonic analysis. For example, the ΔI anomaly can be modeled by an axial dipole with added quadrupolar and octupolar components; the long-term changes in ΔI can then be modeled as changes in the quadrupole-octupole (or primary-secondary family) amplitude ratio. (See Merrill and McElhinny (1983) for more detailed discussion.)

Localized dipole-current-loop models, with either standing or drifting sources, have been applied to individual high-resolution PSV records, as well as to statistical PSV records. An example of a drifting radial dipole model for the Lake St. Croix PSV record (Figure P10; Lund and Banerjee, 1985) is shown in Figure P19. Two drifting radial dipoles plus an axial dipole are able to model almost all of the

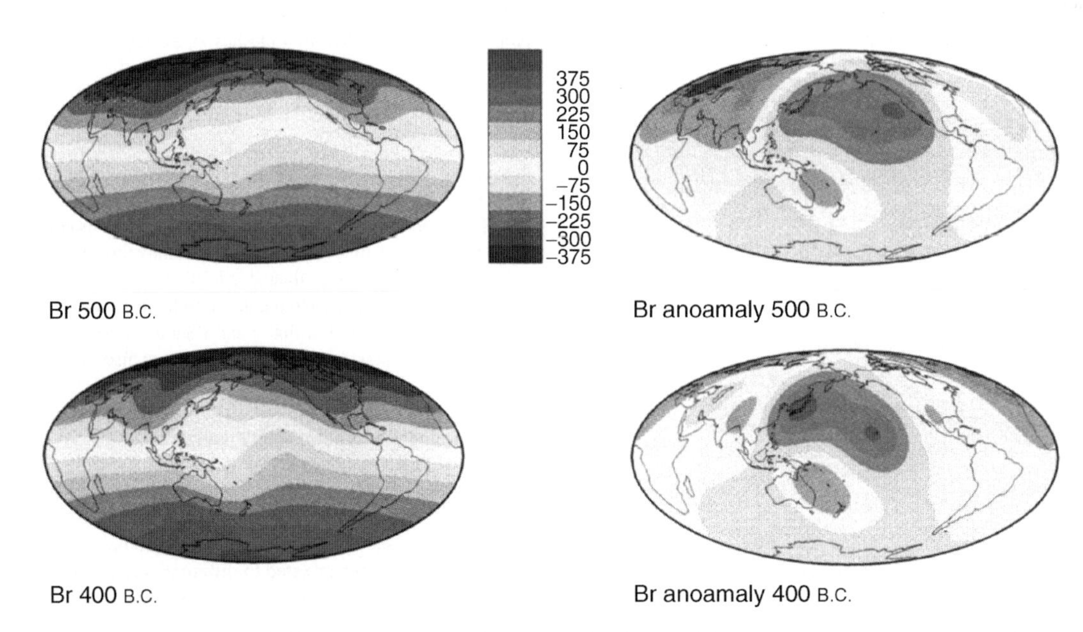

Figure P18 Models of geomagnetic field radial flux (Br) and nonaxial dipole radial flux (Br-nad) at the core-mantle boundary for two prehistoric epochs, 500 B.C. and 400 B.C. The models are the result of downward continuation of spherical harmonic models of PSV data from Constable *et al.* (2000).

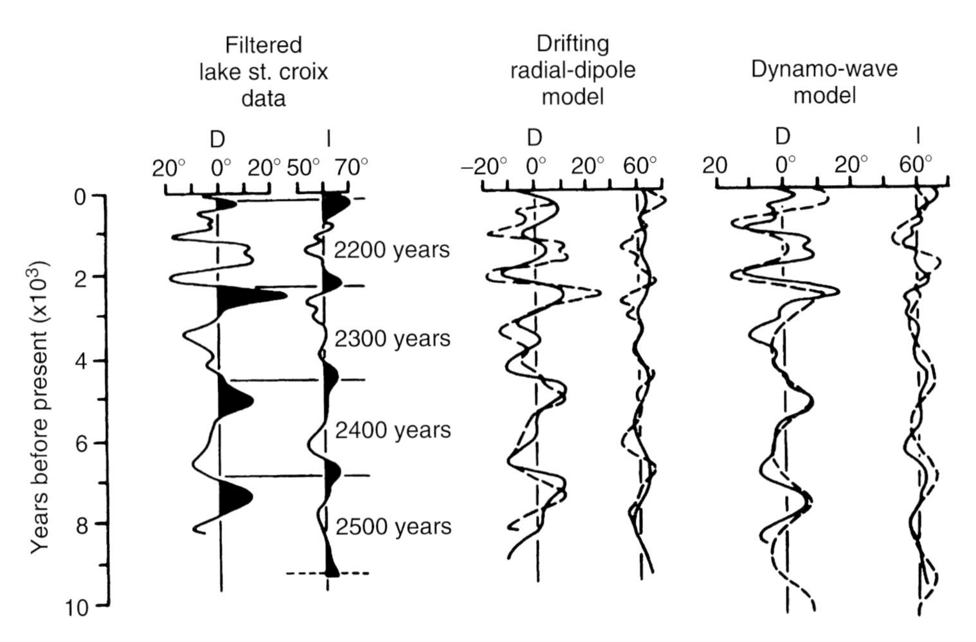

Figure P19 Radial dipole and dynamo wave models of the Lake St. Croix PSV record.

observed variability at Lake St. Croix for the last 9 ka. Unfortunately, more complicated standing radial dipole models could also fit the data. But, these models would require more sources in order to fit the characteristic phase relationships of the Lake St. Croix PSV data. The drifting radial dipole model for Lake St. Croix predicts similar PSV behavior for other sites on Lake St. Croix's latitude; the standing radial dipoles model will only produce regional coherence. The difficulty in correlating the Lake St. Croix PSV record with other global records outside of North America argues that a drifting radial dipole model is probably not appropriate to explain recurring waveforms. However, the complex recurrent waveforms noted in several PSV records are very difficult to model with standing sources due to the number of sources required and the detailed timing of recurrent intensity variations that each source must maintain relative to the other sources.

An alternative model of PSV, based on poleward migrating dynamo waves in the Earth's outer core, has been developed by Olson and Hagee (1987). Figure P19 shows the results of their model applied to the observed Lake St. Croix PSV. It is apparent that the poleward-migrating dynamo-wave model does just as good a job of fitting the observed variability as the drifting radial-dipole model. The dynamo wave model, however, only requires regional coherence in waveform correlations, but not a global-scale correlation. Hagee and Olson (1989) have also applied this dynamo wave model to a global set of Holocene PSV records and shown that the same general pattern of dynamo-wave variability noted to Lake St. Croix, can also explain all other Holocene PSV around the world.

Steve P. Lund

Bibliography

Barton, C.E., 1983. Analysis of paleomagnetic time series-techniques and applications. *Geophysical Surveys*, **5**: 335–368.

Bauer, L.A., 1895. On the distribution and the secular variation of terrestrial magnetism, No. III. *American Journal of Sciences*, **50**: 314.

Bloxham, J., and Gubbins, D., 1985. The secular variation of Earth's magnetic field. *Nature*, **317**: 777–781.

Bloxham, J., and Roberts, P., 1991. The geomagnetic main field and the geodynamo. *Reviews of Geophysics*, **29**: 428–432.

Bonhommct, N., and Zahringer, J., 1969. Paleomagnetism and potassium-argon age determinations of the Laschamp geomagnetic polarity event. *Earth and Planetary Science Letters*, **6**: 43s.

Brock, A., 1971. An experimental study of paleosecular variation. *Geophysical Journal of Royal Astronomical Society*, **24**: 303.

Butler, R., 1992. *Paleomagnetism: Magnetic Domains to Geologic Terranes*, London: Blackwell Publishers, 321 pp.

Cande, S.C., and Kent, D.V., 1995. Revised calibration of the geomagnetic polarity timescale or the late Cretaceous and Cenozoic. *Journal of Geophysical Research*, **100**: 6093–6095.

Champion, D.E., 1980. Holocene geomagnetic secular variation in the western United States: implications for the global geomagnetic field (Open file report 80–824), U.S. Geological Survey, 314 pp.

Channell, J., Hodell, D., McManus, J., and Lehman, B., 1998. Orbital modulation of geomagnetic paleointensity. *Nature*, **394**: 464–468.

Clement, B.M., and Kent, D.V., 1987. Short polarity intervals within the Matuyama: transitional field records from hydraulic piston cored sediments from the North Atlantic. *Earth and Planetary Science Letters*, **81**: 253–264.

Clement, B., and Constable, C., 1991. Polarity transitions, excursions, and paleosecular variation of the Earth's magnetic field. *Reviews of Geophysics*, **29**: 433–442.

Constable, C., Johnson, C., and Lund, S., 2000. Global geomagnetic field models for the last 3000 years. *Philosophical Transactions of the Royal Society of London* A, **358**: 991–1008.

Creer, K.M., 1983. Computer synthesis of geomagnetic paleosecular variations. *Nature*, **304**: 695–699.

Denham, C.E., and Cox, A., 1971. Evidence that the LAschamp polarity event did not occur 13300–30400 years ago. *Earth and Planetary Science Letters*, **13**: 181–190.

Dodson, R.E., 1979. Counterclockwise precession of the geomagnetic field vector and westward drift of the nondipole field. *Journal of Geophysical Research*, **84**: 637.

Dunlop, D., and Ozdemir, O., 2001. *Rock Magnetism*, Cambridge: Cambridge University Press, 596 pp.

Ekman, I., Bakhmutov, V., and Zagniy, G., 1987. Stratification and correlation of varved clays in terms of the fine paleostructures of the Earth's magnetic field, In Kabailiene, M. (ed.), *Methods for the Investigation of Lake Deposits: Paleoecological and Paleoclimatological Aspects,* Academy of the Sciences of the Lithuanian SSR, Vilnius.

Engebretson, D.C., and Beck, M.E., 1978. On the shape of directional data sets. *Journal of Geophysical Research*, **83**: 5979–5982.
Fisher, R.A., 1953. Dispersion on a sphere. *Proceedings of the Royal Society of London*, **A**, **217**: 295.
Gilbert, W., 1600. *De Magnete*, reprinted New York: Dover Publications, 368 pp. 1958.
Guyodo, Y., and Valet, J., 1999. Global changes in intensity of the Earth's magnetic field during the last 800 kyr. *Nature*, **399**: 249.
Holcomb, R., Champion, D., and McWilliams, M., 1986. *Geological Society of America Bulletin*, **97**: 829–839.
Hongre, L., Hulot, G., and Khokhlov, A., 1998. An analysis of the geomagnetic field over the last 3000 years. *Physics of the Earth and Planetary Interiors*, **106**: 311–335.
King, J.K., 1983. Geomagnetic secular variation curves for northeastern North America for the last 9000 years, unpublished PhD dissertation. Minneapolis: University of Minnesota.
Kissel, C., and Laj, C., 2004. Improvements in procedure and paleointensity selection criteria for Thellier-Thellier determinations: application to the Hawaii basaltic long core. *Physics of the Earth and Planetary Interiors*, **45**: 155–169.
Korte, M., and Constable, C., 2003. Continuous global geomagnetic field models for the past 3000 years, *Physics of the Earth and Planetary Interiors;* **140**: 73–89.
Laj, C., Kissel, C., Mazoud, A., Channell, J., and Beer, J., 2000. North Atlantic paleointensity stacksince 75 ka (NAPIS-75) and the duration of the Laschamp event. *Philosophical Transactions of the Royal Society of London, Series* **A**, **358**: 1009.
Liddicoat, J., and Coe, R., 1979. Mono Lake geomagnetic excursion. *Journal of Geophysical Research*, **84**: 261–271.
Lund, S.P., 1985. A comparison of the statistical secular variation recorded in some late Quaternary lava flows and sediments, and its implications. *Geophysical Research Letters*, **12**: 251–254.
Lund, S.P., 1996. A comparison of Holocene paleomagnetic secular variation records from North America. *Journal of Geophysical Research*, **101**(B4): 8007–8024.
Lund, S.P., and Banerjee, S.K., 1985. Late Quaternary paleomagnetic field secular variation from two Minnesota lakes. *Journal of Geophysical Research*, **90**: 803–825.
Lund, S.P., Liddicoat, J.C., Lajoie, K.L., 1988. Henyey and Steve W. Robinson. Paleomagnetic evidence for long-term (10^4 year) memory and periodic behavior in the Earth's core dynamo process. *Geophysical Research Letters*, **15**: 1101–1104.
Lund, S.P., Acton, G. Clement, B. Okada, M. Williams, T., 1998. Initial paleomagnetic results from ODP Leg 172: high resolution geomagnetic field behavior for the last 1.2 Ma. *EOS, Transactions, American Geophysical Union*, **79**: 178–179.
Lund, S., and Schwartz, M., 1999. Environmental factors affecting geomagnetic field intensity estimates from sediments. In Maher, B., and Thompson, R. (eds.), *Quaternary Climates, Environments, and Magnetism*. Cambridge: Cambridge University Press, pp. 323–351.
Lund, S.P., Williams, T., Acton, G., Clement, B., and Okada, M., 2001. Brunhes Epoch magnetic field excursions recorded in ODP Leg 172 sediments. In Keigwin, L., Rio, D., and Acton, G. (eds.), Proceedings of the Ocean Drilling Project, Scientific Results Volume **172**, Ch. 10, 2001.
Lund, S.P., Schwartz, M., Keigwin, L., and Johnson, T., 2005. High-resolution records of the Laschamp geomagnetic field excursion. *Journal of Geophysical Research*, **110**: B04101.
McFadden, P.L., Merrill, R.T., and McElhinny, M.W., 1988. Dipole/quadrapole family modeling of paleosecular variation. *Journal of Geophysical Research*, **93**: 11583–11588.
Merrill, R.T., and McElhinny, M.W., 1977. Anomalies in the time-averaged paleomagnetic field and their implications for the lower mantle. *Reviews of Geophysics and Space Physics*, **15**: 309.
Merrill, R., McElhinny, M., and McFadden, P., 1998. *The Magnetic Field of the Earth*. San Diego, CA: Academic Press, 531 pp.
Merrill, R.T., and McFadden, P.L., 1988. Secular variation and the origin of geomagnetic field reversals. *Journal of Geophysical Research*, **93**: 11589–11598.
Negrini, R., Erbes, D., Roberts, A., Verosub, K., Sarna-Wojcicki, A., and Meyer, C., 1994. Repeating waveform initiated by a 180–190 ka geomagnetic excursion in western North America: implications for field behavior during polarity transitions and subsequent secular variation. *Journal of Geophysical Research*, **99**: 24105–24119.
Olson, P., Sumita, I., and Aurnou, J., 2002. Diffusive magnetic images of upwelling patterns in the core. *Journal of Geophysical Research*, **107**: 2348.
Olson, P., and Hagee, V.L., 1987. Dynamo waves and paleomagnetic secular variation. *Geophysical Journal of the Royal Astronomical Society*, **88**: 139–159.
Runcorn, K., 1959. On the theory of geomagnetic secular variation. *Annales Geophysicae*, **15**: 87.
Schneider, D.A., and Kent, D.V., 1990. The time-averaged paleomagnetic field. *Reviews of Geophysics*, **28**: 71–96.
Skiles, D., 1970. A method of inferring the direction of drift og the geomagnetic field from paleomagnetic data. *Journal of Geomagnetism and Geoelectricity*, **22**: 441.
Smith, J.D., and Foster, J., 1969. Geomagnetic reversal in Brunhes normal polarity epoch. *Science*, **163**: 565–567.
Stoner, J., Channell, J., Hodell, D., and Charles, C., 2003. A 580 kyr paleomagnetic record from the sub-Antarctic South Atlantic (ODPSite 1089). *Journal of Geophysical Research*, **108**: 2244.
Tauxe, L., 1993. Sedimentary records of relative paleointensity of the geomagnetic field: theory and practice. *Reviews of Geophysics*, **31**: 319–354.
Thompson, R., 1984. Geomagnetic evolution: 400 years of change on planet. *Physics of the Earth and Planetary Interiors*, **36**: 61–77.
Thompson, R., and Barraclough, D.R., 1982. Geomagnetic secular variation based on Spherical Harmonic and Cross Validation analyses of historical and archaeomagnetic data. *Journal of Geomagnetism and Geoelectricity*, **34**: 245–263.
Verosub, K.L., and Banerjee, S.K., 1977. Geomagnetic excursions and their paleomagnetic record. *Journal of Geophysical Research*, **15**: 145–155.
Watkins, N., 1976. Polarity group sets up guidelines, *Geotimes*, **21**: 18–20.

PALEOMAGNETISM

Introduction

Most people, certainly mariners and explorers since at least the 15th century, are aware of the value of a compass as a navigational aid. This works because the Earth generates a magnetic field, which, at the Earth's surface, is approximately that of a geocentric axial dipole (GAD). By geocentric we mean that this dipolar field is centered at the center of the Earth and by axial we mean that the axis of the dipolar field aligns with the spin axis of the Earth. This means that a magnetic compass will align approximately in the north-south direction. It also means that a magnetic dip circle will give the inclination of the magnetic field (the angle the direction the magnetic field makes with the horizontal) which, together with a knowledge of the structure of a dipole field, gives the approximate latitude. If the field were precisely that of a GAD then the north-south direction and the latitude could be obtained accurately. The deviation of magnetic north from geographical or true north is called the magnetic declination and was known to the Chinese from about A.D. 720. In 1600, William Gilbert published the results of his experimental studies in the treatise *De Magnete* and confirmed that the geomagnetic field is primarily dipolar.

The properties of lodestone—now known to be magnetite—were known in ancient times to the Chinese, who invented the earliest known form of magnetic compass as early as the 2nd century B.C. This compass consisted of a lodestone spoon rotating on a smooth board. Magnetic inclination (or dip) was discovered by Georg Hartmann in 1544. Hence it has been known for some time that natural rock can behave as a compass needle and align itself along what is now recognized as being the direction of the magnetic field.

At the end of the 18th century, it was recognized that deviation of magnetic compasses could occur because of nearby strongly magnetized rocks. Delesse and Melloni were the first to observe that the magnetization in certain rocks was actually parallel to the Earth's magnetic field. Folgerhaiter extended their work and studied the magnetization of bricks and pottery. He argued that when a brick or pot was fired in the kiln then the remanent magnetization it acquired on cooling provided a record of the direction of the Earth's magnetic field. With the wisdom of hindsight, specifically a better understanding of the physics of magnetization and the mineralogy of rocks, it is fairly obvious that this would be the case. Volcanic rocks are heated well above the Curie point so the magnetization is free to align with the external magnetic field and becomes locked in as the rock cools. This is known as a *thermoremanent magnetization* (*TRM*) (*q.v.*). In sedimentary rocks the magnetic particles will act just like a compass needle and align themselves with the external magnetic field as they settle and then become mechanically locked in as the rock is formed. This is known as a *depositional remanent magnetization* (*DRM*) (*q.v.*).

This is the essence of paleomagnetism that the rock will lock in a fossil record of the ancient (or paleo) magnetic field. The fossil magnetism naturally present in a rock is termed the *natural remanent magnetization* (*NRM*) (*q.v.*). The primary magnetization is the component of the NRM that was acquired when the rock was formed, and may represent all, part, or none of the total NRM. After formation the primary magnetization may decay either partly or completely and additional components, referred to as secondary magnetizations, may be added by several processes. A major task in all paleomagnetic investigations is to identify and separate the magnetic components in the NRM, using a range of demagnetization and analysis procedures. Typically there is also significant effort put in to date the magnetizations so that not only the direction of the magnetic field at that particular sampling site is known but also the time when the field was in that direction. Many of the successful applications of paleomagnetism have derived from an effective partnership with geochronology.

David in 1904 and Brunhes in 1906 reported the first discovery of NRM that was roughly opposite in direction to that of the present field and this led to the speculation that the Earth's magnetic field had reversed its polarity in the past. In 1926 Mercanton pointed out that if the Earth's field had in fact reversed itself in the past, then reversely magnetized rocks should be found in all parts of the world. He demonstrated that this was indeed the case for Quaternary-aged rocks around the world. The speculation gained further support when Matuyama in 1929 observed reversely magnetized lava flows from the past one or two million years in Japan and Manchuria. However, doubts about the validity of the field reversal hypothesis surfaced during the 1950s after Néel presented theory that showed it was possible for samples to acquire a magnetization antiparallel to the external field during cooling, a process referred to as self-reversal. Shortly thereafter Nagata and Uyeda found the first laboratory-reproducible self-reversing rock, the Haruna dacite. Subsequently it was recognized that self-reversal is relatively rare and by the early 1960s it was accepted that the Earth's magnetic field has indeed reversed and that the phenomenon of field reversal has occurred many times. An excellent history of this subject has been given by Glen (1982).

By 1960 the study of the magnetic properties of minerals and rocks and the use of magnetization in rocks to infer the properties of the Earth's past magnetic field had evolved into two separate but related disciplines, respectively referred to as rock magnetism and paleomagnetism. By providing information about the location and orientation of continents relative to the Earth's magnetic pole, paleomagnetism has played a significant role in our understanding of Earth processes, particularly with regard to continental drift and polar wander and the development of plate tectonics.

As currently practiced, paleomagnetism includes topics such as age determination, stratigraphy, magnetic anomaly interpretation and paleoclimatology, as well as the traditional paleomagnetic topics of tectonics, polar wander, and studies of the evolution and history of the Earth's magnetic field.

Determining a paleomagnetic pole

A central assumption for much of paleomagnetism is that if the field directions at any given locality are averaged over an appropriate time interval then the resulting direction is the same as that for a geocentric axial dipole. If that is the case then the inclination I is related to the paleolatitude λ by

$$\tan I = 2 \tan \lambda \qquad \text{(Eq. 1)}$$

Hence it is possible to determine the angular distance (paleocolatitude $p = 90 - \lambda$) of the sampling site from the pole at the time that the magnetization was acquired. Using the declination D and the paleocolatitude of this time-average field direction it is then possible to determine the latitude and longitude (λ_p, ϕ_p) of the time-average magnetic pole, known as the *paleomagnetic pole*, relative to the latitude and longitude (λ_s, ϕ_s) of the sampling site. The spherical geometry needed for this can be visualized with the help of Figure P20. If the continent has moved and/or rotated since the magnetization was acquired, then the paleomagnetic pole will not coincide with the current north pole, leading to the concept of apparent polar wander (see polar wander). Paleomagnetic poles for magnetizations with different ages, but from the same continent, will plot in different positions, providing an apparent polar wander path (APWP) for the continent.

Alternatively, it is a simple matter to place the continent on the globe such that the paleomagnetic pole coincides with the north pole. This then places the continent correctly on the globe with regard to its latitude and orientation at the time the magnetization was acquired. It does not, unfortunately, give any longitudinal information.

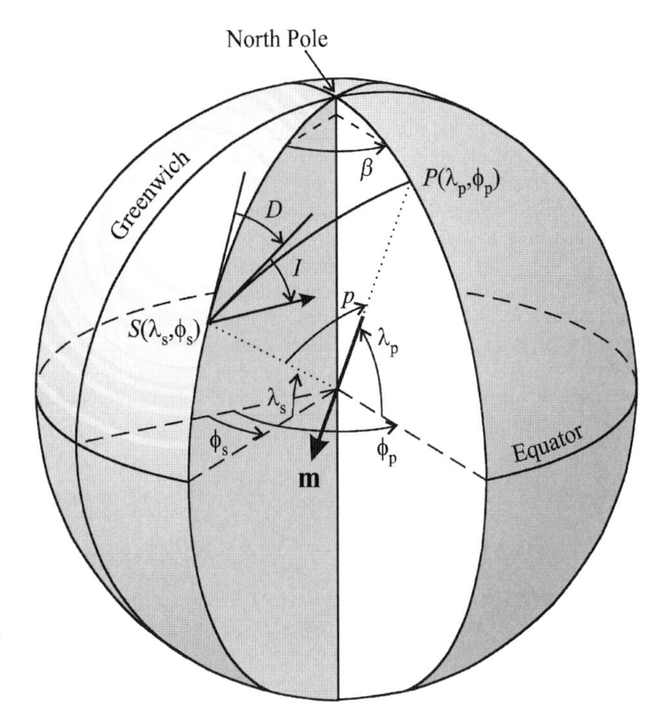

Figure P20 Position of the paleomagnetic pole *P* relative to the sampling site *S*. Modified after Merrill *et al.* (1996).

In addition to geochronology, various geological field tests are used to provide additional evidence on the nature and timing of the acquisition of the magnetization. The three traditional, and most common, tests are known as the fold test (see Magnetization, remanent, fold test), the conglomerate test, and the *baked contact test* (*q.v.*). The field relationships for these classical tests are shown in Figure P21. If the *in situ* directions of magnetization in a folded rock-unit differ from place to place, but agree after "unfolding" the rock-unit to its original pre-folding structure, then the magnetization must predate the folding and must have been stable since that time. Conversely, if the *in situ* directions agree then the magnetization must postdate the folding. If cobbles from the rock formation under investigation can be found in a conglomerate then, because the orientations of the cobbles will have been randomized, the magnetization of these cobbles should have no preferred direction if their original magnetization has been stable. If the direction of magnetization in the baked contact zone surrounding an intrusion is parallel to that observed in the intrusion, but differs from that of the unbaked country rock, then the magnetization of the intrusion has been stable since cooling. It is also important to determine the paleohorizontal of the rock-unit accurately so that the inclination I used in Eq. (1) is meaningful.

It has already been noted that it is necessary to use the time-average field to determine a paleomagnetic pole. This is because the Earth's present magnetic field, and most probably Earth's past field at any particular time, is not that of a simple geocentric axial dipole. Only about 75% of the present field intensity is attributed to a dipole field, a percentage that was about 82% or 83% at the beginning of the 20th century. This dipolar field is not static and is currently tilted about 10.7° from the rotation axis. The remainder of the field is referred to as the nondipole field, which characteristically exhibits more rapid spatial and temporal variations than the dipole field. Consequently it is necessary to average out these variations. Some rocks, for example thin lava flows, can acquire their primary magnetization in less than a year and so that primary magnetization often reflects the instantaneous field and not the time-average field. Thus it becomes necessary to use a sampling scheme that will average out these variations in order to give a reliable paleomagnetic pole.

A hierarchical scheme is typically used to obtain an estimate of a paleomagnetic pole position from a given location. Oriented samples are taken throughout the rock formation of interest. Specimens obtained from these samples are used to investigate rock magnetic properties. Statistical analysis of sample directions provides an estimate of the ancient field direction and its variance for the rock-unit. The mean magnetic direction for this unit is then used in Eq. (1) to determine a *virtual geomagnetic pole* (*VGP*). It is referred to as a "virtual" pole because at this level in the hierarchy it is unlikely that the field would have been averaged to that of a GAD. This makes it possible to average a "sufficient" number of VGP from widely dispersed sampling sites as a final step in the hierarchy to obtain an estimate of the paleomagnetic pole position. Alternatively, if it is apparent from the geology that sampling localities spatially close to each other will provide coverage of a sufficient time interval, then it is possible to average the magnetic field directions and use this mean direction in Eq. (1) to estimate the paleomagnetic pole position. It is presumed that at this level in the hierarchy the geocentric axial dipole assumption is satisfied and that the paleomagnetic pole position provides a reliable estimate of the paleogeographic pole position. There are no hard and fast rules and the procedures used in collecting samples and analyzing them vary depending on such factors as the properties of the site, the nature of the rocks sampled, the goals of the investigation and, to some extent, on the nature of the investigators. Detailed procedures on how this is done are described elsewhere in this volume.

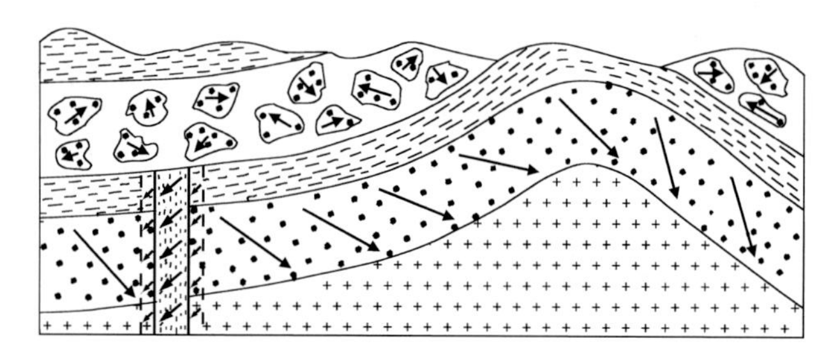

Figure P21 Field relationships for the fold, conglomerate, and baked contact tests. From Merrill and McElhinny (1983).

Buried within this process is the question of how much time is required to obtain a good approximation of the time-average field so that the GAD assumption is in fact reasonable. If a very short interval is used, then the averaging will be incomplete and this can produce significant errors. Conversely, if too long an interval is used then, for example, global tectonic processes can become significant and the sampling locality may have moved a substantial amount relative to the geographic pole. There is no widespread agreement on the optimum time interval for averaging but a minimum of 10 ka is generally viewed as necessary. The GAD assumption appears to be excellent for the time-average magnetic field over the past 5 Ma, although a small nondipole component may also be present. It is difficult to determine whether the GAD assumption has been valid throughout most of geological time but there is reasonable evidence to indicate that the assumption is very good to first order for the last few hundred million years. The evidence is less compelling for times prior to about 300 Ma ago.

Paleomagnetic field

Probably the best-documented fact stemming from paleomagnetism is that Earth's magnetic field has reversed polarity many times. By this it is meant that the north and south geomagnetic poles (the two locations where a line through the geocentric dipole intersects the Earth's surface) have traded places. The duration of a polarity transition is imperfectly known but it is probably in the order of 10^3 to 10^4 years, a time during which the intensity drops to a minimum of around a quarter of its normal value. The field is said to have normal polarity when the north geomagnetic pole is above 45° north latitude, reverse polarity when this pole lies below 45° south latitude, and transitional for all other latitudes. Hundreds of reversals have now been documented and for the past 160 Ma there is now a fairly reliable geomagnetic polarity timescale (GPTS). From this timescale it is evident that reversals do not occur with a simple periodicity but instead appear to occur randomly in time. The underlying rate at which reversals occur seems to vary on a timescale of about 10^8 year, probably reflecting changes in the lower mantle.

The very nature of the generative process of the geomagnetic field—a dynamo operating in the iron-rich liquid outer core—demands that the field is ever changing. The resulting gradual change in the magnetic field with time is referred to as *secular variation* (*q.v.*) which, as already discussed, must be averaged out in order to obtain a paleomagnetic pole. This secular variation occurs over a wide range of timescales, the shortest period that can be observed being around a year. This is because of the shorter period variations screened out by the semiconducting mantle. It is generally felt that the characteristic times of dynamo processes are less than 10^6 years, so variations due to internal processes in the dynamo typically occur in less than 10^6 years. Longer timescale variations, such as in the reversal rate, probably reflect changes in boundary conditions on the core. Apart from reversals, the largest changes in the Earth's magnetic field are known as excursions short intervals of time (of the order 10^4 years) when VGP deviate unusually far from the geographic pole.

The intensity of the field also changes with time. However, because of rock magnetic problems, these changes are more difficult to determine than the changes in direction. Direct measurements of the Earth's magnetic field from magnetic observatories indicate that that the dipole intensity decreased by about 5% during the 20th century.

Indirect measurements from archeomagnetism and paleomagnetism (there are several different absolute and relative paleointensity methods) indicate that the intensity of the Earth's field has varied significantly on a timescale that is geologically short. For example, 2 ka ago the intensity appears to have been 30%–40% greater than the present value but 6 ka ago it was found to be 30% less. Good relative paleointensity data from marine sediments provide a reasonable estimate of the intensity variation for the most recent several hundred thousand years. Intensities from times prior to this come from so-called absolute paleointensity methods, which seemingly should provide more accurate estimates than relative estimates. However, absolute intensity estimates are obtained from lava flows, which provide estimates that represent essentially an instantaneous estimate of the magnetic field rather than a time-average value. Typically there are large spatial and temporal gaps between such estimates. Although there is an incomplete agreement on what constitutes a reliable absolute paleointensity estimate, most paleomagnetists would agree that there are only about 1000 reliable estimates for all of geological time prior to about 5 Ma ago. A consequence is that several models have been advanced to describe the intensity variation over the past few hundred million years and there is little agreement on which model should be preferred.

Tectonics

In the 1950s it was not immediately clear whether apparent polar wander was indeed due to polar wander itself or to movement of the continents. However, because the apparent polar wander paths were different for different continents, paleomagnetists argued that continental drift had occurred. For example, the APWPs from 350 Ma to present for North America and Europe (including Russia west of the Ural Mountains) can effectively be made to coincide by removing the Atlantic Ocean and closing up the continents. This is consistent with Wegener's pioneering suggestion of continental drift to explain various geographic and geologic features—but not magnetic. Nevertheless the claims by paleomagnetists were often dismissed prior to the mid-1960s by other scientists, primarily because of concerns about the reliability of the geocentric axial dipole field assumption, questions concerning the fidelity of rocks as recorders of the primary remanent magnetization and because of the apparent absence of a driving mechanism for continental drift.

Today, paleomagnetic poles are widely used to locate the past positions of continents, particularly when marine magnetic anomaly data are not available. Indeed, as the oldest marine rocks are slightly less than 200 Ma old, paleomagnetic poles and APWPs are the primary quantitative tools for locating land masses prior to 200 Ma. In addition, paleomagnetic poles provide valuable information on tectonics on a smaller scale. For example, a substantial part of north-western North America appears to have been added to the craton in pieces, referred to as displaced terranes, during the Cenozoic.

Marine magnetic anomalies (variations in the intensity of the magnetic field that are usually measured at the sea surface) played a prominent role in the establishment of seafloor spreading and plate tectonics. It was noticed that these magnetic anomalies formed stripes parallel to and symmetric about the mid-ocean ridges. Vine and Matthews and Morley and Larochelle independently recognized that these magnetic stripes were caused by alternating blocks of normally and reversely magnetized volcanic rocks in the upper part of the oceanic crust. According to their model, crust is formed at the mid-ocean ridge, spreads away on both sides of the ridge and cools. As the rock cools at the Curie temperature it acquires a thermoremanent magnetization parallel to the ambient direction of the Earth's magnetic field. Reversals of the Earth's magnetic field will then produce blocks of alternate polarity. Hence the spreading crust acts like a magnetic tape recorder by recording the reversals of Earth's magnetic field, producing the observed stripes. It is now recognized that this magnetization is recorded in Layer 2 of the oceanic crust, an igneous layer of pillow basalts and dikes that lies beneath Layer 1, a sedimentary layer that was deposited later. Because this field varies over Earth's surface, the geocentric axial dipole field assumption is invoked and marine magnetic anomalies are transformed to the poles in order to make comparisons between anomalies from different localities. Because polarity contrast is used, undetected nondipole components will not have a significant effect on the final result. These marine magnetic anomalies have played a major role in the development of a geomagnetic polarity timescale for the last 170 Ma and in our understanding of the motions of tectonic plates during this recent time interval.

In a strict interpretation of plate tectonics, all deformation is confined to three types of plate boundaries: divergent plate boundaries (spreading centers)—where new material is produced; convergent plate boundaries (subduction zones)—where material descends back into the Earth, and transform faults where no new material is produced or removed. Support for this approximation comes from many sources including that the vast majority of energy released by earthquakes occurs at plate boundaries. However, there are places where this strict interpretation must be modified. For example, much of the Tibetan Plateau has experienced uplift of more than 5 km since the collision of India with the rest of Eurasia, which occurred about 50–65 Ma ago. Even such cases as this where interplate tectonics is important, paleomagnetism has proven to be a valuable tool in determining the character of the deformation.

The final tectonic example we give involves an ongoing controversy concerning features known as hot spots. These features are associated with interplate basaltic volcanism characterized by linear volcanic chains that grow older in the direction of plate motion. For more than a quarter of a century, it was believed that hot spots were maintained by material rising from the lower mantle through lower mantle plumes that formed part of the upward flow of a long-lived stable pattern of whole-mantle convention. Furthermore, it was supposed that these plumes do not change their relative positions. As already noted, given a single APWP, it is not immediately apparent whether this is due to polar wander or continental motion, and this is because it is difficult to find a fixed reference frame. Consequently a fixed hot spot reference frame was an enormously attractive concept. The fixed hot spot model implies that all islands and seamounts along a particular hot spot chain are formed at the same latitude and longitude. However, recent inclination data obtained by drilling seamounts from the northern end of the Hawaiian Emperor Seamount chain suggests that this is not the case. The magnetic inclinations of drilled samples from the Emperor Seamounts at the northern end of this chain seem to indicate the seamounts formed at significantly higher latitudes than that of the present hot spot, which lies beneath the island of Hawaii. If the paleomagnetic data from the Emperor Seamounts are sufficient to average out secular variation, if the paleohorizontal has been accurately obtained, and if the primary magnetization has been properly obtained, then the fixed hot spot model is not obeyed, at least for the Hawaiian-Emperor hot spot chain. In 1998 Steinberger and O'Connell showed how it was possible for hot spot plumes under a single plate to show less relative motion while at the same time it was moving relative to plumes under a different plate. This suggests it is unlikely that plumes can be effectively anchored in a convecting mantle and it seems that the weight of evidence is currently against the idea of a fixed hot spot reference frame.

Magnetostratigraphy and paleoclimate studies

Because there is an excellent first-order reversal chronology for most of the Mesozoic and all of the Cenozoic and because reversals occur randomly in time, sequences of reversals can be used for age dating and stratigraphic control. For example, a distinctive reversal pattern found in sediments in one area can sometimes be correlated with the same pattern in a different area. Sedimentation rates will typically be different in the two areas and so some stretching of the record from one area is required to make a correlation. This makes the match nonunique and so the correlation is more convincing the more distinctive the local pattern of reversals is. Absolute ages can be assigned to the sediment sections involved when the pattern of reversals can be identified in the reversal chronology. In addition, rock magnetic properties, such as magnetic susceptibility are sometimes used for correlation of marine cores taken in the same general location. Details on how magnetostratigraphy is done in practice are provided in books such as Opdyke and Channell (1996).

Often the detailed characteristics of a sediment will vary with depth because the source of the sediments has changed with time or because of physical, biological, or chemical alteration of the sediments after deposition. This provides a record that can be used for paleoclimate and environmental studies. Typically, although the magnetic properties will often vary with depth in sediments, this does not usually provide primary information on paleoclimatic changes. However, magnetic properties such as magnetic susceptibility can frequently be correlated with primary proxies for paleoclimate such as the variation in oxygen isotopes. Thus susceptibility, which can be measured relatively quickly, provides valuable data when there are gaps in the isotope record.

Paleomagnetism also plays a key role in testing global paleoclimatic hypotheses. For example, in 1969 Mikhail Budyko proposed a model for ice sheets that could lead to a runaway situation in which the whole Earth becomes covered in ice. The possibility of one or more glaciations that led to an entirely, or mostly, ice-covered Earth around 600 Ma ago is referred to as the snowball earth hypothesis. Paleomagnetism has provided significant support to this hypothesis by showing that, for example, Precambrian igneous rocks associated with glacial deposits in Namibia exhibit shallow magnetic inclinations, providing a clear implication that glacial deposits were formed at tropical latitudes in the Late Precambrian.

Concluding remarks

In essence, paleomagnetism provides three types of information: first, it provides a record of the Earth's magnetic field, which is a proxy for processes deep within the Earth; second, with some assumptions about the structure of the magnetic field, it provides information on the past location of rock-units; and, third, having calibrated the magnetic information, it provides the possibility to use APWPs and the geomagnetic polarity timescale as dating tools. Each of these types of information will contribute to our ongoing quest to understand our planet and its processes.

A major question to be addressed is whether global tectonics early in the Earth's history had a substantially different character from today. Hence paleomagnetism must develop the capacity to provide reliable location information for rocks formed early in the Earth's history.

The geomagnetic polarity timescale for the past 160 Ma is reasonably well developed and provides significant information on processes deep within the Earth with characteristic times in the order of 10^8 years. However, this timescale continues to evolve and it is necessary to invest the effort to obtain a complete and accurate record. The plate tectonic process has ensured that seafloor older than about 160 Ma has been recycled and so there is no well-ordered "tape recording" of reversals older than this. Instead, it is necessary to develop the reversal chronology for times before this through a painstaking process of dating, correlation, and assembly of information from continental rocks. Although the reversal chronology has been extended back into the Paleozoic, large gaps in coverage remain and there are a few data for the Precambrian. This timescale will provide valuable information on internal Earth processes and will also itself become a valuable dating tool.

Reliable paleointensity data for almost all times are needed. The resulting information is required to understand the long-term evolution of the Earth's magnetic field. In turn, this is likely to provide better understanding of the evolution of the Earth's inner core.

At this stage there is insufficient evidence regarding the structure of the time-average field for times earlier than about 300 Ma. This is an impediment to accurate reconstruction of earlier land masses. To some extent this creates a circular problem: a GAD field is assumed and paleoreconstructions are performed; inconsistencies in the paleoreconstructions are then attributed to non-GAD structure in the ancient geomagnetic field. If the structure of the field is actually known then it is possible to obtain an accurate paleoreconstruction. Conversely, if the paleolocations of the continental masses are known then it is possible to get information about the structure of the paleofield. It is going to require substantial innovation to achieve both of these without independent evidence.

Acknowledgments

This article is published with the permission of the Chief Executive Officer, Geoscience Australia.

Ronald T. Merrill and Phillip L. McFadden

Bibliography

Cox, A., and Doell, R.R., 1960. Review of paleomagnetism. *Geological Society of America Bulletin*, **71**: 645–768.

Evans, M., and Heller, F., 2003. *Environmental Magnetism: Principles and Applications of Enviromagnetics*, San Diego, CA: Academic Press.

Glen, W., 1982. *The Road to Jaramillo.* Stanford: Stanford University Press.

McElhinny, M.W., and McFadden, P.L., 2000. *Paleomagnetism: Continents and Oceans.* San Diego CA: Academic Press.

Merrill, R.T., and McElhinny, M.W., 1983. *The Earth's Magnetic Field: its History, Origin and Planetary Perspective.* London: Academic Press.

Merrill, R.T., McElhinny, M.W., and McFadden, P.L., 1996. *The Magnetic Field of the Earth: Paleomagnetism, the Core, and the Deep Mantle.* San Diego, CA: Academic Press.

Morley, L.W., and Larochelle, A., 1964. *Paleomagnetism as a Means of Dating Geological Events.* In Osborne, F.F. (ed.), Geochronology in Canada., The Royal Society of Canada Special Publications, Toronto: University of Toronto Press, **8**: 39–51.

Opdyke, N.D., and Channell, J.E.T., 1996. *Magnetic Stratigraphy.* San Diego, CA: Academic Press.

Steinberger, B., and O'Connell, R.J., 1997. Changes in the Earth's rotation axis owing to advection of mantle density heterogeneities. *Nature*, **387**: 169–173.

Vine, F.J., and Matthews, D.H., 1963. Magnetic anomalies over oceanic ridges. *Nature*, **199**: 947–949.

Cross-references

Baked Contact Test
Dipole Moment Variation
Geocentric Axial Dipole Hypothesis
Geodynamo
Geomagnetic Dipole Field
Geomagnetic Polarity Timescales
Geomagnetic Secular Variation
Harmonics, Spherical
Inner Core
Magnetic Susceptibility
Magnetization, Natural Remanent (NRM)
Magnetization, Remanent, Fold Test
Magnetization, Thermoremanent (TRM)
Nondipole Field
Paleointensity
Reversals, Theory
Statistical Methods for Paleovector Analysis
Time-averaged Paleomagnetic Field
True Polar Wander

PALEOMAGNETISM, DEEP-SEA SEDIMENTS

Introduction

Deep-sea (pelagic) sediments, deposited remotely from sources of continental detritus, have been a very important source for learning about the direction and intensity of the ancient geomagnetic field (paleomagnetism) because they often carry a primary natural remanent magnetization (NRM) acquired at the time of deposition, or shortly thereafter. The age of a primary magnetization can be determined from the accompanying biostratigraphy or isotope stratigraphy. The magnetization directions of known age have been used to assess plate motion (continental drift) or to record the characteristics of the ancient geomagnetic field, such as the sequence of polarity reversal in the geologic past. The record of geomagnetic polarity reversal (magnetostratigraphy) in deep-sea sediments, first practiced by Opdyke *et al.* (1966), has become important in paleoceanography and biostratigraphy, and in geologic timescale construction. The geomagnetic polarity time scale (GPTS), based on the sequence of polarity reversal through time, is the central thread to which the other facets of geologic time (radiometric, isotopic, biostratigraphic) are correlated in the construction of geologic timescales for the last 150 Ma (see Opdyke and Channell, 1996). Prior to ~150 Ma, the GPTS is less well defined, due to the lack of *in situ* oceanic crust and hence marine magnetic anomaly records, and is not adequately correlated to the biozonations that define geologic stages. The global synchronicity of polarity reversals (of the main axial dipole field) means that magnetic polarity stratigraphies can be used to correlate environmental (isotopic) and biostratigraphic events among contrasting environments and remote locations. The stochastic (unpredictable) occurrence of polarity reversal, and our inability to distinguish one normal (reverse) polarity chron from another, means that precise correlation through magnetic polarity stratigraphy is only possible at polarity reversals, with interpolation between these tie points. Within individual polarity chrons (time intervals between polarity reversals) magnetic records can be used for correlation if geomagnetic directional "secular" variation and/or paleointensity are adequately recorded. The conditions for adequate recording of secular variation and geomagnetic paleointensity are more stringent than for the recording of polarity reversals as contamination (overprint) of primary magnetization is less crippling for polarity records (~180° directional changes) than for the more subtle changes that define secular variation and paleointensity.

Origin of primary magnetizations

What makes deep-sea sediments efficient recorders of the geomagnetic field at time of deposition? Sediments can acquire a detrital remanent magnetization (DRM) at the time of deposition by mechanical orientation of fine grained magnetite (Fe_3O_4) or titanomagnetite (xFeTiO$_4$ $[1-x]$ Fe_3O_4) into line with the ambient geomagnetic field at the sediment-water interface. The natural remanent (permanent) magnetization (NRM) of ferrimagnetic (titano) magnetite results in a torque that statistically orients the magnetic moment of the grain population into line with the ambient geomagnetic field. The mechanical orientation of grains may be achieved either at the sediment-water interface (DRM) or in the uppermost few centimeters or decimeters of the sediment in which case the resulting remanence is referred to as pDRM (post-depositional detrital remanent magnetization).

Following introduction of the concept by Irving and Major (1964), progressive acquisition of postdepositional detrital remanent magnetization (pDRM) has been modeled as an exponential or cubic function of progressive lock-in with depth (Hyodo, 1984; Mazaud, 1996; Meynadier and Valet, 1996; Roberts and Winklhofer, 2004). Teanby and Gubbins (2000) used an 8 cm uniform mixed layer (magnetization = 0) to simulate the bioturbated surface layer, underlain by an exponential lock-in function. Channell and Guyodo (2004) have used a sigmoidal pDRM function based on tanh(x). The function can be defined by a surface mixing layer depth (M) at which 5% of the magnetization is acquired, and a lock-in depth (L) that is the depth below M at which 50% of the magnetization is acquired (see Figure P22). The assumption of a well-mixed layer (M), in which the sediment is thoroughly consumed by sipunculid or echiuran worms and other benthos, is valid in deep-sea sediments where the mixing coefficient exceeds the product of mixed layer depth and sedimentation rate (Guinasso and Schink, 1975). In deep-sea sediments, isotopic tracers indicate mean mixed layer thicknesses of about 10 cm (values vary by an order of magnitude from 3–30 cm) that is largely independent of sedimentation rate (Boudreau, 1994). Using ^{14}C of the bulk carbonate fraction as the isotopic tracer, Thomson *et al.* (2000) estimated mixed layer thicknesses of 10–20 cm in box cores collected close to the Rockhall Plateau. These values are greater than the 2–13 cm mixed layer thicknesses obtained from further south in the North Atlantic using ^{14}C in foraminifera (Trauth *et al.*, 1997; Smith and Rabouille, 2002). Estimates of mixed layer thickness are grain size sensitive (see Bard, 2001) and would be expected to be lower for the coarse fraction (foraminifera) than for the bulk carbonate (nannofossils). For this reason, mixed layer thickness estimates based on ^{14}C and other isotopic tracers should be considered as minimum estimates for the fine (PSD) grains that carry stable magnetization. The main control on the mixed layer thickness appears to be organic carbon flux derived from surface water productivity (Trauth *et al.*, 1997; Smith and Rabouille, 2002). A recent redeposition study has suggested that, at least for some lithologies, intergranular interactions overcome the magnetic aligning torque so that bioturbation would not enhance, but rather disrupt, the remanent magnetization (Katari *et al.*, 2000). This proposition is in agreement with Tauxe *et al.* (1996) who, based on an analysis of the position of the Matuyama-Brunhes boundary relative to oxygen isotope records, concluded that magnetization lock-in depth is insignificant in marine sediments. Other studies (DeMenocal *et al.*, 1990; Lund and Keigwin, 1994; Kent and Schneider, 1995; Channell and Guyodo, 2004) invoked pDRM, and hence a finite (decimeter-scale) lock-in depth, to explain reversal-isotope correlations and observed attenuation of secular variation records.

Magnetite and titanomagnetite grains carrying DRM or pDRM in marine sediments may be of detrital or biogenic origin. Titanomagnetite is likely to have a detrital origin from the weathering of igneous rocks (such as mid-ocean ridge or oceanic island basalts) or from volcanic ash falls. A magnetic remanence known as thermal remanent magnetization (TRM) is acquired as the (titano)magnetite grain cools through a blocking temperature spectrum within its igneous host. The detrital grains retain this TRM during erosion, transport, and subsequent incorporation into the deep-sea sediment. The interaction of the TRM of individual grains with the ambient geomagnetic field

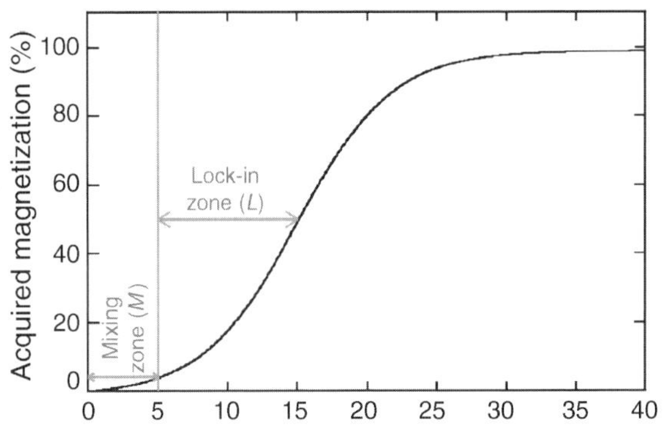

Figure P22 The sigmoidal function based on tanh(x) used to model the mixing zone (thickness M) and the underlying lock-in zone (thickness L) (after Channell and Guyodo, 2004).

at the time of sediment deposition generates the orienting torque that produces DRM or pDRM.

Magnetite (Fe_3O_4), with low or insignificant Ti (titanium) substitution, is commonly associated with biogenic sources. Magnetotactic bacteria are ubiquitous in freshwater and marine environments (e.g., Kirschvink and Chang, 1984; Petersen *et al.*, 1986; Vali *et al.*, 1987). They inhabit aerobic and anaerobic sediments and form intracellular chains of fine-grained (single domain) "magnetosomes" of magnetite (Blakemore and Blakemore, 1990) or, less commonly, goethite (Mann *et al.*, 1990; Heywood *et al.*, 1990). Magnetite and goethite magnetosomes have been observed to coexist within the same bacterium (Bazylinski *et al.*, 1995). The intracellular magnetite and goethite crystals are formed by a process referred to as biologically controlled mineralization (BCM) (Lowenstam and Weiner, 1989; Bazylinski and Frankel, 2003). Intracellular magnetite and goethite are usually structurally well ordered and have a narrow (single domain) size distribution that is optimal for the retention of magnetic remanence in ferrimagnetic minerals such as magnetite and goethite (Figure P23).

Other Fe(III)-reducing bacteria secrete magnetite, and a range of other iron minerals, outside the cell by a process referred to as biologically induced mineralization (BIM) (Lowenstam and Weiner, 1989; Frankel and Bazylinski, 2003). Magnetite and other iron minerals produced by BIM are often poorly crystalline, fine-grained, and not structurally ordered. In the case of *Geobacter metallireducens* (also referred to as bacterium strain GS-15), magnetite is produced outside the cell as fine (superparamagnetic) grains produced by the oxidation of organic matter and Fe(III) reduction (Lovley, 1990). Unlike magnetotactic bacteria, these iron reducers are nonmagnetotactic (nonmotile), unconcerned about the size, shape, or composition of the iron mineral product, and can produce a range of iron minerals depending on the nature of the surrounding medium.

Magnetotactic bacteria apparently utilize the internal magnetite-goethite chains to navigate along geomagnetic field lines (magnetotaxis) to find optimal redox conditions in the near-surface sediment. In the absence of sensitivity to gravity due to neutral buoyancy in seawater, the ambient magnetic field provides up-down orientation, either in the Northern or Southern hemisphere (Kirschvink, 1980). The single-domain grain size typical for magnetite and goethite produced by BCM indicates that magnetic remanence is central to magnetosome function. Living magnetotactic bacteria tend to be concentrated at depths of few tens of centimeters below the sediment-water interface at the transition from iron-oxidizing to iron-reducing conditions (Karlin *et al.*, 1987). The conditions under which BIM and BCM of magnetite takes place, and the depth in the sediment column to which these microbes are active (see Liu *et al.*, 1997), are clearly of great importance in understanding the origin of the magnetic signature in sediments. Magnetic methods for the recognition of magnetosome chains, and individual magnetosomes, have been proposed by Moskowitz *et al.* (1988, 1994).

Apart from bacteria, other biogenic marine sources of magnetite are chiton (mollusk) teeth, fish, whales, and turtles. The role of magnetite in fish and marine mammals is thought to be navigational (Kirschvink and Lowenstam, 1979; Kirschvink *et al.*, 2001; Walker *et al.*, 2003). As for bacterial magnetite, these biogenic magnetite grains are often in the SD grain size range, the optimal grain size range for a stable magnetic remanence and are usually pure magnetite. The grains acquire a so-called chemical remanent magnetization (CRM) as they grow through their "critical" volume within the host organism.

Biogenic magnetite has been commonly observed in marine sediments on the basis of particle shape and size (e.g., Kirschvink and Chang, 1984; Petersen *et al.*, 1986; Vali *et al.*, 1987). The small elongate (SD) grain size of biogenic magnetite, typically $\sim 0.2 \pm 0.05$ μm across (Figure P23), and the resulting high surface area to volume ratio makes these grains susceptible to diagenetic dissolution. A large proportion of the biogenic magnetite in surface sediment may not survive sediment burial. Detrital magnetite (often titanomagnetite) appears to have greater survivability perhaps due to larger initial grain size and/or less intimate contact with pore waters. Stable primary NRMs in deep-sea sediments are commonly associated with PSD, rather than SD or MD, magnetite. Larger multidomain (MD) magnetite, with dimensions in excess of a few microns, carry a lower coercivity magnetization (than PSD or SD grains) that is more prone to remagnetization during the history of the sediment or sedimentary rock.

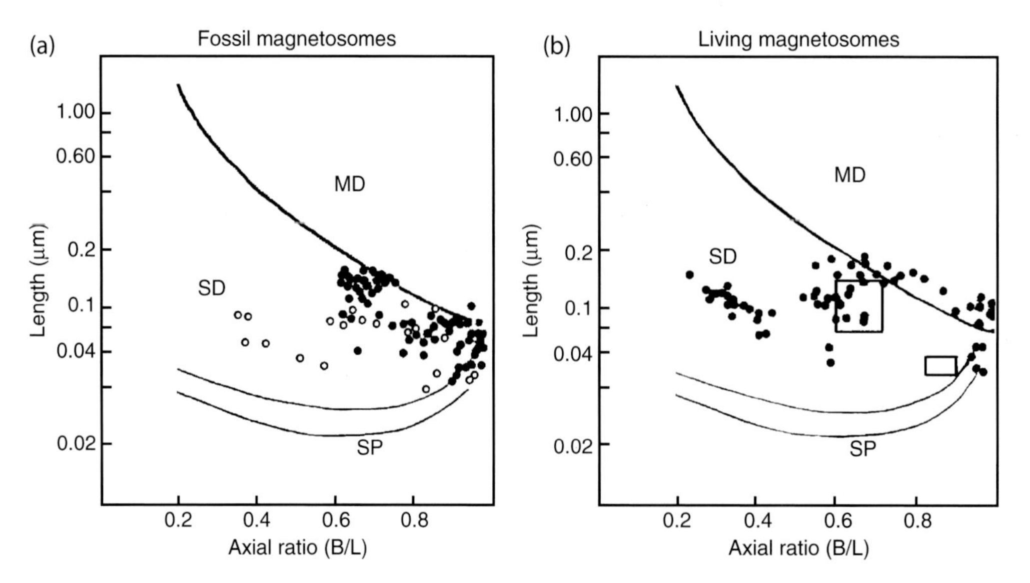

Figure P23 (a) Size and axial ratio (breadth/length) of fossil magnetite magnetosomes observed in Mesozoic and Tertiary sediments (open circles) and Quaternary sediments (closed circles). (b) Size and axial ratio (breadth/length) of magnetite magnetosomes from living bacteria (modified after Vali *et al.*, 1987). Boxed regions in (b) from observations of Kirschvink (1983). Size distribution of multidomain (MD), single-domain (SD) and superparamagnetic (SP) magnetite grains after Butler and Banerjee (1975).

Origin of secondary magnetizations

Reduction diagenesis

CRM acquired during sediment diagenesis is the principal process that generates secondary magnetizations in marine sediments. The other important secondary magnetization is viscous remanent magnetization (VRM), the time-dependent acquisition of magnetization by the ambient (usually Brunhes Chron) geomagnetic field. A CRM postdates sediment deposition (usually by an unknown amount of time) and must be differentiated from the primary magnetization if sedimentary magnetic data are to be correctly interpreted. The primary and secondary magnetizations are differentiated in the laboratory by stepwise, progressive, demagnetization using either temperature or alternating fields. The differentiation can be successful if the two magnetization components respond differently to the demagnetization process (see Dunlop and Ozdemir, 1997).

A wide range of reactions, many of which are mediated by microbes, occur during deep-sea sediment diagenesis. Studies of pelagic pore waters indicate a sequence of electron acceptors used in the oxidation of sedimentary organic matter: O_2, nitrate NO_3^-, manganese (Mn^{3+}), iron (Fe^{3+}) then sulfate SO_4^{2-} (see Burdige, 1993). The utilization of sulfate has been found to be particularly important in the dissolution of magnetite grains in pelagic sediments. Microbial reduction of pore water sulfate (usually derived from seawater) yields sulfide ions that combine with the most reactive iron phase often magnetite to produce iron sulfide (Canfield and Berner, 1987). In most marine sedimentary environments, authigenic iron sulfides are dispersed throughout the sediment or associated with organic-rich burrows. In visual core descriptions of deep-sea sediments recovered during the 20-year duration of the Ocean Drilling Program, "pyrite" and amorphous iron sulfides are widespread and these iron sulfides probably originated through microbial sulfate reduction. The initial products are iron monosulfides (FeS), such as mackinawite, and further diagenesis may produce greigite (Fe_3S_4), pyrite (FeS_2) and pyrrhotite (FeS_{1+x} where $x = 0$–0.14). Pyrite is paramagnetic and therefore does not carry magnetic remanence. Pyrrhotite (e.g., Fe_7S_8) and greigite, on the other hand, are ferrimagnetic and capable of carrying a stable magnetization.

A secondary magnetization (CRM) carried by authigenic pyrrhotite often develops at the expense of a primary magnetization carried by magnetite. The dissolution of magnetite will be accentuated where sufficient pore water sulfate and labile organic matter support microbial sulfate-reducing communities. In organic rich, reducing, diagenetic conditions, such as the Japan Sea, pyrite and pyrrhotite are readily formed (Kobayashi and Nomura, 1972), at the expense of magnetite and other forms of reactive iron. Iron sulfide formation in marine sediments is limited either by the availability of pore-water sulfate (usually derived from sea-water) or the availability of organic carbon to sustain sulfate-reducing bacteria (Canfield and Berner, 1987). On the Ontong-Java Plateau, increased C_{org} in glacial isotopic stages has led to enhanced magnetite dissolution (Tarduno, 1994). In Taiwan, Pleistocene sediments contain detrital magnetite, authigenic greigite and pyrrhotite (Horng *et al.*, 1998). Authigenic gregite has been detected in lake sediments (Giovanoli, 1979; Snowball and Thompson, 1988; Roberts *et al.*, 1996) and in marine environments (Tric *et al.*, 1991; Reynolds *et al.*, 1994; Lee and Jin, 1995; Roberts and Weaver, 2005).

The dissolution of magnetite within the upper few decimeters of hemipelagic sediments has been documented by a reduction in coercivity and magnetization intensity as finer grains undergo preferential dissolution (Karlin and Levi, 1983; Leslie *et al.*, 1990). In pelagic, high sedimentation rate, "drift" sediments from the Sub-Antarctic South Atlantic (see Figure P24a) and Iceland Basin (see Figure P24b), the pore water profiles indicate steady sulfate depletion with depth. At both sites, a primary magnetization records the polarity stratigraphy (Channell and Lehman, 1999; Channell, 1999; Channell and Stoner, 2002; Stoner *et al.*, 2003) although there is abundant visual evidence that authigenic iron sulfide formation accompanies pore-water sulfate depletion. In the case of the Sub-Antarctic South Atlantic site (Figure P24a), NRM intensity decreases with depth, together with reduction in the ratio of anhysteretic susceptibility (k_{arm}) to susceptibility (k). This ratio (k_{arm}/k) is a magnetite grain-size proxy where higher values indicate finer grains. In the Iceland Basin record, the depletion of sulfate is not accompanied by marked changes in these magnetic parameters (Figure P24b). This may be due to the presence of a more reactive iron phase than magnetite at the Iceland Basin site and /or to the presence of a particularly reactive *magnetite* phase (e.g., fine-grained biogenic magnetite) at the Sub-Antarctic South Atlantic site.

Two additional profiles from the sub-Antarctic South Atlantic indicate a marked decrease in NRM intensity and k_{arm}/k at 150–200 cm depth (see Figures P25 and P26). This denotes a marked *increase* in grain size of magnetite at this depth relative to surface sediment (Figure P25) due to preferential dissolution of fine-grained magnetite due to the higher surface area to volume ratio. This dissolved fine-grained magnetite fraction carries NRM but is not an important contributor to volume susceptibility (Figure P25). Its fine grain size (k_{arm}/k values >6 are consistent with magnetite grain sizes of less than 0.1 μm) imply that the magnetite may be partly biogenic in origin. We conclude that the fine-grained (biogenic) magnetite is often lost during early diagenesis due to its small grain size and hence high reactivity. Note that k_{arm}/k values $>\sim 3$ are absent in the Iceland Basin record (Figure P24b) implying that biogenic magnetite is a less important component of the total magnetic budget at this site.

At some deep-sea sites, the concentration of sulfate in pore waters remains high to the base of the recovered section indicating (in the absence of a sulfate source at depth) that sulfate reduction is not a continuing process, even in the presence of a few percent organic carbon. This has been observed in siliceous sediments from the South Atlantic (ODP Leg 177), NW Pacific (ODP Leg 145), equatorial Pacific (ODP Leg 138) where magnetostratigraphic records indicate preservation of a primary magnetization (Channell and Stoner, 2002; Weeks *et al.*, 1995; Schneider, 1995). We speculate that the organic carbon associated with diatomaceous (siliceous) oozes may be too refractory, or otherwise unavailable due to adsorption or encapsulation in/on siliceous surfaces, to be utilized by sulfate reducing microbes. Under these conditions of arrested sulfate reduction, a primary magnetization carried by magnetite will often be preserved.

Dissolution of magnetite by combination with sulfide ions, released by microbial reduction of pore water sulfate, is the most important process that accounts for the degradation of the magnetic signal in deep-sea sediments. In the presence of an iron phase that is more reactive than magnetite (e.g., goethite), the formation of iron sulfides may occur without appreciable magnetite dissolution.

Oxidation diagenesis

In contrast to the role of reduction diagenesis on magnetite dissolution, the primary magnetic signal in deep-sea sediments can be destroyed by oxidation of magnetite to maghemite. Kent and Lowrie (1974) and Johnson *et al.* (1975) documented this process in the so-called red clay facies that occurs over a large part of the mid-latitude Pacific (Davies and Gorsline, 1976). Henshaw and Merrill (1980) showed that the magnetic signal in these sediments is also affected by the authigenic growth of ferromanganese oxides and oxyhydroxides. The red clays are devoid of calcareous and siliceous microfossils and accumulated below the CCD (calcium compensation depth) at rates of ~25 cm/Ma, more than an order of magnitude slower than "average" pelagic sedimentation rates. The primary magnetization in the red clay facies is lost at depths of a few meters below the sediment-water interface, and the loss of magnetization often coincides with the later part of the Gauss Chron, close to the time of onset of Northern Hemisphere glaciation. The reduced grain size of eolian detrital magnetite, due to lower prevailing wind velocities prior to the onset of Northern Hemisphere glaciation, may influence the degradation of the magnetic signal in these sediments (Yamazaki and Katsura, 1990).

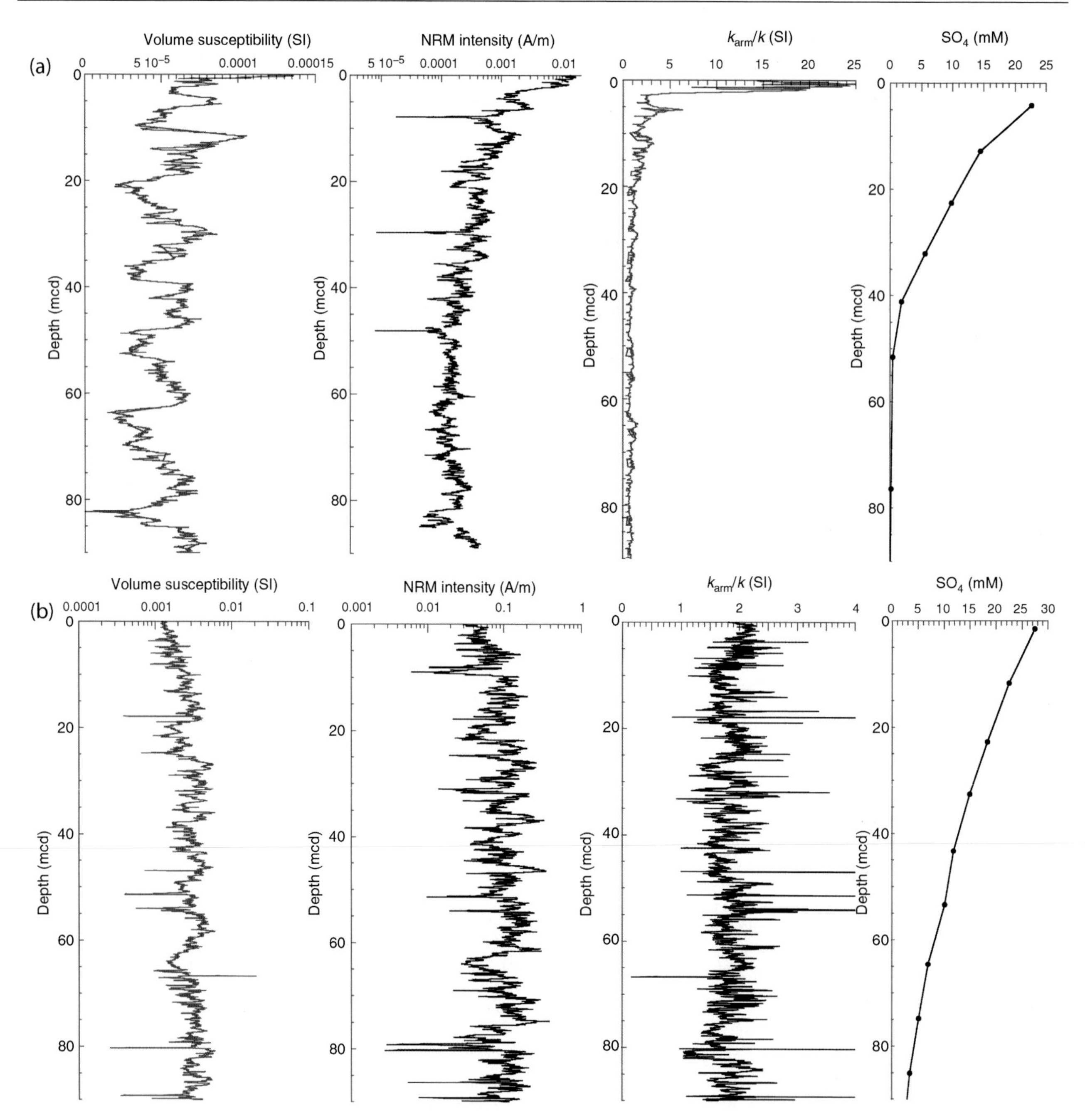

Figure P24 Volume susceptibility, NRM intensity after demagnetization at peak fields of 25 mT, anhysteretic susceptibility divided by susceptibility (k_{arm}/k), and pore water sulfate concentrations: (a) ODP Site 1089 (sub-Antarctic South Atlantic at 40.9°S, 9.9°E, 4620 m water depth) and (b) ODP Site 984 (Iceland Basin at 61.4°N, 24.1°W, 1648 m water depth). Data from Stoner *et al.* (2003), Channell (1999), and Shipboard Scientific Party (1996, 1999). At both sites, the 90 m composite depths (mcd) shown here represent ~600 ka, at mean sedimentation rates of 15 cm/ka.

Other origins of secondary magnetizations

Florindo *et al.* (2003) attribute low susceptibilities at some deep-sea drilling sites to dissolution of magnetite to form smectite in the presence of high concentrations of dissolved silica. Although authigenic smectite is widespread in deep-sea sediments, there are a number of different pathways for its formation and magnetite is one of a number of different possible precursors.

Degradation of the primary magnetic signal in deep-sea sediment cores can be attributed to factors associated with drilling and recovery. For example, magnetite in calcareous oozes from ODP Leg 154 (Ceara Rise, equatorial Atlantic) carry a secondary magnetization imposed by drilling that is apparently oriented radially in the core cross-section (Curry *et al.*, 1997). No primary magnetization was resolved in these sediments. Magnetite grains are, however, not apparently greatly affected by sulfate reduction as pore water sulfate remains high (>20 mM) to ~200 m depth and magnetic susceptibility does not decrease with depth (Curry *et al.*, 1997; Richter *et al.*, 1997). Low activity of sulfate reducing microbes may be due to low levels of (labile) organic matter in these sediments. The magnetic susceptibility record is not affected by diagenetic dissolution or authigenesis of

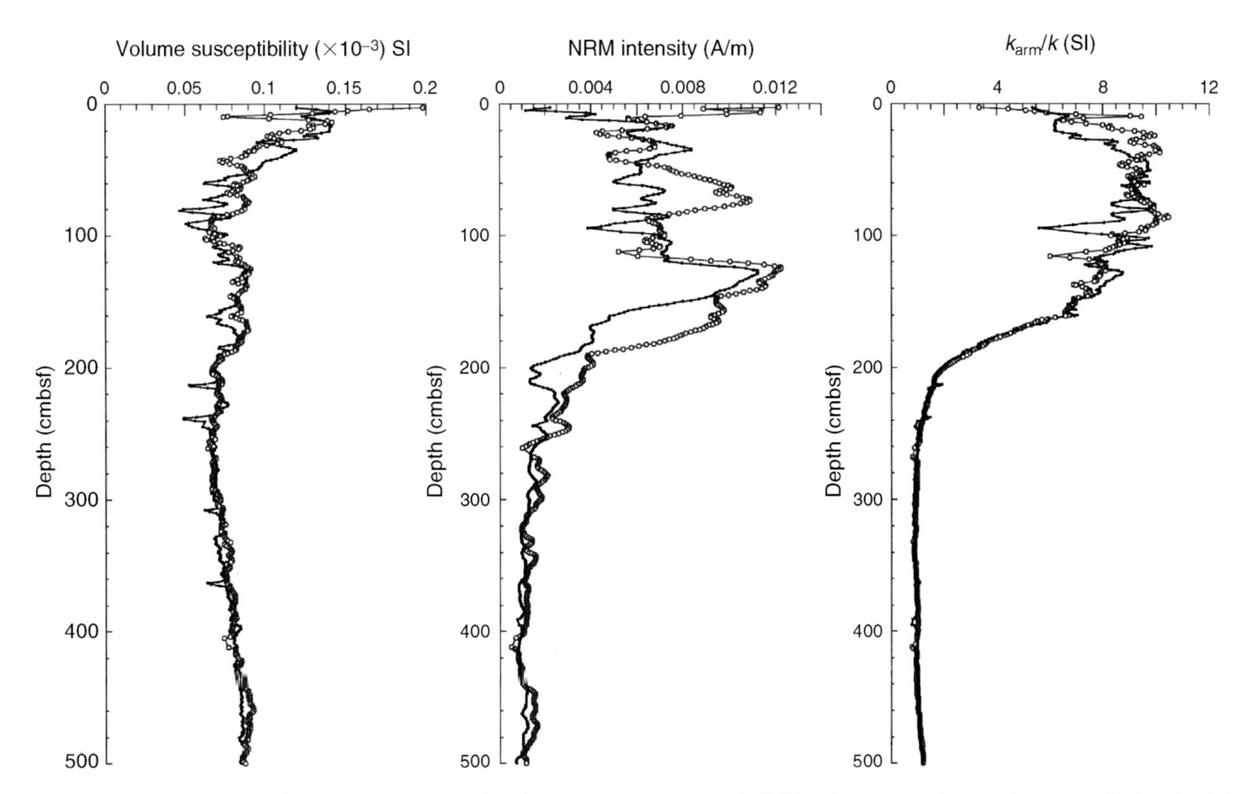

Figure P25 Volume susceptibility, NRM intensity after demagnetization at peak fields of 25 mT, anhysteretic susceptibility divided by susceptibility (k_{arm}/k) for two piston cores (4-PC03 and 5-PC01) collected during cruise TTN-057 in the sub-Antarctic South Atlantic at 40.9°S, 9.9°E and 41.0°S, 9.6°E, respectively, and 4620 m water depth. The 500 cm record shown here represents ~25 ka at a mean sedimentation rate of 20 cm/ka. Data from Channell *et al.* (2000).

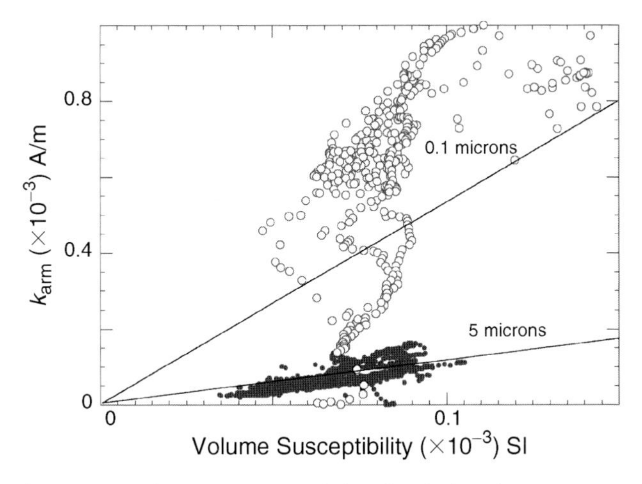

Figure P26 Anhysteretic susceptibility (k_{arm}) plotted against susceptibility (k) for Cores 4-PC03 and 5-PC01, from the sub-Antarctic South Atlantic. The measurements represent the last ~25 ka. Open symbols indicate Holocene samples. Data from Channell *et al.* (2000). The lines corresponding to magnetite grain size estimates of 0.1 μm and 5 μm after King *et al.* (1983).

magnetic minerals, and the cycles in susceptibility, produced by variations in surface water productivity, have provided a robust astrochronology (Shackleton *et al.*, 1999). The fidelity of the cyclostratigraphy supports other rock magnetic data (Richter *et al.*, 1997), indicating that magnetite has remained unaltered in ODP Leg 154 sediments. The magnetite grains are at least partially in the pseudo-single domain (PSD) size range (Richter *et al.*, 1997) and therefore capable of carrying stable primary magnetization. The observed radial (re)magnetization is attributed to an isothermal remanence (IRM) imposed by the coring procedure (Curry *et al.*, 1997).

In pelagic and volcaniclastic sediments of ODP Leg 157, from close to the Canary Islands, PSD magnetite appears to carry an inward-directed radial magnetization (Herr *et al.*, 1998; Fuller *et al.*, 1998). The coercivity of this radial remagnetization is greater than the more commonly observed downward-directed remagnetization associated with drilling, that can often be eliminated in peak alternating fields of ~20 mT. The coercivity of the radial remagnetization is greater than that of a simple IRM acquired in the few tens of mT fields associated with the steel core barrels and the remagnetization is more pronounced in poorly lithified sediments (Fuller *et al.*, 1998). We speculate that the nonmagnetic matrix in these sediments, when disturbed or shocked by drilling, allows physical re-orientation of magnetite grains in ambient magnetic fields emanating from the core barrels, bottom hole assembly or cutting shoe.

At any time during the history of the sediment or sedimentary rock, secondary magnetizations may be acquired (as CRMs) by chemical alteration of existing magnetic minerals, or by growth of new magnetic minerals.. Apart from the diagenetic changes noted above, events such as uplift, weathering, and deformation can all trigger the development of secondary CRMs. For example, much of the Paleozoic sequence of North America and Europe was remagnetizated coeval with the Hercynian (Late Carboniferous) orogenic pulse. This "orogenic" remagnetization is very widespread in North America, extending thousands of kilometers into the continental interior from the Appalachian margin. In carbonate rocks in North America and Europe, the Hercynian remagnetization is a CRM carried by magnetite. Authigenesis of magnetite may be triggered by orogenic activity in the Appalachians and the migration of hydrocarbon rich fluids into the hinterland (McCabe *et al.*, 1983, 1989; McCabe and Channell, 1994). Weathering and uplift can result in the growth of magnetic iron oxides and oxyhydroxides (such as hematite and goethite) from preexisting Fe-bearing minerals such as clays or iron sulfides. The high coercivity of CRMs carried by hematite and goethite makes them difficult to remove by alternating field demagnetization techniques.

Conclusions

On the basis of a survey of the global distribution of high-quality magnetic stratigraphies in deep-sea sediments, Clement *et al.* (1996) concluded that terrigeneous sediment input is a factor that contributes to data quality. Sediments from the North Atlantic and the North Pacific, and the Indian Ocean, often contain terrigeneous material that appears to have aided preservation of a primary magnetization in these sediments and hence enhanced the magnetostratigraphic records from these regions. Although biogenic magnetite is ubiquitous in modern deep-sea sediments, fine SD particles are particularly susceptible to dissolution due to their high surface area to volume ratio. Much of this biogenic magnetite fraction does not appear to survive early diagenesis. The detrital magnetite component is more likely to survive, and record the direction and intensity of the geomagnetic field at the time of deposition. Diagenetic dissolution of magnetite in deep-sea sediments often occurs by reaction with sulfide ions to form iron sulfides. The formation of sulfide ions is generally a result of microbially mediated reduction of seawater sulfate in pore waters. The resulting sulfide ions in pore water react with magnetite or, if present, a more reactive iron phase, to produce iron sulfides that may carry a secondary CRM. The process is limited by the availability of sulfate and labile organic matter required by sulfate-reducing bacteria. Even in the presence of abundant pore water sulfate, high-quality magnetic records (preservation of primary magnetization) in siliceous sediments may be due to the low activity of sulfate-reducing microbes due to the role of tests of siliceous organisms in shielding labtile organic matter from sulfate-reducing microbes.

James E.T. Channell

Bibliography

Bard, E., 2001. Paleoceanographic implications of the difference in deep-sea sediment mixing between large and fine particles. *Paleoceanography*, **16**: 235–239.

Bazylinski, D.A., Frankel, R.B., Heywood, B.R., Mann, S., King, J.W., Donaghay, P.L., and Hanson, P.K., 1995. Controlled biomineralization of magnetite (Fe_3O_4) and greigite (Fe_3S_4 in a magnetotactic bacterium. *Applied and Environmental Microbiology*, **61**: 3232–3239.

Bazylinski, D.A., and Frankel, R.B., 2003. Biologically controlled mineralization in Prokaryotes. In Biomineralization, P.M., Dove, J.J., De Yoreo, and Weiner, S. (eds.), *Reviews in Mineralogy and Geochemistry*, **54**: 217–247. Washington, DC: The Mineralogical Society of America.

Blakemore, R.P., and Blakemore, N.A. Magnetotactic magnetogens. In Frankel, R.B., and Blakemore, R.P. (eds.), *Iron Biominerals*. New York: Plenum Press, pp. 51–67.

Boudreau, B.P., 1994. Is burial velocity a master parameter for bioturbation? *Geochimica et Cosmochimica Acta*, **58**: 1243–1249.

Burdige, D.J., 1993. The biogeochemistry of manganese and iron reduction in marine sediments. *Earth Science Reviews*, **35**: 249–284.

Butler, R.F., and Banerjee, S.K., 1975. Theoretical single-domain grain-size range in magnetite and titanomagnetite. *Journal of Geophysical Research*, **80**: 4049–4058.

Canfield, D.E., and Berner, R.A., 1987. Dissolution and pyritization of magnetite in anoxic marine sediments. *Geochimica et Cosmochimica Acta*, **51**: 645–659.

Channell, J.E.T., 1999. Geomagnetic paleointensity and directional secular variation at Ocean Drilling Program (ODP) Site 984 (Bjorn Drift) since 500 ka: Comparisons with ODP Site 983 (Gardar Drift). *Journal of Geophysical Research*, **104**: 22937–22951.

Channell, J.E.T., and Lehman, B., 1999. Magnetic stratigraphy of Leg 162 North Atlantic Sites 980–984. In Jansen, E., Raymo, M.E., Blum, P., and Herbert, T. (eds.), Proceedings of the Ocean Drilling Program-Scientific Results, College Station, TX (Ocean Drilling Program), **162**: 113–130.

Channell, J.E.T., and Guyodo, Y., 2004. Magnetic stratigraphy of the Matuyama Chronozone at ODP Site 982 (Rockall Bank), evidence for finite magnetization lock-in depths. In *Timescales of the Internal Geomagnetic Field. American Geophysical Union* Geophysical Monograph, **145**: 205–219.

Channell, J.E.T., and Stoner, J.S., 2002. Plio-Pleistocene magnetic polarity stratigraphies and diagenetic magnetite dissolution at Ocean Drilling Program Leg 177 Sites (1089, 1091, 1093 and 1094). *Marine Micropaleontology*, **45**: 269–290.

Channell, J.E.T., Stoner, J.S., Hodell, D.A., and Charles, C., 2000. Geomagnetic paleointensity for the last 100 kyr from the subantarctic South Atlantic: a tool for interhemispheric correlation. *Earth and Planetary Science Letters*, **175**: 145–160.

Clement, B.M., Kent, D.V., and Opdyke, N.D., 1996. A synthesis of magnetostratigraphic results from Pliocene-Pleistocene sediments cored using the hydraulic piston corer. *Paleoceanography*, **11**: 299–308.

Curry, W.B., Shackleton, N.J., and Richter, C., *et al.*, 1997. Proceedings of the Ocean Drilling Program, Initial Reports, **154**: College Station, Texas (Ocean Drilling Program).

Davies, T.A., and Gorsline, D.S., 1976. Oceanic sediments and sedimentary processes. In Riley, J.P., and Chester, R., (eds.), *Chemical Oceanography*. New York: Academic Press, pp. 1–80.

DeMenocal, P.B., Ruddiman, W.F., and Kent, D.V., 1990. Depth of post-depositional remanence acquisition in deep-sea sediments: a case study of the Brunhes-Matuyama reversal and oxygen isotopic stage 19.1. *Earth and Planetary Science Letters*, **99**: 1–13.

Dunlop, D.J., and Ozdemir, O., 1997. *Rock Magnetism*. Cambridge: Cambridge University Press, 573 pp.

Florindo, F., Roberts, A.P., and Palmer, M.R., 2003. Magnetite dissolution in siliceous sediments. *Geochemistry, Geophysics, Geosystems*, **4**(7), 1053, doi:10.1029/2003GC000516.

Frankel, R.B., and Bazylinski, D.A., 2003. Biologically induced mineralization by bacteria. In Biomineralization. Dove, P.M., De Yoreo, J.J., and Weiner, S. (eds.), *Reviews in Mineralogy and Geochemistry*, **54**: 95–114. Washington, DC: The Mineralogical Society of America.

Fuller, M., Hastedt, M., and Herr, B., 1998. Coring-induced magnetization of recovered sediment. In Weaver, P.E.E., Schmincke, H-U., Firth, J.V., and Duffield, W. (eds.), Proceedings of the Ocean Drilling Program-Scientific Results, 157: College station, Texas (Ocean Drilling Program), pp. 47–56.

Giovanoli, F., 1979. A comparison of the magnetization of detrital and chemical sediments from Lake Zurich. *Geophysical Research Letters*, **6**: 233–235.

Guinasso, N.L., and Schink, D.R., 1975. Quantitative estimates of biological mixing rates in abyssal sediments. *Journal of Geophysical Research*, **80**: 3032–3043.

Henshaw, P.C., and Merrill, R.T., 1980. Magnetic and chemical changes in marine sediments. *Reviews of Geophysics and Space Physics*, **18**: 483–504.

Herr, B., Fuller, M., Haag, M., and Heider, F., 1998. Influence of drilling on two records of the Matuyama/Brunhes polarity transition in marine sediment cores near Gran Canaria. In Weaver, P.E.E., Schmincke, H-U., Firth, J.V., and Duffield, W. (eds.), Proceedings of the Ocean Drilling Program-Scientific Results, **157**: College station, Texas (Ocean Drilling Program), pp. 57–69.

Heywood, B.R., Bazylinski, D.A., Garratt-Reed, A.J., Mann, S., and Frankel, R.B., 1990. Controlled biosynthesis of greigite (Fe_3S_4) in magnetotactic bacteria. *Naturwiss*, **77**: 536–538.

Horng, C.-S., Torii, M., Shea, K.-S., and Kao, S.-J., 1998. Inconsistent magnetic polarities between greigite- and pyrrhotite/magnetite-bearing marine sediments from the Tsailiao-chi section, southwestern Taiwan. *Earth and Planetary Science Letters*, **164**: 467–481.

Hyodo, M., 1984. Possibility of reconstruction of the past geomagnetic field from homogeneous sediments, *Journal of Geomagnetism and Geoelectricity*, **36**: 45–62.

Irving, E., and Major, A., 1964. Post-depositional detrital remanent magnetization in a synthetic sediment. *Sedimentology*, **3**: 135–143.

Jansen, E., Raymo, M.E., Blum, P., *et al.*, 1996. Proceedings of The Ocean Drilling Program, Initial Reports, **162**: College Station, Texas (Ocean Drilling Program).

Johnson, H.P., Lowrie, W., and Kent, D.V., 1975. Stability of anhysteretic remanent magnetization in fine and coarse magnetite and maghemite particles. *Geophysical Journal of the Royal Astronomical Society*, **41**: 1–10.

Karlin, R., and Levi, S., 1983. Diagenesis of magnetic minerals in recent hemipelagic sediments. *Nature*, **303**: 327–330.

Karlin, R., Lyle, M., and Heath, G.R., 1987. Authigenic magnetite formation in suboxic marine sediments. *Nature*, **6**: 490–493.

[illegible] depositional remanent magnetism: preliminary experiments with natural sediments. *Earth and Planetary Science Letters*, **183**: 147–160.

Kent, D.V., and Lowrie, W., 1974. Origin of magnetic instability in sediment cores from the central north Pacific. *Journal of Geophysical Research*, **79**: 2987–3000.

Kent, D.V., and Schneider, D.A., 1995. Correlation of paleointensity variation records in the Brunhes/Matuyama polarity transition interval. *Earth and Planetary Science Letters*, **129**: 135–144.

King, J.W., Banerjee, S.K., and Marvin, J., 1983. A new rock-magnetic approach to selecting sediments for geomagnetic paleointensity studies: application to paleointensity for the last 4000 years. *Journal of Geophysical Research*, **88**: 5911–5921.

Kirschvink, J.L., 1980. South-seeking magnetic bacteria. *Journal of Experimental Biology*, **86**: 345–347.

Kirschvink, J.L., 1983. Biogenic ferrimagnetism: A new biomagnetism. In Williamson, S.J., Romani, G.-L., Kaufman, L., and Modena, I. (eds.), *Biomagnetism*. New York: Plenum Press, pp. 501–531.

Kirschvink, J.L., and Chang, S.B.R., 1984. Ultrafine-grained magnetite in deep-sea sediments: possible bacterial magnetofossils. *Geology*, **12**: 559–562.

Kirschvink, J.L., and Lowenstam, H.A., 1979. Mineralization and magnetization of chiton teeth: paleomagnetic, sedimentologic, and biologic implications of organic magnetite. *Earth and Planetary Science Letters*, **44**: 193–204.

Kirschvink, J.L., Walker, M.M., and Deibel, C.E., 2001. Magnetite-based magnetoreception. *Current Opinion in Neurobiology*, **11**: 462–467.

Kobayashi, K., and Nomura, M., 1972. Iron sulfides in the sediment cores from the Sea of Japan and their geophysical implications. *Earth and Planetary Science Letters*, **16**: 200–208.

Lee, C.H., and Jin, J.-H., 1995. Authigenic greigite in mud from the continental shelf of the Yellow Sea, off the southwestern Korean peninsula. *Marine Geology*, **128**: 11–15.

Leslie, B.W., Lund, S.P., and Hammond, D.E., 1990. Rock magnetic evidence for the dissolution and authigenic growth of magnetic minerals within anoxic marine sediments of the California continental borderland. *Journal of Geophysical Research*, **95**: 4437–4452.

Liu, S.V., Zhou, J., Zhang, C., Coile, D.R., Gajdarziska-Josifovska, M., and Phelps, T.J., 1997. Thermophilic Fe (III)-reducing bacteria from the deep subsurface: the evolutionary implications. *Science*, **277**: 1106–1109.

Lovley, D.R., 1990. Magnetite formation during microbial dissimilatory iron reduction. In Frankel, R.B., and Blakemore, R.P. (eds.), *Iron Biominerals*, New York: Plenum Press, pp. 151–166.

Lowenstam, H.A., and Weiner, S., 1989. *On Biomineralization*. New York: Oxford University Press.

Lund, S., and Keigwin, L., 1994. Measurement of the degree of smoothing in sediments paleomagnetic secular variation records: an example from late Quaternary deep-sea sediments of the Bermuda rise, western North Atlantic Ocean. *Earth and Planetary Science Letters*, **122**: 317–330.

Mann, S., Sparks, N.H.C., Frankel, R.B., Bazylinski, D.A., and Jannasch, H.W., 1990. Biomineralization of ferrimagnetic greigite (Fe_3S_4) and iron pyrite (FeS_2) in a magnetotactic bacterium. *Nature*, **343**: 258–260.

Mazaud, A., 1996. Sawtooth variation in magnetic intensity profiles and delayed acquisition of magnetization in deep sea cores. *Earth and Planetary Science Letters*, **139**: 379–386.

McCabe, C., and Channell, J.E.T., 1994. Late Paleozoic remagnetization in limestones of the Craven Basin (northern England) and the rock magnetic fingerprint of remagnetization secondary carbonates. *Journal of Geophysical Research*, **99**: 4603–4612.

McCabe, C., Van der Voo, R., Peacor, D.R., Scotese, C.R., and Freeman, R., 1983. Diagenetic magnetite carries ancient yet [illegible] *Geology*, **11**: 221–223.

McCabe, C., Jackson, M., and Saffer, B., 1989. Regional pattern of magnetic authigenesis in the Appalachian Basin: implications for the mechanism of late Paleozoic remagnetization. *Journal of Geophysical Research*, **94**: 10429–10443.

Meynadier, L., and Valet, J.-P., 1996. Post-depositional realignment of magnetic grains and asymetrical saw-toothed pattern of magnetization intensity. *Earth and Planetary Science Letters*, **140**: 123–132.

Moskowitz, B.M., Frankel, R.B., Flanders, P.J., Blakemore, R.P., and Schwartz, B.B., 1988. Magnetic properties of magnetotactic bacteria. *Journal of Magnetism and Magnetic Materials*, **73**: 273–288.

Moskowitz, B.M., Frankel, R.B., and Bazylinski, D.A., 1994. Rock magnetic criteria for the detection of biogenic magnetite. *Earth and Planetary Science Letters*, **120**: 283–300.

Opdyke, N.D., Glass, B., Hays, J.P., and Foster, J., 1966. Paleomagnetic study of Antarctica deep-sea cores. *Science*, **154**: 349–357.

Opdyke, N.D., and Channell, J.E.T., 1996. *Magnetic Stratigraphy*. London: Academic Press, 346 pp.

Petersen, N., von Dobeneck, T., and Vali, H., 1986. Fossil bacterial magnetite in deep-sea sediments from the South Atlantic Ocean. *Nature*, **320**: 611–615.

Reynolds, R.L., Tuttle, M.L., Rice, C.A., Fishman, N.S., Karachewski, J.A., and Sherman, D.M., 1994. Magnetization and geochemistry of greigite-bearing Cretaceous strata, north slope basin. Alaska, *American Journal of Science*, **294**: 485–528.

Richter, C., Valet, J.-P., and Solheid, P.A., 1997. Rock magnetic properties of sediments from Ceara Rise (Site 929): implications for the origin of the magnetic susceptibility signal. In Shackleton, N.J., Curry, W., and Bralower, T.J. (eds.), Proceedings of the Ocean Drilling Program-Scientific Results, 154: College station, Texas (Ocean Drilling Program), pp. 169–179.

Roberts, A.P., Reynolds, R.L., Verosub, K.L., and Adam, D.P., 1996. Environmental magnetic implications of greigite (Fe_3S_4) formation in a 3 Ma lake sediment record from Butte Valley, northern California. *Geophysical Research Letters*, **23**: 2859–2862.

Roberts, A.P., and Winklhofer, M., 2004. Why are geomagnetic excursions not always recorded in sediments? Constraints from post-depositional remanent magnetization lock-in modeling. *Earth and Planetary Science Letters*, **227**: 345–359.

Roberts, A.P., and Weaver, R., 2005. Multiple mechanisms of remagnetization involving sedimentary greigite (Fe_3S_4). *Earth and Planetary Science Letters*, **231**: 263–277.

Schneider, D.A., 1995. Paleomagnetism of some Leg 138 sediments: Detailing Miocene magnetostratigraphy, In Pisias, N.G., Mayer, L.A., Janecek, T.R., Palmer-Julson, A., and van Andel, T.H. (eds.), Proceedings of The Ocean Drilling Program—Scientific Results, 138: College Station Texas (Ocean Drilling Program), pp. 59–72.

Shackleton, N.J., Crowhurst, S.J., Weedon, G., and Laskar, L., 1999. Astronomical calibration of Oligocene-Miocene time. *The Royal Society London Philosophical Transactions*, **357**: 1909–1927.

Shipboard Scientific Party, Site 984, In Jansen, E., M. Raymo, P. Blum, *et al.* (eds.), 1996. Proceedings of The Ocean Drilling Program, Initial Reports, College Station, Texas (Ocean Drilling Program), **162**: 169–222.

Shipboard Scientific Party, Site 1089. In Gersonde, R., Hodell, D.A., Blum, P. *et al.*, 1999. Proceedings of The Ocean Drilling Program, Initial Reports, **177**, 1–97, [CD ROM]. Available from Ocean Drilling Program, Texas A&M University, College Station, Texas 77845–9547, USA.

Smith, C.R., and Rabouille, C., 2002. What controls the mixed-layer depth in deep-sea sediments? The importance of POC flux. *Limnology and Oceanography*, **47**: 418–426.

Snowball, I., and Thompson, R., 1988. The occurrence of greigite in sediments from Loch Lomond. *Journal of Quaternary Science*, **3**: 121–125.

Stoner, J.S., Channell, J.E.T., Hodell, D.A., and Charles, C., 2003. A 580 kyr paleomagnetic record from the sub-Antarctic South Atlantic (ODP Site 1089). *Journal of Geophysical Research*, **108**: 2244, doi:10.1029/2001JB001390.

Tarduno, J.A., 1994. Temporal trends of magnetic dissolution in the pelagic realm: Gauging paleoproductivity. *Earth and Planetary Science Letters*, **123**: 39–48.

Tauxe, L., Herbert, T., Shackleton, N.J., and Kok, Y.S., 1996. Astronomical calibration of the Matuyama-Brunhes boundary: consequences for the magnetic remanence acquisition in marine carbonates and Asian loess sequences. *Earth and Planetary Science Letters*, **140**: 133–146.

Teanby, N., and Gubbins, D., 2000. The effect of aliasing and lock-in processes on paleosecular variation records from sediments. *Geophysical Journal International*, **142**: 563–570.

Thomson, J., Brown, L., Nixon, S., Cook, G.T., and McKenzie, A.B., 2000. Bioturbation and Holocene sediment accumulation fluxes in the north-east Atlantic Ocean (Benthic Boundary Layer experiment sites). *Marine Geology*, **169**: 21–39.

Trauth, M.H., Sarnthein, M., and Arnold, M., 1997. Bioturbational mixing depth and carbon flux at the seafloor. *Paleoceanography*, **12**: 517–526.

Tric, E., Laj, C., Jéhanno, C., Valet, J-P., Kissel, C., Mazaud, A., and Iaccarino, S.,1991. High resolution record of the Olduvai transition from Po valley (Italy) sediments: support for dipolar transition geometry. *Physics of the Earth and Planetary Interiors* **65**: 319–336.

Vali, H., Förster, O., Amarantidis, G., and Petersen, N., 1987. Magnetotactic bacteria and their magnetofossils in sediments. *Earth and Planetary Science Letters*, **86**: 389–400.

Walker, M.M., Diebel, C.E., and Kirschvink, J.L., 2003. Detection and use of the Earth's magnetic field by aquatic vertebrates. In Collin, S.P., and Marshall, N.J. (eds.), *Sensory Processing in Aquatic Environments*. New York: Springer-Verlag, pp. 53–74.

Weeks, R.J., Roberts, A.P., Verosub, K.L., Okada, M., and Dubuisson, G.J., 1995. Magnetostratigraphy of upper Cenozoic sediments from Leg 145, North Pacific Ocean. In Rea, D.K., Basov, I.A., Scholl, D.W., and Allan, J. (eds.), Proceedings of The Ocean Drilling Program—Scientific Results, 145: College Station, Texas (Ocean Drilling Program), pp. 209–302.

Yamazaki, T., and Katsura, T., 1990. Magnetic grain size and viscous remanent magnetization of pelagic clay. *Journal of Geophysical Research*, **95**: 4373–4382.

Cross-references

Biomagnetism
Geomagnetic Polarity Timescales
Iron Sulfides
Magnetization, Chemical emanent (CRM)
Magnetization, Depositional Remanent (DRM)
Magnetization, Isothermal Remanent (IRM)
Magnetization, Natural Remanent (NRM)
Magnetization, Thermoremanent (TRM)
Magnetization, Viscous Remanent (VRM).
Magnetostratigraphy
Paleomagnetism

PALEOMAGNETISM, EXTRATERRESTRIAL

With the advent of the space age, a rich paleomagnetic record has been discovered in the inner solar system. The various bodies of the solar system differ in their interactions with the solar wind, depending upon their size, the presence or absence of active dynamos, remanent magnetic fields, and atmospheres. Observations of these interactions have shown that like Earth, Mercury appears to have an active dynamo, whereas Venus, the Moon, Mars, and those asteroids studied do not.

In a planet with an active dynamo, magnetic fields arise from magnetic material because of the induced magnetism in the ambient field and their earlier acquired remanent magnetism. Magnetic fields observed on the Moon and Mars in the absence of an active dynamo must be due to remanent magnetism only. These fields due to remanent magnetization, or paleomagnetic fields, have been investigated on the Moon and Mars with satellite-based surveys. In addition, samples brought to Earth by the Apollo missions from the Moon and those that have come to Earth from the Moon, Mars, and asteroids in the form of meteorites provide additional paleomagnetic data. Other meteorites include primitive material from the early solar system and even presolar grains.

This extraterrestrial paleomagnetic record may offer a fascinating glimpse of the magnetic fields in the early solar system, providing we are clever enough to interpret it correctly.

Lunar paleomagnetism

Observations from spacecraft prior to Apollo established limits on any lunar dipole moment should be at least five orders smaller than the geomagnetic dipole, but to the surprise of many, the Apollo-returned basalts and breccias carried stable natural remanent magnetization (NRM). With the advent of surface magnetometers on Apollo 12, 14, 15 and 16, subsatellites in low orbits on Apollo 15 and 16, and the more recent Lunar Prospector, it also became clear that the substantial volumes of the lunar crust are magnetized.

Apollo samples

The Apollo samples were collected from the lunar regolith and consisted of igneous rocks, breccias, and soil. Their rock magnetism is very different from that of terrestrial samples. The dominant magnetic phases in the lunar samples are metallic iron and iron-nickel as shown in Figure P27 (Strangway *et al.*, 1970; Nagata *et al.*, 1972). The Fe-Ni is in the form of the Ni poor alpha phase kamacite. Kamacite has an irreversible thermomagnetic curve with a Curie point between 740°C–770°C, whereas taenite has a range of Curie points dependent upon Ni content. On heating, kamacite converts to taenite with a sluggish return transition on cooling, whose temperature is dependent upon Ni content. This inversion is responsible for the irreversible thermomagnetic curve illustrated in Figure P27b. The mare basalts recovered from the regolith had iron as the dominant remanence carrying phase, whereas the breccias contained the Fe-Ni. In some breccias

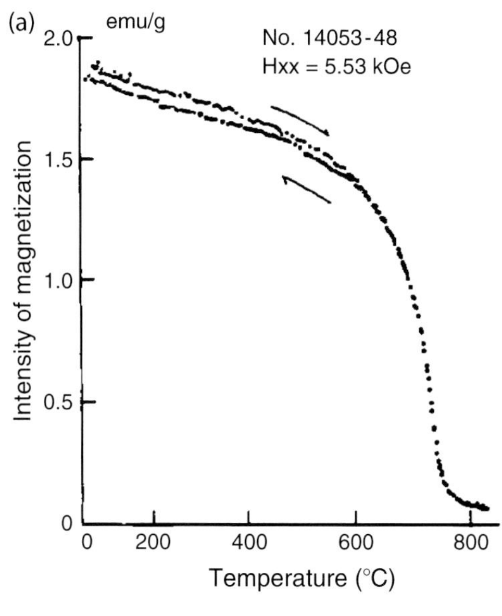

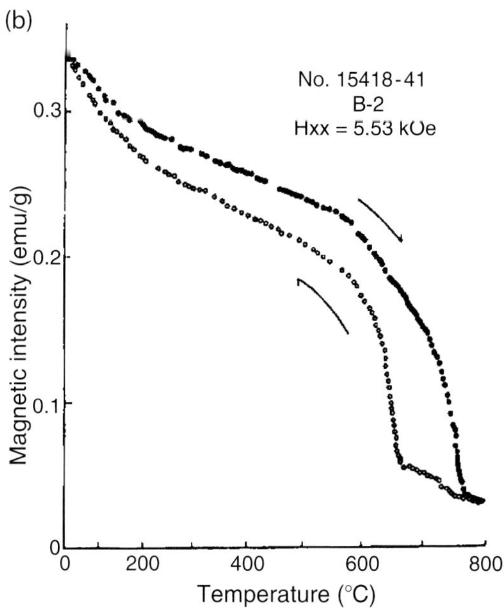

Figure P27 Magnetic phases in lunar rocks: (a) Apollo 14 mare basalt showing reversible thermomagnetic curve demonstrating presence of metallic iron and (b) Apollo 15 breccia, illustrating the effect of kamacite-taenite transition (Nagata *et al.*, 1972).

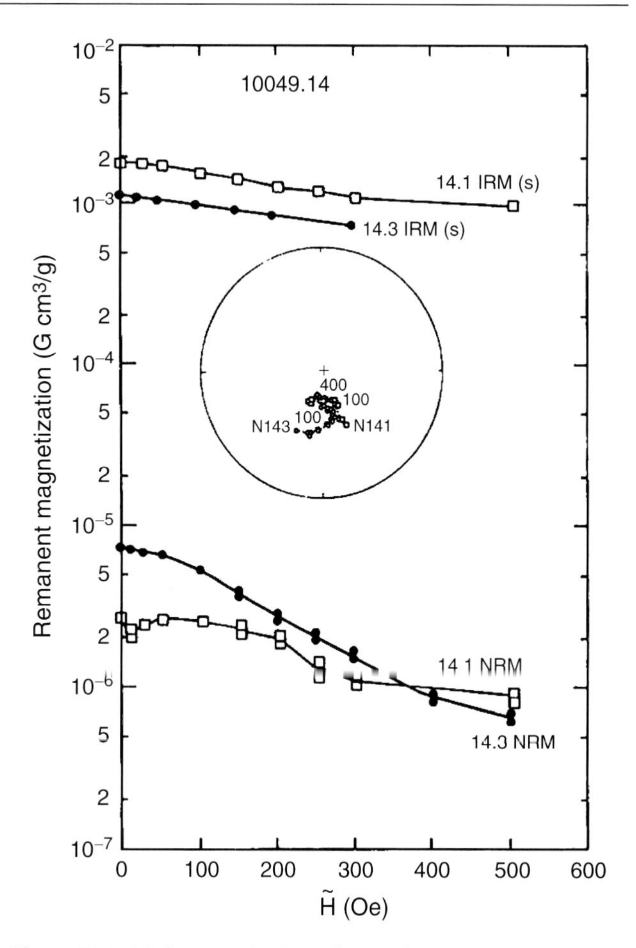

Figure P28 AF demagnetization of natural remanent magnetization (NRM) and saturation isothermal remanent magnetization (IRMs) of an Apollo 11 mare basalt. Note the directional similarity of the two samples after demagnetization to 100 Oe (10 mT).

with extreme coercivity the mineral "tetrataenite" may be present (Wasilewski, 1988).

The rock magnetism of the soils presented an immediate problem because they contained more ferromagnetic iron than the basalts from which they were principally derived. This iron is predominantly superparamagnetic as demonstrated by Nagata (1972). Three possible explanations of the origin of the excess iron were suggested: (1) by reduction of paramagnetic iron-rich silicates in the presence of solar wind hydrogen in the uppermost surface layer of the soil (Houseley *et al.*, 1973), (2) by reduction in the thermal blankets of ejecta material (Pearce *et al.*, 1972), and (3) by shock decomposition of iron rich silicates (Cisowski *et al.*, 1973). To some extent these processes overlap and all may contribute to the excess metallic iron, but the first mechanism appears to be dominant and may have relevance elsewhere in the solar system, for example on asteroids. The paramagnetic and ferromagnetic iron content of soils forms a useful classification scheme for soils and serves an indicative of provenance (Wasilewski and Fuller, 1975). The magnetic properties of the breccias vary according to their subgroups. The regolith breccias are similar to the soil from which they are derived by varying degrees of shock metamorphism. The fragmental breccias contain clasts in a porous matrix. The melt breccias, which consist of clasts in an igneous textured matrix, have properties similar to the igneous rocks.

The lunar paleomagnetic record comes primarily from the mare basalts and the melt breccias (see Figure P28). The plutonic rocks that are largely unsuitable for paleomagnetism are in the form of clasts in breccias. The paleomagnetism of the breccias is frequently hard to interpret, but in the melt breccias thermoremanent magnetization (TRM) appears to dominate, so that their mode of magnetization is similar to that of the mare basalts. The samples had experienced some degree of shock, exposure to radiation, and thermal cycling on the lunar surface.

The role of shock as a possible source of magnetization in the presence of a magnetic field, or demagnetization in the absence of such a field, was investigated in a number of experimental studies (Cisowski *et al.*, 1973, 1976). Flying plate experiments in which soil samples were encapsulated in the plate demonstrated that shock levels of a few GPa (tens of kilobars) in fields of tens of microteslas were sufficient to induce magnetization in lunar soil comparable in intensity with that found in the regolith breccias samples. In the absence of the magnetic field, shock was shown to harden thermoremanent magnetization by preferentially demagnetizing the softer magnetic phases. A flying plate impact at a velocity of 17 km/s into a block of terrestrial basalt in an ambient field of 1 mT field demonstrated shock magnetization and the generation of a plasma that displaced and compressed the ambient field (Figure P29, Srnka *et al.*, 1986). Effects of radiation and thermal cycling were also investigated and are unlikely to be important factors in the origin of lunar magnetism.

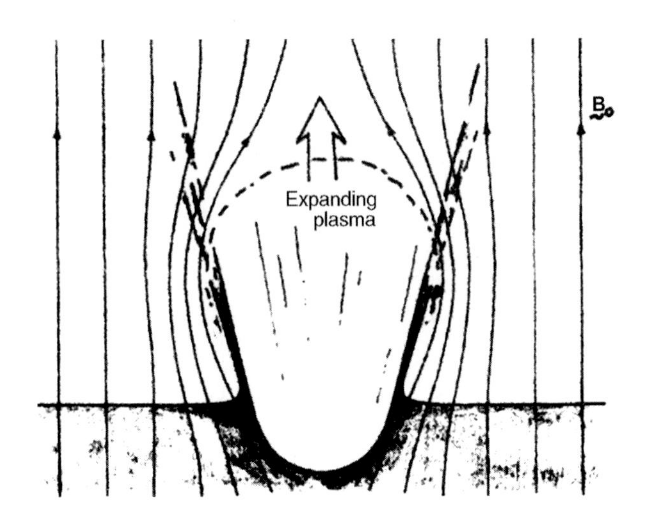

Figure P29 Model of field compression by impact in flying plate experiment. The plate was traveling at 17 km/s and impacted a block of basalt. The ambient field was 10 Oe (1 mT). The block was magnetized by the event and pick up coils demonstrated the formation of a plasma and field compression consistent with the model shown.

Later it became clear that at least some stable primary magnetization was preserved from the initial cooling of samples on the moon as igneous rocks. In the absence of any knowledge of the orientation,—in which the samples acquired their magnetization—the interest focused in on the intensity of magnetization they carry and the possibility of finding the history of the intensity of ancient lunar magnetic fields. Despite some success, classical methods of determining intensity involving heating the sample to give a TRM in a known field in the laboratory, were confounded by irreversible changes brought about by heating the samples. Various other methods including using ARM as a proxy for TRM (Shaw Stephenson and Collinson, 1974), heating by microwave radiation (Hale *et al.*, 1978) and IRM normalization (Cisowski and Fuller, 1986) were utilized. The IRM normalization method is admittedly a poor substitute for classical intensity determination methods, but because of the availability of data from a large number of samples it had a value as a rough indicator. Such a distinctive result followed that even the poor accuracy likely for IRM normalization was useful. From all of the intensity work it emerged that there was a high field with a surface intensity comparable to the Earth's surface field between 3.85 and 3.65 Ga (Figure P30). Two discussions of the subsequent history were that the field intensities gradually decreased over a period of about 1 Ga (Collinson, 1984; Runcorn, 1994). In contrast, Cisowski and Fuller (1986) suggested that there was a relatively brief period of high intensity from about 3.85 to 3.65 Ga and that outside of this period intensities could not be distinguished from background noise. Additional work is needed to clarify this point.

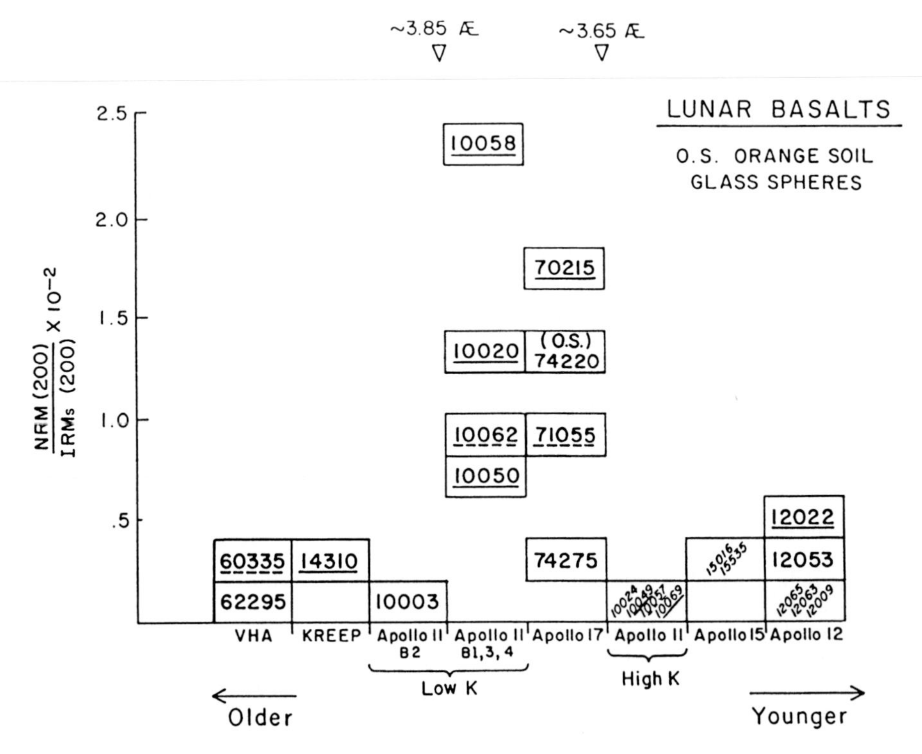

Figure P30 Paleointensity estimates from Apollo mare basalts by the IRMs normalization method. Calibration of the method by thermoremanent magnetism (TRM) experiments suggest that values of NRM(200)/IRMs(200) are equivalent to field intensities of 0.5 Oe. Note that only some of the Apollo 11 and 17 basalts with ages between approximately 3.85 and 3.65 Ga show values of NRM(200)/IRMs(200) greater than 0.5 suggesting that only in this age range were fields comparable to or greater than the geomagnetic surface field at present.

Magnetization of the lunar crust observed with surface and subsatellite magnetometers and electron reflectance experiment

The Apollo surface magnetometers (e.g., Dyall, 1972) revealed fields from a few tens of gammas (nT) at the Mare sites to 100 nT at the Fra Mauro Apollo 14 site and hundreds of nanoteslas at the Apollo 16 highland site. The stronger magnetizations could be explained by the strong magnetization of basin ejecta of the Fra Mauro and Cayley formations (Strangway *et al.*, 1973). The scale length of coherence appeared to be relatively short –from 100 m to km. The results from the subsatellite magnetometers (Coleman *et al.*, 1972; Russell *et al.*, 1975) and the electron reflectance experiments (Lin *et al.*, 1988) confirmed the results from the surface magnetometers in finding stronger fields over the highlands than over the Mare. They also showed stronger fields over the older eastern Mare than the younger western Mare. The high resolution of the electron reflectance experiment demonstrated that cratered regions were not systematically more strongly, or weakly, magnetized than surrounding regions (Lin, 1979). It is also confirmed from the observation of the subsatellite magnetometer surveys that the strongest magnetic anomalies were antipodal to the younger large impact basins. This suggested that a lunar field was required at the time of the younger giant impacts to account for the field compression and magnetization at the antipodes proposed by (Hood and Huang, 1991) as well as to account for the strong magnetization of mare basalts of this age. The sites of the strongest anomalies include regions of swirls similar to the Reiner gamma feature. These features are probably caused by the strong fields of the magnetic anomalies that stand off the radiation that otherwise darkens the albedo of the lunar surface.

The magnetometer/electron reflectometer experiment on Lunar Prospector has greatly improved the resolution and coverage of the magnetic mapping of the Moon, so that the entire surface is now mapped. The results were not only confirmed but also extended the earlier results that showed that the fields over the highlands were strong compared with those over the Mare despite the fact that the dominant occurrence of iron is of course in the Mare (Lucey *et al.*, 1995). The electron reflectance results (Mitchell *et al.*, 2004) also confirmed the ubiquitous presence of small-scale anomalies with dimensions of order tens of kilometers. This is consistent with the small scale of surface anomalies observed with the Apollo surface magnetometers. The absence of fringing anomalies around craters of up to 50 km in diameter (Halekas *et al.*, 2003) indicates that there cannot be coherent magnetization in the upper tens of kilometers of the crust. Moreover, since the Curie point of iron is soon reached in deeper regions, it appears that the crust does not contain large volumes of uniform magnetization. Thus the elegant theorem of Runcorn (1975), which proved that the field recorded in a spherical shell from cooling in the field of an internal dipole source is not be detectable from the outside, may not be relevant to the interpretation of lunar crustal magnetism. Following the assumption of shallow sources, Mitchell *et al.* (2003) developed a simple model for the magnetism of the lunar crust. They reconstructed the field by sequentially applying the magnetizing and demagnetizing effects of the known basin forming events to give the field illustrated in Figure P31. They found that the model was much improved by adding 1 nT background field. It therefore appears that the magnetization of the lunar crust is basically due to the effects of large basin forming events and that a relatively stronger field is required at the age of the proposed strong field era to explain the anomalies over the antipodes of the young basins.

Origin of lunar magnetism

After attempts to explain the magnetization of the Moon and of returned samples by exotic models had been investigated and found wanting, Runcorn's initial suggestion of a lunar dynamo (e.g., Runcorn, 1975) was generally accepted. The history of this dynamo was less clear and two rival ideas emerged corresponding to the two views of the field intensity records. One invoked a lunar dynamo that operated from about 3.9 Ga, giving a surface field comparable to the Earth's surface field and gradually lost intensity over about a billion years (Collinson, 1984; Runcorn 1994). In contrast, Cisowski and Fuller (1986) argued for a relatively brief period of high intensity between about 3.9 and 3.6 Ga with intensity outside of this interval indistinguishable from background noise. But still, it is not clear which of these suggestions were correct. With either model, the strongly magnetized mare and highland samples would have acquired their NRM as they get cooled in the dynamo field and the strongest anomalies antipodal to the younger giant basins could have acquired their magnetization in an enhanced antipodal field by shock magnetization as suggested by Hood *et al.* (1991). Recently, a short period of dynamo action has been proposed to be associated with a major overturn of the lunar mantle that made the core cool rapidly. This could have generated a magnetic field between 3.85 and 3.65 Ga, during a brief period of convection (Stegman *et al.*, 2003).

In conclusion, it should be remembered that the lunar dipole moment required to give a particular surface field on the Moon is smaller than the geomagnetic dipole moment required to give a comparable surface field on Earth because the radius of the Moon is smaller than the Earth's. Moreover, since we do not know how dynamo fields scale with the size of the region in which they are generated, we should not assume that the ratio of lunar to terrestrial dipole moments varies as the volume (Stevenson, 2002), or as the cube of the radii of the cores. Only in this case, does the cube of the core dimension cancel with the inverse cube of the dipole field fall off, so that the decrease in moment of the core is canceled by the decrease in the distance to source. In contrast, if the lunar dynamo had given the same moment as the geodynamo, which is admittedly unlikely given the small size of the lunar core, the lunar surface field would be two orders stronger than the surface field on the Earth. Admittedly, this is a rough and ready argument, but dynamo action in the lunar core with a dipole moment weaker than that of the geodynamo could still have generated surface fields similar to those interpreted from the paleomagnetic studies.

Martian paleomagnetism

Studies of Martian magnetism consist of direct observations of the magnetism of the soil by Viking, the magnetic surveys of the crustal field carried out by Mars Surveyor, and the paleomagnetic record of the Shergotite, Nahklite, and Chassignite (SNC) meteorites.

Viking and Pathfinder: soil and dust magnetic experiments

The Viking Magnetic Properties Experiment (Hargraves *et al.*, 1977, 1979) provided an estimate of the amount of magnetic material in the martian soil at a few percent. Moreover, the magnet experiments demonstrated that some 80% of the magnetic material was strongly magnetic suggesting that it was magnetite, titanomagnetite, or their more oxidized equivalents maghemite and titanomaghemite. Their preferred interpretation was that the mineral was maghemite and that its occurrence was in the form of a pigment dispersed throughout the soil.

Pollack *et al.* (1977) analyzed the airborne dust from the Martian images and found that the best match they achieved to their derived imaginary refractive indices was for fine-grained magnetite. Additional analysis of the airborne dust particles including Pathfinder data indicated that fine magnetite rather than titanomagnetite was required to explain the high-saturation magnetization.

The Mars Pathfinder lander carried similar magnetic experiments to those on Viking with the important addition of a multispectral imager (Hargaves *et al.*, 2000). Despite difficulties, the results of these analyses confirmed the earlier suggestion that there is a strongly magnetic phase present either as coatings, or nanoparticles. The phase is either magnetite or maghemite. Again, maghemite was the preferred choice.

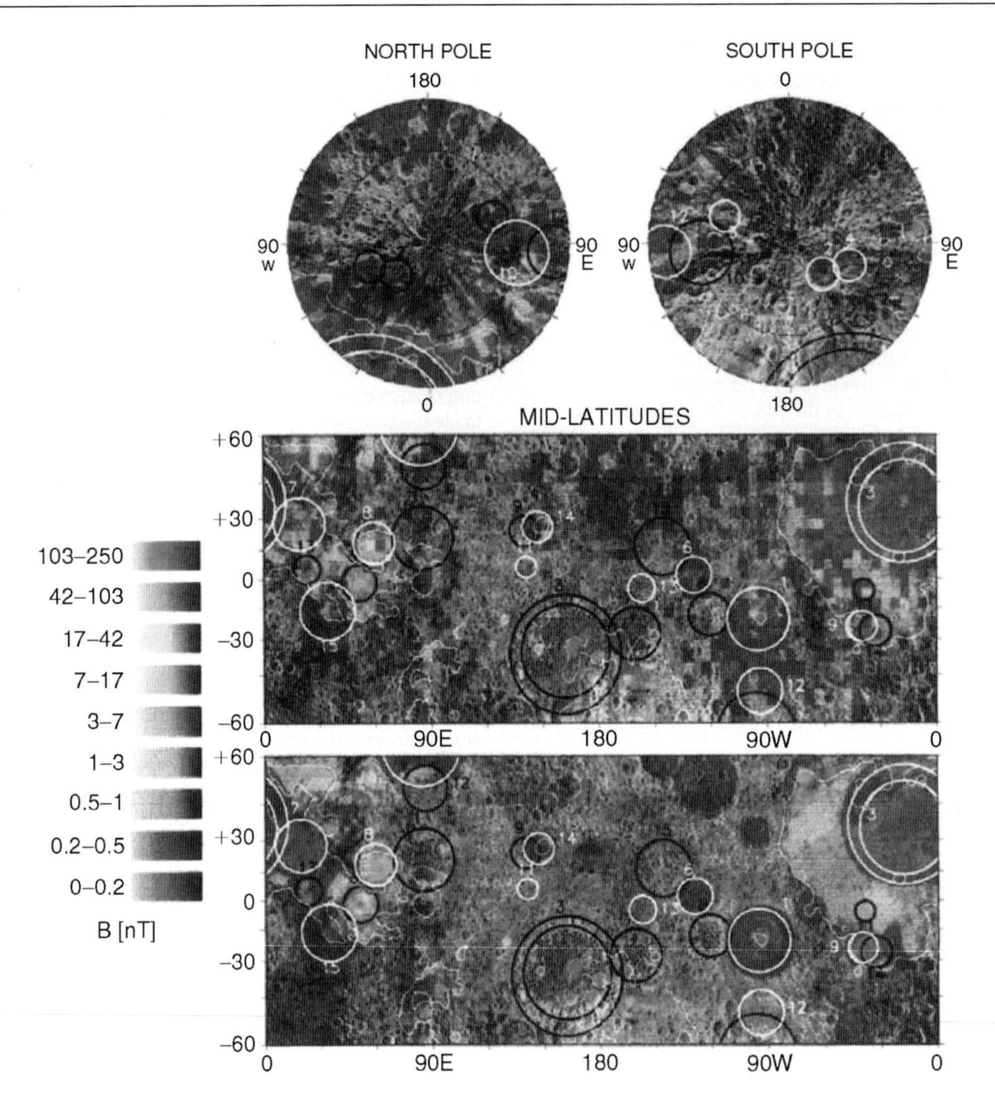

Figure P31 Model for lunar crustal magnetization; the top panel for mid-latitudes shows the magnetic anomaly data and the bottom panel shows the model reconstruction (With permission, Halekas, 2003).

On the missions presently at Mars, there are Mössbauer experiments that should identify the phases present. Such identification of the magnetic phases in the soil will be important in distinguishing an origin by direct weathering from surface basalts, or something more exotic. In the former case, titanium would be expected to be present, whereas pure magnetite, or maghemite, would suggest the latter.

Viking had also measured the composition of the martian atmosphere, which turned out to be a key piece of information demonstrating that the meteorites Shergottites, Nahklites, and Chassignites (SNC) did indeed come from Mars. The main discussion of them will therefore be given in this martian section.

Shergottites, Nahklites, and Chassignites (SNC) meteorites: the Martian meteorites

As noted above, the primary evidence for the martian origin of the SNC meteorites is that they contain noble gases that match the present martian atmosphere. In particular, it was found that the trapped noble gases in impact glass melt in EETA79001 were a perfect match for those measured by the Viking lander. Moreover, the ages of all except ALH84001 are young enough (200–1300 Ma) that given their volcanic nature they must have come from an evolved body with recent volcanic activity. Histories have been established for the various meteorites with estimates of the age of their initial formation on Mars, the age of the event that blasted them from the martian surface, their time in transit, and finally the age of arrival on Earth.

The rock magnetism of the SNC meteorites is much more like that of terrestrial samples than is that of the lunar samples, or indeed of other meteorites. The dominant carriers of the paleomagnetic record in these martian samples are magnetite and low Ti titanomagnetites and pyrrhotite. The paleomagnetism of the SNC meteorites has yielded a possible record of the surface magnetic field of Mars. Estimates of the order of 1 μT have come from microwave intensity determinations with Nakhla (Shaw *et al.*, 2001).

ALH84001 has played such a special role because of the claims of the presence of life forms in it (McKay *et al.*, 1997) that a separate discussion is given here. ALH84001 is a cataclastic orthopyroxenite. It initially crystallized at 4.5 Ga, was involved in a major impact event at 4.0 Ga, was excavated from the Martian surface by another event at approximately 16 Ma, and eventually landed in Antarctica 13 ka ago. It contains secondary carbonate of controversial origin within which magnetite also of controversial origin is found. The carbonate has been dated at approximately 4.0 Ga. This magnetite played a critical role in the controversy over the evidence for life because it was interpreted to be biogenic having been formed in martian bacteria analogously to the magnetite formed in terrestrial bacteria (Thomas-Keprta, 2000, 2001). However, the demonstration of a topotactic relationship of both the magnetite in iron-rich carbonate and periclase in magnesium-rich carbonate strongly suggests that most of the magnetite was formed by breakdown of the iron-rich carbonate during the impact event at 4.0 Ga (Barber and Scott, 2002). Whether there is other magnetite of biogenic origin remains to be seen, although Barber and Scott (2002) see no requirement for it.

The natural remanent magnetization (NRM) of ALH84001 is predominantly carried by the fine magnetite (Collinson, 1997; Cisowski, 1986). Collinson's initial study of mutually oriented samples of ALH84001 (Collinson 1997) indicated that individual subsamples had soft magnetizations whose directions were dissimilar, but that a harder magnetization, which demagnetized between 20 and 40 mT, was similar in different samples. His interpretation of the magnetization gave an estimate of the Martian field of one order smaller than the geomagnetic field.

Isolated NRM in two neighboring pyroxene grains of ALH84001 was reported by Kirschvink *et al.* (1997) to differ by some 70°. The magnetization was interpreted to have been acquired during cooling after the 4.0 Ga impact event and now differed due to subsequent differential rotation of the grains during brittle deformation. Given this interpretation, Kirschvink *et al.* (1997) argued that the differing magnetizations in the neighboring grains leads to an upper temperature limit, since magnetization on Mars of 110C (the maximum temperature the sample was exposed to during sample preparation). Using exquisite technique with high-resolution SQUID magnetometers, Weiss *et al.* (2000) discovered that the magnetization was inhomogeneous on the scale of the individual carbonate blebs. They suggested that the magnetization was probably carried by pyrrhotite and magnetite in association with the carbonate. In this work, observations of partial thermal demagnetization of small regions of millimeter size with the high-resolution SQUIDs further reduced the constraints on heating, since the time that the magnetization was acquired on the Martian surface, to 40C (Figure P32).

Antretter *et al.* (2003) confirmed the observation by Collinson (1997) that despite soft magnetization, which differed from one subsample to another, at higher intermediate harder magnetization from different subsamples agreed in direction. Assuming that the magnetite was formed from the carbonate in the major impact event at 4.0 Ga, as petrological evidence suggests, the NRM will record this field.

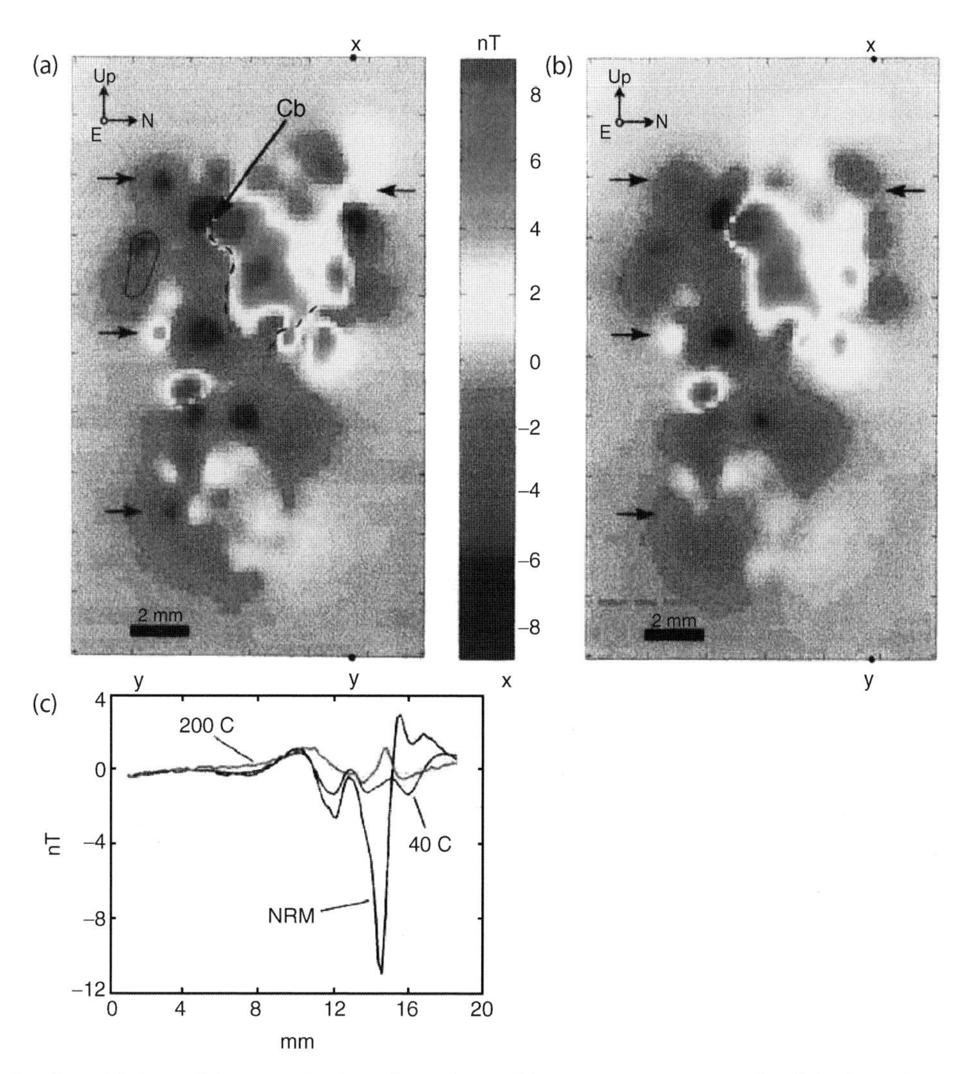

Figure P32 Results of partial thermal demagnetization of a region within ALH 84001 measured with high-resolution SQUID measurements. Note that the magnetization is changed upon heating to 40°C. Hence assuming that the NRM was acquired entirely on Mars, the material cannot have been heated subsequently above 40°C (With permission; Weiss *et al.*, 2000).

Normalizing the intensity of the NRM by the saturation isothermal remanence (IRMs) then gives an estimate for the 4.0 Ga martian field one order smaller than the present geomagnetic field. If the magnetization is heterogeneous due to magnetization at different times, or due to subsequent misalignment as suggested by Kirschvink *et al.* (1997), this may be a minimum field value.

Crustal magnetic field of Mars

Just as the paleomagnetic record discovered by the Apollo missions was one of the bigger surprises of the lunar exploration, so the magnitude and distribution of the martian magnetic anomalies has proved a major surprise in martian exploration. The magnetic crustal field on Mars differs from the terrestrial crustal field in its intensity and distribution (Acuna *et al.*, 1999; Connerney *et al.*, 2001). On Earth, the magnetic features are distributed more or less uniformly over the planet. On Mars, the strong features are largely confined to a band that covers two-thirds of the southern highlands (Figure P33). They also reach an order of magnitude stronger than substantial terrestrial anomalies. An $n = 90$ spherical harmonic model (Cain *et al.*, 2003) gave an upper limit for the martian dipole six orders smaller than the geomagnetic dipole moment and close to the noise in the martian coefficients. However, between $n = 20$ and 40 the martian power spectrum is 40 times stronger than that of the Earth.

There have been discussions of the linear nature of the anomalies and the possibility that they record a seafloor-spreading-like mechanism on Mars (Acuna *et al.*, 1999) or terrane accretions (Fairen *et al.*, 2002). However, the evidence for seafloor-like anomalies has been questioned by Harrison (Harrison, 2000) and even the linear nature of the features has been disputed by Arkhani-Hamed (2001a). The possibility of reversals of the Martian dynamo is supported by calculations of paleomagnetic poles from isolated anomalies that turned out to be in similar locations, but of both polarities (Arkhani-Hamed, 2001b). More recent analyses have increased the resolution of the surveys and drawn attention to the presence of smaller anomalies in the northern plains and the possibility of the role of shock demagnetization associated with the major basin of Utopia (Halekas, 2003).

Discussion

The origin of the martian anomalies and the explanation of their distribution and strength remain unclear. However, it appears that there must have been a relatively strong surface magnetic field, at least comparable with that of the geomagnetic surface field, when the strong anomalies of the southern highlands were formed. The caveat discussed in reference to the lunar dynamo and the strength of the surface fields it may generate must also be remembered here. Hence, if the dipole moments generated by the martian and geodynamos are similar, say to order of magnitude, then the surface martian field would be stronger than terrestrial surface field simply due to the ratio of the radii of the planets. The strongest observed anomalies are about an order larger than those on Earth.

Numerous magnetic phases have been suggested to help in understanding the high remanent magnetization of the Martian crust: e.g., hematite (Dunlop and Kletetschka, 2001, Kletetschka *et al.*, 2000a, Özedmir and Dunlop, 2002), hematite-ilmenite solid solution (Kletetschka *et al.*, 2002, Robinson *et al.*, 2000), and low-temperature oxidation, or weathering to give maghemite, or hematite (Özdemir, 2000). Cooling experiments with iron-rich basalts were carried out and revealed the presence of titanium-rich cruciform titanomagnetites that were modeled as the source of the martian crustal anomalies (Hammer *et al.*, 2003). The role of pyrrhotite has been emphasized by Rochette *et al.* (2001). The most potent magnetic material that might give rise to these strongly magnetized rocks is generally recognized to be SD magnetite (e.g., Kletetschka *et al.*, 2000b).

Given the saturation magnetization of magnetite of 4.9×10^5 A/m, the saturation remanent magnetization of a 0.5% dispersion of uniaxial, SD magnetite will be between 1×10^3 A/m and 2×10^3 A/m. Assuming a surface field intensity an order of magnitude greater than the geomagnetic field, the TRM, or indeed the chemical remanent magnetization

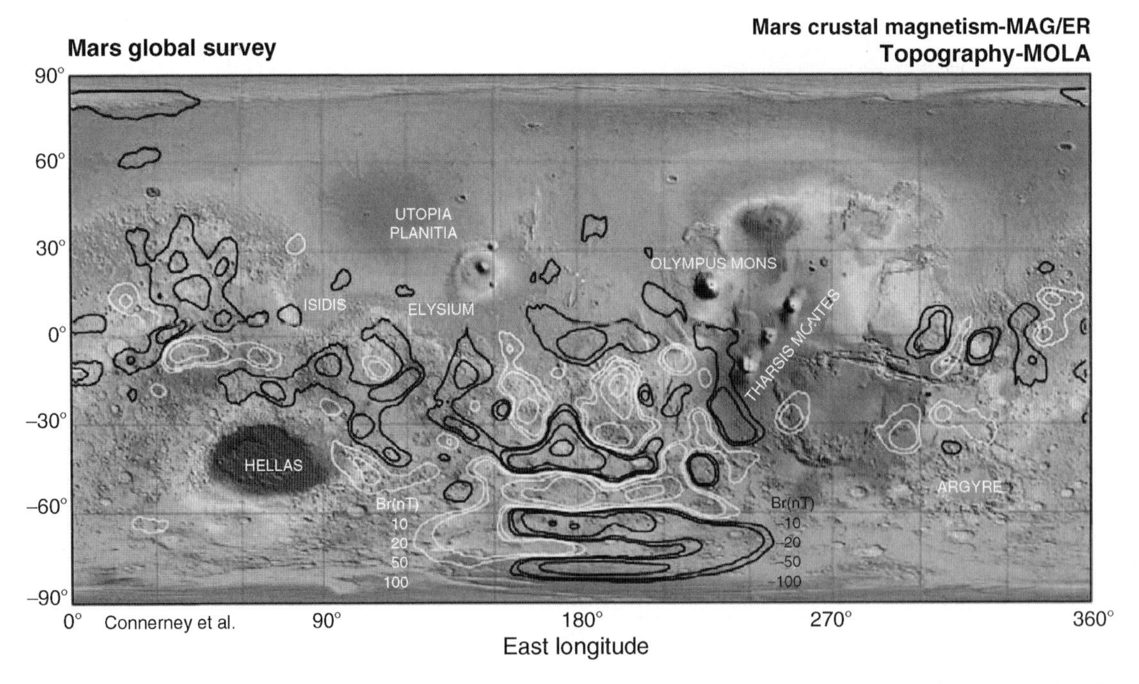

Figure P33 Mars crustal magnetic anomalies from Connerney *et al.* (2001). Note that the strongest anomalies are in a band stretching some 60° of latitude, which is centered equatorially near 0° longitude and expands to cover most of the Southern Hemisphere near 180° longitude. Within the Southern Hemisphere, the anomalies are weak in Hellas and Argyre, and over the Tharsis region (With permission, Purucker).

(CRM), will give the ~10–20 A/m required by the models for the Martian crustal fields. With a surface field comparable to the geomagnetic field, 0.5% of SD magnetite may be sufficient to give the required to depths of ~10 km in parts of the Southern Hemisphere to meet the requirements of the models. The key point is that this magnetite must be in a SD state.

No matter what proves to be the explanation of the intensity of the martian anomalies, their distribution and in particular their confinement to a band largely in the Southern Hemisphere remains a puzzle (Figure P31). Moreover, there is strong evidence that beneath the younger surface material of the northern plains, there is cratered older material similar to that found in the Southern Highlands (Frey, 2003), so that the superficial difference in appearance of the two regions is not a sufficient explanation.

To explain the lack of strong anomalies in the north, hydrothermal activity has been invoked to demagnetize the features in the northern basin (Solomon *et al.*, 2002). Another possibility is that the strong magnetic anomalies are never formed in the Northern Hemisphere, but only in the Southern Highlands (Scott and Fuller, 2004). This idea is based on: (1) an early carbon dioxide rich atmosphere in which weak acids formed, dissolved iron from igneous rocks, depositing iron rich carbonates in the upper crust and (2) the fine magnetite that was then formed by decomposition of the siderite on subsequent heating due to intrusions. Given the observed correlation between water channels and magnetic anomalies in the Southern Highlands (Harrison and Grimm, 2000), this suggestion is consistent with the disposition of the anomalies. Only in these regions water was found to be present in the appropriate amounts to permit the necessarily intermittent dissolution and deposition processes to take place.

Another factor that has emerged with the newer analyses is the possible role of demagnetization caused by the impacts that formed the large basins in the Northern Hemisphere. In this view, it appears that just as in the absence of strong magnetization in the major basins on the Moon permits a simple model of magnetization of the lunar crust, so on Mars, the same absence of strong anomalies in Argyre, Hellas, and Utopia basin suggest a similar role of shock demagnetization on Mars. Indeed Hood *et al.* (2000) have argued that the size of the demagnetized region around Hellas corresponds to the region that would have experienced 2 GPa, the shock value required for pyrrhotite to pass through a transition into a nonmagnetic phase. If shock does play a dominant role of in determining not only the lack of anomalies associated with Argyre and Hellas, but also with even bigger features of Utopia in the Northern Hemisphere, then important aspects of the Martian crustal magnetic field may be modeled by similar demagnetizing effects to those utilized in the lunar model of Mitchell *et al.* (2003).

Conclusion

The martian crustal anomalies, the magnetization of ALH840001, and other SNC meteorites suggest that a martian dynamo generated a global magnetic field early in martian history, but that it shut off sometime soon after 4.0 Ga. The strength of the surface anomalies may be partially explained by the smaller distance from the sources in the martian core to the surface, but their distribution on the martian surface requires an explanation. Clearly the role of shock demagnetization associated with large basin-forming events has played an important role in determining the observed distribution, but there appears to be a need for some additional explanation of the absence of anomalies in the Northern Hemisphere.

Asteroid paleomagnetism

Prior to the 1990s our knowledge of the asteroids was from telescopic observations and interpretations of the sources of certain meteorites. With advent of the space era of exploration, observations by spacecraft with magnetometers onboard permitted estimates of the magnetic fields and ultimately of the magnetization of a sampling of asteroids. Beginning with the flyby of 9969 Braille in 1991 by Deep Space 1 (DS1) (Richter *et al.*, 2001), there have been flybys of 951 Gaspra (Kivelson *et al.*, 1993) and 243 Ida by Galileo on its way to Jupiter, and of 433 Eros and of 253 Mathilde by the Near Earth Rendezvous Mission (NEAR) which subsequently landed on 433 Eros (Acuna *et al.*, 2001). With the exception of the observations on 433 Eros, estimates of the magnetic fields of the various asteroids rely on the interpretation of the interaction of the solar wind with these bodies. The nature of this interaction can reveal whether it is strongly and coherently magnetized, weakly and incoherently magnetized, or whether it is without remanent magnetization and simply has an induced magnetization dependent on the solar wind field. The results of such measurements constrain the origin and history of the asteroid and potentially have important bearing on the origin of the solar system.

The closest approach of DS1 of 28 km was achieved at 1.3 AU. Despite the strong magnetic noise from the ion propulsion unit of DS1, magnetic field measurements were made of 9969 Braille and yielded an estimate of its magnetic moment of 2.1×10^{11} A m^2 (Richter *et al.*, 2001). The magnetometer on Galileo observed a disturbance of the solar wind field during the flyby of Gaspra consistent with the asteroid having a remanent magnetization (Figure P34). The intensity of magnetization was comparable with that of chondritic meteorites (Kivelson *et al.*, 1993; Baumgartel *et al.*, 1994). A similar effect was observed on the flyby of 243 Ida, although it was weaker. In contrast to these results, the magnetometer on the NEAR spacecraft on its approach to 433 Eros and after landing failed to detect a global magnetic field (Acuna *et al.*, 2002). The upper limit on the field value was reported 0.005 A/m, giving a remanent magnetization of 1.9×10^{-6} A m^2/kg. This is orders of magnitude smaller than those reported for 951 Gaspra and 9969 Braille and for meteorites likely to have come from this type of asteroid.

Taken at face value these results suggest that Gaspra and Ida have relatively strong coherent magnetization comparable in intensity to that seen in strongly magnetized chondrites (see below) to which they are most closely linked spectroscopically. This would be consistent with their S-type classification. Such a magnetization might have been acquired by initial cooling in a substantial magnetic field, or as a result of some subsequent event. On the other hand, the results from Eros, also an S-type, suggest cooling in the absence of a field, or random magnetization, such as might be acquired in a cold accretion process. The puzzling results from the asteroid magnetic field observations led to suggestions of random magnetization within 433 Eros (Wasilewski *et al.*, 2002) and to reconsideration of the possibility that the magnetization of chondritic meteorites might be randomly oriented. The latter will be discussed below.

Conclusion

The results from 9969 Braille, 951 Gaspra, and to a lesser extent 243 Ida suggest homogenous magnetization and hence magnetization recording the field of the planetesimal in which it was formed or some subsequent event in a field. In contrast, the results from 433 Eros suggest that its constituent magnetic phases either carried little intrinsic magnetization, or that they were not strongly heated in the accretion process.

Meteorite paleomagnetism

Introduction

Meteorites provide a sampling of the oldest material in the solar system and even pre-solar grains. In addition they sample the moon, planets, and asteroids. Recently, the number of meteorites available for analysis has more than doubled because of the remarkable Antarctic meteorite collections obtained by recent Japanese and US expeditions, in which Prof. William Cassidy of the University of Pittsburgh played a key role. We thus have the possibility of studying the paleomagnetism,

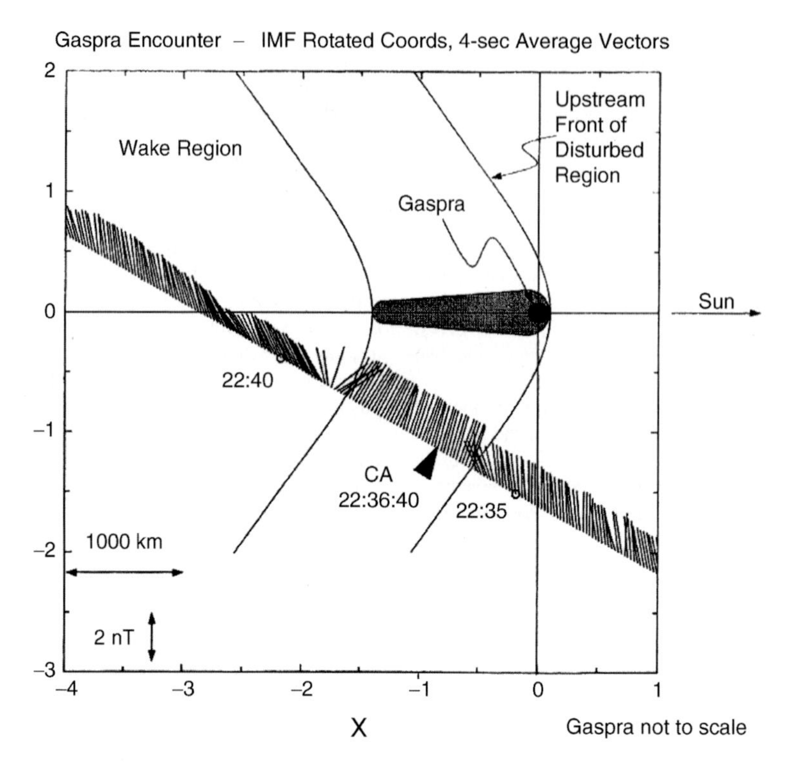

Figure P34 Disturbance of solar wind field during Gaspra flyby and a model for its magnetosphere (With permission; Kivelson *et al.*, 1993).

as well as the petrology, chemical, and isotopic composition and age of this material, so fortuitously brought to Earth and preserved in a relatively pristine state. Meteorites are divided into chondrites, achondrites, stoney-irons, and irons. The irons are separated because they are metal rich. The chondrites and achondrites are separated according to the presence, or absence of chondrules, which are small spherules, 0.1–2 mm in size. The achondrites are differentiated, in contrast to the chondrites. McSween (1999) provides an excellent modern introduction to meteorites.

Chondrites are heterogeneous aggregates with radiogenic ages of 4.56 Ga, which is consistent with the presence in chondrites of products of short-lived radioactive isotopes such as ^{26}Al, and the similarity of their composition to that of the sun. Among the inclusions in chondrites are the refractory calcium aluminum inclusions (CAI). They formed between 1400 and 1100°C as condensates, or evaporative residues, or more complex mixtures of thermally processed material. As such, they are of great interest as a means of glimpsing early solar system magnetic fields, although the magnetic material in them formed somewhat later than the earliest condensates. The chondrule population in the chondrites, is the next most primitive material available to us, which probably formed a few million years after the CAIs, although their origin remains somewhat controversial. Unfortunately, the CAIs and chondrules have suffered either the varying degrees of aqueous alteration especially in the various carbonaceous chondrites, or the thermal metamorphism especially in the ordinary chondrites.

The achondrites in principle are simpler to understand than the chondrites because they are derived from the products of igneous differentiation similar to processes seen on the Earth. However, achondrites are also frequently brecciated, so that we find rarely anything like a pristine igneous rock. They crystallized at approximately 4.5 Ga.

The paleomagnetism of meteorites has been a major research area for more than half a century since the pioneering work of Stacey *et al.* (1961) and a comprehensive review appeared in the 1980s (Cisowski, 1987), as well as a very useful discussion in the *Rock Magnetism* text by Dunlop and Özdemir (1999). It is a rich field with the possibility of giving information about early solar system magnetic fields, and in addition key magnetic data for asteroids, the moon, and Mars. There are however major difficulties in the interpretation of the paleomagnetic record of meteorites that must be addressed before the data can be used with confidence (e.g., Wasilewski *et al.*, 2002). Magnetic studies have focused upon the chondrites, their inclusions, and achondrites. Yet, the complicated histories of these meteorites make the interpretation of their paleomagnetic record particularly tricky.

Magnetic properties of different categories of meteorites

Irons contain dominantly Fe-Ni, in the form of the Ni poor alpha phase kamacite, and the Ni-rich gamma phase taenite. Irons have been classified according to their composition and the dimensions of the Widmanstatten pattern, which is related to their Ni content and cooling rate. Kamacite has an irreversible thermomagnetic curve with a Curie point of between 740 C and 770 C, whereas taenite has a range of Curie points dependent upon Ni content. As we noted in the discussion of the magnetic properties of the Apollo samples, on heating kamacite inverts to taenite with a sluggish return transition on cooling. This temperature is dependent upon Ni content. A plot of the fraction of the saturation magnetization due to the alpha Fe-Ni phase kamacite against the temperature of the gamma to alpha transition separates the various subgroups (Figure P35a) (Strangway *et al.*, 1970; Nagata *et al.*, 1972).

Because many chondrites contain metallic Fe-Ni, whereas achondrites do not, the subdividision into chondrites, or achondrites, has a natural magnetic expression. This is best seen if the fraction of the total saturation magnetization due to the alpha phase, or kamacite, is plotted against the total saturation magnetization of the meteorite (Figure P35b, P35c). Kamacite and taenite, alpha and gamma phases of NiFe dominate the magnetic properties of most meteorites. It is tetrataenite, the stable form of NiFe below 320°C that is seen for the most part in ordinary chondrites as was first pointed out by Wasilewski

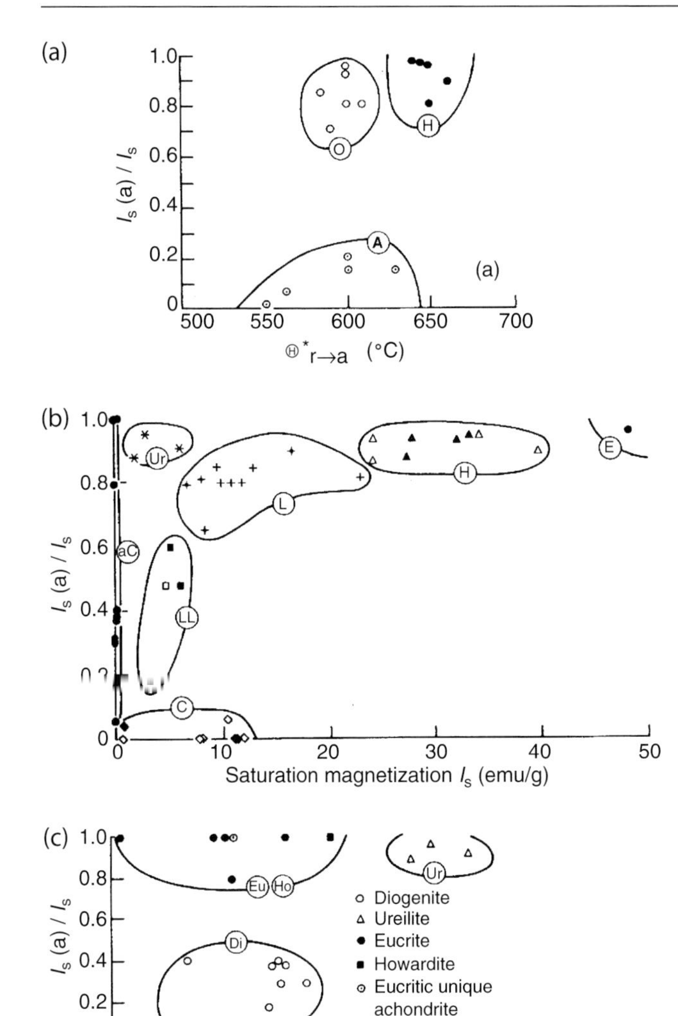

Figure P35 Magnetic classification of meteorites (a) the ratio of the saturation magnetization of the alpha phase to the total saturation magnetization plotted against the temperature of the gamma to alpha transition defines the various group of the irons H—hexahedrites, O—octahedrites, and A—ataxites. (b) the same ratio plotted against saturation magnetization defines the various groups of chondrites and (c) also separates the various groups of achondrites (Strangway *et al.*, 1970; Nagata *et al.*, 1972; Cisowski, 1987).

(e.g., 1988). Tetrataenite is remarkable for its extreme anisotropy and coercivity that can reach as much as 100 mT. The carbonaceous chondrites have multiple magnetic phases that are not yet completely understood.

Achondrites are the products of igneous differentiation and have in general much less metallic content than the chondrites. As Cisowski (1987) notes, they are somewhat analogous to lunar samples with eucrites and diogenites being similar to monomict lunar breccias. They too are conveniently divided in the same parameters as those used for the chondrites (Figure P35c).

The SNCs (Shergotites, Nahklites, and Chassignites) are a special group of meteorites from an evolved body with relatively recent volcanic activity that has been shown to be Mars. They contain magnetite, titanomagnetites, and pyrrhotite and have been discussed above in the section on Mars.

A lunar origin has been demonstrated for certain meteorites. These have the characteristic mixes of metallic Fe and Fe-Ni found in lunar samples.

Paleomagnetic record of meteorites

The prime interest in meteorite paleomagnetic studies is in the intensity of the fields in which the NRM was acquired and the possibility of gaining insight to early solar system magnetic fields, or in the martian and lunar meteorites to the history of the magnetic field of Mars and the Moon. Meteorites may be exposed to shock effects in their assembly or in their subsequent history, to radiation, and to temperature changes in space, and finally to heating as they come through the atmosphere. All these factors can have an effect on their paleomagnetic record, but they do not eliminate the possibility of some primary magnetization acquired at the time of their formation.

Iron meteorites can carry stable NRM, but the microstructure of the kamacite taenite intergrowths has been shown to control the direction of TRM, while the intensity of the TRM was not found to depend upon the field in which it was acquired (Brecher and Albright, 1977). For these reasons the paleomagnetic record of the irons has not been pursued strongly.

The paleomagnetic record of the chondrites and achondrites has been energetically investigated. Unfortunately, many of the meteorites collected in earlier times were contaminated by magnets used to recognize them as meteorites. Recently, and particularly with the Antarctic collections, care has been taken to maintain the meteorites in as pristine a condition as possible. In earlier work, plots of NRM versus IRMs were used to aid in the detection of contamination and in the interpretation of the origin of the NRM (Sugiura and Strangway, 1988). Wasilewski and Dickinson (2000) have recently suggested using the ratio of NRM:IRMs (or REM as they call it) as a preliminary indication of contamination, with values greater than 0.1 being indicative of exposure to a strong magnet.

The carbonaceous chondrites are some of the most intensely investigated of all meteorites with studies of individual CAIs, chondrules, and of bulk samples. The individual refractory inclusions and other chondrules are of particular interest because they may contain a record of solar system magnetic fields extant prior to the assembly of the meteorite. A necessary test for this is the demonstration that their magnetization was acquired before aggregation of the parent meteorites randomly oriented in the sample. This is an application of the classical conglomerate test of paleomagnetism, in which the clasts within a conglomerate are shown to be magnetized randomly within the conglomerate, whereas the matrix is coherently magnetized. Thus their magnetization may have been acquired prior to assembly into the conglomerate because they do not exhibit the coherent magnetization that might have been acquired during, or since the conglomerate was assembled. Unfortunately, this very direct test has been confounded by the random nature of the magnetization of the matrix and its virtual absence as a significant carrier of homogeneous NRM in some chondrites.

Paleomagnetic analyses of CAIs in Allende have been reported (Smethurst and Herrero-Bervera, 2002) and gave consistent values of 5.2–5.4 μT by the Shaw method (Shaw, 1974) for the intensity of the field they record. Results from Allende chondrules are inconsistent: values range from hundreds of millitesla (Lanoix *et al.*, 1978) to 70 and 130 mT (Nagata and Sugiura, 1977). Moreover, Nagata and Sugiura (1977) obtained similar results from the matrix and from bulk samples as they had from chondrules in double heating experiments from 20 to 300°C. This of course strongly suggests that the magnetization was a low-temperature overprint. Such information may be of interest as a means of estimating the temperature to which the meteorite was heated at the time of acquisition of the low-temperature overprint. In ALH 76009, 13 chondrules were found to have random orientation despite the matrix samples showing an overprint (Nagata and Funaki, 1981). Stepwise thermal demagnetization suggested that the chondrules may have acquired their magnetization in a steady field within which they were rotating slowly.

The interpretation of the magnetization of carbonaceous chondrites, as opposed to the inclusions that they contain, is complicated by the discovery that the paleomagnetic record of the matrix is randomly oriented within some of these meteorites (e.g., Brecher, Stein, and Fuhrman, 1977). The magnetization of some carbonaceous chondrites such as Coolidge, Murray, and Yamato fall off slowly with AF demagnetization and exhibit directional stability. Investigations of multiple oriented samples from these meteorites to check for random magnetization would be of great interest. The presence of tetrataenite is suggested by extreme stability against AF demagnetization. Other carbonaceous chondrites show quite different behavior with NRM that falls off rapidly with the NRM decreasing by as much as an order of magnitude with AF to 10 mT. e.g., Tysnhes Island and Bushof (Gus'kova, 1963). Results are also available for samples that have been cycled through the low-temperature transition of magnetite at ~120 K (Brecher and Arrhenius, 1974). Classical double heating intensity determinations have been carried out for Orgueil, Murchison, Leoville, with multiple determinations for Allende. The results fall in the range of tens and hundreds of microtesla. However, the bulk of the magnetization is blocked below 200°C and one must question how well such magnetization is likely to reflect fields more than 4 Ga old.

The magnetization of ordinary chondrites suffers from the same problem of the heterogeneity of the magnetization that is encountered with the carbonaceous chondrites. The phenomenon was investigated systematically by Morden and Collinson (1992). They demonstrated that the magnetism of subsamples of 8LL chondrites and 3L chondrites was randomly oriented down to the scale of millimeters, which was the smallest for which they could be confident of maintaining accurate orientation of subsamples (Figure P36). They also discovered a magnetic fabric, which they considered to be a primary structure formed during the final lithification of the various chondrites. The random paleomagnetic result was therefore interpreted as primary. However, this interpretation is not consistent with the strong metamorphic heating of some of these meteorites, unless magnetization took place before final aggregation. Subsamples of Bjurbole including were also found to be randomly magnetized (Wasilewski *et al.*, 2002). This once again raises a general question of how homogeneous the magnetization of chondrite is and hence what size of sample can be regarded as giving an indication of any possible primary magnetization. Nevertheless numerous dual heating intensity determinations on bulk samples have been carried out. They have yielded values a little lower than the carbonaceous chondrites with values falling between a few μT and nearly 200 μT. The blocking temperatures are rather higher than in the carbonaceous chondrites with the pTRM acquired and the NRM demagnetized linearly related up to steps near to 400°C.

The interpretation of the magnetization of achondrites is in principle simpler than that of chondrites because achondrites originated by crystallization from a magma, or in the case of primitive achondrites, as the residues of partial melting. Therefore they are formed above the Curie points of any magnetic minerals they contain and should have acquired a primary NRM of thermal origin on the evolved body in which they were formed. However, unfortunately only very few achondrites appear to be unbrecciated igneous rocks. Rather achondrites mostly compare with the lunar suite of breccias with igneous clasts, whereas just a few may be true igneous rocks. The crystallization ages of the howardite, eucrite, diogenite clan (HED) are between 4.40 and 4.55 Ga

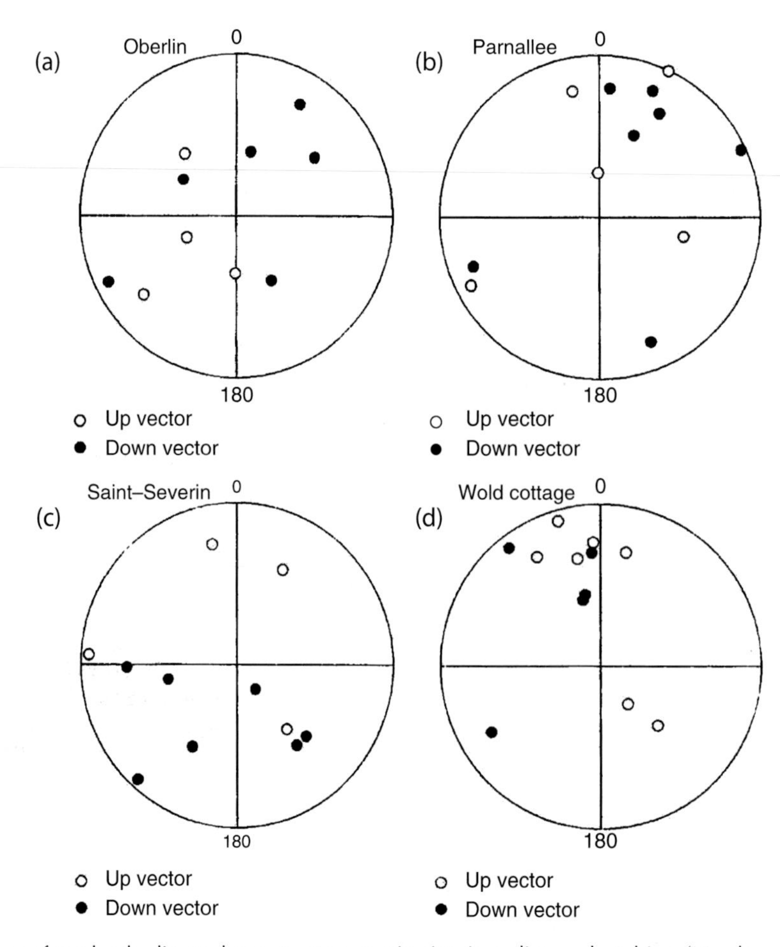

Figure P36 Demonstration of randomly directed remanent magnetization in ordinary chondrites (Morden and Collinson, 1992).

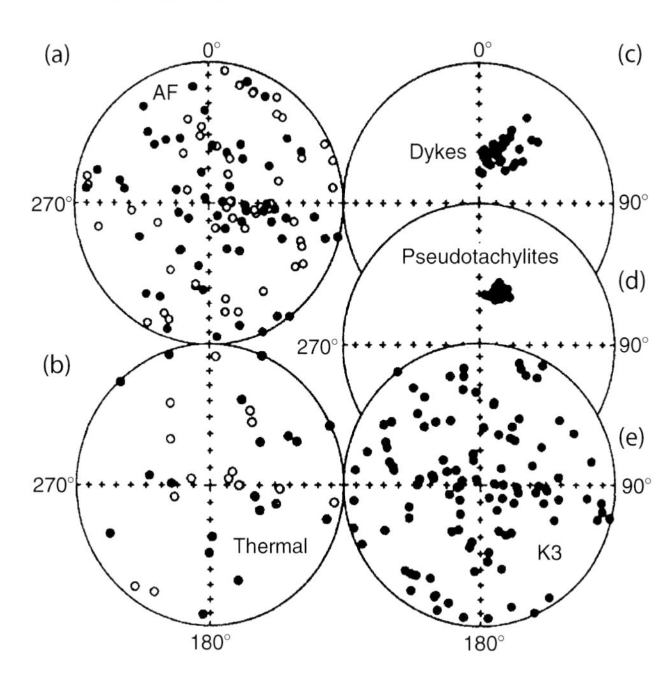

Figure P37 Magnetization of Vredefort impact crater samples. Strong random magnetization of shocked basement after (a) AF demagnetization, and (b) thermal demagnetization. (c) and (d) TRM like magnetization in dykes and melt rock recording earth's field, (e) orientation of minimum susceptibility axes (Carporzen *et al.*, 2005).

and thus appear to have come from a parent body that had a brief and active history. The shergottites, nahklites, and chassignites (SNC) meteorites, although achondrites were discussed above because of their demonstrated martian origin.

Not a lot of paleomagnetism has been done on the achondrites other than the SNC meteorites. Some exhibit directionally stable NRM that is resistant to AF demagnetization and to some degree against thermal demagnetization. Some have ratios of NRM:IRMs so high that according to the criterion suggested by Wasilewski *et al.* (2002) contamination must be suspected. One dual heating intensity determination has been done on an Antarctic achondrite (Yamato 7038) and yielded a field intensity of less than 10 μT.

Recent work has produced important results, which bear on the studies of randomly directed remanent magnetization in meteorites and on the role of shock in many meteorites. The first result comes from a study of the Vredefort meteorite impact crater by Carporzen *et al.* (2005). Rocks such as the dykes and pseudotachylites carrying TRM are acquired during cooling after the impact was magnetized coherently with small scatter between samples (Figure P37c,d). In contrast, shocked granitoid basement rocks are magnetized randomly (Figure P37a, b). These shocked rocks also have anomalously high intensities of magnetization. This magnetization of the shocked rocks is interpreted to have been acquired in the intense fields of the plasmas generated by the shock event. The anisotropy of these samples was also determined and the minimum susceptibility axes found to be randomly oriented. In individual samples the anisotropy was related to the remanence direction, indicating that both were caused by the shock events. In other related work, it has been demonstrated that the foliation of chondrites does indeed have an impact origin. Gattacecca *et al.* (2005) showed that the magnitude of the susceptibility anisotropy is correlated with shock level (Figure P38). These two results call in to doubt again the use of randomly directed magnetization as an indication of magnetization acquired prior to meteorite assembly, demonstrate that foliation can be shock generated, and make the interpretation of paleointensities obtained from meteorites more problematical.

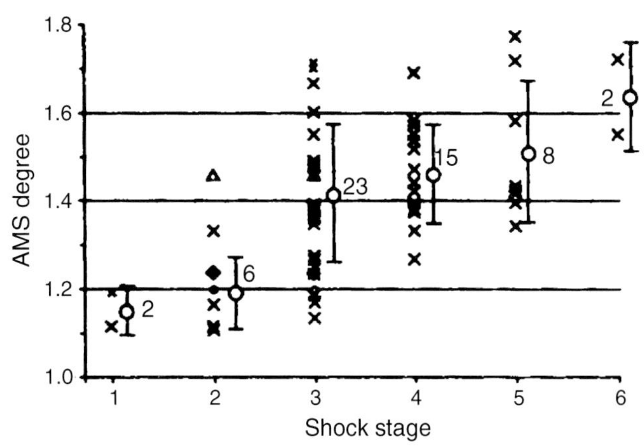

Figure P38 The relation between fabric and shock demonstrated by the correlation of magnetic anisotropy with shock level (Gattacecca *et al.*, 2005).

Discussion

The paleomagnetic record of meteorites is probably the most difficult of all paleomagnetic records to interpret. In particular, the record in CAIs and chondrules is difficult because they are later incorporated in chondrites and so one must separate the history prior to accretion from postaccretion history in the meteorite. Nevertheless there are standard techniques for doing this and progress is being made, despite the difficulties discussed above. In some cases, CAIs and chondrules may carry a record of the fields experienced prior to accretion, but much of the magnetization is probably a postaccretional low-temperature overprint. The magnetization of carbonaceous chondrites and ordinary chondrites may well have been acquired on the surface of the parent body and yielded, respectively, fields of hundreds and tens of microtesla. The NRM of achondrites and primitive achondrites may have been acquired at different depths on an evolved parent body. Only a single double heating experiment was available that yielded a value in the single microtesla range.

The interpretation of these results in terms of early solar system fields is not easy. The fields in which CAIs and chondrules were formed must have been very early solar system fields and Strangway *et al.* (1988) suggested that these may have been strong fields associated with T-Tauri winds. However, as we have seen that care needs to be taken in while interpreting these field estimates. The fields recorded by carbonaceous and ordinary chondrites are presumably due to surface processes on the parent bodies and modern studies are increasingly recognizing the difficulty of interpreting these results (Gattacecca *et al.*, 2003). Achondrites and primitive achondrites may yield values for different depths in short-lived parent bodies in the form of planets or planetesimals, that might had active dynamos. The question of parent bodies of meteorites a major research area discussed in McSween (1999). For the present purposes, we simply note that S-type asteroids have similar spectra, but not identical to ordinary chondrites, whereas C-type asteroids have features in their spectra characteristic of hydrated minerals typical of the aqueous alteration exhibited by carbonaceous chondrites. It is therefore interesting that S-type asteroid Gaspra appears to have an intensity of magnetization comparable with chondrites, but puzzling that Eros, which is also an S-type asteroid, has a very low magnetization.

Conclusion

A remarkable extraterrestrial paleomagnetic record has been found to be associated with the Moon, Mars, the asteroids, and the meteorites. Observations from spacecraft provide evidence that suggests that just as the Earth and Mercury have dynamos generating magnetic fields at the present, the Moon and Mars may have had similar dynamos in the past that generated comparable fields. They also suggest that there were magnetic fields in the solar system at the time of origin of chondrites and achondrites, but the records of the chondrules in carbonaceous chondrites and of the primitive CAIs remain hard to interpret in these observations.

Michael D. Fuller

Bibliography

Acuña, M.H., Anderson, B.J., Russell, C.T., Wasilewski, P., Kletetschka, G., Zanetti, L., and Omidi, N., 2002. NEAR Magnetic field observations at eros: first measurements from the surface of an asteroid. *Icarus*, **155**: 220–228.

Acuña, M.H., Connerney, J.E.P., Ness, N.F., Lin, R.P., Mitchell, D., Carlson, C.W., McFadden, J., Anderson, K.A., Reme, H., Mazelle, C., Vignes, D., Wasilewski, P., and Cloutier, P., 1999. Global distribution of crustal Mars Global Surveyor MAG/ER experiment. *Science*, **284**: 790–793.

Arkani-Hamed, J., 2001a. A 50-degree spherical harmonic model of the magnetic field of Mars. *Journal of Geophysical Research*, **106**: 23197–23208.

Arkani-Hamed, J., 2001b. Paleomagnetic pole positions and pole reversals of Mars. *Geophysical Research Letters*, **28**: 3409–3412.

Arkani-Hamed, J., 2002. Magnetization of the martian crust, JGR,107E5, 10, 1029/2001JE001496.

Barber, D.J., and Scott, E.R.D., 2003. Transmission electron microscopy of minerals in the martian meteorite Allan Hills 84001. *Meteoritics and Planetary Science*, **38**(6): 831–848.

Baumgartel, K., Sauer, K., and Bogdanov, A., 1994. A magnetohydrodynamic model of solar wind interaction with asteroid Gaspra. *Science*, **263**: 653–655.

Brecher, A., and Arrhenius, G., 1974. The paleomagnetic record of carbonaceous chondrites: natural remanence and magnetic properties. *Journal of Geophysical Research*, **79**: 2081–2106.

Brecher, A., Stein, J., and Fuhrman, M., 1977. The magnetic effects of brecciation and shock in meteorites. *The Moon*, **17**: 205–216.

Cain, J.C., Ferguson, B.B., and Mozzoni, D., 2003. An $n = 90$ internal potential function of the Martian crustal field, *Journal of Geophysical Research*, **108E2**: 5008, doi 10, 1029/2000JEE001487, 2003.

Carporzen, L., Gilder, S.A., and Hart, R.J., 2005. Paleomagnetism of the Vredefort meteorite crater and implications f or craters on Mars. *Nature*, **435**: 198–201.

Cisowski, S.M., 1987. Magnetism of meteorites. In Jacobs, J.A. (ed.) *Geomagnetism*, **2**: 525–560.

Cisowski, S.M., and Fuller, M., 1986. Lunar paleointensities via the IRM(s) normalization method and the early magnetic history of the moon. In Hartmann, W.K., Phillips, R.J., and Taylor, G.J. (eds.), *The Origin of the Moon.* Houston: Lunar and Planetary Science Institute, pp. 411–424.

Cisowski, S.M., Fuller, M., Rose, M., and Wasilewski, P.J., 1973. Magnetic effects of experimental shocking of lunar soil. *Proceedings of The Fourth Lunar Science Conference*, 3003–3017.

Cisowski, S.M., Fuller, M., Rose, M.F., and Wasilewski, P.J., 1976. Magnetic effects of shock and their implications for magnetism of lunar samples. *Proceedings of the Sixth Lunar Science Conference*, 3123–3141.

Cisowski, S.M., Collinson, D.W., Runcorn, S.K., Stephenson, A., and Fuller, M.,1986. A review of lunar paleointensity data and implications for the origin of lunar magnetism, *Proceedings of the Thirteenth Lunar and Planetary Science Conference. Journal of Geophysical Research*, **88**: A691–A704.

Collinson, D.W., 1997. Magnetic properties of Martian meteorites: Implications for an ancient Martian magnetic field. *Meteoritics and Planetary Science*, **32**: 803–811.

Connerney, J.E.P., Acuña, M.H., Wasilewski, P.J., Kletetschka, G., Ness, N.F., Reme, H., Lin, R.P., and Mitchell, D.L., 2001. The global magnetic field of Mars and implications for crustal evolution. *Geophysical Research Letters*, **28**: 4015–4018.

Dunlop, D.J., and Ozdemir, O., 1997. *Rock Magnetism.* Cambridge: Cambridge University Press.

Dunlop, D.J., and Kletetschka, G., 2001. Multi-domain hematite: A source of planetary magnetic anomalies? *Geophysical Research Letters*, **28**: 3345–3348.

Dyall, P., and Parkin 1971. The magnetism of the Moon. *Scientific American*, **225**: 62–73.

Fairén, A.G., Ruiz, J., and Anguita, F., 2002. An origin for the linear magnetic anomalies on Mars through accretion of terranes; Implications for dynamo timing. *Icarus*, **160**: 220–223.

Frey, H.V., 2003. Large diameter visible and buried impact basin on Mars: Implications for the age of the highlands and (buried) lowlands and turnoff of the global magnetic field, Abstract, *Lunar and Planetary Sciences*, XXXIV, 1838pdf.

Fuller, M., Rose, F., and Wasilewski, P.J., 1973. Preliminary results of an experimental study on the magnetic effects of shocking lunar soil. *Proceedings of the Fourth Lunar Science Conference*, pp 58–61.

Fuller, M., 1974. Lunar magnetism. *Reviews of Geophysics*, **12**: 23–70.

Fuller, M., 1987. Lunar paleomagnetism. In Jacobs, J.A. (ed.), *Geomagnetism*, **2**: 307–456.

Fuller, M., 1998. Lunar magnetism—a retrospective view of the Apollo sample magnetic studies. *Physics, Chemistry and Earth Sciences*, **23**(78): 725–735.

Gattacceca, J., Rochette, P., and Bourot-Denise, M., 2003. Magnetic properties of a freshly fallen LL ordinary chondrite: the Bensour meteorite. *Physics of the Earth and Planetary Interiors*, **140**: 345–358.

Gattacceca, J., Rochette, P., Denise, M., Cosolmagno, G., and Folco, L., 2005. An impact origin for the foliation of chondrites. *Earth and Planetary Science Letters*, **234**: 351–368.

Gus'kova, Y.G., 1963. Investigation of natural remanent magnetization of stoney meteorites. *Journal of Geomagnetism and Aeronomy*, **3**: 308–313.

Hale, C.J., Fuller, M., and Bailey, R.C., 1978. On the application of microwave heating to lunar paleointensity determination. *Proceedings of the Ninth Lunar Science Conference*, 3165–3179.

Halekas, J.S., 2003. The origins of Lunar Crustal Magnetic Fields, PhD Thesis, Berkeley: University of California.

Halekas, J.S., Lin. R.P., and Mitchell, D.L., 2003. Magnetic effects of impacts on the moon (2003), ABS, *Eos, Transcations, American Geophysical Union* **84**(46): GP22A-04.

Harrison, C.G.A., 2000. Questions about magnetic lineations in the ancient crust of Mars. *Science*, **287**: 547a.

Harrison, K.P., and Grimm, R.E., 2002. Controls on Martian hydrothermal systems: Application to valley network and magnetic anomaly formation. *Journal of Geophysical Research*, **107**(E5) 10.1029/2001JE001616.

Hammer, J.E., Brachfield, S., and Rutherford, M.J., 2003. An igneous origin for martian magnetic anomalies?, Abstract, *Lunar Planetary Science*, XXXIV #1918, CD-ROM.

Hargraves, R.B., Collinson, D.W., Arvidsen, R.E., and Spitzer, C.R., 1977. Viking magnetic properties experiment: primary mission results. *Journal of Geophysical Research*, **82**: 4547–4558.

Hargraves, R.B., Collinson, D.W., Arvidsen, R.E., 1979. Viking magnetic properties experiment: extended mission results. *Journal of Geophysical Research*, **84**: 8379–8384.

Hargraves, R.B., Knudsen, J.M., Bertelsen, P., Goetz, W., Gunnlaugsson, H.P., Hviid, S.F., Madsen, M.B., and Olsen, M., 2000. Magnetic enhancement on the surface of Mars. *Journal of Geophysical Research*, **105**(E1): 1819–1827.

Hood, L., and Huang, Z., 1991. Formation of anomalies antipodal to lunar impact basins: Two dimensional model calculations, *Journal of Geophysical Research*, **96**: 9837–9846.

Hood, L.L., Richmond, N.C., Pierazzo, E., Rochette, P., 2003. Distribution of crustal magnetic fields on Mars: Shock effects of basin-forming impacts. *Geophysical Research Letters*, **30**(6): 1281, doi:10.1029/2002GL016657.

Houseley, R.M., Grant, R.W., and Patton, N.E., 1973. Origin and characteristics of Fe metal in lunar glass welded aggregates. *Geochimica et Cosmochimica Acta*, **35**, Supplement(4): 2737–2749.

Kirschvink, J.L., Maine, A.T., and Vali, H., 1997. Paleomagnetic evidence of a low temperature origin of carbonate in the Martian meteorite ALH84001. *Science*, **275**: 1629–1638.

Kivelsen, M., Bargatgze, L.F., Khurana, K.K., Southwood, D.J., Walker, R.J., and Coleman, P.J., 1993. Magnetic field signatures near Galileo's closest approach to Gaspra. *Science*, **271**: 331–334.

Kletetschka, G., Wasilewski, P.J., and Taylor, P.T., 2000a. Hematite vs. magnetite as the signature for planetary magnetic anomalies? *Physics of the Earth and Planetary Interiors*, **119**: 259–267.

Kletetschka, G., Wasilewski, P.J., and Taylor, P.T., 2000b. Mineralogy of the sources for magnetic anomalies on Mars. *Meteoritics and Planetary Science*, **35**: 895–899.

Kletetschka, G., Wasilewski, P.J., and Taylor, P.T., 2002. The role of hematite-ilmenite solid solution in the production of magnetic anomalies in ground- and satellite-based data. *Tectonophysics*, **347**: 167–177.

Lanoix, M., Strangway, D.W., and Pearce, G.W., 1978. The primordial magnetic field preserved in chondrules of the Allende meteorite. *Geophysical Research Letters*, **5**: 73–76.

Lin, R.P., Anderson, K.A., and Hood, L.L., 1988. Lunar surface magnetic field concentrations antipodal to young large impact basins. *Icarus*, **74**: 529–541,

Lucey, P.G., Taylor, G.J., and Malareet, E., 1995. Abundance and distribution of iron on the moon. *Science*, **268**: 1150–1153.

McSween, H.Y., 2000. *Meteorites and their Parent Planets*. Cambridge: Cambridge University Press.

Mitchell, D.L., Lillis, R., Lin. R.P., Connerney, J., and Acuna, M., 2003. Evidence for Demagnetization of the Utopia Impacet Basin on Mars, *Eos, Transactions, American Geophysical Union*, **84**, 46, ABS., GP22A-08.

Morden, S.J., and Collinson, D.W., 1992. The implications of the magnetism of ordinary chondrite meteorites. *Earth and the Planetary Science Letters*, **109**(1–2): 185–204.

Nagata, T., Fisher, R.M., and Schwerer, F.C., 1972. Lunar rock magnetism. *The Moon*, **4**: 160–186.

Nagata, T., and Funaki, M., 1981. The comparison of natural remanent magnetization of an Antarctic chondrite ALH 76009 L6. *Proceedings of the Lunar Planetary Science Conference*, **12B**: 1229–1241.

Nagata, T., and Sugiura, N., 1977. Paleomagnetic field intensity derived from meteorite magnetization. *Physics of the Earth and Planetary Interiors*, **13**: 373–379.

Özdemir, O., 2000. Chemical remanent magnetization-possible source for the magnetization of Martian crust (abstract), American Geophysical Union Spring Meeting GP31A-05.

Özdemir, O., and Dunlop, D.J., 2002. Thermoremanence and stable memory of single-domain hematites. *Geophysical Research Letters*, **29**, No. 18 10.1029/2002GL015597.

Pearce, G.W., Williams, R.J., and McKay, D.S., 1972. The magnetic properties and morphology of metallic iron produced by sub-solidus reduction of synthetic Apollo 11 composition glasses. *Earth and Planetary Science Letters*, **17**: 95–104.

Richter, L., Brinza, D.E., Cassel, M., Glassmeir, K.-H., Kunhke, F., Musman, G., Ohtmer, C., Schwingenschu, and Tsurutani, 2001. First direct field measurements of an asteroidal magnetic field: DSI at Braille. *Geophysical Research Letters*, **28**(10): 1913–1916.

Robinson, P., Harrison, R.J., McEnroe, S.A., and Hargraves, R.B., 2000. Lamellar magnetism in the hematite-ilmenite series as an explanation for strong remanent magnetization. *Nature*, **418**: 517–520.

Rochette, P., Lorand, G., Fillion, G., and Sautter, V., 2001. Pyrrhotite and the remanent magnetization of SNC meteorites: a changing perspective on martian magnetism. *Earth and Planetary Science Letters*, **190**: 1–12.

Runcorn, S.K.R., 1975. An ancient lunar magnetic dipole field. *Nature*, **253**: 701–703.

Russell, C.T., Weiss, H., Colema, P.J., Soderblom. L.A., Stuart-Alexander, D.E., and Wilhelms, D.E., 1975. Geologic-magnetic correlations on the Moon: sub-satellite results, *Proceeedings of the eight Lunar Conference. Geochimica et Cosmochimica Acta*, **41**, Supplement 8: 1171–1185.

Scott, E., and Fuller, M., 2004. A possible source for the martian crustal magnetic field, EPSL., 220, 83–90.

Shaw, J., 1974. Anew method of determining the magnitude of the paleomagnetic field. *Geophysical Journal of the Royal Astronomical Society*, **15**: 205–211.

Shaw, J., Hill, M.J., and Openshaw, S.J., 2001. Investigating the martian magnetic field using microwaves. *Earth and Planetary Science Letters*, **190**: 103–109.

Smethurst, M. T., and Herrero-Bervera, E., 2002. Paleomagnetic analysis of Calcium-Aluminium Inckusions (CAI's) from the Allende meteorite EOS Trans. AGu, 83(47) Fall meet. Suppl., Abstract GP72A-0989.

Srnka, L.J., Martelli, G., Newton, G., Cisowski, S.M., Fuller, M., and Schaal, R.B., 1979. Magnetic field and shock effects and remanent magnetization in a hypervelocity impact experiment. *Earth and Planetary Science Letters*, **42**: 127–137.

Solomon, S. C., Aharonson, O., Banerdt, W. B., Dombard, A. J., Frey, H. V., Golombek, M. P., Hauck, III, S. A., Head, III, J. W., Johnson, C. L., McGovern, P. J. Phillips, R. J., Smith, D. E., and Zuber, M. T., 2003. Why are there so few magnetic anomalies in Martian Lowlands and Basins? Abstract *Lunar Planetary Science*, XXXIV #1382, CD-ROM.

Stacey, F.D., Lovering, J.F., and Parry L.G., 1961. Thermomagnetic properties natural magnetic moments and magnetic anisotropies of some chondritic meteorites. *Journal of Geophysical Research*, **66**: 1523–1534.

Stegman, D.R., Jellinek, A.M., Zatman, S.A., Baumgardnerk, J.R., and Richards, M.A., 2003. An early lunar Core dynamo driven by thermochemical mantle evolution. *Nature*, **451**: 146.

Stevenson, D.J., 2001. Mars' core and magnetism. *Nature*, **412**: 214–219.

Strangway, D.W., Larson, E.E., and Perce, G.W., 1970. Magnetic studies of Lunar samples, breccias and fines, *Proceedings of the Apollo 11, Lunar Science Conference. Geochimica et Cosmochimica Acta*, Supplement **1**: 2435–2451.

Strangway, D.W., Pearce, G.W., Gose, W.A., and McConnell, R.K., 1973. Lunar magnetic anomalies and the Cayley Formation. *Nature*, **246**: 112–114.

Thomas-Keprta, K.L., Bazylinski, D.A., Kirschvink, J.L., Clement, S.J., McKay, D.S., Wentworth, S.J., Vali, H., Gibson, Jr. E. K., and Romanek, C.S., 2000. Elongated prismatic magnetite crystals in ALH84001 carbonate globules: Potential Martian magnetofossils. *Geochimica et Cosmochimica Acta*, **64**: 4049–4081.

Wasilewski, P.J., 1988. Magnetic characterization of the new mineral tetrataenite and its contrast with isochemical taenite. *Physics of the Earth and Planetary Interiors*, **52**(1–2): 150–158.

Wasilewski, P.J., and Fuller, M., 1975. Magnetochemistry of the Apollo landing sites. *The Moon*, **14**: 79–101.

Wasilewski, P.J., Acuna, M.H., and Kletetschka, G., 2002. 433 Eros, problems with the meteorite magnetism record in attempting an asteroid match. *Meteoritics and Planetary Science*, **37**: 937–950.

Weiss, B.P., Kirschvink, J.L., Baudenbacher, F.J., Vali, H., Peters, N.T., Macdonald, F.A., and Wikswo, J.P., 2000. A low temperature transfer of ALH84001 from Mars to Earth. *Science*, **290**: 791–795.

PALEOMAGNETISM, OROGENIC BELTS

Orogenic belts are the key signature of Plate Tectonics at the sites of plate convergence. At these margins the consumption of ancient ocean basins culminates in the collision of blocks of continental crust and this is responsible for deforming broad tracts of crust on either side of the site of collision (*suture*). The deformed rocks are mostly marine sediments and the products of the preceding phases of ocean development. After collision they are squeezed into increasingly tight folds that may ultimately become overfolded and develop shear planes along which wedges of crust are transported horizontally for large distances as *nappes*. Slices of young, light, and buoyant ocean crust may also become incorporated in this process and emplaced (*obducted*) into the deformed crust as *ophiolites* instead of being subducted back into the mantle.

Orogenic deformation is concentrated mainly within the rock wedges accreted to the peripheries of the continents during the preceding phase of ocean growth, which deform in a semiplastic way. The ancient, relatively cold metamorphic and igneous basements to the continental crust act as the rigid indenters and the inherited shapes of these blocks of old hard crust ultimately control the shape of the orogenic belt. Further from the site of the suture, the sedimentary covers on the continents may be deformed by collision into open folds with amplitudes and wavelengths declining away from the collisional zone. This cover is typically folded independently of the underlying basement by slipping along detachment zones (*decollements*). Such detachments are also developed within the orogenic belt where rocks deform by contrasting mechanisms controlled by their ductility and bedding. This deformation is constrained between the limiting *parallel* case where the folding layers in the rock section preserve their thickness but not their shape, and the *similar* case where they preserve their shape but not their thickness.

The compressional phases of orogenic deformation lead in turn, to crustal thickening, isostatic rebound and uplift. The thickened wedge of deformed rocks becomes topographically unstable and flows outwards as a viscous medium on a geological timescale in a secondary phase of plate collision known as *orogenic collapse*. Compressional deformation then gives away to extensional deformation that may be regarded as a later phase of a single Plate Tectonic cycle. The development of the Basin and Range Province in western North America during Tertiary times for example, followed compression and crustal thickening during the Laramide Orogeny in Late Cretaceous times. Extensional deformation in the continental crust can occur by two mechanisms. Pure shear attenuates a viscous crust (McKenzie 1978) leading to a change in shape with little or no rotation. In contrast, deformation by *simple shear* develops a low-angle detachment zone through the crust (Wernicke *et al.*, 1987). As this crust attenuates, the changing thickness balance of crust to mantle lithosphere causes partial uplift of the basement and rotation of blocks above the detachment about horizontal axes; the upper zone (*hanging wall*) may dissect into fault blocks that rotate about horizontal axes in a domino fashion and, if the detachment extends right through the crustal section, both upper and lower (*footwall*) sections may rotate with respect to each other.

Following continental collision, the rate of convergence of the two converging continents decelerates but relative movements may not cease altogether. One continental margin may then drive under, and uplift, the orogenic belt as in the case of India beneath Tibet. If the indenter has an irregular shape, or there is ocean basin to one side, blocks of the orogen defined by major vertical faults are extruded laterally by *tectonic escape*. This latter phenomenon also occurs when the subduction of ocean crust is oblique to the orogenic margin. It is responsible for the tectonics of the Cordilleran margin of western North America where terranes are slipping differentially northwards along the continental margin until they become locked into a dumping ground in Alaska. In the orthogonal continent-continent collision of the Alpine-Himalayan Belt the tectonic escape defines the last phase of the orogenic cycle.

The ability of rocks to record the direction of the ancient magnetic field provides the key quantitative information for unraveling tectonic deformation in orogenic belts. Geological indicators such as bedding directions, stratigraphic variations and dyke intersections can be used only rarely to provide data of comparable significance, and they are unconstrained to a paleogeographic orientation. The orogenic environment, however, causes specific problems to the analysis of the magnetism in rocks due to the effects of strong deformation and temperature increase. Rocks in this environment are more likely to have their primary magnetism partially or completely overprinted by magnetizations linked to the deformation. Partially set against this disadvantage, the deformed nature of the rocks means that the paleomagnetic fold test is more readily applied to constrain the geological ages of the magnetizations than would be possible in less deformed settings. The study of rock-units may include fault flakes up to a few kilometers in size rotating within fault zones, thrust sheets in interior zones of strong deformation and blocks of crust on a 10–100 km scale bounded by major near vertical faults called *terranes*; only if the terrane extends to the base of the continental lithosphere can it properly be referred to as a *microplate*.

Paleomagnetic analysis assumes that the mean direction of magnetization recovered from the rock-unit is a record of the time averaged dipole field source at the time of magnetization. The cumulative deformation can then be resolved by comparing this direction with the predicted direction at the study location at the time of magnetization. These predicted directions are usually calculated from the *apparent polar wander paths* (*APWPs*) of the adjoining continental indenter. In the case of the Alpine-Himalayan orogenic belt for example, they would be calculated from APWPs of continental Africa, Arabia, India, or Eurasia. Usually the difference is the resultant of more than one episode of deformation and in favorable circumstances it can be decomposed using geological evidence or from studying the paleomagnetism of rocks of different ages.

The angular deformation is usually considered in terms of differences of declination and inclination with respect to the reference direction from the stable plate. The difference between the observed and expected directions is described in terms of rotation and flattening. A difference in declination implies rotation (R) of a crustal block about a quasivertical axis and is usually calculated by the method of Beck (1980). Both observed and reference directions have confidence limits (ΔD and ΔD_{ref}, respectively) and Beck (1980) proposed a confidence limit on R' defined by $\Delta R' = \sqrt{(\Delta D^2 + \Delta D_{\mathrm{ref}}^2)}$. From a statistical analysis of this proposed confidence limit, Demarest (1983) concluded that this value overestimates errors and showed that, provided α_{95} is small and preferably less than 10°, a standard correction factor is applicable. For a direction derived from $n \geq 6$ or more samples, the correction factor lies between 0.78 and 0.80. Then $\Delta R'$ is derived from the equation $\Delta R' = 0.8\sqrt{(\Delta D^2 + \Delta D_{\mathrm{ref}}^2)}$ and only, if $R' > \Delta R'$ can tectonic rotation be regarded as significant.

A difference in inclination(I), or flattening(F), is also subject to error limits (ΔI and ΔI_{ref}, respectively). Following Demarest (1983) it is given by $\Delta F = 0.8\sqrt{(\Delta I^2 + \Delta I_{\mathrm{ref}}^2)}$ but provided that $F > \Delta F$ the flattening may be regarded as significant. F has two possible explanations. Firstly, it could result from rotation of crustal blocks about near horizontal axes. This contribution can usually be isolated if references to the paleohorizontal, such as sedimentary bedding, are present. Secondly, the continental block could have moved through latitude since the time of magnetization. This latter interpretation of F has proved more contentious than the interpretation of R, in part because there is often structural or paleogeographic evidence with which the paleomagnetic analysis can be tested independently. It has been found that inclination differences within the Alpine-Himalayan Belt are often too large to be attributed to continental movement through latitude alone. The anomalies are more important in sedimentary rocks and imply that inclination shallowing during compaction and lithification is a factor here (Bazhenov and Mikolaichuk, 2002). However, Tertiary inclinations from igneous rocks also appear to be too shallow to be accommodated solely by latitudinal movement (Beck *et al.*, 2001). The difference is probably to be explained in terms of small departures from the geocentric axial dipole assumption and a full explanation for this anomaly currently limits the usefulness of flattening in tectonic analysis.

The values of F and R need to be interpreted in terms of the scale of the crustal blocks undergoing deformation and temporal setting of the deformation within the orogenic cycle. Two examples from the

Anatolian sector of the Alpine-Himalayan orogenic belt illustrate contrasts in scale (see Figures P39, P40). Figure P39 shows an example of rotation across an intracontinental transform, the North Anatolian Fault Zone, separating the Eurasian and Anatolian plates. The crust on either side of this fault zone shows comparable regional anticlockwise rotation. This occurred before the initiation of the transform plate boundary in Late Pliocene times. Within the transform fault zone, small fault blocks 10 km in size are being rotated clockwise in ball bearing fashion between master faults at rates of the order of 1° per 10 ka. Figure P40 shows a large-scale example of deformation in which blocks on a scale of ~50–200 km are being rotated as Anatolian crust is being extruded outwards to the west by the impingement of the Arabian indenter into a collage of terranes accreted to the Eurasian margin during closure of the Tethyan Ocean in Mesozoic and Early Tertiary times. Most of these rotations are concentrated within the last few millions of years and are thus a record of the last phase of orogenic development (tectonic escape) following continental collision and crustal thickening. Both these examples illustrate deformation during late stages of orogenic deformation following collision, folding and crustal thickening. Within the Alpine-Himalayan Belt, they fall into the timescale of postcollisional events described as *neotectonic* to distinguish it from the pre and syn collisional history described as *paleotectonic*.

In principle there are two aspects to the earlier history of orogenesis that paleomagnetism may aim to quantify. Firstly, there is the phase of ocean closure in which latitudinal convergence of the two continental margins is recorded by a decline in the difference between paleomagnetic inclinations from either margin. In the Anatolian case, for example, this is identified by progressive reduction in flattening as the Afro-Arabian crust has moved northwards to close the Tethyan Ocean (Kaymakci *et al.*, 2003). Secondly, the plastic deformation associated with continental collision is responsible for nappe emplacement and *R* is a measure of the differential rotation between the nappe and the footwall beneath the thrust. Figure P41 shows directions of magnetization resolved from Late Cretaceous rocks, mostly passive margin limestones, in the overthrust front of the Apenninic and Silcilian nappes of southern Italy (Channell *et al.*, 1980). The distribution of declinations is a composite of rotation during nappe emplacement and of differential rotations during the opening of the Tyrrhenian Sea to form the Calabrian arc of southern Italy.

The large-scale deformation shown in Figures P40, P41 producing a radial distribution of paleomagnetic directions over hundreds of kilometers of continental crust is often referred to as "orogenic bending." One of the best documented examples is the Japanese Arc (see Figure P42) where magnetic declinations in north east Japan are rotated anticlockwise and directions in south west Japan are rotated clockwise. The relative rotation between these two regions appears to have occurred over a relatively short interval of time (~16–20 Ma) during the opening of the Japan Sea by back-arc spreading (Otofuji *et al.*, 1985). However, the actual deformation mechanisms operating in such areas are certainly complex (Randall, 1998). They no doubt vary from region to region and are likely to involve both small block rotations

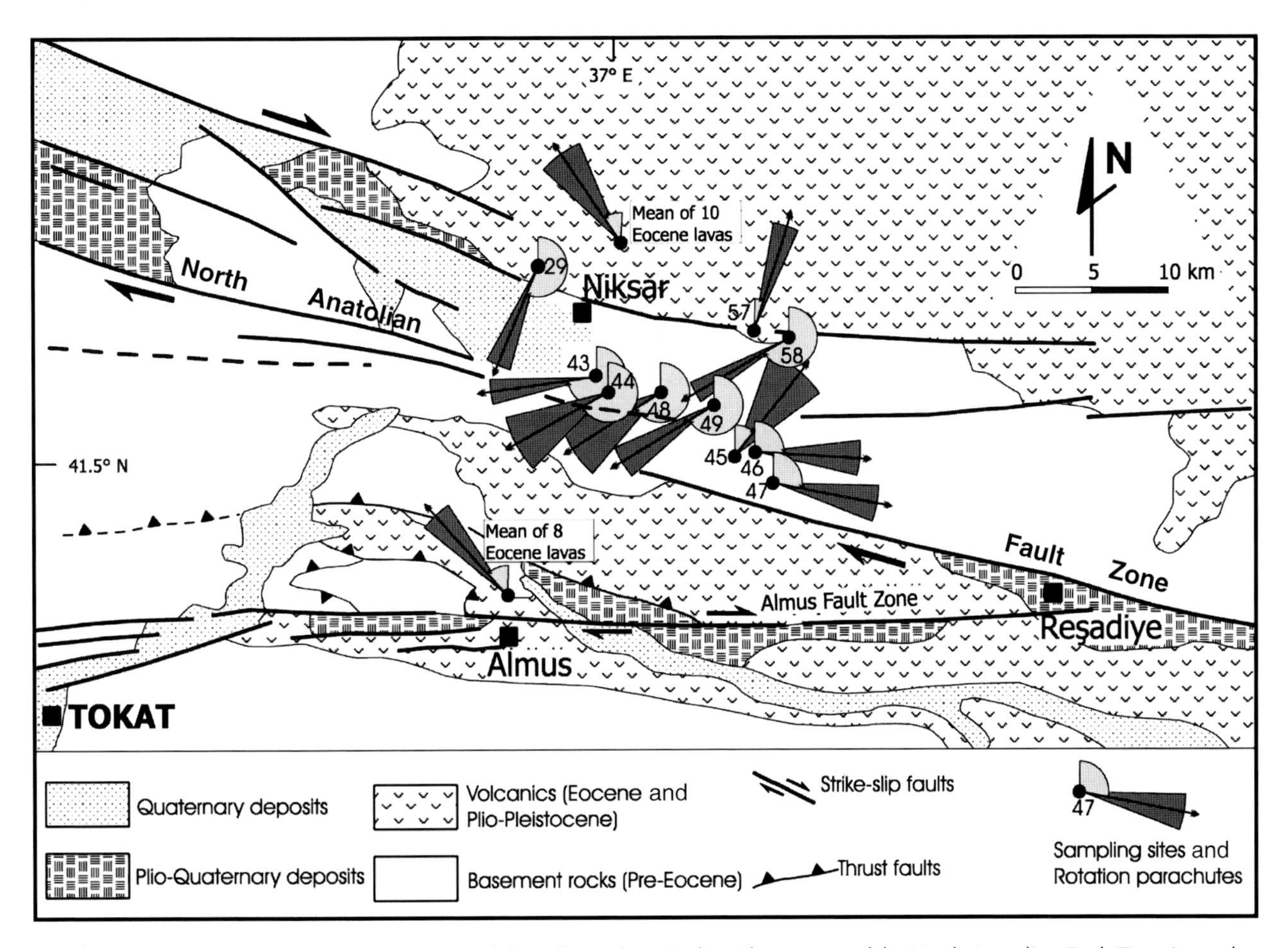

Figure P39 Deformation across an intracontinental dextral transform in the Niksar sector of the North Anatolian Fault Zone in northern Turkey. The arrows with 95% confidence cones are normal polarity directions. Rapid clockwise rotation of small blocks between the master faults is identified in Brunhes chron lava units with ages defined by high-precision K-Ar dating at ca. 520 ka. After Piper *et al.* (1997).

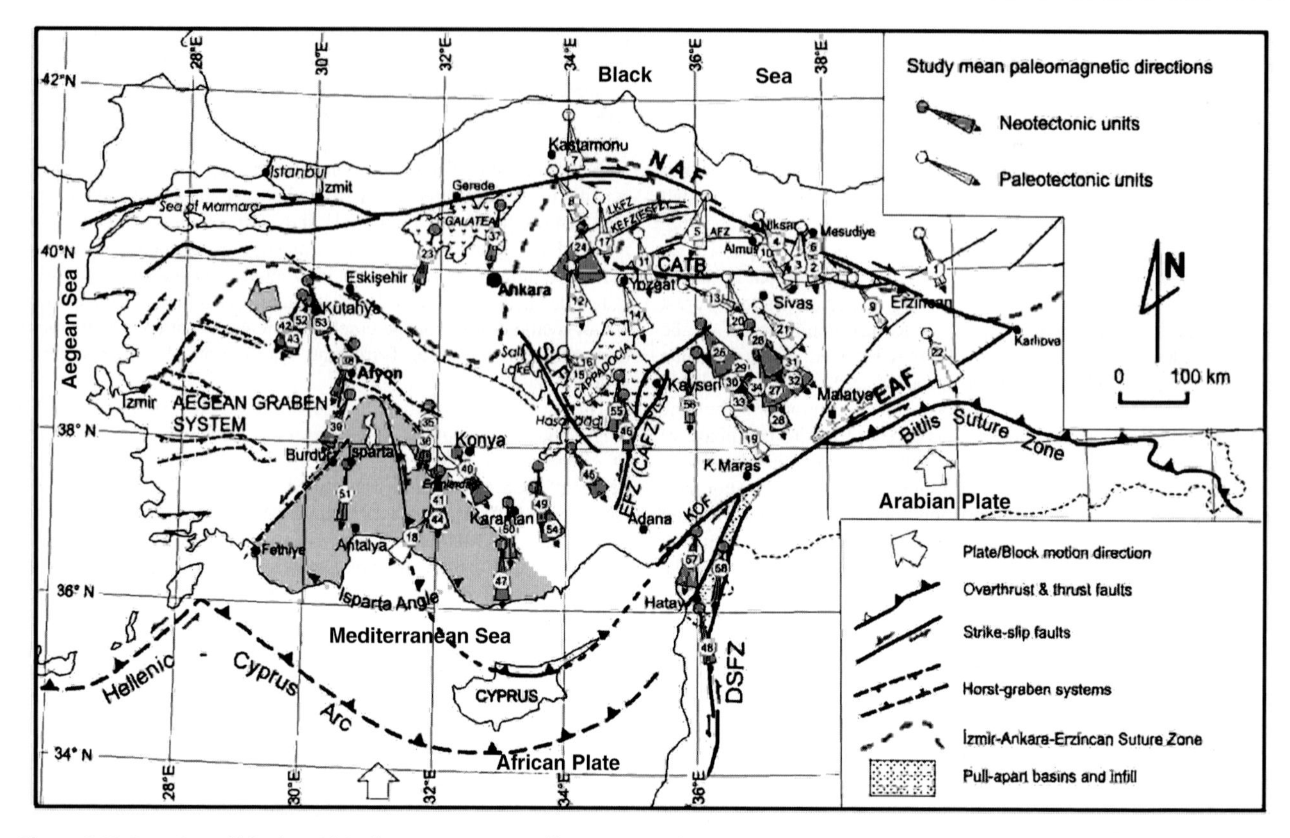

Figure P40 Rotation of blocks within the accretionary collage in central Anatolia formed by closure of the Neotethys Ocean prior to the termination of suturing ~12 Ma ago and defined by paleomagnetic directions (here shown with common reversed polarity). The neotectonic magnetizations are less than 12 Ma in age and the rotations are an expression of block rotation during tectonic escape to the west as the Arabian indenter is continuing to impinge into the weak continental crust of Anatolia and the Eurasian margin in the east. After Piper *et al.* (2003).

and differential slip along major lineaments such as the "D" zones in the Calabrian example shown in Figure P41a.

Rotation can occur within an orogenic belt either during translation of nappes over a low-angle thrust (Figure P41) or in response to strike slip fault movement within the seismogenic upper crust. The stress conditions for strike slip faulting occur when the maximum (σ_1) and minimum (σ_3) principal stresses are in the horizontal plane. Whilst this ideally generates two sets of subvertical faults that are conjugate to one another, one fault set always becomes dominant over time. Vertical axis rotation is demonstrably a major component of crustal deformation within orogenic belts although the kinematics of *in situ* vertical axis rotations in the upper crust is still controversial due to uncertainties relating to the ways in which continental crust deforms. Limiting models consider the deformation to be *continuum* or *discrete* with the division between them based on the driving mechanism for the rotation rather than the type of deformation produced. In continuum models, the rotating blocks are considered to be significantly smaller than the size of the deforming zone and occur passively in response to motion of a ductile lower crust. Shear is distributed across a wide zone with creep and diffusive mass transfer accommodating the deformation. Paleomagnetic directions will be rotated by an amount that diminishes away from the main fault (McKenzie and Jackson, 1983).

In the discrete case the strain is localized into bands and deformation involves the rotation of rigid blocks with slip between them accommodated by fault movement. The blocks themselves are internally undeformed although space constraints imply that they may be expected to undergo tectonic erosion at some peripheries and *spheochasm* formation to open conduits to the lower crust at others. Whilst the evidence indicates that continental lithosphere as a whole deforms on a large scale as a fluid medium with a behavior that can be modeled as a thin viscous sheet (England and Jackson, 1989), pervasive deformation in the upper crust can only occur if there is a sufficient high-level heat source to promote ductile deformation and permit diffusive mass transfer so that strain is not focused. As Figure P39 demonstrates, this condition does not pertain in the brittle upper crust, even adjacent to intracontinental transform faults. Deformation here takes the form of intense rotation confined between master faults and realistic models to explain rotations must therefore invoke a discrete mechanism. The base of this upper crust is evidently decoupled from a viscous lower crust obeying power law behavior. This detachment is defined by an abrupt reduction in seismicity in the depth range 10–20 km over most of the continental crust. It correlates with the combined effects of increase in temperature and in relatively weak pyroxene and plagioclase feldspar mineral assemblages, which combine to produce a rapid reduction in the strength of the crust below this level.

In discrete models the deformation is taken up on rigid crustal blocks with a length comparable to the width of the deforming zone (Figure P43). The rotation and deformation is the response to shear applied along the edges of the blocks by strike slip faults. All blocks and block bounding faults within the same fault domain will rotate by the same amount in the same direction. An exception to this rule will occur when the strike slip faults form two domains and the sense of rotation will then be opposite within each domain (Ron *et al.*, 1984). It is usually considered that rotation will be clockwise in regions of net right lateral shear and anticlockwise in regions of net left lateral shear (Nelson and Jones, 1987) as shown in Figure P43a. However, the sense of fault motion on the block bounding faults controlling the rotation need not necessarily be the same as the motion sense on the system

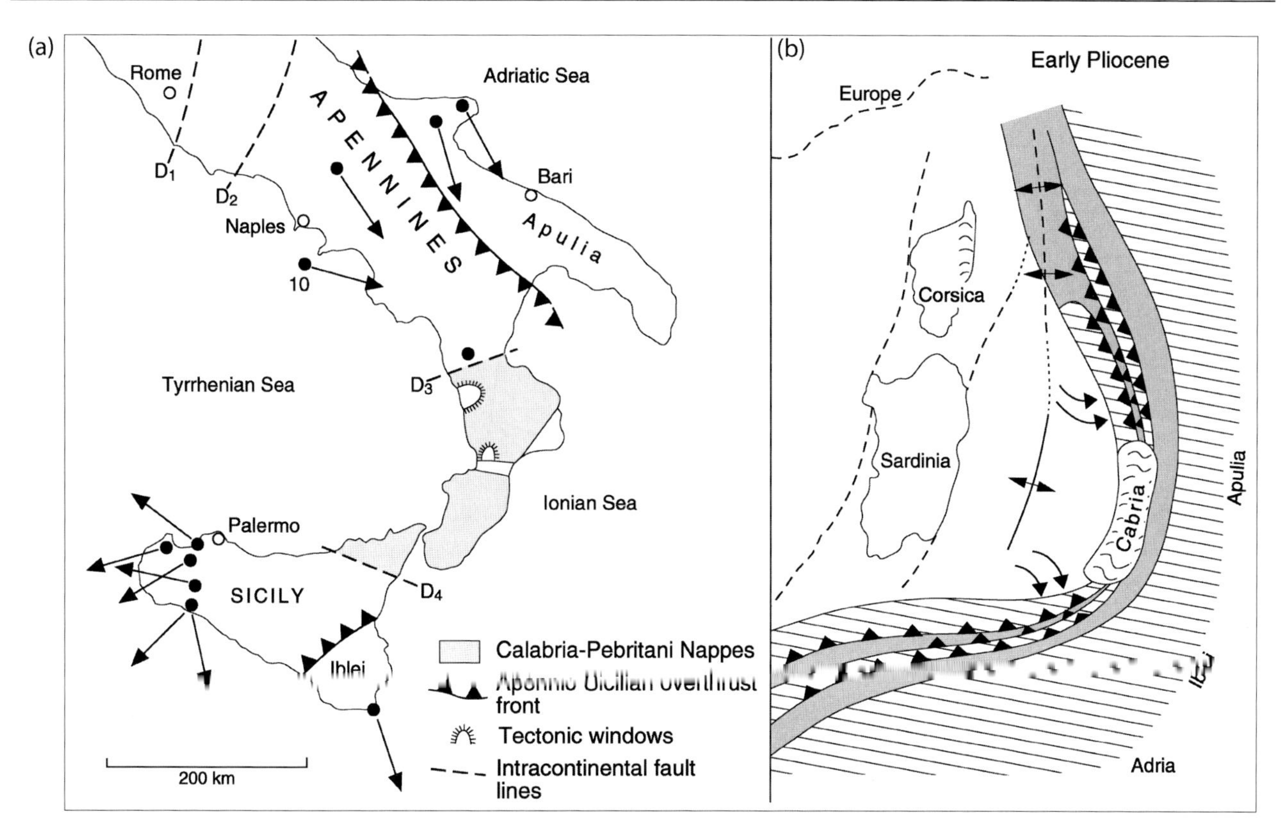

Figure P41 (a): Mean directions (reversed polarity) in Late Cretaceous rocks of the Apenninic and Sicilian nappes and the bordering autochthonous zones of Apulia and Iblei in southern Italy. (b) Schematic reconstruction showing the opening of the Tyrrhenian Sea in Early Pliocene times responsible for the large-scale regional differences in magnetic declination in (a) Compiled from Channell *et al.* (1980).

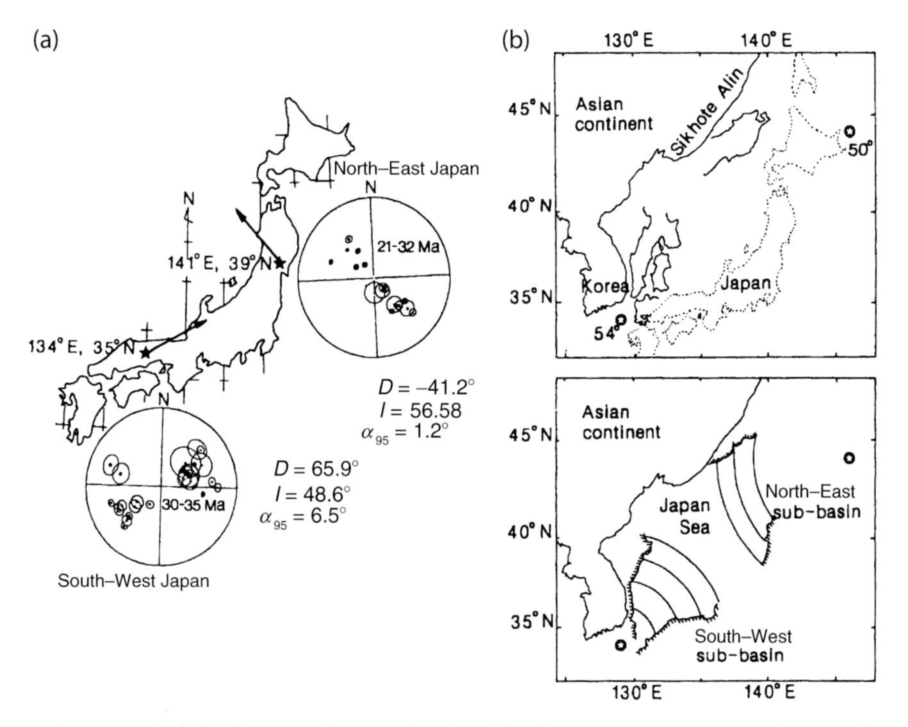

Figure P42 (a): Mean paleomagnetic field directions from rock-units older than 20 Ma in age in northeast and southwest Japan. (b) Explanation for the difference between the regional directions shown in (a) in terms of rotation of the two sectors of Japan away from the Asian margin as a result of the fan-shaped opening of the Japan Sea. After Otofuji *et al.* (1985).

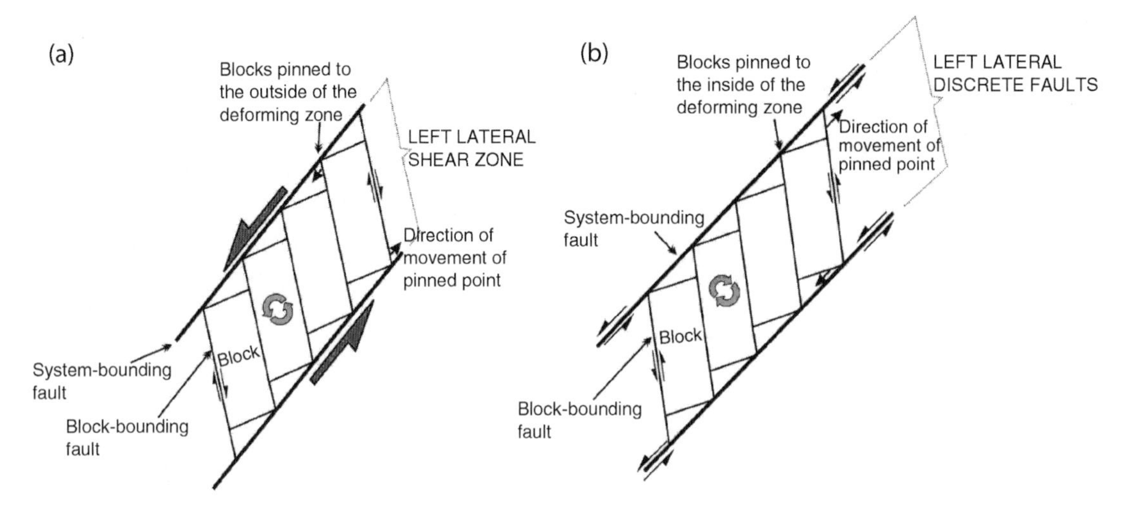

Figure P43 The origin of clockwise and anticlockwise rotations defined by paleomagnetic study of small blocks influenced by major vertical faults cutting through the thickness of the seismogenic upper crust when the deformation is (a) distributed and (b) discrete. Adapted from Randall (1998) and Tatar *et al.* (2004).

bounding faults (Randall, 1998). The sense of rotation is controlled by the initial orientation of the block bounding faults with respect to the system bounding faults, and whether the system bounding faults are acting as the margins of a wide shear zone (Figure P43a) or as discrete faults either side of the rotating blocks (Figure P43b). If the block bounding faults have the same sense of motion to the discrete system bounding faults (Figure P43b) the blocks rotate in the direction opposite to that suggested by the bulk shear on the deformation zone. Examples of both types of rotation have been reported from application of paleomagnetism to orogenic belts and are evidently a signature of the distributed or discrete character of the causative major strike slip zones.

In general there are two scales of tectonic rotation observed in orogenic belts. Large and rapid rotations are present within, or adjacent to, the intracontinental transform faults defining the conservative boundaries of the plates; Figure P39 is an example of this kind of deformation. More modest and slower rotations are present where terranes are being pushed laterally by oblique subduction or tectonic escape, or nappes are being emplaced by an advancing fold belt. Figures P39 (remote form the intracontinental transform) and P40 are examples of this scale of deformation.

A further important aspect of orogenic paleomagnetism is the study of remagnetization events directly related to orogenesis. A burial related heating that partially or completely remagnetizes preexisting ferromagnetic minerals may cause these overprints. More commonly they reside in chemical remanent magnetizations (CRMs) carried by new minerals that have either been transformed from the preexisting minerals (*diagenesis*) or have been precipitated from fluids (*authigenesis*). Comparison of the pole position with the APWP of the adjoining indenters may then allow diagenetic and authigenic events to be dated. In addition progressive unfolding of remagnetized rocks within fold structures can constrain this event within a pre, syn or postfolding timeframe. The most commonly invoked mechanism for regional CRMs in the vicinity of orogenic belts is lateral migration of preexisting connate brines within the deforming rock assemblage as it is loaded and heated within the orogenic belt. Forced migration of these orogenic fluids along ancient aquifers under a temperature and pressure gradient is considered to be a plausible explanation for the widespread distribution of CRMs away from some orogenic belts such as the Appalachians, although evidence for the link is still in part circumstantial (Elmore *et al.*, 2000). Precipitation of new ferromagnetic minerals will be controlled by the availability of ions, and the temperature, pH and Eh of the ambient environment. Oxidizing conditions will favor precipitation of hematite whereas alkaline and mildly reducing environments favor precipitation of magnetite. In more strongly reducing conditions, the ferromagnetic pyrrhotite is precipitated and can produce important late stage CRMs within the orogenic belt where abundant mudrocks originally containing pyrite are present or where mineralizing fluids have been able to permeate. Thus, in addition to applications of the primary or early tectonic paleomagnetic record for resolving the scale of deformation, the study of chemical magnetic overprints can constrain the age and regional extent of important later events in the development of orogenic belts.

John D.A. Piper

Bibliography

Bazhenov, M.L., and Mikolaichuk, A.V., 2002. Palaeomagnetism of Palaeogene basalts from the Tien Shan, Kyrgystan: rigid Eurasia and dipole geomagnetic field. *Earth and Planetary Science Letters*, **195**: 155–166.

Beck, M.E., 1980. Palaeomagnetic record of plate margin processes along the western edge of North America. *Journal of Geophysical Research*, **87**: 7115–7131.

Beck, M.E., Burmester, R.F., Kondopoulou, P., and Atzemoglou, A., 2001. The palaeomagnetism of Lesbos, NE Aegean, and the eastern Mediterranean inclination anomaly. *Geophysical Journal International*, **145**: 233–245.

Channell, J.E.T., Catalano, R., and D'Argenio, B., 1980. Palaeomagnetism and the deformation of the Mesozoic continental margin in Sicily. *Tectonophysics*, **61**: 391–407.

Demarest, H.H., 1983. Error analysis for the determination of tectonic rotation from palaeomagnetic data. *Journal Geophysical Research*, **88**: 4321–4328.

Elmore, R.D., Kelley, J., Evans, M., and Lewchuk, M., 2000. Remagnetization and Orogenic Fluids: Testing the hypothesis in the central Appalachians. *Geophysical Journal International*, **144**: 568–576.

England, P.C., and Jackson, J., 1989. Active deformation of the continents. *Annual Review of Earth and Planetary Science*, **17**: 197–226.

Kaymakci, N., Duermeijer, C.E., Langereis, C., White, S.H., and Van Dijk, P.M., 2003. Palaeomagnetic evolution of the Cankiri Basin (Central Anatolia, Turkey): implications for oroclinal bending due to indentation. *Geological Magazine*, **140**: 343–355.

King, G., Oppenheimer, D., and Amelung, F., 1994. Block versus continuum deformation in the western United States. *Earth and Planetary Science Letters*, **128**: 55–64.
McKenzie, D.P., 1978. Some remarks on the development of sedimentary basins. *Earth and Planetary Science Letters*, **40**: 25–32.
McKenzie, D.P., and Jackson, J.A., 1983. The relationship between strain rates, crustal thickening, palaeomagnetism, finite strain and fault movements within a deforming zone. *Earth and Planetary Science Letters*, **65**: 182–202.
Nelson, M.R., and Jones, C.H., 1987. Palaeomagnetism and crustal rotations along a shear zone, Las Vegas Range, Southern Nevada. *Tectonics*, **6**: 13–33.
Otofuji, Y.-O., Matsuda, T., and Nohda, S., 1985. Opening mode of the Japan Sea inferred from the palaeomagnetism of the Japan Arc. *Nature*, **317**: 603–604.
Piper, J.D.A., Tatar, O., and Gürsoy, H., 1997. Deformational behaviour of continental lithosphere deduced from block rotations across the North Anatolian fault zone in Turkey. *Earth and Planetary Science Letters*, **150**: 191–203.
Piper, J.D.A., Gürsoy, H., and Tatar, O., 2001. Palaeomagnetism and magnetic properties of the Cappadocian ignimbrite succession, central Turkey and Neogene tectonics of the Anatolian collage. *Journal of Volcanology and Geothermal Research*, **117**: 237–262.
Randall, D.E., 1998. A new Jurassic-Recent apparent polar wander path for South America and a review of central Andean tectonic models. *Tectonophysics*, **299**: 49–74.
Ron, H., Freund, R., Garfunkel, Z., and Nur, A., 1984. Block rotation by strike slip faulting: structural and palaeomagnetic evidence. *Journal of Geophysical Research*, **89**: 6256–6270.
Tatar, O., Piper, J.D.A., Gürsoy, H., Heimann, A., and Kocbulut, F., 2004. Neotectonic deformation in the transition zone between the Dead Sea Transform and the East Anatolian Fault Zone, Southern Turkey: a palaeomagnetic study of the Karasu Rift Volcanism, *Tectonophysics*, **385**: 17–43.
Wernicke, B.P., Christiansen, R.L., England, P.C., and Sonder, L.J., 1987. Tectonomagmatic evolution of Cenozoic extension in the North American Cordillera. In Coward, M.P., Dewey, J.F., and Hancock, P.L. (eds.), *Continental Extensional Tectonics*. Geological Society of London Special Publication, **28**: 203–221.

PARKINSON, WILFRED DUDLEY

Twentieth century scientist

Wilfred Dudley Parkinson (see Figure P44) was an internationally prominent geomagnetician of the 20th century (Barton and Banks, 2002). His origins in geomagnetism were classical, and he was active through the remarkable period that saw observing apparatus change from mechanical to modern electronic, and recording methods change from paper chart and photographic paper to electronic memory. Data analysis methods changed from graphical and hand-calculated to the full range of numerical, modeling and inversion methods, which were possible with electronic computers by the end of the century. Parkinson was a dedicated and effective teacher at the University of Tasmania, and his book *Introduction to Geomagnetism* (Parkinson, 1983) reflected the benefits of being based on an excellent lecture course.

His research contributions, together, thus very much appear as a period piece of his time. The International Geophysical Year (IGY), which spanned the 18 months from July 1, 1957 to December 31, 1958 (Bates *et al.*, 1982) promoted the observation of the geomagnetic time-varying field at an enhanced density of observatories relative to the then standard global network, and focussed attention on the spatial pattern of these time-varying fields. It was from an analysis of the fluctuating fields observed at Australian stations that Parkinson first observed the "coast effect" for which he became well known, and which he characterized by graphically constructed "Parkinson vectors" (also called "Parkinson arrows" see induction arrows).

The time of his activity, plus his natural skills, meant Parkinson led in pioneering much Australian geophysics. In pursuits promulgated by his national government employer, the Bureau of Mineral Resources, he was a pioneer in the construction of magnetic maps of the Australian continent, in developing the aeromagnetic survey method for Australia, and in instituting geomagnetic measurements in Antarctica.

However he is best known for his widely used "Parkinson arrow" contribution.

Brief biography

Parkinson was born in 1919 into a family active in geomagnetism, as his father worked for the Carnegie Institution of Washington (CIW), and was assistant observer at the Watheroo Observatory in Western Australia. Parkinson thus traveled extensively with his family in connection with geomagnetic activities. He received his first degree (BSc hons in mathematics) from the University of Western Australia, and then following service with the CIW at Huancayo Observatory in Peru he studied for his PhD at Johns Hopkins University in the United States of America. After some time in USA, he returned to Australia in 1954 to join the then Bureau of Mineral Resources. In 1967 he joined the Department of Geology at the University of Tasmania. He was promoted to Reader in Geophysics and spent the rest of his career at that university, traveling internationally to other places active in geomagnetism during his sabbatical leave periods.

Hobart and environs had been significant in the history of geomagnetism and Parkinson played an important role both in the celebration of the bicentenary of the year 1792 D'Entrecasteaux expedition (Lilley and Day, 1993), and in ensuring appropriate recognition of the site and history of the 1840 Rossbank Magnetic Observatory.

As a leader in Australian geomagnetism, he was an important figure at four Australian Geomagnetic Workshops which were convened in Canberra by C.E. Barton and F.E.M. Lilley in the years 1985, 1987, 1993, and 2000. It was appropriate that he was the invited guest speaker at the last workshop he was to attend, in 2000—the year before his death in 2001.

Dudley Parkinson, as he was widely known, was survived by his wife Mary, two sons Charles and Richard, and four grandchildren.

Figure P44 Wilfred Dudley Parkinson 1919–2001 (photo taken 1990, by F.E.M. Lilley).

Recognition of the coast effect, Parkinson arrows, and crustal conductivity structure

From his familiarity with magnetic observatory data, in the 1950s Parkinson had noticed that commonly for coastal observatories the vertical component of the fluctuating field had an evident correlation, best seen during magnetic storm activity, with the onshore horizontal component. Parkinson realized that from some starting point, the vectors of magnetic field change, reckoned at intervals of say half-an-hour, tended to lie on a plane in space. Parkinson called this plane the "preferred plane."

In investigating the phenomenon, Parkinson developed a graphical method for determining the plane at a particular observatory. To characterize the effect he plotted the horizontal projection of the (unit-length) downwards normal to the plane on a map, at the observatory site. The plane became known as the "Parkinson plane," and the horizontal projection of the downwards normal became known as the "Parkinson vector."

Subsequently, usage of the term Parkinson arrows was adopted to avoid any implication that Parkinson vectors of structures considered in isolation could necessarily be added vectorially to produce the effect of the structures in combination. Later Parkinson supported use of the term "Induction arrow", out of respect for others who had developed similar ideas.

Remote from a coastline, such arrows generally point to the high conductivity side of any conductivity structure present in the regional geology. (Note that in some conventions, such as that attributed to Wiese, the arrows are plotted in the opposite direction.) Plotted for an array of stations, the arrows demonstrated that electrical conductivity contrasts within continents, as well as at continent-ocean boundaries, could thus be mapped. The realization that the phenomenon was controlled by Earth electrical conductivity structure guided much of Parkinson's research for the rest of his career.

Chronicle of publications

Parkinson's published work provides a representative chronicle in geomagnetic research for the later half of the 20th century. Thus Parkinson (1959) describes recognition of the "preferred plane," and Parkinson (1962) introduces the Parkinson vector technique. In Parkinson (1964) a laboratory model comprising copper sheeting for the seawater of the world's oceans is described—Parkinson's "terrella"—in an attempt to test whether induction in the seawater alone is sufficient to account for the coast effect. The coast effect is reviewed further in Parkinson and Jones (1979).

Parkinson (1971) addresses an analysis of the Sq variations recorded during the IGY. These observations had been a prime objective of the IGY global network of observatories, and Parkinson's analysis was one of the first to exploit the power of electronic computers, which were developing rapidly in the 1960s. This thread is picked up again in Parkinson (1977), Parkinson (1980) and Parkinson (1988).

Dosso *et al.* (1985), Parkinson *et al.* (1988), and Parkinson (1989) see attention given again to local induction, and the discovery of the Tamar conductivity anomaly in northeast Tasmania. The Australian island State proved a productive field area for Parkinson and his students, based in Hobart. In Parkinson (1999) several earlier pursuits are brought together, in addressing the influence of time-variations on aeromagnetic surveying. Parkinson and Hutton (1989) is a major review, bringing together threads of earlier research contributions.

F.E.M. Lilley

Bibliography

Barton, C., and Banks, M., 2002. Wilfred Dudley Parkinson: Obituary. *Preview (Newsletter of the Australian Society of Exploration Geophysicists)*, **96**: 11.

Bates, C.C., Gaskell, T.F., and Rice, R.R., 1982. *Geophysics in the Affairs of Man.* New York: Pergamon Press.

Dosso, H.W., Nienaber, W., and Parkinson, W.D., 1985. An analogue model study of electromagnetic induction in the Tasmania region. *Physics of the Earth and Planetary Interiors*, **39**: 118–133.

Lilley, F.E.M., and Day, A.A., 1993. D'Entrecasteaux 1792: celebrating a bicentennial in Geomagnetism. *Eos, Transactions, American Geophysical Union*, **74**: 97 and 102–103.

Parkinson, W.D., 1959. Directions of rapid geomagnetic fluctuations. *Geophysical Journal of the Royal Astronomical Society*, **2**: 1–14.

Parkinson, W.D., 1962. The influence of continents and oceans on geomagnetic variations. *Geophysical Journal of the Royal Astronomical Society*, **6**: 441–449.

Parkinson, W.D., 1964. Conductivity anomalies in Australia and the ocean effect. *Journal of Geomagnetism and Geoelectricity*, **15**: 222–226.

Parkinson, W.D., 1971. An analysis of the geomagnetic diurnal variation during the International Geophysical Year. *Gerlands Beitrage Zur Geophysik*, **80**: 199–232.

Parkinson, W.D., 1977. An analysis of the geomagnetic diurnal variation during the International Geophysical Year. *Bureau of Mineral Resources, Geology and Geophysics, (Australia)*, **173**.

Parkinson, W.D., 1980. Induction by *Sq*. *Journal of Geomagnetism and Geoelectricity*, **32**(Supplement I): SI 79–SI 88.

Parkinson, W.D., 1983. *Introduction to Geomagnetism.* Edinburgh: Scottish Academic Press.

Parkinson, W.D., 1988. The global conductivity distribution. *Surveys in Geophysics*, **9**: 235–243.

Parkinson, W.D., 1989. The analysis of single site induction data. *Physics of the Earth and Planetary Interiors*, **53**: 360–364.

Parkinson, W.D., 1999. The influence of time variations on aeromagnetic surveying. *Exploration Geophysics*, **30**: 113–114.

Parkinson, W.D., and Hutton, V.R.S., 1989. The electrical conductivity of the Earth. In Jacobs, J.A. (ed.) *Geomagnetism.* London: Academic Press, vol. 3, pp. 261–321.

Parkinson, W.D., and Jones, F.W., 1979. The geomagnetic coast effect. *Reviews of Geophysics and Space Physics*, **17**: 1999–2015.

Parkinson, W.D., Hermanto, R., Sayers, J., and Bindoff, N.L., 1988. The Tamar conductivity anomaly. *Physics of the Earth and Planetary Interiors*, **52**: 8–22.

Cross-references

Coast Effect of Induced Currents
Electromagnetic Induction (EM)
Geomagnetic Deep Sounding

PEREGRINUS, PETRUS (FLOURISHED 1269)

Virtually nothing is known of the life or wider circumstances of the Frenchman Peter, Pierre, or Petrus Peregrinus, beyond what he wrote in his *Epistola Petri Peregrini de Maricourt*, from the camp of the army besieging Lucera in Apulia, Italy, and dated August 8, 1269. Most probably, he was a native of Méharicourt, Picardy, in north-east France, and from his fascination with all kinds of machines and self-acting devices, he could have been a military engineer. His honorific title "Peregrinus," no doubt stemming from the Latin *peregrinator*, or wanderer, could have derived from his having been on pilgrimage or Crusade, though there is no evidence to back the legend that he was a monk or priest. Whether he was the same "Master Peter" referred to as a mathematician of brilliance by his English contemporary Friar Roger Bacon is uncertain but possible.

Peter's historical importance, however, derives from his being the first significant author on magnetism. His *Epistola* of 1269 was widely circulated across Europe in manuscript form, and in the 16th century was acknowledged by no less a figure than William Gilbert in *De Magnete* (1600). The *Epistola* recounts that magnetic phenomena was already familiar to Europeans by 1269 (for knowledge of the compass, originally obtained from China, had already been used for direction finding in the West for about 100 years by Peter's time), along with new experiments and insights, which could well have been Peter's own. What Peter does, however, is producing a coherent treatise that surveys the current state of magnetic knowledge in Europe by 1269, before going on to describe his own experiments.

Peter experimented with different types of magnets, including elongated pieces of magnetite stone, with what seem to have been spherical magnets (probably carved from blocks of magnetite) and with the transfer of magnetism from the natural stone to pieces of iron or steel. While there was some conjecture amongst scholars as to whether he was the first person to use the term "pole" with relation to magnetism, he was certainly familiar with the concept of the north and south terminations of all magnetic bodies. When using what seems to have been a spherical magnet, for instance, he describes employing the natural orientation of needles upon its surface to trace lines that would converge in polar points.

He also conducted experiments in the breaking in two of natural magnets, and found that when a piece of magnetic stone was so divided, the ends of the two broken pieces suddenly acquired north and south polar characteristics of their own. Yet if one rejoins the broken pieces and cemented them together, then the two magnetic fields seemed to recombine and form one magnet again. Peter's experiments showed that like poles repel and unlike poles attract. Heavy pieces of magnetite with an apparently homogenous composition were more powerful attractors of iron than the light-weighted and less pure pieces of the stone.

Needles and pieces of iron (or steel), under the right conditions and by bringing them into contact with natural magnets, themselves become magnetic, and display all the characteristics of the natural magnets from which they were derived.

Peter Peregrinus described several experiments with floating magnets which he had placed in cups that were then made to float in water—the pieces of magnetite, indeed, being like passengers in a boat. He found that when placed in an unrestricted environment of this kind, the magnets always oriented themselves north-south by the Earth's own astronomical poles. An advancement on this experiment was made when he encased an elongated piece of magnetite in wood, so as to make it float. First, he proposed that if the rim of the water-containing vessel was divided into the cardinal points and the whole vessel into 360°, then one had an instrument which would allow an observer or navigator at sea to determine the exact rising and setting points of astronomical bodies on the horizon, with reference to the north pole. Next Peter proposed a dry compass, where instead of water, the balanced needle rotated upon a vertical pin, but which still allowed astronomical bearings to be read off against a graduated edge. This would have been a very early version of a marine azimuth compass.

But in addition to his more practical experimental work, Peter was interested in the source of the magnet's power. First, he dismissed what seem to have been a series of folk myths about magnetism. For instance, magnetism could *not* have been a uniquely north-polar phenomenon because the north pole is too cold to be inhabited and hence there can be no people there to mine the magnetite familiar to scholars. Peter reminds us that magnetite deposits occur in many places in Europe that are many degrees south of the north polar regions, while—let us not forget—the needle also points south. And while Peter, like all educated medieval men who were heirs to the classical tradition, knew that the Earth itself was a sphere. In 1269 no European knew anything for certain about the nature, inhabitants, or geography of the Southern Hemisphere.

As an educated man who had clearly received some training in the classical sciences of astronomy, geometry, and arithmetic, and had no doubt encountered the increasingly influential ideas of Aristotle (all of which were part of the undergraduate *Quadrivium* of Europe's universities in the 13th century), he would have taken it as axiomatic that the Earth rested motionless at the center of the universe and that the heavens revolved about it. In this way of thinking, the Earth was associated with cosmological stability, whereas the heavens were imbued with motion. So it was probably this line of reasoning which led Peter to attribute the motion inducing capacity of the magnet to the heavens. This also prompted him to suggest the devising of a piece of apparatus to demonstrate the phenomenon. He proposed that if a spherical magnet were mounted upon its polar points in some sort of bearing, and the north and south poles exactly oriented to correspond with those of the Earth (in what would now be called an Equatorial Mount), then the sphere would rotate of its own accord. It would do so in accordance with the natural rotation of the heavens, the sphere's own magnetism simply responding to its celestial source. Peter never reported any success with this experiment as the reasons seem to be obvious.

But Peter's obvious interest in inventions next led him to join that band of medieval men who were fascinated with perpetual motion. Yet whereas his 12th-century fellow-Frenchman, Villard de Honnecourt, and others like him saw their machines as powered by arrangements of falling weights, Peter's proposed perpetual motion machine was set in train by an oval magnet the north-seeking properties of which actuated the teeth of a denticulated wheel.

What is so tantalizing about Peter is that we know nothing of him beyond his *Epistola* of August 8, 1269. His very fleeting passage across the pages of history, therefore, begs more questions than it answers. For how many educated men were there in medieval Europe who traveled, possibly crusaded, worked as military engineers, and were fascinated by mechanical invention? And let us not forget that the *Epistola* was sent not to a Bishop or university corporation, but to "*Sygerum de Foucaucourt, Militem*", or "Sygerus... the *Soldier*". Did ingenious military engineers conduct experiments into magnetism and investigate astronomy and other branches of science on those long tedious days in camp, when they were not supervising storming engines? Indeed, 13th- and 14th-century Europe was vibrant with mechanical invention: improved wind and water mills, iron founding machinery, daring Gothic cathedrals, early gun founding and explosives manufacture, the ship's rudder, and soon after 13th century the first geared mechanical clocks, many of which possessed elaborate automata. To see medieval Europe as technologically primitive is just as much a myth of our own time as was Peter's refutation of the existence of a great lump of ironstone at the north pole. And it was this ingenious culture, which produced the first experimentally based treatise on magnetism.

Allan Chapman

Bibliography

For English edition of Epistola, see Silvanus P. Thompson, Epistle of Petrus Peregrinus of Maricourt, to Sygerus of Foucaucourt, Soldier, Concerning the Magnet (London, 1902).

Grant, E., 1972. Peter Peregrinus, *Dictionary of Scientific Biography* New York: Scribner's. Grant's bibliographical essay on Peter is exhaustive.

Hellmann, G., Rara Magnetica, Neudrucke von Schriften und Karten über Meteorologie und Erdmagnetismus, no. 10 (Berlin, 1898).

PERIODIC EXTERNAL FIELDS

Definition

The lines of force of the Earth's magnetic field extend upwards from the surface of the Earth to the magnetopause, which is the boundary at great altitudes within which the solar wind confines the Earth's magnetic field. At an altitude of 100 km the atmospheric pressure is low and molecules dissociate to form a conducting layer called the ionosphere. When lower layers of the atmosphere are heated by sunlight and caused to move, the motion is propagated in all directions. Electromotive forces are formed everywhere in the moving atmosphere, where from Lenz's law, $E = \mathbf{v} \times B$. In the electrically conducting ionosphere, electrical currents flow, and the magnetic field of these currents forms the external periodic magnetic variation. On magnetically quiet days this variation is called the solar quiet day variation, denoted Sq.

In every month of the calendar, five days are denoted as disturbed days, and another five as quiet days. The difference between the magnetic field, hour after hour, on the quiet and disturbed days, is called the disturbance daily variation, denoted SD, and it is another periodic field of external origin.

The gravitational force of the Sun and the Moon causes tidal movements in the oceans, the atmosphere and the solid Earth. The tidal movements of the ionosphere caused by the Moon lead to the lunar magnetic daily variations. Solar tides are much smaller than lunar tides and the corresponding solar tidal magnetic effect is difficult to discern from the magnetic variation arising from solar heating of the atmosphere.

The deep oceans in tidal movement also give rise to magnetic variations. The separation of oceanic and ionospheric components in the analysis of solar and lunar daily magnetic variations is based on the assumption that there is no ionospheric current flow at local midnight.

Because of their ionospheric origin, the regular daily variations at the surface of the Earth have an external component that is greater than the corresponding induced internal component, and the results are used for the modeling of the distribution of electrical conductivity within the Earth's interior land eigenmodes of atmospheric oscillation.

Magnetograms

In geomagnetism, the declination, denoted D, refers to the angle between the direction of the compass needle and true north, positive towards the east. The horizontal intensity, H, of the magnetic field is that component of the field in the direction of true north, and the vertical intensity, Z, is the vertically downwards component. Magnetic variations are recorded in these three elements at magnetic observatories throughout the world. Satellite magnetic data from low-Earth-orbiting satellites at an altitude of approximately 400 km, above the ionosphere, record the main field components and the determination of daily variations from satellite magnetic data has only just begun.

Magnetograms of the declination of the Earth's magnetic field recorded at Canberra magnetic observatory are given in Figure P45 with one calendar month to each row and 12 monthly rows for the year 1999. The most prominent feature of the daily magnetograms is the regular daily change that occurs on each day throughout the year, with a daily range that is smaller in local winter months.

The monthly mean values of the magnetic elements show a small annual, one cycle per year, external, periodic variation of the order of 10 nT, and it is necessary to use several years of data to get a significant result. This variation is thought to arise from the seasonal displacement of the ring current from the equatorial plane. The accurate analysis of long-period terms such as the annual variation and the 11 year cycle, is plagued by a number of difficulties, amongst which are baseline problems arising from small errors in instrument calibration.

Hourly mean values

For many years the daily variations in the magnetic field were given as hourly mean values that had been prepared manually from magnetogram photographic traces, and given in Universal Time centered midway between the hours. The results were published in yearbooks, usually three elements, month by month, after a long delay, often of some years. With modern three component vector magnetometers (see *Magnetometers*) providing samples every 10 s with a resolution of 0.1 nT, it is now possible to have 1-min values transmitted within minutes via the Intermagnet program. In 2001 about half of the world's magnetic observatories contributed to the Intermagnet program, and since that time the number of observatories contributing has increased.

Fourier analysis

The Fourier analysis of solar and lunar magnetic tides from observatory hourly mean values has a number of technical difficulties, such as noncyclic variation, but in every case, harmonic analyses of either 24-hourly mean values or 1440 one-minute values are required. The fast Fourier transform can be used, although for relatively small calculations the Goertzel algorithm (Goertzel, 1958) is useful and easily programmed. For the analysis of lunar magnetic tides, each day is "tagged" with an integer, usually from 1 to 12, called a "character number," based on a linear combination of the astronomical parameters denoted s, h, and p, by Bartels (Bartels, 1957, 747, see *Bartels, Julius*). These represent the east longitudes of the Moon, the Sun and the Moon's perigee respectively, from January 1, 1900. For the lunar variations, the character number is determined from the angular measure of $2s - 2h$, and seasonal sidebands are determined using character numbers $2s - h$ and $2s - 3h$. Hourly values from days with the same tag provide an average daily sequence, and the 12 groups are called "group sum sequences."

The Chapman-Miller (Chapman and Miller, 1940), (see *Chapman, Sydney*) method analyses the 12 groups of 24 values, by a modified two-dimensional Fourier analysis, with adjustments for the daily noncyclic variation. By using the variability of Fourier coefficients determined for the 12 groups, reasonable estimates of standard deviations for computed coefficients are obtained. The Chapman-Miller method has the great advantage of being able to deal effectively with missing days of data, or the rejection of days because of magnetic disturbance.

The Chapman-Miller method can be adapted (Malin, 1970) to separate the lunar magnetic variations into parts of ionospheric and oceanic origin.

The Fourier analysis of the elements X, Y, and Z, being the northward, eastward, and vertically downward components, respectively, of the Earth's magnetic field, over a sphere of radius a, leads to the following Fourier series:

$$X(a,\theta,\phi) = \sum_{M=1}^{4} [C_{XM}(\theta,\phi)\cos Mt + D_{XM}(\theta,\phi)\sin Mt],$$
$$Y(a,\theta,\phi) = \sum_{M=1}^{4} [C_{YM}(\theta,\phi)\cos Mt + D_{YM}(\theta,\phi)\sin Mt],$$
$$Z(a,\theta,\phi) = \sum_{M=1}^{4} [C_{ZM}(\theta,\phi)\cos Mt + D_{ZM}(\theta,\phi)\sin Mt].$$

(Eq. 1)

Spherical harmonic analysis of the Fourier coefficients C_{XM}, C_{YM}, and C_{ZM} gives internal field coefficients $g_{NCi}^{mM}, h_{nCi}^{mM}$, and external field coefficients $g_{NCe}^{mM}, h_{nCe}^{mM}$, in the potential function $V_C^M(r,\theta,\phi)$, where

$$V_C^M(r,\theta,\phi) = a\sum_n\sum_m \left[\left(\frac{a}{r}\right)^{n+1}\left(g_{nCi}^{mM}\cos m\phi + h_{nCi}^{mM}\sin m\phi\right) + \left(\frac{r}{a}\right)^n\left(g_{nCe}^{mM}\cos m\phi + h_{nCe}^{mM}\sin m\phi\right)\right]P_n^m(\cos\theta).$$

(Eq. 2)

It is the usual practice to do a spherical harmonic analysis of the Fourier coefficients D_{XM}, D_{YM}, and D_{ZM}, giving the two sets of spherical

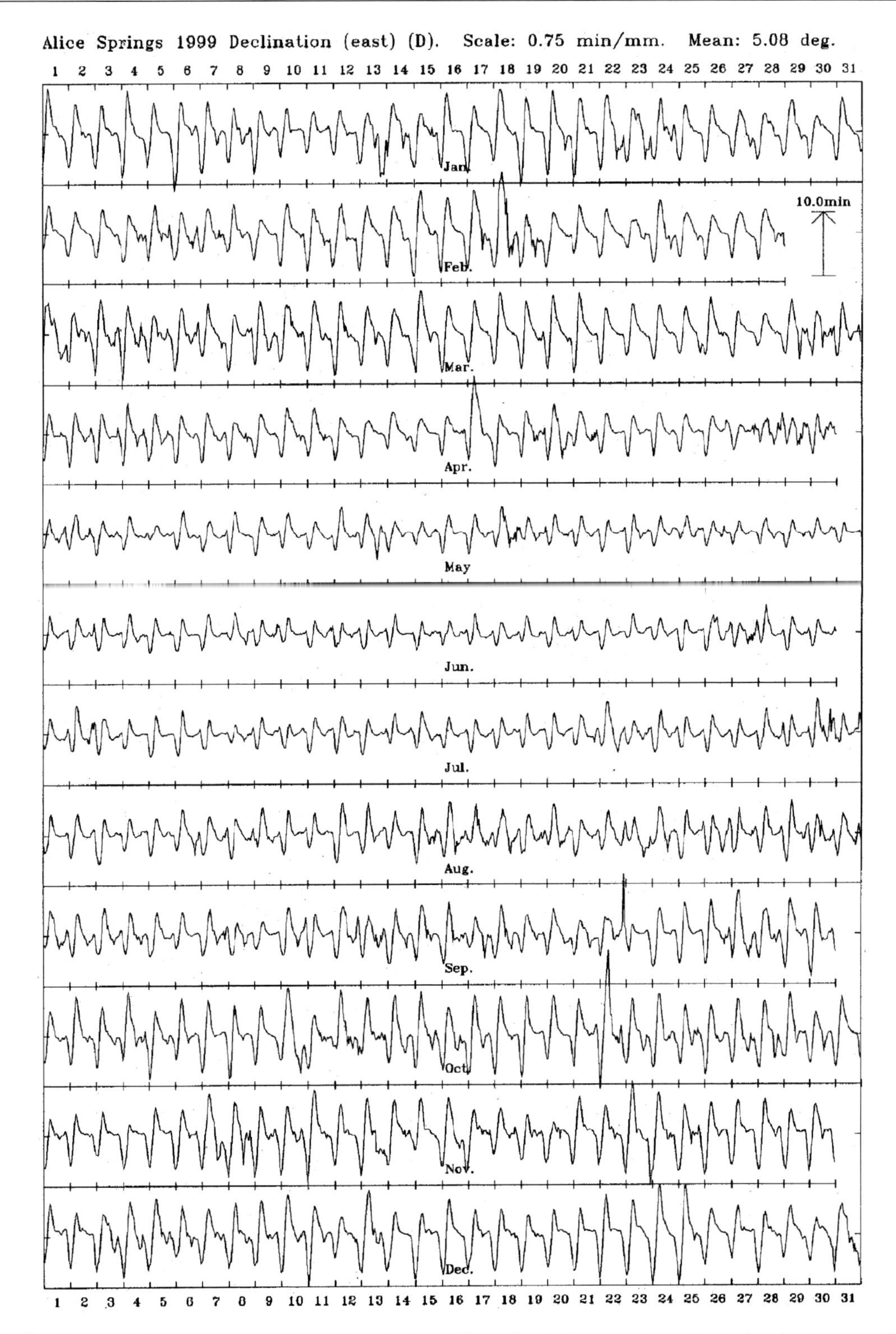

Figure P45 Daily change of the compass needle at Alice Springs, 1999. The daily range is smaller in local winter months, the lunar changes appear as a small, fortnightly modulation. (From the Australian Geomagnetism Report, 1999. Magnetic Observatories, vol. 47, Part 1, 2001).

harmonic coefficients, but leaving the researcher with the difficult task of deciding which coefficient is which, and of separating the westward and eastward moving terms.

However, by using the g and h coefficients from $V_C^M(r,\theta,\phi)$ to represent the potential $V_D^M(r,\theta,\phi)$ for the D coefficients in the form:

$$V_D^M(r,\theta,\phi) = a\sum_n\sum_m\left[\left(\frac{a}{r}\right)^{n+1}\left(h_{nCi}^{mM}\cos m\phi - g_{nCi}^{mM}\sin m\phi\right) + \left(\frac{r}{a}\right)^n\left(h_{nCe}^{mM}\cos m\phi - g_{nCe}^{mM}\sin m\phi\right)\right]P_n^m(\cos\theta) \quad \text{(Eq. 3)}$$

the potential function based on both $V_C^M(r,\theta,\phi)$ and $V_D^M(r,\theta,\phi)$, namely

$$V_C^M(r,\theta,\phi)\cos Mt + V_D^M(r,\theta,\phi)\sin Mt$$
$$= a\sum_{n=1}^{N}\sum_{m=0}^{n}\left\{\left(\frac{a}{r}\right)^{n+1}\left[g_{nCi}^{mM}\cos(m\phi+Mt)+h_{nCi}^{M}\sin(m\phi+Mt)\right]\right.$$
$$\left.+\left(\frac{r}{a}\right)^{n}\left[g_{nCe}^{mM}\cos(m\phi+Mt)+h_{nCe}^{M}\sin(m\phi+Mt)\right]\right\}P_n^m(\cos\theta) \quad \text{(Eq. 4)}$$

consists entirely of westward moving terms. The actual numerical analysis is based on Eqs. (2) and (3) together. The terms for which $m=M$ are local time terms. It is a simple matter to construct an expression corresponding to Eq. (3), which will give the results for eastward moving terms, which analysis shows to be much smaller than the westward moving terms.

Amplitudes and phase angles for internal and external fields are needed for conductivity studies, so that results of Eq. (4) should be provided in the form

$$a\sum_{n=1}^{N}\sum_{m=0}^{n}\left\{\left(\frac{a}{r}\right)^{n+1}A_{nCi}^{mM}\cos(m\phi+Mt+\epsilon_{nCi}^{mM})\right.$$
$$\left.+\left(\frac{r}{a}\right)^{n}A_{nCe}^{mM}\cos(m\phi+Mt+\epsilon_{nCe}^{mM})\right\}P_n^m(\cos\theta). \quad \text{(Eq. 5)}$$

Note particularly that the phase angles ϵ_{nCi}^{mM} and ϵ_{nCe}^{mM} in Eq. (5) are given for cosine functions.

No general method of analysis for periodic external variations from satellite magnetic data is available. Significant orbital drift through local time is required, and the more rapid the drift, the better.

Ionospheric current systems

It is very convenient to represent the vector field of external (ionospheric) origin by a scalar "stream function" whose contours run parallel to the lines of flow of electrical currents in the thin conducting shell used to represent the ionosphere. This scalar function $W(\theta,\phi)$, A, in Eq. (6), is determined from the *external* field coefficients only. The surface current density $\mathbf{K}(\theta,\phi)$, A/m, is

$$\mathbf{K}(\theta,\phi)=\nabla\times[\mathbf{e}_r\Psi(\theta,\phi)]=-\mathbf{e}_r\times\nabla W(\theta,\phi),$$
$$=\frac{1}{R}\left(\frac{1}{\sin\theta}\frac{\partial W}{\partial\phi}\mathbf{e}_\theta-\frac{\partial W}{\partial\theta}\mathbf{e}_\phi\right),\ \text{A/m}. \quad \text{(Eq. 6)}$$

The current function $W(\theta,\phi)$, A, for a shell of radius R, relative to a reference sphere of radius a, is given by

$$W(\theta,\phi)=\sum_{n=1}^{N}\sum_{m=0}^{n}\left[W_{gn}^{m}\cos(m\phi+Mt)\right.$$
$$\left.+W_{hn}^{m}\sin(m\phi+Mt)\right]P_n^m(\cos\theta) \quad \text{(Eq. 7)}$$

with coefficients W_{gn}^m and W_{hn}^m as given originally by Maxwell (Maxwell, 1873, p. 672),

$$W_{gn}^{m}=-a_{km}\frac{10}{4\pi}\frac{2n+1}{n+1}\left(\frac{R_{km}}{a_{km}}\right)^{n+1}(g_{nCe}^{mM})_{\text{nT}},$$
$$W_{hn}^{m}=-a_{km}\frac{10}{4\pi}\frac{2n+1}{n+1}\left(\frac{R_{km}}{a_{km}}\right)^{n+1}(h_{nCe}^{mM})_{\text{nT}}. \quad \text{(Eq. 8)}$$

Contours in kiloamperes of the current function for the solar quiet day variation are given in Figure P46.

Ionospheric dynamo theory

An article by Stewart (1883) in the 9th edition of the *Encyclopedia Britannica*—in a subsection of the article, "Meteorology"—proposed a dynamo in the upper atmosphere as the only plausible theory for the regular solar daily magnetic variation. Schuster (1889) analyzed Sq from four Northern Hemisphere observatories, and then used the theory of electromagnetic induction (see *Electromagnetic induction* and *Price, Albert Thomas*) in a uniformly conducting sphere as derived by Lamb (in an appendix to Schuster's 1889 paper) to determine the electrical conductivity models of the Earth's interior. The theory, for a uniformly conducting sphere, uses phase angle differences and amplitude ratios of internal and external components. Schuster (1908) took the process a step further, inferring upper atmosphere wind velocities from global representations of atmospheric pressure. In particular he represented the components of electric force by a potential function $S(\theta,\phi)$ and a stream function $T(\theta,\phi)$,

$$E_\theta=-\frac{1}{a}\left(\frac{\partial S}{\partial\theta}-\frac{1}{\sin\theta}\frac{\partial T}{\partial\phi}\right),$$
$$E_\phi=-\frac{1}{a}\left(\frac{1}{\sin\theta}\frac{\partial S}{\partial\phi}+\frac{\partial T}{\partial\theta}\right). \quad \text{(Eq. 9)}$$

Schuster included an analysis of the seasonal change of *Sq*. His theory provided a solid foundation for ionospheric dynamos, adapted in more recent times by using the eigenfunctions of the Laplace tidal equation to represent the forced and free atmospheric oscillations and wind velocities.

Chapman (1919) (see *Chapman, Sydney*) developed the ionospheric dynamo theory (see *Ionosphere*) of the magnetic daily variations, of solar and lunar origin, using atmospheric velocities given as the gradient of a scalar potential. Mathematical expressions for the lunar magnetic variations, and the Chapman-Miller method for their analysis developed gradually over many years.

In the following sections, r,θ,ϕ, will denote radial distance, colatitude, and east longitude in a spherical polar coordinate system. Subscripts will be used to denote the r,θ,ϕ, components of vectors. For wind movements in the high atmosphere, the radial component v_r is assumed to be negligibly small, relative to the horizontal components, v_θ and v_ϕ. In a magnetic field with magnetic flux density components (B_r,B_θ,B_ϕ), the dynamo electric field $\mathbf{v}\times\boldsymbol{B}$ has spherical polar components

$$E_r=v_\theta B_\phi-v_\phi B_\theta,$$
$$E_\theta=v_\phi B_r,$$
$$E_\phi=-v_\theta B_r. \quad \text{(Eq. 10)}$$

Consider the dipole term only in the geomagnetic potential, V, where

$$V=a\left(\frac{a}{r}\right)^2 g_1^0\cos\theta, \quad \text{(Eq. 11)}$$

so that the components of magnetic flux density from $\boldsymbol{B}=-\nabla V$ at $r=R$, are,

$$B_r=-\left.\frac{\partial V}{\partial r}\right|_{r=R}=2G_1^0\cos\theta,$$
$$B_\theta=-\left.\frac{1}{r}\frac{\partial V}{\partial\theta}\right|_{r=R}=G_1^0\sin\theta,$$
$$B_\phi=-\left.\frac{1}{r\sin\theta}\frac{\partial V}{\partial\phi}\right|_{r=R}=0, \quad \text{(Eq. 12)}$$

where $G_1^0=(a/R)^3g_1^0$.

With the Schuster representation of the electric field, in Eq. (9), the currents driven in the direction of the gradient of the function $S(\theta,\phi)$ will go towards maxima or minima of the function, which act as sinks or sources. In a conducting thin shell, these currents have nowhere to go and an electrostatic field is set up so that no current

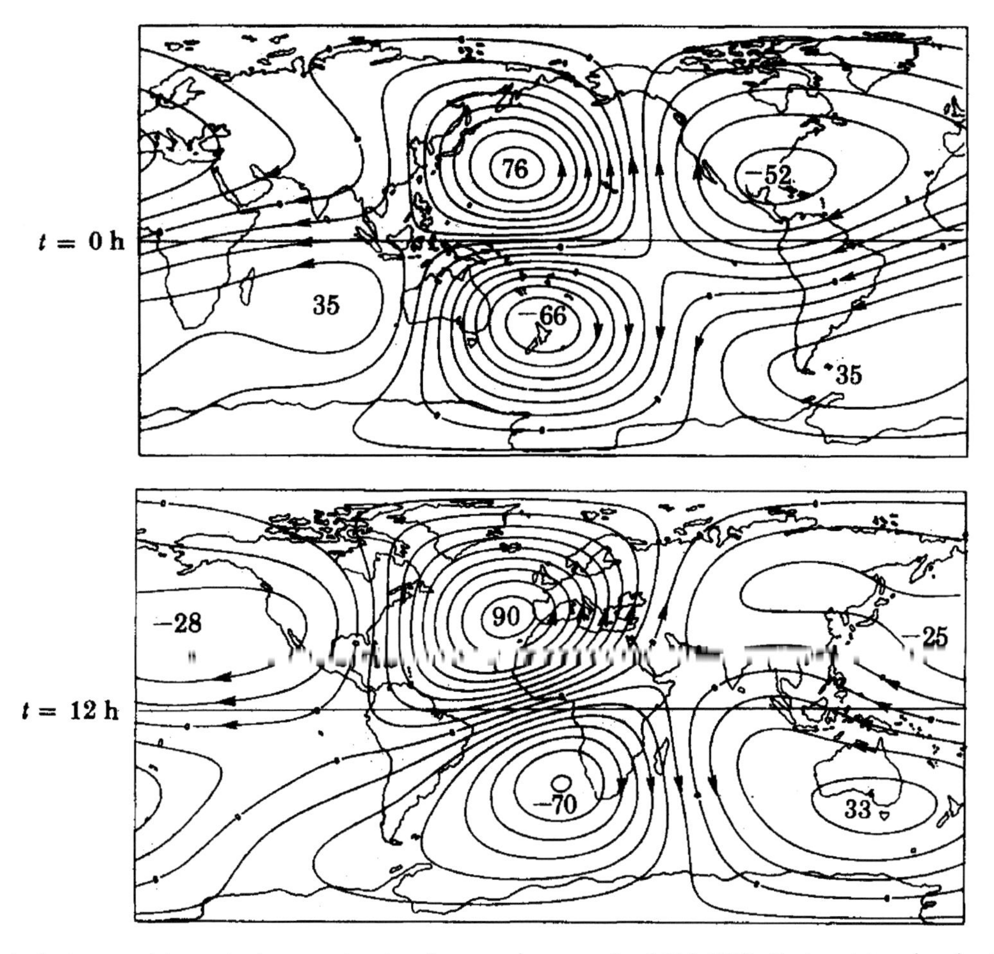

Figure P46 Equivalent external ionospheric current system for annual average Sq, 1964–1965. Contour interval and extrema in kiloamperes (from Winch, 1981).

flows. The stream function, or toroidal component $T(\theta, \phi)$, drives a stream function electrical current parallel to the contours of $T(\theta, \phi)$ in the thin spherical shell representing the ionosphere. Thus, backing off the electrostatic field from the dynamo field,

$$E_\theta + \frac{1}{R}\frac{\partial S}{\partial \theta} = \frac{1}{R\sin\theta}\frac{\partial T}{\partial \phi}, \quad E_\phi + \frac{1}{R\sin\theta}\frac{\partial S}{\partial \phi} = -\frac{1}{R}\frac{\partial T}{\partial \theta}. \quad \text{(Eq. 13)}$$

From the current function $W(\theta, \phi)$, determined from observatory data, the components of surface current density are J_θ, J_ϕ, where

$$J_\theta = \frac{1}{R\sin\theta}\frac{\partial W}{\partial \phi}, \quad J_\phi = -\frac{1}{R}\frac{\partial W}{\partial \theta}. \quad \text{(Eq. 14)}$$

Therefore, by Ohm's law in the form $\boldsymbol{E} = \boldsymbol{J}/\sigma(\theta, \phi)$,

$$\frac{1}{\sigma(\theta,\phi)R\sin\theta}\frac{\partial W}{\partial \phi} = \frac{1}{R\sin\theta}\frac{\partial T}{\partial \phi},$$
$$\frac{1}{\sigma(\theta,\phi)R}\frac{\partial W}{\partial \theta} = -\frac{1}{R}\frac{\partial T}{\partial \theta}. \quad \text{(Eq. 15)}$$

Therefore, from Eqs. (10) to (15), ignoring self-induction,

$$2v_\phi G_1^0 \cos\theta - \frac{1}{R}\frac{\partial S}{\partial \theta} = \frac{1}{\sigma(\theta,\phi)R\sin\theta}\frac{\partial W}{\partial \phi},$$
$$2v_\theta G_1^0 \cos\theta + \frac{1}{R\sin\theta}\frac{\partial S}{\partial \phi} = \frac{1}{\sigma(\theta,\phi)R}\frac{\partial W}{\partial \theta}. \quad \text{(Eq. 16)}$$

Eliminating the electrostatic function $S(\theta, \phi)$ from Eq. (16) gives

$$\frac{2G_1^0}{\sin\theta}\left[\frac{\partial v_\phi}{\partial \phi}\cos\theta + \frac{\partial}{\partial \theta}(v_\theta \sin\theta\cos\theta)\right]$$
$$= -\frac{1}{\sigma(\theta,\phi)R}\left[\frac{1}{\sin^2\theta}\frac{\partial^2 W}{\partial \phi^2} + \frac{1}{\sin\theta}\frac{\partial}{\partial \theta}\left(\sin\theta\frac{\partial W}{\partial \theta}\right)\right]$$
$$- \frac{1}{\sigma^2(\theta,\phi)R}\left[\frac{\partial \sigma}{\partial \phi}\frac{1}{\sin\theta}\frac{\partial W}{\partial \phi} + \frac{\partial \sigma}{\partial \theta}\frac{\partial W}{\partial \theta}\sin\theta\right]. \quad \text{(Eq. 17)}$$

In the special case of constant conductivity $\sigma(\theta, \phi) = \sigma$, Eq. (17) can be written

$$\frac{2G_1^0}{\sin\theta}\left[\frac{\partial v_\phi}{\partial \phi}\cos\theta + \frac{\partial}{\partial \theta}(v_\theta \sin\theta\cos\theta)\right] = -\frac{1}{\sigma R}\nabla_H^2 W(\theta,\phi), \quad \text{(Eq. 18)}$$

where ∇_{H}^2 is the horizontal Laplacian operator,

$$\nabla_{\mathrm{H}}^2 W = \frac{1}{\sin\theta}\frac{\partial}{\partial \theta}\left(\sin\theta\frac{\partial W}{\partial \theta}\right) + \frac{1}{\sin^2\theta}\frac{\partial^2 W}{\partial \phi^2}. \quad \text{(Eq. 19)}$$

Wind velocities in the upper atmosphere will be governed by the Laplace tidal equation, and the components of the equation can be used to simplify the expression on the left of Eq. (18).

Laplace tidal equation

Taylor, in his paper on oscillations of the atmosphere (Taylor, 1936), showed that when the temperature of the atmosphere is a function only of height above the ground, free oscillations are possible which are identical with those of a sea of uniform depth H, except that the amplitude of the oscillations is a function of the height.

Movements of the upper atmosphere are governed by the Laplace tidal equation, which includes Coriolis forces and conservative forces. In vectorial form

$$\frac{\partial \boldsymbol{v}}{\partial t} + 2\Omega \times \boldsymbol{v} = -\nabla F \qquad \text{(Eq. 20)}$$

If v_θ and v_ϕ denote the southward and eastward components of velocity relative to the surface of the rotating Earth of radius a, the horizontal components of Eq. (20) are

$$\frac{\partial v_\theta}{\partial t} - 2\Omega v_\phi \cos\theta = -\frac{1}{R}\frac{\partial F}{\partial \theta},$$
$$\frac{\partial v_\phi}{\partial t} + 2\Omega v_\theta \cos\theta = -\frac{1}{R\sin\theta}\frac{\partial F}{\partial \phi}, \qquad \text{(Eq. 21)}$$

The function F is the scalar potential of conservative forces,

$$F = \delta p/\rho + V. \qquad \text{(Eq. 22)}$$

With e^{iMt} time dependence, Eq. (21) can be written

$$\mathrm{i}Mv_\theta - 2\Omega v_\phi \cos\theta = -\frac{1}{R}\frac{\partial F}{\partial \theta},$$
$$\mathrm{i}Mv_\phi + 2\Omega v_\theta \cos\theta = -\frac{1}{R\sin\theta}\frac{\partial F}{\partial \phi}. \qquad \text{(Eq. 23)}$$

With $\mathrm{e}^{im\phi}$ longitude dependence, the Laplace tidal equation in the form of Eq. (23) can be solved for v_θ and v_ϕ,

$$v_\theta = \frac{iM}{4R\Omega^2(f^2 - \cos^2\theta)}\left(\frac{\partial F}{\partial \theta} + \frac{m}{f}F\cot\theta\right), \qquad \text{(Eq. 24)}$$

$$v_\phi = \frac{-M}{4R\Omega^2(f^2 - \cos^2\theta)}\left(\frac{\cos\theta}{f}\frac{\partial F}{\partial \theta} + \frac{m}{\sin\theta}F\right) \qquad \text{(Eq. 25)}$$

where $f = M/2\Omega$.

From Eqs. (24) and (25), the divergence of the velocity is

$$\frac{1}{\sin\theta}\left[\frac{\partial v_\phi}{\partial \phi} + \frac{\partial}{\partial \theta}(v_\theta \sin\theta)\right] = \frac{iM}{4R\Omega^2}LF \qquad \text{(Eq. 26)}$$

where the operator L is defined by

$$LF = \frac{1}{\sin\theta}\frac{\partial}{\partial\theta}\left(\frac{\sin\theta}{f^2 - \cos^2\theta}\frac{\partial F}{\partial \theta}\right) - \frac{1}{f^2 - \cos^2\theta}\left[\frac{m}{f}\frac{f^2 + \cos^2\theta}{f^2 - \cos^2\theta} + \frac{m^2}{\sin^2\theta}\right]F, \qquad \text{(Eq. 27)}$$

where $f = M/2\Omega$. Eigenfunctions $F(\theta)$ of the operator L, are denoted $\Theta^M(\theta)$ and have $\mathrm{e}^{im\phi}$ longitudinal dependence, and e^{iMt} time dependence,

$$L\Theta^M = -(4R^2\Omega^2/gh^M)\Theta^M. \qquad \text{(Eq. 28)}$$

For specific values of M, the eigenfunctions Θ^M form an orthogonal set. The eigenvalue $4R^2\Omega^2/gh^M$ is chosen to correspond to the theory of the oscillations of a shallow ocean of depth h^M on a rotating sphere of radius R.

Eliminating the scalar potential $F(\theta,\phi)$ in Eq. (23) leads to the remarkable simplification required in Eq. (18), namely

$$\frac{1}{\sin\theta}\left[\frac{\partial v_\phi}{\partial \phi}\cos\theta + \frac{\partial}{\partial\theta}(v_\theta \sin\theta\cos\theta)\right] = \frac{iM}{2\Omega}\frac{1}{\sin\theta}\left[\frac{\partial v_\theta}{\partial\phi} - \frac{\partial}{\partial\theta}(v_\phi\sin\theta)\right] \qquad \text{(Eq. 29)}$$

Longuet-Higgins calculation

Longuet-Higgins (1968) used spectral analysis of tidal modes of "stream function" and "potential" function type for a shallow ocean on a rotating sphere. Tarpley (1970) pointed out the need to use wind velocities based on the Laplace tidal equation, and not just the gradient of a scalar potential. He showed the influence of the various tidal modes in the combinations of spherical harmonic coefficients obtained by analysis of global distributions of hourly mean values.

The horizontal components of velocity with potential and stream function components, Φ and Ψ, respectively, are

$$v_\theta = \frac{\partial\Phi}{\partial\theta} + \frac{1}{\sin\theta}\frac{\partial\Psi}{\partial\phi},$$
$$v_\phi = \frac{1}{\sin\theta}\frac{\partial\Phi}{\partial\phi} - \frac{\partial\Psi}{\partial\theta}. \qquad \text{(Eq. 30)}$$

In particular, the eigenvalues of the Laplace tidal equation yield eigenfunctions for Φ and Ψ simultaneously. It follows from Eq. (30) that

$$\frac{1}{\sin\theta}\left[\frac{\partial v_\phi}{\partial\phi} + \frac{\partial}{\partial\theta}(v_\theta\sin\theta)\right] = \nabla_{\mathrm{H}}^2\Phi, \qquad \text{(Eq. 31)}$$

$$\frac{1}{\sin\theta}\left[\frac{\partial v_\theta}{\partial\phi} - \frac{\partial}{\partial\theta}(v_\phi\sin\theta)\right] = \nabla_{\mathrm{H}}^2\Psi, \qquad \text{(Eq. 32)}$$

where ∇_{H}^2 is the horizontal Laplacian, defined in Eq. (19).

When $\Phi(\theta,\phi)$ and $\Psi(\theta,\phi)$ are represented as a series of associated Legendre functions,

$$\Phi(\theta,\phi) = \sum_{n=m}^{\infty} A_n^m P_n^m(\cos\theta)\mathrm{e}^{i(m\phi+Mt)},$$
$$\Psi(\theta,\phi) = \sum_{n=m}^{\infty} iB_n^m P_n^m(\cos\theta)\mathrm{e}^{i(m\phi+Mt)},$$

then Eqs. (31) and (32) lead to the equations for the eigenvalues h^M, and eigenfunctions, namely, in which the coefficients A_n^m and B_n^m are calculated simultaneously.

From Eqs. (26), (28), and (31), for the horizontal divergence of the flow, the horizontal Laplacian of the potential flow is proportional to the eigenfunctions Θ^M

$$\nabla_H^2\Phi = -\frac{iM}{4R\Omega^2}L\Theta^M = -\frac{iMR}{gh^M}\Theta^M. \qquad \text{(Eq. 33)}$$

From Eqs. (18), (29), and (32), the velocity stream function term $\Psi(\theta,\phi)$, is given in terms of the current function $W(\theta,\phi)$, which has been calculated by spherical harmonic analysis of geomagnetic data:

$$\nabla_{\mathrm{H}}^2\Psi = \frac{i\kappa\Omega}{aMG_1^0}\nabla_{\mathrm{H}}^2 W(\theta,\phi). \qquad \text{(Eq. 34)}$$

Equations (33) and (34) are valid only for an ionosphere with constant conductivity, whereas the ionospheric conductivity will be small during nighttime hours and large during daylight hours. For this particular approximation, Eq. (34) indicates that from the known current function $W(\theta, \phi)$, the coefficients B_n^m of the stream function component $\Psi(\theta, \Phi)$ of wind velocity can be obtained. These coefficients are determined by eigenvalue calculation at the same time as coefficients A_n^m of the potential function component $\Phi(\theta, \phi)$ of wind velocity and allow a direct determination of wind velocity modes from the magnetic field daily variation current function. See, for example, Winch (1981).

Ionospheric conductivity

The electrical conductivity of an ionized gas in the presence of a magnetic field is no longer uniform, but depends on the relative orientation of the applied electric field and the magnetic flux density $\boldsymbol{B}$. The electric field can be resolved into three components. Thus component of $\boldsymbol{E}$ parallel to $\boldsymbol{B}$ is denoted $\boldsymbol{E}_0$; the component of $\boldsymbol{E}$ perpendicular to $\boldsymbol{B}$ in the plane containing both $\boldsymbol{E}$ and $\boldsymbol{B}$ is denoted $\boldsymbol{E}_1$; the component of $\boldsymbol{E}$ in the direction $\boldsymbol{B} \times \boldsymbol{E}$ is denoted $\boldsymbol{E}_2$. The current density $\boldsymbol{J}$ is given by

$$\boldsymbol{J} = \sigma_0 \boldsymbol{E}_0 + \sigma_1 \boldsymbol{E}_1 + \sigma_2 \boldsymbol{E}_2. \qquad \text{(Eq. 35)}$$

in which σ_0 is the longitudinal conductivity, σ_1 the Pedersen conductivity, and σ_2 the Hall conductivity.

Thus, if $\boldsymbol{b}$ is used to denote a unit vector in the direction of the magnetic flux density $\boldsymbol{B}$, then if χ is used to denote the magnetic dip, being the angle that the magnetic needle dips below the horizontal, the spherical polar components of the unit vector $\boldsymbol{b}$ are

$$\boldsymbol{b} = -\sin\chi\, \boldsymbol{e}_r - \cos\chi\, \boldsymbol{e}_\theta. \qquad \text{(Eq. 36)}$$

Spherical polar components of the electric field $\boldsymbol{E}_0, \boldsymbol{E}_1$and $\boldsymbol{E}_1$ are

$$\begin{aligned} \boldsymbol{E}_0 &= (\boldsymbol{E} \cdot \boldsymbol{b})\boldsymbol{b}, \\ &= (E_r \sin\chi + E_\theta \cos\chi)\sin\chi\, \boldsymbol{e}_r + (E_r \sin\chi \\ &\quad + E_\theta \cos\chi)\cos\chi\, \boldsymbol{e}_\theta \\ \boldsymbol{E}_1 &= \boldsymbol{E} - \boldsymbol{E}_0, \\ &= (E_r \cos\chi - E_\theta \sin\chi)\cos\chi\, \boldsymbol{e}_r - (E_r \cos\chi - E_\theta \sin\chi) \\ &\quad \times \sin\chi\, \boldsymbol{e}_\theta + E_\phi \boldsymbol{e}_\phi, \\ \boldsymbol{E}_2 &= \boldsymbol{b} \times \boldsymbol{E}, \\ &= -E_\phi \cos\chi\, \boldsymbol{e}_r + E_\phi \sin\chi\, \boldsymbol{e}_\theta + (E_r \cos\chi - E_\theta \sin\chi)\boldsymbol{e}_\phi. \end{aligned} \qquad \text{(Eq. 37)}$$

The spherical polar components of current are

$$\begin{aligned} J_r &= (\sigma_0 \sin^2\chi + \sigma_1 \cos^2\chi)E_r + (\sigma_0 - \sigma_1)\sin\chi\cos\chi E_\theta \\ &\quad - \sigma_2 \cos\chi E_\phi, \\ J_\theta &= (\sigma_0 - \sigma_1)\sin\chi\cos\chi E_r + (\sigma_0 \cos^2\chi + \sigma_1 \sin^2\chi)E_\theta \\ &\quad + \sigma_2 \sin\chi E_\phi, \\ J_\phi &= \sigma_2 \cos\chi E_r - \sigma_2 \sin\chi E_\theta + \sigma_1 E_\phi. \end{aligned} \qquad \text{(Eq. 38)}$$

At the dip equator, $\chi = 0$, and Eq. 38 reduces to

$$\begin{aligned} J_r\big|_{\chi=0} &= \sigma_1 E_r\big|_{\chi=0} - \sigma_2 E_\phi\big|_{\chi=0}, \\ J_\theta\big|_{\chi=0} &= \sigma_0 E_\theta\big|_{\chi=0}, \\ J_\phi\big|_{\chi=0} &= \sigma_2 E_r\big|_{\chi=0} + \sigma_1 E_\phi\big|_{\chi=0}. \end{aligned} \qquad \text{(Eq. 39)}$$

At the dip equator, the fluid velocity $\boldsymbol{v}$ and the magnetic flux density $\boldsymbol{B}$ are both horizontal, and the dynamo electric field $\boldsymbol{v} \times \boldsymbol{B}$ will be a purely radial field $E_r\big|_{\chi=0}$. By the first of Eq. (39), and electrostatic field $E_\phi\big|_{\chi=0}$ appears,

$$E_r\big|_{\chi=0} = \frac{1}{\sigma_1} J_r\big|_{\chi=0} + \frac{\sigma_2}{\sigma_1} E_\phi\big|_{\chi=0}. \qquad \text{(Eq. 40)}$$

Substitution of Eq. (40) into $J_\phi\big|_{\chi=0}$ of Eq. (39) gives

$$J_\phi\big|_{\chi=0} = \frac{\sigma_2}{\sigma_1} J_r\big|_{\chi=0} + \sigma_3 E_\phi\big|_{\chi=0}$$

in which σ_3 is the Cowling conductivity, defined at the dip equator by

$$\sigma_3 = \sigma_1 + \frac{\sigma_2^2}{\sigma_1} = \frac{\sigma_1^2 + \sigma_2^2}{\sigma_1}, \qquad \text{(Eq. 41)}$$

which is clearly greater than the Pedersen conductivity σ_1.

Equatorial electrojet

At observatories within one or two degrees of the magnetic equator, particularly at Huancayo in Peru, the regular daily variations in horizontal intensity are larger than at observatories further away from the magnetic equator. The section of the equivalent ionospheric current system that is responsible for this variation was first called the equatorial electrojet by Chapman (Chapman, 1951), and it is a narrow band of current flowing eastward along the magnetic dip equator in the sunlit section of the ionosphere. Baker and Martyn (Baker and Martyn, 1953) showed that "thin-shell" approximation, in which the radial component of current generated by the ionospheric dynamo is inhibited, leads to an effective enhancement of conductivity in a narrow band along the dip equator.

Vector component data from the Magsat satellite of 1980, in an orbit that remained on the dawn and dusk terminators in the ionosphere-magnetosphere region, allowed modeling of the morning and afternoon structure of the equatorial electrojet. The more recent CHAMP and Ørsted satellites, moving steadily through a range of local times, provide information on the longitudinal distribution and local time variation of the equatorial electrojet. Studies also continue at observatories in India, Africa, and Vietnam that are along, and in lines directly across the magnetic equator.

Thin-shell approximation

From the first of Eq. (38) use

$$\begin{aligned} &(\sigma_0 \sin^2\chi + \sigma_1 \cos^2\chi)E_r \\ &\quad = J_r - (\sigma_0 - \sigma_1)\sin\chi\cos\chi\, E_\theta + \sigma_2 \cos\chi. \end{aligned} \qquad \text{(Eq. 42)}$$

to eliminate E_r in Eq. (38) to obtain

$$\begin{aligned} J_\theta &= \frac{(\sigma_0 - \sigma_1)\sin\chi\cos\chi}{\sigma_0 \sin^2\chi + \sigma_1 \cos^2\chi} J_r + \sigma_{\theta\theta} E_\theta + \sigma_{\theta\phi} E_\phi, \\ J_\phi &= \frac{\sigma_2 \cos\chi}{\sigma_0 \sin^2\chi + \sigma_1 \cos^2\chi} J_r - \sigma_{\theta\phi} E_\theta + \sigma_{\phi\phi} E_\phi. \end{aligned} \qquad \text{(Eq. 43)}$$

The tensor conductivity components are

$$\begin{aligned} \sigma_{\theta\theta} &= \frac{\sigma_0\sigma_1}{\sigma_0 \sin^2\chi + \sigma_1 \cos^2\chi}, \\ \sigma_{\theta\phi} &= \frac{\sigma_0\sigma_2 \sin\chi}{\sigma_0 \sin^2\chi + \sigma_1 \cos^2\chi}, \\ \sigma_{\phi\phi} &= \frac{\sigma_0 \sin^2\chi + \sigma_3 \cos^2\chi}{\frac{\sigma_0}{\sigma_1}\sin^2\chi + \cos^2\chi}. \end{aligned} \qquad \text{(Eq. 44)}$$

At the dip equator, $\chi = 0$, the conductivity$\sigma_{\phi\phi}$ reduces to the Cowling conductivity,

$$\sigma_{\phi\phi}\big|_{\chi=0} = \sigma_3. \quad \text{(Eq. 45)}$$

In the ionosphere, $\sigma_0 \approx 800\sigma_1$, and this tends to keep the conductivity $\sigma_{\phi\phi}$ much smaller than the Cowling conductivity σ_3 in all regions except those where the magnetic dip is small. In a narrow band along the dip equator, corresponding to the equatorial electrojet, $\sigma_{\phi\phi}$ is approximately equal to the Cowling conductivity.

Disturbance daily variation

The disturbance daily variation is simply the daily variation that occurs when the Earth rotates about the geographic axis when it is surrounded by a ring current system symmetric about the geomagnetic axis.

Annual variation

Walker, in his Adam's Prize essay of 1866 (Walker, 1866) includes a chapter on the annual variation, recording that Cassini noted an annual variation in the direction of the compass needle in 1786. Modern analyses show that there are both annual and semiannual components. Spherical harmonic analysis shows that the annual term is dominantly a zonal quadrupolar term, and the semiannual term is a zonal dipolar term. Determination of these long period terms is difficult due to aliasing by shorter period variations. For example, a calculation based on midnight values only will detect the K1 tide, a sidereal time variation, as an annual term. The precise origin of these terms is still debated, but annual changes in the Sun-Earth geometry have been proposed, together with the corresponding changes in the influences of the solar wind on the Earth's magnetosphere and ring current.

Denis Winch

Bibliography

Baker, W.G., and Martyn, D.F., 1953. Electric currents in the ionosphere. I. The conductivity. *Philosophical Transactions of the Royal Society of London*, **246**A: 281–294.

Bartels, J., 1957. Geizeitenkräfte. *Handbuch der Physik*, **48**: 734–774.

Chapman, S., 1919. The solar and lunar diurnal variations of terrestrial magnetism. *Philosophical Transactions of the Royal Society of London*, **218**A: 1–118.

Chapman, S., and Bartels, J., 1940. *Geomagnetism.* Oxford: Clarendon Press.

Chapman, S., 1951. The equatorial electrojet as detected from the abnormal electric current distribution above Huancayo, Peru and elsewhere. *Archiv für Meteorologie Geophysik und Bioklimatologie, Serie A, Meteorologie und Geophysik*, **4**: 368–390.

Chapman, S., and Miller, J.C.P., 1940. The statistical determination of lunar daily variations in geomagnetic and meteorological elements. *Monthly Notices of the Royal Astronomical Society, Geophysical Supplement*, **4**: 649–659.

Goertzel, G., 1958. An algorithm for the evaluation of finite trigonometric series. *American Mathematical Monthly*, **65**(1): 34–35.

Longuet-Higgins, M.S., 1968. The eigenfunctions of Laplace's tidal equations over a sphere. *Philosophical Transactions of the Royal Society of London.* **262**A: 511–607.

Malin, S.R.C., 1970. Separation of lunar daily geomagnetic variations into parts of ionospheric and oceanic origin. *Geophysical Journal of the Royal Astronomical Society*, **21**: 447–455.

Maxwell, J.C., 1873. *A Treatise on Electricity and Magnetism.* Oxford: Clarendon Press.

Schuster, A., 1889. The diurnal variation of terrestrial magnetism. *Philosophical Transactions of the Royal Society of London*, **180**A: 467–518.

Schuster, A., 1908. The diurnal variation of terrestrial magnetism. *Philosophical Transactions of the Royal Society of London*, **208**A: 163–204.

Stewart, B., 1883. Terrestrial magnetism. *The Encyclopædia Britannica*, 9th ed, pp. 159–184.

Tarpley, J.D., 1970. The ionospheric wind dynamo–1. Lunar Tide. *Planetary and Space Science*, **18**: 2075–1090.

Taylor, G.I., 1936. The oscillations of the atmosphere. *Proceedings of the Royal Society of London*, **156**A: 318–326.

Walker, E., 1866. *Terrestrial and cosmical magnetism.* The Adams prize essay for 1865. Cambridge: Deighton, Bell and Co.

Winch, D.E., 1981. Spherical harmonic analysis of geomagnetic tides, 1964–1965. *Philosophical Transactions of the Royal Society of London*, **303**A: 1–104.

Cross-references

Bartels, Julius (1899–1964)
Chapman, Sydney (1888–1970)
Electromagnetic Induction (EM)
Ionosphere
Magnetometers
Ocean, Electromagnetic Effects
Price, Albert Thomas (1903–1978)

PLATE TECTONICS, CHINA

Our Earth is a layered planet with a dense iron core, a surficial crust of light rock, and between them, a solid silicate, convecting mantle. Plate tectonics is the surface manifestation of mantle convection. This activity gives rise to seafloor spreading, continental drift, earthquakes, and volcanic eruptions. Specifically, plate tectonics is the theory and study of plate formation, movement, interactions, and destruction, with seafloor spreading ridges, transform faults, and subduction zones as the plate boundaries. Today, many displaced continental blocks have been found in the world, and the original locations of some have yet to be worked out. Nevertheless, the concept of continental drift is now accepted as a necessary consequence of plate tectonics, and the complexity of mountain belts is well recognized as the product of the great mobility of lithospheric plates.

Paleomagnetism, which is the study of the ancient magnetism of the Earth, has played a central role in this developing view of the dynamic Earth. In the 1950s and early 1960s, scientists collected rocks on land all over the world and determined their magnetism and ages in order to reconstruct the history of the Earth's magnetic field. This worldwide effort led to the discovery of reversals of the geomagnetic field, and thus the realization that the seafloor acts as a tape recorder of the Earth's magnetic reversals, and polar wandering (McElhinny and McFadden, 2000). In particular, polar wandering is an interpretation given to the observation that the magnetic pole—and the geographic pole—has moved extensively in the geologic past. In the absence of any external influences, however, the direction of the Earth's axis of rotation cannot shift but remains fixed in space for all the time due to the law of conservation of angular momentum. The only way out of this dilemma is—the continents have to be moved relative to the axis of rotation. Thus, polar wandering is actually a manifestation of the mobility of the lithospheric plates. There are two principal ways of summarizing paleomagnetic data for a given region. One approach is to construct paleogeographic maps of the region for different geologic periods. These maps are mostly useful for comparison with relevant information, but not easy to view overall variations over different geologic time intervals. A much simpler and convenient way is to plot successive positions of paleomagnetic pole for a given continent from epoch to epoch on the present latitude-longitude grid. Such a path, traced out by the paleomagnetic poles and relative to the continent, is called the apparent polar wander path (APWP).

Fundamental principles in paleomagnetism

The first assumption of paleomagnetism is that the time-averaged geomagnetic field is produced by a single magnetic dipole at the center of the Earth and aligned with the Earth's rotation axis (McElhinny and McFadden, 2000). Thus, the calculated paleomagnetic pole coincides with the geographic pole. This geocentric axial dipole (GAD) hypothesis means that, when averaged over time, magnetic north is geographic north and there are simple relationships between the geographic latitude and the inclination of the field, which are the cornerstones of paleomagnetic methods applied to plate reconstruction and paleogeography. Paleomagnetic data can be used to find latitude and north-south orientation of the paleocontinents. Paleolatitude can be also checked with paleoclimate records (Irving, 1964). Although it is not possible to assign longitudinal position to the paleocontinents, the relative positions of the continents around the globe can often be pieced together by matching the shapes of apparent polar wander paths.

The configuration of the Earth's field today is modeled fairly closely by a geocentric dipole inclined at 11.5° to the rotation axis, but the field for most of the Tertiary, when averaged over periods of several thousand years, agrees well with the GAD model (Merrill *et al.*, 1996). During the 1950s and early 1960s, many questioned the validity of the GAD hypothesis during the Paleozoic and Mesozoic. With the expansion of paleomagnetic data and development of plate tectonics studies, the validity of the GAD hypothesis as a good working approximation now appears to be on solid ground (McElhinny and McFadden, 2000). Rigorously selected paleomagnetic data are in quite good agreement with plate tectonic reconstructions and also internally self-consistent under the assumption of a GAD field (Besse and Courtillot, 2002). Moreover, paleolatitude changes calculated from paleomagnetic data are consistent with paleoclimatic changes: the distribution of various paleoclimatic indicators is latitude dependent (Irving, 1964; Ziegler *et al.*, 1996).

The second assumption of paleomagnetism applied to plate tectonics is that one can use rocks as fossil compasses that record the Earth's ancient field. Study and experiments have proved that igneous and sedimentary rocks acquire primary magnetization when they were formed (McElhinny and McFadden, 2000). But determining whether the natural remanent magnetization of rocks is primary is often problematical and challenging. The original magnetization may be unstable to later physical or chemical processes and partially or completely overprinted over geologic time. Some sedimentary rocks have been shown to record paleomagnetic directions significantly shallower than the ancient field direction, a phenomenon termed inclination error (Tauxe, 1998, 2005). An important factor in making meaningful tectonic interpretations from volcanic rocks is the need to sample many separate flows that span many thousand years, a requirement that cannot always be fulfilled. The reason for this is the necessity of averaging out secular variation of the magnetic field, which typically produces a 10–20° angular dispersion of instantaneous directions, so as to obtain the time-averaged, GAD field direction (McElhinny and McFadden, 2000).

Supercontinent Gondwanaland and its dispersion

The past existence of the Gondwanaland supercontinent, a huge landmass comprised of over half of the world's present continental crust, is now clearly confirmed by paleomagnetic and paleobiogeographic data (Li, 1998). Late Precambrian to Early Paleozoic paleomagnetic data suggest that the Gondwanaland supercontinent had assembled just before the end of Early Cambrian (520 million years ago, Ma) as a result of collision between proto East Gondwanaland (Antarctica, India, Australia) and proto West Gondwanaland (Africa, Arabia, South America, and possibly Madagascar). For at least 100 Ma during the Late Devonian to Middle Carboniferous, Gondwanaland is thought to have remained over the south pole, as indicated by the available paleomagnetic data (Li and Powell, 2001) and southern cold-water fauna (Yin, 1988) and cool-climate flora (Runnegar and Cambell, 1976; Ziegler, 1990). By 320–310 Ma, Gondwanaland had collided with Laurussia to form a single supercontinent named Pangea, which stretched from pole to pole. The fragmentation of Pangea occurred as a result of plate tectonics and continental drift over Late Paleozoic, Mesozoic, and Cenozoic time to form the modern continents and oceans (Metcalfe, 1999).

Starting in Late Paleozoic time, fragmentation of Pangea (and former Gondwanaland continents) and the concomitant expansion of Eurasia have occurred repeatedly. The latest example is the addition of India to Eurasia in Early Cenozoic time (McElhinny and McFadden, 2000). Indeed, this global geographical reorganization is still in progress, as evidenced in part by the continued movement of Australia northward toward Asia (White, 1995). The ability of modern science to reconstruct this paleo-supercontinent is truly impressive. Earth scientists are continuing to sort out more details of this complex jigsaw puzzle, whose individual pieces change shape over geologic time.

Tectonic framework of China

China, which covers 9.6 million square kilometers (6.5% of the Earth's land surface) in eastern Asia, is complex and interesting from both tectonic and paleogeographic standpoints. In its simplest form, China consists of three large Precambrian continental cratons, namely, the Tarim, North China Blocks, and South China Blocks (NCB and SCB), separated by Paleozoic and Mesozoic accretionary belts, as well as several smaller blocks or terranes (Tibet, Junggar and Qaidam basins, and the Alashan/Hexi Corridor, see Figure P47). Several recent studies favor the hypothesis that these continental blocks and terranes were derived from Gondwanaland (Wang *et al.*, 1999). A close biogeographic association of South China with Australia has been recognized for both the Cambrian and Devonian, suggesting that South China may have been part of Gondwanaland in Early and Middle Paleozoic time (Burrett and Richardson, 1980; Long and Burrett, 1989). Some paleontological data from Tarim show that Cambrian and Ordovician Bradoriida fossils are very similar to those of South China, which in turn would suggest that Tarim was also part of Gondwanaland in the Early Paleozoic (McKerrow and Scotese, 1990; Shu and Chen, 1991). But it is less clear whether North China belonged to Gondwanaland or not during this same period. Early Cambrian fauna from North China show affinity with Australia, but Ordovician benthic trilobite assemblages in North China show more similarity with those of Siberia and North America (Burrett *et al.*, 1990). Comparative studies of the tectono-stratigraphy, paleontology, and structural geology of the various terranes in the Tibetan Plateau suggest that they were all derived directly or indirectly from Gondwanaland (Metcalfe, 1999) and added successively to the Eurasian plate during the Mesozoic and Cenozoic eras (Burg and Chen, 1984).

If, indeed, these Chinese blocks and terranes were associated with Gondwanaland during the Early Paleozoic, then the key questions would be, when and in what sequence did these blocks rift off from Gondwanaland and stitch as they are at present. As outlined below, geologic data germane to these questions are sparse and subject to conflicting interpretations (e.g., Klimetz, 1983; Mattauer *et al.*, 1991; Hsu *et al.*, 1988).

North China block

The North China block is bounded on the north by the Central Asian Fold Belt and on the south by the Qilian-Qingling-Dabie Shan ("shan" = mountain in Chinese). The Tancheng-Lujiang (more often referred to by the Chinese contraction "Tanlu") fault runs through the eastern part of the NCB, and sharply cuts the Qinling belt at its eastern end. There is considerable debate as to whether or not the fault represents the eastern boundary of the NCB, and its origin and history of development. The oldest basement rock was dated as 3670 ± 230 Ma, which is possibly the oldest age in China (Li and Cong, 1980). The oldest unmetamorphosed sedimentary cover has been dated at 1700 Ma (Wang, 1993).

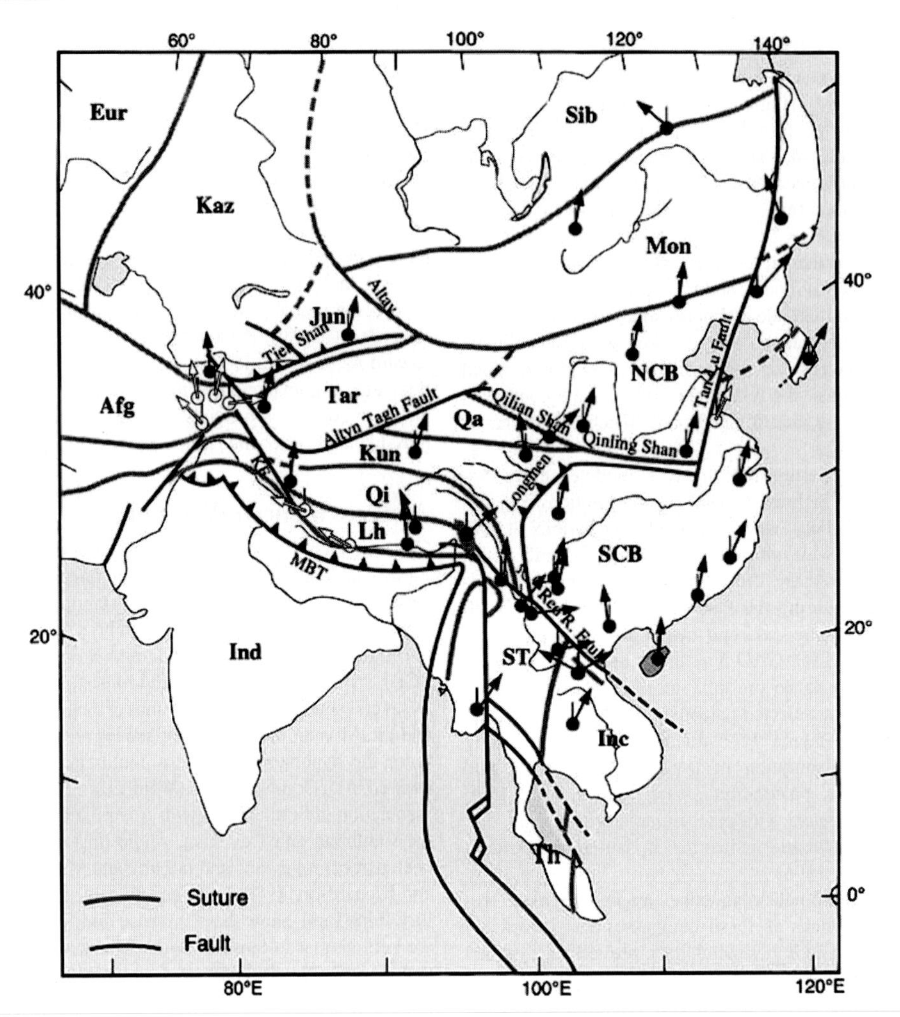

Figure P47 Simplified map (after Halim *et al.*, 2003) of Eastern Asia showing the main sutures and faults (KF, Karakorum fault; MBT, Main Boundary Thrust). The major blocks are: Afg, Afganistan; Eur, Eurasia; Inc, Indochina; Ind, India; Jun, Junggar; Kaz, Kazakhtan; Kun, Kunlun; Mon, Mongolia; NCB, North China block; Qa, Qaidam; Qi, Qiangtang; Sib, Siberia; SCB, South China block; ST, Shan Tai; and Tar, Tarim. Also indicated are the main locations where Cretaceous paleomagnetic data are available. Declination is indicated by arrows (a thin line indicates geographic north, it is vertical because of the projection selected). Open arrows are for areas suspected of major local scale deformation or unreliable paleomagnetic data. Solid arrows are thought to be representative of the individual blocks at a larger scale.

The Paleozoic sections of the NCB are dominated by Lower Paleozoic marine carbonates and characterized by a great sedimentary hiatus from the Late Ordovician through Early Carboniferous, although rare strata within this age span apparently exist on its western boundary. This remarkable hiatus has led many geologists to the conclusion that the NCB was separated from currently adjacent blocks during much of Paleozoic time. From the Late Permian onwards, its sediments are of continental origin.

The Korean Peninsula is situated east of the NCB. Traditionally, Korea has been incorporated with the NCB into a single craton (Sino-Korea) based on the close similarity of Paleozoic stratigraphy. Depending on the significance of Tanlu fault movement, it is possible that the Korean Peninsula (or part of it) was formerly located close to the South China Block, a question that is still intensively debated.

Geological estimates of the age of accretion of the NCB with Siberia to the north are also controversial. It is assigned a Middle Paleozoic age (end of Devonian and Hercynian Orogeny), while Klimetz (1983) ascribes it to the Mesozoic. By analyzing paleobiogeographic data, Laveine *et al.* (1989) argued that the Chinese blocks must have been close to Europe in the Late Carboniferous, as shown by plant migration routes.

South China block

The South China Block is bounded on the north by the Qinling and Dabie Shan, which mark the suture between the SCB and NCB (see Figure P47). Exposures of basement rocks are limited, but a U-Pb ziron date of 2860 Ma was reported for migmatitic granite in Guangxi province, indicating that the block may have formed in the late Archean. The basement of the SCB was finally consolidated around 850–800 Ma. The Paleozoic and Mesozoic strata consist mainly of marine carbonate and clastic sediments, with continental lava flows in the Late Permian. The Cenozoic strata are mainly red beds.

In the eastern part of the Qinling-Dabie Mountains, ultrahigh-pressure, coesite- and diamond-bearing metamorphic rocks crop out. Ages from U-Pb dating of zirons from eclogites range from 250 to 220 Ma, which correspond to the time when sedimentation in the SCB changed from marine to continental. This change was heralded by the voluminous eruption of the Emeishan basalts in southwestern China around 250 Ma. Great quantities of Jurassic red beds in the Sichuan Basin indicate that large neighboring mountains were being eroded at that time. Northeast of the Dabie Shan in eastern Shandong Peninsula lies another ultrahigh-pressure coesite-bearing metamorphic suite. Similarities in mineralogy of the Shandong and Dabie Shan metamorphic assemblages suggest they were once part of the same belt that was later displaced about 500 km by the Tanlu fault in a left-lateral sense. There are widely different estimates on the timing of collision between the SCB and NCB based on geologic data. These estimates vary from the Precambrian and Middle Paleozoic to the end of the Paleozoic and earliest Mesozoic.

The SCB is commonly segregated into the Yangtze block to the northwest and mobile belts to the southeast. In general, the division between the Yangtze block and exotic terranes is demarcated south of the exposed Meso- to Neoproterozoic low-grade metamorphic rocks called the Jiangnan old land. Jiangnan rocks include ophiolite and thick argillaceous turbidites that extend for over 1000 km. Using Sm (samarium) and Nd (neodymium) isotopes, two laboratories have dated two of the ophiloite suites at 1024 ± 24 Ma and 935 ± 34 Ma, and 1034 ± 24 Ma and 935 ± 10 Ma, respectively. Li *et al*. (1995) and Li (1998) suggested that the Yangtze Block and fold belts to the southeast were amalgamated in the late Proterozoic at around 1000 Ma and are separated by a failed intracontinental rift activated during the breakup of Rodinia at around 700 Ma. There is no well-documented evidence of a Phanerozoic suture in the SCB. Hsu *et al*. (1988) proposed a Mesozoic age collision model based on the discovery of Mesozoic thrust structures in south China. This new interpretation of south China tectonics has since led intensive debate.

Tarim and Junggar basins

Tarim is a rhombic-shaped basin, bounded by the Tianshan, ("Celestial Mountains"), Kunlun and Altyn Tagh mountain ranges, with Precambrian basement (3.3 Ga) exposed along its periphery (Hu and Rogers, 1992), attesting to an extensive geologic history for the block. Tarim became an inland basin since the beginning of the Mesozoic and it is presently covered by the Takalamagan (Takla Makan) desert, with Paleozoic rocks occurring mostly along its northern and southern margins. Cambrian and Ordovician strata are distributed mainly along the northern periphery of the basin, whereas the Silurian, Devonian, Carboniferous and Permian are mainly distributed in the northwestern part of the basin (Keping area). The lithologies of these Paleozoic sequences are mainly dolomitic limestone and red and yellow siltstones. The Mesozoic and Cenozoic strata consist mainly of green sandstone and red mudstone and are distributed in the northeast part of the basin (Kuche depression). A late Paleozoic (Hercynian) orogeny was responsible for a general unconformity between Paleozoic and Mesozoic formations.

Of the three major blocks that compose the tectonic framework of China, Tarim, being the most inboard must have arrived at its present position at least as early as any of the more outboard blocks and terranes. There is no clear suture zone between Tarim and the NCB. Indeed, some geologists considered the NCB and Tarim to have formed a rigid block since the Neoproterozoic because an almost continuous—"Hexizoulang", equivalent to "Hexi Corridor"—of Precambrian and Paleozoic formations connects the NCB and Tarim. This idea has been abandoned because available paleomagnetic data suggest that these two blocks did not assume their present configuration for sometime in the Mesozoic.

The Junggar Basin, often considered to be the southeast extremity of the Kazakhstan Block, sits in a complicated region squeezed between the Tarim Basin on the south, the Kazakhstan Block on the west, and Siberia to the northeast. Unlike the Tarim Basin, there are no known exposures of Precambrian rocks along the Junggar Basin margins. However, some researchers have suggested that Junggar is underlain by Precambrian crystalline basement, based entirely upon gravimetrically determined depths to the Moho (e.g., Ren *et al*., 1987). Other workers believe that it is underlain by a basement of structurally imbricated Ordovician to Lower Carboniferous accretionary metasedimentary rocks and oceanic crustal fragments. The oldest rocks in this region are of Late Cambrian age (508 ± 60 Ma). Middle Devonian strata consist of spilitic pillow lavas interlayered with radiolarian cherts and associated limestone horizons. Carboniferous units consist mainly of marine to continental volcanic rocks interbedded with clastics. Potassium-rich granitic plutons and their extrusive equivalents, which have Permian K-Ar ages and Middle to Late Carboniferous U-Pb ages are common in west Junggar. Permian rocks are mainly continental clastics and volcanic extrusive. At many localities igneous dikes of Permian age intrude the Carboniferous formations. Triassic and Jurassic sequences are all continental clastics.

Collision between Tarim and Junggar along the Tianshan Range has been inferred to be Early Permian in age, based on geological features such as the termination of andesitic volcanism, the development of unconformitites between late Early Permian and underlying units, the intrusion of widespread syncollisional to postcollisional granites in the Tianshan and the mixture of Angaran and Cathaysian flora across the Tianshan.

Alashan/Hexi Corridor terrane

Situated west of Ordos, the Alashan/Hexi Corridor terrane is bounded on the north by Central Asian Fold Belt and is separated from the Qaidam terrane by the Qilian Shan fold belt to the south. The western boundary with the Tarim is not clearly delineated but is thought to be within the northern Qilian Shan.

The oldest rock of the Alashan/Hexi Corridor terrane is Precambrian (about 1850 Ma), as the Precambrian Longshoushan Formation in the southern part is intruded by granites 1789 Ma in age (Yang *et al*., 1986, p. 244). The first most recognized regional uncomformity in southern Alashan area is between Middle and Late Cambrian (the Xianshan Formation), which can be correlated to the coeval unconformity the Ordos basin, one of the most stable parts of the NCB. In contrast, this geologic pattern never penetrated into the Central Asian Fold Belt, indicating the latter indeed is the geologic/block boundary. Because of these geological observations, Alashan/Hexi Corridor terrane is believed to be part of the NCB at least since the Late Cambrian and was still receiving deposition while most other parts of the NCB were standing higher during Late Ordovician to Early Carboniferous time.

Tibet and Qaidam basin

Tibet is a collage of continental terranes that accreted onto Eurasia ahead of India. From north to south, the Tibetan terranes are the Kunlun, Songpan-Ganze, Qiangtang, Lhasa, and Himalaya terranes, the latter being actually part of the Indian subcontinent (see Figure P47). Geologic studies suggest that these terranes collided with Eurasia at times progressively younging southward (Dewey *et al*., 1988; Yin and Harrison, 2000; Davis *et al*., 2002). The amalgamation of the Kunlun and Songpan-Ganze terranes is marked by a broad Early Paleozoic arc on which was later superposed a Triassic arc (Burchfield *et al*., 1992). The Qiangtang terrane accreted to the Songpan-Ganze terrane along the Jinsha suture during the Late Triassic or Early Jurassic (Xiao *et al*., 2002), followed by the docking of the Lhasa terrane against the Qiangtang block along the Bangong-Nujiang suture during the Late Jurassic (Dewey *et al*., 1988), although the precise age of suturing and subsequent reactivation by other collisions are still debated. The Indian subcontinent in turn collided with these terranes along the Yarlung-Zangbo suture during Late Cretaceous to Eocene time (Murphy *et al*., 1997).

The Qaidam Basin is relatively a lower elevation region of the northeast Tibetan Plateau bounded by the Altyn Tagh fault to the northwest, the Kunlun fault to the south, and the Nan Shan fold-and-thrust belt to the northeast. The nature of Qaidam's basement is not known because it is buried under thick cover strata and still no sufficient borehole or geophysical data are available. Cover strata consist of Cenozoic clastics of terrestrial facies. Tertiary continental sediments in the Qaidam Basin reach several kilometers and serve as the main oil producing layers in the basin. Qaidam's mountain ranges probably formed as ramp anticlines stacked upon a crustal-scale thrust fault. However, compared to complexly deformed surrounding areas, the Qaidam Basin shows only limited deformation of Cretaceous to Tertiary sedimentary cover.

In summary, these geological observations provide fundamental constraints on tectonics models of Chinese blocks. Although the geology of the fold belts between these major Chinese blocks and terranes has the potential of constraining the times of collisions more tightly than the geology of the blocks themselves, the complexity of these fold belts has led to widely different views of the tectonic history of the region. With Gondwanaland in high southern paleolatitudes during the Late Paleozoic and the Chinese blocks nowadays situated at mid-northern latitudes, paleomagnetism is obviously an ideal tool by which to test various Paleozoic paleogeographic models and to unravel the later northward relative motions of these blocks and terranes if they were part of Gondwanaland. As explored below, paleomagnetism offers an independent—and complementary—source of constraints in terms of the style of convergence and timing of amalgamation of individual continental blocks.

Paleomagnetic constraints on tectonic history of major chinese blocks

The tectonic relationship between crustal blocks can be constrained by comparison of their paleomagnetic poles. If reliable poles of the same age from two blocks are significantly different, one must conclude that the two blocks have moved relative to one another sometime after remanent magnetization was acquired. The converse is not necessarily true, since it is not possible to assign longitudinal position to the paleocontinents according to the GAD model.

Paleomagnetic investigation in China began in the early 1960s. Great development in the acquisition of paleomagnetic data has been made since early 1980s. In particular, the Cretaceous is well-studied paleomagnetically both in terms of data quality and area distribution (see Figure P47). The Global Paleomagnetic Data Base nowadays has included studies from the Chinese literature, but there is still a large variation in quality and reliability of the poles. Detailed reviews of paleomagnetic poles ("paleopoles" for short) for Chinese blocks have been given in several papers and books (e.g., Lin *et al.*, 1985; Zheng *et al.*, 1991; Enkin *et al.*, 1992; Zhao and Coe, 1987; Van der Voo, 1993; Zhao *et al.*, 1996, 1999; McElhinny and McFadden, 2000; Zhu and Tschu, 2001; Yang and Besse, 2001). Several additional paleopoles have become available since 2001. Up to 2003, a total of 415 paleomagnetic poles have been published for Chinese blocks and terranes. Some of these poles are still subject to debate and interpretation. Any serious use of paleomagnetic data from China should always involve independent evaluation and assessment of each study. Thus, our intention is not to list the detailed references for these 415 paleopoles and discuss exhaustively the significance of data. Readers are referred to the Global Paleomagnetic Data Base Version 4.6 (www.tsrc.uwa.edu.au/data_bases) for more information. These paleomagnetic poles, from both Chinese and English language journals and books, are briefly outlined below.

Paleomagnetic studies in the North China block (+Hexi Corridor and Korea): 118 Poles

Quaternary (3 poles)

The Quaternary is represented by only three paleomagnetic poles: two studies on the Chinese Loess Plateau in Shaanxi province and one study on basalt flows in Shandong province.

Tertiary (7 poles)

Little work has been done on Tertiary rocks in North China as well: Only a total of seven paleopoles have been reported from volcanic and sedimentary rocks from Inner Mongolia, Hebei, Shandong, Shanxi, and Gansu provinces. Most of the results passed reversal tests, although the reliability of some of the studies can still be argued. The scarcity of reliable paleomagnetic poles for these younger geologic time intervals seems to be a result of the low research priority given to them.

Cretaceous (16 poles)

Twelve Cretaceous paleopoles have been reported from Ningxia, Gansu, Shaanxi, Shanxi, Henan, Shandong, and Inner Mongolia of the NCB. In addition, four studies on Cretaceous rocks from South Korea have been published in the literature. These Cretaceous-aged poles obtained from widely separated regions are consistent with each other, suggesting that these areas occupied the same relative positions in terms of paleolatitudes since the Cretaceous. Clockwise vertical-axis block rotations of 30° of Korea relative to the NCB, however, are implied by the Cretaceous data for Korea.

Jurassic (12 poles)

Only one Early Jurassic pole is available from Shaanxi province. Four Late Jurassic poles have been published from three areas of Inner Mongolia and Shanxi provinces and six paleopoles are available for the Middle Jurassic that were derived from sedimentary rocks in Shaanxi and Shanxi provinces. One Late Jurassic pole was reported from Korea, again showing the region has rotated clockwise with respect to China. There is still a worldwide deficiency of Early Jurassic paleomagnetic poles, and North China is no exception.

Triassic (23 poles)

The Late Triassic in the NCB has only been studied in one region of Shanxi province (2 poles). Nine paleopoles were reported from Middle Triassic formations in Shaanxi and Shanxi provinces. Twelve paleopoles are available for the NCB from the Early Triassic sections in Liaoning, Shandong, Shanxi, and Shaanxi provinces, but five of them are clearly questionable.

Permian (26 poles)

A total of 19 Late Permian poles have been obtained from 14 sections distributed over Hebei, Shanxi, Shaanxi, and Inner Mongolia. There are also four Late Permian paleopoles available for Korea. In contrast, only three Early Permian paleopoles are available from Shaanxi and Shanxi province, which are significantly different from each other. Like the Early Triassic, the distribution of the Late Permian paleomagnetic poles is streaked along a small circle centered on the NCB, indicating significant internal deformation that has caused local rotations.

Carboniferous (6 poles)

Two studies have reported Middle-Late Carboniferous results from Shanxi and Shaanxi province for the NCB. None of them was considered reliable (Zhu and Tschu, 2001), underscoring that more paleomagnetic study of Carboniferous rocks should be a future research focus. Four preliminary Late Carboniferous paleomagnetic poles were reported from Korea.

Devonian and Silurian (5 poles)

Because of the great sedimentary hiatus during the Devonian and Silurian for the NCB, three studies were conducted in the Alashan/Hexi Corridor in the Ningxia Hui Autonomous region on Middle to Late Devonian and Silurian red beds. Two Middle to Late Devonian and three Middle to Late Silurian paleopoles have been published from these studies.

Ordovician (10 poles)

Nine paleopoles have been obtained from Liaoning, Hebei, Shandong, Shaanxi provinces and the Ningxia Hui Autonomous region that are distributed along a small circle centered on the NCB, indicating local rotations. One Early Ordovician pole was obtained from Korea. There was a dispute on the magnetic polarity of the stable magnetization from the Ordovician rocks in China, which was resolved by a recent magnetostratigraphic study on rocks spanning the Cambrian/Ordovician boundary and into early Ordovician.

Cambrian (10 poles)

Ten Cambrian paleopoles have been reported for the NCB that were derived from rock formations from Liaoning, Shandong, Shanxi, Shaanxi, and Ningxia. These poles can be grouped into three clusters. One group (3 poles) is close to the Triassic pole of the NCB and without stability tests. These poles are suspicious because they may have been remagnetized in the Triassic. Poles in the second group fall near the equatorial area of the western Atlantic while those in the third group are in the eastern Atlantic. The scatter of paleopoles may result from the local rotation of sampling areas or unsatisfactory removal of a second magnetization component.

Paleomagnetic study in the South China block: 204 poles

Quaternary (5 poles)

Early Quaternary poles were obtained from a basalt formation in Jiangsu (1 pole) and clays from Yunnan (4 poles). No details concerning sampling sites and laboratory work were given.

Tertiary (24 poles)

Tertiary paleomagnetic data are numerous for the SCB. A total of 24 paleopoles are available for the SCB, which come from red beds and basalt formations in Yunnan, Guangdong, Sichuan, Hubei, Hunan, Zhejiang, and Jiangsu provinces.

Cretaceous (26 poles)

Of the 26 paleopoles determined for the Cretaceous, 12 come from sedimentary rocks in Sichuan province. The rest are obtained from red beds and mudstones in Guangxi, Fujian, Yunnan, and Guangxi provinces. There are two poles obtained from volcanic rocks in Zhejiang and Hong Kong, respectively.

Jurassic (31 poles)

Fourteen poles have been published for the Late Jurassic of the SCB, which were obtained from sandstone and mudstone formations in Zhejiang, Sichuan, Anhui, and Yunnan provinces. The Middle Jurassic is represented by 11 paleopoles that come from sedimentary and volcanic rocks from Guangxi, Hubei, Anhui, Sichuan, Guizhou, and Hong Kong. Six paleopoles from sandstone formations in Sichuan, Hubei, Anhui, Guangxi, and Yunnan are available for the Early Jurassic.

Triassic (49 poles)

Fifteen Late Triassic poles have been isolated from siltstone sites in Sichuan and Yunnan provinces. Nine Middle Triassic poles were reported from limestone and red sandstone in Hubei, Jiangsu, Guizhou, Sichuan, Hunan, and Guangxi provinces. Most of poles from the Early Triassic limestone formations are from Sichuan. It is important to note that Early Triassic poles from limestone and sandstone formations of Jiangsu and Zhejiang in the eastern extreme of the block are coherent with the poles from west (Sichuan), suggesting that the SCB has behaved more or less rigidly since at least this period.

Permian (30 poles)

Many high-quality paleomagnetic data for the Late Permian have been presented since McElhinny *et al.*'s (1981) pioneering paleomagnetic paper on the Emeishan Basalt from Sichuan. Those poles are now from both the Emeishan basalt from Yunnan and Guizhou provinces and red limestone and sandstone from Hubei, Zhejiang, Guangdong, and Sichuan. In contrast, only one Early Permian pole is reported from a limestone section in Hubei province. It is worth mentioning that, like many other fields in Earth sciences, progress in the paleomagnetic data of China has grown through repeated trials, criticisms, and improvements. Some studies may encompass entire lifetimes of devoted work. The debate on the age of magnetization of the Late Permian Emeishan Basalt from the SCB is a case in point. Debate has continued for the last 15 years as to whether the observed magnetic direction is primary or secondary. Although a reported positive conglomerate test by Huang and Opdyke (1995) effectively refuted the suspicion that the directions from the Emeishan Basalt may represent a remagnetization of Triassic or even Jurassic age, it may still require more extensive investigations to unequivocally settle the question.

Carboniferous (5 poles)

Out of the 5 Carboniferous poles published from limestone and mudstone formations in Zhejiang, Yunnan, and Fujian provinces, none of them, however, can demonstrate the age of magnetization with certainty, reflecting the still insufficient inventory of Chinese paleomagnetic data.

Devonian (5 poles)

Five poles of Devonian age for the SCB are reported from sandstone formations in Zhejiang, Jiangsu, Yunnan, and Sichuan. The data set is relatively small and also subject to debate. Only one pole was selected by McElhinny and McFadden (2000) to be included in their book.

Silurian (8 poles)

Silurian poles for the SCB were reported from red beds, sandstone, and limestone in Guizhou, Zhejiang, Hubei, Yunnan, and Sichuan provinces. These poles have been all determined using the modern methods and analytical techniques.

Ordovician (8 poles)

Eight Ordovician paleomagnetic poles have been reported for South China that were derived from limestone and sandstone in Zhejiang, Hubei, Yunnan, and Sichuan province. These poles generally have poor quality.

Cambrian (5 poles)

The collected Cambrian poles were derived from rocks from Zhejiang, Hubei, and Yunnan provinces. The poles are very different from Cambrian poles of the NCB, suggesting that North and South China were two different continental blocks during Cambrian.

Paleomagnetic study in the Tarim and Junggar basins: 52 poles

Quaternary (1 pole)

Only one study of Quaternary basalt from the southeast of the Tarim Basin has been published. The age of the basalt is still in dispute, but it seems reasonable for it to represent the Quaternary.

Tertiary (11 poles)

Tertiary paleomagnetic poles for Tarim were mainly obtained from the Aertushi region of western Tarim, the Kuche depression of northwest Tarim, and the Maza Tagh range in the central Tarim Basin. Up to 1990, the paleomagnetic database for Tarim was very sparse. The last 13 years saw a major increase in the number of paleomagnetic data for Tarim, which is rather fortunate because tectonic rotations caused by the Indian collision might be constrained by such data.

Cretaceous (10 poles)

Eight Cretaceous poles were reported for Tarim, all of them from the Kuche depression in northwestern Tarim. Of the eight poles, six were derived from red beds and two were from basalt formations in the Tuoyun area of western Tarim. Two Cretaceous poles were published for the Junggar basin, which are close to Cretaceous poles for Tarim.

Jurassic (4 poles)

One Late Jurassic and three Middle Jurassic pole are available for Tarim. Both were derived from red bed formations in southwestern Tarim. No coeval pole is available for the Junggar basin.

Triassic (3 poles)

The Triassic of Tarim is represented by three poles (one each in the Late, Middle, and Early Triassic). Among them, only the Late Triassic was well studied through a magnetostratigraphic investigation.

Permian (14 poles)

Permian paleomagnetic poles for Tarim are mainly from volcanic rocks in the Aksu, Keping, and Kuche areas of western Tarim, with 7 poles for Late Permian and 5 poles for the Early Permian. Two Late Permian poles have been reported for the Junggar Basin, which is concord with the Permian poles for Tarim. The number and quality of Permian paleopoles for Tarim is quite impressive, suggesting one should treat these poles seriously. No paleopoles older than Permian were reported from the Junggar basin.

Carboniferous (4 poles)

Four Carboniferous paleopoles are available for Tarim, which were all derived from sedimentary rocks (limestone and sandstone) from South western Tarim.

Devonian (3 poles)

Three Devonian paleomagnetic poles were reported for the Tarim, obtained from both the north and south rims of Tarim.

Silurian (3 poles)

There are three middle Silurian poles for Tarim. The magnetization is dominated by an extremely weak, recent-field component of magnetization that was carried by goethite. Clearly, these poles are unacceptable for tectonic study.

Ordovician (1 pole) and Cambrian (1 pole)

Although the number of poles for these two periods is extremely sparse, the reported poles each pass fold and reversal tests and place Tarim in low paleolatitudes during Cambrian and Ordovician.

Paleomagnetic studies in the Tibet and Qaidam basin: 38 poles

Since the early 1980s, paleomagnetic investigation has been an important aspect of all the major geologic expeditions conducted on the Tibetan plateau. While paleomagnetic investigations of the Cretaceous and younger rocks from the plateau have yielded useful constraints for the Lhasa and Qiangtang terranes (Achache *et al.*, 1984; Lin and Watts, 1988; Otofuji *et al.*, 1989; Huang *et al.*, 1992; Chen *et al.*, 1993), studies on the Paleozoic formations were often hampered by lack of favorable lithology, multiphase deformation, and metamorphism that has obliterated the primary magnetization (Lin and Watts, 1988; Otofuji *et al.*, 1989; Huang *et al.*, 1992). Triassic and Jurassic paleomagnetic directions also tend to be remagnetized (Chen *et al.*, 1993). As a result, there are a total of 8 Tertiary and 13 Cretaceous paleomagnetic poles for Tibet.

Among the Late Cretaceous paleomagnetic results from Tibet, those reported from south Tibet (Achache *et al.*, 1984), east Qiangtang (Huang *et al.*, 1992), and west Qiangtang (Otofuji *et al.*, 1989; Chen *et al.*, 1993) appear to have high quality, each with positive fold test result. However, there is a serious problem in interpreting these Late Cretaceous data in terms of paleolatitude. The study on the Late Cretaceous Takena red beds and Paleocene andesites of the Lhasa Block yielded a paleolatitude of 15–20° N for Lhasa. It should be pointed out that the Paleocene Lingzizong volcanics in the Lhasa area give consistent result (with positive fold and reversal test) with the underlying Cretaceous strata (Achache *et al.*, 1984) and provide a notable example that both red beds and volcanic rock in the Lhasa area recorded the same ancient magnetization. Lin and Watts (1988) sampled the Takena formation in the Linzhou area as well as Late Cretaceous volcanic units in Nagqu and Qelico. Their data pass a fold test and give slightly lower paleolatitudes ($7.6 \pm 3.5°$ N), but still similar to the results of Achache *et al.* (1984). If we use the mean of these three results, the paleolatitude of Lhasa in the Late Cretaceous would be about 12.5° N. A comparison of this value with that predicted by the reference pole (Besse and Courtillot, 1991) would confirm the conclusion of Achache *et al.* (1984, p. 10335) that crustal shortening of the order of 2000 km has occurred between Lhasa and Eurasia since the Late Cretaceous. Calculations of shortening using the Early Tertiary Linzizong volcanic rocks yield a similar amount of northward displacement of the Lhasa terrane. This is compatible with other geophysical and geological estimation that India has moved about 2500 km northward about 50 Ma (e.g., Dewey *et al.*, 1988). However, when comparing the paleolatitude of Lhasa with that of west Qiangtang (Chen *et al.*, 1993) and Tarim/Junggar (Li *et al.*, 1988; Chen *et al.*, 1991), the data suggest that there is no net crustal shortening between Lhasa and west Qiangtang/Tarim/Junggar from the Late Cretaceous to present. In other words, the entire 2000 km of northward displacement of the Lhasa block suggested by the Late Cretaceous paleomagnetic data would have to be taken up in the region to the north of Junggar. This would conflict with existing understanding of the formation of the Tibetan plateau by either the crustal shortening model (Dewey *et al.*, 1988) or the tectonic extrusion model (Tapponnier *et al.*, 2001), or a combination of the two (Harrison *et al.*, 1992). In addition, there is no geologic evidence to suggest a candidate place between Siberia platform and Junggar/Tarim to accommodate crustal shortening of this magnitude in the Tertiary. If the configuration for most or all of the Tibetan plateau has been created by the India/Asia collision, then the Late Cretaceous paleomagnetic results from Lhasa and west Qiangtang /Tarim/Junggar cannot both be correct. Obviously, further work is badly needed to solve this problem, and a reliable paleomagnetic pole from central Qiangtang would provide a direct test.

For the Qaidam Basin, there are 5 Jurassic, 3 Cretaceous, and 9 Tertiary poles reported from paleomagnetic investigations of the basin. These studies were designed to detect possible clockwise rotation of Qaidam basin, as implied by extrusion models. Rotation within the Qaidam-Altyn Tagh region is very controversial, and existing paleomagnetic data are contradictory. Halim *et al.* (2003) presented one Late Jurassic pole from red beds in Huatugou of Qaidam basin. This pole indicates negligible northward convergence of the Qaidam block with respect to Tarim since Late Jurassic times, but implies a significant relative clockwise rotation of the studied area with respect to Tarim (~16°). Similarly, Chen *et al.* (2002) reported that three widely spread areas in Qaidam exhibit 20° clockwise rotations during the Late Oligocene-Late Miocene time. However, recent study from the east end of the Altyn Tagh fault by Gilder *et al.* (2001) shows that area has rotated 27° ± 5° counterclockwise with respect to the Eurasian reference pole during the last 19 Ma. Otofuji *et al.* (1995) also suggest that their sampling sites along the Altyn Tagh fault showed 7.4° ± 5.2° counterclockwise rotation during the last 1.0 Ma. On the other hand, Dupont-Nivet *et al.* (2002, 2003) and Rumelhart *et al.* (1999) argue for no significant Neogene vertical-axis rotation in northern Tibet since Eocene. These results would contradict extrusion models which imply clockwise rotation of Qaidam basin at rates approaching 1°/Ma.

Paleo-positions of Chinese blocks

Although a great number of Late Mesozoic and Cenozoic paleomagnetic results are available, the current paleomagnetic data base for China is still limited to allow a satisfactory interpretation of the questions concerning the paleogeography of China and the implications for Gondwanaland. Because many studies are performed at a single locality or over a small region and are of variable quality, paleomagnetists must critically assess and update the previously published data from China and Eurasia in order to better understand the tectonic relationship between the Chinese blocks and their relative motions with respect to Eurasia (i.e., Europe, Siberia, and Kazakhstan) during the Phanerozoic. Basic criteria for an acceptable paleomagnetic pole determination should include (i) age of the magnetization; (ii) structural control; and (iii) paleomagnetic laboratory treatment (demagnetization) of sufficient samples.

After such evaluations and assessments of each published study, the current state of paleomagnetic knowledge can be synthesized as follows: during Early (~530 Ma) and early-middle Late Cambrian (505 Ma), the paleolatitudes of South China, Tarim, and Australia permit them to be next to each other. North China can be placed between Siberia and Australia. All these blocks were in the area around the equator. Between late Middle to early Late Cambrian (~505 Ma), Tarim, North and South China Blocks lay closer to Australia and shared close trilobite affinities until Late Cambrian (~490 Ma). The counterclockwise rotation of Gondwanaland during the Cambrian was reversed during the Ordovician. The change in sense of rotation occurred at the Cambro-Ordovican boundary. The North and South China Blocks, and Tarim, remained in low latitudes, and rotated sinistrally along the Cimmerian margin of Gondwanaland towards positions they are inferred to have occupied later on the basis of faunal relationships.

By Early Silurian time (430 Ma) Siberia remained in the vicinity of the equator but has moved slightly northward to a position north of the equator. Gondwanaland appears to have moved southward and rotated clockwise. North China may also have rotated clockwise and appears to be compatible with a position next to the northern margin of Australia, which also fits the distribution of freshwater fish fauna on the two continents. While South China remains in roughly the same position, Tarim appears to have rifted off the SCB by rapid clockwise rotation and northward displacement.

At Middle Devonian time (~380 Ma) the location of South China suggests that it may have rifted off Australia. Tarim is continuing its northward journey. Siberia has also drifted northward and rotated clockwise. North China, on the other hand, is still on the equator. The development of highly endemic vertebrate faunas on both the North and South China Blocks indicates that these two blocks were separated from Gondwanaland during Early-Middle Devonian.

Later at the Late Devonian time (~354 Ma), Gondwanaland began to drift rapidly across the south pole, and by mid-Carboniferous (320–310 Ma), it has collided with Laurussia to start assemblage the supercontinent Pangea. It was for the first time that Australia had been drifted close to the south pole during the Paleozoic. The major Chinese blocks remained in low and medium paleolatitudes and started to develop their distinctive Cathaysia flora. The Late Carboniferous poles for Tarim and Eurasia are in agreement, suggesting that these regions could have started the process of accretion by Late Carboniferous time.

The paleogeography of the Early Permian (~280 Ma) remained more or less the same as the mid-Carboniferous, except that Siberia collided with Laurussia along the Urals. Mongolian terranes accreted to the North China, the first phase of continental collision between the NCB and SCB at their eastern ends started soon. Supercontinent Pangea was complete at this time.

Pangea endured from ~280 Ma to ~175 Ma when the first seafloor spreading began in the North Atlantic Ocean. During this time, several Tibetan terranes (Kulun, Songpan-Ganze, and Qiangtang) accreted to Eurasia. The apparent polar wander paths for Eurasia and North and South China converge during the Late Jurassic (~150 Ma) and are tightly grouped throughout the Cretaceous and Tertiary, implying that large relative motion (translation and/or rotation) between the blocks was accomplished by Late Jurassic. On the other hand, although the paleolatitudes for Tarim and Eurasia have been similar since Carboniferous time, a close inspection reveals that there is a significant difference between the Late Permian, Early Triassic and Late Jurassic through Cretaceous poles for Eurasia/Siberia and Tarim, indicating relative motions between Tarim and Eurasia.

Tectonic models inferred by the paleomagnetic data

The paleomagnetic results available from China allow plausible tectonic models to be constructed for the amalgamation of the Chinese blocks with themselves and with Eurasia (Zhao and Coe, 1987; Zhao *et al.*, 1996). The collision between Siberia/Eurasia and the Mongolia-NCB initially occurred in the Permian near the western end of their boundary in the Irtysh-Zaysan region (50° N, 75° E), which is near the edge of the Kazakhstan Block (in today's geographical frame). Suturing progressed eastward as the Mongolia-NCB rotated counterclockwise about 90° in relation to Eurasia during the Early Mesozoic with a scissors-like motion (see Figure P48). The amalgamation of North and South China took place in the Late Permian by a similar but antithetic mechanism, with collision first occurring near the eastern end (about 30° N, 120° E) of the Qinling Fold belt and diachronously suturing from east to west due to a clockwise rotation of 65° of South China relative to North China.

Early Triassic paleomagnetic poles for all continental blocks are very similar to those for the Late Permian, consistent with the same scissor model discussed above and reinforcing the tectonic model. Reliable Middle and Late Triassic paleomagnetic data for the NCB, SCB, and Eurasia are still relatively sparse. Little change appears to have taken place by Middle Triassic time, with about 65° of rotation still left between Eurasia and the NCB to reconcile their poles.

In the Early Jurassic, more than half of the Qinling Sea between the NCB and SCBs was subducted, and so was the Mongol-Okhotsk Ocean between Eurasia and the NCB. Therefore, paleomagnetic data suggest that most of the relative rotation between the Chinese blocks took place during the Late Triassic and Early Jurassic. By the Middle Jurassic, the Qinling Sea is almost completely closed, although this may not be entirely correct due to the relatively large uncertainty in the SCB poles. The Late Jurassic and Early Cretaceous paleomagnetic poles for the NCB and SCB are statistically indistinguishable at this time, suggesting that accretion between the two blocks was accomplished some time between the Middle and Late Jurassic. However, the Chinese blocks were still significantly south of Siberia, with ~1000 km of the Mongol-Okhotsk Ocean to be closed north of Mongolia. The final closure of the Mongol-Okhotsk Ocean took place sometime in Cretaceous.

The tectonic model mentioned above contradicted widely held views at the time that the NCB and SCB were joined by the Devonian at the latest and that the Central Asian Fold Belt, which runs across North China just north of 40° N, is the zone where China was sutured to Siberia in Permian time. Rather, the paleomagnetic evidence suggested that in both cases suturing was completed much later, in Mid-Mesozoic time,

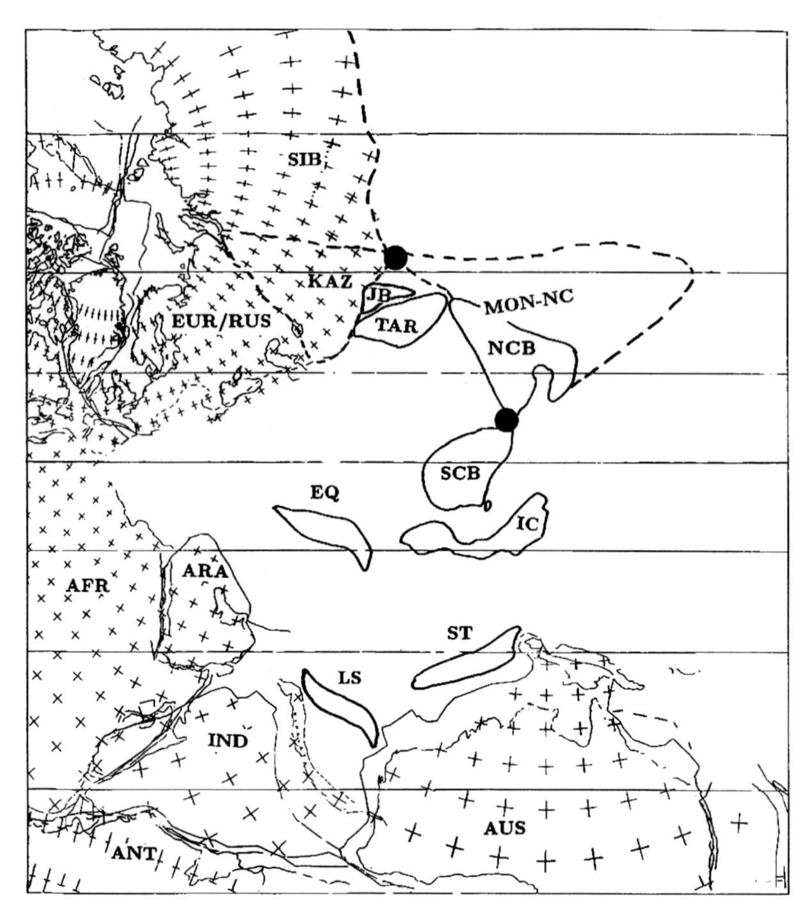

Figure P48 Schematic reconstruction of Mongolia-North China (MON-NC), Europe-Siberia-Kazakhstan (EUR-SIB-KAZ), Tarim (TAR), Junggar Basin (JB), South China block (SCB), Africa (AFR), Arabian Shield (ARA), India (IND), Australia (AUS), Antarctic (ANT), the East Qiangtang (EQ), Lhasa (LS), Indochina (IC), and Shan-Thai (ST) blocks in Late Permian time. Mecator Projection. After Zhao *et al.* (1990).

and that it was a composite Mongolia-North China plate, not the NCB, that became sutured to Siberia Mesozoic time in a zone much further north along what is termed the South Siberian Fold Belt. Furthermore, the paleomagnetic data suggested in both cases a very simple mechanism involving a scissors-like closing of the intervening ocean basins by antithetic senses of rotation about the points of initial collision at the eastern and western ends of the convergence zones of the southern and northern suture zones, respectively. These paleomagnetic models have been reinforced by many later studies and backed up by reinterpretation of relevant geological information, such as the propagation of abrupt regional and local changes from marine to continental basin deposition and of syncollisional to postcollisional alkalic granites in the collision zone. Therefore, here is a case where paleomagnetism has again provided first-order contributions in deciphering continental collision and amalgamation.

Paleomagnetic implications concerning tectonic controversies in China

Implications for the duration of continent-continent collisions

As noted above, the scissor model of NCB-SCB collision based on paleomagnetic data matches the change in sedimentary style from marine to continental on the northern margin of the SCB. Both paleomagnetic and geologic data agree that the onset of collision occurred in Permian to Triassic time. The data sets diverge, however, regarding the end (duration) of the collision, with the paleomagnetic data suggesting younger estimates (Middle Jurassic) than geological indicators (Late Triassic) by 20–30 Ma. Part of the discrepancy could be the lack of more reliable Late Triassic to Middle Jurassic paleomagnetic data from both blocks. Further studies could show that the paleomagnetic poles of Early Jurassic or Late Triassic age are indistinguishable, thereby bringing the paleomagnetic and geological data sets closer into agreement. On the other hand, it is possible that the rotation suggested by the paleomagnetic data also includes postcollisional shortening between the two blocks. Taking the ongoing India-Asia collision as an example of typical continent-continent collisions, deformation and shortening could continue at least 50 Ma after initial contact of the two plates. Such postcollision shortening could also potentially lead the age of initial India-Asia collision based on paleomagnetic data to be somewhat younger than those based on stratigraphic data.

Tertiary low-paleolatitude dilemma of western China

As the Chinese paleomagnetic database grows, several research teams are continuing to discover the outstanding regional disagreement between the observed and predicted Cenozoic paleomagnetic directions from red bed formations in western China. The inclinations of stable magnetization in Cretaceous red beds from Tarim and surrounding region's are somewhat shallower than expected using Besse and Courtillot (2002) APWP Cretaceous reference pole for Eurasia, whereas the Tertiary red beds are even more shallow using the corresponding Tertiary reference pole. The inclinations translate to anomalously low paleolatitudes. If all these paleomagnetic data, the ages, and the GAD hypothesis were correct, then in the Cretaceous Tarim would have been somewhat further south with respect to Eurasia compared to its present situation, relatively even further south when the Tertiary red beds were magnetized, and then had to move to its present more northerly relative position after that time. This implies a southward (extensional) relative displacement away from Eurasia in the period between the times when the Cretaceous and Tertiary redbeds were magnetized, and significantly more northerly (compressional) relative displacement toward Eurasia after the Tertiary red beds were magnetized, a scenario that is geologically indigestible. Recent results from Cretaceous red beds in the Hexi-Corridor (central China) show that there is no evidence that red beds there suffered the same inclination shallowing as the neighborhood Cenozoic red beds to the west (Sun *et al.*, 2001; Chen *et al.*, 2002). A review of published poles also confirms previous analyses that red beds from South and North China do not reveal the shallow inclination and these blocks (plus Mongolia) and Siberia have not undergone very significant internal deformation since the Cretaceous (Besse and Courtillot, 2002). It appears that some red beds suffer from inclination error, and others do not. The task is thus to develop ways to identify which of the Tertiary red beds do and do not undergo inclination shallowing.

The most direct way would be to obtain reliable Tertiary paleomagnetic results from volcanic rocks and compare them with inclinations of coeval red beds in the same study area. Two such studies have yielded contrasting results: (1) the Paleocene Lingzizong volcanics in the Lhasa area give consistent result (with positive fold and reversal test) with the underlying Cretaceous strata (Achache *et al.*, 1984), providing a notable example that both red beds and volcanic rock in the Lhasa area recorded the same ancient magnetization; and (2) Early Cretaceous red beds and underlying basalt flows from Tarim and central Asia fold belt (Gilder *et al.*, 2003) have directions that pass both fold and reversal tests, but the mean paleolatitude of the Early Cretaceous red beds is 11° lower than that of the Early Cretaceous basalts. Obviously, more studies should be conducted on Tertiary volcanic and red sedimentary rocks from the same area.

Tectonic affiliation of the Korean peninsula

The question about the Korean Peninsula's affinities with both the NCB and SCB is one of the highly charged controversies in Asian tectonics. The lack of knowledge about this question hampers our understanding concerning the models of collision of eastern Asian blocks and their paleogeographic settings. Pre-Cretaceous paleomagnetic poles for Korea are all different from each other and make it difficult to compare the coeval poles from the NCB and SCB. At first glance, Triassic and Jurassic paleomagnetic poles for Korea published prior to 1994 are indeed somewhat similar to coeval ones from the SCB, and this was perhaps the reason to lead some workers to suggest that Korea and South China may have behaved as a single tectonic block since the Triassic. However, all these poles that apparently correspond to the SCB poles were derived from rocks within the Okchon zone, which is known to be a site of severe deformation in the Mesozoic. In fact, Cretaceous remagnetization of Carboniferous rocks has been concluded for several previous published results. Although detailed structural analysis and mapping are still insufficient to unravel the kinematic history of the Okchon fold belt and assess its effects on these paleomagnetic poles, the proximity of these poles to the SCB poles may be a coincidence.

If one leaves out these poles derived from the Okchon zone and merely retains those poles derived from areas bordering the Okchon, a striking feature between the Korean and NCB poles emerges: the Late Permian, Early Triassic, and Late Jurassic poles for Korea are systematically displaced some 30° eastward with respect to the coeval poles of the NCB. A 30° clockwise rotation of Korea about a vertical axis brings these poles into general coincidence with those coeval poles for the NCB. Thus, paleomagnetic study suggests that the whole region was part of the North China landmass at least as early as Late Permian and rotated clockwise by about 30° with respect to the NCB during the Cretaceous (Zhao *et al.*, 1999).

It is apparent that the rotated (or corrected) poles show large apparent polar wander motion of Korea between Early Triassic and Late Jurassic, replicating those found in the paleomagnetic results from the NCB. Late Jurassic paleomagnetic poles for the NCB, SCB, and Korea are statistically indistinguishable, reinforcing the hypothesis that the accretion of the NCB and SCB was finished at this time. Thus, the NCB-Korea connection is not only consistent with the majority geologic observations, suggesting affinities between the two blocks, but also consistent with the collisional tectonic history of the eastern Asian margin derived from paleomagnetic data.

Terrane accretion between east and west Qiangtang

Comparison of paleomagnetic data obtained from east and west Qiangtang shows a problem that obstructs our understanding of terrane accretion in Tibet. The Cretaceous paleolatitudes of Otofuji *et al.* (1989) and Huang *et al.* (1992) in the Markam area of east Qiangtang are similar to the expected paleolatitudes for the sampling locality. The data would suggest that east Qiangtang was 21° ± 10° farther north than expected for Lhasa and 22° ± 11° farther north than expected for western Qiangtang. Taken at face value, these data would suggest that no significant crustal shortening has taken place north of east Qiangtang since at least the Late Cretaceous. The 2000 km shortening of Lhasa must have been accommodated between the Qiangtang and Lhasa blocks in the Tertiary. On the other hand, west Qiangtang must be close to the Lhasa trerrane in Late Cretaceous time, so there should be a suture between east and west Qiangtang. Huang *et al.* (1992) interpreted their paleomagnetic observation in terms of a giant right-lateral transpressional zone advocated by Dewey *et al.* (1998). Chen *et al.* (1993), on the other hand, suggested that east Qiangtang may have belonged to South China or to a block that was separated from western Qiangtang.

Tectonic rotational history of northern Tibet

A closely related question is the tectonic rotational history of northern Tibet, which is often related to uplift and deformation. Magnetic declinations, which are not typically affected by depositional compaction, can be compared to expected magnetic directions to assess the extent to which regions within the northern Tibetan plateau have rotated about vertical axes with respect to stable Eurasia. In eastern Tibet, Cretaceous and Tertiary paleomagnetic data indicate large amounts—in the order of 40°—of clockwise rotation in the eastern Himalayan syntaxis (e.g., Achache *et al.*, 1984; Huang *et al.*, 1992), which is consistent with right-lateral simple shear and associated oroclinal bending. However, paleomagnetic data obtained from Cretaceous and Tertiary red sedimentary rocks of western Qiangtang (Chen *et al.*, 1993) show no significant Neogene vertical-axis rotation in northern Tibet since the Late Cretaceous. Thus, all paleomagnetic data for tectonic rotations are not in agreement; some of them are apparently contradicting extrusion models that imply clockwise rotation of the region. These existing results, however, should be incorporated with forthcoming new results to advance our understanding of the growth history of the Tibetan plateau.

Summary

Plate tectonics views the whole Earth as a dynamic system in which the internal heat drives lithospheric plates in relative motion with respect to each other. Paleomagnetism enables geologists to view important aspects of continental movements in the past namely, changes in latitude and orientation relative to the geographic poles. It is important to realize the crucial role that paleomagnetic studies in China have played in providing the tectonic framework that guides geological investigation and interpretation. Until McElhinny *et al.*'s pioneering paleomagnetic paper in 1981 revealed discrepancies in Late

Permian paleolatitudes on the order of 20°, literally all geological interpretation concurred (incorrectly) that North and South China had been in place as part of the Eurasian landmass since the Early Permian or before. For some years after, with a notable exception (Klimetz, 1983), few geologists took these paleomagnetic results seriously enough to question the prevailing view. Meanwhile, additional paleomagnetic data began to accumulate in support of McElhinny *et al.*'s findings: discrepancies not only in Late Permian and but also in Early Mesozoic poles between South China and Siberia (Opdyke *et al.*, 1986), between North and South China, and between North China (including Mongolia) and Eurasia (see Enkin *et al.*, 1992 for references). It was on the basis of paleomagnetic evidence that the proposal was first made that South and North China collided at the eastern end of the Qinling belt at the close of the Permian and sutured progressively westward by relative rotation in the Mesozoic (Zhao and Coe, 1987), then backed up by reinterpretation of relevant geological information from the literature. Likewise, paleomagnetic evidence was the original basis of the similar mechanism of opposing sense (collision in the west followed by progressive rotation and suturing to the east) proposed for the incorporation of northern China into Eurasia in the Mesozoic (Zhao and Coe, 1989; Zhao *et al.*, 1990).

Although paleomagnetism has already made a dramatic contribution to the tectonics of China, this is just the beginning. By carefully gathering the available paleomagnetic, geological, and geophysical data and integrating them, Earth scientists are testing new hypotheses about smaller scale, intracontinental deformation. The tectonics of China will continue as a fruitful focus in the study of the dynamics of the Earth.

Xixi Zhao and Robert S. Coe

Bibliography

Achache, J., Courtillot, V., and Zhou, Y.X., 1984. Paleogeographic and tectonic evolution of southern Tibet since middle Cretaceous time: New paleomagnetic data and synthesis. *Journal of Geophysical Research-Solid Earth*, **89**: 311–339.

Besse, J., and Courtillot, V., 1991. Revised and synthetic apparent polar wander paths of the African, Eurasian, North American and Indian plates, and true polar wander since 200 Ma. *Journal of Geophysical Research*, **96**: 4029–4050.

Besse, J., and Courtillot, V., 2002. Apparent and true polar wander and the geometry of the geomagnetic field in the last 200 Myr. *Journal of Geophysical Research*, **107**: 2300 doi:10.1029/2000JB000050.

Burchfiel, B.C., Chen, Z., Hodges, K.V., Liu, Y., Royden, L.H., *et al.*, 1992. The south Tibetan detachment system, Himalayan orogen: extension contemporaneous with and parallel to shortening in a collisional mountain belt. *Geological Society of America Special Paper*, **269**: 1–41.

Burg, J.P., and Chen, G.M., 1984. Tectonics and structural zonation of southern Tibet, China. *Nature*, **311**: 219–223.

Burrett, C., and Richardson, R., 1980. Trilobite biogeography and Cambrian tectonic models. *Tectonophysics*, **63**: 155–192.

Burrett, C., Long, J., and Stait, B.,1990. Early-Middle Palaeozoic biogeography of Asian terranes derived from Gondwana. In Mckerrow, W.S. and Scotese, C.R. (eds.), *Palaeozoic Palaeogeography and Biogeography, Geological Society Memoir*, Bath, England. **12**: 163–174.

Chen, Y., Cogné, J.P., Courtillot, V., Avouac, J.P., Tapponnier, P., Buffetaut, E., Wang, G., Bai, M., You, H., Li, M., and Wei, C., 1991. Paleomagnetic study of Mesozoic continental sediments along the northern Tien Shan (China) and heterogeneous strain in central Asia. *Journal of Geophysical Research*, **96**: 4065–4082.

Chen, Y., Cogne, J.P., Courtillot, V., Tapponnier, P., and Zhu, X.Y. 1993. Cretaceous paleomagnetic results from western Tibet and tectonic implications. *Journal of Geophysical Research*, **98**(B10): 17981–17999.

Chen, Y., Gilder, S., Halim, N., Cogné, J.-P., and Courtillot, V. 2002. New Mezosoic and Cenozoic paleomagnetic data help constrain the age of motion on the Altyn Tagh fault and rotation of the Qaidam basin. *Tectonics*, **21**(5): 1042, doi:10.1029/2001 TC901030, 2002.

Davis, A.M., Aitchison, J.C., Zhu, B.D., and Huo, L., 2002. Paleogene Island arc collision related conglomerates, Yarlung Tsangpo suture zone, Tibet. *Sedimentary Geology*, **150**: 247–273.

Dewey, J.F., Shackleton, R.M., Chang, C.F., and Sun, Y.Y., 1988. The tectonic evolution of the Tibetan plateau, *Philosophical Transactions of the Royal Society of London Series A, Mathematical Physical and Engineering Sciences*, **327**: 379–413.

Dewey, J.F., Shackleton, R.M., Chang, C.F., and Sun, Y.Y., 1990. The tectonic evolution of the Tibetan plateau. In Sino-British Comprehensive Geological Expedition Team of the Qinghai-Tibet Plateau (ed.), *The Geological Evolution of the Qinghai-Tibet* (in Chinese). Beijing: Science Press, pp. 384–415.

Dupont-Nivet, G., Butler, R.F., Yin, A., and Chen, X., 2002. Paleomagnetism indicates no Neogene rotation of the Qaidam basin in north Tibet during Indo-Asian collision. *Geology*, **30**(3): 263–266.

Dupont-Nivet, G., Butler, R.F., Yin, A., and Chen, X., 2003. Paleomagnetism indicates no Neogene rotation of the Northeastern Tibetan Plateau. *Journal of Geophysical Research*, **108**(B8): 2386, doi:10.1029/2003JB002399, 2003.

Enkin, R.J., Yang, Z., Chen, Y., and Courtillot, V., 1992. Paleomagnetic constraints on the geodynamic history of the major blocks of China from the Permian to the Present. *Journal of Geophysical Research*, **97**: 13953–13984.

Gilder, S., Chen, Y., and Sen, S., 2001. Oligo-Miocene magnetostratigraphy and environmental magnetism of the Xishuigou section, Subei (Gansu Province, western China): Further implication on the shallow inclination of central Asia. *Journal of Geophysical Research-Solid Earth*, **106**: 30505–30521.

Gilder, S., Chen, Y., Cogne, J.P., Tan, X.D., Courtillot, V., Sun, D.J., and Li, Y.G., 2003. Paleomagnetism of Upper Jurrassic to Lower Cretaceous volcanic and sedimentary rocks from the western Tarim Basin and implications for inclination shallowing and absolute dating of the M-0 (ISEA?) chron. *Earth and Planetary Science Letters*, **206**: 587–600.

Halim, N., Chen, Y., and Cogné, J.P., 2003. A first palaeomagnetic study of Jurassic formations from the Qaidam basin, northeastern Tibet, China-tectonic implications. *Geophysical Journal International*, **153**: 20–26.

Harrison, T.W., Copeland, P., Kidd, W.S.F., and Yi, A., 1992. Raising Tibet. *Science* **255**: 1663–1670.

Hsu, K.J., Sun, S., Chen, H.H., Pen, H.B., and Sengor, AM.C., 1988. Mesozoic overthrust tectonics in south China. *Geology*, **16**: 418–421.

Hu, A.Q., and Rogers, G., 1992. Discovery of 3.3 Ga Archean rocks in north Tarim block of Xinjiang, western China. *Chinese Science Bulletin*, **37**: 1546–1549.

Huang, K., and Opdyke, N., 1995. A positive conglomerate test for the characteristic remanent magnetization of the Emeishan basalt from Southwest China. *Geophysical Research Letters*, **22**: 2769–2772.

Huang, K.N., Opdyke, N.D., Li, J., and Peng, X., 1992. Paleomagnetism of Cretaceous from eastern Qiangtang terrane of Tibet. *Journal of Geophysical Research*, **97**: 1789–1800.

Irving, E., 1964. *Paleomagnetism and its Application to Geological and Geophysical Problems*, New York: Wiley, 399 pp.

Klimetz, M.P., 1983. Speculations on the Mesozoic plate tectonic evolution of eastern China. *Tectonics*, **2**: 139–166.

Laveine, J.P., Zhang, S., and Lemoigne, Y., 1989. Global paleobotany, as exemplied by some Upper Carboniferous Pteridosperms. *Bulletin Societe Belge Geologique*, **98**: 115–125.

Li, Y.P., Zhang, Z., McWilliams, M., Nur, A., Li, Y., Li, Q., and Cox, A., 1988. Mesozoic paleomagnetic results of the Tarim Cration: Tertiary relative motion between China and Siberia. *Geophysical Research Letters*, **15**: 217–220.

Li, J.L., and Cong, B.L., 1980. A preliminary study on the early stage evolution of the Earth's crust in the North China fault block region.

In: *Formation and Development on the North China Fault Block Region*, Beijing, China: Publishing House of Academic, pp. 23–25.

Li, Z.X., Zhang, L., and Powell, C.McA., 1995. South China in Rodinia: part of the missing link beween Australia-East Antarctica and Laurentia? *Geology*, **23**: 407–410.

Li, Z.X.,1998. Tectonic history of the major east Asian lithospheric blocks since the mid Proterozoic—A synthesis. In Martin, F.J., Chung, S.-L., Lo, C.-J., and Lee, T.-Y. (eds.), *Mantle Dynamics and plate interactions in East Asia*, AGU Geodynamics Series **27**: Washington, DC: American Geophysical Union, pp. 221–243.

Li, Z.X., and Powell, C. McA., 2001. An outline of the palaeogeographic evolution of the Australasian region since the beginning of the Neoproterozoic. *Earth Science Reviews*, **53**: 237–277.

Lin, J.L., Fuller, M., and Zhang, W.Y., 1985. Preliminary Phanerozoic polar wander paths for the North and South China blocks. *Nature*, **313**: 444–449.

Lin, J., and Watts, D.R., 1988. Palaeomagnetic constraints on Himalayan-Tibetan tectonic evolution. *Philosophical Transactions of the Royal Society of London, Series A*, **326**: 177–188.

Long, J., and Burrett, C., 1989. Fish from the Upper Devonian of the Shan-Thai terrance indicate proximity to east Gondwana and south China terranes. *Geology*, **17**: 811–813.

Mattauer, M., Matte, P., Maluski, H., Qin, Z.Q., Zhang, Q.W., and Yong, Y.M., 1991. Paleozoic and Triassic plate boundary between North and South China, New structural and radiometric data on the Dabie-Shan (eastern China). *Comptes Rendus de l' Academie des Sciences—Series II*, **312**: 1227–1233.

McElhinny, M.W., Embleton, B.J., Ma, X.H., and Zhang, Z.K., 1981. Fragmentation of Asia in the Permian. *Nature*, **293**: 212–215.

McElhinny, M.W., and McFadden, P.L., 2000. *Paleomagnetism: Continents and Oceans International Geophysics Series* **73**, San Diego: Academic Press, 386 pp.

McKerrow, W.S., and Scotese, C.R., (eds.), 1990. Palaezoic paleogeography and biogeography. *Geology Society London Memoir* **12:** 435 pp.

Merrill, R.T., McElhinny, M.W., and McFadden, P.L., 1996. *The Magnetic Field of the Earth: Paleomagnetism, the Core, and the Deep Mantle.* San Diego: Academic Press, 541 pp.

Metcalfe, I., (ed.), 1999. *Gondwana Dispersion and Asian Accretion*, Final Results Volume for IGCP Project 321. Rotterdam: A.A. Balkema, VIII + 361 pp. ISBN 90 5410 446 5.

Murphy, M.A., Yin, A., Harrison, T.M., Durr, S.B., Chen, Z., Ryerson, F.J., Kidd, W.S.F., Wang, X., and Zhou, X., 1997. Did the Indo-Asian collision alone create the Tibetan plateau? *Geology*, **25**: 719–722.

Opdyke, N.D., Huang, K., Xu, G., Zhang, W.Y., and Kent, D.V., 1986. Paleomagnetic results from the Triassic of the Yangtze platform. *Journal of Geophysical Research*, **91**: 9553–9568.

Otofuji, Y., Funahara, S., Matsuo, J., Murata, F., Nishiyama, T., Zheng, X.L., and Yaskawa, K., 1989. Paleomagnetic study of western Tibet: Deformation of a narrow zone along the Indus Zangbo suture between India and Asia. *Earth and Planetary Science Letters*, **92**: 307–316.

Otofuji, Y., Matsuda, T., Itaya, T., Shibata, T., Matsumoto, M., Yamamoto, T., Morimoto, C., Kulinich, R.G., Zimin, P.S., Matunin, A.P., Sakhno, V.G., and Kimura, K., 1995. Late Cretaceous to early Paleogene paleomagnetic results from Sikhote-Alin, far eastern Russia: implications for deformation of East Asia. *Earth and Planetary Science Letters*, **130**: 95–108.

Otofuji, Y, Itaya, T., Wang, H., and Nohda, S., 1995. Paleomagnetism and K-Ar Dating of Pleistocene Volcanic-Rocks along the Altyn-Tagh Fault, Northern Border of Tibet. *Geophysical Journal International*, **120**(2): 367–374.

Ren, J.S., Jiang, C.F., Zhang, Z.K., and Qin, D.Y., 1987. *The Geotectonic Evolution of China.* Science Press. Springer-Verlag,

Rumelhart, P.E., Yin, A., Cowgill, E., Bulter, R., Zhang, Q., and Wang, X., 1999. Cenozoic vertical-axis rotation of the Altyn Tagh fault system. *Geology*, **27**: 819–822.

Runnegar, B., and Cambell, K.S.W., 1976. Late Palaeozoic faunas of Australia. *Earth Science Reviews*, **12**: 235–257.

Shu, D., and Chen, L., 1991. Paleobiogeography of Bradoriida and break-up of Gondwana. In Ren, J., and Xie, G. (eds.), *Proceedings of the 1st International Symposium on Gondwana Dispersion and Asia Accretion*. 314 pp., China University of Geosciences, Beijing, China, pp. 16–21.

Tapponier, P., Xu, Z.Q., Roger, F., Meyer, B., Arnaud, N.O., Wittlinger, G., and Yang, J.S., 2001. Oblique stepwise rise and growth of the Tibet Plateau. *Science*, **294**: 1671–1677.

Tauxe, L., 1998. *Paleomagnetic Principles and Practices.* Dordrecht: Kluwer Academic, 299 pp.

Tauxe, L., 2005. Inclination flattening and the geocentric axial dipole hypothesis. *Earth and Planetary Science Letters*, **233**: 247–261.

Van der Voo, R., 1993. *Paleomagnetism of the Atlantic, Tethys and Iapetus Oceans.* Cambridge: Cambridge University Press, 411 pp.

Wang, X.F., Xu, T., and Wei, C.S., 1999. Geodynamic and tectonic evolution of China and related Gondwana crustal fragments: Preface. *Gondwana Research*, **2**(4): 509–509.

Wang, S., 1993. The relationship between Ar and Cl in cherts and vein quartz and its significance on geochronology. *Acta Petrologica Sinica*, **9**: 319–328 (in Chinese).

White, M.C., 1995. Finding documents split of Indo-Australian plate. *EOS*, **76**: 337–343.

Xiao, W.J., Windley, B.F., Hao, J., and Li, J.L., 2002. Arc-ophiolite obduction in the Western Kunlun Range (China): implications for the Palaeozoic evolution of central Asia. *Journal of the Geological Society*, **159**: 517–528.

Yang, Z., and Besse, J., 2001. New Mesozoic apparent polar wandering path for south China: Tectonic consequences.. *Journal of Geophysical Research*, **106**: 8493–8520.

Yang, Z.Y., Cheng, Y.Q., and H.Z., Wang 1986. *The Geology of China.* Oxford: Clarendon Press.

Yin, A., and Harrison, T.M., 2000. Geologic Evolution of the Himalayan-Tibetan Orogen. *Annual Review of Earth and Planetary Science*, **28**: 211–280.

Yin, H.F., (ed.) 1988. *Paleobiogeography of China.* Beijing: Chinese University of Earth Sciences Press, 329 pp. (in Chinese).

Zhao, X., and Coe, R.S., 1987. Palaeomagnetic constraints on the collision and rotation of North and South China. *Nature*, **327**: 141–144.

Zhao, X., and Coe, R.,1989. Tectonic Implications of Permo-Triassic Paleomagnetic Results From North and South China. In Hillhouse, J.W. (ed.), Deep Structure and Past Kinematics of Accreted Terranes, Geophysical Monograph/International Union of Geodesy and Geophysics, **5**: 267–283.

Zhao, X., Coe, R.S., Zhou, Y.X., Wu, H.R., and Wang, J., 1990. New paleomagnetic results from northern China: collision and suturing with Siberia and Kazakhstan. *Tectonophysics*, **181**: 43–81.

Zhao, X., Coe, R.S., Gilder, S., and Frost, G., 1996. Palaeomagnetic constraints on palaeogeography of China: Implications for Gondwanaland, In Breakup of Rodinia and Gonwanaland and Assembly of Asia. *Special Issue of the Australian Journal of Earth Sciences*, **43**: 643–672.

Zhao, X., Coe, R.S., Chang, K.H., Park, S.O., Omarzai, S.K., Zhu, R. X., Gilder, S., and Zheng, Z., 1999. Clockwise rotations recorded in cretaceous rocks of South Korea: Implications for tectonic affinity between Korea Peninsula and North China. *Geophysical Journal International*, **139**: 447–463.

Zheng, Z., Kono, M., Tsunakawa, H., Kimura, G., Wei, Q.Y., Zhu, X.Y., and Hao, T., 1991. The apparent polar wander path for North China Block since the Jurassic. *Geophysical Journal International*, **104**: 29–40.

Zhu, R.X., and Tschu, K.K., 2001. *Studies on Paleomagnetism and Reversals of Geomagnetic Field in China.* Beijing: Science Press, 168 pp.

Zhu, X.Y., Liu, C., Ye, S.J., and Lin, J.L., 1977. Remanence of red beds from Linzhou, Xizang and the northward movement of the Indian plate (in Chinese with English abstract). *Scientia Geologica Sinica*, **1**: 44–51.

Ziegler, A.M., Rees, P.M., Rowley, D.B., Bekker, A., Li, Q., and Hulver, M.L., 1996. Mesozoic assembly of Asia: constraints from fossil floras, tectonics, and paleomagnetism, In An Yin and Harrison, M. (eds.), *The Tectonic Evolution of Asia.* Cambridge: Cambridge University Press, pp. 371–400.
Ziegler, A.M.,1990. Phytogeographic paterns and continental configurations during the Permian period. In McKerrow, W.S., and Scotese, C.R. (eds.), *Palaeozoic Palaeogeography and Biogeography.* Geological Society of London Memoir **12**, pp. 363–379.

Cross-references

Geomagnetic Polarity Reversals
Magnetosphere of the Earth
Paleomagnetism

POGO (OGO-2, -4 AND -6 SPACECRAFT)

The Polar Orbiting Geophysical Observatories (POGO) were the low altitude half of the Orbiting Geophysical Observatories (OGO) intended to carry a large number of instruments to observe the physical environment of the Earth's magnetosphere and outer ionosphere (Jackson and Vette, 1975). The characteristics of the three POGO spacecraft are shown in Table P1 and the spacecraft design is shown in Figure P49. The magnetometers were each a pair of optically pumped rubidium vapor units whose output signal strength had a sin θ cos θ dependence, where θ is the angle between the optical axis and the observed magnetic field vector (Farthing and Folz, 1967). Each then had equatorial and polar "null zones" relative to their optical axes where the signals were too weak to measure. The equatorial ones were less than 7° half angle, and the polar zones less than 15°. The two instruments were mounted with their optical axes at 55° to each other so the combined output signal was available except for the two small angular areas where these zones overlapped.

The maximum altitude of the OGO-2 spacecraft was planned to be only 900 km but the tracking devices had lost its contact at launch due to the early morning fog, so it burned its full load of fuel to attain the high apogee seen in the table. The spacecraft also experienced problems soon after the operation began because its horizon sensors were found to be too sensitive; confusing temperature gradients in clouds with those at the horizon. It had utilized its full load of control gas within ten days of launch. It was then put into a slow spin mode. In April 1966 both batteries failed; so operation thereafter was limited to sunlit periods of the orbit. Nevertheless, there were some 306 days of observations. The OGO-4 spacecraft experienced far fewer problems and operated almost continuously until the tape recorders failed in January 1969. Because the orbits of both of these spacecraft drifted earlier in local time (5.5 min/day for OGO-2 and 6.1 for OGO-4) it was possible for the first time during September and early October 1967 to obtain magnetic field data simultaneously in two differing local time planes. OGO-6 was equally successful as OGO-4 though the data collection after August 1970 was sporadic. Having a lower inclination, the orbital drift with local time was higher—earlier by 8 min/day.

The magnetometer data from these three spacecrafts were the most accurate and continuous of any to that date, though the analysis was limited to use of the scalar field. A proposal to add vector instruments to OGO-6 by J.P. Heppner was not accepted by NASA, thus missing an opportunity to observe field-aligned currents in the polar regions for the first time. Initially, it was thought that accurate scalar observations would be sufficient to map the internal field as well as to study the *in situ* electrical currents. However, it was shown mathematically by Backus (1970) that there was one example of different vector

Table P1 The POGO spacecraft

Spacecraft	Dates of operation	Inclination (degrees)	Perigee (km)	Apogee (km)	Sample interval (s)
OGO-2	Oct 14, 1965-Oct 2, 1967	87.3	410	1510	0.5
OGO-4	July 28, 1967-Jan 19, 1969	86.0	410	910	0.5
OGO-6	June 5, 1969-June 1971	82.0	400	1100	0.288

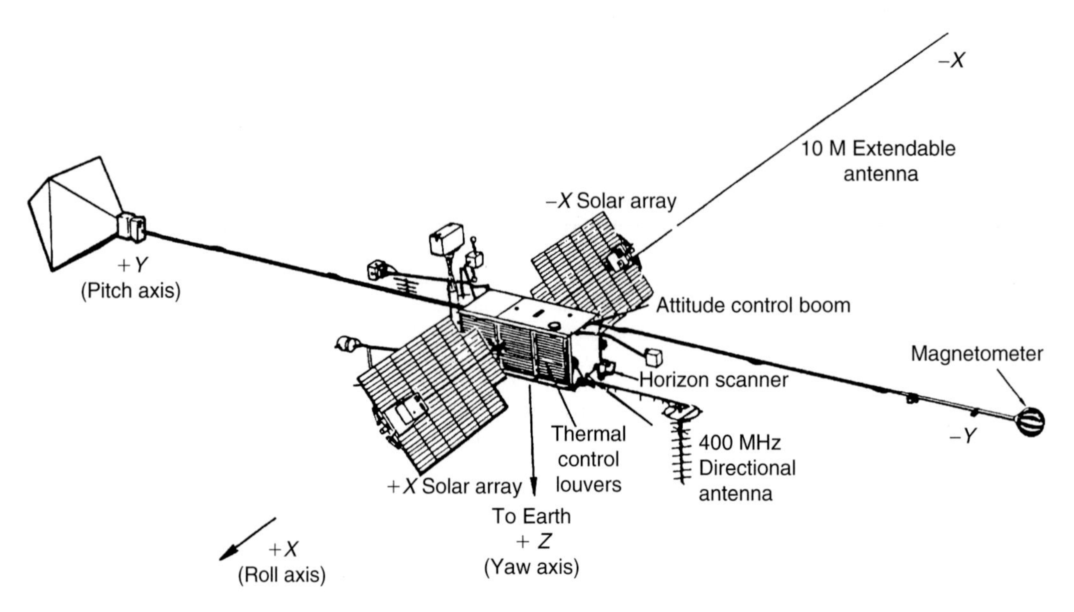

Figure P49 Typical OGO spacecraft in deployed configuration. The spacecrafts were of the order of 18 m long on the *Y*-axis. They were maintained in Earth orientation using gyroscopic reaction wheels with corrections as needed with gas jets. The *Z*-axis was toward the Earth and the direction of travel was along *X*-axis.

fields that could not be distinguished by observations of the scalar magnitude. Hurwitz and Knapp (1974) first demonstrated that models based only on scalar magnitude have larger vertical intensity errors in equatorial regions. Nevertheless, the data, especially for OGO-4, have been shown to be accurate to better than 4 nT, including the effects of orbital errors of position (Sabaka *et al.*, 2004).

The data from OGO-2 were used as a standard in establishing the first International Geomagnetic Reference Field (IGRF) (Cain and Cain, 1971), and those from all three spacecrafts are the best control over the main field for the 1965–1971 epochs (Sabaka *et al.*, 2004). Although Benkova *et al.* (1973) believed that Cosmos 49 measurements showed the presence of broad-scale magnetic anomalies, the noise in the data and orbital positions were too great to bring them out clearly. Until the POGO magnetic data were analyzed, the first accurate global magnetic anomaly map (Regan *et al.*, 1975) could not be constructed.

The POGO data also allowed study of the field sources external to the Earth. Cain and Sweeney (1973) were the first to observe the equatorial electrojet from above the ionosphere. OGO-2 and OGO-4 were the first spacecrafts to operate simultaneously in different local time planes. Olsen *et al.* (2002) used data from a brief period in 1967 to show how such multispacecraft observations could help in obtaining a detailed picture of ionospheric current systems. Important contributions were also made to the understanding of high latitude external magnetic variations (Langel, 1974a) and their relation to the interplanetary field (Langel, 1974b) as well as the low energy plasma environment about the Earth (Langel and Sweeney, 1971).

Information about the POGO series spacecraft, experiments, and data sets can be obtained from the Master Catalog at the National Space Science Data Center Web site http://nssdc.gsfc.nasa.gov/nmc/sc-query.html by entering a query for the spacecraft name "pogo" and then following the response. Other internet sources include: http://www.spacecenter.dk/data/magnetic-satellites/ and http://core2.gsfc.nasa.gov/CM/.

Joseph C. Cain

Bibliography

Backus, G.E., 1970. Nonuniqueness of the external geomagnetic field determined by surface intensity measurements. *Journal of Geophysical Research*, **75**: 6339–6341.

Benkova, N.P., Dolginov, Sh., and Simonenko, T.N., 1975. *Journal of Geophysical Research*, **80**: 794–802.

Cain, J.C., and Cain, S.J., 1971. Derivation of the International Geomagnetic Reference Field [IGRF(1068)]. In Zmuda, A.J. (ed.), *The World Magnetic Survey* 1957–1969. pp. 163–166, 190–199, and 202–203, IAGA Bulletin, 28, Paris. Also published as *NASA Technical Note D-4527*, April, 1968.

Cain, J.C., and Sweeney, R.E., 1973. The POGO Data. *Journal of Atmospheric and Terrestrial Physics*, **35**: 1231–1247.

Farthing, W.H., and Folz, W.C., 1967. Rubidium vapor magnetometer for near Earth orbiting spacecraft. *Review of Scientific Instruments*, **38**: 1023–1030.

Hurwitz, L., and Knapp, D.G., 1974. Inherent vector discrepancies in geomagnetic main field models based on scalar F. *Journal of Geophysical Research*, **79**: 3009–3013.

Langel, R.A., 1974a. Near-earth magnetic disturbance in total field at high latitudes 1. Summary of data from OGO 2, 4, and 6. *Journal of Geophysical Research*, **79**: 2363–2371.

Langel, R.A., 1974b. Variation with interplanetary sector of the total magnetic field measured at the OGO 2, 4 and 6 satellites. *Planetary and Space Sciences*, **22**: 1413–1425.

Langel, R.A., and Sweeney, R.A., 1971. Asymmetric ring current at twilight local time. *Journal of Geophysical Research*, **76**: 4420–4427.

Jackson, J.E., and Vette, J.I., 1975. *OGO Program Summary*, NASA, SP-7601.

Olsen, N., Moretto, T., and Friis-Christensen, E., 2002. New approaches to explore the Earth's magnetic field. *Journal of Geodynamics*, **33**: 29–41.

Regan, R.D., Cain, J.C., and Davis, W.M., 1975. A global magnetic anomaly map. *Journal of Geophysical Research*, **80**: 794–802.

Sabaka, T.J., Olsen, N., and Purucker, M.E., 2004. Extending comprehensive models of the Earth's magnetic field with Ørsted and CHAMP. *Geophysical Journal International*, **159**: 521–547, 10111/j.1365-246X.2004.942421.x.

Cross-references

CHAMP, Ørsted, Magsat
Langel, Robert A. (1937–2000)
Magnetic anomalies, Long-wavelength
Magnetometers, Laboratory

POLARITY TRANSITION, PALEOMAGNETIC RECORD

Early work on the Matuyama-Brunhes transition field

The last change of geomagnetic polarity, bringing about an end to the Matuyama Reverse Chron and the onset of the Brunhes Normal Chron, occurred between [illegible] and [illegible] ka. This age was determined by both orbital tuning of marine sediment records (e.g., Tauxe *et al.*, 1996) and by $^{40}Ar/^{39}Ar$ age determinations on transitionally-magnetized lavas (e.g., Singer and Pringle, 1996). Being the most youthful field reversal, it has always been associated with the largest number of available paleomagnetic records (see e.g., Love and Mazaud, 1997). One reason for this is that sedimentary records, often obtained from marine cores, penetrate this reverse-to-normal (R-N) polarity change before all others. It is no wonder then that the Matuyama-Brunhes (M-B) reversal was the first polarity transition to be associated with the availability of synchronous records from distant sites. These multiple recordings facilitated an initial attempt to better understand global characteristics of a transitioning field and the dynamo process that causes field reversal.

In 1976 the path of the virtual geomagnetic pole (VGP)—that is, the north geomagnetic pole position calculated by assuming each recorded paleodirection to be associated with a geocentric dipole–from a M-B record recovered from the dry sediments at Lake Tecopa, California, was compared by Hillhouse and Cox to the detailed, exposed marine sediment record published five years earlier (Niitsuma, 1971) from the Boso Peninsula, Japan. What was found was that the two VGP paths traced quite distinct tracks over Earth's surface from the geographic south polar region to the north polar region (see Figure P50), an observational finding that suggested for the first time that the transitioning field becomes largely nondipolar during, at least, this particular polarity change. On the other hand, if the two VGP tracks were quite similar, such a finding would be compatible with a transitioning field that remained dominantly dipolar. However, for this case, the VGP paths were found in the Atlantic and the west Pacific, respectively. Much later, the Lake Tecopa sediments became the focus of a second study by Valet *et al.* (1989) that attempted to confirm the earlier findings. This study found a far more complex VGP path, one that encompassed a significant portion of the central-to-east Pacific. Although these later findings still suggested a significant nondipole contribution to the transitioning M-B field, this study made clear that a complete understanding of even the most basic aspects of field geometry during such events would require a much larger database of contemporaneous records.

As the number of reversal records associated with the M-B transition steadily grew, so did the number of analyses that attempted to better define the structure of the transitional field. In 1981 a phenomenological model (Hoffman, 1981) was applied to available M-B datasets, an approach that assumed a reversal to begin at a particular locality within the core and then proliferate, or "flood," this "reversed" magnetic flux throughout the dynamo until completion of the process. This flooding model appeared to generally account for the M-B VGP paths that were available at the time.

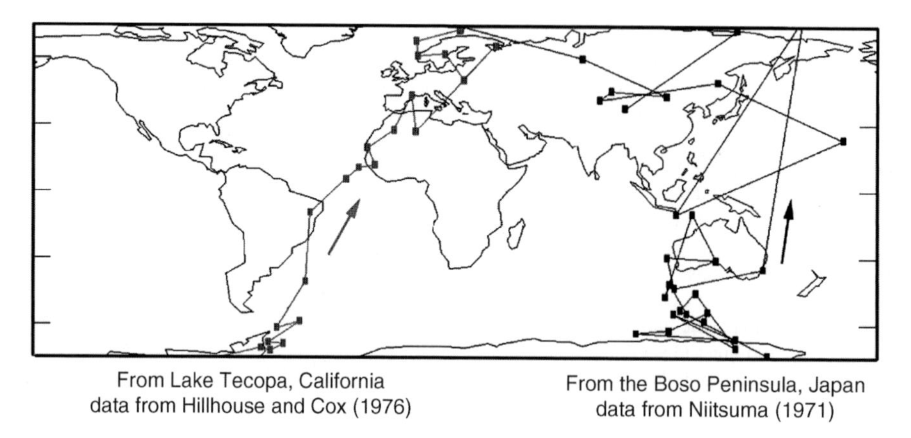

Figure P50 The path of the virtual geomagnetic pole (VGP) corresponding to the first two acquired paleomagnetic records of the Matuyama-Brunhes transition. Given the very different pathways, Hillhouse and Cox (1976) concluded that the M-B transition field was predominantly nondipolar.

Questions regarding the role of Earth's mantle

A decade later research regarding the last reversal took a new and exciting turn: Clement (1991) found that VGP paths associated with M-B records contained in a more complete database of published results displayed a significant bimodal distribution of transitional poles in geographic longitude. More specifically, the M-B VGPs appeared to preferentially lie within two particular antipodal bands, one running through the Americas, the other through East Asia and Australia (Figure P51a). This finding nearly coincided with another study—the results of which can be seen on the cover of a 1991 issue of *Nature* magazine—which found the same geographical bandings of transitional VGPs associated with several records of paleomagnetic reversals going back in time into the Miocene (Laj *et al.,* 1991).

This rather surprising result implied that, for the last several millions of years, and, in particular, during the time of the M-B event, transitional VGPs ran along "preferred" pathways. This, for the first time,

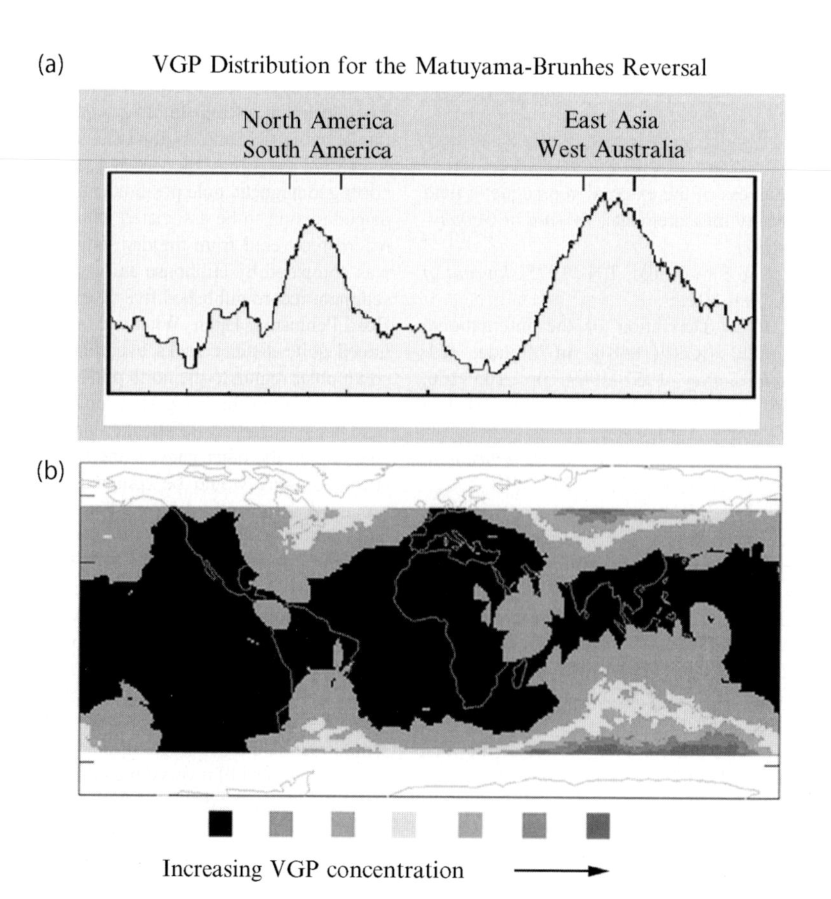

Figure P51 Geographical distribution of M-B transitional VGPs. (a) *after* Clement's (1991) finding of a bimodal distribution in longitude. (b) *after* Hoffman's (2000) global histogram.

suggested that Earth's mantle was intimately involved in the process of field reversal. That is, it was argued that such recurrence of field behavior had to be tied to physical conditions about the lowermost mantle, the variations of which imposing a control on the manner in which magnetic flux can emerge from the fluid outer core during reversals. Since the time of these two seminal papers by Clement (1991) and Laj *et al.* (1991), the question as to whether or not field behavior during successful reversals as well as during other transitional events—like, say, an aborted attempt—is largely influenced by such lower mantle heterogeneities, has been hotly debated (see e.g., Jacobs, 1994; Merrill and McFadden, 1999).

An alternative explanation put forward was that the apparent preferential behavior—rather than having a geomagnetic basis—was, in fact, an artifact related to limitations in the way sediments acquire their remanent magnetization (e.g., Rochette, 1990; Valet *et al.,* 1992; Langereis *et al.*, 1992). However, during the following year an analysis (Hoffman, 1992) of transitional field behavior recorded in volcanic sequences having ages running back to the Miocene—rocks that, typically, provide the most accurate spot recordings of the paleofield vector—showed, not preferred banding of VGP paths, but rather preferred groupings, or clusters. Interestingly, these clusters resided within the hypothesized preferred longitudinal bands. Hence, this "preferred patch" finding implied virtually the same geomagnetic basis as proposed for sediment records. Yet, the correspondences and differences between what was observed in distinct rock types broadened the debate, while raising it to a new level. If preferred VGP behavior is assumed to be a manifestation of a dynamo process controlled by the lower mantle, could the apparent discrepancy observed in sediment and lava records—that is, preferred bands and preferred patches, respectively—be explained through the known differences in the manner these two types of rocks become magnetized? And, if so, could long-lived transitional field states during which time the field direction at a given site, although transitional, remained essentially stationary, produce both observations?

The MBD97 database

In 1997, the first rigorous attempt was made (Love and Mazaud, 1997) to distinguish which records of the full M-B database may be considered reliable, and, hence, worthy of inclusion in a robust analysis. Nearly 62 M-B records from sites about the globe were found in the literature, and each was subjected to a number of strict reliability criteria. Of these records only 11 satisfied this scrutiny and were admitted to the so-called *MBD97* database. These records now separated from less reliable records, were analyzed in an attempt to answer the primary question regarding the existence or not of preferential VGP behaviour during the last geomagnetic reversal. The findings by the authors of the *MBD97* database suggested a statistically significant answer in the affirmative. Specifically, they argued that the distribution of transitional M-B VGPs showed some degree of preferred longitudinal behavior compatible with the hypothesis that the bias was, in fact, a real geomagnetic phenomenon.

Other analyses of the *MBD97* database followed. One study (Gubbins and Love, 1998) raised the point that the VGP behavior, when viewed in a global sense, was associated with a dependence on the site of observation, and that each individual VGP path was compatible with inferred symmetry conditions within the core dynamo. Still another investigation of the *MBD97* database (Hoffman, 2000) produced a global histogram of transitional M-B VGPs (Figure P51b). This analysis, although again confirming a longitudinal bimodal nature of virtual poles, demonstrated that very few VGPs existed in the database at low latitudes near the equator. From this result it was argued that during the M-B transition considerable time was spent by the reversing field in particular configurations. Specifically, clusters of sequential VGPs near Australia, and in the South Atlantic, were found to exist in the database-data that was observed in records from sequences of lavas as well as sediments. Yet, the question regarding whether or not the VGP migrated across the equator along preferred longitudinally-confined pathways, remained unresolved.

Further work

Since the construction of the *MBD97* database, records of the Matuyama-Brunhes obtained from deep-sea sediments in the West Pacific (Oda *et al.*, 2000) and the North Atlantic (Channell and Lehman, 1997; Channell *et al.*, 2004), have found their way to publication. In particular, the several parallel records from North Atlantic high deposition-rate drift deposits are particularly noteworthy (see examples shown in Figure P52). These data sets, all displaying remarkably similar,

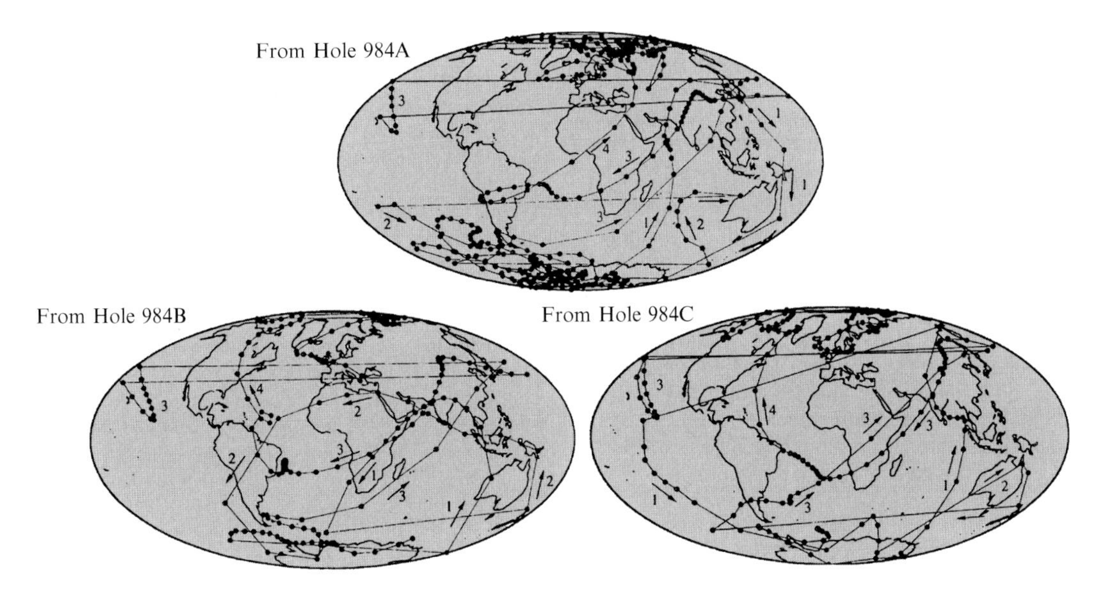

Figure P52 Three examples of VGP paths associated with parallel M-B marine sediment core records from the North Atlantic. Note the remarkable similarity of detail. After Channell and Lehman (1997).

though complex, field behavior, are arguably the most detailed and complete records of the M-B reversal presently available. In some contrast to the most reliable records obtained from mid-latitudes, these high latitude North Atlantic records show little in the way of VGP clustering near Australia, yet contain clusters both in the South Atlantic and northeast Asia (Channell and Lehman, 1997; Channell *et al.*, 2004). If it is the case that VGP cluster localities are dependent on the site of observation, then the associated long-held field states must be largely nondipolar.

Constable (1992) was the first to show a link between preferred VGP path behavior and the modern-day geomagnetic field. Specifically,

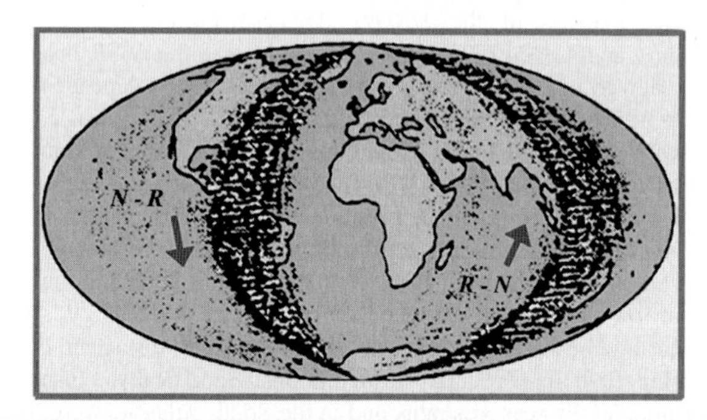

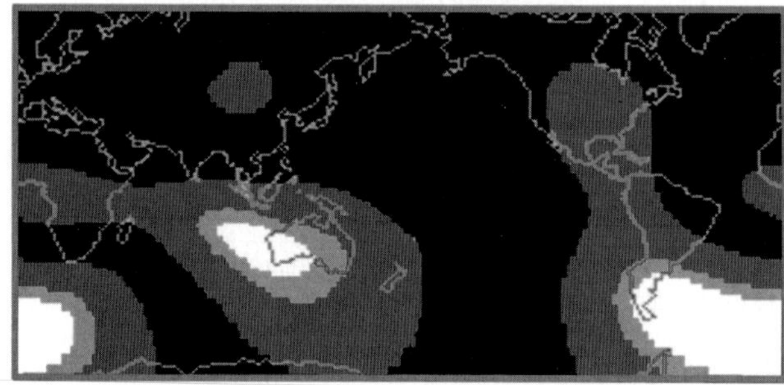

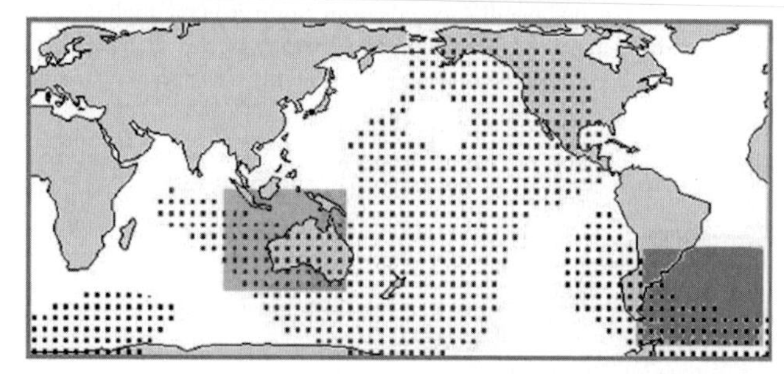

Figure P53 Aspects of the modern-day field. *(Top)* VGP pathways associated with numerous sites from about the globe for simulated normal-to-reverse (N-R) and reverse-to-normal (R-N) polarity transitions in which only the axial dipole is assumed to be involved in the process. After Constable (1992). *(Middle)* Intensity of the vertical component of the field at Earth's surface following synthetic removal of the axial dipole from the modern-day field (hence, the *NAD*-field). The two locations in white—near west Australia and in the South Atlantic—are the most intense, having at least 90% of the most intense value found. *(Bottom)* Two grids of sites defining regions that, when subjected to the modern-day *NAD*-field, would possess paleomagnetic directions corresponding to VGPs found in either of the two shaded localities (also found to be areas of the most intense *NAD*-field). The smaller and larger grids correspond to north-VGPs in the South Atlantic and south VGPs in the west Australian shaded regions, respectively *after* Hoffman and Singer (2004).

it was shown (see Figure P53, *top*) that if a polarity reversal simply involved the decay, vanishing, and reemergence of the axial dipole having opposite sign—keeping the remainder of the field constant—VGP pathways for sites from around the globe would preferentially lie along the same antipodal longitudinal bands proposed by Laj *et al.* (1991) and Clement (1991). The field configuration for which only the axial dipole has been removed is commonly termed the nonaxial dipole or, *NAD*-field (although it is sometimes also called the *sans*-axial dipole, or *SAD*-field, translated from French as to be "without" the axial dipole). Such a field state must occur during a successful change in polarity. That is, the axial dipole must vanish before it can change sign.

Interestingly, the primary locations of observed M-B clustered VGPs from available paleomagnetic records near western Australia, in the southwest Atlantic, and within Siberia, are not only sites within the two proposed preferred bands, but also correspond to the three locations at Earth's surface which possess the most intense vertical component of the modern-day field after synthetic removal of the axial dipole (i.e., the *NAD*-field) from the calculated spherical harmonic analysis (Figure P53 *middle*). Moreover, if VGPs are calculated for sites from around the globe when assumed to be experiencing the modern-day *NAD*-field, one finds that in the Southern Hemisphere, the very same Australian and South Atlantic localities are the most preferred regions (Figure P53 *bottom*). Thus, available paleomagnetic transition data appear to be linked to the field of today, suggesting that primary features of the modern-day *NAD*-field remain virtually stationary over times in order of 10^6 or, perhaps, 10^7 years (Hoffman and Singer, 2004). If so, this is a definite sign of a controlling influence by the lowermost mantle on flux emerging from the fluid outer core.

Duration of the Matuyama-Brunhes transition

Recently, the duration of the M-B transition has been under study. The investigation of paleodirectional records obtained from several marine sediment cores (Clement, 2004) found a clear dependence of reversal time on the latitude of the site of observation. Specifically, the duration varied from about 2000 years at low latitudes to 10000 years at high latitude sites. However, the complete temporal process of field reversal appears from other studies to be rather complex and require a significantly longer span of time: an earlier analysis by Hartl and Tauxe (1996) of several marine sediment cores that recorded the M-B found a significant decrease in field strength some 15 ka prior to a second weakening associated with the actual change in polarity. The first decrease of this "double-dip" in intensity was explained by the authors as a precursory geomagnetic event to the subsequent successful reversal.

Following this finding, a recent geochronological study in which several precise $^{40}Ar/^{39}Ar$ ages were determined for transitionally magnetized M-B lavas from volcanic sequences at four mutually distant sites found a bimodal age distribution consistent with such a precursor (Singer *et al.*, 2005). The initial instability, associated with the first of the two decreases in field intensity, was found to have occurred 18 ka prior to the polarity switch. The Australian and South Atlantic VGP clustering (Figure P54) appears now to be associated with the precursor, and thus suggests that the beginning of the period of dynamo instability may be simply explained by a process which essentially destroys the axial dipole, leaving the transitional field to have the configuration of the *NAD*-field at that time. Singer *et al.* (2005) further suggest that it is such a two-stage process that may be required for a given reversal attempt to be successful. Whether these characteristics of the M-B, the last reversal of Earth's magnetic field, are common to earlier polarity transitions (see e.g., Coe and Glen, 2004), and hence, typical of a fundamentally systematic process, requires the availability of considerably more paleomagnetic reversal data.

Kenneth A. Hoffman

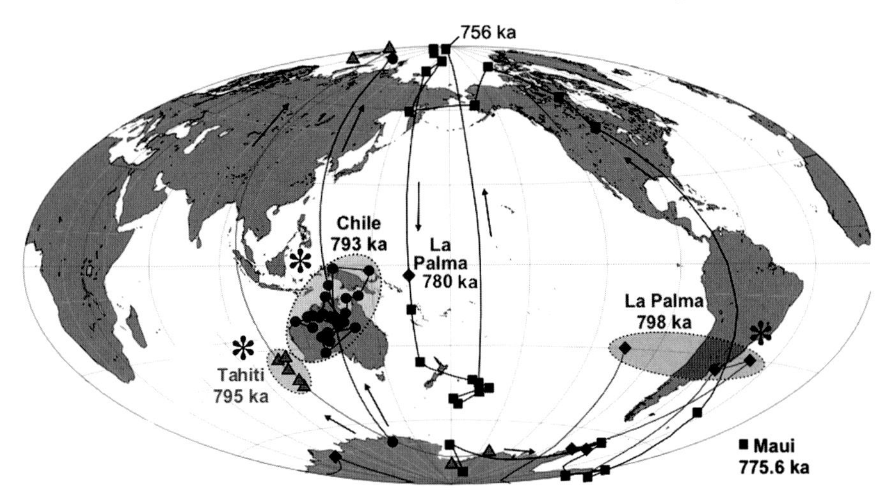

Figure P54 Matuyama-Brunhes VGPs and precise $^{40}Ar/^{39}Ar$ ages for sequences of transitionally-magnetized lavas from each of four mutually distant sites. Note that those sites associated with VGP clusters near Australia and the South Atlantic/South American regions (indicated with an asterisk) correspond to lavas that were erupted some 10 ka prior to the actual change in polarity that was recorded on Maui, Hawaii. After Singer *et al.* (2005).

Bibliography

Channell, J.E.T., and Lehman, B., 1997. The last two geomagnetic polarity reversals recorded in high-deposition-rate sediment drifts. *Nature*, **389**: 712–715.

Channell, J.E.T., Curtis J.H., and Flower, B.P., 2004. The Matuyama-Brunhes boundary interval (500–900 ka) in North Atlantic drift sediments. *Geophysical Journal International*, **158**: 489–505.

Clement, B.M., 1991. Geographical distribution of transitional VGPs: evidence for non-zonal equatorial symmetry during the Matuyama-Brunhes geomagnetic reversal. *Earth and the Planetary Science Letters*, **104**: 48–58.

Clement, B., 2004. Dependence on the duration of geomagnetic polarity reversals on site latitude. *Nature*, **428**: 637–640.

Constable, C., 1992. Link between geomagnetic reversal paths and secular variation of the field over the last 5 Ma *Nature*, **358**: 230–233.

Gubbins, D., and Love, J.J., 1998. Preferred VGP paths during geomagnetic polarity reversals: Symmetry conditions. *Geophysical Research Letters*, **25**: 1079–1082.

Hartl, P., and Tauxe, L., 1996. A precursor to the Matuyama/Brunhes transition-field instability as recorded in pelagic sediments. *Earth and the Planetary Science Letters*, **138**: 121–135.

Hillhouse, J., and Cox, A., 1976. Brunhes-Matuyama polarity transition. *Earth and the Planetary Science Letters*, **29**: 51–64.

Hoffman, K.A., 1981. Quantitative description of the geomagnetic field during the Matuyama-Brunhes polarity transition. In: *Proceedings of Symposium 10 "Dynamics of Core and Mantle"*, XVII General Assembly IUGG, Canberra, Australia, *Physics of the Earth and Planetary Interiors*, **24**: 229–235.

Hoffman, K.A., 1992. Dipolar reversal states of the geomagnetic field and core-mantle dynamics. *Nature*, **359**: 789–794.

Hoffman, K.A., 2000. Temporal aspects of the last reversal of Earth's magnetic field. *Philosophical Transactions of the Royal Society of London series A*, **358**: 1181–1190.

Hoffman, K.A., and Singer, B.S., 2004. Regionally recurrent paleomagnetic transitional fields and mantle processes. In Channell, J.E.T., Kent, D.V., Lowrie, W., and Meert, J. (eds.), *Timescales of the Internal Geomagnetic Field.*, Geophysical Monograph, Washington, DC: American Geophysical Union, pp. 233–243.

Jacobs, J.A., 1994. *Reversals of the Earth's Magnetic Field*, 2nd edn, Cambridge: Cambridge University Press.

Langereis, C.G., van Hoof, A.A.M., and Rochette, P., 1992. Longitudinal confinement of geomagnetic reversal paths as a possible sedimentary artefact. *Nature*, **358**: 226–229.

Laj, C., *et al.*,1991. Geomagnetic reversal paths. *Nature*, **351**: 447.

Love, J.J., and Mazaud, A., 1997. A database for the Matuyama-Brunhes magnetic reversal. *Physics of the Earth and Planetary Interiors*, **103**: 207–245.

Merrill, R.T., and McFadden, P.L., 1999. Geomagnetic polarity transitions. *Reviews of Geophysics* **37**: 201–226.

Niitsuma, N., 1971. Detailed study of the sediments recording the Matuyama-Brunhes geomagnetic reversal., *Science Reports of the Tohuku University Second Series (Geology)*, **43**: 1–39.

Oda, H., Shibuya, H., and Hsu, V., 2000. Paleomagnetic records of Brunhes/Matuyama polarity transition from Ocean Drilling Program Leg 124 (Celebes and Sulu Seas). *Geophysical Journal International*, **142**: 319–338.

Rochette, P., 1990. Rationale of geomagnetic reversals versus remanence recording processes in rocks: a critical review. *Earth and the Planetary Science Letters*, **98**: 33–39.

Singer, B.S., *et al.*, 2005. Structural and temporal requirements for geomagnetic field reversal deduced from lava flows. *Nature*, **434**: 633–636.

Singer, B.S., and Pringle, M.S., 1996. The age and duration of the Matuyama-Brunhes geomagnetic polarity reversal from $^{40}Ar/^{39}Ar$ incremental heating analyses of lavas. *Earth and the Planetary Science Letters*, **139**: 47–61.

Tauxe, L., *et al.*,1996. Astronomical calibration of the Matuyama-Brunhes boundary: consequences for magnetic remanence acquisition in marine carbonates and the Asian loess sequences. *Earth and the Planetary Science Letters*, **140**: 133–146.

Valet, J.P., Tauxe, L., and Clark, D.R., 1988. The Matuyama-Brunhes transition recorded from Lake Tecopa sediments (California). *Earth and the Planetary Science Letters*, **87**: 463–472.

Valet, J.P., *et al.* Palaeomagnetic constraints on the geometry of the geomagnetic field during reversals. *Nature*, **356**: 400–407.

Cross-references

Geomagnetic Polarity Reversals
Geomagnetic Polarity Reversals, Archives
Geomagnetic Polarity Reversals, Observations
Polarity Transitions: Radioisotopic Dating

POLARITY TRANSITIONS: RADIOISOTOPIC DATING

The Plio-Pleistocene Geomagnetic Polarity Time Scale (GPTS) developed beginning in the early 1960s as Cox *et al.* (1963) and McDougall and Tarling (1963) used K-Ar ages and magnetic remanence data from volcanic rocks to propose a sequence of normal and reverse polarity "epochs" now called Chrons (Opdyke and Channell, 1996). A large set of K-Ar ages were acquired, mainly from lavas, that defined the lengths of Chrons, but if few, these lavas showed transitional directions of remanent magnetization, the timing of the polarity reversals between the Chrons had to be interpolated. This was done using the chronogram method (Cox and Dalrymple, 1967) to minimize the error function of K-Ar ages near the polarity boundary relative to assumed ages for the boundary. In addition to the polarity Chrons, that are periods on the order of 10^6 year in length, K-Ar ages and magnetic data from widely dispersed locations began to delineate several $10^4 - 10^5$ year long subchrons including the Jaramillo, Olduvai, and Reunion subchrons (Opdyke and Channell, 1996).

Still shorter-lived geomagnetic events or Cryptochrons, that include excursions, aborted reversal attempts, or rapid back-to-back reversals, were also detected magnetically in sediments and volcanic rocks. K-Ar dating of normally magnetized basalt flows in Iceland (McDougall and Wensink, 1966) and a transitionally magnetized rhyolite in California (Mankinen *et al.*,1978) provided the first radioisotopic ages for events within the Matuyama reversed Chron, named the Gilsa and Cobb Mountain events. In the Brunhes normal chron, several transitionally magnetized lavas in the Massif central, France that recorded the Laschamp event have been K-Ar dated, but these ages are imprecise and scatter between 60 and 15 ka (Guillou *et al.*, 2004). In a remarkable study, Hall and York (1978) obtained both K-Ar and $^{40}Ar/^{39}Ar$ ages from these lavas and concluded that the Laschamp event occurred about 47 ka; it marked the first use of the $^{40}Ar/^{39}Ar$ method to date lavas this young. The K-Ar age of 565 ± 28 ka (uncertainties herein are $\pm 2\sigma$ analytical errors) determined from two of four transitionally magnetized lava flows found in Idaho by Champion *et al.* (1988) provided a precise radioisotopic date for an event during the Brunhes normal chron which was named the Big Lost event for a nearby river. Notably, Champion *et al.* (1988) discussed the evidence from sediments and volcanic rocks globally that at least 10 geomagnetic events of short duration had occurred during the last 1 Ma.

More recently, as directional excursions associated with brief periods of low relative paleointensity were revealed in sediments worldwide, many of these geomagnetic events have been verified, and new ones proposed (Langereis *et al.*, 1997; Guyodo and Valet, 1999; Channell, 1999; Channell *et al.*, 2002). It is the promise of an accurate, high resolution correlation of these events globally that sparked considerable interest and debate concerning their number and precise timing. There are at least two reasons for this: First, marine and terrestrial sediments contain a rich, quasi continuous, archive of major climate changes that have shaped the Earth's surface over the past several Ma. An accurate calendar of changes in climate is crucial to determining the underlying driving mechanisms in the oceans and atmosphere. Second, theoretical arguments and numerical models have used the frequency and duration of reversals and excursions to infer magnetic or thermal interactions between the liquid outer core and either the solid inner core, or the lowermost mantle, that act to control the stability of the geodynamo (e.g., Gubbins, 1999; see *Geodynamo, numerical simulations; Core-mantle coupling, thermal*). Thus, an accurate accounting of spatial temporal instabilities of the magnetic field, from excursions to reversals, is critical in advancing the next generation of geodynamo models (Singer *et al.*, 2002, 2005).

In sediments, the timing of an event may be estimated by astronomical tuning of oxygen isotope, or other variations that proxy for orbitally-driven changes in climate (e.g., Langereis *et al.*, 1997; Channell *et al.*, 2002). However, not all marine core is amenable to oxygen isotope measurements, few geomagnetic records have been astronomically tuned, and even the sediment that accumulates at the highest rates may contain hiatuses leading to underestimates of the amount of time actually recorded. In detail, magnetic recordings in sediment are further complicated by postdepositional processes, including lock-in of the magnetic remanence, such that an accurate chronology may be elusive (Merrill and McFadden, 1999). Alternatively, $^{40}Ar/^{39}Ar$ dating of lava flows or tuffs that rapidly lock in a thermal magnetic remanence during cooling has recently made it possible to obtain precise radioisotopic ages for at least 14 reversals and excursions, including previously unrecognized events that occurred during the past 2.2 Ma (Singer *et al.*, 2002, 2004).

$^{40}Ar/^{39}Ar$ variant of K-Ar dating

Until the early 1990's K-Ar dating was the preferred method for determining ages of volcanic rocks. The $^{40}Ar/^{39}Ar$ variant offers many advantages over K-Ar dating and coupled with more sensitive measurement capabilities in such a way it has eclipsed K-Ar dating as the method of choice. The history, theory, and application of these techniques are summarized by McDougall and Harrison (1999), Renne (2000), and Kelley (2002); hence only a brief outline, drawn from these sources, is presented here. Both methods rely on the radioactive decay of naturally occurring ^{40}K to ^{40}Ar, which is proportional at any time t to the N number of ^{40}K atoms present:

$$\frac{dN}{dt} = -\lambda N \qquad \text{(Eq. 1)}$$

where λ is the total decay of ^{40}K. A large fraction, 89.52%, of ^{40}K decays to ^{40}Ca via beta emission ($\lambda_\beta = 4.962 \times 10^{-10}$ year), whereas 10.48% decays to ^{40}Ar via electron capture ($\lambda_\epsilon = 5.808 \times 10^{-11}$ year) so that the total decay constant ($\lambda = 5.543 \times 10^{-10}$ year) corresponds to a half-life of 1.25×10^9 year, Solving the differential equation (1) yields the age equation for the K-Ar isotope system:

$$t = \left(\frac{1}{\lambda}\right) \cdot \ln\left[1 + \left(\frac{\lambda}{\lambda_\epsilon}\right) \cdot \left(\frac{^{40}Ar^*}{^{40}K}\right)\right] \qquad \text{(Eq. 2)}$$

where $^{40}Ar^*$ is radiogenic argon, distinguished from other sources such as the atmosphere. The ratio of modern $^{40}K/K$ is constant, hence measurement of K and Ar concentrations (in mol/g), and isotope ratios of Ar, allow the age to be calculated. In practice, the sample is completely fused, the gas released is purified and spiked with a known amount of ^{38}Ar tracer, then its isotopic composition is measured in a mass spectrometer. K is a salt and must be measured independently on a separate aliquot of the same sample. Alternatively, $^{40}Ar^*$ may be measured directly by using a unique mass spectrometer which functions as a sensitive manometer able to distinguish minute quantities of radiogenic argon from the large atmospheric component found in most lavas (e.g., Guillou *et al.*, 2004).

One of the more important assumptions of any radioisotopic dating system, including the K-Ar and $^{40}Ar/^{39}Ar$ methods, is that the rock or mineral has remained closed to loss or gain of parent and daughter nuclides. For example, loss of Ar or gain of K can yield spuriously young dates, whereas gain of Ar or loss of K may give spuriously old dates. From a K-Ar age determination alone, it is impossible to assess the validity of this assumption. Another limitation of K-Ar dates is their precision, which includes the compound uncertainty of two different analytical procedures used to measure separately the absolute contents in mol/g of Ar and K.

The $^{40}Ar/^{39}Ar$ method is based on the same radioisotopic decay system as the K-Ar method, but instead of measuring K in a separate aliquot of the sample, it is measured by creating ^{39}Ar from ^{39}K via neutron activation. The ^{39}Ar is produced by fast neutron bombardment in a nuclear reactor, and because the ratio of ^{39}K to ^{40}K is constant, it becomes a proxy for ^{40}K that allows the $^{40}Ar^*/^{40}K$ ratio to be calculated simply from the measurement of the ratio of two Ar isotopes $^{40}Ar/^{39}Ar$.

Production of ^{39}Ar atoms is given by

$$^{39}\mathrm{Ar} = {^{39}\mathrm{K}}\Delta \int \phi(\mathrm{E})\sigma(\mathrm{E})\mathrm{d}E \quad \text{(Eq. 3)}$$

where ^{39}K is the number of atoms originally present, Δ the duration of irradiation, $\phi(E)$ the neutron flux density at energy, E, and $\sigma(\mathrm{E})$ the neutron capture cross section of ^{39}K for neutrons of energy E in the $^{39}K(n,p)^{39}Ar$ reaction. Solving Eq. (2) for $^{40}Ar^*$ and combining with Eq. (3) yields:

$$\left(\frac{^{40}\mathrm{Ar}^*}{^{39}\mathrm{K}}\right) = \frac{\left(\frac{^{40}\mathrm{K}}{^{39}\mathrm{K}}\right)\left(\frac{\lambda_\epsilon}{\lambda}\right)\left(\frac{1}{\Delta}\right)(e^{\lambda t}-1)}{\Delta \int \phi(\mathrm{E})\sigma(\mathrm{E})\mathrm{d}E} \quad \text{(Eq. 4)}$$

This can be simplified by defining a dimensionless irradiation parameter J:

$$J = \left(\frac{^{39}\mathrm{K}}{^{40}\mathrm{K}}\right)\left(\frac{\lambda}{\lambda_\epsilon}\right)\Delta \int \phi(\mathrm{E})\sigma(\mathrm{E})\mathrm{d}E \quad \text{(Eq. 5)}$$

which allows Eq. (4) to be written as

$$\left(\frac{^{40}\mathrm{Ar}^*}{^{39}\mathrm{Ar}}\right) = \frac{(e^{\lambda t}-1)}{J} \quad \text{(Eq. 6)}$$

and rearranged to yield the age equation for $^{40}Ar/^{39}Ar$ dating:

$$t = \left(\frac{1}{\lambda}\right) ln\left[1 + J \cdot \left(\frac{^{40}\mathrm{Ar}^*}{^{39}\mathrm{Ar}}\right)\right] \quad \text{(Eq. 7)}$$

The value for the neutron fluence parameter J is determined by irradiating a standard mineral of known age t together with the samples, measuring its $^{40}Ar^*/^{39}Ar$ ratio, and solving Eq. (6) for J. In this sense, the $^{40}Ar/^{39}Ar$ method is a relative dating technique that relies on high-quality, homogenous standards that are available widely. There are many neutron fluence standards in use, however, not all laboratories use the same standards, nor does each use identical ages for some standards. Therefore, it is critical when comparing $^{40}Ar/^{39}Ar$ ages determined in various laboratories to normalize the ages to a common value, based on intercalibration of several standards (see Renne, 2002). For example, all the $^{40}Ar/^{39}Ar$ ages discussed below are calculated or normalized relative to an age of 28.02 Ma for the widely used and intercalibrated Fish Canyon Tuff sanidine standard.

The amount of $^{40}Ar^*$ measured must be corrected for the presence of atmospheric argon which has a $^{40}Ar/^{36}Ar$ ratio of 295.5. This is done using the relationship $(^{40}\mathrm{Ar}^*/^{39}\mathrm{Ar}) = (^{40}\mathrm{Ar}^*/^{39}\mathrm{Ar})_m - 295.5(^{36}\mathrm{Ar}/^{39}\mathrm{Ar})_m$, where the subscript m denotes the measured ratio. Other corrections for Ar isotopes produced by undesirable nuclear reactions during irradiation are also important, particularly for Plio-Pleistocene samples (Renne, 2000).

The fundamental difference between the K-Ar and $^{40}Ar/^{39}Ar$ dating methods is that the latter allows calculation of an age based solely on the isotopic composition of Ar in the sample, rather than absolute abundances of Ar and K, thereby eliminating major sources of potential error. The $^{40}Ar^*/^{39}Ar$ ratio is determined using highly sensitive mass spectrometry, thus the total fusion age for a Plio-Pleistocene sample of 10 mg of basalt groundmass or a single 1 mg phenocryst of sanidine can be determined to a precision of ~1–2%. Systematic uncertainties, including a potential uncertainty in the ^{40}K decay constant currently in use, may be important when comparing $^{40}Ar/^{39}Ar$ ages with those determined by other methods. This contribution to the total uncertainty associated with a particular age determination may be small, amounting to only a few hundred to a few thousand years for Pleistocene lava flows, but may be important when comparing $^{40}Ar/^{39}Ar$ ages of geomagnetic excursions or reversals to those obtained astronomically from marine sediment (Guillou *et al.*, 2004).

Age spectra

Because measurement of the $^{40}Ar^*/^{39}Ar$ ratio does not require absolute abundances, another powerful aspect of the $^{40}Ar/^{39}Ar$ method is that rather than totally fusing an irradiated sample, it can be incrementally degassed by stepwise heating to higher temperatures using a controllable furance or laser. The apparent ages of each successive gas increment are monitored for consistency by plotting the ages vs. the cumulative release of ^{39}Ar, as a proxy for the parent isotope ^{40}K (see Figure P55). If successive gas increments yield consistent ages, the spectrum is concordant Figure P55a, reflecting an internally homogenous distribution of $^{40}Ar^*/^{39}Ar$, thereby providing evidence against mobility of K and Ar. The "plateau" age is commonly defined as the

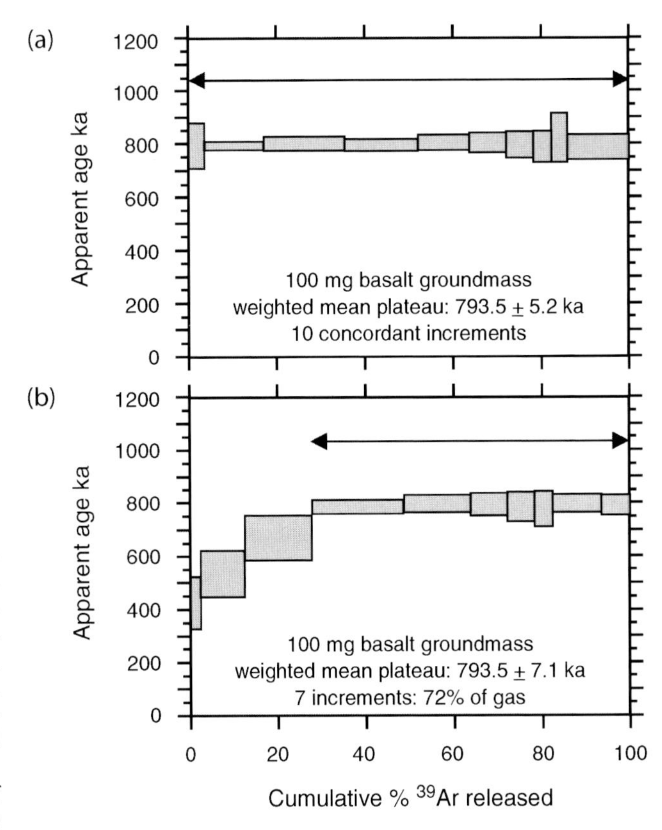

Figure P55 Age spectrum diagrams for two basaltic lava flows erupted during the Matuyama-Brunhes transition. Vertical height of each gas step gives the 2σ analytical uncertainty. (a) Concordant age spectrum. The ten gas steps were released by successively raising the temperature of a resistance furnace in 50°C increments between 700 and 1200°C. (b) Discordant age spectrum produced by the same procedure used in A. The plateau age is calculated from seven contiguous steps that comprise 72% of the gas released. The gas released at lower temperatures gives ages younger than the plateau, possibly reflecting argon loss during weathering. In both A and B, the isotopic composition of the plateau steps may be used to calculate an isochron (Figure P56).

mean of the concordant step ages, each step being weighted by its inverse variance. Though arbitrary, most definitions of a plateau specify that it must include at least three successive increments that comprise $\geq 50\%$ of the ^{39}Ar released that are concordant in age at the 95% confidence level. Weighting each plateau increment (or individual total fusion age) by its inverse variance $(1/\sigma_i^2)$, reduces the uncertainty of the mean age t according to

$$\sigma_t = \sqrt{\sum \left(\frac{1}{\sigma_i^2} \right)} \qquad \text{(Eq. 8)}$$

In this manner, the plateau age may approach a precision of 0.2%, depending on the number of concordant increments and the $^{40}Ar^*$ content of the material.

It is common, particularly with Plio-Pleistocene lavas, that age spectra are discordant. Notwithstanding, discordant spectra may yield a plateau, and depending on the nature of the disturbance, may provide important evidence for mobility of Ar or K. Figure P55b illustrates an example from a Pleistocene basalt interpreted to have lost a small fraction of $^{40}Ar^*$ due to low temperature weathering of the matrix.

Isochrons

It is routine in the $^{40}Ar/^{39}Ar$ method to measure five isotopes of Ar including ^{36}Ar, ^{39}Ar, ^{40}Ar, ^{37}Ar, ^{38}Ar; the latter two are required to correct for undesirable quantities of ^{40}Ar and ^{36}Ar produced artificially from K and Cl in the sample during irradiation. The other three isotopes facilitate isochron analysis; the most commonly used isochron is the plot of $^{36}Ar/^{40}Ar$ vs. $^{39}Ar/^{40}Ar$ where the intercepts yield reciprocals of $^{40}Ar^*/^{39}Ar$, hence age (from Eq. (7)), and $(^{40}Ar/^{36}Ar)_i$ which gives the initial or trapped component (see Figure P56). Isochrons are calculated by a least squares method that weights individual measurements by their inverse variance. Incremental heating experiments on Plio-Pleistocene volcanic rocks are particularly amenable to isochron analysis. In some instances isochrons are more powerful than the age spectrum approach, because no *a priori* assumption regarding the $(^{40}Ar/^{36}Ar)_i$ is made, which can be important for the small number of lavas or tuffs that may contain initially a nonatmospheric (or excess) component of argon (e.g., Singer and Brown, 2002). Indeed, the ability to use incremental heating data to test the assumptions of (1) closed system behavior and (2) an initial $(^{40}Ar/^{36}Ar)_i$ ratio corresponding to the atmosphere, makes it the preferred method for precisely dating lavas and tuffs thought to record geomagnetic reversals and events.

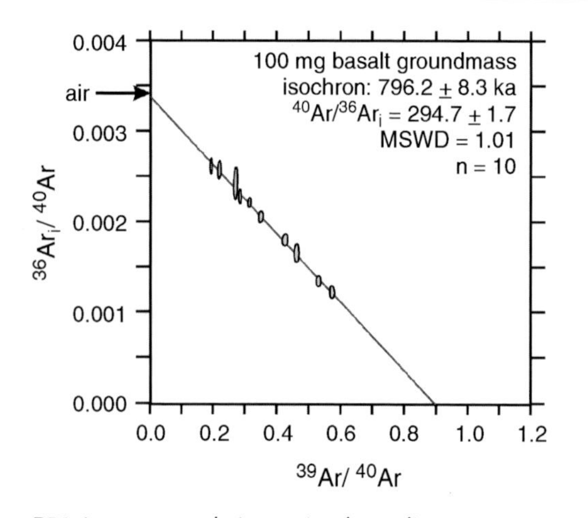

Figure P56 Isotope correlation or isochron diagram. Intercepts of the isochron in this diagram give the reciprocal of $^{40}Ar/^{39}Ar$ and $^{40}Ar/^{36}Ar_i$. Hence the isochron is a mixing line between the radiogenic argon and the initial or trapped component. The data points shown correspond to the ten gas increments in Figure P55a. In this example, the trapped component is indistinguishable from air argon. MSWD is the mean square weighted deviation, which indicates the magnitude of scatter about the isochron. A value of 1.0 means that the analytical uncertainties account for most of the error in the age which is calculated using least squares regression that weights each data point by the inverse of its variance.

Direct $^{40}Ar/^{39}Ar$ dating of polarity transitions and excursional events

Matuyama-Brunhes transition

The first direct radioisotopic date for the Matuyama-Brunhes polarity transition corresponding to the last reversal of Earth's magnetic field is based on $^{40}Ar/^{39}Ar$ isochrons from four incremental heating experiments by Baksi *et al.* (1992) on three lavas at Haleakala volcano on the island of Maui that possess transitional directions of magnetic remanence. Baksi *et al.*'s age of 795 ± 16 ka is ca. It was 65 ka older than the K-Ar based age for this reversal, but within error of the astronomical estimate of 780 ka. Singer and Pringle (1996) incrementally heated groundmass separates from eight basaltic to andesitic lavas that erupted during the M-B transition at volcanoes in Chile, and on the islands of Tahiti, La Palma, and Maui, including several replicate experiments to improve precision of the isochrons. Although Singer and Pringle (1996) determined the weighted mean age of these eight lavas at 791 ± 4 ka, they also emphasized that the 12 ka difference in age between the Chilean lavas and those from Maui may reflect the duration of this reversal. To further investigate the duration and structure of the M-B transition, Singer *et al.* (2005a) obtained 71 $^{40}Ar/^{39}Ar$ incremental heating experiments on 23 of the transitionally magnetized lavas from the four volcanoes above. Surprisingly, the mean $^{40}Ar/^{39}Ar$ ages of the lava sequences on La Palma, Tatara San Pedro, and Tahiti were found to cluster at 793 ± 3 ka, whereas the age of the lavas at Haleakala, 776 ± 2 ka, is about 18 ka younger. Singer *et al.* (2005a) propose that the older lavas record the onset of geodynamo instability and nondipolar field behavior, which persisted for a long enough period of time that magnetic flux diffused out of the solid inner core, thereby weakening its stabilizing effect and allowing the field to reverse itself at 776 ± 2 ka. This interpretation is bolstered, in part, by the paleointensity records from marine sediments, in which are observed a pronounced drop in intensity about 15 ka prior to the intensity low associated with the field reversal itself. Singer *et al.* (2005a) argue that these lava flows provide the first observational evidence in support of Gubbins' (1999) hypothesis that the solid inner core may control the frequency of excursions and probability of full field reversals.

Dating other reversals and events toward a geomagnetic instability timescale (GITS)

The K-Ar age of the Cobb Mountain event, 1120 ± 40 ka (Mankinen *et al.*, 1978), was revised to 1194 ± 12 ka on the basis of $^{40}Ar/^{39}Ar$ laser fusion and furnace incremental heating experiments on sanidine by Turrin *et al.* (1994). The 6% to 7% increase in age was attributed to incomplete degassing of sanidine during the K-Ar experiments, a problem that may have biased many K-Ar dates used to develop the original GPTS, including those used in the chronogram estimate of 730 ka for the M-B transition.

Isochron ages for 18 lavas within long basalt flow sequences at Punaruu Valley, Tahiti, and Haleakala volcano, Maui, known to record several polarity transitions and events were determined using the $^{40}Ar/^{39}Ar$ incremental heating method on groundmass separated by Singer *et al.* (1999). The results indicate that an event recorded 1122 ± 10 ka at Tahiti, which was originally correlated with the Cobb Mountain event on the basis of K-Ar dating, is actually a previously

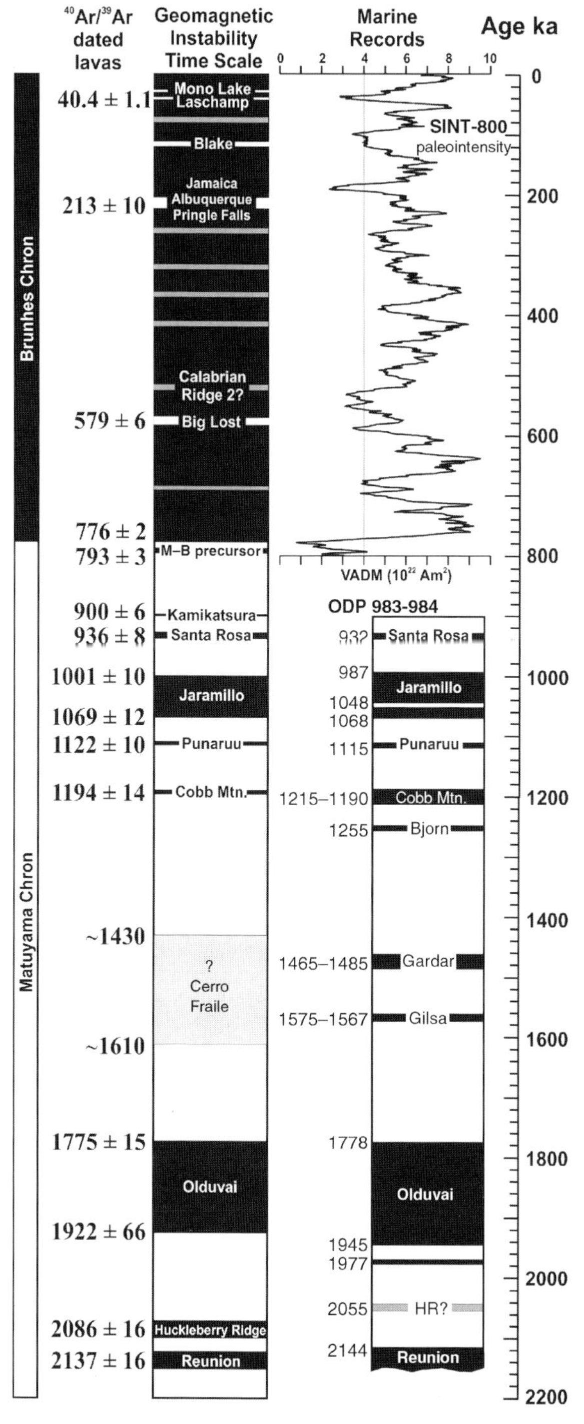

Figure P57 A Geomagnetic Instability Time Scale (GITS) for the Matuyama reversed and Brunhes normal Chrons. The ^{40}Ar/^{39}Ar ages of transitionally magnetized lava flows or tuffs recording reversals or excursional events are from Singer *et al.* (1999, 2002, 2004, 2005a,b), Singer and Brown (2002), and references therein. Postulated events in grey are undated, imprecisely dated, or less well documented magnetically (Singer *et al.*, 2002). The astronomically-dated record of reversals and excursions includes SINT-800, a global compilation of geomagnetic field intensity from 33 marine sediment records over the last 800 ka (Guyodo and Valet, 1999), and the composite record from sediment cored at Ocean Drilling Program sites 983 and 984 (Channell *et al.*, 2002). In the latter, paleointensity lows that correspond with equator crossing shifts in the direction of the virtual geomagnetic pole are dated using an astronomical age model of oxygen isotope variations in the cores. Note that several paleointensity lows in SINT-800 do not readily correlate to events that have been radioisotopically dated or assigned astronomical ages elsewhere. Similarly, although many radioisotopically dated events match those found in the ODP 983 and 984 cores, several paleointensity lows, including some not shown here, have yet to be found in lavas that have been ^{40}Ar/^{39}Ar dated. Note that all ^{40}Ar/^{39}Ar ages are calculated relative to an age of 28.02 Ma for the Fish Canyon tuff sanidine standard.

unrecognized event, now named the Punaruu event. The results of Singer *et al.* (1999) also provided for the first time,the direct radioisotopic ages for the onset and termination of the Jaramillo normal subchron at 1069 ± 12 and 1001 ± 10 ka, respectively, which are in agreement with astronomical estimates (see Figure P57). Between the Jaramillo subchron and the M-B boundary, eight transitionally magnetized lavas from Tahiti and Halekala yielded a common weighted mean ^{40}Ar/^{39}Ar age of 900 ± 6 ka, corresponding to Champion *et al.*'s (1988) Kamikatsura event (Singer *et al.*, 1999; Figure P57).

Singer *et al.* (1999) suggested that the Kamikatsura event shortly postdated another previously unrecognized event recorded by the Santa Rosa I rhyolite dome in the Valles Caldera, New Mexico. To verify this, Singer and Brown (2002) undertook ^{40}Ar/^{39}Ar laser fusion and incremental heating experiments on sanidine from the transitionally magnetized Santa Rosa I rhyolite dome and determined its age to be 936 ± 8 ka; that is, the Santa Rosa event is 37 ka older than the Kamikatsura event and 65 ka younger than the reversal that terminated the Jaramillo Subchron Figure P57. At Cerro del Fraile in southern Argentina, Singer *et al.* (2004) have ^{40}Ar/^{39}Ar dated a sequence of 10 basaltic lavas that, in concert with dates from lavas elsewhere, provide precise radioisotopic ages for geomagnetic events marking termination of the Reunion Subchron at 2137 ± 16 ka, plus the onset and termination of the Olduvai Subchron at 1922 ± 66 and 1175 ± 15 ka. A single transitionally magnetized lava gave a discordant age spectrum that, together with a K-Ar age determination, indicate an age between about 1610 and 1430 ka (Figure P57). The termination of the Reunion Subchron is temporally distinguished by about 50 ka from the Huckleberry Ridge event recorded in tuff at Yellowstone national park, Wyoming (Singer *et al.*, 2004).

The paleomagnetic and ^{40}Ar/^{39}Ar data of Singer *et al.* (2002) revealed that two geomagnetic events, including the one that preceded the M-B reversal by 18 ka, and a younger event dated at 580 ± 8 ka are recorded in lavas on the northeast flank of La Palma island. The latter correspond to the Big Lost event that was K-Ar dated in Idaho at 565 ± 28 ka by Champion *et al.* (1988). Moreover, Hoffman and Singer (2004) discovered a third recording of the Big Lost event in lavas at Tahiti, clearly demonstrating the global nature of this excursion at 579 ± 6 ka (Figure P57). Lava flows of the Albuquerque Volcanoes, New Mexico along with a sequence of lacustrine sediments and volcanic ash at Pringle Falls, Oregon, that both record excursional field behavior have been ^{40}Ar/^{39}Ar dated at 217 ± 17 and 211 ± 13 ka, respectively (Singer *et al.*, 2005b). The common age from the two sites suggests that they record the same period of geomagnetic instability with a weighted mean age of 213 ± 10 ka. Using both ^{40}Ar/^{39}Ar and unspiked K-Ar methods, two lava flows in the Massif Central, France that record the Laschamp excursion have recently been dated to 40.4 ± 1.1 ka, which is about 10% younger than previous estimates discussed earlier (Guillou *et al.*, 2004). This new age is indistinguishable from that of a prominent paleointensity low at 40.9 ka found in marine sediments of the North Atlantic Ocean whose history has been calibrated using an age model based on oxygen isotopes and annual varve counting in the Greenland Ice Sheet Project II (GISP2) ice core. Thus, despite uncertainties in the decay constant for ^{40}K, which may be as large as 2.4%, the ^{40}Ar/^{39}Ar method is capable of determining ages with an accuracy and precision of between 1% and 2% for virtually the entire Pleistocene (Guillou *et al.*, 2004).

Comparison of these radioisotopically dated events with more continuous records of paleointensity and paleodirection recovered from marine sediments has helped to (1) verify that excursions mark frequent instability of the geodynamo, (2) test the astronomical age estimates of many events, and (3) better interpret long-term temporal changes in the field. Perhaps the highest resolution recording of the field during the Matuyama reversed Chron comes from drift sediment deposited at high rates at Ocean Drilling Program (ODP) sites 983 and 984 in the Iceland basin (Channell *et al.*, 2002). Hydraulic piston cores from these sites revealed changes in the direction of the magnetic field corresponding to at least seven excursional events and six reversals bounding subchrons (Figure P57). Each of these directional shifts is associated with a low in the relative intensity of the magnetic field, but there are also other paleointensity lows not associated with prominent directional excursions (Channell *et al.*, 2002). ODP 983 and 984 cores record paleointensity lows corresponding to the Santa Rosa and Punaruu events, and to reversals bounding the Jaramillo, Cobb Mountain, Olduvai, and Reunion subchrons, each with an astronomical age that is within an error of the radioisotopic ages (Figure P57). Alternatively, a few events found in the ODP 983 and 984 cores, for example the Bjorn, Gardar, and Gilsa events, have yet to be $^{40}Ar/^{39}Ar$ dated in lavas, although the transitionally magnetized lava from Cerro del Fraile that is poorly dated between 1610 and 1430 ka may correspond to one of the latter two events (Figure P57). In the site 983 core, a paleointensity low accompanying an equator crossing of the virtual geomagnetic pole direction was not highlighted by Channell *et al.* (2002), but the astronomically estimated age of 2055 ka is close to the $^{40}Ar/^{39}Ar$ age of the Huckleberry Ridge event (Figure P57).

For the Brunhes normal Chron we have the global compilation of paleointensity data known as SINT-800 that indicates at least eight periods when the virtual axial dipole moment (VADM) is dropped below about 4×10^{22} Am^2 (Guyodo and Valet, 1999). Three of these events, astronomically dated at ca. 590, 190, and 40 ka, broadly correspond to the Big Lost, Jamaica, and Laschamp excursions. However, closer inspection reveals that the $^{40}Ar/^{39}Ar$ ages for the Big Lost and Albuquerque Volcanoes/Pringle Falls excursions are several ka older than first-order intensity lows recorded in the sediment (Figure P57). The SINT-800 paleointensity record is based on 33 sediment cores, no two of which preserve identical patterns of paleointensity (Guyodo and Valet, 1999). Given uncertainties associated with generating astrochronologic age models, and a limited understanding of how magnetic remanence acquisition may be delayed or distorted in marine sediments, these temporal mismatches suggest that caution be used when attempting to correlate the major paleointensity lows across the globe. Alternatively, dating of transitionally magnetized lava flows relies upon well understood process of radioisotopic decay and thermal magnetic remanence.

$^{40}Ar/^{39}Ar$ dating of lava flows which directly record brief periods of geomagnetic field instability, coupled with recognition of these events in marine sediment, indicate that both excursions and reversals reflect intrinsic behavior of the geodynamo. Singer *et al.* (2002) proposed that when the stability of the geodynamo is considered, rather than the lengths of polarity chrons, an alternative approach to the GPTS is needed. Hence, a new $^{40}Ar/^{39}Ar$-based Geomagnetic Instability Time-Scale (GITS) for the last 2.2 Ma is under construction (Figure P57). Experience suggests that many geomagnetic events remain to be discovered or verified as more lava flow sequences are studied in detail.

Brad S. Singer

Bibliography

Baksi, A.K., Hsu, V., McWilliams, M.O., and Farrar, E., 1992. $^{40}Ar/^{39}Ar$ dating of the Brunhes-Matuyama geomagnetic field reversal. *Science*, **256**: 356–357.

Champion, D.E., Lanphere, M.A., and Kuntz, M.A., 1988. Evidence for a new Geomagnetic reversal from lava flows in Idaho: Discussion of short polarity reversals in the Brunhes and Late Matuyama Polarity Chrons. *Journal of Geophysical Research*, **93**: 11667–11680.

Channell, J.E.T., 1999. Geomagnetic paleointensity and directional secular variation at Ocean Drilling Program (ODP) Site 984 (Bjorn Drift) since 500 ka: Comparisons with ODP Site 983 (Gardar Drift). *Journal of Geophysical Research*, **104**: 22937–22951.

Channell, J.E.T., Mazaud, A., Sullivan, P., Turner, S., and Raymo, M.E., 2002. Geomagnetic excursions and paleointensities in the Matuyama Chron at Ocean Drilling Program Sites 983 and 984 (Iceland Basin). *Journal of Geophysical Research*, **107**, doi: 10.1029/2001JB000491.

Cox, A., Doell, R.R., and Dalrymple, G.B., 1963. Geomagnetic polarity epochs and Pleistocene geochronometry. *Nature*, **198**: 1049–1051.

Cox, A., and Dalrymple, G.B., 1967. Statistical analysis of geomagnetic reversal data and the precision of potassium-argon dating. *Journal of Geophysical Research*, **72**: 2603–2614.

Gubbins, D., 1999. The distinction between geomagnetic excursions and reversals. *Geophysical Journal International*, **137**: F1–F3

Guillou, H., Singer, B.S., Laj, C., Kissell, C., Scalliet, S., and Jicha, B.R., 2004. On the age of the Laschamp Event. *Earth and Planetary Science Letters*, **227**: 331–343.

Guyodo, Y., and Valet, J.-P., 1999. Global changes in intensity of the Earth's magnetic field during the past 800 ka. *Nature*, **399**: 249–252.

Hall, C.M., and York, D. 1978. K-Ar and $_{40}Ar/_{39}Ar$ age of the Laschamp geomagnetic polarity reversal. *Nature*, **274**: 462–464.

Hoffman, K.A., and Singer, B.S., 2004. Regionally recurrent paleomagnetic transitional fields and mantle processes. In Channell, J.E.T., Kent, D.V., Lowrie, W., and Meert, J., (eds.), *Timescales of the Paleomagnetic Field, American Geophysical Union,* Geophysical Monograph, **145**: 233–243.

Kelley, S., 2002. K-Ar and Ar-Ar Dating. In Porcelli, D.P., Ballentine, C.J., and Wiele, R. (eds.), *Noble Gases. Reviews in Mineralogy and Geochemistry*, The Mineralogical Society of America, Washington, DC, **47**: 785–818.

Langereis, C.G., Dekkers, M.J., de Lange, G.J., Paterne, M., and van Santvoort, P.J.M., 1997. Magnetostratigraphy and astronomical calibration of the last 1.1 Ma from an eastern Mediterranean piston core and dating of short events in the Brunhes. *Geophysical Journal International*, **129**: 75–94.

Mankinen, E.A., Donnelly, J.M., and Gromme, C.S., 1978. Geomagnetic polarity event recorded at 1.1 Ma B.P. on Cobb Mountain, Clear Lake volcanic field, California. *Geology*, **6**: 653–656.

McDougall, I., and Tarling, D.H., 1963. Dating of reversals of the Earth's magnetic fields. *Nature*, **198**: 1012–1013.

McDougall, I., and Wensink, J., 1966. Paleomagnetism and geochronology of Pliocene-Pleistocene lavas in Iceland. *Earth and Planetary Science Letters*, **1**: 232–236.

McDougall, I., and Harrison, T.M., 1999. *Geochronology and thermochronology by the $^{40}Ar/^{39}Ar$ method.* New York: Oxford University Press.

Merrill, R.T., and McFadden, P.T., 1999. Geomagnetic polarity transitions. *Reviews of Geophysics*, **37**: 201–226.

Opdyke, N.D., and Channell, J.E.T., 1996. *Magnetic Stratigraphy.* San Diego:Academic Press.

Renne, P.R., 2000. K-Ar and $^{40}Ar/^{39}Ar$ dating. In Noller, J.S., Sowers, J.M., and Lettis, W.R., (eds.) *Quaternary Geochronology: Methods and Applications.* American Geophysical Union Reference Shelf 4. Washington, DC: American Geophysical Union, pp. 77–100.

Singer, B.S., and Pringle, M.S., 1996. Age and duration of the Matuyama-Brunhes geomagnetic polarity reversal from $^{40}Ar/^{39}Ar$ incremental heating analyses of lavas. *Earth and Planetary Science Letters*, **139**: 47–61.

Singer, B.S., Hoffman, K.A, Chauvin, A., Coe, R.S., and Pringle, M.S., 1999. Dating transitionally magnetized lavas of the late Matuyama Chron: Toward a new $^{40}Ar/^{39}Ar$ timescale of reversals and events. *Journal of Geophysical Research*, **104**: 679–693.

Singer, B.S., Relle, M.R., Hoffman, K.A., Battle, A., Guillou, H., Laj, C., and Carracedo, J.C., 2002. Ar/Ar ages from transitionally magnetized lavas on La Palma, Canry Islands, and the Geomagnetic Instability Timescale. *Journal of Geophysical Research*, **107**: 2307, doi: 10.1029/2001JB001613.

Singer, B., and Brown, L.L., 2002. The Santa Rosa Event: $^{40}Ar/^{39}Ar$ and paleomagnetic results from the Valles Rhyolite near Jaramillo Creek, Jemez Mountains, New Mexico. *Earth and Planetary Science Letters*, **197**: 51–64.

Singer, B.S., Brown, L.L., Rabassa, J.O., and Guillou, H., 2004. $^{40}Ar/^{39}Ar$ chronology of Late Pliocene and early Pleistocene geomagnetic and glacial events in southern Argentina. In Channell, J.E.T., Kent, D.V., Lowrie, W., and Meert, J. (eds.), *Timescales of the Paleomagnetic Field, American Geophysical Union*, Geophysical Monograph **145**: 175–190.

Singer, B.S., Hoffman, K.A., Coe, R.S., Brown, L.L., Jicha, B.R., Pringle M.S., and Chauvin, A., 2005a. Structural and temporal requirements for geomagnetic field reversal deduced from lava flows. *Nature*, **434**: 633–636.

Singer, B.S., Jicha, B.R., Kirby, B.T., Zhang, X., Geissman, J.W., and Herrero-Bervera, E., 2005b. An $^{40}Ar/^{39}Ar$ age for geomagnetic instability recorded at the Albuquerque Volcanoes and Pringle Falls, Oregon. *EOS, Transactions of the American Geophysical Union* 86 (52), abstract GP21A–0019.

Turrin, B.D., Donnelly-Nolan, J.M., and Hearn, B.C. 1994. $^{40}Ar/^{39}Ar$ ages from the rhyolite of Alder Creek, California: Age of the Cobb Mountain Normal Polarity Subchron revisited. *Geology*, 22: 251–254.

Cross-references

Core-Mantle Coupling, Thermal
Geodynamo, Numerical Simulations
Geomagnetic Excursion

POLE, KEY PALEOMAGNETIC

Introduction

Using paleomagnetism to reconstruct continents in the Precambrian has proven to be exceedingly difficult despite a database of many hundreds of paleomagnetic poles from around the world. The most serious problem involves the large age uncertainties associated with most Precambrian paleopoles. This is due to a number of factors, most notably the poor age constraints on the rock-units from which the paleopoles are derived and magnetic overprinting during metamorphic events.

Roy (1983) first pointed out that most Precambrian paleopoles were too poorly dated, with uncertainties of tens or even hundreds of millions of years, to be of use in defining apparent polar wander paths (APWPs). Buchan and Halls (1990) suggested rigorous criteria to identify individual paleopoles that are sufficiently well defined and well dated that they are useful for defining apparent polar wander paths or reconstructing paleocontinents. Such paleopoles were termed "key paleopoles" by Buchan *et al.* (1994).

Criteria of a key paleomagnetic pole

The following basic criteria for a key paleomagnetic pole are adapted from Buchan and Halls (1990) and Buchan *et al.* (2001).

(a) *Age of the paleopole*: The paleopole should be demonstrated to be primary and the rock-unit precisely and accurately dated.
(b) *Quality of the paleopole*: The primary paleomagnetic remanence should be properly isolated, secular variation averaged out, and, where necessary, correction made for crustal tilting.

The critical importance of the age criterion cannot be overemphasized; *only* well dated paleopoles should be considered as key paleopoles. Field tests are required to demonstrate that the remanence is primary. For example, igneous rocks can be shown to carry a primary remanent magnetization using the baked contact test (see *Baked contact test*). A Precambrian rock-unit must have a radiometric age with an uncertainty of $< \pm 20$ Ma (Buchan *et al.*, 2001). At present, only U-Pb and Ar-Ar dating meet this requirement. In the future, with improved analytical techniques it should be possible to strengthen this criterion and require age uncertainties $< \pm 5$ Ma.

It is important to ensure that the age and the paleomagnetic remanence are actually derived from the same geological unit. Thus, the geochronological and paleomagnetic studies should be integrated and carried out at the same sampling localities wherever possible (e.g., Buchan *et al.*, 1994).

The direction of the paleomagnetic remanence used to calculate the paleomagnetic pole must accurately represent the direction of the Earth's magnetic field. In particular, the paleopole itself must be properly isolated to ensure that secondary magnetizations have been eliminated. This requires the use of stepwise alternating field and thermal demagnetization techniques. Secular variation must be averaged out. For example, in the case of rapidly cooling igneous units such as volcanic flows, several different units must be sampled. The paleopole must be corrected if crustal tilting has occurred since remanence acquisition.

Apparent polar wander path method vs. comparison of key paleomagnetic poles

The standard practice of comparing APWPs for different continental blocks in order to determine their relative movements often breaks down in the Precambrian. At present, there are simply too few key paleopoles from most continental blocks to allow meaningful APWPs to be constructed. In an attempt to circumvent this problem, it has been standard practice to make use of large numbers of poorly dated paleopoles. In some cases, the large uncertainties in the age of most of the paleopoles are disregarded. In other cases, "grand mean" paleopoles are calculated from the poorly constrained data and used as tie-points along the APWP. In still other cases, the APWP is simply interpolated over relatively long time intervals between existing individual paleopoles or grand mean paleopoles.

However, as noted above, most Precambrian paleopoles are too poorly dated to allow them to be properly sequenced along APWPs. Long time gaps in the paleopole record from a given continental block make it difficult to interpolate between them. In addition, interpolation between widely separated paleopoles is hindered by an ambiguity in magnetic polarity that results from the lack of continuous APWPs from the Precambrian to the Present. The unreliability of Precambrian APWPs based on poorly dated paleopoles was illustrated by Buchan *et al.* (1994), who used key paleopoles to demonstrate that the widely used Paleoproterozoic APWP for the Superior Province of the Canadian Shield was invalid.

The extent of the dating problem has been described in recent reviews by Buchan *et al.* (2000) and Buchan *et al.* (2001), who examined the Baltica-Laurentia database for the Proterozoic and the global database for the period 1700–500 Ma, respectively. They concluded that only a very few Proterozoic paleopoles are sufficiently well dated to pass the age criterion for a key paleopole. Although many paleopoles in the database are well defined, their ages are uncertain. Either they do not have a rigorous field test to determine that the remanence is primary, or the rock-unit from which they have been derived is undated or poorly dated. Most of the key paleopoles that were identified come from a single continental block, Laurentia. A few are available from Baltica. More recent work has yielded key paleopoles for Australia (e.g., Wingate and Giddings, 2000; Pisarevsky *et al.*, 2003). Unfortunately, there are as yet no key Proterozoic paleopoles that can be used for reliable APWP construction from most other continental blocks.

The comparison of APWPs derived from different continental blocks is the ideal method of determining if continental blocks were drifting as a unit or separately. However, Buchan and Halls (1990) and Buchan *et al.* (1994) suggested that key paleopoles only be connected to form a segment of an APWP if their ages are within 30 Ma.

Instead, Buchan *et al.* (2000) proposed that, until enough key paleopoles are available to construct reliable APWPs, individual key paleopoles of the same age from different continental blocks should be directly compared in order to determine the relative position of the blocks. Of course, comparing key paleopoles of a single age would not yield a unique reconstruction, because of the longitudinal uncertainty inherent in the paleomagnetic method. However, as noted by Buchan *et al.* (2000, p. 185–186), a unique reconstruction may be possible if two (or more) ages of key paleopoles are compared and if the continental blocks in question moved as a unit and rotated through a significant angle during the period under study.

Conclusion

Reliable APWPs and continental reconstructions cannot be obtained from poorly dated paleomagnetic poles, no matter how many are used or how well defined are the paleopole themselves. Key paleopoles that are both well dated and well defined are a prerequisite for establishing reliable APWPs and producing robust paleocontinental reconstructions.

Kenneth L. Buchan

Bibliography

Buchan, K.L., and Halls, H.C., 1990. Paleomagnetism of Proterozoic mafic dyke swarms of the Canadian Shield. In Parker, A.J., Rickwood, P.C., and Tucker, D.H. (eds.), *Mafic Dykes and Emplacement Mechanisms.* Rotterdam: Balkema, pp. 209–230.

Buchan, K.L., Mortensen, J.K., and Card, K.D., 1994. Integrated paleomagnetic and U-Pb geochronologic studies of mafic intrusions in the southern Canadian Shield. *Precambrian Research*, **69**: 1–10.

Buchan, K.L., Mertanen, S., Park, R.G., Pesonen, L.J., Elming, S.-Å., Abrahamsen, N., Bylund, G., 2000. Comparing the drift of Laurentia and Baltica in the Proterozoic: the importance of key paleomagnetic poles. *Tectonophysics*, **319**: 167–198.

Buchan, K.L., Ernst, R.E., Hamilton, M.A., Mertanen, S., Pesonen, L.J., and Elming, S.-Å., 2001. Rodinia: the evidence from integrated palaeomagnetism and U-Pb geochronology. *Precambrian Research*, **110**: 9–32.

Pisarevsky, S.A., Wingate, M.T.D., and Harris, L.B., 2003. Late Mesoproterozoic (ca. 1.2 Ga) palaeomagnetism of the Albany-Fraser orogen: no pre-Rodinia Australia-Laurentia connection. *Geophysical Journal International*, **155**: 6–11.

Roy, J.L., 1983. Paleomagnetism of the North American Precambrian: a look at the data base. *Precambrian Research*, **19**: 319–348.

Wingate, M.T.D., and Giddings, J.W., 2000. Age and paleomagnetism of the Mundine Well dyke swarm, Western Australia: implications for an Australia-Laurentia connection at 755 Ma. *Precambrian Research*, **100**: 335–357.

Cross-references

Baked Contact Test

POTENTIAL VORTICITY AND POTENTIAL MAGNETIC FIELD THEOREMS

Theoretical studies of motions in the Earth's fluid (liquid) metallic outer core, where the main geomagnetic field is produced by self-exciting *dynamo* (*q.v.*) action, are based on the nonlinear partial differential equations (PDEs) of *magnetohydrodynamics (MHD)* (*q.v.*) that govern the flow of electrically conducting fluids. The equations are mathematical expressions of the laws of *mechanics*, *thermodynamics* and *electrodynamics* applied to a continuous medium.

The equations of electrodynamics are not needed when dealing with fluids of low electrical conductivity, for effects due to Lorentz forces associated with the flow of electric currents are then negligible, as in dynamical meteorology and oceanography. In these highly developed areas of geophysical fluid dynamics a key role is played by the concept of *potential vorticity* (PV). This pseudoscalar quantity—defined as the reciprocal of the density of the fluid multiplied by the scalar product of the vorticity vector and the gradient of any differentiable scalar quantity, such as the potential temperature or specific entropy of the fluid (see Eqs. (2), (17) below)—satisfies an elegant theorem (see Eq. (3)) due to Ertel (1942; see also Gill, 1982; Pedlosky, 1987).

In the words of Pedlosky (1987, p. 38) extolling the virtues of Ertel's theorem: "although the vorticity equation [which expresses the effects of torques associated with forces acting on a moving fluid element] is illuminating because it deals directly with the vector character of vorticity, it is more descriptive of how vorticity is changed than a useful constraint on that change. Kelvin's (circulation) theorem is more powerful, but (it) is an integral theorem dealing with a scalar and requires knowledge of the detailed evolution of material surfaces in the fluid. (The) beautiful and unusually useful theorem due to Ertel (governing the behavior of potential vorticity) provides a constraint on vorticity which is free from many of (these) difficulties "

An analogous constraint (Hide, 1983) on the behavior of the magnetic field (rather than vorticity) in a moving electrically conducting medium follows directly from the equations of electrodynamics (see Eqs. (6), (7) below). The constraint involves the concept of *potential magnetic field* (PMF) (defined by Eq. (6)), the behavior of which is governed by the general expression given by Eq. (7), the electrodynamic analogue of the Ertel PV theorem (for recent references see Hide, 2002, 2004). The full set of mathematical equations obtained by incorporating into extended versions of these PV and PMF theorems (see Eqs. (8) and (9) below) the constraints that arise from thermodynamic considerations on *potential temperature* or *specific entropy* should facilitate future theoretical and numerical investigations of the *geodynamo* (*q.v.*) and other MHD phenomena encountered in theoretical geophysics and astrophysics; see Hide (1996). The rest of present article closely follows Hide (1996), material from which is reproduced here by kind permission of the Royal Astronomical Society and Blackwell Publishers.

Potential vorticity

Consider a continuous medium in which the mass density is $\rho(\boldsymbol{r},t)$ and Eulerian velocity relative to an inertial frame is $\boldsymbol{u}(\boldsymbol{r},t)$ at a general point P with position vector $\mathbf{r}$ at time t. Conservation of mass requires that

$$\frac{\mathrm{D}\rho}{\mathrm{D}t} + \rho\nabla\cdot\boldsymbol{u} = 0 \qquad \text{(Eq. 1)}$$

(where $\mathrm{D}/\mathrm{D}t \equiv \partial/\partial t + \boldsymbol{u}\cdot\nabla$), which reduces to $\nabla\cdot\boldsymbol{u} = 0$ when the medium is incompressible.

If by $\xi \equiv \nabla\times\boldsymbol{u}$ we denote the (absolute) vorticity and by

$$\rho^{-1}\xi\cdot\nabla H^{*} \qquad \text{(Eq. 2)}$$

the so-called "potential vorticity", where $H^{*} = H^{*}(\boldsymbol{r},t)$ is any continuous and differentiable function, then by a very slight extension of Ertel's theorem (see Gill, 1982; Pedlosky, 1987), we have

$$\frac{\mathrm{D}}{\mathrm{D}t}\left(\frac{\xi\cdot\nabla H^{*}}{\rho}\right) = \frac{\xi}{\rho}\cdot\nabla\frac{\mathrm{D}H^{*}}{\mathrm{D}t} + \nabla H^{*}\cdot\Psi \qquad \text{(Eq. 3)}$$

where

$$\Psi \equiv \rho^{-2}[\mathbf{g}\times\nabla\rho + \nabla\times(\boldsymbol{j}\times\boldsymbol{B}) + \nabla\times\boldsymbol{F}] \qquad \text{(Eq. 4)}$$

Here $-\mathbf{g}$ is the acceleration due to gravity and $\boldsymbol{j} \times \boldsymbol{B}$ the, Lorentz ponderomotive force, if $\boldsymbol{j}(\boldsymbol{r},t)$ is the electric current density at P and $\boldsymbol{B}(\boldsymbol{r},t)$ is the magnetic field, which satisfies

$$\nabla \cdot \boldsymbol{B} = 0 \qquad \text{(Eq. 5)}$$

The term $\boldsymbol{F}$ in Eq. (4) represents the visco-elastic force acting on an element of material of unit volume which at time t is situated at P, reducing in the case of a fluid to the usual term representing the viscous force.

Potential magnetic field

Analogous to the derivation of Eq. (3) from the equations of dynamics, is the derivation from the equations of electrodynamics of an expression governing the behavior of the "potential magnetic field", defined as

$$\rho^{-1}\boldsymbol{B} \cdot \nabla G^* \qquad \text{(Eq. 6)}$$

(where G^*, like H^*, is any continuous and differentiable function of $\boldsymbol{r}$ and t), namely

$$\frac{\mathrm{D}}{\mathrm{D}t}\left(\frac{\boldsymbol{B} \cdot \nabla G^*}{\rho}\right) = \frac{\boldsymbol{B}}{\rho} \cdot \nabla \frac{\mathrm{D}G^*}{\mathrm{D}t} + \nabla G^* \cdot \Phi \qquad \text{(Eq. 7)}$$

(Hide, 1983). Here Φ comprises several terms, all of which vanish when the medium is a perfect conductor of electricity and thermoelectric, Hall, Nernst-Ettinghausen and other "nonohmic" effects are negligible. (Consistent with the neglect of relativistic effects, no account is taken of Maxwell displacement currents and electrostatic forces can be ignored in $\boldsymbol{F}$, see Eq. (4).)

Now make the successive substitutions $H^* = H$ and then $H^* = q'H$ in the potential vorticity Eq. (3) and combine the resulting two equations to obtain

$$\frac{\mathrm{D}}{\mathrm{D}t}\left[\frac{H}{\rho}\xi \cdot \nabla q'\right] = \left(\frac{\xi \cdot \nabla q'}{\rho}\right)\frac{\mathrm{D}H}{\mathrm{D}t} + \frac{H}{\rho}\left(\xi \cdot \nabla \frac{\mathrm{D}q'}{\mathrm{D}t}\right) + H\Psi \cdot \nabla q' \qquad \text{(Eq. 8)}$$

In the same way, substitute $G^* = G$ and then $G^* = qG$ in the potential magnetic field equation (7); hence

$$\frac{\mathrm{D}}{\mathrm{D}t}\left[\frac{G}{\rho}\mathbf{B} \cdot \nabla q\right] = \left(\frac{\mathbf{B} \cdot \nabla q}{\rho}\right)\frac{\mathrm{D}G}{\mathrm{D}t} + \frac{G}{\rho}\left(\mathbf{B} \cdot \nabla \frac{\mathrm{D}q}{\mathrm{D}t}\right) + G\Phi \cdot \nabla q \qquad \text{(Eq. 9)}$$

(The functions H, G, q, and q' are also arbitrary. see Hide, 1996.)

Equations (8) and (9) are useful extensions of the general results expressed by Eqs. (3) and (7) respectively, for each equation contains two arbitrary scalars, rather than just one. Suitable choices of these scalars lead directly to other general results, such as a relationship between forms of potential vorticity, helicity and superhelicity in hydrodynamics and its counterpart in electrodynamics (see Hide, 1989, 2002).

Some geophysical applications

As an application of these equations we set $G = \xi \cdot \nabla q'$ in Eq. (9) and H as a comparable function of $\boldsymbol{B}$ and q in Eq. (8) and subtract the resulting equations. Thus:

$$\frac{\mathrm{D}}{\mathrm{D}t}\left[\log\frac{\mathbf{B} \cdot \nabla q}{\xi \cdot \nabla q'}\right] = (\mathbf{B} \cdot \nabla q)^{-1}\left[\mathbf{B} \cdot \nabla \frac{\mathrm{D}q}{\mathrm{D}t} + \rho\Phi \cdot \nabla q\right] - (\xi \cdot \nabla q')^{-1}\left[\xi \cdot \nabla \frac{\mathrm{D}q'}{\mathrm{D}t} + \rho\Psi \cdot \nabla q'\right] \qquad \text{(Eq. 10)}$$

Equation (10) is particularly useful when q and q' are simply related to the coordinates of the general point P. Consider for example the case when spherical polar coordinates (r, θ, ϕ) are used and $q = q' = r$. Equation (10) then reduces to

$$\frac{\mathrm{D}}{\mathrm{D}t}\left[\log\left(\frac{\mathbf{B}_r}{\xi_r}\right)\right] = \frac{1}{B_r}(\mathbf{B} \cdot \nabla u_r + \rho\Phi_r) - \frac{1}{\xi_r}\left[\xi \cdot \nabla u_r + \rho\Psi_r\right] \qquad \text{(Eq. 11)}$$

where the subscript r denotes the r-component of the corresponding vector and the operators

$$\mathbf{B} \cdot \nabla \equiv \mathbf{B} \cdot \nabla - B_r\partial\ \partial r \quad \text{and} \quad \xi \cdot \nabla \equiv \xi \cdot \nabla - \xi_r\partial\ \partial r \qquad \text{(Eq. 12)}$$

Comparable relationships can be found by setting q and q' equal to θ and ϕ.

In geophysical and astrophysical fluid dynamics we are often interested in fluids for which the electrical conductivity is very high and "non-ohmic" effects (see Eq. (7)) are negligible, so that Alfvén's "frozen magnetic flux" (*q.v*) theorem holds. Then the vector $\Phi = 0$ in Eq. (7) (see also Eqs. (9), (10) (11)) which when $G^* = r$ and the fluid is incompressible (so that $\mathrm{D}\rho/\mathrm{D}t$, see Eq. (1)) gives

$$\frac{\mathrm{D}}{\mathrm{D}t}B_r = \mathbf{B} \cdot \nabla u_r \qquad \text{(Eq. 13)}$$

This equation has been put to good use by geophysicists in work on the determination of the flow just below the Earth's core-mantle interface from observations of the *geomagnetic secular variation* (*q.v.*); see Eq. (16) below.

Also of interest are studies of flows in rapidly rotating systems, where it is convenient to work in a frame of reference that rotates relative to an inertial frame with angular velocity $\Omega = (\Omega\cos\theta, -\Omega\sin\theta, 0)$ about the polar axis. Equations (1)–(11) hold in the new frame if we re-define ξ as being equal to $\nabla \times \boldsymbol{u} + 2\Omega$, include centripetal effects in $\mathbf{g}$ and add the term $\rho\boldsymbol{r} \times \mathrm{d}\Omega/\mathrm{d}t$ to $\boldsymbol{F}$ (see Eq. (4)). Such flows include those where to a first approximation buoyancy forces act in the radial direction and other forces and torques acting on individual fluid elements are in magnetostrophic balance which, when Lorentz forces are also negligible, reduces to geostrophic balance. Magnetostrophic balance is characterized here by setting

$$(\xi_r\ \xi_\theta) = 2\Omega(\cos\theta\ - \sin\theta) \quad \text{and} \quad \Psi = \rho^{-2}\boldsymbol{j} \times \boldsymbol{B} \qquad \text{(Eq. 14)}$$

with $\Psi = 0$ in the geostrophic case. When combined with Eq. (14), Eq. (11) gives

$$\frac{\mathrm{D}}{\mathrm{D}t}\log\left(\frac{B_r}{\cos\theta_r}\right) = \left(\frac{\boldsymbol{B} \cdot \nabla}{B_r} - \frac{\xi \cdot \nabla}{2\Omega\cos\theta}\right)u_r - \frac{[\nabla \times (\boldsymbol{j} \times \boldsymbol{B})]_r}{2\rho\ \Omega\cos\theta} \qquad \text{(Eq. 15)}$$

In the case of geostrophic flow over a spherical surface where u_r is negligibly small in magnitude in comparison with u_θ and u_ϕ, it follows immediately from Eq. (15) that

$$\frac{\mathrm{D}}{\mathrm{D}t}(B_r\sec\theta) = 0 \qquad \text{(Eq. 16)}$$

implying the conservation of the quantity $B_r\sec\theta$ on moving fluid elements. This result is a familiar one in work on the determination of motions just below the Earth's core-mantle boundary from *geomagnetic secular variation* data (*q.v.*), where near-uniqueness can be secured by combining the assumption of geostrophy with Alfvén's frozen magnetic flux theorem (see Eq. (13)). Conditions under which the result can be expected to apply are clearly exposed by its novel derivation here from the powerful general theorems governing the behavior of potential vorticity and potential magnetic field in magnetohydrodynamic flows.

The full equations of magnetohydrodynamics express not only the laws of mechanics (the basis of Eq. (3)) and of electrodynamics (the basis of Eq. (7)) but also the laws of thermodynamics governing, inter alia, the behavior of specific entropy $\Theta = \Theta(\boldsymbol{r},t)$. This quantity satisfies

$$\frac{D\Theta}{Dt} = Q \qquad \text{(Eq. 17)}$$

where $Q = Q(\boldsymbol{r},t)$ represents thermal conduction, radiation and other diabatic effects, including ohmic heating. We conclude this short note by observing that Eq. (10) with $q = q' = \Theta$ shows that isentropic flows, for which $Q = 0$ by definition, satisfy

$$\frac{D}{Dt}\left(\frac{\mathbf{B}\cdot\nabla\Theta}{\xi\cdot\nabla\Theta}\right) = 0 \qquad \text{(Eq. 18)}$$

in regions where Ψ^* and Φ are also negligibly small. According to Eq. (18), within such flows the ratio of the components of $\boldsymbol{B}$ and ξ in the direction of $\nabla\Theta$, the gradient of specific entropy, is (like Θ itself) conserved on moving fluid elements. A further result of geophysical and astrophysical interest is that Eq. (18) reduces to Eq. (16) in cases when approximate geostrophic balance obtains and the nonradial components of $\nabla\Theta$ are much weaker than its radial component.

Raymond Hide

Bibliography

Ertel, H., 1942. Ein neuer hydrodynamischer Wirbelsatz. *Meteorologische Zeitscrift*, **59**: 271–281.

Gill, A.E., 1982. *Atmosphere-Ocean Dynamics*. New York: Academic Press.

Hide, R., 1983. The magnetic analogue of Ertel's potential vorticity theorem. *Annales Geophysicae*, **1**: 59–60.

Hide, R., 1989. Superhelicity, helicity and potential vorticity. *Geophysical and Astrophysical Fluid Dynamics*, **48**: 69–79.

Hide, R., 1996. Potential vorticity and potential magnetic field in magnetohydrodynamics. *Geophysical Journal International*, **125**: F1–F3.

Hide, R., 2002. Helicity, superhelicity and weighted relative potential vorticity (and their electrodynamic counterparts): Useful diagnostics pseudoscalars? *Quarterly Journal of the Royal Meteorological Society*, **128**: 1759–1762.

Hide, R., 2004. Reflections on the analogy between the equations of electrodynamics and hydrodynamics. In Schröder, W. (ed.), *Hans Ertel: Gedenkschrift zum 100. Geburtstag*. Bremen: Deutscher Arbeitskreis Geschichte, Geophysik und Kosmische Physik, pp. 25–33.

Pedlosky, J., 1987 *Geophysical Fluid Dynamics*, 2nd edn. New York: Springer-Verlag.

Cross-references

Alfvén's Theorem and the Frozen Flux Approximation
Core Motions
Geodynamo
Geomagnetic Secular Variation
Magnetohydrodynamics
Proudman-Taylor Theorem

PRECESSION AND CORE DYNAMICS

The Earth is forced to precess by the gravitational action of the Sun, the Moon and the other planets of the solar system on its equatorial budge. Like a spinning top, the axis of rotation of the planet moves on a cone of semiangle 23.27° with a period of 26000 years, which is known from early astronomy as the precession of the Earth. The motions of the axis on smaller cones with shorter periods are known as nutations. The amplitudes and periods of the precession nutations are directly connected to the shape (ellipticity) and structure of the Earth. The presence of a liquid layer (the core) at the center of the planet introduces a differential precession between liquid and solid parts and their coupling sets the amplitude of this difference. After a first attempt by Kelvin, Hough (1895) and Poincaré (1910) were the first to model this problem in order to determine whether the interior of the Earth is liquid or solid. Considering an inviscid fluid, Poincaré suggested that the response of the fluid is rather simple as it is a solid body rotation and a gradient flow associated with the ellipticity of the solid boundary. More generally, Busse (1968) revealed an analogy between the core flow induced by precession and the one due to tidal bulge. Malkus (1968) envisioned the core motion associated with precession as a possible source for the geodynamo.

Position of the axis of rotation of the Core

The axis of rotation of the liquid core differs from the axis of rotation of the mantle and the torque balance on the fluid core gives the relative position of the two axes (Poincaré (1910) for the inviscid problem and Busse (1968) for the viscous correction). In Figure P58, the photograph illustrates this situation as the rotation axis of fluid is clearly off the axis of the container. It is worth noting that there is no differential rotation along the axis of rotation of the fluid while the flow

Figure P58 Precessional flow in a laboratory spheroid. Kalliroscope flakes dispersed in the fluid are illuminated by a beam of light in a plan containing both the rotation axis of the container and of the fluid. Any shear in the flow aligns the flakes. The obliquity angle is 20°. Experimental details can be found in Noir *et al.* (2003). Meridional cross sections of geostrophic cylinders aligned with the axis of the Poincaré mode are clearly seen. Oscillations (or instabilities?) of the axis are also visible.

induced by precession is not associated with a spin-up process, which would not preserve the geometry of the Poincaré mode (Greenspan, 1968). As the Earth's rotation decreases, the precessional and nutational forcings vary and the fluid response may match an eigenmode flow in the rotating core known as the "tilt-over mode" (Greenspan, 1968), generating a resonance in the core which may be related to the reversals frequency of the geomagnetic field or major geologic events (Greff and Legros, 1999). Noir *et al.* (2003) have observed this resonance in their experiment and showed its nonlinear implications from the torque balance; very large excursions of the axis of rotation of the fluid are predicted. It remains unclear how the presence of an inner core (with a different ellipticity) would influence these results.

Boundary layer singularities

A thin boundary layer develops at the core-mantle boundary to ensure continuity of the velocity field. Both magnetic (electromagnetic skin layer) and viscous (Ekman layer) effects form a layer a few hundreds meter in depth at the liquid core boundaries (see Core, boundary layers, Loper, 1975; Deleplace and Cardin, 2005). Because of their diurnal oscillations, the viscous boundary layer is singular at the critical latitudes ($\pm 30°$) and diurnal jets or shear layers erupt in the bulk fluid along characteristic cones associated with inertial waves in the fluid core (Hollerbach and Kerswell, 1995; Noir *et al.*, 2001). Asymptotic scaling (with no magnetic field) lead to diurnal motions in the core of amplitude $\approx 10^{-5}$ m/s, depending on the effective viscosity of the fluid outer core.

Geostrophic motions

Cylindrical motions have been observed in precession experiments (Malkus, 1968; Vanyo *et al.*, 1995). In Figure P58, we clearly see axisymmetric geostrophic shear flows (Black and white lines parallel to the axis of rotation). Their source is associated with a nonlinear effect in the boundary layer at the critical latitude. These geostrophic motions have been retrieved in nonlinear calculations of precessing flows (Noir *et al.*, 2001). Malkus (1968) determined the amplitude of the velocities of the geostrophic cylinders and the recent numerical results agree qualitatively with his experimental findings. In the absence of magnetic field, an amplitude of 10^{-5} m/s of the geostrophic cylinders is predicted in the Earth's core.

Instabilities

In Figure P58 the central axis of rotation of the fluid shows a variation of the brightness. This modulation may be the signature of an elliptical or shear instability, which would result from the nonlinear interaction of two inertial waves and the Poincaré flow according to symmetry rules (Kerswell, 2002). Such an instability may lead to intermittency of small-scale turbulence (Malkus, 1989; Lorenzani and Tilgner, 2003; Lacaze *et al.*, 2004). It is difficult to apply these results to planetary interiors as they are only partially understood in simple systems.

Dynamo action

In the past, energetic arguments based on the Poincaré mode alone, have been used to rule out precession as a source for the geodynamo (Loper, 1975). This result is invalid if we consider that the Poincaré mode transfers energy from the kinetic energy stored in the Earth's equatorial rotation (Malkus, 1968) to secondary motions within the core (geostrophic cylinders, instabilities) through nonlinear effects which have not been taken into account in Loper's approach.

The Poincaré flow cannot produce any dynamo action as it is mainly a solid body rotation. The erupted layers and connected inertial waves are diurnal and confined to very small regions; they may participate to a permanent dynamo only through a nonlinear and alpha-effect but no results have been reported according to our knowledge. Geostrophic motions and associated instabilities can produce a kinematic dynamo as Tilgner (2005) just proved, at least for a very viscous fluid. The importance of elliptical instabilities for the dynamo remains an open question even though Aldridge and Baker (2003) used them as a basis of an interpretation of the paleomagnetic reversals frequency.

Philippe Cardin

Bibliography

Aldridge, K., and Baker, R., 2003. Paleomagnetic intensity data: a window on the dynamics of Earth's fluid core? *Physics of the Earth and Planetary Interiors*, **140**: 91–100.

Busse, F.H., 1968. Steady fluid flow in a precessing spheroidal shell. *Journal of Fluid Mechanics*, **33**: 739–751.

Deleplace, B., and Cardin, P., 2005. Viscomagnetic torque at the core mantle boundary. submitted to *Geophysical Journal International*, GJI, 2006, **167**: 557–566.

Greenspan, H.P., 1968. *The Theory of Rotating Fluids.* Cambridge: Cambridge University Press.

Greff, M., and Legros, H., 1999. Core rotational dynamics and geological events. *Science*, **286**: 1707–1709.

Hollerbach, R., and Kerswell, R.R., 1995. Oscillatory internal shear layers in rotating and precessing flows. *Journal of Fluid Mechanics*, **298**: 327–339.

Hough, S.S., 1895. The oscillations of a rotating ellipsoidal shell containing fluid. *Philosophical Transactions of the Royal Society of London*, **186**: 469.

Kerswell, R.R., 2002. Elliptical instability. *Annual Review of Fluid Mechanics*, **34**: 83–113.

Lacaze, L., Le Gal, P., and Le Dizs, S., 2004. Elliptical instability in a rotating spheroid. *Journal of Fluid Mechanics*, **505**: 1–22.

Loper, D.E., 1975. Torque balance and energy budget for the precessionally driven dynamo. *Physics of the Earth and Planetary Interiors*, **11**: 43–60.

Lorenzani, S., and Tilgner, A., 2003. Inertial instabilities of fluid flow in precessing spheroidal shells. *Journal of Fluid Mechanics*, **492**: 363–379.

Malkus, W.V.R., 1968. Precession of the Earth as the cause of geomagnetism. *Science*, **160**: 259–264.

Malkus, W.V.R., 1989. An experimental study of global instabilities due to tidal (elliptical) distortion of a rotating elastic cylinder. *Geophysical and Astrophysical Fluid Dynamics*, **48**: 123–134.

Noir, J., Cardin, P., Jault, D., and Masson, J.-P., Experimental evidence of nonlinear resonance effects between retrograde precession and the tilt-over mode within a spheroid. *Geophysical Journal International*, **154**: 407–416.

Noir, J., Jault, D., and Cardin, P., 2001. Numerical study of the motions within a slowly precessing sphere at low Ekman number. *Journal of Fluid Mechanics*, **437**: 283–299.

Poincaré, H., 1910. Sur la précession des corps déformables. *Bulletin of Astronomical Society*, **27**: 321–356.

Tilgner, A., 2005. Precession driven dynamos. *Physics of Fluid*, **17**: 034104.

Vanyo, J., Wilde, P., Cardin, P., and Olson, P., 1995. Experiments on precessings flows in the Earth's liquid core. *Geophysical Journal International*, **121**: 136–142.

Cross-references

Core, Boundary Layers
Core, Magnetic Instabilities
Fluid Dynamics Experiments
Geodynamo, Energy Sources
Gravity-Inertio Waves and Inertial Oscillations

PRICE, ALBERT THOMAS (1903–1978)

An applied mathematician, born in Nantwich, Cheshire, Price is regarded as one of the founding figures in the development of geomagnetism, and was particularly concerned with the determination of the electrical properties of the Earth's interior. He pioneered methods for analyzing magnetic variations and for separating them into parts arising externally and internally to the Earth's surface. Treating the external part as an inducing field, he sought to explain the internal induced part by establishing theory for electromagnetic induction in spheres, in thin sheets and shells and in half-spaces. These had direct relevance to global problems of induction in the Earth, the oceans and the ionosphere, and to local problems of electromagnetic exploration. He sought mathematical theories to explain existing data and drove data acquisition programs to test mathematical theories.

A fundamental paper by Price and Chapman (1928) (see *Chapman, Sydney*) established that diurnal variations in the Earth's magnetic field are derivable from a scalar potential. In the 1930s, Price derived formal solutions to induction in uniformly conducting spheres by aperiodic inducing fields (Price, 1930) and included permeability (Price, 1931), complementing solutions of Lamb for periodic fields. This made him possible to investigate the conductivity of the Earth using both the periodic daily variation of the magnetic field on quiet days (S_q) and the aperiodic storm time variations (D_{st}) (see *Storms and substorms, magnetic* (*q.v.*)). He applied these theories, again with Chapman, establishing the validity of global electromagnetic induction and determining conductivity models for D_{st} (Chapman and Price, 1930) (see *Mantle, electrical conductivity, mineralogy*). The deeper penetrating storm time fields required a higher conductivity than was the case for S_q and this was early evidence that the Earth's electrical conductivity increases with depth. In a classic paper (Lahiri and Price, 1939), spheres in which the conductivity varied as a function of radius were considered. Analytic solutions were obtained for the special case of conductivity varying as a power of the normalized radius. These remain the only analytic solutions determined to date. Application of the theory jointly to both S_q and D_{st} revealed a rapid increase in conductivity around 600–700 km depth, a result of major importance in the study of the Earth. An additional requirement was the inclusion of a thin conducting shell at the Earth's surface, which was assumed to represent an effect of the conducting oceans.

Price provided theory for induction in nonuniform thin sheets and shells electrically insulated from their surroundings (Price, 1949), establishing what became known as Price's Equation (Parkinson, 1983). He provided analytic solutions for specific conductivity distributions and iterative schemes for obtaining numerical approximations for the general case. Through this he initiated two lines of research-induction in the ionosphere and induction in the oceans. Price's reappraisal of ionospheric current flow with Cocks (Cocks and Price, 1969) concluded that the prevalent theory of two-dimensional current flow was inadequate and that the external part of the S_q variations required a component of vertical current flow, now a well-established premiss in ionospheric studies (see *Ionosphere*). With Hobbs (Hobbs and Price, 1970) he derived a suite of surface integral formulae which found direct application in his iterative schemes for solving problems of induction in the nonuniformly distributed oceans. An intense period of thin-sheet oceanic induction studies followed until computing power in the 1980s was able to cope with a thin surface sheet electrically coupled to the underlying mantle (see *Ocean, electromagnetic effects*).

Price's continuing interest in S_q led to a new method of separating parts arising externally to the Earth from those arising from within the earth due to induction. The two-step method, proposed with Wilkins (Price and Wilkins, 1963), involved first determining values of a potential function on the Earth's surface representing the total magnetic variation field and then separating this total field into its external and internal parts using surface integrals. The potential function was determined by iteration, minimizing residuals of line integrals of the magnetic field around closed contours on the Earth's surface. Application of the surface integrals required establishing tables of values of integrands over worldwide tesseral elements. The new method was applied, using banks of mechanical calculators, to observations made during the International Polar Year (Price and Wilkins, 1963) and to the International Geophysical Year (1964). Price was instrumental, again with Chapman, in proposing and establishing S_q as a subject of study during the International Quiet Sun Year.

Price also provided theory for so-called local induction problems, whereby the Earth is represented as a half-space. In 1950 he gave complete solutions for induced and freely decaying modes and showed that induction in a half-space by a uniform field is an indeterminate problem (Price, 1950). In 1962 he presented formal theory for the magnetotelluric method including consideration of the dimensions of the source field (Price, 1962) (see *Magnetotellurics*). Work with Jones (Jones and Price, 1970) involved a detailed analysis of boundary conditions applicable to two-dimensional induction problems and showed how numerical solutions could be obtained for models that included inhomogeneous conducting regions. This paved the way for the interpretation of anomalies found in the rapidly expanding field of magnetotelluric surveys.

Summaries of much of his work, and that of others, are given in two valuable reviews (Price, 1967a,b). These reviews include observational material, theory, and applications together with physical and mathematical insights into geomagnetic phenomena. They are still useful works of reference.

In an academic career spanning of 43 years, Price held appointments at Queen's University, Belfast (1925); Imperial College, London (1926–1951); Royal Technical College, Glasgow (1951–1952) and University of Exeter (1952–1968). He was an IGY Research Associate at the National Academy of Sciences, Washington (1961–1962) and Chairman of Commission IV of IAGA on Magnetic Activity and Disturbances (1964–1968). The Gold Medal of the Royal Astronomical Society was awarded to Price in 1969 for his work on geomagnetism and especially for his studies of the electrical conductivity in the interior of the Earth. The Royal Astronomical Society awards the Price Medal for geomagnetism and aeronomy, in his honor.

Bruce A. Hobbs

Bibliography

Chapman, S., and Price, A.T., 1930. The electric and magnetic state of the interior of the Earth as inferred from terrestrial magnetic variations. *Philosophical Transactions of the Royal Society of London*, A**229**: 427–460.

Cocks, A.C., and Price, A.T., 1969. Sq currents in a 3-dimensional ionosphere. *Planetary and Space Science*, **17**: 471–482.

Hobbs, B.A., and Price, A.T., 1970. Surface integral formulae for geomagnetic studies. *Geophysical Journal of the Royal Astronomical Society*, **20**: 49–63.

Jones, F.W., and Price, A.T., 1970. Perturbations of alternating geomagnetic fields by conductivity anomalies. *Geophysical Journal of the Royal Astronomical Society*, **20**: 317–334.

Lahiri, B.N., and Price, A.T., 1939. Electromagnetic induction in nonuniform conductors and the determination of the conductivity of the Earth from terrestrial magnetic variations. *Philosophical Transactions of the Royal Society of London*, A**237**: 509–540.

Parkinson, W.D., 1983. *Introduction to Geomagnetism.* Edinburgh: Scottish Academic Press.

Price, A.T., 1930. Electromagnetic induction in a conducting sphere. *Proceedings of the London Mathematical Society*, [2], **31**: 217–224.

Price, A.T., 1931. Electromagnetic induction in a permeable conducting sphere. *Proceedings of the London Mathematical Society*, [2], **33**: 233–245.

Price, A.T., 1949. The induction of electric currents in nonuniform thin sheets and shells. *Quarterly Journal of Mechanics and Applied Mathematics*, **2**: 283–310.
Price, A.T., 1950. Electromagnetic induction in a semi-infinite conductor with a plane boundary. *Quarterly Journal of Mechanics and Applied Mathematics*, **3**: 385–410.
Price, A.T., 1962. Theory of magnetotelluric methods when the source field is considered. *Journal of Geophysical Research*, **67**: 1907–1918.
Price, A.T., 1967a. Electromagnetic Induction within the Earth. In Matsushita, S., and Campbell, W.H. (eds.), *Physics of Geomagnetic Phenomena*. London and New York: Academic Press, pp. 235–295.
Price, A.T., 1967b. Magnetic Variations and Telluric Currents. In Gaskell, T.F. (ed.), *The Earth's Mantle*. London and New York: Academic Press, pp. 125–170.
Price, A.T., and Chapman, S., 1928. On line-integrals of the diurnal magnetic variations. *Proceedings of the Royal Society of London*, A**119**: 182–196.
Price, A.T., and Stone, D.J., 1964. The quiet day magnetic variations during the IGY. In *Annals of the International Geophysical Year*, **35** Part III.
Price, A.T., and Wilkins, G.A., 1963. New methods for the analysis of geomagnetic fields and their application to the Sq field of 1932–1933, *Philosophical Transactions of the Royal Society of London*, A**256**: 31–98.

Cross-references:

Chapman, Sydney (1888–1970)
Ionosphere
Magnetotellurics
Mantle, Electrical Conductivity, Mineralogy
Ocean, Electromagnetic Effects
Storms and Substorms, Magnetic

PRINCIPAL COMPONENT ANALYSIS IN PALEOMAGNETISM

When studying the mean and variance of paleomagnetic data it is a common practice to employ principal component analysis (Jolliffe, 2002). The theory of this method is related to the mathematics quantifying the moment of inertia of a set of particles of mass about some reference point of interest. For the purposes of data analysis, principal component analysis was first promoted by Pearson (1901) and Hotelling (1933), and it also often associated with Karhunen (1947) and Loéve (1977). Principal component analysis is widely applied in crystallography (e.g., Schomaker *et al.*, 1959). In paleomagnetism (e.g., Mardia, 1972; Kirschvink, 1980), it finds application in studies of the average paleofield, paleosecular variation, demagnetization, and magnetic susceptibility. Here we discuss and demonstrate principal component analysis in application to full paleomagnetic vectorial data and, separately, to paleomagnetic directional data.

Vectorial analysis

Consider a set of paleomagnetic vectors, with the *i*th vector $\mathbf{x}(i)$ having intensity, inclination, and declination values ($F(i)$, $I(i)$, $D(i)$). Their equivalent Cartesian expression is just

$$\begin{aligned} x_1(i) &= F(i)\cos I(i)\cos D(i) \\ x_2(i) &= F(i)\cos I(i)\sin D(i) \\ x_3(i) &= F(i)\sin I(i) \end{aligned} \qquad \text{(Eq. 1)}$$

where $\mathbf{x} = (x_1\ x_2\ x_3)$ represents the usual geographic components of (north, east, down). With N such vectorial data we can calculate their values relative to some fixed reference point $\mathbf{r} = (r_1\ r_2\ r_3)$

$$\delta\mathbf{x}(i) = \mathbf{x}(i) - \mathbf{r} \qquad \text{(Eq. 2)}$$

and with which we can calculate their covariance matrix,

$$\mathrm{C} = \frac{1}{N}\begin{pmatrix} \sum \delta x_1(i)^2 & \sum \delta x_1(i)\delta x_2(i) & \sum \delta x_1(i)\delta x_3(i) \\ \sum \delta x_2(i)\delta x_1(i) & \sum \delta x_2(i)^2 & \sum \delta x_2(i)\delta x_3(i) \\ \sum \delta x_3(i)\delta x_1(i) & \sum \delta x_3(i)\delta x_2(i) & \sum \delta x_3(i)^2 \end{pmatrix} \qquad \text{(Eq. 3)}$$

This matrix can be reduced to diagonal form by an appropriate choice of axes, obtained by solving the eigenvalue problem

$$\mathrm{C}\mathbf{e}^m = \lambda_m \mathbf{e}^m \qquad \text{(Eq. 4)}$$

where $\mathbf{e}^m$ is an eigenvector and λ_m is an eigenvalue (e.g., Strang, 1980). Three eigenvectors $\mathbf{e}^m$, for $m = 1\ 2\ 3$, and their corresponding eigenvalues λ_m can be found using standard numerical packages, such as LAPACK. The eigenvectors are orthogonal, and therefore

$$\mathbf{e}^m \cdot \mathbf{e}^n = 0 \quad \text{for} \quad m \neq n \qquad \text{(Eq. 5)}$$

The eigenvectors only define directions; they are of arbitrary length. Here we choose to normalize the eigenvectors, so that

$$\mathbf{e}^m \cdot \mathbf{e}^m = 1 \qquad \text{(Eq. 6)}$$

The eigenvalues λ_m are real, but they have no particular ordering; we choose an ascending order here $\lambda_1 \leq \lambda_2 \leq \lambda_3$ for specificity.

Let us now examine transformations between geographic space $\mathbf{x}$ and eigen space, which we designate $\mathbf{z}$. A transformation matrix $\mathbf{E}$ can be constructed from a columnar arrangement of the eigenvectors $\mathbf{e}^m$. We choose to arrange the eigenvector columns in the order of their ascending eigenvalues,

$$\mathbf{E} = \begin{pmatrix} e_1^1 & e_1^2 & e_1^3 \\ e_2^1 & e_2^2 & e_2^3 \\ e_3^1 & e_3^2 & e_3^3 \end{pmatrix} \qquad \text{(Eq. 7)}$$

The matrix $\mathbf{E}$ is orthonormal, and therefore its inverse is equal to its transpose,

$$\mathbf{E}^{-1} = \mathbf{E}^{\mathrm{T}} \qquad \text{(Eq. 8)}$$

The matrix $\mathbf{E}$ rotates data from the eigenspace $\mathbf{z} = (z_1\ z_2\ z_3)$ into the original geographic space $\mathbf{x} = (x_1\ x_2\ x_3)$ through the transformation

$$\mathbf{x}(i) = \mathbf{E}\mathbf{z}(i) \qquad \text{(Eq. 9)}$$

The inverse transformation is given by

$$\mathbf{z}(i) = \mathbf{E}^{\mathrm{T}}\mathbf{x}(i) \qquad \text{(Eq. 10)}$$

The matrix $\mathbf{E}$ also diagonalizes the covariance matrix, so that

$$\mathbf{E}^{\mathrm{T}}\mathrm{C}\mathbf{E} = \Lambda \qquad \text{(Eq. 11)}$$

where

$$\Lambda = \begin{pmatrix} \lambda_1 & 0 & 0 \\ 0 & \lambda_2 & 0 \\ 0 & 0 & \lambda_3 \end{pmatrix} \qquad \text{(Eq. 12)}$$

All of these transformations are orthonormal, they are special cases of a more general similarity transformation, for which the trace is invariant, and so

$$\mathrm{tr}(C) = \mathrm{tr}(\Lambda) = \sum_m \lambda_m. \quad \text{(Eq. 13)}$$

The eigenvectors and values have an important geometric interpretation. The variance of the data about the reference point **r** is an ellipsoid, which has the following simple expression in eigenspace

$$\frac{z_1^2}{\lambda_1} + \frac{z_2^2}{\lambda_2} + \frac{z_3^2}{\lambda_3} = 1. \quad \text{(Eq. 14)}$$

Depending on the relative sizes of the eigenvalues, different symmetries of the data distribution are revealed; a summary is given in Table P2. Note that each eigenvalue λ_m is the variance of the data along the direction defined by its corresponding eigenvector $\mathbf{e}^m$. If λ_3 is the largest eigenvalue, then the major axis is defined as the line segment joining the two points $\pm\sqrt{\lambda_3}\mathbf{e}^3$; the two other shorter minor axes are defined similarly. The eccentricity of each ellipsoidal equator is measured by

$$\epsilon_{mn} = \sqrt{1 - \frac{\lambda_m}{\lambda_n}}, \quad \text{for} \quad m < n. \quad \text{(Eq. 15)}$$

The overall anisotropy of the variance can be roughly quantified by the most eccentric ellipsoidal equator ϵ_{13}. Finally, it is important to recognize the fact that the line parallel to $\mathbf{e}^3(\mathbf{r})$ is, in a least-squares sense, the best fitting line to the data that passes through the reference point **r**, with $\lambda_3(\mathbf{r})$ being a measure of the misfit to this line. Moreover, the plane normal to $\mathbf{e}^1(\mathbf{r})$, and which contains $\mathbf{e}^2(\mathbf{r})$ and $\mathbf{e}^3(\mathbf{r})$, is the least-squares, best fitting plane to the data that contains the reference point **r**, with $\lambda_2(\mathbf{r}) + \lambda_3(\mathbf{r})$ functioning as a measure of misfit.

It is of interest to translate the eigenvalues into more conventionally interpretable quantities. For a prolate variance ellipsoid, Kirschvink (1980) has defined an approximate maximum angular deviation ($\mathrm{MAD_p}$) from the major axis $\mathbf{e}^3(0)$ by the conic angle determined by a projection of the minor ellipse onto the unit sphere

$$\mathrm{MAD_p} = \tan^{-1}\left[\sqrt{\frac{\lambda_1(0) + \lambda_2(0)}{\lambda_3(0)}}\right], \text{for} \quad \lambda_1 \simeq \lambda_2 \ll \lambda_3. \quad \text{(Eq. 16)}$$

For an oblate variance ellipsoid, we can also define a corresponding maximum angular deviation ($\mathrm{MAD_o}$) from the plane normal to the most minor axis along $\mathbf{e}^1(0)$,

$$\mathrm{MAD_o} = \tan^{-1}\left[\sqrt{\frac{\lambda_1(0)}{\lambda_2(0) + \lambda_3(0)}}\right], \quad \text{for} \quad \lambda_1 \ll \lambda_2 \simeq \lambda_3, \quad \text{(Eq. 17)}$$

which is somewhat different from the definition offered by Kirschvink. With respect to the maximum intensity deviation, if the variance ellipsoid about $\mathbf{r} = \bar{\mathbf{x}}$ is more or less spherical, then the maximum intensity deviation (MID) can be estimated as

Table P2 Eigenvalue classifications

Eigenvalues	Eigenvectors	Variance ellipsoid
$\lambda_1 = \lambda_2 = \lambda_3$	No preferred oriention	Spherical
$\lambda_1 = \lambda_2 < \lambda_3$	$\mathbf{e}^1$ and $\mathbf{e}^2$ have no preferred orientation	Prolate ellipsoid with rotational symmetry about $\mathbf{e}^3$
$\lambda_1 < \lambda_2 = \lambda_3$	$\mathbf{e}^2$ and $\mathbf{e}^3$ have no preferred orientation	Oblate ellipsoid with rotational symmetry about $\mathbf{e}^1$
$\lambda_1 \neq \lambda_2 \neq \lambda_3$	All eigenvectors have definite orientation	Scalene ellipsoid, with no axis of symmetry

Table P3 Principal component analysis of the Hawaiian data

Reference	Eigenvalue (μT^2)	Eigen direction I(°)	D(°)	Eccentricities ϵ_{12}	ϵ_{23}	ϵ_{13}	MID (μT)	$\mathrm{MAD_p}$ (°)
			Past 5 Ma, Full vectors					
	34.4 (μT^2)	10.0	−96.4					
$r = 0$	65.6 (μT^2)	55.7	158.8	0.69				15.0
	1384.2 (μT^2)	32.5	0.1					
			Brunhes only, Full vectors					
	26.2 (μT^2)	12.5	−94.6					
$r = 0$	52.1 (μT^2)	54.2	157.5	0.71				12.8
	1505.4 (μT^2)	32.9	3.7					
	24.1 (μT^2)	11.4	−83.9					
$r = \bar{x}$	47.2 (μT^2)	44.3	174.7	0.70	0.81	0.91	11.8	
	138.4 (μT^2)	43.5	17.1					
			Past 5 Ma, Directions only					
	0.0304	−14.6	87.6					
$r = 0$	0.0601	−58.9	−48.4	0.71				17.6
	0.9088	31.1	−1.5					

Note: Results are shown for data recording a mixture of normal and reverse polarities over the past 5 Ma, as well as for normal Brunhes data only. Covariance is measured relative to the origin $\boldsymbol{r} = (0, 0, 0)$ and, for Brunhes data, relative to the vectorial mean $\bar{\boldsymbol{x}} = (31.3, 15.4, 19.7)\mu$T. The direction-only analysis utilizes unit vectors of the 5 Ma data set.

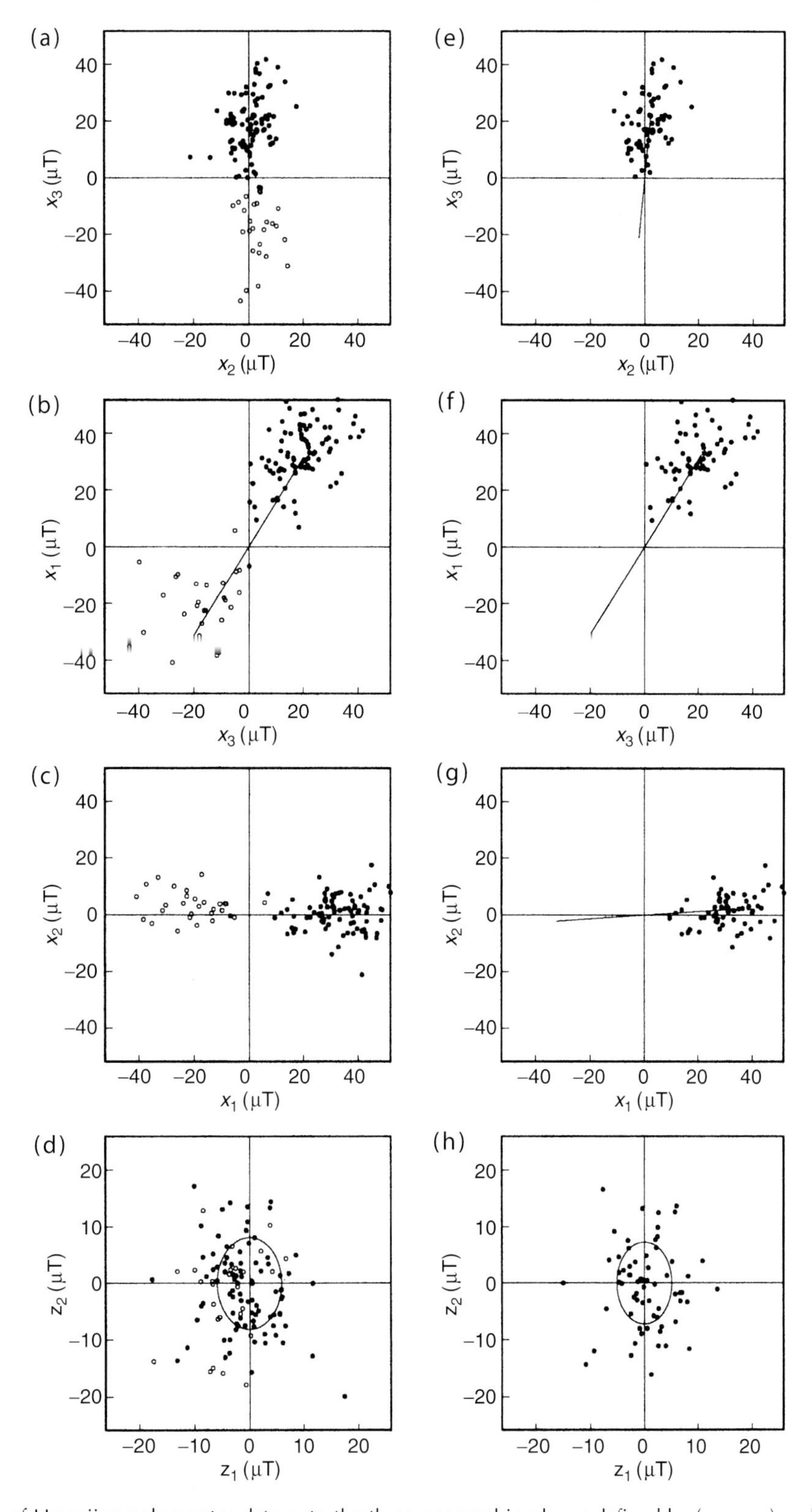

Figure P59 Projection of Hawaiian paleovector data onto the three geographic planes defined by (x_1,x_2,x_3) and onto the eigenplane $z_1 - z_2$ normal to the direction of the major axis defined by $\mathbf{e}^3$. (a-d) show data covering the past 5 Ma. (e-f) show data covering the normal Brunhes. In (a-c) and (e-f) we plot the major axis, and in (d) and (h) we plot the projection of the variance ellipse. Solid circles (disks) represent normal data, open circles represent reversed data.

$$\mathrm{MID} = \sqrt{\frac{\lambda_1(\bar{\mathbf{x}}) + \lambda_2(\bar{\mathbf{x}}) + \lambda_3(\bar{\mathbf{x}})}{3}}, \quad \text{for} \quad \lambda_1 \simeq \lambda_2 \simeq \lambda_3. \tag{Eq. 18}$$

Otherwise, for aspherical dispersion, we need to consider the projection of the variance onto $\mathbf{e}^3(0)$,

$$\mathrm{MID} = \sqrt{\sum_m \lambda_m(\bar{\mathbf{x}})[\mathbf{e}^m(\bar{\mathbf{x}}) \cdot \mathbf{e}^3(0)]^2}. \tag{Eq. 19}$$

Hawaiian bimodal vectors

Let us now illustrate the utility of principal component analysis with paleomagnetic data. For this example, we consider Hawaiian paleomagnetic vector data coming from lava flows emplaced over the past 5 Ma (Love and Constable, 2003), a period of time that encompasses several periods of normal and reverse field polarity. We consider only those flows that, upon sampling and subsequent measurement, have yielded complete triplets of intensity, inclination, and declination ($F(i)$, $I(i)$, $D(i)$), and which, therefore, represent the full ambient magnetic vector at the time of deposition. Calculating the covariance matrix about the origin, using $\mathbf{r} = 0$ in Eq. (3), we perform a principal component analysis to obtain the eigenvalues and corresponding eigenvectors; see Table P3. The major axis along $\mathbf{e}^3$ is orientated almost parallel to the mean direction found by others using other methodologies. This axis is shown in Figure P59a-P59c, where we see that it passes through both the zero point origin and the cloud of points defining the paleosecular variation. As a physical interpretation, it is this axis about which the geomagnetic field varies over time and, even, occasionally reverses its polarity. In Figure P59d we show the projection of the data onto the eigenplane $z_1 - z_2$, where we also plot the variance ellipse defined by the projection of the variance ellipsoid.

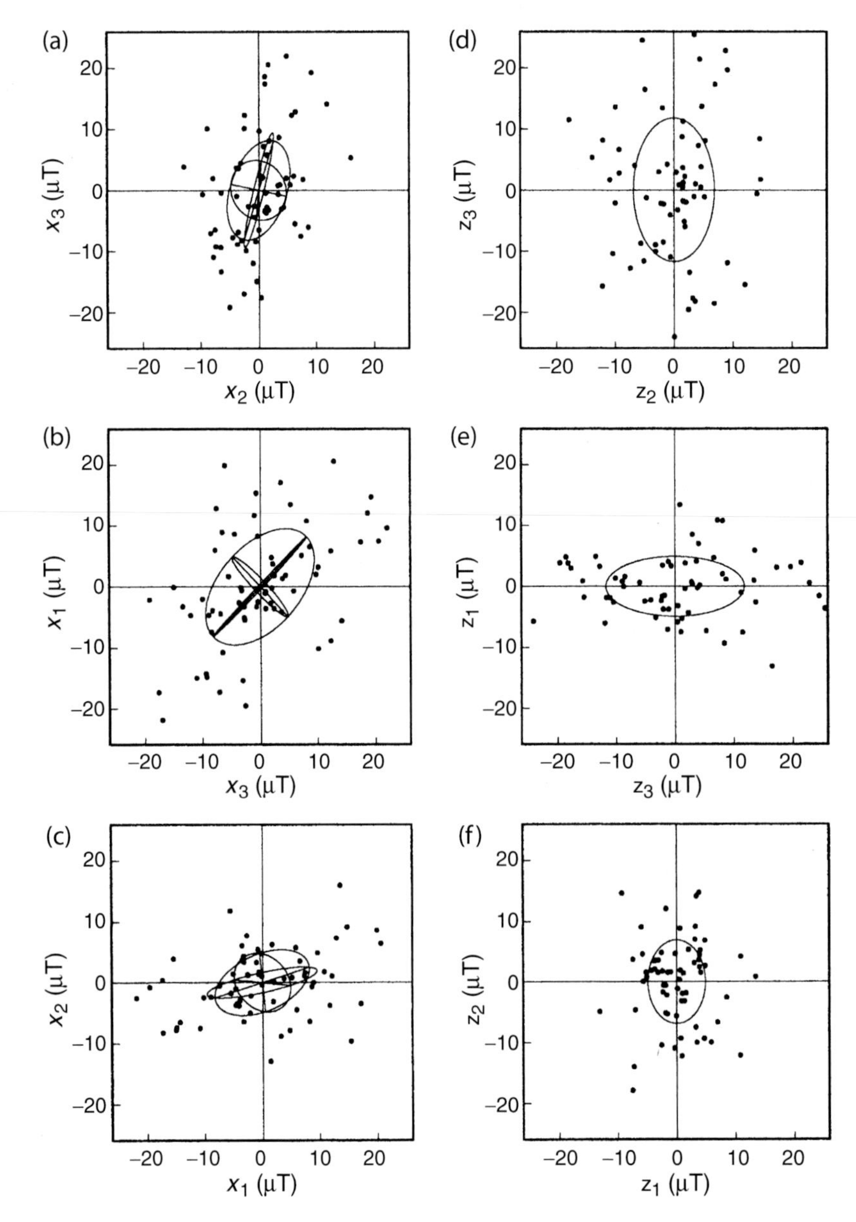

Figure P60 Projection of Brunhes Hawaiian paleovector secular-variation data (with mean vector having been subtracted) onto (a-c) the three geographic planes defined by (x_1, x_2, x_3) and onto (d-f) the three eigenplanes defined by (z_1, z_2, z_3). Also shown are the projections of the three equators of the variance ellipsoid in each coordinate system.

The slightly asymmetric form of the paleosecular variation about the major axis is to be noted.

Hawaiian unimodal vectors

Next, let us consider Brunhes normal data from Hawaii. Results of a principal component analysis of these data for reference point r = 0 are given in Table P3. In Figure P59e-P59g we plot the major axis defined by $\mathbf{e}^3$, and in Figure P59h we show the projection of the data onto the eigenplane $z_1 - z_2$, along with the corresponding variance ellipse. The results here are not dramatically different from the previous result, where we used data of mixed polarities, although some slight differences in the orientation of the major axis and variance size are noted. In order to better inspect the nature of the paleosecular variation at Hawaii during the Brunhes, we perform a principal component analysis of the covariance about the mean vector $\mathbf{r} = \bar{\mathbf{x}}$. In Table P3 we see that much of the secular variation is roughly parallel with the mean vector; the angular difference between the orientation of the two vectors is only 14.9°. The geometric form of the variance of the secular variation is shown in Figure P60, where we plot both the data and the variance ellipse in geographic and eigencoordinates. The utility of measuring the variance about the vectorial mean $\mathbf{r} = \bar{\mathbf{x}}$ and in the eigenspace $\mathbf{z}$ should now be obvious.

Directional analysis

Usually, paleomagnetists do not have complete vectorial data. Instead, directional data, consisting of inclination and declination values ($I(i)$, $D(i)$), are most commonly available. Therefore, let us consider a principal component analysis for directional-only data. The equivalent Cartesian expression of the data is

$$\begin{aligned}\hat{x}_1(i) &= \cos I(i) \cos D(i),\\ \hat{x}_2(i) &= \cos I(i) \sin D(i),\\ \hat{x}_3(i) &= \sin I(i).\end{aligned} \qquad \text{(Eq. 20)}$$

With N such directional data we can calculate their covariance about the defined origin $\mathbf{r} = (0, 0, 0)$

$$\hat{\mathbf{C}} = \frac{1}{N}\begin{pmatrix} \sum \hat{x}_1(i)^2 & \sum \hat{x}_1(i)\hat{x}_2(i) & \sum \hat{x}_1(i)\hat{x}_3(i) \\ \sum \hat{x}_2(i)\hat{x}_1(i) & \sum \hat{x}_2(i)^2 & \sum \hat{x}_2(i)\hat{x}_3(i) \\ \sum \hat{x}_3(i)\hat{x}_1(i) & \sum \hat{x}_3(i)\hat{x}_2(i) & \sum \hat{x}_3(i)^2 \end{pmatrix}. \qquad \text{(Eq. 21)}$$

As before, eigenvalues λ_m and eigenvectors $\mathbf{e}^m$ can be obtained for this matrix. Because all the data are unit vectors, the total variance measured relative to the origin is one and the trace of $\hat{\mathbf{C}}$ is unity:

$$\mathrm{tr}(\hat{\mathbf{C}}) = 1, \qquad \text{(Eq. 22)}$$

and so the sum of the three eigenvalues is determined,

$$\lambda_1 + \lambda_2 + \lambda_3 = 1. \qquad \text{(Eq. 23)}$$

This means that the eigenvalues have only two degrees of freedom. For a prolate variance ellipsoid, the approximate maximum angular deviation from the major axis along $\mathbf{e}^3$ is

$$\mathrm{MAD_p} = \tan^{-1}\left[\sqrt{\frac{1-\lambda_3}{\lambda_3}}\right], \qquad \text{(Eq. 24)}$$

and for an oblate variance ellipsoid, the corresponding angular deviation from the plane normal to $\mathbf{e}^1$ is

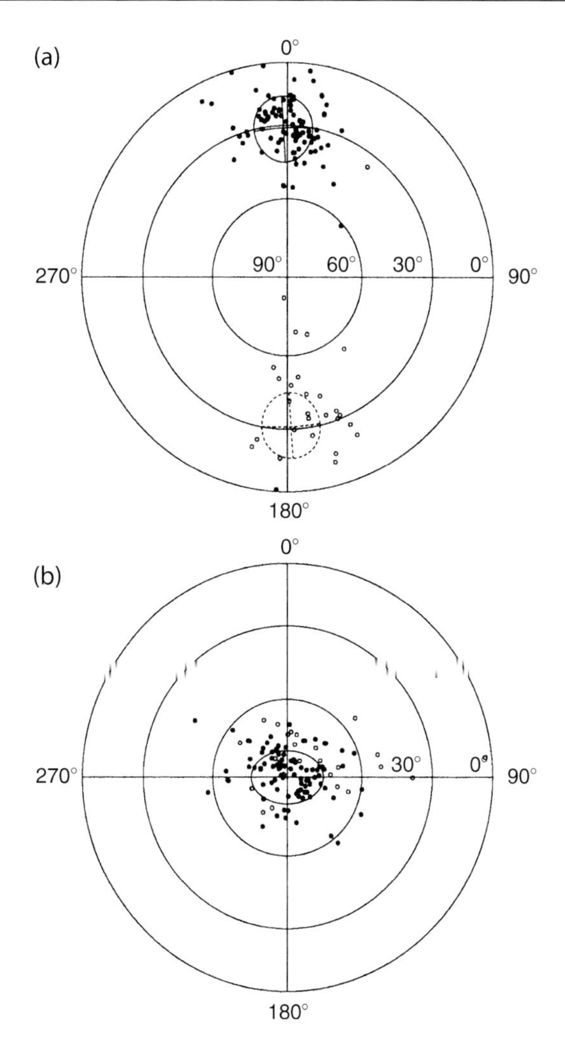

Figure P61 Equal-area projection of Hawaiian directional data, defined in (a) geographic coordinates and (b) eigen coordinates. Also shown are the projections of the variance minor ellipse, defined by λ_1 and λ_2. As is conventional, the azimuthal coordinate is declination (clockwise positive, 0°–360°), and the radial coordinate is inclination (from 90° in the center to 0° on the circular edge).

$$\mathrm{MAD_o} = \tan^{-1}\left[\sqrt{\frac{\lambda_1}{1-\lambda_1}}\right]. \qquad \text{(Eq. 25)}$$

Hawaiian bimodal directions

As a final example, we return to the mixed polarity Hawaiian data covering the past 5 Ma, but this time we only consider the paleodirections $(I(i), D(i))$. Results of a principal component analysis of the corresponding unit vectors for reference point $r = 0$ are given in Table P3. In Figure P61a we plot the data, and the minor ellipse defined by λ_1 and λ_2, in an equal-area projection of geographic coordinates. After rotation into eigenspace (Figure P61b), we see, quite clearly, the asymmetric variance of paleodirections, thus, demonstrating the utility of inspecting directional data in the eigenspace $\mathbf{z}$.

Jeffrey J. Love

Bibliography

Hotelling, H., 1933. Analysis of a complex of statistical variables into principal components. *Journal of Educational Psychology*, **24**: 417–441, 498–520.

Jolliffe, I.T., 2002. *Principal Component Analysis*, 2nd edn. New York: Springer-Verlag.

Karhunen, K., 1947. Über lineare Methoden in der Wahrscheinlichkeitsrechnung. *Annales Academiae Scientiarum Fennocaeseries A1*, **37**: 3–79.

Kirschvink, J.L., 1980. The least-squares line and plane and the analysis of palaeomagnetic data. *Geophysical Journal International*, **62**: 699–718.

Loéve, M., 1977. *Probability Theory*, 4th edn. New York: Springer-Verlag.

Love, J.J. and Constable, C.G., 2003. Gaussian statistics for palaeomagnetic vectors. *Geophysical Journal International*, **152**: 515–565.

Mardia, K.V., 1972. *Statistics of Directional Data.* New York: Academic Press.

Pearson, K., 1901. On lines and planes of closest fit to systems of point in space. *Philosophical Magazine, series 6,* **2**: 559–572.

Schomaker, V., Wasser, J., Marsh, R.E. and Bergman, G., 1959. To fit a plane or a line to a set of points by least squares. *Acta crystallographica*, **12**: 600–604.

Strang, G., 1980. *Linear Algebra and Its Applications*, New York: Academic Press.

Cross-references

Bingham Statistics
Fisher Statistics
Magnetic Remanence, Anisotropy
Paleomagnetic Secular Variation
Statistical Methods for Paleovector Analysis

PROJECT MAGNET

Project Magnet is a comprehensive vector aeromagnetic surveying enterprise that spanned much of the last five decades. Under the direction of the United States Navy and management by the Naval Oceanographic Office, test flights began in 1951 and full operational capabilities were established in 1953. The project ran through 1994, ultimately contributing many thousands of track miles (see Figure P62) of geomagnetic data (Coleman, 1992). The data are available on CD-ROM through the National Geophysical Data Center (Hittelman *et al.*, 1996).

The primary purpose of Project Magnet was to supply data in support of the World Magnetic Modeling (WMM) and charting program, which in turn supported civilian and military navigation requirements. The WMM has been incorporated into many global positioning system (GPS) receivers manufactured in the United States and has been used to control drift rates in inertial navigation systems. As a geophysical tool, the WMM has been useful as a reference measurement for Earth's core–mantle boundary field and as an aid in geophysical prospecting and resource evaluation.

Over the years, Project Magnet surveys have been performed using five different aircrafts, with each successive one improving in range, speed or altitude, navigational capabilities, and geophysical instrumentation. From 1953 to 1970 survey aircraft flew at 4615 m and most surveys were confined to remote ocean areas. Navigation methods in use at the time were periodic celestial fixes, LORAN, and dead reckoning. As a result, navigational accuracy was rather poor and was limited to about 5 nautical miles. Altitudes were determined using a baroclinic altimeter, with an uncertainty of ± 30 m. Observations of declination, inclination, and intensity (to an accuracy of ± 15 nT) were made using a self-orienting fluxgate magnetometer, while a towed, optically pumped metastable helium magnetometer measured field intensity to ± 4 nT. Until 1970, data acquisition systems consisted primarily of strip chart recorders and navigation logs. The majority of this data has been manually digitized.

The introduction of a new aircraft in 1970 permitted high level (over 4615 m) vector aeromagnetic surveying, usually conducted at altitudes between 6200 and 7700 m. During this era, the use of inertial navigation systems improved navigational accuracy to about 1 nautical mile, and in 1987, after the appearance of GPS, accuracies were further increased to several tens of meters, but only when a reliable signal was available. A baroclinic altimeter similar to the previous one was again the only source of altitude data. Improved magnetic measurements were facilitated by a fluxgate magnetometer, providing three vector components X, Y, and Z in the local reference frame to accuracies of ± 40 nT. An optically pumped metastable helium magnetometer mounted on a stinger extending from the rear of the aircraft measured intensity to ± 1 nT.

Major technological improvements arrived with the 1990s and by 1992 the Project Magnet aircraft had been fitted with an ASG-81 scalar magnetometer, a NAROD ring-core fluxgate vector magnetometer, and a ring-laser gyro (RLG) inertial system. The latter two instruments were mounted on a rigid beam in a magnetically clean area at the rear of the aircraft. The RLG was used primarily for attitude determination, while GPS, a radar system and a precision barometer were employed for altitude measurements, which were determined to a precision of less than 2 m. The GPS system was the primary navigational tool, with a circular error probability of 15 m. However, the accuracy of the magnetic measurements remained the same. Technological advancements also led to the incorporation of a number of other surveying capabilities, including gravity, ocean acoustics, and ocean temperature.

Survey data were calibrated by several low level "airswings" prior to and during the high level surveying. The airswings included straight and level passes as well as roll, pitch and yaw maneuvers in each cardinal direction at 1000 feet over a ground based observatory. The calibration data were used to model aircraft intrinsic magnetic fields that perturb the magnetometer data. The contaminating fields are due to: the permanently magnetized part of the aircraft; magnetization induced by the core field; and eddy currents driven by changes in the crustal and core fields as the aircraft passed through them. A compensation model, based on developments by Leliak (1961), involves a two-step iterative least squares solution for 21 coefficients, which characterize the contaminating fields, and three bias angles, which account for differences in the orientations of the vector magnetometer and the inertial navigation system. Measurements taken over all airswing maneuvers are then included in a least squares minimization of the difference between the magnitudes of the decontaminated aeromagnetic observations and the transformed ground based observations. This transformation involves upward continuation to aircraft altitude and rotation into the aircraft reference frame using roll, pitch and heading data.

The compensation model is imprecise in that it is based on the assumption that the coefficients are constant with respect to latitude and frequency of the airswing maneuvers. There have been attempts to account for these variations and the Project Magnet database includes compensation coefficients from repeated calibration flights at observatories over the globe. More importantly, Project Magnet orientation accuracy is insufficient to prevent contamination of the magnetic field vector components due to misalignment of the aircraft coordinate system (Parker and O'Brien, 1997). Because field intensity is immune from orientation errors, Parker and O'Brien (1997) show that simple spectral analysis of intensity and the three vector magnetic components as functions of along track distance may be used as a diagnostic for this contamination.

Although one of the primary goals of Project Magnet was to characterize the magnetic field generated in Earth's core, the majority of published studies exploiting Project Magnet data have been focused on the crustal part of the field, including numerous regional magnetic

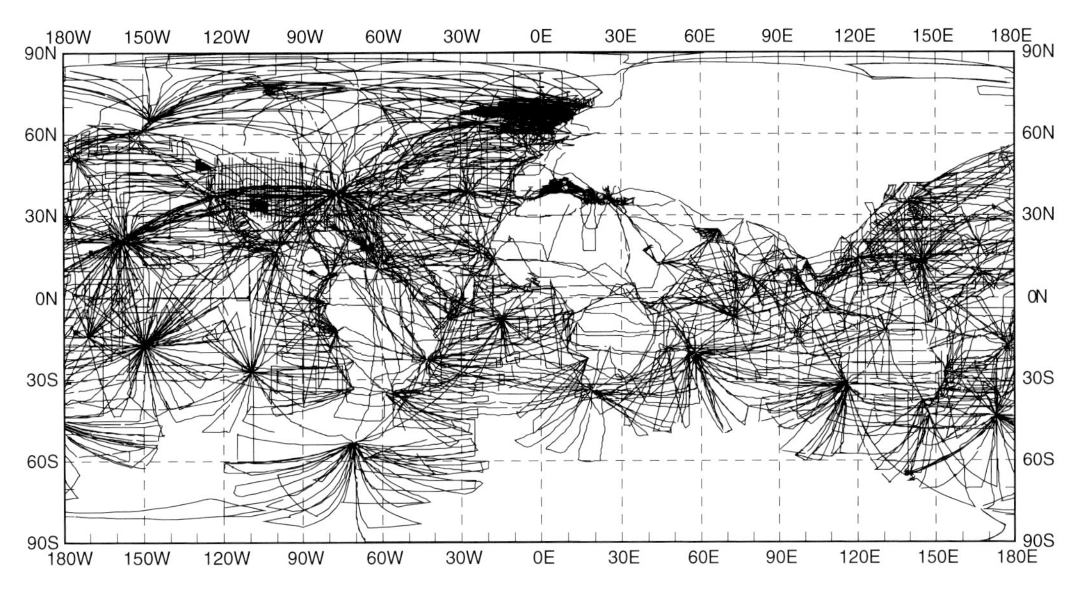

Figure P62 Coverage of Project Magnet aeromagnetic surveys, 1953–1994 (cylindrical equidistant projection). Reproduced by permission of the National Geophysical Data Center.

mapping efforts spread across the globe. These include sub-Saharan Africa (Green, 1976), South Georgia Island (Simpson and Griffiths, 1982), Iceland (Jonsson *et al.*, 1991), and New Zealand (McKnight, 1996). A detailed study by Horner-Johnson and Gordon (2003) of short wavelength features (20–150 km) over the equatorial Pacific, suggests that magnetic anomalies associated with seafloor spreading are more clearly delineated by vector aeromagnetic data than by the traditional shipboard total field observations. O'Brien *et al.* (1999) developed a method, based on spectral analysis of Project Magnet data along great circle paths, to estimate the geomagnetic spatial spectrum of the oceanic crust for spherical harmonic degrees from $l \approx 60$ to 1200. Observational models based on satellite data had previously been limited to $l \approx 100$. Korte *et al.* (2002) observed that the agreement of the new model with satellite based observations was unsatisfactory for degrees $l < 100$, and found that O'Brien *et al.* had overcompensated for external magnetic field variations in their analysis. The improved model of Korte *et al.* (2002) achieves better agreement with theoretical and satellite based models of the spatial spectrum.

The advent of satellite based magnetic observations, in particular Magsat in the 1980s raised the question of agreement between aeromagnetic and spaced based geomagnetic observations. An early study by Won and Son (1982), using Project Magnet data spanning the continental US and upward continued to Magsat altitude, revealed agreement in gross features of the crustal field. Using an improved method for removing the core field (spherical harmonic degrees $l \leq 13$) and long wavelength external contributions, Wang *et al.* (2000) demonstrated excellent agreement between aeromagnetic and satellite data over the continental US, and suggest that an anomaly field derived from a combined dataset provides accurate estimates of the long wavelength crustal field.

David G. McMillan

Bibliography

Coleman, R.J., 1992. Project magnet high-level vector survey data reduction. In Langel, R.A., and Baldwin, R.T. (eds.), *Types and Characteristics of Data for Geomagnetic Field Modelling*. NASA Conference Publication 3153, pp. 215–248.

Green, A.G., 1976. Interpretation of project magnet aeromagnetic profiles across Africa. *The Geophysical Journal of the Royal Astronomical Society*, **44**: 203–228.

Hittelman, A.M., Buhmann, R.W., and Racey, S.W., 1996. *Project Magnet Data (1953–1994) CD-ROM User's Manual.* Boulder: National Geophysical Data Center.

Horner-Johnson, B.C., and Gordon, R.G., 2003. Equatorial pacific magnetic anomalies identified from vector aeromagnetic data. *Geophysical Journal International*, **155**: 547–556.

Jonsson, G., Kristjansson, L., and Sverrisson, M., 1991. Magnetic surveys of Iceland. *Tectonophysics*, **189**: 229–247.

Korte, M., Constable, C.G., and Parker, R.L., 2002. Revised magnetic power spectrum of the Oceanic crust. *Journal of Geophysical Research*, 107, EPM 6.

Leliak, P., 1961. Identification of magnetic field sources of magnetic airborne detector equipped aircraft. *IRE Transactions on Aerospace and Navigational Electronics*, **8**: 95–105.

McKnight, J.D., 1996. An updated regional geomagnetic field model for New Zealand. Institute of Geological and Nuclear Sciences Science Report: Report 96/4; M24/773; M24/774.

O'Brien, M.S., Parker, R.L, and Constable, C.G., 1999. Magnetic power spectrum of the Ocean crust on large scales. *Journal of Geophysical Research*, **104**: 29189–29201.

Parker, R.L., and O'Brien, M.S., 1997. Spectral analysis of vector magnetic profiles. *Journal of Geophysical Research*, **100**: 20111–20136.

Simpson, P., and Griffiths, D.H., 1982. The structure of the South Georgia Continental block. In Craddock, C. (ed.) *Antarctic Geoscience.* Madison, WI: University of Wisconsin Press, pp. 185–191.

Wang, B., Ravat, D., Sabaka, T., and Hildenbrand, T.G., 2000. Long-wavelength magnetic field for the conterminous US. from project magnet and magsat data. *Eos,Transactions of the American Geophysical Union*, **81**, GP71B-09.

Won, I.J., and Son, K.H., 1982. A preliminary comparison of the MAGSAT data and aeromagnetic data in the Continental US. *Geophysical Research Letters*, **9**: 296–298.

Cross-references

Aeromagnetic Surveying
Core-mantle Boundary
Electromagnetic Induction (EM)
Geomagnetic Spectrum, Spatial
Instrumentation, History of
Magsat
Upward and Downward Continuation

PROUDMAN-TAYLOR THEOREM

Geophysicists accept a suggestion traceable back to a paper by Elsasser (1939) that the near-alignment of the main geomagnetic field (see *Main field*) with the Earth's geographic polar axis is a manifestation of the influence on convective motions in the liquid, metallic outer *core* (*q.v.*) that is exerted by gyroscopic (Coriolis) forces associated with the Earth's diurnal rotation. Such forces render core motions anisotropic and highly sensitive to the mechanical and other (thermal and electromagnetic) boundary conditions imposed on the motions by the underlying solid inner *core* (*q.v.*) and the overlying *mantle* (*q.v.*) (for further references see Hide, 2000).

The influence of Coriolis forces on motions in spinning fluids is most pronounced in mechanically-driven flows satisfying the so-called Proudman-Taylor (PT) theorem. Stated in words, the theorem (see Eq. (5)) shows that "slow and steady hydrodynamical motion of an inviscid (and electrically insulating) fluid of constant density that is otherwise in steady "rigid-body" rotation will be the same in all planes perpendicular to the axis of rotation". Given in a paper by J. Proudman in 1916, the PT theorem was successfully tested by means of a laboratory experiment in which G.I. Taylor investigated the relative flow produced in a tank of water in rapid rotation about a vertical axis by the slow and steady horizontal motion of a solid body through the water (see Greenspan, 1968). In accordance with the theorem, the flow was highly two-dimensional nearly everywhere and the moving solid object carried with it a "Taylor column" of water extending axially throughout the whole depth of the tank.

We note here in passing (a) that the theorem had appeared much earlier in the literature, in a paper on tidal theory by S.S. Hough (Gill, 1982), and (b) that some writers find it convenient to refer to it as the "Proudman theorem" (Hide, 1977), to avoid confusion with another theorem, due to G.I. Taylor, concerning effects of Coriolis forces on fluid motions (see Greenspan, 1968; Hide, 1997).

Convective motions in the *core* (*q.v.*) cannot satisfy the PT theorem exactly because they are driven by buoyancy forces due to the action of gravity on density inhomogeneities, which give rise to axial variations of the horizontal components of the relative flow velocity (see Eq. (6)). Core motions are also influenced by Lorentz forces associated with electric currents and magnetic fields in the core, which in regions well below the *core-mantle boundary* (*q.v.*) may be comparable in strength with Coriolis forces (see Eqs. (1) and (2)). But—as in theoretical work in dynamical meteorology and oceanography, see e.g., Gill, 1982; Pedlosky, 1987—certain basic dynamical processes and phenomena can be elucidated by means of simplified theoretical models in which axial variations of the nonaxial components of the relative flow velocity are neglected in the first instance.

Characteristic shear waves that occur in fluids subject to Coriolis and/or Lorentz forces—such as Rossby waves, Alfvén (*magnetohydrodynamic waves* (*q.v.*)) and related hybrid waves (one important type of which may underlie the *geomagnetic secular variation* (*q.v.*))—have been investigated in this way. And so have other phenomena, such as torsional oscillations of the core-mantle system associated with angular momentum transfer within the Earth's core (Hide *et al.*, 2000) and differential rotation produced in a rapidly rotating spherical fluid annulus by *potential vorticity* (*q.v.*) mixing (Hide and James, 1983).

Geostrophic and magnetostrophic flows

Flows satisfying the PT theorem belong to a wider class of "geostrophic" flows (see Gill, 1982; Pedlosky, 1987), for which the horizontal component of the local pressure gradient, **grad** p, is closely balanced by the horizontal component of the Coriolis force (per unit volume), $2\rho\boldsymbol{\Omega}\mathbf{x}\boldsymbol{u}$ (see Eq. (1)), but torques produced by buoyancy forces associated with horizontal density gradients give rise to systematic axial variations in the horizontal components of $\boldsymbol{u}$ (see Eqs. (5) and (6)).

Consider a moving element of fluid of density ρ which at time t is located at a general point P with vector position $\boldsymbol{r}$ in a frame of reference that rotates with angular velocity $\boldsymbol{\Omega}$ relative to an inertial frame. Newton's second law is conveniently expressed as follows (see Hide, 1971):

$$2\rho\boldsymbol{\Omega}\mathbf{x}\boldsymbol{u} + \mathbf{grad}\,p + \rho\,\mathbf{grad}\,V + A = 0 \qquad \text{(Eq. 1)}$$

where the Eulerian flow velocity at P and the corresponding pressure and the potential due to gravitational plus centripetal effects are denoted by $\boldsymbol{u}$, p, and V respectively. The "ageostrophic" term, $\boldsymbol{A}$, in Eq. (1) is defined as follows:

$$\boldsymbol{A} = \rho[\partial\boldsymbol{u}\ \partial t + (\boldsymbol{u}\,\mathbf{grad})\boldsymbol{u} - \boldsymbol{r}\mathrm{x}\mathrm{d}\boldsymbol{\Omega}\ \mathrm{d}t] + \mathbf{curl}\,(\nu\rho\boldsymbol{\omega}) - \boldsymbol{j}\mathbf{x}\boldsymbol{B} \qquad \text{(Eq. 2)}$$

if $\boldsymbol{\omega} = \mathbf{curl}\,\boldsymbol{u}$, the relative vorticity vector at P, and ν denotes the coefficient of kinematic viscosity. The term $\boldsymbol{j}\mathbf{x}\boldsymbol{B}$ in Eq. (2) is needed when dealing with electrically conducting fluids; it represents the Lorentz force per unit volume acting on the fluid element if $\boldsymbol{j}$ is the electric current density and $\boldsymbol{B}$ the magnetic field at P.

The flow is said to be *geostrophic* in regions where $\boldsymbol{A}$ is negligible in comparison with other terms in Eq. (1). Within the Earth's *core* (*q.v.*), such regions probably arise in its upper reaches (LeMouël, 1984; Hide, 1995) where the toroidal part of the *geomagnetic field* (*q.v.*) is no stronger than the poloidal part and the corresponding Lorentz term is typically no more than a few parts percent of the Coriolis term in magnitude. On the other hand, the flow is said to be in "*magnetostrophic*" (rather than geostrophic) balance in any regions where the Lorentz force provides the main contribution to $\boldsymbol{A}$ and is comparable in magnitude with $2\rho\boldsymbol{\Omega}\mathbf{x}\boldsymbol{u}$. Such regions probably arise at depth within the core, where the toroidal part of the magnetic field may be an order of magnitude stronger than the poloidal part.

Following Hide, 1971, now consider the balance of torques acting on the moving fluid element by taking the **curl** of Eq. (1), thereby obtaining the so-called "vorticity" equation in the useful form

$$2\boldsymbol{\Omega}\ \mathbf{grad}\,(\rho\boldsymbol{u}) + \mathbf{grad}\,V\,\mathrm{x}\,\mathbf{grad}\,\rho + \mathbf{curl}\,\boldsymbol{A} + 2\boldsymbol{\Omega}C = 0 \qquad \text{(Eq. 3)}$$

Here $C = \partial\rho/\partial t$ which, by the mass continuity equation, satisfies

$$\mathbf{div}\,(\rho\boldsymbol{u}) + C = 0 \qquad \text{(Eq. 4)}$$

The PT theorem follows immediately from Eq. (3), for when the conditions $\boldsymbol{A} = 0$, $\mathbf{grad}\ \rho = 0$ and $C = 0$ are satisfied, the equation reduces to

$$(2\boldsymbol{\Omega}\ \mathbf{grad})\ \boldsymbol{u} = 0 \qquad \text{(Eq. 5)}$$

The above equation implies that all three components of $\boldsymbol{u}$ are independent of the axial coordinate. When the restriction to cases of fluids of constant density is relaxed we have

$$2\boldsymbol{\Omega}\ \mathbf{grad}\,(\rho\boldsymbol{u}) + \mathbf{grad}\,V\,\mathbf{x}\,\mathbf{grad}\,\rho = 0 \qquad \text{(Eq. 6)}$$

in place of Eq. (5).

Flow patterns satisfying Eqs. (5) or (6) are constrained by Coriolis forces to have no circulation in (meridian) planes containing the axis of rotation. But by Eq. (6) the horizontal component of $\boldsymbol{u}$ is not independent of the axial direction, except in regions where **grad** ρ has no horizontal component. We note here in passing (a) that the *specific helicity,* $\boldsymbol{u}.\boldsymbol{\omega}$,—a quantity of importance in *dynamo theory* (*q.v.*)—of flows satisfying Eq. (6) is generally nonzero (Hide, 1976), and (b) that with certain geometrical simplifications Eq. (6) leads to the meteorologist's "*thermal wind*" equation (*q.v.*) relating the rate of increase of the speed and direction of the geostrophic wind with height in the atmosphere to the horizontal gradient of temperature.

Ageostrophic effects and detached shear layers

The full expression of the laws of mechanics given by Eq. (1) is, of course, *prognostic,* in the sense that when solved simultaneously with the full equations of thermodynamics and electrodynamics under all the relevant boundary conditions the equation gives the fields of all the dependent variables, p, $\boldsymbol{u}$, ρ, $\boldsymbol{j}$, $\boldsymbol{B}$ etc. But the expression for geostrophic flow to which Eq. (1) reduces in regions where the ageostrophic term, $\boldsymbol{A}$, is negligible is *diagnostic* (rather than prognostic), in the sense that it provides useful relationships between $\boldsymbol{u}$, p and ρ, but it cannot give full solutions satisfying all the relevant boundary conditions.

It follows that flows in real systems cannot be geostrophic everywhere. Regions of ageostrophic flow involving length scales so short that $\boldsymbol{A}$ is comparable in magnitude with $2\rho\boldsymbol{\Omega}\mathbf{x}\boldsymbol{u}$ must be present not only on bounding surfaces (in Ekman-Hartmann and Stewartson boundary layers (see core, boundary layers)) but also within the main body of the fluid, in "detached shear layers". It is within such regions, where effects due to Coriolis forces are countered by strong ageostrophic effects, that meridional circulation can occur. Detached shear layers formed the "walls" of the so-called "Taylor column" of fluid that remained attached to the moving solid body in G. I. Taylor's experiment. Physical examples of ageostrophic detached shear layers embedded in geostrophic flows are the jet streams and their associated frontal systems seen in the Earth's atmosphere and "western boundary" currents such as the Gulf Stream in the Atlantic Ocean and the Kuroshio Current in the Pacific Ocean (see Gill, 1982; Pedlosky, 1987).

Amongst the various laboratory investigations that were stimulated directly by the original "Taylor column" experiment were several studies of the effect of the inner spherical boundary on patterns of mechanically-driven (Greenspan, 1968; Hide, 1977) and buoyancy-driven flows in a rotating spherical annulus of fluid. These studies clearly demonstrated what general arguments based on Eq. (5) or (6) predict, namely that near the imaginary "tangent cylindrical surface" (see Inner core tangent cylinder) in contact with the equator of the inner spherical surface and extending axially throughout the fluid, a detached shear layer would form inhibiting mixing of fluid in the "polar" regions inside the cylinder with fluid in the "equatorial" region outside the cylinder. Such behaviour is thought to bear on core dynamics and the influence of the solid inner core on structure of the geomagnetic field (for further references see Hide, 2000 and Hide, 1966a,b).

Raymond Hide

Bibliography

Elsasser, W.M., 1939. On the origin of the Earth's magnetic field. *Physical Reviews*, **55**: 489–497.

Gill, A.E., 1982. *Atmosphere-Ocean Dynamics.* New York: Academic Press.

Greenspan, H.P., 1968. *The Theory of Rotating Fluids.* Cambridge: Cambridge University Press.

Hide, R., 1966a. The dynamics of rotating fluids and related topics in geophysical fluid dynamics. *Bulletin of the American Meteorological Society*, **47**: 873–885.

Hide, R., 1966b. Free hydromagnetic oscillations of the Earth's core and the theory of the geomagnetic secular variation. *Philosophical Transactions of the Royal Society of London*, **A259**: 615–647.

Hide, R., 1971. On geostrophic motion of a nonhomogeneous fluid. *Journal of Fluid Mechanics*, **49**: 745–751.

Hide, R., 1976. A note on helicity. *Geophysical (Astrophysical) Fluid Dynamics*, **7**: 157–161.

Hide, R., 1977. Experiments with rotating fluids; Presidential address. *Quarterly Journal of the Royal Meteorological Society*, **103**: 1–28.

Hide, R., 1995. The topographic torque on the rigid bounding surface of a rotating gravitating fluid and the excitation of decadal fluctuations in the Earth's rotation. *Geophysical Research Letters*, **22**: 1057–1059.

Hide, R., 1997. On the effects of rotation on fluid motions in containers of various shapes and topological characteristics. *Dynamics of Atmospheres and Oceans*, **27**: 243–256.

Hide, R., 2000. Generic nonlinear processes in self-exciting dynamos and the long-term behaviour of the main geomagnetic field, including superchrons. *Philosophical Transactions of the Royal Society of London*, **A358**: 943–955.

Hide, R., Boggs, D.H., and Dickey, J.O., 2000. Angular momentum fluctuations within the Earth's core and torsional oscillations of the core-mantle system. *Geophysical Journal International*, **143**: 777–786.

Hide, R., and James, I.N., 1983. Differential rotation produced by large-scale potential vorticity mixing in a rapidly-rotating fluid. *Geophysical Journal of the Royal Astronomical Society*, **74**: 301–312.

Le Mouël, J.-L., 1984. Outer core geostrophic flow and the secular variation of the Earth's magnetic field. *Nature*, **311**: 734–735.

Pedlosky, J., 1987. *Geophysical Fluid Dynamics*, 2nd edn. New York: Springer-Verlag.

Cross-references

Core, Boundary Layers
Core Convection
Core-Mantle Boundary
Core Motions
Equilibration of Magnetic Field, Weak and Strong Field Dynamos
Geodynamo
Geomagnetic Secular Variation
Inner Core Tangent Cylinder
Length of the Day Variations, Decadal
Magnetohydrodynamic Waves
Oscillations, Torsional
Potential Vorticity and Potential Magnetic Field Theorems
Thermal Wind

R

RADIOACTIVE ISOTOPES, THEIR DECAY IN MANTLE AND CORE

Introduction

The Earth is a thermal engine driven largely by heat produced from the decay of naturally occurring radioactive isotopes in its interior. At present, the main radioactive isotopes are ^{40}K, ^{235}U, ^{238}U, and ^{232}Th, whose atomic percentages, radioactive decay constants, half-life, and the heat production characteristics are given in Table R1. ^{26}Al, an extinct radioactive isotope may have been an important heat source in the early history of the Earth. Many other radioactive isotopes are also present in the Earth but they play a negligible role in heat production, either because of their low abundance or their low heat producing capacity.

The abundances of K, U, and Th in common rocks from the crust and some upper mantle materials are available from direct measurements. The compilation by Van Schmus (1995) lists their abundances and also shows some primitive mantle (present mantle+crust) estimates. Since then, new estimates of these elements in the primitive mantle (bulk silicate Earth—BSE) and the Earth have appeared, as also new experimental and theoretical studies bearing on radioactivity in the core.

Radioactivity of the mantle

Direct measurements of the radioactivity of rocks from the mantle are limited to its outermost layers just a few hundred kilometers deep in the upper mantle. Much of the lower mantle is not accessible for direct measurements and geophysical parameters such as seismic velocities and mineral physics data cannot provide strong constraints to its composition (Mattern *et al.*, 2005). Current estimates of the abundances of K, U, and Th in BSE, and the Earth as a whole, are derived from several geochemical models. These combine cosmochemical considerations and composition of meteorites with some coherent trace element ratios observed in upper mantle samples and meteorites. The composition of volatile-rich C1 carbonaceous chondrites (a primitive class of meteorites with elemental abundances that closely match the composition of the solar photosphere) is used as a reference to compare the terrestrial abundances and deduce chemical fractionation patterns in the Earth (C1 models). The most comprehensive modeling of this type by McDonough and Sun (1995) provides a complete description of the methods and references to prior work. The abundance data of K, U, and Th in the primitive mantle from some of these studies (Table R2) are now included in the GERM database (GERM-A Geochemical Earth Reference Model at http://earthref.org/GERM/index.html). Implicit in these models is the assumption that the lower mantle is compositionally similar to the upper mantle and that the Fe-metallic core of the Earth has no radioactivity. Whether or not the mantle is compositionally layered is still unresolved (see for example, Anderson, 1989a,b). The K/U ratio in rocks from the crust and upper mantle is $\sim 1 \times 10^4$ but in chondrite meteorites is $\sim 8 \times 10^4$ (see Lassiter, 2004 for an up-to-date discussion of K/U ratios in terrestrial materials and meteorites). In the geochemical models, the low K/U ratio in the Earth, and the estimated low K abundance in the Earth relative to meteorites is attributed to the loss of moderately volatile K from the Earth.

Some recent experiments have raised questions about the C1-chondrite comparisons and the postulated loss due to volatility. Oxygen, the most abundant major element in the Earth (>30% by mass) has a different isotopic composition from that in carbonaceous and ordinary chondrites (Clayton, 1993) casting doubt on their suitability for the precursor Earth material. The identical K isotopic composition of terrestrial and meteoritic material indicates that the loss of K could not be due to partial vaporization from the Earth but must have preceded the process of planetary formation (Humayun and Clayton, 1995). Furthermore, C1 meteorites cannot provide for the ~30% Fe abundance in the Earth (Javoy, 1995).

Table R1 Important heat producing radioactive isotopes in the Earth: their decay constants, half-lives and heat production values (Van Schmus, 1995)

Isotope	Atomic percentage	Decay constant, λ (yr^{-1})	Half-life (yr)	Specific isotopic heat production ($\mu W kg^{-1}$)
^{40}K	0.01167	5.54×10^{-10}	1.251×10^{9}	29.17
^{235}U	0.7200	9.85×10^{-10}	7.038×10^{8}	568.7
^{238}U	99.2743	1.551×10^{-10}	4.468×10^{9}	94.65
^{232}Th	100.00	4.95×10^{-11}	1.401×10^{10}	26.38

Table R2 Some geochemical estimates of K, U, and Th in Earth's primitive mantle (mantle plus crust or bulk silicate Earth (BSE))

K	U	Th	References
231	0.021	–	Wanke *et al.* (1984)
180	0.018	0.064	Taylor and McClennan (1985)
266	0.0208	–	Hart and Zindler (1986)
258	0.0203	0.0813	Hofmann (1988)
240	0.0203	0.0795	McDonough and Sun (1995)
235	0.0202	0.0764	Van Schmus (1995)[a]

All values are in parts per million.
[a]Suggested average values.

Drake and Righter (2002) have discussed the general difficulty of using chondritic meteorites for the initial starting composition of the Earth. Given these considerations, and the debate about whether the mantle is compositionally layered, the geochemical estimates of K, U, and Th in Table R2 seem not well constrained.

Geophysical models use terrestrial heat-flow data or other aspects of geodynamics and plate tectonics to constrain the radioactivity of the mantle. Global average heat-flow estimates range from 30 to 44 TW (TW $= 10^{12}$ W). The higher value (Pollack *et al.*, 1993; Stein, 1995) is greater than twice the radiogenic heat production of various C1 models (see compilation by Lodders and Fegley, 1998). Even at the recently suggested heat flux of 31 $\pm$ 1 TW (Hofmeister and Criss, 2005), the C1 models fall short by about 10–12 TW. This short-fall may be satisfied by considering a significant secular cooling delay of $>$1–2 Ga (Ga $= 10^9$ years) where the surface heat flux represents radiogenic heat produced in the past, or by including residual primordial heat from the time of accretion. Definitive conclusions about either of these possibilities cannot be reached at present. Alternatively, additional radioactive heat sources in the interior can also meet the global surface heat flux requirement.

The only meteorite materials with O-isotopic composition similar to the Earth (and the Moon) are the enstatite chondrites (EH, EL) and the enstatite achondrites (aubrites). Models of the Earth with EH chondrites as precursor materials have been considered in some recent studies (Javoy, 1995; Lodders, 1995, 2000; Hofmeister and Criss, 2005). These models have the advantage of simultaneously satisfying the stringent requirement of the O-isotopic composition, the high iron content of the Earth, and the global heat-flow data. Models of this type result in a BSE with higher radioactive contents than those listed in Table R2. For example, Hofmeister and Criss (2005) find that suitable BSE compositions that satisfy the surface heat flux of 31 TW can be represented by compositions or mixtures with EH meteorites as one end member. If the terrestrial K/U of 10^4 is a constraint, for no secular cooling delay or delay of $<$1 Ga, a BSE composition with 300–350 ppm K, about 30–35 ppb U, and 110–130 ppb Th is required. Higher heat fluxes lead to even higher K, U, and Th contents. Using essentially EH chondrites as precursor materials, Lodders (1995) estimated the model BSE to contain over 1 000 ppm K. It thus appears that estimates of the radioactivity of the silicate portion on the Earth ($\sim$68% by mass of the planet) cannot be uniquely constrained at present; rather, we have some model-dependent estimates that need further evaluation.

Radioactivity of the core

Radioactivity in the core has been a controversial topic for a long time. The traditional view is that there is no radioactivity in the core (see for example, McDonough, 1999). In contrast, some recent developments in experiments and theory suggest the possible presence of radioactivity in the core. Of the radioactive elements discussed here, K is the only element for which there is some basis both in theory and experiment (Murthy *et al.*, 2003, Lee *et al.*, 2004 and the references cited in the papers). Table R3 summarizes the theoretical and experimental estimates of K from these studies and the corresponding radiogenic heat production in the core. U and Th have been hypothesized to be in the core (Herndon, 1980), but the experimental investigations so far are inconclusive or contradictory.

The estimated core-mantle boundary (CMB) heat flux ranges from 2 to 12 TW (Labrosse and Macouin, 2003 and references cited therein). The CMB heat flux controls the rate of core cooling. Nimmo *et al.* (2004) have studied the question of how best to reconcile the CMB heat flux with the small size of inner core together and the $>$3.5 Ga age of the magnetic field. These authors note that the presence of K in the core at about 400 ppm is in best accord with the present size of the inner core and the power needs of a geomagnetic dynamo for the past 3.5 Ga. Using somewhat different parameters, similar conclusions have been reached by others (Labrosse, 2003 and the references cited therein). Purely thermal models of the core call for an additional heat source in the core and in view of the recent experimental data, it is reasonable to attribute this to the radioactivity of K in the core.

Table R3 Estimates of potassium abundance in the core from geochemical models, theoretical calculations, and recent experiments and the corresponding heat production in terawatts (TW) today

Method	Abundance (ppm)	Radiogenic heat production (TW)	References
Geochemical models	0	0	McDonough and Sun (1995), McDonough (1999), and GERM database
Theoretical	550±260	~4–5	Lodders (1995)[a]
	200–400		Buffett (2003)
	250–750		Labrosse (2003)
	Up to 1420	9	Roberts *et al.* (2003)[b]
	400	~3	Nimmo *et al.* (2004)
Experimental	<1	0.01	Chabot and Drake (1999)
	100–250	~0.8–2.0	Gessman and Wood (2002)
	60–130	0.4–0.8	Murthy *et al.* (2003)
	Up to 7000	Up to 45	Lee and Jeanloz (2003)[c]
	35	0.23	Hirao *et al.* (2005)

The GERM database values are shown for comparison.
[a]With chondritic U and Th.
[b]As quoted in Labrosse and Macouin (2003).
[c]See discussion in Lee and Jeanloz (2003). This is an upper limit; for realistic conditions of core formation in the Earth, the authors note that the value is likely to be much less.

We have made major strides in deciphering the chemistry and physics of the Earth's interior in the past few decades but the specific radioactive content of the mantle is not well constrained yet. The new evidence for radioactivity and radiogenic heat in the core is receiving much increased attention from geophysical and geochemical theorists and experimentalists because of its impact on the thermal and chemical evolution of the core, the planet, and the geomagnetic field.

V. Rama Murthy

Bibliography

Anderson, D.L., 1989a. Composition of the Earth. *Science*, **243**: 367–370.

Anderson, D.L., 1989b. *Theory of the Earth*. Boston: Blackwell Scientific Publications (http://resolver.caltech.edu/CaltechBOOK:1989.001).

Buffett, B.A., 2003. The thermal state of the earth's core. *Science*, **299**: 1675–1677.

Chabot, N.L., and Drake, M.J., 1999. Potassium solubility in metal: the effects of composition at 15kbar and 1900°C on partitioning between iron alloys and silicate melts. *Earth and Planetary Science Letters*, **172**: 323–335.

Clayton, R.M., 1993. Oxygen isotopes in meteorites. *Annual Review of Earth and Planetary Sciences*, **21**: 115–149.

Drake, M.J., and Righter, K., 2002. Determining the composition of the Earth. *Nature*, **416**: 39–44.

GERM, *A Geochemical Earth Reference Model* (http://earthref.org/GERM/index.html).

Gessman, C.K., and Wood, B.J., 2002. Potassium in the Earth's core? *Earth and Planetary Science Letters*, **200**: 63–78.

Hart, S.R., and Zindler, A., 1986. In search of a bulk-Earth composition. *Chemical Geology*, **57**: 247–267.

Herndon, J.M., 1980. The chemical composition of the interior shells of the Earth. *Proceedings of the Royal Society of London Series A*, **372**: 149–154.

Hirao, N., Ohtani, E., Kondo, T., Endo, N., Kuba, T., Suzuki, T., and Kikegawa, T., 2005. Partitioning of potassium between iron and silicate at the core-mantle boundary. *Geophysical Research Letters*, **33**: L08303.

Hofmann, A.W., 1988. Chemical differentiation of the Earth. *Earth and Planetary Science Letters*, **90**: 297–314.

Hofmeister, A.M., and Criss, R.E., 2005. Earth's heat flow revised and linked to chemistry. *Tectonophysics*, **395**: 159–170.

Humayun, M., and Clayton, R.N., 1995. Potassium isotope cosmochemistry: genetic implications of volatile element depletion. *Geochimica et Cosmochimica Acta*, **59**: 2131–2148.

Javoy, M., 1995. The integral enstatite chondrite model of the Earth. *Geophysical Research Letters*, **22**: 2219–2222.

Labrosse, S., 2003. Thermal and magnetic evolution of the Earth's core. *Physics of the Earth and Planetary Interiors*, **140**: 127–143.

Labrosse, S., and Macouin, M., 2003. The inner core and the geodynamo. *Comptes Rendus Geoscience*, **335**: 37–50.

Lassiter, J.C., 2004. Role of recycled oceanic crust in the potassium and argon budget of the Earth: toward a resolution of the "missing argon" problem. *Geochemistry Geophysics Geosystems*, **5**: Q11012 (doi: 10.1029/2004GC000711).

Lee, K.K.M., and Jeanloz, R., 2003. High-pressure alloying of potassium and iron: Radioactivity in the Earth's core? *Geophysical Research Letters*, **30**: 2312 (doi: 10.1029/2003GL018515).

Lee, K.K.M., Steinle-Neumann, G., and Jeanloz, R., 2004. *Ab-initio* high-pressure alloying of iron and potassium: implications for the Earth's core. *Geophysical Research Letters*, **31**: L11603 (doi: 10.1029/2004GL019839).

Lodders, K., 1995. Alkali elements in the Earth's core: evidence from enstatite chondrites. *Meteoritics*, **30**: 93–101.

Lodders, K., 2000. An oxygen isotope mixing model for the accretion and composition of rocky planets. *Space Science Reviews*, **92**: 341–354.

Lodders, K., and Fegley, B.J., Jr., 1998. *The Planetary Scientist's Companion*. Oxford: Oxford University Press.

Mattern, E., Matas, J., Ricard, Y., and Bass, J., 2005. Lower mantle composition and temperature from mineral physics and thermodynamic modeling. *Geophysical Journal International*, **160**: 973–990.

McDonough, W.F., 1999. Earth's core. In Marshall, C.P., and Fairbridge, R.W. (eds.), *Encyclopedia of Geochemistry*. Dordrecht: Kluwer Academic Publishers.

McDonough, W.F., and Sun, S.-S., 1995. The composition of the Earth. *Chemical Geology*, **120**: 223–253.

Murthy, V.R., van Westrenen, W., and Fei, Y., 2003. Experimental evidence that potassium is a substantial radioactive heat source in planetary cores. *Nature*, **423**: 163–165.

Nimmo, F., Price, G.D., Brodholt, J., and Gubbins, D., 2004. The influence of potassium on core and geodynamo. *Geophysical Journal International*, **156**: 363–376.

Pollack, H.N., Hunter, S.J., and Johnson, J.R., 1993. Heat flow from the Earth's interior: analysis of the global data set. *Reviews of Geophysics*, **31**: 267–280.

Roberts, P.H., Jones, C.A., and Calderwood, C.A., 2003. Energy fluxes and ohmic dissipation in the Earth's core. In Jones, C.A., Soward, A.M., and Zhang, K. (eds.), *Earth's Core and Lower Mantle*. London: Taylor and Francis.

Stein, C.A., 1995. Heat flow of the Earth. In Ahrens, T.J. (ed.) *A Handbook of Physical Constants: Global Earth Physics,* AGU Reference Shelf 1. Washington, DC: American Geophysical Union, pp. 144–158.

Taylor, S.R., and McClennan, S.M., 1985. *The Continental Crust: Its Composition and Evolution*. Oxford: Blackwell Scientific Publications, 312 pp.

Van Schmus, W.R., 1995. Natural radioactivity of the crust and mantle. In Ahrens, T.J. (ed.) *Global Earth Physics: A Handbook of Physical Constants,* AGU Reference Shelf 1. Washington, DC: American Geophysical Union, pp. 283–291.

Wanke, H.G., Dreibus, G., and Jagoutz, E., 1984. Mantle chemistry and the accretion history of the Earth. In Kroner, A., Hanson, G.N., and Goodwin, A.M. (eds.) *Archean Geochemistry*. New York: Springer-Verlag, pp. 1–24.

Cross-references

Core Composition
Core Origin
Core-Mantle Boundary, Heat Flow Across
Geodynamo, Energy Sources

REDUCTION TO POLE

Introduced by Baranov (1957) (see also, Baranov and Naudy, 1964), the reduction-to-pole transformation of total field magnetic anomalies (see crustal magnetic field) is intended to remove the skewness of the anomalies (see Figure R1). The transformation makes the anomalies overlie the sources, makes it possible to correlate the magnetic anomalies with other types of geophysical anomalies (e.g., gravity) and geological information, and aids their interpretation. In reality, even the amplitude of the anomaly is affected (increased) when sources of induced magnetization are observed at poles in comparison to lower magnetic latitudes because the Earth's field intensity increases from equator to poles; some of the reduction-to-pole methods can take this change in amplitude into account (e.g., equivalent source method) while the others typically do not (e.g., rectangular coordinate wavenumber domain methods). The expression of a magnetic anomaly, ΔT, due to a localized spherical source of uniform magnetization is helpful in understanding the transformation

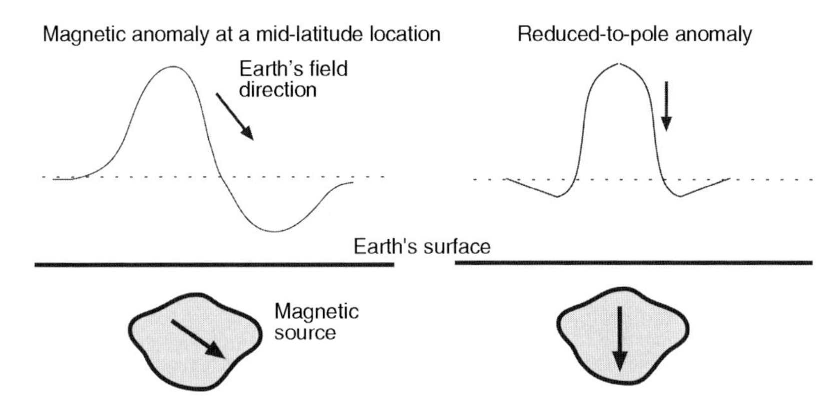

Figure R1 Skewness of a magnetic anomaly due to a uniform arbitrarily magnetized source below Earth's surface in an obliquely oriented Earth's magnetic field (left) and its reduced-to-pole expression in the vertical magnetization and vertical field conditions (right).

$$\Delta T(r) = -\frac{\partial}{\partial \beta}\Delta V_{\alpha}(r) = \Delta J \frac{\partial^2}{\partial \beta \partial \alpha}\frac{1}{r},$$

where r is source to observation distance, ΔV_{α} is anomalous potential due to the uniform anomalous magnetization direction α, ΔJ is the intensity of anomalous magnetization [illegible], and β is the direction of the Earth's main field (assumed uniform). To derive the anomalous source function ($\Delta J/r$), one integrates the equation twice, once with respect to β (to find the anomalous potential), and once with respect to α. The magnetization direction of the source is usually not known and, therefore, induced magnetization is assumed, leading to $\alpha = \beta$. The reduced-to-pole magnetic anomaly (vertical intensity anomaly due to vertical magnetization) can then be computed by twice differentiating this source function in the vertical direction (to find first the potential due to the vertically magnetized source, and then its anomaly in the vertical direction) as

$$\Delta T_z(r) = \Delta Z(r) = \frac{\partial^2}{\partial z^2}\int_{-\infty}^{\infty}\int_{-\infty}^{\infty}\Delta T(r)\partial\alpha\partial\beta,$$

where both ΔZ and ΔT_z represent the vertical intensity magnetic anomaly.

In practice, these computations are significantly easier to perform in the wavenumber domain, where the process of integration involves division by a factor and differentiation involves multiplication by a factor. Under induced magnetization conditions (i.e., $\alpha = \beta$), the reduced-to-pole anomaly in the wavenumber domain is given by

$$\Delta T_z(k)^* = |k|^2 \frac{\Delta T(k)^*}{B^2},$$

where asterisks denote wavenumber domain representation of the respective anomalies, k is the radial wavenumber; in Cartesian coordinates, $k = (k_x^2 + k_y^2)^{1/2}$, where k_x and k_y are the wavenumbers in x and y directions; and $B = 1/[\mathrm{i}k_x \cos I \cos D + \mathrm{i}k_y \cos I \sin D + k \sin I]$, where I and D are the main field inclination and declination, respectively, and the trigonometric quantities represent direction cosines in the north, east, and down directions, respectively. An iterative wavenumber domain method for variable directions of magnetization and Earth's field intensity appropriate for a large region is described by Arkani-Hamed (1988).

Another customary approach of achieving reduction to pole is through the equivalent source technique (Dampney, 1969; Emilia, 1973; von Frese *et al.*, 1981, 1988; Silva, 1986), where a configuration of equivalent sources is first assumed. Magnetization is generally assumed in the direction of the inducing field, but it can also be different if known for particular sources. Using inverse methods, and taking advantage of Green's principle of the equivalent layer (see Blakely, 1995), one can map magnetization variation for the region where anomalies are available. Using the derived magnetization distribution, it is possible to compute the reduced-to-pole anomaly under vertical magnetization and vertical Earth's field conditions.

The equivalent source method is subject to instabilities due to a variety of reasons (e.g., spacing of sources, altitude difference between observations and the sources, low magnetic inclinations, etc.), but most of the instabilities can be reduced or eliminated using damped least-squares or ridge regression approach (Silva, 1986; von Frese *et al.*, 1988). The wavenumber domain reduction-to-pole operations also encounter instabilities in low magnetic latitudes ($<30°$ inclination, i.e., when the terms involving vertical component of direction cosines are close to zero). Under these circumstances, when the other two direction cosines nearly negate one another, which happens along a line in a k_x–k_y plane due to the shape of the reduction-to-pole filter (see, e.g., Blakely, 1995), the quantity B^2 is nearly zero and small errors in the anomaly field are significantly enlarged in the reduction-to-pole process. Hansen and Pawlowski (1989) describe methods to overcome these artifacts by designing a Wiener filter for this purpose.

Dhananjay Ravat

Bibliography

Arkani-Hamed, J., 1988. Differential reduction-to-the-pole of regional magnetic anomalies. *Geophysics*, **53**: 1592–1600.

Baranov, V., 1957. A new method for interpretation of aeromagnetic maps: pseudo-gravimetric anomalies. *Geophysics*, **22**: 359–383.

Baranov, V., and Naudy, H., 1964. Numerical calculation of the formula of reduction to the magnetic pole. *Geophysics*, **29**: 67–79.

Blakely, R.J., 1995. *Potential Theory in Gravity and Magnetic Applications*. Cambridge: Cambridge University Press.

Dampney, C.N.G., 1969. The equivalent source technique. *Geophysics*, **34**: 39–53.

Emilia, D.A., 1973. Equivalent sources used as an analytic base for processing total magnetic field profiles. *Geophysics*, **38**: 339–348.

von Frese, R.R.B., Hinze, W.J., and Braile, L.W., 1981. Spherical earth gravity and magnetic anomaly analysis by equivalent point source inversion. *Earth and Planetary Science Letters*, **53**: 69–83.

von Frese, R.R.B., Ravat, D., Hinze, W.J., and McGue, C.A., 1988. Improved inversion of geopotential field anomalies for lithospheric investigations. *Geophysics*, **53**: 375–385.

Hansen, R.O., and Pawlowski, R.S., 1989. Reduction to the pole at low latitudes by Wiener filtering. *Geophysics*, **54**: 1607–1613.

Silva, B.C.J., 1986. Reduction to the pole as an inverse problem and its application to low latitude anomalies. *Geophysics*, **51**: 369–382.

Cross-references

Crustal Magnetic Field
Magnetic Anomalies

REPEAT STATIONS

Definition

Repeat stations are permanently marked sites where it is possible to make accurate observations of the Earth's magnetic field vector for a period of a few hours (sometimes a few days) every few years. Their main purpose is to track secular variation (see *Geomagnetic secular variation* and *Time-dependent models of the geomagnetic field*) and, if accurate observational techniques and careful reduction procedures are followed, they can be a cost-effective way of supplementing observatory data for secular-variation modeling. Figure R2 shows a plot of repeat station locations. The distribution is somewhat uneven, being determined more by local need and resources rather than requirements to achieve global coverage.

History

One of the earliest repeat station networks was in the UK where 190 stations were established and measurements were made at each by Sabine and colleagues in the 1830s (see *Sabine, Edward*). Many of these sites were reoccupied between 1857 and 1861. However, as the distribution of these stations was somewhat uneven and records enabling their further reoccupation do not seem to have survived, Rücker and Thorpe established a new network between 1884 and 1888 comprising some 200 stations (Barraclough, 1995 and references therein). Some of these stations are still in use today although probably none are exact reoccupations. Elsewhere in the world, the Carnegie Institution of Washington, Department of Terrestrial Magnetism (see *Carnegie Institution of Washington, Department of Terrestrial Magnetism*) permanently marked sites to enable reoccupation during their excursions around the world starting in the early 20th century (see *Bauer, Louis Agricola*) and many countries began to establish repeat station networks at around about this time.

Equipment

The instruments are usually the same as those used at magnetic observatories; a fluxgate theodolite to measure declination and inclination and a proton precession magnetometer (PPM) to measure field strength (see *Observatories, instrumentation*). At observatories fixed marks are usually already surveyed in but this is not always the case at repeat stations. Therefore additional instrumentation may be necessary for the determination of true north. This may be a gyro-attachment for the theodolite, eyepieces or sun filters and accurate timing equipment for sun or star observations, or separate geodetic-quality Global Positioning System (GPS) units. Also often necessary is a tent to provide shelter for the equipment and observer from the weather.

Procedures

Great care is taken to ensure that exact reoccupation of the site is made at each visit because, if there is an appreciable local gradient in the crustal field, any error in positioning the instruments will contaminate the resulting secular-variation data. At each visit a site survey with a PPM is usually done to check for magnetic contamination of the repeat station. An auxiliary station is established a few meters from the marked site and the difference in the total intensity of the field between these two sites is established using two PPMs running simultaneously. Depending on the method being used to determine true north the fluxgate theodolite or GPS unit may then be set up over the marked site. The fluxgate theodolite for observing declination and inclination is then set up over the marked site. The PPM at the auxiliary station logs data for the duration of the declination and inclination observations. Several rounds of observations are made over a number of hours, each round involving four circle readings to eliminate collimation errors between the theodolite telescope and the fluxgate sensor and within the theodolite itself.

An important aspect of the data reduction procedures is correction for the regular daily variation (see *Periodic external fields*) and magnetic storms (see *Storms and substorms, magnetic*). When repeat stations are close to observatories, these variations are observed and corrected for using continuous observatory data. Elsewhere, especially for repeat stations in areas of complex external fields such as the auroral zones or remote from geomagnetic observatories, on-site variometers are sometimes run to monitor these variations.

A useful publication for guidance on repeat station survey procedures is that published under the auspices of IAGA (Newitt *et al.*, 1996, see *IAGA*).

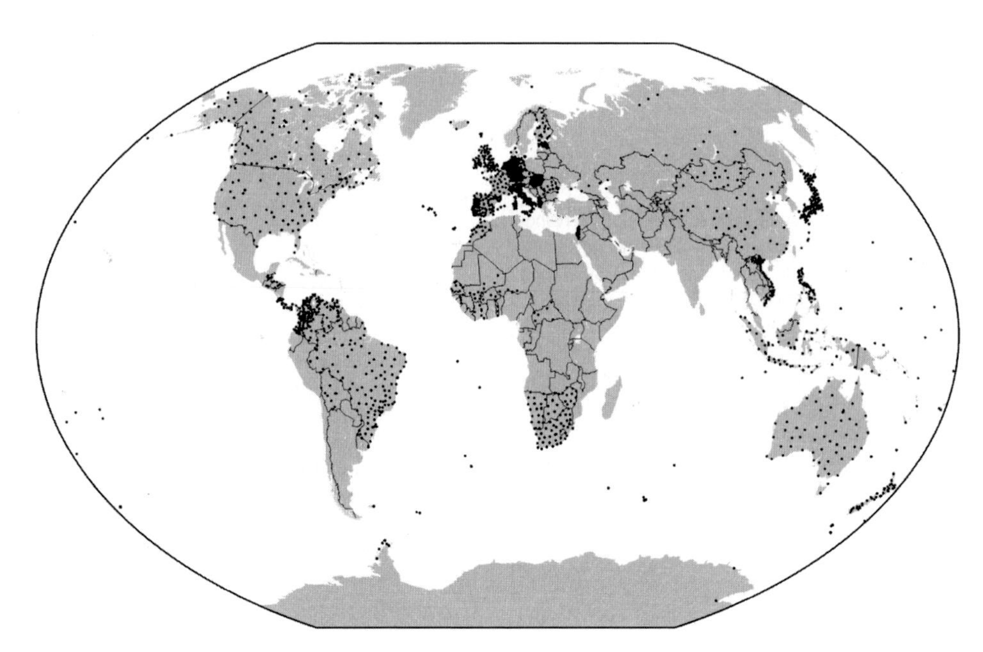

Figure R2 Locations of all known repeat stations visited more than once since 1975.

Use in modeling

Efforts to model the field over restricted regions of the Earth for the purposes of mapping rely on good local coverage of vector data such as that provided by repeat stations. Different approaches are used around the world depending on the extent of the area and coverage of data, but the most common is some form of spherical cap harmonic analysis (SCHA) (see *Harmonics, spherical cap*). As well as being applied in many individual countries to map the magnetic field, for example Canada, Spain, South Africa, SCHA has also been applied to European repeat station data to search for secular-variation anomalies. However, the degree of success has been limited partly because of the high uncertainties associated with the data (Korte and Haak, 2000). Repeat station data have also been used in global modeling and help fill in some gaps in the observatory distribution.

Susan Macmillan

Bibliography

Barraclough, D.R., 1995. Observations of the Earth's magnetic field made in Edinburgh from 1670 to the present day. *Transactions of the Royal Society of Edinburgh: Earth Sciences*, **85**: 239–252.
Korte, M., and Haak, V., 2000. Modelling European magnetic repeat station data by SCHA in search of time-varying anomalies. *Physics of the Earth and Planetary Interiors*, **122**: 205–220.
Newitt, L.R., Barton, C.E., and Bitterly, J., 1996. *Guide for Magnetic Repeat Station Surveys*. International Association of Geomagnetism and Aeronomy.

Cross-references

Bauer, Louis Agricola (1865–1932)
Carnegie Institution of Washington, Department of Terrestrial Magnetism
Geomagnetic Secular Variation
Harmonics, Spherical Cap
IAGA, International Association of Geomagnetism and Aeronomy
Observatories, Instrumentation
Periodic External Fields
Sabine, Edward (1788–1883)
Storms and Substorms, Magnetic
Time-Dependent Models of the Geomagnetic Field

REVERSALS, THEORY

Once it was appreciated that reversals of the geomagnetic field occurred (see *Geomagnetic polarity reversals*), it was natural to seek a theoretical understanding of these events. The dominant, dynamo theory for the generation of the Earth's field has no difficulty accounting for configurations in which the field is reversed from the present "normal" polarity; but detailed mechanisms, explaining how and why the field comes to reverse, remain the topics of active research, with many fundamental questions still to be definitively answered. Are reversals isolated phenomena, triggered by exceptional events within the core, or are they simply part of the normal range of the dynamical variations of the geodynamo? Does a reversal transition involve a fundamentally different regime of dynamo action (e.g., a period of field decay), or is it a relatively minor variation in the normal behavior? How does the transition field compare with the normal field? Despite the current lack of final answers, dynamo theory has been used to address all of these questions. (Reversal mechanisms have also been proposed for some of the alternative theories of geomagnetism; see *Nondynamo theories*.)

Equivalence of normal and reverse polarity states

Within the general theory of *Magnetohydrodynamics* (*q.v.*), the overall sign of the magnetic field is not dynamically important; the governing equations remain unchanged, if a field in the opposite direction is substituted (see *Geodynamo, symmetry properties*). There is therefore no problem in accounting for fields of both "normal" and "reverse" polarities (with dipoles oriented parallel and antiparallel to the present-day field, respectively). Indeed, if a state such as the present-day field exists, then the reverse state must be equally possible, theoretically; it is therefore gratifying that observations suggest that both polarities have occurred equally frequently, over the lifetime of the geomagnetic field.

The dynamo can therefore be viewed as a "bistable" system, which may reside in one of two equivalent states of opposite polarity; it is relatively stable within either of these states (or else reversals would be more common than they are), and yet is prone to occasionally flipping from one state to the other. The theoretical problem is to explain how, and why, the field moves between these two states in the irregular manner which it does. This remains a rather broad question, which may be addressed in many different ways. Some authors have focused on the study of individual reversals, considering in more detail how one particular reversal is "triggered", and exactly what happens to the field as it undergoes the process; others have considered the problem over longer timescales, developing models which seek to explain the statistical properties of sequences of reversals.

Kinematic mechanisms

The first successful models of dynamo action were kinematic models (see *Dynamos, kinematic*), which prescribed a flow, and investigated the magnetic field generated by this flow. It is therefore natural that some of the first attempts to explain reversals related to such models; these were typically formulated in terms of the *mean field* or *Braginsky dynamos* (*q.v.*) which were being developed at about the same time.

Within the kinematic problem, a stationary flow can generate fields of either stationary or oscillatory character (which may both ultimately grow or decay, depending on the overall strength of the flow). If the flow is made up of a number of components whose relative magnitude may be varied (e.g., differential rotation, "cyclonic" convection, and meridional circulation; the most common ingredients of mean field dynamo models), then solutions of different character are typically preferred for different forms of the velocity. Changes in the underlying flow can therefore result in changes in the nature of the field being excited (or decaying); and if the velocity moves through a suitable sequence of states, then a field reversals may be effected. The various kinematic mechanisms all propose some such sequence of velocity variations.

It is worth explicitly noting that some variation in the velocity is required, to obtain an irregularly reversing field; a stationary flow can only produce stationary or periodically reversing fields, as described above. The types of velocity variations envisaged in kinematic mechanisms are therefore entirely plausible; it is only the arbitrary manner in which the variations are assumed, without reference to the coupled dynamics of the full dynamo problem, that makes kinematic mechanisms somewhat unsatisfactory. Nevertheless, these mechanisms may still effectively isolate the important physics of reversal transitions, and thus provide insights often lost in more complex dynamical simulations.

One such mechanism was suggested by Braginsky (1964), when he first considered the consequences of dynamo action in his "nearly axisymmetric" system (see *Dynamo, Braginsky*). He found that, in the absence of a component of meridional circulation (i.e., an axisymmetric flow whose streamlines lie on planes of constant azimuth), his flows preferred to excite oscillatory fields; with some meridional circulation present, however, stationary fields were preferred. (The same situation often pertains to solutions of the closely related mean field dynamo systems.) He therefore argued that the geodynamo may normally function in the latter regime, with some meridional circulation stabilizing the solution, to give the predominantly steady geomagnetic field. If a fluctuation takes the velocity into a state with weaker meridional circulation, however, it is easy to see how an oscillation might commence, from the initially steady field. (And if the original flow was near the boundary in preferences, then only an infinitesimal change of velocity might be needed.) Indeed, such an oscillation might be expected to take the form of a simple *dynamo wave* (*q.v.*). If the flow has returned to its original state after a single change of sign of the field (i.e., about half a period of the oscillating solution), then a stationary solution will again be preferred; but now, the stationary solution will be based on the post-oscillation field, in its new, reversed state. The same mechanism was later highlighted by Gubbins and Sarson (1994), who carried out three-dimensional kinematic calculations; and Sarson and Jones (1999) later found the effect to be important within a dynamical calculation, described below.

An alternative series of kinematic mechanisms was developed from original papers by Parker (1969) and Levy (1972). All of these are based on some form of *mean field dynamo* (*q.v.*), and rely on fluctuations in the strength or location of the eddy flows, or "cyclones", responsible for producing dipolar field from azimuthal field via the so-called "alpha"-effect. Parker (1969) noted that an occasional excess of cyclones at high latitudes (or a relative dearth at low latitudes) led to the fields generated at low latitudes being of opposite sign to the original, stable configuration. The new state therefore acts "degeneratively", with respect to the original field; if this situation persists for long enough for the low-latitude field to reverse sign, before the normal flow regime is restored, then the subsequent generation will be based on the new field sign, and a reversal will have been achieved. In a similar scheme, applied to a related mean field model, Levy (1972) showed that an occasional period of high-latitude cyclone activity, in a dynamo which normally generates field via low-latitude activity (or *vice versa*), would likewise lead to the field in some regions being "flooded" with fields of the "reverse" sign; this again can lead to a reversal, if the episodes are suitably timed. In contrast with the meridional circulation mechanism, these schemes are based entirely on stationary kinematic solutions, and rely upon the different morphologies of field excited before and after the change in velocity, to effect the reversal.

In all such kinematic mechanisms, the change in velocity responsible for triggering the reversal—and the change responsible for subsequently stabilizing the system in the new polarity—is simply imposed, and is not subject to self-consistent dynamical forces (including the feedback of the field on the flow). It may also seem rather convenient that the change of flow persists for just long enough to undergo a single transition, leading to a reversal. But in defense of the mechanism, shorter deviations in flow would plausibly give rise to excursions, which may occur more frequently than full reversals. Such fluctuations in the flow may be happening all the time, with only the more extreme cases leading to reversals (e.g., Gubbins, 1999). Additionally, there are some dynamical reasons to accept that the flow might plausibly be restored on the timescale of the field oscillation itself.

Other kinematic mechanisms are also possible, of course. It has been suggested by more than one author, for example, that reversals may begin when the field and flow have evolved to a state too close to being axisymmetric. Then, since *Cowling's theorem* (*q.v.*) states that such an axisymmetric state cannot be maintained, a period of field decay must commence; and depending upon the subsequent development of field and flow, a reversal may occur before the normal generation process is resumed. (A number of similar "free decay" models have been proposed, where interruptions to the normal state of convection result in flows unable to sustain dynamo action, and thus also leading to periods of field decay, and the possibility of reversals.)

Dynamic mechanisms

The obvious limitation in the kinematic mechanisms outlined above—as with kinematic theories in general—is the lack of a self-consistent treatment of the varying velocity responsible for the reversal; the equations of motion are not explicitly addressed. This is particularly unsatisfactory given that the magnetic field is thought to play a dynamically important role in the dynamo process; so that a reversal is rather likely to influence the velocity.

A strong case can be made for the geodynamo being in a so-called "strong field" state, with the magnetic Lorentz force being a dominant force. Since the magnetic field often acts to make convection easier (see, e.g., *Magnetoconvection*), then such a state may be dynamically distinct from nonmagnetic or weakly magnetic states, in the sense that it requires some "boot-strapping" to attain: the dynamo is stable, as long as the field is present; but if the field is lost, e.g., as a result of natural fluctuations, then the entire convective state may decay. In such a scenario, a reversal may occur as a catastrophic event, with the dynamo losing the strong field solution branch, and both field and flow initially decaying. Subsequently, convection must reestablish itself, and the strong field state may ultimately be reattained. Given the drastic nature of this field collapse, however, it is quite reasonable to allow that fluctuations might result in the new field having been seeded with either polarity, so that a reversal may have occurred. Such a scenario was envisaged by Zhang and Gubbins (2000), who noticed the extreme sensitivity of magnetoconvection calculations to the strength of the dipolar magnetic field, in particular.

Such dynamic mechanisms differ from kinematic mechanisms through their consideration of the evolving equations of motion (particularly with respect to the Lorentz force) during the reversal process. This makes such scenarios formidably complex, however, and detailed reversal mechanisms along these lines remain relatively uncommon. In terms of appreciating the dynamics of reversals, more work has perhaps been invested in studying the reversals observed in numerical geodynamo simulations. Dynamic mechanisms are also inherent in the dynamical systems models discussed further below.

Reversals in numerical simulations

Another approach to understanding reversals theoretically—possible since the advent of detailed numerical simulations (see *Geodynamo, numerical simulations*)—is to study the reversals produced by such simulations phenomenologically, to try to understand why these particular systems, at least, undergo reversals. Given the inability of these simulations to attain the regime of the geodynamo, reversal mechanisms deduced in this way may or may not ultimately prove relevant to geomagnetic reversals; yet such studies remain clearly of interest. The output of such simulations is highly complex, however, with field and flow fluctuating together in ways that can be difficult to track, so that clear-cut mechanisms for reversals remain difficult to isolate.

Glatzmaier and Roberts (1995) noted that their reversal was accompanied by oscillations in the quadrupole component of the field, and also noted that the reversal proceeded through a relatively complex sequence of events, with the azimuthal field, and then the dipole field in the interior, reversing before the dipole field visible at the surface. They also identified an effect earlier noted by Hollerbach and Jones (1993), that the solid inner core can act to stabilize the system, providing a "reservoir" of field that cannot easily be overturned by short-term oscillations (and thus stabilizing the field to the sort of fluctuations envisaged by Gubbins, 1999). Nevertheless the overall picture remains extremely complicated, and a clear causal process is not obvious.

In a calculation with highly truncated azimuthal resolution, Sarson and Jones (1999) observed a reversal to occur shortly after a fluctuation in the meridional circulation of their solution. They could therefore relate their reversal to one of kinematic mechanisms outlined above, an interpretation that was endorsed by subsequent kinematic calculations. In this case, however, the fluctuation in meridional circulation occurred as a natural part of the full dynamic system. (It appeared to be related to a surge in the buoyancy driving.)

In perhaps the most detailed study of simulated reversals to date, Wicht and Olson (2004) analyzed a sequence of reversals obtained from a single calculation, highlighting a relatively clear sequence of effects. They also found meridional circulation to play an important role in this reversal process (and also verified this role via related kinematic calculations). Unfortunately, these reversals did not occur in a very Earth-like solution; e.g., the flow contained a significant component of transequatorial flow, which is not expected to be significant within *rotating convection* (*q.v.*). Furthermore, as the authors themselves point out, the solution behaves almost kinematically, with the magnetic field playing no important dynamical role; again, rather far from the case expected for the Earth. (It is arguably true that all of the current numerical dynamo models behave rather too kinematically, with the magnetic field never yet playing as strong a dynamical role as it should in the geodynamo.)

Li *et al.* (2002) obtain irregular reversals in a weakly compressible, ideal gas, dynamo system. Interestingly, they find their system to possess more than one steady state (in either polarity), differentiated by different relative signs of the dipole and octupole terms. In the "high energy" state, characterized by antialigned dipole and octupole moments, the system is susceptible to reversals. This state is also associated with more vigorously fluctuating convection, and by significant transequatorial flows (as in Wicht and Olson, 2004). As in the reversal of Glatzmaier and Roberts (1995), a significant component of axial quadrupole field is observed during the transitions.

It is somewhat encouraging that the current set of numerical simulations seem to share at least some aspects of their reversal behavior. On the other hand, the detailed mechanisms occurring within any of the simulations remain imperfectly understood; and the ultimate relevance of these simulations to the geodynamo remains unproven. While promising, therefore, the analysis of reversals from numerical simulations remains at rather an early stage, in terms of deriving new models for geomagnetic reversals.

Dynamical systems approaches

The preceding sections discussed attempts to analyze individual reversals in terms of detailed deterministic processes. Another, parallel line of enquiry, however, is concerned with the analysis of sequences of reversals. Sufficiently many reversals have been observed that reasonable sequence statistics are available (e.g., describing the statistical distribution of the durations of the stable polarity intervals). By taking sequences of "synthetic" reversals from numerical models, and calculating comparable statistics for these, theoreticians can check the long-term time-behavior of their dynamo models.

Unfortunately, obtaining reversal sequences of the length required, from the detailed numerical simulations described above, is computationally prohibitive. For this approach, a rather simplified model is required instead; various authors have chosen a variety of ways to isolate simple sets of equations which allow reversals, while, hopefully, retaining much of the essential physics of the dynamo process. In studying these systems, the dynamics omitted from the earlier kinematic mechanisms is restored to central place Yet within this approach, the question "why does the field reverse?" is not often addressed. It is accepted that such dynamical systems are subject to occasional reversals (and the systems studied are, essentially, constructed to exhibit such behavior); reversals are just one part of the normal range of behavior, together with secular variation or excursions.

This approach has a long history, going back to the coupled disk dynamos of Rikitake (1958), based on the homopolar *disk dynamo* (*q.v.*) earlier studied by Bullard. The original disk dynamo had been proposed as a heuristic model of dynamo action; by coupling two such disks together, and obtaining a system of equations exhibiting irregular reversals, Rikitake derived a heuristic model for dynamical systems reversals. This model, notably, predates the 1963 Lorenz equations—derived in a similar spirit to model atmospheric convection—which have become something of a paradigm for deterministic chaos. Although the coupled disk dynamos do not model the geodynamo more than schematically—the disk dynamo is perhaps best viewed simply as an analogue of the true geodynamo system, rather than being essentially related to it—they have retained interest as a simple dynamical model of reversals, and been subject to many studies and refinements; some early work on these systems is summarized in Moffatt (1978). While the coupled disk dynamos exhibit behavior suggestive of reversals and excursions, however, the statistics of such behavior is not particularly "Earth-like", and alternative models have been proposed, following a variety of approaches.

Melbourne *et al.* (2001) constructed a model based on the interaction of various symmetries of magnetic field—axial dipole, equatorial dipole, and axial quadrupole—that separate in the kinematic problem, for geophysically plausible symmetries of the velocity (see *Geodynamo, symmetry properties*). For certain specializations from the generic form for the dynamical interaction of such symmetries, this system has structurally stable "heteroclinic cycles," which allow intermittent excursions and reversals between dominantly axial dipole states; and for some choices of parameters, the sequences of reversals display relatively realistic properties. While somewhat abstract in construction, the explicit use of symmetry interactions in this model is an attractive feature, potentially allowing more detailed comparisons with the paleomagnetic record, or with the output of detailed numerical simulations.

Although most models studied in this vein have been of "low order"—evolving typically just a few scalar variables, as proxies for the whole complex process of dynamo action—models of increasing complexity can now also be studied in this way. Hoyng *et al.* (2001) considered a nonlinear, axisymmetric *mean field dynamo* (*q.v.*), with the addition of random fluctuations to the "alpha"-effect term responsible for the generation of dipole field. The fluctuations model the variations in this flow on rapid timescales (cf. the magnetic timescale, not explicitly modeled here); in a sense they implement, in a random way, the type of fluctuations envisaged in the kinematic mechanisms of Parker and Levy. The resulting system exhibits plausible reversal behavior, with a Poissonian distribution of inter-reversal intervals, as is often claimed for the paleomagnetic data.

A complementary approach is taken by Narteau *et al.* (2000), who couple a simple mean field model for the magnetic field to a rather complex "multiscale" model for the evolution of the flow producing the "alpha"-effect. This model tries to simulate the evolution of eddy flows over a variety of length scales, albeit in a rather phenomenological way (with the cascades of energy both up and down the range of length scales being controlled in a rather arbitrary manner). The result, however, is a system containing some memory of past behavior, allowing for a wide variety of reversal behavior.

All of the above models show how the intermittent behavior observed of reversals can arise, from relatively simple dynamical systems modeling the nonlinear interactions of interest in the full geodynamo system. The challenges for future work in this direction are to identify which interactions appear most important, in modeling the true reversal behavior, and to develop this type of analysis towards more realistic simulations.

Long-term changes in reversal behavior

The preceding analyses have assumed the problem of reversal behavior to be a static problem; they have assumed that the Earth's core is in some stationary base state, and that the observed reversal behavior reflects this underlying state (with a single statistical distribution, for example, being applicable across the whole span). Given the timescales for the evolution of the Earth's core (see, e.g., *Core origin*), this is a reasonable approximation for the analysis of individual transitions, or of reversal sequences of even moderate length. Yet considered over the full history of the geomagnetic field, it need not obviously remain valid. Another approach, also amenable to theoretical enquiry, considers mechanisms whereby variations in the mean state of the core over such timescales can affect the reversal behavior. Note that this approach assumes that there truly are significant variations in reversal behavior which can be investigated, and this is not yet definitively resolved. Even relatively "obvious" changes in long-term reversal behavior, such as superchrons (see *Superchrons, changes in reversal frequency*), might alternatively arise from "intermittency" in the dynamical system responsible for the reversals. Yet there remain some clear theoretical reasons why the long-term reversal behavior might vary.

Receiving most attention to date has been the thermal influence of the overlying mantle, which evolves on a much slower timescale than the core, effectively applying very slowly varying boundary conditions to the latter system (see *Core-mantle boundary coupling, thermal*). Different regimes of reversal behavior can then be anticipated for different states of this boundary; one configuration might produce flows conducive to reversals, while another may produce very stable flows. This possible effect has already been illustrated via a suite of dynamo simulations by Glatzmaier *et al.* (1999), incorporating a series of different thermal boundary conditions, and indeed obtaining different patterns of reversal behavior. While the exact relation between the simulations and the true Earth remains open to debate (as with all current simulations), the potential importance of such coupling on reversals remains clear.

Another long-term effect of potential importance is the slow growth of the Earth's inner core, which has solidified within the fluid outer core since the original differentiation of the Earth's interior (see, e.g., *Core origin*. For a number of reasons—including the important role of the inner core in permitting *chemical convection* (*q.v.*) in addition to thermal convection; the nonnegligible geometrical effect of a moderately sized inner core; the importance of the *Inner core tangent cylinder* (*q.v.*) on outer core fluid dynamics; and the magnetically stabilizing role of the inner core noted by Hollerbach and Jones (1993)—it is very likely that different regimes of dynamo behavior existed during different stages of inner core growth. Although some simulations have begun to address this topic (e.g., Roberts and Glatzmaier, 2001), the net effect of all these factors remains far from clear. Nevertheless, the potential importance to long-term reversal behavior is apparent.

Both of the above mechanisms are discussed in more detail under *Superchrons: changes in reversal frequency* (*q.v.*).

Graeme R. Sarson

Bibliography

Braginsky, S.I., 1964. Kinematic models of the Earth's hydrodynamic dynamo. *Geomagnetism and Aeronomy*, **4**: 572–583 (English translation).

Glatzmaier, G.A., and Roberts, P.H., 1995. A three-dimensional self-consistent computer simulation of a geomagnetic field reversal. *Nature*, **377**: 203–209.

Glatzmaier, G.A., Coe, R.S., Hongre, L., and Roberts, P.H., 1999. The role of the Earth's mantle in controlling the frequency of geomagnetic reversals. *Nature*, **401**: 885–890.

Gubbins, D., 1999. The distinction between geomagnetic excursions and reversals. *Geophysical Journal International*, **137**: F1–F3.

Gubbins, D., and Sarson, G., 1994. Geomagnetic field morphologies from a kinematic dynamo model. *Nature*, **368**: 51–55.

Hollerbach, R., and Jones, C.A., 1993. Influence of the Earth's inner-core on geomagnetic fluctuations and reversals. *Nature*, **365**: 541–543.

Hoyng, P., Ossendrijver, M.A.J.H., and Schmitt, D., 2001. The geodynamo as a bistable oscillator. *Geophysical and Astrophysical Fluid Dynamics*, **94**: 263–314.

Levy, E.H., 1972. Kinematic reversal schemes for the geomagnetic dipole. *Astrophysical Journal*, **171**: 635–642.

Li, J., Sato, T., and Kageyama, A., 2002. Repeated and sudden reversals of the dipole field generated by a spherical dynamo action. *Science*, **295**: 1887–1890.

Melbourne, I., Proctor, M.R.E., and Rucklidge, A.M., 2001. A heteroclinic model of geodynamo reversals and excursions. In Chossat, P., Armbruster, D., and Oprea, I. (eds.) *Dynamo and Dynamics: A Mathematical Challenge.* Dordrecht: Kluwer, pp. 363–370.

Moffatt, H.K., 1978. *Magnetic Field Generation in Electrically Conducting Fluids.* Cambridge: Cambridge University Press.

Narteau, C., Blanter, E., Le Mouël, J.-L., Shirnman, M., and Allègre, C.J., 2000. Reversal sequences in a multiple scale dynamo mechanism. *Physics of the Earth and Planetary Interiors*, **120**: 271–287.

Parker, E.N., 1969. The occasional reversal of the geomagnetic field. *Astrophysical Journal*, **158**: 815–827.

Rikitake, T., 1958. Oscillations in a system of disk dynamos. *Proceedings of the Cambridge Philosophical Society*, **54**: 89–105.

Roberts, P.H., and Glatzmaier, G.A., 2001. The geodynamo, past, present and future. *Geophysical and Astrophysical Fluid Dynamics*, **94**: 47–84.

Sarson, G.R., and Jones, C.A., 1999. A convection driven geodynamo reversal model. *Physics of the Earth and Planetary Interiors*, **111**: 3–20.

Wicht, J., and Olson, P., 2004. A detailed study of the polarity reversal mechanism in a numerical dynamo model. *Geochemistry Geophysics Geosystems*, **5**: Q03H10.

Zhang, K., and Gubbins, D., 2000. Is the geodynamo process intrinsically unstable? *Geophysical Journal International*, **140**: F1–F4.

Cross-references

Convection, Chemical
Convection, Nonmagnetic Rotating
Core-Mantle Coupling, Thermal
Core Origin
Cowling's Theorem
Dynamo Waves
Dynamo, Braginsky
Dynamo, Disk
Dynamos, Kinematic
Dynamos, Mean Field
Equilibration of Magnetic Field, Weak and Strong Field Dynamos
Geodynamo, Numerical Simulations
Geodynamo, Symmetry Properties
Geomagnetic Polarity Reversals
Inner Core Tangent Cylinder
Magnetoconvection
Magnetohydrodynamics
Nondynamo Theories
Superchrons, Changes in Reversal Frequency

RIKITAKE, TSUNEJI (1921–2004)

The late Tsuneji Rikitake (1921–2004), professor emeritus of the University of Tokyo and the Tokyo Institute of Technology, contributed greatly to the advancement of the following areas in geomagnetism through his extensive research activity, mostly at the Earthquake Research Institute, University of Tokyo, Japan.

1. Electromagnetic induction by geomagnetic variations and the electrical state of the Earth's interior
2. Short-period geomagnetic variations and electrical conductivity anomalies in the upper *mantle* (*q.v.*)
3. Dynamo theory as a mechanism of Earth's magnetic field generation
4. Fluid motion in the Earth's *core* (*q.v.*)
5. A model of polarity reversal of the magnetic field (*q.v.*)
6. *Magnetohydrodynamic waves* (*q.v.*) in the Earth's core
7. Geomagnetic and geoelectric changes associated with earthquakes
8. Electrical resistivity changes of rocks associated with strain and their application to earthquake prediction
9. Magnetic anomalies of volcanoes and their changes before and after volcanic eruptions (see *Volcano-electromagnetic effects*)

Figure R3 Professor Tsuneji Rikitake (1921–2004).

Rikitake (Figure R3) proposed a global model of electrical conductivity distribution in the Earth's interior by his own analyses of various types of geomagnetic variations since the latter half of the 1940s and a theory of electromagnetic response of a spherical conductor. He then found an anomaly in the vertical component of short-period geomagnetic variations in central Japan in the early 1950s. Subsequent intensive observations revealed a systematic short-period geomagnetic variation anomaly, called the Central Japan Anomaly, which was one of the earliest findings of a series of similar anomalies over the globe. He interpreted such an anomaly in terms of a conductivity anomaly, possibly undulation of the surface of the conducting *mantle* (*q.v.*). Conductors under oceanic areas seemed to be depressed under the Japanese island arc; this was understood as representing subduction of the oceanic plate beneath the island arc. Such pioneer work stimulated, in 1972, an *IAGA* (*q.v.*) workshop on electromagnetic induction; these workshops have been continued every two years, the latest (18th) having been held in Spain. Rikitake is also well known as one of the pioneers in *Magnetotellurics* (*q.v.*) for investigating the crust and upper mantle.

Rikitake was also one of the pioneers in the so-called dynamo problem in the 1950s, but without the use of high-speed electronic computers. The epoch-making work of Sir *E.C. Bullard* (*q.v.*), widely known for the *Bullard-Gellman dynamo* model (*q.v.*), showed how the magnetic field might be generated in the Earth. However, no kinematic dynamo model can give rise to polarity reversal. One day an idea suddenly came to the mind of Rikitake when he was in the train in Tokyo. He must have been thinking about the *disk dynamo* (*q.v.*), for which a nonreversing analytical solution had been found by Sir Edward Bullard. In an analogy to possible turbulence in the Earth's core, he considered two disks coupled together electrically. This coupled-disk model cannot be solved analytically because of nonlinearity, so Rikitake solved the nonlinear equations numerically using a mechanical calculator. He found a spontaneous polarity *reversal* (*q.v.*). This result was published by Rikitake (1958), 5 years before the finding of a famous example of *chaos* by a meteorologist, E.N. Lorentz. It should be remembered that recent numerical results for MHD dynamo models, derived from high-speed supercomputers, have shown the occurrence of such a spontaneous reversal. This coupled-disk model has been called the Rikitake model.

Rikitake was a professor at the Earthquake Research Institute and hence was naturally involved in earthquake and volcano studies. In the 1960s an earthquake swarm hit a small town in central Japan. Rikitake undertook electric and magnetic measurements in an attempt to find anomalous phenomena associated with earthquakes. He became one of the most distinguished experts in this research field. In particular, he found from *in situ* observations that the electrical resistivity of rock changes in response to small strains. He then devised and installed a very sensitive field instrument and discovered many examples of resistivity steps associated with earthquakes that obviously corresponded to strain steps. His interest expanded further to include various kinds of phenomena precursory to earthquakes and to systematic understanding of precursory phenomena—for example, he established empirical relations for precursory times and earthquake magnitudes.

Rikitake's research achievements were published in more than 200 original papers and more than 50 books. Although most of the books were written in Japanese, some books are written in English, such as "Electromagnetism and the Earth's Interior" (1966), or "Earthquake Prediction" (1976) which have proved invaluable to graduate students and researchers over the world.

Y. Honkura

Bibliography

Rikitake, T., 1958. Oscillations of a system of disk dynamos. *Proceedings of the Cambridge Philological Society*, **54**: 89–105.

Cross-references

Bullard, Edward Crisp (1907–1980)
Core Motions
Dynamo, Bullard-Gellman
Dynamo, Disk
EM, Regional Studies
IAGA, International Association of Geomagnetism and Aeronomy
Magnetohydrodynamic Waves
Magnetotellurics
Reversals, Theory
Seismo-Electromagnetic Effects
Volcano-Electromagnetic Effects

RING CURRENT

The Earth's ring current consists of millions of Amperes of electrical current, encircling the Earth in space near and within the geosynchronous orbit (6.6 Earth radii). It is a feature of the interaction between the magnetized conducting solar wind and the Earth, with its geomagnetic field and conducting ionosphere.

The geomagnetic field is a peculiar feature of Earth when compared to our planetary neighbors in the inner solar system. Mercury has a weak dipolar magnetic field, but Mars' field consists of scattered patches of remnant crustal magnetization, and Venus has no sensible planetary field. It seems plausible that the geomagnetic field has played an important role in the habitability of Earth, but this is so far unproven. We do know that the geomagnetic field creates a well-defined bubble in space, called a *magnetosphere* (*q.v.*). The Earth's ionosphere expands into and fills the magnetosphere with a low density conducting plasma (the fourth state of matter, consisting of free electrons and their parent ions) that is to some degree confined within it but eventually escapes.

The solar atmosphere, also in the plasma state, similarly expands into a well-defined bubble called the *heliosphere*, consisting of the solar wind plasma and magnetic field and their extension to the limits of the solar system. Just as the Earth's magnetospheric bubble is embedded in the solar wind, the heliospheric bubble is embedded within the interstellar medium, a partially ionized plasma having its own magnetic field, flowing generally away from the galactic center, but otherwise of so far indeterminate characteristics. Our galaxy has a bubble of its own, with a boundary at the edge of intergalactic space.

Owing to solar variations and activity, the solar wind inside the heliosphere is highly variable in intensity and magnetic structure, on timescales much shorter than the time required for a particular parcel of the solar wind to expand to the boundary of the heliosphere.

Many types of motion of conducting gases or fluids generate electrical currents that in turn produce magnetic fields. Much as surface tension acts to confine water in droplets or air in bubbles, magnetic fields act to confine plasmas in magnetic cells. When two cells of plasma collide, or when a single magnetic cell of plasma divides into two smaller cells that go separate ways, their magnetic field lines are reconfigured accordingly by a process called *reconnection*, which links or unlinks magnetic field lines between the two cells at their boundaries. The behavior of magnetic cells is loosely analogous to the behavior of surface tension bubbles, but the magnetic field and its cohesive influence are distributed throughout a plasma cell and are not concentrated at the boundaries, as surface tension is for water droplets or air bubbles. Magnetic field lines act more like a connective tissue of fibers that thread the entire cell of plasma, rather than as a membrane at the outer boundary.

When two magnetic cells collide, magnetic field lines that were initially limited to the respective cells become linked from one cell to the other. Cohesive electromagnetic forces are created that act to accelerate each of the cells toward the velocity of the other, tending toward a merger and the formation of a single unified plasma cell with properties weighted by the relative contributions of the two merged cells.

Conversely, if any section of a magnetic plasma cell should acquire a large velocity relative to the overall cell of which it is a part, the magnetic field may not be strong enough to maintain the integrity of the initial cell. In such a case, the cell will stretch out as the errant plasma attempts to escape from the cell proper. Depending upon the amount of momentum acquired by the plasma and the strength of the magnetic field, the cell magnetic field may become so highly distorted by the stretching motion that a separate blob of plasma is formed and the overstretched field is pinched off between the two. Field lines connecting the cell with the errant subcell are then reconnected so that they no longer link the two and are confined to their respective cells. Readers may recognize this behavior as being analogous to the behavior of fluid cells confined by surface tension, a familiar example being found in the "lava lamp" that became popular in the 1960s.

We have been discussing discrete cells of plasma and their magnetic fields. But what if one cell is much larger than another and the smaller is embedded within the larger, as the larger one streams by at high velocity. This is the situation of Earth's magnetosphere, embedded within the solar wind. The large bubble will tend to engulf and acquire the smaller cell; to pick it up and carry it off downstream, and assimilate it. The solar wind and geomagnetic fields reconnect so that they are interlinked, and the magnetic forces that are created seek to accelerate Earth's ionospheric plasma up to solar wind speed while simultaneously exerting drag on the solar wind plasma. The electrical currents that form link the solar wind to the roots of the interlinked field lines, in the auroral zone around each magnetic pole.

The larger cell (solar wind flow in the heliosphere) seeks continuously to link to and entrain the small cell into itself. Conversely, the smaller cell seeks to slow down the solar wind and entrain it into itself, but can succeed only to a limited degree. The outermost contents of the smaller cell are accelerated downstream, driving a return flow through the interior of the small cell. When this circulation is sufficiently strong, the slowed solar wind and accelerated ionospheric plasmas form an errant plasma cell in the tail of the magnetosphere, which episodically pinches off and escapes from the main cell, forming new cells of mixed plasma called "plasmoids." These are carried off downstream in the solar wind. The result is a continual ablation of the plasma in the smaller cell, which is fed by the sunlit atmosphere and auroral zones. A substantial amount of solar wind is slowed down and incorporated into the magnetosphere, with excess energy being either thermalized or transferred to the ionospheric plasmas. This increases the ionospheric plasma contribution to the cell and the rate of loss downstream.

So finally, we come to the formation of the ring current (Figure R4). As the solar wind seeks to erode away the plasma and fields of the Earth and carry them off downstream, the magnetosphere responds by forming a ring of current around the Earth that increases the total dipole moment of the Earth and inflates the magnetic field lines. This is a result of the energy dissipation associated with the work or effort on the part of the solar wind to accelerate and assimilate the magnetosphere. The situation is analogous to a water droplet suspended in a supersonic gas flow, for which the frictional interaction is so intense that it heats the droplet contents, turns them to vapor, and only then carries them off downstream.

In the magnetosphere, solar wind energy dissipation goes partly into the ionosphere and partly into the plasma clouds of the magnetosphere. Energy going into the ionosphere and atmosphere causes them to expand and inflate upward against gravity into the magnetosphere. The energy that goes into the magnetospheric plasma clouds (which

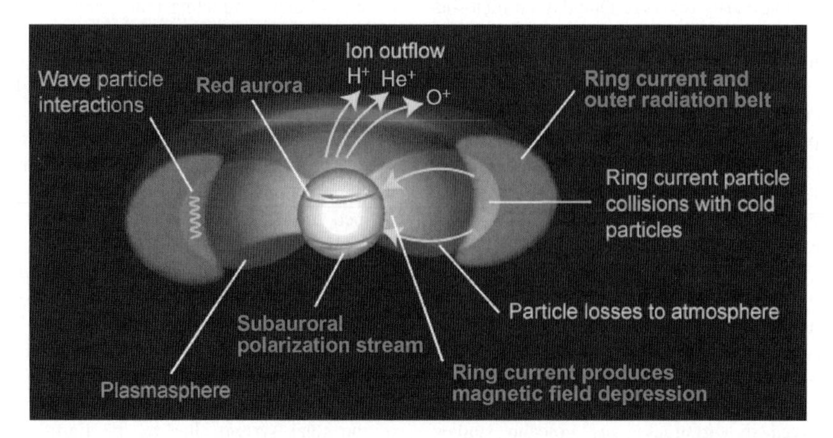

Figure R4 The Earth's ring current is depicted as a red toroidal or donut-shaped region encircling the Earth near the equator inside geosynchronous orbit (6.6 Earth radii). Also shown are the cold extended ionospheric region known as the *plasmasphere;* the ionospheric outflow regions at latitudes higher than the plasmaspheric boundary, and ionospheric features such as the subauroral red (SAR) arcs and polarization streams. Coulomb collisions and wave particle interactions are indicated as loss mechanisms operating on the ring current particles.

then contain more material from the ionosphere) raises the pressure of those plasmas to the point that they inflate the magnetic field that confines them. This is just another way of saying that the plasmas carry currents in the same sense as the Earth's core, adding to the total dipole moment of the planet and inflating the geomagnetic field. Since the pressure and current exists in space around the Earth, it reduces the magnetic field intensity near Earth's surface, while increasing it outside the ring of current.

This can also be viewed as a Faraday's law response: any action that changes a magnetic field induces a current with a sense to oppose that change. The harder the solar wind blows and the more it tries to erode the magnetosphere, the more current is generated in magnetospheric plasmas in such a sense as to oppose that erosion and inflate the magnetosphere. In fact, the solar wind is a tempestuous medium, and the magnetosphere is buffeted by its variations. A side effect of this buffeting is the acceleration of a fraction of the charged particles to very high energies, creating large variations of the Van Allen radiation belts around the Earth.

The ring current was originally discovered by 1917 as the reduction of the magnetic field near the equator of the Earth, by as much as 1–2%. It was inferred by Chapman and Ferraro that this was consistent with a large scale current flowing around the Earth in space (see Bibliography), adding to the dipole moment of the Earth. Theoretical work refined this inference quantitatively, and the advent of spacecraft measurements led to direct observations of the responsible particles, which are mainly ions in the energy range of 50–500 keV. Energetic electrons carry at most 20% of the current. Beginning in the 1970s, ion composition observations showed that the ring current acquires a substantial component of oxygen ions when it becomes strong. These O^+ ions have certainly come from the ionosphere, illustrating the point made above that solar and terrestrial plasmas mix when the solar wind interaction is very strong. The largest ring currents consist almost entirely of ionospheric O^+ plasmas, making it clear that the ring current results in large part from the ablation of the ionosphere into space by solar wind energy deposition.

In recent years, the IMAGE mission has enabled us to globally image and visualize the dynamics of both the outflow of ionospheric plasmas into the magnetosphere, and the creation of high pressure clouds of ionospheric and solar plasmas that produce the ring current. Imaging of plasmas is made possible by a fundamental interaction between atoms and ions in which an electron is exchanged between a fast ion and a slow atom; leading to a fast atom and a slow ion. The fast atom, no longer bound by electromagnetic forces, flies off in a straight line. Suitable cameras record the "glow" of fast atoms coming from any hot plasma that coexists within a neutral gas. For Earth, the hot ring current plasmas coexist with the so-called "geocorona" of the Earth, a spherical region of declining hydrogen density that extends well beyond geosynchronous orbit and a fraction of the way to the moon. Global imaging has revealed some surprises about the ring current, including the fact that it is stronger on the night side than on the day side when it intensifies, and only relaxes into a symmetric ring shape as it fades. This and other more subtle aspects of the ring current can be understood through study of the motions of individual charged particles in the disturbed electromagnetic fields of the storm time magnetosphere.

Other planets that have substantial internal magnetic fields also have ring currents. These include Mercury, Jupiter, Saturn, Neptune, and Uranus in our own planetary system. Presumably other magnetized planets would share this feature, provided that a mechanical energy source dissipates energy in the plasma confined with each magnetosphere. Mercury has a magnetosphere so small that there is not much room for a ring current, and it has very little sensible atmosphere, so the internal source of plasma is weak. Mariner 10 observations of Mercury will soon be complemented by new observations by the Messenger spacecraft, which will greatly increase our understanding of this case. Jupiter is a special case in that it is so large and rotates so rapidly (every 10 h) that more energy comes from braking the rapid rotation than from braking the solar wind flow. Nevertheless, a strong ring current results as plasma from the atmospheres of both the planet and its moons (especially Io) fill the magnetosphere and are spun up by the planetary rotation. Saturn rotates less rapidly and has a solar wind interaction more like that of Earth. It also has a strong ring current inflation of its magnetic field. The magnetosphere of Neptune is also more like that of Earth in terms of ring current, but that of Uranus is peculiar in that the spin and dipole axes are tilted to a large angle from the ecliptic plane normal, so that the magnetosphere lies nearly "pole-on" to the solar wind for half of the Uranian "year." The Earth may resemble Uranus in this respect during a geomagnetic field reversal. This leads to a magnetosphere whose tail is nearly aligned with its magnetic poles, yielding significant differences in the shape of many magnetospheric features. Nevertheless, Uranus has a substantial ring current, and for the same reasons: internal plasmas from the planet, its moons, and the solar wind are heated by energy dissipation, building up pressure and inflating the dipolar magnetic field region.

Ring currents are thus a common feature of magnetospheres filled with hot plasma. That includes in some sense the magnetized astrospheres of the sun and other stars, whose stellar plasma winds inflate their magnetic fields enormously, distending them throughout their astrospheres and generating extended ring currents. In these cases the energy to create and inflate the plasma currents comes from the object itself, rather than being derived from an enveloping medium, as is the case for the Earth's magnetosphere. On the other hand, a rotation-dominated magnetosphere like Jupiter's may be understood as an intermediate case in which energy to create and heat the plasma originates both internally and externally. In summary, ring currents should be thought of as natural extensions of the intrinsic magnetic fields of astrophysical objects, which are sustained when the supply of energy is sufficient to create and maintain a plasma atmosphere of sufficient pressure.

Thomas E. Moore

Bibliography

Burch, J.L., 2001. The Fury of space storms. *Scientific American*, **284**: 86–94.

Cladis, J.B., and Francis, W.E., 1985. The polar ionosphere as a source of the storm time ring current. *Journal of Geophysical Research*, **90**: 3465.

Daglis, I., 2003. Magnetic storm—still an adequate name? *Eos Transactions American Geophysical Union*, **84**(22): 207–208.

Fok, M.-C., Wolf, R.A., Spiro, R.W., and Moore, T.E., 2001. Comprehensive computational model of the Earth's ring current. *Journal of Geophysical Research*, **106**(A5): 8417.

Hamilton, D.C., Gloeckler, G., Ipavich, F.M., Stüdemann, W., Wilken, B., and Kremser, G., 1988. Ring current development during the great geomagnetic storm of February 1986. *Journal of Geophysical Research*, **93**: 14343.

Kistler, L.M., Ipavich, F.M., Hamilton, D.C., Gloeckler, G., Wilken, B., Kremser, G., and Stüdemann, W., 1989. Energy spectra of the major ion species in the ring current during geomagnetic storms. *Journal of Geophysical Research*, **94**: 3579–3599.

Kozyra, J.U., Shelley, E.G., Comfort, R.H., Brace, L.H., Cravens, T.E., and Nagy, A.F., 1987. The role of ring current O+ in the formation of stable auroral red arcs. *Journal of Geophysical Research*, **92**: 7487.

Moore, T.E., Chandler, M.O., Fok, M.-C., Giles, B.L., Delcourt, D.C., Horwitz, J.L., and Pollock, C.J., 2001. Ring currents and internal plasma sources. *Space Science Reviews*, **95**(1/2): 555–568.

Williams, D.J., 1985. Dynamics of the Earth's ring current: theory and observations. *Space Science Reviews*, **42**: 375.

Cross-reference

Magnetosphere of the Earth

ROBUST ELECTROMAGNETIC TRANSFER FUNCTIONS ESTIMATES

Deep probing electromagnetic (EM) induction methods use naturally occurring temporal variations of EM fields observed at the surface of the earth to map variations in electrical conductivity within the Earth's crust and mantle. In the magnetotelluric (MT) method, which is most commonly used for studies of crustal and upper mantle conductivity structure, both electric and magnetic fields are measured at a series of sites. Under the generally reasonable assumption that external sources are spatially uniform, the two horizontal components of the electric field variations can be linearly related to the two horizontal magnetic field components through a 2×2 frequency-dependent impedance tensor. Estimation of these impedances from the raw electromagnetic time series (Figures R5 and R6) is the first step in the interpretation of MT data, followed by inversion of the estimated impedances for Earth's conductivity, and then mapping of electrical conductivity to parameters more directly related to geological processes. The MT impedance tensor is a particular case of a more general electromagnetic response or transfer function (TF). This article focuses primarily on MT impedances and their estimation, but other closely related examples of TFs are also discussed briefly.

The earliest approaches to MT TF estimation were based on classical time series methods (e.g., Bendat and Piersol, 1971), applying a simple linear least squares (LS) fitting procedure to a series of Fourier transformed data segments (Figure R5). However, MT data quality can be highly variable, with both signal and noise amplitudes varying by orders of magnitude over the data record. Furthermore, contrary to the standard LS assumption, there is typically noise in both the "output" or predicted electric field channels, and the "input" or independent magnetic channels, leading to biases in TF estimates. As a consequence the simplest LS approach often fails to yield physically reasonable or reproducible results. Two developments have improved the situation considerably: remote reference, in which data from a simultaneously operating second MT site is used for noise cancellation (Gamble *et al.*, 1979), and data adaptive robust TF estimation schemes which automatically downweight or eliminate poor quality data (Egbert and Booker, 1986; Chave *et al.*, 1987).

Electromagnetic induction transfer functions

The basic assumption underlying the EM TF approach is that the external source variations which induce currents in the earth can be approximated well as random linear combinations of a small number of modes of fixed known geometry. In the most important case, relevant to MT, the external sources are assumed to be spatially uniform and hence linear combinations of two simple modes: unit magnitude sources linearly polarized North-South and East-West, respectively. Physically this assumption is justified if the spatial scale of external magnetic fields at the Earth's surface is large compared to the depth of penetration of the EM fields in the conducting Earth. Except at long periods ($>10^4$ s), for which penetration depths in the Earth exceed several hundred kilometers, and in the auroral zone where ionospheric current systems vary over short-length scales, this assumption generally holds quite well. For very long-period global studies other sorts of simplifying source assumptions are used to justify TFs.

Uniform source transfer functions

For periodic sources at a fixed frequency the uniform source assumption implies that the horizontal electric and magnetic field vectors measured at a single site are linearly related in the frequency domain as

$$\begin{pmatrix} E_x \\ E_y \end{pmatrix} = \begin{pmatrix} Z_{xx}(\omega) & Z_{xy}(\omega) \\ Z_{yx}(\omega) & Z_{yy}(\omega) \end{pmatrix} \begin{pmatrix} H_x \\ H_y \end{pmatrix}. \quad \text{(Eq. 1)}$$

The 2×2 TF $\mathbf{Z}$, which depends on frequency ω, is referred to as the impedance tensor; see Figure R6 for an example. Equation(1) can be justified by the linearity of Maxwell's equations, and the assumption that all sources can be expressed as linear combinations of two linearly independent polarizations. Under these circumstances, two independent field components (e.g., H_x and H_y at one location) uniquely determine all other EM field components everywhere in the domain. The uniform source assumption also justifies other sorts of TFs. In particular, vertical magnetic fields can be linearly related to the two horizontal components at the local site

$$H_z = \begin{pmatrix} T_x(\omega) & T_y(\omega) \end{pmatrix} \begin{pmatrix} H_x \\ H_y \end{pmatrix}. \quad \text{(Eq. 2)}$$

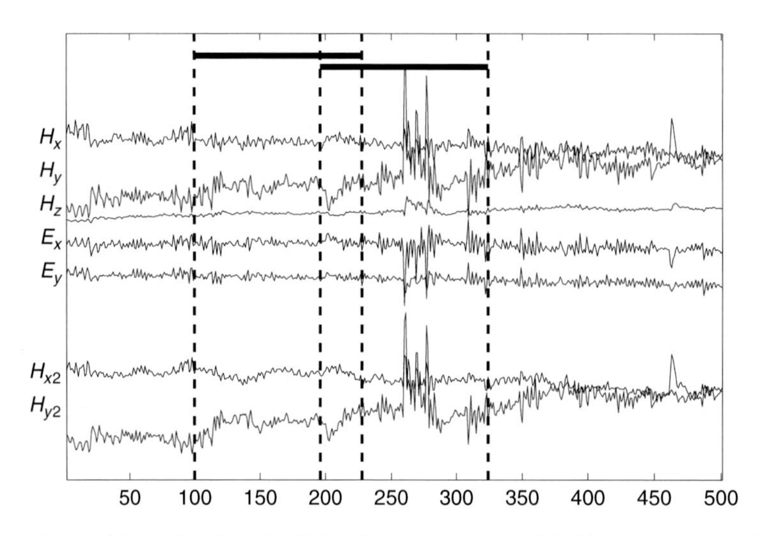

Figure R5 Example of time series used for estimation of MT TFs, from a site near Parkfield, CA. From top, the three vector components of the magnetic field (measured by an induction coil), two colocated orthogonal horizontal electric field components (measured as the potential difference between pairs of electrodes separated by 100m), and two horizontal components of the magnetic field at a distant remote site. Two short (128 point) overlapping data segments are indicated by the vertical dashed lines. Segments such as these are Fourier transformed to yield a series of complex Fourier coefficients for each data channel, which are then used for the various transfer function estimates discussed in the text.

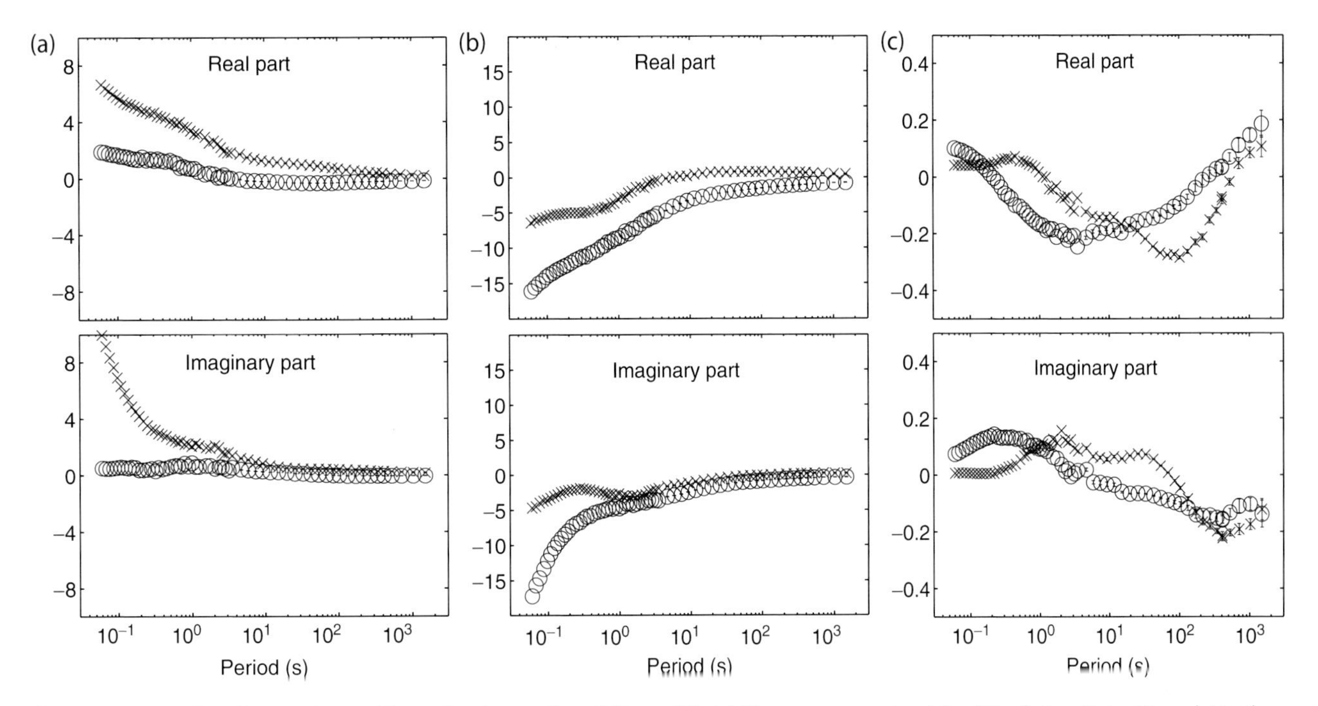

Figure R6 Examples of TFs estimated from the time series of Figure R5. (a) Two components of the TF relating E_x to H_x and H_y, the first row of the impedance tensor: Z_{xx} (circles) and Z_{xy} (crosses). Upper plots give real parts, lower plots imaginary. (b) TF for E_y, the second row of the impedance tensor: Z_{xy} (circles) and Z_{yy} (crosses). Note that the off-diagonal components of the impedance (Z_{xy}, Z_{yx}) are dominant. (c) Vertical field TF T_x (circles), T_y (crosses).

Vertical field TFs (Figure R6c), which are commonly estimated along with the impedance in most MT surveys, are referred to variously as the Tipper vector, Parkinson vector, or Wiese vector (see *Induction Arrows*). Vertical field TFs have frequently been used to map lateral conductivity variations in the crust qualitatively, and they provide useful additional constraints (along with the impedance) in quantitative two- and three-dimensional inversions for conductivity.

When two or more sites are run simultaneously, interstation magnetic TFs can also be defined. These relate magnetic fields at a remote site to the horizontal components at a reference site

$$\begin{pmatrix} H_{x2} \\ H_{y2} \end{pmatrix} = \begin{pmatrix} T_{xx}(\omega) & T_{xy}(\omega) \\ T_{yx}(\omega) & T_{yy}(\omega) \end{pmatrix} \begin{pmatrix} H_{x1} \\ H_{y1} \end{pmatrix}. \qquad \text{(Eq. 3)}$$

Interstation TFs are used to map concentrations of electric current in the crust resulting from lateral variations of conductivity, and can in principle also be used as data for inversion.

Transfer functions for global induction and large arrays

The TFs of (1)–(3) derived from the assumption that external sources are approximately spatially uniform. For global studies and for large arrays other sorts of source assumptions are appropriate, leading to additional examples of induction TFs. The best known of these is the *Z*/*H* TF used for long-period global studies of mantle conductivity. Here the assumption is that external magnetic sources can be approximated as a single zonal dipole (P_1^0). With this one-dimensional source space an appropriate TF is

$$H_z = T_{zh}(\omega)H_x, \qquad \text{(Eq. 4)}$$

i.e., the scalar TF $T_{zh}(\omega)$ is the ratio of vertical (*Z*) to meridional (*H*) geomagnetic variation components. Under the assumption that conductivity in the Earth depends only on depth, and that sources are purely P_1^0, $T_{zh}(\omega)$ can be transformed to an equivalent MT impedance. Variants on (4) have also been applied to studies at daily variation periods, with a fixed geometry for sources (dominated by the P_{m+1}^m spherical harmonic for a frequency of *m* cpd) assumed.

With geomagnetic arrays of large enough size spatial gradients of the magnetic fields can be directly computed, and gradient TFs can be estimated. In the standard approach, the TF function relationship takes the form

$$H_z = C(\omega)[\partial_x H_x + \partial_y H_y]. \qquad \text{(Eq. 5)}$$

Equation (5) strictly applies only to the case where conductivity varies only with depth, and in this case $C(\omega)$ can be converted into an equivalent one-dimensional impedance $Z(\omega) = \mathrm{i}\omega C(\omega)$. Extension to the more general case of multidimensional conductivity is possible, as reviewed in Egbert (2002).

Robust transfer function estimation

Least squares estimates

To be explicit consider estimating the MT TF between one component of the electric field (e.g., E_x) and the two horizontal magnetic field components H_x and H_y at a single fixed frequency ω. This TF corresponds to the first *row* of the usual MT impedance tensor. All of the discussion here and in subsequent sections applies equally to the second row of the impedance, and to the other transfer functions outlined above. Because all the TFs are most succinctly described in the frequency domain as in (1)–(5) the first step in processing is generally to Fourier transform the data. After possibly despiking and/or prewhitening, time series for each component are divided into *M* short time-windows, tapered, and Fourier transformed (Figure R5). Fourier coefficients for *N* periods in a band centered around ω are used for the TF estimate, for a total of $I = MN$ complex data. Estimates $\hat{\mathbf{Z}}$ of the TF (i.e., impedance elements) are then obtained for frequency ω by LS fitting of the linear model

$$\mathbf{E} = \mathbf{HZ} + \mathbf{e}. \quad \text{(Eq. 6)}$$

or in matrix notation

$$\begin{pmatrix} E_1 \\ \vdots \\ E_I \end{pmatrix} = \begin{pmatrix} H_{x1} & H_{y1} \\ \vdots & \vdots \\ H_{xI} & H_{yI} \end{pmatrix} \begin{pmatrix} Z_{xx} \\ Z_{xy} \end{pmatrix} + \begin{pmatrix} \varepsilon_1 \\ \vdots \\ \varepsilon_I \end{pmatrix}, \quad \text{(Eq. 7)}$$

With standard LS this is accomplished by minimizing the sum of the squares of the residuals

$$\sum_i |E_{xi} - (H_{xi}\hat{Z}_{xx} + H_{yi}\hat{Z}_{xy})|^2 = \sum_i |r_i|^2 \to \text{Min}, \quad \text{(Eq. 8)}$$

yielding

$$\hat{\mathbf{Z}} = (\mathbf{H}^\dagger\mathbf{H})^{-1}(\mathbf{H}^\dagger\mathbf{E}), \quad \text{(Eq. 9)}$$

where the superscript † denotes the conjugate transpose of the complex matrix.

The regression M-estimate

The simple LS estimator implicitly assumes a Gaussian distribution for the errors ϵ in (6). This assumption often fails for MT data due to the nonstationarity of both signal and noise, which can result in a marginal error distribution in the frequency domain which is heavy tailed, or contaminated by outliers. As a result, the simple LS estimate all too frequently leads to biased and noisy TF estimates with large error bars (Figure R7a). A number of MT processing methods have been proposed to overcome these difficulties, generally using some sort of automated screening or weighting of the data. Early efforts in this direction used *ad hoc* schemes, for example weighting data segments based on broadband coherence between input and output channels. A more rigorously justifiable approach is based on the regression M-estimate (RME; Huber, 1981), a variant on LS that is robust to violations of distributional assumptions and resistant to outliers (Egbert and Booker, 1986; Chave *et al.*, 1987).

For the RME minimization of the quadratic loss functional of (8) is replaced by the more general form

$$\sum_i \rho(|E_{xi} - (H_{xi}\hat{Z}_{xx} + H_{yi}\hat{Z}_{xy})|/\hat{\sigma}) = \sum_i \rho(|r_i|/\hat{\sigma}) \to \text{Min}, \quad \text{(Eq. 10)}$$

where $\hat{\sigma}$ is some estimate of the scale of typical residuals. LS is a special case of the general form of (10), with $\rho(r) = r^2$. By instead choosing ρ to penalize large residuals less heavily than with the quadratic used for LS, the influence of outliers on the estimate can be substantially reduced. The Huber (1981) loss function

$$\rho(r) = \begin{cases} r^2/2 & |r| < r_0 \\ r_0|r| - r_0^2/2 & |r| \geq r_0 \end{cases} \quad \text{(Eq. 11)}$$

is commonly used with $r_0 = 1.5$ for robust estimation of induction TFs. To find the minimizer of (10) an iterative weighted LS procedure can be used. Let $\psi(r) = \rho'(r)$ be the derivative of the loss function (referred to as the influence function), and set $w(r) = \psi(r)/r$. Then it is easily shown that the minimizer of (10) satisfies

$$\hat{\mathbf{Z}} = (\mathbf{H}^\dagger\mathbf{WH})^{-1}(\mathbf{H}^\dagger\mathbf{WE}), \quad \text{(Eq. 12)}$$

where $\mathbf{W} = \text{diag}(w_1, \ldots, w_I) = \text{diag}(w(|r_1|), \ldots, w(|r_I|))$ is a diagonal matrix of weights. The RME thus corresponds approximately to the weighted LS problem $\sum_i w(|r_i|)|r_i|^2 \to \text{Min}$. However the weights depend on the residuals r_i (and hence on the TF estimate), so an iterative procedure is required. Given an estimate of the TF (and of the error scale $\hat{\sigma}$) residuals can be used to calculate weights, and the weighted LS problem can be solved for a new TF estimate. This procedure can be started from a standard LS estimate of the TF (and some robust estimate of error scale) and then repeated until convergence.

For convex loss functions (e.g., the Huber function of (11)) convergence of this procedure to the unique minimizer of (10) is guaranteed (Huber, 1981). For the loss function in (11) the weights are

$$w(r) = \begin{cases} 1 & |r| < r_0 \\ r_0/|r| & |r| \geq r_0 \end{cases}, \quad \text{(Eq. 13)}$$

i.e., data corresponding to large residual vectors get smaller weights. To allow for a sharp cutoff, with data completely discarded if residuals

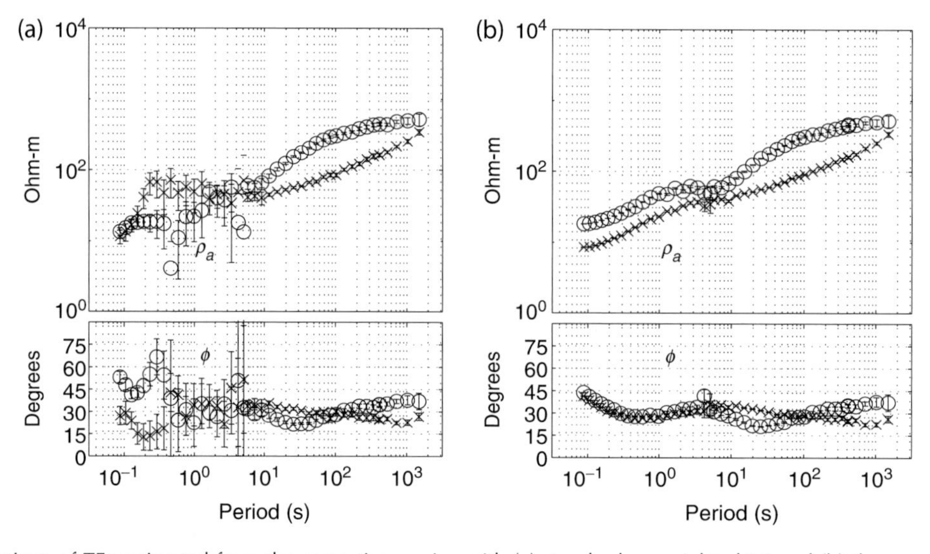

Figure R7 Comparison of TFs estimated from the same time series with (a) standard unweighted LS and (b) the regression M-estimate (RME) described in the text. Here apparent resistivities $\rho_a = (\omega\mu_0)^{-1}|Z|^2$ and phases $\phi = \text{a}\tan(Z)$ are plotted, computed from the dominant off-diagonal impedances Z_{xy} (circles) and Z_{yx} (crosses). The LS estimates are too noisy to be useful over much of the frequency range, while the robust estimates vary smoothly with frequency and have small error bars.

exceed a prescribed threshold, a nonconvex loss function must be used. Convergence of the iterative minimization algorithm cannot be guaranteed in this case, so standard practice is to iterate with a convex loss function such as (11) to convergence, followed by a final step with a hard cutoff to completely discard extreme outliers. This completely automated procedure frequently results in significant improvements in TF estimates, as indicated by improvements in smoothness and physical realizability of apparent resistivity and phase curves (Figure R7b), and reproducibility of results.

More *ad hoc* schemes are also frequently used, sometimes in conjunction with the RME, for down-weighting noisy or inconsistent data. These can be viewed as special cases of the weighted LS estimate of (12), but with weights now determined by some criteria other than residual magnitude. For example, in coherence weighting, broadband coherence of input and output channels is used to downweight time segments with low signal-to-noise ratios.

Leverage

The RME can be excessively influenced by a small number of very large amplitude data sections, resulting in breakdown of the estimator. In the terminology of linear statistical models, these large amplitude data are *leverage points*. For the LS estimate the predicted data are

$$\hat{\mathbf{E}} = [\mathbf{H}(\mathbf{H}^{\dagger}\mathbf{H})^{-1}\mathbf{H}^{\dagger}]\mathbf{E}, \qquad \text{(Eq. 14)}$$

with the matrix expression in brackets (which maps observed data to be predicted) referred to as the "hat matrix." It is readily shown that the sum of the diagonal elements of the hat matrix satisfies, $\sum_{i=1}^{I} h_{ii} = 2$, and that the individual diagonal elements h_{ii} can be interpreted as the fraction of total magnetic field signal power in the ith Fourier coefficient. Large values of h_{ii} indicate data points with inordinate influence on the TF estimates. Furthermore, in extreme cases leverage points are used heavily to predict themselves, making the iterative RME ineffective. To deal with leverage points in routine MT processing, it is thus useful to include an additional weighting term (a function of the magnitude of h_{ii}) to reduce the influence of any observations of unusually large amplitude.

Remote reference

The linear statistical model (6) is strictly appropriate to the case where noise is restricted to the output, or predicted electric field channels. Violation of this assumption results in the downward bias of estimated impedance amplitudes. These biases are proportional to the ratio of noise power to signal power, and can be quite severe in the so-called MT "dead band" at periods of around 1–10 s (Figure R8). To avoid these bias errors horizontal magnetic fields recorded simultaneously at a remote reference site are correlated with the EM fields at the local site. Letting R_{xi} and R_{yi} be the Fourier coefficients for the two remote site components for the ith data segment, and $\mathbf{R}$ the corresponding $I \times 2$ matrix, then the analogue of the LS estimate is

$$\hat{\mathbf{Z}} = (\mathbf{R}^{\dagger}\mathbf{H})^{-1}(\mathbf{R}^{\dagger}\mathbf{E}). \qquad \text{(Eq. 15)}$$

To generalize the RME to remote reference, one can iterate the weighted analogues of (15), i.e., $\hat{\mathbf{Z}} = (\mathbf{R}^{\dagger}\mathbf{W}\mathbf{H})^{-1}(\mathbf{R}^{\dagger}\mathbf{W}\mathbf{E})$, with the weights on the diagonal of $\mathbf{W}$ determined from the residual magnitudes, exactly as for the single site robust estimator. It is also useful to add some additional weighting to allow for outliers at the remote site, and as with the single site estimates, for leverage.

When arrays of simultaneously operating EM instruments are available more complex procedures, which use data from multiple sites to define the reference fields, are possible (Egbert, 2002).

Error estimates

With a statistical approach to TF estimation, one also obtains estimation error variances which define the precision of the TF estimates. These error bars are ultimately required at the modeling or inversion stage to assess the adequacy of fit of a derived model to the measured data, and thus play an important role in the overall interpretation of EM data. The covariance of the linear LS TF estimate of (9) is readily derived from standard theory, using linear propagation of errors

$$\mathrm{Cov}(\hat{\mathbf{Z}}) = \hat{\sigma}^2(\mathbf{H}^{\dagger}\mathbf{H})^{-1}, \quad \hat{\sigma} = (I-2)^{-1}\sum_i |r_i|^2, \qquad \text{(Eq. 16)}$$

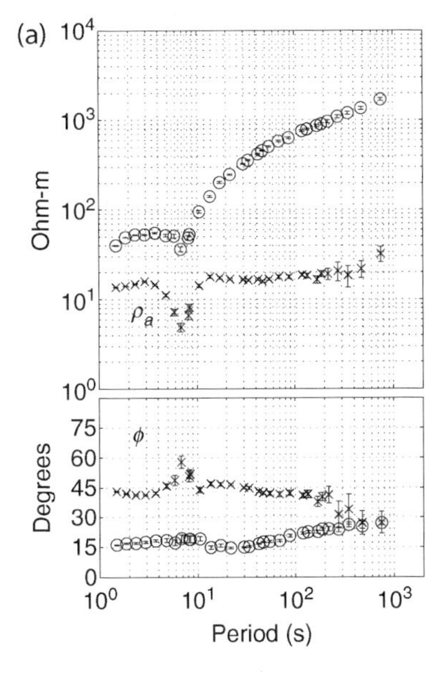

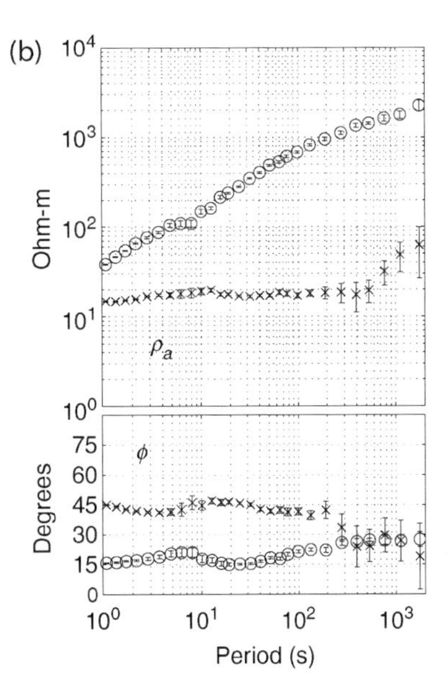

Figure R8 Comparison of TFs (from a different site) estimated with (a) the RME applied to data from a single station and (b) the robust remote reference estimate. As in Figure R7, results are plotted as apparent resistivity and phase. Although both curves are smooth, and have small error bars, a significant (nonphysical) bias is evident in the single site results of (a).

where r_i are again the residuals. An analogous expression for the error covariance for the remote reference estimate is

$$\mathrm{Cov}(\hat{\mathbf{Z}}) = \hat{\sigma}_{\mathrm{r}}^2(\mathbf{R}^\dagger\mathbf{H})^{-1}(\mathbf{R}^\dagger\mathbf{R})(\mathbf{H}^\dagger\mathbf{R})^{-1}, \quad \hat{\sigma}_{\mathrm{r}} = (I-2)^{-1}\sum_i |r_i|^2. \tag{Eq. 17}$$

For both (16) and (17) the diagonal components of the 2 × 2 covariance matrices are the estimation error variances for the two impedance elements.

For the RME variances of the estimates are complicated by nonlinearity, but asymptotic expressions (valid in the limit of large sample sizes) can be obtained from standard theory (Huber, 1981)

$$\mathrm{Cov}(\hat{\mathbf{Z}}) = \frac{(I-2)^{-1}\sum_i w_i^2|r_i|^2}{\left[I^{-1}\sum_i \psi'(|r_i|/\hat{\sigma})\right]^2}(\mathbf{H}^\dagger\mathbf{H})^{-1}, \tag{Eq. 18}$$

where r_i, w_i, $i=1, I$, and $\hat{\sigma}$ are the residuals, weights, and error scale from the final iteration, and the prime denotes the derivative of the influence function. An analogous asymptotic covariances for the robust remote reference estimates can be given with $(\mathbf{H}^\dagger\mathbf{H})^{-1}$ in (18) replaced by $(\mathbf{R}^\dagger\mathbf{H})^{-1}\mathbf{R}^\dagger\mathbf{R}(\mathbf{H}^\dagger\mathbf{R})^{-1}$.

As an alternative to asymptotic error estimates such as (18), the nonparametric jackknife method (Efron, 1982) has also been frequently used for computing TF error estimates (Thomson and Chave, 1991). The jackknife approach can also be applied to compute error bars for complicated nonlinear functions of TFs, as arise for example in some forms of distortion analysis. The jackknife approach significantly increases computational effort required for TF estimation, unless some approximations are used.

Gary D. Egbert

Bibliography

Bendat, J.S., and Piersol, A.G., 1971. *Random Data: Analysis and Measurement Procedures.* New York: John Wiley and Sons.

Chave, A.D., Thomson, D.J., and Ander, M., 1987. On the robust estimation of power spectra: coherence and transfer functions. *Journal of Geophysical Research*, **92**: 633–648.

Efron, B., 1982. *The Jackknife: The Bootstrap and Other Resampling Methods.* Philadelphia: Society for Industrial and Applied Mathematics.

Egbert, G.D., 2002. Processing and interpretation of electromagnetic induction array data: a review. *Survey of Geophysics*, **23**: 207–249.

Egbert, G.D., and Booker, J.R., 1986. Robust estimation of geomagnetic transfer functions. *Geophysical Journal of the Royal Astronomical Society*, **87**: 173–194.

Gamble, T., Goubau, W., and Clarke, J., 1979. Magnetotellurics with a remote reference. *Geophysics*, **44**: 53–68.

Huber, P., 1981. *Robust Statistics.* New York: John Wiley and Sons.

Thomson, D., and Chave, A., 1991. Jackknifed error estimates for spectra, coherence, and transfer functions. In Haykin, S. (ed.), *Advances in Spectrum Analysis and Array Processing.* Englewood Cliffs: Prentice Hall.

Cross-references

Coast Effect of Induced Currents
Conductivity, Ocean Floor Measurements
EM Modeling, Inverse
Geomagnetic Deep Sounding
Induction Arrows
Magnetotellurics
Natural Sources for EM Induction Studies
Transfer Functions

ROCK MAGNETISM

Introduction

Rock magnetism is the study of the magnetic properties of rocks and it includes the study of magnetic minerals. Rock magnetism is one of the oldest sciences known to man. Historians still debate when the magnetic properties of lodestone (leading stone or compass) were first discovered. Some historians suggest that the Greek philosopher Thales made observations in the 6th century B.C. on lodestone, known today as magnetite. Beginning as early as the 3rd century B.C. the Chinese literature makes reference to the magnetic properties *ci shi*, the loving stone, still another name for magnetite. One early school of thought (the animists), which included Thales, argued that the attraction powers of lodestone occurred because it possessed a soul. Even the commonly referred to father of magnetism, William Gilbert of Colchester (who was born in 1544 and died of the plague in 1603), offered the following explanation for lodestone's magnetic attraction: "the Loadstone hath a soul."

The separation of rock magnetism from the broader subject of magnetism of materials occurred in the first half of the 20th century following particularly influential experimental work of Koenigsberger, Thellier, and Nagata. The theoretical foundations of rock magnetism were later established by Néel. The first book on rock magnetism, written by Nagata in 1953, preceded the first book on paleomagnetism written in 1964 by Irving. Rock magnetism primarily deals with the records of magnetic fields recorded in rocks. The inferences drawn from these records usually fall under the subject of paleomagnetism. An excellent modern comprehensive coverage of rock magnetism has been given by Dunlop and Özdemir (1997). In this brief overview article several topics are introduced that are more fully discussed elsewhere in this volume.

Physics of magnetism

Essentially all materials are magnetic because they possess electrons that have a spin (magnetic moments) and because the electron's motion results in currents with their associated magnetic fields. Magnetism is often subdivided into induced magnetization and "permanent" magnetization. Induced magnetization is described by the equation

$$\mathbf{M} = \chi\mathbf{H},$$

where $\mathbf{M}$ is the magnetization, $\mathbf{H}$ is the magnetic field, and χ is a second order tensor referred to as the susceptibility. If χ is negative, the material is diamagnetic (halite, NaCl, is an example) and if χ is positive, the material is paramagnetic (for example, iron-rich olivine or fayalite, Fe_2SiO_4).

Scientists no longer attribute the cause of permanent magnetism to a soul, but instead attribute it to exchange energy. A permanent magnetic order results from exchange energy, which combines Coulomb interaction energy with the Pauli exclusion principle. Thus, exchange energy can only properly be understood through quantum mechanics. Exchange energy is often calculated by evaluating the nature of overlap of electron orbitals. Depending on the details of the electron overlap, a minimum in exchange energy sometimes requires that adjacent atoms have parallel magnetic moments and the material is referred to as ferromagnetic. With different overlap, the minimum in energy occurs when the adjacent magnetic moments have opposite sign. Then the material is described as antiferromagnetic when the moments are of equal magnitude and ferrimagnetic when the adjacent magnetic moments differ in magnitude. The best-known example of a ferrimagnetic mineral is magnetite (Fe_3O_4). Magnetite is said to have an indirect exchange energy because the coupling of adjacent iron atoms occurs through intervening oxygen atoms.

As the temperature of a material is increased the thermal energy (primarily recorded in quantized lattice vibrations referred to as phonons) increases and eventually overwhelms the exchange energy. The Curie temperature is the highest temperature a ferromagnetic material can possess a magnetic structure in the absence of an external magnetic field. The analogous permanent loss of magnetic order in antiferromagnetic and ferrimagnetic structures occurs at the Néel temperature. (Some paleomagnetists also refer to this as the Curie temperature.) The saturation magnetization monotonically increases on cooling from the Curie or Néel temperature to absolute zero providing no phase change occurs on cooling. The Curie temperature at ambient pressure of iron, a ferromagnetic material, is 1043 K and the Néel temperature of ferrimagnetic magnetite is 853 K. Ilmenite ($FeTiO_3$) is an example of an antiferromagnetic substance and it has a Néel temperature of 40 K. It is useful to recognize that the melting temperature of a magnetic material always exceeds the Curie or Néel temperature. (So-called magnetic fluids contain suspended solid magnetic material.) Because the electron overlap is a function of pressure, the Néel and Curie temperatures are also functions of pressure. Physics and engineering aspects of magnetism are summarized in books such as Mattis (1988) and Hubert and Schäfer (1998).

Remanent magnetizations

Remanent magnetization (RM) is what is meant by most people when they refer to "permanent magnetization." An RM is defined as the magnetization that is present in a material in the absence of an external magnetic field. One of the prime goals of rock magnetists is to explain the properties and origins of various forms of RM. Of particular interest is the origin of RM that was acquired under natural conditions, natural RM (NRM).

A particularly useful way to gain insight into RM is through an oversimplified model involving a large ensemble of identical uniformly magnetized particles that are noninteracting. (These particles would correspond to individual magnetic mineral grains in a rock.) We assume there are only two assessable minimum energy states, E_u and E_d, of equal magnitude (in the absence of an external field) in which the magnetization is, respectively, in the up or down direction. There is an energy barrier separating these two states of magnitude E_B. In equilibrium and in the absence of an external field one expects both states to be occupied by the same number of magnetic particles. The material, which we will subsequently refer to as rock, is then said to be demagnetized. Let us apply a very large external magnetic field in the up direction. We take this field to be so large that all the particles become magnetized up. When the external field is removed the majority of the particles will remain in E_u. (All the particles would be in E_u if the experiment were carried out at absolute zero temperature where thermal fluctuations could be ignored.) The sample now has an RM referred to as a saturated isothermal RM (SIRM). If we had used a smaller, but still strong, external field so that there were some particles still in E_d after the external field was applied, then the resulting RM would be referred to as an IRM. Consider yet a different thought experiment in which we start with a demagnetized rock and apply a weak field such that $\mathbf{m} \cdot \mathbf{H}$ (where $\mathbf{m}$ is the magnetic moment of a particle) is smaller than E_B. With time thermal fluctuations may displace some of the particles from E_d to the minimum energy state E_u. The magnetization of the rock acquired over time in a magnetic field is referred to as a viscous RM (VRM) acquired at temperature T. If T is not explicitly given, the VRM is assumed to have been acquired at ambient temperature.

There are many types of RM, far more numerous that can be reviewed here. However, rock magnetists are often interested in distinguishing between a primary RM, one acquired when the rock formed, from a secondary RM formed later. The most common forms of primary RM of interest to paleomagnetists are TRM, DRM, and postdepositional DRM. Thermal RM (TRM) is the primary RM acquired by an igneous rock when it cools from the Curie or Néel temperature to room temperature in a weak field, such as that of Earth's magnetic field. Detrital RM (DRM) is acquired by sediments as they settle out of a quiet fluid environment, such as a lake. Postdepositional DRM seems to be the predominant mechanism by which a primary RM is acquired by marine sediments. Mixing, often by marine organisms, and compaction typically occurs in the sediments closest to the surface in a marine environment. The primary RM, a postdepositional DRM in this case, is acquired at a depth in the sediments. The depth of acquisition varies depending on the type of sediment and its physical and biological environment. Postdepositional DRM also can, and sometimes does, occur in lakes.

Similarly there are many types of secondary RM such as a VRM or an IRM. One of the most common forms of secondary RM is a chemical RM (CRM) usually defined as any RM acquired during chemical change below the Curie or Néel temperature. Some scientists use a more restricted definition of CRM to mean that RM acquired during the growth of a mineral in a magnetic field (CRM then refers to crystalline RM). Unfortunately, this ambiguity in definition of CRM is not always contextually obvious. There are more than 30 different types of RM that have been introduced into the literature, a dozen or so of which are commonly used.

There are two other terms that are useful to introduce and are common to those scientists who have worked with magnetic hysteresis. The bulk coercive force is the magnitude of a field that is applied in the opposite direction to the magnetization to reduce the magnetization to zero. The remanent coercive force is equal to or larger than the bulk coercive force; it is the magnitude of the field applied to a sample in the opposite direction to its RM that will leave the sample demagnetized after the external field is removed. The bulk coercive force and the remanent coercive force provide somewhat different measures of the stability of the RM with respect to an external magnetic field.

Anisotropy

It is important to recognize that the presence of an RM in a sample is a nonequilibrium process that requires magnetic anisotropy. This is manifested in the thought experiment used above. In the absence of an external magnetic field, the particles in the rock example considered possess uniaxial anisotropy: there is an easy axis of magnetization parallel to the magnetization of the grain and a plane of hard directions perpendicular to the easy axis. In the above example discrete energies were used, but in an actual grain it would be possible for the magnetization to be in a direction other than an easy axis in the presence of an external field but not otherwise. (We exclude the possibility of the magnetization residing in the metastable hard direction because thermal fluctuations would quickly lead to the magnetization changing to the easy axis, except close to absolute zero.) Note that the equilibrium state in zero external field is the demagnetized state. If there is no anisotropy, there is no energy barrier separating minimum energy states and thermal fluctuations would quickly reduce the magnetization to zero.

Although magnetic anisotropy at the grain level is a necessary condition to have a remanence, it is also important to recognize that a rock can be magnetically isotropic and carry an RM. For example, if the particles discussed above (each of which is by itself magnetically anisotropic) were randomly orientated, the rock would be isotropic. The most common extrusive rock, basalt, is typically isotropic while other igneous rocks sometimes exhibit magnetic fabrics and thus are anisotropic.

The basic underlying mechanisms that produce magnetic anisotropy in a grain at the microscopic level include dipole-dipole interaction and most importantly the coupling of unquenched orbital moments with spin. However, a continuum approach is applied in rock magnetic calculations that use phenomenological equations to describe anisotropy. Magnetostatic, magnetostriction, shape, and exchange are explicitly described by equations that are not strictly valid at the microscopic level. They probably are valid down to a scale size comparable to that applicable to reconstructed surfaces in solids; i.e., a size of ten to several tens of angstroms. Scientists that claim accuracy below a 100 Å or so using a continuum approach might wish to investigate their assumptions closely. The external part of the magnetostatic energy is the energy difference of a material in zero external field and in the presence of an external field. The internal part of the energy is most commonly referred to as the demagnetization energy. It can be derived from the constitutive equation

$$\mathbf{B} = \mu_0 \mathbf{H} + \mathbf{M},$$

where **B** is the magnetic induction field and μ_0 is the free-air permeability. Using the fact that the divergence of **B** and the curl of **H** (in the absence of electric current and electric displacement current) are zero, it is straightforward to use the constitutive equation to derive Poisson's equation for a scalar potential in which the divergence of **M** acts as a magnetic charge density. This means that wherever the magnetization changes in direction or magnitude there is a bound magnetic charge. In particular, these charges are present on all grain boundaries for which there is some perpendicular component of the magnetization. Then the bound magnetic charges (analogous to electric charges on a capacitor) produce an internal field that has a component opposite to the magnetization. Because of this, the internal field is often referred to as the demagnetization field. The demagnetization field can result in a strong magnetic anisotropy. For example, the presence of a demagnetization field explains why it is typically much easier to magnetize a needle-shaped grain along its long axis than perpendicular to this axis.

The magnetocrystalline anisotropy energy is represented in the uniaxial case by

$$E_K = K \sin^2 \theta,$$

where K is the anisotropy constant and θ is the angle between the easy anisotropy axis and the magnetization. The larger K the more difficult it is to reverse the magnetization. Additional anisotropy constants are used for most materials, such as cubic magnetite for which two anisotropy constants are commonly employed. These constants are usually empirically determined rather than calculated from first principles.

The ordering of magnetic moments in a material results in a stress that produces strain, referred to as magnetostriction. Because stress and strain are second order tensors, the equations for magnetostriction can be complicated. Not only do they depend on crystallography, but they also must be able to accommodate additional sources of stress, such as associated with dislocations. One of the simplest examples is the equation for energy applicable to isotropic magnetostriction in which the magnetization and strain are measured in the same direction

$$E_\lambda = (3/2)\lambda\sigma V \sin^2 \alpha,$$

where the stress σ is a tension applied in the direction of the magnetization and λ is the magnetostriction constant, V is the volume, and α is the angle between the easy magnetostrictive anisotropy axis and the magnetization. Note that in this case the equation for magnetostriction energy is of the same form as that given for the uniaxial magnetocrystalline anisotropy. Most situations are more complicated and additional magnetostriction constants are introduced to describe experimental results.

Although the exchange energy is purely isotropic, it can result in anisotropy when two or more phases are present. An example is when an antiferromagnetic phase with a relatively high Néel temperature is coupled through exchange to a ferromagnetic phase with a lower Curie temperature. In this case there is a uniaxial anisotropy in the ferromagnetic material that has an easy axis aligned with the directions of the antiferromagnetic moments.

Self-reversal

An isotropic rock will usually acquire an RM parallel to the external field. However there are circumstances for which a rock can acquire an RM antiparallel to the field and when such circumstances occur the rock is said to have self-reversed. Néel presented the first mechanisms for self-reversal in the mid-1950s and shortly thereafter Nagata and Uyeda discovered the first reproducible self-reversing rock, the Haruna dacite. Self-reversal always involves two magnetic phases. The first phase acquires an RM parallel to the external field. An antiparallel coupling to a second phase produces an RM opposite to the external field. If the magnetic moment of the second phase exceeds that of the first phase at room temperature, then a self-reversal will have occurred. The negative coupling between the two phases can be cause by exchange interaction (as is the case for the Haruna dacite) or by magnetostatic interaction. Another process that can lead to self-reversal involves an order-disorder transition. For example, suppose the collective magnetic moment of cations at lattice site A is greater than at lattice site B at elevated temperature where a material is disordered. Further suppose that the exchange coupling between the two sites is negative (antiferromagnetic). If on cooling the material becomes an ordered one with the dominant magnetic moment now being that of site B, then a self-reversal will have occurred.

Demagnetization

There are several different demagnetization procedures that can be employed to bring a rock to a demagnetized state (defined above in the section entitled remanent magnetizations). An RM has a relaxation time τ that is useful to describe the decay of an initial magnetization M_0 with time t:

$$M = M_0 e^{-t/\tau}.$$

In actual rock samples there typically is a range of relaxation times that correspond to the fact that there are many contributions to anisotropy energy. These relaxation times are functions of many intrinsic variables (associated with the anisotropy energies) and extrinsic variables, such as magnetic field, stress, and temperature. Demagnetization procedures provide information on the sensitivity of magnetic relaxation to these extrinsic factors. A magnetic component is said to be stable or locked in to a sample when it has a long relaxation time with respect to the time of interest to the investigator.

Only the two most commonly employed demagnetization methods will be mentioned here. The first is thermal demagnetization that measures the sensitivity of the sample to temperature. A complete thermal demagnetization of a sample can occur by heating the sample to a temperature exceeding its Curie or Néel temperature and cooling the sample back to room temperature in the absence of an external magnetic field. Partial thermal demagnetization occurs when the sample is heated to some temperature T that is lower than the Curie or Néel temperature and then cooled to room temperature in zero external field. The magnetization that is demagnetized by the last process is referred to as being unblocked at T. In igneous rocks it is often found that a TRM consists of independent components of magnetization that are locked into the rock during cooling at different temperatures. These components can be unlocked by progressively thermally demagnetization of a sample. This is the process by which a sample is partially demagnetized in a series of experiments in which the peak demagnetization temperature is sequentially increased. (This procedure is sometimes referred to as the progressive method and it is the one most commonly employed. There also is a continuous method in which the RM is measured while the sample is still hot.) The unblocking temperature is typically assumed to be the blocking temperature, that temperature at which the component of magnetization was locked into an igneous rock during cooling. Although it is well known by modern rock magnetists that the unblocking and blocking temperatures are not usually identical, the difference is often negligible for practical purposes (especially for single domain grains defined in the next section). Thermal demagnetization is a measure of the instability of the magnetization as a function of temperature and it often provides valuable information of a sample even when the magnetization is not a TRM.

The second method discussed here is alternating field (AF) demagnetization. This method involves the gradual decrease of an alternating (often at 60 Hz) magnetic field from some peak value H_p to zero. The preferred AF demagnetization method also randomly tumbles the sample with respect to the external field. Those components of magnetization with relaxation times that are small with respect to some characteristic time of the experiment (associated with the rate the alternating field is decreased) are thought to be randomized by this process. Usually AF demagnetization employs a sequence of increasing values for H_p.

Most paleomagnetists use thermal demagnetization, which provides information on the stability of magnetization with respect to temperature, and AF demagnetization, which provides information on the stability of the magnetization with respect to alternating magnetic fields, to infer stability of magnetization in a rock sample with time. The magnetic components that are most stable with respect to temperature or alternating magnetic fields are believed to be the most probable carriers of the primary RM. However, paleomagnetists are well aware that this is not always the case, for example when the sample has acquired a particularly stable CRM. Thus numerous experiments are undertaken to determine the primary RM. There are no hard and fast rules to do this and undoubtedly mistakes are made. For such reasons consistency of a variety of data and interpretations are sought and rock magnetism remains a vibrant field of intellectual endeavor.

Magnetic structure

The structure of magnetic grains depends on many factors including mineralogy, grain size, grain shape, and defects. Grains are classified for a fixed mineralogy as superparamagnetic (SP), single domain (SD), pseudo-single domain (PSD), and multidomain (MD). The smallest are the uniformly magnetized SP grains that have short relaxation times (usually less than 10 s or so). These grains do not contribute to the remanent magnetization but they do affect the susceptibility. SD grains are also uniformly magnetized and can exhibit relaxation times from the largest SP grains to times that exceed the age of Earth (4.6 billion years). PSD grains have the poorest defined structures that range from slight deviations from a SD structure to small multidomain grains. MD grains are grains that contain uniformly magnetized regions, called domains that are separated from each other by transition regions referred to as domain walls. A domain wall is referred to as an x wall, where x is the number of degrees required by the gradual rotation of magnetic moments in the wall to accommodate the magnetization in the adjacent domains. For example, a 180° domain wall separates domains with opposite magnetizations, while a 70° wall is a transition of magnetic moments between two magnetic domains that have magnetic directions that differ by 70°. MD grains are often classified on the basis of the number of domains they have. For example, a two-domain grain has two oppositely magnetized domains separated by a 180° wall.

Micromagnetic calculations minimize the exchange and anisotropy energies to determine the equilibrium states available to a grain. The structures found this way can be compared to those found by various imaging techniques, such as magnetic force microscopy. One finds that there is often more than one magnetic structure that a grain can have depending on its magnetic history. These different structures reflect different local energy minimum or LEM states. Let us consider what happens to an initial SD grain of magnetite as its size is increased. We will assume the grain is a cube and use one side to indicate its size. The SP-SD threshold at room temperature is near 0.045 μm, the SD-PSD is near 0.05 μm, and there is no agreed upon PSD-MD threshold size. This lack of agreement on the PSD-MD transition probably reflects that the transition is a gradual one, although some investigators using a hysteresis parameter definition put this size around 15 μm. The largest PSD domains have several magnetic domains, but they exhibit different stability properties from large MD grains. A reference to a PSD grain or PSD behavior can be confusing because there are incompatible definitions of them that are still commonly used.

It needs to be pointed out that one might find a SD magnetite with a size of say 0.06 μm at room temperature. However, this would presumably not reflect a global minimum energy state. In the case of magnetite the curling, also called vortex, state would have a lower energy at this size. Vortex and flower states, names that reflect the magnetic structure, are the most common states calculated for the smallest PSD grains. A given size grain can exhibit more than one domain structure depending on the circumstances under which the grain became magnetized. Further discussion of this subject can be found in Dunlop and Özdemir (1997).

Magnetic mineralogy

Although the elements Fe, Co, and Ni are ferromagnetic, only Fe plays a prominent role for magnetic properties in minerals and rocks on Earth. Pure iron is the primary phase responsible for remanence on the moon and Fe-Ni alloys are common in meteorites. However, on Earth the most common forms of remanence are found in compounds of iron such as iron oxides, iron-titanium oxides, iron oxyhydroxides, and iron sulfides. There are also other less common systems, such as the solid solution series Mn_3O_4-Fe_3O_4 (end members hausmannite and magnetite and intermediate members collectively called jacobsite) in which the peak in saturation magnetization occurs when Mn_3O_4/Fe_3O_4 is around 16.7%. The most commonly studied solid solution series are those of titanomagnetites, FeO-Fe_3O_4 with end members wüstite and magnetite, and titanohematites, $FeTiO_3$-Fe_2O_3, with end members ilmenite and hematite. Low temperature oxidation of titanomagnetites produce γ phases (titanomaghemites) while high temperature oxidation includes titanohematites as products. For instance, Fe_3O_4 is oxidized to cubic maghemite, γFe_2O_3, at room temperature and to rhombohedral hematite, αFe_2O_3, at say 900 K. The magnetic properties vary significantly with changes in composition. The magnetic properties of hematite will be briefly discussed to illustrate some of the complexities associated with magnetic mineralogy.

Hematite is an antiferromagnetic substance that contains a parasitic ferromagnetism and thus it possesses a different magnetic order than given earlier in the oversimplified description of the origins of magnetic order. It is well known that alternating basal planes in hematite are antiferromagnetically coupled. There is a slight canting of the magnetic moments in these planes such that the moments in alternating planes are not precisely 180° apart. This implies that there is a weak, or parasitic, ferromagnetism that is perpendicular to the ternary axis. In addition, hematite sometimes displays a defect moment in which defects occur more commonly on alternating planes and thereby produces a ferrimagnetic-like moment that is nearly perpendicular to the spin-canted moment. The magnetocrystalline anisotropy of hematite changes sign near 258 K, the Morin transition, and below this temperature the ternary axis (perpendicular to the basal plane) becomes the easy axis. Only the defect moment is present below the Morin temperature. Because of the defect moment the saturation magnetization of hematite is somewhat variable with magnitude around 1/200 of that of magnetite at room temperature (the latter is $480 \times 10^3 \mathrm{A\,m^{-1}}$). Interestingly some intermediate values of titanohematites reach saturation magnetization values that are a significant fraction of that for magnetite. Although this brief discussion of hematite is far from exhaustive, it should be sufficient to illustrate the complexity of magnetic mineralogy and the prominent role such mineralogy plays in the understanding of magnetic properties of rocks.

Some applications

Magnetic minerals are very sensitive to oxidation conditions. Because of this, iron titanium oxide assemblages in igneous rocks are sometimes used to estimate the oxygen fugacity and temperature in Earth's mantle. In partial contrast, it is the pE and pH that is often responsible for the equilibrium assemblages in marine sediments. Because of their sensitivity to climatic conditions, magnetic minerals in sediments are sometimes used as proxies for paleoclimatic variation. For similar reasons magnetic minerals are sensitive to man-induced changes and this has led to the field of environmental magnetism. Such studies are complicated by the fact that microbes can also mobilize some cations, including Fe, in sediments. In addition, some microbes such as magnetotatic bacteria produce chains of SD magnetite within their cells and appear to use this magnetite to determine the vertical direction in mud where they are neutrally buoyant. The microbe examples reflect the use of rock magnetism in the fields of geobiology and biomagnetism. Magnetic properties such as susceptibility, remanence, and magnetic phase transitions are often used to correlate sedimentary layers in magnetostratigraphy. Paleoflow directions of fluids are sometimes estimated from the magnetic fabric of rocks. Finally, the most common uses of rock magnetism are in paleotectonic and paleogeomagnetic studies.

Ronald T. Merrill

Bibliography

Dunlop, D., and Özdemir, Ö., 1997. *Rock Magnetism: Fundamentals and Frontiers.* Cambridge: Cambridge University Press, 573 pp.
Hubert, A., and Schäfer, R., 1998. *Magnetic Domains: The Analysis of Magnetic Microstructures.* New York: Springer-Verlag, 696 pp.
Mattis, D., 1988. *The Theory of Magnetism 1: Statics and Dynamics.* New York: Springer-Verlag, 300 pp.

Cross-references

Biomagnetism
Demagnetization
Iron Sulfides
Magnetic Anisotropy, Sedimentary Rocks and Strain Alteration
Magnetic Domain
Magnetization, Chemical Remanent (CRM)
Magnetization, Isothermal Remanent (IRM)
Magnetization, Natural Remanent (NRM)
Magnetization, Remanent, Application
Paleomagnetism

ROCK MAGNETISM, HYSTERESIS MEASUREMENTS

Use and scope of hysteresis measurements

The term hysteresis in general describes that an effect is lagging behind its cause (from greek *υστερειν*: to lag behind). More specifically, magnetic hysteresis is a property of all ferromagnetic materials (in sensu lato) causing the magnetization of such a material to be strongly dependent on its magnetic history, i.e., exposure to an external magnetic field.

Practical description of magnetic hysteresis

In order to obtain a complete hysteresis loop, the magnetization **M** of a sample is measured as a function of an external inducing field **B** which is cycled from zero to the maximum positive value, back to zero, up to the maximum negative value and then back again to the maximum positive value. Magnetization is generally measured only in the direction of the inducing field so that both quantities are treated as scalars.

The most simple hysteresis measurement of a rock is performed with a set of two concentric coils of which one is inducing the external field B (Figure R9). The magnetic moment of the rock is generating a magnetic stray field which, when the sample is pulled out from the set of coils, in turn induces a voltage pulse in the second (measurement) coil. This induction pulse is a direct measure for the magnetization

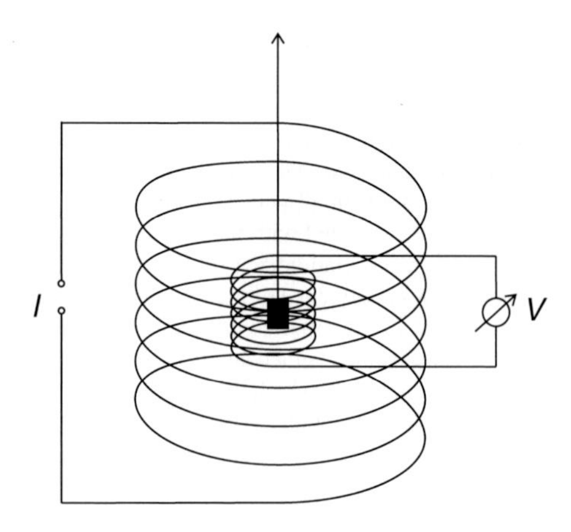

Figure R9 Schematic drawing of a simple hysteresis measurement.

of the sample for a given external field step. The whole procedure can then be repeated for a sufficient number of field steps to obtain the hysteresis loop.

Rocks generally present a mixture of different minerals. Some of these minerals (in most cases ore minerals containing iron) form a long range ferro-, ferri-, or antiferromagnetic spin order below their critical Curie or Néel temperature. In this ordered state the volume dV at r carries a spontaneous magnetization $M(r)$. In most materials $M(r)^2 = M_s^2$ is constant. Within a given external field $B(r)$ and stress tensor $\sigma(r)$ the magnetization structure $M(r)$ adjusts itself into a local minimum of the total magnetic energy $E(M(r), B(r), \sigma(r))$ which encompasses exchange, anisotropy, demagnetizing, magnetoelastic, magnetostrictive, and external field energy.

During an isothermal external field variation the magnetization change of such a system usually lags behind the field change which is expressed in the term "hysteresis." If the field amplitude B_{max} in measuring a hysteresis loop is sufficiently large, this loop defines four basic quantities (Figure R10) (1) the saturation magnetization M_s is the maximal magnetization of the ordered phase within any external field, (2) the saturation remanence M_{rs} is the maximal magnetization of the sample in zero field, (3) the coercive field B_c is the maximal field where zero magnetization of the ordered phase is possible, and (4) the total hysteresis work E_{hys} corresponds to the area inside the hysteresis loop. The coercivity of remanence B_{cr} is the counter field necessary to remove the saturation remanence. It cannot be directly read from the hysteresis loop but requires an additional measurement.

Most rocks also contain paramagnetic minerals, i.e., minerals containing ions with permanent magnetic moments which are not forming a long range spin order. As these minerals cannot carry a remanent magnetization, they do not contribute to M_{rs}, B_c, or E_{hys}. However, they do contribute to the induced magnetization $\mathbf{M}_i = \chi_p(1/\mu_0)\mathbf{B}$, where χ_p is the paramagnetic susceptibility. χ_p is independent of B in the field range typically used for hysteresis measurements. It thus defines the slope of the hysteresis loop above the field where the ordered phase saturates. The paramagnetic contribution has to be subtracted to determine the saturation magnetization of the ordered phase (Figure R10). Another contributor to the field independent susceptibility in rocks are diamagnetic minerals, e.g., quartz.

Definitions and units for magnetic quantities

Much confusion in rock magnetism arises from the fact that the older system of cgs units is still used in many publications besides the now recommended SI units. Table R4 shows the basic magnetic quantities, their units in the cgs and SI system, and the conversion factors. In the SI system the following applies for the relation between magnetic induction **B** and magnetic field **H**

$$\mathbf{B} = \mu_0(\mathbf{H} + \mathbf{M}) \quad \text{(Eq. 1)}$$

with $\mu_0 = 4\pi \times 10^{-7}\,\mathrm{V\,s\,A^{-1}\,m^{-1}}$ being the magnetic permeability of vacuum.

Note, that for convenience in the case of $M = 0$, i.e., outside a sample, the term "magnetic field" is also used for $\mathbf{B} = \mu_0 \cdot \mathbf{H}$.

Most measurement techniques for hysteresis curves determine the magnetic moment of a given sample which is then normalized either by sample volume V or sample mass m. $M = \mu/V$ is called the (volume) magnetization, whereas $\sigma = \mu/m$ is called the specific magnetization.

Use of hysteresis measurements

The shape of hysteresis loops measured on rocks is determined by several properties of the contained minerals: Among these are the type of spin ordering (i.e., ferro-, ferri-, and antiferromagnetic) or the absence of spin order (para- and diamagnetic), the grain size, grain shape, concentration, and also distribution of each contributing mineral. Due to the multitude of influencing factors, generally some additional information about the magnetic mineralogy is needed for the interpretation of hysteresis loops in rock magnetism. If such information is available, hysteresis loops allow a rapid characterization of samples regarding compositional variations, the stability of paleomagnetic information carried by the rocks, change of depositional regime in the case of sediments or alteration processes affecting rocks in general.

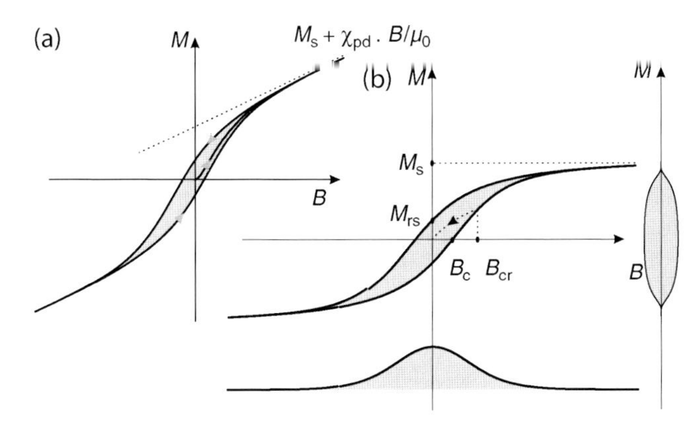

Figure R10 Measured hysteresis loops (a) contain para- or diamagnetic susceptibilities χ_{pd} from sample holder or admixtures which first must be removed to obtain the ferromagnetic loop (b). The coercive force B_c is defined by the zero crossing ($M = 0$) of the lower hysteresis branch, the saturation remanence M_{rs} is the magnetization of the upper branch at $B = 0$. The saturation magnetization M_s is obtained by estimating the limit value of the asymptotic approach to saturation. The coercivity of remanence B_{cr} is defined to be the field value at which switching off the field in course of the lower hysteresis branch would lead to zero remanence. Note that the susceptibility correction leads to different B_c values for (a) and (b), whereas M_{rs}, M_s, and B_{cr} are unchanged. Subtracting the hysteresis branches either in vertical direction or subtracting their average in horizontal direction leads to two representations of the coercivity distribution with respect to B or M. The shaded area in all cases is equal and represents the magnetic energy dissipation E_{hys} during one hysteresis run.

Physical mechanisms of hysteresis

Reversible and irreversible magnetization processes

Magnetic hysteresis is a result of irreversible magnetization changes. An abstract physical process between two states $A \rightarrow B$ is reversible if each intermediate state is arbitrarily close to an equilibrium state. In this case the cycle $A \rightarrow B \rightarrow A$ dissipates no energy. A simple mechanical example of an irreversible process is a switch (Figure R11). It is typical for any system which allows for irreversible changes. (1) It possesses two (or more) metastable positions separated by an energy barrier that must be overcome to change from one to the other. (2) The systems current state depends upon previous physical influences: it serves as a memory.

Natural magnetic systems usually contain a huge number of different metastable magnetization states which give rise to irreversible magnetization processes whenever external variables like magnetic field, temperature, or pressure change. These processes are usually very complex, but as was firstly shown in Stoner and Wohlfarth (1948), there is one important magnetic analog to the mechanical switch: a coherently rotating single-domain (SD) particle.

Single-domain particles

In sufficiently small magnetic particles, quantum mechanical exchange coupling keeps all spins aligned. The most simple case where the particle contains a single axis of preferred spin direction is discussed in many textbooks (Dunlop and Özdemir, 1997; Bertotti, 1998; Hubert and Schäfer, 1998). This is one of the few situations where a complete analytical treatment of magnetic hysteresis is possible. Most of our ideas on physical interpretation of hysteresis implicitly or explicitly are founded on this ideal situation. As sketched in Figure R12a, the magnetic energy of the particle varies during rotation of the magnetization. When a magnetic field $\mathbf{B}$ is applied, the particle behaves as a magnetic switch. When K denotes its uniaxial anisotropy constant

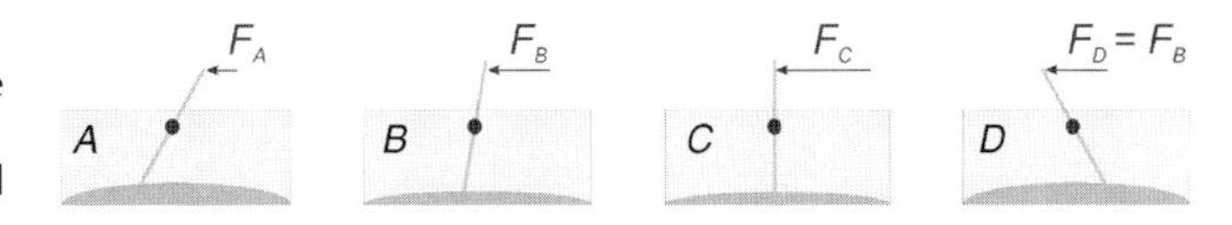

Figure R11 Reversible and irreversible physical processes in a switch. Slowly increasing the force between A and B is a reversible process through a series of well-defined equilibrium states. In contrast, the state transition from B over C to D is irreversible. There are no equilibrium states between C and D. Slowly decreasing the force in position D to the value at B does not reinstall the position of the switch.

Table R4 Units and conversion of magnetic quantities

Quantity	Symbol	SI units	cgs units	Conversion cgs $\leftrightarrow$ SI
Magnetic induction	B	T(Tesla) $= \mathrm{V\,s\,m^{-2}}$	G (Gauss)	$1\,\mathrm{T} = 10^4\,\mathrm{G}$
Magnetic field	H	$\mathrm{A\,m^{-1}}$	Oe (Oersted)	$1\,\mathrm{A\,m^{-1}} = 4\pi \times 10^{-3}\,\mathrm{Oe}$
Magnetic moment	μ	$\mathrm{A\,m^2}$	$\mathrm{G\,cm^3}$	$1\,\mathrm{A\,m^2} = 10^3\,\mathrm{G\,cm^3}$
Magnetization	$M = \mu/V$	$\mathrm{A\,m^{-1}}$	G	$1\,\mathrm{A\,m^{-1}} = 10^{-3}\,\mathrm{G}$
Susceptibility	χ	Dimensionless	Dimensionless	$1(\mathrm{SI}) = 1/4\pi(\mathrm{cgs})$

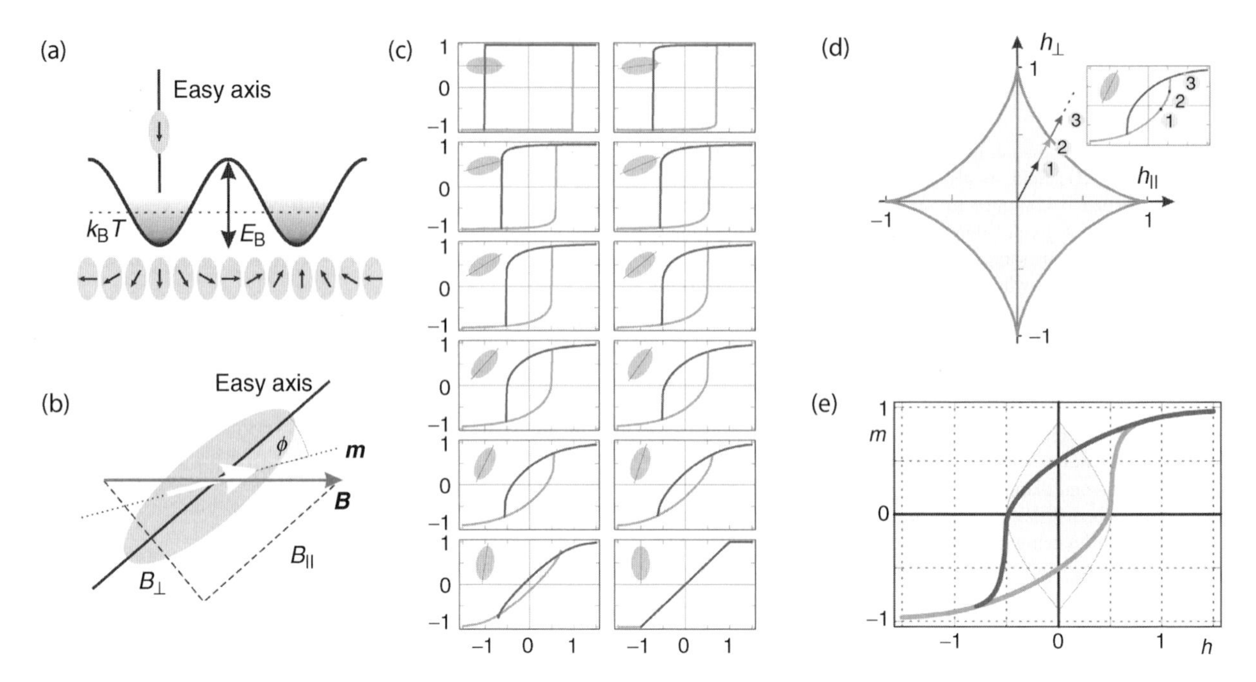

Figure R12 Stoner-Wohlfarth theory of hysteresis loops in coherently rotating single-domain particles. (a) Each particle has a preferred easy axis of magnetization and an energy barrier E_B opposes the rotation. It can result from magnetostatic-, crystal-, or stress-induced anisotropy. Thermal activation k_BT is assumed to be small in comparison to E_B. (b) Application of an external field with components $B_{||}$ and $B_\perp$ with respect to the easy axis results in a magnetization deflection ϕ. (c) Plotting the magnetization component m_B versus B results in single particle hysteresis loops. Their shape depends upon the angle between easy axis and external field B. (d) The switching instability traces an asteroid in the $h_{||} - h_\perp$ space. For field vectors inside the asteroid the system is bistable, outside the asteroid the magnetization state is unique. (e) Averaging over an isotropic ensemble of identical single-domain particles results in the shown Stoner-Wohlfarth hysteresis loop. More realistic ensembles average in addition over a distribution of anisotropy constants.

and M_s its spontaneous magnetization, the particles' energy as a function of the deflection angle φ is given by

$$e(\varphi) = K \sin^2 \varphi - \mu_0 B_{||} M_s \cos \varphi - \mu_0 B_\perp M_s \sin \varphi. \qquad \text{(Eq. 2)}$$

Solving the system of equations $e'(\varphi) = e''(\varphi) = 0$, yields a parametrization of the critical line in field space where the magnetization becomes unstable and switches direction

$$h^*_{||} = -\cos^3 \varphi, \quad h^*_\perp = \sin^3 \varphi,$$

where $h = BM_s/(2K)$ is the normalized field strength and φ serves as parameter. This curve is the Stoner-Wohlfarth asteroid (Bertotti, 1998; Hubert and Schäfer, 1998) (Figure R12d).

Plotting the magnetization component parallel to the field versus the field yields the Stoner-Wohlfarth hysteresis loops in Figure R12. For simple numerical mixing experiments, a useful approximation for a Stoner-Wohlfarth particle ensemble is $m_{SW}(h) = \tanh(0.8h) \pm 0.5\text{sech}(1.9h)$.

Hysteresis in multidomain particles

In magnetic particles beyond the critical SD size, magnetostatic self-interaction overcomes the strong quantum mechanical exchange coupling and deflects a part of the spins to form an inhomogeneous internal magnetization structure which minimizes the energy by partial magnetic flux closure. In large particles the optimal magnetization configuration is a subdivision of the grain in different relatively large magnetic domains, nearly homogeneously magnetized along some preferred axes and separated by relatively small domain walls wherein the magnetization changes continuously. To a certain extent, it is easy to change the magnetization of a multidomain grain by moving its domain walls such as to enlarge one domain and diminish another (Figure R13). One mechanism of hysteresis in multidomain grains originates from impeding the continuous motion of a domain wall by pinning it to a local defect or stress source which creates an energy barrier which must be overcome to proceed with the motion. By larger field variation, the energetically optimal number of domains changes and a global reorganization of domain structure occurs by nucleation or denucleation of domains. In all these cases, one can envisage the self-demagnetizing field as the driving force which adjusts the internal structure to the prescribed external field. Only beyond the saturation field B_{sat}, the internal magnetization structure is homogeneous and essentially rotates like an SD particle into the field direction. Because this rotation is a reversible process, the branches of the hysteresis curve coincide for $B > B_{sat}$.

While a quantitative description of the physical processes in multidomain particles is extremely difficult, a general understanding of multidomain hysteresis in terms of the self-demagnetizing field H_d, as given by Néel (1955), is very easy. In this simplified picture the MD particles' average magnetization M creates an average internal demagnetizing field

$$H_d = -NM,$$

where the numerical demagnetizing factor $0 < N < 1$ nearly does not depend on the domain configuration. The total internal field H_i inside the particle then is the sum $H + H_d$ of external and demagnetizing field. Assuming that anhysteretic magnetization and internal field are linked by an internal susceptibility χ_i one obtains

$$M(H) = \chi_i H_i = \chi_i (H + H_d) = \chi_i (H - NM(H)).$$

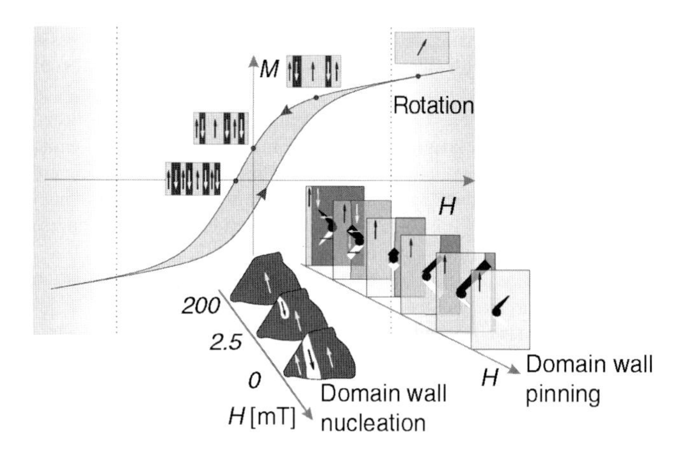

Figure R13 Hysteresis in multidomain particles is effected by changes of the domain structure. Domains grow or shrink by domain wall movement which can be impeded by pinning at lattice defects. An example is irreversible pinning of a 180° domain wall as observed in a single crystal iron film (after Hubert and Schäfer 1998). Changing the number of domains requires global reorganization of the magnetization pattern (nucleation). For example nucleation of domains as observed in a 15-μm titanomagnetite (after Halgedahl and Fuller 1983). In high fields the sample becomes saturated with essentially homogeneous magnetization which in increasing field rotates toward the field direction.

Solving for $M(H)$ yields

$$M(H) = \frac{\chi_i}{1 + N\chi_i} H \approx H/N,$$

where the last approximation assumes large χ_i. For given H_c, the two hysteresis branches then are $M^{\pm}(H) \approx (H \mp H_c)/N$, from which for $H = 0$ Néel (1955) obtains

$$\frac{M_{rs}}{M_s} = \frac{1}{NM_s} H_c. \quad \text{(Eq. 3)}$$

Overview of typical measurements

Instrumentation

There are two commonly used techniques to measure the magnetization of a sample in dependence of an applied field.

Firstly, magnetization can be determined by measuring the stray field generated by the magnetic moment of a sample. Commonly used instruments in rock magnetism applying this principle are vibrating sample magnetometers (VSMs). During the measurement the sample is vibrating and its magnetization is registered by measuring the flux change of the magnetic field through a system of pickup coils. The inducing external field is homogeneous and constant through time and therefore does not contribute to the measurement signal. One of the advantages of VSMs is the quick magnetization measurement, which allows to sample the hysteresis loop with typically up to 1 000 evenly distributed data points. Despite this large number, measurement of a complete hysteresis loop typically does not take longer than one minute. Other instruments use superconducting quantum interference devices (SQUIDs) to measure the stray field of a magnetized sample. These devices are more common in the measurement of remanent magnetization but can also be used for measuring magnetization in applied fields. This technique is realized in the magnetic property measurement system (MPMS).

The second class of instruments measures the force which is exerted by a magnetic gradient field on the magnetized sample in order to determine its magnetization. The classical instrument using this technique is the magnetic balance. Here, the sample is suspended from a force measuring system and is positioned between the pole pieces of the electromagnet generating the inducing field. In order to generate a measurable force, the sample is not positioned in the area of homogeneous magnetic field but rather at the edges of the pole pieces where a large field gradient is present. Another frequently used configuration involves pole pieces which are shaped in a special way so that the inducing field is not homogeneous but has a well-defined gradient which is exerting a translational force on the sample (Collinson, 1983). The field gradient is generally dependent on the strength of the inducing field **B** so that the measurement of **M** is not independent of **B**.

This disadvantage of the magnetic balance can be overcome by separating the source of the gradient field from the inducing homogeneous field. The variable field translation balance (VFTB), a modified magnetic balance, uses this approach by supplying the gradient field through an additional set of coils. In addition, this gradient field is oscillating, forcing the sample to oscillate as well. Thus, the force acting onto the sample and eventually the magnetization can be determined dynamically from the sample's oscillation amplitude. This yields a higher sensitivity in comparison to the static force measurement of the classical magnetic balance. A similar approach is used in the alternating gradient field magnetometer (AGFM) where the sample is mounted on a quartz rod, which itself is attached to a piezoelectric transducer recording the force acting on the sample.

Measurement procedures

The basic interpretation of hysteresis properties requires determination of the four parameters saturation magnetization M_s, saturation remanence M_{rs}, coercive field B_c, and coercivity of remanence B_{cr}. The first three quantities are determined from the hysteresis loop, whereas B_{cr} can only be determined from the so-called backfield curve.

Measurement procedures for hysteresis loops depend on the speed of data acquisition and thus the number of data points. For instruments like the VSM and AGFM with typically 600–1 000 data points per loop, field steps are chosen equidistantly to obtain a quasicontinuous loop. In the case of instruments with slower data acquisition and typically 20–40 field steps (e.g., VFTB, MPMS), nonequidistant spacing is advisable in order to properly sample all features of the loop shape.

von Dobeneck (1996) proposed the following series yielding increasing field increments toward the field maxima

$$B(i) = \frac{|i|}{i} \frac{B_{max}}{\lambda} [(\lambda + 1)^{|i|/n} - 1]$$

with $i = -n, \ldots, -1, 0, 1, \ldots, n$ ($2n$ number of data points per branch) and $\lambda > 0$ defining the accumulation of field steps toward the origin. This provides a good sampling of the bends in the region of irreversible magnetization changes and a less dense spacing in the region of the approach to saturation.

As magnetic hysteresis is temperature dependent, a proper temperature control during hysteresis loop measurement is crucial to obtain high-quality data.

For measurement of the backfield curve the sample is first brought into magnetic saturation by applying the maximum magnetic field in positive field direction. The sample's remanent magnetization is then measured in zero field being equal to M_{rs}. Subsequently, successively higher fields are imparted in negative field (or backfield) direction and after each field-on step, the remanent magnetization is again measured in zero field. The field at which the remanent magnetization switches from positive to negative values is called the coercivity of remanence B_{cr}. As the backfield measurement always consists of discrete steps, the crossing of the abscissa is usually determined by interpolation.

For instruments where the inducing field is generated by an electromagnet with pole pieces, it is necessary to compensate the residual magnetic field of these pole pieces in order to conduct a proper zero field measurement. Usually, this is accomplished by an additional set of coils.

Typical measurement results

Some results of hysteresis loop measurements and associated backfield curves are shown in Figure R14. The loop of synthetic MD magnetite (Figure R14a) shows the typical ramp-like shape and a quick saturation below 500mT. Note also the comparatively low B_c of 3.5mT and the small area inside the loop corresponding to a small E_{hys}. The high B_{cr}/B_c ratio is characteristic for multidomain particles. The young ocean basalt sample (Figure R14b) has a fairly low B_c as well. For the herein contained titanomagnetite, however, much higher fields are necessary for magnetic saturation. Although irreversible processes (causing the difference between the two loop branches) do not occur above $B \approx 0.4$ T, the loop shows significant curvature up to fields of 1.4T due to reversible rotation of magnetic moments. Older ocean basalts, where titanomagnetite is transformed into titanomaghemite by low-temperature oxidation often show almost perfect single-domain behavior as represented by the hysteresis loop in Figure R14c. Both the M_{rs}/M_s and B_{cr}/B_c ratio are approaching the ideal values for noninteracting uniaxial single-domain particles (see Table R5). Natural titanomaghemites as found in ocean basalts have even higher saturation fields than titanomagnetites.

The grain size dependence of hysteresis loops and parameters will be discussed in the following section. More hysteresis loops of rock forming minerals can be found in the literature, e.g., Chevallier and Mathieu (1943) (Hematite) and Keller and Schmidbauer (1999) (Titanomagnetite).

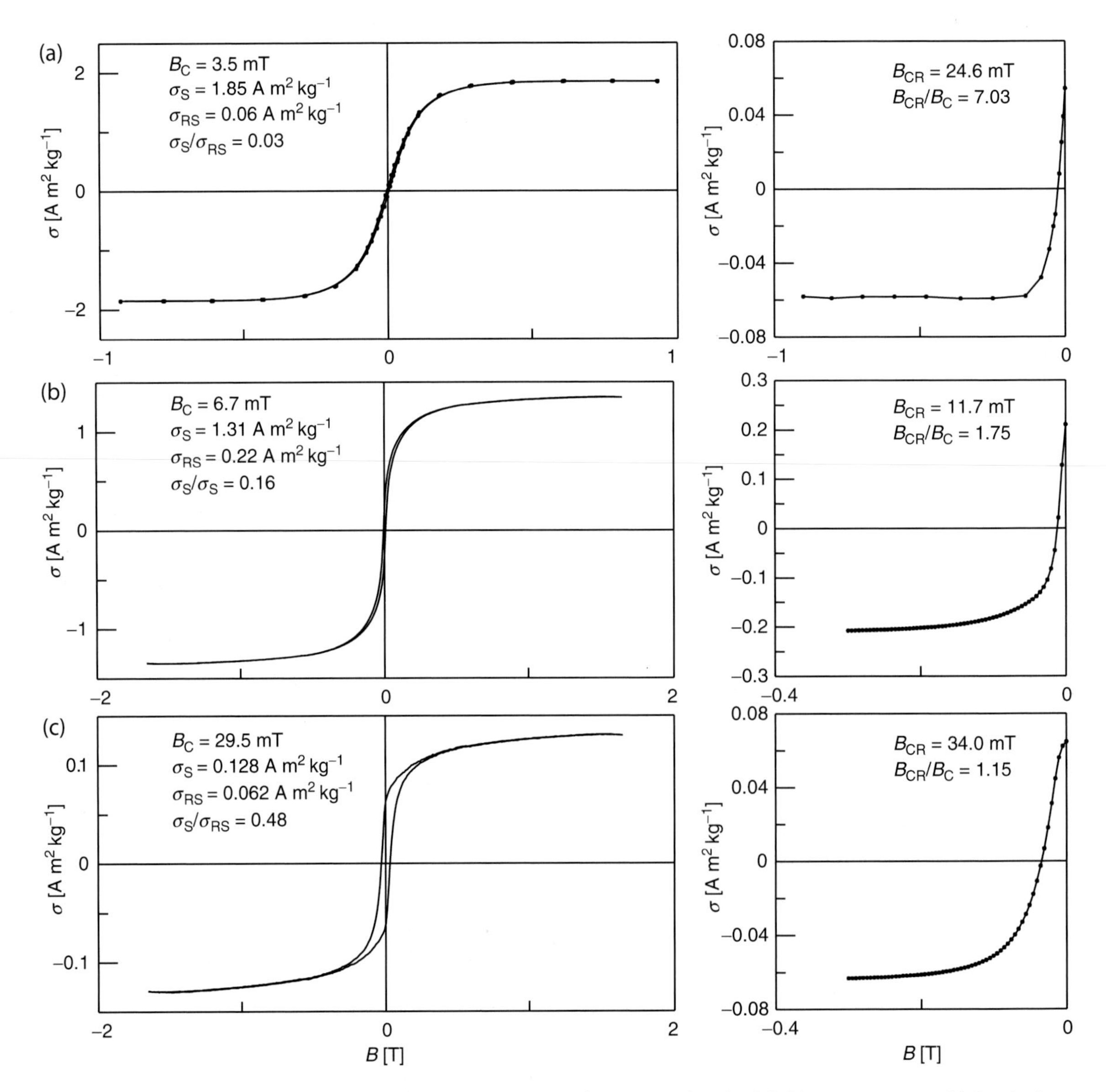

Figure R14 Three examples of room temperature hysteresis curves and corresponding backfield measurements. (a) Synthetic multidomain mangetite (average grain size 12μm) dispersed in CaF_2 and measured with a VFTB. (b) Young (0.7 Ma) ocean basalt sample containing titanomagnetite. (c) 32-Ma old ocean basalt sample containing single-domain titanomaghemite. Both ocean basalts were measured with a Princeton Measurements VSM.

Table R5 Theoretical values of hysteresis parameters for isotropic noninteracting single-domain particle ensembles (after Joffe and Heuberger 1974)

	Prolate	Oblate	Cubic: K 0	Cubic: K 0
B_{cr}/B_c	1.09	1.15	1.08	1.04
M_{rs}/M_s	0.5	0.638	0.866	0.831

Interpretation of hysteresis loops

Constituents of the hysteresis loop

The initial curve of a hysteresis loop starts at zero field and zero remanence (Figure R10a). There exist many different magnetization states having zero remanence and accordingly many different initial curves are possible. While the most symmetric choice is the thermally demagnetized initial state, usually the alternating field (AF) demagnetized state is regarded in the literature. In sufficiently small fields the initial curve of multidomain particle ensembles can be described by the Rayleigh law:

$$M(B) = \kappa_0 B + \alpha B^2$$

The first term accounts for reversible changes (wall movements) whereas the second term is associated with irreversible processes. In rock magnetic applications, the initial susceptibility κ_0 is often used as a relative proxy for magnetic mineral concentration.

The part of the hysteresis curve measured between maximum + and maximum − field is called upper hysteresis branch, the subsequent measurement between maximum − and maximum + field is termed lower hysteresis branch. With only a few exceptions (associated with a certain type of magnetic interaction; Meiklejohn and Carter, 1960), the two branches are point symmetric with respect to the origin. The intersections of the hysteresis branches with the ordinate and abscissa give the values for saturation remanence M_{rs} and B_c, respectively. The area between the two branches defines the total hysteresis work E_{hys}. One of the factors influencing the shape of the hysteresis branches is the grain size of magnetic minerals. Figure R12e shows the idealized loop shape for an ensemble of uniaxial single-domain grains, whereas Figure R14a shows the typical ramp-like shape of a sample containing multidomain grains.

An important difficulty in hysteresis measurements is the determination of the saturation magnetization M_s from the loop. It requires to estimate the limit value of magnetization toward high fields. The *approach to saturation* occurs in the high field region where irreversible magnetization changes are completed. The loop is a single branch which in case of pure rotation theoretically is described by:

$$M(B) = \chi_{pd} B + M_s - \frac{a_2}{B^2} + \cdots$$

Experimental evidence rather indicates an expansion

$$M(B) = \chi_{pd} B + M_s - \frac{a_1}{B} - \frac{a_2}{B^2} + \cdots$$

Deviations from the ideal case result mainly from small stress centers of point defects, dislocation dipoles, or inclusions (Kronmüller and Fähnle, 2003). In rock magnetic applications it is sometimes difficult to ensure that a sample is sufficiently saturated since some natural minerals saturate only in high fields and especially titanomagnetites may be extremely sensitive to stress.

A natural generalization of the hysteresis measurement procedure is the variation of start and end fields. Any zigzag sequence of fields $B_1\ B_2\ \ldots\ B_n$ may be applied to the sample to produce a sequence of higher order hysteresis branches. All these branches lie inside the region outlined by the main hysteresis curve between the maximal field values $\pm B_{max}$. The additional information contained in higher order curves may be helpful for a detailed sample characterization. In the framework of Preisach theory it is possible to infer theoretical relations between different higher order curves. Especially, it is possible to show that the complete information of any Preisach system can be retrieved from its first-order curves, which result from the field sequences B_{max}, B, B_{max} (Mayergoyz, 1991).

The most basic interpretation of the backfield curve yields a single value for the coercivity of remanence B_{cr} of the bulk sample. The curve shape, however, does also contain information about the distribution of coercivities due to a mixture of different magnetominerals or a distribution of grain sizes of the contained magnetic phase. It is important to know that the B_{cr} value of a mixture of different magnetic minerals or grain sizes cannot be explained by simple linear mixing theory (e.g., Dunlop 2002a).

Hysteresis parameters

Saturation magnetization M_s is a material constant being independent of grain size. In rock magnetic applications M_s is thus dependent on the type of magnetic mineral and its concentration. In contrast to M_s, saturation remanence M_{rs} is a grain size sensitive parameter. Meaningful interpretation of this parameter therefore almost always requires normalization to M_s as it is also dependent on mineral type and concentration. It can then be compared to theoretical domain state threshold values to determine the magnetic grain size as discussed in the next section. The coercivity parameters B_{cr} and B_c are independent of concentration and can be interpreted in terms of the stability of magnetic remanence. Coercivity is controlled by magnetomineralogy, grain size and the anisotropies controlling the magnetic energy. In minerals being dominated by magnetostrictive anisotropy, e.g., titanomagnetite, the coercivity parameters are strongly dependent on internal and external stresses acting onto the mineral grains. Other important factors controlling coercivity are grain shape, crystal impurities, and lattice imperfections. B_{cr} is generally normalized by B_c in order to yield independent granulometric information.

The Day plot

It was observed by Nagata (1953) that B_c varies linearly with M_{rs}/M_s in igneous rocks. As seen in (3), Néel (1955) explained this linear dependency as a direct effect of the self-demagnetizing field. His theory implies that M_{rs}/M_s and B_c do not represent independent grain size information and that the slope of M_{rs}/M_s versus B_c strongly depends upon M_s and thus on mineralogy (Wasilewski, 1973). In Day *et al.* (1977) it was reported for titanomagnetites that besides M_{rs}/M_s also the ratio B_{cr}/B_c well reflects grain size variations, but is independent of mineral composition. They therefore suggested to plot M_{rs}/M_s versus B_{cr}/B_c and delineated regions containing predominantly SD or MD remanence carriers (Day *et al.*, 1977) (Figure R15). This *Day plot* has since been extensively used in rock magnetism to characterize magnetic grain size. Between the SD and MD region lies a broad transitional interval which is attributed to pseudo-single-domain (PSD) behavior. Several reasons for this peculiar behavior are discussed in the literature (see, e.g., Dunlop and Özdemir, 1997). Among these are SD-like moments in larger MD grains caused by dislocations or surface defects, the magnetic moments of domain walls, unusual domain states such as vortex structures or mixtures of MD and SD grains. Recent theoretical and experimental studies investigate this influence of grain size mixtures upon the Day plot (Dunlop, 2002a,b; Carter-Stiglitz *et al.*, 2001; Lanci and Kent, 2003).

Table R5 after Joffe and Heuberger (1974) gives the theoretical SD values of B_{cr}/B_c and M_{rs}/M_s for isotropic noninteracting particle ensembles. With increasing temperature the value of B_{cr}/B_c for uniaxial random SD ensembles approaches 1 (Joffe, 1969).

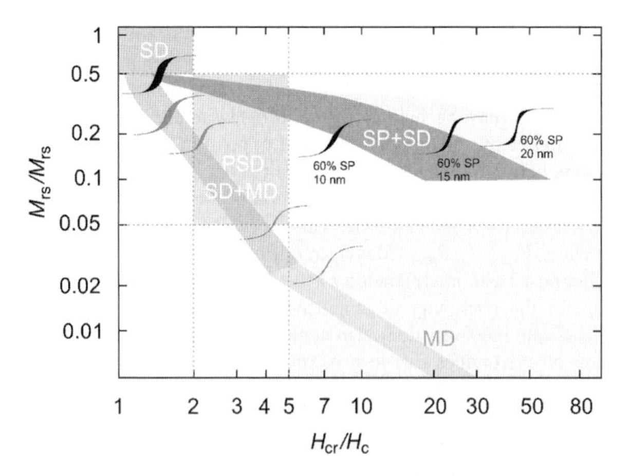

Figure R15 The Day plot classifies hysteresis curves in terms of the grain size sensitive quantities M_{rs}/M_s and B_{cr}/B_c. For magnetite and titanomagnetites the shown grain size classes have been originally defined in Day *et al.* (1977) and recently revised by Dunlop (2002a). The indicated theoretical trends for mixtures are taken from Dunlop (2002a) and Lanci and Kent (2003). While in case of the extensive quantities M_{rs} and M_s normalization is necessary to remove the volume dependence, using the ratio of the intensive quantities B_{cr} and B_c intends to remove dependencies of these quantities upon mineral composition. Exemplary hysteresis loops are shown inset. Loops are centered on the respective M_{rs}/M_s and B_{cr}/B_c values. Black: calculated synthetic loops. Gray: hysteresis loops of synthetic magnetite samples measured with an alternating gradient field magnetometer (maximum field 300mT).

Hysteresis shape

The possible shapes of magnetic hysteresis loops cover a wide range and only few general rules can be stated. (a) If saturation is reached, all magnetic memory is erased and the loop is point symmetric $M^+(B) = -M^-(-B)$. In most materials the coupling between magnetic system and lattice is very weak and the magnetic free energy obeys the thermodynamic laws. (b) After correcting for para- or diamagnetic slope, each hysteresis branch increases monotonically: $B_1 \leq B_2$ implies $M(B_1) \leq M(B_2)$. (c) The area E_{hys} of the loop corresponds to the energy dissipated by the magnetic system which implies $E_{hys} \geq 0$.

In rock magnetism the general outline of hysteresis loops has been classified into constricted (wasp-waisted) rectangular (normal) and rounded (pot-bellied) shapes (Wasilewski, 1973; Muttoni, 1995; Roberts *et al.*, 1995; Hodych, 1996; Tauxe *et al.*, 1996). Extreme shape variations usually result from mixtures of samples with anhysteretic loops of very different saturation behavior and coercivity (Bean, 1955; Wasilewski, 1973; Hejda *et al.*, 1994). Figure R16 demonstrates shape changes for a simple system of noninteracting SD particles mixed with an SP fraction of varying anhysteretic slope (e.g., grain size). While the vertical loop difference is unchanged by this process, the horizontal difference reflects the degree of constriction from pot-bellied to wasp-waisted loops. To quantitatively describe constriction the ratio between maximal and minimal horizontal difference has been used (Xu *et al.*, 2002; Gee and Kent, 1999). Another possibility is the parameter

$$\sigma_{hys} = \log\left(\frac{E_{hys}}{4M_sB_c}\right), \qquad \text{(Eq. 4)}$$

where log denotes the natural logarithm (Fabian, 2003). For constricted loops $\sigma_{hys} > 0$ whereas for round loop shapes $\sigma_{hys} < 0$.

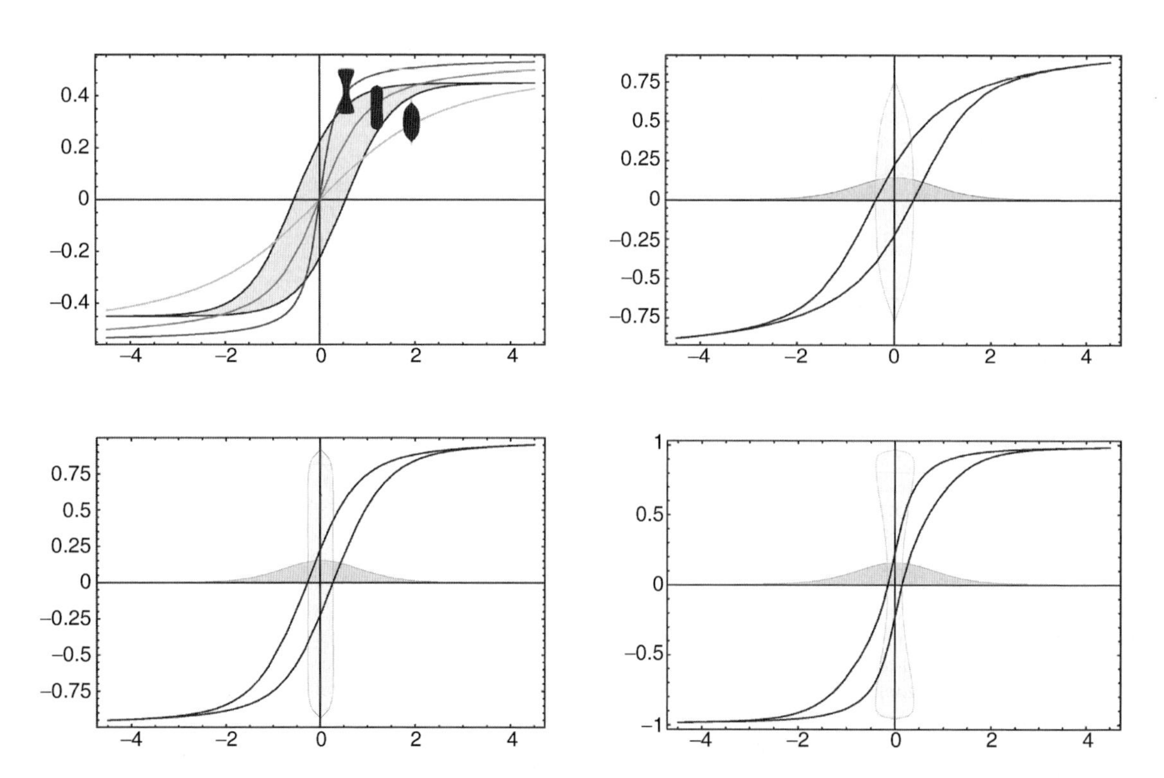

Figure R16 Three hysteresis shapes result from linear mixture of the same primary hysteresis loop with Langevin functions of different susceptibility. The vertical loop difference is unchanged by this process, whereas the horizontal difference clearly reflects the rounded or constricted shape of the loops.

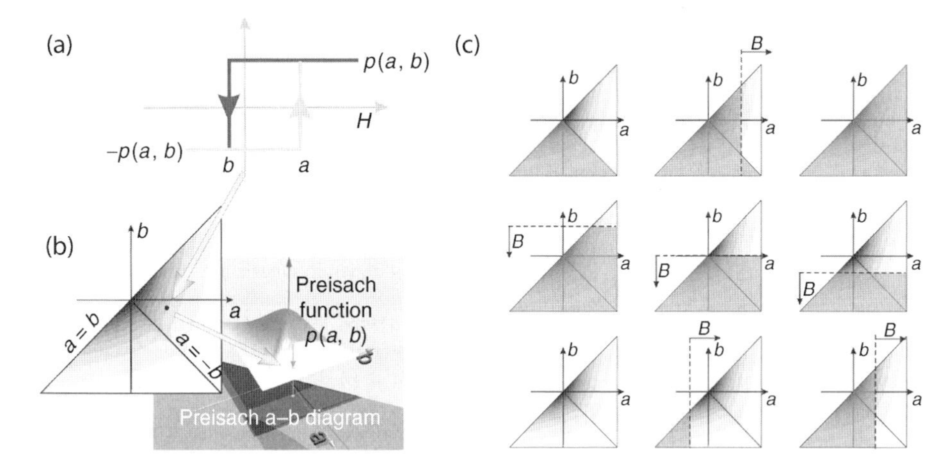

Figure R17 A noninteracting ensemble of rectangular switching curves as in (a) is a very general system with hysteresis. If a and b denote downward and upward switching fields, respectively, the systems hysteresis properties are completely characterized by the Preisach function $p(a, b)$ (b), which is defined for $a \geq b$. In (c) the magnetization change of the system is determined by integration over the Preisach plane. Shaded areas correspond to upward switched loops and count positively, while unshaded areas count negatively (see (Eq. 5)).

Stress control and annealing

An important factor influencing the hysteresis properties of natural and synthetic samples is stress. It originates either from external sources (hydrostatic or shear stress) or from internal sources like defects, dislocations, twin boundaries, or changes in lattice constants due to oxidation or ion diffusion. A further intrinsic stress source for magnetic materials is magnetostriction. It denotes the lattice deformation in reaction to the local magnetic moment vector. Any inhomogeneous magnetization structure creates a magnetostrictive stress field which compensates for the incompatible local deformations (Hubert and Schäfer, 1998).

The inverse effect is that the magnetization structure adapts to external stress patterns. In a material with isotropic magnetostriction constant λ_s an external stress σ leads to an additional uniaxial anisotropy along the stress axis with anisotropy constant $K = \frac{3}{2}\lambda_s\sigma$.

Stress effects both, reversible and irreversible magnetization patterns. Reversible processes which govern the approach to saturation are influenced by the increased anisotropy (Dankers and Sugiura, 1981; Hodych, 1990; Özdemir and Dunlop, 1997), while irreversible processes, reflected by coercive fields as B_c and B_{cr} or remanences as M_{rs}, are enhanced by stress-induced pinning sites. Since characteristic domain sizes scale with $\sqrt{K}$ if all other material parameters are constant (Hubert and Schäfer, 1998), the apparent magnetic grain size, as observed, e.g., in the Day plot, changes with $1/\sqrt{\sigma}$ when stress anisotropy prevails.

Decomposition of hysteresis curves

Conceptually, a hysteresis loop traces a continuous magnetization process and contains much more information about the sample than represented by a few remanence, coercivity or shape parameters. Especially in environmental magnetism one wishes to use all available information to analyze compositional variations within large sample collections. Several procedures have been put forward to exactly describe the hysteresis loop by large parameter sets. Among them are Fourier decomposition (Jackson *et al.*, 1990) and a representation by hyperbolic basis functions (von Dobeneck, 1996). Ideally, such methods attempt to use hysteresis loops and other magnetization curves (Robertson and France, 1994) to decompose a set of given mixed loops into their end members which can be attributed to different sources. In case of linear mixing this kind of decomposition is possible (Thompson, 1986; Carter-Stiglitz *et al.*, 2001). However it is not possible to uniquely identify end members from only mixed samples. Moreover, unmixing procedures which synchronously use different magnetization curves should take into account mutual interdependencies between these curves (Fabian and von Dobeneck, 1997). The extreme case of complete sample description is based on the measurement of the samples Preisach function or—equivalently—its FORC distribution.

Preisach model

Preisach (1935) put forward a simple mathematical concept which allows to understand many fundamental properties of hysteresis loops and other magnetization curves. His model replaces the magnetic sample by a collection of rectangular switching processes characterized by their respective downward and upward switching fields (a, b) (Figure R17a). The weight of (a, b) for the sample is given by the Preisach function $p(a, b)$ (Figure R17b). Whether a loop is in its *up* or *down* position depends on the previous field history. In Figure R17c shaded a-b areas represent loops in *up* position. Increasing the field B switches loops with $a < B$ in the *up* position, decreasing B switches loops with $b > B$ in their *down* position. When $p(a, b)$ is known, the sample magnetization after an arbitrary field history is

$$m = \int_{S^+} p(a,b)\mathrm{d}a\,\mathrm{d}b - \int_{S^-} p(a,b)\mathrm{d}a\,\mathrm{d}b, \qquad \text{(Eq. 5)}$$

where S^+ and S^- denote the subsets of the a-b plane with, respectively, upward or downward switched loops. The Preisach model has been used in rock magnetism to investigate interaction (Dunlop, 1969; Dunlop *et al.*, 1990), viscosity (Mullins and Tite, 1973), and grain size (Ivanov *et al.*, 1981). Mapping of the Preisach function yields detailed information about the sample (Hejda and Zelinka, 1990). A special measurement scheme for mapping of the Preisach function, originally developed and studied by Girke (1960), is now routinely used in rock magnetism (Pike *et al.*, 1999, 2001). Using the Preisach model it is also possible to obtain general theoretical relations between different magnetization processes (Fabian and von Dobeneck, 1997). There is extensive literature on using this or similar models in engineering and mathematics (Mayergoyz, 1991; Visintin, 1991).

David Krása and Karl Fabian

Bibliography

Bean, C., 1955. Hysteresis loops of mixtures of ferromagnetic micropowders. *Journal of Applied Physics*, **26**: 1381–1383.

Bertotti, G., 1998. *Hysteresis in Magnetism.* San Diego: Academic Press.

Carter-Stiglitz, B., Moskowitz, B., and Jackson, M., 2001. Unmixing magnetic assemblages and the magnetic behavior of bimodal mixtures. *Journal of Geophysical Research*, **106**(B11): 26397–26411.

Chevallier, R., and Mathieu, S., 1943. Propriétés magnétiques des poudres d'hématite—influence des dimensions des grains. *Annales de Physique*, **18**: 258–288.

Collinson, D.W., 1983. *Methods in Rock Magnetism and Palaeomagnetism: Techniques and Instrumentation.* New York: Chapman and Hall.

Dankers, P., and Sugiura, N., 1981. The effects of annealing and concentration on the hysteresis properties of magnetite around the PSD-MD transition. *Earth and Planetary Science Letters*, **56**: 422–428.

Day, R., Fuller, M., and Schmidt, V.A., 1977. Hysteresis properties of titanomagnetites: grain-size and compositional dependence. *Physics of the Earth and Planetary Interiors*, **13**: 260–267.

von Dobeneck, T., 1996. A systematic analysis of natural magnetic mineral assemblages based on modelling hysteresis loops with coercivity-related hyperbolic basis functions. *Geophysical Journal International*, **124**: 675–694.

Dunlop, D., 1969. Preisach diagrams and remanent properties of interacting monodomain grains. *Philosophical Magazine*, **19**: 369–378.

Dunlop, D., 2002a. Theory and application of the Day plot (M_{rs}/M_s versus H_{cr}/H_c) 1. Theoretical curves and tests using titanomagnetite data. *Journal of Geophysical Research*, **107**(B3): 2056 (doi: 10.1029/2001JB000486).

Dunlop, D., 2002b. Theory and application of the Day plot (M_{rs}/M_s versus H_{cr}/H_c) 2. Application to data for rocks, sediments, and soils. *Journal of Geophysical Research*, **107**(B3): 2057 (doi: 10.1029/2001JB000487).

Dunlop, D., and Özdemir, Ö., 1997. *Rock Magnetism: Fundamentals and Frontiers. Cambridge studies in Magnetism*. Cambridge: Cambridge University Press.

Dunlop, D.J., Westcott-Lewis, M.F., and Bailey, M.E., 1990. Preisach diagrams and anhysteresis: do they measure interactions? *Physics of the Earth and Planetary Interiors*, **65**: 62–77.

Fabian, K., 2003. Some additional parameters to estimate domain state from isothermal magnetization measurements. *Earth Planetary Science Letters*, **213**: 337–345.

Fabian, K., and von Dobeneck, T., 1997. Isothermal magnetization of samples with stable Preisach function: a survey of hysteresis, remanence and rock magnetic parameters. *Journal of Geophysical Research*, **102**: 17659–17677.

Gee, J., and Kent, D.V., 1999. Calibration of magnetic granulometric trends in oceanic basalts. *Earth Planetary Science Letters*, **170**: 377–390.

Girke, H., 1960. Der Einfluß innerer magnetischer Kopplungen auf die Gestalt der Preisach-funktionen hochpermeabler Materialien. *Zeitschrift für Angewandte Physik*, **11**: 502–508.

Halgedahl, S., and Fuller, M., 1983. The dependence of magnetic domain structure upon magnetization state with emphasis upon nucleation as a mechanism for pseudo-single-domain behavior. *Journal of Geophysical Research*, **88**: 6505–6522.

Hejda, P., and Zelinka, T., 1990. Modelling of hysteresis processes in magnetic rock samples using the Preisach diagram. *Physics of the Earth and Planetary Interiors*, **63**: 32–40.

Hejda, P., Kapička, A., Petrovský, E., and Sjöberg, B.A., 1994. Analysis of hysteresis curves of samples with magnetite and hematite grains. *IEEE Transactions on Magnetics*, **30**: 881–883.

Hodych, J., 1990. Magnetic hysteresis as a function of low temperature in rocks: evidence for internal stress control of remanence in multi-domain and pseudo-single-domain magnetite. *Physics of the Earth and Planetary Interiors*, **64**(1): 21–36.

Hodych, J., 1996. Inferring domain state from magnetic hysteresis in high coercivity dolerites bearing magnetite with ilmenite lamellae. *Earth and Planetary Science Letters*, **142**: 523–533.

Hubert, A., and Schäfer, R., 1998. *Magnetic Domains.* Berlin Heidelberg New York: Springer.

Ivanov, V.A., Khaburzaniya, I.A., and Sholpo, L.Y., 1981. Use of Preisach diagram for diagnosis of single- and multi-domain grains in rock samples. *Izvestiya Earth Physics*, **17**: 36–43.

Jackson, M.J., Worm, H.-U., and Banerjee, S.K., 1990. Fourier analysis of digital hysteresis data: rock magnetic applications. *Physics of the Earth and Planetary Interiors*, **65**: 78–87.

Joffe, I., 1969. The temperature dependence of the coercivity of a random array of uniaxial single domain particles. *Journal of Physics C*, **2**: 1537–1541.

Joffe, I., and Heuberger, R., 1974. Hysteresis properties of distributions of cubic single-domain ferromagnetic particles. *Philosophical Magazine*, **29**: 1051–1059.

Keller, R., and Schmidbauer, E., 1999. Magnetic hysteresis properties and rotational hysteresis losses of synthetic stress-controlled titanomagnetite ($Fe_{2.4}Ti_{0.6}O_4$) particles—I. Magnetic hysteresis properties. *Geophysical Journal International*, **138**(2): 319–333.

Kronmüller, H., and Fähnle, M., 2003. *Micromagnetism and the Microstructure of Ferromagnetic Solids.* Cambridge: Cambridge University Press.

Lanci, L., and Kent, D.V., 2003. Introduction of thermal activation in forward modeling of hysteresis loops for single-domain magnetic particles and implications for the interpretation of the Day diagram. *Journal of Geophysical Research*, **108**(B3): 2142.

Mayergoyz, I.D., 1991. *Mathematical Models of Hysteresis.* New York: Springer-Verlag.

Meiklejohn, W.H., and Carter, R.E., 1960. Exchange anisotropy in rock magnetism. *Journal of Applied Physics*, **31**: 164–165.

Mullins, C., and Tite, M., 1973. Preisach diagrams and magnetic viscosity phenomena for soils and synthetic assemblies of iron oxide grains. *Journal of Geomagnetism and Geoelectricity*, **25**: 213–229.

Muttoni, G., 1995. "Wasp-waisted" hysteresis loops from a pyrrhotite and magnetite-bearing remagnetized Triassic limestone. *Geophysical Research Letters*, **22**: 3167–3170.

Nagata, T., 1953. *Rock Magnetism.* Tokyo: Maruzen.

Néel, L., 1955. Some theoretical aspects of rock-magnetism. *Advances in Physics*, **4**: 191–243.

Özdemir, Ö., and Dunlop, D.J., 1997. Effect of crystal defects and internal stress on the domain structure and magnetic properties of magnetite. *Journal of Geophysical Research*, **102**(B9): 20211–20224.

Pike, C.R., Roberts, A.P., and Verosub, K.L., 1999. Characterizing interactions in fine magnetic particle systems using first order reversal curves. *Journal of Applied Physics*, **85**(9): 6660–6667.

Pike, C.R., Roberts, A.P., Dekkers, M.J., and Verosub, K.L., 2001. An investigation of multi-domain hysteresis mechanisms using force diagrams. *Physics of the Earth and Planetary Interiors*, **126**(1–2): 11–25.

Preisach, F., 1935. Über die magnetische Nachwirkung. *Zeitschrift für Physik*, **94**: 277–302.

Roberts, A., Cui, Y., and Verosub, K., 1995. Wasp-waisted hysteresis loops: minerals magnetic characteristics and discrimination of components in mixed magnetic systems. *Journal of Geophysical Research*, **100**(B9), 17909–17924.

Robertson, D.J., and France, D.E., 1994. Discrimination of remanence-carrying minerals in mixtures, using isothermal remanent magnetization acquisition curves. *Physics of the Earth and Planetary Interiors*, **82**(3–4): 223–234.

Stoner, E., and Wohlfarth, E., 1948. A mechanism of magnetic hysteresis in heterogeneous alloys. *Philosophical Transactions of the Royal Society of London Series A*, **240**: 599–642.

Tauxe, L., Mullender, T., and Pick, T., 1996. Potbellies, wasp-waists, and super-paramagnetism in magnetic hysteresis. *Journal of Geophysical Research*, **101**: 571–583.

Thompson, R., 1986. Modeling magnetization data using simplex. *Physics of the Earth and Planetary Interiors*, **42**(1–2): 113–127.
Visintin, A., 1991. *Differential Models of Hysteresis*. Berlin: Springer-Verlag.
Wasilewski, P., 1973. Magnetic hysteresis in natural materials. *Earth Planetary Science Letter*, **20**: 67–72.
Xu, L., Van der Voo, R., Peacor, D.R., and Beaubouef, R.T., 2002. Alteration and dissolution of fine grained magnetite and its effects on magnetization of the ocean floor. *Earth Planetary Science Letters*, **151**: 279–288.

Cross-references

Magnetization, Isothermal Remanent
Magnetization, Thermoremanent
Paleomagnetism
Rock Magnetism

ROCK MAGNETOMETER, SUPERCONDUCTING

Superconducting Rock Magnetometers, often abbreviated as SRM, are the most commonly used magnetometers in modern Paleomagnetism. The SRM uses superconducting quantum interference devices (SQUIDs), superconducting magnetic flux transformers, superconducting magnetic shielding, and a liquid helium temperature environment of 4.2 K to measure the remanent and induced magnetic moment in samples inserted into the instrument. These samples are usually measured at room temperature. There have been many different configurations of superconducting sensors used for paleomagnetic measurements but, we will focus on the instrument designs that are, or have been commercially available and that have been the most widely used in these magnetic studies.

History

The SRM had its beginning in 1967 with the first development made by the author after meeting Dr. Richard Doell and Dr. Brent Dalrymple of the US Geological Survey at Menlo Park, CA and Prof. Alan Cox of Stanford University. They were conducting research on the reversals of the earth's magnetic field (Cox *et al.*, 1964). The author was at Stanford Research Institute in Menlo Park, CA studying the recently discovered quantum interference properties of superconducting devices. These devices were being used as very sensitive detectors of magnetic fields (Deaver and Goree, 1967). In discussions with Doell and Cox it became clear that the use of these superconducting devices could be helpful in the measurement of magnetic properties of geophysical samples.

The standard paleomagnetic measurement systems in use in the 1960s were the astatic, the induction coil spinner, and the fluxgate spinner. See Collinson (1983) and Fuller (1987) for complete descriptions of these instruments. The meetings with Cox, Doell, and Dalrymple in 1967 and 1968, and later discussions with Prof. Michael Fuller at the University of Pittsburgh led to the design and construction of the first SRM by the authors group, then at Develco, Inc., Mountain View, CA. Fuller envisioned many applications of the SRM in the study of magnetic properties of rocks and obtained funding from the US National Science Foundation to buy the first SRM. Fuller's scientific and technical expertise was fundamental in defining the instrument features so that this instrument could be used for many different measurements important to paleo- and rock magnetism (Goree and Fuller, 1976). Figure R18 is a photograph of this first SRM design installed at the University of Pittsburgh in 1970. This instrument was the foundation for design evolution that has resulted in over 200 SRMs used in laboratories throughout the world.

This first SRM was horizontally orientated with a straight through, room temperature access. This design proved to be somewhat complex for the existing state of the art in SQUID and thermal insulation technology, and only two of these systems were constructed. For the next 13 years all SRMs were a more conventional vertical cryogenic design with sample access from the top end, as shown in Figure R19. In this design the superconducting components were installed in a probe that was lowered into a conventional liquid helium dewar. The advantage of this design was primarily that the probe could be easily removed to gain access to the SQUIDs, superconducting shield and cryogenic vacuum seals. The primary disadvantages were the high liquid helium loss rate due to thermal load down the large dewar and probe neck tubes, and the fact that the room temperature access could not be straight through the magnetometer. Therefore the design was not suitable for long core measurements. The typical SRM of this design had a helium loss rate of about 3 l day^{-1}. Several vertical systems were constructed with cryocoolers mounted on the dewar to intercept much of the thermal load, and the loss rate was reduced to about 1 l day^{-1}.

The author reintroduced the horizontal straight through access system in 1981 and incorporated the cryogenic components with

Figure R18 First SRM design.

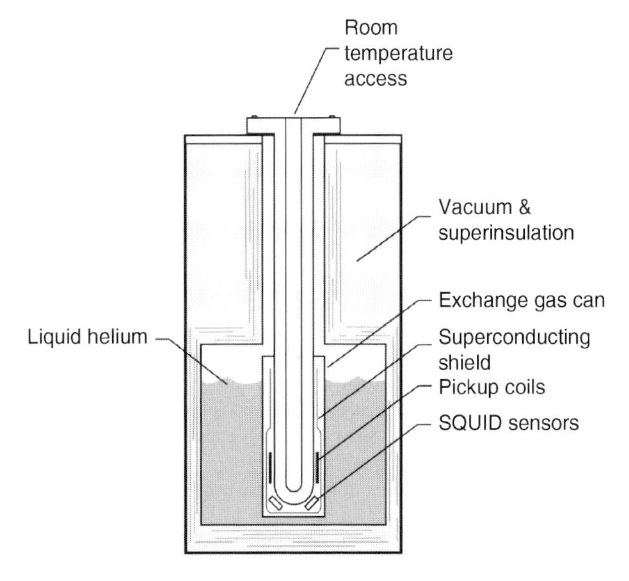

Figure R19 Conventional liquid helium dewar with vertical SRM insert.

Table R6 SRM manufacturers

Company	SRMs delivered
Develco, Inc., 1969–1973	30
Superconducting Technology, Inc., 1973–1975	45
Cryogenic Consultants, Ltd, 1977–1994	15
CTF Systems, Inc., 1977–1990	15
CEA-LETI, France, 1978	3
2G Enterprises, 1981–August 2005	104

the SQUIDs and superconducting shield to provide an integrated design that eliminated the large neck tube heat load between room temperature and liquid helium. This new design has been sold through a marketing arrangement between the authors company (William S. Goree, Inc.) and William Goodman's company (Applied Physics Systems, Inc.) as 2G Enterprises. The new design eliminated all cryogenic fiberglass-epoxy vacuum parts, thereby removing the diffusion of helium gas into the high vacuum insulation volume. These vacuum changes combined with improved superinsulation and mechanical supports and a cryocooler to intercept much of the thermal load resulted in systems with liquid helium loss rates below 0.11 day^{-1}. These very low helium loss rates provided operating times exceeding 3 years on a single fill of helium and made it practical to use the SRM in laboratories where liquid helium was very expensive and/or difficult to obtain.

There have been six different companies, starting in 1969 that have built various styles of the basic SRM. Table R6 list the companies and the approximate number of instruments built by each as of August 2005.

Instrument description

The SRM is an instrument that uses SQUIDs (Kleiner *et al.*, 2004) to measure the magnetic field of a sample typically at room temperature. The SQUID is the world's most sensitive detector of magnetic fields and, with the use of superconducting flux transformers, large samples can be coupled to the very small sensor (Goree and Fuller, 1976). One of the first design parameters used in the SRM was the concept of a very homogeneous measurement region. For a "standard" paleomagnetic sample of 2.5-cm diameter by 2.5-cm long it was desired to measure the three orthogonal components of the samples magnetic moment with a single insertion of the sample into the measurement volume. The measurement should be done quickly, in a few seconds, and should be independent of the exact location of the magnetic sources within the sample. A complete magnetic measuring system as used in most paleomagnetic laboratories consists of the SRM for the accurate measurement of the magnetic moment of the sample, sample handling equipment to place the sample into the center of the measurement region of the SRM, alternating field demagnetization coils to produce precise demagnetization of the sample at any desired field strength up to approximately 300 mT, ARM coils used in conjunction with the demagnetization coils, IRM coils to produce large induced moments in the sample, and thermal demagnetization furnaces. Each of these measurement techniques is described in other sections of this encyclopedia.

Many of the SRM systems in use today incorporate the AF demagnetization, ARM and IRM in-line with the SRM and all computer controlled so that very detailed studies of samples can be done without changing samples. The thermal demagnetization is most often done in a separate furnace where up to 100 samples are heated and cooled, then measured with the SRM. This thermal demagnetization is repeated for sequential temperature steps.

The use of the SRM for the measurement of long cores was first envisioned by Fuller and shown to be almost as accurate as cutting a core into discrete samples and measuring each sample independently (Goree and Fuller, 1976). This technique was incorporated in the first long core system specifically designed for the Ocean Drilling Program (ODP, 1985). The ODP system, shown in Figure R20, was the first use of an SRM on an ocean going ship. This SRM was used continuously by ODP from 1985 until 1992 when it was replaced with an improved design SRM that used DC SQUIDs for better sensitivity. This SRM is still in use on the ODP ship as of 2005. It has been estimated that ODP scientists with the two SRMs have measured 200 000 m of ocean core. Most of the measurements were done on semicircular half cores 150-cm long, and the cores were often passed through the magnetometer many times with sequential AF demagnetization. As of 2005 there have been ten long core SRMs installed.

Figure R20 Ocean Drilling Program ship and SRM.

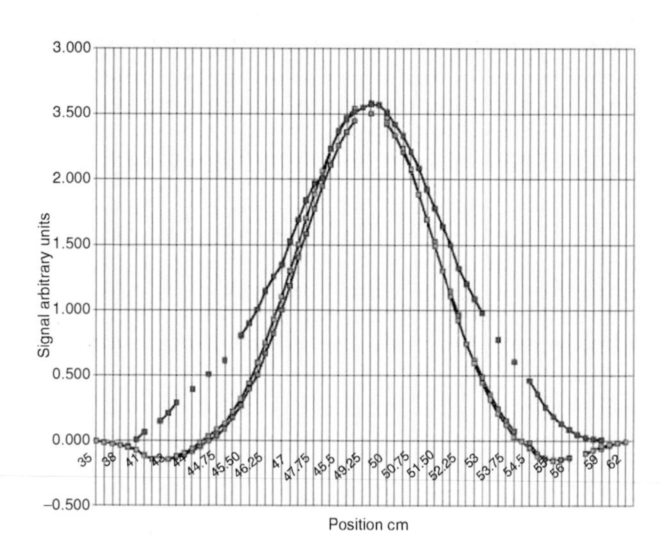

Figure R21 Sensitivity profile for high-resolution pickup coils.

Carlo Laj and colleagues at GIF Sur Yevette, France became interested in measuring long sediment samples with better axial resolution of the magnetic moment. Tauxe *et al.* (1983) suggested a U-channel sample configuration to obtain a clean section of sediment cores. These samples were 2 cm by 2 cm cross-section by 150-cm long cut from near the center of a circular core. The first U-channel SRM was delivered to GIF in 1991 (Weeks *et al.*, 1993). The pickup coil construction for U-channel systems was optimized for sampling a smaller volume of the sample and is called a high-resolution (HR) pickup coil. (Figure R21 shows sensitivity profiles for HR pickup coils.) A second U-channel system was installed at IPGP, Paris in 1992 to measure discrete samples as well as U-channels (Nagy and Valet, 1993). Many laboratories have adopted the use of the high-resolution pickup coil system to measure both U-channels and discrete samples since 1992. There have been seventy-two 2G SRMs built since the GIF system was delivered in 1991 and 14 of these have used the high-resolution U-channel design.

If a 150-cm long core or U-channel is measured every cm of its length, with AF, ARM, and IRM sequences, it may take on the order of 10h to complete a full measurement cycle where thousands of individual measurements are made. A study with this detail using an astatic magnetometer of the 1960s would have taken years of measurement time.

Superconducting quantum interference device

The SQUID is a very small device with a magnetic field sensitive area of the order of 10^{-6} m^2 (Deaver and Goree, 1967) and is the heart of an SRM. The SQUID is basically a superconducting loop with a diameter of less than 1 mm. The optimum energy sensitivity of an RF SQUID is obtained when the energy of a quantum of flux in the SQUID, $\varnothing_0^2/2L$ is larger than the thermal energy, kT, where $\varnothing_0$ is the flux quantum (2.07×10^{-15}weber), L is the SQUID loop inductance, k is the Boltzmann constant, and T is the absolute temperature of the SQUID (Zimmerman and Campbell, 1975). The optimum energy sensitivity for a DC SQUID scales as $T\ (LC)^{1/2}$ (Kleiner *et al.*, 2004). Thus, the SQUID must have a very small capacitance, C, and a very small inductance, L. The SQUID capacitance is set by the junction technology and is about 1 pF for thin film niobium junctions. The inductance is about 100 pH, which gives a SQUID loop diameter of less than 1 mm. This very small SQUID area requires that most samples to be measured must be coupled to the SQUID using a superconducting flux transformer that will be discussed in the next section, rather than being placed within the SQUID. The SQUID responds to changes in the magnetic flux that links the body of the SQUID. A unique quantum mechanical property of a SQUID is that the magnetic flux linking the SQUID area is quantized in units of $h/2e$, where h is Planck's constant and e is the charge of an electron. This fundamental property of superconductors gives the SQUID a digital output in units of the flux quanta $\varnothing_0$. In an SRM each measurement axis is calibrated by determining the magnetic moment applied at the center of the pickup coil that will produce one flux quanta change at the SQUID. A typical calibration number for an SRM with 4.2-cm diameter room temperature access is 5×10^{-8}Am^2 per flux quanta. The noise level of a SQUID is often measured in terms of the fraction of single flux quanta that can be resolved. This can approach one part in 10^6 for DC SQUID instruments.

The first SRMs used radio frequency driven single junction SQUIDs (Kleiner *et al.*, 2004) beginning with a cylindrical SQUID approximately 1-mm diameter. Later this type was replaced with a toroidal niobium body SQUID. These SQUIDs were driven at radio frequencies of about 20MHz. The noise level of the RF SQUID was typically 10^{-4} flux quanta and, in the early instruments this noise was approximately equal to the noise from the magnetometer produced by vibration of magnetic impurities within the pickup coil region and Johnson noise in the thermal insulation. After about 10 years of SRM development the instrument noise was reduced to less than the RF SQUID noise.

A major improvement in the SRM sensitivity came with the use of DC SQUIDs in an SRM in 1994. The DC SQUID uses two Josephson junctions connected in parallel (Kleiner *et al.*, 2004). It is operated by applying a DC voltage across the two junctions, which then oscillate at the Josephson frequency of 500 MHz μV^{-1}. The applied voltage is normally tens of microvolts so the SQUID frequency is about 10 GHz. This very high operating frequency results in a greatly improved SQUID noise level. A typical DC SQUID has an intrinsic noise level of a few by 10^{-6} flux quanta. This is about 100 times lower than the RF SQUIDs used in the earlier SRMs. After many changes in the internal construction of the SRM a usable sensitivity of less 10^{-5} flux quanta has been obtained. This gives a magnetic moment noise level of about 3×10^{-13}Am^2 dipole moment for a 4.2-cm access DC SQUID SRM compared to about 8×10^{-12}Am^2 for the RF SQUID SRM or an improvement of over 20 to 1. Further improvements in the DC SQUID SRM sensitivity can be expected as the noise from the superinsulation is reduced.

Pickup coils and flux transformers

The desire to measure large samples with relatively uniform sensitivity over the sample volume can be achieved by using the zero resistance property of a superconductor. A two axis superconducting magnetic flux transformer is used in the SRM as shown in Figure R22 (Goree and Fuller, 1976). The SRM normally uses three orthogonal superconducting pickup coils cocentered about the sample position. The axial pickup coils are typically constructed in a Helmholtz configuration (Helmholtz, 1849), and the transverse pickup coils are saddle-shaped coils. The large superconducting pickup coil is connected to the SQUID input coil. Efficient energy transfer from this large coil to the small SQUID is obtained if the inductance of the pickup coil is equal to the inductance of the SQUID input coil. This inductive matching transfers one-half of the magnetic field energy applied to the pickup coil to the SQUID input coil (Deaver and Goree, 1967).

Superconducting shields

Another crucial element of an SRM is the superconducting magnetic shield. The magnetic field sensitivity of an SRM is of the order of 10^{-14} T where the earth's magnetic field is approximately 5×10^{-4} T. In order to use the extreme sensitivity of the SQUID system, very effective magnetic shielding is a necessity. A superconducting shield is the best magnetic shield known. In order to provide sample access inside the shield the typical SRM uses a superconducting shield that is open at one or both ends. If an external magnetic field is applied to the open end of a superconducting cylinder the net field at any cross-section of the cylinder will be zero, but the distribution of the field change will be such that the field along the cylinder axis is

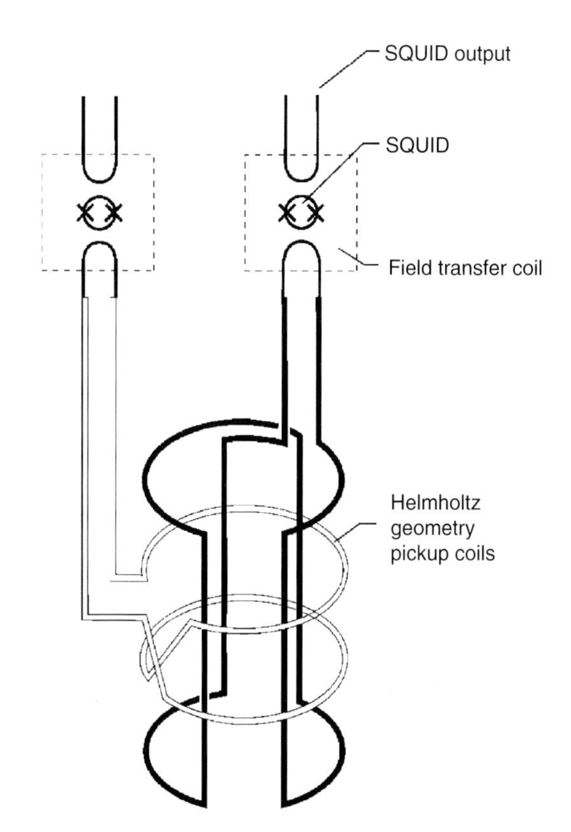

Figure R22 Two axis pickup coils.

in the same direction as the applied field and the field near the inner walls of the cylinder will oppose the applied field. In the design of a magnetic shield for the SRM the field change measured along the shield axis is normally used to determine the required shield length. The on-axis attenuation of axially applied fields is given by $31^{(x/r)}$ where x is the on-axis distance into the cylinder from an open end and r is the radius of the superconducting shield. The on-axis attenuation of fields applied transverse to the shield axis is given by $36^{(x/2r)}$ (Deaver and Goree, 1967). A typical design attenuation for transverse fields is 10^6 and this gives a minimum shield length of 7.7 times shield radii. For this shield length and transverse attenuation of 10^6 the axial attenuation will be $31^{(7.7r)/r} = 3.0 \times 10^{11}$. Thus, the axial attenuation of a cylindrical shield is orders of magnitude better than its transverse attenuation, and it is important to place the SRM where the ambient magnetic transverse field noise is a minimum.

Cryogenics

Another important performance variable in an SRM is the cryogenic environment. The SRM normally uses liquid helium as the thermal environment. A few SRMs have used liquid nitrogen as a secondary coolant, but most have used superinsulated dewar construction (Kropschot *et al.*, 1968). In a superinsulated dewar the liquid helium reservoir and SQUID volume are enclosed in a vacuum region surrounded by many layers of aluminized mylar placed within this vacuum. Thermally conducting shields are also placed within this region and are cooled by utilizing the heat capacity of the evolving helium gas. The first SRMs used about 31 of liquid helium per day. Since the initial development of the SRM cryocoolers have been incorporated into the thermal design. The cryocooler is used to provide external cooling to the thermal shields placed within the superinsulation. A typical two stage cryocooler as used with the 2G Enterprises SRM will absorb about 0.5 W of heat at 15 K and 15 W at 80 K (CTI model 350, 2004). The extra cooling provided by the cryocooler will reduce the helium loss rate to below 0.11 day^{-1} in a properly designed dewar. This greatly reduced loss rate has made it practical to use SRMs in laboratories where liquid helium is difficult to obtain and/or is very expensive.

It is also possible to use cryocoolers that will either reduce the helium loss rate to essentially zero, or produce the required low temperature without a reservoir of liquid helium. Cryocoolers that will produce 4K temperatures and/or liquid helium have been available for many years but the cost and poor long-term reliability have made them unacceptable for use with SRMs. Recent developments with pulse tube cryocoolers appear to change this situation. The author, in October 2004, began the development of a new magnetometer that would not use any liquid helium. The first SRM using this new pulse tube cryocooler technology (Cryomech, Inc., 2005) was completed and delivered to Dr. J. Hus at Dourbes, Belgium in May 2005 (Institut Royal Metrologique, Belgium). SQUID and superconducting shield temperatures of approximately 4.0K were obtained from room temperature to operation in about 3days and the magnetic measurement performance is identical to that achieved with the liquid helium cooled systems. The pulse tube cryocooler has no moving mechanical parts at low temperatures and is predicted to have a very long mean time between failure, probably in excess of 15years for the cold head, 10 years for the easily replaced drive motor, and about 10years for the compressor. The projected excellent reliability of the pulse tube cryocooler has made it practical to use with this new SRM.

High temperature superconductors

The discovery of materials that become superconducting at "high" temperatures in 1986 (Bednorz and Muller 1986) caused much excitement in all fields of superconductivity. If instruments could be built with these new materials and operate at 70–90 K the fabrication and operating cost could be dramatically reduced. Unfortunately, the high temperature materials that have been discovered to date (2004) are very brittle and it has been difficult to make high temperature superconducting wires and SQUIDs. In particular, wires have not been produced where superconducting pickup coils and transformers can be constructed (Kleiner *et al.*, 2004).

William S. Goree

Bibliography

2G Enterprises, Goree, W.S., at William S. Goree, Inc., 2040 Sunset Drive, Pacific Grove, CA, 93950 (wgoree@wsgi.us), and Goodman, W.L., at Applied Physics Systems, Inc., 1245 Space Park Way, Mountain View, CA 94043 (goodman@appliedphysics.com).

Bednorz, J.G., and Muller, K.A., 1986. Z. Phys. B. **64**: p. 189.

Collinson, D.W., 1983. *Methods in Rock and Paleomagnetism: Techniques and Instrumentation.* London: Chapman and Hall, 503 pp.

Cox, A., Doell, R.R., and Dalrymple, G.B., 1964. Reversals of the earth's magnetic field. *Science*, **144**: 1537–1543.

Cryomech, Inc., 2005. 113 Falso Drive, Syracuse, NY 13211 (http://www.cryomech.com).

CTI, 2004. Cryogenic Technology, Inc., a division of Helix Technology (http://www.helixtechnology.com).

Deaver, B., and Goree, W.S., 1967. Some techniques for sensitive magnetic measurements using superconducting circuits and magnetic shields. *Review of Scientific Instruments*, **38**: 311–318.

Fuller, M., 1987. Experimental methods in rock magnetism and paleomagnetism. *Methods of Experimental Physics*, **24**: 303–471.

Galbrun, B., At CEA, Grenoble, France.

Good, J., Cryogenic Limited (http://www.cryogenic.co.uk).

Goree, W.S., and Fuller, M.D., 1976. Magnetometers using RF-driven SQUIDs and their applications in rock magnetism and paleomagnetism. *Reviews of Geophysics and Space Physics*, **14**: 591–608.

Helmholtz, H.V., 1849. *Vortrag in der Sitzung*, Volume 16, III. Berlin: Physikalische Gesellschaft; Loeb, L.B., 1947. *Fundamentals of Electricity and Magnetism*, 3rd edition. New York: John Wiley & Sons.

Kleiner, R., Koelle, D., Ludwig, F., and Clarke, J., 2004. *Superconducting Quantum Interference Devices: State-of-the-Art and Applications.* Proceedings of the IEEE, Special Issue on Applications of Superconductivity, Volume 92, October 2004, pp. 1534–1548.

Kropschot, R.H., *et al.*,1968. Technology of liquid helium. *National Bureau of Standards Monograph*, **111**: 170.

Nagy, E.A., and Valet, J.P., 1993. New advances in paleomagnetic studies of sediment cores using U-channels. *Geophysics Research Letters*, **20**: 671–674.

ODP, 1985. Please see the ODP website: http://www.odp.edu/.

Tauxe, L., Labrecque, J.L., Dodson, R., and Fuller, M., 1983. "U" channels—a new technique for paleomagnetic analysis of hydraulic piston cores. *Eos, Transactions, American Geophysical Union*, **64**: 219.

Vrba, J., CTF Systems, Inc., 9 Burbidge Street, Coquitlam, BC, Canada Y3K7B2 (corp@vsmmedtech.com).

Weeks, R., Laj, C., Edignoux, L., Fuller, M., Roberts, A., Manganne, R., Blanchard, E., and Goree, W.S., 1993. Improvements in long-core measurement techniques: applications in paleomagnetism and paleoceanography. *Geophysical Journal International*, **114**: 651–662.

Zimmerman, J.E., and Campbell, W., 1975. Test of cryogenic Squid for geomagnetic field measurements. *Geophysics*, **40**: 269–284.

RUNCORN, S. KEITH (1922–1995)

I first met S.K. Runcorn in June of 1955. I had just graduated from Columbia College with a degree in Geology. He reached me at my grandparent's home on a Saturday morning and asked if I would like to be his field assistant working in Arizona for the summer and by the way I had to leave with him on Monday morning. I thought it was a joke but the longer we talked the more it seemed to be a real offer. So after a lot of discussion in the family I decided to take the risk on Runcorn so off we went on Monday to the wild west. Runcorn was supremely self-confident and somehow always expected things to turn out the way he wanted them to. By the end of the summer I was on my way to Cambridge England instead of Wyoming where I had expected to do my graduate studies. We first met when Keith was 33-years old having been born in 1922 in Southport Lancashire. He loved to swim and played rugby until he was past 40. He had a great interest in religion and during a trip to France, Italy, and Switzerland, which we took in late 1955, we made a tour of the great Cathedrals of France and northern Italy. In his later years he was to serve on the Vatican Council even though he was a lifelong member of the Church of England.

Without a doubt however his great love was science. His undergraduate education was obtained at Cambridge University where he was a member of Gonville and Caius College. He studied engineering, receiving his degree in 1943 and spent the rest of the war researching at the Royal Radar Establishment. In 1946 he was appointed to a lectureship in the Physics Department at Manchester University were P.M.S. Blackett was Department Chairman. Research in the Physics Department was dominated by cosmic ray research. However, Blackett was becoming interested in the generation of Earth's magnetic field and proposed a theory that rotating matter would generate a magnetic field and suggested that Runcorn get involved in this research. The competing theory was proposed by Elsasser and expanded by Bullard that the Earth's magnetic field arises from electric currents in the Earth's core. The simple test to differentiate between the two is to measure the field at the surface and at depth in the earth, if the field increases then the field is generated within the Earth. This experiment was carried out by Runcorn and others in the coalmines of Lancashire and Yorkshire. The field was shown to increase with depth showing that the field originated within the earth. This study served as Runcorn's PhD thesis.

Blackett had produced a very sensitive astatic magnetometer in his attempt to measure the magnetic field of a rotating gold ball, this failed but the magnetometer was put to work trying to find a record of secular variation in sediments and Runcorn became interested in this problem. Runcorn accepted a position at the department of Geodesy and Geophysics at Cambridge at about the same time (1950) that Blackett moved to Imperial College London, paleomagnetic research began at both institutions, which of course became a competition. When Runcorn arrived at Cambridge he found that paleomagnetic research had already begun on lavas from Iceland by Jan Hospers who found superimposed lavas of normal and reverse polarity, which were aligned along the axis of Earth's rotation however there was no laboratory. Runcorn set about to construct a new Paleomagnetic laboratory with the help of his students, K.M. Creer, E. Irving, D.W. Collinson, and J.H. Parry. The emphasis was still on secular variation but it soon became apparent that directions obtained by Irving in the pre-Cambrian and by Creer in Triassic red sediments were widely divergent from the present field and it became apparent that red sediments and igneous rocks gave internally consistent paleomagnetic results. The group used statistics developed by Fisher and the first apparent polar wander curve was produced by 1954. These results were interpreted as true polar wander an idea that was proposed by T. Gold. The early results from Runcorn's group were mostly from British rocks. However by 1954 Runcorn was traveling to the United States to collect rock samples and the first results had been obtained. By 1954 the original group began to disperse; Irving to Australia and Creer to the British Geological Survey.

Runcorn did not limit his research to paleomagnetism and he built a research group at Cambridge that studied various aspects of magnetic field behavior, e.g., R. Hide on convection in rotating fluids, P.H. Roberts on magnetohydrodynamics, F. Lowes on detecting earth currents which he believed might be leaking from the core. After the departure of Irving and Creer the paleomagnetic group at Cambridge consisted of J.C. Belshe, A.E.M. Nairn, D.W. Collinson, and J. Perry. This group was joined in the autumn of 1955 by N.D. Opdyke.

Runcorn was offered the position of Professor of Physics in the Department of Physics at the College of Newcastle on Tyne, which at the time was King's College of Durham University. He transferred the paleomagnetics laboratory to Newcastle and it was up and running in 1956. At the time of the move to Newcastle he was still maintaining that the data from North America and Europe could still be explained by Polar Wander alone. However the data was becoming better and a separation of the polar wander paths of Europe and North America persisted with the North American curve falling systematically to the West of the European curve. In the spring of 1956 he decided that the data could not be explained solely by Polar Wandering and he embraced Continental Drift. It is also true that new data was coming in from Australia that certainly could not be explained by Polar Wander alone. He became a fervent supporter of Continental Drift and lectured widely on the subject in Europe and North America where the Earth Science community was hostile to the idea. He and all paleomagnetists would see these ideas vindicated in the late 1960s by the success of the Plate Tectonic Theory. Runcorn adopted a strategy to convince the scientific community that crustal mobility was real. Ken Creer had rejoined him at Newcastle and was given the task of producing a polar wander curve for South America and Alan Nairn was sent to Africa were Anton Hales and Ian Gough were already working. Ted Irving had gone to Australia where his results were beginning to appear in print. The last of the easily accessible Gondwana continents was India where Blackett's group from Imperial College was working hard. It is interesting that Blackett was a supporter of Continental Drift in the early 1950s but did not say so in print until 1960. North American paleomagnetists were very skeptical of Continental Drift, in particular John Graham was a strong opponent of Continental Drift and he and Runcorn had some rather heated public debates, which on Graham's part became rather personal. Cox and Doell were not convinced that paleomagnetic data supported Continental Drift as late as 1960. Runcorn was always cool under verbal attack and I cannot ever recall him responding in kind. He was always interested by the idea of convection in the mantle and proposed that solid-state creep under stress from thermal and gravitational effects was responsible for the motion of continents.

He stopped doing active research in paleomagnetism in the mid-1960s and turned his attention to other scientific problems. He became fascinated by the origin and evolution of the Moon and the possibility of convection during the early history of the moon. He took advantage of the fact that the USA was going to the moon, which meant that funds would become available for lunar research. Moon rocks were returned from the moon in 1969 and in subsequent years till the end of the Apollo program. The magnetic properties of these rocks were studied at Newcastle and in other laboratories and all were found to be magnetic. Runcorn concluded that the Moon must have had a magnetic field early in its history and that this field must have been generated by magnetohydrodynamic motions in the Moon's core. Satellites orbiting the moon detected small magnetic anomalies, which Runcorn attributed to magnetizations being acquired at different times during the early history of the moon prior to 3.9-Ga years ago and resulting from the moon having a dipole field. The different pole positions he attributed to polar wander on the moon, which he suggested was caused by impacts on the Moon's surface. His last paper published in 1996 was titled "The Formation of the Lunar core."

Runcorn had a inquiring mind and during his career he published on many different subjects for example wind directions determined from eolian sandstone, the changing length of the day through time as recorded in coral growth, earth currents using abandoned cable in the Pacific, earthquakes and polar motion, excitation of the Chandler wobble, and convection in the Planets.

Runcorn was regarded highly by the scientific Community and this high regard resulted in many honors. He was elected a fellow of the Royal Society in 1965, the Indian National Academy of Science (1980), and he was an honorary member of the Royal Netherlands Academy of Arts and Sciences (1985), The Royal Norwegian Academy of Science and letters (1985), the Bavarian Academy of Science (1990), and the Royal Society of New South Wales (1993). He won a series of awards and medals beginning with the Napier Shaw prize of the Royal Meteorological Society (1959), the Charles Chree prize from the Institute of Physics (1969), the Vetlesen prize of Columbia University (1971), the John Adams Fleming Medal from the AGU (1983), the Gold Medal From the RAS (1984), and the Wegener Medal from the EGU (1987). He was appointed to Papal Academy of Sciences and helped to get Galileo reinstated by the Catholic Church.

Runcorn retired as the Professor of Physics at Newcastle in 1988 but he never stopped working. He was appointed to the Chapman Chair of Physical Sciences at the University of Alaska at Fairbanks, spending three months a year in Alaska teaching and doing research. He was found dead in his hotel room in San Diego on 5th December 1995 being the victim of a violent attack. He is mourned by many to whom he was a friend and mentor. He was a unique individual who was married to science and whose home was the Universe.

Neil Opdyke

RUNCORN'S THEOREM

This remarkable theorem is concerned with the exterior magnetic field generated by magnetized materials at the surface of a planet. Usually magnetized materials above their Curie temperature lead to patterns of *magnetic anomalies* (*q.v.*); however, under idealized circumstances it transpires that some arrangements of magnetization lead to no external magnetic field, because of extremely subtle and fortuitous cancellation of the magnetic fields generated by one area of magnetization by the magnetic fields generated by another area of magnetization. The proof of this fact is quite complex, and the reader uninterested in the details should skip to the last two paragraphs to discover the application of this theorem.

We fix attention on the exterior magnetic field $\mathbf{B}(\mathbf{r}_j)$ (at observation point $\mathbf{r}_j$) generated by a magnetization distribution $\mathbf{M}(\mathbf{r})$ enclosed in a spherical shell whose volume we denote by V and surfaces by ∂V. The field $\mathbf{B}(\mathbf{r}_j)$ can be derived from a potential ϕ_m as

$$\mathbf{B}(\mathbf{r}_j) = -\nabla\phi_m(\mathbf{r}_j),$$

where ϕ_m is given by

$$\phi_m(\mathbf{r}_j) = \frac{\mu_0}{4\pi}\int_V \mathbf{G}(\mathbf{r},\mathbf{r}_j)\cdot\mathbf{M}(\mathbf{r})\mathrm{d}^3\mathbf{r}, \qquad \text{(Eq. 1)}$$

where the vector Green's function $\mathbf{G}(\mathbf{r},\mathbf{r}_j)$ is given by

$$\mathbf{G}(\mathbf{r},\mathbf{r}_j) = \nabla\frac{1}{|\mathbf{r}_j-\mathbf{r}|}. \qquad \text{(Eq. 2)}$$

The potential and thus the magnetic field is in general nonzero for arbitrary $\mathbf{M}(\mathbf{r})$. There are, however, a few distributions of $\mathbf{M}(\mathbf{r})$ (termed *annihilators*) that generate no exterior, a simple example is the arrangement $\mathbf{M}(\mathbf{r}) = A(r)\hat{\mathbf{r}}$ for arbitrary function $A(r)$, a purely radial magnetization. This spherical arrangement is the equivalent of the infinite plane layer with constant magnetization in a Cartesian geometry, which also generates no external field. In the spherical case it generates no magnetic field because the radial part of the vector Green's function contains no spherical harmonic of degree zero (see (7) below), whereas the magnetization $\mathbf{M}(\mathbf{r}) = A(r)\hat{\mathbf{r}}$ contains only degree zero; consequently the orthogonality of *spherical harmonics* (*q.v.*) leads to a zero result.

In 1975 Keith Runcorn proved that a more complex arrangement of magnetization was also an annihilator, namely that arising when a homogeneous material is magnetized by an arbitrary internal source. The theorem applies equally well to the case of remanent or induced magnetization, but we treat the case of induced magnetization here. Assuming that a linear relation exists between magnetization and applied magnetic field via a constant α, we have an equation of the form

$$\mathbf{M}(\mathbf{r}) = \alpha\mathbf{B}(\mathbf{r})$$

(in SI the constant $\alpha = \chi/\mu_0$; see Crustal magnetic field). It is the case that the inducing field is itself a potential field satisfying Laplace's equation, so we can write $\mathbf{B} = -\nabla V_\mathrm{i}$ where V_i is the internal potential (see Main field modeling); then the integral (1) reads

$$\phi_m(\mathbf{r}_j) = -\alpha\frac{\mu_0}{4\pi}\int_V \nabla\frac{1}{|\mathbf{r}_j-\mathbf{r}|}\cdot\nabla V_\mathrm{i}\mathrm{d}^3\mathbf{r}, \qquad \text{(Eq. 3)}$$

where the "self-magnetization" due to terms of order α^2 have been dropped.

We expand V_i in terms of fully normalized spherical harmonics with coefficients β_l^m, analogous to the conventional Gauss coefficients $\{g_l^m; h_l^m\}$

$$V_\mathrm{i} = c\sum_{l=1}^{\infty}\sum_{m=-l}^{l}\left(\frac{a}{r}\right)^{l+1}\beta_l^m Y_l^m(\theta,\phi), \qquad \text{(Eq. 4)}$$

where c is the core radius; by assuming that the mantle is an insulator, V_i is harmonic everywhere outside of c. The spherical harmonics satisfy an orthogonality relation when integrated over the unit sphere Ω on which $|r| = 1$ as follows:

$$4\pi\int_\Omega Y_l^m Y_p^q \mathrm{d}\Omega = \delta_{lp}\delta_{mq}. \qquad \text{(Eq. 5)}$$

For constant susceptibility, (3) can be written as

$$\phi_m(\mathbf{r}_j) = -\alpha\frac{\mu_0}{4\pi}\int_{\partial V}\hat{\mathbf{n}}\cdot\nabla\frac{1}{|\mathbf{r}_j-\mathbf{r}|}\mathbf{V}_\mathrm{i}(\mathbf{r})\mathrm{d}^2\mathbf{r} \qquad \text{(Eq. 6)}$$

by an application of Gauss' Theorem, where $\hat{\mathbf{n}}$ is a unit normal and ∂V represents both the inner and outer surfaces of the shell. We use the expansion of the reciprocal distance in spherical harmonics for $|\mathbf{r}_j| > |\mathbf{r}|$

$$|\mathbf{r}_j-\mathbf{r}| = \frac{1}{r_j}\sum_{l=0}^{\infty}\sum_{m=-l}^{l}\frac{1}{2l+1}\left(\frac{r}{r_j}\right)^l Y_l^m(\theta,\phi)Y_l^m(\theta_j,\phi_j) \qquad \text{(Eq. 7)}$$

to obtain

$$\phi_m(\mathbf{r}_j) = -\alpha\frac{\mu_0}{4\pi}\int_{\partial V}\sum_{l,m}\frac{l}{2l+1}\left(\frac{r}{r_j}\right)^{l+1}Y_l^m(\theta,\phi)Y_l^m(\theta_j,\phi_j)$$
$$\times c\sum_{p,q}\left(\frac{c}{r}\right)^{p+1}\beta_p^q Y_p^q(\theta,\phi)\mathrm{d}\Omega \qquad \text{(Eq. 8)}$$

We invoke equation(s) and find that the orthogonality of the Y_l^m couples l and p in the sums, leading to

$$\phi_m(\mathbf{r}_j) = -\alpha\mu_0\int_{\partial V}\sum_{l,m}c\beta_l^m\frac{l}{2l+1}\left(\frac{c}{r_j}\right)^{l+1}Y_l^m(\theta_j,\phi_j)\mathrm{d}\Omega. \qquad \text{(Eq. 9)}$$

Since the integrand consists of two spherical surfaces with opposing outward normals, then each surface integral gives a contribution of the same absolute value but of opposite sign, and therefore the external potential generated is zero. This is Runcorn's (1975a,b) theorem. Note that the theorem also allows α to be a function of r simply by noting that the theorem applies equally well to infinitesimally thin shells; each shell generates no field even when α is different in each one.

The importance of this theorem was in explaining the absence of any appreciable exterior field outside Earth's moon, despite the presence of apparently significantly magnetized rocks on the Moon's surface. Runcorn (1975a,b) hypothesized that the mantle rocks had been magnetized by an internally generated dynamo field (now switched off) as they cooled through their Curie temperature; provided the Curie isotherm is everywhere spherical and the remanent susceptibility is everywhere only a function of radius, this even allows a time-dependent internal source field.

For a higher order analysis in α and the treatment of an ellipsoidal Earth, see Jackson *et al.* (1999) and Lesur and Jackson (2000). Further description of the general form of annihilators can be found in Arkani-Hamed and Dyment (1996) and Maus and Haak (2003).

A. Jackson

Bibliography

Arkani-Hamed, J., and Dyment, J., 1996. Magnetic potential and magnetization contrasts of Earth's lithosphere. *Journal of Geophysical Research*, **101**(B5): 11401–11426 (doi: 10.1029/95JB03537).

Jackson, A., Winch, D.E., and Lesur, V., 1999. Geomagnetic effects of the Earth's ellipticity. *Geophysical Journal International*, **138**: 285–289.

Lesur, V., and Jackson, A., 2000. Exact nonlinear solutions for internally induced magnetisation in a shell. *Geophysical Journal International*, **140**: 453–459.

Maus, S., and Haak, V., 2003. Magnetic field annihilators: invisible magnetisation at the magnetic equator. *Geophysical Journal International*, **155**: 509–513.

Runcorn, S.K., 1975a. An ancient lunar magnetic dipole field. *Nature*, **253**: 701–703.

Runcorn, S.K., 1975b. On the interpretation of lunar magnetism. *Physics of the Earth and Planetary Interiors*, **10**: 327–335.

Cross-references

Crustal Magnetic Field
Harmonics, Spherical
Magnetic Anomalies, Long Wavelength
Main Field Modeling

S

SABINE, EDWARD (1788–1883)

General Sir Edward Sabine (Figure S1), soldier-scientist, made many of the first magnetic surveys to include magnetic force using Gauss' newly devised method for determining absolute intensity, was instrumental in setting up magnetic observatories and Ross' magnetic survey of the Antarctic regions, and completed the definitive survey of the geomagnetic field for epoch 1818–1876.

Born in Dublin to a distinguished military family, he was educated at Marlow and Woolwich where he excelled in mathematics. He obtained his first commission in 1803 at the age of 15, rising to General in 1870. After service in Canada, which involved capture by the American privateer the *Yorktown* after a running fight of 24 h and a close engagement of an hour with a greatly superior force, he began his work on terrestrial magnetism and gravity in 1816. In 1827 he obtained from the Duke of Wellington general leave of absence from military duties as long as he could usefully be employed in scientific pursuits. He was knighted in 1869 and served as President of the Royal Society of London from 1861–1871.

Figure S1 Bust of Edward Sabine by Joseph Durham. It currently stands in the cafeteria of the Royal Society of London (reproduced with permission of The Royal Society, © The Royal Society).

His work on gravity, influenced by Henry Kater, dominated over his work on magnetism early in his career (Anon, 1892) but we shall not dwell on it here. His first expedition was as astronomer to John Ross' expedition to find a northwest passage in 1818; this was followed by Parry's expedition in the following year. Gravity occupied most of his time on these early expeditions, but he also observed magnetic inclination, for the first time in many places, thus forming the basis for later measurements of secular variation.

In 1834, while posted in Ireland, he began the first systematic magnetic survey of the British Isles, starting in Ireland before moving on to Scotland and finally completing England and Wales in 1837, working with Rev. *Humphrey Lloyd* (*q.v.*) and John Ross. They repeated the survey twenty three years later, from 1858–1861. A meeting with Baron Alexander von Humboldt in Berlin in 1836 led to von Humboldt's urging the British Government to set up magnetic observatories throughout the Empire. Sabine played a major role in recommending the establishment of observatories in both hemispheres (Toronto, St Helena, Cape of Good Hope, and Hobart Tasmania) and in recommending a magnetic survey of the Antarctic regions. The magnetic observatories were run by Lieutenants of the Royal Artillery trained by Lloyd in Dublin, and the magnetic survey was carried out by John Ross's expedition of 1840.

Sabine's campaign for global magnetic surveys is regarded by historians of science as the first scientific endeavor requiring substantial government support, both financial and in kind in the form of navy ships and army logistics. It is known as the Magnetic Crusade and is described in detail by Cawood (1979).

Sabine was almost fanatical in his collection and archiving of observations, which he believed to be the essential foundation of advancement of knowledge. His main scientific legacy is his survey of all magnetic observations made from about 1818 until 1876, when his scientific activity ceased. The compilation is thorough and accurate; it is published in fifteen contributions to the Philosophical Transactions of the Royal Society, a gigantic task. His compilation forms the major part of modern models of the geomagnetic field for the first half of the 19th century (Bloxham, 1986; Bloxham *et al.*, 1989; Jackson *et al.*, 2000); so carefully was it constructed that it could be used virtually unedited in a modern inversion. Sabine's work was certainly not restricted to data compilation: by careful statistical analysis he demonstrated a relationship between sunspots and magnetic storms (Sabine, 1851, 1852), thus heralding the start of a completely new discipline of solar-terrestrial physics.

David Gubbins

Bibliography

Anon, 1892. Obituary notices of fellows deceased. *Proceedings of the Royal Society of London*, **51**: xliii–lii.

Bloxham, J., 1986. Models of the magnetic field at the core-mantle boundary. *Journal of Geophysical Research*, **91**: 13954–13966.

Bloxham, J., Gubbins, D., and Jackson, A., 1989. Geomagnetic secular variation. *Philosophical Transactions of the Royal Society of London*, **329**, 415–502.

Cawood, J., 1979. The Magnetic Crusade: science and politics in early Victorian Britain. *Isis*, **70**: 493–519.

Jackson, A., Jonkers, A.R.T., and Walker, M.R., 2000. Four centuries of geomagnetic secular variation from historical records. *Philosophical Transactions of the Royal Society of London*, **358**: 957–990.

Sabine, E., 1851. On periodical laws discoverable in the mean effects of the larger magnetic disturbances. *Philosophical Transactions of the Royal Society of London*, **141**: 123–139.

Sabine, E., 1852. On periodical laws discoverable in the mean effects of the larger magnetic disturbances. II. *Philosophical Transactions of the Royal Society of London*, **142**: 103–129.

Cross-references

Gauss, Carl Friedrich (1777–1855)
Gauss' Determination of Absolute Intensity
Humboldt, Alexander von (1759–1859)
Humboldt, Alexander von and Magnetic Storms
Lloyd, Humphrey
Time-Dependent Models of the Geomagnetic Field
Voyages Making Geomagnetic Measurements

SEAMOUNT MAGNETISM

It has been estimated that there are more than a million seamounts in the oceans of the world (Smith and Jordan, 1988). These rise from all depths in the oceans and may reach the ocean surface or even rise above the surface. If they rise above the surface they are called islands. Although many oceanic islands have the same origin as seamounts, they will not be discussed here. Shallow water corals are frequently found on the top of seamounts. Because the seamount is denser than the surrounding seafloor they may sink with time so the coral tops are now deeply underwater.

Seamount shape and location

Most seamounts are approximately conical in shape, many with a base diameter of 100 km or more and sides with slopes of about 5°. In recent decades, many shipboard surveys and even submarine investigations have been made of seamounts. Many seamounts are isolated features; however, they frequently lie in a chain, with the direction of the chain extending away from a present or past mid-ocean spreading center. Some, however, exist in clusters, do not display a conical shape, and may appear more like a small underwater plateau. Most are believed to have been formed as a rapidly cooled volcanic eruption near the axis of mid-ocean spreading centers. While this would make a linear progression in age along a chain of seamounts, there are some chains that do not have a simple increase in age with distance along the chain. Why seamounts form in some areas and not in others is unknown. When the magma forming the seamount cools it is magnetized in the direction of the Earth's magnetic field at that time. Because it is on a moving plate, both the location of the seamount and the direction of the Earth's magnetic field change with time. Studies of seamount magnetism attempt to learn about each or both of these two variables. Such studies are aided by knowing the age of the seamount from radiometric or other absolute dating means; however, this absolute age is frequently not known.

If one takes the direction of the past geomagnetic field as known, then finding the direction of magnetization of the seamount will give its age. That can be compared to the age of the adjacent seafloor, as determined from magnetic anomalies, and provide information about tectonic plate motions.

Why to study seamount magnetization?

To a first approximation the two-dimensional magnetic anomaly pattern over a seamount is like that of a magnetic dipole; however, for many seamounts, especially those of an irregular shape, this is a poor approximation. When a reference field, such as the IGRF, is subtracted from the observed magnetic field reading, the anomaly pattern generally shows a high area and a low area. If the anomaly *high* is on the equator side of the seamount, then it is approximately normally magnetized. If the high is on the polar side it is largely reversely magnetized. Model calculations can be made to better determine the strength and direction of magnetization of different parts of the seamount.

Alternatively, the magnetization of a seamount may be studied from the magnetization of rocks recovered by dredging or deep-sea drilling cores. Dredging allows the intensity of magnetization to be determined, but not its direction, since the *in situ* orientation of the dredged rock is unknown. Drilled rocks allow the intensity and inclination (or dip) of magnetization to be determined, but the azimuth is less well known because the azimuth of the core is not accurately determined.

In principle both the magnetic anomaly pattern and the rocks could be studied for the same seamount, but that has rarely been done. Opinion is divided as to which method gives the more accurate information about the direction of magnetization of the seamount. Furthermore, in seamount magnetic studies it is usually assumed that the seamount was formed quickly, i.e., all in the same place and with little change, or secular variation of the geomagnetic field during formation. Studies have not clearly shown that the body of the seamount has different magnetizations in its different parts, except some models that indicate a nonmagnetic cap on the top.

Some seamount magnetization studies

In the 1960s, when marine magnetometers came into use, the first magnetic surveys were made over simple, conical seamounts. A review of early geomagnetic studies, especially for Pacific Ocean seamounts, has been given by Grossling (1970). One early study was made of Vema Seamount, which lies in the South Atlantic Ocean about 1000 km west of Cape Town. It was discovered in 1959 that Vema Seamount rises from a depth of 4600 m, coming to within 40 m of the sea surface. This seamount was found to be approximately reversely magnetized. Model studies showed it has a magnetization of about 0.0015 emu, which is characteristic of basalt (Heirtzler and Hadley, 1966). This study had no ages for this seamount.

Several magnetic field studies have been made over the New England seamounts in the North Atlantic. Mayhew (1986) treated the magnetic field over several of these seamounts as due to a dipole. From Bear Seamount, nearer to shore, to Nashville Seamount, further off shore, the radiometric ages vary from 103 to 83 my and are of the same age as the adjacent seafloor as determined by magnetic anomalies and the basal age of sediments. The direction of the equivalent geomagnetic dipole was about what is expected for this Cretaceous normal epoch and showed that this simple means of determining the magnetization direction may be useful.

In the eastern central Pacific Ocean, studies have been made of the northern Cocos Seamounts (McNutt and Batiza, 1981), of the magnetism of the Line Island Seamounts (Sager and Keating, 1984), and a rather different type of study of the Emperor Seamounts (Tarduno *et al.*, 2003).

McNutt and Batiza (1981) studied the magnetic survey data from six scattered seamounts, ranging in page from 1 to 7 my in age in the Cocos plate in the eastern Central Pacific. Over this time the magnetization of the seamounts showed a counterclockwise rotation of 30° agreeing with models of plate motion. Model studies showed a magnetization of up to 10 $\mathrm{A\,m^{-1}}$ is required, with better fits to observations if the upper layer of the seamount is not included in the body shape.

The Line Islands form a 4200 km chain in the central equatorial Pacific, south of the Hawaiian Islands. Several investigators have studied these features. Sager and Keating (1984) have studied eight of these seamounts. Paleomagnetic directions of four, with ages of 72–85 my, agree with other geomagnetic data. Some central seamounts are much younger (Eocene) in age. Model studies show that they have magnetizations of 3–5 $\mathrm{A\,m^{-1}}$.

After the Ocean Drilling Program drilled several seamounts of the Emperor Seamount chain in the northwest Pacific Ocean, Tarduno *et al.* (2003) studied the magnetic properties of recovered rocks with radiometric ages of 81–47 my after reasoning that secular variation of the geomagnetic field did not influence their measurements. Paleolatitude determinations indicated a progressive change and that the seamounts were formed by a rapid motion over the Hawaiian hotspot.

Model calculations

The calculations of magnetic anomalies due to magnetic bodies of irregular shape have advanced considerably since the 1960s, largely due to the use of digital computers. Special computer codes have been written for seamounts. One of the earlier codes calculated the anomaly due to a thin "pancake" with its individual magnetization and approximated the seamount structure as a stack of pancakes, and summing the anomaly components of the entire stack (Talwani, 1965). In contrast to this forward modeling technique, it is possible to start with the observed anomalies and calculate the approximate magnetization constrained by the shape of the body (Parker, 1991). This can yield information about how magnetization is distributed over the body structure although the solutions are not unique.

J.R. Heirtzler and K.A. Nazarova

Bibliography

Grossling, B.F., 1970. Seamount magnetization. In Maxwell, A.E. (ed.), *The Sea,* Volume 4, Part 1. New York: Wiley-Interscience, Chap. 4, pp. 129–156.

Heirtzler, J.R., and Hadley, M.L., 1966. Magnetic anomaly over Vema Seamount. *Nature*, **212**: 912–913.

Mayhew, M.A., 1986. Approximate paleomagnetic poles for some of the New England Seamounts. *Earth and Planetary Science Letters*, **79**: 183–194.

McNutt, M., and Batiza, R., 1981. Paleomagnetism of northern Cocos seamounts: constraints on absolute plate motion. *Geology*, **9**: 148–154.

Parker, R., 1991. A theory of ideal bodies for seamount magnetism. *Journal of Geophysical Research*, **96**: 16101–16112.

Sager, W.W., 1992. Seamount age estimates from paleomagnetism and their implications for the history of volcanism on the Pacific plate. In Keating, B.H., and Bolton, B.R. (eds.), *Geology and Offshore Mineral Resources of the Central Pacific Basin.* Circum-Pacific Council for Energy and Mineral Resources Earth Science Series. New York: Springer.

Sager, W.W., and Keating, B.H., 1984. Paleomagnetism of Line Island Seamounts: evidence for late Cretaceous and early Tertiary volcanism. *Journal of Geophysical Research*, **89**: 11135–11151.

Smith, D.K., and Jordan, T.H., 1988. Seamount statistics in the Pacific Ocean. *Journal of Geophysical Research*, **93**: 2899–2918.

Talwani, M., 1965. Computation with the help of a digital computer of magnetic anomalies caused by bodies of arbitrary shape. *Geophysics*, **30**: 797–817.

Tarduno, J.A., Duncan, R.A., Schell, D.W., Cottrell, R.D., Steinberger, B., Thordarson, T., Kerr, B.C., Neal, C.R., Frey, F.A., Torii, M., and Carvallo, C., 2003. The Emperor Seamounts: southward motion of the Hawaiian hotspot plume in Earth's mantle. *Science*, **301**: 1064–1069.

Cross-references

Magnetization
Oceanic Crust

SECULAR VARIATION MODEL

The secular variation of Earth's magnetic field has been studied for almost as long as paleomagnetic measurements have been made. Many models have been proposed to explain the latitudinal variation of the secular variation. It was found simpler to study the scatter of the virtual geomagnetic poles (VGPs) rather than the directions because VGPs usually have a uniform scatter around a mean pole whereas directions can have an elongated distribution for observations made close to the equator. A short history of secular variation models will be presented, where it will be shown that the present day field has been an important factor in determining many model parameters. It will be argued that one of the drawbacks of most models is the practice of ignoring data that give low-latitude VGPs falling within a certain band, centered on the equator, whose width varied from model to model. No rigorous explanation is produced for the specific choice of the width of the band. The model proposed here uses information from VGPs falling in all latitude bands following the Camps and Prévot (1996) model which used VGPs from all latitudes. The model has a central dipole and a number of off-centered dipoles arranged within the core, to represent the nondipole field. The strength of all the dipoles is allowed to vary statistically in a predetermined manner, and the scatter of VGPs is calculated for various observational latitudes. The model fits the data as well as the Camps and Prévot model. It has a larger number of off-centered dipoles close to the geographic poles than to the equator, this being necessary to give the higher scatter of VGPs at high observation latitudes compared to low observation latitudes.

Introduction

The study of Earth's magnetic field has been materially influenced by the measurements of Earth orbiting satellites, starting by MAGSAT, which was launched in 1979 and measured the field for about 6 months before its orbit decayed. Satellites launched more recently are Ørsted, Champ, and SAC-C. A good source of information about recent geopotential field satellites is given at http://denali.gsfc.nasa.gov/terr_mag/e_missions.html. Although these satellites allow a very accurate picture of Earth's magnetic field to be seen, they do not give the full picture of the field because of the long time scale of field changes. For this we have to turn to paleomagnetic methods.

The advantage of paleomagnetic methods is that they cover a much wider range of time than do direct observations. The disadvantage is that the coverage of measurements over Earth's surface is usually quite poor, and so it is difficult to obtain viable spherical harmonic representations of the field. We are, therefore, left with spot readings of the geomagnetic field at certain locations on Earth's surface, from which we need to infer something about the geomagnetic field. The quantity that paleomagnetists have chosen to look at is the scatter of directions or VGPs as a function of observation latitude. It turns out that this is frequently a useful parameter to work with because it does tell us something interesting about the generation of the geomagnetic field.

Edmund Gunter (the inventor of a precursor to the slide rule and also of Gunter's chain, still used in surveying today) was the first person to notice that the variation or declination of a compass needle from true north changed with time. Gellibrand, who was Gunter's colleague at Gresham's College, published the first paper attempting to describe the secular variation measured using compass needle directions (Gellibrand, 1635).

A parade of secular variation models

Model A (Irving and Ward, 1964) supposed that the field was due to an axial dipole field plus randomly directed fields at the surface. These surface fields are expected to be spherically symmetric with respect to the axial dipole field, and in the first use of this model the surface perturbing fields were uniform as a function of latitude and constant. This resulted in a total field dispersion (as measured by the angular standard deviation) that decreased with absolute latitude. Irving and Ward (1964) showed that this dispersion was very close to the dispersion obtained when the present day field was rotated around the geographic pole. These results are shown in Figure S2. The data used for demonstration of this model are not those used by Irving and Ward (1964), who gave no list of the data. They come from Creer (1962) who proposed model B and who generated the results from the rotation of the 1945 field around the spin axis of Earth that is shown in Figure S2 too. Model A proposes that the scatter in directions becomes lower by a factor of 2 on going from equator to pole, and as can be seen this is roughly what the 1945 field does and what the paleomagnetic data suggest. This is a direct result of the strength of the dipole field, which increases by a factor of 2 on going from equator to poles, and of the fact that the perturbing field was of constant size.

Model B (Creer *et al.*, 1959) simply uses a dipole wobble to explain the angular scatter. The data studied in this paper were from the Paleozoic Great Whin Sill which has roughly horizontal directions of magnetization. These directions form an elongated pattern but when used to calculate VGPs, give circularly symmetric patterns, thus suggesting that they are caused by dipole wobble with a Fisherian (1953) distribution of dipole directions. This observation has meant that many of the SV models proposed have been based on an analysis of VGP scatter rather than on direction scatter. One other notable thing is that this model gives no latitudinal variation of VGP scatter. Creer *et al.* (1959) also showed that angular dispersion of field data was lower in the northern hemisphere than in the southern hemisphere.

Creer (1962) used a number of different geomagnetic field models and rotated them around the spin axis to determine angular standard deviation of field directions. He compiled four data sets from 1882 to 1955 and suggested that the hemispheric difference was likely to be temporary. So he combined the northern and southern hemispheres and showed that the angular dispersion of the field was similar for each of the data sets. A curve drawn through the data showed a drop from about 21° at the pole to 10° at a latitude of 60°. This was compared with scatter data from igneous rocks (shown here in Figure S2) and there was considerable agreement.

When going from Fisherian poles to directions or Fisherian directions to poles, there are equations that determine the ratio of scatter (Merrill *et al.*, 1996). Two different equations have been suggested for the transformation of Fisherian VGPs to directions (Creer, 1962; Cox, 1970). The solution to this dilemma is shown in Figure S3. The Monte Carlo distributions were done using 4000 randomly generated VGPs with a precision parameter of 60. The upper of these two curves was calculated using the input precision parameter of 60 and the lower curve was calculated using the calculated VGP precision parameter for the 4000 generated VGPs. It can be seen that the Cox (1970) version of the ratio is closer to the experimental results than is the equation of Creer (1962).

Model C was proposed by Cox (1962). He allowed for both dipole wobble and variations in the nondipole field. He also advocated the use of VGPs rather than directions because dipole wobble is simple to deal with when using VGPs as there is then no latitudinal variation.

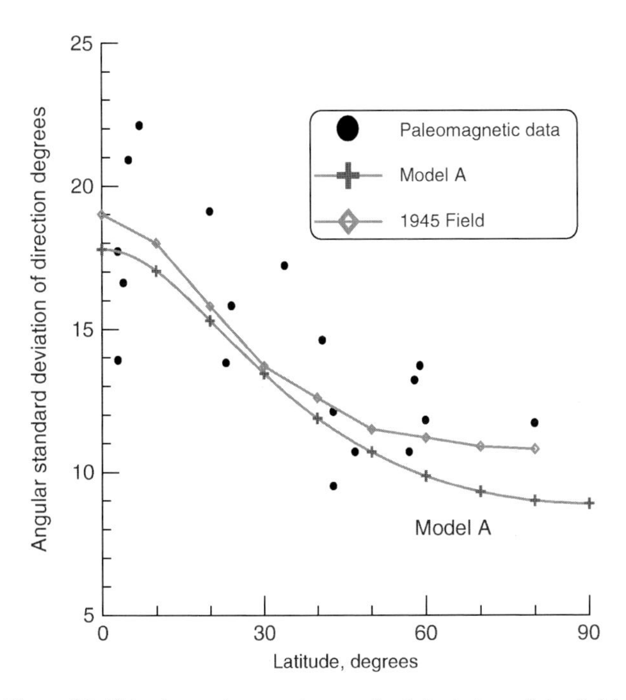

Figure S2 This shows the angular standard deviation of the field as a function of latitude. Model A is from Irving and Ward (1964). The line marked by diamonds is the scatter observed at a specific point when the 1945 field is rotated through 360°. The black dots are results obtained from igneous rocks collected at different latitudes.

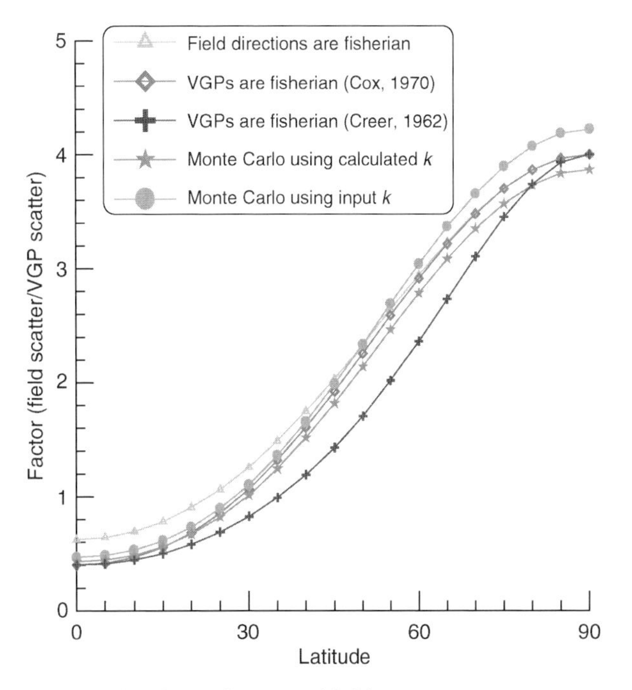

Figure S3 This shows the ratio of field scatter to VGP scatter (as calculated by the ratios of the precision parameters) as a function of latitude. The curve marked by triangles uses the equation for Fisherian directions to poles from Cox (1970), for comparison purposes. The two curves marked by diamonds and plusses are from Cox (1970) and Creer (1962). The Monte Carlo curves were calculated as described in the text.

He showed that the VGPs calculated from the field as it rotates around the geomagnetic pole, thus showing the variation of the nondipole field only, has a low angular standard deviation (ASD) at the equator of about 8° which rises as the pole is approached, so that close to the pole the ASD is about 16°. He also showed that there was a hemispheric difference, the high-latitude reading for the South Pole being 8° greater than that close to the North Pole. Cox used an equation proposed by Runcorn (1957) to express the precision parameter of VGP scatter from both dipole and nondipole sources, which is

$$\frac{1}{k_T} = \frac{1}{k_D} + \frac{1}{k_N}, \quad \text{(Eq. 1)}$$

where k is Fisher's precision parameter for polar scatter, T shows total contributions, N shows nondipole contributions and D shows dipole contributions. The corresponding equation for angular standard deviation is

$$\delta_T^2 = \delta_D^2 + \delta_N^2. \quad \text{(Eq. 2)}$$

Cox did not compare his model of the present day VGP dispersion with paleomagnetic data. The results of his calculations are shown in Figure S4. The bottom curve shows the result of adding a dipole field scatter equivalent to half of the poles lying within 11.5° of the geographic pole, giving a k_D of 34.5. The scatter, therefore, rises from 16° at the equator to 20° at the poles.

Model D was also proposed by Cox (1970). Cox allowed for changes in intensity and direction of the nondipole field, and also changes in the direction and intensity of the dipole field. The nondipole field intensity was described by Rayleigh (1919) function that was calculated to explain the directions of magnetization in small rock samples when there was a random component added to some stable component. It is not clear whether this is a reasonable distribution for the nondipole field portion of Earth's magnetic field. This distribution is given by:

$$p(f) = \left(\frac{4f^2}{\sqrt{\pi} f_R^3}\right) \exp\left(\frac{-f^2}{f_R^2}\right), \quad \text{(Eq. 3)}$$

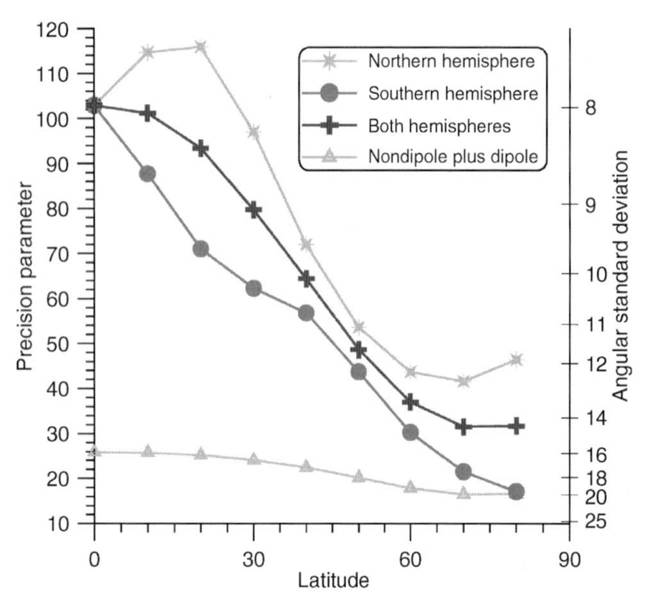

Figure S4 Cox's (1962) model C. Nondipole field scatter shown by the top three curves, and dipole plus nondipole scatter shown on the bottom curve.

where f_R completely specifies the distribution and is the mode. This is not the same as the more usual Rayleigh function that describes things like the distribution of wave heights given by:

$$p(f) = \left(\frac{2f}{f_R^2}\right) \exp\left(\frac{-f^2}{f_R^2}\right). \quad \text{(Eq. 4)}$$

This distribution function produces a value for the angular variance of the population of nondipole (f) and dipole (F) fields:

$$\sigma^2 = \left(\frac{f_R(\lambda)}{F_0}\right)^2 \left(\frac{1}{(1-r^2)^{3/2}}\right)\left(\frac{1}{1+3\sin^2\lambda}\right). \quad \text{(Eq. 5)}$$

Assuming that the VGPs are distributed according to Fisherian distribution and not the field directions, Cox determined that the variance in pole positions should follow the equation below.

$$S^2 = S_D^2 + \frac{2}{5}\left(\frac{f_R(\lambda)}{F_0}\right)\left(\frac{1}{(1-r^2)^{3/2}}\right) W_{NP}^2. \quad \text{(Eq. 6)}$$

S_D is the contribution from dipole wobble and so has no latitudinal variance. In the variable W_{NP}, N stands for the nondipole part of the field and P stands for the poles that are Fisherian in distribution. W_{NP} is given by:

$$W_{NP} = \left(5\frac{1+3\sin^2\lambda}{5+3\sin^2\lambda}\right)^{1/2}. \quad \text{(Eq. 7)}$$

This function increases by a factor of 1.6 on going from equator to pole. Since the rotated 1945 field produces poles whose scatter rises by a factor of more than 2 on going from equator to pole, it is natural to suppose that the function $f_R(\lambda)$ must produce a scatter that rises modestly by a factor of 1.3 on going from equator to pole. No attempt was made by Cox to compare his models with paleomagnetic results. Figure S5 shows some of the results from Cox's model recalculated for this paper. The red line marked by diamonds is the "observed" scatter, obtained by rotating the present day field around the geomagnetic pole axis and adding an 11° dipole wobble. The green curve marked by circles is the model result using a dipole wobble of 11° plus a nondipole field that is constant, i.e., it does not depend on the dipole field. The blue curve marked with plusses uses the same dipole wobble but assumes that the nondipole field is directly related to the dipole field (i.e., it increases by a factor of 2 on going from equator to pole).

Model E was proposed by Baag and Helsley (1974). They wished to explain why the green dotted curve in Figure S5 does not increase enough with latitude to match the red diamond curve. They suggested that the nondipole and dipole fields were correlated by some amount, as expressed by the following:

$$S_T^2 = S_D^2 + S_N^2 + 2\rho_{DN} S_D S_N, \quad \text{(Eq. 8)}$$

where ρ_{DN} is variable correlation coefficient between the dipole field and the nondipole field. This means that the following holds:

$$S_D^2 + S_N^2 \leq S_T^2 \leq (S_D + S_N)^2, \quad \text{(Eq. 9)}$$

thus allowing us to have lower and upper bounds to the angular standard deviation. Figure S6 shows all five models discussed so far using the angular standard deviation of VGP distribution as the common ordinate, so that comparisons may be made (data from Baag and Helsley, 1974). The data shown by Baag and Helsley (1974) show a general increase of scatter with latitude, with several results falling above most of the curves except E_U.

Many of the models discussed above have not considered the growing number of paleomagnetic results that might contribute to

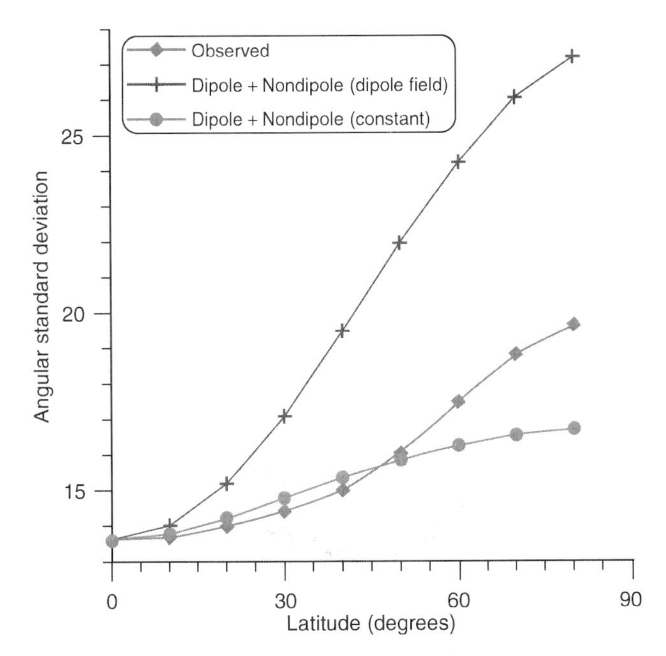

Figure S5 Model [illegible] Angular standard deviation of VGPs as a function of latitude. The red line marked by rhombi is the present day nondipole field variation to which has been added an 11° dipole wobble. The green line (dots) is model D using a nondipole field that does not vary with latitude. The blue line (plusses) is model D using a nondipole field that varies in intensity in the same way that a dipole field does. This is equivalent to model M.

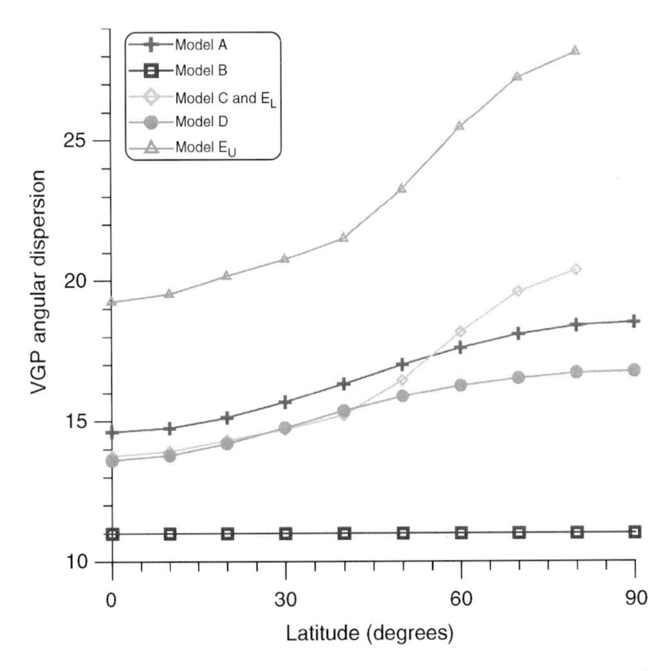

Figure S6 VGP angular dispersion as a function of latitude plotted for models A-E. Model A is in red (plusses), model B is in blue (squares), model C is in yellow (rhombus) (same as model E_L), model D is in green (circles), and model E_U is in gray (triangles).

understanding the variation of angular standard deviation as a function of latitude. The paper introducing model M (McElhinny and Merrill, 1975) made the first serious attempt to analyze paleomagnetic data.

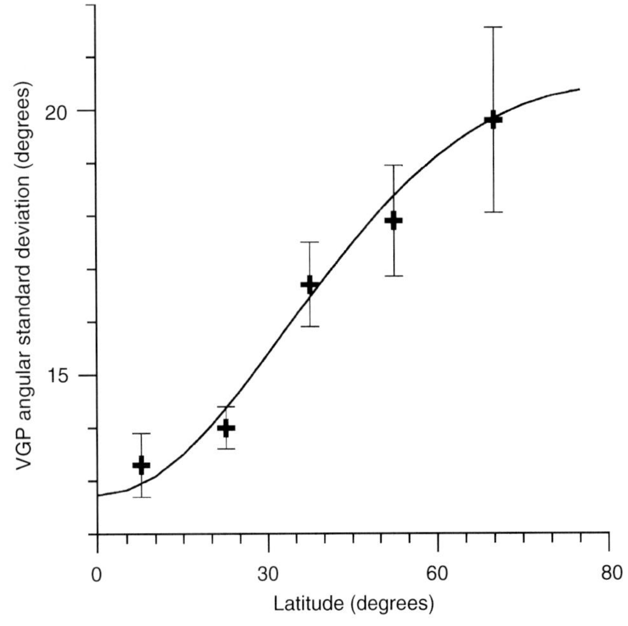

Figure S7 Model proposed by Harrison (1980) compared with data compiled by McElhinny and Merrill (1975).

They used information for the past 5 Ma and rejected data whose average VGP was further than 40° away from the poles to remove results showing transitional directions that might be recording a reversal. This is a more extreme position that other workers who have often used directions more than 45° away from the poles to signify transitional behavior. Their data showed that the VGP scatter rose from about 12° at the equator to almost 20° at high latitudes. Data from the Brunhes epoch showed similar results, with slightly less scatter at high latitudes but with significantly greater uncertainty, rendering the two results indistinguishable. A minor error (Harrison, 1980) rendered model M incorrect in that it predicted a much greater latitudinal variation of VGP scatter than actually existed in the data that they presented. Model M is basically similar to the blue curve (marked by plusses) in Figure S5, in which the latitudinal variation of the nondipole field was equivalent to the latitudinal variation of an axial dipole field. Harrison (1980) suggested a modification to model M in which the nondipole field had two components, one of which increased with latitude like a dipole field and another component that depended on the rate of change of angular momentum in a rising convection cell, a function that is responsible for creating core surface current rings that are probably responsible for the nondipole field. Figure S7 shows an excellent agreement between model and data. With the low latitude cutoff proposed by McElhinny and Merrill (1975) VGP scatter rose from about 13° at the equator to 20° at high latitudes.

Model F was proposed by McFadden and McElhinny (1984). It assumes that the VGPs are symmetrically distributed about the mean, as was done for some earlier models. This agrees with observation and also is convenient because dipole wobble in this system is latitude-independent. The VGP scatter is usually written as:

$$S^2 = S_D^2 + S_N^2 W^2(\lambda). \quad \text{(Eq. 10)}$$

They point out that using paleosecular variation data cannot solve for the unknowns in this equation because they cannot separate out dipole from nondipole effects. The authors, therefore, suggest that the model predict a value of $W(\lambda)$ that is consistent with the latitudinal variation of the present day nondipole field. They then have an extended discussion concerning the distribution of field directions that will result in a

Fisherian distribution of VGP positions. Numerical mapping was then used to determine the angular variance in field directions as a function of latitude, which is in radians:

$$\delta_{\mathrm{d}}^2(\lambda) = \left(\frac{2}{3}\right)\frac{1 + a\,\exp(-b\lambda^2)}{\sqrt{1 + 3\sin^2\lambda}}\left(\frac{f}{F}\right)^2 \qquad \text{(Eq. 11)}$$

where f and F are nondipole and dipole field intensities, respectively, and a, b are numerically determined numbers. The top part of the latitudinal term is plotted in Figure S8 (the top three curves, asymptotic to 1). The whole of the latitudinal term is plotted as the bottom three curves asymptotic to 0.5. What these curves mean is that if the ratio of f/F is constant, then the variance in direction will fall by a factor of somewhat greater than 2 on going from the equator to the poles. This latitudinal term was found numerically. McFadden and McElhinny (1984) believed that f/F was in fact constant, i.e., the nondipole field varied with latitude just like the dipole field. Numerical techniques also have to be used to transfer field variance into VGP variance. The result is that the nondipole latitudinal variation of VGP scatter increases by a factor of about 1.93 ($W(\lambda)$) on going from equator to pole. A fit to the 1965 IGRF nondipole field is shown in Figure S9. Figure S10 shows the fit of model F to the paleomagnetic data aged less than 5 Ma, and it can be seen that there are no serious deviations. But it should be stressed that this fit was achieved by the omission of low-latitude VGPs, to allow the nondipole part of model F to fit in general with the 1965 IGRF (Figure S9).

Model G was proposed by McFadden *et al.* (1988), and was described by Merrill *et al.* (1996) as the only phenomenological model of the paleosecular variation. It is based on separation of the field into a dipole family and a quadrupole family. The dipole family consists of the axial dipole plus all spherical harmonics whose degree and order add up to an odd number. The quadrupole family consists of those spherical harmonics whose sum of degree and order is even. They found out that, using the 1965 IGRF, if the field generated by the dipole family is rotated around the spin axis, the resulting scatter of VGPs, as a function of latitude, is zero at the equator and grows linearly with latitude up to 70° where it is about 15°. When the quadrupole family plus the axial dipole are rotated around the spin axis, the VGP scatter scarcely varies and is about 13°. Since the scatters from the quadrupole and dipole families are independent due to the large size of the axial dipole, model G is just the quadrature sum of the dipole and quadrupole families and is given by the following equation:

$$S = \sqrt{(A^2\lambda^2 + B^2)}, \qquad \text{(Eq. 12)}$$

where $S(\lambda)$ is the VGP scatter, A is the constant showing rise of dipole scatter with latitude, λ, and B is the constant quadrupole scatter. With $A = 0.21 \pm 0.01$ and $B = 13.6 \pm 0.3°$ (both showing 95% confidence) the fit to the 1965 IGRF is excellent as is the fit to the VGP scatter obtained from paleomagnetic samples over the past 5 Ma, provided that high latitudes are not considered (Figure S11).

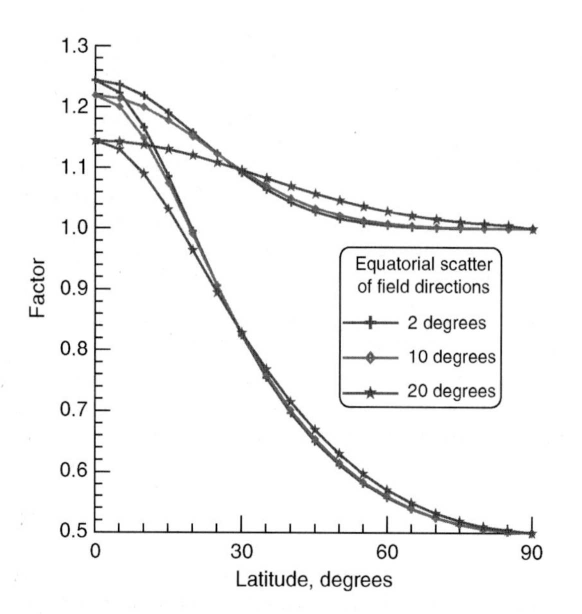

Figure S8 Top latitude term in (11) is given by upper curves asymptotic to 1, as a function of three equatorial field scatters. Bottom curves asymptotic to 0.5 show total latitudinal term from (11).

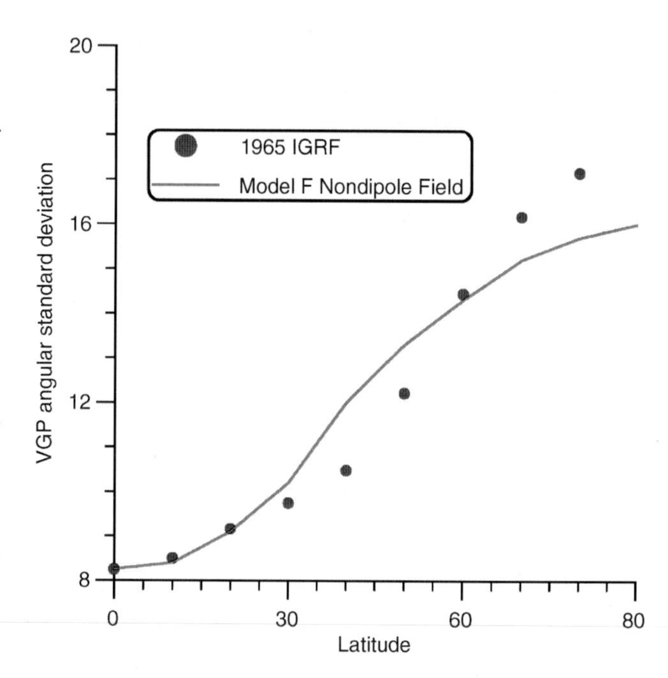

Figure S9 Fit of model F to 1965 IGRF. The nondipole parts of these are shown here.

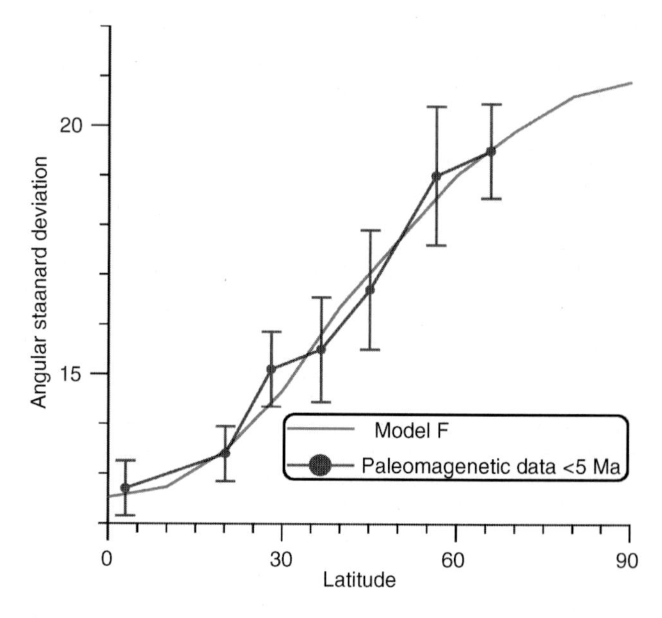

Figure S10 Model F fit to the paleomagnetic data aged <5 Ma.

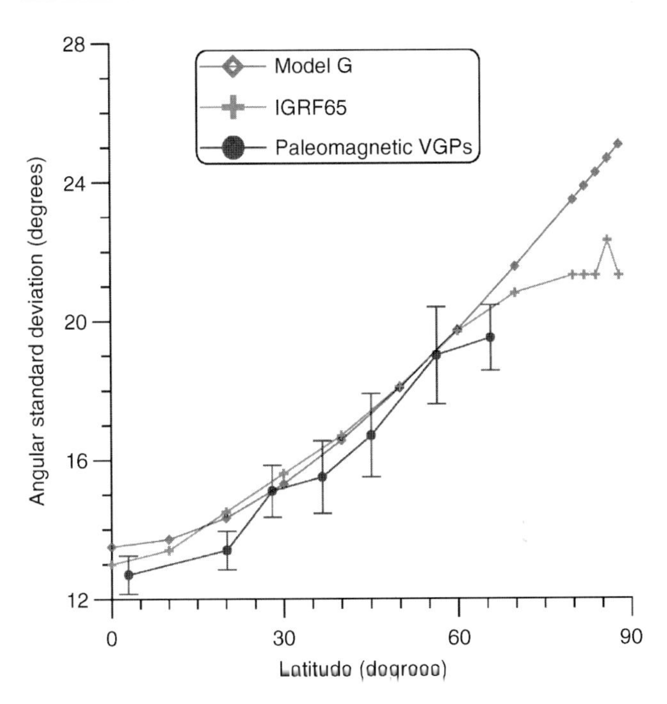

Figure S11 Model G fit to IGRF65 rotated around geographic axis and to paleomagnetic data <5 Ma old from which low latitude results have been removed.

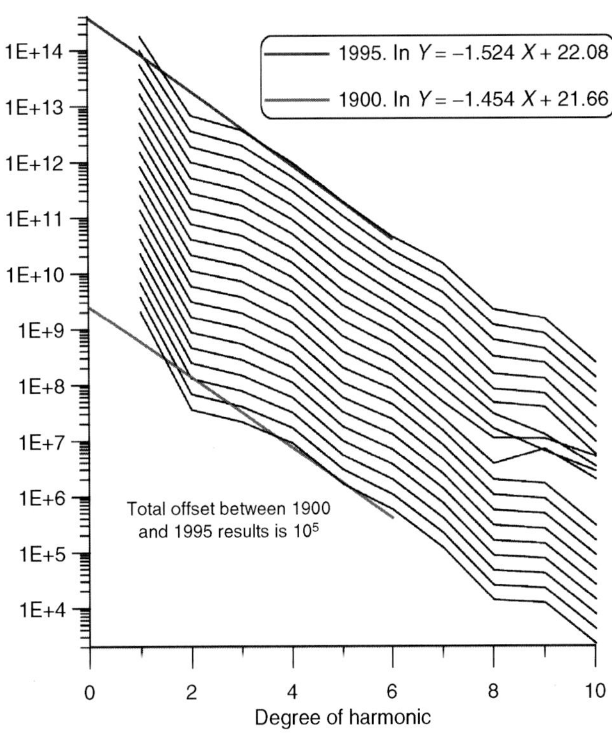

Figure S12 Lowes-Mauersberger function for the IGRF between 1900 (bottom) and 1995 (top). Each graph is displaced by a common multiple (equal to $10^{5/19}$) from its nearest neighbors for clarity. The red and blue straight lines are best fitting lines through degrees 1 to 6 for the 1900 and 1995 fields.

Unfortunately if other periods of time when the field is well known are looked at, the shape of the scatter produced by the field changes considerably (Harrison, 1995; Hulot and Gallet, 1996). Basically, the latitudinal variation has grown continuously for well over 200 years. In 1770, the scatter rose from 12° (equator) to 13.5° at high latitudes. In 1840, the change was from 12.5° to 17°, and in 1990, the change was from 13° to 22°. So it is just coincidental that the paleomagnetic field and the 1965 field gave similar results in 1965. At other epochs the similarity is not there.

There are, however, aspects of the recent measurements of the Earth's magnetic field that do not change very much over 300 years. The equivalent of the power spectrum in two-dimensional signals is replaced by a different function when measurements are taken over a spherical surface. In this case it has been shown by Mauersberger (1956) and Lowes (1966) that the mean square value of the field over the sphere as a function of degree of harmonic is given by the following function:

$$R_n = (n+1)\sum_{m=0}^{n}\left[(g_n^m)^2+(h_n^m)^2\right]. \qquad \text{(Eq. 13)}$$

The degree of harmonic is roughly equivalent to the wavenumber in a linear system. This function is plotted for the IGRF for 5-year intervals in Figure S12. There are some obvious errors in the data, i.e., there are some peculiarities in the higher degrees of harmonic in the middle years of this presentation. But in general the power falls off as a function of degree of harmonic quite uniformly from one representation of the field to the next. Analyses of the field back to 1690 (Harrison, 1994) show similar behavior, with some possible small changes due to the fact that the earlier field models lacked data to define the higher degree harmonics accurately.

Models based on spherical harmonics

The Giant Gaussian Distribution model of Constable and Parker (1988) attempted to derive a spherical harmonic spectrum for the geomagnetic field based on the supposition that the power spectrum (Lowes-Mauersberger function) for degrees 2–12 evaluated at the core-mantle boundary is statistically flat. The Gauss coefficients are chosen from a zero-mean Gaussian distribution with the appropriate standard deviation for each degree. However, no statistical variation as a function of order is allowed. Because of this, this model, despite its elegance, does not produce any variation of VGP scatter as a function of latitude, and since this is one of the main findings of paleomagnetic work on younger lava flows, it fails as a model. Later attempts to modify this model, while giving a more satisfactory latitudinal variation of VGP scatter, make the model much less elegant, one of the attractive features of the original model. For instance, Constable and Johnson (1999) produced a model in which the standard deviation of Gauss coefficients changed as a function of harmonic order, but only for degree 2. Quidelleur and Courtillot (1996) and Kono and Tanaka (1995) have also suggested modifications to the Giant Gaussian Distribution to explain the VGP scatter variation as a function of latitude. We shall see that modifications to the Gauss coefficients are necessary to obtain a reasonable model.

Another model of the geomagnetic field

One other model that deserves mention is that suggested by Camps and Prévot (1996). They used a field model (i.e., they specified the statistical properties field components at two latitudes). Their dipole field was drawn from that produced by a normal distribution of dipole strength, in which the standard deviation was 45% (high latitude) or 50% (low latitude) of the mean. The nondipole field was modeled by Maxwell Boltzmann probability density function (in which each Cartesian field coordinate is drawn from a zero-mean Gaussian distribution, all with the same standard deviation). In contrast to almost all models already discussed they did not use the present day field to constrain their model, neither did they restrict their data to lava flows that only gave high latitude VGPs (to quote from their paper "... implies that paleosecular variation is physically distinct from the reversing regime. We know of no observation that has confirmed this view."). However, their model did not provide a complete description of the field because they essentially divided up the observations into high latitude and low latitude, and did not try and join the two.

Source models of the geomagnetic field

Several people have produced source models of the geomagnetic field. Dodson (1980) used a series of dipoles buried within the core to represent the nondipole field of Earth. He used up to 20 dipoles located at the CMB for this purpose and found that he had to concentrate the off-centered dipoles close to the geographic axes to achieve the necessary variance of VGP scatter as a function of observation latitude. Kono (1971) developed a model in which the nondipole field was represented by small dipoles oriented either parallel or antiparallel to the central dipole with uniform moment. One of his models had 80 dipoles whose intensity was equal to the central axial dipole divided by 80. If the central axial dipole has a strength of 30 μT (like that of today), then the noncentral dipoles will have a strength of 375 nT.

A new source model of the geomagnetic field

Model H (after model G of McFadden *et al.*, 1988), is presented here. A complete explanation is found in Harrison (2004, 2006). The model is summarized in Table S1. The model is a variation of the one proposed by Dodson (1980) and consists of dipolar sources placed within the core to simulate the electric currents that cause the geomagnetic field. The central dipole source is assumed to be axial and is allowed to vary about its mean value that is taken to be a simple number close to the axial component today (30 μT). Since directions are determined by ratios of Gauss coefficients there is no loss of generality in choosing the central dipole source in this way. No allowance is made for far sided effects. The off-centered dipoles are buried sufficiently deeply within the core so that the correct slope to the Lowes-Mauersberger function is obtained (Figure S12). This is at a depth of $0.4R_e$. If the dipole sources are placed at the CMB (Dodson, 1980) the absolute slope is far too small. Deep burial well below the CMB achieves the correct observed slope of −0.55. Note that this aspect of the Lowes-Mauersberger function is about constant for the past 300 years, in contrast to the VGP scatter as a function of latitude that varies considerably over this time period. Benton and Alldredge (1987) suggested a depth of $0.404R_e$. The present day field variation of VGP scatter as a function of latitude is not used to constrain the model, because it has changed considerably over the past few hundred years. With this constraint gone, there seemed to be no reason to try and separate the field behavior into a "secular variation" mode and a "reversal or excursion" mode, and so VGPs from all latitudes were considered, following Camps and Prévot (1996).

How does the latitudinal variation of VGP scatter come about? This is determined by the fact that analysis of the present day field reveals that there is a variation in strength of Gauss coefficients as a function of order within each degree. This is shown in Figure S13. Figure S13, which shows the relative strength of the squared Gauss coefficients which are normalized so that mean squared values of the reduced coefficients in each degree are equal to 1. This is equivalent to the following:

$$\sum_{m=0}^{n}\left[\left(G_n^m\right)^2+\left(H_n^m\right)^2\right]=n+1, \quad \text{(Eq. 14)}$$

where

$$G_n^m = c_n \cdot g_n^m, \quad \text{(Eq. 15)}$$

$$H_n^m = c_n \cdot h_n^m, \quad \text{(Eq. 16)}$$

and c_n is a constant for degree n.

Figure S13 demonstrates that the high-order harmonics within each degree tend to be smaller. The zeroth degree harmonics ($m = 0$) tend to be the highest of all, while the sectoral harmonics ($m = n$) are mostly small, especially in 1995.

Figure S14 shows the difference between low-order harmonics and high-order harmonics in all harmonics up to degree 8. The harmonics are normalized in the same way as in Figure S13. The

Table S1 Methods of determining model parameter values

Item	Parameter	Method of determination
1	Mean central axial dipole intensity	Free parameter, made about equal to present day dipole, or 30 μT
2	Central axial dipole intensity variation (standard deviation of central dipole)	Governed by comparison with latitudinal variation of Icelandic VGPs. Different standard deviations are each tried a number of times and compared with observed distribution of VGP latitudes using random numbers to generate specific models for a given SD. Results show that 0–10% variation is permissible, but that higher variation does not give as good a fit
3	Off-centered dipole depth (constant for a specific model)	Chosen to fit the Lowes-Mauersberger function for geomagnetic field over past 300 years (Figures S12). Depth is at $0.4R_e$
4	Number of off-centered dipoles	Chosen to fit the latitudinal distribution of Icelandic VGPs. For each number of dipoles their intensity, as given by the standard deviation of a zero-mean Gaussian distribution, was varied so as to give the best fit to the Icelandic data. The number of dipoles which gave the best fit was 90, the number used in later calculations
5	Polar concentration of off-centered dipoles	Chosen by comparison with VGP scatter calculated for different latitudinal observations (Figure S10). Random numbers used to choose hemisphere (equal probability), absolute latitude (according to a Fisher distribution). The Fisher precision parameter, κ, is 3
6	Longitude of off-centered dipoles	Chosen by random number so that probability of any longitude is uniform
7	Variation of off-centered dipoles	Zero-mean Gaussian distribution with standard deviation fixed by comparison with Icelandic VGP data. See item 4. With 90 dipoles, the standard deviation is 1040 nT. To fit more restrictive data sets the standard deviation has to be reduced to 740 nT (see Figure S10)

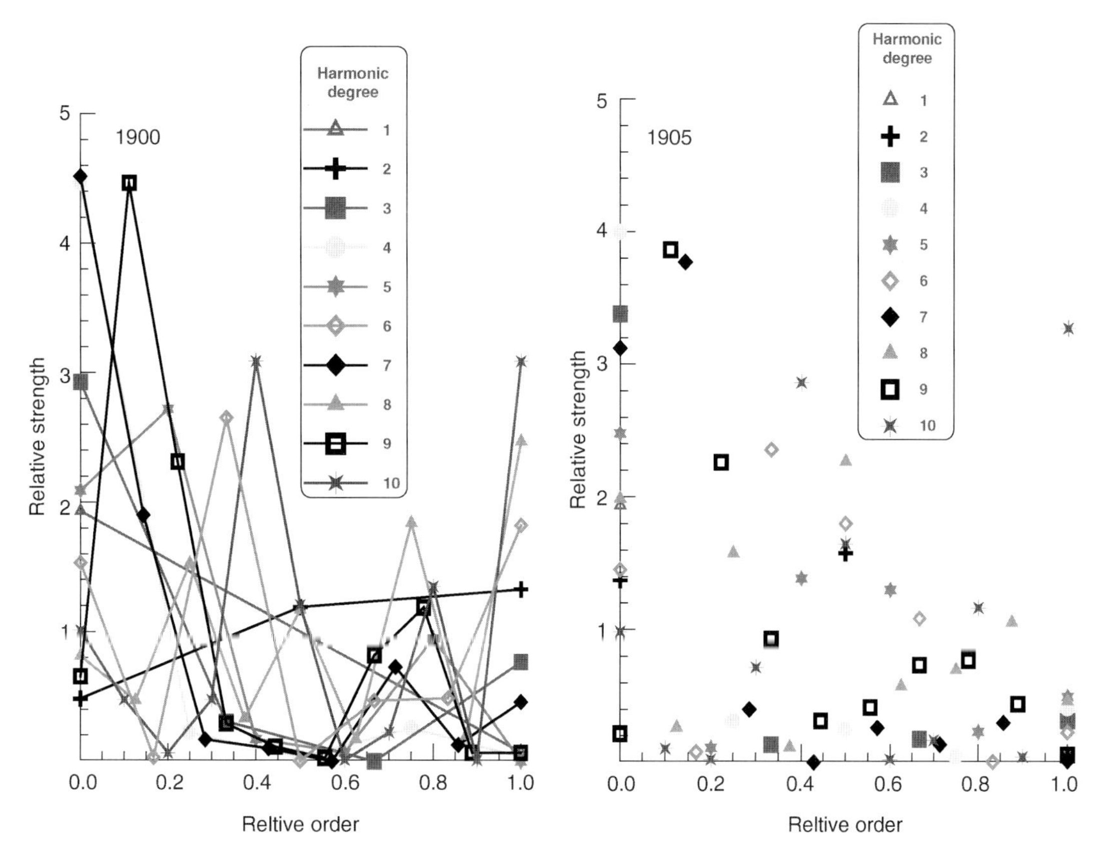

Figure S13 Squared values of reduced harmonics are plotted against relative order (m/n). Harmonics are reduced so that in each degree, the average squared harmonic is 1. Left, field for 1900 with harmonics in the same degree connected by lines; Right, field for 1995. Each degree is plotted using a different color and symbol.

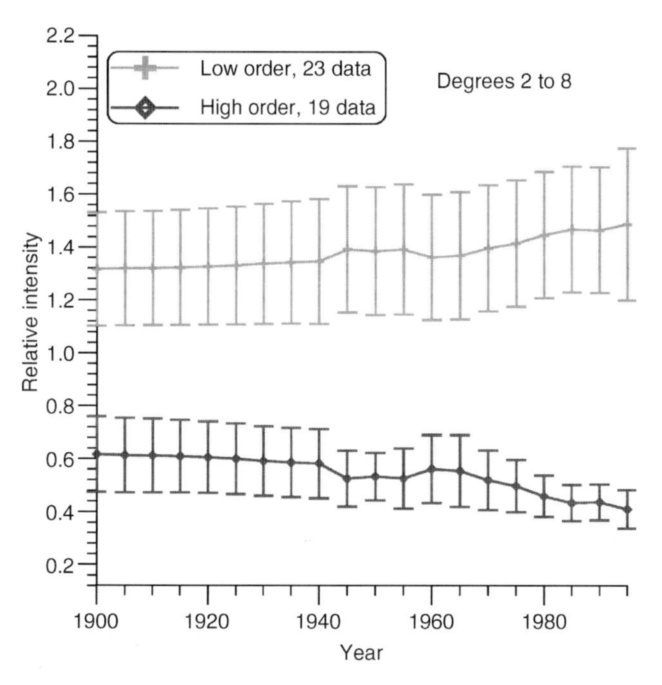

Figure S14 Low-order harmonics and high-order from degrees 2 to 8, and normalized so that in each degree the harmonics average unity, plotted against epoch. Standard errors are shown.

low-order harmonics ($m \leq n/2$) were averaged and the high-order harmonics ($m > n/2$) were separately averaged and plotted as a function of time. It can be seen that the average of the low-order harmonics is consistently more than twice as large as the average of the high-order harmonics. During the last few decades the low-order harmonics are on an average greater than three times as large as the high-order harmonics. This effect is seen in models of the field going back more than 300 years (Harrison, 1994), if the analysis is restricted to degrees lower than 6. This will include almost all of the information necessary for paleomagnetic observations of direction, since harmonics higher than five make almost no difference to directional observations. This indicates that the difference between high and low harmonics is a more permanent aspect of the geomagnetic field than the difference between northern and southern VGP scatters.

The characteristic of the low-order harmonics being in general greater than the high-order harmonics is closely connected with the necessity of having greater scatter in VGPs observed at high latitudes compared with the scatter at low latitudes. The gradual increase in the difference between low-order and high-order harmonics is mirrored by the gradual increase on the difference in VGP scatter from low-latitude observations to high-latitude observations. By having low orders stronger than high orders, low-latitude observation points have in general smaller nondipole fields than do high-latitude observation points. This is because in the expression for spherical harmonics of order m there is a term in $\sin^m \theta$, which is concentrated more and more at the equator as m is increased. This is illustrated in Figure S15 with one example, which shows all of the degree 6 associated Legendre polynomials as a function of colatitude, θ.

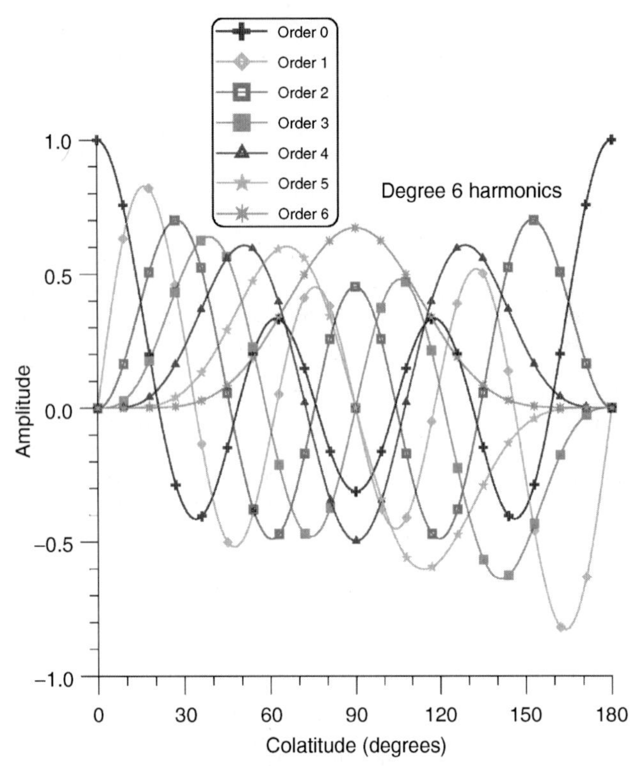

Figure S15 Associated Legendre polynomials for degree 6. As the order is increased the harmonics take on very low values for latitudes close to the poles.

In order to achieve the higher scatter of VGPs at high latitudes, it was found that the nondipole sources (radial or vertical dipoles placed at a depth of $0.4R_e$) had to be concentrated towards the poles and this was done by distributing them according to Fisher (1953) distribution. This distribution was used by Dodson (1980) in his model. The Fisher distribution is given by the following equation:

$$P_\theta \, d\theta = \frac{\kappa}{2 \sinh \kappa} e^{\kappa \cos \theta} \sin \theta \, d\theta. \qquad \text{(Eq. 17)}$$

$P_\theta d\theta$ is the probability of finding a dipole within an angular distance between θ and $\theta + d\theta$ from the mean (in this case the poles). κ is the Fisher precision parameter and is equivalent to 1/variance in a normal distribution. Large values of κ give tight groupings. A double-sided distribution is used. The best value of κ was found to be 3. This is smaller than the value used by Dodson (1980) and smaller than the value suggested by the scatter in the present day field. But it does give a factor of 10 difference between frequency of sources at the equator and sources at the poles if the double-sided distribution is used where both hemispheres of the distribution are added together. In other words, the value of $(e^3 + e^{-3})/2e^0$ is about 10.

Each separate representation of the field is determined in the following way. Ninety off-centered dipoles are placed at a radius of $0.4R_e$. Each longitude is random and determined using a random number generator. The hemisphere of each dipole is randomly determined by another random number such that each hemisphere has an equal chance of being chosen. The strength of each dipole is taken from a zero-mean Gaussian distribution using another random number. The latitude of each dipole is chosen from a distribution given above, using another random number. The central dipole is chosen from a Gaussian distribution whose mean is 30 μT and whose standard deviation is relatively small (less than 10% of average). All of these values are chosen to fit specific distributions of VGPs from lava flow data from Iceland. Because of my belief in Camps and Prévot's statement that there is no rational way of choosing the boundary between "normal" field directions and "abnormal" field directions or VGPs, I used VGPs of all latitudes. Once a model has been completely specified the field so produced is rotated around the spin axis to generate a field variation at any specified observation latitude, from which a VGP scatter is obtained, as well as other parameters such as field strength, etc. The whole process is repeated sufficiently often to enable statistical characteristics to be obtained. The model calculations are then compared with observations.

Basic characteristics are obtained by comparison with the frequency plot of VGP latitudes observed in Icelandic lavas. This enabled us to fix the central dipole standard deviation to be small compared with its mean, and the number and standard deviation of the off-centered dipoles (90 dipoles and an SD of 1040 nT). Depth of off-centered dipoles is governed by the desirability of producing the correct slope to the Lowes-Mauersberger function plotted against degree of harmonic. Polar concentration of off-centered dipoles is chosen to give the best fit to the VGP scatter as a function of observation latitude determined from paleomagnetic information.

This model agrees with other data, as shown in Table S2. This model then, predicts several aspects of the Earth's magnetic field as well as or better than any other model. But are there other things that the model predicts which will make it stand out from its competitors? It turns out that there is one further property in which the model agrees with Icelandic data. Figure S16 shows how the magnetization intensity varies as a function of VGP latitude. If magnetizations are separated into those that come from VGPs lying within the longitudinal quadrant centered on Iceland, the reduction in intensity as the VGP latitude is decreased is smaller than for those results whose VGPs lie within the other three quadrants (Harrison, 2004). An alternative more sophisticated model is to measure the fraction of variance of magnetization that is explained by the distance of the VGP from the pole. This is 0.178 for the observed data and 0.355 for 2000 model data. Then further calculations are done where the distance from a point between the geographic pole and Iceland is used instead of colatitude. As this point is moved from the geographic pole toward Iceland, the fraction of variance explained increases and then decreases. Results are shown in Figure S17. The fact that the model data show a higher degree of variance, explained by distance from pole or other location close to the pole, is because in contrast to the Icelandic data, there are no spurious effects causing magnetization to vary in the model data compared to the Icelandic data. This experiment is another prediction from the model that is mirrored in the observations from Icelandic lava flows.

Table S2 Agreement with observations

Observation	Reference
Symmetrical distribution of directions at high latitudes	Camps and Prévot (1996)
Assymetric distribution of directions at low latitudes	Tanaka (1999)
Symmetric distribution of VGPs at high latitudes	Camps and Prévot (1996)
Symmetric distribution of VGPs at low latitudes	Tanaka (1999)
Agreement with paleofield intensity in Iceland as a function of VGP latitude	Harrison (2004)
Agreement with paleofield intensity distribution from Iceland	Harrison (2004)

Implications

Model H predicts that the nondipole field should have stronger sources at high latitudes. The ratio of source strength at the poles to the source strength at the equator is given by:

$$\frac{e^{\kappa}+e^{-\kappa}}{2e^{0}}, \quad \text{(Eq. 18)}$$

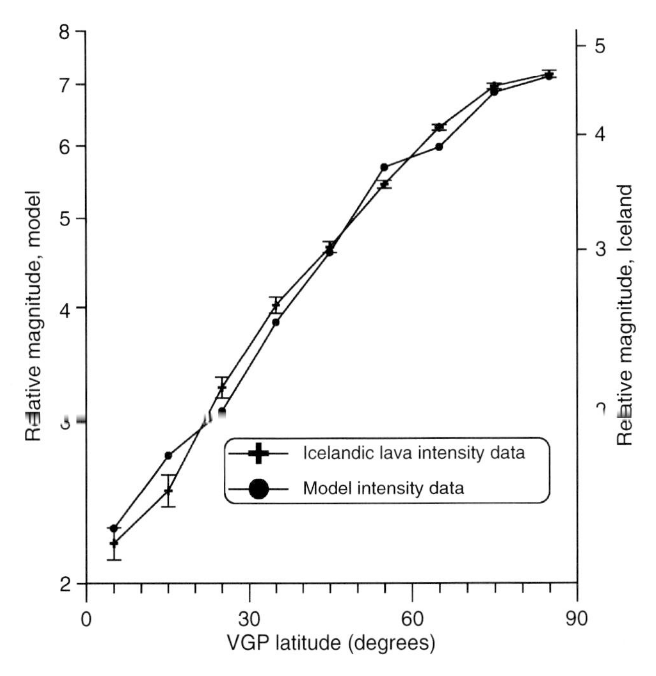

Figure S16 Relative intensity of Icelandic lava flow magnetization plotted against VGP latitude and compared with results generated by model. Both ordinates are logarithmic.

where $\kappa = 3$ is the value of Fisher's precision parameter necessary to give the correct polar concentration of nondipole sources. This is a "double" Fisher distribution, with the distribution from the back half of the sphere added to the front half, to mimic what is done to VGPs, which are represented by their antipoles if they fall in the southern hemisphere. This expression is equal to 10. Modelers who have studied the generation of Earth's magnetic field have frequently remarked on the importance of the tangent cylinder to the inner core. This tangent cylinder (whose axis is parallel to the rotation axis) impinges on the surface of the outer core at a latitude of 70°. Busse (1975) emphasized the importance of convective rolls running parallel to the spin axis and tangent to the inner core in his discussion of Earth dynamos. Gubbins and Bloxham (1987) pointed out that patches of high flux were located close to the tangent cylinder. Others have shown that there is more activity within the tangent cylinder. Kuang and Bloxham (1997) showed that there was more differential circulation within the tangent cylinder than elsewhere, and Glatzmaier and Roberts (1995) showed that there were strong toroidal fields and differential rotation within the tangent cylinder, suggesting that these might be strong sources for the nondipole field preferentially happening at high latitudes.

The radial dipoles placed within the outer core used in this model represent the sources of the nondipole field. They are deeply buried within the outer core because the slope of the Lowes-Mauersberger function requires this. These dipoles may represent core-surface sources that are spread out over a finite area. For instance, core surface current loops of finite radius will produce Lowes-Mauersberger functions that slope at about the right amount if they have a radius of about 1600 km (Harrison, 1994). A loop of this radius also gives a low power for degree 8, as is observed in the present day field (Figure S12). The off-centered radial dipoles could on the other hand represent sources deep within the core, unless conductivity considerations prevent sources at this depth from influencing the relatively rapidly changing core surface nondipole field. Further details of this model are to be found in Harrison (2006).

Christopher G.A. Harrison

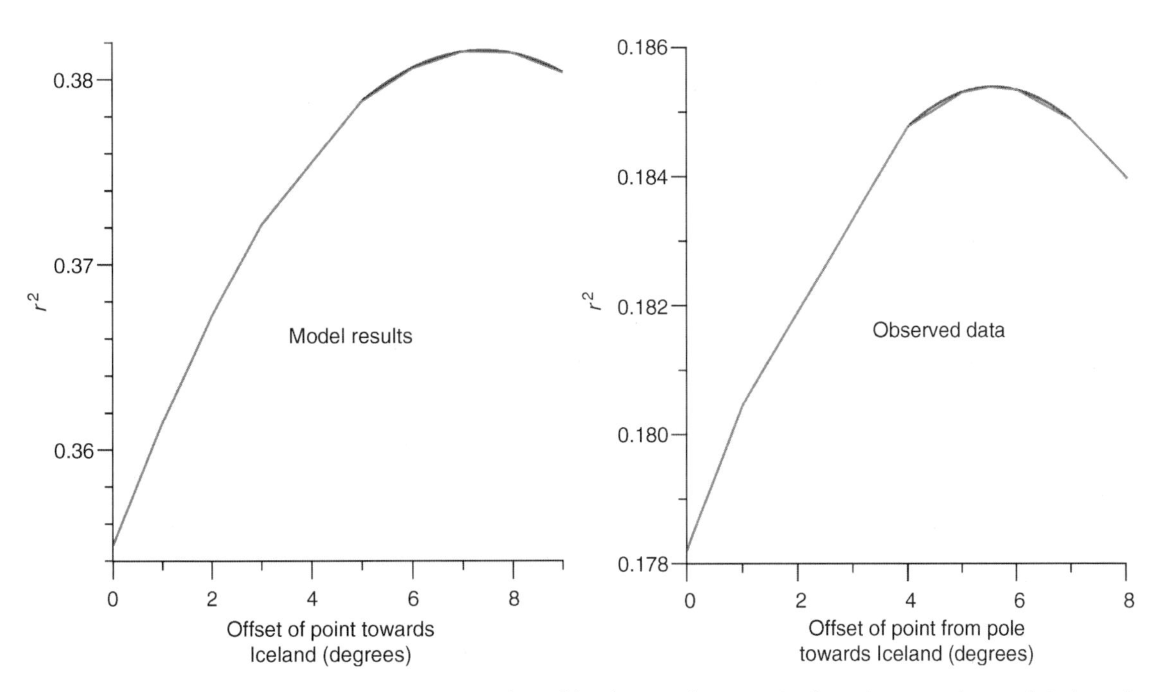

Figure S17 Fractional variance of logarithmic intensity explained by distance from a point lying between the North Pole and Iceland. The left-hand graph shows results from 2000 model calculations. The maximum is observed at a displacement of 7.4°. The right-hand graph shows magnetization results from Icelandic lavas. The maximum is observed at a displacement of 5.6°. Parabolas in blue are used to determine maximum values.

Bibliography

Baag, C., and Helsley, C.E., 1974. Geomagnetic secular variation model E. *Journal of Geophysical Research*, **79**: 4918–4922.

Benton, E.R., and Alldredge, L.R., 1987. On the interpretation of the geomagnetic energy spectrum. *Phyics of the Earth and Planetary Interiors*, **48**: 265–278.

Busse, F.H., 1975. A model of the geodynamo. *Geophysical Journal of Royal Astronomical Society*, **42**: 437–459.

Camps, O., and Prévot, M., 1996. A statistical model of the fluctuation in the geomagnetic field from paleosecular variation to reversal. *Science*, **273**: 776–779.

Constable, C.G., and Johnson, C.L., 1999. Anisotropic paleosecular variation models: implications for geomagnetic field observables. *Physics of the Earth and Planetary Interiors*, **115**: 35–51.

Constable, C.G., and Parker, R.L., 1988. Statistics of the geomagnetic secular variation for the past 5 m.y. *Journal of Geophysical Research*, **93**: 11569–11581.

Cox, A., 1962. Analysis of present geomagnetic field for comparison with paleomagnetic results. *Journal of Geomagnetism and Geoelectricity*, **13**: 101–112.

Cox, A., 1970. Latitude dependence of the angular dispersion of the geomagnetic field. *Geophysical Journal of Royal Astronomical Society*, **20**: 253–269.

Creer, K.M., 1962. The dispersion of the geomagnetic field due to secular variation and its determination for remote times from paleomagnetic data. *Journal of Geophysical Research*, **67**: 3461–3476.

Creer, K.M., Irving, E., and Nairn, A.E.M., 1959. Palaeomagnetism of the Great Whin Sill. *Geophysical Journal of Royal Astronomical Society*, **2**: 306–323.

Dodson, R.E., 1980. Late tertiary secular variation of the geomagnetic field in the North Atlantic. *Journal of Geophysical Research*, **85**: 3606–3622.

Fisher, R.A., 1953. Dispersion on a sphere. *Proceedings of the Royal Society of London, Series A*, **217**: 295–305.

Gellibrand, H., 1635. *A Discourse Mathematical on the Variation of the Magnetical Needle.* London: William Jones.

Glatzmaier, G.A., and Roberts, P.H., 1995. A three dimensional convective dynamo solution with rotating and finitely conducting inner core and mantle. *Physics of the Earth and Planetary Interiors*, **91**: 63–75.

Gubbins, D., and Bloxham, J., 1987. Morphology of the geomagnetic field and implications for the geodynamo. *Nature*, **325**: 509–511.

Harrison, C.G.A., 1980. Secular variation and excursions of the Earth's magnetic field. *Journal of Geophysical Research*, **85**: 3511–3522.

Harrison, C.G.A., 1994. An alternative picture of the geomagnetic field. *Journal of Geomagnetism and Geoelectricity*, **46**: 127–142.

Harrison, C.G.A., 1995. Secular variation of the Earth's magnetic field. *Journal of Geomagnetism and Geoelectricity*, **47**: 131–147.

Harrison, C.G.A., 2004. An equivalent source model for the geomagnetic field. In Channell, J.E.T., Kent, D.V., Lowrie, W., and Meert, J.G. (eds.), Timescales of the paleomagnetic Field. *Geophysical Monograph*, AGU, 145.

Harrison, C.G.A., 2006. Variation of spherical harmonic power as a function of order for Earth's core and crustal field and for Mars' crustal field. *Geochem. Geophys. Geosyst.*, **7#10**: Q100014, doi:10.1029/2006GC001334.

Hulot, G., and Gallet, Y., 1996. On the interpretation of virtual geomagnetic pole (VGP) scatter curves. *Physics of the Earth and Planetary Interiors*, **95**: 37–53.

Irving, E., and Ward, M.A., 1964. A statistical model of the geomagnetic field. *Pure and Applied Geophysics*, **57**: 47–52.

Kono, M., 1971. Mathematical models of the Earth's magnetic field. *Physics of the Earth and Planetary Interiors*, **5**: 140–150.

Kono, M., and Tanaka, H., 1995. Mapping the Gauss coefficients to the pole and the models of paleosecular variation. *Journal of Geomagnetism and Geoelectricity*, **47**: 115–130.

Kuang, W., and Bloxham, J., 1997. An Earth-like numerical dynamo model. *Nature*, **389**: 371–374.

Lowes, F.J., 1966. Mean square values of the spear of spherical harmonic vector fields, *J. Geophys. Res.*, **71**: 2179.

Mauersberger, D., 1956. Das Mittel der Energiedichte des Geomagnetischen Hauptfeldes an der Erdoberfläche und seine Säkulare Änderung, Gerlands Beitr. *Geophys.*, **65**: 207–215.

McElhinny, M.W., and Merrill, R.T., 1995. Geomagnetic secular variation over the past 5 m.y. *Reviews of Geophysics and Space Physics*, **13**: 687–708.

McFadden, P.L., and M.W. McElhinny, 1984. A physical model for paleosecular variation, *Geophys. J. Roy. Astron. Soc.*, **78**: 809–830.

McFadden, P.L., Merrill, R.T., and McElhinny, M.W., 1988. Dipole/quadrupole family modelling of paleosecular variation. *Journal of Geophysical Research*, **93**: 11583–11588.

Merrill, R.T., and McFadden, P.L., 1988. Secular variation and the origin of geomagnetic field reversals. *Journal of Geophysical Research*, **93**: 11589–11597.

Merrill, R.T., McElhinny, M.W., and McFadden, P.L., 1996. *The Magnetic Field of the Earth.* New York: Academic Press, 532 pp.

Quidelleur, X., and Courtillot, V., 1996. On low degree spherical harmonic models of paleosecular variation. *Physics of the Earth and Planetary Interiors*, **95**: 55–77.

Rayleigh, R.J., Strutt, (*Lord*), 1919. On the problem of random vibrations, and of random flights in one, two, ro three dimensions, *Phil. Mag.*, **37**: 321.

Runcorn, S.K., 1957. The sampling of rocks for paleomagnetic comparisons between continents, *Adv. Phys.*, **6**: 169–176.

Tanaka, H., 1999. Circular asymmetry of the paleomagnetic directions observed at low latitude volcanic sites. *Earth Planets Space*, **51**: 1279–1286.

Cross-references

IGRF, International Geomagnetic Reference Field
Inner Core Tangent Cylinder
Paleomagnetic Secular Variation
Paleomagnetism

SEDI

SEDI is an international scientific organization dedicated to the study of the Earth's deep interior. The ultimate goal of SEDI is an enhanced understanding of the past evolution and current thermal, dynamical, and chemical state of the Earth's deep interior, and of the effect that the interior has on the structures and processes observed at the surface of the Earth. The "deep interior" is generally considered to be the core and lower mantle, but interest may extend to the surface, for example, in the study of mantle plumes or dynamics of descending lithospheric slabs. The scientific questions and problems of interest to SEDI include the geomagnetic dynamo and secular variation, paleomagnetism and the evolution of Earth's magnetic field, composition, structure and dynamics of the outer core, dynamo energetics, structure of the inner core, core cooling and the core-mantle boundary region, core-mantle boundary shape, coupling and the rotation of the Earth, lower mantle: structure, convection and plumes, nature and location of deep geochemical reservoirs, etc.

Since its inception in 1987, SEDI has been a Union Committee of the International Union of Geodesy and Geophysics (IUGG). It provides a bridge across the traditional discipline-oriented bounds of the associations of the IUGG (particularly the International Associations of Geodesy, Geomagnetism and Aeronomy, Seismology and Physics of the Earth's Interior, and Volcanology and Chemistry of the Earth's Interior), which normally study the Earth from a particular point of view. The intent of SEDI is to amalgamate all sources of data and all points of view to generate the most coherent and consistent picture of the workings of the Earth's deep interior.

SEDI is guided by an Executive Committee, headed by a Chairman, a Vice Chairman, and a Secretary General. The officers and committee members are chosen at open business meetings held at the quadrennial IUGG General Assemblies. The term of these three offices is 4 years, with the current term running from September 1, 2003 to August 31, 2007. The officers for the present term are Chairman: Bruce Buffett (University of Chicago), Vice Chairman: Gauthier Hulot (Institut de Physique du Globe, France), and Secretary General: Michael Bergman (Simon's Rock College of Bard, Great Barrington, MA). In contrast to the associations of IUGG, which are formally structured and which have national membership dues, membership in SEDI is open to all individuals and there are no membership dues. The member list is essentially an e-mail list maintained by the secretary. To apply for membership, inform the current Secretary General of your desire.

SEDI serves the interests of science in three ways (1) serving as an information conduit for scientific activity related to the deep interior of the Earth, (2) organizing scientific symposia and sessions, and (3) providing an organizational framework for scientific projects. Symposia, which are held every two years, include:

1988: Structure and Dynamics of the Core and Adjacent Mantle; Blanes, Spain
1990: Reversals, Secular Variation and Dynamo Theory; Santa Fe, NM
1992: The Core-Mantle Boundary Region: Structure and Dynamics; Mizusawa, Japan
1994: The Earth's Deep Interior; Whistler Mountain, BC, Canada
1996: Untitled; Brisbane, Australia
1998: Voyage to the Center of the Earth; Tours, France
2000: SEDI 2000; Exeter, UK
2002: Geophysical and Geochemical Evolution of the Deep Earth; Tahoe City, CA
2004: SEDI 2004; Garmisch-Partenkirchen, Germany
2006: 10th Symposium of SEDI, Prague, Czech Republic

The Doornbos Memorial Prize, in memory and honor of the Dutch seismologist Durk Doornbos, is presented to several young scientists at the biennial symposia in recognition of their outstanding work on the Earth's deep interior. Past winners are:

1994: R.E. Cohen, R. Hollerbach, J. Tromp
1996: R. van der Hilst, X. Song, S. Widiyantoro
1988: D. Andrault, A. Tilgner, L. Volcado
2000: M. Bergman, E. Dormy, I. Sumita
2002: D. Alfie, R. Holme, S. Labrosse
2004: A. Deuss, C. Farentani
2006: J. Aubert, K. Koper, J. Mound

The activities of SEDI for its first 12 years of existence are documented in a set of annual newsletters (the SEDI Dialogs), edited by David Loper, found at www.sedigroup.org.

David Loper

Cross-references

Benton, Edward R. (1934–1992)
Earth Structure, Major Divisions

SEISMIC PHASES

The body wave portion of a seismogram is marked by the arrival of distinct bursts of energy which we associate with different classes of propagation path through the Earth, as illustrated in Figure S18, where the arrivals are marked with their phase code. The individual seismic arrivals sample different parts of the Earth in their passage between the source and the receiver and their properties are dictated by the structure they encounter. Thus the time of arrival of *PcP* which is reflected from the core-mantle boundary is a strong function of the radius of the core. Information from many different phases, with sensitivity to structure in different parts of the Earth, is used in the construction of models of seismic wavespeed.

The phase code describing each propagation path is built up by combining the different elements of the path though the major zones in the structure of the Earth (see *Earth structure, major divisions*):

1. A leg in the mantle is denoted by *P* or *S* depending on wave type.
2. A compressional leg in the outer core is labeled as *K*.
3. A compressional leg in the inner core is indicated with *I*, the corresponding shear wave leg would be denoted *J*.
4. Waves leaving upward from the source are indicated by lowercase letters so that *pP* represents a surface reflection for *P* near the source and *sP* a reflection with conversion above the source.
5. Reflected waves from major interfaces are indicated by lowercase letters: *m* for the Mohorovičić discontinuity (Moho), *c* at the core-mantle boundary and *i* at the boundary between the inner and outer cores.

A *P* wave which returns to the surface after propagation through the mantle and is then reflected at the surface to produce a further mantle leg will be represented by *PP*. *PKIKP* is a *P* wave which has traveled through the mantle and both the inner and outer cores, whilst *PKiKP* is reflected back from the surface of the inner core. Similarly an *S* wave reflected at the core-mantle boundary is indicated by *ScS*, and if conversion occurs in reflection we have *ScP*.

PKJKP (*q.v.*) would correspond to a wave that traversed the inner core as a shear wave, a number of claims have been made for the observation of this phase but none have been confirmed. However, a very careful analysis by Duess *et al.* (2000) indicates that the arrival *SKJKP* has been detected using recordings from many stations for a deep in the Flores sea, Indonesia.

The "depth phases" (*pP*, *sP* and *pS*, *sS*) can be very distinct for deep events, and their separation in time from the main phases *P*, *S* provides a useful measure of depth.

An understanding of the way in which seismic energy travels through the Earth to emerge at the surface can be gained by looking at the nature of the ray paths and wavefronts associated with different classes of arrivals. Figures S19-S22 show the behavior for major *P* and *S* phases for a surface source in the model AK 135 of Kennett *et al.* (1995). *P* wave legs are shown in black and *S* wave legs are indicated by using grey tone. The time progression of the waves through the Earth are shown by marking the wavefronts with ticks on the ray paths at one minute intervals; the spreading and concentration of the ray paths provides a simple visual indication of the local amplitude associated with each phase.

An extended treatment of the nature of seismic wave propagation within the Earth with illustrations of the seismograms for the major phases can be found in Kennett (2002). Shearer (1999) shows stacks of seismograms recorded around the globe that reveal the complexity of the range of propagation paths through the Earth.

The direct *P* waves refracted back from the velocity gradients in the mantle extend to about 100° away from the source but then reach the core-mantle boundary at grazing incidence. *P* waves with steeper take-off angles at the source are either reflected from the core-mantle boundary (*PcP*) or refracted into the core (*PKP*). The refracted *PKP* waves have a relatively complex pattern of propagation in the core. The absence of shear strength in the fluid outer core means that the *P* wavespeed at the top of the core is markedly less than at the base of the mantle. From Snell's law *P* waves entering the core therefore have a much steeper inclination than in the mantle. The refraction into the lower velocity medium combined with the wavespeed gradients in the outer core produce a pronounced *PKP* caustic with a concentration

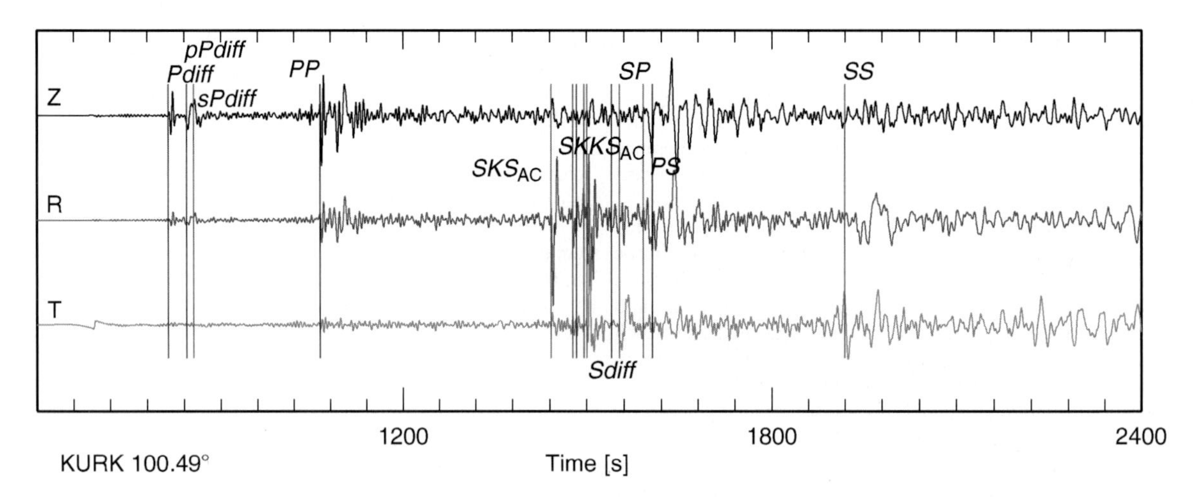

Figure S18 Three-component seismogram from an intermediate depth earthquake in Vanuatu recorded in Kazakhstan at approximately 100° from the source, showing a rich set of seismic phase arrivals.

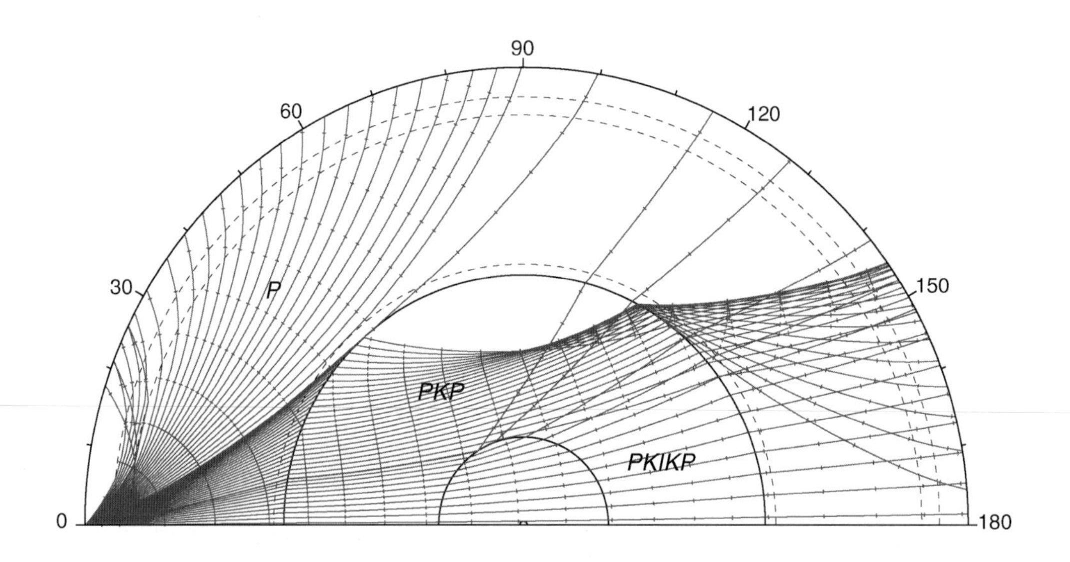

Figure S19 Ray paths for major *P* phases. Wavefronts are indicated by ticks on the ray paths at 1 min intervals.

of arrivals near 145° from the source. The apparent "shadow zone" with no *P* arrivals from 100° to 145° led to the discovery of the core by Oldham at the end of the 19th century. Reflection from the boundary between the inner and outer cores (*PKiKP*) and refraction through outermost part of the inner core (*PKIKP*) produce small amplitude arrivals which help to fill in the gap between direct *P* out to 100° and the *PKP* caustic. The identification of these arrivals lead to the discovery of the inner core by Inge Lehmann. Note that the presence of the *PKP* caustic means that there are no turning points for *P* waves in the upper part of the core.

For *S* waves (Figure S20) the pattern of propagation in the mantle is similar to that for *P*. However, since the *P* wavespeed in the outer core is slightly higher than the *S* wavespeed at the base of the core, it is possible for an *SKS* wave to travel faster in the core and overtake the direct *S* wave traveling just in the mantle. For distances beyond 82°, the *SKS* phases arrives before *S*. Unlike the *PKP* waves, the *SKS* family of waves (including underside reflections at the core-mantle boundary as in *SKKS*) sample the whole of the core.

The *P* reflections from the major internal boundaries of the Earth can often be seen as distinct phases. In Figure S21 we show the ray paths for *PcP* reflected from the core-mantle boundary and *PKiKP* reflected from the inner-core boundary. The reflections can sometimes be seen relatively close to the source but their time trajectories cut across a number of other phases and so they can often be obscured by other energy. The *P* wave refracted in the mantle and the *PcP* wave reflected from the core-mantle boundary have very similar ray paths for near grazing incidence, and so their travel time curves converge close to 90° epicentral distance.

The strong contrast in physical properties at the core-mantle boundary has the effect of inducing conversion between *S* and *P* waves. An incident *S* wave at the core-mantle boundary can give rise to the converted reflection *ScP* which can be quite prominent out to 60° from the source. In addition conversion on transmission into the core generates *SKP* as illustrated in Figure S22.

The Earth's surface and the underside of the core-mantle boundary can give rise to multiply reflected phases (Figure S23) with very clear internal caustics. The surface multiples can be tracked for *P* and *S* to great distances and can often be recognized as distinct phases for the third or higher multiples (*PPP*, *SSS*, etc.). The internal core reflections also carry energy to substantial distance, and retain the character of the original wave system. Thus *PKKP* only has strong sensitivity to *P* wave structure in the upper part of the core near the bounce point at

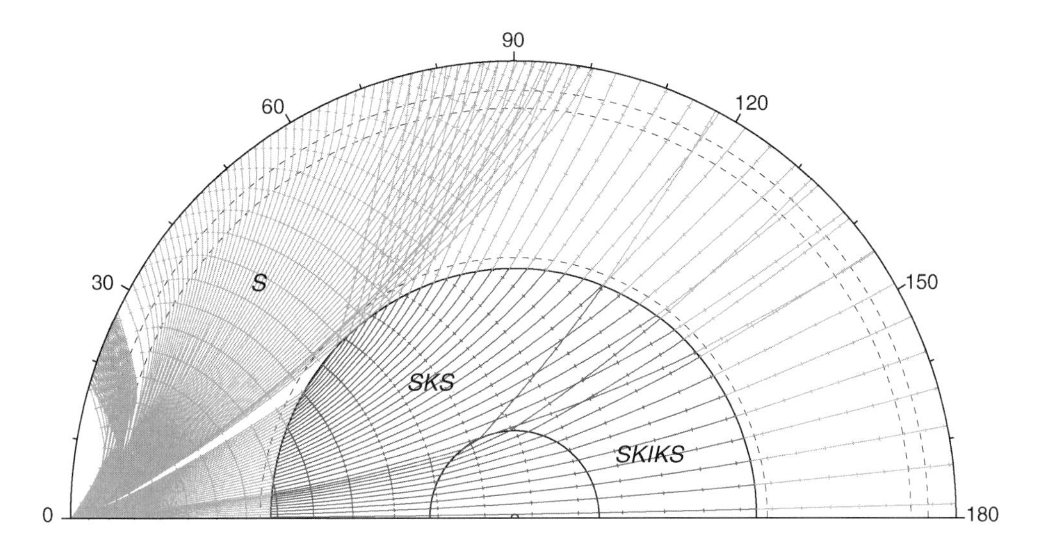

Figure S20 Ray paths for major *S* phases for the AK 135 model of seismic wavespeeds.

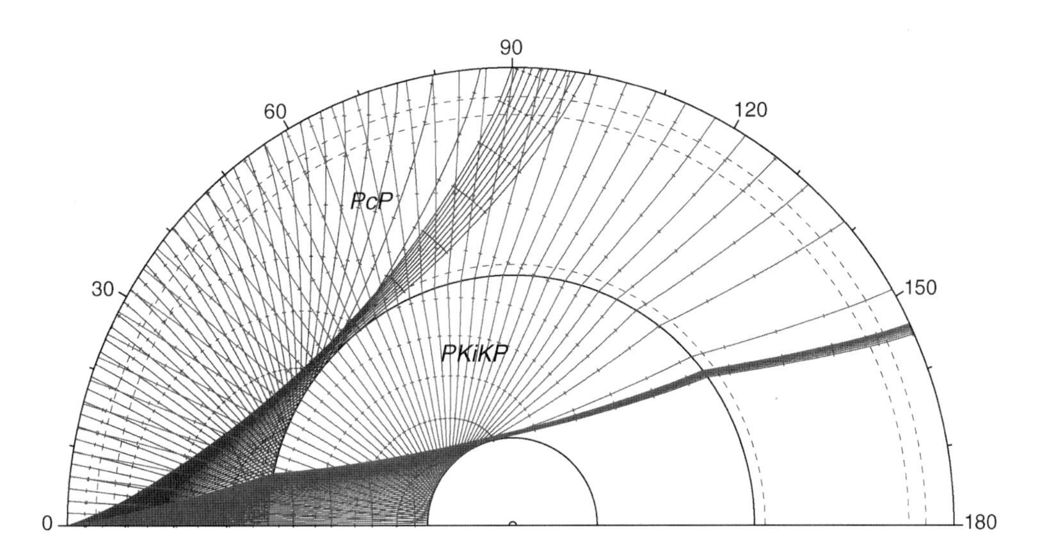

Figure S21 Ray paths for the reflected phases *PcP* and *PKiKP*.

the core-mantle boundary, whereas *SKKS* samples the *P* wavespeed distribution through the whole core. The higher order multiples *SKKKS*, etc. provide the closest probing of the structure in the outermost parts of the core.

Multiply reflected waves from the core-mantle boundary are of particular significance for horizontally polarized shear waves (*SH*), since the solid-fluid interface is close to a perfect reflector. With efficient reflection at the Earth's surface, a long train of multiple $(ScS)_H$ can be established that carry with them information on the internal structure of the Earth, in terms of the influence of internal discontinuities.

Branches of core phases

The propagation of *P* waves into the Earth's core in the *PKP* wave group includes refraction back from the velocity gradients in the outer core and reflection (*PKiKP*) and refraction (*PKIKP*) from the inner core. The components of the wave group are also commonly designated by a notation based on the different branches of the travel-time curve (Figure S24).

The patterns of the travel times and the associated ray paths can be followed in Figure S24, where the critical slowness points corresponding to the transition between branches are clearly labeled. The details of the positions of these critical slownesses vary slightly between different models for the Earth's core, but the general pattern is maintained.

The PKP_{AB} branch corresponds to the waves entering the core at the shallowest angles and PKP_{BC} to waves refracted back from the lower part of the outer core. The CD branch corresponds to wide-angle reflection from the inner boundary (the extension of the *PKiKP* phase). The PKP_{DF} phases, which is equivalent to *PKIKP* is refracted through the inner core. Rays penetrating into the upper part are strongly bent and emerge near 110° at the D point, steeper entry leads to more direct propagation through the inner core with the F point at 180° corresponding to transmission without deflection.

The concentration of rays near the *PKP* caustic at B is reflected in localized large amplitude arrivals for *PKP* near 145°. The observations

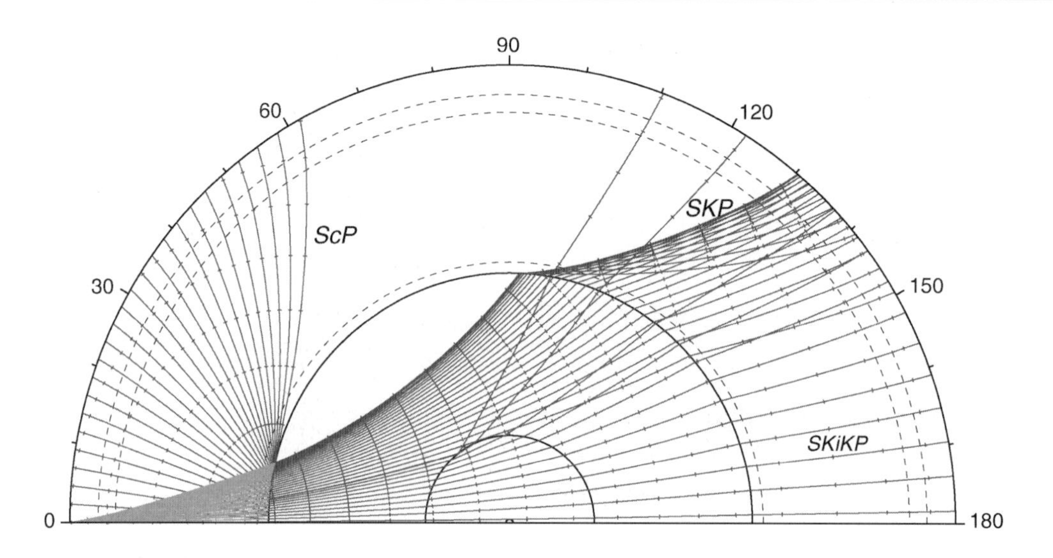

Figure S22 Ray paths for the converted phases *ScP* and *SKiKP*.

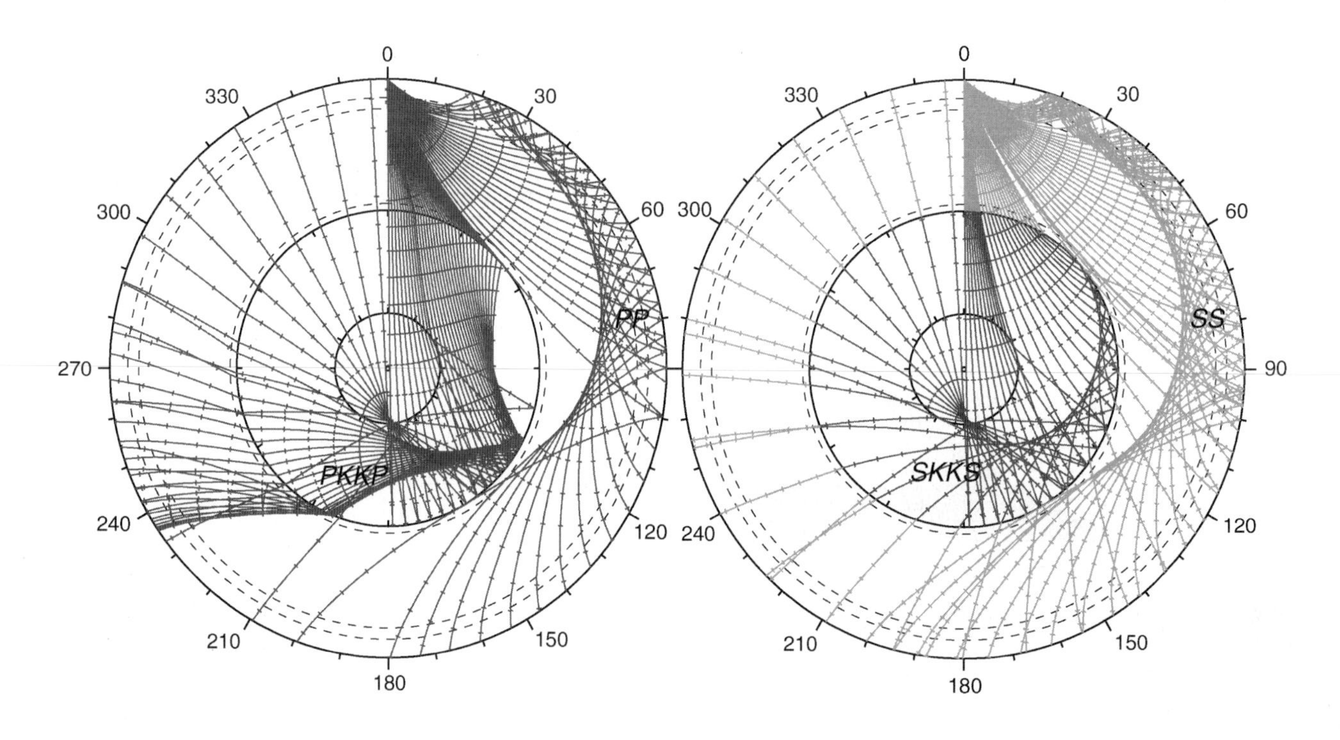

Figure S23 Ray paths for multiply reflected phases (a) *PP, PKKP,* (b) *SS, SKKS.*

of the BC branch tend to extend beyond the ray theoretical predictions. For epicentral distances beyond the C point, there is the possibility of diffraction around the inner core. At the B caustic the real branch does not just stop and there will be frequency dependent decay into the shadow side of the caustic. In addition scattering in the mantle from *PKP* produces short-period arrivals as precursors to *PKIKP* that can be seen because they arrive in a quiet portion of the seismic record. The envelope of possible precursors is indicated in Figure S24.

The pattern of branches for the *SKS* phase is somewhat different because the *P* wavespeed at the top of the core is higher than the *S* at the base of the mantle.

The AC branch extends from *S* incident on the core-mantle boundary that can just propagate as *P* at the top of the outer core and emerge at A – 63°, through to grazing incidence at the inner core boundary at C. Diffracted waves around the inner core can extend the branch beyond the formal C point. The DF branch (*SKIKS*) again corresponds to refracted waves in the inner core. The postcritical reflections from the inner core boundary (*SKiKS*) form the CD branch and connect directly into the precritical reflection at shorter distances than the D point at 104° (Figure S25).

When the refraction just begins at A, the *S* wave path to the same epicentral distance is shorter and *SKS* is about 75 s behind *S*. However, as the proportion of faster *P* wave path in the core increases, the discrepancies in *S* and *SKS* travel time are reduced. Eventually, the travel time of *SKS* becomes less than that for *S* at the same epicentral distance. Beyond 83° *SKS* becomes the onset of the shear wave group

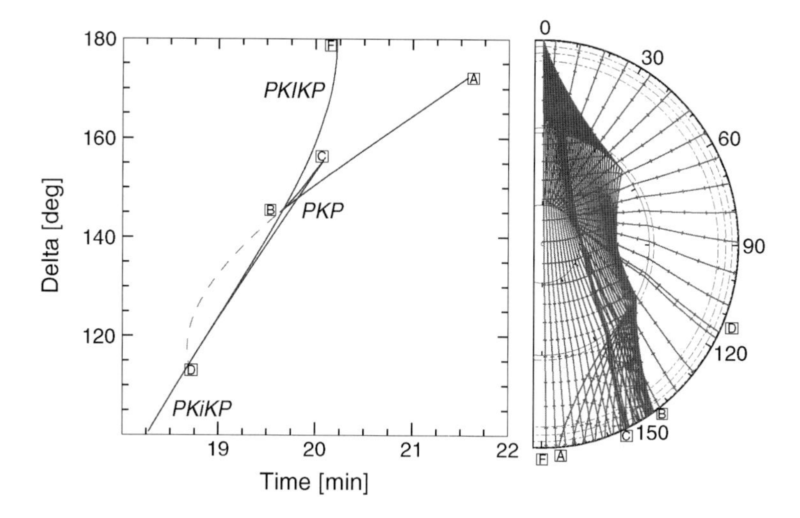

Figure S24 Rays and travel times for *PKP* for the AK135 model, wavefronts are indicated by tick marks at 60 s intervals. The critical points for the various *PKP* branches are indicated on both the travel time curve and the ray pattern. The dashed segment indicates the locus of precursors to *PKIKP* from scattering at the core-mantle boundary.

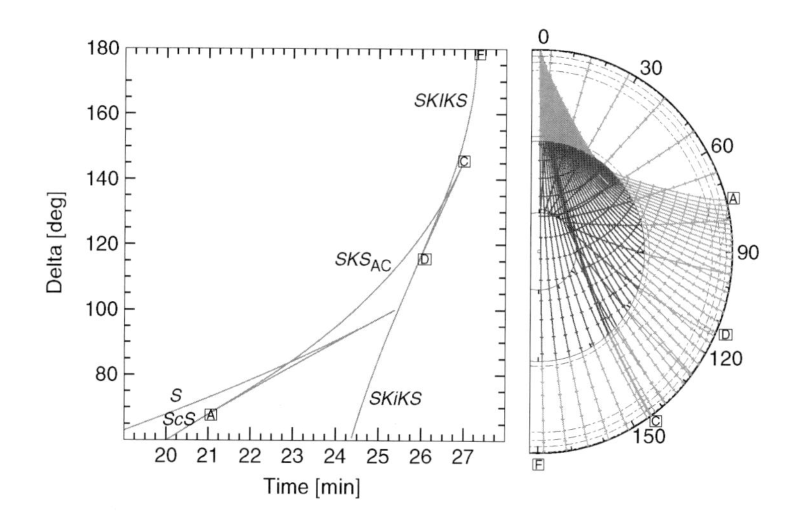

Figure S25 Rays and travel times for *SKS* for the AK 135 model. *S* legs are plotted in grey and wavefronts are indicated by tick marks at 60 s intervals. The critical points for the various *SKS* branches are indicated on both the travel time curves and the ray pattern.

and vertically polarized *S*, *Sdiff* have to be sought in the *SKS* coda. The transversely polarized *S* wave is very distinct and small precursory *SKS* contributions (as in Figure S18) can arise from either anisotropy of heterogeneity in passage through the mantle.

Brian Kennett

Bibliography

Duess, A., Woodhouse, J.H., Paulssen, H., and Trampert, J., 2000. The observation of inner core shear phases. *Geophysical Journal International*, **142**: 67–73.

Kennett, B.L.N., 2002. *The Seismic Wavefield II: Interpretation of seismograms on regional and global scales.* Cambridge: Cambridge University Press.

Kennett, B.L.N., Engdahl, E.R., and Buland, R., 1995. Constraints on seismic velocities in the Earth from travel times. *Geophysical Journal International*, **122**: 108–124.

Shearer, P.M., 1999. *An Introduction to Seismology.* Cambridge: Cambridge University Press.

Stein, S., and Wysession, M., 2003. *An Introduction to Seismology, Earthquakes, and Earth Structure.* Oxford: Blackwell Publishing.

Cross-references

Earth Structure, Major Divisions
Inner Core, PKJKP
Inner Core Seismic Velocities
Lehmann, Inge (1888–1993)
Oldham, Richard Dixon (1858–1936)

SEISMO-ELECTROMAGNETIC EFFECTS

Seismo-electromagnetic effects refer to electromagnetic (EM) signals generated by fault failure processes in the Earth's crust. These may occur slowly (when associated with plate tectonic loading, slow earthquakes, postseismic slip, etc.) or rapidly preceding, during and following earthquakes. Several different physical processes related to crustal failure can contribute to the generation of seismo-electromagnetic (SEM) effects. Unambiguous observations of SEM effects provide new independent information about the physics of fault failure. Causal relations between co-seismic magnetic field changes and earthquake stress drops have been clearly documented. However, despite several decades of high quality monitoring, clear demonstration of the existence of precursory EM signals has not been achieved.

Brief history

Suggestions that electromagnetic field disturbances are a consequence of crustal failure processes have been made throughout recorded history. Unfortunately, much of the earliest work was recognized as spurious by Reid (1914) who showed that transients recorded by magnetographs located close to earthquake epicenters resulted from earthquake shaking, not earthquake source processes. This invalidated earlier reports from magnetic variometers (suspended magnets) and other instruments sensitive to ground displacement, acceleration, and rotation common in epicentral regions during the propagation of seismic waves. Other early problems resulted from inadequate rejection of ionospheric, magnetospheric, and man-made noise (Rikitake *et al.*, 1966).

Since the mid-1960s, these problems have been avoided through the use of absolute magnetometers installed in regions of low magnetic field gradient to reduce sensitivity to earthquake shaking and by the application of new noise reduction techniques. As a consequence, unambiguous observations of EM variations related to earthquakes and tectonic stress/strain loading, have now been obtained near active faults in many countries (Japan, China, Russia, USA, and other locations). However, careful work still needs to be done to convincingly demonstrate causality between "precursory" EM signals and earthquakes and consistency with other geophysical data reflecting the state of stress, strain, material properties, fluid content, and approach to failure of the Earth's crust in seismically active regions.

Physical mechanisms involved

The loading and rupture of water-saturated crustal rocks during earthquakes, together with fluid/gas movement, stress redistribution, change in material properties, has long been expected to generate associated magnetic and electric field perturbations. The primary mechanisms for generation of electric and magnetic fields with crustal deformation and earthquake related fault failure include piezomagnetism, stress/conductivity, electrokinetic effects, charge generation processes, charge dispersion, magnetohydrodynamic effects, and thermal remagnetization and demagnetization effects. Physical limitations, constraints, and frequency limitations placed on these processes are discussed in Johnston (2002).

Basic measurement limitations

The precision of local magnetic and electric field measurements on active faults varies as a function of frequency, spatial scale, instrument type, and site location. Most measurement systems on the Earth's surface are limited more by noise generated by ionosphere, magnetosphere, and cultural noise than by instrumental noise. Thus, systems for quantifying these noise sources are of crucial importance if changes in electromagnetic fields are to be uniquely identified. For spatial scales of a few kilometers to a few tens of kilometers comparable to moderate magnitude earthquake sources, geomagnetic, and electric noise power decreases with frequency as $1/f^2$, similar to the "red" spectrum behavior of most geophysical parameters. Against this background noise, transient magnetic fields can be measured to several nanotesla over months, to 1 nT over days, to 0.1 nT over minutes, and 0.01 nT over seconds. Long term changes and field offsets can be determined if their amplitudes exceed about a nanotesla. Comparable electric field noise limits are 10 mV km^{-1} over months, several mV km^{-1} over days, 1 mV km^{-1} over minutes and 0.1 mV km^{-1} over seconds. EM noise increases approximately linearly with site separation. Cultural noise further complicates measurement capability because of its inherent unpredictability. This largely precludes measurements in urban areas. At lower frequencies (microhertz to hertz) for both electric and magnetic field measurements, the most common technique involves the use of reference sites with synchronized data sampling in arrays using site spacing comparable to the expected source sizes of a few tens of kilometers. Adaptive filtering, use of multiple variable-length sensors in the same and nearby locations further reduce noise by about a factor of 3.

These same techniques can be applied to electromagnetic field measurements at higher frequencies (100 Hz to MHz) but much less is known about the scale and temporal variation of noise. These frequencies may be less important since basic physics precludes simple propagation of high-frequency EM signals from seismogenic depths (5–100 km) on active faults in the Earth's crust where the electrical conductivity is more than 0.1 S m^{-1}.

Recent results: general constraints

If reliable magnetic and electric field observations are indeed source related, clear signals should occur at the time of large local earthquakes because the primary energy release occurs at this time. These signals should scale with the earthquake moment (size) and source geometry. In fact, co-event observations provide a determination of stress sensitivity since the stress redistribution and the source geometry of earthquakes are well determined. With this "calibration," SEM effects can be quantified and spurious signals identified. Observations without consistent and physically sensible co-seismic effects are generally considered suspect.

High-resolution strain data at the epicenters of moderate to large earthquakes show that precursive moment release during the months to minutes before rupture is less than 0.1% of that occurring coseismically (Johnston and Linde, 2002). This strongly limits the scale of precursive failure and the expected "size" of precursive effects.

Examples of seismomagnetic effects

The primary features of seismomagnetic effects are shown in Figure S26 from Mueller and Johnston (1998). It is apparent that maximum signals are not more than a nanotesla or so and these signals occur only for larger earthquakes ($M > 6$) for which corresponding strain changes are about a microstrain or so. An example of a magnetic record observed at the epicenter of the 1986 M5.9 North Palm Springs earthquake and 17 km from the 1992 M7.4 Landers earthquake is shown in Figure S27. For this, and some 40 other earthquakes with magnitude between 5.5 and 7.4, no significant precursory magnetic signals were observed.

Seismoelectric effects

Seismoelectric observations that show expected scaling with both earthquake moment release and inverse distance cubed are difficult to make because of the sensitivity of electrode contact potential to earthquake shaking. Measurements of electrical resistivity to better than 1% have been made since 1988 in a well designed experiment installed near Parkfield, California (Park, 1997). An expected M6 earthquake together with several M5 earthquakes have occurred beneath this array since 1990. None of these earthquakes generated any observable changes in resistivity above the measurement resolution (Park, 1997; Langbein *et al.*, 2005).

Indirect observations of possible SE signals might be obtained using the magnetotelluric (MT) technique to monitor apparent resistivity in seismically active regions. Even with the best designed systems using remote referencing systems to reduce noise and obtain stable impedance tensors, it is difficult to reduce errors below 5% for good soundings and 10–40% for poor soundings.

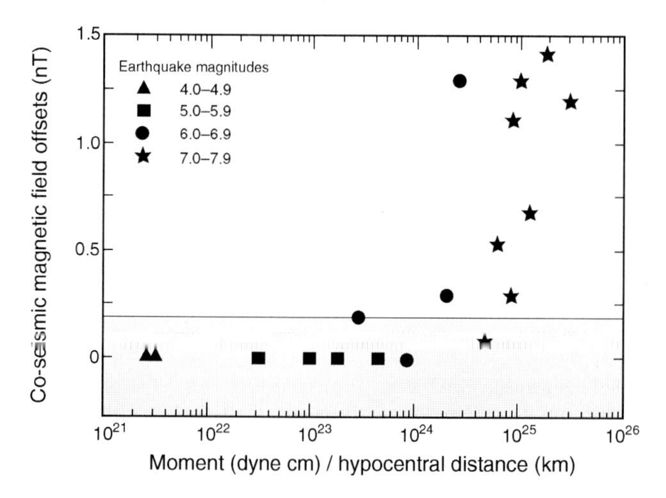

Figure S26 Co-seismic magnetic field offsets as a function of seismic moment scaled by hypocentral distance. The shaded region shows the 2-sigma measurement resolution (from Mueller and Johnston, 1998). Geodetically based seismomagnetic models (Sasai, 1991) fit each offset.

Possible high-frequency precursory effects

A number of observations purported to be high-frequency SEM effects have been recently reported (Hayakawa, 1999). Interest in these higher ULF frequencies primarily resulted from the fortuitous observation of elevated ULF noise power on a single 3-component magnetometer near the epicenter of the M7.1 Loma Prieta earthquake of October 18, 1989. However, similar records were not obtained with the 1992 M7.4 Landers earthquake, the 1994 M6.7 Northridge earthquake, the 1999 M7.1 Hector Mine earthquake, the 1999 M7.4 Izhmet, Turkey earthquake and the 2004 M6 Parkfield, Ca, earthquake.

Though controversial, increased interest in tectonoelectric (TE) phenomena related to earthquakes has resulted from suggestions in Greece and Japan that short-term geoelectric field transients (SES) of particular form and character precede earthquakes with $M > 5$ at distances up to several hundreds of kilometers. These transients appear to have a spatially uniform source field on the scale of the array but no clear corresponding magnetic field transients and no sensible coseismic effects. The SES have been empirically associated with subsequent distant earthquakes and claimed as precursors (Varotsos *et al.*, 1996).

Careful study of the SES recordings indicates that the SES signals have the form expected from rectification/saturation effects of local radio transmissions from high-power transmitters on nearby military bases. Without any clear physical explanations describing how the SES signals are earthquake generated yet coseismic effects related to the much larger earthquake source are not observed, these observations have been extremely controversial (Debate on VAN, 1996).

Another enigma concerns the generation of high-frequency (>1 kHz) electromagnetic emissions associated with subsequent moderate earthquakes but, again, with no coseismic effects. Such emissions are reported to have been detected at great distances from these earthquakes (see summary by Hayakawa and Fujinawa, 1994) and by magnetometers onboard satellites. However, the statistical significance of these observations is under dispute.

The generation of high-frequency electromagnetic radiation can be easily demonstrated in controlled laboratory experiments involving

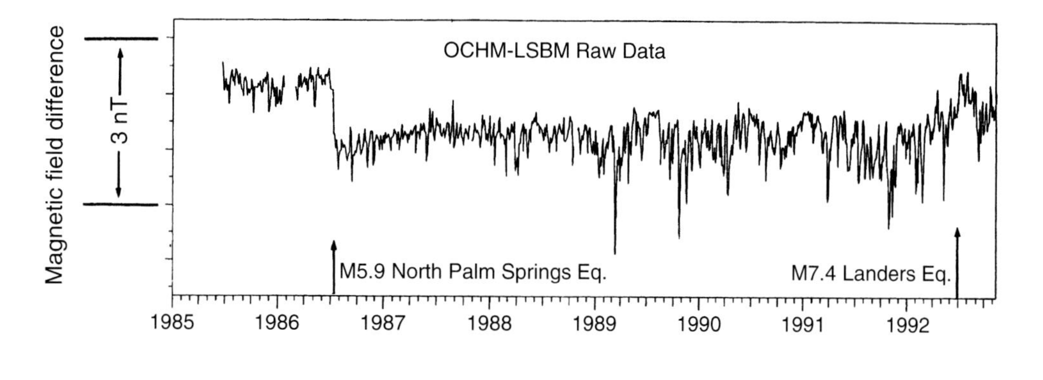

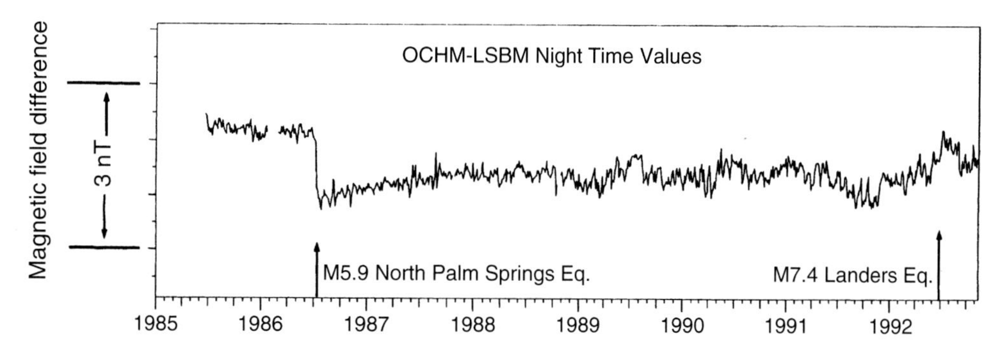

Figure S27 Magnetic field differences between stations OCHM and LSBM before, during and after the July, 1986 M5.9 North Palm Springs and the June, 1992 M7.4 Landers earthquakes (from Johnston *et al.*, 1994).

rock fracture in dry rocks. However, the Earth's crust in seismically active areas is quite conducting (0.3–0.001 S m^{-1}) and propagation of very high-frequency (VHF) electromagnetic waves even short distances through the crust is difficult to justify physically. Propagation from earthquake source regions (10–100 km in depth), and in some cases through oceans with conductivities of 1 S m^{-1}, is physically implausible. More significantly, the amount of allowable rock fracturing prior to earthquakes is strongly constrained by high sensitivity crustal strain measurements in the near-field of many earthquakes. These measurements indicate moment release ($\mu \times$ slip $\times$ area) prior to earthquakes is at least three orders of magnitude smaller than that released at the time of an earthquake (Johnston and Linde, 2002). Appeal to secondary sources at the Earth's surface may avoid this difficulty but the expected associated near-field crustal strain and displacement fields are not observed.

High-frequency disturbances are generated in the ionosphere as a result of coupled infrasonic waves generated by earthquakes and are readily detected with routine ionospheric monitoring techniques and global position system (GPS) measurements. In essence, displacement of the Earth's surface by an earthquake acts like a huge piston, generating propagating pressure waves in the atmosphere/ionosphere waveguide. Thus, traveling waves in the ionosphere (traveling ionospheric disturbances or TIDs) are a consequence of earthquakes (and volcanic eruptions). EM data at VHF frequencies recorded on ground receivers or by satellite require correction for TID and other disturbances before any association can be made to source processes or earthquake precursors.

Malcolm J.S. Johnston

Bibliography

Debate on VAN, 1996. Special issue. *Geophysical Research Letters*, **23**: 1291–1452.

Hayakawa, M. (ed.), 1999. *Atmospheric and Ionospheric Electromagnetic Phenomena Associated with Earthquakes.* Tokyo: Terra Scientific Publishing Company, p. 996.

Hayakawa, M., and Fujinawa, F. (eds.), 1994. *Electromagnetic Phenomena Related to Earthquake Prediction.* Tokyo: Terra Scientific Publishing Company, p. 677.

Johnston, M.J.S., 2002. Electromagnetic fields generated by earthquakes. *International Handbook of Earthquake and Engineering Seismology,* Volume 81A. New York: Academic Press, pp. 621–635.

Johnston, M.J.S., and Linde, A.T., 2002. Implications of crustal strain during conventional, slow and silent earthquakes. *International Handbook of Earthquake and Engineering Seismology,* Volume 81A. New York: Academic Press, pp. 589–605.

Johnston, M.J.S., Mueller, R.J., and Sasai, Y., 1994. Magnetic field observations in the near-field of the 28 June 1992 M7.3 Landers, California, earthquake. *Bulletin of the Seismological Society of America*, **84**: 792–798.

Langbein, J., Borcherdt, R., Dreger, D., Fletcher, J., Hardebeck, J.L., Hellweg, M., Ji, C., Johnston, M., Murray, J.R., Nadeau, R., Rymer, M., and Treiman, J., 2005. Preliminary report on the 28 September 2004, M 6.0 Parkfield, California earthquake. *Seismological Research Letters*, **76**: 10–26.

Mueller, R.J., and Johnston, M.J.S., 1998. Review of magnetic field monitoring near active faults and volcanic calderas in California: 1974–1995. *Physics of the Earth and Planetary Interiors*, **105**: 131–144.

Park, S.K., 1997. Monitoring resistivity change in Parkfield, California: 1988–1995, *Journal of Geophysical Research*, **102**: 24545–24559.

Reid, H.F., 1914. Free and forced vibrations of a suspended magnet. *Terrestrial Magnetism*, **19**: 57–189.

Rikitake, T., 1966. Elimination of non-local changes from total intensity values of the geomagnetic field. *Bulletin of Earthquake Research Institute*, **44**: 1041–1070.

Sasai, Y., 1991. Tectonomagnetic modeling on the basis of linear piezomagnetic effect. *Bulletin of the Earthquake Research Institute*, **66**: 585–722.

Varotsos, P., Eftaxias, K., Lazaridou, M., Nomicos, K., Sarlis, N., Bogris, N., Makris, J., Antonopoulos, G., and Kopanas, J., 1996. Recent earthquake predictions results in Greece based on the observation of seismic electric signals. *Acta Geophysica Polonica*, **44**: 301–307.

Cross-references

Geomagnetic Spectrum, Temporal
Gravity-Inertio Waves and Inertial Oscillations
Magnetotellurics
Volcano-Electromagnetic Effects

SHAW AND MICROWAVE METHODS, ABSOLUTE PALEOINTENSITY DETERMINATION

The Shaw technique is used to determine the strength of the Earth's magnetic field from igneous rocks and man fired artifacts. In this technique (Shaw, 1974) the natural remanent magnetization (NRM) of the sample is demagnetized using alternating field (AF) demagnetization but is remagnetized by heating to above the Curie temperature and cooling in a known field to form a full thermoremanent magnetization (TRM) in a known field strength. By comparing the NRM and TRM values it is possible to calculate the strength of the field in which the NRM was formed. There are checks for alteration of the sample and also several variations of the original technique that are described below.

The original Shaw technique

This technique used anhysteretic remanent magnetizations (ARMs) to check for any alteration in the sample due to the laboratory heating. There are four demagnetization stages to the experiment:

1. AF demagnetize the NRM in steps to a maximum AF value.
2. Give the sample an ARM in a known constant field in the maximum AF used in 1 (ARM1) and AF demagnetize in the same steps used in 1.
3. Give the sample a full TRM in a known constant laboratory field strength and AF demagnetize in the same steps used in 1.
4. Give the sample an ARM in the same known field as used in 2 (ARM2) and AF demagnetize in the same steps used in 1.

The basic premise is simply that if a sample can be demagnetized between two AF values then it can also be remagnetized with an ARM between those values. The value of the ARM is not simply related to the value of the NRM (or TRM) lost between those AF values but it is a function of the samples' ability to retain magnetization between those two AF values. If the sample becomes altered during heating to form the TRM in such a way that the ability to retain magnetization (TRM or ARM) between those two values changes, then the value of ARM2 (given after heating) between the two AF values will be different from the value of ARM1 (which was given before heating and, therefore, before alteration).

By plotting the demagnetization values of ARM1 against ARM2 and using the value of the AF demagnetizing field as a parameter, the slope of the line should be unity if there has been no alteration of the sample. If there has been alteration, then the slope will not be unity. By selecting the continuous AF demagnetization region where the slope is unity (if such a region exists) and plotting NRM against TRM it is possible to calculate the ratio NRM/TRM (Figure S28)

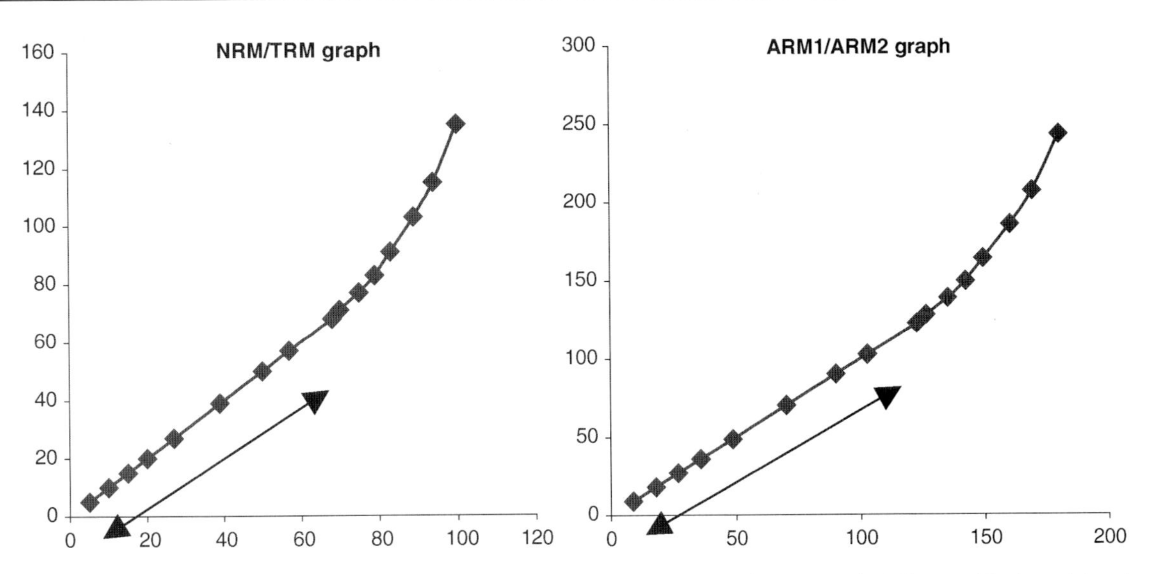

Figure S28 Shaw technique paleointensity analysis, the high AF region of the ARM1/ARM2 graph is linear with slope 1 (marked with double ended arrow). The same AF region on the NRM/TRM graph is also linear and as this AF region is unaltered the paleointensity value can be calculated from the slope.

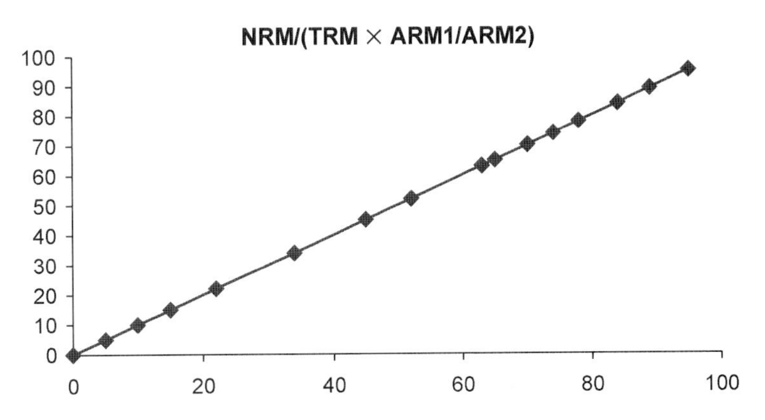

Figure S29 The data in this figure analyzed using the Rolph and Shaw correction where changes in ARM are used to correct for changes in TRM. In the Rolph and Shaw technique only data with peak AF values above 0.1 T were used in order to restrict the analysis to single or pseudosingle domain grains.

which is the same ratio as the ancient field strength/laboratory field strength (Nagata, 1943).

Kono, and Rolph and Shaw corrections

Kono (1978) proposed that samples with altered TRMs could be used for paleointensity analysis provided that the NRM/TRM graph has a linear segment and that the equivalent ARM1/ARM2 graph is also linear. He proposed multiplying the paleointensity obtained from the NRM/TRM linear section by the inverse of the slope of the equivalent ARM1/ARM2 section in order to correct for laboratory thermal alteration.

Rolph and Shaw (1985) proposed that the degree of alteration of the TRM was reflected in the degree of alteration of ARM2 and so it should be possible to use the ARM data to continuously correct for alteration in the TRM. In order to achieve this, the final highest AF field value of NRM, TRM, ARM1, and ARM2 are subtracted from their respective data. Each TRM value is then multiplied by the ratio of ARM1/ARM2 for the same AF value. A new graph plotting NRM against TRM × ARM1/ARM2 should correct for any alteration of the TRM due to laboratory heating (Figure S29).

Double heating technique

Following on from the Rolph and Shaw correction, Tsunakawa and Shaw (1994) applied a second heating to form a second TRM (TRM2) in the same field as the first TRM (TRM1). If both the first and second heatings cause thermal alteration, then it is possible to check if the ARM correction proposed by Rolph and Shaw works by plotting TRM1 against ARM corrected TRM2. The slope of the line should be unity if the correction works. This adaptation of the Shaw technique is lengthier because of the need for two heatings and a third set of ARM measurements but it gives confidence to final result if the TRM2 correction works.

Low temperature demagnetization double heating technique

Tsunakawa *et al.* (1997) and Yamamoto *et al.* (2003) used low temperature demagnetization (LTD) in conjunction with the double heating technique. The low temperature demagnetization (to liquid nitrogen temperature) removed the magnetization formed during high temperature oxidation of the magnetic grains. Although the technique is lengthy, it has provided accurate paleointensity determinations from recent Hawaiian lavas.

John Shaw

Bibliography

Kono, M., 1978. Reliability of palaeointensity methods using alternating field demagnetization and anhysteretic remanence. *Geophysical Journal of Royal Astronomical Society*, **54**: 241–261.

Nagata, T., 1943. The natural remnant magnetism of volcanic rocks and its relation to geomagnetic phenomena. *Bulletin of the Earthquake Research Institute*, **21**: 1–196.

Rolph, T.C., and Shaw, J., 1985. A new method of palaeofield magnitude correction for thermally altered samples and its application to Lower Carboniferous lavas. *Geophysical Journal of Royal Astronomical Society*, **80**: 773–781.

Shaw, J., 1974. A new method of determining the magnitude of the palaeomagnetic field. Application to five historic lavas and five archaeological samples. *Geophysical Journal of Royal Astronomical Society*, **39**: 133–144.

Tsunakawa, H., and Shaw, J., 1994. The Shaw method of palaeointensity determination and its application to recent volcanic rocks. *Geophysical Journal International*, **118**: 781–787.

Tsunakawa, H., Shimura, K., and Yamamoto, Y., 1997. *Application of Double Heating Technique of the Shaw Method to the Brunhes Epoch Volcanic Rocks*. 8th Scientific Assemble of IAGA.

Yamamoto, Y., Tsunakawa, H., and Shibuya, H., 2003. Palaeointensity study of the Hawaiian 1960 lava: implications for possible causes of erroneously high intensities. *Geophysical Journal International*, **153**: 263–276.

Cross-references

Demagnetization
Magnetization, Natural Remanent (NRM)
Magnetization, Thermoremanent (TRM)

SHOCK WAVE EXPERIMENTS

Introduction and background

Shock wave experiments have a special place in the study of the Earth's interior and the interiors of the terrestrial planets because virtually the entire pressure (to 470 GPa) and temperature (to 6000 K) ranges existing with these objects can be achieved, via this technique, in the laboratory. The study of the properties of iron and alloys, solid and molten, relevant to core composition via shock compression to high pressures and temperatures of the Earth and planetary cores has continued to challenge mineral physicists.

The geomagnetic field is thought to be generated by complex convective flows of electrically conducting Fe-rich fluids in the fluid outer core. The author has measured physical properties under shock compression which provide information on where possible Fe-rich materials are fluid and, hence, where convection and generation of the geomagnetic field occurs.

Dynamic compression of cosmochemically candidate metal-like compounds which, in the molten state, are soluble in molten iron such as FeS, FeO, FeC, FeSi, and FeH_2 continue to be of interest to theories of the formation and evolution of the molten conductive cores that generates the Earth's and other planetary magnetic fields (see *Core composition*). The electrical resistivity of iron alloys and compounds under shock compression are compared to approximate bounds for the Earth that are derived theoretically. Recent compositional and thermodynamic models of the Earth's core (see *Core, adiabatic gradient*) have been strongly revised (Anderson and Ahrens, 1994; Poirier, 1994; Stacey, 2001; Anderson and Isaak, 2002) in part, on account of new shock wave data for iron and its alloys.

Here we present a summary of modern techniques for shock wave equation of state (EOS) (pressure-volume-energy) measurements which are applicable to condensed matter, in general, and specific to iron and its alloys. The study of compression behavior of other standard materials and phase transitions in selected materials have provided standards for the more widely used diamond and multianvil high pressure apparatuses (Holzapfel, 1996). Improvements in techniques for measuring sound, or rarefaction velocity in shock-compressed material (Brown and McQueen, 1986; McQueen, 1992; Chen and Ahrens, 1998) have continued to complement the EOS measurements for these materials and provide data specifying whether materials in various shock states are in solid, partially or completely molten form. Such data also provide firm constraints on the high-pressure and high-temperature Grüneisen parameter of iron and iron alloys (see *Grüneisen's parameter for iron and Earth's core*). As discussed below, knowledge of the Grüneisen ratio of iron is crucial to reducing shock wave data to pressure-volume isotherms and isentropes so that it may be compared to data from other sources, as well as to constrain the thermodynamic models of the Earth's core.

Shock wave generation and the Rankine-Hugoniot equation

The shock waves used for dynamic compression research must be maintained for time intervals of $\sim 10^3$ times the intrinsic rise-time τ_s of the shock front (Figure S30).

The impact of gun and explosively launched flyer plates both provide the requisite nearly steady waves in materials. For steady waves, a shock, with velocity U with respect to the laboratory, independent of time, can be defined and conservation of mass, momentum, and energy across a shock front can be expressed as:

$$\rho_1 = \rho_0 U/(U - u_1), \quad \text{(Eq. 1)}$$

$$P_1 = \rho_0 u_1 U, \quad \text{(Eq. 2)}$$

$$E_1 - E_0 = P_1(1/\rho_0 - 1/\rho_1)/2 = 1/2u_1^2, \quad \text{(Eq. 3)}$$

where ρ, u, P, and E are density, particle velocity, shock pressure, and internal energy and, (as indicated in Figure S31), the subscripts 0 and 1 refer to the state in front of and behind the shock front, respectively. It should be understood that in this section, pressure is used in place of stress in the indicated wave propagation direction. Actually stress, in the wave propagation direction, is what is specified by (2). A detailed derivation of (1)–(3) is given in Melosh (1989). Equation (3) also indicates that the material achieves an increase in internal energy (per unit mass) which is exactly equal to one-half of the kinetic energy per unit mass.

Upon driving a shock of pressure P_1 into a material, a final shock state is achieved which is described by (1)–(3). This shock state is shown in relation to other thermodynamic paths in Figure S32 in the pressure-volume plane. Here $V_0 = 1/\rho_0$ and $V = 1/\rho$. In the case of the isotherm and isentrope, it is possible to follow, as a thermodynamic path, the actual isothermal or isentropic curve to achieve a state on the isotherm or isentrope. A shock or Hugoniot state is different, however.

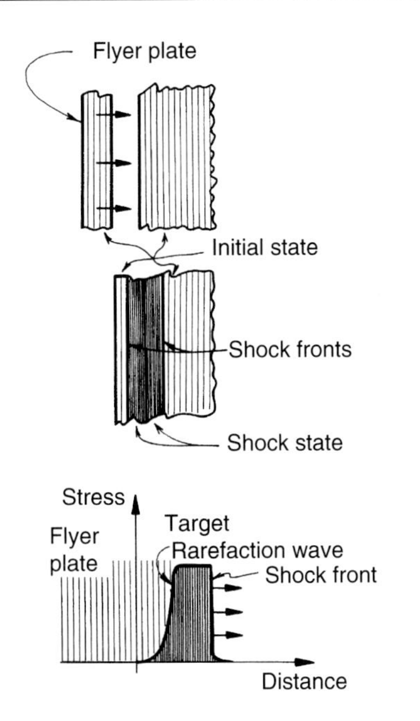

Figure S30 Generation of shock waves by flyer plate impact. Impact induces steady flat-topped shock wave and this is followed by a rarefaction wave (from Ahrens (1987)).

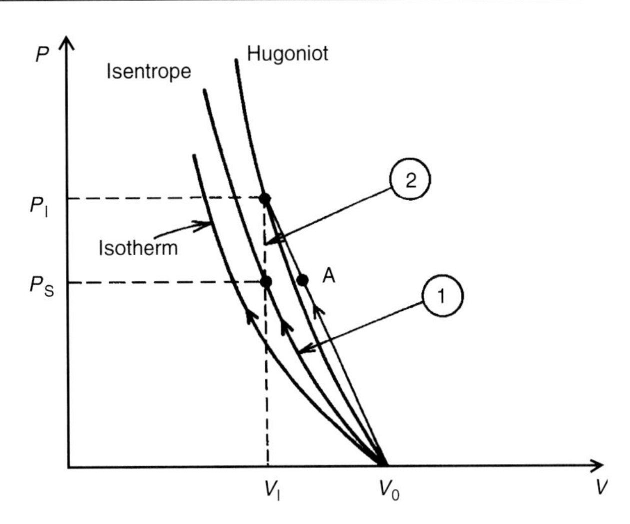

Figure S32 Pressure-volume compression curves. For isentrope and isotherm, the thermodynamic path coincides with the locus of states, whereas for the shock, the thermodynamic path is a straight line (from point 0, V_0) to point (P_1, V_1) on the Hugoniot curve (from Ahrens (1987)).

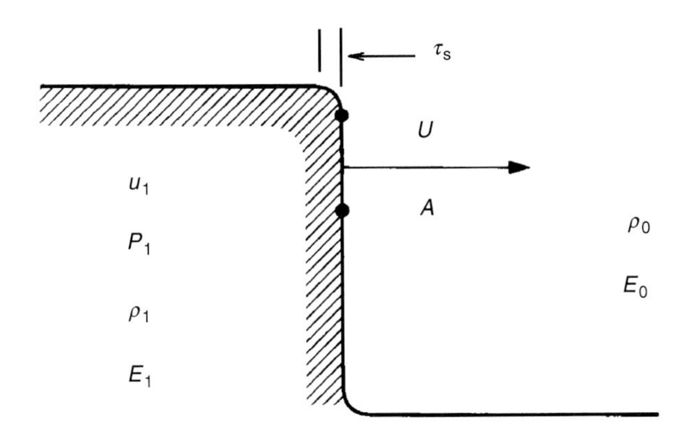

Figure S31 Profile of a steady shock wave, rise time τ_s, imparting a particle velocity u_1, pressure P_1, density ρ_1, and internal energy density E_1, propagating with velocity U into material that is at rest at density ρ_0 and internal energy density E_0 (from Ahrens (1987)).

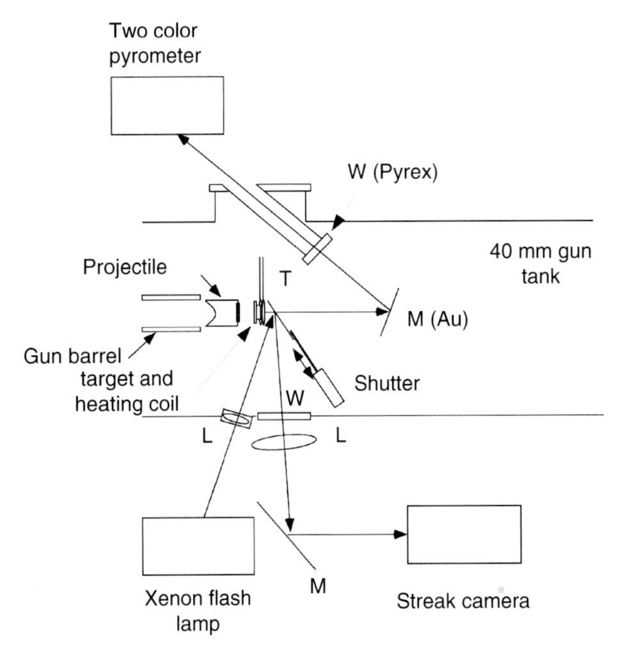

Figure S33 Diagrams of propellant gun launching system to impact metallic targets up to 2.6 km s^{-1}. The iron target is preheated with radiofrequency induction coils. Resulting target temperature is monitored using the optical signal reflected from gold mirror (M). Abbreviations: W: window; L: lens; M: mirror; M (Au): gold mirror; T: EOS turning mirror. About 1 s before launching projectile, EOS turning mirror is inserted in front of the target free-surface (after Chen and Ahrens (1998)).

The Hugoniot state P_1, V_1 is achieved via a thermodynamic path given by the straight line called Rayleigh line (Figure S32). Thus successive states along the Hugoniot curve cannot be achieved, one from another, by a shock process. The Hugoniot curve itself then just represents the locus of final shock states.

To achieve such shock waves with flat-topped stress pulses, flat plates called flyer plates are launched with guns (Figure S33) or explosive devices (Figures S34 and S35). The flyer plate impacts the sample or a "driver plate" upon which the sample is mounted and drives a shock into the driver plate and the sample. Although the various plate launching systems depicted in Figures S33–S35 operate over an enormous velocity range, all actual systems have shortcomings. Corrections for converging flow must be made in the case of explosive imploding systems of Figure S35. Tilt and flyer plate distortion, must be taken into account in the case of all explosive and gun launched impactors, as for example, indicated in Figure S36.

Comparison of electrical resistivity of shocked iron, iron alloys, and compounds to Earth

Our understanding of the conditions required to generate the Earth's magnetic field provides some constraints on the electrical conductivity of the Earth which may be usefully compared to the conductivity of

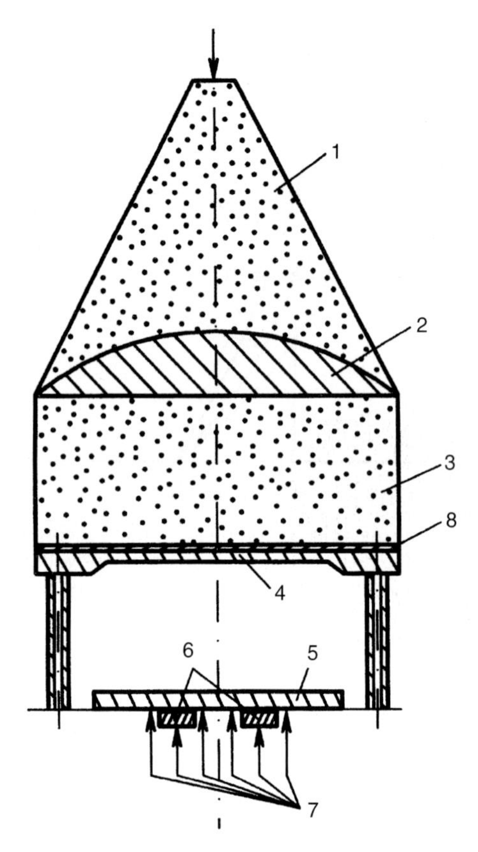

Figure S34 Explosive flyer plate launching (to ~5 km s^{-1}) apparatus. (1) and (2) Explosive plane-wave lens is point detonated at upper (downwards pointing) arrow. (1) Fast detonating explosive, (2) slow detonating explosive, (3) main explosive. (8) Plexiglas buffer plate, (4) metal flyer plate, (5) driver plate, (6) specimens under study, (7) electrical contactor pin switches that detect shock wave arrival (after Al'tshuler *et al.* (1999)).

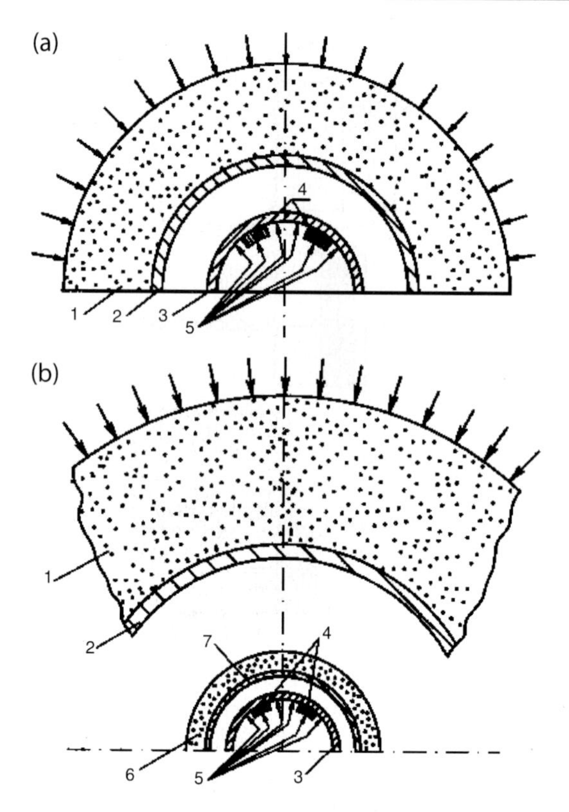

Figure S35 Hemispheric implosive flyer plate launching devices for ultra-high pressure shock wave research. (a) Single stage apparatus launches metal flyer plates to velocities of 9–11 km s^{-1}. (b) Explosive charge is multiply point detonated at outer surface of explosive hemisphere. (1) Main charge, (2) hemispherical primary flyer plate. Upon acceleration of primary flyer plate to ~5–7 km s^{-1} impact onto (6) secondary explosive charge induces an overdriven detonation wave. Overdriven detonating wave in secondary charge accelerates (7) secondary flyer plate to velocities of up to 14 km s^{-1}. Upon impact onto (3) driver plate, shock waves with pressures in 1400–1800 GPa are driven into (4) specimens. Resulting shock velocities are measured via (5) electric contactor pin switches (after Al'tshuler *et al.* (1999)).

solid and molten iron and alloys as a function of temperature and pressure.

The resistivity of the Earth's core is thus constrained to be sufficiently low, such that magnetic field lines and unable to diffuse out of the core on the timescale of ~10^4 years approximately obtained by assuming radius of the core (usually assumed to be the inner core) (~1200 km), divided by the westward drift velocity (of features of the magnetic field such as mapped by Bloxham and Gubbins (1985) from the years 1715 to present). Using this loose constraint, Matassov (1977) first showed that the core should have a resistivity of

$$\rho < 10^{-5} - 10^{-4}\,\Omega\,\text{m}. \qquad \text{(Eq. 4)}$$

These low resistivity values assume that the magnetic field will not diffuse out of the core over the 10^4 year time scale field deflation.

By assuming a fixed value of 8–12% of the total Earth heat flow (44 TW) across core-mantle boundary, a lower limit resistivity of the core region of

$$\rho < 10^{-6}\,\Omega\,\text{m} \qquad \text{(Eq. 5)}$$

is obtained. This rather simple constraint limit is obtained by taking the extreme assumption that the heat flow across the CMB is totally resulting from ohmic heating by the core of the geodynamo.

As can be seen in Figure S37, upon extrapolation, the mostly lower pressure shock data for iron, and its alloys and compounds to 135–450 GPa, the resistivity of these media lies in the 10^{-6}–10^{-5} Ωm range in agreement with the very loose theoretical expectations described above.

At least three sources of energy for powering the geodynamo are recognized:

1. Hydrodynamic flow of core fluid induced by the difference in dynamic ellipticity of the whole Earth versus that of the core.
2. Thermal convection of core fluid driving the geomagnetic dynamo also appears possible. Such thermal convection could be driven by radioactivity of the core, such as for example, supplied by ^{40}K in the core.
3. Another source of sufficient thermal energy to operate the geodynamic dynamo is the latent heat of crystallization of the inner, solid core.

Shock wave equation of state (EOS) measurements

In the simplest case when a single shock state is achieved via a shock front, the Rankine-Hugoniot equations involved six variables (U, u_1, ρ_0, ρ_1, E_1–E_0, and P_1); thus, measuring three, usually U, u_1, and ρ_0, determines the shock state, ρ_1, E_1–E_0, and P_1.

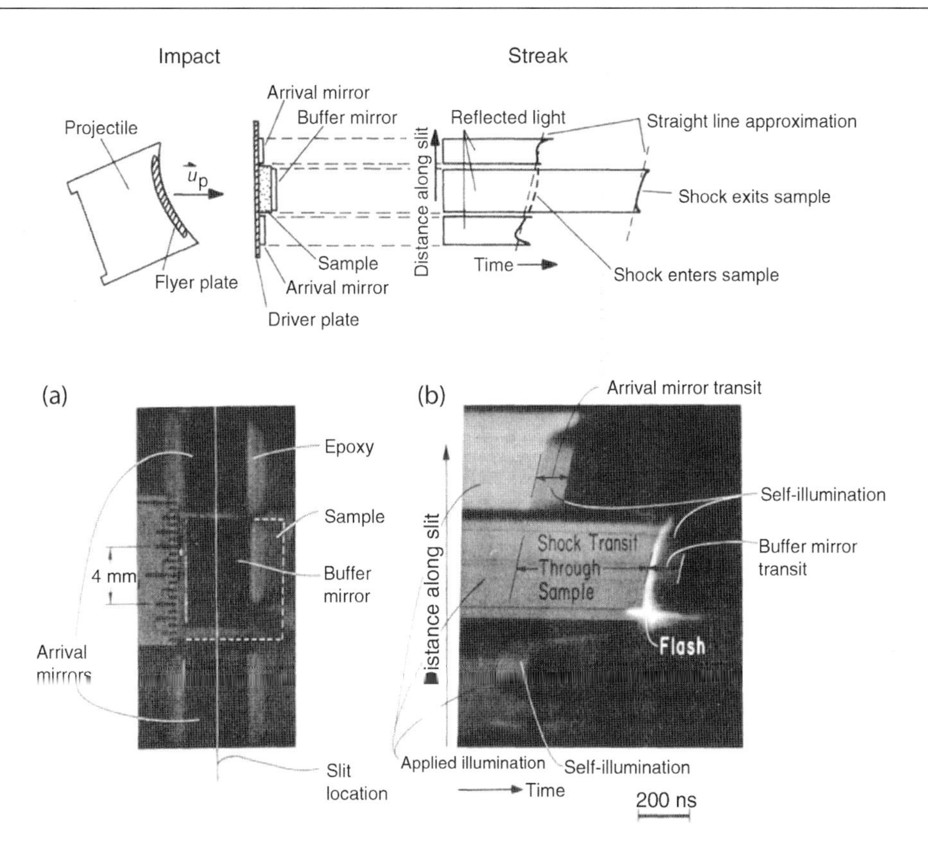

Figure S36 (Upper) Schematic illustration (highly exaggerated) of projectile tilt and bowing, and their effect on the streak record. Particle-velocity vector (up) is normal to target (seen edge-on) despite tilt. The arrival on the streak record corresponding to the shock-wave entering the sample could significantly be curved due to the bowing of the projectile (after Jeanloz and Ahrens (1980)). (Lower) (a) Experimental sample image taken through image converter camera, viewing the back-side of the target. The sample and mirror can be seen, as well as the location of the slit. (b) Streak photograph taken by sweeping the image of the slit towards the right (at nearly constant velocity) as the projectile impacted the target. Note the intense flash of light as the shock-wave crosses the sample/buffer mirror interface (after Jeanloz and Ahrens (1978)).

Careful measurement of sample volumes and mass can result in ρ_0 to be determined within $\sim\pm0.3\%$. Projectile or flyer plate velocity is measured via successive closure of a series of contactor pin switches or via optical or flash X-ray techniques (Ahrens, 1987). In the simplest case where the sample and impactor are made of the same material, the shock-induced particle velocity is

$$u_1 = u_{fp}/2, \qquad \text{(Eq. 6)}$$

where u_{fp} is the flyer plate velocity which is measured to $\sim\pm0.5\%$.

Shock velocity is usually measured to $\sim$1% via streak camera photography (Figures S33 and S36) or via contactor pin closures (Figures S34 and S35) as indicated.

Shock compression data for candidate core constituents are tabulated by Ahrens and Johnson (1995). Upon shock compression to greater than 13 GPa, the Hugoniot data for iron demonstrates the onset of a 5% volume decrease upon transition from the low pressure, bcc, α-phase to the high pressure hcp, ε-phase. Moreover, the phase change demonstrates hysteretic unloading (for example, as seen from 23 GPa shown in Figure S38). The shock wave data for iron (up to 200 GPa were originally obtained using explosively launched flyer plates, Figure S34) and extended to 900, and later, 1400 GPa using the techniques sketched in Figures S35a and b by Al'tschuler and co-workers in the former Soviet Union (now Russia) (Ahrens and Johnson, 1995). Shock wave data for Fe-Si (Matassov, 1977), pyrrhotite (FeS) (Brown *et al.*, 1984), pyrite (FeS_2) (Ahrens and Jeanloz, 1987), and wüstite (FeO) (Jeanloz and Ahrens, 1980) may be used to estimate the fraction of light elements in the core by correcting Hugoniot temperatures to those of the Earth (as for example as shown in Figures S39 and S44) (see *Inner Core composition* and *Core temperature*).

Strategy for applying shock wave data to modeling the Earth's core

Theoretical temperature models of the Earth's core

Thermal models of the outer core of the Earth have been anchored to estimates of the pure iron melting point at 330 GPa, the pressure of the inner core-outer core (IC-OC) boundary (see *Melting temperature of iron in the core, theory* and *Melting temperature of iron in the core, experimental*). Since the temperature variation $T_{oc} = T_{oc}(\gamma/V_{oc})$ in the outer core (which is assumed to be in approximately adiabatic equilibrium) is usually modeled as an isentropic temperature distribution which is described by (see *Core, adiabatic gradient*):

$$T_{oc} = T_{ic-oc} \exp\left[-\int_{V_{ic-oc}}^{V_{oc}} \frac{\gamma}{V} dV\right]. \qquad \text{(Eq. 7)}$$

Here γ is the Grüneisen parameter of the molten material of the outer core (usually assumed to be approximately equal to that of molten

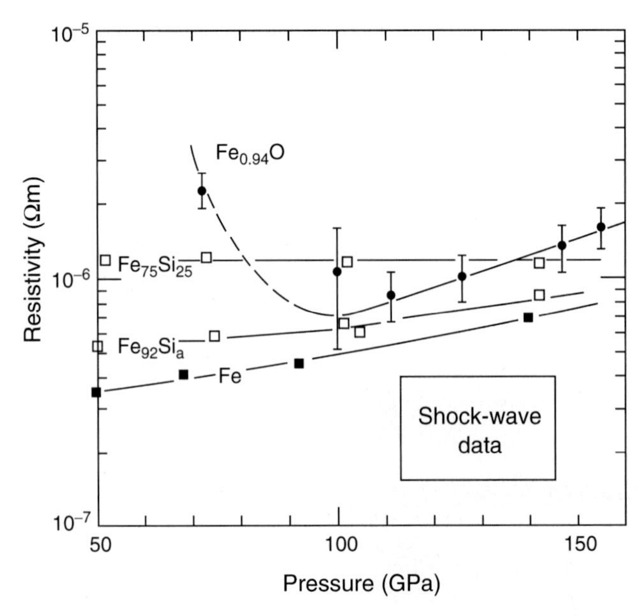

Figure S37 Measurements of the electrical resistivity of $Fe_{0.94}O$ under shock loading are compared with previous data for iron and two iron-silicon alloys, iron, iron alloys, and compounds versus shock pressure (after Knittle *et al.* (1986)). Upon extrapolating these data to the pressure range of core (135–450 GPa), it can be concluded that these iron-bearing minerals have resistivities of 10^{-6}–10^{-5} Ωm at core pressures and temperatures.

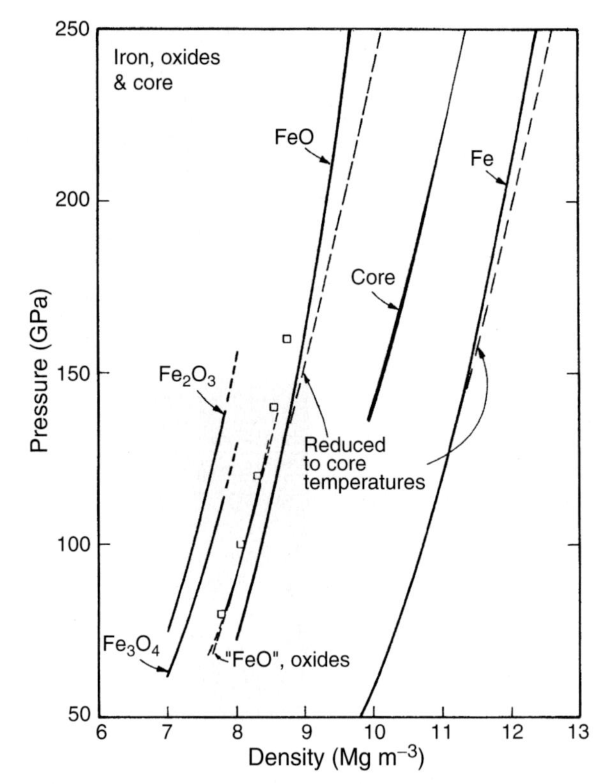

Figure S39 Shock-wave data for FeO (Jeanloz and Ahrens, 1980), Fe (McQueen *et al.*, 1970), Fe_2O_3 and Fe_3O_4 (McQueen and Marsh, quoted in Birch (1966); not corrected for porosity) (heavy curves) compared with the seismologically determined compression curve of the core. The Fe and FeO data, corrected to core temperatures, are shown as thin dashed curves. Hugoniots of mixtures of oxides and Fe corresponding to FeO are given as thin curves; $\frac{1}{3}(Fe + Fe_2O_3)$, $\frac{1}{4}(Fe + Fe_3O_4)$ and $(Fe_3O_4 - Fe_2O_3)$ are plotted but overlap; open squares are from Al'tshuler and Sharipdzhanov (1971). Hugoniots for only the high-pressure phase are shown, and the wüstite data are corrected for initial porosity and nonstoichiometry (after Jeanloz and Ahrens (1980)).

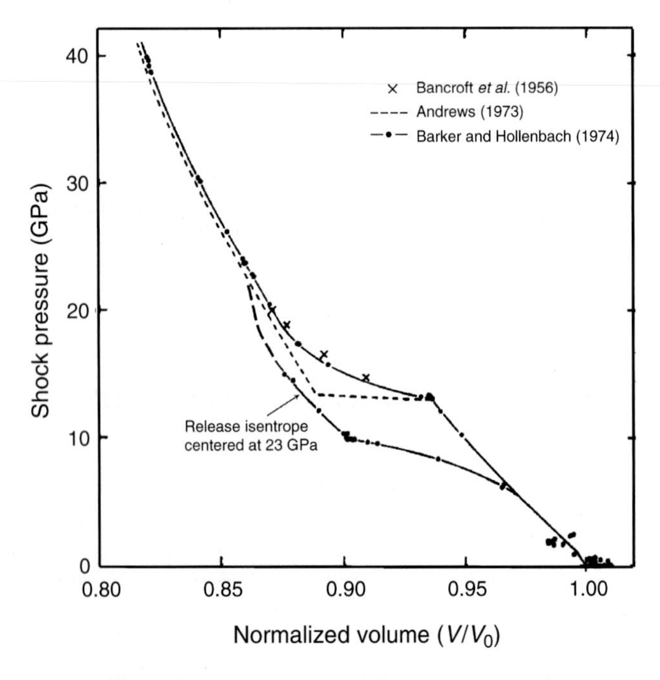

Figure S38 Hugoniot shock pressure versus reduced volume (V/V_0) data for γ-iron. Upon shock loading, the onset of phase change which α-iron (bcc) transforms to ε-iron (hcp) with a 5% decrease in volume occurs at 13 GPa. Upon unloading from 23 GPa dynamic transformation from hcp to bcc-iron occurs at 10 GPa. The data are from Bancroft *et al.* (1956) and Barker and Hollenbach (1974). Dashed curve (Andrews, 1973) is theoretical curve.

iron). The Grüneisen parameter is given by (see *Grüneisen's parameter for iron and Earth's core*):

$$\gamma = V\left(\frac{\partial P}{\partial E}\right)_V = \frac{\alpha K_s}{C_p \rho_0} = \frac{\alpha K_T}{\rho_0 C_V}, \quad \text{(Eq. 8)}$$

where, α, K_s, K_T, C_p, and C_v are the thermal expansion coefficient, isentropic and isothermal bulk modulus, and specific heat at constant pressure and volume. Equation (8) is used to evaluate γ at ambient pressures.

Determining Grüneisen parameter

The Grüneisen parameter at high pressure is obtained from the measured bulk sound velocity, C, behind the shock front and is given as (see *Grüneisen's parameter for iron and Earth's core*):

$$\gamma = \frac{\rho_0}{\rho s \eta^2}\left[1 + s\eta - R^{*2}(1 - s\eta)\right], \quad \text{(Eq. 9)}$$

when

$$\eta = u_1/U = (V_0 - V)/V_0 \quad \text{(Eq. 10)}$$

and

$$R^* = (\rho/\rho_0)(c/U). \qquad \text{(Eq. 11)}$$

The derivation of (9) assumes that the Hugoniot shock and particle relation are described by:

$$U = c_0 + s u_1, \qquad \text{(Eq. 12)}$$

where

$$c_0 = \sqrt{K_{os}/\rho_0} \qquad \text{(Eq. 13)}$$

and

$$s = \left(K'_{os} + 1\right)/4. \qquad \text{(Eq. 14)}$$

By measuring the free-surface velocity profiles of shock states being overtaken by rarefactions originating at the rear of the iron flyer plate, the sound velocity c (and hence γ) of different states along the principal Hugoniot and 1373 K preheated Hugoniot of γ-iron may be obtained as a function of density or specific volume (Figure S40). From a series of free-surface profiles, such as shown in Figure S41, data on the value of c, and hence γ (Ahrens *et al.*, 2002a) can be obtained, for example, for preheated iron, using apparatus depicted in Figure S42.

Detection of shock-induced melting

As described above, the bulk sound velocities can be measured for shock states achieved in molten materials,

$$c = \sqrt{K_s/\rho_0}. \qquad \text{(Eq. 15)}$$

In the solid regime, the rarefaction wave propagates with the longitudinal elastic velocity which is given by:

$$V_p = \sqrt{\frac{K_s + \frac{4}{3}\mu}{\rho_0}}. \qquad \text{(Eq. 16)}$$

Thus by measuring the sound velocity as a function of shock pressure, the shock pressure upon which the sound velocity decreases sharply,

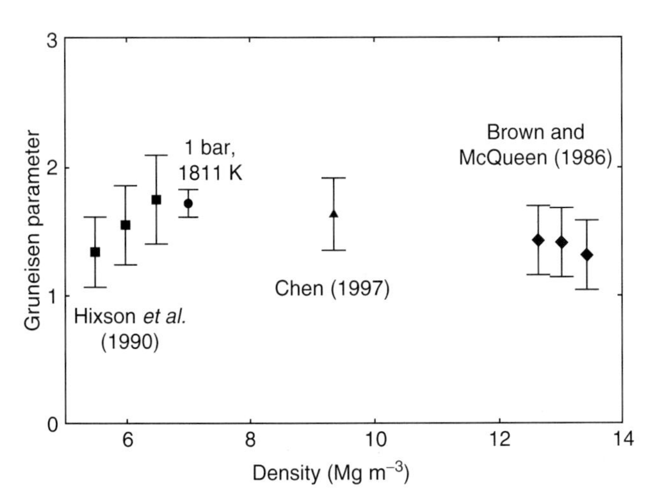

Figure S40 Thermodynamic Grüneisen parameter γ of liquid iron versus density. The 1 bar, 1811 K datum is from Anderson and Ahrens (1994). "This study" with "Chen (1997)."

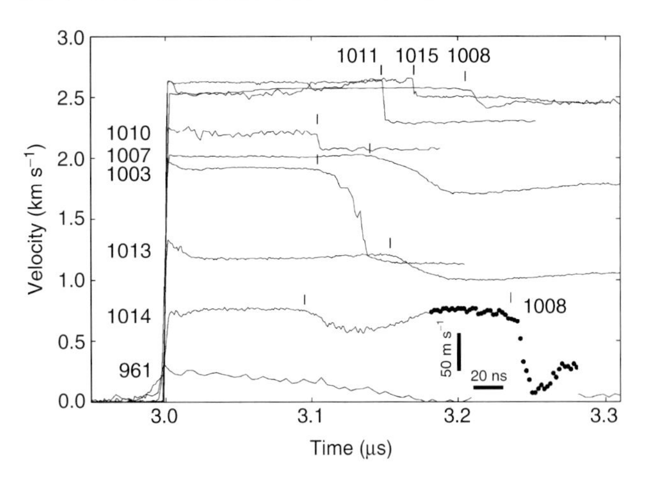

Figure S41 Free-surface velocity profiles from VISAR experiments in geometry of Figures S33 and S42 on iron preheated to 1373 K. Shot numbers are shown near each wave profile. Vertical lines indicate initial unloading arrivals. Peak particle velocities are within 4% of impedance match solutions. Origin of time axis is arbitrary. Inset: detailed plot of first release wave arrival for shot 1008 (after Chen (1997)).

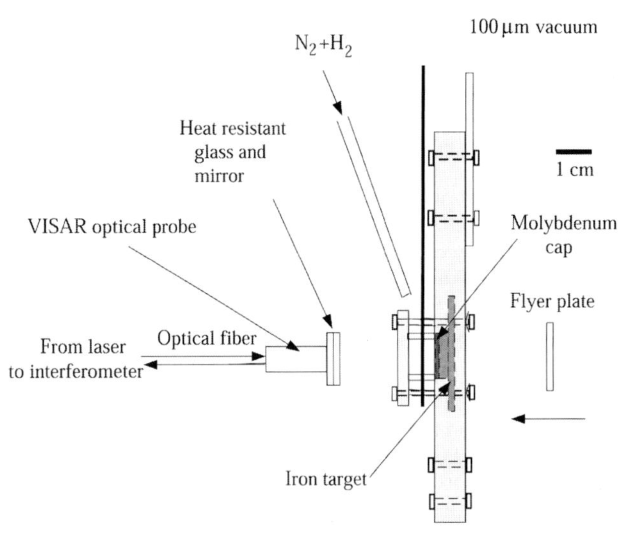

Figure S42 Experimental arrangement for conducting VISAR sound speed measurements in preheated 1373 K iron (after Chen and Ahrens (1998)).

indicates where the Hugoniot curve enters the regime of transformation from the solid, into the partially molten and then completely liquid regime. Knowledge of the Grüneisen parameter at high pressure and temperature (9–11) and thermodynamic calculations allow construction of the high pressure-temperature phase diagram (Brown and McQueen, 1986; Ahrens *et al.*, 2002b; Nguyen and Holmes, 2004) (see *Melting temperature of iron in the core, theory* and *Melting temperature of iron in the core, experiment*).

Very extensive data for iron, centered at normal temperatures (Figure S43), demonstrates that shock-induced melting begins at 220 GPa (Brown and McQueen, 1986; Nguyen and Holmes, 2004). The sound velocity of the Hugoniot of other metals is also available for Al, Cu, Ta, Pb, Mo, W, V, and stainless steel (Dai *et al.*, 2002; Luo and Ahrens, 2004). Shock temperatures upon melting, for all these

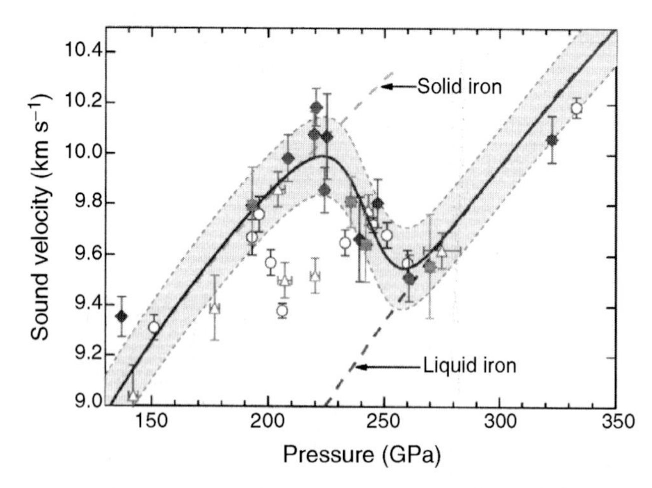

Figure S43 Sound velocity upon unloading rarefaction upon shock compression of α-iron. Solid symbols exhibit a single solid-liquid transition along Hugoniot at 222 ± 3 GPa (according to Nguyen and Holmes (2004)). Brown and McQueen's (1986) data are shown by open symbols. Open squares (triangles) represent (explosive driven) shock data (after Nguyen and Holmes (2004)).

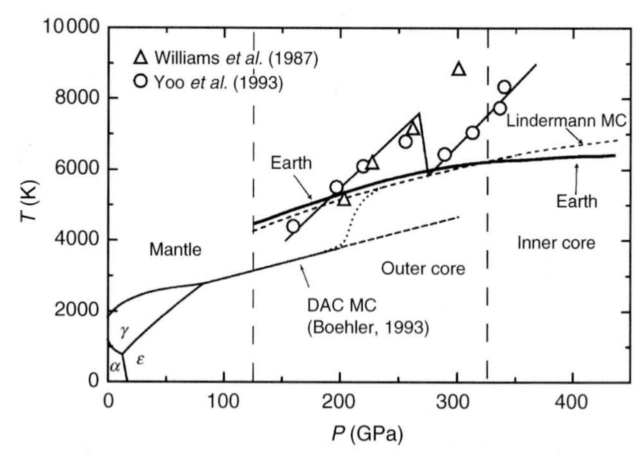

Figure S44 Shock temperature data for iron relative to the temperatures in the outer and inner core. States between 200 and 250 GPa are assigned to solid ε-iron driven to temperatures above melting line (Lindemann MC, melting curve). States above 270 GPa are believed to represent totally melted material. DAC MC is diamond anvil melting curve (after Luo and Ahrens (2004)).

metals and other materials of geophysical interest were calculated. These calculations are substantially verified by the, at present, limited shock temperature and high pressure melting data which are measured for iron (Williams *et al.*, 1987; Yoo *et al.*, 1993), V (Dai *et al.*, 2001), FeS and FeS_2 (Anderson and Ahrens, 1996), and stainless steel (Ahrens *et al.*, 1990). Moreover, recent studies of melting in the Fe-O and Fe-S system in diamond anvil apparatus indicate that eutectic melting temperatures of plausible iron alloys are lowered from those of iron by <500 K at core pressures (Boehler, 1992, 2000). The above referenced iron and iron alloys data have indicated the IC-OC melting temperature of approximately 6000 ± 500 K. These values have profoundly influenced models of the core as discussed in the next section.

Influence of shock wave data on models of the core

Thermal models of the Earth's core of the 1970s (e.g., Stacey (1977)) were, in part, based on extrapolation of 10 GPa melting data for the FeS eutectic (Usselman, 1975a, b). Stacey and others modeled the outer core as having temperature of ~3200 K at CMB rising to 4200 K at the IC-OC boundary. The shock temperatures first calculated by Brown and McQueen (1986) on the basis of their pioneering study of the shock temperature upon melting along the principal Hugoniot of iron at 220 GPa (5000–5700 K) extrapolated to 5800 ± 500 K for melting of iron at the IC-OC boundary. Later shock temperature measurements (Figure S44) supported these considerably higher temperatures. The shock temperatures for iron when applied to the core indicates higher (by ~1000 K) melting points than static high pressure diamond anvil melting data to 200 GPa (Boehler, 1993). Present models of the Earth's core assume IC-OC temperature of ~5000–6000 K (e.g., Poirier (1994), Anderson and Ahrens (1994), Boehler (2000), Anderson (2002)) are summarized by the temperature-pressure relation labeled Earth (representing the outer and inner core) in Figure S44, were influenced by the shock-induced melting data for iron. These rather higher values lead to rather larger predicted heat flows from the core of 6–10 Tw, compared to the previous assumed values of CMB heat flow of 4–5 Tw of Matassov (1977). High temperature core values lead to implausibly short model existence times of the solid Earth core (Labrosse *et al.*, 2001). This then leads to possible, but still very poorly constrained models, of radioactive element contents of the core. Such models of radiogenic heat production lead to higher sustained core temperatures over the history of the Earth (Buffett, 2003).

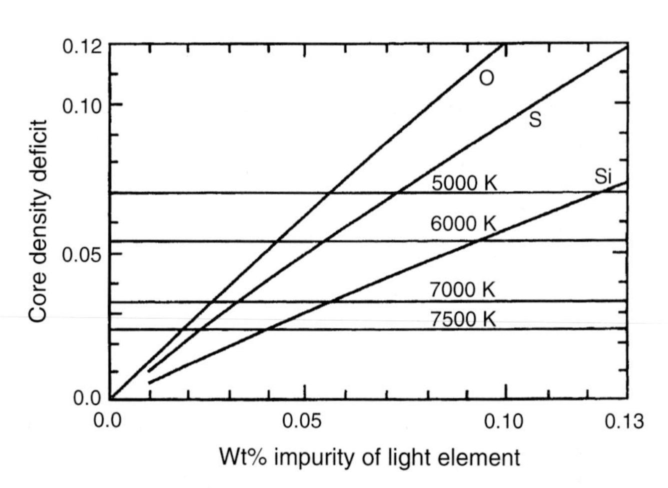

Figure S45 Core density deficit in terms of wt% of three cosmochemically acceptable elements. Four parallel lines represent the value of T_m at IC-OC (330 GPa) corresponding to temperatures indicated. Intersection of these parallel lines with curves for Si, S, and O indicate wt% values of element (after Anderson and Isaak (2002)).

As can be seen in Figure S45, as the model values of outer core temperatures are increased from 5000 to 7000 K, the calculated density deficit (with respect to molten iron at that temperature largely on the basis of shock wave data for FeS, FeO, and FeSi) goes down from 7 to 3.5% and the wt% of impurity of light element decreases drastically (see *Core, adiabatic gradient* and *Core composition*).

Acknowledgments

I am grateful to William Nellis for a number of suggestions that improved this manuscript. Research supported by NSF EAR-0207934. Contribution No. 9039, Division of Geological and Planetary Sciences, California Institute of Technology, Pasadena, CA.

Thomas J. Ahrens

Bibliography

Ahrens, T.J., 1987. Shock wave techniques for geophysics and planetary physics. In Sammis, C.G., and Henyey, T.L. (eds.), *Methods of Experimental Physics,* Volume 24, Part A. New York: Academic Press, pp. 185–235.

Ahrens, T.J., and Jeanloz, R., 1987. Pyrite: shock compression, isentropic release, and composition of the Earth's core. *Journal of Geophysical Research*, **92**: 10363–10375.

Ahrens, T.J., and Johnson, M.L., 1995. Shock wave data for minerals. In Ahrens, T.J. (ed.), *Mineral Physics and Crystallography, A Handbook of Physical Constants,* Volume 2. Washington, DC: American Geophysical Union, pp. 143–183.

Ahrens, T.J., Bass, J.D., and Abelson, J.R., 1990. Shock temperatures in metals. In Schmidt, S.C., Johnson, J.N., and Davison, L.W. (eds.), *Shock Compression of Condensed Matter—1989.* Amsterdam: Elsevier Publishers, pp. 851–857.

Ahrens, T.J., Holland, K.G., and Chen, G.Q., 2002a. Phase diagram of iron, revised-core temperatures. *Geophysical Research Letters*, **29**: 54-1–54-4. doi:10.1029/2001GL014350.

Ahrens, T.J., Xia, K., and Coker, D., 2002b. Depth of cracking beneath impact craters: new constraint for impact velocity. In Furnish, M.D., Thadhani, N.N., and Horie, Y. (eds.), *Shock-Compression of Condensed Matter—2001.* New York: American Institute of Physics, pp. 1393–1396.

Al'tshuler, L.V., and Sharipdzhanov, I.I., 1971. Additive equations of state of silicates at high pressures. *Izvestiya, Earth Physics, English Translation*, **3**: 167–177.

Al'tshuler, L.V., Trunin, R.F., Urlin, V.D., *et al.,*1999. Development of dynamic high-pressure techniques in Russia. *Physics-Uspekhi*, **42**: 261–280.

Anderson, O.L., 2002. The power balance at the core-mantle boundary. *Physics of the Earth and Planetary Interiors*, **131**: 1–17.

Anderson, W.W., and Ahrens, T.J., 1994. An equation of state for liquid iron and implications for the Earth's core. *Journal of Geophysical Research*, **99**: 4273–4284.

Anderson, W.W., and Ahrens, T.J., 1996. Shock temperatures and melting in iron sulfides at core pressures. *Journal of Geophysical Research*, **101**: 5627–5642.

Anderson, O.L., and Isaak, D.G., 2002. Another look at the core density deficit of Earth's outer core. *Physics of the Earth and Planetary Interiors*, **131**: 19–27.

Andrews, D.J., 1973. Equation of state of the alpha and epsilon phase of iron. *Journal of Physics in Chemical Solids*, **34**: 825–840.

Bancroft, D., Peterson, E.L., and Minshall, S., 1956. Polymorphism of iron at high pressure. *Journal of Applied Physics*, **27**: 291–298.

Barker, L.M., and Hollenbach, R.E., 1974. Shock wave study of the α–ε transition in iron. *Journal of Applied Physics*, **45**: 4872–4887.

Birch, F., 1966. Compressibility: elastic constants. In Clark, S.P., Jr. (ed.), *Handbook of Physical Constants,* revised edition. New York: The Geological Society of America, pp. 153–159. Bloxham, J., and Gubbins, D., 1985. *Nature*, **317**: 777.

Boehler, R., 1992. Melting of the Fe-FeO and Fe-FeS systems at high pressure: constraints on core temperatures. *Earth and Planetary Science Letters*, **111**: 217–227.

Boehler, R., 1993. Temperatures in the Earth's core from melting-point measurements of iron at high pressures. *Nature*, **363**: 534–536.

Boehler, R., 2000. High-pressure experiments and the phase diagram of lower mantle and core materials. *Reviews of Geophysics*, **38**: 221–245.

Brown, J.M., and McQueen, R.G., 1986. Phase transitions, Grüneisen parameter, and elasticity for shocked iron between 77 GPa and 400 GPa. *Journal of Geophysical Research*, **91**: 7485–7494.

Brown, J.M., Ahrens, T.J., and Shampine, D.L., 1984. Hugoniot data for pyrrhotite and the Earth's core. *Journal of Geophysical Research*, **89**: 6041–6048.

Buffett, B.A., 2003. The thermal state of Earth's core. *Science*, **299**: 1675–1676.

Chen, G., 1997. I. High pressure melting of γ-iron and the thermal profile in the Earth's core, II. High pressure, high temperature equation of state of fayalite (Fe_2SiO_4). Ph.D. (major: physics) thesis, Division of Physics, Mathematics, and Astronomy, California Institute of Technology, Pasadena, California.

Chen, G.Q., and Ahrens, T.J., 1998. High pressure and high temperature equation-of-state of gamma and liquid iron. In Wentzcovitch, R.M., Hemley, R.J., Nellis, W.J., and Yu, P.Y. (eds.), *High-Pressure Materials Research. Materials Research Society Symposium Proceedings, Warrendale, PA, 499*, pp. 41–61.

Dai, C., Jin, X., Zhou, X., *et al.,* 2001. Sound velocity variations and melting of vanadium under shock compression. *Journal of Physics D: Applied Physics*, **34**: 3064–3070.

Dai, C., Tan, H., and Geng, H., 2002. Model for assessing the melting on Hugoniots of metals: Al, Pb, Cu, Mo, Fe, and U. *Journal of Applied Physics*, **92**: 5019–5026.

Holzapfel, W.B., 1996. Physics of solids under strong compression. *Reports on Progress in Physics*, **59**: 29–90.

Jeanloz, R., and Ahrens, T.J., 1978. *The equation of state of a lunar anorthosite: 60025.* Lunar and Planetary Science Conference 9th. Houston, TX: Pergamon Press, pp. 2789–2803.

Jeanloz, R., and Ahrens, T.J., 1980. Equations of state of FeO and CaO. *Geophysical Journal of the Royal Astronomical Society*, **62**: 505–528.

Knittle, E., Jeanloz, R., Mitchell, A.C., *et al.,* 1986. Metallization of $Fe_{0.94}O$ at elevated pressures and temperatures observed by shock wave electrical resistivity measurements. *Solid State Communication*, **59**: 513–515.

Labrosse, S., Poirier, J.-P., and Le Mouël, J.-L., 2001. The age of the inner core. *Earth and Planetary Science Letters*, **190**: 111–123.

Luo, S.-N., and Ahrens, T.J., 2004. Shock-induced superheating and melting curves of geophysically important minerals. *Physics of the Earth and Planetary Interiors*, **143–144**: 369–386.

Matassov, G., 1977. The electrical conductivity of iron-silicon alloy at high pressures and the Earth's core. Lawrence Livermore Laboratory, University of California.

McQueen, R.G., 1992. The velocity of sound behind strong shocks in SiO_2. In Tasker, D.G. (ed.), *Shock Compression of Condensed Matter 1991.* Amsterdam: Elsevier, pp. 75–78.

McQueen, R.G., Marsh, S.P., Taylor, J.W., *et al.,*1970. The equation of state of solids from shock wave studies. In Kinslow, R. (ed.), *High-Velocity Impact Phenomena.* New York: Academic Press, pp. 293–417.

Melosh, H.J., 1989. *Impact Cratering, A Geologic Process.* New York: Oxford University Press, 245 pp.

Nguyen, J.H., and Holmes, N.C., 2004. Melting of iron at the physical conditions of the Earth's core. *Nature*, **427**: 339–342.

Poirier, J.-P., 1994. Light elements in the Earth's outer core: a critical review. *Physics of the Earth and Planetary Interiors*, **85**: 319–337.

Stacey, F.D., 1977. A thermal model of the Earth. *Physics of the Earth and Planetary Interiors*, **15**: 341–348.

Stacey, F.D., 2001. Finite strain, thermodynamics and the earth's core. *Physics of the Earth and Planetary Interiors*, **128**: 179–193.

Usselman, T.M., 1975a. Experimental approach to the state of the core: Part I. The liquidus relations of the Fe-rich portion of the Fe-Ni-S system from 30 to 100 kb. *American Journal of Science*, **275**: 278–290.

Usselman, T.M., 1975b. Experimental approach to the state of the core: Part II. Composition and thermal regime. *American Journal of Science*, **275**: 291–303.

Williams, Q., Jeanloz, R., Bass, J., *et al.,*1987. The melting curve of iron to 250 Gigapascals: a constraint on the temperature at Earth's center. *Science*, **236**: 181–182.

Yoo, C.S., Holmes, N.C., Ross, M., *et al.,*1993. Shock temperatures and melting of iron at Earth core conditions. *Physical Review Letters*, **70**: 3931–3934.

Cross-references

Core Composition
Core Temperature
Core, Adiabatic Gradient
Grüneisen's Parameter for Iron and Earth's Core
Inner Core Composition
Melting Temperature of Iron in the Core, Experimental
Melting Temperature of Iron in the Core, Theory

SPINNER MAGNETOMETER

Introduction

The spinner magnetometer is an instrument for measuring remanent magnetization (RM) of rock specimens in studies of the magnetic properties of rocks.

Natural remanent magnetization (NRM) is the magnetic quantity, which was preserved in the rock during its formation in the presence of the prevailing magnetic field and then during its long-life existence when it was exposed to many physical and chemical factors. Thus, various sorts of RM may exist simultaneously and in different ratio in the same rock. RM is a vector quantity. If the rock is magnetically isotropic, the direction of the RM is identical with the direction of the magnetic field at the time of rock formation. Strong anisotropy usually causes a difference between the directions of the RM and the magnetizing field. The spinner magnetometer measures the total NRM, which is the sum of several RM vectors. The remanent magnetizations differ by their formation and have different stabilities. Magnetization strengths are in the order of 10^{-6} Am^{-1} for weakly magnetized sediments to 10^{2} Am^{-1} for strongly magnetized ores. Typical spinner magnetometer applications are:

Paleomagnetism: Changes in Earth's magnetic field in geological history can be investigated through the measurement of the rock's remanent magnetization and in the investigation of its stability, age-dating the rocks, solving some tectonic problems in particular terrains, dating the development of mineralization of ore deposits, and many other geological problems.
Archeomagnetism: Changes of Earth's magnetic field in human history. These investigations are mostly applicable to dating archeological materials.
Mineralogy: Impurities of ferromagnetic grains in para- or diamagnetic minerals can be investigated.
Magnetic fabric studies: Measurement of the anisotropy of isothermal remanent magnetization (IRM) can help in the separation of ferromagnetic and paramagnetic fractions of a rock.
Magnetometry: In the interpretation of ground or airborne magnetic measurements it is useful to know whether the rock's magnetization is due to its induced or remanent component. Investigation of RM can help to solve this problem.

Measuring principles

The principle of measurement is very simple, based on electromagnetic induction law formulated by Michael Faraday as early as 1831. Rock specimen of defined size and shape, fixed in a specimen holder rotates at a constant angular speed in the vicinity of a detector.

Provided that the spinning specimen has nonzero magnetic moment, an AC voltage is induced in the detector. The detector is shielded with a multilayer permalloy shield. At constant speed of rotation and constant specimen size, the amplitude and phase of the induced signal depend on the magnitude and direction of RM vector of the specimen. The voltage is amplified, filtered, digitized, and analyzed. In a specimen rotating about a single axis inside a pair of detector coils, the output voltage amplitude is proportional to the projection of the RM vector to the plane perpendicular to the rotation axis; the component of the RM vector parallel to the rotation axis induces no signal. The phase of the signal depends on the direction of projection of the RM vector into the plane perpendicular to the rotation axis. To measure the third RM component, the specimen must be at least once repositioned by turning it by 90° and re-measured. In practice, the standard measurement of the RM vector consists of successive measurements in three, four, or six positions to reduce the measurement errors. The specimen position change is performed manually or using special automatic manipulator, which turns the specimen between successive spinning steps around the imaginary body diagonal of a cube which has threefold symmetry. Thus, the automatic manipulator eliminates the need of manual specimen positioning during the measuring process and shortens the measuring time. The other solution is to use a special specimen holder simultaneously spinning the specimen about two or three axes. However, this solution usually results in decreasing the sensitivity of the instrument due to the occurrence of additional mechanical vibrations caused by the two axes spinning mechanism.

The induced signal is amplitude sensitive proportional to speed of rotation and phase sensitive due to the band pass filtration of the signal by the instrument hardware. For this reason a very precise stabilization of rotation speed is necessary. A reference signal synchronous to the specimen rotation is needed for phase calibration, speed control, filtration, induced signal scanning, and digital processing. The reference signal is generated by an optocoupler built right on specimen holder shaft or in some cases into the motor unit. The magnetometer calibration is a special measurement procedure of the standard, which yields the gain and phase for calculating the remanence vector components. Since the specimen is fixed in the specimen holder, the measurement procedure and specimen position design should eliminate the residual RM value of the holder. If this is not possible due to principal reasons, correction for the holder in the case of a weak specimen must be done by subtracting the empty holder values from the data measured.

In general, the properties of the instrument depend upon the geometric arrangement of the detector relative to the spinning specimen, the type of detector and its size, homogeneity and geometric precision, noise of detector, noise of electronic unit, speed of rotation, size and homogeneity (magnetic and physical) of the specimen, the type of specimen holder, the influence of external disturbing fields, the resistance to mechanical shock, and temperature.

Sensitivity of the spinner magnetometer

The signal detector is the heart of the instrument. The design of the detector together with the preamplifier of the induced signal significantly determines the sensitivity of the instrument. Since some rocks have very low amplitude of the RM vector, of the order of 10^{-5}–10^{-6} Am^{-1}, the sensitivity of the instrument should be at least in the order 10^{-6} Am^{-1}.

Spinner magnetometer measurements are disturbed by noise, which can be in three categories: First the thermal noise of the detector (pick-up coils), second the noise of the electric circuits (preamplifier of the induced signal and the control unit), and third the noise whose source may be by various causes (impurities on the specimen holder, vibration of the pick-up coils, strange external magnetic fields, etc.). The sum of all of these is called the operating noise, and the sum of the first two categories is called the basic noise. Other sources of errors occur during removing and inserting of the specimen in cases of manual position change, or during specimen position change by the automatic manipulator. The sensitivity, in case we are able to remove all additional disturbing effects (except for basic noise), is called the limiting sensitivity or minimum detectable field. For practical use, it is significant to work with the standard deviation of repeated measurements of a weak specimen (at the lower limit of measuring range) using a standard measuring routine for a particular instrument. This standard deviation characterizes the sensitivity of instrument. The value of the noise is usually lower than the sensitivity.

The sensitivity of instrument depends on the time of integration, the prolongation of measurement time (four times results in increasing the sensitivity two times). The limitation factor is the operating noise. Evaluation of the noise and sensitivity of the instrument is an essential step for instrument selection suitable for a particular application. For measurement of strong rocks specimens, ores for example, the sensitivity of the instrument is not critical but in the case of weakly magnetized sediments, it is the most important parameter to determine the suitability of the intended method of measurement.

Deviations in specimen size and shape will cause errors in position, and/or possibly increase of vibration during measurement and thus decrease the precision of measurement.

The detectors used in classical spinner magnetometers are a pair of Helmholtz coils, or a fluxgate sensor. Recently, the SQUID detector working at the temperature of liquid nitrogen, the so-called high temperature HT-SQUID has been used. HT-SQUID based spinner magnetometer is a compromise between the classical spinner and the low-temperature liquid helium cryogenic magnetometer.

The sensitivity of spinner magnetometer with a fluxgate sensor is about 1.0×10^{-4} Am^{-1}, with a spinning rate of 6 Hz, an integration time of 24 s (Leaflet Molspin).

The sensitivity of the HT-SQUID spinner magnetometer expressed in terms of intrinsic noise is 3.0×10^{-7} Am^{-1} (Leaflet F.I.T. Messtechnik). The detector output voltage is due to a different measuring principle in comparison with classical spinner independent of the spinning rate.

The measurements with the instrument using the two axes spinning specimen method and HT-SQUID sensor show a sensitivity of 1.0×10^{-5} Am^{-1} (standard deviation) in measurement time of 13 s (Leslie *et al.*, 2001).

The sensitivity achieved with the standard Helmholtz coils is 2.4×10^{-6} Am^{-1} for a spinning rate of 88 Hz, and 10 s integration time and about 1.0×10^{-5} Am^{-1} for a slow spinning rate of 17 Hz (Agico, 2001).

Description of Agico spinner magnetometer

The JR-6/JR-6A dual speed spinner magnetometer is an innovated version of the instrument series based on classical spinner magnetometer design with pick-up unit Helmholtz coils. The instrument is equipped with up to date microelectronic components. Two microprocessors control and test the speed of specimen rotation, signal gain, acquisition of the data, carry out digital filtration, and control the autoposition manipulator. The magnetometer is fully controlled by an external computer via a serial channel RS232C.

The spinner magnetometer JR-6A, equipped with a specimen autoposition manipulator and an automatic specimen holder, enables the automatic measuring of all components of the RM vector.

The low speed of rotation increases the possibility of measuring fragile specimens, soft specimens placed in a perspex container and/or specimens with considerable deviations in size and shape. The integration time may be shortened or prolonged by a factor of 2. The repeat mode enables the repetition of measurement in the current specimen position without stopping the motor.

A nominal specimen is either a cube with an edge of 20 mm or a cylinder 25.4 mm (1 in.) in diameter and 22 mm long. Calibration relates to the nominal volume of the specimen. One side of a cubic specimen or the base of a cylindrical specimen is marked with an arrow, which defines the coordinate system of the specimen in which the RM vector is measured. The components of RM vector are calculated using Fourier harmonic analysis. The system of the measurement positions is identical for cubic and for cylindrical specimens. The sensitivity is 2.4×10^{-6} Am^{-1} for a spinning rate of 87.7 rps and an integration time of 10 s, and for low-speed rotation of 16.7 rps, the sensitivity is 1.0×10^{-5} Am^{-1}. The accuracy of measurement of the RM components is 1%, $\pm$ 2.4 $\mu\mathrm{Am}^{-1}$. The measuring range of the instrument in full auto range is from 0 (2.4 $\mu\mathrm{Am}^{-1}$) to 12500 Am^{-1} in seven decimal ranges.

JR-6/JR-6A spinner magnetometer measurement procedure

Standard measurement of the RM vector consists of successive measurements in four positions of the specimen with regard to the holder using the manual specimen holder for four or six positions. The complete measurement yields four values for z component of the RM vector and two values of the x and y components from which the average values are calculated. This process eliminates any residual non-compensated value of the holder RM and reduces the measuring errors caused by inaccurate shape of the specimen and by instrument noise. The residual value of the remanent magnetization of the manual holder for measurement in four or six positions is fully eliminated even if the correction for the holder was not done. It is due to position design of manual positioning with respect to the holder.

If less than four positions are used, the residual components of the holder are not eliminated automatically. We must rely upon the correction of the holder by subtracting the empty holder values. No correction is made for errors due to irregular shapes of the specimens. It is therefore recommended to reduce the measuring cycle to two positions only, for approximate measurements.

Sometimes it may be useful to expand the measuring cycle to six measuring positions, especially in case of specimens of an irregular shape. If six measuring positions are used, the residual value of the RM vector of the holder is fully eliminated. All three components are always measured four times, so the statistics may be better.

Measurement with an automatic holder is available with model JR-6A. The specimen is fixed in the holder only once; the automatic manipulator changes the measuring position without the need for the operator intervention. The further advantage is that the specimen is not exposed to external magnetic fields during position change, as the specimen remains closed inside the triple permalloy shield during the three-position measurement routine. The automatic specimen holder consists of a partly spherical outer shell and a rotatable inner spherical core in which the specimen is secured by a plastic screw. The system of specimen orientation occurs by specimen rotation around the imaginary body diagonal of a cube, which has threefold symmetry. For this reason it is impossible to find a combination of positions which would eliminate the residual value of the remanent magnetization of the holder as with the traditional four- or six-orientation magnetometer measurement. It is necessary to rely entirely upon the correction done by the instrument. This fact may negatively influence the result, especially in case of very soft and weak specimens when small impurities, moving obligatory on spherical core, can generate unrealistic data. Specimens, which differ in size or shape from nominal values, could cause strong vibrations or imbalance while spinning. Such specimens cannot be measured using automatic holder. But in this case it is usually possible to use standard manual holder.

Jiří Pokorný

Bibliography

Agico s.r.o., 2001. *Instruction Manual for Spinner Magnetometer JR-6/ JR-6A*. Brno, Czech Republic.

F.I.T. Messtechnik GmbH. 2001. *Leaflet of the HSM2 SQUID-Based Spinner Magnetometer.* Hildesheim, Germany.

Leslie, K.E., Binks, R.A., Lewis, C.J., Scott, M.D., Tilbrook, D.L., and Du, J., 2001. Three component spinner magnetometer featuring rapid measurement times. *IEEE Transactions on Applied Superconductivity*, **11**(1): 252–255.

Natsuhara Gigen Ltd, 2003. *Leaflet of the Aspin Spinner Magnetometer.* Osaka, Japan.
Molspin Ltd. *Leaflet of the Minispin Fluxgate Magnetometer.* Newcastle on Tyne, England.

Cross-references

Magnetization, Natural Remanent (NRM)
Magnetization, Remanent

STATISTICAL METHODS FOR PALEOVECTOR ANALYSIS

Our concern is with the statistical description of paleomagnetic vectors and the estimation of their mean and variance. These vectors may come from a number of different rock units or archeological samples, representing a range of acquisition times, and be useful for studies of the mean paleomagnetic field and *paleosecular variation*; alternatively, the vectors may come from individual measurements taken from a given rock unit or archeological sample, representing the same moment of acquisition, and be useful for studying the acquisition process itself. Directional data of a particular polarity are usually analyzed with a *Fisher distribution* (1953), and data of mixed polarities are usually analyzed with a *Bingham distribution* (1964). Occasionally, other directional distributions are used. For example, Bingham (1983) considered the projection of a three-dimensional (3D), scalar-variance Gaussian distribution onto the unit sphere, something he called the "angular-Gaussian" distribution. More recently, Khokhlov *et al.* (2001) considered a generalization of the angular-Gaussian distribution, one with a covariance matrix, which they used to analyze directional data from a number of sites. With respect to intensity data, they have traditionally been treated separately from paleodirections, analyzed with normal, log-normal, or gamma distributions. Here, for data of either a particular polarity or of mixed polarities, we summarize these works, and that of Love and Constable (2003), who developed a full-vector, scalar-variance, Gaussian-statistical framework for treating directional and intensity data simultaneously and self-consistently.

The distributions

In our statistical treatment, each paleomagnetic vector $\mathbf{x}$ is considered to be an independent realization occurring in probability according to a statistical distribution. In Cartesian coordinates (X, Y, Z) the probability $P(\mathbf{x})$ that $\mathbf{x}$ lies within the infinitesimal differential volume

$$d^3\mathbf{x} = dX\, dY\, dZ \qquad \text{(Eq. 1)}$$

is

$$P(\mathbf{x}) = \int p(\mathbf{x}) d^3\mathbf{x}, \qquad \text{(Eq. 2)}$$

where $p(\mathbf{x})$ is the density function,

$$p(\mathbf{x}) = \frac{d^3}{d\mathbf{x}^3} P(\mathbf{x}). \qquad \text{(Eq. 3)}$$

With respect to the many different distributions of probability theory, the Gaussian occupies the most prominent position. This is due to the central limit theorem, which, roughly speaking, asserts that the distribution of the sum of independent, identically distributed random variables is approximately Gaussian. This theoretical underpinning is appealing, and, therefore, for the analysis of paleomagnetic vectors, we consider probability-density functions in a Cartesian three-space of orthogonal magnetic-field components consisting of nonzero mean Gaussian distributions. We model vectors recording data of a particular polarity with a unimodal, Gaussian probability-density function defined in terms of a mean paleomagnetic vector $\mathbf{x}_\mu$ and an associated scalar variance σ^2:

$$p_{g_1}(\mathbf{x}|\mathbf{x}_\mu, \sigma^2) = \frac{1}{(2\pi)^{3/2}\sigma^3} \exp\left[-\frac{1}{2\sigma^2}(\mathbf{x} - \mathbf{x}_\mu)\cdot(\mathbf{x} - \mathbf{x}_\mu)\right]. \qquad \text{(Eq. 4)}$$

We model vectors of mixed polarities with a bimodal, bi-Gaussian probability-density function, which, in Cartesian coordinates, is

$$p_{g_2}(\mathbf{x}|\mathbf{x}_\mu, \sigma^2) = \frac{1}{2}\left[p_{g_1}(\mathbf{x}|\mathbf{x}_\mu, \sigma^2) + p_{g_1}(\mathbf{x}| - \mathbf{x}_\mu, \sigma^2)\right]. \qquad \text{(Eq. 5)}$$

The relevant paleomagnetic coordinates are spherical, being the familiar quantities of magnetic intensity, inclination, and declination (F, I, D). In this case, the differential volume element is transformed according to

$$dX\, dY\, dZ \rightarrow F^2 \cos I\, dF\, dI\, dD, \qquad \text{(Eq. 6)}$$

and the Gaussian probability-density function is

$$p_{g_1}(\mathbf{x}|\mathbf{x}_\mu, \sigma^2) = p_{g_1}(F, I, D|F_\mu, I_\mu, D_\mu, \sigma^2) = F^2 \cos I q(\mathbf{x}|\mathbf{x}_\mu, \sigma^2), \qquad \text{(Eq. 7)}$$

where

$$\begin{aligned} q(\mathbf{x}|\mathbf{x}_\mu, \sigma^2) = & \frac{1}{(2\pi)^{\frac{3}{2}}\sigma^3} \\ & \times \exp\left[-\frac{1}{2\sigma^2}(F\cos I \cos D - F_\mu \cos I_\mu \cos D_\mu)^2\right] \\ & \times \exp\left[-\frac{1}{2\sigma^2}(F\cos I \sin D - F_\mu \cos I_\mu \sin D_\mu)^2\right] \\ & \times \exp\left[-\frac{1}{2\sigma^2}(F \sin I - F_\mu \sin I_\mu)^2\right]. \end{aligned} \qquad \text{(Eq. 8)}$$

The quantity F_μ is the magnetic intensity, or the Euclidean length, of the mean vector x_μ, and I_μ and D_μ are the inclination and declination of the mean vector. In spherical coordinates the bi-Gaussian probability-density function is

$$\begin{aligned} p_{g_2}(\mathbf{x}|\mathbf{x}_\mu, \sigma^2) &= p_{g_2}(F, I, D|F_\mu, I_\mu, D_\mu, \sigma^2) \\ &= \frac{1}{2}F^2 \cos I\left[q(\mathbf{x}|\mathbf{x}_\mu, \sigma^2) + q(\mathbf{x}| - \mathbf{x}_\mu, \sigma^2)\right]. \end{aligned} \qquad \text{(Eq. 9)}$$

If we define θ to be the off-axis angle between a particular unit paleomagnetic vector and the mean unit vector,

$$\hat{\mathbf{x}} = \frac{\mathbf{x}}{|\mathbf{x}|} \quad \text{and} \quad \hat{\mathbf{x}}_\mu = \frac{\mathbf{x}_\mu}{|\mathbf{x}_\mu|}, \qquad \text{(Eq. 10)}$$

then

$$\cos\theta = \hat{\mathbf{x}} \cdot \hat{\mathbf{x}}_\mu. \qquad \text{(Eq. 11)}$$

For both the Gaussian and bi-Gaussian cases the vectorial variance is taken to be spherically symmetrical. That is, the three Cartesian vectorial components are assumed to be independent and to have equal scalar variance. Such a situation is sometimes described as being one of "isotropic" variance. More generally, however, a Gaussian distribution can be defined in terms of a covariance matrix, where the Cartesian vectorial variance is ellipsoidal, the components having possibly different variances and, even, correlation. Such a situation that is sometimes described as being one of "anisotropic" variance. Of course, because the anisotropic distribution has a larger number of degrees of freedom, it will always fit a given dataset at least as well as the isotropic distribution. In either case, however, it is worth remarking that the Gaussian distributions are idealizations. We do not expect that they will fit all paleovector datasets, since the data themselves result from a myriad of physical processes that in all likelihood cannot be completely distilled down to simple mathematical descriptions. Instead, the utility of statistical distributions is as benchmarks for comparison, and in that sense, the isotropic unimodal and bimodal Gaussian distributions, being relatively mathematically simple, are the most practically attractive.

Marginal forms

For paleomagnetic datasets consisting wholly of coincident intensity and directional measurements, the Gaussian distribution (7) and the bi-Gaussian distribution (9) are of obvious utility. However, in most circumstances, paleomagnetic data consist of only parts, or mixtures of different parts of the full paleomagnetic vector. Thus, what we need are the appropriately marginalized probability-density functions corresponding to the underlying Gaussian distributions. So, for example, most paleomagnetic data are only directional, they consist of inclination-declination pairs with no associated absolute paleointensity. To analyze such data we need the joint probability-density function for inclination and declination, obtained by integrating (7) and (9) over all intensities,

$$p_g\left(I, D|I_\mu, D_\mu, (\sigma/F_\mu)^2\right) = \int_0^\infty p_g(F, I, D)\mathrm{d}F. \qquad \text{(Eq. 12)}$$

Alternatively, if we are analyzing data from an azimuthally unoriented borecore, providing (say) intensity-inclination data, then we need the marginal density function

$$p_g(F, I|F_\mu, I_\mu, \sigma^2) = \int_0^{2\pi} p_g(F, I, D)\mathrm{d}D. \qquad \text{(Eq. 13)}$$

If the data consist only of inclinations then we integrate (13) over all intensities, etc. In each case, we integrate over the vectorial components that are either not available or are not needed. For reference, all required integrations are given in Love and Constable (2003). In Figure S46 we show the intensity, inclination, declination, and off-axis angular distributions corresponding to the Gaussian distribution (7) for a variety of different dispersions and mean inclinations.

Intensity

In recent years the analysis of paleointensity data, be they from within a particular epoch or spanning a much longer geological period of time, has become the subject of increasing interest to researchers. It is useful, therefore, to compare such data to the intensity distribution corresponding to the 3D Gaussian distributions. For both unimodal and bimodal cases the intensity density function is the same, obtained by integrating either (7) or (9) over all angles,

$$p_g(F|F_\mu, \sigma^2) = \sigma^{-1}\left(\frac{2}{\pi}\right)^{1/2}\left(\frac{F}{F_\mu}\right) \times \exp\left[-\frac{1}{2}\left(\frac{F}{\sigma}\right)^2 - \frac{1}{2}\left(\frac{F_\mu}{\sigma}\right)^2\right]\sinh\left[\frac{FF_\mu}{\sigma^2}\right], \qquad \text{(Eq. 14)}$$

which is a special case of the n-dimensional Rayleigh-Rician distribution. This function is invariant with respect to change in sign of intensity, although, of course, intensity is, by convention, taken to be a positive quantity. As an aside, we note that distributions of this type have application to digital communications and the radar identification of targets surrounded by Gaussian clutter.

Next, it is enlightening to consider the limiting form of the intensity distribution where $\sigma \ll F_\mu$. It is approximately that for a one-dimensional normal distribution,

$$p_n(F|F_\mu, \sigma^2) = \frac{1}{\sqrt{2\pi}\sigma}\exp\left[-\frac{1}{2}\left(\frac{F - F_\mu}{\sigma}\right)^2\right], \qquad \text{(Eq. 15)}$$

which McFadden and McElhinny (1982) have suggested might be appropriate for paleointensity studies, after truncation of negative intensities. Note that (15) is not a log-normal distribution and it is not a gamma distribution, each of which have been employed on occasion in the analysis of paleointensity data. However, because it is directly linked to the 3D Gaussian distributions (7) and (9), which can be used for directional analyses as well, and because it applies to a complete and proper range of mean intensities and dispersions, (14) is suitable for paleointensity studies, even (say) during periods of reversal when one can expect that the mean intensity would be small, but the vectorial dispersion would be large.

Off-axis angle

For the Gaussian distribution, the marginal density function for off-axis angle is

$$p_{g_1}\left(\theta|(\sigma/F_\mu)^2\right) = \frac{1}{2}\sin\theta\exp\left[-\frac{1}{2}\left(\frac{F_\mu}{\sigma}\right)^2\right] \times \left\{\left[1 + \left(\frac{F_\mu}{\sigma}\right)^2\cos^2\theta\right]\exp\left[\frac{1}{2}\left(\frac{F_\mu}{\sigma}\right)^2\cos^2\theta\right] \times \left[1 + \operatorname{erf}\left[\frac{1}{\sqrt{2}}\left(\frac{F_\mu}{\sigma}\right)\cos\theta\right]\right] + \left(\frac{2}{\pi}\right)^{\frac{1}{2}}\left(\frac{F_\mu}{\sigma}\right)\cos\theta\right\}, \qquad \text{(Eq. 16)}$$

and the marginal density function for off-axis angle corresponding to the bi-Gaussian distribution is just

$$p_{g_2}\left(\theta|(\sigma/F_\mu)^2\right) = \frac{1}{2}\sin\theta\exp\left[-\frac{1}{2}\left(\frac{F_\mu}{\sigma}\right)^2\sin^2\theta\right] \times \left[1 + \left(\frac{F_\mu}{\sigma}\right)^2\cos^2\theta\right]. \qquad \text{(Eq. 17)}$$

For the limiting case where $\sigma/F_\mu \ll 1$, the unimodal off-axis angular probability-density function (16) is approximately

$$p_f\left(\theta|(F_\mu/\sigma)^2\right) \propto \sin\theta\exp\left[\left(\frac{F_\mu}{\sigma}\right)^2\cos\theta\right], \qquad \text{(Eq. 18)}$$

this corresponding to the Fisher distribution so often used by the paleomagnetic community for unimodal directional. For the bi-Gaussian off-axis angular probability-density function (16), and in the same limit, the off-axis angular density function is approximately

$$p_b\left(\theta|(F_\mu/\sigma)^2\right) \propto \sin\theta\exp\left[\frac{1}{2}\left(\frac{F_\mu}{\sigma}\right)^2\cos^2\theta\right], \qquad \text{(Eq. 19)}$$

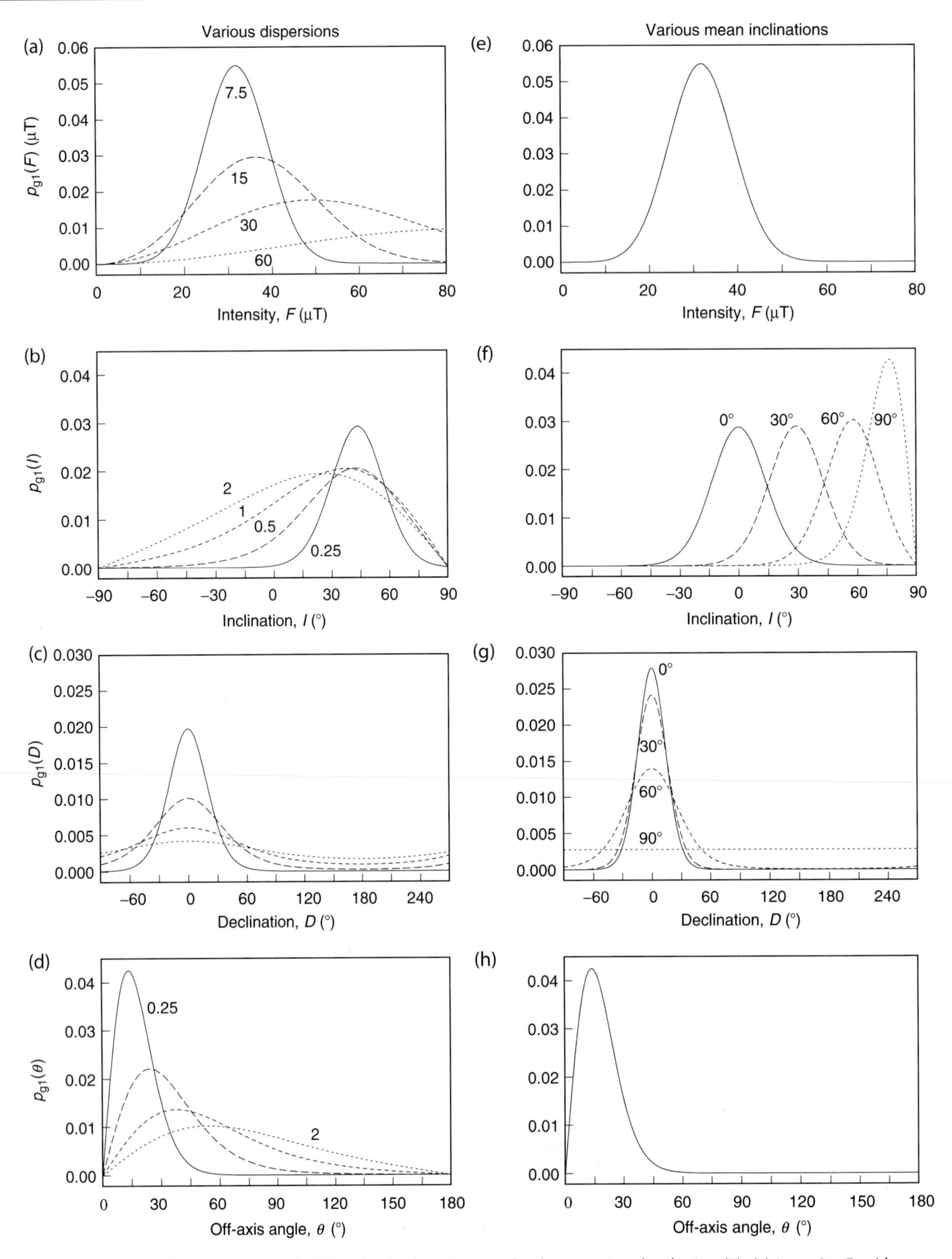

Figure S46 Examples of the marginal probability-density functions p_{g_1} for the Gaussian distribution (7). (a) Intensity F, with vectorial-mean intensity F_μ, and with vectorial dispersions σ of 7.5, 15, 30, and 60 μT, shown, respectively, by solid, long-dashed, short-dashed, and dotted lines. (b) Inclination I, (c) declination D, and (d) off-axis angle θ with vectorial mean direction $(I_\mu, D_\mu) = (45°, 0°)$, and with relative vectorial dispersions σ/F_μ of 0.25, 0.5, 1, and 2, shown, respectively, by solid, long-dashed, short-dashed, and dotted lines. Examples (e-h) are for different mean inclinations, but only the (f) inclination and (g) declination density functions are affected; for vectorial-mean values of $(F_\mu, D_\mu, \sigma) = (30\ \mu T, 0°, 7.5\ \mu T)$ the solid, long-dashed, short-dashed, and dotted lines are for vectorial-mean inclinations I_μ of 0°, 30°, 60°, and 90°, respectively.

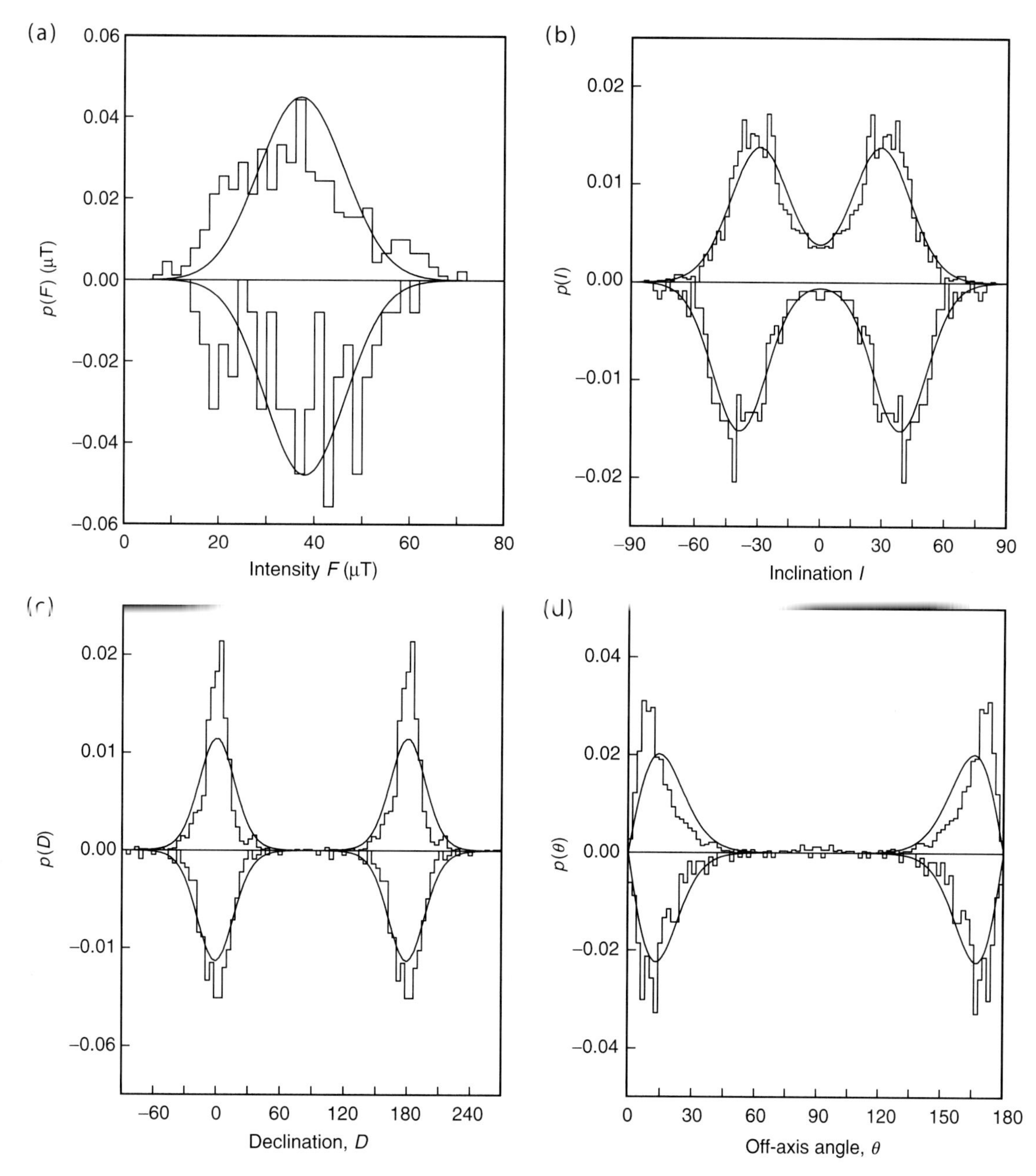

Figure S47 Comparison of maximum-likelihood fits of the 3D bi-Gaussian distribution to the (positive, top) Hawaiian data and (negative, bottom) Réunion data covering the past 5 Ma. Both the probability-density functions and the histograms of the data are shown for (a) intensity, (b) inclination, (c) declination, and (d) off-axis angle. Note that the Réunion data, particularly the declination and off-axis angle data, are fitted better than the Hawaiian data, also note the sizable difference in mean inclination between these two sites.

this corresponding to the Bingham distribution so often used by the paleomagnetic community for bimodal directional analysis. As with our comment about intensity distributions, because they are directly linked to the 3D Gaussian distributions, and because they apply to the complete range of possible vectorial dispersions, (16) and (17) are suitable for most paleodirection studies.

Maximum-likelihood estimation

In fitting paleomagnetic data to a particular Gaussian distribution, thereby yielding a measure of mean and variance, a convenient method is that of maximum-likelihood; for a general review see Stuart *et al.* (1999). With this formalism, the likelihood function is constructed from the joint probability-density function for the existing dataset. In our case we use the Gaussian density functions and/or their appropriate marginalizations to construct the likelihood, which, in its most general form, for all normally encountered types of data groups, is just

$$L = \prod_{j=1}^{N_{FID}} p_g(F_j, I_j, D_j) \prod_{k=1}^{N_{ID}} p_g(I_k, D_k) \times \prod_{l=1}^{N_{FI}} p_g(F_l, I_l) \prod_{m=1}^{N_F} p_g(F_m) \prod_{n=1}^{N_I} p_g(I_n). \quad \text{(Eq. 20)}$$

Here, N_{FID} is the number of intensity-inclination-declination triplets; N_{ID} is the number of inclination-declination pairs, etc. Maximizing L is accomplished numerically (Press *et al.*, 1992), an exercise yielding a particular paleomagnetic vectorial mean and variance. Insofar as the Gaussian distribution is an appropriate description of the paleovector field, then in the limit of large number of data, the maximum-likelihood method yields unbiased estimates of the vectorial mean intensity and the vectorial mean direction (Love and Constable, 2003), which is not usually the case with the traditional method of making separate numerical averages of intensity data and unit directional vectors (Creer, 1983).

Application example

In Figure S47 we show the results (after Love and Constable, 2003) of a maximum-likelihood analysis using the 3D bi-Gaussian distribution and fitting paleomagnetic data covering the past 5 Ma from both Hawaii and Réunion. Comparison of data from these two sites is of interest since they are on almost opposite latitudes, and therefore the asymmetry seen in the data, most prominently in inclination, is indicative of mean-field ingredients other than a simple *geocentric axial dipole*. The Réunion data are fitted much better than the Hawaiian data by the bi-Gaussian distribution with scalar variance, and we can say, therefore, that the Réunion data are relatively "isotropic" in their vectorial variance, while the Hawaiian data display an "anisotropy" in vectorial variance. Better fits to the Hawaiian data would require the introduction of covariance into the underlying Gaussian distribution functions; nonetheless this comparative analysis is enlightening. Software for fitting the 3D Gaussian distributions to paleomagnetic data can be obtained at http://geomag.usgs.gov.

Jeffrey J. Love

Bibliography

Bingham, C., 1964. Distributions on the sphere and on the projective plane. Ph.D. thesis, Yale University, New Haven.

Bingham, C., 1983. A series expansion for the angular Gaussian distribution. In Watson, G.S. (ed.) *Statistics on Spheres*. New York: John Wiley and Sons, pp. 226–231.

Creer, K.M., 1983. Computer synthesis of geomagnetic palaeosecular variations. *Nature*, **304**: 695–699.

Fisher, R.A., 1953. Dispersion on a sphere. *Proceedings of the Royal Society of London, Series A*, **217**: 295–305.

Khokhlov, A., Hulot, G., and Carlut, J., 2001. Towards a self-consistent approach to paleomagnetic field modelling. *Geophysical Journal International*, **145**: 157–171.

Love, J.J., and Constable, C.G., 2003. Gaussian statistics for palaeomagnetic vectors. *Geophysical Journal International*, **152**: 515–565.

McFadden, P.L., and McElhinny, M.W., 1982. Variations in the geomagnetic dipole 2: Statistical analysis of VDMs for the past 5 million years. *Journal of Geomagnetism and Geoelectricity*, **34**: 163–189.

Press, W.H., Teukolsky, S.A., Vetterling, W.T., and Flannery, B.P., 1992. *Numerical Recipes*. Cambridge, UK: Cambridge University Press.

Stuart, A., Ord, K., and Arnold, S., 1999. *Kendall's Advanced Theory of Statistics*, Volume 2A, Classical Inference and the Linear Model. London: Arnold.

Cross-references

Bingham Statistics
Dipole Moment Variation
Fisher Statistics
Geocentric Axial Dipole Hypothesis
Magnetization, Remanent
Nondipole Field
Paleomagnetic Secular Variation
Paleomagnetism

STORMS AND SUBSTORMS, MAGNETIC

Introduction

Magnetic storms were first defined in the mid-19th century when large variations in the horizontal intensity of the magnetic field were measured at a variety of locations on the surface of the Earth. Although much work was subsequently initiated on this topic, it has been the space age that has led to a more detailed understanding of the phenomena involved in magnetic storms. It is now established that magnetic storms are an element of the interaction between processes that occur on the Sun, the coupling between the solar wind and the Earth's magnetosphere, and the subsequent energization of particles within the Earth's magnetosphere (Tsurutani *et al.*, 1997). Furthermore, it is clear that storms occur as a result of abnormal conditions at the Sun and in the solar wind. As a result, the effects during magnetic storms on the space environment surrounding the Earth can be serious in terms of human activities in space and on the ground.

The term "magnetospheric substorm" is used to describe a range of associated phenomena that occur in the magnetosphere. The word substorm was initially used in the early part of the 1960s to portray rapid, repeatable variations of the polar magnetic field during magnetic storms. In order to characterize the overall phenomenology of auroral disturbances, the term was modified to the auroral substorm (Akasofu, 1964), before becoming more widely incorporated as the magnetospheric substorm in the 1970s. Substorms can be considered to be part of the normal solar wind magnetosphere interaction.

Magnetic storms

Magnetic storms are most easily observed in equatorial or low-latitude magnetograms as a depression in the magnetic field. In order to produce a simple means of identifying a magnetic storm, the Dst index was derived, which is based on the change in the horizontal component of the magnetic field measured at a number of low-latitude stations, which are separated in longitude. A schematic representation of the Dst index for an individual storm is given in Figure S48, which illustrates the three specific intervals or phases of a storm, each of which has different timescales.

The initial phase, often but not always, follows a rapid enhancement in the Dst index, over a timescale of a few minutes, which is referred to as a storm sudden commencement (SSC) and is caused by a rapid increase in the solar wind pressure incident on the Earth's magnetosphere.

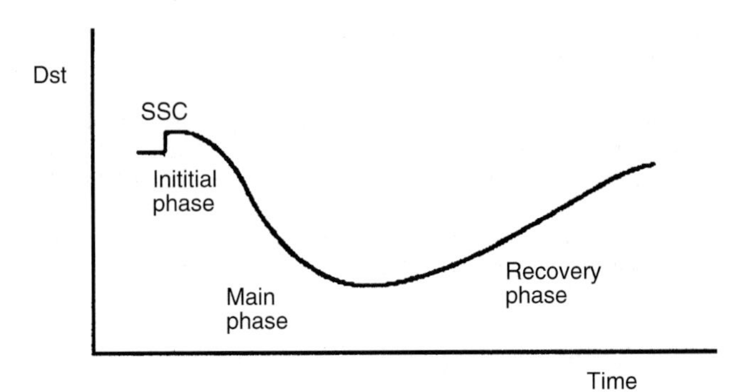

Figure S48 A schematic representation of the Dst index during a typical magnetic storm.

The initial phase lasts up to a few hours and is the period during which the Dst remains at a high level following the SSC. The second phase is the main phase of the storm, which is the time frame over which the Dst index reduces to its minimum value and represents the period of the main energy transfer and storage into the Earth's magnetosphere. The main phase of the storm typically lasts for 5–10 h. The final phase of the storm is the recovery phase, during which the Dst returns to its prestorm values and this phase lasts for about 10–15 h. Thus, the timescale for the whole storm is typically 24–30 h. Minimum values of Dst which can be reached are typically several hundred nanoteslas.

The Dst index, however, is just a means to identify the storm. The cause of these changes in the magnetic field is in the magnetosphere and is referred to as the ring current. This current flows in a westward direction in the equatorial plane of the magnetosphere and during storms ultimately surrounds the whole Earth at a distance between about 2 and $7R_e$ from the center of the Earth (where R_e is the radius of the Earth and is approximately 6380 km). The ring current is carried by ions in the energy range 20–200 keV and the main phase of the storm is the time during which the ring current increases in intensity while during the recovery phase the ring current decays.

The initiators of magnetic storms are solar phenomena such as coronal mass ejections, solar flares, and coronal holes. These phenomena result in unusual conditions in the solar wind, for example, high values of the solar wind density and velocity (and hence pressure) and the interplanetary magnetic field (IMF). The orientation of the IMF is central to the energy transfer from the solar wind to the magnetosphere such that when the IMF has a southward component, the energy transfer is most efficient and occurs through the process of magnetic reconnection between the IMF and the geomagnetic field.

Magnetospheric substorms

A magnetospheric substorm comprises three separate phases: the growth, expansion, and recovery phases (e.g., Baker *et al.*, 1996). The growth phase is the interval during which energy transfer from the solar wind is the main process and this results in storage of that energy in the tail of the magnetosphere together with a small amount of dissipation in the ionosphere. The expansion phase is the interval during which the energy that has been stored in the tail is explosively released, with some energy appearing in the ionosphere, some in the ring current, and some being lost to the solar wind. Finally, the recovery phase is an interval when the magnetosphere-ionosphere system relaxes back to a quiescent state. The overall timescale of a magnetospheric substorm is typically 2–4 h.

The growth phase of the magnetospheric substorm begins with the onset of a period of southward IMF at the dayside magnetopause, which leads to magnetic reconnection between the IMF and the geomagnetic field, thereby creating open magnetic flux, which is connected at one end to the Earth and the other to the solar wind plasma. This newly opened magnetic flux is transported by the motion of the solar wind into the magnetotail where it is added to the lobes. In this way, magnetic energy, $B^2/2\mu_0$ per volume, is stored in the tail as the lobe magnetic field intensity increases. A consequence of the addition of this magnetic flux to the tail lobes is the reorientation of the magnetic field in the near-Earth tail, for example at geostationary orbit, where the field becomes less dipolar-like and more like the stretched mid- and far-tail field. This reorientation results from the formation and subsequent intensification of a thin current sheet in the inner magnetosphere termed "crosstail current."

The expansion phase of the magnetospheric substorm is normally accepted as starting with the brightening of the equatorward most auroral arc in the nightside ionosphere. This brightening begins in a localized region near midnight and propagates along the length of the arc both westward and eastward at phase velocities of up to 10 km s^{-1}. This is then followed by a poleward expansion of the auroral luminosity at speeds of typically a few hundred m s^{-1}. This is known as the auroral breakup because of the rapid deformation of the original quiet arc. Accompanying the auroral breakup is a host of phenomena of which the most important with regard to expansion phase onset identification are the Pi2 pulsation and the westward electrojet, both observed in the magnetic field. Pi2 pulsations are MHD waves with periods between 40 and 200 s, which are impulsive and last for a few cycles. The westward electrojet is the current that flows in the ionosphere primarily in the enhanced conductivity region associated with the auroral activity. This electrojet is sensed on the ground by magnetometers and has characteristic spatial variations in the horizontal and vertical intensities of the magnetic field. In the near-Earth tail, the enhanced crosstail current decreases, resulting in a change in the orientation of the field back to more dipolar like. The cause of this change in the crosstail current remains uncertain and is the source of much controversy. The energy stored in the tail is released in a bursty fashion, with the trigger for the release as yet unknown, and there may be more than one burst of energy release, resulting in secondary substorm intensifications after the expansion phase onset.

The recovery phase is the phase during which the system returns to a quiescent state. Overall, this results in the magnetic field becoming less disturbed and the auroral activity declining. Despite this, however, there are still dynamic variations in the field during the recovery phase, which illustrate that the process is by no means simple.

Outstanding questions

Regarding magnetic storms, the outstanding questions concern the growth and decay of the ring current. The mechanisms by which particles can be energized to increase the ring current intensity remain unclear. Furthermore, once the ring current is formed, the mechanisms, which lead to the loss of particles from the ring current and the consequent decay of the ring current, are still debated. It is likely that waves over a range of frequencies play central roles in both these processes. Debate also continues over the solar cause of magnetic storms in terms of which phenomenon is the most important.

The outstanding questions regarding magnetospheric substorms center on the expansion phase onset. Several models exist for this and can be summarized thus. Reconnection at a near-Earth neutral line in the tail (approximately $20R_e$ downstream from Earth) leads to a disruption of the enhanced tail current formed during the growth phase. This disruption propagates earthward, resulting in signatures in the ionosphere and at geosynchronous orbit (Baker *et al.*, 1996). Alternatively, the current disruption could be caused nearer to the Earth (approximately $6–10R_e$) as a result of some instability, which is responsible for the signatures in the ionosphere and geosynchronous orbit, and then propagates tailward, causing the reconnection at a near-Earth neutral line (Lui, 1996). In addition, there is a question over what triggers the expansion phase onset (Lyons, 1996). It is unclear if a threshold has to be reached in the state of the tail or whether external triggers are sufficient to drive the process. Finally, the linkage between storms and substorms is more complex than previously thought and requires further attention (Kamide *et al.*, 1998). These and many other questions concerning magnetic storms and substorms will be answered over the next decade of space research.

Mark Lester

Bibliography

Akasofu, S.I., 1964. The development of the auroral substorm. *Planetary and Space Science*, **12**: 273.

Baker, D.N., Pulkkinen, T.I., Angelopoulos, V., Baumjohann, W., and McPherron, R.L., 1996. Neutral line model of substorms: past results and present view. *Journal of Geophysical Research*, **101**: 12975.

Kamide, Y., *et al.*, 1998. Current understanding of magnetic storms: storm-substorm relationships. *Journal of Geophysical Research*, **103**: 17705.

Lui, A.T.Y., 1996. Current disruption in the earth's magnetosphere: observations and models. *Journal of Geophysical Research*, **101**: 13067.

Lyons, L.R., 1996. Substorms: fundamental observational features, distinction from other disturbances, and external triggering. *Journal of Geophysical Research*, **101**: 13011.

Tsurutani, B.T., Gonzalez, W.D, Kamide, Y., and Arballo, J.K., 1997. *Magnetic Storms*. Washington, DC: American Geophysical Union.

Cross-references

Archeomagnetism
Magnetization, Natural Remanent (NRM)
Paleomagnetism
Rock Magnetism

SUPERCHRONS, CHANGES IN REVERSAL FREQUENCY*

The geomagnetic reversal timescale appears to contain intervals of four distinct durations. They are called cryptochrons, subchrons, chrons, and superchrons, with durations of the order of thousands of years, tens of thousands, hundreds of thousands, and tens of millions. Cryptochrons are observed as excursions and are interpreted as aborted reversals; empirically they are said to differ from secular variation if the geomagnetic pole moves more than 45° from the geographic pole. Theoretically an excursion differs from a pair of reversals if the intermediate state fails to completely reverse, that is change from $\boldsymbol{B}$ to $-\boldsymbol{B}$ (Gubbins, 1999). There is probably no theoretical distinction between subchrons and chrons. Superchrons are distinct periods of one polarity lasting longer than any natural timescale for the geodynamo. Since this implies a cause related to some longer term evolution of the Earth, such as mantle convection or inner core growth, any theory aimed at explaining superchron behavior must also address long-term changes in the frequency of reversals, unless it invokes some sudden catastrophic change in the state of the Earth's core.

Theories for long-term changes in reversal rate can be listed under five categories:

1. Natural nonlinear behavior of the geodynamo
2. Growth and changes of the inner core
3. Changes in lower mantle electrical conductivity
4. Changes in total heat flow from the core
5. Changes in the geographical pattern of heat flow from the core

All theories are in a very rudimentary state as it has only recently become possible to test speculations against dynamo theory.

There are two clear periods in the past when the field remained the same polarity for a considerable length of time: from about 119 to 84 million years ago (Ma) the field was normal (the *Cretaceous Normal Superchron—CNS*) while from about 312 to 262 Ma the field was reversed (the *Permo-Carboniferous Superchron—PCRS*). The CNS is established beyond doubt because it is recorded on the ocean floors; any but the shortest of reversed intervals would have been discovered by now. The PCRS is older than the oceans and therefore normal intervals may remain to be discovered.

312 Myr still represents a small fraction of the age of the Earth and of the geomagnetic field. A number of authors have suggested the possible existence of another superchron during the Ordovician (see, e.g., Gallet *et al.* (1992)). Using a global paleomagnetic database, Johnson *et al.* (1995) derived a model of long-term reversal behavior for the last 570 Myr. In addition to the two established superchrons (the CNS and PCRS), they found a long period of very low reversal rate in the Ordovician (502–470 Ma), which they suggested may be a previously unidentified superchron.

Johnson *et al.* (1995) also found a short period of anomalously low reversal rate around the Jurassic-Triassic boundary (212–185 Ma). However, Kent and Olsen (1999) in their analysis of cores from the Newark rift basin in North America found no evidence for an interval of low reversal rate around this time. They suspect Johnson *et al.*'s results may be an artifact of sampling bias. Algeo (1996) carried out a study of geomagnetic bias patterns through the Phanerozoic. In addition to the CNS and PCRS, he found four other first order polarity features—an interval of reversed polarity bias in the middle Cambrian-middle Ordovician and three intervals of normal polarity bias in the late Ordovician-late Silurian, in the early Jurassic, and in the middle Jurassic.

Gallet and Pavlov (1996), in an analysis of data from the Moyera river section of northwestern Siberia, found very few reversals, the field being dominantly reversed from the upper Cambrian to the middle Ordovician, and normal from the upper Ordovician to the lower Silurian. However only the Avenig (140 samples) has an acceptable time resolution (~100 Myr???) and they suggested the possibility of a single reversed interval lasting at least 15 Myr during this time. This is in agreement with the result from a composite sequence for the same period by Trench *et al.* (1991), but not with those of Piper (1978). In a later paper, Pavlov and Gallet (1998), using data from the Kulumbe river section in the same area suggested a 25–30 Ma-long period during the lower and middle Ordovician, dominated by reversed polarity with very few reversals. Algeo (1996) also found strong polarity bias during the first half of the Ordovician. Pavlov and Gallet (2001) later returned to the Kulumbe river section, sampling older rocks, and found a sequence of 28 magnetic polarity intervals, eight of which were defined by only one sample. Since the section had a duration of ~5 Myr, they proposed that the magnetic reversal frequency was high (~4–6 reversals per Myr) during part of the middle Cambrian, which is amongst the highest rates known within the Phanerozoic. This reversal frequency may also have existed during the lower Cambrian. They further suggested a drastic decrease in the reversal frequency between the lower-middle Cambrian and the end of the Tremodoc (Ordovician) when a superchron may have occurred.

The reversal record suggests a gradual increase in reversal frequency since the CNS, perhaps a result of slow changes on the timescale of mantle convection. McFadden and Merrill (1984) found an approximately linear trend in the mean rate of reversals, increasing from zero at 84 Ma at the end of the CNS to over four per Myr in recent times. Visual inspection of the reversal sequence suggests that, superimposed on this linear trend, there are oscillations with a period of ~30 Myr There has been much controversy over the reality of a 30 Ma periodicity (or 15 Myr proposed by some authors) but on the whole the claims of periodicities are not convincing (see Jacobs (1994) for a review of the debate). Merrill and McFadden (1990) later showed the pattern of reversals to have a linear decrease in reversal frequency from about 165 to 119 Ma, the commencement of the CNS. In a further paper, McFadden and Merrill (1997) showed that the reversal rate prior to the CNS is larger than the rate after the CNS.

Gallet and Courtillot (1995) adopted a different approach, plotting the successive lengths of polarity intervals as a function of their order of occurrence in the sequence. The longer duration polarity intervals become shorter, fewer and further part to indicate a change in behavior at 15–30 Ma. Flat "white noise" characterizes the behavior from this time to the present, while a quasiperiodic structure in magnetic interval lengths is observed between 85 and 25 Ma. Such behavior is similar to that proposed by Dubois and Pambrun (1990)—low-dimensional deterministic chaos from 85–30 Ma and a random process from 25 Ma to the present.

Gallet and Hulot (1997) offered an alternative interpretation to that of McFadden and Merrill (1997) for the evolution of the geomagnetic

*This article was left by Jack Jacobs in manuscript form and completed by the editor D. Gubbins, who takes responsibility for all errors and omissions in the final article.

field over the last 160 Myr. They plotted the duration of individual polarity intervals as a function of their order of occurrence (as Gallet and Courtillot (1995) had done) and suggested that the geomagnetic polarity timescale could be split into three segments with different characteristics. For the time period 160–130 Ma, they proposed that the duration of magnetic polarity intervals was constant, i.e., during this period, the reversal process, unlike that of McFadden and Merrill was stationary. They suggested that at $\sim$130 Ma, a change occurred leading to a discontinuous decrease in reversal frequency up to the CNS. The CNS is followed by a period of gradually increasing reversal frequency until 25 Ma, when the reversal process again became steady.

McFadden and Merrill (2000) derived a new statistic to distinguish between the two extreme alternatives of themselves (1997) and of Gallet and Hulot (1997). They showed that the data between 160 and 130 Ma require a linear trend towards longer polarity intervals, this trend continuing to the beginning of the CNS, i.e., a discontinuity at 130 Ma (as proposed by Gallet and Hulot) is neither required nor supported by the data. Moreover the trend towards shorter polarity intervals from 85 Ma continues to at least 12 Ma, there being no need for a change in the reversal process at 25 Ma as demanded by Gallet and Hulot (1997). Constable (2000) obtained a new estimate for the reversal rate as a function of time and was able to obtain confidence bounds on the temporal probability density function for geomagnetic reversals. For the period 0–158 Ma, she found that the derivative of the rate functions of reversals must have changed sign at least once. The timing of this sign change is constrained to be between 152 and 22 Ma at the 95% confidence level. Unfortunately she was unable to accept or reject either of the models of McFadden and Merrill and Gallet and Hulot.

Reversal frequency estimates for earlier times are hampered by the lack of precise dating. There are very few records showing a sequence of several reversals in rocks older than 1 Ga. Opdyke and Channel (1996) compiled a histogram of the lengths of polarity intervals from 330 Ma to the present. Apart from the CNS and PCRS, the majority of polarity intervals lie in the time range from 0.1 to 1 Myr, and there are very few with durations longer than 2 Myr.

A related question is the relationship (if any) between the strength of the Earth's magnetic field and the frequency of reversals—in particular with superchrons. Fuller and Weeks (1992) have given a summary of the two opposing views—either the field is strong, or it is weak during a superchron. Pick and Tauxe (1993) found extremely low average intensity for the CNS, and low values of the geomagnetic field during the PCRS were found by Prévot and Perrin (1992) from a list of paleointensity data determined from magmatic rocks up to 3.5 Ga, and by Harcombe-Smee *et al.* (1996) from lavas in S.W.

The latest contribution to this controversial question has been given by Tarduno *et al.* (2001), who found high geomagnetic intensities during the mid-Cretaceous. They used single crystals of plagioclase feldspar from eight independent lava flows that erupted at different times within 0.1–1 Ma from basalts of the Raymahal Traps of India (113–116 Ma). The average value from all the flows and the 56 crystals with reproducible data are taken to represent the true dipole field during the CNS. The averaging process removes possible sources of error in the magnitude of the virtual dipole moment from the highly variable (in space and time) nondipole components of the field, which vary on a timescale shorter than 10 kyr and hence cancel out unless data are averaged over 0.1–1 Myr. Their value of the mean intensity of the field is significantly higher than other values for the past 180 Myr, and the times of the mean paleointensity for the mean Cenozoic and early Cretaceous-late Jurassic when reversals were frequent. The question is still open.

Turning to theory, we must first ask whether the geodynamo could move spontaneously between reversing and nonreversing states, whether in fact no special external changes are required at all to produce superchrons. The longest natural mescale is the magnetic diffusion time (see *Geodynamo, dimensional analysis and timescales*), 25 kyr. If this Ordovician low reversal rate is substantiated, the 200 Ma spacing between the CNS, PCRS, and the Ordovician superchron would be a fundamental time constant for long-term geomagnetic field behavior, simulation for anything like this time, but studies of simpler systems suggest it is highly unlikely (Jones and Wallace, 1992). Barring some dramatic new findings, we can be fairly confident that some change is needed external to the liquid core.

The inner core is known to stabilize numerical dynamo simulations (Hollerbach and Jones, 1993). It is growing slowly as the Earth cools (see *Geodynamo, energy sources*), and was therefore smaller in the past. This may have made superchrons less likely in the distant past (Gubbins, 1999). Some dynamo calculations have more stable solutions without an inner core (Morrison and Fearn, 2000) because the two hemispheres are better connected, but this may be a special feature of this weakly driven model. For the inner core to account for the changes since the Ordovician superchron it would have to change size very significantly in 100 Myr, and repeatedly, with implied warming of the entire core. There is as yet no evidence for this, so the theory must remain a long shot.

This leaves changes in the lower mantle, which controls the boundary conditions for the geodynamo. The electrical conductivity of the lower mantle could have an effect if it were large enough, for example in substantial regions of entrained iron. This idea appears in theories of short-term core-mantle coupling (see Core-mantle coupling) and of reversal transition fields (Herrero-Bervera and Runcorn, 1997), but to affect the geodynamo seriously the highly conductive layer would have to be of 10–100 km thick, and this would affect the short term secular variation in a way that would surely have been observed by now.

This leaves thermal effects, which have received most attention to date. One idea is that a change in the style of mantle convection draws more heat from the core, as might follow for example from a major reorganization of the plate boundaries at the Earth's surface (it is interesting to note that Venus had a resurfacing event at 500 Ga and no longer has plate tectonics or magnetic field, but it may have had both in the past). Naïvely one might expect the geomagnetic field to be stronger and also to reverse in unstable fashion if the geodynamo is being driven hard, and weak and stable if the driving is weak. However, Pal and Roberts (1988) and Larson and Olson (1991) argued persuasively that the field is strong during a superchron, arguing that a thin D″ layer at the bottom of the mantle would increase heat flow from the outer core (OC) to the mantle and so create more vigorous convection in the OC. The associated heat loss would lead to a strong, stable, nonreversing field (a superchron). Loper and McCartney (1986), McFadden and Merrill (1986) and Courtillot and Besse (1987), on the other hand, had argued earlier against a strong magnetic field during a superchron, which they claimed represents a low energy state with infrequent instabilities corresponding to inactive periods in the D″ later when it is thick and heat transfer across the CMB relatively small. This view receives some support from dynamo studies (Jones and Wallace, 1992), which now include some sophisticated 3D numerical simulations (Kutzner and Christensen, 2002), although their reversing solutions do not resemble the geomagnetic field very much. Further support for low field strength during a superchron has been given by Hide (2000), who carried out a mathematical investigation of nonlinear processes in self-exciting dynamos. In his models the maximum strength of the magnetic field could be systematically higher when reversals are frequently than when they are infrequent. He suggested that a superchron is associated with a long quiescent period when convection in the lower mantle is comparatively feeble.

The second idea invokes changes not in the average heat flux from the core but the lateral variations in heat flux into the lower mantle (see *Core-mantle coupling, thermal*). These lateral variations almost certainly affect convection in the core, and there have been a number of geodynamo simulations that incorporate inhomogeneous thermal boundary conditions. A change in this pattern, which at present has high heat flow around the Pacific and low below Africa and the central Pacific, may change the style of convection from one that allows reversals to one that does not. There is some evidence for this in recent simulations (Glatzmaier *et al.*, 1999). The theory has the appeal of a natural process of evolution of the Earth with the right timescale. This study also indicates correlations between the frequency of reversals, the direction, and the strength of the magnetic field. The most stable geodynamo (i.e., nonreversing, low SV) is the most efficient with the highest dipole moment. The near future will surely bring some exciting new insights into this problem.

Jack A. Jacobs

Bibliography

Algeo, T.J., 1996. Geomagnetic polarity bias patterns through the Phanerozoic. *Journal of Geophysical Research-Solid Earth*, **101**: 2785–2814.

Constable, C., 2000. On rates of occurrence of geomagnetic reversals. *Physics of the Earth and Planetary Interiors*, **118**: 181–193.

Courtillot, V., and Besse, J., 1987. Magnetic-field reversals, polar wander, and core-mantle coupling. *Science*, **237**: 1140–1147.

Dubois, J., and Pambrun, C., 1990. Distribution of reversals of the earths magnetic-field from 165 Ma to the present time—research of an attractor of the reversal process. *Comptes Rendus de l'Academie des sciences, Série II*, **311**: 643–650.

Fuller, M., and Weeks, R., 1992. Geomagnetism—superplumes and superchrons. *Nature*, **356**: 16–17.

Gallet, Y., and Courtillot, V., 1995. Geomagnetic reversal behavior since 100 Ma. *Physics of the Earth and Planetary Interiors*, **92**: 235–244.

Gallet, Y., and Hulot, G., 1997. Stationary and nonstationary behaviour within the geomagnetic polarity time scale. *Geophysical Research Letters*, **24**: 1875–1878.

Gallet, Y., and Pavlov, V., 1996. Magnetostratigraphy of the Moyero river section (north-western siberia): Constraints on geomagnetic reversal frequency during the early palaeozoic. *Geophysical Journal International*, **125**: 95–105.

Gallet, Y., Besse, J., Krystyn, L., Marcoux, J., and Theveniaut, H., 1992. Magnetostratigraphy of the late triassic bolucektasi tepe section (southwestern turkey)—implications for changes in magnetic reversal frequency. *Physics of the Earth and Planetary Interiors*, **73**: 85–108.

Glatzmaier, G.A., Coe, R.S., Hongre, L., and Roberts, P.H., 1999. The role of the Earth's mantle in controlling the frequency of geomagnetic reversals. *Nature*, **401**: 885–890.

Gubbins, D., 1999. The distinction between geomagnetic excursions and reversals. *Geophysical Journal International*, **137**: F1–F3.

Harcombe-Smee, B.J., Piper, J.D.A., Rolph, T.C., and Thomas, D.N., 1996. A palaeomagnetic and palaeointensity study of the Mauchline lavas, south-west Scotland. *Physics of the Earth and Planetary Interiors*, **94**: 63–73.

Herrero-Bervera, E., and Runcorn, S.K., 1997. Transition fields during geomagnetic reversals and their geodynamic significance. *Philosophical Transactions of the Royal Society of London*, **355**: 1713–1742.

Hide, R., 2000. Generic non-linear processes in self-exciting dynamos and the long-term behaviour of the main geomagnetic field, including polarity superchrons. *Philosophical Transactions*, **358**: 943.

Hollerbach, R., and Jones, C.A. 1993. Influence of the Earth's inner core on geomagnetic fluctuations and reversals. *Nature*, **365**: 541–543.

Jacobs, J.A., 1994. *Reversals of the Earth's Magnetic Field.* Cambridge: Cambridge University Press.

Johnson, H.P., Vanpatten, D., Tivey, M., and Sager, W.W., 1995. Geomagnetic polarity reversal rate for the phanerozoic. *Geophysical Research Letters*, **22**: 231–234.

Jones, C.A., and Wallace, S.G., 1992. Periodic, chaotic and steady solutions in $\alpha\omega$ dynamos. *Geophysical and Astrophysical Fluid Dynamics*, **67**: 37–64.

Kent, D.V., and Olsen, P.E., 1999. Astronomically tuned geomagnetic polarity timescale for the Late Triassic. *Journal of Geophysical Research-Solid Earth*, **104**: 12831–12841.

Kutzner, C., and Christensen, U.R., 2002. From stable dipolar towards reversing numerical dynamos. *Physics of the Earth and Planetary Interiors*, **131**: 29–45.

Larson, R.L., and Olson, P., 1991. Mantle plumes control magnetic reversal frequency. *Earth and Planetary Science Letters*, **107**: 437–447.

Loper, D.E., and McCartney, K., 1986. Mantle plumes and the periodicity of magnetic-field reversals. *Geophysical Research Letters*, **13**: 1525–1528.

McFadden, P.L., and Merrill, R.T., 1984. Lower mantle convection and geomagnetism. *Journal of Geophysical Research*, **89**: 3354–3362.

McFadden, P.L., and Merrill, R.T., 1986. Geodynamo energy-source constraints from paleomagnetic data. *Physics of the Earth and Planetary Interiors*, **43**: 22–33.

McFadden, P.L., and Merrill, R.T., 1997. Asymmetry in the reversal rate before and after the cretaceous normal polarity superchron. *Earth and Planetary Science Letters*, **149**: 43–47.

McFadden, P.L., and Merrill, R.T., 2000. Evolution of the geomagnetic reversal rate since 160 ma: Is the process continuous? *Journal of Geophysical Research-Solid Earth*, **105**: 28455–28460.

Merrill, R.T., and McFadden, P.L., 1990. Paleomagnetism and the nature of the geodynamo. *Science*, **248**: 345–350.

Morrison, G., and Fearn, D.R., 2000. The influence of rayleigh number, azimuthal wavenumber and inner core radius on 2-1/2 d hydromagnetic dynamos. *Physics of the Earth and Planetary Interiors*, **117**: 237–258.

Opdyke, N.D., and Channel, J.E.T., 1996. *Magnetic Stratigraphy.* San Diego: Academic Press.

Pal, P.C., and Roberts, P.H., 1988. Long-term polarity stability and strength of the geomagnetic dipole. *Nature*, **331**: 702–705.

Pavlov, V., and Gallet, Y., 1998. Upper Cambrian to Middle Ordovician magnetostratigraphy from the kulumbe river section (northwestern siberia). *Physics of the Earth and Planetary Interiors*, **108**: 49–59.

Pavlov, V., and Gallet, Y., 2001. Middle Cambrian high magnetic reversal frequency (Kulumbe River section, northwestern Siberia) and reversal behaviour during the early palaeozoic. *Earth and Planetary Science Letters*, **185**: 173–183.

Pick, T., and Tauxe, L., 1993. Geomagnetic palaeointensities during the Cretaceous Normal Superchron measured using submarine basaltic glass. *Nature*, **366**: 238–242.

Piper, J.D.A., 1978. Paleomagnetic survey of (Paleozoic) Sheive inlier and Berwvn Hills, Welsh Borderlands. *Geophysical Journal of the Royal Astronomical Society*, **53**: 355–371.

Prévot, M., and Perrin, M., 1992. Intensity of the Earth's magnetic field since Precambrian from Thellier-type paleointensity data and inferences on the thermal history of the core. *Geophysical Journal International*, **108**: 613–620.

Tarduno, J.A., Cottrell, R.D., and Smirnov, A.V., 2001. High geomagnetic intensity during the mid-Cretaceous from Thellier analyses of single plagioclase crystals. *Science*, **291**: 1779–1783.

Trench, A., McKerrow, W.S., and Torsvik, T.H., 1991. Ordovician magnetostratigraphy—a correlation of global data. *Journal of the Geological Society*, **148**: 949–957.

Cross-references

Core-Mantle Boundary, Heat Flow Across
Core-Mantle Coupling, Electromagnetic
Core-Mantle Coupling, Thermal
D″ as a Boundary Layer
Dipole Moment Variation
Geodynamo

Geodynamo, Dimensional Analysis and Timescales
Geodynamo, Energy Sources
Geomagnetic Polarity Reversals, Observations
Geomagnetic Polarity Timescales
Reversals, Theory

SUSCEPTIBILITY

What is magnetic susceptibility? This is a question I use to ask students, and I have got many different answers. One even said that it is the number on the display of instrument used. It is true that one can find several definitions of this physical parameter, more or less correct. For example, Encyclopedia Britannica gives quite nice description, saying that magnetic susceptibility is "quantitative measure of the extent to which a material may be magnetized in relation to a given applied magnetic field." Let us improve this definition, saying that magnetic susceptibility is a parameter characterizing the nature and intensity of response of material to external magnetic field.

Since magnetic properties of materials are mostly studied by applying a magnetic field (**H**) and measuring the magnetization (**M**) induced in these materials, magnetic susceptibility is defined by equation relating **M** and **H**:

$$\mathbf{M} = \chi_v \mathbf{H},$$

which yields

$$\chi_v = \mathbf{M}/\mathbf{H},$$

where the subscript v denotes "volume." Thus, χ_v is a material parameter—transfer function, linking induced magnetization and applied magnetic field.

Magnetization induced by the applied magnetic field is due to the effect of the field upon moving charged particles. Magnetism is characteristic of all materials containing charged particles. Since contribution of nucleons is very small and can be neglected, magnetism can be considered as practically entirely of electronic origin. Thus, magnetic susceptibility can be used to classify materials in terms of their response to external magnetic field as *diamagnetics*, *paramagnetics*, and a group of *ferrimagnetics* (for more details, reader is referred to, e.g., Dekker (1957) or Dunlop and Özdemir (1997)).

However, the above definition of magnetic susceptibility is only valid for linear behavior of induced magnetization in dependence of the applied magnetic field. In other words, it can only be applied to diamagnetic or paramagnetic materials, or ferrimagnetics in the range of magnetic saturation. In all other cases, and these are more common in natural rocks and minerals, more general definition applies

$$\chi_v = \left.\frac{\partial \mathbf{M}}{\partial \mathbf{H}}\right|_{\mathbf{H},\mathbf{M}} \quad \text{or } \chi_v = \lim_{\Delta H \to 0} \left.\frac{\Delta \mathbf{M}}{\Delta \mathbf{H}}\right|_{\mathbf{H},\mathbf{M}}.$$

Mathematically speaking, it represents an angle between the tangent to the *magnetization curve* $M(\mathbf{H})$ at given **H** and the H axis, and in physics literature a term reversible susceptibility is sometimes used, or differential for $\mathbf{H} = \mathbf{M} = 0$. From this it is clear, that magnetic susceptibility depends not only on the material and the applied field, but also on the way it is measured.

Diamagnetic susceptibility rises from the distortion of electronic orbit due to applied magnetic field, forcing the electrons to additional precession motion. It is very small, negative, and independent of temperature and magnetic field

$$\chi_v = -\frac{z\mu_0 e^2 \overline{a^2} n}{6m_0},$$

where z is the number of electrons in atom, n is the number of atoms in unit volume, $\overline{a^2}$ is the median of square of distance of electrons from the nucleus. The magnitude of diamagnetic susceptibility is very small, in the range of 10^{-6} cgs Most organic compounds are diamagnetic only. Among minerals, diamagnetism is expressed in quartz, plagioclase, calcite, and apatite.

In paramagnetic substances, the constituent atoms have unpaired electrons, resulting in a net magnetic moment. At room temperature, the individual atomic moments are randomly oriented. If an external magnetic field is applied, they will tend to become aligned in a direction parallel to the external field. The resulting induced magnetization is called paramagnetic, and the corresponding susceptibility is defined by Curie law (first empirically derived by P. Curie in 1895):

$$\chi_v = \frac{C}{T},$$

where $C = \frac{M_\infty^2}{3n\mu_0 k}$ and $M_\infty = nM$, k being the Boltzman constant.

The variation of the inverse susceptibility with temperature is linear, with slope $1/C$, which is related to the number of unpaired electrons in the substance (n). The magnitude of paramagnetic susceptibility is in the range of 10^{-3}–10^{-6} cgs and is temperature dependent (Figure S49).

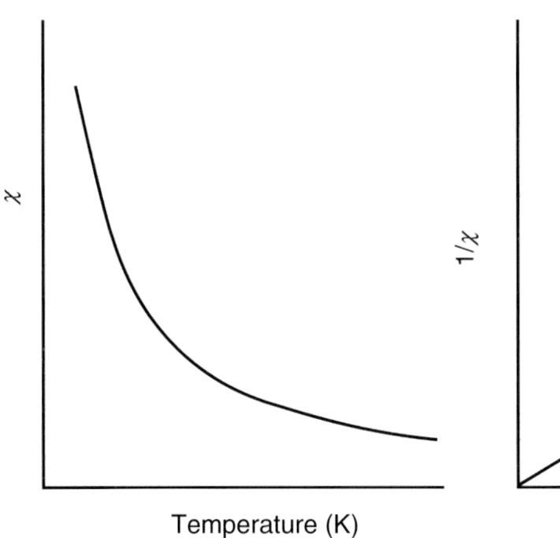

Figure S49 Temperature behavior of paramagnetic susceptibility.

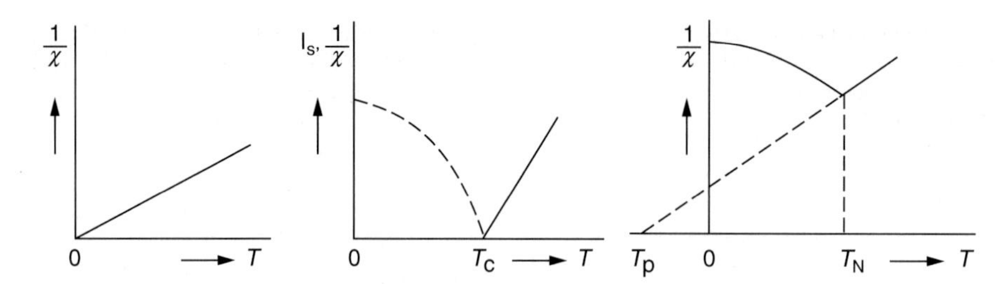

Figure S50 Temperature dependence of inverse susceptibility for paramagnetics (left), ferrimagnetics (middle) and antiferromagnetics (right).

In nature, paramagnetic minerals may include clay minerals (chlorite, smectite, and glauconite), iron and manganese carbonates (siderite, rhodochrosite), ferromagnesian silicates (olivine, amphiboles, pyroxenes, etc.), as well as a variety of ferric-oxyhydroxide mineraloids.

In paramagnetic substances, individual atomic magnetic moments do not interact with each other. However, in many materials, interactions (of quantum-mechanic nature) between atomic moments result in spontaneous parallel (or antiparallel) alignment of atomic magnetic moments in the absence of external magnetic field, typical for a group of ferromagnetic, ferrimagnetic, and antiferromagnetic materials (all called herein as ferrimagnetics). In these materials, Curie's law is not obeyed and magnetization curve $M(\mathbf{H})$ is not linear, but shows *hysteresis*, instead. In this case, magnetic susceptibility cannot be expressed on the basis of classical theory, as in the case of dia- or paramagnetism. However, above the transition *Curie* (*Néel*) *temperature*, material behaves paramagnetically, and the inverse value of magnetic susceptibility follows a linear line defined by Curie-Weiss law (Figure S50):

$$\chi_{\mathrm{v}} = \frac{C}{(T - T_P)},$$

where T_P is Weiss constant, often called paramagnetic Curie temperature. Note that mathematically T_P can be <0 K, which is physically impossible, but typical for antiferromagnetics. Typical ferrimagnetics are, for instance, magnetite and maghemite, while hematite is antiferromagnetic. There are no true ferromagnetic minerals in nature.

Magnetic response of ferrimagnetics becomes stronger as the number of unpaired electrons increases. Therefore, it is customary to measure magnetic susceptibility of a substance as a function of its weight, yielding mass-specific susceptibility. In order to distinguish the volume and mass-specific susceptibility, the former one is commonly denoted with κ (unlike most physics literature), while χ is used for the latter one:

$$\chi = \frac{\kappa}{\rho},$$

where κ replaced χ_{v} and ρ is density.

Mass-specific susceptibility is related to molar susceptibility χ_{M} by

$$\chi_{\mathrm{M}} = \frac{\chi W_{\mathrm{M}}}{Z},$$

where W_{M} is molar weight of the substance and Z is the number of moles. After some more calculations and simplifications, one can arrive at an expression for the effective magnetic moment of an ion:

$$\mu_{\mathrm{eff}} = 2.828\left[\chi_{\mathrm{M}}(T - T_P)\right]^{1/2}.$$

Regarding the fact that magnetization curve of ferrimagnetics is not linear, experimental determination of magnetic susceptibility depends also on the intensity of magnetic field (or point of the $M(\mathbf{H})$ curve, at which it is measured), and on the way how it is measured. Therefore, in physics literature one can find terms initial, reversible, and differential magnetic susceptibility (Figure S51).

In terms of the applied magnetic field, frequency of the AC field has significant effect in case of presence of ultrafine super-paramagnetic

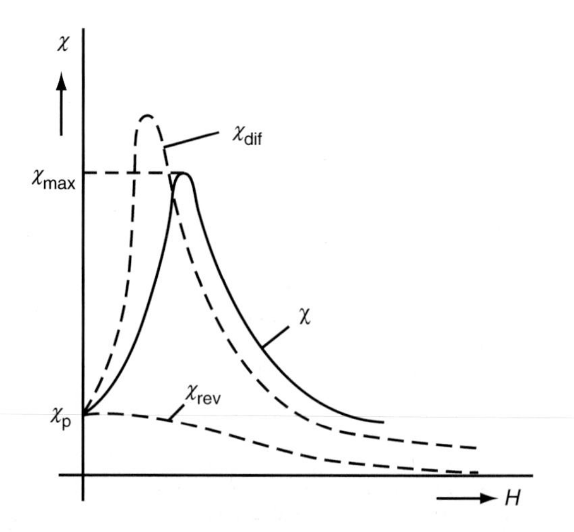

Figure S51 Reversible and differential susceptibility as a function of the magnetic field at which the slope of magnetization curve is determined. χ_P stands for initial susceptibility.

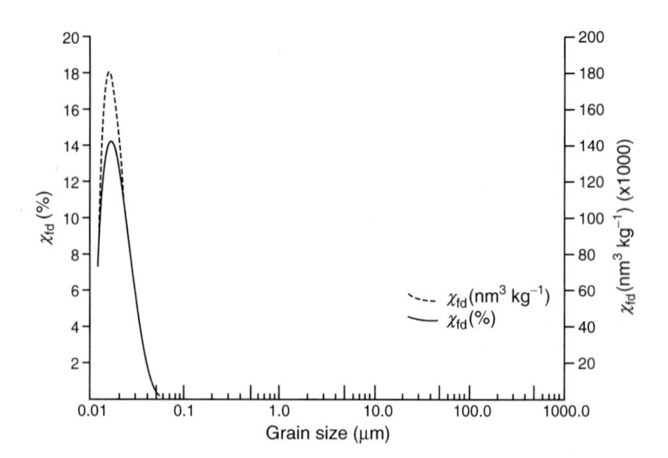

Figure S52 Schematic dependence of frequency-dependent magnetic susceptibility on grain size (after Dearing, 1994).

particles. At higher frequencies, energy of thermal excitations overcomes the alignment effect of the applied field and, consequently, magnetic moments of such particles are not aligned parallel to the applied field and do not contribute to magnetic susceptibility. Therefore, magnetic susceptibility measured at higher frequencies is always equal or lower than that measured at lower frequencies. Difference, expressed as percentage, is called frequency-dependent magnetic susceptibility (Figure S52),

$$\chi_{FD} = \frac{\chi_{LF} - \chi_{HF}}{\chi_{LF}}.$$

While χ_{FD} depends strongly on grain size (e.g., Eyre, 1997; Dearing *et al.*, 1996), and is used as indicator of significance of SP particles, magnetic susceptibility itself depends mostly on mineralogy, and for a given mineral on its concentration.

Eduard Petrovsky

Bibliography

Dearing, J.A., 1994. *Environmental Magnetic Susceptibility—Using the MS2 Bartington System*. Chi Publishing, Kenilworth, UK.

Dearing, J.A., Dann, R.J.L., Hay, K., Lees, J.A., Loveland, P.J., and O'Grady, K., 1996. Frequency-dependent susceptibility measurements of environmental materials. *Geophysical Journal International*, **124**: 228–240.

Dekker, A.J., 1957. *Solid State Physics*. Englewood Cliffs, NJ: Prentice-Hall.

Dunlop, D., and Özdemir, Ö., 1997: *Rock Magnetism—Fundamentals and Frontiers*. Cambridge: Cambridge University Press.

Eyre, J.K., 1997. Frequency dependence of magnetic susceptibility for populations of single-domain grains. *Geophysical Journal International*, **129**: 209–211.

SUSCEPTIBILITY, MEASUREMENTS OF SOLIDS

Introduction

Magnetism is a property characteristic of all materials that contain electrically charged particles. A moving electrical charge (i.e., electric current) gives rise to a magnetic field in a material. The contribution of nucleons to the magnetic field in substances is small and requires special instrumentation to observe. For the purpose of this laboratory the effect from nucleons is neglected, thus magnetism may be considered to be entirely of electronic origin. In an atom, the magnetic field is due to the coupled orbital and spin magnetic moments associated with the motion of electrons. The orbital magnetic moment is due to the motion of electrons around the nucleus whereas the spin magnetic moment is due to the precession of the electrons about their own axes. The resultant of the orbital and spin magnetic moments of the constituent atoms of a material gives rise to the observed magnetic properties.

Knowledge about the magnetic properties of materials is an important aspect in the study and characterization of substances. Unlike many others, commonly employed analytical methods, magnetic property characterization is a nondestructive technique.

Theory

Often magnetic properties of materials are studied by applying a magnetic field and measuring the induced magnetization in these materials. The magnetic induction, B, that a substance experiences when placed in an applied external magnetic field, H, is given by the expression

$$B = H + 4\pi M, \qquad \text{(Eq. 1)}$$

where M is the magnetic moment of the compound per unit volume or the Magnetization. The volume susceptibility of the compound, χ_v, is defined as the ratio of the induced magnetic moment per unit volume to the applied field:

$$\chi_v = M/H. \qquad \text{(Eq. 2)}$$

The two most important responses observed are characterized as diamagnetic and paramagnetic moment (Figure S53). As shown in this figure, the applied magnetic field remains unaffected in a vacuum but is attracted by a paramagnetic, and repelled by an ideal diamagnetic media.

In diamagnetic materials, all the electrons in the atoms are paired (e.g., $H_{2(g)}$, $NaCl_{(s)}$) and the resultant magnetic moment is zero due to the absence of unpaired electrons. The external magnetic field induces a current whose associated magnetic field is directed opposite to the applied field. The induced magnetic moment associated with this field is called a diamagnetic moment and is negative relative to the applied magnetic field and independent of temperature. The magnitude of the diamagnetic susceptibility is usually small, in the range of 10^{-6} cgs units. All inert gases and most organic compounds are examples of diamagnetic materials.

In paramagnetic materials, the constituent atoms have unpaired electrons resulting in a net magnetic moment. Generally at room temperature, the individual magnetic moments are randomly oriented as shown in Figure S54 above. If an external magnetic field is applied to these randomly oriented intrinsic magnetic moments, they will tend to become aligned in a direction parallel to the external magnetic field.

The magnetic moment induced in such materials by the application of an external magnetic field is called a paramagnetic moment. The magnitude of the susceptibility of paramagnetic substances is in the range of 10^{-3}–10^{-6} cgs units and is positive. Typical paramagnetic compounds include gaseous compounds like molecular oxygen (O_2) and nitric oxide (NO), vapors of alkali metals and certain salts of transition and rare-earth metals.

The alignment of the magnetic moments of a paramagnetic substance under an applied magnetic field is opposed by the thermal motion of the magnetic ions which tends to randomize the moments. Hence the observed paramagnetic susceptibility of a substance increases with decreasing temperature since the effect of thermal

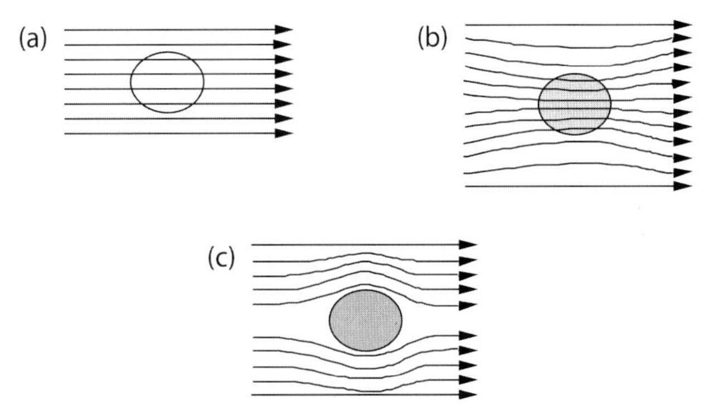

Figure S53 Behavior of applied magnetic field in different media: (a) vacuum, (b) paramagnet, and (c) ideal diamagnet.

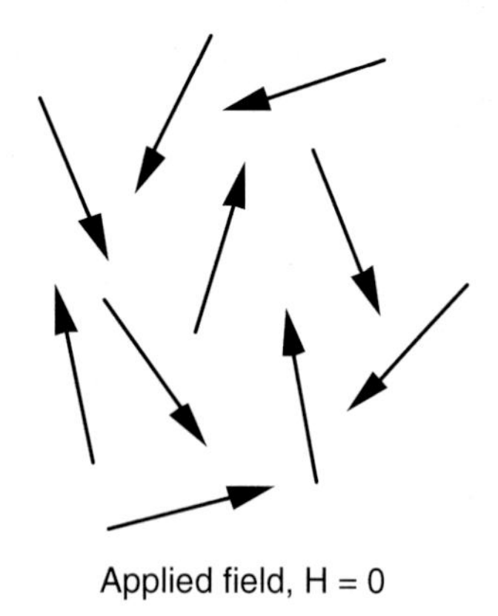

Figure S54 Random orientation of magnetic moments in paramagnetic material.

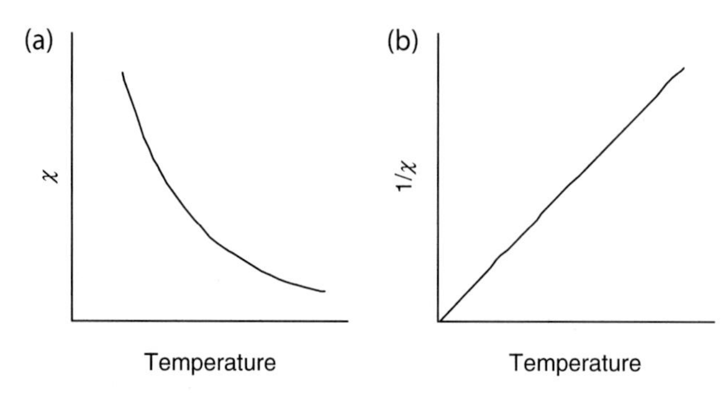

Figure S55 (a) Magnetic susceptibility and (b) inverse magnetic susceptibility dependence on temperature of a paramagnetic material.

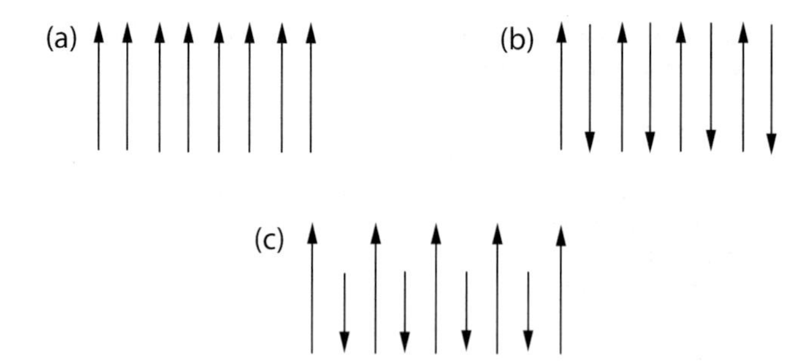

Figure S56 Types of magnetic ordering: (a) ferromagnetic, (b) antiferromagnetic, and (c) ferrimagnetic.

motion is minimized at lower temperatures (Figure S55). According to Curie's Law,

$$\chi_M = \frac{C}{T} \qquad \text{(Eq. 3)}$$

the susceptibility of a paramagnetic substance is inversely proportional to temperature where C is the Curie constant. The variation of the inverse magnetic susceptibility (χ_M^{-1}) with temperature is linear. The slope of this line is equal to $1/C$. The value of C is related to the number of unpaired electrons present in the compound by the following expression

$$(8C)^{1/2} = g[S(S+1)]^{1/2} = [n(n+2)]^{1/2}, \qquad \text{(Eq. 4)}$$

where Lande's factor $g \approx 2.0$ for an electron or ion with no orbital contribution to the magnetic moment and S is the resultant spin quantum number (it is the total spin angular momentum of all the unpaired electrons in the system), and n is the number of unpaired electrons in the system (i.e., $S = n/2$). Equation (4) is generally valid for compounds of most of the first row transition metals, where the orbital contribution to the magnetic moment is completely quenched and only the spin contribution, S need be considered. This will be discussed in more detail subsequently.

Figure S55 shows the plot of χ versus T and $1/\chi$ versus T (T in absolute temperature, K) typical for a paramagnetic substance in which orbital contribution to the susceptibility is negligible.

Paramagnetic materials, besides having unpaired electrons, contain paired electrons in the inner (closed) shells of the constituent atoms. The presence of these paired electrons makes diamagnetism an inherent property of all materials. Thus the magnetic moment measured is in fact the sum of both paramagnetic (positive quantity) and the associated diamagnetic (negative quantity) moment. Since the presence of an intrinsic magnetic moment in a substance results in a large paramagnetic moment, the diamagnetic effects are often neglected in the calculations. However, if the paramagnetic and diamagnetic moments of the substance under investigation are of comparable magnitude, and where accurate measurements are desired, corrections due to the diamagnetic contributions are made to the measured magnetic moments.

So far we have considered paramagnetic substances in which there are no significant interactions between the magnetic moments of its atoms. However, magnetic moments of individual atoms in many substances do interact with each other in several different ways. In these materials, Curie's law is not obeyed; often, the magnetic behavior of these substances is best described by a Curie-Weiss law which takes into account the interactions among the individual magnetic moments

$$\chi_M = \frac{C}{(T-\theta)}, \qquad \text{(Eq. 5)}$$

where θ is the Weiss constant (or paramagnetic Curie temperature).

Interactions of magnetic moments in condensed systems (i.e., solids) will in most cases lead to different types of magnetic ordering, characteristic of the substance. This is shown in Figure S56 where the net spin on each atom is represented by an arrow aligned "with" or "against" the applied external magnetic field. The type of magnetic interactions present in a particular substance is primarily determined by the nature of the constituent ions and chemical bonds. The magnetic interactions between the individual magnetic moments result in a net stabilization energy. The two most common types of magnetic ordering or interactions are called ferromagnetism and antiferromagnetism. In ferromagnetic substances, the magnetic moments of adjacent atoms are aligned parallel to one another below a certain critical temperature (known as the Curie temperature, T_c) as shown in Figure S56a. This type of magnetic ordering is characterized by a spontaneous magnetization of the substance below T_c even in the absence of an external magnetic field. Iron, nickel, and cobalt and some rare-earth metals, as well as some compounds and alloys of these elements are typical examples that show ferromagnetic ordering.

In antiferromagnetic substances, the individual magnetic moments are aligned antiparallel to one another below a certain critical temperature (known as Néel temperature, T_N). This interaction gives no net magnetization when the neighboring atoms have magnetic moments of identical magnitude as shown in Figure S56b. Most compounds of the transition elements such as MnO, MnF_2, $FeCl_2$, and elements such as chromium, α-manganese and some rare-earth metals such as cerium, praseodymium, neodymium, samarium, and europium display antiferromagnetic ordering. Compounds that exhibit ferromagnetic or antiferromagnetic type of magnetic ordering often obey the

Curie-Weiss law (5) at temperatures well above the transition temperatures (ordering temperature, T_c or T_N).

Ferrimagnetism is another prevalent type of ordering of the magnetic moments. In a ferrimagnetic substance, although the individual magnetic moments of adjacent atomic particles are aligned antiparallel, they are not of equal magnitude (i.e., when the type of interacting neighboring atoms are different), e.g., $Fe_3O_4(FeO \cdot Fe_2O_3)$.

As expected, the magnetic response becomes stronger, as the number of unpaired electrons present increases. Hence it is customary to measure experimentally the magnetic susceptibility of a substance as a function of its weight, χ_g

$$\chi_g = \chi_v / \rho, \quad \text{(Eq. 6)}$$

where χ_g, the mass susceptibility of the compound, is the induced magnetic moment per unit mass per unit applied magnetic field, and ρ is the density of the material.

The mass susceptibility of the compound is related to the molar susceptibility, χ_M

$$\chi_M = \chi_g \cdot \mathrm{MW}/Z, \quad \text{(Eq. 7)}$$

where MW is the molecular weight of the compound and Z is the number of moles of magnetic ions per formula weight of the compound. The molar magnetic susceptibility of a compound having isolated magnetic ions with no magnetic interactions is given by (assuming total quenching of orbital moments)

$$\chi_M = \frac{N_A \beta^2 g^2 S(S+1)}{3k_B T}. \quad \text{(Eq. 8a)}$$

In the above expression, N_A is Avogadro's number, k_B is the Boltzmann constant, and β is a constant unit called the Bohr magneton (BM). Note that (8a) is similar to that shown by Curie's law which was derived from experimental observations, giving the Curie constant, C

$$C = \frac{N_A \beta^2 g^2 S(S+1)}{3k_B}. \quad \text{(Eq. 8b)}$$

As discussed above, in condensed systems, e.g., liquids and solids, generally there are some magnetic interactions between the magnetic ions such that a modified Curie's law known as the Curie-Weiss law is applied

$$\chi_M = \frac{N_A \beta^2 g^2 S(S+1)}{3k_B (T-\theta)}. \quad \text{(Eq. 9)}$$

In substances with interacting magnetic moments and where the orbital contribution to the magnetic moments is significant, the molar susceptibility is given by

$$\chi_M = \frac{N_A \beta^2 g^2 J(J+1)}{3k_B (T-\theta)}, \quad \text{(Eq. 10)}$$

here the resultant total angular momentum, J, is given by

$$J(J+1) = L(L+1) + S(S+1), \quad \text{(Eq. 11)}$$

and the value of g factor is given by Landé's equation:

$$g = 1 + \frac{S(S+1) - L(L+1) + J(J+1)}{2J(J+1)}. \quad \text{(Eq. 12)}$$

The quantum number L is the resultant orbital angular momentum of all unpaired electrons present in the atom (i.e., L is zero for a completely filled orbital). However for most compounds of the first row transition metal series, the orbital contribution to the magnetic moment is negligible, such that $J = S$ and (10) is identical to (9). In these and other compounds of the transition metal series the metal ions are typically surrounded by strong ligands that quench the orbital contribution to the magnetic moment.

Equation (9) can be further simplified by defining μ_{eff}^2 as $g^2 S(S+1)$ and combining all of the constants to give:

$$\chi_M = \left(\frac{N_A \beta^2}{3k_B}\right) \frac{\mu_{\mathrm{eff}}^2}{(T-\theta)} \approx 0.125 \frac{\mu_{\mathrm{eff}}^2}{(T-\theta)}. \quad \text{(Eq. 13)}$$

This equation allows us to calculate the effective magnetic moment of an ion, μ_{eff}, which can be rewritten as

$$\mu_{\mathrm{eff}} = 2.828[\chi_M (T-\theta)]^{1/2}, \quad \text{(Eq. 14)}$$

where χ_M, the molar susceptibility, is obtained from (7) and T is the temperature in Kelvin.

Instrumentation

In this experiment, the method to be used in the measurement of the magnetic susceptibility of substances is based on the principle of stationary sample but moving magnet. Here, the balance (i.e., Johnson Matthey Magnetic Susceptibility Balance, MSB), measures the force that the sample exerts on a suspended permanent magnet, unlike the Gouy balance which measures the equal and opposite force which a magnet exerts on the sample.

In the experiment that follows, the mass susceptibility of a compound, χ_g, to be used in (7) is calculated from the quantities obtained using the MSB by the following formula

$$\chi_g = \frac{L C_{\mathrm{bal}} (R - R_0)}{(m \times 10^9)}, \quad \text{(Eq. 15)}$$

where C_{bal} is the calibration constant of the balance, L is the length in centimeters (cm) of a well-packed sample in the tube, m is sample mass in grams (g), R is the reading obtained for the tube plus sample, R_0 is the empty tube reading (normally a negative value, i.e., diamagnetic).

The MSB must be calibrated with the standard compound, $HgCo(SCN)_4$ which has a $\chi_g = 16.44 \times 10^{-6}$ cgs at 20°C. This first measurement is required for the calculation of the calibration constant of the balance, C_{bal}, in (15).

Experimental procedure

1. Turn the *Range* Knob to $\times 1$ scale and let the MSB warm up for.
2. Adjust the MSB to read 000 using the *Zero* knob when no tube is inside the tube guide.

Many of the following steps may not be required if we supply preprepared samples.

3. Place the empty tube into the tube guide of MSB, and record the reading, R_0 (normally a negative value, i.e., diamagnetic).
4. Remove the empty tube from the MSB and weigh it carefully on a balance to the nearest milligram; record the mass of the empty tube in grams, m_t.
5. Prepare your sample for measurement by grinding (if not already in a powdered form). Sample must be ground to a fine powder in a mortar and pestle.
6. Introduce your sample into the preweighed empty tube and tap (the sample tubes are expensive, made of brittle quartz materials; so tap them gently to avoid breakage) on a wooden bench several times to achieve densely packed (i.e., minimum empty space

between grains) sample. The length of the well packed sample in the tube should be between 2.5 and 3.5 cm (or a length appropriate to the diameter of the tube being used, as suggested by the manufacturer of the MSB).

7. Place the packed tube into the tube guide in the MSB, wait for stable reading before recording this first reading, R_1. If the reading is off-scale, remove the packed tube, set the *Range* Knob to ×10 scale and then *rezero* the MSB before putting back the packed sample. Record your reading, but remember to multiply this reading by 10.
8. Remove the packed tube from the MSB and tap it several times to further compress the solid in the tube and remove any trapped air.
9. Place the packed tube back into the tube guide in the MSB, and record a second reading, R_2. If this second reading (R_2) is not the same as the first reading (R_1), then repeat further the tapping/measuring sequence until *two* successive readings are the same. Even packing gives constant reading hence tap the sample tube gently till a constant reading is indicated.
10. Take the temperature reading, T, of the area around the MSB; remove the packed tube from the MSB and measure the sample length in the tube in centimeters, L.
11. Immediately, measure the mass of the packed tube on a balance and record its mass in grams, $m_{t,s}$; find the actual mass of the sample (m) in the packed tube in grams by subtracting the mass of the empty tube (m_t) from this value; i.e.,

$$m = m_{t,s} - m_t.$$

12. Do another measurement of the same sample by repacking the same tube and following the procedures above. (Note: data for the same empty tube, i.e., R_0 and m_t should remain the same.)
13. Clean the tube by tapping out the powder, followed by several washings with a squirt bottle filled with soapy water. Finally rinse and dry with acetone.

Calculations and worked examples

1. Calculation of the calibration constant, C_{bal}:

(a) $HgCo(SCN)_4$ has a $\chi_g = 16.44 \times 10^{-6}$ cgs at 20°C

$$\chi_g \text{ at } T\,^\circ\text{C} = \frac{4\,981}{(283 + T)} \times 10^{-6} \text{ cgs},$$

then at $T = 21.0\,^\circ$C, χ_g is

$$\chi_g = \frac{4\,981}{(283 + 21)} \times 10^{-6} \text{ cgs} = 16.38 \times 10^{-6} \text{ cgs}.$$

(b) Refer to Table S3, data No. 1 for $HgCo(SCN)_4$;

$$\chi_g = \frac{LC_{bal}(R - R_0)}{(m \times 10^9)},$$

on rearranging;

$$C_{bal} = \frac{\chi_g(m \times 10^9)}{L(R - R_0)}.$$

Hence for $T = 21.0\,^\circ$C, C_{bal} becomes;

$$C_{bal} = \frac{16.38 \times 10^{-6}(0.5082 \times 10^9)}{3.25(2\,120 - (-31))} = 1.19076.$$

(c) χ_g for $Ni(en)_3S_2O_3$ refer to Table S3, data No. 2;

$$\chi_g = \frac{LC_{bal}(R - R_0)}{(m \times 10^9)} = \frac{3.30 \times 1.19076(750 - (-31))}{(0.2787 \times 10^9)},$$

Table S3 The list of chemicals to be used in this experiment

No.	Chemical	MW (g mole^{-1})
Standard	$HgCo(SCN)_4$	491.85816
1	$MnSO_4{\cdot}2H_2O$	187.03221
2	MnO_2	86.93685
3	$Cu(HCOO)_2{\cdot}H_2O$	171.59676
4	Mn_2O_3	157.8743
5	$Ni(en)_3S_2O_3$	351.1206
6	$KMnO_4$	158.03395
7	$K_4Fe(CN)_6{\cdot}3H_2O$	422.39248
8	$Co(NH_4)_2(SO_4)_2{\cdot}6H_2O$	395.22908
9	$Fe(NH_4)_2(SO_4)_2{\cdot}6H_2O$	392.14288
10	$K_2Cr_2O_7$	294.18476

i.e., $\chi_g = 11.01 \times 10^{-6}$ cgs at $T = 21.0\,^\circ$C.

Compare this value of χ_g for $Ni(en)_3S_2O_3$ with the value given in literature, i.e., χ_g at $20.0\,^\circ\text{C} = 11.04 \times 10^{-6}$ cgs and since

$$\chi_g \text{ at } T\,^\circ\text{C} = \frac{2\,759}{(230 + T)} \times 10^{-6} \text{ cgs};$$

then

$$\chi_g \text{ at } 21.0\,^\circ\text{C} = \frac{2\,759}{(230 + 21)} \times 10^{-6} \text{ cgs} = 10.99 \times 10^{-6} \text{ cgs}.$$

2. Worked examples: refer to Table S3, data for α-MnS (data No. 4)

$$\chi_g = \frac{LC_{bal}(R - R_0)}{(m \times 10^9)} = \frac{3.15 \times 1.19076(8\,840 - (-27))}{(0.5358 \times 10^9)},$$

i.e., $\chi_g = 62.07 \times 10^{-6}$ cgs at $T = 23.0\,^\circ$C.

Hence χ_M for this compound can be calculated using (7), i.e., $\chi_M = \chi_g \times MW/Z = 62.07 \times 10^{-6} \times 87.002$ g mol^{-1}, since $Z = 1$ for α-MnS,

therefore, $\chi_M = 5.40 \times 10^{-3}$ emu mol^{-1}.

To calculate the effective magnetic moment, μ_{eff}, we make use of (14), i.e.,

$$\begin{aligned}\mu_{eff} &= 2.828(\text{emu mol}^{-1}\,\text{K})^{-1/2}[\chi_M(T - \theta)]^{1/2} \\ &= 2.828(\text{emu mol}^{-1}\,\text{K})^{-1/2}[5.40 \times 10^{-3}\,\text{emu mol}^{-1} \\ &\quad (296 - (-490))\,\text{K}]^{1/2} \\ &= 5.83\,\text{BM}.\end{aligned}$$

The students will study a series of first row transition metal compounds to determine the oxidation state of the ion in each compound and the number of unpaired electrons, from the measurement of the magnetic susceptibility. The compounds will include various transition metal complexes varying in different number of d-electrons. The students will directly measure the mass susceptibility of each compound using a Magnetic Susceptibility Balance (MSB, Johnson Matthey model). Given the values of Curie constant (θ) for each compound, the students will be able to find the effective magnetic moment of the manganese ion from the measured values of the molar susceptibility of each compound using (14), hence will be able to calculate the number of unpaired electrons.

Z.S. Teweldemedhin, R.L. Fuller and M. Greenblatt

Bibliography

Carlin, R.L., 1986. *Magnetochemistry.* Berlin: Springer-Verlag.

Drago, R.S., 1977. *Physical Methods in Chemistry.* Philadelphia: W. B. Saunders Co.

Goodenough, J.B., 1963. *Magnetism and the Chemical Bond.* New York: John Wiley & Sons.

Kittel, C., 1962. *Elementary Solid State Physics: A Short Course.* New York: John Wiley & Sons.

Konig, E., and Landolt-Bornstein, 1966. In Hellwege. K.-H., and Hellwege, A.M., (eds.), *Magnetic properties of Coordination and Organo-Metallic Transition Metal Compounds*, Group II, Volume 2. Berlin: Springer-Verlag.

Magnetic Susceptibility Balance Instructional Manual. Wayne, PA: Johnson Matthey, 1991.

O'Connor, C.J., 1982. In Stiefel, E. (ed.), *Magnetochemistry—Advances in Theory and Experimentation*. New York: John Wiley and Sons; pp. 203–283.

Shoemaker, D.P., *et al.,* 1996. *Customized Experiments in Physical Chemistry,* 6th Edition. Lawrence, KS: Primus/McGraw Hill, p. 109.

Vonsovskii, S.V., 1974. *Magnetism*, Volume 1. Israel: Ketter Publishing House Jerusalem Ltd (Translated by R. Hardin).

Woolcock, J., and Zafar, A., 1992. Microscale techniques for determination of magnetic susceptibility. *Journal of Chemical Education*, **69**: A176–A179.

SUSCEPTIBILITY, PARAMETERS, ANISOTROPY

One of the main interests in the determination of the low field anisotropy of magnetic susceptibility (AMS) is its value as a petrofabric indicator. An important part of the interpretation of AMS measurements in this context involves the orientation of the principal susceptibilities. Several methods of calculating mean directions of the principal susceptibilities, and associated regions of confidence have been devised, but the tensor-averaging technique will provide a better estimate in most circumstances (Ernst and Pearce, 1989). Also, different methods have been proposed to estimate regions of confidence around the mean principal susceptibilities. When the individual measurements are not very different from the calculated mean (in which case it is said that the uncertainty is not large), the regions of confidence calculated with different methods are similar to each other and exert a small influence in the interpretation of results. When the uncertainty in the mean susceptibility is large, however, the regions of confidence found with different methods can be very different (Tauxe, 1998). Nevertheless, instead of entering in a pointless discussion concerning which statistical method is better suited for every specific case, when a large scatter of directions of principal susceptibilities is found it is more significant to check whether the individual data used for the calculation of the average directions of susceptibility might include systematic variations of principal susceptibilities with a well-defined physical meaning. The identification of different statistical populations made in this form not only provides a more meaningful interpretation of AMS results than an interpretation based in a very rigorous statistical treatment that does not acknowledge physical controls on the mineral fabric, but also leads to situations in which the uncertainty associated with the average of each AMS population is relatively small. Consequently, by adequately identifying systematic fluctuations in the AMS of a rock unit and associating them with changes in the physical conditions controlling the magnetic fabric (like for example at opposite sides of a tabular intrusive where shear direction is expected to be different in orientation), it turns out that the method used in the calculation of the associated regions of confidence is relatively unimportant.

On the other hand, sometimes is necessary to concentrate our attention in the numeric components of the susceptibility tensor. This part of the analysis may be somewhat problematic not only because of the large number of AMS parameters that have been proposed over time (Tables S4 and S5), but also because the parameters used in a particular analysis generally are selected based more on previous use than on some objective criteria. Historically, five classes of AMS parameters have been defined: Bulk susceptibility, foliation, lineation, shape, and degree of anisotropy. Among these, the bulk susceptibility is the only one that has a well-defined physical meaning. Whether it is calculated as the arithmetic or the geometric average of the three principal susceptibilities, the bulk susceptibility is one of the three invariants associated with any second rank tensor, and consequently it is a measure of susceptibility independent of the coordinate system used as a reference frame.

In contrast, the physical meaning of all the other AMS parameters is less clear, and even it may be controversial. For instance, the lineation parameters were proposed to supposedly yield a measure of the intensity of alignment of elongated particles as detected through AMS

Table S4 Shape parameters

	Parameters	Name	Range	Group	Source
1.	$(k_1 - k_2)/(k_2 - k_3)$	Prolateness	0 to ...		(Khan, 1962)
2.	$(k_2 - k_3)/(k_1 - k_2)$	Oblateness	0 to ...		(Khan, 1962)
3.	$a \sin\{(k_2 - k_3)/(k_1 - k_3)\}^{1/2}$	Angle V	0 to 90		(Graham, 1966)
4.	$(2k_2 - k_1 - k_3)/(k_1 - k_3)$	Difference shape factor	−1 to 1		(Jelinek, 1981)
5.	$(k_1 - k_2)/\{(k_1 + k_2)/2 - k_3\}$	*q* factor	0 to 2	S1	(Granar, 1957)
6.	$(k_2(k_1 - k_2))/(k_1(k_2 - k_3))$		0 to ...		(Janak and Hrouda, 1969)
7.	$(k_3(k_1 - k_2))/(k_1(k_2 - k_3))$	Shape indicator	0 to ...		(Stacey *et al.*, 1960)
8.	$(k_2/k_3 - 1)/(k_1/k_2 - 1)$	*D* factor	0 to ...		(Hrouda, 1976)
9.	$(2n_2 - n_1 - n_3)/(n_1 - n_3)$	Shape parameter *T*	−1 to 1		(Jelinek, 1981)
10.	k_1/k_2	Lineation	1 to ...	S2a	(Balsey and Buddington, 1960)
11.	$(k_1 - k_2)/k_m$	Magnetic lineation	0 to 3		(Khan, 1962)
12.	$(k_2 - k_3)/k_m$	Magnetic foliation	0–1.5		(Khan, 1962)
13.	k_2/k_3	Foliation	1 to ...	S2b	(Stacey *et al.*, 1960)
14.	$1 - (k_3/k_2)$	Foliation	0 to 1		(Porath, 1971)
15.	k_1k_3/k_2^2		0 to ...		(Stacey *et al.*, 1960)
16.	k_2^2/k_1k_3	*E* factor		S2	(Hrouda *et al.*, 1971)
17.	$k_3/k_1 - 2k_2/k_1 + 1$	*B* parameter	−1 to 1		(Cañón-Tapia, 1992)

$n_i = \ln(k_i)$, $i = 1,2,3$, and $k_m = (k_1 + k_2 + k_3)/3$.

Table S5 Parameters quantifying the degree of anisotropy

	Parameters	Name	Range	Group	Source
1.	k_1/k_3	Anisotropy degree P	1–...		(Nagata, 1961)
2.	$100(k_1 - k_3)/k_1$	Percentage anisotropy	0–100		(Graham, 1966)
3.	$(k_1 - k_3)/k_m$	Total anisotropy H	0–3	A1	(Owens, 1974)
4.	$(k_1 - k_3)/2k_m$	Percentage anisotropy	0–1.5		(Khan, 1962)
5.	$(k_1 - k_3)/k_2$	Absolute anisotropy	0–...		(Rees, 1966)
6.	$(k_1 + k_2)/2k_3$	Foliation	1–...	A2	(Balsey and Buddington, 1960)
7.	$((k_1 - k_2)/2) - k_3$	Magnetic excess	0–1 ($\times k_1$)		(Granar, 1957)
8.	$\exp\{(2[(a_1)^2 + (a_2)^2 + (a_3)^2]\}^{1/2}$	Corrected anisotropy degree (P')	1–...	A3	(Jelinek, 1981)
9.	$k_1/(k_3k_2)^{1/2}$	Emplacement factor	1–...		(Ellwood, 1975)
10.	$2k_1/(k_3 + k_2)$	Lineation degree	1–...	A4	(Hrouda *et al.*, 1971)
11.	$100(1-k_3/2k_1-k_2/2k_1)$	A parameter	0–100		(Cañón-Tapia, 1992)

$a_i = \ln(k_i/k_m')$, $i = 1,2,3$, and $k_m = (k_1 + k_2 + k_3)/3$, $k_m' = (k_1k_2k_3)^{1/3}$.

measurements. This interpretation is only correct if the AMS is controlled by elongated objects of similar (and previously known) aspect ratios and for which the AMS bears a direct relation with the shape of the object (normal shape-effect), but it becomes misleading under other conditions. In particular, if the AMS is controlled by minerals in which the physical dimensions of the grain are inversely related to the intensity of susceptibility (inverse shape-effect Rochette *et al.*, 1991), the lineation becomes a measure of the grouping of particles over one plane rather than reflecting the degree of clustering of the largest dimensions of the minerals. Alternatively, if the AMS of two rocks is controlled by minerals with a normal shape-effect but different average aspect ratios, the magnetic lineation of those two rocks will be different even when the degree of clustering of the minerals is the same. In an extreme case, a rock formed by very elongated minerals that are not perfectly aligned might yield a larger lineation than a second rock formed by a perfect alignment of not so elongated minerals, therefore potentially leading to erroneous petrofabric interpretations.

These problems are not exclusive of the lineation parameter, as ambiguities are found when the physical meaning of all the other types of AMS parameters is examined in detail. Until now, there have been two different approaches to alleviate this situation, both of which have been unsuccessful in finding a general agreement within the scientific community. The first attempt in suggesting criteria for the selection of AMS parameters was made by Ellwood *et al.* (1988). They pointed out that some methods of measurement favor the use of parameters that include the differences of the principal susceptibilities while others call for the use of their ratios (see also Hrouda and Jelinek, 1990). Undoubtedly, this distinction has some merit within an instrumental basis, but it overlooks the fact that most of the available parameters use both ratios and differences of the principal susceptibilities in their definitions. Furthermore, this approach has the problem that actual calculation of different parameters often reduces in practice to a simple arithmetical combination of the principal susceptibilities regardless of the method used for the measurement of the susceptibility tensor (Ellwood *et al.*, 1988; Tarling and Hrouda, 1993). The second attempt to devise useful criteria to guide the selection of AMS parameters was made by Cañón-Tapia (1994). He used a mathematically oriented point of view to compare the most important parameters proposed until that time, and to estimate the influence that such parameters could have in the final interpretation of results. Based on a theoretical comparison in which individual parameters could be conveniently expressed as families of lines, it was shown that many of the parameters proposed in the literature are equivalent to each other and that the only difference between two apparently independent parameters can be reduced to a difference in a numeric range. In particular, using this approach it was shown that the foliation and lineation parameters are special cases of the shape parameters, and therefore little is gained with their use as separate categories. In contrast with the experimentally derived criteria, the geometrically based comparison is applicable to any parameter defined as a combination of two or three of the principal susceptibilities. Consequently, this set of criteria can be used in a general context to guide the selection of AMS parameters in a more ordered form than possible with other criteria available until now. Unfortunately, costume derived from continual use of a particular type of AMS parameters seems to be the dominant criteria in some cases. Consequently, foliation and lineation parameters have continued to be used as independent types and, in some instances, these two parameters are reported together with a third shape parameter in the same work.

Although lacking an objective criteria behind their usage, in practice it is found that the two most commonly used parameters are those called P' and T (Tables S4 and S5, respectively). Another pair of parameters that has been used with increasing frequency, especially because of its use in studies of lava flows, is formed by the A and B parameters. These two pairs of AMS parameters yield different numeric ranges in the analysis of AMS measurements (hence requiring caution when comparing different works). Qualitatively, however, these two pairs of AMS parameters yield equivalent results and therefore the main aspects in the interpretation will remain unaltered irrespective of which pair of parameters was used (Cañón-Tapia, 1994).

In summary, despite the various methods available for the calculation of average orientations of principal susceptibilities and associated regions of confidence, when attention is given to the physical factors controlling the acquisition of the AMS of a rock, the differences between alternative methods of calculation becomes negligible for the interpretation of results. In contrast, little agreement has been reached concerning the selection of AMS parameters. More details concerning the physical value of individual AMS parameters will be required before reaching a clearly satisfactory set of criteria that can be used as a general guide for their selection.

Edgardo Cañón-Tapia

Bibliography

Balsey, J.R., and Buddington, A.F., 1960. Magnetic susceptibility anisotropy and fabric of some Adirondack granites and orthogneisses. *American Journal of Science*, **258A**: 6–20.

Cañón-Tapia, E., 1992. Applications to volcanology of paleomagnetic and rock magnetic techniques. MSc thesis, University of Hawaii, Honolulu, p. 146.

Cañón-Tapia, E., 1994. AMS parameters: guidelines for their rational selection. *Pure and Applied Geophysics*, **142**: 365–382.

Ellwood, B.B., 1975. Analysis of emplacement mode in basalt from deep-sea drilling project holes 319A and 321 using anisotropy of magnetic susceptibility. *Journal of Geophysical Research*, **80**: 4805–4808.

Ellwood, B.B., Hrouda, F., and Wagner, J.J., 1988. Symposia on magnetic fabrics: introductory comments. *Physics of the Earth and Planetary Interiors*, **51**: 249–252.

Ernst, R.E., and Pearce, G.W., 1989. Averaging of anisotropy of magnetic susceptibility data. *Geological Survey of Canada, Paper 89-9*: 297–305.

Graham, J.W., 1966. Significance of magnetic anisotropy in Appalachian sedimentary rocks. In Steinhart, J.S., and Smith, T.J. (eds.), *The Earth Beneath the Continents.* Washington, DC: American Geophysical Union, pp. 627–648.

Granar, L., 1957. Magnetic measurements on Swedish varved sediments. *Arkiv Geofysik*, **3**: 1–40.

Hrouda, F., 1976. The origin of cleavage in the light of magnetic anisotropy investigations. *Physics of the Earth and Planetary Interiors*, **13**: 132–142.

Hrouda, F., and Jelinek, V., 1990. Resolution of ferrimagnetic and paramagnetic anisotropies in rocks, using combined low-field and high-field measurements. *Geophysical Journal International*, **103**: 75–84.

Hrouda, F., Janak, F., Rejl, L., and Weiss, J., 1971. The use of magnetic susceptibility anisotropy for estimating the ferromagnetic mineral fabrics of metamorphic rocks. *Geologische Rundschau*, **60**: 1124–1142.

Janak, F., F., and Hrouda, 1969. Research of magnetic susceptibility and its anisotropy, Geofyzika, Brno.

Jelinek, V., 1981. Characterization of the magnetic fabric of rocks. *Tectonophysics*, **79**: T63–T67.

Khan, M.A., 1962. The anisotropy of magnetic susceptibility of some igneous and metamorphic rocks. *Journal of Geophysical Research*, **67**: 2873–2885.

Nagata, T., 1961. *Rock magnetism.* Tokyo: Maruzen.

Owens, W.H., 1974. Mathematical model studies on factors affecting the magnetic anisotropy of deformed rocks. *Tectonophysics*, **24**: 115–131.

Porath, H., 1971. Anisotropie der magnetischen Suszeptibilitat und Sättigungsmagnetisierung als Hilfsmittel der Gefügekunde. *Geologische Rundschau*, **60**: 1088–1062.

Rees, A.I., 1966. The effect of depositional slopes on the anisotropy of magnetic susceptibility of laboratory deposited sands. *Journal of Geology*, **74**: 856–867.

Rochette, P., Jackson, M., and Aubourg, C., 1991. Rock magnetism and the interpretation of anisotropy of magnetic susceptibility. *Reviews of Geophysics*, **30**: 209–226.

Stacey, F.D., Joplin, G., and Lindsay, J., 1960. Magnetic anisotropy and fabric of some foliated rocks from SE Australia. *Pure and Applied Geophysics*, **47**: 30–40.

Tarling, D., and Hrouda, F., 1993. *The Magnetic Anisotropy of Rocks.* London: Chapman & Hall, p. 217.

Tauxe, L., 1998. *Paleomagnetic Principles and Practice.* Dordrecht: Kluwer, p. 299.

T

TAYLOR'S CONDITION

Taylor's condition

Taylor's condition, first derived by J.B. Taylor in 1963, is a statement about the electromagnetic torque within the Earth's core, namely that it must vanish when integrated over cylindrical shells parallel to the axis of rotation. It is one of the most important results in geodynamo theory.

Basic equations

The two equations we will need are the induction equation:

$$\frac{\partial \mathbf{B}}{\partial t} = \nabla \times (\mathbf{U} \times \mathbf{B}) + \nabla^2 \mathbf{B}, \qquad \text{(Eq. 1)}$$

governing the evolution of the magnetic field $\mathbf{B}$ in the core, and the Navier-Stokes equation (in its simplest, Boussinesq form):

$$Ro\left[\frac{\partial}{\partial t} + \mathbf{U}\cdot\nabla\right]\mathbf{U} + 2\hat{\mathbf{e}}_z \times \mathbf{U} = -\nabla p + E\nabla^2\mathbf{U} + (\nabla \times \mathbf{B}) \times \mathbf{B} + \mathbf{F}_\mathrm{T}, \qquad \text{(Eq. 2)}$$

governing the fluid flow $\mathbf{U}$. In general there would also be an equation for the thermal and/or compositional buoyancy that ultimately drives the whole system, but for our discussion here we can take the buoyancy force $\mathbf{F}_\mathrm{T}$ to be prescribed. The only feature that will matter is that it is purely radial.

In these equations, length has been scaled by r_0, time by r_0^2/η, $\mathbf{U}$ by η/r_0, and $\mathbf{B}$ by $\sqrt{\Omega\rho\mu_0\eta}$, where r_0 is the outer core radius, η the magnetic diffusivity, Ω the Earth's rotation rate, ρ the density, and μ_0 the magnetic permeability. The two nondimensional parameters then appearing in (2) are the Rossby and Ekman numbers $Ro = \eta/\Omega r_0^2$ and $E = \nu/\Omega r_0^2$, where ν is the viscosity. Inserting the numbers, we find that $Ro = O(10^{-9})$ and $E = O(10^{-15})$, indicating the extreme dominance of rotation over inertia and viscosity.

One very basic question then is where does the Lorentz force fit into this balance? Is it dominant, like the Coriolis force, or very small, like inertia and viscosity? Formally it is certainly comparable to the Coriolis force in (2), but that merely reflects our scalings for $\mathbf{B}$ and $\mathbf{U}$. So the question is, are these the right scalings, that is, do $\mathbf{B}$ and $\mathbf{U}$ come out to be $O(1)$ when measured in these $\sqrt{\Omega\rho\mu_0\eta}$ and η/r_0 units? As we will see, Taylor's condition touches upon precisely this point, and suggests that $\mathbf{B}$ (and possibly $\mathbf{U}$ as well, see *dynamo, model-Z*) may not necessarily come out to be $O(1)$, but may scale as some power of E instead.

Taylor's analysis

If Ro and E are so small, we might try to simplify the Navier-Stokes equation by dropping the inertial and viscous terms entirely, yielding

$$2\hat{\mathbf{e}}_z \times \mathbf{U} = -\nabla p + (\nabla \times \mathbf{B}) \times \mathbf{B} + \mathbf{F}_\mathrm{T}. \qquad \text{(Eq. 3)}$$

Taking the curl (and using also $\nabla \cdot \mathbf{U} = 0$) yields

$$-2\frac{\partial}{\partial z}\mathbf{U} = \nabla \times [(\nabla \times \mathbf{B}) \times \mathbf{B} + \mathbf{F}_\mathrm{T}], \qquad \text{(Eq. 4)}$$

where (z, s, ϕ) are standard cylindrical coordinates. One might suppose then that the solution is simply $\mathbf{U} = \mathbf{U}_\mathrm{M} + \mathbf{U}_\mathrm{T}$, with the so-called magnetic and thermal winds given by

$$\mathbf{U}_\mathrm{M} = -\frac{1}{2}\int \nabla \times [(\nabla \times \mathbf{B}) \times \mathbf{B}]\mathrm{d}z, \quad \mathbf{U}_\mathrm{T} = -\frac{1}{2}\int \nabla \times \mathbf{F}_\mathrm{T}\,\mathrm{d}z. \qquad \text{(Eq. 5)}$$

Instead, it turns out that in general (3) has no solution at all. To see why, consider its ϕ-component:

$$2\mathbf{U}\cdot\hat{\mathbf{e}}_s = -\frac{1}{s}\frac{\partial p}{\partial \phi} + [(\nabla \times \mathbf{B}) \times \mathbf{B}]\cdot\hat{\mathbf{e}}_\phi \qquad \text{(Eq. 6)}$$

(it is at this stage that we use the fact that $\mathbf{F}_\mathrm{T}$ has no ϕ-component). Next, integrate (6) over the so-called geostrophic surfaces $C(s)$ shown in Figure T1, consisting of concentric cylindrical shells parallel to the z-axis. The pressure-gradient term then vanishes, since

$$\oint \frac{\partial p}{\partial \phi}\mathrm{d}\phi = 0. \qquad \text{(Eq. 7)}$$

The Coriolis term also vanishes, since $\int \mathbf{U}\cdot\hat{\mathbf{e}}_s\,\mathrm{d}S$ is the net flux across $C(s)$, which must be zero if the fluid is incompressible, as in the Boussinesq approximation here. We are thus left with Taylor's celebrated result, stating that $\mathbf{B}$ must satisfy

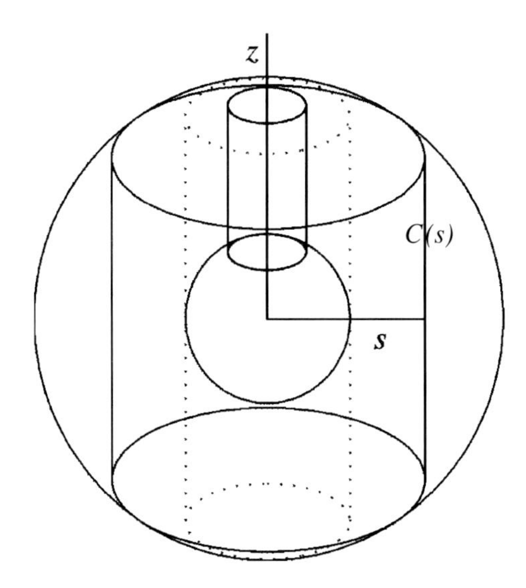

Figure T1 The geostrophic surfaces $C(s)$ over which (8) is to be integrated. Note also how these surfaces change abruptly across the so-called tangent cylinder, the particular surface (indicated by the dotted lines) just touching the inner core.

$$\int_{C(s)} [(\nabla \times \mathbf{B}) \times \mathbf{B}] \cdot \hat{\mathbf{e}}_\phi \, \mathrm{d}S = 0. \qquad \text{(Eq. 8)}$$

That is, when considered pointwise, all four forces contribute to (3), but when the corresponding torques are integrated over these particular surfaces $C(s)$, the buoyancy, pressure-gradient and Coriolis terms all drop out, so the only remaining term must then also be zero. We see therefore that unless **B** satisfies (8), and on *each* such surface $C(s)$, (3) simply has no solution at all.

Now suppose that **B** does satisfy (8), so a solution exists. There is then the additional complication that this solution is not unique, since one can add to it an arbitrary geostrophic flow $U_g(s)\hat{\mathbf{e}}_\phi$, which after all satisfies $\partial\mathbf{U}/\partial z = 0, \nabla \cdot \mathbf{U} = 0$, and the no normal flow boundary conditions. Equation (3) thus has the awkward feature that it has either no solution at all if (8) is not satisfied, or else an infinite number of solutions if it is.

This arbitrariness in the geostrophic flow can fortunately be removed, by requiring that **B** should not only satisfy (8) at some particular instant, but that it should evolve in time in such a way that (8) continues to be satisfied. That is, we want not only (8), but also $\mathrm{d}/\mathrm{d}t(8)$, or

$$\int_{C(s)} \left[(\nabla \times \mathbf{B}) \times \frac{\partial \mathbf{B}}{\partial t} + \left(\nabla \times \frac{\partial \mathbf{B}}{\partial t}\right) \times \mathbf{B}\right] \cdot \hat{\mathbf{e}}_\phi \, \mathrm{d}S = 0. \qquad \text{(Eq. 9)}$$

To see how this determines U_g, we recall first (1), giving $\partial\mathbf{B}/\partial t$ in terms of **B** and **U**. We remember further that except for U_g, the rest of **U** is known, from (5). If we therefore write

$$\mathbf{U} = U_g(s)\hat{\mathbf{e}}_\phi + \mathbf{U}_\mathrm{M} + \mathbf{U}_\mathrm{T}, \qquad \text{(Eq. 10)}$$

then inserting (10) into (1), and then (1) into (9), we end up with an (horribly complicated) equation in which the only unknown quantity is U_g. In this way (3) has a unique solution after all, and in the process (8) continues to be satisfied into the future. Taylor therefore envisioned the geodynamo evolving through such a sequence of states.

Alternatives to Taylor's analysis

Unfortunately, no one has ever succeeded in evolving a numerical geodynamo model according to Taylor's prescription. Instead, most workers have proceeded more cautiously, not setting *Ro* and *E* identically to zero. If one restores viscosity, at least in the Ekman boundary layers, then an asymptotic analysis (Tough and Roberts 1968) of these layers yields:

$$\int_{C(s)} [(\nabla \times \mathbf{B}) \times \mathbf{B}] \cdot \hat{\mathbf{e}}_\phi \, \mathrm{d}S = E^{1/2} \frac{4\pi s}{(1-s^2)^{1/4}} U_g(s), \qquad \text{(Eq. 11)}$$

indicating that the integrated electromagnetic torque does not have to vanish identically, but can instead be balanced by this small but nonzero viscous torque. (One might suppose that $\mathbf{U}_\mathrm{M}$ and $\mathbf{U}_\mathrm{T}$ would have associated viscous torques as well. However, the constants of integration in (5) can be chosen such that their contributions vanish.)

Equation (11) can then be solved for U_g in terms of **B**. According to this prescription therefore, the field evolves according to (1), and at each instant in time the flow is still given by (10), but with U_g now given by (11) rather than Taylor's more complicated formula. The advantage of this approach is that it is not only easier to implement numerically (although still nontrivial), it is also more general than Taylor's approach. In particular, by postulating an expansion of the form

$$\mathbf{B} = \mathbf{B}_0 + E^{1/2}\mathbf{B}_1 + E\,\mathbf{B}_2 + \cdots, \qquad \text{(Eq. 12)}$$

and assuming that $\mathbf{B}_0$ satisfies (8), we can recover Taylor's solution (and in the process also obtain the higher order viscous corrections to it). This is not the only possibility though. We can also postulate an expansion of the form

$$\mathbf{B} = E^{1/4}\mathbf{B}_0 + E^{3/4}\mathbf{B}_1 + E^{5/4}\mathbf{B}_2 + \cdots, \qquad \text{(Eq. 13)}$$

where now $\mathbf{B}_0$ does *not* satisfy (8). In doing so, we obtain the so-called Ekman states, which do not exist at all in Taylor's analysis, because there $E \equiv 0$. Malkus and Proctor (1975) then conjectured that in general the field would start out in the Ekman regime, but would undergo a transition to the Taylor regime for sufficiently strong forcing. That such a transition can indeed occur was demonstrated numerically by Hollerbach and Ierley (1991).

There are also additional possibilities beyond these two states; for example, Hollerbach (1997) showed that an intermediate state can perhaps exist in which the leading order field scales as $E^{1/8}$. If one allows the existence of boundary layers other than the Ekman layers implicitly contained in (11), one can also obtain the so-called model-Z states (see *dynamo, model-Z*).

The geophysical significance of this plethora of possible states is that it is believed that the geodynamo is not necessarily in a single state throughout its entire evolution. Most of the time it is probably in a Taylor state (or possibly a model-Z state), since those are the only two states in which $\mathbf{B} = O(1)$, as we would like it to be. During reversals or excursions though, it could conceivably be in one of the weaker states. Indeed, what triggers these events may well be a breakdown of the previously existing Taylor state.

Finally, we note that fully three-dimensional numerical simulations are unfortunately not yet in a parameter regime where one could test these various ideas; numerically accessible Ekman numbers are so large that one cannot clearly distinguish between these different states. It is only in mean-field models such as that of Hollerbach and Ierley (1991) that one can reduce E sufficiently to obtain clear distinctions between different states. Because Taylor's condition is inherently axisymmetric though, these results should carry over to the fully three-dimensional models, if only we could reduce E further there as well.

Rainer Hollerbach

Bibliography

Hollerbach, R., 1997. The dynamical balance in semi-Taylor states. *Geophysical and Astrophysical Fluid Dynamics*, **84**: 85–98.

Hollerbach, R., and Ierley, G.R., 1991. A modal α^2-dynamo in the limit of asymptotically small viscosity. *Geophysical and Astrophysical Fluid Dynamics*, **56**: 133–158.
Malkus, W.V.R., and Proctor, M.R.E., 1975. The macrodynamics of α-effect dynamos in rotating fluids. *Journal of Fluid Mechanics*, **67**: 417–443.
Taylor, J.B., 1963. The magnetohydrodynamics of a rotating fluid and the Earth's dynamo problem. *Proceedings of the Royal Society of London, Series A*, **274**: 274–283.
Tough, J.G., and Roberts, P.H., 1968. Nearly symmetric hydromagnetic dynamos. *Physics of the Earth and Planetary Interiors*, **1**: 288–296.

Cross-references

Anelastic and Boussinesq Approximations
Core, Boundary Layers
Dynamo, Model-Z
Geodynamo, Numerical Simulations
Inner Core Tangent Cylinder
Thermal Wind

THELLIER, ÉMILE (1904–1987)

Introduction

Émile Thellier (Figure T2), a pioneer in rock magnetism and archeomagnetism, was born February 11, 1904 at Mont-en-Ternois (Pas-de-Calais), France and died May 11, 1987 in Paris. Following studies at the École Normale Supérieure de Saint-Cloud (1924–1926), he taught at the École Primaire Supérieure de Bourges (1927–1930). In spite of a heavy teaching load, 21 h per week, he successfully completed the Licence ès-sciences physiques of the Faculté des Sciences de Paris, where he would spend the remainder of his career. His outstanding performance led his professors to urge him to embark on research. At the Sorbonne in the laboratory of Prof. Charles Maurain, he completed a Diplôme d'Études Supérieures (1931), studying the magnetism of baked clay, and the Agrégation de Sciences physiques (1932).

Figure T2 Thellier, Émile (1904–1987).

It had been known since at least the time of Brunhes (1906) that the thermoremanent magnetization (TRM) acquired by baked clays and volcanic rocks paralleled the direction of the geomagnetic field acting when they cooled and survived later reversals of the field. Maurain proposed to Thellier the idea of using the TRM of ancient pottery to deduce not only the direction but also the strength of the Earth's field at the time of firing. This program Thellier later carried out brilliantly during a long career, but his first desire was to reproduce TRM in the laboratory and understand its properties. His success in this undertaking makes him the true "discoverer of magnetic memory," in the words of his great confrere, Louis Néel. Working with clays from the famed Sèvres pottery factories, and later with volcanic rocks, Thellier minutely and exhaustively determined how TRM forms, producing a remarkable Docteur ès-Sciences thesis (Thellier, 1938).

This landmark paper, which has lost none of its immediacy today, launched his academic career at the Institut de Physique du Globe de Paris, where he became successively Physicien-adjoint (1943), Maître de Conférences (1945), Professeur (1948) and Director (1954–1966). From 1967, he moved his research to the Observatoire du Parc Saint-Maur, in the suburbs of Paris, where he established a CNRS Laboratoire de Géomagnétisme, which he molded into one of the world's leading rock magnetic and paleomagnetic research centers. Thellier received many honors in his lifetime. He was named an Officer of the Legion of Honour (1957), elected to the Académie des Sciences (1967), and awarded four prizes by the academy and one by the Université de Paris. His first love, to the end of his life, remained his research and the laboratory he had created.

Thellier's laws of thermoremanent magnetization

Thellier worked exclusively with samples containing very fine particles of minerals like magnetite (Fe_3O_4) and hematite (α-Fe_2O_3) which we now know contain only one magnetic domain (single-domain particles) or at most a very few domains. For such particles he established a series of universal properties (Thellier, 1938, 1941; Thellier and Thellier, 1941). A master experimentalist, he perfected astatic, translation, and rotating-sample magnetometers of unprecedented sensitivity. Without such instruments, it would have been impossible to satisfy himself or others of the universality of laws such as the additivity of partial TRMs because the clays he used in his earliest work and the hematites used by his student Roquet (Roquet and Thellier, 1946; Roquet, 1954) were weakly magnetic compared to the volcanic rocks favored by others (Koenigsberger, 1938; Nagata, 1943).

The Thellier "laws" are most clearly stated in Thellier (1946), a masterpiece of mature reflection about results obtained in the preceding one and a half decades and their implications for theories of magnetism. [For alternative statements, see Roquet (1954, p. 21–23) and Aubouin and Coulomb (1987).] Quoting Thellier (translated from the French original), we have the following general properties.

1. TRM (intensity) is proportional to field (strength) for weak fields.
2. All partial TRMs are parallel to the field $\boldsymbol{H}$ in which they were acquired.
3. A partial TRM acquired by cooling in $\boldsymbol{H}$ through the temperature interval (T_2, T_1) and in zero field through all other temperature intervals is unaffected by reheating in zero field to a temperature $\leq T_1$ and completely disappears after reheating to T_2.
4. A partial TRM (T_2, T_1, $\boldsymbol{H}$) is independent of other partial TRMs acquired in temperature intervals outside (T_2, T_1), which may be due to fields differing from $\boldsymbol{H}$ in strength and direction.
5. All these partial TRMs add geometrically but each of them retains true autonomy and an exact memory of the temperatures and field in which they were acquired.

The first two properties were known or assumed before Thellier's time (e.g., Folgerhaiter, 1899). The latter three are entirely novel and are usually referred to as Thellier's laws. Thellier's statements recognize the vectorial nature of partial TRM additivity and independence, and form the basis for using thermal demagnetization to separate natural remanent magnetization (NRM) vectors of different ages in paleomagnetism. These statements go well beyond the usual scalar laws given in textbooks (e.g., Dunlop and Özdemir, 1997). Collectively, the five laws form a firm fundamental justification for determining the direction and strength of the paleomagnetic field from the NRM of rocks and baked materials, provided the NRM fraction utilized was produced originally as one or more partial TRMs carried by single-domain or nearly single-domain size grains. Number 3 is the basis for magnetic paleothermometry (e.g., Pullaiah *et al.*, 1975).

Thellier and the theory of thermoremanent magnetization

Louis Néel was intrigued by the apparent generality of the Thellier laws, clearly related to the fine size of the magnetic carriers and not to their chemistry. In 1949, he published his celebrated theory of thermal relaxation of single-domain grains, which very elegantly and completely explained Thellier's observations. Partial TRM was seen to be acquired sharply at a blocking temperature T_B during cooling, as a grain passed from a superparamagnetic to a thermally blocked state, and demagnetized ("unblocked") at the identical temperature T_B during reheating. From this single property flow the three Thellier laws of reciprocity (blocking and unblocking as reciprocal processes), independence and additivity. This powerful and far-reaching theoretical picture remains the foundation of magnetic theory today, in both geophysics and magnetic recording.

What is not generally recognized is that Thellier himself anticipated Néel's ideas, albeit in qualitative rather than quantitative form, three years earlier. Again quoting from Thellier (1946): "These facts, which do not as yet seem to have attracted the attention of theoreticians, seem to me to demand a mechanism of immobilization of elementary magnetic moments below a temperature θ, this temperature having a strong variation from point to point within a body. One can imagine the body to be constituted of elementary domains, perhaps very tiny, each with spontaneous magnetization. Above θ, their moments will be free (rotation, wall displacement or reversal) and the magnetic state of the body will establish itself by the combined action of the field and of thermal agitation (a sort of paramagnetism). Below θ, the elementary moments will be bound to equilibrium positions from which they undergo only reversible displacements in weak fields. The temperature θ will vary at each point in the body, perhaps with the dimensions and the shape of the crystalline grains, and will be broadly distributed between the Curie point and room temperature. One can thus explain thermoremanence by the progressive fixing, in the course of cooling, of moments, which find themselves held fast when they pass through their individual temperature θ. One can also account for the shape of thermomagnetic curves, in particular the strong increase in susceptibility below the Curie point."

In this remarkable paragraph are all the essential ingredients of Néel's picture: blocking temperature θ; unblocked or superparamagnetic state above θ in which thermal agitation establishes an equilibrium; freezing in of the equilibrium in cooling with only reversible changes possible below θ; and distribution of θ values due to variable sizes and shapes of grains, explaining the spectrum of partial TRMs comprising the total TRM of the body. Thellier even anticipates the explanation of the Hopkinson peak in susceptibility at high temperature. Although Thellier's ideas were not crystallized into a quantitative theory, his intuition was correct in every essential.

Néel did not directly acknowledge these early theoretical ideas but his appreciation of Thellier's enormous experimental contributions is clear in his commentary on a paper summarizing all of Thellier's results (Thellier, 1951, p. 217; translated from French): "Among the phenomena demonstrated in the magnetization of baked clays and lavas, we should distinguish carefully (total) TRM on the one hand and the independent magnetizations acquired in nonoverlapping temperature intervals (i.e., partial TRMs) on the other... The second is a much more remarkable phenomenon and it is to Thellier that we owe our experimental knowledge of it."

Anhysteretic magnetization and AF demagnetization

Thellier's earliest scientific achievements were made in collaboration with three remarkable women, whose contributions tend to be overlooked. First was his wife Odette Thellier, who participated in experiments showing that thermal demagnetization of TRM to a temperature T_1 resulted in precisely the same intensity as suppressing the field at T_1 during initial cooling from the Curie point T_C (i.e., pTRM $(T_C, T_1, \boldsymbol{H})$ (Thellier and Thellier, 1941). From these experiments evolved the pTRM reciprocity law. Odette Thellier was also her husband's full partner in archeomagnetic research on both paleodirections and paleointensities throughout the 1940s and 1950s, culminating in their famous paper, Thellier and Thellier (1959) (see following section).

Thellier's second collaborator, Juliette Roquet, extended investigations to synthetic minerals of both fine and coarse grain sizes and natural materials containing different size fractions. Her thesis (Roquet, 1954) is the first systematic rock magnetic study of grain-size trends in isothermal and thermal properties of partial TRM and other remanences. It set a standard that has yet to be surpassed, thanks to her dedication and the exacting standards set by her mentor.

Third was Francine Rimbert, who constructed one of the first AF demagnetizers and used it to demonstrate that anhysteretic remanent magnetization (ARM) is almost as strong and resistant to alternating fields as TRM (Thellier and Rimbert, 1954). They pointed out the need to scrupulously eliminate any extraneous fields during AF demagnetization, including the ambient geomagnetic field, to avoid contaminating NRM by unwanted ARMs. This was the pioneering study in AF demagnetization, which was then developing as a standard paleomagnetic cleaning method.

Rimbert's thesis (Rimbert, 1959), another monumental piece of work testifying to her ability and her supervisor's stringent expectations, is a complete experimental and theoretical study of ARM, partial ARMs, and AF demagnetization. It is the ultimate justification of AF bias methods for encoding magnetic information, for example in analog audio and video recording, and of AF cleaning of successive generations of primary and secondary NRMs in rocks. Laws analogous to those of Thellier for partial TRM have since been verified for fine-grained magnetic oxides (Dunlop and West, 1969) and ferromagnetic thin films (Papusoi and Apostol, 1979), ARM playing the role of TRM and AF substituting for temperature.

Archeomagnetism and the Thellier-Thellier paleointensity method

Archeomagnetism is the study of the direction and intensity of the geomagnetic field during historical times. Directional work is on the face of it more straightforward than determining paleointensity because loss of part of the primary NRM does not affect the direction of the remainder. Loss of NRM intensity, on the other hand, translates directly into a diminished estimate of paleointensity; a problem the Thellier-Thellier method was designed to overcome.

In spite of the comparative ease of directional measurements and the early beginning made by Émile and Odette Thellier, they published during the 1940s and 1950s only sporadic determinations of inclination I (for pots and bricks for which only paleohorizontal at the time of firing was known) or both I and declination D (for fixed objects such as the kilns in which firing took place). This was not for want of experimental work but because of the Thelliers desire to have a complete and trustworthy picture of the variations of D and I through time before publishing a master curve that would certainly be widely used by archeologists as a dating tool. Thus it was only after investigations were complete for about 100 sites in France, Turkey, Cambodia, and North Africa that Émile Thellier revealed in a presentation at the 1971 IUGG assembly in Moscow the fruits of all their labors, curves of secular variation of the field throughout the past 2000 years.

The Thelliers moved more quickly to establish the historical record of field intensity. For paleodirectional work, Émile Thellier had perfected a sensitive spinner magnetometer on a grand scale, still in use today, capable of measuring the NRMs of intact pots with dimensions as large as 50 cm without the need to subsample these priceless archeological treasures. Paleointensity work required a methodological invention of equal ingenuity, the Thellier-Thellier protocol, still the standard method used today.

The Thellier-Thellier method compares NRM, produced thermally in an unknown ancient field $\boldsymbol{H}_A$, with a TRM produced in a known laboratory field $\boldsymbol{H}_L$. This basic idea had been put forward earlier by Folgerhaiter (1899) and Koenigsberger (1938) but no trustworthy results emerged. Folgerhaiter, quoted by Thellier (1938, 287), says (translated from French) "One could also arrive at some conclusion about the intensity of the terrestrial field by reheating ancient vases and comparing the ancient and presently acquired intensities of magnetization; but measurements made on vases heated and reheated in repeated experiments have shown me that this method leads to too uncertain results." The reason for the uncertainty is alteration of the chemistry and physical state of the magnetic minerals resulting from heating, demonstrated very clearly by Koenigsberger's progressive heating experiments on igneous rocks.

The novelty of the Thelliers' procedure is in the interweaving of pairs of heating-cooling steps to successively higher temperatures, instead of a single heating-cooling to T_C. In their original version (Thellier and Thellier, 1959), both heatings to a particular temperature T_i were carried out in the presence of a laboratory field (the ambient Earth's field in their experiments) but the sample was rotated 180° between heatings. In the currently most used version (Coe, 1967), the first heating-cooling is in zero field and the second in $\boldsymbol{H}_L$. In the Coe version, the first heating serves to demagnetize that part of the NRM with $T_B \leq T_i$, while the second heating replaces this loss with a partial TRM ($T_i, T_o, \boldsymbol{H}_L$). The NRM and partial TRM are not generally in the same direction, so that they must be obtained by vector subtraction of the results of the two heatings. In the original version, equal partial TRMs in opposite directions are acquired in the two heatings, but the vector subtractions are still straightforward. Although the modified version gives a neater segregation between NRM loss and partial TRM gain, the original version has some bonuses: the two heatings have perfect symmetry and no null field is needed.

The Thellier-Thellier method is firmly rooted in Thellier's three laws of partial TRM. This protocol has three tremendous advantages, not matched by other techniques:

1. There is a built-in test of the TRM origin of the NRM, namely constancy of the ratio of NRM lost/partial TRM gained over successive heating steps.
2. Portions of the NRM that are unreliable can be recognized and discarded. A common contaminant at low T_i is viscous remanent magnetization (VRM) produced by the present Earth's field. Alteration of mineral microstructure or chemistry tends to occur at high T_i.
3. Linear least-squares fitting to the set of acceptable NRM and partial TRM data can be used to obtain the mean paleointensity ratio H_A/H_L and its associated error. This procedure, the Arai plot, was introduced later, by Nagata *et al.* (1963) but it is a natural consequence of the linear replications in the Thellier-Thellier method.

The name of Koenigsberger has been associated by some with this method and given precedence over the Thelliers themselves (so-called KTT method). This misrepresents the facts. In four of Koenigsberger's papers published in German journals between 1930 and 1936 and in his summary work in English (Koenigsberger, 1938), there is no indication that he was aware of partial TRMs or their properties. The only paper by Thellier he cites is on determining directions. He did carry out stepwise heatings with $\boldsymbol{H}_L$ parallel or antiparallel to NRM but these were separate, not interwoven, experiments, and there was no attempt to use the results to estimate field strength. Most of his samples altered so much in the first set of heatings that there was no symmetry between $+\boldsymbol{H}_L$ and $-\boldsymbol{H}_L$ curves.

Koenigsberger espoused the idea that NRM spontaneously decayed with time (what we would nowadays call viscous decay) so that the ratio Q_n = NRM/kH would be systematically lower than Q_t = TRM/$k'H$ for rocks of increasing age. Here k, k' are susceptibilities measured before and after TRM acquisition and H is the local present Earth's field. He viewed this as an age determination, not a paleointensity, method. Thellier (1938, p. 293) correctly ascribed the differences between Q_n and Q_t to increased susceptibility resulting from alteration during heating and not to spontaneous decay of NRM. Thellier went on to suggest that the ratio Q_n/Q_t = (NRM/TRM) (k'/k) might serve as a rough estimate of the paleointensity ratio H_A/H_L. This is a slight improvement on Folgerhaiter's prescription, H_A/H_L = NRM/TRM, in that it takes some account of alteration of a sample through the ratio k'/k, but it is far from being an earlier incarnation of the powerful and sophisticated Thelliers' method.

Viewing the considerable alteration evidenced by the differences between k' and k in Koenigsberger's igneous samples, Thellier soon came to the reasonable conclusion that the suggested correction procedure was unjustified, particularly since the quantitative relation between remanence and susceptibility depends strongly on mineralogy and grain size. He says (Thellier, 1938, p. 293) "The susceptibility has changed markedly as a result of heating for many of these rocks; they should be rejected for the purpose of studying the intensity of the terrestrial field." Since then, many intricate and ingenious schemes have been proposed for "undoing" the effects of alteration during heating but none in the final analysis gives results that the most paleomagnetists would trust. There is really no substitute for the Thellier-Thellier method (under which we include microwave heating methods that heat the magnetic minerals but not the rock matrix), nor for the uncompromising standards set by Émile Thellier.

David J. Dunlop

Bibliography

Aubouin, J., and Coulomb, J., 1987. La vie et l'oeuvre d'Émile Thellier. La Vie des Sciences. *Comptes rendus de l'Académie des sciences (Paris), Série génerale*, **4**(6): 607–610.

Brunhes, B., 1906. Recherches sur la direction d'aimantation des roches volcaniques. *Journal of Physique, Série 4*, **5**: 705–724.

Coe, R.S., 1967. Paleointensities of the Earth's magnetic field determined from Tertiary and Quaternary rocks. *Journal of Geophysical Research*, **72**: 3247–3262.

Dunlop, D.J., and Özdemir, Ö., 1997. *Rock Magnetism: Fundamentals and Frontiers*. Cambridge and New York: Cambridge University Press.

Dunlop, D.J., and West, G.F., 1969. An experimental evaluation of single-domain theories. *Reviews of Geophysics*, **7**: 709–757.

Folgerhaiter, G., 1899. Sur les variations séculaires de l'inclinaison magnétique dans l'antiquité. *Journal de Physique, Série 3*, **8**: 5–16.

Koenigsberger, J.G., 1938. Natural residual magnetism of eruptive rocks. *Terrestrial Magnetism and Atmospheric Electricity*, **43**: 119–130, 299–320.

Nagata, T., 1943. The natural remanent magnetism of volcanic rocks and its relation to geomagnetic phenomena. *Bulletin of Earthquake Research Institute, University of Tokyo*, **21**: 1–196.

Nagata, T., Arai, Y., and Momose, K., 1963. Secular variation of the geomagnetic total force during the last 5000 years. *Journal of Geophysical Research*, **68**: 5277–5281.

Néel, L., 1949. Théorie du traînage magnétique des ferromagnétiques en grains fins avec applications aux terres cuites. *Annales de Géophysique*, **5**: 99–136.
Papusoi, C., and Apostol, P., 1979. Propriétés d'indépendance des rémanences magnétiques hystérétiques et anhystérétiques dans les couches électrolytiques de Co-Ni-P. *Physica Status Solidi (a)*, **54**: 477–486.
Pullaiah, G., Irving, E., Buchan, K.L., and Dunlop, D.J., 1975. Magnetization changes caused by burial and uplift. *Earth and Planetary Science Letters*, **28**: 133–143.
Rimbert, F., 1959. Contribution à l'étude de l'action de champs alternatifs sur les aimantation rémanentes des roches. *Revue de l'Institut Francais du Petrole et Annales des Liquides Combustibles*, **14**: 17–54, 123–155.
Roquet, J., 1954. Sur les rémanences magnétiques des oxydes de fer et leur intérêt en géomagnétisme. *Annales de Géophysique*, **10**: 226–247, 282–325.
Roquet, J., and Thellier, É., 1946. Sur des lois numériques simples, relatives à l'aimantation thermorémanente du sesquioxyde de fer rhomboédrique. *Comptes rendus de l'Académie des sciences (Paris)*, **222**: 1288–1290.
Thellier, É., 1938. Sur l'aimantation des terres cuites et ses applications géophysiques. Annales de. l'Institut de Physique du Globe, Université de Paris, **16**: 157–302.
Thellier, É., 1941. Sur les propriétés de l'aimantation thermorémanente des terres cuites. *Comptes rendus de l'Académie des sciences (Paris)*, **213**: 1019–1022.
Thellier, É., 1946. Sur la thermorémanence et la théorie du métamagnétisme. *Comptes rendus de l'Académie des sciences (Paris)*, **223**: 319–321.
Thellier, É., 1951. Propriétés magnétiques des terres cuites et des roches. *Journal de Physique et le Radium*, **12**: 205–218.
Thellier, É., 1967. *Notice sur les titres et travaux scientifiques de M. Émile Thellier, Professeur à la Faculté des Sciences de Paris.* 28 pp.
Thellier, É., and Rimbert, F., 1954. Sur l'analyse d'aimantations fossiles par action de champs magnétiques alternatifs. *Comptes rendus de l'Académie des sciences (Paris)*, **239**: 1399–1401.
Thellier, É., and Thellier, O., 1941. Sur les variations thermiques de l'aimantation thermorémanente des terres cuites. *Comptes rendus de l'Académie des sciences (Paris)*, **213**: 59–61.
Thellier, É., and Thellier, O., 1959. Sur l'intensité du champ magnétique terrestre dans le passé historique et géologique, *Annales Géophysique*, **15**: 285–376.

Cross-references

Archeomagnetism
Demagnetization
Magnetic Susceptibility
Magnetization, Anhysteretic Remanent (ARM)
Magnetization, Natural Remanent (NRM)
Magnetization, Thermoremanent (TRM)
Magnetization, Viscous Remanent (VRM)
Nagata, Takesi (1913–1991)
Neél, Louis (1904–2000)

THERMAL WIND

The term *thermal wind* refers to steady or slowly time-varying shear flow, driven by buoyancy forces due to lateral gradients in density and modified by planetary rotation through the Coriolis acceleration. Thermal winds in the Earth's fluid outer core are important in the geodynamo. These flows generate toroidal magnetic field in the outer core, drive anomalous rotation of the solid inner core, and contribute to the observed secular variation of the geomagnetic field, particularly the westward drift. Thermal winds may also play a key role in geomagnetic polarity reversals. This article reviews the basic properties of thermal winds in the core and describes how these flows affect the geomagnetic field.

Examples of thermal wind flows

Examples of thermal wind flows abound in Nature. The general circulation of Earth's atmosphere is a thermal wind to a first approximation, and is driven by lateral variations in density associated with the pole-to-equator zonal temperature gradient. Some major ocean current systems such as the Gulf Stream are driven partly by lateral density gradients and qualify as thermal wind flows, the density gradients associated with these currents arising from both temperature and salinity variations. Lateral density variations are certainly present in the outer core, due to the combined effects of thermal and compositional variations. Because rotation influences fluid motion in the core as much as it does in the ocean and atmosphere, thermal winds are likely to be important parts of core flow. However, since the outer core is a much deeper fluid and is affected by convection and the forces associated with Earth's magnetic field, thermal winds in the core are expected to differ from their atmospheric and oceanic counterparts in certain respects.

The thermal wind equation

A thermal wind equation appropriate for Earth's outer core can be obtained from the vorticity conservation equation in the limit where the fluid inertia and the viscous force are negligible in comparison with effects of the Coriolis acceleration, density gradients, and the Lorentz force due to Earth's magnetic field. Here the focus is on the zonal (the longitude-averaged) component of motion in the longitude (east-west) direction, which is usually the largest part of the thermal wind. Assuming a Boussinesq fluid, the equation for this component of motion near the surface of the core is (Davidson, 2001)

$$2\Omega\frac{\partial u_\phi}{\partial z} = -\frac{g}{\rho R}\frac{\partial \rho}{\partial \theta} - \left(\nabla \times \frac{\vec{J}\times\vec{B}}{\rho}\right)_\phi,$$

where u_ϕ is the zonal velocity in the ϕ-direction (measured eastward), ρ is fluid density, θ is co-latitude, Ω is the angular velocity of rotation in the z-direction, $\vec{J}$ and $\vec{B}$ are electric current density and magnetic field intensity, respectively, and g is the acceleration of gravity at the core's outer radius R. The thermal wind equation is written here in terms of density, but as its name suggests, it is often applied to fluids where the density is a function of temperature only. In these cases

$$\frac{\partial \rho}{\partial \theta} = -\alpha\rho\frac{\partial T}{\partial \theta},$$

where T is temperature and α is the thermal expansion coefficient of the fluid.

Before describing thermal winds in the core, it is useful to consider a rotating fluid in which lateral density gradients and electric currents are negligible. In this case the thermal wind equation reduces to the Proudman-Taylor constraint:

$$\frac{\partial u_\phi}{\partial z} = 0.$$

Fluid motions satisfying this constraint are sometimes called columnar, because they have zero shear in the axial direction, parallel to the rotation axis. In a rotating fluid with strong lateral density gradients or strong Lorentz forces, the Proudman-Taylor constraint does not hold. The above equation dictates that the flow contains axial shear in proportion to these effects. Other terminology is sometimes used for these flows. In Meteorology and Oceanography, the term thermal wind usually refers specifically to the shear, not the flow itself

(Cushman-Roisin, 1994). In rotating, electrically conducting fluids, Lorentz force-driven shear flows are sometimes called *magnetic winds*.

Thermal wind in the core

The general pattern of zonal flow we expect in the outer core is illustrated in Figure T3, which shows cross sections of the time-averaged zonal temperature, fluid motion, and magnetic field structure from a numerical dynamo model. This dynamo is driven by thermal convection and the lateral density variations are proportional to the lateral temperature variations. The model is a highly idealized (and probably over-simplified) representation of the geodynamo, but it illustrates the basic structure of thermal winds in a rotating, convecting dynamo. Although some of the input parameters are not accurately scaled (in particular, the rate of rotation is far too slow for the Earth and the fluid has higher electrical conductivity than the outer core) the model gives realistic results for several important properties. For example, its external magnetic field is dipole-dominant, like the geomagnetic field, and it tends toward axial symmetry when averaged over long periods of time. At any instant, however, both the flow and the magnetic field are not as symmetric, as shown in Figure T3. The magnetic secular variation in this model exhibits a preference for westward drift, especially near the poles, which is also Earth-like. These general similarities with observed geomagnetic field behavior result from fundamental constraints on convection in a rotating spherical shell, and are not particular to this model. It is therefore reasonable to assume that the pattern of zonal flow in the model may approximate the time-averaged pattern of zonal flow within the Earth's core.

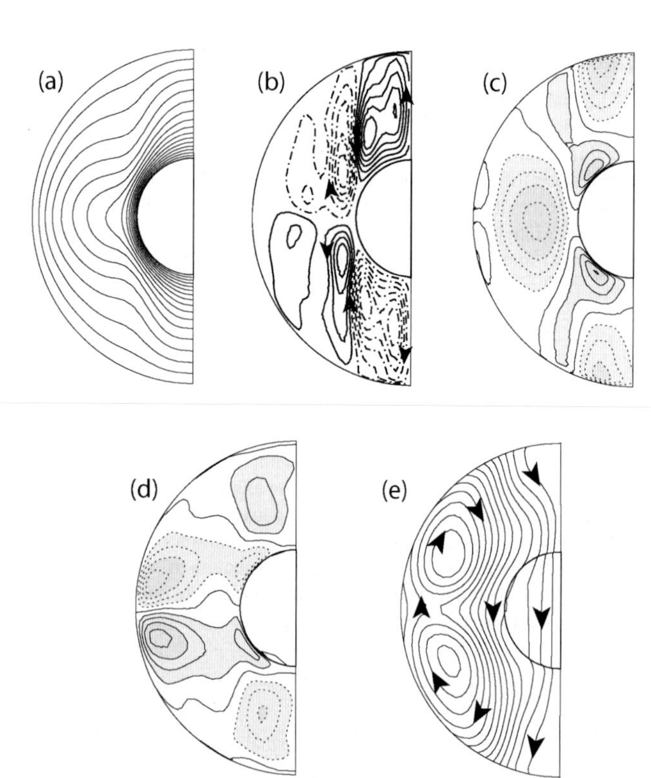

Figure T3 Cross-sections of time- and azimuthally averaged structure of a numerical dynamo model driven by thermal convection in a rotating, electrically conducting spherical fluid shell. The parameters of this model are the same as the thermal convection benchmark dynamo of Christensen *et al.* (2001), but with twice the rotation and heating rates. (a) Contours of temperature; (b) streamlines of meridional circulation with flow direction arrows; (c) contours of zonal velocity showing the thermal wind; (d) contours of toroidal magnetic field intensity; (e) poloidal magnetic field lines with direction arrows. Contour shading is proportional to magnitude; solid and dashed lines indicate positive and negative values, respectively. Zonal velocity and toroidal magnetic field intensity are positive eastward. Magnetic polarity in this model is normal relative to present-day geomagnetic polarity; this has no effect on the flow.

The large amount of axial shear in Figure T3c and its spatial relationship to the lateral temperature variations in Figure T3a and the meridional circulation in Figure T3b indicate the time-averaged zonal velocity is mostly thermal wind. In contrast, there is very little columnar motion in the time-averaged flow in this dynamo. In the Earth's core, there is evidence from the geomagnetic secular variation for time-varying zonal columnar flow with periodicity of order one century. Oscillating columnar core flow is thought to be important in the exchange of momentum between the core and the mantle associated with decade time-scale fluctuations in the length of the day. But the existence of columnar flow in the core over longer periods of time has yet to be established.

The time-average zonal temperature shows a basic octupole pattern. Elevated temperatures occur near the equatorial plane in Figure T3a, a consequence of efficient heat transport by small-scale convection, which is suppressed by the zonal averaging and is not evident in the velocity fields in Figure T3b and T3c. The thermal wind is particularly strong within the axial cylinder containing the inner core, the inner core tangent cylinder (*q.v.*). Here it is closely related to the axisymmetric meridional circulation, which consists of a polar plume with return flow along the tangent cylinder. Zonal flow inside the tangent cylinder is directed westward near the core-mantle boundary, but reverses direction with depth and becomes eastward near the inner core boundary. The high-latitude westward motion near the core-mantle boundary in both hemispheres are polar vortices, analogs to the ozone-trapping polar vortices in the stratosphere. The eastward motion of the thermal wind deep in the outer core results in viscous and electromagnetic stresses on the inner core boundary. The drag produced by these stresses tends to make the inner core super-rotate, that is, spin slightly faster than the mantle. This offers an explanation for the anomalous rotation of the inner core inferred by some seismic studies. Figure T3c shows that the belts of eastward motion responsible for inner core super-rotation extend out to the core-mantle boundary at midlatitudes. At lower latitudes, the thermal wind flow immediately below the core-mantle boundary is weak, but at greater depth there is a submerged westward jet centered on the equator, an equatorial undercurrent.

The combination of the westward undercurrent and the eastward flow near the inner core boundary generates much of the toroidal component of the magnetic field shown in Figure T3d. The shear contained in these flows stretches the poloidal magnetic field lines shown in Figure T3e, winding them into pairs of toroidal field coils with opposite polarity in each hemisphere. This is an example of toroidal field induction by the so-called ω-dynamo effect, and is considered to be a critical part of the geodynamo process in the core. The thermal wind also influences the poloidal magnetic field, although in a different way. Comparison of Figure T3e and T3c shows that the poloidal field lines tend to align with the contours of the thermal wind velocity, a configuration that allows the field to nearly co-rotate with the thermal wind. The tendency of poloidal field lines to align with the thermal wind is an example of Ferraro's (1937) law of co-rotation, a general magnetohydrodynamic principle that has been used to infer planetary rotation rates from magnetic variations.

In rotating fluids where the lateral density variations are produced by convection, the amplitude of the thermal wind obeys the following relationship:

$$U \approx \left(\frac{F}{\Omega}\right)^{1/2},$$

where U is the r.m.s. thermal wind amplitude and F is the flux of buoyancy associated with the convection. The same formula has been found to hold in convection-driven dynamos with relatively strong magnetic fields, another consequence of Ferraro's law of corotation (Aubert, 2005). According to this relationship, a thermal wind velocity of $U = 10^{-3}\ \mathrm{m\,s^{-1}}$ in the core, which is consistent with the observed geomagnetic secular variation, is associated with a convective buoyancy flux of about $F = 7 \times 10^{-11}\ \mathrm{m^2\,s^{-3}}$. This is within the range of the buoyancy flux estimates based on the core's present-day energy budget (Aurnou *et al.*, 2003). A scaling analysis of the thermal wind equation reveals that the lateral density variations needed to maintain a thermal wind flow with this amplitude in the core are less than one part in a million (Jackson and Bloxham, 1991), too small, unfortunately, to be observed in the gravity field or by seismic methods.

The high degree of symmetry in Figure T3 is a consequence of time and azimuthal averaging. Significant departures from axial symmetry are present in the thermal wind and the other variables at any instant, because of transient chaotic fluctuations in the model. In the core, similar transient chaotic fluctuations are also likely, and these might account for the hemispheric asymmetry observed in the geomagnetic field during historical times, and may be the cause of dipole excursions and polarity reversals. Even longer-lasting deviations from symmetry are also possible in the geodynamo, due to interaction of core flow with heterogeneity in the lower mantle. It is well established that spatial heterogeneity in boundary heat flow can produce localized thermal winds in a rotating fluid (Sumita and Olson, 1999), and it has been proposed that thermal winds in the core are tied to lower mantle heterogeneity by this mechanism. It has also been proposed that slow changes in the core flow, responding to changes in the structure of the lower mantle heterogeneity, can causes the long-term variations observed in magnetic reversal frequency record (Gubbins, 1987).

In summary, multiple lines of evidence indicate thermal winds are present in the core and affect the geodynamo in a variety of ways. Much remains to be learned about the nature of these flows and their relationship to the other parts of the Earth.

Peter Olson

Bibliography

Aubert, J., 2005. Steady zonal flows in spherical shell dynamos. *Journal of Fluid Mechanics*, **542**: 53–67.

Aurnou, J., Andreadis, S., Zhu, L., and Olson, P., 2003. Experiments on convection in Earth's core tangent cylinder. *Earth and Planetary Science Letters*, **212**: 119–134.

Christensen, U.R., *et al.*, 2001. A numerical dynamo benchmark. *Physics of Earth and Planetary Interiors*, **128**: 51–65.

Cushman-Roisin, B., 1994. *Introduction to Geophysical Fluid Dynamics.* Englewood Clifts, NJ: Prentice-Hall.

Davidson, P.A., 2001. *An Introduction to Magnetohydrodynamics.* New York: Cambridge University Press.

Ferraro, V.C.A., 1937. The non-uniform rotation of the sun and its magnetic field. *Monthly Notices of the Royal Astronomical Society*, **97**: 458.

Gubbins, D., 1987. Mechanism for geomagnetic polarity reversals. *Nature*, **326**: 167–169.

Jackson, A., and Bloxham, J., 1991. Mapping the fluid flow and shear near the core surface using the radial and horizontal components of the magnetic field. *Geophysical Journal International*, **105**: 199–212.

Sumita, I., and Olson, P., 1999. A laboratory model for convection in Earth's core driven by a thermally heterogeneous mantle. *Science*, **286**: 1547–1549.

Cross-references

Convection, Chemical
Core Convection
Core Motions
Dynamos, Mean Field
Geodynamo, Numerical Simulations
Inner Core Rotation
Inner Core Tangent Cylinder
Proudman-Taylor Theorem

TIME-AVERAGED PALEOMAGNETIC FIELD

The *geocentric axial dipole hypothesis* (*q.v.*), or GAD, is a central tenet of paleomagnetism: the geomagnetic field is assumed to average over time to a simple dipolar form aligned with the Earth's rotation axis. This enables measurement of past plate motion, continental drift, and regional geological deformation. The dipole assumption allows any magnetic direction determined at a site to be transformed to a virtual geomagnetic pole (VGP), defining the axis of the assumed dipole. Many VGPs determined within a single rock unit are found to cluster about a center, called the paleomagnetic pole. The paleomagnetic pole is treated as an estimate of the location of the geographic pole at the time of formation of the rock unit; the scatter of VGPs is supposed to be caused by geomagnetic secular variation. The latitude of the rock unit at the time of its formation, and the angle through which the rock unit has rotated since its formation, can be obtained from the paleomagnetic pole position. The scatter of VGPs around the paleomagnetic pole provides an estimate of the error on the paleomagnetic pole and hence the errors on the latitude and rotation of the rock unit.

Merrill and McElhinny (1996) discuss the establishment of the GAD in detail. Simple clustering of the VGPs is not enough to demonstrate dipolar structure: a definitive determination requires similar paleomagnetic pole positions for a range of contemporary locations. This is possible in the recent geological past when plate motions are negligible or are well determined by other means; for the more distant past it is possible to compare latitudes determined paleomagnetically with those determined by paleoclimate. Statistical distributions of paleomagnetic inclination can also be used to establish the dipole structure. Adequate time sampling of the rock unit is also essential in determining an accurate paleomagnetic pole, which must be long enough to average out the secular variation. In practice the VGP cluster can give confidence of adequate time sampling but dating of individual samples often fails to define the time interval.

Theory has had little or no input into the development of the GAD because dynamo theory has only very recently been developed to the point where it can be used to explore the nature of the time average. There are three separate issues. The first is the time span required to make an average. This must exceed the advection time, or time for fluid to traverse the core, which is about a thousand years. It probably needs to exceed the diffusion time, or time for the magnetic field to decay in the absence of dynamo action, which is about twenty-five thousand years. The magnetic field is quite chaotic—excursions are now known to occur every few tens of thousands of years, at least in the Brunhes—which suggests an even longer time span. Some paleomagnetists advocate much longer times of millions of years, and the issue is not resolved.

The second is the dipolar structure, which requires the inclination to vary with distance from the pole according to the dipole formula, and the vertical field to vary as $\cos\theta$, where θ is co-latitude. The dipole form falls off more slowly than any other configuration with distance from the source, and is therefore favored by our great distance from the Earth's core. It is very unlikely to result from dynamo action, and we now suspect preferred generation of magnetic field just outside the *tangent cylinder* as occurs in many geodynamo models and is seen in the modern field. Paleomagnetic data has been modeled by a variety of axisymmetric configurations limited to low degree axisymmetric spherical harmonics without any consensus. A review is given in Merrill and McElhinny (1996).

The third is whether the time-average contains any departures from axial symmetry. The simplest dynamo model would have the magnetic field rotating freely with respect to the solid mantle, leaving an axisymmetric time-average. Any departure from axial symmetry would therefore be evidence of the influence on core convection of anomalies in the solid mantle. Lateral variations in heat flux, topography, and electrical conductivity have all been suggested, but only heat flux has been investigated quantitatively (see *Core-mantle coupling, thermal*). Some numerical geodynamo models have laterally varying heat flux or temperature on the outer boundary defined by the shear wave velocity in the lowermost mantle, and some of these exhibit nonaxisymmetric time averages bearing some resemblance to the modern field (Bloxham, 2000). Paleodeclination, which is zero for all axisymmetric field configurations, is best for detecting departures from axial symmetry. Unfortunately it is a difficult measurement to make with confidence because the rock may have rotated. Geological mapping is capable of detecting tilt, or rotation about a horizontal axis, but cannot directly detect rotation about a vertical axis. Some authors therefore reject paleodeclination information altogether, while others claim that only an axisymmetric time-average can be estimated (McElhinny *et al.*, 1996). However, when paleomagnetic data are subjected to the same analysis as is used in making models of the modern field (see *Main field modeling*) they produce departures from axial symmetry that resemble, at least in the northern hemisphere, those of the modern field (Gubbins and Kelly, 1993; Johnson and Constable, 1995), as expected from recent geodynamo simulations.

In summary, the time-averaged field encompasses both the oldest and the best-established hypothesis of paleomagnetism, the GAD, and the newest and most controversial meeting point between paleomagnetism and dynamo modeling, the question of whether the solid mantle influences the core. Current research programs into both the geodynamo and the paleomagnetic time average will undoubtedly answer some of the remaining questions in the near future.

David Gubbins

Bibliography

Bloxham, J., 2000. The effect of thermal core-mantle interactions on the paleomagnetic secular variation. *Philosophical Transactions of the Royal Society of London*, **358**: 1171–1179.

Gubbins, D., and Kelly, P., 1993. Persistent patterns in the geomagnetic field during the last 2.5 myr. *Nature*, **365**: 829–832.

Johnson, C., and Constable, C., 1995. The time-averaged geomagnetic field as recorded by lava flows over the past 5 myr. *Geophysical Journal International*, **122**: 489–519.

McElhinny, M.W., McFadden, P.L., and Merrill, R.T., 1996. The time averaged paleomagnetic field 0–5 ma. *Journal of Geophysical Research*, **101**: 25007–25027.

Merrill, R.T., and McElhinny, M.W., 1996. *The Magnetic Field of the Earth.* San Diego, CA: Academic.

Cross-references

Core-Mantle Coupling, Electromagnetic
Core-Mantle Coupling, Thermal
Core-Mantle Coupling, Topographic
D″, Seismic Properties
Geocentric Axial Dipole Hypothesis
Geodynamo, Dimensional Analysis and Timescales
Geodynamo, Numerical Simulations
Geodynamo, Symmetry Properties
Geomagnetic Excursion
Geomagnetic Field, Asymmetries
Geomagnetic Secular Variation
Geomagnetic Spectrum, Temporal
Harmonics, Spherical
Inner Core Tangent Cylinder
Main Field Modeling
Nondipole Field
Paleomagnetic Field Collection Methods
Paleomagnetic Secular Variation
Statistical Methods for Paleovector Analysis
Westward Drift

TIME-DEPENDENT MODELS OF THE GEOMAGNETIC FIELD

In this article we will review a selection of the different time-dependent models of the magnetic field that have been produced. We generally restrict attention to models that have been produced specifically as time-dependent models, typically using a dataset spanning a wide range in time; only a little reference is made to models designed to describe either the static magnetic field or its rate of change (secular variation) at a particular point in time. Useful descriptions of these types of model can be found in Barraclough (1978) or Langel (1987).

Our description centers then on models of the magnetic field **B** which are simultaneously models of its spatial ((r, θ, ϕ) in spherical coordinates) dependence and the temporal dependence (t denotes time). The standard technique which is common to all the analyses we will describe is to employ the *spherical harmonic* (*q.v.*) expansion of the field in terms of Gauss coefficients $\{g_l^m\ h_l^m\}$ for the internal field; some of the most recent models also incorporate coefficients representing the external field. All the models will employ the Schmidt quasinormalization common in geomagnetism.

A time dependent model of the field necessarily must be built using a dataset spanning a period of time, denoted herein $[t_s, t_e]$. In order that a spherical harmonic analysis can be performed a parametrization is required for the temporal variation of the field. The unifying idea, common to all analyses, is to use an expansion for the Gauss coefficients of the form:

$$g_l^m(t) = \sum_i {}^i g_l^m \phi_i(t), \qquad \text{(Eq. 1)}$$

where ϕ_i are a set of basis functions and the ${}^i g_l^m$ are a set of unknown coefficients. (A similar expansion is of course used for h_l^m.) The different models that have been produced over the last few decades differ in their choice of the $\phi_i(t)$. With an expansion of the form (1) the unknown coefficients $\{{}^i g_l^m\ {}^i h_l^m\}$ are denoted as a model vector **m**, and when linear data such as the elements (X, Y, Z) are required to be synthesized (denoted by vector **d**) the resulting forward problem is linear and of the form:

$$\mathbf{d} = A\mathbf{m}, \qquad \text{(Eq. 2)}$$

where A is often termed the equations of condition or design matrix. It is generally the case that the inverse problem of finding the coefficients **m** is solved using the method of least squares (see Main field modeling).

It is straightforward to treat single observations of the field (such as made by surveys or satellites) as being independent measurements that can be fitted simultaneously in a least-squares process. Some words are in order regarding the treatment of observatory data in time-dependent field modeling. Observatories obviously supply critical data on the secular variation, and indeed the accuracy of many of the modern field models rests on the observatory time series. A problem that must be recognized, however, is the fact that the observatories are subject to a (quasi-) constant field associated with the magnetization of the crust in the region that they are located. If observatory data are mixed with other types of data (survey, satellite data), this so-called observatory bias must be recognized, otherwise it will bias the solution for the main field because an observatory time-series essentially records it many times. Two approaches have developed for dealing with this. The first, developed by Langel *et al.* (1982), is to solve for the observatory biases (three per observatory in the *X*, *Y*, and *Z* directions) as unknowns at the same time as solving for the magnetic field. This technique continues to be adopted in the comprehensive series of field models (see below), and works very effectively. The second approach is to desensitize the observatory data to the presence of the bias. An effective way of doing this is to work with the rate-of-change of the field from the observatory, and hence first differences of observatory data are used in the ufm and gufm series of models (see below). There appears to be very little difference in the results of the two approaches.

Taylor series models

The earliest analyses simply used a Taylor expansion for the Gauss coefficients of the form:

$$g_l^m(t) = g_l^m(t_0) + \dot{g}_m^m(t_0)(t - t_0) + \ddot{g}_l^m(t_0)\frac{(t - t_0)^2}{2!} + \cdots \quad \text{(Eq. 3)}$$

about some central epoch here denoted t_0. This expansion is of the form (1), with the identification $\phi_i(t) = (t - t_0)^n/n!$ and ${}^ig_l^m = (\partial_t)^n g_l^m(t_0)$, the nth time derivative at the central epoch.

In the case of the Taylor expansion A is a dense matrix. The first models to be produced this way were those of Cain *et al.* (1965, 1967), who produced models GSFC(4/64) and GSFC(12/66) with temporal expansions truncated at first derivative and second derivative terms, respectively. The truncation level was subsequently raised to third derivative terms in the model GSFC(9/80) of Langel *et al.* (1982).

When it is desired to produce a model of the field spanning a long time period, it is clear that a large number of terms will be required in (3), and it no longer remains an attractive method because of numerical instabilities and lack of flexibility of the parametrization. It is known that in the case of an equidimensional inverse problem (the same number of data as unknowns) with perfect data, the inner-product matrix associated with the problem becomes the classically ill-conditioned Hilbert matrix. As a result the expansion (3) becomes unsuitable as a temporal expansion with many parameters.

Two-step models

A variety of models have been made by a two-step process of first making a series of spatial models at particular epochs, followed by some form of interpolation. For example, the international geomagnetic reference fields and definitive geomagnetic reference fields are strictly snapshot models of the field for particular epochs, but they can be used to calculate the magnetic field at times intermediate between two epochs by linear interpolation between the models. As a result it is possible to evaluate the DGRFs at any point in time between 1900 and the present day, though from a purist point of view they are not strictly time-dependent models of the magnetic field.

Finally, the idea of using splines as a temporal expansion was first employed in the two-step approach by Langel *et al.* (1986), an idea that proved influential on subsequent authors.

Models of SV over century timescales

Beginning in the mid-1980s, more flexible representations of the time dependency were introduced. Beginning with Bloxham (1987), who used Legendre polynomials, a variety of functions have been used—see Table T1. The methods have gradually moved toward the use of splines as basis functions, beginning with the work of Bloxham and Jackson (1992), very much in the pioneering spirit of Langel *et al.* (1986). [illegible] global basis functions such as Legendre or Chebychev expansions are used, the design matrix remains dense and requires considerable memory for its storage, whereas a B-spline basis is a *local* basis, meaning that the basis functions are zero outside a small range (see Figure T4). This fact leads to a design matrix which is sparse (in fact it is banded), and its storage is minimized. Secondly, the B-splines provide a flexible basis for smoothly varying descriptions of data. In fact one can show that of all the interpolators passing through a time-series of points (say $f(t_i), i = 1, N$), an expansion in cubic B-spline of order 4 ($\hat{f}(t)$ say) is the unique interpolator which minimizes a particular measure of roughness

$$\int_{t_s}^{t_e} \left[\frac{\partial^2 \hat{f}(t)}{\partial t^2}\right]^2 \mathrm{d}t. \quad \text{(Eq. 4)}$$

Table T1 Characteristics of some models of the time-varying magnetic field. L is the maximum degree of the internal secular variation, N is the number of temporal basis functions used for each Gauss coefficient

Model	L	N	Time period	Expansion	Regularized?	Author
GSFC(4/64)	5	2	1940–1963	Taylor	No	Cain *et al.* (1965)
GSFC(12/66)	10	3	1900–1966	Taylor	No	Cain *et al.* (1967)
GSFC(9/80)	13	4	1960–1980	Taylor	No	Langel *et al.* (1982)
MFSV/1900/1980/OBS	14	8	1900–1980	Legendre	Yes	Bloxham (1987)
	14	10	1820–1900, 1900–1980	Chebychev	Yes	Bloxham and Jackson (1989)
ufm1, ufm2	14	63	1690–1840, 1840–1990	B-spline	Yes	Bloxham and Jackson (1992)
gufm1	14	163	1690–1990	B-spline	Yes	Jackson *et al.* (2000)
CM3	13	14	1960–1985	B-spline integrals	Yes	Sabaka *et al.* (2002)
CM4	13	24	1960–2002.5	B-spline integrals	Yes	Sabaka *et al.* (2004)

The idea of attempting to construct a smooth representation in time is an application of "Occam's Razor," that there should be no extra detail in the representation than that truly demanded by the data. This idea of "regularization" has been employed in the models of Table T1 from

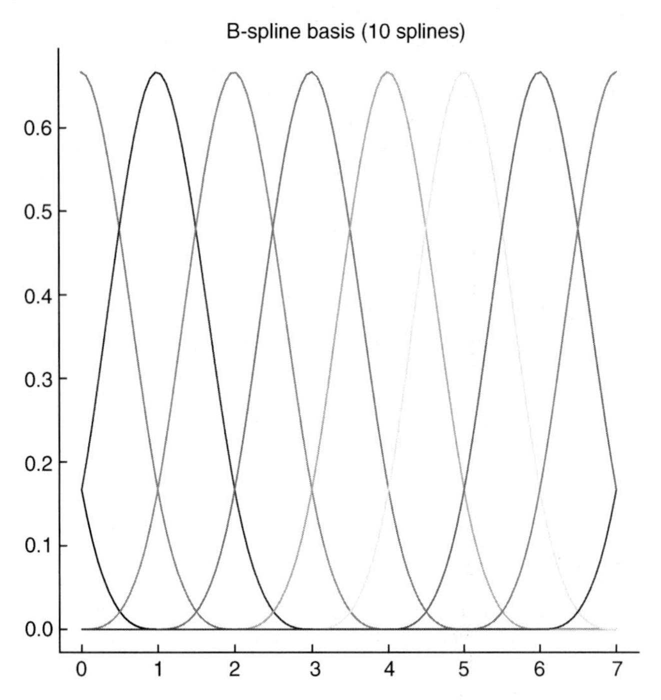

Figure T4 B-splines of order 4.

that of Bloxham (1987) onward. All the models minimize a combination of norms on the core-mantle boundary (CMB) of the form:

$$\int_{t_s}^{t_e} \left[\nabla_h^{(n_1)} \partial_t^{(n_2)} B_r \right]^2 \mathrm{d}\Omega \, \mathrm{d}t, \qquad \text{(Eq. 5)}$$

where B_r is the radial field on the CMB. The models produced by Bloxham, Jackson and coworkers use $n_1 = 0$ and $n_2 = 2$ in one norm, and $n_1 = 1$ and $n_2 = 0$ (approximately) in a second norm; this is slightly different to that of the comprehensive models (see below).

The ufm1/ufm2 and gufm1 field models share a common aim, namely to model the long-term secular variation at the core surface as accurately as possible. They were built using the B-spline basis with knots every 2.5 years, and from the largest datasets possible at the time: ufm1/ufm2 used over 250000 data originating from old ships' logs, survey data, observatories and satellite missions; a description of the oldest data can be found in Bloxham (1986) and Bloxham *et al.* (1989). The gufm1 model was built from similar data from the 20th century, but a vastly expanded historical dataset, described in Jonkers *et al.* (2003)—the model contains over 365000 data and 36512 parameters. No account is explicitly taken of external fields in these models.

The comprehensive models

An effort began in the early 1990s to build a comprehensive series of field models which took account of many effects which are recorded in geomagnetic data in addition to the core secular variation. The first model was reported by Sabaka and Baldwin (1993). We will specifically report on the latest model CM4 (Sabaka *et al.*, 2004). In general terms the model includes representations of the main field, its secular variation, and both local-time (Sun-synchronous) and seasonal modes of the magnetospheric and ionospheric fields, as well as describing

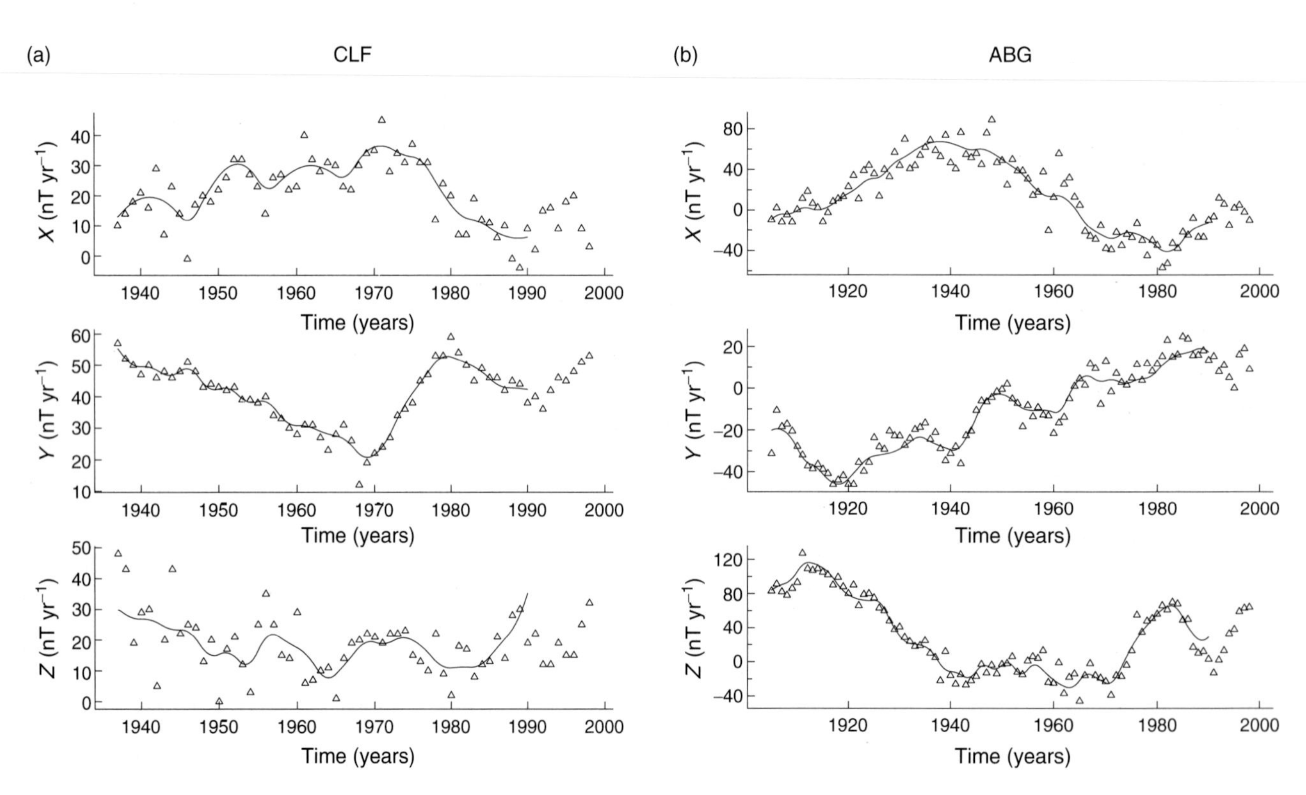

Figure T5 Comparison of secular variation models and annual means at (a) Chambon-la-Forêt, France, (b) Alibag, India. Because the post-1990 data were not used in the creation of *gufm1*, there is a small mismatch at the end of the data series—this shows the difficulty in predicting the secular variation.

ring-current variations through the *D*st index and internal fields induced by time-varying external fields. The data used in creating the model consists of POGO, Magsat, Ørsted, and CHAMP satellite data (totaling over 1.6 million observations) and over 500000 observatory data; the latter consist of either a 1.00 a.m. observation (actually an hourly mean) on the quietest day of the month during the 1960–2002.5 period, plus observations every 2 h on quiet days during the POGO and Magsat missions.

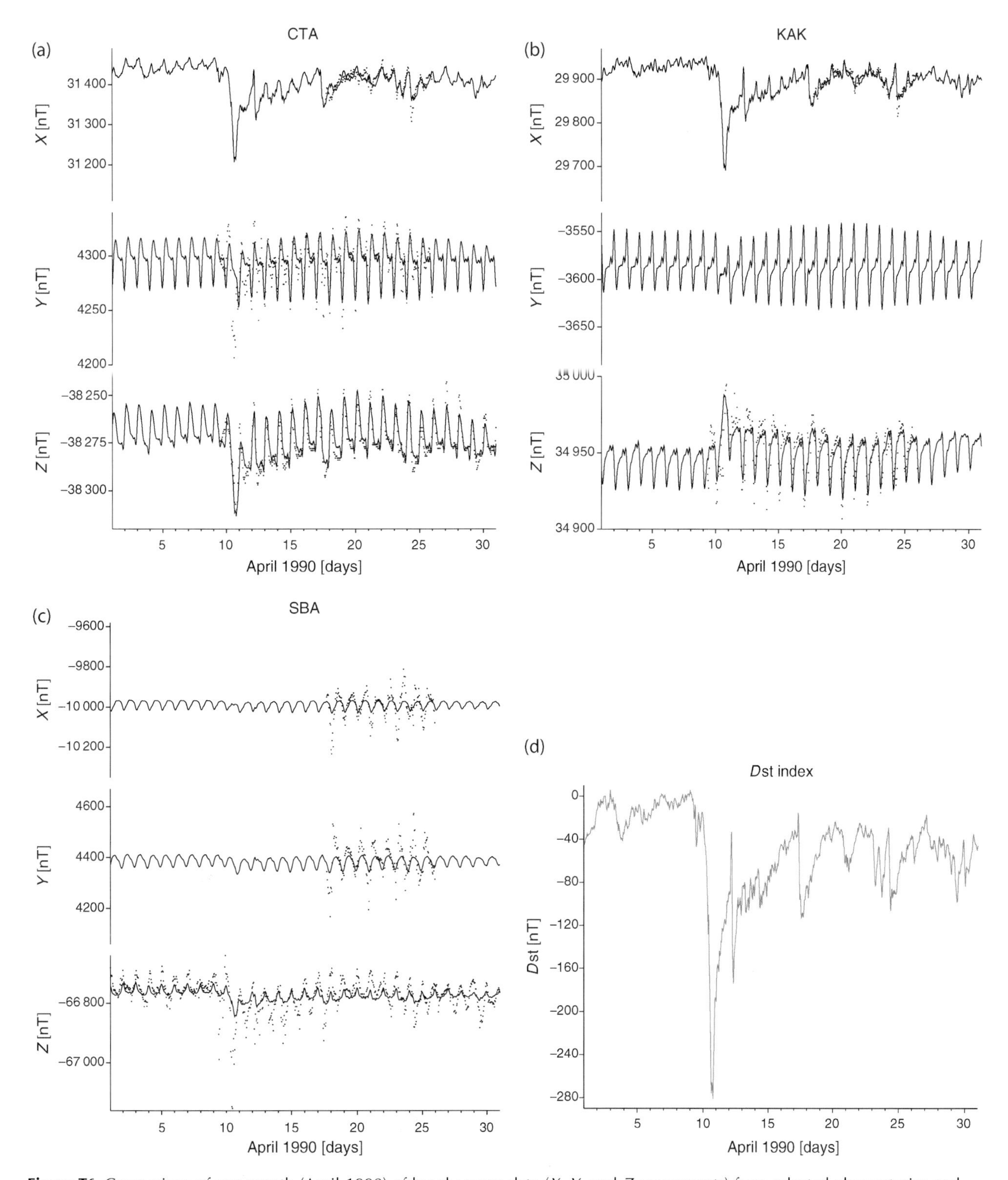

Figure T6 Comparison of one month (April 1990) of hourly mean data (*X, Y,* and *Z* components) from selected observatories and predictions from CM4. (a) Charters Towers, (b) Kakioka, (c) Scott Base, (d) The *D*st index for April. Note the commencement of a magnetic storm on the tenth day. The *D*st index is used in the synthesis of predictions at individual observatories.

The comprehensive model takes into account not only the time-varying core magnetic field (out to degree 13) but also the static crustal field from degree 14 to degree 65. Because a model of the lithospheric field to this degree captures only a small proportion of the total lithospheric signal, it is necessary to also solve for 1635 observatory biases, generally three components at each observatory. The novel features of the model arise in its very sophisticated treatment of the external magnetic fields, and we will discuss these in some detail.

The ionospheric field is modeled as currents flowing in a thin shell at an altitude of 110 km. This leads to magnetic fields which are derived from potentials below and above this layer, which influence the observatory and satellite data, respectively (since all the satellites fly above this layer). In quasidipole coordinates, the currents are allowed to vary with 24, 12, 8, and 6 h periods, as well as annually and semiannually. Induced fields are accounted for by assuming that the conductivity distribution of the Earth varies only in radius, which means that an external spherical harmonic can only excite its corresponding internal spherical harmonic. The magnetospheric field is also parameterized in a similar way, with both daily and seasonal periodicities, but also a modulation is allowed based on the *D*st index.

In order to take into account the poloidal *F*-region currents through which the satellites fly, a parametrization is made in terms of a toroidal magnetic field, which also has periodic time variations.

The model is estimated by an iteratively reweighted least-squares method, using Huber weights, and the core contribution is regularized as in (5) using $n_1 = 2$ and $n_2 = 1$ in one norm and $n_1 = 0$ and $n_2 = 2$ in another. This difference from the ufm1/gufm1 method simply represents a different approach; the fundamental quantity in the comprehensive models is the secular variation $\partial_t B_r$, which has an expansion in B-splines, and the main field B_r is found as the integral of this using the 1980 value as the offset or integration constant. All the other parameters are regularized in a similar way, by smoothing on spheres at different altitudes, representing the physical locations of the sources. In total CM4 consists of 25243 free parameters.

Discussion

To illustrate the fidelity with which the various field models are able to model observatory data, we show in Figure T5 a comparison of model *gufm1*'s predictions with some observatory annual mean datasets. To show CM4's performance on very short timescales, Figure T6 compares the model to hourly mean values for the month of April 1990,

Table T2 Comparison of r.m.s differences between observatory annual means and predictions from the models gufm1 and CM3, the latter with or without its external contributions

Component	No. of data	gufm1	CM3 (all)	CM3 (no external)
X	4047	17.71	17.48	18.09
Y	4047	21.27	21.45	21.47
Z	4047	24.55	24.49	24.53

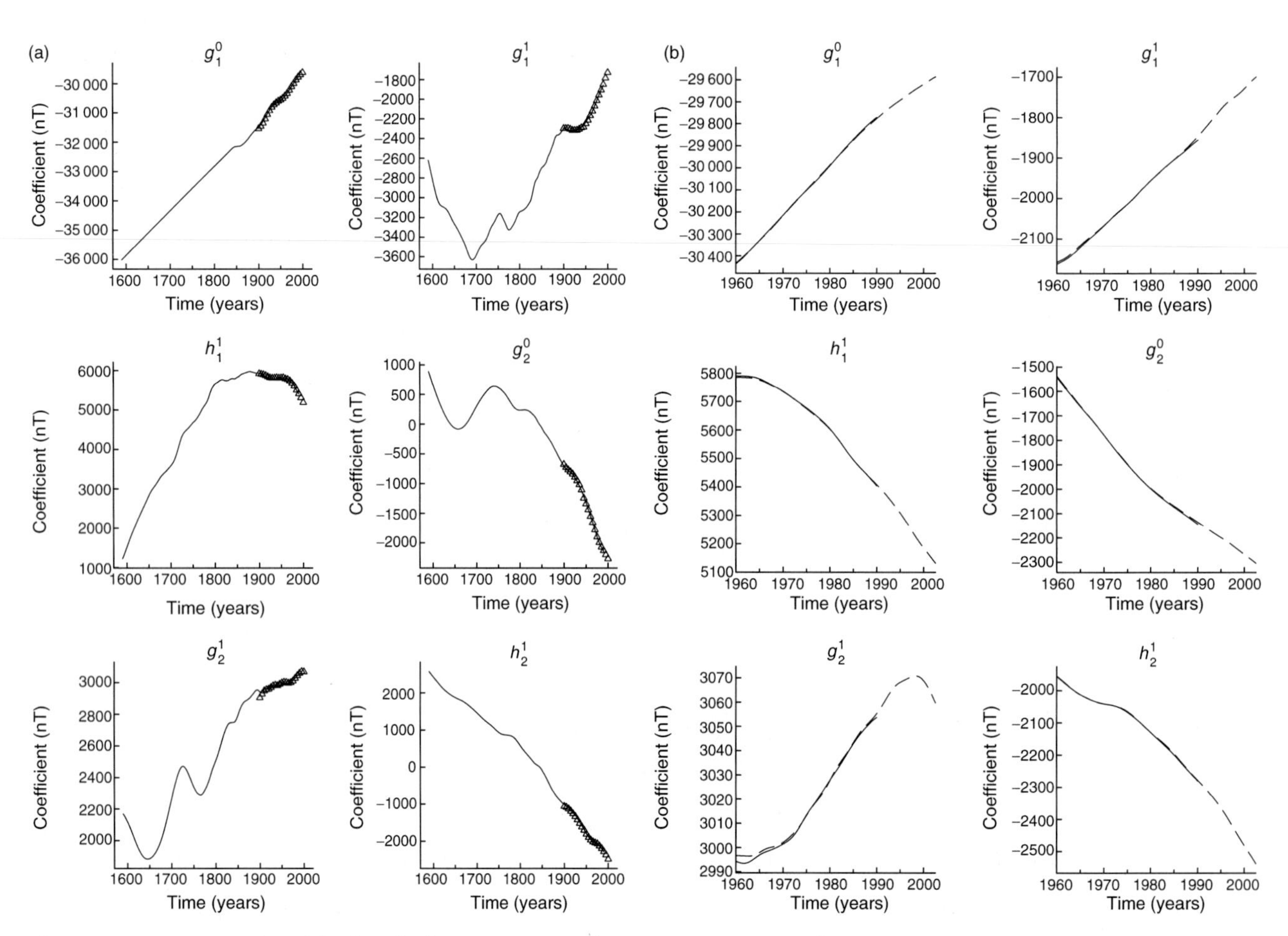

Figure T7 Comparison of model values for first six Gauss coefficients. (a) 1590–1990, (b) 1960–2002.5. Solid is gufm1, dashed is CM4 and the triangles are DGRFs. In (a) g_1^0 has been fixed to decrease at a rate of 15nT yr^{-1} prior to 1840; in the absence of intensity data it is necessary to fix the amplitude of the solutions.

data that were not used in deriving the model. It is clear that the model is capable of predicting variations rather well, though with more difficulty at the Antarctic station SBA (Scott Base). Table T2 compares the performance of models gufm1 and CM3 against observatory data, showing almost identical performance. This comes about principally because of the large intrinsic variance of the data at some observatories, which neither field model is able to capture. Figure T7 shows a comparison of the model predictions for the variation in the first six Gauss coefficients over century and decade timescales. Although small differences exist, particularly in estimates of the instantaneous secular variation, it is apparent that modeling has reached a stage where there is considerable consensus between the models.

Andrew Jackson

Bibliography

Barraclough, D.R., 1978. Spherical harmonic analysis of the geomagnetic field. *Geomagnetic Bulletin Institute of Geological Science*, **8**.

Bloxham, J., 1986. Models of the magnetic field at the core-mantle boundary for 1715, 1777 and 1842. *Journal of Geophysical Research*, **91**: 13954–13966.

Bloxham, J., 1987. Simultaneous stochastic inversion for geomagnetic main field and secular variation 1. A large-scale inverse problem. *Journal of Geophysical Research*, **92**: 11597–11608.

Bloxham, J., and Jackson, A., 1989. Simultaneous stochastic inversion for geomagnetic main field and secular variation 2: 1820–1980. *Journal of Geophysical Research*, **94**: 15753–15769.

Bloxham, J., and Jackson, A., 1992. Time-dependent mapping of the magnetic field at the core-mantle boundary. *Journal of Geophysical Research*, **97**: 19537–19563.

Bloxham, J., Gubbins, D., and Jackson, A., 1989. Geomagnetic secular variation. *Philosophical Transactions of the Royal Society of London*, **329**: 415–502.

Cain, J.C., Daniels, W.E., Hendricks, S.J., and Jensen, D.C., 1965. An evaluation of the main geomagnetic field, 1940–1962. *Journal of Geophysical Research*, **70**: 3647–3674.

Cain, J.C., Hendricks, S.J., Langel, R.A., and Hudson, W.V., 1967. A proposed model for the International Geomagnetic Reference Field—1965. *Journal of Geomagnetism and Geoelectricity*, **19**: 335–355.

Jackson, A., Jonkers, A., and Walker, M., 2000. Four centuries of geomagnetic secular variation from historical records. *Philosophical Transactions of the Royal Society of London*, **358**: 957–990. doi:10.1098/rsta.2000.0569

Jonkers, A.R.T., Jackson, A., and Murray, A., 2003. Four centuries of geomagnetic data from historical records. *Reviews of Geophysics*, **41**: 1006. doi:10.1029/2002RG000115. 2002 6.0

Langel, R.A., 1987. The main field. In Jacobs, J.A. (ed.) *Geomagnetism*, Volume 1. London: Academic Press, pp. 249–512.

Langel, R.A., Estes, R.H., and Mead, G.D., 1982. Some new methods in geomagnetic field modelling applied to the 1960–1980 epoch. *Journal of Geomagnetism and Geoelectricity*, **34**: 327–349.

Langel, R.A., Kerridge, D.J., Barraclough, D.R., and Malin, S.R.C., 1986. Geomagnetic temporal change: 1903–1982, a spline representation. *Journal of Geomagnetism and Geoelectricity*, **38**: 573–597.

Sabaka, T.J., and Baldwin, R.T., 1993. Modeling the Sq magnetic field from POGO and Magsat satellite and contemporaneous hourly observatory data, HSTX/G&G-9302, Hughes STX Corp., 7701 Greenbelt Road, Greenbelt, MD.

Sabaka, T.J., Olsen, N., and Langel, R.A., 2002. A comprehensive model of the quiet-time, near-Earth magnetic field: phase 3, *Geophysical Journal International*, **151**: 32–68.

Sabaka, T.J., Olsen, N., and Purucker, M.E., 2004. Extending comprehensive models of the Earth's magnetic field with Oersted and CHAMP data, *Geophysical Journal International*, **159**(2): 521–547. doi:10.1111/j.1365-246X.2004.02421.x

Cross-references

IGRF, International Geomagnetic Reference Field
Main Field Modeling
Harmonics, Spherical

TRANSFER FUNCTIONS

Introduction

Transfer functions are used in *magnetotellurics* and *geomagnetic depth sounding* to determine subsurface electrical conductivity from surface measurements of natural electric and magnetic fields (Simpson and Bahr, 2005). A transfer function describes the mathematical relationship between surface measurements of variations of electric and magnetic fields.

Magnetotelluric transfer functions

Magnetotellurics uses natural electromagnetic field variations to image subsurface electrical resistivity structure through electromagnetic induction. The surface electric and magnetic fields at each frequency are related through

$$\begin{bmatrix} E_x \\ E_y \end{bmatrix} = \begin{bmatrix} Z_{xx} & Z_{xy} \\ Z_{yx} & Z_{yy} \end{bmatrix} \begin{bmatrix} H_x \\ H_y \end{bmatrix},$$

where Z is defined as the magnetotelluric impedance, a transfer function. The components E_x and E_y are orthogonal components of the electric field and H_x and H_y are orthogonal components of the magnetic field. These electromagnetic fields are nonstationary and the transfer functions are computed as the average of many measurements. Robust statistical methods are used to reliably estimate the transfer functions (Jones *et al.*, 1989; Larsen *et al.*, 1996; Egbert, 1997) and remote-reference processing should always be used (Gamble *et al.*, 1979).

Magneto-variational studies

The vertical and horizontal magnetic fields are related through a magnetic field transfer function **T** defined as

$$[H_z] = [\, T_{xz} \;\; T_{yz} \,] \cdot \begin{bmatrix} H_x \\ H_y \end{bmatrix}.$$

In the simple 2D geometry shown in Figure T8, the electric currents flow along a conductivity anomaly and generate vertical magnetic fields that are oriented upward on one side and downward on the other. The magnetic field transfer function is often called the tipper. The transfer function **T** can also be displayed as *induction arrows* which are horizontal vectors with north and east components T_{zx} and T_{zy}, respectively. Each component of the induction arrow is a complex number with a real (in-phase) and imaginary (out-of-phase) part. In the convention of Parkinson (1962), the real *induction arrows* point toward a conductor (see *Dudley Parkinson*) as shown in Figure T8. In the Wiese convention the arrows point away from conductors (Wiese, 1962). The direction of the induction arrows is further discussed by Lilley and Arora (1982). Induction arrows exhibit a reversal above a conductivity anomaly as illustrated in Figure T8. The ocean is the largest conductor on the surface of the Earth. Thus in surveys in coastal areas, the real Parkinson arrows will point at the ocean and the seawater must be included in conductivity models (see *Coast effect of induced currents*). Edward Bullard is alleged to have commented that "*geomagnetic induction was a good way to find coastlines, but there are easier ones.*" Large-scale magneto-variational studies generate

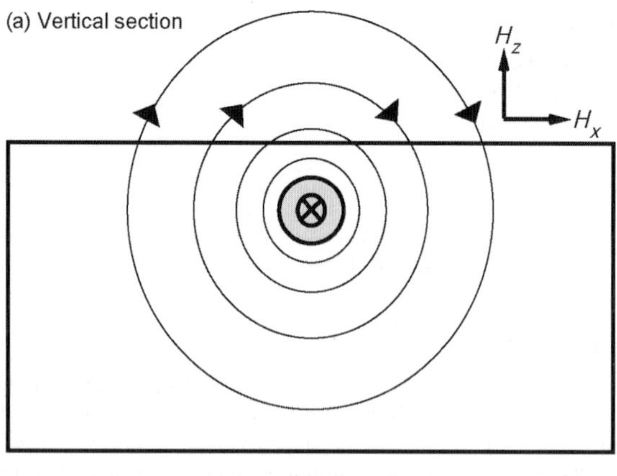

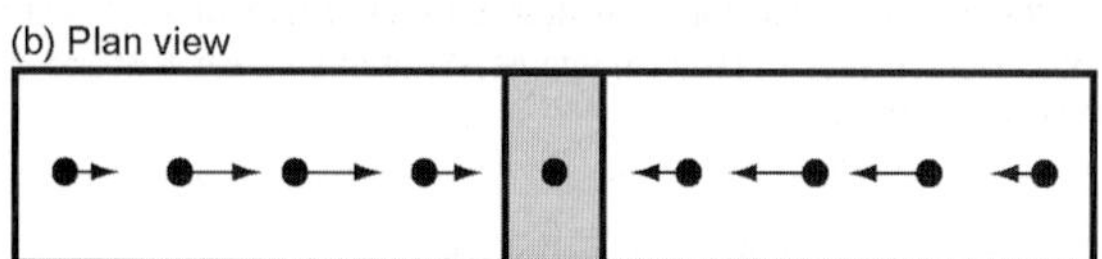

Figure T8 The geometry of the vertical magnetic fields associated with naturally induced electric currents flowing in a buried conductor. In the Parkinson convention the real induction arrows point at the conductor, become stronger as the conductor is approached, and then reverse direction above the conductor.

maps of induction arrows over a wide range of frequencies and have been used to map the structure of the lithosphere on a number of continents (Gough, 1989).

Interstation transfer functions

If magnetotelluric data are recorded simultaneously at an array of stations, then interstation transfer functions can be defined. If a magnetic field $H_x^1(\omega)$, is measured at one station and $H_x^2(\omega)$, is measured at another station, then the interstation transfer function can be written as

$$I_{21}(\omega) = \frac{H_x^2(\omega)}{H_x^1(\omega)}$$

at a frequency ω. All stations in an array are related to the same base station. One of the first applications of this method was described by Schmucker (1970) in a study of the southwestern United States. Interstation magnetic field transfer functions can be inverted in a similar way to apparent resistivity and phase to produce a resistivity model of the Earth.

Martyn Unsworth

Bibliography

Egbert, G.D., 1997. Robust multiple-station magnetotelluric data processing. *Geophysical Journal International*, **130**: 475–496.

Gamble, T.B., Goubau, W.M., and Clarke, J., 1979. Magnetotellurics with a remote reference. *Geophysics*, **44**: 53–68.

Gough, D.I., 1989. Magnetometer array studies, earth structure and tectonic processes. *Reviews of Geophysics*, **27**: 141–157.

Jones, A.G., Chave, A.D., Egbert, G.D., Auld, D., and Bahr, K., 1989. A comparison of techniques for magnetotelluric response function estimates. *Journal of Geophysical Research*, **94**: 14201–14213.

Larsen, J.C., Mackie, R.L., Manzella, A., Fiordelisi, A., and Rieven, S., 1996. Robust smooth magnetotelluric transfer functions. *Geophysical Journal International*, **124**: 801–819.

Lilley, F.E.M., and Arora, B.R., 1982. The sign convention for quadrature Parkinson arrows in geomagnetic induction studies. *Reviews of Geophysics and Space Physics*, **20**: 513–518.

Parkinson, W.D., 1962. The influence of continents and oceans on geomagnetic variations. *Geophysical Journal of the Royal Astronomical Society*, **6**: 441.

Schmucker, U., 1970. *Anomalies of Geomagnetic Variations in the Southwestern United States.* Berkeley: University of California Press.

Simpson, F., and Bahr, K., 2005. *Practical Magnetotellurics.* Cambridge: Cambridge University Press, p. 270.

Wiese, H., 1962. Geomagnetische Tiefentellurik Teil II: Die Streichrichtung der Untergrundstrukturen des elektrischen Widerstandes, erschlossen aus geomagnetischen Variationen [Strike direction of underground structures of electric resistivity, inferred from geomagnetic variations.]. *Pure and applied geophysics*, **52**: 83–103.

Cross-references

Coast Effect of Induced Currents
Geomagnetic Deep Sounding
Induction Arrows
Magnetotellurics
Parkinson, Wilfred Dudley (1919–2001)

TRANSIENT EM INDUCTION

Geophysicists measure the electrical conductivity of subsurface materials for a variety of purposes including geological mapping, mining investigation, groundwater resource evaluation and environmental assessment (Meju, 2002). One of the methods used for mapping subsurface conductivity distributions is the transient electromagnetic (TEM) method that employs artificially controlled source fields. The TEM method is sensitive to electrical conductivity averaged over the volume of ground in which induced electric currents are caused to flow.

Transient electromagnetic induction

The TEM method is founded on Maxwell's equations that govern electromagnetic phenomena. Combining Ohm, Ampere, and Faraday laws (Wangsness, 1986; Jackson, 1998) results in the damped wave equation

$$\frac{\partial^2 B}{\partial t^2} - \mu_0\sigma\frac{\partial B}{\partial t} - \mu_0\epsilon\frac{\partial^2 B}{\partial t^2} = \mu_0\nabla \times J_S, \qquad \text{(Eq. 1)}$$

where B is magnetic field, μ_0 is magnetic permeability, σ is electrical conductivity, ε is dielectric permittivity, J_S is the source current distribution, and t is time. For most geophysical applications of TEM, the earth is generally considered to be nonmagnetic, such that $\mu_0 = 4\pi \times 10^{-7}\ \mathrm{Hm}^{-1}$, the magnetic permeability of free space. The third term on the left-hand side of (1) is the energy storage term describing wave propagation. The second term on the left-hand side of (1) is the energy dissipation term describing electromagnetic diffusion. In the TEM method, $\partial^2 B/\partial t^2$ is relatively insignificant compared to $\partial B/\partial t$, so that the wave propagation term is safely ignored. TEM induction is thus a diffusive phenomenon.

A physical understanding of this phenomenon can be obtained by recognizing that the induction process is equivalent to the diffusion of an image of the transmitter (TX) loop into a conducting medium. The similarity of the equations governing electromagnetic induction and hydrodynamic vortex motion, first noticed by Helmholtz, leads directly to the association of the image current with a smoke ring (Lamb, 1945, p. 210). The latter is not "blown," as commonly thought, but instead moves by self-induction with a velocity that is generated by the smoke ring's own vorticity and described by the familiar

Biot-Savart law (Arms and Hama, 1965). An electromagnetic smoke ring dissipates in a conducting medium much as the strength of a hydrodynamic eddy is attenuated by the viscosity of its host fluid (Taylor, 1958, pp. 96–101). The medium property that dissipates the electromagnetic smoke ring is electrical conductivity.

An inductively coupled TEM system is shown in Figure T9a. A typical TX current waveform $I(t)$ is a slow rise to a steady-on value I_0 followed by a rapid shut-off, as exemplified by the linear ramp in Figure T9b (top plot). Passing a disturbance through the TX loop generates a primary magnetic field that is in-phase with, or proportional to, the TX current. According to Faraday's law of induction, an electromotive force (emf) that scales with the time rate of change of the primary magnetic field is also generated. The emf drives electromagnetic eddy currents in the conductive earth, notably in this case during the ramp-off interval, as shown in Figure T9b (middle plot). After the ramp is terminated, the emf vanishes and the eddy currents start to decay via Ohmic dissipation of heat. A weak, secondary magnetic field is produced in proportion to the waning strength of the eddy currents. The receiver (RX) coil voltage measures the time rate of change of the decaying secondary magnetic field as depicted in Figure T9b (bottom plot). In typical TEM systems, the RX voltage measurements are made during the TX off-time when the primary field is absent. The advantage of measuring during TX off-time is that the relatively weak secondary signal is not swamped by the much stronger primary signal.

During the ramp-off, the induced current assumes the shape of the horizontal projection of the TX loop onto the surface of the conducting ground. The sense of the circulating induced currents is such that the secondary magnetic field they create tends to maintain the total magnetic field at its original steady-on value prior to the TX ramp-off. Thus, the induced currents flow in the same direction as the TX current, i.e., opposing the TX current decrease that served as the emf source. The image current then diffuses downward and outward while diminishing in amplitude. A mathematical treatment of the electromagnetic smoke ring phenomenon appears in Nabighian (1979).

Transient electromagnetic forward modeling

Calculation of the TEM field generated by a magnetic or electric dipole source situated on, above or within a layered conducting medium is a well-known boundary value problem of classical physics. One standard solution technique is to calculate the mutual impedance, Z between the transmitter and receiver loops as a function of measurement (or delay) times t. For a TX loop of radius a and RX loop of radius b that are co-axial and located on the ground's surface, we have that (Knight and Raiche, 1982)

$$Z(t) = -\pi\mu_0 ab \int_0^{\infty} L_p^{-1}\{I(p)\mathrm{p}A_0(\mathbf{m},p,\lambda)\}J_1(\lambda a)J_1(\lambda b)\mathrm{d}\lambda.$$

In the above equation, $A_0(\mathbf{m},p,\lambda)$ is the layered-earth impedance function, $\mathbf{m}$ represents the parameters of the hypothetical earth-model (namely electrical conductivities, σ_j and thicknesses, h_j of the layers), p is the Laplace transform variable corresponding to $-\mathrm{i}\omega$ where ω is the angular frequency, "i" is the imaginary unit, λ is the integration variable for the inverse Hankel transform, $I(p)$ represents the Laplace transform of the normalized current waveform and is equal to $-p^{-1}$ for step function turn-off of TX current, J_1 is the Bessel function of order 1, and L_p^{-1} is the inverse Laplace transform operator with respect to p.

Similar analytic solutions are available for other TX-RX configurations and a variety of special-purpose algorithms are available for evaluating such Hankel transforms or integrals over Bessel functions. Note that the manner in which the TX current is switched off influences the TEM response and needs to be accounted for in the forward models (Raiche, 1984).

The TEM response of 3D subsurface conductivity distributions requires the application of numerical methods for solving the governing Maxwell partial differential equations (see Hohmann, 1987). Full 3D numerical simulations involve a complete description of the physics of electromagnetic induction including galvanic and vortex contributions. The galvanic term appears when induced currents encounter spatial gradients in electrical conductivity. Electric charges accumulate across interfaces in electrical conductivity. In 1D layered media excited by a purely inductive electromagnetic source such as a loop, induced currents flow horizontally and do not cross layer boundaries. In that case, only the vortex contribution is present.

Transient electromagnetic method of geophysical prospecting

In TEM surveying, a primary field is generated by a rectangular or square TX loop, with dimensions of ten to a few hundred meters (Figure T10). A long straight wire grounded at both ends can also serve as the transmitter. For shallow depth probes, a 20 m-sided TX loop could suffice for imaging about 60–100 m depth depending on ground conditions (Meju *et al.*, 2000). The primary field is not continuous but consists of a series of pulses between which the

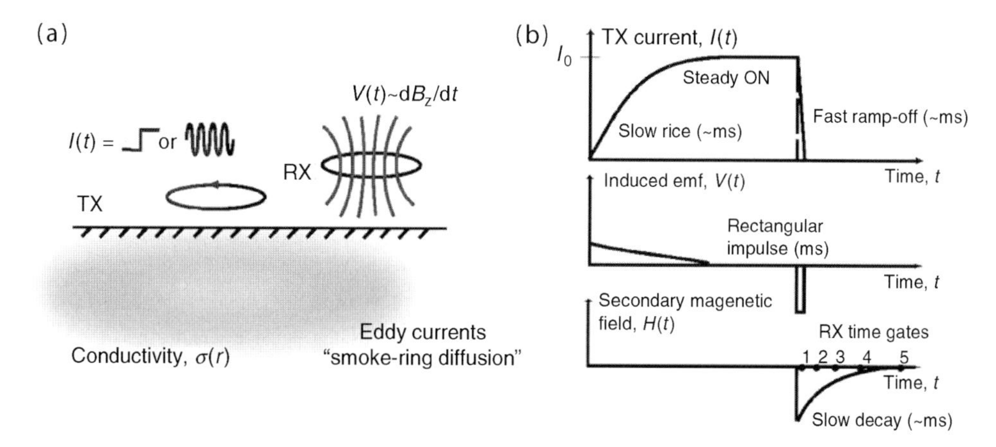

Figure T9 (a) TX loop lying on the surface of an isotropic uniform half-space of electrical conductivity, σ. A disturbance $I(t)$ in the source current immediately generates an electromagnetic eddy current in the ground just beneath the TX loop. An RX coil measures the induced voltage $V(t)$, which is a measure of the time-derivative of the magnetic flux generated by the diffusing eddy current. (b) Typical TX current waveform $I(t)$ with slow rise time and fast ramp-off; induced emf $V(t)$, proportional to the time rate of change of the primary magnetic field; decaying secondary magnetic field $H(t)$ due to the dissipation of currents induced in the ground (Everett and Meju, 2005).

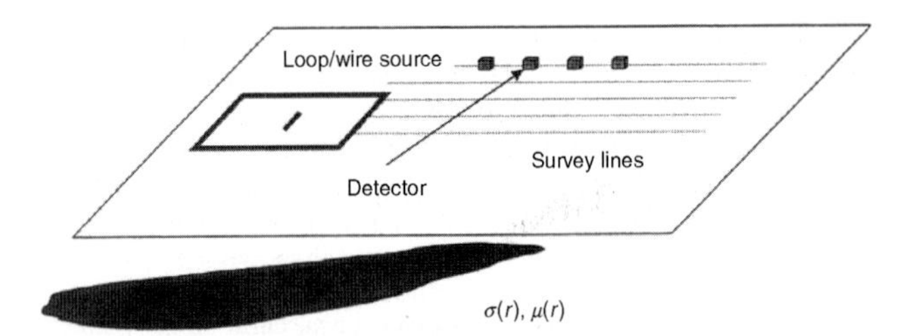

Figure T10 Basic set-up for TEM surveying. The ground is inductively energized using a grounded wire or transmitter loop. The electric or magnetic fields induced in a subsurface conductor ($\sigma(r)$) are systematically measured along survey lines using a suitable detector coil.

primary field is switched off and measurements made. The rapid termination of the primary field in the TX causes Eddy currents to be produced in the subsurface, and the associated secondary field is measured on the surface using a detector (RX) coil (Figure T10). RX may be a small multiturn coil (about 1 m diameter) or a relatively large loop. The decaying transient signal is sampled at various time gates to yield a sounding curve. The transient signal can last from a few microseconds to several hundreds of millisecond, and an observational bandwidth of about 0.005–30 ms is desirable for near-surface investigations (10–400 m). The character of the transient signal is a function of the subsurface resistivity structure, with the shallow structure influencing the TEM response at early times soon after TX turn-off and measurement at increasing times allowing the deeper electrical structure to be interpreted. For shallow structure, TEM systems use high base-frequencies (i.e., pulse repetition rates), small transmitter loops (5–50 m), need only low power source (12 V battery), but in particular requires a very rapid primary field turn-off (currently ≤ 1.5 μs). The accuracy of the narrow early sampling times is also an important factor. For deeper structure, larger transmitter loops, higher power and of course longer transient measuring times at the receiver are required.

There are a number of different configurations as regards the position of the receiver relative to the transmitter (Figure T10). The receiver can be located in the center of the TX loop (in-loop or central-loop technique) or outside it (separated-loop or offset-loop arrangements). In the single-loop or coincident-loop configuration, the TX loop serves as the RX during transmitter-off times or both the TX and RX loops are separate but coincident.

Maxwell A. Meju and Mark E. Everett

Bibliography

Arms, R.J., and Hama, F.R., 1965. Localized-induction concept on a curved vortex and motion of an elliptic vortex ring. *Physics of Fluids*, **8**: 553–559.

Everett, M.E., and Meju, M.A., 2005. Near-surface controlled-source electromagnetic induction: background and recent advance. In Yoram, R., and Hubbard, S. (eds.), *Hydrogeophysics.* Heidelberg: Springer (Chap. 6).

Hohmann, G.W., 1987. Numerical modeling for electromagnetic methods of geophysics. In Nabighian, M.N. (ed.), *Electromagnetic Methods in Applied Geophysics,* Volume 1. Tulsa Oklahoma: Society of Exploration Geophysics, pp. 313–363.

Jackson, J.D., 1998. *Classical Electrodynamics*, 3rd Edition. New York: John Wiley & Sons, 808 pp.

Knight, J.H., and Raiche, A.P., 1982. Transient electromagnetic calculations using the Gaver-Stehfest inverse Laplace transform method. *Geophysics*, **47**: 47–50.

Lamb, H., 1945. *Hydrodynamics,* by Sir Horace Lamb, 6th Edition. New York: Dover Publications, 738 pp.

Meju, M.A., 2002. Geoelectromagnetic exploration for natural resources: models, case studies and challenges. *Surveys of Geophysics*, **23**: 133–205.

Meju, M.A., Fenning, P.J., and Hawkins, T.R.W., 2000. Evaluation of small-loop transient electromagnetic soundings to locate the Sherwood Sandstone aquifer and confining formations at well sites in the Vale of York. *England Journal of Applied Geophysics*, **44**: 217–236.

Nabighian, M.N., 1979. Quasi-static transient response of a conducting half-space: an approximate representation. *Geophysics*, **44**: 1700–1705.

Raiche, A.P., 1984. The effect of ramp function turn-off on the TEM response of layered earth. *Exploration Geophysics*, **15**: 37–41.

Taylor, G.I., 1958. On the dissipation of eddies. In Batchelor, G.K. (ed.) *Scientific Papers,* Volume 2. Cambridge: Cambridge University Press, pp. 96–101.

Wangsness, R.K., 1986. *Electromagnetic Fields*, 2nd Edition. New York: John Wiley & Sons, 608 pp.

TRUE POLAR WANDER

Definitions and early views

The idea that polar wander (PW) must have occurred dates back to the 19th century. Geologists, paleontologists, and paleoclimatologists suggested that the Earth's equator must have been located far from its present position in the distant geological past, at least viewed from certain continents, given the present day locations of climatological belts derived from certain types of rocks or fossil assemblages. Darwin (1877) made the first attempt at quantitative modeling but this suffered from several errors (see Steinberger and O'Connell, 2002, for an analysis). In the early 1950s, paleomagnetists provided quantitative evidence that the geographical latitudes of individual continents had indeed changed with time, hence that the instantaneous geographic or rotation pole had moved with respect to most continents. The paths followed by the poles in the geological past were termed "apparent" polar wander paths (or APWPs), because it was not clear whether it was the pole or the continent that had moved. It would soon become apparent that continents (more accurately tectonic plates) had moved with respect to each other and that a significant part of APW was actually due to these relative motions. The question was: is there a remaining fraction in polar wander, which would be a characteristic of "Earth as a whole," and which would not be accounted for by plate tectonics. This fraction is what is generally called "true" polar wander (TPW).

TPW refers to the large scale motions of the Earth's rotation axis with respect to the "bulk silicate Earth" or "solid Earth" through geological time (rather than "Earth as a whole," indeed, much of the core is a fluid with small viscosity and the parts which will be important to us in this article are the solid though slowly deforming mantle and crust, i.e., the silicate earth; see below). External forces driving shorter-term changes, such as precession, are not considered here. TPW could result from centrifugal forces acting on mass anomalies distributed on or inside the Earth: excess mass, for instance, will lead to slow deformation of the Earth and change its rotation axis, causing the mass to move towards the equator without being displaced with respect to the solid Earth. The problem is to define precisely which part wanders with respect to which reference frame in a deformable Earth. We seek to identify the axis of rotation of a reference frame RF which is relevant to the solid Earth relative to an inertial (or absolute astronomical) frame of reference IFR (i.e., determined by the angular momentum vector of Earth; see f.i. Steinberger and O'Connell, 2002). The reference frame RF can be selected as the one which has zero net rotation when motions are integrated over the entire mantle, called the "mean mantle" reference frame. Or one can define a "mean lithosphere" frame, in which the lithospheric plates have no net rotation. Because of lateral variations in viscosity of the deformable Earth layers, these frames in general do not coincide (Ricard *et al.*, 1991; O'Connell *et al.*, 1991). Although the "mean mantle" frame can be computed in numerical simulations in which the density distribution is given in the entire mantle, when attempting to constrain it with actual (paleomagnetic) observations, we will have to assume that the "mean mantle" frame is identical to some frame based on observations: the "hotspot reference frame" will most often be the one involved. Steinberger and O'Connell (2000) calculate that the approximation is good even if hotspots move in the convecting mantle (see also Davies, 1988; Steinberger, 2000). This is discussed further later in this article.

Is the concept of Earth as a whole meaningful? If the Earth were perfectly rigid, there would be no polar wander. Of course we now know that the Earth is not rigid, and that its crust and mantle actually deform actively at the typical velocities of plate tectonics ($\sim$10–100 mm yr^{-1}). Thus, PW involves (relative) motions of the rotation axis with respect to the Earth's mantle. Lambeck (1980) points out that from the time of Kelvin, in 1863, up to the work published by Gold in 1955, a number of qualified physicists believed that polar wander was not only possible, but even inevitable if the long-term rheology of the planet was inelastic.

However, in a landmark treatise on the rotation of the Earth, Munk and Macdonald (1960) were led to question both the observational and theoretical bases for large-scale polar wandering. They considered that evidence from either paleomagnetism or paleontology available at that time was at best tenuous, and proposed that the "excess" bulge (the present bulge is indeed larger than what would be expected if the Earth behaved as a fluid, called the hydrostatic bulge) was a fossil remainder from a time when the Earth rotated faster.

A decade later, plate tectonics and continental drift had basically been established. Another landmark paper by Goldreich and Toomre (1969) reflected the new view of the Earth. They hypothesized that (large scale) polar wandering and continental drift shared a common explanation through mantle convection and resulting redistribution of masses within the mantle. Goldreich and Toomre (1969) started their analysis of the motions of the rotation axis of a quasirigid (slowly deforming) Earth with the now classical figure (inspired by Gold, 1955) where a colony of beetles (the continents or other mass anomalies) move slowly at the surface of a rigid rotating sphere (Figure T11a). They used this to demonstrate that, once set spinning about the axis with the greatest nonhydrostatic moment of inertia, a slowly deforming body would always continue to spin almost exactly about the same principal axis, as that axis moved through it. Viewed from space, the rotation and principal axes are fixed with respect to the stars, and the crust and mantle move relative to it (again, the shorter-term effects of precession are neglected here). The speed of motion is $N^{1/2}$ times the average speed of the N beetles of mass m; hence if N (and m) is large, large polar wandering is possible. In a numerical experiment, Goldreich and Toomre (1969) simulated very large polar wander, with particularly intense and rapid swings.

The physics of TPW can be explained in a (hopefully) simple way (e.g., Ricard *et al.*, 1993a; Steinberger and O'Connell, 2002). TPW is seen as the result of conservation of angular momentum of a rotating, deformable body (i.e., whose inertia tensor slowly changes with time). In simple terms, the inertia tensor comprises three components (1) the *inertia of a spherical Earth*, (2) the *hydrostatic equilibrium part*, that is the *inertia of its equatorial bulge* (which, as mentioned above, is slightly larger than what it would be if the planet were a fluid), and (3) the remainder, the *"total" nonhydrostatic contribution to the inertia tensor*, which itself comprises a contribution "imposed" by mantle convection (linked to subducting slabs, upwelling plumes, and more generally thermal and chemical mass heterogeneities resulting from the long history of mantle convection) and one caused by the change of the rotation axis and the immediate elastic and delayed viscoelastic adjustment to the new equilibrium shape. Each of the three terms is very small compared to the previous one. Changing the rotation of a spherical object (term 1) does not require energy on geological time scales; the bulge (term 2), itself a function of rotation, results in a viscous drag which controls only the speed at which rotation can

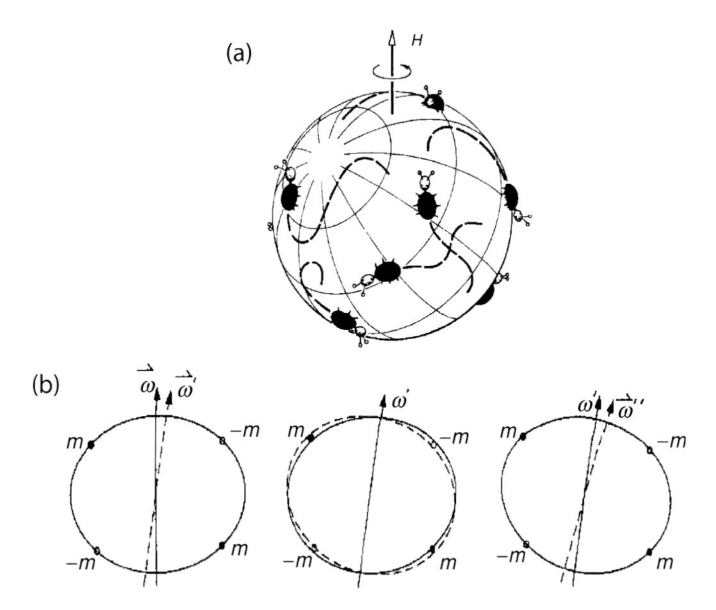

Figure T11 (a) Diagram illustrating polar wander on a quasirigid planet (after Goldreich and Toomre, 1969; Gordon, 1987). The beetles represent mass anomalies which change with time (drifting plates or mantle heterogeneities), inducing changes in the principal nonhydrostatic moments of inertia. The coordinate system attached to the (slowly deforming) solid planet moves with respect to the rotation (or spin) axis. (b) Example of interplay between deformation and change of the rotation axis of the Earth in the hypothetical case when four mass changes (two positive or excess, two negative or deficit) are imposed such that the center of mass does not change and only one nondiagonal inertia tensor element changes (from Steinberger and O'Connell, 2002). In an absolute astronomical reference frame the direction of the rotation axis is fixed; positive masses drift to the equator (and negative ones away). Therefore the relative position of the rotation axis changes with respect to the mass distribution, the change in rotation axis causes the Earth to deform, the change in shape causes the axis to move further, until a new equilibrium is asymptotically approached.

change; as a result, the direction of rotation is fixed by the smallest term (term 3), i.e., the nonhydrostatic component. The rotation axis tracks the maximum eigenvector of the nonhydrostatic inertia tensor as fast as adjustment of the hydrostatic bulge allows (Figure T11b illustrates this for a simple case): the hydrostatic bulge is a response to the current rotation axis, and it will move in accord with the motion of that rotation axis, which is determined by the nonhydrostatic inertia tensor. Seen from an observation point on the surface of the Earth, TPW appears as the wave-like propagation of the bulge. Ultimately, the viscosity of the mantle will control the rate of this motion (Ricard *et al.*, 1993a,b; for a recent synthesis and numerical results showing the time scales involved, see Steinberger and O'Connell, 2002).

With estimates of degree 2 components of the geopotential derived from satellite data, Goldreich and Toomre (1969) showed that the nonhydrostatic part of the earth's inertia ellipsoid is triaxial, the present rotation axis coinciding with the largest of the nonhydrostatic moments of inertia. Therefore, there was no need to resort to the idea of a fossil bulge: the nonhydrostatic Earth simply departs from perfect sphericity. Goldreich and Toomre's (1969) Earth had no bulge to control the rotation speed (on a rotating planet, the faster the beetles the more random their paths and the less the viscous bulge has time to flow, impeding TPW); also isostatic compensation implies that only objects far from interfaces (such as subductions and upwellings, not surface plates or beetles…) which are not isostatically compensated can act (e.g., Ricard *et al.*, 1993b). But their presentation had a goal of pedagogical efficiency and in that sense was quite successful.

How is TPW measured (on short time scales)?

Now, how can we obtain measures of the position of the rotation axis of the Earth? This depends on the time scale. On the short side of the scale, astronomers and geodesists are the appropriate experts. According to these scientists, polar wandering is occurring at present. One of the most recent determinations uses the star catalogue established from the Hipparcos astrometric satellite (determinations of the positions, proper motions, and parallaxes of 100000 stars). Optical astrometric measurements of star positions made during the past century have been rereduced using that catalogue by Gross and Vondrak (1999), producing the longest available homogeneous series of polar motion. The annual, Chandler, and Markowitz wobbles (all periodic or quasi-periodic with annual to decadal periods) have been separated from the longer-term trend of the mean pole (which is of interest to us). The result (see arrow in Figure T12) is a mean drift of the Earth's rotation pole over the period 1900–1992 of 3.5 milliarcsec yr^{-1} (or ~107 mm a^{-1} or 107 km Ma^{-1} for later comparison with geological values), along the direction of the 79°W meridian (uncertainties on absolute values are a few per mil). Data reduction involves modeling of tectonic plate motions, tidal variations and ocean loading and this estimate of present-day TPW is considered as the best unbiased one. This is actually not far from earlier determinations, for instance by Markowitz (1970) (3.5 milliarcsec yr^{-1} towards 65°W longitude).

It has been proposed since the late 1970s that this long-term polar wander is largely due to postglacial rebound, i.e., lithospheric and mantle response to melting of ice (e.g., Sabadini and Peltier, 1981; Mitrovica and Milne, 1998), which has mainly a vertical expression at the surface. The pole is moving toward Hudson Bay, where a large ice mass melted and the lithosphere is currently rising. Possibly contributing sources include mantle convection (Steinberger and O'Connell, 1997), and secular change in ice sheet mass and sea-level (e.g., James and Ivins, 1997). McCarthy and Luzum (1996) compare theoretical estimates and conclude that snow loading, changes in Antarctic ice and Pleistocene deglaciation all provide estimates in reasonable agreement with observations. Melting of glaciers or changes in major hydrological reservoirs are at least a factor of 10 less efficient. Overall, the hydrostatic bulge adjusts to changes in the position of the Earth's rotation axis in about 10^4 years, which is the characteristic time for upper mantle deformation (glacio-isostatic rebound).

How is TPW measured (long time scales)?

How long have the recent rate and direction been maintained, and what values applied in the geological past? On these longer time scales, say longer than 10^6 years, paleomagnetism has become the primary source of information. Although interesting information comes from geological field observations (reconstruction of paleoclimate belts based on occurrences of fossils or certain rock types, or wind

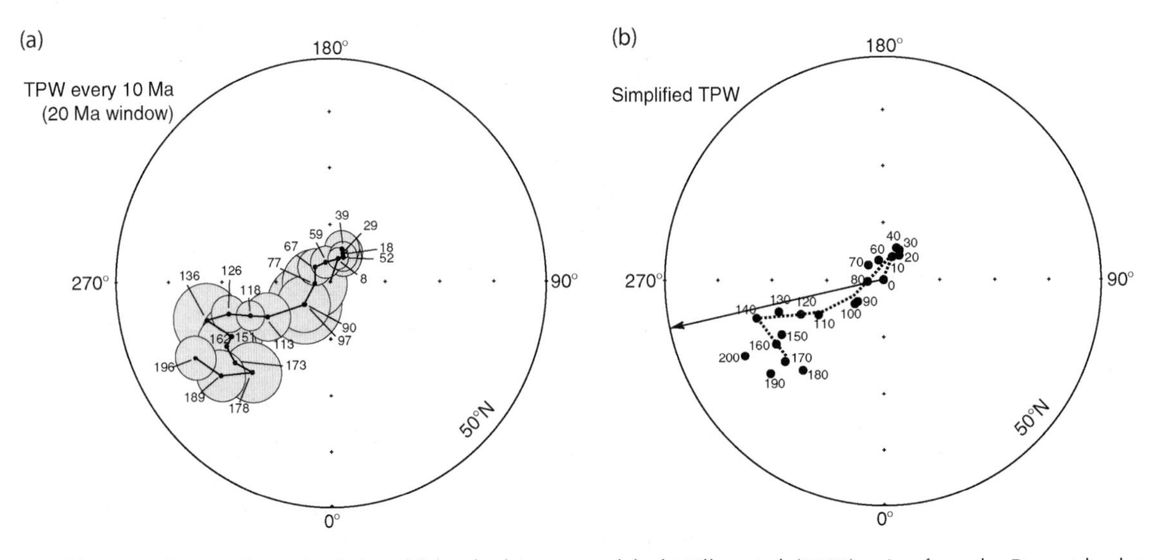

Figure T12 (a) True polar wander paths deduced from the hotspot model of Müller *et al.* (1993) going from the Present back to 130 Ma, and that of Morgan (1983) going from 130 back to 200Ma, with associated 95% confidence ellipses (shaded light gray); 20 Ma sliding window. Ages are in Ma and correspond to the actual mean age of the data in the corresponding window, which is why they are not regularly distributed when the original time windows were. Note that the external circle is at 50° latitude. (b) A simplified outline of the TPW path of (a) outlining successive episodes and standstills. The long arrow shows the mean direction of pole motion between 1900 and 1990 (79°W with a large uncertainty); modulus is ~107 mm a (see text, section "How is TPW measured (on short time scales)?"). (After Besse and Courtillot, 2002).

directions), the key source of quantitative information on polar wander originated from the first paleomagnetic works in the 1950s (with the seminal paper of Creer *et al.*, 1954), and paleomagnetism has remained the main provider of such data. Although clearly doubtful, Munk and MacDonald (1960) already spent a first long chapter on paleomagnetic evidence of polar wander, noting the importance of early syntheses by Runcorn (1956) and several others who demonstrated that continents could not have remained in the same relative positions throughout geological time. A recent review of paleomagnetism is given by McElhinny and McFadden (2000).

The apparent polar wander paths from all the major plates are being defined with increasing accuracy, and their differences reflect continental drift and plate tectonics. This demonstration of the reality of "apparent" polar wander seen from any plate has led some to propose that the concept of "true" polar wandering might be irrelevant or unnecessary (see review in Gordon, 1987). McKenzie (1972) noted that the concept of PW could be useful if the motion of the pole relative to any one plate was much larger than the motion between pairs of plates (Figure T13). The most often accepted view is that the concept is worthwhile if a reference frame external to the plates themselves can be defined, in which there was significant motion of the pole. Various attempts, namely the "vector-sum" and "mean-lithosphere frame" methods, have been reviewed for instance by Jurdy (1981) and Gordon (1987). The "mean-lithosphere frame," first proposed by Jurdy and Van der Voo (1974, 1975), minimizes the difference between a displacement (vector) field and a rigid rotation of the entire Earth's surface. Jurdy and Van der Voo found no significant change in the Cenozoic and Jurdy (1981) concluded in favor of hotspot motion, not TPW. Hargraves and Duncan (1973) had earlier proposed the idea of "mantle roll" in which the nonlithospheric mantle "rolls" beneath the mean lithosphere to which the rotation axis is fixed. McElhinny and McFadden (2000, p. 328) repeated some of these calculations with updated kinematic and paleomagnetic data and concluded that TPW was not discernible from zero over the past 50 myr, as had been found by Jurdy and Van der Voo (1974, 1975).

However, most studies performed in the last two decades follow an alternate approach, based on the possibility, originally suggested by Morgan (1972, 1981), that hotspots could provide a valid reference frame for the mantle. The meaning and existence of that reference frame has been the subject of continued debate, up to this day.

Apparent and TPW, and the hotspot reference frame

We recall that hotspots are active volcanoes, apparently unrelated to (and often but not always far from) plate boundaries, which have been active for tens of millions of years. Hotspots are supposed to be the surface manifestation of plumes, which have left traces in the form of (quasi-) linear chains of extinct volcanoes on the plate(s), which passed over them. Morgan (1972, 1981) proposed that these plumes have a "deep" mantle origin: these convective instabilities may be anchored at least down to the transition zone and possibly all the way down to the core-mantle boundary (yet another topic of heated debate). As a plate moves over such a plume, the relative motion can be described by a rotation about a pole, which is often called the Euler pole: if the rotation axis remains unchanged, the trace of the plume (the hotspot track) will form a small circle about the Euler pole, as do transform faults about the relative rotation axis of two plates (Figure T14). If the traces can be identified and the extinct volcanoes dated, one can define a "hotspot apparent polar wander" (Gordon, 1987) in the same way one defines the apparent polar wander path of any plate. If hotspots are indeed fixed with respect to each other and are fixed within the mantle (we will return to these essential assumptions), then hotspot apparent polar wander describes the wander of the Earth's rotation axis with respect to the hotpots (or vice versa). It is this motion of the pole with respect to the hotspots, assumed to act as a reference frame for the mantle, which has been taken as an estimate of TPW. Motion of plates with respect to the hotspots is sometimes (improperly) called "absolute motion." Note that "APW" (corresponding to a rotation operator Ω_1), i.e., the rotation of the Earth's rotation axis with respect to a given plate, "HS" or "absolute motion" (corresponding to a rotation operator Ω_2), i.e., the rotation of the plate with respect to the hotspot reference frame, and "TPW" (corresponding to a rotation operator Ω_3), i.e., the rotation of the Earth's rotation axis with respect to the hotspot reference frame are linked through the simple equation: $\Omega_1 = \Omega_2 \times \Omega_3$ (i.e., as soon as two of them are given the third can be derived, see, e.g., Besse and Courtillot, 2002).

Determining TPW in that sense requires the best possible knowledge of APWPs of all plates (i.e., the best possible data base of paleomagnetic poles to determine the paleolatitudes and orientations of plates with respect to the rotation axis), of their relative motions (e.g., kinematics from oceanic data such as marine magnetic anomalies and azimuths of transform faults), and positions and ages of individual volcanoes along hotspot tracks on as many plates and for as many hotspots as possible. Paleolatitudes of these seamounts are used for comparison to test model predictions. Note that (except in the final part of this article) we will restrict ourselves to the last 200 million years (i.e., the time for which some ocean kinematic information is preserved).

APWPs and kinematic models were developed from the 1960s to the 1980s under the key assumption that, when averaged over a sufficient amount of time, in excess of a few thousand years, the Earth's magnetic field could be described accurately by a geocentric axial dipole (so-called GAD hypothesis). Following early attempts by Livermore *et al.* (1984) and Andrews (1985), Besse and Courtillot (1988, 1991) proposed to blend the two data sets (i.e., the paleomagnetic poles used

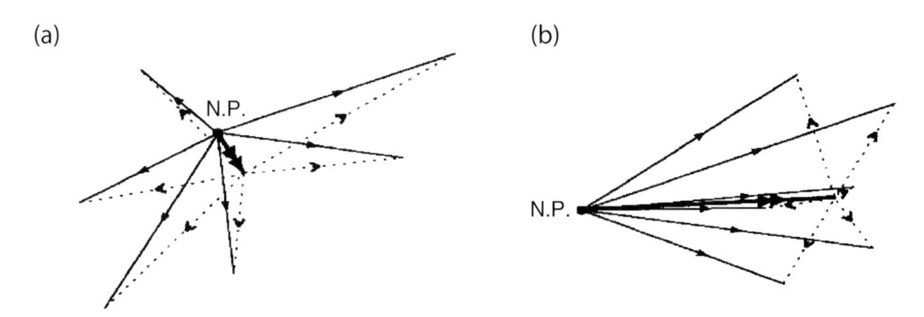

Figure T13 True polar wander in the mean lithosphere reference frame. Each thin solid line with arrow represents motion of the north pole (NP) ${}_n\mathbf{V}_{NP}$ with respect to a plate n (or vice versa). The thick vector $\mathbf{V}_m$ links NP with the center of gravity of the end-points of all individual velocity vectors (weighed by the fractional area occupied by each plate). True polar wander is considered useful when the modulus (scalar magnitude) of $\mathbf{V}_m$ is large compared to those of the dashed vectors representing velocities with respect to the center of gravity, ${}_n\mathbf{V}_{NP}-\mathbf{V}_m$. (after McKenzie, 1972; McElhinny and McFadden, 2000).

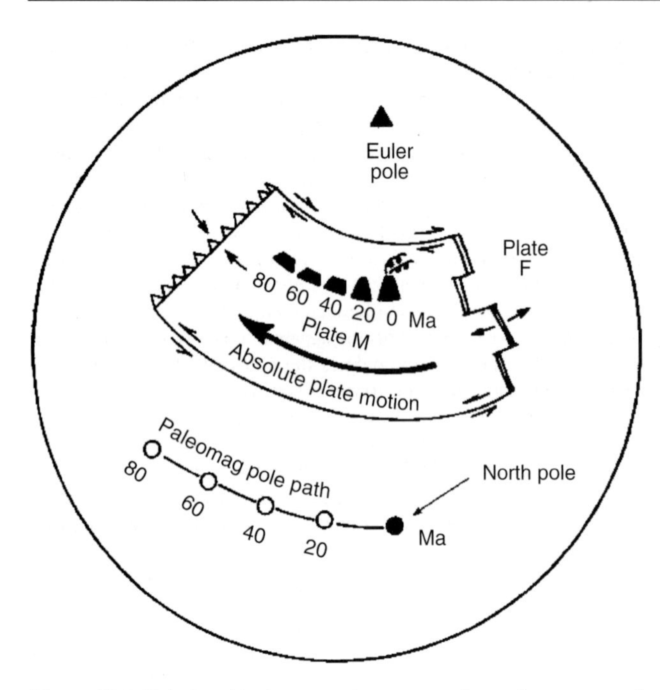

Figure T14 Relationship between hotspot tracks and segments of APW paths if the spin axis has been fixed with respect to the hotspots. Plate F is fixed relative to the hotspots, while plate M rotates about the Euler pole relative to plate F, the hotspots and the paleomagnetic axis (assumed to coincide with the spin axis). The hotspot trace on plate M, the transform faults on the plate boundaries between F and M, and the segment of the APW path all lie on small circles centered on the Euler pole (Gordon, 1987).

to derive the APWPs and the oceanic data used to derive plate kinematic models). They used then available paleomagnetic data (from North America, Africa, Eurasia and India) and kinematic models (from the Indian, Central Atlantic, and North Atlantic oceans) and produced a "synthetic" APWP, which was defined with respect to South Africa and then transferred to all other plates. Motion of hotspots with respect to the plates was then integrated to derive an estimate of TPW. As a consequence of publication of a Global Paleomagnetic computerized Database and of much improved reconstructions of past plate motions, this work was updated recently (Besse and Courtillot, 2002).

A recent TPW curve for the last 200 million years

The resulting TPW curve, based on 242 independent data from the African, Antarctic, Australian, European, Greenland, Indian, and North and South American plates, is shown in Figure T12a with a 20 myr time resolution. Points are shown with their 95% confidence circles (radii A_{95}) every 10 myr and every other point is statistically independent (estimates at 5 myr intervals are also available in that study). We recall that the 95% confidence intervals depend on many parameters linked to paleomagnetic data collection and laboratory treatment (see, e.g., McElhinny and McFadden, 2000, p. 91), but fail to explicitly include uncertainties on plate motions: they are supposed to be our best estimate of the cone within which the "true" direction is expected to lie. In other terms, there is a remaining 5% chance that our determination is more than A_{95} away from the true direction.

The first points, corresponding to the period 8–59 Ma, are all in the same quadrant, between 3.3° and 5.9° away from the present rotation pole, with 95% uncertainties ranging from 2.0° to 3.8°. They are not statistically distinct from each other and therefore could correspond to a standstill. A mean position can be calculated from all (102) data points in that time window: it is found to lie at 85.1°N, 153.3°E ($A_{95} = 1.5°$), significantly displaced from the pole. The youngest mean pole at ~5 Ma (mean age 3.0 ± 2.4 Ma) is at 86.4°N, 166.8°E ($A_{95} = 2.5°$), also significantly displaced from the pole but not statistically different from the 8–59 Ma overall mean. It seems that TPW may have been negligible for an extended period of about 50 myr, but accelerated a few myr ago, with a velocity on the order of 100 km myr^{-1}. The path displays a succession of a standstill at 160–130 Ma, a quasicircular track from 130 to 70 Ma, then the standstill at 50–10 Ma and the faster motion up to the present (Figure T12b). The TPW rate between 130 and 70 Ma averages 30 km my^{-1}. This would confirm earlier findings (e.g., Andrews, 1985; Besse and Courtillot, 1991) that TPW appears episodic in nature, with periods of (quasi-) standstill alternating with periods of faster TPW. The typical duration of these standstill periods is on the order of a few tens of millions of years (50 myr). Because of all the uncertainties in models of hotspot kinematics prior to ~130 Ma, it is not safe to place too much weight on behavior prior to that time. The major event since then is therefore the end of the 130–60 Ma period of relatively fast polar wander, with the standstill (i.e., no or little TPW) from 60 Ma (actually because of larger 95% confidence circles, possibly 80 Ma) to 10 Ma. Marcano *et al.* (1999) find TPW rates on the order of or less than 40 km myr^{-1} in the Permotriassic, i.e., typical of times since then, extending the period of moderate TPW back to more than 250 myr.

However uncertain, evidence for the fact that Earth emerged from a long standstill to enter a new period of faster polar wander in a different direction 10 Ma ago (~5 Ma at the higher resolution) is particularly interesting. That period would then still be going on. The youngest pole at ~5 Ma, taken at face value, would imply a TPW rate of 130 km myr^{-1} in the direction ~15°W (with a large azimuthal uncertainty from 10°E to 40°W, due to proximity of the 5 Ma pole to the present rotation axis). The first-order agreement with recent astronomical determinations (107 km myr^{-1} in the direction 79°W, with a possible range from 65°W to 85°W; see section "How is TPW measured (on short time scales)?") has reinforced the idea that this might be a valid estimate and that TPW might have been rather uniform over the last few millions of years, corresponding to significant, large scale changes in ice cover. Ice is not considered to have played a significant role before, and it was absent prior to ~30 Ma.

Does super fast TPW occur?

Prévot *et al.* (2000) argue for an episode of super fast TPW between 114 and 118 Ma (in excess of 500 km myr^{-1}). Besse and Courtillot (2002; see also Cottrell and Tarduno (2000), and Tarduno and Smirnov, 2001, 2002) critically discuss this and other suggestions at other epochs (e.g., Petronotis and Gordon (1999) at 80–70 Ma; Sager and Koppers (2000) at about 84 Ma). They conclude that none of the several suggested super fast events (with velocities of hundreds of km myr^{-1}) is based on sufficiently robust sets of observations. For instance, Cottrell and Tarduno (2000) tested the hypothesis of Sager and Koppers (2000) using data from Italian sections and found that it failed the test. On the other hand, Besse and Courtillot (2002) and Tarduno and Smirnov (2002) showed that the episode found by Prévot *et al.* (2000; see also Camps *et al.*, 2002) was not statistically robust due to the use of too short time windows. Gordon *et al.* (2004) recently conclude in favor of a super fast episode seen on the Pacific plate between 88 and 81 Ma (see below). However, it remains reasonable (in the author's view) to assume that many of these features correspond to erroneous individual data or other sources of error. Only the recent phase of TPW since 5 Ma may prompt us to accept that TPW velocities up to 100 km myr^{-1} can be maintained over periods of millions of years (Figure T12).

Two of the nagging questions which must be fully addressed before the concept of global TPW and the velocities mentioned above can be accepted are geometry of the mean geomagnetic field and hotspot fixity.

Is the mean field sufficiently close to an axial dipole?

Slight departures from a purely centered ancient dipole field have been noted as early as 1970, when Wilson argued for a far-sided and right-handed distribution of virtual geomagnetic poles during the Cenozoic: average paleomagnetic poles tend to fall beyond the geographical pole and to the east of it, seen from the site where the data have been obtained. A number of analyses concluded that, when averaged over the last few million years, an axial quadrupole component is detectable, with an amplitude on the order of 3–6% of the axial dipole (e.g., Constable and Parker, 1988; Schneider and Kent, 1990b; Quidelleur *et al.*, 1994; Carlut and Courtillot, 1998). Livermore *et al.* (1983, 1984) tried to extract such a quadrupolar term in a worldwide paleomagnetic database going back 200 myr^{-1}. More recently, Besse and Courtillot (2002) find moderately far-sided poles, consistent with a persistent quadrupole moment (on the order of 3 ± 2% of the dipole over the last 200 myr), but its amplitude can be neglected for many applications. The GAD hypothesis has recently been further challenged by suggestions that significant long-term octupolar contributions of between 6 and 25% of the GAD may have existed for Precambrian through early Tertiary time (Kent and Smethurst, 1998; Van der Voo and Torsvik, 2001; Si and Van der Voo, 2001; Torsvik *et al.*, 2002). For instance, Si and Van der Voo (2001) propose that a value of 6% would account for the low inclinations observed in central Asia in the Cretaceous and early Tertiary. A symposium in honor of Neil Opdyke has reviewed the pros and cons of octupolar contributions to the long-term mean field (the reader is referred to the corresponding AGU monograph and to other articles in this encyclopedia). Courtillot and Besse (2004) attempt to find evidence for octupolar contributions in the 0–200 myr period. In that new analysis, data from sites believed to have possibly undergone a tectonic rotation about a local vertical axis are included, contributing 174 out of 465 data. The positions of mean poles are analyzed in 20 myr windows in common-site longitude, respectively, for the northern mid-latitudes, equatorial and southern mid-latitudes, searching for the distinctive antisymmetrical pattern expected for a dipole plus octupole field. They next analyze the distribution of "latitude anomaly" (derived from the inclination anomaly, i.e., observed minus expected in case of a pure dipole) versus dipole latitude. Based on these various data manipulations, they find no robust evidence for an octupole and estimate that values on the order of 5% are unlikely to have been exceeded in the last 200 myr. A preliminary 200 myr overall mean field has an octupole of 3 ± 8% (i.e., not significant; note that the present day value is on the order of 4%, but this is an "instantaneous value" with large secular variation, which in large part could average out). Thus field geometry is unlikely to severely alter TPW estimates, for which it is not the main source of "noise."

Is the estimate of TPW truly global?

A legitimate concern regarding the above conclusions is due to the fact that the analysis is not truly global, in that it fails to encompass the Pacific plate. Petronotis and Gordon (1999) have compiled an APW path (nine poles from 125 to 26 Ma) for the Pacific plate, with four poles based on the analysis of skewness of ocean crust magnetic anomalies, three on seamount magnetic anomaly modeling (Sager and Pringle, 1988) and two undetermined. Using the Pacific plate versus hotspot kinematic model of Engebretson *et al.* (1985), based on dating of volcanoes which are part of hotspot tracks such as Hawaii, Louisville, or MacDonald (see, e.g., Figure T15 in the case of Hawaii), Besse and Courtillot (2002) have determined a corresponding 125–26 Ma "Pacific hotspot only" TPW curve. The "Pacific" and "Indo-Atlantic" TPW curves are compared in Figure T16. This comparison is interesting because the data sets they are based on are

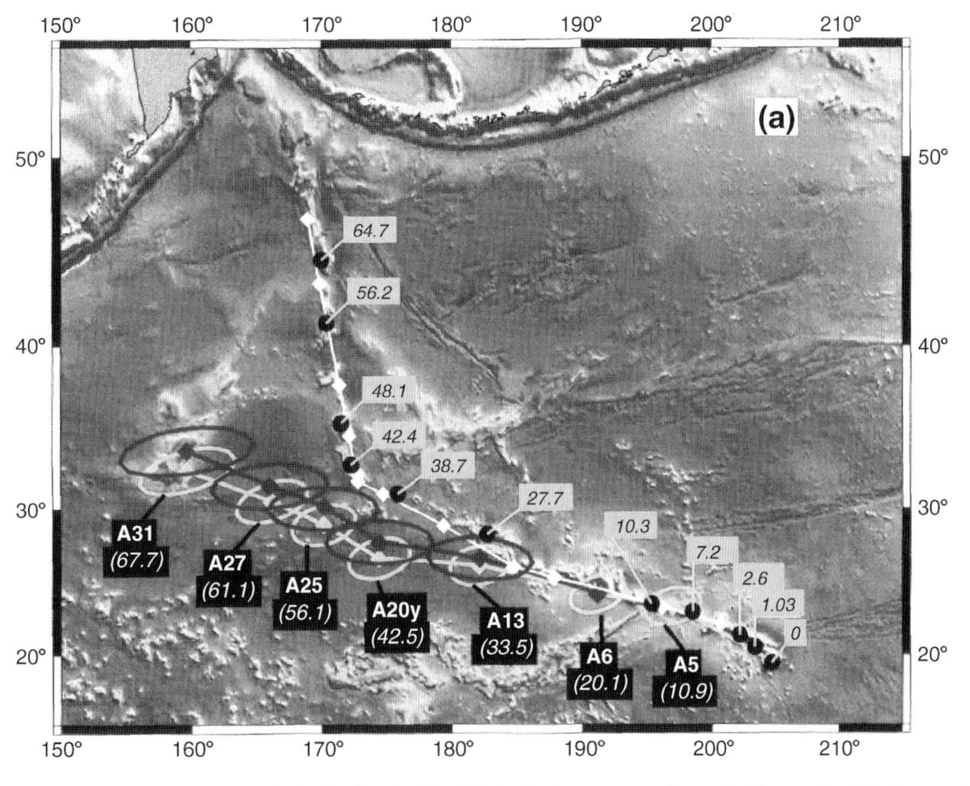

Figure T15 The Hawaii-Emperor seamount chain in the Pacific (ETOPO5 bathymetry image). The track with diamonds, which closely follows the observed one, is for the best-fit Pacific plate motion. Two reconstructed tracks of Pacific-hotspot motion in the Indo-Atlantic reference frame are also shown with quantitative plate reconstruction error ellipses. The reconstructed track, which is closer to the observed one, includes East-West Antarctic motion. Chrons and ages (in Ma) are shown (after Raymond *et al.*, 2000).

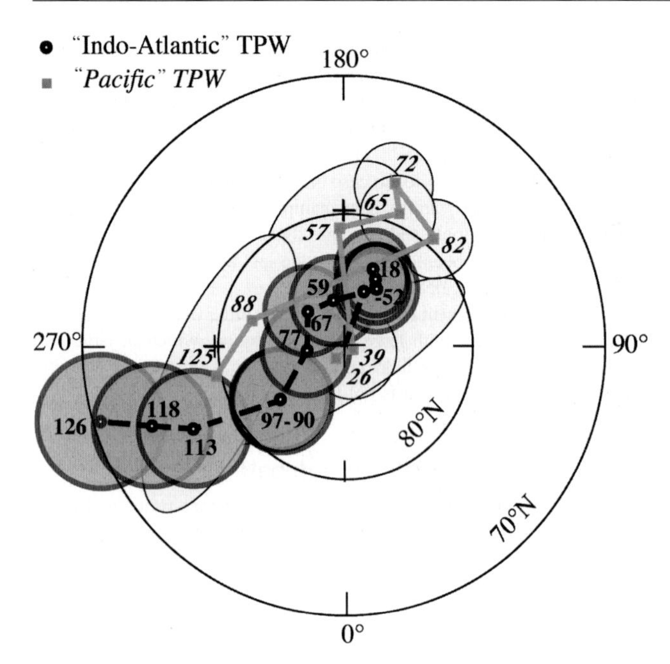

Figure T16 Comparison between the "Indo-Atlantic" (open dots and dashed line) and "Pacific" (squares and solid line) TPW paths (after Besse and Courtillot, 2002).

entirely different and independent. Despite some significant differences to which we return shortly, it is worth emphasizing that the two curves are altogether similar in shape (tracks, amplitudes, azimuths), particularly the 300° longitude trending track from 130 to 70 Ma, though the two are offset (in the same general direction) by about 7°. More precisely, the confidence intervals intersect (though points are not in the intersection) near 125, 90, 60, 40 and 30 Ma. The main differences occur between 80 and 70 Ma: the 82 and 65–72 poles derived from Petronotis and Gordon (1999) and the 77 and 67 Ma poles derived by Besse and Courtillot (2002) are clearly distinct. In as yet unpublished work, Gordon *et al.* (2004 and R. Gordon, personal communication March 2004) have updated the APW path of the Pacific plate. It is now defined by 11 paleomagnetic poles from nonoverlapping age windows going back to 125 Ma. Nine of these poles (32–81 Ma) are determined from skewness analysis. Comparing their APWP to Pacific hotspot motion with respect to the spin axis, Gordon *et al.* (2004) find a TPW curve which agrees fairly well with their previous analysis, except for a significant change in the 32–40 Ma pole. The new curve implies rather small motion between 81 and 32 Ma, with larger swings before and after. The most recent shift, during the past 30 myr, is about 10° (or 35 km myr^{-1} on average). A bigger shift by about 20° would have occurred between 88 and 81 Ma (i.e., more than 300 km myr^{-1}, an episode of super fast motion which in our view remains to be demonstrated). Comparing the new Pacific TPW curve to the Indo-Atlantic one derived by Besse and Courtillot (2002), Gordon *et al.* (2004) conclude that there is no significant net motion between Pacific and Indo-Atlantic hotspots in the past 125 myr^{-1} (see section "Are hotspots fixed with respect to each other?"). This conclusion depends both on data quality and on the validity of the plate motion models.

There are ongoing debates on the validity of the data used by Gordon and collaborators to construct their Pacific APWP, whether the uncertainty regions they use are realistic, and also on the question of fixity of Pacific hotspots with respect to each other and to the "Indo-Atlantic" hotspots (see below). For instance, Tarduno and Cottrell (1997; see also Tarduno *et al.*, 2003) have determined the paleolatitude (based on inclination-only data from cores) of the 81 Ma old Detroit Seamount, which is part of the Emperor chain, not far from its northern termination in the Kuril Trench. The paleolatitude (36.2 + 6.9/−7.2°) is distinct from that based on the 81 Ma pole of the Pacific APWP (Sager and Pringle, 1988) which is on the order of 20°N, based on seamount anomaly poles. Tarduno and Cottrell (1997) exclude the possibilities of inadequate sampling of secular variation, bias due to unremoved overprints or off-vertical drilling. They point out the uncertainties encountered when building an APWP for an oceanic plate, such as the Pacific, solely from inversions of magnetic surveys over seamounts and/or analysis of skewness of marine magnetic profiles. Following several authors (e.g., Parker, 1991), Di Venere and Kent (1999) and Cottrell and Tarduno (2000) argue that the reliability of the Pacific paleopoles based on either modeling of seamount magnetic anomalies or determination of skewness of marine magnetic anomalies should be considered suspect. Both are prone to numerous biases and could yield errors in excess of 10° in the position of the mean poles derived from them. Seamounts suffer from induced and viscous magnetizations, and potential unresolved multiple polarities of magnetization. Skewness suffers from anomalous skewness as recognized by all users of such data. If one uses only the highest resolution, standard paleomagnetic data from the Pacific, much of Mesozoic TPW for that plate is seen as motion between hotspot groups (e.g., Tarduno and Gee, 1995).

Are hotspots fixed with respect to each other?

We must therefore now review briefly the question of hotspot fixity. Norton (1995) has suggested that the famous 43 Ma Hawaiian bend in the Hawaiian-Emperor hotspot track (see Figure T15) was actually a "nonevent," i.e., indicated a change in motion of the Hawaiian hotspot with respect to the mantle rather than a change in Pacific plate motion. Koppers *et al.* (2001) have tested the fixed hotspot hypothesis for Pacific seamount trails. They use seamount locations to first determine stage Euler poles, which they then test against observed $^{40}Ar/^{39}Ar$ age progressions. The main stages tested are 0–43, 43–80, and 80–100 Ma. Koppers *et al.* find that the 0–43 Ma Hawaiian and Foundation seamount trail pair is the only one compatible with the fixed hotspot hypothesis. Most other trail pairs would require relative motions of order (or in excess) of 10 km myr^{-1}. The 43–80 Ma Emperor/Line pair shows particularly large discrepancies, requiring motions of at least 30 km myr^{-1}. Based on careful analysis of 14 Pacific seamount tracks, using updated age determinations and bathymetry, Clouard and Bonneville (2001) show that Pacific seamounts are actually created by different processes, many being short-lived. They conclude that only the Hawaii and Louisville chains qualify as long-lived hotspots that can robustly be tested for fixity.

Di Venere and Kent (1999) have addressed the problem of relative motion between the Pacific and Indo-Atlantic groups of hotspots, using paleomagnetic results from Marie Byrd Land in West Antarctica. Though they demonstrate that some motion must have taken place since 100 Ma between West (Marie Byrd Land) and East Antarctica, they conclude that this motion cannot account for more than 20%, and possibly as little as 4% of the 14.5° offset between the observed and predicted positions of the 65 Ma Suiko seamount on the Emperor continuation of the Hawaiian hotspot track (Figure T15). They also discuss the integrity of the Pacific plate and the role of missing plate boundaries and errors in kinematic plate circuits, and find that they play a small role. Di Venere and Kent (1999) conclude that most of the apparent motion between the two main groups of hotspots is real, with an average drift of about 25 km myr^{-1} since 65 Ma. Torsvik *et al.* (2002) summarize paleomagnetic tests, which in their view raise increasing doubts on the fixity of hotspots. Though Indo-Atlantic hotspots appear to have remained stable, the Pacific hotspots, most notably Hawaii, appear to have drifted southward with respect to the spin axis in the 65–45 Ma period. Torsvik *et al.* (2002) conclude that global analyses fail to support TPW since the late Cretaceous, and suggest controversial, slow TPW in the early Cretaceous. Their preferred explanation for this is southward drift of Atlantic hotspots prior to ~95 Ma.

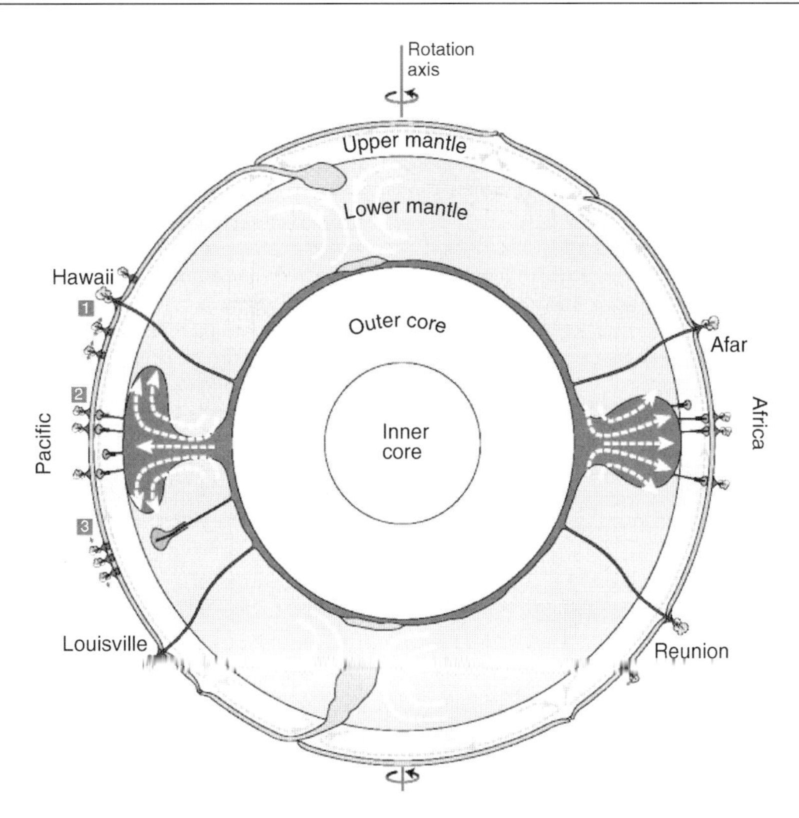

Figure T17 A schematic cross section of the dynamic Earth going through its rotation axis, outlining the possible sources of three types of plumes/hotspots: "primary," deeper plumes possibly coming from the lowermost mantle boundary layer; "secondary" plumes possibly coming from the top of domes near the depth of the transition zone at the locations of the superswells; "tertiary" hotpots may have a superficial origin (after Courtillot *et al.*, 2003).

Courtillot *et al.* (2003) recently reanalyzed the characteristics of the world catalogue of some 50 hotspots. Based on geophysical (buoyancy, tomography), geochemical (He isotope ratio) and geological (presence of a seamount track, of a flood basalt) criteria, Courtillot *et al.* (2003) suggest that surface hotspots on Earth may have three distinct origins (Figure T17): at least seven "primary" hotspots (Hawaii, Easter, Louisville, Iceland, Afar, Reunion, and Tristan da Cunha) would originate from the deepest part of the lower mantle, possibly anchored on chemical heterogeneities deposited in the D" layer. Some ~20 "secondary" hotspots might originate from the bottom of the transition zone in relation with the large transient domes that correspond to the two antipodal superswells under the central Pacific and Africa. And the remainder (~20) could be upper mantle features, as suggested for instance by Anderson (e.g., 2000). Very recently, Montelli *et al.* (2004) have traced the deep roots of some primary plumes into the lowermost mantle, based on a new type of "finite-frequency" tomography. The list of their primary plumes is similar to though not identical with that of Courtillot *et al.* (2003).

Restraining themselves to potential primary hotspots, Courtillot *et al.* (2003) find no evidence for inter-hotspot motion significantly larger than 5 km myr^{-1} either within the Pacific hemisphere or the Indo-Atlantic hemisphere hotspots (however, in very recent work, Steinberger *et al.* (2004) and Koppers *et al.* (2004) do evidence some motion between Hawaii and Louisville). Such rms velocities of 5 km myr^{-1} or less, i.e., an order of magnitude less than rms plate velocities are to first order "small." So, some primary hotspots indeed seem to provide a quasifixed frame in each hemisphere over the last 80–100 myr. Was there then any motion between the two hotspot ensembles? This raises the well-known difficulty of establishing a reliable kinematic connection between the two hemispheres through Antarctica. This has recently been addressed by Raymond *et al.* (2000). Based on updated kinematics, these authors predict the location of the Hawaiian hotspot back in time, under the hypothesis that Reunion and Hawaii have remained fixed with respect to each other; for this, they use the dated tracks left on the African and Indian plates by the Reunion hotspot since it started as the Deccan traps 65 Ma ago. The plot of misfit between the predicted and observed positions for Hawaii as a function of time (Figure T15) indicates that the two hotspots have actually drifted slowly (Figure T18a), at ~10 km myr^{-1}, for the last 45 myr, but at a much faster rate (~50 km myr^{-1}) prior to that (assuming that there is no missing plate boundary or unaccounted for motion between E and W Antarctica), vindicating earlier conclusions reached by Tarduno and Cottrell (1997). Courtillot *et al.* (2003) conclude that the primary hotspots may form two distinct subsets in each one of the two geodynamically distinct hemispheres. Each subset would deform an order of magnitude slower than typical plate velocities. The two subsets would have been in slow motion for the last 45 myr, but in faster motion prior to that.

Figure T18b,c displays paleomagnetically derived paleolatitudes for the Hawaii and Reunion hotspots, which can be taken as the best documented representatives from each hemisphere. However sparse, the data are compatible with the same simple two-phase history, in which there was little latitudinal motion in the last 45 myr, but significant equatorward motion prior to this, at about 50 km myr^{-1} for Hawaii and 30 km myr^{-1} for Reunion. There is an uncertainty of a few myr (up to 5) on the timing of the change from one phase to the next at 40–50 Ma. The ~45 Ma date is most accurately fixed by the age of the bend of the Hawaiian-Emperor chain, if this is indeed the common time of change of all processes described in Figure T18, which we assume to be the case to a first approximation. Tarduno *et al.* (2003) report new data from the Emperor seamounts that confirm an age-progressive paleolatitude trend. The southward drift rate is about

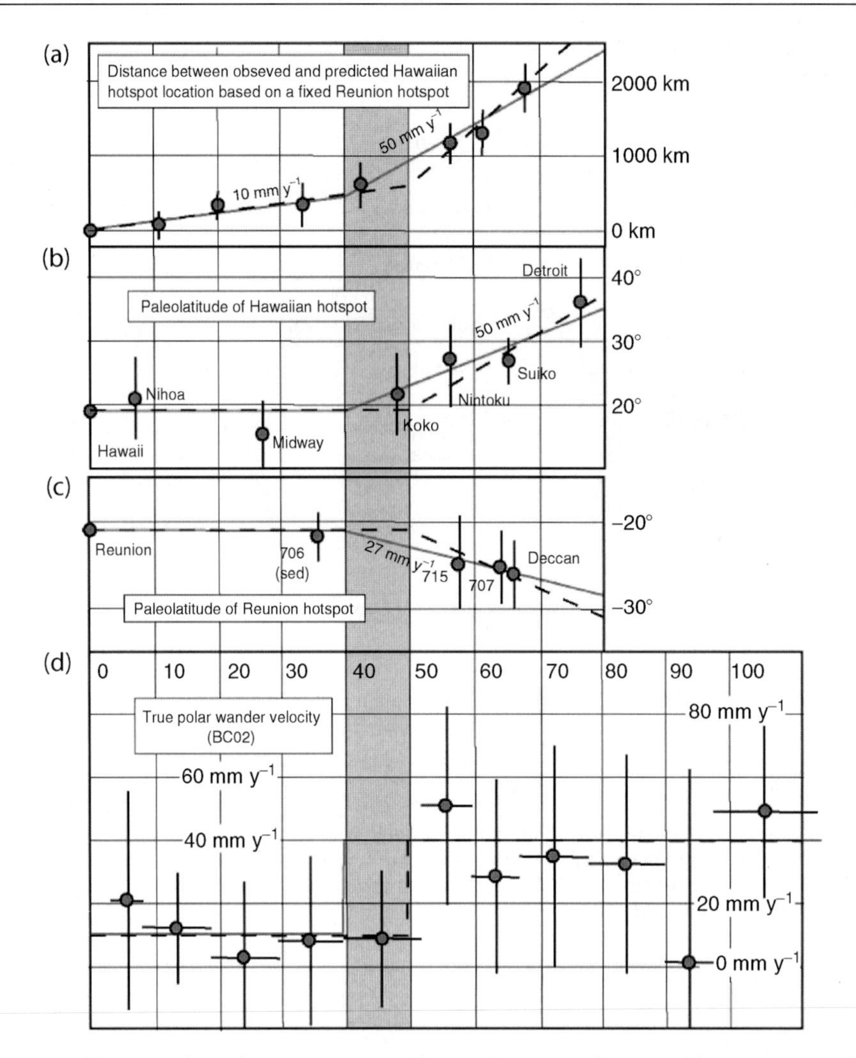

Figure T18 Time variations of four significant kinematic, geographic or dynamic indicators of hotspot motion (all on the same time scale in million years B.P.). All display a steplike change in velocity at ~40–50 Ma. Velocity patterns are shown as full or black lines, depending on whether the change happened at 40 or 50 Ma. (a) Distance between observed and predicted positions for the Hawaiian hotspot. (b) and (c) Latitudinal evolution of the Hawaiian and Reunion hotspots. (d) Along track TPW velocity at 10Ma intervals (after Courtillot *et al.*, 2003; Tarduno *et al.*, 2003).

40 km myr^{-1} between 81 and 47 Ma. Tarduno *et al.* (2003) also conclude that TPW was smaller than suggested in earlier studies.

The recently re-evaluated estimate of TPW over the last 200 myr (Figure T12) is uncertain, as recalled above, being based only on hotspots from the Indo-Atlantic hemisphere. The TPW curve estimated using Pacific-only data is similar to the Indo-Atlantic TPW, seemingly validating to first order the concept that TPW is a global phenomenon (Figure T16). We have seen that Gordon *et al.* (2004) believe that motion between the two subsets of primary hotspots is small enough that the hotspot reference frame remains useful for estimating TPW over the past 125 myr On the other hand, Besse and Courtillot (2002) suggest that TPW pole positions for the two hemispheres are significantly displaced between ~50 and ~90 Ma (Figure T16).

Due to the high viscosity of the bulk of the lower mantle (see e.g., Lambeck and Johnston, 1998; Steinberger and O'Connell, 2000), the primary (and probably also the secondary) plumes follow the large-scale motion of the lower mantle. The two groups of primary hotspots might indicate that the two separate mega-cells of quadrupolar convection, centered on the Pacific and African super-upwellings, have moved little (~10 km myr^{-1}) in the last ~40–50 myr with respect to each other, but underwent some significant (~50 km myr^{-1}), rather uniform, relative motion in the previous tens of millions of years. This motion would have already been going on prior to the oldest preserved trace of the Hawaiian hotspot (e.g., prior to 80 Ma). Hotspot tracks become fewer and more uncertain as one goes back in the past and pre-100 Ma TPW estimates should be regarded with caution. However, there are indications (Besse and Courtillot, 2002) that the phase of TPW which ended around 50 Ma may have started ~130 Ma ago (Figure T12). What could have triggered it? The geometry of density anomalies associated with upwellings (notably with the two superswells) may not have a large enough effect on the principal axes of inertia of the Earth and hence on TPW (Richards *et al.*, 1999). However, because they are antipodal, the direction of maximum principal axis of inertia is very sensitive to time variations in superswell structure and some TPW should occur on the same time scale as the superswells (with alternating quiet and more rapid episodes) and with an amplitude which could be similar to that due to subducting slabs (Greff-Lefftz, 2004). Indeed, cold subducted material may have a larger effect on TPW (e.g., Ricard *et al.*, 1993a,b). It has been proposed that this material may accumulate along the great circle of dominant downwelling in the lower mantle, at the base of subduction zones near the transition zone, and trigger an avalanche in the lower mantle

(Weinstein, 1993; Steinbach *et al.*, 1994; Brunet and Machetel, 1998). Such an avalanche starting some 130 Ma ago could have set Earth on the episode of TPW which lasted until ~40–50 Ma ago. An alternate interpretation would be the disappearance of a major subduction zone system, after which both heat flow and mean temperature would have been rapidly and significantly altered (S. Labrosse and L.-E. Ricou, personal communication, 2002). The more recent halting of TPW at 40–50 Ma could also be related to the closure of the huge Tethys subduction zone, following the generalization of Indian collision, as has been suggested for a long time in order to interpret the Hawaii-Emperor bend (Patriat and Achache, 1984). Episodes of TPW could be the result of such (rare) events, with alternate episodes of quiescence lasting tens of millions of years. And primary hotspots would be our main source of information on their time history, if they can be considered as tracers of readjustments in the two-cell geometry of the lower-mantle reservoirs.

Modeling of true polar wander

Modeling of TPW has made significant progress in the last decade. In an early attempt, Ricard *et al.* (1993b) derived a model of mantle density heterogeneities based on plate motions recorded by the ocean floor, including slabs crossing the upper/lower mantle transition zone slowing down by a factor between 2 and 5. Based on this, they computed a synthetic geoid, which agreed rather well with observations, provided viscosity increased by a factor 40 going from the upper to the lower mantle. Changes in the degree two geoid should reflect the history of the inertia tensor and therefore TPW as revealed by paleomagnetism. Ricard *et al.* (1993a) predicted TPW velocities on the order of 300 km myr^{-1} in the direction 130°W, much larger than observed rates. Their three-layer model could not account simultaneously for the present-day geoid and observed TPW. Slowing TPW, i.e., stabilizing the Earth, can be accomplished in two ways. Either by increasing viscosity at depth and slowing motion of the bulge (Spada *et al.*, 1992): increasing lower mantle viscosity by a factor 100 rather than 40 slows TPW to a more reasonable 40 km myr^{-1} but worsens the fit to the geoid. Or by slowing down time changes in the nonhydrostatic tensor: Richards *et al.* (1997) explain the observed long-term stability of the rotation axis by the slow rate of change in large-scale plate tectonic motions in the last 100 myr. Subducted lithosphere is for them the main component of mantle density heterogeneity and invoking readjustment of the rotational bulge is not necessary (see also early work by Chase, e.g., 1979). They find that the degree 2 geoid coefficient changes little in the last 60 myr, and that a lower/upper mantle viscosity ratio of 30 leads to TPW velocity estimates less than 50 km myr^{-1}. The computed synthetic TPW path they obtain for a model with layered viscosity has some features which qualitatively resemble observations (Figure T19).

In more recent studies, seismic tomography is used to infer 3D maps of density heterogeneities that drive flow in the viscous mantle. Steinberger and O'Connell (2002; see also Steinberger, 1996) have developed an algorithm to calculate changes in the Earth's rotation axis for rather general viscoelastic Earth models. They combine hotspot motion, polar wander and plate motions in a single dynamically consistent model (though plate velocities are imposed on the model, not derived from it), allowing all effects to be simultaneously determined and their relative effects sorted out. Changes in rotation are due to advection of "realistic" density anomaly distributions inferred from a number of seismic tomography models in a "realistic" flow field constrained by observed plate kinematics at the surface. The axis of maximum hydrostatic moment of inertia is by definition parallel to the rotation axis, and one of the principal axes of the "total" nonhydrostatic inertia tensor (see section "Definitions and early views") always follows the rotation axis closely, as originally pointed out by Spada *et al.* (1996a,b). This alignment does not imply an alignment with the axis of maximum "total" (Ricard *et al.*, 1993a) or "imposed" nonhydrostatic moment of inertia; because of delayed adjustment of the

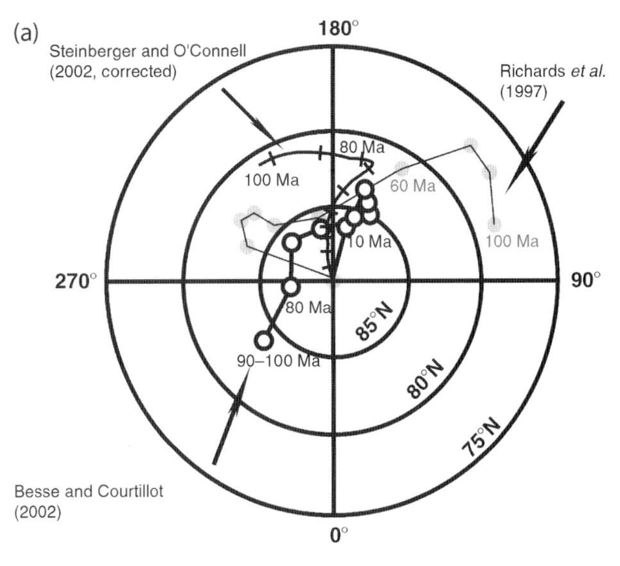

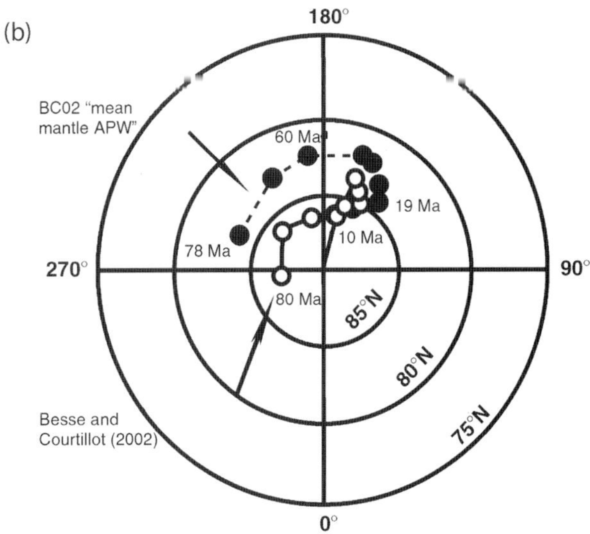

Figure T19 (a) A comparison of TPW observations and models over the past 100myr. Open circles are for observation derived determinations by Besse and Courtillot (2002; see Figures T12 and T16); gray circles are a model calculation with 2-layer mantle viscosity structure—viscosity ratio of 30—by Richards *et al.* (1997); the solid line with tick marks is from a model calculation by Steinberger and O'Connell (2002), with an 8-layer mantle viscosity structure and driving density heterogeneities calculated from a tomographic model of Grand (2001 personal communication to the authors of that paper). Note that this is a corrected curve (see text). (b) A comparison of the TPW curve of Besse and Courtillot (2002) for the past 80Ma and a "mean mantle" apparent polar wander path derived by B. Steinberger (personal communication, 2004) from the BC02 synthetic apparent polar wander path for South Africa (Table 4, with correction given in Besse and Courtillot, 2003) and finite African plate rotation in the mean mantle reference frame for the "preferred" model of Steinberger *et al.* (2004).

bulge, a misalignment may result (Steinberger and O'Connell, 2002). But this is negligible unless the viscosity of the lower mantle is much larger than 10^{23} Pas (and except in the hypothetical case of inertial interchange, see section "Inertial interchange and true polar wander"). Steinberger and O'Connell (2002) also find that the maximum speed of polar wander driven by mantle convection is about 100 km myr^{-1} (and incidentally that a fast rising plume head cannot cause more than a few degrees of TPW during the few myr of its ascent). A recent study by Greff-Lefftz (2004) investigates the influence of mantle plumes varying at different spatial wavelengths on the time variations of TPW. In a first model, she shows that an upwelling plume analytically represented as a sphere whose radius varies as a function of the flux of material in the conduit and which traverses the mantle at the Stokes velocity, produces very little wander of the rotational axis. Greff-Lefftz then studies the effects of two superswells which mimic the ones observed with seismic tomography, and concludes that a doming regime within the mantle involves significant polar wander. Some of the features of this TPW which are directly linked to the periodicity of doming are reminiscent of observed phases of slow and fast TPW, with similar peak velocities.

In all cases, isoviscous mantle models predict TPW rates much larger than observed, and a significant viscosity increase in the lower mantle is required to stabilize the large scale pattern of convection and bring TPW rates closer to observed values. This is also what is needed for hotspot stability as well as matching the geoid. Steinberger and O'Connell (2002) compare their model results for a number of tomographic models and discuss the effects of compressibility, deep mantle viscosity, thickness of a high viscosity layer in the deep mantle, and chemical boundaries on rates of TPW. Their TPW estimates, one of which going back to 100 Ma is shown in Figure T19a, are in qualitative agreement with observations (note that Figures 9 and 10 in Steinberger and O'Connell (2002) are incorrect, corrections being available at http://www.geodynamics.no/STEINBERGER/papers/giavol_fig9.pdf and http://www.geodynamics.no/STEINBERGER/papers/giavol_fig10.pdf; B. Steinberger, personal communication, 2004).

Inertial interchange and true polar wander

If the principal directions of the nonhydrostatic imposed inertia tensor remain unchanged, but the eigenvalues change and the maximum nonhydrostatic imposed moment of inertia decreases whereas the intermediate one increases, until the latter becomes larger than the former, then a quasidiscontinuous 90° rotation becomes possible so that the rotation axis will align with the new axis of maximum nonhydrostatic moment of inertia (Fisher, 1974). More precisely, the preferred axis of rotation moves instantaneously by up to 90° and the rotation axis will move to its new stable position at a rate limited by adjustment of the bulge. The rate also depends on the driving force, linked to the difference in maximum and intermediate eigenvalues. In such an inertial interchange event, only the nonhydrostatic part is interchanged and there is propagation of the equatorial bulge, which must always remain equatorial. Such an occurrence is referred to as an inertial interchange true polar wander (IITPW) event. Actually, the amplitude of an IITPW event may not reach 90°. At the time when the former maximum and intermediate nonhydrostatic imposed moments of inertia become equal, the nonhydrostatic moment of inertia becomes equal along a whole great circle (the tensor is prolate or flattened with cylindrical symmetry) and the new axis of maximum imposed moment of inertia can grow anywhere on this great circle. The fact that the Earth may in the past have catastrophically exchanged two of its principal axes of inertia in that way, i.e., that the whole lithosphere (and mantle) may have rotated by up to 90° in only a few million years, bringing for instance parts of the Earth which were close to the poles at the equator, has become a new area of questioning in the geosciences. It is clear from the available TPW curves going back to the early Mesozoic that such an event has not occurred in the past 250 myr. For certain parameter values (such as lower mantle viscosity) IITPW is encountered in some numerical model runs. This was almost seen in Goldreich and Toomre's earliest calculations. Richards *et al.* (1999) find occasional though infrequent inertial interchanges of polar axes with duration of 20 myr, due to "avalanching" (discontinuous rearrangements in geometry of subducting slabs) in the lower mantle (only one such event occurs in a 600 myr numerical run).

A recent review of evidence for IITPW has been given by Evans (2003). A series of paleomagnetic poles spanning an IITPW event from any continent should fall on a great circle with the pole to that great circles approximating the minimum nonhydrostatic imposed moment of inertia (Evans, 1998). Paleomagnetic indications which would support rather fast TPW in pre-Mesozoic time have been put forward by Van der Voo (1994) and for very fast TPW by Kirschvink *et al.* (1997). For pre-Jurassic time, ocean floor has been destroyed (together with evidence of plate kinematics and many hotspot traces) and TPW cannot be estimated in the way indicated above. Evans proposes to obtain a crude estimate of TPW in the case (to be tested) when TPW velocity would have been much faster than between-plate motions (as in Figure T13b). This requires matching APW paths from a majority of continental fragments, and extracting a large common component of latitudinal drift (if it exists). Geological evidence of subduction and collision and fossil hotspot traces can help in interpreting the common APW record. Kirschvink *et al.* (1997) proposed in this way that 90° of TPW had occurred in a brief interval of Early Cambrian time (possibly 15 myr or less, i.e., TPW velocities in excess of 600 km myr^{-1}!), coinciding with the famous biological Cambrian explosion. Recall that numerical models place an upper limit of about 100 km myr^{-1} on TPW velocity; a full 90° IITPW event would take 3–20 myr for a modern Earth viscosity structure (Steinberger and O'Connell, 2002). To this day, the issue remains contentious because of the dearth of paleomagnetic records and possibility of problems with unremoved overprints, dating uncertainties, etc. (see Evans, 2003). Various authors tend to make different data selections, and each new datum published can still throw the balance in the opposite way. For one school of thought, the terminal Proterozoic to late Paleozoic APW for Gondwanaland (Figure T20) displays up to five large angular swings, some of which reach 90° in amplitude. This oscillatory polar wander is attributed to TPW about a common, long-lived minimum inertial axis near eastern Australia, which would be the prolate axis of Earth figure, inherited from the previous super-continent of Rodinia. In this way, Evans (2003) links TPW episodes to supercontinental cycles (Wilson-like but with durations on the order of 500 myr).

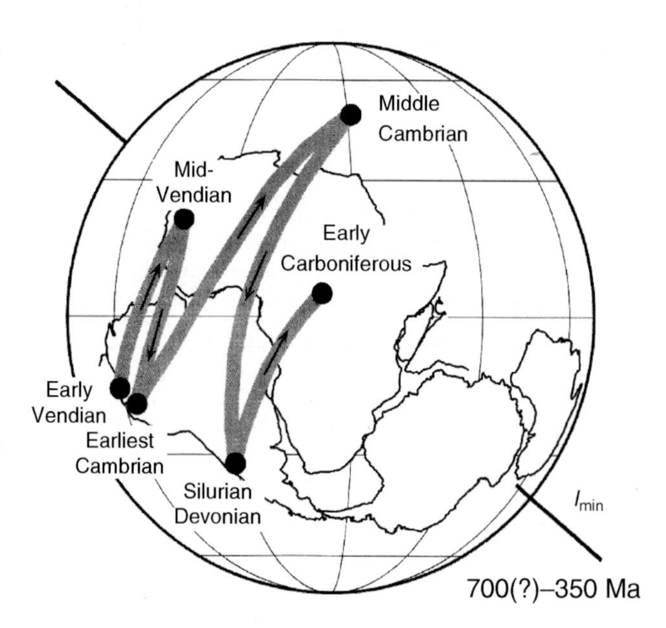

Figure T20 Terminal Proterozoic to late Paleozoic apparent polar wander path for Gondwanaland. Oscillatory APW rotations are interpreted as inertial interchange (IITPW) about a common, long-lived, minimum inertial axis (I_{min}) near eastern Australia (after Evans, 2003).

This is an exciting if controversial chapter of Precambrian paleomagnetism that will require much more data than is currently available, hence a concerted effort of the paleomagnetic community.

Most recent developments and conclusion

For the time being, the hypothesis that TPW is a go-stop-go phenomenon, with long standstills in the tens of millions of years, and periods of faster, rather uniform polar wander at velocities often not exceeding 30 km myr^{-1}, cannot be discounted. Evidence for super fast events at any time in the past remains to be demonstrated. Much recent modeling still failed to some extent to account for the slow values of typical TPW velocity (0–30, and rarely up to 100 km myr^{-1}), and even more so to account for the prolonged (~50 myr) periods with almost no TPW (standstills). Steinberger and O'Connell (2002) gave reasonable velocities but predicted smooth evolutions, rather than the alternating episodes, which some paleomagnetists feel they uncover from the data (Figure T19a).

The similarity between TPW estimates for the Pacific plate and the rest of the world (Besse and Courtillot, 2002), which are based on completely different and independent data sets, may be taken as supporting the idea that small yet significant TPW, on the order of 10°, occurred since the Cretaceous, though it still requires careful verification. Relatively fast rates before 50 Ma could be due to the variable difference between the largest and intermediate nonhydrostatic principal moments of inertia and to combinations of peculiar configurations of downgoing slabs and super-upwellings (Greff-Lefftz, 2004; B. Steinberger, personal communication, 2004). As seen since the time of Goldreich and Toomre, small differences may lead to faster rates. If the two superswells remain more or less in the same locations, the pole may move on the great circle perpendicular to the axis they form. Figures T12 and T16 may indicate such a tendency for the segment from 130 to 50 Ma, though as stated above the pre-80 Ma data have increasing uncertainties as one goes back into the past. The importance of the Pacific plate and severe limitations on presently available data from that plate, the fact that the global paleomagnetic database still is quite scant for certain continents and time windows, the fact that many hotspot traces are not yet sufficiently constrained in terms of age and paleolatitude, all point to the need for many more direct (paleomagnetic) measurements as opposed to indirect/remote sensing determinations of magnetization direction (i.e., "skewness" or "seamount" data).

In a very recent paper, Steinberger *et al.* (2004) combine hotspot motion in the large scale flow field of the mantle and intraplate deformation to account for most of the post-65 Ma misfits between computed and observed hotspot tracks. Motion of initially vertical plume conduits is dominated by advection in the high viscosity lower mantle and buoyant rising in the low viscosity upper mantle. Given the mantle flow field and a preferred viscosity distribution, Steinberger *et al.* (2004) find that computed hotspot motion fits observed tracks well for the last 50 myr, i.e., since the time of the Hawaii-Emperor bend. But the change in track direction at the bend is explained only when an alternative plate circuit through East Antarctica, Australia, and the Lord Howe rise is used rather than the traditional circuit through West Antarctica. This leads to East versus West Antarctic motion with a rotation pole close to the boundary and deformations which the authors believe are compatible with geologic evidence. Remaining pre-65 Ma misfits are attributed to unquantified motion internal to New Zealand. Only a combination of the two effects (hotspot motion and intraplate deformation) successfully predicts the geometry of the four selected seamount chains in the Pacific and Indo-Atlantic hemispheres. If correct, their analysis at the same time invalidates the notion of a hotspot-mantle reference frame or of separate hemispheres (for Steinberger and colleagues plumes indicate local flow and not motion of hemispheric cells) and replaces it with limited, predictable inter-hotpot motion. But TPW can still be calculated, i.e., a "mean mantle" reference frame can still be evaluated. The Hawaiian hotspot would have moved faster between 80 and 50 Ma, and at the same time TPW would have been faster prior to 50 Ma. TPW would account alone for paleolatitudinal changes of the Reunion hotspot, whereas the two effects would be required for the Hawaiian hotspot.

Combining Steinberger *et al.* (2004) preferred model for the finite rotation of the African plate in the "mean mantle" reference frame with the synthetic apparent polar wander path for South Africa (Besse and Courtillot, 2002, Table 5, with correction given in Besse and Courtillot, 2003), one can derive a TPW curve in the "mean mantle" reference frame (or if one prefers, the "apparent polar wander path of the mean mantle") that takes hotspot motions into account. This is shown in Figure T19b (B. Steinberger, personal communication, 2004). The curve captures most of the features of the original BC02 observed TPW (standstill between 10 and 50 Ma, faster TPW in the correct direction prior to that back to 80 Ma) though with somewhat larger amplitudes.

We have seen that a number of scientists suggest possible links between (1) TPW episodes, or major changes between TPW episodes, or IITPW events if indeed they have occurred in the past, (2) reorganizations in the geometry of subduction zones or plume and superplume generation (or any other major geodynamical events occurring in the mantle), and (3) major biotic changes. These tantalizing hypotheses will likely keep analyses of polar wander very much alive in the coming decade, despite the somewhat bothering feeling that TPW still remains an elusive geophysical phenomenon.

Acknowledgments

I am thankful to Emilio Herrero-Bervera and David Gubbins for inviting this contribution. I am particularly thankful to J. Besse, A. Cazenave, R. Gordon, M. Greff-Lefftz, K. Lambeck, J. Laskar, J.L. Le Mouël, M. McElhinny, R. O'Connell, J.P. Poirier, Y. Ricard, J. Tarduno and R. Van der Voo for very useful comments on an earlier version of this paper. Special thanks to B. Steinberger for particularly detailed and helpful comments and help with several parts of Figure T19. IPGP Contribution NS 2094.

Vincent Courtillot

Bibliography

Anderson, D.L., 2000. The thermal state of the upper mantle: no role for mantle plumes. *Geophysical Research Letters*, **27**: 3623–3626.

Andrews, J.A., 1985. True polar wander: an analysis of Cenozoic and Mesozoic paleomagnetic poles. *Journal of Geophysical Research*, **90**: 7737–7750.

Besse, J., and Courtillot, V., 1988. Paleogeographic maps of the Indian Ocean bordering continents since the Upper Jurassic. *Journal of Geophysical Research*, **93**: 11791–11808.

Besse, J., and Courtillot, V., 1991. Revised and synthetic polar wander paths of the African, Eurasian, North American and Indian plates, and true polar wander since 200 Ma. *Journal of Geophysical Research*, **96**: 4029–4050.

Besse, J., and Courtillot, V., 2002. Apparent and true polar wander and the geometry of the geomagnetic field in the last 200 million years. *Journal of Geophysical Research*, **107**. doi:10.1029/2000JB000050.

Besse, J., and Courtillot, V., 2003. Correction to "Apparent and true polar wander and the geometry of the geomagnetic field in the last 200 million years. *Journal of Geophysical Research*, **107**. doi:10.1029/2000JB000050, 2002", *Journal of Geophysical Research*, 108. doi:10.1029/2003JB002684.

Brunet, D., and Machetel, P., 1998. Large-scale tectonic features induced by mantle avalanches with phase, temperature and pressure lateral variations in viscosity. *Journal of Geophysical Research*, **103**: 4929–4945.

Camps, P., Prévot, M., Daignières, M., and Machetel, P., 2002. Comment on stability of the earth with respect to the spin axis for the last 130 million years by J.A. Tarduno and A.Y. Smirnov (*Earth and Planetary Science Letters* **184**(2001): 549–553), *Earth and Planetary Science Letters*, **198**: 529–532.

Carlut, J., and Courtillot, V., 1998. How complex is the time-averaged geomagnetic field over the past 5 Myr? *Geophysical Journal International*, **134**: 527–544.

Chase, C.G., 1979. Subduction, the geoid, and lower mantle convection. *Nature*, **282**: 464–468.

Clouard, V., and Bonneville, A., 2001. How many Pacific hotspots are fed by deep-mantle plumes? *Geology*, **21**: 695–698.

Constable, C.G., and Parker, R.L., 1988. Statistics of the geomagnetic secular variation for the past 5 Myr. *Journal of Geophysical Research*, **93**: 11569–11581.

Cottrell, R.D., and Tarduno, J.A., 2000. Late cretaceous True polar wander: not so fast. *Science*, **288**: 2283a.

Courtillot, V., and Besse, J., 2004. A long-term octupolar component in the geomagnetic field? (0–200 Million Years B.P.). *Geophysical Monograph*, **145**, 59–74.

Courtillot, V., Davaille, A., Besse, J., and Stock, J., 2003. Three distinct types of hotspots in the Earth's mantle. *Earth and Planetary Science Letters*, **205**: 295–308.

Creer, K., Irving, E., and Runcorn, S.K., 1954. The direction of the geomagnetic field in remote epochs in Great Britain. *Journal of Geomagnetism and Geoelectricity*, **6**: 163–168.

Darwin, G., 1877. On the influence of geological changes on the earth's axis of rotation. *Philosophical Transactions of the Royal Society of London, Series A*, **167**: 271–312.

Davies, G.F., 1988. Ocean bathymetry and mantle convection. 1. Large-scale flow and hotspots. *Journal of Geophysical Research*, **90**: 10467–10480.

Di Venere, V., and Kent, D.V., 1999. Are the Pacific and Indo-Atlantic hotspots fixed? Testing the plate circuit through Antarctica. *Earth and Planetary Science Letters*, **170**: 105–117.

Engebretson, D.C., Cox, A., and Gordon, R.G., 1985. Relative motions between oceanic and continental plates in the Pacific Basin. *Geological Society of America Special Paper*, **206**: 59.

Evans, D., 1998. True polar wander, a supercontinental legacy. *Earth and Planetary Science Letters*, **157**: 1–8.

Evans, D., 2003. True polar wander and supercontinents. *Tectonophysics*, **362**: 303–320.

Fisher, D., 1974. Some more remarks on polar wandering. *Journal of Geophysical Research*, **79**: 4041–4045.

Gold, T., 1955. Instability of the Earth's axis of rotation. *Nature*, **175**: 526–529.

Goldreich, P., and Toomre, A., 1969. Some remarks on polar wandering. *Journal of Geophysical Research*, **74**: 2555–2567.

Gordon, R.G., 1987. Polar wandering and paleomagnetism. *Annual Reviews of Earth and Planetary Science*, **15**: 567–593.

Gordon, R.G., Horner-Johnson, B.C., Petronotis, K., and Acton, G.D., 2004. Apparent polar wander of the Pacific plate and Pacific hotspots: implications for true polar wander and hotspot fixity. *EOS, Transactions of the American Geophysical Union*, Spring.

Greff-Lefftz, M., 2004. Upwelling plumes, superswells and true polar wander. *Geophysical Journal International*, **159**: 1125–1137.

Gross, R.S., and Vondrak, J., 1999. Astrometric and space-geodetic observations of polar wander. *Geophysical Research Letters*, **26**: 2085–2088.

Hargraves, R.B., and Duncan, R.A., 1997. Does the mantle roll? *Nature*, **245**: 361–363.

James, T.S., and Ivins, E.R., 1997. Global geodetic signatures of the Antarctic ice sheet. *Journal of Geophysical Research*, **102**: 605–633.

Jurdy, D.M., 1981. True polar wander. *Tectonophysics*, **74**: 1–16.

Jurdy, D.M., and Van der Voo, R., 1974. A method for the separation of polar wander and continental drift. *Journal of Geophysical Research*, **79**: 2945–2952.

Jurdy, D.M., and Van der Voo, R., 1975. True polar wander since the Early Cretaceous. *Science*, **187**: 1193–1196.

Kent, D.V., and Smethurst, M.A., 1998. Shallow bias of paleomagnetic inclinations in the Paleozoic and Precambrian. *Earth and Planetary Science Letters*, **160**: 391–402.

Kirschvink, J.L., Ripperdan, R.L., and Evans, D.A., 1997. Evidence for a large-scale early Cambrian reorganization of continental masses by inertial interchange true polar wander. *Science*, **277**: 541–545.

Koppers, A.A.P., Morgan, J.P., Morgan, J.W., and Staudigel, H., 2001. Testing the fixed hotspot hypothesis using $^{40}Ar/^{39}Ar$ age progressions along seamount trails. *Earth and Planetary Science Letters*, **185**: 237–252.

Koppers, A.A.P., Duncan, R.A., and Steinberger, B., 2004. Implications of a non-linear $^{40}Ar/^{39}Ar$ age progression along the Louisville seamount trail for models of fixed and moving hotspots. *Geochemistry, Geophysics, Geosystems*, **5**. doi:10.1029/2003GC000671.

Lambeck, K., 1980. *The Earth's Variable Rotation: Geophysical Causes and Consequences.* Cambridge: Cambridge University Press, 449 pp.

Lambeck, K., and Johnston, P., 1998. The viscosity of the mantle: evidence from analyses of glacial rebound phenomena. In Jackson I. (ed.) *The Earth's Mantle.* Cambridge: Cambridge University Press, pp. 461–502.

Livermore, R.A., Vine, F.J., and Smith, A.G., 1983. Plate motions and the geomagnetic field. I. Quaternary and late Tertiary. *Geophysical Journal of the Royal Astronomical Society*, **73**: 153–171.

Livermore, R.A., Vine, F.J., and Smith, A.G., 1984. Plate motions and the geomagnetic field. II. Jurassic to Tertiary. *Geophysical Journal of the Royal Astronomical Society*, **79**: 939–961.

Marcano, M.C., Van der Voo, R., and Mac Niocaill, C., 1999. True polar wander during the Permo-Triassic. *Journal of Geodynamics*, **28**: 75–95.

Markowitz, W., 1970. Sudden changes in rotational acceleration of the Earth and secular motion of the pole. In Mansinha, L., Smylie, D.E., and Beck, A.E. (eds.) *Earthquake Displacement Fields and the Rotation of the Earth.* New York: Springer, pp. 69–91.

McCarthy, D.D., and Luzum, B.J., 1996. Path of the mean rotational pole from 1899 to 1994. *Geophysical Journal International*, **125**: 623–629.

McElhinny, M.W., and McFadden, P.L., 2000. *Paleomagnetism: continents and oceans.* San Diego: Academic Press, 386 pp.

McKenzie, D.P., 1972. Plate tectonics. In Robertson, E.C. (ed.) *The Nature of the Solid Earth.* New York: McGraw-Hill, pp. 323–360.

Mitrovica, J.X., and Milne, G.A., 1998. Glaciation-induced perturbations in the Earth's rotation: a new appraisal. *Journal of Geophysical Research*, **103**: 985–1005.

Montelli, R., Nolet, G., Dahlen, F.A., Masters, G., Engdahl, E.R., and Hung, S.H., 2004. Finite-frequency tomography reveals a variety of plumes in the mantle. *Science*, **303**: 338–343.

Morgan, W.J., 1972. Plate motions and deep mantle convection. *Geological Society of America Memoir*, **132**: 7–22.

Morgan, J.W., 1981. Hotspot tracks and the opening of the Atlantic and Indian oceans. In Emiliani, C. (ed.) *The Sea,* Volume 7. New York: Wiley, pp. 443–487.

Morgan, W.J., 1983. Hotspot tracks and the early rifting of the Atlantic. *Tectonophysics*, **94**: 123–139.

Müller, D.M., Royer, J.Y., and Lawver, L.A., 1993. Revised plate motions relative to the hotspots from combined Atlantic and Indian Ocean hotspot tracks. *Geology*, 275–278.

Munk, W.H., and MacDonald, G.J.F., 1960. *The Rotation of the Earth.* Cambridge: Cambridge University Press, 323 pp.

Norton, I.O., 1995. Plate motion in the North Pacific: the 43 Ma nonevent. *Tectonics*, **14**: 1080–1094.

O'Connell, R.J., Gable, C.W., and Hager, B.H., 1991. Toroidal-poloidal partitioning of lithospheric plate motions. In Sabadini, R., and Lambeck, K. (eds.) *Glacial Isostasy, Sea Level and Mantle Rheology.* Dordrecht: Kluwer Academic Publishers, pp. 535–551.

Parker, R.L., 1991. A theory of ideal bodies for seamount magnetism. *Journal of Geophysical Research*, **96**: 16101–16112.

Patriat, P., and Achache, J., 1984. India-Asia collision chronology has implications for crustal shortening and driving mechanism of plates. *Nature*, **311**: 615–621.

Petronotis, K.E., and Gordon, R.G., 1999. A Maastrichtian paleomagnetic pole for the Pacific plate from a skewness analysis of marine magnetic anomaly 32. *Geophysical Journal International*, **139**: 227–247.

Prévot, M., Mattern, E., Camps, P., and Daignières, M., 2000. Evidence for a 20° tilting of the Earth's rotation axis 110 million years ago. *Earth and Planetary Science Letters*, **179**: 517–528.

Quidelleur, X., Valet, J.P., Courtillot, V., and Hulot, G., 1994. Long-term geometry of the geomagnetic field for the last five million years: an updated secular variation database. *Geophysical Research Letters*, **21**: 1639–1642.

Raymond, C.A., Stock, J.M., and Cande, S.C., 2000. Fast Paleogene motion of the Pacific hotspots from revised global plate circuit constraints. *The History and Dynamics of Global Plate Motions, Geophysical Monographs*, **121**: 359–375.

Ricard, Y., Doglioni, C., and Sabadini, R., 1991. Differential rotation between lithosphere and mantle: a consequence of lateral mantle viscosity variations. *Journal of Geophysical Research*, **96**: 8407–8416.

Ricard, Y., Spada, G., and Sabadini, R., 1993a. Polar wandering of a dynamic Earth. *Geophysical Journal International*, **113**: 284–298.

Ricard, Y., Richards, M., Lithgow-Bertelloni, C., and Le Stunff, Y., 1993b. A geodynamic model of mantle density heterogeneity. *Journal of Geophysical Research*, **98**: 21895–21909.

Richards, M.A., Ricard, Y., Lithgow-Bertelloni, C., Spada, G., and Sabadini, R., 1997. An explanation of the Earth's long-term rotational stability. *Science*, **275**: 372–375.

Richards, M.A., Bunge, H.P., Ricard, Y., and Baumgardner, J.R., 1999. Polar wandering in mantle convection models. *Geophysical Research Letters*, **26**: 1777–1780.

Runcorn, S.K., 1956. Paleomagnetic comparisons between Europe and North America. *Proceedings of the Geological Association of Canada*, **8**: 77–85.

Sabadini, R., and Peltier, W.R., 1981. Pleistocenic deglaciation and the Earth's rotation: implications for mantle viscosity. *Geophysical Journal of the Royal Astronomical Society*, **66**: 553–578.

Sager, W.W., and Koppers, A.A.P., 2000. Late cretaceous polar wander of the Pacific plate: evidence of a rapid true polar wander event. *Science*, **287**: 455–459.

Sager, W.W., and Pringle, S., 1988. Mid cretaceous to early tertiary apparent polar wander path of the Pacific Plate. *Journal of Geophysical Research*, **93**: 11753–11771.

Schneider, D.A., and Kent, D.V., 1990a. Paleomagnetism of Leg 115 sediments: implications for Neogene magnetostratigraphy and paleolatitude of the Reunion Hotspot. *Proceedings of the Ocean Drilling Program, Scientific Results*, **115**: 717–36.

Schneider, D.A., and Kent, D.V., 1990b. The time-averaged paleomagnetic field. *Reviews of Geophysics*, **28**: 71–96.

Si, J., and Van der Voo, R., 2001. Too-low magnetic inclinations in central Asia: an indication of a long-term Tertiary non-dipole field? *Terra Nova*, **13**: 471–478.

Spada, G., Ricard, Y., and Sabadini, R., 1992. True polar wander for a dynamic Earth. *Nature*, **360**: 452–454.

Spada, G., Sabadini, R., and Boschi, E., 1996a. The spin and inertia of Venus. *Geophysical Research Letters*, **23**: 1997–2000.

Spada, G., Sabadini, R., and Boschi, E., 1996b. Long-term rotation and mantle dynamics of the earth, Mars and Venus. *Journal of Geophysical Research*, **101**: 2253–2266.

Steinbach, V., and Yuen, D.A., 1994. Effects of depth dependent properties on the thermal anomalies produced in flush instabilities from phase transitions. *Physics of the Earth and Planetary Interiors*, **86**: 165–183.

Steinberger, B., 1996. Motions of hotspots and changes of the earth's rotation axis caused by a convecting mantle, Ph.D. thesis, 203 pp., Harvard University, Cambridge, MA.

Steinberger, B., 2000. Plumes in a convecting mantle: models and observations for individual hotspots. *Journal of Geophysical Research*, **105**: 11127–11152.

Steinberger, B.M., and O'Connell, R.J., 1997. Changes of the Earth's rotation axis inferred from advection of mantle density heterogeneities. *Nature*, **387**: 169–173.

Steinberger, B., and O'Connell, R.J., 2000. Effects of mantle flow on hotspot motions. In Richards, M., Gordon, R., and van der Hilst, R. (eds.), *The History and Dynamics of Global Plate Motions, Geophysical Monograph,* Volume 121, pp. 377–398.

Steinberger, B., and O'Connell, R.J., 2002. The convective mantle flow signal in rates of true polar wander. In Mitrovica, J., and Vermeersen, L. (eds.) *Ice Sheets, Sea-Level and the Dynamic Earth Geodynamic Series,* Volume 29. Washington, DC: AGU, pp. 233–256.

Steinberger, B., Sutherland, R., and O'Connell, R., 2004. Prediction of Emperor-Hawaii seamount locations from a revised model of plate motion and mantle flow. *Nature*, **430**: 167–173.

Tarduno, J.A., and Cottrell, R.D., 1997. Paleomagnetic evidence for motion of the Hawaiian hotspot during formation of the Emperor Seamounts. *Earth and Planetary Science Letters*, **153**: 171–180.

Tarduno, J.A., and Gee, J., 1995. Large-scale motion between Pacific and Atlantic hotspots. *Nature*, **378**: 477–480.

Tarduno, J.A., and Smirnov, A.V., 2001. Stability of the Earth with respect to the spin axis for the last 130 million years. *Earth and Planetary Science Letters*, **184**: 549–553.

Tarduno, J.A., and Smirnov, A.V., 2002. Response to comment on "Stability of the Earth with respect to the spin axis for the last 130 Million Years" by Camps, P., Prévot, M., Daignieres, M., and Machetel, P., *Earth and Planetary Science Letters*, **198**: 533–539.

Tarduno, J.A., Duncan, R.A., Scholl, D.W., Cottrell, R., Steinberger, B., *et al.,* 2003. The Emperor seamounts: southward motion of the Hawaiian hotspot plume in Earth's mantle. *Science*, **301**: 1064–1069.

Torsvik, T.H., and Van der Voo, R., 2002. Refining Gondwana and Pangea paleogeography: estimates of Phanerozoic non-dipole (octupole) fields. *Geophysical Journal International*, **151**: 771–794.

Torsvik, T.H., Van der Voo, R., and Redfield, T.F., 2002. Relative hotspot motions versus true polar wander. *Earth and Planetary Science Letters*, **202**: 185–200.

Van der Voo, R., 1994. True polar wander during the mid-Paleozoic? *Earth and Planetary Science Letters*, **122**: 239–243.

Van der Voo, R., and Torsvik, T.H., 2001. Evidence for late Paleozoic and Mesozoic non-dipole fields provides an explanation for the Pangea reconstruction problems. *Earth and Planetary Science Letters*, **187**: 71–81.

Weinstein, S.A., 1993. Catastrophic overturn of the Earth's mantle driven by multiple phase changes and internal heat generation. *Geophysical Research Letters*, **20**: 101–104.

Wilson, R.L., 1970. Permanent aspects of the Earth's non-dipole magnetic field over Upper Tertiary Times. *Geophysical Journal of the Royal Astronomical Society*, **19**: 417–37.

U

ULVZ, ULTRA-LOW VELOCITY ZONE

Nearly half way to Earth's center, the boundary between the solid silicate rock mantle and the liquid iron-alloy outer core was long thought to be a sharp discontinuity between the two vastly different regimes. Recently, detailed seismological analyses have depicted the core-mantle boundary (CMB) as being far from simple, and in fact shows evidence for an additional thin veneer of anomalous properties in certain geographical regions. Imaged as thin as a couple of km, and up to 50-km thick, these unique zones are characterized by strong reductions in the speeds of seismic waves relative to the overlying mantle. These areas have thus been dubbed "ultra-low velocity zones" (ULVZs).

Seismic probes

Seismology remains the most direct remote sensing tool for deciphering the subtleties of Earth's inaccessible deep interior. This is most commonly accomplished through the use of elastic energy that propagates away from earthquakes, traveling through the entire interior of the planet; some energy propagates continuously through the Earth, some reflects from local or global boundaries between contrastingly different materials, and some of it, in special cases, diffracts along boundaries between strongly contrasting media. Each of these types has provided evidence for extremely sluggish patches at the CMB.

Four of the most commonly utilized seismic probes (or "phases") to date are *SPdKS* referenced to *SKS*, and precursors (seismic energy that just slightly precedes a later more dominant phase) to the waves *PcP*, *ScP*, or *PKP* (Figure U1). Over the past decade a variety of research groups have documented anomalies in these arrivals and attributed them to CMB structure (see, e.g., Garnero *et al*., 1998). *ScP* is a seismic phase that departs from the earthquake as an *S* wave, and upon reflection at the CMB, converts to a *P* wave (Figure U1a). If a low velocity boundary layer is present at the CMB, several additional arrivals are possible (Figure U1a, second panel). The relative timing and amplitude of these arrivals is apparent in a computer generated "synthetic" seismogram 60° in arc away from a 500-km deep hypothetical earthquake (third panel). If the top of the layer is diffuse, then these arrivals diminish in amplitude, which remains an active direction for current research.

Figure U1b shows *PcP*, which also contains pre- and postcursors, analogously to *ScP*, as well as some of the additional arrival geometries, and a synthetic seismogram showing the additional arrivals. One challenge in data analyses is to identify the later arrivals, since they are commonly obscured by the coda of the main *PcP* phase, i.e., additional arrivals due to (for example) reverberations in Earth's crust due to the *PcP* wave.

Waves that travel into the Earth's core can contain important information about anomalous properties at the CMB, which they traverse at least twice. The phase *PKP* has been used to detect CMB structures that scatter energy resulting in precursory arrivals to *PKP*. Figure U1c shows *PKP* paths, along with an associated reference arrival *PKiKP* that reflects from the inner core boundary. Anomalous topography to the ULVZ or inclusions of low velocity material can give rise to precursory arrivals, as shown in the theoretical predictions in the bottom panel.

Another important probe of the CMB is an *S* wave that encounters the CMB at a critical angle to produce a *P* wave that diffracts along the CMB (*Pd*), then continues into the core as a *P* wave, then back through the mantle as an *S* wave (Figure U1d). This phase, *SPdKS*, has short segments of *P* wave diffraction at the core entry and exit locations. Additional internal reflections within the layer are also emerging as important (e.g., *SPuPKS*). Certainly other possible seismic probes of ULVZ structure are possible, as long as seismic energy either refracts, diffracts, or reflects at the CMB.

An important next step in ULVZ research will be to find geographical regions that permit analysis of more than one particular seismic phase, since different waves are sensitive to different ULVZ structural components. For example, the precursor analyses (Figure U1a-c) utilize short period energy, which is particularly sensitive to sharp contrasts in properties, and not able to well-resolve gradational changes in properties. *SPdKS*, on the other hand, is less sensitive to such contrasts, but very sensitive to the velocity structure within the layer, particularly at distances where the *Pd* segment in *SPdKS* is short.

Ultra-low velocity boundary layer possibilities

The seismic phases introduced in Figure U1 have played a critical role in revealing several possibilities for transitional structure between the core and mantle. Three main possibilities are highlighted here: a layer on the mantle side of the CMB (which is most commonly referred to as ULVZ), a layer on the core-side of the CMB (essentially a core-rigidity zone, CRZ), and some thickness over which the mantle changes into the core (hereafter denoted as a core-mantle transition zone, CMTZ) (Figure U2). It is important to recognize these as end-member models, and that any combination of these is equally possible.

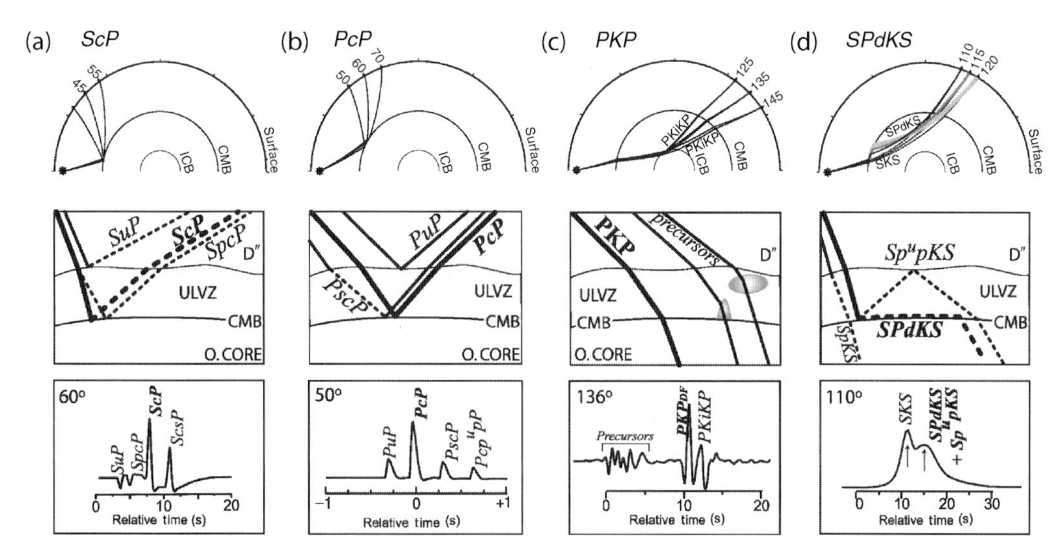

Figure U1 Ray paths, seismic arrivals due to a ULVZ, and synthetic seismograms for (a) *ScP*, (b) *PcP*, (c) *PKP*, and (d) *SPdKS*. Synthetic seismogram predictions of (a) and (c) are from Garnero and Vidale (1999) and Wen and Helmberger (1998), respectively.

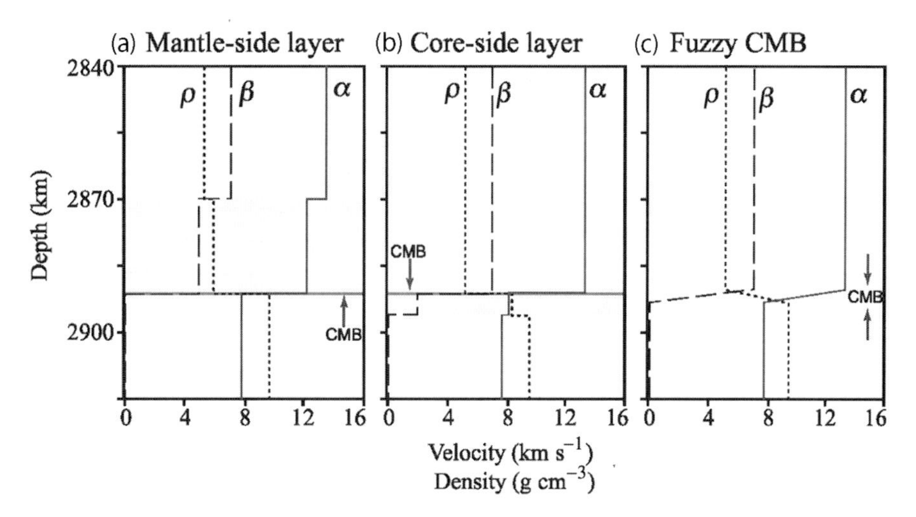

Figure U2 P-wave (α) and S-wave (β) velocity and density (ρ) versus depth for ultra-low velocity boundary layering (shaded regions) on (a) the mantle-side of the core-mantle boundary (CMB) (a ULVZ) (b) the core-side of the CMB (a CRZ), and (c) a finite thickness transition between the mantle and core (CMTZ).

Mantle ultra-low velocity zone

ULVZ thickness has been imaged between 5 and 50km, with strong lateral variations. The most commonly explored model parameters are those compatible with partial melt of the lowermost mantle, which results in a three times larger reduction in shear velocity (e.g., 30%) than that for compressional waves. Density increases are also possible (Figure U2a).

Core-rigidity zone

If large density increases and shear velocity decreases are considered in ULVZ modeling, one must allow for the possibility of the layer residing on the core-side of the CMB (Figure U2b). The liquid outer core of the Earth is predominantly iron, along with a minor constituency of some lighter element(s). As the Earth cools, the solid inner core of the Earth grows, releasing the lighter elements into the outer core, which may result in "underplating" of the CMB in a sedimentation process (see Buffett *et al.*, 2000). Thus, isolated regions of nonzero rigidity may exist beneath positive topography, or, "hills" on the CMB, where such sediments can accumulate and concentrate (up to a couple km thick; Rost and Revenaugh, 2001). If electrically conductive, the CRZ may affect Earth's magnetic field, nutations, and possibly even magnetic field reversal paths.

Core-mantle transition zone

Finally, we consider the possibility of a transitional zone between the mantle and core over some finite thickness (Figure U2c). Chemical reactions between the silicate rock mantle and liquid iron-alloy outer core (see Knittle and Jeanloz, 1989) can result in a thin mixing zone—an effective blurring of the CMB. The CMTZ can appropriately cause precursors to the short period waves (Figure U1a-c) as well as delay the SPdKS relative to SKS (Figure U1d; Garnero and Jeanloz, 2000).

ULVZ geographic distribution

To gain insight into the possible origin of ultra-low velocity layering at the CMB, it is instructive to compare results to other phenomena, such as gross properties of the deep mantle as revealed by shear wave tomography. Figure U3 compares the strongest variations in shear wave velocity with ULVZ distributions. There is suggestion of a connection between ULVZ and low seismic wave speeds (Figure U3a,b). This ULVZ distribution also correlates strongly with hot spot volcanism at Earth's surface (Williams *et al.*, 1998). However, an examination of a larger data set of higher quality broadband data reveals shorter scale variations, with some CMB patches showing evidence for both support and lack of an anomalous layer. Figure U3c summarizes ULVZ likelihood using broadband digital *SPdKS* data. The correlation with large-scale lower mantle shear velocity is more difficult to assess. Greater geographical coverage in future studies will allow more confidence in such comparisons. A main limitation in achieving greater spatial coverage is uneven earthquake and seismometer distribution on the globe, limiting where the deep interior can be probed.

Discussion and summary

While confident correlations between ULVZ layering and bulk mantle properties may be premature at present, the existence of the layer is clear, from a variety of studies and methods. While not constrained, it is instructive to briefly consider possible scenarios relating CMB structure to the thermal, chemical, and dynamical environment. Figure U4 displays a multitude of possibilities beneath upwelling and downwelling mantle regimes. The point here is to recognize the variety of structures and their scale lengths that are possible, yet unresolved at present, and hence future work should seek to sharpen our focusing ability for such possibilities.

Beneath upwelling regions, possibilities include (but certainly are not limited to) (a) a combination of ULVZ, CMTZ, and CRZ structures in the hottest lowermost mantle regions; (b) convection within a partially molten ULVZ which can sweep chemical heterogeneity into localized piles within the ULVZ; (c) large- and fine scale CMB (as well as ULVZ) topography, resulting in (d) multiple scale CRZs of variable strength; (e) some ULVZ melt entrainment into overlying convection currents, which may result in (f) mantle plume genesis, and (g) aligned melt pockets from strong boundary layer shear flow, yielding seismic anisotropy; and (h) chemical mixing between the ULVZ and outer core (or CRZ) material, giving rise to local (or widespread) CMB blurring, including chemical coupling (or interactions).

Beneath downwelling regions, similar phenomena exist, but perhaps suppressed in the vertical dimension (a) spatially organized ULVZ to the front of downwelling motions, where plume instabilities (thus local warmer zones) have been shown to exist (Tan *et al.*, 2002); (b) a thin (undetectable?) ULVZ throughout region; (c) small- and large-scale CMB topography, which may provide localized basins or wells for material with density intermediate to that of the mantle and core; (d) possible chemical contamination from melt from either CMB chemical reaction product entrainment or ponding of former oceanic crust; and (e) anisotropy due to high strains resulting from overlying subduction stresses (McNamara *et al.*, 2002). While provocative, Figure U4 depicts the likely scenario of CMB topography that is intimately coupled to ULVZ, CMTZ, and CRZ chemistry and dynamics. Furthermore, these structures probably play an important role with electromagnetic, gravitational, thermal, and topographic coupling between the mantle and core.

Future analyses should incorporate more realistic 3D wave propagation tools for predictions to compare to data (see, e.g., Helmberger *et al.*, 1998). Analyses that utilize more that one of the wave types presented in Figure U1 for the same patch of the CMB will also greatly reduce uncertainties.

Lastly, it is clear that ULVZ structure is an intimate part of the core-mantle transition, and likely reflects core processes that may be related to the geodynamo. For example, if the ULVZ is enriched in iron from the core, it may affect geomagnetic reversal path geometries, which to

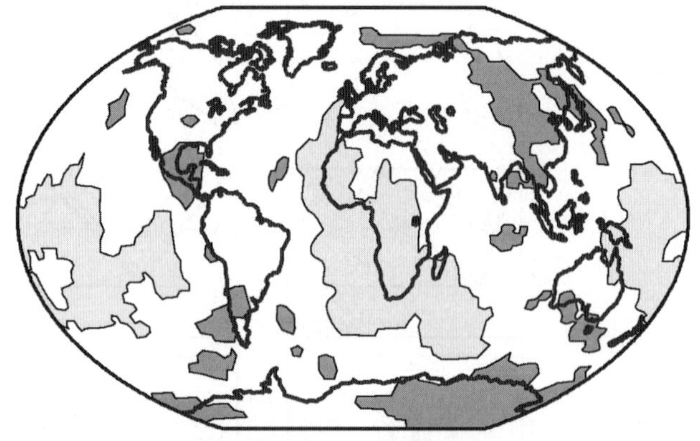

(a) High (dark) and low (light) D″ shear velocity

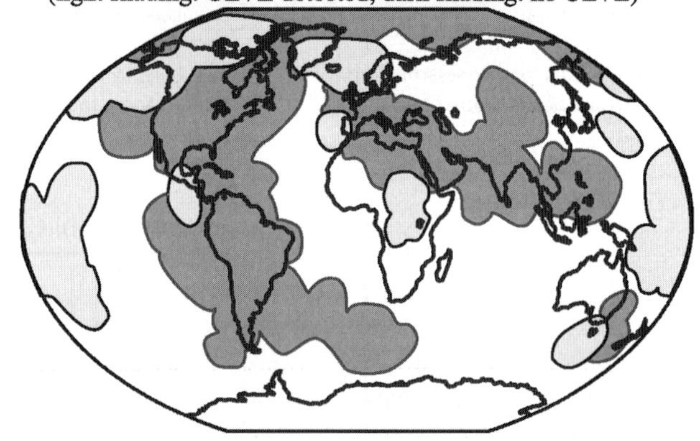

(b) ULVZ distribution from LP WWSSN SPdKS Fresnel zones
(light shading: ULVZ detected, dark shading: no ULVZ)

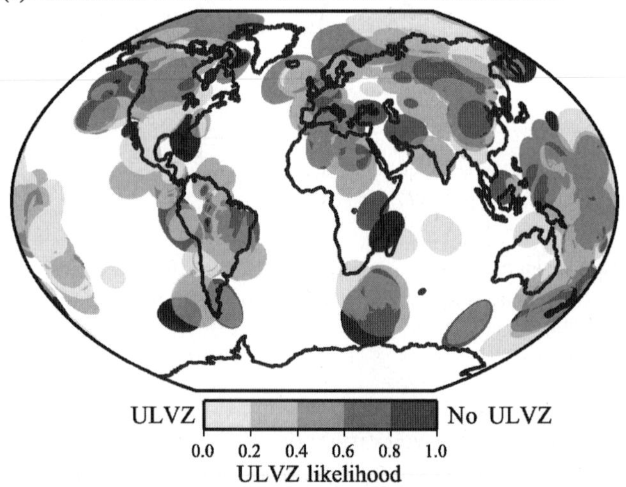

Figure U3 (a) Shear velocity distribution in the lowermost 250 km of the mantle from the model of Grand (2002). Light shaded areas have velocities equal to or lower than −1% relative to the global average, dark areas are at or great than +1%. (b) Fresnel zones of *SPdKS* sampling at the CMB from long period analog data. Light shading is for suggested ULVZ presence; dark shading for ULVZ absence, and no shading represents no data sampling. (c) Same as (b), except using modern broadband digital data, and shading represents the likelihood of ULVZ presence. For each $1° \times 1°$ section of the CMB with data coverage, the following ratio is constructed and plotted: (# of records requiring an ultra-low velocity layer)/(total # of records for that cell). Thus, a value of 1 indicates all data that traversed that cell are anomalous; a value of 0 represents the case where no data sampling a particular region are anomalous (after Thorne and Garnero, 2004).

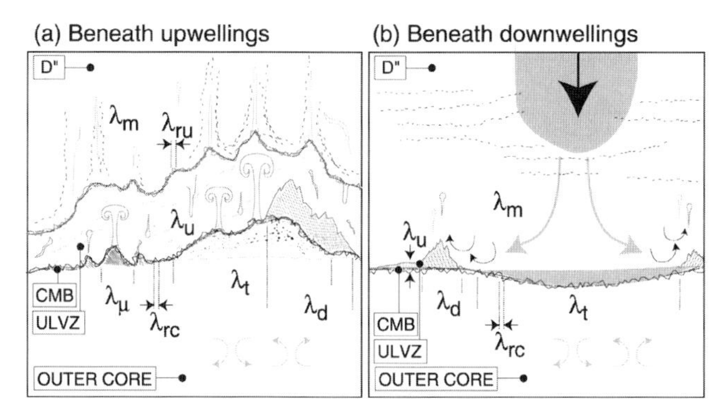

Figure U4 Possible CMB scenarios beneath regions of (a) upwelling, and (b) downwelling (see text for details). Significant (or total) uncertainty exists for many wavelengths of interest, which include: scale-length over which ULVZ phenomena affects the lowermost mantle (λ_m), scale lengths of roughness of the top of the ULVZ (λ_{ru}), ULVZ thickness distribution (λ_u), dimension of isolated core-rigidity zones (λ_μ), scale of roughness of the CMB, including the thickness of transition from pure core-to-mantle (λ_{rc}), lateral and vertical scale of long wavelength CMB topography, and hence possible anomalous zones contained within the topographic depressions/elevations (λ_t), and isolated thermochemical domes, scatterers, or anomalous shapes within the ULVZ (λ_d).

the first order appear anticorrelated with ULVZ distributions (Garnero *et al.*, 1998). Certainly, uneven geographic distribution of patches of partially molten and/or chemically unique ULVZ material can result in variability in core heat flow that may affect core fluid motions, possibly relating to Earth's magnetic field generation and variability.

Ed J. Garnero and M. Thorne

Bibliography

Buffett, B.A., Garnero, E.J., and Jeanloz, R., 2000. Sediments at the top of the Earth's core. *Science*, **290**: 1338–1342.

Garnero, E.J., and Jeanloz, R., 2000. Earth's enigmatic interface. *Science*, **289**: 70–71.

Garnero, E.J., and Vidale, J., 1999. ScP; a probe of ultralow-velocity zones at the base of the mantle. *Geophysical Research Letters*, **26**: 377–380.

Garnero, E.J., Revenaugh, J.S., Williams, Q., Lay, T., and Kellogg, L.H., 1998. Ultralow velocity zone at the core-mantle boundary. In Gurnis, M., Wysession, M., Knittle, E., and Buffet, B. (eds.) *The Core-Mantle Boundary.* Washington, DC: American Geophysical Union, pp. 319–334.

Grand, S.P., 2002. Mantle shear-wave tomography and the fate of subducted slabs. *Philosophical Transactions of the Royal Society of London A*, **360**: 2475–2491.

Helmberger, D.V., Wen, L., and Ding, X., 1998. Seismic evidence that the source of the Iceland hotspot lies at the core-mantle boundary. *Nature*, **396**: 251–255.

Knittle, E., and Jeanloz, R., 1989. Simulating the core-mantle boundary: an experimental study of high-pressure reactions between silicates and liquid iron. *Geophysical Research Letters*, **16**: 609–612.

McNamara, A.K., van Keken, P.E., and Karato, S.I., 2002. Development of anisotropic structure in the Earth's lower mantle by solid-state convection. *Nature*, **416**: 310–314.

Rost, S., and Revenaugh, J., 2001. Seismic detection of rigid zones at the top of the core. *Science*, **294**: 1911–1914.

Tan, E., Gurnis, M., and Han, L., 2002. Slabs in the lower mantle and their modulation of plume formation. *Geochemistry, Geophysics, Geosystems*, **3**(11): 1067 (doi: 10.1029/2001GC000238).

Thorne, M.S., and Garnero, E.J., 2004. Inferences on ultralow-velocity zone structure from a global analysis of SPdKS waves. *Journal of Geophysical Research*, **109**: B08301 (doi: 10.1029/2004JB003010).

Wen, L., and Helmberger, D.V., 1998. Ultra-low velocity zones near the core-mantle boundary from broadband PKP precursors. *Science*, **279**: 1701–1703.

Williams, Q., Revenaugh, J.S., and Garnero, E.J., 1998. A correlation between ultra-low basal velocities in the mantle and hot spots. *Science*, **281**: 546–549.

Cross-references

Core Composition
Core Properties, Physical
Core, Boundary Layers
Core-Mantle Boundary
Core-Mantle Boundary Topography, Implications for Dynamics
Core-Mantle Boundary Topography, Seismology
Core-Mantle Boundary, Heat Flow Across
Core-Mantle Coupling, Electromagnetic
Core-Mantle Coupling, Thermal
Core-Mantle Coupling, Topographic
D″ as a Boundary Layer
D″, Anisotropy
D″, Composition
D″, Seismic Properties
Earth Structure, Major Divisions
Geodynamo, Numerical Simulations
Seismic Phases

UNITS

Scientific units were standardized to the current Système Internationale d'unités (SI) in about 1970. For the most part conversion involves applying simple factors of 10 to convert centimeters and grams to meters and kilograms, but in magnetism conversion is more complicated and confusing because the old cgs system had, for historical reasons, two sets of units for electricity and magnetism: electrostatic and electromagnetic units. The latter, abbreviated to emu, concern us here. Understanding of emu is necessary to understand the old, and some not so old, literature. The paleomagnetic community has been particularly slow to convert to SI, and even today some papers give results in emu.

SI achieved the worthy goal of unifying the system of units for both electricity and magnetism; it also removed some factors of 4π from the basic equations (Table U1). The critical difference between the two systems arises from the distinction between magnetic induction or flux density $\boldsymbol{B}$ and magnetic field intensity $\boldsymbol{H}$. The dimensions of $\boldsymbol{H}$ differ in the two systems of units, as does the magnetic moment per unit

Table U1 Relevant equations in SI and emu systems of units

SI	Emu
$B = \mu_0 H + M$	$B = H + 4\pi M$
$P = \mu_0 M$	$P = M$
$M = \chi_{SI} H$	$M = \chi_{emu} H/4\pi$

Further discussion is given in the appendix of Jackson's (1999), 3rd edition. Butler (1992) and Blakely (1996) also have appendices on systems of units.

Table U2 Dimensions and units of physical quantities used in geomagnetism

Physical quantity	Dimension	SI	Emu	Conversion factor
Length	L	m	cm	10^2
Mass	M	kg	G	10^{-3}
Time	T	s	s	1
Charge	Q	coulomb (C)	coulomb	1
Electric current	QT^{-1}	ampere (A)	abamp	10
Potential difference	$L^2MT^{-2}Q$	volt (V)	emu	10^{-8}
Electric field	$LMT^{-2}Q$	$V\,m^{-1}$	emu	10^{-6}
Resistance	$L^2MT^{-1}Q^{-2}$	ohm	emu	10^{-9}
Resistivity	$L^3MT^{-1}Q^{-2}$	ohmm	emu	10^{-11}
Conductivity	$L^{-2}M^{-1}TQ^2$	siemens$\,m^{-1}$	emu	10^{11}
Magnetic flux	$L^2MT^{-1}Q^{-1}$	weber (W)	maxwell	10^{-8}
Magnetic induction ***B***	$MT^{-1}Q^{-1}$	tesla (T)	gauss	10^{-4}
Magnetic field intensity	$L^{-1}T^{-1}Q$	$A\,m^{-1}$	oersted ($MT^{-1}Q^{-1}$)	$10^3/4\pi$
Inductance	L^2MQ^{-2}	henry (H)	emu	10^{-9}
Permeability	LMQ^{-2}	$H\,m^{-1}$	Dimensionless	$4\pi \times 10^{-7}$
Magnetic moment density	$L^{-1}T^{-1}Q$	$A\,m^{-1}$	emu ($MT^{-1}Q^{-1}$)	$10^3/4\pi$
Magnetic polarization	$MT^{-1}Q^{-1}$	T	gauss	10^{-4}
Susceptibility	Dimensionless	χ_{SI}	χ_{emu}	4π

LMTQ denote length, mass, time, and charge. The conversion factor in the right column should be used to multiply a value in emu to yield the SI value. Note the difference in definition for H, M, and χ between the two systems. The siemen is sometimes called the mho.

volume of a material ***M***. Furthermore, the definition of ***M*** differs by a numerical factor of 4π between the two systems, which has the undesirable effect that the dimensionless susceptibility χ differs. The magnetic polarization ***P*** (usually denoted as ***J*** in paleomagnetism but this is used in *MHD* (*q.v.*) exclusively for electric current density), has the same dimensions as ***B*** in both systems. Confusion propagates because of sloppy terminology: it is standard practice in geomagnetism and paleomagnetism to refer to ***B*** as the magnetic field rather than magnetic induction, and magnetization is used to mean either ***M*** or ***P*** (Table U2).

David Gubbins

Bibliography

Blakely, R.J., 1996. *Potential Theory in Gravity and Magnetic Applications.* Cambridge: Cambridge University Press.

Butler, R.F., 1992. *Paleomagnetism: Magnetic Domains to Geologic Terranes.* Boston: Blackwell Scientific.

Jackson, J.D., 1999. *Classical Electrodynamics,* 3rd edition. New York: Wiley.

Cross-references

Magnetohydrodynamics

UPWARD AND DOWNWARD CONTINUATION

Potential fields known at a set of points can be expressed at neighboring higher or lower spatial locations in a source free region using the continuation integral that results from one of Green's theorems (see, e.g., Blakely, 1995). The principal uses of this concept are to adjust altitude of observations to a datum as an aid to the interpretation of a survey (see Crustal magnetic field), reduce short-wavelength data noise by continuing the field upward, and increasing the horizontal resolution of anomalies and their sources by continuing the field downward. It is possible to continue the field upward or downward in a number of different ways depending on the application at hand; for example, designing continuation operators in spatial or wavenumber space (Henderson and Zietz, 1949; Dean, 1958), using harmonic functions (Courtillot *et al.*, 1978; Shure *et al.*, 1982; Fedi *et al.*, 1999), and deriving physical property variations of sources causing the fields (Dampney, 1969; Emilia, 1973; Langel and Hinze, 1998). Applications also vary widely: from environmental and exploration applications involving short-wavelength anomaly fields over small height differences (a few meters to kilometers) to global distribution of anomalies measured by satellites in which anomalies are downward continued from satellite altitudes (300–700km) to Earth's surface and also downward continuing the core part of the Earth's field all the way to the top of the core to decipher features of core circulation over time.

The effect of upward/downward continuation process on the fields can be understood by examining the continuation operator in the wavenumber domain. The operator has the form $e^{\pm|k|z}$, where $|k|$ is the wavenumber ($|k| = 2\pi\lambda$ where λ is the full wavelength) and z is the continuation level (Dean, 1958). The negative sign in the exponent indicates upward continuation (away from the sources of the field) and the positive sign implies the downward continuation (toward the sources of the field). The response of the continuation operator with respect to wavelength is illustrated in Figure U5, which shows that shorter wavelengths are attenuated and smoothed in the process of upward continuation, whereas in downward continuation the shorter wavelengths are amplified and sharpened. Both operations are susceptible to errors in the data and their results can be rendered invalid or at least severely compromised due to the quality of data. For example, if measurement errors are primarily short-wavelength, then the nature of downward continuation operator which amplifies primarily the short-wavelength components of the data can severely distort the downward continued result. On the other hand, if the long-wavelength portion of the field is contaminated, for example, by inaccurate compilation of different surveys having different base levels, then the retention of the corrupt long wavelengths in the process of upward continuation can render the result unusable (Ravat *et al.*, 2002).

The most straightforward upward/downward continuation of a field is performed from one level surface to another level surface (Henderson

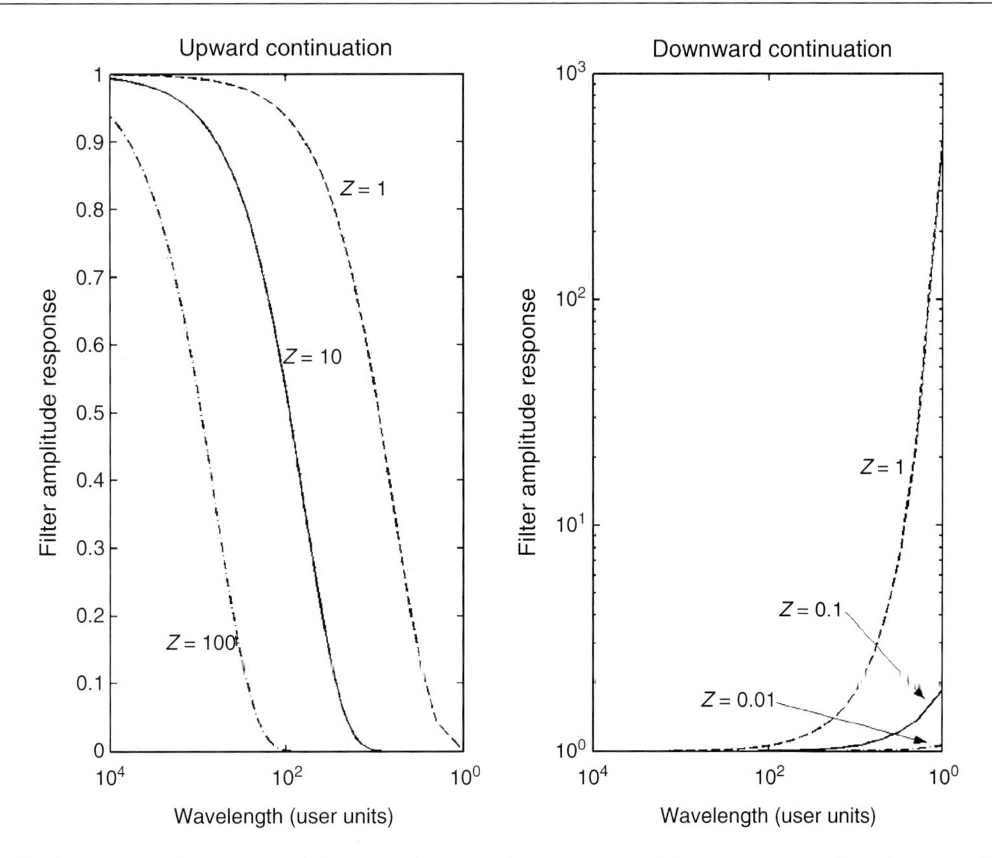

Figure U5 Amplitude response of upward and downward continuation operators with respect to wavelength for certain heights (z) of continuation.

and Zietz, 1949; Henderson, 1970). This is often useful for interpretation and joining two adjacent surveys carried out at different altitudes. As aid in interpretation, upward continuation allows one to assess the effect of deeper sources because in this process the effect of shallower, short-wavelength features is attenuated. Preferential upward and downward continuation operators have been designed that can help attenuate only the shallow, short-wavelength part of the spectrum, leaving the deeper, long-wavelength part unaltered or, alternatively, preferentially amplify only the deeper part of the spectrum without the deleterious effects of amplifying short-wavelength noise (Pawlowski, 1995). Thus, under certain situations, it is possible to isolate a magnetic anomaly signal from different depth layers of the crust. Downward continuation into the region of sources leads the continuation integral to diverge even in the case of noise-free data; in the case of high data density noise-free data the depths at which the continuation integral blows up (data begin to vary wildly) can be used to infer the depth to the top of the shallow magnetic sources in the region.

When airborne magnetic surveys (see Aeromagnetic surveying) are conducted in rugged terrain made up of magnetic formations, it is not advisable to view the data at constant altitude because effects of topographic variations can lead to anomaly artifacts. In such situations, one might prefer to "continue" level survey data at some constant distance away from topography (on a constant terrain clearance or "draped" surface). Challenges of maintaining the constant terrain clearance of aircrafts in a rugged topography may require one to adjust the data further until the survey is accurately draping. Conversely, flying conditions can lead to unintentional altitude variations in surveys originally intended to be flown at constant barometric altitude (level survey), and such surveys need datum corrections as well. Two types of procedures have been commonly used in accomplishing datum transformations from level-to-drape and drape-to-level surfaces: Taylor's series approximation and equivalent source concept. Taylor's series allows extrapolation of a function to nearby points and, given vertical derivatives of the field and certain approximations regarding behavior of the field, the series yields adequate values of level-to-drape transformation. Similarly, an iterative Taylor's series can be used for drape-to-level transformation (Cordell and Grauch, 1985). The equivalent source method (Dampney, 1969) employs Green's equivalent layer concept and uses a set of sources with arbitrary magnetization (often induced dipoles because of their simplicity; Emilia, 1973) to approximate the field. This process is equivalent to finding the potential that satisfies the observed field. The inverted magnetization of the sources is then used to predict the field in the neighborhood of observations. Use of local harmonic functions (Fedi *et al.*, 1999) can also be useful for these purposes.

Dhananjay Ravat

Bibliography

Blakely, R.J., 1995. *Potential Theory in Gravity and Magnetic Applications.* Cambridge:Cambridge University Press.

Cordell, L., and Grauch, V.J.S., 1985. Mapping basement magnetization zones from aeromagnetic data in the San Juan basin, New Mexico. In Hinze, W.J. (ed.) *The Utility of Regional Gravity and Magnetic Anomaly Maps.* Tulsa: Society of Exploration Geophysicists, pp. 181–197.

Courtillot, V., Ducruix, J., and Le Mouël, J.L., 1978. Inverse methods applied to continuation problems in geophysics. In Sabatier, P.C. (ed.) *Applied Inverse Problems.* Berlin: Springer-Verlag, pp. 48–82.

Dampney, C.N.G., 1969. The equivalent source technique. *Geophysics*, **34**: 39–53.

Dean, W.C., 1958. Frequency analysis for gravity and magnetic interpretation. *Geophysics*, **23**: 97–127.
Emilia, D.A., 1973. Equivalent sources used as an analytic base for processing total magnetic field profiles. *Geophysics*, **38**: 339–348.
Fedi, M., Rapolla, A., and Russo, G., 1999. Upward continuation of scattered potential field data. *Geophysics*, **64**: 443–451.
Henderson, R.G., 1970. On the validity of the use of the upward continuation integral for total magnetic intensity data. *Geophysics*, **35**: 916–919.
Henderson, R.G., and Zietz, I., 1949. The upward continuation of anomalies in total magnetic intensity fields. *Geophysics*, **14**: 517–534.
Langel, R.A., and Hinze, W.J., 1998. *The Magnetic Field of the Earth's Lithosphere: The Satellite Perspective.* Cambridge: Cambridge University Press.
Pawlowski, R.S., 1995. Preferential continuation for potential-field anomaly enhancement. *Geophysics*, **60**: 390–398.
Ravat, D., Whaler, K.A., Pilkington, M., Purucker, M., and Sabaka, T., 2002. Compatibility of high-altitude aeromagnetic and satellite-altitude magnetic anomalies over Canada. *Geophysics*, **67**: 546–554.
Shure, L., Parker, R.L., and Backus, G.E., 1982. Harmonic splines for geomagnetic modeling. *Physics of the Earth and Planetary Interiors*, **28**: 215–229.

Cross-references

Aeromagnetic Surveying
Crustal Magnetic Field

V

VARIABLE FIELD TRANSLATION BALANCE

The variable field translation balance (VFTB) is an instrument for measuring isothermal magnetizations in variable fields (e.g., hysteresis loops) as well as the temperature dependence of the associated magnetic parameters. It is specifically designed to measure the weak magnetizations commonly encountered in rock magnetism.

Measurement principles and main instrument components

The VFTB is a modification of the horizontal magnetic translation balance (e.g., Collinson, 1983). In contrast to the normal horizontal translation balance, the magnetic gradient in the VFTB is not produced by a special shape of the pole pieces of the electromagnet, but by a separate set of four gradient coils (AC; Figure V1). This separation of inducing field and gradient field was for the first time proposed by McKeehan (1934). The gradient field of the VFTB is not kept constant, but is oscillating with a certain frequency (2–4 Hz). The VFTB can thus be considered a one-dimensional harmonic oscillator with damping (D), operated in forced oscillation mode. The oscillating part of the instrument is a pendulum with bifilar suspension (BS) with the specimen (S) fixed to it in a nonmagnetic quartz glass sample holder (Q). The motion of the system is excited by a periodic force acting on the sample and generated by the set of gradient coils. The system is normally operated at resonance frequency to obtain highest sensitivity, but can also be operated in a nonresonance mode for the measurement of very strong magnetic samples. The motion of the sample is monitored by a linear variable differential transformer (LVDT). This signal—after processing in the PC—is then proportional to the magnetic moment of the sample.

The temperature is controlled by a heating/cooling unit (H/C) surrounding the sample chamber, thus making possible measurements at variable temperatures. During measurement the specimen is located in the center of either an electromagnet (EM, with pole pieces) or an iron-free solenoid generating a homogeneous inducing magnetic field. In case of the pole piece configuration, two compensation coils (CCs) are attached to compensate the residual field generated by the remanent magnetization of the pole pieces in the case of zero field measurements. All instrument functions like inducing field strength, gradient field amplitude, and temperature are computer controlled.

As the sample is moving back and forth inside the gradient field its motion is detected by the LVDT vibration sensor (Figure V2). The LVDT generates a voltage proportional to the position of the sample holder inside the sensor. This voltage is amplified and read by the A/D converter. The VFTB control software records the digitized voltages from the converter board. The gradient field actuating the sample movement varies sinusoidally. The response movement of the sample is also shaped like a sine wave with the same frequency as the original signal but with differing phase and amplitude. These indicate the actual magnetization of the sample. To compute amplitude and phase of the signal the recorded voltages are analyzed via fast Fourier transformation to find the spectral amplitude at the measurement frequency. The phase is used to compute the sign of the gained amplitude thus yielding the desired magnetization value. This value is normalized by the input gain of the A/D converter and the strength of the actuating field.

Types of measurements

VFTB measurements can be subdivided into field-dependent, temperature-dependent, and time-dependent magnetization measurements. The first group consists of isothermal magnetization measurements, i.e., IRM acquisition curves, hysteresis loops, and backfield curves. The number and spacing of field steps for these measurements can be freely chosen depending on the sample material and the desired measurement speed. It takes about 15 min to measure a hysteresis loop defined by 80 data points. Examples are shown in Figure V3.

The main purpose of temperature-dependent experiments is the determination of crystallographical or physical phase transitions, e.g., Verwey transition and Curie temperatures. The temperature can be varied between −190°C and room temperature with liquid nitrogen as furnace coolant and between room temperature and 800°C with

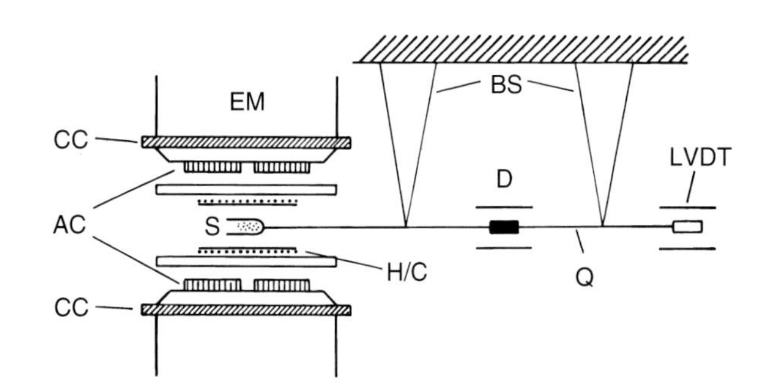

Figure V1 Schematic drawing of the main instrument components. Drawing shows the configuration with an electromagnet with pole pieces (see text for explanation of acronyms).

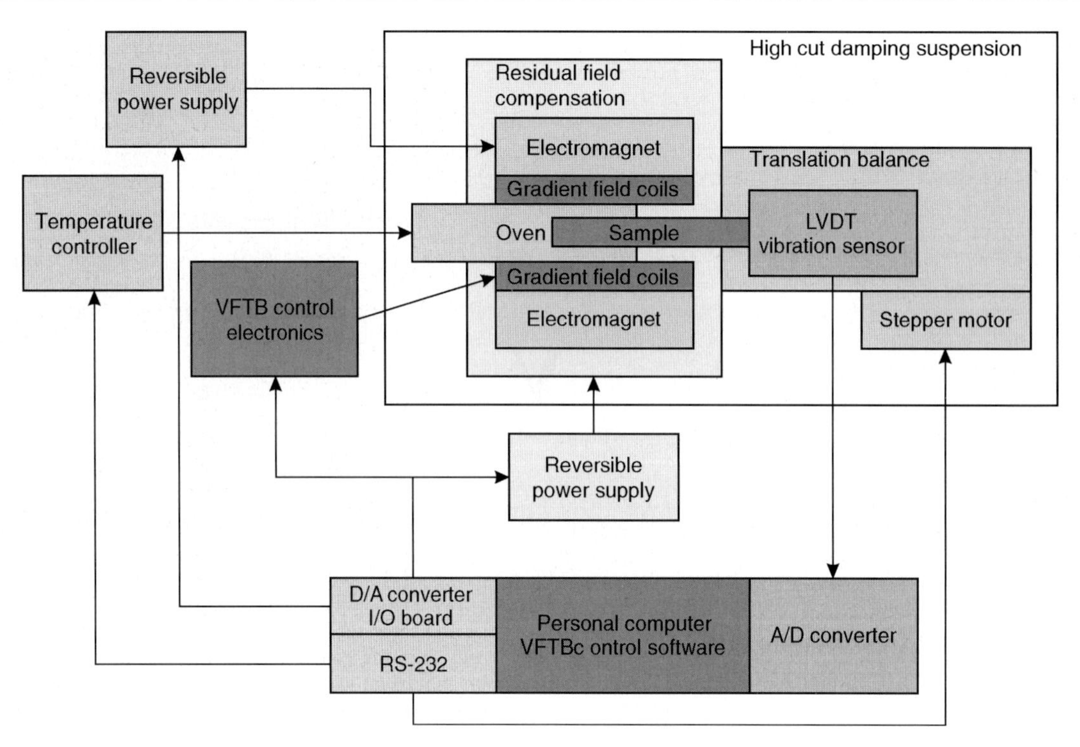

Figure V2 Simplified connection scheme of the VFTB.

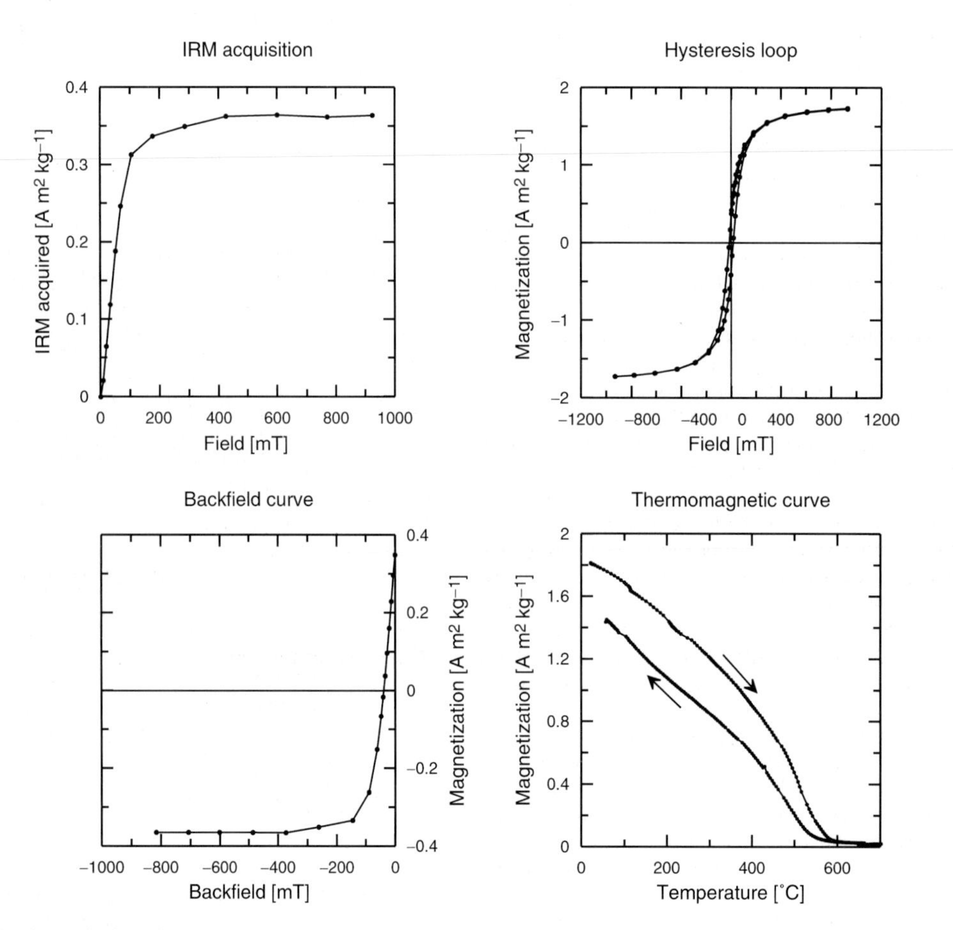

Figure V3 Example data of a basaltic specimen measured with a VFTB.

water cooling. IRM acquisition curves as well as hysteresis loops and backfield curves can also be measured in the whole temperature range between −190 and 800°C. The VFTB can also measure the time dependence of magnetization at given temperatures, useful, e.g., for determining the decay of isothermal remanent magnetization or for observing alteration processes at elevated temperatures.

Magnetic moment sensitivity of the VFTB amounts to 5×10^{-8} A m^2. Sample holders can accommodate samples with a diameter of either 5 or 10 mm and a length of 9mm. It is thus possible to measure fairly large samples with a maximum weight of ≈1.5 g which yields a sensitivity of ≈5×10^{-5} A m^2 kg^{-1} in terms of specific magnetization. Depending on the electromagnet a maximum magnetic field of 1.0 T can be attained and the residual field during zero field measurements does not exceed 50 μT.

David Krása, Klaus Petersen, and Nikolai Petersen

Bibliography

Collinson, D.W., 1983. *Methods in Rock Magnetism and Palaeomagnetism: Techniques and Instrumentation.* New York: Chapman & Hall.

McKeehan, L., 1934. Pendulum magnetometer for crystal ferromagnetism. *The Review of Scientific Instruments*, **5**: 265–268.

[illegible]

Magnetization, Isothermal Remanent
Rock Magnetism
Rock Magnetism, Hysteresis Measurement

VERHOOGEN, JOHN (1912–1993)

John Verhoogen (Figure V4) was an engineer, geologist, geophysicist, and a pioneering advocate of the role of the solid inner core in the geodynamo. He was also an inspirational mentor to a generation of students, including the Earth Scientists who first established the timescale for magnetic polarity reversals.

Verhoogen was born in Brussels in 1912 in an educated, professional family. He became afflicted with poliomyelitis while in high school. Although this disease would progressively limit his physical mobility, it did not deter him from pursuing a career in the geological sciences. He earned degrees in mining engineering at the University of Brussels in 1933, engineering geology at the University of Liege in 1934, and a Ph.D. in geology at Stanford University in 1936. This was his first introduction to the United States, where he would later make his scientific mark. After a decade of fieldwork in equatorial Africa, he joined the faculty of the University of California, Berkeley in 1947 and taught there until his retirement in 1976.

Verhoogen's first major research accomplishment was to measure the gas composition of the active volcano Nyamuragira in Zaire (then the Belgian Congo). This experience convinced him that igneous processes and volcanism were indisputable evidence for large-scale flow deep in Earth's interior. One outcome of these studies proposed an explanation for the thermal evolution of the Earth in terms of heat transfer by fluid motions in the deep interior. Although the significance of this idea was not widely appreciated by his contemporaries at the time, it guided his own research for the ensuing three decades and, ultimately, it has to be accepted throughout the geoscience community.

In the late 1950s and early 1960s, Verhoogen played a seminal role in the development of paleomagnetism, particularly the paleomagnetic reversal time scale. At that time there were many reports of rocks with ferromagnetic directions reversed to the present-day geomagnetic field direction. Two competing explanations were offered to explain this puzzling phenomenon: either these rocks were magnetized during epochs when the main geomagnetic field was reversed, or their original magnetic directions had been reversed by local remagnetization effects.

Figure V4 John Verhoogen (courtesy of Margaret Gennaro, University of California Archives).

The existing techniques in geochronology were inadequate to distinguish between the two competing explanations. Verhoogen initially favored remagnetization, but several of his own graduate students, including Allan Cox and Richard Doell, were measuring the paleomagnetic directions in lavas from Hawaii and other volcanic hotspots and convinced him that the global reversal explanation was far more likely. Within a few years, new developments in mass spectrometer technology provided accurate radiometric dating methods, confirming that the main geomagnetic field had indeed reversed its polarity many times in the past. The resulting geomagnetic polarity time scale is now a cornerstone in geophysics; for example, it provides the interpretation of magnetic stripes in the ocean crust in terms of sea-floor spreading (see Glen 1982 for a description of Verhoogen's role in these developments).

As post-polio injuries curtailed his field and laboratory work, Verhoogen progressively turned his attention toward the energy balance of the Earth as a whole, and in particular, the energy needed to maintain the geodynamo. One of his major contributions to igneous geology had been a quantitative thermodynamic treatment of igneous petrogenesis, which combined the energetics of crystallization with the thermal environment of the deep crust and upper mantle where the magmas originate. He recognized that the core was like a giant magma chamber, in the sense that physical and chemical processes similar to those in magmatic systems must be operating inside the core. This represented a new perspective; previously, only thermal convection had been considered important in core energetics. Verhoogen reasoned that, because the Earth as a whole is cooling, the solid inner core must be growing by crystallization at the expense of the fluid outer core. He realized that crystallization at the inner core boundary would provide a source of energy for the geodynamo, through latent heat release and other effects associated with the solidification. In 1961, he estimated the energy release by this process, and found it adequate to maintain the geodynamo if the rate of inner core solidification is sufficiently high, about 25 m^3 s^{-1} or more. Subsequent investigations by Braginsky (1963), Gubbins (1977), Buffett *et al.* (1992), and others have developed these concepts further, adding an important ingredient to the model, the gravitational potential energy supplied by chemical differentiation at the crystallizing inner-outer core boundary. This model has come to be called the "gravitational dynamo" and is now the favored mechanism for powering the geodynamo.

In his 1980 monograph *Energetics of the Earth*, Verhoogen systematically analyzed the various energy sources available to drive the geodynamo, including inner core solidification, heat loss to the mantle, and possible radioactive heat sources in the outer core. He concluded that, although the gravitational dynamo mechanism evidently provides enough energy, the uncertainties in the rate of heat loss to the mantle, combined with the uncertainties in radioactive abundance and important physical properties of the core, preclude us from making a definitive model of the geodynamo. Researchers today would, no doubt, come to a similar conclusion. However, in an address to the US National Academy of Science, Verhoogen expressed his scientific optimism in the following way:

> Perhaps the greatest step forward since the discovery of radioactivity has been the recognition of convection as the dominant mode of heat and mass transfer in most of the mantle and core. Here at long last is something to hang onto.

This statement is as applicable today as it was two decades ago, and remains the paradigm for understanding the deep Earth.

Peter Olson

Bibliography

Braginsky, S.I., 1963. Structure of the F layer and reasons for convection in the earth's core. *Doklady Akademiya Nauk SSSR*, **149**: 1311–1314 (English Translation).

Buffett, B.A., Huppert, H.H., Lister, J.R., and Woods, A.W., 1992. Analytical model for solidification of Earth's core. *Nature*, **356**: 329–331.

Glen, W., 1982. *The Road to Jaramillo.* Palo Alto, CA: Stanford University Press.

Gubbins, D., 1977. Energetics of the Earth's core. *Journal of Geophysics*, **43**: 453–464.

Verhoogen, J., 1961. Heat balance of the Earth's core. *Geophysical Journal of the Royal Astronomical Society*, **4**: 276–281.

Verhoogen, J., 1980. *Energetics of the Earth.* Washington, DC: National Academy of Science.

Cross-references

Convection, Chemical
Cox, Allan V. (1926–1987)
Geodynamo, Energy Sources
Geomagnetic Polarity Timescales
Magnetization, Thermoremanent (TRM)

VINE-MATTHEWS-MORLEY HYPOTHESIS

Definition

The Vine-Matthews-Morley (VMM) hypothesis states that, when ocean crust forms at a midocean ridge (i.e., a spreading center), the cooling crust becomes magnetized in the direction of Earth's prevailing magnetic field as it cools below the Curie temperature of the magnetic minerals (Morley and Larochelle, 1964; Vine and Matthews, 1963). Typically this Curie temperature is between 150 and 300°C for titanomagnetite, the primary magnetic mineral in the upper oceanic crust, and ~580°C for pure magnetite found in the lower ocean crust. The ocean crust moves away from the spreading center through the process of "seafloor spreading" and in this way the magnetic signal recorded in the newly formed crust is preserved (Figure V5). A good analogy is that of a tape recorder where the ocean crust is the magnetic tape and Earth's magnetic field is the signal, which is recorded onto the moving tape. In this way, changes in Earth's magnetic field polarity through time are preserved in the crustal rocks. The alternating bands of magnetic polarity give rise to magnetic anomalies when measured at the sea surface, the so-called "magnetic stripes" (Figure V6) (see Magnetic surveys, marine; Magnetic anomalies, marine).

The VMM hypothesis elegantly explains the striped patterns of magnetic anomalies measured at the sea surface without the need to call upon petrological, thermal, or some other process to be aligned parallel to the spreading axis. It also links many of the other observations about midocean ridges such as their apparent youth and thin crust. It also explains why the ocean basins are relatively young (<200 My) compared to continents. The magnetic anomalies document the progression in age of the ocean basins from young crust at midocean ridges to the oldest and deepest crust near subduction zones and in doing so strongly supports the global concept of continental drift that ocean basins have opened and closed in the past. The VMM hypothesis forms one of the cornerstones of plate tectonic theory and began a revolution in thinking about how the Earth works and how it has evolved through geologic time.

Historical context

Fred Vine and Drummond Matthews were part of the geophysics group at Cambridge University in England at the time of their discoveries under the leadership of Teddy Bullard and they were very receptive to the ideas of American scientists Harry Hess and Robert Dietz (Dietz, 1961; Hess, 1962) about continental drift and the possibility of seafloor spreading in contrast to the mainstream US science community at that time (Glen, 1982). Vine began his work as a graduate student working for Drummond Matthews and their ideas grew out of work by Vine on magnetic profiles collected by Matthews over the Carlsberg Ridge in the Indian Ocean. Vine found that there were reversely magnetized seamounts indicating that the ocean crust could record both normal and reverse polarity periods in Earth's magnetic field history. A similar conclusion had been reached earlier by Girdler and Peter (1960) who suggested the possibility of reversely magnetized crust in the Gulf of Aden. Vine and Matthews (1963) suggested that this process may be more pervasive that previously thought. The Carlsberg Ridge profiles showed positive and negative magnetic anomalies ~20 km in wavelength, and Vine and Matthews suggested that these anomalies could be easily explained by polarity changes in the crustal magnetization rather than petrological or thermal differences. It was this observation that allowed Vine and Matthews to make the daring statement that perhaps 50% of the ocean floor was reversely magnetized. They borrowed on the ideas of Dietz (1961) and Hess (1962) that seafloor spreading was occurring at midocean ridges and made the key observation that ocean crust could record Earth's magnetic field as it formed and that this record would then be preserved as the crust moved away from the spreading center.

The paper initially received little fanfare, comment, or recognition. Most scientists dismissed the paper as being too speculative. What was not immediately appreciated was the prediction of symmetrical anomalies about a midocean ridge crest and the fact that seafloor spreading rates could be obtained from such magnetic records. Lawrence Morley recognized these key ideas but his initial attempts at publishing were declined. One editor reputedly commented that the paper was more suitable as a subject for discussion at a cocktail party rather than a serious scientific idea (Glen, 1982). Morley was an administrator for the Geological Survey of Canada at the time and had little time to vigorously pursue these ideas. Morley was eventually able to publish his article in an obscure journal in Canada (Morley and Larochelle, 1964). None of the significance of the VMM ideas permeated the scientific community at the time of the publications. It was not until Vine and Wilson (1965) looked at the newly published maps of Raff and Mason (1961) (Figure V6) that they recognized the location of the Juan de Fuca midocean ridge spreading center and its position at the center of symmetry

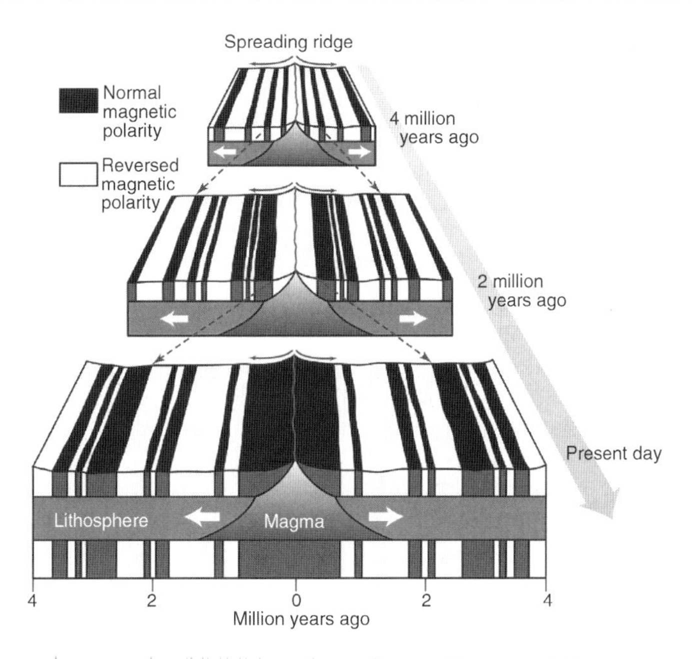

Figure V5 Cartoon of the Vine-Matthews-Morley (VMM) hypothesis showing the process of seafloor spreading at a midocean ridge and the recording Earth's magnetic field to create the magnetic stripes. Reproduced by permission of Woods Hole Oceanographic Institution (Tivey, 2004).

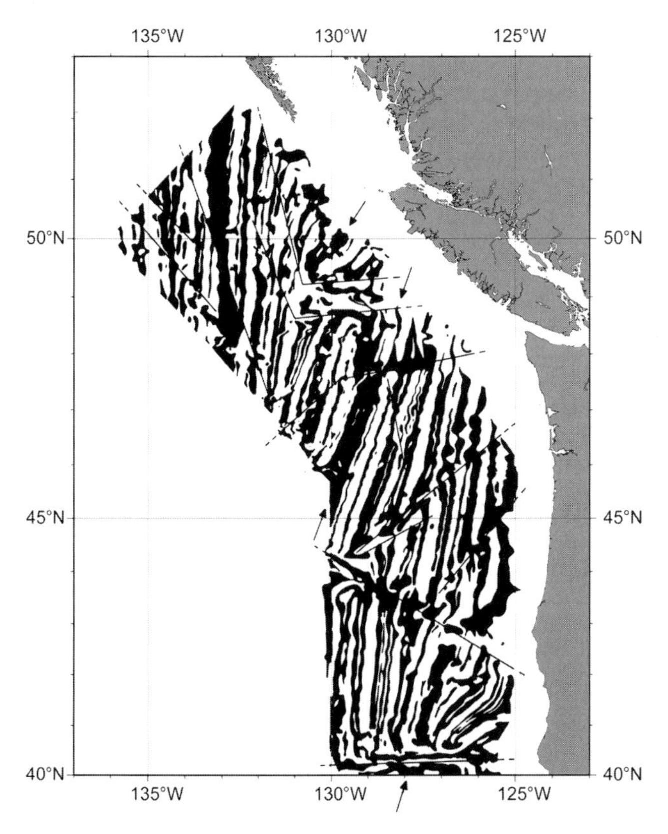

Figure V6 Magnetic anomaly "stripe" map of northeast Pacific Ocean, where positive anomalies are shaded black and reverse polarity is shown in white. It was the symmetrical pattern of anomalies about the newly discovered Juan de Fuca Ridge spreading center that began the acceptance of the VMM hypothesis. Figure modified from Raff and Mason (1961). Reproduced by permission of the Geological Society of America.

in the northeast Pacific magnetic anomalies that the idea began to gain ground and become seriously debated in academic circles. Vine and Wilson (1965) also matched the magnetic profiles with the newly compiled geomagnetic polarity reversal timescale of Cox *et al.* (1963) based on precisely age dated terrestrial lavas.

A subsequent paper by Vine (1966) eventually put all the ideas together using a long magnetic profile over the East Pacific Rise (Eltanin Leg-19) collected by Walter Pitman of Lamont-Doherty Geological Observatory (Pitman and Heirtzler, 1966) (Figure V7). Vine showed that there was universal symmetry and correlation between magnetic profiles from several midocean ridges and that these could be modeled with the geomagnetic polarity reversal timescale of Cox *et al.* (1963). This essentially sealed the case that magnetic anomalies recorded in ocean crust preserved the history of Earth's magnetic field and that this reversal pattern could be correlated with dated reversals on land.

Discoveries stemming from VMM hypothesis

Heirtzler and colleagues (1968) took the observations of Vine and Matthews (Glen, 1982) and Vine (1966) to the logical conclusion by producing a geomagnetic polarity timescale (GPTS) based on the marine magnetic anomaly patterns found over the ocean basins and extended the timescale to 80 My. Larson and Pitman (1972) and Larson and Chase (1972) extended the GPTS with studies of Pacific magnetic anomalies and eventually these were reconciled with anomalies of a similar age in the Atlantic (Vogt *et al.*, 1971) into the Mesozoic or M-series anomalies. The present incarnation of the GPTS was published by Cande and Kent (1995) with extension to the Mesozoic anomalies by Channell *et al.* (1995). The marine magnetic record allows for a detailed history of ocean basin formation and continental drift to be documented in terms of plate tectonics. A detailed history of plate motion and spreading rates has been compiled by various researchers primarily based on the marine magnetic anomaly patterns (DeMets *et al.*, 1990; Minster and Jordan, 1978; Scotese, 1997). A compilation of ship-based mapping of the marine magnetic anomaly patterns has allowed a digital database of seafloor age to be produced (Mueller *et al.*, 1996).

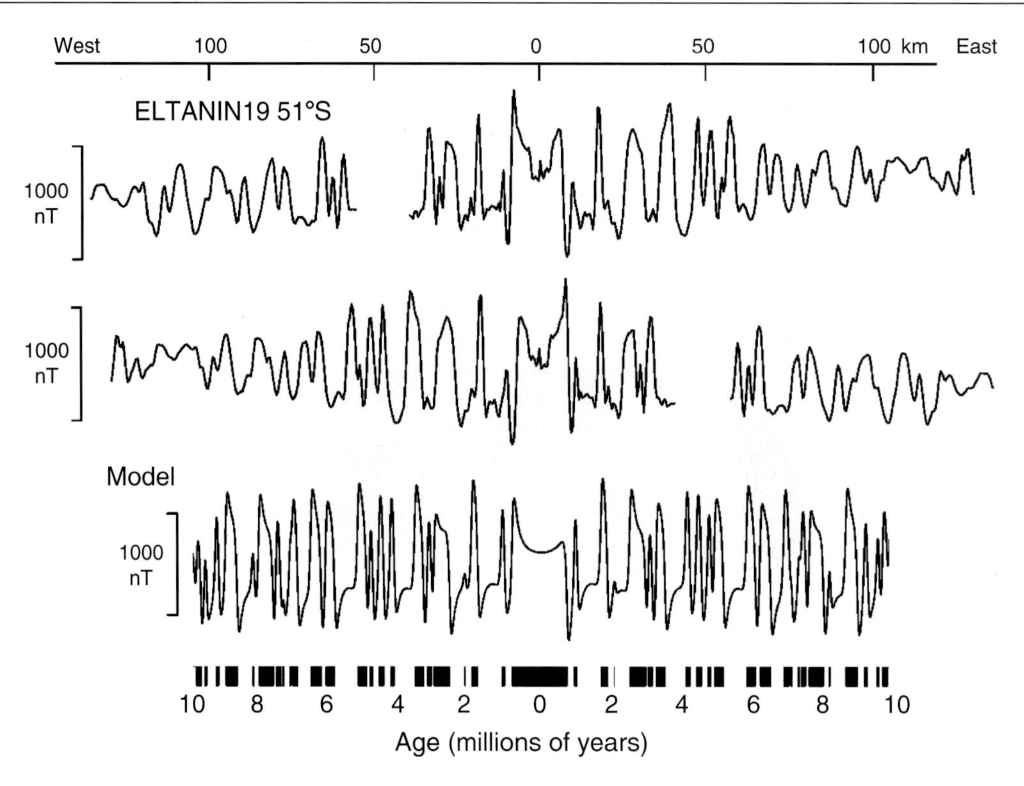

Figure V7 Eltanin-19 magnetic anomaly profile from Pitman and Heirtzler (1966) shown in the middle with the profile reversed shown at top and a computed model profile shown below. The geomagnetic polarity timescale is shown at the bottom with the normal polarity shown in black and the reverse polarity in white.

The shape of magnetic anomalies (i.e., their phase) contains information about their latitude of formation which can be used to determine the paleolatitude when the crust formed (Acton and Gordon, 1991; Petronotis *et al.*, 1994). Magnetic anomaly patterns have also led to the discovery of new types of seafloor spreading in the form of propagating rifts (Hey, 1977; Hey *et al.*, 1980) and rotating microplates (Courtillot, 1980; Hey, 1985; Schouten *et al.*, 1993).

Marine magnetic surveys using deep-towed sensors and autonomous underwater vehicles are now providing increased resolution in mapping magnetic anomalies, which has allowed for short-wavelength magnetic anomalies (<10 km wavelength) to be recognized (Tivey *et al.*, 1997). It now is apparent that the ocean crust not only records the polarity of Earth's magnetic field, but also provides a record of Earth's field intensity through time (Bowers *et al.*, 2001; Pouliquen *et al.*, 2001). Magnetic profiles over ridge crests and show a good correlation of short-wavelength magnetic anomalies (<10 km wavelength) with terrestrial and sediment core derived paleointensity records (Gee *et al.*, 2000; Guyodo and Valet, 1999). It may, therefore, be possible to use the paleointensity record as a fine-scale signal with which to date the seafloor in addition to the polarity reversal signal. Earth's magnetic field intensity apparently undergoes relatively rapid changes compared to the polarity reversal signal and thus the crustal accretion process will ultimately control the resolution of the recorded signal (Schouten *et al.*, 1999).

Some outstanding issues

One of the outstanding and unsolved issues of the VMM hypothesis is the source of the magnetic anomaly stripes. What part of the ocean crust is responsible for the generating the magnetic anomalies? Vine and Matthews (1963) made only vague stipulations as to the possible source of the magnetic anomalies. For a crustal unit to be considered as a viable source of the magnetic anomalies, it must satisfy a number of criteria (Harrison, 1981; Johnson, 1979). First, the crust has to be sufficiently magnetic in order to create the anomalies. Second, the magnetization has to be stable and robust enough to withstand millions of years of alteration without degrading. Third, the crust has to cool quickly in order to capture magnetic reversals, which occur at an average frequency of 350000 years. Finally, the geometry of the reversal boundaries must be sufficiently vertical in order to create magnetic anomalies.

Ocean crust can be crudely thought of being composed of three layers (Figure V8). The upper extrusive lava layer consists of lava that has erupted onto the seafloor and thus has cooled rapidly. This layer is highly magnetic although relatively thin, typically <1 km thick. This layer overlies the intrusive dike layer, which are the subsurface volcanic conduits that feed the overlying lava flows. The dikes cool much more slowly than the lavas and are markedly less magnetic because of their intrinsically larger and less magnetic grains and high temperature alteration associated with their emplacement. The dike layer is on the order of 1–2 km thick. The third crustal layer is the plutonic gabbro section, which can be thought of as the cooling magma chamber that is the magmatic source of the overlying dikes and lavas. Gabbros cool relatively slowly and have large crystals and would therefore be intrinsically less magnetic. Nevertheless, some gabbros can be quite magnetic with microscopic grains of pure magnetite found within plagioclase crystals. The gabbros as a layer are quite thick, up to 2–4 km thick and thus while perhaps less magnetic than the extrusive lavas could be of sufficient volume to generate a magnetic anomaly. Finally, underlying the ocean crust is the mantle, which is composed of peridotite and may also be magnetic, especially if it has undergone alteration to serpentinite through the introduction of water. The contribution of these crustal layers and mantle to the overlying magnetic anomalies is difficult to quantify, especially with in situ measurements. Early magnetic surveys determined that the upper extrusive lavas were the primary source layer based on topographic arguments (Atwater and

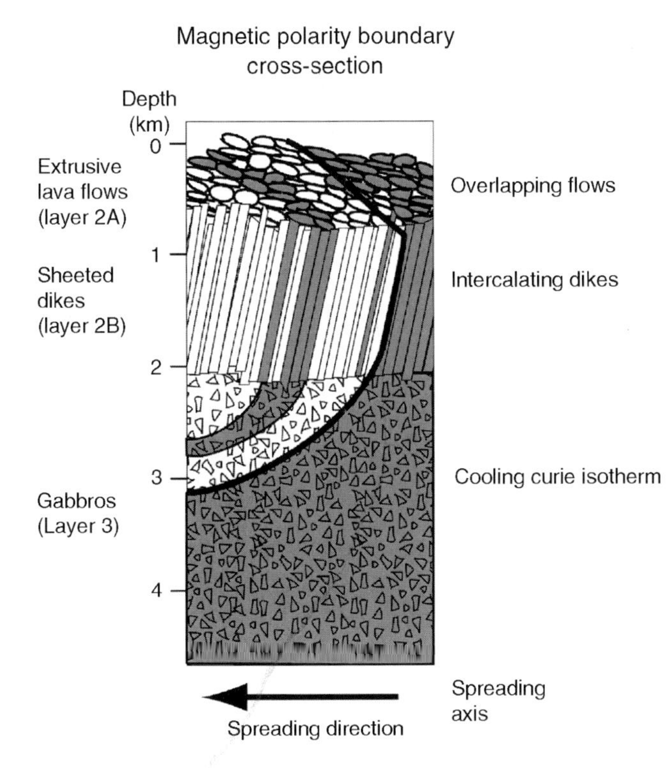

Figure V8 Diagram showing a cross-section through ocean crust with the primary crustal units and thicknesses. Superimposed on the lithology is the presumed geometry of a magnetic polarity reversal boundary (thick black line). The boundary dips toward the spreading axis in the extrusive lavas, but is thought to dip away from the axis in the deeper gabbro layer following a cooling isotherm. This portion of the geometry is hypothetical. Drawing based on Macdonald *et al.* (1983) and Kidd (1977).

Mudie, 1973; Talwani *et al.*, 1971). Oceanic crustal drilling provided ambiguous results showing that the magnetization of the extrusive lavas was on average not enough to generate the anomalies and that deeper layers must be contributing as well (Harrison, 1981). While some gabbros have been drilled and are magnetic (Pariso and Johnson, 1993), ocean crustal drilling to date has not penetrated a complete section of crust from extrusive lava to gabbro so that a complete answer to this question remains.

The dip of a polarity boundary can also reveal useful information about the source and contribution of the crustal layers to the magnetic anomaly. A conceptual model of the polarity boundary (Kidd, 1977) suggests that within the extrusive lava layer the progressive overlap of lavas, as they are formed and are rafted away from the spreading center, generates a boundary that dips toward the spreading center. In the dike layer the boundary is likely to be near vertical and relatively sharp, while in the gabbro section a polarity boundary is likely to be a cooling isotherm, which would dip away from the spreading center (Figure V8). A couple of detailed near-bottom magnetic surveys showed the importance of the dip of the polarity boundary within the extrusive layer. Macdonald *et al.* (1983) used the submersible ALVIN to measure the magnetic polarity of seafloor outcrops of lava at a reversal boundary and found a sharply defined polarity boundary between normal and reversal magnetized lavas. This boundary location was compared to that obtained from a towed near-bottom magnetic survey that is an average of the entire layer contribution. The deep-tow boundary was a kilometer closer to the spreading axis than the surficial boundary, indicating that the polarity boundary dips toward the spreading center. A submarine magnetic survey carried out at a fracture zone wall (Tivey, 1996; Tivey *et al.*, 1998) mapped the magnetic boundaries within the extrusive layer in cross-section and also found that the polarity boundaries dipped toward the spreading center. In this case, the authors also were able to calculate the contribution of the layer to the overlying anomaly and determined that the extrusive lavas produced about 75% of the observed anomaly, indicating that additional contribution was needed from deeper layers such as the dikes and gabbros (Tivey *et al.*, 1998). Only one survey has unambiguously determined the dip of polarity boundaries in the gabbros to date (Allerton and Tivey, 2001). In this survey, seafloor drilling determined the surficial polarity boundary location while a near-bottom survey determined the average boundary position and found that the polarity boundary dips away from the spreading center consistent with the idea of cooling isotherms within the gabbro section (Allerton and Tivey, 2001). The contribution and geometry of the polarity boundaries within the lower ocean crust remain poorly documented and continue to be a topic of research.

Maurice A. Tivey

Bibliography

Acton, G.D., and Gordon, R.G., 1991. A 65 Ma paleomagnetic pole for the Pacific Plate from the skewness of magnetic anomalies 27r-31. *Geophysical Journal International*, [illegible]

Allerton, S., and Tivey, M.A., 2001. Magnetic polarity structure of lower oceanic crust. *Geophysical Research Letters*, **28**: 423–426.

Atwater, T.M., and Mudie, J., 1973. Detailed near-bottom geophysical study of the Gorda Rise. *Journal of Geophysical Research*, **78**: 8665–8686.

Bowers, N.E., Cande, S.C., Gee, J.S., Hildebrand, J.A., and Parker, R. L., 2001. Fluctuations of the paleomagnetic field during chron C5 as recorded in near-bottom marine magnetic anomaly data. *Journal of Geophysical Research*, **106**: 26379–26396.

Cande, S.C., and Kent, D.V., 1995. Revised calibration of the geomagnetic polarity timescale for the late Cretaceous and Cenozoic. *Journal of Geophysical Research*, **100**: 6093–6095.

Channell, J.E.T., Erba, E., Nakanishi, M., and Tamaki, K., 1995. Late Jurassic-Early Cretaceous time scales and oceanic magnetic anomaly block models. In Berggren, W.A., Kent, D.V., Aubry, M.-P., and Hardenbol, J. (eds.), *Geochronology, Time Scales and Global Stratigraphic Correlation,* Volume 54: Special Publication. Tulsa, OK: SEPM (Society for Sedimentary Geology), pp. 51–63.

Courtillot, V., 1980. Plaques, microplaques et dechirures lithospheriques: une hierarchie de structures tectoniques de l'echelle du globe a celle du terrain. *Bulletin de la Societe Geologique de France*, **12**: 981–984.

Cox, A., Doell, R.R., and Dalrymple, G.B., 1963. Geomagnetic polarity epochs and Pleistocene geochronology. *Nature*, **198**: 1049–1051.

DeMets, C., Gordon, R.G., Argus, D.F., and Stein, S., 1990. Current plate motions. *Geophysical Journal International*, **101**: 425–478.

Dietz, R.S., 1961. Continent and ocean basin evolution by spreading of the seafloor. *Nature*, **190**: 854–857.

Gee, J.S., Cande, S.C., Hildebrand, J.A., Donnelly, K., and Parker, R.L., 2000. Geomagnetic intensity variations over the past 780 kyr obtained from near-seafloor magnetic anomalies. *Nature*, **408**: 827–832.

Girdler, R.W., and Peter, G., 1960. An example of the importance of natural remanent magnetization in the interpretation of magnetic anomalies. *Geophysical Prospecting*, **8**: 474–483.

Glen, W., 1982. *The Road to Jaramillo. Critical Years of the Revolution in Earth Science*. Stanford, CA: Stanford University Press, 459 pp.

Guyodo, Y., and Valet, J.-P., 1999. Global changes in intensity of the Earth's magnetic field during the past 800 kyr. *Nature*, **399**: 249–252.

Harrison, C.G.A., 1981. Magnetism of the oceanic crust. In Emiliani, C. (ed.), *The Sea,* Volume 7. New York: Wiley, pp. 219–237.
Heirtzler, J.R., Dickson, G.O., Herron, E.M., Pittman, W.C., III, and Le Pichon, X., 1968. Marine magnetic anomalies, geomagnetic field reversals, and motions of the ocean floor and continents. *Journal of Geophysical Research*, **73**: 2119–2136.
Hess, H.H., 1962. History of ocean basins. In Engel, A.E.J., James, H.L., and Leonard, B.F. (eds.) *Petrologic Studies: The Buddington Volume.* Boulder, CO: Geological Society of America, pp. 599–620.
Hey, R., 1977. A new class of "pseudofaults" and their bearing on plate tectonics: a propagating rift model. *Earth and Planetary Science Letters*, **37**: 321–325.
Hey, R., Duennebier, F.K., and Morgan, W.J., 1980. Propagating rifts on midocean ridges. *Journal of Geophysical Research*, **85**: 3647–3658.
Hey, R.N., Naar, D.F., Kleinrock, M.C., Phipps Morgan, W.J., Morales, E., and Schilling, J.-G., 1985. Microplate tectonics along a superfast seafloor spreading system near Easter Island. *Nature*, **317**: 320–325.
Johnson, H.P., 1979. Magnetization of the oceanic crust. *Reviews of Geophysics and Space Physics*, **17**: 215–225.
Kidd, R.G.W., 1977. The nature and shape of the sources of marine magnetic anomalies. *Earth and Planetary Science Letters*, **33**: 310–320.
Larson, R.L., and Chase, C.G., 1972. Late Mesozoic evolution of the western Pacific Ocean. *Geological Society of America Bulletin*, **83**: 3627–3644.
Larson, R.L., and Pitman, W.C., 1972. Worldwide correlation of Mesozoic magnetic anomalies and its implications. *Geological Society of America Bulletin*, **83**: 3645–3662.
Macdonald, K.C., Miller, S.P., Luyendyk, B.P., Atwater, T.M., and Shure, L., 1983. Investigation of a Vine-Matthews magnetic lineation from a submersible: the source and character of marine magnetic anomalies. *Journal of Geophysical Research*, **88**: 3403–3418.
Minster, B.J., and Jordan, T.H., 1978. Present-day plate motions. *Journal of Geophysical Research*, **83**: 5331–5354.
Morley, L.W., and Larochelle, A., 1964. Paleomagnetism as a means of dating geological events. *Royal Society of Canada Special Publication*, **8**: 39–50.
Mueller, R.D., Roest, W.R., Royer, J.-Y., Gahagan, L.M., and Sclater, J.G., 1996. *Age of the Ocean Floor.* Report MGG-12, Data Announcement 96-MGG-04. Boulder, CO: National Geophysical Data Center.
Pariso, J.E., and Johnson, H.P., 1993. Do layer 3 rocks make a significant contribution to marine magnetic anomalies? In situ magnetization of gabbros at ocean drilling program hole 735B. *Journal of Geophysical Research*, **98**: 16033–16052.
Petronotis, K.E., Gordon, R.G., and Acton, G., 1994. A 57Ma Pacific plate paleomagnetic pole determined from a skewness analysis of crossings of marine magnetic anomaly 25r. *Geophysical Journal International*, **118**: 529–544.
Pitman, W.C., and Heirtzler, J.R., 1966. Magnetic anomalies over the Pacific-Antarctic ridge. *Science*, **154**: 1164.
Pouliquen, G., Gallet, Y., Patriat, P., Dyment, J., and Tamura, C., 2001. A geomagnetic record over the last 3.5 million years from deep-tow magnetic anomaly profiles across the Central Indian Ridge. *Journal of Geophysical Research*, **106**: 10941–10960.
Raff, A.D., and Mason, R.G., 1961. A magnetic survey off the west coast of North America, 40-N to 52-N. *Bulletin of Geological Society of America*, **72**: 1267–1270.
Schouten, H., Klitgord, K., and Gallo, D., 1993. Edge-driven microplate kinematics. *Journal of Geophysical Research*, **98**: 6689–6702.
Schouten, H., Tivey, M.A., Fornari, D.J., and Cochran, J.R., 1999. The central anomaly magnetic high: constraints on volcanic construction of seismic layer 2A at a fast-spreading midocean ridge, the East Pacific Rise at 9°30–50′N. *Earth and Planetary Science Letters*, **169**: 37–50.
Scotese, C.R., 1997. *Continental Drift*, 7th Edition. Arlington, TX: PALEOMAP Project, p. 79.
Talwani, M., Windisch, C.C., and Langseth, M.G., 1971. Reykjanes ridge crest: a detailed geophysical study. *Journal of Geophysical Research*, **76**: 473–517.
Tivey, M., 1996. Vertical magnetic structure of ocean crust determined from near-bottom magnetic field measurements. *Journal of Geophysical Research*, **101**: 20275–20296.
Tivey, M.A., 2004. Paving the seafloor—brick by brick. *Oceanus*, **42**: 44–47.
Tivey, M.A., Bradley, A., Yoerger, D., Catanach, R., Duester, A., Liberatore, S., and Singh, H., 1997. Autonomous underwater vehicle maps seafloor. *EOS*, **78**: 229–230.
Tivey, M., Johnson, H.P., Fleutelot, C., Hussenoeder, S., Lawrence, R., Waters, C., and Wooding, B., 1998. Direct measurement of magnetic reversal polarity boundaries in a cross-section of oceanic crust. *Geophysical Research Letters*, **25**: 3631–3634.
Vine, F.J., 1966. Spreading of the ocean floor: new evidence. *Science*, **154**: 1405–1425.
Vine, F.J., and Matthews, D.H., 1963. Magnetic anomalies over oceanic ridges. *Nature*, **199**: 947–949.
Vine, F.J., and Wilson, J.T., 1965. Magnetic anomalies over a young oceanic ridge off Vancouver Island. *Science*, **150**: 485–489.
Vogt, P.R., Anderson, C.N., and Bracey, D.R., 1971. Mesozoic magnetic anomalies, sea floor spreading, and geomagnetic reversals in the southwestern North Atlantic. *Journal of Geophysical Research*, **76**: 4796–4823.

Cross-references

Magnetic Anomalies, Marine
Magnetic Surveys, Marine

VOLCANO-ELECTROMAGNETIC EFFECTS

Volcano-electromagnetic effects—electromagnetic (EM) signals generated by volcanic activity—derive from a variety of physical processes. These include piezomagnetic effects, electrokinetic effects, fluid vaporization, thermal demagnetization/remagnetization, resistivity changes, thermochemical effects, magnetohydrodynamic effects, and blast-excited traveling ionospheric disturbances (TIDs). Identification of different physical processes and their interdependence is often possible with multiparameter monitoring, now common on volcanoes, since many of these processes occur with different timescales and some are simultaneously identified in other geophysical data (deformation, seismic, gas, ionospheric disturbances, etc.). EM monitoring plays an important part in understanding these processes.

Brief history

The identification of electromagnetic field disturbances associated with volcanic activity has a relatively young history. Initial measurements in the 1950s on Mihara volcano in Japan focused on monitoring the inclination and declination of the magnetic field while on Kilauea in Hawaii static electric fields (self-potential, SP) were shown to reflect the intrusion activity. More robust magnetic and electric field observations since the 1960s show changes of several tens of nanotesla and more than a hundred mV km^{-1} accompany eruptions from volcanoes (Johnston, 1997; Zlotnicki, 1995). Following eruptions, much larger changes result from remagnetization processes.

Physical mechanisms involved

Short-term magnetic and electric field changes during volcanic eruptions result primarily from stress changes in volcanic rocks and injection of hydrothermal/magmatic fluids and gasses. Of particular importance for research in eruption prediction are changes generated in the early stages of eruptions by deformation and/or fluid/gas flow. Detection of these fields provides an independent window into the eruption process.

The dominant physical processes in play during the final stages before eruptions are stress-induced modification of magnetization (piezomagnetism), electric currents produced by fluid flow (electrokinetic effects), and thermal demagnetization and remagnetization of volcanic rocks. Other EM signals result from charge generation during fluid vaporization and the creation of ash. Changes in EM response also result from modification of the electrical conductivity structure. During and following eruptions, dramatic EM signals can result from electric charge generation from rock disintegration, lightning within the eruption ash cloud, magnetohydrodynamic effects, thermal remagnetization, stress driven magnetization changes, and coupled TIDs.

It has become increasingly clear, because of the complexity and interdependence of volcanic processes, that simultaneous multiparameter monitoring (e.g., strain, fluid pressure, tilt, ground displacement, electric fields, magnetic fields, gas emissions, etc.) on volcanoes is necessary to identify and separate contributions from the different physical processes. Interpretation of electric and magnetic fields can thus be best done within the constraints placed by other geophysical data and vise versa.

Piezomagnetic effects

The magnetic properties of rocks have been shown under laboratory conditions to depend on the state of applied stress, and theoretical models for this phenomenon have been developed in terms of single domain, pseudosingle domain rotation and multidomain wall translation. The stress sensitivity of magnetization K, defined as the change in magnetization per unit magnetization per unit stress, can be expressed in the form

$$\Delta\mathbf{I} \approx K\sigma \cdot \mathbf{I}, \qquad \text{(Eq. 1)}$$

where $\Delta\mathbf{I}$ is the change in magnetization in a body with net magnetization $\mathbf{I}$ due to a deviatoric stress σ, K the stress sensitivity, typically has values of about 3×10^{-3} MPa^{-1}. The stress sensitivity of induced and remanent magnetization from theoretical and experimental studies has been combined with stress estimates from dislocation models of fault rupture and elastic pressure loading in active volcanoes to calculate magnetic field changes expected to accompany volcanic activity. These volcanomagnetic models show that moderate scale magnetic anomalies of several nanoteslas (nT) should accompany moderate to large volcanic eruptions for rock magnetizations and stress sensitivities of 1 Amperemeter^{-1} (A m^{-1}) and 10^{-3} MPa^{-1}, respectively (Johnston, 1997).

Electrokinetic effects

Electrokinetic electric and magnetic fields result from fluid flow through layered volcanic rocks as ions anchored to solid material induce equivalent ionic charge of opposite sign in the fluids. The fluid flow itself is thermally and gravitationally driven. Fluid flow in this system transports the ions in the fluid in the direction of flow, producing electric potential of several tens of mV km^{-1} and magnetic fields of a few nanoteslas.

The current density $\mathbf{j}$ and fluid flow rate $\mathbf{v}$ in porous media are found from the coupled equations

$$\mathbf{j} = -s\nabla E - \frac{\xi\zeta\nabla P}{\eta}, \qquad \text{(Eq. 2)}$$

$$\frac{\phi\xi\zeta\nabla E}{\eta} \quad \frac{\kappa\nabla P}{\eta} \qquad \text{(Eq. 3)}$$

where E is streaming potential, s is the electrical conductivity of the fluid, ξ is the dielectric constant of water, η is fluid viscosity, ζ is the zeta potential, ϕ is the porosity, κ is the hydraulic permeability, and P is pore pressure. Unfortunately, many of these parameters are poorly known within volcanoes.

Electrokinetic models combined with estimates of electrical conductivity structure within volcanoes have been proposed to explain the curious "W" form of electric potential commonly observed on active volcanoes (Ishido, 2004).

Electrical resistivity

Changes in electrical resistivity of volcanic rocks result from changes in strain, temperature, chemistry, and fluid content. While little theoretical research has been done on the details of resistivity changes within volcanoes and methods to separate the different contributions, there is no question that resistivity changes occur and are observed during eruptions (e.g., Yukutake *et al.*, 1990; Zlotnicki *et al.*, 2003). Measurement methods include magnetotellurics (MT) and Schlumberger, Wenner, DC/DC, and dipole/dipole systems.

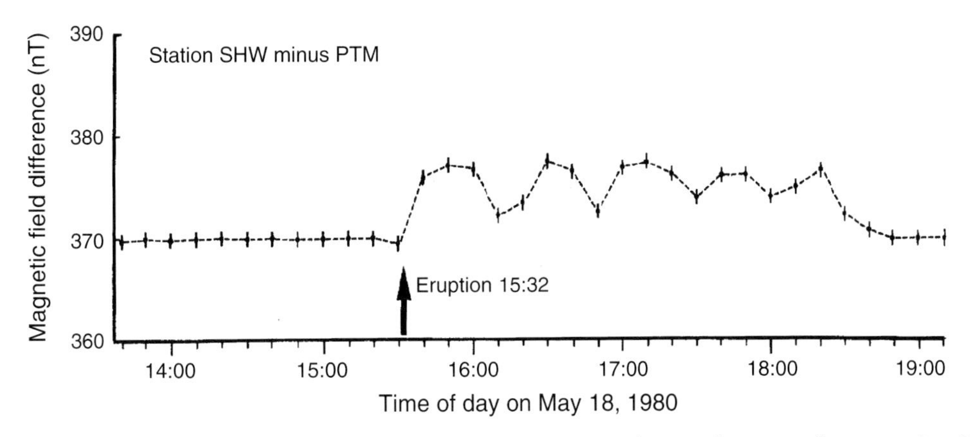

Figure V9 Total magnetic field differences between stations SHW, on Mount St. Helens, and remote reference station, PTM, at Portland Oregon, for the several hours preceding and following the 18 May 1980 catastrophic eruption. Time in UTC (from Johnston *et al.*, 1981).

Thermal remagnetization and demagnetization

Rocks lose their magnetization when temperatures exceed the Curie Point (≈580 °C for magnetite) and become remagnetized again as the temperature drops below this value. At shallow depths in volcanic regions, particularly in recently emplaced extrusions and intrusions, cracking from thermal expansion allows gas and fluid movement to rapidly transport heat. Large local anomalies of hundreds of nanoteslas from remagnetization of the upper few tens of meters can be generated quickly (Dzurisin *et al.*, 1990). At greater depth these effects occur on very long timescales because of the small thermal conductivity of crustal rocks.

Observations

Volcanomagnetic effects

Unambiguous observations of magnetic field changes have been obtained on many active volcanoes. The best examples clearly related to volcanic activity were obtained during (1) the 18 May 1980, eruption from Mount St. Helens, (2) the 1987 eruption sequence from Piton de la Fournaise, Reunion Island, (3) the 16–22 November 1986, eruption of Izu-Oshima volcano, Japan, (4) the 1980–1990 intrusion episode in Long Valley caldera, California, (5) the 2000 eruption of Miyake-Jima, and (6) the 2003 eruption of Mt. Etna, Italy. Figure V9 shows an example of data from Mount St. Helens during the first stage of the eruption on 18 May 1980. A final magnetic field offset of 9 ± 2 nT was permanent and consistent with pressure decrease within the mountain at the time of the eruption. The net positive signal contrasts with the negative change expected from the removal of magnetic material from the mountain (Johnston, 1997). The oscillatory signals are magneto-gas dynamic effects that couple into the ionosphere and propagate many thousands of kilometers round the world. Figure V10 shows this signal propagating across a 20-station magnetometer array between 1000 and 1500 km from the volcano. These TIDs are commonly generated by explosive volcanic eruptions and earthquakes.

Volcanoelectric effects

A number of important new electric field experiments have focused on obtaining measurements on, or near, active volcanoes. In particular, impressive self-potential observations have been made that delineate the changing electric potentials around erupting volcanoes. SP anomalies show apparent correlation with episodes of intrusive activity, active venting, and eruptions. The pertinent physical mechanisms in this case are electrokinetic effects, fluid vaporization and thermoelectric effects. These are all generated by the injection of hot magma, gas, and hydrothermal fluids into the volcano.

The best documented examples are recorded on:

1. Unzen, Japan, during 1990–1992 where large positive self-potential anomalies were identified in the vicinity of the new lava dome. Potentials generated by fluid flow in the hydrothermal system beneath the array is suggested as the most likely cause for this anomaly.

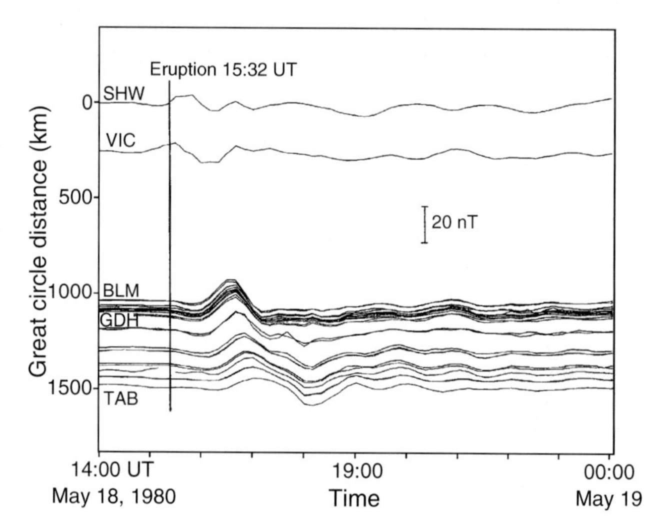

Figure V10 Magnetic field as a function of time and great circle distance following the 18 May 1980 eruption of Mount St. Helens (from Mueller and Johnston, 1989).

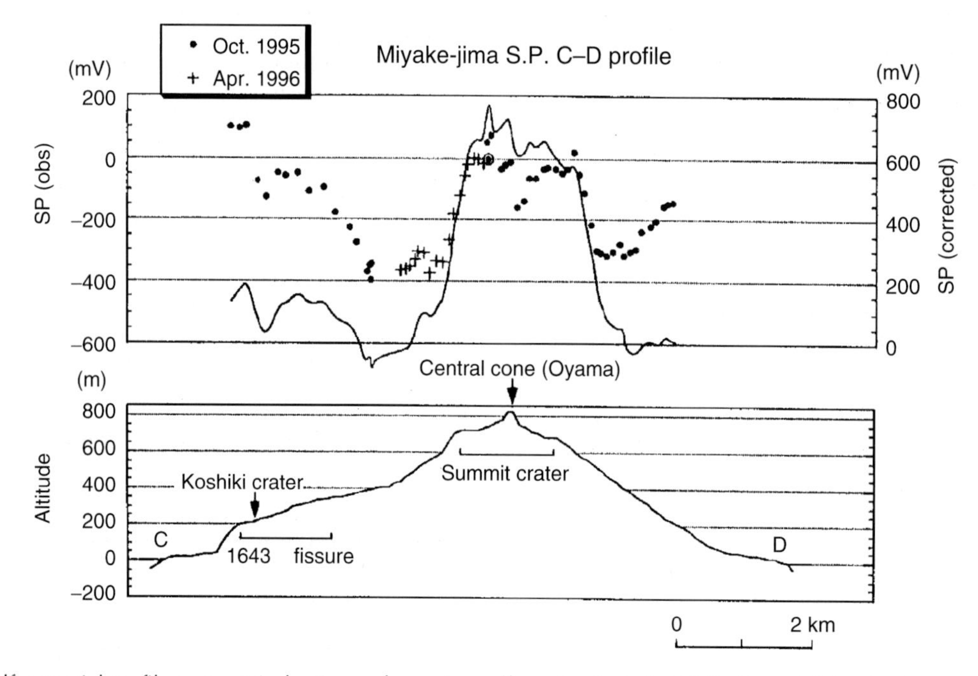

Figure V11 Self-potential profile across Miyake-Jima volcano Japan (from Sasai *et al.*, 1997). The solid line in the upper plot shows the data corrected for elevation effects.

2. Piton de la Fournaise volcano on Reunion Island, where similar positive self-potential anomalies have been observed above the active magma chamber.
3. Oshima Eruption, Japan, in 1986, where continuous measurements of vertical electric fields in the frequency range 0–3 KHz before, during and after a minor eruption from Mt. Mihara show a series of rapidly rising slow-decaying pulses. These effects are attributed to electrokinetic phenomena in the hydrothermal system.
4. Miyake-Jima, Japan, in 2000, where resistivity on scales of 0.5–1.5 km show up to a 300% change before eruptions. Immediately before the onset of caldera formation on 8 July 2000, self-potential near the summit reversed sign from positive to as much as −225 mV. Figure V11 shows the SP record from Miyake-Jima about 3 years before the eruption from Sasai *et al.* (1997). Zlotnicki *et al.* (2003) suggest the preeruption observations resulted from electrokinetic effects in a changing hydrothermal system. Remeasurements in 2003 indicate the SP anomalies are much reduced in amplitude and are localized to the edges of the new crater.

Malcolm J.S. Johnston

Bibliography

Dzurisin, D., Denlinger, R.P., and Rosenbaum, J.G., 2005. Cooling rate and thermal structure determined from progressive magnetization of the dacite dome at Mount St. Helens, Washington. *Bulletin of the Seismological Society of America*, **95**: 2763–2780.

Ishido, T., 2004. Electrokinetic mechanism for the “W”-shaped self-potential profile on volcanoes. *Geophysical Research Letters*, **31**: L22601 (doi: 10.1029/2004GL020964).

Johnston, M.J.S., 1997. Review of electrical and magnetic fields accompanying seismic and volcanic activity. *Surveys in Geophysics*, **18**: 441–475.

Johnston, M.J.S., Mueller, R.J., and Dvorak, J., 1981. Volcano-magnetic observations during eruptions May-August 1980. *US Geological Survey Professional Papers*, **1250**: 183–189.

Mueller, R.J., and Johnston, M.J.S., 1989. Large-scale magnetic field perturbations arising from the 18 May 1980, eruption from Mount St. Helens, Washington. *Physics of the Earth and Planetary Interiors*, **57**: 23–31.

Sasai, Y., Zlotnicki, J., Nishida, Y., Yvetot, P., Morat, P., Murakami, H., Tanaka, Y., Ishikawa, Y., Koyama, S., and Sekiguchi, W., 1997. Electromagnetic monitoring of Miyake-Jima volcano, Izu-Bonin Arc, Japan: a preliminary report. *Journal of Geomagnetism and Geoelectricity*, **49**: 1293–1316.

Yukutake, T., Yoshino, T., Utada, H., Watanabe, H., Hamano, Y., and Shimomura, T., 1990. Changes in the electrical resistivity of the central cone, Mihara-yama, of Oshima volcano observed by a direct current method. *Journal of Geomagnetism and Geoelectricity*, **42**: 151–168.

Zlotnicki, J., 1995. Geomagnetic surveying methods. In McGuire, W., Kilburn, C., and Murray, J. (eds.), *Monitoring Active Volcanoes*. London: University College London Press, pp. 275–296.

Zlotnicki, J., Sasai, Y., Nishida, Y., Uyeshima, M., Takahashi, Y., and Donnadieu, G., 2003. Resistivity and self-potential changes associated with volcanic activity: the 8 July 2000 Miyake-Jima eruption (Japan). *Earth and Planetary Science Letters*, **205**: 139–154.

Cross-references

Depth to Curie Temperature
Geomagnetic Spectrum, temporal
Gravity-Inertio Waves and Inertial Oscillations
Ionosphere
Magnetotellurics
Seismo-Electromagnetic Effects

VOYAGES MAKING GEOMAGNETIC MEASUREMENTS

Introduction

The magnetic *compass (q.v.)* has been used in Western Europe since the early or middle 12th century (see History of Geomagnetism). By 1187 a form with pivoted needle was described by Alexander Neckam, a monk at St. Albans, in his *De naturis rerum*. At this time, however, and for the next two and a half centuries, it was believed that the compass pointed toward true north. As long as this was the case, there was no reason to make or record any form of magnetic measurement.

Even when (probably sometime about the middle of the 15th century) it was realized that the magnetic needle did not, in general, point true north the situation with regard to making measurements did not change. The discrepancy in pointing was regarded more as a fault of the needle itself or of the method of magnetizing it rather than as being caused by some external agency.

And yet, some glimmering of the truth seems to have begun to dawn. Instrument makers in different parts of Europe began to compensate for the difference between true north and magnetic north (the magnetic variation or declination) by offsetting the needle from the true north point of the compass card by the appropriate angle for the place of manufacture. This worked pretty well near where the instrument had been made but caused problems when used further away. On his second voyage to the New World Christopher Columbus took compasses manufactured in Genoa and in Flanders. Flemish compasses had their needles aligned approximately 11.5° east of true north, this being the approximate value of declination at the time in Flanders; Genoese instruments had theirs aligned with true north since the declination was close to zero in the Mediterranean.

Columbus has been credited with the discovery of magnetic declination. Mitchell (1937) made a very detailed study of the surviving versions of the journals of Columbus's four transatlantic voyages and concluded that (a) Columbus was aware of the existence of declination before setting out on his first voyage, and (b) he may have discovered that it varies with position on the Earth's surface. The journals are so vague and have been edited so extensively that these conclusions are by no means certain. It would perhaps be stretching a point to say that Columbus's journals contain geomagnetic measurements; they do however contain comments and some rather obscure discussion of the behavior of his compasses.

As it became obvious that, for accurate navigation, declination had to be allowed for, so the practice developed of measuring and recording values of declination during a voyage. These measurements were useful at the time to navigators (usually of the same nationality or trading company) following the same or similar routes. Later, such knowledge was systematized by drawing declination contour charts using the measurements as a basis. Later still measurements of declination (and, eventually, other magnetic elements such as inclination and field strength) were collected by those wishing to produce charts and/or mathematical models of the historical geomagnetic field. Among such collections are those of Hansteen (1819), *van Bemmelen* (*q.v.*) (1899), *Sabine* (*q.v.*) (1868, 1872, 1875, 1877), Veinberg (1929–1933), and Veinberg and Shibaev (1969).

The most recent, and by far the most comprehensive, data collection is described by Jonkers *et al.* (2003). It incorporates the collections mentioned above as well as others. In this database, the measurements have been derived where possible from the original sources and errors in longitude, which could be very large before the late 18th century, have been corrected. Many of the sources are logbooks and journals (published and unpublished) kept on board ship during the voyages. Others are collections of voyages of a broader nature than the data collections discussed above: collections that summarize or give edited accounts of voyages. Two important examples are Hakluyt (1598–1600) and Purchas (1625).

Since the database of Jonkers *et al.* (2003) contains over 177000 measurements, of declination, inclination and field strength, from over 2000 voyages, space does not allow us to discuss every voyage that made magnetic measurements. We shall instead select a few particularly interesting voyages from each century from the 16th to the 20th, both inclusive.

Sixteenth century

The earliest voyage from which we have geomagnetic measurements is that of the brothers Jean and Raoul Parmentier. They sailed from Dieppe in March 1529, rounded the Cape of Good Hope and arrived in Sumatra in November, having made a brief stop in Madagascar. Soon after their arrival in Sumatra the brothers and many of their crew perished, possibly murdered, and the ships returned, with great difficulty, to France. During the outward voyage declination measurements were made in the South Atlantic and six of these were collected by van Bemmelen (1899), the earliest surviving measurements made during a sea voyage.

The earliest voyage to have scientific work as one of its planned objectives left Lisbon in the spring of 1538 bound for Goa in India. In command of one of the eleven ships that made up the fleet was Dom João de Castro who had been given the task of testing a new form of magnetic compass and its associated method of observation. Harradon (1943) describes the instrument and Harradon (1944) discusses de Castro's declination measurements, 38 made on the outward voyage and five on the return voyage. On the outward voyage de Castro noted that iron objects such as cannon in the vicinity of his measuring position adversely affected some of his measurements. Later, he was the first person to investigate the local attraction of magnetic rocks (what would now be called an intense local magnetic anomaly). He was also able to cast severe doubt on the idea that different lodestones, used in the magnetization of compass needles, could result in different values of declination being measured.

The Portuguese had pioneered and, by the early 16th century, controlled the route to the Far East via the Cape of Good Hope; the Spaniards from their settlements in Central and South America controlled access across the Pacific. The seafarers of northern Europe tried to find routes around the northern fringes of North America and Eurasia, expending vast amounts of energy, and many lives, in searching for the North-West and North-East Passages. An early voyage to the north-east yielded six declination measurements. This voyage, in the years 1556 and 1557 was under the command of Stephen Borough, later Chief Pilot under Queen Elizabeth I. He had sailed with Richard Chancellor in 1553 as part of the earliest English attempt to find the North-East Passage, an attempt that failed to find the Passage but did pioneer a new route to Moscow via the White Sea.

One of the earliest voyages to search for a North-West Passage was that of Martin Frobisher in 1576. He entered Frobisher Bay and followed it inland sufficiently far to convince himself (wrongly) that this was the Passage. Because of the lateness of the season he returned home, taking with him some samples of rock that he hoped might contain gold. This voyage yielded three declination measurements. On his next two voyages, in 1577 and 1578, the collection of rocks took precedence over exploration and magnetic measurements. The ore, alas, proved to be pyrites and did not contain any gold.

The early circumnavigations led by Magellan and Drake have not, apparently, provided any magnetic measurements, but that of Thomas Cavendish (or Candish) between 1586 and 1588 did (five declination measurements). This voyage was planned as a repeat of Drake's and, like Drake's, had an element of piracy about it.

Another English voyage that furnished declination measurements was essentially a privateering expedition in 1589 commanded by the third Earl of Cumberland with the intention of capturing any Spanish, Portuguese, or French ships it might meet. The fleet of seven ships ended up in the Azores and returned with more prizes than they could really manage. Edward Wright, a mathematician and expert on navigation, wrote an account of the voyage in which he included the 30 measurements that he managed to make, presumably in the intervals between the various fights that the ships were engaged in. Wright was the first person to develop the mathematical basis of Mercator's projection. Mercator had devised this map projection, of great use in navigation because a course at a constant bearing becomes a straight line on the chart, in an empirical way in about 1569.

The Dutch also tried to find the North-East Passage. Willem Barentsz made three voyages, in 1594, 1595, and 1596. On the last, in company with Jacob van Heemskerck, he discovered Bjørnøya (Bear Island) and Svalbard. He and his crew wintered on Novaya Zemlya but Barentsz died during the voyage home. The Dutch also began to challenge the Portuguese control of the East India trade via the direct route. A Dutch fleet under the command of Cornelis de Houtman sailed in 1595 and returned, after trading successfully, in 1597. This was followed in 1598 by a second Dutch fleet under Jacob van Neck. Magnetic declination measurements were made on these two Dutch East India voyages and on the third voyage of Barentsz. They, along with the measurements made by Borough, Frobisher, Cavendish and Wright, are included in the collections of van Bemmelen (1899) and Jonkers *et al.* (2003).

Seventeenth century

The English had broken the Portuguese monopoly in the East Indies a little earlier than the Dutch. Sir James Lancaster sailed in 1591 and returned laden with booty in 1594. As a result of this voyage the English East India Company was formed in 1600 and Lancaster was given command of the first Company fleet that sailed in 1601 and returned in 1603. Declination measurements were presumably made on these voyages but only one value has survived, from the 1601 to 1603 voyage. A potentially beneficial discovery was made on the latter voyage. Lancaster took with him bottles of lemon juice and administered three teaspoonfuls every morning to all his crew. His ship remained relatively free of scurvy compared with the other vessels in the fleet. The antiscorbutic effect of lemon juice, due to its vitamin C content, was, however, largely forgotten until the time of James Cook almost 200 years later.

The search continued for the North-West Passage. Two English expeditions visited Hudson Bay at about the same time, believing (wrongly) that there was an exit to the west. Luke Foxe (or Fox) sailed from London at the end of April 1631 and made a thorough exploration of the shores of Hudson Bay, the islands to the north and of Foxe Basin to the south of Baffin Island. He returned safely at the beginning of November without the loss of a single member of his crew. Thomas James sailed from Bristol a month later than Foxe, met and dined with him off the southern shore of Hudson Bay at the end of August and entered and explored James Bay at its southeast extremity. There, on the mainland near Charlton Island, James and his crew were forced to spend a very difficult winter, but they managed to repair their damaged ship and returned to Bristol in October 1632. The voyages of Foxe and of James both provided declination measurements in a part of the world where few, if any, had previously been obtained.

The East India Companies of England and Holland sent out fleets annually. These have provided large numbers of declination measurements, helped by the fact that the English East India Company ordered each captain, master, mate, and purser on their vessels to submit a journal at the end of each voyage. Two Dutch voyages that provided declination measurements were those of Willem Corneliszoon Schouten and Jakob Le Maire between 1615 and 1616, the first European voyage to round Cape Horn, and of Abel Janszoon Tasman between 1642 and 1643, the first European to visit Tasmania and New Zealand.

The discovery by *Henry Gellibrand* (*q.v.*) in 1634 that declination at a particular place varied with time (a phenomenon known as secular variation) made the collection of geomagnetic measurements even more important. It was now crucial to record the date of observation, something that had not always been done before.

It may not have escaped the attentive reader that all the magnetic observations mentioned so far have been of declination, despite the fact that *Robert Norman* (*q.v.*) first measured the angle of inclination or dip (the angle a freely suspended magnetic needle makes with the horizontal plane) in 1576. One reason is that inclination is more difficult to measure accurately than declination, and this is particularly so on board a moving ship. The earliest surviving inclination measurements made at sea are seven values measured by Benjamin Harry in the Atlantic and Indian Oceans during a voyage to the East Indies in 1680. This was followed by a more extensive set of 34 measurements made by James Cunningham, a Scotsman more famous as a botanist, on a voyage to China for the English East India Company in 1700.

We cannot leave the 17th century without mentioning the voyages of *Edmond Halley* (*q.v.*). These were the first voyages planned with scientific endeavors as the main part of their program. With the backing of the Royal Society Halley persuaded the British Admiralty to build a ship in which he could survey the Atlantic Ocean making measurements of declination and, as a further aid to navigation, of the longitude of his ports of call. The ship was named the *Paramore*, Halley was commissioned as a Captain in the Royal Navy and was given command of the vessel. The first voyage, from October 1698 to July 1699, was cut short because of trouble with Halley's officers. A second voyage was therefore ordered. This lasted from September 1699 to September 1700 and surveyed the Atlantic as far south as latitude 52°41′S. During the two Atlantic voyages Halley made 187 measurements of declination (though, strangely, none of inclination) and used them as the basis of the first published contour chart, for the Atlantic Ocean, published in 1701. In the summer of this year Halley made a third voyage in the *Paramore*, sailing around and across the English Channel studying the tides. He also, when opportunity arose, measured declination, 13 values in all. The logbooks of Halley's three voyages have been edited by Thrower (1980).

Eighteenth century

Although by no means commonplace, voyages of circumnavigation became more numerous during the 18th century. Among those that provided geomagnetic measurements were those of Woodes Rogers between 1708 and 1711; Jacob Roggeveen (1721–1722); George Anson (1740–1744); John Byron, grandfather of the poet, (1764–1766); Samuel Wallis (1766–1768) and Philip Carteret (1766–1769); Louis de Bougainville (1766–1769); the three voyages of James Cook undertaken between 1768 and 1780; Jean-François de la Pérouse (1785–1788); Antoine d'Entrecasteaux (1791–1794); and George Vancouver (1791–1795).

The voyages of Rogers and Anson were repetitions of Drake's circumnavigation in that one of their primary purposes was the harrying of Spanish possessions in South and Central America and the disruption of their sea-borne trade. Rogers' voyage was financed by a group of Bristol merchants; Anson's was an official naval project. Both were successful financially and politically. Rogers managed, in large measure, to avoid the scourge of scurvy but the disease almost wrecked Anson's ill-equipped expedition.

The voyages of Roggeveen, Byron, Wallis and Carteret, and Bougainville all had as one of their main objectives a search for Terra Australis Incognita, the great southern continent postulated by armchair geographers. All failed to find the continent but instead added to the list of Pacific islands known to Europeans.

Cook's first and second voyages also searched for the southern continent and their failure to find it after meticulous searching proved that, if it existed, it must lie entirely south of the Antarctic Circle. During the first voyage observations of a transit of the planet Venus across the Sun's disc were made from Tahiti. Cook's third voyage included a search for the western end of the North-West Passage north of the Bering Strait. All three voyages were noted for the accuracy of their positional data (the second and third carried chronometers made by Kendall and Arnold). As well as declination measurements inclination data were also collected.

The voyage of la Pérouse ended in tragedy. He left New South Wales in February 1788 and was never seen alive again. The relief expedition of d'Entrecasteaux helped to solve the mystery of the disappearance and also made some of the earliest magnetic intensity measurements, though these were in relative rather than absolute units. Beaglehole (1934) summarizes the voyages of discovery in the Pacific from Magellan's circumnavigation to the end of the 18th century.

Nineteenth century

Polar exploration flourished during the 19th century, spurred on by commercial interests such as the whale fishery and seal hunting. The search for the North-West Passage also continued.

Naval voyages led by John Ross in 1818 and by William Edward Parry in 1819–1820, 1821–1823, 1824–1825, and 1827 (the latter an attempt to reach the North Pole across the ice from Svalbard) all provided geomagnetic observations, at sea and on land, inclination as well as declination. None of them achieved their major objective but greatly increased geographical knowledge of the Arctic. From magnetic observations made at Parry's winter quarters during his first three expeditions it became obvious that the North Magnetic Dip-pole (where the dip attains its maximum value of 90°) was to the south of the archipelago of islands that lie to the north of the Canadian mainland.

John Ross, having failed to persuade the Admiralty to give him command of another attempt to find the North-West Passage, found a private sponsor: the gin distiller Felix Booth. Accompanied, as second-in-command, by his nephew James Clark Ross, who had been on the elder Ross's previous expedition and on all four of Parry's, Ross established winter quarters on a peninsula that he named, in honor of his sponsor, Boothia Felix. The expedition ship, a paddle steamer named *Victory*, was beset by ice and the explorers had to spend four winters on or near Boothia Felix. During June 1831 the younger Ross led a sledging party to the North Dip-pole, which he claimed for King William IV. Other, more useful, magnetic measurements were also made.

Antarctic waters also began to attract attention, largely because they were rich in sea mammals such as seals and whales. Voyages were also sent out with more scientific aims. The Russian expedition under Baron von Bellingshausen, during his circumnavigation of Antarctica between 1819 and 1821, was the first to sight the continent. He also made geomagnetic measurements.

During the years 1837–1843 three expeditions were exploring the Antarctic: a French Expedition under Captain Dumont d'Urville between 1837 and 1840, the US Exploring Expedition under Lieutenant Charles Wilkes (1838–1842) and the British National Expedition to Antarctic Seas under James Clark Ross (1839–1843). D'Urville discovered Adélie Land and Wilkes also charted the coast of the continent in this same region. Both made geomagnetic observations but neither expedition landed on Antarctica.

Ross's expedition was largely devoted to geomagnetic studies. Three permanent and three temporary magnetic observatories were established during the voyage. Ross's personal objective was to be the first person to set foot on both North and South Magnetic Dip-poles. He was thwarted in this ambition because the magnetic measurements made at sea clearly indicated that the South Dip-pole lay well inland, most probably beyond the mountains of Victoria Land. He did, however, discover the Ice Shelf that now bears his name.

Another unsuccessful attempt to find the North-West Passage was led by Sir John Franklin. It sailed in 1845 but was beset in the ice off King William Island and every member of the expedition perished. Although no results of geomagnetic measurements survived, the expedition is important to our story because of the large number of expeditions that were sent out over the next three decades to try to find news or relics of Franklin; many of these search expeditions did make geomagnetic measurements in regions where such data were scarce or nonexistent.

Two important oceanographic expeditions included geomagnetic measurements in their observational programs. Both the British *Challenger* Expedition between 1872 and 1876 and the German Expedition on board SMS *Gazelle* (1874–1876) made hundreds of measurements of declination, inclination and total intensity during their circumnavigations. And now, thanks to Gauss's pioneering work (Gauss, 1833), the intensity measurements were in absolute units (see Gauss' determination of absolute intensity).

Twentieth century

The North-West Passage was finally discovered and traversed by Roald Amundsen between 1903 and 1907. During the winters of 1903/1904 and 1904/1905 a magnetic observatory was operated at Gjöahavn on King William Island. The measurements made there served to fix the position of the North Magnetic Dip-pole for the first time since Ross's visit in 1831. A second observatory was established at King Point near the Canada-Alaska border during the winter of 1905/1906.

Interest in exploring the Antarctic continent began to increase during the final years of the 19th century, but the early years of the 20th century saw a veritable assault on the South Pole in particular and Antarctica in general. Over a dozen major expeditions visited the southern polar regions between 1900 and 1918 and the majority made geomagnetic measurements during the voyage south or on Antarctica or both. Somewhat surprisingly Amundsen's 1910–1912 expedition, which was the first to attain the South Pole, was not one of these. On Shackelton's British Antarctic Expedition of 1907–1909, David, Mawson, and Mackay became the first to visit the South Magnetic Dip-pole. Mawson retained his interest in the dip-pole and during his Australasian Antarctic Expedition of 1911–1914 detailed geomagnetic measurements by Webb enabled a revised position to be calculated, although adverse conditions prevented the party from reaching it.

Of much greater practical importance was the global magnetic survey undertaken by the Carnegie Institution of Washington under the leadership of *Louis A. Bauer* (*q.v.*) from 1905 onward. The oceanic part of this was begun using a converted brigantine, the *Galilee*, and was continued by the specially built nonmagnetic ship *Carnegie* (*q.v.*) from 1909 until her destruction by fire in 1929.

Plans by the UK Government to build a nonmagnetic ship, the RRS *Research*, to carry on the work of the *Carnegie* were delayed by World War II and were eventually abandoned after the War. Geomagnetic measurements of the complete field vector were continued, however, by the Russian nonmagnetic ship *Zarya* in a series of cruises beginning in 1956.

The development of the proton precession magnetometer revolutionized marine magnetic surveying. This type of magnetometer does not need to be accurately oriented with respect to the geomagnetic field and can be towed in a container behind a ship so that it is not affected by the intrinsic magnetic field of the ship. A disadvantage is that it is only able to measure the total intensity of the geomagnetic field rather than the complete vector. Proton magnetometer surveys have therefore been mainly used for mineral exploration and for studying magnetic anomalies associated with plate tectonics. Measurements of total intensity can be used to supplement vector measurements in producing charts and models of the main geomagnetic field and they have proved to be a valuable source of data, particularly over large areas of deep ocean were few other measurements exist.

So sparse have such measurements been in the past that routine observations made to check the reading of the magnetic compass on board merchant vessels have also been used in the production of global magnetic charts and models. These measurements, although of lower accuracy than those made on board specially designed magnetic survey vessels, have been invaluable in ensuring that the charts and models were of acceptable accuracy in remote ocean regions such as the Southern Ocean and the central Pacific Ocean.

Discussion

The Portuguese and Spanish were the first to venture across the major oceans in search of new trading opportunities and the conquest of newly discovered territories. It is therefore strange that very few Iberian voyages feature in the datasets of magnetic measurements that have been collected. van Bemmelen (1899) pointed out the potential riches that he thought should be available in the Spanish and Portuguese archives but was not able to confirm that this was the case.

Jonkers and his colleagues (2003) have made comprehensive searches in seven of the main Spanish archives. Apart from material relating to Malaspina's expedition in the Atlantic and Pacific between 1789 and 1794, which supplemented observations already published, they found disappointingly few logbooks. In particular, they had hoped to find a manuscript legacy of the Manila galleons that, over many decades, sailed annually across the Pacific between the Philippines and South America. No logbooks from these voyages were found in the Spanish archives. Jonkers *et al.* (2003) speculate that manuscript sources containing magnetic measurements may exist in former Spanish colonies in South America or the Philippines.

In their discussion of possible future work on collecting magnetic measurements made on board ship Jonkers *et al.* (2003) mention the Portuguese archives as one of the most promising of hitherto untapped sources "even though much Portuguese material is known to be no longer extant." The remarks made above concerning possible archives in former overseas possessions also apply here. If found such sources may provide valuable early measurements in the Indian Ocean.

Mention has already been made of the improved accuracy of the measurements made by Cook on his three voyages in the third quarter of the 18th century compared with those made on other contemporaneous voyages. The biggest improvement was in the determination of position at sea, particularly that of longitude and this was primarily due to the employment of some of the first marine chronometers. Cook's use of these was experimental, part of the tests that proved the worth of this method for finding the longitude at sea, and it was some time before their use became general. The Hydrographic Office of the British Admiralty was founded in 1795 and, together with the Royal Observatory at Greenwich, began to test and issue marine chronometers. By the end of John Pond's time as Astronomer Royal in the 1830s complaints were being made that too much of his time and that of his Observatory's staff was being taken up with chronometer testing, to the detriment of the astronomical work. By the middle of the 19th century most major navies and merchant marines were using chronometers and the accuracy of positional data approached modern values.

Conclusion

Since the earliest European voyages of exploration the importance of the magnetic compass as an aid to navigation has been realized along with the necessity to collect information about the behavior of the geomagnetic field that controls it. Beginning with measurements of the most obvious geomagnetic element, declination, observations of the other geomagnetic elements have gradually been added. Because about 70% of the Earth's surface is covered by oceans, measurements made at sea are of the greatest importance in making mathematical models and charts of the global geomagnetic field. This article has summarized the history of such measurements by highlighting a selection (no doubt very subjective) of the more important and interesting voyages that have made geomagnetic measurements since the beginning of the 16th century.

David R. Barraclough

Bibliography

Beaglehole, J.C., 1934. *The Exploration of the Pacific.* London: A & C Black.

van Bemmelen, W., 1899. Die Abweichung der Magnetnadel; Beobachtungen, Säcularvariation, Wert-und Isogonensysteme bis zur Mitte des XVIIIten Jahrhunderts. *Observations of the Royal Magnetic and Meteorological Observatory, Batavia*, **21**(Suppl): 109 pp.

Gauss, C.F., 1833. *Intensitas vis magneticae terrestris ad mensuram absolutam revocata.* Göttingen: Dieterich.

Hakluyt, R., 1598–1600. *The Principal Navigations Voyages Traffiques & Discoveries of the English Nation Made by Sea or Over-Land in the Remote and Farthest Distant Quarters of the Earth at any Time Within the Compasse of these 1600 Years.* London: George Bishop, Ralph Newberie, and Robert Barker (Reprinted: 1903–1905 in 12 Volumes. Glasgow: J. Maclehose & Sons).

Hansteen, C., 1819. *Untersuchungen über den Magnetismus der Erde.* Christiania (Oslo): Lehmann & Gröndahl.

Harradon, H.D., 1943. Some early contributions to the history of geomagnetism—V. *Terrestrial Magnetism and Atmospheric Electricity*, **48**: 197–199.

Harradon, H.D., 1944. Some early contributions to the history of geomagnetism—VII. *Terrestrial Magnetism and Atmospheric Electricity*, **49**: 185–198.

Jonkers, A.R.T., Jackson, A., and Murray, A., [illegible] of geomagnetic data from historical records. *Reviews of Geophysics*, **41**(2): 1006 (doi: 10.1029/2002RG000115).

Mitchell, A.C., 1937. Chapters in the history of terrestrial magnetism. II. The discovery of the magnetic declination. *Terrestrial Magnetism and Atmospheric Electricity*, **42**: 241–280.

Purchas, S., 1625. *Purchas his Pilgrimes. In five books. The first, containing the voyages (...) made by ancient kings (...) and others, to and throw the remoter parts of the known world, etc.* 4 Parts. London: W. Stansby, for H. Fetherstone (Reprinted: 1905–1907 as *Hackluytus Posthumus (or) Purchas His Pilgrimes: Containing a History of the World in Sea Voyages and Land Travels by Englishmen and Others.* 20 Volumes. Glasgow: J. Maclehose & Sons).

Sabine, E., 1868. Contributions to Terrestrial Magnetism, No. XI. *Philosophical Transactions of the Royal Society of London*, **158**: 371–416.

Sabine, E., 1872. Contributions to terrestrial magnetism, No. XIII. *Philosophical Transactions of the Royal Society of London*, **162**: 353–433.

Sabine, E., 1875. Contributions to terrestrial magnetism, No. XIV. *Philosophical Transactions of the Royal Society of London*, **165**: 161–203.

Sabine, E., 1877. Contributions to terrestrial magnetism, No. XV. *Philosophical Transactions of the Royal Society of London*, **167**: 461–508.

Thrower, N.J.W. (ed.), 1980. *The Three Voyages of Edmond Halley in the "Paramore" 1698–1701*, Second Series, Nos. 156, 157. London: The Hakluyt Society.

Veinberg, B.P., 1929–1933. *Catalogue of Magnetic Determinations in USSR and in Adjacent Countries from 1556 to 1926, Parts 1, 2 & 3.* Leningrad (St. Petersburg): Central Geophysical Observatory.

Veinberg, B.P., and Shibaev, V.P. (Editor-in-Chief: Pushkov, A.N.), 1969. *Catalogue. The Results of Magnetic Determinations at Equidistant Points and Epochs, 1500–1940.* Moscow: IZMIRAN (English translation: No. 0031 by the Canadian Department of the Secretary of State, Translation Bureau, 1970).

Cross-references

Bauer, Louis Agricola (1865–1932)
Bemmelen, Willem van (1868–1941)
Carnegie, Research Vessel
Compass
Gauss' Determination of Absolute Intensity
Gellibrand, Henry (1597–1636)
Geomagnetism, History of
Halley, Edmond (1656–1742)
Norman, Robert (1560–1585)
Sabine, Edward (1788–1883)

W

WATKINS, NORMAN DAVID (1934–1977)

Norman Watkins (Figure W1) was an innovative and dynamic paleomagnetist who in just a few years attained international recognition and high praise for his work. As a student in geophysics and paleomagnetism during the period when the polarity timescale was first being developed he experienced first hand the plate tectonics revolution in the earth sciences. He was among the first to develop a modern paleomagnetism laboratory, initially with spinner magnetometers developed during the middle-late 1960s, and then adding one of the first cryogenic magnetometers to his laboratory in 1972. In his short lifetime he published over 100 refereed publications (and hundreds of other works), received six academic degrees, rose to the rank of professor of oceanography at the Graduate School of Oceanography, University of Rhode Island, was continuously funded by NSF from 1963 until his death in 1977 (grant funding starting before he received his Ph.D.), and served as the division director of the Earth Science Division of the National Science Foundation.

Figure W1 Norman Watkins, on board the USNS Eltanin in 1970.

Born in Sunbury-on-Thames in Surrey, England, in 1934, of Welsh parents, Watkins later became a naturalized citizen of the United States. His first college degrees were from the University of London where he received a B.Sc. in mathematics, geology, and geography in 1956, and a second B.Sc. in mathematics, physics, and "geology special (Part 1 only)" in 1957. He then attended the University of Birmingham, England, where he received an M.Sc. in applied geophysics in 1958. Watkins moved to Alberta, Canada, in 1958, where he enrolled at the University of Alberta, eventually receiving another M.Sc., this time in geology with a thesis titled *Studies in Paleomagnetism*. During his time in Alberta, he took a job with Shell Oil Company of Canada where he worked in various areas of geophysics (gravity, magnetic, and seismic interpretation) as well as geological mapping. In 1961, he moved to California where he accepted a position at Menlo College in Menlo Park, California, as a lecturer in physics. He served in that position until 1964 when he was named the acting dean of science at the college. He enrolled as a graduate student at Stanford University in 1962, serving there as a research assistant. At Stanford he worked with Allan Cox and Richard Doell in paleomagnetism, Richard Doell nominally being his advisor. In 1963, he received an appointment upgrade at Stanford to acting assistant professor in the Department of Geophysics, and in 1964, he became a research associate there. Because of very strong personal conflicts with Doell, Watkins moved his dissertation research problem back to the University of London where he defended and was granted his Ph.D. in geophysics in 1964. The title of his dissertation was *Paleomagnetism of the Miocene Lavas of Southeastern Oregon*. After receiving his Ph.D. he took a position as visiting senior fellow in the Chadwick Physics Laboratory at the University of Liverpool in England. In 1965, he returned to the United States as an associate professor in the Department of Geology at Florida State University in Tallahassee, Florida, where he received tenure in 1968. In 1970, he accepted the position of professor of oceanography at the Graduate School of Oceanography, University of Rhode Island (URI). In 1972, he was granted tenure there. Later still, he received a doctor of science degree from the University of Birmingham in recognition for his contributions to geophysics. He took a leave of absence from URI in 1976 to serve as the director of the Earth Science Division of the National Science Foundation, but after 8 months, illness forced his resignation from NSF and a return to URI.

During his career, Watkins mainly taught courses in geophysics and paleomagnetism. While he was at Florida State University he supervised six students who received their MS degrees and one Ph.D. student who finished after Watkins moved to Rhode Island. While at URI he supervised three Ph.D. students, one who graduated after his death, and five MS students. He also served as chief scientist on a cruise of the USNS Eltanin to the Indian Ocean and sent students out on cruises of the RV Trident (URI), RV Chain (Woods Hole Oceanographic Institution of MIT), and other ships as his representatives during various cruises. The primary aim of this work was to collect piston cores of deep-sea sediment, to dredge for marine basalts, and for other related oceanographic studies. Sediment samples from piston cores were returned to his laboratory for paleomagnetic measurement. These samples were also shared with James Kennett (a foraminifera specialist) and his students for biostratigraphic analysis. His relationship with Kennett started when they were professors at Florida State University. Both men moved to the Graduate School of Oceanography at URI in 1970, as a "package deal."

Watkins' contributions to research in paleomagnetism began with a paper published in *Nature* in 1961, where he was a co-author with Ernest Deutsch. This work examined the direction of the geomagnetic field recorded in Triassic rocks in Siberia, and was published at that exciting time in paleomagnetism during the early development of the Plio-Pleistocene polarity timescale. Another paper in *Nature* in 1963 examined the behavior of the Miocene geomagnetic field recorded in Columbia Plateau basalts and led to his important paper on the polarity transition identified at Steens Mountain. He was very concerned with the effects of oxidation on the magnetic properties of rocks. This concern was reflected in a number of papers with Stephen Haggerty and others published in *Nature*, *Science*, and other journals between 1965 and 1968.

In 1968, he published his first paper on the magnetic properties of deep-sea sediments recovered from piston cores. While he and his students continued to work actively on the magnetic properties of basaltic materials, this shift to marine sediments reflected a new direction in the overall field of paleomagnetism. It was facilitated by the availability at about this time of easy-to-use and reliable spinner magnetometers and adequate instruments that allowed rapid demagnetization of samples. While measurements on deep-sea sediments were important, the magnetometers were still only marginally sensitive enough for many samples. It was not until the availability of cryogenic magnetometers that many of these materials could be adequately studied. Therefore, he continued to concentrate primarily on basaltic rocks and polarity studies.

Much of his work in the 1970s involved measurements of basaltic rocks from oceanic islands and characterization of secular variation and polarity of Earth's magnetic field during the last 10–15 Ma. Watkins and his students collected samples from islands all over the world including Reunion Island, Kerguelen, and Amsterdam Island in the Indian Ocean, and the Azores Islands in the Atlantic. Much of this work was concentrated in Iceland with Leo Kristjansson (an Icelandic physicist and paleomagnetist) as well as with the late well-recognized physical volcanologist, professor George P. L. Walker, where they had revolutionized the understanding of the growth of volcanoes in Iceland by means of volcanic and magnetic stratigraphy, and with Ian McDougal (a good friend and potassium-argon dater from Australia), where his group collected thousands of cores from stratigraphic sequences of lavas all over the island. Many of these samples and his notes are archived at the Graduate School of Oceanography, URI.

Watkins was a very aggressive scientist who enjoyed stimulating debate. For example, he would always sit in the front row during presentations at national and international meetings and ask one question of every speaker that was designed to elicit discussion. He would do this while casually resting his arm on the chair next to him and turning sideways toward the audience, as if to get their opinion. His students observed that at these meetings he would usually be dressed in a three-piece suit. That is, except when it was time for him to give a presentation. Then he would appear in his field "traveling" cloths, a light blue dungaree shirt, tan pants, and a white ship-captain's hat with a black bill. Minutes later he would be seen again wearing his three-piece suit.

He was a very big man. One of his students once commented that when sitting at the back of a classroom on the first day of geophysics class he noticed there was something strange about the appearance of Prof. Watkins. The student suddenly realized that Watkins was standing with his left forearm on the top of the blackboard, bending down so that he could write on the board. "This guy is really big!" commented the student to a classmate. Watkins enjoyed impressing his students at Florida State University with the medal he earned throwing the Javelin for Wales at the 15th Commonwealth Games.

Besides being big, Watkins was also powerful. Basaltic paleomagnetic samples were collected by his group using chain saws that he had converted to rock drills. This was before such conversions were commercially available, and these drills were heavy. To impress his students and to demonstrate how the drilling method worked, he would drill several, perfectly straight, six-inch long cores, upward at a 45° angle using only one hand! On his best, in a single day he drilled and collected samples from all 55 lavas contained in a very steep, 1250 m thick stratigraphic sequence. His students were never able to keep up with him in the field.

When working in the field he was a very careful scientist and was always well prepared. He stressed back-up systems and redundant parts so that there would be no delays that might limit the number of samples returned to the laboratory. Early in his work he used a Sun compass to correct problems associated with local or anomalous magnetic field directions. To protect the collection and avoid problems he demanded that every evening the notes be recopied into a separate notebook, thus producing a redundant copy. He insisted that all samples be checked for marking errors or other problems. As a result there were very few ambiguities associated with these field collections.

In summary, Norman D. Watkins made important contributions to paleomagnetism during the early years of that developing field and helped to establish the techniques we use today. He actively stimulated thought and interaction among all who really knew him, and is missed by all of them. During his lifetime, Watkins and his wife Patricia had two sons, the oldest, Paul Watkins, currently living in the United States, became a successful teacher and author of many well-received novels. The youngest, Clive Watkins, now living in Luxembourg, received an MBA and became a successful businessman in Europe.

Brooks B. Ellwood

WESTWARD DRIFT

The time variation (secular variation) of the magnetic field was first recognized in the late 17th century. In attempting to provide a theory to explain it, Halley (1693) (qv) noticed that a large part of the secular variation could be explained in terms of a "westward drift" of the field. If maps of declination are produced over the Earth's surface at different epochs, then there is a clear westward movement of field patterns, particularly lines of zero declination (so-called *agonic lines*) (e.g., Langel, 1987). Using his map of Atlantic secular variation, Halley estimated that a full revolution of the field would take about 700 years, giving a rotation rate of just over 0.5° per year. To explain the drift, he posited a model of the interior of the Earth consisting of concentric shells of magnetic material rotating at varying rates with respect to the Earth's surface. In many ways, his theory has remarkable similarities to our current understanding. The Earth does indeed consist of different layers, and it is one of these layers (the fluid core) which gives rise to the secular variation, of which a large part can indeed be represented by westward motions at the surface of the core.

As understanding of the Earth's interior improved, it became clear that Halley's theory was inadequate both observationally and theoretically, not least because the thermal state of the Earth does not allow permanent magnetization at depth. With the advent of dynamo theory, and the understanding that the magnetic field originates from fluid motions in the molten iron core, the first "modern" theory of westward drift was proposed by Bullard *et al.* (1950). They argued that cooling of the Earth would lead to a pattern of large-scale convection in the core. As a parcel of fluid rose, it would move further from the Earth's rotation axis. Therefore, to conserve angular momentum, the angular velocity of the parcel would reduce. Similarly, a sinking fluid parcel would reduce its moment arm, and so need to increase its angular velocity. Bullard *et al.* suggested that this would lead to a net eastward flow with respect to the solid Earth at the base of the core near the inner core boundary, and westward flow at the core surface, which if advection by flow dominates diffusion (the *frozen-flux approximation* (qv)) would then carry the magnetic field in a westerly direction, giving rise to westward drift. While the basic physical ideas of this theory are very attractive, modeling of rotating convection has shown that the interaction of convection and rotation is much more complicated, and this pleasingly simple explanation for the drift does not work (although, interestingly, an eastward *inner core rotation* (qv) has more recently been suggested both from numerical dynamo models (qv) and seismological observations).

An alternative to Bullard's mechanism was provided by Hide (1966). While Bullard *et al.* had argued for an origin of westward drift in large scale flow, Hide instead suggested that it could arise as a result of wave motion. Magnetic waves called *Alfvén waves* (qv) were known to be supported in a fluid penetrated by a magnetic field. Adding the effect of rotation, additional families of diffusionless *magnetohydrodynamic waves* (qv) were shown to exist. Some have periods of order days (similar to inertial waves), but others, named magnetic Rossby waves or planetary waves, have much longer periods, for reasonable estimates of the magnetic field strength perhaps 300 years. Initial analysis based on propagation in a thin shell (similar to analysis for the atmosphere) unfortunately suggested that these waves would propagate eastward, not westward. Hide provided an intuitive argument, later confirmed by more detailed analysis, as to why in the core (a thick shell) such waves would propagate westward instead. Further examination has suggested that westward motion will dominate when the toroidal magnetic field is stronger than the poloidal magnetic field, as is thought to be the case within the core (Hide and Roberts, 1979).

However, such simple models, while attractive, can in no way explain the full complexity of secular variation. Estimates of westward drift obtained show considerable variation depending on just how it is defined—should it involve a fit to the whole field, just the equatorial dipole, the secular variation, or some combination? The value of Bullard *et al.* (1950) of 0.2° per year has entered the geomagnetism consciousness as a standard, but estimates have varied from 0.08° eastward per year, up to 0.733° westward (for a summary, see Langel 1987). What this range of values demonstrates is that the picture of westward drift is much too simplistic. Even in the 18th century, it was realized some features of the field drift northwards rather than westward, and once Gauss had developed a method for measuring magnetic intensity, the observed decay of the dipole could clearly not be explained by drift alone. Further, westward drift is much more clearly visible in Europe and North America than in Asia and the Pacific hemisphere. Yukutake and Tachinaka (1969) argued that other features in the field did not move at all, and proposed a division of the field into drifting and standing components. Yukutake (1969) further suggested that the equatorial dipole field could be separated into two components, drifting in opposite directions; such a process is reminiscent of wave motion, consistent with Hide's ideas. Wave theory further allows drift rate to vary with location, as the wave velocity is a function of the background field strength, which will vary over the core's surface.

However, this confused picture was largely swept away by the advent of the *Magsat satellite* (qv), and the impetus it gave to a program of high-resolution mapping of the field at the core-mantle boundary. If we want to understand the secular variation, we must really examine it at its source; upward continuation filters the field, complicating considerably the interpretation of the field at the Earth's surface. Field models constructed for different epochs, culminating in the *time-dependent main field model* (qv) "ufm" of Bloxham and Jackson (1992), allowed detailed features of field evolution to be examined at source. Their model from 1690 to 1990 shows apparent upwelling of field under Africa, followed by a westward motion under the Atlantic. However, such features are not observed under the Pacific, leading to the weak westward drift observed in the surface field in the Pacific hemisphere. Other features, particularly two pairs of high-latitude flux patches of opposite signs in the northern and southern hemisphere, are observed to remain stationary over historical time, contributing to Yukatake's standing field. Models of surface *core motion* (qv) have been generated, using the magnetic field as a tracer of fluid motion. While such models uniformly demonstrate a westward flow under the Atlantic, elsewhere the pattern is more complex, and many suggest eastward flow under the Pacific. An alternative view of drift has also recently been presented: instead of the drift of field or secular variation, the underlying flow pattern is considered steady but drifting either westward or eastward (Holme and Whaler, 2001). Such a model can explain the secular variation well with either a westward or eastward drift, again suggesting that the problem may be best posed in terms of wave motion.

In conclusion, the concept of westward drift has largely been superceded by more complex models of the secular variation, with maps of the field evolution at the core-mantle boundary allowing the secular variation to be examined at source. However, the simplicity of westward drift means that it is still used for interpretation of archeo- and paleomagnetic data. Because insufficient data have been available to enable calculation of high-resolution field models as has been done for historical time, we are limited to considering the change of the field at the Earth's surface. Therefore, interpreting changes in the field in terms of drift, and relating such drift to our understanding of the historical field, is still extremely valuable. Furthermore, the underlying controversy as to whether secular variation is dominantly generated by large scale core convection (as proposed by Bullard *et al.*) or wave motion at the surface of the core (as proposed by Hide) remains alive to this day.

Richard Holme

Bibliography

Bloxham, J., and Jackson, A., 1992. Time-dependent mapping of the magnetic field at the core-mantle boundary. *Journal of Geophysical Research*, **97**: 19537–19563.

Bullard, E.C., Freeman, C., Gellman, H., and Nixon, J., 1950. The westward drift of the Earth's magnetic field. *Philosophical Transactions of the Royal Society of London A*, **243**: 61–92.

Halley, E., 1693. On the cause of the change in the variation of the magnetic needle; with an hypothesis of the structure of the internal parts of the Earth. *Philosophical Transactions of the Royal Society of London A*, **17**: 470–478.

Hide, R., 1966. Free hydromagnetic oscillations of the Earth's core and the theory of the geomagnetic secular variation. *Philosophical Transactions of the Royal Society of London A*, **259**: 615–650.

Hide, R., and Roberts, P., 1979. How strong is the magnetic field in the Earth's liquid core? *Physics of the Earth and Planetary Interiors*, **20**: 124–126.

Holme, R., and Whaler, K., 2001. Steady core flow in an azimuthally drifting reference frame. *Geophysical Journal International*, **145**: 560–569.

Langel, R.A., 1987. The main field. In Jacobs, J.A. (ed.), *Geomagnetism,* Volume. 1. New York: Academic Press (Chap. 4).

Yukutake, T., 1979. Review of the geomagnetic secular variations on the historical time scale. *Physics of the Earth and Planetary Interiors*, **20**: 83–95.

Yukutake, T., and Tachinaka, H., 1969. Separation of the earth's magnetic field into drifting and standing parts. *Bulletin of the Earthquake Research Institute*, **47**: 65.

Cross-references

Alfvén Waves
Alfvén's Theorem and the Frozen Flux Approximation
Core Motions
Dipole Moment Variation
Halley, Edmond (1656–1742)
Inner Core Rotation
Magnetohydrodynamic Waves
Time-Dependent Models of the Geomagnetic Field
Upward and Downward Continuation

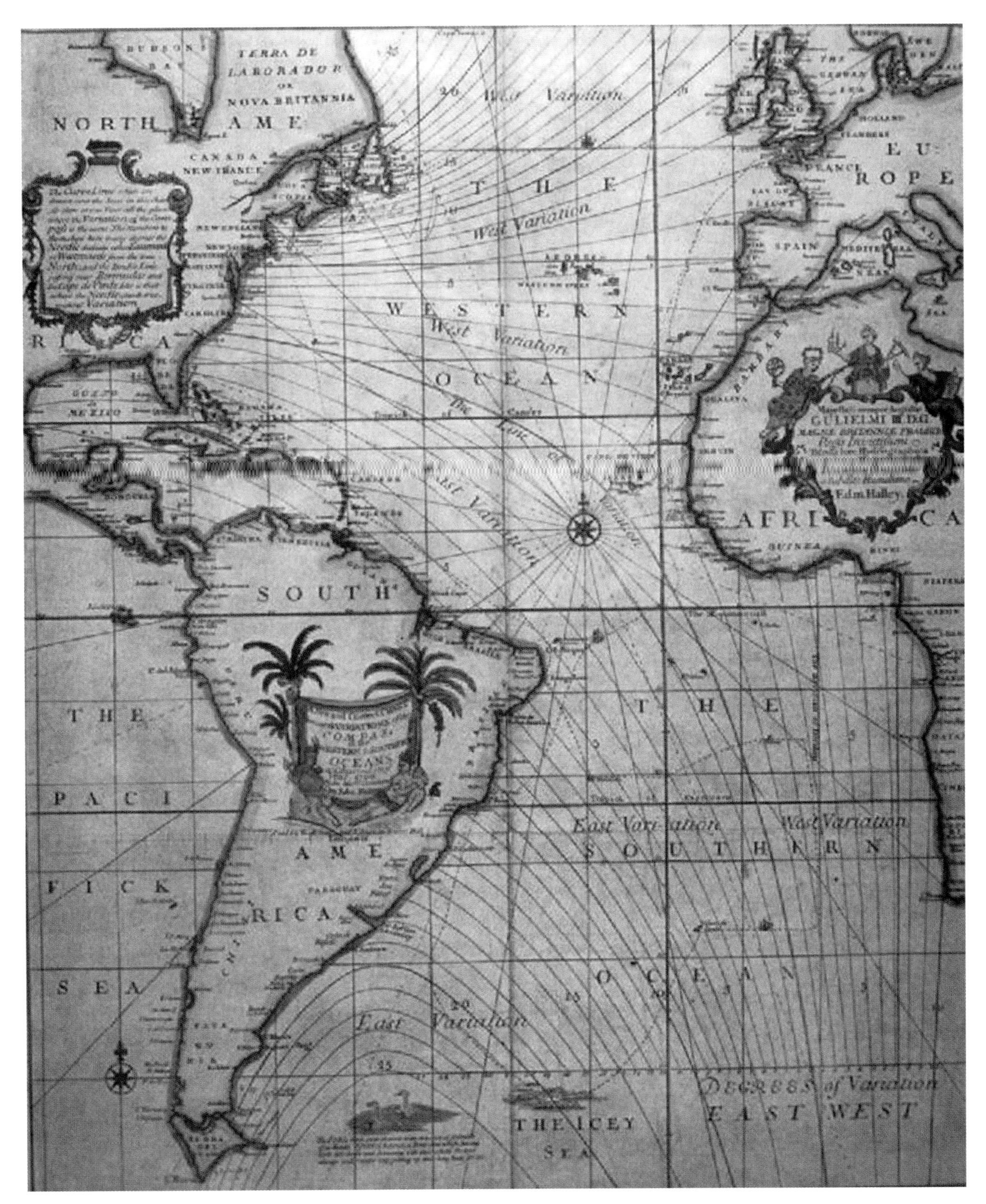

Plate 1 Halley's Atlantic chart showing contours of declination based largely on his voyages at the end of the 17th century. This is the version of the chart with the black cartouche over Africa containing a dedication to King William III. (*Figure M173, Main field maps*).

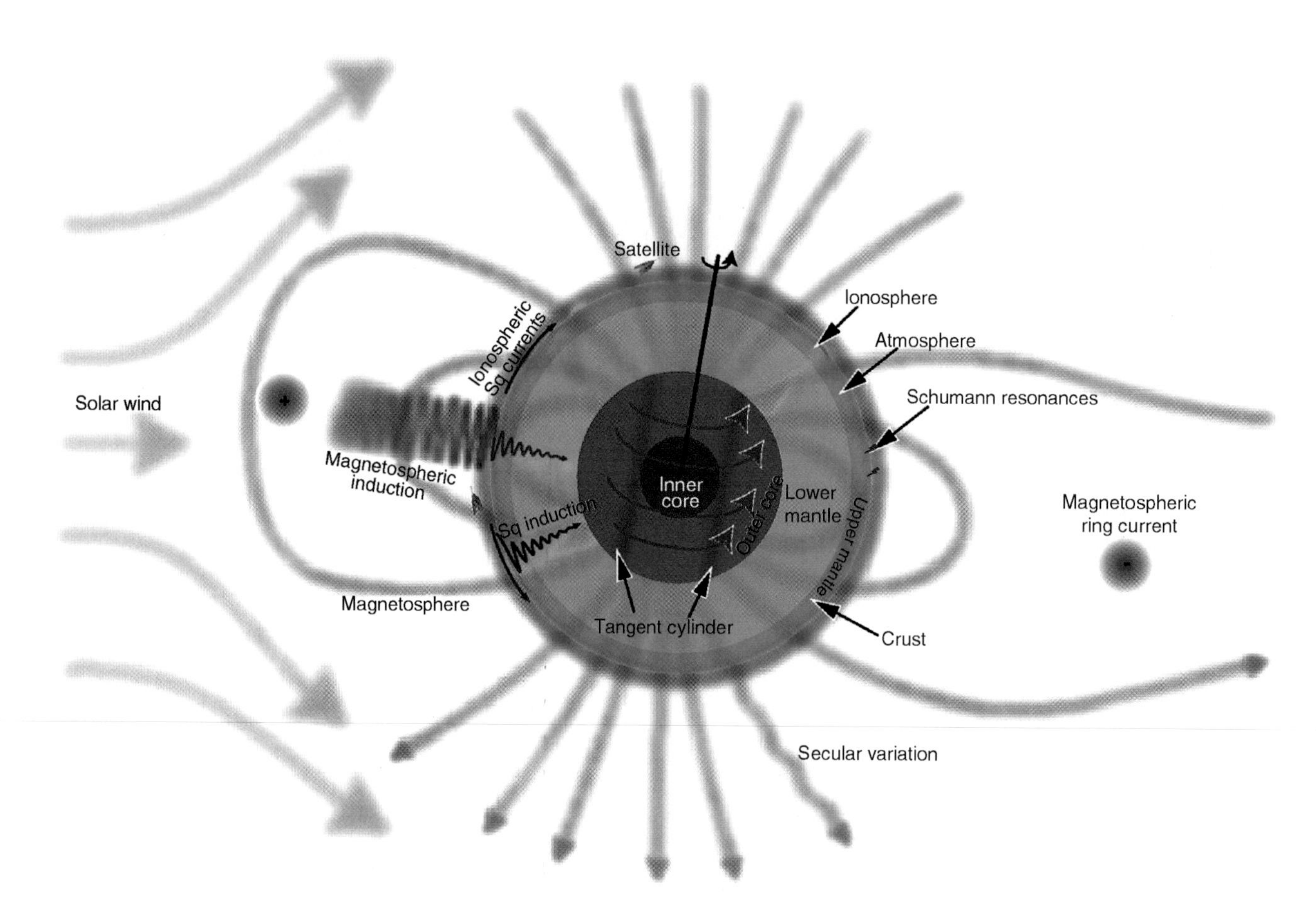

Plate 2 Earth's magnetic field is generated in the liquid *outer core,* where fluid flow is influenced by Earth rotation and the *inner core* geometry (which defines the *tangent cylinder*). Core fluid flow produces a *secular variation* in the magnetic field, which propagates upward through the relatively electrically insulating *mantle* and *crust.* The crust makes a small static contribution to the overall field. Above the insulating *atmosphere* is the electrically conducting *ionosphere,* which supports *Sq currents* as a result of dayside solar heating. Lightning generates high frequency *Schumann resonances* in the Earth/ionsophere cavity. Outside the solid Earth the *magnetosphere,* the manifestation of the core dynamo, is deformed and modulated by the solar wind, compressed on the sunside and elongated on the nightside. At a distance of about 3 earth radii, the *magnetospheric ring current* acts to oppose the main field and is also modulated by solar activity. Magnetic fields generated in the magnetosphere and ionosphere propagate by *induction* into the conductive Earth, providing information on electrical conductivity variations in the crust and mantle. *Magnetic satellites* fly above the ionosphere, but below the magnetospheric induction sources. (*Figure G50, Geomagnetic spectrum, temporal*).

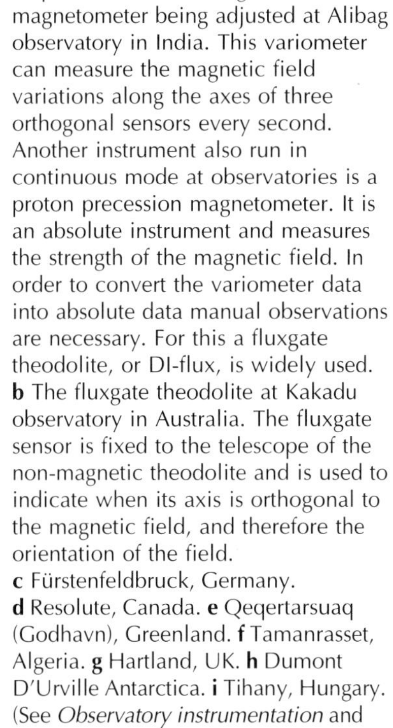

Plate 3 MAGNETIC OBSERVATORIES.
a The Danish Meteorological Institute suspended tri-axial fluxgate magnetometer being adjusted at Alibag observatory in India. This variometer can measure the magnetic field variations along the axes of three orthogonal sensors every second. Another instrument also run in continuous mode at observatories is a proton precession magnetometer. It is an absolute instrument and measures the strength of the magnetic field. In order to convert the variometer data into absolute data manual observations are necessary. For this a fluxgate theodolite, or DI-flux, is widely used.
b The fluxgate theodolite at Kakadu observatory in Australia. The fluxgate sensor is fixed to the telescope of the non-magnetic theodolite and is used to indicate when its axis is orthogonal to the magnetic field, and therefore the orientation of the field.
c Fürstenfeldbruck, Germany.
d Resolute, Canada. **e** Qeqertarsuaq (Godhavn), Greenland. **f** Tamanrasset, Algeria. **g** Hartland, UK. **h** Dumont D'Urville Antarctica. **i** Tihany, Hungary. (See *Observatory instrumentation* and *Observatories, overview.*) (*Figure O2, Observatories, overview*).

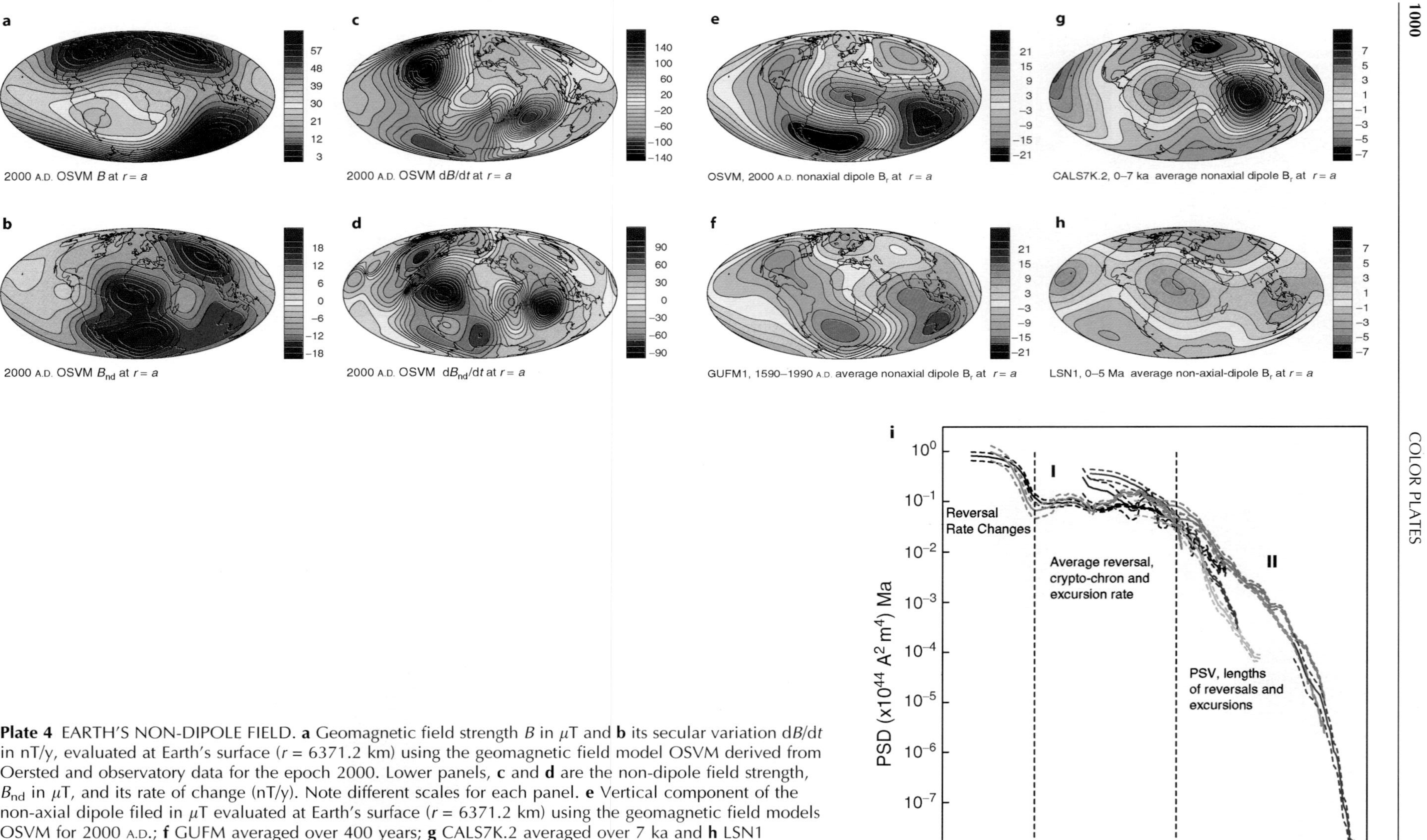

Plate 4 EARTH'S NON-DIPOLE FIELD. **a** Geomagnetic field strength B in μT and **b** its secular variation dB/dt in nT/y, evaluated at Earth's surface (r = 6371.2 km) using the geomagnetic field model OSVM derived from Oersted and observatory data for the epoch 2000. Lower panels, **c** and **d** are the non-dipole field strength, B_{nd} in μT, and its rate of change (nT/y). Note different scales for each panel. **e** Vertical component of the non-axial dipole filed in μT evaluated at Earth's surface (r = 6371.2 km) using the geomagnetic field models OSVM for 2000 A.D.; **f** GUFM averaged over 400 years; **g** CALS7K.2 averaged over 7 ka and **h** LSN1 (Johnson & Constable, 1997), from normal polarity lava and marine sediment directions averaged over 5 Ma. Note scales for g and h differ by factor of 3 from e and f. **i** Composite spectrum for the geomagnetic dipole moment constructed from the magnetostratigraphic reversal record with and without cryptochrons (frequencies 10^{-2}–20 Ma^{-1}), various sedimentary records of relative paleointensity (10^{0}–10^{3} Ma^{-1}), and from the dipole moment of a 0–7 ka paleomagnetic field model (10^{3}–10^{4} Ma^{-1}). Figure redrawn from Constable and Johnson (2005). (*Figure N4, Non-dipole field; Figure N6, Non-dipole field; Figure G52, Geomagnetic spectrum, temporal*).

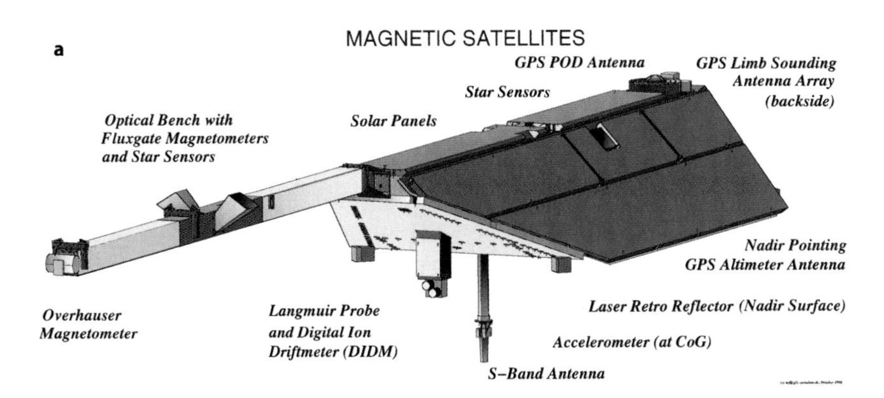

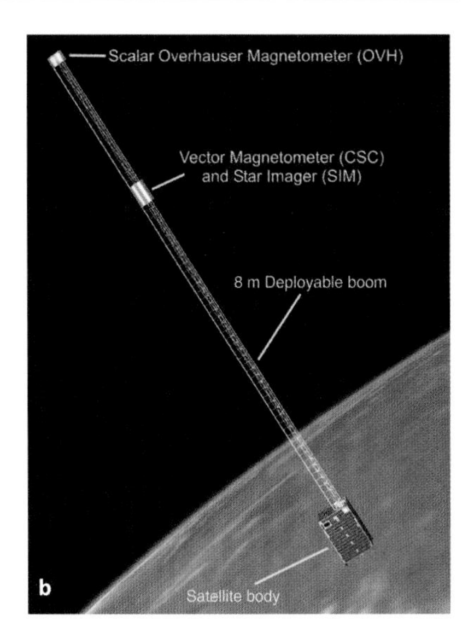

Plate 5 MAGNETIC SATELLITES. Magnetic satellites in operation in 2005. **a** CHAMP and **b** Ørsted. Satellites give the best current information about the Earth's main magnetic field. They are the only way we can determine the intermediate wavelength anomalies in the magnetic field, those larger than covered by conventional surveys and smaller than the main field from the core. (See *CHAMP*; *Ørsted*; *Magnetometers*; *Main field maps*).

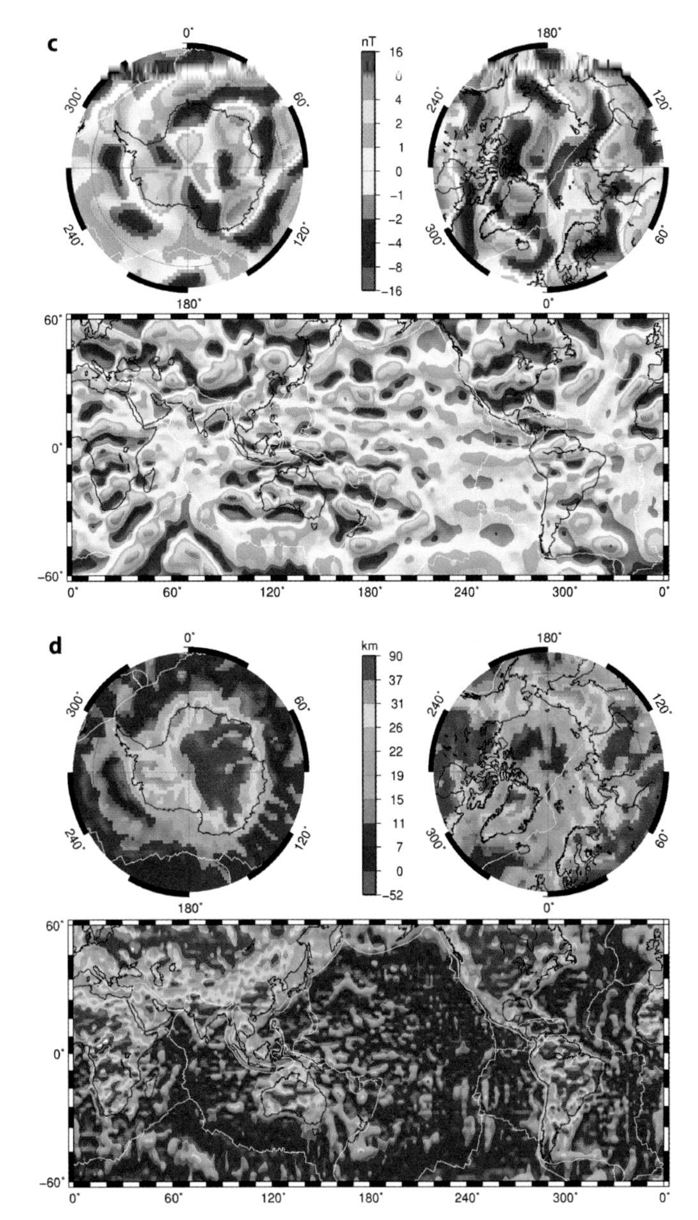

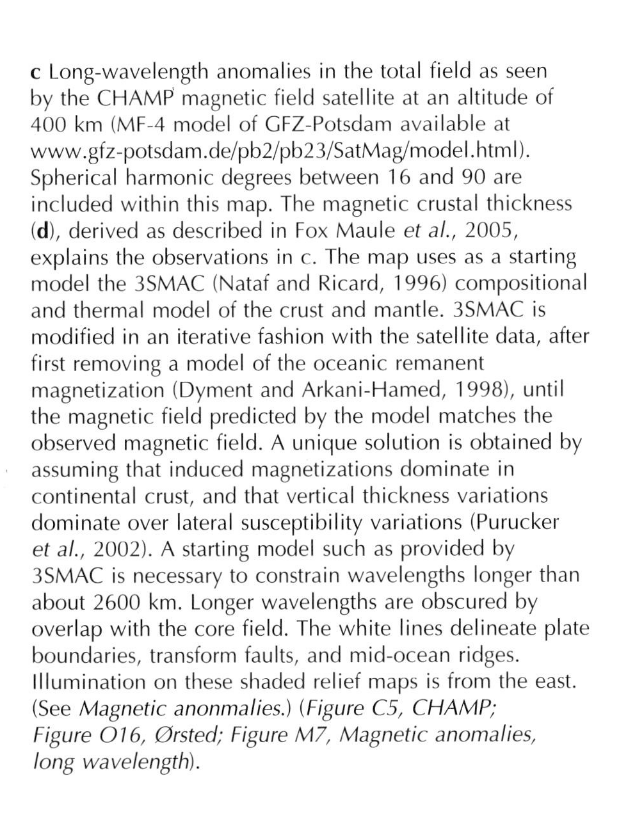

c Long-wavelength anomalies in the total field as seen by the CHAMP magnetic field satellite at an altitude of 400 km (MF-4 model of GFZ-Potsdam available at www.gfz-potsdam.de/pb2/pb23/SatMag/model.html). Spherical harmonic degrees between 16 and 90 are included within this map. The magnetic crustal thickness (**d**), derived as described in Fox Maule *et al.*, 2005, explains the observations in c. The map uses as a starting model the 3SMAC (Nataf and Ricard, 1996) compositional and thermal model of the crust and mantle. 3SMAC is modified in an iterative fashion with the satellite data, after first removing a model of the oceanic remanent magnetization (Dyment and Arkani-Hamed, 1998), until the magnetic field predicted by the model matches the observed magnetic field. A unique solution is obtained by assuming that induced magnetizations dominate in continental crust, and that vertical thickness variations dominate over lateral susceptibility variations (Purucker *et al.*, 2002). A starting model such as provided by 3SMAC is necessary to constrain wavelengths longer than about 2600 km. Longer wavelengths are obscured by overlap with the core field. The white lines delineate plate boundaries, transform faults, and mid-ocean ridges. Illumination on these shaded relief maps is from the east. (See *Magnetic anonmalies.*) (*Figure C5, CHAMP; Figure O16, Ørsted; Figure M7, Magnetic anomalies, long wavelength*).

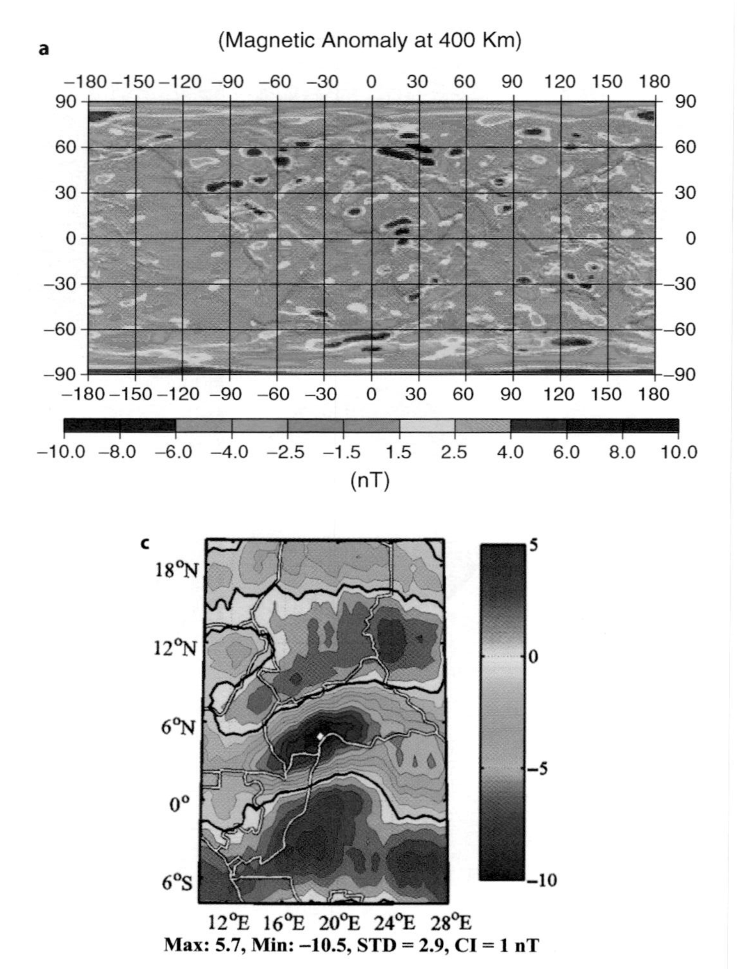

Plate 6 LITHOSPHERIC MAGNETISATION. **a** The magnetic anomaly, or residual magnetic intensity after subtraction of the main geomagnetic field, at a satellite altitude of 400 km derived from POGO and Magsat data. **b** The susceptibility distribution required within an assumed 40 km-thick spherical shell (the assumed upper layer of the lithosphere) required to produce the anomalies in the top figure. The surface topography is given in outline. Note the very large Bangui anomaly in the Central African Republic. Although larger magnetic anomalies are evident at high latitude, these are enhanced because the main field is strongest at high latitudes. The susceptibility map shows the Bangui anomaly is the strongest at this wavelength. **c** Close-up of the scalar magnetic anomaly at 400 km altitude using data from the recent CHAMP satellite (courtesy of Hyung Rae Kim, UMBC and NASA/GSFC). (See *Modelling magnetic anomalies; Bangui anomaly; Crustal magnetic field; POGO; Magsat; CHAMP*).

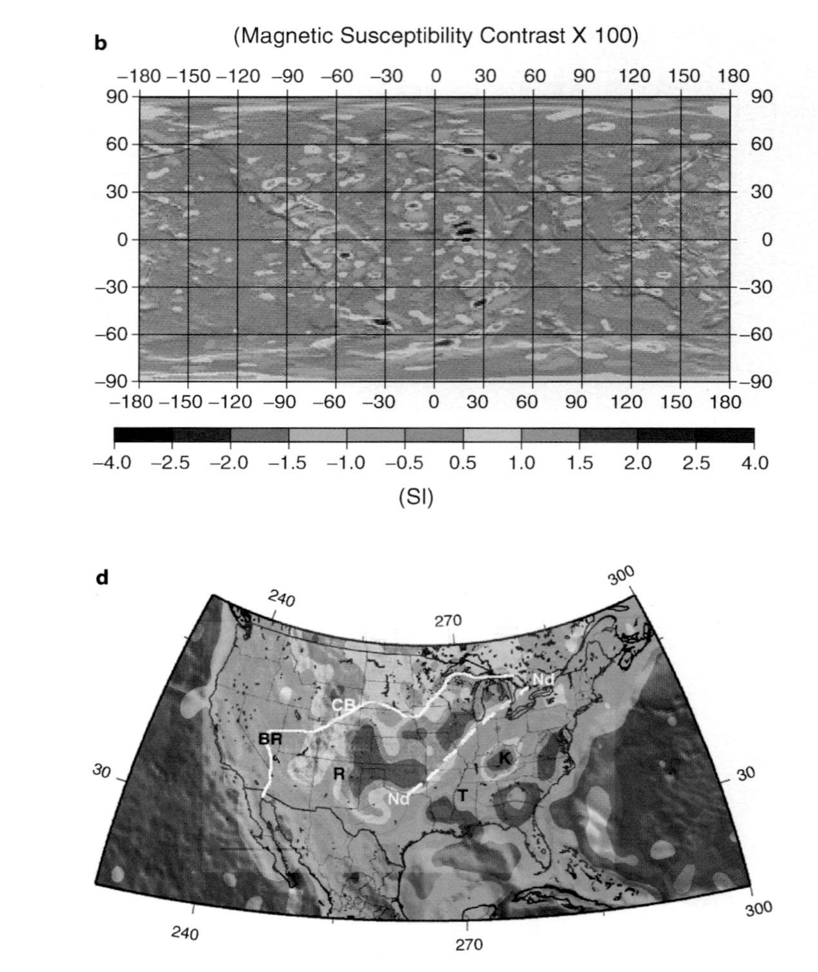

d Depth integrated magnetic contrasts over the United States. Orange/reds are the relative highs in the susceptibility variation and blue/light greens depict relative susceptibility lows. White continuous and dashed lines (Nd) show the Middle Proterozoic provinces in the United States. Mv is excluded from the Middle Proterozoic region of (Figure C47a) because presently this region is governed by the high heat flow related to the Basin and Range province (BR) which has decreased the magnetic thickness of the crust and hence it does not appear as a magnetic high. R- Rio Grande Rift; CB- Cheyenne Belt; K- Kentucky magnetic anomaly; T- Oklahoma-Alabama Transform. (See *Crustal magnetic field*). (*Figures M11, M12 Magnetic anomalies, modeling; Figure B6, Bangui anomaly; Figure C47, Crustal magnetic field*).

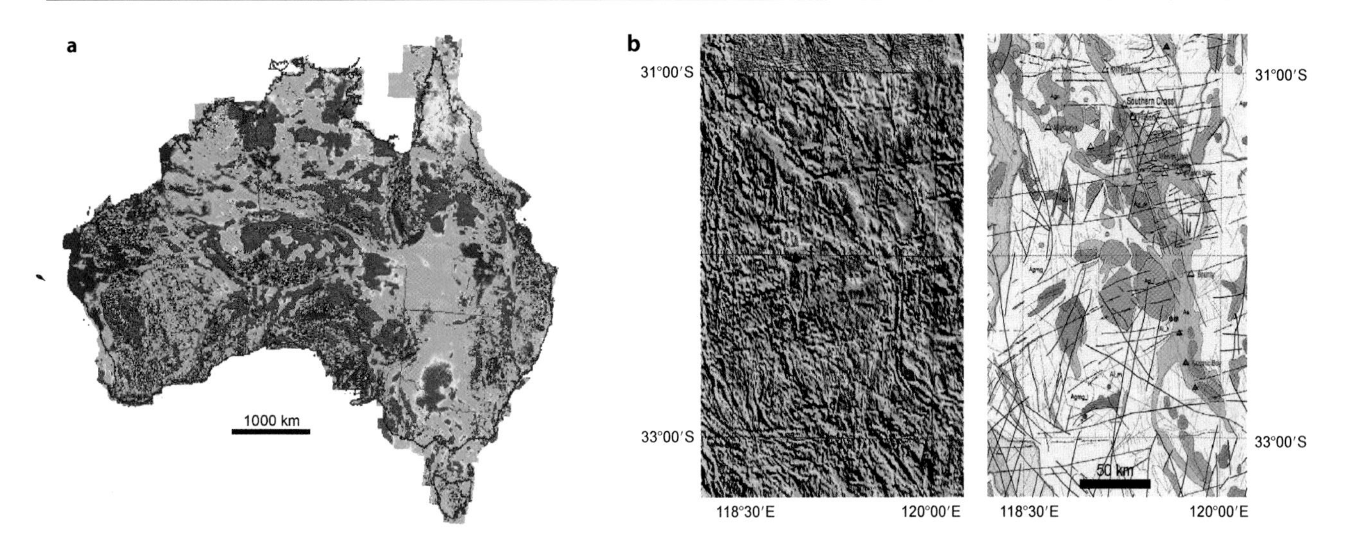

Plate 7 a Magnetic anomaly map of Australia, 3rd edition (Milligan and Tarlowski, 1999), courtesy of Geoscience Australia. **b** Left: Magnetic anomaly patterns over part of Western Australia recorded in various aeromagnetic surveys. Right: Geological interpretation. Courtesy of Geoscience Australia and the Geological Survey of Western Australia. (See *Magnetic anomalies for geology and resources; Crustal magnetic field; Magnetic anomalies, modeling*).

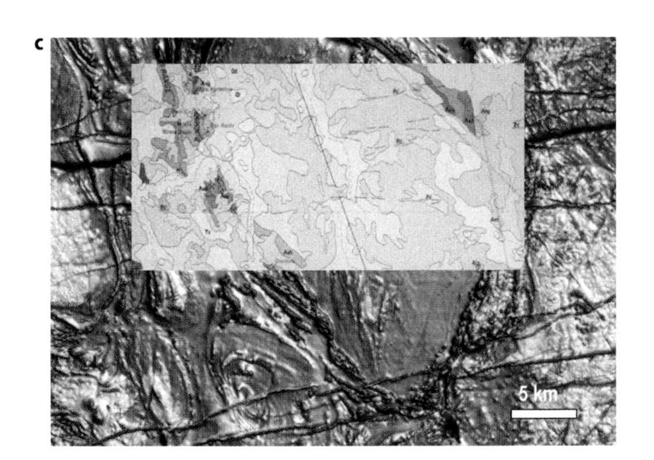

c A low-altitude, high resolution aeromagnetic map of a part of the Archaen craton in Western Australia depicting the power of remotely-sensed magnetic anomalies in peeling off the sedimentary cover. Pixel size is 25 m which makes it possible to observe subtle variations in the textures of the buried craton. The aeromagnetic image is courtesy of Fugro Airborne Surveys Limited. The example is courtesy of Colin Reeves. Blues and greens depict magnetic lows and yellows, orange and brick reds are magnetic highs. (See *Aeromagnetic surveying; Crustal magnetic field; Magnetic anomalies, modeling; Magnetic anomalies for geology and resources*).

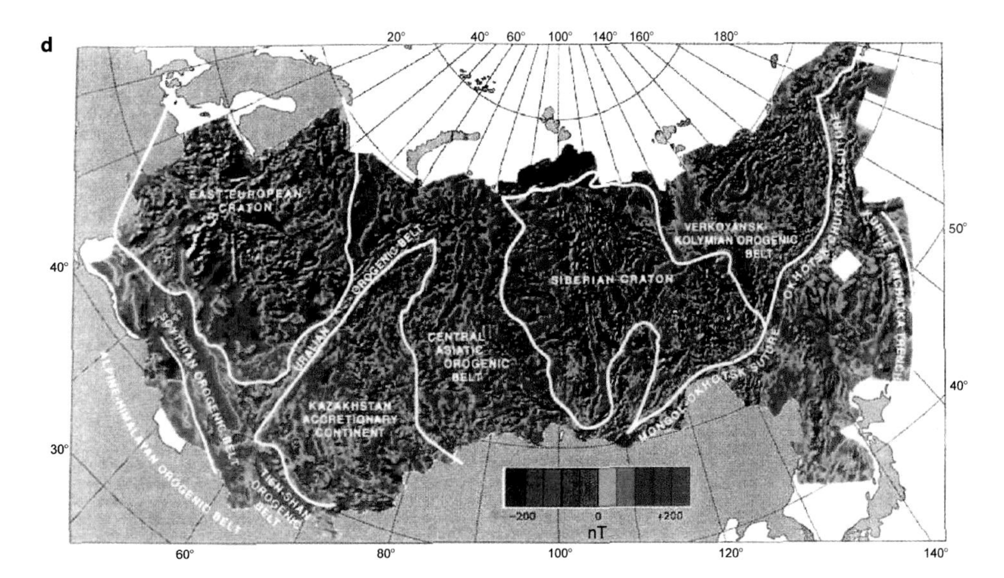

d Magnetic anomaly patterns recorded in aeromagnetic surveys over the former Soviet Union with the outlines of some major tectonic domains added. Russia. From Zonenshain *et al.*, 1991, courtesy of the American Geophysical Union. (See *Magnetic anomalies for geology and resources.*) (*Figures M1, M3, M4, Magnetic anomalies for geology and resources; Figure C46, Crustal magnetic field*).

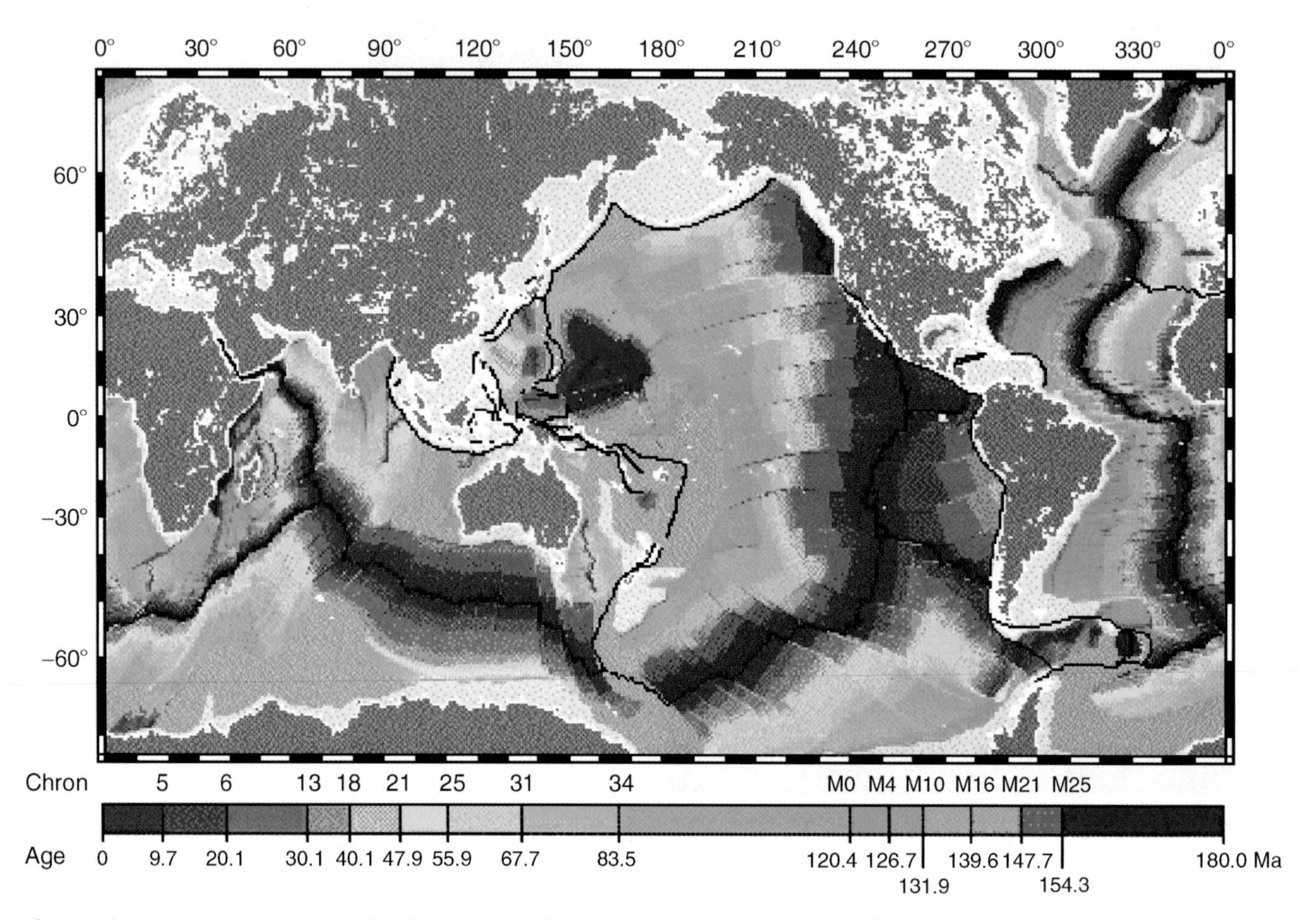

Plate 8 The magnetic stripes are used to date the ocean floor. As new plate forms at the mid-ocean ridges it becomes magnetized in the direction of the ambient magnetic field. As the geomagnetic field reverses the resulting magnetic anomaly changes abruptly at each reversal boundary. The ocean floor can be dated quite precisely by correlating the magnetic anomalies worldwide, identifying each reversal, and comparing it with an age scale for reversals. World Map of Isochrons of the Ocean Floor, after Muller, R.D. *et al.*, 2005. (See *Magnetic anomalies, marine*; *Magnetic surveys, marine*; *Magnetization, ocean crust*; *Geomagnetic polarity timescales*.) (*Figure M9, Magnetic anomalies, marine*).

Plate 9 MARINE MAGNETICS. **a** Photograph showing the deployment of a marine proton precession magnetometer from the stern of a research vessel, taken by Maurice Tivey, Woods Hole Oceanographic Institution. **b** Contour map of the western flank of the mid-Atlantic ridge near 26°N showing the magnetic anomaly stripes that characterize the seafloor over most of the ocean basins. Contour interval 50 nT. Thin black lines indicate ship's tracklines. Map based on data from Tivey and Tucholke (1998). **c** Calculated crustal magnetization based on seafloor depth and magnetic anomaly data (b). Black dots and adjoining lines indicate the normal polarity isochrons while white triangle symbols and adjoining lines indicate reverse polarity chrons. Selected major chrons are identified and can be converted to age using the geomagnetic polarity timescale shown at the bottom of the figure. Map based on data from Tivey and Tucholke (1998).

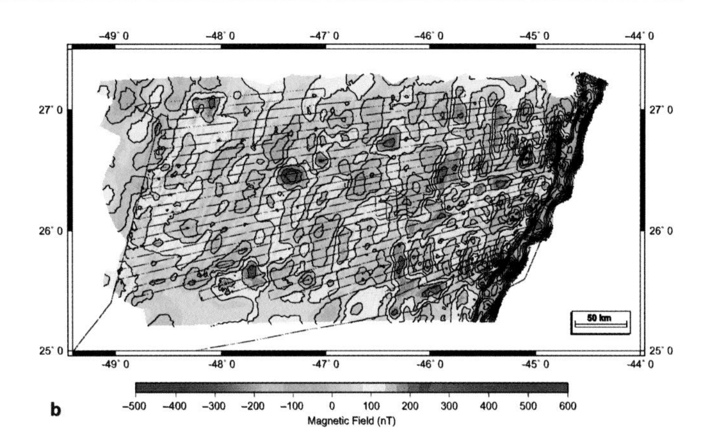

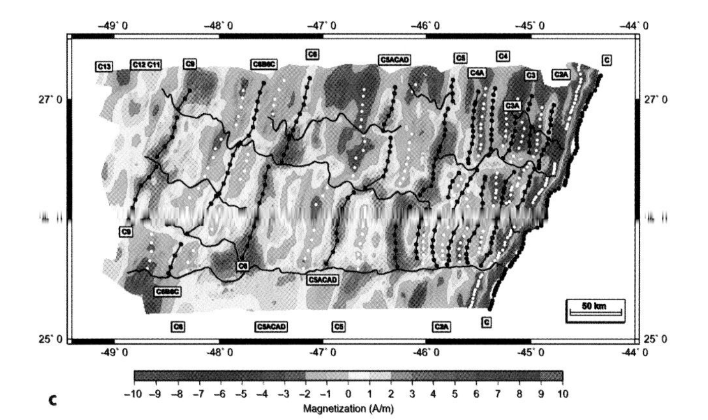

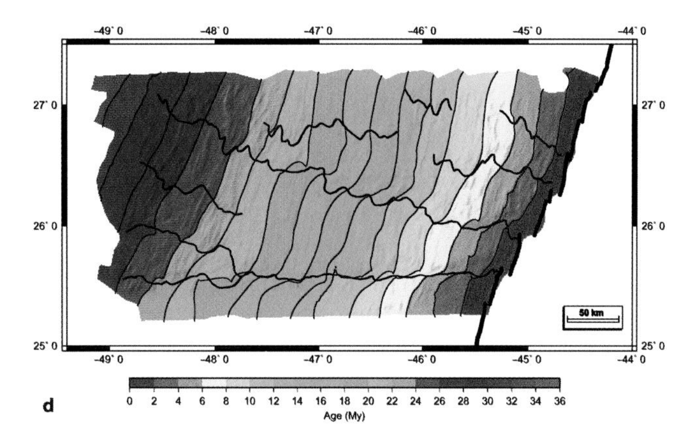

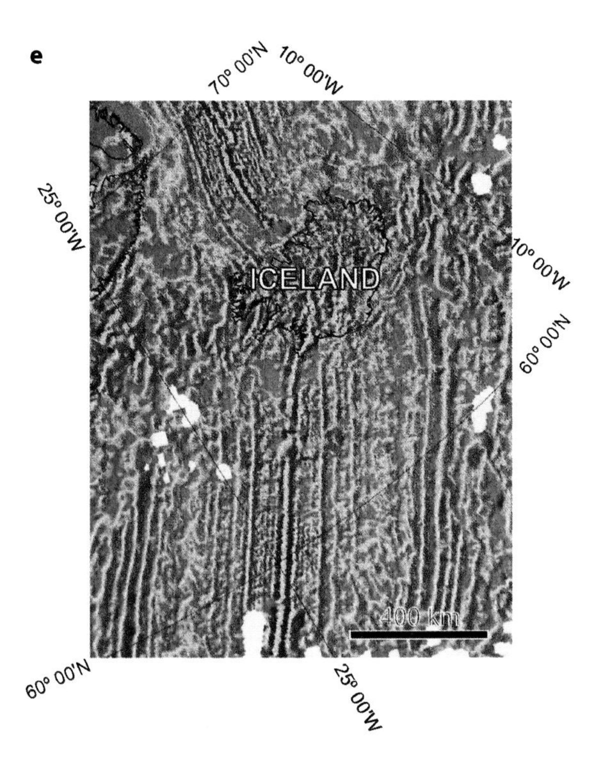

d Age map based on the magnetic anomaly age picks showing the progressive aging of the sea floor away from the midocean ridge spreading center shown by the solid thick black lines. Thin black lines indicate the boundaries of individual spreading ridge segments through time. (See *Magnetic surveys, marine*; *Magnetic anomalies for geology and resources*). **e** Magnetic anomaly patterns in the North Atlantic Ocean showing the symmetry of the anomalies either side of the Mid-Atlantic Ridge in the vicinity of Iceland. Detail from Verhoef *et al.*, 1996, courtesy of the Geological Survey of Canada. (See *Magnetic anomalies for geology and resources; Magnetic anomalies, marine; Magnetic surveys, marine.*) (*Figure M2, Magnetic anomalies for geology and resources*).

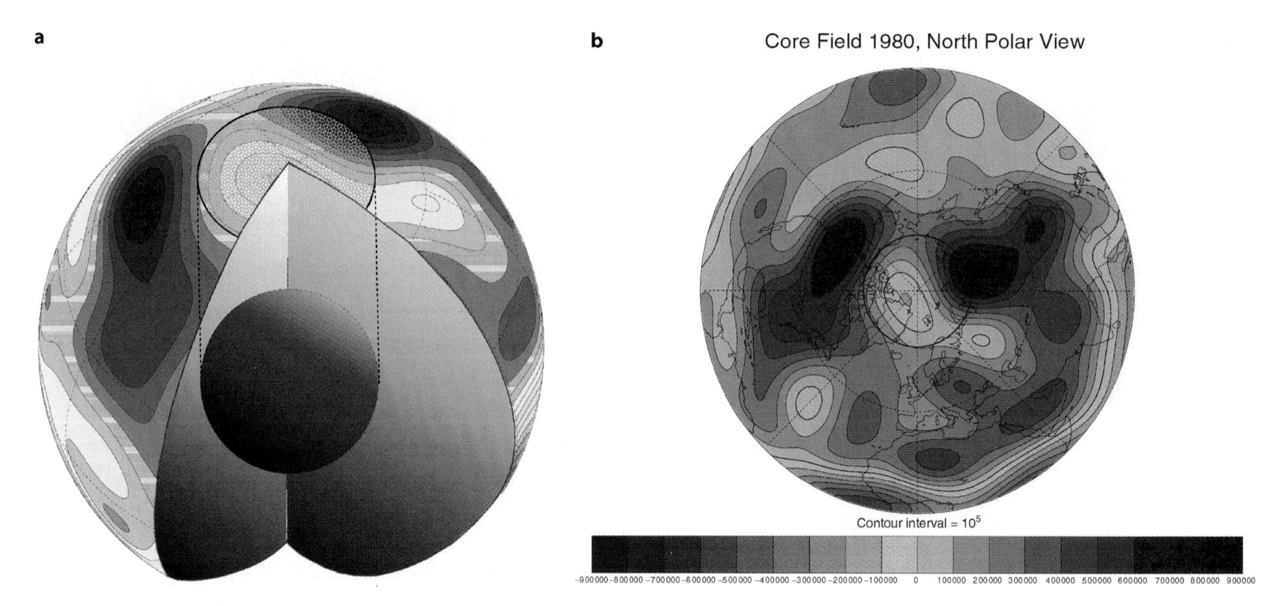

Plate 10 Shown is the vertical component of magnetic field at the core surface, contour interval 0.1 mT blue inwards (**a** satellite view, **b** Lambert equal area northern hemisphere). The tangent cylinder enclosing the inner core with axis parallel to the geographic axis is thought to exert a strong influence on the fluid dynamics of the core and this is seen in the magnetic field, which is weak inside the tangent cylinder and strongest just outside it, where shear layers are thought to develop. The southern hemisphere shows a similar structure but is more disturbed by secular variation. (See *Tangent cylinder; Geodynamo, numerical simulations*).

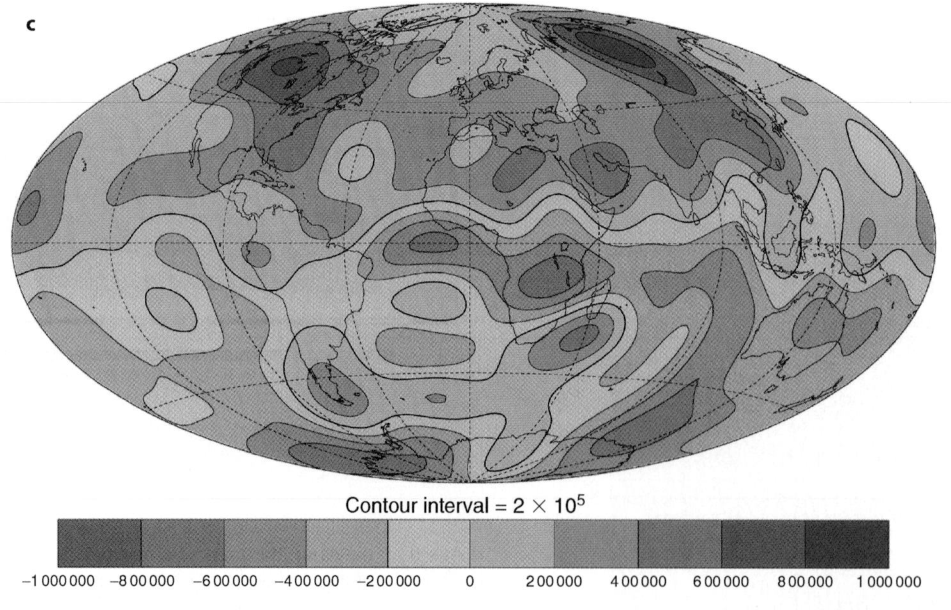

c Radial component of magnetic field at the core surface in 1980, contour interval 0.2 mT. Note the symmetry of the four main concentrations of flux in high latitudes, the zero contours near the poles, and the large blue area of reversed flux in the southern hemisphere from the tip of South America through to the Indian Ocean. High latitude flux is fairly stable while in equatorial regions the anomalies drift west from the Philippines through to the west coast of the Americas. The Pacific region is relatively featureless and is changing little. From the model of Jackson *et al.* (2000); *see time dependent models of the main magnetic field, main field modelling, geomagnetic secular variation, tangent cylinder.* Compare the regions of strong flux with those of low shear wave velocity at the base of the mantle given in Plate 8(a) for an indication of the correlation between field and possible cold lower mantle. (See *Core-mantle interactions: thermal.*) (*Figures I13, I14, I15, Inner core tangent cylinder*).

Plate 11 Maps of S velocity **a** and P velocity **b** variations relative to the global average values in D″ at a depth near 200 km above the CMB in representative global mantle seismic tomography models. S velocity variations are from Mégnin and Romanowicz (2000), and P velocity variations are from Boschi and Dziewonski (2000). Positive values (blue) indicate higher than average velocities and negative values (red) indicate lower than average velocities. The scale bars differ: S velocity variations are 2–3 times stronger than P velocity variations. Large-scale patterns dominate, suggesting the presence of coherent thermal and chemical structures. (See *D″ seismic properties; D″ composition; Earth structure, major divisions; CMB topography*).

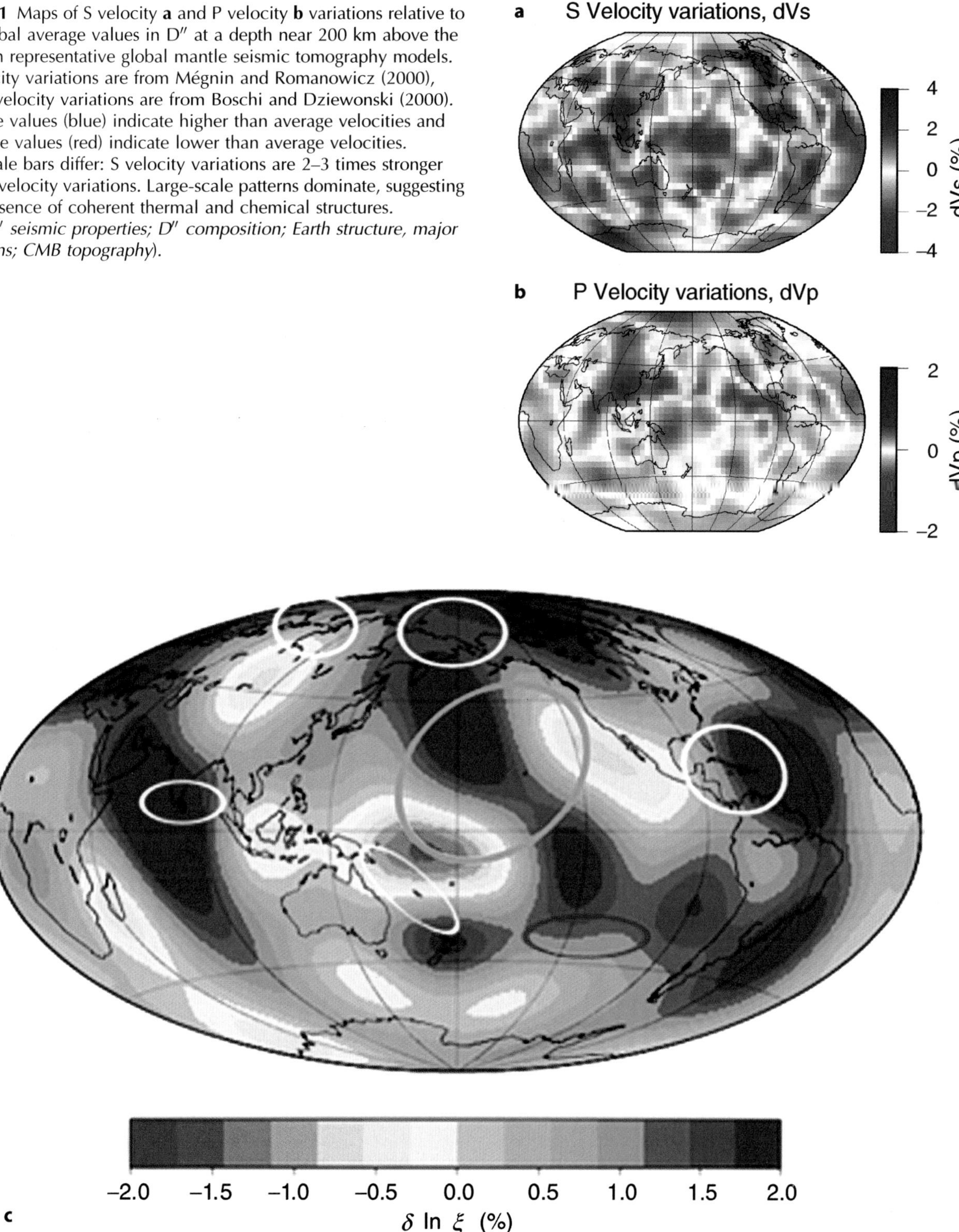

c Map of seismic S-wave anisotropy ($\xi = V_{SH}^2/V_{SV}^2$) in D″, based on a global waveform tomography inversion for a 3D model of radial anisotropy (VTI) (from Panning and Romanowicz, 2004). Also shown are areas of detailed regional analysis of D″ anisotropy: white indicates regions where in D″ horizontally polarised S-waves (SH) are faster than a vertically polarised S-waves (SV); red shows regions where D″ appears to be isotropic; green shows a broad region of D″ beneath the Pacific where the style of anisotropy appears to be complex and quite variable. (See *D″ anisotropy*.) (*Figure D2, D″, anisotropy; Figure D3, D″, seismic properties*).

Plate 12 Most of the planets and their larger satellites have a dynamo-generated magnetic field now, or have had one in the past. Dynamo action occurs in the liquid cores of rocky planets and the atmospheres of gas giants. Cutaway views of what lies beneath the surfaces of many of the planets and moons of the solar system. These models of the planetary interiors are based on a variety of geophysical observations as discussed in *Interiors of Planets and Satellites*. Starting at the top and proceeding down to the right and then to the left are Neptune, Uranus, Saturn, Jupiter, Mars, Earth, Venus and Mercury. Jupiter's four Galilean moons are arranged in line to the left of the planet (in order from right to left are Io, Europa, Ganymede and Callisto). Earth's Moon lies to its lower right. The bodies are not shown to proper relative scale. The figure was prepared by Calvin J. Hamilton and included here with permission. (See *Interiors of Planets and Satellites, Planetary and Satellite Dynamos*.) (*Figure 123, Interiors of planets and satellites*).

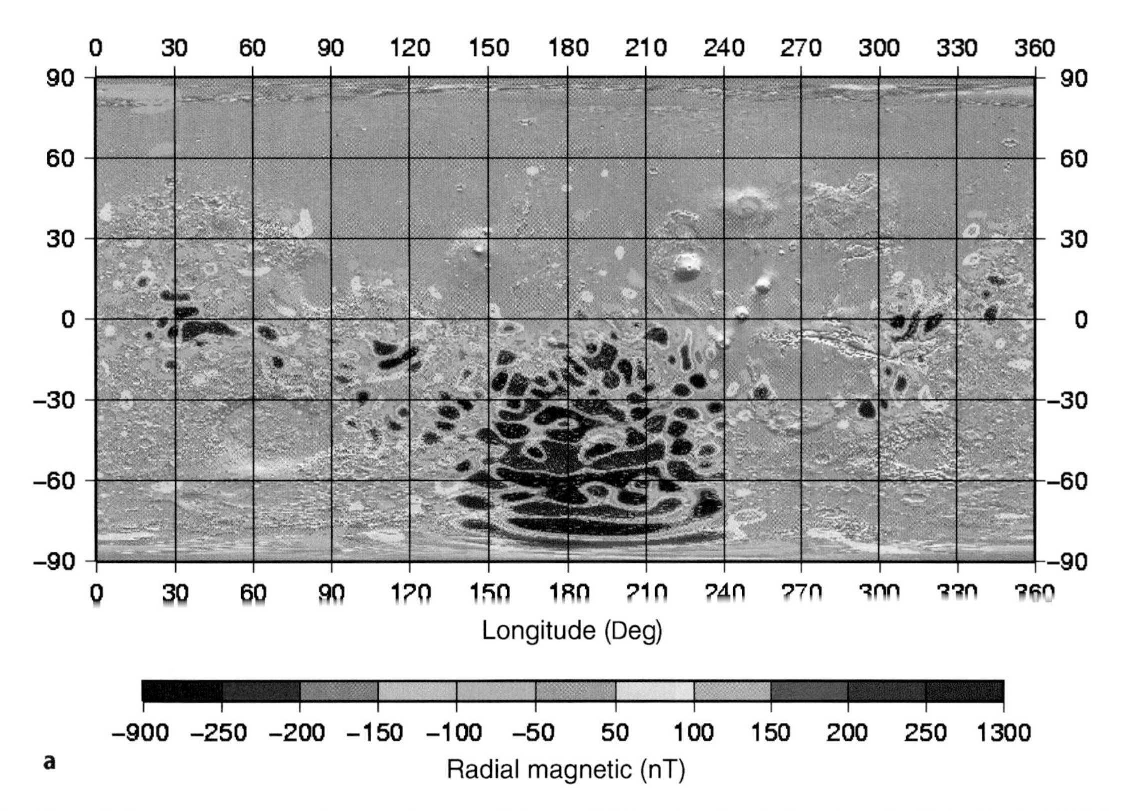

Plate 13 **a** The radial component magnetic anomaly map of Mars at 100 km elevation derived from the high-altitude MGS data using the spherical harmonics of degree up to 50. The color bar is saturated to better illustrate the weak anomalies. Units are in nT. (See *Magnetic field of Mars*).

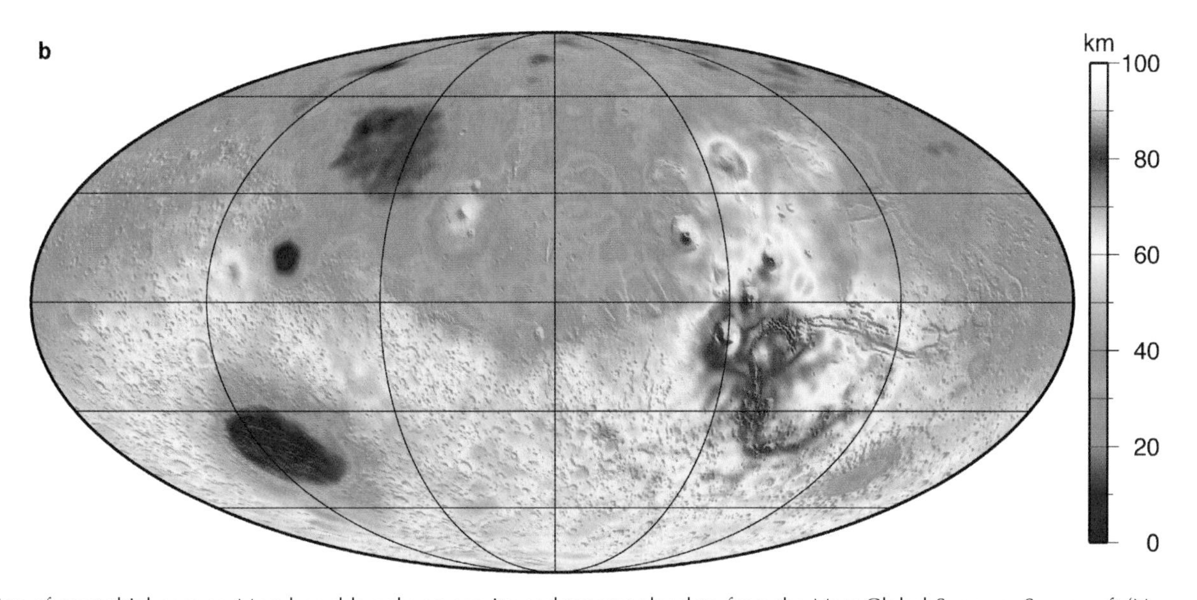

b Map of crust thickness on Mars based largely on gravity and topography data from the Mars Global Surveyor Spacecraft (Neumann *et al.*, 2004). (See *Interiors of planets; Magnetic field of Mars.*) (*Figure I26, Interiors of planets and satellites; Figure M25, Magnetic field of Mars*).

Plate 14 **a** Compound extreme ultraviolet (171 Å) full-disk image of the Sun taken by the TRACE (Transition Region and Coronal Explorer) satellite. The two bands either side of the Sun's equator contain bright features indicating magnetic activity. (See *Sun's magnetic field*).

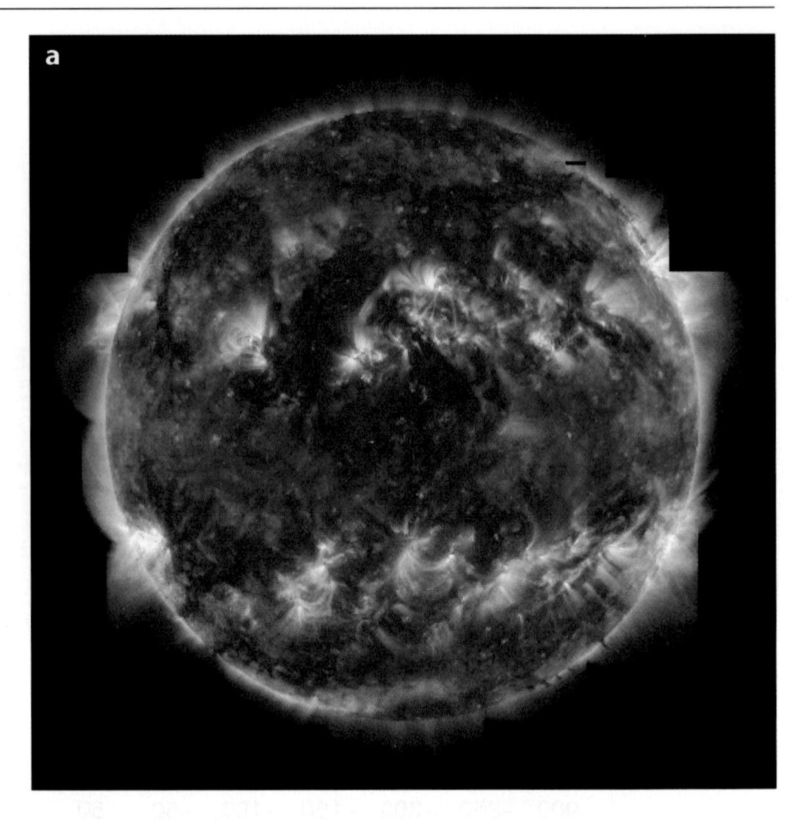

b The rotational variation as a function of time and latitude at radius $r = 0.98\ R_\odot$ are shown for the first nearly 6 years, and thereafter the time series is continued by exhibiting the 11-year harmonic fit. Shown to the right are the residuals from the fit, on the same colour scale. Bottom Panel; As for above, but showing the rotation variation as a function of depth instead of latitude, at latitude 20°. From Vorontsov *et al.*, 2002. (See *Sun's magnetic field, solar dynamo, helioseismology*.) (*Figures M26, M30, Sun's magnetic field*).

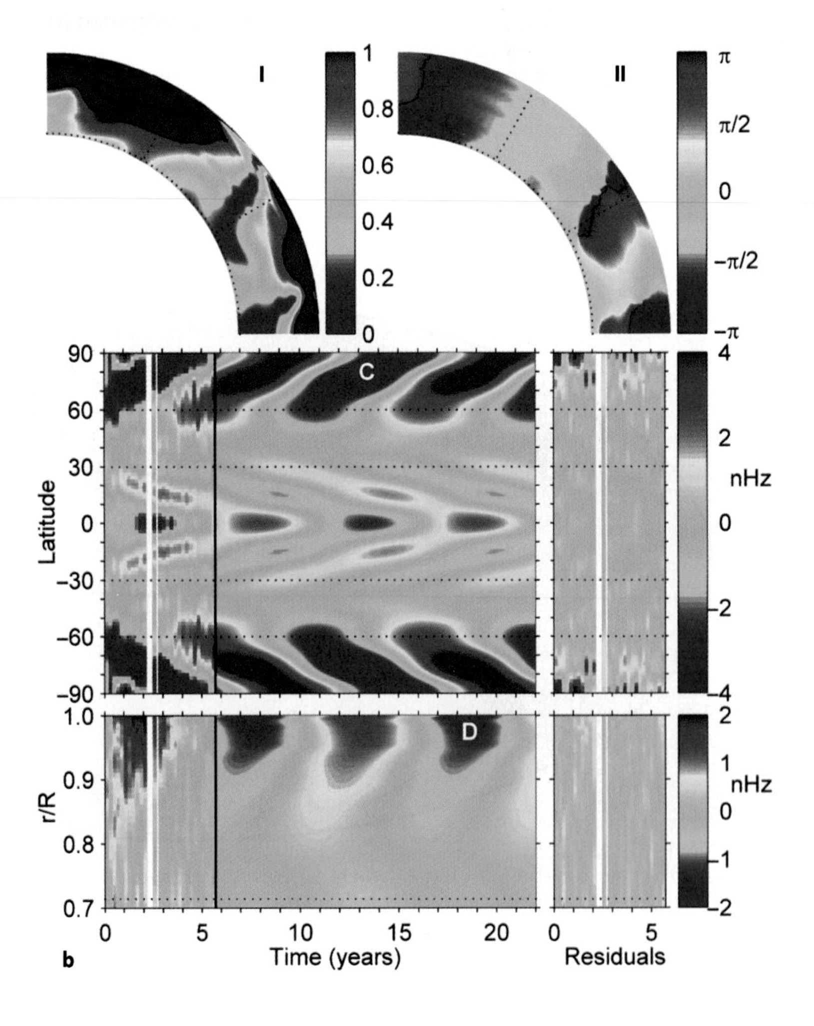

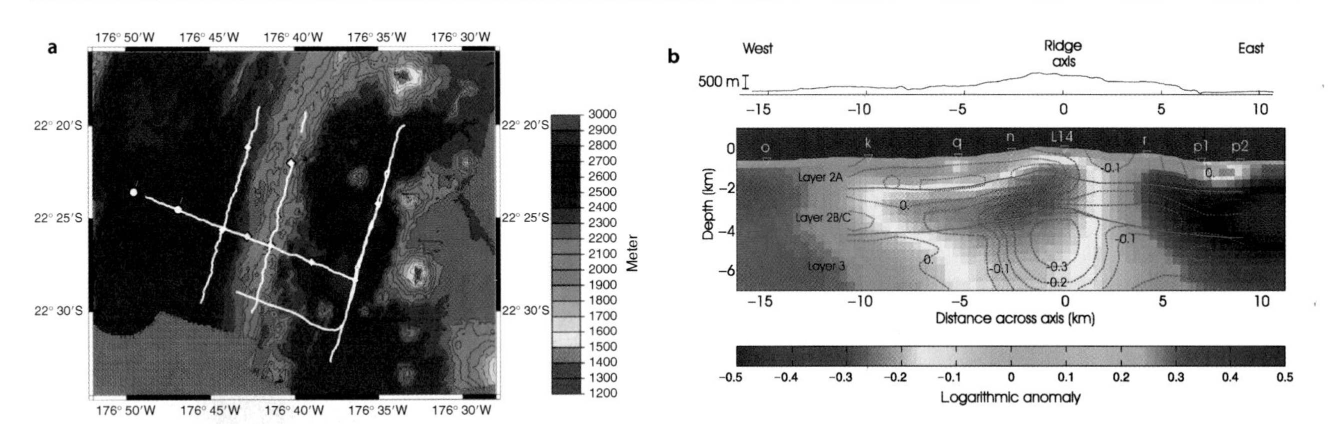

Plate 15 **a** Marine controlled source EM (CSEM) transmitter tracks superimposed on a bathymetry map of the Valu Fa Ridge. The receiver locations are shown as white dots. **b** A section through the ridge showing the logarithmic variation in electrical conductivity found by CSEM and the anomaly in seismic P-wave velocity as colour shading and black contours respectively (courtesy M.C. Sinha). (See *Marine controlled source electromagnetics*).

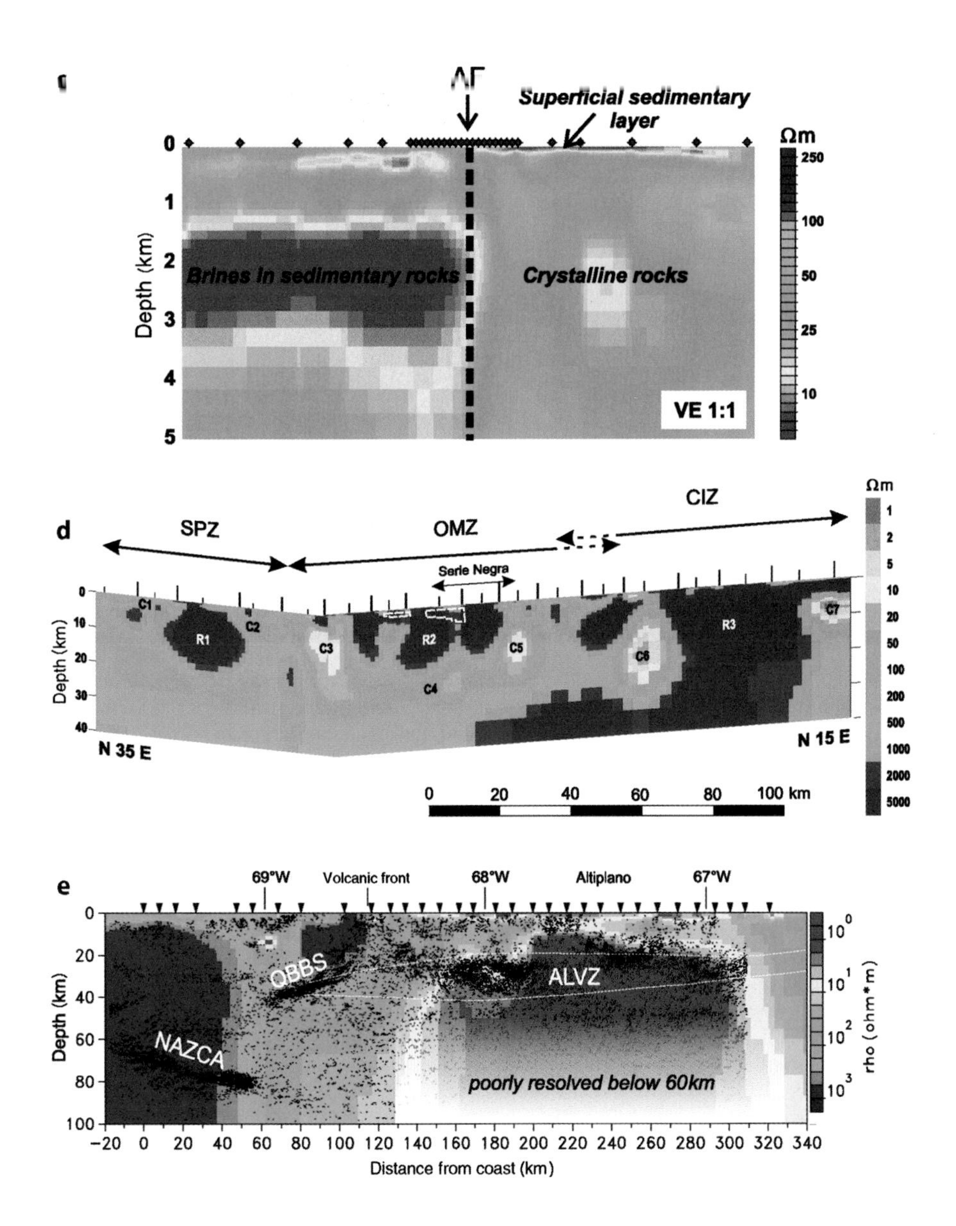

Three examples of regional EM studies. The resistivity values are colour-coded; red and yellow colours indicate zones of high conductivity: **c** Magnetotelluric profile crossing the Dead Sea Transform Fault, locally known as the Arava Fault (AF), modified after Ritter *et al.* (2003). The most prominent feature in the resistivity model are the sharp lateral contrasts under the surface trace of the AF. **d** Two-dimensional electrical resistivity model of a profile across the western part of the Iberian Peninsula (Iberian Massif), modified after Pous *et al.* (2004). The high conductivity zones coincide with the transitions of suture zones of the Variscan orogen. Labels are R: high-resistivity zones, C: high-conductivity zones, SPZ: South Portuguese Zone, OMZ: Ossa Morena Zone and CIZ: the Central Iberian Zone. **e** MT model across the Andes, modified after Brasse *et al.* (2002). The most consistent explanation for the broad and deep-reaching highly conductive zone is granitic partial melt. Reflection seismic data and the location of the Andean Low-Velocity Zone (ALVZ) are superimposed on the MT model. The QBBS (Quebrada Blanca Bright Spot) marks a highly reflective zone in the middle crust of the forearc. (See *EM, regional studies*.) (*Figure E9, EM, regional studies; Figures E26, E27, EM, marine controlled source*).

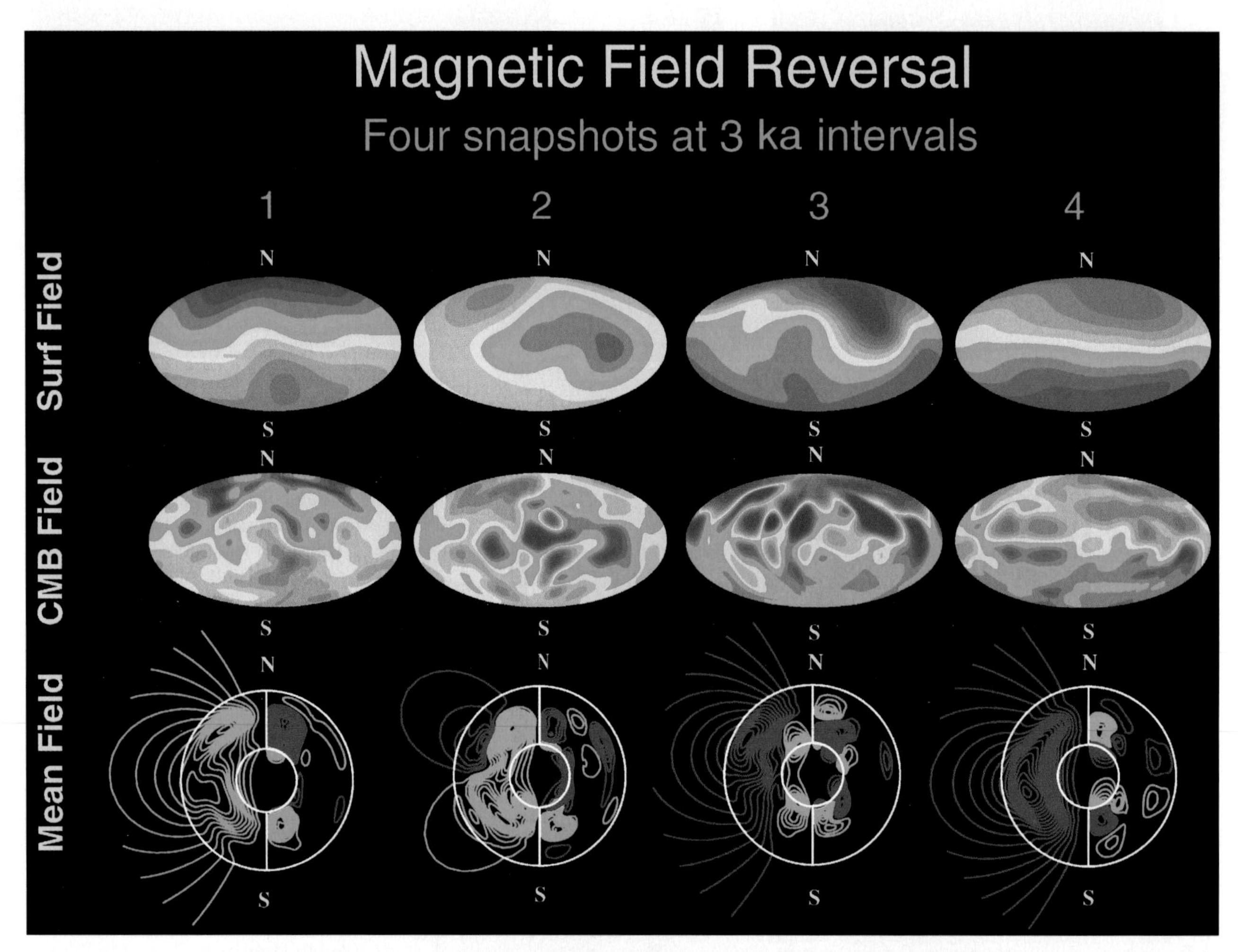

Plate 16 A sequence of snapshots of the longitudinally averaged magnetic field of a numerical geodynamo model through the interior of the core and of the radial component of the field at the core-mantle boundary and at what would be the surface of the Earth, displayed at roughly 3 ka intervals spanning a dipole reversal. In the plots of the average field, the small circle represents the inner core boundary and the large circle is the core-mantle boundary. The poloidal field is shown as magnetic field lines on the left-hand sides of these plots (blue is clockwise and red is counter-clockwise). The toroidal field direction and intensity are represented as contours (not magnetic field lines) on the right-hand sides (red is eastward and blue is westward). Aitoff-Hammer projections of the entire core-mantle boundary and surface are used to display the radial component of the field (with the two different surfaces displayed as the same size). Reds represent outward directed field and blues represent inward field; the surface field, which is typically an order of magnitude weaker, was multiplied by 10 to enhance the color contrast. (See *Geodynamo: numerical simulations.*) (*Figure G14, Geodynamo, numerical simulations*).

Index

A

A/D converter 977
aa index 509, 719, 720
AARM (see *anisotropy of anhysteretic remanent magnetization*)
AAS (see *anisotropy of anhysteretic susceptibility*)
ab initio
  –, calculation 94, 99, 421
  –, method 693
Abelian group of symmetry operations 306
Abinger Observatory 720
Abisko Observatory 737
absolute
  –, age 320
  –, intensity 890
    –, unit 358
  –, plate motion (APM) 20
abundance, isotopic 79
AC
  –, field 932
  –, voltage 920
Académie Royale des Sciences 357
acceleration
  –, centrifugal 76, 135, 275
  –, nontidal 471
accelerometer 59
acceptor, electron 249
accretion
  –, heterogeneous 79
  –, planetary 149, 151
achondrite 796–798
  –, enstatite 855
acquisition, fabric 565
activation
  –, energy 685
  –, enthalpy 106, 687
activity
  –, geomagnetic 318
  –, intrusive 986
  –, magnetic
    –, solar 507
    –, storm-time 509
  –, orogenic 785
  –, solar 316
Adams-Williamson equation 82–84
Addis Ababa Observatory 461
additivity 574, 612
  –, law 754
adiabatic 8[illegible]
  –, bulk modulus 82
  –, compression 212
  –, derivative 92
  –, gradient 80, 106
  –, heat flux 15
  –, heating 128
  –, reference state 14, 16
Adie, Patrick 436
Admiralty Research Laboratory 54
Adolf Schmidt Observatory 730
advection 84
  –, time 297
advective timescale 644
AE index 511, 512
aeronomy 407
aether and matter 468
AF (see *alteniating field*)
African
  –, plate 960
  –, super-upwelling 964
age
  –, absolute 320, 665, 780
  –, C-14 669
  –, determination 829
  –, K-Ar 819
  –, radiometric 767
  –, U-Pb 819
ageostrophic, effect 853
AGFM (see *alternating gradient field magnetometer*)
aggregate, framboidal 456
Agincourt Observatory 726
AGM (see *alternating gradient magnetometer*)
agonic line 993
Agricola, Georgius 3
AGU 274
Ahrrenius relation 105
Aichi Prefecture, Japan 696
air wave 233, 236
AIRM (see *anisotropy of isothermal remanent magnetization*)
airswings 850
Airy compensation 447
AIST (see *National Institute of Advanced Industrial Science and Technology*)
Aitken Basin 116
AK135 model 210, 213, 903
akagenéite 526
Aksu area 822
AL index 512
alabandite 79
Alashan/Hexi Corridor 817, 819
albedo 791
Albuquerque Volcanoes 837
  –, Pringle Falls excursion 838
Alfvén
  –, Hannes 6, 7
  –, number 299
  –, theorem 7, 112
  –, timescale 650
  –, velocity 5, 6
  –, waves 3–7, 746
    –, velocity 334
algorithm
  –, genetic 220
  –, optimization 220
  –, search 220
ALH84001 792
Alibag 710
  –, Observatory 732
Alice Springs Observatory 717
Allende meteorite 798
  –, chondrules 797
  –, refractory calcium aluminum inclusions (CAI) 797
Allgemeine Theorie des Erdmagnetismus 359
alloy, metallic 74
alpha
  –, effect 138, 161, 164, 174, 192, 199, 290, 643, 843, 860, 861
    –, anisotropic 194
    –, turbulent 645
  –, iron 691
  –, omega 164
    –, dynamo 179, 181
    –, mechanism 196, 197
  –, particles, ionizing 53
  –, quenching 198, 262

alpha^2
–, dynamo 162
–, mechanism 196
Alpine Fault 246
Alpine-Himalayan Belt 802
–, Anatolian sector 803
alteration
–, aqueous 796
–, chemical 37, 490, 583, 785
–, diagenetic 258
–, hydrothermal 597
–, low-temperature 584
–, zone 253
–, mineralogical 512–514
–, phase 79
–, postdepositional diagenetic 253
–, process 979
–, severe 695
–, thermal 911
–, thermochemical 575
alternating field (AF) 759, 910
–, demagnetization 498, 759, 873, 884, 943
–, spectra 577
–, gradient field magnetometer (AGFM) 877
–, gradient magnetometer (AGM) 655
Altiplano 244
altitude, orbital 60
Altyn Tagh mountain range 819
aluminum–26 90, 854
Amberley Observatory 735
am-index 719
ammonite 665
AMO (see *automatic magnetic observatory*)
Ampère, André Marie 358
–, law 114, 287, 347, 640
amphiboles 932
amphidrome effect 65
amplitude 67, 977
–, compass 67
AMR (see *anisotropy of magnetic remanence*)
AMS (see *anisotropy of magnetic susceptibility*)
Amsterdam Island 993
AMT (see *audiomagnetotellurics*)
AMTM60 500
Amundsen, Roald 990
amygdalohippocampectomy 50
analysis
–, biostratigraphic 993
–, dimensional 297
–, Fourier 351
–, harmonic 921
–, least squares 86, 356, 359
–, linearized resolution 222
–, periodogram 353
–, principal component 45, 46, 845, 848
–, resolution 84
–, skewness 962
–, spectral 122, 770
–, spherical harmonic 280
–, thermomagnetic 755
–, time series 766
–, waveform 768
–, wavelet 355
analytic signal method 142
Anatolian plate 803
Ancón Observatory 734
Andes 243
anelastic approximation 11
angle, Euler 393
angular
–, momentum 130, 135
–, exchange 469
–, oceanic 469
–, operator 388
–, vector 956
–, standard deviation (ASD) 894
anhysteretic remanent magnetization (ARM) 527, 535, 572, 759, 910, 943
–, acquisition curve 574
–, partial 528, 590
–, spurious 573
–, susceptibility 528, 572
animal navigation 48
animists 870
anisometry, intergranular 554
anisotropy 427, 546, 560, 564, 589, 618, 799, 846, 871
–, bulk conductivity 20
–, conductive 219, 222
–, constant 621
–, corrected 476
–, crystalline 493, 618
–, uniaxial 492
–, cylindrical 419
–, D″ 146
–, degree 561, 937
–, electrical 20, 671
–, ellipsoid 536
–, energy 492, 872, 874
–, exchange 548
–, factor 20
–, intrinsic 20
–, macroscopic 20
–, magnetic 535, 871
–, bulk 590
–, susceptibility 514
–, magnetocrystalline 454, 491, 492, 515–517, 537, 546, 548, 566, 614, 872, 873
–, magnetostrictive, axis 872
–, mechanisms 146
–, media 146
–, multiaxial 590
–, of anhysteretic remanent magnetization (AARM) 477, 577
–, ellipsoid 538
–, of anhysteretic susceptibility (AAS) 536
–, of isothermal remanent magnetization (AIRM) 537
–, of magnetic remanence (AMR) 535
–, of magnetic susceptibility (AMS) 475, 535
–, carrier 557
–, ellipsoid 476, 538, 547, 553, 555
–, low field 937
–, -to-strain relationship 556
–, remanent 536
–, seismic 146
–, shape 546
–, source 564
–, stress 491
–, stress-induced 548
–, susceptibility 799
–, S-wave 148
–, uniaxial 500, 572, 590, 881
annihilator 888, 889
–, magnetic 481
annual variation 308, 816
annulus model 76, 125
anomaly 772, 773
–, attenuation 143
–, Bangui 39
–, Bouguer 39, 504
–, Br- 774
–, Brunhes central 542
–, conductivity 20, 131
–, latitude 961
–, local 676
–, magnetic 140, 143, 320, 477, 888
–, compilation 478
–, correlation with gravity anomaly 856
–, definition 478
–, field 543
–, inversion nonuniqueness 488
–, long wavelength 481
–, low amplitude 666
–, marine 324, 328, 336, 483, 665, 779, 962
–, modeling 485
–, short wavelength 982
–, skewness 857
–, stripe 478
–, vertical intensity 857
–, mapping 478
–, North German 412
–, resistive 309
–, resistivity 235
anorthosite 629
Anson, George 989
Antananarive Observatory 461
Antarctic
–, expedition 696
–, ice surface 696
–, meteorite collections 795
–, plate 960
Antarctica 359, 376, 792, 962
anthropogenic 253
antidynamo theorem 137, 163, 173, 176, 192, 468, 643
antiferromagnet 590
antiferromagnetic 513, 871, 932
–, spin 874
antiferromagnetism 519, 525, 700
AO index 511, 512
apatite 931
Apia 43
–, Observatory 735, 736
APM (see *absolute plate motion*)
Apollo 788
–, exploration 440
–, samples 788
apparent
–, phases 221
–, polar wander path (APWP) 777, 779, 802, 840, 956, 959, 960
–, resistivities 221, 869
application, paleomagnetic 256
approach
–, 2½-dimensional 293
–, bootstrap 608
–, least-squares 284
–, Monte Carlo 608

approach *Contd.*
–, nonlinear conjugate gradient (NLCG) 221
–, reciprocity 221
approximation
–, anelastic 11
–, Born scattering 221
–, Born-Rytov 216
–, Boussinesq 11
–, first-order smoothing 194, 201
–, quasistatic 223
–, second-order correlation 194
–, thin-shell 815
APWP (see *apparent polar wander path*)
aqueous fluid 247
aquifer, saline 224
Arago, Fraçois 436
Arai diagram 754, 755
Arava Fault 243, 413
arc
–, auroral 376
–, transpolar 34
Archean 141, 478
–, craton 141
archeological material 766
archeologist 944
archeology 23
archeomagnetic dating 29
archeomagnetism 31, 920, 942, 943
Archie law 232
archive
–, Portuguese 990
–, Spanish 990
argon
–, –40-argon–39 age 832, 834, 835, 962
–, incremental heating experiment 836
–, isochrons 836
–, isotopes 836
ARGOS (see *Automatic Remote Geomagnetic Observatory System*)
Argyre 795
–, impact basin 503
Aristotle 361
ARM (see *anhysteretic remanent magnetization*)
array
–, Schlumberger 229
–, Wenner 229
Ars Magnesia 463
Arsia Volcano 504
arthropods 49
artifact
–, archaeological 569
–, man fired 910
Ascension Island Observatory 720
Ascreaus Volcano 504
ASD (see *angular standard deviation*)
ASG–81 scalar magnetometer 850
Askania 437
ASMO (see *Automatic Standard Magnetic Observatory*)
assemblage, mixed magnetic 271
Association of Terrestrial Magnetism and Electricity 274, 382
assumption
–, kinematic 188
–, steady flow 88
astatic magnetometer 654
asteroid 788, 795, 796, 800
–, belt 90
–, C-type 799
–, S-type 799
asteroseismology 398
asthenosphere 73, 208
asthenospheric conducting layer 73
astrochronology 785
astronaut protection 316
Astronomer Royal 61, 375
asymmetric rectangular hysteresis loop 575
asymptotics 111
Atlantic 993
–, chart 675
–, hotspot 962
Atlantis, R/V 542
atmosphere
–, martian 792
–, solar 119
atom transmutation 53
ATS (see *attitude transfer system*)
attenuation parameter 209
attitude 59
–, transfer system (ATS) 675
attraction, polar 356
AU index 511, 512
audiomagnetotellurics (AMT) 671
–, controlled-source (CSAMT) 671
–, method 224, 230
aurora 7, 61, 334, 376, 404, 696
–, australis 33
–, borealis 33
–, cusp 33
–, polaris 358
–, rayed 33
–, record 508
–, substorm 926
–, break-up 34
–, bulge 34
–, theta 34
auroral
–, arc 33, 376
–, breakup 927
–, corona 33
–, curl 33
–, curtain 33
–, electrojet index 34, 509
–, oval 33
–, processes 33
–, wavelengths
–, forbidden green oxygen line 33
–, Lyman-Birge-Hopfield lines 33
–, red doublet 33
–, zone 864
Australia 62, 967
Australian plate 960
authigenesis 785, 806
autocollimator 715
automatic
–, computing engine 166
–, magnetic observatory (AMO) 713
Automatic Remote Geomagnetic Observatory System (ARGOS) 714
Automatic Standard Magnetic Observatory (ASMO) 714
autoposition 921
average
–, azimuthal 193
–, geostrophic 648
–, Lagrangian 103
–, local 192
–, space 193
–, statistical or ensemble 192
–, time 193
averaging
–, azimuthal 199
–, Lagrangian 102
–, low-level 199
axial quadrupole 961
axis
–, crystallographic 599
–, pattern 550
–, geographic 311
–, geomagnetic 311
–, maximum 561
–, minimum 561
–, paleovertical 561
azimuth
–, compass 67
–, visor 435
azimuthal
–, averaging 199
–, flow 746
Azorean prime meridian 356
Azores Islands 993

B

Bac Lieu Observatory 734
Bache, A.D. 436
backfield 591
–, curve 877, 979
backup variometer 711
Backus
–, ambiguity 466
–, dynamo 163
–, series 466
Backus-Gilbert theory 83, 221
Bacon, Roger 808
Bacon, Sir Francis 361
bacteria
–, anerobic 24
–, magnetogenic 249
–, magnetotactic 26, 48, 516, 782
–, martian 793
Bahia Province 40
Bajocian stage 666
baked contact test 35, 608, 778
balance
–, horizontal magnetic translation 977
–, hydrostatic 14
–, Lloyd's 435
–, magnetostrophic 841
–, Schmidt's 435
–, variable field translation (VFTB) 977
–, vertical 437
band
–, longitudinal preferred 326, 341
–, pass filtration 920
–, preferred 831
Bangui anomaly 39
bar magnet 310
Baranov 856

Barentsz, Willem 988
Barkhausen
 –, discontinuity 616
 –, jump 613, 623
Barlow, Peter 40
 –, plate 68
Baroque culture 463
Bartels, Julius 42
basalt 128, 596, 871
 –, Emeishan 819
 –, flood 150
 –, layer 597
 –, mare 788
 –, marine 993
 –, oceanic 583
 –, submarine 518, 583
basaltic
 –, magma 72
 –, volcanics 72
Base Roi Baudouin Observatory 725
basement
 –, magnetic 477
 –, structure 224
basestation, magnetometer 140
Basin and Range Province 141, 802
basin, sedimentary 242
basis
 –, function 395
 –, geostrophic 86
Bastille-day magnetic storm 742
Batavia 44
bathymetry 64, 962
battery, celestial 316
Bauer, Georg (see *Agricola, Georgius*)
Bauer, Louis Agricola 42, 438
Bavarian Academy of Science 887
bay, magnetic 662
Bayesian
 –, formulation 123
 –, statistics 29
bcc, iron 916
BCMT (see *Bureau Central de Magnétisme Terrestre*)
Bear Island Observatory 737
Bear Seamount 891
bedrock lithology 477
behavior
 –, piezoremanent 599
 –, pseudo-single-domain (PSD) 879
 –, superparamagnetic 516
Beijing Mingtomb Observatory 728
Bellarmatus, Künstler 721
Belshe, J.C. 887
Belsk Observatory 728
belt, orogenic 801, 806
Bemmelen, Willem van 44
Bénard layer 650
bench, optical 59
bending, orogenic 803
Benton, E.R. 44
Berkner, Lloyd 407
Berlin Observatory 729
bernallite 526
beta
 –, effect 194
 –, quenching 198
Beta Regio 446
BGS (see *British Geological Survey*)
bias field 528
biconjugate gradient 221
bifurcation
 –, Hopf 417
 –, subcritical 262
 –, supercritical 262
Big Bang theory 7
Big Lost event 834, 837, 838
BIM (see *biologically induced mineralization*)
Bingham
 –, distribution 45
 –, statistics 45
Binza Observatory 725
biogenic 783, 793
 –, magnetite 50, 51
biologically induced mineralization (BIM) 782
biomagnetism 48, 874
biostratigraphy 320, 329, 568, 781
Biot-Savart law 955
bioturbation 476
Birch law 77
Birkeland currents 34
Birmingham Solar Oscillation Network (BiSON) 398
Bitter
 –, line 499
 –, pattern method 495
Bjerkreim-Sokndal layered intrusion 141
Bjorn event 838
Bjurbole meteorite 798
Blackett, Patrick Maynard Stuart 53, 887
Blake Event 770
blanket, thermal 789
Bloch
 –, line 497
 –, wall 492
 –, wave 201
block, continental 839
blocking
 –, temperature 493, 610, 614, 615, 872, 943
  –, spectrum 611
 –, volume 584
Bobrov magnetometer 712
Bobrov-system quartz variometer 729
Bodes law 706
body
 –, gray 688
 –, ideal black 688
 –, wave 82
body-centered cubic (bcc) iron 94, 367
Boltzmann
 –, constant 69, 100, 105, 109, 367, 610, 885, 931
 –, equation 137
 –, statistics 500
Bond, Henry 357
Bonnet formula 379
Booth, Felix, 989
bootstrap
 –, approach 608
 –, technique 553, 608
Boreal Hemisphere 464
boresight 59
Born scattering approximation 221
Born-Rytov approximation 216
Borough, Stephen 988
Borough, William 280, 357, 435, 674, 707
Boso Peninsula, Japan 829
bottom-simulating reflector (BSR) 234
Bouguer gravity anomaly 39, 447, 504
Bouma sequence 553
boundary
 –, conditions
  –, impermeable 75
  –, inhomogeneous 416
 –, continental, Precambrian 142
 –, core-mantle 14, 84, 337
  –, topography 124, 125
 –, Cretaceous-Tertiary 666
 –, inner core 14, 74, 83
 –, insulating 191
 –, Jurassic–Triassic 928
 –, layer 76, 85, 111, 145, 152, 174, 186, 642, 657
  –, superadiabatic 82
  –, theory 423
  –, thermal 18, 76, 132
 –, Matuyama-Brunhes 781
 –, outer core 84
 –, reversal 982
 –, Santonian-Campanian stage 666
 –, stage 329
 –, stress-free 76
 –, tectonic 64
 –, tilt 75
bounding theorem 288
Boussinesq
 –, approximation 11
 –, fluid 74, 630
bow shock 657
Bowie Medal 274, 469
boxcar 122
B-polarization mode 221
Braginsky dynamo 164
Braille 795
brain, human 49
breakup, auroral 927
breccia 788, 789, 798
 –, high-grade 629
 –, mature 623
 –, regolith 789
bremsstrahlung X-ray emission 33
Briggs, Henry 357
British Association for the Advancement of Science 468
British Board of Longitude 435
British Geological Survey (BGS) 720, 887
broadside
 –, electric dipole-dipole geometry 233
 –, geometry 236
 –, system geometry 232
Brooks, Charles 436
Brorfelde Observatory 737
Brownian force 588
Brunhes 320, 849, 942
 –, central anomaly 542
 –, chron 311, 829
 –, epoch 626
Bruno, Christoforo (or Borro, C.) 357
Brunt-Väisälä frequency 649

B-spline
–, basis 949
–, cubic 87, 347
BSR (see *bottom-simulating reflector*)
Buda Observatory 728
Budkov Observatory 728
Bulawayo Observatory 461
bulk
–, composition of Earth 213
–, conductivity, anisotropic 20
–, magnetization 141, 487
–, modulus 92, 95, 153
–, isothermal 367
–, susceptibility 937
Bullard, Edward Crisp 54, 887
Bullen parameter 82, 83
Bullseye vent 235
bump 124, 135
bundles, static flux 134
buoyancy 291
–, compositional 206
–, force 12
–, frequency 649
–, number 297
Buoyant rise time 298
Bureau Central de Magnétisme Terrestre (BCMT) 721
burial 588
–, depth 603, 605
–, temperature 628
Bushof 798
Busse column 291
Byron, John 989

C

C response 308
Cabeo, Nicolò 357, 460
cable telecommunication 227
Cabot, Sebastian 674
caesium-vapor magnetometer 2
CAI (see *refractory calcium aluminum inclusion*)
Calabrian 804
calcite 931
calcium, intracellular 49
calculation
–, ab initio 94, 99, 107, 421
–, first principles 370
–, micromagnetic 873
–, molecular dynamics 421
calibration
–, curve 29
–, data 29
Callisto 203, 204, 207, 439, 445
–, differentiation 445
Cambrian 928
–, explosion 966
Cambridge, England 887
Canadian Geomagnetic Reference Field (CGRF) 397
Canadian Geospace Monitoring Program 727
Canadian Magnetic Observatory Network (CANMON) 726
Canadian Magnetic Observatory System (CANMOS) 714
Canberra Observatory 717, 810
CANMON (see *Canadian Magnetic Observatory Network*)
CANMOS (see *Canadian Magnetic Observatory System*)
CANOPUS network 726
cap, spherical 395
Cape Chelyuskin Observatory 738
Cape Denison Observatory 717, 723
Cape Evans Observatory 723
carbon 150
–, –14 781
–, age 669
carbonaceous chondrite (C1) 854
carbonados 40
carbonate, iron-rich 793
Cardús 461
CARISMA network 726
Carixian stage 666
Carnegie Institution of Washington (CIW) 56, 58, 807
Carnegie, RV 57, 58
Carnot engine 128, 290, 301
Carteret, Philip 989
Cartesian
–, components 451
–, coordinate 547
–, unit vector 272
cartridge belt 649
Carty Lake 227
Cascadia margin 238
Cassini 207, 721, 816
Cassini, Jean Jacques 435
Castellaccio Observatory 733
Cathaysia flora 823
Cathedrals 887
Catholic Church 888
Cavallo, Tiberius 435
Cavendish, Thomas 988
c-axis 546, 590
Cayley formation 791
Cayley-Klein parameters 392
CEA (see *China Earthquake Administration*)
cell
–, convective cyclonic 214
–, diamond anvil 94, 99, 421, 503, 689, 690, 692
–, magnetic 864
–, Taylor 649
Cenozoic 929
Central African Republic 39, 40
Central Asian Fold Belt 817, 823
Central Institute of Meteorology and Geomagnetism (ZAMG) 728
Central Japan Anomaly 863
centrifugal acceleration 135, 956
Ceres 90, 279
cesium magnetometer 27, 542
CGRF (see *Canadian Geomagnetic Reference Field*)
cgs unit 874
Cha Pa Observatory 734
Chadwick Physics Laboratory 992
Challenger expedition 359, 990
Chambon la Forêt Observatory 714, 721
CHAMP satellite 59, 60, 141, 349, 351, 352, 676, 743, 892
Chancellor, Richard 988
Chandler wobble 104, 363, 958
Chandrasekhar number 299, 630, 649
Changchun Observatory 728
change
–, angular momentum 320
–, biotic 967
–, geomagnetic 343
–, large amplitude 312
–, magnetomineralogical 754
–, paleoclimatic 669
channel, buried 236
chaos 863
–, deterministic 861
Chapman, Sydney 61
Chapman-Ferraro
–, current 658
–, theory 137
Chapman-Miller method 810
character number 810
characteristics, electric 361
charge, electric rotating 705
Charles II, King 720
Charles Chree Prize 274, 888
charnockite 40
Charon 439, 446
chart
–, Atlantic 675
–, isarhythmic 375
–, isogonic 358
–, magnetic 674
–, nautical 66
Charters Towers Observatory 717, 951
chassignite meteorite 791, 797, 799
Chebyshev polynomial expansion 303
Cheltenham Maryland Observatory 723
chemical
–, demagnetization 157
–, diffusivity 101
–, equilibrium 109
–, layering 151
–, potential 99
–, remanent magnetization (CRM) 457, 580, 609, 696, 782, 806
–, parasitic 513
Chevron Oil Company 263
Cheyenne Belt 143
Chichijima Observatory 733
Chief Magnetician 273
Chilean earthquake, May 22, 1960 423
chimney 74
–, convection 74
China Earthquake Administration (CEA) 727
China Seismological Bureau (CSB) 727
Chinese Academy of Sciences (IGGCAS) 727
Chinese
–, block 823
–, loess 516, 571
–, Plateau 820
Chiripa Observatory 734
chi-squared 47
Chiton 50, 782
chlorite 932
–, paramagnetic 476
Cholesky decomposition 216

chondrite 77, 796–798
–, 3L 798
–, 8LL 798
–, carbonaceous 128, 796–799
–, enstatite 855
chondrules 796
Christchurch Observatory 735
Christoffel formula 379
chromite 526
chron 311, 664, 834, 928
–, Brunhes 311
–, Matuyama 311, 542, 838
–, notation 485
–, polarity 704
chronology, reversal 338
chronometer 989
Church of England 361, 887
Churchman, John 358
CI chondrite 77
ci shi 870
circuit, global electrical 448
circularity 768, 770
circulation
–, Kelvin's theorem 103
–, meridional 162, 181, 190, 294, 859, 946
circumferentor 434
Circum-pan Pacific Magnetometer Network (CPMN) 738
Cirera, Ricardo 461
CIW (see *Carnegie Institution of Washington*)
Clapeyron slope 151
classification 795
clay
–, baked 612, 942
–, archeological 758
–, sequences 570
climate
–, archive 531
–, change 472
–, cycle 571
–, proxy 340
–, variation 531
–, wetter 531
cloud
–, chamber 53
–, magnetic 404
CM4 model 352
CMB (see *core-mantle boundary*)
CME (see *coronal mass ejection*)
CMHD (see *Coriolis magnetohydrodynamics*)
CMTZ (see *core-mantle transition zone*)
CNPS (see *Cretaceous normal polarity superchron*)
CNS (see *Cretaceous normal superchron*)
Coast and Geodetic Survey 43, 723
coast effect 64, 807
coating 791
cobalt 873
Cobb Mountain
–, event 836
–, subchron 838
Cocos Seamounts 891
CODATA 712
coefficient
–, coupling 387
–, covarying 503
–, diffusion 96
coefficient *Contd.*
–, Gauss 9, 10, 84, 311, 346, 681
–, self-diffusion 106
–, thermal expansion 93
–, volume expansion 92
coercive force 157
–, intrinsic 582
coercivity 268, 589, 783, 785
–, high 786
–, low 537
–, microscopic 612
–, parameter 668, 879
–, remanent 591
–, spectrum 591
–, uniaxial 572
coil 541, 874
–, alternating field demagnetization 884
–, compensation 977
–, gradient 977
–, Helmholtz 714, 921
–, Helmholtz-Gaugain 438
–, high-resolution pickup 884
Colaba Observatory 732
colatitude 283
collapse, orogenic 802
College of Newcastle on Tyne 887
College, Alaska 424
Collegio Romano 460
Collinson, D.W. 887
collision, continental 802
Coloumb gauge condition 217
Columbia College 886
Columbia Plateau 993
columbite 148
Columbus, Christopher 674, 987
column
–, fluid 432
–, geostrophic 124
columnar rolls 81
Commission for Terrestrial Magnetism and Atmospheric Electricity 407
Committee on Space Research (COSPAR) 408
communication cable 316
compaction 588
–, process 553
compass 66
–, amplitude 67
–, azimuth 67
–, card 435
–, design 66
–, deviation 67
–, dry 435
–, dry-pivoted 356
–, fish 66
–, fossil 817
–, liquid 67
–, marine 707
–, mariner's 434
–, observation 66
–, oscillating needle 435
–, saturable-core 438
–, ship 40
–, solar 765
–, steering 66
–, Sun 993
–, surveying 66
–, wet 435
compensation
–, Airy 447
–, coil 977
component
–, cryogenic 883
–, horizontal 49
–, in-phase 62
–, microelectronic 921
–, nondipole 342
–, nonzonal 774
–, out-of-phase 62
–, primary 586
–, terrigenous 566
–, viscous 760
composition
–, chemical 796
–, inner core 420
–, isotopic 796
comprehensive model 141
compressibility, isentropic 402
compression
–, adiabatic 212
–, effect 705
computational mineral physics 94
computing, parallel 103
concentration 248
concept, mean-field 192
condensation temperature 78, 89
condition
–, Dirichlet 466
–, field-blocking 575
–, Neumann 466
–, no-slip 76, 112, 646
–, number 216
–, Stewartson 650
conductance 72, 131
conduction
–, electrical 685
–, extrinsic 69
–, heat 75
–, intrinsic 69
–, ionic 685
–, proton 686
–, thermal 116
–, core 120
conductive anisotropy 219
conductivity 71, 901
–, anisotropic bulk 20
–, anomaly 20, 131, 309
–, bulk 244
–, distribution, global 309
–, electrical 62, 69, 72, 92, 93, 98, 109, 149, 192, 200, 204, 215, 223, 540, 641, 672, 812, 985
–, core 116
–, effect of impurities 117
–, inverse problem 671
–, mantle 288, 684
–, subseabed 224
–, Hall 453
–, high 541
–, ionospheric 317, 700, 814, 815
–, iron 116
–, land-ocean contrast 62
–, lateral variation 309
–, lattice 120, 688
–, mantle 867

conductivity *Contd.*
–, mean-field 194
–, melt 244
–, mineral 70
–, model 215
–, ocean 740
–, Pedersen 453
–, seafloor 72, 73, 234
–, solid rock 244
–, structure
–, oceanic 72
–, upper mantle 866
–, thermal 92, 93, 98, 107, 108, 110, 158, 605, 688
–, lateral variations 133
conductor
–, imperfect 644
–, perfect 186, 643
confidence region 938
configuration
–, multidomain 574
–, stable single 574
–, superparamagnetic 574
conglomerate test 608, 778, 797
conservation
–, mass 646
–, momentum 646
consistency test 608, 765
constant
–, barometric altitude 975
–, Boltzmann 69, 100, 105, 109, 367
–, crystalline 619
–, Debye 107
–, magnetoelastic 619
–, solar 508
–, Stefan-Boltzmann 688
–, terrain clearance 975
–, Weiss 932
constituent, ferrimagnetic 567
constrained inversion 221
continent 956
continental
–, boundary, Precambrian 142
–, drift 139, 607, 887, 980
–, shelf 63
–, slope 63
continuation, downward 131
continuum, deformation 804
contour, geostrophic 124, 135
contribution
–, diamagnetic 528
–, octupolar, long-term 961
–, paramagnetic 528
controlled source
–, audiomagnetotellurics (CSAMT) 224, 230, 671
–, electromagnetic (CSEM)
–, mapping 231
–, system 232
–, frequency-domain electromagnetics 230
convection 966
–, chaotic 291
–, chemical 73
–, chimney 74
–, compositional 73, 81, 89
–, core 80
convection *Contd.*
–, cyclonic 179, 859
–, free 18
–, in a plane layer 202
–, inner core 121
–, nonmagnetic rotating 74
–, quasigeostrophic 124
–, rapidly rotating 291
–, Rayleigh-Bénard 117
–, roll 76, 315
–, rotating experiment 275
–, solar zone 137
–, subsolidus 16
–, thermal 11
–, vacillating 76
–, zone 399
–, solar 506
Cook, James 988
Coolidge chondrite 798
cooling
–, rate 327
–, secular 128, 301
coordinates
–, geocentric 450, 683
–, geodetic 683
–, quasidipole 952
–, stretched 111
Copernican 361
Copernicus, Nicholas 397
Copley medal 61
core 884
–, accretion 89
–, adiabatic 82
–, gradient 106
–, boundary
–, crystallization 979
–, layer 111
–, outer 84
–, central 179
–, composition 77
–, conductivity 93
–, convection 80
–, dynamics 81
–, cooling 146
–, correlation 253
–, density 82
–, deficit 78, 371
–, differentiation 89
–, dynamics 290, 842
–, electrical conductivity 116
–, entropy balance 289
–, evolution 89
–, formation 79
–, heat loss 128
–, impurity, resistivity 117
–, inner
–, age 91
–, boundary 14
–, evolution 91
–, light elements 78, 421
–, magnetic
–, diffusion time 298
–, instabilities 117
–, melting temperature of iron 689, 692
–, motions 84
–, Ni 371
–, origin 89
core *Contd.*
–, planetary 274
–, potassium 79
–, properties 94
–, physical 91
–, radioactivity 288, 289
–, radius 124
–, semicircular half 884
–, sequestration 79
–, stably stratified 86
–, surface constraint 122
–, temperature 98
–, thermal conduction 120
–, turbulence 101
–, undertone 364
–, viscosity 104
core-mantle
–, boundary (CMB) 14, 84, 337, 898, 929
–, heat flux 752, 855
–, coupling 135, 348
–, exchange 79
–, reactions 150
–, transition zone (CMTZ) 970, 971
core-rigidity zone (CRZ) 970, 971
coring equipment 766
Coriolis
–, force 5, 6, 74, 81, 105, 180, 194, 291, 430, 630, 746, 852, 940
–, magnetohydrodynamics (CMHD)
–, ideal fluid for 650
–, rise time 298
corner frequency 771
coronal
–, hole 927
–, mass ejection (CME) 317, 505
corotating interaction region 663
corpuscularism, geomagnetic 357
correlation
–, approximation, second-order 194
–, inter-core 256
–, length 127
cosmic radiation 702
cosmochemistry 77
cosmogenic isotope 161
cosmology, Aristotelian 356
COSMOS satellite 59
COSMOS 49 satellite 39
COSPAR (see *Committee on Space Research*)
cost function 682
Couette flow 177
Coulomb
–, gauge 642
–, interaction energy 870
Coulomb, Charles 435
country rock 602
coupling
–, core-mantle 130
–, thermal 132
–, topographic 135
–, dissipative 132
–, electromagnetic 149, 651
–, gravitational 132
–, topographic 651
–, viscous 113, 651
covariance matrix 845, 848, 870

Cowling
–, conductivity 815
–, number 299
–, theorem 21
Cowling, Thomas George 137
Cox, Allan V. 139, 992
CPMN (see *Circum-pan Pacific Magnetometer Network*)
craton, Archean 141
creep
–, dislocation 105
–, Harper-Dorn 105
–, Nabarro-Herring 105
Creer, K.M. 887
C-response 277, 697
Cretaceous 134
–, normal polarity superchron (CNPS) 161, 294, 328, 329, 337, 339, 664, 749, 751, 928
–, Quiet Zone 328, 664
Cretaceous-Tertiary boundary 666
criteria
–, buoyancy 963
–, geochemical 963
–, geophysical 963
–, tomography 963
criterion, Schwarzschild 137
CRM (see *chemical remanent magnetization*)
crosstail current 927
crust
–, anorthositic 446
–, continental 128, 801
–, lunar 443
–, magnetization 791
–, magnetic 142
–, magnetization 350
–, oceanic 150, 208, 596, 779
–, magnetization 596
–, spreading 779
crustal field model 60
cryocooler 884, 886
–, pulse tube 886
cryogenic magnetometer 341, 992
cryogenics 886
cryptochron 664, 928
cryptocontinent 129
crystal
–, defect-free 494
–, lattice 554
–, preferred
–, alignment 148
–, orientation 549
–, twinning 455
crystallinity, degree of 515
crystallization
–, at inner core boundary 979
–, remanence 580
crystallography 535
CRZ (see *core-rigidity zone*)
CSAMT (see *controlled-source audiomagnetotellurics*)
CSB (see *China Seismological Bureau*)
CSEM (see *controlled source electromagnetic*)
C-sequence 664
Cunningham Medal 474
Cunningham, James 989
Curie
–, isotherm 889
–, law 480, 931
–, point 1, 116, 157, 581, 583, 626, 705
–, isotherm 596
–, temperature 23, 140, 157, 454, 481, 485, 514, 597, 609, 753, 871, 874, 889, 932, 980
Curie, Pierre 214
Curie-Weiss law 932
current
–, Birkeland 34
–, channeling 224, 277
–, Chapman-Ferraro 658
–, crosstail 927
–, displacement 21, 287, 640, 841
–, eddy 224, 225, 231, 308, 956
–, electric 192, 339
–, field-aligned 659, 700
–, Hall 453
–, induced 61, 171
–, Pedersen 453, 660
–, primordial 705
–, ring 863
–, discovery 865
–, system, ionospheric 812
–, thermoelectric 206
–, wedge 662
–, western boundary 853
curve
–, first-orde 879
–, magnetization 931
–, master 943
–, null-flux 85
–, thermomagnetic 537
cusp mantle 657
cycle
–, Dungey 33
–, heteroclinic 861
–, magnetic 179
–, solar 507
–, Milankovitch 332
–, seasonal 69
–, solar 160, 319, 405, 452, 700
–, substorm 34
–, sunspot 178, 347, 353, 505, 663
–, supercontinental 966
cycling
–, performance measure 61
–, thermal 789
cyclonic convection 859
cyclostratigraphy 665, 785
cylinder
–, geostrophic 6, 647
–, tangent 115, 430, 647, 946

D
d'Entrecasteaux, Antoine 989
d'Urville, Dumont 989
D region 452
D″
–, anisotropy 146
–, composition 149
–, layer 145, 929, 963
–, seismic properties 151
–, ultralow velocity zone 154
D'Entrecasteaux expedition 358, 435, 807
D^+ inversion 220
DAC (see *diamond anvil cell*)
daily variation 63, 308, 449, 450, 698, 858
–, disturbance 810
–, modulation 700
Dalat Observatory 734
damping 123
–, magnetic 80
–, time, magnetic 645
–, viscous 76
Danian-Selandian (Paleocene) stage boundary 567
Danzig 346
Darwin Observatory 717
Das ist der Rom Weg 674
data
–, global paleomagnetic 820
–, mineral-magnetic 525
–, multivariate magnetic 253
–, paleomagnetic 845
–, Thellier-Thellier 577
database
–, Chinese paleomagnetic 824
–, global paleomagnetic 928
–, MBD97 831
dating 944
–, –40-argon–39 age 832, 834, 835, 962
–, archeomagnetic 23, 29, 30
–, magnetic 31
–, paleomagnetic 23
–, radioisotopic 665, 834
–, relative 571
datum level 665
David 990
day length 320
–, variation 348, 419, 469
Day plot 266, 881
DC resistivity 71
–, response 233
de Acosta, José 460
De Bilt Observatory 725
de Borda, Jean-Charles 403
de Bougainville, Louis 989
de Broglie wavelength 215
de Castro, João 435, 988
de Houtman, Cornelis 988
de la Perouse, Jean-Franqois 989
De Magnete 357, 360, 404, 679, 776
de Moidrey, Joseph 461
De Mundo 361
de Nautonnier, Guillaume 674
De Rossel, E.P.E. 435
Dead Sea transform fault 243, 413
Debye
–, constant 107
–, length 640
–, solid 368
–, temperature 367
decadal variations 131
Decade of Geopotential Research 408
decay
–, mode 163
–, radioactive 205
–, time 128, 138
–, viscous 649
–, viscous 621, 944

Deccan traps 963
declination 66
 –, and inclination magnetometer (DIM) 717, 726
 –, magnetic 776
declinometer 434
decollement 802
decomposition
 –, Cholesky 216
 –, Helmholtz 217
 –, Lancsoz 216, 218
 –, toroidal-poloidal 85
decompression 601
deconvolution
 –, Euler 263
 –, Werner 263
dedolomitization 586
Deep Sea Drilling Project (DSDP) 597
deep sounding
 –, geomagnetic 307
 –, geophysical 242
Deep Space 1 795
deep-sea sediment 253, 781, 993
Definitive Geomagnetic Reference Field (DGRF) 411
deformation
 –, angular 802
 –, ductile 554, 555
 –, extensional 802
 –, plastic 803
 –, postintrusive 554
 –, tidal 365
deforming zone 804
degaussing 54
degeneracy, geostrophic 650
degree variance 351
degrees of freedom 47
dehydration reaction 246
Deimos 439
Delambre 384
Delaunay triangulation 26
Della Porta, Giambattista 156
demagnetization 156, 599, 600, 603, 759, 872, 943
 –, alternating field (AF) 498, 573, 665, 873, 884, 910, 943
 –, chemical 157
 –, curve 516
 –, low temperature (LTD) 912
 –, microwave 694
 –, shock 795
 –, thermal 512, 552, 604, 615, 839, 872, 943
  –, continuous 627
  –, remagnetization 986
demagnetizing
 –, energy 874
 –, field 620
  –, internal 615
density 932
 –, deficit 78
 –, discontinuity 99
 –, electron 60
 –, free charge 640
 –, function 46, 272
 –, gradient 82
 –, jump 82
density *Contd.*
  –, inner core boundary 99
 –, scale-height 13
 –, stratification 649
denucleation 500, 613
Denza, Francesco 733
Department of Terrestrial Magnetism (DTM) 58
 –, Universal Magnetometer 437
deposit
 –, high deposition-rate drift 831
 –, mineral 223
depositional remanent magnetization 758, 777
depth phase 903
derivative
 –, adiabatic 92
 –, Lagrangian 8
Descartes, René 357, 375
detachment zone 802
detector, SQUID 921
detrital remanent magnetization (DRM) 285, 535, 588, 781
 –, postdepositional 871
Detroit Seamount 962
deviation 66
 –, angular standard (ASD) 894
 –, compass 67
 –, maximum angular (MAD) 846, 849
 –, maximum intensity (MID) 846
device, superconducting quantum interference 885
dewar, superinsulated 886
D-function 385
DGRF (see *Definitive Geomagnetic Reference Field*)
diagenesis 554, 783, 806
 –, oxidative 531
diamagnetic 870, 931
diamagnetism 157, 525, 579
diamond
 –, anvil cell (DAC) 94, 99, 421, 503, 689, 690, 692
 –, anvil melting curve 918
DIDD (delta I-delta D) 715
dielectric permittivity 215, 954
difference, finite 303
differential rotation 166, 859
differentiation, planetary 77
diffraction, electron 215
diffusion 84
 –, after-effect 621
 –, coefficient 96, 105
  –, effective 181
 –, magnetic 161
  –, time 297, 298
 –, Ohmic 5, 120, 186
 –, slow 621
 –, Soret 691
 –, thermal 14
  –, time 298
 –, timescale 644
 –, turbulent 181
 –, viscous, time 298
diffusionless limit 85
diffusivity 74, 605
 –, chemical 101
 –, magnetic 8, 84, 92, 94, 117, 287, 641
diffusivity *Contd.*
  –, tensor 139
 –, molecular 17, 80
 –, thermal 12, 74, 80, 92, 121, 688
 –, turbulent 18, 305, 645
 –, viscous 305
DI-flux 710, 711
dike 596
 –, complexes 597
 –, contact 603, 604
 –, intrusion 604
 –, width 604
DIM (see *declination and inclination magnetometer*)
dimension, fractal 158
diogenite 797, 798
dip 602
 –, circle 711
  –, magnetic 776
 –, equator 700, 815
dipole
 –, array method 486
 –, central axial 898
 –, direction, Fisherian distribution 893
 –, equatorial 901
 –, family 307
 –, field 310, 897
  –, centered ancient 961
  –, geomagnetic 310
 –, geocentric axial (GAD) 681, 776, 817, 959
 –, hypothesis 281
 –, inclined 311
 –, moment 311
  –, paleomagnetic 159
  –, variation 159
  –, virtual 159, 159, 160
 –, off-centered 900
 –, radial 901
  –, off-centered 901
 –, tilted, offset 204
 –, window, Pacific 133, 314, 417
 –, wobble 773
dip-pole
 –, North Magnetic 989
 –, south 989
direction 48
 –, intermediate 327, 342
 –, minimum strain 556
 –, paleomagnetic dispersion 272
 –, principal 547
 –, true 960
Dirichlet condition 217, 466
discontinuity
 –, 410-km 687
 –, 660-km 687
 –, Mohorovičić 158, 208
Discovery
 –, Bay Observatory 723
 –, RV 723
discrete deformation 804
disease, neurodegenerative 51
disk
 –, dynamo 167, 202
 –, Faraday 167
dislocation creep 105
disorder, neurological 51

dispersion
–, angular 773
–, relation 76, 292
–, relationship 176
displacement
–, current 21, 287, 640, 841
–, thickness 112
dissipation
–, Ohmic 22, 128, 289, 641
–, viscous 22, 128, 290
dissolution, diagenetic 786
distortion
–, ellipsoidal 443
–, galvanic 277
–, magnetic 277
distribution
–, Bingham 45
–, Fisher 273
–, Gaussian 45
–, log-normal 161
–, magnetization 888
disturbance
–, daily variation 816
–, field 720
–, storm-time index 404
diurnal period 101
divergence, static 217
–, correction 217
Division of Terrestrial Magnetism 723
Dixon Observatory 738
djerfisherite 79
Dodson, James 358
Doell, Richard 992
Dolland, George 435
domain
–, lamellar 499
–, magnetic 490, 548
–, metastable 537
–, multi- 526
–, nucleation 613
–, pseudosingle 526
–, single 157, 526
–, spatial 769
–, structure 493, 591
–, superparamagnetic 526
–, transition 498
–, wall 490, 621, 873, 876
–, migration 601
–, pinned 613
Dombås Observatory 737
DomeC Observatory 724
Doornbos Memorial Prize 903
Doppler
–, velocimetry, ultrasonic 275
–, velocity measurement 398
double
–, disk dynamo 168
–, hexagonal close-packed structure (dhcp) 94
Doubting Thomas 138
Dourbes Observatory 725
downward continuation 122, 131
DP variations 700
drape 2
drift
–, continental 139, 957
–, mirror instability 334
–, westward 133, 347, 773
drilling 784
driver plate 913
driving force 966
DRM (see *detrital remanent magnetization*)
DSDP (see *Deep Sea Drilling Project*)
DSIR (see *New Zealand Department of Scientific and Industrial Research*)
Dst geomagnetic index 414, 739, 926, 951
DTM (see *Department of Terrestrial Magnetism*)
Dublin Observatory 720
duct flow 114
Duke Albert of Prussia 397
Duke of Sussex 403
Dulong-Petit value 93
Dumont D'Urville Observatory 710, 723
Dungey
–, cycle 33, 659
–, equations 333
Dunn, Samuel 358
Dutch Meteorological Institute (KNMI) 725
dyke 799
–, contact zone 603
–, intersection 802
dynamical systems theory 186
dynamics 772
–, molecular 105
–, rotational, inner core 425
Dynamics Explorer 33
dynamo
–, action 313
–, active 788
–, alpha^2 162
–, alpha-omega 165, 179, 181
–, Backus 163
–, Braginsky 164
–, Bullard-Gellman 166
–, condition 643
–, disk 167, 202
–, coupled 861
–, homopolar 861
–, double 168
–, Faraday 186
–, equation 214, 641
–, experimental 183
–, fast 186, 290
–, fluid, homogeneous 166
–, Gailitis 169
–, Herzenberg 170
–, homogeneous 164, 195
–, hydromagnetic 766
–, intermediate model 290
–, Karlsruhe 184, 195, 294
–, kinematic 188, 643
–, Lowes-Wilkinson 173
–, lunar 791, 794
–, martian 795
–, mean-field 192
–, MHD 175, 643
–, model-Z 174, 293
–, omega × j 197
–, periodic 200
–, spatially 200
–, planetary 203
–, Ponomarenko 175
–, precession-driven 288
–, quadrupolar 207
dynamo *Contd.*
–, Riga 184
–, Rikitake 173
–, satellite 203
–, screw 177
–, self-excited 643, 929
–, slow 186
–, solar 165, 178
–, spatially periodic, second order 170
–, strong-field 262
–, subcritical 652
–, theory 928
–, ionospheric 812
–, wave 161
–, model 775
–, weak field 202

E

E region 452
E&EM (see *electrical and electromagnetic*)
Earl of Cumberland 988
Early Cretaceous 929
Early Hesperian 503
Early Jurassic, GPTS 331
Early Permian 826
Earth 203, 204, 207, 439, 930
–, -air current 450
–, bulk
–, composition 213
–, silicate 956
–, core, establishment 742
–, crust 601
–, equator 956
–, figure of 208
–, habitability 863
–, hollow 357
–, inductor 435, 438
–, inner core 312
–, superrotation 304
–, ionosphere waveguide 308
–, magnetic field 37, 603, 656, 778, 897
–, strength 910
–, model
–, 1066A 422
–, PREM 422
–, viscoelastic 965
–, orbit 353
–, orbital
–, geometry 258
–, parameters 332
–, precession rate 442
–, response 308
–, rotation 54, 471
–, axis 959
–, pole 958
–, rotational energy 288
–, snowball hypothesis 780
–, solid 956
–, soul 361
–, structure 208
–, thermal evolution 127, 300
–, viscosity structure 966
earthquake
–, Chile 423
–, Kupatal 208
–, Loma Prieta 909
–, North Palm Springs 908

Earthquake Research Institute, University of Tokyo 862
East Antarctica 967
East India Company 357, 732, 988
East Pacific Rise 981
Eastindiamen 67
Ebre (Ebro) Observatory 461, 739, 740
eccentric dipole model 679
eccentricity 548
eclipse 471
 –, lunar 471
 –, solar 471
 –, timing of 469
Eddington, Arthur 61, 137, 468
eddy
 –, current 224, 225, 231, 308, 956
 –, diffusion 102
 –, diffusivity 101
 –, simulation 102
 –, turbulent 183
 –, viscosity 94, 102, 305
EDOS (see *electronic density of states*)
effect
 –, ageostrophic 853
 –, compression 705
 –, diamagnetic 60
 –, electrokinetic 985
 –, far-sided 283, 703, 772
 –, gyromagnetic 705
 –, Hall 705
 –, magneto-optical Kerr (MOKE) 495
 –, Nernst-Ettinghausen 705
 –, off-diagonal, condensed matter physics 705
 –, pinning 609
 –, rolling-over 565
 –, seismo-electromagnetic (SEM) 908
 –, seismomagnetic 602
 –, tectonomagnetic 696
 –, temperature-induced 560
 –, thermoelectric 214, 288, 705
effective viscosity 104
efficiency 617
 –, equation 301
 –, factor 290, 301
 –, thermoremanent 617
eGY (see *Electronic Geophysical Year*)
eigen analysis 608
eigenvalue 190, 535
 –, problem 845
eigenvector 535, 846
 –, column 845
 –, orthogonal 845
Einstein function 367
ejecta 789
Ekaterinburg Observatory 737
Ekman
 –, layer 76, 112, 113, 648, 843, 941
  –, equatorial 115
 –, number 75, 81, 111, 124, 175, 291, 297, 299, 365, 430, 647, 940
 –, pumping 76, 113, 124, 648
 –, spiral 113
 –, state 174
 –, suction 113
Ekman-Hartmann layer 112, 114, 653, 853
 –, stability 115
elasticity 91
E-layer 697
Electric and Magnetic Studies on Earthquakes and Volcanoes (EMSEV) 408
electrical
 –, and electromagnetic (E&EM) method 228
 –, anisotropy 671
 –, characteristics 361
 –, circuit, global 448
 –, conductivity 62, 69, 72, 92, 98, 109, 149, 192, 200, 204, 215, 223, 540, 641, 672, 685, 812, 985
  –, core 116
  –, effect of impurities 117
  –, inverse problem 671
  –, mantle 684
  –, subseabed 224
 –, current
  –, density 192
  –, gyre 63
 –, field, sensor 242
 –, potential, W form 985
 –, power transmission grid 316
 –, resistance tomography (ERT) 229
 –, resistivity 62, 92, 242, 985
electrode
 –, grounded 224
 –, metal salt 227
 –, noise 227
electrodynamics 640
 –, mean-field 192
electrogeotherm 70
 –, olivine 70
electrojet 34, 927
 –, equatorial 543, 697, 734, 815
 –, index 34
  –, auroral 509
 –, polar 700
electromagnet 977
electromagnetic (EM)
 –, coupling 149, 651
 –, exploration 844
 –, field, polarization 277
 –, frequency-domain 223, 230
 –, induction 863, 866
  –, technique 223, 954
 –, industrial use 223
 –, lake-bottom measurement 227
 –, land uses 228
 –, marine controlled source 231
 –, modeling 215, 219
 –, plane wave 230
 –, refraction 233
 –, regional study 242
 –, spectrum 308
 –, stress tensor 641
 –, tectonic interpretation 245
 –, time domain 223, 225, 231
 –, transient 954
electromagnetism, nonrelativistic 640
electron
 –, acceptor 249
 –, charge 109
 –, density 60, 452
 –, diffraction 215
 –, exchange 599
 –, holography 500
 –, hopping 685
electron *Contd.*
 –, ion recombination 452
 –, mobility 453
 –, overlap 870
 –, reflectance experiment 791
 –, reflectometer 502
 –, specific heat 93
 –, spin 870
 –, temperature 60
 –, total content (TEC) 452
electronic
 –, contribution to specific heat 370
 –, density of states (EDOS) 370
 –, semiconduction 69
 –, transition 79
Electronic Geophysical Year (eGY) 408
electrostatic
 –, force 841
 –, function 813
element
 –, abundance 77
 –, finite 103
 –, lithophile 77
 –, radioactive 79, 128
 [illegible]
 –, refractory 77
 –, siderophile 90, 421
 –, volatile 77, 149
ELF (see *extremely low-frequency*)
Elizabethville Observatory 725
ellipsoid
 –, finite strain 476
 –, oblate 562
ellipticity 842
Elsasser 887
 –, integral 86
 –, number 114, 118, 206, 263, 291, 297
Elsasser, Walter M. 214
Eltanin Leg–19 981
EM (see *electromagnetic*)
Emeishan basalts 819
emission, optical 693
Emperor chain 962
Emperor Seamounts 891, 963
Emperor/Line pair 962
EMSEV (see *Electric and Magnetic Studies on Earthquakes and Volcanoes*)
EMSLAB 245
Encyclopedia Britannica 931
energy 772
 –, absolute minimum 492
 –, activation 685
 –, anisotropy 492
 –, Coulomb interaction 870
 –, density, magnetic 351
 –, equation 300
 –, exchange 491, 870
 –, free 408, 421, 692, 693
 –, geothermal 158
 –, Gibbs free 95, 99
 –, gravitational 14, 128, 300, 362
  –, released 289
 –, Helmholtz free 99
 –, internal 11
 –, magnetic 339
 –, magnetocrystalline anisotropy 872
 –, magnetostatic 491, 493

energy *Contd.*
–, low 496
–, minimum, local (LEM) 494, 873
–, pinning 625
–, rotational, Earth 288
–, source 645
–, surface 491
–, thermal 870
English Heritage 26
enhancement, thermal 561
enthalpy, activation 106
entropy 11
–, balance, core 289
–, equation 12, 301
–, flux, turbulent 17
–, maximum method 353
–, of melting 100
–, specific 840
environment
–, alteration 597
–, cryogenic 886
–, suboxic 585
environmental magnetic signal 259
EOS (see *equation of state*)
Eötvös Loránd Geophysical Institute 715
ephemeral active region 506
Epicurus 156
episode, super fast 960
Epistle on the Magnet 356
Epistola Petri Peregrini de Maricourt 361, 808
E-polarization mode 221
epsilon iron 368
equation
–, Adams-Williamson 82, 84
–, azimuthally averaged 164
–, Boltzmann 137
–, constitutive 872
–, Dungey 333
–, dynamo 214, 641
–, efficiency 301
–, energy 300
–, entropy 301
–, Euler 263
–, Gibbs-Duhem 409
–, induction 118, 176, 178, 186, 188, 190, 214, 290, 307, 634, 940
–, integral, method 218
–, Laplace 84, 204, 377, 443, 449, 680, 888
–, uniqueness of solutions 466
–, magnetic Archimedean Coriolis (MAC) wave 634
–, Maxwell 21, 84, 192, 215, 229, 335, 954
–, mean-field 193
–, Mie-Grüneisen of state 96
–, momentum 634
–, Navier-Stokes 118, 191, 290, 335, 430
–, of state (EOS) 77, 91, 912
–, K-primed 107, 108
–, Poisson 229, 643
–, pre-Maxwell 287, 640
–, Rankine–Hugoniot 912
–, Schrödinger 94
–, Stokes-Einstein 105
–, temperature 634
–, thermal wind 852, 945
equation *Contd.*
–, vector diffusion 223
–, vorticity 75
equator, magnetic 10, 39, 60, 511
equatorial
–, dipole 994
–, electrojet 543, 697, 734
equilibrium
–, chemical 109
–, hydrostatic 107, 443, 957
equinox, September 452
equipartition field 181
equipotential surface 132, 136
equivalent
–, source method 158, 857
–, layer 857
erasure
–, diagenetic 253
–, postdepositional 253
Erebus, HMS 717
ergodicity 130
Eros 795
error
–, Gaussian 32
–, inclination 285
–, pitch and yaw 673
ERT (see *electrical resistance tomography*)
Ertel theorem 840
Esashi Observatory 734
escape, tectonic 802
Eschenhagen magnetograph 735
Eschenhagen, J.F.A.M. 436
Eskdalemuir Observatory 720
establishment, Beno 743
estimation
–, maximum-likelihood 46
–, scheme, data adaptive 866
–, spectral 353
Ettaiyapuram Observatory 732
eucrite 797, 798
EULDPH 263
Euler 679
–, angles 384, 393
–, deconvolution 40, 263
–, equation 263
–, pole 959, 962
–, Rodrigues parameters 390
Euler, Leonhard 358
Eurasia 823, 825
Eurasian
–, landmass 826
–, plate 803
Europa 203, 204, 207, 444
–, internal ocean 445
European plate 960
European Space Agency 59
eustacy, plate-driven 567
eutectic
–, solidus 154
–, temperature 99
event
–, Big Lost 834, 837
–, Bjorn 838
–, Blake 770
–, Cobb Mountain 836
–, deformation 607
–, Flores Sea 434
event *Contd.*
–, Gardar 838
–, Gilsa 838
–, Heinrich 531
–, Huckleberry Ridge 837, 838
–, hypothetical 310
–, impact 792
–, Kamikatsura 837
–, Laschamp 758, 761, 834
–, Mono Lake 761
–, precursory 344
–, Punaruu 837, 838
–, reheating 37
–, resurfacing 207
–, Santa Rosa 838
–, solar flare 404
evolution, thermal 127, 300
excess mass 956
exchange 871
–, energy 491, 870, 874
–, interaction 525, 872
–, quantum mechanical 875
excitation, thermal 933
excursion 311, 344, 928
–, Albuquerque Volcanoes/Pringle Falls 838
–, Big Lost 838
–, geomagnetic 311, 761, 835
–, Jamaica 838
–, Laschamp 311, 837, 838
–, magnetic field 770
–, Mono Lake 770
expansion
–, coefficient
–, compositional 73
–, thermal 73
–, phase 927
–, polynomial, Chebyshev 303
–, spectral 166
–, spherical harmonic 86, 303
–, thermal 74
–, coefficient 93
expansivity, thermal 95, 372
expedition
–, Antarctic 696
–, D'Entrecasteaux 435
experiment
–, electron reflectance 791
–, flying plate 789
experimental
–, dynamo 183
–, law 612
exploration 224
–, Arctic 359
–, electromagnetic 844
–, mineral 1, 477
–, oil 1
–, petroleum 477
–, polar 696
–, resources 477
explosion, Cambrian 966
expression, Cartesian 849
extremely low-frequency (ELF) 700
Eyrewell Observatory 735

F

f mode 399
F region 452

fabric
–, acquisition 565
–, composite 561
–, cryptic 563
–, inverse 514, 563
–, magnetic 579, 798, 920
–, mimetic 554
–, syndeformational 476
–, multicomponent 561
–, sedimentary 475
–, structure 565
face-centered cubic (fcc) iron 94
factor
–, anisotropy 20
–, efficiency 301
–, Lamb-Mössbauer 691
family
–, dipole 307, 896
–, quadrupole 307, 896
Faraday
–, disk 167
–, dynamo 186
–, law 186, 203, 287, 347, 640, 865, 955
Faraday, Michael 358, 920
fast
–, dynamo [illegible]
–, wave 651
fault
–, Arava 243, 413
–, fossil 244
–, North Anatolian 247
–, San Andreas 244
–, seal 244
–, transform 779
–, West, Chile 413
–, zone conductor (FZC) 247
fayalite 522
FCN (see *free core nutation*)
Felix, Boothia 989
FEM (see *frequency-domain electromagnetic*)
Fennoscandian shield 479
Fényi, Gyula 461
feroxyhite 526
Ferraro law of co-rotation 946
Ferraro, V.C.A. 61
Ferrers
–, normalization 381
–, normalized functions 380
ferric 522
ferrihydrite 513, 520, 526
ferrimagnet 249
ferrimagnetic 50, 513, 870, 931, 932
–, geological formation 480
–, rock 480
–, spin 874
ferrimagnetism 525
ferritin 51
ferrogabbro 597
ferromagnetic 116, 932
–, spin 874
ferromagnetism 525
–, parasitic 518
ferromanganese
–, oxide 783
–, oxyhydroxide 783
ferrosilite 522
ferrous 522
fiber
–, magnetic 357
–, suspension instrument 435
fibril 179
–, magnetic 182
FICN (see *free inner core nutation*)
field
–, AC 932
–, alternating (AF) 582, 596, 759
–, demagnetization 759
–, ambient geomagnetic 596
–, ancient 911
–, anomalous 140
–, coercive 874, 877, 881
–, controlled source 223
–, demagnetizing 613
–, dipole 171, 310
–, centered ancient 961
–, inclined 341
–, disturbance 720
–, downward continued 84
–, electric
–, rotating 705
–, recorder 72
–, electromagnetic, polarization 277
–, equipartition 181
–, external 448, 874, 877
–, energy 874
–, inducing 874
–, weak 498
–, geomagnetic 48
–, ancient 612
–, asymmetry 313, 337
–, hemispheric asymmetry 947
–, intensity 140
–, long-term behavior 929
–, mean 960
–, morphology 133
–, orbital energization 259
–, reversals 54, 139
–, spatial spectrum 771
–, time-averaged 817
–, time-dependent model 948
–, transition 140
–, gradient 49, 977
–, growth, runaway 652
–, homogeneous 264
–, inducing, aperiodic 844
–, intensity 797, 944
–, ancient 753
–, internal 448, 876
–, interplanetary 33, 829
–, line
–, closed 33
–, magnetic 7, 8
–, open 33
–, lithospheric 485
–, Lorenz 72
–, magnetic 776, 778, 931
–, alternation 31, 328
–, ambient 617
–, anomaly 543
–, applied 560, 931
–, crustal 140
–, Earth 37
–, excursion 770
–, external 871, 872
field *Contd.*
–, galactic 7
–, helicity 506
–, induction 872
–, instrument-generated 541
–, interplanetary 658
–, Mars 481
–, poloidal 22
–, potential 841
–, primary 225
–, primordial dipole 179
–, Saturn 198
–, secondary 225
–, self-consistent model 123
–, sensing 49
–, source 541
–, spherical harmonic model 502
–, stellar 508
–, Sun 179, 335, 505
–, toroidal 23, 173, 302
–, very-low-inducing 566
–, magnetizing 547
–, intensity 546
–, main
–, [illegible]
–, modeling 679
–, mean, definition 192
–, minimum detectable 920
–, nonaxial-dipole (NAD) 832
–, nondipole 311, 341, 344, 701, 901
–, standing 773
–, standing components 774
–, normal-polarity 626
–, octupolar 171
–, paleomagnetic 778
–, collection 765
–, time-averaged 281, 947
–, poloidal 174, 180, 302
–, potential 448
–, quadrupolar 171, 311
–, reference 396
–, regime, strong 652
–, region, open 659
–, reverse-polarity 626
–, SAD- 832
–, secondary 69
–, induced 224
–, self-demagnetizing 623, 876, 879
–, separation 448
–, solar magnetic 179, 335, 505
–, source, varying 541
–, station 739
–, steady 540
–, stepwise alternating 839
–, strength
–, peak alternating 533
–, geomagnetic 31
–, terrestrial surface 794
–, thermal 596
–, time-averaged 133, 285, 778
–, toroidal 174, 901
–, transitional 321, 326
–, weak, state 652
fifth element 361
filtering 857
filtration, band pass 920
fine-scale layering 148

finite-difference
–, method 216
–, scheme 190
finite-element method 217
Finnish Meteorological Institute 736
first principles calculation 370
first-order reversal curve (FORC) 253, 266, 533
–, diagram 267, 620
–, distribution 881
–, Preisach 267
fish 782
Fisher
–, direction 893
–, distribution 273, 552, 900
–, pole 893
–, precision parameter 894
–, statistics 272, 552, 608
Flamsteed, John 720
flare, solar 404, 505
flattening 802
–, degree 539
–, inclination 539
–, preferential 539
F-layer 145
Fleming, John Adam 273, 723
–, Medal 274, 888
Flinders bar 68
floating total-field magnetometer 65
flood basalt 150
flooding model 321
Floquet
–, parameter 201
–, theorem 201
Flores Sea event 434
flow (see also *flux*)
–, azimuthal 746
–, construction 87
–, convergence 86
–, Couette 177
–, divergence 86
–, duct 114
–, geostrophic 75, 76, 124, 174, 293, 746, 852, 941
–, helical 184
–, instability, mean 76
–, spiraling 165
–, tangentially geostrophic 86
–, time-varying 320
–, twin-vortex 659
fluctuation 192
–, thermal 575, 621
fluid
–, Boussinesq 74
–, column 432
–, dynamics 274
–, geophysical 840
–, dynamo 166
–, flow, helical 302
–, hydrothermal 986
–, imperfect 652
–, pressure torque 425
–, real 7, 177
–, rotating 646
–, rapidly 647
–, saline 244
flux (see also *flow*)
–, creation 10
–, curve 9
–, frozen 6, 645
–, heat 12
–, magnetic, frozen 841
–, Poynting 641
–, quanta 885
–, static 133
–, bundles 133
–, transformer, superconducting 884
–, tube 644
–, magnetic 7, 8
–, turbulent
–, entropy 17
–, heat 17
fluxgate
–, gradiometer 25, 28
–, magnetometer 72, 438, 483, 672, 710, 713, 714
–, sensor 26, 71, 921
–, spinner 883
–, theodolite 710, 858
fly ash 533
flyer plate 912
flying plate experiment 789
fold
–, test 608, 778
–, tight 802
fold, stretch-twist 187
foliation 538, 561, 799, 937
–, magnetic 548
–, parameter 938
footwall 802
foraminifera 665, 762
FORC (see *first-order reversal curve*)
force
–, buoyancy 12
–, centrifugal 135, 956
–, coercive 157, 518
–, Coriolis 74, 81, 105, 180, 194, 291, 430, 630, 746, 852
–, driving 966
–, electrostatic 841
–, gravitational 565
–, gyroscopic 852
–, hydrodynamic 565
–, Lorentz 10, 74, 85, 114, 187, 630, 641, 860
–, mean electromotive 194
–, ponderomotoric 198
–, remanent coercive 527
–, telluric 359
–, viscous 630
forcing
–, orbital 259
–, external 259
–, precessional 289
–, tidal 289
formation
–, ferrimagnet 248
–, geological, ferrimagnetic 480
–, planetismal 77
formula
–, Bonnet 379
–, Christoffel 379
–, Lawson 688
formula *Contd.*
–, Neumann 379
–, Rodrigues 385
formulation, Bayesian 123
Forster, Georg 402
forward modeling 142, 486
fossil
–, assemblage 956
–, compass 817
Fourier
–, analysis 351, 810
–, domain method 486
–, harmonic analysis 921
–, law 12
–, transformation 977
four-pole model 679
Foxe, Luke 988
Fra Mauro Apollo 14 site 791
fractal dimension 158
fracture zone 983
frame, mean-lithosphere 959
framework, Lagrangian 165
Fredericksburg Observatory 509, 723
free
–, core nutation (FCN) 135
–, electron model 109
–, energy, Helmholtz 367
–, inner core nutation (FICN) 426
–, shear layer 114
–, stream 111
freedom, degrees of 47
freezing point, depression 371
F-region 209
French Academy of Sciences 435
French National Magnetic Observatory 721
frequency
–, Brunt-Väisälä 649
–, buoyancy 649
–, Larmor 468
–, reversal 134, 320
frequency-domain electromagnetic (FEM) method 224, 225
Fresnel zone 126
Frezier 358
friction
–, tidal 128, 471
–, viscous 12
–, wheel 435
Friedrich Wilhelm III, King 403
Frobisher, Martin 674, 988
frozen-flux 645, 841
–, hypothesis 85, 320
–, theorem 6
fugacity, oxygen 71
fulgurite 593
function
–, basis 395
–, cost 682
–, D- 385
–, density 46, 272
–, Einstein 367
–, electrostatic 813
–, generating 379
–, Green 218, 888
–, Green-Kubo 105
–, impedance 955

function *Contd.*
  –, Legendre 395, 642
    –, associated 377
  –, log-Gaussian model 533
  –, Lowes–Mauersberger 898, 900
  –, orthogonal 396
  –, poloidal scalar 22
  –, potential 359
  –, probability-density 272
  –, response 219
  –, Schmidt 680
  –, temporal basis 949
  –, toroidal scalar 22
  –, transfer 670, 866, 953
    –, magnetotelluric 953
  –, Z-transfer 310
functional penalty 220
furnace, demagnetization 884
Fürstenfeldbruck Observatory 710, 729
fusion, latent heat 693
FZC (see *fault zone conductor*)

G

g-mode 399
gabbro 73, 596, 597, 983
  –, plutonic 982
GAD (see *geocentric axial dipole*)
Gailitis dynamo 169
Galilean
  –, satellites
    –, Jupiter 203
    –, Mars 203
  –, transformation 640
Galilee, RV 57–59
Galilei, Galileo 357
Galileo 506, 795, 888
  –, spacecraft 443
gallium 275
galvanometer, sine 438
Gambey, H.P. 436
gamma 542
  –, iron 691
    –, face-centered-cubic (fcc) structure 691
  –, process 336
gamma-epsilon-liquid triple point 691
gamma-ray spectrometer survey 478
Gansu province 820
Ganymede 203, 204, 207, 439, 445
  –, magnetic field 287
Gardar event 838
garlic 156
Garzoni, Leonardo 460
gas
  –, giant 77, 203, 204
  –, hydrate 234, 457
  –, pipeline 316
Gaspra 795
Gassendi, Pierre 357
Gaubil, Antoine 460
gauge, Coulomb 642
Gaunt integral 86
Gauss
  –, chron 783
  –, coefficient 9, 10, 84, 311, 346, 681, 888, 898, 948
  –, epoch 667
Gauss *Contd.*
  –, spherical harmonics coefficients 680
  –, theorem 378, 888
Gauss, Carl Friedrich 279, 435
  –, instruments 436
Gauss, RV 723
Gaussian
  –, distribution 45, 897, 898, 900
  –, error 32
Gay-Lussac, J.L. 403, 435
Gazelle expedition 359
GCM (see *ground conductivity meter*)
Gellibrand, Henry 280, 346, 357, 435, 707
GEM (see *global energy minimum*)
generating function 379
genetic algorithm 220
Geobacter metallireducens 782
geobiology 874
geocentric
  –, axial dipole (GAD) 681, 776, 959
    –, hypothesis 281–283, 817, 961
    –, model 820
  –, axial quadrupole 284
  –, octupole 681
[illegible]
Geochemical Earth Reference Model (GERM) 854
geochemistry 586
geocorona 657
geodynamo 287, 791, 838, 930
  –, dimensional analysis 297
  –, energy source 300
  –, numerical simulation: 302
  –, stability 432
  –, symmetry 306
GeoForschungsZentrum Potsdam 512, 729
geographic axis 311
Geographical Survey Institute of Japan 734
geoid 966
  –, coefficient 965
geological
  –, mapping 228, 230, 478
  –, timescale 253, 281
  –, unit 478
Geological Survey of Canada 236, 726
geologist 956
geology 477
geomagnetic
  –, activity 318
  –, axis 311
  –, change, fast 343
  –, corpuscularism 357
  –, field 48
    –, ancient 281
    –, asymmetry 313, 337
    –, core-based inversion 122
    –, hemispheric asymmetry 947
    –, intensity 140
    –, morphology 133
    –, orbital energization 259
    –, reversal 54
    –, spatial spectrum 771
    –, strength 31
    –, time-dependent model 948
    –, transition 140
  –, hazard 316
geomagnetic *Contd.*
  –, inclination 68
  –, information nodes (GIN) 716
  –, index 408, 508
  –, instability time-scale (GITS) 838
  –, jerk 88, 319
  –, metrology 711
  –, observatory data
    –, monthly 319
    –, yearly 319
  –, paleointensity, relative 257
  –, polarity
    –, history 666
    –, reversal 311, 324
    –, timescale (GPTS) (see beyond)
  –, potential 466
  –, reversal 161, 306
    –, archive 339
    –, frequency 133
    –, rate 354
    –, sequence 335
  –, secular variation 68
  –, spectrum
    –, natural 308
    –, spatial 350
    –, temporal 353
    –, variation 308
  –, storm 137, 316, 663
    –, frequency 316
Geomagnetic Network of China 728
geomagnetic polarity timescale (GPTS) 320, 328, 328, 329, 332, 544, 664, 778, 834
  –, Cenozoic 329
  –, Early Cretaceous 330
  –, Early Jurassic 331
  –, Late Cretaceous 329
  –, Late Jurassic 330
  –, M-sequence 330
  –, Triassic 331
geomagnetically induced currents (GIC) 316
geomagnetism 407
  –, history 355
geomancy 356
geometry
  –, broadside 232, 236
    –, electric dipole-dipole 233
  –, in-line 232, 236
    –, electric dipole-dipole 233
  –, two-cell 965
geophysical method 228
geophysics 53, 943, 992
  –, longest historical time series 509
Geostationary Meteorological Satellite (GMS) 716
geostrophic 85
  –, average 648
  –, basis 86
  –, column 124
  –, contour 124, 135
  –, cylinder 647
  –, degeneracy 650
  –, flow 75, 76, 124, 174, 293, 746, 852, 941
  –, mode 125, 647
  –, motion 136, 843
  –, pressure 87

geostrophic *Contd.*
–, surface 940
–, velocity 746
geostrophy, tangential 131
geotherm 227
geothermal
–, investigation 230
–, resources 245
geothermometer, conductivity 69
GERM (see *Geochemical Earth Reference Model*)
German Samoan Islands 736
German South-Polar Expedition 736
GGP (see *Global Geodynamics Project*)
giant
–, gas 203, 204, 439
–, Gaussian distribution model 897
–, ice 203, 204, 439
Gibbs
–, free energy 14, 95, 99, 408, 421, 692
–, relation 12
Gibbs-Duhem
–, equation 409
–, relation 100
GIC (see *geomagnetically induced currents*)
Gilbert, William 360, 435
Gilsa event 838
GIN (see *geomagnetic information nodes*)
GITS (see *geomagnetic instability timescale*)
glacial isotopic stage 783
glacial-interglacial variations 569
glacier 958
glass
–, submarine basaltic 756
–, volcanic 756
glauconite 932
global
–, electrical circuit 448
–, stacks 160
Global Boundary Stratotype Sections and Points (GSSP) 568
global energy minimum (GEM) 493
–, domain state 498
Global Geodynamics Project (GGP) 365
Global Oscillation Network Group (GONG) 398
Global Paleomagnetic Computerized Database 960
Global Paleomagnetic Database 820
Global Positioning System (GPS) 317, 469
GMS (see *Geostationary Meteorological Satellite*)
Gnangara Observatory 719
Godhavn Observatory 736
Goertzel algorithm 810
goethite 251, 516, 520, 526, 560, 760, 783, 786
–, magnetosomes 782
–, nanocrystalline 520
Goetz, Edmund 461
Gold Medal of the Royal Astronomical Society 844
GOLF, helioseismology instrument 398
Gondwana 817, 820, 823, 966
GONG (see *Global Oscillation Network Group*)
goodness-of-fit (GoF) 536
Göttingen Magnetic Union 359, 720, 731, 736
Göttingen Observatory 729
Göttinger Magnetischer Verein 42, 280, 403
Gouin, Pierre 461
GPR (see *ground-penetrating radar*)
GPS (see *Global Positioning System*)
GPTS (see *geomagnetic polarity time-scale*)
gradient
–, adiabatic 80, 106
–, biconjugate 221
–, coil 977
–, field 49, 977
–, geothermal 603
–, local 48
–, magnetic 977
–, paleogeothermal 606
–, superadiabatic 18
gradiometer 27
–, fluxgate 25, 28
Graham fold test 607
Graham, George 435
grain
–, anisotropy 549
–, ferrimagnetic 566, 571
–, multidomain 571
–, flattening 539
–, magnetic 542
–, morphology 566
–, multidomain (MD) 266, 571, 573, 755, 873
–, nonequidimensional 548
–, orientation 535
–, paramagnetic 566, 571
–, pre-solar 795
–, pseudo-single domain (PSD) 527, 571, 573, 617, 755, 873
–, shape 875
–, single domain (SD) 266, 527, 535, 571, 598, 782, 873, 943
–, stable 249
–, size 248, 526, 875, 879, 933
–, dependence 571
–, distribution- 525
–, proxy 577
–, spheroidal 610
–, strain state of the 526
–, superparamagnetic (SP) 268, 528, 535, 571, 873
–, ultrafine 518
–, volume 612
grand minima 507
Grandami, Jacques 357, 460
granite 554
–, sedimental type 554
–, S-type 563
Granite-Rhyolite province 141
granulite 40
granulometry 624
–, magnetic 577
graphite 69
gravitation law 443
gravitational
–, coupling 132
–, dynamo 980
–, energy 362
–, potential 362
gravity
–, meter, superconducting 365
–, torques 136
–, wave 364
gray body 688
Green
–, function 218, 888
–, theorem 466, 974
–, third identity 451
Green-Kubo functions 105
greenstone
–, belt 40
–, formation 141
Greenwich meridian 311
Greenwich Royal Observatory 61, 713, 720
greigite 248, 260, 455, 457, 526
–, authigenic 783
–, magnetic properties 455
–, prismatic 456
–, sedimentary 586
–, synthetic 520
Gresham College, London 707
Gresham, Professor of Astronomy 280, 357
GRM (see *gyromagnetic remanent magnetization*)
Grocka Observatory 729
ground conductivity meter (GCM) 230
ground-penetrating radar (GPR) 226, 229, 230
–, survey 226
grounded wire method 225
groundwater 1, 223
group sum sequences 810
group table 306
growth
–, dendritic 419
–, phase 927
–, rate 176
Grubb, Thomas 473
Grüneisen
–, parameter 13, 91–93, 95, 100, 107, 366
–, ratio 83
GSFC model 346
GSSP (see *Global Boundary Stratotype Sections and Point*)
GTPS (see *Geomagnetic Polarity Timescales*)
Guangdong province 821
Guangzhou Observatory 728
Gubbio, Italy 669
Guérard, Jean 674
Güimar Observatory 740
Gulf Company 263
Gulf Research and Development 438
Gulf Stream 945
Gulmarg Observatory 732
Gunter, Edmund 280, 357, 707
Gutenberg, Beno 469
gyrocompass 68
gyromagnetic
–, effect 705
–, remanent magnetization (GRM) 455, 456
gyroscope 425, 715
–, Sperry 438
gyroscopic force 852

H

habitability, Earth 863
hafnium-tungsten isotopic observations 90
Hale
–, law 506
–, polarity laws 506
Hale, G.E. 468, 506
Haleakala volcano 836

half-life 79
Hall
–, conductivity 453
–, current 453
–, effect 705, 841
–, layer, ionospheric 660
–, sensor 656
Halley
–, Atlantic chart 675
–, four-pole model 679
–, magnetic core 357
Halley Bay 376
Halley, Edmond 357, 375
Halleyan lines 375, 676
Halliburton Observatory 721
Halloween storm 405, 509
hanging wall 802
Hankel transform 233
Hansteen, Christopher 358, 376, 435
harmonics
–, high-order 898
–, low-order 898
–, rectangular 397
–, potential 467
–, sectorial 680
–, spherical 103, 204, 214, 377, 680, 899
–, analysis 449
–, cap 395
–, coefficients 680
–, expansion 86, 303, 356
–, finite series 467
–, model 315
–, sectorial surface 382
–, solid 380
–, surface 378
–, tesseral surface 382
–, transformation 383
–, vector 382, 387
–, spline 123
–, surface 351
–, tesseral 680
–, zonal 314, 680
Harper-Dorn creep 105
Harrison, Edward 357
Harrison, John 674
Harry, Benjamin 989
Hartebeesthoek Observatory 739
Hartland Observatory 710, 720
Hartmann
–, layer 644
–, number 114, 299, 649
Hartmann, Georg 357, 397
Haruna dacite 696, 872
Harvey, William 361
hat matrix 869
Hatizyo Observatory 734
Hawaii
–, chain 962
–, hotspot 246, 892, 961, 964
–, lava 912
–, lineation 330
Hawaiian-Emperor chain 779, 963
Haynald Observatory 461
hazard, geomagnetic 316
HCL 687
hcp iron 916
Heard Island Observatory 717
heat
–, capacity, mantle 128
–, Carnot engine 290
–, conduction 75
–, equation 12
–, flux 12, 929
–, adiabatic 15
–, core-mantle boundary (CMB) 855
–, global average 855
–, tomographic 129
–, turbulent 17
–, internal sources 12
–, latent 74, 92, 128, 206
–, radiogenic 121
–, specific 12, 80, 92, 93, 95
–, electron 93
–, electronic contribution 370
–, transfer
–, radiative 688
–, supercritical 276
heating
–, adiabatic 128
–, conductive 605
–, dielectric 694
–, effects 360
–, laser- 690
–, radioactive 300
–, stepwise 563
–, wire 690
heating/cooling unit 977
heating-cooling steps 944
heavy metal concentration 533
Hebei province 820, 821
Heinrich
–, event 531
–, layer 762
Hel Observatory 728
helicity 55, 87, 170, 175, 190, 290, 307
–, mean 201
–, specific 852
helioseismology 137, 182, 398
heliosphere 863
helium
–, abundance 400
–, liquid 886
–, magnetometer 850
Hell, Maximilian 460
Hellas 795
–, impact basin 503
Helmholtz
–, coil 203, 714, 756, 921
–, decomposition 217
–, free energy 99, 367, 408, 693
–, potentials 217
Helmholtz-Gaugain coil 438
hematite 24, 32, 251, 267, 495, 526, 536, 560, 618, 668, 759, 760, 786, 794, 873, 932, 942
–, crystalline lattice 519
–, ilmenite solid solution 794
heme iron 51
hemisphere
–, boreal 464
–, Indo-Atlantic 964
–, Pacific 994
Henkel plot 591
Hercynian orogenic pulse 785
hercynite 526
Hermanus Observatory 739
Hermitian operator 139
Herzenberg dynamo 170
Hess, Harry 55
heterogeneity
–, inner core 429
–, lower mantle 831
Hevelius, Johannes 375
hexagonal close packed (hcp) iron 94, 367
Hexi-Corridor 825
Hexizoulang 819
hidden ocean 81
Hide, R. 887
Higgins-Kennedy paradox 401
highlands, lunar 446
high-resolution deep-sea record 312
Hilbert matrix 949
Himalaya terrane 819
Hipparcos astrometric satellite 958
hippocampus, human 50
historic secular variation (HSV) 766
historical field observation 773
history
–, accretion 799
–, geological 595
–, metamorphic 480
–, postaccretion 799
holder, automatic 921
hole, coronal 663
Hollandia Observatory 725
hollow Earth 357
holography 500
homogeneous inverse problem 467
Hong Kong 821
Honolulu Observatory 723
Hooke, Robert 357
Hopf bifurcation 417
Hopkinson peak 513, 943
Hornsund Observatory 728, 737
host component 604
hotspot 128, 150, 779, 959
–, apparent polar wander 959
–, Atlantic 962
–, fixity 962
–, Hawaiian 246, 892, 962
–, long-lived 962
–, mantle 245
–, motion 959, 963, 965, 967
–, primary 963
–, reference frame 21, 957
–, track 961, 964
Hough, S.S. 852
howardite 798
HSV (see *historic secular variation*)
Huancayo Observatory 43, 734, 807
Hubei province 821
Huber function 868
Huckleberry Ridge event 837, 838
Hugoniot state 912
Humboldt, Alexander von 402
Hunan province 821
Hurbanovo Observatory 728
Hyderabad Observatory 732
hydrocarbon
–, reserves 478
–, reservoir 72

hydrogen 150
–, geocorona 657
Hydrographic and Oceanographic Department 734
Hydro-Quebec power grid 317
hydrostatic
–, equilibrium 107
–, state 300
hydroxide 513
hyperdiffusion 101, 102
hypothesis
–, dipole 281
–, frozen-flux 85, 320
–, GAD 961
–, paleoclimatic 780
–, standing field 321
–, Taylor 103
hysteresis 497, 589, 874
–, cycle 701
–, loop 266, 527, 540, 874, 875, 878–881, 979
–, analysis 514
–, outlines 880
–, Stoner-Wohlfarth 876
–, magnetic 518
–, measurement 874
–, parameter 879
–, shape 880
–, total 874

I

IA (see *induction arrow*)
IAGA (see *International Association of Geomagnetism and Aeronomy*)
IATME (see *International Association of Terrestrial Magnetism and Electricity*)
Iberian Massif 244
Iberian Peninsula 243
ice
–, cover 960
–, giant 204
Iceland 993
ICM (see *induction coil magnetometer*)
ICME (see *interplanetary coronal mass ejection*)
ICW (see *inner core wobble*)
Ida 795
ideal
–, black body 688
–, body theory 143
identity, third 451
IEEY (see *International Equatorial Electrojet Year*)
IGGCAS (see *Chinese Academy of Sciences*)
Ignatius of Loyola 460
igneous intrusion 40
IGRF (see *International Geomagnetic Reference Field*)
IGY (see *International Geophysical Year*)
IIG (see *Indian Institute of Geomagnetism*)
IITPW (see *inertial interchange true polar wander*)
ilmenite 519, 526
ILP (see *International Lithosphere Program*)
IMAGE mission 865
IMF (see *interplanetary magnetic field*)
IMGA (see *Indonesian Meteorological and Geophysical Agency*)
impact
–, event 792
–, extraterrestrial 40
impedance 671
–, contrast 82
–, function 955
–, magnetotelluric 277
–, tensor 223, 242, 670, 671, 866
Imperial College London 887
impulse
–, response 236
–, sudden 663
impurity
–, scattering 688
–, silicon 110
–, sulfur 110
inclination 802, 943
–, anomaly curve 285
–, error 285, 588
–, geomagnetic 68
–, magnetic 777
–, magnetization 588
–, shallowing 584
inclinometer 435
inclusion 751
–, refractory 797
independence 574, 612
–, Norwegian 377
index
–, aa 509
–, auroral electrojet 509
–, equatorial storm 511
–, geomagnetic 408, 414
–, AA 508
–, activity 318
–, AP 508
–, magnetic 509
–, range 509
–, solar cycle 508
–, structural 263, 264
India-Asia collision 824
Indian plate 960
Indian Institute of Geomagnetism (IIG) 732
Indian National Academy of Science 887
Indian Ocean 993
indicator
–, magnetostratigraphic 664
–, petrofabric 564
Indo-Atlantic hemisphere 964
Indonesian Meteorological and Geophysical Agency (IMGA) 734
Indonesian Observatory 717
induced magnetization 140, 856, 888
inducing field, aperiodic 844
inductance
–, mutual 186
–, self- 186
–, SQUID loop 885
induction
–, arrow (IA) 412, arrow 808
–, coil 71
–, magnetometer (ICM) 72, 713
–, sensor 72
–, electromagnetic 863
–, equation 118, 176, 178, 186, 188, 190, 214, 290, 307, 634, 940
–, global 867
induction *Contd.*
–, magnetic 974
–, field 872
–, satellite data from 413
inductor 435
–, earth 438
inertia 291, 957
–, axes 966
–, moment of 77, 957, 966, 967
–, tensor 957, 966
inertial
–, interchange true polar wander (IITPW) 966
–, wave 365, 843
inertio-gravity wave 364
ING (see *Istituto Nazionale di Geofisica*)
in-line
–, electric dipole-dipole geometry 233
–, geometry 236
–, system geometry 232
inner accretionary belt 141
inner core 312, 901, 929
–, age 91
–, anisotropy 418
–, attenuation 427
–, boundary 14, 74, 83
–, density discontinuity 82, 99
–, composition 420
–, convection 121
–, dynamo operation 425
–, Earth's superrotation 304
–, geodynamo 432
–, gravitationally 426
–, growth 928
–, heterogeneities 429
–, magnetic diffusion time 298
–, oscillation 422
–, PKJKP 433
–, rotation 294, 423
–, dynamics 425
–, seismic velocity 427
–, sensitive mode 428
–, structure 419
–, tangent cylinder 430
–, viscosity 105
–, wobble (ICW) 426
Inner Mongolia province 820
in-phase component 62
input, detrital 248
Inquisition, Spanish 156
instability 119, 120
–, baroclinic 648
–, classes of 118
–, critical level 119
–, ideal 119
–, magnetic 118, 293
–, parametric 365
–, Rayleigh-Taylor 649
–, resistive 119
instrument
–, absolute 710
–, fiber-suspension 435
–, geomagnetic 56
–, La Cour 438
–, magnetic, second generation 438
–, medical diagnostic 541
–, photographically self-registering 436
instrumentation 434

insulation, thermal 885
insurance, mathematics 280
integral
–, Elsasser 86
–, Gaunt 86
integration, thermodynamic 99
intensity 589
–, absolute 890
–, Gauss determination 278
–, magnetic 601
interaction
–, dipolar 541
–, electrostatic 579
–, exchange 872
–, magnetostatic 268, 574, 872
–, inter-grain 268
–, parameter 649
interbedding 233
interferometry, very long baseline (VLBI) 426, 469
INTERMAGNET (see *International Real-time Magnetic Observatory Network*)
International Association of Geomagnetism and Aeronomy (IAGA) 407
International Association of Terrestrial Magnetism and Electricity (IATME) 407
International Council of Scientific Unions 407
International Decade of Geopotential Research 711, 743
International Equatorial Electrojet Year (IEEY) 408
International Geodetic and Geophysical Union 407
International Geological Congress 568
International Geomagnetic Reference Field (IGRF) 411
International Geophysical Year (IGY) 42, 61, 333, 461, 708, 717, 720, 807, 844
International Lithosphere Program (ILP) 408
International Meteorological Organization 407
International Polar Year (IPY) 359, 708, 723, 723–726, 736, 737, 844
–, First 407
–, Second 407
–, Third 407
International Quiet Sun Year 844
International Real-time Magnetic observatory Network (INTERMAGNET) 711, 715
International Seismological Centre 209
International Service of Geomagnetic Indices 512
International Union of Geodesy and Geophysics (IUGG) 407, 902
interplanetary
–, coronal mass ejections (ICME) 404
–, field 829
–, magnetic field (IMF) 658, 676, 927
intrusion
–, cooling 603
–, igneous 40
–, layered 141
inverse
–, fabric 514
–, modeling 142, 487
–, theory 467
inversion
–, constrained 221
–, D^+ 220
–, helioseismic 508
–, Schmucker 220
investigation, geothermal 230
Io 203, 204, 439, 444
–, eccentricity 444
ion
–, cyclotron wave 333
–, drift meter 59
–, propulsion unit 795
ionic conduction 685
ionization potential 33
ionosphere 60, 449, 452, 696
ionospheric
–, conductivity 317, 700, 814, 815
–, current system 812
–, dynamo theory 812
–, physics 61
IPY (see *International Polar Year*)
Ireland, magnetic survey 473
Irkutsk Observatory 737
IRM (see *isothermal remanent magnetization*)
iron 526, 618, 873
–, body-centered cubic (bcc) 367, 421, 691
–, carbonate and manganese 932
–, conductivity 116
–, epsilon 368
–, face-centered-cubic 421
–, ferromagnetic 789
–, formation 141
–, Kursk 141
–, heme 51
–, hexagonal close-packed (hcp) 367, 421, 691, 692
–, high pressure melting curve 693
–, hydrophosphate 521
–, melting temperature 98, 689, 692
–, metallic 526, 629, 788
–, meteorite 77, 796, 797
–, molten 313
–, oxide 248, 525
–, ferrimagnetic 23, 569
–, fine-grained 533
–, hydrous 519
–, paramagnetic 789
–, phase diagram 421, 691
–, silicate 522
–, alloy 916
–, storage 50
–, sulfide 249, 253, 454, 457, 520, 586, 786
–, authigenic 260, 783
–, fine-grained 533
–, greigite 455
–, superparamagnetic 623, 789
iron-nickel 618, 788
–, alloy 873
–, ordered phase 701
–, phase 796
iron-titanium oxide 526, 583
irreducibility 378
Irtysh-Zaysan region 823
Irving, E. 887
isentropy 300
Ishtar Terra 446
Isidis impact basin 503
isochron analysis 836
isoclinics 358
isogonics 357, 676
isotherm, Curie 889
isothermal remanent magnetization (IRM) 51, 455, 535, 572, 589, 596, 621, 668, 758, 760, 979
–, acquisition curve 979
–, anisotropy 536
–, component analysis 532
–, grain-size dependencea 536
–, normalization 790
–, saturation 536
–, triaxial 591
isotope
–, abundance 79
–, cosmogenic 161, 760
–, radioactive, short-lived 796
–, radiogenic 605
–, stratigraphy 320
isotropic point 500, 516
isotropy, vertical transverse 147, 155
Istituto Nazionale di Geofisica (ING) 733
itabrite 40
iteration, Krylov 216
IUGG (see *International Union of Geodesy and Geophysics*)
Izu-Oshima volcano 986

J

jackknife method 870
jacobsite 526
Jaipur Observatory 732
Jamaica excursion 838
James, Thomas 988
Japanese lineation 330, 331
Jaramillo subchron 834, 837, 838
Jassy Observatory 728
Jeans, James 137
Jeffreys, Harold 468, 743
jerk, geomagnetic 88, 319
Jesuits 460
jet stream 294
Jiangsu province 821
John Adam Fleming Medal 274, 888
Jones, Thomas 435
Joule
–, dissipation 288
–, losses 641
Joy, law 506
Juan de Fuca Ridge 981
Juan de Fuca subduction zone 245
Junggar basin 817, 819
Jupiter 90, 203, 204, 207, 439, 446, 865
–, Galilean satellites 203
–, tilted dipole 206
Jurassic Quiet Zone 331, 664
Jurassic–Triassic boundary 928

K

K1 tide 816
Kakadu Observatory 710, 717
Kakioka Observatory 715, 733, 951
Kakioka Automatic Standard Magnetometer (Kasmmer) 715
Kalliroscope flakes 842
kamacite 788, 796

Kamikatsura event 837
Kanoya Observatory 733
Kanozan Observatory 734
Karelian Craton 751
Karlsruhe dynamo 184, 195, 294
Kater, Henry 890
Kazakhstan block 819, 823
Kazan Observatory 737
Keathley lineation 330, 331
Kelvin 214
–, theorem 8, 103, 840
Kelvin-Helmholtz instability 334
Kentucky magnetic anomaly 143
Keping area 822
Kepler, Johannes 357
Kerguelen Island 993
Kerr rotation 499
Kew 359
–, magnetometer 279
–, pattern 436
Kew Observatory 333, 720
kewness, magnetic anomaly 857
key paleopole 839, 840
Kiaman reversed polarity superchron 339, 752
Kilauea 984
Kilns 24
kimberlite pipe 477
K-index 42, 717, 729
kinematic 7
–, assumption 188
–, dynamo 188
–, viscosity 74, 102, 104, 111
kinematics 963
King Charles II 720
King Friedrich Wilhelm III 403
King's College of Durham University 887
Kircher, Athanasius 357, 463
Kiruna magmatic iron ore 477
KNMI (see *Dutch Meteorological Institute*)
Kodaikanal Observatory 732
Koenigsberger 870
–, parameter 597
–, ratio 481, 597
Korean Peninsula 818
Kp-index 719, 720
Kreil, Karl 728
Kroger-Vink notation 685
Kronecker
–, delta 379
–, tensor 193
Krylov
–, iteration 216
–, space solver 221
Kuche depression 822
Kunlun Mountain range 819
Kunlun Terrane 819
Kupatal earthquake 208
Kursk formation 141, 477
Kyoto World Data Center 512

L

L'Aquila Observatory 733
La Cour instrument 438
La Habana Observatory 734
La Hire 721
La Pérouse expedition 358
La Pérouse, Jean-Françoise 435
laboratory
–, heating 911
–, TRM 612
Lagrange multiplier 87, 123
Lagrangian
–, average 102, 103
–, derivative 8
Lake Carty 227
Lake sediment 577
Lake St. Croix PSV data 775
Lake Tecopa, California 829
Lake Washington 227
Lake Woelfersheimer 228
Lamb, Horace 61
Lamb-Mössbauer factor 691
Lambert, Johann Heinrich 358
Lamé parameter 147
lamellae structure 598
Lancaster, Sir James 988
Lancsoz decomposition 216, 218
Landau-Lifshitz-type structure 499
Langel, Robert A. 465
Langevin alignment function 541
Langmuir probe 59
Lanzhou Observatory 728
Laplace
–, equation 84, 204, 377, 443, 449, 680, 888
–, uniqueness of solutions 466
–, tidal equation 813, 814
–, transform 176
Large Aperture Seismic Array (LASA) 433
large-eddy simulation 102
Larmor
–, frequency 468
–, precession 468
Larmor, Joseph 468
LASA (see *Large Aperture Seismic Array*)
Laschamp
–, excursion 161, 311, 837, 838
–, event 758, 761, 834
laser
–, heating technique 690
–, ranging, lunar 471
Late Jurassic 929
Late Permian 818, 821
Late Precambrian 780
Late Silurian 928
Late Triassic 332
latent heat 92, 128
latitude
–, anomaly 961
–, paleomagnetic time-averaged 281
lattice 201
–, cell 201
–, conductivity 688
–, defect 621
–, dynamic 94, 367
–, preferred orientation 148
Laurussia 817
lava 597
–, Hawaiian 912
–, Icelandic 900
–, lamp 864
–, radiometrically dated 666
law
–, Ampère 114, 347, 640
–, Archie 232
law *Contd.*
–, Biot-Savart 955
–, Birch 77
–, Bodes 706
–, Curie 480
–, Curie-Weiss 932
–, Faraday 186, 203, 287, 347, 640, 955
–, Fourier 12
–, gravitation 443
–, Hale 506
–, polarity 506
–, Joy 506
–, Lenz 425, 643, 746
–, Lindemann 401
–, of additivity 754
–, Ohm 21, 84, 114, 192, 287, 335, 641
–, reciprocity 943
–, Wiedemann-Franz 116, 120, 205
Lawson formula 688
Lawson, A.W. 688
layer
–, asthenospheric conducting 73
–, Bénard 650
–, boundary 145, 152, 174, 186, 642
–, thermal 132
–, D″ 145
–, E- 697
–, Ekman 76, 112, 113, 648
–, equatorial 115
–, Ekman-Hartmann 112, 114, 653
–, stability 115
–, F- 145
–, free shear 114
–, Hartmann 644
–, loess 667
–, magnetic 141
–, magnetized 158
–, mushy 421
–, paleosol 667
–, resistive 72
–, seismic 597
–, shear 431
–, stably stratified 87
–, Stewartson 115, 432, 649
–, stratified 81, 632
–, stably 191
layering
–, chemical 151
–, fine-scale 148
Le Maire, Jakob 988
Le Monnier 358
leakage, torque 131
Learmonth Observatory 717
least-squares
–, analysis 86, 356, 359
–, approach 284
–, computation 536
–, fit 680
leeway 66
Legendre
–, function 395, 642
–, associated 377
–, polynomials 350, 364, 378, 899
Lehmann, Inge 208, 468
Lehnert wave 651
Leiden Observatory 725
Leirvogur Observatory 736, 737

LEM (see *local energy minimum*)
length of day (LOD) 88, 149, 320
–, variation 348
–, decadal 419, 469, 470
–, long-term 471
length, Debye 640
Lenoir, Étienne 435
Lenz law 4, 5, 425, 540, 643, 746, 810
Leoville meteorite 798
lepidocrocite 24, 513, 520, 526, 560
Lerwick Observatory 720
level
–, surface 974
–, survey 975
leverage 869
–, point 869
Levi-Civita tensor 193
Lewis number 297
Lhasa
–, area 825
–, terrane 819
Lhasa Observatory 728
Liaoning province 821
Liapunov exponents 187
light elements, core 421
lightning 353, 592, 596
–, overprint 593
–, remanence 593
–, strike 224, 592
limit
–, diffusionless 85
–, static 216
Lindemann law 401
Lindemann, F.A. 116
line
–, agonic 993
–, bedding intersection 555
–, cleavage intersection 555
–, isogonic 375
–, neutral 927
–, semiannual 308
Line Island Seamounts 891
linear variable differential transfonmer (LVDT) 977
linearization, thermodynamic 13, 16, 18
lineation 538, 561, 937
–, Hawaiian 330
–, Japanese 330, 331
–, Keathley 330, 331
–, magnetic 476, 485, 548, 553, 596
–, Pacific 330
–, parameter 938
–, Phoenix 330, 331
Lingzizong 825
Linzhou area 822
liquid
–, compass 67
–, helium 886
–, nitrogen 979
liquidus temperature 371
lithology 480
–, bedrock 477
lithosphere 73, 140, 145
–, mean 956
–, planetary 486
lithospheric field 485
Little America Observatory 723
Livingston Island Observatory 740
Lloyd's balance 435, 436
Lloyd, Humphrey 472
LMT (see *long-period magnetotelluric*)
loadstone (see *lodestone*)
lobe
–, reconnection 661
–, symmetrical pattern 433
local
–, energy minimum (LEM) 494, 613
–, gradient 48
–, turbulence theory 18
LOD (see *length of day*)
lodestone 66, 156, 214, 435, 776, 870
–, rock 356
–, spherical 356
loess 27
–, layer 667
–, paleosol climate archive 531
–, windblown 251
Loess Plateau 252
Loma Prieta earthquake 909
London, historical measurements 55
longitude 283
–, problem 675
longitudinal, preferred band 326
long-period magnetotelluric (LMT) 672
Longuet-Higgins calculation 814
loop, superconducting 885
Loran-C system 317
Lord Howe rise 967
Lorentz
–, equations 861
–, field 72
–, force 4, 10, 74, 85, 114, 187, 630, 641, 746, 841, 852, 860
–, number 120
–, torque 746
–, geostrophic cylinders 6
Lorentz, E.N. 863
Louisville
–, chain 962
–, hotspot 961
Love number 444
Lovø Observatory 737
low temperature demagnetization (LTD) 912
low-Curie temperature 499
lower mantle 211
Lowes spectrum 352
Lowes-Mauersberger
–, function 898, 900
–, spectrum 465, 481
low-level averaging 199
Loyola, Ignatius of 460
LTD (see *low temperature demagnetization*)
Lubber's line 66
Lukiapang Observatory 461
luminosity, solar 508
lunar
–, breccia, monomict 797
–, dynamo 791, 794
–, eclipse 471
–, laser ranging 471
–, paleomagnetic record 789
–, regolith 788
–, sample 788
Lundqvist number 7, 299, 645
Lundqvist, Stig 7
LVDT (see *linear variable differential transfonmer*)
Lwiro Observatory 725
Lyell Medal 742
Lyman-Birge-Hopfield lines 33

M

MAC (see *magnetic Archimedean Coriolis*)
MACCS network 726
MacDonald hotspot 961
Mach number, magnetic 649
Mackay 990
Macquarie Island Observatory 717
MAD (see *maximum angular deviation*)
Madras Observatory 731
Madrid Observatory 739
maghemite 23, 24, 50, 248, 513, 516, 517, 526, 560, 570, 668, 783, 791, 794, 932
–, nanocrystalline 249
–, single-domain 623
–, superparamagnetic 624
maghemitization 749
magma
–, basaltic 72
–, chamber 72, 238, 247
–, ocean 90, 151
–, lunar 446
–, pulse 603
magnes rotundus 356
Magnes, sive de Arte Magnetica 460, 463
magnesiowüstite 73, 148, 688
magnet
–, bar 310
–, project 352
magnetic
–, activity, storm-time 509
–, airborne detector 542
–, anisotropy 535
–, annihilator 481
–, anomaly 140, 143, 477, 888
–, compilations 478
–, correlation with gravity anomaly 856
–, definition 478
–, field 543
–, global map 829
–, inversion nonuniqueness 488
–, long wavelength 481
–, low amplitude 664
–, marine 324, 328, 336, 483, 664, 665, 779, 962
–, modeling 485
–, short-wavelength 982
–, skewness 857
–, stripe 478
–, thermoremanent 619
–, vertical intensity 857
–, Archimedean Coriolis (MAC) (see beyond)
–, assemblage, mixed 271
–, Atlas 358
–, basement 477
–, bay 662
–, cell 864
–, chart 674
–, cloud 404
–, Crusade 376
–, crust 142

magnetic *Contd.*
–, cycle 179
–, solar 507
–, damping 80
–, time 645
–, declination 776
–, diffusion 161
–, time 297
–, diffusivity 8, 84, 92, 94, 117, 204, 287, 641
–, tensor 139
–, dip 707, 776, 777
–, distortion 277
–, domain 490, 548
–, energy 339
–, density 351
–, equator 10, 39, 60, 511
–, fabric 579, 798
–, mimetic 554
–, study 578
–, fiber 357
–, fibril 182
–, field 776, 778, 872, 931
–, alternating 31
–, ambient 617
–, applied 560
–, Earth 37
–, excursion 770
–, galactic 7
–, helicity 506
–, instrument-generated 541
–, interplanetary 658
–, line 7, 8
–, potential 841
–, primary 225
–, primordial dipole 179
–, Saturn 198
–, secondary 225
–, self-consistent model 123
–, sensing 49
–, solar 178
–, source 541
–, spherical harmonic model 502
–, stellar 508
–, Sun 179, 505
–, toroidal 173, 302
–, very-low-inducing 566
–, flux
–, frozen 841
–, tube 7, 8
–, foliation 548
–, force microscopy (MFM) 495
–, gradient 977
–, grain 542
–, multidomain 266
–, single domain 266
–, granulometry 577
–, technique 577
–, hysteresis 518
–, index 509
–, induction 974
–, instability 118, 293
–, intensity 601
–, layer 141
–, lineation 476, 485, 548, 553, 596
–, Mach number 649
–, map 141
–, memory 344

magnetic *Contd.*
–, discoverer 942
–, meridian 66
–, mine 54
–, mineral 140, 248, 526, 546, 870
–, growth 260
–, mineralogy 479, 512, 533, 754
–, mixed 532
–, moment 601, 694, 870, 877, 933
–, atomic 932
–, sensitivity 979
–, spontaneous 758
–, monopole 287, 351, 705
–, mountain 356
–, needle 356
–, north pole 339
–, observatory 460, 715
–, particles, synthetic 48
–, permeability 118, 173, 215, 541, 670, 875, 954
–, phase 552, 600, 794
–, transformation 512
–, transition 515
–, philosophy 357
–, polarization 361, 974
–, pole 310
–, potential 122, 492
–, field (PMF) 840
–, Prandtl number 291, 297
–, pressure 641
–, property
–, map 141
–, measurement system (MPMS) 877
–, Rayleigh number 651
–, reconnection 33, 316, 658
–, recording 943
–, region, bipolar 179
–, relaxation 753
–, remanence 535
–, spontaneous 599
–, response 540
–, Reynolds number 8, 85, 164, 171, 173, 176–178, 183–186, 188, 190, 191, 194, 195, 200, 201, 205, 275, 288, 297, 299, 641–645, 652
–, Rossby number 650
–, Schmidt number 297
–, sheet 541
–, shield 540
–, signal
–, environmental 259
–, ocean flow 60
–, storm 61, 376, 696, 741
–, relationship to sunspots 890
–, stripe 542, 979, 980
–, surface 8
–, survey 56
–, Ireland 473
–, marine 542
–, susceptibility 94, 475, 513, 546, 560, 564, 667, 780
–, anisotropy 514, 535, 546, 578
–, differential 932
–, low-field 566
–, tension 633, 644
–, torque 341
–, variation 310

magnetic *Contd.*
–, mapping 310
–, wave 118, 994
–, wind 174, 946
magnetic Archimedean Coriolis (MAC) 206
–, balance 292, 634
–, regime 205
–, wave 87, 298, 348, 632, 635, 652
–, bifurcation 653
–, buoyancy-driven 635
–, convection-driven 633
–, diffusive 635
–, equation 634
–, instability 651
–, time 298
–, timescale 293
Magnetic Crusade 359, 436, 890
magnetic-Coriolis (MC)
–, Rossby wave 633
–, timescale 633
–, wave, convection-driven 633
Magneticum Naturae Rerum 463
Magnetische Ungewitter 404
Magnetischer Sturm 403
magnetism 931
–, environmental 248, 256, 527, 577, 758, 874, 881
–, induced 24
–, permanent 203, 870
–, physics 870
–, remanent 23
–, rock 54, 139, 701, 942
magnetite 23, 24, 32, 157, 495, 516, 526, 536, 560, 570, 618, 668, 759, 781, 782, 793, 870, 932, 942
–, authigenic 584
–, biogenic 50, 51, 516, 782, 786
–, crystal, synthetic 516
–, detrital 786
–, ferrimagnetic 781
–, glass-ceramic 497
–, low-temperature monoclinic phase of 517
–, nanocrystalline 249
–, botanical 249
–, pedogenically formed 249
–, physiological role 51
–, single-domain 623, 794
–, superparamagnetic 624
–, synthetic 611, 878
–, titanium-poor 539
magnetization 31, 535, 546, 560, 600, 931, 974
–, anhysteretic 876
–, remanent (ARM) 527, 759, 910, 943
–, anisotropy of anhysteretic remanent (AARM) 477, 577
–, bulk 141, 487
–, chemical remanent (CRM) 580, 609, 696, 782, 806
–, parasitic 513
–, crust 350
–, curve 931
–, depositional remanent 588, 758, 777
–, detrital remanent (DRM) 285, 588, 781
–, direction, stable 665
–, distribution 888
–, gyromagnetic 455, 456
–, remanent (GRM) 455, 456

magnetization *Contd.*
–, homogenous 795
–, inclination 588
–, independent 943
–, induced 140, 480, 519, 535, 598, 856, 870, 874, 888, 931
–, intensity 596
–, intrinsic 580
–, irreversible 875
–, isothermal 977
–, remanent (IRM) 51, 455, 572, 589, 596, 621, 668, 758, 760
–, locked-in 342
–, low coercivity 270
–, measurement 931
–, multiple components 594
–, natural remanent (NRM) 335, 515, 594, 607, 777, 910, 911, 920
–, oceanic crust 596
–, partial thermoremanent (pTRM) 37, 596
–, post-metamorphic 456
–, permanent 871
–, piezoremanence 599
–, post-depositional remanent (PDRM) 767
–, pressure-induced remanent (PRM) 696
–, primary 595, 616, 664, 781, 783, 786
–, natural remanent (NRM) 781
–, radial 785
–, random 795
–, remanent (RM) 140, 528, 535–537, 596, 603, 871, 920
–, anhysteretic (ARM) 535, 572
–, application 38
–, chemical 38
–, detrital (DRM) 535
–, fold test 607
–, isothermal (IRM) 535, 613
–, Moon 443
–, multicomponent 156
–, natural 156
–, postdetrital (pDRM) 535
–, saturation 266
–, thermal (TRM) 535
–, viscous 29
–, reversed 339
–, rock 35
–, rotational remanent (RRM) 455, 456
–, saturation 589, 874, 877
–, isothermal remanent (SIRM) 515, 527
–, secondary
–, origin 783
–, viscous 597
–, shock 791
–, remanent (SRM) 696
–, spontaneous 491, 535, 621, 876
–, structure 874
–, thermoremanent (TRM) 35, 37, 325, 514, 528, 609, 616, 696, 749, 777, 781, 910, 911, 942
–, thermoviscous 621
–, vector 546
–, vertically integrated 486
–, viscous 623, 962
–, remanent (VRM) 577, 621, 696
–, weak 608
–, spontaneous 496
magnetocardiagram 48
magnetoconvection 137, 630
magnetoelastic energy 874
magnetoencephalogram 48
magnetofluid mechanics 639
magnetofossil, bacterial 249
magnetogram 506, 810
magnetograph 179, 712
–, Eschenhagen 735
magnetohydrodynamics (MHD) 639
–, classical 649
–, dynamo 175, 643
–, mean-field 198
–, rotating, experiment 276
magnetometer 2, 54, 596, 602, 654
–, alternating gradient (AGM) 655
–, alternating gradient field (AGFM) 877
–, ASG–81 scalar 850
–, astatic 654, 942
–, bifilar 280, 435
–, Bobrov 712
–, cesium 2, 27, 542
–, cryogenic 341, 561, 655, 992
–, declination and inclination (DIM) 717, 726
–, floating total-field 65
–, fluxgate 59, 72, 242, 438, 483, 672, 713, 714
–, sensor 26
–, helium 850
–, induction 435
–, coil (ICM) 72, 242, 713
–, JR–6/JR–6A dual speed spinner 921
–, Kew 279
–, NAROD ring-core fluxgate vector 850
–, ocean-bottom 741
–, optically pumped 438, 712
–, Overhauser 59, 542, 712, 727, 744
–, parastatic 655
–, precession, proton-free 26
–, proton precession (PPM) 2, 60, 438, 468, 484, 542, 710–712, 717, 858, 990
–, quartz horizontal (QHM) 438, 717
–, rotating-sample 942
–, rubidium 714
–, saturable-core 435
–, seafloor 71
–, ship-towed 741
–, spinner 29, 655, 920, 992, 993
–, SQUID 793
–, subsatellite 791
–, superconducting (SRM) 883
–, superconducting quantum interference device (SQUID) 438
–, surface 791
–, survey 25
–, theodolite 435
–, translation 942
–, unifilar 280
–, universal 437
–, vector proton (VPM) 714, 715
–, vibrating sample (VSM) 655, 756, 877
magnetometric resistivity (MMR)
–, method 230
–, sounding 72
magnetometry 920
–, superconducting quantum interference device (SQUID) 51
magnetomineralogy 879
magneto-optical Kerr effect (MOKE) 495
magnetopause 333, 449, 657, 659, 927
magnetosheath 657
magnetosome 782
–, chain 782
–, goethite 782
magnetosphere
–, dynamics 660
–, Earth 656
magnetospheric substorm 662, 700, 926
magnetostatic 871
magnetostratigraphy 331, 664, 760, 780
magnetostriction 515, 535, 614, 618, 871
magnetostrictive
–, anisotropy 879
–, energy 874
magnetostrophic 852
magnetotail 33
magnetotaxis 50, 782
magnetotellurics (MT) 219, 230, 316, 670, 866
–, dead band 869
–, method 71, 227, 844
–, survey 844
magnetozone 664
–, normal polarity 665
magnon 694
MAGPROX 533
Magsat 59, 673, 892
–, mission 465
–, satellite 349, 352
main
–, field
–, ellipticity correction 683
–, model 60
–, modeling 679
–, phase 663
mainstream 111
Maisach Observatory 729
Makerstoun Observatory 720
Malaspina's expedition 990
Mallet, Frederick 435
manganese carbonate 932
Manhay Observatory 725
Manila Observatory 461
mantle
–, conductivity 696, 867
–, electrical 288, 684
–, thermal 688
–, continental 72
–, convection 19, 294, 928, 958, 966
–, geochemistry 77
–, heat capacity 128
–, hot spot 245
–, lateral temperature variations 133
–, lower 211, 929, 964
–, high viscosity 967
–, mean 956, 957, 967
–, mineralogy 684
–, model, isoviscous 966
–, nutation of the 207
–, plume 146, 228, 902
–, primitive 77, 854
–, radioactivity 854
–, rheology 206
–, roll 959
–, upper 211

mantle *Contd.*
–, low viscosity 967
–, upwelling 148
–, viscosity 965
map
–, magnetic 141
–, main field 674
mapping
–, anomaly 478
–, controlled source electromagnetic (CSEM) 231
–, geological 1, 228, 230, 478
–, magnetic variation 310
Maraldi 721
mare
–, basalt 788
–, magnetized 791
margin
–, continental 485
–, passive 64
Marie Byrd Land 962
marine magnetic survey 542
Mariner 10 865
Markowitz wobble 363, 426, 958
Marr, John 357
Mars 89, 203, 204, 207, 439, 788, 800
–, crustal dichotomy 447
–, Galilean satellites 203
–, magnetic field 287, 481, 502
–, magnetization crustal 443
–, moment of inertia 442
–, paleomagnetic pole position of 504
–, polarity reversal had 504
–, southern highlands crust of 443
–, two-layer model 442
Mars Global Surveyor (MGS) 443, 444, 502, 791
Mars Pathfinder 444, 791
martian
–, anomaly 794
–, core 502
–, crust
–, anomalies 795
–, magnetic carrier 503
–, magnetic field 795
–, north-south dichotomy 503
–, dynamo 794, 795
–, field 793, 794
–, meteorites 502
–, power spectrum 794
–, rock 603
Martini, Martin 460
mass
–, anomaly 956
–, conservation 646
–, coronal ejection 505
–, diffusion time 298
–, ejection, coronal 927
–, excess 956
–, spectrometer 834, 835
master curve 943
matching principle 111
material
–, archeological 31, 766
–, heated 694
–, environmental 248
–, ferromagnetic 874
–, high temperature 886
material *Contd.*
–, isotropic is 535
–, magnetic remanent 541
–, synthetic 519
Mathilde 795
matrix
–, covariance 870
–, equation 948
matter, organic 784
Matuyama reversed chron 311, 542, 829, 838
Matuyama, Motonori 689
Matuyama-Brunhes
–, boundary 781
–, polarity transition 311, 836
–, reversal 161, 829, 837
–, record 830
Mauersberger-Lowes spectrum 352
Maunder Minimum 182, 507
Mawson 990
Mawson Observatory 717
maximum
–, angular deviation (MAD) 846, 849
–, entropy method 353
–, intensity deviation (MID) 846
–, Medieval 182
Max-Planck-Institut für Aeronomie 42
Maxwell
–, equation 21, 84, 192, 215, 229, 335, 954
–, relationship 367
–, stress 103, 114
Maxwell-Boltzmann probability density function 897
Mayaud, Pierre Noel 461
Maza Tagh range 822
MC (see *magnetic-Coriolis*)
MD (see *multidomain*)
MDF (see *median destructive field*)
mean
–, field
–, concept 192
–, conductivity 194
–, definition 192
–, dynamo 192
–, electrodynamics 192
–, equation 193
–, flow instability 76
–, lithosphere 956
–, mantle 956, 957, 967
–, pole 46
Meanook Observatory 726
measurement
–, absolute 711
–, airborne 224
–, paleomagnetic, astatic 883
–, relative 435
–, solar surface 508
–, vector magnetic 673
mechanics, magnetofluid 639
medal
–, Bowie 274, 469
–, Copley 61
–, Cunningham 474
–, Fleming 274, 888
–, Lyell 742
–, Price 844
–, Royal 61
–, Royal Astronomical Society 844
medal *Contd.*
–, Wegener 888
media, anisotropic 147
median destructive field (MDF) 533
Medieval Maximum 182
Melbourne Observatory 717
melt
–, breccia 789
–, granitic 244
melting
–, curve 117
–, density change on 92
–, entropy 100
–, eutectic 691
–, latent heat 92
–, partial 150, 247
–, temperature 372
–, iron 98
–, suppression 99
Memambetsu Observatory 733
memory, magnetic 344
–, discoverer 942
Mendelson-Bartholdy, A. 403
Meng Chhi Pi Than 435
Mercator, Gerard 357, 988
merchant marine 139
Mercury 89, 203, 204, 206, 276, 375, 439, 706, 788, 865
–, two-layer model 441
meridian
–, magnetic 66
–, prime 356
meridional circulation 190, 859, 946
Merret, Christopher 707
Mesozoic 826
–, serie 485
MESSENGER mission 206, 865
Met Éireann 720
metal
–, ferromagnetic 581
–, nickel-rich 541
–, salt electrode 227
metal-silicate segregation 77
metamorphism
–, low-grade 555
–, thermal 796
meteorite 77, 446, 696, 788, 796, 799, 800
–, C1 carbonaceous chondrite 854
–, Chassignite 791
–, chondritic 77, 89, 795
–, collections 795
–, composition of 854
–, crater 602
–, Nahklite 791
–, Shergotite 791
–, Shergotty-Nakhla-Chassigny (SNC) 446
Meteorological Agency of Japan 733
meteorology 61
method
–, ab initio 693
–, absolute 436
–, analytic signal 142
–, audiomagnetotelluric (AMT) 224, 230
–, Bitter pattern 495
–, bootstrap 553
–, continuous 872

method *Contd.*
–, controlled-source audiomagnetotellurics (CSAMT) 224, 230
–, dipole array 486
–, electrical and electromagnetic (E&EM) 228
–, finite-difference 216
–, finite-element (FEM) 217, 225
–, Fourier domain 486
–, frequency-domain electromagnetic (FEM) 224, 225
–, geophysical 228
–, grounded wire 225
–, integral equation 218
–, jackknife 870
–, magnetometric resistivity (MMR) 230
–, magnetotelluric (MT) 71, 227, 310, 844
–, maximum entropy 353
–, microwave 31, 910
–, Naudy 263
–, paleointensity 583
–, alternative 575
–, plane wave source 224
–, progressive 872
–, pseudo-Thellier 759
–, Shaw 373, 910
–, Slingram 230
–, space-domain 486
–, surface integral 449, 451
–, Thellier double heating 583, 749
–, Thellier-Thellier 575, 611, 943, 944
–, time-domain electromagnetic 225
–, transient electromagnetic 231
–, transmission electron microscopy 496
–, universal stage 550
–, very low frequency (VLF) 224
–, wavenumber domain centroid 143
–, Z/H- 697
metrology, geomagnetic 711
MFM (see *magnetic force microscopy*)
MGS (see *Mars Global Surveyor*)
MHD (see *magnetohydrodynamics*)
microbe 783, 784, 874
microcoercivity 617, 620
microplate 802
micropulsation 308
microscopy
–, magnetic force (MFM) 495
–, transmission electron 50, 495
microstate 609
microtesla 799
microwave
–, demagnetization 694
–, intensity, determination 792
–, method 31
–, technique 694
MID (see *maximum intensity deviation*)
Middle Ordovician 928
MIDM (see *modified iterative-dissipative method*)
mid-ocean ridge 72, 208, 237, 446, 779, 980
–, basalt 80
Mie-Grüneisen
–, equation of state 96
–, relationship 366
migration, domain wall 601
Mihara volcano 984
Milankovitch cycle 332
Milankovitch periodicities 759
Milne's Observatory 743
Milne, John 742
mine
–, magnetic 54
–, Silesian 403
mineral
–, assemblage, natural magnetic 577
–, authigenic iron sulfide 260
–, conductivity 70
–, diamagnetic 874
–, distribution 566
–, exploration 1, 224, 477
–, fabric 564, 565
–, ferrimagnetic 535
–, ferromagnetic 157, 567
–, secondary 665
–, growth, magnetic 260
–, lithogenic components 256
–, magnetic 140, 248, 258, 526, 546, 604, 870
–, magnetically hard 516
–, orientation 535
–, preferred 546
–, paramagnetic 874, 932
–, physics 94
–, synthetic 943
–, thermoremanent magnetization 616
–, transformation 561
mineralization, biologically induced (BIM) 782
mineralogical alteration 513
mineralogy 248, 684, 920
–, magnetic 479, 512, 754
–, mixed 532, 533
mineraloid, ferric-oxyhydroxide 932
minima, grand 507
minimum
–, energy
–, absolute 492
–, state 871
–, Maunder 182
–, strain direction 556
–, structure 220
–, model 243
minnesotaite 522
mircopulsation 333
mirror symmetry- 643
mixed-polarities 47
mixing length theory 181
Miyake-Jima volcano 986
Mizusawa Observatory 734
MMR (see *magnetometric resistivity*)
mode
–, B-polarization 221
–, core-sensitive 418
–, E-polarization 221
–, f 399
–, freely decaying 844
–, g 399
–, geostrophic 125, 647
–, normal 176, 210
–, p 399
–, resonant 398
–, transverse
–, electric (TE) 221, 671
–, magnetic (TM) 221
model
–, 1066A 83
–, 1066B 83
–, 3D micromagnetic 611
–, AK135 83, 210, 213, 903
–, annulus 76, 125
–, comprehensive 141
–, conductivity 215
–, cosmochemical 128
–, crustal field 60
–, dynamo wave 775
–, dynamo, intermediate 290
–, Earth 422
–, eccentric dipole 679
–, equivalent source 158
–, flooding 321
–, four-pole 679
–, free decay 860
–, free electron 109
–, GAD 820
–, Giant Gaussian Distribution 897
–, GSFC 346
–, gufm1 637
–, interatomic potential 94
–, localized dipole-current-loop 774
–, magnetic field 502
–, main field 60
–, mantle, isoviscous 966
–, micromagnetic 490, 494
–, minimum structure 243
–, Néel-Preisach 575
–, parameterization 220
–, PREM 83, 91
–, PREM2 213
–, reference 208
–, Rikitake 863
–, scale-similarity 102
–, secular variation 892–898, 901
–, self-exciting 173
–, spherical harmonic 315, 774
–, subgrid scale 305
–, terrella 64
–, tomographic 152
–, two-layer 440–442
–, velocity 83
modeling
–, EM 215, 219
–, forward 142, 486
–, inverse 142, 487
–, magnetic anomaly 485
–, main field 679
model-Z
–, dynamo 174, 293
–, state 941
modified iterative-dissipative method (MIDM) 218
modulus
–, bulk 153
–, shear 153
Mohe Observatory 728
Mohorovičić discontinuity 158, 208
Mojave Province 143
MOKE (see *magneto-optical Kerr effect*)
molecular
–, diffusivity 80
–, dynamics 94
–, ab initio 105

molecular *Contd.*
  –, calculation 421
–, viscosity 104, 112
moment
–, atomic 931, 932
–, magnetic 601, 694, 870, 877, 931–933
  –, sensitivity 979
  –, spontaneous 758
–, of inertia 77, 363, 441, 442, 957, 966
  –, axial 443
  –, measurement 442
  –, nonhydrostatic principal 967
momentum
–, angular 6, 130, 135, 179, 994
  –, change 320
  –, exchange 469
  –, oceanic 469
–, conservation 646
–, equation 634
–, vector 956
Mongolia-North China plate 824
Mongol-Okhotsk Ocean 823
Mono Lake
–, excursion 770
–, event 761
monopole, magnetic 287, 351, 705
monsoon
–, relative intensity 531
–, summer 531
–, winter 531
Monte Carlo approach 608
Moon 89, 203, 204, 207, 439, 788, 800, 887
–, crustal types 446
–, exterior field outside Earth 889
–, farside 446
–, highlands 446
–, interior 442
–, magma ocean 446
–, moment of inertia 442
–, remanent magnetization 443
Morgan Stanley 240
Morin transition 519
Mössbauer spectroscopy 600
motion
–, absolute 959
–, geostrophic 136, 843
–, helical 194, 643
–, inter-hotspot 963
–, paleolatitudinal 140
–, speed 957
–, subgrid-scale 276
motor, piezoelectric 715
Mount Etna 986
Mount Haruna 339
Mount St. Helens 986
Mount Wilson Survey 508
mountain, magnetic 356
Mountaine, William 358
MPMS (see *magnetic property measurement system*)
MS standard sequence 571
M-sequence 664
–, GPTS 330
MT (see *magnetotellurics*)
multidomain (MD) 518, 591
–, state 455, 575
–, theory 575
multiplier, Lagrange 87, 123
–, theory 575
Munich Observatory 729
Murchison meteorite 798
Murray chondrite 798
Museum of the History of Science in Florence 711
mush 74
mushy layer 421

N

Nabarro-Herring creep 105
NAD (see *nonaxial-dipole*)
Nagata 870
Nagpur Observatory 732
Nagqu 822
Nagycenk Observatory 728
Nahklite meteorite 791
nahklites 797, 799
Nairn, A.E.M. 887
Nakhla 792
nannofossil 665
nanoparticle 791
nanotesla 350
Napier 384
Napier-Shaw prize 888
nappe 802
Narsarsuaq Observatory 736, 737
NASA 60, 465
Nashville Seamount 891
National Geophysical Data Center 850
National Institute of Advanced Industrial Science and Technology (AIST) 734
National Physical Laboratory (NPL) 55, 438
National Science Foundation 992
natural remanent magnetization (NRM) 156, 335, 515, 535, 594, 607, 777, 910, 911, 920
–, multivectorial nature 156
Naudy method 263
Navier-Stokes equation 118, 191, 290, 335, 430, 940
navigation
–, animal 48
–, oceanic 356
–, system 317
  –, inertial 850
navigator, Arabian 66
NEAR spacecraft 795
Neckam, Alexander 435, 987
needle
–, magnetic 356
–, oscillating 435
–, weighting 435
Néel
–, point temperature 526
–, temperature 516, 519, 871, 872, 874, 932
Néel, Louis 700
Néel-Preisach model 575
Neogene 825
neotectonic 803
Neptune 203, 204, 207, 439, 865
–, magnetic field 287, 865
Nernst-Einstein relation 686
Nernst-Ettinghausen effect 705, 841
Neumann
–, condition 217, 466
–, formula 379
neurological disorder 51
neutral line 927
neutron bombardment 701
New Zealand Department of Scientific and Industrial Research (DSIR) 736
Newe Attractive 707
Newton
–, law of gravitation 443
–, second law 4
nickel 873
–, core 371
–, deposits 477
Niemegk 42, 699
Niemegk Observatory 729, 730
Ningxia Hui Autonomous region 821
Ningxia province 820
niningerite 79
niobium, toroidal body 885
nitrogen, liquid 979
–, temperature 157
NLCG (see *nonlinear conjugate gradient*)
Noachian 443, 503
noble gas 792
noise
–, basic 920
–, cancellation 866
–, operating 920
–, white 928
nonaxial-dipole (NAD) field 832
nondimensionalization 297
nondipole 897
–, component 342
–, field 311, 701, 901
  –, drifting 773
  –, present-day 326
  –, standing components 774
nondynamo
–, generation 288
–, theory 704
nonferrimagnetic 513
non-GAD 780
nonlinear conjugate gradient (NLCG) approach 221
nonmagnetic quartz glass sample holder 977
nonmagnetotactic 782
nonohmic effect 841
normal
–, mode 176
–, polarity 313
normalization
–, Ferrers 381
–, Schmidt 381
Norman, Robert 435, 707
North American plate 960
North Anatolian Fault 247, 803
North China 823, 826
North China Block 817
North German anomaly 412
North Palm Springs earthquake 908
northeasting 356
northwest passage 890
northwesting 356

no-slip condition 112, 646
notation, Kroger-Vink 685
Nouvelle hypothèse sur les variations de l 'aiguille aimantée 460
Nova demonstratio inmobilitatis terrae petita ex virtute magnetica 460
Novum Organon 361
NPL (see *National Physical Laboratory*)
NRM (see *natural remanent magnetization*)
nuclear magnetic resonance magnetometer 435
nucleation 498
 –, failure 613
nucleon 931
null flux curve 9, 85
number
 –, Alfvén 299
 –, buoyancy 297
 –, Chandrasekhar 299, 630, 649
 –, condition 216
 –, Cowling 299
 –, Ekman 75, 81, 111, 124, 175, 291, 297, 299, 365, 430, 647
 –, Elsasser 114, 118, 206, 263, 291, 297
 –, Hartmann 114, 299, 649
 –, Lewis 297
 –, Lorentz 120
 –, Love 444
 –, Lundqvist 299, 645
 –, Nusselt 19, 297, 299
 –, Péclet 299
  –, mass 297
  –, thermal 297
 –, Prandtl 75, 291, 297, 299, 646
  –, magnetic 291, 645
 –, Rayleigh 74, 75, 80, 82, 117, 128, 262, 291, 297, 417, 630, 646
  –, critical 631
  –, magnetic 651
  –, modified 297, 652
 –, Reynolds 85, 101, 111
  –, magnetic 8, 164, 171, 173, 176–178, 183–186, 188, 190, 191, 194, 195, 200, 201, 205, 275, 288, 297, 299, 641–645, 652
 –, Roberts 120
  –, magnetic 297
 –, Rossby 75, 291, 297, 299, 430
  –, magnetic 650
  –, modified 648
 –, Schmidt 297
  –, magnetic 297
 –, Strouhal 194
 –, Taylor 630
Nunes, Pedro 435
Nurmijärvi Observatory 736, 737
Nusselt number 19, 297, 299
nutation 104, 362

O

O'Gyalla Observatory 728
obducted 802
oblate 548, 553
 –, ellipsoid 562
observation
 –, compass 66
 –, field, historical 773
 –, spectroscopic 400
observator
 –, Fürstenfeldbruck 710
 –, Tihany 710
observatory 708
 –, Abinger 720
 –, Abisko 737
 –, Adolf Schmidt 730
 –, Agincourt 726
 –, Alibag 732
 –, Alice Springs 717
 –, Amberley 735
 –, Ancón 734
 –, Apia 735, 736
 –, Ascension Island 720
 –, automatic 712, 713
 –, automation, 713
 –, Bac Lieu 734
 –, Base Roi Baudouin 725
 –, Bear Island 737
 –, Beijing Mingtomb 728
 –, Belsk 728
 –, Berlin 729
 –, Binza 725
 –, [illegible]
 –, Buda 728
 –, Budkov 728
 –, Canberra 717, 810
 –, Cape Chelyuskin 738
 –, Cape Denison 717, 723
 –, Cape Evans 723
 –, Castellaccio 733
 –, Cha Pa 734
 –, Chambon-la-Forêt 714, 721
 –, Changchun 728
 –, Charters Towers 717, 951
 –, Cheltenham Maryland 723
 –, Chichijima 733
 –, Chiripa 734
 –, Christchurch 735
 –, Colaba 732
 –, colonial 359
 –, Dalat 734
 –, Darwin 717
 –, data 708, 720
  –, monthly 319
  –, yearly 319
 –, De Bilt 725
 –, Discovery Bay 723
 –, Dixon 738
 –, Dombås 737
 –, DomeC 724
 –, Dourbes 725
 –, Dublin 720
 –, Dumont D'Urville 710, 723
 –, Ebre 739, 740
 –, Ekaterinburg 737
 –, Elizabethville 725
 –, Esashi 734
 –, Eskdalemuir 720
 –, Ettaiyapuram 732
 –, Eyrewell 735
 –, Fredericksburg 509, 723
 –, Fürstenfeldbruck 729
 –, geophysical
  –, Huancayo 274
  –, Peru 274
  –, Watheroo 274
observatory *Contd.*
 –, Gnangara 719
 –, Godhavn 736
 –, Göttingen 729
 –, Greenwich 61, 720
 –, Grocka 729
 –, Guangzhou 728
 –, Güimar 740
 –, Gulmarg 732
 –, Halliburton 721
 –, Hartebeesthoek 739
 –, Hartland 710, 720
 –, Hatizyo 734
 –, Heard Island 717
 –, Hel 728
 –, Hermanus 739
 –, Hollandia 725
 –, Honolulu 723
 –, Hornsund 728, 737
 –, Huancayo 734, 807
 –, Hurbanovo 728
 –, Hyderabad 732
 –, Indonesian 717
 –, instrumentation 711
 –, Irkutsk 737
 –, Jaipur 732
 –, Jassy 728
 –, Kakadu 710, 717
 –, Kakioka 715, 733, 951
 –, Kanoya 733
 –, Kanozan 734
 –, Kazan 737
 –, Kew 333, 720
 –, Kodaikanal 732
 –, L'Aquila 733
 –, La Habana 734
 –, Lanzhou 728
 –, Learmonth 717
 –, Leiden 725
 –, Leirvogur 736, 737
 –, Lerwick 720
 –, Lhasa 728
 –, Little America 723
 –, Livingston Island 740
 –, Lovø 737
 –, Lwiro 725
 –, Macquarie Island 717
 –, Madras 731
 –, Madrid 739
 –, magnetic 460, 715
 –, Maisach 729
 –, Makerstoun 720
 –, Manhay 725
 –, Mawson 717
 –, Meanook 726
 –, Melbourne 717
 –, Memambetsu 733
 –, Milne's 743
 –, Mizusawa 734
 –, Mohe 728
 –, Munich 729
 –, Nagpur 732
 –, Nagycenk 728
 –, Narsarsuaq 736, 737
 –, Niemegk 729, 730
 –, Nurmijärvi 736, 737
 –, O'Gyalla 728

observatory *Contd.*
–, Orcadas del Sur 723
–, Panagyurishte 728
–, Paramaribo 725
–, Paris 721
–, Pavlovsk 737
–, Phu Thuy 734
–, Pola 728, 733
–, Pondicherry 732
–, Port Martin 723
–, Port Moresby 717
–, Port Stanley 720
–, Potsdam 729
–, Prague 728
–, Prudhoe Bay 721
–, Pruhonice 728
–, Qaanaaq 737
–, Qeqertarsuaq 710, 736, 737
–, Resolute 710
–, Rossbank 717, 807
–, Sabhawala 732
–, Sable Island 721
–, San Fernando 739
–, San Pablo 740
–, Sanya 728
–, Scott Base 735, 736, 951
–, Seddin 729
–, Sheshan 728
–, Shillong 732, 733
–, Shimla 731
–, Silchar 732
–, Sitka Alaska 723
–, Sodankylä 737
–, Srednekan 738
–, Stará Dala 728
–, subauroral 509
–, Surlari 729
–, Swider 728
–, Tamanrasset 710
–, Tangerang 725, 734
–, Terra Nova Bay 733
–, Tiflis 737
–, Tihany 728
–, Tikhaya Bay 738
–, time series 948
–, Tirunelveli 732, 733
–, Toledo 739
–, Tondano 734
–, Toolangi 717
–, Toronto 726
–, Trelew 734
–, Trivandrum 731
–, Tromsø 737
–, Tsumeb 739
–, Tuntungan 734
–, unstaffed 714
–, Uppsala 737
–, Urumqi 728
–, Utrecht 725
–, Valentia Island 720
–, Vienna 728
–, Vishakhapatnam 732, 733
–, Watheroo 717, 807
–, Wellen 738
–, Wien Auhof 728
–, Wilhelmshaven 729
–, Wilkes 717
observatory *Contd.*
–, Wingst 714, 729
–, Witteveen 725
–, Wuhan 728
–, Yakutsk 738
Occam's Razor 950
ocean 61
–, conductivity 740
–, floor
–, conductivity 71
–, dating 484
–, isochrons 484
–, flow 741
–, magnetic signal 60
–, hidden 81, 637
–, induction 742
–, Mongol-Okhotsk 823
–, shallow 814
–, water-ammonia 207
Ocean Drilling Program (ODP) 234, 479, 762
–, drillcore 666
oceanic crust 208, 596
oceanography 992
octupole 284
–, geocentric 681
ODP (see *Ocean Drilling Program*)
offset, frequency-independent 277
Ogg, Alexander 739
OGO (see *Orbiting Geophysical Observatory*)
Ohm law 21, 84, 114, 192, 287, 335, 641, 813
Ohmic
–, diffusion 5, 120, 186
–, dissipation 128, 641, 955
–, time 186
OhmMapper system 229, 230
oil exploration 1
Oldham, Richard Dixon 208, 742
oldhamite 79
Olduvai subchron 834, 837, 838
olivine 20, 21, 560, 685, 686, 932
–, electrical conductivity 72
–, electrogeotherm 70
–, hot-pressed 70
Olympus volcano 504
omega
–, effect 9, 10, 162, 164, 171, 946
–, × j dynamo 197
–, × j effect 194
–, loop 179, 182
–, system 317
Ontong-Java Plateau 783
ooze, calcareous 784
operating noise 920
operator
–, angular momentum 388
–, Hermitian 139
–, reduction-to-pole 487
–, roughening 220
ophiolite 596, 802, 819
optimization algorithms 220
optocoupler 920
orbit 59
–, dawn/dusk 59
–, Earth 353
–, elliptical 59
orbital
–, altitude 60
–, geometry, Earth's 258
–, parameter 161
Orbiting Geophysical Observatory (OGO) 828
Orcadas del Sur Observatory 723
Ordovician 928
–, superchron 929
ore body 62
organic, matter 784
–, labile 786
organism, marine 566
Orgueil meteorite 798
orientation
–, animal 48
–, distribution, isotropic 590
–, lattice preferred 148
orienting devise 765
Ørsted Initial Field Model (OIFM) 745
ØRSTED satellite 349, 351, 352, 676, 743, 892
Ørsted, Hans Christian 358, 376
orthogonality 46
orthopyroxenite, cataclastic 792
oscillation
–, free 82
–, frequencies 77
–, geomagnetic 333
–, inertial 101, 364
–, inner core 422
–, relaxation 76
–, solar 398
–, torsional 6, 87, 508, 746
oscillator, harmonic, damped 132
osmium, isotopic composition 79, 150
outcrop 766
outer core
–, liquid-iron 766
–, viscosity 424
out-of-phase component 62
oval, auroral 33
overburden 224
Overhauser magnetometer 59, 542, 712, 727, 744
overprint 577
–, component 604
–, intensity 593
–, lightning 593
–, low-temperature 797
overprinting 592
–, thermal 35
overturn time 297, 298
oxidation 993
–, degree 629
–, deuteric 597
–, low-temperature 749, 752, 794, 873, 878
–, partial 584
–, state 580
oxide
–, ferromanganese 783
–, magnetic 561
oxyexsolution 629
oxygen 371
–, fugacity 71, 480, 873
–, green line 33
–, isotope stratigraphy 771
–, state 70

oxygen–18 record 762
oxyhydroxide 248
–, ferromanganese 783

P

p mode 399
pace weather observatory 720
PACES system 229, 230
Pacific
–, dipole window 133, 134, 314, 417
–, hemisphere 994
–, lineation 330
–, Ocean 129
–, paleopole 962
–, plate 961, 962, 967
–, super-upwelling 964
PAIGH (see *Pan-American Institute of Geography and History*)
Palazzo, Luigi 733
paleochannel 27, 240
paleoclimatologist 956
paleocontinent 839
paleodirection 313, 943
paleofield, intensity 611
paleoflow 871
paleogeomagnetic 874
paleointensity 32, 514, 589, 770, 799, 943
–, absolute 321
–, techniques 753
–, determination 574
–, absolute 749, 910
–, fluctuation 666
–, geomagnetic 257
–, method 583
–, record 259
–, relative 354, 758
–, reliable 780
–, study 31, 32, 271
–, relative 271
–, Thellier 749
–, variation 353
paleolatitude 283, 777, 959, 962, 967
paleomagnetic 966
–, dipole moment (PDM) 159
–, direction dispersion 45
–, field 281
–, collection 765
–, time-averaged 947
–, mean direction 282
–, microwave technique 694
–, recorder 325
–, secular variation (PSV) 766
–, data 775
–, record 769
–, signal 260
–, signature 273
–, study, baked contact test 35
–, vector 654
paleomagnetism 54, 548, 776, 781, 795, 797, 801, 817, 825, 839, 845, 870, 883, 920, 957–959, 993
–, extraterrestrial 788
–, sedimentary 579
paleomagnetist 992
paleomeridian 283
paleontologist 956
paleontology 957
paleopole 839
–, Cambrian 821
–, Cretaceous 820
–, key 839, 840
–, Pacific 962
–, Precambrian 839
–, Tertiary 821
paleorotation 472
paleoslab region 148
paleosol
–, layer 667
–, pre-Quaternary 252
paleosusceptibility 251
paleotectonic 803, 874
paleothermometry 943
Paleozoic 966
Panagyurishte Observatory 728
Pan-American Institute of Geography and History (PAIGH) 734
Pangea 817
Papal Academy of Sciences 888
paramagnetic 116, 870, 931
–, rock 480
paramagnetism 157
Paramaribo Observatory 725
parameter
–, attenuation 209
–, Bullen 83
–, Cayley-Klein 392
–, coercivity 668
–, Euler Rodrigues 390
–, Floquet 201
–, Grüneisen 13, 91–93, 95, 100, 107, 366
–, interaction 649
–, Lamé 147
–, mineral-magnetic 527
–, orbital 161
–, precision 273
–, proxy 525, 527, 528
–, regime 290
–, response 647
–, rock-magnetic proxy 525
–, seismic 82, 100, 107
–, structural 647
parameterization model 220
parametric bootstrap technique 608
Paramore 675, 989
Paris Observatory 721
Parkinson
–, arrow 807
–, plane 808
–, preferred plane 65
–, terrella 808
–, vector 807, 867
Parkinson, Wilfred Dudley 807
pARM 537, 579
Parmentier, Jean 988
Parmentier, Raoul 988
Parry, J.H. 887
partial
–, melting 150
–, thermal remanent magnetization (pTRM) 596, 612
–, additivity 696
–, check 754
–, law of additivity 754
partial *Contd.*
–, reciprocity law 943
–, relaxation 614
particle
–, ellipsoidal 565
–, ferrimagnetic 571
–, magnetic, synthetic 48
–, motion 147
–, noninteracting SD 574
–, path 187
–, pollutant 251
–, single-domain 589
–, noninteracting uniaxial 268
–, super-paramagnetic 581, 759, 932
passive margin 64
patch, preferred 831
path
–, apparent polar wander (APWP) 779, 802, 956
–, electrically conductive 245
–, polar 428
–, reversal
–, geomagnetic 134
–, preferred 132
pathway, preferred 830
Pauli exclusion principle 870
Pavlovsk Observatory 737
Pavonis Volcano 504
PcP 903, 970
–, ray 125
PDM (see *paleomagnetic dipole moment*)
PDOS (see *phonon density of states*)
pDRM (see *postdetrital remanent magnetization*)
Péclet number 299
–, mass 297
–, thermal 297
Pedersen
–, conductivity 453, 815
–, current 453, 660
pedogenesis 569
penalty functional 220
pencil structure 476, 538
pendulum 977
Penrose crust 598
percolation 91
–, threshold 20
Peregrinus, Petrus 435, 808
periclase 793
peridotite 982
–, serpentinized 597
period
–, diurnal 101
–, magnetically quiet 141
–, tidal 72
periodic dynamo 200
–, spatially 200
periodogram analysis 353
Perkins, Peter 357
permeability 63, 72
–, free-air 872
–, hydraulic 985
–, magnetic 118, 173, 192, 215, 540, 541, 670, 875, 954
Permian 568
Permian-Triassic boundary 568
Perminvar 173

permittivity 641
 –, dielectric 215, 226, 954
Permo-Carboniferous superchron 928
Permotriassic 960
perovskite 148, 687, 688
Perry, Stephen J. 460, 887
perturbation, finite-amplitude 262
petrofabric 560
petroleum
 –, exploration 477
 –, target 224
petrology 796
Pfaff, J.F. 279
Phanerozoic 568, 928
phase 672, 977
 –, calibration 920
 –, compressional 802
 –, curve 869
 –, expansion 927
 –, growth 927
 –, magnetic 600, 794
 –, main 663
 –, recovery 663, 927
 –, seismic 82
 –, transition 515, 620
 –, velocity 76
phenomen, tectonoelectric 909
Phillippes, Henry 357
Philosophia Magnetica 357, 460
Phobos 439
Phoenix lineation 330, 331
phonon 116, 694
 –, density of states (PDOS) 368, 369
 –, scattering 688
photosphere 4, 178
 –, solar 89
Phu Thuy Observatory 734
phyllosilicate, paramagnetic 563
physics
 –, ionospheric 61
 –, solar-terrestrial 61
 –, solid-state 701
Pi2 pulsation 927
Piazzi, G. 279
piezoelectric motor 715
piezomagnetism 985
piezoremanence 599
 –, magnetization 599
pillow lava 597
pin domain wall 621
pinning
 –, effect 609
 –, energy 625
 –, strength 613
piston core 993
 –, gravity-driven 665
pitch 176
 –, error 673
Piton de la Fournaise 986
PKIKP 213, 427, 903, 904
 –, wave 125, 424
PKJKP 427, 433, 903
PKKP wave 125, 126
PKP 970
PKPab 418, 427
PKPbc 427
PKPdf 418, 427
plagioclase 931
 –, c-axes 540
 –, crystal 749
 –, feldspar 752
Plancius, Petrus 357
plane
 –, paleohorizontal 561
 –, Parkinson preferred 65
 –, preferred 808
 –, wave
  –, electromagnetics 230
  –, source method 224
planet
 –, cooling rate 206
 –, formationary 854
 –, icy 204
 –, interior 439, 441
 –, moment of inertia 441
 –, properties 440
 –, rotation 442
 –, terrestrial 89, 203, 439
  –, crust 446
 –, two-layer model 440
planetary
 –, accretion 151
 –, dynamo 203
 –, thermoremanent magnetic anomaly 619
 –, wave 994
planetesimals 90
plasma 7, 60, 790
 –, bubble 316
 –, cavity 60
 –, density distribution 334
 –, instability 333
 –, interstellar 7
 –, mantle 657
 –, sheet 33, 657, 658
 –, solar wind 657
plasmasphere 657, 864
plasmoid 864
plate
 –, convergence 801
 –, motion 20, 891, 965
 –, oceanic 128
 –, tectonics 139, 779, 801, 957, 992
  –, reconstruction 817
  –, theory 887
plateau age 835
platform tilt 708
plume 128, 146, 689, 959
 –, generation 967
 –, mantle 146, 902
 –, polar 946
Pluto 204, 207, 439, 446
PMF (see *potential magnetic field*)
pNRM (see *primary natural remanent magnetization*)
Poggendorff, J.C. 436
POGO (see *Polar Orbiting Geophysical Observatory*)
Poincaré
 –, flow 843
 –, mode 843
point
 –, isotropic 517
 –, antipodal 122
Poisson
 –, distribution 336, 497
 –, equation 229, 643
Pola Observatory 728, 733
polar
 –, electrojet 700
 –, null zone 828
 –, path 428
 –, plume 946
 –, vector 306
 –, wander 504, 959, 965
  –, curve 887
  –, large-scale 957
  –, long-term 958
  –, true 956, 957
 –, wandering 959
 –, wind 657
Polar Orbiting Geophysical Observatory (POGO) 39, 828
 –, spacecraft 828
polarity
 –, asymmetries 314
 –, bias 314
 –, chron
  –, C-sequence 329
  –, M-sequence 329
  –, short-lasting 665
 –, geomagnetic 664
  –, history 666
  –, timescale 320, 328, 544, 664
 –, interval, statistics 140
 –, magnetic 361
 –, normal 313, 664, 778
 –, reversal 324
  –, geomagnetic 311, 320, 781
  –, Mars 504
 –, reverse 313, 664, 778
 –, state, equivalence 859
 –, stratigraphy, magnetic 664
 –, study 993
 –, superchron, Cretaceous 329
 –, timescale 992
  –, geomagnetic 664, 780, 929
  –, Plio-Pleistocene 993
 –, transition 321, 325, 829, 834
polarization 155
 –, induced 230
 –, magnetic 974
polaron 685
pole
 –, Early Quaternary 821
 –, Euler 959
 –, Fisherian 893
 –, geographic 350
 –, geomagnetic 33
  –, virtual 150, 321, 947, 961
 –, Jurassic 821
 –, key paleomagnetic 839
 –, Late Carboniferous 820
 –, Late Jurassic 820
 –, Late Permian 820
 –, Late Triassic 820
 –, magnetic 310
  –, north 339
 –, mean 46
 –, Ordovician 821
 –, paleomagnetic 282, 777–779

pole *Contd.*
 –, Permian 821
 –, reduction to 856
 –, Silurian 820
 –, virtual geomagnetic (VGP) 282, 778, 829, 892
  –, low-latitude 892
Polestar 356
pollutant particle 251
pollution
 –, anthropogenic 533
 –, load 533
poloidal 163
 –, field 22, 174, 180
 –, mode 163
 –, scalar functions 22
polynomial
 –, homogeneous 377
 –, Legendre 350, 364, 378
POMME (model) 676
ponderomotoric force 198
Pondicherry Observatory 732
pore pressure 985
porosity 77, 240, 525, 985
Port Martin Observatory 723
Port Moresby Observatory 717
Port Stanley Observatory 720
postdepositional 588
 –, remanent magnetization 767
postdetrital remanent magnetization (pDRM) 535
postmetamorphic partial thermomagnetic remanent magnetization (pTRM) 456
post-perovskite 148, 151, 153
potassium in core 79
potassium–39-potassium–40 ratio 835
potassium–40 89
potassium-argon age 834
potassium-argon dating 335, 834
potassium-uranium ratio 854
potential
 –, chemical 99, 408, 421
 –, electric, W form 985
 –, field 448
 –, function 359
 –, geomagnetic 466
 –, gravitational 362, 379
 –, harmonics 467
 –, Helmholtz 217
 –, interatomic 94
 –, magnetic 122, 492
  –, field (PMF) 840, 841
 –, scalar 194, 350
 –, self-gravitation 362
 –, temperature 840
 –, theory 466
 –, vector 194
 –, vorticity 840, 852
Potsdam 42
Potsdam Observatory 729
power
 –, spectrum 350, 353, 771, 897
 –, transmission grid 316
Poynting flux 641
PPM (see *proton precession magnetometer*)
Prague Observatory 728
Prandtl number 75, 291, 297, 299, 646
 –, magnetic 291, 645
Precambrian 839
 –, craton 817
 –, paleopole 839
 –, shield 478, 479
precession 185, 842, 956
 –, frequency, proton 468
 –, Larmor 468
 –, luni-solar 645
precision parameter 273
preconditioner 216
precursor, transitional 323
precursory EM signal 908
preferred band 322
Preisach
 –, diagram 267, 610
 –, function 881
 –, theory 879
Preisach-Néel
 –, interpretation 268
 –, theory of hysteresis 267
Preliminary Reference Earth Model (PREM) 91, 107, 140, 152, 210, 426, 434
Preliminary Reference Earth Model 2 (PREM2) 213
pre-Maxwell
 –, equation 287
  –, EM 640
 –, theory 640
present-day nondipole field 326
pressure
 –, dynamic 135
 –, effective 15
 –, geostrophic 87
 –, kinetic 644
 –, magnetic 641
 –, thermal 366
 –, vessel apparatus 600
pressure-induced remanent magnetization (PRM) 696
pre-superchron 338
Price Medal 844
Price, Albert Thomas 844
primary
 –, magnetization 595
 –, natural remanent magnetization (pNRM) 781
 –, primary
 –, remanence 595
 –, signal 69
principal
 –, axe, crystallographic 476
 –, component analysis 45, 46, 845, 848
Pringle Falls, Oregon 837
prism, accretionary 246
PRM (see *pressure-induced remanent magnetization*)
probability-density function 272
probe, seismic 970
problem
 –, inverse homogeneous 467
 –, linear stability 176
 –, Proudman-Stewartson 115
process
 –, auroral 33
 –, compaction 553
process *Contd.*
 –, diagenetic 475, 584
 –, self-reversing 339
 –, tectonic 245
 –, transitional 340
prognostic 853
prograde 76
project, magnet 850
prokaryotes 516
prolate 548
 –, variance ellipsoid 846
proliferation, nuclear 7
property
 –, dielectric 223
 –, magnetic 527
  –, low-temperature 515
  –, nondirectional 248
 –, statistical 772
prospection
 –, archeological 24, 248
 –, magnetic 23
Proterozoic 966
protoEarth 151
proton
 –, conduction 686
 –, gyromagnetic ratio 710
 –, magnetometer 468
 –, precession
  –, frequency 468
  –, magnetometer (PPM) 2, 60, 542, 711, 712, 717, 858, 990
Proudman, J. 852
Proudman-Stewartson problem 115
Proudman-Taylor (PT)
 –, constraint 291, 945
 –, theorem 631, 852, 852
provenance area 533
proxies 253
proxy 567, 790
 –, mineral-magnetic 525, 531
 –, parameter 525, 527, 528
Prudhoe Bay Observatory 721
Pruhonice Observatory 728
Prussian Meteorological Institute 437
Prussian state 403
PSD (see *pseudo-single domain*)
pseudoscalar 306, 840
pseudo-single domain (PSD) 492, 785
 –, grain 755, 873
pseudosusceptibility 536
pseudotachylite 602, 799
pseudo-Thellier method 759
pseudounconformity 567
pseudovector 306
PSV (see *paleomagnetic secular variation*)
PT (see *Proudman-Taylor*)
pTRM (see *partial thermoremanent magnetization*)
Pugwash Conference 7
pulsation 63, 698
 –, classification scheme 333
 –, continuous 333, 700
 –, geomagnetic 333
 –, irregular 333, 700
 –, ULF 333
pumping, Ekman 113, 124, 648

Punaruu event 837, 838
P-wave 147
pyrite 260, 454, 520, 560, 783
–, paramagnetic 457
pyroxenes 560, 932
pyrrhotite 157, 454, 457, 495, 520, 526, 536, 614, 668, 792–795
–, antiferromagnetic hexagonal 561
–, authigenic 783
–, ferrimagnetic monoclinic 561
–, ferromagnetic 806
–, monoclinic 260, 457, 520
–, natural 496
–, polycrystalline 498
–, single-domain 626
–, stochiometric 454

Q

Qaanaaq Observatory 737
Qaidam 823
–, basin 817, 820, 822
Qelico 822
Qeqertarsuaq Observatory 710, 736, 737
QHM (see *quartz horizontal magnetometer*)
Qiangtang 822, 825
–, terrane 819
Qinling-Dabie Mountains 818
Q-ratio 480
quadrupole
–, geocentric axial 284
–, family 307
quantum
–, interference properties 883
–, mechanics 700
quartz 931
–, horizontal magnetometer (QHM) 717
quasidipole coordinates 952
quasigeostrophic 75
–, convection 124
Quaternary climate 571
Quebrada Blanca Bright Spot 243
quenching
–, alpha 198
–, beta 198
Quetelet, Adolphe 725
quiescent state 74

R

Racah square 386
radar, ground-penetrating (GPR) 223, 229, 230
–, survey 226
Radau equation 443
radiation 789
–, cosmic 702
–, synchrotron 690
radio
–, communication 317
–, wave propagation 452
radioactivity, core 288, 289
radioastronomy 54
radioisotope production 507
radiometry, spectral 690
radionuclide signature 252
radius, solar 179
Rajmahal Traps 751
range
–, dynamic 234
–, index 509
Rankine-Hugoniot equation 912
ray
–, cosmic 53, 507, 887
–, direction 147
–, gamma 478
–, PcP 125
–, seismic 125, 126, 147, 210, 418, 426, 428, 434, 905–907
–, X- 33, 51, 317, 369, 452, 508, 546, 550, 691, 701, 914
Rayleigh
–, law 701
–, number 14, 74, 75, 80, 82, 117, 128, 262, 291, 297, 417, 630, 646
–, critical 631
–, magnetic 651
–, modified 297, 652
Rayleigh-Bénard convection 117
Rayleigh-Taylor instability 649
Raymahal Traps, India 929
real fluid 7
rebound
–, glacio-isostatic 958
–, postglacial 958
reciprocity 574, 612
–, approach 221
reconnection 33, 644, 864, 927
–, magnetic 33, 316, 658
reconstruction, continental 840
record
–, cosmogenic isotopic 343
–, deep-sea, high-resolution 312
–, estuarine 762
–, excursion, high-resolution 770
–, gravimetric 423
–, lacustrine 762
–, lunar paleomagnetic 789
–, magnetostratigraphic 354
–, marine oxygen isotopic 571
–, paleointensity 259
–, paleomagnetic 564, 788, 799, 829
recorder, paleomagnetic 248, 325
recording, magnetic 943
recovery phase 509, 663, 927
recrystallization, dynamic 148
rectifier 66
recurrence relations 379
red
–, clay facies 783
–, doublet 33
reference
–, field 396
–, frame 956
–, model 208
–, state 300
–, adiabatic (isentropic) 14, 16
reflectometer, electron 502
refraction
–, atmospheric 67
–, electromagnetic 233
refractory
–, inclusion 797
–, calcium aluminum inclusions (CAI) 796
–, Allende 797
region
–, bipolar 179, 181
–, corotating interaction 663
–, ephemeral 179
–, active 182
–, subadiabatic 83
region1 FAC system 660
region2 FAC system 660
regional 242
regolith
–, breccia 789
–, lunar 788
regression 568
–, M-estimate 868
–, minimization 868
–, ridge 857
regularization 950
–, constraints 87
Reiner gamma feature 791
relation
–, Ahrrenius 105
–, constitutive 640
–, Gibbs-Duhem 100
–, Nernst-Einstein 686
–, orthogonality 888
–, recurrence 379
–, thermodynamic 12
relationship
–, Maxwell 367
–, Mie-Grüneisen 366
relaxation
–, magnetic 753
–, oscillation 76
–, thermal 121, 533, 943
–, time 157, 269, 513, 610, 622, 628, 872
–, long 872
remagnetization 343, 910, 984
–, radial 785
–, thermal 986
remanence
–, acquisition 607
–, coercivity 266, 877
–, crystallization 580
–, detrital 29
–, lightning 593
–, lock-in 325
–, magnetic 535, 540
–, anisotropy 535
–, spontaneous 599
–, measurements, low temperature 253
–, natural 536
–, net 573
–, paleomagnetic 839
–, primary 595, 596
–, saturation 590, 874, 879
–, isothermal 794
–, secondary 581, 596
–, tensor, anisotropic detrital 579
–, viscous 701
remanent magnetization (RM) 140, 920
–, chemical 38
–, isothermal 613
–, multicomponent 156
–, residual 920
–, saturation 266
–, vector 921
–, viscous 871

repeat station 359, 719, 858
research, archeomagnetic 943
reservoir, hydrocarbon 72
resetting, thermal 603
resistive layer 72
resistivity
–, apparent 222, 233, 277, 670, 869
–, bulk 245
–, core impurity 117
–, crustal 232
–, electrical 62, 92, 242, 985
–, marine sediment 232
–, seawater 231, 246
Resolute Observatory 710
resolution
–, analysis 84
–, linearized 222
–, subcellular 51
resonance, Schumann 308
resources 477
–, economic 223
–, exploration 477
–, geothermal 245
[illegible]
–, curve 233
–, DC resistivity 233
–, diamagnetic 60
–, electromagnetic 309
–, estimation 309
–, function 219, 308, 866
–, impulse 236
–, magnetic 540
–, step-on 233, 234
return stroke 592
Reunion Island 963, 993
Reunion subchron 834, 837, 838
reversal 835
–, behavior, changes 861
–, boundary 329
–, chronology, first-order 779
–, duration 326, 344
–, free decay model 860
–, frequency 133, 134, 320, 843, 929
–, variations 133
–, geomagnetic 161, 306, 335
–, archive 339
–, paths 134
–, rate 354
–, mechanism 859, 860
–, numerical simulation in 860
–, process 929
–, rate 337
–, statistics 199
–, successive 340
–, test 608
–, theory, 859
–, time scale 979
reverse polarity 313
Reykjanes Ridge 483
Reynolds
–, magnetic number 21, 101, 111, 8, 22, 85, 164, 171, 173, 176–178, 183–186, 188, 190, 191, 194, 195, 200, 201, 205, 275, 288, 297, 299, 641–645, 652
–, rule 198, 199
–, stress 103, 181
–, tensor 17, 198
rhodochrosite 521, 932
Riddell, C.J.B. 436
–, magnetical instructions 436
ridge
–, mid-ocean 72, 208, 237, 446
–, basalt 80
–, regression 857
–, Reykjanes 483
–, slowly spreading 72
–, Valu Fa 237
Riga dynamo 184
rigidity modulus 92
Rikitake
–, coupled-dynamo (1958) 173
–, dynamo model 863
Rikitake, Tsuneji 862
ring current 160, 404, 657, 863
–, discovery 865
–, substorm-enhanced 334
–, variation 415
ringwoodite 685, 687
RM (see *remanent magnetization*)
Roberts number 120
–, magnetic 297
rock
–, basaltic 993
–, conductivity
–, electrical 62, 69
–, solid 244
–, cratonic 62
–, extrusive 871
–, fabric 564
–, ferrimagnetic 480
–, hemo-ilmenitic 141
–, isotropic 872
–, lithology 480
–, magnetism 54, 139, 696, 701, 870, 874, 883, 942
–, magnetite-bearing 516, 614
–, magnetization 35
–, magnetometer, superconducting (SRM) 883
–, Martian 603
–, metamorphic 555
–, low-grade 555, 819
–, paramagnetic 480
–, plutonic 554, 629
–, pyrrhotite-bearing 537
–, sample 751
–, sedimentary 782
–, Tertiary 820
–, Triassic 993
–, ultramafic 597
–, VRM 629
Rodinia 819, 966
Rodrigues formula 385
Rogers, Woodes 989
Roggeveen, Jacob 989
roll
–, convection 165, 901
–, tilted convection 190
Ross' magnetic Antarctic survey 890
Ross, John 890, 989
Ross, Sir James Clark 359, 403, 717
Rossbank Observatory 717, 807
Rossby
–, number 75, 291, 297, 299, 430, 940
–, magnetic 650
Rossby *Contd.*
–, modified 648
–, wave 75, 124, 136, 432, 852, 994
–, buoyancy-driven magneto 635
–, magnetic-Coriolis (MC) 633
–, thermal 75
rotating convection experiment 275
rotation 804
–, about an axis 389
–, axis 965
–, differential 118, 136, 164, 166, 189, 213, 294, 302, 859
–, diurnal 365
–, Earth 54, 471
–, inner core 294, 423
–, matrix 383
–, nonuniform 180
–, pole 956
–, rigid body 391
–, solar 353, 415
–, Sun 308
–, tectonic 806, 961
rotational remanent magnetization (RRM) 455, 456
rotor, spherical 171
roughening, operator 220
Royal Danish Academy of Sciences 376
Royal Danish Geodetic Institute 469
Royal Hungarian Meteorological Institute 728
Royal Irish Academy 474
Royal medal 61
Royal Meteorological Society 888
Royal Netherlands Academy of Arts and Sciences 887
Royal Observatory at Greenwich 990
Royal Radar Establishment 887
Royal School of Mines 742
Royal Society 357, 887
Royal Society of New South Wales 887
RRM (see *rotational remanent magnetization*)
rubidium magnetometer 714
ruby fluorescence 690
Runcorn theorem 791, 888
Runcorn, S. Keith 886, 887
Russian Academy of Science 738
Russian Hydrometeorological Service 738
RV Chain 993
RV Trident 993
RX coil 955

S
Sabhawala Observatory 732
Sabine, Edward 890
Sable Island Observatory 721
SAC-C 60, 892
Safegarde of Saylers, or, Great Rutter 707
sailing
–, direction 66
–, instruction 356
saline
–, aquifer 224
–, fluid 244
sample
–, dredged 597
–, drilled 597
–, lunar 788

sample *Contd.*
–, oceanic 596
–, paleomagnetic 604, 884
sampling 604
San Andreas Fault 244
San Fernando Observatory 739
San Pablo Observatory 740
sand, wind-deposited 475
Santa Rosa event 838
Santonian-Campanian stage boundary 666
Sanya Observatory 728
sapropel environment 531
SAR (see *subauroral red*)
Sarrabat, Nicolas 460
SAS (see *Small Astronomical Satellite*)
satellite 59
–, dynamo 203
–, icy 204
–, interior 439
–, low Earth orbit 318
saturation
–, isothermal remanent magnetization (SIRM) 515, 527
–, magnetization 589, 874, 877
–, remanence 590, 617, 874, 877, 879
Saturn 203, 204, 439, 446, 865
–, magnetic field 198, 287
Savery, Servington 358
Scaglia Rossa limestone 669
scalar
–, functions 22
–, potential 194, 350
scale
–, height 689
–, length 277
–, similarity 102
–, subgrid- 102
–, model 305
scaling rule 264
SCAR (see *Scientific Committee on Antarctic Research*)
Schlumberger array 229
Schmidt 735
–, balance 435
–, vertical 437
–, functions 680
–, normalization 381
–, number 297
–, magnetic 297
–, seminormalized 351
Schmidt, Adolf 42, 437, 438, 729, 736
Schmidt-normalized 395
Schmucker inversion 220
Schouten, Willem Corneliszoon 988
Schrödinger equation 94
Schumann resonance oscillation 308, 353, 698
Schuster 812
–, representation 812
Schuster, Arthur 353, 438
Schwarzschild criterion 80, 137
Scientific Committee on Antarctic Research (SCAR) 408
Scientific Committee on Solar Terrestrial Physics (SCOSTEP) 408
Scott Base Observatory 735, 736, 951
Scott, R.F. expedition 723
ScP 970
screw dynamo 177
Scripps Institute of Oceanography 237, 483
SD (see *single domain*)
seafloor
–, anomaly 794
–, conductivity 72, 234
–, profile 73
–, electric field recorder 72
–, magnetic anomaly 320
–, magnetometer 71
–, spreading 140, 484, 542, 664, 779, 794, 979, 980
seamount 959
–, magnetism 891
search algorithms 220
seasonal cycle 69
seawater 62, 71
–, resistivity 231, 246
Secchi, Angelo 460
secondary field 69
sectorial harmonics 680
secular
–, cooling 128
–, variation 86, 588, 778, 779, 839, 892, 929
–, historic (HSV) 766
–, impulses 651
–, model 892–898, 901
–, paleomagnetic 133
Seddin Observatory 729
SEDI 902
sediment 560
–, anoxic 521, 585
–, deep-sea 253, 758, 761, 781, 993
–, lacustrine 353
–, magnetic 24
–, marine 72, 353
–, resistivity 232
–, paleointensity
–, relative 577
–, pelagic 785
–, red 584, 629
–, relative paleointensity 758
–, suboxic 249
–, terrigeneous 786
–, unconsolidated 665
–, unlithified 567
–, volcaniclastic 785
–, -water interface 781
sedimentation rate 779
–, pelagic 783
segregation
–, liquid 79
–, metal-silicate 77
seismic
–, anisotropy 146
–, blank zone 239
–, layer 597
–, 2A 597
–, 2B 597
–, 3 597
–, migration 264
–, parameter 100, 107
–, probe 970
–, tomography 129, 416
–, baseline uncertainty 153
–, velocity pattern 152
–, wavespeed 209
seismo-electromagnetic (SEM) 908
self-demagnetization 497
self-diffusion coefficient 106
self-gravitating 76
self-gravitation potential 362
self-induction 954
self-magnetization 888
self-potential (SP) 71, 984
self-reversal 335, 777, 872
self-reversing processes 339
self-sustaining 302
SEM (see *seismo-electromagnetic*)
semiconduction, electronic 69
sensitivity 525, 921
–, limiting 920
sensor
–, deep-towed 982
–, electric field 242
–, fluxgate 71, 921
–, induction coil 72
separation, transmitter-receiver 234
September equinox 452
sequence
–, lacustrine 567
–, Quaternary 766
series, Mesozoic 485
serpentinite 598
serpentinization 598
SF (see *smoothing factor*)
Shaanxi province 821
Shackelton's British Antarctic Expedition 990
shadow zone 904
shallow ocean 814
Shandong province 820, 821
Shanxi province 820
shape 871, 937
–, anisotropy 564, 610
–, factor 143
–, inverse effect 938
–, parameter 476, 938
Shaw
–, method 575
–, technique 910
shear
–, flow, parallel 118
–, layer 431
–, detached 853
–, modulus 153
–, movement 556
–, plane 802
–, simple 802
–, total 565
–, wave splitting 147, 155
–, zone 244
sheet
–, flow 597
–, high-permeability 541
–, magnetic 541
–, remanent 541
–, plasma 658
–, thin 63
–, solutions 215
sheeted
–, dike layer 597
–, lava 597
shelf, continental 63
shell structure, nuclear 215

Shen Kua 435
Shergotite meteorite 791
shergottites 797, 799
Shergotty-Nakhla-Chassigny (SNC) meteorite 446
Sheshan Observatory 728
shield
–, dual material 541
–, Fennoscandian 479
–, magnetic 540
–, superconducting 885
–, magnetostatic 541
–, Precambrian 479
shielding
–, internal 541
–, magnetic 540, 541
–, stage 541
shift
–, factor 277
–, static 222, 671
Shillong Observatory 732, 733
Shimla Observatory 731
ship
–, compass 40
–, logbook 339
shock 789
–, decomposition 789
–, demagnetization 600, 603, 795
–, experiment 692
–, magnetization 600, 791
–, remanent magnetization (SRM) 696
–, system 884
–, wave experiment 912
shunt 167
SI (see *Système Internationale d'unités*)
Siberia 993
Sichuan province 821
siderite 521, 526, 560, 932
Sidgreaves, Walter 461
signal
–, detector 920
–, magnetic, environmental 259
–, paleomagnetic 260
–, syndepositional 260
–, precursory EM 908
–, primary 69
signature
–, paleomagnetic 273
–, radionuclide 252
Silchar Observatory 732
silica, dissolved 784
silicate
–, ferromagnesian 932
–, perovskite 73
silicon 371
–, impurity 110
–, steel alloy 541
similarity transform 264
simulation
–, large-eddy 102
–, numerical 292
sinc function 122
sine galvanometer 438
single-domain (SD)
–, grain 249, 873
–, magnetic grain 266
–, state 455
singularity, logarithmic 382
SINT800 record 160
SIRM (see *saturation isothermal remanent magnetization*)
site 765
–, archeological 23
Sitka Alaska Observatory 723
skewness analysis 962
skin
–, depth 63, 69, 224, 308, 670, 699, 742
–, electromagnetic 206
–, effect 242, 642
–, layer, electromagnetic 843
SKJKP 434, 903
slab
–, downgoing 967
–, graveyard 212
–, subducted 149
Slave craton 227
Slichter triplet 422
–, rotational splitting of the 423
Slingram method 230
slope
–, continental 63
–, diamagnetic 880
–, paramagnetic 880
slow
–, dynamo 186
–, wave 651
slowness 145
Small Astronomical Satellite (SAS) 673
smectite 784, 932
Smith, Frank 438
Smithsonian Institution 437
smooth, optimally 122
smoothing factor (SF) 266
SMS (see *synchrotron Mössbauer spectroscopy*)
SNC (see *Shergotty-Nakhla-Chassigny meteorite*)
Snell law 210, 903
snowball earth hypothesis 780
Society of Jesus 460
Socorro 699
Sodankylä Observatory 737
SOHO (see *Solar and Heliospheric Orbiter*)
soil 789
–, lunar 629
–, magnetic 25
solar
–, activity 316, 353
–, atmosphere 119
–, compass 765
–, constant 508
–, convection zone 137, 506
–, coronal mass ejection (CME) 316
–, cycle 33, 160, 319, 405, 452, 698, 700
–, index 508
–, dynamo 165, 178
–, eclipse 471
–, flare 316, 404, 505, 927
–, luminosity 508
–, magnetic
–, activity 507
–, cycle 507
–, field 178, 335
–, maximum 316, 346, 506
–, minimum 316, 506
solar *Contd.*
–, oscillations 398
–, photosphere 89
–, quiet 63, 810
–, radius 179
–, rotation 353, 415
–, surface, Doppler measurement 508
–, system 77, 89, 788, 795
–, magnetic field 796
–, terrestrial physics 61
–, tide 471, 810
–, wind 33, 60, 318, 333, 505, 657, 658, 864, 926, 927
–, plasma 657
–, field 795
Solar and Heliospheric Orbiter (SOHO) 398
solenoid 41
solid
–, anisotropic polycrystalline 419
–, isotropic 82
–, solution series 873
solidus, eutectic 154
solution
–, disjointed quadrupole 358
–, separable 306
–, theory 408
Songpan-Ganze terrane 819
Soret diffusion 691
sound speed
–, adiabatic 12
–, isothermal 12
sounding
–, depth 308
–, electromagnetic 277
–, geophysical 242
–, magnetometric resistivity (MMR) 72
source
–, body 486
–, polarization 223
South American plate 960
South Atlantic magnetic anomaly 734
South China 823
South China Block 817, 818
South Magnetic Pole 723
South Portuguese Zone 243
South Sandwich Islands 424
Southern Alps, New Zealand 246, 666
Southern Highlands 795
SP (see *superparamagnetic*)
space
–, station 318
–, weather 408
–, forecast 317
space-domain method 486
spacecraft shielding 316
Spanish civil war 461, 739
SPdKS 970
species, magnetogenic
–, anaerobic 248
–, microaerobic 248
specific
–, entropy 840
–, heat 92, 93, 95
–, electron 93
specimen 977
–, holder 920
–, multiple 765

spectral
–, analysis 122, 770
–, estimation 353
–, Lancsoz decomposition method 218
–, radiometry 690
spectrometer, gamma-ray 478
spectroscopy, synchrotron Mössbauer (SMS) 691
spectrum
–, electromagnetic 308
–, geomagnetic
–, natural 308
–, spatial 350
–, temporal 353
–, power 350
–, Sturm-Liouville 220
–, vibrational 368
Sperry gyroscope 438
spheochasm 804
sphere, uniformly magnetized 486
spherical
–, harmonics 103, 214, 377, 680, 772, 899
–, analysis 449
–, coefficients 680
–, expansion 303
–, finite series 467
–, sectorial surface 382
–, solid 380
–, surface 378
–, tesseral surface 382
–, transformation 383
–, vector 382, 387
spheroid, oblate 208
spherules 796
spin 870
–, antiferromagnetic 874
–, axis 901
–, behavior 525
–, canting 518
–, ferrimagnetic 874
–, ferromagnetic 874
–, glass transition temperature 519
–, moment 525
–, ordering 875
–, wave 694
spinel structure 687
spinner
–, induction coil 883
–, magnetometer 29, 920, 921, 992, 993
spinning rate 921
spinors 383, 392
spiral, Ekman 113
spiralling 190
spline, harmonic 123
splitting 147
–, shear wave 147
spreading
–, center 485, 980
–, seafloor 484, 542
springtime 248
Sputnik 3 673
Sq 700
–, variation 808
SQUID (see *superconducting quantum interference device*)
Srednekan Observatory 738
SRM (see *shock remanent magnetization*)
SSC (see *sudden storm commencement*)
stability, neutral 14
stacks, global 160
stage
–, boundary 329
–, paleontological 666
stagnation, point 186
standing field hypothesis 321, 994
Stanford Research Institute 883
star
–, camera 59
–, imager 744
–, main-sequence 137
Stará Dala Observatory 728
state
–, Ekman 174
–, GEM domain 498
–, hydrostatic 300
–, micromagnetic 574
–, model-Z 941
–, multidomain 575
–, SD-like 497
–, superparamagnetic 943
–, Taylor 174, 175, 941
–, transitional 323
–, weak field 652
static
–, divergence 217
–, correction 217
–, flux bundles 133, 134
–, limit 216
–, shift factor 277
station
–, magnetotelluric 69
–, repeat 719
statistics
–, Bayesian 29
–, Bingham 45
–, Boltzmann 500
–, Fisher 272, 608
–, reversal 199
steady
–, field 540
–, flow assumption 88
steel alloy 600
Steens Mountain, Oregon 327, 343, 993
steering compass 66
Stefan-Boltzmann constant 688
Stevin, Simon 357
Stewart, Balfour 333
Stewartson
–, conditions 650
–, layer 115, 432, 649, 853
stirring, mechanical 207
stoichiometry 515
Stokes theorem 8, 112
Stokes-Einstein equation 105
Stoner-Wohlfarth hysteresis loop 876
stoney-iron meteorite 796
Stonyhurst College Observatory 460
storm 308, 926
–, Bastille-day 742
–, geomagnetic 137, 316, 663
–, frequency 316
–, Halloween 405
–, index, equatorial 511
storm *Contd.*
–, magnetic 27, 61, 376, 696, 741
–, relationship to sunspots 890
–, sudden commencement (SSC) 663, 926
–, time field 844
storm-time
–, disturbance index D_{st} 511
–, variation 698
strain
–, analysis 556
–, indicator, quantitative 558
–, principal 556
stratification
–, chemical 128
–, stably 150
–, density 119
–, thermal 129, 130, 206, 302
stratigraphic variation 256
stratigraphy, magnetic 786
streak record 915
stream
–, function 85
–, free 85
streaming, potential 985
stress 535, 599, 872
–, anisotropy 491
–, control 881
–, demagnetization 599, 600
–, deviatoric 985
–, external 491
–, field, magnetostrictive 881
–, hydrostatic 599
–, internal 493
–, Maxwell 103, 114
–, Reynolds 103
–, sensitivity 985
–, source, external 881
–, tensor 641, 874
–, uniaxial 600
stretch-twist-fold 187
strike 602
–, electromagnetic 20
–, geoelectric, regional 222
stripe
–, magnetic 542
–, magnetized rock 140
Strouhal number 194
structure
–, basement 224
–, columnar 166
–, crystal's spin 609
–, crystalline 564
–, crystallographic 561, 563
–, fibril 182
–, geological 69, 223
–, hydrothermal 596
–, inner core 419
–, Landau-Lifshitz-type 499
–, minimum 220
–, multidomain 589
study
–, directional 31, 32
–, downhole 226
–, environmental 226
–, high-resolution 256
–, magnetic fabric 578, 920

study *Contd.*
–, marine magnetotelluric 73
–, paleointensity 31, 32, 271
–, absolute 266
–, relative 271
–, paleomagnetic, baked contact test 35
–, polarity 993
–, sequence stratigraphic 568
Sturm-Liouville spectrum 220
subadiabatic region 83
subauroral red (SAR) arcs 864
subchron 312, 928
–, Cobb Mountain 838
–, Jaramillo 834, 837, 838
–, Olduvai 834, 837, 838
–, Reunion 834, 837, 838
subduction 150
–, zone 64, 72, 129
–, active 244
subgrid scale 101, 102
–, field 17
–, model 305
subgroup 306
submarine, communication with 221
subsatellite, magnetometer 791
subsolidus convection 16
substage, boundary 666
substance
–, diamagnetic 60
–, paramagnetic 525
substorm 308, 662, 926
–, auroral 926
–, bulge 662
–, cycle 34
–, electrojet 662
–, westward 34
–, expansion phase of the 662
–, magnetospheric 662, 700, 926
–, recovery phase 662
suction, Ekman 113
sudden storm commencement (SSC) 663, 926
Suiko seamount 962
sulfide 248
–, ion 786
sulfur 371
–, impurity 110
sum rule 385
Sun
–, compass 993
–, dipole magnetic field of the 179
–, magnetic field of 505
–, rotation 308
–, torsional oscillations of the 179
sundial 66, 356, 435
sunspot 10, 138, 468, 505
–, cycle 178, 347, 353, 414, 505, 663
–, maximum 316
–, minimum 316
–, number 507
Sun-synchronous 950
superadiabatic 82
–, temperature 152
–, gradient 80
superchron 320, 338, 339, 928
–, Cretaceous 294
superchron *Contd.*
–, Cretaceous normal (CNS) 339, 928
–, Cretaceous normal polarity (CNPS) 328, 749, 751
–, Kiaman Reversed Polarity 752
–, Ordovician 929
–, Permo-Carboniferous 928
superconducting quantum interference device (SQUID) 655, 877, 883
–, detector 921
–, high temperature 921
–, high-resolution 793
–, magnetometer 51, 793
superconductor, high temperature 886
supergranule 179
Superior craton 227
superparamagnetic (SP) 50, 782
–, grain 268, 873
–, anomaly 986
superparamagnetism 610, 943
Superplume 154
–, generation 967
superrotation, inner core 304
super upwelling 967
surface
–, bioturbated 781
–, charge, density 640
–, current 640
–, distortion parameter 222
–, draped 975
–, drape-to-level 975
–, energy 491
–, equipotential 132, 136
–, geostrophic 940
–, harmonics 351
–, integral method 449, 451
–, level-to-drape 975
–, magnetic 8
–, magnetometer 791
–, mixing layer 781
–, wave 334
Surlari Observatory 729
survey
–, continental 141
–, gamma-ray spectrometer 478
–, geotechnical 236
–, magnetic 56
–, Ireland 473
–, marine 542
–, susceptibility 26
–, magnetotelluric 844
–, satellite-based 788
–, World Magnetic 56
surveying
–, aeromagnetic 1
–, compass 66
susceptibility 513, 870, 931, 974
–, anhysteretic 783
–, anisotropy 799
–, anisotropy of magnetic (AMS) 937
–, bulk 937
–, contrast map 489
–, diamagnetic 931
–, distribution 487
–, ellipsoid 547, 561
–, fabric 538
susceptibility *Contd.*
–, ferrimagnetic range of 480
–, ferromagnetic 552
–, frequency-dependent 592, 933
–, induced 755
–, initial 548, 624
–, interaction 592
–, intrinsic 549
–, low-field 517, 528, 533
–, magnetic 24, 94, 251, 475, 513, 546, 560, 564, 667, 780
–, anisotropy 514, 535, 546, 578
–, differential 932
–, low-field 566
–, survey 26
–, variations 256
–, mass-specific 932
–, mean bulk 552
–, measurement 933
–, paramagnetic 931
–, range 480
–, parameter 937
–, principal 547, 556, 937
tensor 549, 938
–, variation 513
–, volume 25
suspension
–, bifilar 977
–, cardanic 66
suture 801, 818
–, zone 242
swarm 59, 743
S-wave 147
–, anisotropy 148
–, reflection 153
Swider Observatory 728
swinging the ship 68
symmetry
–, breaking 306, 643
–, mirror 643
–, properties 306
synchrotron
–, Mössbauer spectroscopy (SMS) 691
–, Gazelle 990
–, radiation 690
Syowa Base 696
system
–, bistable 859
–, controlled source electromagnetic (CSEM) 232
–, core-mantle 132
–, hydrothermal 247
–, natural 525
–, nearly axisymmetric 164
–, nonlinear chaotic 304
–, solar 77, 89, 788
–, three-rotor 172
Système Internationale d'unités (SI) 874, 973

T

Tacchini, Pietro 733
tachocline 179, 182, 400
taenite 788, 796
tail lobe 661
Takalamagan 819
Takena red beds 822

talc, ferrous 522
tall azimuth visor 435
Tamanrasset Observatory 710
tangent cylinder 76, 115, 430, 901, 946
tangential geostrophy 131
Tangerang Observatory 725, 734
taper 122
tapering 353
target curve, astronomical 258
Tarim
–, basin 819, 822
–, block 817
Task Group WDMAM 479
Tasman, Abel Janszoon 988
taxonomy for magnetic phenomena 361
Taylor
–, cell 649
–, column 365, 400, 852
–, condition 940
–, constraint 292, 293
–, hypothesis 103
–, number 630
–, state 174, 175, 650, 941
Taylor, G.I. 852
Taylorization 293
TE (see *transverse electric*)
TEC (see *total electron content*)
Technical University of Braunschweig 279
technique
–, electrical galvanic 223
–, EM induction 223
–, tensor-averaging 937
tectonic
–, active belt 208
–, boundary 64
–, process 245
, rotation 961
tectonics 779
tectonoelectric phenomen 909
teeth 782
telecommunications cable 227
–, submarine 131
telegraph
–, network 316
–, station 280
telluric force 359
temperature 526, 872
–, absolute 116, 581
–, adiabatic 74, 301
–, blocking 156, 493, 610, 622, 943
–, burial 628
–, condensation 78, 89
–, cryogenic 515
–, Curie 23, 157, 454, 481, 485, 889
–, Debye 367
–, depression 109
–, due to impurities 110
–, difference, superadiabatic 16
–, electron 60
–, equation 634
–, eutectic 99
–, fluctuation, superadiabatic 292
–, gradient 93
–, subadiabatic 289
–, superadiabatic 80
temperature *Contd.*
–, homologous 105
–, liquid nitrogen 157
–, low-Curie 499
–, maximum 603
–, melting 372
–, Néel 519
–, potential 74, 840
–, spin glass transition 519
–, superadiabatic 152
–, transition 517, 519
–, unblocking 36, 626
–, Verwey 517
temporal basis function 949
tension, magnetic 633, 644
tensor
–, decomposition 672
–, magnetotelluric 278
–, impedance 866
–, inertia 966
–, Kronecker 193
–, Levi-Civita 193
–, nonhydrostatic 965
–, second-rank 578
tensor-averaging technique 937
tephra, volcanic 253
Terra Nova Bay Observatory 733
terrane 477, 802
–, tectonostratigraphic 140
terrella 361, 464
–, model 64
Terror, HMS 717
Tertiary 961
tesseral harmonics 680
test
–, baked contact 35, 608
–, conglomerate 608
–, consistency 608
–, fold 608
–, reversal 608
tetrataenite 701, 789, 796–798
texture, authigenic 585
Thales 870
Tharsis bulge 503
The Formation of the Lunar core 887
The Royal Norwegian Academy of Science 887
Thellier 870
–, double heating method 583, 749
–, law 574, 942, 943
–, paleointensity 749
–, technique, conventional 694
Thellier, Emile 942
Thellier-Thellier
–, data 577
–, method 575, 611
–, paleointensity method 943, 944
–, technique 754
THEMIS network 726
thenaoremanent magnetization (TRM) 910, 911
theodolite, fluxgate 710, 858
theorem
–, Alfvén's 112
–, antidynamo 21, 22, 137, 163, 173, 176, 192, 468, 643
–, binomial 383
theorem *Contd.*
–, bounding 21, 22
–, Cowling's 21
–, Floquet's 201
–, Gauss 378, 888
–, Green 466, 974
–, Kelvin 8
–, Proudman-Taylor 631, 852
–, Runcorn 888
–, Stokes 8, 112
–, toroidal 21
theory
–, Backus-Gilbert 221
–, Big Bang 7
–, boundary layer 423
–, Chapman-Ferraro 137
–, continental drift 607
–, dynamical systems 186
–, dynamo 928
–, ideal body 143
–, ideal solution 408
–, inverse 467
–, mixing length 181
–, multidomain 575
–, nondynamo 704
–, Plate Tectonic 887
–, potential 466
–, pre-Maxwell 640
–, relaxation 575, 609, 701
–, reversals 859
–, turbulence 192
–, X-ray 369
thermal
–, boundary layer 18
–, conduction 116
–, conductivity 92, 98, 107, 108, 110, 688
–, lateral variations 133
–, demagnetization 604
–, continuous 627
–, diffusivity 74, 80, 92, 121, 688
–, expansion 74
–, coefficient 93
–, expansivity 95, 372
–, Péclet number 297
–, relaxation 121
–, stratification 129, 130
–, vibration 688
–, wind 76, 174, 648, 945
–, equation 852, 945
thermodynamic
–, integration 99
–, linearization 13, 16, 18
–, relation 12
thermoelectric 841
–, effect 705
thermopower 686
thermoremanence 23, 701, 943
thermoremanent
–, efficiency 617
–, magnetization (TRM) 35, 37, 161, 325, 493, 514, 535, 609, 616, 696, 777, 781, 942
–, acquisition 616
–, field-blocked 614
–, multidomain 613

thermoremanent *Contd.*
–, partial (pTRM) 37, 590, 753, 943
–, self-reversal 696
–, weak-field 499
thermoviscous magnetization 621
theta aurora 34
thickness, displacement 112
thin sheet 63
thin-shell approximation 815
thorium–232 854
threshold, percolation 20
thunderstorm activity 308
Tibet 825
Tibet basin 817
Tibetan plateau 208, 247, 822
TID (see *traveling ionospheric disturbance*)
tidal
–, friction 128, 471
–, mode, spectral analysis 814
tide
–, semidiurnal 365
–, solar 471
Tiflis Observatory 737
Tihany Observatory 710, 720
Tikhaya Bay Observatory 738
tillite, Late Precambrian 561
tilt-over mode 843
time
–, annealing 582
–, -average field 778
–, domain electromagnetic 225, 231
–, domain transmitter 225
–, magnetic diffusion 298
–, Ohmic 186
–, relaxation 269, 628
–, sequence 336
–, series 509
–, analyses 766
–, equi-spaced 767
–, translation 306
timescale
–, advective 644
–, Alfvénic 650
–, convective turnover 133
–, diffusion 644
–, diffusive 348
–, dynamical 323
–, geological 253, 281
–, geomagnetic polarity (GPTS) 778
–, reversal 484
–, spin-up 651
tiny wiggle 599, 664
Tipper vector 867
Tirunelveli Observatory 732, 733
tissue 50
–, diamagnetic 51
Titan 204, 207, 439, 445
titanohematite 518, 519, 873
titanomaghemite 597, 791, 878
titanomagnetite 495, 517, 536, 560, 597, 781, 873, 878, 879
–, intermediate 499
–, iron-rich 626
–, low Ti 792
–, oxidation-exsolution 597
–, stoichiometric 583
titanomagnetite *Contd.*
–, Ti-rich 517, 518
–, cruciform 794
TM (see *transverse magnetic*)
Toledo Observatory 739
tomographic heat flux 129
tomography 146, 148
–, electrical resistance (ERT) 229
–, finite-frequency 963
–, seismic 129, 416, 965
–, baseline uncertainty 153
Tondano Observatory 734
tool, paleomagnetic 564
Toolangi Observatory 717
topographic, variations 975
toroidal 163
–, field 23, 174
–, mode 163
–, scalar functions 22
–, theorem 21
Toronto Observatory 726
torque
–, advective 131
–, electromagnetic 425
–, fluid pressure 425
–, gravitational 136, 362
–, leakage 131
–, Lorentz 746
–, magnetic 341
–, poloidal 131
–, viscous 425
torsional
–, oscillation 746
–, wave 650
total electron content (TEC) 318, 452
TPW (see *true polar wander*)
TRACE (see *Transition Region and Coronal Explorer*)
trace mineral, ferromagnetic 475
track, quasicircular 960
transdomain 623
transfer
–, function 277, 670, 697, 866, 953
–, magnetotelluric 953
–, matrix 223
–, vertical field 221
transformer, linear variable differential (LVDT) 977
transform
–, fault 779, 959
–, intracontinental 806
–, Hankel 233
–, similarity 264
transformation
–, Galilean 640
–, magnetic phase 512
–, mineralogical 513
transient electromagnetic 954
–, method 231
–, sounding 277
transition
–, electronic 79
–, magnetic phase 515
–, metamagnetic 522
–, Morin 519
–, order-disorder 872
transition *Contd.*
–, paramagnetic 515
–, phase 515
–, spin glass 519
–, temperature 517, 519
–, zone 145, 211
Transition Region and Coronal Explorer (TRACE) 506
transitional
–, long-lived 342
–, precursor 323
–, state 323
translation
–, balance 656
–, nappes 804
transmission electron microscopy 50, 495
transmitter
–, loop 224, 954
–, time-domain 225
transmutation of atoms 53
transport properties 104
transverse
–, electric (TE) mode 221, 671
–, magnetic (TM) mode 221
traveling ionospheric disturbance (TID) 910, 984
treatment, thermal 561
–, progressive 563
Trelew Observatory 734
triangulation, Delaunay 26
Triassic 568
–, GPTS 331
Triaxe 756
triaxial 554
trigonometry, hyperspherical 391
Trinity College, Dublin 472, 742
Triton 204, 207, 439, 446
Trivandrum Observatory 731
TRM (see *thermoremanent magnetization*)
Trodos Ophiolite 756
Tromsø Observatory 737
true polar wander (TPW) 956
–, Indo-Atlantic 961
–, curve 962
–, Pacific curve 962
–, synthetic path 965
truncation point 122
Tsumeb Observatory 739
T-Tauri wind 799
tunneling process 334
Tuntungan Observatory 734
turbulence 17, 101, 169, 645
–, entropy flux 17
–, isotropic 193
–, magnetohydrodynamic 193
–, reflectionally 194
–, theory 18, 192
turbulent
–, diffusivity 18, 305
–, eddies 183
–, heat flux 17
–, viscosity 104
turtles 782
twin-vortex flow 659
Tycho Brahe 361
Tysnhes Island 798

U

u-channel 665, 760, 884
UK National Physics Laboratory 166
ultralow
  –, frequency (ULF) 700
    –, plasma wave 333
  –, velocity zone (ULVZ) 154, 208, 970
ulvöspinel 517, 526, 597
ULVZ (see *ultra-low velocity zone*)
Umbrian Apennines 666
unblocking temperature 626, 872
unconformity 567
undersea communication cable 316
undertone, core 364
unit 973
  –, absolute intensity 358
  –, geological 478
  –, igneous 35
  –, universal system of 278
United States Navy 850
universal
  –, stage method 550
  –, system of units 278
  –, time 810
University of Göttingen 279
University of Liverpool 992
University of London 992
upper, mantle 211
Uppsala Observatory 737
upstream wave 333
Urals 823
uranium–235 854
uranium–238 854
Uranus 203, 204, 207, 439, 865
  –, magnetic field 287
Urumqi Observatory 728
US Coast and Geodetic Survey 436
US Coast and Geodetic Survey's Division Magnetism 273
US National Academy of Science 980
US Naval Research Laboratory 438
USNS Eltanin 993
Utopia basin 795
Utrecht Observatory 725

V

Vacquier, Victor 438
vacuum seals, cryogenic 883
VADM (see *virtual axial dipole moment*)
Valentia Island Observatory 720
Valles Marineris 504
Valu Fa Ridge 237
value
  –, chi-squared 47
  –, Dulong-Petit 93
Van Allen radiation belt 865
van Heemskerck, Jacob 988
van Musschenbroek, Pieter 358, 725
van Neck, Jacob 988
Vanadis expedition 359
Vancouver, George 989
Vardö Island 460
variable
  –, effective 164
  –, field translation balance (VFTB) 877, 977
variable *Contd.*
    –, measurement 977
  –, separation 378
variance 846
  –, ellipsoid 846
variation
  –, annual 308, 698, 816
  –, daily 3, 27, 63, 308, 359, 449, 450, 858
    –, disturbance 816
    –, quiet-time 509
    –, regular 698
  –, decadal 131
  –, DP 700
  –, geomagnetic
    –, secular 68
    –, spectrum 308
  –, lateral 129
  –, latitudinal 892, 897
  –, magnetic 310
    –, mapping 310
  –, paleoclimatic 874
  –, paleointensity 353
  –, paleomagnetic secular (PSV) 766
  –, paleosecular 256
  –, secular 86, 280, 340, 346, 588, 778, 779, 892, 929, 993
    –, impulses 651
    –, model 892–898, 901
    –, paleomagnetic 133
  –, semiannual 698
  –, storm-time 698
  –, stratigraphic 256
variometer 27, 435, 436, 708
  –, Bobrov-system quartz 729
  –, declination 712
Variscan orogen 243
Vatican Council 887
VDM (see *virtual dipole moment*)
vector
  –, algebra 391
  –, Cartesian unit 272
  –, diffusion equation 223
  –, Hawaiian paleomagnetic 848
  –, looping 768, 770
  –, magnetic measurement 673
  –, paleomagnetic 654
  –, Parkinson 867
  –, polar 306
  –, potential 180, 194
  –, proton magnetometer (VPM) 715
  –, sum 595, 959
  –, Tipper 867
  –, waveform 768
  –, Wiese 867
velocity
  –, effective cyclonic 182
  –, geostrophic 746
  –, gradient 565
  –, measurement 398
  –, pattern, seismic 152
  –, phase 76
  –, potential 85
  –, zone, ultra-low (ULVZ) 154, 208, 970
Vema Seamount 891
Venus 89, 203, 204, 439, 788
Venus *Contd.*
  –, transit 720
  –, two-layer model 441
Venusian lowland plains 447
Verhoogen, John 979
vernier scale 435
versorium 66, 361
vertical transverse isotropy (VTI) 147, 155
Verwey
  –, temperature 517
  –, transition 500, 516, 552, 749, 751
very early time EM (VETEM) system 231
very long baseline radio interferometry (VLBI) 426, 469
very low frequency (VLF) 224
  –, method 224
Vesta 90
VFTB (see *variable field translation balance*)
VGP (see *virtual geomagnetic pole*)
vibrating sample magnetometer (VSM) 877
vibration, thermal 688
Victory 989
Vienna Observatory 728
Vietnam, National Centre for Natural Science and Technology 734
Viking 791
  –, magnetic properties experiment 791
  –, spacecraft 441
VIRGO, helioseismology instrument 398
virtual
  –, axial dipole moment (VADM) 159, 160
  –, dipole moment (VDM) 159
  –, geomagnetic pole (VGP) 282, 321, 778, 829, 892, 947
    –, path 322
    –, low-latitude 892
    –, path 829
    –, scatter 895, 899
VISAR sound speed measurement 917
viscosity 81, 92, 94, 96
  –, coefficient 622
  –, definition of 104
  –, eddy 94, 102, 305
  –, effective 104, 275
  –, inner core 105
  –, kinematic 12, 74, 80, 102, 104, 111
  –, magnetic 513
  –, molecular 104, 112
  –, temperature-dependent 128
  –, turbulent 104
  –, variation with temperature 146
viscous
  –, coupling 113
  –, dissipation 128
  –, force 630
  –, remanent magnetization (VRM) 577, 621, 696
    –, ambient-temperature 626
    –, elevated-temperature 626
    –, hard 629
    –, intensity 627
    –, soft 629
Vishakhapatnam Observatory 732, 733
vivianite 521
VLBI (see *very long baseline radio interferometry*)

VLF (see *very low frequency*)
vocation, religious 462
volatiles 89
volcanics
–, basalt 72
–, tephra 253
volcanoe
–, active 959
–, extinct 959
–, linear chain 959
volcano-electromagnetic effects 984
voltage
–, AC 920
–, amplitude 920
volume 931
–, expansion coefficient 92
–, magnetization 875
–, susceptibility 25
von Bellingshausen, Baron 989
von Humboldt, Alexander 358
von Lamont, Johann 729
Voronoi cell 26
vortex 170, 172
vorticity
–, absolute 75
–, equation 75
–, potential 840, 852
–, relative 75
Vostoks 359
VPM (see *vector proton magnetometer*)
Vredefort meteorite impact crater 799
VRM (see *viscous remanent magnetization*)
VSM (see *vibrating sample magnetometer*)
VTI (see *vertical transverse isotropy*)

W

wadsleyite 685, 687
Wallis, Samuel 989
Walton-Middleton-Schmidt contour 627
wander
–, inertial interchange true polar (IITPW) 966
–, polar 504, 959, 965
–, true polar (TPW) 956, 959
warming, climatic 669
Watheroo Observatory 43, 717, 807
Watkins, Norman David 992
wave
–, acoustic 398
–, ageostrophic 651
–, air 233
–, Alfvén 746
–, velocity 334
–, Bloch 201
–, body 82
–, dynamo 161
–, extremely low-frequency (ELF) 700
–, fast 651
–, frequency 651
–, gravitational 398
–, gravity 364
–, inertial 365, 633, 647, 843
–, inertio-gravity 364
–, ion-cyclotron 333
–, Lehnert 651
–, magnetic 118, 994

wave *Contd.*
–, magnetic Archimedean Coriolis (MAC) 87, 298, 348, 632, 652
–, bifurcation 653
–, buoyancy-driven 635
–, diffusive 635
–, instability 651
–, magnetohydrodynamic 333, 632
–, dispersion relations 632
–, MC- 124
–, number 76
–, domain centroid method 143
–, P- 147
–, PKIKP 125, 424
–, PKKP 125, 126
–, PKP 210
–, planetary 994
–, propagation 147
–, radio 452
–, Rossby 75, 124, 136, 432, 852, 994
–, buoyancy-driven magneto 635
–, thermal 75
–, S- 147
–, shear splitting 147
–, SKS 210
–, slow 651
–, surface 334
–, torsional 650
–, ultralow frequency (ULF) plasma 333
–, upstream 333
waveform 768, 770
–, analysis 768
–, recurring 774
–, vector 768
wavefront 147
wavelength
–, de Broglie 215
–, repeat 200
–, short 664
wavelet analysis 355
wavenumber 897
–, domain method 856
wavespeed, seismic 209
weak field
–, state 652
–, TRM 499
weathering 794
Weber, Wilhelm E. 278, 280, 359, 403, 679, 729
–, instruments 436
Wegener Medal 888
Weiss constant 932
Wellen Observatory 738
Wellington, Duke of 890
Welsh, John 720
Wenner array 229
Werner deconvolution 263
West Fault, Chile 413
western boundary current 853
westward drift 347, 993
wetting angle 687
whales 782
whistler 698
–, tone 700
Whiston, William 676

white noise 928
Widmanstatten pattern 796
Wiechert, Emil 743
Wiedemann-Franz
–, law 116, 120, 205
–, ratio 109
Wien Auhof Observatory 728
Wiener filter 857
Wiese
–, convention 413
–, vector 867
Wigner 3-j coefficients 386
Wilcke, Johann Karl 358
Wilhelmshaven Observatory 729
Wilkes Observatory 717
Wilkes, Charles 989
wind
–, magnetic 174, 946
–, polar 657
–, solar 60, 318, 333, 505, 657, 658, 864, 926, 927
–, plasma 657
–, thermal 76, 174, 304, 648, 945
window, sliding 338
Wingst Observatory 714, 729
wire heating 690
Witteveen Observatory 725
WMM (see *World Magnetic Model*)
WMS (see *World Magnetic Survey*)
wobble
–, Chandler 363
–, Markowitz 363, 426
Woelfersheimer Lake 228
Woods Hole Oceanographic Institution 236, 981
World Data Center 44, 407, 512, 728
World Magnetic Model (WMM) 359, 676, 850
World Magnetic Survey (WMS) 43, 56, 57, 274, 359, 411
Wright, Edward 674, 707, 988
Wuhan Observatory 728

X

Xingtai Earthquake 728
X-ray theory 369

Y

Yakutsk Observatory 738
Yamato 7038 799
Yamato chondrite 798
Yorktown 890
Yunnan province 821

Z

Z transfer function 310
Z/H-method 697
ZAMG (see *Central Institute of Meteorology and Geomagnetism*)
Zarya, RV 59, 359, 990
Zeeman
–, frequency 712
–, splitting 506
zenith 67
zeta potential 985
Zhejiang province 821

Zikawei Observatory 461
zirconia 685
zonal harmonics 680
zone
–, auroral 864
–, baked 603
–, cataclastic 247
–, conductive 247
zone *Contd.*
–, convective 179, 181
–, Fresnel 126
–, iron-reducing 585
–, isodynamic 404, 676
–, mushy 145
–, radiative 179
–, reacting 150
zone *Contd.*
–, resistive 234
–, seismic blank 239
–, subduction 64, 72
–, suture 242
–, transition 211
–, ultralow velocity 154
Zose Observatory 461